HISTORICAL FIGURES IN CHEMISTRY AND

Paul Dirac

English physicist
Born: Bristol, England, August 8, 1902
Died: Tallahassee, Florida, United States, October 20, 1984
Professor of Physics at the University of Cambridge, University of Miami, and Florida State University

- Papers published in *Proceedings of the Royal Society* titled "The Quantum Theory of the Electron" (1928) and "The Quantum Theory of the Electron Part II" (1929) formulated the first complete and accurate merging of quantum mechanics and relativity, through what is now known as the Dirac equation
- The Dirac equation both predicted the anti-electron (positron) and explained the occurrence of spin
- Received the 1933 Nobel Prize in Physics (with Erwin Schrödinger) for the "discovery of new productive forms of atomic theory"
- Proposed the concept of vacuum polarization and helped develop the field of quantum electrodynamics
- Demonstrated that the existence of a magnetic monopole would explain the existence of the electric charge
- Wrote the 1930 textbook *The Principles of Quantum Mechanics* that is still in use today
- Trained initially as an electrical engineer, he became one of the foremost theoretical scientists of the 20th century

"Scientific progress is measured in units of courage, not intelligence."

"In science one tries to tell people, in such a way as to be understood by everyone, something that no one ever knew before. But in poetry, it's the exact opposite."

"I do not see how a man can work on the frontiers of physics and write poetry at the same time. They are in opposition."

"God used beautiful mathematics in creating the world."

"A great deal of my work is just playing with equations and see what they give."

"If you are humble and receptive, mathematics will lead you by the hand."

"Pick a flower on earth and you move the farthest star."

Jacobus Henricus van't Hoff

Dutch chemist
Born: Rotterdam, Netherlands, August 30, 1852
Died: Steglitz, Germany, March 1, 1911
Professor of Chemistry at the University of Amsterdam and University of Berlin

- Published "Voorstel tot Uitbreiding der Tegenwoordige" in *de Scheikunde gebruikte Structuurformules in de Ruimte* (Proposal for the Development of Three-Dimensional Chemical Structural Formulae) in 1874, which led to the development of stereochemistry
- Proposed the idea of an "asymmetrical carbon atom" that provided an understanding of organic isomers and provided an explanation of optical activity of organic compounds
- First recipient of the Nobel Prize in Chemistry (1901) "in recognition of the extraordinary services he has rendered by the discovery of the laws of chemical dynamics and osmotic pressure in solutions"
- Developed a general thermodynamic relationship between the heat of conversion and changes in equilibrium due to temperature variation – The Principle of Mobile Equilibrium (1885) – that led to the van't Hoff-Le Chatelier principle
- Showed osmotic pressure in solutions sufficiently dilute is proportional to the solute concentration and that thermodynamic laws are not only valid for gases, but also for dilute solutions
- Proposed classifying the kinetics of molecular reactions according to the number of molecules involved

"It is sometimes easier to circumvent prevailing difficulties [in science] rather than to attack them."

"A famous name has this peculiarity that it becomes gradually smaller especially in natural sciences where each succeeding discovery invariably overshadows what precedes."

"In Leiden, all was prose – the town, the country, the people. In Bonn, all is poetry."

Johannes Diderik van der Waals

Dutch physicist
Born: Leiden, Netherlands, November 23, 1837
Died: Amsterdam, Netherlands, March 8, 1923
Professor of Physics at the Municipal University of Amsterdam

- Introduced the van der Waals equation that modifies ideal gas behavior by taking into account intramolecular forces and the finite volume of individual molecules
- Provided the first successful description of real gases that also provided insights into the need for atomistic theory
- Awarded the Nobel Prize in Physics in 1910 for "his work on equation of state of gases and liquids"
- Formulated the Law of Corresponding States stating all gases behave alike at the same reduced conditions (a reduced variable is the normal variable divided by the critical value of that variable $P_r = P/P_c$, etc.)
- In 1890, wrote the first treatise on the "Theory of Binary Solutions" that appeared in the *Archives Néerlandaises*; this gave rise to a numerous series of experiments, including those by J. P. Kuenen, who found characteristics of critical phenomena fully predicted by the theory
- Proposed his thermodynamic theory of capillarity in 1893, in which he accepted the existence of a gradual, though very rapid, change of density at the boundary layer between liquid and vapor; van der Waals also held the view that molecules are in permanent, rapid motion
- Was a high school teacher for 12 years before becoming a professor and made his most famous discoveries in that position

Erwin Schrödinger

Austrian physicist and natural science philosopher
Born: Vienna, Austria, August 12, 1887
Died: Vienna, Austria, January 4, 1961
Professor of Physics at the University of Breslau, University of Zürich, Frederick William University, Humboldt-Universität zu Berlin, University of Graz, Dublin Institute for Advanced Studies, and University of Vienna

- Paper published in *Annalen der Physik* (1926) titled "Quantisierung als Eigenwertproblem" [Eng. Tr. Quantization as an Eigenvalue Problem] on wave mechanics that established the Schrödinger equation as the fundamental equation of quantum mechanics
- Interpreted the square of the eigenfunction as related to observable physical quantities, leading to the basis for most quantum chemistry and quantum physical calculations in the past 70 years
- Other papers he wrote at about the same time showed the equivalence of the Schrödinger equation with Heisenberg's matrix formulation of quantum mechanics and treated time-dependent problems (i.e., scattering)
- Received the 1933 Nobel Prize in Physics (with Paul Dirac) for the "discovery of new productive forms of atomic theory"
- Proposed thought experiment (Schrödinger's Cat) to illustrate problems with the Copenhagen School interpretation of quantum superposition
- Wrote the influential book *What is Life* that explored the relationship of modern physical theory with biology and evolution, in part influencing James Watson and Francis Crick to explore gene theory and the structure of DNA

"The task is ... not so much to see what no one has yet seen; but to think what nobody has yet thought, about that which everybody sees."

"The scientist only imposes two things, namely truth and sincerity, imposes them upon himself and upon other scientists."

"What we observe as material bodies and forces are nothing but shapes and variations in the structure of space."

"Nirvana is a state of pure blissful knowledge ... It has nothing to do with the individual."

"It is by avoiding the rapid decay into the inert state of 'equilibrium' that an organism appears so enigmatic."

"If a man never contradicts himself, the reason must be that he virtually never says anything at all."

CRC Handbook of Chemistry and Physics

98th Edition

CRC Handbook of Chemistry and Physics

A Ready-Reference Book of Chemical and Physical Data

HANDBOOK OF CHEMISTRY AND PHYSICS
2017-2018
98th
EDITION
CRC PRESS

Editor-in-Chief

John R. Rumble, Jr., Ph.D.
R&R Data Services

Associate Editors

David R. Lide, Ph.D.
Former Director, Standard Reference Data
National Institute of Standards and Technology

Thomas J. Bruno, Ph.D.
Group Leader
National Institute of Standards and Technology

CRC Press
Taylor & Francis Group
Boca Raton London New York

CRC Press is an imprint of the
Taylor & Francis Group, an **informa** business

CRC Press
Taylor & Francis Group
6000 Broken Sound Parkway NW, Suite 300
Boca Raton, FL 33487-2742

Printed on acid-free paper

International Standard Book Number-13: 978-1-4987-8454-2 (Hardback)

Visit the Taylor & Francis Web site at
http://www.taylorandfrancis.com

and the CRC Press Web site at
http://www.crcpress.com

INTRODUCTION

It is with some pride and considerable humility that I introduce myself as the next Editor-in-Chief of the *CRC Handbook of Chemistry and Physics*, following the pioneering efforts of W. M. (Mickey) Haynes, Editor-in-Chief from 2010 to 2016 and who passed away in 2016, and David R. Lide, Editor-in-Chief from 1988 to 2009. Before describing my background, let me first write a few words about the *CRC Handbook* and its immediate past editors.

In an era when large Web-based search engines seemingly provide answers to every question and supply requested data on every subject, it is vital to recognize the importance and critical value of quality in a resource for scientific data. Most users of scientific data are not experts on how those data are measured and reported. They need data as input to their modeling and experiments, and those data must be reliable and the best available to ensure their results are valid and useful. In this universe of easy access to data of unknown quality, the *CRC Handbook of Chemistry and Physics* remains a trusted source of reliable data. Using subject matter experts from the international scientific community, David Lide and Mickey Haynes have maintained the integrity and quality in the book edition and created a powerful database version to meet the needs of e-science. Their vision and leadership has enabled users to continue to have access to high-quality scientific data.

While I am new to the *CRC Handbook*, my many years of experience in the scientific data community has prepared me for the tasks ahead, as outlined below. I have worked extensively with chemistry, physics, materials, and other types of scientific data, both during my years in industry, with the NIST Standard Reference Data Program, and more recently in the private sector. I worked on data compilations, especially in physics, materials science, and nanotechnology as well as in many related biological and engineering areas. This experience has highlighted for me the importance of accessibility to accurate data for chemists and physicists, from their earliest educational experiences to their advanced research efforts.

What lies in the future, especially as the 100th Edition of the *CRC Handbook* approaches? Let me offer a few thoughts of the tasks and challenges ahead.

- The printed version of the *CRC Handbook* has reached its maximum number of pages. As we add new data tables, we must carefully remove some existing tables and move them exclusively to the *CRC Online Handbook of Chemistry and Physics* hbcponline.com. This dual publication approach allows us to provide access to all *CRC Handbook* data online while keeping the printed edition manageable.
- The emergence of new chemistry disciplines such as materials chemistry, astro-chemistry, bio-medical chemistry, and green chemistry has created a need for reliable data to support experimental and theoretical progress in these areas. We are actively identifying new research fields to cover and recruiting the necessary experts to provide high-quality data to them.
- Nanotechnology has created new science and technology in every area it touches: chemistry, physics, materials, solid state science, bio-medicine, environmental impact, and toxicity. Slowly but surely, sets of reliable data are being generated and will be covered in future editions of the *CRC Handbook*.
- The incredible richness of modern electronic-based technology continually grows, and industrial physics is constantly moving the boundaries in terms of device scale, size, and complexity. These advances in turn require very reliable data to support the modeling and metrology associated with 21st century electronics. The *CRC Handbook* will explore how to meet the needs for additional solid state physics data.
- Expanding our interactions with users is a crucial step to ensuring the *CRC Handbook* stays relevant in this era of information overload. We are looking at a variety of methods for learning more about what users want in the future. The online edition for the first time gives us quantitative feedback on what data are being accessed. We want to go further and find out what additional data subjects need to be covered. We will be talking to you at meetings, conferences, and by other means. We want to hear from you.

As a long time user of the *CRC Handbook* and a data zealot for most of my professional career, I am delighted to be in a position to help continue its excellence in the future and to make it even more relevant as science advances. I look forward to working with everyone as we move forward.

John Rumble
Gaithersburg, MD
john.rumble@crchbcp.com
April 2017

EDITORIAL ADVISORY BOARD

PREFACE

The 98th Edition of the *CRC Handbook of Chemistry and Physics* continues our progress on transforming the full content into a 21st century modern electronic data resource from which both the print version and the Internet edition are generated. This work, about complete, provides our user community with many advantages.

- Consistency throughout the *CRC Handbook* with respect to chemical names, formulas, and basic physical properties
- More current data, especially in the Internet edition, from our ability to update property values as new measurements are reported
- Supplemental data tables in the Internet edition that cannot easily be included in the printed *CRC Handbook*
- The ability to use data directly in a wide variety of analytical and visualization tools

As we reach completion of this transformation, our dedication remains to provide the highest data quality standards and comprehensive data currency in the chemical and information world. In the production of the 98th Edition of the *CRC Handbook*, we have made several major updates to existing data tables and have included a number of new tables of important chemical and physical data, including the following.

- The Elements – Updated with a new summary table of key properties
- Table of the Isotopes – The first major update to this important data collection in several years with hundreds of new and more accurate data points for more than 3000 isotopes of all the elements
- Gas Clathrates – New data tables for important mixed clathrate systems
- Ionic Liquids – Expanded and revised information on over 150 important ionic liquids
- Protein-Ligand Binding Thermodynamics – The first major compilation of high-quality data on these important biochemical reactions
- Common Drugs – Completely revised with property and functionality data on over 250 drugs
- ^{1}H Chemical Shifts of Contaminants in Deuterated Solvents – An important new collection of NMR data critical for understanding outcomes of organic syntheses
- Electron Affinities – New data on over 200 atoms and molecules
- Atomic and Molecular Polarizabilities – Updated data value for virtually every element and many important molecules
- IUPAC Guide to Nomenclature of Inorganic Chemistry – This completely revised, four-page guide provides easy-to-use access to the nomenclature rules for inorganic compounds

A number of other tables have received updates as listed below.

- Bond Dissociation Energies – The latest experimental values have been added for a number of molecules
- Standard Atomic Weights – 2016 changes made by IUPAC
- Physical Properties of Organic Compounds – Results of the first phase of a complete review, with new data of densities
- Atomic Masses and Abundances – Update of isotopic abundances
- Periodic Table of the Elements – Updated to include new element names
- Interstellar Molecules – Complete list of molecules observed in space to date
- Major World Earthquakes – Updated through 2016
- Atmospheric Concentration of CO_2 – Updated through 2016
- Global Temperature Trend – Updated through 2016
- Threshold Limits for Airborne Contaminants – Updated to 2016 recommendations
- Chemical Carcinogens – Inclusion of 2016 recommendations of the National Toxicology Program
- Nanomaterial Safety Guidelines – Definition of terms in nanoscience; guidelines for safe handling and disposal

The Index to the *CRC Handbook* covers properties, processes, general concepts, and classes of substances. Space does not permit an index to individual chemical compounds, but the Internet edition provides searching for compounds on the basis of name, synonym, molecular formula, CAS Registry Number, and structure. For the book edition, indexes to the tables "Physical Constants of Organic Compounds" and "Physical Constants of Inorganic Compounds" are available in PDF form by e-mail request to john.rumble@crchbcp.com.

The success of the *CRC Handbook* is very dependent on feedback from its users. The Editor-in-Chief appreciates any suggestions from readers on proposed new topics or comments on how the usefulness may be improved in future editions. Please send your comments to john.rumble@crchbcp.com.

Numerous international experts make key contributions to the *CRC Handbook*. These contributors are listed on pages immediately following the Preface. Their efforts play a key role in the quality and diversity of the subject matter covered in the *CRC Handbook*. The sound advice and guidance of the Editorial Advisory Board members, who are listed in the front matter, is very much appreciated. Fiona Macdonald, Senior Publisher – Chemical & Life Sciences, CRC Press/Taylor & Francis Group has been of great assistance and support in providing oversight to ensure that the *CRC Handbook* continues to satisfy the needs of the user community. Thanks are also due to Linda Manis Leggio and Pam Morrell for their detailed, cooperative work and extreme care in the production of the *CRC Handbook*. Special thanks are due to David R. Lide and Thomas Bruno for their assistance in support of the Editor-in-Chief.

John Rumble
Gaithersburg, MD
John.rumble@crchbcp.com
April 2017

CURRENT CONTRIBUTORS

Ian H. Bell
Applied Chemicals and Materials Division
National Institute of Standards and Technology
Boulder, Colorado 80305

Lev I. Berger
California Institute of Electronics and Materials Science
2115 Flame Tree Way
Hemet, California 92545

Peter E. Bradley
Applied Chemicals and Materials Division
National Institute of Standards and Technology
Boulder, Colorado 80305

Thomas J. Bruno
Applied Chemicals and Materials Division
National Institute of Standards and Technology
Boulder, Colorado 80305

Jessica L. Burger
Applied Chemicals and Materials Division
National Institute of Standards and Technology
Boulder, Colorado 80305

Charles E. Carraher
Department of Chemistry and Biochemistry
Florida Atlantic University
Boca Raton, Florida 33431

Jin-Pei Cheng
Ministry of Science and Technology
Beijing 100862, China

Robert D. Chirico
Thermodynamics Research Center
Applied Chemicals and Materials Division
National Institute of Standards and Technology
Boulder, Colorado 80305

Ivan Cibulka
Department of Physical Chemistry
Institute of Chemical Technology
CZ-166 28 Prague, Czech Republic

Christopher J. Cramer
Department of Chemistry
University of Minnesota
Minneapolis, Minnesota 55455

Vladimir Diky
Thermodynamics Research Center
Applied Chemicals and Materials Division
National Institute of Standards and Technology
Boulder, Colorado 80305

Michael Frenkel
14809 Glen Eden Drive
Naples, Florida 34110

Jeffrey R. Fuhr
Quantum Measurement Division
National Institute of Standards and Technology
Gaithersburg, Maryland 20899

Robert N. Goldberg
Biosystems and Biomaterials Division
National Institute of Standards and Technology
Gaithersburg, Maryland 20899

Thomas W. Grove
Boulder Safety, Health, and Environmental Division
National Institute of Standards and Technology
Boulder, Colorado 80305

Allan H. Harvey
Applied Chemicals and Materials Division
National Institute of Standards and Technology
Boulder, Colorado 80305

Steven R. Heller
Chemical and Biochemical Reference Data Division
National Institute of Standards and Technology
Gaithersburg, Maryland 20899

Norman E. Holden
National Nuclear Data Center
Brookhaven National Laboratory
Upton, New York 11973

Marcia L. Huber
Applied Chemicals and Materials Division
National Institute of Standards and Technology
Boulder, Colorado 80305

Andrei Kazakov
Thermodynamics Research Center
Applied Chemicals and Materials Division
National Institute of Standards and Technology
Boulder, Colorado 80305

Daniel E. Kelleher
Quantum Measurement Division
National Institute of Standards and Technology
Gaithersburg, Maryland 20899

M. Naveed Khan
Center for Hydrate Research
Colorado School of Mines
Golden, Colorado 80401
Petroleum Institute, Abu Dhabi, U.A.E.

Carolyn A. Koh
Center for Hydrate Research
Colorado School of Mines
1600 Illinois Street
Golden, Colorado 80401

Eric W. Lemmon
Applied Chemicals and Materials Division
National Institute of Standards and Technology
Boulder, Colorado 80305

David R. Lide
13901 Riding Loop Dr.
North Potomac, Maryland 20878

Frank J. Lovas
Sensor Sciences Division
National Institute of Standards and Technology
Gaithersburg, Maryland 20899

Yu-Ran Luo (deceased)
Department of Chemistry
Shanghai University
Shanghai 200444, China

Serguei N. Lvov
Department of Energy and Mineral Engineering
Pennsylvania State University
University Park, Pennsylvania 16802

Manjeera Mantina
Department of Chemistry
University of Minnesota
Minneapolis, Minnesota 55455

Alan D. McNaught
8 Cavendish Avenue
Cambridge CB1 7US
England

Thomas M. Miller
Air Force Research Laboratory/RVBXT
3550 Aberdeen Av, SE, Bldg 570.
Kirtland AFB, New Mexico 87117-5776

Nasser Moazzen-Ahmadi
Department of Physics and Astronomy
University of Calgary
2500 University Drive NW
Calgary, Alberta T2N 1N4, Canada

Peter J. Mohr
Quantum Measurement Division
National Institute of Standards and Technology
Gaithersburg, Maryland 20899

Chris D. Muzny
Thermodynamics Research Center
Applied Chemicals and Materials Division
National Institute of Standards and Technology
Boulder, Colorado 80305

David B. Newell
Quantum Measurement Division
National Institute of Standards and Technology
Gaithersburg, Maryland 20899

Irving Ozier
Department of Physics and Astronomy
University of British Columbia
6224 Agricultural Road
Vancouver, British Columbia V6T 1Z1, Canada

Eugene Paulechka
Applied Chemicals and Materials Division
National Institute of Standards and Technology
Boulder, Colorado 80305

Larissa I. Podobedova
Biomolecular Measurement Division
National Institute of Standards and Technology
Gaithersburg, Maryland 20899

Cedric J. Powell
Materials Measurement Science Division
National Institute of Standards and Technology
Gaithersburg, Maryland 20899

Ray Radebaugh
Applied Chemicals and Materials Division
National Institute of Standards and Technology
Boulder, Colorado 80305

Joseph Reader
Quantum Measurement Division
National Institute of Standards and Technology
Gaithersburg, Maryland 20899

Anthony J. Remijan
National Radio Astronomy Observatory
520 Edgemont Road
Charlottesville, Virginia 22903

E. Dendy Sloan
Center for Hydrate Research
Colorado School of Mines
1600 Illinois Street
Golden, Colorado 80401

Beverly L. Smith
3835 E Mineral Pl.
Centennial, CO 80122

Lewis E. Snyder
Astronomy Department
University of Illinois
Urbana, Illinois 61801

Paris D. N. Svoronos
Queensborough Community College
City University of New York
Bayside, NY 11364

Barry N. Taylor
Quantum Measurement Division
National Institute of Standards and Technology
Gaithersburg, Maryland 20899

Donald G. Truhlar
Department of Chemistry
University of Minnesota
Minneapolis, Minnesota 55455

Rosendo Valero
Chemistry Department
University of Coimbra
Coimbra, Portugal

Wolfgang L. Wiese
Quantum Measurement Division
National Institute of Standards and Technology
Gaithersburg, Maryland 20899

Mark A. Williams
Institute of Structural and Molecular Biology
Department of Biological Sciences
Birkbeck, University of London
Malet Street, London, England WC1E 7HX

Christian Wohlfarth
Martin Luther University
Institute of Physical Chemistry
Mühlpforte 1
06108 Halle (Saale), Germany

Daniel Zwillinger
Mathematics Department
Rensselaer Polytechnic Institute
Troy, New York 12180

TABLE OF CONTENTS

SECTION 1: BASIC CONSTANTS, UNITS, AND CONVERSION FACTORS

SECTION 2: SYMBOLS, TERMINOLOGY, AND NOMENCLATURE

SECTION 3: PHYSICAL CONSTANTS OF ORGANIC COMPOUNDS

SECTION 4: PROPERTIES OF THE ELEMENTS AND INORGANIC COMPOUNDS

SECTION 5: THERMOCHEMISTRY, ELECTROCHEMISTRY, AND SOLUTION CHEMISTRY

SECTION 6: FLUID PROPERTIES

SECTION 11: NUCLEAR AND PARTICLE PHYSICS

SECTION 12: PROPERTIES OF SOLIDS

SECTION 13: POLYMER PROPERTIES

SECTION 14: GEOPHYSICS, ASTRONOMY, AND ACOUSTICS

SECTION 15: PRACTICAL LABORATORY DATA

Section 1
Basic Constants, Units, and Conversion Factors

CODATA RECOMMENDED VALUES OF THE FUNDAMENTAL PHYSICAL CONSTANTS: 2014

Peter J. Mohr, Barry N. Taylor, and David B. Newell

This report gives the 2014 self-consistent set of values of the constants and conversion factors of physics and chemistry recommended by the Committee on Data for Science and Technology (CODATA). These values are based on a least-squares adjustment that takes into account all data available up to 31 December 2014. The recommended values may also be found on the Web at physics.nist.gov/constants.

This report was prepared by the authors under the auspices of the CODATA Task Group on Fundamental Constants. The members of the task group are:

F. Cabiati, Istituto Nazionale di Ricerca Metrologica, Italy
J. Fischer, Physikalisch-Technische Bundesanstalt, Germany
J. Flowers (deceased), National Physical Laboratory, United Kingdom
K. Fujii, National Metrology Institute of Japan, Japan
S. G. Karshenboim, Pulkovo Observatory, Russian Federation
E. de Mirandés, Bureau international des poids et mesures
P. J. Mohr, National Institute of Standards and Technology, United States of America
D. B. Newell, National Institute of Standards and Technology, United States of America
F. Nez, Laboratoire Kastler-Brossel, France
K. Pachucki, University of Warsaw, Poland
T. J. Quinn, Bureau international des poids et mesures
C. Thomas, Bureau international des poids et mesures
B. N. Taylor, National Institute of Standards and Technology, United States of America
B. M. Wood, National Research Council, Canada
Z. Zhang, National Institute of Metrology, China (People's Republic of)

References

1. Mohr, P. J., Newell, D. B., and Taylor, B. N., *J. Phys. Chem. Ref. Data* 45, 043102, 2016.
2. Mohr, P. J., Newell, D. B., and Taylor, B. N., *Rev. Mod. Phys.* 88, 035009, 2016.
3. Available on the Web at physics.nist.gov/constants.

Table I. An Abbreviated List of the CODATA Recommended Values of the Fundamental Constants of Physics and Chemistry Based on the 2014 Adjustment

Quantity	Symbol	Numerical value	Unit	Relative std. uncert. u_r
speed of light in vacuum	c, c_0	299 792 458	$m\ s^{-1}$	exact
magnetic constant	μ_0	$4\pi\times10^{-7}$	$N\ A^{-2}$	
		$=12.566\,370\,614\ldots\times10^{-7}$	$N\ A^{-2}$	exact
electric constant $1/\mu_0 c^2$	ϵ_0	$8.854\,187\,817\ldots\times10^{-12}$	$F\ m^{-1}$	exact
Newtonian constant of gravitation	G	$6.674\,08(31)\times10^{-11}$	$m^3\ kg^{-1}\ s^{-2}$	4.7×10^{-5}
Planck constant	h	$6.626\,070\,040(81)\times10^{-34}$	J s	1.2×10^{-8}
$h/2\pi$	$\hbar$	$1.054\,571\,800(13)\times10^{-34}$	J s	1.2×10^{-8}
elementary charge	e	$1.602\,176\,6208(98)\times10^{-19}$	C	6.1×10^{-9}
magnetic flux quantum $h/2e$	Φ_0	$2.067\,833\,831(13)\times10^{-15}$	Wb	6.1×10^{-9}
conductance quantum $2e^2/h$	G_0	$7.748\,091\,7310(18)\times10^{-5}$	S	2.3×10^{-10}
electron mass	m_e	$9.109\,383\,56(11)\times10^{-31}$	kg	1.2×10^{-8}
proton mass	m_p	$1.672\,621\,898(21)\times10^{-27}$	kg	1.2×10^{-8}
proton-electron mass ratio	m_p/m_e	1836.152 673 89 (17)		9.5×10^{-11}
fine-structure constant $e^2/4\pi\epsilon_0\hbar c$	α	$7.297\,352\,5664(17)\times10^{-3}$		2.3×10^{-10}
inverse fine-structure constant	α^{-1}	137.035 999 139 (31)		2.3×10^{-10}
Rydberg constant $\alpha^2 m_e c/2h$	R_∞	10 973 731.568 508 (65)	m^{-1}	5.9×10^{-12}
Avogadro constant	N_A, L	$6.022\,140\,857(74)\times10^{23}$	mol^{-1}	1.2×10^{-8}
Faraday constant $N_A e$	F	96 485.332 89 (59)	$C\ mol^{-1}$	6.2×10^{-9}
molar gas constant	R	8.314 4598 (48)	$J\ mol^{-1}\ K^{-1}$	5.7×10^{-7}
Boltzmann constant R/N_A	k	$1.380\,648\,52(79)\times10^{-23}$	$J\ K^{-1}$	5.7×10^{-7}
Stefan-Boltzmann constant $(\pi^2/60)k^4/\hbar^3c^2$	σ	$5.670\,367(13)\times10^{-8}$	$W\ m^{-2}\ K^{-4}$	2.3×10^{-6}
Non-SI units accepted for use with the SI				
electron volt (e/C) J	eV	$1.602\,176\,6208(98)\times10^{-19}$	J	6.1×10^{-9}
(unified) atomic mass unit $\frac{1}{12}\,m(^{12}C)$	u	$1.660\,539\,040(20)\times10^{-27}$	kg	1.2×10^{-8}

Table II. The CODATA Recommended Values of the Fundamental Constants of Physics and Chemistry Based on the 2014 Adjustment

Quantity	Symbol	Numerical value	Unit	Relative std. uncert. u_r
Universal				
speed of light in vacuum	c, c_0	299 792 458	$m\ s^{-1}$	exact
magnetic constant	μ_0	$4\pi \times 10^{-7}$	$N\ A^{-2}$	
		$=12.566\ 370\ 614... \times 10^{-7}$	$N\ A^{-2}$	exact
electric constant $1/\mu_0 c^2$	ϵ_0	$8.854\ 187\ 817... \times 10^{-12}$	$F\ m^{-1}$	exact
characteristic impedance of vacuum $\mu_0 c$	Z_0	376.730 313 461...	Ω	exact
Newtonian constant of gravitation	G	$6.674\ 08(31) \times 10^{-11}$	$m^3\ kg^{-1}\ s^{-2}$	4.7×10^{-5}
	$G/\hbar c$	$6.708\ 61(31) \times 10^{-39}$	$(GeV/c^2)^{-2}$	4.7×10^{-5}
Planck constant	h	$6.626\ 070\ 040(81) \times 10^{-34}$	$J\ s$	1.2×10^{-8}
$h/2\pi$	$\hbar$	$1.054\ 571\ 800(13) \times 10^{-34}$	$J\ s$	1.2×10^{-8}
		$6.582\ 119\ 514(40) \times 10^{-16}$	$eV\ s$	6.1×10^{-9}
	$\hbar c$	197.326 9788(12)	MeV fm	6.1×10^{-9}
Planck mass $(\hbar c/G)^{1/2}$	m_P	$2.176\ 470(51) \times 10^{-8}$	kg	2.3×10^{-5}
energy equivalent	$m_P c^2$	$1.220\ 910(29) \times 10^{19}$	GeV	2.3×10^{-5}
Planck temperature $(\hbar c^5/G)^{1/2}/k$	T_P	$1.416\ 808(33) \times 10^{32}$	K	2.3×10^{-5}
Planck length $\hbar/m_P c = (\hbar G/c^3)^{1/2}$	l_P	$1.616\ 229(38) \times 10^{-35}$	m	2.3×10^{-5}
Planck time $l_P/c = (\hbar G/c^5)^{1/2}$	t_P	$5.391\ 16(13) \times 10^{-44}$	s	2.3×10^{-5}
Electromagnetic				
elementary charge	e	$1.602\ 176\ 6208(98) \times 10^{-19}$	C	6.1×10^{-9}
	e/h	$2.417\ 989\ 262(15) \times 10^{14}$	$A\ J^{-1}$	6.1×10^{-9}
magnetic flux quantum $h/2e$	Φ_0	$2.067\ 833\ 831(13) \times 10^{-15}$	Wb	6.1×10^{-9}
conductance quantum $2e^2/h$	G_0	$7.748\ 091\ 7310(18) \times 10^{-5}$	S	2.3×10^{-10}
inverse of conductance quantum	G_0^{-1}	12 906.403 7278(29)	Ω	2.3×10^{-10}
Josephson constant[1] $2e/h$	K_J	$483\ 597.8525(30) \times 10^9$	$Hz\ V^{-1}$	6.1×10^{-9}
von Klitzing constant[2] $h/e^2 = \mu_0 c/2\alpha$	R_K	25 812.807 4555(59)	Ω	2.3×10^{-10}
Bohr magneton $e\hbar/2m_e$	μ_B	$927.400\ 9994(57) \times 10^{-26}$	$J\ T^{-1}$	6.2×10^{-9}
		$5.788\ 381\ 8012(26) \times 10^{-5}$	$eV\ T^{-1}$	4.5×10^{-10}
	μ_B/h	$13.996\ 245\ 042(86) \times 10^9$	$Hz\ T^{-1}$	6.2×10^{-9}
	μ_B/hc	46.686 448 14(29)	$m^{-1}\ T^{-1}$	6.2×10^{-9}
	μ_B/k	0.671 714 05(39)	$K\ T^{-1}$	5.7×10^{-7}
nuclear magneton $e\hbar/2m_p$	μ_N	$5.050\ 783\ 699(31) \times 10^{-27}$	$J\ T^{-1}$	6.2×10^{-9}
		$3.152\ 451\ 2550(15) \times 10^{-8}$	$eV\ T^{-1}$	4.6×10^{-10}
	μ_N/h	7.622 593 285(47)	$MHz\ T^{-1}$	6.2×10^{-9}
	μ_N/hc	$2.542\ 623\ 432(16) \times 10^{-2}$	$m^{-1}\ T^{-1}$	6.2×10^{-9}
	μ_N/k	$3.658\ 2690(21) \times 10^{-4}$	$K\ T^{-1}$	5.7×10^{-7}
Atomic and Nuclear				
General				
fine-structure constant $e^2/4\pi\epsilon_0\hbar c$	α	$7.297\ 352\ 5664(17) \times 10^{-3}$		2.3×10^{-10}
inverse fine-structure constant	α^{-1}	137.035 999 139(31)		2.3×10^{-10}
Rydberg constant $\alpha^2 m_e c/2h$	R_∞	10 973 731.568 508(65)	m^{-1}	5.9×10^{-12}
	$R_\infty c$	$3.289\ 841\ 960\ 355(19) \times 10^{15}$	Hz	5.9×10^{-12}
	$R_\infty hc$	$2.179\ 872\ 325(27) \times 10^{-18}$	J	1.2×10^{-8}
		13.605 693 009(84)	eV	6.1×10^{-9}
Bohr radius $\alpha/4\pi R_\infty = 4\pi\epsilon_0\hbar^2/m_e e^2$	a_0	$0.529\ 177\ 210\ 67(12) \times 10^{-10}$	m	2.3×10^{-10}
Hartree energy $e^2/4\pi\epsilon_0 a_0 = 2R_\infty hc = \alpha^2 m_e c^2$	E_h	$4.359\ 744\ 650(54) \times 10^{-18}$	J	1.2×10^{-8}
		27.211 386 02(17)	eV	6.1×10^{-9}
quantum of circulation	$h/2m_e$	$3.636\ 947\ 5486(17) \times 10^{-4}$	$m^2\ s^{-1}$	4.5×10^{-10}
	h/m_e	$7.273\ 895\ 0972(33) \times 10^{-4}$	$m^2\ s^{-1}$	4.5×10^{-10}
Electroweak				
Fermi coupling constant[3]	$G_F/(\hbar c)^3$	$1.166\ 3787(6) \times 10^{-5}$	GeV^{-2}	5.1×10^{-7}
weak mixing angle[4] θ_W (on-shell scheme) $\sin^2\theta_W = s_W^2 \equiv 1-(m_W/m_Z)^2$	$\sin^2\theta_W$	0.2223(21)		9.5×10^{-3}

Quantity	Symbol	Numerical value	Unit	Relative std. uncert. u_r
Electron, e⁻				
electron mass	m_e	$9.109\,383\,56(11)\times10^{-31}$	kg	1.2×10^{-8}
		$5.485\,799\,090\,70(16)\times10^{-4}$	u	2.9×10^{-11}
energy equivalent	$m_e c^2$	$8.187\,105\,65(10)\times10^{-14}$	J	1.2×10^{-8}
		$0.510\,998\,9461(31)$	MeV	6.2×10^{-9}
electron-muon mass ratio	m_e/m_μ	$4.836\,331\,70(11)\times10^{-3}$		2.2×10^{-8}
electron-tau mass ratio	m_e/m_τ	$2.875\,92(26)\times10^{-4}$		9.0×10^{-5}
electron-proton mass ratio	m_e/m_p	$5.446\,170\,213\,52(52)\times10^{-4}$		9.5×10^{-11}
electron-neutron mass ratio	m_e/m_n	$5.438\,673\,4428(27)\times10^{-4}$		4.9×10^{-10}
electron-deuteron mass ratio	m_e/m_d	$2.724\,437\,107\,484(96)\times10^{-4}$		3.5×10^{-11}
electron-triton mass ratio	m_e/m_t	$1.819\,200\,062\,203(84)\times10^{-4}$		4.6×10^{-11}
electron-helion mass ratio	m_e/m_h	$1.819\,543\,074\,854(88)\times10^{-4}$		4.9×10^{-11}
electron to alpha particle mass ratio	$m_e/m\alpha$	$1.370\,933\,554\,798(45)\times10^{-4}$		3.3×10^{-11}
electron charge to mass quotient	$-e/m_e$	$-1.758\,820\,024(11)\times10^{11}$	C kg^{-1}	6.2×10^{-9}
electron molar mass $N_A m_e$	$M(e)$, M_e	$5.485\,799\,090\,70(16)\times10^{-7}$	kg mol^{-1}	2.9×10^{-11}
Compton wavelength $h/m_e c$	λ_C	$2.426\,310\,2367(11)\times10^{-12}$	m	4.5×10^{-10}
$\lambda_C/2\pi = \alpha a_0 = \alpha^2/4\pi R_\infty$	$ƛ_C$	$386.159\,267\,64(18)\times10^{-15}$	m	4.5×10^{-10}
classical electron radius $\alpha^2 a_0$	r_e	$2.817\,940\,3227(19)\times10^{-15}$	m	6.8×10^{-10}
Thomson cross section $(8\pi/3)r_e^2$	σ_e	$0.665\,245\,871\,58(91)\times10^{-28}$	m^2	1.4×10^{-9}
electron magnetic moment	μ_e	$-928.476\,4620(57)\times10^{-26}$	J T^{-1}	6.2×10^{-9}
to Bohr magneton ratio	μ_e/μ_B	$-1.001\,159\,652\,180\,91(26)$		2.6×10^{-13}
to nuclear magneton ratio	μ_e/μ_N	$-1838.281\,972\,34(17)$		9.5×10^{-11}
electron magnetic moment anomaly $\lvert\mu_e\rvert/\mu_B-1$	a_e	$1.159\,652\,180\,91(26)\times10^{-3}$		2.3×10^{-10}
electron g-factor $-2(1+a_e)$	g_e	$-2.002\,319\,304\,361\,82(52)$		2.6×10^{-13}
electron-muon magnetic moment ratio	μ_e/μ_μ	$206.766\,9880(46)$		2.2×10^{-8}
electron-proton magnetic moment ratio	μ_e/μ_p	$-658.210\,6866(20)$		3.0×10^{-9}
electron to shielded proton magnetic moment ratio (H_2O, sphere, 25 °C)	μ_e/μ'_p	$-658.227\,5971(72)$		1.1×10^{-8}
electron-neutron magnetic moment ratio	μ_e/μ_n	$960.920\,50(23)$		2.4×10^{-7}
electron-deuteron magnetic moment ratio	μ_e/μ_d	$-2143.923\,499(12)$		5.5×10^{-9}
electron to shielded helion magnetic moment ratio (gas, sphere, 25 °C)	μ_e/μ'_h	$864.058\,257(10)$		1.2×10^{-8}
electron gyromagnetic ratio $2\lvert\mu_e\rvert/\hbar$	γ_e	$1.760\,859\,644(11)\times10^{11}$	s^{-1} T^{-1}	6.2×10^{-9}
	$\gamma_e/2\pi$	$28\,024.951\,64(17)$	MHz T^{-1}	6.2×10^{-9}
Muon, μ⁻				
muon mass	m_μ	$1.883\,531\,594(48)\times10^{-28}$	kg	2.5×10^{-8}
		$0.113\,428\,9257(25)$	u	2.2×10^{-8}
energy equivalent	$m_\mu c^2$	$1.692\,833\,774(43)\times10^{-11}$	J	2.5×10^{-8}
		$105.658\,3745(24)$	MeV	2.3×10^{-8}
muon-electron mass ratio	m_μ/m_e	$206.768\,2826(46)$		2.2×10^{-8}
muon-tau mass ratio	m_μ/m_τ	$5.946\,49(54)\times10^{-2}$		9.0×10^{-5}
muon-proton mass ratio	m_μ/m_p	$0.112\,609\,5262(25)$		2.2×10^{-8}
muon-neutron mass ratio	m_μ/m_n	$0.112\,454\,5167(25)$		2.2×10^{-8}
muon molar mass $N_A m_\mu$	$M(\mu)$, M_μ	$0.113\,428\,9257(25)\times10^{-3}$	kg mol^{-1}	2.2×10^{-8}
muon Compton wavelength $h/m_\mu c$	$\lambda_{C,\mu}$	$11.734\,441\,11(26)\times10^{-15}$	m	2.2×10^{-8}
$\lambda_{C,\mu}/2\pi$	$ƛ_{C,\mu}$	$1.867\,594\,308(42)\times10^{-15}$	m	2.2×10^{-8}
muon magnetic moment	μ_μ	$-4.490\,448\,26(10)\times10^{-26}$	J T^{-1}	2.3×10^{-8}
to Bohr magneton ratio	μ_μ/μ_B	$-4.841\,970\,48(11)\times10^{-3}$		2.2×10^{-8}
to nuclear magneton ratio	μ_μ/μ_N	$-8.890\,597\,05(20)$		2.2×10^{-8}
muon magnetic moment anomaly $\lvert\mu_\mu\rvert/(e\hbar/2m_\mu)-1$	a_μ	$1.165\,920\,89(63)\times10^{-3}$		5.4×10^{-7}
muon g-factor $-2(1+a_\mu)$	g_μ	$-2.002\,331\,8418(13)$		6.3×10^{-10}
muon-proton magnetic moment ratio	μ_μ/μ_p	$-3.183\,345\,142(71)$		2.2×10^{-8}
Tau, τ⁻				
tau mass[5]	m_τ	$3.167\,47(29)\times10^{-27}$	kg	9.0×10^{-5}
		$1.907\,49(17)$	u	9.0×10^{-5}

Quantity	Symbol	Numerical value	Unit	Relative std. uncert. u_r
energy equivalent	$m_\tau c^2$	$2.846\,78(26)\times10^{-10}$	J	9.0×10^{-5}
		1776.82 (16)	MeV	9.0×10^{-5}
tau-electron mass ratio	m_τ/m_e	3477.15 (31)		9.0×10^{-5}
tau-muon mass ratio	m_τ/m_μ	16.8167 (15)		9.0×10^{-5}
tau-proton mass ratio	m_τ/m_p	1.893 72 (17)		9.0×10^{-5}
tau-neutron mass ratio	m_τ/m_n	1.891 11 (17)		9.0×10^{-5}
tau molar mass $N_A m_\tau$	$M(\tau)$, M_τ	$1.907\,49(17)\times10^{-3}$	kg mol^{-1}	9.0×10^{-5}
tau Compton wavelength $h/m_\tau c$	$\lambda_{C,\tau}$	$0.697\,787(63)\times10^{-15}$	m	9.0×10^{-5}
$\lambda_{C,\tau}/2\pi$	$\bar{\lambda}_{C,\tau}$	$0.111\,056(10)\times10^{-15}$	m	9.0×10^{-5}
***Proton,* p**				
proton mass	m_p	$1.672\,621\,898(21)\times10^{-27}$	kg	1.2×10^{-8}
		1.007 276 466 879 (91)	u	9.0×10^{-11}
energy equivalent	$m_p c^2$	$1.503\,277\,593(18)\times10^{-10}$	J	1.2×10^{-8}
		938.272 0813 (58)	MeV	6.2×10^{-9}
proton-electron mass ratio	m_p/m_e	1836.152 673 89 (17)		9.5×10^{-11}
proton-muon mass ratio	m_p/m_μ	8.880 243 38 (20)		2.2×10^{-8}
proton-tau mass ratio	m_p/m_τ	0.528 063 (48)		9.0×10^{-5}
proton-neutron mass ratio	m_p/m_n	0.998 623 478 44 (51)		5.1×10^{-10}
proton charge to mass quotient	e/m_p	$9.578\,833\,226(59)\times10^{7}$	C kg^{-1}	6.2×10^{-9}
proton molar mass $N_A m_p$	$M(\mathrm{p})$, M_p	$1.007\,276\,466\,879(91)\times10^{-3}$	kg mol^{-1}	9.0×10^{-11}
proton Compton wavelength $h/m_p c$	$\lambda_{C,p}$	$1.321\,409\,853\,96(61)\times10^{-15}$	m	4.6×10^{-10}
$\lambda_{C,p}/2\pi$	$\bar{\lambda}_{C,p}$	$0.210\,308\,910\,109(97)\times10^{-15}$	m	4.6×10^{-10}
proton rms charge radius	r_p	$0.8751(61)\times10^{-15}$	m	7.0×10^{-3}
proton magnetic moment	μ_p	$1.410\,606\,7873(97)\times10^{-26}$	J T^{-1}	6.9×10^{-9}
to Bohr magneton ratio	μ_p/μ_B	$1.521\,032\,2053(46)\times10^{-3}$		3.0×10^{-9}
to nuclear magneton ratio	μ_p/μ_N	2.792 847 3508 (85)		3.0×10^{-9}
proton *g*-factor $2\mu_p/\mu_N$	g_p	5.585 694 702 (17)		3.0×10^{-9}
proton-neutron magnetic moment ratio	μ_p/μ_n	-1.459 898 05 (34)		2.4×10^{-7}
shielded proton magnetic moment (H_2O, sphere, 25 °C)	μ_p'	$1.410\,570\,547(18)\times10^{-26}$	J T^{-1}	1.3×10^{-8}
to Bohr magneton ratio	μ_p'/μ_B	$1.520\,993\,128(17)\times10^{-3}$		1.1×10^{-8}
to nuclear magneton ratio	μ_p'/μ_N	2.792 775 600 (30)		1.1×10^{-8}
proton magnetic shielding correction $1-\mu_p'/\mu_p$ (H_2O, sphere, 25 °C)	σ_p'	$25.691(11)\times10^{-6}$		4.4×10^{-4}
proton gyromagnetic ratio $2\mu_p/\hbar$	γ_p	$2.675\,221\,900(18)\times10^{8}$	s^{-1}T^{-1}	6.9×10^{-9}
	$\gamma_p/2\pi$	42.577 478 92 (29)	MHz T^{-1}	6.9×10^{-9}
shielded proton gyromagnetic ratio $2\mu_p'/\hbar$ (H_2O, sphere, 25 °C)	γ_p'	$2.675\,153\,171(33)\times10^{8}$	s^{-1}T^{-1}	1.3×10^{-8}
	$\gamma_p'/2\pi$	42.576 385 07 (53)	MHz T^{-1}	1.3×10^{-8}
***Neutron,* n**				
neutron mass	m_n	$1.674\,927\,471(21)\times10^{-27}$	kg	1.2×10^{-8}
		1.008 664 915 88 (49)	u	4.9×10^{-10}
energy equivalent	$m_n c^2$	$1.505\,349\,739(19)\times10^{-10}$	J	1.2×10^{-8}
		939.565 4133 (58)	MeV	6.2×10^{-9}
neutron-electron mass ratio	m_n/m_e	1838.683 661 58 (90)		4.9×10^{-10}
neutron-muon mass ratio	m_n/m_μ	8.892 484 08 (20)		2.2×10^{-8}
neutron-tau mass ratio	m_n/m_τ	0.528 790 (48)		9.0×10^{-5}
neutron-proton mass ratio	m_n/m_p	1.001 378 418 98 (51)		5.1×10^{-10}
neutron-proton mass difference	$m_n - m_p$	$2.305\,573\,77(85)\times10^{-30}$	kg	3.7×10^{-7}
		0.001 388 449 00 (51)	u	3.7×10^{-7}
energy equivalent	$(m_n - m_p)c^2$	$2.072\,146\,37(76)\times10^{-13}$	J	3.7×10^{-7}
		1.293 332 05 (48)	MeV	3.7×10^{-7}
neutron molar mass $N_A m_n$	$M(\mathrm{n})$, M_n	$1.008\,664\,915\,88(49)\times10^{-3}$	kg mol^{-1}	4.9×10^{-10}
neutron Compton wavelength $h/m_n c$	$\lambda_{C,n}$	$1.319\,590\,904\,81(88)\times10^{-15}$	m	6.7×10^{-10}

Quantity	Symbol	Numerical value	Unit	Relative std. uncert. u_r
$\lambda_{C,n}/2\pi$	$ƛ_{C,n}$	$0.210\,019\,415\,36(14)\times10^{-15}$	m	6.7×10^{-10}
neutron magnetic moment	μ_n	$-0.966\,236\,50(23)\times10^{-26}$	J T^{-1}	2.4×10^{-7}
to Bohr magneton ratio	μ_n/μ_B	$-1.041\,875\,63(25)\times10^{-3}$		2.4×10^{-7}
to nuclear magneton ratio	μ_n/μ_N	$-1.913\,042\,73(45)$		2.4×10^{-7}
neutron *g*-factor $2\mu_n/\mu_N$	g_n	$-3.826\,085\,45(90)$		2.4×10^{-7}
neutron-electron magnetic moment ratio	μ_n/μ_e	$1.040\,668\,82(25)\times10^{-3}$		2.4×10^{-7}
neutron-proton magnetic moment ratio	μ_n/μ_p	$-0.684\,979\,34(16)$		2.4×10^{-7}
neutron to shielded proton magnetic moment ratio (H_2O, sphere, 25 °C)	μ_n/μ'_p	$-0.684\,996\,94(16)$		2.4×10^{-7}
neutron gyromagnetic ratio $2\lvert\mu_n\rvert/\hbar$	γ_n	$1.832\,471\,72(43)\times10^{8}$	s^{-1}T^{-1}	2.4×10^{-7}
	$\gamma_n/2\pi$	$29.164\,6933(69)$	MHz T^{-1}	2.4×10^{-7}
***Deuteron*, d**				
deuteron mass	m_d	$3.343\,583\,719(41)\times10^{-27}$	kg	1.2×10^{-8}
		$2.013\,553\,212\,745(40)$	u	2.0×10^{-11}
energy equivalent	$m_d c^2$	$3.005\,063\,183(37)\times10^{-10}$	J	1.2×10^{-8}
		$1875.612\,928(12)$	MeV	6.2×10^{-9}
deuteron-electron mass ratio	m_d/m_e	$3670.482\,967\,85(13)$		3.5×10^{-11}
deuteron-proton mass ratio	m_d/m_p	$1.999\,007\,500\,87(19)$		9.3×10^{-11}
deuteron molar mass $N_A m_d$	$M(d)$, M_d	$2.013\,553\,212\,745(40)\times10^{-3}$	kg mol^{-1}	2.0×10^{-11}
deuteron rms charge radius	r_d	$2.1413(25)\times10^{-15}$	m	1.2×10^{-3}
deuteron magnetic moment	μ_d	$0.433\,073\,5040(36)\times10^{-26}$	J T^{-1}	8.3×10^{-9}
to Bohr magneton ratio	μ_d/μ_B	$0.466\,975\,4554(26)\times10^{-3}$		5.5×10^{-9}
to nuclear magneton ratio	μ_d/μ_N	$0.857\,438\,2311(48)$		5.5×10^{-9}
deuteron *g*-factor μ_d/μ_N	g_d	$0.857\,438\,2311(48)$		5.5×10^{-9}
deuteron-electron magnetic moment ratio	μ_d/μ_e	$-4.664\,345\,535(26)\times10^{-4}$		5.5×10^{-9}
deuteron-proton magnetic moment ratio	μ_d/μ_p	$0.307\,012\,2077(15)$		5.0×10^{-9}
deuteron-neutron magnetic moment ratio	μ_d/μ_n	$-0.448\,206\,52(11)$		2.4×10^{-7}
***Triton*, t**				
triton mass	m_t	$5.007\,356\,665(62)\times10^{-27}$	kg	1.2×10^{-8}
		$3.015\,500\,716\,32(11)$	u	3.6×10^{-11}
energy equivalent	$m_t c^2$	$4.500\,387\,735(55)\times10^{-10}$	J	1.2×10^{-8}
		$2808.921\,112(17)$	MeV	6.2×10^{-9}
triton-electron mass ratio	m_t/m_e	$5496.921\,535\,88(26)$		4.6×10^{-11}
triton-proton mass ratio	m_t/m_p	$2.993\,717\,033\,48(22)$		7.5×10^{-11}
triton molar mass $N_A m_t$	$M(t)$, M_t	$3.015\,500\,716\,32(11)\times10^{-3}$	kg mol^{-1}	3.6×10^{-11}
triton magnetic moment	μ_t	$1.504\,609\,503(12)\times10^{-26}$	J T^{-1}	7.8×10^{-9}
to Bohr magneton ratio	μ_t/μ_B	$1.622\,393\,6616(76)\times10^{-3}$		4.7×10^{-9}
to nuclear magneton ratio	μ_t/μ_N	$2.978\,962\,460(14)$		4.7×10^{-9}
triton *g*-factor $2\mu_t/\mu_N$	g_t	$5.957\,924\,920(28)$		4.7×10^{-9}
***Helion*, h**				
helion mass	m_h	$5.006\,412\,700(62)\times10^{-27}$	kg	1.2×10^{-8}
		$3.014\,932\,246\,73(12)$	u	3.9×10^{-11}
energy equivalent	$m_h c^2$	$4.499\,539\,341(55)\times10^{-10}$	J	1.2×10^{-8}
		$2808.391\,586(17)$	MeV	6.2×10^{-9}
helion-electron mass ratio	m_h/m_e	$5495.885\,279\,22(27)$		4.9×10^{-11}
helion-proton mass ratio	m_h/m_p	$2.993\,152\,670\,46(29)$		9.6×10^{-11}
helion molar mass $N_A m_h$	$M(h)$, M_h	$3.014\,932\,246\,73(12)\times10^{-3}$	kg mol^{-1}	3.9×10^{-11}
helion magnetic moment	μ_h	$-1.074\,617\,522(14)\times10^{-26}$	J T^{-1}	1.3×10^{-8}
to Bohr magneton ratio	μ_h/μ_B	$-1.158\,740\,958(14)\times10^{-3}$		1.2×10^{-8}
to nuclear magneton ratio	μ_h/μ_N	$-2.127\,625\,308(25)$		1.2×10^{-8}
helion *g*-factor $2\mu_h/\mu_N$	g_h	$-4.255\,250\,616(50)$		1.2×10^{-8}
shielded helion magnetic moment (gas, sphere, 25 °C)	μ'_h	$-1.074\,553\,080(14)\times10^{-26}$	J T^{-1}	1.3×10^{-8}
to Bohr magneton ratio	μ'_h/μ_B	$-1.158\,671\,471(14)\times10^{-3}$		1.2×10^{-8}
to nuclear magneton ratio	μ'_h/μ_N	$-2.127\,497\,720(25)$		1.2×10^{-8}

Quantity	Symbol	Numerical value	Unit	Relative std. uncert. u_r
shielded helion to proton magnetic moment ratio (gas, sphere, 25 °C)	μ'_h/μ_p	-0.761 766 5603 (92)		1.2×10^{-8}
shielded helion to shielded proton magnetic moment ratio (gas/H_2O, spheres, 25 °C)	μ'_h/μ'_p	-0.761 786 1313 (33)		4.3×10^{-9}
shielded helion gyromagnetic ratio $2\|\mu'_h\|/\hbar$ (gas, sphere, 25 °C)	γ'_h	$2.037\,894\,585(27)\times10^{8}$	$s^{-1}\,T^{-1}$	1.3×10^{-8}
	$\gamma'_h/2\pi$	32.434 099 66 (43)	MHz T^{-1}	1.3×10^{-8}
Alpha particle, α				
alpha particle mass	m_α	$6.644\,657\,230(82)\times10^{-27}$	kg	1.2×10^{-8}
		4.001 506 179 127 (63)	u	1.6×10^{-11}
energy equivalent	$m_\alpha c^2$	$5.971\,920\,097(73)\times10^{-10}$	J	1.2×10^{-8}
		3727.379 378 (23)	MeV	6.2×10^{-9}
alpha particle to electron mass ratio	m_α/m_e	7294.299 541 36 (24)		3.3×10^{-11}
alpha particle to proton mass ratio	m_α/m_p	3.972 599 689 07 (36)		9.2×10^{-11}
alpha particle molar mass $N_A m_\alpha$	$M(\alpha)$, M_α	$4.001\,506\,179\,127(63)\times10^{-3}$	kg mol^{-1}	1.6×10^{-11}
Physicochemical				
Avogadro constant	N_A, L	$6.022\,140\,857(74)\times10^{23}$	mol^{-1}	1.2×10^{-8}
atomic mass constant $m_u = \frac{1}{12}m(^{12}C) = 1$ u	m_u	$1.660\,539\,040(20)\times10^{-27}$	kg	1.2×10^{-8}
energy equivalent	$m_u c^2$	$1.492\,418\,062(18)\times10^{-10}$	J	1.2×10^{-8}
		931.494 0954 (57)	MeV	6.2×10^{-9}
Faraday constant[6] $N_A e$	F	96 485.332 89 (59)	C mol^{-1}	6.2×10^{-9}
molar Planck constant	$N_A h$	$3.990\,312\,7110(18)\times10^{-10}$	J s mol^{-1}	4.5×10^{-10}
	$N_A hc$	0.119 626 565 582 (54)	J m mol^{-1}	4.5×10^{-10}
molar gas constant	R	8.314 4598 (48)	J mol^{-1} K^{-1}	5.7×10^{-7}
Boltzmann constant R/N_A	k	$1.380\,648\,52(79)\times10^{-23}$	J K^{-1}	5.7×10^{-7}
		$8.617\,3303(50)\times10^{-5}$	eV K^{-1}	5.7×10^{-7}
	k/h	$2.083\,6612(12)\times10^{10}$	Hz K^{-1}	5.7×10^{-7}
	k/hc	69.503 457 (40)	m^{-1} K^{-1}	5.7×10^{-7}
molar volume of ideal gas RT/p [T = 273.15 K, p = 100 kPa]	V_m	$22.710\,947(13)\times10^{-3}$	m^3 mol^{-1}	5.7×10^{-7}
Loschmidt constant N_A/V_m	n_0	$2.651\,6467(15)\times10^{25}$	m^{-3}	5.7×10^{-7}
molar volume of ideal gas RT/p [T = 273.15 K, p = 101.325 kPa]	V_m	$22.413\,962(13)\times10^{-3}$	m^3 mol^{-1}	5.7×10^{-7}
Loschmidt constant N_A/V_m	n_0	$2.686\,7811(15)\times10^{25}$	m^{-3}	5.7×10^{-7}
Sackur-Tetrode (absolute entropy) constant[7] $\frac{5}{2}+\ln[(2\pi m_u kT_1/h^2)^{3/2}kT_1/p_0]$				
T_1 = 1 K, p_0 = 100 kPa	S_0/R	-1.151 7084 (14)		1.2×10^{-6}
T_1 = 1 K, p_0 = 101.325 kPa		-1.164 8714 (14)		1.2×10^{-6}
Stefan-Boltzmann constant $(\pi^2/60)k^4/\hbar^3c^2$	σ	$5.670\,367(13)\times10^{-8}$	W m^{-2} K^{-4}	2.3×10^{-6}
first radiation constant $2\pi hc^2$	c_1	$3.741\,771\,790(46)\times10^{-16}$	W m^2	1.2×10^{-8}
first radiation constant for spectral radiance $2hc^2$	c_{1L}	$1.191\,042\,953(15)\times10^{-16}$	W m^2 sr^{-1}	1.2×10^{-8}
second radiation constant hc/k	c_2	$1.438\,777\,36(83)\times10^{-2}$	m K	5.7×10^{-7}
Wien displacement law constants:				
$b = \lambda_{max}T = c_2/4.965\,114\,231...$	b	$2.897\,7729(17)\times10^{-3}$	m K	5.7×10^{-7}
$b' = \nu_{max}/T = 2.821\,439\,372...\,c/c_2$	b'	$5.878\,9238(34)\times10^{10}$	Hz K^{-1}	5.7×10^{-7}

1 See Table IV for the conventional value adopted internationally for realizing representations of the volt using the Josephson effect.
2 See Table IV for the conventional value adopted internationally for realizing representations of the volt using the quantum Hall effect.
3 Value recommended by the Particle Data Group (Olive et al., 2014).
4 Based on the ratio of the masses of the W and Z bosons m_W/m_Z recommended by the Particle Data Group (Olive et al., 2014). The value for $\sin^2\theta_W$ they recommend, which is based on a particular variant of the modified minimal subtraction (MS) scheme, is $\sin^2\hat{\theta}_W(M_Z) = 0.231\,26(5)$.
5 This and all other values involving m_τ are based on the value of $m_\tau c^2$ in MeV recommended by the Particle Data Group (Olive et al., 2014).
6 The numerical value of F to be used in coulometric chemical measurements is 96 485.3251(12) [1.2×10^{-8}] when the relevant current is measured in terms of representations of the volt and ohm based on the Josephson and quantum Hall effects and the internationally adopted conventional values of the Josephson and von Klitzing constants K_{J-90} and R_{K-90} given in Table IV. 7.
7 The entropy of an ideal monoatomic gas of relative atomic mass A_r is given by $S = S_0 + 3/2\,R\ln A_r - R\ln(p/p_0) + 5/2\,R\ln(T/K)$.

Reference

Olive, K. A., et al., and Particle Data Center, 2014, *Chin. Phys. C* 38, 090001.

Table III. The Variances, Covariances, and Correlation Coefficients of the Values of a Selected Group of Constants Based on the 2014 CODATA Adjustment. The Numbers in Bold above the Main Diagonal Are 10^{16} Times the Numerical Values of the Relative Covariances; the Numbers in Bold on the Main Diagonal Are 10^{16} Times the Numerical Values of the Relative Variances; and the Numbers in Italics below the Main Diagonal Are the Correlation Coefficients[1]

	α	h	e	m_e	N_A	m_e/m_μ	F
α	**0.0005**	**0.0005**	**0.0005**	**-0.0005**	**0.0005**	**-0.0010**	**0.0010**
h	*0.0176*	**1.5096**	**0.7550**	**1.5086**	**-1.5086**	**-0.0010**	**-0.7536**
e	*0.0361*	*0.9998*	**0.3778**	**0.7540**	**-0.7540**	**-0.0010**	**-0.3763**
m_e	*-0.0193*	*0.9993*	*0.9985*	**1.5097**	**-1.5097**	**0.0011**	**-0.7556**
N_A	*0.0193*	*-0.9993*	*-0.9985*	*-1.0000*	**1.5097**	**-0.0011**	**0.7557**
m_e/m_μ	*-0.0202*	*-0.0004*	*-0.0007*	*0.0004*	*-0.0004*	**4.9471**	**-0.0021**
F	*0.0745*	*-0.9957*	*-0.9939*	*-0.9985*	*0.9985*	*-0.0015*	**0.3794**

[1] The relative covariance is $u_r(x_i, x_j) = u(x_i, x_j)/(x_i x_j)$, where $u(x_i, x_j)$ is the covariance of x_i and x_j; the relative variance is $u_r^2(x_i) = u_r(x_i, x_i)$; and the correlation coefficient is $r(x_i, x_j) = u(x_i, x_j)/[u(x_i)u(x_j)]$.

Table IV. Internationally Adopted Values of Various Quantities

Quantity	Symbol	Numerical value	Unit	Relative std. uncert. u_r
relative atomic mass[1] of ^{12}C	$A_r(^{12}C)$	12		exact
molar mass constant	M_u	1×10^{-3}	kg mol^{-1}	exact
molar mass of ^{12}C	$M(^{12}C)$	12×10^{-3}	kg mol^{-1}	exact
conventional value of Josephson constant[2]	K_{J-90}	483 597.9	GHz V^{-1}	exact
conventional value of von Klitzing constant[3]	R_{K-90}	25 812.807	Ω	exact
standard-state pressure		100	kPa	exact
standard atmosphere		101.325	kPa	exact

[1] The relative atomic mass $A_r(X)$ of particle X with mass $m(X)$ is defined by $A_r(X) = m(X)/m_u$, where $m_u = m(^{12}C)/12 = M_u/N_A = 1$ u is the atomic mass constant, M_u is the molar mass constant, N_A is the Avogadro constant, and u is the unified atomic mass unit. Thus the mass of particle X is $m(X) = A_r(X)$ u and the molar mass of X is $M(X) = A_r(X)M_u$.
[2] This is the value adopted internationally for realizing representations of the volt using the Josephson effect.
[3] This is the value adopted internationally for realizing representations of the ohm using the quantum Hall effect.

Table V. Values of Some X-ray-Related Quantities Based on the 2014 CODATA Adjustment of the Values of the Constants

Quantity	Symbol	Numerical value	Unit	Relative std. uncert. u_r
Cu x unit: $\lambda(\mathrm{CuK}\alpha_1)/1\,537.400$	xu$(\mathrm{CuK}\alpha_1)$	$1.002\,076\,97(28)\times10^{-13}$	m	2.8×10^{-7}
Mo x unit: $\lambda(\mathrm{MoK}\alpha_1)/707.831$	xu$(\mathrm{MoK}\alpha_1)$	$1.002\,099\,52(53)\times10^{-13}$	m	5.3×10^{-7}
ångstrom star: $\lambda(\mathrm{WK}\alpha_1)/0.209\,010\,0$	Å*	$1.000\,014\,95(90)\times10^{-10}$	m	9.0×10^{-7}
lattice parameter[1] of Si (in vacuum, 22.5 °C)	a	$543.102\,0504(89)\times10^{-12}$	m	1.6×10^{-8}
{220} lattice spacing of Si $a/\sqrt{8}$ (in vacuum, 22.5 °C)	d_{220}	$192.015\,5714(32)\times10^{-12}$	m	1.6×10^{-8}
molar volume of Si $M(\mathrm{Si})/\rho(\mathrm{Si}) = N_A a^3/8$ (in vacuum, 22.5 °C)	$V_m(\mathrm{Si})$	$12.058\,832\,14(61)\times10^{-6}$	m^3 mol^{-1}	5.1×10^{-8}

[1] This is the lattice parameter (unit cell edge length) of an ideal single crystal of naturally occurring Si free of impurities and imperfections, and is deduced from measurements on extremely pure and nearly perfect single crystals of Si by correcting for the effects of impurities.

Table VI. The Values in SI Units of Some Non-SI Units Based on the 2014 CODATA Adjustment of the Values of the Constants

Quantity	Symbol	Numerical value	Unit	Relative std. uncert. u_r
Non-SI units accepted for use with the SI				
electron volt: (e/C) J	eV	$1.602\,176\,6208(98)\times10^{-19}$	J	6.1×10^{-9}
(unified) atomic mass unit: $\frac{1}{12}m(^{12}C)$	u	$1.660\,539\,040(20)\times10^{-27}$	kg	1.2×10^{-8}
Natural units (n.u.)				
n.u. of velocity	c, c_0	299 792 458	m s^{-1}	exact
n.u. of action: $h/2\pi$	$\hbar$	$1.054\,571\,800(13)\times10^{-34}$	J s	1.2×10^{-8}

Quantity	Symbol	Numerical value	Unit	Relative std. uncert. u_r
		6.582 119 514 (40)×10^{-16}	eV s	6.1×10^{-9}
	$\hbar c$	197.326 9788 (12)	MeV fm	6.1×10^{-9}
n.u. of mass	m_e	9.109 383 56 (11)×10^{-31}	kg	1.2×10^{-8}
n.u. of energy	$m_e c^2$	8.187 105 65 (10)×10^{-14}	J	1.2×10^{-8}
		0.510 998 9461 (31)	MeV	6.2×10^{-9}
n.u. of momentum	$m_e c$	2.730 924 488 (34)×10^{-22}	kg m s^{-1}	1.2×10^{-8}
		0.510 998 9461 (31)	MeV/c	6.2×10^{-9}
n.u. of length: $\hbar/m_e c$	$\bar{\lambda}_C$	386.159 267 64 (18)×10^{-15}	m	4.5×10^{-10}
n.u. of time	$\hbar/m_e c^2$	1.288 088 667 12 (58)×10^{-21}	s	4.5×10^{-10}
Atomic units (a.u.)				
a.u. of charge	e	1.602 176 6208 (98)×10^{-19}	C	6.1×10^{-9}
a.u. of mass	m_e	9.109 383 56 (11)×10^{-31}	kg	1.2×10^{-8}
a.u. of action: $h/2\pi$	$\hbar$	1.054 571 800 (13)×10^{-34}	J s	1.2×10^{-8}
a.u. of length: Bohr radius (bohr) $\alpha/4\pi R_\infty$	a_0	0.529 177 210 67 (12)×10^{-10}	m	2.3×10^{-10}
a.u. of energy: Hartree energy (hartree) $e^2/4\pi\epsilon_0 a_0 = 2R_\infty hc = \alpha^2 m_e c^2$	E_h	4.359 744 650 (54)×10^{-18}	J	1.2×10^{-8}
a.u. of time	$\hbar/E_h$	2.418 884 326 509 (14)×10^{-17}	s	5.9×10^{-12}
a.u. of force	E_h/a_0	8.238 723 36 (10)×10^{-8}	N	1.2×10^{-8}
a.u. of velocity: αc	$a_0 E_h/\hbar$	2.187 691 262 77 (50)×10^{6}	m s^{-1}	2.3×10^{-10}
a.u. of momentum	$\hbar/a_0$	1.992 851 882 (24)×10^{-24}	kg m s^{-1}	1.2×10^{-8}
a.u. of current	$e E_h/\hbar$	6.623 618 183 (41)×10^{-3}	A	6.1×10^{-9}
a.u. of charge density	e/a_0^3	1.081 202 3770 (67)×10^{12}	C m^{-3}	6.2×10^{-9}
a.u. of electric potential	E_h/e	27.211 386 02 (17)	V	6.1×10^{-9}
a.u. of electric field	$E_h/e a_0$	5.142 206 707 (32)×10^{11}	V m^{-1}	6.1×10^{-9}
a.u. of electric field gradient	$E_h/e a_0^2$	9.717 362 356 (60)×10^{21}	V m^{-2}	6.2×10^{-9}
a.u. of electric dipole moment	$e a_0$	8.478 353 552 (52)×10^{-30}	C m	6.2×10^{-9}
a.u. of electric quadrupole moment	$e a_0^2$	4.486 551 484 (28)×10^{-40}	C m^2	6.2×10^{-9}
a.u. of electric polarizability	$e^2 a_0^2/E_h$	1.648 777 2731 (11)×10^{-41}	C^2 m^2 J^{-1}	6.8×10^{-10}
a.u. of 1*st* hyperpolarizability	$e^3 a_0^3/E_h^2$	3.206 361 329 (20)×10^{-53}	C^3 m^3 J^{-2}	6.2×10^{-9}
a.u. of 2*nd* hyperpolarizability	$e^4 a_0^4/E_h^3$	6.235 380 085 (77)×10^{-65}	C^4 m^4 J^{-3}	1.2×10^{-8}
a.u. of magnetic flux density	$\hbar/e a_0^2$	2.350 517 550 (14)×10^{5}	T	6.2×10^{-9}
a.u. of magnetic dipole moment: $2\mu_B$	$\hbar e/m_e$	1.854 801 999 (11)×10^{-23}	J T^{-1}	6.2×10^{-9}
a.u. of magnetizability	$e^2 a_0^2/m_e$	7.891 036 5886 (90)×10^{-29}	J T^{-2}	1.1×10^{-9}
a.u. of permittivity: $10^7/c^2$	$e^2/a_0 E_h$	1.112 650 056…×10^{-10}	F m^{-1}	exact

Table VII. The Values of Some Energy Equivalents Derived from the Relations $E = mc^2 = hc/\lambda = h\nu = kT$, and Based on the 2010 CODATA Adjustment of the Values of the Constants; 1 eV = (e/C) J, 1 u = $m_u = (1/12)m(^{12}\mathrm{C}) = 10^{-3}$ kg mol$^{-1}/N_A$, and $E_h = 2R_\infty hc = \alpha^2 m_e c^2$ Is the Hartree Energy (hartree)

	J	kg	m^{-1}	Hz
1 J	(1 J) = 1 J	(1 J)/c^2 = 1.112 650 056…×10^{-17} kg	(1 J)/hc = 5.034 116 651 (62)×10^{24} m^{-1}	(1 J)/h = 1.509 190 205 (19)×10^{33} Hz
1 kg	(1 kg)c^2 = 8.987 551 787…×10^{16} J	(1 kg) 1 kg	(1 kg)c/h = 4.524 438 411 (56)×10^{41} m^{-1}	(1 kg)c^2/h = 1.356 392 512 (17)×10^{50} Hz
1 m^{-1}	(1 m^{-1})hc = 1.986 445 824 (24)×10^{-25} J	(1 m^{-1})h/c = 2.210 219 057 (27)×10^{-42} kg	(1 m^{-1}) = 1 m^{-1}	(1 m^{-1})c = 299 792 458 Hz
1 Hz	(1 Hz)h = 6.626 070 040 (81)×10^{-34} J	(1 Hz)h/c^2 = 7.372 497 201 (91)×10^{-51} kg	(1 Hz)/c = 3.335 640 951…×10^{-9} m^{-1}	(1 Hz) = 1 Hz
1 K	(1 K)k = 1.380 648 52 (79)×10^{-23} J	(1 K)k/c^2 = 1.536 178 65 (88)×10^{-40} kg	(1 K)k/hc = 69.503 457 (40) m^{-1}	(1 K)k/h = 2.083 6612 (12)×10^{10} Hz
1 eV	(1 eV) = 1.602 176 6208 (98)×10^{-19} J	(1 eV)/c^2 = 1.782 661 907 (11)×10^{-36} kg	(1 eV)/hc = 8.065 544 005 (50)×10^{5} m^{-1}	(1 eV)/h = 2.417 989 262 (15)×10^{14} Hz
1 u	(1 u)c^2 = 1.492 418 062 (18)×10^{-10} J	(1 u) = 1.660 539 040 (20)×10^{-27} kg	(1 u)c/h = 7.513 006 6166 (34)×10^{14} m^{-1}	(1 u)c^2/h = 2.252 342 7206 (10)×10^{23} Hz
1 E_h	(1 E_h) = 4.359 744 650 (54)×10^{-18} J	(1 E_h)/c^2 = 4.850 870 129 (60)×10^{-35} kg	(1 E_h)/hc = 2.194 746 313 702 (13)×10^{7} m^{-1}	(1 E_h)/h = 6.579 683 920 711 (39)×10^{15} Hz

Table VIII. The Values of Some Energy Equivalents Derived from the Relations $E = mc^2 = hc/\lambda = h\nu = kT$, and Based on the 2010 CODATA Adjustment of the Values of the Constants; 1 eV = (e/C) J, 1 u = $m_u = (1/12)m(^{12}\mathrm{C}) = 10^{-3}$ kg mol$^{-1}/N_A$, and $E_h = 2R_\infty hc = \alpha^2 m_e c^2$ Is the Hartree Energy (hartree)

	K	eV	u	E_h
1 J	$(1\ \mathrm{J})/k$ = 7.242 9731 (42)×10^{22} K	$(1\ \mathrm{J})$ = 6.241 509 126 (38)×10^{18} eV	$(1\ \mathrm{J})/c^2$ = 6.700 535 363 (82)×10^{9} u	$(1\ \mathrm{J})$ = 2.293 712 317 (28)×10^{17} E_h
1 kg	$(1\ \mathrm{kg})c^2/k$ = 6.509 6595 (37)×10^{39} K	$(1\ \mathrm{kg})c^2$ = 5.609 588 650 (34)×10^{35} eV	$(1\ \mathrm{kg})$ = 6.022 140 857 (74)×10^{26} u	$(1\ \mathrm{kg})c^2$ = 2.061 485 823 (25)×10^{34} E_h
1 m^{-1}	$(1\ \mathrm{m}^{-1})hc/k$ = 1.438 777 36 (83)×10^{-2} K	$(1\ \mathrm{m}^{-1})hc$ = 1.239 841 9739 (76)×10^{-6} eV	$(1\ \mathrm{m}^{-1})h/c$ = 1.331 025 049 00 (61)×10^{-15} u	$(1\ \mathrm{m}^{-1})hc$ = 4.556 335 252 767 (27)×10^{-8} E_h
1 Hz	$(1\ \mathrm{Hz})h/k$ = 4.799 2447 (28)×10^{-11} K	$(1\ \mathrm{Hz})h$ = 4.135 667 662 (25)×10^{-15} eV	$(1\ \mathrm{Hz})h/c^2$ = 4.439 821 6616 (20)×10^{-24} u	$(1\ \mathrm{Hz})h$ = 1.519 829 846 0088 (90)×10^{-16} E_h
1 K	$(1\ \mathrm{K})$ = 1 K	$(1\ \mathrm{K})k$ = 8.617 3303 (50)×10^{-5} eV	$(1\ \mathrm{K})k/c^2$ = 9.251 0842 (53)×10^{-14} u	$(1\ \mathrm{K})k$ = 3.166 8105 (18)×10^{-6} E_h
1 eV	$(1\ \mathrm{eV})/k$ = 1.160 452 21 (67)×10^{4} K	$(1\ \mathrm{eV})$ = 1 eV	$(1\ \mathrm{eV})/c^2$ = 1.073 544 1105 (66)×10^{-9} u	$(1\ \mathrm{eV})$ = 3.674 932 248 (23)×10^{-2} E_h
1 u	$(1\ \mathrm{u})c^2/k$ = 1.080 954 38 (62)×10^{13} K	$(1\ \mathrm{u})c^2$ = 931.494 0954 (57)×10^{6} eV	$(1\ \mathrm{u})$ = 1 u	$(1\ \mathrm{u})c^2$ = 3.423 177 6902 (16)×10^{7} E_h
1 E_h	$(1\ E_h)/k$ = 3.157 7513 (18)×10^{5} K	$(1\ E_h)$ = 27.211 386 02 (17) eV	$(1\ E_h)/c^2$ = 2.921 262 3197 (13)×10^{-8} u	$(1\ E_h)$ = 1 E_h

STANDARD ATOMIC WEIGHTS

This table of atomic weights includes the changes made through 2016 by the International Union of Pure and Applied Chemistry (IUPAC) Commission on Isotopic Abundances and Atomic Weights (Ref. 1,5,6,7).

IUPAC made a significant policy change in 2009 (Ref. 2-4). Each atomic weight had previously been given as a single value with an uncertainty that took into account both the measurement uncertainty and the variation in isotopic abundance in samples of the element from different terrestrial sources. For a variety of reasons (Ref. 3), this fails to give complete information on the natural variability in isotopic abundance of several elements. Therefore, beginning in 2009 the recommendations expressed the atomic weights of several elements as intervals rather than single numbers plus uncertainties. The symbol for these intervals is [*a*, *b*], where *a* is the lower bound of values found in normal materials and *b* the upper bound. For the other elements in the table, a single recommended atomic weight value is given; the number in parentheses following the value gives the uncertainty in the last digit.

Table 1 gives the atomic weights of the elements listed in alphabetical order by name. Table 2 gives reference atomic weights for the 11 elements whose entries in Table 1 are intervals rather than single numbers. These conventional reference values are suggested for use on samples of unspecified origin and for calculation of molecular weights in tables intended to be broadly applicable. They have been selected such that most or all natural terrestrial atomic-weight variation is covered in an interval of plus or minus one in the last digit. It should be emphasized that the conventional values are not simply midpoints of the intervals, but rather represent the best judgment of the data evaluators.

References

1. Wieser, M. E., et al., *Pure Appl. Chem.* 85, 1047, 2013.
2. Wieser, M. E., and Coplen, T. D., *Pure Appl. Chem.* 83, 359, 2011.
3. Coplen, T. B., and Holden, N. E., *Chemistry International*, Vol. 33, No. 2, p. 10, 2011.
4. Berglund, M., and Wieser, M. E., *Pure Appl. Chem.* 83, 397, 2011.
5. *Chemistry International*, Vol. 35, No. 6, p. 17, 2013; <www.ciaaw.org>.
6. *Chemistry International*, Vol. 37, No. 5-6, p. 26, 2015.
7. Meija, J., et al., *Pure Appl. Chem.* 88, 265, 2016.

Table 1. Standard Atomic Weights 2013

Element	Symbol	Atomic number	Atomic weight
Actinium[u]	Ac	89	
Aluminum	Al	13	26.9815385(7)
Americium[u]	Am	95	
Antimony[g]	Sb	51	121.760(1)
Argon[g, r]	Ar	18	39.948(1)
Arsenic	As	33	74.921595(6)
Astatine[u]	At	85	
Barium	Ba	56	137.327(7)
Berkelium[u]	Bk	97	
Beryllium	Be	4	9.0121831(5)
Bismuth[u]	Bi	83	208.98040(1)
Bohrium[u]	Bh	107	
Boron[m, *]	B	5	[10.806, 10.821]
Bromine[*]	Br	35	[79.901, 79.907]
Cadmium[g]	Cd	48	112.414(4)
Calcium[g]	Ca	20	40.078(4)
Californium[u]	Cf	98	
Carbon[*]	C	6	[12.0096, 12.0116]
Cerium[g]	Ce	58	140.116(1)
Cesium[g]	Cs	55	132.90545196(6)
Chlorine[m, *]	Cl	17	[35.446, 35.457]
Chromium	Cr	24	51.9961(6)
Cobalt	Co	27	58.933194(4)
Copernicium[u]	Cn	112	
Copper[r]	Cu	29	63.546(3)
Curium[u]	Cm	96	
Darmstadtium[u]	Ds	110	
Dubnium[u]	Db	105	
Dysprosium[g]	Dy	66	162.500(1)
Einsteinium[u]	Es	99	
Erbium[g]	Er	68	167.259(3)
Europium[g]	Eu	63	151.964(1)
Fermium[u]	Fm	100	
Flerovium[u]	Fl	114	
Fluorine	F	9	18.998403163(6)
Francium[u]	Fr	87	
Gadolinium[g]	Gd	64	157.25(3)
Gallium	Ga	31	69.723(1)
Germanium	Ge	32	72.630(8)
Gold	Au	79	196.966569(5)
Hafnium	Hf	72	178.49(2)
Hassium[u]	Hs	108	
Helium[g, r]	He	2	4.002602(2)
Holmium	Ho	67	164.93033(2)
Hydrogen[m, *]	H	1	[1.00784, 1.00811]
Indium	In	49	114.818(1)
Iodine	I	53	126.90447(3)
Iridium	Ir	77	192.217(3)
Iron	Fe	26	55.845(2)
Krypton[g, m]	Kr	36	83.798(2)
Lanthanum[g]	La	57	138.90547(7)
Lawrencium[u]	Lr	103	
Lead[g, r]	Pb	82	207.2(1)
Lithium[m, *]	Li	3	[6.938, 6.997]
Livermorium[u]	Lv	116	
Lutetium[g]	Lu	71	174.9668(1)
Magnesium[*]	Mg	12	[24.304, 24.307]
Manganese	Mn	25	54.938044(3)
Meitnerium[u]	Mt	109	
Mendelevium[u]	Md	101	
Mercury	Hg	80	200.592(3)
Molybdenum[g]	Mo	42	95.95(1)
Moscovium[u]	Mc	115	
Neodymium[g]	Nd	60	144.242(3)
Neon[g, m]	Ne	10	20.1797(6)
Neptunium[u]	Np	93	

Element	Symbol	Atomic number	Atomic weight
Nickel	Ni	28	58.6934(4)
Nihonium[u]	Nh	113	
Niobium	Nb	41	92.90637(2)
Nitrogen*	N	7	[14.00643, 14.00728]
Nobelium[u]	No	102	
Oganesson[u]	Og	118	
Osmium[g]	Os	76	190.23(3)
Oxygen*	O	8	[15.99903, 15.99977]
Palladium[g]	Pd	46	106.42(1)
Phosphorus	P	15	30.973761998(5)
Platinum	Pt	78	195.084(9)
Plutonium[u]	Pu	94	
Polonium[u]	Po	84	
Potassium[g]	K	19	39.0983(1)
Praseodymium	Pr	59	140.90766(2)
Promethium[u]	Pm	61	
Protactinium[z]	Pa	91	231.03588(2)
Radium[u]	Ra	88	
Radon[u]	Rn	86	
Rhenium	Re	75	186.207(1)
Rhodium	Rh	45	102.90550(2)
Roentgenium[u]	Rg	111	
Rubidium[g]	Rb	37	85.4678(3)
Ruthenium[g]	Ru	44	101.07(2)
Rutherfordium[u]	Rf	104	
Samarium[g]	Sm	62	150.36(2)
Scandium	Sc	21	44.955908(5)
Seaborgium[u]	Sg	106	
Selenium	Se	34	78.971(8)
Silicon*	Si	14	[28.084, 28.086]
Silver[g]	Ag	47	107.8682(2)
Sodium	Na	11	22.98976928(2)
Strontium[g, r]	Sr	38	87.62(1)
Sulfur*	S	16	[32.059, 32.076]
Tantalum	Ta	73	180.94788(2)
Technetium[u]	Tc	43	
Tellurium[g]	Te	52	127.60(3)
Tennessine[u]	Ts	117	
Terbium	Tb	65	158.92535(2)
Thallium*	Tl	81	[204.382, 204.385]
Thorium[g, u]	Th	90	232.0377(4)
Thulium	Tm	69	168.93422(2)
Tin[g]	Sn	50	118.710(7)
Titanium	Ti	22	47.867(1)
Tungsten	W	74	183.84(1)
Uranium[g, m, u]	U	92	238.02891(3)
Vanadium	V	23	50.9415(1)
Xenon[g, m]	Xe	54	131.293(6)
Ytterbium[g]	Yb	70	173.045(10)
Yttrium	Y	39	88.90584(2)
Zinc[r]	Zn	30	65.38(2)
Zirconium[g]	Zr	40	91.224(2)

[g] Geological specimens are known in which the element has an isotopic composition outside the limits for the normal material. The difference between the atomic weight of the element in such specimens and that given in the table may exceed the stated uncertainty.

[m] Modified isotopic compositions may be found in commercially available material because the material has been subjected to an undisclosed or inadvertent isotopic fractionation. Substantial deviations in atomic weight of the element from that given in the table can occur.

[r] Range in isotopic composition of normal terrestrial material prevents a more precise atomic weight being given; the tabulated value and uncertainty should be applicable to any normal material.

[u] Element has no stable isotopes. See "Table of the Isotopes" in Section 11 for individual isotopic masses. However, four such elements (Bi, Th, Pa, and U) do have a characteristic terrestrial isotopic composition, and for these elements standard atomic weights are tabulated.

* See Table 2

Table 2. Conventional Atomic Weights 2013

Element	Symbol	Atomic number	Reference atomic weight[a]
Boron	B	5	10.81
Bromine	Br	35	79.904
Carbon	C	6	12.011
Chlorine	Cl	17	35.45
Hydrogen	H	1	1.008
Lithium	Li	3	6.94
Magnesium	Mg	12	24.305
Nitrogen	N	7	14.007
Oxygen	O	8	15.999
Silicon	Si	14	28.085
Sulfur	S	16	32.06
Thallium	Tl	81	204.38

[a] For users needing an atomic-weight value for an unspecified sample, such as for trade or commerce. See text.

ATOMIC MASSES AND ABUNDANCES

This table lists the mass (in atomic mass units, symbol u) and the natural abundance (in percent) of the stable nuclides and a few important radioactive nuclides. The atomic masses were taken from the AME 2012 evaluation of the Atomic Mass Data Center, now located at the Institute of Modern Physics in Lanzhou, China (Ref. 1,2). The number in parentheses following the mass value is the uncertainty in the last digit(s) given. The mass values for elements with $Z = 102$ and higher were derived from a combination of experimental data and systematic trends.

Natural abundance values were taken from the IUPAC Technical Report "Isotopic compositions of the elements 2009" (Ref. 3); these entries are also followed by uncertainties in the last digit(s) of the stated values. This uncertainty includes both the estimated measurement uncertainty and the reported range of variation in different terrestrial sources of the element (see Ref. 3 for full details and caveats regarding elements whose abundance is variable). The absence of an entry in the Abundance column indicates a radioactive nuclide not present in nature or an element whose isotopic composition varies so widely that a meaningful natural abundance cannot be defined.

Reference 1 contains mass data on over 3000 nuclides and describes the evaluation procedure in detail. Masses and other properties of nuclides may also be found in Section 11, "Table of the Isotopes" (Ref. 4).

References

1. Wang, M., Audi, G., Wapstra, A. H., Kondev, F. G., MacCormick, M., Xu, X., and Pfeiffer, B., *Chinese Physics* C 36, 1603-2014, 2012.
2. <http//amdc.impcas.ac.cn>
3. Berglund, M., and Wieser, M. E., *Pure Appl. Chem.* 83, 397, 2011.
4. Holden, N. E., "Table of the Isotopes", in Haynes, W. M., Ed., *CRC Handbook of Chemistry and Physics, 97th Ed.*, CRC Press, Boca Raton, FL, 2016.

Z	Isotope	Mass in u	Abundance in %
1	^{1}H	1.00782503223(9)	99.9885(70)
	^{2}H	2.01410177812(12)	0.0115(70)
	^{3}H	3.0160492779(24)	
2	^{3}He	3.0160293201(25)	0.000134(3)
	^{4}He	4.00260325413(6)	99.999866(3)
3	^{6}Li	6.0151228874(15)	7.59(4)
	^{7}Li	7.016003437(5)	92.41(4)
4	^{9}Be	9.01218307(8)	100
5	^{10}B	10.0129369(4)	19.9(7)
	^{11}B	11.0093054(4)	80.1(7)
6	^{11}C	11.0114336(10)	
	^{12}C	12.0000000(0)	98.93(8)
	^{13}C	13.00335483507(23)	1.07(8)
	^{14}C	14.003241988(4)	
7	^{14}N	14.00307400443(20)	99.636(20)
	^{15}N	15.0001088989(6)	0.364(20)
8	^{16}O	15.99491461957(17)	99.757(16)
	^{17}O	16.9991317565(7)	0.038(1)
	^{18}O	17.9991596129(8)	0.205(14)
9	^{18}F	18.0009373(5)	
	^{19}F	18.9984031627(9)	100
10	^{20}Ne	19.9924401762(17)	90.48(3)
	^{21}Ne	20.99384669(4)	0.27(1)
	^{22}Ne	21.991385114(18)	9.25(3)
11	^{22}Na	21.99443741(18)	
	^{23}Na	22.9897692820(19)	100
	^{24}Na	23.99096295(4)	
12	^{24}Mg	23.985041697(14)	78.951(12)
	^{25}Mg	24.98583698(5)	10.020(8)
	^{26}Mg	25.98259297(3)	11.029(10)
13	^{27}Al	26.98153853(11)	100
14	^{28}Si	27.9769265346(4)	92.223(19)
	^{29}Si	28.9764946649(5)	4.685(8)
	^{30}Si	29.973770136(23)	3.092(11)
15	^{31}P	30.9737619984(7)	100
	^{32}P	31.97390764(4)	
16	^{32}S	31.9720711744(14)	94.99(26)
	^{33}S	32.9714589098(15)	0.75(2)
	^{34}S	33.96786700(5)	4.25(24)
	^{35}S	34.96903231(4)	
	^{36}S	35.96708071(20)	0.01(1)
17	^{35}Cl	34.96885268(4)	75.76(10)
	^{37}Cl	36.96590260(6)	24.24(10)
18	^{36}Ar	35.967545105(28)	0.3365(30)
	^{38}Ar	37.96273211(21)	0.0632(5)
	^{40}Ar	39.9623831237(24)	99.6003(30)
19	^{39}K	38.963706486(5)	93.2581(44)
	^{40}K	39.96399817(6)	0.0117(1)
	^{41}K	40.961825258(4)	6.7302(44)
	^{42}K	41.96240231(11)	
	^{43}K	42.9607347(4)	
20	^{40}Ca	39.962590863(22)	96.941(156)
	^{42}Ca	41.95861783(16)	0.647(23)
	^{43}Ca	42.95876644(24)	0.135(10)
	^{44}Ca	43.9554816(3)	2.086(110)
	^{45}Ca	44.9561864(4)	
	^{46}Ca	45.9536890(24)	0.004(3)
	^{47}Ca	46.9545424(24)	
	^{48}Ca	47.95252277(13)	0.187(21)
21	^{45}Sc	44.9559083(8)	100
22	^{46}Ti	45.9526277(3)	8.25(3)
	^{47}Ti	46.9517588(4)	7.44(2)
	^{48}Ti	47.9479420(4)	73.72(3)
	^{49}Ti	48.9478657(4)	5.41(2)
	^{50}Ti	49.9447869(4)	5.18(2)
23	^{50}V	49.9471560(9)	0.250(4)
	^{51}V	50.9439570(9)	99.750(4)
24	^{50}Cr	49.9460418(9)	4.345(13)
	^{51}Cr	50.9447650(9)	
	^{52}Cr	51.9405062(6)	83.789(18)
	^{53}Cr	52.9406481(6)	9.501(17)
	^{54}Cr	53.9388792(6)	2.365(7)
25	^{54}Mn	53.9403576(12)	
	^{55}Mn	54.9380439(5)	100
26	^{52}Fe	51.948113(7)	
	^{54}Fe	53.9396090(5)	5.845(35)
	^{55}Fe	54.9382920(5)	
	^{56}Fe	55.9349363(5)	91.754(36)

Z	Isotope	Mass in u	Abundance in %
	^{57}Fe	56.9353928(5)	2.119(10)
	^{58}Fe	57.9332744(5)	0.282(4)
	^{59}Fe	58.9348743(5)	
27	^{57}Co	56.9362906(7)	
	^{58}Co	57.9357521(13)	
	^{59}Co	58.9331943(6)	100
	^{60}Co	59.9338163(6)	
28	^{58}Ni	57.9353424(5)	68.077(19)
	^{59}Ni	58.9343462(5)	
	^{60}Ni	59.9307859(5)	26.223(15)
	^{61}Ni	60.9310556(5)	1.1399(13)
	^{62}Ni	61.9283454(6)	3.6345(40)
	^{63}Ni	62.9296696(6)	
	^{64}Ni	63.9279668(6)	0.9255(19)
29	^{63}Cu	62.9295977(6)	69.15(3)
	^{64}Cu	63.9297643(6)	
	^{65}Cu	64.9277897(7)	30.85(3)
30	^{64}Zn	63.9291420(7)	49.17(75)
	^{65}Zn	64.9292408(7)	
	^{66}Zn	65.9260338(9)	27.73(98)
	^{67}Zn	66.9271277(10)	4.04(16)
	^{68}Zn	67.9248446(10)	18.45(63)
	^{70}Zn	69.9253192(21)	0.61(10)
	^{67}Ga	66.9282025(13)	
	^{68}Ga	67.9279805(16)	
	^{69}Ga	68.9255735(13)	60.108(9)
	^{71}Ga	70.9247026(9)	39.892(9)
32	^{68}Ge	67.9280953(20)	
	^{70}Ge	69.9242488(9)	20.52(19)
	^{72}Ge	71.92207583(8)	27.45(15)
	^{73}Ge	72.92345896(6)	7.76(8)
	^{74}Ge	73.921177761(13)	36.52(12)
	^{76}Ge	75.921402726(19)	7.75(12)
33	^{75}As	74.9215946(9)	100
34	^{74}Se	73.922475934(15)	0.86(3)
	^{75}Se	74.92252287(8)	
	^{76}Se	75.919213704(17)	9.23(7)
	^{77}Se	76.91991415(7)	7.60(7)
	^{78}Se	77.91730928(20)	23.69(22)
	^{79}Se	78.91849929(24)	
	^{80}Se	79.9165218(13)	49.80(36)
	^{82}Se	81.9166995(15)	8.82(15)
35	^{79}Br	78.9183376(14)	50.69(7)
	^{81}Br	80.9162897(14)	49.31(7)
36	^{78}Kr	77.9203649(8)	0.355(3)
	^{80}Kr	79.9163781(7)	2.286(10)
	^{82}Kr	81.9134827(9)	11.593(31)
	^{83}Kr	82.9141272(3)	11.500(19)
	^{84}Kr	83.911497728(4)	56.987(15)
	^{86}Kr	85.910610627(4)	17.279(41)
37	^{85}Rb	84.911789738(5)	72.17(2)
	^{86}Rb	85.91116743(21)	
	^{87}Rb	86.909180531(6)	27.83(2)
38	^{84}Sr	83.9134191(13)	0.56(1)
	^{85}Sr	84.912932(3)	
	^{86}Sr	85.9092606(12)	9.86(1)
	^{87}Sr	86.9088775(12)	7.00(1)
	^{88}Sr	87.9056125(12)	82.58(1)
	^{89}Sr	88.9074511(12)	
	^{90}Sr	89.9077300(28)	
39	^{89}Y	88.9058403(24)	100
40	^{90}Zr	89.9046977(20)	51.45(40)
	^{91}Zr	90.9056396(20)	11.22(5)
	^{92}Zr	91.9050347(20)	17.15(8)
	^{94}Zr	93.9063108(20)	17.38(28)
	^{96}Zr	95.9082714(21)	2.80(9)
41	^{93}Nb	92.9063730(20)	100
42	^{92}Mo	91.9068080(8)	14.649(106)
	^{94}Mo	93.9050849(5)	9.187(33)
	^{95}Mo	94.9058388(5)	15.873(30)
	^{96}Mo	95.9046761(5)	16.673(8)
	^{97}Mo	96.9060181(5)	9.582(15)
	^{98}Mo	97.9054048(5)	24.292(80)
	^{99}Mo	98.9077085(5)	
	^{100}Mo	99.9074718(11)	9.744(65)
43	^{97}Tc	96.906367(4)	
	^{98}Tc	97.907212(4)	
	^{99}Tc	98.9062508(10)	
44	^{96}Ru	95.9075903(5)	5.54(14)
	^{98}Ru	97.905287(7)	1.87(3)
	^{99}Ru	98.9059341(11)	12.76(14)
	^{100}Ru	99.9042143(11)	12.60(7)
	^{101}Ru	100.9055769(12)	17.06(2)
	^{102}Ru	101.9043441(12)	31.55(14)
	^{104}Ru	103.9054275(28)	18.62(27)
	^{106}Ru	105.907329(6)	
45	^{103}Rh	102.9054980(26)	100
46	^{102}Pd	101.9056022(28)	1.02(1)
	^{104}Pd	103.9040305(14)	11.14(8)
	^{105}Pd	104.9050796(12)	22.33(8)
	^{106}Pd	105.9034804(12)	27.33(3)
	^{108}Pd	107.9038916(12)	26.46(9)
	^{110}Pd	109.9051722(7)	11.72(9)
47	^{107}Ag	106.9050916(26)	51.839(8)
	^{109}Ag	108.9047553(14)	48.161(8)
48	^{106}Cd	105.9064599(12)	1.25(6)
	^{108}Cd	107.9041834(12)	0.89(3)
	^{110}Cd	109.9030066(6)	12.49(18)
	^{111}Cd	110.9041829(6)	12.80(12)
	^{112}Cd	111.9027629(6)	24.13(21)
	^{113}Cd	112.9044081(4)	12.22(12)
	^{114}Cd	113.9033651(4)	28.73(42)
	^{116}Cd	115.90476315(17)	7.49(18)
49	^{111}In	110.905108(4)	
	^{113}In	112.9040618(9)	4.281(52)
	^{115}In	114.903878776(12)	95.719(52)
50	^{112}Sn	111.9048239(6)	0.97(1)
	^{113}Sn	112.9051757(18)	
	^{114}Sn	113.9027827(10)	0.66(1)
	^{115}Sn	114.903344699(16)	0.34(1)
	^{116}Sn	115.90174280(10)	14.54(9)
	^{117}Sn	116.9029540(5)	7.68(7)
	^{118}Sn	117.9016066(5)	24.22(9)
	^{119}Sn	118.9033112(8)	8.59(4)
	^{120}Sn	119.9022016(10)	32.58(9)
	^{122}Sn	121.9034438(26)	4.63(3)
	^{124}Sn	123.9052766(11)	5.79(5)
51	^{121}Sb	120.903812(3)	57.21(5)
	^{123}Sb	122.9042132(23)	42.79(5)
52	^{120}Te	119.904059(3)	0.09(1)
	^{122}Te	121.9030435(16)	2.55(12)
	^{123}Te	122.9042698(16)	0.89(3)

Z	Isotope	Mass in u	Abundance in %
	^{124}Te	123.9028179(16)	4.74(14)
	^{125}Te	124.9044299(16)	7.07(15)
	^{126}Te	125.9033109(16)	18.84(25)
	^{128}Te	127.9044613(9)	31.74(8)
	^{130}Te	129.906222748(12)	34.08(62)
53	^{123}I	122.905589(4)	
	^{125}I	124.9046294(16)	
	^{127}I	126.904472(4)	100
	^{129}I	128.904984(3)	
	^{131}I	130.9061263(7)	
54	^{124}Xe	123.9058920(19)	0.0952(3)
	^{126}Xe	125.904298(4)	0.0890(2)
	^{128}Xe	127.9035310(11)	1.9102(8)
	^{129}Xe	128.904780861(6)	26.4006(82)
	^{130}Xe	129.903509349(10)	4.0710(13)
	^{131}Xe	130.90508406(24)	21.2324(30)
	^{132}Xe	131.904155086(6)	26.9086(33)
	^{134}Xe	133.9053947(9)	10.4357(21)
	^{136}Xe	135.907214484(11)	8.8573(44)
55	^{129}Cs	128.906066(5)	
	^{133}Cs	132.905451961(8)	100
	^{134}Cs	133.906718503(17)	
	^{136}Cs	135.9073114(20)	
	^{137}Cs	136.9070892(4)	
56	^{130}Ba	129.9063207(28)	0.106(1)
	^{132}Ba	131.9050611(11)	0.101(1)
	^{133}Ba	132.9060074(11)	
	^{134}Ba	133.90450818(30)	2.417(18)
	^{135}Ba	134.90568838(29)	6.592(12)
	^{136}Ba	135.90457573(29)	7.854(24)
	^{137}Ba	136.9058271(3)	11.232(24)
	^{138}Ba	137.9052470(3)	71.698(42)
	^{140}Ba	139.910606(9)	
57	^{138}La	137.907115(4)	0.090(1)
	^{139}La	138.9063563(24)	99.910(1)
58	^{136}Ce	135.9071292(4)	0.185(2)
	^{138}Ce	137.905991(10)	0.251(2)
	^{140}Ce	139.9054431(23)	88.450(51)
	^{141}Ce	140.9082807(23)	
	^{142}Ce	141.9092504(29)	11.114(51)
	^{144}Ce	143.913653(3)	
59	^{141}Pr	140.9076576(23)	100
60	^{142}Nd	141.9077290(20)	27.2(5)
	^{143}Nd	142.9098200(20)	12.2(2)
	^{144}Nd	143.9100930(20)	23.8(3)
	^{145}Nd	144.9125793(20)	8.3(1)
	^{146}Nd	145.9131226(20)	17.2(3)
	^{148}Nd	147.9168993(26)	5.7(1)
	^{150}Nd	149.9209022(18)	5.6(2)
61	^{145}Pm	144.912756(3)	
	^{147}Pm	146.9151450(19)	
62	^{144}Sm	143.9120065(21)	3.07(7)
	^{147}Sm	146.9149044(19)	14.99(18)
	^{148}Sm	147.9148292(19)	11.24(10)
	^{149}Sm	148.9171921(18)	13.82(7)
	^{150}Sm	149.9172829(18)	7.38(1)
	^{152}Sm	151.9197397(18)	26.75(16)
	^{154}Sm	153.9222169(20)	22.75(29)
63	^{151}Eu	150.9198578(18)	47.81(6)
	^{153}Eu	152.9212380(18)	52.19(6)
64	^{152}Gd	151.9197995(18)	0.20(1)
	^{154}Gd	153.9208741(17)	2.18(3)
	^{155}Gd	154.9226305(17)	14.80(12)
	^{156}Gd	155.9221312(17)	20.47(9)
	^{157}Gd	156.9239686(17)	15.65(2)
	^{158}Gd	157.9241123(17)	24.84(7)
	^{160}Gd	159.9270624(18)	21.86(19)
65	^{159}Tb	158.9253547(19)	100
66	^{156}Dy	155.9242847(17)	0.056(3)
	^{158}Dy	157.924416(3)	0.095(3)
	^{160}Dy	159.9252046(20)	2.329(18)
	^{161}Dy	160.9269405(20)	18.889(42)
	^{162}Dy	161.9268056(20)	25.475(36)
	^{163}Dy	162.9287383(20)	24.896(42)
	^{164}Dy	163.9291819(20)	28.260(54)
67	^{165}Ho	164.9303288(21)	100
68	^{162}Er	161.9287884(20)	0.139(5)
	^{164}Er	163.9292088(20)	1.601(3)
	^{166}Er	165.9302995(22)	33.503(36)
	^{167}Er	166.9320546(22)	22.869(9)
	^{168}Er	167.9323767(22)	26.978(18)
	^{170}Er	169.9354702(26)	14.910(36)
69	^{169}Tm	168.9342179(22)	100
70	^{168}Yb	167.9338896(22)	0.13(1)
	^{169}Yb	168.9351825(22)	
	^{170}Yb	169.9347664(22)	3.04(15)
	^{171}Yb	170.9363302(22)	14.28(57)
	^{172}Yb	171.9363859(22)	21.83(67)
	^{173}Yb	172.9382151(22)	16.13(27)
	^{174}Yb	173.9388664(22)	31.83(92)
	^{176}Yb	175.9425764(24)	12.76(41)
71	^{175}Lu	174.9407718(23)	97.41(2)
	^{176}Lu	175.9426897(20)	2.59(2)
	^{174}Hf	173.9400461(28)	0.16(1)
	^{176}Hf	175.9414076(22)	5.26(7)
	^{177}Hf	176.9432277(20)	18.60(9)
	^{178}Hf	177.9437058(20)	27.28(7)
	^{179}Hf	178.9458232(20)	13.62(2)
	^{180}Hf	179.9465570(20)	35.08(16)
73	^{180}Ta	179.9474648(24)	0.012(2)
	^{181}Ta	180.9479958(20)	99.988(2)
74	^{180}W	179.9467108(20)	0.12(1)
	^{182}W	181.9482039(9)	26.50(16)
	^{183}W	182.9502227(9)	14.31(4)
	^{184}W	183.9509309(9)	30.64(2)
	^{186}W	185.9543628(17)	28.43(19)
75	^{185}Re	184.9529545(13)	37.40(2)
	^{187}Re	186.9557501(16)	62.60(2)
76	^{184}Os	183.9524885(14)	0.02(1)
	^{186}Os	185.9538350(16)	1.59(3)
	^{187}Os	186.9557474(16)	1.96(2)
	^{188}Os	187.9558352(16)	13.24(8)
	^{189}Os	188.9581442(17)	16.15(5)
	^{190}Os	189.9584437(17)	26.26(2)
	^{192}Os	191.9614770(29)	40.78(19)
77	^{191}Ir	190.9605893(21)	37.3(2)
	^{193}Ir	192.9629216(21)	62.7(2)
78	^{190}Pt	189.959930(6)	0.014(1)
	^{192}Pt	191.961039(3)	0.782(7)
	^{194}Pt	193.9626809(10)	32.967(99)
	^{195}Pt	194.9647917(10)	33.832(10)
	^{196}Pt	195.9649521(10)	25.242(41)

Z	Isotope	Mass in u	Abundance in %
	^{198}Pt	197.9678949(23)	7.163(55)
79	^{197}Au	196.9665688(7)	100
	^{198}Au	197.9682424(7)	
80	^{196}Hg	195.965833(3)	0.155(12)
	^{197}Hg	196.967213(3)	
	^{198}Hg	197.9667686(5)	10.038(16)
	^{199}Hg	198.9682806(5)	16.938(39)
	^{200}Hg	199.9683266(5)	23.138(65)
	^{201}Hg	200.9703028(7)	13.170(66)
	^{202}Hg	201.9706434(7)	29.743(89)
	^{203}Hg	202.9728728(18)	
	^{204}Hg	203.9734940(5)	6.818(35)
81	^{201}Tl	200.970822(15)	
	^{203}Tl	202.9723446(14)	29.52(1)
	^{205}Tl	204.9744278(14)	70.48(1)
82	^{204}Pb	203.9730440(13)	1.4(1)
	^{206}Pb	205.9744657(13)	24.1(1)
	^{207}Pb	206.9758973(13)	22.1(1)
	^{208}Pb	207.9766525(13)	52.4(1)
	^{210}Pb	209.9841889(16)	
83	^{207}Bi	206.9784710(26)	
	^{209}Bi	208.9803991(16)	100
84	^{209}Po	208.9824308(20)	
	^{210}Po	209.9828741(13)	
85	^{210}At	209.987148(8)	
	^{211}At	210.9874963(30)	
86	^{211}Rn	210.990601(7)	
	^{220}Rn	220.0113941(23)	
	^{222}Rn	222.0175782(25)	
87	^{223}Fr	223.0197360(25)	
88	^{223}Ra	223.0185023(27)	
	^{224}Ra	224.0202120(23)	
	^{226}Ra	226.0254103(25)	
	^{228}Ra	228.0310707(26)	
89	^{227}Ac	227.0277523(25)	
90	^{228}Th	228.0287413(23)	
	^{230}Th	230.0331341(19)	
	^{232}Th	232.0380558(21)	100
91	^{231}Pa	231.0358842(24)	100
92	^{233}U	233.0396355(29)	
	^{234}U	234.0409523(19)	0.0054(5)
	^{235}U	235.0439301(19)	0.7204(6)
	^{236}U	236.0455682(19)	
	^{238}U	238.0507884(20)	99.2742(10)
93	^{237}Np	237.0481736(19)	
	^{239}Np	239.0529392(22)	
94	^{238}Pu	238.0495601(19)	
	^{239}Pu	239.0521636(19)	
	^{240}Pu	240.0538138(19)	
	^{241}Pu	241.0568517(19)	
	^{242}Pu	242.0587428(20)	
	^{244}Pu	244.064205(6)	
95	^{241}Am	241.0568293(19)	
	^{243}Am	243.0613813(24)	
96	^{243}Cm	243.0613893(22)	
	^{244}Cm	244.0627528(19)	
	^{245}Cm	245.0654915(22)	
	^{246}Cm	246.0672238(22)	
	^{247}Cm	247.070354(5)	
	^{248}Cm	248.072350(6)	
97	^{247}Bk	247.070307(6)	
	^{249}Bk	249.0749877(27)	
98	^{249}Cf	249.0748539(23)	
	^{250}Cf	250.0764062(22)	
	^{251}Cf	251.079589(5)	
	^{252}Cf	252.081627(6)	
99	^{252}Es	252.08298(5)	
100	^{257}Fm	257.095106(7)	
101	^{256}Md	256.09389(13)	
	^{258}Md	258.098431(5)	
102	^{259}No	259.10103(11)	
103	^{262}Lr	262.10961(22)	
104	^{261}Rf	261.10877(5)	
105	^{262}Db	262.11407(15)	
106	^{266}Sg	266.1220(3)	
107	^{272}Bh	272.1383(6)	
108	^{277}Hs	277.1519(6)	
109	^{276}Mt	276.1516(6)	
110	^{281}Ds	281.1645(6)	
111	^{280}Rg	280.1651(6)	
112	^{285}Cn	285.1771(6)	
114	^{287}Fl	287.1868(7)	
116	^{291}Lv	291.2011(7)	

ELECTRON CONFIGURATION AND IONIZATION ENERGY OF NEUTRAL ATOMS IN THE GROUND STATE

William C. Martin

The ground-state electron configuration, ground level, and ionization energy of the elements hydrogen through rutherfordium are listed in this table. The electron configurations of elements heavier than neon are shortened by using rare-gas element symbols in brackets to represent the corresponding electrons. See the references for details of the notation for Pa, U, and Np. Ionization energies to higher states (and more precise values of the first ionization energy for certain elements) may be found in the table "Ionization Energies of Atoms and Atomic Ions" in Section 10 of this *Handbook*.

References

1. Martin, W. C., Musgrove, A., Kotochigova, S., and Sansonetti, J. E., NIST Physical Reference Data Web Site, <http://www.nist.gov/pml/data/ion_energy.cfm>, June 2013.
2. Martin, W. C., and Wiese, W. L., "Atomic Spectroscopy", in *Atomic, Molecular, & Optical Physics Handbook*, ed. by G.W.F. Drake (AIP, Woodbury, NY, 1996) Chapter 10, pp. 135-153.

Z	Symbol	Element	Ground-state configuration	Ground level	Ionization energy (eV)
1	H	Hydrogen	1s	$^2S_{1/2}$	13.5984
2	He	Helium	1s^2	$^1S_0$	24.5874
3	Li	Lithium	1s^2 2s	$^2S_{1/2}$	5.3917
4	Be	Beryllium	1s^2 2s^2	$^1S_0$	9.3227
5	B	Boron	1s^2 2s^2 2p	$^2P^o_{1/2}$	8.2980
6	C	Carbon	1s^2 2s^2 2p^2	$^3P_0$	11.2603
7	N	Nitrogen	1s^2 2s^2 2p^3	$^4S^o_{3/2}$	14.5341
8	O	Oxygen	1s^2 2s^2 2p^4	$^3P_2$	13.6181
9	F	Fluorine	1s^2 2s^2 2p^5	$^2P^o_{3/2}$	17.4228
10	Ne	Neon	1s^2 2s^2 2p^6	$^1S_0$	21.5645
11	Na	Sodium	[Ne] 3s	$^2S_{1/2}$	5.1391
12	Mg	Magnesium	[Ne] 3s^2	$^1S_0$	7.6462
13	Al	Aluminum	[Ne] 3s^2 3p	$^2P^o_{1/2}$	5.9858
14	Si	Silicon	[Ne] 3s^2 3p^2	$^3P_0$	8.1517
15	P	Phosphorus	[Ne] 3s^2 3p^3	$^4S^o_{3/2}$	10.4867
16	S	Sulfur	[Ne] 3s^2 3p^4	$^3P_2$	10.3600
17	Cl	Chlorine	[Ne] 3s^2 3p^5	$^2P^o_{3/2}$	12.9676
18	Ar	Argon	[Ne] 3s^2 3p^6	$^1S_0$	15.7596
19	K	Potassium	[Ar] 4s	$^2S_{1/2}$	4.3407
20	Ca	Calcium	[Ar] 4s^2	$^1S_0$	6.1132
21	Sc	Scandium	[Ar] 3d 4s^2	$^2D_{3/2}$	6.5615
22	Ti	Titanium	[Ar] 3d^2 4s^2	$^3F_2$	6.8281
23	V	Vanadium	[Ar] 3d^3 4s^2	$^4F_{3/2}$	6.7462
24	Cr	Chromium	[Ar] 3d^5 4s	$^7S_3$	6.7665
25	Mn	Manganese	[Ar] 3d^5 4s^2	$^6S_{5/2}$	7.4340
26	Fe	Iron	[Ar] 3d^6 4s^2	$^5D_4$	7.9024
27	Co	Cobalt	[Ar] 3d^7 4s^2	$^4F_{9/2}$	7.8810
28	Ni	Nickel	[Ar] 3d^8 4s^2	$^3F_4$	7.6399
29	Cu	Copper	[Ar] 3d^{10} 4s	$^2S_{1/2}$	7.7264
30	Zn	Zinc	[Ar] 3d^{10} 4s^2	$^1S_0$	9.3942
31	Ga	Gallium	[Ar] 3d^{10} 4s^2 4p	$^2P^o_{1/2}$	5.9993
32	Ge	Germanium	[Ar] 3d^{10} 4s^2 4p^2	$^3P_0$	7.8994
33	As	Arsenic	[Ar] 3d^{10} 4s^2 4p^3	$^4S^o_{3/2}$	9.7886
34	Se	Selenium	[Ar] 3d^{10} 4s^2 4p^4	$^3P_2$	9.7524
35	Br	Bromine	[Ar] 3d^{10} 4s^2 4p^5	$^2P^o_{3/2}$	11.8138
36	Kr	Krypton	[Ar] 3d^{10} 4s^2 4p^6	$^1S_0$	13.9996
37	Rb	Rubidium	[Kr] 5s	$^2S_{1/2}$	4.1771
38	Sr	Strontium	[Kr] 5s^2	$^1S_0$	5.6949
39	Y	Yttrium	[Kr] 4d 5s^2	$^2D_{3/2}$	6.2173
40	Zr	Zirconium	[Kr] 4d^2 5s^2	$^3F_2$	6.6339
41	Nb	Niobium	[Kr] 4d^4 5s	$^6D_{1/2}$	6.7589
42	Mo	Molybdenum	[Kr] 4d^5 5s	$^7S_3$	7.0924
43	Tc	Technetium	[Kr] 4d^5 5s^2	$^6S_{5/2}$	7.28
44	Ru	Ruthenium	[Kr] 4d^7 5s	$^5F_5$	7.3605
45	Rh	Rhodium	[Kr] 4d^8 5s	$^4F_{9/2}$	7.4589
46	Pd	Palladium	[Kr] 4d^{10}	$^1S_0$	8.3369
47	Ag	Silver	[Kr] 4d^{10} 5s	$^2S_{1/2}$	7.5762
48	Cd	Cadmium	[Kr] 4d^{10} 5s^2	$^1S_0$	8.9938
49	In	Indium	[Kr] 4d^{10} 5s^2 5p	$^2P^o_{1/2}$	5.7864
50	Sn	Tin	[Kr] 4d^{10} 5s^2 5p^2	$^3P_0$	7.3439
51	Sb	Antimony	[Kr] 4d^{10} 5s^2 5p^3	$^4S^o_{3/2}$	8.6084
52	Te	Tellurium	[Kr] 4d^{10} 5s^2 5p^4	$^3P_2$	9.0096
53	I	Iodine	[Kr] 4d^{10} 5s^2 5p^5	$^2P^o_{3/2}$	10.4513
54	Xe	Xenon	[Kr] 4d^{10} 5s^2 5p^6	$^1S_0$	12.1298
55	Cs	Cesium	[Xe] 6s	$^2S_{1/2}$	3.8939
56	Ba	Barium	[Xe] 6s^2	$^1S_0$	5.2117
57	La	Lanthanum	[Xe] 5d 6s^2	$^2D_{3/2}$	5.5769
58	Ce	Cerium	[Xe] 4f 5d 6s^2	$^1G^o_4$	5.5387
59	Pr	Praseodymium	[Xe] 4f^3 6s^2	$^4I^o_{9/2}$	5.473
60	Nd	Neodymium	[Xe] 4f^4 6s^2	$^5I_4$	5.5250
61	Pm	Promethium	[Xe] 4f^5 6s^2	$^6H^o_{5/2}$	5.582
62	Sm	Samarium	[Xe] 4f^6 6s^2	$^7F_0$	5.6437
63	Eu	Europium	[Xe] 4f^7 6s^2	$^8S^o_{7/2}$	5.6704
64	Gd	Gadolinium	[Xe] 4f^7 5d 6s^2	$^9D^o_2$	6.1498
65	Tb	Terbium	[Xe] 4f^9 6s^2	$^6H^o_{15/2}$	5.8638
66	Dy	Dysprosium	[Xe] 4f^{10} 6s^2	$^5I_8$	5.9389
67	Ho	Holmium	[Xe] 4f^{11} 6s^2	$^4I^o_{15/2}$	6.0215
68	Er	Erbium	[Xe] 4f^{12} 6s^2	$^3H_6$	6.1077
69	Tm	Thulium	[Xe] 4f^{13} 6s^2	$^2F^o_{7/2}$	6.1843
70	Yb	Ytterbium	[Xe] 4f^{14} 6s^2	$^1S_0$	6.2542
71	Lu	Lutetium	[Xe] 4f^{14} 5d 6s^2	$^2D_{3/2}$	5.4259
72	Hf	Hafnium	[Xe] 4f^{14} 5d^2 6s^2	$^3F_2$	6.8251
73	Ta	Tantalum	[Xe] 4f^{14} 5d^3 6s^2	$^4F_{3/2}$	7.5496
74	W	Tungsten	[Xe] 4f^{14} 5d^4 6s^2	$^5D_0$	7.8640
75	Re	Rhenium	[Xe] 4f^{14} 5d^5 6s^2	$^6S_{5/2}$	7.8335
76	Os	Osmium	[Xe] 4f^{14} 5d^6 6s^2	$^5D_4$	8.4382
77	Ir	Iridium	[Xe] 4f^{14} 5d^7 6s^2	$^4F_{9/2}$	8.9670
78	Pt	Platinum	[Xe] 4f^{14} 5d^9 6s	$^3D_3$	8.9588
79	Au	Gold	[Xe] 4f^{14} 5d^{10} 6s	$^2S_{1/2}$	9.2255
80	Hg	Mercury	[Xe] 4f^{14} 5d^{10} 6s^2	$^1S_0$	10.4375
81	Tl	Thallium	[Xe] 4f^{14} 5d^{10} 6s^2 6p	$^2P^o_{1/2}$	6.1082
82	Pb	Lead	[Xe] 4f^{14} 5d^{10} 6s^2 6p^2	$^3P_0$	7.4167
83	Bi	Bismuth	[Xe] 4f^{14} 5d^{10} 6s^2 6p^3	$^4S^o_{3/2}$	7.2855
84	Po	Polonium	[Xe] 4f^{14} 5d^{10} 6s^2 6p^4	$^3P_2$	8.414
85	At	Astatine	[Xe] 4f^{14} 5d^{10} 6s^2 6p^5	$^2P^o_{3/2}$	
86	Rn	Radon	[Xe] 4f^{14} 5d^{10} 6s^2 6p^6	$^1S_0$	10.7485
87	Fr	Francium	[Rn] 7s	$^2S_{1/2}$	4.0727
88	Ra	Radium	[Rn] 7s^2	$^1S_0$	5.2784
89	Ac	Actinium	[Rn] 6d 7s^2	$^2D_{3/2}$	5.3807
90	Th	Thorium	[Rn] 6d^2 7s^2	$^3F_2$	6.3067
91	Pa	Protactinium	[Rn] 5f^2($^3H_4$) 6d 7s^2	$(4,3/2)_{11/2}$	5.89
92	U	Uranium	[Rn] 5f^3($^4I^o_{9/2}$) 6d 7s^2	$(9/2,3/2)^o_6$	6.1939
93	Np	Neptunium	[Rn] 5f^4($^5I_4$) 6d 7s^2	$(4,3/2)_{11/2}$	6.2657
94	Pu	Plutonium	[Rn] 5f^6 7s^2	$^7F_0$	6.0260
95	Am	Americium	[Rn] 5f^7 7s^2	$^8S^o_{7/2}$	5.9738
96	Cm	Curium	[Rn] 5f^7 6d 7s^2	$^9D^o_2$	5.9914
97	Bk	Berkelium	[Rn] 5f^9 7s^2	$^6H^o_{15/2}$	6.1979
98	Cf	Californium	[Rn] 5f^{10} 7s^2	$^5I_8$	6.2817
99	Es	Einsteinium	[Rn] 5f^{11} 7s^2	$^4I^o_{15/2}$	6.3676
100	Fm	Fermium	[Rn] 5f^{12} 7s^2	$^3H_6$	6.50
101	Md	Mendelevium	[Rn] 5f^{13} 7s^2	$^2F^o_{7/2}$	6.58
102	No	Nobelium	[Rn] 5f^{14} 7s^2	$^1S_0$	6.65
103	Lr	Lawrencium	[Rn] 5f^{14} 7s^2 7p*	$^2P^o_{1/2}{}^*$	4.96
104	Rf	Rutherfordium	[Rn] 5f^{14} 6d^2 7s^{2*}	$^3F_2{}^*$	6.0*

* Uncertain

INTERNATIONAL TEMPERATURE SCALE OF 1990 (ITS-90)

B. W. Mangum

A new temperature scale, the International Temperature Scale of 1990 (ITS-90), was officially adopted by the Comité International des Poids et Mesures (CIPM), meeting 26–28 September 1989 at the Bureau International des Poids et Mesures (BIPM). The ITS-90 was recommended to the CIPM for its adoption following the completion of the final details of the new scale by the Comité Consultatif de Thermométrie (CCT), meeting 12–14 September 1989 at the BIPM in its 17th Session. The ITS-90 became the official international temperature scale on 1 January 1990. The ITS-90 supersedes the previous scales, the International Practical Temperature Scale of 1968 (IPTS-68) and the 1976 Provisional 0.5 to 30 K Temperature Scale (EPT-76).

The ITS-90 (Ref. 1, 2) extends upward from 0.65 K, and temperatures on this scale are in much better agreement with thermodynamic values than are those on the IPTS-68 and the EPT-76. The new scale has subranges and alternative definitions in certain ranges that greatly facilitate its use. Furthermore, its continuity, precision, and reproducibility throughout its ranges are much improved over that of the previous scales. The replacement of the thermocouple with the platinum resistance thermometer at temperatures below 961.78 °C resulted in the biggest improvement in reproducibility.

The ITS-90 is divided into four primary ranges:

1. Between 0.65 and 3.2 K, the ITS-90 is defined by the vapor pressure-temperature relation of ^{3}He, and between 1.25 and 2.1768 K (the λ point) and between 2.1768 and 5.0 K by the vapor pressure–temperature relations of ^{4}He. T_{90} is defined by the vapor pressure equations of the form:

$$T_{90}/\mathrm{K} = A_0 + \sum_{i=1}^{9} A_i \left[(\ln(p/\mathrm{Pa}) - B)/C \right]^i$$

The values of the coefficients A_i, and of the constants A_o, B, and C of the equations are given below.

2. Between 3.0 and 24.5561 K, the ITS-90 is defined in terms of a ^{3}He or ^{4}He constant volume gas thermometer (CVGT). The thermometer is calibrated at three temperatures — at the triple point of neon (24.5561 K), at the triple point of equilibrium hydrogen (13.8033 K), and at a temperature between 3.0 and 5.0 K, the value of which is determined by using either ^{3}He or ^{4}He vapor pressure thermometry.
3. Between 13.8033 K (−259.3467 °C) and 1234.93 K (961.78 °C), the ITS-90 is defined in terms of the specified fixed points given below, by resistance ratios of platinum resistance thermometers obtained by calibration at specified sets of the fixed points, and by reference functions and deviation functions of resistance ratios which relate to T_{90} between the fixed points.
4. Above 1234.93 K, the ITS-90 is defined in terms of Planck's radiation law, using the freezing-point temperature of either silver, gold, or copper as the reference temperature.

Since the adoption of ITS-90, the isotopic composition of the water and hydrogen whose fixed points appear in the table has been specified (Ref. 3). A Provisional Low Temperature Scale (PLTS-2000) has been developed, covering the region from 0.9 mK to 1 K (Ref. 2, 4). This scale is based on the melting temperature of ^{3}He.

References

1. The International Temperature Scale of 1990, Metrologia 27, 3, 1990; errata in Metrologia 27, 107, 1990.
2. *Mise en pratique* for definition of the kelvin, <www.bipm.org/utils/en/pdf/MeP_K.pdf>, 2011.
3. Technical annex for the International Temperature Scale of 1990, <www.bipm.org/utils/en/pdf/MeP_K_Technical_Annex.pdf>, 2005.
4. The Provisional Low Temperature Scale from 0.9 mK to 1 K, <www.bipm.org/utils/en/pdf/PLTS-2000.pdf>, 2000.

Defining Fixed Points of the ITS-90

Material[a]	Equilibrium state[b]	Temperature T_{90} (K)	t_{90} (°C)
He	VP	3 to 5	−270.15 to −268.15
e-H_2	TP	13.8033	−259.3467
e-H_2 (or He)	VP (or CVGT)	≈17	≈ −256.15
e-H_2 (or He)	VP (or CVGT)	≈20.3	≈ −252.85
Ne[c]	TP	24.5561	−248.5939
O_2	TP	54.3584	−218.7916
Ar	TP	83.8058	−189.3442
Hg[c]	TP	234.3156	−38.8344
H_2O	TP	273.16	0.01
Ga[c]	MP	302.9146	29.7646
In[c]	FP	429.7485	156.5985
Sn	FP	505.078	231.928
Zn	FP	692.677	419.527
Al[c]	FP	933.473	660.323
Ag	FP	1234.93	961.78
Au	FP	1337.33	1064.18
Cu[c]	FP	1357.77	1084.62

[a] e-H_2 indicates equilibrium hydrogen, that is, hydrogen with the equilibrium distribution of its ortho and para states. Normal hydrogen at room temperature contains 25% para hydrogen and 75% ortho hydrogen.

[b] VP indicates vapor pressure point; CVGT indicates constant volume gas thermometer point; TP indicates triple point (equilibrium temperature at which the solid, liquid, and vapor phases coexist); FP indicates freezing point, and MP indicates melting point (the equilibrium temperatures at which the solid and liquid phases coexist under a pressure of 101 325 Pa, one standard atmosphere). The isotopic composition is that naturally occurring.

[c] Previously, these were secondary fixed points.

Values of Coefficients in the Vapor Pressure Equations for Helium

Coef. or constant	^{3}He 0.65–3.2 K	^{4}He 1.25–2.1768 K	^{4}He 2.1768–5.0 K
A_0	1.053 447	1.392 408	3.146 631
A_1	0.980 106	0.527 153	1.357 655
A_2	0.676 380	0.166 756	0.413 923
A_3	0.372 692	0.050 988	0.091 159
A_4	0.151 656	0.026 514	0.016 349
A_5	−0.002 263	0.001 975	0.001 826
A_6	0.006 596	−0.017 976	−0.004 325
A_7	0.088 966	0.005 409	−0.004 973
A_8	−0.004 770	0.013 259	0
A_9	−0.054 943	0	0
B	7.3	5.6	10.3
C	4.3	2.9	1.9

CONVERSION OF TEMPERATURES FROM THE 1948 AND 1968 SCALES TO ITS-90

This table gives temperature corrections from older scales to the current International Temperature Scale of 1990 (see the preceding table for details on ITS-90). The first table may be used for converting Celsius temperatures in the range –180 to 4000 °C from IPTS-68 or IPTS-48 to ITS-90. Within the accuracy of the corrections, the temperature in the first column may be identified with either t_{68}, t_{48}, or t_{90}. The second table is designed for use at lower temperatures to convert values expressed in kelvins from EPT-76 or IPTS-68 to ITS-90.

The references give analytical equations for expressing these relations. Note that Ref. 1 and Ref. 4 supersede Ref. 2 with respect to corrections in the 630 to 1064 °C range.

References

1. Burns, G. W. et al., in *Temperature: Its Measurement and Control in Science and Industry*, Vol. 6, Schooley, J. F., Ed., American Institute of Physics, New York, 1993.
2. Goldberg, R. N., and Weir, R. D., *Pure and Appl. Chem.*, 64, 1545, 1992.
3. Bedford, R. E., Bonnier, G., Maas, H., and Pavese, F., Techniques for approximating the ITS-90, <www.bipm.org/en/publications/mep_kelvin/its-90_techniques.html>, 1997.
4. Rusby, R. L., Hudson, R. P., and Durieux, M., Revised values for (t_{90} – t_{68}) from 630 °C to 1064 °C, *Metrologia* 31, 149, 1994.

Conversion from IPTS-68 or IPTS-48 to ITS-90

t/°C	t_{90}-t_{68}	t_{90}-t_{48}
-180	0.008	0.020
-170	0.010	0.017
-160	0.012	0.007
-150	0.013	0.000
-140	0.014	0.001
-130	0.014	0.008
-120	0.014	0.017
-110	0.013	0.026
-100	0.013	0.035
-90	0.012	0.041
-80	0.012	0.045
-70	0.011	0.041
-60	0.010	0.042
-50	0.009	0.038
-40	0.008	0.032
-30	0.006	0.024
-20	0.004	0.016
-10	0.002	0.008
0	0.000	0.000
10	-0.002	-0.006
20	-0.005	-0.012
30	-0.007	-0.016
40	-0.010	-0.020
50	-0.013	-0.023
60	-0.016	-0.026
70	-0.018	-0.026
80	-0.021	-0.027
90	-0.024	-0.027
100	-0.026	-0.026
110	-0.028	-0.024
120	-0.030	-0.023
130	-0.032	-0.020
140	-0.034	-0.018
150	-0.036	-0.016
160	-0.037	-0.012
170	-0.038	-0.009
180	-0.039	-0.005
190	-0.039	-0.001
200	-0.040	0.003
210	-0.040	0.007
220	-0.040	0.011
230	-0.040	0.014
240	-0.040	0.018
250	-0.040	0.021
260	-0.040	0.024
270	-0.039	0.028
280	-0.039	0.030
290	-0.039	0.032
300	-0.039	0.034
310	-0.039	0.035
320	-0.039	0.036
330	-0.040	0.036
340	-0.040	0.037
350	-0.041	0.036
360	-0.042	0.035
370	-0.043	0.034
380	-0.045	0.032
390	-0.046	0.030
400	-0.048	0.028
410	-0.051	0.024
420	-0.053	0.022
430	-0.056	0.019
440	-0.059	0.015
450	-0.062	0.012
460	-0.065	0.009
470	-0.068	0.007
480	-0.072	0.004
490	-0.075	0.002
500	-0.079	0.000
510	-0.083	-0.001
520	-0.087	-0.002
530	-0.090	-0.001
540	-0.094	0.000
550	-0.098	0.002
560	-0.101	0.007
570	-0.105	0.011
580	-0.108	0.018
590	-0.112	0.025
600	-0.115	0.035
610	-0.118	0.047
620	-0.122	0.060
630	-0.125	0.075
640	-0.11	0.12
650	-0.10	0.15
660	-0.09	0.19
670	-0.07	0.24
680	-0.05	0.29
690	-0.04	0.32
700	-0.02	0.37
710	-0.01	0.41
720	0.00	0.45
730	0.02	0.49
740	0.03	0.53
750	0.03	0.56
760	0.04	0.60
770	0.05	0.63
780	0.05	0.66
790	0.05	0.69
800	0.05	0.72
810	0.05	0.75
820	0.04	0.76
830	0.04	0.79
840	0.03	0.81
850	0.02	0.83
860	0.01	0.85
870	0.00	0.87
880	-0.02	0.87
890	-0.03	0.89
900	-0.05	0.90
910	-0.06	0.92
920	-0.08	0.93
930	-0.10	0.94
940	-0.11	0.96
950	-0.13	0.97
960	-0.15	0.97
970	-0.16	0.99
980	-0.18	1.00
990	-0.19	1.02
1000	-0.20	1.04
1010	-0.22	1.07
1020	-0.23	1.09
1030	-0.23	1.11
1040	-0.24	1.13
1050	-0.25	1.15
1060	-0.25	1.17
1070	-0.25	1.19
1080	-0.26	1.20
1090	-0.26	1.20
1100	-0.26	1.2
1200	-0.30	1.4

t/°C	t_{90}-t_{68}	t_{90}-t_{48}
1300	-0.35	1.5
1400	-0.39	1.6
1500	-0.44	1.8
1600	-0.49	1.9
1700	-0.54	2.1
1800	-0.60	2.2
1900	-0.66	2.3
2000	-0.72	2.5
2100	-0.79	2.7
2200	-0.85	2.9
2300	-0.93	3.1
2400	-1.00	3.2
2500	-1.07	3.4
2600	-1.15	3.7
2700	-1.24	3.8
2800	-1.32	4.0
2900	-1.41	4.2
3000	-1.50	4.4
3100	-1.59	4.6
3200	-1.69	4.8
3300	-1.78	5.1
3400	-1.89	5.3
3500	-1.99	5.5
3600	-2.10	5.8
3700	-2.21	6.0
3800	-2.32	6.3
3900	-2.43	6.6
4000	-2.55	6.8

Conversion from EPT-76 or IPTS-68 to ITS-90 at Low Temperatures

T/K	T_{90}-T_{76}	T_{90}-T_{68}
5	-0.0001	
6	-0.0002	
7	-0.0003	
8	-0.0004	
9	-0.0005	
10	-0.0006	
11	-0.0007	
12	-0.0008	
13	-0.0010	
14	-0.0011	-0.006
15	-0.0013	-0.003
16	-0.0014	-0.004
17	-0.0016	-0.006
18	-0.0018	-0.008
19	-0.0020	-0.009
20	-0.0022	-0.009
21	-0.0025	-0.008
22	-0.0027	-0.007
23	-0.0030	-0.007
24	-0.0032	-0.006
25	-0.0035	-0.005
26	-0.0038	-0.004
27	-0.0041	-0.004
28		-0.005
29		-0.006
30		-0.006
31		-0.007
32		-0.008
33		-0.008
34		-0.008
35		-0.007
36		-0.007
37		-0.007
38		-0.006
39		-0.006
40		-0.006
41		-0.006
42		-0.006
43		-0.006
44		-0.006
45		-0.007
46		-0.007
47		-0.007
48		-0.006
49		-0.006
50		-0.006
51		-0.005
52		-0.005
53		-0.004
54		-0.003
55		-0.002
56		-0.001
57		0.000
58		0.001
59		0.002
60		0.003
61		0.003
62		0.004
63		0.004
64		0.005
65		0.005
66		0.006
67		0.006
68		0.007
69		0.007
70		0.007
71		0.007
72		0.007
73		0.007
74		0.007
75		0.008
76		0.008
77		0.008
78		0.008
79		0.008
80		0.008
81		0.008
82		0.008
83		0.008
84		0.008
85		0.008
86		0.008
87		0.008
88		0.008
89		0.008
90		0.008
91		0.008
92		0.008
93		0.008
94		0.008
95		0.008
96		0.008
97		0.009
98		0.009
99		0.009
100		0.009
110		0.011
120		0.013
130		0.014
140		0.014
150		0.014
160		0.014
170		0.013
180		0.012
190		0.012
200		0.011
210		0.010
220		0.009
230		0.008
240		0.007
250		0.005
260		0.003
270		0.001
273.16		0.000
300		-0.006
400		-0.031
500		-0.040
600		-0.040
700		-0.055
800		-0.089
900		-0.124

INTERNATIONAL SYSTEM OF UNITS (SI)

The International System of Units, abbreviated as SI (from the French name *Le Système International d'Unitès*), was established in 1960 by the 11th General Conference on Weights and Measures (CGPM) as the modern metric system of measurement. The core of the SI is the seven base units for the physical quantities length, mass, time, electric current, thermodynamic temperature, amount of substance, and luminous intensity. These base units are:

Table 1. SI Base Units

Base quantity	Name	Symbol
length	meter	m
mass	kilogram	kg
time	second	s
electric current	ampere	A
thermodynamic temperature	kelvin	K
amount of substance	mole	mol
luminous intensity	candela	cd

The SI base units are defined as follows:

ampere: The ampere is that constant current which, if maintained in two straight parallel conductors of infinite length, of negligible circular cross-section, and placed 1 meter apart in vacuum, would produce between these conductors a force equal to $2{\cdot}10^{-7}$ newton per meter of length.

candela: The candela is the luminous intensity, in a given direction, of a source that emits monochromatic radiation of frequency $540{\cdot}10^{12}$ hertz and that has a radiant intensity in that direction of 1/683 watt per steradian.

kelvin: The kelvin, unit of thermodynamic temperature, is the fraction 1/273.16 of the thermodynamic temperature of the triple point of water.

kilogram: The kilogram is the unit of mass; it is equal to the mass of the international prototype of the kilogram.

meter: The meter is the length of the path travelled by light in vacuum during a time interval of 1/299 792 458 of a second.

mole: The mole is the amount of substance of a system which contains as many elementary entities as there are atoms in 0.012 kilogram of carbon 12. When the mole is used, the elementary entities must be specified and may be atoms, molecules, ions, electrons, other particles, or specified groups of such particles.

second: The second is the duration of 9 192 631 770 periods of the radiation corresponding to the transition between the two hyperfine levels of the ground state of the cesium 133 atom.

SI derived units

Derived units are units which may be expressed in terms of base units by means of the mathematical symbols of multiplication and division (and, in the case of °C, subtraction). Certain derived units have been given special names and symbols, and these special names and symbols may themselves be used in combination with those for base and other derived units to express the units of other quantities. The next table lists some examples of derived units expressed directly in terms of base units:

Table 2. Examples of SI Derived Units

Physical quantity	Name	Symbol
area	square meter	m^2
volume	cubic meter	m^3
speed, velocity	meter per second	m/s
acceleration	meter per second squared	m/s^2
wave number	reciprocal meter	m^{-1}
density, mass density	kilogram per cubic meter	kg/m^3
specific volume	cubic meter per kilogram	m^3/kg
current density	ampere per square meter	A/m^2
magnetic field strength	ampere per meter	A/m
concentration (of amount of substance)	mole per cubic meter	mol/m^3
luminance	candela per square meter	cd/m^2
refractive index	(the number) one	1[a]

[a] The symbol "1" is generally omitted in combination with a numerical value.

For convenience, certain derived units, which are listed in the next table, have been given special names and symbols. These names and symbols may themselves be used to express other derived units. The special names and symbols are a compact form for the expression of units that are used frequently. The final column shows how the SI units concerned may be expressed in terms of SI base units. In this column, factors such as m^0, kg^0 ..., which are all equal to 1, are not shown explicitly.

Table 3. SI Derived Units with Special Names and Symbols

Physical quantity	Name	Symbol	Other SI units	SI base units
plane angle	radian[(a)]	rad	$m \cdot m^{-1} = 1$[(b)]	
solid angle	steradian[(a)]	sr[(c)]	$m^2 \cdot m^{-2} = 1$[(b)]	
frequency	hertz	Hz	s^{-1}	
force	newton	N	$m \cdot kg \cdot s^{-2}$	
pressure, stress	pascal	Pa	N/m^2	$m^{-1} \cdot kg \cdot s^{-2}$
energy, work, quantity of heat	joule	J	$N \cdot m$	$m^2 \cdot kg \cdot s^{-2}$
power, radiant flux	watt	W	J/s	$m^2 \cdot kg \cdot s^{-3}$
electric charge, quantity of electricity	coulomb	C	$s \cdot A$	
electric potential difference, electromotive force	volt	V	W/A	$m^2 \cdot kg \cdot s^{-3} \cdot A^{-1}$
capacitance	farad	F	C/V	$m^{-2} \cdot kg^{-1} \cdot s^4 \cdot A^2$
electric resistance	ohm	Ω	V/A	$m^2 \cdot kg \cdot s^{-3} \cdot A^{-2}$
electric conductance	siemens	S	A/V	$m^{-2} \cdot kg^{-1} \cdot s^3 \cdot A^2$
magnetic flux	weber	Wb	$V \cdot s$	$m^2 \cdot kg \cdot s^{-2} \cdot A^{-1}$
magnetic flux density	tesla	T	Wb/m^2	$kg \cdot s^{-2} \cdot A^{-1}$
inductance	henry	H	Wb/A	$m^2 \cdot kg \cdot s^{-2} \cdot A^{-2}$
Celsius temperature	degree Celsius[(d)]	°C		K
luminous flux	lumen	lm	$cd \cdot sr$[(c)]	$m^2 \cdot m^{-2} \cdot cd = cd$
illuminance	lux	lx	lm/m^2	$m^2 \cdot m^{-4} \cdot cd = m^{-2} \cdot cd$

Physical quantity	Name	Symbol	Other SI units	SI base units
activity (of a radionuclide)	becquerel	Bq		s^{-1}
absorbed dose, specific energy (imparted), kerma	gray	Gy	J/kg	$m^2 \cdot s^{-2}$
dose equivalent, ambient dose equivalent, directional dose equivalent, personal dose equivalent, organ equivalent dose	sievert	Sv	J/kg	$m^2 \cdot s^{-2}$
catalytic activity	katal	kat		$s^{-1} \cdot mol$

(a) The radian and steradian may be used with advantage in expressions for derived units to distinguish between quantities of different nature but the same dimension. Some examples of their use in forming derived units are given in the next table.

(b) In practice, the symbols rad and sr are used where appropriate, but the derived unit "1" is generally omitted in combination with a numerical value.

(c) In photometry, the name steradian and the symbol sr are usually retained in expressions for units.

(d) It is common practice to express a thermodynamic temperature, symbol T, in terms of its difference from the reference temperature T_0 = 273.15 K. The numerical value of a Celsius temperature t expressed in degrees Celsius is given by t/°C = T/K-273.15. The unit °C may be used in combination with SI prefixes, e.g., millidegree Celsius, m°C. Note that there should never be a space between the ° sign and the letter C, and that the symbol for kelvin is K, not °K.

The SI derived units with special names may be used in combinations to provide a convenient way to express more complex physical quantities. Examples are given in the next table:

Table 4. Examples of Using SI Derived Units with Special Names to Express Complex Physical Quantities

Physical quantity	Name	Symbol	As SI base units
dynamic viscosity	pascal second	Pa · s	$m^{-1} \cdot kg \cdot s^{-1}$
moment of force	newton meter	N · m	$m^2 \cdot kg \cdot s^{-2}$
surface tension	newton per meter	N/m	$kg \cdot s^{-2}$
angular velocity	radian per second	rad/s	$m \cdot m^{-1} \cdot s^{-1} = s^{-1}$
angular acceleration	radian per second squared	rad/s²	$m \cdot m^{-1} \cdot s^{-2} = s^{-2}$
heat flux density, irradiance	watt per square meter	W/m²	$kg \cdot s^{-3}$
heat capacity, entropy	joule per kelvin	J/K	$m^{-3} \cdot kg \cdot s^{-2} \cdot K^{-1}$
specific heat capacity, specific entropy	joule per kilogram kelvin	J/(kg · K)	$m^2 \cdot s^{-2} \cdot K^{-1}$
specific energy	joule per kilogram	J/kg	$m^2 \cdot s^{-2}$
thermal conductivity	watt per meter kelvin	W/(m · K)	$m \cdot kg \cdot s^{-3} \cdot K^{-1}$
energy density	joule per cubic meter	J/m³	$m^{-1} \cdot kg \cdot s^{-2}$
electric field strength	volt per meter	V/m	$m \cdot kg \cdot s^{-3} \cdot A^{-1}$
electric charge density	coulomb per cubic meter	C/m³	$m^{-3} \cdot s \cdot A$
electric flux density	coulomb per square meter	C/m²	$m^{-2} \cdot s \cdot A$
permittivity	farad per meter	F/m	$m^{-3} \cdot kg^{-1} \cdot s^4 \cdot A^2$
permeability	henry per meter	H/m	$m \cdot kg \cdot s^{-2} \cdot A^{-2}$
molar energy	joule per mole	J/mol	$m^2 \cdot kg \cdot s^{-2} \cdot mol^{-1}$
molar entropy, molar heat capacity	joule per mole kelvin	J/(mol · K)	$m^2 \cdot kg \cdot s^{-2} \cdot K^{-1} \cdot mol^{-1}$
exposure (x and γ rays)	coulomb per kilogram	C/kg	$kg^{-1} \cdot s \cdot A$
absorbed dose rate	gray per second	Gy/s	$m^2 \cdot s^{-3}$
radiant intensity	watt per steradian	W/sr	$m^4 \cdot m^{-2} \cdot kg \cdot s^{-3} = m^2 \cdot kg \cdot s^{-3}$
radiance	watt per square meter steradian	W/(m² ·sr)	$m^2 \cdot m^{-2} \cdot kg \cdot s^{-3} = kg \cdot s^{-3}$
catalytic (activity) concentration	katal per cubic meter	kat/m³	$m^{-3} \cdot s^{-1} \cdot mol$

In practice, with certain quantities preference is given to the use of certain special unit names, or combinations of unit names, in order to facilitate the distinction between different quantities having the same dimension. For example, the SI unit of frequency is designated the hertz, rather than the reciprocal second, and the SI unit of angular velocity is designated the radian per second rather than the reciprocal second (in this case retaining the word radian emphasizes that angular velocity is equal to 2π times the rotational frequency). Similarly, the SI unit of moment of force is designated the newton meter rather than the joule.

In the field of ionizing radiation, the SI unit of activity is designated the becquerel rather than the reciprocal second, and the SI units of absorbed dose and dose equivalent the gray and sievert, respectively, rather than the joule per kilogram. In the field of catalysis, the SI unit of catalytic activity is designated the katal rather than the mole per second. The special names becquerel, gray, sievert, and katal were specifically introduced because of the dangers to human health which might arise from mistakes involving the units reciprocal second, joule per kilogram, and mole per second.

Units for dimensionless quantities, quantities of dimension one

Certain quantities are defined as the ratios of two quantities of the same kind, and thus have a dimension which may be expressed by the number one. The unit of such quantities is necessarily a derived unit coherent with the other units of the SI and, since it is formed as the ratio of two identical SI units, the unit also may be expressed by the number one. Thus the SI unit of all quantities having the dimensional product one is the number one. Examples of such quantities are refractive index, relative permeability, and friction factor. Other quantities having the unit 1 include "characteristic numbers" like the Prandtl number and numbers which represent a count, such as a number of molecules, degeneracy (number of energy levels), and partition function in statistical thermodynamics. All of these quantities are described as being dimensionless, or of dimension one, and have the coherent SI unit 1. Their values are simply expressed as numbers and, in general, the unit 1 is not explicitly shown. In a few cases, however, a special name is given to this unit, mainly to avoid confusion between some compound derived units. This is the case for the radian, steradian, and neper.

SI prefixes

The following prefixes have been approved by the CGPM for use with SI units. Only one prefix may be used before a unit. Thus 10^{-12} farad should be designated pF, not μμF.

Table 5. SI Prefixes

Factor	Name	Symbol
10^{24}	yotta	Y
10^{21}	zetta	Z
10^{18}	exa	E

Factor	Name	Symbol
10^{15}	peta	P
10^{12}	tera	T
10^{9}	giga	G
10^{6}	mega	M
10^{3}	kilo	k
10^{2}	hecto	h
10^{1}	deka	da
10^{-1}	deci	d
10^{-2}	centi	c
10^{-3}	milli	m
10^{-6}	micro	µ
10^{-9}	nano	n
10^{-12}	pico	p
10^{-15}	femto	f
10^{-18}	atto	a
10^{-21}	zepto	z
10^{-24}	yocto	y

The kilogram

Among the base units of the International System, the unit of mass is the only one whose name, for historical reasons, contains a prefix. Names and symbols for decimal multiples and submultiples of the unit of mass are formed by attaching prefix names to the unit name "gram" and prefix symbols to the unit symbol "g".

Example: 10^{-6} kg = 1 mg (1 milligram) *but not* 1 µkg (1 microkilogram).

Units used with the SI

Many units that are not part of the SI are important and widely used in everyday life. The CGPM has adopted a classification of non-SI units: (1) units accepted for use with the SI (such as the traditional units of time and of angle); (2) units accepted for use with the SI whose values are obtained experimentally; and (3) other units currently accepted for use with the SI to satisfy the needs of special interests.

Table 6. Non-SI Units Accepted for Use with the International System

Name	Symbol	Value in SI units
minute	min	1 min = 60 s
hour	h	1 h = 60 min = 3600 s
day	d	1d = 24 h = 86 400 s
degree	°	1° = (π/180) rad
minute	′	1′ = (1/60)° = (π/10 800) rad
second	″	1″ = (1/60)′ = (π/648 000) rad
liter	l, L	1L = 1 dm³ = 10^{-3} m³
metric ton	t	1 t = 10^{3} kg
neper[a]	Np	1 Np = 1
bel[b]	B	1 B = (1/2) ln 10 Np

[a] The neper is used to express values of such logarithmic quantities as field level, power level, sound pressure level, and logarithmic decrement. Natural logarithms are used to obtain the numerical values of quantities expressed in nepers. The neper is coherent with the SI, but is not yet adopted by the CGPM as an SI unit. In using the neper, it is important to specify the quantity.

[b] The bel is used to express values of such logarithmic quantities as field level, power level, sound-pressure level, and attenuation. Logarithms to base ten are used to obtain the numerical values of quantities expressed in bels. The submultiple decibel, dB, is commonly used.

Table 7. Non-SI Units Accepted for Use with the International System, Whose Values in SI Units Are Obtained Experimentally

Name	Symbol	Value in SI units
electronvolt[b]	eV	1 eV = 1.602 176 565(35) ·10^{-19} J[a]
dalton[c]	Da	1 Da = 1.660 538 921(73) · 10^{-27} kg[a]
unified atomic mass unit[c]	u	1 u = 1 Da
astronomical unit[d]	au	1 au = 149 597 870 700 m (exact)[a]

[a] For the electronvolt and the dalton (unified atomic mass unit), values are quoted from the 2014 CODATA set of the Fundamental Physical Constants (p. 1-1 of this *Handbook*). The value given for the astronomical unit is from Resolution B2 of the XXVIII General Assembly of the International Astronomical Union (Captaine, N., Klioner, S., and McCarthy, D. [2012], "The re-definition of the astronomical unit of length: reasons and consequences", IAU Joint Discussion 7: Space-Time Reference Systems for Future Research, Bibcode: 2012IAUJD 7E 40C). Note the definition of the au in terms of the meter is now exact.

[b] The electronvolt is the kinetic energy acquired by an electron in passing through a potential difference of 1 V in vacuum.

[c] The dalton and unified atomic mass unit are alternative names for the same unit, equal to 1/12 of the mass of an unbound atom of the nuclide ^{12}C, at rest and in its ground state. The dalton may be combined with SI prefixes to express the masses of large molecules in kilodalton, kDa, or megadalton, MDa.

[d] The mean Earth-Sun distance is approximately equal to the astronomical unit.

Table 8. Other Non-SI Units Currently Accepted for Use with the International System

Name	Symbol	Value in SI units
nautical mile		1 nautical mile = 1852 m
knot		1 nautical mile per hour = (1852/3600) m/s
are		1 a = 1 dam² = 10^{2} m²
hectare	ha	1 ha = 1 hm² = 10^{4} m²
bar	bar	1 bar = 0.1 MPa = 100 kPa = 10^{5} Pa
ångström	Å	1 Å = 0.1 nm = 10^{-10} m
barn	b	1 b = 100 fm² = 10^{-28} m²

Other non-SI units

The SI does not encourage the use of cgs units, but these are frequently found in old scientific texts. The following table gives the relation of some common cgs units to SI units.

Table 9. Relation of cgs Units to SI Units

Name	Symbol	Value in SI units
erg	erg	1 erg = 10^{-7} J
dyne	dyn	1 dyn = 10^{-5} N
poise	P	1P = 1dyn · s/cm² = 0.1 Pa · s
stokes	St	1 St = 1 cm²/s = 10^{-4} m²/s
gauss	G	1G ≙ 10^{-4} T
oersted	Oe	1 Oe ≙ (1000/4π) A/m
maxwell	Mx	1Mx ≙ 10^{-8} Wb
stilb	sb	1 sb = 1 cd/cm² = 10^{4} cd/m²
phot	ph	1 ph = 10^{4} lx
gal	Gal	1 Gal = 1 cm/s² = 10^{-2} m/s²

Note: The symbol ≙ should be read as "corresponds to"; these units cannot strictly be equated because of the different dimensions of the electromagnetic cgs and the SI.

Examples of other non-SI units found in the older literature and their relation to the SI are given below. Use of these units in current texts is discouraged.

Table 10. Relation of Other Older Units to SI Units

Name	Symbol	Value in SI units
curie	Ci	1 Ci = $3.7 \cdot 10^{10}$ Bq
roentgen	R	1 R = $2.58 \cdot 10^{-4}$ C/kg
rad	rad	1 rad = 1 cGy = 10^{-2} Gy
rem	rem	1 rem = 1 cSv = 10^{-2} Sv
X unit		1 X unit ≈ $1.002 \cdot 10^{-4}$ nm
gamma	γ	1 γ = 1 nT = 10^{-9} T
jansky	Jy	1 Jy = 10^{-26} W · m^{-2} · Hz^{-1}
fermi		1 fermi = 1 fm = 10^{-15} m
metric carat		1 metric carat = 200 mg = $2 \cdot 10^{-4}$ kg
torr	Torr	1 Torr = (101325/760) Pa
standard atmosphere	atm	1 atm = 101325 Pa
calorie[(a)]	cal	1 cal = 4.184 J
micron	μ	1 μ = 1 μm = 10^{-6} m

[(a)] Several types of calorie have been used; the value given here is the so-called "thermochemical calorie".

Prefixes for binary multiples

In December 1998 the International Electrotechnical Commission (IEC), the leading international organization for worldwide standardization in electrotechnology, approved as an IEC International Standard names and symbols for prefixes for binary multiples for use in the fields of data processing and data transmission. The prefixes are as follows:

Table 11. Prefixes for Binary Multiples

Factor	Name	Symbol	Origin	Derivation
2^{10}	kibi	Ki	kilobinary: $(2^{10})^1$	kilo: $(10^3)^1$
2^{20}	mebi	Mi	megabinary: $(2^{10})^2$	mega: $(10^3)^2$
2^{30}	gibi	Gi	gigabinary: $(2^{10})^3$	giga: $(10^3)^3$
2^{40}	tebi	Ti	terabinary: $(2^{10})^4$	tera: $(10^3)^4$
2^{50}	pebi	Pi	petabinary: $(2^{10})^5$	peta: $(10^3)^5$
2^{60}	exbi	Ei	exabinary: $(2^{10})^6$	exa: $(10^3)^6$

Table 12. Examples and Comparisons with SI Prefixes

Multiple	SI equivalent
one **kibibit**	1 Kibit = 2^{10} bit = **1024 bit**
one **kilobit**	1 kbit = 10^3 bit = **1000 bit**
one **mebibyte**	1 MiB = 2^{20} B = **1 048 576 B**
one **megabyte**	1 MB = 10^6 B = **1 000 000 B**
one **gibibyte**	1 GiB = 2^{30} B = **1 073 741 824 B**
one **gigabyte**	1 GB = 10^9 B = **1 000 000 000 B**

It is suggested that in English, the first syllable of the name of the binary-multiple prefix should be pronounced in the same way as the first syllable of the name of the corresponding SI prefix, and that the second syllable should be pronounced as "bee."

It is important to recognize that the new prefixes for binary multiples are not part of the International System of Units (SI), the modern metric system. However, for ease of understanding and recall, they were derived from the SI prefixes for positive powers of ten. As can be seen from the above table, the name of each new prefix is derived from the name of the corresponding SI prefix by retaining the first two letters of the name of the SI prefix and adding the letters "bi," which recalls the word "binary." Similarly, the symbol of each new prefix is derived from the symbol of the corresponding SI prefix by adding the letter "i," which again recalls the word "binary." (For consistency with the other prefixes for binary multiples, the symbol Ki is used for 2^{10} rather than ki.)

References

1. Taylor, B. N., and Thompson, A., *The International System of Unit (SI)*, NIST Special Publication 330, National Institute of Standards and Technology, Gaithersburg, MD, 2008.
2. Bureau International des Poids et Mesures, *Le Système International d'Unités (SI)*, 8th French and English Edition, BIPM, Sèvres, France, 2006.
3. Thompson, A., and Taylor, B. N., *Guide for the Use of the International System of Unit (SI)*, NIST Special Publication 811, National Institute of Standards and Technology, Gaithersburg, MD, 2008.
4. NIST Physical Reference Data Web site, http://physics.nist.gov/cuu/Units/index.html, October 2004.
5. Amendment 2 to IEC International Standard IEC 60027-2, 1999-01, Letter symbols to be used in electrical technology – Part 2: Telecommunications and electronics.
6. IEC 60027-2, Second edition, 2000-11, Letter symbols to be used in electrical technology - Part 2: Telecommunications and electronics.
7. Barrow, B., "A Lesson in Megabytes," *IEEE Stand. Bearer*, January 1997, p. 5.

UNITS FOR MAGNETIC PROPERTIES

Quantity	Symbol	Gaussian & cgs emu [a]	Conversion factor, C [b]	SI & rationalized mks [c]
Magnetic flux density, magnetic induction	B	gauss (G)[d]	10^{-4}	tesla (T), Wb/m²
Magnetic flux	Φ	maxwell (Mx), G · cm²	10^{-8}	weber (Wb), volt second (V · s)
Magnetic potential difference, magnetomotive force	U, F	gilbert (Gb)	$10/4\pi$	ampere (A)
Magnetic field strength, magnetizing force	H	oersted (Oe)[e], Gb/cm	$10^3/4\pi$	A/m[f]
(Volume) magnetization[g]	M	emu/cm³ [h]	10^3	A/m
(Volume) magnetization	$4\pi M$	G	$10^3/4\pi$	A/m
Magnetic polarization, intensity of magnetization	J, I	emu/cm³	$4\pi \times 10^{-4}$	T, Wb/m² [i]
(Mass) magnetization	σ, M	emu/g	1	A · m²/kg
			$4\pi \times 10^{-7}$	Wb · m/kg
Magnetic moment	m	emu, erg/G	10^{-3}	A · m², joule per tesla (J/T)
Magnetic dipole moment	j	emu, erg/G	$4\pi \times 10^{-10}$	Wb · m[i]
(Volume) susceptibility	χ, κ	dimensionless, emu/cm³	4π	dimensionless
			$(4\pi)^2 \times 10^{-7}$	henry per meter (H/m), Wb/(A · m)
(Mass) susceptibility	χ_ρ, κ_ρ	cm³/g, emu/g	$4\pi \times 10^{-3}$	m³/kg
			$(4\pi)^2 \times 10^{-10}$	H · m²/kg
(Molar) susceptibility	χ_{mol}, κ_{mol}	cm³/mol, emu/mol	$4\pi \times 10^{-6}$	m³/mol
			$(4\pi)^2 \times 10^{-13}$	H · m²/mol
Permeability	μ	dimensionless	$4\pi \times 10^{-7}$	H/m, Wb/(A · m)
Relative permeability[j]	μ_r	not defined		dimensionless
(Volume) energy density, energy product[k]	W	erg/cm³	10^{-1}	J/m³
Demagnetization factor	D, N	dimensionless	$1/4\pi$	dimensionless

a Gaussian units and cgs emu are the same for magnetic properties. The defining relation is $B = H + 4\pi M$.

b Multiply a number in Gaussian units by C to convert it to SI (e.g., 1 G × 10^{-4} T/G = 10^{-4} T).

c SI (*Système International d'Unités*) has been adopted by the National Institute of Standards and Technology. Where two conversion factors are given, the upper one is recognized under, or consistent with, SI and is based on the definition $B = \mu_0(H + M)$, where $\mu_0 = 4\pi \times 10^{-7}\ H/M$. The lower one is not recognized under SI and is based on the definition $B = \mu_0 H + J$, where the symbol I is often used in place of J.

d 1 gauss = 10^5 gamma (γ).

e Both oersted and gauss are expressed as $cm^{-1/2} \cdot g^{1/2} \cdot s^{-1}$ in terms of base units.

f A/m was often expressed as "ampere–turn per meter" when used for magnetic field strength.

g Magnetic moment per unit volume.

h The designation "emu" is not a unit.

i Recognized under SI, even though based on the definition $B = \mu_0 H + J$. See footnote *c*.

j $\mu_r = \mu/\mu_0 = 1+\chi$, all in SI. μ_r is equal to Gaussian μ.

k $B \cdot H$ and $\mu_0 M \cdot H$ have SI units J/m³; $M \cdot H$ and $B \cdot H/4\pi$ have Gaussian units erg/cm³.

Reference

R. B. Goldfarb and F. R. Fickett, U.S. Department of Commerce, National Bureau of Standards, Boulder, Colorado 80303, March 1985, NBS Special Publication 696. Superintendent of Documents, U.S. Government Printing Office, Washington, DC 20402, 1985.

CONVERSION FACTORS

The following table gives conversion factors from various units of measure to SI units. It is reproduced from NIST Special Publication 811, *Guide for the Use of the International System of Units (SI).* The table gives the factor by which a quantity expressed in a non-SI unit should be multiplied in order to calculate its value in the SI. The SI values are expressed in terms of the base, supplementary, and derived units of SI in order to provide a coherent presentation of the conversion factors and facilitate computations (see the table "International System of Units" in this section). If desired, powers of ten can be avoided by using SI prefixes and shifting the decimal point if necessary.

Conversion from a non-SI unit to a different non-SI unit may be carried out by using this table in two stages, e.g.,

$$1\ \mathrm{cal_{th}} = 4.184\ \mathrm{J}$$

$$1\ \mathrm{Btu_{IT}} = 1.055056\ \mathrm{E{+}03\ J}$$

Thus,

$$1\ \mathrm{Btu_{IT}} = (1.055056\ \mathrm{E{+}03} \div 4.184)\ \mathrm{cal_{th}} = 252.164\ \mathrm{cal_{th}}$$

Conversion factors are presented for ready adaptation to computer readout and electronic data transmission. The factors are written as a number equal to or greater than one and less than ten with six or fewer decimal places. This number is followed by the letter E (for exponent), a plus or a minus sign, and two digits that indicate the power of 10 by which the number must be multiplied to obtain the correct value. For example:

$$3.523\,907\ \mathrm{E{-}02}\ \text{is}\ 3.523\,907 \times 10^{-2}$$

or

$$0.035\,239\,07$$

Similarly:

$$3.386\,389\ \mathrm{E{+}03}\ \text{is}\ 3.386\,389 \times 10^{3}$$

or

$$3\,386.389$$

A factor in boldface is exact; i.e., all subsequent digits are zero. All other conversion factors have been rounded to the figures given in accordance with accepted practice. Where less than six digits after the decimal point are shown, more precision is not warranted.

It is often desirable to round a number obtained from a conversion of units in order to retain information on the precision of the value. The following rounding rules may be followed:

1. If the digits to be discarded begin with a digit less than 5, the digit preceding the first discarded digit is not changed.

Example: 6.974 951 5 rounded to 3 digits is 6.97

2. If the digits to be discarded begin with a digit greater than 5, the digit preceding the first discarded digit is increased by one.

Example: 6.974 951 5 rounded to 4 digits is 6.975

3. If the digits to be discarded begin with a 5 and at least one of the following digits is greater than 0, the digit preceding the 5 is increased by 1.

Example: 6.974 851 rounded to 5 digits is 6.974 9

4. If the digits to be discarded begin with a 5 and all of the following digits are 0, the digit preceding the 5 is unchanged if it is even and increased by one if it is odd. (Note that this means that the final digit is always even.)

Examples:

6.974 951 5 rounded to 7 digits is 6.974 952

6.974 950 5 rounded to 7 digits is 6.974 950

Reference

Thompson, A., and Taylor, B. N., *Guide for the Use of the International System of Units (SI)*, NIST Special Publication 811, 2008 Edition, Superintendent of Documents, U.S. Government Printing Office, Washington, D.C. 20402, 2008.

Factors in **boldface** are exact

To convert from	to	Multiply by	
abampere	ampere (A)	**1.0**	**E+01**
abcoulomb	coulomb (C)	**1.0**	**E+01**
abfarad	farad (F)	**1.0**	**E+09**
abhenry	henry (H)	**1.0**	**E−09**
abmho	siemens (S)	**1.0**	**E+09**
abohm	ohm (Ω)	**1.0**	**E−09**
abvolt	volt (V)	**1.0**	**E−08**
acceleration of free fall, standard (g_n)	meter per second squared (m/s^2)	**9.806 65**	**E+00**
acre (based on U.S. survey foot)[a]	square meter (m^2)	4.046 873	E+03
acre foot (based on U.S. survey foot)[a]	cubic meter (m^3)	1.233 489	E+03
ampere hour (A · h)	coulomb (C)	**3.6**	**E+03**
ångström (Å)	meter (m)	**1.0**	**E−10**
ångström (Å)	nanometer (nm)	**1.0**	**E−01**
apostilb (asb)	candela per meter squared (cd/m^2)	3.183 098	E−01
are (a)	square meter (m^2)	**1.0**	**E+02**
astronomical unit (au)	meter (m)	**1.495 978 70700**	**E+11**
atmosphere, standard (atm)	pascal (Pa)	**1.013 25**	**E+05**
atmosphere, standard (atm)	kilopascal (kPa)	**1.013 25**	**E+02**
atmosphere, technical (at)[b]	pascal (Pa)	**9.806 65**	**E+04**
atmosphere, technical (at)[b]	kilopascal (kPa)	**9.806 65**	**E+01**

[a] The U.S. survey foot equals (1200/3937) m. 1 international foot = 0.999998 survey foot.

[b] One technical atmosphere equals one kilogram-force per square centimeter (1 at = 1 kgf/cm^2).

To convert from	to	Multiply by	
bar (bar)	pascal (Pa)	**1.0**	**E+05**
bar (bar)	kilopascal (kPa)	**1.0**	**E+02**
barn (b)	square meter (m^2)	**1.0**	**E−28**
barrel [for petroleum, 42 gallons (U.S.)](bbl)	cubic meter (m^3)	1.589 873	E−01
barrel [for petroleum, 42 gallons (U.S.)](bbl)	liter (L)	1.589 873	E+02
biot (Bi)	ampere (A)	**1.0**	**E+01**
British thermal unit$_{IT}$ (Btu$_{IT}$)[c]	joule (J)	1.055 056	E+03
British thermal unit$_{th}$ (Btu$_{th}$)[c]	joule (J)	1.054 350	E+03
British thermal unit (mean) (Btu)	joule (J)	1.055 87	E+03
British thermal unit (39 °F) (Btu)	joule (J)	1.059 67	E+03
British thermal unit (59 °F) (Btu)	joule (J)	1.054 80	E+03
British thermal unit (60 °F) (Btu)	joule (J)	1.054 68	E+03
British thermal unit$_{IT}$ foot per hour square foot degree Fahrenheit [Btu$_{IT}$ · ft/(h · ft^2 · °F)]	watt per meter kelvin [W/(m · K)]	1.730 735	E+00
British thermal unit$_{th}$ foot per hour square foot degree Fahrenheit [Btu$_{th}$ · ft/(h · ft^2 · °F)]	watt per meter kelvin [W/(m · K)]	1.729 577	E+00
British thermal unit$_{IT}$ inch per hour square foot degree Fahrenheit [Btu$_{IT}$ · in/(h · ft^2 · °F)]	watt per meter kelvin [W/(m · K)]	1.442 279	E−01
British thermal unit$_{th}$ inch per hour square foot degree Fahrenheit [Btu$_{th}$ · in/(h · ft^2 · °F)]	watt per meter kelvin [W/(m · K)]	1.441 314	E−01
British thermal unit$_{IT}$ inch per second square foot degree Fahrenheit [Btu$_{IT}$ · in/(s · ft^2 · °F)]	watt per meter kelvin [W/(m · K)]	5.192 204	E+02
British thermal unit$_{th}$ inch per second square foot degree Fahrenheit [Btu$_{th}$ · in/(s · ft^2 · °F)]	watt per meter kelvin [W/(m · K)]	5.188 732	E+02
British thermal unit$_{IT}$ per cubic foot (Btu$_{IT}$/ft^3)	joule per cubic meter (J/m^3)	3.725 895	E+04
British thermal unit$_{th}$ per cubic foot (Btu$_{th}$/ft^3)	joule per cubic meter (J/m^3)	3.723 403	E+04
British thermal unit$_{IT}$ per degree Fahrenheit (Btu$_{IT}$/°F)	joule per kelvin (J/k)	1.899 101	E+03
British thermal unit$_{th}$ per degree Fahrenheit (Btu$_{th}$/°F)	joule per kelvin (J/k)	1.897 830	E+03
British thermal unit$_{IT}$ per degree Rankine (Btu$_{IT}$/°R)	joule per kelvin (J/k)	1.899 101	E+03
British thermal unit$_{th}$ per degree Rankine (Btu$_{th}$/°R)	joule per kelvin (J/k)	1.897 830	E+03
British thermal unit$_{IT}$ per hour (Btu$_{IT}$/h)	watt (W)	2.930 711	E−01
British thermal unit$_{th}$ per hour (Btu$_{th}$/h)	watt (W)	2.928 751	E−01
British thermal unit$_{IT}$ per hour square foot degree Fahrenheit [Btu$_{IT}$/(h · ft^2 · °F)]	watt per square meter kelvin [W/(m^2 · K)]	5.678 263	E+00
British thermal unit$_{th}$ per hour square foot degree Fahrenheit [Btu$_{th}$/(h · ft^2 · °F)]	watt per square meter kelvin [W/(m^2 · K)]	5.674 466	E+00
British thermal unit$_{th}$ per minute (Btu$_{th}$/min)	watt (W)	1.757 250	E+01
British thermal unit$_{IT}$ per pound (Btu$_{IT}$/lb)	joule per kilogram (J/kg)	**2.326**	**E+03**
British thermal unit$_{th}$ per pound (Btu$_{th}$/lb)	joule per kilogram (J/kg)	2.324 444	E+03
British thermal unit$_{IT}$ per pound degree Fahrenheit [Btu$_{IT}$/(lb · °F)]	joule per kilogram kelvin (J/(kg · K)]	**4.1868**	**E+03**
British thermal unit$_{th}$ per pound degree Fahrenheit [Btu$_{th}$/(lb · °F)]	joule per kilogram kelvin [J/(kg · K)]	**4.184**	**E+03**
British thermal unit$_{IT}$ per pound degree Rankine [Btu$_{IT}$/(lb · °R)]	joule per kilogram kelvin [J/(kg · K)]	**4.1868**	**E+03**
British thermal unit$_{th}$ per pound degree Rankine [Btu$_{th}$/(lb · °R)]	joule per kilogram kelvin [J/(kg · K)]	**4.184**	**E+03**
British thermal unit$_{IT}$ per second (Btu$_{IT}$/s)	watt (W)	1.055 056	E+03
British thermal unit$_{th}$ per second (Btu$_{th}$/s)	watt (W)	1.054 350	E+03

[c] The Fifth International Conference on the Properties of Steam (London, July 1956) defined the International Table calorie as 4.1868 J. Therefore the exact conversion factor for the International Table Btu is 1.055 055 852 62 kJ. Note that the notation for the International Table used in this listing is subscript "IT". Similarly, the notation for thermochemical is subscript "th." Further, the thermochemical Btu, Btu$_{th}$, is based on the thermochemical calorie, cal$_{th}$, where cal$_{th}$ = 4.184 J exactly.

To convert from	to	Multiply by	
British thermal unit$_{IT}$ per second square foot degree Fahrenheit [Btu$_{IT}$/(s · ft^2 · °F)]	watt per square meter kelvin [W/(m^2 · K)]	2.044 175	E+04
British thermal unit$_{th}$ per second square foot degree Fahrenheit [Btu$_{th}$/(s · ft^2 · °F)]	watt per square meter kelvin [W/(m^2 · K)]	2.042 808	E+04
British thermal unit$_{IT}$ per square foot (Btu$_{IT}$/ft^2)	joule per square meter (J/m^2)	1.135 653	E+04
British thermal unit$_{th}$ per square foot (Btu$_{th}$/ft^2)	joule per square meter (J/m^2)	1.134 893	E+04
British thermal unit$_{IT}$ per square foot hour [(Btu$_{IT}$/(ft^2 · h)]	watt per square meter (W/m^2)	3.154 591	E+00
British thermal unit$_{th}$ per square foot hour [Btu$_{th}$/(ft^2 · h)]	watt per square meter (W/m^2)	3.152 481	E+00
British thermal unit$_{th}$ per square foot minute [Btu$_{th}$/(ft^2 · min)]	watt per square meter (W/m^2)	1.891 489	E+02
British thermal unit$_{IT}$ per square foot second [(Btu$_{IT}$/(ft^2 · s)]	watt per square meter (W/m^2)	1.135 653	E+04
British thermal unit$_{th}$ per square foot second [Btu$_{th}$/(ft^2 · s)]	watt per square meter (W/m^2)	1.134 893	E+04
British thermal unit$_{th}$ per square inch second [Btu$_{th}$/(in^2 · s)]	watt per square meter (W/m^2)	1.634 246	E+06
bushel (U.S.) (bu)	cubic meter (m^3)	3.523 907	E–02
bushel (U.S.) (bu)	liter (L)	3.523 907	E+01
calorie$_{IT}$ (cal$_{IT}$)[c]	joule (J)	**4.1868**	**E+00**
calorie$_{th}$ (cal$_{th}$)[c]	joule (J)	**4.184**	**E+00**
calorie (cal) (mean)	joule (J)	4.190 02	E+00
calorie (15 °C) (cal$_{15}$)	joule (J)	4.185 80	E+00
calorie (20 °C) (cal$_{20}$)	joule (J)	4.181 90	E+00
calorie$_{IT}$, kilogram (nutrition)[d]	joule (J)	**4.1868**	**E+03**
calorie$_{th}$, kilogram (nutrition)[d]	joule (J)	**4.184**	**E+03**
calorie (mean), kilogram (nutrition)[d]	joule (J)	4.190 02	E+03
calorie$_{th}$ per centimeter second degree Celsius [cal$_{th}$/(cm · s · °C)]	watt per meter kelvin [W/(m · K)]	**4.184**	**E+02**
calorie$_{IT}$ per gram (cal$_{IT}$/g)	joule per kilogram (J/kg)	**4.1868**	**E+03**
calorie$_{th}$ per gram (cal$_{th}$/g)	joule per kilogram (J/kg)	**4.184**	**E+03**
calorie$_{IT}$ per gram degree Celsius [cal$_{IT}$/(g · °C)]	joule per kilogram kelvin [J/(kg · K)]	**4.1868**	**E+03**
calorie$_{th}$ per gram degree Celsius [cal$_{th}$/(g · °C)]	joule per kilogram kelvin [J/(kg · K)]	**4.184**	**E+03**
calorie$_{IT}$ per gram kelvin [cal$_{IT}$/(g · K)]	joule per kilogram kelvin [J/(kg · K)]	**4.1868**	**E+03**
calorie$_{th}$ per gram kelvin [cal$_{th}$/(g · K)]	joule per kilogram kelvin [J/(kg · K)]	**4.184**	**E+03**
calorie$_{th}$ per minute (cal$_{th}$/min)	watt (W)	6.973 333	E–02
calorie$_{th}$ per second (cal$_{th}$/s)	watt (W)	**4.184**	**E+00**
calorie$_{th}$ per square centimeter (cal$_{th}$/cm^2)	joule per square meter (J/m^2)	**4.184**	**E+04**
calorie$_{th}$ per square centimeter minute [cal$_{th}$/(cm^2 · min)]	watt per square meter (W/m^2)	6.973 333	E+02
calorie$_{th}$ per square centimeter second [cal$_{th}$/(cm^2 · s)]	watt per square meter (W/m^2)	**4.184**	**E+04**
candela per square inch (cd/in^2)	candela per square meter (cd/m^2)	1.550 003	E+03
carat, metric	kilogram (kg)	**2.0**	**E–04**
carat, metric	gram (g)	**2.0**	**E–01**
centimeter of mercury (0 °C)[e]	pascal (Pa)	1.333 22	E+03
centimeter of mercury (0 °C)[e]	kilopascal (kPa)	1.333 22	E+00
centimeter of mercury, conventional (cmHg)[e]	pascal (Pa)	1.333 224	E+03

[d] The kilogram calorie or "large calorie" is an obsolete term used for the kilocalorie, which is the calorie used to express the energy content of foods. However, in practice, the prefix "kilo" is usually omitted.

[e] Conversion factors for mercury manometer pressure units are calculated using the standard value for the acceleration of gravity and the density of mercury at the stated temperature. Additional digits are not justified because the definitions of the units do not take into account the compressibility of mercury or the change in density caused by the revised practical temperature scale, ITS-90. Similar comments also apply to water manometer pressure units. Conversion factors for conventional mercury and water manometer pressure units are based on ISO 31-3.

To convert from	to	Multiply by	
centimeter of mercury, conventional (cmHg)[e]	kilopascal (kPa)	1.333 224	E+00
centimeter of water (4 °C)[e]	pascal (Pa)	9.806 38	E+01
centimeter of water, conventional (cmH_2O)[e]	pascal (Pa)	**9.806 65**	**E+01**
centipoise (cP)	pascal second (Pa · s)	**1.0**	**E−03**
centistokes (cSt)	meter squared per second (m^2/s)	**1.0**	**E−06**
chain (based on U.S. survey foot) (ch)[a]	meter (m)	2.011 684	E+01
circular mil	square meter (m^2)	5.067 075	E−10
circular mil	square millimeter (mm^2)	5.067 075	E−04
clo	square meter kelvin per watt ($m^2 \cdot K/W$)	1.55	E−01
cord (128 ft^3)	cubic meter (m^3)	3.624 556	E+00
cubic foot (ft^3)	cubic meter (m^3)	2.831 685	E−02
cubic foot per minute (ft^3/min)	cubic meter per second (m^3/s)	4.719 474	E−04
cubic foot per minute (ft^3/min)	liter per second (L/s)	4.719 474	E−01
cubic foot per second (ft^3/s)	cubic meter per second (m^3/s)	2.831 685	E−02
cubic inch (in^3)[f]	cubic meter (m^3)	1.638 706	E−05
cubic inch per minute (in^3/min)	cubic meter per second (m^3/s)	2.731 177	E−07
cubic mile (mi^3)	cubic meter (m^3)	4.168 182	E+09
cubic yard (yd^3)	cubic meter (m^3)	7.645 549	E−01
cubic yard per minute (yd^3/min)	cubic meter per second (m^3/s)	1.274 258	E−02
cup (U.S.)	cubic meter (m^3)	2.365 882	E−04
cup (U.S.)	liter (L)	2.365 882	E−01
cup (U.S.)	milliliter (mL)	2.365 882	E+02
curie (Ci)	becquerel (Bq)	**3.7**	**E+10**
darcy[g]	meter squared (m^2)	9.869 233	E−13
day (d)	second (s)	**8.64**	**E+04**
day (sidereal)	second (s)	8.616 409	E+04
debye (D)	coulomb meter (C · m)	3.335 641	E−30
degree (angle) (°)	radian (rad)	1.745 329	E−02
degree Celsius (temperature) (°C)	kelvin (K)	$T/K = t/°C + \mathbf{273.15}$	
degree Celsius (temperature interval) (°C)	kelvin (K)	**1.0**	**E+00**
degree centigrade (temperature)[h]	degree Celsius (°C)	$t/°C \approx t/\text{deg.cent.}$	
degree centigrade (temperature interval)[h]	degree Celsius (°C)	1.0	E+00
degree Fahrenheit (temperature) (°F)	degree Celsius (°C)	$t/°C = (t/°F - \mathbf{32})/\mathbf{1.8}$	
degree Fahrenheit (temperature) (°F)	kelvin (K)	$T/K = (t/°F + \mathbf{459.67})/\mathbf{1.8}$	
degree Fahrenheit (temperature interval)(°F)	degree Celsius (°C)	5.555 556	E−01
degree Fahrenheit (temperature interval) (°F)	kelvin (K)	5.555 556	E−01
degree Fahrenheit hour per British thermal unit$_{IT}$ (°F · h/Btu$_{IT}$)	kelvin per watt (K/W)	1.895 634	E+00
degree Fahrenheit hour per British thermal unit$_{th}$ (°F · h/Btu$_{th}$)	kelvin per watt (K/W)	1.896 903	E+00
degree Fahrenheit hour square foot per British thermal unit$_{IT}$ (°F · h · ft^2/Btu$_{IT}$)	square meter kelvin per watt ($m^2 \cdot K/W$)	1.761 102	E−01
degree Fahrenheit hour square foot per British thermal unit$_{th}$ (°F · h · ft^2/Btu$_{th}$)	square meter kelvin per watt ($m^2 \cdot K/W$)	1.762 280	E−01
degree Fahrenheit hour square foot per British thermal unit$_{IT}$ inch [°F · h · ft^2/(Btu$_{IT}$ · in)]	meter kelvin per watt (m · K/W)	6.933 472	E+00
degree Fahrenheit hour square foot per British thermal unit$_{th}$ inch [°F · h · ft^2/(Btu$_{th}$ · in)]	meter kelvin per watt (m · K/W)	6.938 112	E+00
degree Fahrenheit second per British thermal unit$_{IT}$ (°F · s/Btu$_{IT}$)	kelvin per watt (K/W)	5.265 651	E−04
degree Fahrenheit second per British thermal unit$_{th}$ (°F · s/Btu$_{th}$)	kelvin per watt (K/W)	5.269 175	E−04
degree Rankine (°R)	kelvin (K)	$T/K = (T/°R)/\mathbf{1.8}$	
degree Rankine (temperature interval) (°R)	kelvin (K)	5.555 556	E−01
denier	kilogram per meter (kg/m)	1.111 111	E−07
denier	gram per meter (g/m)	1.111 111	E−04
dyne (dyn)	newton (N)	**1.0**	**E−05**
dyne centimeter (dyn · cm)	newton meter (N · m)	**1.0**	**E−07**
dyne per square centimeter (dyn/cm^2)	pascal (Pa)	**1.0**	**E−01**

[f] The exact conversion factor is 1.638 706 4 E−05.
[g] The darcy is a unit for expressing the permeability of porous solids, not area.
[h] The centigrade temperature scale is obsolete; the degree centigrade is only approximately equal to the degree Celsius.

To convert from	to	Multiply by	
electronvolt (eV)	joule (J)	1.602 177	E−19
EMU of capacitance (abfarad)	farad (F)	**1.0**	**E+09**
EMU of current (abampere)	ampere (A)	**1.0**	**E+01**
EMU of electric potential (abvolt)	volt (V)	**1.0**	**E−08**
EMU of inductance (abhenry)	henry (H)	**1.0**	**E−09**
EMU of resistance (abohm)	ohm (Ω)	**1.0**	**E−09**
erg (erg)	joule (J)	**1.0**	**E−07**
erg per second (erg/s)	watt (W)	**1.0**	**E−07**
erg per square centimeter second [erg/(cm² · s)]	watt per square meter (W/m²)	**1.0**	**E−03**
ESU of capacitance (statfarad)	farad (F)	1.112 650	E−12
ESU of current (statampere)	ampere (A)	3.335 641	E−10
ESU of electric potential (statvolt)	volt (V)	2.997 925	E+02
ESU of inductance (stathenry)	henry (H)	8.987 552	E+11
ESU of resistance (statohm)	ohm (Ω)	8.987 552	E+11
faraday (based on carbon 12)	coulomb (C)	9.648 531	E+04
fathom (based on U.S survey foot)[a]	meter (m)	1.828 804	E+00
fermi	meter (m)	**1.0**	**E−15**
fermi	femtometer (fm)	**1.0**	**E+00**
fluid ounce (U.S.) (fl oz)	cubic meter (m³)	2.957 353	E−05
fluid ounce (U.S.) (fl oz)	milliliter (mL)	2.957 353	E+01
foot (ft)	meter (m)	**3.048**	**E−01**
foot (U.S. survey ft)[a]	meter (m)	3.048 006	E−01
footcandle	lux (lx)	1.076 391	E+01
footlambert	candela per square meter (cd/m²)	3.426 259	E+00
foot of mercury, conventional (ftHg)[e]	pascal (Pa)	4.063 666	E+04
foot of mercury, conventional (ftHg)[e]	kilopascal (kPa)	4.063 666	E+01
foot of water (39.2 °F)[e]	pascal (Pa)	2.988 98	E+03
foot of water (39.2 °F)[e]	kilopascal (kPa)	2.988 98	E+00
foot of water, conventional (ftH_2O)[e]	pascal (Pa)	2.989 067	E+03
foot of water, conventional (ftH_2O)[e]	kilopascal (kPa)	2.989 067	E+00
foot per hour (ft/h)	meter per second (m/s)	8.466 667	E−05
foot per minute (ft/min)	meter per second (m/s)	**5.08**	**E−03**
foot per second (ft/s)	meter per second (m/s)	**3.048**	**E−01**
foot per second squared (ft/s²)	meter per second squared (m/s²)	**3.048**	**E−01**
foot poundal	joule (J)	4.214 011	E−02
foot pound-force (ft · lbf)	joule (J)	1.355 818	E+00
foot pound-force per hour (ft · lbf/h)	watt (W)	3.766 161	E−04
foot pound-force per minute (ft · lbf/min)	watt (W)	2.259 697	E−02
foot pound-force per second (ft · lbf/s)	watt (W)	1.355 818	E+00
foot to the fourth power (ft⁴)[i]	meter to the fourth power (m⁴)	8.630 975	E−03
franklin (Fr)	coulomb (C)	3.335 641	E−10
gal (Gal)	meter per second squared (m/s²)	**1.0**	**E−02**
gallon [Canadian and U.K. (Imperial)] (gal)	cubic meter (m³)	**4.546 09**	**E−03**
gallon [Canadian and U.K. (Imperial)] (gal)	liter (L)	**4.546 09**	**E+00**
gallon (U.S.) (gal)	cubic meter (m³)	3.785 412	E−03
gallon (U.S.) (gal)	liter (L)	3.785 412	E+00
gallon (U.S.) per day (gal/d)	cubic meter per second (m³/s)	4.381 264	E−08
gallon (U.S.) per day (gal/d)	liter per second (L/s)	4.381 264	E−05
gallon (U.S.) per horsepower hour [gal/(hp · h)]	cubic meter per joule (m³/J)	1.410 089	E−09
gallon (U.S.) per horsepower hour [gal/(hp · h)]	liter per joule (L/J)	1.410 089	E−06
gallon (U.S.) per minute (gpm)(gal/min)	cubic meter per second (m³/s)	6.309 020	E−05
gallon (U.S.) per minute (gpm)(gal/min)	liter per second (L/s)	6.309 020	E−02
gamma (γ)	tesla (T)	**1.0**	**E−09**
gauss (Gs, G)	tesla (T)	**1.0**	**E−04**
gilbert (Gi)	ampere (A)	7.957 747	E−01

[i] This is a unit for the quantity second moment of area, which is sometimes called the "moment of section" or "area moment of inertia" of a plane section about a specified axis.

To convert from	to	Multiply by	
gill [Canadian and U.K. (Imperial)] (gi)	cubic meter (m^3)	1.420 653	E–04
gill [Canadian and U.K. (Imperial)] (gi)	liter (L)	1.420 653	E–01
gill (U.S.) (gi)	cubic meter (m^3)	1.182 941	E–04
gill (U.S.) (gi)	liter (L)	1.182 941	E–01
gon (also called grade) (gon)	radian (rad)	1.570 796	E–02
gon (also called grade) (gon)	degree (angle) (°)	**9.0**	**E–01**
grain (gr)	kilogram (kg)	**6.479 891**	**E–05**
grain (gr)	milligram (mg)	**6.479 891**	**E+01**
grain per gallon (U.S.) (gr/gal)	kilogram per cubic meter (kg/m^3)	1.711 806	E–02
grain per gallon (U.S.) (gr/gal)	milligram per liter (mg/L)	1.711 806	E+01
gram-force per square centimeter (gf/cm^2)	pascal (Pa)	**9.806 65**	**E+01**
gram per cubic centimeter (g/cm^3)	kilogram per cubic meter (kg/m^3)	**1.0**	**E+03**
hectare (ha)	square meter (m^2)	**1.0**	**E+04**
horsepower (550 ft · lbf/s) (hp)	watt (W)	7.456 999	E+02
horsepower (boiler)	watt (W)	9.809 50	E+03
horsepower (electric)	watt (W)	**7.46**	**E+02**
horsepower (metric)	watt (W)	7.354 988	E+02
horsepower (U.K.)	watt (W)	7.4570	E+02
horsepower (water)	watt (W)	7.460 43	E+02
hour (h)	second (s)	**3.6**	**E+03**
hour (sidereal)	second (s)	3.590 170	E+03
hundredweight (long, 112 lb)	kilogram (kg)	5.080 235	E+01
hundredweight (short, 100 lb)	kilogram (kg)	4.535 924	E+01
inch (in)	meter (m)	**2.54**	**E–02**
inch (in)	centimeter (cm)	**2.54**	**E+00**
inch of mercury (32 °F)[e]	pascal (Pa)	3.386 38	E+03
inch of mercury (32 °F)[e]	kilopascal (kPa)	3.386 38	E+00
inch of mercury (60 °F)[e]	pascal (Pa)	3.376 85	E+03
inch of mercury (60 °F)[e]	kilopascal (kPa)	3.376 85	E+00
inch of mercury, conventional (inHg)[e]	pascal (Pa)	3.386 389	E+03
inch of mercury, conventional (inHg)[e]	kilopascal (kPa)	3.386 389	E+00
inch of water (39.2 °F)[e]	pascal (Pa)	2.490 82	E+02
inch of water (60 °F)[e]	pascal (Pa)	2.4884	E+02
inch of water, conventional (inH_2O)[e]	pascal (Pa)	2.490 889	E+02
inch per second (in/s)	meter per second (m/s)	**2.54**	**E–02**
inch per second squared (in/s^2)	meter per second squared (m/s^2)	**2.54**	**E–02**
inch to the fourth power (in^4)[i]	meter to the fourth power (m^4)	4.162 314	E–07
kayser (K)	reciprocal meter (m^{-1})	**1.0**	**E+02**
kelvin (K)	degree Celsius (°C)	$t/°C = T/K$ – **273.15**	
kilocalorie$_{IT}$ (kcal$_{IT}$)	joule (J)	**4.1868**	**E+03**
kilocalorie$_{th}$ (kcal$_{th}$)	joule (J)	**4.184**	**E+03**
kilocalorie (mean) (kcal)	joule (J)	4.190 02	E+03
kilocalorie$_{th}$ per minute (kcal$_{th}$/min)	watt (W)	6.973 333	E+01
kilocalorie$_{th}$ per second (kcal$_{th}$/s)	watt (W)	**4.184**	**E+03**
kilogram-force (kgf)	newton (N)	**9.806 65**	**E+00**
kilogram-force meter (kgf · m)	newton meter (N · m)	**9.806 65**	**E+00**
kilogram-force per square centimeter (kgf/cm^2)	pascal (Pa)	**9.806 65**	**E+04**
kilogram-force per square centimeter (kgf/cm^2)	kilopascal (kPa)	**9.806 65**	**E+01**
kilogram-force per square meter (kgf/m^2)	pascal (Pa)	**9.806 65**	**E+00**
kilogram-force per square millimeter (kgf/mm^2)	pascal (Pa)	**9.806 65**	**E+06**
kilogram-force per square millimeter (kgf/mm^2)	megapascal (MPa)	**9.806 65**	**E+00**
kilogram-force second squared per meter ($kgf \cdot s^2/m$)	kilogram (kg)	**9.806 65**	**E+00**
kilometer per hour (km/h)	meter per second (m/s)	2.777 778	E–01
kilopond (kilogram-force) (kp)	newton (N)	**9.806 65**	**E+00**
kilowatt hour (kW · h)	joule (J)	**3.6**	**E+06**
kilowatt hour (kW · h)	megajoule (MJ)	**3.6**	**E+00**

To convert from	to	Multiply by	
kip (1 kip=1000 lbf)	newton (N)	4.448 222	E+03
kip (1 kip=1000 lbf)	kilonewton (kN)	4.448 222	E+00
kip per square inch (ksi) (kip/in^2)	pascal (Pa)	6.894 757	E+06
kip per square inch (ksi) (kip/in^2)	kilopascal (kPa)	6.894 757	E+03
knot (nautical mile per hour)	meter per second (m/s)	5.144 444	E−01
lambert[j]	candela per square meter (cd/m^2)	3.183 099	E+03
langley (cal_{th}/cm^2)	joule per square meter (J/m^2)	**4.184**	**E+04**
light year (l.y.)[k]	meter (m)	9.460 73	E+15
liter (L)[l]	cubic meter (m^3)	**1.0**	**E−03**
lumen per square foot (lm/ft^2)	lux (lx)	1.076 391	E+01
maxwell (Mx)	weber (Wb)	**1.0**	**E−08**
mho	siemens (S)	**1.0**	**E+00**
microinch	meter (m)	**2.54**	**E−08**
microinch	micrometer (μm)	**2.54**	**E−02**
micron (μ)	meter (m)	**1.0**	**E−06**
micron (μ)	micrometer (μm)	**1.0**	**E+00**
mil (0.001 in)	meter (m)	**2.54**	**E−05**
mil (0.001 in)	millimeter (mm)	**2.54**	**E−02**
mil (angle)	radian (rad)	9.817 477	E−04
mil (angle)	degree (°)	**5.625**	**E−02**
mile (mi)	meter (m)	**1.609 344**	**E+03**
mile (mi)	kilometer (km)	**1.609 344**	**E+00**
mile (based on U.S. survey foot) (mi)[a]	meter (m)	1.609 347	E+03
mile (based on U.S. survey foot) (mi)[a]	kilometer (km)	1.609 347	E+00
mile, nautical[m]	meter (m)	**1.852**	**E+03**
mile per gallon (U.S.) (mpg) (mi/gal)	meter per cubic meter (m/m^3)	4.251 437	E+05
mile per gallon (U.S.) (mpg) (mi/gal)	kilometer per liter (km/L)	4.251 437	E−01
mile per gallon (U.S.) (mpg) (mi/gal)[n]	liter per 100 kilometer (L/100 km)..	divide 235.215 by number of miles per gallon	
mile per hour (mi/h)	meter per second (m/s)	**4.4704**	**E−01**
mile per hour (mi/h)	kilometer per hour (km/h)	**1.609 344**	**E+00**
mile per minute (mi/min)	meter per second (m/s)	**.2.682 24**	**E+01**
mile per second (mi/s)	meter per second (m/s)	**.1.609 344**	**E+03**
millibar (mbar)	pascal (Pa)	**.1.0**	**E+02**
millibar (mbar)	kilopascal (kPa)	**.1.0**	**E−01**
millimeter of mercury, conventional (mmHg)[e]	pascal (Pa)	.1.333 224	E+02
millimeter of water, conventional (mmH_2O)[e]	pascal (Pa)	**9.806 65**	**E+00**
minute (angle) (')	radian (rad)	.2.908 882	E−04
minute (min)	second (s)	**.6.0**	**E+01**
minute (sidereal)	second (s)	.5.983 617	E+01
nit	candela per meter squared (cd/m^2)	**1.0**	**E+00**
nox	lux (lx)	**1.0**	**E−03**
oersted (Oe)	ampere per meter (A/m)	.7.957 747	E+01
ohm centimeter (Ω · cm)	ohm meter (Ω · m)	**.1.0**	**E−02**
ohm circular-mil per foot	ohm meter (Ω · m)	.1.662 426	E−09
ohm circular-mil per foot	ohm square millimeter per meter (Ω · mm^2/m)	.1.662 426	E−03
ounce (avoirdupois) (oz)	kilogram (kg)	.2.834 952	E−02
ounce (avoirdupois) (oz)	gram (g)	.2.834 952	E+01
ounce (troy or apothecary) (oz)	kilogram (kg)	.3.110 348	E−02
ounce (troy or apothecary) (oz)	gram (g)	.3.110 348	E+01
ounce [Canadian and U.K. fluid (Imperial)] (fl oz)	cubic meter (m^3)	.2.841 306	E−05

[j] The exact conversion factor is $10^4/\pi$.

[k] This conversion factor is based on 1 d = 86 400 s; and 1 Julian century = 36 525 d. (See *The Astronomical Almanac for the Year 1995*, page K6, U.S. Government Printing Office, Washington, DC, 1994.)

[l] In 1964 the General Conference on Weights and Measures reestablished the name "liter" as a special name for the cubic decimeter. Between 1901 and 1964 the liter was slightly larger (1.000 028 dm^3); when one uses high-accuracy volume data of that time, this fact must be kept in mind.

[m] The value of this unit, 1 nautical mile = 1852 m, was adopted by the First International Extraordinary Hydrographic Conference, Monaco, 1929, under the name "International nautical mile."

[n] For converting fuel economy, as used in the U.S., to fuel consumption.

To convert from	to	Multiply by	
ounce [Canadian and U.K. fluid (Imperial)] (fl oz)	milliliter (mL)	2.841 306	E+01
ounce (U.S. fluid) (fl oz)	cubic meter (m^3)	2.957 353	E−05
ounce (U.S. fluid) (fl oz)	milliliter (mL)	2.957 353	E+01
ounce (avoirdupois)-force (ozf)	newton (N)	2.780 139	E−01
ounce (avoirdupois)-force inch (ozf · in)	newton meter (N · m)	7.061 552	E−03
ounce (avoirdupois)-force inch (ozf · in)	millinewton meter (mN · m)	7.061 552	E+00
ounce (avoirdupois) per cubic inch (oz/in^3)	kilogram per cubic meter (kg/m^3)	1.729 994	E+03
ounce (avoirdupois) per gallon [Canadian and U.K. (Imperial)] (oz/gal)	kilogram per cubic meter (kg/m^3)	6.236 023	E+00
ounce (avoirdupois) per gallon [Canadian and U.K. (Imperial)] (oz/gal)	gram per liter (g/L)	6.236 023	E+00
ounce (avoirdupois) per gallon (U.S.)(oz/gal)	kilogram per cubic meter (kg/m^3)	7.489 152	E+00
ounce (avoirdupois) per gallon (U.S.)(oz/gal)	gram per liter (g/L)	7.489 152	E+00
ounce (avoirdupois) per square foot (oz/ft^2)	kilogram per square meter (kg/m^2)	3.051 517	E−01
ounce (avoirdupois) per square inch (oz/in^2)	kilogram per square meter (kg/m^2)	4.394 185	E+01
ounce (avoirdupois) per square yard(oz/yd^2)	kilogram per square meter (kg/m^2)	3.390 575	E−02
parsec (pc)	meter (m)	3.085 678	E+16
peck (U.S.) (pk)	cubic meter (m^3)	8.809 768	E−03
peck (U.S.) (pk)	liter (L)	8.809 768	E+00
pennyweight (dwt)	kilogram (kg)	1.555 174	E−03
pennyweight (dwt)	gram (g)	1.555 174	E+00
perm (0 °C)	kilogram per pascal second square meter [kg/(Pa · s · m^2)]	5.721 35	E−11
perm (23 °C)	kilogram per pascal second square meter [kg/(Pa · s · m^2)]	5.745 25	E−11
perm inch (0 °C)	kilogram per pascal second meter [kg/(Pa · s · m)]	1.453 22	E−12
perm inch (23 °C)	kilogram per pascal second meter [kg/(Pa · s · m)]	1.459 29	E−12
phot (ph)	lux (lx)	**1.0**	**E+04**
pica (computer) (1/6 in)	meter (m)	4.233 333	E−03
pica (computer) (1/6 in)	millimeter (mm)	4.233 333	E+00
pica (printer's)	meter (m)	4.217 518	E−03
pica (printer's)	millimeter (mm)	4.217 518	E+00
pint (U.S. dry) (dry pt)	cubic meter (m^3)	5.506 105	E−04
pint (U.S. dry) (dry pt)	liter (L)	5.506 105	E−01
pint (U.S. liquid) (liq pt)	cubic meter (m^3)	4.731 765	E−04
pint (U.S. liquid) (liq pt)	liter (L)	4.731 765	E−01
point (computer) (1/72 in)	meter (m)	3.527 778	E−04
point (computer) (1/72 in)	millimeter (mm)	3.527 778	E−01
point (printer's)	meter (m)	3.514 598	E−04
point (printer's)	millimeter (mm)	3.514 598	E−01
poise (P)	pascal second (Pa · s)	**1.0**	**E−01**
pound (avoirdupois) (lb)[o]	kilogram (kg)	4.535 924	E−01
pound (troy or apothecary) (lb)	kilogram (kg)	3.732 417	E−01
poundal	newton (N)	1.382 550	E−01
poundal per square foot	pascal (Pa)	1.488 164	E+00
poundal second per square foot	pascal second (Pa · s)	1.488 164	E+00
pound foot squared (lb · ft^2)	kilogram meter squared (kg · m^2)	4.214 011	E−02
pound-force (lbf)[p]	newton (N)	4.448 222	E+00
pound-force foot (lbf · ft)	newton meter (N · m)	1.355 818	E+00
pound-force foot per inch (lbf · ft/in)	newton meter per meter (N · m/m)	5.337 866	E+01
pound-force inch (lbf · in)	newton meter (N · m)	1.129 848	E−01
pound-force inch per inch (lbf · in/in)	newton meter per meter (N · m/m)	4.448 222	E+00
pound-force per foot (lbf/ft)	newton per meter (N/m)	1.459 390	E+01
pound-force per inch (lbf/in)	newton per meter (N/m)	1.751 268	E+02
pound-force per pound (lbf/lb) (thrust to mass ratio)	newton per kilogram (N/kg)	**9.806 65**	**E+00**

[o] The exact conversion factor is 4.535 923 7 E−01. All units that contain the pound refer to the avoirdupois pound unless otherwise specified.

[p] If the local value of the acceleration of free fall is taken as g_n=9.806 65 m/ s^2 (the standard value), the exact conversion factor is 4.448 221 615 260 5 E+00.

To convert from	to	Multiply by	
pound-force per square foot (lbf/ft^2)	pascal (Pa)	4.788 026	E+01
pound-force per square inch (psi) (lbf/in^2)	pascal (Pa)	6.894 757	E+03
pound-force per square inch (psi) (lbf/in^2)	kilopascal (kPa)	6.894 757	E+00
pound-force second per square foot ($lbf \cdot s/ft^2$)	pascal second ($Pa \cdot s$)	4.788 026	E+01
pound-force second per square inch ($lbf \cdot s/in^2$)	pascal second ($Pa \cdot s$)	6.894 757	E+03
pound inch squared ($lb \cdot in^2$)	kilogram meter squared ($kg \cdot m^2$)	2.926 397	E−04
pound per cubic foot (lb/ft^3)	kilogram per cubic meter (kg/m^3)	1.601 846	E+01
pound per cubic inch (lb/in^3)	kilogram per cubic meter (kg/m^3)	2.767 990	E+04
pound per cubic yard (lb/yd^3)	kilogram per cubic meter (kg/m^3)	5.932 764	E−01
pound per foot (lb/ft)	kilogram per meter (kg/m)	1.488 164	E+00
pound per foot hour [$lb/(ft \cdot h)$]	pascal second ($Pa \cdot s$)	4.133 789	E−04
pound per foot second [$lb/(ft \cdot s)$]	pascal second ($Pa \cdot s$)	1.488 164	E+00
pound per gallon [Canadian and U.K. (Imperial)] (lb/gal)	kilogram per cubic meter (kg/m^3)	9.977 637	E+01
pound per gallon [Canadian and U.K. (Imperial)] (lb/gal)	kilogram per liter (kg/L)	9.977 637	E−02
pound per gallon (U.S.) (lb/gal)	kilogram per cubic meter (kg/m^3)	1.198 264	E+02
pound per gallon (U.S.) (lb/gal)	kilogram per liter (kg/L)	1.198 264	E−01
pound per horsepower hour [$lb/(hp \cdot h)$]	kilogram per joule (kg/J)	1.689 659	E−07
pound per hour (lb/h)	kilogram per second (kg/s)	1.259 979	E−04
pound per inch (lb/in)	kilogram per meter (kg/m)	1.785 797	E+01
pound per minute (lb/min)	kilogram per second (kg/s)	7.559 873	E−03
pound per second (lb/s)	kilogram per second (kg/s)	4.535 924	E−01
pound per square foot (lb/ft^2)	kilogram per square meter (kg/m^2)	4.882 428	E+00
pound per square inch (*not* pound-force) (lb/in^2)	kilogram per square meter (kg/m^2)	7.030 696	E+02
pound per yard (lb/yd)	kilogram per meter (kg/m)	4.960 546	E−01
psi (pound-force per square inch) (lbf/in^2)	pascal (Pa)	6.894 757	E+03
psi (pound-force per square inch) (lbf/in^2)	kilopascal (kPa)	6.894 757	E+00
quad (10^{15} Btu_{IT})[c]	joule (J)	1.055 056	E+18
quart (U.S. dry) (dry qt)	cubic meter (m^3)	1.101 221	E−03
quart (U.S. dry) (dry qt)	liter (L)	1.101 221	E+00
quart (U.S. liquid) (liq qt)	cubic meter (m^3)	9.463 529	E−04
quart (U.S. liquid) (liq qt)	liter (L)	9.463 529	E−01
rad (absorbed dose) (rad)	gray (Gy)	**1.0**	**E−02**
rem (rem)	sievert (Sv)	**1.0**	**E−02**
revolution (r)	radian (rad)	6.283 185	E+00
revolution per minute (rpm) (r/min)	radian per second (rad/s)	1.047 198	E−01
rhe	reciprocal pascal second [$(Pa \cdot s)^{-1}$]	**1.0**	**E+01**
rod (based on U.S. survey foot) (rd)[a]	meter (m)	5.029 210	E+00
roentgen (R)	coulomb per kilogram (C/kg)	**2.58**	**E−04**
rpm (revolution per minute) (r/min)	radian per second (rad/s)	1.047 198	E−01
second (angle) (″)	radian (rad)	4.848 137	E−06
second (sidereal)	second (s)	9.972 696	E−01
shake	second (s)	**1.0**	**E−08**
shake	nanosecond (ns)	**1.0**	**E+01**
skot	candela per meter squared (cd/m^2)	3.183 098	E−04
slug (slug)	kilogram (kg)	1.459 390	E+01
slug per cubic foot ($slug/ft^3$)	kilogram per cubic meter (kg/m^3)	5.153 788	E+02
slug per foot second [$slug/(ft \cdot s)$]	pascal second ($Pa \cdot s$)	4.788 026	E+01
square foot (ft^2)	square meter (m^2)	**9.290 304**	**E−02**
square foot per hour (ft^2/h)	square meter per second (m^2/s)	**2.580 64**	**E−05**
square foot per second (ft^2/s)	square meter per second (m^2/s)	**9.290 304**	**E−02**
square inch (in^2)	square meter (m^2)	**6.4516**	**E−04**
square inch (in^2)	square centimeter (cm^2)	**6.4516**	**E+00**
square mile (mi^2)	square meter (m^2)	2.589 988	E+06
square mile (mi^2)	square kilometer (km^2)	2.589 988	E+00

To convert from	to	Multiply by	
square mile (based on U.S. survey foot) (mi^2)[a]	square meter (m^2)	2.589 998	E+06
square mile (based on U.S. survey foot) (mi^2)[a]	square kilometer (km^2)	2.589 998	E+00
square yard (yd^2)	square meter (m^2)	8.361 274	E−01
statampere	ampere (A)	3.335 641	E−10
statcoulomb	coulomb (C)	3.335 641	E−10
statfarad	farad (F)	1.112 650	E−12
stathenry	henry (H)	8.987 552	E+11
statmho	siemens (S)	1.112 650	E−12
statohm	ohm (Ω)	8.987 552	E+11
statvolt	volt (V)	2.997 925	E+02
stere (st)	cubic meter (m^3)	**1.0**	**E+00**
stilb (sb)	candela per square meter (cd/m^2)	**1.0**	**E+04**
stokes (St)	meter squared per second (m^2/s)	**1.0**	**E−04**
tablespoon	cubic meter (m^3)	1.478 676	E−05
tablespoon	milliliter (mL)	1.478 676	E+01
teaspoon	cubic meter (m^3)	4.928 922	E−06
teaspoon	milliliter (mL)	4.928 922	E+00
tex	kilogram per meter (kg/m)	**1.0**	**E−06**
therm (EC)[q]	joule (J)	**1.055 06**	**E+08**
therm (U.S.)[q]	joule (J)	**1.054 804**	**E+08**
ton, assay (AT)	kilogram (kg)	2.916 667	E−02
ton, assay (AT)	gram (g)	2.916 667	E+01
ton-force (2000 lbf)	newton (N)	8.896 443	E+03
ton-force (2000 lbf)	kilonewton (kN)	8.896 443	E+00
ton, long (2240 lb)	kilogram (kg)	1.016 047	E+03
ton, long, per cubic yard	kilogram per cubic meter (kg/m^3)	1.328 939	E+03
ton, metric (t)	kilogram (kg)	**1.0**	**E+03**
tonne (called "metric ton" in U.S.) (t)	kilogram (kg)	**1.0**	**E+03**
ton of refrigeration (12 000 Btu_{IT}/h)	watt (W)	3.516 853	E+03
ton of TNT (energy equivalent)[r]	joule (J)	**4.184**	**E+09**
ton, register	cubic meter (m^3)	2.831 685	E+00
ton, short (2000 lb)	kilogram (kg)	9.071 847	E+02
ton, short, per cubic yard	kilogram per cubic meter (kg/m^3)	1.186 553	E+03
ton, short, per hour	kilogram per second (kg/s)	2.519 958	E−01
torr (Torr)	pascal (Pa)	1.333 224	E+02
unit pole	weber (Wb)	1.256 637	E−07
watt hour (W · h)	joule (J)	**3.6**	**E+03**
watt per square centimeter (W/cm^2)	watt per square meter (W/m^2)	**1.0**	**E+04**
watt per square inch (W/in^2)	watt per square meter (W/m^2)	1.550 003	E+03
watt second (W · s)	joule (J)	**1.0**	**E+00**
yard (yd)	meter (m)	**9.144**	**E−01**
year (365 days)	second (s)	**3.1536**	**E+07**
year (sidereal)	second (s)	3.155 815	E+07
year (tropical)	second (s)	3.155 693	E+07

[q] The therm (EC) is legally defined in the Council Directive of 20 December 1979, Council of the European Communities (now the European Union, EU). The therm (U.S.) is legally defined in the Federal Register of July 27, 1968. Although the therm (EC), which is based on the International Table Btu, is frequently used by engineers in the United States, the therm (U.S.) is the legal unit used by the U.S natural gas industry.

[r] Defined (not measured) value.

CONVERSION OF TEMPERATURES

From	To	Conversion formula
Celsius	Fahrenheit	$t_F/°F = (9/5)t/°C + 32$
	Kelvin	$T/K = t/°C + 273.15$
	Rankine	$T/°R = (9/5)(t/°C + 273.15)$
Fahrenheit	Celsius	$t/°C = (5/9)[(t_F/°F) - 32]$
	Kelvin	$T/K = (5/9)[(t_F/°F) - 32] + 273.15$
	Rankine	$T/°R = t_F/°F + 459.67$
Kelvin	Celsius	$t/°C = T/K - 273.15$
	Rankine	$T/°R = (9/5)T/K$
Rankine	Fahrenheit	$t_F/°F = T/°R - 459.67$
	Kelvin	$T/K = (5/9)T/°R$

Definition of symbols:

T: thermodynamic (absolute) temperature

t: Celsius temperature (the symbol θ is also used for Celsius temperature)

t_F: Fahrenheit temperature

Designation of Large Numbers

Power of 10	U.S.A.	Other countries
10^6	million	million
10^9	billion	milliard
10^{12}	trillion	billion
10^{15}	quadrillion	billiard
10^{18}	quintillion	trillion
100^{100}	googol	
10^{googol}	googolplex	

CONVERSION FACTORS FOR ENERGY UNITS

If greater accuracy is required, use the Energy Equivalents section of the "CODATA Recommended Values of the Fundamental Physical Constants" table.

Unit	Wavenumber $\bar{\nu}$ cm⁻¹	Frequency ν MHz	Energy E aJ	Energy E eV	Energy E E_h	Molar energy E_m kJ/mol	Molar energy E_m kcal/mol	Temperature T K
$\bar{\nu}$: 1 cm^{-1}	1	2.997925×10^4	1.986447×10^{-5}	1.239842×10^{-4}	4.556335×10^{-6}	11.96266×10^{-3}	2.85914×10^{-3}	1.438769
ν: 1 MHz	3.33564×10^{-5}	1	6.626076×10^{-10}	4.135669×10^{-9}	1.519830×10^{-10}	3.990313×10^{-7}	9.53708×10^{-8}	4.79922×10^{-5}
E: 1 aJ	50341.1	1.509189×10^9	1	6.241506	0.2293710	602.2137	143.9325	7.24292×10^4
E: 1 eV	8065.54	2.417988×10^8	0.1602177	1	3.674931×10^{-2}	96.4853	23.0605	1.16045×10^4
E: E_h	219474.63	6.579684×10^9	4.359748	27.2114	1	2625.500	627.510	3.15773×10^5
E_m: 1 kJ/mol	83.5935	2.506069×10^6	1.660540×10^{-3}	1.036427×10^{-2}	3.808798×10^{-4}	1	0.239006	120.272
E_m: 1 kcal/mol	349.755	1.048539×10^7	6.947700×10^{-3}	4.336411×10^{-2}	1.593601×10^{-3}	4.184	1	503.217
T: 1 K	0.695039	2.08367×10^4	1.380658×10^{-5}	8.61738×10^{-5}	3.16683×10^{-6}	8.31451×10^{-3}	1.98722×10^{-3}	1

Examples of the use of this table:

1 aJ ≙ 50341.1 cm^{-1}

1 ev ≙ 96.4853 kJ mol^{-1}

The symbol ≙ should be read as meaning corresponds to or is equivalent to.

$E = h\nu = hc\bar{\nu} = kT$

$E_m = N_A E$ where N_A is the Avogadro constant

E_h is the Hartree energy

CONVERSION FACTORS FOR PRESSURE UNITS

	Pa	kPa	MPa	bar	atm	Torr	μmHg	psi
Pa	1	0.001	0.000001	0.00001	9.8692×10^{-6}	0.0075006	7.5006	0.0001450377
kPa	1000	1	0.001	0.01	0.0098692	7.5006	7500.6	0.1450377
MPa	1000000	1000	1	10	9.8692	7500.6	7500600	145.0377
bar	100000	100	0.1	1	0.98692	750.06	750060	14.50377
atm	101325	101.325	0.101325	1.01325	1	760	760000	14.69594
Torr	133.322	0.133322	0.000133322	0.00133322	0.00131579	1	1000	0.01933672
μmHg	0.133322	0.000133322	1.33322×10^{-7}	1.33322×10^{-6}	1.31579×10^{-6}	0.001	1	1.933672×10^{-5}
psi	6894.757	6.894757	0.006894757	0.06894757	0.068046	51.7151	51715.1	1

To convert a pressure value from a unit in the left-hand column to a new unit, multiply the value by the factor appearing in the column for the new unit. For example:

$$1 \text{ kPa} = 9.8692 \times 10^{-3} \text{ atm}$$
$$1 \text{ Torr} = 1.33322 \times 10^{-4} \text{ MPa}$$

Notes: μmHg is often referred to as "micron"
Torr is essentially identical to mmHg
psi is an abbreviation for the unit pound–force per square inch
psia (as a term for a physical quantity) implies the true (absolute) pressure
psig implies the true pressure minus the local atmospheric pressure

CONVERSION FACTORS FOR THERMAL CONDUCTIVITY UNITS

MULTIPLY ↓ by appropriate factor to OBTAIN→	Btu_{IT} h^{-1} ft^{-1} °F^{-1}	Btu_{IT} in. h^{-1} ft^{-2} °F^{-1}	Btu_{th} h^{-1} ft^{-1} °F^{-1}	Btu_{th} in. h^{-1} ft^{-2} °F^{-1}	cal_{IT} s^{-1} cm^{-1} °C^{-1}	cal_{th} s^{-1} cm^{-1} °C^{-1}	$kcal_{th}$ h^{-1} m^{-1} °C^{-1}	J s^{-1} cm^{-1} K^{-1}	W cm^{-1} K^{-1}	W m^{-1} K^{-1}	mW cm^{-1} K^{-1}
Btu_{IT} h^{-1} ft^{-1} °F^{-1}	1	12	1.00067	12.0080	$4.13379{\times}10^{-3}$	$4.13656{\times}10^{-3}$	1.48916	$1.73073{\times}10^{-2}$	$1.73073{\times}10^{-2}$	1.73073	17.3073
Btu_{IT} in h^{-1} ft^{-2} °F^{-1}	$8.33333{\times}10^{-2}$	1	$8.33891{\times}10^{-2}$	1.00067	$3.44482{\times}10^{-4}$	$3.44713{\times}10^{-4}$	0.124097	$1.44228{\times}10^{-3}$	$1.44228{\times}10^{-3}$	0.144228	1.44228
Btu_{th} h^{-1} ft^{-1} °F^{-1}	0.999331	11.9920	1	12	$4.13102{\times}10^{-3}$	$4.13379{\times}10^{-3}$	1.48816	$1.72958{\times}10^{-2}$	$1.72958{\times}10^{-2}$	1.72958	17.2958
Btu_{th} in. h^{-1} ft^{-2} °F^{-1}	$8.32776{\times}10^{-2}$	0.999331	$8.33333{\times}10^{-2}$	1	$3.44252{\times}10^{-4}$	$3.44482{\times}10^{-4}$	0.124014	$1.44131{\times}10^{-3}$	$1.44131{\times}10^{-3}$	0.144131	1.44131
cal_{IT} s^{-1} cm^{-1} °C^{-1}	$2.41909{\times}10^{2}$	$2.90291{\times}10^{3}$	$2.42071{\times}10^{2}$	$2.90485{\times}10^{3}$	1	1.00067	$3.60241{\times}10^{2}$	4.1868	4.1868	$4.1868{\times}10^{2}$	$4.1868{\times}10^{3}$
cal_{th} s^{-1} cm^{-1} °C^{-1}	$2.41747{\times}10^{2}$	$2.90096{\times}10^{3}$	$2.41909{\times}10^{2}$	$2.90291{\times}10^{3}$	0.999331	1	$3.6{\times}10^{2}$	4.184	4.184	$4.184{\times}10^{2}$	$4.184{\times}10^{3}$
$kcal_{th}$ h^{-1} m^{-1} °C^{-1}	0.671520	8.05824	0.671969	8.06363	$2.77592{\times}10^{-3}$	$2.77778{\times}10^{-3}$	1	$1.16222{\times}10^{-2}$	$1.16222{\times}10^{-2}$	1.16222	11.6222
J s^{-1} cm^{-1} K^{-1}	57.7789	$6.93347{\times}10^{2}$	57.8176	$6.93811{\times}10^{2}$	0.238846	0.239006	86.0421	1	1	$1{\times}10^{2}$	$1{\times}10^{3}$
W cm^{-1} K^{-1}	57.7789	$6.93347{\times}10^{2}$	57.8176	$6.93811{\times}10^{2}$	0.238846	0.239006	86.0421	1	1	$1{\times}10^{2}$	$1{\times}10^{3}$
W m^{-1} K^{-1}	0.577789	6.93347	0.578176	6.93811	$2.38846{\times}10^{-3}$	$2.39006{\times}10^{-3}$	0.860421	$1{\times}10^{-2}$	$1{\times}10^{-2}$	1	10
mW cm^{-1} K^{-1}	$5.77789{\times}10^{-2}$	0.693347	$5.78176{\times}10^{-2}$	0.693811	$2.38846{\times}10^{-4}$	$2.39006{\times}10^{-4}$	$8.60421{\times}10^{-2}$	$1{\times}10^{-3}$	$1{\times}10^{-3}$	0.1	1

CONVERSION FACTORS FOR ELECTRICAL RESISTIVITY UNITS

To convert FROM ↓ multiply by appropriate factor to OBTAIN →	abΩ cm	μΩ cm	Ω cm	StatΩ cm	Ω m	Ω cir. mil ft^{-1}	Ω in.	Ω ft
abohm centimeter	1	1×10^{-3}	10^{-9}	1.113×10^{-21}	10^{-11}	6.015×10^{-3}	3.937×10^{-10}	3.281×10^{-11}
microohm centimeter	10^{3}	1	10^{-6}	1.113×10^{-18}	10^{-8}	6.015	3.937×10^{-7}	3.281×10^{-6}
ohm centimeter	10^{9}	10^{6}	1	1.113×10^{-12}	1×10^{-2}	6.015×10^{6}	3.937×10^{-1}	3.281×10^{-2}
statohm centimeter (esu)	8.987×10^{20}	8.987×10^{17}	8.987×10^{11}	1	8.987×10^{9}	5.406×10^{18}	3.538×10^{11}	2.949×10^{10}
ohm meter	10^{11}	10^{8}	10^{2}	1.113×10^{-10}	1	6.015×10^{8}	3.937×10^{1}	3.281
ohm circular mil per foot	1.662×10^{2}	1.662×10^{-1}	1.662×10^{-7}	1.850×10^{-19}	1.662×10^{-9}	1	6.54×10^{-6}	5.45×10^{-9}
ohm inch	2.54×10^{9}	2.54×10^{6}	2.54	2.827×10^{-12}	2.54×10^{-2}	1.528×10^{7}	1	8.3×10^{-2}
ohm foot	3.048×10^{10}	3.048×10^{7}	3.048×10^{-1}	3.3924×10^{-11}	3.048×10^{-1}	1.833×10^{8}	12	1

CONVERSION FORMULAS FOR CONCENTRATION OF SOLUTIONS

A: Weight percent of solute
B: Molecular weight of solvent
E: Molecular weight of solute
F: Grams of solute per liter of solution
G: Molality
M: Molarity
N: Mole fraction
R: Density of solution in grams per milliliter

Relation between Concentration Expressed in a Desired Unit (First Column) and Concentration Given in Other Units (Subsequent Columns)

Desired unit	*A*	*N*	*G*	*M*	*F*
A	—	$\frac{100N \times E}{N \times E + (1-N)B}$	$\frac{100G \times E}{1000 + G \times E}$	$\frac{M \times E}{10R}$	$\frac{F}{10R}$
N	$\frac{\frac{A}{E}}{\frac{A}{E} + \frac{100-A}{B}}$	—	$\frac{B \times G}{B \times G + 1000}$	$\frac{B \times M}{M(B-E) + 1000R}$	$\frac{B \times F}{F(B-E) + 1000R \times E}$
G	$\frac{1000A}{E(100-A)}$	$\frac{1000N}{B - N \times B}$	—	$\frac{1000M}{1000R - (M \times E)}$	$\frac{1000F}{E(1000R - F)}$
M	$\frac{10R \times A}{E}$	$\frac{1000R \times N}{N \times E + (1-N)B}$	$\frac{1000R \times G}{1000 + E \times G}$	—	$\frac{F}{E}$
F	$10AR$	$\frac{1000R \times N \times E}{N \times E + (1-N)B}$	$\frac{1000R \times G \times E}{1000 + G \times E}$	$M \times E$	—

DESCRIPTIVE TERMS FOR SOLUBILITY

This table lists approximate solubility as indicated by one of the following descriptive terms as defined by USP32-NF27 [1].

Descriptive term	Parts of solvent required for 1 part of solute
Very soluble	Less than 1
Freely soluble	From 1 to 10
Soluble	From 10 to 30
Sparingly soluble	From 30 to 100
Slightly soluble	From 100 to 1000
Very slightly soluble	From 1000 to 10000
Practically insoluble, or Insoluble	Greater than or equal to 10000

Reference

1. *The Pharmacopeia of the United States of America,* Thirty-Second Revision and *The National Formulary,* Twenty-Seventh Edition (General Notices and Requirements), pp. 1-12, 2009.

CONVERSION FACTORS FOR CHEMICAL KINETICS

Equivalent Second Order Rate Constants

A \ B	$cm^3\ mol^{-1}s^{-1}$	$dm^3\ mol^{-1}s^{-1}$	$m^3\ mol^{-1}s^{-1}$	cm^3 $molecule^{-1}s^{-1}$	$(mmHg)^{-1}s^{-1}$	$atm^{-1}s^{-1}$	$ppm^{-1}min^{-1}$	$m^2kN^{-1}s^{-1}$
1 $cm^3\ mol^{-1}\ s^{-1}$ =	1	10^{-3}	10^{-6}	1.66×10^{-24}	$1.604 \times 10^{-5}\ T^{-1}$	$1.219 \times 10^{-2}\ T^{-1}$	2.453×10^{-9}	$1.203 \times 10^{-4}\ T^{-1}$
1 $dm^3\ mol^{-1}\ s^{-1}$ =	10^3	1	10^{-3}	1.66×10^{-21}	$1.604 \times 10^{-2}\ T^{-1}$	$12.19\ T^{-1}$	2.453×10^{-6}	$1.203 \times 10^{-1}\ T^{-1}$
1 $m^3\ mol^{-1}\ s^{-1}$ =	10^6	10^3	1	1.66×10^{-18}	$16.04\ T^{-1}$	$1.219 \times 10^4\ T^{-1}$	2.453×10^{-3}	$120.3\ T^{-1}$
1cm^3 $molecule^{-1}s^{-1}$ =	6.023×10^{23}	6.023×10^{20}	6.023×10^{17}	1	$9.658 \times 10^{18}T^{-1}$	$7.34 \times 10^{21}\ T^{-1}$	1.478×10^{15}	$7.244 \times 10^{19}T^{-1}$
1 $(mmHg)^{-1}\ s^{-1}$ =	$6.236 \times 10^4\ T$	$62.36\ T$	$6.236 \times 10^{-2}\ T$	$1.035 \times 10^{-19}\ T$	1	760	4.56×10^{-2}	7.500
1 $atm^{-1}\ s^{-1}$	$82.06\ T$	$8.206 \times 10^{-2}\ T$	$8.206 \times 10^{-5}\ T$	$1.362 \times 10^{-22}\ T$	1.316×10^{-3}	1	6×10^{-5}	9.869×10^{-3}
1 $ppm^{-1}\ min^{-1}$ = at 298 K, 1 atm total pressure	4.077×10^8	4.077×10^5	407.7	6.76×10^{-16}	21.93	1.667×10^4	1	164.5
1 $m^2kN^{-1}s^{-1}$ =	$8314\ T$	$8.314\ T$	$8.314 \times 10^{-3}\ T$	$1.38 \times 10^{-20}\ T$	0.1333	101.325	6.079×10^{-3}	1

To convert a rate constant from one set of units A to a new set B find the conversion factor for the row A under column B and multiply the old value by it, e.g.. to convert cm^3 $molecule^{-1}$ s^{-1} to m^3 mol^{-1} s^{-1} multiply by 6.023×10^{17}.

Table adapted from High Temperature Reaction Rate Data No. 5, The University, Leeds (1970).

Equivalent Third Order Rate Constants

A \ B	$cm^6\ mol^{-2}s^{-1}$	$dm^6\ mol^{-1}s^{-1}$	$m^6\ mol^{-2}\ s^{-1}$	cm^6 $molecule^{-2}\ s^{-1}$	$(mmHg)^{-2}\ s^{-1}$	$atm^{-2}\ s^{-1}$	$ppm^{-2}\ min^{-1}$	$m^4\ kN^{-2}s^{-1}$
1 $cm^6\ mol^{-2}\ s^{-1}$=	1	10^{-6}	10^{-12}	2.76×10^{-48}	$2.57 \times 10^{-10}T^{-2}$	$1.48 \times 10^{-4}\ T^{-2}$	1.003×10^{-19}	$1.477 \times 10^{-8}T^{-2}$
1 $dm^6\ mol^{-2}\ s^{-1}$ =	10^6	1	10^{-6}	2.76×10^{-42}	$2.57 \times 10^{-4}T^{-2}$	$148\ T^{-2}$	1.003×10^{-13}	$1.477 \times 10^{-2}T^{-2}$
1 $m^6\ mol^{-2}\ s^{-1}$=	10^{12}	10^6	1	2.76×10^{-36}	$257\ T^{-2}$	$1.48 \times 10^8T^{-2}$	1.003×10^{-7}	$1.477 \times 10^4T^{-2}$
1cm^6 $molecule^{-2}s^{-1}$=	3.628×10^{47}	3.628×10^{41}	3.628×10^{35}	1	$9.328 \times 10^{37}T^{-2}$	$5.388 \times 10^{43}T^{-2}$	3.64×10^{28}	$5.248 \times 10^{39}T^{-2}$
1 $(mmHg)^{-2}\ s^{-1}$ =	$3.89 \times 10^9\ T^2$	$3.89 \times 10^3\ T^2$	$3.89 \times 10^{-3}\ T^2$	$1.07 \times 10^{-38}T^2$	1	5.776×10^5	3.46×10^{-5}	56.25
1 $atm^{-2}\ s^{-1}$ =	$6.733 \times 10^3\ T^2$	$6.733 \times 10^{-3}\ T^2$	$6.733 \times 10^{-9}\ T^2$	$1.86 \times 10^{-44}T^2$	1.73×10^{-6}	1	6×10^{-11}	9.74×10^{-5}
1 $ppm^{-2}\ min^{-1}$= at 298K, 1 atm total pressure	9.97×10^{18}	9.97×10^{12}	9.97×10^6	2.75×10^{-29}	2.89×10^4	1.667×10^{10}	1	1.623×10^6
1 $m^4kN^{-2}s^{-1}$=	$6.91 \times 10^7\ T^2$	$6.91\ T^2$	$69.1 \times 10^{-5}\ T^2$	$1.904 \times 10^{-40}\ T^2$	0.0178	1.027×10^4	6.16×10^{-7}	1

From *J. Phys. Chem. Ref. Data*, 9, 470, 1980, by permission of the authors and the copyright owner, the American Institute of Physics.

CONVERSION FACTORS FOR IONIZING RADIATION

Conversion between SI and Other Units

Quantity	Symbol for quantity	Expression in SI units	Expression in symbols for SI units	Special name for SI units	Symbols using special names	Conventional units	Symbol for conventional unit	Value of conventional unit in SI units
Activity	A	1 per second	s^{-1}	becquerel	Bq	curie	Ci	3.7×10^{10} Bq
Absorbed dose	D	joule per kilogram	$J\ kg^{-1}$	gray	Gy	rad	rad	0.01 Gy
Absorbed dose rate	$\dot{D}$	joule per kilogram second	$J\ kg^{-1}\ s^{-1}$		$Gy\ s^{-1}$	rad	rad s^{-1}	0.01 Gy s^{-1}
Average energy per ion pair	W	joule	J			electronvolt	eV	1.602×10^{-19} J
Dose equivalent	H	joule per kilogram	$J\ kg^{-1}$	sievert	Sv	rem	rem	0.01 Sv
Dose equivalent rate	$\dot{H}$	joule per kilogram second	$J\ kg^{-1}\ s^{-1}$		$Sv\ s^{-1}$	rem per second	rem s^{-1}	0.01 Sv s^{-1}
Electric current	I	ampere	A			ampere	A	1.0 A
Electric potential difference	U, V	watt per ampere	$W\ A^{-1}$	volt	V	volt	V	1.0 V
Exposure	$\dot{X}$	coulomb per kilogram	$C\ kg^{-1}$			roentgen	R	2.58×10^{-4} C kg^{-1}
Exposure rate	X	coulomb per kilogram second	$C\ kg^{-1}\ s^{-1}$			roentgen per second	R s^{-1}	2.58×10^{-4} C $kg^{-1}\ s^{-1}$
Fluence	ϕ	1 per meter squared	m^{-2}			1 per centimeter squared	cm^{-2}	$1.0 \times 10^{4}\ m^{-2}$
Fluence rate	Φ	1 per meter squared second	$m^{-2}\ s^{-1}$			1 per centimeter squared second	$cm^{-2}\ s^{-1}$	$1.0 \times 10^{4}\ m^{-2}\ s^{-1}$
Kerma	K	joule per kilogram	$J\ kg^{-1}$	gray	Gy	rad	rad	0.01 Gy
Kerma rate	$\dot{K}$	joule per kilogram second	$J\ kg^{-1}\ s^{-1}$		$Gy\ s^{-1}$	rad per second	rad s^{-1}	0.01 Gy s^{-1}
Lineal energy	y	joule per meter	$J\ m^{-1}$			kiloelectron volt per micrometer	keV μm^{-1}	1.602×10^{-10} J m^{-1}
Linear energy transfer	L	joule per meter	$J\ m^{-1}$			kiloelectron volt per micrometer	keV μm^{-1}	1.602×10^{-10} J m^{-1}
Mass attenuation coefficient	μ/ρ	meter squared per kilogram	$m^2\ kg^{-1}$			centimeter squared per gram	$cm^2\ g^{-1}$	0.1 $m^2\ kg^{-1}$
Mass energy transfer coefficient	μ_{tr}/ρ	meter squared per kilogram	$m^2\ kg^{-1}$			centimeter squared per gram	$cm^2\ g^{-1}$	0.1 $m^2\ kg^{-1}$
Mass energy absorption coefficient	μ_{en}/ρ	meter squared per kilogram	$m^2\ kg^{-1}$			centimeter squared per gram	$cm^2\ g^{-1}$	0.1 $m^2\ kg^{-1}$
Mass stopping power	S/ρ	joule meter squared per kilogram	$J\ m^2\ kg^{-1}$			MeV centimeter squared per gram	MeV $cm^2\ g^{-1}$	1.602×10^{-14} J $m^2\ kg^{-1}$
Power	P	joule per second	$J\ s^{-1}$	watt	W	watt	W	1.0 W
Pressure	p	newton per meter squared	$N\ m^{-2}$	pascal	Pa	torr	torr	(101325/760) Pa
Radiation chemical yield	G	mole per joule	$mol\ J^{-1}$			molecules per 100 electron volts	molecules $(100\ eV)^{-1}$	1.04×10^{-7} mol J^{-1}
Specific energy	z	joule per kilogram	$J\ kg^{-1}$	gray	Gy	rad	rad	0.01 Gy

Conversion of Radioactivity Units from Becquerel to Curie Units

Bq units		Ci units	
7000	Bq	189	mCi
6000	Bq	162	mCi
5000	Bq	135	mCi
4000	Bq	108	mCi
3000	Bq	81	mCi
2000	Bq	54	mCi
1000	Bq	27.0	mCi
900	Bq	24.3	mCi
800	Bq	21.6	mCi
700	Bq	18.9	mCi

Bq units		Ci units	
600	Bq	16.2	mCi
500	Bq	13.5	mCi
400	Bq	10.8	mCi
300	Bq	8.11	mCi
200	Bq	5.41	mCi
100	Bq	2.70	mCi
90	Bq	2.43	mCi
80	Bq	2.16	mCi
70	Bq	1.89	mCi
60	Bq	1.62	mCi
50	Bq	1.35	mCi
40	Bq	1.08	mCi
30	Bq	810	μCi

Bq units		Ci units	
20	Bq	540	μCi
10	Bq	270	μCi
9	Bq	240	μCi
8	Bq	220	μCi
7	Bq	189	μCi
6	Bq	162	μCi
5	Bq	135	μCi
4	Bq	108	μCi
3	Bq	81	μCi
2	Bq	54	μCi
1	Bq	27.0	μCi
0.9	Bq	24.3	μCi
0.8	Bq	21.6	μCi

Bq units		Ci units	
0.7	Bq	18.9	μCi
0.6	Bq	16.2	μCi
0.5	Bq	13.5	μCi
0.4	Bq	10.8	μCi
0.3	Bq	8.1	μCi
0.2	Bq	5.4	μCi
0.1	Bq	2.7	μCi

Conversion of Radioactivity Units from mCi to MBq

mCi	MBq
200	7400
150	5550
100	3700
90	3330
80	2960
70	2590
60	2220
50	1850
40	1480
30	1110
20	740
10	370
9	333
8	296
7	259
6	222
5	185
4	148
3	111
2	74
1	37

Conversion of Radioactivity Units from μCi to MBq

μCi	MBq
1000	37.0
900	33.3
800	29.6
700	25.9
600	22.2
500	18.5
400	14.8
300	11.1
200	7.4
100	3.7
90	3.33
80	2.96
70	2.59
60	2.22
50	1.85
40	1.48
30	1.11
20	0.74
10	0.37
5	0.185
2	0.074
1	0.037

Conversion of Radioactivity Units

SI units	Ci units
100 TBq (10^{14} Bq)	2.7 kCi (2.7×10^3 Ci)
10 TBq (10^{13} Bq)	270 Ci (2.7×10^2 Ci)
1 TBq (10^{12} Bq)	27 Ci (2.7×10^1 Ci)
100 GBq (10^{11} Bq)	2.7 Ci (2.7×10^0 Ci)
10 GBq (10^{10} Bq)	270 mCi (2.7×10^{-1} Ci)
1 GBq (10^9 Bq)	27 mCi (2.7×10^{-2} Ci)
100 MBq (10^8 Bq)	2.7 mCi (2.7×10^{-3} Ci)
10 MBq (10^7 Bq)	270 μCi (2.7×10^{-4} Ci)
1 MBq (10^6 Bq)	27 μCi (2.7×10^{-5} Ci)
100 kBq (10^5 Bq)	2.7 μCi (2.7×10^{-6} Ci)
10 kBq (10^4 Bq)	270 nCi (2.7×10^{-7} Ci)
1 kBq (10^3 Bq)	27 nCi (2.7×10^{-8} Ci)
100 Bq (10^2 Bq)	2.7 nCi (2.7×10^{-9} Ci)
10 Bq (10^1 Bq)	270 pCi (2.7×10^{-10} Ci)
1 Bq (10^0 Bq)	27 pCi (2.7×10^{-11} Ci)
100 mBq (10^{-1} Bq)	2.7 pCi (2.7×10^{-12} Ci)
10 mBq (10^{-2} Bq)	270 fCi (2.7×10^{-13} Ci)
1 mBq (10^{-3} Bq)	27 fCi (2.7×10^{-14} Ci)

Conversion of Absorbed Dose Units

SI units	Conventional
100 Gy (10^2 Gy)	10,000 rad (10^4 rad)
10 Gy (10^1 Gy)	1,000 rad (10^3 rad)
1 Gy (10^0 Gy)	100 rad (10^2 rad)
100 mGy (10^{-1} Gy)	10 rad (10^1 rad)
10 mGy (10^{-2} Gy)	1 rad (10^0 rad)
1 mGy (10^{-3} Gy)	100 mrad (10^{-1} rad)
100 μGy (10^{-4} Gy)	10 mrad (10^{-2} rad)
10 μGy (10^{-5} Gy)	1 mrad (10^{-3} rad)
1 μGy (10^{-6} Gy)	100 μrad (10^{-4} rad)
100 nGy (10^{-7} Gy)	10 μrad (10^{-5} rad)
10 nGy (10^{-8} Gy)	1 μrad (10^{-6} rad)
1 nGy (10^{-9} Gy)	100 nrad (10^{-7} rad)

Conversion of Dose Equivalent Units

SI units	Conventional
100 Sv (10^2 Sv)	10,000 rem (10^4 rem)
10 Sv (10^1 Sv)	1,000 rem (10^3 rem)
1 Sv (10^0 Sv)	100 rem (10^2 rem)
100 mSv (10^{-1} Sv)	10 rem (10^1 rem)
10 mSv (10^{-2} Sv)	1 rem (10^0 rem)
1 mSv (10^{-3} Sv)	100 mrem (10^{-1} rem)
100 μSv (10^{-4} Sv)	10 mrem (10^{-2} rem)
10 μSv (10^{-5} Sv)	1 mrem (10^{-3} rem)
1 μSv (10^{-6} Sv)	100 μrem (10^{-4} rem)
100 nSv (10^{-7} Sv)	10 μrem (10^{-5} rem)
10 nSv (10^{-8} Sv)	1 μrem (10^{-6} rem)
1 nSv (10^{-9} Sv)	100 nrem (10^{-7} rem)

VALUES OF THE GAS CONSTANT IN DIFFERENT UNIT SYSTEMS

In SI units the value of the gas constant, R, is:

$$R = 8.314460 \text{ Pa m}^3 \text{ K}^{-1} \text{ mol}^{-1}$$
$$= 8314.460 \text{ Pa L K}^{-1} \text{ mol}^{-1}$$
$$= 0.08314460 \text{ bar L K}^{-1} \text{ mol}^{-1}$$

This table gives the appropriate value of R for use in the ideal gas equation, $PV = nRT$, when the variables are expressed in other units. The following conversion factors for pressure units were used in generating the table:

1 atm = 101325 Pa
1 psi = 6894.757 Pa
1 torr (mm Hg) = 133.322 Pa [at 0 °C]
1 in Hg = 3386.38 Pa [at 0 °C]
1 in H_2O = 249.082 Pa [at 4 °C]
1 ft H_2O = 2988.98 Pa [at 4 °C]

Reference

Mohr, P. J., Taylor, B. N., and Newell, D. B., "CODATA Recommended Values of the Fundamental Physical Constants: 2014," Sec. 1 of this *Handbook.*

Values of the Gas Constant *R* When *V*, *T*, *n*, and *P* Are Expressed in the Indicated Units

V unit	*T* unit	*n* unit	*R* (*P*/psi)	*R* (*P*/kPa)	*R* (*P*/atm)	*R* (*P*/mm Hg)	*R* (*P*/in Hg)	*R* (*P*/in H_2O)	*R* (*P*/ft H_2O)
ft^3	K	mol	0.0425863	0.2936224	0.00289784	2.20236	0.0867069	1.17881	0.0982350
ft^3	K	lb-mol	19.3168	133.1849	1.31443	998.972	39.3295	534.703	44.5586
ft^3	°R	mol	0.0236591	0.1631236	0.00160990	1.22353	0.0481705	0.654899	0.0545750
ft^3	°R	lb-mol	10.7316	73.99159	0.730241	554.983	21.8498	297.058	24.7548
cm^3	K	mol	1205.91	8314.460	82.0573	62363.7	2455.27	33380.4	2781.71
cm^3	K	lb-mol	546992	3771375	37220.5	282878000	1113690	15141100	1261760
cm^3	°R	mol	669.950	4619.144	45.5874	34646.5	1364.03	18544.7	1545.39
cm^3	°R	lb-mol	303885	2095208	20678.1	15715400	618716	8411729	700978
L	K	mol	1.20591	8.314460	0.0820573	62.3637	2.45527	33.3804	2.78171
L	K	lb-mol	546.992	3771.376	37.2205	28287.8	1113.69	15141.1	1261.76
L	°R	mol	0.669950	4.619144	0.0455874	34.6465	1.36403	18.5447	1.54539
L	°R	lb-mol	303.885	2095.208	20.6781	15715.4	618.716	8411.72	700.978
m^3	K	mol	0.00120591	0.008314460	0.0000820573	0.0623637	0.00245527	0.0333804	0.00278171
m^3	K	lb-mol	0.546992	3.771376	0.0372205	28.2878	1.11369	15.1411	1.26176
m^3	°R	mol	0.000669950	0.004619144	0.0000455874	0.0346465	0.00136403	0.0185447	0.00154539
m^3	°R	lb-mol	0.303885	2.095208	0.0206781	15.7154	0.618716	8.41172	0.700978

Section 2
Symbols, Terminology, and Nomenclature

SYMBOLS AND TERMINOLOGY FOR PHYSICAL AND CHEMICAL QUANTITIES

The International Organization for Standardization (ISO), International Union of Pure and Applied Chemistry (IUPAC), and the International Union of Pure and Applied Physics (IUPAP) have jointly developed a set of recommended symbols for physical and chemical quantities. Consistent use of these recommended symbols helps assure unambiguous scientific communication. The list below is reprinted from Reference 1 with permission from IUPAC; this list includes updates from the Third Edition of Reference 1. Full details may be found in the following references:

1. Mills, Ian, Ed., *Quantities, Units, and Symbols in Physical Chemistry,* Blackwell Scientific Publications, Oxford, 1988. Third Edition: RSC Publishing, Cambridge, UK, 2007.
2. Cohen, E. R., and Giacomo, P., *Symbols, Units, Nomenclature, and Fundamental Constants in Physics,* Document IUPAP–25, 1987; also published in *Physica* 146A, 1–68, 1987.
3. *ISO Standards Handbook 2: Units of Measurement,* International Organization of Standardization, Geneva, 1982.

General Rules

The value of a physical quantity is expressed as the product of a numerical value and a unit, e.g.:

$T = 300$ K
$V = 26.2$ cm^3
$C_p = 45.3$ J mol^{-1} K^{-1}

The symbol for a physical quantity is always given in italic (sloping) type, while symbols for units are given in roman type. Column headings in tables and axis labels on graphs may conveniently be written as the physical quantity symbol divided by the unit symbol, e.g.:

T/K
V/cm^3
C_p/J mol^{-1} K^{-1}

The values in the table or graph axis are then pure numbers. Subscripts to symbols for physical quantities should be italic if the subscript refers to another physical quantity or to a number, e.g.:

C_p – heat capacity at constant pressure
B_n – nth virial coefficient

Subscripts that have other meanings should be in roman type:

m_{p} – mass of the proton
E_{k} – kinetic energy

The following tables give the recommended symbols for the major classes of physical and chemical quantities. The expression in the Definition column is given as an aid in identifying the quantity but is not necessarily the complete or unique definition. The SI Unit gives one (not necessarily unique) expression for the coherent SI unit for the quantity. Other equivalent unit expressions, including those that involve SI prefixes, may be used.

Symbols, Units, and Definitions of Basic Physical Quantities

Name	Symbol	Definition	SI unit
Space and Time			
cartesian space coordinates	x, y, z		m
spherical polar coordinates	r, θ, ϕ		m, 1, 1
generalized coordinate	q, q_i		(varies)
position vector	$\mathbf{r}$	$\mathbf{r} = x\mathbf{i} + y\mathbf{j} + z\mathbf{k}$	m
length	l		m
special symbols:			
height	h		
breadth	b		
thickness	d, δ		
distance	d		
radius	r		
diameter	d		
path length	s		
length of arc	s		
area	A, A_s, S		m^2
volume	$V, (v)$		m^3
plane angle	$\alpha, \beta, \gamma, \theta, \phi$...	$\alpha = s/r$	rad, 1
solid angle	ω, Ω	$\omega = A/r^2$	sr, 1
time	t		s
period	T	$T = t/N$	s
frequency	ν, f	$\nu = 1/T$	Hz
circular frequency, angular frequency	ω	$\omega = 2\pi\nu$	rad s^{-1}, s^{-1}
characteristic time interval, relaxation time, time constant	τ, T	$\tau = \lvert \mathrm{d}t/\mathrm{d}\ln x \rvert$	s
angular velocity	ω	$\omega = \mathrm{d}\phi/\mathrm{d}t$	rad s^{-1}, s^{-1}

Name	Symbol	Definition	SI unit
velocity	$\boldsymbol{v}, \boldsymbol{u}, \boldsymbol{w}, \boldsymbol{c}, \dot{\boldsymbol{r}}$	$\boldsymbol{v} = \mathrm{d}\boldsymbol{r}/\mathrm{d}t$	m s^{-1}
speed	v, u, w, c	$v = \lvert\boldsymbol{v}\rvert$	m s^{-1}
acceleration	$\boldsymbol{a}, (g)$	$\boldsymbol{a} = \mathrm{d}\boldsymbol{v}/\mathrm{d}t$	m s^{-2}
Classical Mechanics			
mass	m		kg
reduced mass	μ	$\mu = m_1 m_2/(m_1 + m_2)$	kg
density, mass density	ρ	$\rho = m/V$	kg m^{-3}
relative density	d	$d = \rho/\rho$	1
surface density	ρ_A, ρ_S	$\rho_A = m/A$	kg m^{-2}
specific volume	v	$v = V/m = 1/\rho$	m^3 kg^{-1}
momentum	$\boldsymbol{p}$	$\boldsymbol{p} = m\boldsymbol{v}$	kg m s^{-1}
angular momentum, action	$\boldsymbol{L}$	$\boldsymbol{L} = \boldsymbol{r} \times \boldsymbol{p}$	J s
moment of inertia	I, J	$I = \Sigma m_i r_i^2$	kg m^2
force	$\boldsymbol{F}$	$\boldsymbol{F} = \mathrm{d}\boldsymbol{p}/\mathrm{d}t = m\boldsymbol{a}$	N
torque, moment of a force	$\boldsymbol{T}, (\boldsymbol{M})$	$\boldsymbol{T} = \boldsymbol{r} \times \boldsymbol{F}$	N m
energy	E		J
potential energy	E_p, V, Φ	$E_p = \int \boldsymbol{F} \cdot \mathrm{d}\boldsymbol{s}$	J
kinetic energy	E_k, T, K	$E_k = 1/2mv^2$	J
work	W, w	$W = \int \boldsymbol{F} \cdot \mathrm{d}\boldsymbol{s}$	J
Hamilton function	H	$H(q, p) = T(q, p) + V(q)$	J
Lagrange function	L	$L(q, \dot{q}) = T(q, \dot{q}) - V(q)$	J
pressure	p, P	$p = F/A$	Pa, N m^{-2}
surface tension	γ, σ	$\gamma = \mathrm{d}W/\mathrm{d}A$	N m^{-1}, J m^{-2}
weight	$G, (W, P)$	$G = mg$	N
gravitational constant	G	$F = Gm_1m_2/r^2$	N m^2 kg^{-2}
normal stress	σ	$\sigma = F/A$	Pa
shear stress	τ	$\tau = F/A$	Pa
linear strain, relative elongation	ε, e	$\varepsilon = \Delta l/l$	1
modulus of elasticity, Young's modulus	E	$E = \sigma/\varepsilon$	Pa
shear strain	γ	$\gamma = \Delta x/d$	1
shear modulus	G	$G = \tau/\gamma$	Pa
volume strain, bulk strain	θ	$\theta = \Delta V/V_0$	1
bulk modulus, compression modulus	K	$K = -V_0(\mathrm{d}p/\mathrm{d}V)$	Pa
viscosity, dynamic viscosity	η, μ	$\tau_{x,z} = \eta(\mathrm{d}v_x/\mathrm{d}z)$	Pa s
fluidity	ϕ	$\phi = 1/\eta$	m kg^{-1} s
kinematic viscosity	ν	$\nu = \eta/\rho$	m^2 s^{-1}
friction coefficient	$\mu, (f)$	$F_{\mathrm{frict}} = \mu F_{\mathrm{norm}}$	1
power	P	$P = \mathrm{d}W/\mathrm{d}t$	W
sound energy flux	P, P_a	$P = \mathrm{d}E/\mathrm{d}t$	W
acoustic factors:			
reflection factor	ρ	$\rho = P_r/P_0$	1
acoustic absorption factor	$\alpha_a, (\alpha)$	$\alpha_a = 1 - \rho$	1
transmission factor	τ	$\tau = P_{tr}/P_0$	1
dissipation factor	δ	$\delta = \alpha_a - \tau$	1
Electricity and Magnetism			
quantity of electricity, electric charge	Q		C
charge density	ρ	$\rho = Q/V$	C m^{-3}
surface charge density	σ	$\sigma = Q/A$	C m^{-2}
electric potential	V, ϕ	$V = \mathrm{d}W/\mathrm{d}Q$	V, J C^{-1}
electric potential difference	$U, \Delta V, \Delta\phi$	$U = V_2 - V_1$	V
electromotive force	E	$E = \int(\boldsymbol{F}/Q) \cdot \mathrm{d}\boldsymbol{s}$	V
electric field strength	$\boldsymbol{E}$	$\boldsymbol{E} = \boldsymbol{F}/Q = -\mathrm{grad}\ V$	V m^{-1}
electric flux	Ψ	$\Psi = \int \boldsymbol{D} \cdot \mathrm{d}\boldsymbol{A}$	C
electric displacement	$\boldsymbol{D}$	$\boldsymbol{D} = \varepsilon\boldsymbol{E}$	C m^{-2}
capacitance	C	$C = Q/U$	F, C V^{-1}
permittivity	ε	$\boldsymbol{D} = \varepsilon\boldsymbol{E}$	F m^{-1}
permittivity of vacuum	ε_0	$\varepsilon_0 = \mu_0^{-1}c_0^{-2}$	F m^{-1}
relative permittivity	ε_r	$\varepsilon_r = \varepsilon/\varepsilon_0$	1
dielectric polarization (dipole moment per volume)	$\boldsymbol{P}$	$\boldsymbol{P} = \boldsymbol{D} - \varepsilon_0\boldsymbol{E}$	C m^{-2}

Name	Symbol	Definition	SI unit
electric susceptibility	χ_e	$\chi_e = \varepsilon r - 1$	1
electric dipole moment	$\boldsymbol{p}, \boldsymbol{\mu}$	$\boldsymbol{p} = Q\boldsymbol{r}$	C m
electric current	I	$I = \mathrm{d}Q/\mathrm{d}t$	A
electric current density	$\boldsymbol{j}, \boldsymbol{J}$	$I = \int \boldsymbol{j} \cdot \mathrm{d}\boldsymbol{A}$	A m^{-2}
magnetic flux density, magnetic induction	$\boldsymbol{B}$	$\boldsymbol{F} = Q\boldsymbol{v}\times\boldsymbol{B}$	T
magnetic flux	Φ	$\Phi = \int \boldsymbol{B} \cdot \mathrm{d}\boldsymbol{A}$	A m^{-2}
magnetic field strength	$\boldsymbol{H}$	$\boldsymbol{B} = \mu\boldsymbol{H}$	A m^{-2}
permeability	μ	$\boldsymbol{B} = \mu\boldsymbol{H}$	N A^{-2}, H m^{-1}
permeability of vacuum	μ_0		H m^{-1}
relative permeability	μ_r	$\mu_r = \mu/\mu_0$	1
magnetization (magnetic dipole moment per volume)	$\boldsymbol{M}$	$\boldsymbol{M} = \boldsymbol{B}/\mu_0 - H$	A m^{-1}
magnetic susceptibility	$\chi, \kappa, (\chi_m)$	$\chi = \mu_r - 1$	1
molar magnetic susceptibility	χ_m	$\chi_m = V_m\chi$	m^3 mol^{-1}
magnetic dipole moment	$\boldsymbol{m}, \boldsymbol{\mu}$	$E_p = -\boldsymbol{m}\cdot\boldsymbol{B}$	A m^2, J T^{-1}
electrical resistance	R	$R = U/I$	Ω
conductance	G	$G = 1/R$	S
loss angle	δ	$\delta = (\pi/2) + \phi_I - \phi_U$	1, rad
reactance	X	$X = (U/I)\sin\delta$	Ω
impedance (complex impedance)	Z	$Z = R + iX$	Ω
admittance (complex admittance)	Y	$Y = 1/Z$	S
susceptance	B	$Y = G + iB$	S
resistivity	ρ	$\rho = E/j$	Ω m
conductivity	κ, γ, σ	$\kappa = 1/\rho$	S m^{-1}
self-inductance	L	$E = -L(\mathrm{d}I/\mathrm{d}t)$	H
mutual inductance	M, L_{12}	$E_1 = L_{12}(\mathrm{d}I_2/\mathrm{d}t)$	H
magnetic vector potential	$\boldsymbol{A}$	$\boldsymbol{B} = \nabla\times\boldsymbol{A}$	Wb m^{-1}
Poynting vector	$\boldsymbol{S}$	$\boldsymbol{S} = \boldsymbol{E}\times\boldsymbol{H}$	W m^{-2}
Quantum Mechanics			
momentum operator	$\dot{\boldsymbol{p}}$	$\dot{\boldsymbol{p}} = -\mathrm{i}h\nabla$	m^{-1}J s
kinetic energy operator	$\hat{T}$	$\hat{T} = -(h^2/2m)\nabla^2$	J
Hamiltonian operator	$\hat{H}$	$\hat{H} = \hat{T} + V$	J
wave function*, state function	Ψ, ψ, ϕ	$\hat{H}\psi = E\psi$	($m^{-3/2}$)
probability density	P	$P = \psi^*\psi$	(m^{-3})
charge density of electrons	ρ	$\rho = -eP$	(C m^{-3})
probability current density	$\boldsymbol{S}$	$\boldsymbol{S} = -i\hbar(\psi^*\nabla\psi - \psi\nabla\psi^*)/2m_e$	($m^{-2}s^{-1}$)
electric current density of electrons	$\boldsymbol{j}$	$\boldsymbol{j} = -e\boldsymbol{S}$	(A m^{-2})
matrix element of operator Â	$A_{ij}, \langle i\|\hat{A}\|j\rangle$	$A_{ij} = \int\psi_i{}^*\hat{A}\psi_j\mathrm{d}\tau$	(varies)
expectation value of operator Â	$\langle A\rangle, \bar{A}$	$\langle A\rangle = \int\psi^*\hat{A}\Psi\mathrm{d}\tau$	(varies)
hermitian conjugate of Â	$\hat{A}^\dagger$	$(\hat{A}^\dagger)_{ij} = (A_{ji})^*$	(varies)
commutator of $\hat{A}$ and $\hat{B}$	$[\hat{A},\hat{B}], [\hat{A},\hat{B}]_-$	$[\hat{A},\hat{B}] = \hat{A}\hat{B} - \hat{B}\hat{A}$	(varies)
anticommutator	$[\hat{A},\hat{B}]_+$	$[\hat{A},\hat{B}]_+ = \hat{A}\hat{B} + \hat{B}\hat{A}$	(varies)
spin wave function*	$\alpha; \beta$		1
coulomb integral	H_{AA}	$H_{AA} = \int\psi_A{}^*\hat{H}\psi_A\mathrm{d}\tau$	J
resonance integral	H_{AB}	$H_{AB} = \int\psi_A{}^*\hat{H}\psi_B\mathrm{d}\tau$	J
overlap integral	S_{AB}	$S_{AB} = \int\psi_A{}^*\psi_B\mathrm{d}\tau$	1
Atoms and Molecules			
nucleon number, mass number	A		1
proton number, atomic number	Z		1
neutron number	N	$N = A - Z$	1
electron rest mass	m_e		kg
mass of atom, atomic mass	m_a, m		kg
atomic mass constant	m_u	$m_u = m_a(^{12}C)/12$	kg
mass excess	Δ	$\Delta = m_a - Am_u$	kg
elementary charge, proton charge	e		C
Planck constant	h		J s
Planck constant/2π	$\hbar$	$\hbar = h/2\pi$	J s
Bohr radius	a_0	$a_0 = 4\pi\varepsilon_0\hbar^2/m_e e^2$	m
Hartree energy	E_h	$E_h = \hbar^2/m_e a_0{}^2$	J
Rydberg constant	R_∞	$R_\infty = E_h/2hc$	m^{-1}
fine structure constant	α	$\alpha = e^2/4\pi\varepsilon_0\hbar c$	1

Name	Symbol	Definition	SI unit
ionization energy	E_i		J
electron affinity	E_{ea}		J
dissociation energy	E_d, D		J
from the ground state	D_0		J
from the potential minimum	D_e		J
principal quantum number (H atom)	n	$E = -hcR/n^2$	1
angular momentum quantum numbers	see under Spectroscopy		
magnetic dipole moment of a molecule	$\boldsymbol{m}, \boldsymbol{\mu}$	$E_p = -\boldsymbol{m}\cdot\boldsymbol{B}$	$J\ T^{-1}$
magnetizability of a molecule	ξ	$\boldsymbol{m} = \xi B$	$J\ T^{-2}$
Bohr magneton	μ_B	$\mu_B = e\hbar/2m_e$	$J\ T^{-1}$
nuclear magneton	μ_N	$\mu_N = (m_e/m_p)\mu_B$	$J\ T^{-1}$
magnetogyric ratio (gyromagnetic ratio)	γ	$\gamma = \mu/L$	$C\ kg^{-1}$
g factor	g		1
Larmor circular frequency	ω_L	$\omega_L = (e/2m)B$	s^{-1}
Larmor frequency	ν_L	$\nu_L = \omega_L/2\pi$	Hz
longitudinal relaxation time	T_1		s
transverse relaxation time	T_2		s
electric dipole moment of a molecule	$\boldsymbol{p}, \boldsymbol{\mu}$	$E_p = -\boldsymbol{p}\cdot\boldsymbol{E}$	C m
quadrupole moment of a molecule	$\boldsymbol{Q}; \boldsymbol{\Theta}$	$E_p = 1/2\boldsymbol{Q}{:}V'' = 1/3\boldsymbol{\Theta}{:}V''$	$C\ m^2$
quadrupole moment of a nucleus	eQ	$eQ = 2\cdot\langle\Theta_{ZZ}\rangle$	$C\ m^2$
electric field gradient tensor	$\boldsymbol{q}$	$q_{\alpha\beta} = -\partial^2V/\partial\alpha\partial\beta$	$V\ m^{-2}$
quadrupole interaction energy tensor	χ	$\chi_{\alpha\beta} = eQq_{\alpha\beta}$	J
electric polarizability of a molecule	α	p (induced) $= \alpha E$	$C\ m^2\ V^{-1}$
activity (of a radioactive substance)	A	$A = -dN_B/dt$	Bq
decay (rate) constant, disintegration (rate) constant	λ	$A = \gamma N_B$	s^{-1}
half-life**	$t_{1/2}$, $T_{1/2}$		s
mean life	τ		s
level width	Γ	$\Gamma = \hbar/\tau$	J
disintegration energy	Q		J
cross section (of a nuclear reaction)	σ		m^2
Spectroscopy			
total term	T	$T = E_{tot}/hc$	m^{-1}
transition wavenumber	$\tilde{\nu}$, (ν)	$\tilde{\nu} = T' - T''$	m^{-1}
transition frequency	ν	$\nu = (E' - E'')/h$	Hz
electronic term	T_e	$T_e = E_e/hc$	m^{-1}
vibrational term	G	$G = E_{vib}/hc$	m^{-1}
rotational term	F	$F = E_{rot}/hc$	m^{-1}
spin orbit coupling constant	A	$T_{s.o.} = A\langle\hat{\mathbf{L}}\cdot\hat{\mathbf{S}}\rangle$	m^{-1}
principal moments of inertia	$I_A; I_B; I_C$	$I_A \le I_B \le I_C$	$kg\ m^2$
rotational constants, in wavenumber	$\tilde{A}; \tilde{B}; \tilde{C}$	$\tilde{A} = h/8\pi^2cI_A$	m^{-1}
rotational constants, in frequency	$A; B; C$	$A = h/8\pi^2I_A$	Hz
inertial defect	Δ	$\Delta = I_C - I_A - I_B$	$kg\ m^2$
asymmetry parameter	κ	$\kappa = \dfrac{(2B - A - C)}{(A - C)}$	1
centrifugal distortion constants:			
S reduction	$D_J; D_{JK}; D_K; d_1; d_2$		m^{-1}
A reduction	$\Delta_J; \Delta_{JK}; \Delta_K; \delta_J; \delta_K$		m^{-1}
harmonic vibration wavenumber	$\omega_e; \omega_r$		m^{-1}
vibrational anharmonicity constant	$\omega_e x_e; x_{rs}; g_{tt'}$		m^{-1}
vibrational quantum numbers	$v_r; l_t$		1
Coriolis zeta constant	ζ_{rs}^{α}		1
angular momentum quantum numbers	see additional information below		
degeneracy, statistical weight	g, d, β		1
electric dipole moment of a molecule	$\boldsymbol{p}, \boldsymbol{\mu}$	$E_p = -\boldsymbol{p}\cdot\boldsymbol{E}$	C m
transition dipole moment of a molecule	$\boldsymbol{M}, \boldsymbol{R}$	$\boldsymbol{M} = \int\psi'\boldsymbol{p}\psi''d\tau$	C m
molecular geometry, interatomic distances:			
equilibrium distance	r_e		m
zero–point average distance	r_z		m

Name	Symbol	Definition	SI unit
ground state distance	r_0		m
substitution structure distance	r_s		m
vibrational coordinates:			
internal coordinates	R_i, r_i, θ_i, *etc.*		(varies)
symmetry coordinates	S_i		(varies)
normal coordinates:			
mass adjusted	Q_r		$kg^{1/2}$ m
dimensionless	q_r		1
vibrational force constants:			
diatomic	f, (k)	$f = \partial^2 V/\partial r^2$	J m^{-2}
polyatomic,			
internal coordinates	f_{ij}	$f_{ij} = \partial^2 V/\partial r_i \partial r_j$	(varies)
symmetry coordinates	F_{ij}	$F_{ij} = \partial^2 V/\partial S_i \partial S_j$	(varies)
dimensionless normal coordinates	$\varphi_{rst...}$, $k_{rst...}$		m^{-1}
nuclear magnetic resonance (NMR):			
magnetogyric ratio	γ	$\gamma = \mu/I\hbar$	C kg^{-1}
shielding constant	σ_A	$B_A = (1 - \sigma_A)B$	1
chemical shift, δ scale	δ	$\delta = 10^6(\nu - \nu_0)/\nu_0$	1
(indirect) spin–spin coupling constant	J_{AB}	$\hat{H}/h = J_{AB}\hat{I}_A \cdot \hat{I}_B$	Hz
direct (dipolar) coupling constant	D_{AB}		Hz
longitudinal relaxation time	T_1		s
transverse relaxation time	T_2		s
electron spin resonance, electron paramagnetic resonance (ESR, EPR):			
magnetogyric ratio	γ	$\gamma = \mu/s\hbar$	C kg^{-1}
g-factor	g	$h\nu = g\mu_B B$	1
hyperfine coupling constant,			
in liquids	a, A	$\hat{H}_{hfs}/h = a\hat{S} \cdot \hat{I}$	Hz
in solids	T	$\hat{H}_{hfs}/h = \hat{S} \cdot T \cdot \hat{I}$	Hz

* IUPAC recommends "wavefunction"
** IUPAC recommends "half life"

Quantum Numbers Describing Angular Momentum

Angular momentum	Operator symbol	Quantum number (total)	Quantum number (Z–axis)	Quantum number (z-axis)
electron orbital	$\hat{L}$	L	M_L	Λ
one electron only	$\hat{l}$	l	m	λ
electron spin	$\hat{S}$	S	M_S	Σ
one electron only	$\hat{s}$	s	m_s	σ
electron orbital + spin	$\hat{L} + \hat{S}$			$\Omega = \Lambda + \Sigma$
nuclear orbital (rotational)	$\hat{R}$	R		K_R, k_R
nuclear spin	$\hat{I}$	I	M_I	
internal vibrational				
spherical top	$\hat{l}$	$l(l\zeta)$		K_l
other	$\hat{j}$, $\hat{\pi}$			$l(l\zeta)$
sum of *R*+*L*(+*j*)	$\hat{N}$	N		K, k
sum of *N*+*S*	$\hat{J}$	J	M_J	K, k
sum of *J*+*I*	$\hat{F}$	F	M_F	

Symbols and Teminology in Physical Chemistry

Name	Symbol	Definition	SI unit
Electromagnetic Radiation			
wavelength	λ		m
speed of light			
in vacuum	c_0		m s^{-1}
in a medium	c	$c = c_0/n$	m s^{-1}

Symbols

Name	Symbol	Definition	SI unit
wavenumber in vacuum	$\tilde{\nu}$	$\tilde{\nu} = \nu/c_0 = 1/n\lambda$	m^{-1}
wavenumber (in a medium)	σ	$\sigma = 1/\lambda$	m^{-1}
frequency	ν	$\nu = c/\lambda$	Hz
circular frequency, pulsatance	ω	$\omega = 2\pi\nu$	s^{-1}, rad s^{-1}
refractive index	n	$n = c_0/c$	1
Planck constant	h		J s
Planck constant/2π	$\hbar$	$\hbar = h/2\pi$	J s
radiant energy	Q, W		J
radiant energy density	ρ, w	$\rho = Q/V$	J m^{-3}
spectral radiant energy density			
in terms of frequency	ρ_ν, w_ν	$\rho = d\rho/d\nu$	J m^{-3} Hz^{-1}
in terms of wavenumber	$\rho_{\tilde{\nu}}$, $\omega_{\tilde{\nu}}$	$\rho_{\tilde{\nu}} = d\rho / d\tilde{\nu}$	J m^{-2}
in terms of wavelength	ρ_λ, w_λ	$\rho_\lambda = d\rho/d\lambda$	J m^{-4}
Einstein transition probabilities,			
spontaneous emission	A_{nm}	$dN_n/dt = -A_{nm}N_n$	s^{-1}
stimulated emission	B_{nm}	$dN_n/dt = -\Sigma\, \rho\tilde{\nu}(\tilde{\nu}_{nm}) \times B_{nm}N_n$	s kg^{-1}
stimulated absorption	B_{mn}	$dN_n/dt = -\Sigma\rho\tilde{\nu}(\tilde{\nu}_{nm})B_{mn}N_m$	s kg^{-1}
radiant power, radiant energy per time	Φ, P	$\Phi = dQ/dt$	W
radiant intensity	I	$I = d\Phi/d\Omega$	W sr^{-1}
radiant exitance (emitted radiant flux)	M	$M = d\Phi/dA_{source}$	W m^{-2}
irradiance, (radiant flux received)	E, (I)	$E = d\Phi/dA$	W m^{-2}
emittance	ε	$\varepsilon = M/M_{bb}$	1
Stefan–Boltzmann constant	σ	$M_{bb} = \sigma T^4$	W m^{-2} K^{-4}
first radiation constant	c_1	$c_1 = 2\pi hc_0^2$	W m^2
second radiation constant	c_2	$c_2 = hc_0/k$	K m
transmittance, transmission factor	τ, T	$\tau = \Phi_{tr}/\Phi_0$	1
absorptance, absorption factor	α	$\alpha = \Phi_{abs}/\Phi_0$	1
reflectance, reflection factor	ρ	$\rho = \Phi_{refl}/\Phi_0$	1
(decadic) absorbance	A	$A = -\lg(1 - \alpha_i)$	1
napierian absorbance	B	$B = -\ln(1 - \alpha_i)$	1
absorption coefficient,			
(linear) decadic	a, K	$a = A/l$	m^{-1}
(linear) napierian	α	$\alpha = B/l$	m^{-1}
molar (decadic)	ε	$\varepsilon = a/c = A/cl$	m^2 mol^{-1}
molar napierian	κ	$\kappa = \alpha/c = B/cl$	m^2 mol^{-1}
absorption index	k	$k = \alpha/4\pi\tilde{\nu}$	1
complex refractive index	$\hat{n}$	$\hat{n} = n + ik$	1
molar refraction	R, R_m	$R = \frac{(n^2-1)}{(n^2+2)}V_m$	m^3 mol^{-1}
angle of optical rotation	α		1, rad
Solid State			
lattice vector	$\boldsymbol{R}$, $\boldsymbol{R}_0$		m
fundamental translation vectors for the crystal lattice	$\boldsymbol{a}_1$; $\boldsymbol{a}_2$; $\boldsymbol{a}_3$, $\boldsymbol{a}$; $\boldsymbol{b}$; $\boldsymbol{c}$	$\boldsymbol{R} = n_1\boldsymbol{a}_1 + n_2\boldsymbol{a}_2 + n_3\boldsymbol{a}_3$	m
(circular) reciprocal lattice vector	$\boldsymbol{G}$	$\boldsymbol{G}\cdot\boldsymbol{R} = 2\pi m$	m^{-1}
(circular) fundamental translation vectors for the reciprocal lattice	$\boldsymbol{b}_1$; $\boldsymbol{b}_2$; $\boldsymbol{b}_3$, $\boldsymbol{a}^*$; $\boldsymbol{b}^*$; $\boldsymbol{c}^*$	$\boldsymbol{a}i\cdot\boldsymbol{b}k = 2\pi\delta_{ik}$	m^{-1}
lattice plane spacing	d		m
Bragg angle	θ	$n\lambda = 2d \sin\theta$	1, rad
order of reflection	n		1
order parameters			
short range	σ		1
long range	s		1
Burgers vector	$\boldsymbol{b}$		m
particle position vector	$\boldsymbol{r}$, $\boldsymbol{R}_j$		m
equilibrium position vector of an ion	$\boldsymbol{R}_0$		m
displacement vector of an ion	$\boldsymbol{u}$	$\boldsymbol{u} = \boldsymbol{R} - \boldsymbol{R}_0$	m
Debye–Waller factor	B, D		1
Debye circular wavenumber	q_D		m^{-1}
Debye circular frequency	ω_D		s^{-1}
Grüneisen parameter	γ, Γ	$\gamma = \alpha V/\kappa C_v$	1

Name	Symbol	Definition	SI unit
Madelung constant	α, M	$E_{coul} = \frac{\alpha N_A z_+ z_- e^2}{4\pi\varepsilon_0 R_0}$	1
density of states	N_E	$N_E = dN(E)/dE$	$J^{-1}\ m^{-3}$
(spectral) density of vibrational modes	N_ω, g	$N_\omega = dN(\omega)/d\omega$	$s\ m^{-3}$
resistivity tensor	ρ_{ik}	$\boldsymbol{E}=\boldsymbol{\rho}\cdot\boldsymbol{j}$	$\Omega\ m$
conductivity tensor	σ_{ik}	$\sigma = \rho^{-1}$	$S\ m^{-1}$
thermal conductivity tensor	λ_{ik}	$J_q = -\boldsymbol{\lambda} \cdot \text{grad}\ T$	$W\ m^{-1}\ K^{-1}$
residual resistivity	ρ_R		$\Omega\ m$
relaxation time	τ	$\tau = l/v_F$	s
Lorenz coefficient	L	$L = \lambda/\sigma T$	$V^2\ K^{-2}$
Hall coefficient	$A_H, \boldsymbol{R}_H$	$\boldsymbol{E} = \boldsymbol{\rho} \cdot \boldsymbol{j} + R_H (\boldsymbol{B} \times \boldsymbol{j})$	$m^3\ C^{-1}$
thermoelectric force	E		V
Peltier coefficient	Π		V
Thomson coefficient	$\mu, (\tau)$		$V\ K^{-1}$
work function	Φ	$\Phi = E_\infty - E_F$	J
number density, number concentration	$n, (p)$		m^{-3}
gap energy	E_g		J
donor ionization energy	E_d		J
acceptor ionization energy	E_a		J
Fermi energy	E_F, ε_F		J
circular wave vector, propagation vector	$k, \boldsymbol{q}$	$k = 2\pi/\lambda$	m^{-1}
Bloch function	$u_k(r)$	$\psi(\boldsymbol{r}) = u_k(\boldsymbol{r})\exp(i\boldsymbol{k}\cdot\boldsymbol{r})$	$m^{-3/2}$
charge density of electrons	ρ	$\rho(\boldsymbol{r}) = -e\psi^*(\boldsymbol{r})\psi(\boldsymbol{r})$	$C\ m^{-3}$
effective mass	m^*		kg
mobility	μ	$\mu = v_{drift}/E$	$m^2\ V^{-1}\ s^{-1}$
mobility ratio	b	$b = \mu_n/\mu_p$	1
diffusion coefficient	D	$dN/dt = -DA(dn/dx)$	$m^2\ s^{-1}$
diffusion length	L	$L = \sqrt{D\tau}$	m
characteristic (Weiss) temperature	θ, θ_W		K
Curie temperature	T_C		K
Néel temperature	T_N		K
Statistical Thermodynamics			
number of entities	N		1
number density of entities, number concentration	n, C	$n = N/V$	m^{-3}
Avogadro constant	L, N_A		mol^{-1}
Boltzmann constant	k, k_B		$J\ K^{-1}$
gas constant (molar)	R	$R = Lk$	$J\ K^{-1}\ mol^{-1}$
molecular position vector	$\boldsymbol{r}(x, y, z)$		m
molecular velocity vector	$\boldsymbol{c}(c_x, c_y, c_z), \boldsymbol{u}(u_x, u_y, u_z)$	$\boldsymbol{c} = d\boldsymbol{r}/dt$	$m\ s^{-1}$
molecular momentum vector	$\boldsymbol{p}(p_x, p_y, p_z)$	$\boldsymbol{p} = m\boldsymbol{c}$	$kg\ m\ s^{-1}$
velocity distribution function (Maxwell)	$f(c_x)$	$f(c_x) = (m/2\pi kT)^{1/2} \times \exp(-mc_x^2/2kT)$	$m^{-1}\ s$
speed distribution function (Maxwell–Boltzmann)	$F(c)$	$F(c) = (m/2\pi kT)^{3/2} \times 4\pi c^2 \exp(-mc^2/2kT)$	$m^{-1}\ s$
average speed	$\bar{c}, \bar{u}, \langle c \rangle, \langle u \rangle$	$\bar{c} = \int cF(c)dc$	$m\ s^{-1}$
generalized coordinate	q		(m)
generalized momentum	p	$p = \partial L/\partial\dot{q}$	$(kg\ m\ s^{-1})$
volume in phase space	Ω	$\Omega = (1/h)\int p dq$	1
probability	P		1
statistical weight, degeneracy	g, d, W, ω, β		1
density of states	$\rho(E)$	$\rho(E) = dN/dE$	J^{-1}
partition function, sum over states, for a single molecule	q, z	$q = \sum_i g_i \exp(-\varepsilon_i / kT)$	1
for a canonical ensemble (system, or assembly)	Q, Z		1
microcanonical ensemble	Ω		1
grand (canonical ensemble)	Ξ		1
symmetry number	σ, s		1
reciprocal temperature parameter	β	$\beta = 1/kT$	J^{-1}
characteristic temperature	Θ		K

Name	Symbol	Definition	SI unit
General Chemistry			
number of entities (e.g. molecules, atoms, ions, formula units)	N		1
amount (of substance)	n	$n_B = N_B/L$	mol
Avogadro constant	L, N_A		mol^{-1}
mass of atom, atomic mass	m_a, m		kg
mass of entity (molecule, or formula unit)	m_f, m		kg
atomic mass constant	m_u	$m_u = m_a(^{12}C)/12$	kg
molar mass	M	$M_B = m/n_B$	kg mol^{-1}
relative molecular mass (relative molar mass, molecular weight)	M_r	$M_{r,B} = m_B/m_u$	1
molar volume	V_m	$V_{m,B} = V/n_B$	m^3mol^{-1}
mass fraction	w	$w_B = m_B/\Sigma m_i$	1
volume fraction	ϕ	$\phi_B = V_B/\Sigma V_i$	1
mole fraction, amount fraction, number fraction	x, y	$x_B = n_B/\Sigma n_i$	1
(total) pressure	p, P		Pa
partial pressure	p_B	$p_B = y_B p$	Pa
mass concentration (mass density)	γ, ρ	$\gamma_B = m_B/V$	kg m^{-3}
number concentration, number density of entities	C, n	$C_B = N_B/V$	m^{-3}
amount concentration, concentration	c	$c_B = n_B/V$	mol m^{-3}
solubility	s	$s_B = c_B$ (saturated solution)	mol m^{-3}
molality (of a solute)	$m, (b)$	$m_B = n_B/m_A$	mol kg^{-1}
surface concentration	Γ	$\Gamma_B = n_B/A$	mol m^{-2}
stoichiometric number	ν		1
extent of reaction, advancement	ξ	$\Delta\xi = \Delta n_B/\nu_B$	mol
degree of dissociation	α		1
Chemical Thermodynamics			
heat	q, Q		J
work	w, W		J
internal energy	U	$\Delta U = q + w$	J
enthalpy	H	$H = U + pV$	J
thermodynamic temperature	T		K
Celsius temperature	θ, t	$\theta/°C = T/K - 273.15$	°C
entropy	S	$dS \geq dq/T$	J K^{-1}
Helmholtz energy (Helmholtz function)	A	$A = U - TS$	J
Gibbs energy (Gibbs function)	G	$G = H - TS$	J
Massieu function	J	$J = -A/T$	J K^{-1}
Planck function	Y	$Y = -G/T$	J K^{-1}
surface tension	γ, σ	$\gamma = (\partial G/\partial A_s)_{T,p}$	J m^{-2}, N m^{-1}
molar quantity X	X_m	$X_m = X/n$	(varies)
specific quantity X	x	$x = X/m$	(varies)
pressure coefficient	β	$\beta = (\partial p/\partial T)\nu$	Pa K^{-1}
relative pressure coefficient	α_p	$\alpha_p = (1/p)(\partial p/\partial T)_V$	K^{-1}
compressibility,			
isothermal	κ_T	$\kappa_T = -(1/V)(\partial V/\partial p)_T$	Pa^{-1}
isentropic	κ_S	$\kappa_S = -(1/V)(\partial V/\partial p)_S$	Pa^{-1}
linear expansion coefficient	α_l	$\alpha_l = (1/l)(\partial l/\partial T)$	K^{-1}
cubic expansion coefficient	α, α_V, γ	$\alpha = (1/V)(\partial V/\partial T)_p$	K^{-1}
heat capacity,			
at constant pressure	C_p	$C_p = (\partial H/\partial T)_p$	J K^{-1}
at constant volume	C_V	$C_V = (\partial U/\partial T)_V$	J K^{-1}
ratio of heat capacities	$\gamma, (\kappa)$	$\gamma = C_p/C_V$	1
Joule–Thomson coefficient	μ, μ_{JT}	$\mu = (\partial T/\partial p)_H$	K Pa^{-1}
second virial coefficient	B	$pV_m = RT(1 + B/V_m + \ldots)$	m^3 mol^{-1}
compression factor (compressibility factor)	Z	$Z = pV_m/RT$	1
partial molar quantity X	$X_B, (X'_B)$	$X_B = (\partial X/\partial n_B)_{T,p,nj \neq B}$	(varies)
chemical potential (partial molar Gibbs energy)	μ	$\mu_B = (\partial G/\partial n_B)_{T,p,nj \neq B}$	J mol^{-1}
absolute activity	λ	$\lambda_B = \exp(\mu_B/RT)$	1
standard chemical potential	$\mu^\ominus, \mu^o$		J mol^{-1}
standard partial molar enthalpy	$H_B^\ominus$	$H_B^\ominus = \mu_B^\ominus + TS_B^\ominus$	J mol^{-1}
standard partial molar entropy	$S_B^\ominus$	$S_B^\ominus = -(\partial \mu_B^\ominus/\partial T)_p$	J mol^{-1} K^{-1}

Name	Symbol	Definition	SI unit
standard reaction Gibbs energy (function)	$\Delta_r G^\ominus$	$\Delta_r G^\ominus = \sum_B \nu_B \mu_B^\ominus$	J mol^{-1}
affinity of reaction	A, (A)	$A = -(\partial G / \partial \xi)_{p,T} = -\sum_B \nu_B \mu_B$	J mol^{-1}
standard reaction enthalpy	$\Delta_r H^\ominus$	$\Delta_r H^\ominus = \sum_B \nu_B H_B^\ominus$	J mol^{-1}
standard reaction entropy	$\Delta_r S^\ominus$	$\Delta_r S^\ominus = \sum_B \nu_B S_B^\ominus$	J mol^{-1} K^{-1}
equilibrium constant	$K^\ominus$, K	$K^\ominus = \exp(-\Delta_r G^\ominus/RT)$	1
equilibrium constant, pressure basis	K_p	$K_p = \prod_B p_B^{\nu_B}$	Pa$^{\Sigma\nu}$
concentration basis	K_c	$K_c = \prod_B c_B^{\nu_B}$	(mol m^{-3})$^{\Sigma\nu}$
molality basis	K_m	$K_m = \prod_B m_B^{\nu_B}$	(mol kg^{-1})$^{\Sigma\nu}$
fugacity	f, $\tilde{p}$	$f_B = \lambda_B \lim_{p\to 0}(p_B / \lambda_B)_T$	Pa
fugacity coefficient	ϕ	$\phi_B = f_B/p_B$	1
activity and activity coefficient referenced to Raoult's law, (relative) activity	a	$a_B = \exp\left[\frac{\mu_B - \mu_B^\varnothing}{RT}\right]$	1
activity coefficient	f	$f_B = a_B/x_B$	1
activities and activity coefficients referenced to Henry's law, (relative) activity, molality basis	a_m	$a_{m,B} = \exp\left[\frac{\mu_B - \mu_B^\varnothing}{RT}\right]$	1
concentration basis	a_c	$a_{c,B} = \exp\left[\frac{\mu_B - \mu_B^\varnothing}{RT}\right]$	1
mole fraction basis	a_x	$a_{x,B} = \exp\left[\frac{\mu_B - \mu_B^\varnothing}{RT}\right]$	1
activity coefficient, molality basis	γ_m	$a_{m,B} = \gamma_{m,B} m_B/m^\ominus$	1
concentration basis	γ_c	$a_{c,B} = \gamma_{c,B} c_B/c^\ominus$	1
mole fraction basis	γ_x	$a_{x,B} = \gamma_{x,B} x_B$	1
ionic strength, molality basis	I_m, I	$I_m = ½ \Sigma m_B z_B^2$	mol kg^{-1}
concentration basis	I_c, I	$I_c = ½ \Sigma c_B z_B^2$	mol m^{-3}
osmotic coefficient, molality basis	ϕ_m	$\phi_m = (\mu_A^* - \mu_A)/(RTM_A \Sigma m_B)$	1
mole fraction basis	ϕ_x	$\phi_x = (\mu_A - \mu_A^*)/(RT \ln x_A)$	1
osmotic pressure	Π	$\Pi = c_B RT$ (ideal dilute solution)	Pa

Symbols Used in Kinetics, Electrochemistry, Surface Chemistry, and Transport Properties

Name	Symbol	Definition	SI unit
Chemical Kinetics			
rate of change of quantity X	$\dot{X}$	$\dot{X} = dX/dt$	(varies)
rate of conversion	$\dot{\xi}$	$\dot{\xi} = d\xi/dt$	mol s^{-1}
rate of concentration change (due to chemical reaction)	r_B, v_B	$r_B = dc_B/dt$	mol m^{-3} s^{-1}
rate of reaction (based on amount concentration)	v	$v = \dot{\xi}/V = v_B^{-1} dc_B/dt$	mol m^{-3} s^{-1}
partial order of reaction	n_B	$v = k\Pi c_B^{nB}$	1
overall order of reaction	n	$n = \Sigma n_B$	1
rate constant, rate coefficient	k	$v = k\Pi c_B^{nB}$	(mol^{-1} m^3)$^{n-1}$ s^{-1}
Boltzmann constant	k, k_B		J K^{-1}

Name	Symbol	Definition	SI unit
half-life	$t_{½}$	$c(t_{½}) = c_0/2$	s
relaxation time	τ	$\tau = 1/(k_1+k_{-1})$	s
energy of activation, activation energy	E_a, E	$E_a = RT^2\, \mathrm{d}\ln k/\mathrm{d}T$	J mol^{-1}
pre-exponential factor	A	$k = A \exp(-E_a/RT)$	(mol^{-1} m^3)$^{n-1}$ s^{-1}
volume of activation	$\Delta^{\ddagger}V$	$\Delta^{\ddagger}V = -RT \times (\partial \ln k/\partial p)_T$	m^3 mol^{-1}
collision diameter	d	$d_{AB} = r_A + r_B$	m
collision cross-section	σ	$\sigma_{AB} = \pi d_{AB}^2$	m^2
collision frequency	Z_A		s^{-1}
collision number	Z_{AB}, Z_{AA}		m^{-3} s^{-1}
collision frequency factor	z_{AB}, z_{AA}	$z_{AB} = Z_{AB}/Lc_Ac_B$	m^3 mol^{-1} s^{-1}
standard enthalpy of activation	$\Delta^{\ddagger}H^{\ominus}$, $\Delta H^{\ddagger}$		J mol^{-1}
standard entropy of activation	$\Delta^{\ddagger}S^{\ominus}$, $\Delta S^{\ddagger}$		J mol^{-1} K^{-1}
standard Gibbs energy of activation	$\Delta^{\ddagger}G^{\ominus}$, $\Delta G^{\ddagger}$		J mol^{-1}
quantum yield, photochemical yield	ϕ		1
Electrochemistry			
elementary charge (proton charge)	e		C
Faraday constant	F	$F=eL$	C mol^{-1}
charge number of an ion	z	$z_B = Q_B/e$	1
ionic strength	I_c, I	$I_c = ½\Sigma c_i z_i^2$	mol m^{-3}
mean ionic activity	$a_{\pm}$	$a_{\pm} = m_{\pm}\gamma_{\pm}/m^{\ominus}$	1
mean ionic molality	$m_{\pm}$	$m_{\pm}^{(\nu_+ + \nu_-)} = m_+^{\nu+} m_-^{\nu-}$	mol kg^{-1}
mean ionic activity coefficient	$\gamma_{\pm}$	$\gamma_{\pm}^{(\nu_+ + \nu_-)} = \gamma_+^{\nu+}\gamma_-^{\nu-}$	1
charge number of electrochemical cell reaction	n, (z)		1
electric potential difference (of a galvanic cell)	ΔV, E, U	$\Delta V = V_R - V_L$	V
emf, electromotive force	E	$E = \lim_{I\to 0} \Delta V$	V
standard emf, standard potential of the electrochemical cell reaction	$E^{\ominus}$	$E^{\ominus} = -\Delta_r G^{\ominus}/nF = (RT/nF)\ln K^{\ominus}$	V
standard electrode potential	$E^{\ominus}$		V
emf of the cell, potential of the electrochemical cell reaction	E	$E = E^{\ominus} - (RT/nF) \times \Sigma \nu_i \ln a_i$	V
pH	pH	$\mathrm{pH} \approx -\lg\left[\dfrac{c(\mathbf{H}^+)}{\mathrm{mol\ dm}^{-3}}\right]$	1
inner electric potential	ϕ	$\nabla\phi = -E$	V
outer electric potential	ψ	$\psi = Q/4\pi\varepsilon_0 r$	V
surface electric potential	χ	$\chi = \phi - \psi$	V
Galvani potential difference	$\Delta\phi$	$\Delta_{\alpha}^{\beta}\phi = \phi^{\beta} - \phi^{\alpha}$	V
volta potential difference	$\Delta\psi$	$\Delta_{\alpha}^{\beta}\psi = \psi^{\beta} - \psi^{\alpha}$	V
electrochemical potential	$\tilde{\mu}$	$\tilde{\mu}_B^{\alpha} = (\partial G/\partial n_B^{\alpha})$	J mol^{-1}
electric current	I	$I = \mathrm{d}Q/\mathrm{d}t$	A
(electric) current density	j	$j = I/A$	A m^{-2}
(surface) charge density	σ	$\sigma = Q/A$	C m^{-2}
electrode reaction rate constant	k	$k_{ox} = I_a/(nFA\prod_i c_i^{n_i})$	(varies)
mass transfer coefficient, diffusion rate constant	k_d	$k_{d,B} = \lvert\nu_B\rvert I_{l,B}/nFcA$	m s^{-1}
thickness of diffusion layer	δ	$\delta_B = D_B/k_{d,B}$	m
transfer coefficient (electrochemical)	α	$\alpha_c = \dfrac{-\lvert\nu\rvert RT}{nF}\dfrac{\partial \ln\lvert I_c\rvert}{\partial E}$	1
overpotential	η	$\eta = E_I - E_{I=0} - IR_u$	V
electrokinetic potential (zeta potential)	ζ		V
conductivity	κ, (σ)	$\kappa = j/E$	S m^{-1}
conductivity cell constant	K_{cell}	$K_{cell} = \kappa R$	m^{-1}
molar conductivity (of an electrolyte)	Λ	$\Lambda_B = \kappa/c_B$	S m^2 mol^{-1}
ionic conductivity, molar conductivity of an ion	λ	$\lambda_B = \lvert z_B\rvert F u_B$	S m^2 mol^{-1}
electric mobility	u, (μ)	$u_B = \nu_B/E$	m^2 V^{-1} s^{-1}
transport number	t	$t_B = j_B/\Sigma j_i$	1
reciprocal radius of ionic atmosphere	κ	$\kappa = (2F^2 I/\varepsilon RT)^{½}$	m^{-1}
Colloid and Surface Chemistry			
specific surface area	a, a_s, s	$a = A/m$	m^2 kg^{-1}
surface amount of B, adsorbed amount of B	n_B^s, n_B^a		mol

Name	Symbol	Definition	SI unit
surface excess of B	n_B^σ		mol
surface excess concentration of B	$\Gamma_B, (\Gamma_B^\sigma)$	$\Gamma_B = n_B^\sigma/A$	mol m^{-2}
total surface excess concentration	$\Gamma, (\Gamma^\sigma)$	$\Gamma = \sum_i \Gamma_i$	mol m^{-2}
area per molecule	a, σ	$a_B = A/N_B^\sigma$	m^2
area per molecule in a filled monolayer	a_m, σ_m	$a_{m,B} = A/N_{m,B}$	m^2
surface coverage	θ	$\theta = N_B^\sigma/N_{m,B}$	1
contact angle	θ		1, rad
film thickness	t, h, δ		m
thickness of (surface or interfacial) layer	τ, δ, t		m
surface tension, interfacial tension	γ, σ	$\gamma = (\partial G/\partial A_s)_{T,p}$	N m^{-1}, J m^{-2}
film tension	Σ_f	$\Sigma_f = 2\gamma_f$	N m^{-1}
reciprocal thickness of the double layer	κ	$\kappa = [2F^2 I_c/\varepsilon RT]^{1/2}$	m^{-1}
average molar masses			
number–average	M_n	$M_n = \Sigma n_i M_i/\Sigma n_i$	kg mol^{-1}
mass–average	M_m	$M_m = \Sigma n_i M_i^2/\Sigma n_i M_i$	kg mol^{-1}
Z–average	M_Z	$M_Z = \Sigma n_i M_i^3/\Sigma n_i M_i^2$	kg mol^{-1}
sedimentation coefficient	s	$s = v/a$	s
van der Waals constant	λ		J
retarded van der Waals constant	β, B		J
van der Waals–Hamaker constant	A_H		J
surface pressure	π^s, π	$\pi^s = \gamma^0 - \gamma$	N m^{-1}
Transport Properties			
flux (of a quantity X)	J_X, J	$J_X = A^{-1}\, dX/dt$	(varies)
volume flow rate	$q_V, \dot{V}$	$q_v = dV/dt$	m^3 s^{-1}
mass flow rate	$q_m, \dot{m}$	$q_m = dm/dt$	kg s^{-1}
mass transfer coefficient	k_d		m s^{-1}
heat flow rate	ϕ	$\phi = dq/dt$	W
heat flux	J_q	$J_q = \phi/A$	W m^{-2}
thermal conductance	G	$G = \phi/\Delta T$	W K^{-1}
thermal resistance	R	$R = 1/G$	K W^{-1}
thermal conductivity	λ, k	$\lambda = J_q/(dT/dl)$	W m^{-1} K^{-1}
coefficient of heat transfer	$h, (k, K, \alpha)$	$h = J_q/\Delta T$	W m^{-2} K^{-1}
thermal diffusivity	a	$a = \lambda/\rho c_p$	m^2 s^{-1}
diffusion coefficient	D	$D = J_n/(dc/dl)$	m^2 s^{-1}

Symbols Used to Denote a Chemical Process, Reaction, or Condition

Process	Symbol
Symbols used as subscripts	
vaporization, evaporation (liquid → gas)	vap
sublimation (solid → gas)	sub
melting, fusion (solid → liquid)	fus
transition (between two phases)	trs
mixing of fluids	mix
solution (of solute in solvent)	sol
dilution (of a solution)	dil
adsorption	ads
displacement	dpl
immersion	imm
reaction in general	r
atomization	at
combustion reaction	c
formation reaction	f

Process	Symbol
Symbols used as superscripts	
standard	$\ominus$, o
pure substance	*
infinite dilution	∞
ideal	id
activated complex, transition state	‡, ≠
excess quantity	E
apparent	app

The following symbols are used in the definitions of the dimensionless quantities: mass (m), time (t), volume (V), area (A), density (ρ), speed (v), length (l), viscosity (η), pressure (p), acceleration of free fall (g), cubic expansion coefficient (α), temperature (T), surface tension (γ), speed of sound (c), mean free path (λ), frequency (f), thermal diffusivity (a), coefficient of heat transfer (h), thermal conductivity (k), specific heat capacity at constant pressure (c_p), diffusion coefficient (D), mole fraction (x), mass transfer coefficient (k_d), permeability (μ), electric conductivity (κ), and magnetic flux density (B).

Dimensionless Quantities

Name	Symbol	Definition	SI unit
Reynolds number	Re	$Re = \rho vl/\eta$	1
Euler number	Eu	$Eu = \Delta p/\rho v^2$	1
Froude number	Fr	$Fr = v/(lg)^{1/2}$	1
Grashof number	Gr	$Gr = l^3 g\alpha\Delta T\rho^2/\eta^2$	1
Weber number	We	$We = \rho v^2 l/\gamma$	1
Mach number	Ma	$Ma = v/c$	1
Knudsen number	Kn	$Kn = \lambda/l$	1
Strouhal number	Sr	$Sr = lf/v$	1
Fourier number	Fo	$Fo = at/l^2$	1
Péclet number	Pe	$Pe = vl/a$	1
Rayleigh number	Ra	$Ra = l^3 g\alpha\Delta T\rho/\eta a$	1
Nusselt number	Nu	$Nu = hl/k$	1
Stanton number	St	$St = h/\rho vc_p$	1
Fourier number for mass transfer	Fo^*	$Fo^* = Dt/l^2$	1
Péclet number for mass transfer	Pe^*	$Pe^* = vl/D$	1
Grashof number for mass transfer	Gr^*	$Gr^* = l^3 g\left(\frac{\partial\rho}{\partial x}\right)_{T,p}\left(\frac{\Delta x\rho}{\eta}\right)$	1
Nusselt number for mass transfer	Nu^*	$Nu^* = k_d l/D$	1
Stanton number for mass transfer	St^*	$St^* = k_d/v$	1
Prandtl number	Pr	$Pr = \eta/\rho a$	1
Schmidt number	Sc	$Sc = \eta/\rho D$	1
Lewis number	Le	$Le = a/D$	1
magnetic Reynolds number	Rm, Re_m	$Rm = v\mu\kappa l$	1
Alfvén number	Al	$Al = v(\rho\mu)^{1/2}/B$	1
Hartmann number	Ha	$Ha = Bl(\kappa/\eta)^{1/2}$	1
Cowling number	Co	$Co = B^2/\mu\rho v^2$	1

EXPRESSION OF UNCERTAINTY OF MEASUREMENTS

In general, the result of a measurement is only an approximation or estimate of the true value of the quantity subject to measurement, and thus the result is of limited value unless accompanied by a statement of its uncertainty. Much (but not all) of the scientific data appearing in the literature does include some indication of the uncertainty, but this may be stated in many different ways and is often explained poorly. In an effort to encourage consistency in uncertainty statements, the International Committee for Weights and Measures (CIPM) initiated a project, in collaboration with several other international organizations, to prepare a set of guidelines expressing international consensus on the recommended method of stating uncertainties. This project resulted in the publication of the *Guide to the Expression of Uncertainty in Measurement* (Reference 1), which is often referred to as *GUM*. The recommendations of *GUM* have been summarized by the National Institute of Standards and Technology in *NIST Technical Note 1297, Guidelines for Evaluating the Uncertainty of NIST Measurement Results* (Reference 2).

In the notation of *GUM*, we are concerned with the **measurand**, i.e., the quantity that is being measured. In physics and chemistry this is usually called a **physical quantity** and represents some inherent characteristic of a material, system, or process that can be expressed in numerical terms — specifically as the product of a number and a reference, commonly called a **unit.** Thus the density of water at room temperature is (approximately) 0.998 g/mL (grams per milliliter) or, alternatively 998 kg m^{-3} (kilograms per meter cubed). This statement gives the most likely value of the measurand, to this level of precision, but gives no information on how much the stated value might differ from the true value.

It is important to differentiate between the terms **error** and **uncertainty**. The error in a measurement is the difference between the measured value and the true value; the error can be stated if the true value is known (to some level of accuracy). The uncertainty is an estimate of the maximum reasonable extent to which the measured value is believed to deviate from the true value, in a situation where the true value is not known (most often the case). The result of a measurement can unknowably be very close to the true value, and thus have negligible error, even though its uncertainty is large.

The uncertainty of the result of a measurement generally consists of several components, which may be grouped in two types according to the method used to estimate their numerical values:

Type A. Those which are evaluated by statistical methods
Type B. Those which are evaluated by other means

The terms "random uncertainty" and "systematic uncertainty" are often used, but these terms do not always correspond in a simple way to the A and B categories. This is because the nature of an uncertainty component is conditioned by how the quantity appears in the mathematical model that describes the current measurement process. An uncertainty component arising from a systematic effect may in some cases be evaluated by methods of Type A while in other cases by methods of Type B.

In the *GUM* formulation, each component of uncertainty, whether in the A or B category, is represented by an estimated standard deviation, termed **standard uncertainty**, symbol u_i, and equal to the positive square root of the estimated variance u_i^2.

For an uncertainty component of Type A, $u_i = s_i$, where s_i is the statistically estimated standard deviation, as determined from a series of observations by appropriate statistical analysis. Any valid statistical method may be used. Examples are calculating the standard deviation of the mean of a series of independent observations; using the method of least squares to fit a curve to data in order to estimate parameters of the curve and their standard deviations; and carrying out an analysis of variance (ANOVA) in order to identify and quantify random effects in certain types of measurements. Details of statistical analysis are given in References 4–7 and many other places.

In a similar manner, each uncertainty component of Type B is represented by a quantity u_j, which is obtained from an assumed probability distribution based on all the available information about the measurement process. Since u_j is treated like a standard deviation, the standard uncertainty in each Type B component is simply u_j. The evaluation of u_j is usually based on scientific judgment using all the relevant information available, which may include

- previous measurement data
- experience with, or general knowledge of, the behavior and properties of relevant materials and instruments
- manufacturer's specifications
- data provided in calibrations and other reports
- uncertainties assigned to reference data taken from handbooks.

The specific approach to evaluating the standard uncertainty u_j of a Type B uncertainty will depend on the detailed model of the measurement process. The following are examples of steps that may be used:

1. Convert a quoted uncertainty (for example, in a calibration factor) that is a stated multiple of an estimated standard deviation to a standard uncertainty by dividing the quoted uncertainty by the multiplier.
2. Convert a quoted uncertainty that defines a "confidence interval" having a stated level of confidence, such as 95% or 99%, to a standard uncertainty by treating the quoted uncertainty as if a normal distribution had been used to calculate it (unless otherwise indicated) and dividing it by the appropriate factor for such a distribution. These factors are 1.960 and 2.576 for the two levels of confidence given.
3. Model knowledge of the quantity in question by a normal distribution and estimate lower and upper limits a_- and a_+ such that the best estimated value of the quantity is $(a_+ + a_-)/2$ (i.e., the midpoint of the limits) and there is 1 chance out of 2 (i.e., a 50 percent probability) that the value of the quantity lies in the interval a_- to a_+. Then $u_j \approx 1.48\,a$, where $a = (a_+ - a_-)/2$ is the half-width of the interval.
4. Model knowledge of the quantity in question by a normal distribution and estimate lower and upper limits a_- and a_+ such that the best estimated value of the quantity is $(a_+ + a_-)/2$ and there is about a 2 out of 3 chance (i.e., a 67 percent probability) that the value of the quantity lies in the interval a_- to a_+. Then $u_j \approx a$, where $a = (a_+ - a_-)/2$.
5. Estimate lower and upper limits a_- and a_+ for the value of the quantity in question such that the probability that the value lies in the interval a_- to a_+ is, for all practical purposes, 100 percent. Provided that there is no contradictory information, treat the quantity as if it is equally probable for its value to lie anywhere within the interval a_- to a_+;

that is, model it by a uniform or rectangular probability distribution. The best estimate of the value of the quantity is then $(a_+ + a_-)/2$ with $u_j = a/\sqrt{3}$ where $a = (a_+ - a_-)/2$. If the distribution used to model the quantity is triangular rather than rectangular, then $u_j = a/\sqrt{6}$. The rectangular distribution is a reasonable default model in the absence of any other information. But if it is known that values of the quantity in question near the center of the limits are more likely than values close to the limits, a triangular or a normal distribution may be a better model.

When all the standard uncertainties of Type A and Type B have been determined in this way, they should be combined to produce the **combined standard uncertainty** (suggested symbol u_c), which may be regarded as the estimated standard deviation of the measurement result. This process, often called the *law of propagation of uncertainty* or "root-sum-of-squares," involves taking the square root of the sum of the squares of all the u_i. In many practical measurement situations, the probability distribution characterized by the measurement result y and its combined standard uncertainty $u_c(y)$ is approximately normal (Gaussian). When this is the case, $u_c(y)$ defines an interval $y - u_c(y)$ to $y + u_c(y)$ about the measurement result y within which the value of the measurand Y estimated by y is believed to lie with a level of confidence of approximately 68 percent. That is, it is believed with an approximate level of confidence of 68 percent that $y - u_c(y) \leq Y \leq y + u_c(y)$, which is commonly written as $Y = y \pm u_c(y)$.

In fundamental metrological research (involving physical constants, calibration standards, and the like) the combined standard uncertainty u_c is normally used as the statement of uncertainty in a measurement. In most cases, however, it is desirable to use a measure of uncertainty that defines an interval about the measurement result y within which the value of the measurand Y is confidently believed to lie. The measure of uncertainty intended to meet this requirement is termed **expanded uncertainty,** suggested symbol U, and is obtained by multiplying $u_c(y)$ by a **coverage factor,** suggested symbol k. Thus $U = ku_c(y)$ and it is believed with high confidence that $y - U \leq Y \leq y + U$, which is commonly written as $Y = y \pm U$. The value of the coverage factor k is chosen on the basis of the desired level of confidence to be associated with the interval defined by $U = ku_c$. Typically, k is in the range 2 to 3. When the normal distribution applies, $U = 2u_c$ (i.e., $k = 2$) defines an interval having a level of confidence of approximately 95 percent, and $U = 3u_c$ defines an interval having a confidence level greater than 99 percent. In current international practice it is most common to use $k = 2$, corresponding to about 95 percent confidence, but the value of k should be stated in each case to avoid confusion. See References 1 and 2 for methods of calculating k when a value other than $k = 2$ is needed for a specific requirement.

Summary of Key Steps

- Group the uncertainty components into Type A (can be evaluated by statistical methods) and Type B (must be evaluated by other means).
- Determine the standard uncertainty for each component of Type A by statistical methods and for each component of Type B by other suitable methods, based on modeling the measurement process.
- Take the square root of the sum of the squares of all the standard uncertainties to get the combined standard uncertainty u_c.
- Specify a coverage factor k which, when multiplied by u_c, gives the expanded uncertainty U. In fundamental metrological research $k = 1$ is usually chosen; in other cases, $k = 2$ (corresponding to a confidence level of about 95%) is the most common choice.

References

1. ISO, *Guide to the Expression of Uncertainty in Measurement*, International Organization for Standardization, Geneva, Switzerland, 1993. Several supplements have been published; see Bich, W., Cox, M. C., and Harris, P. M., "Evolution of the *Guide to the Expression of Uncertainty in Measurement*," *Metrologia* 43, S161, 2006.
2. Taylor, B. N., and Kuyatt, C. E., *Guidelines for Evaluating and Expressing the Uncertainty of NIST Measurement Results*, NIST Technical Note 1297, National Institute of Standards and Technology, Gaithersburg, MD, 1994; available for free download at <physics.nist.gov/cuu/Uncertainty/bibliography.html>.
3. Bell, S., *A Beginner's Guide to Uncertainty of Measurement*, National Physical Laboratory, Teddington, Middlesex, UK, 2001; available on the Internet through <www.npl.co.uk/server.php?show=ConWebDoc.1785>.
4. Eisenhart, C., "Realistic Evaluation of the Precision and Accuracy of Instrument Calibration Systems," *J. Res. Natl. Bur. Stand.* (U.S.) 67C, 161, 1963.
5. Mandel, J., *The Statistical Analysis of Experimental Data*, Dover Publishers, New York, 1984.
6. Nantrella, M. G., *Experimental Statistics*, NBS Handbook 91, U.S. Government Printing Office, Washington, DC, 1966.
7. Box, G. E. P., Hunter, J. S., and Hunter, W. G., *Statistics for Experimenters: Design, Innovation, and Discovery, 2nd Edition*, John Wiley & Sons, Hoboken, NJ, 2005.

NOMENCLATURE FOR CHEMICAL COMPOUNDS

The International Union of Pure and Applied Chemistry (IUPAC) maintains several commissions that deal with the naming of chemical substances. In general, the approach of IUPAC is to present rules for arriving at names in a systematic manner, rather than recommending a unique name for each compound. Thus there are often several alternative "IUPAC names," depending on which nomenclature system is used, each of which may have advantages in specific applications. However, each of these names will be unambiguous.

Organizations such as the Chemical Abstracts Service that prepare indexes to the chemical literature must adopt a system for selecting unique names in order to avoid excessive cross referencing. Chemical Abstracts Service uses a system which groups together compounds derived from a single parent compound. Thus most index names are inverted (e.g., Benzene, bromo rather than bromobenzene; Acetic acid, sodium salt rather than sodium acetate).

Recommended names for the most common substituent groups, ligands, ions, and organic rings are given in the two following tables, "Nomenclature for Inorganic Ions and Ligands" and "Organic Substituent Groups and Ring Systems." For the basics of macromolecular nomenclature, see "Nomenclature for Organic Polymers" in Section 13.

Some of the most useful recent guides to chemical nomenclature, prepared by IUPAC and other organizations such as the International Union of Biochemistry and Molecular Biology (IUBMB) and the American Chemical Society are listed below. These books contain citations to the more detailed nomenclature documents in each area. Two very useful links to nomenclature documents are:

iupac.org/what-we-do/nomenclature/
www.chem.qmul.ac.uk/iupac/

Inorganic Chemistry

1. Block, B. P., Powell, W. H., and Fernelius, W. C., *Inorganic Chemical Nomenclature, Principles and Practice*, American Chemical Society, Washington, DC, 1990.
2. *Nomenclature of Inorganic Chemistry - IUPAC Recommendations 2005.* Connelly, N. G., Damhus, T., Hartshorn, R. M., and Hutton, A. T., The Royal Society of Chemistry, 2005.
3. Hartshorn, R. M., Hellwich, K. H., Yerin, A., Damhus, T., and Hutton, A. T., *Brief Guide to the Nomenclature of Inorganic Chemistry, Pure Appl. Chem.*, 87, 1039, 2015.

Organic Chemistry

1. International Union of Pure and Applied Chemistry, *A Guide to IUPAC Nomenclature of Organic Compounds, Recommendations 1993,* Panico, R., Powell, W. H., and Richer, J.-C., Eds., Blackwell Scientific Publications, Oxford, 1993.
2. International Union of Pure and Applied Chemistry, *Glossary of Class Names of Organic Compounds and Reactive Intermediates Based on Structure*, Moss, G. P., Smith, P. A. S., and Tavernier, D., Eds., *Pure Appl. Chem.* 67, 1307, 1995.
3. International Union of Pure and Applied Chemistry, *Basic Terminology of Stereochemistry*, Moss, G. P., Ed., *Pure Appl. Chem.* 68, 2193, 1996.
4. Rhodes, P. H., *The Organic Chemist's Desk Reference*, Chapman & Hall, London, 1995.

Macromolecular Chemistry

1. International Union of Pure and Applied Chemistry, *Compendium of Polymer Nomenclature*, Hess, M., Jones, R. G., Kahovec, J., Kitayama, T., Metanomski, W. V., Stepto, R., and Wilks, E. S., RSC Publishing, 2008.
2. International Union of Pure and Applied Chemistry, *Glossary of Basic Terms in Polymer Science*, Jenkins, A.D., Kratochvil, P., Stepto, R. F. T., and Suter, U. W., Eds., *Pure Appl. Chem.* 68, 2287, 1996.

Biochemistry

1. International Union of Biochemistry and Molecular Biology, *Biochemical Nomenclatureand Related Documents, 2nd Edition, 1992,* Portland Press, London, 1993; includes recommendations of the IUPAC-IUBMB Joint Commission on Biochemical Nomenclature.
2. International Union of Biochemistry and Molecular Biology, *Enzyme Nomenclature, 1992*, Academic Press, Orlando, FL, 1992.
3. IUPAC-IUBMB Joint Commission on Biochemical Nomenclature, *Nomenclature of Carbohydrates, Recommendations 1996,* McNaught, A. D., Ed., *Pure & Appl. Chem.* 68, 1919, 1996.

General

Principles of Chemical Nomenclature, Leigh, G. J., RSC Publishing, 2011.

NOMENCLATURE OF INORGANIC CHEMISTRY

R. M. Hartshorn, K.-H. Hellwich, A. Yerin, T. Damhus, and A. T. Hutton

Preamble

The universal adoption of an agreed chemical nomenclature is a key tool for communication in the chemical sciences, for computer-based searching in databases, and for regulatory purposes, such as those associated with health and safety or commercial activity. The International Union of Pure and Applied Chemistry (IUPAC) provides recommendations on the nature and use of chemical nomenclature.[1] The basics of this nomenclature are shown here, and in companion documents on the nomenclature systems for organic chemistry[2] and polymers,[3] with hyperlinks to the original documents. An overall summary of chemical nomenclature can be found in *Principles of Chemical Nomenclature*.[4] Greater detail can be found in the *Nomenclature of Inorganic Chemistry*, colloquially known as the Red Book,[5] and in the related publications for organic compounds (the Blue Book)[6] and polymers (the Purple Book).[7] It should be noted that many compounds may have non-systematic or semi-systematic names (some of which are not accepted by IUPAC for several reasons, for example because they are ambiguous) and IUPAC rules allow for more than one systematic name in many cases. IUPAC is working towards identification of single names which are to be preferred for regulatory purposes (Preferred IUPAC Names, or PINs). *Note*: In this document, the symbol '=' is used to split names that happen to be too long for the column format, unless there is a convenient hyphen already present in the name.

The boundaries between 'organic' and 'inorganic' compounds are blurred. The nomenclature types described in this document are applicable to compounds, molecules and ions that do not contain carbon, but also to many structures that do contain carbon (Section 2), notably those containing elements of Groups 1–12. Most boron-containing compounds are treated using a special nomenclature.[8]

References

1. Freely available at: (a) http://www.degruyter.com/pac; (b) http://www.chem.qmul.ac.uk/iupac/.
2. K.-H. Hellwich, R. M. Hartshorn, A. Yerin, T. Damhus, A. T. Hutton, Brief Guide to the Nomenclature of Organic Chemistry, *Pure Appl. Chem.*, in preparation.
3. R. C. Hiorns, R. J. Boucher, R. Duhlev, K.-H. Hellwich, P. Hodge, A. D. Jenkins, R. G. Jones, J. Kahovec, G. Moad, C. K. Ober, D. W. Smith, R. F. T. Stepto, J.-P. Vairon, J. Vohlídal, *Pure Appl. Chem.* 84(10), 2167–2169 (2012).
4. *Principles of Chemical Nomenclature – A Guide to IUPAC Recommendations, 2011 Edition*, G. J. Leigh (Ed.), Royal Society of Chemistry, Cambridge, U.K., ISBN 978-1-84973-007-5.
5. *Nomenclature of Inorganic Chemistry – IUPAC Recommendations 2005*, N. G. Connelly, T. Damhus, R. M. Hartshorn, A. T. Hutton (Eds.), Royal Society of Chemistry, Cambridge, U.K., ISBN 0-85404-438-8.
6. *Nomenclature of Organic Chemistry – IUPAC Recommendations and Preferred Names 2013*, H. A. Favre, W. H. Powell (Eds.), Royal Society of Chemistry, Cambridge, U.K., ISBN 978-0-85404-182-4.
7. *Compendium of Polymer Terminology and Nomenclature – IUPAC Recommendations 2008*, R. G. Jones, J. Kahovec, R. Stepto, E. S. Wilks, M. Hess, T. Kitayama, W. V. Metanomski (Eds.), Royal Society of Chemistry, Cambridge, U.K., ISBN 978-0-85404-491-7.
8. Reference 4, Chapter 10.
9. Reference 5, Table IX.
10. Reference 4, Table P10.
11. Reference 5, Chapter IR-6.
12. Reference 5, Table X.
13. Reference 5, Section IR-9.2.4.
14. Reference 5, Table IR-10.4.
15. Reference 5, Section IR-10.3.
16. Reference 5, Chapter IR-8.
17. Reference 4, Table P5; Reference 5, Tables IR-9.2 and IR-9.3.
18. Reference 5, Section IR-9.3.3.
19. R. S. Cahn, C. Ingold, V. Prelog, *Angew. Chem., Int. Ed. Engl.*, 5, 385–415 and 511 (1966); V. Prelog, G. Helmchen, *Angew. Chem., Int. Ed. Engl.*, 21, 567–583 (1982).
20. J. Brecher, K. N. Degtyarenko, H. Gottlieb, R. M. Hartshorn, G. P. Moss, P. Murray-Rust, J. Nyitrai, W. Powell, A. Smith, S. Stein, K. Taylor, W. Town, A. Williams, A. Yerin, *Pure Appl. Chem.*, 78(10), 1897–1970 (2006); J. Brecher, K. N. Degtyarenko, H. Gottlieb, R. M. Hartshorn, K.-H. Hellwich, J. Kahovec, G. P. Moss, A. McNaught, J. Nyitrai, W. Powell, A. Smith, K. Taylor, W. Town, A. Williams, A. Yerin, *Pure Appl. Chem.*, 80(2), 277–410 (2008).

1 Stoichiometric or Compositional Names

A stoichiometric or compositional name provides information only on the composition of an ion, molecule, or compound, and may be related to either the empirical or molecular formula for that entity. It does not provide any structural information.

For homoatomic entities, where only one element is present, the name is formed (Table 1) by combining the element name with the appropriate multiplicative prefix (Table 2). Ions are named by adding charge numbers in parentheses, e.g., (1+), (3+), (2−), and for (most) homoatomic anion names 'ide' is added in place of the 'en', 'ese', 'ic', 'ine', 'ium', 'ogen', 'on', 'orus', 'um', 'ur', 'y', or 'ygen' endings of element names.[9] Exceptions include Zn and Group 18 elements ending in 'on', where the 'ide' ending is added to the element names. For some elements (e.g., Fe, Ag, Au) a Latin stem is used before the 'ide' ending (cf. Section 2.3).[9] Certain ions may have acceptable traditional names (used without charge numbers).

Table 1: Examples of Homoatomic Entities

Formula	Name	Formula	Name
O_2	dioxygen	Cl^-	chloride(1-) or chloride
S_8	octasulfur	I_3^-	triiodide(1-)
Na^+	sodium(1+)	O_2^{2-}	dioxide(2-) or peroxide
Fe^{3+}	iron(3+)	N_3^-	trinitride(1-) or azide

Table 2: Multiplicative Prefixes for Simple and Complicated Entities

No.	Simple	Complicated	No.	Simple	Complicated
2	di	bis	8	octa	octakis
3	tri	tris	9	nona	nonakis
4	tetra	tetrakis	10	deca	decakis
5	penta	pentakis	11	undeca	undecakis
6	hexa	hexakis	12	dodeca	dodecakis
7	hepta	heptakis	20	icosa	icosakis

Binary compounds (those containing atoms of two elements) are named stoichiometrically by combining the element names and treating, by convention, the element reached first when following the arrow in the element sequence (Figure 1) as if it were an anion. Thus the name of this formally 'electronegative' element is given an 'ide' ending and is placed after the name of the formally 'electropositive' element followed by a space (Table 3).

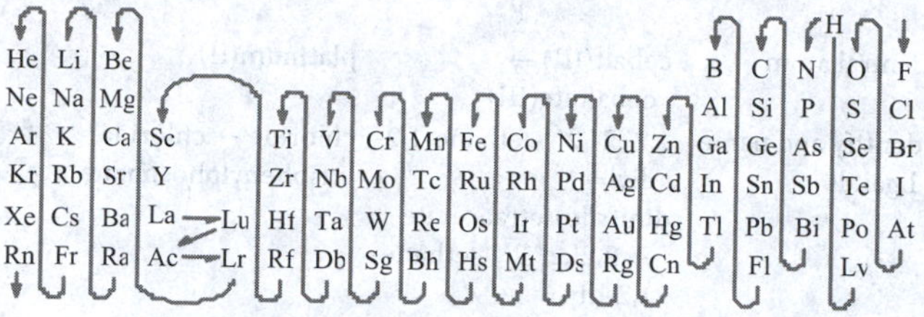

Figure 1: Element Sequence

Table 3: Examples of Binary Compounds

Formula	Name	Formula	Name
GaAs	gallium arsenide	$FeCl_2$	iron dichloride or iron(II) chloride
CO_2	carbon dioxide	$FeCl_3$	iron trichloride or iron(III) chloride
CaF_2	calcium difluoride or calcium fluoride	H_2O_2	dihydrogen dioxide or hydrogen peroxide

Again, multiplicative prefixes (Table 2) are applied as needed, and certain acceptable alternative names[10] may be used. Stoichiometry may be implied in some cases by the use of oxidation numbers, but is often omitted for common cases, such as in calcium fluoride.

Heteropolyatomic entities in general can be named similarly using compositional nomenclature, but often either substitutive[11] or additive nomenclature (Section 2) is used. In the latter case information is also provided about the way atoms are connected. For example, $POCl_3$ (or PCl_3O, compositional name phosphorus trichloride oxide) is given an additive name in Table 10.

Certain ions have traditional short names, which are commonly used and are still acceptable (e.g., ammonium, NH_4^+; hydroxide, OH^-; nitrite, NO_2^-; phosphate, PO_4^{3-}; diphosphate, $P_2O_7^{4-}$).

Inorganic compounds in general can be combinations of cations, anions, and neutral entities. By convention, the name of a compound is made up of the names of its component entities: cations before anions and neutral components last (see examples in Table 4).

The number of each entity present has to be specified in order to reflect the composition of the compound. For this purpose, multiplicative prefixes (Table 2) are added to the name of each entity. The prefixes are 'di', 'tri', 'tetra', etc., for use with names for simple entities, or 'bis( )', 'tris( )', 'tetrakis( )', etc., for names for most entities which themselves contain multiplicative prefixes or locants. Care must also be taken in situations when use of a simple multiplicative prefix may be misinterpreted, e.g., tris(iodide) must be used for $3I^-$ rather than triiodide (which is used for I_3^-), and bis(phosphate) rather than diphosphate (which is used for $P_2O_7^{4-}$). Examples are shown in Table 4. There is no elision of vowels (e.g., tetraaqua, pentaoxide), except in the special case of monoxide.

Names of neutral components are separated by 'em' dashes without spaces. Inorganic compounds may themselves be components in (formal) **addition compounds** (last four examples in Table 4). The ratios of component compounds can be indicated, in general, using a stoichiometric descriptor in parentheses after the name (see the last three examples in Table 4). In the special case of hydrates, multiplicative prefixes can be used with the term 'hydrate'.

Table 4: Use of Multiplicative Prefixes in Compositional Names

Formula	Name
$Ca_3(PO_4)_2$	tricalcium bis(phosphate)
$Ca_2P_2O_7$	dicalcium diphosphate
BaO_2	barium(2+) dioxide(2−) or barium peroxide
$MgSO_4{\cdot}7H_2O$	magnesium sulfate heptahydrate
$CdSO_4{\cdot}6NH_3$	cadmium sulfate—ammonia (1/6)
$AlK(SO_4)_2{\cdot}12H_2O$	aluminum potassium bis(sulfate)—water (1/12) or aluminum potassium bis(sulfate) dodecahydrate
$Al_2(SO_4)_3{\cdot}K_2SO_4{\cdot}24H_2O$	dialuminum tris(sulfate)—dipotassium sulfate—water (1/1/24)

2 Complexes and Additive Nomenclature

2.1 Overall approach

Additive nomenclature was developed in order to describe the structures of coordination entities, or complexes, but this method is readily extended to other molecular entities as well. Mononuclear complexes are considered to consist of a central atom, often a metal ion, which is bonded to surrounding small molecules or ions, which are referred to as ligands. The names of complexes are constructed (Table 5) by adding the names of the ligands before those of the central atoms, using appropriate multiplicative prefixes. Formulae are constructed by adding the symbols or abbreviations of the ligands *after* the symbols of the central atoms (Section 2.7).

Table 5: Producing Names for Complexes: Simple Ligands

Structure to be named	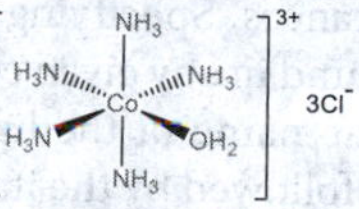	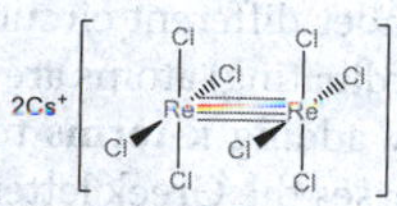
Central atom(s)	cobalt(III)	2 × rhenium
Identify and name ligands	ammonia → ammine water → aqua	chloride → chlorido
Assemble name	pentaammineaqua=cobalt(III) chloride	cesium bis(tetrachlorido=rhenate)(*Re—Re*)(2-)

2.2 Central atom(s) and ligands

The first step is to identify the central atom(s) and thereby also the ligands. By convention, the electrons involved in bonding between the central atom and a ligand are usually treated as belonging to the ligand (and this will determine how it is named).

Each ligand is named as a separate entity, using appropriate nomenclature[4] – usually substitutive nomenclature for organic ligands[2,4,6] and additive nomenclature for inorganic ligands. A small number of common molecules and ions are given special names when present in complexes. For example, a water ligand is represented in the full name by the term 'aqua'. An ammonia ligand is represented by 'ammine', while carbon monoxide bound to the central atom through the carbon atom is represented by the term 'carbonyl' and nitrogen monoxide bound through nitrogen is represented by 'nitrosyl'. Names of **anionic ligands** that end in 'ide', 'ate', or 'ite' are modified within the full additive name for the complex to end in 'ido', 'ato', or 'ito', respectively. Note that the 'ido' ending is now used for halide and oxide ligands as well. By convention, a single coordinated hydrogen atom is always considered anionic and it is represented in the name by the term 'hydrido', whereas coordinated dihydrogen is usually treated as a neutral two-electron donor entity.

2.3 Assembling additive names

Once the ligands have been named, the name can be assembled. This is done by listing the ligand names in alphabetical order before the name of the central atom(s), *without* regard to ligand charge.

If there is more than one ligand of a particular kind bound to a central atom in the same way, the number of such identical ligands is indicated using the appropriate multiplicative prefix for simple or complicated ligands (Table 2), not changing the already established alphabetical order of ligands. The nesting order of enclosing marks, for use in names where more than one set of enclosing marks is required, is: ( ), [( )], {[( )]}, ({[( )]}), etc.

Any **metal-metal bonds** are indicated by placing the central atom symbols in parentheses, in italics and connected by an 'em' dash, after the name of the complex (without spaces). The **charge number** of the complex or the **oxidation number** of the central atom is appended to the name of the complex. For **anions** that are named additively, the name of the central atom is given the 'ate' ending in a similar way to the 'ide' endings of homoatomic anions (Section 1). In some cases, by tradition, the Latin stem is used for the 'ate' names, such as in ferrate (for iron), cuprate (for copper), argentate (for silver), stannate (for tin), aurate (for gold), and plumbate (for lead).[12] Finally, the rules of compositional nomenclature (Section 1) are used to combine the additive names of ionic or neutral coordination entities with the names of any other entities that are part of the compound.

2.4 Specifying connectivity

Some ligands can bind to a central atom through different atoms under different circumstances. Specifying just which ligating (coordinating) atoms are bound in any given complex can be achieved by adding **κ-terms** to the name of the ligand. The κ-term comprises the Greek letter κ followed by the italicized element symbol of the ligating atom. For more complicated ligands the κ-term is often placed within the ligand name following the group to which the κ-term refers. Multiple identical links to a central atom can be indicated by addition of the appropriate numeral as a superscript between the κ and element symbols (see Table 6). These possibilities are discussed in more detail in the Red Book.[13] If the ligating atoms of a ligand are contiguous (i.e., directly bonded to one another), then an **η-term** is used instead, for example, for many organometallic compounds (Section 2.6) and the peroxido complex in Table 6.

A κ-term is required for ligands where more than one coordination mode is possible. Typical cases are thiocyanate, which can be bound through either the sulfur atom (thiocyanato-κ*S*) or the nitrogen atom (thiocyanato-κ*N*), and nitrite, which can be bound through either the nitrogen atom (M–NO_2, nitrito-κ*N*), or an oxygen atom (M–ONO, nitrito-κ*O*). The names pentaammine(nitrito-κ*N*)cobalt(2+) and pentaammine(nitrito-κ*O*)cobalt(2+) are used for each of the isomeric nitrito complex cations. More examples of constructing names using κ-terms to specify the connectivity of ligands are shown in Table 6. A κ-term may also be used to indicate to which central atom a ligand is bound if there is more than one central atom (Section 2.5).

Table 6: Producing Names for Complexes: Complicated Ligands

Structure to be named	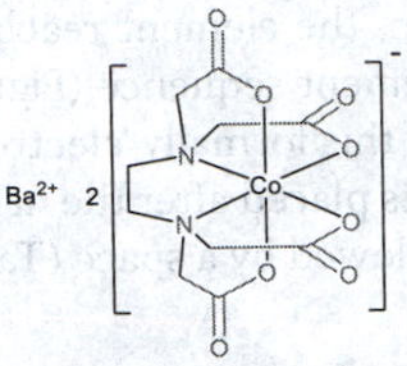	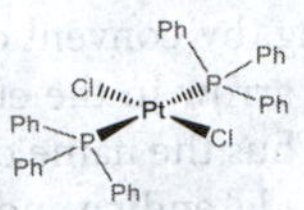
Central atom	cobalt(III) → cobaltate(III)	platinum(II)
Identify and name ligands	2,2′,2″,2‴-(ethane-1,2-diyl= dinitrilo)tetraacetate → 2,2′,2″,2‴-(ethane-1,2-diyl= dinitrilo)tetraacetato	chloride → chlorido triphenylphosphane
Specify ligating atoms	2,2′,2″,2‴-(ethane-1,2-diyl= dinitrilo-κ²*N*) tetraacetato-κ⁴*O*	*not required for chloride* triphenylphosphane-κ*P*
Assemble name	barium [2,2′,2″,2‴-(ethane-1,2-diyldinitrilo-κ²*N*) tetra=acetato-κ⁴*O*] cobaltate(III)	dichloridobis(triphenyl= phosphane-κ*P*)platinum(II)
Structure to be named	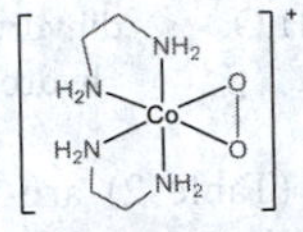	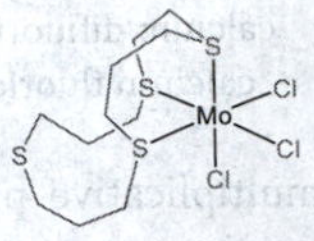
Central atom	cobalt(III)	molybdenum(III)
Identify and name ligands	ethane-1,2-diamine peroxide → peroxido	chloride → chlorido 1,4,8,12-tetrathiacyclopenta-decane
Specify ligating atoms	ethane-1,2-diamine-κ²*N* η²-peroxido	*not required for chloride* 1,4,8,12-tetrathiacyclo= pentadecane-κ³*S*¹,*S*⁴,*S*⁸
Assemble name	bis(ethane-1,2-diamine-κ²*N*)= (η²-peroxido) cobalt(III)	trichlorido(1,4,8,12-tetrathiacyclopentadecane-κ³*S*¹,*S*⁴,*S*⁸)molybdenum(III)

2.5 Bridging ligands

Bridging ligands are those bound to more than one central atom. They are differentiated in names by the addition of the prefix 'μ' (Greek mu), with the prefix and the name of the bridging ligand being separated from each other, and from the rest of the name, by hyphens. This is sufficient if the ligand is monoatomic, but if the ligand is more complicated it may be necessary to specify which ligating atom of the ligand is attached to which central atom. This is certainly the case if the ligating atoms are of different kinds, and κ-terms can be used for this purpose.

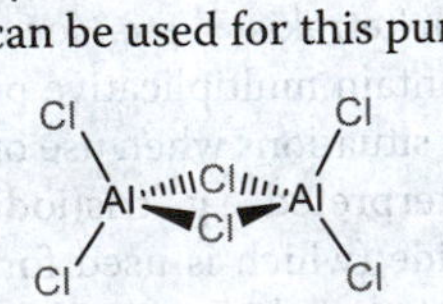

di-μ-chlorido-bis[di= chloridoaluminium(III)]
$[Cl_2Al(\mu\text{-}Cl)_2AlCl_2]$

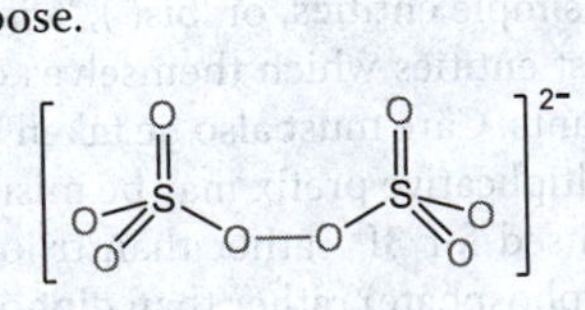

μ-peroxido-1κ*O*¹,2κ*O*²-bis(tri= oxidosulfate(2−)
$[O_3S(\mu\text{-}O_2)SO_3]^{2-}$

2.6 Organometallic compounds

Organometallic compounds contain at least one bond between a metal atom and a carbon atom. They are named as coordination compounds, using the additive nomenclature system (see above).

The name for an organic ligand **binding through one carbon atom** may be derived either by treating the ligand as an anion or as a neutral substituent group. The compound $[Ti(CH_2CH_2CH_3)Cl_3]$ is thus named as trichlorido(propan-1-ido)titanium or as trichlorido(propyl)titanium. Similarly, 'methanido' or 'methyl' may be used for the ligand $-CH_3$.

When an organic ligand forms **two or three metal-carbon single bonds** (to one or more metal centers), the ligand may be treated as a di- or tri-anion, with the endings 'diido' or 'triido' being used, with no removal of the terminal 'e' of the name of the parent hydrocarbon. Again, names derived by regarding such ligands as substituent groups and using the suffixes 'diyl' and 'triyl' are still commonly encountered. Thus, the bidentate ligand $-CH_2CH_2CH_2-$ would be named propane-1,3-diido (or propane-1,3-diyl) when chelating a metal centre, and μ-propane-1,3-diido (or μ-propane-1,3-diyl) when bridging two metal atoms.

Organometallic compounds containing a **metal-carbon multiple bond** are given substituent prefix names derived from the parent hydrides which end with the suffix 'ylidene' for a metal-carbon double bond and with 'ylidyne' for a triple bond. These suffixes either replace the ending 'ane' of the parent hydride, or, more generally, are added to the name of the parent hydride with insertion of a locant and elision of the terminal 'e', if present. Thus, the entity $CH_3CH_2CH=$ as a ligand is named propylidene and $(CH_3)_2C=$ is called propan-2-ylidene. The 'diido'/'triido' approach, outlined above, can also be used in this situation. The terms 'carbene' and 'carbyne' are not used in systematic nomenclature.

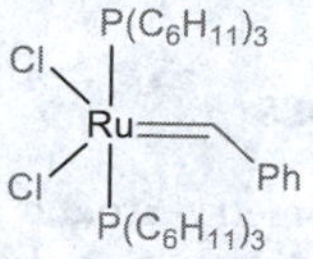

dichlorido(phenylmethylidene)bis(tricyclohexylphosphane-κ*P*)ruthenium, dichlorido(phenylmethanediido)bis(tricylohexylphosphane-κ*P*)ruthenium, **or** (benzylidene)dichloridobis(tricyclohexylphosphane-κ*P*)ruthenium

The special nature of the bonding to metals of unsaturated hydrocarbons in a 'side-on' fashion via their π-electrons requires the **eta (η) convention**. In this 'hapto' nomenclature, the number of *contiguous* atoms in the ligand coordinated to the metal (the hapticity of the ligand) is indicated by a right superscript on the eta symbol, e.g., η^3 ('eta three' or 'trihapto'). The η-term is added as a prefix to the ligand name, or to that portion of the ligand name most appropriate to indicate the connectivity, with locants if necessary.

A list of many **π-bonding unsaturated ligands**, neutral and anionic, can be found in the Red Book.[14]

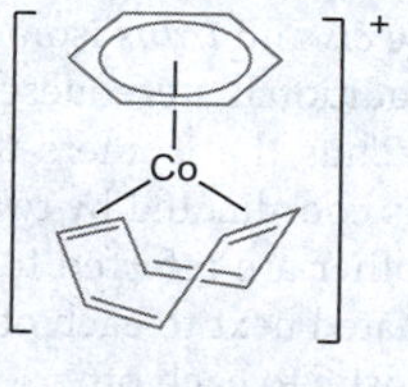

(η^6-benzene)[(1,2,5,6-η)-cycloocta-1,3,5,7-tetraene]cobalt(1+)

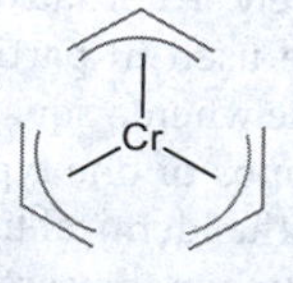

tris(η^3-prop-2-en-1-ido)chromium, tris(η^3-prop-2-en-1-yl)chromium or tris(η^3-allyl)chromium

Note that the ubiquitous ligand η5-C_5H_5, strictly η5-cyclopenta-2,4-dien-1-ido, is also acceptably named η5-cyclopentadienido or η5-cyclopentadienyl. When cyclopenta-2,4-dien-1-ido coordinates through one carbon atom via a σ bond, a κ-term is added for explicit indication of that bonding. The symbol η^1 should not be used, as the eta convention applies only to the bonding of contiguous atoms in a ligand.

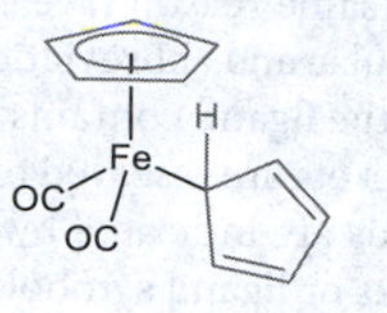

dicarbonyl(η^5-cyclopentadienido)(cyclopenta-2,4-dien-1-ido-κC^1)iron **or** dicarbonyl(η^5-cyclopentadienyl)(cyclopenta-2,4-dien-1-yl-κC^1)iron

Discrete molecules containing two parallel η5-cyclopentadienido ligands in a 'sandwich' structure around a transition metal, as in bis(η5-cyclopentadienido)iron, $[Fe(\eta^5\text{-}C_5H_5)_2]$, are generically called metallocenes and may be given 'ocene' names, in this case ferrocene. These 'ocene' names may be used in the same way as parent hydride names are used in substitutive nomenclature, with substituent group names taking the forms 'ocenyl', 'ocenediyl', 'ocenetriyl' (with insertion of appropriate locants).

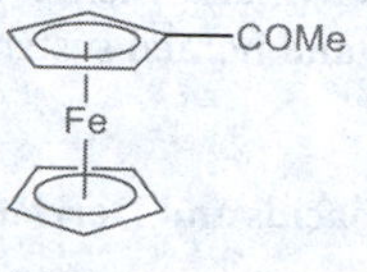

1-ferrocenylethan-1-one

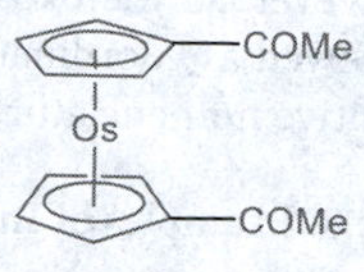

1,1′-(osmocene-1,1′-diyl)di=(ethan-1-one)

By convention, 'organoelement' compounds of the **main group elements** are named by substitutive nomenclature if derived from the elements of Groups 13–16, but by additive nomenclature if derived from the elements of Groups 1 and 2. In some cases compositional nomenclature is used if less structural information is to be conveyed. More detail is provided in the Red Book.[15]

Table 7: Producing Line Formulae for Complexes

Structure	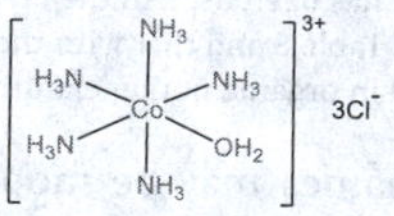	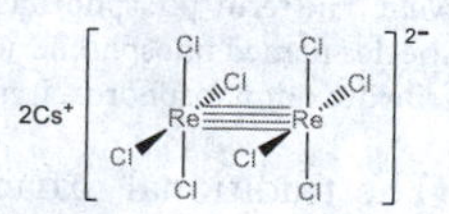
Central atom(s)	Co	2 × Re
Ligands	NH_3, OH_2	Cl
Assemble formula	$[Co(NH_3)_5(OH_2)]Cl_3$	$Cs_2[Cl_4ReReCl_4]$
Structure	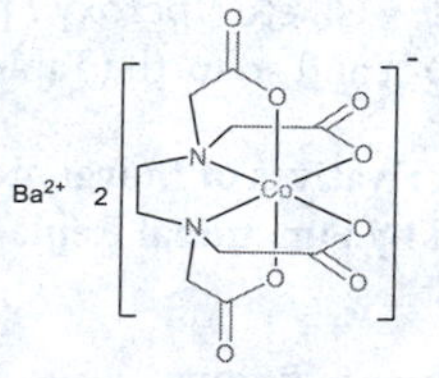	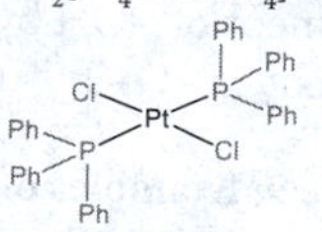
Central atom(s)	Co	Pt
Abbreviate ligands	2,2′,2″,2‴-(ethane-1,2-diyl)dinitrilotetraacetate → edta	Cl triphenylphosphane → PPh_3
Assemble formula	$Ba[Co(edta)]_2$	$[PtCl_2(PPh_3)_2]$

2.7 Formulae of coordination compounds

Line formulae for coordination entities are constructed within square brackets to specify the composition of the entity. The overall process is shown in Table 7. The symbol for the central atom is placed first and is then followed by the symbols or abbreviations for the ligands (in alphabetical order according to the way they are presented in the formula). Where possible the coordinating (ligating) atom should be placed nearer the central atom in order to provide more information about the structure of the complex. If possible, bridging ligands should be placed between central atom symbols for this same reason (see examples in Section 2.5). Generally, ligand formulae and abbreviations are placed within enclosing marks (unless the ligand contains only one atom), remembering that square brackets are reserved to define the coordination sphere. Multiple ligands are indicated by a right subscript following the enclosing marks or ligand symbol.

2.8 Inorganic oxoacids and related compounds

Inorganic oxoacids, and the anions formed by removing the acidic **hydrons** (H^+) from them, have traditional names, many of which are well-known and can be found in many textbooks: sulfuric acid, sulfate; nitric acid, nitrate; nitrous acid, nitrite; phosphoric acid, phosphate; arsenic acid, arsenate; arsinous acid, arsinite; silicic acid, silicate; etc. These names are retained in IUPAC nomenclature, first because they almost invariably are the names used in practice, and second because they play a special role in organic nomenclature when names are needed for organic derivatives. However, all the oxoacids themselves and their derivatives may be viewed as coordination entities and named systematically using additive nomenclature (Table 8).[16]

Table 8: Examples of Inorganic Oxoacids and Derivatives

Formula	Traditional or organic name	Additive name
H_2SO_4 or $[S(O)_2(OH)_2]$	sulfuric acid	dihydroxidodioxidosulfur
$(CH_3)_2SO_4$ or $[S(O)_2(OMe)_2]$	dimethyl sulfate	dimethoxidodioxidosulfur or dimethanolatodioxidosulfur
H_2PHO_3 or $[P(H)(O)(OH)_2]$	phosphonic acid*	hydridodihydroxidooxido-phosphorus
$PhP(O)(OH)_2$	phenylphosphonic acid	dihydroxidooxido(phenyl) phosphorus

* *Note:* The term 'phosphorous acid' has been used in the literature for both the species named phosphonic acid in Table 8 and that with the formula $P(OH)_3$, trihydroxidophosphorus. It is used in organic nomenclature in the latter sense.

The traditional oxoacid names may be modified according to established rules for naming derivatives formed by **functional replacement**[16]: thus 'thio' denotes replacement of =O by =S; prefixes 'fluoro', 'chloro', etc., and infixes 'fluorid', 'chlorid', etc., denote replacement of –OH by –F, –Cl, etc.; 'peroxy'/'peroxo' denote replacement of –O– by –OO–; and so forth (Table 9).

Table 9: Examples of Derivatives of Inorganic Oxoacids and Anions Formed by Functional Replacement

Formula	Name indicating functional replacement	Additive name
H_3PS_4 or $[P(S)(SH)_3]$	tetrathiophosphoric acid or phosphorotetrathioic acid	tris(sulfanido)sulfido= phosphorus
H_2PFO_3 or $[PF(O)(OH)_2]$	fluorophosphoric acid or phosphorofluoridic acid	fluoridodihydroxido= oxidophosphorus
$S_2O_3^{2-}$ or $[S(O)_3(S)]^{2-}$	thiosulfate or sulfurothioate	trioxidosulfido= sulfate(2−)
$[O_3S(\mu\text{-}O_2)SO_3]^{2-}$	peroxydisulfate	see Section 2.5

If all hydroxy groups in an oxoacid are replaced, the compound is no longer an acid and is not named as such, but will have a traditional **functional class name**[16] as, e.g., an acid halide or amide. Such compounds may again be systematically named using additive nomenclature (Table 10).

Table 10: Examples of Functional Class Names and Corresponding Additive Names

Formula	Functional class name	Additive name
PCl_3O	phosphoryl trichloride	trichloridooxido= phosphorus
SCl_2O_2	sulfuryl dichloride	dichloridodioxidosulfur
$S(NH_2)_2O_2$	sulfuric diamide	diamidodioxidosulfur

A special construction is used in **hydrogen names**, which allows the indication of hydrons bound to an anion without specifying exactly where. In such names, the word 'hydrogen' is placed at the front of the name with a multiplicative prefix (if applicable) and with no space between it and the rest of the name, which is placed in parentheses. For example, dihydrogen(diphosphate)(2−) denotes $H_2P_2O_7^{2-}$, a diphosphate ion to which two hydrons have been added, with the positions not known or at least not being specified.

One may view the common names for partially dehydronated oxoacids, such as hydrogenphosphate, HPO_4^{2-}, and dihydrogenphosphate, $H_2PO_4^-$, as special cases of such hydrogen names. In these simplified names, the charge number and the parentheses around the main part of the name are left out. Again, these particular anions may be named systematically by additive nomenclature. The word 'hydrogen' is placed separately in forming analogous names in organic nomenclature, for example, dodecyl hydrogen sulfate, $C_{12}H_{25}OS(O)_2OH$. This difference between the two systems has the consequence that the important carbon-containing ion HCO_3^- can be named equally correctly as 'hydrogen carbonate' and as 'hydrogencarbonate' (but not as bicarbonate).

3 Stereodescriptors

The approximate geometry around the central atom is described using a **polyhedral symbol** placed in front of the name. The symbol is made up of italicized letter codes for the geometry and a number that indicates the coordination number. Frequently used polyhedral symbols are *OC*-6 (octahedral), *SP*-4 (square-planar), *T*-4 (tetrahedral), *SPY*-5 (square-pyramidal), and *TBPY*-5 (trigonal-bipyramidal). More complete lists are available.[17]

The relative positions of ligating groups[19] around a central atom can be described using a **configuration index** that is determined in a particular way for each geometry,[18] based on the Cahn-Ingold-Prelog priorities of the ligating groups,[19] and it may change if the ligands change, even if the geometry remains the same. The absolute configuration can also be described. Generally, configuration indices are used only if there is more than one possibility and a particular stereoisomer is to be identified. The full stereodescriptors for the particular square-planar platinum complexes shown below are (*SP*-4-2) and (*SP*-4-1), for the *cis* and *trans* isomers, respectively. Alternatively, a range of traditional stereodescriptors may be used in particular situations. Thus the isomers that are possible when a square-planar centre is coordinated by two ligating groups of one type and two of another are referred to as *cis*- (when the identical ligands are coordinated next to each other) or *trans*- (when they are coordinated opposite to each other).

cis-diamminedichloridoplatinum(II)

trans-diamminedichlorido-platinum(II)

Octahedral centers with four ligands of one kind and two of another can also be referred to as *cis*- (when the two identical ligands are coordinated next to each other) or *trans*- (when they are coordinated opposite each other). Octahedral centers with three of each of two kinds of ligand can be described as *fac*- (facial), when the three ligands of a particular kind are located at the corners of a face of the octahedron, or *mer*- (meridional), when they are not.

4 Summary

This document provides an outline of the essential nomenclature rules for producing names and formulae for inorganic compounds, coordination compounds, and organometallic compounds. The complementary document for nomenclature systems of organic chemistry[2] will also be useful to the reader.

Names and formulae have only served half their role when they are created and used to describe or identify compounds, for example, in publications. Achieving their full role requires that the reader of a name or formula is able to interpret it successfully, for example, to produce a structural diagram. The present document is also intended to assist in the interpretation of names and formulae.

Finally, we note that IUPAC has produced recommendations on the graphical representation of chemical structures and their stereochemical configurations.[20]

ORGANIC SUBSTITUENT GROUPS AND RING SYSTEMS

The first part of this table lists substituent groups and their line formulas. A substituent group is defined by IUPAC as a group that replaces one or more hydrogen atoms attached to a parent structure. Such groups are sometimes called radicals, but IUPAC now reserves the term radical for a free molecular species with unpaired electrons. IUPAC does not recommend some of these names, which are marked here with asterisks (e.g., amyl*), but they are included in this list because they are often encountered in the older literature. Substituent group names that are formed by systematic rules (e.g., methyl from methane, ethyl from ethane, etc.) are included here only for the first few members of a homologous series.

In the second part of the table a number of common organic ring compounds are shown, with the conventional numbering of the ring positions indicated.

The help of Warren H. Powell in preparing this table is greatly appreciated. Pertinent references may be found in the table "Nomenclature for Chemical Compounds."

Substituent Groups

Name	Formula
acetamido (acetylamino)	CH_3CONH-
acetoacetyl	CH_3COCH_2CO-
acetonyl	CH_3COCH_2-
acetyl	CH_3CO-
acryloyl* (1-oxo-2-propenyl)	$CH_2=CHCO-$
alanyl (from alanine)	$CH_3CH(NH_2)CO-$
β-alanyl	$H_2N(CH_2)_2CO-$
allyl (2-propenyl)	$CH_2=CHCH_2-$
allylidene (2-propenylidene)	$CH_2=CHCH=$
amidino (aminoiminomethyl)	$H_2NC(=NH)-$
amino	H_2N-
amyl* (pentyl)	$CH_3(CH_2)_4-$
anilino (phenylamino)	C_6H_5NH-
anisidino	$CH_3OC_6H_4NH-$
anthranoyl (2-aminobenzoyl)	$2\text{-}H_2NC_6H_4CO-$
arsino	AsH_2-
azelaoyl (from azelaic acid)	$-OC(CH_2)_7CO-$
azido	N_3-
azino	$=N-N=$
azo	$-N=N-$
azoxy	$-N(O)=N-$
benzal* (benzylidene)	$C_6H_5CH=$
benzamido (benzoylamino)	C_6H_5CONH-
benzhydryl (diphenylmethyl)	$(C_6H_5)_2CH-$
benzoxy* (benzoyloxy)	C_6H_5COO-
benzoyl	C_6H_5CO-
benzyl	$C_6H_5CH_2-$
benzylidene	$C_6H_5CH=$
benzylidyne	$C_6H_5C\equiv$
biphenylyl	$C_6H_5C_6H_5-$
biphenylene	$-C_6H_4-C_6H_4-$
butoxy	C_4H_9O-
sec-butoxy (1-methylpropoxy)	$C_2H_5CH(CH_3)O-$
tert-butoxy (1,1-dimethylethoxy)	$(CH_3)_3CO-$
butyl	$CH_3(CH_2)_3-$
sec-butyl (1-methylpropyl)	$CH_3CH_2CH(CH_3)-$
tert-butyl (1,1-dimethylethyl)	$(CH_3)_3C-$
butyryl (1-oxobutyl)	$CH_3(CH_2)_2CO-$
caproyl* (hexanoyl)	$CH_3(CH_2)_4CO-$
capryl* (decanoyl)	$CH_3(CH_2)_8CO-$
capryloyl* (octanoyl)	$CH_3(CH_2)_6CO-$
carbamido (carbamoylamino)	$H_2NCONH-$
carbamoyl (aminocarbonyl)	H_2NCO-
carbamyl (aminocarbonyl)	H_2NCO-
carbazoyl (hydrazinocarbonyl)	$H_2NNHCO-$
carbethoxy (ethoxycarbonyl)	C_2H_5OCO-
carbonyl	$=C=O$
carboxy	$HOOC-$
cetyl* (hexadecyl)	$CH_3(CH_2)_{15}-$
chloroformyl (chlorcarbonyl)	$ClCO-$
cinnamoyl	$C_6H_5CH=CHCO-$
cinnamyl (3-phenyl-2-propenyl)	$C_6H_5CH=CHCH_2-$
cinnamylidene	$C_6H_5CH=CHCH=$
cresyl* (hydroxymethylphenyl)	$HO(CH_3)C_6H_4-$
crotonoyl	$CH_3CH=CHCO-$
crotyl (2-butenyl)	$CH_3CH=CHCH_2-$
cyanamido (cyanoamino)	$NCNH-$
cyanato	$NCO-$
cyano	$NC-$
decanedioyl	$-OC(CH_2)_8CO-$
decanoyl	$CH_3(CH_2)_8CO-$
diazo	$N_2=$
diazoamino	$-NHN=N-$
disilanyl	H_3SiSiH_2-
disiloxanyloxy	$H_3SiOSiH_2O-$
disulfinyl	$-S(O)S(O)-$
dithio	$-SS-$
enanthoyl* (heptanoyl)	$CH_3(CH_2)_5CO-$
epoxy	$-O-$
ethenyl (vinyl)	$CH_2=CH-$
ethynyl	$HC\equiv C-$
ethoxy	C_2H_5O-
ethyl	CH_3CH_2-
ethylene	$-CH_2CH_2-$
ethylidene	$CH_3CH=$
ethylthio	C_2H_5S-
formamido (formylamino)	$HCONH-$
formyl	$HCO-$
furmaroyl (from fumaric acid)	$-OCCH=CHCO-$
furfuryl (2-furanylmethyl)	$OC_4H_3CH_2-$
furfurylidene (2-furanylmethylene)	$OC_4H_3CH=$
glutamoyl (from glutamic acid)	$-OC(CH_2)_2CH(NH_2)CO-$
glutaryl (from glutaric acid)	$-OC(CH_2)_3CO-$
glycylamino	H_2NCH_2CONH-
glycoloyl; glycolyl (hydroxyacetyl)	$HOCH_2CO-$
glycyl (aminoacetyl)	H_2NCH_2CO-
glyoxyloyl; glyoxylyl (oxoacetyl)	$HCOCO-$
guanidino	$H_2NC(=NH)NH-$
guanyl (aminoiminomethyl)	$H_2NC(=NH)-$
heptadecanoyl	$CH_3(CH_2)_{15}CO-$
heptanamido	$CH_3(CH_2)_5CONH-$
heptanedioyl	$-OC(CH_2)_5CO-$
heptanoyl	$CH_3(CH_2)_5CO-$

Name	Formula
hexadecanoyl	$CH_3(CH_2)_{14}CO-$
hexamethylene (1,6-hexanediyl)	$-(CH_2)_6-$
hexanedioyl	$-OC(CH_2)_4CO-$
hippuryl (N-benzoylglycyl)	$C_6H_5CONHCH_2CO-$
hydrazino	H_2NNH-
hydrazo	$-HNNH-$
hydrocinnamoyl	$C_6H_5(CH_2)_2CO-$
hydroperoxy	$HOO-$
hydroxyamino	$HONH-$
hydroxy	$HO-$
imino	$HN=$
iodoso* (iodosyl)	$OI-$
iodyl	O_2I-
isoamyl* (isopentyl; 3-methylbutyl)	$(CH_3)_2CH(CH_2)_2-$
isobutenyl (2-methyl-1-propenyl)	$(CH_3)_2C=CH-$
isobutoxy (2-methylpropoxy)	$(CH_3)_2CHCH_2O-$
isobutyl (2-methylpropyl)	$(CH_3)_2CHCH_2-$
isobutylidene (3-methylpropylidene)	$(CH_3)_2CHCH=$
isobutyryl (2-methyl-1-oxopropyl)	$(CH_3)_2CHCO-$
isocyanato	$OCN-$
isocyano	$CN-$
isohexyl (4-methylpentyl)	$(CH_3)_2CH(CH_2)_3-$
isoleucyl (from isoleucine)	$C_2H_5CH(CH_3)CH(NH_2)CO-$
isonitroso* (hydroxyamino)	$HON=$
isopentyl (3-methylbutyl)	$(CH_3)_2CH(CH_2)_2-$
isopentylidene (3-methylbutylidene)	$(CH_3)_2CHCH_2CH=$
isopropenyl (1-methylethenyl)	$CH_2=C(CH_3)-$
isopropoxy (1-methylethoxy)	$(CH_3)_2CHO-$
isopropyl (1-methylethyl)	$(CH_3)_2CH-$
isopropylidene (1-methylethylidene)	$(CH_3)_2C=$
isothiocyanato (isothiocyano)	$SCN-$
isovaleryl* (3-methyl-1-oxobutyl)	$(CH_3)_2CHCH_2CO-$
lactoyl (from lactic acid)	$CH_3CH(OH)CO-$
lauroyl (from lauric acid)	$CH_3(CH_2)_{10}CO-$
lauryl (dodecyl)	$CH_3(CH_2)_{11}-$
leucyl (from leucine)	$(CH_3)_2CHCH_2CH(NH_2)CO-$
levulinoyl (from levulinic acid)	$CH_3CO(CH_2)_2CO-$
malonyl (from malonic acid)	$-OCCH_2CO-$
mandeloyl (from mandelic acid)	$C_6H_5CH(OH)CO-$
mercapto	$HS-$
mesityl	$2,4,6-(CH_3)_3C_6H_2-$
methacryloyl (from methacrylic acid)	$CH_2=C(CH_3)CO-$
methallyl (2-methyl-2-propenyl)	$CH_2=C(CH_3)CH_2-$
methionyl (from methionine)	$CH_3SCH_2CH_2CH(NH_2)CO-$
methoxy	CH_3O-
methyl	H_3C-
methylene	$H_2C=$
methylthio	CH_3S-
myristoyl (from myristic acid)	$CH_3(CH_2)_{12}CO-$
myristyl (tetradecyl)	$CH_3(CH_2)_{13}-$
naphthyl	$(C_{10}H_7)-$
naphthylene	$-(C_{10}H_6)-$
neopentyl (2,2-dimethylpropyl)	$(CH_3)_3CCH_2-$
nitramino (nitroamino)	O_2NNH-
nitro	O_2N-
nitrosamino (nitrosoamino)	$ONNH-$
nitrosimino (nitrosoimino)	$ONN=$
nitroso	$ON-$
nonanoyl (from nonanoic acid)	$CH_3(CH_2)_7CO-$
oleoyl (from oleic acid)	$CH_3(CH_2)_7CH=CH(CH_2)_7CO-$
oxalyl (from oxalic acid)	$-OCCO-$
oxo	$O=$
palmitoyl (from palmitic acid)	$CH_3(CH_2)_{14}CO-$
pentamethylene (1,5-pentanediyl)	$-(CH_2)_5-$
pentyl	$CH_3(CH_2)_4-$
tert-pentyl	$CH_3CH_2C(CH_3)_2-$
phenacyl	$C_6H_5COCH_2-$
phenacylidene	$C_6H_5COCH=$
phenethyl (2-phenylethyl)	$C_6H_5CH_2CH_2-$
phenoxy	C_6H_5O-
phenyl	C_6H_5-
phenylene (benzenediyl)	$-C_6H_4-$
phosphino* (phosphanyl)	H_2P-
phosphinyl* (phosphinoyl)	$H_2P(O)-$
phospho	O_2P-
phosphono	$(HO)_2P(O)-$
phthaloyl (from phthalic acid)	$1,2-C_6H_4(CO-)_2$
picryl (2,4,6-trinitrophenyl)	$2,4,6-(NO_2)_3C_6H_2-$
pimeloyl (from pimelic acid)	$-OC(CH_2)_5CO-$
piperidino (1-piperidinyl)	$C_5H_{10}N-$
pivaloyl (from pivalic acid)	$(CH_3)_3CCO-$
prenyl (3-methyl-2-butenyl)	$(CH_3)_2C=CHCH_2-$
propargyl (2-propynyl)	HC´CCH$_2$-
1-propenyl	$-CH=CHCH_2$
2-propenyl (allyl)	$CH_2=CHCH_2-$
propionyl* (propanyl)	CH_3CH_2CO-
propoxy	$CH_3CH_2CH_2O-$
propyl	$CH_3CH_2CH_2-$
propylidene	$CH_3CH_2CH=$
pyrryl (pyrrolyl)	C_3H_4N-
salicyloyl (2-hydroxybenzoyl)	$2-HOC_6H_4CO-$
selenyl* (selanyl; hydroseleno)	$HSe-$
seryl (from serine)	$HOCH_2CH(NH_2)CO-$
siloxy	H_3SiO-
silyl	H_3Si-
silylene	$H_2Si=$
sorboyl (from sorbic acid)	$CH_3CH=CHCH=CHCO-$
stearoyl (from stearic acid)	$CH_3(CH_2)_{14}CO-$
stearyl (octadecyl)	$CH_3(CH_2)_{17}-$
styryl (2-phenylethenyl)	$C_6H_5CH=CH-$
suberoyl (from suberic acid)	$-OC(CH_2)_6CO-$
succinyl (from succinic acid)	$-OCCH_2CH_2CO-$
sulfamino (sulfoamino)	$HOSO_2NH-$
sulfamoyl (sulfamyl)	H_2NSO_2-
sulfanilyl [(4-aminophenyl)sulfonyl]	$4-H_2NC_6H_4SO_2-$
sulfeno	$HOS-$
sulfhydryl (mercapto)	$HS-$
sulfinyl	$OS=$
sulfo	HO_3S-
sulfonyl (sulfuryl)	$-SO_2-$
terephthaloyl	$1,4-C_6H_4(CO-)_2$
tetramethylene	$-(CH_2)_4-$
thienyl (from thiophene)	$(C_4H_3S)-$
thiocarbonyl (carbothionyl)	$=CS$
thiocarboxy	$HOSC-$
thiocyanato (thiocyano)	$NCS-$
thionyl* (sulfinyl)	$-SO-$
threonyl (from threonine)	$CH_3CH(OH)CH(NH_2)CO-$
toluidino [(methylphenyl)amino]	$CH_3C_6H_4NH-$
toluoyl (methylbenzoyl)	$CH_3C_6H_4CO-$
tolyl (methylphenyl)	$CH_3C_6H_4-$
α-tolyl (benzyl)	$C_6H_5CH_2-$

Name	Formula
tolylene (methylphenylene)	-$(CH_3C_6H_3)$-
tosyl [(4-methylphenyl) sulfonyl)]	4-$CH_3C_6H_4SO_2$-
triazano	H_2NNHNH-
trimethylene (1,3-propanediyl)	-$(CH_2)_3$-
trityl (triphenylmethyl)	$(C_6H_5)_3C$-
valeryl* (pentanoyl)	$CH_3(CH_2)_3CO$-
valyl (from valine)	$(CH_3)_2CHCH(NH_2)CO$-
vinyl (ethenyl)	CH_2=CH-
vinylidene (ethenylidene)	CH_2=C=
xylidino [(dimethylphenyl)amino]	$(CH_3)_2C_6H_3NH$-
xylyl (dimethylphenyl)	$(CH_3)_2C_6H_3$-
xylylene [phenelenebis(methylene)]	-$CH_2C_6H_4CH_2$-

Organic Ring Compounds

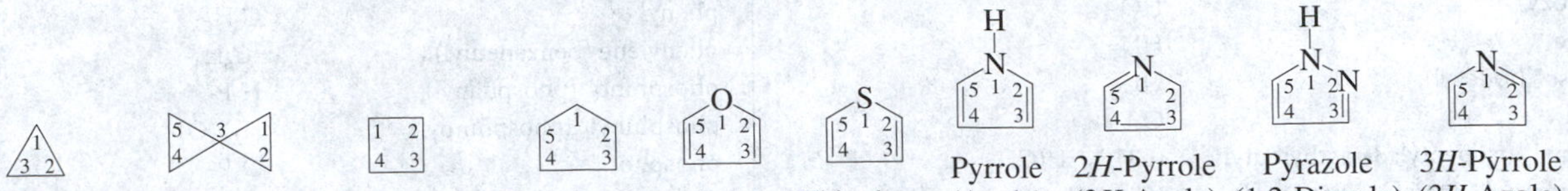

Cyclopropane, Spiropentane, Cyclobutane, Cyclopentane, Furan, Thiophene, Pyrrole (Azole), 2*H*-Pyrrole (2*H*-Azole), Pyrazole (1,2-Diazole), 3*H*-Pyrrole (3*H*-Azole)

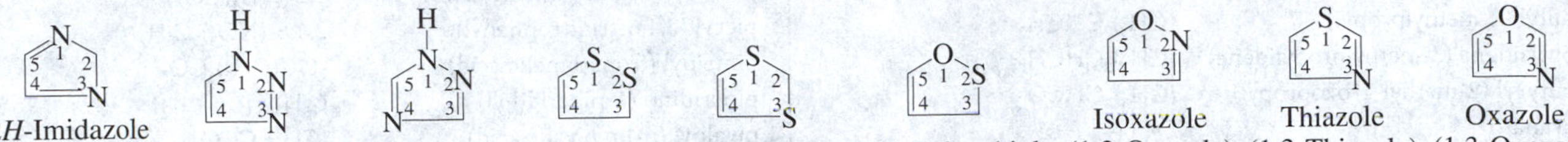

2*H*-Imidazole (1,3-Diazole), 1,2,3-Triazole, 1,2,4-Triazole, 1,2-Dithiole, 1,3-Dithiole, 3*H*-1,2-Oxathiole, Isoxazole (1,2-Oxazole), Thiazole (1,3-Thiazole), Oxazole (1,3-Oxazole)

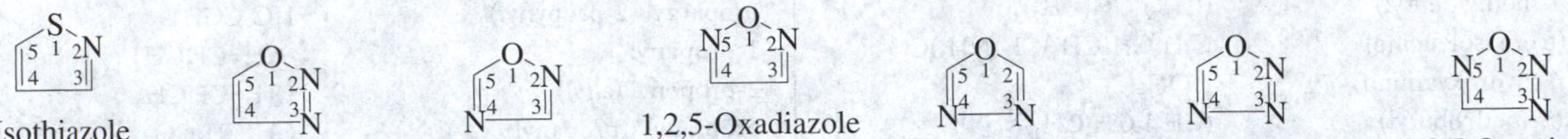

Isothiazole (1,2-Thiazole), 1,2,3-Oxadiazole, 1,2,4-Oxadiazole, 1,2,5-Oxadiazole (Furazan), 1,3,4-Oxadiazole, 1,2,3,4-Oxatriazole, 1,2,3,5-Oxatriazole

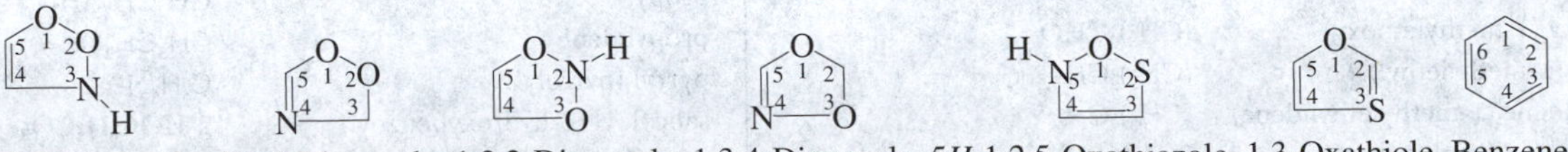

3*H*-1,2,3-Dioxazole, 1,2,4-Dioxazole, 1,3,2-Dioxazole, 1,3,4-Dioxazole, 5*H*-1,2,5-Oxathiazole, 1,3-Oxathiole, Benzene

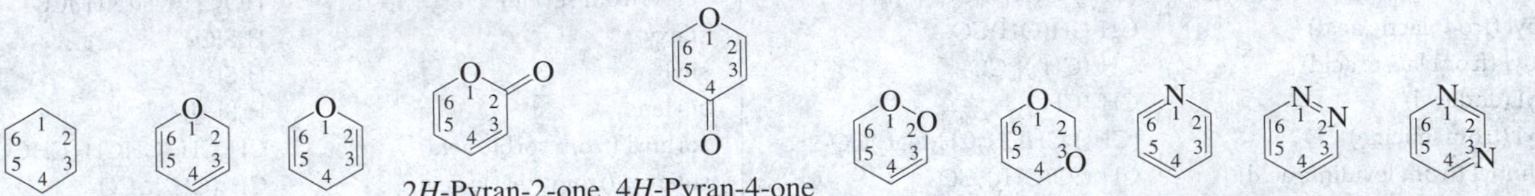

Cyclohexane, 2*H*-Pyran, 4*H*-Pyran, 2*H*-Pyran-2-one (2-Pyrone), 4*H*-Pyran-4-one (4-Pyrone), 1,2-Dioxin, 1,3-Dioxin, Pyridine, Pyridazine, Pyrimidine

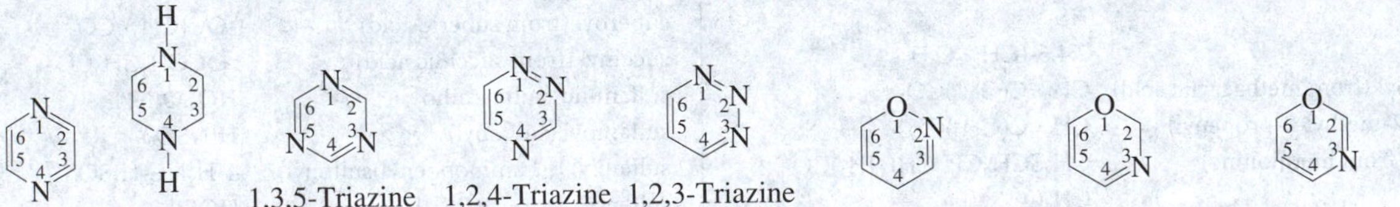

Pyrazine, Piperazine, 1,3,5-Triazine (*s*-Triazine), 1,2,4-Triazine (*as*-Triazine), 1,2,3-Triazine (*v*-Triazine), 4*H*-1,2-Oxazine, 2*H*-1,3-Oxazine, 6*H*-1,3-Oxazine

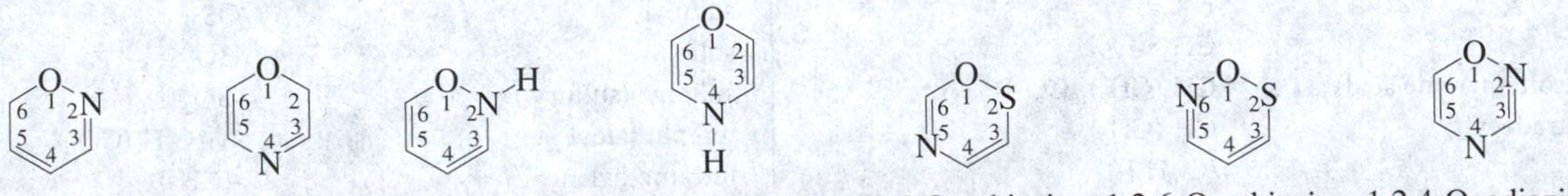

6*H*-1,2-Oxazine, 1,4-Oxazine, 2*H*-1,2-Oxazine, 4*H*-1,4-Oxazine, 1,2,5-Oxathiazine, 1,2,6-Oxathiazine, 1,2,4-Oxadiazine

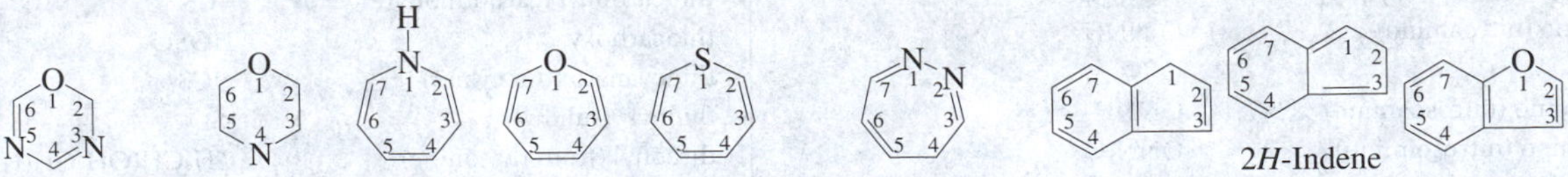

1,3,5-Oxadiazine, Morpholine, Azepine, Oxepin, Thiepin, 4*H*-1,2-Diazepine, Indene, 2*H*-Indene (Isoindene), Benzofuran

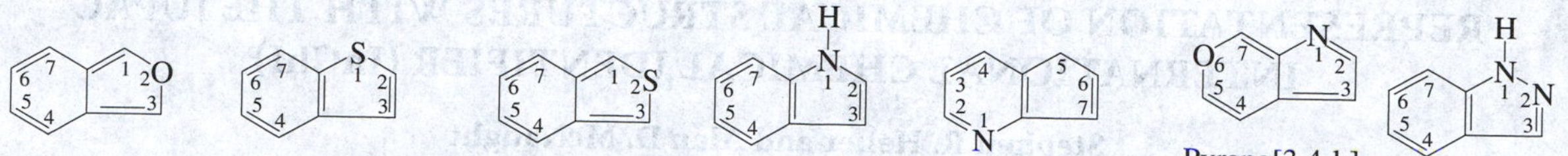

Isobenzofuran | Benzo[b]thiophene | Benzo[c]thiophene | Indole | Cyclopenta[b]pyridine | Pyrano[3,4-b]-pyrrole | Indazole

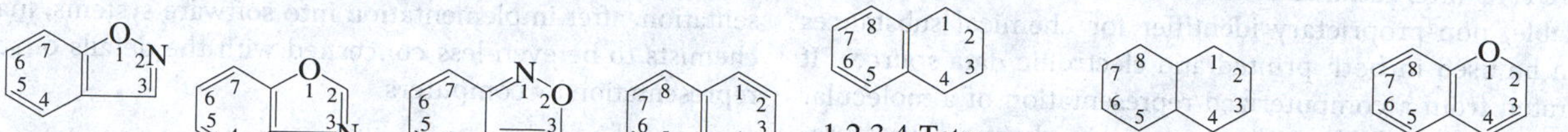

Benzisoxazole (Indoxazene) | Benzoxazole | 2,1-Benzisoxazole | Naphthalene | 1,2,3,4-Tetrahydronaphthalene (Tetralin) | Octahydronaphthalene (Decalin) | 2*H*-1-Benzopyran (2*H*-Chromene)

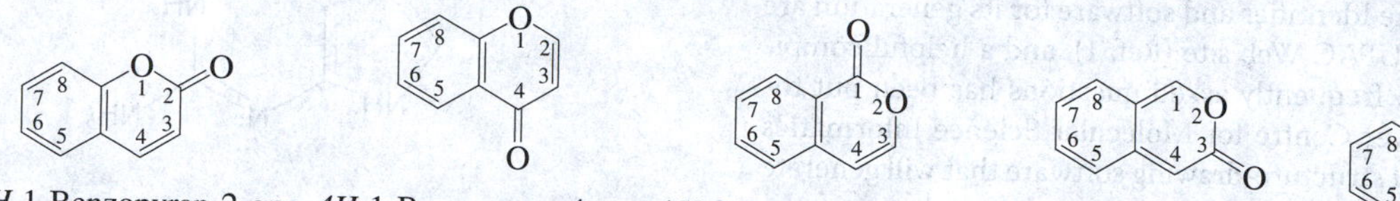

2*H*-1-Benzopyran-2-one (Coumarin) | 4*H*-1-Benzopyran-4-one (Chromen-4-one) | 1*H*-2-Benzopyran-1-one (Isocoumarin) | 3*H*-2-Benzopyran-1-one (Isochromen-3-one) | Quinoline

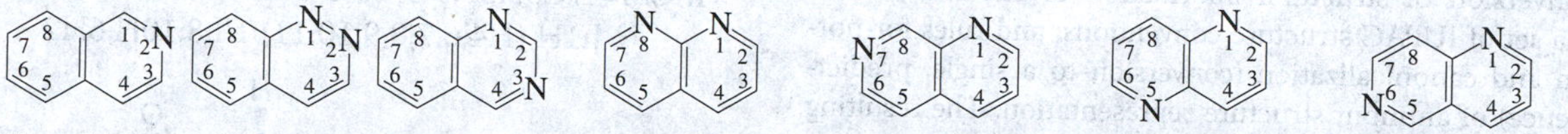

Isoquinoline | Cinnoline | Quinazoline | 1,8-Naphthyridine | 1,7-Naphthyridine | 1,5-Naphthyridine | 1,6-Naphthyridine

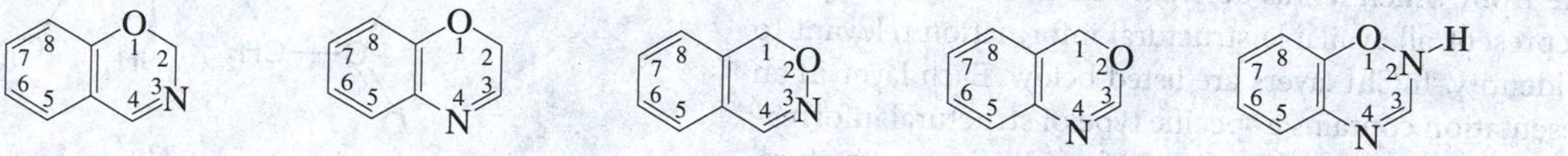

2*H*-1,3-Benzoxazine | 2*H*-1,4-Benzoxazine | 1*H*-2,3-Benzoxazine | 4*H*-3,1-Benzoxazine | 2*H*-1,2-Benzoxazine

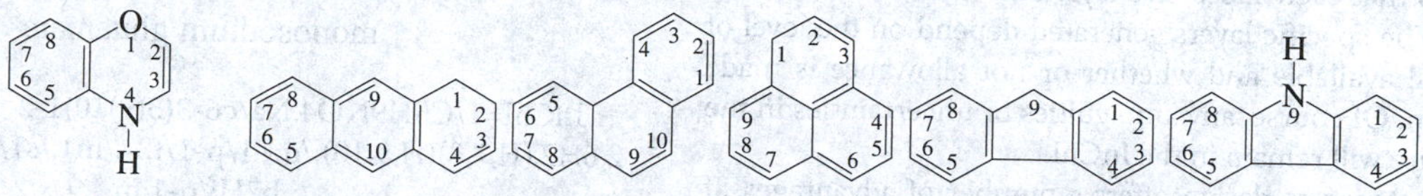

4*H*-1,4-Benzoxazine | Anthracene | Phenanthrene | Phenalene | Fluorene | Carbazole

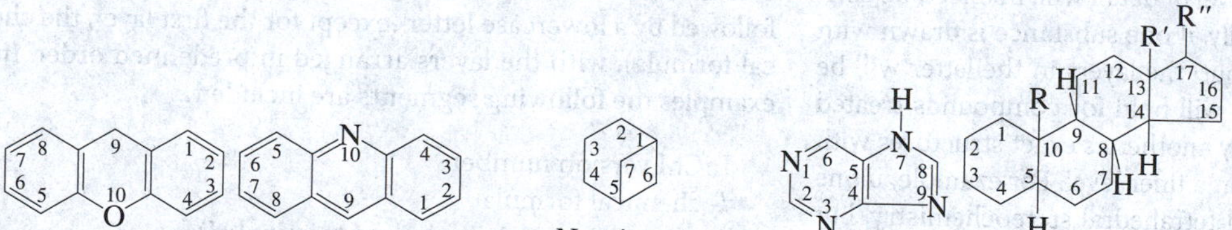

R = Nearly always methyl
R′ = Usually methyl
R″ = Various groups

Xanthene | Acridine | Norpinane (Bicyclo[3.1.1]heptane) | 7*H*-Purine | Steroid ring system

Symbols

REPRESENTATION OF CHEMICAL STRUCTURES WITH THE IUPAC INTERNATIONAL CHEMICAL IDENTIFIER (InChI)

Stephen R. Heller and Alan D. McNaught

The IUPAC International Chemical Identifier (InChI) is a freely available, non-proprietary identifier for chemical substances that can be used in both printed and electronic data sources. It is generated from a computerized representation of a molecular structure diagram, which can be produced by chemical structure-drawing software. Its use enables linking of diverse data compilations and unambiguous identification of chemical substances. A full description of the Identifier and software for its generation are available from the IUPAC Web site (Ref. 1), and a helpful compilation of answers to frequently asked questions has been put together at the Unilever Centre for Molecular Science Informatics (Ref. 2). Commercial structure-drawing software that will generate the Identifier is available from several organizations, listed on the IUPAC Web site.

The conversion of structural information to the Identifier is based on a set of IUPAC structure conventions, and rules for normalization and canonicalization (conversion to a single, predictable sequence) of an input structure representation. The resulting InChI is simply a series of characters that serve to uniquely identify the structure from which it was derived. The InChI uses a layered format to represent all available structural information relevant to compound identity. InChI layers are listed below. Each layer in an InChI representation contains a specific type of structural information. These layers, automatically extracted from the input structure, are designed so that each successive layer adds additional detail to the Identifier. The specific layers generated depend on the level of structural detail available and whether or not allowance is made for tautomerism. Of course, any ambiguities or uncertainties in the original structure will remain in the InChI.

This layered structure design offers a number of advantages. If two structures for the same substance are drawn at different levels of detail, the one with the lower level of detail will, in effect, be contained within the other. Specifically, if one substance is drawn with stereo-bonds and the other without, the layers in the latter will be a subset of the former. The same will hold for compounds treated by one author as tautomers and by another as exact structures with all H-atoms fixed. This can work at a finer level. For example, if one author includes double bond and tetrahedral stereochemistry, but another omits stereochemistry, the latter InChI will be contained in the former.

The InChI layers are

1. Formula
2. Connectivity (no formal bond orders)
    a. disconnected metals
    b. connected metals
3. Isotopes
4. Stereochemistry
    a. double bond (*Z*/*E*)
    b. tetrahedral (sp^3)
5. Tautomers (on or off)

Charges are not part of the basic InChI, but rather are added at the end of the InChI string.

Two examples of InChI representations are given below. It is important to recognize, however, that InChI strings are intended for use by computers and end users need not understand any of their details. In fact, the open nature of InChI and its flexibility of representation, after implementation into software systems, may allow chemists to be even less concerned with the details of structure representation by computers.

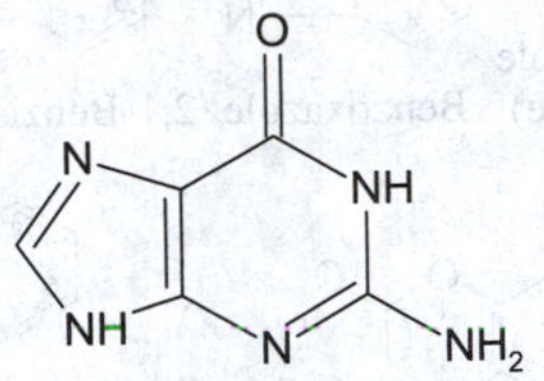

guanine

InChI=1/C5H5N5O/c6-5-9-3-2(4(11)10-5)7-1-8-3/h1H,(H4,6,7,8,9,10,11)/f/h8,10H,6H2

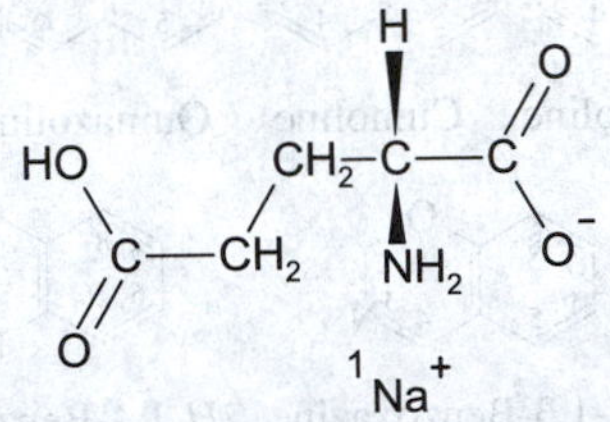

monosodium glutamate

InChI=1/C5H9NO4.Na/c6-3(5(9)10)1-2-4(7)8;/h3H,1-2,6H2,(H,7,8)(H,9,10);/q;+1/p-1/t3-;/m1./s1/fC5H8NO4.Na/h7H;/q-1;m

The layers in the InChI string are separated by the '/' character followed by a lowercase letter (except for the first layer, the chemical formula), with the layers arranged in predefined order. In the examples the following segments are included

InChI version number
/- chemical formula
/c connectivity-1.1 (excluding terminal H)
/h connectivity-1.2 (locations of terminal H, including mobile H attachment points)
/q charge
/p proton balance
/t sp^3 (tetrahedral) parity
/m parity inverted to obtain relative stereo (1 = inverted, 0 = not inverted)
/s stereo type (1 = absolute, 2 = relative, 3 = racemic)
/f chemical formula of the fixed-H structure if it is different
/h connectivity-2 (locations of fixed mobile H)
/q charge
/t sp^3 (tetrahedral) parity
/m parity inverted to obtain relative stereo (1 = inverted, 0 = not inverted, . = inversion does not affect the parity)
/s stereo type (1 = absolute, 2 = relative, 3 = racemic)

One of the most important applications of InChI is the facility to locate mention of a chemical substance using Internet-based search engines. This is made easier by using a shorter (compressed) form of InChI, known as InChIKey. The InChIKey is a 27-character representation that, because it is compressed, cannot be reconverted into the original structure, but it is not subject to the undesirable and unpredictable breaking of longer character strings by some search engines. The usefulness of the InChIKey as a search tool is enhanced by its derivation from a "standard" InChI. i.e., an InChI produced with standard option settings for features such as tautomerism and stereochemistry. An example is shown below; the "standard" InChI is denoted by the letter "S" after the version number.

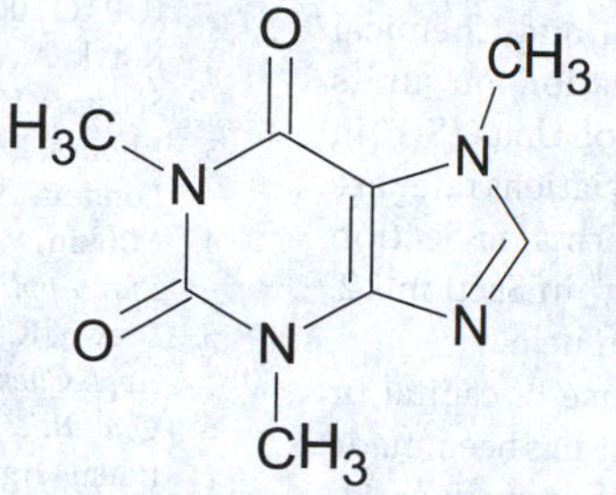

InChI=1S/C8H10N4O2/c1-10-4-9-6-5(10)7(13)12(3)8(14)11(6)2/h4H,1-3H3 (caffeine)

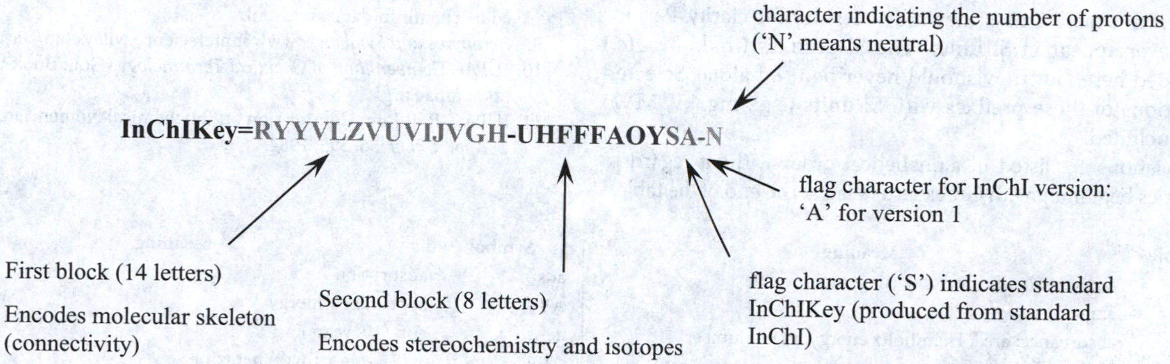

Use of InChIKey also allows searches based solely on atomic connectivity (first 14 characters). Software for generating InChIKey is available from the IUPAC Web site (Ref. 1).

The enormous databases compiled by organizations such as PubChem (Ref. 3), the U.S. National Cancer Institute (NCI), and ChemSpider (Ref. 4) contain millions of InChIs and InChIKeys, which allow sophisticated searching of these collections. PubChem provides InChI-based structure-search facilities for both identical and similar structures (Ref. 5), and ChemSpider offers both search facilities and Web services enabling a variety of InChI and InChIKey conversions (Ref. 6). The NCI Chemical Structure Lookup Service (Ref. 7) provides InChI-based search access to over 39 million chemical structures from over 80 different public and commercial data sources.

References

1. http://www.iupac.org/inchi
2. http://wwmm.ch.cam.ac.uk/inchifaq/
3. http://pubchem.ncbi.nlm.nih.gov
4. http://www.chemspider.com
5. http://pubchem.ncbi.nlm.nih.gov/search
6. http://www.chemspider.com/InChI.asmx
7. http://cholla.chemnavigator.com/cgi-bin/lookup/new/search

SCIENTIFIC ABBREVIATIONS, ACRONYMS, AND SYMBOLS

This table lists some abbreviations, acronyms, and symbols encountered in the physical sciences. Most entries in italic type are symbols for physical quantities; for more details on these, see the table "Symbols and Terminology for Physical and Chemical Quantities" in this section. Additional information on units may be found in the table "International System of Units (SI)" in Section 1. Many of the terms to which these abbreviations refer are included in the tables "Definitions of Scientific Terms" in Section 2 and "Techniques for Materials Characterization" in Section 12. Useful references for further information are given below.

Publication practices vary with regard to the use of capital or lower case letters for many abbreviations. An effort has been made to follow the most common practices in this table, but much variation is found in the literature. Likewise, policies on the use of periods in an abbreviation vary considerably. Periods are generally omitted in this table unless they are necessary for clarity. Periods should never appear in SI units. The SI prefixes (m, k, M, etc.) are included here, but they should never be used alone. Selected combinations of these prefixes with SI units (e.g., mg, kV, MW) are also included.

Abbreviations are listed in alphabetical order without regard to case. Entries beginning with Greek letters fall at the end of the table.

References

1. *Quantities, Units, and Symbols in Physical Chemistry, Third Edition*, IUPAC 2007, RSC Publishing, 2007.
2. Kotyk, A., *Quantities, Symbols, Units, and Abbreviations in the Life Sciences*, Humana Press, Totawa, NJ, 1999.
3. Rhodes, P. H., *The Organic Chemist's Desk Reference*, Chapman & Hall, London, 1995.
4. Minkin, V., Glossary of Terms used in Theoretical Organic Chemistry, *Pure Appl. Chem.* 71, 1919–1981, 1999.
5. Brown, R. D., Ed., Acronyms Used in Theoretical Chemistry, *Pure Appl. Chem.* 68, 387–456, 1996.
6. *Quantities and Units, ISO Standards Handbook, Third Edition*, International Organization for Standardization, Geneva, 1993.
7. Cohen, E. R., and Giacomo, P., Symbols, Units, Nomenclature, and Fundamental Constants in Physics, *Physica* 146A, 1–68, 1987.
8. *Chemical Acronyms Database*, Indiana University, <www.indiana.edu/~cheminfo/databases.html>.
9. *Acronyms and Symbols*, <www3.interscience.wiley.com/stasa/>.
10. *IUPAC Compendium of Chemical Terminology* (Gold Book), <goldbook.iupac.org>.
11. IUPAC-IUB Joint Commission on Biochemical Nomenclature, *Pure & Appl. Chem.* 56, 595, 1984.

Symbol	Meaning
A	ampere; alanine; adenine (in genetic code)
Å	ångström
A	absorbance; area; Helmholtz energy; mass number
A_H	Hall coefficient
A_r	atomic weight (relative atomic mass)
a	atto (SI prefix for 10^{-18})
a	absorption coefficient; acceleration; activity; van der Waals constant
a_0	Bohr radius
AAA	acetoacetanilide
Aad	2-aminoadipic acid
AAF	2-(acetylamino)fluorene
AAN	aminoacetonitrile
AAO	acetaldehyde oxime
AAS	atomic absorption spectroscopy
ABA	abscisic acid; acrylonitrile-butadiene acrylate
Abe	abequose
ABL	α-acetylbutyrolactone
ABS	acrylonitrile-butadiene-styrene copolymer
abs	absolute
Abu	2-aminobutanoic acid
Ac	acetyl; acetate
ac, AC	alternating current
ACAC	acetylacetone
Aces	2-[(2-amino-2-oxoethyl)amino]ethanesulfonic acid
ACS	acrylonitrile-chlorinated polyethylene-styrene copolymer
ACT	activated complex theory
ACTH	adrenocorticotropic hormone
A/D	analog to digital
Ad	adamantyl
Ada	[(carbamoylmethyl)imino]diacetic acid
Ade	adenine
ADI	acceptable daily intake
Ado	adenosine
ADP	adenosine diphosphate; ammonium dihydrogen phosphate
ads	adsorption
AE	appearance energy
ae	eon (10^9 years)
AEP	1-(2-aminoethyl)piperazine
AEPD	2-amino-2-ethyl-1,3-propanediol
AES	atomic emission spectroscopy; Auger electron spectroscopy
AF	audio frequency
AFM	atomic force microscopy
Ahx	2-aminohexanoic acid
AI	artificial intelligence
AIBN	2,2'-azobis[isobutyronitrile]
AICA	5-amino-1*H*-imidazole-4-carboxamide
AIM	atoms in molecules (method)
AIP	aluminum isopropoxide
Al	Alfén number
Ala	alanine
alc	alcohol
ALE	atomic layer epitaxy
aliph.	aliphatic
alk.	alkaline
All	allose
Alt	altrose
AM	amplitude modulation
Am	amyl
am	amorphous solid
AMA	acrylate maleic anhydride terpolymer
AMMA	acrylate-methyl methacrylate copolymer
AMP	adenosine monophosphate
AMPD	2-amino-2-methyl-1,3-propanediol
AMS	accelerator mass spectrometry
AMTCS	amyltrichlorosilane [trichloropentylsilane]
amu	atomic mass unit (recommended symbol is u)
AN	acetonitrile; acrylonitrile
anh, anhyd	anhydrous
ANOVA	analysis of variance

Symbol	Meaning
antilog	antilogarithm
ANTU	1-naphthalenylthiourea
AO	atomic orbital
AOM	angular overlap model
AP	ethylene-propylene copolymer
APAD	3-acetylpyridine adenine dinucleotide
APAP	acetyl *p*-amino phenol (acetaminophen)
Ape	2-aminopentanoic acid
API	atmospheric pressure ionization
Api	apiose
APM	atomic probe microanalysis
Apm	2-aminopimelic acid
APO	amorphous polyolefin
APPI	atmospheric pressure photoionization
APS	appearance potential spectroscopy; adenosine phosphosulfate
APW	augmented plane wave
aq	aqueous
Ar	aryl
Ara	arabinose
Ara-ol	arabinitol
Arg	arginine
ARPES	angular resolved photoelectron spectroscopy
AS	acrylonitrile styrene copolymer
ASA	acetylsalicylic acid; acrylonitrile-styrene-acrylonitrile block copolymer
ASC	4-(acetylamino)benzenesulfonyl chloride
ASCII	American National Standard Code for Information Interchange
ASE	aromatic stabilization model
Asn	asparagine
Asp	aspartic acid
at	atomization
ATCP	4-amino-3,5,6-trichloro-2-pyridinecarboxlic acid
ATEE	*N*-acetyl-L-tyrosine ethyl ester
ATLC	adsorption thin layer chromatography
atm	standard atmosphere
ATP	adenosine triphosphate
ATR	attenuated total internal reflection
at.wt.	atomic weight
AU	astronomical unit (ua is also used); polyurethane
AUC	area under the time-concentration curve
av	average
avdp	avoirdupois
B	bel; asparagine or aspartic acid (unspecified)
B	magnetic flux density; second virial coefficient; susceptance
b	barn
b	van der Waals constant; molality
BA	benzyladenine
BAL	British anti-Lewisite [2,3-dimercapto-1-propanol]
BAP, BaP	benzo[a]pyrene
bar	bar (pressure unit)
bbl	barrel
BBP	benzyl butyl phthalate
BCB	bromocresol blue
bcc	body centered cubic
BCF	bioconcentration factor
BCG	bromocresol green
BCME	bis(chloromethyl) ether
BCNU	*N,N'*-bis(2-chloroethyl)-*N*-nitrosourea
BCP	bromocresol purple
BCPB	bromochlorophenol blue
BCPE	1,1-bis(4-chlorophenyl)ethanol
BCS	Bardeen-Cooper-Schrieffer (theory)
BDE	bond dissociation energy
BDEA	butyldiethanolamime
BDMA	benzyldimethylamine
Bé	Baumé
BEBO	bond energy bond order (method)
BEI	biological exposure index
BEM	biological effect monitoring
BEP	2-butyl-2-ethyl-1,3-propanediol
Bes	2-[bis(2-hydroxyethyl)amino]ethanesulfonic acid
BET	Brunauer-Emmett-Teller (isotherm)
BeV	billion electronvolt
BGE	butyl glycidyl ether
BHA	*tert*-butyl-4-hydroxyanisole
BHC	benzene hexachloride [hexachlorobenzene]
Bhn	Brinell hardness number
BHT	butylated hydroxytoluene [2,6-di-*tert*-butyl-4-methylphenol]
Bi	biot
Bicine	*N,N*-bis(2-hydroxyethyl)glycine
BIRD	blackbody infrared radiative dissociation
Bistris	2-[bis(2-hydroxyethyl)amino]-2-(hydroxymethyl) propane-1,3-diol
Bistris-propane	1,3-bis[tris(hydroxymethyl)methylamino]propane
BLO	γ-butyrolactone
BN	bond number; benzonitrile
BNS	nuclear backscattering spectroscopy
BO	Born-Oppenheimer (approximation); bond order
BOD	biochemical oxygen demand
BON	β-hydroxynaphthoic acid
BP	base peak (in mass spectrometry); benzo[a]pyrene
bp	boiling point; base pair
BPB	bromophenol blue
BPG	2,3-bis(phospho)-*D*-glycerate
BPL	β-propiolactone
BPO	benzoyl peroxide
bpy	2,2'-bipyridine
Bq	becquerel
Br	butyryl
BRE	bond resonance energy
BrUrd	5-bromouridine
BS	Birge-Sponer extrapolation
BSE	back scattered electron(s)
BSSE	basis set superposition error
BTMSA	1,2-bis(trimethylsilyl)acetylene
Btu	British thermal unit
BTX	benzene, toluene, and xylene
Bu	butyl
bu	bushel
BVE	butyl vinyl ether
Bz	benzoyl
Bzl	benzyl
C	coulomb; cysteine; cytosine (in genetic code)
°C	degree Celsius
C	capacitance; heat capacity; number concentration
c	centi (SI prefix for 10^{-2}); combustion reaction
c	amount concentration; specific heat; velocity
c_0	speed of light in vacuum
CA	collisional activation; cellulose acetate
ca.	approximately

Symbols

Symbol	Meaning
CAB	cellulose acetate butyrate
CADD	computer-assisted drug design
cal	calorie
calc	calculated
cAMP	adenosine cyclic 3',5'-(hydrogen phosphate)
CAN	ceric ammonium nitrate
CAR	carbon fiber
CARS	coherent anti-Stokes Raman spectroscopy
CAS	complete active space
CASRN	Chemical Abstracts Service Registry Number
CAT	computerized axial tomography; clear air turbulence
CBE	chemical beam epitaxy
CBS	complete basis set (of orbitals)
CC	coupled cluster; combustion calorimetry
cc	cubic centimeter
CCD	charge-coupled device
CD	circular dichroism
cd	candela; condensed (phase)
CDAA	2-chloro-*N*,*N*-diallylacetamide
CDNO	complete neglect of differential overlap
CDP	cytidine 5'-diphosphate
CDT	1,5,9-cyclododecatriene
CDTA	(1,2-cyclohexylenedinitrilo)tetraacetic acid monohydrate
CDW	charge density waves
CED	cohesive energy density
CEM	channel electron multiplier
CEP	counter electrophoresis
CEPA	coupled electron-pair approximation
cf.	compare
CFC	chlorofluorocarbon compound
cfm	cubic feet per minute
CFRP	carbon reinforced plastics
cgs	centimeter-gram-second system
Chaps	3-[3-(cholamidopropyl)dimethylammonio]-1-propanesulfonic acid
Ches	2-(*N*-cyclohexylamino)ethanesulfonic acid
CHF	coupled Hartree-Fock (method)
Chl	chlorophyll
Cho	choline
CHT	1,3,5-cycloheptatriene
Ci	curie
CI	configuration interaction; chemical ionization; color index
CID	charge-injection device; collision-induced dissociation
CIDEP	chemically induced dynamic electron polarization
CIDNP	chemically induced dynamic nuclear polarization
CIE	countercurrent immunoelectrophoresis
cir	circular
CKFF	Cotton-Kraihanzel force field
CL	cathode luminescence (spectroscopy)
CLT	central limit theorem
cm	centimeter
c.m.	center of mass
CMC	carboxymethylcellulose
c.m.c.	critical micelle concentration
CMO	canonical molecular orbital
CMP	cytidine 5'-monophosphate; chemical measurement process
CN	coordination number; cellulose nitrate
CNDO	complete neglect of differential overlap
Co	Cowling number
COC	cycloolefin copolymer
COD	chemical oxygen demand; 1,4-cyclooctadiene
conc	concentrated; concentration
const	constant
COOP	crystal orbital overlap population
cos	cosine
cosh	hyperbolic cosine
COSY	correlation spectroscopy
COT	1,3,5,7-cyclooctatetraene
cot	cotangent
coth	hyperbolic cotangent
CP	chemically pure
Cp	cyclopentadienyl
Cp*	pentamethylcyclopentadienyl
cP	centipoise
cp	candle power
CPA	coherent potential approximation
CPC	centrifugal partition chromatography
cpd	contact potential difference
CPE	chlorinated polyethylene
CPL	circular polarization of luminescence
CPR	chlorophenol red
cps	cycles per second
CPT	charge conjugation/space inversion/time inversion (theorem)
CPU	central processing unit
CPVC	chlorinated poly(vinyl chloride)
CR	chloroprene rubber (neoprene)
cr, cryst	crystalline (phase)
CRF	charge remote fragmentation
CRU	constitutional repeating unit (in polymer nomenclature)
CSA	camphorsulfonic acid
csc	cosecant
CSR	charge stripping reaction
CT	charge transfer
ct	carat
CTA	cellulose triacetate
CTEM	conventional transmission electron microscopy
CTFE	chlorotrifluoroethylene
CTP	cytidine 5'-triphosphate
CTR	controlled thermonuclear reaction
cu	cubic
CV	cyclic voltammetry
CVD	chemical vapor deposition
cw	continuous wave
cwt	hundredweight (112 pounds)
Cy	cyclohexyl
Cya	cysteic acid
Cyd	cytidine
cyl	cylinder
Cys	cysteine
Cyt	cytosine
D	debye unit; aspartic acid
D	diffusion coefficient; dissociation energy; electric displacement
d	day; deuteron; deci (SI prefix for 10^{-1})
d	distance; density; dextrorotatory
2,4-D	2,4-dichlorophenoxyacetic acid
D/A	digital to analog
Da	dalton
DA	donor-acceptor (complex)
da	deka (SI prefix for 10^{1})
DAA	diacetone alcohol
DAB	4-(dimethylamino)azobenzene
Dab	2,4-diaminobutanoic acid

Symbol	Meaning
DACH	*trans*-1,2-diaminocyclohexane
DAIP	diallyl isophthalate plasticizer
DAP	diammonium phosphate
DART	direct analysis in real time mass spectrometry
dB	decibel
DBA	dibenz[a,h]anthracene; dibenzylamine
DBCP	1,2-dibromo-3-chloropropane
DBED	dibenzyl ethylene diamine
DBM	dibutyl maleate
DBMC	2,4-di-*tert*-butyl-5-methylphenol
DBMS	database management system
DBP	dibutyl phthalate; 2,3-dibromo-1-propanol
DBPC	2,6-di-*tert*-butyl-*p*-cresol
dc, DC	direct current
DCB	dicyanobenzene
DCBP	4,4'-dichlorobenzophenone
DCEE	dichloroethyl ether
DCHA	dicyclohexylamine
DCM	dichloromethane
DCNP	2,6-dichloro-4-nitrophenol
DCP	2,4-dichlorophenol
DCPD	dicyclopentadiene
DDM	4,4'-diaminodiphenylmethane
DDT	dichlorodiphenyltrichloroethane
DE	delocalization energy; delayed extraction
DEA	*N,N*-diethylaniline; diethanolamine
Dec	decyl
dec	decomposes
DEET	diethyltoluamide [*N,N*-diethyl-3-methylbenzamide]
deg	degree
DEK	diethyl ketone
den	density
DEP	2,2-diethyl-1,3-propanediol
DES	diethyl sulfate
DESI	desorption electrospray ionization (in mass spectrometry)
det	determinant
dev	deviation
DFT	density functional theory
dGlc	2-deoxyglucose
DHBA	2,3-dihydroxybenzoic acid
DHH	dehydroheliotridine
DHR	dehydroretronecine
DHU	dihydrouridine
DI	desorption ionization
diam	diameter
DIBA	diisobutyl adipate
DIBK	diisobutyl ketone
dil	dilute; dilution
DIM	diatomics in molecules (method); digital imaging microscopy
DIPA	diisopropanolamine
dm	decimeter
DMA	*N,N*-dimethylaniline
DMAB	4-(dimethylamino)azobenzene
DMAC	*N,N*-dimethylacetamide
DMAE	*N,N*-dimethylethanolamine
DMBA	7,12-dimethylbenz[a]anthracene
DME	1,2-dimethoxyethane
DMF	*N,N*-dimethylformamide
DMP	dimethyl phthalate
DMS	dimethyl sulfide
DMSO	dimethyl sulfoxide
DMT	dimethyl terephthalate; dimethyl tartrate
DN	donor number
DNA	deoxyribonucleic acid
DNase	deoxyribonuclease
DNB	1,3-dinitrobenzene
DNMR	dynamic NMR spectroscopy
DNP	dinitropyrene
Dod	dodecyl
DOP	dioctyl phthalate
DOS	density of states; digital operating system; dioctyl sebacate
doz	dozen
DP, d.p.	degree of polymerization
DPA	diphenylamine
DPG	*N,N'*-diphenylguanidine
dpl	displacement
Dpm	2,6-diaminopimelic acid
dpm	disintegrations per minute
dps	disintegrations per second
DPU	*N,N'*-diphenylurea
dr	dram
DRE	Dewar resonance energy
dRib	2-deoxyribose
DRIFT	diffuse reflectance infrared Fourier transform
DRP	dynamic reaction path
DRS	diffuse reflectance spectroscopy
DS	degree of substitution
DSC	differential scanning calorimetry
DTA	differential thermal analysis
DTBP	di-*tert*-butyl peroxide
DVB	divinylbenzene
dyn	dyne
DZ	double-zeta (type of basis set)
E	exa (SI prefix for 10^{18}); glutamic acid
E	electric field strength; electromotive force; energy; Young's modulus of elasticity; entgegen (*trans*configuration)
E_h	Hartree energy
e	electron; base of natural logarithms
e	elementary charge; linear strain
EA	electron affinity
EAA	ethylene acrylic acid copolymer; ethyl acetoacetate
EAK	ethyl amyl ketone (3-octanone)
EAN	effective atomic number
EC	ethyl cellulose
ECD	electron capture dissociation
ECP	effective core potential
ECR	electron cyclotron resonance
ECTFE	ethylene-chlorotrifluoroethylene copolymer
ED	electron diffraction
EDAX	energy dispersive analysis by x-rays
EDB	ethylene dibromide [1,2-dibromoethane]
EDC	ethylene dichloride [1,2-dichloroethane]
EDI	estimated daily intake
EDS	energy-dispersive x-ray spectroscopy
EDTA	ethylenediaminetetraacetic acid
EEA	ethylene-ethyl acetate copolymer
EEDQ	ethyl 2-ethoxy-1(2*H*)-quinolinecarboxylate
EEL	environmental exposure level
EELS	electron energy loss spectroscopy
EES	excitation emission spectrum
EFF	empirical force field
EFFF	energy factored force field
EG	equilibrium in the gas phase

Symbol	Meaning
EGA	evolved gas analysis
EGG	Einstein-Guth-Gold equation
EHMO, EHT	extended Hückel molecular orbital (theory)
EIMS	electron impact mass spectrometry
EIS	electron impact spectroscopy; electrochemical impedance spectroscopy
ELISA	enzyme-linked immunosorbent assay
ELS	energy loss spectroscopy
EM	extended molarity; electron microscopy
EMAC	ethylene-methyl acrylate copolymer
emf	electromotive force
EMPA, EMA	electron probe microanalysis
emu	electromagnetic unit system
en	ethylenediamine
ENDOR	electron-nuclear double resonance
EOS	equation of state
EP	epoxy resin
EPDS	electron photodetachment spectroscopy
EPM	ethylene-propylene copolymer
EPR	electron paramagnetic resonance; ethylene propylene rubber
EPS	expanded polystyrene
EPT-76	provisional low temperature scale of 1976
EPTC	dipropylcarbamothioic acid, *S*-ethyl ester
EPXMA	electron probe x-ray microanalysis
eq, eqn	equation
eqQ	quadrupole coupling constant
erf	error function
erg	erg (energy unit)
ES	equilibrium in solution
ESA	electrostatic energy analyzer
ESCA	electron spectroscopy for chemical analysis
ESD	electron stimulated desorption
e.s.d.	estimated standard deviation
ESI	electrospray ionization
ESR	electron spin resonance
est	estimated
esu	electrostatic unit system
ET	ephemeris time; electron transfer
Et	ethyl
ETA	electrothermal analysis
ETFE	ethylene tetrafluoroethylene polymer
Etn	ethanolamine
ETO	ethylene oxide
ETS	electron tunneling spectroscopy
ETU	ethylene thiourea
EU	polyether polyurethane
Eu	Euler number
e.u.	entropy unit
eV	electronvolt
EVA	ethylene-vinyl acetate copolymer
EVE	ethyl vinyl ether
EXAFS	extended x-ray absorption fine structure (spectroscopy)
EXELFS	extended energy loss fine structure
exp	exponential function
expt	experimental
ext	external
F	farad; phenylalanine
°F	degree Fahrenheit
F	Faraday constant; force; angular momentum
f	formation reaction; femto (SI prefix for 10^{-15})
f	activity coefficient; aperture ratio; focal length; force constant; frequency; fugacity
FAB	fast atom bombardment
FAD	flavine adenine dinucleotide
FAIMS	high-field asymmetric waveform ion mobility spectrometry
FA-SIFT	flowing afterglow – selected ion-flow tube
fcc	face centered cubic
FD	field desorption
FEL	free electron laser
FEM	field emission microscopy
FEMO	free electron molecular orbital
FEP	fluorinated ethylene propylene
FET	field effect transistor
FI	field ionization
fid	free induction decay
FIM	field ion microscopy
FIR	far infrared
fl	fluid (phase)
FM	frequency modulation
Fo	Fourier number
fp	freezing point
fpm	feet per minute
fps	feet per second; foot-pound-second system
Fr	franklin
Fr	Froude number
FRP	fibrous glass reinforced polyester; fiber reinforced plastic
Fru	fructose
FSGO	floating spherical gaussian orbitals
FT	Fourier transform
ft	foot
ft-lb	foot pound
FTIR	Fourier transform infrared spectroscopy
FTMS	Fourier transform mass spectrometry
FTNMR	Fourier transform nuclear magnetic resonance
fus	fusion (melting)
FVP	flash vacuum pyrolysis
FWHM	full width at half maximum
G	gauss; guanine (in genetic code); giga (SI prefix for 10^9); glycine
G	electrical conductance; Gibbs energy; gravitational constant; sheer modulus
g	gram; gas (phase)
g	acceleration due to gravity; degeneracy; Landé g-factor; statistical weight
GABA	γ-aminobutyric acid
Gal	gal; galactose
gal	gallon
GalN	galactosamine
GB	gas-phase basicity
GC	gas chromatography
GC-MS	gas chromatography-mass spectroscopy
GDMS	glow discharge mass spectroscopy
GDP	guanosine 5′-diphosphate
gem	geminal (on the same carbon atom)
GeV	gigaelectronvolt
GF	glass reinforced
GIAO	gauge invariant atomic orbital
GIBMS	guided ion beam mass spectrometry
gl	glacial
Gla	4-carboxyglutamic acid
GLC	gas-liquid chromatography
Glc	glucose
GlcA	gluconic acid
GlcN	glucosamine

Symbol	Meaning
GlcNAc	*N*-acetylglucosamine
GlcU	glucuronic acid
Gln	glutamine
GLP	good laboratory practice
Glu	glutamic acid
Glx	glutamine or glutamic acid (unspecified)
Gly	glycine
GMP	guanosine 5'-monophosphate
GMT	Greenwich mean time
GPC	gel-permeation chromatography
gpm	gallons per minute
gps	gallon per second
Gr	Grashof number
gr	grain
Gra	glyceraldehyde
Gri	glyceric acid
Grn	glycerone [dihydroxyacetone]
Gro	glycerol
GTO	gaussian-type orbital
GTP	guanosine 5'-triphosphate
Gua	guanine
Gul	gulose
Guo	guanosine
GUT	grand unified theory
GVB	generalized valence bond (method)
GWS	Glashow-Weinberg-Salam (theory)
Gy	gray; gigayear
H	henry; histidine
H	enthalpy; Hamiltonian function; magnetic field
H_0	Hubble constant
h	helion; hour; hecto (SI prefix for 10^2)
h	Planck constant
Ha	Hartmann number
ha	hectare
HAM	hydrogenic atoms in molecules
hav	haversine
Hb	hemoglobin
HCA	heterocyclic amine; hexachloroacetone
HCB	hexachlorobenzene
hcp	hexagonal closed packed
Hcy	homocysteine
HCZ, HCTZ	hydrochlorothiazide
HDL	high-density lipoprotein
HDPE	high-density polyethylene
HDS	hydrodesulfurization
HEIS	high-energy ion scattering
HEP	high energy physics
Hepes	4-(2-hydroxyethyl)-1-piperazineethanesulfonic acid
Hepps	4-(2-hydroxyethyl)-1-piperazinepropanesulfonic acid
HF	high frequency; Hartree-Fock (method)
HFA	hexafluoroacetone
HFO	Hartree-Fock orbital
hfs	hyperfine structure
HHPA	hexahydrophthalic anhydride
HIPS	high-impact polystyrene
His	histidine
HMC	high strength molding compound
HMDA	hexamethylenediamine
HMO	Hückel molecular orbital
HMT	hexamethylenetetramine
HMX	cyclotetramethylenetetranitramine
HN1	2-chloro-*N*-(2-chloroethyl)-*N*-ethylethanamine
HOAc	acetic acid
HOC	halogenated organic compound(s)
HOMAS	harmonic oscillator model of aromatic stabilization
HOMO	highest occupied molecular orbital
HOSE	harmonic oscillator stabilization energy
Hp	heptyl
hp	horsepower
HPLC	high-performance liquid chromatography
HPMS	high pressure mass spectrometry
HQ	*p*-hydroquinone
hr	hour
HRE	Hückel resonance energy
HREELS	high resolution electron energy loss spectroscopy
HREM	high resolution electron microscopy
HSAB	hard-soft acid-base (theory)
HSE	homodesmotic stabilization energy
Hse	homoserine
HVA	homovanillic acid
Hx	hexyl
Hyl	5-hydroxylysine
Hyp	hypoxanthine; 4-hydroxyproline
Hz	hertz
I	isoleucine; inositol; ionomer
I	electric current; ionic strength; moment of inertia; nuclear spin angular momentum; radiant intensity
i	square root of minus one
i	electric current
I/O	input/output
IAT	international atomic time
IC	integrated circuit
ICD	induced circular dichroism
ICP	inductive-coupled plasma
ICR	ion cyclotron resonance
ICVTST	improved canonical variational transition-state theory
ID	inside diameter
id	ideal (solution)
Ido	iodose
IdoA	iduronic acid
IDP	inosine 5'-diphosphate
IE	ionization energy
i.e.p.	isoelectric point
IEPA	independent electron pair approximation
IF	intermediate frequency
IGLO	individual gauge for localized orbitals
IIR	isobutylene-isoprene rubber (butyl rubber)
IKES	ion kinetic energy spectrometry
Ile	isoleucine
Im	imaginary part
IMFP	inelastic mean free path (of electrons)
imm	immersion
IMP	inosine 5'-monophosphate
IMPATT	impact ionization avalanche transit time
IMS	ion mobility spectrometry
in.	inch
InChI	IUPAC International Chemical Identifier
INDO	immediate neglect of differential overlap
Ino	inosine
INS	inelastic neutron scattering; ion neutralization spectroscopy
Ins	*myo*-inositol
int	internal
IP	ionization potential
IPA	isopropyl alcohol

Symbol	Meaning
IPMA	ion probe microanalysis
IPN	interpenetrating polymer network
IPR	isotope perturbation of resonance
IPTS	International Practical Temperature Scale
IQ	2-amino-3-methyl-3*H*-imidazo(4,5-f)quinoline
IR	infrared
IRAS	infrared reflection-absorption spectroscopy
IRC	intrinsic reaction coordinate
IRMPD	infrared multiphoton dissociation
IRMS	isotope ratio mass spectrometry
IRS	infrared spectroscopy
isc	intersystem crossing
ISE	ion-selective electrode; isodesmic stabilization energy
ISS	ion scattering spectroscopy
IT	ion trap; information technology
ITP	inosine 5’-triphosphate
ITS	International Temperature Scale (1990)
IU	international unit
IVE	isobutyl vinyl ether
J	joule; leucine or isoleucine (unspecified)
J	angular momentum; electric current density; flux; Massieu function
j	angular momentum; electric current density
JT	Jahn-Teller (effect)
K	kelvin; lysine
K	absorption coefficient; bulk modulus; equilibrium constant; kinetic energy
k	kilo (SI prefix for 10^3)
k	absorption index; Boltzmann constant; rate constant; thermal conductivity; wave vector
kat	katal (unit of catalytic activity)
kb	kilobar; kilobases (DNA or RNA)
KC-MS	Knudson cell mass spectrometry
kcal	kilocalorie
KDP	potassium dihydrogen phosphate
KE	kinetic energy
KERD	kinetic energy release distributions
keV	kiloelectronvolt
KG	kinetics in the gas phase
kg	kilogram
kgf	kilogram force
KIE	kinetic isotope effect
kJ	kilojoule
km	kilometer
Kn	Knudsen number
kPa	kilopascal
KS	kinetics in solution
kt	karat
KTP	potassium titanium phosphate
kV	kilovolt
kva	kilovolt ampere
kW	kilowatt
kwh	kilowatt hour
L	liter; lambert; leucine
L	Avogadro constant; inductance; Lagrange function; angular momentum
l	liter; liquid (phase)
l	angular momentum; length; mean free path; levorotatory
Lac	lactose
LAH	lithium aluminum hydride
lat.	latitude
lb	pound

Symbol	Meaning
lbf	pound force
LC	liquid chromatography; liquid crystal
LC-MS	liquid chromatography-mass spectrometry
lc	liquid crystal (phase)
LCAO	linear combination of atomic orbitals
LD	lethal dose; laser desorption
LDA	local density approximation; lithium diisopropylamide
LDL	low-density lipoprotein
LDPE	low-density polyethylene
LDV	laser-Doppler velocimetry
Le	Lewis function
LE	localization energy
LEC	liquid exchange chromatography
LED	light emitting diode
LEED	low-energy electron diffraction
LEIS	low-energy ion scattering
Leu	leucine
LFER	linear free energy relationships
LFL	lower flammable limit
LI	laser ionization
lim	limit
LIMS	laser ionization mass spectroscopy; laboratory information management system
liq	liquid
LIT	linear ion trap
LLCT	ligand to ligand charge transfer
lm	lumen
LMCT	ligand to metal charge transfer
LMMS	laser microprobe mass spectrometry
LMO	localized molecular orbital
LMR	laser magnetic resonance
ln	logarithm (natural)
LNDO	local neglect of differential overlap
log	logarithm (common)
LOMO	lowest occupied molecular orbital
long.	longitude
LPE	linear polyethylene
LPG	liquid petroleum gas
LPHP	laser-powered homogeneous pyrolysis
LPU	law of propagation of uncertainty
LSFE	linear field stabilization energy
LSI	liquid secondary ionization
LST	local sidereal time
LT	local time
LTE	local thermodynamic equilibrium
LUMO	lowest unoccupied molecular orbital
lx	lux
ly	langley
l.y.	light year
Lys	lysine
Lyx	lyxose
M	molar (as in 0.1 M solution); mega (SI prefix for 10^6); methionine
M	magnetization; molar mass; mutual inductance; torque; angular momentum component; median
M_r	molecular weight (relative molar mass)
m	meter; molal (as in 0.1 m solution); metastable (isotope); milli (SI prefix for 10^{-3})
m	magnetic dipole moment; mass; molality; angular momentum component; *meta* (locant on aromatic ring)
Ma	Mach number
MA	maleic anhydride

Symbol	Meaning
MAAc	methyl amyl acetate
Mal	maltose
Man	mannose
MASNMR	magic angle spinning nuclear magnetic resonance
max	maximum
Mb	myoglobin
MBE	molecular beam epitaxy
MBER	molecular beam electron resonance
MBK	methyl butyl ketone
MBOCA	4,4'-methylenebis[2-chloroaniline]
MBPT	many body perturbation theory
MBS	methyl methacrylate butadiene styrene terpolymer
MC	Monte Carlo (method)
MCAA	monochloroacetic acid
MCD	magnetic circular dichroism
MCP	microchannel plate
MCPA	(4-chloro-2-methylphenoxy)acetic acid
MCPF	modified coupled pair functional
MCS	Monte Carlo simulation
MCSCF	multiconfigurational self-consistent field (approximation)
MD	molecular dynamics (method)
MDI	methylene diphenylisocyanate
MDPE	medium density polyethylene
Me	methyl
MeCCNU	1-(2-chloroethyl)-3-(4-methylcyclohexyl)-1-nitrosourea
MeIQ	2-amino-3,4-dimethylimidazo[4,5-f]quinoline
MeIQx	2-amino-3,8-dimethylimidazo[4,5-f]quinoxaline
MEK	methyl ethyl ketone
MEP	molecular electrostatic potential
MERP	minimum energy reaction path
Mes	4-morpholineethanesulfonic acid
MESFET	metal-semiconductor field-effect transistor
Met	methionine
MeV	megaelectronvolt
meV	millielectronvolt
MF	molecular formula; melamine-formaldehyde resin
mg	milligram
MHD	magnetohydrodynamics
mi	mile
MIAK	methyl isoamyl ketone
MIBK	methyl isobutyl ketone
MIC	methyl isocyanate
MIK	methyl isobutyl ketone
MIKES	mass-analyzed ion kinetic energy spectrometry
min	minimum; minute
MINDO	modified INDO (method)
MIPK	methyl isopropyl ketone
MIR	mid infrared
misc	miscible
MKS	meter-kilogram-second system
MKSA	meter-kilogram-second-ampere system
mL, ml	milliliter
MM	molecular mechanics
mm	millimeter
MMDR	microwave-microwave double resonance
mmf	magnetomotive force
mmHg	millimeter of mercury
MNA	*m*-nitroaniline
MNDO	modified neglect of diatomic overlap
MNT	*m*-nitrotoluene
MNU	*N*-methyl-*N*-nitrosourea
MO	molecular orbital; methyl orange
MODR	microwave-optical double resonance
mol	mole
mol.wt.	molecular weight
mon	monomeric form
Mops	4-morpholinepropanesulfonic acid
MOS	metal-oxide semiconductor
MOSFET	metal-oxide semiconductor field-effect transistor
mp	melting point
MPa	megapascal
MPA	Mulliken population analysis
Mpc	megaparsec
MPD	2-methyl-2,4-pentanediol
MPI	multiphoton ionization
MPTP	1,2,3,6-tetrahydro-1-methyl-4-phenylpyridine
MR	methyl red
MRD	multireference double substitution (method)
MRI	magnetic resonance imaging
mRNA	messenger RNA
MS	mass spectroscopy
ms	millisecond
MSA	methanesulfonic acid
MSDS	Material Safety Data Sheet
MSF	methanesulfonyl fluoride
MS-K	mass spectroscopy – kinetic method
MSL	mean sea level
MTBE	methyl *tert*-butyl ether
MTD	maximum tolerable dose
Mur	muramic acid
mV	millivolt
MVK	methyl vinyl ketone
MW	megawatt; microwave; molecular weight
mW	milliwatt
MWD	molecular weight distribution
Mx	maxwell
N	newton; asparagine
N	angular momentum; neutron number; number density
N_A	Avogadro constant
n	neutron; nano (SI prefix for 10^{-9})
n	amount of substance; number density; principal quantum number; refractive index; normal (in chemical formulas)
NAA	nuclear activation analysis; 1-naphthaleneacetic acid
NAAD	nicotinic acid adenine dinucleotide
NAD	nicotinamide adenine dinucleotide
NADH	reduced NAD
NADP	NAD phosphate
NANA	*N*-acetylneuraminic acid
NAO	natural atomic orbital
NBO	natural bond orbital
nbp	normal boiling point
NBR	nitrile butadiene rubber [poly(butadiene-co-acrylonitrile)]
NDELA	*N*-nitrosodiethanolamine
NEDOR	nuclear electron double resonance
NEM	*N*-ethylmorpholine
Neu	neuraminic acid
NEXAFS	near-edge x-ray absorption fine structure
ng	nanogram
NHO	natural hybrid orbital
NHOMO	next-to-highest occupied molecular orbital
NICI	negative ion chemical ionization
NICS	nuclear independent chemical shift
NIR	near infrared; ribosylnicotinamide
nm	nanometer

Symbol	Meaning
NMN	β-nicotinamide mononucleotide
NMR	nuclear magnetic resonance
Nn	nonyl
NNDO	neglect of nonbonded differential overlap
NO	natural orbital
NOE	nuclear Overhauser effect
NOEL	no-observed-effect level
NOx	nitrogen oxides
NP	nitropyrene
NPA	natural population analysis
NQR	nuclear quadrupole resonance
NR	natural rubber
NRA	nuclear reaction analysis
ns	nanosecond
NSE	neutron spin echo
NTA	nitrilotriacetic acid
NTP	normal temperature and pressure
Nu	nucleophile
Nu	Nusselt number
o	*ortho* (locant on aromatic ring)
OAA	oxaloacetic acid
obs, obsd	observed
Oc	octyl
OD	optical density; outside diameter
ODMR	optically detected magnetic resonance
Oe	oersted
OFGF	outer valence Green's function (method)
ONA	*o*-nitroaniline
ORD	optical rotatory dispersion
Oro	orotate; orotidine
oz	ounce
P	poise; peta (SI prefix for 10^{15}); proline
P	power; pressure; probability; sound energy flux
p	proton; pico (SI prefix for 10^{-12})
p	dielectric polarization; electric dipole moment; momentum; pressure; bond order; *para* (as aromatic ring locant)
Pa	pascal
PA	proton affinity; pyrrolizidine alkaloid; polyamide (nylon)
PAA	poly(acrylic acid)
PABA	*p*-aminobenzoic acid
PABS	*p*-aminobenzenesulfonamide
PAC	photoacoustic calorimetry
PAH	polycyclic aromatic hydrocarbon(s)
PAI	polyamide-imide
PAL	polyaniline
PAM	polyacrylamide
PAN	1-(2-pyridylazo)-2-naphthol; polyacrylonitrile
PAR	4-(2'-pyridylazo)resorcinol
PARA	polyaryl amide
PAS	photoacoustic spectroscopy; polyarylsulfone
PB	polybutylene
PBA	poly(butyl acrylate)
PBAN	polybutylene-acrylonitrile copolymer
PBB	polybrominated biphenyl
PBD	poly(1,3-butadiene)
PBI	polybenzimidazole
PBMA	poly(butyl methacrylate)
PBS	polybutadiene-styrene copolymer
PBT	poly(butylene terephthalate)
PC	paper chromatography; photocalorimetry; polycarbonate
pc	parsec
PCB	polychlorinated biphenyl
PCHO	paraldehyde (2,4,6-trimethyl-1,3,5-trioxane)
PCL	polycaprolactone
PCM	polarizable continuum model
PCNB	pentachloronitrobenzene
PCP	pentachlorophenol
PCR	polymerase chain reaction
PCT	poly(cyclohexylene terephthalate)
PCTFE	polymonochlorotrifluoroethylene
PD	potential difference
PDB	*p*-dichlorobenzene
pdl	poundal
PDMS	poly(dimethylsiloxane)
PE	polyethylene
Pe	pentyl
Pe	Péclet number
pe	probable error
PEA	poly(ethyl acrylate)
PEEK	poly(ether ether ketone)
PEG	poly(ethylene glycol)
PEI	polyetherimide
PEK	polyetherketone
PEL	permissible exposure limit
PEO	poly(ethylene oxide)
PES	photoelectron spectroscopy; potential energy surface; polyethersulfone
PET	positron emission tomography; poly(ethylene terephthalate); pentaerythritol tetranitrate
peth	petroleum ether
PEX	crosslinked polyethylene
PF	phenol-formaldehyde resin
pf	power factor
PFOA	perfluorooctanoic acid
pg	picogram
Ph	phenyl
pH	negative log of hydrogen ion concentration
Phe	phenylalanine
PhIP	2-amino-1-methyl-6-phenylimidazo[4,5-b]pyridine
PHPMS	pulsed high pressure mass spectrometry
PI	polyimide
pI	isoelectric point
PIB	polyisobutylene
PIMS	photoionization mass spectrometry
PIN	p-intrinsic-n (diode)
Pipes	1,4-piperazinediethanesulfonic acid
PIV	particle-image velocimetry
PIXE	particle induced x-ray emission
p*K*	negative log of ionization constant
PLM	principle of least motion
PLOT	porous-layer open-tabular (column)
PLS	partial least squares
pm	picometer
PMA	poly(methyl acrylate)
PMAC	phenylmercuric acetate
PMMA	poly(methyl methacrylate)
PMO	perturbation MO (theory)
PMP	polymethylpentene
PMS	polymethylstyrene; *p*-methylstyrene
PNA	*p*-nitroaniline
PNDO	partial neglect of differential overlap
PNO	pair natural orbitals
PNRA	prompt nuclear reaction analysis
PNT	*p*-nitrotoluene

Symbol	Meaning
PO	polyolefin
POAV	π-orbital axis vector
pol	polymeric form
POM	polyoxymethylene
POx	phosphorus oxides
PP	polypropylene
ppb	parts per billion
PPC	chlorinated polypropylene
PPE	poly(phenylene ether)
ppm	parts per million
PPO	poly(phenylene oxide)
PPOX	polypropylene oxide
PPP	Pariser-Parr-Pople (method)
PPS	poly(phenylene sulfide)
PPSU	poly(phenylene sulfone)
PPT	poly(propylene terephthalate)
ppt	parts per thousand; precipitate
Pr	propyl
Pr	Prandtl number
PRDDO	partial retention of diatomic differential overlap
Pro	proline
PS	photoelectron spectroscopy; polystyrene
ps	picosecond
PSD	photon stimulated desorption
psi	pounds per square inch
psia	pounds per square inch absolute
psig	pounds per square inch gage
PT	perturbation theory
pt	pint
PTFE	poly(tetrafluoroethylene)
PTME	poly(tetramethylene terephthalate)
PTMS	propyltrimethoxysilane
PTP	*p*-terphenyl
PTU	phenylthiourea
PU	polyurethane
Pu	purine
PVA	poly(vinyl alcohol)
PVAc	poly(vinyl acetate)
PVC	poly(vinyl chloride)
PVD	physical vapor deposition
PVDC	poly(vinylidene chloride)
PVDF	poly(vinylidene fluoride)
PVF	poly(vinyl fluoride)
PVK	poly(vinyl carbazole)
PVME	poly(methyl vinyl ether)
PVOH	poly(vinyl alcohol)
PVP	poly(vinyl pyrrolidone)
PVT	pressure-volume-temperature
Py	pyrimidine
PyMS	pyrolysis mass spectrometry
p.z.c.	point of zero charge
Q	electric charge; heat; partition function; quadrupole moment; radiant energy; vibrational normal coordinate; glutamine
q	electric field gradient; flow rate; heat; wave vector (phonons)
QCD	quantum chromodynamics
QCI	quadratic configuration interaction
QCT	quasi-classical trajectory (method)
QED	quantum electrodynamics
Q.E.D.	quod erat demonstrandum (which was to be proved)
QIT	quadrupole ion trap
QMRE	quantum mechanical resonance energy
QMS	quadrupole mass spectrometry
QSAR	quantitative structure-activity relations
QSO	quasi-stellar object
qt	quart
quad	quadrillion BTU (=$1.055 \cdot 10^{18}$ joules)
Qui	quinovose
q.v.	quod vide (which you should see)
R	roentgen; arginine; alkyl radical (in chemical formulas)
°R	degree Rankine
R	electrical resistance; gas constant; molar refraction; Rydberg constant; coefficient of multiple correlation
r	reaction (as in $\Delta_r H$)
r	position vector; radius
RA	right ascension
rad	radian
RAIRS	reflection-absorption infrared spectroscopy
RAM	random access memory
RBS	Rutherford back scattering
Rbu, Rul	ribulose
RCI	ring current index
RDA	rubidium dihydrogen arsenate
RDS	rate determining step
RDX	Royal Demolition Explosive (hexahydro-1,3,5-trinitro-1,3,5-triazine)
Re	real part
RE	resonance energy
RED	radial electron distribution
REELS	reflection electron energy loss spectroscopy
REM	reflection electron microscopy
rem	roentgen equivalent man
REMPI	resonance-enhanced multiphoton ionization
REPE	resonance energy per electron
RF	radiofrequency
RGA	residual gas analyzer
Rha	rhamnose
RHEED	reflection high-energy electron diffraction
RHF	restricted Hartree-Fock (theory)
RI	resonance ionization
RIA	radioimmunoassay
Rib	ribose
Ribulo	ribulose
rms	root-mean-square
RNA	ribonucleic acid
RNase	ribonuclease
ROHF	restricted open shell Hartree-Fock
ROM	read only memory
ROMP	ring opening metathesis polymerization
ROP	ring opening polymerization
RPA	random phase approximation
RPH	reaction path Hamiltonian
RPLC	reversed-phase liquid chromatography
rpm	revolutions per minute
rps	revolutions per second
RRK	Rice-Ramsperger-Kassel (theory)
RRKM	Rice-Ramsperger-Kassel-Marcus (theory)
rRNA	ribosomal RNA
RRS	resonance Raman spectroscopy
RS	Raman spectroscopy
RSC	reaction-solution calorimetry
Ry	rydberg
S	siemens; serine

Symbol	Meaning
S	area; entropy; probability current density; Poynting vector; symmetry coordinate; spin angular momentum
s	second; solid (phase)
s	path length; spin angular momentum; symmetry number; sedimentation coefficient; solubility; symmetrical (as stereochemical descriptor)
SAED	selected area electron diffraction
SALC	symmetry adapted linear combinations
SALI	surface analysis by laser ionization
SAM	scanning Auger microscopy
SAMS	self-assembled monolayers
SANS	small angle neutron scattering
SAR	structure-activity relationship
Sar	sarcosine
sat, satd	saturated
SAXS	small angle x-ray scattering
SB	styrene butadiene copolymer
SBS	styrene butadiene styrene block copolymer
Sc	Schmidt number
SC	spin-coupled (method)
SCD	state correlation diagram
SCE	saturated calomel electrode
SCF	self-consistent field (method); supercritical fluid
SCP	single cell protein
SCR	silicon-controlled rectifier
SCRF	self-consistent reaction field (method)
sd	standard deviation
SDA	sulfadiazine
SDW	spin density wave
SE	strain energy
SEBS	styrene ethylene butylene styrene block copolymer
SEC	size exclusion chromatography
sec	secant; second
sec	secondary (in chemical name)
SECSY	spin-echo correlated spectroscopy
Sed	sedoheptulose
SEELFS	surface extended energy loss fine structure
SEM	scanning electron microscopy; standard error of the mean
sepn	separation
Ser	serine
SERS	surface-enhanced Raman spectroscopy
SET	single electron transfer
SEXAF	surface extended x-ray absorption fine structure
SFC	supercritical fluid chromatography
Sh	Sherwood number
Shy	thiohypoxanthine
SI	International System of Units; surface ionization
SID	surface-induced dissociation
SILAR	successive ionic layer adsorption and reaction
SIM	selected ion monitoring
SIMS	secondary-ion spectroscopy
sin	sine
sinh	hyperbolic sine
SIPN	semi-interpenetrating polymer network
SIS	styrene isoprene styrene block copolymer
SLAM	scanning laser acoustic microscopy
SLUMO	second lowest unoccupied molecular orbital
SMILES	simplified molecular input line entry system
SMMA	styrene methyl methacrylate copolymer
SMO	semiempirical molecular orbital
SMOW	Standard Mean Ocean Water (Vienna)
SNMS	sputtered neutral mass spectroscopy

Symbol	Meaning
Sno	thiouridine
SNU	solar neutrino unit
SOJT	second-order Jahn-Teller (effect)
sol	soluble; solution
soln, sln	solution
SOMO	singly occupied molecular orbital
Sor	sorbose
sp gr	specific gravity
SPM	scanned probe microscopy
SPST	single-pulse shock tubes
sq	square
Sr	Strouhal number
sr	steradian
Srd	6-thioinosine
SSMS	source spark mass spectroscopy
St	stoke
St	Stanton number
std, stnd	standard (state)
STEL	short-term exposure limit
STEM	scanning transmission electron microscope
STM	scanning tunneling microscopy
STO	Slater-type orbital
STP	standard temperature and pressure
sub, subl	sublimes; sublimation
Suc, Sac	sucrose
Sur	thiouracil
Sv	sievert
SWIFT	stored waveform inverse Fourier transform
T	tesla; tera (SI prefix for 10^{12}); threonine
T	kinetic energy; period; term value; temperature (thermodynamic); torque; transmittance
t	metric tonne; triton
t	Celsius temperature; thickness; time; transport number
TAC	time-to-amplitude converter
TAI	International Atomic Time
Tal	talose
tan	tangent
tanh	hyperbolic tangent
Taps	3-{[2-hydroxy-1,1-bis(hydroxymethyl)ethyl]amino}-1-propanesulfonic acid
TBE	1,1,2,2-tetrabromoethane
TBP	tributyl phosphate
TC	titration calorimetry
TCA	trichloroacetic acid
TCB, TCBA	2,3,6-trichlorobenzoic acid
TCE	trichloroethylene
TCG	Geocentric Coordinated Time
TCNE	tetracyanoethylene
TCNQ	tetracyanoquinodimethane
TCP	tricresyl phosphate
TCSCF	two configuration self-consistent field
TDA	toluene-2,4-diamine
TDI	toluene diisocyanate
tDNA	transfer DNA
TE	transverse electric
TEA	triethanolamine; triethylamine
TED	transferred electron device; transmission electron diffraction
TEDA	triethylenediamine
TEELS	transmission electron energy loss spectroscopy
TEM	transverse electromagnetic; transmission electron microscope
temp	temperature

Symbol	Meaning
TEO	thermoplastic elastic olefin
TEPP	tetraethyl pyrophosphate
tert	tertiary (in chemical name)
Tes	2-{[2-hydroxy-1,1-bis(hydroxymethyl)ethyl]amino}-1-propanesulfonic acid
TFD	Thomas-Fermi-Dirac (method)
TFE	tetrafluoroethylene
TGA	thermogravimetric analysis
Thd	ribosylthymine
THEED	transmission high energy electron diffraction
theor	theoretical
thf, THF	tetrahydrofuran
THQ	1,2,3,4-tetrahydroquinoline
Thr	threonine
Thy	thymine
TI	thermal ionization
TIPA	triisopropanolamine
TL	thermoluminescence
TLC	thin-layer chromatography
TLV	threshold limit value
TM	transverse magnetic
TMAB	tetrabutylammonium bromide
TMAO	trimethylamine oxide
TMCP	tri-*m*-cresyl phosphate
TMEDA	*N,N,N',N'*-tetramethyl-1,2-ethanediamine
TMMV	threshold molecular weight value
TMS	tetramethylsilane
TNA	2,4,6-trinitroaniline
TNB	1,3,5-trinitrobenzene
TNM	tetranitromethane
TNT	2,4,6-trinitrotoluene
TOCP	tri-*o*-cresyl phosphate
TOF	turnover frequency
TOF-MS	time-of-flight mass spectrometer
tol	tolyl
TON	turnover number
TOPO	trioctylphosphine oxide
Torr	torr (pressure unit)
TOTP	tri-*o*-tolyl phosphate
TPE	thermoplastic elastomer
TPTA	triphenyltin acetate
TPTC	triphenyltin chloride
TRE	topological resonance energy
Tre	trehalose
Tricine	*N*-[2-hydroxy-1,1-bis(hydroxymethyl)ethyl]glycine
Tris	2-amino-2-(hydroxymethyl)-1,3-propanediol
TRMC	time-resolved microwave conductivity
tRNA	transfer RNA
Trp	tryptophan
trs	transition
TS	transition state
TSS	transition state spectroscopy
TST	generalized transition-state theory
TTF	tetrathiofulvalene
Tyr	tyrosine
U	uracil (in genetic code)
U	electric potential difference; internal energy
u	unified atomic mass unit
u	Bloch function; electric mobility; velocity
ua	astronomical unit (AU is also used)
UBFF	Urey-Bradley force field
UDMH	1,1-dimethylhydrazine

Symbol	Meaning
UDP	uridine 5'-diphosphate
UHF	ultrahigh frequency; unrestricted Hartree-Fock (method)
UHMWPE	ultrahigh molecular weight polyethylene
ULDPE	ultra low density polyethylene
ULPE	ultra linear polyethylene
UMP	uridine 5'-monophosphate
uns, unsym	unsymmetrical (as chemical descriptor)
UPS, UPES	ultraviolet photoelectron spectroscopy
Ura	uracil
Urd	uridine
USP	United States Pharmacopeia
UT	universal time
UTC	coordinated universal time
UTP	uridine 5'-triphosphate
UV	ultraviolet
V	volt; valine
V	electric potential; potential energy; volume
v	reaction rate; specific volume; velocity; vibrational quantum number; vicinal (as chemical descriptor)
v/v	volume per volume (volume of solute divided by volume of solution, expressed as percent)
VA	vinyl acetate, vanillic acid
Val	valine
vap	vaporization
VAT	vibration assisted tunneling
VB	valence band; valence bond (theory)
VCD	vibrational circular dichroism
VDW	van der Waals interaction
VHF	very high frequency
vic	vicinal (on adjacent carbon atom)
VIS	visible region of the spectrum
vit	vitreous (phase)
VLDPE	very low density polyethylene
VLPP	very low pressure pyrolysis
VMA	vanilmandelic acid
VOC	volatile organic compound(s)
VOFF	valence orbital force field
VPC	vapor phase chromatography
VSEPR	valence shell electron-pair repulsion (method)
VSIP	valence state ionization potential
VSLI	very large scale integrated (circuit)
VSMOW	Vienna Standard Mean Ocean Water
VTCS	vinyltrichlorosilane
VUV	vacuum ultraviolet
W	watt; tryptophan
W	radiant energy; statistical weight; work
w	energy density; mass fraction; velocity; work
w/v	weight per volume (mass of solute divided by volume of solution, usually expressed as g/100 mL)
w/w	weight per weight (mass of solute divided by mass of solution, expressed as percent)
WAXS	wide angle x-ray scattering
Wb	weber
We	Weber number
WKB	Wentzel-Kramers-Brillouin (approximation)
WLF	Williams-Landel-Ferry (equation)
WLN	Wiswesser line notation
wt	weight
X	X unit; halogen (in chemical formula)
X	reactance
x	mole fraction
X, Xaa	unspecified amino acid

Symbol	Meaning
XAFS	x-ray absorption fine structure
Xan	xanthine
XANES	x-ray absorption near-edge structure
Xao	xanthosine
Xle	leucine or isoleucine (unspecified)
XLPE	crosslinked polyethylene
Xlu, Xul	xylulose
XPS, XPES	x-ray photoelectron spectroscopy
XRD	x-ray diffraction
XRF	x-ray fluorescence
XRS	x-ray spectroscopy
Xyl	xylose
Y	yotta (SI prefix for 10^{24}); tyrosine
Y	admittance; Planck function; Young's modulus
y	yocto (SI prefix for 10^{-24})
y	mole fraction for gas (when *x* refers to liquid phase)
y, yr	year
YAG	yttrium aluminum garnet
yd	yard
YIG	yttrium iron garnet
Z	zetta (SI prefix for 10^{21}); glutamine or glutamic acid (unspecified)
Z	atomic number; compression factor; collision number; impedance; partition function; zusammen (*cis*-configuration)
z	zepto (SI prefix for 10^{-21})
z	charge number (of an ion); collision frequency factor
ZDO	zero differential overlap
ZINDO	Zerner's INDO method
ZPE, ZPVE	zero point vibrational energy
ZULU	Greenwich mean time
α	alpha particle
α	absorption coefficient; degree of dissociation; electric polarizability; expansion coefficient; fine structure constant
β	beta particle
β	reciprocal temperature parameter (= $1/kT$)
γ	photon; gamma (obsolete mass unit = μg)
γ	activity coefficient; conductivity; magnetogyric ratio; mass concentration; ratio of heat capacities; surface tension
Γ	Grüneisen parameter; level width; surface concentration
Δ	inertial defect; mass excess
δ	chemical shift; Dirac delta function; Kronecker delta; loss angle
ε	emittance; Levi-Civita symbol; linear strain; molar absorption coefficient; permittivity
ζ	Coriolis coupling constant; electrokinetic potential
η	overpotential; viscosity
κ	compressibility; conductivity; magnetic susceptibility; molar absorption coefficient
λ	absolute activity; radioactive decay constant; thermal conductivity; wavelength
Λ	angular momentum; ionic conductivity
μ	muon; micro (SI prefix for 10^{-6})
μ	chemical potential; electric dipole moment; electric mobility; friction coefficient; Joule-Thompson coefficient; magnetic dipole moment; mobility; permeability
μF	microfarad
μg	microgram
μm	micrometer
μs	microsecond
ν	frequency; kinematic velocity; stoichiometric number
ν_e	neutrino
$\tilde{\nu}$	wavenumber
Π	osmotic pressure; Peltier coefficient
π	pion
ρ	density; reflectance; resistivity
σ	electrical conductivity; cross section; normal stress; shielding constant (NMR); Stefan-Boltzmann constant; surface tension; standard deviation
τ	transmittance; chemical shift; shear stress; relaxation time
Φ	magnetic flux; potential energy; radiant power; work function
φ	electrical potential; fugacity coefficient; osmotic coefficient; quantum yield; wave function
χ	magnetic susceptibility
χ_e	electric susceptibility
ψ	wave function
Ω	ohm
Ω	axial angular momentum; solid angle
ω	circular frequency; angular velocity; harmonic vibration wavenumber; statistical weight

GREEK, HEBREW, AND RUSSIAN ALPHABETS

The following table presents the Greek, Hebrew, and Russian alphabets, their letters, the names of the letters, and the English equivalents.

Greek Alphabet[1]

Letter	Name	English equivalent
Α α	alpha	a
Β β	beta	b
Γ γ	gamma	g, n
Δ δ	delta	d
Ε ε	epsilon	e
Ζ ζ	zeta	z
Η η	eta	ē
Θ θ	theta	th
Ι ι	iota	i
Κ κ	kappa	k
Λ λ	lambda	l
Μ μ	mu	m
Ν ν	nu	n
Ξ ξ	xi	x
Ο ο	omicron	o
Π π	pi	p
Ρ ρ	rho	r, rh
Σ σ ς	sigma	s
Τ τ	tau	t
Υ υ	upsilon	y, u
Φ φ	phi	ph
Χ χ	chi	ch
Ψ ψ	psi	ps
Ω ω	omega	ō

[1] The letter gamma is transliterated "n" only before velars; the letter upsilon is transliterated "u" only as the final element in diphthongs.

Hebrew Alphabet[1,2]

Letter	Name	English equivalent
א	aleph	’[3]
ב	beth	b, bh
ג	gimel	g, gh
ד	daleth	d, dh
ה	he	h
ו	waw	w
ז	zayin	z
ח	heth	ḥ
ט	teth	ṭ
י	yodh	y
כ ך	kaph	k, kh
ל	lamedh	l
מ ם	mem	m
נ ן	nun	n
ס	samekh	s
ע	ayin	‘
פ ף	pe	p, ph
צ ץ	sadhe	ṣ
ק	qoph	q
ר	resh	r
ש	sin	ś
ש	shin	sh
ת	taw	t, th

[1] Where two forms of a letter are given, the second one is the form used at the end of a word.
[2] The Hebrew letters are primarily consonants; a few of them are also used secondarily to represent certain vowels, when provided at all, by means of a system of dots or strokes adjacent to the consonated characters.
[3] Not represented in transliteration when initial.

Russian Alphabet

Letter	English equivalent
А а	a
Б б	b
В в	v
Г г	g
Д д	d
Е е	e
Ж ж	zh
З з	z
И и Й й	i, ĭ
К к	k
Л л	l
М м	m
Н н	n
О о	o
П п	p
Р р	r
С с	s
Т т	t
У у	u
Ф ф	f
Х х	kh
Ц ц	ts
Ч ч	ch
Ш ш	sh
Щ щ	shch
Ъ ъ[1]	”
Ы ы	y
Ь ь[2]	’
Э э	e
Ю ю	yu
Я я	ya

[1] This sign indicates that the immediately preceding consonant is not palatized even though immediately followed by a palatized vowel.
[2] This sign indicates that the immediately preceding consonant is palatized even though not immediately followed by a palatized vowel.

DEFINITIONS OF SCIENTIFIC TERMS

Symbols

Brief definitions of selected terms of importance in chemistry, physics, and related fields of science are given in this section. The selection process emphasizes the following types of terms:

- Physical quantities
- Units of measure
- Classes of chemical compounds and materials
- Important theories, laws, and basic concepts.

Individual chemical compounds are not included.

Definitions have taken wherever possible from the recommendations of international or national bodies, especially the International Union of Pure and Applied Chemistry (IUPAC) and International Organization for Standardization (ISO). For physical quantities and units, the recommended symbol is also given. The source of such definitions is indicated by the reference number in brackets following the definition. In many cases these official definitions have been edited in the interest of stylistic consistency and economy of space. The user is referred to the original source for further details.

An asterisk (*) following a term indicates that further information can be found by consulting the index of this *Handbook* under the entry for that term.

References

1. *ISO Standards Handbook 2, Units of Measurement,* International Organization for Standardization, Geneva, 1992.
2. *Quantities, Units, and Symbols in Physical Chemistry, Second Edition,* International Union of Pure and Applied Chemistry, Blackwell Scientific Publications, Oxford, 1993.
3. *Compendium of Chemical Terminology,* International Union of Pure and Applied Chemistry, Blackwell Scientific Publications, Oxford, 1987.
4. *A Guide to IUPAC Nomenclature of Organic Compounds,* International Union of Pure and Applied Chemistry, Blackwell Scientific Publications, Oxford, 1993.
5. *Glossary of Class Names of Organic Compounds and Reactive Intermediates Based on Structure, Pure and Applied Chemistry,* 67, 1307, 1995.
6. *Compendium of Analytical Nomenclature,* International Union of Pure and Applied Chemistry, Blackwell Scientific Publications, Oxford, 1987.
7. *Nomenclature of Inorganic Chemistry,* International Union of Pure and Applied Chemistry, Blackwell Scientific Publications, Oxford, 1990.
8. *Glossary of Basic Terms in Polymer Science, Pure and Applied Chemistry,* 68, 2287, 1996.
9. *The International Temperature Scale of 1990, Metrologia,* 27, 107, 1990.
10. *Compilation of ASTM Standard Definitions,* American Society of Testing and Materials, Philadelphia, 1990.
11. *ASM Metals Reference Book,* American Society for Metals, Metals Park, OH, 1983.

***Ab initio* method** - An approach to quantum-mechanical calculations on molecules which starts with the Schrödinger equation and carries out a complete integration, without introducing empirical factors derived from experimental measurement.

Absorbance (A) - Defined as $-\log(1-\alpha) = \log(1/\tau)$, where α is the absorptance and τ the transmittance of a medium through which a light beam passes. [2]

Absorbed dose (D) - For any ionizing radiation, the mean energy imparted to an element of irradiated matter divided by the mass of that element. [1]

Absorptance (α) - Ratio of the radiant or luminous flux in a given spectral interval absorbed in a medium to that of the incident radiation. Also called absorption factor. [1]

Absorption coefficient (a) - The relative decrease in the intensity of a collimated beam of electromagnetic radiation, as a result of absorption by a medium, during traversal of an infinitesimal layer of the medium, divided by the length traversed. [1]

Absorption coefficient, molar (ε) - Absorption coefficient divided by amount-of-substance concentration of the absorbing material in the sample solution ($\varepsilon = a/c$). The SI unit is m^2/mol. Also called extinction coefficient, but usually in units of $mol^{-1}\ dm^3\ cm^{-1}$. [2]

Acceleration - Rate of change of velocity with respect to time.

Acceleration due to gravity (g)* - The standard value (9.80665 m/s^2) of the acceleration experienced by a body in the earth's gravitational field. [1]

Acenes - Polycyclic aromatic hydrocarbons consisting of fused benzene rings in a rectilinear arrangement. [5]

Acid - Historically, a substance that yields an H^+ ion when it dissociates in solution, resulting in a pH<7. In the Brönsted definition, an acid is a substance that donates a proton in any type of reaction. The most general definition, due to G.N. Lewis, classifies any chemical species capable of accepting an electron pair as an acid.

Acid dissociation constant (K_a)* - The equilibrium constant for the dissociation of an acid HA through the reaction $HA + H_2O \rightleftharpoons A^- + H_3O^+$. The quantity $pK_a = -\log K_a$ is often used to express the acid dissociation constant.

Actinides - The elements of atomic number 89 through 103, e.g., Ac, Th, Pa, U, Np, Pu, Am, Cm, Bk, Cf, Es, Fm, Md, No, Lr. [7]

Activation energy* - In general, the energy that must be added to a system in order for a process to occur, even though the process may already be thermodynamically possible. In chemical kinetics, the activation energy is the height of the potential barrier separating the products and reactants. It determines the temperature dependence of the reaction rate.

Activity - For a mixture of substances, the absolute activity λ of substance B is defined as $\lambda_B = \exp(\mu_B/RT)$, where μ_B is the chemical potential of substance B, R the gas constant, and T the thermodynamic temperature. The relative activity a is defined as $a_B = \exp[(\mu_B-\mu_B^\circ)/RT]$, where μ_B° designates the chemical potential in the standard state. [2]

Activity coefficient (γ)* - Ratio of the activity a_B of component B of a mixture to the concentration of that component. The value of γ depends on the method of stating the composition. For mole fraction x_B, the relation is $a_B = \gamma_B x_B$; for molarity c_B, it is $a_B = \gamma_B c_B/c^\circ$, where c° is the standard state composition (typically chosen as 1 mol/L); for molality m_B, it is $a_B = \gamma_B m_B/m^\circ$, where m° is the standard state molality (typically 1 mol/kg). [2]

Activity, of radioactive substance (A) - The average number of spontaneous nuclear transitions from a particular energy state occurring in an amount of a radionuclide in a small time interval divided by that interval. [1]

Acyl groups - Groups formed by removing the hydroxy groups from oxoacids that have the general structure RC(=O)(OH) and replacement analogues of such acyl groups. [5]

Adiabatic process - A thermodynamic process in which no heat enters or leaves the system.

Admittance (*Y*) - Reciprocal of impedance. $Y = G + iB$, where G is conductance and B is susceptance. [1]

Adsorption - A process in which molecules of gas, of dissolved substances in liquids, or of liquids adhere in an extremely thin layer to surfaces of solid bodies with which they are in contact. [10]

Albedo* - The ratio of the light reflected or scattered from a surface to the intensity of incident light. The term is often used in reference to specific types of terrain or to entire planets.

Alcohols - Compounds in which a hydroxy group, -OH, is attached to a saturated carbon atom. [5]

Aldehydes - Compounds RC(=O)H, in which a carbonyl group is bonded to one hydrogen atom and to one R group. [5]

Aldoses - Aldehydic parent sugars (polyhydroxyaldehydes $H[CH(OH)]_nC(=O)H$, n>1) and their intramolecular hemiacetals. [5]

Aldoximes - Oximes of aldehydes: RCH=NOH. [5]

Alfvén number (*Al*) - A dimensionless quantity used in plasma physics, defined by $Al = v(\rho\mu)^{1/2}/B$, where ρ is density, v is velocity, μ is permeability, and B is magnetic flux density. [2]

Alfvén waves - Very low frequency waves which can exist in a plasma in the presence of a uniform magnetic field. Also called magnetohydrodynamic waves.

Alicyclic compounds - Aliphatic compounds having a carbocyclic ring structure which may be saturated or unsaturated, but may not be a benzenoid or other aromatic system. [5]

Aliphatic compounds - Acyclic or cyclic, saturated or unsaturated carbon compounds, excluding aromatic compounds. [5]

Alkali metals - The elements lithium, sodium, potassium, rubidium, cesium, and francium.

Alkaline earth metals - The elements calcium, strontium, barium, and radium. [7]

Alkaloids - Basic nitrogen compounds (mostly heterocyclic) occurring mostly in the plant kingdom (but not excluding those of animal origin). Amino acids, peptides, proteins, nucleotides, nucleic acids, and amino sugars are not normally regarded as alkaloids. [5]

Alkanes - Acyclic branched or unbranched hydrocarbons having the general formula C_nH_{2n+2}, and therefore consisting entirely of hydrogen atoms and saturated carbon atoms. [5]

Alkenes - Acyclic branched or unbranched hydrocarbons having one carbon-carbon double bond and the general formula C_nH_{2n}. Acyclic branched or unbranched hydrocarbons having more than one double bond are alkadienes, alkatrienes, etc. [5]

Alkoxides - Compounds, ROM, derivatives of alcohols, ROH, in which R is saturated at the site of its attachment to oxygen and M is a metal or other cationic species. [5]

Alkyl groups - Univalent groups derived from alkanes by removal of a hydrogen atom from any carbon atom: C_nH_{2n+1}-. The groups derived by removal of a hydrogen atom from a terminal carbon atom of unbranched alkanes form a subclass of normal alkyl (*n*-alkyl) groups. The groups RCH_2-, R_2CH-, and R_3C- (R not equal to H) are primary, secondary, and tertiary alkyl groups, respectively. [5]

Alkynes - Acyclic branched or unbranched hydrocarbons having a carbon-carbon triple bond and the general formula C_nH_{2n-2}, RC≡CR′. Acyclic branched or unbranched hydrocarbons having more than one triple bond are known as alkadiynes, alkatriynes, etc. [5]

Allotropy - The occurrence of an element in two or more crystalline forms.

Allylic groups - The group CH_2=$CHCH_2$- (allyl) and derivatives formed by substitution. The term 'allylic position' or 'allylic site' refers to the saturated carbon atom. A group, such as - OH, attached at an allylic site is sometimes described as "allylic." [5]

Amagat volume unit - A non-SI unit previously used in high pressure science. It is defined as the molar volume of a real gas at one atmosphere pressure and 273.15 K. The approximate value is 22.4 L/mol.

Amides - Derivatives of oxoacids R(C=O)(OH) in which the hydroxy group has been replaced by an amino or substituted amino group. [5]

Amine oxides - Compounds derived from tertiary amines by the attachment of one oxygen atom to the nitrogen atom: R_3N^+-O^-. By extension the term includes the analogous derivatives of primary and secondary amines. [5]

Amines - Compounds formally derived from ammonia by replacing one, two, or three hydrogen atoms by hydrocarbyl groups, and having the general structures RNH_2 (primary amines), R_2NH (secondary amines), R_3N (tertiary amines). [5]

Amino acids* - Compounds containing both a carboxylic acid group (-COOH) and an amino group (-NH_2). The most important are the α-amino acids, in which the -NH_2 group in attached to the C atom adjacent to the -COOH group. In the β-amino acids, there is an intervening carbon atom. [4]

Ampere (A)* - The SI base unit of electric current. [1]

Ampere's law - The defining equation for the magnetic induction $\boldsymbol{B}$, viz., $\mathrm{d}\boldsymbol{F} = I\mathrm{d}\boldsymbol{l} \times \boldsymbol{B}$, where $\mathrm{d}\boldsymbol{F}$ is the force produced by a current I flowing in an element of the conductor $\mathrm{d}\boldsymbol{l}$ pointing in the direction of the current.

Ångström (Å) - A unit of length used in spectroscopy, crystallography, and molecular structure, equal to 10^{-10} m.

Angular momentum (*L*) - The angular momentum of a particle about a point is the vector product of the radius vector from this point to the particle and the momentum of the particle; i.e., $\boldsymbol{L} = \boldsymbol{r} \times \boldsymbol{p}$. [1]

Angular velocity (*ω*) - The angle through which a body rotates per unit time.

Anilides - Compounds derived from oxoacids R(C=O)(OH) by replacing the -OH group by the -NHPh group or derivative formed by ring substitution. Also used for salts formed by replacement of a nitrogen-bound hydrogen of aniline by a metal. [5]

Anion - A negatively charged atomic or molecular particle.

Antiferroelectricity* - An effect analogous to antiferromagnetism in which electric dipoles in a crystal are ordered in two sublattices that are polarized in opposite directions, leading to zero net polarization. The effect vanishes above a critical temperature.

Antiferromagnetism* - A type of magnetism in which the magnetic moments of atoms in a solid are ordered into two antiparallel aligned sublattices. Antiferromagnets are characterized by a zero or small positive magnetic susceptibility. The susceptibility increases with temperature up to a critical value, the Néel temperature, above which the material becomes paramagnetic.

Antiparticle - A particle having the same mass as a given elementary particle and a charge equal in magnitude but opposite in sign.

Appearance potential* - The lowest energy which must be imparted to the parent molecule to cause it to produce a particular specified parent ion. This energy, usually stated in eV, may be imparted by electron impact, photon impact, or in other ways. More properly called appearance energy. [3]

Appearance potential spectroscopy (APS) - See Techniques for Materials Characterization, Section 12.

Are (a) - A unit of area equal to 100 m^2. [1]

Arenes - Monocyclic and polycyclic aromatic hydrocarbons. See aromatic compounds. [5]

Aromatic compounds - Compounds whose structure includes a cyclic delocalized π-electron system. Historical use of the term implies a ring containing only carbon (e.g., benzene, naphthalene), but it is often generalized to include heterocyclic structures such as pyridine and thiophene. [5]

Arrhenius equation - A key equation in chemical kinetics which expresses the rate constant k as $k = A\exp(-E_a/RT)$, where E_a is the activation energy, R the molar gas constant, and T the temperature. A is called the preexponential factor and, for simple gas phase reactions, may be identified with the collision frequency.

Arsines - AsH_3 and compounds derived from it by substituting one, two or three hydrogen atoms by hydrocarbyl groups. $RAsH_2$, R_2AsH, R_3As (R not equal to H) are called primary, secondary and tertiary arsines, respectively. [5]

Aryl groups - Groups derived from arenes by removal of a hydrogen atom from a ring carbon atom. Groups similarly derived from heteroarenes are sometimes subsumed in this definition. [5]

Astronomical unit (AU)* - The mean distance of the earth from the sun, equal to $1.49597870 \times 10^{11}$ m.

Atomic absorption spectroscopy (AAS) - See Techniques for Materials Characterization, Section 12.

Atomic emission spectroscopy (AES) - See Techniques for Materials Characterization, Section 12.

Atomic force microscopy (AFM) - See Techniques for Materials Characterization, Section 12.

Atomic mass* - The mass of a nuclide, normally expressed in unified atomic mass units (u).

Atomic mass unit (u)* - A unit of mass used in atomic, molecular, and nuclear science, defined as the mass of one atom of ^{12}C divided by 12. Its approximate value is 1.66054×10^{-27} kg. Also called the unified atomic mass unit. [1]

Atomic number (*Z*) - A characteristic property of an element, equal to the number of protons in the nucleus.

Atomic weight (A_r)* - The ratio of the average mass per atom of an element to 1/12 of the mass of nuclide ^{12}C. An atomic weight can be defined for a sample of any given isotopic composition. The standard atomic weight refers to a sample of normal terrestrial isotopic composition. The term relative atomic mass is synonymous with atomic weight. [2]

Attenuated total reflection (ATR) - See Techniques for Materials Characterization, Section 12.

Auger effect - An atomic process in which an electron from a higher energy level fills a vacancy in an inner shell, transferring the released energy to another electron which is ejected.

Aurora - An atmospheric phenomenon in which streamers of light are produced when electrons from the sun are guided into the thermosphere by the earth's magnetic field. It occurs in the polar regions at altitudes of 95—300 km.

Avogadro constant (N_A)* - The number of elementary entities in one mole of a substance.

Azeotrope - A liquid mixture in a state where the variation of vapor pressure with composition at constant temperature (or, alternatively, the variation of normal boiling point with composition) shows either a maximum or a minimum. Thus when an azeotrope boils the vapor has the same composition as the liquid.

Azides - Compounds bearing the group $-N_3$, viz. $-N{=}N^+{=}N^-$; usually attached to carbon, e.g. PhN_3, phenyl azide or azidobenzene. Also used for salts of hydrazoic acid, HN_3, e.g. NaN_3, sodium azide. [5]

Azines - Condensation products, $R_2C{=}NN{=}CR_2$, of two moles of a carbonyl compound with one mole of hydrazine. [5]

Azo compounds - Derivatives of diazene (diimide), HN=NH, wherein both hydrogens are substituted by hydrocarbyl groups, e.g., PhN=NPh, azobenzene or diphenyldiazene. [5]

Balmer series - The series of lines in the spectrum of the hydrogen atom which corresponds to transitions between the state with principal quantum number $n = 2$ and successive higher states. The wavelengths are given by $1/\lambda = R_H(1/4 - 1/n^2)$, where $n = 3,4,...$ and R_H is the Rydberg constant for hydrogen. The first member of the series ($n = 2 \rightleftharpoons 3$), which is often called the H_α line, falls at a wavelength of 6563 Å.

Bar (bar) - A unit of pressure equal to 10^5 Pa.

Bardeen-Cooper-Schrieffer (BCS) theory - A theory of superconductivity which is based upon the formation of electron pairs as a result of an electron-lattice interaction. The theory relates the superconducting transition temperature to the density of states and the Debye temperature.

Barn (b) - A unit used for expressing cross sections of nuclear processes, equal to 10^{-28} m^2.

Barrel - A unit of volume equal to 158.9873 L.

Baryon - Any elementary particle built up from three quarks. Examples are the proton, neutron, and various short-lived hyperons. Baryons have odd half-integer spins.

Base - Historically, a substance that yields an OH^- ion when it dissociates in solution, resulting in a pH>7. In the Brönsted definition, a base is a substance capable of accepting a proton in any type of reaction. The more general definition, due to G.N. Lewis, classifies any chemical species capable of donating an electron pair as a base.

Becquerel (Bq)* - The SI unit of radioactivity (disintegrations per unit time), equal to s^{-1}. [1]

Beer's law - An approximate expression for the change in intensity of a light beam that passes through an absorbing medium, viz., $\log(I/I_0) = -\varepsilon cl$, where I_0 is the incident intensity, I is the final intensity, ε is the molar (decadic) absorption coefficient, c is the molar concentration of the absorbing substance, and l is the path length. Also called the Beer-Lambert law.

Binding energy* - A generic term for the energy required to decompose a system into two or more of its constituent parts. In nuclear physics, the binding energy is the energy difference between a nucleus and the separated nucleons of which it is composed (the energy equivalent of the mass defect). In atomic physics, it is the energy required to remove an electron from an atom.

Biot (Bi) - A name sometimes used for the unit of current in the emu system.

Birefringence - A property of certain crystals in which two refracted rays result from a single incident light ray. One, the ordinary ray, follows the normal laws of refraction, while the other, the extraordinary ray, exhibits a variable refractive index which depends on the direction in the crystal.

Blackbody radiation* - The radiation emitted by a perfect black body, i.e., a body which absorbs all radiation incident on it and reflects none. The wavelength dependence of the radiated energy density ρ (energy per unit volume per unit wavelength range) is given by the Planck formula

$$\rho = \frac{8\pi hc}{\lambda^5 (e^{hc/\lambda kt} - 1)}$$

where λ is the wavelength, h is Planck's constant, c is the speed of light, k is the Boltzmann constant, and T is the temperature.

Black hole - A very dense object, formed in a supernova explosion, whose gravitational field is so large that no matter or radiation can escape from the object.

Bloch wave function - A solution of the Schrödinger equation for an electron moving in a spatially periodic potential; used in the band theory of solids.

Bohr magneton (μ_B)* - The atomic unit of magnetic moment, defined as $eh/4\pi m_e$, where h is Planck's constant, m_e the electron mass, and e the elementary charge. It is the moment associated with a single electron spin.

Bohr, bohr radius (a_0)* - The radius of the lowest orbit in the Bohr model of the hydrogen atom, defined as $\varepsilon_o h^2/\pi m_e e^2$, where ε_o is the permittivity of a vacuum, h is Planck's constant, m_e the electron mass, and e the elementary charge. It is customarily taken as the unit of length when using atomic units.

Boiling point - The temperature at which the liquid and gas phases of a substance are in equilibrium at a specified pressure. The normal boiling point is the boiling point at normal atmospheric pressure (101.325 kPa).

Boltzmann constant (k)* - The molar gas constant R divided by Avogadro's constant.

Boltzmann distribution - An expression for the equilibrium distribution of molecules as a function of their energy, in which the number of molecules in a state of energy E is proportional to $\exp(-E/kT)$, where k is the Boltzmann constant and T is the temperature.

Bond strength - See Dissociation energy.

Born-Haber cycle* - A thermodynamic cycle in which a crystalline solid is converted to gaseous ions and then reconverted to the solid. The cycle permits calculation of the lattice energy of the crystal.

Bose-Einstein distribution - A modification of the Boltzmann distribution which applies to a system of particles that are bosons. The number of particles of energy E is proportional to $[e^{(E-\mu)/kT}-1]^{-1}$, where μ is a normalization constant, k is the Boltzmann constant, and T is the temperature.

Boson - A particle that obeys Bose-Einstein Statistics; specifically, any particle with spin equal to zero or an integer. This includes the photon, pion, deuteron, and all nuclei of even mass number.

Boyle's law - The empirical law, exact only for an ideal gas, which states that the volume of a gas is inversely proportional to its pressure at constant temperature.

Bragg angle (θ) - Defined by the equation $n\lambda = 2d\sin\theta$, which relates the angle θ between a crystal plane and the diffracted x-ray beam, the wavelength λ of the x-rays, the crystal plane spacing d, and the diffraction order n (any integer).

Bravais lattices* - The 14 distinct crystal lattices that can exist in three dimensions. They include three in the cubic crystal system, two in the tetragonal, four in the orthorhombic, two in the monoclinic, and one each in the triclinic, hexagonal, and trigonal systems.

Breakdown voltage - The potential difference at which an insulating substance undergoes a physical or chemical change that causes it to become a conductor, thus allowing current to flow through the sample.

Bremsstrahlung - Electromagnetic radiation generated when the velocity of a charged particle is reduced (literally, "braking radiation"). An example is the x-ray continuum resulting from collisions of electrons with the target in an x-ray tube.

Brewster angle - The angle of incidence for which the maximum degree of plane polarization occurs when a beam of unpolarized light is incident on the surface of a medium of refractive index n. At this angle, the angle between the reflected and refracted beams is 90°. The value of the Brewster angle is $\tan^{-1}n$.

Brillouin scattering - The scattering of light by acoustic phonons in a solid or liquid.

Brillouin zone - A region of allowed wave vectors and energy levels in a crystalline solid, which plays a part in the propagation of waves through the lattice.

British thermal unit (Btu) - A non-SI unit of energy, equal to approximately 1055 J. Several values of the Btu, defined in slightly different ways, have been used.

Brownian motion - The random movements of small particles suspended in a fluid, which arise from collisions with the fluid molecules.

Brunauer-Emmett-Teller method (BET) - See Techniques for Materials Characterization, Section 12.

Buffer* - A solution designed to maintain a constant pH when small amounts of a strong acid or base are added. Buffers usually consist of a fairly weak acid and its salt with a strong base. Suitable concentrations are chosen so that the pH of the solution remains close to the pK_a of the weak acid.

Calorie (cal) - A non-SI unit of energy, originally defined as the heat required to raise the temperature of 1 g of water by 1 °C. Several calories of slightly different values have been used. The thermochemical calorie is now defined as 4.184 J.

Candela (cd)* - The SI base unit of luminous intensity. [1]

Capacitance (C) - Ratio of the charge acquired by a body to the change in potential. [1]

Symbols

Carbamates - Salts or esters of carbamic acid, $H_2NC(=O)OH$, or of N-substituted carbamic acids: $R_2NC(=O)OR'$, (R' = hydrocarbyl or a cation). The esters are often called urethanes or urethans, a usage that is strictly correct only for the ethyl esters. [5]

Carbenes - The electrically neutral species H_2C: and its derivatives, in which the carbon is covalently bonded to two univalent groups of any kind or a divalent group and bears two nonbonding electrons, which may be spin-paired (singlet state) or spin-non-paired (triplet state). [5]

Carbinols - An obsolete term for substituted methanols, in which the name carbinol is synonymous with methanol. [5]

Carbohydrates - Originally, compounds such as aldoses and ketoses, having the stoichiometric formula $C_n(H_2O)_n$ (hence "hydrates of carbon"). The generic term carbohydrate now includes mono-, oligo-, and polysaccharides, as well as their reaction products and derivatives. [5]

Carboranes - A contraction of carbaboranes. Compounds in which a boron atom in a polyboron hydride is replaced by a carbon atom with maintenance of the skeletal structure. [5]

Carboxylic acids - Oxoacids having the structure RC(=O)OH. The term is used as a suffix in systematic name formation to denote the -C(=O)OH group including its carbon atom. [5]

Carnot cycle - A sequence of reversible changes in a heat engine using a perfect gas as the working substance, which is used to demonstrate that entropy is a state function. The Carnot cycle also provides a means to calculate the efficiency of a heat engine.

Catalyst - A substance that participates in a particular chemical reaction and thereby increases its rate but without a net change in the amount of that substance in the system. [3]

Catenanes, catena compounds - Hydrocarbons having two or more rings connected in the manner of links of a chain, without a covalent bond. More generally, the class catena compounds embraces functional derivatives and hetero analogues. [5]

Cation - A positively charged atomic or molecular particle.

Centipoise (cP) - A common non-SI unit of viscosity, equal to mPa s.

Centrifugal distortion - An effect in molecular spectroscopy in which rotational levels are lowered in energy, relative to the values of a rigid rotor, as the rotational angular momentum increases. The effect may be understood classically as a stretching of the bonds in the molecule as it rotates faster, thus increasing the moment of inertia.

Ceramic - A nonmetallic material of very high melting point.

Cerenkov radiation - Light emitted when a beam of charged particles travels through a medium at a speed greater than the speed of light in the medium. It is typically blue in color.

Cgs system of units - A system of units based upon the centimeter, gram, and second. The cgs system has been supplanted by the International System (SI).

Chalcogens - The Group VIA elements (oxygen, sulfur, selenium, tellurium, and polonium). Compounds of these elements are called chalcogenides. [7]

Chaotic system - A complex system whose behavior is governed by deterministic laws but whose evolution can vary drastically when small changes are made in the initial conditions.

Charge - See Electric charge.

Charles' law - The empirical law, exact only for an ideal gas, which states that the volume of a gas is directly proportional to its temperature at constant pressure.

Charm - A quantum number introduced in particle physics to account for certain properties of elementary particles and their reactions.

Chelate - A compound characterized by the presence of bonds from two or more bonding sites within the same ligand to a central metal atom. [3]

Chemical potential - For a mixture of substances, the chemical potential of constituent B is defined as the partial derivative of the Gibbs energy G with respect to the amount (number of moles) of B, with temperature, pressure, and amounts of all other constituents held constant. Also called partial molar Gibbs energy. [2]

Chemical shift* - A small change in the energy levels (and hence in the spectra associated with these levels) resulting from the effects of chemical binding in a molecule. The term is used in fields such as NMR, Mössbauer, and photoelectron spectroscopy, where the energy levels are determined primarily by nuclear or atomic effects.

Chiral molecule - A molecule which cannot be superimposed on its mirror image. A common example is an organic molecule containing a carbon atom to which four different atoms or groups are attached. Such molecules exhibit optical activity, i.e., they rotate the plane of a polarized light beam.

Chlorocarbons - Compounds consisting solely of chlorine and carbon. [5]

Chromatography* - A method for separation of the components of a sample in which the components are distributed between two phases, one of which is stationary while the other moves. In gas chromatography the gas moves over a liquid or solid stationary phase. In liquid chromatography the liquid mixture moves through another liquid, a solid, or a gel. The mechanism of separation of components may be adsorption, differential solubility, ion-exchange, permeation, or other mechanisms. [6]

Clapeyron equation - A relation between pressure and temperature of two phases of a pure substance that are in equilibrium, viz., $dp/dT = \Delta_{trs}S/\Delta_{trs}V$, where $\Delta_{trs}S$ is the difference in entropy between the phases and $\Delta_{trs}V$ the corresponding difference in volume.

Clathrates - Inclusion compounds in which the guest molecule is in a cage formed by the host molecule or by a lattice of host molecules. [5]

Clausius (Cl) - A non-SI unit of entropy or heat capacity defined as cal/K = 4.184 J/K. [2]

Clausius-Clapeyron equation - An approximation to the Clapeyron equation applicable to liquid-gas and solid-gas equilibrium, in which one assumes an ideal gas with volume much greater than the condensed phase volume. For the liquid-gas case, it takes the form $d(\ln p)/dT = \Delta_{vap}H/RT^2$, where R is the molar gas constant and $\Delta_{vap}H$ is the molar enthalpy of vaporization. For the solid-gas case, $\Delta_{vap}H$ is replaced by the molar enthalpy of sublimation, $\Delta_{sub}H$.

Clausius-Mosotti equation - A relation between the dielectric constant ε_r at optical frequencies and the polarizability α:

$$\frac{\varepsilon_r - 1}{\varepsilon_r + 2} = \frac{\rho N_A \alpha}{3M\varepsilon_0}$$

where ρ is density, N_A is Avogadro's number, M is molar mass, and ε_0 is the permittivity of a vacuum.

Clebsch-Gordon coefficients - A set of coefficients used to describe the vector coupling of angular momenta in atomic and nuclear physics.

Codon - A set of three bases, chosen from the four primary bases found in the DNA molecule (uracil, cytosine, adenine, and guanine), which specifies the production of a particular amino acid or carries some other genetic instruction. For example, the codon UCA specifies the amino acid serine, CAG specifies glutamine, etc. There are a total of 64 codons.

Coercive force - The magnetizing force at which the magnetic flux density is equal to zero. [10]

Coercivity* - The maximum value of coercive force that can be attained when a magnetic material is symmetrically magnetized to saturation induction. [10]

Coherent anti-Stokes Raman spectroscopy (CARS) - See Techniques for Materials Characterization, Section 12.

Colloid - Molecules or polymolecular particles dispersed in a medium that have, at least in one direction, a dimension roughly between 1 nm and 1 μm. [3]

Color center - A defect in a crystal that gives rise to optical absorption, thus changing the color of the material. A common type is the F-center, which results when an electron occupies the site of a negative ion.

Compressibility (κ)* - The fractional change of volume as pressure is increased, viz., $\kappa = -(1/V)(dV/dp)$. [1]

Compton wavelength (λ_C)* - In the scattering of electromagnetic radiation by a free particle (e.g., electron, proton), $\lambda_C = h/mc$ is the increase in wavelength, at a 90° scattering angle, corresponding to the transfer of energy from radiation to particle. Here h is Planck's constant, c the speed of light, and m the mass of the particle.

Conductance (G)* - For direct current, the reciprocal of resistance. More generally, the real part of admittance. [1]

Conductivity, electrical (σ)* - The reciprocal of the resistivity. [1]

Conductivity, thermal - See Thermal conductivity.

Congruent transformation - A phase transition (melting, vaporization, etc.) in which the substance preserves its exact chemical composition.

Constitutional repeating unit (CRU) - In polymer science, the smallest constitutional unit, the repetition of which constitutes a regular macromolecule, i.e., a macromolecule with all units connected identically with respect to directional sense. [8]

Copolymer - A polymer derived from more than one species of monomer. [8]

Coriolis effect - The deviation from simple trajectories when a mechanical system is described in a rotating coordinate system. It affects the motion of projectiles on the earth and in molecular spectroscopy leads to an important interaction between the rotational and vibrational motions. The effect may be described by an additional term in the equations of motion, called the Coriolis force.

Cosmic rays* - High energy nuclear particles, electrons, and photons, originating mostly outside the solar system, which continually bombard the earth's atmosphere.

Coulomb (C)* - The SI unit of electric charge, equal to A s. [1]

Coulomb's law - The statement that the force F between two electrical charges q_1 and q_2 separated by a distance r is $F = (4\pi\varepsilon_0)^{-1}q_1q_2/r^2$, where ε_0 is the permittivity of a vacuum.

Covalent bond - A chemical bond between two atoms whose stability results from the sharing of two electrons, one from each atom.

Cowling number (Co) - A dimensionless quantity used in plasma physics, defined by $Co = B^2/\mu\rho v^2$, where ρ is density, v is velocity, μ is permeability, and B is magnetic flux density. [2]

CPT theorem - A theorem in particle physics which states that any local Lagrangian theory that is invariant under proper Lorentz transformations is also invariant under the combined operations of charge conjugation, C, space inversion, P, and time reversal, T, taken in any order.

Critical point* - In general, the point on the phase diagram of a two-phase system at which the two coexisting phases have identical properties and therefore represent a single phase. At the liquid-gas critical point of a pure substance, the distinction between liquid and gas vanishes, and the vapor pressure curve ends. The coordinates of this point are called the critical temperature and critical pressure. Above the critical temperature, it is not possible to liquefy the substance.

Cross section (σ)* - A measure of the probability of collision (or other interaction) between a beam of particles and a target which it encounters. In rough terms it is the effective area the target particles present to the incident ones; however, the precise definition depends on the nature of the interaction. A general definition of σ is the number of encounters per unit time divided by nv, where n is the concentration of incident particles and v their velocity.

Crosslink - In polymer science, a small region in a macromolecule from which at least four chains emanate, and formed by reactions involving sites or groups on existing macromolecules or by interactions between existing macromolecules. [8]

Crown compounds - Macrocyclic polydentate compounds, usually uncharged, in which three or more coordinating ring atoms (usually oxygen or nitrogen) are or may become suitably close for easy formation of chelate complexes with metal ions or other cationic species. [5]

Crust* - The outer layer of the solid earth, above the Mohorovicic discontinuity. Its thickness averages about 35 km on the continents and about 7 km below the ocean floor.

Cryoscopic constant (E_f)* - The constant that expresses the amount by which the freezing point T_f of a solvent is lowered by a non-dissociating solute, through the relation $\Delta T_f = E_f m$, where m is the molality of the solute.

Curie (Ci) - A non-SI unit of radioactivity (disintegrations per unit time), equal to 3.7×10^{10} s^{-1}.

Curie temperature (T_C)* - For a ferromagnetic material, the critical temperature above which the material becomes paramagnetic. Also applied to the temperature at which the spontaneous polarization disappears in a ferroelectric solid. [1]

Cyanohydrins - Alcohols substituted by a cyano group, most commonly, but not limited to, examples having a CN and an OH group attached to the same carbon atom. They are formally derived from aldehydes or ketones by the addition of hydrogen cyanide. [5]

Cycloalkanes - Saturated monocyclic hydrocarbons (with or without side chains). See alicyclic compounds. Unsaturated monocyclic hydrocarbons having one endocyclic double or one triple bond are called cycloalkenes and cycloalkynes, respectively. [5]

Symbols

Symbols

Cyclotron resonance - The resonant absorption of energy from a system in which electrons or ions that are orbiting in a uniform magnetic field are subjected to radiofrequency or microwave radiation. The resonance frequency is given by $\nu = eH/2\pi m^*c$, where e is the elementary charge, H is the magnetic field strength, m^* is the effective mass of the charged particle, and c is the speed of light. The effect occurs in both solids (involving electrons or holes) and in low pressure gasses (involving ions)

Dalton (Da) - A name sometimes used in biochemistry for the unified atomic mass unit (u).

De Broglie wavelength - The wavelength associated with the wave representation of a moving particle, given by h/mv, where h is Planck's constant, m the particle mass, and v the velocity.

De Haas-Van Alphen effect - An effect observed in certain metals and semiconductors at low temperatures and high magnetic fields, characterized by a periodic variation of magnetic susceptibility with field strength.

Debye equation* - The relation between the relative permittivity (dielectric constant) ε_r, polarizability α, and permanent dipole moment μ in a dielectric material whose molecules are free to rotate. It takes the form

$$\frac{\varepsilon_r - 1}{\varepsilon_r + 2} = \frac{\rho N_A}{3M\varepsilon_0}\left(\alpha + \frac{\mu^2}{3kT}\right)$$

where ρ is density, N_A is Avogadro's number, M is molar mass, and ε_0 is the permittivity of a vacuum.

Debye length - In the Debye-Hückel theory of ionic solutions, the effective thickness of the cloud of ions of opposite charge which surrounds each given ion and shields the Coulomb potential produced by that ion.

Debye temperature (θ_D)* - In the Debye model of the heat capacity of a crystalline solid, $\theta_D = h\nu_D/k$, where h is Planck's constant, k is the Boltzmann constant, and ν_D is the maximum vibrational frequency the crystal can support. For $T \ll \theta_D$, the heat capacity is proportional to T^3.

Debye unit (D) - A non-SI unit of electric dipole moment used in molecular physics, equal to 3.335641×10^{-30} C m.

Debye-Waller factor (D) - The factor by which the intensity of a diffraction line is reduced because of lattice vibrations. [1]

Defect - Any departure from the regular structure of a crystal lattice. A Frenkel defect results when an atom or ion moves to an interstitial position and leaves behind a vacancy. A Schottky defect involves either a vacancy where the atom has moved to the surface or a structure where a surface atom has moved to an interstitial position.

Degree of polymerization - The number of monomeric units in a macromolecule or an oligomer molecule. [8]

Dendrite - A tree-like crystalline pattern often observed, for example, in ice crystals and alloys in which the crystal growth branches repeatedly.

Density (ρ)* - In the most common usage, mass density or mass per unit volume. More generally, the amount of some quantity (mass, charge, energy, etc.) divided by a length, area, or volume.

Density of states (NE, ρ) - The number of one-electron states in an infinitesimal interval of energy, divided by the range of that interval and by volume. [1]

Dew point* - The temperature at which liquid begins to condense as the temperature of a gas mixture is lowered. In meteorology, it is the temperature at which moisture begins to condense on a surface in contact with the air.

Diamagnetism - A type of magnetism characterized by a negative magnetic susceptibility, so that the material, when placed in an external magnetic field, becomes weakly magnetized in the direction opposite to the field. This magnetization is independent of temperature.

Diazo compounds - Compounds having the divalent diazo group, $=N^+=N^-$, attached to a carbon atom, e.g., $CH_2=N_2$ diazomethane. [5]

Dielectric constant (ε)* - Ratio of the electric displacement in a medium to the electric field strength. Also called permittivity. [1]

Dienes - Compounds that contain two fixed double bonds (usually assumed to be between carbon atoms). Dienes in which the two double-bond units are linked by one single bond are termed conjugated. [5]

Differential scanning calorimetry (DSC) - See Techniques for Materials Characterization, Section 12.

Differential thermal analysis (DTA) - See Techniques for Materials Characterization, Section 12.

Diffusion* - The migration of atoms, molecules, ions, or other particles as a result of some type of gradient (concentration, temperature, etc.).

Diopter - A unit used in optics, formally equal to m^{-1}. It is used in expressing dioptic power, which is the reciprocal of the focal length of a lens.

Dipole moment, electric (p,μ)* - For a distribution of equal positive and negative charge, the magnitude of the dipole moment vector is the positive charge multiplied by the distance between the centers of positive and negative charge distribution. The direction is given by the line from the center of negative charge to the center of positive charge.

Dipole moment, magnetic (m,μ) - Formally defined in electromagnetic theory as a vector quantity whose vector product with the magnetic flux density equals the torque. The magnetic dipole generated by a current I flowing in a small loop of area A has a magnetic moment of magnitude IA. In atomic and nuclear physics, a magnetic moment is associated with the angular momentum of a particle; e.g., an electron with orbital angular momentum $\boldsymbol{l}$ exhibits a magnetic moment of $-el/2m_e$ where e is the elementary charge and m_e the mass of the electron. [1]

Disaccharides - Compounds in which two monosaccharides are joined by a glycosidic bond. [5]

Dislocation - An extended displacement of a crystal from a regular lattice. An edge dislocation results when one portion of the crystal has partially slipped with respect to the other, resulting in an extra plane of atoms extending through part of the crystal. A screw dislocation transforms successive atomic planes into the surface of a helix.

Dispersion - Splitting of a beam of light (or other electromagnetic radiation) of mixed wavelengths into the constituent wavelengths as a result of the variation of refractive index of the medium with wavelength.

Dissociation constant* - The equilibrium constant for a chemical reaction in which a compound dissociates into its constituent parts.

Dissociation energy (D_e)* - For a diatomic molecule, the difference between the energies of the free atoms at rest and the minimum in the potential energy curve. The term bond dissociation energy (D_0), which can be applied to polyatomic molecules as well, is used for the difference between the energies of the fragments resulting when a bond is broken and the energy of the original molecule in its lowest energy state. The term bond strength implies differences in enthalpy rather than energy.

Domain - A small region of a solid in which the magnetic or electric moments of the individual units (atoms, molecules, or ions) are aligned in the same direction.

Domain wall - The transition region between adjacent ferromagnetic domains, generally a layer with a thickness of a few hundred ångström units. Also called Bloch wall.

Doppler effect - The change in the apparent frequency of a wave (sound, light, or other) when the source of the wave is moving relative to the observer.

Dose equivalent (*H*) - The product of the absorbed dose of radiation at a point of interest in tissue and various modifying factors which depend on the type of tissue and radiation. [1]

Drift velocity - The velocity of charge carriers (electrons, ions, etc.) moving under the influence of an electric field in a medium which subjects the carriers to some frictional force.

Dyne (dyn) - A non-SI (cgs) unit of force, equal to 10^{-5} N.

Ebullioscopic constant (E_b)* - The constant that expresses the amount by which the boiling point T_b of a solvent is raised by a non-dissociating solute, through the relation $\Delta T_b = E_b m$, where *m* is the molality of the solute.

Eddy currents - Circulating currents set up in conducting bulk materials or sheets by varying magnetic fields.

Effinghausen effect - The appearance of a temperature gradient in a current carrying conductor that is placed in a transverse magnetic field. The direction of the gradient is perpendicular to the current and the field.

Eigenvalue - An allowed value of the constant *a* in the equation $Au = au$, where *A* is an operator acting on a function *u* (which is called an eigenfunction). In quantum mechanics, the outcome of any observation is an eigenvalue of the corresponding operator. Also called characteristic value.

Einstein - A non-SI unit used in photochemistry, equal to one mole of photons.

Einstein temperature (θ_v) - In the Einstein theory of the heat capacity of a crystalline solid, $\theta_v = h\nu/k$, where *h* is Planck's constant, *k* is the Boltzmann constant, and ν is the vibrational frequency of the crystal.

Einstein transition probability - A constant in the Einstein relation $A_{ij} + B_{ij}\rho$ for the probability of a transition between two energy levels *i* and *j* in a radiation field of energy density ρ. The A_{ij} coefficient describes the probability of spontaneous emission, while B_{ij} and B_{ji} govern the probability of stimulated emission and absorption, respectively ($B_{ij} = B_{ji}$).

Elastic limit - The greatest stress which a material is capable of sustaining without any permanent strain remaining after complete release of the stress. [10]

Elastic modulus - See Young's modulus.

Electric charge (*Q*) - The quantity of electricity; i.e., the property that controls interactions between bodies through electrical forces.

Electric current (*I*) - The charge passing through a circuit per unit time. [1]

Electric displacement (*D*) - A vector quantity whose magnitude equals the electric field strength multiplied by the permittivity of the medium and whose direction is the same as that of the field strength.

Electric field strength (*E*) - The force exerted by an electric field on a point charge divided by the electric charge. [1]

Electric potential (*V*) - A scalar quantity whose gradient is equal to the negative of the electric field strength.

Electrical conductance - See Conductance.

Electrical resistance - See Resistance.

Electrical resistivity - See Resistivity.

Electrochemical series* - An arrangement of reactions which produce or consume electrons in an order based on standard electrode potentials. A common arrangement places metals in decreasing order of their tendency to give up electrons.

Electrode potential* - The electromotive force of a cell in which the electrode on the left is the standard hydrogen electrode and that on the right is the electrode in question. [2]

Electrolysis - The decomposition of a substance as a result of passing an electric current between two electrodes immersed in the sample.

Electromotive force (emf) - The energy supplied by a source divided by the charge transported through the source. [1]

Electron* - An elementary particle in the family of leptons, with negative charge and spin of ½.

Electron affinity* - The energy difference between the ground state of a gas-phase atom or molecule and the lowest state of the corresponding negative ion.

Electron cyclotron resonance (ECR) - See Techniques for Materials Characterization, Section 12.

Electron energy loss spectroscopy (EELS) - See Techniques for Materials Characterization, Section 12.

Electron nuclear double resonance (ENDOR) - See Techniques for Materials Characterization, Section 12.

Electron paramagnetic resonance (EPR) - See Techniques for Materials Characterization, Section 12.

Electron probe microanalysis (EPMA) - See Techniques for Materials Characterization, Section 12.

Electron spectroscopy for chemical analysis (ESCA) - See Techniques for Materials Characterization, Section 12.

Electron spin (*s*) - The quantum number, equal to 1/2, that specifies the intrinsic angular momentum of the electron.

Electron stimulated desorption (ESD) - See Techniques for Materials Characterization, Section 12.

Electron volt (eV)* - A non-SI unit of energy used in atomic and nuclear physics, equal to approximately 1.602177×10^{-19} J. The electron volt is defined as the kinetic energy acquired by an electron upon acceleration through a potential difference of 1 V. [1]

Electronegativity* - A parameter originally introduced by Pauling which describes, on a relative basis, the power of an atom or group of atoms to attract electrons from the same molecular entity. [3]

Electrophoresis - The motion of macromolecules or colloidal particles in an electric field. [3]

Emissivity (ε)* - Ratio of the radiant flux emitted per unit area to that of an ideal black body at the same temperature. Also called emittance. [1]

Emu - The electromagnetic system of units, based upon the cm, g, and s plus the emu of current (sometimes called the abampere).

Enantiomers - A chiral molecule and its non-superposable mirror image. The two forms rotate the plane of polarized

Symbols

light by equal amounts in opposite directions. Also called optical isomers.

Energy (*E,U*)* - The characteristic of a system that enables it to do work.

Energy gap* - In the theory of solids, the region between two energy bands, in which no bound states can occur.

Enols, alkenols - The term refers specifically to vinylic alcohols, which have the structure $HOCR'{=}CR_2$. Enols are tautomeric with aldehydes (R′ = H) or ketones (R′ not equal to H). [5]

Enthalpy (*H*)* - A thermodynamic function, especially useful when dealing with constant-pressure processes, defined by $H = E + PV$, where E is energy, P pressure, and V volume. [1]

Enthalpy of combustion* - The enthalpy change in a combustion reaction. Its negative is the heat released in combustion.

Enthalpy of formation, standard* - The enthalpy change for the reaction in which a substance is formed from its constituent elements, each in its standard reference state (normally refers to 1 mol, sometimes to 1 g, of the substance).

Enthalpy of fusion* - The enthalpy change in the transition from solid to liquid state.

Enthalpy of sublimation - The enthalpy change in the transition from solid to gas state.

Enthalpy of vaporization* - The enthalpy change in the transition from liquid to gas state.

Entropy (*S*)* - A thermodynamic function defined such that when a small quantity of heat dQ is received by a system at temperature T, the entropy of the system is increased by dQ/T, provided that no irreversible change takes place in the system. [1]

Entropy unit (e.u.) - A non-SI unit of entropy, equal to 4.184 J/K mol.

Ephemeris time - Time measured in tropical years from January 1, 1900.

Epoxy compounds - Compounds in which an oxygen atom is directly attached to two adjacent or non-adjacent carbon atoms of a carbon chain or ring system; thus cyclic ethers. [5]

Equation of continuity - Any of a class of equations that express the fact that some quantity (mass, charge, energy, etc.) cannot be created or destroyed. Such equations typically specify that the rate of increase of the quantity in a given region of space equals the net current of the quantity flowing into the region.

Equation of state* - An equation relating the pressure, volume, and temperature of a substance or system.

Equilibrium constant (*K*)* - For a chemical reaction $aA + bB \rightleftharpoons cC + dD$, the equilibrium constant is defined by:

$$K = \frac{a_C{}^c \cdot a_D{}^d}{a_A{}^a \cdot a_B{}^b}$$

where a_i is the activity of component i. To a certain approximation, the activities can be replaced by concentrations. The equilibrium constant is related to $\Delta_r G°$, the standard Gibbs energy change in the reaction, by $RT \ln K = -\Delta_r G°$.

Equivalent conductance - See Conductivity, electrical

Erg (erg) - A non-SI (cgs) unit of energy, equal to 10^{-7} J.

Esters - Compounds formally derived from an oxoacid RC(=O)(OH) and an alcohol, phenol, heteroarenol, or enol by linking, with formal loss of water from an acidic hydroxy group of the former and a hydroxy group of the latter. [5]

Esu - The electrostatic system of units, based upon the cm, g, and s plus the esu of charge (sometimes called the statcoulomb or franklin).

Ethers - Compounds with formula ROR , where R is not equal to H. [5]

Euler number (*Eu*) - A dimensionless quantity used in fluid mechanics, defined by $Eu = \Delta p/\rho v^2$, where p is pressure, ρ is density, and v is velocity. [2]

Eutectic - The point on a two-component solid-liquid phase diagram which represents the lowest melting point of any possible mixture. A liquid having the eutectic composition will freeze at a single temperature without change of composition.

Excitance (*M*) - Radiant energy flux leaving an element of a surface divided by the area of that element. [1]

Exciton - A localized excited state consisting of a bound electron-hole pair in a molecular or ionic crystal. The exciton can propagate through the crystal.

Exosphere - The outermost part of the earth's atmosphere, beginning at about 500 to 1000 km above the surface. It is characterized by densities so low that air molecules can escape into outer space.

Expansion coefficient - See Thermal expansion coefficient.

Extended electron energy loss fine structure (EXELFS) - See Techniques for Materials Characterization, Section 12.

Extended x-ray absorption fine structure (EXAFS) - See Techniques for Materials Characterization, Section 12.

Extinction coefficient - See Absorption coefficient, molar.

F-Center - See Color center.

Fahrenheit temperature (°F) - The temperature scale based on the assignment of 32 °F = 0 °C and a temperature interval of °F =(5/9)°C; i.e., t/°F = (9/5)t/°C + 32.

Farad (F)* - The SI unit of electric capacitance, equal to C/V. [1]

Faraday constant (*F*)* - The electric charge of 1 mol of singly charged positive ions; i.e., $F = N_A e$, where N_A is Avogadro's constant and e is the elementary charge. [1]

Faraday effect* - The rotation of the plane of plane-polarized light by a medium placed in a magnetic field parallel to the direction of the light beam. The effect can be observed in solids, liquids, and gases.

Fatty acids - Aliphatic monocarboxylic acids derived from or contained in esterified form in an animal or vegetable fat, oil, or wax. Natural fatty acids commonly have a chain of 4 to 28 carbons (usually unbranched and even-numbered), which may be saturated or unsaturated. By extension, the term is sometimes used to embrace all acyclic aliphatic carboxylic acids. [5]

Fermat's principle - The law that a ray of light traversing one or more media will follow a path which minimizes the time required to pass between two given points.

Fermi (f) - Name sometimes used in nuclear physics for the femtometer.

Fermi level - The highest energy of occupied states in a solid at zero temperature. Sometimes called Fermi energy. The Fermi surface is the surface in momentum space formed by electrons occupying the Fermi level.

Fermi resonance - An effect observed in vibrational spectroscopy when an overtone of one fundamental vibration closely coincides in energy with another fundamental of the same symmetry species. It leads to a splitting of vibrational bands.

Fermi-Dirac distribution - A modification of the Boltzmann distribution which takes into account the Pauli exclusion principle. The number of particles of energy E is proportional to $[e^{(E-\mu)/kT}+1]^{-1}$, where μ is a normalization constant, k the Boltzmann constant, and T the temperature. The distribution is applicable to a system of fermions.

Fermion - A particle that obeys Fermi-Dirac statistics. Specifically, any particle with spin equal to an odd multiple of ½. Examples are the electron, proton, neutron, muon, etc.

Ferrimagnetism* - A type of magnetism in which the magnetic moments of atoms in a solid are ordered into two nonequivalent sublattices with unequal magnetic moments, leading to a nonzero magnetic susceptibility.

Ferrite - A ferrimagnetic material of nominal formula MFe_2O_4, where M is a divalent metal; widely used in microwave switches and other solid state devices.

Ferroelectricity* - The retention of electric polarization by certain materials after the external field that produced the polarization has been removed.

Ferromagnetism* - A type of magnetism in which the magnetic moments of atoms in a solid are aligned within domains which can in turn be aligned with each other by a weak magnetic field. Some ferromagnetic materials can retain their magnetization when the external field is removed, as long as the temperature is below a critical value, the Curie temperature. They are characterized by a large positive magnetic susceptibility.

Fick's law - The statement that the flux J of a diffusing substance is proportional to the concentration gradient, i.e., $J = -D(dc/dx)$, where D is called the diffusion coefficient.

Field - A mathematical construct which describes the interaction between particles resulting from gravity, electromagnetism, or other physical phenomena. In classical physics a field is described by equations. Quantum field theory introduces operators to represent the physical observables.

Field emission microscopy (FEM) - See Techniques for Materials Characterization, Section 12.

Field ion microscopy (FIM) - See Techniques for Materials Characterization, Section 12.

Fine structure - The splitting in spectral lines that results from interactions of the electron spin with the orbital angular momentum.

Fine structure constant (α)* - Defined as $e^2/2hc\varepsilon_0$, where e is the elementary charge, h Planck's constant, c the speed of light, and ε_0 the permittivity of a vacuum. It is a measure of the strength of the electromagnetic interaction between particles.

First radiation constant (c_1)* - Constant (= $2\pi hc^2$) in the equation for the radiant excitance M_λ of a black body:

$$M_\lambda = \frac{c_1\lambda^{-5}\ \lambda}{e^{c_2/\lambda T}-1}$$

where λ is the wavelength, T is the temperature, and $c_2 = hc/k$ is the second radiation constant.

Flash point - The lowest temperature at which vapors above a volatile combustible substance will ignite in air when exposed to a flame. [10]

Fluence (F) - Term used in photochemistry to specify the energy per unit area delivered in a given time interval, for example by a laser pulse. [2]

Fluorocarbons - Compounds consisting solely of fluorine and carbon. [5]

Fluxoid - The quantum of magnetic flux in superconductivity theory, equal to $hc/2e$, where h is Planck's constant, c the velocity of light, and e the elementary charge.

Force (F) - The rate of change of momentum with time. [1]

Force constants (f, k)* - In molecular vibrations, the coefficients in the expression of the potential energy in terms of atom displacements from their equilibrium positions. In a diatomic molecule, $f = d^2V/dr^2$, where $V(r)$ is the potential energy and r is the interatomic distance. [2]

Fourier number (Fo) - A dimensionless quantity used in fluid mechanics, defined by $Fo = at/l^2$, where a is thermal diffusivity, t is time, and l is length. [2]

Fourier transform infrared spectroscopy (FTIR) - A technique for obtaining an infrared spectrum by use of an interferometer in which the path length of one of the beams is varied. A Fourier transformation of the resulting interferogram yields the actual spectrum. The technique is also used for NMR and other types of spectroscopy.

Fractals - Geometrical objects that are self-similar under a change of scale; i.e., they appear similar at all levels of magnification. They can be considered to have fractional dimensionality. Examples occur in diverse fields such as geography (rivers and shorelines), biology (trees), and solid state physics (amorphous materials).

Franck-Condon principle - An important principle in molecular spectroscopy which states that the nuclei in a molecule remain essentially stationary while an electronic transition is taking place. The physical interpretation rests on the fact that the electrons move much more rapidly than the nuclei because of their much smaller mass.

Franklin (Fr) - Name sometimes given to the unit of charge in the esu system.

Fraunhofer diffraction - Diffraction of light in situations where the source and observation point are so far removed that the wave surfaces may be considered planar.

Fraunhofer lines - Sharp absorption lines in the spectrum of sunlight, caused by absorption of the solar blackbody radiation by atoms near the sun's surface.

Free radical - See Radicals. The term "free radical" is often used more broadly for molecules that have a paramagnetic ground state (e.g., O_2) and sometimes for any transient or highly reactive molecular species.

Freezing point - See Melting point.

Frequency (ν)* - Number of cycles of a periodic phenomenon divided by time. [1]

Fresnel diffraction - Diffraction of light in a situation where the source and observation point are sufficiently close together that the curvature of the wave surfaces must be taken into account.

Froude number (Fr) - A dimensionless quantity used in fluid mechanics, defined by $Fr = v/(lg)^{1/2}$, where v is velocity, l is length, and g is acceleration due to gravity. [2]

Fugacity (f_B) - For a gas mixture, the fugacity of component B is defined as the absolute activity λ_B times the limit, as the pressure p approaches zero at constant temperature, of p_B/λ_B. [2]

Fullerenes - Compounds composed solely of an even number of carbon atoms, which form a cage-like fused-ring polycyclic system with twelve five-membered rings and the rest six-membered rings. The archetypal example is [60]fullerene, where the atoms and bonds delineate a truncated icosahedron. The term has been broadened to include any closed cage structure consisting entirely of three-coordinate carbon atoms. [5]

Fulvalenes - The hydrocarbon fulvalene and its derivatives formed by substitution (and by extension, analogues formed by replacement of one or more carbon atoms of the fulvalene skeleton by a heteroatom). [5]

Fulvenes - The hydrocarbon fulvene and its derivatives formed by substitution (and by extension, analogues formed by replacement of one or more carbon atoms of the fulvene skeleton by a heteroatom). [5]

Fundamental vibrational frequencies* - In molecular spectroscopy, the characteristic vibrational frequencies obtained when the vibrational energy is expressed in normal coordinates. They determine the primary features of the infrared and Raman spectra of the molecule.

γ - Name sometimes used for microgram.

γ-rays* - Electromagnetic radiation (photons) with energy greater than about 0.1 MeV (wavelength less than about 1 pm).

g-Factor of the electron* - The proportionality factor in the equation relating the magnetic moment μ of an electron to its total angular momentum quantum number J, i.e., $\mu = -g\mu_B J$, where μ_B is the Bohr magneton. Also called Landé factor.

Gal - A non-SI unit of acceleration, equal to 0.01 m/s. Also called galileo.

Gallon (US) - A unit of volume equal to 3.785412 L.

Gallon (UK, Imperial) - A unit of volume equal to 4.546090 L.

Gauss (G) - A non-SI unit of magnetic flux density (B) equal to 10^{-4} T.

Gaussian system of units - A hybrid system used in electromagnetic theory, which combines features of both the esu and emu systems.

Gel - A colloidal system with a finite, but usually rather small, yield stress (the sheer stress at which yielding starts abruptly). [3]

Genetic code* - The set of relations between each of the 64 codons of DNA and a specific amino acid (or other genetic instruction).

Gibbs energy (*G*)* - An important function in chemical thermodynamics, defined by $G = H\text{-}TS$, where H is the enthalpy, S the entropy, and T the thermodynamic temperature. Sometimes called Gibbs free energy and, in older literature, simply "free energy." [2]

Gibbs phase rule - The relation $F = C - P + 2$, where C is the number of components in a mixture, P is the number of phases, and F is the degrees of freedom, i.e., the number of intensive variables that can be changed independently without affecting the number of phases.

Glass transition temperature* - The temperature at which an amorphous polymer is transformed, in a reversible way, from a viscous or rubbery condition to a hard and relatively brittle one. [10]

Glow discharge mass spectroscopy (GDMS) - See Techniques for Materials Characterization, Section 12.

Gluon - A hypothetical particle postulated to take part in the binding of quarks, in analogy to the role of the photon in electromagnetic interactions.

Glycerides - Esters of glycerol (propane-1,2,3-triol) with fatty acids, widely distributed in nature. They are by long-established custom subdivided into triglycerides, 1,2- or 1,3-diglycerides, and 1- or 2-monoglycerides, according to the number and positions of acyl groups. [5]

Glycols - Dihydric alcohols in which two hydroxy groups are on different carbon atoms, usually but not necessarily adjacent. Also called diols. [5]

Grain (gr) - A non-SI unit of mass, equal to 64.79891 mg.

Grain boundary - The interface between two regions of different crystal orientation.

Grashof number (*Gr*) - A dimensionless quantity used in fluid mechanics, defined by $Gr = l^3 g\alpha\Delta T\rho^2/\eta^2$, where T is temperature, ρ is density, l is length, η is viscosity, α is cubic expansion coefficient, and g is acceleration of gravity. [2]

Gravitational constant (*G*)* - The universal constant in the equation for the gravitational force between two particles, $F = Gm_1m_2/r^2$, where r is the distance between the particles and m_1 and m_2 are their masses. [1]

Gray (Gy)* - The SI unit of absorbed dose of radiation, equal to J/kg. [1]

Gregorian calendar - The modification of the Julian calendar introduced in 1582 by Pope Gregory XII which specified that a year divisible by 100 is a leap year only if divisible by 400.

Grignard reagents - Organomagnesium halides, RMgX, having a carbon–magnesium bond (or their equilibrium mixtures in solution with $R_2Mg + MgX_2$). [5]

Gruneisen parameter (γ) - Defined by $\gamma = \alpha_V/\kappa c_V\rho$, where α_V is the cubic thermal expansion coefficient, κ is the isothermal compressibility, c_V is the specific heat capacity at constant volume, and ρ is the mass density. γ is independent of temperature for most crystalline solids. [1]

Gyromagnetic ratio (γ) - Ratio of the magnetic moment of a particle to its angular momentum. Also called magnetogyric ratio.

Hadron - Any elementary particle that can take part in the strong interaction. Hadrons are subdivided into baryons, with odd half integer spins, and mesons, which have zero or integral spin.

Hall effect* - The development of a transverse potential difference V in a conducting material when subjected to a magnetic field H perpendicular to the direction of the current. The potential difference is given by $V = R_H BJt$, where B is the magnetic induction, J the current density, t the thickness of the specimen in the direction of the potential difference, and R_H is called the Hall coefficient.

Halocarbon - A compound containing no elements other than carbon, hydrogen, and one or more halogens. In common practice, the term is used mainly for compounds of no more than four or five carbon atoms.

Halogens - The elements F, Cl, Br, I, and At. Compounds of these elements are called halogenides or halides. [7]

Hamiltonian (*H*) - An expression for the total energy of a mechanical system in terms of the momenta and positions of constituent particles. In quantum mechanics, the Hamiltonian operator appears in the eigenvalue equation $H\psi = E\psi$, where E is an energy eigenvalue and ψ the corresponding eigenfunction.

Hardness* - The resistance of a material to deformation, indentation, or scratching. Hardness is measured on various scales, such as Mohs, Brinell, Knoop, Rockwell, and Vickers. [10]

Hartmann number (*Ha*) - A dimensionless quantity used in plasma physics, defined by $Ha = Bl(\kappa/\eta)^{1/2}$, where B is magnetic flux density, l is length, κ is electric conductivity, and η is viscosity. [2]

Hartree (E_h)* - An energy unit used in atomic and molecular science, equal to approximately 4.3597482 × 10^{-18} J.

Hartree-Fock method - A iterative procedure for solving the Schrödinger equation for an atom or molecule in which the equation is solved for each electron in an initial assumed potential from all the other electrons. The new potential that results is used to repeat the calculation and the procedure continued until convergence is reached. Also called self-consistent field (SCF) method.

Heat capacity* - Defined in general as dQ/dT, where dQ is the amount of heat that must be added to a system to increase its temperature by a small amount dT. The heat capacity at constant pressure is $C_p = (\partial H/\partial T)_p$; that at constant volume is $C_V = (\partial E/\partial T)_V$, where H is enthalpy, E is internal energy, p is pressure, V is volume, and T is temperature. An upper case C normally indicates the molar heat capacity, while a lower case c is used for the specific (per unit mass) heat capacity. [1]

Heat of formation, vaporization, etc. - See corresponding terms under Enthalpy.

Hectare (ha) - A unit of area equal to 10^4 m^2. [1]

Heisenberg uncertainty principle - The statement that two observable properties of a system that are complementary, in the sense that their quantum-mechanical operators do not commute, cannot be specified simultaneously with absolute precision. An example is the position and momentum of a particle; according to this principle, the uncertainties in position Δq and momentum Δp must satisfy the relation $\Delta p \Delta q \geq h/4\pi$, where h is Planck's constant.

Heitler-London model - An early quantum-mechanical model of the hydrogen atom which introduced the concept of the exchange interaction between electrons as the primary reason for stability of the chemical bond.

Helicon - A low-frequency wave generated when a metal at low temperature is exposed to a uniform magnetic field and a circularly polarized electric field.

Helmholz energy (A) - A thermodynamic function defined by $A = E\text{-}TS$, where E is the energy, S the entropy, and T the thermodynamic temperature. [2]

Hemiacetals - Compounds having the general formula $R_2C(OH)OR'$ (R' not equal to H). [5]

Henry (H)* - The SI unit of inductance, equal to Wb/A. [1]

Henry's law* - An expression which applies to an ideal dilute solution in which one or more gasses are dissolved, viz., $p_i = H_i x_i$, where p_i is the partial pressure of component i above the solution, x_i is its mole fraction in the solution, and H_i is the Henry's law constant (a characteristic of the given gas and solvent, as well as the temperature).

Hermitian operator - An operator A that satisfies the relation $\int u_m{}^*Au_n dx = (\int u_n{}^*Au_m dx)^*$, where * indicates the complex conjugate. The eigenvalues of Hermitian operators are real, and eigenfunctions belonging to different eigenvalues are orthogonal.

Hertz (Hz) - The SI unit of frequency, equal to s^{-1}. [1]

Heterocyclic compounds - Cyclic compounds having as ring members atoms of at least two different elements, e.g., quinoline, 1,2-thiazole, bicyclo[3.3.1]tetrasiloxane. [5]

Heusler alloys - Alloys of manganese, copper, aluminum, nickel, and sometimes other metals which find important uses as permanent magnets.

Holography - A technique for creating a three-dimensional image of a object by recording the interference pattern between a light beam diffracted from the object and a reference beam. The image can be reconstructed from this pattern by a suitable optical system.

Homopolymer - A polymer derived from one species of (real, implicit, or hypothetical) monomer. [8]

Hooke's law - The statement that the ratio of stress to strain is a constant in a totally elastic medium.

Horse power - A non-SI unit of energy, equal to approximately 746 W.

Hubble constant - The ratio of the recessional velocity of an extragalactic object to the distance of that object. Its value is about 2 × 10^{-18} s^{-1}.

Huckel theory - A simple approximation for calculating the energy of conjugated molecules in which only the resonance integrals between neighboring bonds are considered. Also called CNDO method (complete neglect of differential overlap).

Hume-Rothery rules - A set of empirical rules for predicting the occurrence of solid solutions in metallic systems. The rules involve size, crystal structure, and electronegativity.

Hund's rules - A series of rules for predicting the sequence of energy states in atoms and molecules. One of the important results is that when two electrons exist in different orbitals, the state with their spins parallel (triplet state) lies at lower energy than the state with antiparallel spins (singlet).

Hydrazines - Hydrazine (diazane), H_2NNH_2, and its hydrocarbyl derivatives. When one or more substituents are acyl groups, the compound is a hydrazide. [5]

Hydrocarbon - A compound containing only carbon and hydrogen. [5]

Hydrolysis - A reaction occurring in water in which a chemical bond is cleaved and a new bond formed with the oxygen atom of water.

Hyperfine structure - Splitting of energy levels and spectral lines into several closely spaced components as a result of interaction of nuclear spin angular momentum with other angular momenta in the atom or molecule.

Hysteresis* - An irreversible response of a system (parameter A) as a function of an external force (parameter F), usually symmetric with respect to the origin of the A vs. F graph after the initial application of the force. A common example is magnetic induction vs. magnetic field strength in a ferromagnet.

Ideal gas law - The equation of state $pV = RT$, which defines an ideal gas, where p is pressure, V molar volume, T temperature, and R the molar gas constant.

Ideal solution - A solution in which solvent-solvent and solvent-solute interactions are identical, so that properties such as volume and enthalpy are exactly additive. Ideal solutions follow Raoult's law, which states that the vapor pressure p_i of component i is $p_i = x_i p_i{}^*$, where x_i is the mole fraction of component i and $p_i{}^*$ the vapor pressure of the pure substance i.

Ignition temperature* - The lowest temperature at which combustion of a material will occur spontaneously under specified conditions. Sometimes called autoignition temperature, kindling point. [10]

Imides - Diacyl derivatives of ammonia or primary amines, especially those cyclic compounds derived from diacids. Also used for salts having the anion $RN_2{}^-$. [5]

Impedance (Z) - The complex representation of potential difference divided by the complex representation of current. In terms of reactance X and resistance R, the impedance is given by $Z = R + iX$. [1]

Index of refraction (n)* - For a non-absorbing medium, the ratio of the velocity of electromagnetic radiation *in vacuo* to the phase velocity of radiation of a specified frequency in the medium. [1]

Inductance - The ratio of the electromagnetic force induced in a coil by a current to the rate of change of the current.

Inductive coupled plasma mass spectroscopy (ICPMS) - See Techniques for Materials Characterization, Section 12.

Inertial defect - In molecular spectroscopy, the quantity I_c-I_a-I_b for a molecule whose equilibrium configuration is planar, where I_a, I_b, and I_c are the effective principal moments of inertia. The inertial defect for a rigid planar molecule would be zero, but vibration–rotation interactions in a real molecule lead to a positive inertial defect.

Insulator - A material in which the highest occupied energy band (valence band) is completely filled with electrons, while the next higher band (conduction band) is empty. Solids with an energy gap of 5 eV or more are generally considered as insulators at room temperature. Their conductivity is less than 10^{-6} S/m and increases with temperature.

Intercalation compounds - Compounds resulting from reversible inclusion, without covalent bonding, of one kind of molecule in a solid matrix of another compound, which has a laminar structure. The host compound, a solid, may be macromolecular, crystalline, or amorphous. [5]

International System of Units (SI)* - The unit system adopted by the General Conference on Weights and Measures in 1960. It consists of seven base units (meter, kilogram, second, ampere, kelvin, mole, candela), plus derived units and prefixes. [1]

International Temperature Scale (ITS-90)* - The official international temperature scale adopted in 1990. It consists of a set of fixed points and equations which enable the thermodynamic temperature to be determined from operational measurements. [9]

Ion - An atomic or molecular particle having a net electric charge. [3]

Ion exchange - A process involving the adsorption of one or several ionic species accompanied by the simultaneous desorption (displacement) of one or more other ionic species. [3]

Ion neutralization spectroscopy (INS) - See Techniques for Materials Characterization, Section 12.

Ionic strength (I) - A measure of the total concentration of ions in a solution, defined by $I = 1/2\Sigma_i z_i^2 m_i$, where z_i is the charge of ionic species i and m_i is its molality. For a 1-1 electrolyte at molality m, $I = m$.

Ionization constant* - The equilibrium constant for a reaction in which a substance in solution dissociates into ions.

Ionization potential* - The minimum energy required to remove an electron from an isolated atom or molecule (in its vibrational ground state) in the gaseous phase. More properly called ionization energy. [3]

Irradiance (E) - The radiant energy flux incident on an element of a surface, divided by the area of that element. [1]

Isentropic process - A thermodynamic process in which the entropy of the system does not change.

Ising model - A model describing the coupling between two atoms in a ferromagnetic lattice, in which the interaction energy is proportional to the negative of the product of the spin components along a specified axis.

Isobar - A line connecting points of equal pressure on a graphical representation of a physical system.

Isochore - A line or surface of constant volume on a graphical representation of a physical system.

Isoelectric point* - The pH of a solution or dispersion at which the net charge on the macromolecules or colloidal particles is zero. In electrophoresis there is no motion of the particles in an electric field at the isoelectric point.

Isomers - In chemistry, compounds that have identical molecular formulas but differ in the nature or sequence of bonding of their atoms or in the arrangement of their atoms in space. In physics, nuclei of the same atomic number Z and mass number A but in different energy states. [3]

Isomorphs - Substances of different chemical nature but having the same crystal structure.

Isotactic macromolecule - A tactic macromolecule, essentially comprising only one species of repeating unit which has chiral or prochiral atoms in the main chain in a unique arrangement with respect to its adjacent constitutional units. [8]

Isotherm - A line connecting points of equal temperature on a graphical representation of a physical system.

Isothermal process - A thermodynamic process in which the temperature of the system does not change.

Isotones - Nuclides having the same neutron number N but different atomic number Z. [3]

Isotopes - Two or more nuclides with the same atomic number Z but different mass number A. The term is sometimes used synonymously with nuclide, but it is preferable to reserve the word nuclide for a species of specific Z and A. [3]

Jahn-Teller effect - An interaction of vibrational and electronic motions in a nonlinear molecule which removes the degeneracy of certain electronic energy levels. It can influence the spectrum, crystal structure, and magnetic properties of the substance.

Johnson noise - Electrical noise generated by random thermal motion of electrons in a conductor or semiconductor. Also called thermal noise.

Josephson effect - The tunneling of electron pairs through a thin insulating layer which separates two superconductors. When a potential difference is applied to the superconductors, an alternating current is generated whose frequency is precisely proportional to the potential difference. This effect has important applications in metrology and determination of fundamental physical constants.

Joule (J)* - The SI unit of energy, equal to N m. [1]

Joule-Thomson coefficient (μ) - A parameter which describes the temperature change when a gas expands adiabatically through a nozzle from a high pressure to a low pressure region. It is defined by $\mu = (\partial T/\partial p)_H$, where H is enthalpy.

Julian calendar - The calendar introduced by Julius Caesar in 46 B.C. which divided the year into 365 days with a leap year of 366 days every fourth year.

Julian date (JD) - The number of days elapsed since noon Greenwich Mean Time on January 1, 4713 B.C. Thus January 1, 2000, 0h (midnight) will be JD 2,451,543.5.

This dating system was introduced by Joseph Scaliger in 1582.

Kaon - One of the elementary particles in the family of mesons. Kaons have a spin of zero and may be neutral or charged.

Kelvin (K)* - The SI base unit of thermodynamic temperature. [1]

Kepler's laws - The three laws of planetary motion, which established the elliptical shape of planetary orbits and the relation between orbital dimensions and the period of rotation.

Kerr effect* - An electrooptical effect in which birefringence is induced in a liquid or gas when a strong electric field is applied perpendicular to the direction of an incident light beam. The Kerr constant k is given by n_1-n_2 = $k\lambda E^2$, where λ is the wavelength, E is the electric field strength, and n_1 and n_2 are the indices of refraction of the ordinary and extraordinary rays, respectively.

Ketenes - Compounds in which a carbonyl group is connected by a double bond to an alkylidene group: $R_2C{=}C{=}O$. [5]

Ketones - Compounds in which a carbonyl group is bonded to two carbon atoms: $R_1R_2C{=}O$ (neither R may be H). [5]

Kilogram (kg)* - The SI base unit of mass. [1]

Kinetic energy (E_k, T) - The energy associated with the motion of a system of particles in a specified reference frame. For a single particle of mass m moving at velocity v, $E_k = \frac{1}{2}mv^2$.

Kirchhoff's laws - Basic rules for electric circuits, which state (a) the algebraic sum of the currents at a network node is zero and (b) the algebraic sum of the voltage drops around a closed path is zero.

Klein-Gordon equation - A relativistic extension of the Schrödinger equation.

Klein-Nishima formula - An expression for the scattering cross section of a photon by an unbound electron, based upon the Dirac electron theory.

Knight shift - The change in magnetic resonance frequency of a nucleus in a metal relative to the same nucleus in a diamagnetic solid. The effect is due to the polarization of the conduction electrons in the metal.

Knudsen number (Kn) - A dimensionless quantity used in fluid mechanics, defined by $Kn = \lambda/l$, where λ is mean free path and l is length. [2]

Kondo effect - A large increase in electrical resistance observed at low temperatures in certain dilute alloys of a magnetic metal in a nonmagnetic material.

Kramers-Kronig relation - A set of equations relating the real and imaginary parts of the index of refraction of a medium

Lactams - Cyclic amides of amino carboxylic acids, having a 1-azacycloalkan-2-one structure, or analogues having unsaturation or heteroatoms replacing one or more carbon atoms of the ring. [5]

Lactones - Cyclic esters of hydroxy carboxylic acids, containing a 1-oxacycloalkan-2-one structure, or analogues having unsaturation or heteroatoms replacing one or more carbon atoms of the ring. [5]

Lagrangian function (L) - A function used in classical mechanics, defined as the kinetic energy minus the potential energy for a system of particles.

Lamb shift - The small energy difference between the $^2S_{1/2}$ and $^2P_{1/2}$ levels in the hydrogen atom, which results from interactions between the electron and the radiation field.

Laminar flow - Smooth, uniform, non-turbulent flow of a gas or liquid in parallel layers, with little mixing between layers. It is characterized by small values of the Reynolds number.

Landé *g*-factor - See *g*-Factor of the electron

Langevin function - The mathematical function $L(x) = (e^x+e^{-x})/(e^x-e^{-x})-1/x$, which occurs in the expression for the average dipole moment of a group of rotating polar molecules in an electric field: $\mu_{av} = \mu L(\mu E/kT)$, where μ is the electric dipole moment of a single molecule, E is the electric field strength, k is the Boltzmann constant, and T is the temperature.

Lanthanides - The elements of atomic number 57 through 71, which share common chemical properties: La, Ce, Pr, Nd, Pm, Sm, Eu, Gd, Tb, Dy, Ho, Er, Tm, Yb, Lu. [7]

Larmor frequency (ν_L) - The precession frequency of a magnetic dipole in an applied magnetic field. In particular, a nucleus in a magnetic field of strength B has a Larmor frequency of $\gamma B/2\pi$, where γ is the magnetogyric ratio of the nucleus.

Laser* - A device in which an optical cavity is filled with a medium where a population inversion can be produced by some means. When the resonant frequency of the cavity bears the proper relation to the separation of the inverted energy levels, stimulated emission occurs, producing a highly monochromatic, coherent beam of light.

Laser ionization mass spectroscopy (LIMS) - See Techniques for Materials Characterization, Section 12.

Lattice constants* - Parameters specifying the dimensions of a unit cell in a crystal lattice, specifically the lengths of the cell edges and the angles between them.

Lattice energy* - The energy per ion pair required to separate completely the ions in a crystal lattice at a temperature of absolute zero.

Laue diagram - A diffraction pattern produced when an x-ray beam passes through a thin slice of a crystal and impinges on a detector behind the crystal.

Lenz's law - The statement that the current induced in a circuit by a change in magnetic flux is so directed as to oppose the change in flux.

Leonard-Jones potential - A simple but useful function for approximating the interaction between two neutral atoms or molecules separated by a distance r by writing the potential energy as $U(r) = 4\varepsilon\{(r_0/r)^{12} - (r_0/r)^6\}$, where ε and r_0 are adjustable parameters. In this form the depth of the potential well is ε and the minimum occurs at $2^{1/6}r_0$. The $(1/r)^{12}$ term is often replaced by other powers of $1/r$.

Lepton - One of the class of elementary particles that do not take part in the strong interaction. Included are the electron, muon, and neutrino. All leptons have a spin of ½.

Lewis number (Le) - A dimensionless quantity used in fluid mechanics, defined by $Le = a/D$, where a is thermal diffusivity and D is diffusion coefficient. [2]

Ligand field theory - A description of the structure of crystals containing a transition metal ion surrounded by nonmetallic ions (ligands). It is based on construction of molecular orbitals involving the *d*-orbitals of the central metal ion and combinations of atomic orbitals of the ligands.

Light year (l.y.) - A unit of distance used in astronomy, defined as the distance light travels in one year in a vacuum. Its approximate value is 9.46073×10^{15} m.

Lignins - Macromolecular constituents of wood related to lignans, composed of phenolic propylbenzene skeletal units, linked at various sites and apparently randomly. [5]

Ligroin - The petroleum fraction consisting mostly of C_7 and C_8 hydrocarbons and boiling in the range 90-140 °C; commonly used as a laboratory solvent.

Lipids - A loosely defined term for substances of biological origin that are soluble in nonpolar solvents. They consist of saponifiable lipids, such as glycerides (fats and oils) and phospholipids, as well as nonsaponifiable lipids, principally steroids. [5]

Lipoproteins - Clathrate complexes consisting of a lipid enwrapped in a protein host without covalent binding, in such a way that the complex has a hydrophilic outer surface consisting of all the protein and the polar ends of any phospholipids. [5]

Liter (L)* - A synonym for cubic decimeter. [1]

Lithosphere* - The outer layer of the solid earth, extending from the base of the mantle to the surface of the crust.

Lorentz contraction - The reduction in length of a moving body in the direction of motion, given by the factor $(1-v^2/c^2)^{1/2}$, where v is the velocity of the body and c the velocity of light. Also known as the FitzGerald-Lorentz contraction.

Lorentz force - The force exerted on a point charge Q moving at velocity $\boldsymbol{v}$ in the presence of external fields $\boldsymbol{E}$ and $\boldsymbol{B}$. It is given (in SI units) by $\boldsymbol{F} = Q(\boldsymbol{E} + \boldsymbol{v} \times \boldsymbol{B})$.

Loss angle (δ) - For a dielectric material in an alternating electromagnetic field, δ is the phase difference between the current and the potential difference. The function tan δ is a measure of the ratio of the power dissipated in the dielectric to the power stored.

Low energy electron diffraction (LEED) - See Techniques for Materials Characterization, Section 12.

Lumen (lm)* - The SI unit of luminous flux, equal to cd sr. [1]

Luminous flux (Φ) - The intensity of light from a source multiplied by the solid angle. The SI unit is lumen. [1]

Lux (lx)* - The SI unit of illuminance, equal to cd sr m^{-2}. [1]

Lyddane-Sachs-Teller relation - A relation between the phonon frequencies and dielectric constants of an ionic crystal which states that $(\omega_T/\omega_L)^2 = \varepsilon(\infty)/\varepsilon(0)$, where ω_T is the angular frequency of transverse optical phonons, ω_L that of longitudinal optical phonons, $\varepsilon(0)$ is the static dielectric constant, and $\varepsilon(\infty)$ the dielectric constant at optical frequencies.

Lyman series - The series of lines in the spectrum of the hydrogen atom which corresponds to transitions between the ground state (principal quantum number $n = 1$) and successive excited states. The wavelengths are given by $1/\lambda = R_H(1-1/n^2)$, where $n = 2,3,4,...$ and R_H is the Rydberg constant for hydrogen. The first member of the series ($n = 1 \leftrightarrow 2$), which is often called the Lyman-α line, falls at a wavelength of 1216 Å, and the series converges at 912 Å, the ionization limit of hydrogen.

Mach number (Ma) - A dimensionless quantity used in fluid mechanics, defined by $Ma = v/c$, where v is velocity and c is the speed of sound. [2]

Macromolecule - A molecule of high relative molecular mass (molecular weight), the structure of which essentially comprises the multiple repetition of units derived, actually or conceptually, from molecules of low relative molecular mass. [8]

Madelung constant* - A constant characteristic of a particular crystalline material which gives a measure of the electrostatic energy binding the ions in the crystal.

Magnetic field strength (H) - An axial vector quantity, the curl of which is equal to the current density, including the displacement current. [1]

Magnetic induction (B) - An axial vector quantity such that the force exerted on an element of current is equal to the vector product of this element and the magnetic induction. [1]

Magnetic moment - See Dipole moment, magnetic.

Magnetic susceptibility (χ_m, κ)* - Defined by $\chi_m = (\mu-\mu_0)/\mu_0$, where μ is the permeability of the medium and μ_0 the permeability of a vacuum. [1]

Magnetization (M) - Defined by $M = (B/\mu_0)-H$, where B is magnetic induction, H magnetic field strength, and μ_0 the permeability of a vacuum. [1]

Magnetogyric ratio (γ) - Ratio of the magnetic moment of a particle to its angular momentum. Also called gyromagnetic ratio.

Magneton - See Bohr magneton, Nuclear magneton.

Magnetostriction* - The change in dimensions of a solid sample when it is placed in a magnetic field.

Magnon - A quantum of magnetic energy associated with a spin wave in a ferromagnetic or antiferromagnetic crystal.

Mantle - The layer of the earth between the crust and the liquid outer core, which begins about 2900 km below the Earth's surface.

Maser - A device in which a microwave cavity is filled with a medium where a population inversion can be produced by some means. When the resonant frequency of the cavity bears the proper relation to the separation of the inverted energy levels, the device can serve as an amplifier or oscillator at that frequency.

Mass (m)* - Quantity of matter. Mass can also be defined as "resistance to acceleration."

Mass defect (B) - Defined by $B = Zm(^1H) + Nm_n - m_a$, where Z is the atomic number, $m(^1H)$ is the mass of the hydrogen atom, N is the neutron number, m_n is the rest mass of the neutron, and m_a is the mass of the atom in question. Thus Bc^2 can be equated to the binding energy of the nucleus if the binding energy of atomic electrons is neglected. [1]

Mass excess (Δ) - Defined by $\Delta = m_a - Am_u$, where m_a is the mass of the atom, A the number of nucleons, and m_u the unified atomic mass constant ($m_u = 1$ u). [1]

Mass fraction (w_B) - The ratio of the mass of substance B to the total mass of a mixture. [1]

Mass number (A) - A characteristic property of a specific isotope of an element, equal to the sum of the number of protons and neutrons in the nucleus.

Mass spectrometry - An analytical technique in which ions are separated according to the mass/charge ratio and detected by a suitable detector. The ions may be produced by electron impact on a gas, a chemical reaction, energetic vaporization of a solid, etc. [6]

Massieu function - A thermodynamic function defined by $J = -A/T$, where A is the Helmholz energy and T the thermodynamic temperature. [2]

Matthiessen's rule - The statement that the electrical resistivity ρ of a metal can be written as $\rho = \rho_L + \rho_i$, where ρ_L is due to scattering of conduction electrons by lattice vibrations and ρ_i to scattering by impurities and imperfections. If the impurity concentration is small, ρ_i is temperature independent.

Maxwell (Mx)* - A non-SI unit of magnetic field strength (H) equal to 10^{-8} Wb. [1]

Maxwell's equations - The fundamental equations of electromagnetism. In a form appropriate to SI units, they are:

$$\text{curl}\, \boldsymbol{H} = \partial \boldsymbol{D}/\partial t + \boldsymbol{j}$$
$$\text{div}\, \boldsymbol{B} = 0$$
$$\text{curl}\, \boldsymbol{E} = -\partial \boldsymbol{B}/\partial t$$
$$\text{div}\, \boldsymbol{D} = \rho$$

where $\boldsymbol{H}$ is the magnetic field strength, $\boldsymbol{B}$ the magnetic induction, $\boldsymbol{E}$ the electric field strength, $\boldsymbol{D}$ the electric displacement, $\boldsymbol{j}$ the current density, ρ the charge density, and t is time.

Maxwell-Boltzmann distribution - An expression for the fraction of molecules $f(v)$ in a gas that have velocity v within a specified interval. It takes the form

$$f(v) = 4\pi(M/2\pi RT)^{3/2} v^2 e^{-Mv^2/2RT}$$

where M is the molar mass, R the molar gas constant, and T the temperature.

Mean free path* - The average distance a gas molecule travels between collisions.

Meissner effect - The complete exclusion of magnetic induction from the interior of a superconductor.

Melting point* - The temperature at which the solid and liquid phases of a substance are in equilibrium at a specified pressure (normally taken to be atmospheric unless stated otherwise).

Mercaptans - A traditional term abandoned by IUPAC, synonymous with thiols. This term is still widely used. [5]

Meson - Any elementary particle that has zero or integral spin. Mesons are responsible for the forces between protons and neutrons in the nucleus.

Mesosphere - The part of the Earth's atmosphere extending from the top of the stratosphere (about 50 km above the surface) to 80–90 km. It is characterized by a decrease in temperature with increasing altitude.

Metal - A material in which the highest occupied energy band (conduction band) is only partially filled with electrons. The electrical conductivity of metals generally decreases with temperature.

Metallocenes - Organometallic coordination compounds in which one atom of a transition metal such as iron, ruthenium or osmium is bonded to and only to the face of two cyclopentadienyl ligands which lie in parallel planes. [5]

Meter (m)* - The SI base unit of length. [1]

Methine group - In organic compounds, the -C= group. [5]

Mho - An archaic name for the SI unit siemens (reciprocal ohm).

Micelle - A particle formed by the aggregation of surfactant molecules (typically, 10 to 100 molecules) in solution. For aqueous solutions, the hydrophilic end of the molecule is on the surface of the micelle, while the hydrophobic end (often a hydrocarbon chain) points toward the center. At the critical micelle concentration (cmc) the previously dissolved molecules aggregate into a micelle.

Micron (*μ*) - An obsolete name for micrometer.

Mie scattering - The scattering of light by spherical dielectric particles whose diameter is comparable to the wavelength of the light.

Milky way - The band of light in the night sky resulting from the stars in the galactic plane. The term is also used to denote the galaxy in which the sun is located.

Miller indices (*hkl*) - A set of indices used to label planes in a crystal lattice. [2]

Millimeter of mercury (mmHg) - A non-SI unit of pressure, equal to 133.322 Pa. The name is generally considered interchangeable with torr.

Mobility (*μ*)* - In solid state physics, the drift velocity of electrons or holes in a solid divided by the applied electric field strength. The term is used in a similar sense in other fields.

Molality (*m*) - A measure of concentration of a solution in which one states the amount of substance (i.e., number of moles) of solute per kilogram of solvent. Thus a 0.1 molal solution (often written as 0.1 m) has m = 0.1 mol/kg.

Molar mass - The mass of one mole of a substance. It is normally expressed in units of g/mol, in which case its numerical value is identical with the molecular weight (relative molecular mass). [1]

Molar quantity - It is often convenient to express an extensive quantity (e.g., volume, enthalpy, heat capacity, etc.) as the actual value divided by amount of substance (number of moles). The resulting quantity is called molar volume, molar enthalpy, etc.

Molar refraction (*R*) - A property of a dielectric defined by the equation $R = V_m[(n^2-1)/(n^2+2)]$, where n is the index of refraction of the medium (at optical wavelengths) and V_m the molar volume. It is related to the polarizability α of the molecules that make up the medium by the Lorenz-Lorentz equation, $R = N_A\alpha/3\varepsilon_0$, where N_A is Avogadro's constant and ε_0 is the permittivity of a vacuum.

Molarity (*c*) - A measure of concentration of a solution in which one states the amount of substance (i.e., number of moles) of solute per liter of solution. Thus a 0.1 molar solution (often referred to as 0.1 M) has a concentration c = 0.1 mol/L.

Mole (mol)* - The SI base unit of amount of substance. [1]

Mole fraction (x_B) - The ratio of the amount of substance (number of moles) of substance B to the total amount of substance in a mixture. [1]

Molecular orbital - See Orbital.

Molecular weight (M_r)* - The ratio of the average mass per molecule or specified entity of a substance to 1/12 of the mass of nuclide ^{12}C. Also called relative molar (or molecular) mass. [1]

Moment of inertia (*I*) - The moment of inertia of a body about an axis is the sum (or integral) of the products of its elements of mass and the squares of their distances from the axis. [1]

Momentum (*p*) - The product of mass and velocity. [1]

Monomer - A substance consisting of molecules which can undergo polymerization, thereby contributing constitutional units to the essential structure of a macromolecule. [8]

Monosaccharides - A term which includes aldoses, ketoses, and a wide variety of derivatives. [5]

Mössbauer effect - The recoilless emission of γ-rays from nuclei bound in a crystal under conditions where the recoil energy associated with the γ emission is taken up by the crystal as a whole. This results in a very narrow line width, which can be exploited in various types of precise measurements.

Muon* - An unstable elementary particle of spin 1/2 and mass about 200 times that of the electron.

Naphtha - The petroleum fraction consisting mostly of C_6 to C_8 hydrocarbons and boiling in the range 80–120 °C.

Solvents derived from this fraction include ligroin and petroleum ether.

Nautical mile - A non-SI unit of length, equal to exactly 1852 m.

Navier-Stokes equations - A set of complex equations for the motion of a viscous fluid subject to external forces.

Néel temperature (T_N)* - The critical temperature above which an antiferromagnetic substance becomes paramagnetic. [1]

Nernst effect - The production of an electric field in a conductor subject to an applied magnetic field and containing a transverse temperature gradient. The electric field is perpendicular to the magnetic field and the temperature gradient.

Network - In polymer science, a highly ramified macromolecule in which essentially each constitutional unit is connected to each other constitutional unit and to the macroscopic phase boundary by many permanent paths through the macromolecule, the number of such paths increasing with the number of intervening bonds. The paths must on the average be coextensive with the macromolecule. [8]

Neutrino - A stable elementary particle in the lepton family. Neutrinos have zero (or at least near-zero) rest mass and spin 1/2.

Neutron* - An elementary particle on spin 1/2 and zero charge. The free neutron has a mean lifetime of 887 seconds. Neutrons and protons, which are collectively called nucleons, are the constituents of the nucleus.

Neutron activation analysis (NAA) - See Techniques for Materials Characterization, Section 12.

Neutron number (*N*) - A characteristic property of a specific isotope of an element, equal to the number of neutrons in the nucleus.

Newton (N)* - The SI unit of force, equal to m kg s^{-2}. [1]

Nitriles - Compounds having the structure RC≡N; thus C-substituted derivatives of hydrocyanic acid, HC≡N. [5]

Nitrosamines - *N*-Nitroso amines: compounds of the structure R_2NNO. Compounds RNHNO are not ordinarily isolatable, but they, too, are nitrosamines. The name is a contraction of *N*-nitrosoamine and, as such, does not require the *N* locant. [5]

Nuclear magnetic resonance (NMR)* - A widely used technique in which the resonant absorption of radiofrequency radiation by magnetic nuclei in a magnetic field is measured. The results give important information on the local environment of each nucleus.

Nuclear magneton(μ_N)* - The unit of nuclear magnetic moment, defined as $eh/4\pi m_p$, where *h* is Planck's constant, m_p the proton mass, and *e* the elementary charge.

Nuclear quadrupole resonance (NQR) - See Techniques for Materials Characterization, Section 12.

Nuclear reaction analysis (NRA) - See Techniques for Materials Characterization, Section 12.

Nuclear spin (*I*) - The quantum number that specifies the intrinsic angular momentum of a particular nucleus. The magnitude of the angular momentum is given by $[I(I+1)]^{½}h/2\pi$, where *h* is Planck's constant.

Nucleic acids* - Macromolecules, the major organic matter of the nuclei of biological cells, made up of nucleotide units, and hydrolyzable into certain pyrimidine or purine bases (usually adenine, cytosine, guanine, thymine, uracil), D-ribose or 2-deoxy-D-ribose. [5]

Nucleon - A collective term for the proton and neutron.

Nucleosides - Ribosyl or deoxyribosyl derivatives (rarely, other glycosyl derivatives) of certain pyrimidine or purine bases. They are thus glycosylamines or *N*-glycosides related to nucleotides by the lack of phosphorylation. [5]

Nucleotides - Compounds formally obtained by esterification of the 3′ or 5′ hydroxy group of nucleosides with phosphoric acid. They are the monomers of nucleic acids and are formed from them by hydrolytic cleavage. [5]

Nuclide - A species of atoms in which each atom has identical atomic number *Z* and identical mass number *A*. [3]

Nusselt number (*Nu*) - A dimensionless quantity used in fluid mechanics, defined by $Nu = hl/k$, where *h* is coefficient of heat transfer, *l* is length, and *k* is thermal conductivity. [2]

Nyquist theorem - An expression for the mean square thermal noise voltage across a resistor, given by $4RkT\Delta f$ where *R* is the resistance, *k* the Boltzmann constant, *T* the temperature, and Δf the frequency band within which the voltage is measured.

Octanol-water partition coefficient (*P*)* - A measure of the way in which a compound will partition itself between the octanol and water phases in the two-phase octanol-water system, and thus an indicator of certain types of biological activity. Specifically, *P* is the ratio of the concentration (in moles per liter) of the compound in the octanol phase to that in the water phase at infinite dilution. The quantity normally reported is log *P*.

Oersted (Oe) - A non-SI unit of magnetic field (*H*), equal to 79.57747 A/m.

Ohm (Ω)* - The SI unit of electric resistance, equal to V/A. [1]

Ohm's law - A relation among electric current *I*, potential difference *V*, and resistance *R*, viz., $I = V/R$. At constant temperature the resistance for many materials is constant to high precision.

Olefins - Acyclic and cyclic hydrocarbons having one or more carbon-carbon double bonds, apart from the formal ones in aromatic compounds. The class olefins subsumes alkenes and cycloalkenes and the corresponding polyenes. [5]

Oligomer - A substance consisting of molecules of intermediate relative molecular mass (molecular weight), the structure of which essentially comprises the multiple repetition of units derived, actually or conceptually, from molecules of low relative molecular mass. In contrast to a polymer, the properties of an oligomer can vary significantly with the removal of one or a few of its units. [8]

Oligopeptides - Peptides containing from three to nine amino groups. [5]

Onsager relations - An important set of equations in the thermodynamics of irreversible processes. They express the symmetry between the transport coefficients describing reciprocal processes in systems with a linear dependence of flux on driving forces.

Optical rotary power - Angle by which the plane of polarization of a light beam is rotated by an optically active medium, divided by path length and by concentration of the active constituent. Depending on whether mass or molar concentration is used, the modifier "specific" or "molar" is attached. [2]

Orbital - A one-electron wave function. Atomic orbitals are classified as *s*-, *p*-, *d*-, or f-orbitals according to whether the angular momentum quantum number l = 0, 1, 2, or 3. Molecular orbitals, which are usually constructed as linear combinations of atomic orbitals, describe the distribution of electrons over the entire molecule.

Oscillator strength (f) - A measure of the intensity of a spectroscopic transition, defined by

$$f = \frac{8\pi^2 m_e \nu}{3he^2}\left|\mu_{ij}\right|^2$$

where ν is the frequency, μ_{ij} the transition dipole moment, m_e the mass of the electron, e the elementary charge, and h Planck's constant.

Osmosis - The flow of a solvent in a system in which two solutions of different concentration are separated by a semipermeable membrane which cannot pass solute molecules. The solvent will flow from the side of lower concentration to that of higher concentration, thus tending to equalize the concentrations. The pressure that must be applied to the more concentrated side to stop the flow is called the osmotic pressure.

Osmotic coefficient (ϕ) - Defined by $\phi = \ln a_A/(M_A \Sigma m_B)$, where M_A is the molar mass of substance A (normally the solvent), a_A is its activity, and the m_B are molalities of the solutes. [1]

Osmotic pressure (Π) - The excess pressure necessary to maintain osmotic equilibrium between a solution and the pure solvent separated by a membrane permeable only to the solvent. In an ideal dilute solution $\Pi = c_B RT$, where c_B is the amount-of-substance concentration of the solute, R is the molar gas constant, and T the temperature. [1,2]

Ostwald dilution law - A relation for the concentration dependence of the molar conductivity Λ of an electrolyte solution, viz.,

$$\frac{1}{\Lambda} = \frac{1}{\Lambda^\circ} + \frac{\Lambda c}{K(\Lambda^\circ)^2}$$

where c is the solute concentration, K is the equilibrium constant for dissociation of the solute, and Λ° is the conductivity at $c\Lambda = 0$.

Ounce (oz) - A non-SI unit of mass. The avoirdupois ounce equals 28.34952 g, while the troy ounce equals 31.10348 g.

Overpotential (η) - In an electrochemical cell, the difference between the potential of an electrode and its zero-current value.

Oximes - Compounds of structure $R_2C{=}NOH$ derived from condensation of aldehydes or ketones with hydroxylamine. Oximes from aldehydes may be called aldoximes; those from ketones may be called ketoximes. [5]

Oxo compounds - Compounds containing an oxygen atom, =O, doubly bonded to carbon or another element. The term thus embraces aldehydes, carboxylic acids, ketones, sulfonic acids, amides and esters. [5]

Ozonides - The 1,2,4-trioxolanes formed by the reaction of ozone at a carbon-carbon double bond, or the analogous compounds derived from acetylenic compounds. [5]

Pair production - A process in which a photon is converted into a particle and its antiparticle (e.g., an electron and positron) in the electromagnetic field of a nucleus.

Paraffins - Obsolescent term for saturated hydrocarbons, commonly but not necessarily acyclic. Still widely used in the petrochemical industry, where the term designates acyclic saturated hydrocarbons, and stands in contradistinction to naphthenes. [5]

Paramagnetism* - A type of magnetism characterized by a positive magnetic susceptibility, so that the material becomes weakly magnetized in the direction of an external field. The magnetization disappears when the field in removed. In the simplest approximation (Curie's law) the susceptibility is inversely proportional to temperature.

Parity - The property of a quantum-mechanical wave function that describes its behavior under the symmetry operation of coordinate inversion. A parity of +1 (or even) is assigned if the wave function does not change sign when the signs of all the coordinates are changed; the parity is –1 (or odd) if the wave function changes sign under this operation.

Parsec (pc) - A unit of distance defined as the distance at which 1 astronomical unit (AU) subtends an angle of 1 second of arc. It is equal to 206264.806 AU or 3.085678×10^{16} m.

Particle induced x-ray emission (PIXE) - See Techniques for Materials Characterization, Section 12.

Partition function (q, z) - For a single molecule, $q = \Sigma g_i \exp(\varepsilon_i/kT)$, where ε_i is an energy level of degeneracy g_i, k the Boltzmann constant, and T the absolute temperature; the summation extends over all energy states. For a system of N non-interacting molecules which are indistinguishable, as in an ideal gas, the canonical partition function $Q = qN/N!$.

Pascal (Pa)* - The SI unit of pressure, equal to N/m^2. [1]

Paschen series - The series of lines in the spectrum of the hydrogen atom which corresponds to transitions between the state with principal quantum number $n = 3$ and successive higher states. The wavelengths are given by $1/\lambda = R_H(1/9\text{-}1/n^2)$, where n = 4,5,6,... and R_H is the Rydberg constant. The first member of the series ($n = 3{\leftrightarrow}4$), which is often called the P_α line, falls in the infrared at a wavelength of 1.875 μm.

Paschen-Back effect - In atomic spectroscopy, the decoupling of electron spin from orbital angular momentum as the strength of an external magnetic field is increased.

Pauli exclusion principle - The statement that two electrons in an atom cannot have identical quantum numbers; thus if there are two electrons in the same orbital, their spin quantum numbers must be of opposite sign.

Pearson symbol - A code for designating crystallographic information, including the crystal system, the lattice type, and the number of atoms per unit cell.

Péclet number (Pe) - A dimensionless quantity used in fluid mechanics, defined by $Pe = vl/a$, where v is velocity, l is length, and a is thermal diffussivity. [2]

Peltier effect - The absorption or generation of heat (depending on the current direction) which occurs when an electric current is passed through a junction between two materials.

Peptides - Amides derived from two or more amino carboxylic acid molecules (the same or different) by formation of a covalent bond from the carbonyl carbon of one to the nitrogen atom of another with formal loss of water. [5]

Permeability (μ) - Magnetic induction divided by magnetic field strength; i.e. $\mu = B/H$. The relative permeability $\mu_r = \mu/\mu_0$, where μ_0 is the permeability of a vacuum. [1]

Permittivity (ε) - Ratio of the electric displacement in a medium to the electric field strength. Also called dielectric constant. [1]

Symbols

Peroxides - Compounds of structure ROOR in which R may be any organic group. In inorganic chemistry, salts of the anion O_2^{-2}. [5]

Peroxy acids - Acids in which an acidic -OH group has been replaced by an -OOH group; e.g., $CH_3C(=O)OOH$ peroxyacetic acid, $PhS(=O)_2OOH$ benzeneperoxysulfonic acid. [5]

Petroleum ether - The petroleum fraction consisting of C_5 and C_6 hydrocarbons and boiling in the range 35–60 °C; commonly used as a laboratory solvent.

pH* - A convenient measure of the acid-base character of a solution, usually defined by pH = –log [$c(H^+)$/mol L^{-1})], where $c(H^+)$ is the concentration of hydrogen ions. The more precise definition is in terms of activity rather than concentration. [2]

Phenols - Compounds having one or more hydroxy groups attached to a benzene or other arene ring. [5]

Phonon - A quantum of energy associated with a vibrational mode of a crystal lattice.

Phosphines - PH_3 and compounds derived from it by substituting one, two or three hydrogen atoms by hydrocarbyl groups. RPH_2, R_2PH and R_3P (R not equal to H) are called primary, secondary and tertiary phosphines, respectively. [5]

Phosphonium compounds - Salts (and hydroxides) $[R_4P]^+X^-$ containing tetracoordinate phosphonium ion and the associated anion. [5]

Phosphonium ylides - Compounds having the structure R_3P^+-$C^-R_2 \rightleftharpoons R_3P{=}CR_2$. Also known as Wittig reagents. [5]

Phosphorescence - The process by which a molecule is excited by light to a higher electronic state and then undergoes a radiationless transition to a state of different multiplicity from which it decays, after some delay, to the ground state. The emitted light is normally of longer wavelength than the exciting light because vibrational energy has been dissipated.

Photoelectric effect - The complete absorption of a photon by a solid with the emission of an electron.

Photon - An elementary particle of zero mass and spin 1. The photon is involved in electromagnetic interactions and is the quantum of electromagnetic radiation.

Photon stimulated desorption (PSD) - See Techniques for Materials Characterization, Section 12.

Pinacols - Tetra(hydrocarbyl)ethane-1,2-diols, $R_2C(OH)C(OH)R_2$, of which the tetramethyl example is the simplest one and is itself commonly known as pinacol. [5]

Pion - An elementary particle in the family of mesons. Pions have zero spin and may be neutral or charged. They participate in the strong interaction which holds the nucleus together.

p*K** - The negative logarithm (base 10) of an equilibrium constant *K*. For pK_a, see Acid dissociation constant.

Planck constant (*h*)* - The elementary quantum of action, which relates energy to frequency through the equation $E = h\nu$.

Planck distribution - See Black body radiation.

Planck function - A thermodynamic function defined by $Y = -G/T$, where G is Gibbs energy and T thermodynamic temperature. [2]

Plasma - A highly ionized gas in which the charge of the electrons is balanced by the charge of the positive ions, so that the system as a whole is electrically neutral.

Plasmon - A quantum associated with a plasma oscillation in the electron gas of a solid.

Point group* - A group of symmetry operations (rotations, reflections, etc.) that leave a molecule invariant. Every molecular conformation can be assigned to a specific point group, which plays a major role in determining the spectrum of the molecule.

Poise (P) - A non-SI unit of viscosity, equal to 0.1 Pa s.

Poiseuille's equation - A formula for the rate of flow of a viscous fluid through a tube:

$$\frac{dV}{dt} = \frac{(p_1^2 - p_2^2)\pi r^4}{16 l \eta p_0}$$

where V is the volume as measured at pressure p_0; p_1 and p_2 are the pressures at each end of the tube; r is the radius and l the length of the tube; and η is the viscosity.

Poisson ratio (μ) - The absolute value of the ratio of the transverse strain to the corresponding axial strain resulting from uniformly distributed axial stress below the proportional limit (i.e., where Hooke's law is valid). [10]

Polariton - A quantum associated with the coupled modes of photons and optical phonons in an ionic crystal.

Polarizability (α)* - The change in dipole moment of a molecule produced by an external electric field; specifically, $\alpha_{ab} = \partial p_a / \partial E_b$, where p_a is the dipole moment component on the a axis and E_b is the component of the electric field strength along the b axis. [2]

Polymer - A substance composed of molecules of high relative molecular mass (molecular weight), the structure of which essentially comprises the multiple repetition of units derived, actually or conceptually, from molecules of low relative molecular mass. A single molecule of a polymer is called a macromolecule. [8]

Polypeptides - Peptides containing 10 or more amino acid residues. See also Peptides. [5]

Polysaccharides - Compounds consisting of a large number of monosaccharides linked glycosidically. This term is commonly used only for those containing more than ten monosaccharide residues. Also called glycans. [5]

Porphyrins - Natural pigments containing a fundamental skeleton of four pyrrole nuclei united through the α-positions by four methine groups to form a macrocyclic structure (porphyrin is designated porphine in Chemical Abstracts indexes). [5]

Positron - The antiparticle of the electron. It has the same mass and spin as an electron, and an equal but opposite charge.

Positronium - The hydrogen-like "atom" formed from a positron nucleus and an electron. Its lifetime is very short because of annihilation of the positron and electron.

Potential - See Electric potential.

Potential energy (E_p, V, U) - The portion of the energy of a system that is associated with its position in a force field.

Pound (lb) - A non-SI unit of mass, equal to 0.4535924 kg.

Power (P) - Rate of energy transfer. For electrical circuits, this is equal to the product of current and potential difference, $P = IV$. [1]

Poynting vector (S) - For electromagnetic radiation, the vector product of the electric field strength and the magnetic field strength. [1]

Prandtl number (Pr) - A dimensionless quantity used in fluid mechanics, defined by $Pr = \eta/\rho a$, where η is viscosity, ρ is density, and a is thermal diffusivity. [2]

Pressure* - Force divided by area. [1]

Proteins - Naturally occurring and synthetic polypeptides having molecular weights greater than about 10,000 (the limit is not precise). See also Peptides. [5]

Proton* - A stable elementary particle of unit positive charge and spin 1/2. Protons and neutrons, which are collectively called nucleons, are the constituents of the nucleus.

Pulsar - A neutron star which rotates rapidly and emits electromagnetic radiation in regular pulses at a frequency related to the rotation period.

Purine bases* - Purine and its substitution derivatives, especially naturally occurring examples. [5]

Pyrimidine bases* - Pyrimidine and its substitution derivatives, especially naturally occurring examples. [5]

Q-switching - A technique for obtaining very high power from a laser by keeping the *Q* factor of the laser cavity low while the population inversion builds up, then suddenly increasing the *Q* to initiate the stimulated emission.

Quad - A unit of energy defined as 10^{15} Btu, equal to approximately 1.055056×10^{18} J.

Quadrupole moment - A coefficient of the third term (after monopole and dipole) in the power series expansion of the electric potential of an array of charges. A nucleus of spin greater than 1/2 has a non-vanishing nuclear quadrupole moment which can interact with the electric field gradient of the surrounding electrons. Molecular quadrupole moments have an influence on intermolecular forces.

Quality factor (*Q*) - The ratio of the absolute value of the reactance of an electrical system to the resistance; thus a measure of the energy stored per cycle relative to the energy dissipated.

Quantum yield - In photochemistry, the number of moles transformed in a specific process, either physically (e.g., by emission of photons) or chemically, per mole of photons absorbed by the system. [3]

Quark - An elementary entity which has not been directly observed but is considered a constituent of protons, neutrons, and other hadrons.

Quasar - An extragalactic object emitting electromagnetic radiation at a very high power level and showing a very large red shift, thus indicating that the object is receding at a speed approaching the speed of light.

Quasicrystal - A solid having conventional crystalline properties but whose lattice does not display translational periodicity.

Quaternary ammonium compounds - Derivatives of ammonium compounds, $NH_4^+Y^-$, in which all four of the hydrogens bonded to nitrogen have been replaced with hydrocarbyl groups. Compounds having a carbon-nitrogen double bond (i.e. $R_2C{=}N^+R_2Y^-$) are more accurately called iminium compounds. [5]

Quinones - Compounds having a fully conjugated cyclic dione structure, such as that of benzoquinones, derived from aromatic compounds by conversion of an even number of -CH= groups into -C(=O)- groups with any necessary rearrangement of double bonds. [5]

Racemic mixture - A mixture of equal amounts of a pair of enantiomers (optical isomers); such a mixture is not optically active.

Rad - A non-SI unit of absorbed dose of radiation, equal to 0.01 Gy.

Radiance (*L*) - The radiant intensity in a given direction from an element of a surface, divided by the area of the orthogonal projection of this element on a plane perpendicular to the given direction. [1]

Radiant intensity (*I*) - The radiant energy flux leaving an element of a source within an element of solid angle, divided by that element of solid angle. [1]

Radicals - Molecular entities possessing an unpaired electron, such as $\bullet CH_3$, $\bullet SnH_3$, $\bullet Cl$. (In these formulas the dot, symbolizing the unpaired electron, should be placed so as to indicate the atom of highest spin density, if this is possible.) [5]

Raman effect - The inelastic scattering of light by a molecule, in which the incident photon either gives up to, or receives energy from, one of the internal vibrational modes of the molecule. The scattered light thus has either a lower frequency (Stokes radiation) or higher frequency (anti-Stokes radiation) than the incident light. These shifts provide a measure of the normal vibrational frequencies of the molecule.

Rankine cycle - A thermodynamic cycle which can be used to calculate the ideal performance of a heat engine that uses a condensable vapor as the working fluid (e.g., a steam engine or a heat pump).

Rankine temperature - A thermodynamic temperature scale based on a temperature interval °R = (5/9) K; i.e., $T/°R = (9/5)\ T/K = t/°F + 459.67$.

Raoult's law - The expression for the vapor pressure p_i of component i in an ideal solution, viz., $p_i = x_i p_{i0}$, where x_i is the mole fraction of component i and p_{i0} the vapor pressure of the pure substance i.

Rare earth elements - The elements Sc, Y, and the lanthanides (La, Ce, Pr, Nd, Pm, Sm, Eu, Gd, Tb, Dy, Ho, Er, Tm, Yb, Lu). [7]

Rayleigh number (*Ra*) - A dimensionless quantity used in fluid mechanics, defined by $Ra = l^3 g\alpha\Delta T\rho/\eta a$, where l is length, g is acceleration of gravity, α is cubic expansion coefficient, T is temperature, ρ is density, η is viscosity, and a is thermal diffusivity. [2]

Rayleigh scattering - The scattering of light by particles which are much smaller than the wavelength of the light. It is characterized by a scattered intensity which varies as the inverse fourth power of the wavelength.

Rayleigh wave - A guided elastic wave along the surface of a solid; also called surface acoustic wave.

Reactance (*X*) - The imaginary part of impedance. For an inductive reactance L and a capacitive reactance C in series, the reactance is $X = L\omega - 1/(C\omega)$, where ω is 2π times the frequency of the current. [1]

Red shift - A displacement of a spectral line toward longer wavelengths. This can occur through the Doppler effect (e.g., in the light from receding galaxies) or, in the general theory of relativity, from the effects of a star's gravitational field.

Reflectance (*ρ*) - Ratio of the radiant or luminous flux at a given wavelength that is reflected to that of the incident radiation. Also called reflection factor. [1]

Reflection high energy electron diffraction (RHEED) - See Techniques for Materials Characterization, Section 12.

Relative humidity* - The ratio of the partial pressure of water vapor in air to the saturation vapor pressure of water at the same temperature, expressed as a percentage. [10]

Relative molar mass - See Molecular weight.

Rem - A non-SI unit of dose equivalent, equal to 0.01 Sv.

Symbols

Resistance (R) - Electric potential difference divided by current when there is no electromotive force in the conductor. This definition applies to direct current. More generally, resistance is defined as the real part of impedance. [1]

Resistivity (ρ) - Electric field strength divided by current density when there is no electromotive force in the conductor. Resistivity is an intrinsic property of a material. For a conductor of uniform cross section with area A and length L, and whose resistance is R, the resistivity is given by $\rho = RA/L$. [1]

Reynolds number (Re) - A dimensionless quantity used in fluid mechanics, defined by $Re = \rho v l/\eta$, where ρ is density, v is velocity, l is length, and η is viscosity. [2]

Rheology - The study of the flow of liquids and deformation of solids. Rheology addresses such phenomena as creep, stress relaxation, anelasticity, nonlinear stress deformation, and viscosity.

Ribonucleic acids (RNA) - Naturally occurring polyribonucleotides. See also nucleic acids, nucleosides, nucleotides, ribonucleotides. [5]

Ribonucleotides - Nucleotides in which the glycosyl group is a ribosyl group. See also nucleotides. [5]

Roentgen (R) - A unit used for expressing the charge (positive or negative) liberated by x-ray or γ radiation in air, divided by the mass of air. A roentgen is defined as 2.58×10^{-4} C/kg.

Rotational constants - In molecular spectroscopy, the constants appearing in the expression for the rotational energy levels as a function of the angular momentum quantum numbers. These constants are proportional to the reciprocals of the principal moments of inertia, averaged over the vibrational motion.

Rutherford back scattering (RBS) - See Techniques for Materials Characterization, Section 12.

Rydberg constant (R_∞)* - The fundamental constant which appears in the equation for the energy levels of hydrogen-like atoms; i.e., $E_n = hcR_\infty Z^2\mu/n^2$, where h is Planck's constant, c the speed of light, Z the atomic number, μ the reduced mass of nucleus and electron, and n the principal quantum number ($n = 1, 2, ...$).

Rydberg series - A regular series of lines in the spectrum of an atom or molecule, with the spacing between successive lines becoming smaller as the frequency increases (wavelength decreases). The series eventually converges to a limit which usually corresponds to the complete removal of an electron from the atom or molecule.

Sackur-Tetrode equation* - An equation for the molar entropy S_m of an ideal monatomic gas: $S_m = R\ln(e^{5/2}V/N_A\Lambda^3)$, where R is the molar gas constant, V is the volume, and N_A is Avogadro's number. The constant Λ is given by $\Lambda = h/(2\pi mkT)^{1/2}$, where h is Planck's constant, m the atomic mass, k the Boltzmann constant, and T the temperature.

Salinity (S)* - A parameter used in oceanography to describe the concentration of dissolved salts in seawater. It is defined in terms of electrical conductivity relative to a standard solution of KCl. When expressed in units of parts per thousand, S may be roughly equated to the concentration of dissolved material in grams per kilogram of seawater.

Salt - An ionic compound formed by the reaction of an acid and a base.

Scanned probe microscopy (SPM) - See Techniques for Materials Characterization, Section 12.

Scanning electron microscopy (SEM) - See Techniques for Materials Characterization, Section 12.

Scanning laser acoustic microscopy (SLAM) - See Techniques for Materials Characterization, Section 12.

Scanning transmission electron microscopy (STEM) - See Techniques for Materials Characterization, Section 12.

Scanning tunneling microscopy (STM) - See Techniques for Materials Characterization, Section 12.

Schiff bases - Imines bearing a hydrocarbyl group on the nitrogen atom: $R_2C{=}NR'$ (R' not equal to H). Considered by many to be synonymous with azomethines. [5]

Schmidt number (Sc) - A dimensionless quantity used in fluid mechanics, defined by $Sc = \eta/\rho D$, where η is viscosity, ρ is density, and D is diffusion coefficient. [2]

Schottky barrier - A potential barrier associated with a metal-semiconductor contact. It forms the basis for the rectifying device known as the Schottly diode.

Schrödinger equation - The basic equation of wave mechanics which, for systems not dependent on time, takes the form:

$$-(\hbar/2m)\nabla^2\psi + V\psi = E\psi$$

where ψ is the wave function, V is the potential energy expressed as a function of the spatial coordinates, E is an energy eigenvalue, ∇^2 is the Laplacian operator, $\hbar$ is Planck's constant divided by 2π, and m is the mass.

Second (s)* - The SI base unit of time. [1]

Second radiation constant (c_2)* - See First radiation constant.

Secondary ion mass spectroscopy (SIMS) - See Techniques for Materials Characterization, Section 12.

Seebeck effect - The development of a potential difference in a circuit where two different metals or semiconductors are joined and their junctions maintained at different temperatures. It is the basis of the thermocouple.

Selenides - Compounds having the structure RSeR (R not equal to H). They are thus selenium analogues of ethers. Also used for metal salts of H_2Se. [5]

Semicarbazones - Compounds having the structure $R_2C{=}NNHC({=}O)NH_2$, formally derived by condensation of aldehydes or ketones with semicarbazide [$NH_2NHC({=}O)NH_2$]. [5]

Semiconductor - A material in which the highest occupied energy band (valence band) is completely filled with electrons at $T = 0$ K, and the energy gap to the next highest band (conduction band) ranges from 0 to 4 or 5 eV. With increasing temperature electrons are excited into the conduction band, leading to an increase in the electrical conductivity.

Semiquinones - Radical anions having the structure -O-Z-O· where Z is an ortho- or para-arylene group or analogous heteroarylene group; they are formally generated by the addition of an electron to a quinone. [5]

SI units* - The International System of Units adopted in 1960 and recommended for use in all scientific and technical fields. [1]

Siemens (S)* - The SI unit of electric conductance, equal to Ω^{-1}. [1]

Sievert (Sv)* - The SI unit of dose equivalent (of radiation), equal to J/kg. [1]

Silanes - Saturated silicon hydrides, analogues of the alkanes; i.e., compounds of the general formula Si_nH_{2n+2}. Silanes may be subdivided into silane, oligosilanes, and polysilanes. Hydrocarbyl derivatives are often referred to loosely as silanes. [5]

Silicones - Polymeric or oligomeric siloxanes, usually considered unbranched, of general formula $[-OSiR_2-]_n$ (R not equal to H). [5]

Siloxanes - Saturated silicon-oxygen hydrides with unbranched or branched chains of alternating silicon and oxygen atoms (each silicon atom is separated from its nearest silicon neighbors by single oxygen atoms). [5]

Skin effect - The concentration of high frequency alternating currents near the surface of a conductor.

Slater orbital - A particular mathematical expression for the radial part of the wave function of a single electron, which is used in quantum-mechanical calculations of the energy and other properties of atoms and molecules.

Small angle neutron scattering (SANS) - See Techniques for Materials Characterization, Section 12.

Snell's law - The relation between the angle of incidence *i* and the angle of refraction *r* of a light beam which passes from a medium of refractive index n_0 to a medium of index n_1, viz., $\sin i/\sin r = n_1/n_0$.

Solar constant* - The mean radiant energy flux from the sun on a unit surface normal to the direction of the rays at the mean distance of the earth from the sun. The value is approximately 1373 W/m^2.

Solar wind - The stream of high velocity hydrogen and helium ions emitted by the sun which flows through the solar system and beyond.

Soliton - A spatially localized wave in a solid or liquid that can interact strongly with other solitons but will afterwards regain its original form.

Solubility* - A quantity expressing the maximum concentration of some material (the solute) that can exist in another liquid or solid material (the solvent) at thermodynamic equilibrium at specified temperature and pressure. Common measures of solubility include the mass of solute per unit mass of solution (mass fraction), mole fraction of solute, molality, molarity, and others.

Solubility product constant (K_{sp})* - The equilibrium constant for the dissolution of a sparsely soluble salt into its constituent ions.

Space group* - A group of symmetry operations (reflections, rotations, etc.) that leave a crystal invariant. A total of 230 space groups have been identified.

Spark source mass spectroscopy (SSMS) - See Techniques for Materials Characterization, Section 12.

Specific gravity - Ratio of the mass density of a material to that of water. Since one must specify the temperature of both the sample and the water to have a precisely defined quantity, the use of this term is now discouraged.

Specific heat - Heat capacity divided by mass. See Heat capacity.

Specific quantity - It is often convenient to express an extensive quantity (e.g., volume, enthalpy, heat capacity, etc.) as the actual value divided by mass. The resulting quantity is called specific volume, specific enthalpy, etc.

Specific rotation $[\alpha]^\theta\lambda$- For an optically active substance, defined by $[\alpha]^\theta_\lambda = \alpha/\gamma l$, where α is the angle through which plane polarized light is rotated by a solution of mass concentration γ and path length l. Here θ is the Celsius temperature and λ the wavelength of the light at which the measurement is carried out. Also called specific optical rotatory power. [2]

Spin (s, I)* - A measure of the intrinsic angular momentum of a particle, which it possesses independent of its orbital motion. The symbol s is used for the spin quantum number of an electron, while I is generally used for nuclear spin.

Spiro compounds - Compounds having one atom (usually a quaternary carbon) as the only common member of two rings. [5]

Stacking fault - An error in the normal sequence of layer growth in a crystal.

Standard mean ocean water (SMOW) - A standard sample of pure water of accurately known isotopic composition which is maintained by the International Atomic Energy Agency. It is used for precise calibration of density and isotopic composition measurements.

Standard reduction potential ($E°$) - The zero-current potential of a cell in which the specified reduction reaction occurs at the right-hand electrode and the left-hand electrode is the standard hydrogen electrode. Also called Standard electrode potential.

Standard state - A defined state (specified temperature, pressure, concentration, etc.) for tabulating thermodynamic functions and carrying out thermodynamic calculations. The standard state pressure is usually taken as 100,000 Pa (1 bar), but various standard state temperatures are used. [2]

Stanton number (St) - A dimensionless quantity used in fluid mechanics, defined by $St = h/\rho vc_p$, where h is coefficient of heat transfer, ρ is density, v is velocity, and c_p is specific heat capacity at constant pressure. [2]

Stark effect - The splitting of an energy level of an atom or molecule, and hence a splitting of spectral lines arising from that level, as a result of the application of an external electric field.

Statistical weight (g) - The number of distinct states corresponding to the same energy level. Also called degeneracy.

Stefan-Boltzmann constant (σ)* - Constant in the equation for the radiant exitance M (radiant energy flux per unit area) from a black body at thermodynamic temperature T, viz. $M = \sigma T^4$. [1]

Stibines - SbH_3 and compounds derived from it by substituting one, two or three hydrogen atoms by hydrocarbyl groups: R_3Sb. $RSbH_2$, R_2SbH, and R_3Sb (R not equal to H) are called primary, secondary and tertiary stibines, respectively. [5]

Stochastic process - A process which involves random variables and whose outcome can thus be described only in terms of probabilities.

Stoichiometric number (ν) - The number appearing before the symbol for each compound in the equation for a chemical reaction. By convention, it is negative for reactants and positive for products. [2]

Stokes (St) - A non-SI unit of kinematic viscosity, equal to 10^{-4} m^2/s.

Stokes' law - The statement, valid under certain conditions, that the viscous force F experienced by a sphere of radius a moving at velocity v in a medium of viscosity η is given by $F = -6\pi\eta av$.

Strain - The deformation of a body that results from an applied stress.

Stratosphere - The part of the earth's atmosphere extending from the top of the troposphere (typically 10 to 15 km above the surface) to about 50 km. It is characterized by an increase in temperature with increasing altitude.

Stress - Force per unit area (pressure) applied to a body. Tensile stress tends to stretch or compress the body in the

Symbols

direction of the applied force. Sheer stress results from a tangential force which tends to twist the body.

Strong interaction - The short range (order of 1 fm) attractive forces between protons, neutrons, and other hadrons which are responsible for the stability of the nucleus.

Strouhal number (*Sr*) - A dimensionless quantity used in fluid mechanics, defined by $Sr = lf/v$, where l is length, f is frequency, and v is velocity. [2]

Structure factor - In x-ray crystallography, the sum of the scattering factors of all the atoms in a unit cell, weighted by an appropriate phase factor. The intensity of a given reflection is proportional to the square of the structure factor.

Sublimation pressure - The pressure of a gas in equilibrium with a solid at a specified temperature.

Sulfides - Compounds having the structure RSR (R not equal to H). Such compounds were once called thioethers. In an inorganic sense, salts or other derivatives of hydrogen sulfide. [5]

Sulfones - Compounds having the structure, RS(=O)$_2$R (R not equal to H), e.g. $C_2H_5S(=O)_2CH_3$, ethyl methyl sulfone. [5]

Sulfonic acids - HS(=O)$_2$OH, sulfonic acid, and its S-hydrocarbyl derivatives. [5]

Sulfoxides - Compounds having the structure R_2S=O (R not equal to H), e.g., Ph_2S=O, diphenyl sulfoxide. [5]

Superconductor - A material that experiences a nearly total loss of electrical resistivity below a critical temperature T_c. The effect can occur in pure metals, alloys, semiconductors, organic compounds, and certain inorganic solids.

Superfluid - A fluid with near-zero viscosity and extremely high thermal conductivity. Liquid helium exhibits these properties below 2.186 K (the λ point).

Supernova - A star in the process of exploding because of instabilities which follow the exhaustion of its nuclear fuel.

Surface analysis by laser ionization (SALI) - See Techniques for Materials Characterization, Section 12.

Surface tension (γ,σ)* - The force per unit length in the plane of the interface between a liquid and a gas, which resists an increase in the area of that surface. It can also be equated to the surface Gibbs energy per unit area.

Surfactant - A substance which lowers the surface tension of the medium in which it is dissolved, and/or the interfacial tension with other phases, and accordingly is positively adsorbed at the liquid-vapor or other interfaces. [3]

Susceptance (*B*) - Imaginary part of admittance. [1]

Svedberg - A non-SI unit of time, used to express sedimentation coefficients, equal to 10^{-13} s.

Syndiotactic macromolecule - A tactic macromolecule, essentially comprising alternating enantiomeric configurational base units which have chiral or prochiral atoms in the main chain in a unique arrangement with respect to their adjacent constitutional units. In this case the repeating unit consists of two configurational base units that are enantiomeric. [8]

Tacticity - The orderliness of the succession of configurational repeating units of a macromolecule or oligomer molecule. In a tactic macromolecule essentially all the configurational repeating units are identical with respect to directional sense. See Configurational repeating unit, Isotactic, Syndiotactic. [8]

Tautomerism - Isomerism of the general form G-X-Y=Z $\rightleftharpoons$ X=Y-Z-G, where the isomers (called tautomers) are readily interconvertible; the atoms connecting the groups X, Y, Z are typically any of C, H, O, or S, and G is a group which becomes an electrofuge (i.e., a group that does not carry away the bonding electron pair when it leaves its position in the molecule) or nucleofuge (a group that does carry away the bonding electrons when leaving) during isomerization. The commonest case, when the electrofuge is H^+, is also known as prototropy. A common example, written so as to illustrate the general pattern given above, is keto-enol tautomerism, such as

H-O-C(CH_3)=CH-CO_2Et (enol) $\rightleftharpoons$
(CH_3)C(=O)-CH_2-CO_2Et (keto)

In some cases the interconversion rate between tautomers is slow enough to permit isolation of the separate keto and enol forms. [5]

Tensile strength* - In tensile testing, the ratio of maximum load a body can bear before breaking to original cross-sectional area. Also called ultimate strength. [11]

Terpenes - Hydrocarbons of biological origin having carbon skeletons formally derived from isoprene [CH_2=C(CH_3)CH=CH_2]. [5]

Terpenoids - Natural products and related compounds formally derived from isoprene units. They contain oxygen in various functional groups. The skeleton of terpenoids may differ from strict additivity of isoprene units by the loss or shift of a methyl (or other) group. [5]

Tesla (T)* - The SI unit of magnetic flux density (*B*), equal to V s/m^2. [1]

Thermal conductivity* - Rate of heat flow divided by area and by temperature gradient. [1]

Thermal diffusivity - Thermal conductivity divided by density and by specific heat capacity at constant pressure. [1]

Thermal expansion coefficient (α)* - The linear expansion coefficient is defined by $\alpha_l = (1/l)(dl/dT)$; the volume expansion coefficient by $\alpha_V = (1/V)(dV/dT)$. [1]

Thermionic emission - The emission of electrons from a solid as a result of heat. The effect requires a high enough temperature to impart sufficient kinetic energy to the electrons to exceed the work function of the solid.

Thermodynamic laws - The foundation of the science of thermodynamics:

First law: The internal energy of an isolated system is constant; if energy is supplied to the system in the form of heat dq and work dw, then the change in energy $dU = dq + dw$.

Second law: No process is possible in which the only result is the transfer of heat from a reservoir and its complete conversion to work.

Third law: The entropy of a perfect crystal approaches zero as the thermodynamic temperature approaches zero.

Thermoelectric power - For a bar of a pure material whose ends are at different temperatures, the potential difference divided by the difference in temperature of the ends. See also Seeback effect.

Thermogravimetric analysis (TGA) - See Techniques for Materials Characterization, Section 12.

Thermosphere - The layer of the Earth's atmosphere extending from the top of the mesosphere (typically 80–90 km above the surface) to about 500 km. It is characterized by a rapid increase in temperature with increasing altitude up to about 200 km, followed by a leveling off in the 300–500 km region.

Thiols - Compounds having the structure RSH (R not equal to H). Also known by the term mercaptans (abandoned by IUPAC); e.g., CH_3CH_2SH, ethanethiol. [5]

Thomson coefficient (*μ*, *τ*) - The heat power developed in the Thomson effect (whereby heat is evolved in a conductor when a current is flowing in the presence of a temperature gradient), divided by the current and the temperature difference. [1]

Tonne (t) - An alternative name for megagram (1000 kg). [1]

Torque (*T*) - For a force $\boldsymbol{F}$ that produces a torsional motion, $\boldsymbol{T} = \boldsymbol{r} \times \boldsymbol{F}$, where $\boldsymbol{r}$ is a vector from some reference point to the point of application of the force.

Torr - A non-SI unit of pressure, equal to 133.322 Pa. The name is generally considered interchangeable with millimeter of mercury.

Townsend coefficient - In a radiation counter, the number of ionizing collisions by an electron per unit path length in the direction of an applied electric field.

Transducer - Any device that converts a signal from acoustical, optical, or some other form of energy into an electrical signal (or vice versa) while preserving the information content of the original signal.

Transistor - A voltage amplifier using controlled electron currents inside a semiconductor.

Transition metals - Elements characterized by a partially filled *d* subshell. The First Transition Series comprises Sc, Ti, V, Cr, Mn, Fe, Co, Ni, Cu. The Second and Third Transition Series include the lanthanides and actinides, respectively. [7]

Transition probability* - See Einstein transition probability.

Transmittance (*τ*) - Ratio of the radiant or luminous flux at a given wavelength that is transmitted to that of the incident radiation. Also called transmission factor. [1]

Tribology - The study of frictional forces between solid surfaces.

Triple point* - The point in *p*, *T* space where the solid, liquid, and gas phases of a substance are in thermodynamic equilibrium. The corresponding temperature and pressure are called the triple point temperature and triple point pressure.

Troposphere - The lowest part of the earth's atmosphere, extending to 10–15 km above the surface. It is characterized by a decrease in temperature with increasing altitude. The exact height varies with latitude and season.

Tunnel diode - A device involving a p-n junction in which both sides are so heavily doped that the Fermi level on the p-side lies in the valence band and on the n-side in the conduction band. This leads to a current-voltage curve with a maximum, so that the device exhibits a negative resistance in some regions.

Ultraviolet photoelectron spectroscopy (UPS) - See Techniques for Materials Characterization, Section 12.

Umklapp process - A process involving the interaction of three or more waves (lattice or electron) in a solid in which the sum of the wave vectors does not equal zero.

Unified atomic mass unit (u)* - A unit of mass used in atomic, molecular, and nuclear science, defined as the mass of one atom of ^{12}C divided by 12. Its approximate value is 1.66054×10^{-27} kg. [1]

Universal time (t_U, UT) - Mean solar time counted from midnight at the Greenwich meridian. Also called Greenwich mean time (GMT). The interval of mean solar time is based on the average, over one year, of the time between successive transits of the sun across the observer's meridian.

Vacancy - A missing atom or ion in a crystal lattice.

Van Allen belts - Two toroidal regions above the earth's atmosphere containing protons and electrons. The outer belt at about 25,000 km above the surface is probably of solar origin. The inner belt at about 3000 km contains more energetic particles from outside the solar system.

Van der Waals' equation* - An equation of state for fluids which takes the form:

$$pV_m = RT\frac{1}{V_m - b} - \frac{a}{V_m^{\ 2}}$$

where *p* is pressure, V_m is molar volume, *T* is temperature, *R* is the molar gas constant, and *a* and *b* are characteristic parameters of the substance which describe the effect of attractive and repulsive intermolecular forces, respectively.

Van der Waals' force - The weak attractive force between two molecules which arises from electric dipole interactions. It can lead to the formation of stable but weakly bound dimer molecules or clusters.

Van't Hoff equation - The equation expressing the temperature dependence of the equilibrium constant *K* of a chemical reaction:

$$\frac{d\ln K}{dT} = \frac{\Delta_r H^\circ}{RT^2}$$

where $\Delta_r H^\circ$ is the standard enthalpy of reaction, *R* the molar gas constant, and *T* the temperature. Also called van't Hoff isochore.

Vapor pressure* - The pressure of a gas in equilibrium with a liquid (or, in some usage, a solid) at a specified temperature.

Varistor - A device that utilizes the properties of certain metal oxides with small amounts of impurities, which show abrupt nonlinearities at specific voltages where the material changes from a semiconductor to an insulator.

Velocity (*v*) - Rate of change of distance with time.

Verdet constants (*V*)* - Angle of rotation of a plane polarized light beam passing through a medium in a magnetic field, divided by the field strength and by the path length.

Virial equation of state* - An equation relating the pressure *p*, molar volume V_m, and temperature *T* of a real gas in the form of an expansion in powers of the molar volume, viz., $pV_m = RT(1 + BV_m^{-1} + CV_m^{-2} + \ldots)$, where *R* is the molar gas constant. *B* is called the second virial coefficient, *C* the third virial coefficient, etc. The virial coefficients are functions of temperature.

Viscosity (*η*)* - The proportionality factor between sheer rate and sheer stress, defined through the equation $F = \eta A(dv/dx)$, where *F* is the tangential force required to move a planar surface of area *A* at velocity *v* relative to a parallel surface separated from the first by a distance *x*. Sometimes called dynamic or absolute viscosity. The term kinematic viscosity (symbol *ν*) is defined as *η* divided by the mass density.

Volt (V)* - The SI unit of electric potential, equal to W/A. [1]

Volume fraction (ϕ_j) - Defined as $V_j/\Sigma_i V_i$, where V_j is the volume of the specified component and the V_i are the volumes of all the components of a mixture prior to mixing. [2]

Watt (W)* - The SI unit of power, equal to J/s. [1]

Wave function - A function of the coordinates of all the particles in a quantum mechanical system (and, in general, of time) which fully describes the state of the system. The product of the wave function and its complex conjugate is proportional to the probability of finding a particle at a particular point in space.

Weak interaction - The weak forces (order of 10^{-12} of the strong interaction) between elementary particles which are responsible for beta decay and other nuclear effects.

Weber (Wb)* - The SI unit of magnetic flux, equal to V s. [1]

Weber number (*We*) - A dimensionless quantity used in fluid mechanics, defined by $We = \rho v^2 l/\gamma$, where ρ is density, v is velocity, l is length, and γ is surface tension. [2]

Weight - That force which, when applied to a body, would give it an acceleration equal to the local acceleration of gravity. [1]

Wiedeman-Franz law - The law stating that the thermal conductivity k and electrical conductivity σ of a pure metal are related by $k = L\sigma T$, where T is the temperature and L (called the Lorenz ratio) has the approximate value 2.45 × 10^{-8} V^2/K^2.

Wien displacement law - The relation, which can be derived from the Planck formula for black body radiation, that $\lambda_{max} T = 0.0028978$ m K, where λ_{max} is the wavelength of maximum radiance at temperature T.

Wigner-Seitz method - A method of calculating electron energy levels in a solid using a model in which each electron is subject to a spherically symmetric potential.

Wittig reagents - See Phosphonium ylides.

Work (*W*) - Force multiplied by the displacement in the direction of the force. [1]

Work function (*Φ*)* - The energy difference between an electron at rest at infinity and an electron at the Fermi level in the interior of a substance. It is thus the minimum energy required to remove an electron from the interior of a solid to a point just outside the surface. [1]

X unit (X) - A unit of length used in x-ray crystallography, equal to approximately 1.002 × 10^{-13} m.

X-ray photoelectron spectroscopy (XPS) - See Techniques for Materials Characterization, Section 12.

Yield strength - The stress at which a material exhibits a specified deviation (often chosen as 0.2% for metals) from proportionality of stress and strain. [11]

Young's modulus (*E*) - In tension or compression of a body below its elastic limit, the ratio of stress to corresponding strain. Since strain is normally expressed on a fractional basis, Young's modulus has dimensions of pressure. Also called elastic modulus. [11]

Zeeman effect - The splitting of an energy level of an atom or molecule, and hence a splitting of spectral lines arising from that level, as a result of the application of an external magnetic field.

Zener diode - A control device utilizing a p-n junction with a well defined reverse-bias avalanche breakdown voltage.

Zeotrope - A liquid mixture that shows no maximum or minimum when vapor pressure is plotted against composition at constant temperature. See Azeotrope.

Zero-point energy - The energy possessed by a quantum mechanical system as a result of the uncertainty principle even when it is in its lowest energy state; e.g., the difference between the lowest energy level of a harmonic oscillator and the minimum in the potential well.

Zeta potential (ζ) - The electric potential at the surface of a colloidal particle relative to the potential in the bulk medium at a long distance. Also called electrokinetic potential.

Zwitterions - Neutral compounds having formal unit electrical charges of opposite sign. Some chemists restrict the term to compounds with the charges on non-adjacent atoms. Sometimes referred to as inner salts, dipolar ions (a misnomer). [5]

Symbols

THERMODYNAMIC FUNCTIONS AND RELATIONS

p = pressure V = volume T = temperature $x_i = n_i/\Sigma_j n_j$ = mole fraction of substance i

n_i = amount of substance i

Energy	U
Entropy	S
Enthalpy	$H = U + pV$
Helmholtz energy	$A = U - TS$
Gibbs energy	$G = U + pV - TS$
Isobaric heat capacity	$C_p = (\partial H/\partial T)_p$
Isochoric heat capacity	$C_V = (\partial U/\partial T)_V$
Isobaric expansivity	$\alpha = V^{-1}(\partial V/\partial T)_p$
Isothermal compressibility	$\kappa_T = -V^{-1}(\partial V/\partial p)_T$
Isentropic compressibility	$\kappa_S = -V^{-1}(\partial V/\partial p)_S$
	$\kappa_T - \kappa_S = T\alpha^2 V/C_p$
	$C_p - C_V = T\alpha^2 V/\kappa_T$
Gibbs-Helmholtz equation	$H = G - T(\partial G/\partial T)_p$
Maxwell relations	$(\partial S/\partial p)_T = -(\partial V/\partial T)_p$
	$(\partial S/\partial V)_T = -(\partial p/\partial T)_V$
Joule-Thomson expansion	$\mu_{JT} = (\partial T/\partial p)_H = -\{V - T(\partial V/\partial T)_p\}/C_p$
	$\phi_{JT} = (\partial H/\partial p)_T = V - T(\partial V/\partial T)_p$
Partial molar quantity	$X_i = (\partial X/\partial n_i)_{T,p,nj\neq i}$
Chemical potential	$\mu_i = (\partial G/\partial n_i)_{T,p,nj\neq i}$
Perfect gas [symbol^{pg}]	$pV = (\Sigma_i n_i)RT$
	$\mu_i^{pg} = \mu_i^{\theta} + RT\ln(x_i p/p^{\theta})$
Fugacity	$f_i = (x_i p)\exp\{(\mu_i - \mu_i^{pg})/RT\}$
Activity coefficient	$\gamma_i = f_i/(x_i f_i^{\theta})$
Gibbs-Duhem relation	$0 = SdT - Vdp + \Sigma_i n_i d\mu_i$

Notation for chemical and physical changes (*X* = *H,S,G,* etc.):

Chemical reaction	$\Delta_r X$
Formation from elements	$\Delta_f X$
Combustion	$\Delta_c X$
Fusion (cry→liq)	$\Delta_{fus} X$
Vaporization (liq→gas)	$\Delta_{vap} X$
Sublimation (cry→gas)	$\Delta_{sub} X$
Phase transition	$\Delta_{trs} X$
Solution	$\Delta_{sol} X$
Mixing	$\Delta_{mix} X$
Dilution	$\Delta_{dil} X$

Note: Superscript θ in above equations indicates standard state.

NOBEL LAUREATES IN CHEMISTRY AND PHYSICS

Full details on nationality and basis of the awards can be found at <http://www.nobelprize.org>.

Chemistry

Year	
2016	Jean-Pierre Sauvage, J. Fraser Stoddart, Bernard L. Feringa
2015	Tomas Lindahl, Paul Modrich, Aziz Sancar
2014	Eric Betzig, Stefan W. Hell, William E. Moerner
2013	Martin Karplus, Michael Levitt, Arieh Warshel
2012	Robert J. Lefkowitz, Brian K. Kobilka
2011	Dan Shechtman
2010	Richard F. Heck, Ei-ichi Negishi, Akira Suzuki
2009	Venkatraman Ramakrishnan, Thomas A. Steitz, Ada E. Yonath
2008	Martin Chalfie, Osamu Shimomura, Roger Y. Tsien
2007	Gerhard Ertl
2006	Roger D. Kornberg
2005	Yves Chauvin, Robert H. Grubbs, Richard R. Schrock
2004	Aaron Ciechanover, Avram Hershko, Irwin Rose
2003	Peter Agre, Roderick MacKinnon
2002	John B. Fenn, Koichi Tanaka, Kurt Wüthrich
2001	William S. Knowles, Ryoji Noyori, K. Barry Sharpless
2000	Alan Heeger, Alan G. MacDiarmid, Hideki Shirakawa
1999	Ahmed Zewail
1998	Walter Kohn, John Pople
1997	Paul D. Boyer, John E. Walker, Jens C. Skou
1996	Robert F. Curl Jr., Sir Harold Kroto, Richard E. Smalley
1995	Paul J. Crutzen, Mario J. Molina, F. Sherwood Rowland
1994	George A. Olah
1993	Kary B. Mullis, Michael Smith
1992	Rudolph A. Marcus
1991	Richard R. Ernst
1990	Elias James Corey
1989	Sidney Altman, Thomas R. Cech
1988	Johann Deisenhofer, Robert Huber, Hartmut Michel
1987	Donald J. Cram, Jean-Marie Lehn, Charles J. Pedersen
1986	Dudley R. Herschbach, Yuan T. Lee, John C. Polanyi
1985	Herbert A. Hauptman, Jerome Karle
1984	Bruce Merrifield
1983	Henry Taube
1982	Aaron Klug
1981	Kenichi Fukui, Roald Hoffmann
1980	Paul Berg, Walter Gilbert, Frederick Sanger
1979	Herbert C. Brown, Georg Wittig
1978	Peter Mitchell
1977	Ilya Prigogine
1976	William Lipscomb
1975	John Cornforth, Vladimir Prelog
1974	Paul J. Flory
1973	Ernst Otto Fischer, Geoffrey Wilkinson
1972	Christian Anfinsen, Stanford Moore, William H. Stein
1971	Gerhard Herzberg
1970	Luis Leloir
1969	Derek Barton, Odd Hassel
1968	Lars Onsager
1967	Manfred Eigen, Ronald G. W. Norrish, George Porter
1966	Robert S. Mulliken
1965	Robert B. Woodward
1964	Dorothy Crowfoot Hodgkin
1963	Karl Ziegler, Giulio Natta
1962	Max F. Perutz, John C. Kendrew
1961	Melvin Calvin
1960	Willard F. Libby
1959	Jaroslav Heyrovsky
1958	Frederick Sanger
1957	Lord Todd
1956	Sir Cyril Hinshelwood, Nikolay Semenov
1955	Vincent du Vigneaud
1954	Linus Pauling
1953	Hermann Staudinger
1952	Archer J. P. Martin, Richard L. M. Synge
1951	Edwin M. McMillan, Glenn T. Seaborg
1950	Otto Diels, Kurt Alder
1949	William F. Giauque
1948	Arne Tiselius
1947	Sir Robert Robinson
1946	James B. Sumner, John H. Northrop, Wendell M. Stanley
1945	Artturi Virtanen
1944	Otto Hahn
1943	George de Hevesy
1942	*No prize awarded*
1941	*No prize awarded*
1940	*No prize awarded*
1939	Adolf Butenandt, Leopold Ruzicka
1938	Richard Kuhn
1937	Norman Haworth, Paul Karrer
1936	Peter Debye
1935	Frédéric Joliot, Irène Joliot-Curie
1934	Harold C. Urey
1933	*No prize awarded*
1932	Irving Langmuir
1931	Carl Bosch, Friedrich Bergius
1930	Hans Fischer
1929	Arthur Harden, Hans von Euler-Chelpin
1928	Adolf Windaus
1927	Heinrich Wieland
1926	The Svedberg
1925	Richard Zsigmondy
1924	*No prize awarded*
1923	Fritz Pregl
1922	Francis W. Aston
1921	Frederick Soddy
1920	Walther Nernst
1919	*No prize awarded*
1918	Fritz Haber
1917	*No prize awarded*
1916	*No prize awarded*
1915	Richard Willstätter
1914	Theodore W. Richards
1913	Alfred Werner
1912	Victor Grignard, Paul Sabatier
1911	Marie Curie
1910	Otto Wallach
1909	Wilhelm Ostwald
1908	Ernest Rutherford
1907	Eduard Buchner
1906	Henri Moissan
1905	Adolf von Baeyer
1904	Sir William Ramsay
1903	Svante Arrhenius
1902	Emil Fischer
1901	Jacobus H. van't Hoff

Physics

Year	
2016	David J. Thouless, F. Duncan M. Haldane, J. Michael Kosterlitz
2015	Takaaki Kajita, Arthur B. McDonald
2014	Isamu Akasaki, Hiroshi Amano, Shuji Nakamura
2013	François Englert, Peter Higgs
2012	Serge Haroche, David J. Wineland
2011	Saul Perlmutter, Brian P. Schmidt, Adam G. Riess
2010	Andre Geim, Konstantin Novoselov
2009	Charles K. Kao, Willard S. Boyle, George E. Smith
2008	Makoto Kobayashi, Toshihide Maskawa, Yoichiro Nambu
2007	Albert Fert, Peter Grünberg
2006	John C. Mather, George F. Smoot
2005	Roy J. Glauber, John L. Hall, Theodor W. Hänsch
2004	David J. Gross, H. David Politzer, Frank Wilczek
2003	Alexei A. Abrikosov, Vitaly L. Ginzburg, Anthony J. Leggett
2002	Raymond Davis Jr., Masatoshi Koshiba, Riccardo Giacconi
2001	Eric A. Cornell, Wolfgang Ketterle, Carl E. Wieman
2000	Zhores I. Alferov, Herbert Kroemer, Jack S. Kilby
1999	Gerardus't Hooft, Martinus J. G. Veltman
1998	Robert B. Laughlin, Horst L. Störmer, Daniel C. Tsui
1997	Steven Chu, Claude Cohen-Tannoudji, William D. Phillips
1996	David M. Lee, Douglas D. Osheroff, Robert C. Richardson
1995	Martin L. Perl, Frederick Reines
1994	Bertram N. Brockhouse, Clifford G. Shull
1993	Russell A. Hulse, Joseph H. Taylor Jr.
1992	Georges Charpak
1991	Pierre-Gilles de Gennes
1990	Jerome I. Friedman, Henry W. Kendall, Richard E. Taylor
1989	Norman F. Ramsey, Hans G. Dehmelt, Wolfgang Paul
1988	Leon M. Lederman, Melvin Schwartz, Jack Steinberger
1987	J. Georg Bednorz, K. Alex Müller
1986	Ernst Ruska, Gerd Binnig, Heinrich Rohrer
1985	Klaus von Klitzing
1984	Carlo Rubbia, Simon van der Meer
1983	Subramanyan Chandrasekhar, William A. Fowler
1982	Kenneth G. Wilson
1981	Nicolaas Bloembergen, Arthur L. Schawlow, Kai M. Siegbahn
1980	James Cronin, Val Fitch
1979	Sheldon Glashow, Abdus Salam, Steven Weinberg
1978	Pyotr Kapitsa, Arno Penzias, Robert Woodrow Wilson
1977	Philip W. Anderson, Sir Nevill F. Mott, John H. van Vleck
1976	Burton Richter, Samuel C. C. Ting
1975	Aage N. Bohr, Ben R. Mottelson, James Rainwater
1974	Martin Ryle, Antony Hewish
1973	Leo Esaki, Ivar Giaever, Brian D. Josephson
1972	John Bardeen, Leon N. Cooper, Robert Schrieffer
1971	Dennis Gabor
1970	Hannes Alfvén, Louis Néel
1969	Murray Gell-Mann
1968	Luis Alvarez
1967	Hans Bethe
1966	Alfred Kastler
1965	Sin-Itiro Tomonaga, Julian Schwinger, Richard P. Feynman
1964	Charles H. Townes, Nicolay G. Basov, Aleksandr M. Prokhorov
1963	Eugene Wigner, Maria Goeppert-Mayer, J. Hans D. Jensen
1962	Lev Landau
1961	Robert Hofstadter, Rudolf Mössbauer
1960	Donald A. Glaser

Year	
1959	Emilio Segrè, Owen Chamberlain
1958	Pavel A. Cherenkov, Il'ja M. Frank, Igor Y. Tamm
1957	Chen Ning Yang, Tsung-Dao Lee
1956	William B. Shockley, John Bardeen, Walter H. Brattain
1955	Willis E. Lamb, Polykarp Kusch
1954	Max Born, Walther Bothe
1953	Frits Zernike
1952	Felix Bloch, E. M. Purcell
1951	John Cockcroft, Ernest T. S. Walton
1950	Cecil Powell
1949	Hideki Yukawa
1948	Patrick M. S. Blackett
1947	Edward V. Appleton
1946	Percy W. Bridgman
1945	Wolfgang Pauli
1944	Isidor Isaac Rabi
1943	Otto Stern
1942	*No prize awarded*
1941	*No prize awarded*
1940	*No prize awarded*
1939	Ernest Lawrence
1938	Enrico Fermi
1937	Clinton Davisson, George Paget Thomson
1936	Victor F. Hess, Carl D. Anderson
1935	James Chadwick
1934	*No prize awarded*
1933	Erwin Schrödinger, Paul A. M. Dirac
1932	Werner Heisenberg
1931	*No prize awarded*
1930	Sir Venkata Raman
1929	Louis de Broglie
1928	Owen Willans Richardson
1927	Arthur H. Compton, C. T. R. Wilson
1926	Jean Baptiste Perrin
1925	James Franck, Gustav Hertz
1924	Manne Siegbahn
1923	Robert A. Millikan
1922	Niels Bohr
1921	Albert Einstein
1920	Charles Edouard Guillaume
1919	Johannes Stark
1918	Max Planck
1917	Charles Glover Barkla
1916	*No prize awarded*
1915	William Bragg, Lawrence Bragg
1914	Max von Laue
1913	Heike Kamerlingh Onnes
1912	Gustaf Dalén
1911	Wilhelm Wien
1910	Johannes Diderik van der Waals
1909	Guglielmo Marconi, Ferdinand Braun
1908	Gabriel Lippmann
1907	Albert A. Michelson
1906	J. J. Thomson
1905	Philipp Lenard
1904	Lord Rayleigh
1903	Henri Becquerel, Pierre Curie, Marie Curie
1902	Hendrik A. Lorentz, Pieter Zeeman
1901	Wilhelm Conrad Röntgen

Section 3
Physical Constants of Organic Compounds

PHYSICAL CONSTANTS OF ORGANIC COMPOUNDS

The basic physical constants and structure diagrams for about 10,900 organic compounds are presented in this table. An effort has been made to include the compounds most frequently encountered in the laboratory, the workplace, and the environment. Particular emphasis has been given to substances that are considered environmental or human health hazards. In making the selection of compounds for the table, added weight was assigned to the appearance of a compound in various lists or reference sources such as:

- Laboratory reagent lists, e.g., the ACS *Reagent Chemicals* volume (Ref. 1)
- The DIPPR list of industrially important compounds (Ref. 2) and the (much larger) TSCA Inventory of chemicals used in commerce
- The Hazardous Substance Data Bank (Ref. 3)
- The UNEP list of Persistent Organic Pollutants (Ref. 4)
- Chemicals on Reporting Rules (CORR), a database of about 7500 regulated compounds prepared by the Environmental Protection Agency (Ref. 5)
- The EPA Integrated Risk Information System (IRIS), a database of human health effects of exposure to chemicals in the environment (Ref. 6)
- Compendia of chemicals of biochemical or medical importance, such as *The Merck Index* (Ref. 10)
- Specialized tables in this *Handbook*

It should be noted that the above lists vary widely in their choice of chemical names, and even in the use of Chemical Abstracts Registry Numbers. To the extent possible, we have attempted to systematize the names and registry numbers for this table.

Clearly, criteria of this type are somewhat subjective, and compounds considered important by some users have undoubtedly been omitted. Suggestions for additional compounds or other improvements are welcomed.

The data in the table have been derived from many sources, including both the primary literature and evaluated compilations. The *Handbook of Data on Organic Compounds, Third Edition* (Ref. 7) and the *Combined Chemical Dictionary* (Ref. 8) were important sources. Other useful sources of physical property data on organic compounds are listed in Refs. 9-19. The values in the table for the normal boiling point and the melting point that are accompanied with uncertainties (in parentheses) have been critically evaluated using the NIST ThermoData Engine (TDE, Ref. 20), designed to implement the dynamic data evaluation concept (Refs. 21-24). This concept requires large electronic databases capable of storing essentially all relevant experimental data known to date with detailed descriptions of metadata and uncertainties. The combination of these electronic databases with expert-system software, designed to automatically generate recommended property values based on available experimental and predicted data, leads to the ability to produce critically evaluated data dynamically or "to order." The uncertainties listed are combined expanded uncertainties (level of confidence, approximately 95 %) representing the most comprehensive measure of the overall data reliability (Refs. 25-28).

The table is arranged alphabetically by substance name, which generally is either an IUPAC systematic name or, in the case of pesticides, pharmaceuticals, and other complex compounds, a simple trivial name. Names in ubiquitous use, such as acetic acid and formaldehyde, are adopted rather than their systematic equivalents. Synonyms are given in the column following the primary name, and structure diagrams are given on the page facing the data listing. The explanation of the data columns follows:

No.: An identification number used in the indexes.

Name: Primary name of the substance.

Synonym: A synonym in common use. When the primary name is non-systematic, a systematic name may appear here.

Mol. form.: The molecular formula written in the Hill convention.

CAS Reg. No.: The Chemical Abstracts Service Registry Number for the compound.

Mol. wt: Molecular weight (relative molar mass) as calculated with the 2001 IUPAC Standard Atomic Weights.

Physical form: A notation of the physical phase, color, crystal type, or other features of the compound at ambient temperature. Abbreviations are given below.

mp: Normal melting point in °C. A value is sometimes followed by "dec", indicating decomposition is observed at the stated temperature (so that it is probably not a true melting point). The notation "tp" indicates a triple point, where solid, liquid, and gas are in equilibrium. A number in parentheses following the melting point value is the combined expanded uncertainty (see above).

bp: Normal boiling point in °C, if it is available. This is the temperature at which the liquid phase is in equilibrium with the vapor at a pressure of 760 mmHg (101.325 kPa). A number in parentheses following the boiling point value is the combined expanded uncertainty (see above). A notation "sp" following the value indicates a sublimation point, where the vapor pressure of the solid phase reaches 760 mmHg. When a notation such as "dec" (decomposes) or "exp" (explodes) follows the value, the temperature may not be a true boiling point. A simply entry "sub" indicates the solid has a significant sublimation pressure at ambient temperatures. When the normal boiling point is not available, a boiling point at reduced pressure may be listed with a superscript indicating the pressure in mmHg.

den: Density (mass per unit volume) in g/cm^3. The temperature in °C is indicated by a superscript. Values refer to the liquid or solid phase, and all values are true densities, not specific gravities. The number of decimal places gives a rough estimate of the accuracy of the value.

n_D: Refractive index, at the temperature in °C indicated by the superscript. Unless otherwise indicated, all values refer to a wavelength of 589 nm (sodium D line). Values are given only for liquids and solids.

Solubility: Qualitative indication of solubility in common solvents. Abbreviations are:

i	insoluble
sl	slightly soluble
s	soluble
vs	very soluble
msc	miscible
dec	decomposes

Abbreviations for solvents are given below.

In order to facilitate the location of compounds in the table, an index to synonyms follows the main table. Indexes to Molecular Formulas and CAS Registry Numbers are available in the electronic versions of the *Handbook* or as pdf files by request via e-mail (fiona.macdonald@taylorandfrancis.com).

The assistance of members of the Thermodynamics Research Center (TRC) of the National Institute of Standards and Technology (Vladimir Diky, Rob Chirico, Andrei Kazakov) and especially Chris Muzny and Michael Frenkel in the determination of values of the normal-boiling-point and melting-point temperatures with uncertainties is greatly appreciated. The editors of the *Handbook* are much indebted to Chris Muzny who spent countless hours in producing these critically evaluated results. The assistance of Fiona Macdonald in checking names and formulas is gratefully acknowledged, as well as the efforts of Janice Shackleton, Trupti Desai, Nazila Kamaly, Matt Griffiths, and Lawrence Braschi in preparing the structure diagrams.

List of Abbreviations

Ac	acetyl	flr	fluorescent	pow	powder
Ac_2O	acetic anhydride	fum	fumes, fuming	Pr	propyl
AcOEt	ethyl acetate	gl	glacial	PrOH	1-propanol
ac	acid	gr	gray	pr	prisms
ace	acetone	gran	granular	purp	purple
al	alcohol (ethanol)	grn	green	py	pyridine
alk	alkali	hex	hexagonal	pym	pyramids, pyramidal
amor	amorphous	HOAc	acetic acid	reac	reacts
anh	anhydrous	hp	heptane	rhom	rhombic
aq	aqueous	hx	hexane	s	soluble
bipym	bipyramidal	hyd	hydrate	sat	saturated
bl	blue	hyg	hygroscopic	sc	scales
blk	black	i	insoluble	sl	slightly soluble
bp	boiling point	i-	iso-	soln	solution
br	brown	iso	isooctane	sp	sublimation point
bt	bright	lf	leaves	stab	stable
Bu	butyl	lig	ligroin	sub	sublimes
BuOH	1-butanol	liq	liquid	sulf	sulfuric acid
bz	benzene	lo	long	syr	syrup
chl	chloroform	mcl	monoclinic	tab	tablets
col	colorless	Me	methyl	tcl	triclinic
con, conc	concentrated	MeCN	acetonitrile	tetr	tetragonal
cry	crystals	MeOH	methanol	tfa	trifluoroacetic acid
ctc	carbon tetrachloride	misc	miscible	thf, THF	tetrahydrofuran
cy, cyhex	cyclohexane	mp	melting point	tol	toluene
dec	decomposes	n	refractive index	tp	triple point
den	density	nd	needles	trg	trigonal
dil	dilute	oct	octahedra, octahedral	unstab	unstable
diox	dioxane	oran	orange	vap	vapor
dk	dark	orth	orthorhombic	viol	violet
DMF	dimethylformamide	os	organic solvents	visc	viscous
DMSO	dimethyl sulfoxide	pa	pale	vol	volatile
efflor	efflorescent	peth	petroleum ether	vs	very soluble
Et	ethyl	Ph	phenyl	w	water
EtOH	ethanol	PhCl	chlorobenzene	wh	white
eth	diethyl ether	$PhNH_2$	aniline	xyl	xylene
exp	explodes	$PhNO_2$	nitrobenzene	ye	yellow
fl	flakes	pl	plates		

References

1. American Chemical Society, Reagent Chemicals, Tenth Edition, Oxford University Press, New York, 2005.
2. Design Institute for Physical Properties, American Institute of Chemical Engineers, <http://www.aiche.org/dippr/>.
3. National Library of Medicine, Hazardous Substances Data Bank, <http://toxnet.nlm.nih.gov/cgi-bin/sis/htmlgen?HSDB>.
4. United Nations Environmental Program, Persistent Organic Pollutants, <http://www.chem.unep.ch/pops/>.
5. Environmental Protection Agency, Chemicals on Reporting Rules, <http://www.epa.gov/opptintr/CORR>.
6. Environmental Protection Agency, Integrated Risk Information System, <http://www.epa.gov/iris/index.html>.
7. Lide, D. R., and Milne, G. W. A., Editors, Handbook of Data on Organic Compounds, Third Edition, CRC Press, Boca Raton, FL, 1993.
8. Combined Chemical Dictionary, <http://ccd.chemnetbase.com/>.
9. Linstrom, P. J., and Mallard, W. G., Editors, NIST Chemistry WebBook, NIST Standard Reference Database No. 69, February 2010, National Institute of Standards and Technology, Gaithersburg, MD 20899, <http://webbook.nist.gov>.
10. Thermodynamic Research Center, National Institute of Standards and Technology, TRC Thermodynamic Tables, <http://trc.nist.gov>.
11. O'Neil, M. J., Editor, The Merck Index, Fourteenth Edition, Merck & Co., Whitehouse Station, NJ, 2006.
12. Stevenson, R. M., and Malanowski, S., Handbook of the Thermodynamics of Organic Compounds, Elsevier, New York, 1987.
13. Riddick, J. A., Bunger, W. B., and Sakano, T. K., Organic Solvents, Fourth Edition, John Wiley & Sons, New York, 1986.
14. ChemSpider, <http://www.chemspider.com/>.
15. Crossfire Beilstein, <http://accelrys.com/products/>.
16. Springer Materials, The Landolt-Börnstein Database, <http://www.springermaterials.com>.
17. Vargaftik, N.B., Vinogradov, Y. K., and Yargin, V. S., Handbook of Physical Properties of Liquids and Gases, Third Edition, Begell House, New York, 1996.
18. Lide, D. R., and Kehiaian, H. V., Handbook of Thermophysical and Thermochemical Data, CRC Press, Boca Raton, FL, 1994.
19. Lide, D. R., Editor, Properties of Organic Compounds, <http://www.chemnetbase.com/tours/poc/intro.jsf>.
20. Frenkel, M., Chirico, R. D., Diky, V. V., Kazakov, A., and Muzny, C. D., ThermoData Engine, NIST Standard Reference Database 103b, Version 5.0 (Pure Compounds, Binary Mixtures, and Chemical Reactions, TDE-SOURCE Version 5.1), National Institute of Standards and Technology, Gaithersburg, MD – Boulder, CO, 2010, <http://www.nist.gov/srd/nist103b.cfm>.
21. Frenkel, M., Chirico, R. D., Diky, V., Yan, X., Dong, Q., and Muzny, C., J. Chem. Inf. Model. 45, 816, 2005.
22. Diky, V., Muzny, C. D., Lemmon, E. W., Chirico, R. D., and Frenkel, M., J. Chem. Inf. Model. 47, 1713, 2007.
23. Diky, V., Chirico, R. D., Kazakov, A. F., Muzny, C., and Frenkel, M., J. Chem. Inf. Model. 49, 503, 2009.
24. Diky, V., Chirico, R. D., Kazakov, A. F., Muzny, C., and Frenkel, M., J. Chem. Inf. Model. 49, 2883, 2009.
25. Chirico, R. D., Frenkel, M., Diky, V. V., March, K. N., and Wilhoit, R. C., J. Chem. Eng. Data 48, 1344, 2003.
26. Guide to the Expression of Uncertainty in Measurement, International Organization for Standardization, Geneva, Switzerland, 1993.
27. U. S. Guide to the Expression of Uncertainty in Measurement, ANSI/NCSL, Z540-2-1997, ISBN 1-58464-005-7, NCSL Int., Boulder, CO, 1997.
28. Taylor, B. N., and Kuyatt, C. E., Guidelines for the Evaluation and Expression of Uncertainty in NIST Measurement Results, NIST Tech. Note 1297, Natl. Inst. Stand. Technol., Gaithersburg, MD, 1994.

No.	Name	Synonym	Mol. form.	CAS Reg. No.	Mol. wt.	Physical form	mp/°C	bp/°C	den/ g cm⁻³	n_D	Solubility
1	Abate	Temephos	$C_{16}H_{20}O_6P_2S_3$	3383-96-8	466.469	cry	31.6(0.5)		1.32		sl H_2O, hx; s ctc, eth, tol
2	Abietic acid		$C_{20}H_{30}O_2$	514-10-3	302.451	mcl pl (al-w)	173.5	250^{9}			vs ace, bz, eth, EtOH
3	Abscisic acid		$C_{15}H_{20}O_4$	21293-29-8	264.318	cry (chl-peth)	160	120 sub			vs ace, eth, chl
4	Acacetin	5,7-Dihydroxy-2-(4-methoxyphenyl)-4*H*-1-benzopyran-4-one	$C_{16}H_{12}O_5$	480-44-4	284.263	ye nd (95% al)	263				vs EtOH
5	Acebutolol, (±)-		$C_{18}H_{28}N_2O_4$	37517-30-9	336.426	cry	121				
6	Acedapsone		$C_{16}H_{16}N_2O_4S$	77-46-3	332.374	pa ye nd (eth) lf (dil al)	290				sl H_2O
7	Acenaphthene	1,2-Dihydroacenaphthylene	$C_{12}H_{10}$	83-32-9	154.207		93(2)	277.5(0.8)	1.222^{20}	1.6048^{95}	i H_2O; sl EtOH, chl; vs bz; s HOAc
8	Acenaphthylene	Acenaphthalene	$C_{12}H_8$	208-96-8	152.192		89.4(0.3)	280	0.8987^{16}		i H_2O; vs EtOH, eth, bz; sl chl
9	1,2-Acenaphthylenedione		$C_{12}H_6O_2$	82-86-0	182.175	ye nd (HOAc)	259(9)	sub	1.4800^{20}		i H_2O; sl EtOH, bz, HOAc; s lig
10	Acenocoumarol	Nicoumalone	$C_{19}H_{15}NO_6$	152-72-7	353.325	cry (ace aq)	198				i H_2O
11	Acephate	Phosphoramidothioic acid, acetyl-, *O,S*-dimethyl ester	$C_4H_{10}NO_3PS$	30560-19-1	183.166		92.0(0.5)		1.35^{20}		
12	Acepromazine		$C_{19}H_{22}N_2OS$	61-00-7	326.455	oran oil		$230^{0.5}$			
13	Acesulfame		$C_4H_5NO_4S$	33665-90-6	163.153	nd (bz)	123.2				s bz, chl
14	Acetaldehyde	Ethanal	C_2H_4O	75-07-0	44.052	vol liq or gas	-123.4(0.7)	20.8(0.6)	0.7834^{18}	1.3316^{20}	msc H_2O, EtOH, eth, bz; sl chl
15	Acetaldehyde phenylhydrazone		$C_8H_{10}N_2$	935-07-9	134.178		99.5	150^{40}			vs EtOH
16	Acetaldoxime	Acetaldehyde oxime	C_2H_5NO	107-29-9	59.067	nd	25(1)	115.24(0.1)	0.9656^{20}	1.4264^{20}	s H_2O, chl; msc EtOH, eth
17	Acetamide	Ethanamide	C_2H_5NO	60-35-5	59.067	trg mcl (al-eth)	80.16(0.04)	222.0	0.9986^{85}	1.4278	vs H_2O, EtOH
18	Acetanilide	*N*-Phenylacetamide	C_8H_9NO	103-84-4	135.163		114.35(0.04)	292(9)	1.2190^{15}		sl H_2O; vs EtOH, ace; s eth, s bz, tol
19	Acetazolamide	*N*-[5-(Aminosulfonyl)-1,3,4-thiadiazol-2-yl]acetamide	$C_4H_6N_4O_3S_2$	59-66-5	222.246		260.5				sl H_2O
20	Acethion		$C_8H_{17}O_4PS_2$	919-54-0	272.322	liq		$137^{1.5}$	1.18^{20}		
21	Acetic acid	Ethanoic acid	$C_2H_4O_2$	64-19-7	60.052	col liq	17(3)	117.9(0.2)	1.0446^{25}	1.3720^{20}	msc H_2O, EtOH, eth, ace, bz; s chl, CS_2
22	Acetic acid, 2-phenylhydrazide		$C_8H_{10}N_2O$	114-83-0	150.177	hex pr (eth)	130.0				vs H_2O, EtOH; sl eth, chl, tfa; s bz
23	Acetic anhydride	Acetyl acetate	$C_4H_6O_3$	108-24-7	102.089	liq	-73.4(0.8)	139.5(0.3)	1.082^{20}	1.3901^{20}	vs H_2O; s EtOH, bz; msc eth; sl ctc
24	Acetoacetanilide		$C_{10}H_{11}NO_2$	102-01-2	177.200	pr or nd (bz or lig)	86				sl H_2O; s EtOH, eth, bz, chl, acid, lig
25	Acetoacetic acid		$C_4H_6O_3$	541-50-4	102.089	cry (eth)	36.5	100 dec			vs H_2O, eth, EtOH
26	2-Acetoacetoxyethyl methacrylate	2-(Methacryloyloxy)ethyl acetoacetate	$C_{10}H_{14}O_5$	21282-97-3	214.215	liq		$100^{0.8}$	1.122	1.4560^{20}	
27	Acetochlor		$C_{14}H_{20}ClNO_2$	34256-82-1	269.768	ye liq		$134^{0.4}$		1.5272^{20}	sl H_2O
28	Acetohexamide		$C_{15}H_{20}N_2O_4S$	968-81-0	324.396	cry (EtOH aq)	188				i H_2O, eth; sl EtOH, chl; s py
29	Acetohydrazide		$C_2H_6N_2O$	1068-57-1	74.081		67	137^{25}			s H_2O, EtOH; sl eth
30	Acetohydroxamic acid	*N*-Hydroxyacetamide	$C_2H_5NO_2$	546-88-3	75.067	hyg cry	90				
31	1-Acetonaphthone		$C_{12}H_{10}O$	941-98-0	170.206		34	297	1.1171^{21}	1.6280^{22}	i H_2O; s EtOH, eth, ace, chl
32	2-Acetonaphthone		$C_{12}H_{10}O$	93-08-3	170.206	nd (lig, dil al)	52(1)	302			sl EtOH, ctc
33	Acetone	2-Propanone	C_3H_6O	67-64-1	58.079	liq	-94.9(0.4)	56.08(0.07)	0.7845^{25}	1.3588^{20}	msc H_2O, EtOH, eth, ace, bz, chl
34	Acetone cyanohydrin		C_4H_7NO	75-86-5	85.105	liq	-19	180(21)	0.932^{19}	1.3992^{20}	vs H_2O, EtOH, eth; s ace, bz, chl; i peth
35	Acetone (2,4-dinitrophenyl)-hydrazone		$C_9H_{10}N_4O_4$	1567-89-1	238.200	ye nd or pl (al)	128				i H_2O; s EtOH, eth, bz, chl, AcOEt

Physical Constants of Organic Compounds

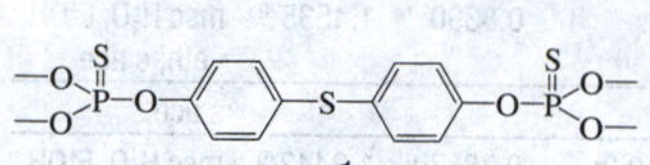
1 Abate

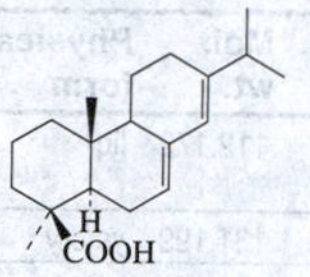
2 Abietic acid

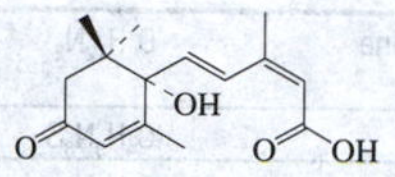
3 Abscisic acid

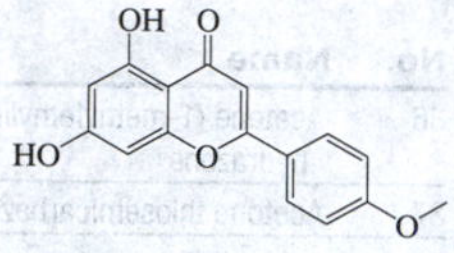
4 Acacetin

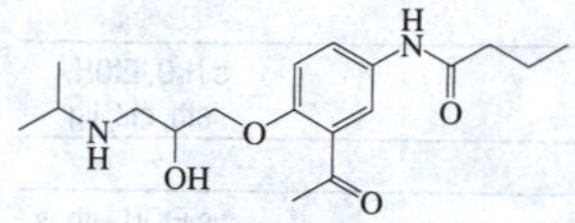
5 Acebutolol, (±)-

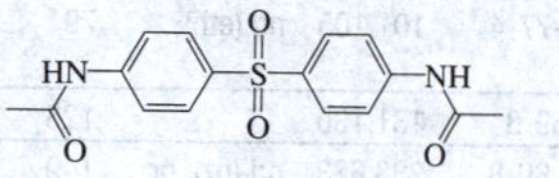
6 Acedapsone

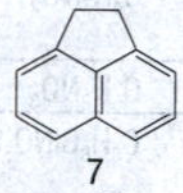
7 Acenaphthene

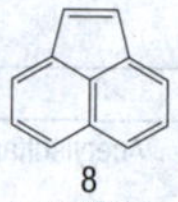
8 Acenaphthylene

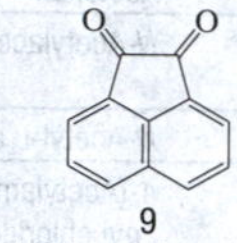
9 1,2-Acenaphthylenedione

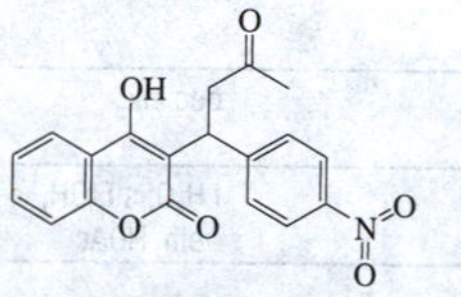
10 Acenocoumarol

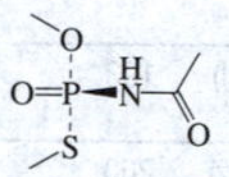
11 Acephate

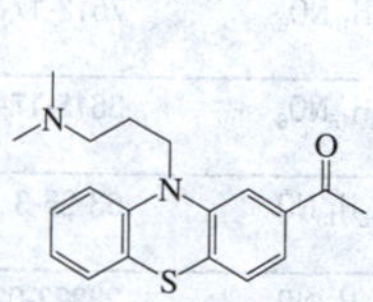
12 Acepromazine

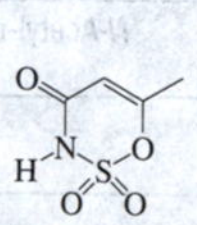
13 Acesulfame

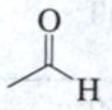
14 Acetaldehyde

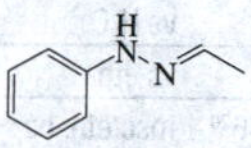
15 Acetaldehyde phenylhydrazone

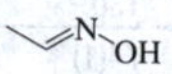
16 Acetaldoxime

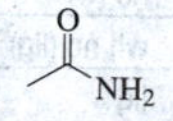
17 Acetamide

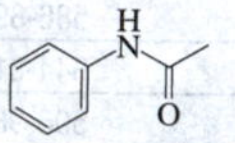
18 Acetanilide

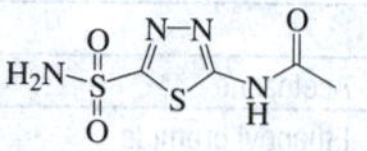
19 Acetazolamide

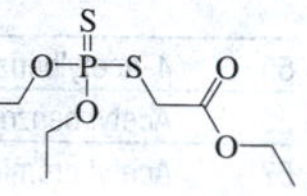
20 Acethion

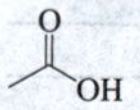
21 Acetic acid

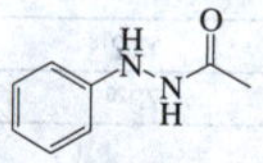
22 Acetic acid, 2-phenylhydrazide

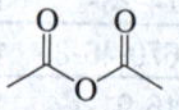
23 Acetic anhydride

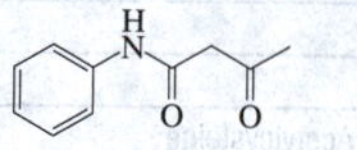
24 Acetoacetanilide

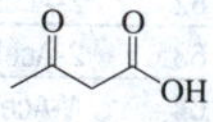
25 Acetoacetic acid

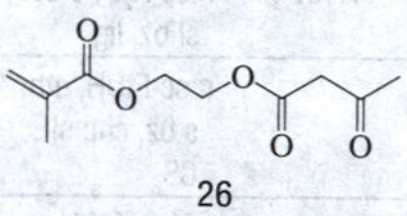
26 2-Acetoacetoxyethyl methacrylate

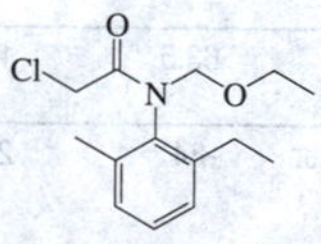
27 Acetochlor

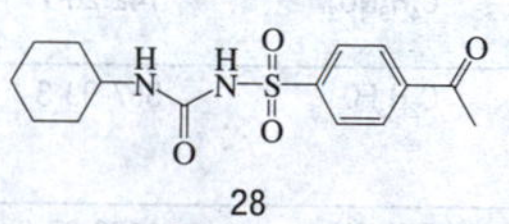
28 Acetohexamide

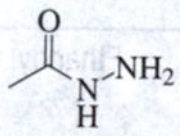
29 Acetohydrazide

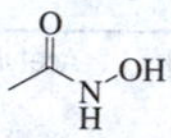
30 Acetohydroxamic acid

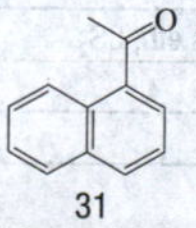
31 1-Acetonaphthone

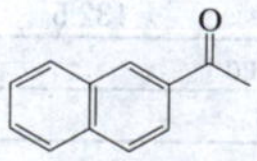
32 2-Acetonaphthone

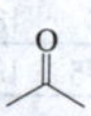
33 Acetone

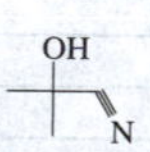
34 Acetone cyanohydrin

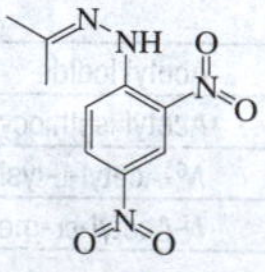
35 Acetone (2,4-dinitrophenyl)hydrazone

Organic

No.	Name	Synonym	Mol. form.	CAS Reg. No.	Mol. wt.	Physical form	mp/°C	bp/°C	den/g cm^{-3}	n_D	Solubility
36	Acetone (1-methylethylidene)-hydrazone	Dimethyl ketazine	$C_6H_{12}N_2$	627-70-3	112.172	liq	-12.5	133	0.8390^{20}	1.4535^{20}	msc H_2O, EtOH, eth; s ace
37	Acetone thiosemicarbazide		$C_4H_9N_3S$	1752-30-3	131.199	ye cry	176				s ace
38	Acetonitrile	Methyl cyanide	C_2H_3N	75-05-8	41.052	liq	-44(1)	81.6(0.2)	0.7857^{20}	1.3442^{30}	msc H_2O, EtOH, eth, ace, bz, ctc
39	Acetophenone	Methyl phenyl ketone	C_8H_8O	98-86-2	120.149	mcl pr or pl	19.4(0.4)	202.1(0.2)	1.0281^{20}	1.5372^{20}	sl H_2O; s EtOH, eth, ace, bz, con sulf, chl
40	Acetophenone azine	Methylphenyl ketazine	$C_{16}H_{16}N_2$	729-43-1	236.311		120				
41	Acetoxon	Acetophos	$C_8H_{17}O_5PS$	2425-25-4	256.257	liq		$73^{0.005}$			
42	*N*-Acetylacetamide		$C_4H_7NO_2$	625-77-4	101.105	nd (eth)	79	223.5			s H_2O, EtOH, eth, chl, lig
43	*N*-Acetyl-L-alanine		$C_5H_9NO_3$	97-69-8	131.130		125				
44	4-(Acetylamino)benzenesulfonyl chloride	Acetylsulfanilyl chloride	$C_8H_8ClNO_3S$	121-60-8	233.673	nd (bz), pr (bz-chl)	149				vs EtOH, eth; s bz, chl
45	2-(Acetylamino)benzoic acid		$C_9H_9NO_3$	89-52-1	179.172	nd (HOAc)	187.5				sl H_2O; s EtOH; vs eth, ace, bz, HOAc
46	4-(Acetylamino)benzoic acid		$C_9H_9NO_3$	556-08-1	179.172	nd (HOAc)	256.5				i H_2O; s EtOH; sl eth, tfa
47	2-(Acetylamino)-2-deoxy-D-glucose	*N*-Acetyl-D-glucosamine	$C_8H_{15}NO_6$	7512-17-6	221.208		205				
48	2-(Acetylamino)-2-deoxy-D-mannose	*N*-Acetyl-D-mannosamine	$C_8H_{15}NO_6$	3615-17-6	221.208	cry (ace aq)	128				dec alk
49	2-(Acetylamino)fluorene		$C_{15}H_{13}NO$	53-96-3	223.270	cry (dil al)	193				i H_2O; s EtOH, eth, HOAc
50	4-(Acetylamino)fluorene		$C_{15}H_{13}NO$	28322-02-3	223.270	br cry (bz)	200				
51	6-(Acetylamino)hexanoic acid	ε-Acetamidocaproic acid	$C_8H_{15}NO_3$	57-08-9	173.210	cry (ace)	104.5				
52	4-Acetylanisole		$C_9H_{10}O_2$	100-06-1	150.174	pl (peth)	38.2(0.6)	254(12)	1.0818^{41}	1.547^{41}	sl H_2O; s EtOH, eth, ace, chl
53	2-Acetylbenzoic acid		$C_9H_8O_3$	577-56-0	164.158	nd (w), pr (bz)	114.5	111^{2}			vs H_2O, eth, EtOH
54	3-Acetylbenzoic acid		$C_9H_8O_3$	586-42-5	164.158		172	111^{2}			s H_2O; msc EtOH
55	4-Acetylbenzoic acid		$C_9H_8O_3$	586-89-0	164.158	nd (w)	208	sub			vs H_2O
56	Acetyl benzoylperoxide	Acetozone	$C_9H_8O_4$	644-31-5	180.158	wh nd (lig)	37	130^{19}			vs eth
57	Acetyl bromide	Ethanoyl bromide	C_2H_3BrO	506-96-7	122.948	liq	-96.5(0.5)	74(1)	1.6625^{16}	1.4486^{20}	msc eth, bz, chl; s ace
58	Acetyl chloride	Ethanoyl chloride	C_2H_3ClO	75-36-5	78.497	liq	-112.7(0.8)	51(2)	1.1051^{20}	1.3886^{20}	msc eth, ace, bz, chl; s ctc
59	Acetylcholine bromide		$C_7H_{16}BrNO_2$	66-23-9	226.112	hyg cry	146				vs H_2O
60	Acetylcholine chloride		$C_7H_{16}ClNO_2$	60-31-1	181.661		150				s H_2O, EtOH; i eth
61	Acetylcholine iodide		$C_7H_{16}INO_2$	2260-50-6	273.112	hyg	163				
62	2-Acetylcyclohexanone		$C_8H_{12}O_2$	874-23-7	140.180		-11	112^{18}	1.0782^{25}	1.5138^{20}	s ctc
63	2-Acetylcyclopentanone		$C_7H_{10}O_2$	1670-46-8	126.153			73^{20}	1.0431^{25}	1.4906^{20}	
64	*N*-Acetyl-L-cysteine	Acetylcysteine	$C_5H_9NO_3S$	616-91-1	163.195	cry (w)	109.5				
65	3-Acetyldihydro-2(3*H*)-furanone	α-Acetylbutyrolactone	$C_6H_8O_3$	517-23-7	128.126			107^{5}	1.1846^{20}	1.4585^{20}	vs H_2O
66	1-Acetyl-2,5-dihydroxybenzene	2,5-Dihydroxyacetophenone	$C_8H_8O_3$	490-78-8	152.148	ye grn nd (dil al or w)	205.3				sl H_2O, eth, bz; s EtOH
67	Acetylene	Ethyne	C_2H_2	74-86-2	26.037	col gas	-81.5(0.9)	-84.7 sp	0.377^{25} (p>1 atm)		sl H_2O, EtOH, CS_2; s ace, bz, chl
68	*N*-Acetylethanolamine		$C_4H_9NO_2$	142-26-7	103.120		63.5	166^{8}	1.1079^{25}	1.4674^{20}	msc H_2O; s ace; sl bz, lig
69	Acetyl fluoride	Ethanoyl fluoride	C_2H_3FO	557-99-3	62.042	vol liq or gas	-84	22(7)	1.032^{25}		msc EtOH, eth; s bz, chl; sl CS_2
70	*N*-Acetylglutamic acid		$C_7H_{11}NO_5$	1188-37-0	189.166	pr (w)	199				s H_2O, EtOH
71	*N*-Acetylglycine	Aceturic acid	$C_4H_7NO_3$	543-24-8	117.104	lo nd (w, MeOH)	206				vs H_2O, ace, EtOH
72	*trans*-1-Acetyl-4-hydroxy-L-proline	Oxaceprol	$C_7H_{11}NO_4$	33996-33-7	173.167	cry (Ac)	132				vs H_2O, MeOH
73	1-Acetyl-1*H*-imidazole		$C_5H_6N_2O$	2466-76-4	110.114		104.5				sl H_2O; s EtOH, eth, chl, THF
74	Acetyl iodide	Ethanoyl iodide	C_2H_3IO	507-02-8	169.948			109(5)	2.0673^{20}	1.5491^{20}	vs eth
75	Acetyl isothiocyanate		C_3H_3NOS	13250-46-9	101.127			132.5	1.1523^{13}	1.5231^{18}	s eth, CS_2
76	N^6-Acetyl-L-lysine		$C_8H_{16}N_2O_3$	692-04-6	188.224		265 dec				
77	*N*-Acetyl-DL-methionine		$C_7H_{13}NO_3S$	1115-47-5	191.248		114.5				

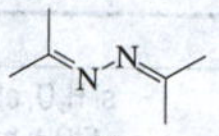
36
Acetone (1-methylethylidene)hydrazone

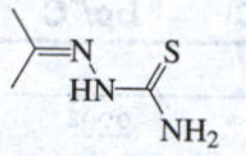
37
Acetone thiosemicarbazide

38
Acetonitrile

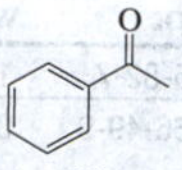
39
Acetophenone

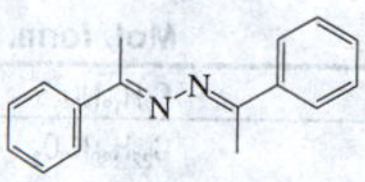
40
Acetophenone azine

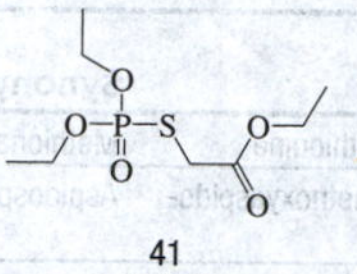
41
Acetoxon

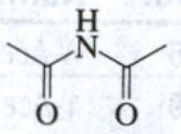
42
N-Acetylacetamide

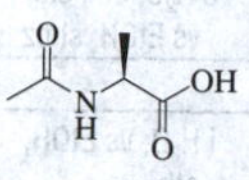
43
N-Acetyl-L-alanine

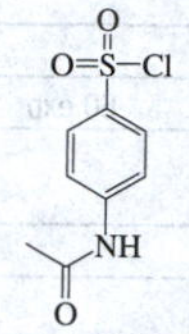
44
4-(Acetylamino)benzenesulfonyl chloride

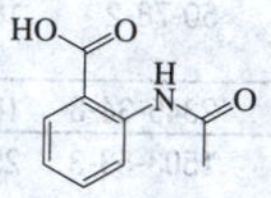
45
2-(Acetylamino)benzoic acid

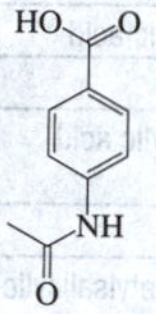
46
4-(Acetylamino)benzoic acid

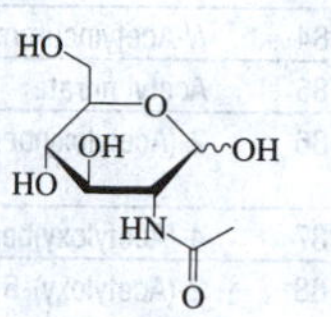
47
2-(Acetylamino)-2-deoxy-D-glucose

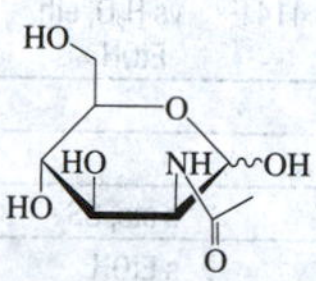
48
2-(Acetylamino)-2-deoxy-D-mannose

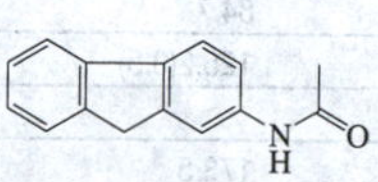
49
2-(Acetylamino)fluorene

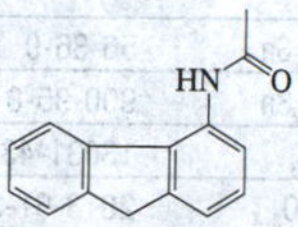
50
4-(Acetylamino)fluorene

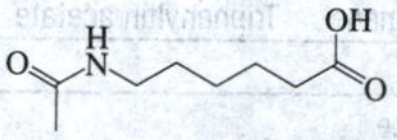
51
6-(Acetylamino)hexanoic acid

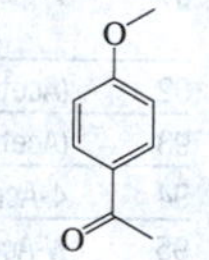
52
4-Acetylanisole

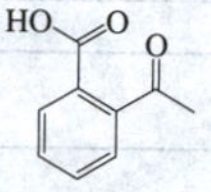
53
2-Acetylbenzoic acid

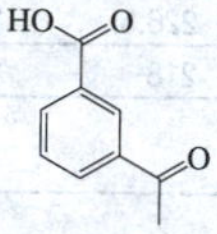
54
3-Acetylbenzoic acid

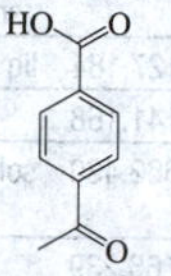
55
4-Acetylbenzoic acid

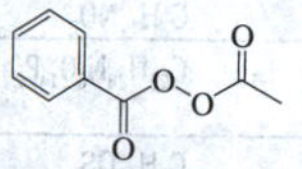
56
Acetyl benzoylperoxide

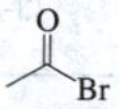
57
Acetyl bromide

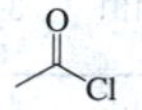
58
Acetyl chloride

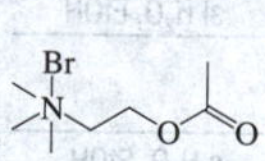
59
Acetylcholine bromide

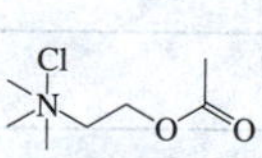
60
Acetylcholine chloride

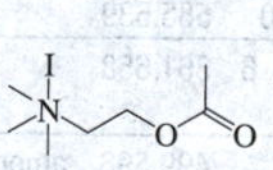
61
Acetylcholine iodide

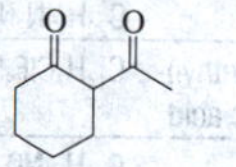
62
2-Acetylcyclohexanone

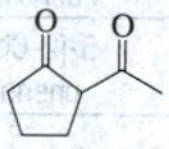
63
2-Acetylcyclopentanone

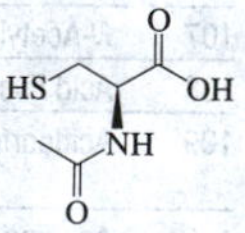
64
N-Acetyl-L-cysteine

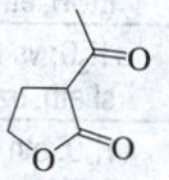
65
3-Acetyldihydro-2(3*H*)-furanone

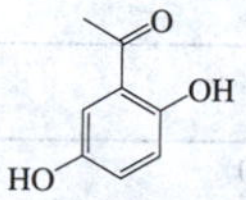
66
1-Acetyl-2,5-dihydroxybenzene

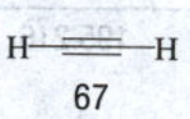
67
Acetylene

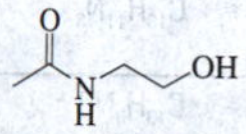
68
N-Acetylethanolamine

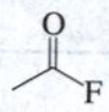
69
Acetyl fluoride

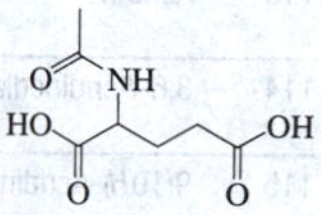
70
N-Acetylglutamic acid

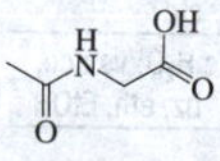
71
N-Acetylglycine

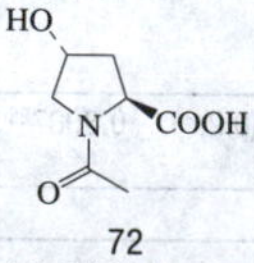
72
trans-1-Acetyl-4-hydroxy-L-proline

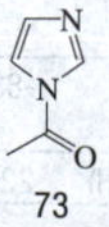
73
1-Acetyl-1*H*-imidazole

74
Acetyl iodide

75
Acetyl isothiocyanate

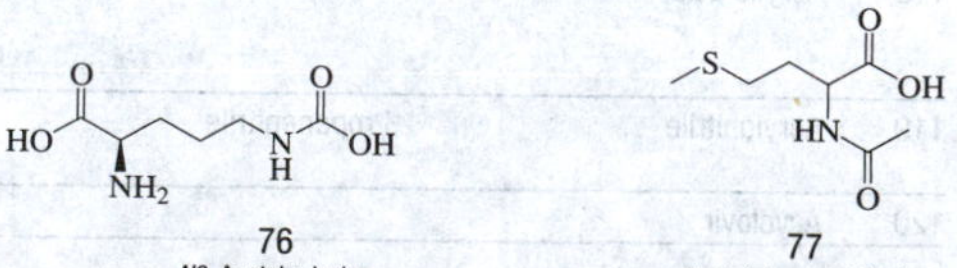
76
*N*6-Acetyl-L-lysine

77
N-Acetyl-DL-methionine

No.	Name	Synonym	Mol. form.	CAS Reg. No.	Mol. wt.	Physical form	mp/°C	bp/°C	den/ g cm^{-3}	n_D	Solubility
78	*N*-Acetyl-L-methionine	Methionamine	$C_7H_{13}NO_3S$	65-82-7	191.248		105.5				
79	1-Acetyl-17-methoxyaspido-spermidine	Aspidospermine	$C_{22}H_{30}N_2O_2$	466-49-9	354.485	nd or pr (al) nd (peth)	208	220^{2}			sl H_2O, eth; s EtOH, bz, chl
80	*N*-Acetyl-*N*-methylacetamide		$C_5H_9NO_2$	1113-68-4	115.131	liq	-25	195	1.0663^{25}	1.4502^{25}	msc H_2O; i eth
81	1-Acetyl-3-methylpiperidine		$C_8H_{15}NO$	4593-16-2	141.211	liq	-13.6	239	0.9684^{25}	1.4731^{25}	vs H_2O
82	3-Acetyl-6-methyl-2*H*-pyran-2,4(3*H*)-dione	Dehydroacetic acid	$C_8H_8O_4$	520-45-6	168.148		109	270			vs H_2O, eth; sl EtOH, chl
83	4-Acetylmorpholine		$C_6H_{11}NO_2$	1696-20-4	129.157		14.5	152^{50}	1.1145^{20}	1.4827^{20}	msc H_2O; s EtOH, ace, ctc
84	*N*-Acetylneuraminic acid	Aceneuramic acid	$C_{11}H_{19}NO_9$	131-48-6	309.271		186				
85	Acetyl nitrate		$C_2H_3NO_4$	591-09-3	105.050			60 exp	1.24^{15}		
86	2-(Acetyloxy)benzoic acid	Acetylsalicylic acid	$C_9H_8O_4$	50-78-2	180.158	nd (w), mcl tab (w)	136(4)				s H_2O, eth, chl; vs EtOH; sl bz
87	4-(Acetyloxy)benzoic acid		$C_9H_8O_4$	2345-34-8	180.158		188.5				
88	2-(Acetyloxy)-5-bromobenzoic acid	5-Bromoacetylsalicylic acid	$C_9H_7BrO_4$	1503-53-3	259.054	nd (al)	60				i H_2O; vs EtOH, eth
89	4-(Acetyloxy)-3-methoxybenz-aldehyde		$C_{10}H_{10}O_4$	881-68-5	194.184		78				sl H_2O; vs EtOH, eth
90	2-(Acetyloxy)-1-phenyletha-none		$C_{10}H_{10}O_3$	2243-35-8	178.184	orth pl	49	270	1.1169^{65}	1.5036^{65}	i H_2O; vs EtOH, eth, chl; sl bz, lig
91	1-(Acetyloxy)-2-propanone	Acetoxyacetone	$C_5H_8O_3$	592-20-1	116.116			171	1.0757^{20}	1.4141^{20}	vs H_2O, eth, EtOH
92	(Acetyloxy)tributylstannane	Tributyltin acetate	$C_{14}H_{30}O_2Sn$	56-36-0	349.097		84.7				
93	(Acetyloxy)triphenylstannane	Triphenyltin acetate	$C_{20}H_{18}O_2Sn$	900-95-8	409.066		125.2(0.5)				
94	4-Acetylphenyl acetate		$C_{10}H_{10}O_3$	13031-43-1	178.184						s ctc, CS_2
95	*N*-Acetyl-L-phenylalanine		$C_{11}H_{13}NO_3$	2018-61-3	207.226		173.5				s EtOH
96	*N*-Acetyl-L-phenylalanine, ethyl ester		$C_{13}H_{17}NO_3$	2361-96-8	235.279	cry (EtOH aq)	93				
97	*N*-Acetyl-L-phenylalanine, methyl ester		$C_{12}H_{15}NO_3$	3618-96-0	221.252	nd (peth) or visc oil (chl)	91				
98	Acetyl phosphate		$C_2H_5O_5P$	590-54-5	140.032	unstab in soln					
99	1-Acetylpiperidine		$C_7H_{13}NO$	618-42-8	127.184	liq	-13.4	226.5	1.011^{9}	1.4790^{25}	vs H_2O, EtOH
100	1-Acetyl-4-piperidinone		$C_7H_{11}NO_2$	32161-06-1	141.168			218	1.146^{25}	1.5026^{20}	
101	3-Acetylpyridine adenine dinucleotide	3-Acetyl NAD	$C_{22}H_{28}N_6O_{14}P_2$	86-08-8	662.436	solid					
102	4-Acetylthioanisole		$C_9H_{10}OS$	1778-09-2	166.239		81.5				
103	Acetyl thiocholine iodide		$C_7H_{16}INOS$	1866-15-5	289.177		205				
104	*N*-Acetyl-L-tryptophan		$C_{13}H_{14}N_2O_3$	1218-34-4	246.261	nd (dil MeOH)	189.5				s H_2O, EtOH, alk
105	*N*-Acetyl-L-tyrosine		$C_{11}H_{13}NO_4$	537-55-3	223.226	cry (w); pl (diox)	153				
106	*N*-Acetyl-L-tyrosine ethyl ester		$C_{13}H_{17}NO_4$	840-97-1	251.279		80.5				
107	*N*-Acetyl-L-valine		$C_7H_{13}NO_3$	96-81-1	159.183		164				
108	Acid Fuchsin	Fuchsin, acid	$C_{20}H_{17}N_3Na_2O_9S_3$	3244-88-0	585.539						sl H_2O, EtOH
109	Acifluorfen	5-[2-Chloro-4-(trifluoromethyl)-phenoxy]-2-nitrobenzoic acid	$C_{14}H_7ClF_3NO_5$	50594-66-6	361.658		164.3(0.5)				
110	Aconine		$C_{25}H_{41}NO_9$	509-20-6	499.596	amor	132				s H_2O, EtOH, chl; sl eth, lig
111	Aconitine		$C_{34}H_{47}NO_{11}$	302-27-2	645.737	orth lf	204				vs bz, EtOH, chl
112	9-Acridinamine	Aminacrine	$C_{13}H_{10}N_2$	90-45-9	194.231	ye nd (ace or al)	241				s EtOH, ace; sl DMSO; vs dil HCl
113	Acridine	Dibenzo[*b*,*e*]pyridine	$C_{13}H_9N$	260-94-6	179.217	orth nd or pr (al)	110.06(0.05)	346.9(1)	1.005^{20}		i H_2O; sl ctc; vs EtOH, eth, bz
114	3,6-Acridinediamine	Proflavine	$C_{13}H_{11}N_3$	92-62-6	209.246	ye nd (al or w)	285				s H_2O; vs EtOH; sl eth, bz
115	9(10*H*)-Acridinone		$C_{13}H_9NO$	578-95-0	195.216	ye lf (al)	>300				i H_2O, eth, bz; sl EtOH; s HOAc, alk
116	Acrolein	2-Propenal	C_3H_4O	107-02-8	56.063	liq	-87.8(0.9)	52.3(0.1)	0.840^{20}	1.4017^{20}	vs H_2O; s EtOH, eth, ace; sl chl
117	Acrylamide	2-Propenamide	C_3H_5NO	79-06-1	71.078	lf (bz)	85(1)	192.6			vs H_2O, chl; s EtOH, eth, ace
118	Acrylic acid	2-Propenoic acid	$C_3H_4O_2$	79-10-7	72.063	acrid liq	13.56(0.05)	142(2)	1.0511^{20}	1.4224^{20}	msc H_2O, EtOH, eth; s ace, bz, ctc
119	Acrylonitrile	Propenenitrile	C_3H_3N	107-13-1	53.063	liq	-83.51(0.05)	77.2(0.2)	0.8007^{25}	1.3911^{20}	s H_2O; vs ace, bz, eth, EtOH
120	Acyclovir		$C_8H_{11}N_5O_3$	59277-89-3	225.205	cry (EtOH)	225				

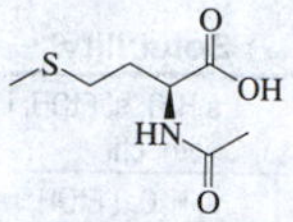
78
N-Acetyl-L-methionine

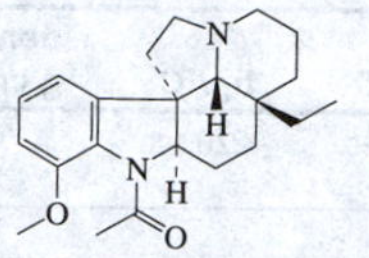
79
1-Acetyl-17-methoxyaspidospermidine

80
N-Acetyl-*N*-methylacetamide

81
1-Acetyl-3-methylpiperidine

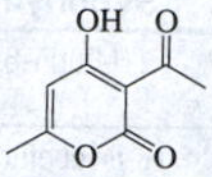
82
3-Acetyl-6-methyl-2*H*-pyran-2,4(3*H*)-dione

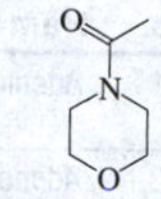
83
4-Acetylmorpholine

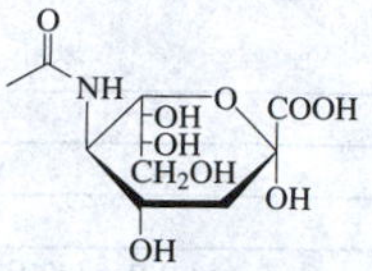
84
N-Acetylneuraminic acid

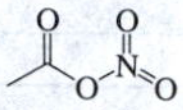
85
Acetyl nitrate

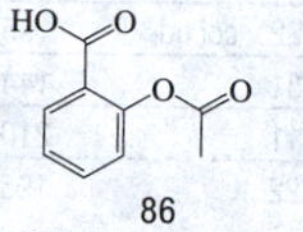
86
2-(Acetyloxy)benzoic acid

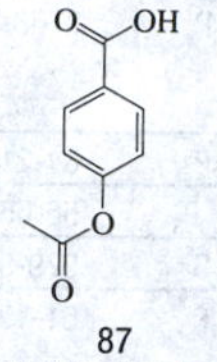
87
4-(Acetyloxy)benzoic acid

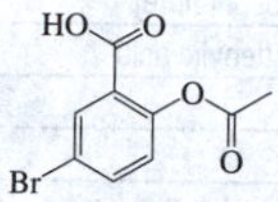
88
2-(Acetyloxy)-5-bromobenzoic acid

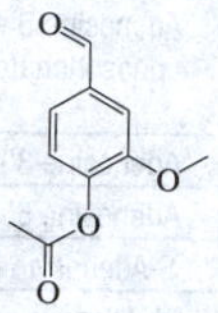
89
4-(Acetyloxy)-3-methoxybenzaldehyde

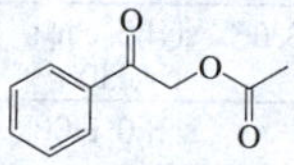
90
2-(Acetyloxy)-1-phenylethanone

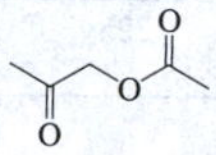
91
1-(Acetyloxy)-2-propanone

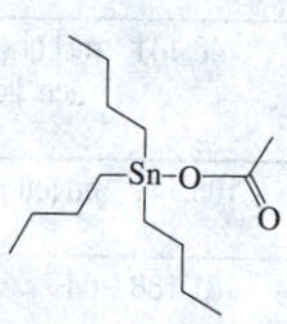
92
(Acetyloxy)tributylstannane

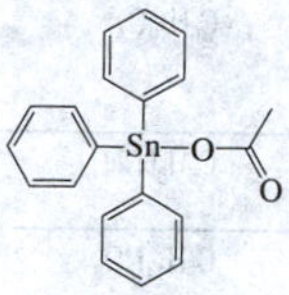
93
(Acetyloxy)triphenylstannane

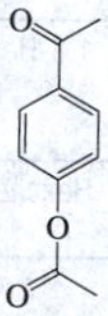
94
4-Acetylphenyl acetate

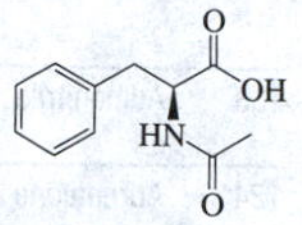
95
N-Acetyl-L-phenylalanine

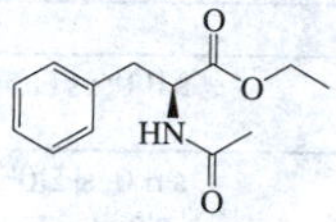
96
N-Acetyl-L-phenylalanine, ethyl ester

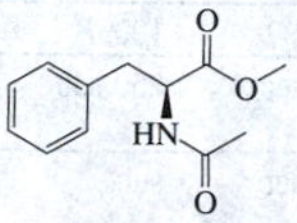
97
N-Acetyl-L-phenylalanine, methyl ester

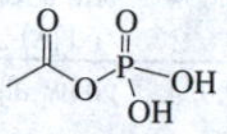
98
Acetyl phosphate

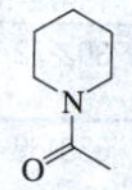
99
1-Acetylpiperidine

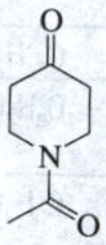
100
1-Acetyl-4-piperidinone

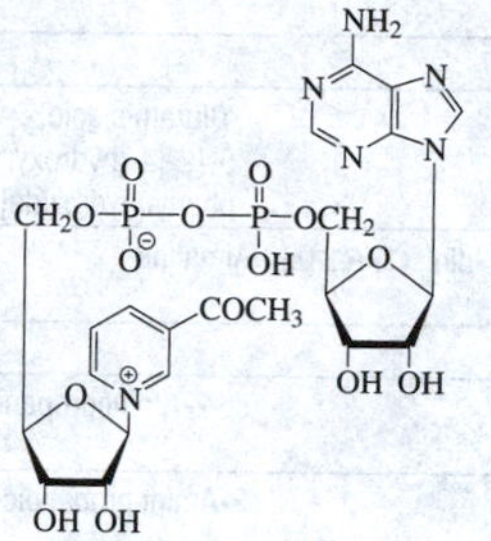
101
3-Acetylpyridine adenine dinucleotide

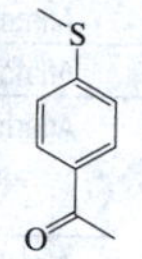
102
4-Acetylthioanisole

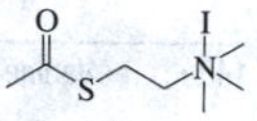
103
Acetyl thiocholine iodide

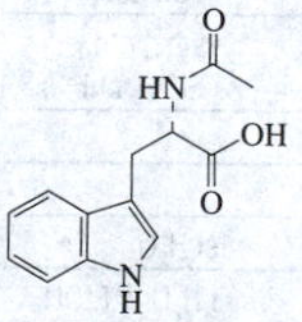
104
N-Acetyl-L-tryptophan

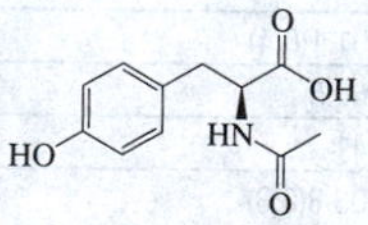
105
N-Acetyl-L-tyrosine

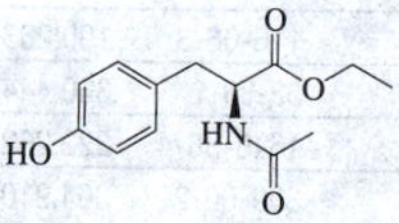
106
N-Acetyl-L-tyrosine ethyl ester

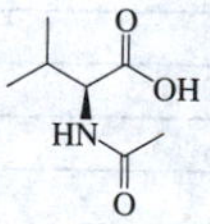
107
N-Acetyl-L-valine

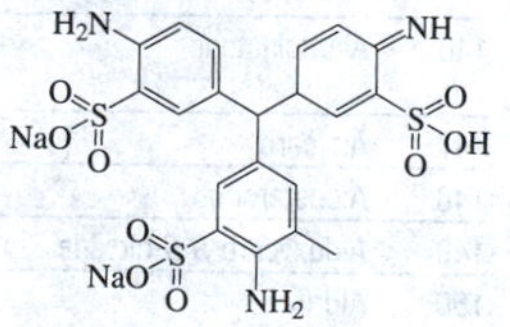
108
Acid Fuchsin

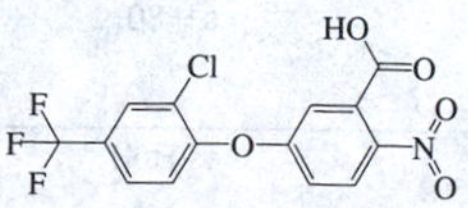
109
Acifluorfen

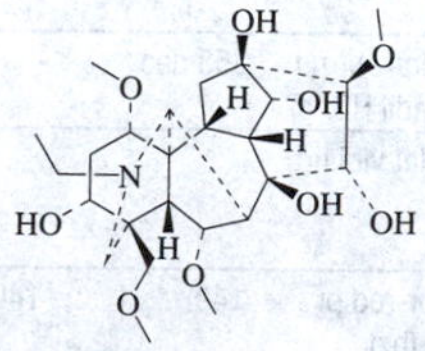
110
Aconine

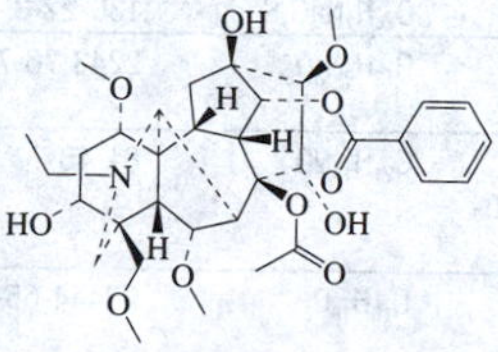
111
Aconitine

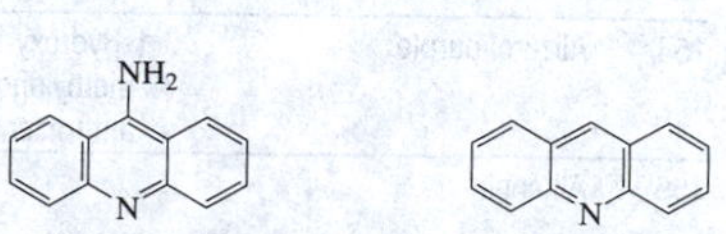
112
9-Acridinamine

113
Acridine

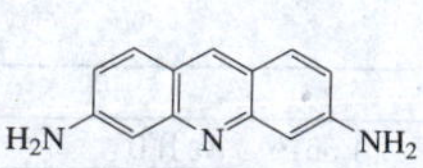
114
3,6-Acridinediamine

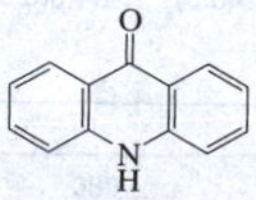
115
9(10*H*)-Acridinone

116
Acrolein

117
Acrylamide

118
Acrylic acid

119
Acrylonitrile

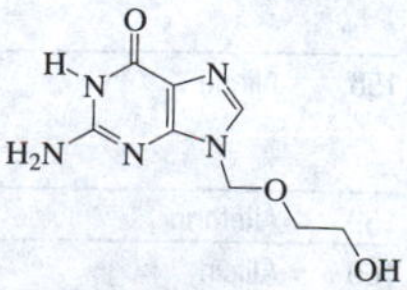
120
Acyclovir

Organic

No.	Name	Synonym	Mol. form.	CAS Reg. No.	Mol. wt.	Physical form	mp/°C	bp/°C	den/ g cm^{-3}	n_D	Solubility
121	Adenine	1*H*-Purin-6-amine	$C_5H_5N_5$	73-24-5	135.128	orth nd (+3w)	360 dec	220 sub			s H_2O; sl EtOH; i eth, chl
122	Adenosine	β-D-Ribofuranoside, adenine-9	$C_{10}H_{13}N_5O_4$	58-61-7	267.242	n(w+3/2)	235.5				sl H_2O; i EtOH
123	Adenosine cyclic 3',5'-(hydrogen phosphate)	cAMP	$C_{10}H_{12}N_5O_6P$	60-92-4	329.206	cry	219				
124	Adenosine 3',5'-diphosphate	3'-Adenylic acid, 5'-(dihydrogen phosphate)	$C_{10}H_{15}N_5O_{10}P_2$	1053-73-2	427.202	amor pow					
125	Adenosine 5'-methylenediphosphonate	Adenosine, 5'-[hydrogen (phosphonomethyl)-phosphonate]	$C_{11}H_{17}N_5O_9P_2$	3768-14-7	425.229	cry (w)	204				s H_2O
126	Adenosine 3'-phosphate	3'-Adenylic acid	$C_{10}H_{14}N_5O_7P$	84-21-9	347.222	col nd	195 dec				
127	Adenosine 5'-triphosphate	ATP	$C_{10}H_{16}N_5O_{13}P_3$	56-65-5	507.181		144 dec				
128	*S*-Adenosyl-L-homocysteine		$C_{14}H_{20}N_6O_5S$	979-92-0	384.411		210 dec				
129	5'-Adenylic acid	Adenosine 5'-monophosphate	$C_{10}H_{14}N_5O_7P$	61-19-8	347.222		195 dec				vs H_2O; s EtOH, 10% HCl
130	Adipamic acid		$C_6H_{11}NO_3$	334-25-8	145.156	nd (w)	161.5				
131	Adiphenine hydrochloride		$C_{20}H_{26}ClNO_2$	50-42-0	347.879	cry	113.5				vs H_2O; sl EtOH, eth
132	Adipic acid	1,6-Hexanedioic acid	$C_6H_{10}O_4$	124-04-9	146.141	mcl pr (w, ace, lig)	151.5(0.6)	337.5	1.360^{25}		sl H_2O; vs EtOH; s eth; i HOAc, lig
133	Adiponitrile	Hexanedinitrile	$C_6H_8N_2$	111-69-3	108.141	nd (eth)	2.2(0.4)	295	0.9676^{20}	1.4380^{20}	sl H_2O, eth; s chl, EtOH
134	Adrenalone		$C_9H_{11}NO_3$	99-45-6	181.188	nd	235 dec				sl H_2O, EtOH, eth
135	Affinin	*N*-(2-Methylpropyl)-2,6,8-decatrienamide	$C_{14}H_{23}NO$	25394-57-4	221.339	ye oil	23	$162^{0.5}$		1.5134^{25}	i H_2O
136	Aflatoxin B1		$C_{17}H_{12}O_6$	1162-65-8	312.273	cry	268				
137	Aflatoxin B2		$C_{17}H_{14}O_6$	7220-81-7	314.289		287.5				
138	Aflatoxin G1		$C_{17}H_{12}O_7$	1165-39-5	328.273	cry	245				
139	Agaritine	L-Glutamic acid, 5-[2-[4-(hydroxymethyl)-phenyl]hydrazide]	$C_{12}H_{17}N_3O_4$	2757-90-6	267.281	cry (dil al)	207 dec				vs H_2O
140	Ajmalan-17,21-diol, (17*R*,21α)	Ajmaline	$C_{20}H_{26}N_2O_2$	4360-12-7	326.432	pl (+3.5w) (aq AcOEt)	206				i H_2O; s EtOH, chl; sl eth, bz
141	Alachlor		$C_{14}H_{20}ClNO_2$	15972-60-8	269.768		42(2)	$100^{0.02}$	1.133^{25}		
142	DL-Alanine	DL-2-Aminopropanoic acid	$C_3H_7NO_2$	302-72-7	89.094	orth pr or nd (w)	300 dec	250 sub	1.424^{25}		s H_2O; vs EtOH
143	D-Alanine	2-Aminopropanoic acid, (*R*)	$C_3H_7NO_2$	338-69-2	89.094	nd (w, al)	314 dec	sub			s H_2O; sl EtOH; i eth
144	L-Alanine	2-Aminopropanoic acid, (*S*)	$C_3H_7NO_2$	56-41-7	89.094	orth (w)	297 dec	250 sub	1.432^{22}		s H_2O; sl EtOH, py; i eth, ace
145	β-Alanine	3-Aminopropanoic acid	$C_3H_7NO_2$	107-95-9	89.094	nd, orth pr (al)	200 dec		1.437^{19}		s H_2O; sl EtOH; i eth, ace
146	Alantolactone		$C_{15}H_{20}O_2$	546-43-0	232.319	nd	76	275			vs bz, eth, EtOH, chl
147	Aldicarb		$C_7H_{14}N_2O_2S$	116-06-3	190.263		101.1(0.4)		1.195^{25}		
148	Aldosterone		$C_{21}H_{28}O_5$	52-39-1	360.444	cry (HOAc)	166.5				
149	Aldoxycarb *S*,*S*-dioxide		$C_7H_{14}N_2O_4S$	1646-88-4	222.262	cry	141				sl H_2O
150	Aldrin		$C_{12}H_8Cl_6$	309-00-2	364.910		103.8(0.3)				i H_2O; s EtOH, eth, ace, bz
151	Alizarin	1,2-Dihydroxy-9,10-anthracenedione	$C_{14}H_8O_4$	72-48-0	240.212	oran or red tcl nd or pr (al)	289.5				sl H_2O; s EtOH, eth, ace, bz; i chl
152	Alizarin Red S	Sodium alizarinesulfonate	$C_{14}H_7NaO_7S$	130-22-3	342.257						vs H_2O; s EtOH
153	Alizarin Yellow R		$C_{13}H_9N_3O_5$	2243-76-7	287.227	oran-br nd (dil HOAc)	253 dec				vs H_2O, EtOH
154	Alizurol purple	1-Hydroxy-4-[(4-methylphenyl)amino]-9,10-anthracenedione	$C_{21}H_{15}NO_3$	81-48-1	329.349	flat viol nd					s H_2SO_4
155	Alkannin		$C_{16}H_{16}O_5$	23444-65-7	288.295	br-red pr (bz)	149	140 sub			vs EtOH
156	Allantoic acid	Bis[(aminocarbonyl) amino]acetic acid	$C_4H_8N_4O_4$	99-16-1	176.132	nd	170 dec				sl H_2O, os, dil acid
157	Allantoin		$C_4H_6N_4O_3$	97-59-6	158.116	mcl pl or	239				sl H_2O; s EtOH, NaOH; i eth, MeOH
158	Allene		C_3H_4	463-49-0	40.064	col gas	-136.4(0.5)	-34.8(0.3)	0.584^{25} (p>1 atm)	1.4168	vs bz, peth
159	Allethrin		$C_{19}H_{26}O_3$	584-79-2	302.407				1.010^{20}		
160	Allicin		$C_6H_{10}OS_2$	539-86-6	162.272			dec	1.112^{20}	1.561^{20}	vs H_2O

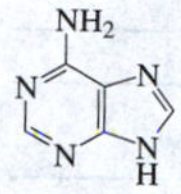
121
Adenine

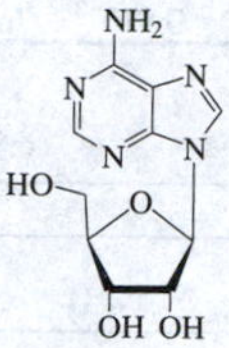
122
Adenosine

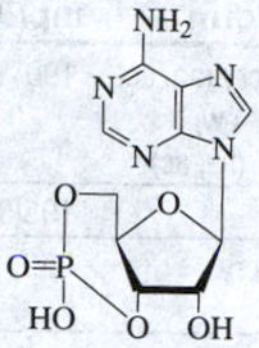
123
Adenosine cyclic 3',5'-(hydrogen phosphate)

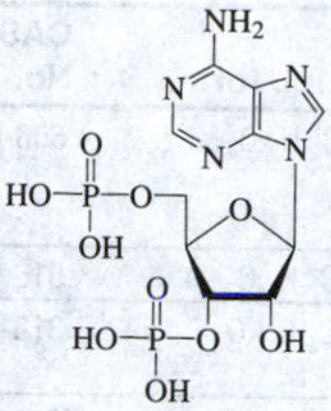
124
Adenosine 3',5'-diphosphate

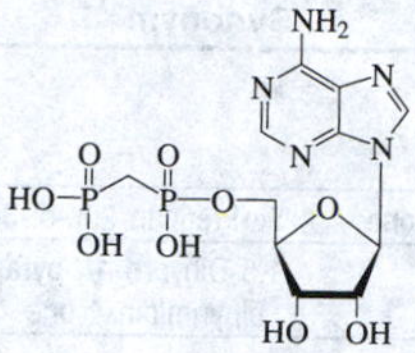
125
Adenosine 5'-methylenediphosphonate

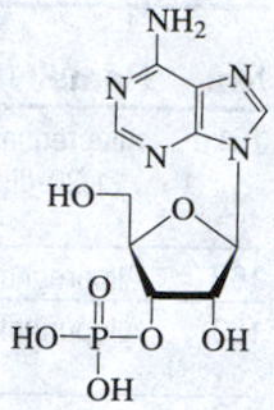
126
Adenosine 3'-phosphate

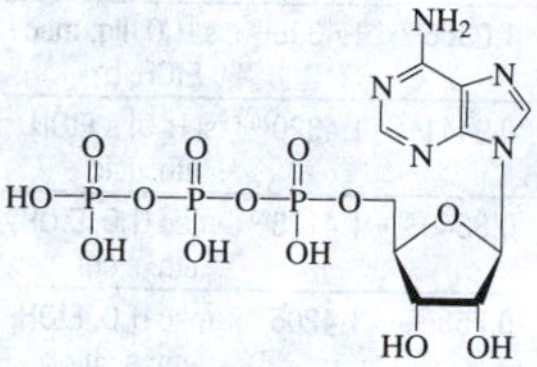
127
Adenosine 5'-triphosphate

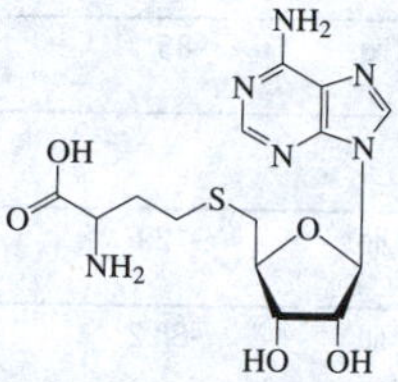
128
S-Adenosyl-L-homocysteine

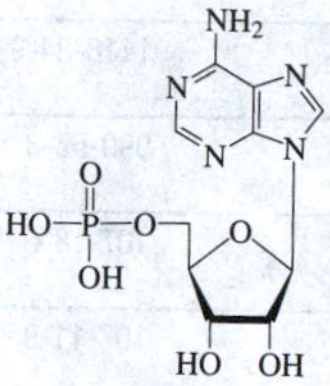
129
5'-Adenylic acid

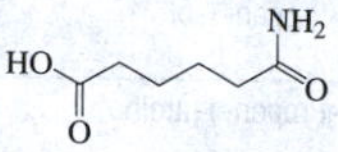
130
Adipamic acid

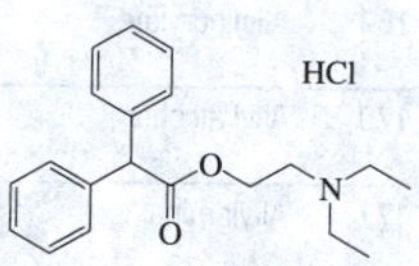
131
Adiphenine hydrochloride

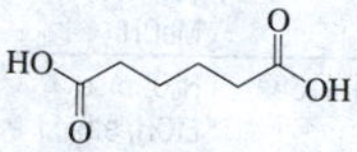
132
Adipic acid

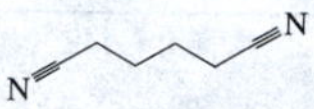
133
Adiponitrile

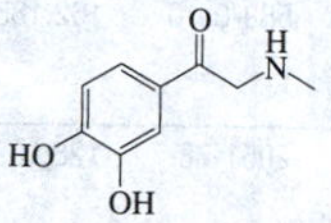
134
Adrenalone

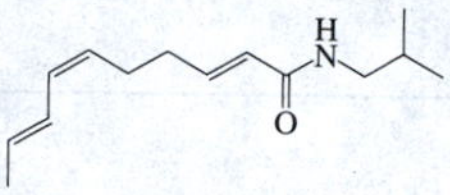
135
Affinin

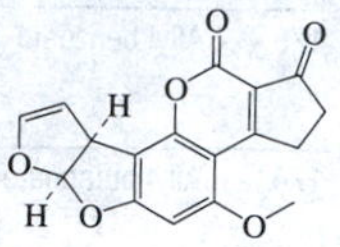
136
Aflatoxin B1

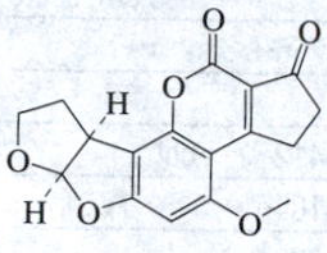
137
Aflatoxin B2

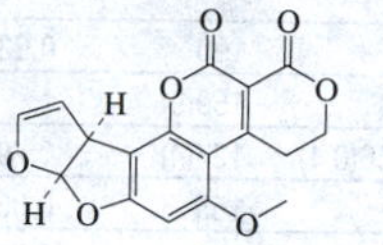
138
Aflatoxin G1

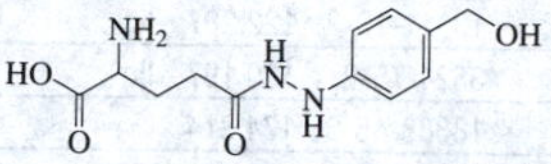
139
Agaritine

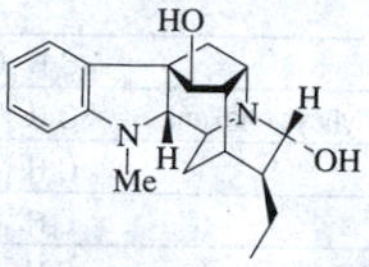
140
Ajmalan-17,21-diol, (17*R*,21α)

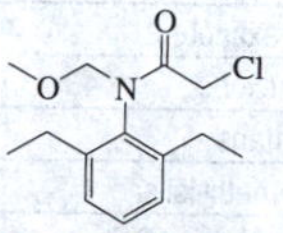
141
Alachlor

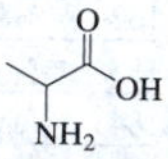
142
DL-Alanine

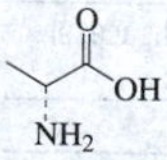
143
D-Alanine

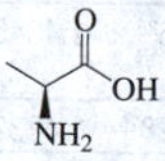
144
L-Alanine

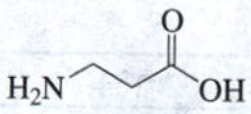
145
β-Alanine

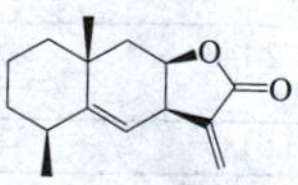
146
Alantolactone

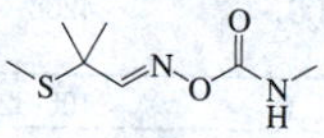
147
Aldicarb

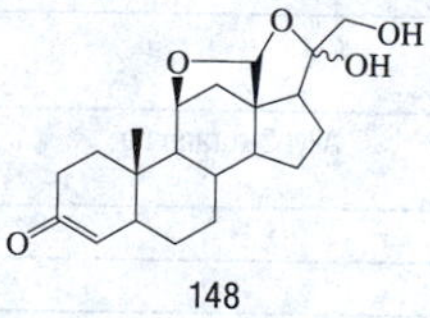
148
Aldosterone

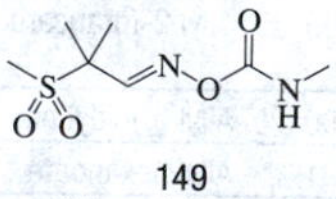
149
Aldoxycarb *S,S*-dioxide

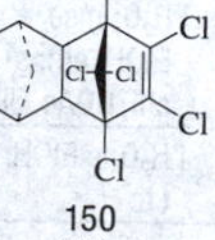
150
Aldrin

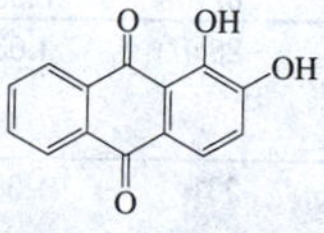
151
Alizarin

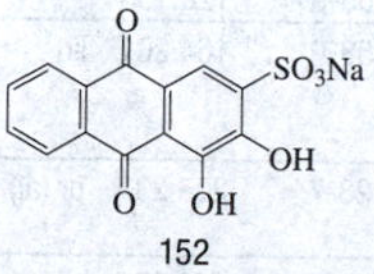
152
Alizarin Red S

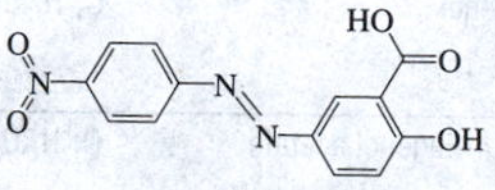
153
Alizarin Yellow R

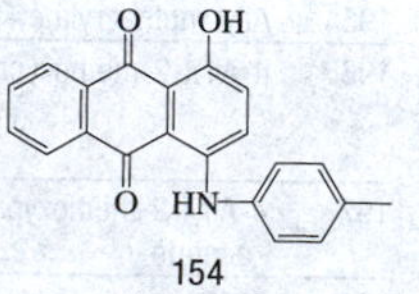
154
Alizurol purple

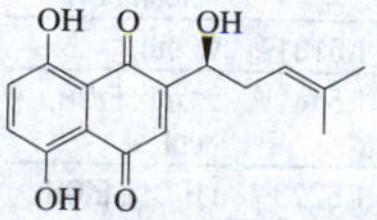
155
Alkannin

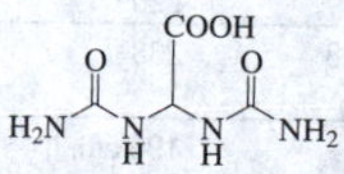
156
Allantoic acid

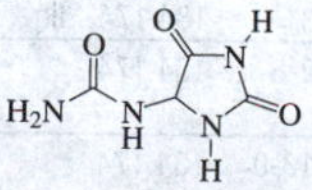
157
Allantoin

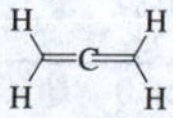
158
Allene

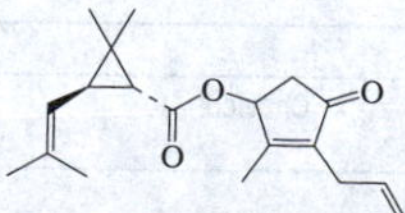
159
Allethrin

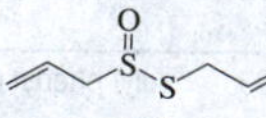
160
Allicin

No.	Name	Synonym	Mol. form.	CAS Reg. No.	Mol. wt.	Physical form	mp/°C	bp/°C	den/ g cm^{-3}	n_D	Solubility
161	Allopregnane-3β,21-diol-11,20-dione		$C_{21}H_{32}O_4$	566-02-9	348.477	cry (aq, ac, +w) nd (bz, ac)	190				
162	Allopregnan-20β-ol-3-one	5α-Pregnan-20β-ol-3-one	$C_{21}H_{34}O_2$	516-58-5	318.494		195(3)				
163	Allopurinol	1,5-Dihydro-4*H*-pyrazolo[3,4-*d*]pyrimidin-4-one	$C_5H_4N_4O$	315-30-0	136.112	cry	350				
164	D-Allose		$C_6H_{12}O_6$	2595-97-3	180.155	cry (w)	128				vs H_2O
165	Alloxanic acid		$C_4H_4N_2O_5$	470-44-0	160.085	tcl pr (eth)	162 dec				vs H_2O, EtOH
166	Alloxantin		$C_8H_6N_4O_8$	76-24-4	286.156	orth pr (w+2)	254 dec				sl H_2O, EtOH, eth
167	Allyl acetate	3-Acetoxypropene	$C_5H_8O_2$	591-87-7	100.117			104(2)	0.9275^{20}	1.4049^{20}	sl H_2O; s ace; msc EtOH, eth
168	Allyl acetoacetate		$C_7H_{10}O_3$	1118-84-9	142.152	liq	-85	198(19)	1.0366^{20}	1.4398^{20}	s H_2O, lig; msc EtOH, bz
169	Allyl acrylate		$C_6H_8O_2$	999-55-3	112.127			121	0.9441^{20}	1.4320^{20}	sl H_2O; s EtOH, eth, acid
170	Allyl alcohol	2-Propen-1-ol	C_3H_6O	107-18-6	58.079	liq	-129	96.9(0.5)	0.8540^{20}	1.4135^{20}	msc H_2O, EtOH, eth; s chl
171	Allylamine	2-Propen-1-amine	C_3H_7N	107-11-9	57.095	liq	-88.2	54(2)	0.758^{20}	1.4205^{20}	msc H_2O, EtOH, eth; s chl
172	*N*-Allylaniline	Allylphenylamine	$C_9H_{11}N$	589-09-3	133.190			219	0.9736^{25}	1.563^{20}	sl H_2O; s EtOH, ace; msc eth
173	Allylbenzene	2-Propenylbenzene	C_9H_{10}	300-57-2	118.175	liq	-40(4)	158(2)	0.8920^{20}	1.5131^{20}	i H_2O; s EtOH, eth, bz, ctc
174	α-Allylbenzenemethanol		$C_{10}H_{12}O$	936-58-3	148.201			228.5	1.004^{18}	1.5289^{21}	
175	Allyl benzoate		$C_{10}H_{10}O_2$	583-04-0	162.185				1.0569^{15}	1.5178^{20}	i H_2O; s EtOH, eth, ace, MeOH
176	Allyl butanoate		$C_7H_{12}O_2$	2051-78-7	128.169			142	0.9017^{20}	1.4158^{20}	i H_2O; msc EtOH, eth; sl ctc
177	Allyl carbamate		$C_4H_7NO_2$	2114-11-6	101.105						sl ctc
178	Allylchlorodimethylsilane		$C_5H_{11}ClSi$	4028-23-3	134.680			111	0.8964^{20}	1.4195^{20}	
179	Allyl chloroformate		$C_4H_5ClO_2$	2937-50-0	120.535	hyg liq		109.5	1.136	1.4220^{20}	
180	Allyl *trans*-cinnamate	Allyl *trans*-3-phenyl-2-propenoate	$C_{12}H_{12}O_2$	1866-31-5	188.222			268 dec	1.048^{23}	1.530^{20}	i H_2O; vs EtOH; msc eth; sl ctc
181	1-Allylcyclohexanol		$C_9H_{16}O$	1123-34-8	140.222			190	0.9341^{22}	1.4756^{22}	
182	1-Allylcyclohexene	1-(2-Propenyl)cyclohexene	C_9H_{14}	13511-13-2	122.207	liq		159(9)			
183	Allylcyclopentane		C_8H_{14}	3524-75-2	110.197	liq	-110.6(0.1)	127(4)	0.793^{25}	1.4412^{20}	s chl
184	Allyldiethoxymethylsilane		$C_8H_{18}O_2Si$	18388-45-9	174.314			155	0.8572^{25}	1.4104^{20}	
185	Allyldiethylamine	*N*,*N*-Diethyl-2-propen-1-amine	$C_7H_{15}N$	5666-17-1	113.201			110	0.7477^{25}	1.4209^{20}	
186	Allyldimethylamine	*N*,*N*-Dimethyl-2-propen-1-amine	$C_5H_{11}N$	2155-94-4	85.148			62(4)	0.7094^{25}	1.4010^{20}	
187	Allyl ethyl ether		$C_5H_{10}O$	557-31-3	86.132			65(4)	0.7651^{20}	1.3881^{20}	i H_2O; msc EtOH, eth; s ace
188	Allyl formate		$C_4H_6O_2$	1838-59-1	86.090			83.6	0.9460^{20}		sl H_2O; s EtOH; msc eth
189	Allyl 2-furancarboxylate	Allyl 2-furanoate	$C_8H_8O_3$	4208-49-5	152.148			207.5	1.115^{25}	1.4945^{20}	s eth, ace; sl ctc
190	Allyl glycidyl ether		$C_6H_{10}O_2$	106-92-3	114.142			154	0.9698^{20}	1.4332^{20}	
191	Allyl hexanoate		$C_9H_{16}O_2$	123-68-2	156.222			186	0.8869^{20}		
192	Allyl (hydroxymethyl)-carbamate		$C_5H_9NO_3$	24935-97-5	131.130	cry (tol)	57				
193	Allyl isocyanate		C_4H_5NO	1476-23-9	83.089			88			
194	Allyl isothiocyanate		C_4H_5NS	57-06-7	99.155	liq	-80	160(6)	1.0126^{20}	1.5306^{20}	vs bz, eth, EtOH
195	Allyl methacrylate		$C_7H_{10}O_2$	96-05-9	126.153			67^{50}	0.9335^{20}	1.4360^{20}	
196	4-Allyl-2-methoxyphenol	Eugenol	$C_{10}H_{12}O_2$	97-53-0	164.201	liq	-7.5	254(7)	1.0652^{20}	1.5405^{20}	i H_2O; msc EtOH, eth; s chl, HOAc, oils
197	4-Allyl-2-methoxyphenyl acetate	1,3,4-Eugenol acetate	$C_{12}H_{14}O_3$	93-28-7	206.237	pr (al)	30.5	281	1.0806^{20}	1.5205^{20}	i H_2O; s EtOH; sl ctc
198	Allyl 3-methylbutanoate		$C_8H_{14}O_2$	2835-39-4	142.196			154			
199	Allylmethyldichlorosilane		$C_4H_8Cl_2Si$	1873-92-3	155.099			119.5	1.0758^{20}	1.4419^{20}	
200	2-(Allyloxy)ethanol	Ethylene glycol monoallyl ether	$C_5H_{10}O_2$	111-45-5	102.132			158.5	0.9580^{20}	1.4358^{20}	msc H_2O; vs EtOH; s bz, ctc, MeOH
201	2-Allylphenol		$C_9H_{10}O$	1745-81-9	134.174	liq	-6	220	1.0246^{15}	1.5181^{20}	vs eth
202	4-Allylphenol	Chavicol	$C_9H_{10}O$	501-92-8	134.174		15.8	238	1.0203^{15}	1.5441^{18}	vs eth, EtOH, chl
203	Allyl phenyl ether		$C_9H_{10}O$	1746-13-0	134.174			190.5(0.8)	0.9811^{20}	1.5223^{20}	i H_2O; s EtOH; msc eth; sl ctc

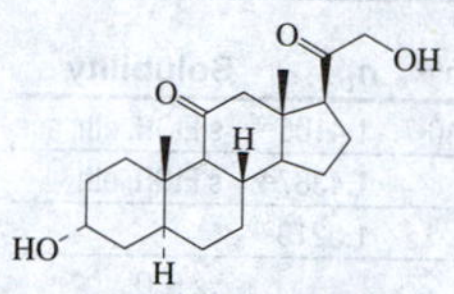
161
Allopregnane-3β,21-diol-11,20-dione

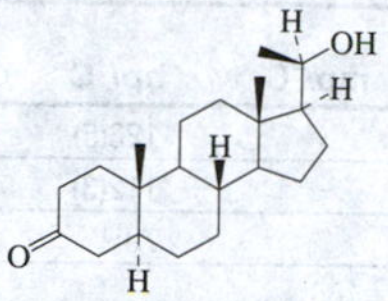
162
Allopregnan-20β-ol-3-one

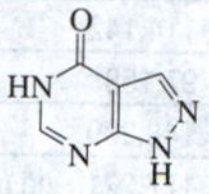
163
Allopurinol

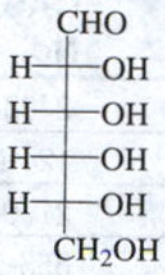
164
D-Allose

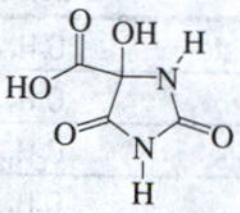
165
Alloxanic acid

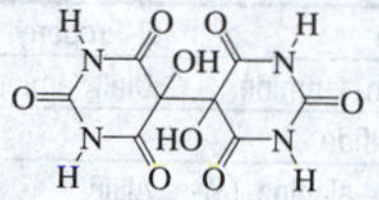
166
Alloxantin

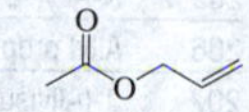
167
Allyl acetate

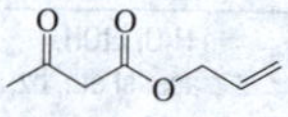
168
Allyl acetoacetate

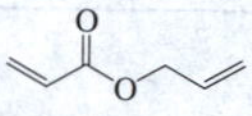
169
Allyl acrylate

170
Allyl alcohol

171
Allylamine

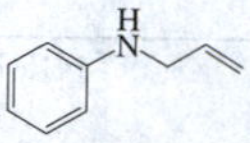
172
N-Allylaniline

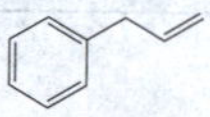
173
Allylbenzene

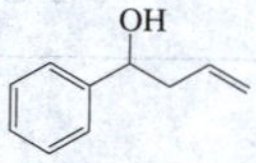
174
α-Allylbenzenemethanol

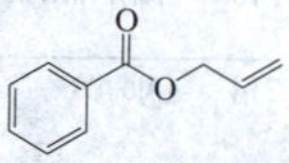
175
Allyl benzoate

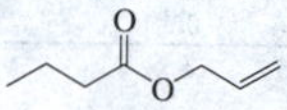
176
Allyl butanoate

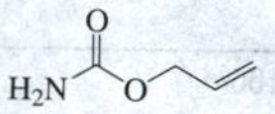
177
Allyl carbamate

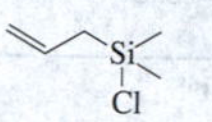
178
Allylchlorodimethylsilane

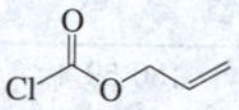
179
Allyl chloroformate

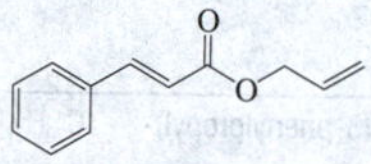
180
Allyl *trans*-cinnamate

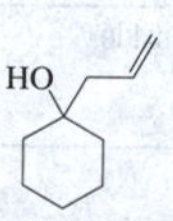
181
1-Allylcyclohexanol

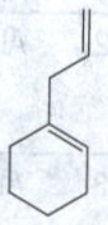
182
1-Allylcyclohexene

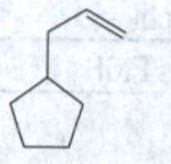
183
Allylcyclopentane

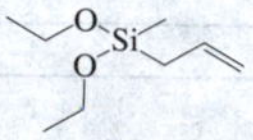
184
Allyldiethoxymethylsilane

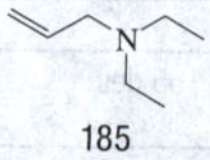
185
Allyldiethylamine

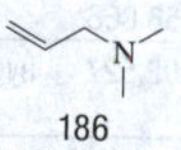
186
Allyldimethylamine

187
Allyl ethyl ether

188
Allyl formate

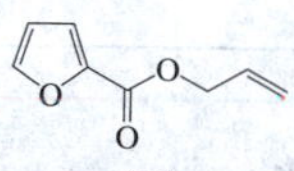
189
Allyl 2-furancarboxylate

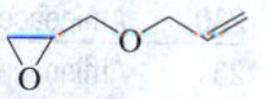
190
Allyl glycidyl ether

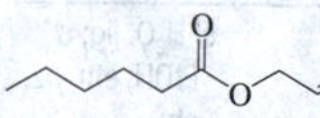
191
Allyl hexanoate

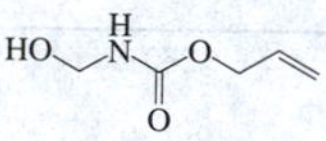
192
Allyl (hydroxymethyl)carbamate

193
Allyl isocyanate

194
Allyl isothiocyanate

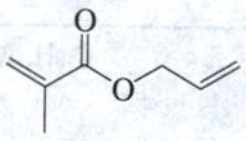
195
Allyl methacrylate

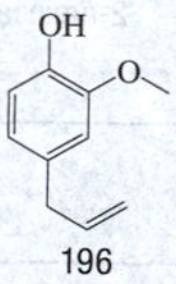
196
4-Allyl-2-methoxyphenol

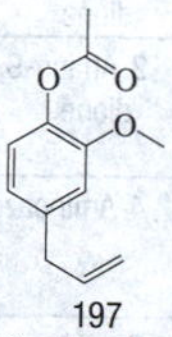
197
4-Allyl-2-methoxyphenyl acetate

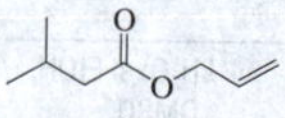
198
Allyl 3-methylbutanoate

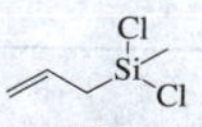
199
Allylmethyldichlorosilane

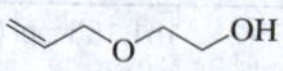
200
2-(Allyloxy)ethanol

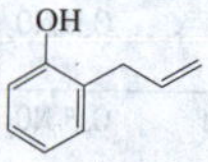
201
2-Allylphenol

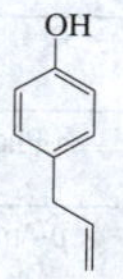
202
4-Allylphenol

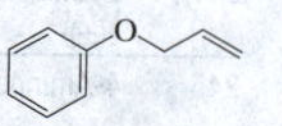
203
Allyl phenyl ether

Organic

No.	Name	Synonym	Mol. form.	CAS Reg. No.	Mol. wt.	Physical form	mp/°C	bp/°C	den/ g cm^{-3}	n_D	Solubility
204	Allyl propanoate	2-Propenyl propanoate	$C_6H_{10}O_2$	2408-20-0	114.142			123(5)	0.9140^{20}	1.4105^{20}	s EtOH, eth, ace
205	*N*-Allyl-2-propen-1-amine	Diallylamine	$C_6H_{11}N$	124-02-7	97.158			112(3)		1.4387^{20}	s EtOH, eth
206	Allyl propyl disulfide		$C_6H_{12}S_2$	2179-59-1	148.289			79^{13}		1.5219^{20}	
207	3-(Allylsulfinyl)-L-alanine, (*S*)-	Alliin	$C_6H_{11}NO_3S$	556-27-4	177.221	nd (dil ac)	165				vs H_2O
208	Allylthiourea	Thiosinamine	$C_4H_8N_2S$	109-57-9	116.185	mcl or orth pr (w)	77(3)		1.217^{20}	1.5936^{78}	s H_2O, EtOH; sl eth; i bz
209	Allyltrichlorosilane	Trichloro-2-propenylsilane	$C_3H_5Cl_3Si$	107-37-9	175.517		35	118(3)	1.2011^{20}	1.4460^{20}	
210	Allyltriethoxysilane		$C_9H_{20}O_3Si$	2550-04-1	204.339			100^{50}	0.9030^{20}	1.4072^{20}	
211	Allyltrimethylsilane		$C_6H_{14}Si$	762-72-1	114.261			85	0.7158^{25}	1.4074^{20}	i H_2O
212	Allylurea		$C_4H_8N_2O$	557-11-9	100.119	nd (al)	85				msc H_2O, EtOH; sl eth, chl; i peth
213	Allyl vinyl ether	3-(Ethenyloxy)-1-propene	C_5H_8O	3917-15-5	84.117			66	0.7900^{20}	1.4062^{20}	i H_2O; s eth, ace, chl
214	Aloin A		$C_{21}H_{22}O_9$	1415-73-2	418.395		149.3				s H_2O, EtOH, ace; sl eth, bz; i chl
215	Alphaprodine		$C_{16}H_{23}NO_2$	15867-21-7	261.360	cry	103				
216	Alstonidine		$C_{22}H_{24}N_2O_4$	25394-75-6	380.437	cry (eth)	189				vs ace, EtOH
217	Alstonine		$C_{21}H_{20}N_2O_3$	642-18-2	348.395	ye nd (ace)	207 dec				
218	D-Altrose		$C_6H_{12}O_6$	1990-29-0	180.155	pr (MeOH,al)	103.5				vs H_2O
219	Aluminum 2-butoxide	2-Butanol, aluminum salt	$C_{12}H_{27}AlO_3$	2269-22-9	246.322			197^{20}			
220	Aluminum distearate	Hydroxyaluminum distearate	$C_{36}H_{71}AlO_5$	300-92-5	610.928	wh pow	145				i H_2O
221	Aluminum ethanolate	Aluminum ethoxide	$C_6H_{15}AlO_3$	555-75-9	162.163	liq/wh solid	140	200^7			dec H_2O; sl xyl
222	Aluminum isopropoxide		$C_9H_{21}AlO_3$	555-31-7	204.243	hyg wh solid	119	135^{10}			reac H_2O; s EtOH, bz, peth, chl
223	Alverine	*N*-Ethyl-bis(3-phenylpropyl)-amine	$C_{20}H_{27}N$	150-59-4	281.435	oil		$166^{0.3}$			
224	α-Amanitin		$C_{39}H_{54}N_{10}O_{14}S$	23109-05-9	918.970	nd	254 dec				
225	Amaranth dye		$C_{20}H_{11}N_2Na_3O_{10}S_3$	915-67-3	604.472	dk red pow					s H_2O
226	Ametryn		$C_9H_{17}N_5S$	834-12-8	227.330		83.6(0.5)				
227	Amminetrimethylboron		$C_3H_{12}BN$	1830-95-1	72.945		73.5				
228	19-Amino-8,11,13-abieta-triene		$C_{20}H_{31}N$	1446-61-3	285.467	cry	44.5				
229	2-Aminoacetamide		$C_2H_6N_2O$	598-41-4	74.081	hyg nd (chl)	67.5				vs H_2O, EtOH; sl eth, bz; s ace, chl
230	Aminoacetonitrile		$C_2H_4N_2$	540-61-4	56.066			58^{15}			vs EtOH
231	Aminoacetonitrile monohydrochloride		$C_2H_5ClN_2$	6011-14-9	92.527	hyg cry (al)	165 dec				
232	α-Aminoacetophenone hydrochloride		$C_8H_{10}ClNO$	5468-37-1	171.624		194 dec				
233	1-Aminoadamantane hydrochloride	Adamantanamine hydrochloride	$C_{10}H_{18}ClN$	665-66-7	187.710	cry (al-eth)	360 dec				vs H_2O, EtOH
234	2-Aminoadipic acid		$C_6H_{11}NO_4$	626-71-1	161.156	pl (w)	207.0				sl H_2O, EtOH, eth
235	3-Aminoalanine	2,3-Diaminopropionic acid	$C_3H_8N_2O_2$	515-94-6	104.108	hyg rosettes	110				vs H_2O
236	1-Amino-9,10-anthracene-dione	1-Aminoanthraquinone	$C_{14}H_9NO_2$	82-45-1	223.227	red nd (al)	253.5	sub			vs ace, bz, EtOH, chl
237	2-Amino-9,10-anthracene-dione	2-Aminoanthraquinone	$C_{14}H_9NO_2$	117-79-3	223.227	red nd (al, HOAc)	304.5	sub			i H_2O, eth; sl EtOH; s ace, bz, chl
238	4-Aminoazobenzene		$C_{12}H_{11}N_3$	60-09-3	197.235	oran mcl nd (al)	125(1)	>360			sl H_2O, lig; s EtOH, eth, bz, chl
239	2-Aminobenzaldehyde		C_7H_7NO	529-23-7	121.137	silv lf	40.5	80^2			sl H_2O; vs EtOH, eth; s bz, chl; i lig
240	3-Aminobenzaldehyde		C_7H_7NO	1709-44-0	121.137	nd (AcOEt)	29				s eth, acid
241	4-Aminobenzaldehyde		C_7H_7NO	556-18-3	121.137	pl (w)	71.5				s H_2O, EtOH, eth, acid
242	2-Aminobenzamide		$C_7H_8N_2O$	88-68-6	136.151		110.5 dec				s H_2O, EtOH; sl eth, bz; vs AcOEt
243	4-Aminobenzamide		$C_7H_8N_2O$	2835-68-9	136.151	ye cry (+1/4w)	183				sl H_2O; s EtOH, eth
244	α-Aminobenzeneacetic acid, (±)-	α-Phenylglycine	$C_8H_9NO_2$	2835-06-5	151.163	pl	292 dec	255 sub			s alk; sl os
245	4-Aminobenzeneacetic acid	*p*-Aminophenylacetic acid	$C_8H_9NO_2$	1197-55-3	151.163	pl (w)	195(1)				i H_2O; sl EtOH, DMSO

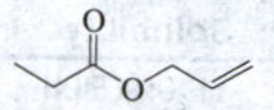
204
Allyl propanoate

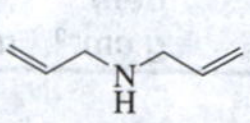
205
N-Allyl-2-propen-1-amine

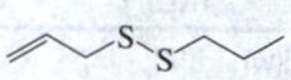
206
Allyl propyl disulfide

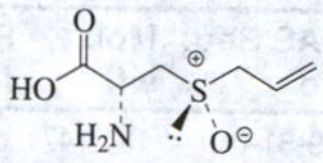
207
3-(Allylsulfinyl)-L-alanine, (*S*)-

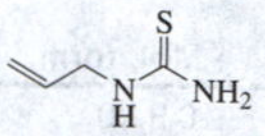
208
Allylthiourea

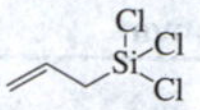
209
Allyltrichlorosilane

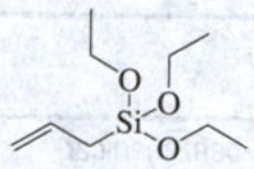
210
Allyltriethoxysilane

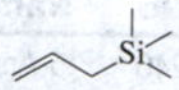
211
Allyltrimethylsilane

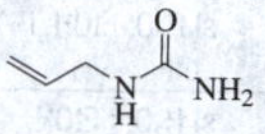
212
Allylurea

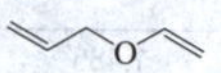
213
Allyl vinyl ether

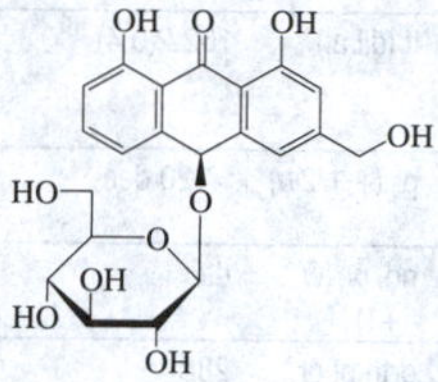
214
Aloin A

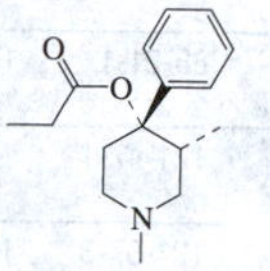
215
Alphaprodine

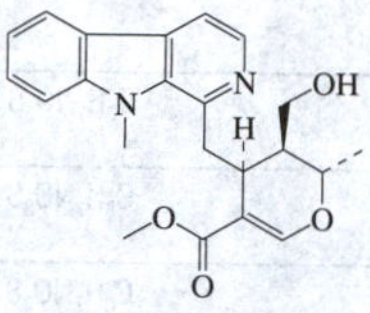
216
Alstonidine

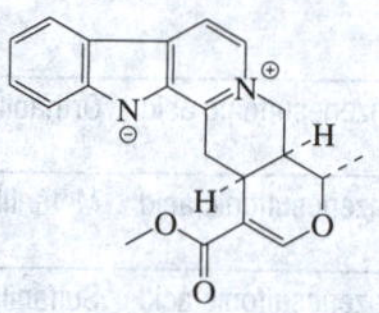
217
Alstonine

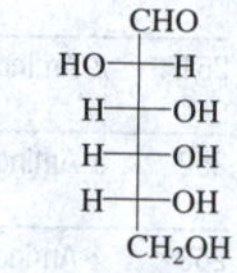
218
D-Altrose

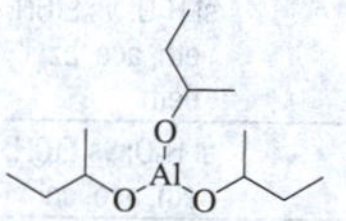
219
Aluminum 2-butoxide

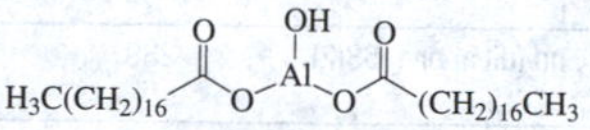
220
Aluminum distearate

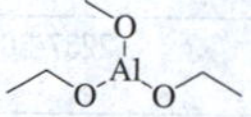
221
Aluminum ethanolate

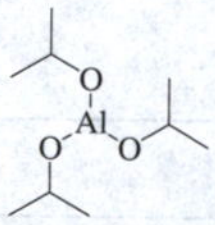
222
Aluminum isopropoxide

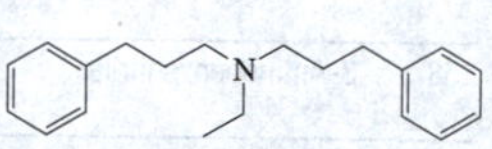
223
Alverine

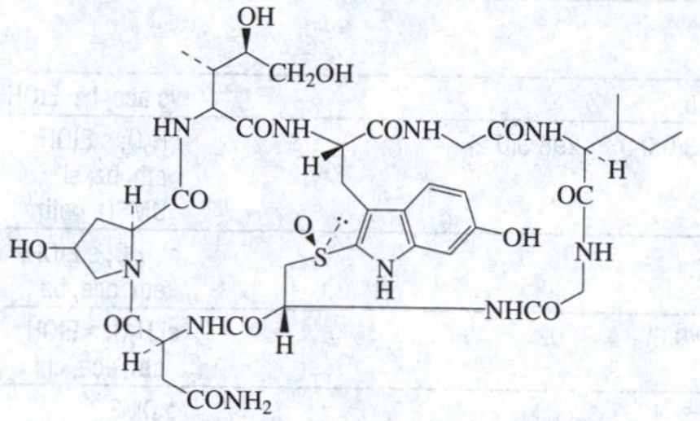
224
α-Amanitin

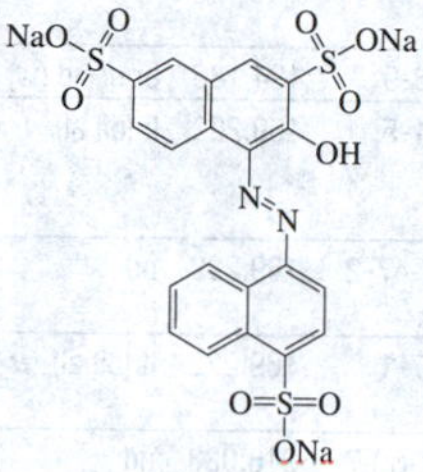
225
Amaranth dye

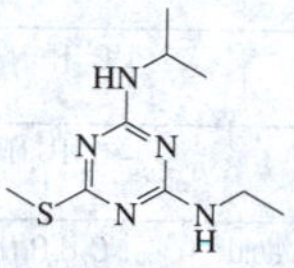
226
Ametryn

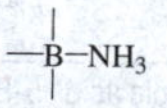
227
Amminetrimethylboron

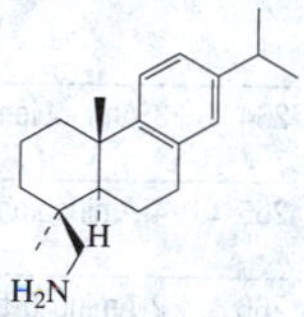
228
19-Amino-8,11,13-abietatriene

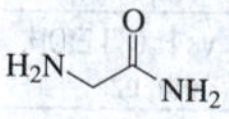
229
2-Aminoacetamide

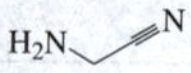
230
Aminoacetonitrile

231
Aminoacetonitrile monohydrochloride

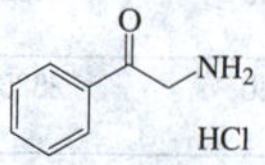
232
α-Aminoacetophenone hydrochloride

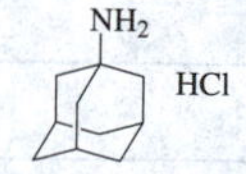
233
1-Aminoadamantane hydrochloride

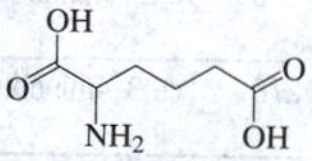
234
2-Aminoadipic acid

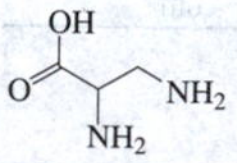
235
3-Aminoalanine

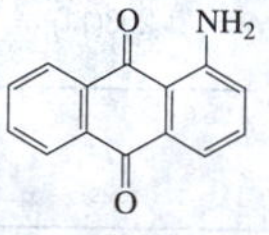
236
1-Amino-9,10-anthracenedione

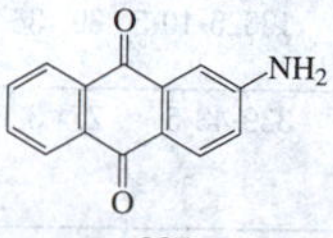
237
2-Amino-9,10-anthracenedione

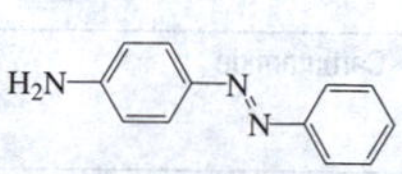
238
4-Aminoazobenzene

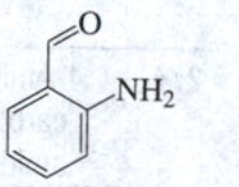
239
2-Aminobenzaldehyde

240
3-Aminobenzaldehyde
241
4-Aminobenzaldehyde

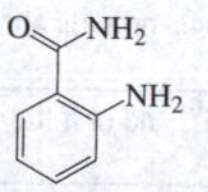
242
2-Aminobenzamide

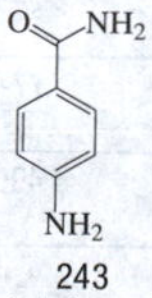
243
4-Aminobenzamide

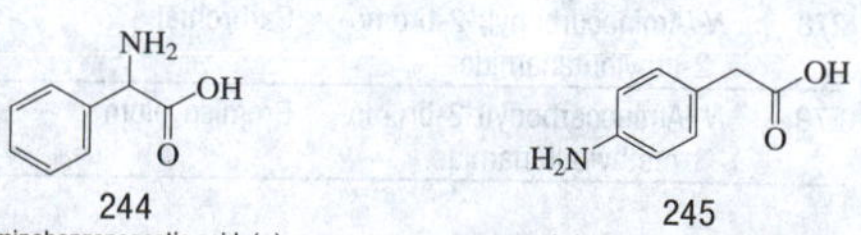
244
α-Aminobenzeneacetic acid, (±)-
245
4-Aminobenzeneacetic acid

No.	Name	Synonym	Mol. form.	CAS Reg. No.	Mol. wt.	Physical form	mp/°C	bp/°C	den/ g cm^{-3}	n_D	Solubility
246	5-Amino-1,3-benzenedicarboxylic acid		$C_8H_7NO_4$	99-31-0	181.147	pr(al), pl(w)	360	sub			i H_2O; sl EtOH
247	4-Aminobenzeneethanol		$C_8H_{11}NO$	104-10-9	137.179	nd (al)	108				
248	2-Aminobenzenemethanamine		$C_7H_{10}N_2$	4403-69-4	122.167		61	269			vs EtOH
249	2-Aminobenzenemethanol		C_7H_9NO	5344-90-1	123.152		83.5	273			s H_2O, EtOH, eth, HOAc; vs bz, chl
250	4-Aminobenzenesulfonamide	Sulfanilamide	$C_6H_8N_2O_2S$	63-74-1	172.205	lf (dil al)	162.2(0.4)		1.08^{25}		s H_2O, EtOH, eth, ace; sl chl, peth
251	2-Aminobenzenesulfonic acid	Orthanilic acid	$C_6H_7NO_3S$	88-21-1	173.190	pr (+ 1/2w)	>320 dec				sl H_2O; i EtOH, eth
252	3-Aminobenzenesulfonic acid	Metanilic acid	$C_6H_7NO_3S$	121-47-1	173.190	nd, pr (w +1)	dec				sl H_2O, EtOH; i eth
253	4-Aminobenzenesulfonic acid	Sulfanilic acid	$C_6H_7NO_3S$	121-57-3	173.190	orth pl or mcl (w+2)	288		1.485^{25}		sl H_2O; i EtOH, eth
254	4-Aminobenzenesulfonyl fluoride	*p*-Sulfanilyl fluoride	$C_6H_6FNO_2S$	98-62-4	175.181		68.5				
255	2-Aminobenzenethiol		C_6H_7NS	137-07-5	125.192		26	234		1.4606^{20}	s EtOH, eth
256	4-Aminobenzenethiol		C_6H_7NS	1193-02-8	125.192		46	143^{17}			s H_2O, EtOH
257	2-Aminobenzonitrile		$C_7H_6N_2$	1885-29-6	118.136	ye pr (CS_2) nd (peth)	51	263			sl H_2O; vs EtOH, eth, ace, bz; i peth
258	3-Aminobenzonitrile		$C_7H_6N_2$	2237-30-1	118.136	nd (dil al or CCl_4)	53(2)	289			sl H_2O; vs EtOH, eth, ace, chl
259	4-Aminobenzonitrile		$C_7H_6N_2$	873-74-5	118.136	pr or pl (w)	86.2(0.5)				sl H_2O, ctc; vs EtOH, eth, ace, bz
260	4-Aminobenzophenone		$C_{13}H_{11}NO$	1137-41-3	197.232	lf (dil al)	123(2)	246^{13}			sl H_2O, tfa; s EtOH, eth, HOAc
261	*N*-(4-Aminobenzoyl)-L-glutamic acid		$C_{12}H_{14}N_2O_5$	4271-30-1	266.249	cry (w)	173				
262	*N*-(4-Aminobenzoyl)glycine	*p*-Aminohippuric acid	$C_9H_{10}N_2O_3$	61-78-9	194.186	pr or nd (w)	198.5				vs ace, bz, EtOH
263	2-Aminobiphenyl		$C_{12}H_{11}N$	90-41-5	169.222	lf (dil al)	49.13(0.04)	298.3(0.2)			i H_2O; s EtOH, eth, bz; sl DMSO, peth
264	3-Aminobiphenyl		$C_{12}H_{11}N$	2243-47-2	169.222	nd	31.5				sl H_2O; s EtOH, eth, ace, bz
265	4-Aminobiphenyl	*p*-Biphenylamine	$C_{12}H_{11}N$	92-67-1	169.222	lf (dil al)	51.0(0.6)	302			sl H_2O; s EtOH, eth, ace, chl
266	2-Amino-5-bromobenzoic acid	5-Bromoanthranilic acid	$C_7H_6BrNO_2$	5794-88-7	216.033	nd	219.5				s DMSO
267	1-Amino-4-bromo-9,10-dihydro-9,10-dioxo-2-anthracenesulfonic acid	1-Amino-4-bromoanthraquinone-2-sulfonic acid	$C_{14}H_8BrNO_5S$	116-81-4	382.187	red nd (w)					
268	DL-2-Aminobutanoic acid		$C_4H_9NO_2$	2835-81-6	103.120	lf (w)	304 dec	sub	1.2300^{20}		vs H_2O; sl EtOH; i eth, bz
269	L-2-Aminobutanoic acid		$C_4H_9NO_2$	1492-24-6	103.120	lf (dil al), cry (al)	292 dec				s H_2O; sl EtOH, eth; i bz
270	DL-3-Aminobutanoic acid		$C_4H_9NO_2$	2835-82-7	103.120	nd (al)	194.3				vs H_2O; i EtOH, eth, bz
271	4-Aminobutanoic acid	γ-Aminobutyric acid	$C_4H_9NO_2$	56-12-2	103.120	pr or nd (al) lf (MeOH-eth)	203 dec				vs H_2O; sl EtOH, ace; i eth, bz
272	2-Amino-1-butanol, (±)-		$C_4H_{11}NO$	13054-87-0	89.136	liq	-1.0	178(9)	0.9162^{20}	1.4489^{25}	msc H_2O, EtOH, eth; sl chl
273	4-Amino-1-butanol		$C_4H_{11}NO$	13325-10-5	89.136			203(11)	0.967^{12}	1.4625^{20}	s H_2O, EtOH; i eth
274	4-Amino-*N*-[(butylamino)-carbonyl]benzenesulfonamide	Carbutamide	$C_{11}H_{17}N_3O_3S$	339-43-5	271.336		144.5				
275	Aminocarb		$C_{11}H_{16}N_2O_2$	2032-59-9	208.257	cry	95.0(0.3)				sl H_2O, bz; s ace
276	*N*-(Aminocarbonyl)acetamide		$C_3H_6N_2O_2$	591-07-1	102.092		218	180 sub			sl H_2O, eth; s EtOH
277	[4-[(Aminocarbonyl)amino]-phenyl]arsonic acid	Carbarsone	$C_7H_9AsN_2O_4$	121-59-5	260.079	nd (w)	174				sl H_2O, DMSO, EtOH; i eth, chl; s alk
278	*N*-(Aminocarbonyl)-2-bromo-2-ethylbutanamide	Carbromal	$C_7H_{13}BrN_2O_2$	77-65-6	237.094	orth (dil al)	118		1.544^{25}		sl H_2O, chl; s ace, bz
279	*N*-(Aminocarbonyl)-2-bromo-3-methylbutanamide	Bromisovalum	$C_6H_{11}BrN_2O_2$	496-67-3	223.067	nd or lf (to)	154	sub	1.56^{15}		vs ace, bz, eth, EtOH

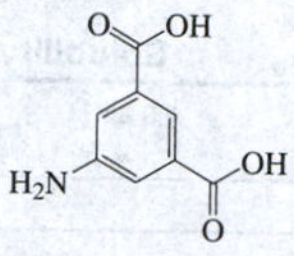
246
5-Amino-1,3-benzenedicarboxylic acid

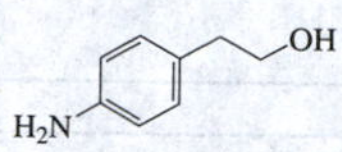
247
4-Aminobenzeneethanol

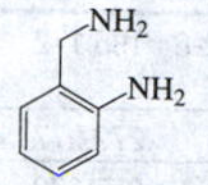
248
2-Aminobenzenemethanamine

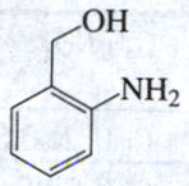
249
2-Aminobenzenemethanol

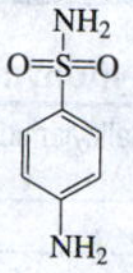
250
4-Aminobenzenesulfonamide

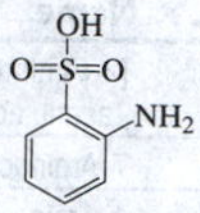
251
2-Aminobenzenesulfonic acid

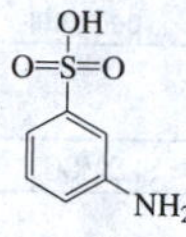
252
3-Aminobenzenesulfonic acid

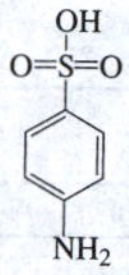
253
4-Aminobenzenesulfonic acid

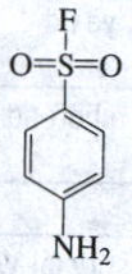
254
4-Aminobenzenesulfonyl fluoride

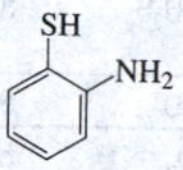
255
2-Aminobenzenethiol

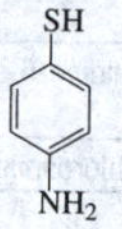
256
4-Aminobenzenethiol

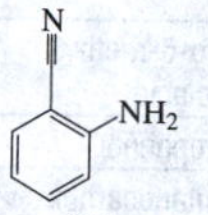
257
2-Aminobenzonitrile

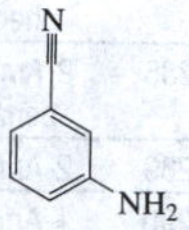
258
3-Aminobenzonitrile

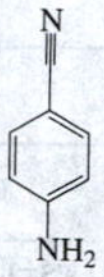
259
4-Aminobenzonitrile

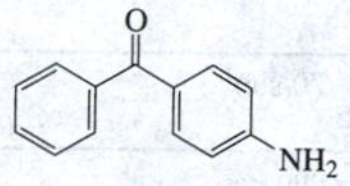
260
4-Aminobenzophenone

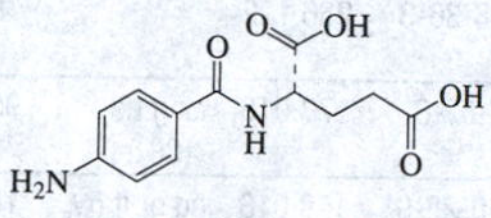
261
N-(4-Aminobenzoyl)-L-glutamic acid

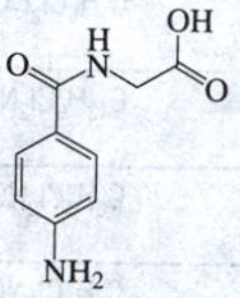
262
N-(4-Aminobenzoyl)glycine

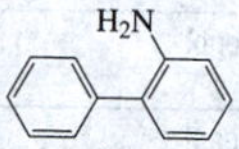
263
2-Aminobiphenyl

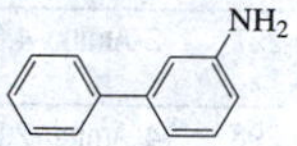
264
3-Aminobiphenyl

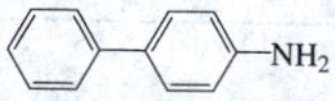
265
4-Aminobiphenyl

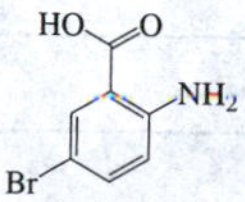
266
2-Amino-5-bromobenzoic acid

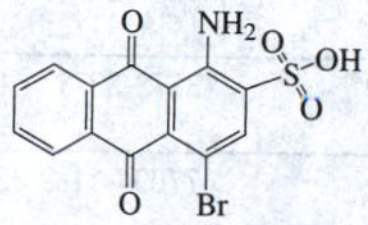
267
1-Amino-4-bromo-9,10-dihydro-9,10-dioxo-2-anthracenesulfonic acid

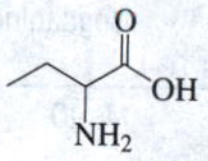
268
DL-2-Aminobutanoic acid

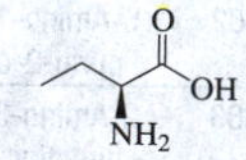
269
L-2-Aminobutanoic acid

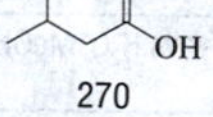
270
DL-3-Aminobutanoic acid

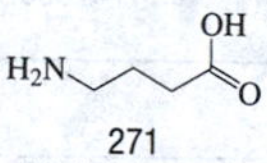
271
4-Aminobutanoic acid

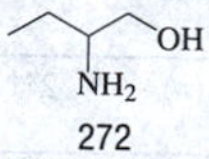
272
2-Amino-1-butanol, (±)-

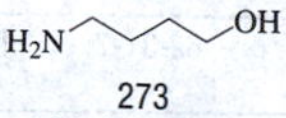
273
4-Amino-1-butanol

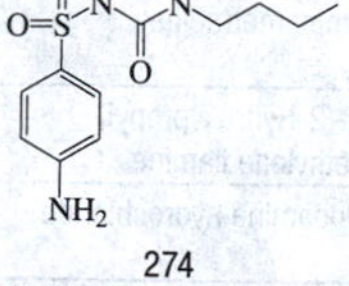
274
4-Amino-*N*-[(butylamino)carbonyl]benzenesulfonamide

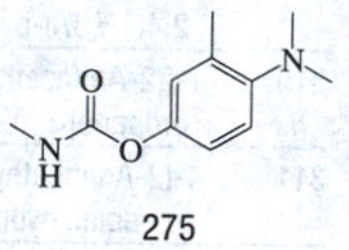
275
Aminocarb

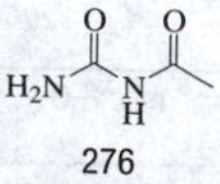
276
N-(Aminocarbonyl)acetamide

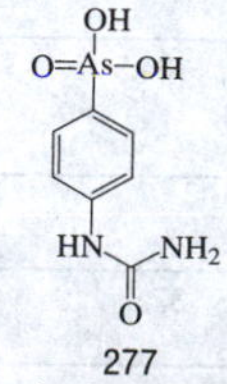
277
[4-[(Aminocarbonyl)amino]phenyl]arsonic acid

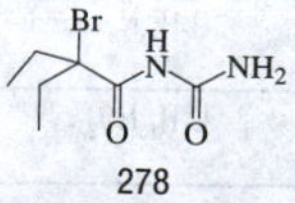
278
N-(Aminocarbonyl)-2-bromo-2-ethylbutanamide

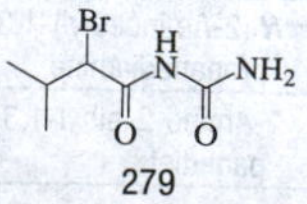
279
N-(Aminocarbonyl)-2-bromo-3-methylbutanamide

No.	Name	Synonym	Mol. form.	CAS Reg. No.	Mol. wt.	Physical form	mp/°C	bp/°C	den/g cm^{-3}	n_D	Solubility
280	[2-(Aminocarbonyl)phenoxy]-acetic acid	Salicylamide *O*-acetic acid	$C_9H_9NO_4$	25395-22-6	195.172		221				s alk
281	7-Aminocephalosporanic acid		$C_{10}H_{12}N_2O_5S$	957-68-6	272.277	cry					
282	1-Amino-5-chloro-9,10-anthracenedione	1-Amino-5-chloroanthraquinone	$C_{14}H_8ClNO_2$	117-11-3	257.673		212				
283	4-Amino-6-chloro-1,3-benzenedisulfonamide	Chloraminophenamide	$C_6H_8ClN_3O_4S_2$	121-30-2	285.729		254.5				
284	5-Amino-2-chlorobenzenesulfonic acid	6-Chlorometanilic acid	$C_6H_6ClNO_3S$	88-43-7	207.635	nd (w)	280 dec				
285	2-Amino-5-chlorobenzoic acid		$C_7H_6ClNO_2$	635-21-2	171.582		211				
286	5-Amino-2-chlorobenzoic acid		$C_7H_6ClNO_2$	89-54-3	171.582		188		1.519^{15}		vs EtOH
287	2-Amino-5-chlorobenzophenone	2-Benzoyl-4-chloroaniline	$C_{13}H_{10}ClNO$	719-59-5	231.677	ye nd	100.5				vs H_2O, EtOH, peth, chl
288	2-Amino-4-chloro-5-methylbenzenesulfonic acid	2-Chloro-*p*-toluidine-5-sulfonic acid	$C_7H_8ClNO_3S$	88-51-7	221.662	short nd (w)					
289	2-Amino-4-chlorophenol	2-Hydroxy-5-chloroaniline	C_6H_6ClNO	95-85-2	143.571		140				sl DMSO
290	1-Aminocyclopentanecarboxylic acid	Cycloleucine	$C_6H_{11}NO_2$	52-52-8	129.157	cry (al-w)	330 dec				
291	7-Aminodeacetoxycephalosporanic acid		$C_8H_{10}N_2O_3S$	22252-43-3	214.241		241 dec				
292	1-Amino-1-deoxy-D-glucitol	Glucamine	$C_6H_{15}NO_5$	488-43-7	181.187	cry (MeOH)	127				vs H_2O, EtOH
293	2-Amino-2-deoxy-D-glucose	D-Glucosamine	$C_6H_{13}NO_5$	3416-24-8	179.171						vs H_2O
294	1-Amino-2,4-dibromo-9,10-anthracenedione		$C_{14}H_7Br_2NO_2$	81-49-2	381.020	red nd (xyl)	226				
295	3-Amino-2,5-dichlorobenzoic acid	Chloramben	$C_7H_5Cl_2NO_2$	133-90-4	206.027		202(1)				sl DMSO
296	2-Amino-2',5-dichlorobenzophenone		$C_{13}H_9Cl_2NO$	2958-36-3	266.122		≈80				
297	2-Amino-4,6-dichlorophenol		$C_6H_5Cl_2NO$	527-62-8	178.016	long nd (CS_2)	95.5	70 sub			
298	4-Amino-2,6-dichlorophenol		$C_6H_5Cl_2NO$	5930-28-9	178.016	nd or lf (w, bz)	168	sub			i H_2O; vs EtOH, eth; s ace; sl bz, HOAc
299	2-Amino-1,7-dihydro-7-methyl-6*H*-purin-6-one	7-Methylguanine	$C_6H_7N_5O$	578-76-7	165.153		370				
300	5-Amino-2,3-dihydro-1,4-phthalazinedione	Luminol	$C_8H_7N_3O_2$	521-31-3	177.161	ye nd (al)	330.5				i H_2O; sl EtOH, eth; vs alk; s HOAc
301	2-Amino-1,7-dihydro-6*H*-purine-6-thione	Thioguanine	$C_5H_5N_5S$	154-42-7	167.193		>360				
302	6-Amino-1,3-dihydro-2*H*-purin-2-one	Isoguanine	$C_5H_5N_5O$	3373-53-3	151.127		>360				i H_2O
303	2-Amino-3,4-dimethylimidazo[4,5-*f*]-quinoline	Me-IQ	$C_{12}H_{12}N_4$	77094-11-2	212.250	cry	297				
304	2-Amino-4,6-dinitrophenol	Picramic acid	$C_6H_5N_3O_5$	96-91-3	199.121	dk red nd (al) pr (chl)	168(1)				vs bz, EtOH
305	2-Aminoethanesulfonic acid	Taurine	$C_2H_7NO_3S$	107-35-7	125.147	mcl pr (w)	328				vs H_2O
306	1-Aminoethanol	Acetaldehyde ammonia	C_2H_7NO	75-39-8	61.083	orth (eth-al)	97	110 dec			s H_2O; sl eth
307	2-(2-Aminoethoxy)ethanol	Diglycolamine	$C_4H_{11}NO_2$	929-06-6	105.136		-12.5	223.1(0.1)	1.0572^{20}		
308	*N*-(2-Aminoethyl)acetamide		$C_4H_{10}N_2O$	1001-53-2	102.134		51				s H_2O, EtOH, bz; i eth
309	6-Amino-3-ethyl-1-allyl-2,4(1*H*,3*H*)-pyrimidinedione	Aminometradine	$C_9H_{13}N_3O_2$	642-44-4	195.218	cry (+1w, w)	143				
310	1-[(2-Aminoethyl)amino]-2-propanol	*N*-(2-Hydroxypropyl)-ethylenediamine	$C_5H_{14}N_2O$	123-84-2	118.177			94^{3}	0.9837^{25}	1.4738^{20}	
311	4-(2-Aminoethyl)-1,2-benzenediol, hydrochloride	Dopamine hydrochloride	$C_8H_{12}ClNO_2$	62-31-7	189.640	nd (w)	241 dec				vs H_2O, MeOH
312	α-(1-Aminoethyl)-benzenemethanol, [*S*-(*R**,*R**)]-		$C_9H_{13}NO$	492-39-7	151.205	pl(MeOH)	77.5				vs eth, EtOH, chl
313	α-(1-Aminoethyl)-benzenemethanol, hydrochloride		$C_9H_{14}ClNO$	53631-70-2	187.666		198.5				s H_2O
314	*N*-(2-Aminoethyl)ethanolamine		$C_4H_{12}N_2O$	111-41-1	104.150			242(5)	1.0286^{20}	1.4863^{20}	msc H_2O, EtOH; s ace; sl bz, lig
315	4-(2-Aminoethyl)phenol	Tyramine	$C_8H_{11}NO$	51-67-2	137.179	pl or nd (bz, w), cry (al)	164.5	206^{25}			sl H_2O, bz, DMSO; s EtOH, xyl; i tol
316	*N*-(2-Aminoethyl)-1,3-propanediamine	*N*-(3-Aminopropyl)-ethylenediamine	$C_5H_{15}N_3$	13531-52-7	117.193			87^{3}		1.4805^{25}	
317	2-Amino-2-ethyl-1,3-propanediol		$C_5H_{13}NO_2$	115-70-8	119.163		37.5	152^{10}	1.099^{20}	1.490^{20}	msc H_2O

Organic

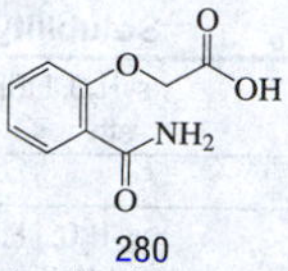
280
[2-(Aminocarbonyl)phenoxy]acetic acid

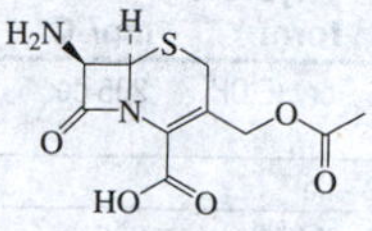
281
7-Aminocephalosporanic acid

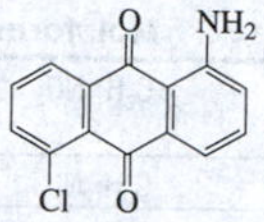
282
1-Amino-5-chloro-9,10-anthracenedione

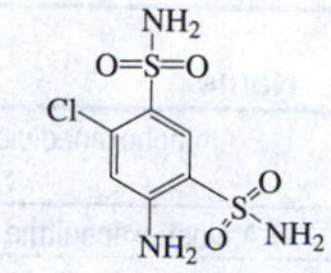
283
4-Amino-6-chloro-1,3-benzenedisulfonamide

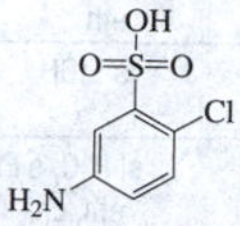
284
5-Amino-2-chlorobenzenesulfonic acid

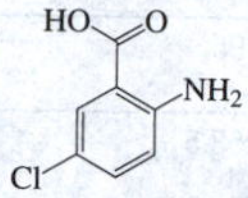
285
2-Amino-5-chlorobenzoic acid

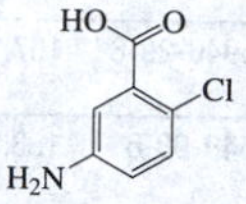
286
5-Amino-2-chlorobenzoic acid

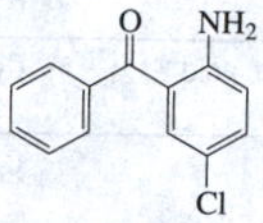
287
2-Amino-5-chlorobenzophenone

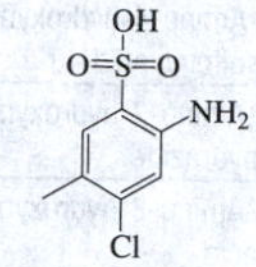
288
2-Amino-4-chloro-5-methylbenzenesulfonic acid

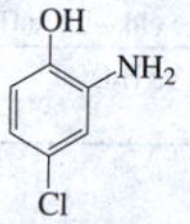
289
2-Amino-4-chlorophenol

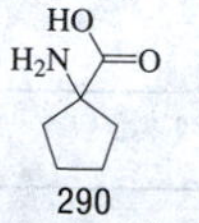
290
1-Aminocyclopentanecarboxylic acid

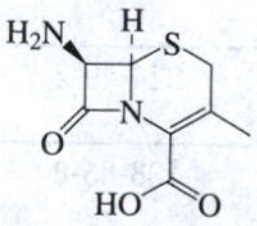
291
7-Aminodeacetoxycephalosporanic acid

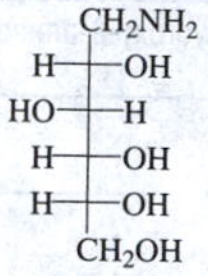
292
1-Amino-1-deoxy-D-glucitol

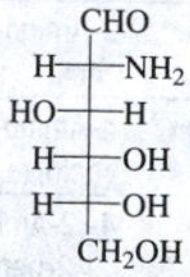
293
2-Amino-2-deoxy-D-glucose

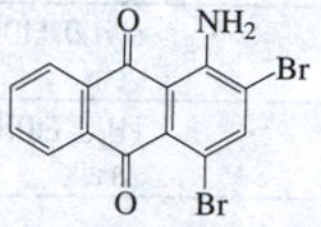
294
1-Amino-2,4-dibromo-9,10-anthracenedione

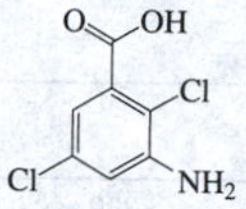
295
3-Amino-2,5-dichlorobenzoic acid

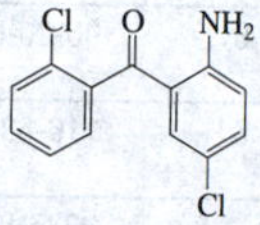
296
2-Amino-2′,5-dichlorobenzophenone

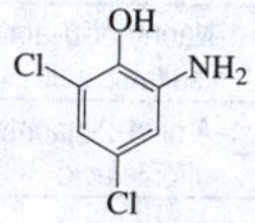
297
2-Amino-4,6-dichlorophenol

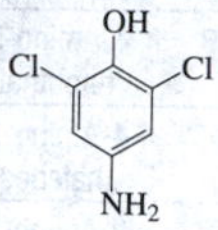
298
4-Amino-2,6-dichlorophenol

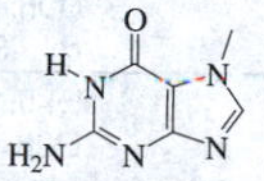
299
2-Amino-1,7-dihydro-7-methyl-6*H*-purin-6-one

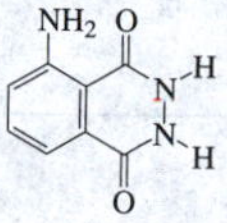
300
5-Amino-2,3-dihydro-1,4-phthalazinedione

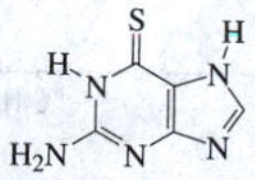
301
2-Amino-1,7-dihydro-6*H*-purine-6-thione

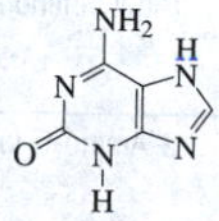
302
6-Amino-1,3-dihydro-2*H*-purin-2-one

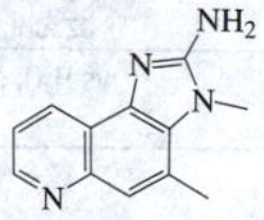
303
2-Amino-3,4-dimethylimidazo[4,5-f]quinoline

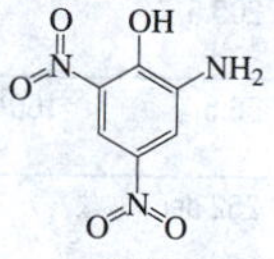
304
2-Amino-4,6-dinitrophenol

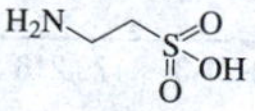
305
2-Aminoethanesulfonic acid

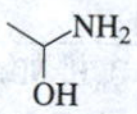
306
1-Aminoethanol

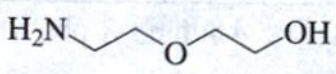
307
2-(2-Aminoethoxy)ethanol

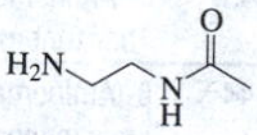
308
N-(2-Aminoethyl)acetamide

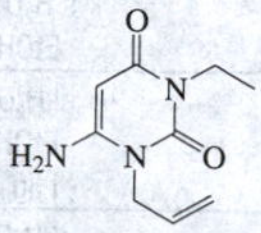
309
6-Amino-3-ethyl-1-allyl-2,4(1*H*,3*H*)-pyrimidinedione

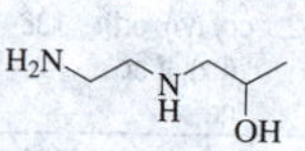
310
1-[(2-Aminoethyl)amino]-2-propanol

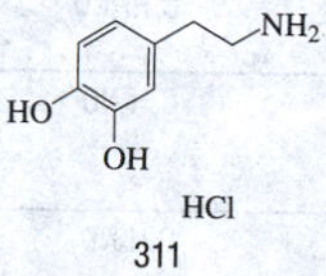
311
4-(2-Aminoethyl)-1,2-benzenediol, hydrochloride

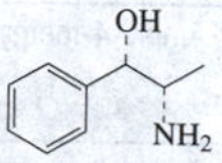
312
α-(1-Aminoethyl)benzenemethanol, [*S*-(*R**,*R**)]-

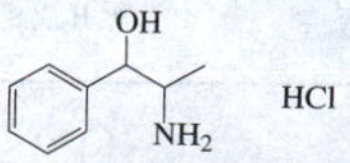
313
α-(1-Aminoethyl)benzenemethanol, hydrochloride

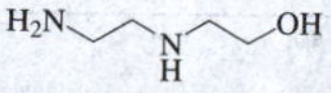
314
N-(2-Aminoethyl)ethanolamine

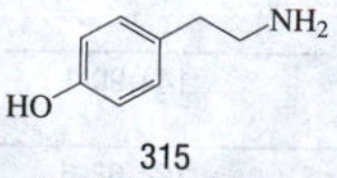
315
4-(2-Aminoethyl)phenol

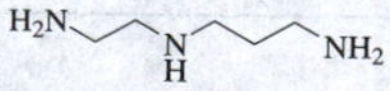
316
N-(2-Aminoethyl)-1,3-propanediamine

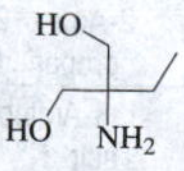
317
2-Amino-2-ethyl-1,3-propanediol

Organic

No.	Name	Synonym	Mol. form.	CAS Reg. No.	Mol. wt.	Physical form	mp/°C	bp/°C	den/ g cm^{-3}	n_D	Solubility
318	L-2-Aminohexanedioic acid	2-Aminoadipic acid	$C_6H_{11}NO_4$	542-32-5	161.156	cry (EtOH, w)	205 dec				sl H_2O, EtOH, eth
319	6-Aminohexanenitrile	5-Cyano-1-pentylamine	$C_6H_{12}N_2$	2432-74-8	112.172	liq		118^{16}			
320	6-Aminohexanoic acid	ε-Aminocaproic acid	$C_6H_{13}NO_2$	60-32-2	131.173	lf (eth)	205				vs H_2O; i EtOH; sl MeOH
321	6-Amino-1-hexanol		$C_6H_{15}NO$	4048-33-3	117.189		57	137^{30}			
322	1-Amino-4-hydroxy-9,10-anthracenedione		$C_{14}H_9NO_3$	116-85-8	239.226		216.5				s EtOH, ace
323	3-Amino-4-hydroxybenzene-sulfonic acid		$C_6H_7NO_4S$	98-37-3	189.190	orth (w+1)	>300				sl H_2O; i EtOH, eth
324	4-Amino-2-hydroxybenzo-hydrazide	*p*-Aminosalicylic acid hydrazide	$C_7H_9N_3O_2$	6946-29-8	167.165	nd (al)	195				vs EtOH
325	2-Amino-3-hydroxybenzoic acid		$C_7H_7NO_3$	548-93-6	153.136	lf (w)	253.5				sl H_2O; s EtOH, eth, chl
326	4-Amino-2-hydroxybenzoic acid	*p*-Aminosalicylic acid	$C_7H_7NO_3$	65-49-6	153.136	nd, pl (al-eth)	150 dec				s H_2O, EtOH, eth, ace; i bz, peth, chl
327	5-Amino-2-hydroxybenzoic acid	Mesalamine	$C_7H_7NO_3$	89-57-6	153.136		281.0(0.5)				sl H_2O; i EtOH
328	3-Amino-4-hydroxybutanoic acid	γ-Hydroxy-β-aminobutyric acid	$C_4H_9NO_3$	589-44-6	119.119	pr	216				vs H_2O; sl EtOH, chl, eth, AcOEt
329	4-Amino-3-hydroxybutanoic acid, (±)-		$C_4H_9NO_3$	924-49-2	119.119	pr (w), cry (dil al)	218				vs H_2O
330	4-(2-Amino-1-hydroxyethyl)-1,2-benzenediol, (±)-		$C_8H_{11}NO_3$	138-65-8	169.178		189 dec				
331	1-Amino-4-hydroxy-2-methoxy-9,10-anthracene-dione		$C_{15}H_{11}NO_4$	2379-90-0	269.253						sl chl
332	4-Amino-5-(hydroxymethyl)-2(1*H*)-pyrimidinone	5-Hydroxymethylcytosine	$C_5H_7N_3O_2$	1123-95-1	141.129		>300 dec				
333	4-Amino-5-hydroxy-2,7-naphthalenedisulfonic acid	1-Naphthol-8-amino-3,6-disulfonic acid	$C_{10}H_9NO_7S_2$	90-20-0	319.311						sl H_2O, EtOH, eth
334	4-Amino-3-hydroxy-1-naph-thalenesulfonic acid	1-Amino-2-naphthol-4-sulfonic acid	$C_{10}H_9NO_4S$	116-63-2	239.248	gray nd					i H_2O, EtOH, bz; s alk
335	2-Amino-4-hydroxypteridine		$C_6H_5N_5O$	2236-60-4	163.137	ye cry	>360				
336	5-Amino-1*H*-imidazole-4-carboxamide		$C_4H_6N_4O$	360-97-4	126.117	cry (EtOH)	170				
337	*O*-[(Aminoiminomethyl) amino]-L-homoserine	Canavanine	$C_5H_{12}N_4O_3$	543-38-4	176.174	cry (al)	172				vs H_2O
338	(Aminoiminomethyl)urea		$C_2H_6N_4O$	141-83-3	102.095	pr	105	160 dec			s H_2O, py; sl EtOH; i eth, bz, chl, CS_2
339	2-Amino-5-iodobenzoic acid		$C_7H_6INO_2$	5326-47-6	263.033		220 dec				sl H_2O, tfa; vs EtOH, eth, ace; s bz
340	4-Amino-1*H*-isoindole-1,3(2*H*)-dione		$C_8H_6N_2O_2$	2518-24-3	162.146		269.5				
341	4-Amino-3-isoxazolidinone, (*R*)-	Cycloserine	$C_3H_6N_2O_2$	68-41-7	102.092		155 dec				s H_2O; sl MeOH
342	1-Amino-2-methyl-9,10-anthracenedione	1-Amino-2-methylanthraqui-none	$C_{15}H_{11}NO_2$	82-28-0	237.254		205.5				i H_2O; s EtOH, bz, chl; sl eth
343	α-(Aminomethyl)-benzenemethanol	Phenylethanolamine	$C_8H_{11}NO$	7568-93-6	137.179		56.5	160^{17}			vs H_2O; s EtOH
344	β-(Aminomethyl)-benzenepropanoic acid	4-Amino-3-phenylbutyric acid	$C_{10}H_{13}NO_2$	1078-21-3	179.216		252 dec				
345	2-Amino-5-methylbenzenesul-fonic acid		$C_7H_9NO_3S$	88-44-8	187.216	lt ye nd	132 dec				vs H_2O
346	*trans*-4-(Aminomethyl)-cyclohexanecarboxylic acid	Tranexamic acid	$C_8H_{15}NO_2$	1197-18-8	157.211		249(4)				vs H_2O
347	4-Amino-4-methyl-2-penta-none	Diacetonamine	$C_6H_{13}NO$	625-04-7	115.173			$25^{0.14}$			s H_2O; msc EtOH, eth
348	2-Amino-4-methylphenol		C_7H_9NO	95-84-1	123.152	cry (w), orth (bz), lf or nd	136	sub			sl H_2O, bz; s EtOH, eth, chl; i lig
349	4-Amino-2-methylphenol		C_7H_9NO	2835-96-3	123.152	nd or lf (bz)	176.5	sub			sl H_2O, bz; s EtOH, eth
350	4-Amino-3-methylphenol		C_7H_9NO	2835-99-6	123.152	pr (dil al) cry (bz)	179				sl H_2O; vs EtOH, eth; s DMSO
351	(Aminomethyl)phosphonic acid		CH_6NO_3P	1066-51-9	111.038	cry	309				
352	2-Amino-2-methyl-1,3-propanediol		$C_4H_{11}NO_2$	115-69-5	105.136		110.93(0.05)	151^{10}			vs H_2O; s EtOH
353	L-3-Amino-2-methylpropanoic acid		$C_4H_9NO_2$	144-90-1	103.120	cry (w)	185				
354	2-Amino-2-methyl-1-propanol	2-Aminoisobutanol	$C_4H_{11}NO$	124-68-5	89.136		25.5	163.8(0.8)	0.934^{20}	1.449^{20}	msc H_2O; s ctc

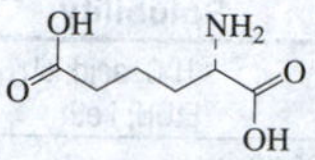
318
L-2-Aminohexanedioic acid

319
6-Aminohexanenitrile

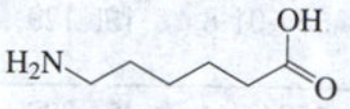
320
6-Aminohexanoic acid

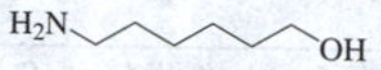
321
6-Amino-1-hexanol

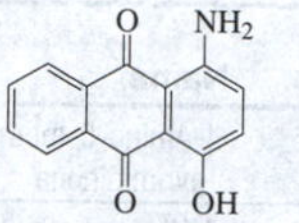
322
1-Amino-4-hydroxy-9,10-anthracenedione

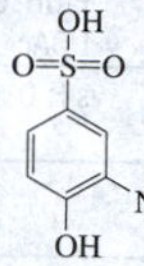
323
3-Amino-4-hydroxybenzenesulfonic acid

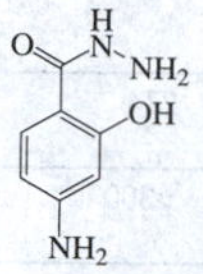
324
4-Amino-2-hydroxybenzohydrazide

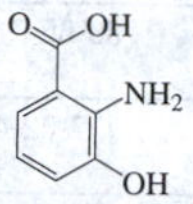
325
2-Amino-3-hydroxybenzoic acid

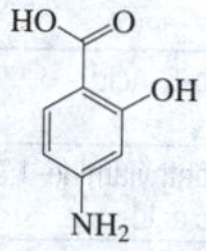
326
4-Amino-2-hydroxybenzoic acid

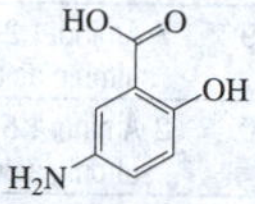
327
5-Amino-2-hydroxybenzoic acid

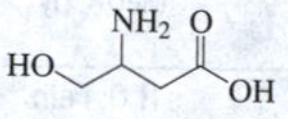
328
3-Amino-4-hydroxybutanoic acid

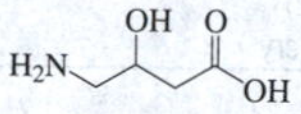
329
4-Amino-3-hydroxybutanoic acid, (±)-

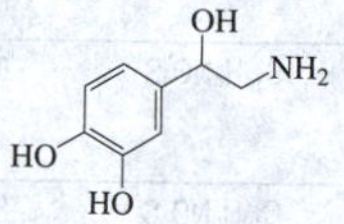
330
4-(2-Amino-1-hydroxyethyl)-1,2-benzenediol, (±)-

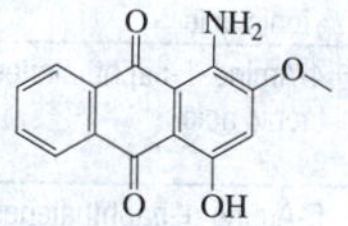
331
1-Amino-4-hydroxy-2-methoxy-9,10-anthracenedione

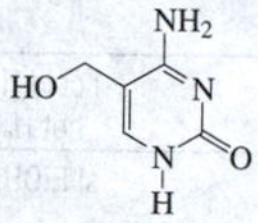
332
4-Amino-5-(hydroxymethyl)-2(1*H*)-pyrimidinone

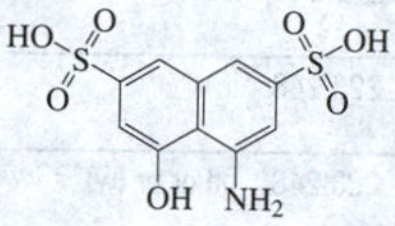
333
4-Amino-5-hydroxy-2,7-naphthalenedisulfonic acid

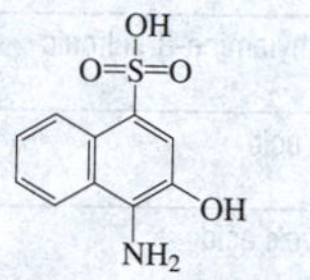
334
4-Amino-3-hydroxy-1-naphthalenesulfonic acid

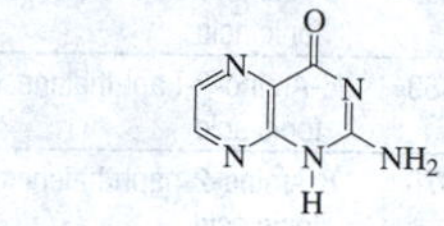
335
2-Amino-4-hydroxypteridine

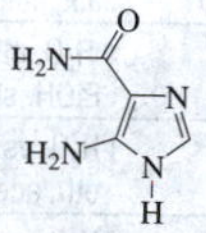
336
5-Amino-1*H*-imidazole-4-carboxamide

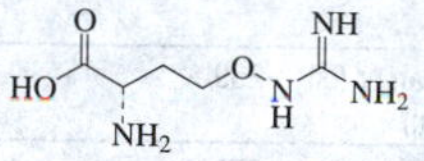
337
O-[(Aminoiminomethyl)amino]-L-homoserine

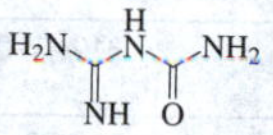
338
(Aminoiminomethyl)urea

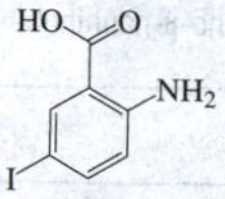
339
2-Amino-5-iodobenzoic acid

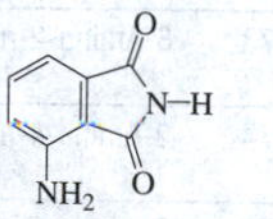
340
4-Amino-1*H*-isoindole-1,3(2*H*)-dione

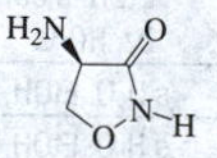
341
4-Amino-3-isoxazolidinone, (*R*)-

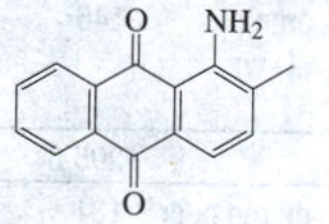
342
1-Amino-2-methyl-9,10-anthracenedione

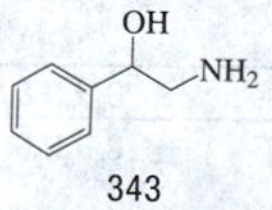
343
α-(Aminomethyl)benzenemethanol

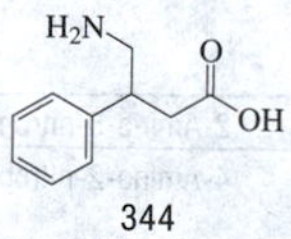
344
β-(Aminomethyl)benzenepropanoic acid

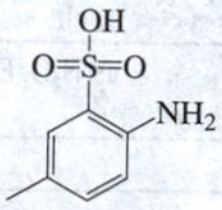
345
2-Amino-5-methylbenzenesulfonic acid

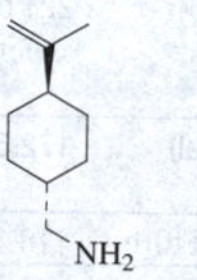
346
trans-4-(Aminomethyl)cyclohexanecarboxylic acid

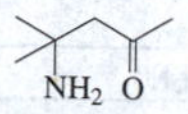
347
4-Amino-4-methyl-2-pentanone

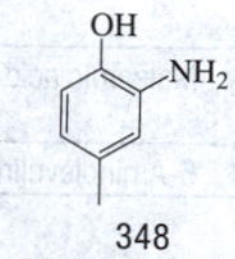
348
2-Amino-4-methylphenol

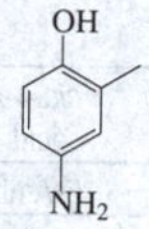
349
4-Amino-2-methylphenol

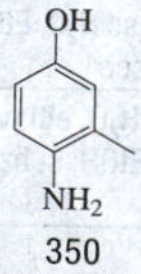
350
4-Amino-3-methylphenol

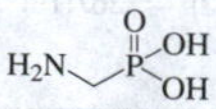
351
(Aminomethyl)phosphonic acid

352
2-Amino-2-methyl-1,3-propanediol

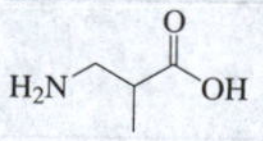
353
L-3-Amino-2-methylpropanoic acid

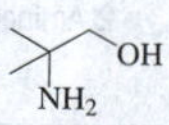
354
2-Amino-2-methyl-1-propanol

No.	Name	Synonym	Mol. form.	CAS Reg. No.	Mol. wt.	Physical form	mp/°C	bp/°C	den/ g cm^{-3}	n_D	Solubility
355	4-Amino-5-methyl-2(1*H*)-pyrimidinone	5-Methylcytosine	$C_5H_7N_3O$	554-01-8	125.129	pr (w+1/2)	270 dec				s H_2O, acid; sl EtOH; i eth
356	3-(Aminomethyl)-3,5,5-trimethylcyclohexanol	1-Hydroxy-3-aminomethyl-3,5,5-trimethylcyclohexane	$C_{10}H_{21}NO$	15647-11-7	171.280		45.5	265	0.969^{25}	1.4904^{20}	
357	3-Amino-2-naphthalene-carboxylic acid	3-Amino-2-naphthoic acid	$C_{11}H_9NO_2$	5959-52-4	187.195	ye lf (dil al)	216.5				s EtOH, eth
358	2-Amino-1,4-naphthalenedi-one		$C_{10}H_7NO_2$	2348-81-4	173.169		207				i H_2O, alk; s EtOH, eth, HOAc
359	7-Amino-1,3-naphthalenedi-sulfonic acid	Amido-G-Acid	$C_{10}H_9NO_6S_2$	86-65-7	303.311	mcl pr or nd (w+4)	274				vs H_2O, EtOH
360	2-Amino-1,5-naphthalenedi-sulfonic acid	2-Naphthylamine-1,5-disul-fonic acid	$C_{10}H_9NO_6S_2$	117-62-4	303.311		>300				
361	4-Amino-1,6-naphthalenedi-sulfonic acid	1-Naphthylamine-4,7-disul-fonic acid	$C_{10}H_9NO_6S_2$	85-75-6	303.311						vs H_2O
362	4-Amino-1,7-naphthalenedi-sulfonic acid	1-Naphthylamine-4,6-disul-fonic acid	$C_{10}H_9NO_6S_2$	85-74-5	303.311						vs H_2O, EtOH
363	2-Amino-1-naphthalenesul-fonic acid	2-Naphthylamine-1-sulfonic acid	$C_{10}H_9NO_3S$	81-16-3	223.248	sc(hot w)					s DMSO
364	4-Amino-1-naphthalenesul-fonic acid	1-Naphthylamine-4-sulfonic acid	$C_{10}H_9NO_3S$	84-86-6	223.248	wh nd (w+1/2) red-br cry	dec		1.6703^{25}		i H_2O; sl EtOH; s MeOH, py
365	5-Amino-1-naphthalenesul-fonic acid	1-Naphthylamine-5-sulfonic acid	$C_{10}H_9NO_3S$	84-89-9	223.248	wh cry					s H_2O; i eth
366	6-Amino-1-naphthalenesul-fonic acid	2-Naphthylamine-5-sulfonic acid	$C_{10}H_9NO_3S$	81-05-0	223.248	nd(w)					i H_2O, EtOH, eth
367	7-Amino-1-naphthalenesul-fonic acid	Badische acid	$C_{10}H_9NO_3S$	86-60-2	223.248	nd (w+1), pl (aq ace)					vs HOAc
368	8-Amino-1-naphthalenesul-fonic acid	1-Naphthylamine-8-sulfonic acid	$C_{10}H_9NO_3S$	82-75-7	223.248	nd					vs gl HOAc
369	6-Amino-2-naphthalenesul-fonic acid	Bronner acid	$C_{10}H_9NO_3S$	93-00-5	223.248	lf					i cold H_2O; sl hot H_2O
370	8-Amino-2-naphthalenesul-fonic acid	1,7-Cleve's acid	$C_{10}H_9NO_3S$	119-28-8	223.248	nd or pr (w)					sl EtOH; s eth
371	5-Amino-1-naphthol	1-Amino-6-hydroxynaphtha-lene	$C_{10}H_9NO$	83-55-6	159.184		170				sl DMSO
372	1-Amino-2-naphthol		$C_{10}H_9NO$	2834-92-6	159.184	silvery lf (bz, eth)	150 dec				sl H_2O, eth; s EtOH; vs dil alk, acid
373	8-Amino-2-naphthol	8-Amino-β-naphthol	$C_{10}H_9NO$	118-46-7	159.184	nd (w, al)	206	sub			s H_2O, eth; vs EtOH; sl bz, lig
374	2-Amino-4-nitrobenzoic acid		$C_7H_6N_2O_4$	619-17-0	182.134	oran pr (dil al)	269				i H_2O; vs EtOH, eth, ace; s xyl
375	2-Amino-5-nitrobenzoic acid		$C_7H_6N_2O_4$	616-79-5	182.134	lf (al), ye nd (w, dil al)	278(5)				i H_2O, bz, chl, xyl; s EtOH, eth
376	2-Amino-5-nitrobenzonitrile		$C_7H_5N_3O_2$	17420-30-3	163.134		203.5				sl DMSO
377	3-Amino-1-nitroguanidine		$CH_5N_5O_2$	18264-75-0	119.084		187.8				sl H_2O
378	2-Amino-4-nitrophenol		$C_6H_6N_2O_3$	99-57-0	154.123	oran pr (+w)	146				sl H_2O, ace; vs EtOH; s eth, bz, HOAc
379	2-Amino-5-nitrophenol		$C_6H_6N_2O_3$	121-88-0	154.123		205.8				s H_2O, EtOH, bz
380	4-Amino-2-nitrophenol		$C_6H_6N_2O_3$	119-34-6	154.123	dk red pl or nd (w, al)	131	110^{12}			s H_2O, EtOH, eth; sl DMSO
381	2-Aminooctanoic acid, (±)-		$C_8H_{17}NO_2$	644-90-6	159.227	lf (w)	270	sub			sl H_2O, EtOH, eth, bz; s HOAc
382	Aminooxoacetohydrazide	Semioxamazide	$C_2H_5N_3O_2$	515-96-8	103.080		221 dec				sl H_2O; i EtOH, eth; vs alk, acid
383	*cis*-4-Amino-4-oxo-2-butenoic acid	Maleamic acid	$C_4H_5NO_3$	557-24-4	115.088	cry (al)	172.5				vs H_2O, EtOH
384	5-Amino-4-oxopentanoic acid	5-Aminolevulinic acid	$C_5H_9NO_3$	106-60-5	131.130	cry (EtOH)	118				
385	(Aminooxy)acetic acid, hydrochloride (2:1)		$C_4H_{11}ClN_2O_6$	2921-14-4	218.592		152.5				
386	6-Aminopenicillanic acid	Penicin	$C_8H_{12}N_2O_3S$	551-16-6	216.257	cry (w)	208				
387	5-Aminopentanoic acid		$C_5H_{11}NO_2$	660-88-8	117.147	lf (dil al)	157 dec	dec			s H_2O; sl EtOH; i eth, bz, lig
388	5-Amino-1-pentanol		$C_5H_{13}NO$	2508-29-4	103.163		38.5	221.5	0.9488^{17}	1.4618^{17}	msc H_2O, EtOH, ace
389	2-Aminophenol		C_6H_7NO	95-55-6	109.126	wh orth bipym nd (bz)	173.5(0.3)	267(19)	1.328^{25}		s H_2O, eth; vs EtOH; sl bz, tfa

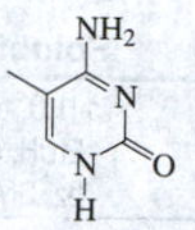

355
4-Amino-5-methyl-2(1*H*)-pyrimidinone

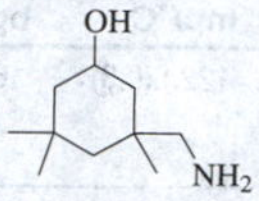

356
3-(Aminomethyl)-3,5,5-trimethylcyclohexanol

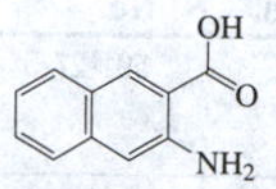

357
3-Amino-2-naphthalenecarboxylic acid

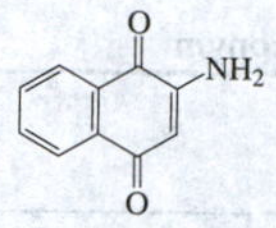

358
2-Amino-1,4-naphthalenedione

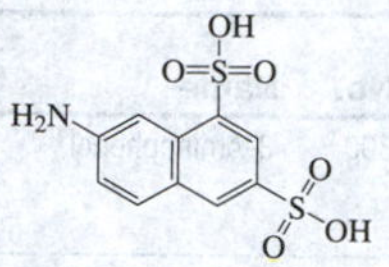

359
7-Amino-1,3-naphthalenedisulfonic acid

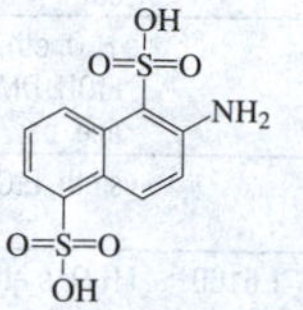

360
2-Amino-1,5-naphthalenedisulfonic acid

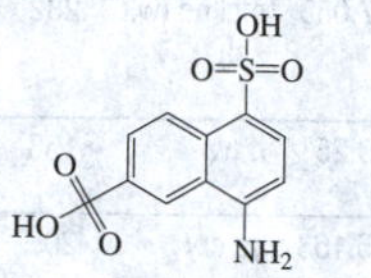

361
4-Amino-1,6-naphthalenedisulfonic acid

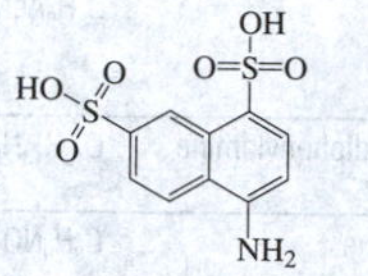

362
4-Amino-1,7-naphthalenedisulfonic acid

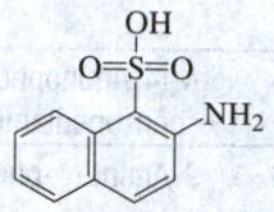

363
2-Amino-1-naphthalenesulfonic acid

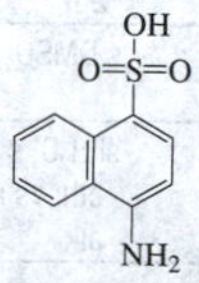

364
4-Amino-1-naphthalenesulfonic acid

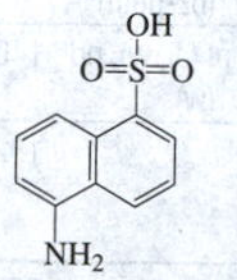

365
5-Amino-1-naphthalenesulfonic acid

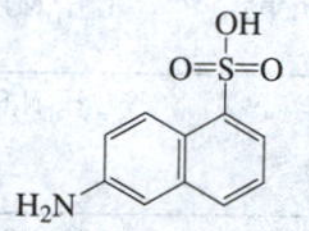

366
6-Amino-1-naphthalenesulfonic acid

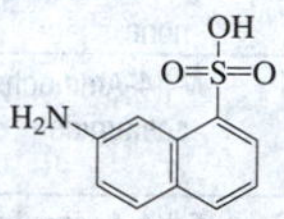

367
7-Amino-1-naphthalenesulfonic acid

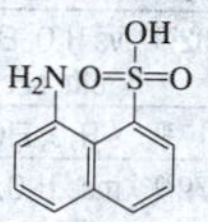

368
8-Amino-1-naphthalenesulfonic acid

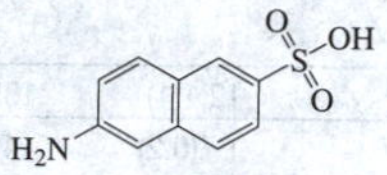

369
6-Amino-2-naphthalenesulfonic acid

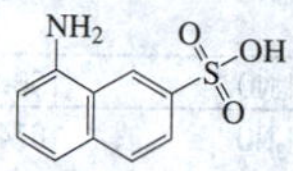

370
8-Amino-2-naphthalenesulfonic acid

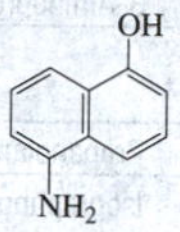

371
5-Amino-1-naphthol

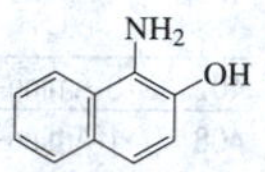

372
1-Amino-2-naphthol

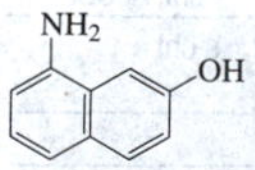

373
8-Amino-2-naphthol

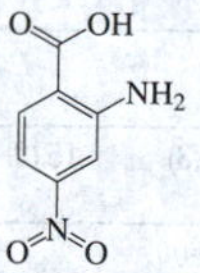

374
2-Amino-4-nitrobenzoic acid

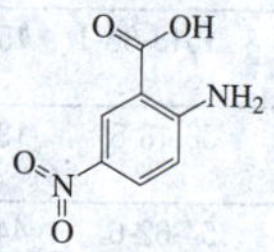

375
2-Amino-5-nitrobenzoic acid

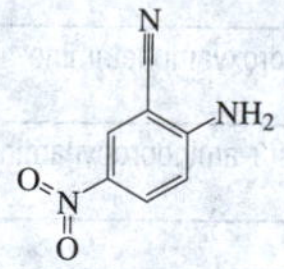

376
2-Amino-5-nitrobenzonitrile

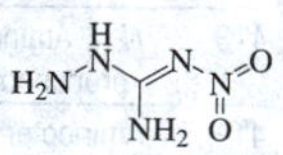

377
3-Amino-1-nitroguanidine

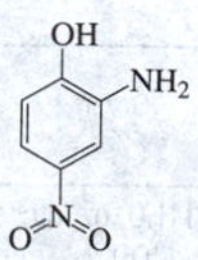

378
2-Amino-4-nitrophenol

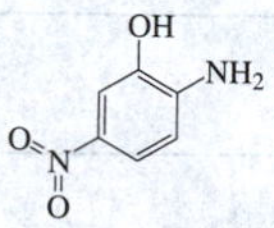

379
2-Amino-5-nitrophenol

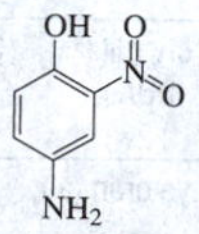

380
4-Amino-2-nitrophenol

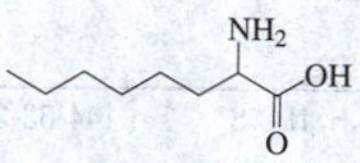

381
2-Aminooctanoic acid, (±)-

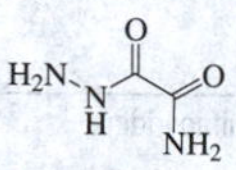

382
Aminooxoacetohydrazide

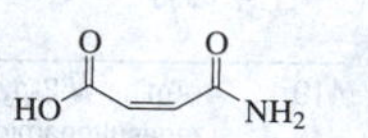

383
cis-4-Amino-4-oxo-2-butenoic acid

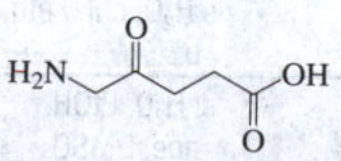

384
5-Amino-4-oxopentanoic acid

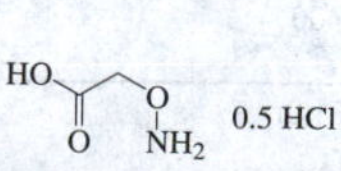

385
(Aminooxy)acetic acid, hydrochloride (2:1)

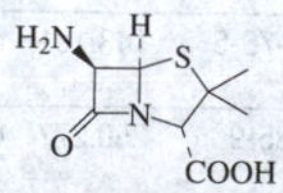

386
6-Aminopenicillanic acid

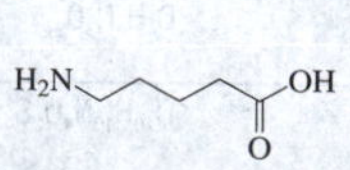

387
5-Aminopentanoic acid

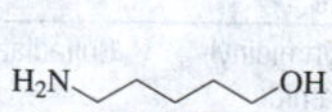

388
5-Amino-1-pentanol

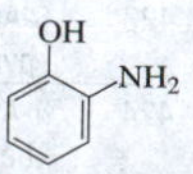

389
2-Aminophenol

Organic

No.	Name	Synonym	Mol. form.	CAS Reg. No.	Mol. wt.	Physical form	mp/°C	bp/°C	den/g cm^{-3}	n_D	Solubility
390	3-Aminophenol		C_6H_7NO	591-27-5	109.126	pr (to)	122.5(0.3)	164^{11}			s H_2O, tol; vs EtOH, eth; sl bz, DMSO
391	4-Aminophenol		C_6H_7NO	123-30-8	109.126	wh pl (w)	186(7)	$110^{0.3}$			sl H_2O, tfa; vs EtOH; i bz, chl; s alk
392	*N*-(3-Aminophenyl)acetamide		$C_8H_{10}N_2O$	102-28-3	150.177	nd or pl (bz)	88				vs H_2O, EtOH, ace; sl eth, bz
393	*N*-(4-Aminophenyl)acetamide	*p*-Aminoacetanilide	$C_8H_{10}N_2O$	122-80-5	150.177	nd (w)	166.5	267			s H_2O; vs EtOH, eth
394	(4-Aminophenyl)arsonic acid	Arsanilic acid	$C_6H_8AsNO_3$	98-50-0	217.055	mcl nd (w, al)	232		1.9571^{10}		s H_2O, eth; sl EtOH, DMSO; i ace, bz
395	*N*-(4-Aminophenyl)-1,4-benzenediamine	4,4'-Diaminodiphenylamine	$C_{12}H_{13}N_3$	537-65-5	199.251	lf (w)	158	dec			vs eth, EtOH
396	2-Amino-1-phenylethanone	Phenacylamine	C_8H_9NO	613-89-8	135.163	ye cry	20	251		1.6160^{20}	i H_2O; s eth; sl ctc
397	1-(3-Aminophenyl)ethanone	*m*-Aminoacetophenone	C_8H_9NO	99-03-6	135.163	pa ye pl (al), lf (eth)	98.5	289.5			sl H_2O; s EtOH
398	1-(4-Aminophenyl)ethanone	*p*-Aminoacetophenone	C_8H_9NO	99-92-3	135.163	ye mcl pr (al)	105(1)	294			vs eth, EtOH
399	1-(4-Aminophenyl)-1-pentanone		$C_{11}H_{15}NO$	38237-74-0	177.243	cry (bz-peth)	74.5	161^{3}			i H_2O; s EtOH, eth
400	1-(4-Aminophenyl)-1-propanone	*p*-Aminopropiophenone	$C_9H_{11}NO$	70-69-9	149.189	pl (al, w), nd (w)	140				s DMSO
401	*N*-[(4-Aminophenyl)sulfonyl]-acetamide	Sulfacetamide	$C_8H_{10}N_2O_3S$	144-80-9	214.241		182.0(0.4)				sl H_2O; s EtOH; i eth; vs ace, alk
402	5-[(4-Aminophenyl) sulfonyl]-2-thiazolamine	Thiazolsulfone	$C_9H_9N_3O_2S_2$	473-30-3	255.316	nd (al)	220 dec				vs ace, eth, EtOH, diox
403	4-Aminophthalimide	5-Amino-1*H*-isoindole-1,3(2*H*)-dione	$C_8H_6N_2O_2$	3676-85-5	162.146			$224^{0.5}$			
404	3-Amino-1,2-propanediol, (±)-		$C_3H_9NO_2$	13552-31-3	91.109			265 dec	1.1752^{20}	1.4910^{25}	s H_2O, EtOH; i eth, bz
405	3-Aminopropanenitrile	3-Aminopropionitrile	$C_3H_6N_2$	151-18-8	70.093			185	0.9584^{20}	1.4396^{20}	
406	2-Amino-1-propanol, (±)-		C_3H_9NO	6168-72-5	75.109			174.5		1.4502^{20}	vs H_2O, EtOH, eth; sl chl
407	3-Amino-1-propanol	Propanolamine	C_3H_9NO	156-87-6	75.109	liq	12.1(1)	185(4)	0.9824^{26}	1.4617^{20}	s H_2O, EtOH, eth
408	1-Amino-2-propanol	Isopropanolamine	C_3H_9NO	1674-56-2	75.109		1.7(0.2)	141(12)	0.9611^{20}	1.4479^{20}	msc H_2O, EtOH, eth, ace, bz, ctc
409	α-(1-Aminopropyl)-benzenemethanol	α-(α-Aminopropyl)benzyl alcohol	$C_{10}H_{15}NO$	5897-76-7	165.232	pl (bz-eth)	79.5				
410	*N*-(3-Aminopropyl)-*N*-methyl-1,3-propanediamine		$C_7H_{19}N_3$	105-83-9	145.246			232(3)	0.9023^{20}	1.4705^{25}	
411	Aminopropylon		$C_{16}H_{22}N_4O_2$	3690-04-8	302.372	pr (bz)	181				vs H_2O
412	4-(2-Aminopropyl)phenol, (±)-	Hydroxyamphetamine	$C_9H_{13}NO$	1518-86-1	151.205	cry (bz)	125.5				s H_2O, EtOH, bz, chl, AcOEt
413	*N*-(3-Aminopropyl)-1,3-propanediamine	Bis(3-aminopropyl)amine	$C_6H_{17}N_3$	56-18-8	131.219		-5.1(0.3)	151^{50}	0.938^{25}	1.4810^{20}	s chl
414	Aminopterin		$C_{19}H_{20}N_8O_5$	54-62-6	440.413	ye cry	262 dec				
415	4-Amino-*N*-pyrazinylbenzene-sulfonamide	Sulfapyrazine	$C_{10}H_{10}N_4O_2S$	116-44-9	250.277	nd ($PhNO_2$)	251				i H_2O, EtOH, eth, bz, chl; s py; sl ace
416	3-Amino-1*H*-pyrazole-4-carbonitrile	3-Amino-4-cyanopyrazole	$C_4H_4N_4$	16617-46-2	108.102	cry (w)	173				
417	2-Amino-3-pyridinecarboxylic acid		$C_6H_6N_2O_2$	5345-47-1	138.124		296 dec				sl H_2O
418	6-Amino-3-pyridinecarboxylic acid	6-Aminonicotinic acid	$C_6H_6N_2O_2$	3167-49-5	138.124	cry (dil HOAc, +2w)	312				
419	4-Amino-*N*-2-pyridinylbenzenesulfonamide	Sulfapyridine	$C_{11}H_{11}N_3O_2S$	144-83-2	249.289	ye oran (al)	190(2)				i H_2O, bz, ctc; s EtOH
420	5-Amino-2,4(1*H*,3*H*)-pyrimidinedione	5-Aminouracil	$C_4H_5N_3O_2$	932-52-5	127.102	nd (w)	dec				i H_2O; s alk, acid
421	6-Amino-2,4(1*H*,3*H*)-pyrimidinedione		$C_4H_5N_3O_2$	873-83-6	127.102	cry (w)	dec				vs H_2O
422	4-Amino-2(1*H*)-pyrimidinethione	2-Thiocytosine	$C_4H_5N_3S$	333-49-3	127.168						sl DMSO
423	5-Amino-2,4,6(1*H*,3*H*,5*H*)-pyrimidinetrione	Uramil	$C_4H_5N_3O_3$	118-78-5	143.101	nd or pl (w)	>400				s H_2O, chl; i eth, bz
424	4-Amino-*N*-2-pyrimidinyl-benzenesulfonamide	Sulfadiazine	$C_{10}H_{10}N_4O_2S$	68-35-9	250.277	cry (w), wh pow	261(3)				sl H_2O, EtOH, ace, DMSO

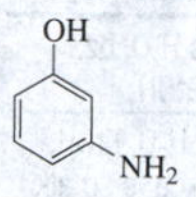
390
3-Aminophenol

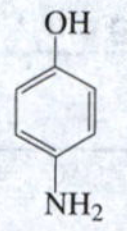
391
4-Aminophenol

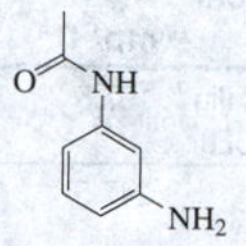
392
N-(3-Aminophenyl)acetamide

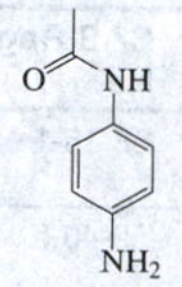
393
N-(4-Aminophenyl)acetamide

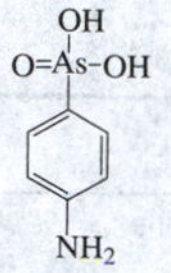
394
(4-Aminophenyl)arsonic acid

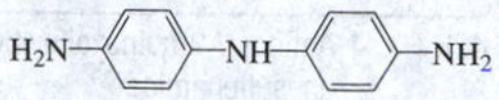
395
N-(4-Aminophenyl)-1,4-benzenediamine

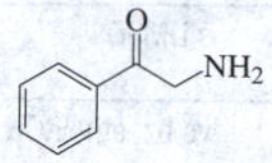
396
2-Amino-1-phenylethanone

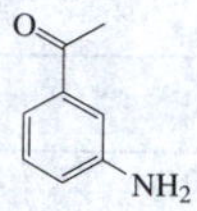
397
1-(3-Aminophenyl)ethanone

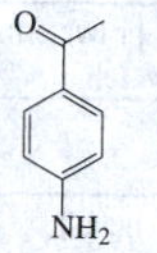
398
1-(4-Aminophenyl)ethanone

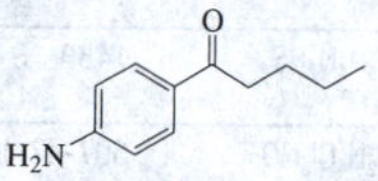
399
1-(4-Aminophenyl)-1-pentanone

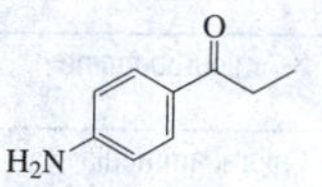
400
1-(4-Aminophenyl)-1-propanone

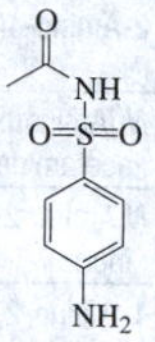
401
N-[(4-Aminophenyl)sulfonyl]acetamide

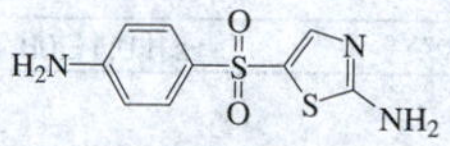
402
5-[(4-Aminophenyl)sulfonyl]-2-thiazolamine

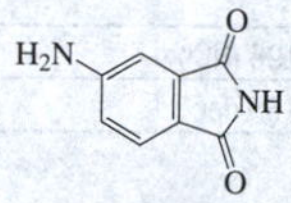
403
4-Aminophthalimide

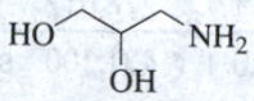
404
3-Amino-1,2-propanediol, (±)-

405
3-Aminopropanenitrile

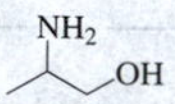
406
2-Amino-1-propanol, (±)-

407
3-Amino-1-propanol

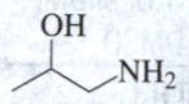
408
1-Amino-2-propanol

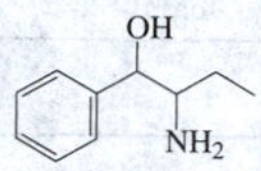
409
α-(1-Aminopropyl)benzenemethanol

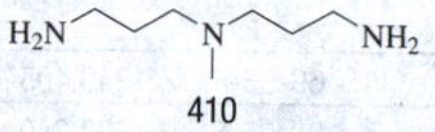
410
N-(3-Aminopropyl)-*N*-methyl-1,3-propanediamine

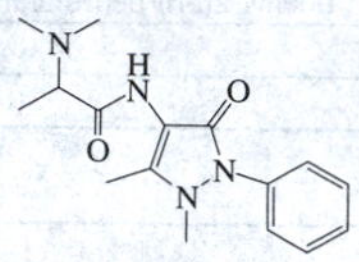
411
Aminopropylon

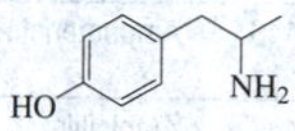
412
4-(2-Aminopropyl)phenol, (±)-

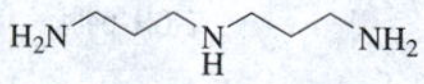
413
N-(3-Aminopropyl)-1,3-propanediamine

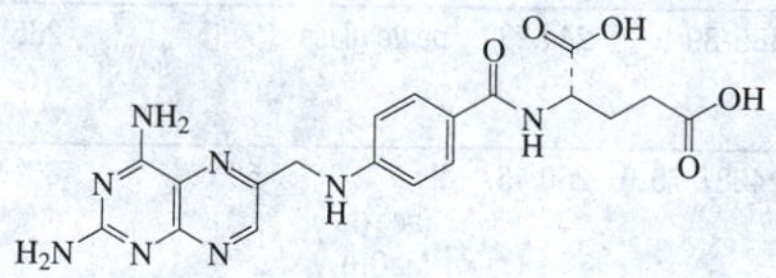
414
Aminopterin

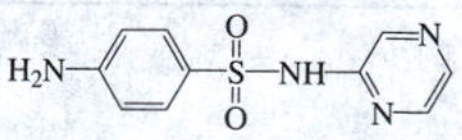
415
4-Amino-*N*-pyrazinylbenzenesulfonamide

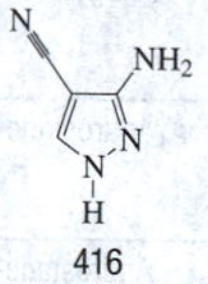
416
3-Amino-1*H*-pyrazole-4-carbonitrile

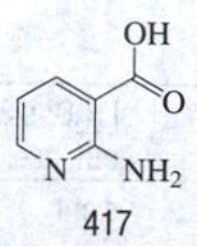
417
2-Amino-3-pyridinecarboxylic acid

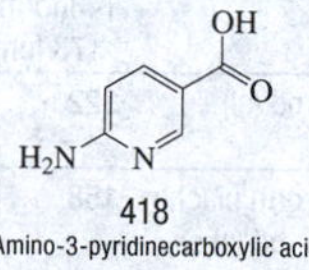
418
6-Amino-3-pyridinecarboxylic acid

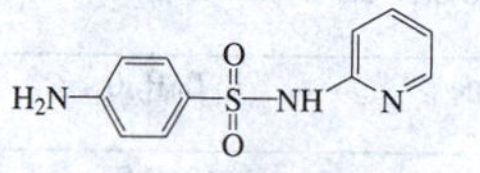
419
4-Amino-*N*-2-pyridinylbenzenesulfonamide

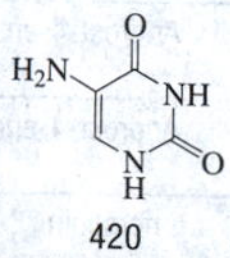
420
5-Amino-2,4(1*H*,3*H*)-pyrimidinedione

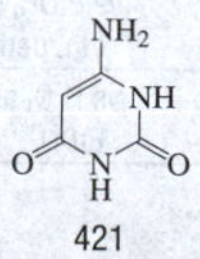
421
6-Amino-2,4(1*H*,3*H*)-pyrimidinedione

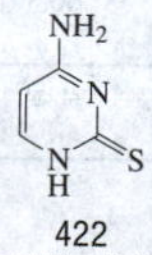
422
4-Amino-2(1*H*)-pyrimidinethione

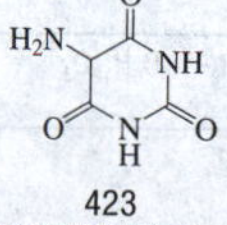
423
5-Amino-2,4,6(1*H*,3*H*,5*H*)-pyrimidinetrione

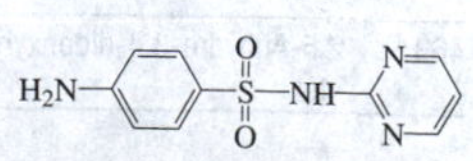
424
4-Amino-*N*-2-pyrimidinylbenzenesulfonamide

No.	Name	Synonym	Mol. form.	CAS Reg. No.	Mol. wt.	Physical form	mp/°C	bp/°C	den/ g cm⁻³	n_D	Solubility
425	Aminopyrine		$C_{13}H_{17}N_3O$	58-15-1	231.293	pr or pl (lig or AcOEt)	107.5				vs H_2O, bz, EtOH
426	4-Amino-*N*-2-quinoxalinylbenzenesulfonamide	Sulfaquinoxaline	$C_{14}H_{12}N_4O_2S$	59-40-5	300.336		247.5				sl H_2O, EtOH, ace; s aq alk
427	4-(Aminosulfonyl)benzoic acid	Carzenide	$C_7H_7NO_4S$	138-41-0	201.201	pr or lf (w)	291 dec				i H_2O; vs EtOH; sl eth; i bz
428	*N*-[4-(Aminosulfonyl)phenyl]-acetamide	Acetylsulfanilamide	$C_8H_{10}N_2O_3S$	121-61-9	214.241	nd (HOAc)	219.5				s H_2O, EtOH, ace
429	5-Amino-1,3,4-thiadiazole-2(3*H*)-thione		$C_2H_3N_3S_2$	2349-67-9	133.195		243.0				
430	2-Amino-4(5*H*)-thiazolone		$C_3H_4N_2OS$	556-90-1	116.141	pr or nd (w)	256 dec				sl H_2O; i EtOH, eth
431	*N*-(Aminothioxomethyl)-acetamide	Acetylthiourea	$C_3H_6N_2OS$	591-08-2	118.157	pr (w), orth (al)	165				sl H_2O, eth; s DMSO, EtOH
432	*N*-Amino-2-thioxo-4-thiazolidinone	3-Aminorhodanine	$C_3H_4N_2OS_2$	1438-16-0	148.206		101.5				s DMSO
433	1-Amino-2,2,2-trichloro-ethanol	Chloral ammonia	$C_2H_4Cl_3NO$	507-47-1	164.418	nd (al)	73	100 dec			vs bz, eth, EtOH
434	4-Amino-3,5,6-trichloro-2-pyridinecarboxlic acid	Picloram	$C_6H_3Cl_3N_2O_2$	1918-02-1	241.459		218.5				
435	11-Aminoundecanoic acid		$C_{11}H_{23}NO_2$	2432-99-7	201.307		189.0				
436	Amiton		$C_{10}H_{24}NO_3PS$	78-53-5	269.342	liq		$76^{0.01}$		1.4655^{27}	
437	Amitraz	*N*-Methylbis(2,4-xylylimino-methyl)amine	$C_{19}H_{23}N_3$	33089-61-1	293.406		86		1.128^{20}		
438	Amitriptyline		$C_{20}H_{23}N$	50-48-6	277.404	cry	196 (HCl)				
439	Ammonium ferric oxalate		$C_6H_{12}FeN_3O_{12}$	14221-47-7	374.017		165 dec		$1.78^{17.5}$		vs H_2O; i EtOH
440	Ammonium perfluorooctanoate		$C_8H_4F_{15}NO_2$	3825-26-1	431.100	solid					
441	Ammonium propanoate		$C_3H_9NO_2$	17496-08-1	91.109	hyg cry	45				s H_2O
442	Amobarbital	5-Ethyl-5-isopentyl-2,4,6(1*H*,3*H*,5*H*)-pyrimidinetrione	$C_{11}H_{18}N_2O_3$	57-43-2	226.272		153(1)				vs bz, EtOH, chl
443	Amolanone	3-[2-(Diethylamino)ethyl]-3-phenyl-2(3*H*)-benzofuranone	$C_{20}H_{23}NO_2$	76-65-3	309.403	cry (peth)	43.4	$193^{2.0}$		1.5614^{25}	
444	Amoxicillin		$C_{16}H_{19}N_3O_5S$	26787-78-0	365.404	cry (w)					s H_2O
445	Amphecloral		$C_{11}H_{12}Cl_3N$	5581-35-1	264.579			$96^{0.5}$		1.530	
446	Amphotericin B		$C_{47}H_{73}NO_{17}$	1397-89-3	924.080	ye pr (DMF)	170 dec				i H_2O; sl DMF; s DMSO
447	Ampicillin		$C_{16}H_{19}N_3O_4S$	69-53-4	349.405	cry	200 dec				sl H_2O
448	Ampyrone		$C_{11}H_{13}N_3O$	83-07-8	203.240	pa ye cry (bz)	109				s H_2O, EtOH, bz, chl; sl eth
449	Amygdalin		$C_{20}H_{27}NO_{11}$	29883-15-6	457.428		224.5				vs H_2O; sl EtOH; i eth, chl
450	Anacardic acid		$C_{22}H_{32}O_3$	11034-77-8	344.487	cry (ace)	35.5				vs eth, EtOH, peth
451	Anagyrine		$C_{15}H_{20}N_2O$	486-89-5	244.332	pe ye glass		265^{12}			s H_2O, eth, bz; vs EtOH, chl; i lig
452	Androstane		$C_{19}H_{32}$	24887-75-0	260.457	lf (ace-MeOH)	50	$60^{0.003}$			vs ace, eth, EtOH, peth
453	Androstane-17-carboxylic acid, (5β,17β)	Etiocholanic acid	$C_{20}H_{32}O_2$	438-08-4	304.467	nd (gl HOAc)	228.5	160 sub			
454	Androstane-3,17-diol, (3α,5α,17β)	Epiandrostanediol	$C_{19}H_{32}O_2$	1852-53-5	292.456	nd (ace aq)	223				
455	5α-Androstane-3,17-dione		$C_{19}H_{28}O_2$	846-46-8	288.424	cry (MeOH)	130(3)				
456	5β-Androstane-3,17-dione		$C_{19}H_{28}O_2$	1229-12-5	288.424	cry (ace-hx)	135				
457	Androst-4-ene-3,17-dione	4-Androstene-3,17-dione	$C_{19}H_{26}O_2$	63-05-8	286.408		143(form a); 173(form b)				
458	Androst-4-ene-3,11,17-trione	Adrenosterone	$C_{19}H_{24}O_3$	382-45-6	300.392	nd (al)	222	sub			sl H_2O; s EtOH, eth, ace, chl
459	Anemonin	*trans*-1,7-Dioxadispiro[4.0.4.2]dodeca-3,9-diene-2,8-dione	$C_{10}H_8O_4$	508-44-1	192.169	orth pl (chl) nd (al or bz)	158				vs chl
460	Anhalamine		$C_{11}H_{15}NO_3$	643-60-7	209.242	nd (al)	187.5				vs eth, EtOH
461	Anhalonidine		$C_{12}H_{17}NO_3$	17627-77-9	223.268	oct cry (bz, eth)	160.5				vs H_2O, EtOH
462	Anhalonine		$C_{12}H_{15}NO_3$	519-04-0	221.252	rhom nd	86	$140^{0.02}$			vs EtOH, bz, chl, eth, peth
463	2,5-Anhydro-3,4-dideoxyhexitol	Tetrahydro-2,5-furandimethanol	$C_6H_{12}O_3$	104-80-3	132.157		<-50	265	1.154^{20}		vs H_2O, ace, bz, EtOH

Organic

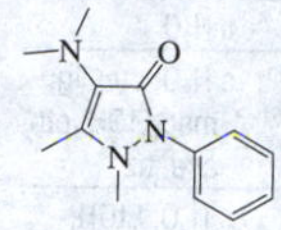
425
Aminopyrine

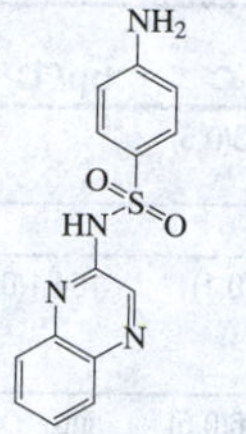
426
4-Amino-*N*-2-quinoxalinylbenzenesulfonamide

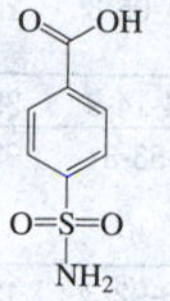
427
4-(Aminosulfonyl)benzoic acid

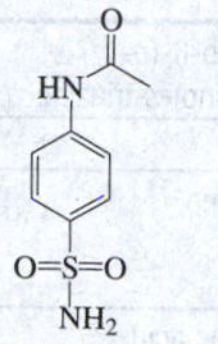
428
N-[4-(Aminosulfonyl)phenyl]acetamide

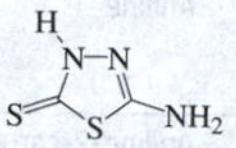
429
5-Amino-1,3,4-thiadiazole-2(3*H*)-thione

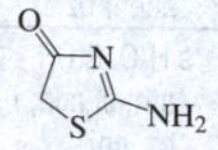
430
2-Amino-4(5*H*)-thiazolone

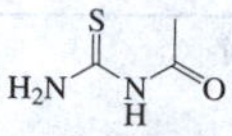
431
N-(Aminothioxomethyl)acetamide

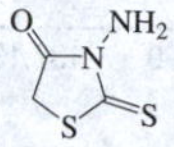
432
N-Amino-2-thioxo-4-thiazolidinone

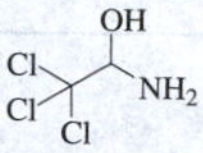
433
1-Amino-2,2,2-trichloroethanol

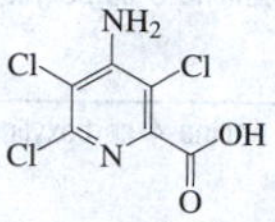
434
4-Amino-3,5,6-trichloro-2-pyridinecarboxlic acid

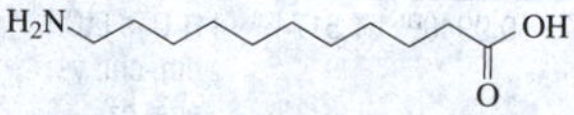
435
11-Aminoundecanoic acid

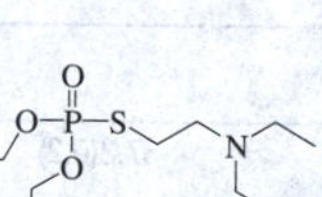
436
Amiton

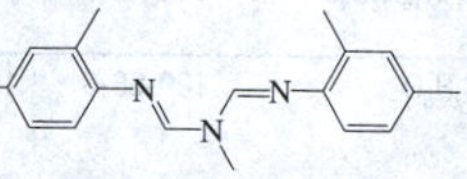
437
Amitraz

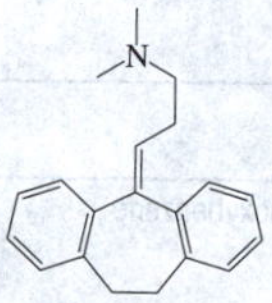
438
Amitriptyline

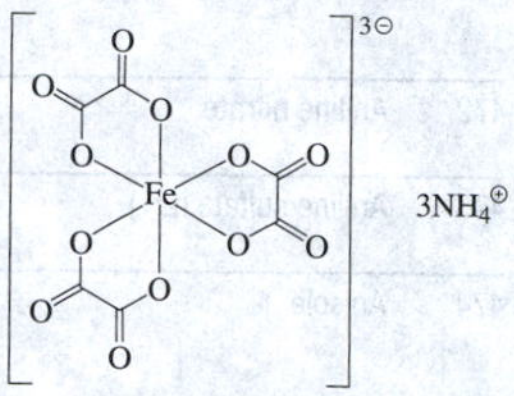
439
Ammonium ferric oxalate

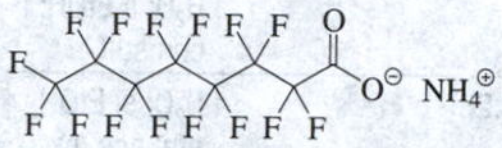
440
Ammonium perfluorooctanoate

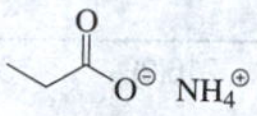
441
Ammonium propanoate

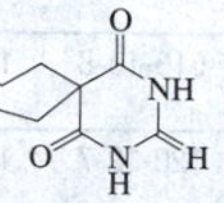
442
Amobarbital

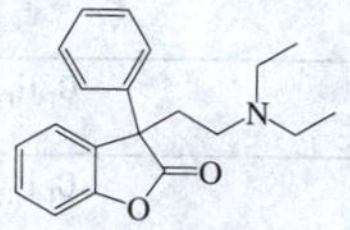
443
Amolanone

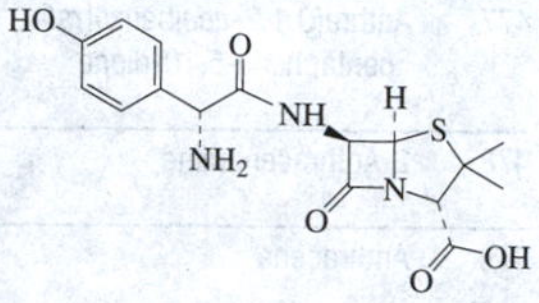
444
Amoxicillin

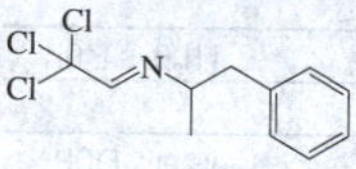
445
Amphecloral

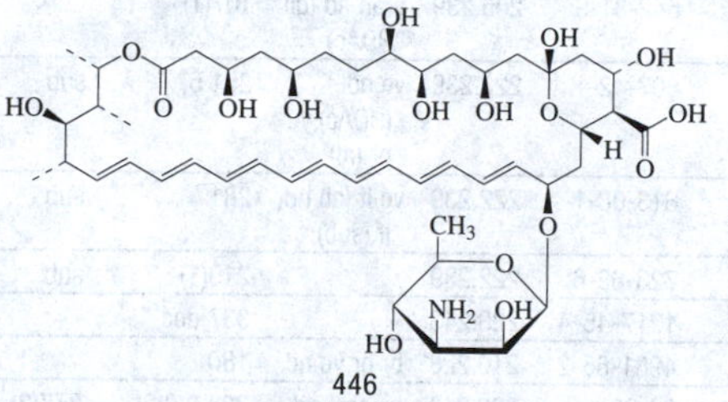
446
Amphotericin B

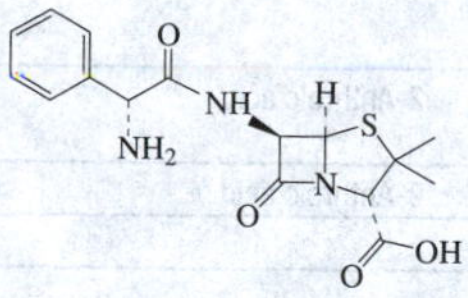
447
Ampicillin

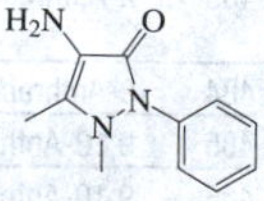
448
Ampyrone

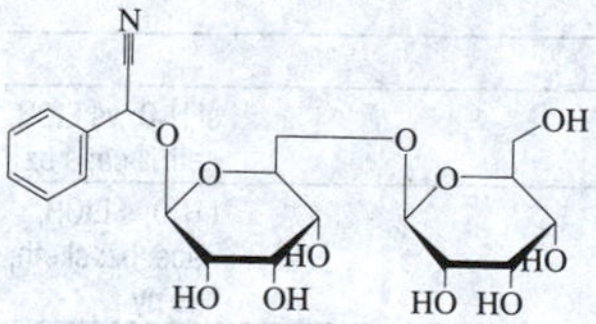
449
Amygdalin

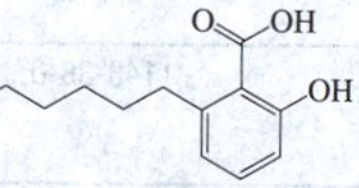
450
Anacardic acid

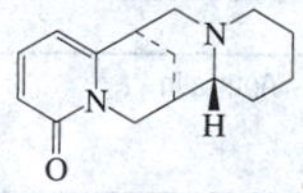
451
Anagyrine

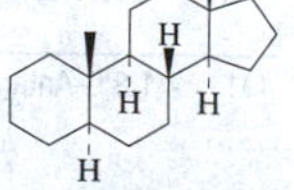
452
Androstane

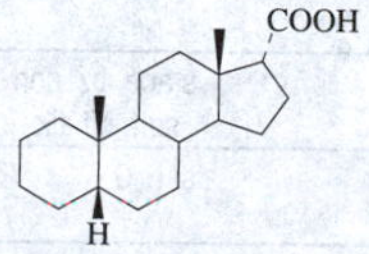
453
Androstane-17-carboxylic acid, (5β,17β)

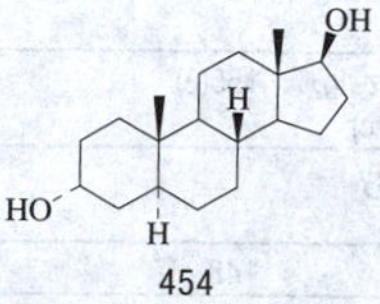
454
Androstane-3,17-diol, (3α,5α,17β)

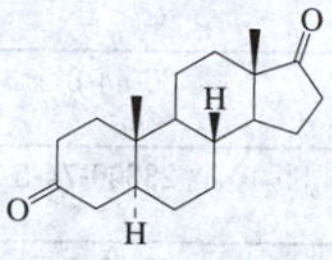
455
5α-Androstane-3,17-dione

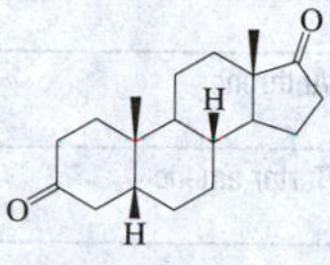
456
5β-Androstane-3,17-dione

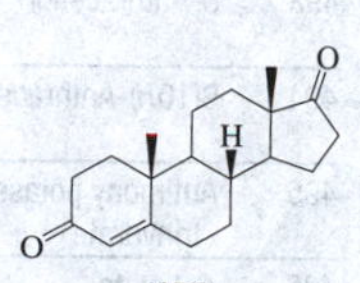
457
Androst-4-ene-3,17-dione

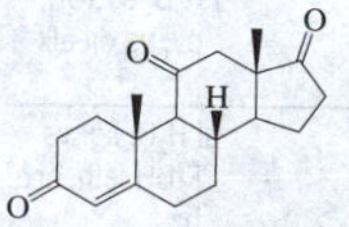
458
Androst-4-ene-3,11,17-trione

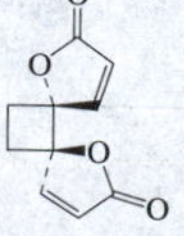
459
Anemonin

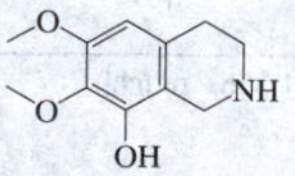
460
Anhalamine

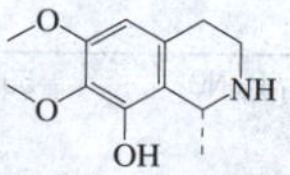
461
Anhalonidine

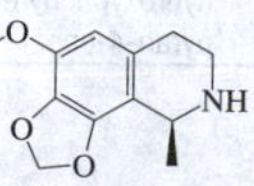
462
Anhalonine

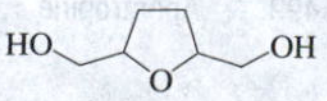
463
2,5-Anhydro-3,4-dideoxyhexitol

No.	Name	Synonym	Mol. form.	CAS Reg. No.	Mol. wt.	Physical form	mp/°C	bp/°C	den/ g cm^{-3}	n_D	Solubility
464	Anilazine	2,4-Dichloro-6-(*o*-chloroanilino)-*s*-triazine	$C_9H_5Cl_3N_4$	101-05-3	275.522		159.3(0.5)		1.8^{20}		
465	Anileridine		$C_{22}H_{28}N_2O_2$	144-14-9	352.469	cry	83				s H_2O
466	Aniline	Benzenamine	C_6H_7N	62-53-3	93.127	oily liq	-6.0(0.1)	184.1(0.4)	1.0217^{20}	1.5863^{20}	s H_2O, ctc, lig; msc EtOH, eth, ace, bz
467	Aniline-2-carboxylic acid	*o*-Anthranilic acid	$C_7H_7NO_2$	118-92-3	137.137	lf (al)	144.6(0.5)	sub	1.412^{20}		s H_2O, EtOH, eth; sl bz, tfa; vs chl, py
468	Aniline-3-carboxylic acid	*m*-Anthranilic acid	$C_7H_7NO_2$	99-05-8	137.137		179.7(0.6)		1.51^{25}		sl H_2O, EtOH; s eth, tfa; vs ace; i bz
469	Aniline-4-carboxylic acid	*p*-Anthranilic acid	$C_7H_7NO_2$	150-13-0	137.137	mcl pr (w)	188.2(0.6)		1.374^{20}		s H_2O, EtOH, eth; sl ace; i bz, chl
470	Aniline hydrobromide		C_6H_8BrN	542-11-0	174.039		286				
471	Aniline hydrochloride	Benzenamine hydrochloride	C_6H_8ClN	142-04-1	129.588	lf or nd	198		1.2215^4		vs H_2O, EtOH; i eth, chl; sl DMSO
472	Aniline nitrate		$C_6H_8N_2O_3$	542-15-4	156.139	orth	190 dec		1.356^4		vs H_2O, eth, EtOH
473	Aniline sulfate (2:1)		$C_{12}H_{16}N_2O_4S$	542-16-5	284.331				1.377^4		s H_2O; sl EtOH, tfa; i eth
474	Anisole	Methoxybenzene	C_7H_8O	100-66-3	108.138	liq	-37.3(0.2)	153.6(0.2)	0.9940^{20}	1.5174^{20}	i H_2O; s EtOH, eth, chl; vs ace, bz
475	Anisotropine methylbromide	Octatropine methylbromide	$C_{17}H_{32}BrNO_2$	80-50-2	362.346	cry (ace)	329				
476	Antazoline		$C_{17}H_{19}N_3$	91-75-8	265.353	cry	122				
477	Anthra[9,1,2-*cde*]benzo[*rst*]pentaphene-5,10-dione		$C_{34}H_{16}O_2$	116-71-2	456.490	viol-bl or blk nd ($PhNO_2$)	492 dec				i EtOH, bz, HOAc; s xyl, py, sulf
478	2-Anthracenamine		$C_{14}H_{11}N$	613-13-8	193.244	ye lf (al)	238.8	sub			i H_2O; s EtOH; i con sulf
479	Anthracene		$C_{14}H_{10}$	120-12-7	178.229	tab or mcl pr (al)	216(2)	341.3(0.4)	1.28^{25}		i H_2O; sl EtOH, eth, ace, bz, chl, ctc
480	9-Anthracenecarbonitrile		$C_{15}H_9N$	1210-12-4	203.239		177.5		1.3000^{20}		
481	9-Anthracenecarboxaldehyde		$C_{15}H_{10}O$	642-31-9	206.239	oran nd (dil HOAc)	107(1)				i H_2O; s bz, HOAc
482	1-Anthracenecarboxylic acid	1-Anthroic acid	$C_{15}H_{10}O_2$	607-42-1	222.239	ye nd (HOAc) ye pr (al)	251.5	sub			i H_2O; s EtOH, eth; sl bz, chl
483	2-Anthracenecarboxylic acid	2-Anthroic acid	$C_{15}H_{10}O_2$	613-08-1	222.239	ye lf (al) nd, lf (sub)	281	sub			vs HOAc
484	9-Anthracenecarboxylic acid	9-Anthroic acid	$C_{15}H_{10}O_2$	723-62-6	222.239		219(1)	sub			i H_2O; s EtOH
485	9,10-Anthracenedicarbonitrile		$C_{16}H_8N_2$	1217-45-4	228.248		337 dec				
486	9,10-Anthracenediol		$C_{14}H_{10}O_2$	4981-66-2	210.228	br or ye nd	180				vs eth, EtOH
487	9,10-Anthracenedione	Anthraquinone	$C_{14}H_8O_2$	84-65-1	208.213	ye orth nd (al, bz)	284.8(0.2)	377(2)	1.438^{20}		i H_2O; sl EtOH, eth, bz, chl
488	9-Anthracenemethanol		$C_{15}H_{12}O$	1468-95-7	208.255		160.5				
489	1,4,9,10-Anthracenetetrol		$C_{14}H_{10}O_4$	476-60-8	242.227		148				
490	1,2,10-Anthracenetriol	Anthrarobin	$C_{14}H_{10}O_3$	577-33-3	226.227	ye lf, nd (al-w)	208				sl H_2O; vs EtOH, eth, ace; s bz
491	1,8,9-Anthracenetriol	Anthralin	$C_{14}H_{10}O_3$	1143-38-0	226.227	ye pl or nd (lig)	179				i H_2O; s EtOH, ace, bz; sl eth; vs py
492	1-Anthracenol		$C_{14}H_{10}O$	610-50-4	194.228	cry (bz), br nd or lf (al)	158	234^{13}			i H_2O; vs EtOH, eth; s NaOH
493	9-Anthracenol	Anthranol	$C_{14}H_{10}O$	529-86-2	194.228	ye red lf (dil al)	152				
494	9(10*H*)-Anthracenone	Anthrone	$C_{14}H_{10}O$	90-44-8	194.228	nd (bz-lig, HOAc)	155(3)				s ace, bz, con sulf, dil alk
495	Antimony potassium tartrate trihydrate	Tartar emetic	$C_8H_{10}K_2O_{15}Sb_2$	28300-74-5	667.873	col cry			2.6		sl H_2O
496	Apholate		$C_{12}H_{24}N_9P_3$	52-46-0	387.300		148				
497	Aphylline		$C_{15}H_{24}N_2O$	577-37-7	248.364	cry	52.5	200^4			vs ace, bz, eth, EtOH
498	Apigenin	5,7-Dihydroxy-2-(4-hydroxyphenyl)-4*H*-1-benzopyran-4-one	$C_{15}H_{10}O_5$	520-36-5	270.237	ye nd (aq py)	347.5				i H_2O; s EtOH, py; vs dil alk
499	Apoatropine		$C_{17}H_{21}NO_2$	500-55-0	271.355	pr (chl)	62				sl H_2O, lig; vs EtOH, eth, ace, bz

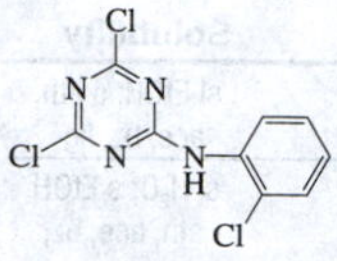
464
Anilazine

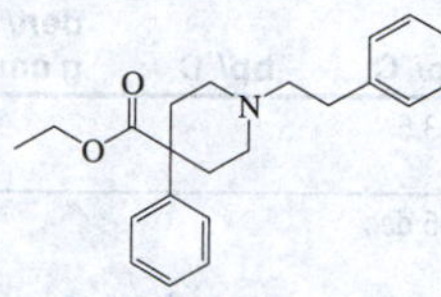
465
Anileridine

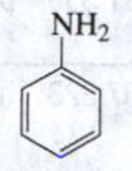
466
Aniline

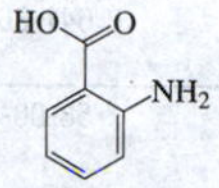
467
Aniline-2-carboxylic acid

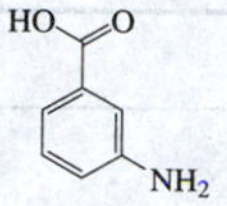
468
Aniline-3-carboxylic acid

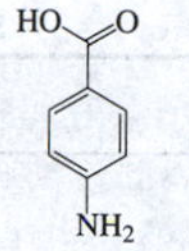
469
Aniline-4-carboxylic acid

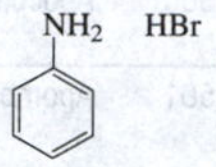
470
Aniline hydrobromide

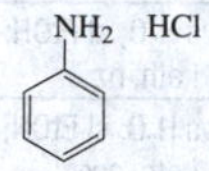
471
Aniline hydrochloride

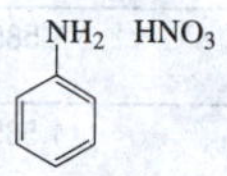
472
Aniline nitrate

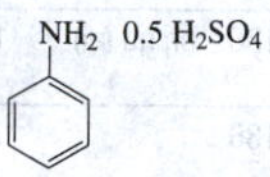
473
Aniline sulfate (2:1)

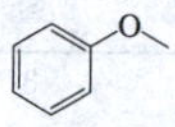
474
Anisole

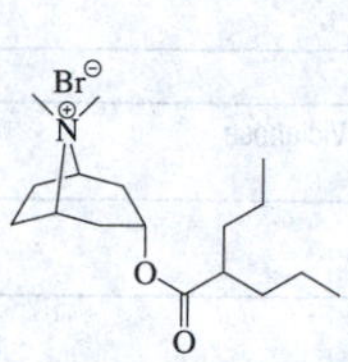
475
Anisotropine methylbromide

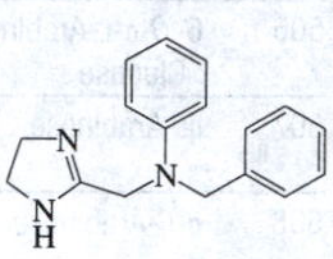
476
Antazoline

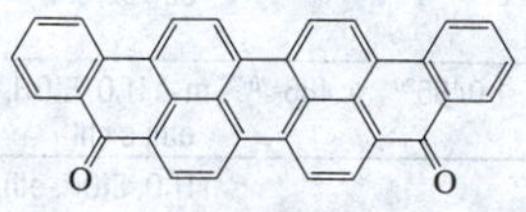
477
Anthra[9,1,2-cde]benzo[rst]pentaphene-5,10-dione

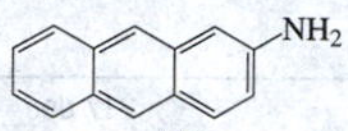
478
2-Anthracenamine

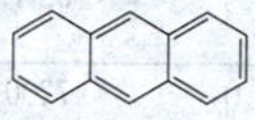
479
Anthracene

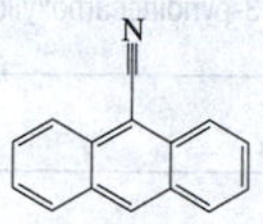
480
9-Anthracenecarbonitrile

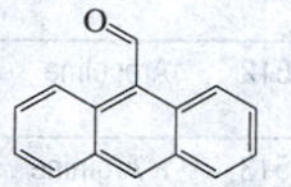
481
9-Anthracenecarboxaldehyde

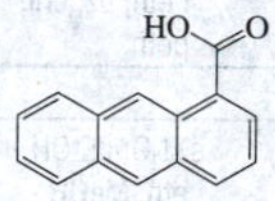
482
1-Anthracenecarboxylic acid

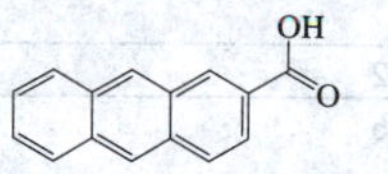
483
2-Anthracenecarboxylic acid

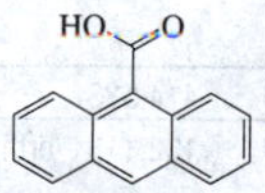
484
9-Anthracenecarboxylic acid

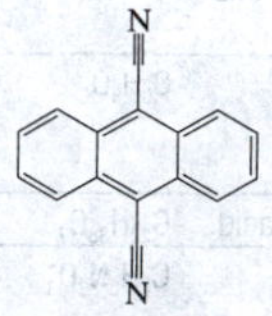
485
9,10-Anthracenedicarbonitrile

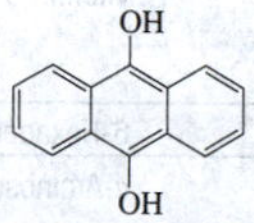
486
9,10-Anthracenediol

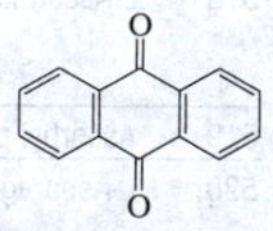
487
9,10-Anthracenedione

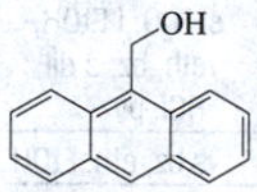
488
9-Anthracenemethanol

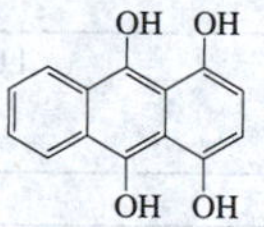
489
1,4,9,10-Anthracenetetrol

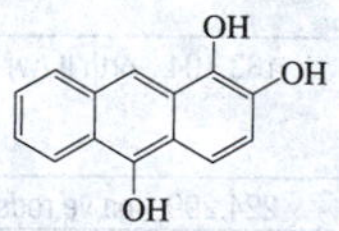
490
1,2,10-Anthracenetriol

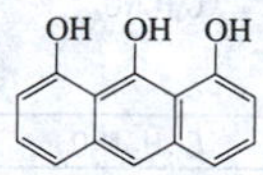
491
1,8,9-Anthracenetriol

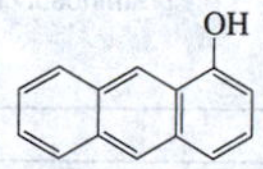
492
1-Anthracenol

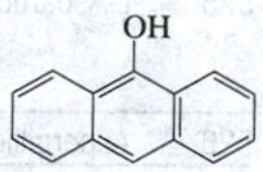
493
9-Anthracenol

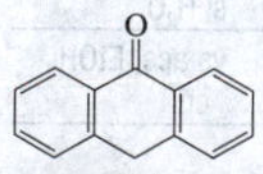
494
9(10*H*)-Anthracenone

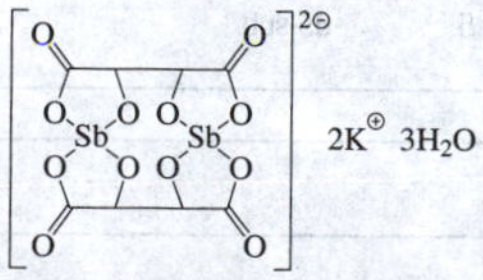
495
Antimony potassium tartrate trihydrate

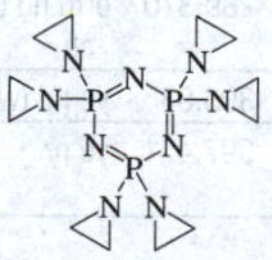
496
Apholate

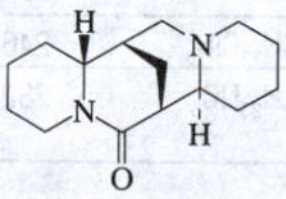
497
Aphylline

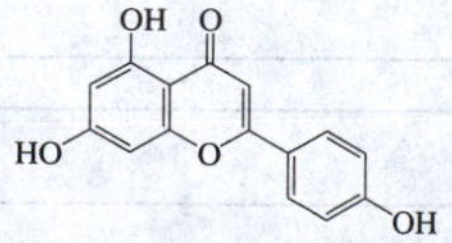
498
Apigenin

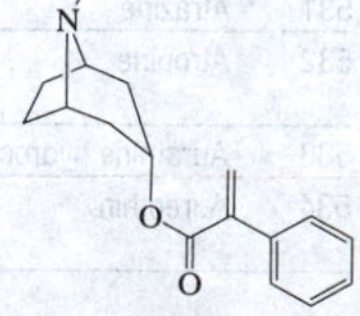
499
Apoatropine

Organic

No.	Name	Synonym	Mol. form.	CAS Reg. No.	Mol. wt.	Physical form	mp/°C	bp/°C	den/g cm^{-3}	n_D	Solubility
500	Apocodeine		$C_{18}H_{19}NO_2$	641-36-1	281.350	pr (MeOH)	123.5				sl EtOH; s eth, ace, bz, lig
501	Apomorphine		$C_{17}H_{17}NO_2$	58-00-4	267.323	hex pl (chl-peth) rods (eth)	195 dec				sl H_2O; s EtOH, eth, ace, bz, alk
502	Apomorphine, hydrochloride		$C_{17}H_{18}ClNO_2$	314-19-2	303.784	grn in air mcl pr	205 dec				
503	Aprobarbital	5-Isopropyl-5-allyl-2,4,6(1*H*,3*H*,5*H*)-pyrimidinetrione	$C_{10}H_{14}N_2O_3$	77-02-1	210.229	cry	141				vs ace, eth, EtOH, chl
504	L-Arabinitol		$C_5H_{12}O_5$	7643-75-6	152.146		102.5				vs H_2O; sl EtOH; i eth
505	α-D-Arabinopyranose		$C_5H_{10}O_5$	608-45-7	150.130	cry (MeOH)	155.5		1.585^{25}		
506	6-*O*-α-L-Arabinopyranosyl-D-Glucose	Vicianose	$C_{11}H_{20}O_{10}$	14116-69-9	312.271	nd (dil al)	210 dec				vs H_2O
507	DL-Arabinose		$C_5H_{10}O_5$	20235-19-2	150.130	pr, nd (al)	164.5		1.585^{20}		vs H_2O; sl EtOH; i eth, bz
508	α-D-Arabinose		$C_5H_{10}O_5$	31178-68-4	150.130		156		1.585^{25}		vs H_2O; sl EtOH; i eth, ace, MeOH
509	β-D-Arabinose		$C_5H_{10}O_5$	31178-69-5	150.130		156		1.625^{25}		vs H_2O; sl EtOH; i eth, ace, MeOH
510	Aramite		$C_{15}H_{23}ClO_4S$	140-57-8	334.860		-37.3	195^{2}	1.143^{20}	1.5100^{20}	vs ace, bz, eth, EtOH
511	Arecaidine	1,2,5,6-Tetrahydro-1-methyl-3-pyridinecarboxylic acid	$C_7H_{11}NO_2$	499-04-7	141.168	pl (dil al) tab (dil al +1w)	232 dec				vs H_2O; i EtOH, eth, bz, chl
512	Arecoline		$C_8H_{13}NO_2$	63-75-2	155.195			209	1.0485^{20}	1.486-^{20}	msc H_2O, EtOH, eth; s chl
513	D-Arginine		$C_6H_{14}N_4O_2$	7200-25-1	174.201		217 dec				i H_2O, EtOH, eth, bz
514	L-Arginine		$C_6H_{14}N_4O_2$	74-79-3	174.201		260(3)				s H_2O; sl EtOH; i eth
515	L-Arginine, monohydrochloride		$C_6H_{15}ClN_4O_2$	1119-34-2	210.662		219				
516	Artemisin	8-Hydroxysantonin	$C_{15}H_{18}O_4$	481-05-0	262.302	cry	203	$260^{0.1}$			sl H_2O, chl; s AcOEt; i peth
517	Ascaridole	1-Methyl-4-isopropyl-2,3-dioxabicyclo[2.2.2]oct-5-ene	$C_{10}H_{16}O_2$	512-85-6	168.233	liq	3.3	exp	1.0103^{20}	1.4769^{20}	i H_2O; s EtOH, ace, bz, tol; sl chl
518	L-Ascorbic acid	Vitamin C	$C_6H_8O_6$	50-81-7	176.124		191(4)		1.65^{25}		vs H_2O; s EtOH; i eth, bz, chl, peth
519	Ascorbyl palmitate	6-Hexadecanoylascorbic acid	$C_{22}H_{38}O_7$	137-66-6	414.533		112				
520	L-Asparagine	α-Aminosuccinamic acid	$C_4H_8N_2O_3$	70-47-3	132.118	orth (w+1)	235		1.543^{15}		s H_2O; i EtOH, eth, MeOH
521	D-Asparagine, monohydrate		$C_4H_{10}N_2O_4$	5794-24-1	150.133		215		1.523^{15}		sl H_2O; i EtOH, eth, bz, MeOH
522	L-Asparagine, monohydrate		$C_4H_{10}N_2O_4$	5794-13-8	150.133		234		1.543^{15}		sl H_2O; i EtOH, eth, bz, MeOH
523	Aspartame	L-α-Aspartyl-L-phenylalanine, 2-methyl ester	$C_{14}H_{18}N_2O_5$	22839-47-0	294.303	nd (w)	246.5				
524	DL-Aspartic acid		$C_4H_7NO_4$	617-45-8	133.104	mcl pr (w)	277.5		1.6622^{13}		sl H_2O; i EtOH, eth, bz, py
525	L-Aspartic acid	L-Aminosuccinic acid	$C_4H_7NO_4$	56-84-8	133.104	orth lf (w)	270		1.6603^{13}		sl H_2O; i EtOH, eth, bz; s dil HCl, py
526	Aspergillic acid		$C_{12}H_{20}N_2O_2$	490-02-8	224.299	pa ye rods	98				vs bz, eth, EtOH
527	Astemizole		$C_{28}H_{31}FN_4O$	68844-77-9	458.570	wh cry	149.1				i H_2O; s os
528	Asulam	Methyl [(4-aminophenyl)-sulfonyl]carbamate	$C_8H_{10}N_2O_4S$	3337-71-1	230.241		144.2(0.5)				
529	Atenolol		$C_{14}H_{22}N_2O_3$	29122-68-7	266.336	cry (AcOEt)	147				sl H_2O, diox, ace; i chl; s MeOH, HOAc
530	Atisine	Anthorine	$C_{22}H_{33}NO_2$	466-43-3	343.503	orth bipym	58.5				vs eth, EtOH, chl
531	Atrazine		$C_8H_{14}ClN_5$	1912-24-9	215.684		177.0(0.5)				
532	Atropine		$C_{17}H_{23}NO_3$	51-55-8	289.370	orth nd (dil al)	118.5	95 sub			vs H_2O, EtOH; i eth; sl chl
533	Auramine hydrochloride		$C_{17}H_{22}ClN_3$	2465-27-2	303.83	ye nd (w)	267				sl H_2O
534	Aureothin		$C_{22}H_{23}NO_6$	2825-00-5	397.421	ye pr	158				vs ace, EtOH, chl

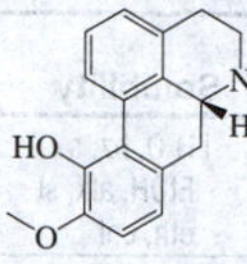
500 Apocodeine

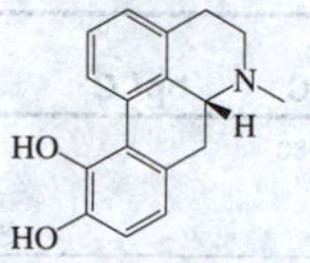
501 Apomorphine

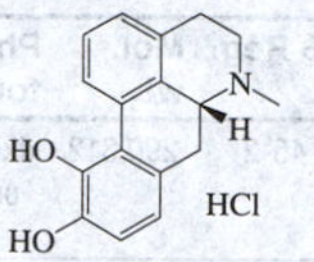
502 Apomorphine, hydrochloride

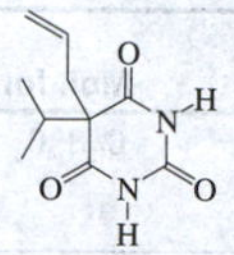
503 Aprobarbital

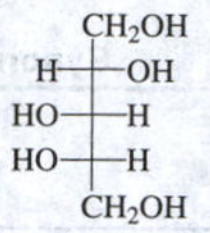
504 L-Arabinitol

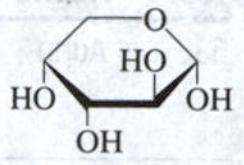
505 α-D-Arabinopyranose

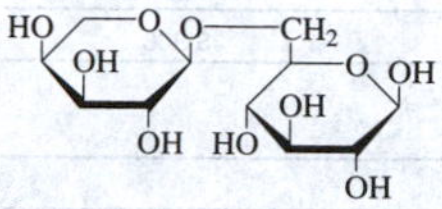
506 6-*O*-α-L-Arabinopyranosyl-D-Glucose

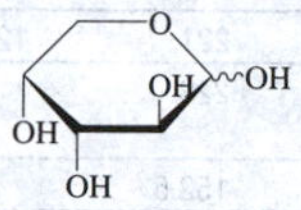
507 DL-Arabinose

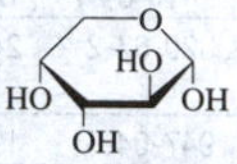
508 α-D-Arabinose

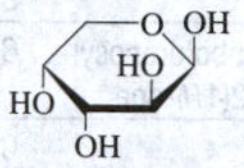
509 β-D-Arabinose

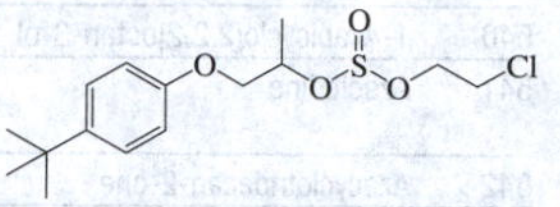
510 Aramite

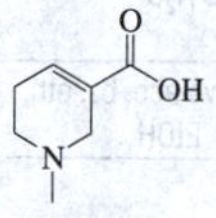
511 Arecaidine

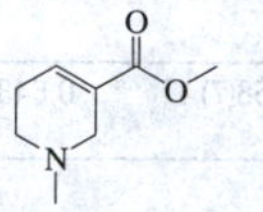
512 Arecoline

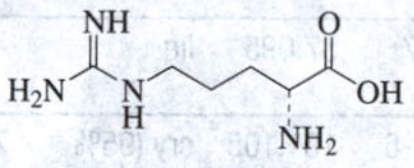
513 D-Arginine

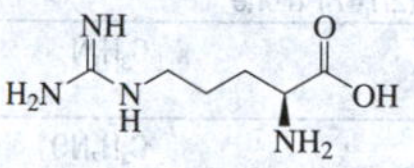
514 L-Arginine

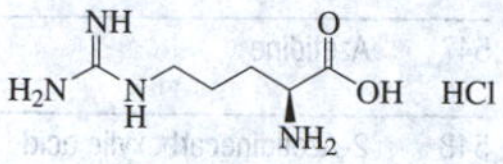
515 L-Arginine, monohydrochloride

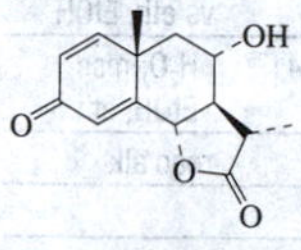
516 Artemisin

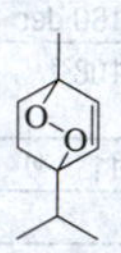
517 Ascaridole

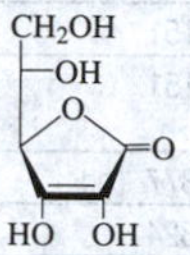
518 L-Ascorbic acid

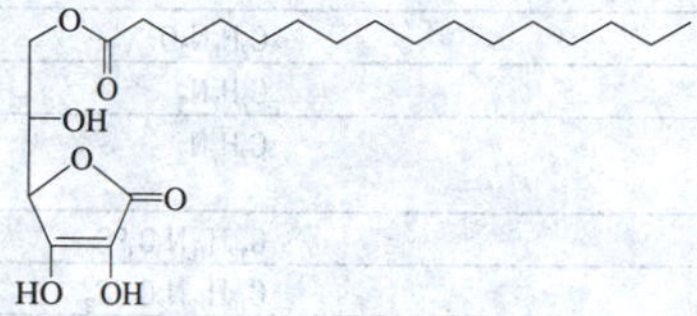
519 Ascorbyl palmitate

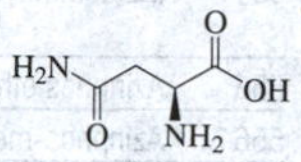
520 L-Asparagine

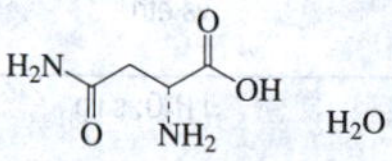
521 D-Asparagine, monohydrate

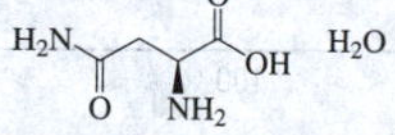
522 L-Asparagine, monohydrate

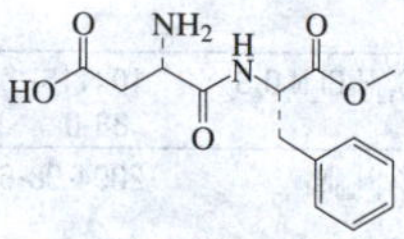
523 Aspartame

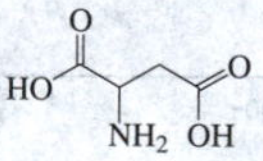
524 DL-Aspartic acid

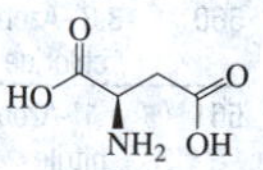
525 L-Aspartic acid

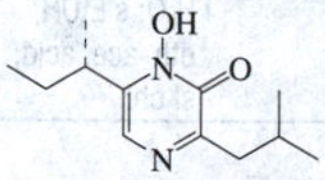
526 Aspergillic acid

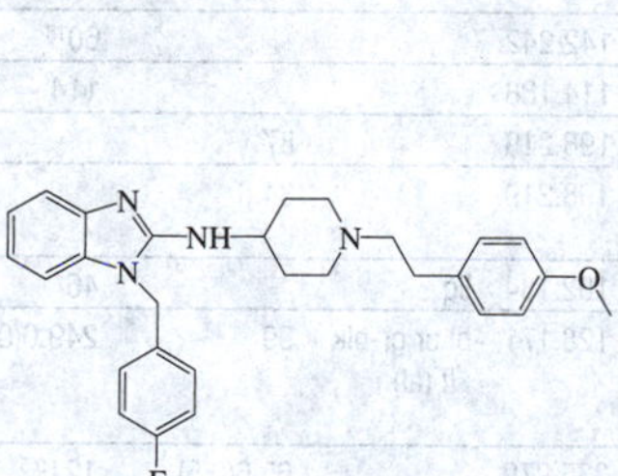
527 Astemizole

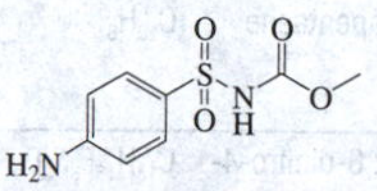
528 Asulam

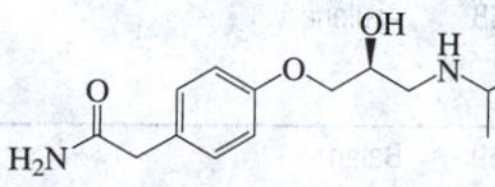
529 Atenolol

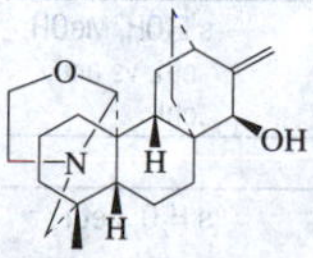
530 Atisine

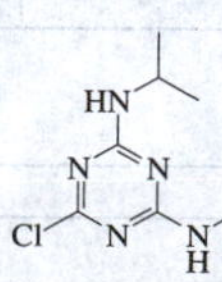
531 Atrazine

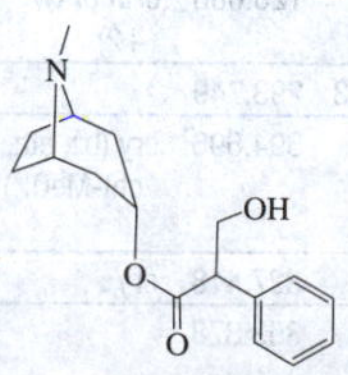
532 Atropine

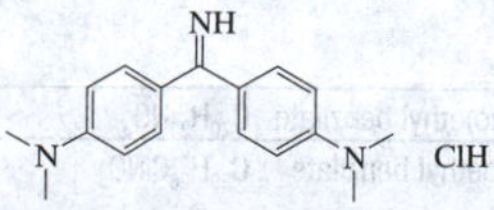
533 Auramine hydrochloride

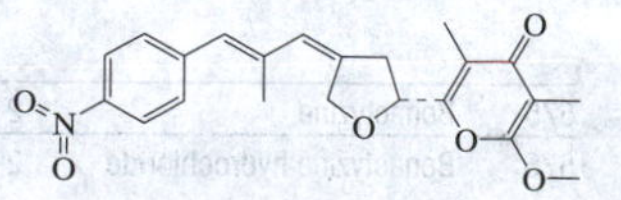
534 Aureothin

No.	Name	Synonym	Mol. form.	CAS Reg. No.	Mol. wt.	Physical form	mp/°C	bp/°C	den/ g cm^{-3}	n_D	Solubility
535	Aurin		$C_{19}H_{14}O_3$	603-45-2	290.312	dk red lf or orth	309 dec				i H_2O, bz; s EtOH, alk; sl eth, chl
536	Aurin tricarboxylic acid, triammonium salt	Aluminon	$C_{22}H_{23}N_3O_9$	569-58-4	473.433	red-br pow					s H_2O; sl EtOH; i peth
537	Avermectin B1a	Abamectin	$C_{48}H_{72}O_{14}$	71751-41-2	873.078		152				
538	3-Azabicyclo[3.2.2]nonane		$C_8H_{15}N$	283-24-9	125.212			166^{500}			
539	1-Azabicyclo[2.2.2]octane	Quinuclidine	$C_7H_{13}N$	100-76-5	111.185	cry (eth)	157(3)				vs H_2O, ace, eth, EtOH
540	1-Azabicyclo[2.2.2]octan-3-ol	3-Quinuclidinol	$C_7H_{13}NO$	1619-34-7	127.184	cry (bz)	221	120 sub			s ace
541	Azacitidine	4-Amino-1-β-D-ribofuranosyl-1,3,5-triazine-2(1*H*)-one	$C_8H_{12}N_4O_5$	320-67-2	244.205	cry	229				
542	Azacyclotridecan-2-one		$C_{12}H_{23}NO$	947-04-6	197.317		152.5				
543	8-Azaguanine		$C_4H_4N_6O$	134-58-7	152.114		300				
544	Azaserine		$C_5H_7N_3O_4$	115-02-6	173.128	ye-grn orth cry	150 dec				vs H_2O; sl EtOH, ace, MeOH
545	Azathioprine	1*H*-Purine, 6-[(1-methyl-4-nitro-1*H*-imidazol-5-yl)thio]-	$C_9H_7N_7O_2S$	446-86-6	277.263	ye cry	243 dec				sl H_2O, EtOH, chl
546	6-Azauridine	2-β-D-Ribofuranosyl-1,2,4-triazine-3,5(2*H*,4*H*)-dione	$C_8H_{11}N_3O_6$	54-25-1	245.189		158				s H_2O
547	Azetidine		C_3H_7N	503-29-7	57.095	liq	-70.0	58(7)	0.8436^{20}	1.4287^{25}	vs ace, bz, eth, EtOH
548	2-Azetidinecarboxylic acid		$C_4H_7NO_2$	2517-04-6	101.105	cry (95% MeOH)	217 dec				
549	2-Azetidinone		C_3H_5NO	930-21-2	71.078		73.5	106^{15}			vs eth, EtOH, chl
550	Azidobenzene		$C_6H_5N_3$	622-37-7	119.124	pa ye oil	-27.5	70^{11}	1.0860^{20}	1.5589^{25}	i H_2O; sl EtOH, eth
551	1-Azido-4-chlorobenzene		$C_6H_4ClN_3$	3296-05-7	153.569		20	96^{20}	1.2634^{25}		i H_2O; s eth
552	2-Azidoethanol		$C_2H_5N_3O$	1517-05-1	87.080			75^{40}	1.146^{24}		vs H_2O
553	1-Azido-4-methylbenzene		$C_7H_7N_3$	2101-86-2	133.151		-29.0	180 dec	1.0527^{23}		vs eth, EtOH
554	(Azidomethyl)benzene		$C_7H_7N_3$	622-79-7	133.151			108^{23}	1.0730^{19}	1.5341^{25}	i H_2O; msc EtOH, eth
555	Azinphos ethyl		$C_{12}H_{16}N_3O_3PS_2$	2642-71-9	345.377	nd	51.3(0.5)	$111^{0.001}$	1.284^{20}		reac alk
556	Azinphos-methyl		$C_{10}H_{12}N_3O_3PS_2$	86-50-0	317.324		72(2)		1.44^{20}		
557	1-Aziridineethanol		C_4H_9NO	1072-52-2	87.120			168	1.088^{25}	1.4560^{20}	
558	*trans*-Azobenzene	*trans*-Diphenyldiazene	$C_{12}H_{10}N_2$	17082-12-1	182.220	oran-red mcl lf (al)	67.88(0.03)	300.0(0.6)	1.203^{20}	1.6266^{78}	sl H_2O; s EtOH, eth, bz, chl; vs py
559	*cis*-Azobenzene	*cis*-Diphenyldiazene	$C_{12}H_{10}N_2$	1080-16-6	182.220	oran-red pl (peth)	71.6(0.4)				sl H_2O; s EtOH, eth, bz, HOAc, lig
560	3,3'-Azobenzenedisulfonyl chloride		$C_{12}H_8Cl_2N_2O_4S_2$	104115-88-0	379.239	red nd (eth)	166.5				vs eth
561	1,1'-Azobiscyclohexanecarbonitrile		$C_{14}H_{20}N_4$	2094-98-6	244.336		100				i H_2O; s lig
562	2,2'-Azobis[isobutyronitrile]	2,2'-Azobis[2-methylpropionitrile]	$C_8H_{12}N_4$	78-67-1	164.208						i H_2O; sl EtOH, eth
563	Azobutane		$C_8H_{18}N_2$	2159-75-3	142.242			60^{18}			
564	Azopropane		$C_6H_{14}N_2$	821-67-0	114.188			114			
565	*cis*-Azoxybenzene	Diphenyldiazene 1-oxide, (*E*)	$C_{12}H_{10}N_2O$	21650-65-7	198.219		87		1.166^{20}	1.633^{20}	
566	*trans*-Azoxybenzene	Diphenyldiazene 1-oxide, (*Z*)	$C_{12}H_{10}N_2O$	20972-43-4	198.219		34.6		1.1590^{26}		i H_2O; s EtOH, eth
567	Azoxyethane	Diethyldiazine 1-oxide	$C_4H_{10}N_2O$	16301-26-1	102.134	liq		46			
568	Azulene	Bicyclo[5.3.0]decapentaene	$C_{10}H_8$	275-51-4	128.171	bl or gr-blk lf (al)	99	249.0(0.5)			i H_2O; s EtOH, eth, ace, acid; sl chl
569	Balan	*N*-Butyl-*N*-ethyl-2,6-dinitro-4-(trifluoromethyl)aniline	$C_{13}H_{16}F_3N_3O_4$	1861-40-1	335.279		65.6(0.5)	$121^{0.5}$			
570	Barban		$C_{11}H_9Cl_2NO_2$	101-27-9	258.101		72.4(0.5)				
571	Barbital	5,5-Diethylbarbituric acid	$C_8H_{12}N_2O_3$	57-44-3	184.192	nd (w)	189(1)		1.220^{25}		sl H_2O; s EtOH, eth, ace, chl, lig, tfa
572	Barbituric acid		$C_4H_4N_2O_3$	67-52-7	128.086	orth pr (w +2)	248	260 dec			s H_2O, eth; sl EtOH
573	Bayleton	Triadimefon	$C_{14}H_{16}ClN_3O_2$	43121-43-3	293.749		77.0(0.5)		1.22^{20}		
574	Bebeerine		$C_{36}H_{38}N_2O_6$	477-60-1	594.696	cry (bz, eth, chl-MeOH)	221				s EtOH, MeOH, eth; vs ace, chl
575	Benactyzine	2-(Diethylamino)ethyl benzilate	$C_{20}H_{25}NO_3$	302-40-9	327.418	cry	51				
576	Benactyzine hydrochloride	2-Diethylaminoethyl benzilate hydrochloride	$C_{20}H_{26}ClNO_3$	57-37-4	363.878		177.5				s H_2O; i eth

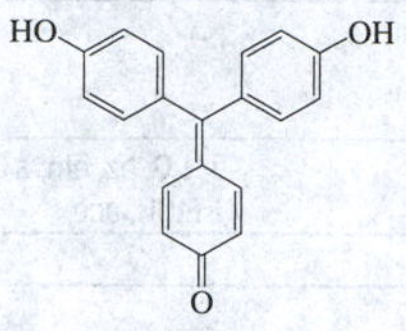
535
Aurin

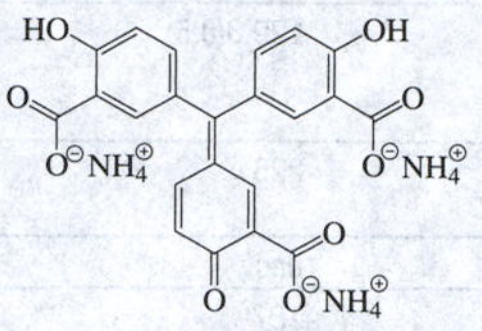
536
Aurin tricarboxylic acid, triammonium salt

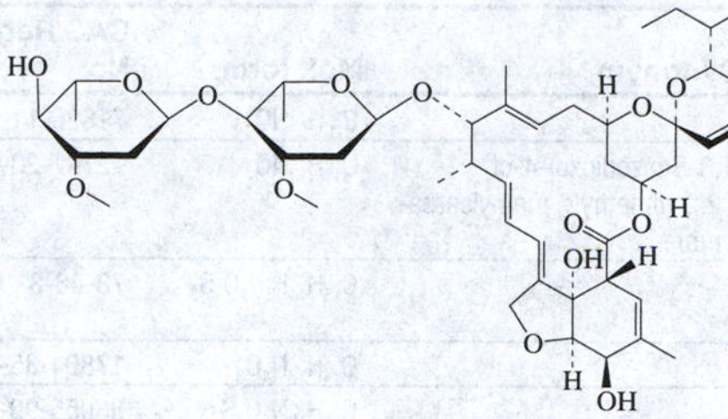
537
Avermectin B1a

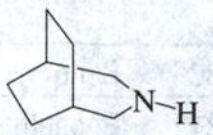
538
3-Azabicyclo[3.2.2]nonane

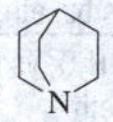
539
1-Azabicyclo[2.2.2]octane

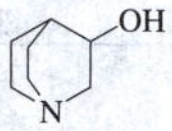
540
1-Azabicyclo[2.2.2]octan-3-ol

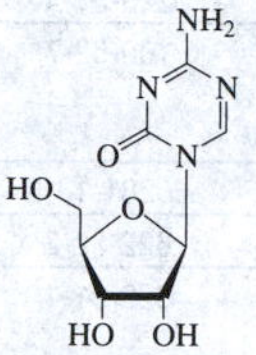
541
Azacitidine

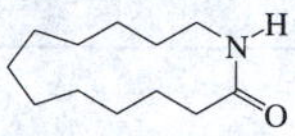
542
Azacyclotridecan-2-one

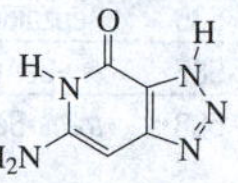
543
8-Azaguanine

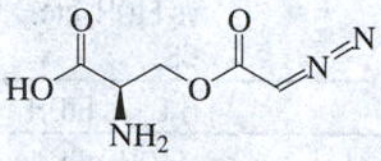
544
Azaserine

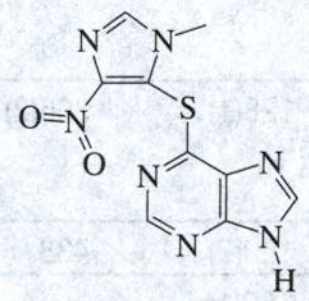
545
Azathioprine

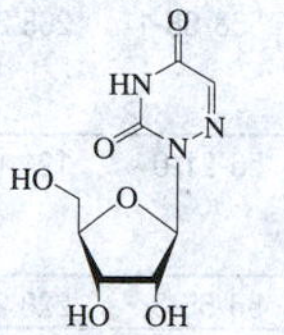
546
6-Azauridine

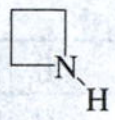
547
Azetidine

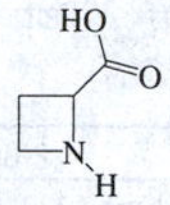
548
2-Azetidinecarboxylic acid

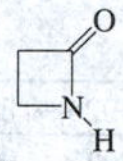
549
2-Azetidinone

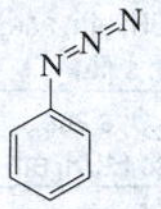
550
Azidobenzene

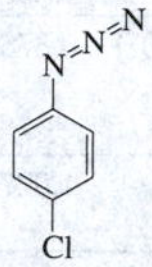
551
1-Azido-4-chlorobenzene

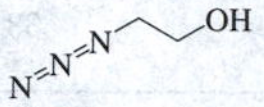
552
2-Azidoethanol

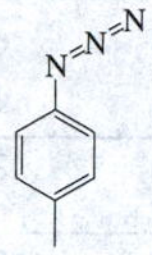
553
1-Azido-4-methylbenzene

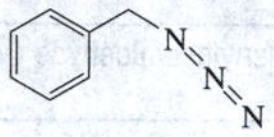
554
(Azidomethyl)benzene

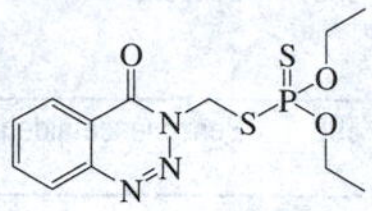
555
Azinphos ethyl

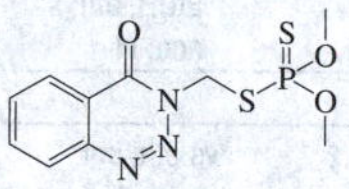
556
Azinphos-methyl

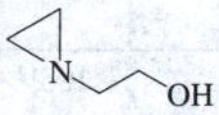
557
1-Aziridineethanol

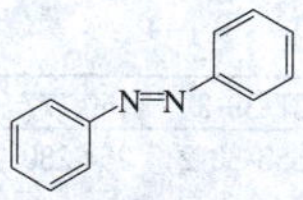
558
trans-Azobenzene

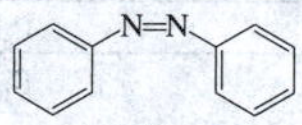
559
cis-Azobenzene

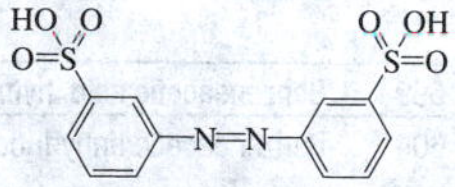
560
3,3'-Azobenzenedisulfonyl chloride

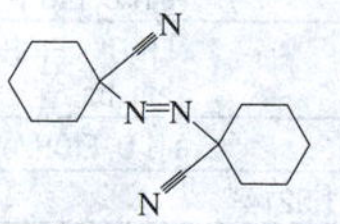
561
1,1'-Azobiscyclohexanecarbonitrile

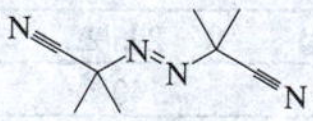
562
2,2'-Azobis[isobutyronitrile]

563
Azobutane

564
Azopropane

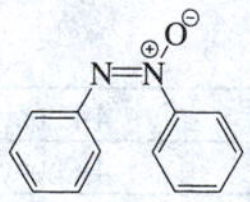
565
cis-Azoxybenzene

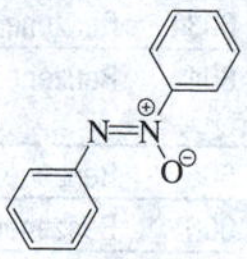
566
trans-Azoxybenzene

567
Azoxyethane

568
Azulene

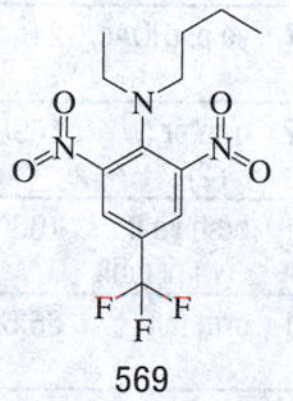
569
Balan

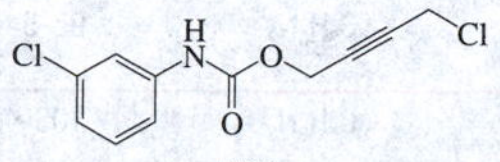
570
Barban

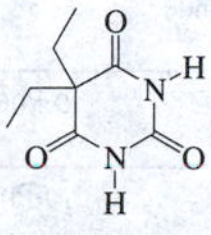
571
Barbital

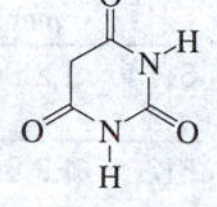
572
Barbituric acid

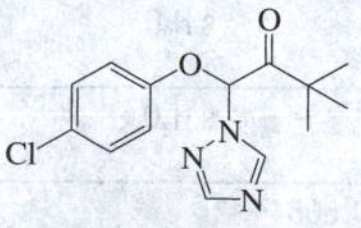
573
Bayleton

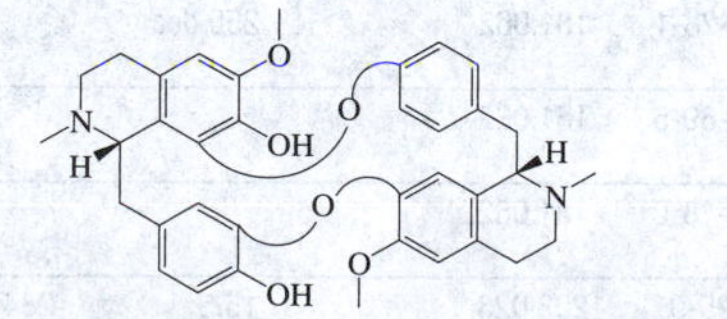
574
Bebeerine

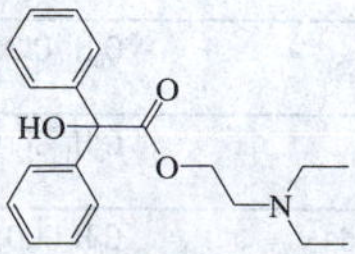
575
Benactyzine

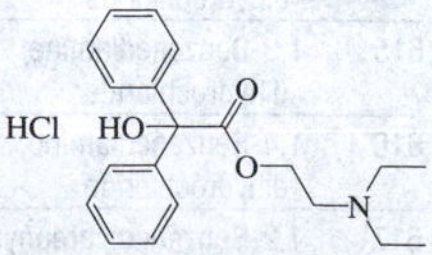
576
Benactyzine hydrochloride

Organic

No.	Name	Synonym	Mol. form.	CAS Reg. No.	Mol. wt.	Physical form	mp/°C	bp/°C	den/ g cm^{-3}	n_D	Solubility
577	Benalaxyl		$C_{20}H_{23}NO_3$	71626-11-4	325.402		79		1.27^{25}		
578	Bendiocarb	1,3-Benzodioxol-4-ol, 2,2-dimethyl-, methylcarbamate	$C_{11}H_{13}NO_4$	22781-23-3	223.226		129.6(0.5)		1.25^{20}		
579	Bendroflumethiazide		$C_{15}H_{14}F_3N_3O_4S_2$	73-48-3	421.415	cry	225				i H_2O, bz, eth; s EtOH, ace
580	Benomyl		$C_{14}H_{18}N_4O_3$	17804-35-2	290.318		dec				
581	Bensulfuron-methyl		$C_{16}H_{18}N_4O_7S$	83055-99-6	410.402		187				
582	Bensulide		$C_{14}H_{24}NO_4PS_3$	741-58-2	397.514		38.4(0.5)		1.224^{20}		
583	Bentazon		$C_{10}H_{12}N_2O_3S$	25057-89-0	240.278		139.7(0.5)				
584	Benz[*c*]acridine	12-Azabenz[*a*]anthracene	$C_{17}H_{11}N$	225-51-4	229.276	nd (dil al)	132				vs bz, eth, EtOH
585	Benzaldehyde	Benzenecarboxaldehyde	C_7H_6O	100-52-7	106.122	liq	-57.12(0.05)	178.7(0.4)	1.0401^{25}	1.5463^{20}	sl H_2O; msc EtOH, eth; vs ace, bz
586	Benzaldehyde hydrazone	Benzylidene hydrazine	$C_7H_8N_2$	5281-18-5	120.152	lf	16	140^{14}			s EtOH
587	*cis*-Benzaldehyde oxime		C_7H_7NO	622-32-2	121.137	pr	36.5	200	1.1111^{20}	1.5908^{20}	vs bz, eth, EtOH
588	*trans*-Benzaldehyde oxime		C_7H_7NO	622-31-1	121.137	nd (eth)	35	119^{10}	1.145^{20}		s H_2O; vs EtOH, eth
589	Benzaldehyde, phenylhydrazone		$C_{13}H_{12}N_2$	588-64-7	196.247	nd (lig), pr	157.0				sl EtOH, eth; s ace, bz, liq NH_3
590	Benzaldehyde, (phenylmethylene)hydrazone		$C_{14}H_{12}N_2$	588-68-1	208.258	ye pr (al)	93(1)				i H_2O; s EtOH, eth, ace, bz, chl; sl ctc
591	Benzamide	Benzoic acid amide	C_7H_7NO	55-21-0	121.137	mcl pr or pl (w)	128(1)	306(2)	1.0792^{130}		sl H_2O, eth, bz; vs EtOH, ctc, CS_2
592	Benz[*a*]anthracene	1,2-Benzanthracene	$C_{18}H_{12}$	56-55-3	228.288	lf (al)	160(2)	438			i H_2O; vs EtOH
593	Benz[*a*]anthracene-7,12-dione		$C_{18}H_{10}O_2$	2498-66-0	258.271		168(1)				sl EtOH, eth, lig; s ace; vs bz, chl
594	Benzanthrone		$C_{17}H_{10}O$	82-05-3	230.260		173.0(0.3)				sl bz
595	Benzene	[6]Annulene	C_6H_6	71-43-2	78.112	orth pr or liq	5.538(0.002)	80.08(0.07)	0.8765^{20}	1.5011^{20}	sl H_2O; msc EtOH, eth, ace, chl; s ctc
596	Benzeneacetaldehyde	Phenylacetaldehyde	C_8H_8O	122-78-1	120.149		33.5	202(5)	1.0272^{20}	1.5255^{20}	sl H_2O; s ace; msc EtOH, eth
597	Benzeneacetamide	α-Phenylacetamide	C_8H_9NO	103-81-1	135.163		157				sl H_2O, eth, bz; s EtOH
598	Benzeneacetic acid	Phenylacetic acid	$C_8H_8O_2$	103-82-2	136.149	lf, pl (peth)	76.7(0.2)	268(2)	1.228^{6}		sl H_2O, chl; vs EtOH, eth; s ace; i lig
599	Benzeneacetic acid, hydrazide		$C_8H_{10}N_2O$	937-39-3	150.177		115.5				
600	Benzeneacetic anhydride		$C_{16}H_{14}O_3$	1555-80-2	254.280	pr or nd (eth)	73.3	195^{12}			vs eth, chl
601	Benzeneacetonitrile	Benzyl cyanide	C_8H_7N	140-29-4	117.149	liq	-22.1(0.5)	232(2)	1.0205^{15}	1.5211^{25}	
602	Benzeneacetyl chloride	Phenylacetyl chloride	C_8H_7ClO	103-80-0	154.594			170^{250}	1.1682^{20}	1.5325^{20}	vs eth
603	Benzenearsonic acid		$C_6H_7AsO_3$	98-05-5	202.040	cry (w)	158 dec				vs H_2O, EtOH
604	Benzeneboronic acid		$C_6H_7BO_2$	98-80-6	121.930		219				sl H_2O; s EtOH, eth, bz
605	Benzenebutanoic acid	4-Phenylbutanoic acid	$C_{10}H_{12}O_2$	1821-12-1	164.201	lf (w)	51(1)	296(2)			s H_2O, EtOH, eth
606	Benzenebutanol	4-Phenyl-1-butanol	$C_{10}H_{14}O$	3360-41-6	150.217			140^{14}		1.5214^{20}	
607	Benzenecarboperoxoic acid	Perbenzoic acid	$C_7H_6O_3$	93-59-4	138.121	mcl pl (peth)	42	100^{14}			vs ace, bz, eth, EtOH
608	Benzenecarbothioamide		C_7H_7NS	2227-79-4	137.203		117				
609	Benzenecarbothioic acid		C_7H_6OS	98-91-9	138.187	ye pl (HOAc)	24	86^{10}	1.28^{20}	1.6040^{20}	vs ace, bz, eth, EtOH
610	Benzenecarboximidamide, monohydrochloride		$C_7H_9ClN_2$	1670-14-0	156.612	orth pr (w +2)	169				s H_2O, EtOH; sl tfa
611	1,2-Benzenediamine	*o*-Phenylenediamine	$C_6H_8N_2$	95-54-5	108.141	brsh ye lf (w) pl (chl)	103(1)	257(9)			s H_2O, eth, bz, chl; vs EtOH
612	1,3-Benzenediamine	*m*-Phenylenediamine	$C_6H_8N_2$	108-45-2	108.141	orth (al)	65.5(0.9)	282(18)	1.0096^{58}	1.6339^{58}	vs H_2O; s EtOH, eth, bz
613	1,4-Benzenediamine	*p*-Phenylenediamine	$C_6H_8N_2$	106-50-3	108.141	wh pl (bz, eth)	140.3(0.6)	267			sl H_2O; s EtOH, eth, bz, chl
614	1,2-Benzenediamine, dihydrochloride		$C_6H_{10}Cl_2N_2$	615-28-1	181.062		250 dec				
615	1,3-Benzenediamine, dihydrochloride		$C_6H_{10}Cl_2N_2$	541-69-5	181.062						s H_2O
616	1,4-Benzenediamine, dihydrochloride		$C_6H_{10}Cl_2N_2$	624-18-0	181.062						s H_2O
617	1,2-Benzenedicarbonyl dichloride	Phthaloyl chloride	$C_8H_4Cl_2O_2$	88-95-9	203.023		15.5	284(10)	1.4089^{20}	1.5684^{20}	

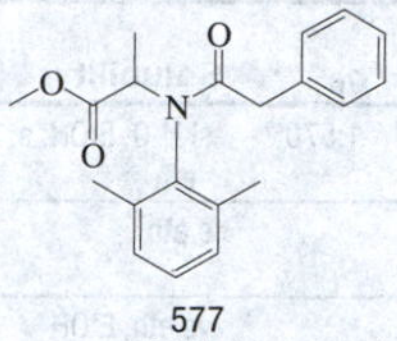
577
Benalaxyl

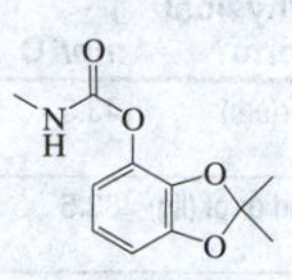
578
Bendiocarb

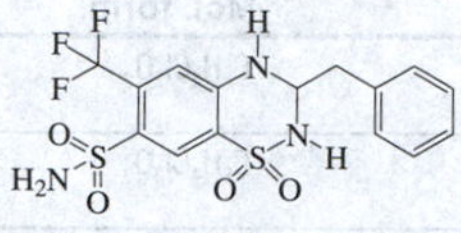
579
Bendroflumethiazide

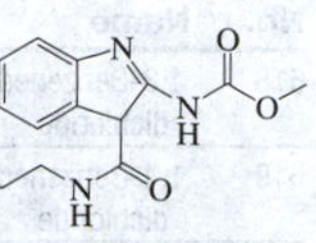
580
Benomyl

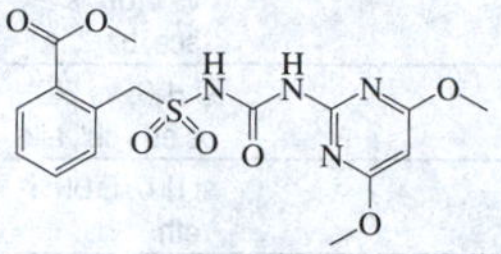
581
Bensulfuron-methyl

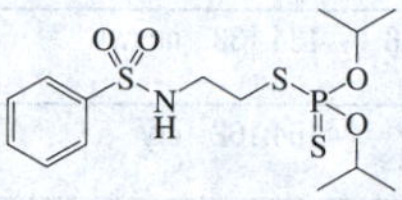
582
Bensulide

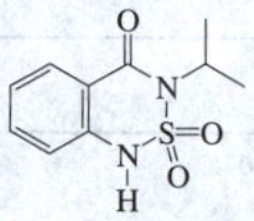
583
Bentazon

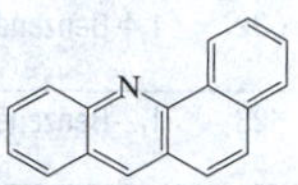
584
Benz[c]acridine

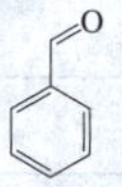
585
Benzaldehyde

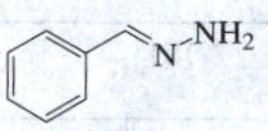
586
Benzaldehyde hydrazone

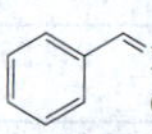
587
cis-Benzaldehyde oxime

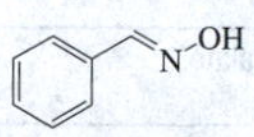
588
trans-Benzaldehyde oxime

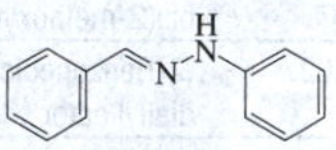
589
Benzaldehyde, phenylhydrazone

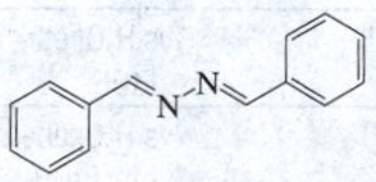
590
Benzaldehyde, (phenylmethylene)hydrazone

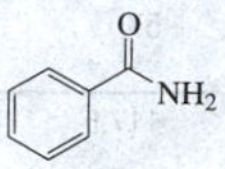
591
Benzamide

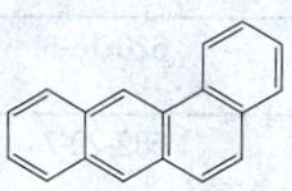
592
Benz[a]anthracene

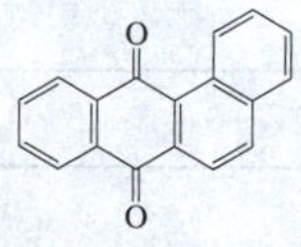
593
Benz[a]anthracene-7,12-dione

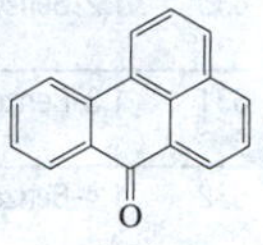
594
Benzanthrone

595
Benzene

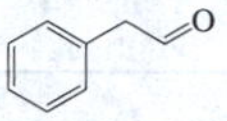
596
Benzeneacetaldehyde

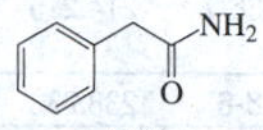
597
Benzeneacetamide

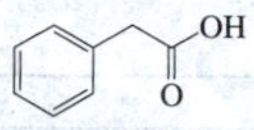
598
Benzeneacetic acid

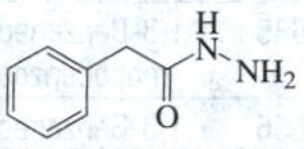
599
Benzeneacetic acid, hydrazide

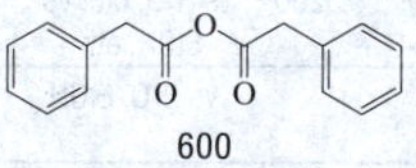
600
Benzeneacetic anhydride

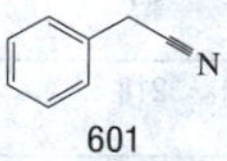
601
Benzeneacetonitrile

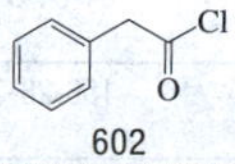
602
Benzeneacetyl chloride

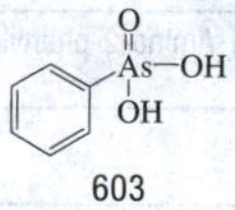
603
Benzenearsonic acid

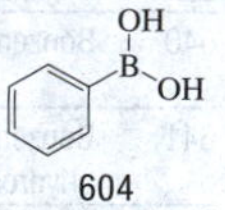
604
Benzeneboronic acid

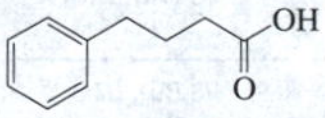
605
Benzenebutanoic acid

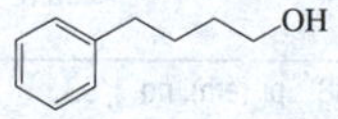
606
Benzenebutanol

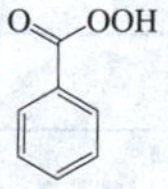
607
Benzenecarboperoxoic acid

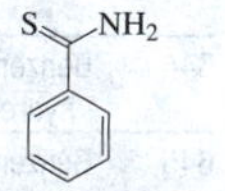
608
Benzenecarbothioamide

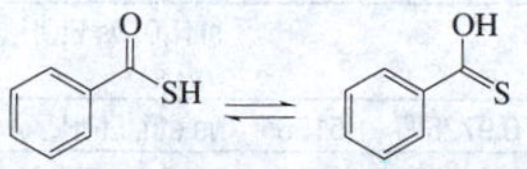
609
Benzenecarbothioic acid

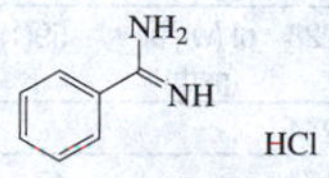
610
Benzenecarboximidamide, monohydrochloride

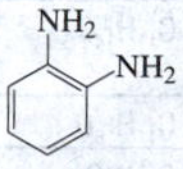
611
1,2-Benzenediamine

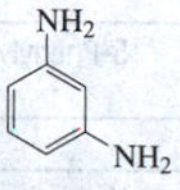
612
1,3-Benzenediamine

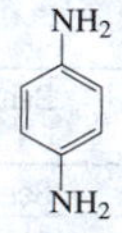
613
1,4-Benzenediamine

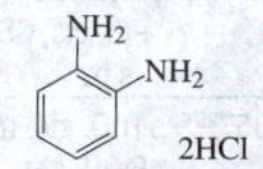
614
1,2-Benzenediamine, dihydrochloride

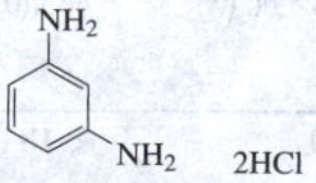
615
1,3-Benzenediamine, dihydrochloride

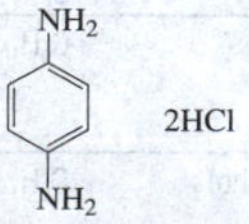
616
1,4-Benzenediamine, dihydrochloride

617
1,2-Benzenedicarbonyl dichloride

No.	Name	Synonym	Mol. form.	CAS Reg. No.	Mol. wt.	Physical form	mp/˚C	bp/˚C	den/ g cm^{-3}	n_D	Solubility
618	1,3-Benzenedicarbonyl dichloride		$C_8H_4Cl_2O_2$	99-63-8	203.023	pr(eth)	43.5	276	1.3880^{17}	1.570^{47}	sl H_2O, EtOH; s eth
619	1,4-Benzenedicarbonyl dichloride		$C_8H_4Cl_2O_2$	100-20-9	203.023	nd or pl (lig)	83.5	258			s eth
620	1,2-Benzenedicarboxaldehyde		$C_8H_6O_2$	643-79-8	134.133	ye cry or nd (lig)	55.8	$83^{0.8}$			vs eth, EtOH
621	1,3-Benzenedicarboxaldehyde		$C_8H_6O_2$	626-19-7	134.133	nd (dil al)	89(2)	246			sl H_2O, eth, chl; vs EtOH; s ace, bz
622	1,4-Benzenedicarboxaldehyde		$C_8H_6O_2$	623-27-8	134.133	nd (w)	117	246			sl H_2O; vs EtOH; s eth, chl, alk
623	1,2-Benzenedicarboxamide	Phthalamide	$C_8H_8N_2O_2$	88-96-0	164.162	cry	222	dec			sl H_2O, EtOH; i eth
624	1,4-Benzenedicarboxamide		$C_8H_8N_2O_2$	3010-82-0	164.162	nd (w), pl (HOAc)	322.3				
625	1,2-Benzenedicarboxylic acid, bis(2-butoxyethyl) ester	Bis(2-butoxyethyl) phthalate	$C_{20}H_{30}O_6$	117-83-9	366.448			409(18)			
626	1,2-Benzenedicarboxylic acid, bis(2-methoxyethyl) ester	Bis(2-methoxyethyl) phthalate	$C_{14}H_{18}O_6$	117-82-8	282.289		-60.0	230^{10}	1.1596^{20}		
627	1,2-Benzenedicarboxylic acid, diallyl ester	Diallyl phthalate	$C_{14}H_{14}O_4$	131-17-9	246.259			161^{4}			
628	1,2-Benzenedicarboxylic acid, dipropyl ester	Dipropyl phthalate	$C_{14}H_{18}O_4$	131-16-8	250.291	liq	-31.0	319(2)	1.0767^{20}		i H_2O; s EtOH, eth
629	1,3-Benzenedimethanamine	*m*-Xylene diamine	$C_8H_{12}N_2$	1477-55-0	136.194			247	1.052^{20}		vs H_2O, eth, EtOH
630	1,2-Benzenedimethanol		$C_8H_{10}O_2$	612-14-6	138.164	pl (eth, peth)	66(3)	145^{3}			s H_2O, EtOH; vs eth; sl bz
631	1,3-Benzenedimethanol		$C_8H_{10}O_2$	626-18-6	138.164	nd (bz)	57	156^{13}	1.1610^{18}		vs H_2O, eth, EtOH
632	1,4-Benzenedimethanol		$C_8H_{10}O_2$	589-29-7	138.164	nd (w)	117.5	140^{1}			vs H_2O, ace, eth, EtOH
633	1,2-Benzenediol, diacetate		$C_{10}H_{10}O_4$	635-67-6	194.184	nd (al)	64.5	142^{9}			i H_2O; vs EtOH, eth, chl; s peth
634	1,4-Benzenediol, diacetate		$C_{10}H_{10}O_4$	1205-91-0	194.184	pl (w, al)	123.5		0.8731^{25}		s H_2O; vs EtOH, eth, chl, lig
635	1,3-Benzenediol, monobenzoate		$C_{13}H_{10}O_3$	136-36-7	214.216		134.5				
636	1,3-Benzenedisulfonic acid		$C_6H_6O_6S_2$	98-48-6	238.238	hyg cry					
637	1,3-Benzenedisulfonyl dichloride		$C_6H_4Cl_2O_4S_2$	585-47-7	275.130		61.8	$195^{10.5}$			
638	1,2-Benzenedithiol		$C_6H_6S_2$	17534-15-5	142.242		28.5	238.5			vs EtOH, eth, bz; s AcOEt
639	1,3-Benzenedithiol		$C_6H_6S_2$	626-04-0	142.242	lf	27	245			vs bz, eth, EtOH
640	Benzeneethanamine	1-Amino-2-phenylethane	$C_8H_{11}N$	64-04-0	121.180	liq	<0	204(4)	0.9640^{25}	1.5290^{25}	s H_2O, ctc; vs EtOH, eth
641	Benzeneethanamine, hydrochloride		$C_8H_{12}ClN$	156-28-5	157.641	pl or lf (al)	218.5				vs H_2O, EtOH
642	Benzeneethanol	Phenethyl alcohol	$C_8H_{10}O$	60-12-8	122.164	liq	-19(2)	220(3)	1.0202^{20}	1.5325^{20}	sl H_2O; msc EtOH, eth
643	Benzenehexacarboxylic acid	Mellitic acid	$C_{12}H_6O_{12}$	517-60-2	342.169	nd (al)	287 dec				vs H_2O; s EtOH, sulf
644	Benzenemethanamine, hydrochloride		$C_7H_{10}ClN$	3287-99-8	143.614		258.3				vs H_2O, EtOH
645	Benzenemethanesulfonyl chloride		$C_7H_7ClO_2S$	1939-99-7	190.648	pr (eth), nd (bz)	93				vs eth, bz
646	Benzenemethanesulfonyl fluoride		$C_7H_7FO_2S$	329-98-6	174.193		92.0				
647	Benzenemethanethiol	Thiobenzyl alcohol	C_7H_8S	100-53-8	124.204	liq	-30	199(1)	1.058^{20}	1.5151^{20}	i H_2O; vs EtOH, eth; sl ctc; s CS_2
648	Benzenepentanoic acid	5-Phenylvaleric acid	$C_{11}H_{14}O_2$	2270-20-4	178.228	pl (w), pr (peth)	59(1)	190^{30}			sl H_2O; vs EtOH; s os
649	Benzenepentanol		$C_{11}H_{16}O$	10521-91-2	164.244			155^{20}	0.9725^{20}	1.5156^{20}	vs eth, EtOH
650	Benzenepropanal	Hydrocinnamic aldehyde	$C_9H_{10}O$	104-53-0	134.174	mcl	47	224	1.0190^{20}		i H_2O; vs EtOH; msc eth
651	Benzenepropanenitrile	Hydrocinnamonitrile	C_9H_9N	645-59-0	131.174	liq	-1	261	1.0016^{20}	1.5266^{28}	s EtOH, eth; sl chl
652	Benzenepropanethiol		$C_9H_{12}S$	24734-68-7	152.256			121^{23}	1.01^{25}	1.5494^{20}	
653	Benzenepropanoic acid	Hydrocinnamic acid	$C_9H_{10}O_2$	501-52-0	150.174	nd (w)	48.4(0.3)	284(2)	1.0712^{49}		s H_2O, EtOH, eth, ctc, CS_2; vs bz
654	Benzenepropanol	Hydrocinnamyl alcohol	$C_9H_{12}O$	122-97-4	136.190		<-18	241(4)	0.995^{25}	1.5357^{25}	s H_2O, ctc; msc EtOH, eth

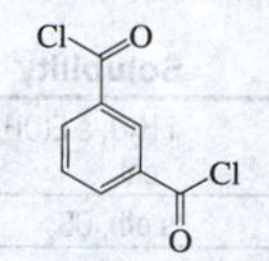
618
1,3-Benzenedicarbonyl dichloride

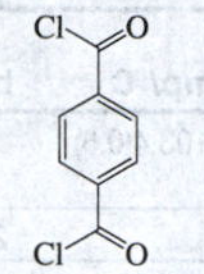
619
1,4-Benzenedicarbonyl dichloride

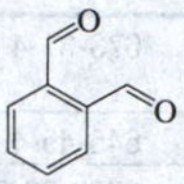
620
1,2-Benzenedicarboxaldehyde

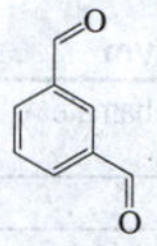
621
1,3-Benzenedicarboxaldehyde

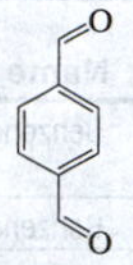
622
1,4-Benzenedicarboxaldehyde

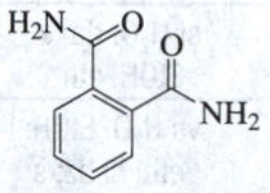
623
1,2-Benzenedicarboxamide

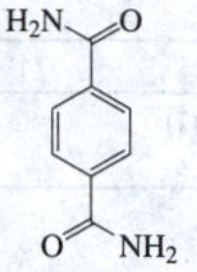
624
1,4-Benzenedicarboxamide

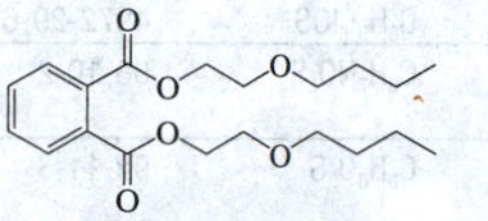
625
1,2-Benzenedicarboxylic acid, bis(2-butoxyethyl) ester

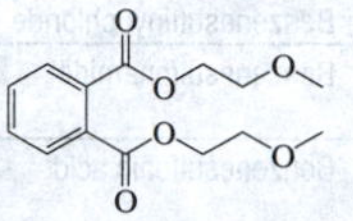
626
1,2-Benzenedicarboxylic acid, bis(2-methoxyethyl) ester

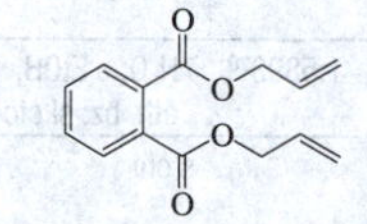
627
1,2-Benzenedicarboxylic acid, diallyl ester

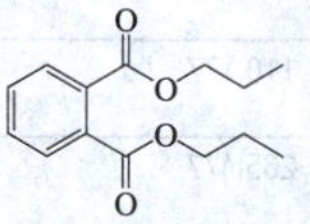
628
1,2-Benzenedicarboxylic acid, dipropyl ester

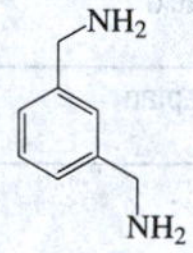
629
1,3-Benzenedimethanamine

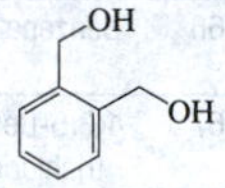
630
1,2-Benzenedimethanol

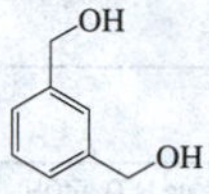
631
1,3-Benzenedimethanol

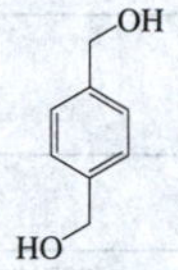
632
1,4-Benzenedimethanol

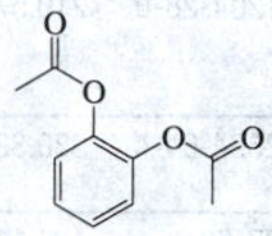
633
1,2-Benzenediol, diacetate

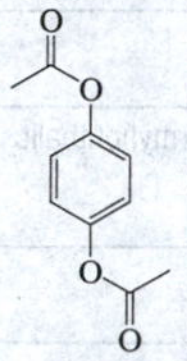
634
1,4-Benzenediol, diacetate

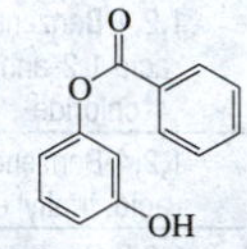
635
1,3-Benzenediol, monobenzoate

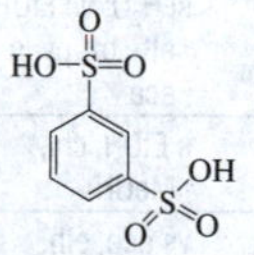
636
1,3-Benzenedisulfonic acid

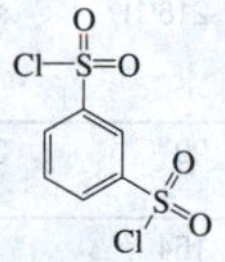
637
1,3-Benzenedisulfonyl dichloride

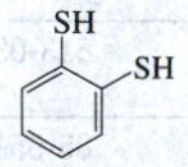
638
1,2-Benzenedithiol

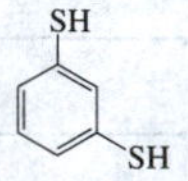
639
1,3-Benzenedithiol

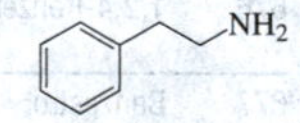
640
Benzeneethanamine

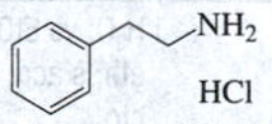
641
Benzeneethanamine, hydrochloride

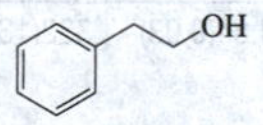
642
Benzeneethanol

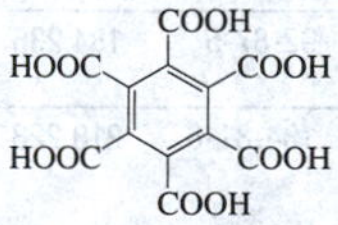
643
Benzenehexacarboxylic acid

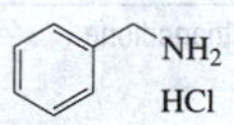
644
Benzenemethanamine, hydrochloride

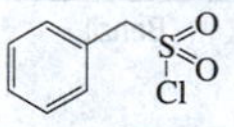
645
Benzenemethanesulfonyl chloride

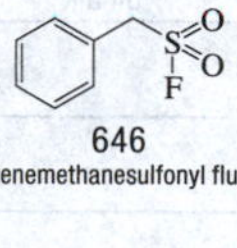
646
Benzenemethanesulfonyl fluoride

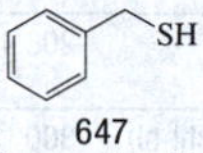
647
Benzenemethanethiol

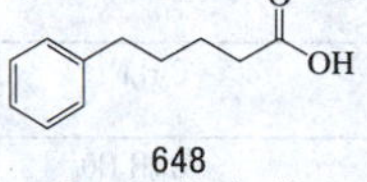
648
Benzenepentanoic acid

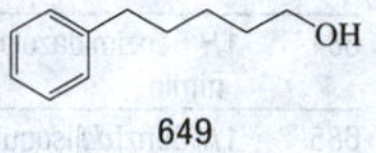
649
Benzenepentanol

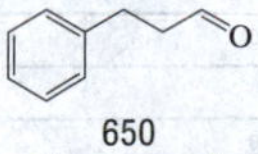
650
Benzenepropanal

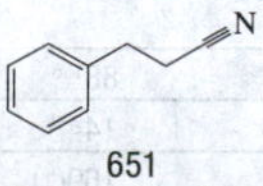
651
Benzenepropanenitrile

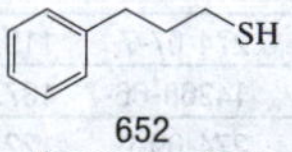
652
Benzenepropanethiol

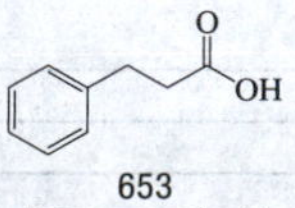
653
Benzenepropanoic acid

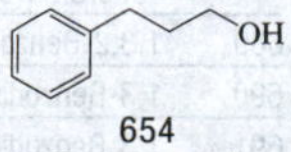
654
Benzenepropanol

No.	Name	Synonym	Mol. form.	CAS Reg. No.	Mol. wt.	Physical form	mp/°C	bp/°C	den/g cm^{-3}	n_D	Solubility
655	Benzenepropanol carbamate	Phenprobamate	$C_{10}H_{13}NO_2$	673-31-4	179.216		103.4(0.5)				i H_2O; s EtOH, chl
656	Benzenepropanoyl chloride		C_9H_9ClO	645-45-4	168.619			225 dec	1.135^{21}		s eth, CS_2
657	Benzeneseleninic acid	Phenylseleninic acid	$C_6H_6O_2Se$	6996-92-5	189.07		124.5		1.93^{20}		sl H_2O; i bz; vs alk
658	Benzeneselenol		C_6H_6Se	645-96-5	157.07			183.6	1.4865^{15}		i H_2O; s EtOH; vs eth, ctc
659	Benzenesulfinic acid		$C_6H_6O_2S$	618-41-7	142.176	pr (w)	84	dec			sl H_2O; s EtOH, eth, bz; i peth
660	Benzenesulfinyl chloride		C_6H_5ClOS	4972-29-6	160.621	pl (peth)	38	$71^{1.5}$	1.3469^{25}	1.3470^{25}	s eth, chl
661	Benzenesulfonamide		$C_6H_7NO_2S$	98-10-2	157.191	lf, nd (w)	156(1)				sl H_2O, tfa; s EtOH, eth
662	Benzenesulfonic acid	Besylic acid	$C_6H_6O_3S$	98-11-3	158.175	nd (bz)	65				vs H_2O, EtOH; i eth; sl bz; s HOAc
663	Benzenesulfonyl chloride	Phenylsulfonyl chloride	$C_6H_5ClO_2S$	98-09-9	176.621	liq	14.5	252(7)	1.3470^{15}		i H_2O; vs EtOH; s eth, ctc
664	Benzenesulfonyl fluoride	Phenylsulfonyl fluoride	$C_6H_5FO_2S$	368-43-4	160.166			203.5	1.3286^{20}	1.4932^{18}	s EtOH, eth
665	1,2,4,5-Benzenetetracarboxylic acid	Pyromellitic acid	$C_{10}H_6O_8$	89-05-4	254.150	tcl pr (w+2)	271(2)				sl H_2O; s EtOH
666	Benzenethiol	Phenyl mercaptan	C_6H_6S	108-98-5	110.177	liq	-14.87(0.05)	169.1(0.2)	1.0775^{20}	1.5893^{20}	i H_2O; s EtOH, eth, bz; sl ctc
667	1,3,5-Benzenetricarbonyl trichloride		$C_9H_3Cl_3O_3$	4422-95-1	265.477		36.3	180^{16}			s chl
668	1,2,3-Benzenetricarboxylic acid	Hemimellitic acid	$C_9H_6O_6$	569-51-7	210.140	pr (al)	209(2)		1.546^{20}		vs eth, EtOH
669	1,2,4-Benzenetricarboxylic acid	Trimellitic acid	$C_9H_6O_6$	528-44-9	210.140	nd (w) cry (al) cry (HOAc)	219				vs H_2O, eth, EtOH
670	1,3,5-Benzenetricarboxylic acid		$C_9H_6O_6$	554-95-0	210.140	pr or nd (w+1)	380				sl H_2O; vs EtOH, eth
671	1,2,4-Benzenetricarboxylic acid 1,2-anhydride, 4-chloride	4-(Chloroformyl)phthalic anhydride	$C_9H_3ClO_4$	1204-28-0	210.571		66				
672	1,2,4-Benzenetricarboxylic acid, triallyl ester		$C_{18}H_{18}O_6$	2694-54-4	330.332		<-30		1.164^{20}		
673	1,2,3-Benzenetriol	Pyrogallol	$C_6H_6O_3$	87-66-1	126.110	lf or nd (bz)	125.5(0.5)	307(4)	1.453^{4}	1.561^{134}	vs H_2O, EtOH, eth, NH_3; s ace; i bz
674	1,2,4-Benzenetriol	Hydroxyhydroquinone	$C_6H_6O_3$	533-73-3	126.110	pl (eth), lf or pl (w)	140.5				vs H_2O, EtOH, eth; i bz, chl
675	1,3,5-Benzenetriol	Phloroglucinol	$C_6H_6O_3$	108-73-6	126.110	lf or pl (w +2)	216(1)	sub	1.46^{25}		sl H_2O; vs EtOH, eth, bz, py; s ace
676	1,2,4-Benzenetriol triacetate		$C_{12}H_{12}O_6$	613-03-6	252.219		99	300			s EtOH, chl, MeOH
677	Benzestrol		$C_{20}H_{26}O_2$	85-95-0	298.419	cry (al)	164				vs ace, eth, EtOH, HOAc
678	Benzethonium chloride		$C_{27}H_{42}ClNO_2$	121-54-0	448.081	pl (chl/eth)	165 (hyd)				vs H_2O; s ace, chl, EtOH
679	Benzidene-3,3'-dicarboxylic acid	3,3'-Dicarboxybenzidine	$C_{14}H_{12}N_2O_4$	2130-56-5	272.256	nd	300 dec				
680	*p*-Benzidine	[1,1'-Biphenyl]-4,4'-diamine	$C_{12}H_{12}N_2$	92-87-5	184.236	nd (w)	127.0(0.5)	401			sl H_2O, eth, DMSO; s EtOH
681	Benzil	Diphenylethanedione	$C_{14}H_{10}O_2$	134-81-6	210.228	ye pr (al)	94.84(0.03)	332(13)	1.084^{102}		i H_2O; vs EtOH, eth; s ace; sl ctc
682	1*H*-Benzimidazol-2-amine		$C_7H_7N_3$	934-32-7	133.151	pl (w)	231.9(0.3)				s H_2O, EtOH, ace; sl eth, bz, DMSO
683	1*H*-Benzimidazole	*N*,*N*-Methenyl-*o*-phenylenediamine	$C_7H_6N_2$	51-17-2	118.136	orth bipym pl (w)	172.2(0.9)	>360			sl H_2O, eth; vs EtOH; i bz; s dil alk
684	1*H*-Benzimidazole-2-acetonitrile		$C_9H_7N_3$	4414-88-4	157.172		208.4				
685	1*H*-Benz[*de*]isoquinoline-1,3(2*H*)-dione		$C_{12}H_7NO_2$	81-83-4	197.190	nd (chl-al)	300				
686	Benzo[*c*]chrysene		$C_{22}H_{14}$	194-69-4	278.346	nd (AcOH)	125.4(0.5)				
687	Benzo[*g*]chrysene	Benzo[*a*]triphenylene	$C_{22}H_{14}$	196-78-1	278.346	nd (AcOH)	114.5				
688	1*H*,3*H*-Benzo[1,2-*c*:4,5-*c'*]difuran-1,3,5,7-tetrone		$C_{10}H_2O_6$	89-32-7	218.119		286(2)				
689	1,3,2-Benzodioxaborole		$C_6H_5BO_2$	274-07-7	119.914		12	88^{156}	1.2700^{20}	1.5070^{20}	
690	1,3-Benzodioxol-5-amine		$C_7H_7NO_2$	14268-66-7	137.137		42	144^{16}			
691	1,3-Benzodioxole		$C_7H_6O_2$	274-09-9	122.122			169(1)	1.271^{25}	1.5398^{20}	

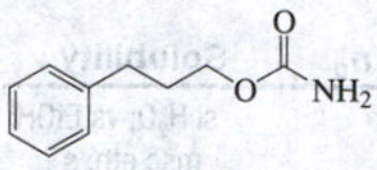
655
Benzenepropanol carbamate

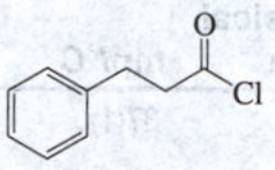
656
Benzenepropanoyl chloride

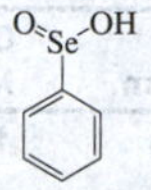
657
Benzeneseleninic acid

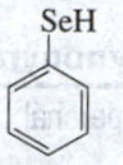
658
Benzeneselenol

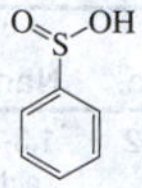
659
Benzenesulfinic acid

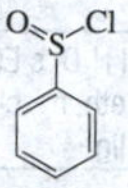
660
Benzenesulfinyl chloride

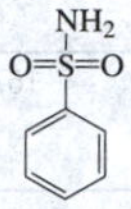
661
Benzenesulfonamide

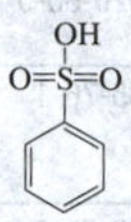
662
Benzenesulfonic acid

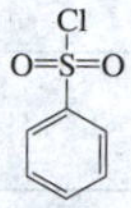
663
Benzenesulfonyl chloride

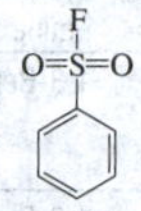
664
Benzenesulfonyl fluoride

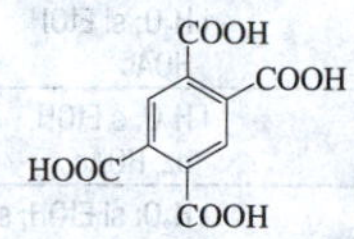
665
1,2,4,5-Benzenetetracarboxylic acid

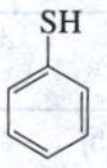
666
Benzenethiol

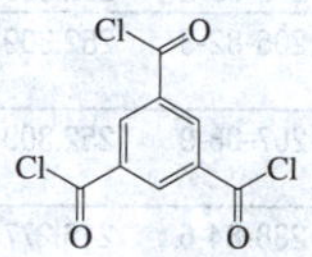
667
1,3,5-Benzenetricarbonyl trichloride

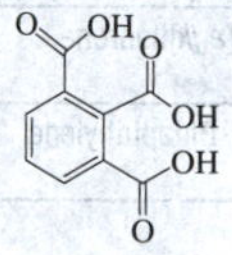
668
1,2,3-Benzenetricarboxylic acid

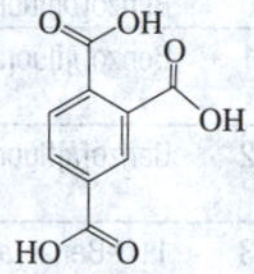
669
1,2,4-Benzenetricarboxylic acid

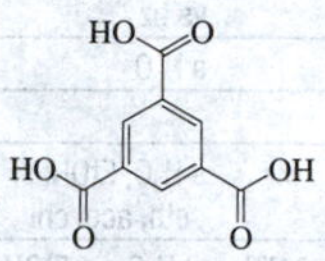
670
1,3,5-Benzenetricarboxylic acid

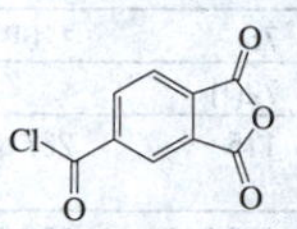
671
1,2,4-Benzenetricarboxylic acid 1,2-anhydride, 4-chloride

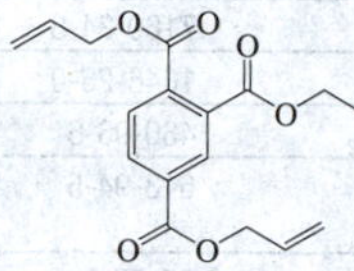
672
1,2,4-Benzenetricarboxylic acid, triallyl ester

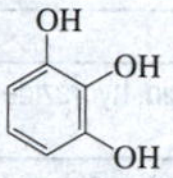
673
1,2,3-Benzenetriol

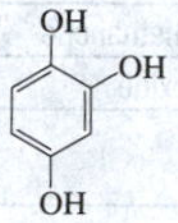
674
1,2,4-Benzenetriol

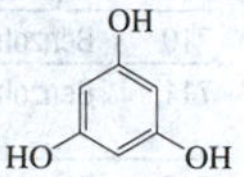
675
1,3,5-Benzenetriol

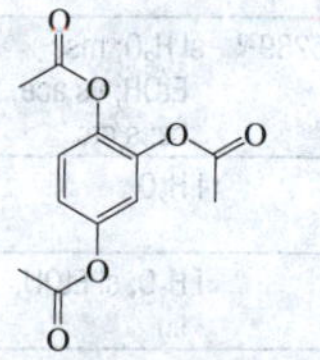
676
1,2,4-Benzenetriol triacetate

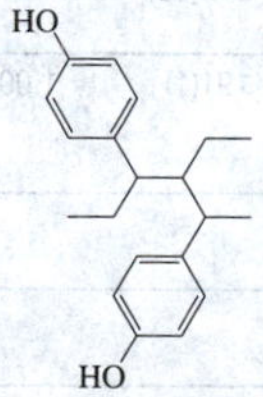
677
Benzestrol

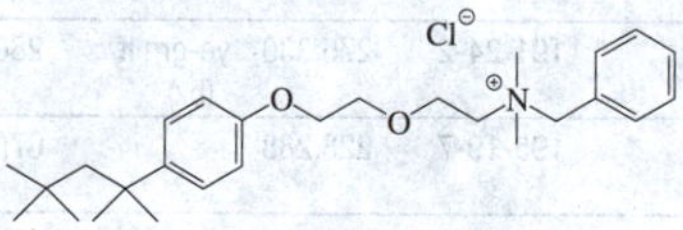
678
Benzethonium chloride

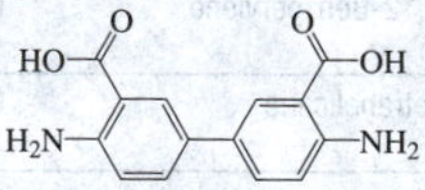
679
Benzidene-3,3'-dicarboxylic acid

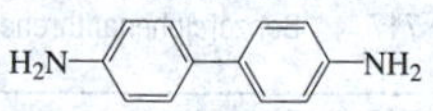
680
p-Benzidine

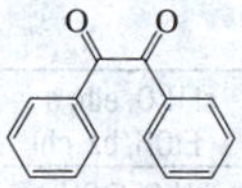
681
Benzil

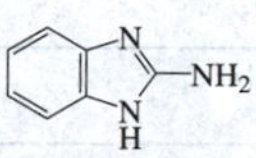
682
1*H*-Benzimidazol-2-amine

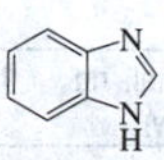
683
1*H*-Benzimidazole

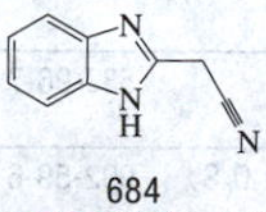
684
1*H*-Benzimidazole-2-acetonitrile

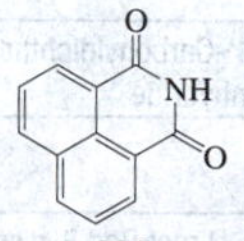
685
1*H*-Benz[de]isoquinoline-1,3(2*H*)-dione

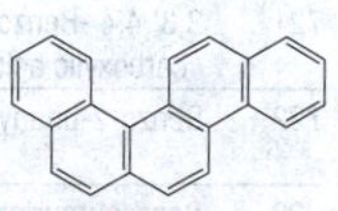
686
Benzo[c]chrysene

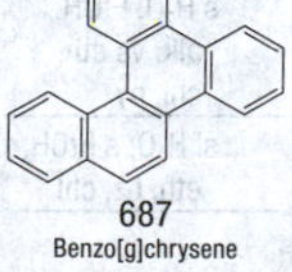
687
Benzo[g]chrysene

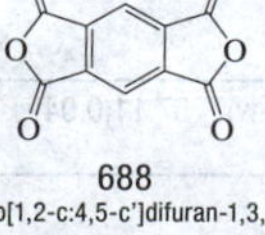
688
1*H*,3*H*-Benzo[1,2-c:4,5-c']difuran-1,3,5,7-tetrone

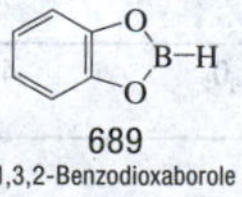
689
1,3,2-Benzodioxaborole

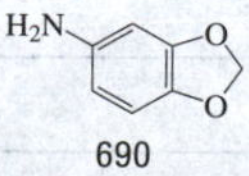
690
1,3-Benzodioxol-5-amine

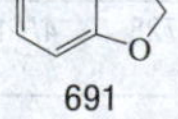
691
1,3-Benzodioxole

No.	Name	Synonym	Mol. form.	CAS Reg. No.	Mol. wt.	Physical form	mp/°C	bp/°C	den/ g cm^{-3}	n_D	Solubility
692	1,3-Benzodioxole-5-carboxaldehyde	Piperonal	$C_8H_6O_3$	120-57-0	150.132		37(1)	263			sl H_2O; vs EtOH; msc eth; s ace, chl
693	1,3-Benzodioxole-5-carboxylic acid	Piperonylic acid	$C_8H_6O_4$	94-53-1	166.132		229				
694	1,3-Benzodioxole-5-ethanamine		$C_9H_{11}NO_2$	1484-85-1	165.189			166^{20}	1.225^{20}	1.5620^{20}	
695	1,3-Benzodioxole-5-methanamine		$C_8H_9NO_2$	2620-50-0	151.163			139^{13}	1.214^{25}	1.5635^{20}	
696	1,3-Benzodioxole-5-methanol		$C_8H_8O_3$	495-76-1	152.148	nd (peth)	58	157^{16}			sl H_2O; s EtOH, eth, bz, chl; i lig
697	1,3-Benzodioxol-5-ol		$C_7H_6O_3$	533-31-3	138.121		64.9				
698	*trans,trans*-5-(1,3-Benzodioxol-5-yl)-2,4-pentadienoic acid	Piperinic acid	$C_{12}H_{10}O_4$	136-72-1	218.205	nd (al), ye nd (sub)	215.8	sub			vs EtOH
699	7,8-Benzoflavone	2-Phenyl-4*H*-naphtho[1,2-*b*]-pyran-4-one	$C_{19}H_{12}O_2$	604-59-1	272.297	ye pl (al)	157				sl EtOH, chl; s sulf
700	Benzo[*b*]fluoranthene	Benz[*e*]acephenanthrylene	$C_{20}H_{12}$	205-99-2	252.309	nd (bz)	168.4(0.7)				i H_2O; msc bz
701	Benzo[*j*]fluoranthene	Dibenzo[*a,jk*]fluorene	$C_{20}H_{12}$	205-82-3	252.309	ye pl (al) nd (HOAc)	165.2(0.7)				i H_2O; sl EtOH, HOAc
702	Benzo[*k*]fluoranthene	2,3,1',8'-Binaphthylene	$C_{20}H_{12}$	207-08-9	252.309	pa ye nd (bz)	217(1)	480			i H_2O; s EtOH, bz, HOAc
703	11*H*-Benzo[*a*]fluorene		$C_{17}H_{12}$	238-84-6	216.277	pl (ace or HOAc)	189.6(0.5)	405			i H_2O; sl EtOH; s eth, bz, chl
704	11*H*-Benzo[*b*]fluorene		$C_{17}H_{12}$	243-17-4	216.277		215(5)	401			i H_2O
705	Benzofuran	Coumarone	C_8H_6O	271-89-6	118.133		<-18	172(6)	1.0913^{25}	1.5615^{17}	i H_2O; s EtOH, eth
706	2-Benzofurancarboxylic acid	Coumarilic acid	$C_9H_6O_3$	496-41-3	162.142	nd (w)	192.5	312.5			vs EtOH
707	2(3*H*)-Benzofuranone		$C_8H_6O_2$	553-86-6	134.133		50	249	1.2236^{14}		
708	3(2*H*)-Benzofuranone		$C_8H_6O_2$	7169-34-8	134.133	red nd (al)	102.5	152^{15}			vs bz
709	1-(2-Benzofuranyl)ethanone		$C_{10}H_8O_2$	1646-26-0	160.170		76	126^{11}			s H_2O
710	Benzofurazan, 1-oxide		$C_6H_4N_2O_2$	480-96-6	136.108		72(1)		1.280^{80}		
711	Benzohydrazide	Benzoic acid, hydrazide	$C_7H_8N_2O$	613-94-5	136.151	pl (w)	115	267 dec			s H_2O, EtOH; sl eth, ace, chl
712	Benzoic acid	Benzenecarboxylic acid	$C_7H_6O_2$	65-85-0	122.122	mcl lf or nd	122.340 (0.005)	250.2(0.6)	1.2659^{15}	1.504^{132}	sl H_2O; vs EtOH, eth; s ace, bz, chl
713	Benzoic anhydride		$C_{14}H_{10}O_3$	93-97-0	226.227	pr (eth)	42.6(0.8)	360	1.989^{15}	1.5767^{15}	i H_2O, lig; s EtOH, eth; sl chl
714	Benzoin	2-Hydroxy-1,2-diphenylethanone, (±)	$C_{14}H_{12}O_2$	579-44-2	212.244		137(2)	343(13)	1.310^{20}		vs EtOH, chl
715	Benzonitrile	Phenyl cyanide	C_7H_5N	100-47-0	103.122	liq	-12.82(0.02)	191(1)	1.0093^{15}	1.5289^{20}	sl H_2O; msc EtOH; vs ace, bz; s ctc
716	Benzo[*ghi*]perylene	1,12-Benzperylene	$C_{22}H_{12}$	191-24-2	276.330	ye-grn lf (bz)	280(1)				i H_2O
717	Benzo[*c*]phenanthrene	Tetrahelicene	$C_{18}H_{12}$	195-19-7	228.288		67(2)				i H_2O; sl EtOH, lig
718	Benzophenone	Diphenyl ketone	$C_{13}H_{10}O$	119-61-9	182.217	(α) orth pr (al); (β) mcl pr	48.0(0.2)	305.9(0.2)	1.111^{18}	1.6077^{19}	i H_2O; vs EtOH, eth, chl, ace; s bz
719	Benzophenone hydrazone		$C_{13}H_{12}N_2$	5350-57-2	196.247		97.3	227^{55}			
720	Benzophenone, oxime	Diphenyl ketoxime	$C_{13}H_{11}NO$	574-66-3	197.232	nd (al)	144				i H_2O; vs EtOH, eth, chl, ace; s bz
721	3,3',4,4'-Benzophenonetetracarboxylic acid dianhydride	4,4'-Carbonyldiphthalic anhydride	$C_{17}H_6O_7$	2421-28-5	322.226		216				
722	Benzo-2-phenylhydrazide		$C_{13}H_{12}N_2O$	532-96-7	212.246	pr (al), nd (w)	168	314			sl H_2O, eth; s EtOH, bz, chl
723	Benzopurpurine 4B	C.I. Direct Red 2, disodium salt	$C_{34}H_{26}N_6Na_2O_6S_2$	992-59-6	724.716	br pow					s H_2O, EtOH, ac, H_2SO_4
724	2*H*-1-Benzopyran	1,2-Chromene	C_9H_8O	254-04-6	132.159			132^{102}	1.0993^{16}	1.5869^{24}	i H_2O
725	[2]Benzopyrano[6,5,4-*def*][2]-benzopyran-1,3,6,8-tetrone	1,4,5,8-Naphthalenetetracarboxylic acid anhydride	$C_{14}H_4O_6$	81-30-1	268.178	nd (al)	450	320 sub			i H_2O; s Na_2CO_3, HOAc
726	1*H*-2-Benzopyran-1-one	Isocoumarin	$C_9H_6O_2$	491-31-6	146.143	pl (bz)	47	286			i H_2O; vs EtOH, eth, bz, CS_2
727	2*H*-1-Benzopyran-2-one	Coumarin	$C_9H_6O_2$	91-64-5	146.143	orth pym (eth)	68(3)	301.7	0.935^{20}		s H_2O, EtOH, alk; vs eth, chl, py
728	4*H*-1-Benzopyran-4-one		$C_9H_6O_2$	491-38-3	146.143	nd (peth w)	57.11(0.04)	sub	1.2900^{20}		sl H_2O; s EtOH, eth, bz, chl

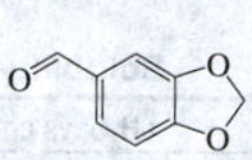
692
1,3-Benzodioxole-5-carboxaldehyde

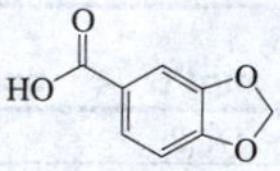
693
1,3-Benzodioxole-5-carboxylic acid

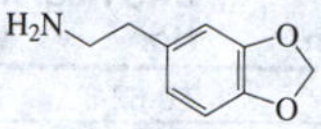
694
1,3-Benzodioxole-5-ethanamine

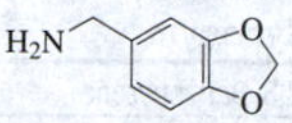
695
1,3-Benzodioxole-5-methanamine

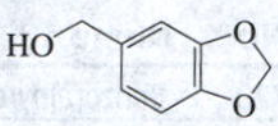
696
1,3-Benzodioxole-5-methanol

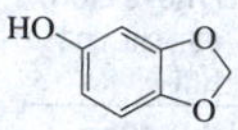
697
1,3-Benzodioxol-5-ol

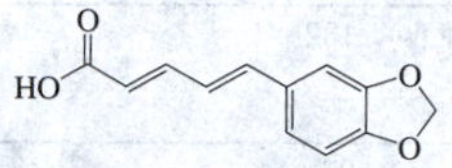
698
trans,trans-5-(1,3-Benzodioxol-5-yl)-2,4-pentadienoic acid

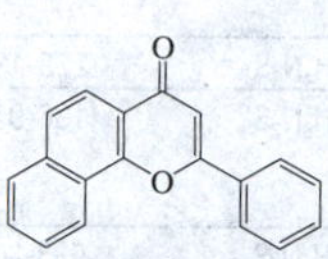
699
7,8-Benzoflavone

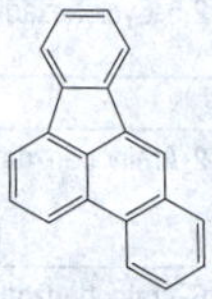
700
Benzo[b]fluoranthene

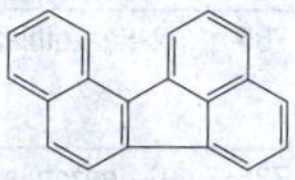
701
Benzo[j]fluoranthene

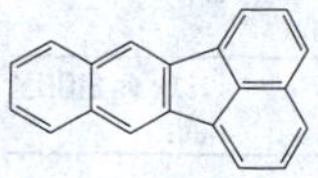
702
Benzo[k]fluoranthene

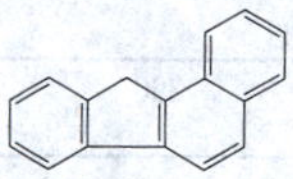
703
11*H*-Benzo[a]fluorene

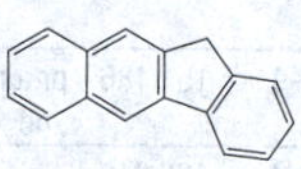
704
11*H*-Benzo[b]fluorene

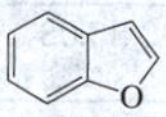
705
Benzofuran

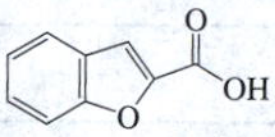
706
2-Benzofurancarboxylic acid

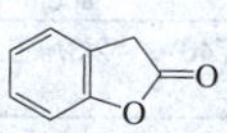
707
2(3*H*)-Benzofuranone

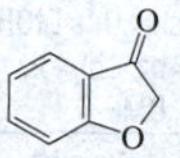
708
3(2*H*)-Benzofuranone

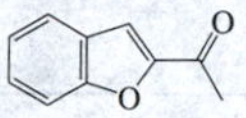
709
1-(2-Benzofuranyl)ethanone

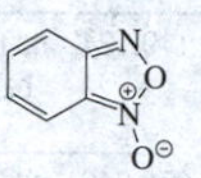
710
Benzofurazan, 1-oxide

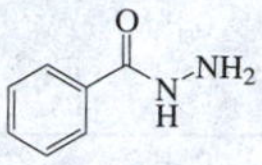
711
Benzohydrazide

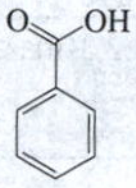
712
Benzoic acid

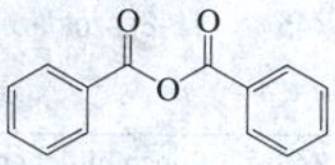
713
Benzoic anhydride

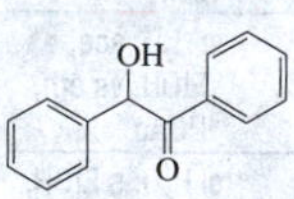
714
Benzoin

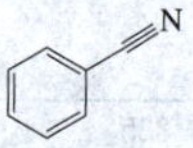
715
Benzonitrile

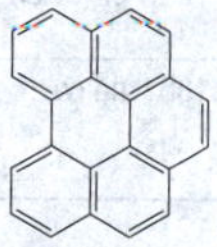
716
Benzo[ghi]perylene

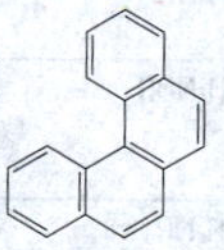
717
Benzo[c]phenanthrene

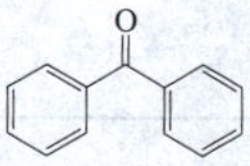
718
Benzophenone

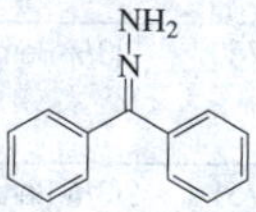
719
Benzophenone hydrazone

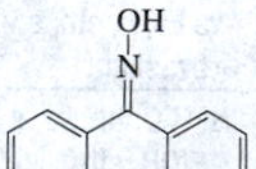
720
Benzophenone, oxime

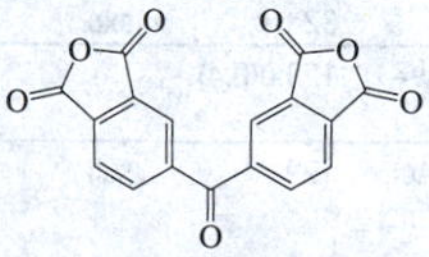
721
3,3',4,4'-Benzophenonetetracarboxylic acid dianhydride

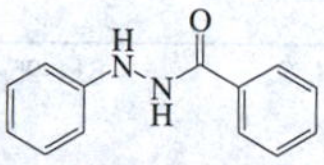
722
Benzo-2-phenylhydrazide

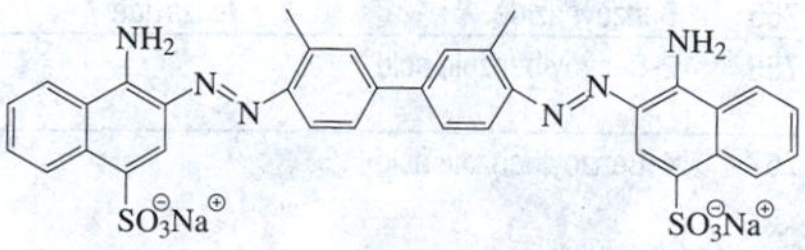
723
Benzopurpurine 4B

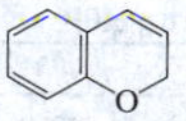
724
2*H*-1-Benzopyran

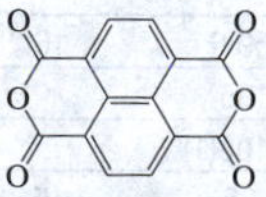
725
[2]Benzopyrano[6,5,4-def][2]benzopyran-1,3,6,8-tetrone

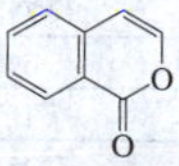
726
1*H*-2-Benzopyran-1-one

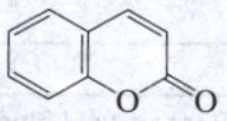
727
2*H*-1-Benzopyran-2-one

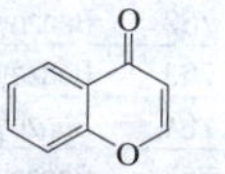
728
4*H*-1-Benzopyran-4-one

No.	Name	Synonym	Mol. form.	CAS Reg. No.	Mol. wt.	Physical form	mp/°C	bp/°C	den/ g cm^{-3}	n_D	Solubility
729	Benzo[*a*]pyrene	2,3-Benzopyrene	$C_{20}H_{12}$	50-32-8	252.309		179(2)				i H_2O; vs chl
730	Benzo[*e*]pyrene	1,2-Benzpyrene	$C_{20}H_{12}$	192-97-2	252.309	pa ye nd (bz-MeOH)	180(3)	311			i H_2O
731	Benzo[*f*]quinoline	β-Naphthoquinoline	$C_{13}H_9N$	85-02-9	179.217	lf (peth or w)	92(2)	352			sl H_2O; vs EtOH, bz, eth; s ace
732	Benzo[*h*]quinoline		$C_{13}H_9N$	230-27-3	179.217	lf (eth), pl (peth)	50.94(0.05)	341.3(0.3)	1.2340^{20}		sl H_2O; s EtOH, eth, ace, bz, ctc
733	*p*-Benzoquinone	2,5-Cyclohexadiene-1,4-dione	$C_6H_4O_2$	106-51-4	108.095	ye mcl pr (w)	113(2)	sub	1.318^{20}		sl H_2O, peth; s EtOH, eth, chl
734	2,1,3-Benzothiadiazole		$C_6H_4N_2S$	273-13-2	136.174		43.7(0.4)	206			
735	2-Benzothiazolamine	2-Aminobenzothiazole	$C_7H_6N_2S$	136-95-8	150.201	pl (w), lf (w)	132				sl H_2O; s EtOH, eth, chl, con HCl
736	6-Benzothiazolamine	6-Aminobenzothiazole	$C_7H_6N_2S$	533-30-2	150.201	pr (w)	87				i H_2O, eth; s EtOH
737	Benzothiazole	Benzosulfonazole	C_7H_5NS	95-16-9	135.187		2.5(0.1)	234.6(0.6)	1.2460^{20}	1.6379^{20}	sl H_2O; vs EtOH, eth, CS_2; s ace
738	2(3*H*)-Benzothiazolethione	2-Mercaptobenzothiazole	$C_7H_5NS_2$	149-30-4	167.252	pa ye mcl nd(al, MeOH)	182(1)		1.42^{20}		i H_2O; s EtOH; sl eth, bz, DMSO
739	2(3*H*)-Benzothiazolethione, sodium salt		$C_7H_4NNaS_2$	2492-26-4	189.234						sl H_2O
740	2(3*H*)-Benzothiazolone		C_7H_5NOS	934-34-9	151.186	pr (dil al), nd	139	360			i H_2O; vs EtOH, eth
741	2(3*H*)-Benzothiazolone, hydrazone		$C_7H_7N_3S$	615-21-4	165.216		202.8				
742	2-(2-Benzothiazolyl)phenol		$C_{13}H_9NOS$	3411-95-8	227.281	nd or lf (al)	131	179^3			s EtOH
743	Benzo[*b*]thiophene	Thianaphthene	C_8H_6S	95-15-8	134.199	lf	31.33(0.03)	220.9(0.4)	1.1484^{32}	1.6374^{37}	i H_2O; vs EtOH; s eth, ace, bz; sl chl
744	Benzo[*b*]thiophene-2-carboxylic acid	Thionaphthene-2-carboxylic acid	$C_9H_6O_2S$	6314-28-9	178.208	nd (w)	240.5				vs eth
745	1*H*-Benzotriazole	1,2,3-Triaza-1*H*-indene	$C_6H_5N_3$	95-14-7	119.124	nd (chl or bz)	100	204^{15}			sl H_2O; s EtOH, bz, chl, tol, DMF
746	Benzo[b]triphenylene		$C_{22}H_{14}$	215-58-7	278.346	nd (al, HOAc)	205				i H_2O; vs bz
747	3*H*-2,1-Benzoxathiol-3-one 1,1-dioxide		$C_7H_4O_4S$	81-08-3	184.170	nd or pr (bz)	129.5	184^{18}			vs bz, chl
748	2*H*-3,1-Benzoxazine-2,4(1*H*)-dione		$C_8H_5NO_3$	118-48-9	163.131	pr (al, gl HOAc) cry (al)	243 dec				sl H_2O, EtOH, ace; i eth, bz, chl
749	Benzoxazole	1-Oxa-3-azaindene	C_7H_5NO	273-53-0	119.121	pr (dil al)	29.36(0.02)	185.3(0.5)	1.1754^{20}	1.5594^{20}	i H_2O; s EtOH, sulf
750	2(3*H*)-Benzoxazolethione		C_7H_5NOS	2382-96-9	151.186	nd (w)	195.1(0.5)				sl H_2O, ace, EtOH; vs eth, HOAc
751	2(3*H*)-Benzoxazolone		$C_7H_5NO_2$	59-49-4	135.121		140.0(0.4)	335			sl H_2O; s EtOH, eth, tfa
752	2-(2-Benzoxazolyl)phenol		$C_{13}H_9NO_2$	835-64-3	211.216	pink nd (al, HOAc)	123.5	338			sl H_2O; vs EtOH; s eth, ace, bz
753	*N*-Benzoyl-DL-alanine		$C_{10}H_{11}NO_3$	1205-02-3	193.199	pl, pr or lf (eth)	165.5	dec			s H_2O, EtOH; sl eth, DMSO
754	4-(Benzoylamino)-2-hydroxybenzoic acid	Benzoylpas	$C_{14}H_{11}NO_4$	13898-58-3	257.242		260.5				
755	Benzoyl azide	Benzazide	$C_7H_5N_3O$	582-61-6	147.134	pl (ace)	32	exp	1.1680^{35}		vs eth, EtOH
756	2-Benzoylbenzoic acid		$C_{14}H_{10}O_3$	85-52-9	226.227	tcl nd (w+1)	129.0(0.4)				vs EtOH, eth; s bz; sl chl
757	4-Benzoylbenzoic acid		$C_{14}H_{10}O_3$	611-95-0	226.227	nd (HOAc), pl (al) mcl lf (w)	199	sub			sl H_2O, tfa, bz; s EtOH, eth, HOAc
758	2-Benzoylbenzoic acid, hydrazide		$C_{14}H_{12}N_2O_2$	787-84-8	240.257	nd (al)	238(3)				sl H_2O; i EtOH, eth, chl; s MeOH
759	4-Benzoylbiphenyl	4-Phenylbenzophenone	$C_{19}H_{14}O$	2128-93-0	258.313		101.5	420			
760	Benzoyl bromide	Benzoic acid, bromide	C_7H_5BrO	618-32-6	185.018	liq	-24	220(10)	1.570^{15}	1.5868^{25}	msc eth
761	Benzoyl chloride	Benzenecarbonyl chloride	C_7H_5ClO	98-88-4	140.567	liq	-0.5(0.2)	201(8)	1.2120^{20}	1.5537^{20}	msc eth; s bz, ctc, CS_2
762	Benzoyl cyclohexane	Cyclohexyl phenyl ketone	$C_{13}H_{16}O$	712-50-5	188.265	nd (peth)	59.5	164^{18}			
763	Benzoylecgonine		$C_{16}H_{19}NO_4$	519-09-5	289.327	nd (w)	195				vs bz, EtOH
764	Benzoylferrocene		$C_{17}H_{14}FeO$	1272-44-2	290.137		108(1)				
765	Benzoyl fluoride	Benzoic acid, fluoride	C_7H_5FO	455-32-3	124.112	liq	-28	154.5	1.1400^{20}		vs EtOH, eth; s ctc

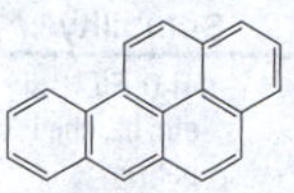
729
Benzo[a]pyrene

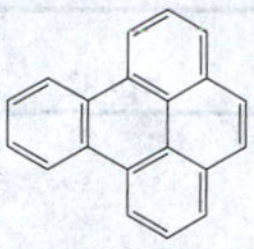
730
Benzo[e]pyrene

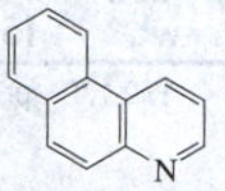
731
Benzo[f]quinoline

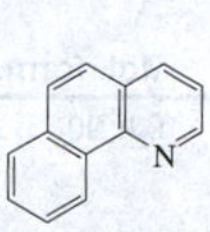
732
Benzo[h]quinoline

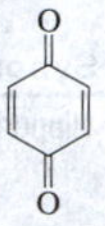
733
p-Benzoquinone

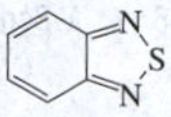
734
2,1,3-Benzothiadiazole

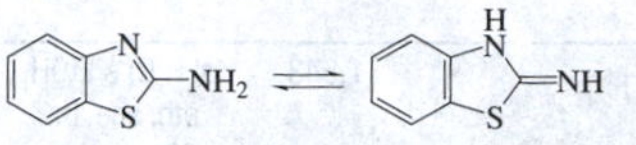
735
2-Benzothiazolamine

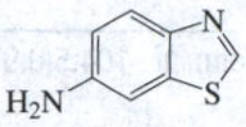
736
6-Benzothiazolamine

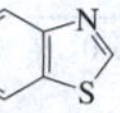
737
Benzothiazole

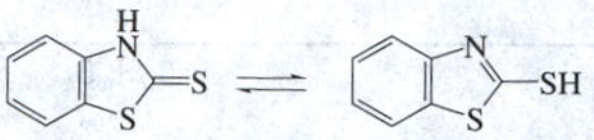
738
2(3*H*)-Benzothiazolethione

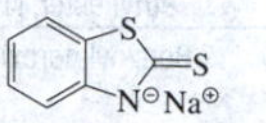
739
2(3*H*)-Benzothiazolethione, sodium salt

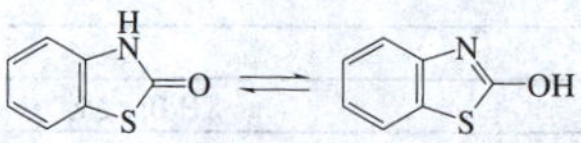
740
2(3*H*)-Benzothiazolone

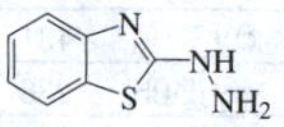
741
2(3*H*)-Benzothiazolone, hydrazone

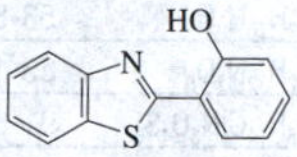
742
2-(2-Benzothiazolyl)phenol

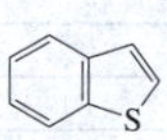
743
Benzo[b]thiophene

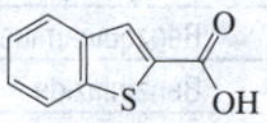
744
Benzo[b]thiophene-2-carboxylic acid

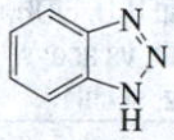
745
1*H*-Benzotriazole

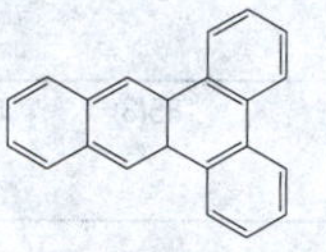
746
Benzo[b]triphenylene

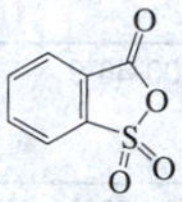
747
3*H*-2,1-Benzoxathiol-3-one 1,1-dioxide

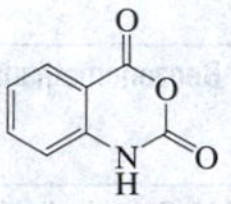
748
2*H*-3,1-Benzoxazine-2,4(1*H*)-dione

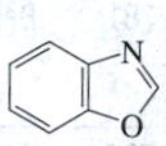
749
Benzoxazole

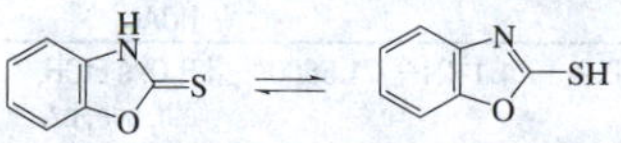
750
2(3*H*)-Benzoxazolethione

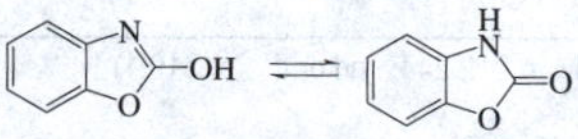
751
2(3*H*)-Benzoxazolone

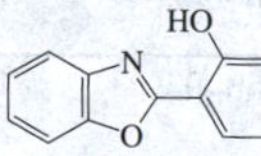
752
2-(2-Benzoxazolyl)phenol

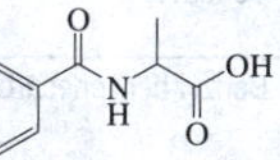
753
N-Benzoyl-DL-alanine

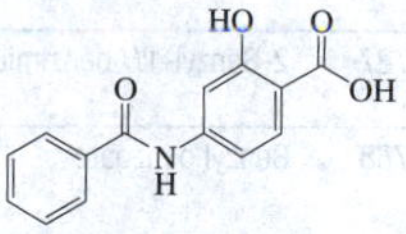
754
4-(Benzoylamino)-2-hydroxybenzoic acid

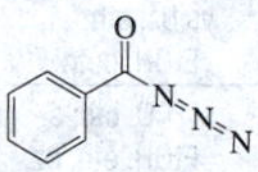
755
Benzoyl azide

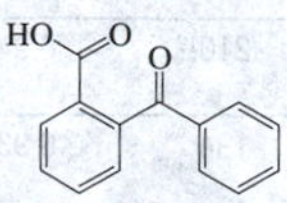
756
2-Benzoylbenzoic acid

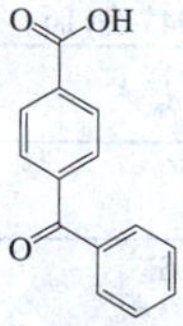
757
4-Benzoylbenzoic acid

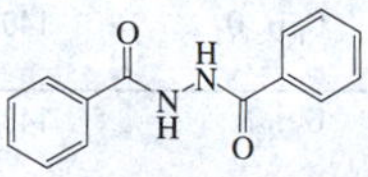
758
2-Benzoylbenzoic acid, hydrazide

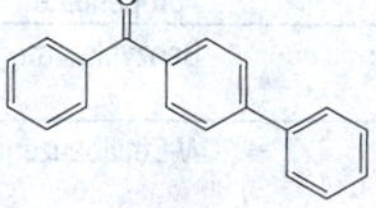
759
4-Benzoylbiphenyl

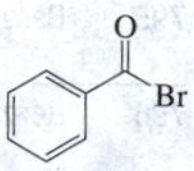
760
Benzoyl bromide

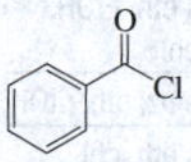
761
Benzoyl chloride

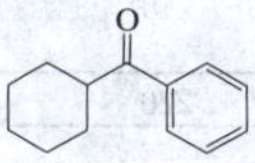
762
Benzoyl cyclohexane

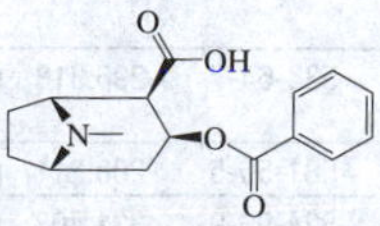
763
Benzoylecgonine

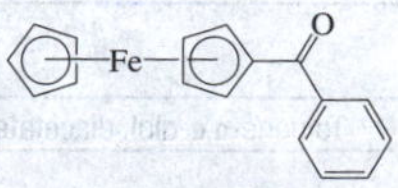
764
Benzoylferrocene

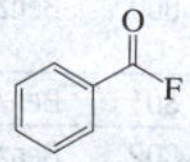
765
Benzoyl fluoride

No.	Name	Synonym	Mol. form.	CAS Reg. No.	Mol. wt.	Physical form	mp/°C	bp/°C	den/ g cm^{-3}	n_D	Solubility
766	*N*-Benzoylglycine	Hippuric acid	$C_9H_9NO_3$	495-69-2	179.172	pr (w or al)	191.5		1.371^{20}		s H_2O, EtOH; sl eth, bz, chl; i peth
767	Benzoyl iodide	Benzoic acid, iodide	C_7H_5IO	618-38-2	232.018	nd	3(2)	128^{20}	1.746^{18}		vs eth, EtOH
768	2-Benzoylmethyl-6(2-hydroxy-2-phenylethyl)-1-methylpiperidine, hydrochloride		$C_{22}H_{28}ClNO_2$	63990-84-1	373.916		183.5				sl H_2O; s EtOH; vs chl
769	3-(Benzoyloxy)-8-methyl-8-azabicyclo[3.2.1] octane-2-carboxylic acid, ethyl ester, [1*R*-(*exo,exo*)]	Cocaethylene	$C_{18}H_{23}NO_4$	529-38-4	317.381	pr (eth)	109				vs eth, EtOH
770	Benzoyl peroxide		$C_{14}H_{10}O_4$	94-36-0	242.227	orth (eth), pr	104.5(0.9)	exp		1.543	sl H_2O; s EtOH, eth, ace, bz, CS_2
771	1-Benzoylpiperidine		$C_{12}H_{15}NO$	776-75-0	189.253	tcl	49	320.5			i H_2O; s EtOH, eth; sl ctc
772	*N*-Benzoyl-L-tyrosine ethyl ester		$C_{18}H_{19}NO_4$	3483-82-7	313.349		119.5				
773	Benzphetamine		$C_{17}H_{21}N$	156-08-1	239.356			$127^{0.02}$		1.5515^{19}	vs eth, EtOH, MeOH, chl
774	Benzpiperylon		$C_{22}H_{25}N_3O$	53-89-4	347.453	cry (al)	182 dec				
775	Benzquinamide		$C_{22}H_{32}N_2O_5$	63-12-7	404.499	cry	131				
776	Benzthiazide		$C_{15}H_{14}ClN_3O_4S_3$	91-33-8	431.938	cry (EtOH)	236				i H_2O; s alk
777	*N*-Benzylacetamide		$C_9H_{11}NO$	588-46-5	149.189		61	157^2			vs EtOH, eth
778	Benzyl acetate	(Acetoxymethyl)benzene	$C_9H_{10}O_2$	140-11-4	150.174	liq	-51.5(0.4)	215(1)	1.0550^{20}	1.5232^{20}	sl H_2O; msc EtOH; s eth, ace, chl
779	Benzyl acrylate		$C_{10}H_{10}O_2$	2495-35-4	162.185			228	1.0573^{20}	1.5143^{20}	i H_2O; s EtOH, eth, ace, ctc
780	Benzyl alcohol	Benzenemethanol	C_7H_8O	100-51-6	108.138	liq	-15.5(0.2)	205.3(0.2)	1.0419^{24}	1.5396^{20}	s H_2O, EtOH, eth, ace, bz, MeOH, chl
781	Benzylamine	Benzenemethanamine	C_7H_9N	100-46-9	107.153	liq		185(3)	0.9813^{20}	1.5401^{20}	msc H_2O, EtOH, eth; vs ace; s bz; sl chl
782	4-(Benzylamino)benzenesulfonamide	N^1-Benzylsulfanilamide	$C_{13}H_{14}N_2O_2S$	104-22-3	262.327		171				
783	2-[Benzylamino]ethanol		$C_9H_{13}NO$	104-63-2	151.205			280	1.065^{25}	1.5430^{20}	
784	4-Benzylaniline		$C_{13}H_{13}N$	1135-12-2	183.249	mcl (lig)	34.5	300	1.038^{25}		vs eth, EtOH, lig
785	*N*-Benzylaniline	*N*-Phenylbenzenemethanamine	$C_{13}H_{13}N$	103-32-2	183.249	pr	33(4)	306.5	1.0298^{65}	1.6118^{25}	vs eth, EtOH
786	α-Benzylbenzenepropanoic acid		$C_{16}H_{16}O_2$	618-68-8	240.297	pl (peth HOAc) nd (w)	90	235^{18}			vs bz, eth, EtOH
787	2-Benzyl-1*H*-benzimidazole	Bendazol	$C_{14}H_{12}N_2$	621-72-7	208.258	nd (bz)	187				vs bz, EtOH, gl HOAc
788	Benzyl benzoate	Benzyl benzenecarboxylate	$C_{14}H_{12}O_2$	120-51-4	212.244	nd or lf	19(1)	321.3(0.9)	1.1121^{25}	1.5680^{20}	i H_2O; s EtOH, eth, ace, bz, MeOH, chl
789	4-Benzyl-1,1′-biphenyl		$C_{19}H_{16}$	613-42-3	244.330	lf	85	285^{110}	1.171^{0}		i H_2O; s EtOH, ctc; vs eth, bz
790	Benzyl butanoate		$C_{11}H_{14}O_2$	103-37-7	178.228			239	1.0111^{20}	1.4920^{20}	i H_2O; vs EtOH, eth; s ctc
791	Benzyl butyl phthalate	Butyl benzyl phthalate	$C_{19}H_{20}O_4$	85-68-7	312.360	liq		370	1.119^{25}		i H_2O
792	Benzyl chloroacetate		$C_9H_9ClO_2$	140-18-1	184.619			147^9	1.2223^4	1.5426^{18}	vs eth, EtOH
793	Benzyl chloroformate	Carbobenzoxy chloride	$C_8H_7ClO_2$	501-53-1	170.594	oily liq		103^{20}	1.195^{25}	1.5190^{20}	s eth, ace, bz
794	Benzyl *trans*-cinnamate	Benzyl *trans*-3-phenyl-2-propenoate	$C_{16}H_{14}O_2$	78277-23-3	238.281	pr	39	350 dec	1.109^{15}		i H_2O; s EtOH, eth; sl bz
795	Benzyl dodecanoate	Benzyl laurate	$C_{19}H_{30}O_2$	140-25-0	290.440		8.5	210^{12}	0.9429^{25}	1.4812^{24}	vs bz, eth, EtOH, peth
796	Benzylethylamine	*N*-Ethylbenzenemethanamine	$C_9H_{13}N$	14321-27-8	135.206			194	0.9342^{17}	1.5117^{20}	sl H_2O, ctc; s EtOH, eth, bz, chl
797	*N*-Benzyl-*N*-ethylaniline	Ethylbenzylaniline	$C_{15}H_{17}N$	92-59-1	211.303	pa ye oil	35	288	1.001^{55}	1.5943^{23}	i H_2O; s EtOH, eth, chl
798	Benzyl ethyl ether	(Ethoxymethyl)benzene	$C_9H_{12}O$	539-30-0	136.190			188(4)	0.9478^{20}	1.4955^{20}	i H_2O; msc EtOH, eth
799	Benzyl formate		$C_8H_8O_2$	104-57-4	136.149			203(8)	1.081^{20}	1.5154^{20}	i H_2O; s EtOH, ace; msc eth; sl ctc
800	Benzyl fumarate		$C_{18}H_{16}O_4$	538-64-7	296.318	cry pow	59	210^5			vs eth, EtOH, chl
801	Benzylidene diacetate	Toluene-α,α-diol, diacetate	$C_{11}H_{12}O_4$	581-55-5	208.211	pl (eth)	45.1(0.5)	220	1.11^{20}		vs bz, eth, EtOH
802	Benzylimidobis(*p*-methoxyphenyl)methane		$C_{22}H_{21}NO_2$	524-96-9	331.408	pa ye cry	90				vs eth, chl

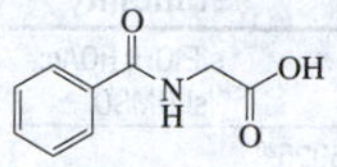
766
N-Benzoylglycine

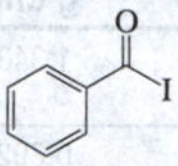
767
Benzoyl iodide

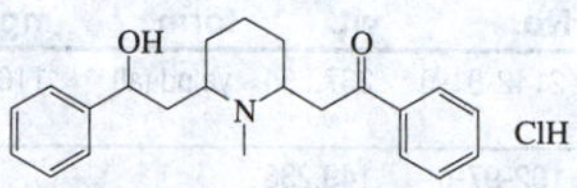
768
2-Benzoylmethyl-6(2-hydroxy-2-phenylethyl)-1-methylpiperidine, hydrochloride

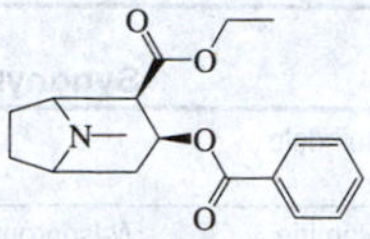
769
3-(Benzoyloxy)-8-methyl-8-azabicyclo[3.2.1]octane-2-carboxylic acid, ethyl ester, [1R-(exo,exo)]

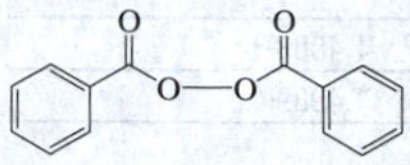
770
Benzoyl peroxide

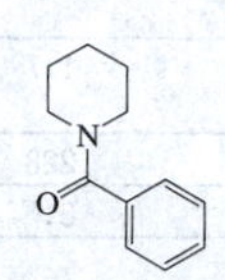
771
1-Benzoylpiperidine

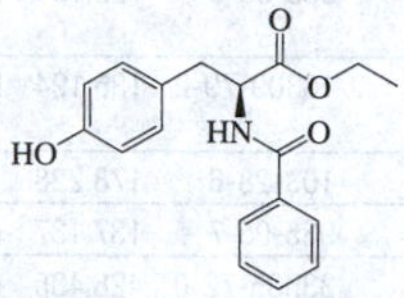
772
N-Benzoyl-L-tyrosine ethyl ester

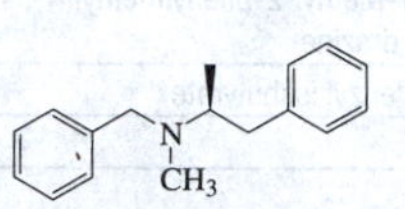
773
Benzphetamine

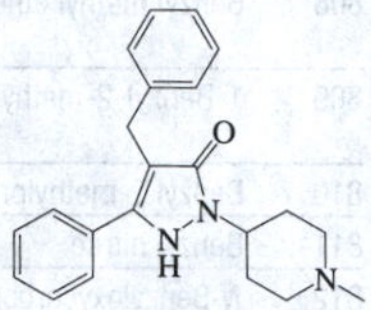
774
Benzpiperylon

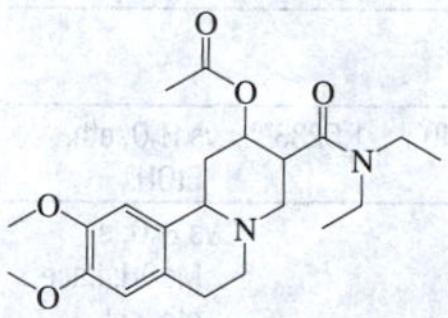
775
Benzquinamide

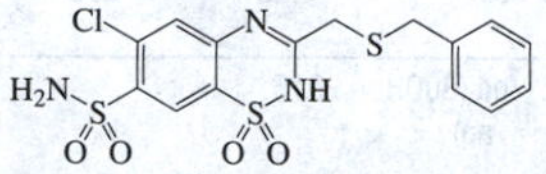
776
Benzthiazide

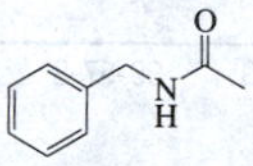
777
N-Benzylacetamide

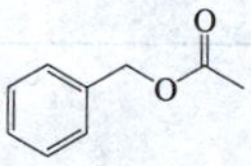
778
Benzyl acetate

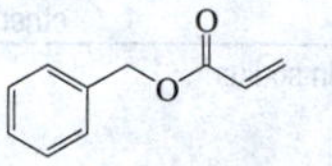
779
Benzyl acrylate

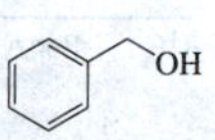
780
Benzyl alcohol

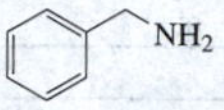
781
Benzylamine

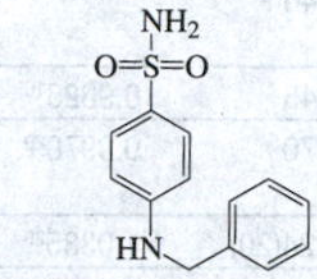
782
4-(Benzylamino)benzenesulfonamide

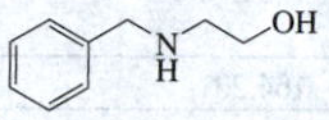
783
2-[Benzylamino]ethanol

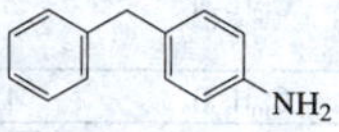
784
4-Benzylaniline

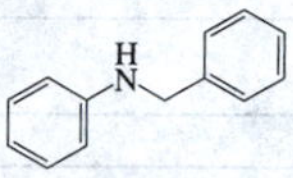
785
N-Benzylaniline

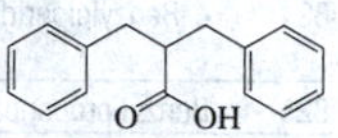
786
α-Benzylbenzenepropanoic acid

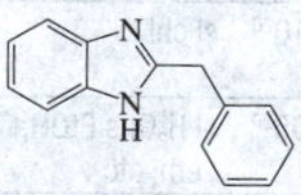
787
2-Benzyl-1H-benzimidazole

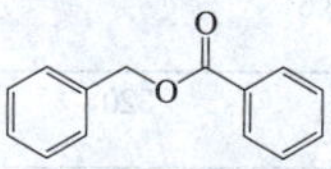
788
Benzyl benzoate

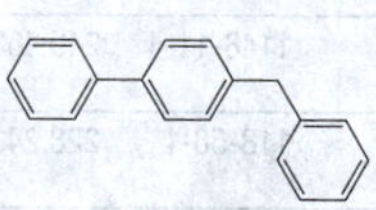
789
4-Benzyl-1,1'-biphenyl

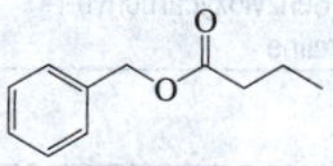
790
Benzyl butanoate

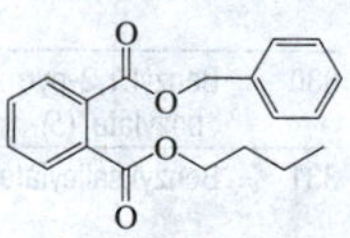
791
Benzyl butyl phthalate

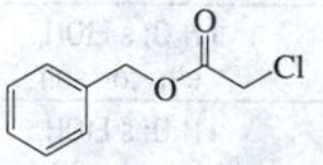
792
Benzyl chloroacetate

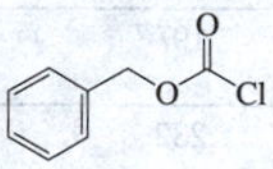
793
Benzyl chloroformate

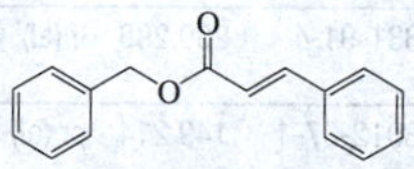
794
Benzyl trans-cinnamate

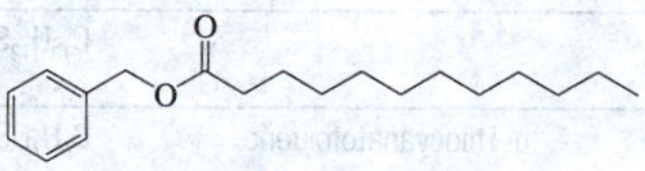
795
Benzyl dodecanoate

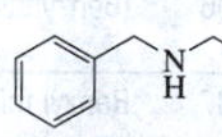
796
Benzylethylamine

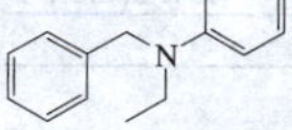
797
N-Benzyl-N-ethylaniline

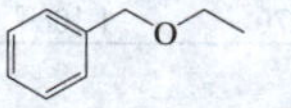
798
Benzyl ethyl ether

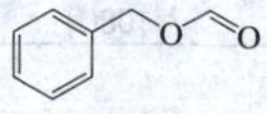
799
Benzyl formate

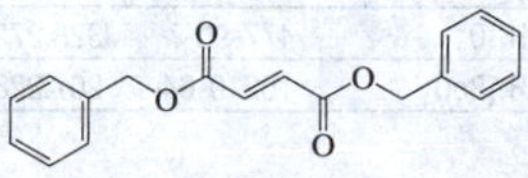
800
Benzyl fumarate

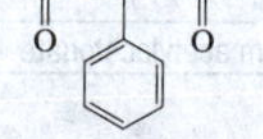
801
Benzylidene diacetate

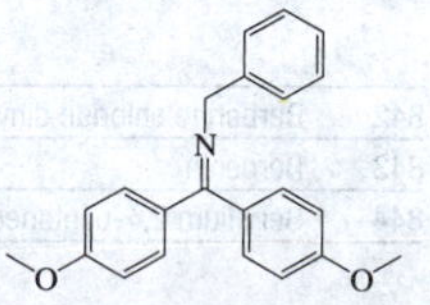
802
Benzylimidobis(p-methoxyphenyl)methane

Organic

No.	Name	Synonym	Mol. form.	CAS Reg. No.	Mol. wt.	Physical form	mp/°C	bp/°C	den/ g cm⁻³	n_D	Solubility
803	2-Benzyl-1*H*-isoindole-1,3(2*H*)-dione		$C_{15}H_{11}NO_2$	2142-01-0	237.254	ye nd (al)	116		1.343^{18}		s EtOH, HOAc; sl DMSO
804	Benzylisopropylamine	*N*-Isopropylbenzenemethanamine	$C_{10}H_{15}N$	102-97-6	149.233			200	0.892^{25}	1.5025^{20}	
805	Benzyl isothiocyanate	(Isothiocyanatomethyl)benzene	C_8H_7NS	622-78-6	149.214	ye oil		243	1.1246^{16}	1.6049^{15}	i H_2O; msc EtOH; s eth
806	Benzyl methacrylate		$C_{11}H_{12}O_2$	2495-37-6	176.212			144^{50}			
807	Benzyl 3-methylbutanoate		$C_{12}H_{16}O_2$	103-38-8	192.254			245	0.9983^{15}	1.4884^{20}	
808	Benzyl methyl ether		$C_8H_{10}O$	538-86-3	122.164	liq	-52.6(0.4)	175(3)	0.9634^{20}	1.5008^{20}	i H_2O, lig; vs EtOH, eth; s bz
809	1-Benzyl-2-methylhydrazine	1-Methyl-2-phenylmethylhydrazine	$C_8H_{12}N_2$	10309-79-2	136.194	liq		117^{20}			
810	Benzyl 2-methylpropanoate	Benzyl isobutyrate	$C_{11}H_{14}O_2$	103-28-6	178.228			228	1.0159^{18}	1.4883^{20}	
811	Benzyl nitrite		$C_7H_7NO_2$	935-05-7	137.137	oil		81^{35}	1.075^{25}	1.4989^{25}	
812	*N*-Benzyloxycarbonylaspartame		$C_{22}H_{24}N_2O_7$	33605-72-0	428.435	cry	122				
813	Benzyloxycarbonyl-L-glutamine		$C_{13}H_{16}N_2O_5$	2650-64-8	280.276		134.5				s DMSO
814	Benzyloxycarbonylglycine		$C_{10}H_{11}NO_4$	1138-80-3	209.199		121				s ace
815	Benzyloxycarbonylglycyl-L-leucine		$C_{16}H_{22}N_2O_5$	1421-69-8	322.356		100				
816	Benzyloxycarbonylglycyl-L-phenylalanine		$C_{19}H_{20}N_2O_5$	1170-76-9	356.372		126				
817	2-(Benzyloxy)ethanol	Ethylene glycol monobenzyl ether	$C_9H_{12}O_2$	622-08-2	152.190	oil	<-75	256	1.0640^{20}	1.5233^{20}	vs H_2O, eth, EtOH
818	Benzylpenicillin sodium		$C_{16}H_{17}N_2NaO_4S$	69-57-8	356.372	nd (BuOH aq)	215		1.41		vs H_2O; s MeOH; i ace, eth, chl
819	2-Benzylphenol	*o*-Benzylphenol	$C_{13}H_{12}O$	28994-41-4	184.233		21	312		1.5994^{20}	vs ace, bz, EtOH
820	4-Benzylphenol	*p*-Benzylphenol	$C_{13}H_{12}O$	101-53-1	184.233		84	322			s H_2O, EtOH, eth, bz, ctc, HOAc, chl
821	Benzyl phenyl ether		$C_{13}H_{12}O$	946-80-5	184.233	lf (al)	39(3)	285(13)			
822	1-Benzylpiperazine		$C_{11}H_{16}N_2$	2759-28-6	176.258			146^{12}		1.5430^{28}	s H_2O, EtOH, eth; sl chl
823	1-Benzylpiperidine		$C_{12}H_{17}N$	2905-56-8	175.270			245	0.9625^{16}	1.5227^{20}	
824	4-Benzylpiperidine		$C_{12}H_{17}N$	31252-42-3	175.270		16.8	270	0.9970^{20}	1.5337^{25}	i H_2O; s EtOH, eth
825	Benzyl propanoate		$C_{10}H_{12}O_2$	122-63-4	164.201			224(20)	1.0335^{20}		
826	2-Benzylpyridine		$C_{12}H_{11}N$	101-82-6	169.222	nd	12.5	277	1.067^{0}	1.5785^{20}	i H_2O; s EtOH, eth, chl
827	4-Benzylpyridine		$C_{12}H_{11}N$	2116-65-6	169.222		12.4	288	1.0612^{20}	1.5818^{20}	i H_2O; s EtOH, ctc; vs eth
828	Benzyl 3-pyridinecarboxylate	Benzyl nicotinate	$C_{13}H_{11}NO_2$	94-44-0	213.232			170^{3}			
829	1-Benzyl-1*H*-pyrrole		$C_{11}H_{11}N$	2051-97-0	157.212		14(2)	247	1.0183^{20}	1.5655^{24}	i H_2O; vs EtOH, eth
830	Benzyl 1,2-pyrrolidinedicarboxylate, (*S*)-	*N*-(Benzyloxycarbonyl)-L-proline	$C_{13}H_{15}NO_4$	1148-11-4	249.263		78.5			1.5310^{20}	sl chl
831	Benzyl salicylate		$C_{14}H_{12}O_3$	118-58-1	228.243			320	1.1799^{20}	1.5805^{20}	sl H_2O; s EtOH, eth, ctc
832	*O*-Benzyl-L-serine	3-(Benzyloxy)-L-alanine	$C_{10}H_{13}NO_3$	4726-96-9	195.215		218 dec				
833	Benzylsulfonic acid		$C_7H_8O_3S$	100-87-8	172.202	hyg cry					
834	4-[(Benzylsulfonyl)amino]-benzoic acid	*p*-(Benzylsulfonamido)benzoic acid	$C_{14}H_{13}NO_4S$	536-95-8	291.323		229.5				vs EtOH
835	(Benzylsulfonyl)benzene		$C_{13}H_{12}O_2S$	3112-88-7	232.298	nd (al)	146(3)		1.1261^{153}		i H_2O; sl EtOH, eth, bz
836	(Benzylthio)benzene		$C_{13}H_{12}S$	831-91-4	200.299	lf (al)	43.5	197^{27}			i H_2O; s EtOH, eth, con sulf
837	Benzyl thiocyanate	α-Thiocyanatotoluene	C_8H_7NS	3012-37-1	149.214	pr (al)	43	232			i H_2O; s EtOH, eth, chl, CS_2
838	Benzyltrimethylammonium chloride		$C_{10}H_{16}ClN$	56-93-9	185.694		243				vs H_2O; s ace
839	Benzylurea		$C_8H_{10}N_2O$	538-32-9	150.177	nd (al)	148	200 dec			vs ace, EtOH
840	Bephenium chloride		$C_{17}H_{22}ClNO$	13928-81-9	291.816	cry (ace)	135				
841	Berberine		$C_{20}H_{19}NO_5$	2086-83-1	353.369	red-ye nd (w+6) cry (chl)	145				vs eth, EtOH
842	Berberine chloride dihydrate		$C_{20}H_{22}ClNO_6$	5956-60-5	407.845	ye cry					
843	Bergenin		$C_{14}H_{16}O_9$	477-90-7	328.272	cry (MeOH)	238				vs H_2O, EtOH
844	Beryllium 2,4-pentanedioate	Beryllium acetylacetonate	$C_{10}H_{14}BeO_4$	10210-64-7	207.228		108(1)	270	1.168^{20}		

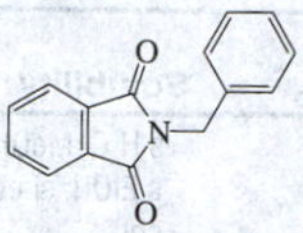
803
2-Benzyl-1*H*-isoindole-1,3(2*H*)-dione

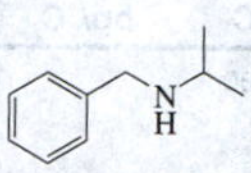
804
Benzylisopropylamine

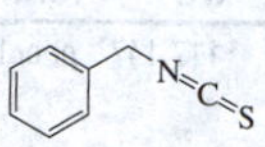
805
Benzyl isothiocyanate

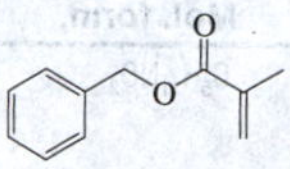
806
Benzyl methacrylate

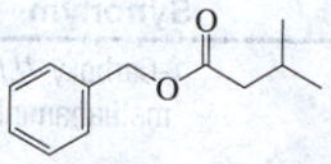
807
Benzyl 3-methylbutanoate

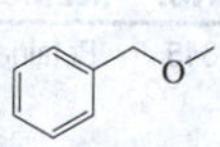
808
Benzyl methyl ether

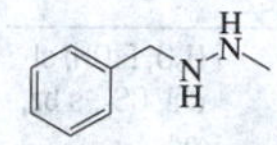
809
1-Benzyl-2-methylhydrazine

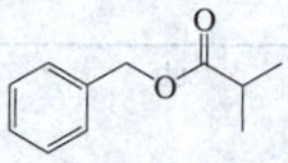
810
Benzyl 2-methylpropanoate

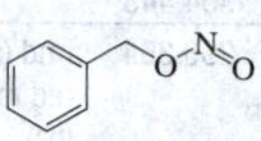
811
Benzyl nitrite

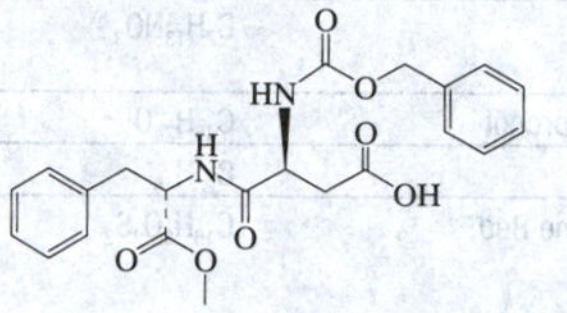
812
N-Benzyloxycarbonylaspartame

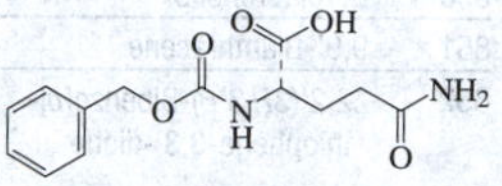
813
Benzyloxycarbonyl-L-glutamine

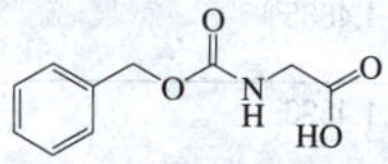
814
Benzyloxycarbonylglycine

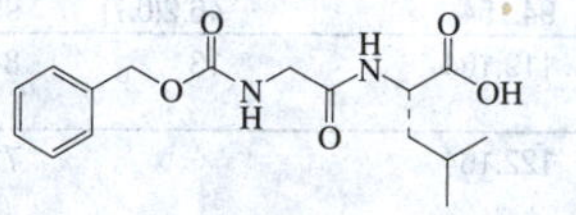
815
Benzyloxycarbonylglycyl-L-leucine

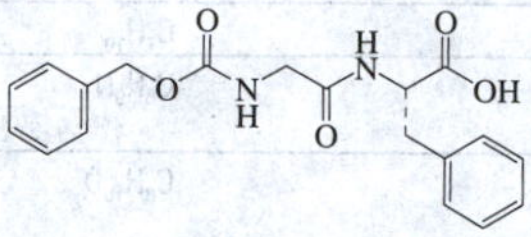
816
Benzyloxycarbonylglycyl-L-phenylalanine

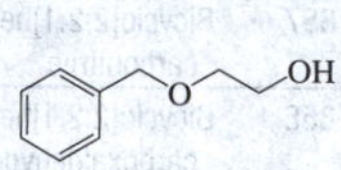
817
2-(Benzyloxy)ethanol

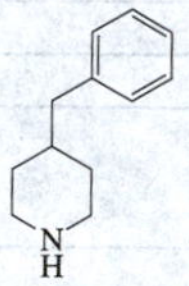
824
4-Benzylpiperidine

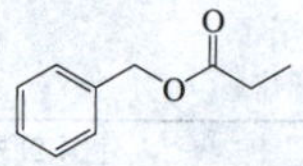
825
Benzyl propanoate

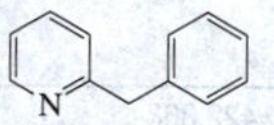
826
2-Benzylpyridine

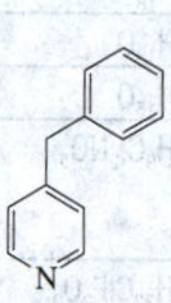
827
4-Benzylpyridine

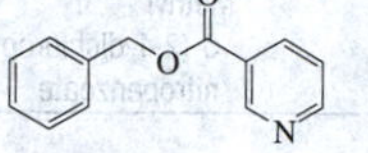
828
Benzyl 3-pyridinecarboxylate

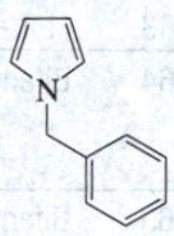
829
1-Benzyl-1*H*-pyrrole

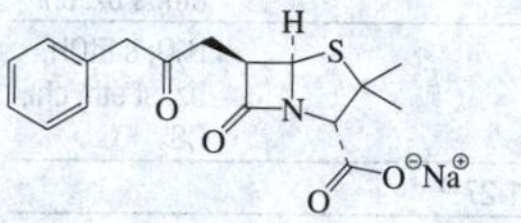
818
Benzylpenicillin sodium

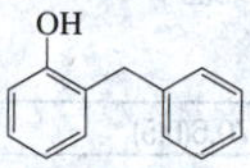
819
2-Benzylphenol

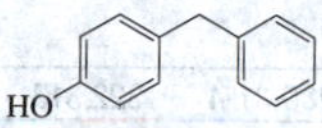
820
4-Benzylphenol

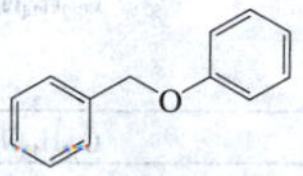
821
Benzyl phenyl ether

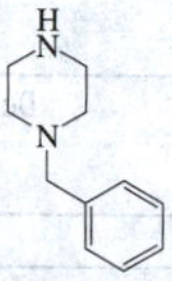
822
1-Benzylpiperazine

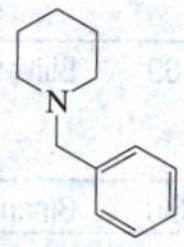
823
1-Benzylpiperidine

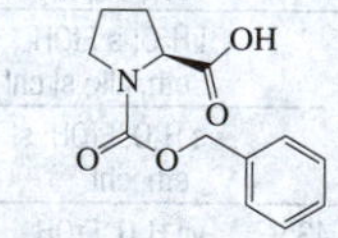
830
Benzyl 1,2-pyrrolidinedicarboxylate, (*S*)-

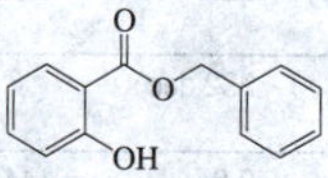
831
Benzyl salicylate

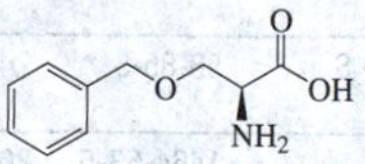
832
O-Benzyl-L-serine

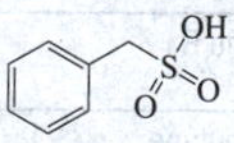
833
Benzylsulfonic acid

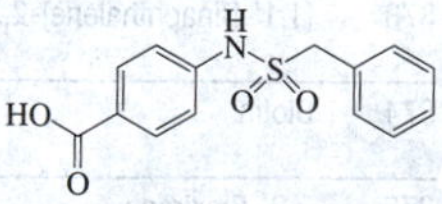
834
4-[(Benzylsulfonyl)amino]benzoic acid

835
(Benzylsulfonyl)benzene

836
(Benzylthio)benzene

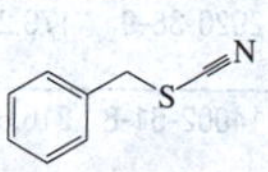
837
Benzyl thiocyanate

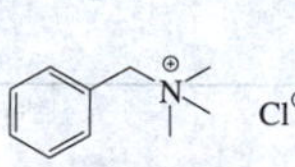
838
Benzyltrimethylammonium chloride

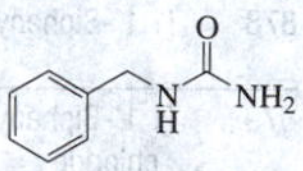
839
Benzylurea

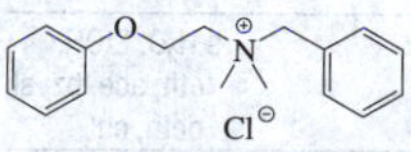
840
Bephenium chloride

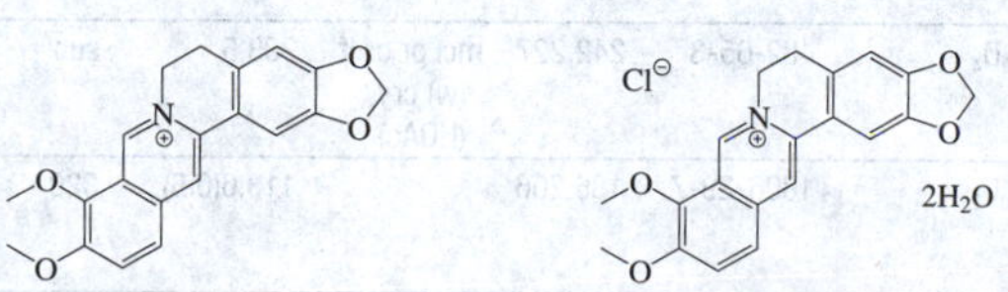
841
Berberine

842
Berberine chloride dihydrate

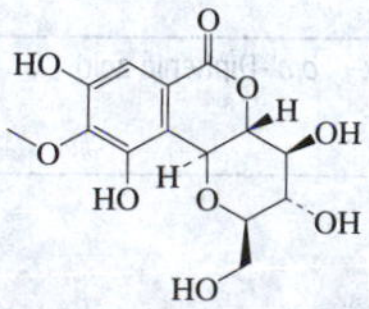
843
Bergenin

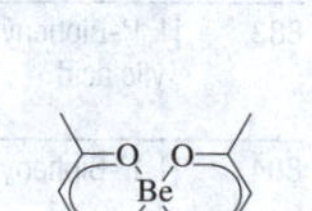
844
Beryllium 2,4-pentanedioate

Organic

No.	Name	Synonym	Mol. form.	CAS Reg. No.	Mol. wt.	Physical form	mp/°C	bp/°C	den/ g cm⁻³	n_D	Solubility
845	Betaine	1-Carboxy-*N,N,N*-trimethyl-methanaminium, inner salt	$C_5H_{11}NO_2$	107-43-7	117.147	pr or lf (al)	293 dec				vs H_2O, MeOH; s EtOH; sl eth, chl
846	Betaine, hydrochloride		$C_5H_{12}ClNO_2$	590-46-5	153.608	mcl cry (al)	227.5				vs H_2O
847	Betamethasone		$C_{22}H_{29}FO_5$	378-44-9	392.460	cry (AcOMe)	232 dec				
848	Bethanidine		$C_{10}H_{15}N_3$	55-73-2	177.246	cry (aq MeOH)	196				
849	Betonicine		$C_7H_{13}NO_3$	515-25-3	159.183	pr (dil al, +1w)	252 dec				vs EtOH
850	Betulaprenol 9	Nonaisoprenol	$C_{45}H_{74}O$	13190-97-1	631.069	oil or cry	41				s chl
851	9,9'-Bianthracene		$C_{28}H_{18}$	1055-23-8	354.443		321.3				
852	Δ2,2'(3*H*,3'*H*)-Bibenzo[*b*]-thiophene-3,3'-dione	Durindone Red	$C_{16}H_8O_2S_2$	522-75-8	296.364	br nd (xyl) red mcl nd (bz)	359	sub			i H_2O, EtOH; sl chl, CS_2; s bz, xyl
853	Bicyclo[2.2.1]heptane		C_7H_{12}	279-23-2	96.170		87.3(0.6)	107(6)			vs ace, bz, eth, EtOH
854	Bicyclo[4.1.0]heptane	Norcarane	C_7H_{12}	286-08-8	96.170			113.8(0.3)	0.853^{25}	1.4564^{20}	
855	Bicyclo[2.2.1]heptan-2-one		$C_7H_{10}O$	497-38-1	110.153		98.0(0.5)	170			
856	Bicyclo[2.2.1]hept-2-ene		C_7H_{10}	498-66-8	94.154		46.2(0.7)	95.69(0.08)			
857	Bicyclo[2.2.1]hept-5-ene-2-carbonitrile		C_8H_9N	95-11-4	119.164		13	84^{10}	0.999^{25}	1.4885^{20}	
858	Bicyclo[2.2.1]hept-5-ene-2-carboxaldehyde		$C_8H_{10}O$	5453-80-5	122.164			71^{20}	1.018^{25}	1.4893^{20}	
859	Bicyclo[2.2.1]hept-5-ene-2-methanol		$C_8H_{12}O$	95-12-5	124.180			103^{20}			
860	[1,1'-Bicyclohexyl]-2-one	2-Cyclohexylcyclohexanone	$C_{12}H_{20}O$	90-42-6	180.286	liq	-32	264	0.9696^{25}	1.4877^{25}	
861	1,1'-Bicyclopentyl		$C_{10}H_{18}$	1636-39-1	138.250			190.46(0.03)			s ctc, CS_2
862	[1,1'-Bicyclopentyl]-2-ol	2-Hydroxybicyclopentyl	$C_{10}H_{18}O$	4884-25-7	154.249		20	233(11)	0.9785^{15}	1.4884^{17}	
863	[1,1'-Bicyclopentyl]-2-one		$C_{10}H_{16}O$	4884-24-6	152.233	liq	-13	232.5	0.9745^{21}	1.4763	
864	Bifenox	Methyl 5-(2,4-dichlorophenoxy)-2-nitrobenzoate	$C_{14}H_9Cl_2NO_5$	42576-02-3	342.131		86.1(0.5)				
865	Bifenthrin		$C_{23}H_{22}ClF_3O_2$	82657-04-3	422.868		69		1.2^{125}		
866	Biguanide	Imidodicarbonimidic diamide	$C_2H_7N_5$	56-03-1	101.111	pr or nd (al)	136	142 dec			vs H_2O; s EtOH; i bz, chl
867	Bikhaconitine	3-Deoxypseudaconitine	$C_{36}H_{51}NO_{11}$	6078-26-8	673.790		164				vs eth, EtOH, chl
868	Bilirubin		$C_{33}H_{36}N_4O_6$	635-65-4	584.662	red mcl pr or pl (chl)					i H_2O; sl EtOH, eth; s bz, chl
869	Biliverdine	Dehydrobilirubin	$C_{33}H_{34}N_4O_6$	114-25-0	582.646	dk grn pl or pr (MeOH)	>300				i H_2O; s EtOH, bz; sl eth, chl, CS_2
870	Binapacryl		$C_{15}H_{18}N_2O_6$	485-31-4	322.313		70.6(0.5)		1.27^{20}		
871	1,1'-Binaphthalene	1,1'-Binaphthyl	$C_{20}H_{14}$	604-53-5	254.325	(i) pl(HOAc) (ii) orth (peth)	159(12)	>360	1.3000^{20}		i H_2O; sl EtOH; s eth, ace, bz, CS_2
872	2,2'-Binaphthalene		$C_{20}H_{14}$	612-78-2	254.325	bl flr pl (al)	188.1(1)	452(26)			i H_2O; sl EtOH; s eth, bz, CS_2
873	[1,1'-Binaphthalene]-2,2'-diol		$C_{20}H_{14}O_2$	602-09-5	286.324	nd (al), cry (w)	220				i H_2O; s EtOH, eth, alk; sl chl
874	Biotin	Coenzyme R	$C_{10}H_{16}N_2O_3S$	58-85-5	244.310	nd (w)	232 dec				s H_2O, EtOH; sl eth, chl
875	2,2'-Bioxirane	Diepoxybutane	$C_4H_6O_2$	1464-53-5	86.090	liq	2.0	144	1.113^{20}	1.435^{20}	vs H_2O, EtOH
876	Biphenyl	Diphenyl	$C_{12}H_{10}$	92-52-4	154.207	lf (dil al)	68.93(0.02)	255.2(0.3)	1.04^{20}	1.588^{77}	i H_2O; s EtOH, eth; vs bz, ctc, MeOH
877	[1,1'-Biphenyl]-4-acetic acid	Felbinac	$C_{14}H_{12}O_2$	5728-52-9	212.244		160.5				
878	[1,1'-Biphenyl]-4-carbonitrile		$C_{13}H_9N$	2920-38-9	179.217		88	190^{20}			i H_2O; vs EtOH, eth
879	[1,1'-Biphenyl]-4-carbonyl chloride		$C_{13}H_9ClO$	14002-51-8	216.662		111	160^{2}			
880	[1,1'-Biphenyl]-2,2'-diamine		$C_{12}H_{12}N_2$	1454-80-4	184.236	pr or nd (al)	81	162^{4}	1.3090^{20}		s H_2O, ace, bz
881	[1,1'-Biphenyl]-2,4'-diamine	2,4'-Biphenyldiamine	$C_{12}H_{12}N_2$	492-17-1	184.236	nd (dil al)	54.5	363			i H_2O; s EtOH, eth
882	[1,1'-Biphenyl]-4,4'-diamine, dihydrochloride		$C_{12}H_{14}Cl_2N_2$	531-85-1	257.158		>300				
883	[1,1'-Biphenyl]-2,2'-dicarboxylic acid	*o,o'*-Diphenic acid	$C_{14}H_{10}O_4$	482-05-3	242.227	mcl pr or lf (w) cry (HOAc)	233.5	sub			i H_2O; s EtOH, eth
884	[1,1'-Biphenyl]-2,2'-diol		$C_{12}H_{10}O_2$	1806-29-7	186.206		113.6(0.5)	320	1.3420^{20}		s H_2O, EtOH, eth, ace, bz; sl peth, chl
885	[1,1'-Biphenyl]-2,5-diol		$C_{12}H_{10}O_2$	1079-21-6	186.206	nd (dil al)	97.5				vs EtOH

845
Betaine

846
Betaine, hydrochloride

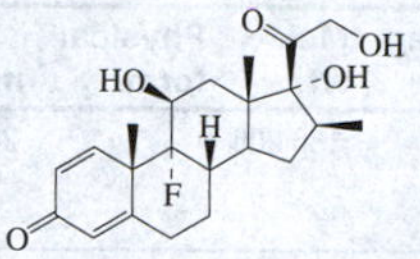

847
Betamethasone

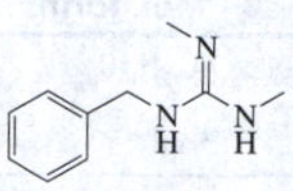

848
Bethanidine

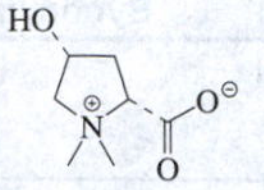

849
Betonicine

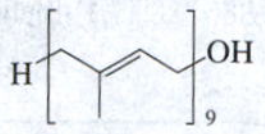

850
Betulaprenol 9

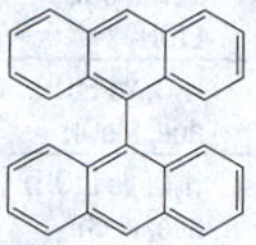

851
9,9'-Bianthracene

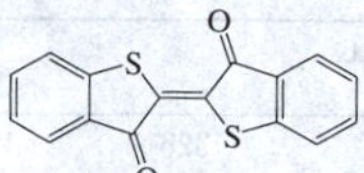

852
Δ2,2'(3*H*,3'*H*)-Bibenzo[b]thiophene-3,3'-dione

853
Bicyclo[2.2.1]heptane

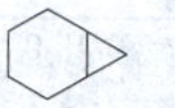

854
Bicyclo[4.1.0]heptane

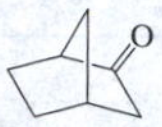

855
Bicyclo[2.2.1]heptan-2-one

856
Bicyclo[2.2.1]hept-2-ene

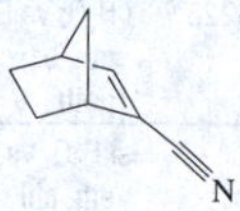

857
Bicyclo[2.2.1]hept-5-ene-2-carbonitrile

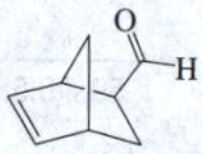

858
Bicyclo[2.2.1]hept-5-ene-2-carboxaldehyde

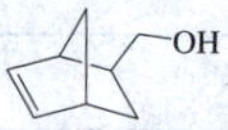

859
Bicyclo[2.2.1]hept-5-ene-2-methanol

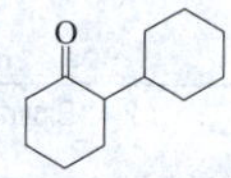

860
[1,1'-Bicyclohexyl]-2-one

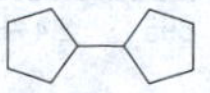

861
1,1'-Bicyclopentyl

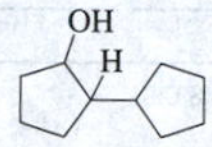

862
[1,1'-Bicyclopentyl]-2-ol

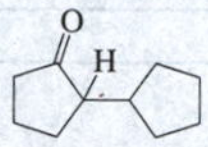

863
[1,1'-Bicyclopentyl]-2-one

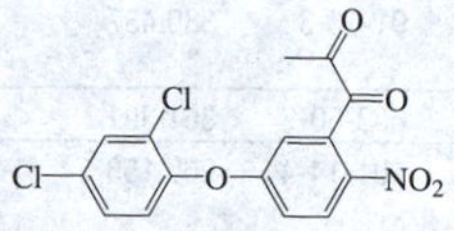

864
Bifenox

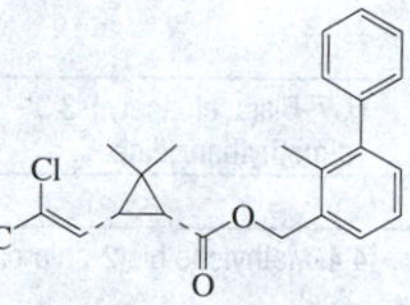

865
Bifenthrin

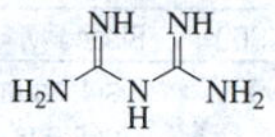

866
Biguanide

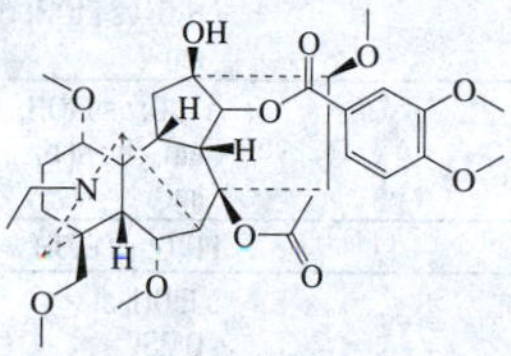

867
Bikhaconitine

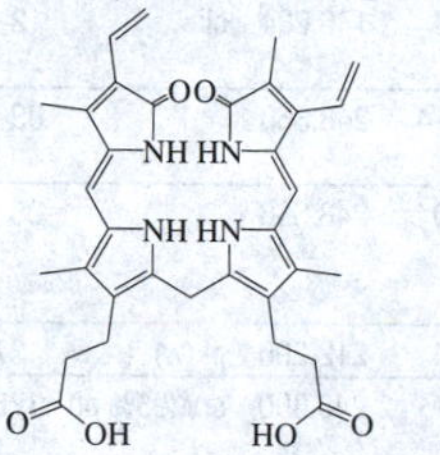

868
Bilirubin

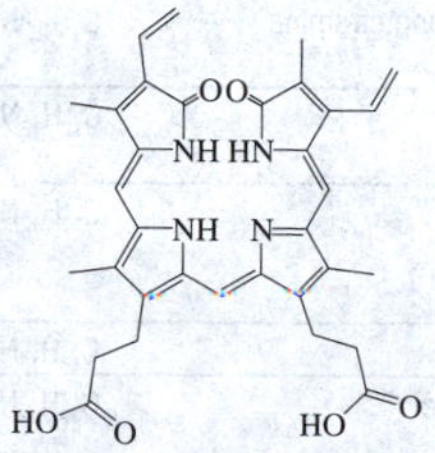

869
Biliverdine

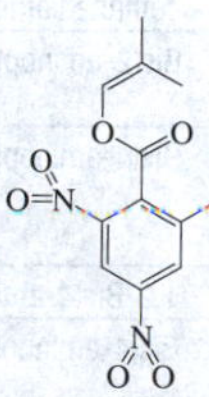

870
Binapacryl

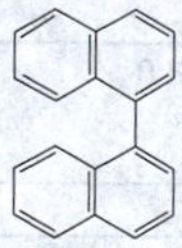

871
1,1'-Binaphthalene

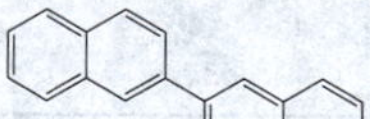

872
2,2'-Binaphthalene

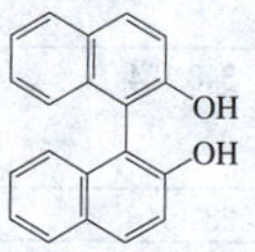

873
[1,1'-Binaphthalene]-2,2'-diol

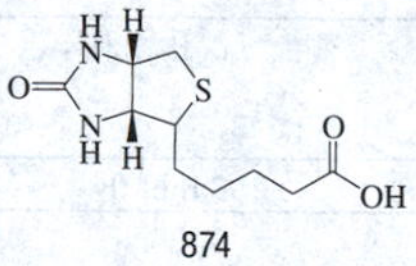

874
Biotin

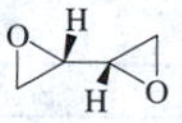

875
2,2'-Bioxirane

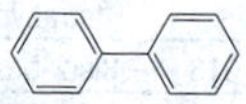

876
Biphenyl

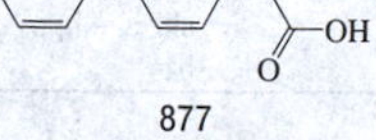

877
[1,1'-Biphenyl]-4-acetic acid

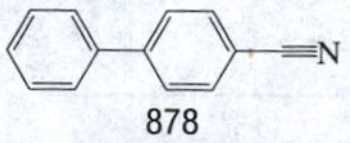

878
[1,1'-Biphenyl]-4-carbonitrile

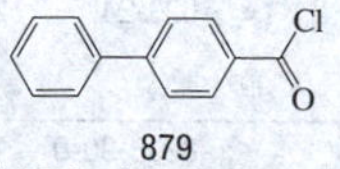

879
[1,1'-Biphenyl]-4-carbonyl chloride

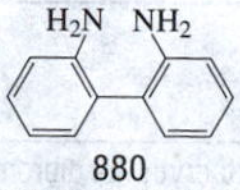

880
[1,1'-Biphenyl]-2,2'-diamine

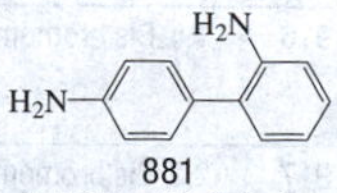

881
[1,1'-Biphenyl]-2,4'-diamine

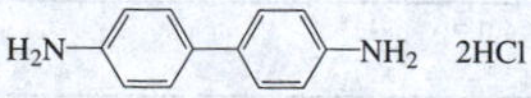

882
[1,1'-Biphenyl]-4,4'-diamine, dihydrochloride

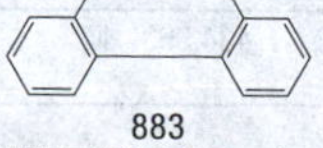

883
[1,1'-Biphenyl]-2,2'-dicarboxylic acid

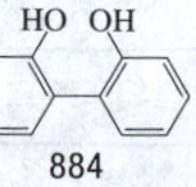

884
[1,1'-Biphenyl]-2,2'-diol

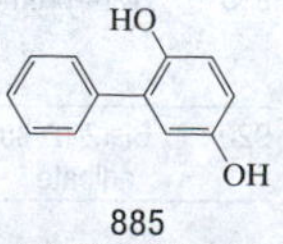

885
[1,1'-Biphenyl]-2,5-diol

No.	Name	Synonym	Mol. form.	CAS Reg. No.	Mol. wt.	Physical form	mp/°C	bp/°C	den/ g cm^{-3}	n_D	Solubility
886	[1,1'-Biphenyl]-4,4'-diol		$C_{12}H_{10}O_2$	92-88-6	186.206		287.6(0.5)				sl H_2O, bz, DMSO; s EtOH, eth
887	[1,1'-Biphenyl]-4,4'-disulfonic acid		$C_{12}H_{10}O_6S_2$	5314-37-4	314.333	pr	72.5	>200			vs H_2O
888	[1,1'-Biphenyl]-3,3',4,4'-tetramine, tetrahydrochloride		$C_{12}H_{18}Cl_4N_4$	7411-49-6	360.110		245 dec				
889	[1,1'-Biphenyl]-3,3',5,5'-tetrol	Diresorcinol	$C_{12}H_{10}O_4$	531-02-2	218.205	pl or nd (w+2)	310				vs H_2O, eth, EtOH
890	*N*-[1,1'-Biphenyl]-4-ylacetamide		$C_{14}H_{13}NO$	4075-79-0	211.259	cry (dil MeOH)	172.8				i H_2O; vs EtOH, ace, MeOH
891	1-[1,1'-Biphenyl]-4-ylethanone		$C_{14}H_{12}O$	92-91-1	196.244	pr (ace), cry (al)	121	326	1.2510^{0}		i H_2O; vs EtOH, ace; sl chl
892	2-[1,1'-Biphenyl]-4-yl-5-phenyl-1,3,4-oxadiazole		$C_{20}H_{14}N_2O$	852-38-0	298.337		168				
893	2,2'-Bipyridine	α,α'-Dipyridyl	$C_{10}H_8N_2$	366-18-7	156.184	pr (peth)	69.9(0.2)	273(10)			sl H_2O; vs EtOH, eth, bz, chl
894	2,3'-Bipyridine	2,3'-Bipyridyl	$C_{10}H_8N_2$	581-50-0	156.184			295.5	1.140^{20}	1.6223^{20}	i H_2O; vs EtOH, eth, bz, chl; sl peth
895	2,4'-Bipyridine	2,4'-Bipyridyl	$C_{10}H_8N_2$	581-47-5	156.184		59.7(0.5)	283(11)			sl H_2O; vs EtOH, eth, chl
896	3,3'-Bipyridine	3,3'-Bipyridyl	$C_{10}H_8N_2$	581-46-4	156.184		68	291.5	1.1614^{20}		vs H_2O, EtOH; sl eth
897	4,4'-Bipyridine	γ,γ'-Dipyridyl	$C_{10}H_8N_2$	553-26-4	156.184	nd (w+2)	104.4(0.5)	305			sl H_2O; vs EtOH, bz, chl; s eth
898	2,2'-Biquinoline		$C_{18}H_{12}N_2$	119-91-5	256.301	pl or lf (al)	196				i H_2O; vs EtOH; s eth, ace, bz
899	4,4'-Bis(acetoacetamido)-3,3'-dimethyl-1,1'-biphenyl	*N,N'*-Bis(acetoacetyl)-3,3'-dimethylbenzidine	$C_{22}H_{24}N_2O_4$	91-96-3	380.437		212				sl DMSO
900	Bisacodyl		$C_{22}H_{19}NO_4$	603-50-9	361.391		134(1)				
901	Bis(4-amino-3-chlorophenyl)-methane	4,4-Methylene bis(2-chloroaniline)	$C_{13}H_{12}Cl_2N_2$	101-14-4	267.153						s ctc
902	Bis(4-aminocyclohexyl)-methane		$C_{13}H_{26}N_2$	1761-71-3	210.358		15	320	0.92^{75}		
903	Bis(2-aminoethyl)amine	Diethylenetriamine	$C_4H_{13}N_3$	111-40-0	103.166	ye hyg liq	-39	206.5(0.3)	0.9569^{20}	1.4810^{25}	msc H_2O, EtOH; i eth; s lig
904	*N,N'*-Bis(2-aminoethyl)-1,2-ethanediamine	Triethylenetetramine	$C_6H_{18}N_4$	112-24-3	146.234	oil	12	266.5		1.4971^{20}	s H_2O, EtOH, acid
905	Bis(2-aminophenyl) disulfide		$C_{12}H_{12}N_2S_2$	1141-88-4	248.366		93				i H_2O; vs EtOH, eth
906	Bis(4-aminophenyl) disulfide		$C_{12}H_{12}N_2S_2$	722-27-0	248.366		85				s H_2O; vs EtOH, eth, chl; sl bz, lig
907	1,2-Bis(4-aminophenyl)ethane		$C_{14}H_{16}N_2$	621-95-4	212.290	pl (w)	137	sub			i H_2O; vs EtOH
908	Bis(4-aminophenyl) sulfone	Dapsone	$C_{12}H_{12}N_2O_2S$	80-08-0	248.300	cry (95% al)	178(1)				s EtOH; sl DMSO
909	Bis(4-aminophenyl) sulfoxide	4,4'-Sulfinyldianiline	$C_{12}H_{12}N_2OS$	119-59-5	232.300	pr (w, al)	175 dec				s H_2O, EtOH
910	1,4-Bis(3-aminopropoxy)-butane	1,4-Butanediol bis(3-aminopropyl) ether	$C_{10}H_{24}N_2O_2$	7300-34-7	204.310	liq		135^{3}	0.96^{20}	1.4619^{20}	
911	*N,N'*-Bis(3-aminopropyl)-1,4-butanediamine	Spermine	$C_{10}H_{26}N_4$	71-44-3	202.340		29	150^{5}			
912	*N,N'*-Bis(3-aminopropyl)-1,4-butanediamine, tetrahydrochloride		$C_{10}H_{30}Cl_4N_4$	306-67-2	348.184		301.5				s H_2O
913	Bis(2-bromoethyl) ether	Bromex	$C_4H_8Br_2O$	5414-19-7	231.914			115^{32}	1.8452^{20}	1.5131^{27}	
914	1,2-Bis(bromomethyl)benzene		$C_8H_8Br_2$	91-13-4	263.958	orth (chl)	95(1)	$129^{4.5}$	1.988^{25}		i H_2O; s EtOH, eth, ctc, chl, peth, lig
915	1,3-Bis(bromomethyl)benzene		$C_8H_8Br_2$	626-15-3	263.958	nd (chl), pr (ace)	77(1)	137^{20}	1.959^{25}		i H_2O; s EtOH, eth, chl, lig
916	1,4-Bis(bromomethyl)benzene		$C_8H_8Br_2$	623-24-5	263.958	mcl pr (al), cry (chl, bz)	144.5	245	2.012^{25}		i H_2O; vs EtOH, chl; sl eth; s bz
917	2,2-Bis(bromomethyl)-1,3-propanediol	Pentaerythritol dibromide	$C_5H_{10}Br_2O_2$	3296-90-0	261.940	nd (bz)	113				
918	1,3-Bis(bromomethyl)-tetramethyldisiloxane		$C_6H_{16}Br_2OSi_2$	2351-13-5	320.169			233	1.3918^{25}	1.4719^{25}	
919	Bis(4-bromophenyl) ether		$C_{12}H_8Br_2O$	2050-47-7	327.999	lf (al)	54(1)	339	1.8^{25}		i H_2O; s EtOH, bz; vs eth; sl chl
920	Bis(2-(2-butoxyethoxy)ethyl) adipate		$C_{22}H_{42}O_8$	141-17-3	434.563	liq			1.1^{25}		

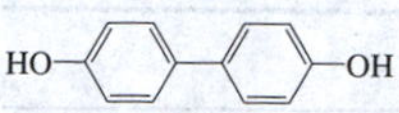
886
[1,1'-Biphenyl]-4,4'-diol

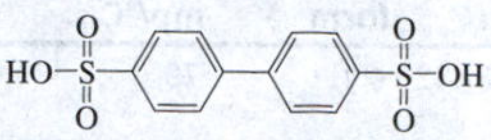
887
[1,1'-Biphenyl]-4,4'-disulfonic acid

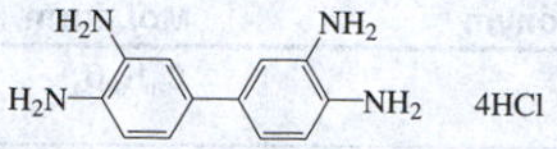
888
[1,1'-Biphenyl]-3,3',4,4'-tetramine, tetrahydrochloride

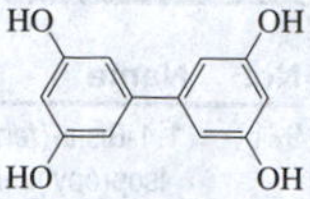
889
[1,1'-Biphenyl]-3,3',5,5'-tetrol

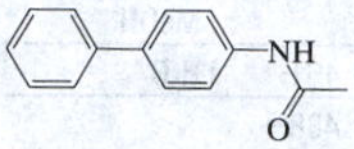
890
N-[1,1'-Biphenyl]-4-ylacetamide

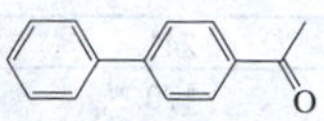
891
1-[1,1'-Biphenyl]-4-ylethanone

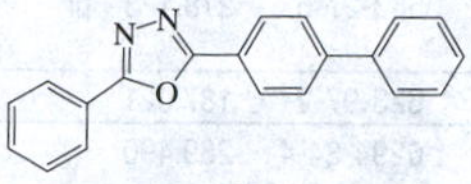
892
2-[1,1'-Biphenyl]-4-yl-5-phenyl-1,3,4-oxadiazole

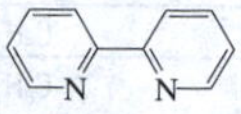
893
2,2'-Bipyridine

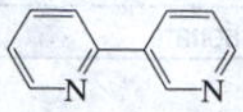
894
2,3'-Bipyridine

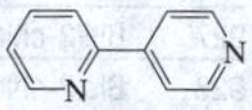
895
2,4'-Bipyridine

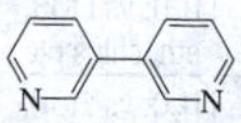
896
3,3'-Bipyridine

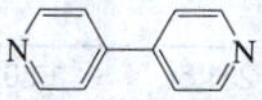
897
4,4'-Bipyridine

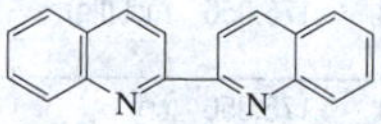
898
2,2'-Biquinoline

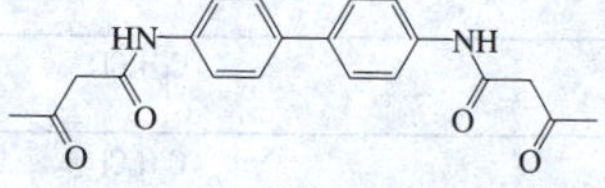
899
4,4'-Bis(acetoacetamido)-3,3'-dimethyl-1,1'-biphenyl

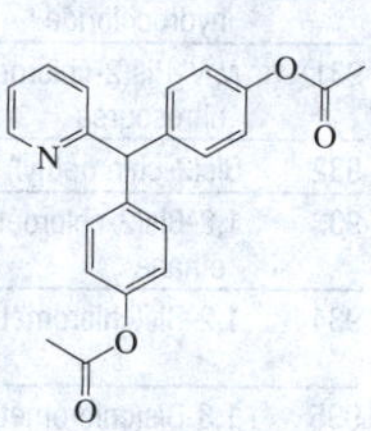
900
Bisacodyl

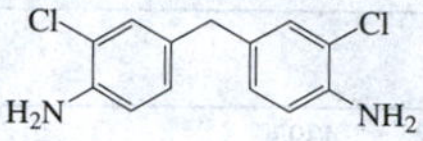
901
Bis(4-amino-3-chlorophenyl)methane

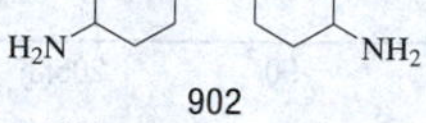
902
Bis(4-aminocyclohexyl)methane

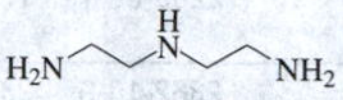
903
Bis(2-aminoethyl)amine

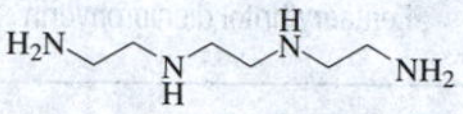
904
N,N'-Bis(2-aminoethyl)-1,2-ethanediamine

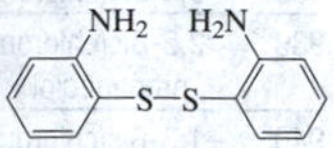
905
Bis(2-aminophenyl) disulfide

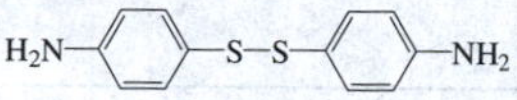
906
Bis(4-aminophenyl) disulfide

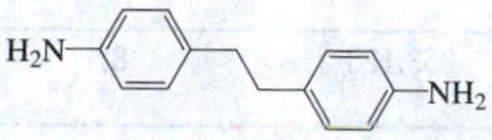
907
1,2-Bis(4-aminophenyl)ethane

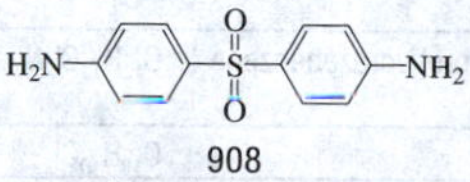
908
Bis(4-aminophenyl) sulfone

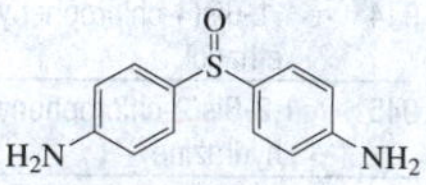
909
Bis(4-aminophenyl) sulfoxide

910
1,4-Bis(3-aminopropoxy)butane

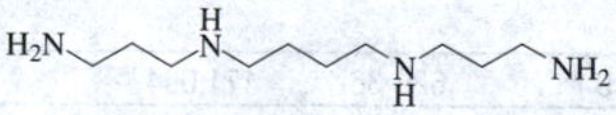
911
N,N'-Bis(3-aminopropyl)-1,4-butanediamine

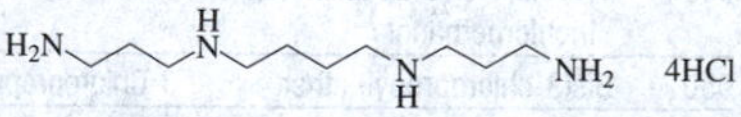
912
N,N'-Bis(3-aminopropyl)-1,4-butanediamine, tetrahydrochloride

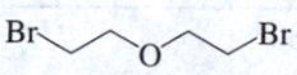
913
Bis(2-bromoethyl) ether

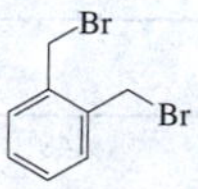
914
1,2-Bis(bromomethyl)benzene

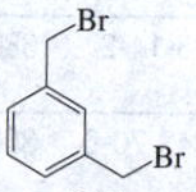
915
1,3-Bis(bromomethyl)benzene

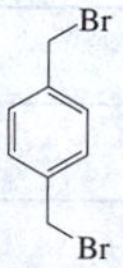
916
1,4-Bis(bromomethyl)benzene

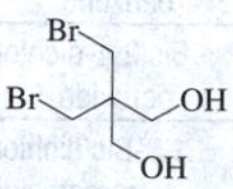
917
2,2-Bis(bromomethyl)-1,3-propanediol

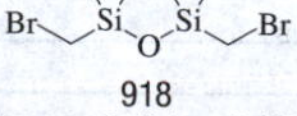
918
1,3-Bis(bromomethyl)tetramethyldisiloxane

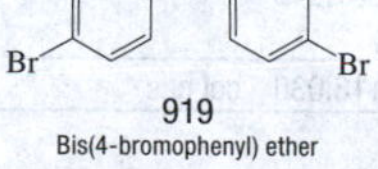
919
Bis(4-bromophenyl) ether

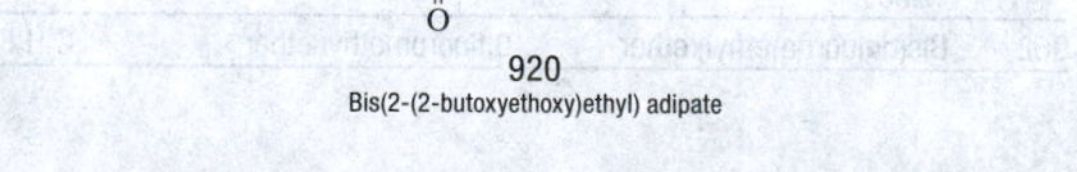
920
Bis(2-(2-butoxyethoxy)ethyl) adipate

No.	Name	Synonym	Mol. form.	CAS Reg. No.	Mol. wt.	Physical form	mp/°C	bp/°C	den/ g cm^{-3}	n_D	Solubility
921	1,4-Bis(α-(*tert*-butyldioxy)-isopropyl)benzene		$C_{20}H_{34}O_4$	2781-00-2	338.482	cry	79				
922	Bis(3-*tert*-butyl-5-ethyl-2-hydroxyphenyl)methane		$C_{25}H_{36}O_2$	88-24-4	368.553	cry	123				
923	Bis(4-chlorobenzoyl) peroxide		$C_{14}H_8Cl_2O_4$	94-17-7	311.118	pr cry (bz)	141				
924	1,2-Bis(2-chloroethoxy)ethane		$C_6H_{12}Cl_2O_2$	112-26-5	187.064			214(18)	1.195^{20}	1.4592^{25}	s ctc
925	Bis(2-chloroethoxy)methane		$C_5H_{10}Cl_2O_2$	111-91-1	173.037			216(3)			
926	*N,N*-Bis(2-chloroethyl)aniline	Aniline mustard	$C_{10}H_{13}Cl_2N$	553-27-5	218.123	pr	45	164^{14}			sl eth; s EtOH, MeOH
927	Bis(2-chloroethyl) carbonate		$C_5H_8Cl_2O_3$	623-97-2	187.021		8	241	1.3506^{20}	1.461^{20}	i H_2O
928	Bis(2-chloroethyl) 2-chloroethylphosphonate		$C_6H_{12}Cl_3O_3P$	6294-34-4	269.490			170.2^5		1.488^{25}	
929	Bis(2-chloroethyl) ether	Dichloroethyl ether	$C_4H_8Cl_2O$	111-44-4	143.012	liq	-46.9(0.5)	178(2)	1.22^{20}	1.451^{20}	i H_2O; s EtOH, eth, ace; msc bz
930	Bis(2-chloroethyl)methylamine hydrochloride	Nitrogen mustard hydrochloride	$C_5H_{12}Cl_3N$	55-86-7	192.515	hyg nd	111.5				
931	*N,N'*-Bis(2-chloroethyl)-*N*-nitrosourea	Carmustine	$C_5H_9Cl_2N_3O_2$	154-93-8	214.049	lt ye pow	31				vs H_2O, EtOH
932	Bis(2-chloroethyl) sulfide	Mustard gas	$C_4H_8Cl_2S$	505-60-2	159.078	liq	13.5	198(9)	1.2741^{20}	1.5313^{20}	
933	1,2-Bis(2-chloroethylsulfonyl)-ethane		$C_6H_{12}Cl_2O_4S_2$	3944-87-4	283.193	cry (MeOH/ HOAc)	205				
934	1,2-Bis(chloromethyl)benzene		$C_8H_8Cl_2$	612-12-4	175.056	mcl (lig)	55(1)	239.5	1.393^{25}		i H_2O; vs EtOH, eth, chl; s ctc
935	1,3-Bis(chloromethyl)benzene		$C_8H_8Cl_2$	626-16-4	175.056	cry	34(1)	251.5	1.302^{20}		i H_2O; vs EtOH, eth; sl chl
936	1,4-Bis(chloromethyl)benzene		$C_8H_8Cl_2$	623-25-6	175.056	pl (al)	100	257(18)	1.417^{25}		i H_2O; vs EtOH, eth, ace, chl; sl HOAc
937	Bis(chloromethyl) ether		$C_2H_4Cl_2O$	542-88-1	114.958	liq	-41.5	104(4)	1.323^{15}	1.435^{21}	msc EtOH, eth
938	3,3-Bis(chloromethyl)oxetane		$C_5H_8Cl_2O$	78-71-7	155.022	liq	19.0(0.1)	101^{27}	1.295^{25}		
939	2,2-Bis(chloromethyl)-1,3-propanediol	Pentaerythritol dichlorohydrin	$C_5H_{10}Cl_2O_2$	2209-86-1	173.037	cry	83	159^{12}			
940	1,3-Bis(chloromethyl)-tetramethyldisiloxane		$C_6H_{16}Cl_2OSi_2$	2362-10-9	231.267	liq	-90	205(3)	1.045^{20}	1.4398^{20}	
941	Bis(4-chlorophenoxy)methane	Di(4-chlorophenoxy)methane	$C_{13}H_{10}Cl_2O_2$	555-89-5	269.123	cry (peth)	70.5	191^6			vs ace, bz
942	Bis(4-chlorophenyl) disulfide		$C_{12}H_8Cl_2S_2$	1142-19-4	287.228		72.8				s chl
943	Bis(4-chlorophenyl)-ethanedione		$C_{14}H_8Cl_2O_2$	3457-46-3	279.119		197.8				
944	1,1-Bis(4-chlorophenyl)-ethanol		$C_{14}H_{12}Cl_2O$	80-06-8	267.150		70				i H_2O, EtOH; s eth, bz
945	1,2-Bis(2-chlorophenyl)-hydrazine	2,2'-Dichlorohydrazobenzene	$C_{12}H_{10}Cl_2N_2$	782-74-1	253.126		87				
946	Bis(4-chlorophenyl)methane		$C_{13}H_{10}Cl_2$	101-76-8	237.124		55.5	188^{18}	1.365^{17}		s EtOH
947	Bis(4-chlorophenyl) sulfone		$C_{12}H_8Cl_2O_2S$	80-07-9	287.162		148.8(1)	250^{10}			sl H_2O; s EtOH, chl
948	*N,N'*-Bis(4-chlorophenyl)-thiourea	Di(*p*-chlorophenyl)thiourea	$C_{13}H_{10}Cl_2N_2S$	1220-00-4	297.202	nd	176				
949	1,1-Bis(4-chlorophenyl)-2,2,2-trichloroethanol		$C_{14}H_9Cl_5O$	115-32-2	370.485	cry (petr)	75.1(0.3)	$180^{0.1}$			i H_2O, os
950	Bis(3-chloropropyl) ether	3-Chloropropyl ether	$C_6H_{12}Cl_2O$	629-36-7	171.064			205(7)	1.136^{20}	1.4158^{20}	s EtOH, eth
951	Bis(2-cyanoethyl) ether		$C_6H_8N_2O$	1656-48-0	124.140			161^5	1.0504^{20}	1.4405^{20}	
952	Bis(2-cyanoethyl) sulfide		$C_6H_8N_2S$	111-97-7	140.206			$163^{1.5}$		1.5047^{20}	
953	Bis(η-cyclopentadienyl)-titanium chloride		$C_{10}H_{10}Cl_2Ti$	1271-19-8	248.959	red cry	289	258^{10}	1.60		sl H_2O, bz; s chl, EtOH, tol
954	Bis(η-cyclopentadienyl)-zirconium chloride		$C_{10}H_{10}Cl_2Zr$	1291-32-3	292.316	col cry	248	$180^{0.5}$			
955	1,2-Bis(dibromomethyl)-benzene		$C_8H_6Br_4$	13209-15-9	421.750	mcl	116.5				sl H_2O; vs chl; i lig
956	Bis(2,4-dichlorobenzoyl) peroxide		$C_{14}H_6Cl_4O_4$	133-14-2	380.008		106				
957	1,3-Bis(dichloromethyl)-tetramethyldisiloxane		$C_6H_{14}Cl_4OSi_2$	2943-70-6	300.157			149^{50}	1.2213^{20}	1.4660^{20}	
958	Bis(2,4-dichlorophenyl)ether	2,2',4,4'-Tetrachlorodiphenyl ether	$C_{12}H_6Cl_4O$	28076-73-5	307.987	cry (eth)	71				
959	4,4'-Bis(diethylamino)-benzophenone	Michler's ethyl ketone	$C_{21}H_{28}N_2O$	90-93-7	324.459	lf (al)	95.3				
960	Bis(diethyldithiocarbamate)-nickel		$C_{10}H_{20}N_2NiS_4$	14267-17-5	355.232			$202^{0.02}$			
961	Bis(diethyldithiocarbamate)-zinc		$C_{10}H_{20}N_2S_4Zn$	14324-55-1	361.948			$178^{0.05}$			
962	Bis(difluoromethyl) ether	Difluoromethyl ether	$C_2H_2F_4O$	1691-17-4	118.030	col gas		5.5(0.4)	1.43^{20}		

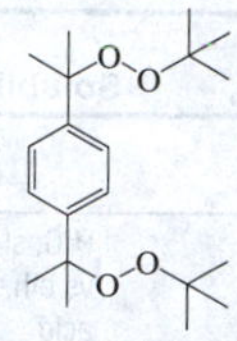
921
1,4-Bis(α-(*tert*-butyldioxy)isopropyl)benzene

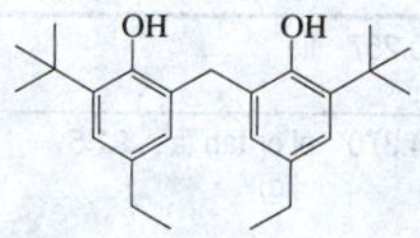
922
Bis(3-*tert*-butyl-5-ethyl-2-hydroxyphenyl)methane

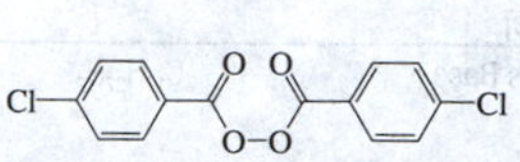
923
Bis(4-chlorobenzoyl) peroxide

924
1,2-Bis(2-chloroethoxy)ethane

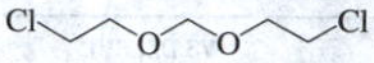
925
Bis(2-chloroethoxy)methane

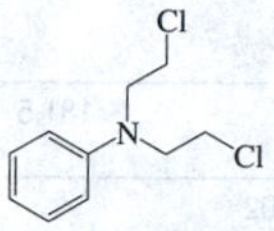
926
N,N-Bis(2-chloroethyl)aniline

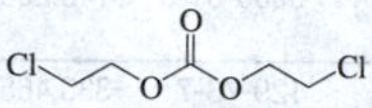
927
Bis(2-chloroethyl) carbonate

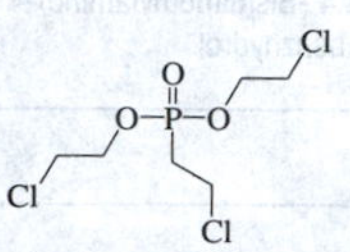
928
Bis(2-chloroethyl) 2-chloroethylphosphonate

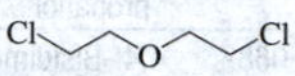
929
Bis(2-chloroethyl) ether

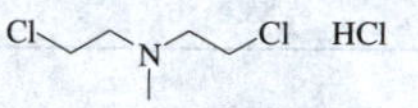
930
Bis(2-chloroethyl)methylamine hydrochloride

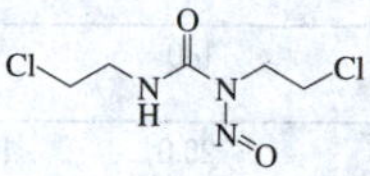
931
N,N'-Bis(2-chloroethyl)-*N*-nitrosourea

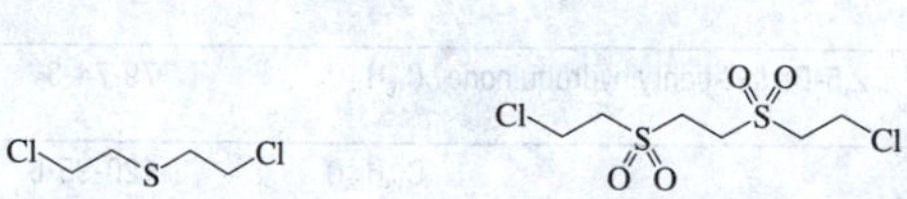
932
Bis(2-chloroethyl) sulfide

933
1,2-Bis(2-chloroethylsulfonyl)ethane

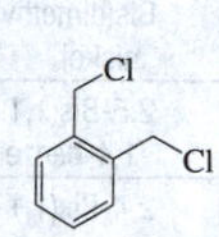
934
1,2-Bis(chloromethyl)benzene

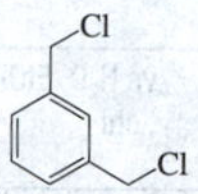
935
1,3-Bis(chloromethyl)benzene

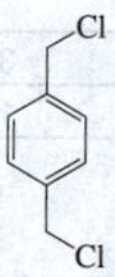
936
1,4-Bis(chloromethyl)benzene

937
Bis(chloromethyl) ether

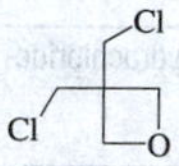
938
3,3-Bis(chloromethyl)oxetane

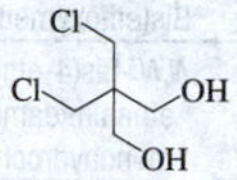
939
2,2-Bis(chloromethyl)-1,3-propanediol

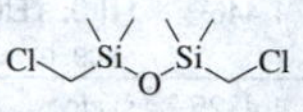
940
1,3-Bis(chloromethyl)tetramethyldisiloxane

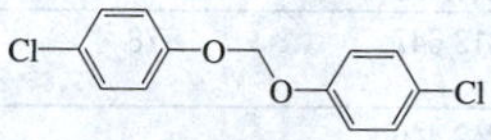
941
Bis(4-chlorophenoxy)methane

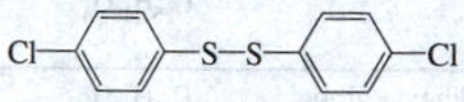
942
Bis(4-chlorophenyl) disulfide

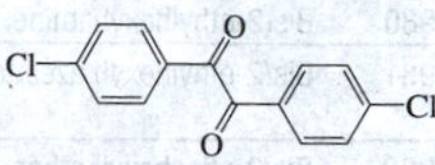
943
Bis(4-chlorophenyl)ethanedione

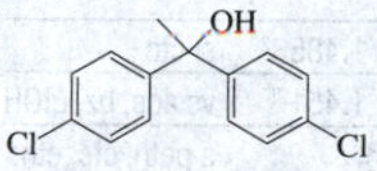
944
1,1-Bis(4-chlorophenyl)ethanol

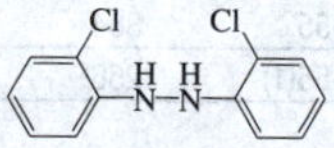
945
1,2-Bis(2-chlorophenyl)-hydrazine

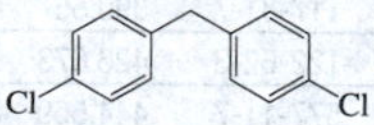
946
Bis(4-chlorophenyl)methane

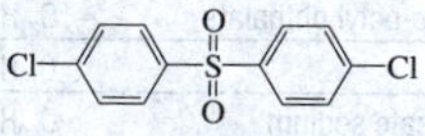
947
Bis(4-chlorophenyl) sulfone

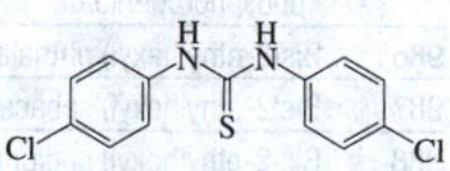
948
N,N'-Bis(4-chlorophenyl)thiourea

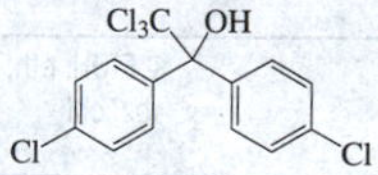
949
1,1-Bis(4-chlorophenyl)-2,2,2-trichloroethanol

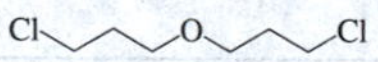
950
Bis(3-chloropropyl) ether

951
Bis(2-cyanoethyl) ether

952
Bis(2-cyanoethyl) sulfide

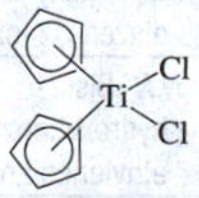
953
Bis(η-cyclopentadienyl)titanium chloride

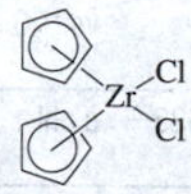
954
Bis(η-cyclopentadienyl)zirconium chloride

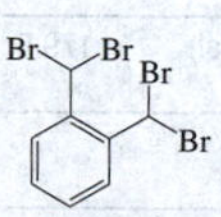
955
1,2-Bis(dibromomethyl)benzene

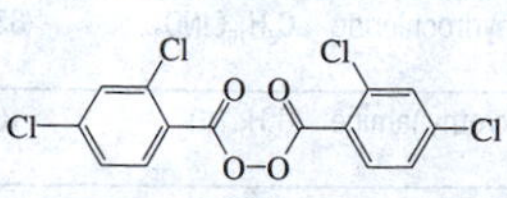
956
Bis(2,4-dichlorobenzoyl) peroxide

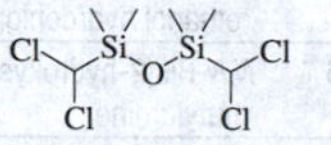
957
1,3-Bis(dichloromethyl)tetramethyldisiloxane

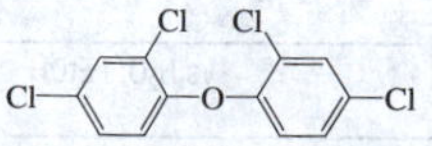
958
Bis(2,4-dichlorophenyl)ether

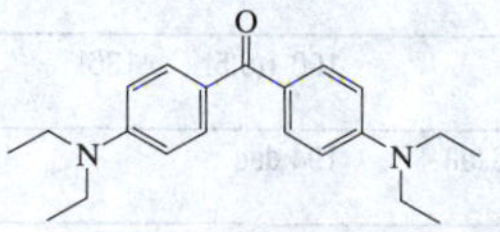
959
4,4'-Bis(diethylamino)benzophenone

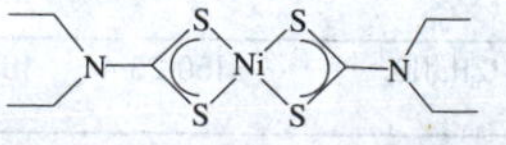
960
Bis(diethyldithiocarbamate)nickel

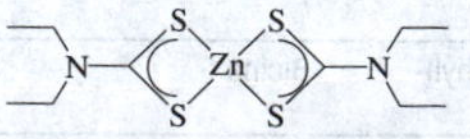
961
Bis(diethyldithiocarbamate)zinc

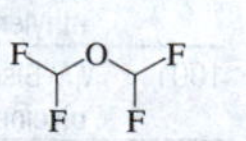
962
Bis(difluoromethyl) ether

No.	Name	Synonym	Mol. form.	CAS Reg. No.	Mol. wt.	Physical form	mp/°C	bp/°C	den/ g cm⁻³	n_D	Solubility
963	Bis(2-dimethylaminoethyl) ether	2,2'-Oxybis[*N,N*-dimethylethanamine]	$C_8H_{20}N_2O$	3033-62-3	160.257	liq		80^{15}			
964	Bis[4-(dimethylamino)phenyl]-methane	Michler's Base	$C_{17}H_{22}N_2$	101-61-1	254.370	pl or tab (al, lig)	91.5	390 dec			i H_2O; sl EtOH; vs eth, bz; s acid
965	Bis[4-(dimethylamino)phenyl]-methanethione	4,4'-Bis(dimethylamino)-thiobenzophenone	$C_{17}H_{20}N_2S$	1226-46-6	284.419	pl	204				i H_2O, EtOH, lig; sl eth; s bz, chl, HOAc
966	Bis(4-dimethylaminophenyl)-methanol	4,4'-Bis(dimethylamino)-benzhydrol	$C_{17}H_{22}N_2O$	119-58-4	270.369		102.0				i H_2O; vs EtOH; s eth, bz, HOAc
967	1,3-Bis(dimethylamino)-2-propanol		$C_7H_{18}N_2O$	5966-51-8	146.230			181.5	0.8788^{20}	1.4418^{20}	vs H_2O
968	4,4'-Bis(dimethylamino)-triphenylmethane		$C_{23}H_{26}N_2$	129-73-7	330.465	nd or lf (al, bz)	102				vs bz, eth
969	Bis(dimethyldithiocarbamate)-copper		$C_6H_{12}CuN_2S_4$	137-29-1	303.978			$206^{0.01}$			
970	Bis(dimethyldithiocarbamate)-nickel		$C_6H_{12}N_2NiS_4$	15521-65-0	299.125			$208^{0.002}$			
971	2,5-Bis(1,1-dimethylpropyl)-1,4-benzenediol	2,5-Di-*tert*-pentylhydroquinone	$C_{16}H_{26}O_2$	79-74-3	250.376		180				
972	2,4-Bis(1,1-dimethylpropyl)-phenol		$C_{16}H_{26}O$	120-95-6	234.376		26.0	169^{22}			
973	1,2-Bis(diphenylphosphino)-ethane	Diphos	$C_{26}H_{24}P_2$	1663-45-2	398.417		143.5				
974	1,3-Bis(2,3-epoxypropoxy)-benzene	Diglycidyl resorcinol ether	$C_{12}H_{14}O_4$	101-90-6	222.237		42.5	$147^{0.4}$	1.2183^{30}	1.5408^{20}	
975	Bis(2-ethoxyethyl) phthalate		$C_{16}H_{22}O_6$	605-54-9	310.342		34	345	1.1229^{21}		
976	Bis(ethoxymethyl) ether		$C_6H_{14}O_3$	5648-29-3	134.173			136(10)			
977	*N,N'*-Bis(4-ethoxyphenyl)-ethanimidamide monohydrochloride	Phenacaine hydrochloride	$C_{18}H_{23}ClN_2O_2$	620-99-5	334.841	cry (w+1)	191				vs H_2O, EtOH, chl
978	Bis(ethylenediamine)copper dichloride	Cupriethylenediamine dichloride	$C_4H_{16}Cl_2CuN_4$	15243-01-3	254.649	dk bl cry					s EtOH
979	Bis(2-ethylhexyl) adipate	Bis(2-ethylhexyl) hexanedioate	$C_{22}H_{42}O_4$	103-23-1	370.566		-67.8	214^5	0.922^{25}	1.4474^{20}	vs ace, eth, EtOH
980	Bis(2-ethylhexyl)amine		$C_{16}H_{35}N$	106-20-7	241.456			161^{21}			
981	Bis(2-ethylhexyl) azelate		$C_{25}H_{48}O_4$	103-24-2	412.647		-78	237^5	0.915^{25}	1.446^{25}	i H_2O; s EtOH, ace, bz; sl ctc
982	Bis(2-ethylhexyl) ether	2,2'-Diethyldihexyl ether	$C_{16}H_{34}O$	10143-60-9	242.440			279(8)		1.4325^{20}	sl ctc
983	Bis(2-ethylhexyl) phosphate		$C_{16}H_{35}O_4P$	298-07-7	322.420	visc liq		$155^{0.015}$	0.975^{25}		sl H_2O; s bz, hx
984	Bis(2-ethylhexyl) phosphonate	Bis(2-ethylhexyl) phosphite	$C_{16}H_{35}O_3P$	3658-48-8	306.421	liq		150^1	0.93^{25}	1.4420^{20}	
985	Bis(2-ethylhexyl) phosphorodithioate		$C_{16}H_{35}O_2PS_2$	5810-88-8	354.552	cry					s bz, hp, chl
986	Bis(2-ethylhexyl) phthalate	Di-*sec*-octyl phthalate	$C_{24}H_{38}O_4$	117-81-7	390.557	liq	-55	384	0.981^{25}	1.4853^{20}	sl ctc
987	Bis(2-ethylhexyl) sebacate		$C_{26}H_{50}O_4$	122-62-3	426.673		-55(1)	256^5	0.912^{25}	1.451^{25}	vs ace, bz, EtOH
988	Bis(2-ethylhexyl) sodium sulfosuccinate	Docusate sodium	$C_{20}H_{37}NaO_7S$	577-11-7	444.559	waxy solid					s peth, ctc, eth, ace
989	Bis(2-ethylhexyl) terephthalate		$C_{24}H_{38}O_4$	6422-86-2	390.557			383			
990	2,2-Bis(ethylsulfonyl)butane	Sulfonethylmethane	$C_8H_{18}O_4S_2$	76-20-0	242.357	pl (w)	76	dec	1.199^{85}		s chl
991	Bis[4-(hexyloxy)phenyl]-diazene, 1-oxide		$C_{24}H_{34}N_2O_3$	2587-42-0	398.538						s chl
992	*N,N'*-Bis(2-hydroxybenzylidene)-1,2-ethylenediamine	Disalicylidene-1,2-ethanediamine	$C_{16}H_{16}N_2O_2$	94-93-9	268.310		125.5				sl EtOH, eth; s bz, chl
993	Bis(2-hydroxy-3-*tert*-butyl-5-methylphenyl)methane		$C_{23}H_{32}O_2$	119-47-1	340.499	nd (peth)	131				
994	Bis(2-hydroxy-5-chlorophenyl) sulfide	Fenticlor	$C_{12}H_8Cl_2O_2S$	97-24-5	287.162		174				i H_2O; s EtOH, eth, gl HOAc
995	2-[Bis(2-hydroxyethyl)amino]-ethanol hydrochloride	Triethanolamine hydrochloride	$C_6H_{16}ClNO_3$	637-39-8	185.649	cry (al)	179.5				vs H_2O
996	*N,N*-Bis(2-hydroxyethyl)-butylamine	Butylbis(2-hydroxyethyl)amine	$C_8H_{19}NO_2$	102-79-4	161.243			279(19)	0.9681^{20}	1.4625^{20}	s chl
997	Bis(2-hydroxyethyl) disulfide		$C_4H_{10}O_2S_2$	1892-29-1	154.251		26	$160^{3.5}$			
998	*N,N*-Bis(2-hydroxyethyl)-dodecanamide		$C_{16}H_{33}NO_3$	120-40-1	287.438	waxy solid	38.7				
999	*N,N*-Bis(2-hydroxyethyl)-ethylamine	*N*-Ethyldiethanolamine	$C_6H_{15}NO_2$	139-87-7	133.189	ye liq	-50	248(19)	1.0135^{20}	1.4663^{20}	vs H_2O, EtOH; sl eth
1000	*N,N'*-Bis(2-hydroxyethyl)-ethylenediamine		$C_6H_{16}N_2O_2$	4439-20-7	148.203		100.1(0.5)	136^1			s H_2O
1001	*N,N*-Bis(2-hydroxyethyl)-glycine	Bicine	$C_6H_{13}NO_4$	150-25-4	163.172	nd (al)	194 dec				vs H_2O; i EtOH

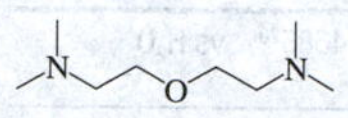
963
Bis(2-dimethylaminoethyl) ether

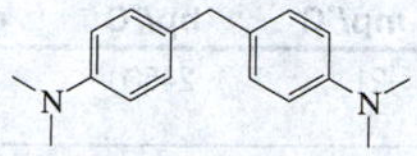
964
Bis[4-(dimethylamino)phenyl]methane

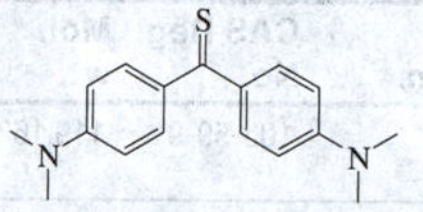
965
Bis[4-(dimethylamino)phenyl]methanethione

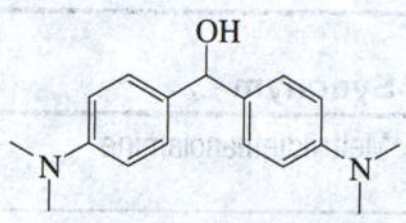
966
Bis(4-dimethylaminophenyl)methanol

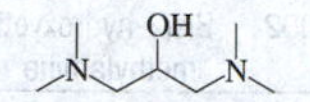
967
1,3-Bis(dimethylamino)-2-propanol

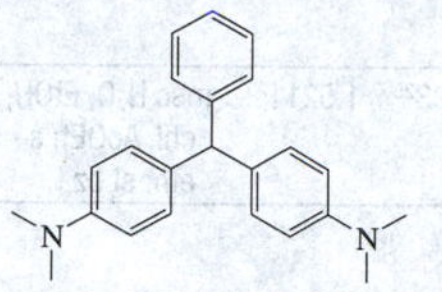
968
4,4'-Bis(dimethylamino)triphenylmethane

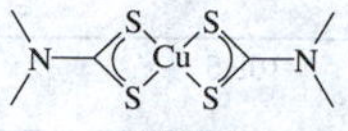
969
Bis(dimethyldithiocarbamate)copper

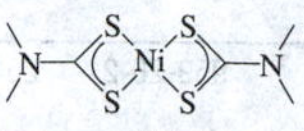
970
Bis(dimethyldithiocarbamate)nickel

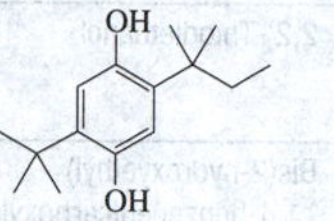
971
2,5-Bis(1,1-dimethylpropyl)-1,4-benzenediol

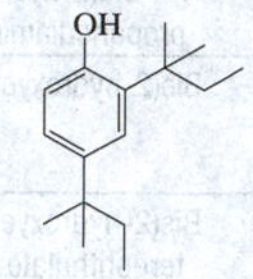
972
2,4-Bis(1,1-dimethylpropyl)phenol

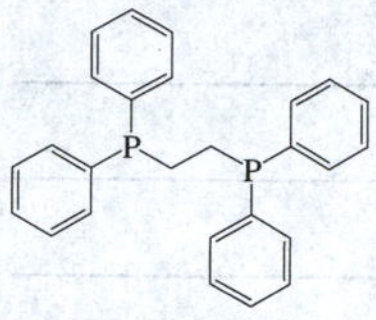
973
1,2-Bis(diphenylphosphino)ethane

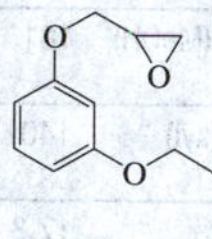
974
1,3-Bis(2,3-epoxypropoxy)benzene

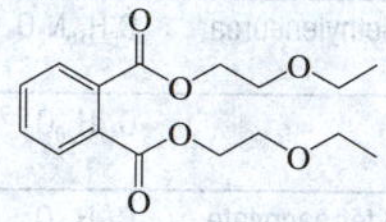
975
Bis(2-ethoxyethyl) phthalate

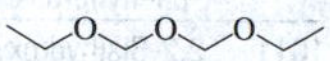
976
Bis(ethoxymethyl) ether

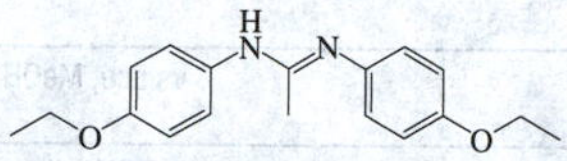
977
N,N'-Bis(4-ethoxyphenyl)ethanimidamide monohydrochloride

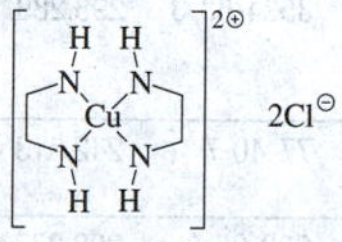
978
Bis(ethylenediamine)copper dichloride

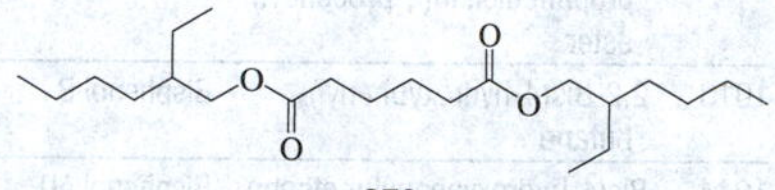
979
Bis(2-ethylhexyl) adipate

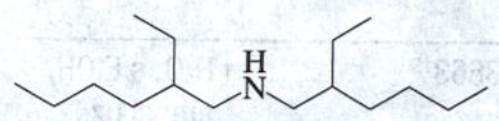
980
Bis(2-ethylhexyl)amine

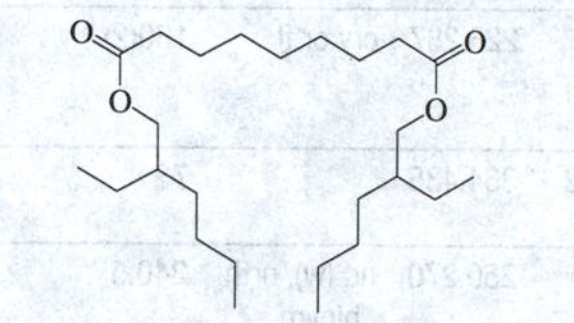
981
Bis(2-ethylhexyl) azelate

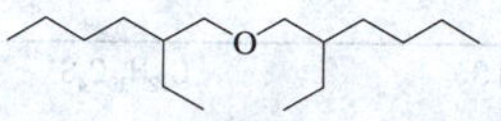
982
Bis(2-ethylhexyl) ether

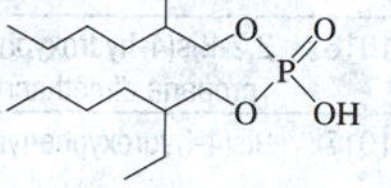
983
Bis(2-ethylhexyl) phosphate

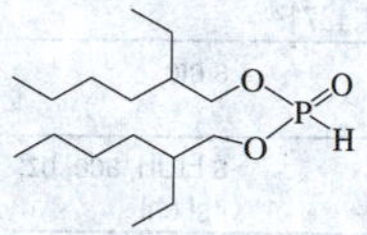
984
Bis(2-ethylhexyl) phosphonate

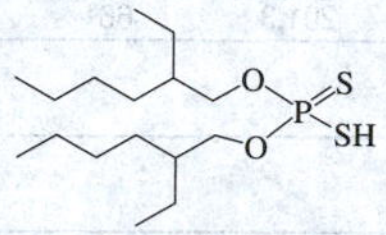
985
Bis(2-ethylhexyl) phosphorodithioate

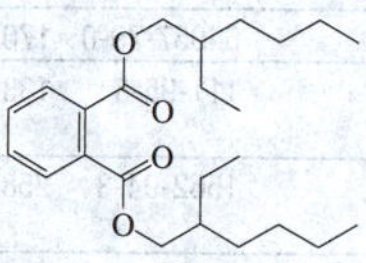
986
Bis(2-ethylhexyl) phthalate

987
Bis(2-ethylhexyl) sebacate

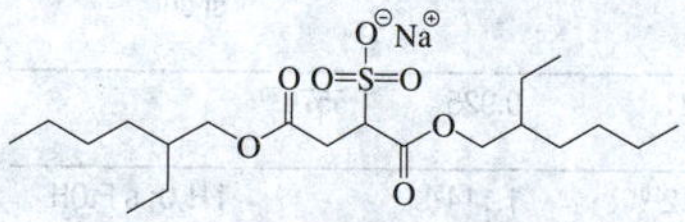
988
Bis(2-ethylhexyl) sodium sulfosuccinate

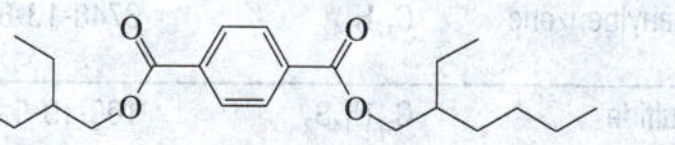
989
Bis(2-ethylhexyl) terephthalate

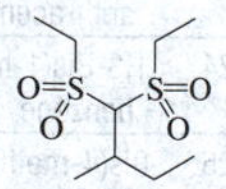
990
2,2-Bis(ethylsulfonyl)butane

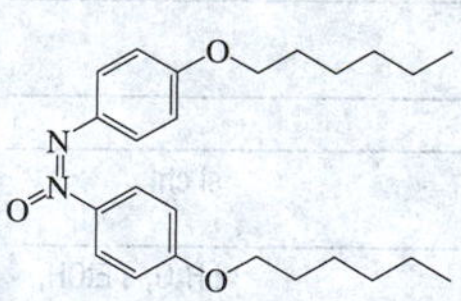
991
Bis[4-(hexyloxy)phenyl]diazene, 1-oxide

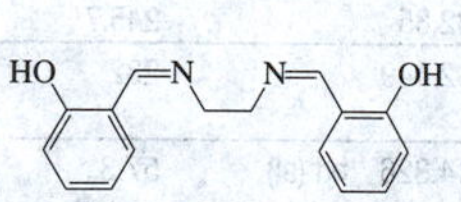
992
N,N'-Bis(2-hydroxybenzylidene)-1,2-ethylenediamine

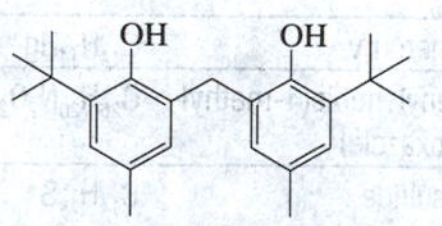
993
Bis(2-hydroxy-3-*tert*-butyl-5-methylphenyl)methane

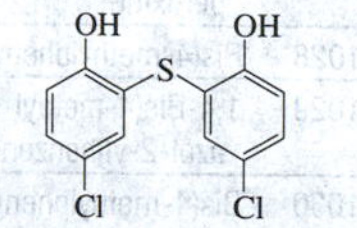
994
Bis(2-hydroxy-5-chlorophenyl) sulfide

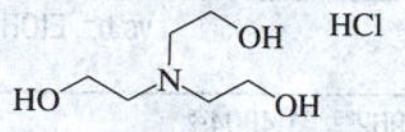
995
2-[Bis(2-hydroxyethyl)amino]ethanol hydrochloride

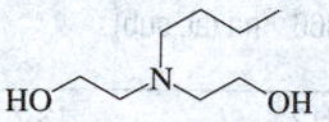
996
N,N-Bis(2-hydroxyethyl)butylamine

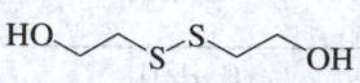
997
Bis(2-hydroxyethyl) disulfide

998
N,N-Bis(2-hydroxyethyl)dodecanamide

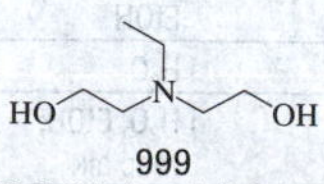
999
N,N-Bis(2-hydroxyethyl)ethylamine

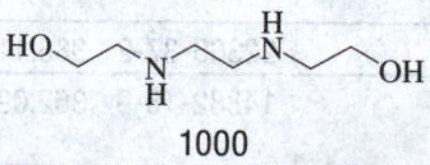
1000
N,N'-Bis(2-hydroxyethyl)ethylenediamine

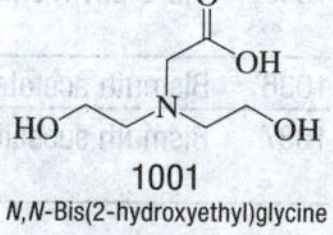
1001
N,N-Bis(2-hydroxyethyl)glycine

No.	Name	Synonym	Mol. form.	CAS Reg. No.	Mol. wt.	Physical form	mp/°C	bp/°C	den/ g cm^{-3}	n_D	Solubility
1002	Bis(2-hydroxyethyl)-methylamine	Methyldiethanolamine	$C_5H_{13}NO_2$	105-59-9	119.163	liq	-21	245(1)	1.043^{25}	1.4685^{20}	vs H_2O
1003	*N,N*-Bis(2-hydroxyethyl)-3-methylaniline	Diethanol-*m*-toluidine	$C_{11}H_{17}NO_2$	91-99-6	195.259		64.5	160^{1}			sl chl
1004	*N,N*-Bis(2-hydroxyethyl)-1,3-propanediamine	3-(Aminopropyl)diethanol-amine	$C_7H_{18}N_2O_2$	4985-85-7	162.230			160^{1}			
1005	Bis(2-hydroxyethyl) sulfide	2,2'-Thiodiethanol	$C_4H_{10}O_2S$	111-48-8	122.186	liq	-10.2	282	1.1793^{25}	1.5211^{20}	msc H_2O, EtOH, chl, AcOEt; s eth; sl bz
1006	Bis(2-hydroxyethyl) terephthalate	Bis(2-hydroxyethyl) 1,4-benzenedicarboxylate	$C_{12}H_{14}O_6$	959-26-2	254.235	cry (w)	109.5				
1007	1,2-Bis(2-hydroxyethylthio)-ethane		$C_6H_{14}O_2S_2$	5244-34-8	182.304		64.8	$170^{0.5}$			s H_2O, EtOH, bz, peth
1008	Bis(2-hydroxy-4-methoxyphe-nyl)methanone	2,2'-Dihydroxy-4,4'-dime-thoxybenzophenone	$C_{15}H_{14}O_5$	131-54-4	274.269		139.1(0.5)				
1009	1,3-Bis(hydroxymethyl)-2-imidazolidone	1,3-Dimethylolethyleneurea	$C_5H_{10}N_2O_3$	136-84-5	146.144	cry (MeOH)	101				
1010	2,2-Bis(4-hydroxy-3-methyl-phenyl)propane	Bisphenol C	$C_{17}H_{20}O_2$	79-97-0	256.340	nd (xyl)	140				
1011	2,2-Bis(hydroxymethyl)-1,3-propanediol, tetra(2-prope-noyl) ester	Pentaerythritol tetraacrylate	$C_{17}H_{20}O_8$	4986-89-4	352.336		17.3		1.185^{25}		
1012	2,2-Bis(hydroxymethyl)-1,3-propanediol, tri(2-propenoyl) ester	Pentaerythritol triacrylate	$C_{14}H_{18}O_7$	3524-68-3	298.289				1.180^{20}		
1013	2,2-Bis(4-hydroxyphenyl)-butane	Bisphenol B	$C_{16}H_{18}O_2$	77-40-7	242.313		120.5				vs ace, MeOH
1014	Bis(4-hydroxyphenyl)methane	Bisphenol AD	$C_{13}H_{12}O_2$	620-92-8	200.233		162.5	sub			s EtOH, eth, chl, alk; sl DMSO; i CS_2
1015	2,2-Bis(4-hydroxyphenyl)-propane	Bisphenol A	$C_{15}H_{16}O_2$	80-05-7	228.287	cry or fl	160(2)	220^{4}			i H_2O; vs EtOH, eth, bz, alk; s HOAc
1016	2,2-Bis(4-hydroxyphenyl)-propane dimethacrylate	Bisphenol A dimethacrylate	$C_{23}H_{24}O_4$	3253-39-2	364.435		73				
1017	Bis(4-hydroxyphenyl) sulfone	Bisphenol S	$C_{12}H_{10}O_4S$	80-09-1	250.270	nd (w), orth bipym	240.5		1.3663^{15}		i H_2O; s EtOH, eth; sl bz, DMSO
1018	Bis(2-mercaptoethyl) sulfide	2,2'-Dimercaptodiethyl sulfide	$C_4H_{10}S_3$	3570-55-6	154.317		-11	135^{18}	1.183^{25}	1.5982^{20}	
1019	Bis(2-methallyl) carbonate		$C_9H_{14}O_3$	64057-79-0	170.205		201.3	66^{3}	0.943^{25}	1.4371^{20}	
1020	Bis(2-methoxyethyl)amine	2-Methoxy-*N*-(2-methoxyethyl)ethanamine	$C_6H_{15}NO_2$	111-95-5	133.189						s ctc
1021	Bis(4-methoxyphenyl)diazene, 1-oxide		$C_{14}H_{14}N_2O_3$	1562-94-3	258.272	ye nd (al)			1.1711^{11}		s EtOH, ace, bz; sl chl
1022	Bis(4-methoxyphenyl)-ethanedione		$C_{16}H_{14}O_4$	1226-42-2	270.280		133				sl EtOH, chl
1023	1,4-Bis(methylamino)-9,10-anthracenedione		$C_{16}H_{14}N_2O_2$	2475-44-7	266.294						sl chl
1024	1,3-Bis(1-methylethenyl)-benzene	1,3-Diisopropenylbenzene	$C_{12}H_{14}$	3748-13-8	158.239	liq		231	0.925	1.5570^{20}	
1025	Bis(4-methylphenyl) disulfide	Di-*p*-Tolyl disulfide	$C_{14}H_{14}S_2$	103-19-5	246.391	nd or lf (al)	47.5	212^{20}	1.114^{51}		i H_2O; s EtOH, ace; vs eth
1026	Bis(4-methylphenyl) ether	*p*-Tolyl ether	$C_{14}H_{14}O$	1579-40-4	198.260		51	285			vs bz, eth, EtOH
1027	Bis(1-methyl-1-phenylethyl) peroxide	Dicumyl peroxide	$C_{18}H_{22}O_2$	80-43-3	270.367	cry (EtOH)	40	$100^{0.2}$			
1028	Bis(4-methylphenyl)mercury	Di-*p*-tolylmercury	$C_{14}H_{14}Hg$	537-64-4	382.85		245.7				
1029	1,4-Bis(4-methyl-5-phenylox-azol-2-yl)benzene	2,2'-*p*-Phenylenebis(4-methyl-5-phenyloxazole)	$C_{26}H_{20}N_2O_2$	3073-87-8	392.449		232				sl chl
1030	Bis(4-methylphenyl) sulfide	Di-*p*-tolyl sulfide	$C_{14}H_{14}S$	620-94-0	214.326	nd (al)	57.3	>300			i H_2O; s EtOH, ace, bz, HOAc; sl chl
1031	Bis(4-methylphenyl) sulfone	Di-*p*-tolyl sulfone	$C_{14}H_{14}O_2S$	599-66-6	246.325	pr(bz), nd(w,al)	150(8)	406			sl H_2O, eth; s EtOH, bz, chl, CS_2
1032	*N,N'*-Bis(2-methylphenyl)-thiourea		$C_{15}H_{16}N_2S$	137-97-3	256.366	nd (al, sub)					vs bz, EtOH, chl
1033	1,3-Bis(1-methyl-4-piperidyl)-propane		$C_{15}H_{30}N_2$	64168-11-2	238.412		13.7	215^{50}	0.8962^{25}	1.4804^{25}	
1034	Bis(methylthio)methane		$C_3H_8S_2$	1618-26-4	108.226			148			
1035	1,2-Bis(*N*-morpholino)ethane		$C_{10}H_{20}N_2O_2$	1723-94-0	200.278	wh-ye (eth,lig)	75	285			vs H_2O, ace, bz, EtOH
1036	Bismuth acetate		$C_6H_9BiO_6$	22306-37-2	386.111	col tablets	250				i H_2O
1037	Bismuth subsalicylate		$C_7H_5BiO_4$	14882-18-9	362.093	pr					i H_2O, EtOH; reac alk

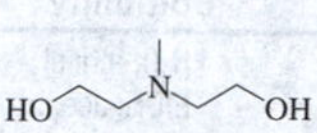
1002
Bis(2-hydroxyethyl)methylamine

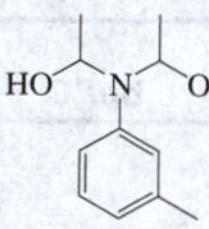
1003
N,N-Bis(2-hydroxyethyl)-3-methylaniline

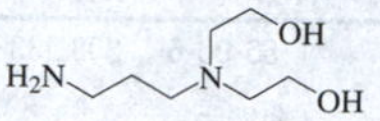
1004
N,N-Bis(2-hydroxyethyl)-1,3-propanediamine

1005
Bis(2-hydroxyethyl) sulfide

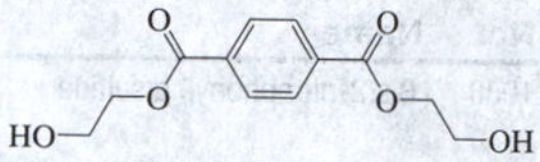
1006
Bis(2-hydroxyethyl) terephthalate

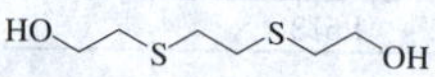
1007
1,2-Bis(2-hydroxyethylthio)ethane

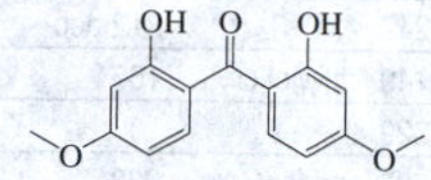
1008
Bis(2-hydroxy-4-methoxyphenyl)methanone

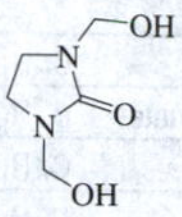
1009
1,3-Bis(hydroxymethyl)-2-imidazolidone

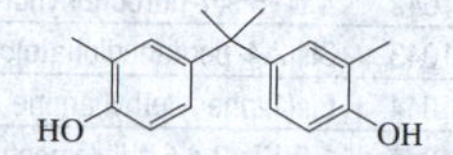
1010
2,2-Bis(4-hydroxy-3-methylphenyl)propane

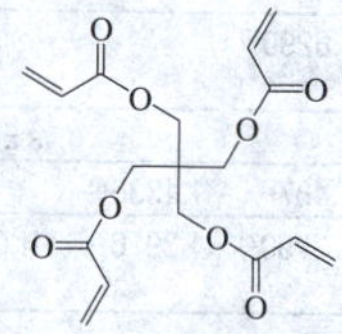
1011
2,2-Bis(hydroxymethyl)-1,3-propanediol, tetra(2-propenoyl) ester

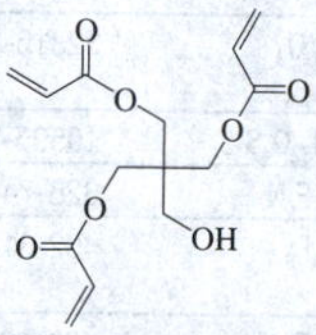
1012
2,2-Bis(hydroxymethyl)-1,3-propanediol, tri(2-propenoyl) ester

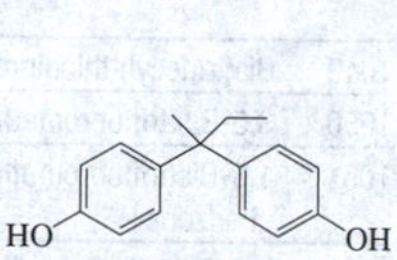
1013
2,2-Bis(4-hydroxyphenyl)butane

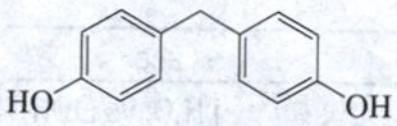
1014
Bis(4-hydroxyphenyl)methane

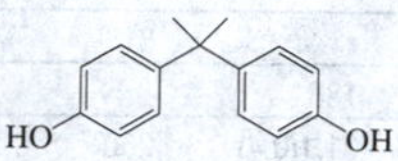
1015
2,2-Bis(4-hydroxyphenyl)propane

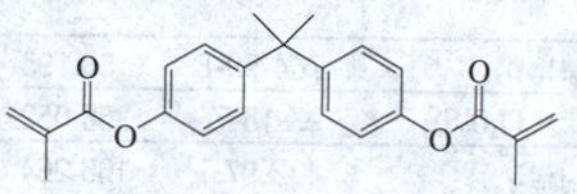
1016
2,2-Bis(4-hydroxyphenyl)propane dimethacrylate

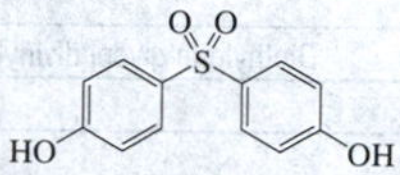
1017
Bis(4-hydroxyphenyl) sulfone

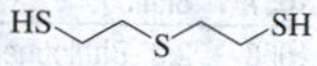
1018
Bis(2-mercaptoethyl) sulfide

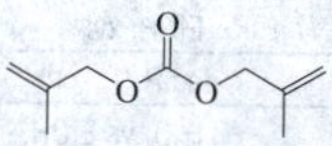
1019
Bis(2-methallyl) carbonate

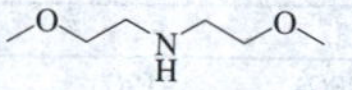
1020
Bis(2-methoxyethyl)amine

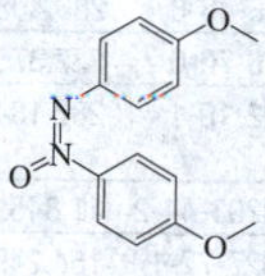
1021
Bis(4-methoxyphenyl)diazene, 1-oxide

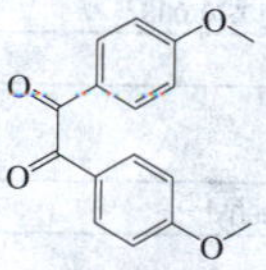
1022
Bis(4-methoxyphenyl)ethanedione

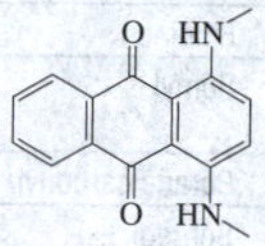
1023
1,4-Bis(methylamino)-9,10-anthracenedione

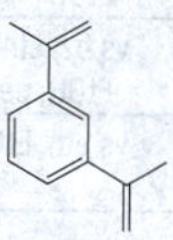
1024
1,3-Bis(1-methylethenyl)benzene

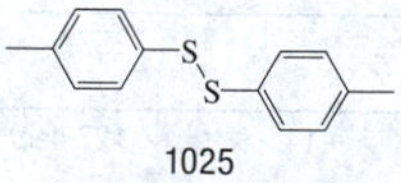
1025
Bis(4-methylphenyl) disulfide

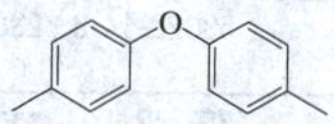
1026
Bis(4-methylphenyl) ether

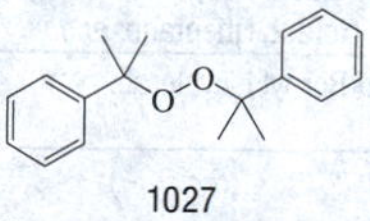
1027
Bis(1-methyl-1-phenylethyl)peroxide

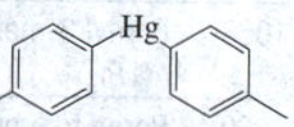
1028
Bis(4-methylphenyl)mercury

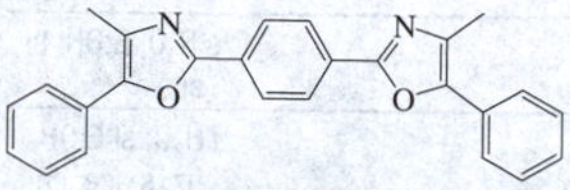
1029
1,4-Bis(4-methyl-5-phenyloxazol-2-yl)benzene

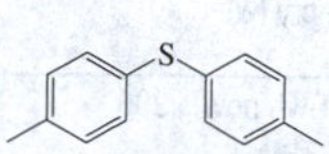
1030
Bis(4-methylphenyl) sulfide

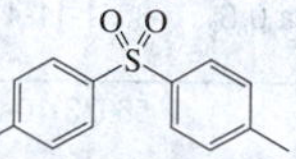
1031
Bis(4-methylphenyl) sulfone

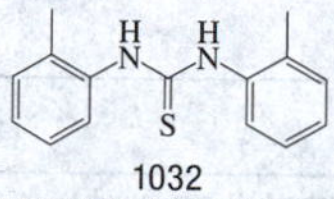
1032
N,N'-Bis(2-methylphenyl)thiourea

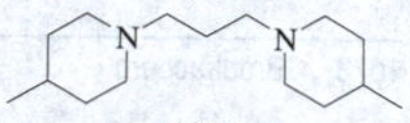
1033
1,3-Bis(1-methyl-4-piperidyl)propane

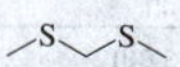
1034
Bis(methylthio)methane

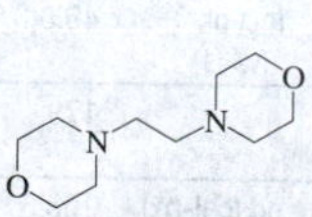
1035
1,2-Bis(*N*-morpholino)ethane

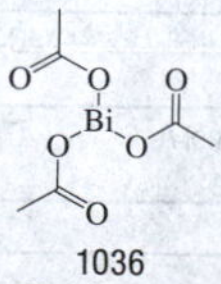
1036
Bismuth acetate

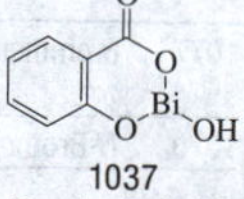
1037
Bismuth subsalicylate

No.	Name	Synonym	Mol. form.	CAS Reg. No.	Mol. wt.	Physical form	mp/°C	bp/°C	den/g cm⁻³	n_D	Solubility
1038	Bis(2-nitrophenyl) disulfide		$C_{12}H_8N_2O_4S_2$	1155-00-6	308.333		198.5				i H_2O, eth; sl EtOH, ace, bz, HOAc
1039	Bis(3-nitrophenyl) disulfide	Nitrophenide	$C_{12}H_8N_2O_4S_2$	537-91-7	308.333		84				sl EtOH, chl; s eth
1040	Bis(4-nitrophenyl) disulfide		$C_{12}H_8N_2O_4S_2$	100-32-3	308.333		182	$255^{0.1}$			sl EtOH, HOAc
1041	1,2-Bis(4-nitrophenyl)ethane	4,4′-Dinitrobibenzyl	$C_{14}H_{12}N_2O_4$	736-30-1	272.256	ye nd (al,bz)	181.8				i EtOH; sl eth, bz, chl, HOAc
1042	*N,N′*-Bis(4-nitrophenyl)urea	4,4′-Dinitrocarbanilide	$C_{13}H_{10}N_4O_5$	587-90-6	302.242		312 dec				
1043	Bis(2,4-pentanedionato)cobalt	Cobalt(II) bis(acetylacetonate)	$C_{10}H_{14}CoO_4$	14024-48-7	257.149	bl-viol cry	167				
1044	Bis(1-phenylethyl)amine		$C_{16}H_{19}N$	10024-74-5	225.329			296.5	1.018^{15}	1.573	
1045	1,2-Bis(2,4,6-tribromophenoxy)ethane		$C_{14}H_8Br_6O_2$	37853-59-1	687.637	nd (bz/EtOH)	222				
1046	*N,N′*-Bis(2,2,2-trichloro-1-hydroxyethyl)urea		$C_5H_6Cl_6N_2O_3$	116-52-9	354.831		196				vs ace, EtOH
1047	1,4-Bis(trichloromethyl)benzene		$C_8H_4Cl_6$	68-36-0	312.836	cry (bz, eth)	109				s chl
1048	Bis(trichloromethyl) carbonate	Triphosgene	$C_3Cl_6O_3$	32315-10-9	296.748	cry (eth, peth)	79	203	1.6290^{80}		
1049	Bis(tridecyl) thiodipropanoate	Ditridecyl thiodipropionate	$C_{32}H_{62}O_4S$	10595-72-9	542.897			$265^{0.25}$			vs EtOH
1050	3,5-Bis(trifluoromethyl)aniline		$C_8H_5F_6N$	328-74-5	229.123			85^{15}	1.487^{25}	1.4335^{20}	
1051	1,3-Bis(trifluoromethyl)benzene		$C_8H_4F_6$	402-31-3	214.108			116	1.3790^{25}	1.3916^{25}	i H_2O
1052	1,4-Bis(trifluoromethyl)benzene		$C_8H_4F_6$	433-19-2	214.108	liq		115			
1053	Bis(trifluoromethyl) disulfide		$C_2F_6S_2$	372-64-5	202.141			34.6			vs EtOH, peth
1054	1,2-Bis(trimethylsilyl)acetylene		$C_8H_{18}Si_2$	14630-40-1	170.400		26	134	0.770^{20}	1.413^{20}	
1055	Bis(2,4,6-trinitrophenyl) sulfide	Dipicryl sulfide	$C_{12}H_4N_6O_{12}S$	2217-06-3	456.258	ye cry	230				
1056	Bis[2-(vinyloxy)ethyl] ether	Diethylene glycol divinyl ether	$C_8H_{14}O_3$	764-99-8	158.195			81^{10}			
1057	Bithionol		$C_{12}H_6Cl_4O_2S$	97-18-7	356.052		188		1.73^{25}		vs ace
1058	2,2′-Bithiophene		$C_8H_6S_2$	492-97-7	166.264		31.1(0.4)	260			i H_2O; vs EtOH; s eth, ctc, HOAc
1059	Bixin		$C_{25}H_{30}O_4$	6983-79-5	394.504	viol pr (ace)	198				i H_2O; s EtOH, ace; sl eth, bz, HOAc
1060	Boldenone	Dehydrotestosterone	$C_{19}H_{26}O_2$	846-48-0	286.408		165				
1061	Boldine		$C_{19}H_{21}NO_4$	476-70-0	327.375	cry (eth)	163				vs EtOH, chl
1062	Bomyl		$C_9H_{15}O_8P$	122-10-1	282.184	ye oil		160^{17}			sl H_2O; vs ace, EtOH, xyl
1063	Borane carbonyl	Borine carbonyl	CH_3BO	13205-44-2	41.845	col gas	-137	-64			dec H_2O
1064	Borneol, (±)-		$C_{10}H_{18}O$	6627-72-1	154.249	lf (lig)	206(7)	213(7)	1.011^{20}		i H_2O; vs EtOH, eth, bz
1065	*l*-Bornyl acetate		$C_{12}H_{20}O_2$	5655-61-8	196.286		27	223.5	0.982^{25}	1.4626^{20}	sl H_2O; s EtOH, eth
1066	Bornylamine		$C_{10}H_{19}N$	32511-34-5	153.265		163				vs ace, bz, eth, EtOH
1067	Bornyl chloride	2-Chloro-1,7,7-trimethylbicyclo[2.2.1]heptane, *endo*	$C_{10}H_{17}Cl$	464-41-5	172.695	nd	128(6)	207.5			vs bz, eth, EtOH, peth
1068	Bornyl 3-methylbutanoate, (1*R*)-	*d*-Bornyl isovalerate	$C_{15}H_{26}O_2$	53022-14-3	238.366			257.5	0.955^{25}		vs eth, EtOH
1069	Boron trifluoride - dimethyl ether complex		$C_2H_6BF_3O$	353-42-4	113.874		-14	127 dec	1.2410^{20}	1.302^{20}	
1070	Boron trifluoride etherate	Trifluoroboron etherate	$C_4H_{10}BF_3O$	109-63-7	141.927	liq	-60.4	125.5	1.125^{25}	1.348^{20}	dec H_2O; vs eth, EtOH
1071	Brilliant Green		$C_{27}H_{34}N_2O_4S$	633-03-4	482.635	small gold cry					vs H_2O, EtOH
1072	Brilliant Yellow		$C_{26}H_{20}N_4Na_2O_8S_2$	3051-11-4	626.569	ye cry (w)					s H_2O, EtOH; sl ace
1073	Brodifacoum		$C_{31}H_{23}BrO_3$	56073-10-0	523.417	off-wh pow	230				i H_2O; sl EtOH, bz; s ace, chl
1074	Bromacil	5-Bromo-3-*sec*-butyl-6-methyluracil	$C_9H_{13}BrN_2O_2$	314-40-9	261.115		157.8(0.5)		1.55^{25}		
1075	Bromadiolone		$C_{30}H_{23}BrO_4$	28772-56-7	527.406	ye-wh pow	205				vs DMF; sl ace, chl, EtOH, eth; i hx
1076	Bromal hydrate		$C_2H_3Br_3O_2$	507-42-6	298.756	mcl pr (w+1)	46.0(0.7)	dec	2.5661^{40}		vs eth, EtOH
1077	Bromdian	Tetrabromobisphenol A	$C_{15}H_{12}Br_4O_2$	79-94-7	543.871		179				s EtOH, eth, bz, chl
1078	*N*-Bromoacetamide		C_2H_4BrNO	79-15-2	137.963	nd (chl-hx)	103.5				vs eth

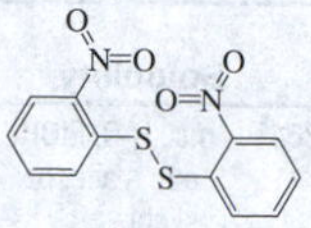
1038
Bis(2-nitrophenyl) disulfide

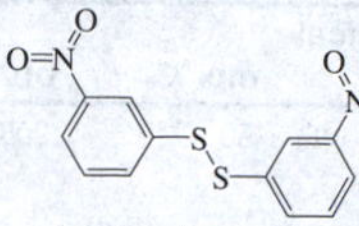
1039
Bis(3-nitrophenyl) disulfide

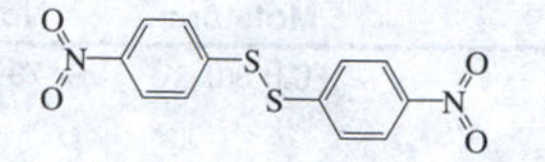
1040
Bis(4-nitrophenyl) disulfide

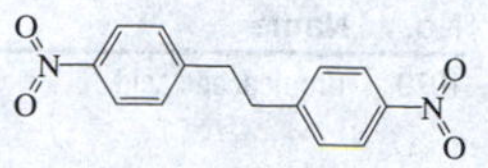
1041
1,2-Bis(4-nitrophenyl)ethane

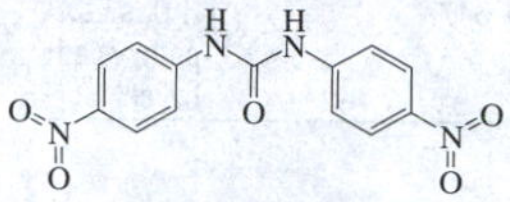
1042
N,N'-Bis(4-nitrophenyl)urea

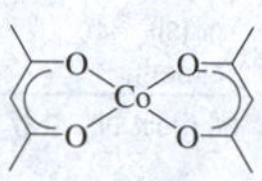
1043
Bis(2,4-pentanedionato)cobalt

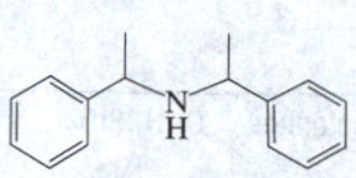
1044
Bis(1-phenylethyl)amine

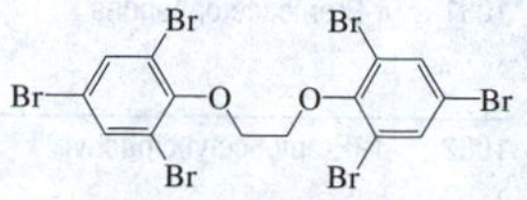
1045
1,2-Bis(2,4,6-tribromophenoxy)ethane

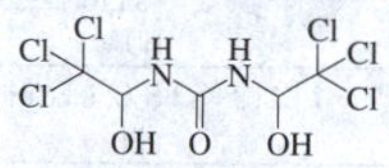
1046
N,N'-Bis(2,2,2-trichloro-1-hydroxyethyl)urea

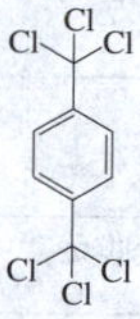
1047
1,4-Bis(trichloromethyl)benzene

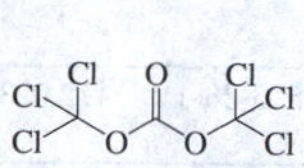
1048
Bis(trichloromethyl) carbonate

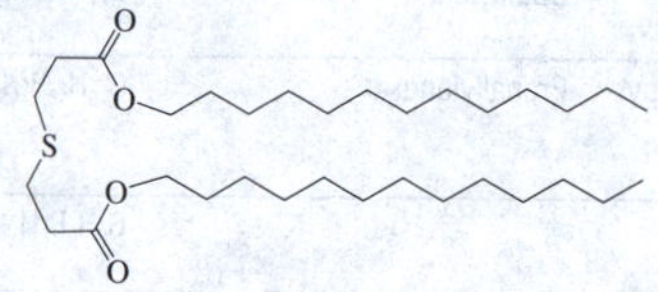
1049
Bis(tridecyl) thiodipropanoate

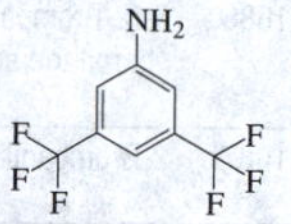
1050
3,5-Bis(trifluoromethyl)aniline

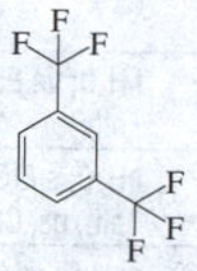
1051
1,3-Bis(trifluoromethyl)benzene

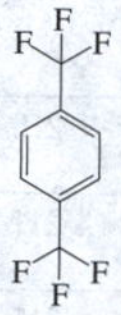
1052
1,4-Bis(trifluoromethyl)benzene

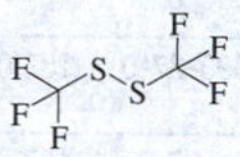
1053
Bis(trifluoromethyl) disulfide

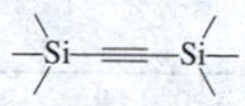
1054
1,2-Bis(trimethylsilyl)acetylene

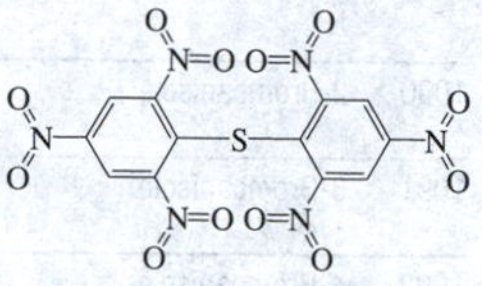
1055
Bis(2,4,6-trinitrophenyl) sulfide

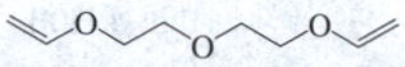
1056
Bis[2-(vinyloxy)ethyl] ether

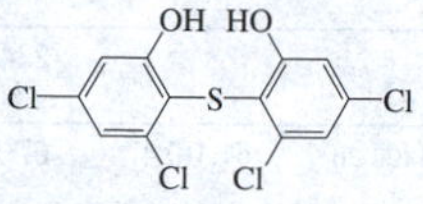
1057
Bithionol

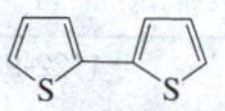
1058
2,2'-Bithiophene

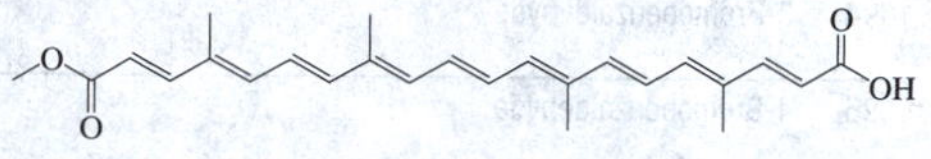
1059
Bixin

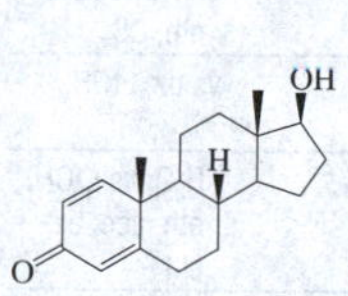
1060
Boldenone

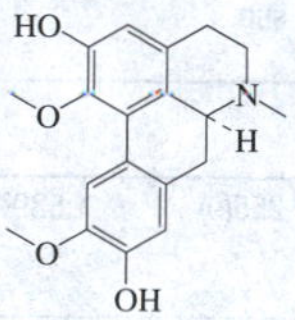
1061
Boldine

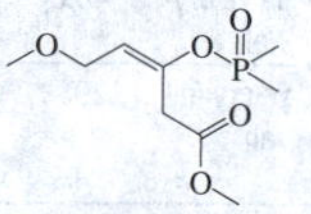
1062
Bomyl

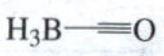
1063
Borane carbonyl

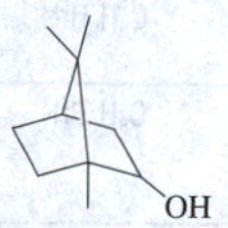
1064
Borneol, (±)-

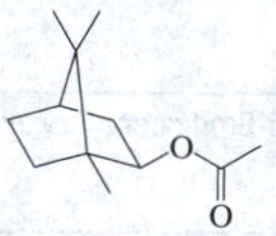
1065
l-Bornyl acetate

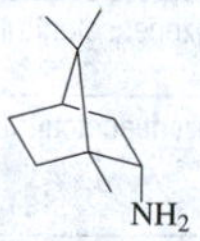
1066
Bornylamine

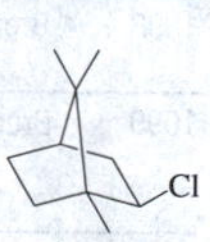
1067
Bornyl chloride

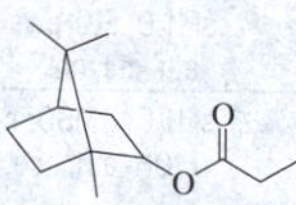
1068
Bornyl 3-methylbutanoate, (1*R*)-

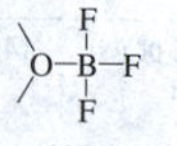
1069
Boron trifluoride - dimethyl ether complex

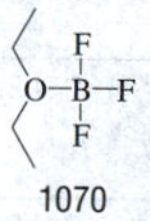
1070
Boron trifluoride etherate

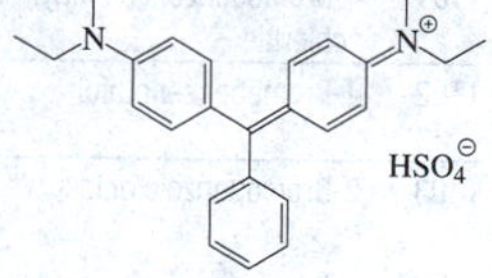
1071
Brilliant Green

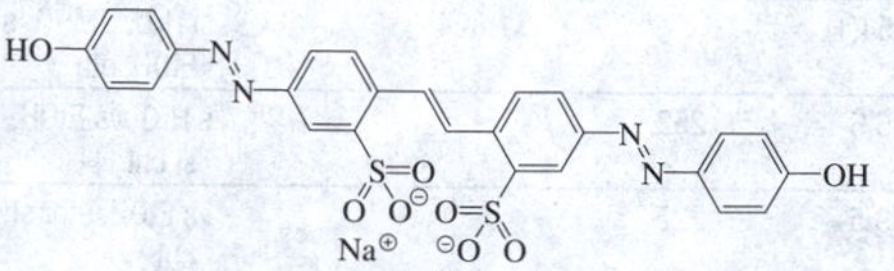
1072
Brilliant Yellow

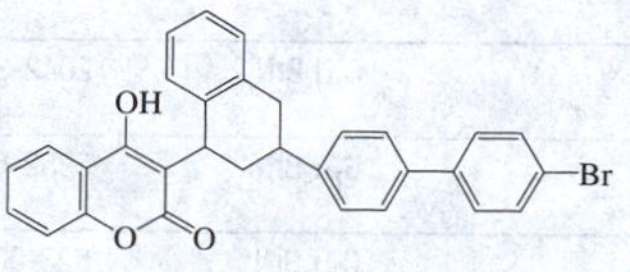
1073
Brodifacoum

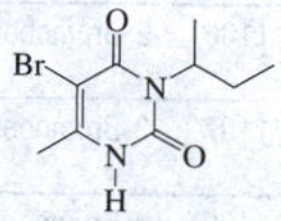
1074
Bromacil

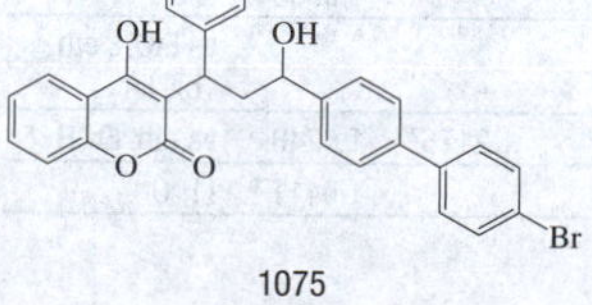
1075
Bromadiolone

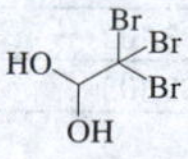
1076
Bromal hydrate

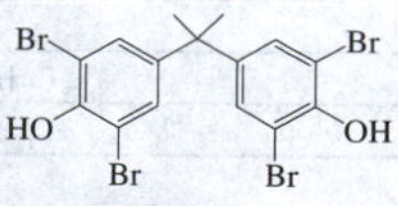
1077
Bromdian

1078
N-Bromoacetamide

No.	Name	Synonym	Mol. form.	CAS Reg. No.	Mol. wt.	Physical form	mp/°C	bp/°C	den/ g cm^{-3}	n_D	Solubility
1079	Bromoacetic acid		$C_2H_3BrO_2$	79-08-3	138.948	hex or orth cry	50	208	1.9335^{50}	1.4804^{50}	msc H_2O, EtOH, eth; s ace, bz; sl chl
1080	Bromoacetone		C_3H_5BrO	598-31-2	136.975	liq	-36.5	138	1.634^{23}	1.4697^{15}	sl H_2O; s EtOH, eth, ace
1081	α-Bromoacetophenone	ω-Bromoacetophenone	C_8H_7BrO	70-11-1	199.045	nd (al) orth pr (al) pl(peth)	51(1)	135^{18}	1.647^{20}		i H_2O; s EtOH, peth; vs eth, bz, chl
1082	4-(Bromoacetyl)biphenyl	2-Bromo-4'-phenylacetophenone	$C_{14}H_{11}BrO$	135-73-9	275.140	nd (95% al)	127				
1083	Bromoacetyl bromide		$C_2H_2Br_2O$	598-21-0	201.844			148.5	2.312^{22}	1.5449^{20}	s ace, ctc
1084	Bromoacetylene		C_2HBr	593-61-3	104.933	col gas		4.7			vs eth
1085	5-(2-Bromoallyl)-5-*sec*-butylbarbituric acid	Butallylonal	$C_{11}H_{15}BrN_2O_3$	1142-70-7	303.152		131.5				vs eth, EtOH
1086	5-(2-Bromoallyl)-5-isopropylbarbituric acid	Propallylonal	$C_{10}H_{13}BrN_2O_3$	545-93-7	289.125	cry (dil HOAc, dil al)	181				sl H_2O, eth, bz; vs EtOH, ace, HOAc
1087	2-Bromoaniline		C_6H_6BrN	615-36-1	172.023		30.9(0.5)	229	1.578^{20}	1.6113^{20}	i H_2O; s EtOH, eth
1088	3-Bromoaniline		C_6H_6BrN	591-19-5	172.023		18.5	250(3)	1.5793^{20}	1.6260^{20}	sl H_2O; s EtOH, eth
1089	4-Bromoaniline		C_6H_6BrN	106-40-1	172.023	orth bipym nd (60% al)	78.2(0.5)	220(6)	1.4970^{100}		i H_2O; s EtOH, eth; sl chl
1090	2-Bromoanisole		C_7H_7BrO	578-57-4	187.034		1.3	216	1.5018^{20}	1.5727^{20}	i H_2O; vs EtOH, eth
1091	3-Bromoanisole		C_7H_7BrO	2398-37-0	187.034			211		1.5635^{20}	i H_2O; s EtOH, eth, bz, CS_2
1092	4-Bromoanisole		C_7H_7BrO	104-92-7	187.034		13.5	215	1.4564^{20}	1.5642^{20}	sl H_2O; vs EtOH, eth, chl; s ctc
1093	2-Bromobenzaldehyde		C_7H_5BrO	6630-33-7	185.018		21.5	230		1.5925^{20}	i H_2O; vs EtOH, bz; sl ctc
1094	3-Bromobenzaldehyde		C_7H_5BrO	3132-99-8	185.018			234		1.5935^{20}	i H_2O; vs EtOH, eth; sl ctc
1095	4-Bromobenzaldehyde		C_7H_5BrO	1122-91-4	185.018	lf (dil al)	61.1(0.4)	67^{2}			i H_2O; vs EtOH, bz; sl chl
1096	Bromobenzene	Phenyl bromide	C_6H_5Br	108-86-1	157.008	liq	-30.74(0.03)	155.9(0.2)	1.4950^{20}	1.5597^{20}	i H_2O; vs EtOH, eth, bz; s ctc
1097	4-Bromobenzeneacetic acid		$C_8H_7BrO_2$	1878-68-8	215.045	nd (w)	116	sub			sl H_2O; vs EtOH, eth, CS_2
1098	4-Bromobenzeneacetonitrile		C_8H_6BrN	16532-79-9	196.045	pa ye cry (al)	48.0				vs bz, EtOH
1099	α-Bromobenzeneacetonitrile	α-Bromobenzyl cyanide	C_8H_6BrN	5798-79-8	196.045	ye cry (dil al)	29	255(9)	1.539^{29}		i H_2O; vs EtOH, eth, ace, bz, chl
1100	2-Bromo-1,4-benzenediol		$C_6H_5BrO_2$	583-69-7	189.007	lf (lig), cry (chl)	111.5	sub			vs H_2O, EtOH, eth, bz; sl chl, lig; s HOAc
1101	4-Bromobenzenesulfonyl chloride	*p*-Brosyl chloride	$C_6H_4BrClO_2S$	98-58-8	255.517	tcl or mcl pl (eth)	76	153^{15}			i H_2O; vs eth; s chl
1102	4-Bromobenzenethiol		C_6H_5BrS	106-53-6	189.073	lf (al)	73	230.5	1.5260^{83}		sl H_2O, EtOH; vs eth, ctc, chl
1103	2-Bromobenzoic acid		$C_7H_5BrO_2$	88-65-3	201.018	mcl pr (w), nd	149.0(0.9)	295(18)	1.929^{25}		sl H_2O, DMSO; s EtOH, eth, ace, chl
1104	3-Bromobenzoic acid		$C_7H_5BrO_2$	585-76-2	201.018	mcl nd (dil al)	156.7(0.5)	285(15)	1.845^{20}		i H_2O; s EtOH, eth
1105	4-Bromobenzoic acid		$C_7H_5BrO_2$	586-76-5	201.018	nd (eth), lf (w), mcl pr	254(1)		1.894^{20}		sl H_2O, DMSO; s EtOH, eth
1106	2-Bromobenzonitrile		C_7H_4BrN	2042-37-7	182.018	nd (w)	55.5	252			s H_2O; vs EtOH; sl chl
1107	3-Bromobenzonitrile		C_7H_4BrN	6952-59-6	182.018		39.5	225			vs EtOH, eth; sl chl
1108	4-Bromobenzonitrile		C_7H_4BrN	623-00-7	182.018	nd (w, al)	114	236			s H_2O, EtOH, eth, chl
1109	6-Bromobenzo[*a*]pyrene		$C_{20}H_{11}Br$	21248-00-0	331.205	cry (ace/MeOH)	223				
1110	2-Bromobenzoyl chloride		C_7H_4BrClO	7154-66-7	219.463	nd	11	238(14)		1.5963^{20}	sl ctc
1111	4-Bromobenzoyl chloride		C_7H_4BrClO	586-75-4	219.463	nd (peth)	37(1)	245(11)			vs EtOH, eth, bz, lig
1112	2-Bromobiphenyl		$C_{12}H_9Br$	2052-07-5	233.103		0.8	297	1.2175^{26}	1.6248^{25}	vs eth, EtOH
1113	3-Bromobiphenyl		$C_{12}H_9Br$	2113-57-7	233.103			300		1.6411^{20}	i H_2O

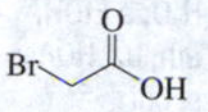
1079
Bromoacetic acid

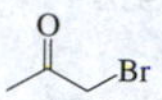
1080
Bromoacetone

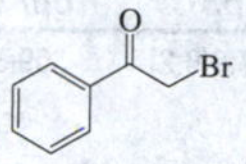
1081
α-Bromoacetophenone

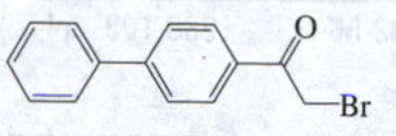
1082
4-(Bromoacetyl)biphenyl

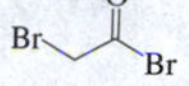
1083
Bromoacetyl bromide

1084
Bromoacetylene

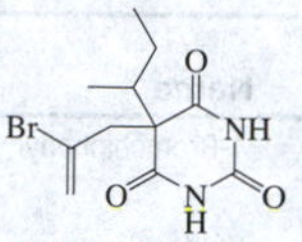
1085
5-(2-Bromoallyl)-5-*sec*-butylbarbituric acid

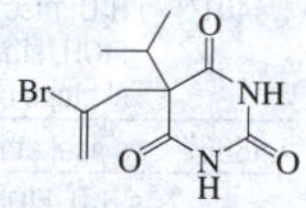
1086
5-(2-Bromoallyl)-5-isopropylbarbituric acid

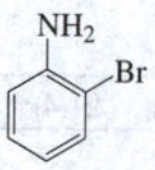
1087
2-Bromoaniline

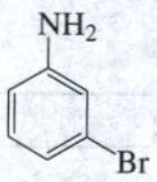
1088
3-Bromoaniline

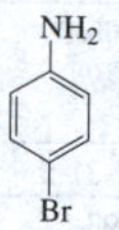
1089
4-Bromoaniline

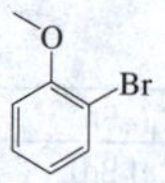
1090
2-Bromoanisole

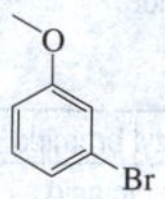
1091
3-Bromoanisole

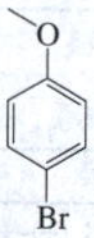
1092
4-Bromoanisole

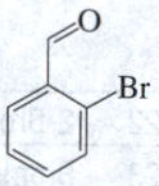
1093
2-Bromobenzaldehyde

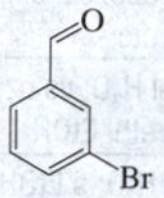
1094
3-Bromobenzaldehyde

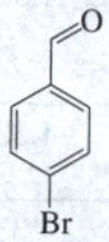
1095
4-Bromobenzaldehyde

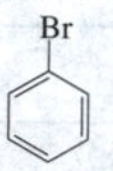
1096
Bromobenzene

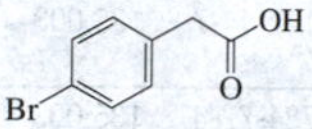
1097
4-Bromobenzeneacetic acid

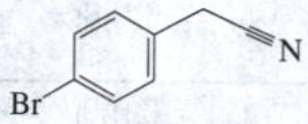
1098
4-Bromobenzeneacetonitrile

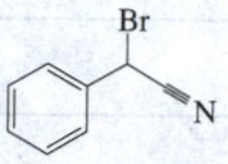
1099
α-Bromobenzeneacetonitrile

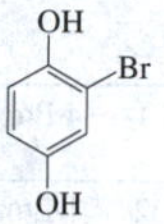
1100
2-Bromo-1,4-benzenediol

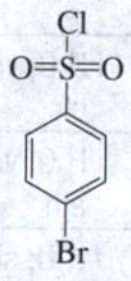
1101
4-Bromobenzenesulfonyl chloride

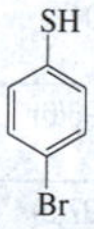
1102
4-Bromobenzenethiol

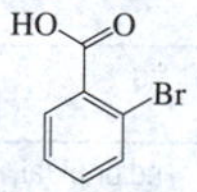
1103
2-Bromobenzoic acid

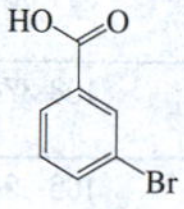
1104
3-Bromobenzoic acid

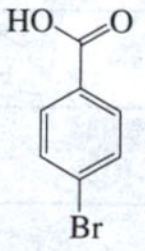
1105
4-Bromobenzoic acid

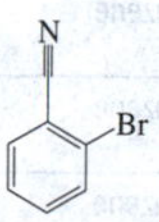
1106
2-Bromobenzonitrile

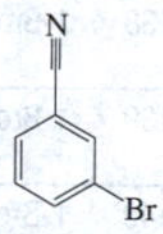
1107
3-Bromobenzonitrile

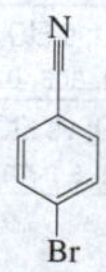
1108
4-Bromobenzonitrile

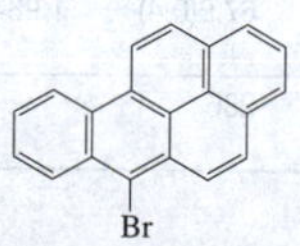
1109
6-Bromobenzo[a]pyrene

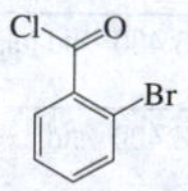
1110
2-Bromobenzoyl chloride

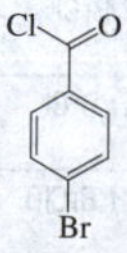
1111
4-Bromobenzoyl chloride

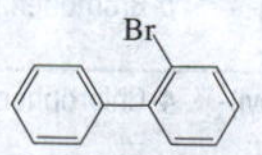
1112
2-Bromobiphenyl

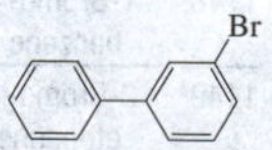
1113
3-Bromobiphenyl

No.	Name	Synonym	Mol. form.	CAS Reg. No.	Mol. wt.	Physical form	mp/°C	bp/°C	den/g cm^{-3}	n_D	Solubility
1114	4-Bromobiphenyl		$C_{12}H_9Br$	92-66-0	233.103	pl (al)	87.0(0.2)	309(3)	0.9327^{25}		i H_2O; s EtOH, eth, bz, HOAc; sl chl
1115	1-Bromo-2-(bromomethyl)-benzene		$C_7H_6Br_2$	3433-80-5	249.931	cry (al, lig)	31	129^{19}			vs eth, EtOH, HOAc
1116	1-Bromo-3-(bromomethyl)-benzene		$C_7H_6Br_2$	823-78-9	249.931	nd or lf	42	122^{12}			s chl
1117	1-Bromo-4-(bromomethyl)-benzene	*p*-Bromobenzyl bromide	$C_7H_6Br_2$	589-15-1	249.931	nd (al)	63				sl H_2O; s EtOH, bz, chl; vs eth, CS_2
1118	2-Bromo-2-(bromomethyl)-pentanedinitrile	1,2-Dibromo-2,4-dicyanobu-tane	$C_6H_6Br_2N_2$	35691-65-7	265.933		52				i H_2O; vs ace, bz, DMF
1119	2-Bromo-1-(4-bromophenyl)-ethanone	*p*-Bromophenacyl bromide	$C_8H_6Br_2O$	99-73-0	277.941	nd (al)	111				i H_2O; s EtOH, eth, chl
1120	2-Bromo-1,3-butadiene		C_4H_5Br	1822-86-2	132.987			42^{165}	1.397^{20}	1.4988^{20}	vs eth, EtOH
1121	1-Bromobutane	Butyl bromide	C_4H_9Br	109-65-9	137.018	liq	-112.5(0.3)	101.4(0.7)	1.2758^{20}	1.4401^{20}	i H_2O; msc EtOH, eth, ace; sl ctc; s chl
1122	2-Bromobutane, (±)-	(±)-*sec*-Butyl bromide	C_4H_9Br	5787-31-5	137.018	liq	-112.6(0.2)	91(4)	1.2585^{20}	1.4366^{20}	vs ace, eth, chl
1123	Bromobutanedioic acid, (±)-	Bromosuccinic acid	$C_4H_5BrO_4$	584-98-5	196.985		161		2.073^{25}		s H_2O, EtOH; sl HOAc
1124	4-Bromobutanenitrile		C_4H_6BrN	5332-06-9	148.002			198(7)	1.4967^{20}	1.4818^{20}	s EtOH, eth, chl
1125	2-Bromobutanoic acid, (±)-	DL-α-Bromobutyric acid	$C_4H_7BrO_2$	2385-70-8	167.002		-2.0	217 dec	1.5641^{20}		s H_2O, EtOH, eth
1126	4-Bromobutanoic acid		$C_4H_7BrO_2$	2623-87-2	167.002		33	142^{25}			
1127	3-Bromo-2-butanone		C_4H_7BrO	814-75-5	151.002			36^{11}			
1128	*cis*-1-Bromo-1-butene		C_4H_7Br	31849-78-2	135.003			86(5)	1.3265^{15}	1.4536^{20}	i H_2O; s eth, ace, bz, chl; sl ctc
1129	*trans*-1-Bromo-1-butene		C_4H_7Br	32620-08-9	135.003	liq	-100.3	95(3)	1.3209^{15}	1.4527^{20}	i H_2O; s eth, ace, bz, chl; sl ctc
1130	2-Bromo-1-butene		C_4H_7Br	23074-36-4	135.003	liq	-133.4	81(3)	1.3209^{15}	1.4527^{20}	i H_2O; s eth, ace, bz, chl; sl ctc
1131	4-Bromo-1-butene		C_4H_7Br	5162-44-7	135.003			98.5	1.3230^{20}	1.4622^{20}	sl H_2O; vs bz, eth, EtOH
1132	1-Bromo-2-butene		C_4H_7Br	4784-77-4	135.003			98(4)	1.3371^{25}	1.4822^{20}	i H_2O; s EtOH, eth, ctc; vs chl, bz
1133	*cis*-2-Bromo-2-butene		C_4H_7Br	3017-68-3	135.003	liq	-111.2(0.5)	89(3)	1.3416^{15}	1.4631^{19}	i H_2O; s EtOH, eth, ctc; vs chl, bz
1134	*trans*-2-Bromo-2-butene		C_4H_7Br	3017-71-8	135.003	liq	-115.4(0.5)	86(3)	1.3323^{15}	1.4602^{16}	i H_2O; s EtOH, eth, ctc; vs chl, bz
1135	(4-Bromobutoxy)benzene		$C_{10}H_{13}BrO$	1200-03-9	229.113	cry (al)	41	154^{18}			sl EtOH, ctc
1136	1-Bromo-4-*tert*-butylbenzene		$C_{10}H_{13}Br$	3972-65-4	213.114		19	231(10)	1.2286^{20}	1.5436^{20}	i H_2O; s eth, bz, chl
1137	2-Bromo-3'-chloroacetophe-none	3-Chlorophenacyl bromide	C_8H_6BrClO	41011-01-2	233.490	nd	40	397.5			vs EtOH
1138	1-Bromo-2-chlorobenzene		C_6H_4BrCl	694-80-4	191.453	liq	-12.6(0.3)	204	1.6387^{25}	1.5809^{20}	i H_2O; vs bz; sl ctc
1139	1-Bromo-3-chlorobenzene		C_6H_4BrCl	108-37-2	191.453	liq	-21.4(0.3)	196(6)	1.6302^{20}	1.5771^{20}	i H_2O; vs EtOH, eth
1140	1-Bromo-4-chlorobenzene		C_6H_4BrCl	106-39-8	191.453	nd or pl (al, eth)	64.78(0.05)	197(3)	1.576^{71}	1.5531^{70}	i H_2O; sl EtOH; s eth, bz, ctc, chl
1141	1-Bromo-4-chlorobutane		C_4H_8BrCl	6940-78-9	171.464			161(11)	1.489^{20}	1.4885^{20}	i H_2O; s EtOH, eth, chl; sl ctc
1142	Bromochlorodifluoromethane	Halon 1211	$CBrClF_2$	353-59-3	165.365	col gas	-159.5	-3.9(0.7)			
1143	3-Bromo-1-chloro-5,5-dimethylhydantoin		$C_5H_6BrClN_2O_2$	126-06-7	241.471		162				
1144	1-Bromo-1-chloroethane		C_2H_4BrCl	593-96-4	143.410			101(12)	1.667^{10}	1.4660^{20}	
1145	1-Bromo-2-chloroethane	2-Chloro-1-bromoethane	C_2H_4BrCl	107-04-0	143.410	liq	-16.7(0.3)	106(2)	1.7392^{20}	1.4908^{20}	sl H_2O; s EtOH, eth, chl
1146	Bromochlorofluoromethane		$CHBrClF$	593-98-6	147.374	liq	-115	39(15)	1.9771^{0}	1.4144^{25}	i H_2O; s eth, ace, chl
1147	Bromochloromethane	Halon 1011	CH_2BrCl	74-97-5	129.384	liq	-87.9(0.2)	67.9(0.4)	1.9344^{20}	1.4838^{20}	i H_2O; s EtOH, eth, ace, bz
1148	1-Bromo-4-(chloromethyl)-benzene	*p*-Bromobenzyl chloride	C_7H_6BrCl	589-17-3	205.480	nd (al, peth)	42(4)	236			i H_2O; vs EtOH, eth; s peth
1149	2-Bromo-1-(4-chlorophenyl)-ethanone	4-Chlorophenacyl bromide	C_8H_6BrClO	536-38-9	233.490	nd	96.5				

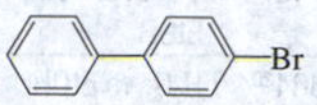
1114 4-Bromobiphenyl

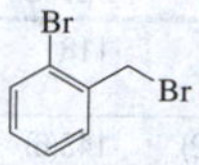
1115 1-Bromo-2-(bromomethyl)benzene

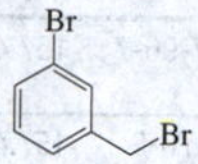
1116 1-Bromo-3-(bromomethyl)benzene

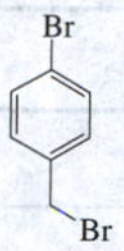
1117 1-Bromo-4-(bromomethyl)benzene

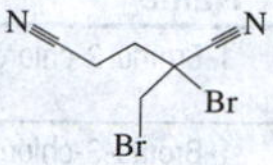
1118 2-Bromo-2-(bromomethyl)pentanedinitrile

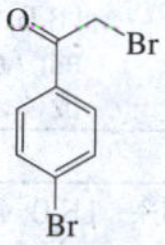
1119 2-Bromo-1-(4-bromophenyl)ethanone

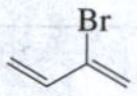
1120 2-Bromo-1,3-butadiene

1121 1-Bromobutane

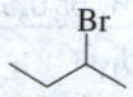
1122 2-Bromobutane, (±)-

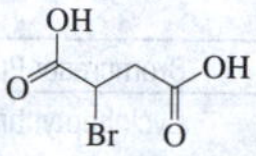
1123 Bromobutanedioic acid, (±)-

1124 4-Bromobutanenitrile

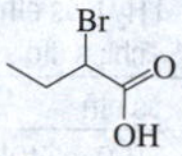
1125 2-Bromobutanoic acid, (±)-

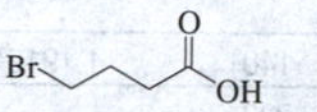
1126 4-Bromobutanoic acid

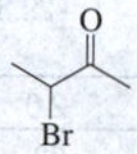
1127 3-Bromo-2-butanone

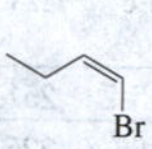
1128 *cis*-1-Bromo-1-butene

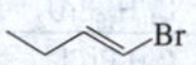
1129 *trans*-1-Bromo-1-butene

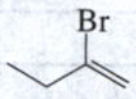
1130 2-Bromo-1-butene

1131 4-Bromo-1-butene

1132 1-Bromo-2-butene

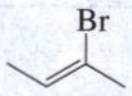
1133 *cis*-2-Bromo-2-butene

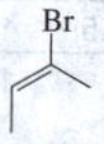
1134 *trans*-2-Bromo-2-butene

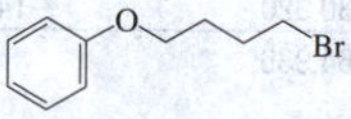
1135 (4-Bromobutoxy)benzene

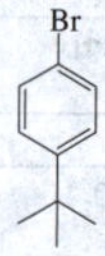
1136 1-Bromo-4-*tert*-butylbenzene

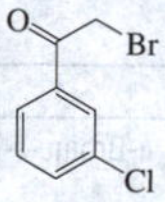
1137 2-Bromo-3'-chloroacetophenone

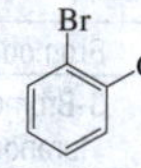
1138 1-Bromo-2-chlorobenzene

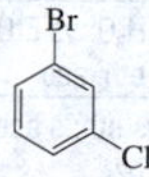
1139 1-Bromo-3-chlorobenzene

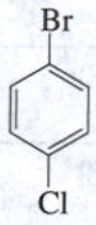
1140 1-Bromo-4-chlorobenzene

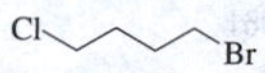
1141 1-Bromo-4-chlorobutane

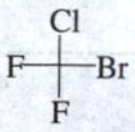
1142 Bromochlorodifluoromethane

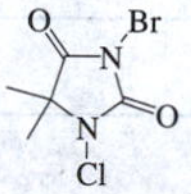
1143 3-Bromo-1-chloro-5,5-dimethylhydantoin

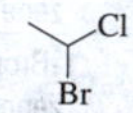
1144 1-Bromo-1-chloroethane

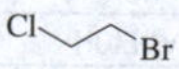
1145 1-Bromo-2-chloroethane

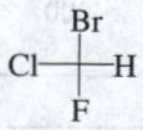
1146 Bromochlorofluoromethane

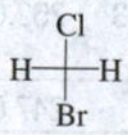
1147 Bromochloromethane

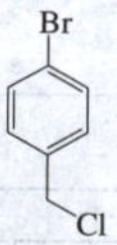
1148 1-Bromo-4-(chloromethyl)benzene

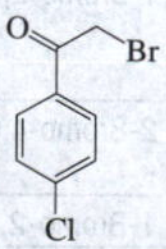
1149 2-Bromo-1-(4-chlorophenyl)ethanone

No.	Name	Synonym	Mol. form.	CAS Reg. No.	Mol. wt.	Physical form	mp/°C	bp/°C	den/ g cm^{-3}	n_D	Solubility
1150	1-Bromo-2-chloropropane		C_3H_6BrCl	3017-96-7	157.437			118	1.531^{20}	1.4745^{20}	vs ace, bz, eth, EtOH
1151	1-Bromo-3-chloropropane		C_3H_6BrCl	109-70-6	157.437	liq	-58.8(0.2)	143(6)	1.5969^{20}	1.4864^{20}	i H_2O; vs EtOH, eth, chl
1152	2-Bromo-1-chloropropane		C_3H_6BrCl	3017-95-6	157.437			117	1.537^{20}	1.4795^{20}	i H_2O; vs EtOH, eth; s ace, bz
1153	2-Bromo-2-chloropropane		C_3H_6BrCl	2310-98-7	157.437			95	1.495^{20}	1.4575^{20}	vs ace, bz, eth, EtOH
1154	1-Bromo-2-chloro-1,1,2-trifluoroethane		$C_2HBrClF_3$	354-06-3	197.381			52.4(0.2)	1.8574^{25}	1.3738^{20}	
1155	2-Bromo-2-chloro-1,1,1-trifluoroethane	Halothane	$C_2HBrClF_3$	151-67-7	197.381			50(1)	1.8563^{25}	1.3697^{0}	sl H_2O; s peth
1156	Bromocresol Green	Bromcresol Green	$C_{21}H_{14}Br_4O_5S$	76-60-8	698.014	wh or red (+7w) ye (HOAc)	218.5				sl H_2O; vs EtOH, eth, AcOEt; s bz
1157	Bromocresol Purple	Bromcresol Purple	$C_{21}H_{16}Br_2O_5S$	115-40-2	540.222		241.5				
1158	Bromocycloheptane	Cycloheptyl bromide	$C_7H_{13}Br$	2404-35-5	177.082			101^{40}	1.3080^{20}	1.4996^{20}	i H_2O; vs eth, chl
1159	Bromocyclohexane	Cyclohexyl bromide	$C_6H_{11}Br$	108-85-0	163.055	liq	-56.28(0.07)	165.9(0.8)	1.3359^{20}	1.4957^{20}	i H_2O; msc EtOH, eth, ace, bz, lig, ctc
1160	*trans*-4-Bromocyclohexanol		$C_6H_{11}BrO$	32388-22-0	179.054	pl (hx)	81.5				
1161	2-Bromocyclohexanone		C_6H_9BrO	822-85-5	177.038			114^{32}	1.340^{25}	1.5085^{25}	
1162	3-Bromocyclohexene		C_6H_9Br	1521-51-3	161.039			81^{40}	1.3890^{20}	1.5320^{20}	i H_2O; s eth, bz, chl
1163	Bromocyclopentane	Cyclopentyl bromide	C_5H_9Br	137-43-9	149.029			137.5	1.3873^{20}	1.4886^{20}	sl ctc
1164	1-Bromodecane	Decyl bromide	$C_{10}H_{21}Br$	112-29-8	221.178	liq	-29.3(0.5)	240.6	1.0702^{20}	1.4557^{20}	i H_2O; vs eth, chl; s ctc
1165	2-Bromodecanoic acid		$C_{10}H_{19}BrO_2$	2623-95-2	251.161		2.0	140^{2}	1.1912^{24}	1.4595^{24}	vs eth
1166	1-Bromo-3,5-dichlorobenzene		$C_6H_3BrCl_2$	19752-55-7	225.898	pr (al)	83	232			i H_2O; s EtOH, eth, chl; vs bz
1167	4-Bromo-1,2-dichlorobenzene		$C_6H_3BrCl_2$	18282-59-2	225.898	pr	25	237			i H_2O; sl EtOH; vs eth, bz, chl
1168	Bromodichlorofluoromethane	Halon 1121	$CBrCl_2F$	353-58-2	181.819	liq		52.8	1.95^{22}		
1169	Bromodichloromethane		$CHBrCl_2$	75-27-4	163.829	liq	-56.0(0.4)	90(2)	1.980^{20}	1.4964^{20}	i H_2O; vs EtOH, eth, ace, bz; sl ctc
1170	4-Bromo-2,5-dichlorophenol		$C_6H_3BrCl_2O$	1940-42-7	241.897	nd	70.8(0.5)				
1171	2-Bromo-1,1-diethoxyethane		$C_6H_{13}BrO_2$	2032-35-1	197.070			170	1.283^{20}	1.4387^{20}	s EtOH, eth
1172	4-Bromo-*N*,*N*-diethylaniline		$C_{10}H_{14}BrN$	2052-06-4	228.129	nd or pr	38	270			i H_2O; vs EtOH, eth
1173	Bromodifluoromethane		$CHBrF_2$	1511-62-2	130.920		-145(4)	-15.6(0.5)	1.832^{16}		s H_2O; vs EtOH
1174	3-Bromo-4,5-dihydro-2(3*H*)-furanone	α-Bromo-γ-butyrolactone	$C_4H_5BrO_2$	5061-21-2	164.986			130^{20}	1.8^{20}	1.5059^{20}	
1175	5-Bromo-*N*,2-dihydroxybenzamide	5-Bromosalicylhydroxamic acid	$C_7H_6BrNO_3$	5798-94-7	232.032	cry (al)	232 dec				
1176	2-Bromo-1,4-dimethoxybenzene		$C_8H_9BrO_2$	25245-34-5	217.060	oil		262	1.445	1.5700^{20}	
1177	4-Bromo-1,2-dimethoxybenzene		$C_8H_9BrO_2$	2859-78-1	217.060			254.5	1.702^{25}	1.5743^{20}	
1178	2-Bromo-1,1-dimethoxyethane		$C_4H_9BrO_2$	7252-83-7	169.017			149	1.430^{20}	1.4450^{20}	s eth, ace, chl
1179	4-Bromo-*N*,*N*-dimethylaniline		$C_8H_{10}BrN$	586-77-6	200.076		55	264	1.3220^{100}		i H_2O; s EtOH; vs eth
1180	1-Bromo-2,4-dimethylbenzene		C_8H_9Br	583-70-0	185.061	liq	-17	204(8)	1.3419^{20}	1.5501^{20}	i H_2O; vs EtOH, eth, ace
1181	1-Bromo-3,5-dimethylbenzene		C_8H_9Br	556-96-7	185.061			203(8)	1.362^{20}	1.5462^{22}	vs eth; s ace, bz
1182	2-Bromo-1,3-dimethylbenzene		C_8H_9Br	576-22-7	185.061			204(6)		1.5552^{20}	vs eth; s ace, bz
1183	2-Bromo-1,4-dimethylbenzene		C_8H_9Br	553-94-6	185.061	lf or pl	9	207(6)	1.3582^{18}	1.5514^{18}	i H_2O; vs EtOH; s bz
1184	4-Bromo-1,2-dimethylbenzene		C_8H_9Br	583-71-1	185.061	liq	-0.2	215(7)	1.3708^{20}	1.5530^{20}	i H_2O; vs EtOH, eth
1185	*trans*-1-Bromo-3,7-dimethyl-2,6-octadiene	*trans*-Geranyl bromide	$C_{10}H_{17}Br$	6138-90-5	217.146			101^{12}	1.0940^{22}	1.5027^{20}	
1186	1-Bromo-2,2-dimethylpropane		$C_5H_{11}Br$	630-17-1	151.045			108(11)	1.1997^{20}	1.4370^{20}	i H_2O; s EtOH, eth, ace, bz; vs chl
1187	2-Bromo-4,6-dinitroaniline		$C_6H_4BrN_3O_4$	1817-73-8	262.018	ye nd (al or HOAc)	153.5	sub			vs EtOH, ace; s HOAc
1188	1-Bromo-2,4-dinitrobenzene		$C_6H_3BrN_2O_4$	584-48-5	247.003	ye nd (al)	75				vs EtOH

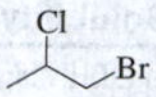
1150
1-Bromo-2-chloropropane

1151
1-Bromo-3-chloropropane

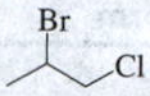
1152
2-Bromo-1-chloropropane

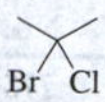
1153
2-Bromo-2-chloropropane

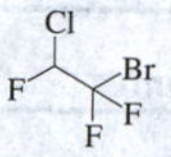
1154
1-Bromo-2-chloro-1,1,2-trifluoroethane

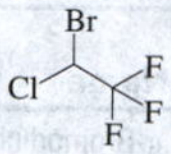
1155
2-Bromo-2-chloro-1,1,1-trifluoroethane

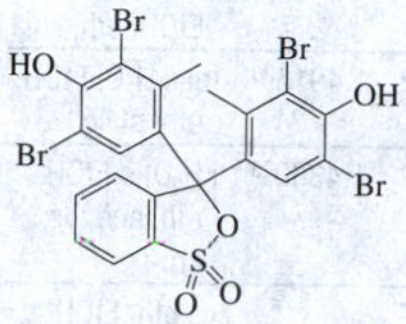
1156
Bromocresol Green

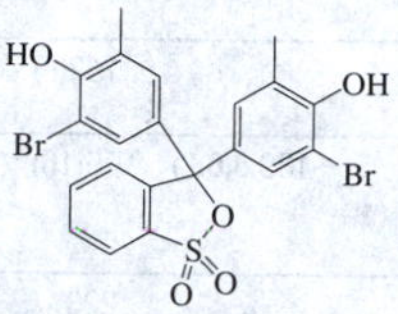
1157
Bromocresol Purple

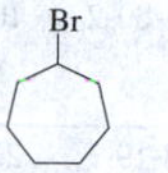
1158
Bromocycloheptane

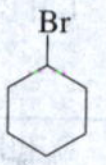
1159
Bromocyclohexane

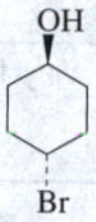
1160
trans-4-Bromocyclohexanol

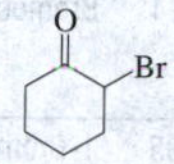
1161
2-Bromocyclohexanone

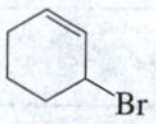
1162
3-Bromocyclohexene

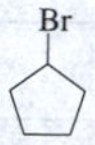
1163
Bromocyclopentane

1164
1-Bromodecane

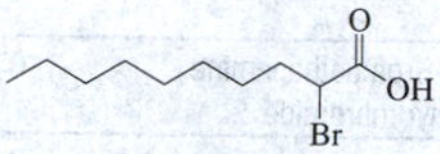
1165
2-Bromodecanoic acid

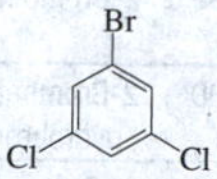
1166
1-Bromo-3,5-dichlorobenzene

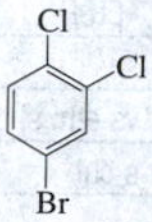
1167
4-Bromo-1,2-dichlorobenzene

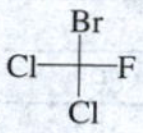
1168
Bromodichlorofluoromethane

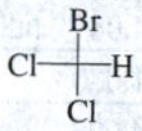
1169
Bromodichloromethane

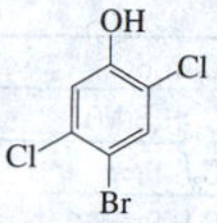
1170
4-Bromo-2,5-dichlorophenol

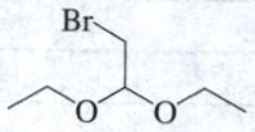
1171
2-Bromo-1,1-diethoxyethane

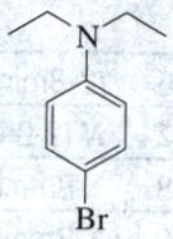
1172
4-Bromo-*N*,*N*-diethylaniline

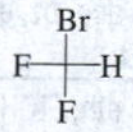
1173
Bromodifluoromethane

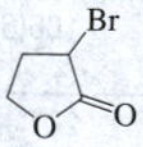
1174
3-Bromo-4,5-dihydro-2(3*H*)-furanone

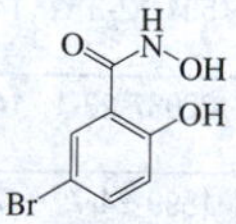
1175
5-Bromo-*N*,2-dihydroxybenzamide

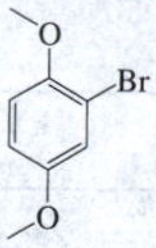
1176
2-Bromo-1,4-dimethoxybenzene

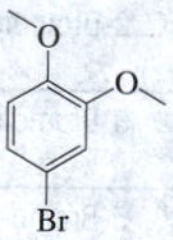
1177
4-Bromo-1,2-dimethoxybenzene

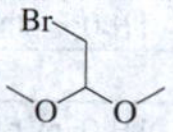
1178
2-Bromo-1,1-dimethoxyethane

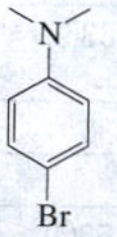
1179
4-Bromo-*N*,*N*-dimethylaniline

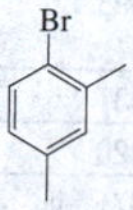
1180
1-Bromo-2,4-dimethylbenzene

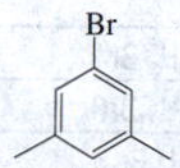
1181
1-Bromo-3,5-dimethylbenzene

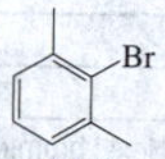
1182
2-Bromo-1,3-dimethylbenzene

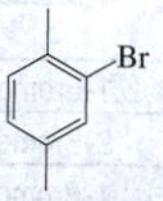
1183
2-Bromo-1,4-dimethylbenzene

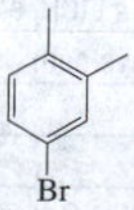
1184
4-Bromo-1,2-dimethylbenzene

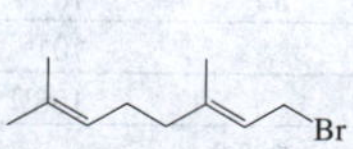
1185
trans-1-Bromo-3,7-dimethyl-2,6-octadiene

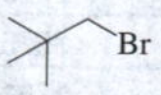
1186
1-Bromo-2,2-dimethylpropane

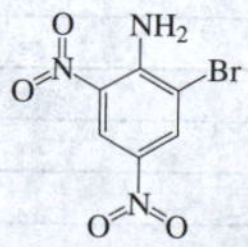
1187
2-Bromo-4,6-dinitroaniline

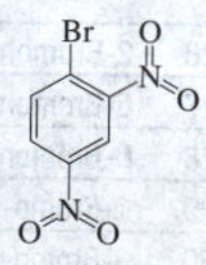
1188
1-Bromo-2,4-dinitrobenzene

No.	Name	Synonym	Mol. form.	CAS Reg. No.	Mol. wt.	Physical form	mp/°C	bp/°C	den/ g cm⁻³	n_D	Solubility
1189	α-Bromodiphenylmethane		$C_{13}H_{11}Br$	776-74-9	247.130		45	184^{20}			s EtOH, chl; vs bz
1190	1-Bromododecane	Lauryl bromide	$C_{12}H_{25}Br$	143-15-7	249.231	liq	-9.6(0.4)	275(20)	1.0399^{20}	1.4583^{20}	i H_2O; s EtOH, eth, ctc; msc ace
1191	2-Bromododecanoic acid		$C_{12}H_{23}BrO_2$	111-56-8	279.214	pl	32	158^{2}	1.1474^{74}	1.4585^{24}	vs bz, eth, EtOH, lig
1192	Bromoethane	Ethyl bromide	C_2H_5Br	74-96-4	108.965	liq	-118.4(1)	38.2(0.6)	1.4604^{20}	1.4239^{20}	sl H_2O; msc EtOH, eth, chl
1193	2-Bromoethanol	Ethylene bromohydrin	C_2H_5BrO	540-51-2	124.964			142(4)	1.7629^{20}	1.4915^{20}	msc H_2O, EtOH, eth; sl lig
1194	Bromoethene	Vinyl bromide	C_2H_3Br	593-60-2	106.949	vol liq or gas	-139.5(0.2)	16(16)	1.4933^{20}	1.4380^{20}	i H_2O; s EtOH, eth, ace, bz, chl
1195	1-Bromo-2-ethoxybenzene		C_8H_9BrO	583-19-7	201.060			223	1.4223^{20}		vs eth, EtOH
1196	1-Bromo-4-ethoxybenzene		C_8H_9BrO	588-96-5	201.060		2.0	231	1.4071^{25}	1.5517^{20}	i H_2O; vs EtOH, eth; s chl
1197	(2-Bromoethoxy)benzene		C_8H_9BrO	589-10-6	201.060		39	240 dec	1.3555^{20}		i H_2O; vs EtOH, eth
1198	1-Bromo-2-ethoxyethane	2-Bromoethyl ethyl ether	C_4H_9BrO	592-55-2	153.017			127(4)	1.3852^{0}	1.4447^{20}	sl H_2O; msc EtOH, eth
1199	2-Bromoethyl acetate		$C_4H_7BrO_2$	927-68-4	167.002	liq	-13.8	159(5)	1.514^{20}	1.457^{23}	vs H_2O, chl; msc EtOH, eth
1200	2-Bromoethylamine hydrobromide	2-Bromoethanamine hydrobromide	$C_2H_7Br_2N$	2576-47-8	204.892		174.0				
1201	(1-Bromoethyl)benzene		C_8H_9Br	585-71-7	185.061			201(8)	1.3535^{25}	1.5543^{25}	
1202	(2-Bromoethyl)benzene		C_8H_9Br	103-63-9	185.061	liq	-55.9(0.2)	216(4)	1.3643^{20}	1.5372^{20}	i H_2O; s eth, bz; sl ctc
1203	1-Bromo-2-ethylbenzene		C_8H_9Br	1973-22-4	185.061	liq	-67.5(0.2)	202(4)	1.3548^{20}	1.5472^{20}	vs ace, bz, eth, EtOH
1204	1-Bromo-3-ethylbenzene		C_8H_9Br	2725-82-8	185.061			203(11)	1.3493^{20}	1.5465^{20}	
1205	1-Bromo-4-ethylbenzene		C_8H_9Br	1585-07-5	185.061	liq	-43.4(0.2)	204(4)	1.3423^{20}	1.5445^{20}	vs ace, bz, eth, EtOH
1206	(2-Bromoethyl)cyclohexane		$C_8H_{15}Br$	1647-26-3	191.109	liq	-57	212(3)	1.2357^{20}	1.4899^{20}	
1207	*N*-(2-Bromoethyl)phthalimide		$C_{10}H_8BrNO_2$	574-98-1	254.081	nd (w)	81.5(0.5)				vs eth; sl chl
1208	1-Bromo-4-ethynylbenzene		C_8H_5Br	766-96-1	181.030		64.5	89^{16}			s chl
1209	1-Bromo-2-fluorobenzene		C_6H_4BrF	1072-85-1	174.998			154	1.6103^{21}	1.5337^{20}	
1210	1-Bromo-3-fluorobenzene		C_6H_4BrF	1073-06-9	174.998			150	1.7081^{20}	1.5257^{20}	s ctc
1211	1-Bromo-4-fluorobenzene		C_6H_4BrF	460-00-4	174.998	liq	-17.4	150(2)	1.593^{15}	1.5310^{15}	i H_2O; s EtOH, eth, chl
1212	1-Bromo-2-fluoroethane		C_2H_4BrF	762-49-2	126.955			58(12)	1.7044^{25}	1.4236^{20}	vs eth, EtOH
1213	Bromofluoromethane		CH_2BrF	373-52-4	112.929	vol liq or gas		23(12)			s EtOH; vs chl
1214	2-Bromofuran		C_4H_3BrO	584-12-3	146.970			103(6)	1.6500^{20}	1.4980^{20}	sl H_2O; s EtOH, eth, ace, bz
1215	3-Bromofuran		C_4H_3BrO	22037-28-1	146.970			106(6)	1.6606^{20}	1.4958^{20}	vs ace, bz, eth, EtOH
1216	5-Bromo-2-furancarboxaldehyde		$C_5H_3BrO_2$	1899-24-7	174.981	cry (50% al)	83.5	201			vs eth, EtOH
1217	1-Bromoheptadecane		$C_{17}H_{35}Br$	3508-00-7	319.364		28.4(0.4)	345(13)	0.9916^{20}	1.4625^{20}	i H_2O; vs chl
1218	1-Bromoheptane	Heptyl bromide	$C_7H_{15}Br$	629-04-9	179.098	liq	-56.1(0.3)	179(5)	1.1400^{20}	1.4502^{20}	i H_2O; vs EtOH, eth; sl ctc; s chl
1219	2-Bromoheptane	2-Heptyl bromide	$C_7H_{15}Br$	1974-04-5	179.098		47	166(6)	1.1277^{20}	1.4503^{20}	i H_2O; vs bz; s ctc, chl
1220	4-Bromoheptane	4-Heptyl bromide	$C_7H_{15}Br$	998-93-6	179.098			163(9)	1.1351^{20}	1.4495^{20}	i H_2O; s bz, ctc, chl
1221	1-Bromohexadecane		$C_{16}H_{33}Br$	112-82-3	305.337		17.5(0.4)	336	0.9991^{20}	1.4618^{25}	i H_2O; s eth
1222	2-Bromohexadecanoic acid		$C_{16}H_{31}BrO_2$	18263-25-7	335.320		52.8				
1223	1-Bromohexane	Hexyl bromide	$C_6H_{13}Br$	111-25-1	165.071	liq	-84.9(0.4)	156(4)	1.1744^{20}	1.4478^{20}	i H_2O; msc EtOH, eth; s ace; vs chl
1224	2-Bromohexane		$C_6H_{13}Br$	3377-86-4	165.071			139(9)	1.1658^{20}	1.4832^{25}	i H_2O; vs EtOH; s eth, ace; sl ctc
1225	3-Bromohexane		$C_6H_{13}Br$	3377-87-5	165.071			143(4)	1.1799^{20}	1.4472^{20}	vs ace, eth, EtOH, chl
1226	2-Bromohexanoic acid, (±)-		$C_6H_{11}BrO_2$	2681-83-6	195.054		2.0	242(11)	1.2810^{33}		s EtOH, eth
1227	6-Bromohexanoic acid		$C_6H_{11}BrO_2$	4224-70-8	195.054	cry (peth)	35	167^{20}			vs peth
1228	6-Bromohexanoyl chloride		$C_6H_{10}BrClO$	22809-37-6	213.499			101^{6}			
1229	1-Bromo-4-(hexyloxy)benzene		$C_{12}H_{17}BrO$	30752-19-3	257.166			156^{13}	1.2306^{20}	1.5262^{20}	
1230	5-Bromo-2-hydroxybenzaldehyde		$C_7H_5BrO_2$	1761-61-1	201.018	nd (al), lf (eth)	105.5				i H_2O; s EtOH, eth; sl chl

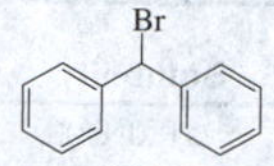
1189
α-Bromodiphenylmethane

1190
1-Bromododecane

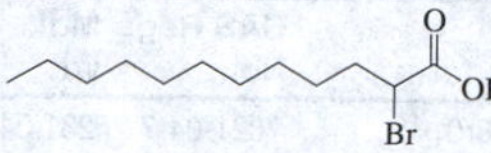
1191
2-Bromododecanoic acid

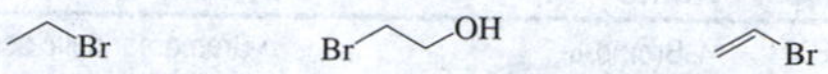
1192
Bromoethane

1193
2-Bromoethanol

1194
Bromoethene

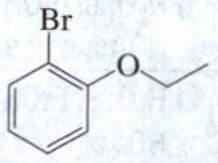
1195
1-Bromo-2-ethoxybenzene

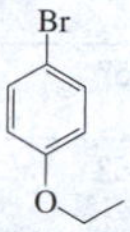
1196
1-Bromo-4-ethoxybenzene

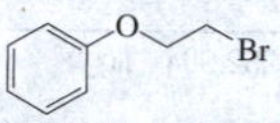
1197
(2-Bromoethoxy)benzene

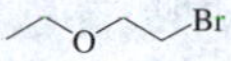
1198
1-Bromo-2-ethoxyethane

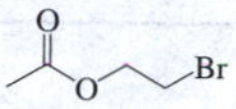
1199
2-Bromoethyl acetate

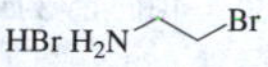
1200
2-Bromoethylamine hydrobromide

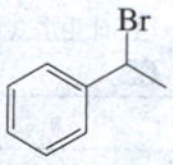
1201
(1-Bromoethyl)benzene

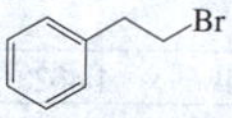
1202
(2-Bromoethyl)benzene

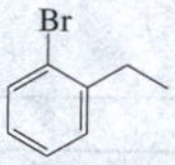
1203
1-Bromo-2-ethylbenzene

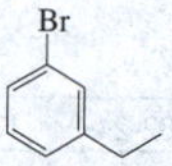
1204
1-Bromo-3-ethylbenzene

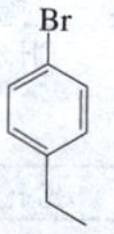
1205
1-Bromo-4-ethylbenzene

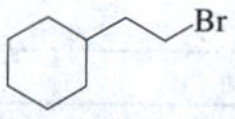
1206
(2-Bromoethyl)cyclohexane

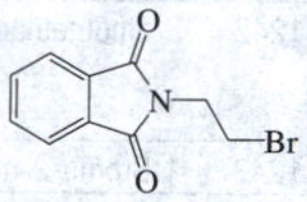
1207
N-(2-Bromoethyl)phthalimide

Organic

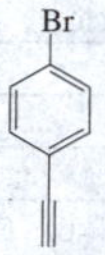
1208
1-Bromo-4-ethynylbenzene

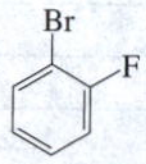
1209
1-Bromo-2-fluorobenzene

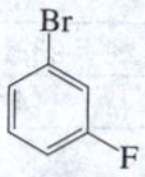
1210
1-Bromo-3-fluorobenzene

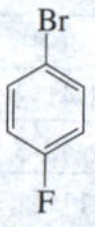
1211
1-Bromo-4-fluorobenzene

1212
1-Bromo-2-fluoroethane

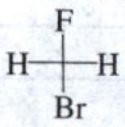
1213
Bromofluoromethane

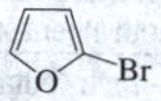
1214
2-Bromofuran

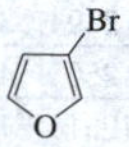
1215
3-Bromofuran

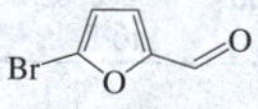
1216
5-Bromo-2-furancarboxaldehyde

1217
1-Bromoheptadecane

1218
1-Bromoheptane

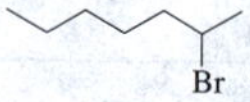
1219
2-Bromoheptane

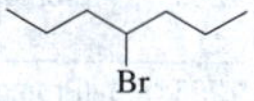
1220
4-Bromoheptane

1221
1-Bromohexadecane

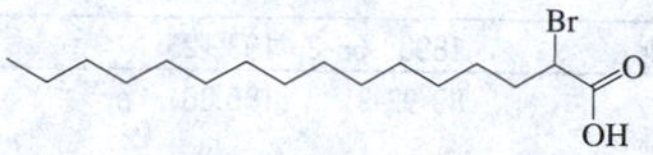
1222
2-Bromohexadecanoic acid

1223
1-Bromohexane

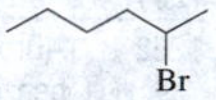
1224
2-Bromohexane

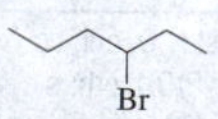
1225
3-Bromohexane

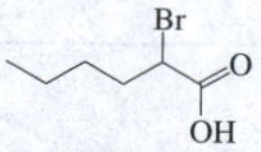
1226
2-Bromohexanoic acid, (±)-

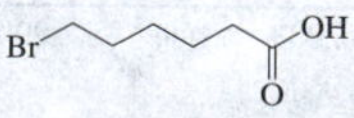
1227
6-Bromohexanoic acid

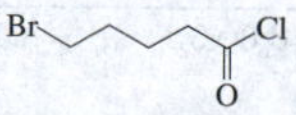
1228
6-Bromohexanoyl chloride

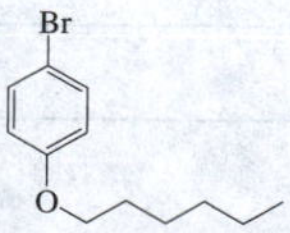
1229
1-Bromo-4-(hexyloxy)benzene

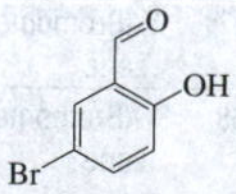
1230
5-Bromo-2-hydroxybenzaldehyde

No.	Name	Synonym	Mol. form.	CAS Reg. No.	Mol. wt.	Physical form	mp/°C	bp/°C	den/ g cm⁻³	n_D	Solubility
1231	4-Bromo-α-hydroxybenzeneacetic acid, (±)-	*p*-Bromomandelic acid	$C_8H_7BrO_3$	7021-04-7	231.044		119				vs H_2O, EtOH, eth, bz, chl
1232	5-Bromo-2-hydroxybenzene-methanol	Bromosaligenin	$C_7H_7BrO_2$	2316-64-5	203.034	lf (bz)	113				vs bz, eth, EtOH, chl
1233	5-Bromo-2-hydroxybenzoic acid		$C_7H_5BrO_3$	89-55-4	217.017	nd (w, dil al)	169.8	100 sub			sl H_2O, ace; vs EtOH, eth
1234	3-Bromo-4-hydroxy-5-methoxybenzaldehyde		$C_8H_7BrO_3$	2973-76-4	231.044	pl (HOAc), nd, pl (al)	167.0				i H_2O; s EtOH, DMSO; sl eth, bz
1235	1-Bromo-2-iodobenzene		C_6H_4BrI	583-55-1	282.904		2.1(0.5)	257	2.2570^{25}	1.6618^{25}	i H_2O; sl EtOH, HOAc; s ace
1236	1-Bromo-3-iodobenzene		C_6H_4BrI	591-18-4	282.904	liq	-9.3(0.6)	252			i H_2O; sl EtOH, HOAc
1237	1-Bromo-4-iodobenzene		C_6H_4BrI	589-87-7	282.904	pr or pl (eth-al)	90.37(0.05)	252			i H_2O; sl EtOH, chl; s eth
1238	Bromoiodomethane		CH_2BrI	557-68-6	220.835			136(14)	2.926^{17}	1.6410^{20}	vs chl
1239	1-Bromo-4-isocyanatobenzene	*p*-Bromophenyl isocyanate	C_7H_4BrNO	2493-02-9	198.017	nd		226			vs eth
1240	1-Bromo-4-isopropylbenzene		$C_9H_{11}Br$	586-61-8	199.087	liq	-22.4(0.2)	219(3)	1.3145^{20}	1.5569^{20}	i H_2O; s eth, bz, chl; sl ctc
1241	4-Bromoisoquinoline		C_9H_6BrN	1532-97-4	208.055	cry (peth)	41.5	282.5			vs eth
1242	Bromomethane	Methyl bromide	CH_3Br	74-83-9	94.939	col gas	-93.7(0.4)	3.4(0.1)	1.6755^{20}	1.4218^{20}	sl H_2O; msc EtOH, eth, chl, CS_2
1243	1-Bromo-2-methoxyethane		C_3H_7BrO	6482-24-2	138.991			112(4)	1.4623^{20}	1.44753^{20}	
1244	Bromomethoxymethane		C_2H_5BrO	13057-17-5	124.964			87	1.5976^{20}	1.4562^{20}	
1245	2-Bromo-4-methylaniline		C_7H_8BrN	583-68-6	186.050	lf	26	240	1.510^{20}	1.5999^{20}	i H_2O; s EtOH, eth
1246	4-Bromo-2-methylaniline		C_7H_8BrN	583-75-5	186.050	cry (al)	59.5	240			sl H_2O, chl; s EtOH; vs eth, HOAc
1247	(Bromomethyl)benzene	Benzyl bromide	C_7H_7Br	100-39-0	171.035	liq	-1.5	191(4)	1.4380^{25}	1.5752^{20}	i H_2O; msc EtOH, eth; s ctc
1248	4-(Bromomethyl)benzoic acid		$C_8H_7BrO_2$	6232-88-8	215.045		226.3				
1249	3-(Bromomethyl)benzonitrile		C_8H_6BrN	28188-41-2	196.045		96.5	130^{4}			
1250	4-(Bromomethyl)benzonitrile		C_8H_6BrN	17201-43-3	196.045		114				
1251	1-Bromo-2-methylbutane, DL		$C_5H_{11}Br$	5973-11-5	151.045			119	1.2205^{20}	1.4452^{20}	i H_2O; s EtOH, eth; vs chl
1252	1-Bromo-3-methylbutane	Isopentyl bromide	$C_5H_{11}Br$	107-82-4	151.045	liq	-112	121(1)	1.2071^{20}	1.4420^{20}	i H_2O; s EtOH, eth; sl ctc; vs chl
1253	2-Bromo-2-methylbutane	*tert*-Pentyl bromide	$C_5H_{11}Br$	507-36-8	151.045			105(8)	1.197^{18}	1.4421	
1254	3-Bromo-3-methylbutanoic acid	β-Bromoisovaleric acid	$C_5H_9BrO_2$	5798-88-9	181.028	nd (lig)	74				vs bz, eth, EtOH
1255	1-Bromo-3-methyl-2-butene		C_5H_9Br	870-63-3	149.029			121(18)	1.2930^{15}	1.4930^{15}	vs ace, bz, eth, EtOH
1256	1-(Bromomethyl)-2-chlorobenzene		C_7H_6BrCl	611-17-6	205.480			109^{10}			
1257	(Bromomethyl)chlorodimethylsilane		$C_3H_8BrClSi$	16532-02-8	187.539			131	1.375^{25}	1.4630^{25}	
1258	1-Bromo-3-methylcyclohexane	3-Methylcyclohexyl bromide	$C_7H_{13}Br$	13905-48-1	177.082			181	1.2676^{15}	1.4979^{20}	i H_2O; vs eth; s bz
1259	(Bromomethyl)cyclohexane		$C_7H_{13}Br$	2550-36-9	177.082			76^{26}	1.283^{20}	1.4907^{30}	vs bz, eth, chl
1260	1-(Bromomethyl)-3-fluorobenzene		C_7H_6BrF	456-41-7	189.025			88^{20}		1.5474^{20}	
1261	3-(Bromomethyl)heptane		$C_8H_{17}Br$	18908-66-2	193.125			67^{10}			
1262	1-(Bromomethyl)-2-methylbenzene		C_8H_9Br	89-92-9	185.061	pr	21	217	1.3811^{23}	1.5730^{20}	i H_2O; s EtOH, eth, ace, bz
1263	1-(Bromomethyl)-3-methylbenzene		C_8H_9Br	620-13-3	185.061			212.5	1.3711^{23}	1.5660^{20}	i H_2O; vs EtOH, eth
1264	1-(Bromomethyl)-4-methylbenzene		C_8H_9Br	104-81-4	185.061	nd (al)	34(2)	220	1.324^{25}		i H_2O; s EtOH; vs eth, chl
1265	1-(Bromomethyl)naphthalene		$C_{11}H_9Br$	3163-27-7	221.093	cry (peth, al)	56(2)	183^{18}			vs ace, bz, eth, EtOH
1266	2-(Bromomethyl)naphthalene		$C_{11}H_9Br$	939-26-4	221.093	lf (al)	56	213^{100}			s EtOH, eth, chl, HOAc
1267	1-(Bromomethyl)-3-nitrobenzene		$C_7H_6BrNO_2$	3958-57-4	216.033	nd or pl (al)	59.3	162^{13}			i H_2O; s EtOH
1268	1-(Bromomethyl)-4-nitrobenzene		$C_7H_6BrNO_2$	100-11-8	216.033	nd (al)	99(2)				sl H_2O, chl; vs EtOH, eth; s HOAc

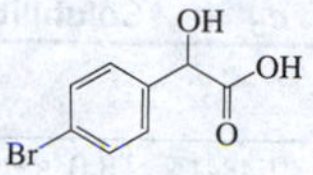
1231
4-Bromo-α-hydroxybenzeneacetic acid, (±)-

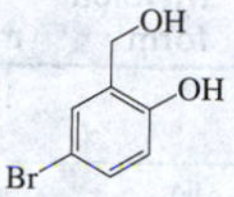
1232
5-Bromo-2-hydroxybenzenemethanol

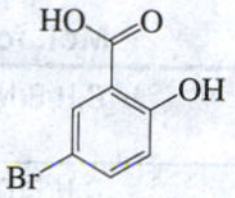
1233
5-Bromo-2-hydroxybenzoic acid

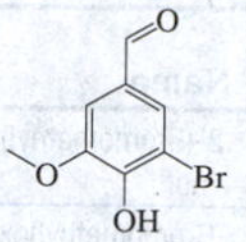
1234
3-Bromo-4-hydroxy-5-methoxybenzaldehyde

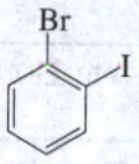
1235
1-Bromo-2-iodobenzene

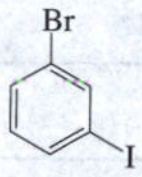
1236
1-Bromo-3-iodobenzene

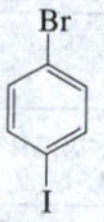
1237
1-Bromo-4-iodobenzene

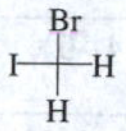
1238
Bromoiodomethane

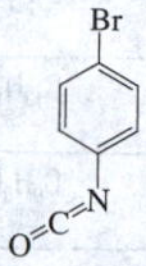
1239
1-Bromo-4-isocyanatobenzene

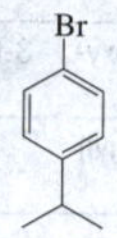
1240
1-Bromo-4-isopropylbenzene

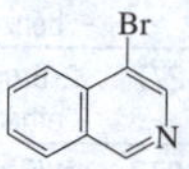
1241
4-Bromoisoquinoline

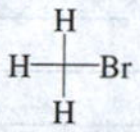
1242
Bromomethane

1243
1-Bromo-2-methoxyethane

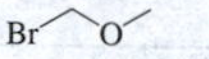
1244
Bromomethoxymethane

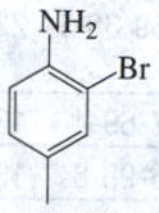
1245
2-Bromo-4-methylaniline

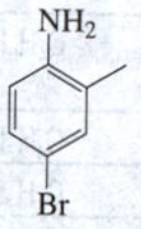
1246
4-Bromo-2-methylaniline

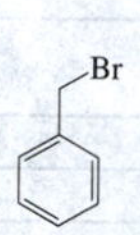
1247
(Bromomethyl)benzene

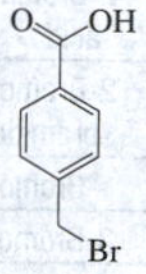
1248
4-(Bromomethyl)benzoic acid

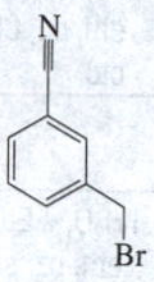
1249
3-(Bromomethyl)benzonitrile

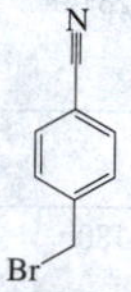
1250
4-(Bromomethyl)benzonitrile

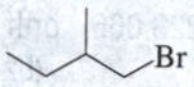
1251
1-Bromo-2-methylbutane, DL

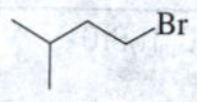
1252
1-Bromo-3-methylbutane

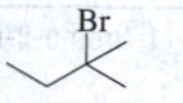
1253
2-Bromo-2-methylbutane

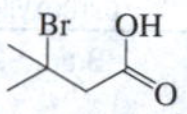
1254
3-Bromo-3-methylbutanoic acid

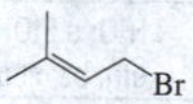
1255
1-Bromo-3-methyl-2-butene

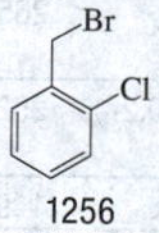
1256
1-(Bromomethyl)-2-chlorobenzene

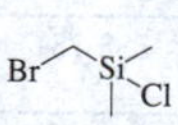
1257
(Bromomethyl)chlorodimethylsilane

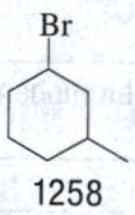
1258
1-Bromo-3-methylcyclohexane

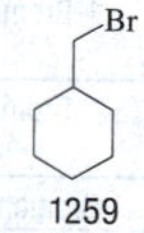
1259
(Bromomethyl)cyclohexane

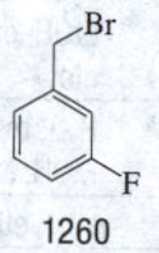
1260
1-(Bromomethyl)-3-fluorobenzene

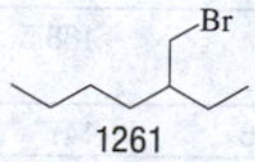
1261
3-(Bromomethyl)heptane

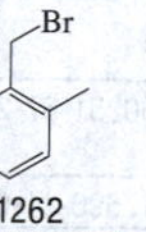
1262
1-(Bromomethyl)-2-methylbenzene

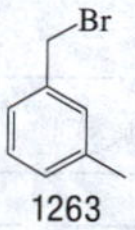
1263
1-(Bromomethyl)-3-methylbenzene

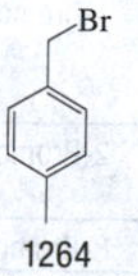
1264
1-(Bromomethyl)-4-methylbenzene

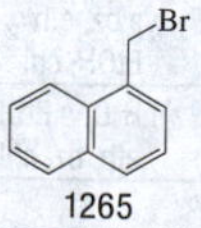
1265
1-(Bromomethyl)naphthalene

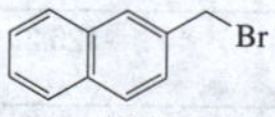
1266
2-(Bromomethyl)naphthalene

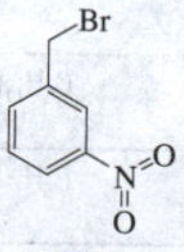
1267
1-(Bromomethyl)-3-nitrobenzene

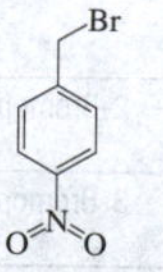
1268
1-(Bromomethyl)-4-nitrobenzene

No.	Name	Synonym	Mol. form.	CAS Reg. No.	Mol. wt.	Physical form	mp/°C	bp/°C	den/g cm^{-3}	n_D	Solubility
1269	2-(Bromomethyl)-4-nitrophenol		$C_7H_6BrNO_3$	772-33-8	232.032		148				
1270	(Bromomethyl)oxirane, (±)-		C_3H_5BrO	82584-73-4	136.975	liq	-40	138(3)	1.615^{14}	1.4841^{20}	i H_2O; s EtOH, eth, bz, chl
1271	1-Bromo-2-methylpentane	2-Methylpentyl bromide	$C_6H_{13}Br$	25346-33-2	165.071			139(8)	1.1624^{20}	1.4495^{20}	vs eth, chl
1272	1-Bromo-4-methylpentane		$C_6H_{13}Br$	626-88-0	165.071			145	1.1683^{20}	1.4490	vs eth, chl
1273	2-Bromo-2-methylpentane		$C_6H_{13}Br$	4283-80-1	165.071			142.5		1.442^{23}	vs eth, chl
1274	3-Bromo-3-methylpentane		$C_6H_{13}Br$	25346-31-0	165.071			130	1.1835^{20}	1.4525^{20}	vs eth, chl
1275	2-Bromo-4-methylphenol		C_7H_7BrO	6627-55-0	187.034	nd (peth)	55.3(0.5)	213.5	1.5422^{25}	1.5772^{20}	sl H_2O; s EtOH, bz, chl
1276	1-(Bromomethyl)-3-phenoxybenzene	3-Phenoxybenzyl bromide	$C_{13}H_{11}BrO$	51632-16-7	263.129	oil					
1277	2-Bromo-1-(4-methylphenyl)-ethanone		C_9H_9BrO	619-41-0	213.070	nd or lf (al)	51	157^{14}			vs eth, EtOH
1278	*N*-(Bromomethyl)phthalimide	2-(Bromomethyl)-1*H*-isoindole-1,3(2*H*)-dione	$C_9H_6BrNO_2$	5332-26-3	240.054	pr (chl, bz)	151.5				s ace; sl bz, chl; vs AcOEt
1279	1-Bromo-2-methylpropane	Isobutyl bromide	C_4H_9Br	78-77-3	137.018	liq	-117.8(0.7)	91.3(0.5)	1.272^{15}	1.4348^{20}	i H_2O; vs EtOH, eth, ace, chl, bz; s ctc
1280	2-Bromo-2-methylpropane	*tert*-Butyl bromide	C_4H_9Br	507-19-7	137.018	liq	-16.8(0.9)	73.3	1.4278^{20}	1.4278^{20}	i H_2O; sl ctc
1281	2-Bromo-2-methylpropanoic acid	α-Bromoisobutyric acid	$C_4H_7BrO_2$	2052-01-9	167.002	cry (peth)	48.5	199	1.4969^{60}		
1282	2-Bromo-2-methylpropanoyl bromide		$C_4H_6Br_2O$	20769-85-1	229.898			163	1.4067^{14}		vs ace, CS_2
1283	1-Bromo-2-methylpropene		C_4H_7Br	3017-69-4	135.003			91(7)	1.336^{20}		
1284	3-Bromo-2-methylpropene		C_4H_7Br	1458-98-6	135.003			92(18)	1.313^{20}		
1285	2-(Bromomethyl)-tetrahydrofuran		C_5H_9BrO	1192-30-9	165.028			170	1.4679^{20}	1.4850^{20}	s EtOH, eth
1286	(Bromomethyl)trimethylsilane		$C_4H_{11}BrSi$	18243-41-9	167.120			116.5	1.170^{25}	1.4460^{20}	
1287	1-Bromonaphthalene	1-Naphthyl bromide	$C_{10}H_7Br$	90-11-9	207.067	oily liq	6.1(0.1)	280(2)	1.4785^{20}	1.658^{20}	s H_2O, ace; msc EtOH, eth, bz; sl ctc
1288	2-Bromonaphthalene		$C_{10}H_7Br$	580-13-2	207.067	pl or orth lf (al)	58(2)	281(9)	1.605^{25}	1.6382^{60}	i H_2O; s EtOH, eth, bz, CS_2; sl ctc
1289	4-Bromo-1,8-naphthalenedicarboxylic anhydride		$C_{12}H_5BrO_3$	81-86-7	277.070		222				
1290	1-Bromo-2-naphthol	1-Bromo-β-naphthol	$C_{10}H_7BrO$	573-97-7	223.066	orth pr (bz-lig) nd (HOAc)	84	130			i H_2O; s EtOH, eth, bz; sl chl; vs HOAc
1291	4-Bromo-2-nitroaniline		$C_6H_5BrN_2O_2$	875-51-4	217.020	oran-ye nd (w)	111.5	sub			vs EtOH
1292	1-Bromo-2-nitrobenzene		$C_6H_4BrNO_2$	577-19-5	202.006	pa ye (al)	38.5(0.2)	253(6)	1.6245^{80}		i H_2O; vs EtOH; s eth, ace, bz; sl chl
1293	1-Bromo-3-nitrobenzene		$C_6H_4BrNO_2$	585-79-5	202.006	orth	54(3)	265	1.7036^{20}	1.5979^{20}	sl H_2O; s EtOH, eth, bz
1294	1-Bromo-4-nitrobenzene	*p*-Nitrobromobenzene	$C_6H_4BrNO_2$	586-78-7	202.006	orth or mcl pr (al)	133.0(0.2)	252(5)	1.948^{25}		i H_2O; s EtOH, eth, bz; sl chl
1295	Bromonitromethane		CH_2BrNO_2	563-70-2	139.937			149		1.4880^{20}	vs EtOH
1296	2-Bromo-2-nitro-1,3-propanediol	Bronopol	$C_3H_6BrNO_4$	52-51-7	199.989		131.5				
1297	1-Bromononane		$C_9H_{19}Br$	693-58-3	207.151	liq	-29.0(0.2)	221.4	1.0845^{25}	1.4522^{25}	
1298	1-Bromooctadecane		$C_{18}H_{37}Br$	112-89-0	333.391	cry (al)	27.6(0.4)	357(17)	0.9848^{20}	1.4631^{20}	i H_2O; s EtOH, eth; sl ctc
1299	1-Bromooctane	Octyl bromide	$C_8H_{17}Br$	111-83-1	193.125	liq	-55.0(0.3)	199(6)	1.1072^{25}	1.4503^{25}	i H_2O; msc EtOH, eth; sl ctc
1300	2-Bromooctane, (±)-		$C_8H_{17}Br$	60251-57-2	193.125			188.5	1.0878^{25}	1.4442^{25}	i H_2O; msc EtOH, eth
1301	8-Bromooctanoic acid		$C_8H_{15}BrO_2$	17696-11-6	223.108	nd (peth)	38.5	147^{2}			vs bz, eth, EtOH
1302	1-Bromopentadecane		$C_{15}H_{31}Br$	629-72-1	291.311		18.6(0.3)	322	1.0675^{20}	1.4611^{20}	i H_2O; s ace; vs chl
1303	Bromopentafluorobenzene		C_6BrF_5	344-04-7	246.960	liq	-31	133(2)	1.981^{25}	1.4490^{20}	
1304	Bromopentafluoroethane		C_2BrF_5	354-55-2	198.917	col gas		-21	1.8098^{25}		
1305	1-Bromopentane	Pentyl bromide	$C_5H_{11}Br$	110-53-2	151.045	liq	-88.0(0.2)	126(3)	1.2182^{20}	1.4447^{20}	i H_2O; s EtOH, bz, chl; sl ctc; msc eth
1306	2-Bromopentane		$C_5H_{11}Br$	107-81-3	151.045	liq	-95.5	117(5)	1.2075^{20}	1.4413^{20}	vs bz, eth, EtOH, chl
1307	3-Bromopentane		$C_5H_{11}Br$	1809-10-5	151.045	liq	-126.2	118(5)	1.214^{20}	1.4441^{20}	i H_2O; s EtOH, eth, bz, chl
1308	5-Bromopentanenitrile		C_5H_8BrN	5414-21-1	162.029			111^{12}	1.3989^{20}	1.4780^{20}	

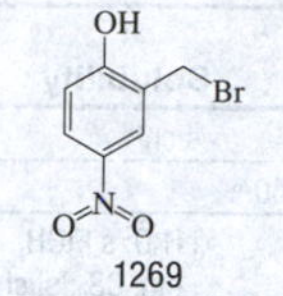
1269
2-(Bromomethyl)-4-nitrophenol

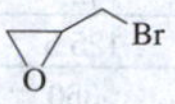
1270
(Bromomethyl)oxirane, (±)-

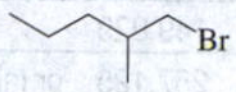
1271
1-Bromo-2-methylpentane

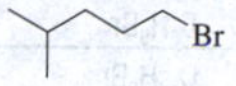
1272
1-Bromo-4-methylpentane

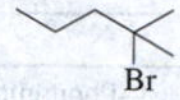
1273
2-Bromo-2-methylpentane

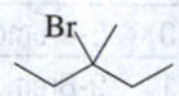
1274
3-Bromo-3-methylpentane

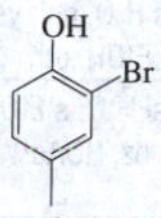
1275
2-Bromo-4-methylphenol

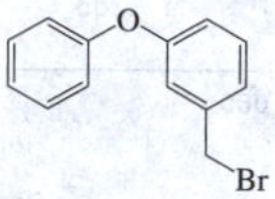
1276
1-(Bromomethyl)-3-phenoxybenzene

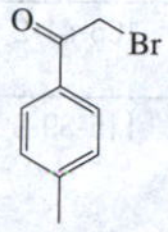
1277
2-Bromo-1-(4-methylphenyl)ethanone

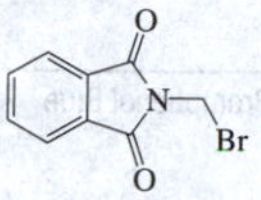
1278
N-(Bromomethyl)phthalimide

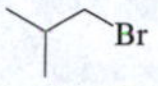
1279
1-Bromo-2-methylpropane

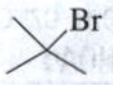
1280
2-Bromo-2-methylpropane

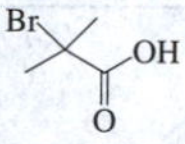
1281
2-Bromo-2-methylpropanoic acid

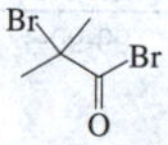
1282
2-Bromo-2-methylpropanoyl bromide

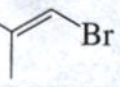
1283
1-Bromo-2-methylpropene

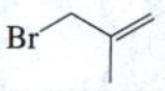
1284
3-Bromo-2-methylpropene

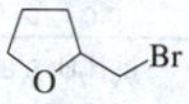
1285
2-(Bromomethyl)tetrahydrofuran

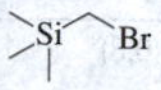
1286
(Bromomethyl)trimethylsilane

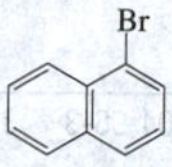
1287
1-Bromonaphthalene

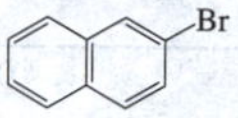
1288
2-Bromonaphthalene

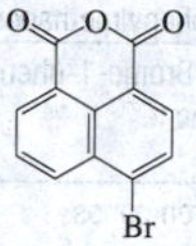
1289
4-Bromo-1,8-naphthalenedicarboxylic anhydride

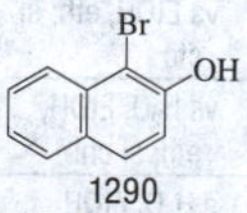
1290
1-Bromo-2-naphthol

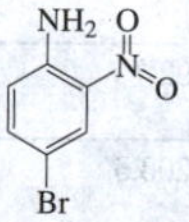
1291
4-Bromo-2-nitroaniline

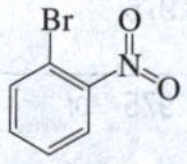
1292
1-Bromo-2-nitrobenzene

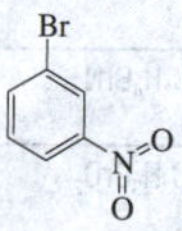
1293
1-Bromo-3-nitrobenzene

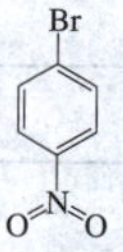
1294
1-Bromo-4-nitrobenzene

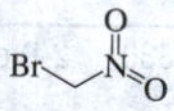
1295
Bromonitromethane

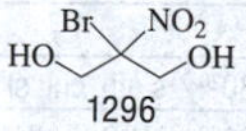
1296
2-Bromo-2-nitro-1,3-propanediol

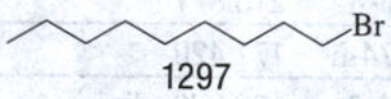
1297
1-Bromononane

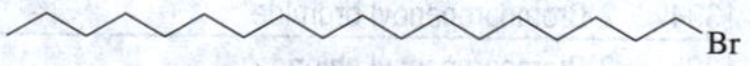
1298
1-Bromooctadecane

1299
1-Bromooctane

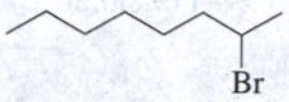
1300
2-Bromooctane, (±)-

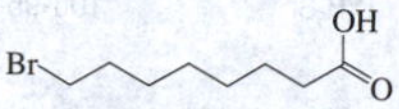
1301
8-Bromooctanoic acid

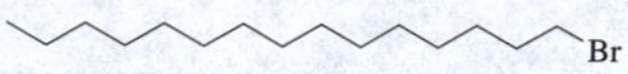
1302
1-Bromopentadecane

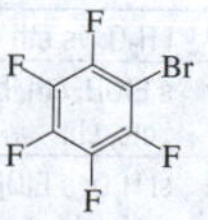
1303
Bromopentafluorobenzene

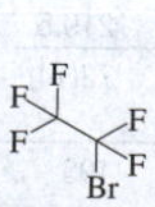
1304
Bromopentafluoroethane

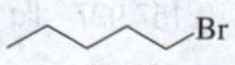
1305
1-Bromopentane

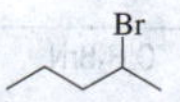
1306
2-Bromopentane

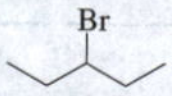
1307
3-Bromopentane

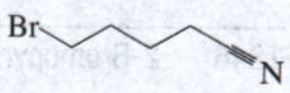
1308
5-Bromopentanenitrile

No.	Name	Synonym	Mol. form.	CAS Reg. No.	Mol. wt.	Physical form	mp/°C	bp/°C	den/ g cm^{-3}	n_D	Solubility
1309	5-Bromopentanoic acid		$C_5H_9BrO_2$	2067-33-6	181.028		40.0	142^{13}			s chl
1310	5-Bromo-1-pentene		C_5H_9Br	1119-51-3	149.029			125.5	1.2581^{20}	1.4640^{20}	
1311	9-Bromophenanthrene	9-Phenanthryl bromide	$C_{14}H_9Br$	573-17-1	257.125	pr (al)	64.5	>360	1.4093^{10}		i H_2O; s EtOH, eth, CS_2; sl chl
1312	2-Bromophenol		C_6H_5BrO	95-56-7	173.007		5.6	194.5	1.4924^{20}	1.589^{20}	sl H_2O, chl; s EtOH, eth, alk
1313	3-Bromophenol		C_6H_5BrO	591-20-8	173.007		33	236.5			sl H_2O, ctc; vs EtOH, eth; s chl, alk
1314	4-Bromophenol		C_6H_5BrO	106-41-2	173.007		63(1)	238	1.840^{15}		s H_2O, chl; vs EtOH, eth
1315	Bromophenol Blue	Bromphenol Blue	$C_{19}H_{10}Br_4O_5S$	115-39-9	669.960	hex pr (HOAc-ace)	279 dec				sl H_2O; s EtOH, bz, HOAc
1316	1-Bromo-4-phenoxybenzene	4-Bromophenyl phenyl ether	$C_{12}H_9BrO$	101-55-3	249.102		18.7(0.2)	$126^{3.5}$	1.6088^{20}	1.6084^{20}	i H_2O; s eth, ctc
1317	(4-Bromophenoxy)-trimethylsilane		$C_9H_{13}BrOSi$	17878-44-3	245.188			126^{25}	1.2619^{20}	1.5145^{20}	
1318	*N*-(4-Bromophenyl)acetamide	*p*-Bromoacetanilide	C_8H_8BrNO	103-88-8	214.060	nd (60% al)	168		1.717^{25}		i H_2O; s EtOH, chl; sl eth, bz
1319	1-(3-Bromophenyl)ethanone		C_8H_7BrO	2142-63-4	199.045		7.5	133^{19}		1.5755^{20}	i H_2O; s ace, bz
1320	1-(4-Bromophenyl)ethanone	*p*-Bromoacetophenone	C_8H_7BrO	99-90-1	199.045	lf (al)	50.5	257	1.647^{25}	1.647	i H_2O; s EtOH, eth, bz, ctc, HOAc
1321	(4-Bromophenyl)hydrazine	(*p*-Bromophenyl)hydrazine	$C_6H_7BrN_2$	589-21-9	187.037	nd (w), lf (lig), cry (al)	108				vs eth, EtOH, lig
1322	2-(4-Bromophenyl)-1*H*-indene-1,3(2*H*)-dione	Bromindione	$C_{15}H_9BrO_2$	1146-98-1	301.135	cry (lig)	138				
1323	(4-Bromophenyl)-phenylmethanone		$C_{13}H_9BrO$	90-90-4	261.113	lf (al)	82.5	350			i H_2O; sl EtOH, eth, bz, peth
1324	2-Bromo-1-phenyl-1-propanone		C_9H_9BrO	2114-00-3	213.070			247.5	1.4298^{20}	1.5720^{20}	i H_2O; s EtOH, eth, ace, bz, ctc
1325	Bromophos		$C_8H_8BrCl_2PS$	2104-96-3	317.999	ye cry	56.3(0.3)	$141^{0.01}$			sl H_2O; s eth, ctc, tol
1326	Bromophos-ethyl		$C_{10}H_{12}BrCl_2O_3PS$	4824-78-6	394.049	pale-ye liq		$122^{0.004}$			
1327	1-Bromopropane	Propyl bromide	C_3H_7Br	106-94-5	122.992	liq	-110.1(0.3)	70.8(0.2)	1.3537^{20}	1.4343^{20}	sl H_2O; s EtOH, eth, ace, bz, chl, ctc
1328	2-Bromopropane	Isopropyl bromide	C_3H_7Br	75-26-3	122.992	liq	-88.9(0.5)	59.34(0.09)	1.3140^{20}	1.4251^{20}	sl H_2O; s ace, bz, chl; msc EtOH, eth
1329	3-Bromopropanenitrile		C_3H_4BrN	2417-90-5	133.975			92^{25}	1.6152^{20}	1.4800^{20}	vs EtOH, eth; sl ctc
1330	2-Bromopropanoic acid, (±)-		$C_3H_5BrO_2$	10327-08-9	152.975	pr	25.7	203.5	1.7000^{20}	1.4753^{20}	vs H_2O, EtOH, eth; sl chl
1331	3-Bromopropanoic acid	β-Bromopropionic acid	$C_3H_5BrO_2$	590-92-1	152.975	pl (CCl_4)	62.5	141^{45}	1.48^{25}		s H_2O, EtOH, eth, bz, chl
1332	3-Bromo-1-propanol		C_3H_7BrO	627-18-9	138.991			105^{185}	1.5374^{20}	1.4834^{25}	s H_2O; msc EtOH, eth
1333	1-Bromo-2-propanol		C_3H_7BrO	19686-73-8	138.991			146.5	1.5585^{30}	1.4801^{20}	s H_2O; vs EtOH, eth
1334	2-Bromopropanoyl bromide		$C_3H_4Br_2O$	563-76-8	215.871			153	2.0611^{16}		
1335	2-Bromopropanoyl chloride		C_3H_4BrClO	7148-74-5	171.420			132	1.697^{11}	1.4780^{20}	s eth, chl; sl ctc
1336	*cis*-1-Bromopropene		C_3H_5Br	590-13-6	120.976	liq	-113	58(3)	1.4291^{20}	1.4560^{20}	i H_2O; s eth, ace, chl
1337	*trans*-1-Bromopropene		C_3H_5Br	590-15-8	120.976			61(5)			
1338	2-Bromopropene		C_3H_5Br	557-93-7	120.976	liq	-126	49(4)	1.3965^{16}	1.4467^{16}	i H_2O; s eth, ace, chl
1339	3-Bromopropene	Allyl bromide	C_3H_5Br	106-95-6	120.976	liq	-119.3(0.5)	70.1(0.5)	1.398^{20}	1.4697^{20}	i H_2O; msc EtOH, eth; s ctc, chl, CS_2
1340	(3-Bromo-1-propenyl)benzene		C_9H_9Br	4392-24-9	197.071	nd (al, eth)	34	130^{10}	1.3428^{30}	1.613^{20}	vs EtOH
1341	(3-Bromopropoxy)benzene		$C_9H_{11}BrO$	588-63-6	215.086		10.7	127^{18}	1.364^{16}		vs eth
1342	3-Bromopropylamine hydrobromide	3-Bromo-1-propanamine hydrobromide	$C_3H_9Br_2N$	5003-71-4	218.918		171.5				
1343	Bromopropylate	4,4'-Dibromobenzilic acid isopropyl ester	$C_{17}H_{16}Br_2O_3$	18181-80-1	428.115		76.0(0.5)		1.59^{20}		
1344	(3-Bromopropyl)benzene		$C_9H_{11}Br$	637-59-2	199.087			219.5	1.3106^{25}	1.5440^{25}	i H_2O; vs eth
1345	3-Bromo-1-propyne	Propargyl bromide	C_3H_3Br	106-96-7	118.960			73(14)	1.579^{19}	1.4922^{20}	s EtOH, eth, bz, ctc, chl
1346	2-Bromopyridine		C_5H_4BrN	109-04-6	157.997	liq	-40.1	193	1.6337^{20}	1.5734^{20}	sl H_2O; s EtOH, eth, ctc

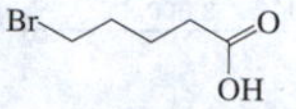
1309
5-Bromopentanoic acid

1310
5-Bromo-1-pentene

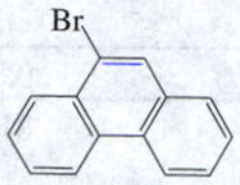
1311
9-Bromophenanthrene

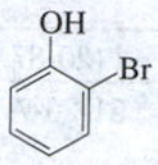
1312
2-Bromophenol

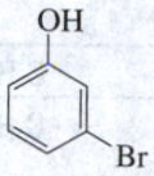
1313
3-Bromophenol

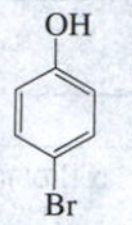
1314
4-Bromophenol

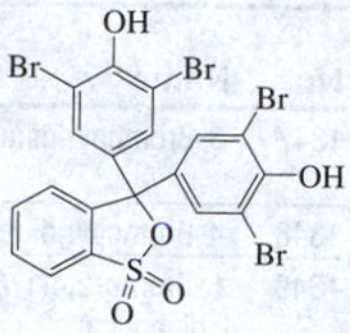
1315
Bromophenol Blue

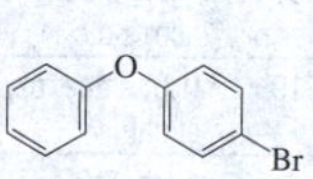
1316
1-Bromo-4-phenoxybenzene

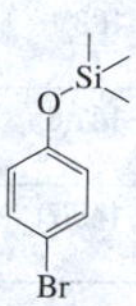
1317
(4-Bromophenoxy)trimethylsilane

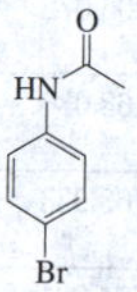
1318
N-(4-Bromophenyl)acetamide

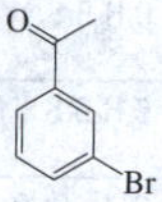
1319
1-(3-Bromophenyl)ethanone

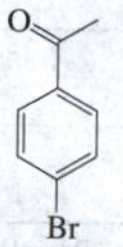
1320
1-(4-Bromophenyl)ethanone

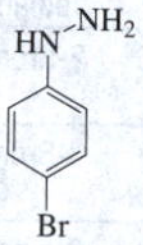
1321
(4-Bromophenyl)hydrazine

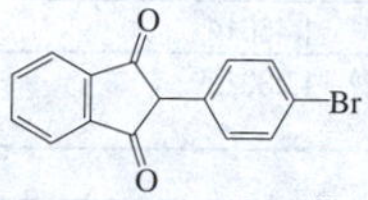
1322
2-(4-Bromophenyl)-1*H*-indene-1,3(2*H*)-dione

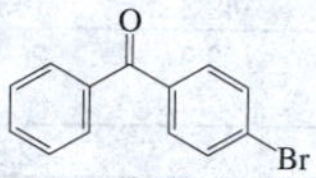
1323
(4-Bromophenyl)phenylmethanone

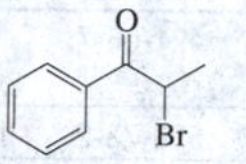
1324
2-Bromo-1-phenyl-1-propanone

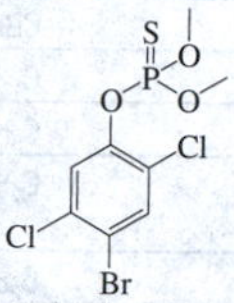
1325
Bromophos

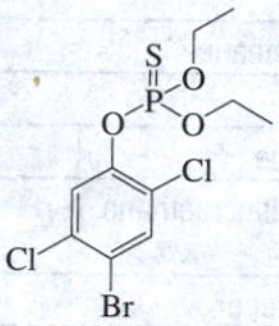
1326
Bromophos-ethyl

1327
1-Bromopropane

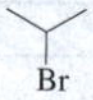
1328
2-Bromopropane

1329
3-Bromopropanenitrile

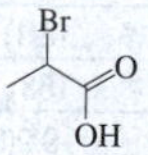
1330
2-Bromopropanoic acid, (±)-

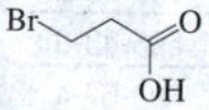
1331
3-Bromopropanoic acid

1332
3-Bromo-1-propanol

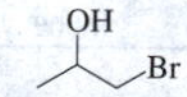
1333
1-Bromo-2-propanol

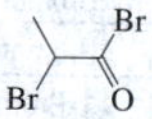
1334
2-Bromopropanoyl bromide

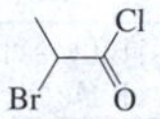
1335
2-Bromopropanoyl chloride

1336
cis-1-Bromopropene

1337
trans-1-Bromopropene

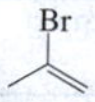
1338
2-Bromopropene

1339
3-Bromopropene

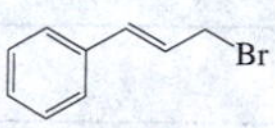
1340
(3-Bromo-1-propenyl)benzene

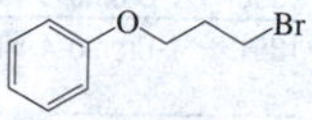
1341
(3-Bromopropoxy)benzene

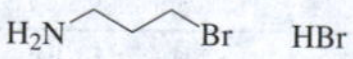
1342
3-Bromopropylamine hydrobromide

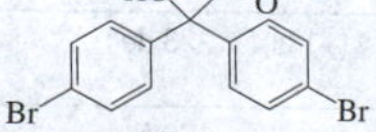
1343
Bromopropylate

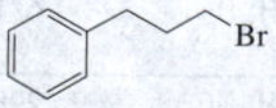
1344
(3-Bromopropyl)benzene

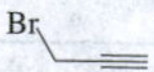
1345
3-Bromo-1-propyne

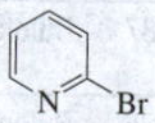
1346
2-Bromopyridine

No.	Name	Synonym	Mol. form.	CAS Reg. No.	Mol. wt.	Physical form	mp/°C	bp/°C	den/ g cm^{-3}	n_D	Solubility
1347	3-Bromopyridine		C_5H_4BrN	626-55-1	157.997	liq	-27.3	172(3)	1.645^{0}	1.5694^{20}	s H_2O; vs EtOH, eth
1348	4-Bromopyridine		C_5H_4BrN	1120-87-2	157.997		0.5	$29^{0.4}$	1.6450^{0}	1.5694^{20}	s ace, bz
1349	5-Bromo-2,4(1*H*,3*H*)-pyrimidinedione	5-Bromouracil	$C_4H_3BrN_2O_2$	51-20-7	190.983		310				
1350	3-Bromoquinoline		C_9H_6BrN	5332-24-1	208.055	ye oil	13.3	275		1.6641^{20}	s chl; vs HOAc
1351	6-Bromoquinoline		C_9H_6BrN	5332-25-2	208.055		24	281			s EtOH, eth, acid
1352	*N*-Bromosuccinimide		$C_4H_4BrNO_2$	128-08-5	177.985	cry (bz)	174		2.098^{25}		sl H_2O, AcOEt, eth; vs ace; i hx
1353	1-Bromotetradecane		$C_{14}H_{29}Br$	112-71-0	277.284		5.7(0.4)	307	1.0170^{20}	1.4603^{20}	vs ace, bz, EtOH
1354	2-Bromothiazole		C_3H_2BrNS	3034-53-5	164.024			171	1.82^{25}	1.5927^{20}	
1355	1-(5-Bromo-2-thienyl)-ethanone		C_6H_5BrOS	5370-25-2	205.072	nd (al)	94.5	103^{4}			sl EtOH; s ctc
1356	2-Bromothiophene	2-Thienyl bromide	C_4H_3BrS	1003-09-4	163.036			141(6)	1.684^{20}	1.5868^{20}	i H_2O; vs eth, ace; s ctc
1357	3-Bromothiophene		C_4H_3BrS	872-31-1	163.036			149(7)	1.735^{20}	1.5919^{20}	i H_2O; s ace, bz; sl chl
1358	Bromothymol Blue	Bromthymol Blue	$C_{27}H_{28}Br_2O_5S$	76-59-5	624.381		201				vs eth, EtOH
1359	2-Bromotoluene		C_7H_7Br	95-46-5	171.035	liq	-27.5(0.8)	182(2)	1.4232^{20}	1.5565^{20}	i H_2O; vs EtOH, eth, bz; msc ctc
1360	3-Bromotoluene		C_7H_7Br	591-17-3	171.035	liq	-38.1(0.2)	184(2)	1.4099^{20}	1.5510^{20}	i H_2O; s EtOH, ace, chl; msc eth; sl ctc
1361	4-Bromotoluene		C_7H_7Br	106-38-7	171.035	cry (al)	26.2(0.7)	184(4)	1.3959^{35}	1.5477^{20}	i H_2O; s EtOH, eth, ace, bz, chl; sl ctc
1362	Bromotrichloromethane		$CBrCl_3$	75-62-7	198.274	liq	-5.6(0.2)	103(5)	2.012^{25}	1.5065^{20}	vs eth, EtOH
1363	1-Bromotridecane		$C_{13}H_{27}Br$	765-09-3	263.257		5.9(0.3)	292	1.0234^{25}	1.4574^{25}	i H_2O; vs chl
1364	Bromotriethylsilane		$C_6H_{15}BrSi$	1112-48-7	195.173	liq	-49.3	163	1.143^{20}	1.4561^{20}	
1365	2-Bromo-1,1,1-trifluoroethane		$C_2H_2BrF_3$	421-06-7	162.936	vol liq or gas	-93.9	26	1.7881^{20}	1.3331^{20}	
1366	Bromotrifluoroethene		C_2BrF_3	598-73-2	160.920	col gas		-2.5			
1367	Bromotrifluoromethane	Halon-1301	$CBrF_3$	75-63-8	148.910	col gas	-174.4(0.2)	-57.8(0.4)	1.5800^{20}		i H_2O; vs chl
1368	1-Bromo-2-(trifluoromethyl)-benzene		$C_7H_4BrF_3$	392-83-6	225.006			167.5	1.652^{25}	1.4817^{20}	
1369	1-Bromo-3-(trifluoromethyl)-benzene		$C_7H_4BrF_3$	401-78-5	225.006		1	151.5	1.613^{25}	1.4716^{20}	
1370	1-Bromo-4-(trifluoromethyl)-benzene		$C_7H_4BrF_3$	402-43-7	225.006			160	1.607^{25}	1.4705^{25}	
1371	2-Bromo-1,3,5-trimethylbenzene		$C_9H_{11}Br$	576-83-0	199.087	liq	-1	225	1.3191^{10}	1.5510^{20}	i H_2O; vs eth; s bz; sl ctc
1372	Bromotrinitromethane		$CBrN_3O_6$	560-95-2	229.931		18(1)	56^{10}	2.0312^{20}	1.4808^{20}	vs EtOH, chl
1373	Bromotriphenylmethane	Triphenylmethyl bromide	$C_{19}H_{15}Br$	596-43-0	323.226		153	230^{15}	1.5500^{20}		
1374	1-Bromoundecane		$C_{11}H_{23}Br$	693-67-4	235.205	liq	-9.9(0.4)	258.8	1.0494^{25}	1.4552^{25}	sl ctc
1375	11-Bromoundecanoic acid		$C_{11}H_{21}BrO_2$	2834-05-1	265.188	nd (liq)	57	188^{18}			vs ace, bz, eth, EtOH
1376	(1-Bromovinyl)benzene		C_8H_7Br	98-81-7	183.046		-44	86^{14}	1.4025^{23}	1.5881^{20}	
1377	(*cis*-2-Bromovinyl)benzene		C_8H_7Br	588-73-8	183.046		-7	55^{2}	1.4322^{10}	1.5990^{22}	
1378	(*trans*-2-Bromovinyl)benzene		C_8H_7Br	588-72-7	183.046		7	219 dec	1.4269^{16}	1.6093^{20}	i H_2O; msc EtOH, eth; s chl
1379	1-Bromo-2-vinylbenzene		C_8H_7Br	2039-88-5	183.046	liq	-52.7(0.2)	215(2)	1.4160^{20}	1.5927^{20}	
1380	1-Bromo-3-vinylbenzene		C_8H_7Br	2039-86-3	183.046			92^{20}	1.4059^{20}	1.5933^{20}	
1381	1-Bromo-4-vinylbenzene		C_8H_7Br	2039-82-9	183.046		5(2)	213(1)	1.3984^{20}	1.5947^{20}	i H_2O; vs chl; s HOAc
1382	Brompheniramine		$C_{16}H_{19}BrN_2$	86-22-6	319.239	ye oily liq		$150^{0.5}$			s dil acid
1383	Brucine		$C_{23}H_{26}N_2O_4$	357-57-3	394.463	mcl pr (w +4)	178				sl H_2O, eth, bz; vs EtOH, chl
1384	Brucine hydrochloride	2,3-Dimethoxystrychnidin-10-one, monohydrochloride	$C_{23}H_{27}ClN_2O_4$	5786-96-9	430.924	pr					vs H_2O, EtOH
1385	Brucine sulfate heptahydrate	2,3-Dimethoxystrychnidin-10-one, sulfate, heptahydrate	$C_{46}H_{68}N_4O_{19}S$	60583-39-3	1013.113	nd (w)					s H_2O; sl EtOH, chl, tfa; vs MeOH; i bz
1386	Bucolome	5-Butyl-1-cyclohexyl-2,4,6(1*H*,3*H*,5*H*)-pyrimidinetrione	$C_{14}H_{22}N_2O_3$	841-73-6	266.336	nd (MeOH)	84	$186^{0.8}$			
1387	Bufotalin		$C_{26}H_{36}O_6$	471-95-4	444.560	cry (+1 al)	223 dec				i H_2O; s EtOH, chl
1388	Bulbocapnine		$C_{19}H_{19}NO_4$	298-45-3	325.359	pr (al)	199.5				i H_2O; s EtOH; vs chl

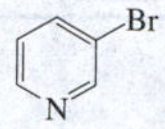
1347
3-Bromopyridine

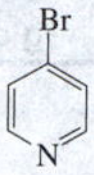
1348
4-Bromopyridine

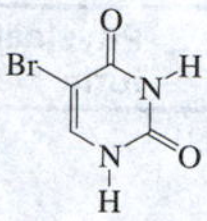
1349
5-Bromo-2,4(1*H*,3*H*)-pyrimidinedione

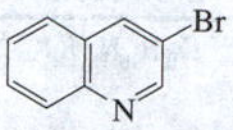
1350
3-Bromoquinoline

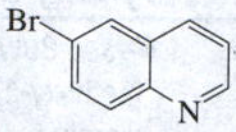
1351
6-Bromoquinoline

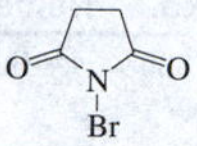
1352
N-Bromosuccinimide

1353
1-Bromotetradecane

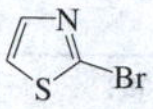
1354
2-Bromothiazole

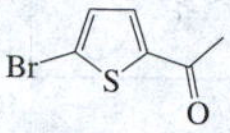
1355
1-(5-Bromo-2-thienyl)ethanone

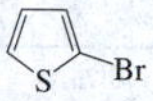
1356
2-Bromothiophene

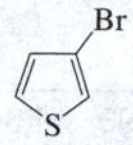
1357
3-Bromothiophene

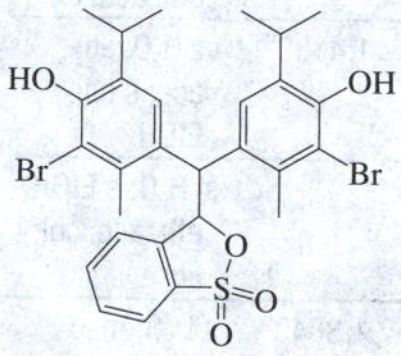
1358
Bromothymol Blue

1359
2-Bromotoluene

1360
3-Bromotoluene

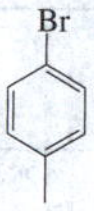
1361
4-Bromotoluene

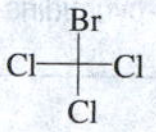
1362
Bromotrichloromethane

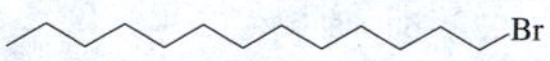
1363
1-Bromotridecane

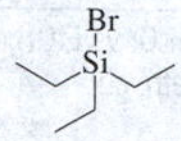
1364
Bromotriethylsilane

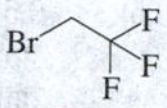
1365
2-Bromo-1,1,1-trifluoroethane

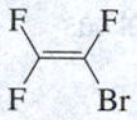
1366
Bromotrifluoroethene

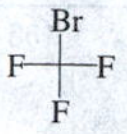
1367
Bromotrifluoromethane

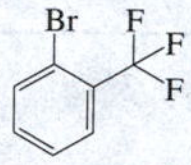
1368
1-Bromo-2-(trifluoromethyl)benzene

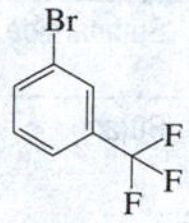
1369
1-Bromo-3-(trifluoromethyl)benzene

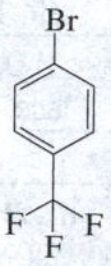
1370
1-Bromo-4-(trifluoromethyl)benzene

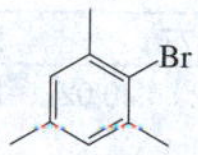
1371
2-Bromo-1,3,5-trimethylbenzene

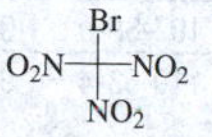
1372
Bromotrinitromethane

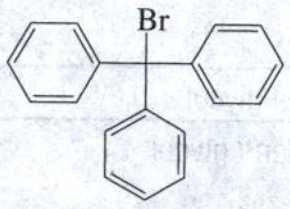
1373
Bromotriphenylmethane

1374
1-Bromoundecane

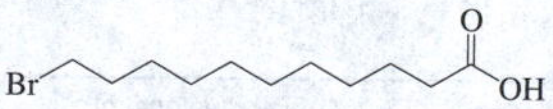
1375
11-Bromoundecanoic acid

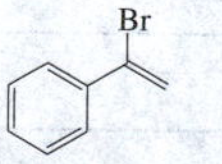
1376
(1-Bromovinyl)benzene

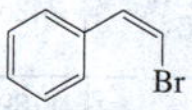
1377
(*cis*-2-Bromovinyl)benzene

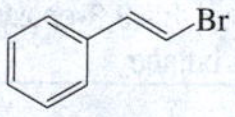
1378
(*trans*-2-Bromovinyl)benzene

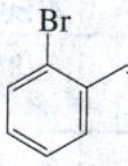
1379
1-Bromo-2-vinylbenzene

1380
1-Bromo-3-vinylbenzene

1381
1-Bromo-4-vinylbenzene

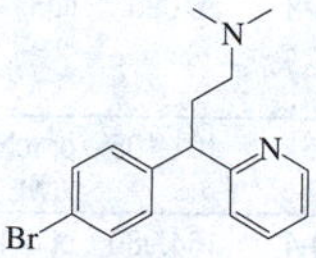
1382
Brompheniramine

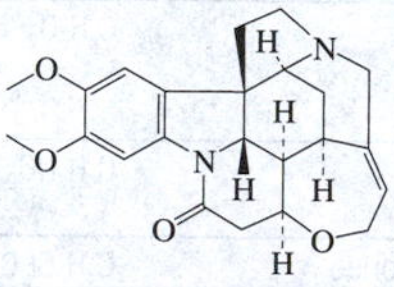
1383
Brucine

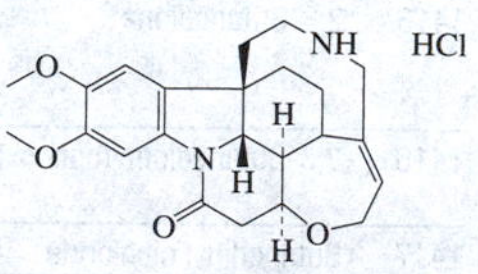
1384
Brucine hydrochloride

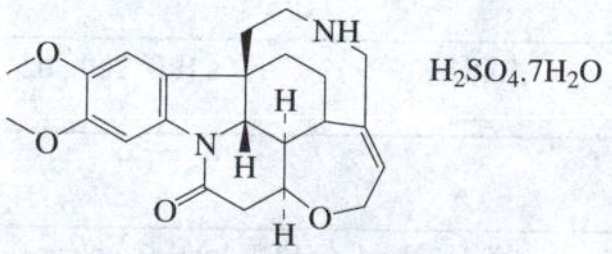
1385
Brucine sulfate heptahydrate

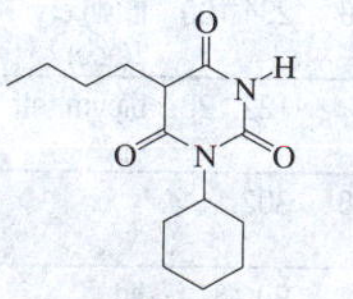
1386
Bucolome

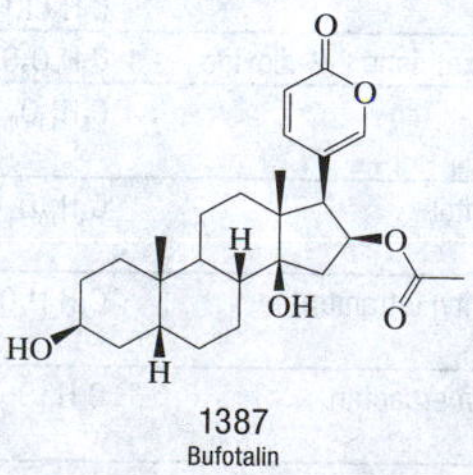
1387
Bufotalin

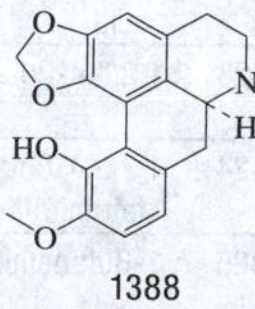
1388
Bulbocapnine

Organic

No.	Name	Synonym	Mol. form.	CAS Reg. No.	Mol. wt.	Physical form	mp/°C	bp/°C	den/ g cm⁻³	n_D	Solubility
1389	*sec*-Bumeton	N^2-*sec*-Butyl-N^4-ethyl-6-methoxy-1,3,5-triazine-2,4-diamine	$C_{10}H_{19}N_5O$	26259-45-0	225.291		87				
1390	BUSAN 72A	(2-Benzothiazolylthio)methyl thiocyanate	$C_9H_6N_2S_3$	21564-17-0	238.352	liq					
1391	Butachlor		$C_{17}H_{26}ClNO_2$	23184-66-9	311.847		<-5	$156^{.5}$	1.070^{25}		
1392	1,2-Butadiene	Methylallene	C_4H_6	590-19-2	54.091	vol liq or gas	-136.20(0.05)	11.0(0.2)	0.676^{0}	1.4205^{1}	i H_2O; msc EtOH, eth; vs bz
1393	1,3-Butadiene	Divinyl	C_4H_6	106-99-0	54.091	col gas	-108.9(0.1)	-4.6(0.2)	0.6149^{25} (p>1 atm	1.4292^{-25}	i H_2O; s EtOH, eth, bz; vs ace
1394	1,3-Butadien-1-ol acetate		$C_6H_8O_2$	1515-76-0	112.127			58^{40}	0.945^{25}	1.4690^{20}	
1395	*trans*-1,3-Butadienylbenzene		$C_{10}H_{10}$	16939-57-4	130.186		2.3	76^{11}	0.9286^{20}	1.6089^{25}	i H_2O; s EtOH, eth, ace, bz
1396	1,3-Butadiyne	Diacetylene	C_4H_2	460-12-8	50.059	vol liq or gas	-35(3)	10(2)	0.7364^{0}	1.4189^{5}	vs H_2O, eth, ace; s chl, EtOH
1397	Butalbital	5-Isobutyl-5-allyl-2,4,6(1*H*,3*H*,5*H*)-pyrimidinetrione	$C_{11}H_{16}N_2O_3$	77-26-9	224.256	pr	138.5				sl H_2O; s EtOH, eth, ace, chl; i lig
1398	Butanal	Butyraldehyde	C_4H_8O	123-72-8	72.106	liq	-96.86(0.02)	74.8(0.2)	0.8016^{20}	1.3843^{20}	s H_2O; msc EtOH; vs ace, bz; sl chl
1399	Butanal oxime		C_4H_9NO	110-69-0	87.120	liq	-29.5	152(9)	0.923^{20}		vs H_2O, ace, bz; msc EtOH, eth; s chl
1400	Butanamide	Butyramide	C_4H_9NO	541-35-5	87.120	lf (bz)	116(1)	231.8(1)	0.8850^{120}	1.4087^{130}	sl H_2O, eth; i bz; s EtOH
1401	Butane		C_4H_{10}	106-97-8	58.122	col gas	-138.3(0.1)	-0.5(0.5)	0.573^{25} (p>1 atm)	1.3326^{20}	i H_2O; vs EtOH, eth, chl
1402	Butanedial		$C_4H_6O_2$	638-37-9	86.090			170 dec	1.065^{20}	1.4262^{18}	vs H_2O, ace, eth, EtOH
1403	1,4-Butanediamine	Putrescine	$C_4H_{12}N_2$	110-60-1	88.151	lf	21.9(0.4)	156(10)	0.877^{25}	1.4969^{20}	s H_2O
1404	1,4-Butanediamine dihydrochloride		$C_4H_{14}Cl_2N_2$	333-93-7	161.073	nd or lf (al, w)	280 dec	sub			vs H_2O, EtOH; i eth, bz, MeOH
1405	1,2-Butanediol, (±)-		$C_4H_{10}O_2$	26171-83-5	90.121			196.42(0.06)	1.0024^{20}	1.4378^{20}	s H_2O, EtOH, ace
1406	1,3-Butanediol	1,3-Butylene glycol	$C_4H_{10}O_2$	107-88-0	90.121	visc liq	-77	208.2(0.1)	1.0053^{20}	1.4401^{20}	
1407	1,4-Butanediol	Tetramethylene glycol	$C_4H_{10}O_2$	110-63-4	90.121		20.43(0.02)	229.5(0.4)	1.0171^{20}	1.4460^{20}	msc H_2O; s EtOH, DMSO; sl eth
1408	2,3-Butanediol		$C_4H_{10}O_2$	6982-25-8	90.121	cry (eth)	7(2)	178(3)	1.0033^{20}	1.4310^{25}	msc H_2O, EtOH; s eth, ace, chl
1409	1,4-Butanediol diacetate		$C_8H_{14}O_4$	628-67-1	174.195		12	233(6)	1.0479^{15}	1.4251^{15}	
1410	1,4-Butanediol diacrylate		$C_{10}H_{14}O_4$	1070-70-8	198.216			$83^{0.3}$	1.105^{25}		
1411	1,4-Butanediol diglycidyl ether	1,4-Bis(2,3-epoxypropoxy)-butane	$C_{10}H_{18}O_4$	2425-79-8	202.248			266	1.1^{25}	1.4611^{20}	
1412	1,3-Butanediol dimethacrylate		$C_{12}H_{18}O_4$	1189-08-8	226.269			290		1.4495^{25}	vs ace, eth, EtOH, lig
1413	1,4-Butanediol dimethacrylate		$C_{12}H_{18}O_4$	2082-81-7	226.269	liq		133^{4}	1.025^{20}	1.4560^{20}	sl H_2O
1414	1,4-Butanediol dimethylsulfonate	Busulfan	$C_6H_{14}O_6S_2$	55-98-1	246.301	cry	116				i H_2O; sl EtOH, ace
1415	2,3-Butanedione	Diacetyl	$C_4H_6O_2$	431-03-8	86.090	liq	-1.2	87.5(0.8)	0.9808^{18}	1.3951^{20}	vs H_2O; msc EtOH, eth; s bz, ctc
1416	2,3-Butanedione monooxime		$C_4H_7NO_2$	57-71-6	101.105	pr (chl), lf (w)	76.8	185.5			sl H_2O; vs EtOH, eth, chl; s alk
1417	Butanedioyl dichloride	Succinyl chloride	$C_4H_4Cl_2O_2$	543-20-4	154.980	pl or lf	16.7(0.6)	190(5)	1.3748^{20}	1.4683^{20}	s eth, ace, bz
1418	1,4-Butanedithiol	Tetramethylenedithiol	$C_4H_{10}S_2$	1191-08-8	122.252	liq	-53.9	195.5	1.0021^{0}	1.5290^{20}	i H_2O; vs EtOH; sl ctc
1419	Butanenitrile	Propyl cyanide	C_4H_7N	109-74-0	69.106	liq	-111.76(0.05)	117.6(0.4)	0.7936^{20}	1.3842^{20}	sl H_2O, ctc; msc EtOH, eth; s bz
1420	1-Butanesulfonyl chloride		$C_4H_9ClO_2S$	2386-60-9	156.631			75^{10}		1.4559^{20}	
1421	1,4-Butane sultone	1,2-Oxathiane 2,2-dioxide	$C_4H_8O_3S$	1633-83-6	136.170	liq	13.5	135^{4}	1.331^{20}	1.4640^{20}	
1422	1,2,3,4-Butanetetracarboxylic acid		$C_8H_{10}O_8$	1703-58-8	234.160	lf (w) cry (ace)	236.5				vs H_2O, EtOH
1423	1,2,3,4-Butanetetrol	Erythritol	$C_4H_{10}O_4$	149-32-6	122.120	bipym tetr pr	118.1(0.7)	330.5	1.451^{20}		s H_2O; i eth, bz
1424	1,2,3,4-Butanetetrol tetranitrate, (*R**,*S**)-	Erythrityl tetranitrate	$C_4H_6N_4O_{12}$	7297-25-8	302.111		61				vs EtOH
1425	1-Butanethiol	Butyl mercaptan	$C_4H_{10}S$	109-79-5	90.187	liq	-115.66(0.06)	98.4(0.5)	0.8416^{20}	1.4440^{20}	sl H_2O, chl; vs EtOH, eth

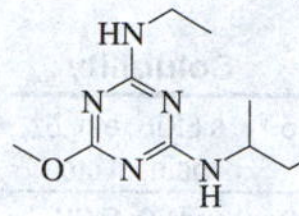
1389
sec-Bumeton

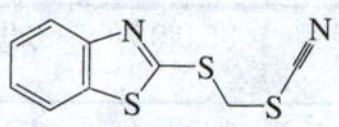
1390
BUSAN 72A

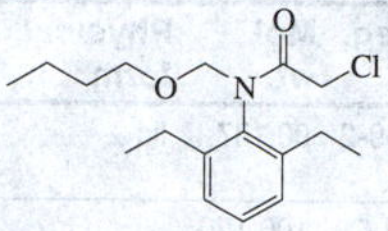
1391
Butachlor

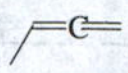
1392
1,2-Butadiene

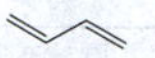
1393
1,3-Butadiene

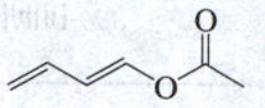
1394
1,3-Butadien-1-ol acetate

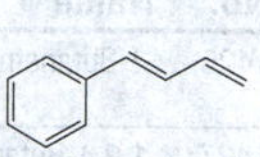
1395
trans-1,3-Butadienylbenzene

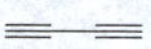

1396
1,3-Butadiyne

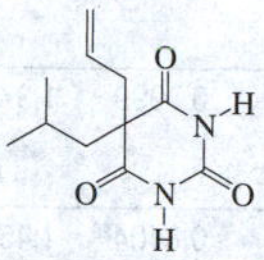
1397
Butalbital

1398
Butanal

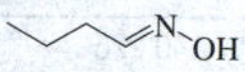
1399
Butanal oxime

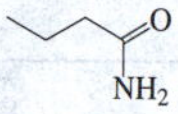
1400
Butanamide

1401
Butane

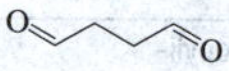
1402
Butanedial

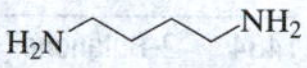
1403
1,4-Butanediamine

1404
1,4-Butanediamine dihydrochloride

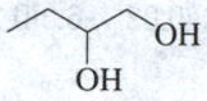
1405
1,2-Butanediol, (±)-

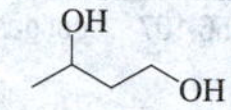
1406
1,3-Butanediol

1407
1,4-Butanediol

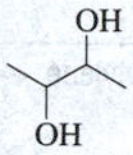
1408
2,3-Butanediol

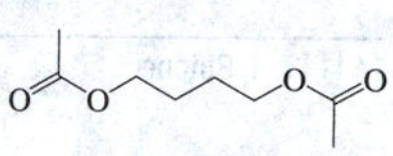
1409
1,4-Butanediol diacetate

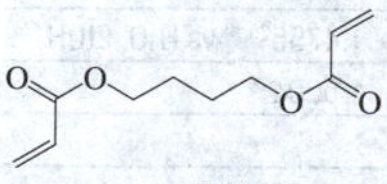
1410
1,4-Butanediol diacrylate

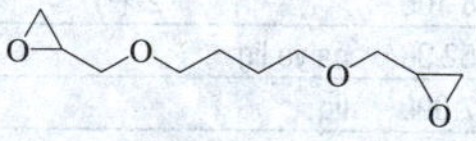
1411
1,4-Butanediol diglycidyl ether

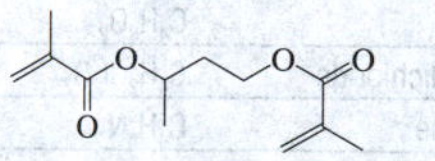
1412
1,3-Butanediol dimethacrylate

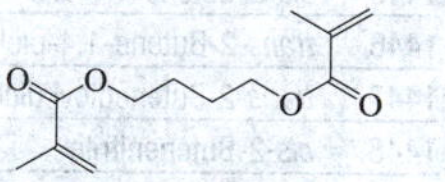
1413
1,4-Butanediol dimethacrylate

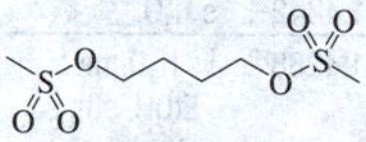
1414
1,4-Butanediol dimethylsulfonate

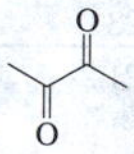
1415
2,3-Butanedione

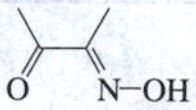
1416
2,3-Butanedione monooxime

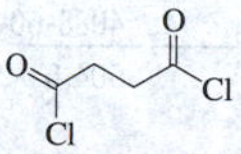
1417
Butanedioyl dichloride

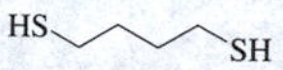
1418
1,4-Butanedithiol

1419
Butanenitrile

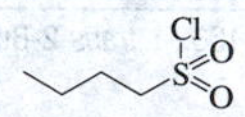
1420
1-Butanesulfonyl chloride

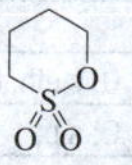
1421
1,4-Butane sultone

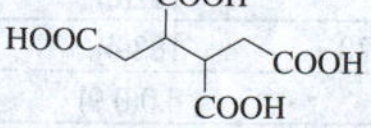
1422
1,2,3,4-Butanetetracarboxylic acid

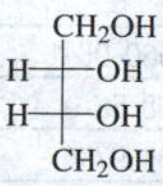
1423
1,2,3,4-Butanetetrol

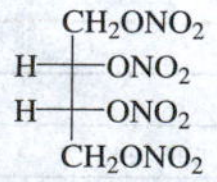
1424
1,2,3,4-Butanetetrol tetranitrate, (*R**,*S**)-

1425
1-Butanethiol

Organic

No.	Name	Synonym	Mol. form.	CAS Reg. No.	Mol. wt.	Physical form	mp/°C	bp/°C	den/ g cm^{-3}	n_D	Solubility
1426	2-Butanethiol	*sec*-Butyl mercaptan	$C_4H_{10}S$	91840-99-2	90.187	liq	-165	85.0(0.7)	0.8295^{20}	1.4366^{20}	s EtOH, eth, bz, peth; sl ctc
1427	1,2,4-Butanetriol		$C_4H_{10}O_3$	3068-00-6	106.120			190^{18}	1.18^{20}	1.4688^{20}	vs H_2O, EtOH
1428	Butanilicaine	2-(Butylamino)-*N*-(2-chloro-6-methylphenyl)acetamide	$C_{13}H_{19}ClN_2O$	3785-21-5	254.755	cry	46	$145^{0.001}$			
1429	Butanoic acid	Butyric acid	$C_4H_8O_2$	107-92-6	88.106	liq	-5.12(0.09)	163.7(0.1)	0.9528^{25}	1.3980^{20}	msc H_2O, EtOH, eth; sl ctc
1430	Butanoic anhydride	Butyric anhydride	$C_8H_{14}O_3$	106-31-0	158.195	liq	-75.0(0.6)	195(1)	0.9668^{20}	1.4070^{20}	s eth; sl ctc
1431	1-Butanol	Butyl alcohol	$C_4H_{10}O$	71-36-3	74.121	liq	-88.60(0.02)	117.6(0.2)	0.8095^{20}	1.3988^{20}	s H_2O, bz; msc EtOH, eth; vs ace
1432	2-Butanol	*sec*-Butyl alcohol	$C_4H_{10}O$	78-92-2	74.121	liq	-88.44(0.07)	99.4(0.2)	0.8063^{20}	1.3978^{20}	vs H_2O; msc EtOH, eth; s bz, ctc
1433	2-Butanone	Methyl ethyl ketone	C_4H_8O	78-93-3	72.106	liq	-86.67(0.01)	79.6(0.2)	0.7999^{25}	1.3788^{20}	vs H_2O; msc EtOH, eth, ace, bz; s chl
1434	2-Butanone (1-methylpropylidene)hydrazone		$C_8H_{16}N_2$	5921-54-0	140.226			170(7)	0.8404^{20}	1.4511^{20}	
1435	2-Butanone oxime		C_4H_9NO	96-29-7	87.120	liq	-29.5(0.5)	151.5(0.6)	0.9232^{20}	1.4410^{20}	s H_2O, chl; msc EtOH, eth
1436	2-Butanone peroxide	Methyl ethyl ketone peroxide	$C_8H_{16}O_4$	1338-23-4	176.211	col liq		exp			sl H_2O; misc os
1437	Butanoyl chloride	*n*-Butyryl chloride	C_4H_7ClO	141-75-3	106.551	liq	-89.0(0.5)	101(3)	1.0277^{20}	1.4121^{20}	msc eth
1438	Butaperazine		$C_{24}H_{31}N_3OS$	653-03-2	409.587			$275^{0.05}$			
1439	Butazolamide	*N*-[5-(Aminosulfonyl)-1,3,4-thiadiazol-2-yl]butanamide	$C_6H_{10}N_4O_3S_2$	16790-49-1	250.298	cry	261 dec				
1440	*trans*-2-Butenal	*trans*-Crotonaldehyde	C_4H_6O	123-73-9	70.090	liq	-76.6(0.3)	102.2(0.3)	0.8516^{20}	1.4366^{20}	s H_2O, chl; vs EtOH, eth, ace; msc bz
1441	1-Butene	1-Butylene	C_4H_8	106-98-9	56.107	col gas	-185.33(0.02)	-6.3(0.2)	0.588^{25} (p>1 atm)	1.3962^{20}	i H_2O; vs EtOH, eth; s bz
1442	*cis*-2-Butene		C_4H_8	590-18-1	56.107	col gas	-138.89(0.02)	3.72(0.08)	0.616^{25} (p>1 atm)	1.3931^{-25}	i H_2O; vs EtOH, eth; s bz
1443	*trans*-2-Butene		C_4H_8	624-64-6	56.107	col gas	-105.52(0.02)	0.88(0.09)	0.599^{25} (p>1 atm)	1.3848^{-25}	s bz
1444	*trans*-2-Butenedinitrile		$C_4H_2N_2$	764-42-1	78.072	nd (bz-peth)	96.0(0.8)	186	0.9416^{111}	1.4349^{111}	s H_2O, EtOH, eth, ace, bz, chl; sl peth
1445	*cis*-2-Butene-1,4-diol		$C_4H_8O_2$	6117-80-2	88.106		11.0(0.5)	235	1.0698^{20}	1.4782^{20}	s H_2O; vs EtOH
1446	*trans*-2-Butene-1,4-diol		$C_4H_8O_2$	821-11-4	88.106		27(1)	131^{13}	1.0700^{20}	1.4755^{20}	vs H_2O, EtOH
1447	*trans*-2-Butenedioyl dichloride	Fumaric acid dichloride	$C_4H_2Cl_2O_2$	627-63-4	152.964	pa ye lig		159	1.408^{20}	1.5004^{18}	
1448	*cis*-2-Butenenitrile	Isocrotononitrile	C_4H_5N	1190-76-7	67.090	liq		106(6)			
1449	*trans*-2-Butenenitrile	Crotononitrile	C_4H_5N	627-26-9	67.090	liq	-51.5	120	0.8239^{20}	1.4225^{20}	s eth, ace
1450	3-Butenenitrile	Allyl cyanide	C_4H_5N	109-75-1	67.090	liq	-87	117(5)	0.8341^{20}	1.4060^{20}	sl H_2O; msc EtOH, eth
1451	*cis*-2-Butenoic acid	Isocrotonic acid	$C_4H_6O_2$	503-64-0	86.090	nd or pr (peth)	15	169	1.0267^{20}	1.4450^{20}	vs H_2O; s EtOH
1452	*trans*-2-Butenoic acid	Crotonic acid	$C_4H_6O_2$	107-93-7	86.090	mcl pr or nd (w, lig)	71.3(0.2)	184.7	0.9604^{77}	1.4249^{77}	vs H_2O, EtOH; s eth, ace, lig
1453	3-Butenoic acid		$C_4H_6O_2$	625-38-7	86.090	liq	-35	169	1.0091^{20}	1.4239^{20}	s H_2O; msc EtOH, eth
1454	2-Butenoic anhydride	Crotonic acid anhydride	$C_8H_{10}O_3$	623-68-7	154.163			247	1.0397^{20}	1.4745^{20}	vs eth
1455	*cis*-2-Buten-1-ol	*cis*-Crotyl alcohol	C_4H_8O	4088-60-2	72.106			123	0.8662^{20}	1.4342^{25}	s H_2O
1456	*trans*-2-Buten-1-ol	*trans*-Crotyl alcohol	C_4H_8O	504-61-0	72.106		<-30	121.2	0.8521^{20}	1.4288^{20}	vs H_2O; msc EtOH, eth; s chl
1457	3-Buten-1-ol		C_4H_8O	627-27-0	72.106			112(5)	0.8424^{20}	1.4224^{20}	s H_2O, ace; msc EtOH, eth; sl chl
1458	3-Buten-2-ol		C_4H_8O	598-32-3	72.106			97(4)			
1459	3-Buten-2-one	Methyl vinyl ketone	C_4H_6O	78-94-4	70.090			81(4)	0.864^{20}	1.4081^{20}	s H_2O, EtOH, bz; vs eth, ace; sl ctc
1460	2-Butenoyl chloride		C_4H_5ClO	10487-71-5	104.535			121(8)	1.0905^{20}	1.460^{18}	vs ace
1461	*trans*-1-Butenylbenzene		$C_{10}H_{12}$	1005-64-7	132.202	liq	-43.1(0.4)	201(4)	0.9019^{20}	1.5420^{20}	i H_2O; s EtOH, eth, bz, ctc
1462	2-Butenylbenzene		$C_{10}H_{12}$	1560-06-1	132.202			182(6)	0.8831^{20}	1.5101^{20}	
1463	3-Butenylbenzene		$C_{10}H_{12}$	768-56-9	132.202	liq	-70	183(4)	0.8831^{20}	1.5059^{20}	i H_2O; s eth, bz
1464	1-Buten-3-yne	Vinylacetylene	C_4H_4	689-97-4	52.075	col gas		6.0(0.9)	0.7094^{0}	1.4161^{1}	i H_2O; s bz

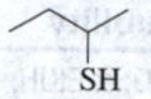
1426
2-Butanethiol

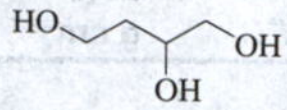
1427
1,2,4-Butanetriol

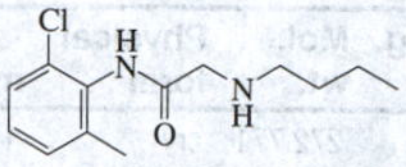
1428
Butanilicaine

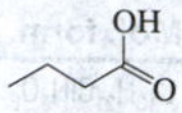
1429
Butanoic acid

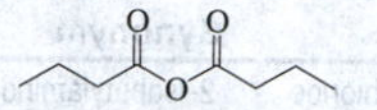
1430
Butanoic anhydride

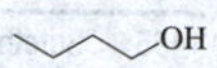
1431
1-Butanol

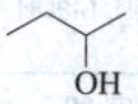
1432
2-Butanol

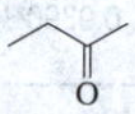
1433
2-Butanone

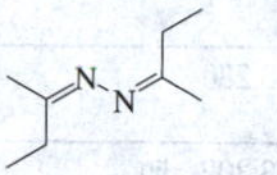
1434
2-Butanone (1-methylpropylidene)hydrazone

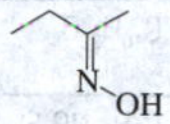
1435
2-Butanone oxime

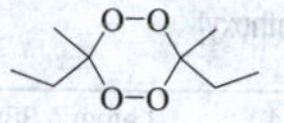
1436
2-Butanone peroxide

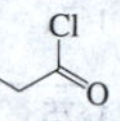
1437
Butanoyl chloride

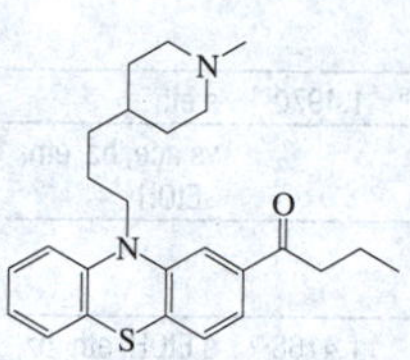
1438
Butaperazine

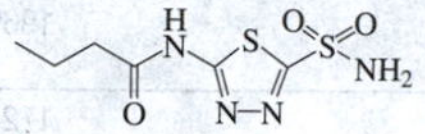
1439
Butazolamide

1440
trans-2-Butenal

1441
1-Butene

1442
cis-2-Butene

1443
trans-2-Butene

1444
trans-2-Butenedinitrile

1445
cis-2-Butene-1,4-diol

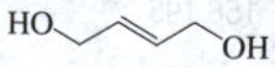
1446
trans-2-Butene-1,4-diol

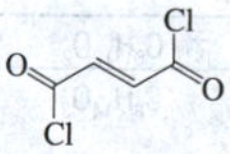
1447
trans-2-Butenedioyl dichloride

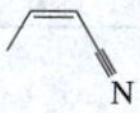
1448
cis-2-Butenenitrile

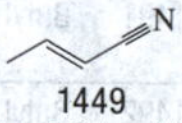
1449
trans-2-Butenenitrile

1450
3-Butenenitrile

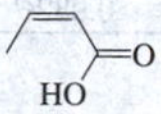
1451
cis-2-Butenoic acid

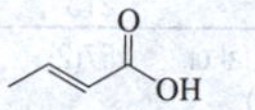
1452
trans-2-Butenoic acid

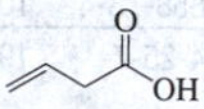
1453
3-Butenoic acid

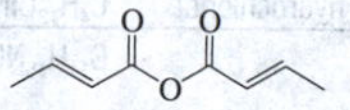
1454
2-Butenoic anhydride

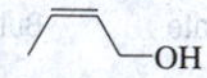
1455
cis-2-Buten-1-ol

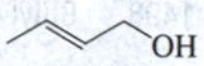
1456
trans-2-Buten-1-ol

1457
3-Buten-1-ol

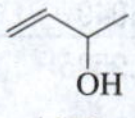
1458
3-Buten-2-ol

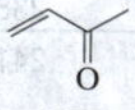
1459
3-Buten-2-one

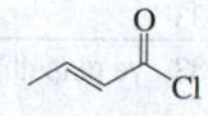
1460
2-Butenoyl chloride

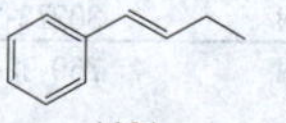
1461
trans-1-Butenylbenzene

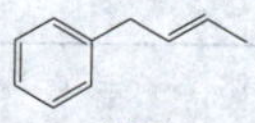
1462
2-Butenylbenzene

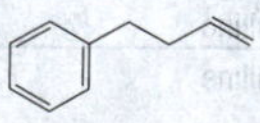
1463
3-Butenylbenzene

1464
1-Buten-3-yne

Organic

No.	Name	Synonym	Mol. form.	CAS Reg. No.	Mol. wt.	Physical form	mp/°C	bp/°C	den/ g cm^{-3}	n_D	Solubility
1465	Butethamine hydrochloride	2-Isobutylaminoethyl 4-aminobenzoate	$C_{13}H_{21}ClN_2O_2$	553-68-4	272.771	cry	194				s H_2O; sl EtOH, bz, chl; i eth
1466	Buthalital sodium		$C_{11}H_{15}N_2NaO_2S$	510-90-7	262.304						vs H_2O; sl EtOH; i eth, bz
1467	Buthiazide		$C_{11}H_{16}ClN_3O_4S_2$	2043-38-1	353.846		221.5				
1468	Buthiobate	Denmert	$C_{21}H_{28}N_2S_2$	51308-54-4	372.590	ye oil	32		1.0865^{25}	1.596^{26}	i H_2O; s os
1469	Butonate		$C_8H_{14}Cl_3O_5P$	126-22-7	327.527			$129^{0.5}$			
1470	Butoxyacetylene		$C_6H_{10}O$	3329-56-4	98.142			104	0.8200^{20}	1.4067	vs eth, EtOH
1471	4-Butoxyaniline		$C_{10}H_{15}NO$	4344-55-2	165.232			132^{4}			
1472	4-Butoxybenzaldehyde		$C_{11}H_{14}O_2$	5736-88-9	178.228			148^{10}			
1473	2-Butoxyethanol	Ethylene glycol monobutyl ether	$C_6H_{14}O_2$	111-76-2	118.174	liq	-74.8	171(2)	0.9015^{20}	1.4198^{20}	msc H_2O, EtOH, eth; sl ctc
1474	2-[2-(2-Butoxyethoxy)ethoxy]-ethanol		$C_{10}H_{22}O_4$	143-22-6	206.280			278	0.9890^{20}	1.4389^{20}	vs EtOH, MeOH
1475	2-(2-Butoxyethoxy)ethyl thiocyanate	Lethane 384	$C_9H_{17}NO_2S$	112-56-1	203.302	liq		$122^{0.25}$			i H_2O; vs os
1476	1-(2-Butoxyethoxy)-2-propanol		$C_9H_{20}O_3$	124-16-3	176.253	col liq	-90	234(14)	0.931^{20}		s H_2O
1477	2-Butoxyethyl acetate	Ethylene glycol monobutyl ether acetate	$C_8H_{16}O_3$	112-07-2	160.211	liq		191.1(0.9)			
1478	2-Butoxyethyl (2,4-dichlorophenoxy)acetate	2,4-D 2-Butoxyethyl ester	$C_{14}H_{18}Cl_2O_4$	1929-73-3	321.197			159^{1}	1.232^{20}		
1479	2-Butoxyethyl (2,4,5-trichlorophenoxy)acetate	2,4,5-T Butoxyethyl ester	$C_{14}H_{17}Cl_3O_4$	2545-59-7	355.642			164^{1}	1.280^{20}		s ctc
1480	4-Butoxy-*N*-hydroxybenzeneacetamide	Bufexamac	$C_{12}H_{17}NO_3$	2438-72-4	223.268	nd (ace)	154				
1481	1-Butoxy-4-methylbenzene		$C_{11}H_{16}O$	10519-06-9	164.244			229.5	0.9205^{25}	1.4970^{20}	s eth
1482	4-Butoxyphenol		$C_{10}H_{14}O_2$	122-94-1	166.217		65.5	125^{4}			vs ace, bz, eth, EtOH
1483	4-[3-(4-Butoxyphenoxy)propyl] morpholine	Pramoxine	$C_{17}H_{27}NO_3$	140-65-8	293.401			196^{6}			
1484	1-Butoxy-2-propanol		$C_7H_{16}O_2$	5131-66-8	132.201			172(3)	0.882^{20}	1.4168^{20}	s EtOH, eth, bz, ctc, MeOH
1485	Butralin	4-*tert*-Butyl-*N*-*sec*-butyl-2,6-dinitroaniline	$C_{14}H_{21}N_3O_4$	33629-47-9	295.335		59.3(0.5)	$135^{0.5}$			
1486	*N*-Butylacetamide		$C_6H_{13}NO$	1119-49-9	115.173			229	0.8960^{25}	1.4388^{25}	
1487	Butyl acetate		$C_6H_{12}O_2$	123-86-4	116.158	liq	-77.0(0.1)	126.0(0.1)	0.8825^{20}	1.3941^{20}	sl H_2O; msc EtOH, eth; s ace, chl
1488	*sec*-Butyl acetate	1-Methylpropyl acetate	$C_6H_{12}O_2$	105-46-4	116.158	liq	-98.9	108(4)	0.8748^{20}	1.3888^{20}	sl H_2O, ctc; s EtOH, eth
1489	*tert*-Butyl acetate		$C_6H_{12}O_2$	540-88-5	116.158	liq		97.9(1)	0.8665^{20}	1.3855^{20}	s EtOH, eth, chl, HOAc
1490	*tert*-Butylacetic acid		$C_6H_{12}O_2$	1070-83-3	116.158		6(2)	184(2)	0.9124^{20}	1.4096^{20}	s EtOH, eth
1491	Butyl acetoacetate		$C_8H_{14}O_3$	591-60-6	158.195		-35.6	127^{50}	0.9671^{25}	1.4137^{20}	sl H_2O; msc EtOH, bz, lig
1492	Butyl acrylate		$C_7H_{12}O_2$	141-32-2	128.169	liq	-63.6(0.5)	146.6(0.6)	0.8898^{20}	1.4185^{20}	i H_2O; s EtOH, eth, ace; sl ctc
1493	*tert*-Butyl acrylate		$C_7H_{12}O_2$	1663-39-4	128.169	liq		120	0.879^{25}	1.4110^{20}	
1494	Butylamine	1-Butanamine	$C_4H_{11}N$	109-73-9	73.137	liq	-49(1)	77.0(0.2)	0.7414^{20}	1.4031^{20}	msc H_2O; s EtOH, eth
1495	*sec*-Butylamine	2-Butanamine, (±)-	$C_4H_{11}N$	33966-50-6	73.137	liq	-104.5(0.6)	62.71(0.08)	0.7246^{20}	1.3932^{20}	s H_2O, chl; msc EtOH, eth; vs ace
1496	*tert*-Butylamine	2-Methyl-2-propanamine	$C_4H_{11}N$	75-64-9	73.137	liq	-66.92(0.06)	44.02(0.07)	0.6958^{20}	1.3784^{20}	msc H_2O, EtOH, eth; s chl
1497	Butylamine hydrochloride	1-Butanamine hydrochloride	$C_4H_{12}ClN$	3858-78-4	109.598		213		0.982^{20}		sl H_2O, EtOH
1498	Butyl 4-aminobenzoate	Butamben	$C_{11}H_{15}NO_2$	94-25-7	193.243	cry (al or bz)	57(2)	173^{8}			i H_2O; s EtOH, eth, bz, chl
1499	2-(Butylamino)ethanol		$C_6H_{15}NO$	111-75-1	117.189			199	0.8907^{20}	1.4437^{20}	vs H_2O, EtOH, eth
1500	2-(*tert*-Butylamino)ethanol		$C_6H_{15}NO$	4620-70-6	117.189		44	180(16)	0.8818^{20}		
1501	*N*-*tert*-Butylaminoethyl methacrylate		$C_{10}H_{19}NO_2$	3775-90-4	185.264			102^{12}			s chl
1502	2-(*tert*-Butylaminothio)-benzothiazole	*N*-*tert*-Butyl-2-benzothiazole-sulfenamide	$C_{11}H_{14}N_2S_2$	95-31-8	238.372		108				
1503	2-*sec*-Butylaniline		$C_{10}H_{15}N$	55751-54-7	149.233			120^{16}	0.9574^{20}		s EtOH, ace, bz; sl ctc
1504	4-Butylaniline		$C_{10}H_{15}N$	104-13-2	149.233	pa ye		261	0.945^{20}		sl ctc
1505	4-*sec*-Butylaniline		$C_{10}H_{15}N$	30273-11-1	149.233			238	0.949^{15}	1.5360^{29}	vs bz, eth
1506	4-*tert*-Butylaniline		$C_{10}H_{15}N$	769-92-6	149.233	ye rd (peth)	17	241	0.9525^{15}	1.5380^{20}	sl H_2O; msc EtOH, eth; vs bz; s ctc

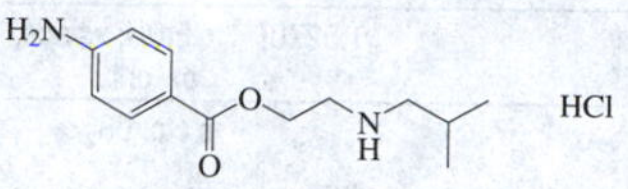

1465
Butethamine hydrochloride

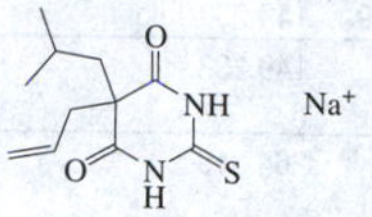

1466
Buthalital sodium

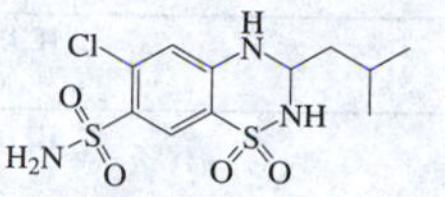

1467
Buthiazide

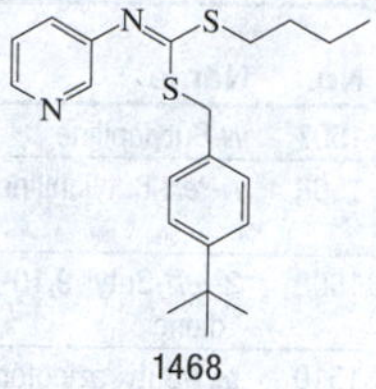

1468
Buthiobate

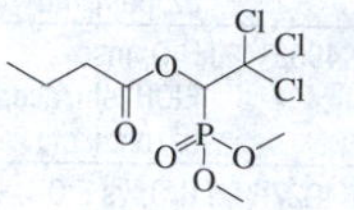

1469
Butonate

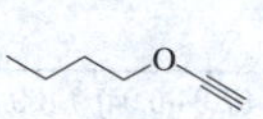

1470
Butoxyacetylene

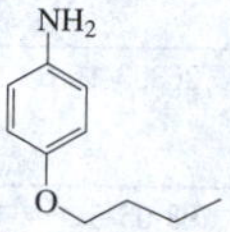

1471
4-Butoxyaniline

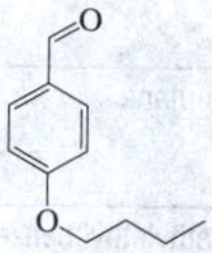

1472
4-Butoxybenzaldehyde

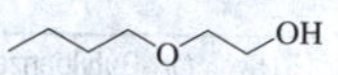

1473
2-Butoxyethanol

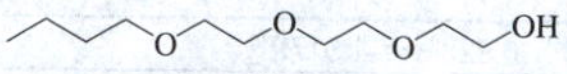

1474
2-[2-(2-Butoxyethoxy)ethoxy]ethanol

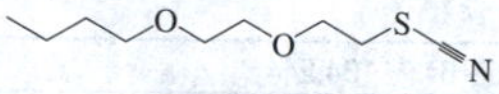

1475
2-(2-Butoxyethoxy)ethyl thiocyanate

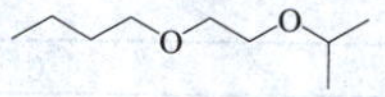

1476
1-(2-Butoxyethoxy)-2-propanol

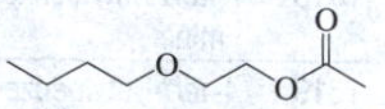

1477
2-Butoxyethyl acetate

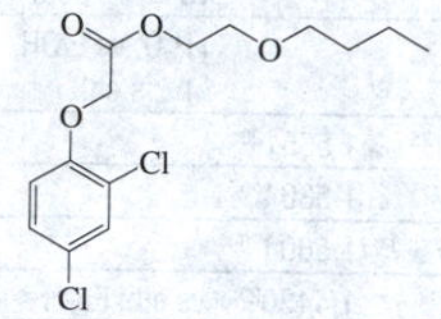

1478
2-Butoxyethyl (2,4-dichlorophenoxy)acetate

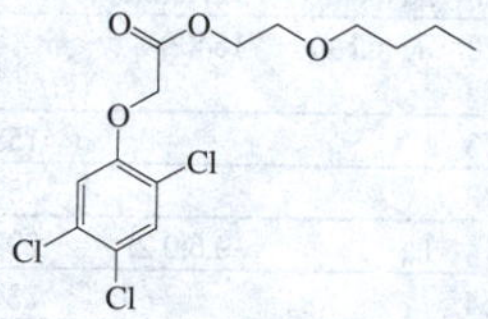

1479
2-Butoxyethyl (2,4,5-trichlorophenoxy)acetate

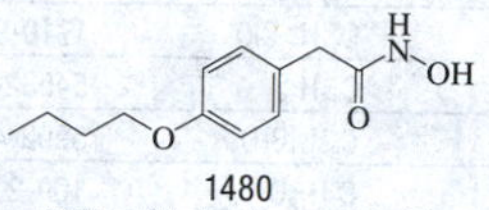

1480
4-Butoxy-*N*-hydroxybenzeneacetamide

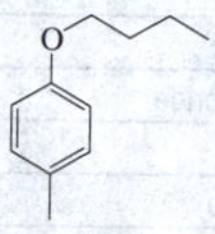

1481
1-Butoxy-4-methylbenzene

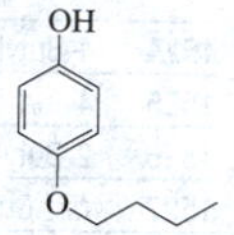

1482
4-Butoxyphenol

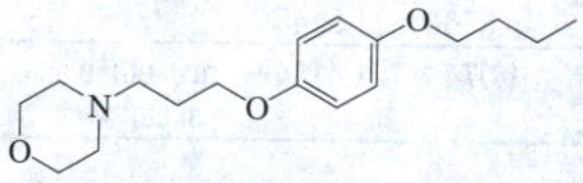

1483
4-[3-(4-Butoxyphenoxy)propyl]morpholine

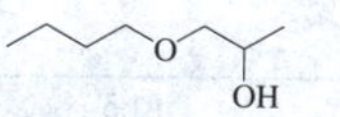

1484
1-Butoxy-2-propanol

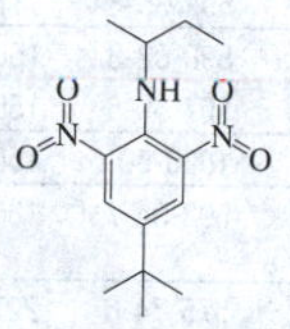

1485
Butralin

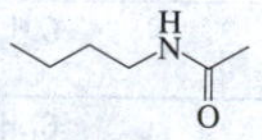

1486
N-Butylacetamide

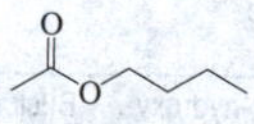

1487
Butyl acetate

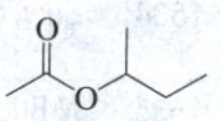

1488
sec-Butyl acetate

1489
tert-Butyl acetate

1490
tert-Butylacetic acid

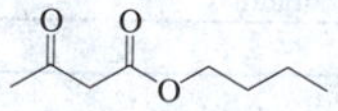

1491
Butyl acetoacetate

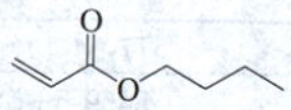

1492
Butyl acrylate

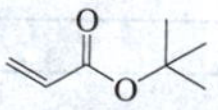

1493
tert-Butyl acrylate

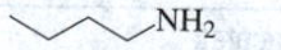

1494
Butylamine

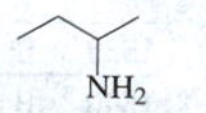

1495
sec-Butylamine

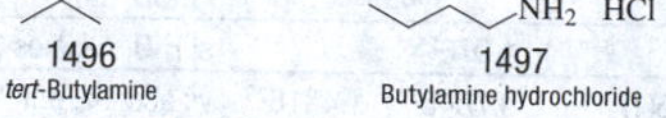

1496
tert-Butylamine

1497
Butylamine hydrochloride

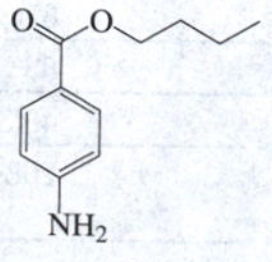

1498
Butyl 4-aminobenzoate

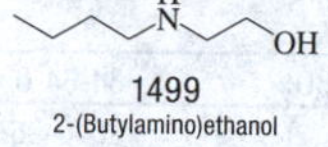

1499
2-(Butylamino)ethanol

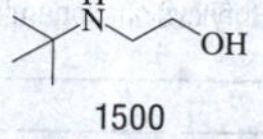

1500
2-(*tert*-Butylamino)ethanol

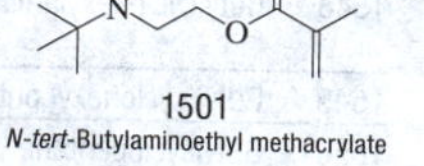

1501
N-*tert*-Butylaminoethyl methacrylate

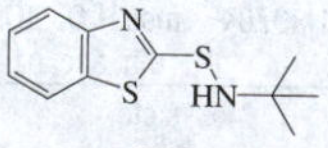

1502
2-(*tert*-Butylaminothio)benzothiazole

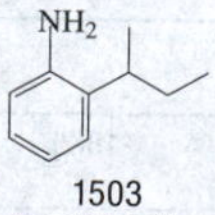

1503
2-*sec*-Butylaniline

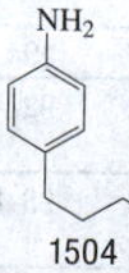

1504
4-Butylaniline

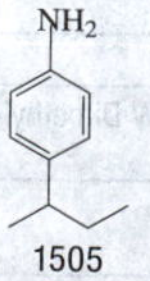

1505
4-*sec*-Butylaniline

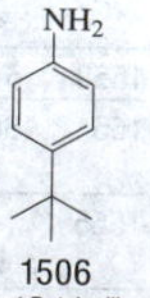

1506
4-*tert*-Butylaniline

No.	Name	Synonym	Mol. form.	CAS Reg. No.	Mol. wt.	Physical form	mp/°C	bp/°C	den/ g cm⁻³	n_D	Solubility
1507	*N*-Butylaniline		$C_{10}H_{15}N$	1126-78-9	149.233	liq	-14.4	254(9)	0.9323^{20}	1.5341^{20}	vs eth, EtOH
1508	*N-tert*-Butylaniline		$C_{10}H_{15}N$	937-33-7	149.233			215		1.5270^{20}	s EtOH; vs ace, bz, chl
1509	2-*tert*-Butyl-9,10-anthracenedione		$C_{18}H_{16}O_2$	84-47-9	264.319		99				s ctc, CS_2
1510	*tert*-Butyl azidoformate	*tert*-Butyl carbonazidate	$C_5H_9N_3O_2$	1070-19-5	143.144	unstab >80		73^{70}			
1511	4-Butylbenzaldehyde		$C_{11}H_{14}O$	1200-14-2	162.228			123^{7}		1.5265	
1512	4-*tert*-Butylbenzaldehyde		$C_{11}H_{14}O$	939-97-9	162.228	liq		107^{11}	0.970	1.5270^{20}	
1513	Butylbenzene	1-Phenylbutane	$C_{10}H_{14}$	104-51-8	134.218	liq	-87.81(0.05)	183.3(0.3)	0.8601^{20}	1.4898^{20}	i H_2O; msc EtOH, eth, ace, bz, peth, ctc
1514	*sec*-Butylbenzene, (±)-	2-Phenylbutane	$C_{10}H_{14}$	36383-15-0	134.218	liq	-75.5(0.3)	173.3(0.4)	0.8621^{20}	1.4902^{20}	i H_2O; msc EtOH, eth, ace, bz, peth, ctc
1515	*tert*-Butylbenzene	(1,1-Dimethylethyl)benzene	$C_{10}H_{14}$	98-06-6	134.218	liq	-57.84(0.04)	169.1(0.3)	0.8665^{20}	1.4927^{20}	i H_2O; vs EtOH, eth; msc ace, bz
1516	4-*tert*-Butyl-1,2-benzenediol		$C_{10}H_{14}O_2$	98-29-3	166.217		54(2)	286(1)			s tfa
1517	2-*tert*-Butyl-1,4-benzenediol		$C_{10}H_{14}O_2$	1948-33-0	166.217		128				
1518	*N-tert*-Butylbenzenemethanamine		$C_{11}H_{17}N$	3378-72-1	163.260			75^{5}		1.4951^{25}	
1519	4-*tert*-Butylbenzenemethanol		$C_{11}H_{16}O$	877-65-6	164.244			236	0.928^{25}	1.5179^{20}	
1520	Butyl benzoate		$C_{11}H_{14}O_2$	136-60-7	178.228	liq	-22.4(0.4)	249(3)	1.000^{20}	1.4940^{25}	i H_2O; msc EtOH, eth; s ace; sl ctc
1521	2-*tert*-Butylbenzoic acid		$C_{11}H_{14}O_2$	1077-58-3	178.228	pl (dil al)	80.5(0.3)				vs EtOH
1522	3-*tert*-Butylbenzoic acid		$C_{11}H_{14}O_2$	7498-54-6	178.228	nd (peth)	127.3(0.5)				vs EtOH, peth
1523	4-*tert*-Butylbenzoic acid	*p-tert*-Butylbenzoic acid	$C_{11}H_{14}O_2$	98-73-7	178.228	nd (dil al)	164(2)				i H_2O; vs EtOH, bz; s chl
1524	4-Butylbenzoyl chloride		$C_{11}H_{13}ClO$	28788-62-7	196.673			155^{26}	1.051^{25}	1.5351^{20}	
1525	4-*tert*-Butylbenzoyl chloride		$C_{11}H_{13}ClO$	1710-98-1	196.673			266	1.007^{25}	1.5364^{20}	
1526	2-Butyl-1,1'-biphenyl		$C_{16}H_{18}$	54532-97-7	210.314	liq	-9.6(0.2)	292(3)	0.9676^{20}	1.5604^{20}	
1527	*tert*-Butyl bromoacetate		$C_6H_{11}BrO_2$	5292-43-3	195.054			73^{25}		1.4430^{20}	vs eth, EtOH
1528	Butyl butanoate		$C_8H_{16}O_2$	109-21-7	144.212	liq	-91.5(0.1)	164.95(0.1)	0.8700^{20}	1.4075^{20}	i H_2O; msc EtOH, eth; s ctc
1529	Butyl *cis*-2-butenedioate	Monobutyl maleate	$C_8H_{12}O_4$	925-21-3	172.179	oil			1.09^{25}		
1530	Butyl carbamate		$C_5H_{11}NO_2$	592-35-8	117.147	pr	53	204 dec			vs EtOH; sl chl
1531	Butyl chloroacetate		$C_6H_{11}ClO_2$	590-02-3	150.603			181(3)	1.0704^{20}	1.4297^{20}	vs eth, EtOH
1532	*tert*-Butyl chloroacetate		$C_6H_{11}ClO_2$	107-59-5	150.603			150		1.4260^{20}	dec H_2O
1533	Butylchlorodimethylsilane		$C_6H_{15}ClSi$	1000-50-6	150.722			139	0.876^{20}	1.5145^{20}	
1534	Butyl chloroformate		$C_5H_9ClO_2$	592-34-7	136.577			142	1.074^{25}	1.4114^{20}	msc eth; s ace; sl ctc
1535	*N*-Butyl-4-chloro-2-hydroxybenzamide	Buclosamide	$C_{11}H_{14}ClNO_2$	575-74-6	227.688		91.5				
1536	Butyl 2-chloropropanoate		$C_7H_{13}ClO_2$	54819-86-2	164.630			184	1.0253^{20}	1.4263^{20}	vs eth
1537	Butyl 3-chloropropanoate		$C_7H_{13}ClO_2$	27387-79-7	164.630			104^{22}	1.0370^{20}	1.4321^{20}	vs H_2O, eth
1538	*tert*-Butyl chromate		$C_8H_{18}CrO_4$	1189-85-1	230.223	red cry (peth)	-5				reac H_2O
1539	Butyl citrate		$C_{18}H_{32}O_7$	77-94-1	360.443		-20	233^{22}	1.043^{20}	1.4460^{20}	
1540	Butyl cyanoacetate		$C_7H_{11}NO_2$	5459-58-5	141.168			231	1.0010^{20}	1.4200^{20}	
1541	Butylcyclohexane		$C_{10}H_{20}$	1678-93-9	140.266	liq	-74.68(0.05)	180.9(0.6)	0.7902^{20}	1.4408^{20}	i H_2O
1542	*sec*-Butylcyclohexane		$C_{10}H_{20}$	7058-01-7	140.266			179.3(0.5)	0.8131^{20}	1.4467^{20}	i H_2O; s ace
1543	*tert*-Butylcyclohexane		$C_{10}H_{20}$	3178-22-1	140.266	liq	-41.2(0.3)	171.6(0.4)	0.8127^{20}	1.4469^{20}	i H_2O
1544	2-*tert*-Butylcyclohexanol		$C_{10}H_{20}O$	13491-79-7	156.265		45	139^{95}	0.902^{25}		
1545	*cis*-4-*tert*-Butylcyclohexanol		$C_{10}H_{20}O$	937-05-3	156.265		82(3)	112^{15}			
1546	*trans*-4-*tert*-Butylcyclohexanol		$C_{10}H_{20}O$	21862-63-5	156.265		83	112^{15}			
1547	4-*tert*-Butylcyclohexanone		$C_{10}H_{18}O$	98-53-3	154.249		48(3)	90^{9}			
1548	Butylcyclohexylamine	*N*-Butylcyclohexanamine	$C_{10}H_{21}N$	10108-56-2	155.281		208.3				sl H_2O, ctc; vs EtOH, eth
1549	Butyl cyclohexyl phthalate		$C_{18}H_{24}O_4$	84-64-0	304.382	col liq			1.076^{25}		sl H_2O; misc os
1550	Butylcyclopentane		C_9H_{18}	2040-95-1	126.239	liq	-107.95(0.05)	156(1)	0.7846^{20}	1.4316^{20}	vs ace, bz, eth, EtOH
1551	Butyl dichloroacetate		$C_6H_{10}Cl_2O_2$	29003-73-4	185.048			193.5	1.1820^{20}	1.4420^{20}	vs eth, EtOH
1552	Butyl (2,4-dichlorophenoxy) acetate	2,4-D Butyl ester	$C_{12}H_{14}Cl_2O_3$	94-80-4	277.143		9	133^{1}			
1553	5-Butyldihydro-2(3*H*)-furanone		$C_8H_{14}O_2$	104-50-7	142.196			132^{20}	0.9796^{19}	1.4451^{19}	s EtOH; sl ctc
1554	Butyldimethylamine	*N,N*-Dimethyl-1-butanamine	$C_6H_{15}N$	927-62-8	101.190			92.2(0.7)	0.7206^{20}	1.3970^{20}	msc H_2O, EtOH, eth, ace, bz
1555	1-*tert*-Butyl-3,5-dimethylbenzene		$C_{12}H_{18}$	98-19-1	162.271	liq	-18(2)	207(1)	0.8668^{20}		s ctc

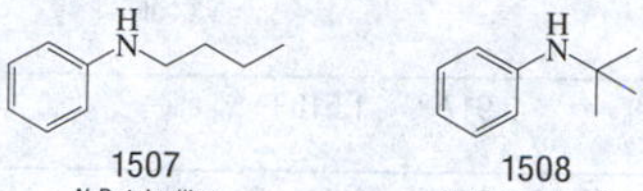
1507 *N*-Butylaniline
1508 *N-tert*-Butylaniline

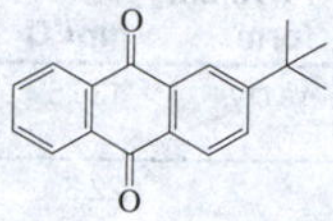
1509 2-*tert*-Butyl-9,10-anthracenedione

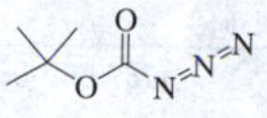
1510 *tert*-Butyl azidoformate

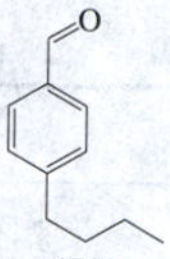
1511 4-Butylbenzaldehyde

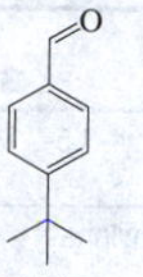
1512 4-*tert*-Butylbenzaldehyde

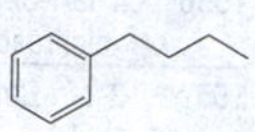
1513 Butylbenzene

1514 *sec*-Butylbenzene, (±)-
1515 *tert*-Butylbenzene

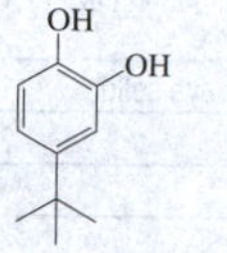
1516 4-*tert*-Butyl-1,2-benzenediol

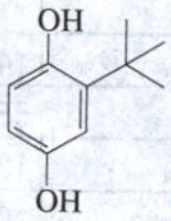
1517 2-*tert*-Butyl-1,4-benzenediol

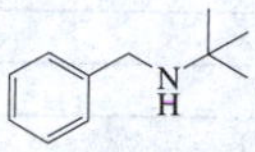
1518 *N-tert*-Butylbenzenemethanamine

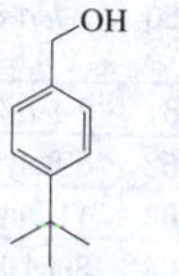
1519 4-*tert*-Butylbenzenemethanol

1520 Butyl benzoate
1521 2-*tert*-Butylbenzoic acid

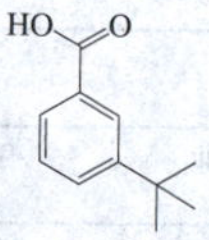
1522 3-*tert*-Butylbenzoic acid

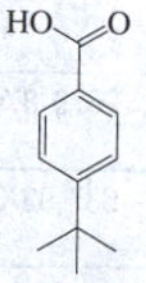
1523 4-*tert*-Butylbenzoic acid

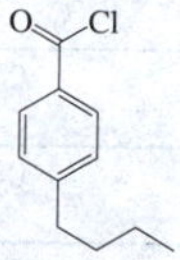
1524 4-Butylbenzoyl chloride

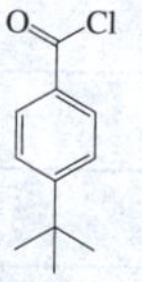
1525 4-*tert*-Butylbenzoyl chloride

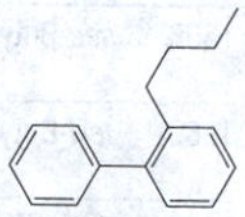
1526 2-Butyl-1,1′-biphenyl

1527 *tert*-Butyl bromoacetate
1528 Butyl butanoate

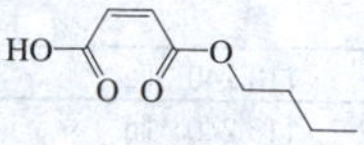
1529 Butyl *cis*-2-butenedioate

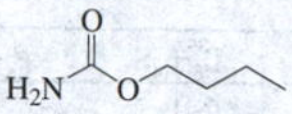
1530 Butyl carbamate

1531 Butyl chloroacetate
1532 *tert*-Butyl chloroacetate

1533 Butylchlorodimethylsilane
1534 Butyl chloroformate

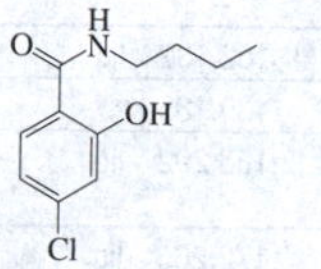
1535 *N*-Butyl-4-chloro-2-hydroxybenzamide

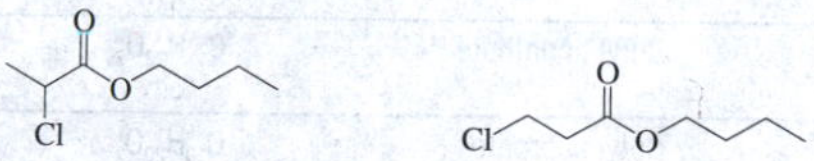
1536 Butyl 2-chloropropanoate
1537 Butyl 3-chloropropanoate

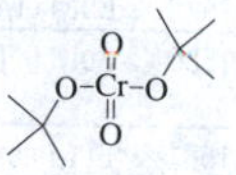
1538 *tert*-Butyl chromate

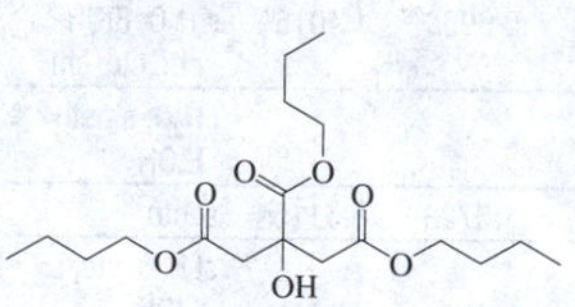
1539 Butyl citrate

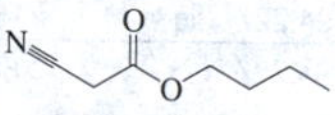
1540 Butyl cyanoacetate

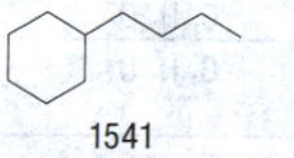
1541 Butylcyclohexane

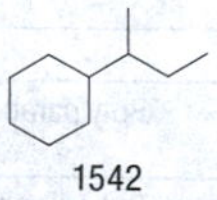
1542 *sec*-Butylcyclohexane

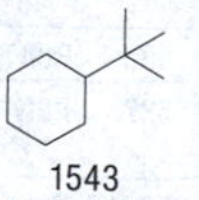
1543 *tert*-Butylcyclohexane

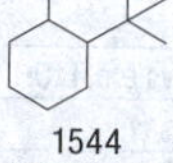
1544 2-*tert*-Butylcyclohexanol

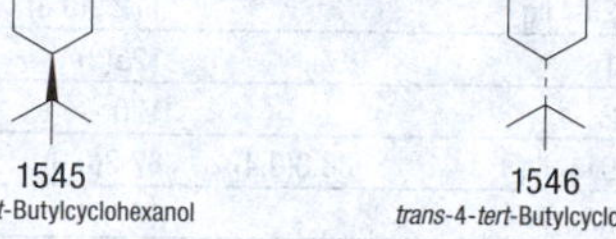
1545 *cis*-4-*tert*-Butylcyclohexanol
1546 *trans*-4-*tert*-Butylcyclohexanol

1547 4-*tert*-Butylcyclohexanone

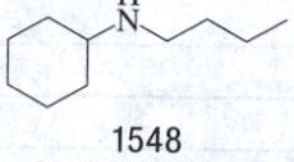
1548 Butylcyclohexylamine

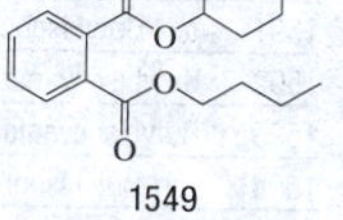
1549 Butyl cyclohexyl phthalate

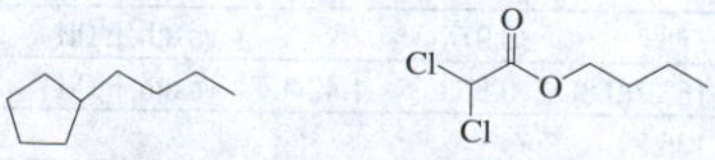
1550 Butylcyclopentane
1551 Butyl dichloroacetate

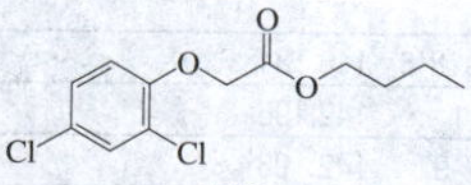
1552 Butyl (2,4-dichlorophenoxy)acetate

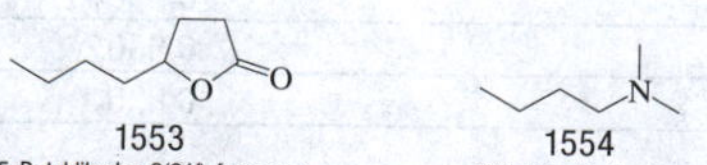
1553 5-Butyldihydro-2(3*H*)-furanone
1554 Butyldimethylamine

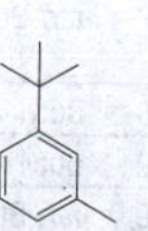
1555 1-*tert*-Butyl-3,5-dimethylbenzene

Organic

No.	Name	Synonym	Mol. form.	CAS Reg. No.	Mol. wt.	Physical form	mp/°C	bp/°C	den/ g cm^{-3}	n_D	Solubility
1556	4-*tert*-Butyl-2,6-dimethyl-3,5-dinitroacetophenone	Musk ketone	$C_{14}H_{18}N_2O_5$	81-14-1	294.303	ye cry	135.5				vs chl
1557	2-*tert*-Butyl-4,6-dimethylphenol		$C_{12}H_{18}O$	1879-09-0	178.270		22.3	247(9)	0.917^{80}	1.5183^{20}	i alk
1558	4-*tert*-Butyl-2,5-dimethylphenol		$C_{12}H_{18}O$	17696-37-6	178.270		71.2	262(27)	0.939^{80}	1.5311^{20}	s alk
1559	4-*tert*-Butyl-2,6-dimethylphenol		$C_{12}H_{18}O$	879-97-0	178.270		82.4	248	0.916^{80}		s alk
1560	1-*tert*-Butyl-3,5-dimethyl-2,4,6-trinitrobenzene		$C_{12}H_{15}N_3O_6$	81-15-2	297.263	pl, nd (al)	111.5(0.2)				i H_2O; sl EtOH; s eth, chl
1561	2-*tert*-Butyl-4,6-dinitrophenol		$C_{10}H_{12}N_2O_5$	1420-07-1	240.212	ye solid	126				
1562	5-Butyldocosane		$C_{26}H_{54}$	55282-16-1	366.707		208	244^{10}	0.8058^{20}	1.4503^{20}	
1563	11-Butyldocosane		$C_{26}H_{54}$	13475-76-8	366.707			242.5^{10}	0.8041^{20}	1.4499^{20}	
1564	Butyl dodecanoate		$C_{16}H_{32}O_2$	106-18-3	256.424			180^{18}			
1565	Butylethylamine	*N*-Ethyl-1-butanamine	$C_6H_{15}N$	13360-63-9	101.190			104.8(0.8)	0.7398^{20}	1.4040^{20}	msc EtOH, eth, ace, bz
1566	1-*tert*-Butyl-4-ethylbenzene		$C_{12}H_{18}$	7364-19-4	162.271	liq	-38.4	211	0.8641^{20}		
1567	Butyl ethyl ether	Ethyl butyl ether	$C_6H_{14}O$	628-81-9	102.174	liq	-124	89(2)	0.7495^{20}	1.3818^{20}	i H_2O; msc EtOH, eth; vs ace
1568	*sec*-Butyl ethyl ether		$C_6H_{14}O$	2679-87-0	102.174			81(4)	0.7503^{20}	1.3802^{20}	i H_2O; vs EtOH, eth
1569	*tert*-Butyl ethyl ether	Ethyl *tert*-butyl ether	$C_6H_{14}O$	637-92-3	102.174	liq	-94.0(0.3)	72.7(0.1)	0.736^{25}	1.3756^{20}	i H_2O; vs EtOH, eth
1570	2-*tert*-Butyl-4-ethylphenol		$C_{12}H_{18}O$	96-70-8	178.270		23	250			
1571	2-Butyl-2-ethyl-1,3-propanediol		$C_9H_{20}O_2$	115-84-4	160.254	wh cry	43(2)	269.0(0.2)	0.927^{50}	1.4587^{25}	sl H_2O, ace; s EtOH
1572	5-Butyl-5-ethyl-2,4,6(1*H*,3*H*,5*H*)-pyrimidinetrione	Butethal	$C_{10}H_{16}N_2O_3$	77-28-1	212.245		123(1)				
1573	Butyl ethyl sulfide		$C_6H_{14}S$	638-46-0	118.240	liq	-95.1(0.2)	144.2(0.8)	0.8376^{20}	1.4492^{10}	vs EtOH; s chl
1574	*tert*-Butyl ethyl sulfide	2-Methyl-2-propanethiol	$C_6H_{14}S$	14290-92-7	118.240	liq	-85.9(0.3)	120.4(0.6)			
1575	*N*-*tert*-Butylformamide		$C_5H_{11}NO$	2425-74-3	101.147	liq	16	202	0.903	1.4330^{20}	
1576	Butyl formate		$C_5H_{10}O_2$	592-84-7	102.132	liq	-90.0(0.4)	106.1(0.1)	0.8958^{20}	1.3887^{20}	sl H_2O; s ace; msc EtOH, eth
1577	*sec*-Butyl formate		$C_5H_{10}O_2$	589-40-2	102.132			93.6(0.3)	0.8846^{20}	1.3865^{20}	sl H_2O; s ace; msc EtOH, eth
1578	*tert*-Butyl formate	1,1-Dimethylethyl formate	$C_5H_{10}O_2$	762-75-4	102.132	liq		83(6)	0.872	1.3790^{20}	
1579	Butyl glycidyl ether		$C_7H_{14}O_2$	2426-08-6	130.185			171(18)	0.918^{20}		
1580	Butyl heptanoate	Butyl enanthate	$C_{11}H_{22}O_2$	5454-28-4	186.292	liq	-68(1)	225(4)	0.8638^{20}	1.4204^{20}	vs ace, bz, eth, EtOH
1581	Butyl hexanoate	Butyl caproate	$C_{10}H_{20}O_2$	626-82-4	172.265	liq	-50.0(0.5)	204(3)	0.8653^{20}	1.4152^{20}	i H_2O; s EtOH; msc eth
1582	*tert*-Butylhydrazine hydrochloride		$C_4H_{13}ClN_2$	7400-27-3	124.612		192.5				
1583	Butyl hydrogen succinate	Monobutyl succinate	$C_8H_{14}O_4$	5150-93-6	174.195		8.6	136.5^{3}	1.0732^{20}	1.4360^{20}	
1584	*tert*-Butyl hydroperoxide		$C_4H_{10}O_2$	75-91-2	90.121	liq	6	89 dec	0.8960^{20}	1.4015^{20}	s H_2O, EtOH, eth, ctc, chl
1585	*tert*-Butyl-4-hydroxyanisole	Butylated hydroxyanisole	$C_{11}H_{16}O_2$	25013-16-5	180.244	wax	51	268			i H_2O; s peth, EtOH
1586	Butyl 2-hydroxybenzoate		$C_{11}H_{14}O_3$	2052-14-4	194.227	liq	-5.9	271	1.0728^{20}	1.5115^{20}	sl ctc
1587	Butyl 4-hydroxybenzoate	Butylparaben	$C_{11}H_{14}O_3$	94-26-8	194.227		68.5				sl H_2O, ctc; s EtOH
1588	Butyl *cis*-12-hydroxy-9-octadecenoate, (*R*)-	Butyl ricinoleate	$C_{22}H_{42}O_3$	151-13-3	354.566			275^{13}	0.9058^{22}	1.4566^{22}	vs eth
1589	*tert*-Butyl hypochlorite		C_4H_9ClO	507-40-4	108.566	ye liq		77.5	0.9583^{18}	1.403^{20}	i H_2O; vs eth, bz; s ace
1590	Butyl isobutyl ether		$C_8H_{18}O$	17071-47-5	130.228	liq		135(8)	0.763^{15}	1.4077^{21}	vs ace, eth, EtOH
1591	*tert*-Butyl isobutyl ether		$C_8H_{18}O$	33021-02-2	130.228	liq		112.9(0.3)			
1592	Butyl isocyanate		C_5H_9NO	111-36-4	99.131			125(3)	0.880^{20}	1.4060^{20}	
1593	Butyl isocyanide		C_5H_9N	2769-64-4	83.132			120	0.78^{20}		vs eth, EtOH
1594	*tert*-Butyl isopropyl ether		$C_7H_{16}O$	17348-59-3	116.201	liq	-88.3(0.4)	87.3(0.3)	0.7365^{25}		s chl
1595	Butyl isothiocyanate	1-Isothiocyanatobutane	C_5H_9NS	592-82-5	115.197			167(7)	0.9546^{20}	1.501^{20}	vs eth, EtOH
1596	*sec*-Butyl isothiocyanate, (±)-	2-Isothiocyanatobutane, (±)	C_5H_9NS	116724-11-9	115.197			159.5	0.944^{12}		vs eth, EtOH
1597	*tert*-Butyl isothiocyanate	2-Isothiocyanato-2-methylpropane	C_5H_9NS	590-42-1	115.197		10.5	140	0.9187^{10}		
1598	Butyl lactate		$C_7H_{14}O_3$	34451-18-8	146.184			77^{10}	0.9744^{27}		vs eth, EtOH
1599	Butyl methacrylate		$C_8H_{14}O_2$	97-88-1	142.196			163.7(0.8)	0.8936^{20}	1.4240^{20}	vs eth, EtOH
1600	*tert*-Butyl methacrylate		$C_8H_{14}O_2$	585-07-9	142.196			135.2			

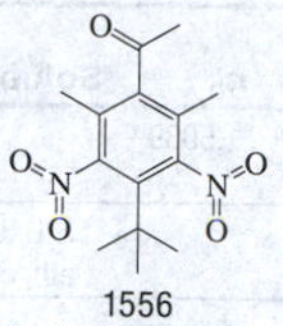
1556
4-*tert*-Butyl-2,6-dimethyl-3,5-dinitroacetophenone

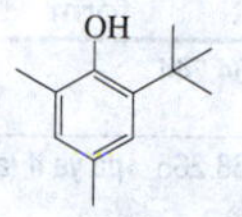
1557
2-*tert*-Butyl-4,6-dimethylphenol

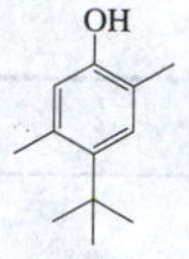
1558
4-*tert*-Butyl-2,5-dimethylphenol

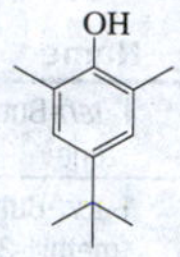
1559
4-*tert*-Butyl-2,6-dimethylphenol

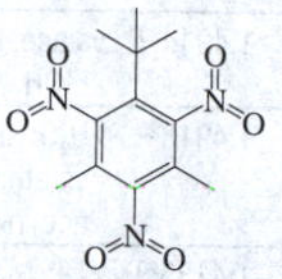
1560
1-*tert*-Butyl-3,5-dimethyl-2,4,6-trinitrobenzene

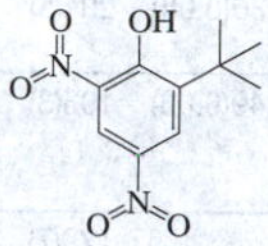
1561
2-*tert*-Butyl-4,6-dinitrophenol

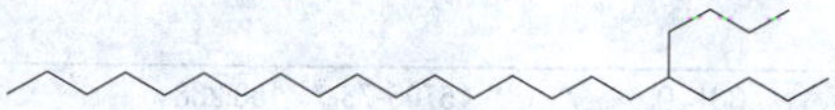
1562
5-Butyldocosane

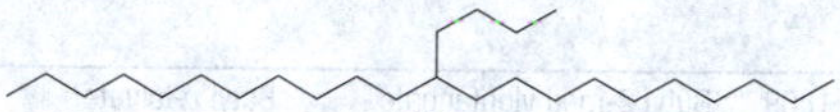
1563
11-Butyldocosane

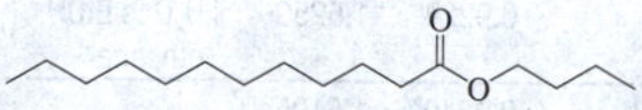
1564
Butyl dodecanoate

1565
Butylethylamine

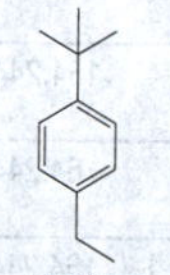
1566
1-*tert*-Butyl-4-ethylbenzene

1567
Butyl ethyl ether

1568
sec-Butyl ethyl ether

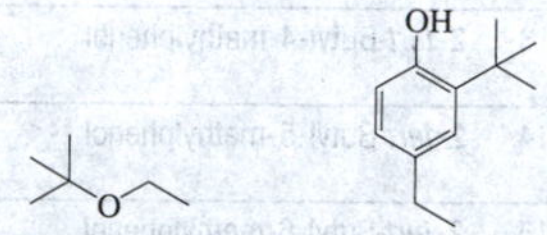
1569
tert-Butyl ethyl ether

1570
2-*tert*-Butyl-4-ethylphenol

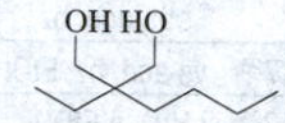
1571
2-Butyl-2-ethyl-1,3-propanediol

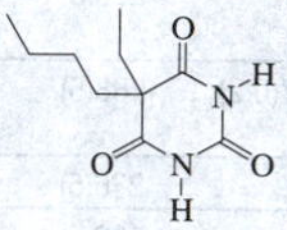
1572
5-Butyl-5-ethyl-2,4,6(1*H*,3*H*,5*H*)-pyrimidinetrione

1573
Butyl ethyl sulfide

1574
tert-Butyl ethyl sulfide

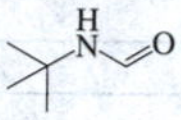
1575
N-*tert*-Butylformamide

1576
Butyl formate

1577
sec-Butyl formate

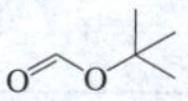
1578
tert-Butyl formate

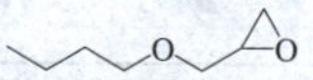
1579
Butyl glycidyl ether

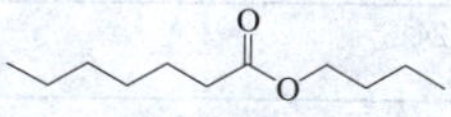
1580
Butyl heptanoate

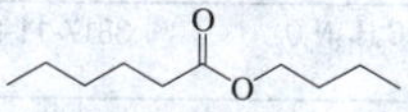
1581
Butyl hexanoate

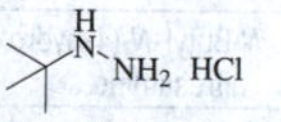
1582
tert-Butylhydrazine hydrochloride

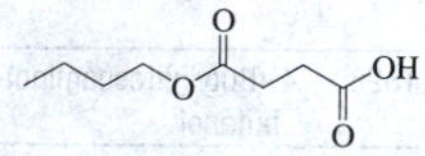
1583
Butyl hydrogen succinate

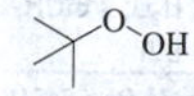
1584
tert-Butyl hydroperoxide

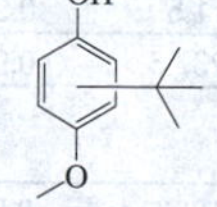
1585
tert-Butyl-4-hydroxyanisole

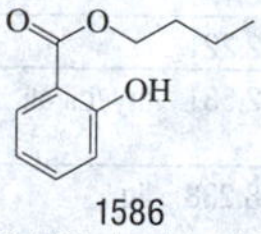
1586
Butyl 2-hydroxybenzoate

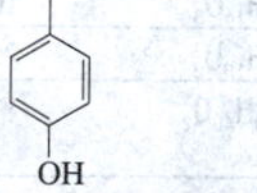
1587
Butyl 4-hydroxybenzoate

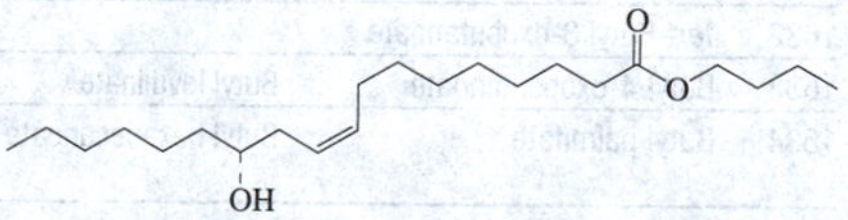
1588
Butyl *cis*-12-hydroxy-9-octadecenoate, (*R*)-

1589
tert-Butyl hypochlorite

1590
Butyl isobutyl ether

1591
tert-Butyl isobutyl ether

1592
Butyl isocyanate

1593
Butyl isocyanide

1594
tert-Butyl isopropyl ether

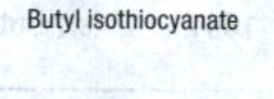
1595
Butyl isothiocyanate

1596
sec-Butyl isothiocyanate, (±)-

1597
tert-Butyl isothiocyanate

1598
Butyl lactate

1599
Butyl methacrylate

1600
tert-Butyl methacrylate

No.	Name	Synonym	Mol. form.	CAS Reg. No.	Mol. wt.	Physical form	mp/°C	bp/°C	den/g cm^{-3}	n_D	Solubility
1601	1-*tert*-Butyl-4-methoxybenzene		$C_{11}H_{16}O$	5396-38-3	164.244		19.1(0.3)	223(5)	0.9383^{20}	1.5039^{20}	
1602	1-*tert*-Butyl-2-methoxy-4-methyl-3,5-dinitrobenzene		$C_{12}H_{16}N_2O_5$	83-66-9	268.265	pa ye lf (al)	85	185^{16}			i H_2O; sl EtOH; s eth, chl
1603	2-*tert*-Butyl-4-methoxyphenol		$C_{11}H_{16}O_2$	121-00-6	180.244			184^{50}			
1604	3-*tert*-Butyl-4-methoxyphenol		$C_{11}H_{16}O_2$	88-32-4	180.244		65				
1605	Butylmethylamine	*N*-Methyl-1-butanamine	$C_5H_{13}N$	110-68-9	87.164			91(2)	0.7637^{15}		
1606	1-*tert*-Butyl-2-methylbenzene	2-*tert*-Butyltoluene	$C_{11}H_{16}$	1074-92-6	148.245	liq	-50.3(0.2)	200(4)	0.8897^{20}	1.5076^{20}	vs ace, bz, eth, EtOH
1607	1-*tert*-Butyl-3-methylbenzene	3-*tert*-Butyltoluene	$C_{11}H_{16}$	1075-38-3	148.245	liq	-41.36(0.08)	204(5)	0.8657^{20}	1.4944^{20}	vs ace, bz, eth, EtOH
1608	1-*tert*-Butyl-4-methylbenzene	4-*tert*-Butyltoluene	$C_{11}H_{16}$	98-51-1	148.245	liq	-52.49(0.08)	193(3)	0.8612^{20}	1.4918^{20}	i H_2O; sl EtOH; vs eth, chl; s ace, bz
1609	Butyl 2-methylbutanoate	Butyl *o*-toluate	$C_9H_{18}O_2$	15706-73-7	158.238			179(7)	0.8620^{20}	1.4135^{20}	
1610	Butyl 3-methylbutanoate	Butyl *p*-toluate	$C_9H_{18}O_2$	109-19-3	158.238					1.4058^{25}	
1611	Butyl methyl ether		$C_5H_{12}O$	628-28-4	88.148	liq	-115.7(0.1)	70.1(0.3)	0.7392^{25}	1.3736^{20}	i H_2O; msc EtOH, eth; s ace
1612	*sec*-Butyl methyl ether		$C_5H_{12}O$	116783-23-4	88.148			59.1	0.7415^{20}	1.3680^{25}	vs ace, eth, EtOH
1613	2-*tert*-Butyl-4-methylphenol		$C_{11}H_{16}O$	2409-55-4	164.244		52.3(0.9)	236(7)	0.9247^{75}	1.4969^{75}	sl H_2O; s ace, bz, chl
1614	2-*tert*-Butyl-5-methylphenol		$C_{11}H_{16}O$	88-60-8	164.244		46.5	127^{11}	0.922^{80}	1.5250^{20}	i H_2O; s EtOH, eth, ace
1615	2-*tert*-Butyl-6-methylphenol		$C_{11}H_{16}O$	2219-82-1	164.244		29(1)	233(7)	0.9240^{80}	1.5195^{20}	
1616	4-*tert*-Butyl-2-methylphenol		$C_{11}H_{16}O$	98-27-1	164.244		27.5	256(4)	0.965^{20}	1.5230^{20}	i H_2O; s eth, ace, bz
1617	Butyl methyl sulfide		$C_5H_{12}S$	628-29-5	104.214	liq	-97.81(0.05)	123.4(0.5)	0.8426^{20}	1.4477^{20}	vs EtOH, MeOH
1618	*tert*-Butyl methyl sulfide		$C_5H_{12}S$	6163-64-0	104.214	liq		98.9(0.3)			
1619	4-Butylmorpholine		$C_8H_{17}NO$	1005-67-0	143.227	liq	-57.1	213.5	0.9068^{20}	1.4451^{20}	vs H_2O, ace, bz, EtOH
1620	1-Butylnaphthalene		$C_{14}H_{16}$	1634-09-9	184.277	liq	-19.7(0.2)	288(5)	0.9738^{20}	1.5819^{20}	i H_2O; s EtOH, eth, ace, bz
1621	2-Butylnaphthalene		$C_{14}H_{16}$	1134-62-9	184.277	liq	-7(4)	286(5)	0.9673^{20}	1.5777^{20}	vs ace, bz, EtOH
1622	Butyl nitrate		$C_4H_9NO_3$	928-45-0	119.119			133	1.0228^{30}	1.4013^{23}	i H_2O; s EtOH, eth; sl ctc
1623	Butyl nitrite		$C_4H_9NO_2$	544-16-1	103.120			78	0.9114^{25}	1.3762^{20}	msc EtOH, eth
1624	*tert*-Butyl nitrite		$C_4H_9NO_2$	540-80-7	103.120	pa ye liq		64(2)	0.8670^{20}	1.368^{20}	sl H_2O; s EtOH, eth, chl, CS_2
1625	*sec*-Butyl nitrite		$C_4H_9NO_2$	924-43-6	103.120			65(2)	0.8726^{20}	1.3710^{20}	vs eth, EtOH, chl
1626	4-(Butylnitrosoamino)-1-butanol	*N*-Butyl-*N*-(4-hydroxybutyl)-nitrosamine	$C_8H_{18}N_2O_2$	3817-11-6	174.241			$115^{0.01}$			
1627	5-Butylnonane		$C_{13}H_{28}$	17312-63-9	184.361			219(5)	0.7635^{18}	1.4273^{18}	
1628	Butyl nonanoate	Butyl pelargonate	$C_{13}H_{26}O_2$	50623-57-9	214.344		-38.0(0.7)	123^{20}	0.8520^{25}	1.4262^{25}	
1629	Butyl octanoate		$C_{12}H_{24}O_2$	589-75-3	200.318	liq	-42.9(0.5)	240(4)	0.8628^{20}	1.4232^{25}	vs ace, eth, EtOH
1630	2-Butyl-1-octanol		$C_{12}H_{26}O$	3913-02-8	186.333			248(23)	0.891^{20}		
1631	Butyl oleate	Butyl *cis*-9-octadecenoate	$C_{22}H_{42}O_2$	142-77-8	338.567	ye cry	-26.4	227^{15}	0.8704^{15}	1.4480^{25}	vs EtOH
1632	*tert*-Butyl 3-oxobutanoate		$C_8H_{14}O_3$	1694-31-1	158.195			71.5^{11}	0.9756^{20}	1.4180^{20}	
1633	Butyl 4-oxopentanoate	Butyl levulinate	$C_9H_{16}O_3$	2052-15-5	172.221			237.5	0.9735^{20}	1.4290^{20}	sl chl
1634	Butyl palmitate	Butyl hexadecanoate	$C_{20}H_{40}O_2$	111-06-8	312.531	cry (dil al)	16(2)			1.4312^{50}	i H_2O; s EtOH, eth
1635	Butyl pentanoate		$C_9H_{18}O_2$	591-68-4	158.238	liq	-83.77(0.02)	186(5)	0.8710^{15}	1.4128^{20}	sl H_2O; s EtOH, eth
1636	*sec*-Butyl pentanoate		$C_9H_{18}O_2$	116836-32-9	158.238			174.5	0.8605^{20}	1.4070^{20}	vs bz, eth, py, EtOH
1637	4-(1-Butylpentyl)pyridine		$C_{14}H_{23}N$	2961-47-9	205.340			265	0.8878^{25}	1.4846^{25}	
1638	*tert*-Butyl peroxybenzoate	Benzoyl *tert*-butyl peroxide	$C_{11}H_{14}O_3$	614-45-9	194.227			$75^{0.2}$	1.021^{25}	1.4990^{20}	
1639	2-Butylphenol		$C_{10}H_{14}O$	3180-09-4	150.217	liq	-20(2)	234(5)	0.975^{20}	1.5180^{25}	i H_2O; s EtOH, eth, alk
1640	2-*sec*-Butylphenol		$C_{10}H_{14}O$	89-72-5	150.217		18(3)	229(3)	0.9804^{25}	1.5200^{25}	
1641	2-*tert*-Butylphenol		$C_{10}H_{14}O$	88-18-6	150.217	liq	-5.6(0.2)	224.3(0.6)	0.9783^{20}	1.5160^{20}	s EtOH, ctc, alk; vs eth
1642	3-Butylphenol		$C_{10}H_{14}O$	4074-43-5	150.217			249(4)	0.974^{20}		vs eth, EtOH
1643	3-*tert*-Butylphenol		$C_{10}H_{14}O$	585-34-2	150.217	nd (peth)	47(1)	240			s EtOH, alk; vs eth
1644	4-Butylphenol		$C_{10}H_{14}O$	1638-22-8	150.217		22	251(4)	0.976^{22}	1.5165^{25}	i H_2O; s EtOH, eth, alk; sl ctc
1645	4-*sec*-Butylphenol	4-(1-Methylpropyl)phenol	$C_{10}H_{14}O$	99-71-8	150.217		60(1)	243(3)	0.986^{20}	1.5182^{21}	i H_2O; s EtOH, alk; vs eth

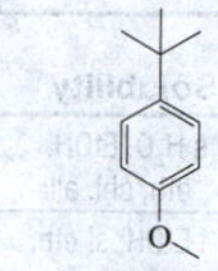
1601
1-*tert*-Butyl-4-methoxybenzene

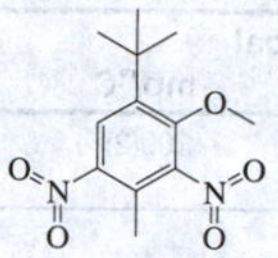
1602
1-*tert*-Butyl-2-methoxy-4-methyl-3,5-dinitrobenzene

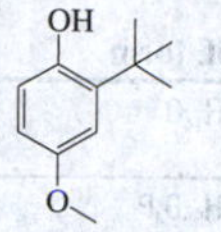
1603
2-*tert*-Butyl-4-methoxyphenol

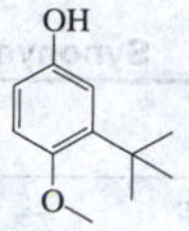
1604
3-*tert*-Butyl-4-methoxyphenol

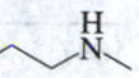
1605
Butylmethylamine

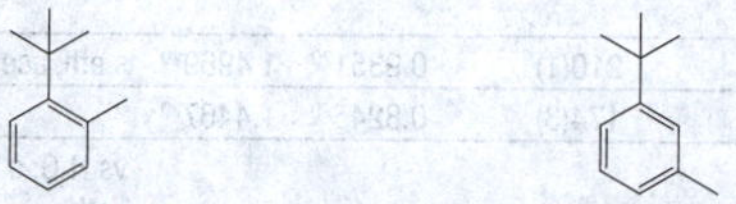
1606
1-*tert*-Butyl-2-methylbenzene

1607
1-*tert*-Butyl-3-methylbenzene

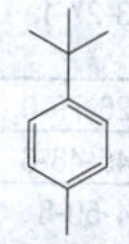
1608
1-*tert*-Butyl-4-methylbenzene

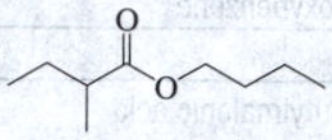
1609
Butyl 2-methylbutanoate

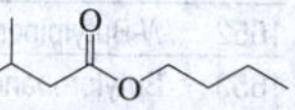
1610
Butyl 3-methylbutanoate

1611
Butyl methyl ether

1612
sec-Butyl methyl ether

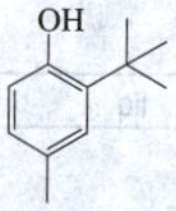
1613
2-*tert*-Butyl-4-methylphenol

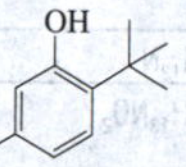
1614
2-*tert*-Butyl-5-methylphenol

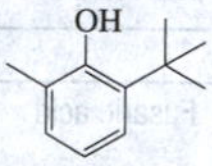
1615
2-*tert*-Butyl-6-methylphenol

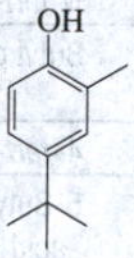
1616
4-*tert*-Butyl-2-methylphenol

1617
Butyl methyl sulfide

1618
tert-Butyl methyl sulfide

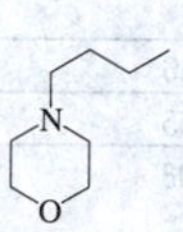
1619
4-Butylmorpholine

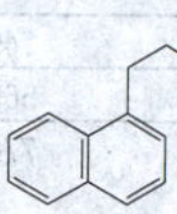
1620
1-Butylnaphthalene

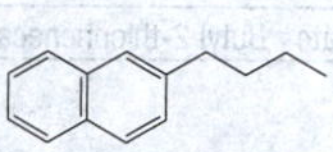
1621
2-Butylnaphthalene

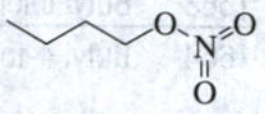
1622
Butyl nitrate

1623
Butyl nitrite

1624
tert-Butyl nitrite

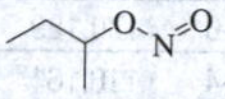
1625
sec-Butyl nitrite

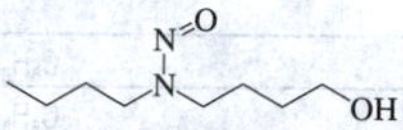
1626
4-(Butylnitrosoamino)-1-butanol

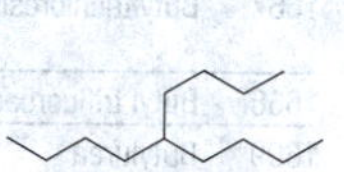
1627
5-Butylnonane

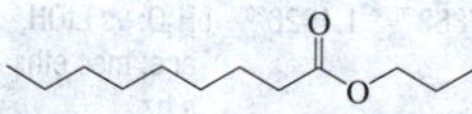
1628
Butyl nonanoate

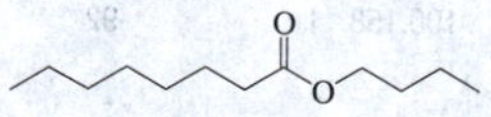
1629
Butyl octanoate

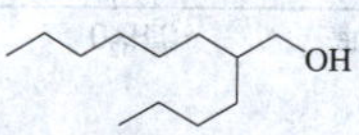
1630
2-Butyl-1-octanol

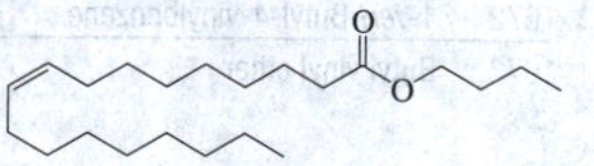
1631
Butyl oleate

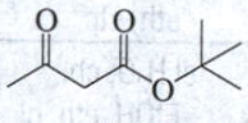
1632
tert-Butyl 3-oxobutanoate

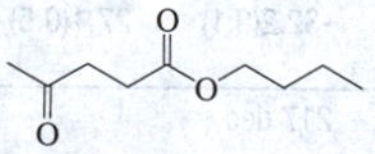
1633
Butyl 4-oxopentanoate

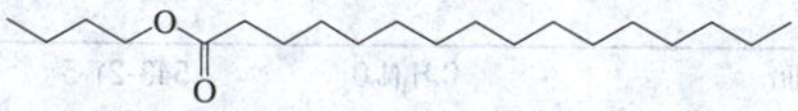
1634
Butyl palmitate

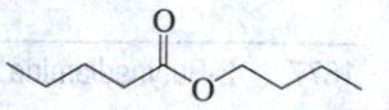
1635
Butyl pentanoate

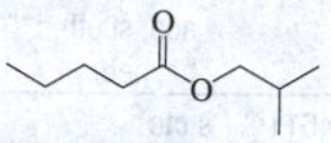
1636
sec-Butyl pentanoate

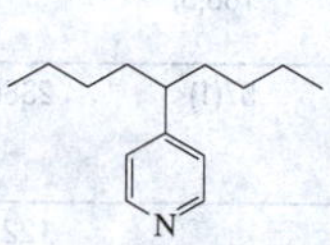
1637
4-(1-Butylpentyl)pyridine

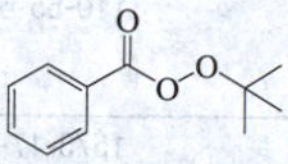
1638
tert-Butyl peroxybenzoate

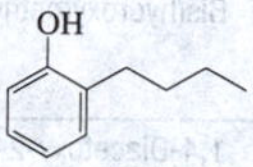
1639
2-Butylphenol

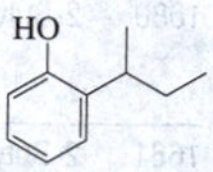
1640
2-*sec*-Butylphenol

1641
2-*tert*-Butylphenol

1642
3-Butylphenol

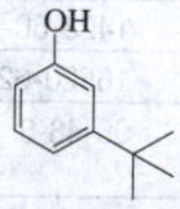
1643
3-*tert*-Butylphenol

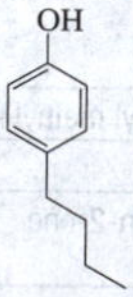
1644
4-Butylphenol

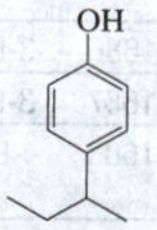
1645
4-*sec*-Butylphenol

No.	Name	Synonym	Mol. form.	CAS Reg. No.	Mol. wt.	Physical form	mp/°C	bp/°C	den/ g cm⁻³	n_D	Solubility
1646	4-*tert*-Butylphenol		$C_{10}H_{14}O$	98-54-4	150.217	nd (lig)	100(2)	244(5)	0.908^{80}	1.4787^{114}	s H_2O, EtOH, eth, chl, alk
1647	4-*tert*-Butylphenol, phosphate (3:1)		$C_{30}H_{39}O_4P$	78-33-1	494.602						i EtOH; sl eth, bz
1648	[(4-*tert*-Butylphenoxy)methyl]-oxirane		$C_{13}H_{18}O_2$	3101-60-8	206.281			167^{14}	1.036^{25}	1.5145^{20}	
1649	*N*-Butyl-*N*-phenylacetamide		$C_{12}H_{17}NO$	91-49-6	191.269		24.5(0.2)	281	0.9912^{20}	1.5146^{20}	sl chl
1650	1-(4-*tert*-Butylphenyl)-ethanone		$C_{12}H_{16}O$	943-27-1	176.254		17.7	263	0.9635^{20}	1.518^{15}	
1651	Butyl phenyl ether	Butoxybenzene	$C_{10}H_{14}O$	1126-79-0	150.217	liq	-19.4	210(1)	0.9351^{20}	1.4969^{20}	s eth, ace
1652	*N*-Butylpiperidine		$C_9H_{19}N$	4945-48-6	141.254			174(3)	0.8245^{20}	1.4467^{20}	
1653	Butylpropanedioic acid	*n*-Butylmalonic acid	$C_7H_{12}O_4$	534-59-8	160.168	pr (w)	104.5				vs H_2O; s EtOH, eth
1654	Butyl propanoate	Butyl propionate	$C_7H_{14}O_2$	590-01-2	130.185	liq	-89.5(0.5)	145.1(0.1)	0.8754^{20}	1.4014^{20}	sl H_2O, ctc; msc EtOH, eth
1655	*sec*-Butyl propanoate		$C_7H_{14}O_2$	591-34-4	130.185			133	0.8657^{20}	1.3952^{20}	s EtOH, eth
1656	*N*-*tert*-Butyl-2-propenamide	*N*-*tert*-Butylacrylamide	$C_7H_{13}NO$	107-58-4	127.184	cry (bz)	128				sl H_2O; i peth
1657	Butyl propyl ether		$C_7H_{16}O$	3073-92-5	116.201			117(4)	0.7772^{0}		i H_2O; vs EtOH, eth
1658	4-*tert*-Butylpyridine		$C_9H_{13}N$	3978-81-2	135.206	liq	-39.7(0.5)	197(5)	0.915^{25}	1.4958^{20}	s ctc, CS_2
1659	5-Butyl-2-pyridinecarboxylic acid	Fusaric acid	$C_{10}H_{13}NO_2$	536-69-6	179.216		97				
1660	Butyl stearate		$C_{22}H_{44}O_2$	123-95-5	340.583		26.56(0.02)	343	0.854^{25}	1.4328^{50}	i H_2O; s EtOH; vs ace
1661	Butyl thiocyanate	1-Thiocyanatobutane	C_5H_9NS	628-83-1	115.197			186	0.9563^{15}	1.4360^{20}	i H_2O; s EtOH, eth
1662	2-Butylthiophene		$C_8H_{12}S$	1455-20-5	140.246			179(17)	0.9537^{20}	1.5090^{20}	
1663	Butyl thiophene-2-carboxylate	Butyl 2-thiophenecarboxylate	$C_9H_{12}O_2S$	56053-84-0	184.255			$58^{0.15}$			
1664	Butyl 4-toluenesulfonate		$C_{11}H_{16}O_3S$	778-28-9	228.308			165^{6}	1.1319^{20}	1.5050^{20}	i H_2O; s eth; sl ctc
1665	Butyl trichloroacetate		$C_6H_9Cl_3O_2$	3657-07-6	219.493			204	1.2778^{20}	1.4525^{25}	s ctc
1666	Butyl (2,4,5-trichlorophenoxy)-acetate	2,4,5-T Butyl ester	$C_{12}H_{13}Cl_3O_3$	93-79-8	311.588		28.5	337			
1667	Butyltrichlorosilane	Trichlorobutylsilane	$C_4H_9Cl_3Si$	7521-80-4	191.559			148.5	1.1606^{20}	1.4363^{20}	s eth, bz, tol, AcOEt
1668	Butyl trifluoroacetate		$C_6H_9F_3O_2$	367-64-6	170.129			104.5(0.4)	1.0268^{22}	1.353^{22}	s chl
1669	Butylurea		$C_5H_{12}N_2O$	592-31-4	116.161	tab (w), nd (bz)	96.3(0.9)				vs H_2O, EtOH; sl chl
1670	*sec*-Butylurea	(1-Methylpropyl)urea	$C_5H_{12}N_2O$	689-11-2	116.161	pr (w)	169				
1671	*tert*-Butylurea		$C_5H_{12}N_2O$	1118-12-3	116.161		176.6(0.7)				s H_2O; vs EtOH; sl bz
1672	1-*tert*-Butyl-4-vinylbenzene	*p*-*tert*-Butylstyrene	$C_{12}H_{16}$	1746-23-2	160.255	liq	-36.9	99^{14}	0.89^{20}		
1673	Butyl vinyl ether	1-(Ethenyloxy)butane	$C_6H_{12}O$	111-34-2	100.158	liq	-92	94(1)	0.7888^{20}	1.4026^{20}	i H_2O; vs EtOH, ace; msc eth; s bz
1674	*tert*-Butyl vinyl ether	2-(Ethenyloxy)-2-methylpropane	$C_6H_{12}O$	926-02-3	100.158	liq	-112	75	0.7691^{20}	1.3922^{20}	
1675	1-Butyne	Ethylacetylene	C_4H_6	107-00-6	54.091	col gas	-125.7(0.2)	8.1(0.3)	0.6783^{0}	1.3962^{20}	i H_2O; s EtOH, eth
1676	2-Butyne	Dimethylacetylene	C_4H_6	503-17-3	54.091	vol liq or gas	-32.2(0.1)	27.1(0.5)	0.6910^{20}	1.3921^{20}	i H_2O; s EtOH, eth, ctc
1677	2-Butynediamide	Cellocidin	$C_4H_4N_2O_2$	543-21-5	112.087	cry (dil MeOH)	217 dec				sl H_2O, chl, EtOH, eth, gl HOAc
1678	2-Butynedinitrile		C_4N_2	1071-98-3	76.056		20(1)	76.5	0.9708^{25}	1.4647^{25}	
1679	2-Butynedioic acid		$C_4H_2O_4$	142-45-0	114.057		166(3)				vs H_2O, EtOH, eth
1680	2-Butyne-1,4-diol	Bis(hydroxymethyl)acetylene	$C_4H_6O_2$	110-65-6	86.090	pl (bz, AcOEt)	57(1)	238(8)		1.4804^{20}	vs H_2O, EtOH, ace; sl eth; i bz, peth
1681	2-Butyne-1,4-diol diacetate	1,4-Diacetoxy-2-butyne	$C_8H_{10}O_4$	1573-17-7	170.163			122^{10}		1.4611^{20}	s ctc
1682	2-Butynoic acid		$C_4H_4O_2$	590-93-2	84.074	pl (eth, peth)	78	203	0.9641^{20}		vs H_2O, eth, EtOH, chl
1683	2-Butyn-1-ol		C_4H_6O	764-01-2	70.090	liq	-1.1	140(5)	0.9370^{20}	1.4530^{20}	vs eth, EtOH
1684	3-Butyn-1-ol		C_4H_6O	927-74-2	70.090	liq	-63.5(0.4)	129(5)	0.9257^{20}	1.4409^{20}	vs H_2O, EtOH
1685	3-Butyn-2-ol		C_4H_6O	2028-63-9	70.090	liq	-1.5	106.5	0.8618^{20}	1.4207^{20}	vs H_2O, eth, EtOH
1686	3-Butyn-2-one	Ethynyl methyl ketone	C_4H_4O	1423-60-5	68.074			84	0.8793^{20}	1.4070^{20}	
1687	3-Butynylbenzene		$C_{10}H_{10}$	16520-62-0	130.186			177(8)	0.9258^{20}	1.5208^{20}	
1688	γ-Butyrolactone	Oxolan-2-one	$C_4H_6O_2$	96-48-0	86.090	liq	-43.36(0.08)	204.6(0.4)	1.1296^{20}	1.4341^{20}	vs ace, bz, eth, EtOH
1689	Cacotheline		$C_{21}H_{21}N_3O_7$	561-20-6	427.408	ye cry	>300				sl H_2O

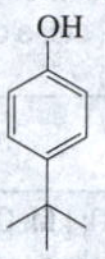
1646
4-*tert*-Butylphenol

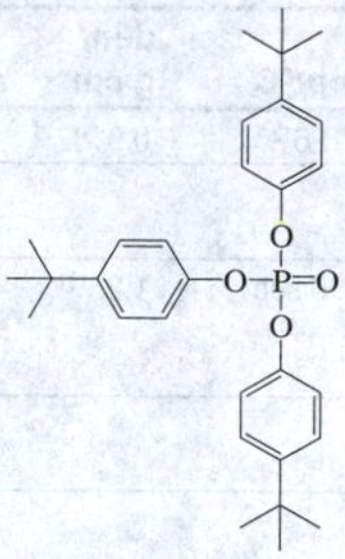
1647
4-*tert*-Butylphenol, phosphate (3:1)

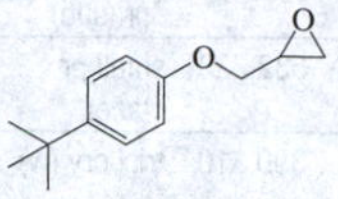
1648
[(4-*tert*-Butylphenoxy)methyl]oxirane

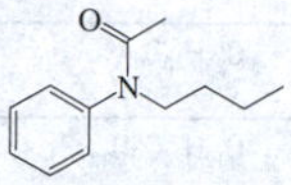
1649
N-Butyl-*N*-phenylacetamide

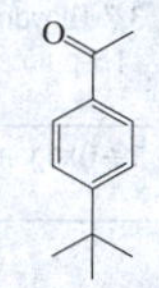
1650
1-(4-*tert*-Butylphenyl)ethanone

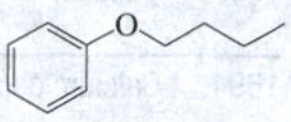
1651
Butyl phenyl ether

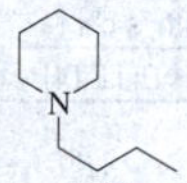
1652
N-Butylpiperidine

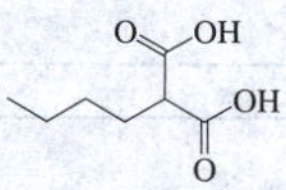
1653
Butylpropanedioic acid

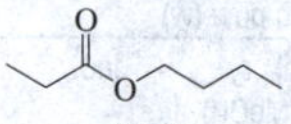
1654
Butyl propanoate

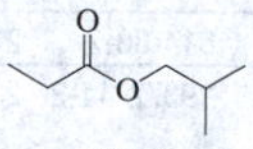
1655
sec-Butyl propanoate

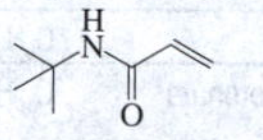
1656
N-*tert*-Butyl-2-propenamide

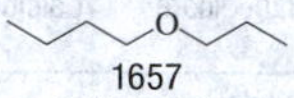
1657
Butyl propyl ether

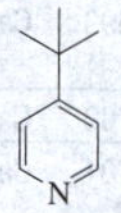
1658
4-*tert*-Butylpyridine

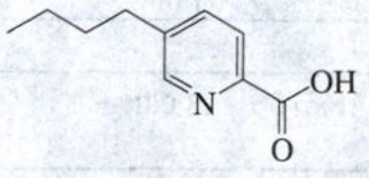
1659
5-Butyl-2-pyridinecarboxylic acid

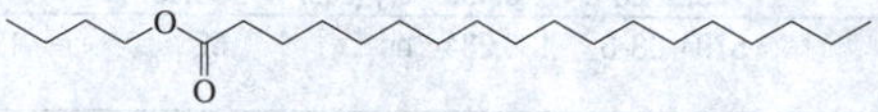
1660
Butyl stearate

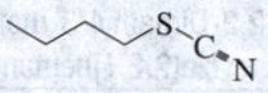
1661
Butyl thiocyanate

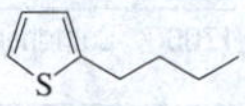
1662
2-Butylthiophene

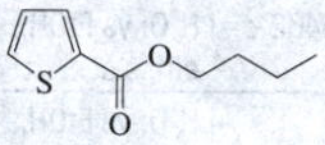
1663
Butyl thiophene-2-carboxylate

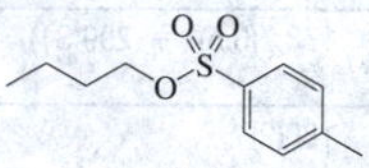
1664
Butyl 4-toluenesulfonate

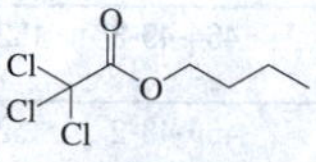
1665
Butyl trichloroacetate

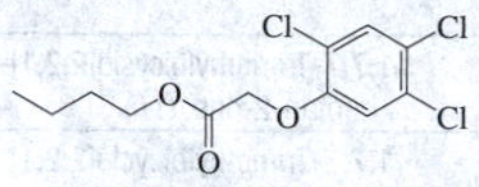
1666
Butyl (2,4,5-trichlorophenoxy)acetate

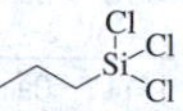
1667
Butyltrichlorosilane

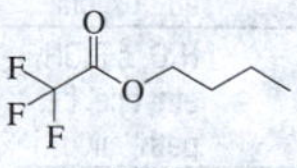
1668
Butyl trifluoroacetate

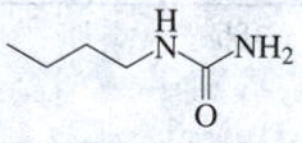
1669
Butylurea

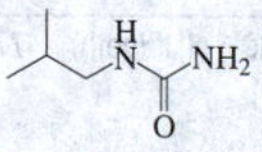
1670
sec-Butylurea

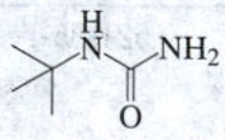
1671
tert-Butylurea

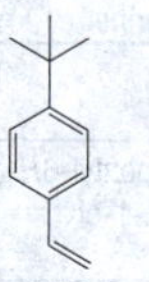
1672
1-*tert*-Butyl-4-vinylbenzene

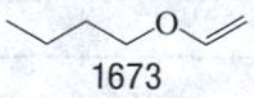
1673
Butyl vinyl ether

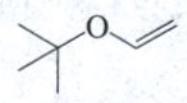
1674
tert-Butyl vinyl ether

1675
1-Butyne

1676
2-Butyne

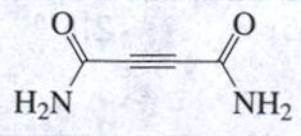
1677
2-Butynediamide

1678
2-Butynedinitrile

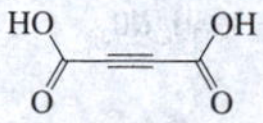
1679
2-Butynedioic acid

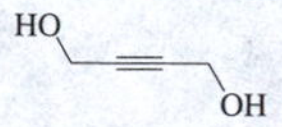
1680
2-Butyne-1,4-diol

1681
2-Butyne-1,4-diol diacetate

1682
2-Butynoic acid

1683
2-Butyn-1-ol

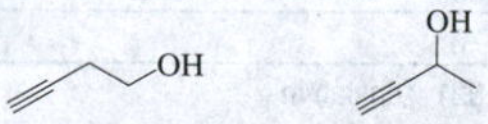
1684
3-Butyn-1-ol

1685
3-Butyn-2-ol

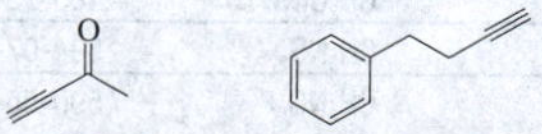
1686
3-Butyn-2-one

1687
3-Butynylbenzene

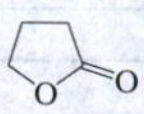
1688
γ-Butyrolactone

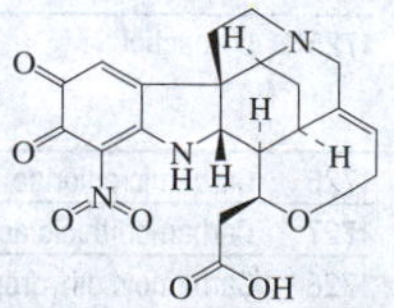
1689
Cacotheline

No.	Name	Synonym	Mol. form.	CAS Reg. No.	Mol. wt.	Physical form	mp/°C	bp/°C	den/ g cm^{-3}	n_D	Solubility
1690	γ-Cadinene		$C_{15}H_{24}$	39029-41-9	204.352			126^{12}	0.9182^{15}	1.3166^{20}	
1691	Cadmium bis(diethyldithiocarbamate)		$C_{10}H_{20}CdN_2S_4$	14239-68-0	408.950	wh cry	255				
1692	Caffeine	3,7-Dihydro-1,3,7-trimethyl-1*H*-purine-2,6-dione	$C_8H_{10}N_4O_2$	58-08-2	194.191	wh nd (w+1), hex pr (sub)	236.1(0.2)	90 sub	1.23^{19}		sl H_2O, EtOH; i eth, ctc; s chl, py
1693	Calactin	19-Oxogomphoside	$C_{29}H_{40}O_9$	20304-47-6	532.623	small pr (ace)	271				
1694	Calcium ascorbate		$C_{12}H_{14}CaO_{12}$	5743-27-1	390.310	tricl cry (w)					s H_2O; i MeOH, EtOH
1695	Calcium citrate	Tricalcium citrate	$C_{12}H_{10}Ca_3O_{14}$	813-94-5	498.433	cry (w)	≈100 dec (hyd)				sl H_2O; i EtOH
1696	Calcium cyanamide	Calcium carbimide	$CCaN_2$	156-62-7	80.102	col hex cry	≈1340	sub	2.29		dec H_2O
1697	Calcium cyclamate		$C_{12}H_{24}CaN_2O_6S_2$	139-06-0	396.536	cry					vs H_2O
1698	Calcium gluconate		$C_{12}H_{22}CaO_{14}$	299-28-5	430.373	cry					i EtOH, os
1699	Calcium iodobehenate	Iododocosanoic acid, calcium salt	$C_{44}H_{84}CaI_2O_4$	1319-91-1	971.023	wh-ye pow					i H_2O, EtOH, eth; s chl
1700	Calcium lactate		$C_6H_{10}CaO_6$	814-80-2	218.217	wh pow (w)					s H_2O; i EtOH
1701	Calcium 2,4-pentanedioate	Calcium acetylacetonate	$C_{10}H_{14}CaO_4$	19372-44-2	238.294	col cry (MeOH)	dec				
1702	Calcium thioglycollate		$C_4H_6CaO_4S_2$	814-71-1	222.297	pr (w)	220 dec				s H_2O, chl; sl EtOH; i eth, bz
1703	Calotoxin	4'β-Hydroxy-19-oxogomphoside	$C_{29}H_{40}O_{10}$	20304-49-8	548.622	cry (EtOH)	268				
1704	Calotropin		$C_{29}H_{40}O_9$	1986-70-5	532.623	pl (EtOH)	221				s H_2O, EtOH; i eth
1705	Calusterone		$C_{21}H_{32}O_2$	17021-26-0	316.483	cry (ace)	157.5				
1706	Camphene, (+)	2,2-Dimethyl-3-methylenebicyclo[2.2.1]heptane, (1R)-	$C_{10}H_{16}$	5794-03-6	136.234	nd	52	161	0.8950^{50}	1.4570^{25}	vs eth
1707	Camphene, (-)	2,2-Dimethyl-3-methylenebicyclo[2.2.1]heptane, (1*S*)-	$C_{10}H_{16}$	5794-04-7	136.234		52	158	0.8446^{50}	1.4564^{54}	vs eth
1708	*d*-Camphocarboxylic acid		$C_{11}H_{16}O_3$	18530-30-8	196.243	pr (eth, 50% al)	127.5				vs bz, eth, EtOH
1709	Camphor, (±)-	1,7,7-Trimethylbicyclo[2.2.1]heptan-2-one, (±)	$C_{10}H_{16}O$	21368-68-3	152.233	wh rhom cry (EtOH)	177.7(0.2)	sub			i H_2O; vs EtOH, eth; s ace, bz, ctc
1710	Camphor, (+)	1,7,7-Trimethylbicyclo[2.2.1]heptan-2-one, (1*R*)	$C_{10}H_{16}O$	464-49-3	152.233	pl	178.7(0.5)	209(31)	0.990^{25}	1.5462	i H_2O; vs EtOH, eth; s ace, bz
1711	Camphor, (-)	1,7,7-Trimethylbicyclo[2.2.1]heptan-2-one, (1*S*)	$C_{10}H_{16}O$	464-48-2	152.233		180(2)		0.9853^{18}		i H_2O; vs EtOH, eth, HOAc; s ace, bz
1712	(±)-Camphoric acid	1,2,2-Trimethyl-1,3-cyclopentanedicarboxylic acid	$C_{10}H_{16}O_4$	5394-83-2	200.232	pr, lf	202		1.186		sl H_2O; s chl, eth, EtOH
1713	*d*-Camphorsulfonic acid		$C_{10}H_{16}O_4S$	3144-16-9	232.297	pr (HOAc)	195 dec				vs H_2O; i eth; sl HOAc
1714	Canadine, (±)-	DL-Tetrahydroberberine	$C_{20}H_{21}NO_4$	29074-38-2	339.386	mcl nd (al)	134				vs EtOH, chl
1715	Cannabidiol		$C_{21}H_{30}O_2$	13956-29-1	314.462	rods (peth)	67	188^2	1.040^{40}	1.5404^{20}	i H_2O; s EtOH, eth, bz, chl
1716	Cannabinol	6,6,9-Trimethyl-3-pentyl-6*H*-dibenzo[*b*,*d*]pyran-1-ol	$C_{21}H_{26}O_2$	521-35-7	310.430	pl, lf (peth)	77	$185^{0.05}$			i H_2O; s EtOH, eth, ace, bz, peth, alk
1717	Canrenone		$C_{22}H_{28}O_3$	976-71-6	340.455	cry (AcOEt)	150				
1718	Cantharidin		$C_{10}H_{12}O_4$	56-25-7	196.200	orth pl	218	84 sub			i H_2O; sl EtOH, eth, ace, bz; s HOAc
1719	Caprolactam	6-Hexanelactam	$C_6H_{11}NO$	105-60-2	113.157	lf (lig)	69.16(0.01)	270.8(0.1)			vs H_2O, bz, EtOH, chl
1720	Capsaicin		$C_{18}H_{27}NO_3$	404-86-4	305.412	mcl pl or sc (peth)	65	$215^{0.01}$			i H_2O; vs EtOH; s eth, bz, peth; sl con HCl
1721	Capsanthin	3,3'-Dihydroxy-β,κ-caroten-6'-one, (3*R*,3'*S*,5'*R*)	$C_{40}H_{56}O_3$	465-42-9	584.871		176				
1722	Captafol		$C_{10}H_9Cl_4NO_2S$	2425-06-1	349.061	cry	159.0(0.9)				
1723	Captan		$C_9H_8Cl_3NO_2S$	133-06-2	300.590	cry (CCl_4)	173.9(0.3)		1.74^{25}		vs chl
1724	Captopril	1-(3-Mercapto-2-methyl-1-oxypropyl)proline	$C_9H_{15}NO_3S$	62571-86-2	217.285	cry (AcOEt)	105				s H_2O, EtOH, chl
1725	Carbachol		$C_6H_{15}ClN_2O_2$	51-83-2	182.648		208(6)				vs H_2O, MeOH; sl EtOH; i eth, chl
1726	Carbamic chloride	Carbamyl chloride	CH_2ClNO	463-72-9	79.486			62 dec			
1727	Carbamodithioic acid		CH_3NS_2	594-07-0	93.172						vs EtOH, eth
1728	Carbamoyl dihydrogen phosphate		CH_4NO_5P	590-55-6	141.021	unstab in soln					

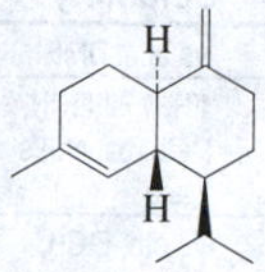
1690
γ-Cadinene

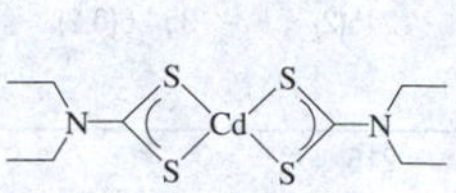
1691
Cadmium bis(diethyldithiocarbamate)

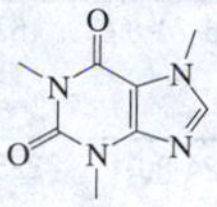
1692
Caffeine

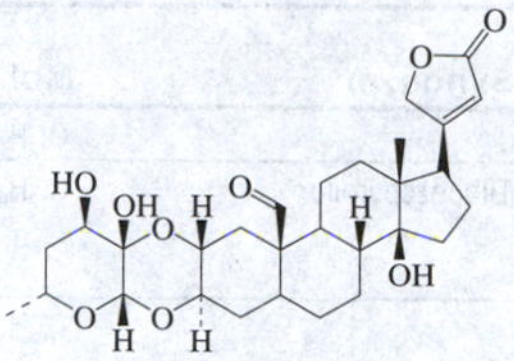
1693
Calactin

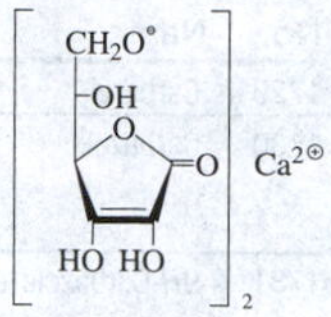
1694
Calcium ascorbate

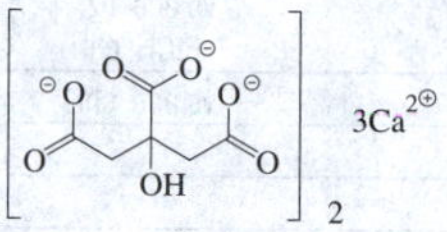
1695
Calcium citrate

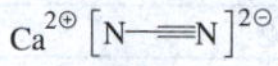
1696
Calcium cyanamide

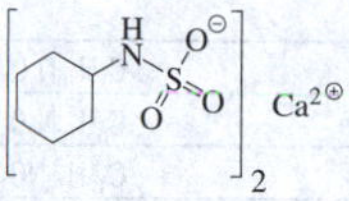
1697
Calcium cyclamate

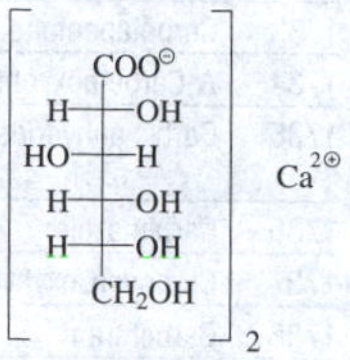
1698
Calcium gluconate

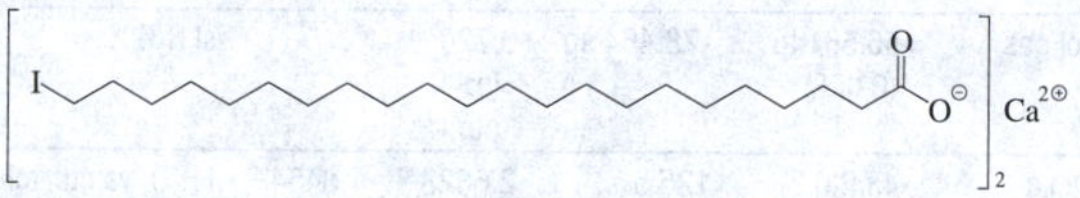
1699
Calcium iodobehenate

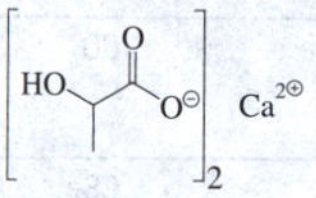
1700
Calcium lactate

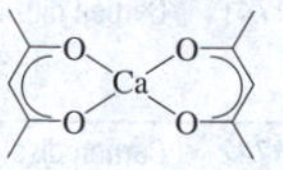
1701
Calcium 2,4-pentanedioate

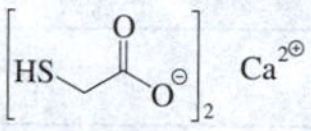
1702
Calcium thioglycollate

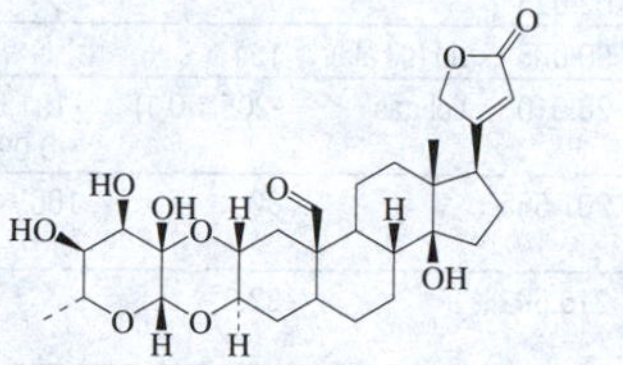
1703
Calotoxin

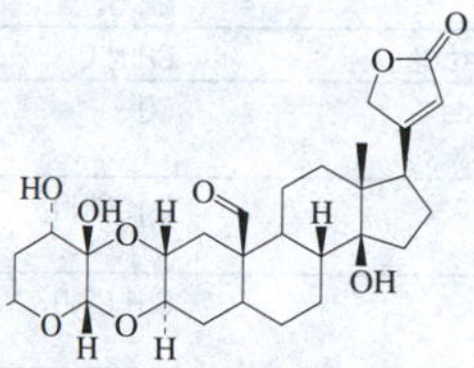
1704
Calotropin

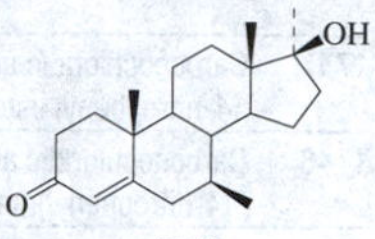
1705
Calusterone

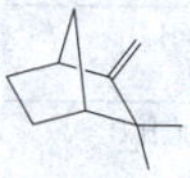
1706
Camphene, (+)

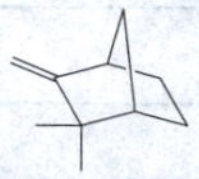
1707
Camphene, (-)

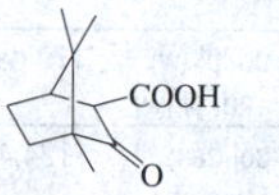
1708
d-Camphocarboxylic acid

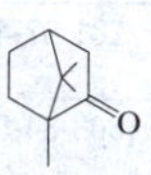
1709
Camphor, (±)-

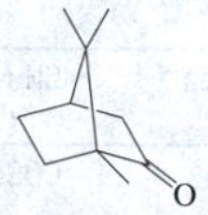
1710
Camphor, (+)

1711
Camphor, (-)

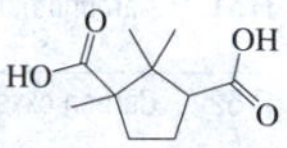
1712
(±)-Camphoric acid

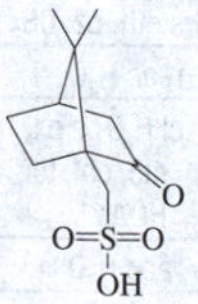
1713
d-Camphorsulfonic acid

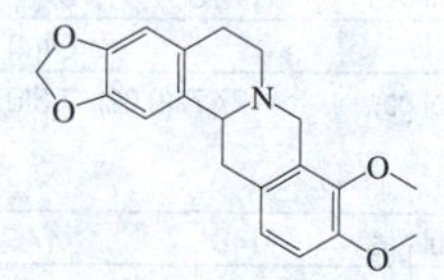
1714
Canadine, (±)-

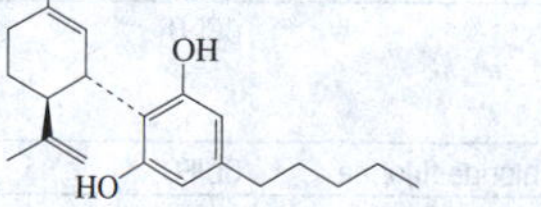
1715
Cannabidiol

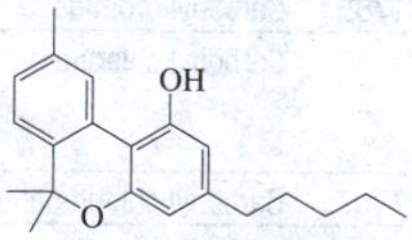
1716
Cannabinol

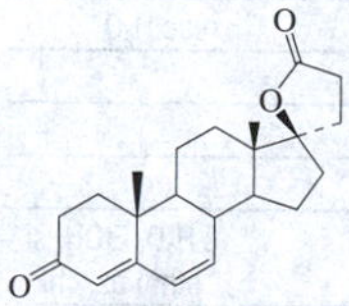
1717
Canrenone

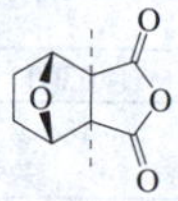
1718
Cantharidin

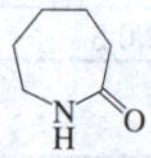
1719
Caprolactam

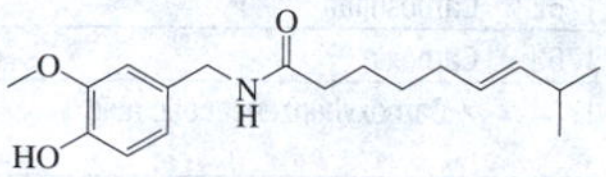
1720
Capsaicin

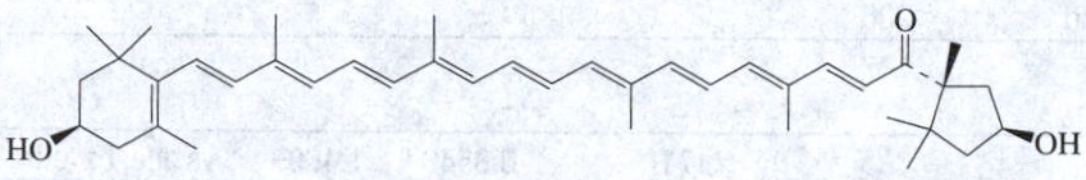
1721
Capsanthin

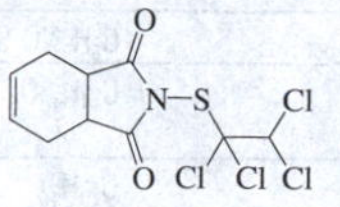
1722
Captafol

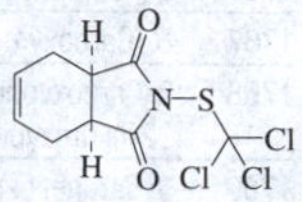
1723
Captan

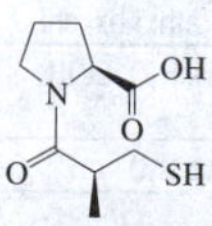
1724
Captopril

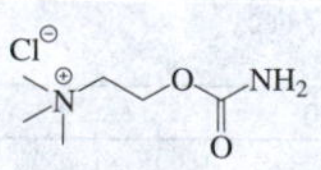
1725
Carbachol

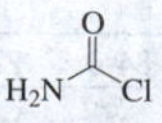
1726
Carbamic chloride

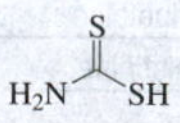
1727
Carbamodithioic acid

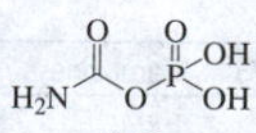
1728
Carbamoyl dihydrogen phosphate

No.	Name	Synonym	Mol. form.	CAS Reg. No.	Mol. wt.	Physical form	mp/°C	bp/°C	den/g cm^{-3}	n_D	Solubility
1729	Carbaryl		$C_{12}H_{11}NO_2$	63-25-2	201.221		142.7(0.7)		1.228^{25}		vs ace, DMF
1730	Carbazole	Dibenzopyrolle	$C_{12}H_9N$	86-74-8	167.206	pl or lf	245(2)	354.6(0.2)	1.297^{25}		i H_2O; sl EtOH, eth, bz, chl; s ace
1731	9*H*-Carbazole-9-acetic acid		$C_{14}H_{11}NO_2$	524-80-1	225.243	lf (AcOEt)	215				vs eth, EtOH, chl, HOAc
1732	Carbendazim	Carbamic acid, 1*H*-benzimidazol-2-yl-, methyl ester	$C_9H_9N_3O_2$	10605-21-7	191.186		300 dec		1.45		
1733	Carbetapentane	Pentoxyverine	$C_{20}H_{31}NO_3$	77-23-6	333.465			$165^{0.01}$			
1734	*N*-Carbethoxyphthalimide	*N*-(Ethoxycarbonyl)phthalimide	$C_{11}H_9NO_4$	22509-74-6	219.194		91				
1735	Carbic anhydride		$C_9H_8O_3$	129-64-6	164.158	orth cry (peth)	163(3)		1.417^{25}		vs ace, bz, EtOH, chl
1736	Carbimazole		$C_7H_{10}N_2O_2S$	22232-54-8	186.231	cry, pow	123.5				vs ace, chl
1737	Carbobenzoxyhydrazine	Benzyl carbazate	$C_8H_{10}N_2O_2$	5331-43-1	166.177		69.5				
1738	Carbofuran		$C_{12}H_{15}NO_3$	1563-66-2	221.252		153.2(0.5)		1.18		
1739	Carboimidic difluoride		CHF_2N	2712-98-3	65.023	gas	-90	-13 dec			
1740	γ-Carboline	5*H*-Pyrido[4,3-*b*]indole	$C_{11}H_8N_2$	244-69-9	168.195	nd	225		1.352		sl H_2O, bz; vs MeOH; s EtOH
1741	Carbon dioxide	Carbonic anhydride	CO_2	124-38-9	44.010	col gas	-56.561 tp (0.008)	-78.464 sp	0.720^{25} (p>1 atm)		sl H_2O
1742	Carbon diselenide	Carbon selenide	CSe_2	506-80-9	169.93	ye liq	-43.6(0.3)	125.5	2.6823^{20}	1.8454^{20}	i H_2O; vs ctc, tol
1743	Carbon disulfide	Carbon bisulfide	CS_2	75-15-0	76.141	col liq	-111.7(0.3)	46.2(0.1)	1.2632^{20}	1.6319^{20}	s H_2O, chl; msc EtOH, eth
1744	Carbonic acid		CH_2O_3	463-79-6	62.025						Aq. soln. of CO_2
1745	Carbonic dihydrazide	Carbohydrazide	CH_6N_4O	497-18-7	90.085	nd (dil al)	154		1.616^{20}		vs H_2O, EtOH
1746	Carbon monoxide	Carbon oxide	CO	630-08-0	28.010	col gas	-205.1(0.1)	-191.51 (0.09)	0.7909^{-19}		sl H_2O; s bz, HOAc
1747	Carbonochloridic acid, 4-nitrophenyl ester		$C_7H_4ClNO_4$	7693-46-1	201.565		80	160^{19}			
1748	Carbonochloridic acid, (4-nitrophenyl)methyl ester		$C_8H_6ClNO_4$	4457-32-3	215.592		32.8				
1749	Carbonochloridic acid, 2,2,2-trichloroethyl ester		$C_3H_2Cl_4O_2$	17341-93-4	211.859			63^{11}			
1750	Carbonothioic dichloride	Thiophosgene	CCl_2S	463-71-8	114.982	red liq		73	1.508^{15}	1.5442^{20}	dec H_2O, EtOH; s eth
1751	Carbonothioic dihydrazide	1,3-Diamino-2-thiourea	CH_6N_4S	2231-57-4	106.151	nd, pl (w) nd, pl (w)	170 dec				vs H_2O
1752	Carbon oxyselenide	Carbonyl selenide	COSe	1603-84-5	106.97	col gas; unstab	-124.4	-21.7			dec H_2O
1753	Carbon oxysulfide	Carbonyl sulfide	COS	463-58-1	60.075	col gas	-138.8(0.1)	-50.2(0.3)	1.028^{17}	1.24^{-87}	sl H_2O; s EtOH; vs KOH
1754	Carbon suboxide	1,2-Propadiene-1,3-dione	C_3O_2	504-64-3	68.031	col gas	-112.5	6.8	1.114^{0}	1.4538^{0}	s eth, bz, CS_2
1755	Carbonyl bromide	Bromophosgene	CBr_2O	593-95-3	187.818			64(4)	2.52^{15}		reac H_2O
1756	Carbonyl chloride	Phosgene	CCl_2O	75-44-5	98.916	col gas	-127.77(0.02)	7.5(0.4)	1.3719^{25} (p>1 atm		sl H_2O; s bz, ctc, chl, tol, HOAc
1757	Carbonyl chloride fluoride	Carbonic chloride fluoride	CClFO	353-49-1	82.461	col gas	-148	-47.2			reac H_2O
1758	Carbonyl dicyanide		C_3N_2O	1115-12-4	80.044	liq	-36	65.5	1.124^{20}	1.3919^{20}	s eth, ace, ctc, chl
1759	*N,N'*-Carbonyldiimidazole		$C_7H_6N_4O$	530-62-1	162.149	cry (bz)	119				
1760	Carbonyl fluoride		CF_2O	353-50-4	66.007	col gas	-111.2	-84.5			reac H_2O
1761	Carbophenothion		$C_{11}H_{16}ClO_2PS_3$	786-19-6	342.866			$82^{0.01}$	1.271^{20}		
1762	Carbosulfan		$C_{20}H_{32}N_2O_3S$	55285-14-8	380.544			126	1.056^{20}		
1763	Carboxin		$C_{12}H_{13}NO_2S$	5234-68-4	235.302		96(7)				
1764	2-Carboxybenzeneacetic acid		$C_9H_8O_4$	89-51-0	180.158		184.5		1.4100^{20}		s H_2O, EtOH; sl eth; i bz, chl
1765	*N*-(D-1-Carboxyethyl)-L-arginine	Octopine	$C_9H_{18}N_4O_4$	34522-32-2	246.264	nd (w)	281				
1766	L-γ-Carboxyglutamic acid		$C_6H_9NO_6$	53861-57-7	191.138	cry	167				
1767	*S*-(Carboxymethyl)-L-cysteine	Carbocysteine	$C_5H_9NO_4S$	638-23-3	179.195	nd	206				
1768	2-Carboxyphenyl 2-hydroxybenzoate	Salsalate	$C_{14}H_{10}O_5$	552-94-3	258.226		147				sl ace
1769	3-Carene, (+)		$C_{10}H_{16}$	498-15-7	136.234			171	0.8549^{30}	1.469^{3}	vs ace, bz, eth
1770	Carisoprodol		$C_{12}H_{24}N_2O_4$	78-44-4	260.330	cry	92				s os
1771	Carminic acid		$C_{22}H_{20}O_{13}$	1260-17-9	492.386	red mcl pr (aq, MeOH)	136 dec				s H_2O, EtOH; sl eth; i bz, chl
1772	Carnitine	4-Amino-3-hydroxybutanoic acid trimethylbetaine	$C_7H_{15}NO_3$	541-15-1	161.199	cry (al-ace), hyg	197 dec				vs H_2O, EtOH
1773	Carnosine	*N*-β-Alanyl-L-histidine	$C_9H_{14}N_4O_3$	305-84-0	226.232		260				vs H_2O

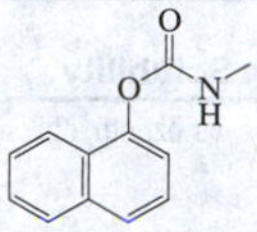
1729
Carbaryl

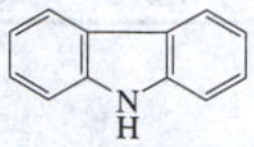
1730
Carbazole

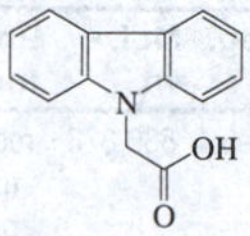
1731
9H-Carbazole-9-acetic acid

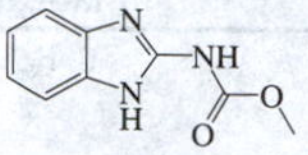
1732
Carbendazim

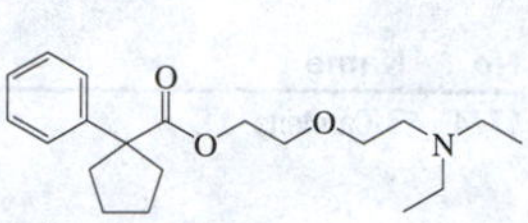
1733
Carbetapentane

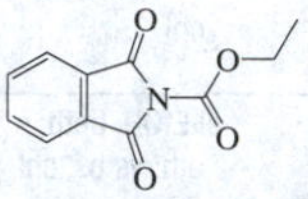
1734
N-Carbethoxyphthalimide

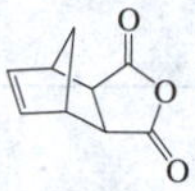
1735
Carbic anhydride

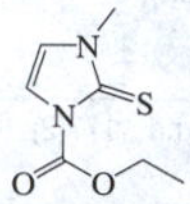
1736
Carbimazole

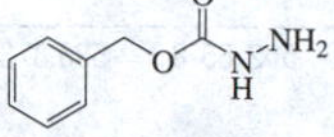
1737
Carbobenzoxyhydrazine

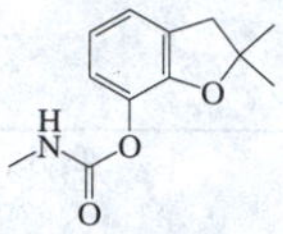
1738
Carbofuran

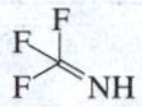
1739
Carboimidic difluoride

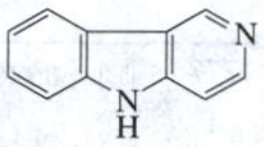
1740
γ-Carboline

O=C=O
1741
Carbon dioxide

Se=C=Se
1742
Carbon diselenide

S=C=S
1743
Carbon disulfide

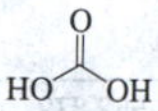
1744
Carbonic acid

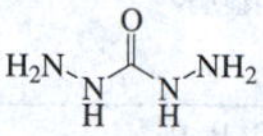
1745
Carbonic dihydrazide

C≡O
1746
Carbon monoxide

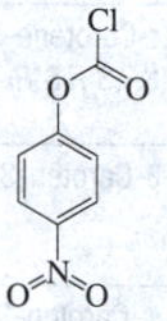
1747
Carbonochloridic acid, 4-nitrophenyl ester

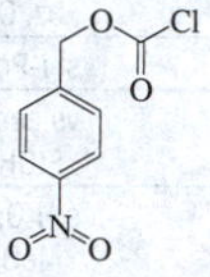
1748
Carbonochloridic acid, (4-nitrophenyl)methyl ester

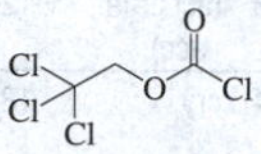
1749
Carbonochloridic acid, 2,2,2-trichloroethyl ester

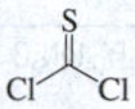
1750
Carbonothioic dichloride

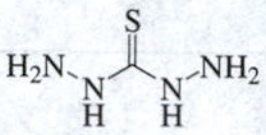
1751
Carbonothioic dihydrazide

O=C=Se
1752
Carbon oxyselenide

O=C=S
1753
Carbon oxysulfide

O=C=C=C=O
1754
Carbon suboxide

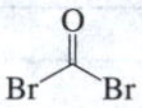
1755
Carbonyl bromide

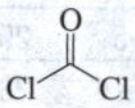
1756
Carbonyl chloride

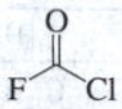
1757
Carbonyl chloride fluoride

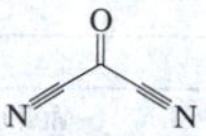
1758
Carbonyl dicyanide

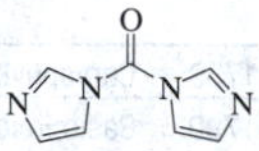
1759
N,N'-Carbonyldiimidazole

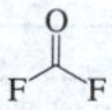
1760
Carbonyl fluoride

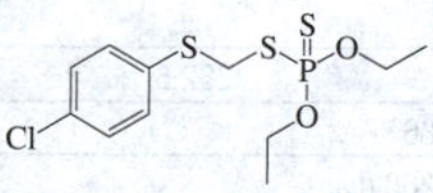
1761
Carbophenothion

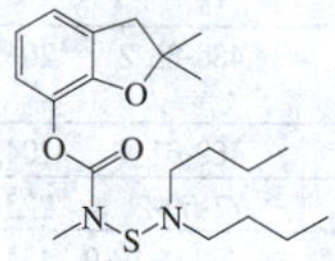
1762
Carbosulfan

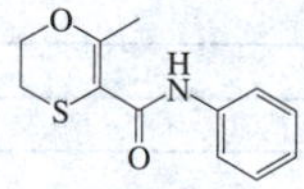
1763
Carboxin

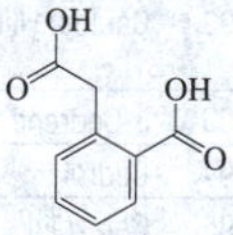
1764
2-Carboxybenzeneacetic acid

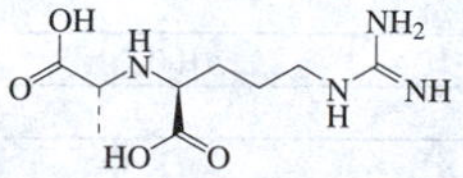
1765
N-(D-1-Carboxyethyl)-L-arginine

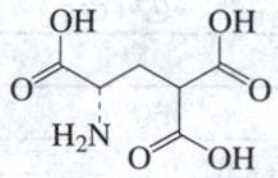
1766
L-γ-Carboxyglutamic acid

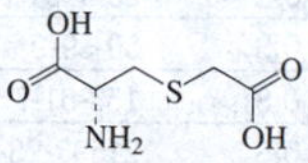
1767
S-(Carboxymethyl)-L-cysteine

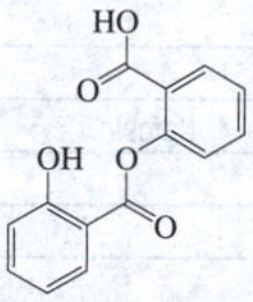
1768
2-Carboxyphenyl 2-hydroxybenzoate

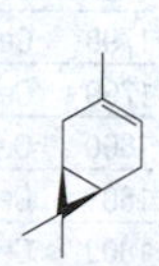
1769
3-Carene, (+)

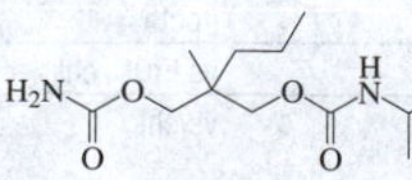
1770
Carisoprodol

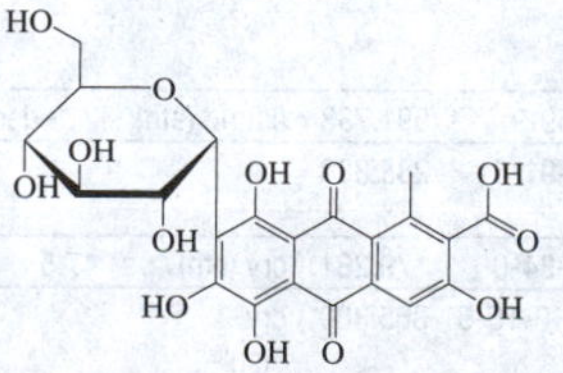
1771
Carminic acid

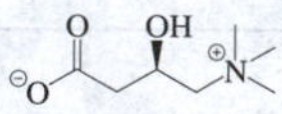
1772
Carnitine

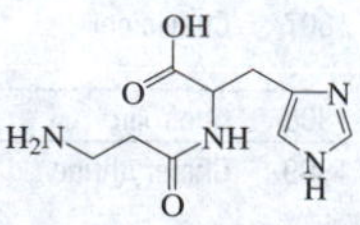
1773
Carnosine

No.	Name	Synonym	Mol. form.	CAS Reg. No.	Mol. wt.	Physical form	mp/°C	bp/°C	den/ g cm^{-3}	n_D	Solubility
1774	α-Carotene		$C_{40}H_{56}$	7488-99-5	536.873	red pl or pr (peth, bz-MeOH)	187.5		1.00^{20}		vs bz, eth, chl
1775	β-Carotene		$C_{40}H_{56}$	7235-40-7	536.873	red br hex pr (bz-MeOH)	183(2)		1.00^{20}		i H_2O; sl EtOH, chl; s eth, ace, bz
1776	β,ψ-Carotene	γ-Carotene	$C_{40}H_{56}$	472-93-5	536.873	red pr (bz-MeOH), viol pr (eth)	153				i H_2O, EtOH; sl eth, peth; s bz, chl
1777	ψ,ψ-Carotene	*trans*-Lycopene	$C_{40}H_{56}$	502-65-8	536.873	red pr or nd (peth)	177(2)				sl EtOH, peth; s eth; vs bz, chl, CS_2
1778	β,β-Carotene-3,3'-diol, (3*R*,3'*R*)-	Zeaxanthin	$C_{40}H_{56}O_2$	144-68-3	568.872	ye pr (MeOH) orth (chl-eth)	215.5	$227^{0.06}$			i H_2O; sl EtOH; s eth, ace, bz, py, chl
1779	β,ε-Carotene-3,3'-diol, (3*R*,3'*R*,6'*R*)-	Xanthophyll	$C_{40}H_{56}O_2$	127-40-2	568.872	ye or viol pr (eth-MeOH)	196				vs bz, eth, EtOH, peth
1780	β,β-Caroten-3-ol, (3*R*)-	Cryptoxanthin	$C_{40}H_{56}O$	472-70-8	552.872	garnet red pr (bz-MeOH)	160				vs bz, chl
1781	β,ψ-Caroten-3-ol, (3*R*)-	Rubixanthin	$C_{40}H_{56}O$	3763-55-1	552.872	dk red nd (bz-MeOH) oran-red (bz-peth)	160				sl EtOH, peth; s bz, chl
1782	ψ,ψ-Caroten-16-ol	Lycoxanthin	$C_{40}H_{56}O$	19891-74-8	552.872	red pl (bz-MeOH)	168				i H_2O; sl EtOH; s bz, CS_2
1783	Caroverine		$C_{22}H_{27}N_3O_2$	23465-76-1	365.468	cry	69	$202^{0.01}$			sl i-PrOH
1784	Carpaine		$C_{28}H_{50}N_2O_4$	3463-92-1	478.708	mcl pr (al, ace)	121				vs ace, bz, eth, EtOH
1785	Cartap hydrochloride		$C_7H_{16}ClN_3O_2S_2$	22042-59-7	273.804	cry	180				s H_2O; sl EtOH, MeOH
1786	Carvenone, (*S*)-		$C_{10}H_{16}O$	10395-45-6	152.233			233	0.9289^{20}	1.4805^{20}	i H_2O; s ace
1787	(*R*)-Carvone	*p*-Mentha-1,8-dien-6-one, (*R*)	$C_{10}H_{14}O$	6485-40-1	150.217		25.2	231	0.9593^{20}	1.4988^{20}	sl H_2O; vs EtOH; s eth, ctc, chl
1788	(*S*)-Carvone	*p*-Mentha-1,8-dien-6-one, (*S*)	$C_{10}H_{14}O$	2244-16-8	150.217		<15	231	0.965^{20}	1.4989^{20}	sl H_2O; vs EtOH; s eth, chl
1789	Caryophyllene		$C_{15}H_{24}$	87-44-5	204.352			$122^{13.5}$	0.9075^{20}	1.4986^{20}	vs bz
1790	Casimiroin	6-Methoxy-9-methyl-1,3-dioxolo[4,5-*h*]quinolin-8(9*H*)-one	$C_{12}H_{11}NO_4$	477-89-4	233.220						sl chl
1791	Cassaine		$C_{24}H_{39}NO_4$	468-76-8	405.572	fl (eth)	142.5				s EtOH, ace, chl, eth, bz, MeOH
1792	Caulophylline		$C_{12}H_{16}N_2O$	486-86-2	204.267	cry (w+2), nd (al, bz)	137				vs H_2O, ace, bz, EtOH
1793	α-Cedrene		$C_{15}H_{24}$	469-61-4	204.352	oil		262.5			
1794	Cedrol		$C_{15}H_{26}O$	77-53-2	222.366		86		0.9479^{90}	1.4824^{90}	
1795	Cefazolin		$C_{14}H_{14}N_8O_4S_3$	25953-19-9	454.508	nd (ace aq)	200 dec				s DMF, py; sl MeOH; i chl, bz, eth
1796	β-Cellobiose		$C_{12}H_{22}O_{11}$	13360-52-6	342.296	cry (dil al)	225 dec				s H_2O; i EtOH, eth, ace, bz
1797	Cellotriose		$C_{18}H_{32}O_{16}$	33404-34-1	504.437		208				
1798	Cephalexin		$C_{16}H_{17}N_3O_4S$	15686-71-2	347.389	cry					
1799	Cephaloglycin	Kafocin	$C_{18}H_{19}N_3O_6S$	3577-01-3	405.425	cry (w)	≈220 dec				
1800	Cephaloridine		$C_{19}H_{17}N_3O_4S_2$	50-59-9	415.486	cry					s H_2O
1801	Cephalothin		$C_{16}H_{16}N_2O_6S_2$	153-61-7	396.437		160				
1802	Cephapirin		$C_{17}H_{17}N_3O_6S_2$	21593-23-7	423.463	cry (ace aq)	155				
1803	Cepharanthine		$C_{37}H_{38}N_2O_6$	481-49-2	606.707	ye amor pow	150				
1804	Cephradine		$C_{16}H_{19}N_3O_4S$	38821-53-3	349.405	col cry (w)	141 dec				
1805	Cerulenin	2,3-Epoxy-4-oxo-7,10-dodecadienamide, (2*R*,3*S*)-	$C_{12}H_{17}NO_3$	17397-89-6	223.268	wh nd	94				sl H_2O; s bz, EtOH, ace; i peth
1806	Cevadine		$C_{32}H_{49}NO_9$	62-59-9	591.733	flat nd (eth)	213 dec				
1807	Chavicine		$C_{17}H_{19}NO_3$	495-91-0	285.338						vs eth, EtOH, peth
1808	Cheirolin		$C_5H_9NO_2S_2$	505-34-0	179.261	cry (eth)	47.5	200^3			vs EtOH, chl
1809	Chelerythrine		$C_{21}H_{19}NO_5$	34316-15-9	365.380	cry (chl-MeOH)		207			vs chl

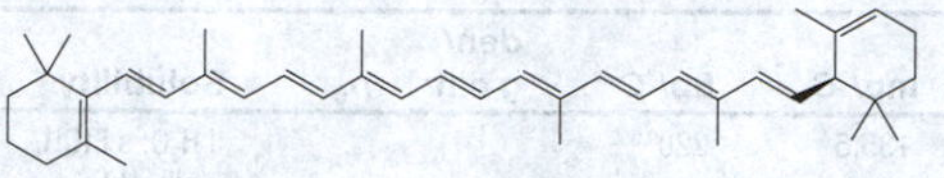
1774
α-Carotene

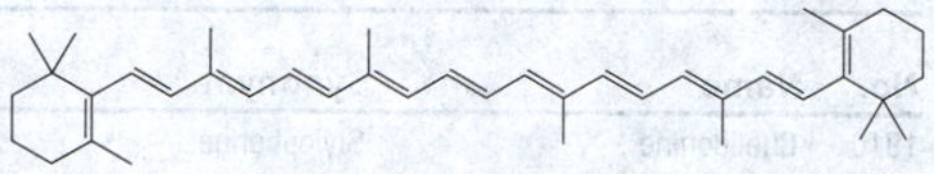
1775
β-Carotene

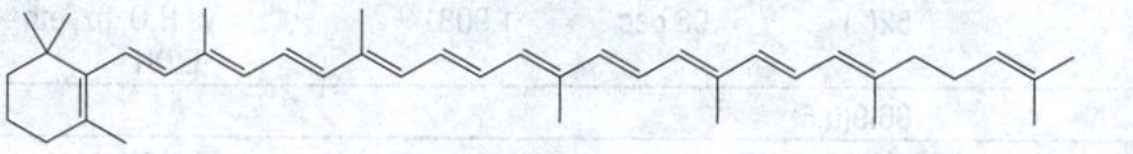
1776
β,ψ-Carotene

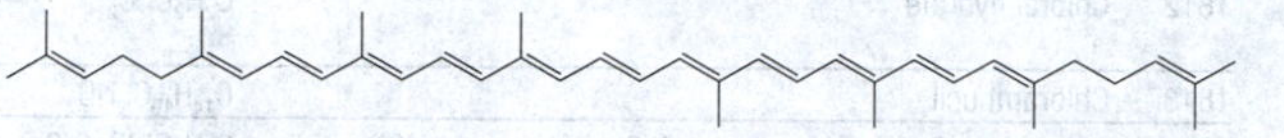
1777
ψ,ψ-Carotene

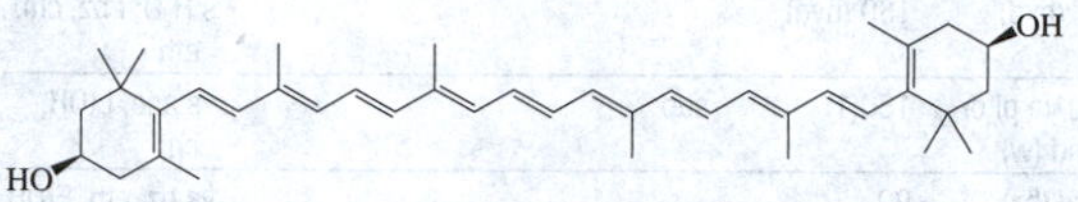
1778
β,β-Carotene-3,3′-diol, (3*R*,3′*R*)-

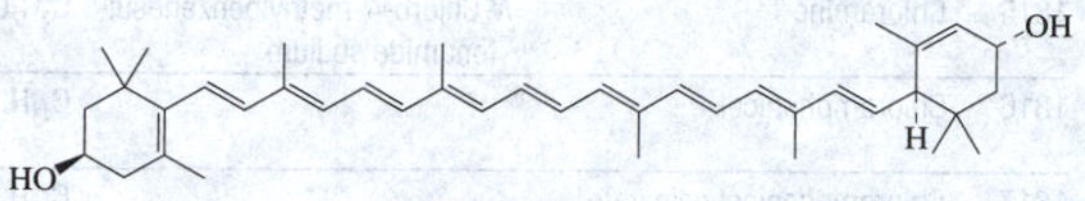
1779
β,ε-Carotene-3,3′-diol, (3*R*,3′*R*,6′*R*)-

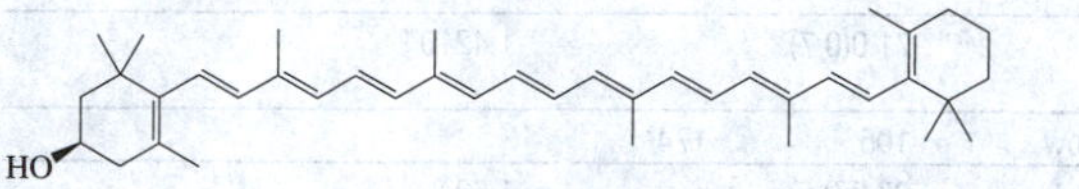
1780
β,β-Caroten-3-ol, (3*R*)-

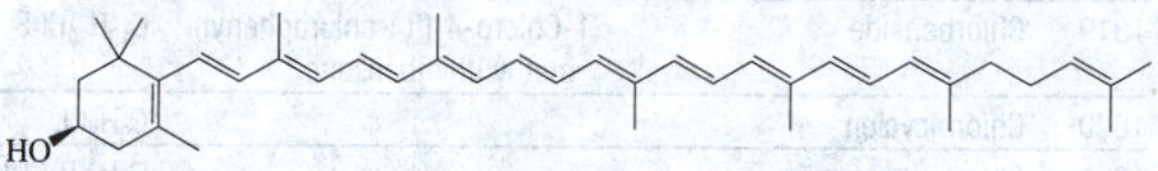
1781
β,ψ-Caroten-3-ol, (3*R*)-

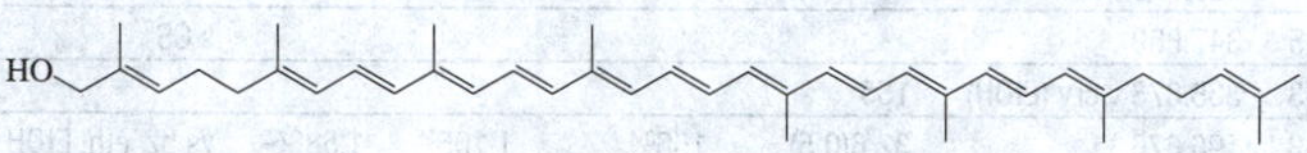
1782
ψ,ψ-Caroten-16-ol

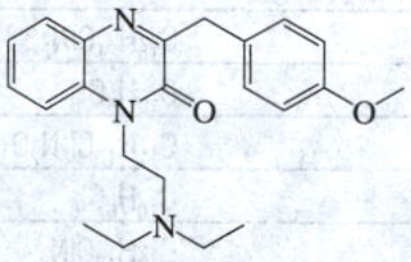
1783
Caroverine

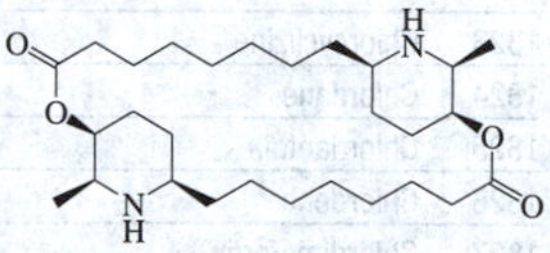
1784
Carpaine

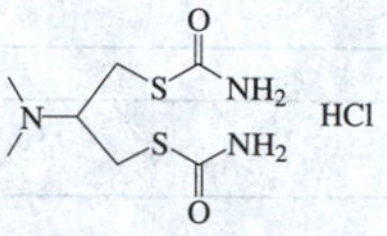
1785
Cartap hydrochloride

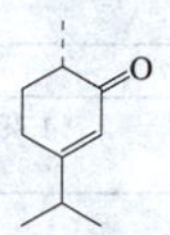
1786
Carvenone, (*S*)-

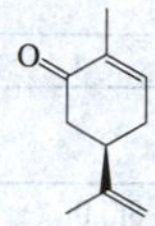
1787
(*R*)-Carvone

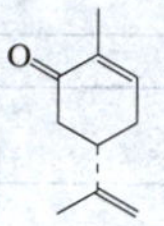
1788
(*S*)-Carvone

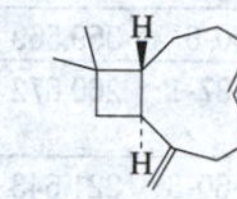
1789
Caryophyllene

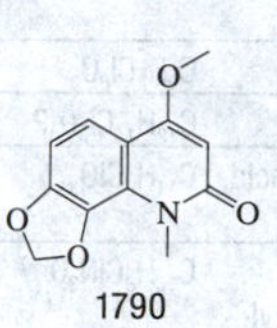
1790
Casimiroin

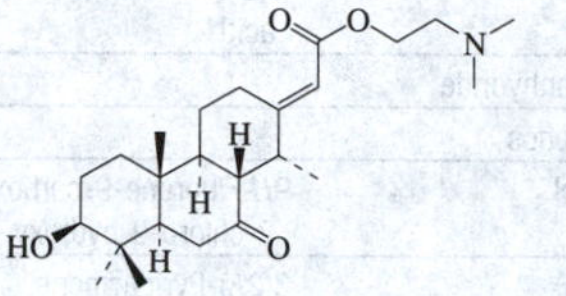
1791
Cassaine

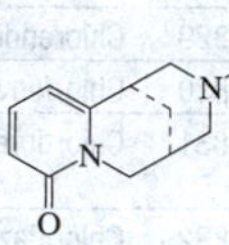
1792
Caulophylline

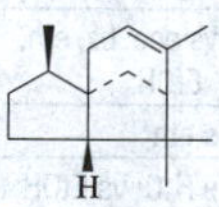
1793
α-Cedrene

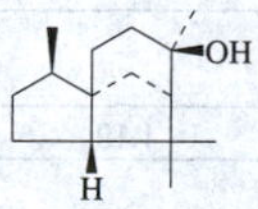
1794
Cedrol

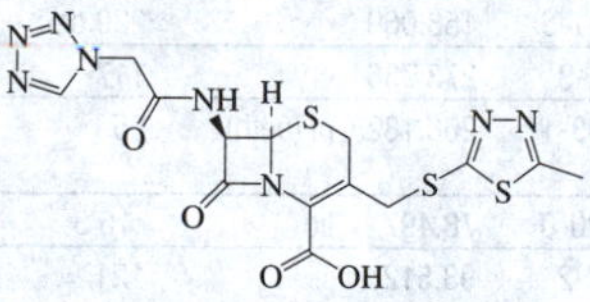
1795
Cefazolin

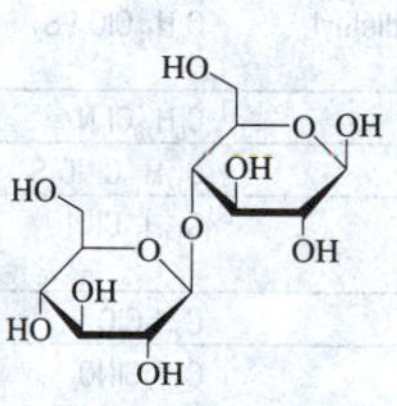
1796
β-Cellobiose

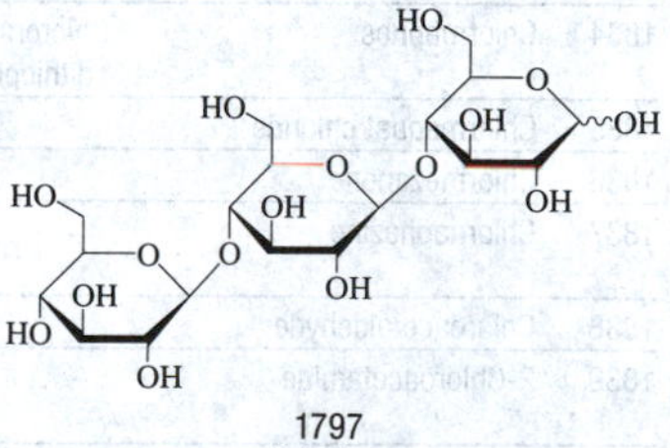
1797
Cellotriose

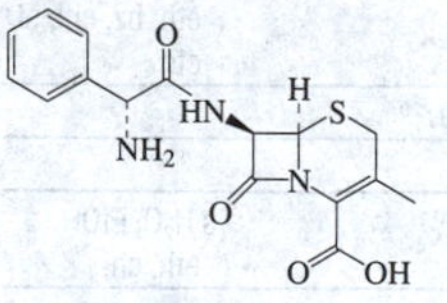
1798
Cephalexin

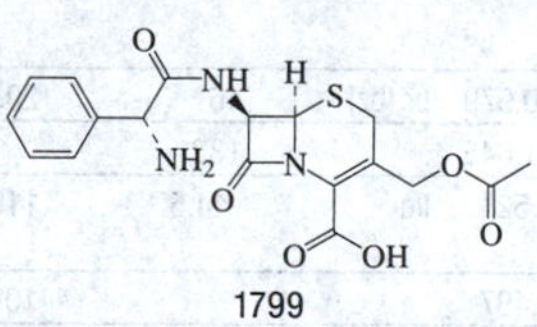
1799
Cephaloglycin

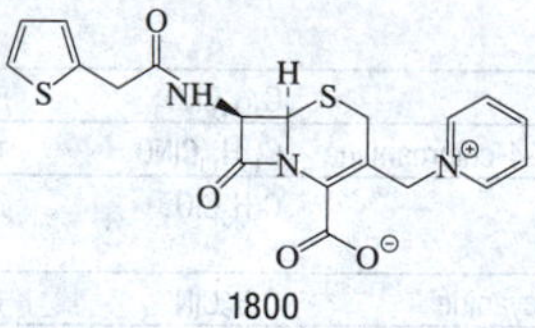
1800
Cephaloridine

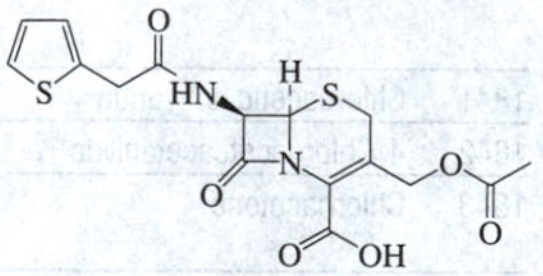
1801
Cephalothin

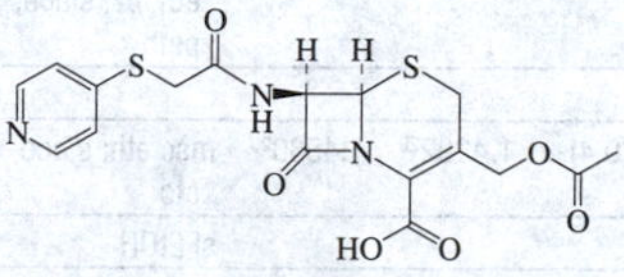
1802
Cephapirin

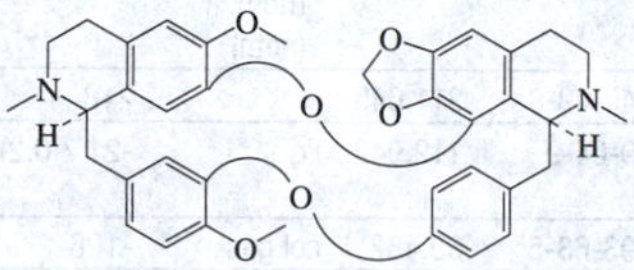
1803
Cepharanthine

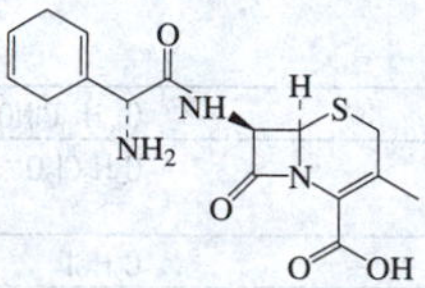
1804
Cephradine

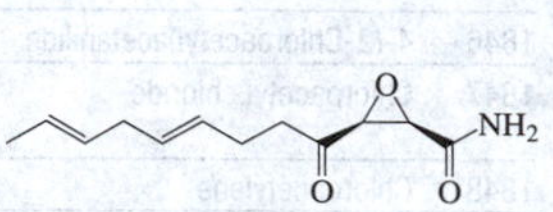
1805
Cerulenin

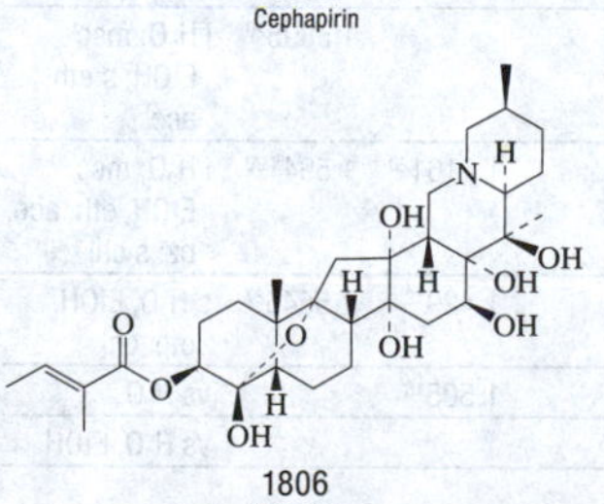
1806
Cevadine

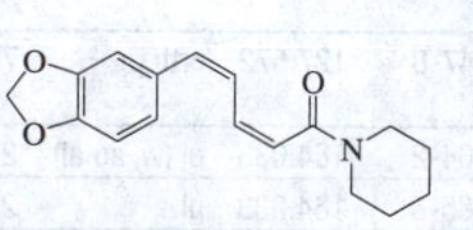
1807
Chavicine

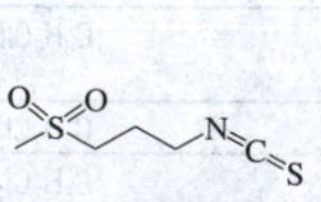
1808
Cheirolin

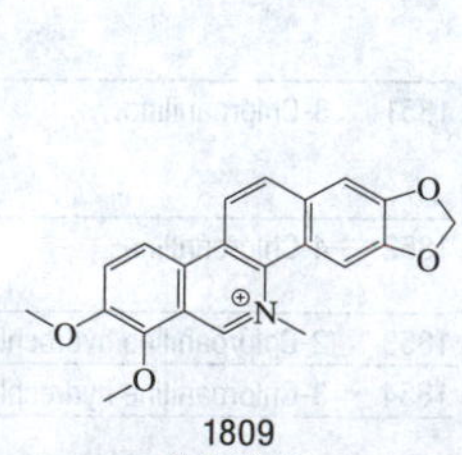
1809
Chelerythrine

No.	Name	Synonym	Mol. form.	CAS Reg. No.	Mol. wt.	Physical form	mp/°C	bp/°C	den/g cm^{-3}	n_D	Solubility
1810	Chelidonine	Stylophorine	$C_{20}H_{19}NO_5$	476-32-4	353.369	mcl pr (al)	135.5	$220^{0.002}$			i H_2O; s EtOH, eth, chl
1811	Chinomethionat		$C_{10}H_6N_2OS_2$	2439-01-2	234.297		170.5(0.5)				
1812	Chloral hydrate		$C_2H_3Cl_3O_2$	302-17-0	165.403		52(2)	96 dec	1.9081^{20}		vs H_2O, bz, eth, EtOH
1813	Chlorambucil		$C_{14}H_{19}Cl_2NO_2$	305-03-3	304.213		66.9(0.5)				
1814	Chloramine B	*N*-Chlorobenzenesulfonamide sodium	$C_6H_5ClNNaO_2S$	127-52-6	213.618	pr (w)	190				sl EtOH; i chl, eth
1815	Chloramine T	*N*-Chloro-4-methylbenzenesulfonamide sodium	$C_7H_7ClNNaO_2S$	127-65-1	227.645	pr (hyd)	180 (hyd)				s H_2O; i bz, chl, eth
1816	Chloramphenicol		$C_{11}H_{12}Cl_2N_2O_5$	56-75-7	323.129	pa ye pl or nd (w)	150(1)	sub			vs ace, EtOH, chl
1817	Chloramphenicol palmitate		$C_{27}H_{42}Cl_2N_2O_6$	530-43-8	561.537	cry (bz)	90				vs bz, eth, EtOH
1818	Chloranilic acid	2,5-Dichloro-3,6-dihydroxy-2,5-cyclohexadiene-1,4-dione	$C_6H_2Cl_2O_4$	87-88-7	208.984	red lf (w+2)	283.5				s H_2O
1819	Chlorbenside	1-Chloro-4-[[(4-chlorophenyl)-methyl]thio]benzene	$C_{13}H_{10}Cl_2S$	103-17-3	269.189		71.0(0.7)		1.4210^{20}		
1820	Chlorbicyclen		$C_9H_6Cl_8$	2550-75-6	397.768	pow	105	174^{2}			
1821	Chlorbromuron		$C_9H_{10}BrClN_2O_2$	13360-45-7	293.544		97.2(0.5)		1.69^{20}		
1822	Chlorbufam	1-Methyl-2-propynyl(3-chloro-phenyl)carbamate	$C_{11}H_{10}ClNO_2$	1967-16-4	223.656	cry	45.5				sl H_2O; s MeOH, EtOH, ace
1823	Chlorcyclizine		$C_{18}H_{21}ClN_2$	82-93-9	300.826	oil		$140^{0.12}$			
1824	Chlordane		$C_{10}H_6Cl_8$	57-74-9	409.779		101.1(0.3)	175^{1}	1.60^{25}		
1825	Chlordantoin		$C_{11}H_{17}Cl_3N_2O_2S$	5588-20-5	347.689						s CS_2
1826	Chlordene		$C_{10}H_6Cl_6$	3734-48-3	338.873	cry (EtOH)	155				
1827	Chlordimeform		$C_{10}H_{13}ClN_2$	6164-98-3	196.676		32.6(0.5)	$156^{0.4}$	1.105^{25}	1.5885^{25}	vs bz, eth, EtOH
1828	Chlorendic acid	1,4,5,6,7,7-Hexachloro-5-norbornene-2,3-dicarboxylic acid	$C_9H_4Cl_6O_4$	115-28-6	388.844	cry (w)	232				
1829	Chlorendic anhydride		$C_9H_2Cl_6O_3$	115-27-5	370.828		235				
1830	Chlorfenvinphos		$C_{12}H_{14}Cl_3O_4P$	470-90-6	359.569			$170^{0.05}$			
1831	Chlorflurecol	9*H*-Fluorene-9-carboxylic acid, 2-chloro-9-hydroxy-	$C_{14}H_9ClO_3$	2464-37-1	260.672				1.496^{20}		
1832	Chloridazon	3(2*H*)-Pyridazinone, 5-amino-4-chloro-2-phenyl-	$C_{10}H_8ClN_3O$	1698-60-8	221.643		206.8(0.9)				
1833	Chlorimuron-ethyl		$C_{15}H_{15}ClN_4O_6S$	90982-32-4	414.821		186				
1834	Chlormephos	Chloromethyl *O,O*-diethyl dithiophosphate	$C_5H_{12}ClO_2PS_2$	24934-91-6	234.705	oil		$83^{0.1}$		1.5244	sl H_2O; misc os
1835	Chlormequat chloride		$C_5H_{13}Cl_2N$	999-81-5	158.069		239 dec				
1836	Chlormezanone		$C_{11}H_{12}ClNO_3S$	80-77-3	273.736	cry	117				sl EtOH
1837	Chlornaphazine		$C_{14}H_{15}Cl_2N$	494-03-1	268.182	pl (peth)	55	210^{5}			vs ace, bz, eth, EtOH
1838	Chloroacetaldehyde		C_2H_3ClO	107-20-0	78.497	liq	-16.3	87(13)	1.19		s eth
1839	2-Chloroacetamide		C_2H_4ClNO	79-07-2	93.512		121	225			s H_2O; vs EtOH; sl eth
1840	Chloroacetic acid		$C_2H_3ClO_2$	79-11-8	94.497	mcl pl	62.0(0.7)	189.11(0.03)	1.4043^{40}	1.4351^{55}	vs H_2O; s EtOH, eth, bz, chl; sl ctc
1841	Chloroacetic anhydride		$C_4H_4Cl_2O_3$	541-88-8	170.979	pr (bz)	46	203	1.5497^{20}		
1842	4-Chloroacetoacetanilide	*N*-Acetoacetyl-4-chloroaniline	$C_{10}H_{10}ClNO_2$	101-92-8	211.645		132				
1843	Chloroacetone		C_3H_5ClO	78-95-5	92.524	liq	-44.5	116(13)	1.15^{20}		s H_2O, EtOH, eth, chl
1844	Chloroacetonitrile	Chloromethyl cyanide	C_2H_2ClN	107-14-2	75.497			108(5)	1.1930^{20}	1.4202^{25}	vs eth, EtOH
1845	α-Chloroacetophenone	ω-Chloroacetophenone	C_8H_7ClO	532-27-4	154.594	pl(dil al), rhom, lf (peth)	56.5	247	1.324^{15}		i H_2O; vs EtOH, eth, bz; s ace, peth
1846	4-(2-Chloroacetyl)acetanilide		$C_{10}H_{10}ClNO_2$	140-49-8	211.645		218				
1847	Chloroacetyl chloride		$C_2H_2Cl_2O$	79-04-9	112.942	liq	-21.7(0.2)	106.0(0.4)	1.4202^{20}	1.4530^{20}	msc eth; s ace, ctc
1848	Chloroacetylene		C_2HCl	593-63-5	60.482	col gas	-126	-30			sl EtOH
1849	9-Chloroacridine		$C_{13}H_8ClN$	1207-69-8	213.663	nd (al)	121	sub			vs H_2O, EtOH
1850	2-Chloroaniline		C_6H_6ClN	95-51-2	127.572	liq	-2.3(0.9)	209(1)		1.5895^{20}	i H_2O; msc EtOH; s eth, ace
1851	3-Chloroaniline		C_6H_6ClN	108-42-9	127.572	liq	-10.3(0.2)	230(1)	1.2161^{20}	1.5941^{20}	i H_2O; msc EtOH, eth, ace, bz; s chl
1852	4-Chloroaniline		C_6H_6ClN	106-47-8	127.572	orth pr	70.4(0.7)	231(4)	1.429^{19}	1.5546^{87}	s H_2O, EtOH, eth, chl
1853	2-Chloroaniline hydrochloride		$C_6H_7Cl_2N$	137-04-2	164.033	pl (w, aq al)	235		1.505^{18}		vs H_2O
1854	3-Chloroaniline hydrochloride		$C_6H_7Cl_2N$	141-85-5	164.033	pl	222				vs H_2O, EtOH

Organic

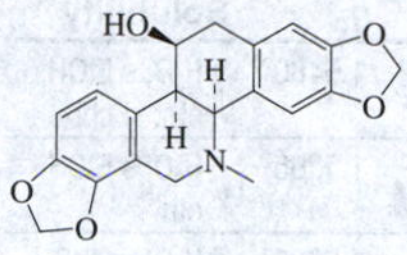
1810
Chelidonine

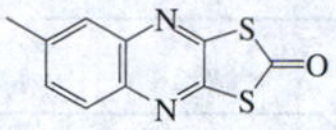
1811
Chinomethionat

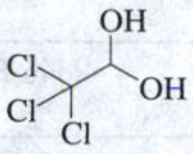
1812
Chloral hydrate

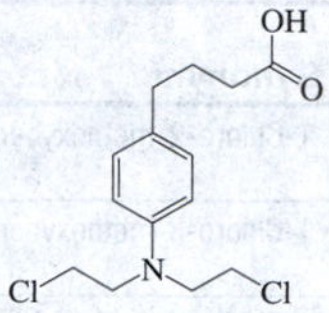
1813
Chlorambucil

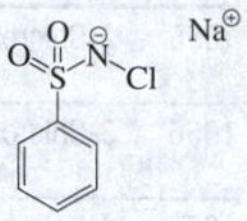
1814
Chloramine B

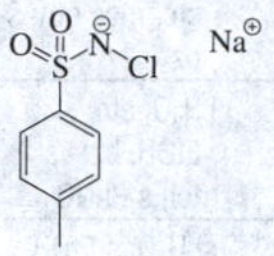
1815
Chloramine T

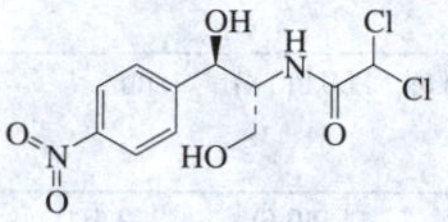
1816
Chloramphenicol

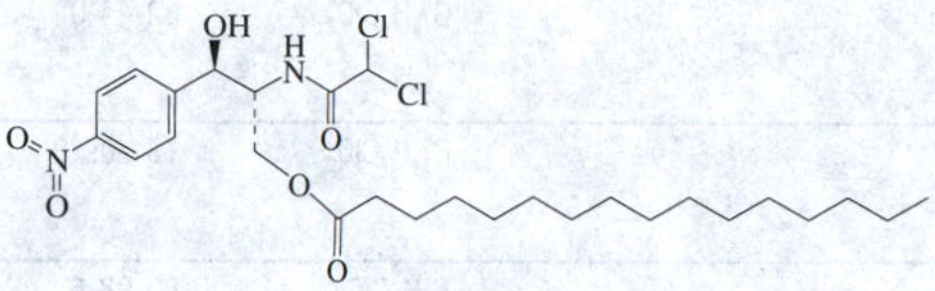
1817
Chloramphenicol palmitate

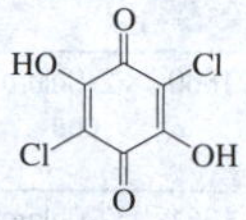
1818
Chloranilic acid

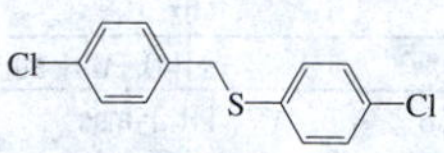
1819
Chlorbenside

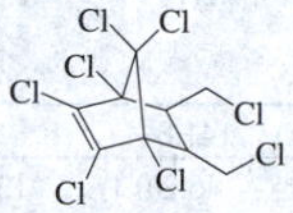
1820
Chlorbicyclen

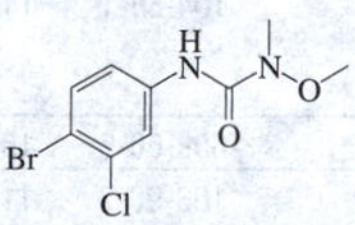
1821
Chlorbromuron

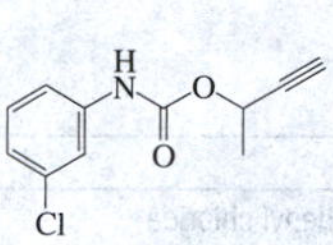
1822
Chlorbufam

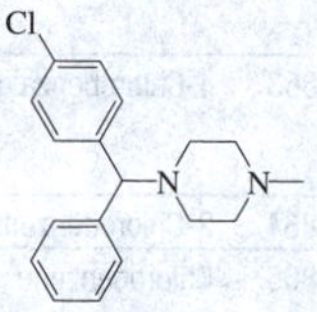
1823
Chlorcyclizine

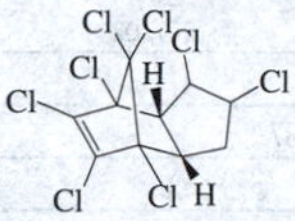
1824
Chlordane

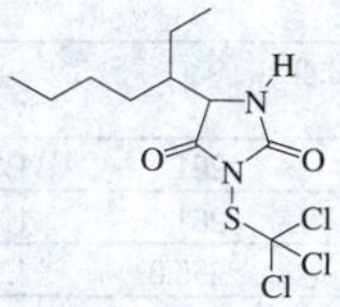
1825
Chlordantoin

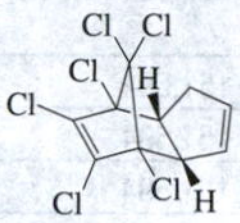
1826
Chlordene

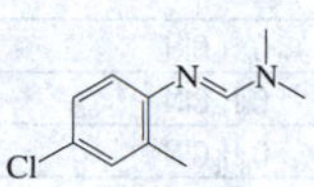
1827
Chlordimeform

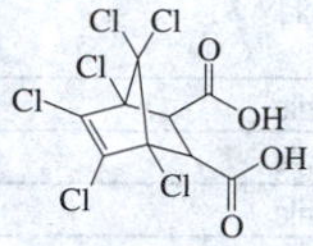
1828
Chlorendic acid

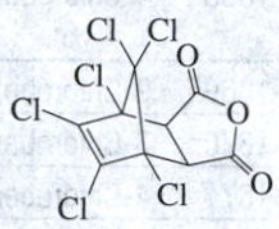
1829
Chlorendic anhydride

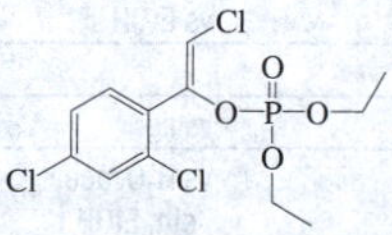
1830
Chlorfenvinphos

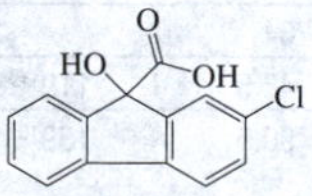
1831
Chlorflurecol

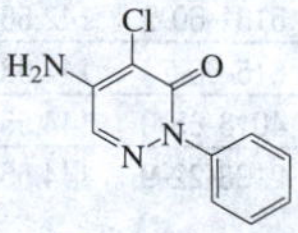
1832
Chloridazon

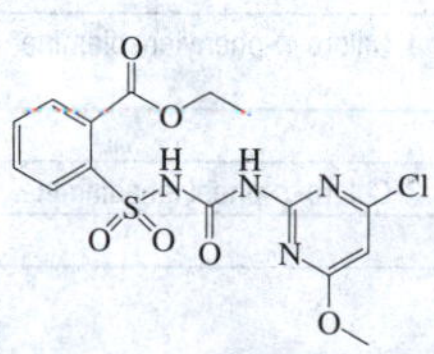
1833
Chlorimuron-ethyl

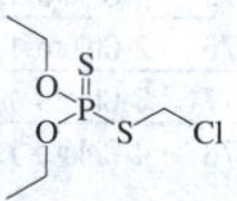
1834
Chlormephos

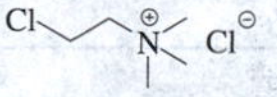
1835
Chlormequat chloride

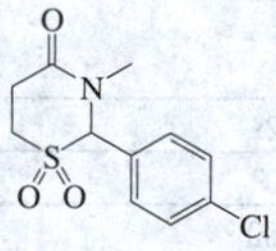
1836
Chlormezanone

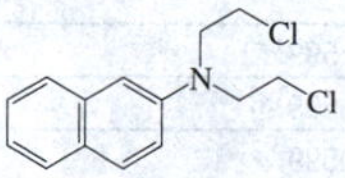
1837
Chlornaphazine

1838
Chloroacetaldehyde

1839
2-Chloroacetamide

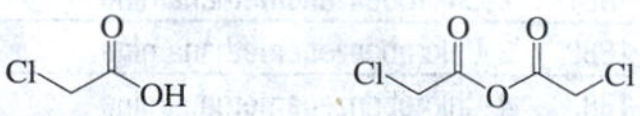
1840
Chloroacetic acid

1841
Chloroacetic anhydride

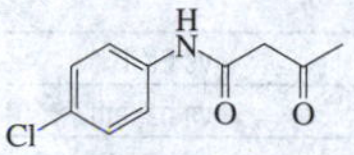
1842
4-Chloroacetoacetanilide

1843
Chloroacetone

1844
Chloroacetonitrile

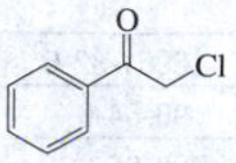
1845
α-Chloroacetophenone

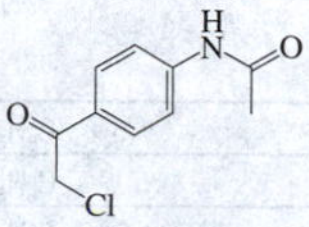
1846
4-(2-Chloroacetyl)acetanilide

1847
Chloroacetyl chloride

1848
Chloroacetylene

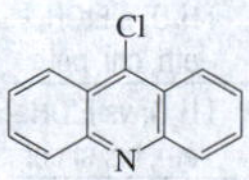
1849
9-Chloroacridine

1850
2-Chloroaniline

1851
3-Chloroaniline

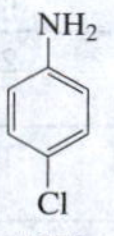
1852
4-Chloroaniline

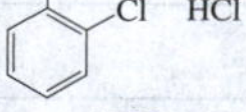
1853
2-Chloroaniline hydrochloride

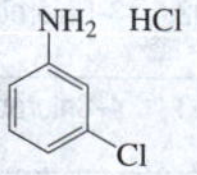
1854
3-Chloroaniline hydrochloride

Organic

No.	Name	Synonym	Mol. form.	CAS Reg. No.	Mol. wt.	Physical form	mp/°C	bp/°C	den/ g cm⁻³	n_D	Solubility
1855	2-Chloroanisole	1-Chloro-2-methoxybenzene	C_7H_7ClO	766-51-8	142.583	liq	-26.5(0.2)	201.8(0.4)	1.1911^{20}	1.5480^{20}	i H_2O; s EtOH, eth; sl chl
1856	3-Chloroanisole	1-Chloro-3-methoxybenzene	C_7H_7ClO	2845-89-8	142.583			193.5	1.1759^{12}	1.5365^{20}	i H_2O; s EtOH, eth
1857	4-Chloroanisole	1-Chloro-4-methoxybenzene	C_7H_7ClO	623-12-1	142.583		<-18	197.5	1.201^{20}	1.5390^{20}	i H_2O; vs EtOH, eth, chl; s ctc
1858	1-Chloroanthracene		$C_{14}H_9Cl$	4985-70-0	212.674	lf (HOAc)	83.5		1.1707^{100}	1.6959^{100}	i H_2O; s EtOH, eth, bz, ctc
1859	1-Chloro-9,10-anthracenedione		$C_{14}H_7ClO_2$	82-44-0	242.658	ye nd (to or al)	163	sub			i H_2O; sl EtOH, ctc; msc eth; s bz
1860	2-Chloro-9,10-anthracenedione		$C_{14}H_7ClO_2$	131-09-9	242.658	pa ye nd (al, HOAc)	209.9(0.4)	sub			i H_2O, eth; sl EtOH, bz; vs tol; s $PhNO_2$
1861	2-Chlorobenzaldehyde		C_7H_5ClO	89-98-5	140.567	nd	11.9(0.6)	212.1(1)	1.2483^{20}	1.5662^{20}	sl H_2O; s EtOH, eth, ace, bz, ctc
1862	3-Chlorobenzaldehyde		C_7H_5ClO	587-04-2	140.567	pr	17.5	213.5	1.2410^{20}	1.5650^{20}	sl H_2O, chl; s EtOH, eth, ace, bz
1863	4-Chlorobenzaldehyde		C_7H_5ClO	104-88-1	140.567	pl	47(1)	213.5	1.196^{61}	1.555^{61}	s H_2O, ace, chl; vs EtOH, eth, bz
1864	2-Chlorobenzamide		C_7H_6ClNO	609-66-5	155.582	orth nd (w)	141.8				s H_2O, EtOH, eth
1865	Chlorobenzene	Phenyl chloride	C_6H_5Cl	108-90-7	112.557	liq	-45.2(0.1)	131.6(0.2)	1.1058^{20}	1.5241^{20}	i H_2O; msc EtOH, eth; vs bz, ctc
1866	2-Chlorobenzeneacetic acid		$C_8H_7ClO_2$	2444-36-2	170.594	nd (w)	94.2(0.4)				sl H_2O; vs EtOH
1867	3-Chlorobenzeneacetic acid		$C_8H_7ClO_2$	1878-65-5	170.594	pl (dil al), nd (lig)	76.6(0.4)				sl H_2O, bz, ctc, EtOH; msc eth
1868	4-Chlorobenzeneacetic acid		$C_8H_7ClO_2$	1878-66-6	170.594	nd (w)	104.8(0.4)				s H_2O, EtOH, eth, bz
1869	2-Chlorobenzeneacetonitrile		C_8H_6ClN	2856-63-5	151.594		24	251	1.1737^{18}		
1870	3-Chlorobenzeneacetonitrile		C_8H_6ClN	1529-41-5	151.594		11.5	261	1.1806^{30}	1.5437^{20}	
1871	4-Chlorobenzeneacetonitrile		C_8H_6ClN	140-53-4	151.594		29	265.0	1.1778^{30}		s ctc
1872	α-Chlorobenzeneacetyl chloride		$C_8H_6Cl_2O$	2912-62-1	189.039			120^{23}	1.196^{25}	1.5440^{20}	
1873	3-Chlorobenzenecarboperoxoic acid		$C_7H_5ClO_3$	937-14-4	172.566		92 dec				
1874	4-Chloro-1,2-benzenediamine	4-Chloro-*o*-phenylenediamine	$C_6H_7ClN_2$	95-83-0	142.586	pl (bz-lig) lf (w)	76				sl H_2O; vs EtOH, eth; s bz, lig
1875	4-Chloro-1,3-benzenediamine		$C_6H_7ClN_2$	5131-60-2	142.586	pl or nd	91				vs EtOH
1876	2-Chloro-1,4-benzenediamine	2-Chloro-*p*-phenylenediamine	$C_6H_7ClN_2$	615-66-7	142.586	nd	64				
1877	3-Chloro-1,2-benzenediol		$C_6H_5ClO_2$	4018-65-9	144.556	cry (lig)	48.5	110^{11}			vs lig
1878	4-Chloro-1,2-benzenediol		$C_6H_5ClO_2$	2138-22-9	144.556	lf (bz-peth)	90.5	$139^{10.5}$			vs H_2O, ace, eth, EtOH
1879	4-Chloro-1,3-benzenediol		$C_6H_5ClO_2$	95-88-5	144.556			257			vs H_2O, EtOH, eth, ace, bz, CS_2
1880	2-Chloro-1,4-benzenediol		$C_6H_5ClO_2$	615-67-8	144.556	red lf (chl), nd (bz)	108	263			vs H_2O, chl; s EtOH, eth; vs bz
1881	2-Chlorobenzenemethanamine		C_7H_8ClN	89-97-4	141.599			72^{2}		1.5594^{25}	
1882	3-Chlorobenzenemethanamine		C_7H_8ClN	4152-90-3	141.599			89^{2}		1.5570^{25}	
1883	4-Chlorobenzenemethanamine		C_7H_8ClN	104-86-9	141.599			109^{13}		1.5566^{25}	
1884	4-Chlorobenzenemethanethiol		C_7H_7ClS	6258-66-8	158.649		19.5	113^{17}	1.202^{25}	1.5893^{20}	
1885	2-Chlorobenzenemethanol		C_7H_7ClO	17849-38-6	142.583	lf or nd (dil al)	73	230			sl H_2O; vs EtOH, eth, lig
1886	4-Chlorobenzenemethanol		C_7H_7ClO	873-76-7	142.583	nd (w), pl (bz or bz-lig)	75	235			vs bz, eth, EtOH
1887	2-Chlorobenzenesulfonamide		$C_6H_6ClNO_2S$	6961-82-6	191.636	lf (al)	188.3(0.5)				vs EtOH
1888	4-Chlorobenzenesulfonamide		$C_6H_6ClNO_2S$	98-64-6	191.636	pr or pl (eth)	146				vs bz, eth
1889	4-Chlorobenzenesulfonic acid	*p*-Chlorobenzenesulfonic acid	$C_6H_5ClO_3S$	98-66-8	192.620	nd (w+1)	67	147^{25}			s H_2O, EtOH; i eth, bz
1890	4-Chlorobenzenesulfonyl chloride		$C_6H_4Cl_2O_2S$	98-60-2	211.066		51	141^{15}			vs eth, bz
1891	2-Chlorobenzenethiol		C_6H_5ClS	6320-03-2	144.622			205.5	1.2752^{10}		sl H_2O, EtOH
1892	3-Chlorobenzenethiol		C_6H_5ClS	2037-31-2	144.622			206	1.2637^{13}		i H_2O; s EtOH, eth, chl, peth
1893	4-Chlorobenzenethiol		C_6H_5ClS	106-54-7	144.622		61	206	1.1911^{20}	1.5480^{20}	i H_2O; vs EtOH, eth, bz; sl chl
1894	Chlorobenzilate		$C_{16}H_{14}Cl_2O_3$	510-15-6	325.186		39.0(0.5)	$157^{0.07}$	1.2816^{20}		

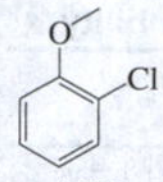
1855
2-Chloroanisole

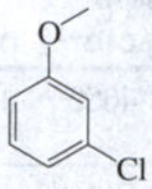
1856
3-Chloroanisole

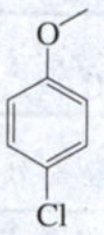
1857
4-Chloroanisole

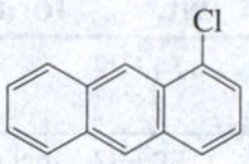
1858
1-Chloroanthracene

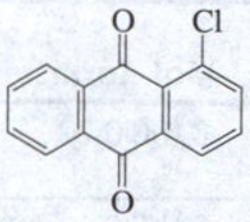
1859
1-Chloro-9,10-anthracenedione

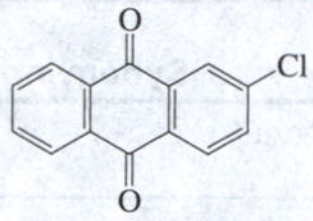
2-Chloro-9,10-anthracenedione

1860

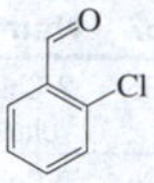
1861
2-Chlorobenzaldehyde

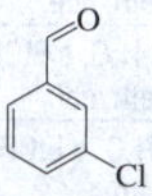
1862
3-Chlorobenzaldehyde

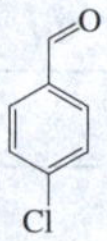
1863
4-Chlorobenzaldehyde

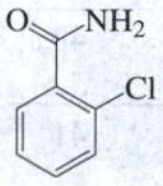
1864
2-Chlorobenzamide

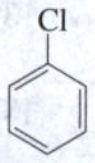
1865
Chlorobenzene

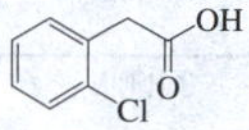
1866
2-Chlorobenzeneacetic acid

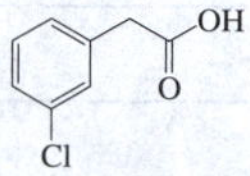
1867
3-Chlorobenzeneacetic acid

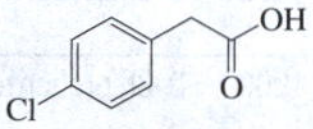
1868
4-Chlorobenzeneacetic acid

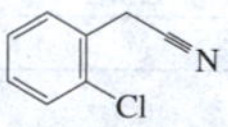
1869
2-Chlorobenzeneacetonitrile

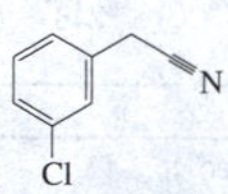
1870
3-Chlorobenzeneacetonitrile

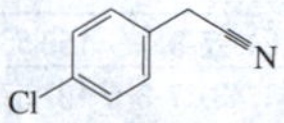
1871
4-Chlorobenzeneacetonitrile

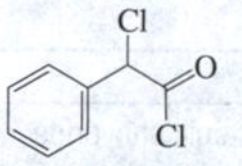
1872
α-Chlorobenzeneacetyl chloride

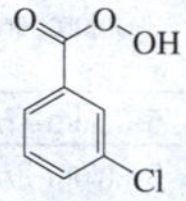
1873
3-Chlorobenzenecarboperoxoic acid

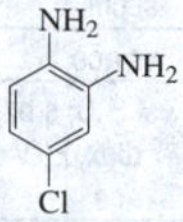
1874
4-Chloro-1,2-benzenediamine

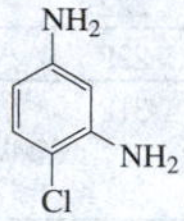
1875
4-Chloro-1,3-benzenediamine

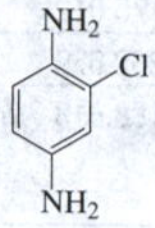
1876
2-Chloro-1,4-benzenediamine

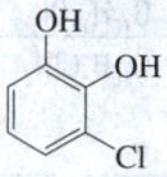
1877
3-Chloro-1,2-benzenediol

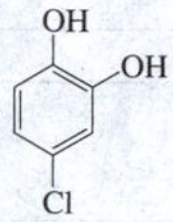
1878
4-Chloro-1,2-benzenediol

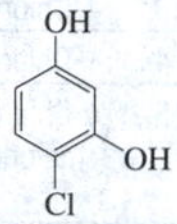
1879
4-Chloro-1,3-benzenediol

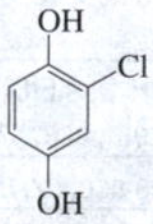
1880
2-Chloro-1,4-benzenediol

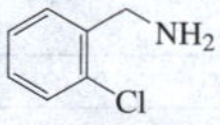
1881
2-Chlorobenzenemethanamine

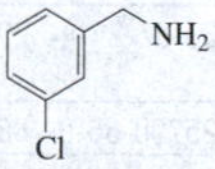
1882
3-Chlorobenzenemethanamine

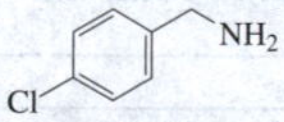
1883
4-Chlorobenzenemethanamine

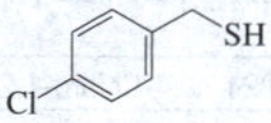
1884
4-Chlorobenzenemethanethiol

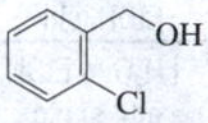
1885
2-Chlorobenzenemethanol

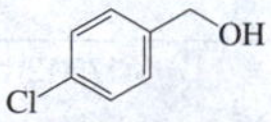
1886
4-Chlorobenzenemethanol

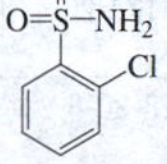
1887
2-Chlorobenzenesulfonamide

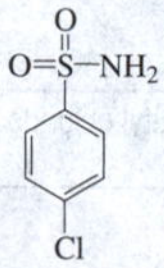
1888
4-Chlorobenzenesulfonamide

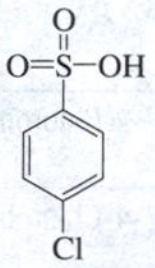
1889
4-Chlorobenzenesulfonic acid

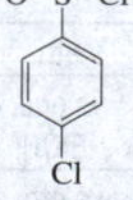
1890
4-Chlorobenzenesulfonyl chloride

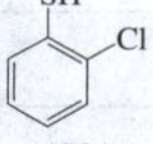
1891
2-Chlorobenzenethiol

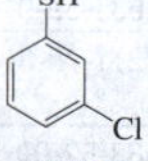
1892
3-Chlorobenzenethiol

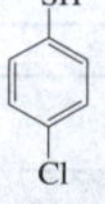
1893
4-Chlorobenzenethiol

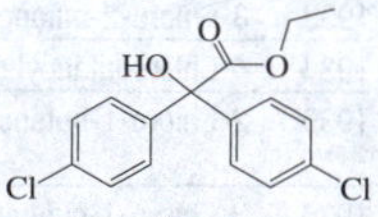
1894
Chlorobenzilate

Organic

No.	Name	Synonym	Mol. form.	CAS Reg. No.	Mol. wt.	Physical form	mp/°C	bp/°C	den/g cm^{-3}	n_D	Solubility
1895	2-Chloro-1,3,2-benzodioxaphosphole		$C_6H_4ClO_2P$	1641-40-3	174.522		30	80^{20}	1.4650^{20}	1.5712^{20}	
1896	2-Chlorobenzoic acid		$C_7H_5ClO_2$	118-91-2	156.567	mcl pr (w)	140.4(0.7)	274(14)	1.544^{20}		s H_2O, bz; vs EtOH, eth, ace; sl CS_2
1897	3-Chlorobenzoic acid		$C_7H_5ClO_2$	535-80-8	156.567	pr (w)	154.2(0.2)	283(17)	1.496^{25}		sl H_2O, bz, ctc, CS_2; s EtOH, eth
1898	4-Chlorobenzoic acid		$C_7H_5ClO_2$	74-11-3	156.567	tcl pr (al-eth)	239.5(0.6)				i H_2O, bz, ctc; vs EtOH; sl eth, ace
1899	2-Chlorobenzonitrile		C_7H_4ClN	873-32-5	137.567	nd	43.5(0.4)	235.1(0.5)			sl H_2O; s EtOH, eth, chl
1900	3-Chlorobenzonitrile		C_7H_4ClN	766-84-7	137.567		41	100^{15}			i H_2O; s EtOH, eth
1901	4-Chlorobenzonitrile		C_7H_4ClN	623-03-0	137.567	nd (al)	91.6(0.4)	223.0(0.4)	1.1133^{17}		sl H_2O, lig; s EtOH, eth, bz, chl
1902	2-Chlorobenzophenone	2-Chlorophenyl phenyl ketone	$C_{13}H_9ClO$	5162-03-8	216.662	pl (chl-lig)	54	330			
1903	4-Chloro-2-benzothiazolamine		$C_7H_5ClN_2S$	19952-47-7	184.646		204				
1904	6-Chloro-2-benzothiazolamine		$C_7H_5ClN_2S$	95-24-9	184.646		200				
1905	2-Chlorobenzothiazole		C_7H_4ClNS	615-20-3	169.632		24	248	1.3715^{10}	1.6338^{10}	vs ace, eth, EtOH
1906	5-Chloro-1*H*-benzotriazole		$C_6H_4ClN_3$	94-97-3	153.569		158				
1907	6-Chloro-2*H*-3,1-benzoxazine-2,4(1*H*)-dione	5-Chloroisatoic anhydride	$C_8H_4ClNO_3$	4743-17-3	197.576		280 dec				
1908	5-Chloro-2-benzoxazolamine	Zoxazolamine	$C_7H_5ClN_2O$	61-80-3	168.580	pl (bz)	184.5				vs EtOH
1909	2-Chlorobenzoxazole		C_7H_4ClNO	615-18-9	153.566		7	201.5	1.3453^{18}	1.5678^{20}	
1910	5-Chloro-2(3*H*)-benzoxazolone	Chlorzoxazone	$C_7H_4ClNO_2$	95-25-0	169.566	cry (ace)	191.5				vs EtOH, MeOH
1911	2-Chlorobenzoyl chloride		$C_7H_4Cl_2O$	609-65-4	175.012	liq	-4	241(12)		1.5726^{16}	s ctc
1912	3-Chlorobenzoyl chloride		$C_7H_4Cl_2O$	618-46-2	175.012			225		1.5677^{20}	
1913	4-Chlorobenzoyl chloride		$C_7H_4Cl_2O$	122-01-0	175.012		16	222	1.3770^{20}	1.5756^{20}	sl chl
1914	1-Chloro-4-benzylbenzene		$C_{13}H_{11}Cl$	831-81-2	202.679		7.5	299	1.1247^{20}		vs ace
1915	*o*-Chlorobenzylidene malononitrile		$C_{10}H_5ClN_2$	2698-41-1	188.613	wh cry	96	312			sl H_2O; s bz, diox, EtOAc, ace
1916	2-Chlorobiphenyl		$C_{12}H_9Cl$	2051-60-7	188.652	mcl (dil al)	31.78(0.08)	273(7)	1.1499^{32}		i H_2O; vs eth, EtOH, lig
1917	3-Chlorobiphenyl		$C_{12}H_9Cl$	2051-61-8	188.652		16	284.5	1.1579^{25}	1.6181^{25}	vs ace, eth, EtOH
1918	4-Chlorobiphenyl		$C_{12}H_9Cl$	2051-62-9	188.652	lf (lig or al)	75.4(0.2)	293(3)			i H_2O; s EtOH, eth, lig
1919	4'-Chloro-[1,1'-biphenyl]-4-amine	4-Amino-4'-chlorodiphenyl	$C_{12}H_{10}ClN$	135-68-2	203.667	cry (peth)	134				vs ace, bz, eth
1920	3-Chloro-[1,1'-biphenyl]-2-ol	2-Phenyl-6-chlorophenol	$C_{12}H_9ClO$	85-97-2	204.651		6	319(4)	1.24^{25}	1.6237^{30}	i H_2O; s EtOH, eth, ace, bz
1921	4-Chloro-1,2-butadiene		C_4H_5Cl	25790-55-0	88.536			88	0.9891^{20}	1.4775^{20}	vs ace, bz, eth
1922	1-Chloro-1,3-butadiene		C_4H_5Cl	627-22-5	88.536			67(6)	0.9606^{20}	1.4712^{20}	vs eth, EtOH, chl
1923	2-Chloro-1,3-butadiene	Chloroprene	C_4H_5Cl	126-99-8	88.536	liq	-130	59(3)	0.956^{20}	1.4583^{20}	sl H_2O; msc eth, ace, bz
1924	4-Chlorobutanal		C_4H_7ClO	6139-84-0	106.551			51^{13}	1.106^{8}	1.4466^{8}	vs ace, eth, EtOH
1925	1-Chlorobutane	Butyl chloride	C_4H_9Cl	109-69-3	92.567	liq	-123.1(0.2)	78.4(0.2)	0.8857^{20}	1.4023^{20}	i H_2O; msc EtOH, eth; sl ctc
1926	2-Chlorobutane	(±)-*sec*-Butyl chloride	C_4H_9Cl	53178-20-4	92.567	liq	-131.3	71(8)	0.8732^{20}	1.3971^{20}	vs bz, eth, EtOH, chl
1927	4-Chlorobutanenitrile		C_4H_6ClN	628-20-6	103.551			175(12)	1.0934^{15}	1.4413^{20}	i H_2O; s EtOH, eth; sl ctc
1928	2-Chlorobutanoic acid		$C_4H_7ClO_2$	4170-24-5	122.551			189^{627}	1.1796^{20}	1.441^{20}	sl H_2O; vs EtOH, eth
1929	3-Chlorobutanoic acid		$C_4H_7ClO_2$	625-68-3	122.551	cry (eth)	16	116^{22}	1.1898^{20}	1.4221^{20}	s EtOH; vs eth; sl ctc
1930	4-Chlorobutanoic acid		$C_4H_7ClO_2$	627-00-9	122.551		16	196^{22}	1.2236^{20}	1.4642^{20}	vs EtOH
1931	4-Chloro-1-butanol		C_4H_9ClO	928-51-8	108.566			84^{16}	1.0883^{20}	1.4518^{20}	vs eth, EtOH
1932	1-Chloro-2-butanol	α-Butylene chlorohydrin	C_4H_9ClO	1873-25-2	108.566			141	1.068^{25}	1.4400^{20}	s EtOH, eth
1933	3-Chloro-2-butanone		C_4H_7ClO	4091-39-8	106.551			129(13)	1.0554^{25}	1.4219^{20}	
1934	4-Chlorobutanoyl chloride		$C_4H_6Cl_2O$	4635-59-0	140.996			173.5	1.2581^{20}	1.4616^{20}	s eth
1935	2-Chloro-1-butene		C_4H_7Cl	2211-70-3	90.552			58(3)	0.9107^{15}	1.4165^{21}	vs ace, bz, eth, EtOH
1936	3-Chloro-1-butene		C_4H_7Cl	563-52-0	90.552			64(3)	0.8978^{20}	1.4149^{20}	vs eth, ace; s chl

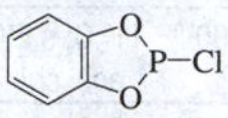
1895
2-Chloro-1,3,2-benzodioxaphosphole

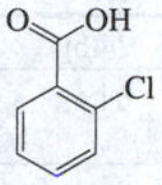
1896
2-Chlorobenzoic acid

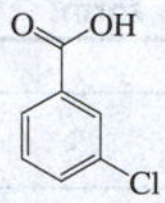
1897
3-Chlorobenzoic acid

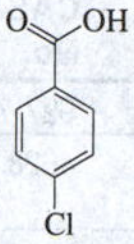
1898
4-Chlorobenzoic acid

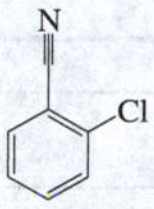
1899
2-Chlorobenzonitrile

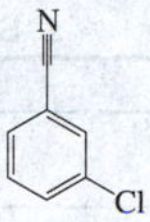
1900
3-Chlorobenzonitrile

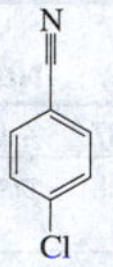
1901
4-Chlorobenzonitrile

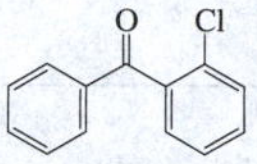
1902
2-Chlorobenzophenone

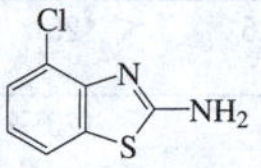
1903
4-Chloro-2-benzothiazolamine

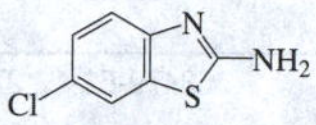
1904
6-Chloro-2-benzothiazolamine

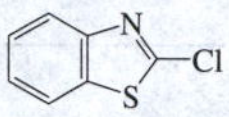
1905
2-Chlorobenzothiazole

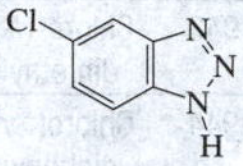
1906
5-Chloro-1*H*-benzotriazole

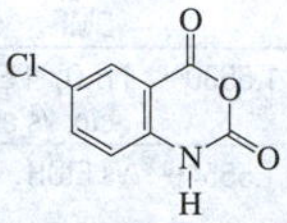
1907
6-Chloro-2*H*-3,1-benzoxazine-2,4(1*H*)-dione

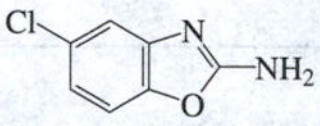
1908
5-Chloro-2-benzoxazolamine

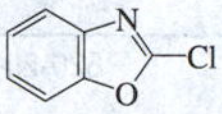
1909
2-Chlorobenzoxazole

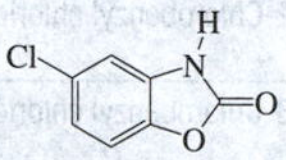
1910
5-Chloro-2(3*H*)-benzoxazolone

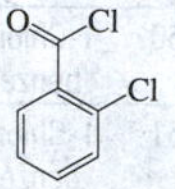
1911
2-Chlorobenzoyl chloride

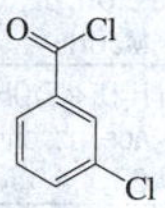
1912
3-Chlorobenzoyl chloride

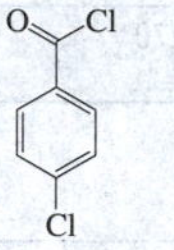
1913
4-Chlorobenzoyl chloride

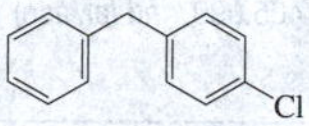
1914
1-Chloro-4-benzylbenzene

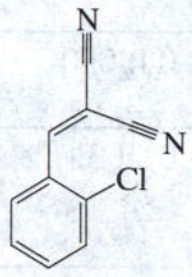
1915
o-Chlorobenzylidene malononitrile

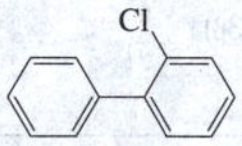
1916
2-Chlorobiphenyl

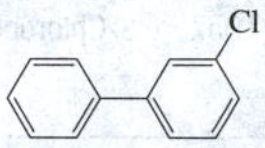
1917
3-Chlorobiphenyl

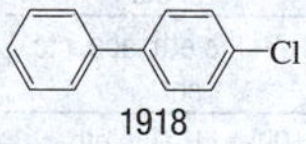
1918
4-Chlorobiphenyl

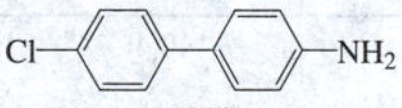
1919
4′-Chloro-[1,1′-biphenyl]-4-amine

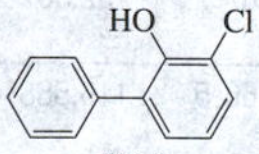
1920
3-Chloro-[1,1′-biphenyl]-2-ol

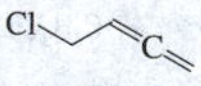
1921
4-Chloro-1,2-butadiene

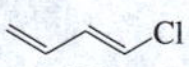
1922
1-Chloro-1,3-butadiene

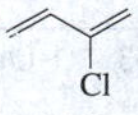
1923
2-Chloro-1,3-butadiene

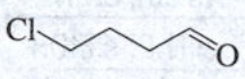
1924
4-Chlorobutanal

1925
1-Chlorobutane

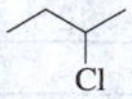
1926
2-Chlorobutane

1927
4-Chlorobutanenitrile

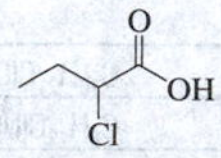
1928
2-Chlorobutanoic acid

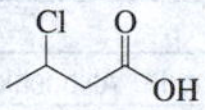
1929
3-Chlorobutanoic acid

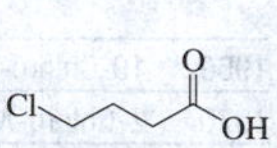
1930
4-Chlorobutanoic acid

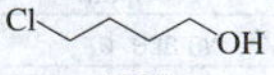
1931
4-Chloro-1-butanol

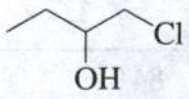
1932
1-Chloro-2-butanol

1933
3-Chloro-2-butanone

1934
4-Chlorobutanoyl chloride

1935
2-Chloro-1-butene

1936
3-Chloro-1-butene

No.	Name	Synonym	Mol. form.	CAS Reg. No.	Mol. wt.	Physical form	mp/°C	bp/°C	den/ g cm⁻³	n_D	Solubility
1937	4-Chloro-1-butene		C_4H_7Cl	927-73-1	90.552			74(5)	0.9211^{20}	1.4233^{20}	vs ace, eth, chl
1938	*cis*-1-Chloro-2-butene		C_4H_7Cl	4628-21-1	90.552			84(4)	0.9426^{20}	1.4390^{20}	i H_2O; s EtOH, ace, chl
1939	*trans*-1-Chloro-2-butene		C_4H_7Cl	4894-61-5	90.552			85(4)	0.9295^{20}	1.4350^{20}	i H_2O; s ace, chl
1940	*cis*-2-Chloro-2-butene		C_4H_7Cl	2211-69-0	90.552	liq	-117.3	65(5)	0.9239^{20}	1.4240^{20}	i H_2O; msc EtOH; s ace, chl
1941	*trans*-2-Chloro-2-butene		C_4H_7Cl	2211-68-9	90.552	liq	-105.8	63(3)	0.9138^{20}	1.4190^{20}	i H_2O; msc EtOH; s ace, chl
1942	1-Chloro-4-*tert*-butylbenzene		$C_{10}H_{13}Cl$	3972-56-3	168.663			214(10)	1.0075^{18}	1.5123^{20}	
1943	Chloro-(*tert*-butyl)-dimethylsilane		$C_6H_{15}ClSi$	18162-48-6	150.722		89.5	125			
1944	Chloro(*tert*-butyl)-diphenylsilane		$C_{16}H_{19}ClSi$	58479-61-1	274.861			$120^{0.06}$	1.07^{20}	1.5675^{20}	
1945	2-Chloro-4-*tert*-butylphenol		$C_{10}H_{13}ClO$	98-28-2	184.662			114^{8}			
1946	3-Chloro-1-butyne		C_4H_5Cl	21020-24-6	88.536			68.5	1.4218^{25}	1.4218^{25}	
1947	2-Chloro-*N*-(2-chloroethyl)-ethanamine, hydrochloride		$C_4H_{10}Cl_3N$	821-48-7	178.488		215.0				
1948	2-Chloro-*N*-(2-chloroethyl)-*N*-ethylethanamine	HN1	$C_6H_{13}Cl_2N$	538-07-8	170.080	col liq	-34	66^{12}	1.0861^{23}	1.4653^{25}	i H_2O
1949	2-Chloro-*N*-(2-chloroethyl)-*N*-methylethanamine	Mechlorethamine	$C_5H_{11}Cl_2N$	51-75-2	156.053		-60	87^{18}			sl H_2O; msc ctc, DMF
1950	1-Chloro-2-(chloromethyl)-benzene	2-Chlorobenzyl chloride	$C_7H_6Cl_2$	611-19-8	161.029	liq	-17	217	1.2699^{0}	1.5530^{20}	i H_2O; sl EtOH, ctc; vs eth, bz
1951	1-Chloro-3-(chloromethyl)-benzene	3-Chlorobenzyl chloride	$C_7H_6Cl_2$	620-20-2	161.029			216	1.2695^{15}	1.5554^{20}	vs EtOH
1952	1-Chloro-4-(chloromethyl)-benzene	4-Chlorobenzyl chloride	$C_7H_6Cl_2$	104-83-6	161.029	nd (dil al)	31	223			sl ctc
1953	Chloro(chloromethyl)-dimethylsilane		$C_3H_8Cl_2Si$	1719-57-9	143.088			115.5	1.0865^{20}	1.4360^{20}	
1954	3-Chloro-2-(chloromethyl)-1-propene		$C_4H_6Cl_2$	1871-57-4	124.997	liq	-14	138	1.1782^{20}	1.4753	vs EtOH, chl
1955	1-Chloro-4-[(chloromethyl)-thio]benzene		$C_7H_6Cl_2S$	7205-90-5	193.094		21.5	128^{12}	1.346^{25}	1.6055^{20}	
1956	2-Chloro-1-(4-chlorophenyl)-ethanone		$C_8H_6Cl_2O$	937-20-2	189.039	nd (al)	101.5	270			s EtOH, bz, MeOH
1957	3-Chlorocholest-5-ene, (3β)		$C_{27}H_{45}Cl$	910-31-6	405.099	nd (al, ace)	96				i H_2O; s EtOH, ace, bz, chl; vs CS_2
1958	*trans*-*o*-Chlorocinnamic acid		$C_9H_7ClO_2$	939-58-2	182.604		212				vs eth, EtOH
1959	*trans*-*m*-Chlorocinnamic acid		$C_9H_7ClO_2$	14473-90-6	182.604		165				s EtOH, eth
1960	*trans*-*p*-Chlorocinnamic acid		$C_9H_7ClO_2$	940-62-5	182.604		249.5				vs ace, eth, EtOH
1961	Chlorocyclohexane	Cyclohexyl chloride	$C_6H_{11}Cl$	542-18-7	118.604	liq	-45(1)	142.6(0.5)	1.000^{20}	1.4626^{20}	i H_2O; msc EtOH, eth, ace, bz; vs chl
1962	2-Chlorocyclohexanone		C_6H_9ClO	822-87-7	132.587		23	82^{15}	1.160^{20}	1.4825^{20}	s eth, bz, diox; sl ctc
1963	1-Chlorocyclohexene		C_6H_9Cl	930-66-5	116.588			140(4)	1.0361^{19}	1.4797^{20}	s eth, ace, ctc, chl
1964	Chlorocyclopentane	Cyclopentyl chloride	C_5H_9Cl	930-28-9	104.578			113.1(0.6)	1.0051^{20}	1.4510^{20}	i H_2O; s eth, ace, bz, ctc
1965	2-Chlorocyclopentanone		C_5H_7ClO	694-28-0	118.562			87^{19}	1.185^{25}	1.4750^{20}	
1966	3-Chlorocyclopentene		C_5H_7Cl	96-40-2	102.563			40^{40}	1.0388^{25}	1.4708^{26}	vs eth, EtOH, chl
1967	4-Chloro-2-cyclopentylphenol	Dowicide 9	$C_{11}H_{13}ClO$	13347-42-7	196.673			183^{18}			
1968	1-Chlorodecane		$C_{10}H_{21}Cl$	1002-69-3	176.727	liq	-31.3	225(3)	0.8696^{20}	1.4380^{20}	i H_2O; vs eth, chl; s ctc
1969	10-Chloro-1-decanol		$C_{10}H_{21}ClO$	51309-10-5	192.726		12.5	187^{15}	0.9630^{25}	1.4578^{20}	vs eth, EtOH
1970	2-Chloro-*N*,*N*-diallylacetamide	Allidochlor	$C_8H_{12}ClNO$	93-71-0	173.640	liq		116^{1}	1.088^{25}	1.4932^{25}	sl H_2O; s EtOH
1971	Chlorodiazepoxide		$C_{16}H_{14}ClN_3O$	58-25-3	299.754		236.2				
1972	Chlorodibromomethane		$CHBr_2Cl$	124-48-1	208.280	liq	-20	120	2.451^{20}	1.5482^{20}	i H_2O; s EtOH, eth, ace, bz
1973	Chloro(dichloromethyl)-dimethylsilane	(Dichloromethyl)dimethylchlorosilane	$C_3H_7Cl_3Si$	18171-59-0	177.533	liq	-48	149	1.2369^{20}	1.461^{20}	
1974	5-Chloro-*N*-(3,4-dichlorophenyl)-2-hydroxy-benzamide	3',4',5-Trichlorosalicylanilide	$C_{13}H_8Cl_3NO_2$	642-84-2	316.568		247				
1975	2-Chloro-1,1-diethoxyethane		$C_6H_{13}ClO_2$	621-62-5	152.619			157(4)	1.0180^{20}	1.4170^{20}	sl H_2O, ctc; msc EtOH, eth
1976	3-Chloro-1,1-diethoxypropane		$C_7H_{15}ClO_2$	35573-93-4	166.646			84^{25}	0.9951^{19}	1.4268^{20}	vs ace, bz

1937
4-Chloro-1-butene

1938
cis-1-Chloro-2-butene

1939
trans-1-Chloro-2-butene

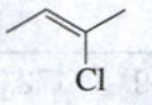
1940
cis-2-Chloro-2-butene

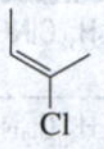
1941
trans-2-Chloro-2-butene

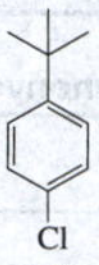
1942
1-Chloro-4-*tert*-butylbenzene

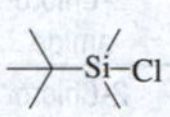
1943
Chloro-(*tert*-butyl)dimethylsilane

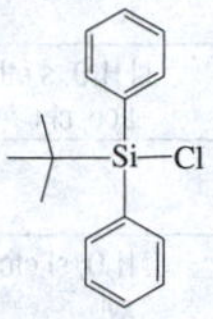
1944
Chloro(*tert*-butyl)diphenylsilane

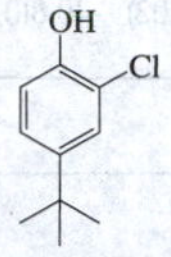
1945
2-Chloro-4-*tert*-butylphenol

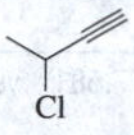
1946
3-Chloro-1-butyne

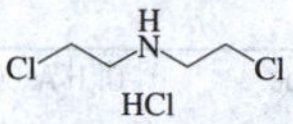
1947
2-Chloro-*N*-(2-chloroethyl)ethanamine, hydrochloride

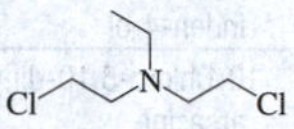
1948
2-Chloro-*N*-(2-chloroethyl)-*N*-ethylethanamine

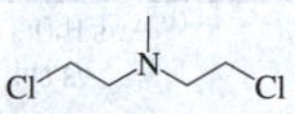
1949
2-Chloro-*N*-(2-chloroethyl)-*N*-methylethanamine

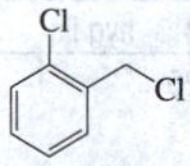
1950
1-Chloro-2-(chloromethyl)benzene

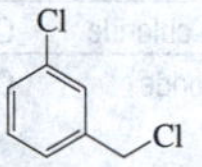
1951
1-Chloro-3-(chloromethyl)benzene

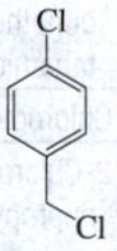
1952
1-Chloro-4-(chloromethyl)benzene

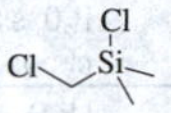
1953
Chloro(chloromethyl)dimethylsilane

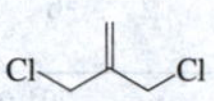
1954
3-Chloro-2-(chloromethyl)-1-propene

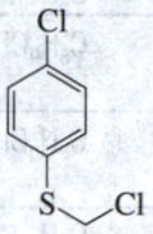
1955
1-Chloro-4-[(chloromethyl)thio]benzene

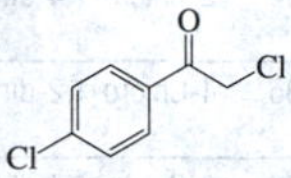
1956
2-Chloro-1-(4-chlorophenyl)ethanone

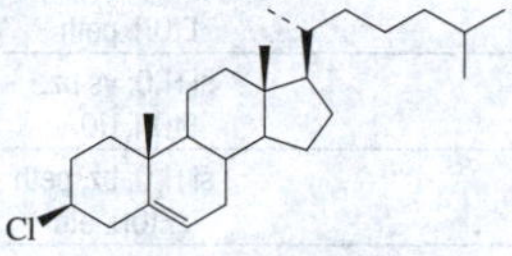
1957
3-Chlorocholest-5-ene, (3β)

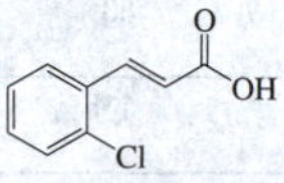
1958
trans-o-Chlorocinnamic acid

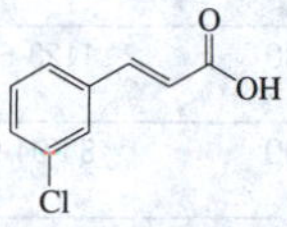
1959
trans-m-Chlorocinnamic acid

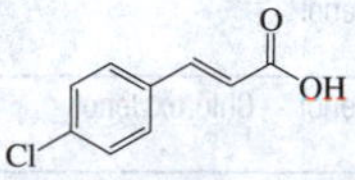
1960
trans-p-Chlorocinnamic acid

1961
Chlorocyclohexane

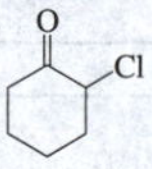
1962
2-Chlorocyclohexanone

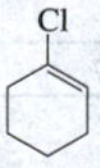
1963
1-Chlorocyclohexene

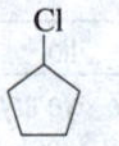
1964
Chlorocyclopentane

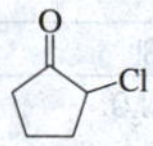
1965
2-Chlorocyclopentanone

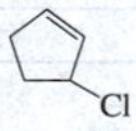
1966
3-Chlorocyclopentene

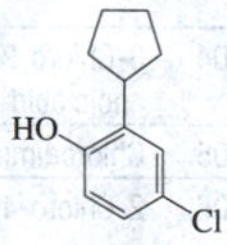
1967
4-Chloro-2-cyclopentylphenol

1968
1-Chlorodecane

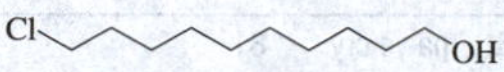
1969
10-Chloro-1-decanol

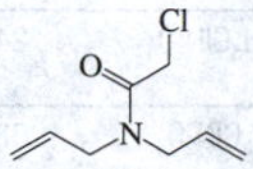
1970
2-Chloro-*N*,*N*-diallylacetamide

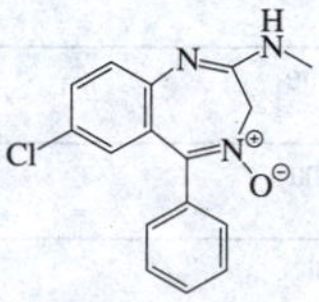
1971
Chlorodiazepoxide

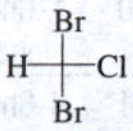
1972
Chlorodibromomethane

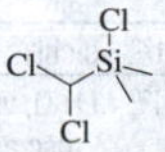
1973
Chloro(dichloromethyl)dimethylsilane

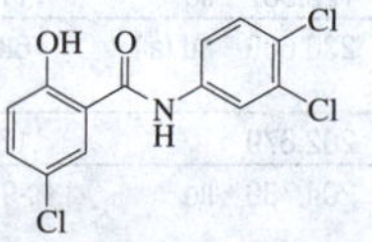
1974
5-Chloro-*N*-(3,4-dichlorophenyl)-2-hydroxybenzamide

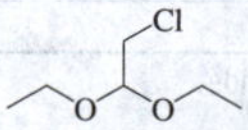
1975
2-Chloro-1,1-diethoxyethane

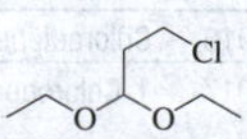
1976
3-Chloro-1,1-diethoxypropane

No.	Name	Synonym	Mol. form.	CAS Reg. No.	Mol. wt.	Physical form	mp/°C	bp/°C	den/ g cm^{-3}	n_D	Solubility
1977	2-Chloro-*N,N*-diethylacetamide		$C_6H_{12}ClNO$	2315-36-8	149.618			192^{25}			
1978	2-Chloro-*N,N*-diethylethanamine, hydrochloride		$C_6H_{15}Cl_2N$	869-24-9	172.096		200				sl H_2O
1979	Chlorodifluoroacetic acid		$C_2HClF_2O_2$	76-04-0	130.478	hyg	25	122		1.3559^{20}	s chl
1980	1-Chloro-1,1-difluoroethane	Refrigerant 142b	$C_2H_3ClF_2$	75-68-3	100.495	col gas	-130.43(0.02)	-9.12(0.07)	1.107^{25}		i H_2O; s bz
1981	1-Chloro-2,2-difluoroethane		$C_2H_3ClF_2$	338-65-8	100.495			35(7)			
1982	1-Chloro-2,2-difluoroethene	1-Chloro-2,2-difluoroethylene	C_2HClF_2	359-10-4	98.479	col gas	-138.5	-18.8(0.5)			
1983	Chlorodifluoromethane	Refrigerant 22	$CHClF_2$	75-45-6	86.469	col gas	-157.41(0.03)	-40.8(0.5)	1.4909^{-69}		sl H_2O; s eth, ace, chl
1984	7-Chloro-2,3-dihydro-1*H*-inden-4-ol	Chlorindanol	C_9H_9ClO	145-94-8	168.619	nd (peth)	92				
1985	10-Chloro-5,10-dihydrophenarsazine	Phenarsazine chloride	$C_{12}H_9AsClN$	578-94-9	277.581	ye cry	195		1.65		i H_2O; sl ctc, bz, xyl
1986	5-Chloro-2,4-dimethoxyaniline		$C_8H_{10}ClNO_2$	97-50-7	187.624		91				
1987	2-Chloro-1,1-dimethoxyethane		$C_4H_9ClO_2$	97-97-2	124.566			127.5	1.068^{20}	1.4150^{20}	sl EtOH, eth, bz, ctc
1988	*N*-(4-Chloro-2,5-dimethoxyphenyl)-3-oxobutanamide		$C_{12}H_{14}ClNO_4$	4433-79-8	271.697		107				s chl
1989	Chlorodimethylaluminum	Dimethylaluminum chloride	C_2H_6AlCl	1184-58-3	92.504	hyg liq	-45	126	0.996		reac H_2O; s hx
1990	2-Chloro-10-(3-dimethylaminopropyl)phenothiazine monohydrochloride	Aminazin hydrochloride	$C_{17}H_{20}Cl_2N_2S$	69-09-0	355.325		198(2)				s H_2O; i eth, bz; vs chl, EtOH
1991	2-Chloro-*N,N*-dimethylaniline		$C_8H_{10}ClN$	698-01-1	155.625			205	1.1067^{20}	1.5578^{20}	vs bz, EtOH
1992	3-Chloro-*N,N*-dimethylaniline		$C_8H_{10}ClN$	6848-13-1	155.625			232			sl H_2O; s EtOH, ace, bz
1993	4-Chloro-*N,N*-dimethylaniline		$C_8H_{10}ClN$	698-69-1	155.625	nd (al)	35.5	231	1.0480^{100}		s EtOH
1994	2-Chloro-1,4-dimethylbenzene		C_8H_9Cl	95-72-7	140.610		0.8	185(8)	1.0589^{15}		i H_2O; s ace, ctc; vs bz
1995	4-Chloro-1,2-dimethylbenzene		C_8H_9Cl	615-60-1	140.610	liq	-6	190(13)	1.0682^{15}		i H_2O; s ace, ctc; vs bz
1996	2-Chloro-*N,N*-dimethylethanamine, hydrochloride		$C_4H_{11}Cl_2N$	4584-46-7	144.043		201.0				sl H_2O
1997	(2-Chloro-1,1-dimethylethyl)-benzene	Neophyl chloride	$C_{10}H_{13}Cl$	515-40-2	168.663			227(12)	1.047^{20}	1.5247^{20}	vs ace, bz, eth, EtOH
1998	4-Chloro-2,5-dimethylphenol		C_8H_9ClO	1124-06-7	156.609	silv-grn nd (lig)	74.5				sl H_2O; vs bz, EtOH, peth
1999	4-Chloro-2,6-dimethylphenol		C_8H_9ClO	1123-63-3	156.609	nd (w)	83				sl H_2O; vs bz, EtOH, HOAc
2000	4-Chloro-3,5-dimethylphenol	Chloroxylenol	C_8H_9ClO	88-04-0	156.609		115	246			sl H_2O, bz, peth; s EtOH, eth
2001	Chlorodimethylphenylsilane		$C_8H_{11}ClSi$	768-33-2	170.712			194(3)	1.032^{20}	1.5082^{20}	
2002	1-Chloro-*N,N*-dimethyl-2-propanamine, hydrochloride		$C_5H_{13}Cl_2N$	17256-39-2	158.069						s chl
2003	1-Chloro-2,2-dimethylpropane		$C_5H_{11}Cl$	753-89-9	106.594	liq	-20(4)	84(1)	0.8660^{20}	1.4044^{20}	vs bz, eth, EtOH, chl
2004	3-Chloro-2,2-dimethylpropanoic acid		$C_5H_9ClO_2$	13511-38-1	136.577		41(4)	110^{10}			vs ctc
2005	Chlorodimethylsilane		C_2H_7ClSi	1066-35-9	94.616	liq	-111	34.7	0.852	1.3830^{20}	
2006	2-Chloro-4,6-dinitroaniline		$C_6H_4ClN_3O_4$	3531-19-9	217.567	ye cry (DMF aq)	157				
2007	4-Chloro-2,6-dinitroaniline		$C_6H_4ClN_3O_4$	5388-62-5	217.567	oran-ye nd (al)	147				s EtOH
2008	1-Chloro-2,4-dinitrobenzene		$C_6H_3ClN_2O_4$	97-00-7	202.552	ye orth (eth) nd (al) ye cry	50.2(0.9)	315	1.4982^{75}	1.5857^{60}	i H_2O; sl EtOH; s eth, bz, CS_2
2009	2-Chloro-1,3-dinitrobenzene		$C_6H_3ClN_2O_4$	606-21-3	202.552	ye nd (al, HOAc)	88	315	1.6867^{16}		i H_2O; s EtOH, eth, tol; sl chl
2010	1-Chloro-2,4-dinitronaphthalene		$C_{10}H_5ClN_2O_4$	2401-85-6	252.611	ye nd (bz)	146.5				
2011	4-Chloro-2,6-dinitrophenol		$C_6H_3ClN_2O_5$	88-87-9	218.551	pa ye cry	81		1.74^{22}		vs eth, EtOH, chl
2012	2-Chloro-3,5-dinitropyridine		$C_5H_2ClN_3O_4$	2578-45-2	203.541		66.5				
2013	2-Chloro-1,3-dinitro-5-(trifluoromethyl)benzene		$C_7H_2ClF_3N_2O_4$	393-75-9	270.550		57				
2014	4-Chloro-1,3-dioxolan-2-one	Chloroethylene carbonate	$C_3H_3ClO_3$	3967-54-2	122.507	liq	110	213	1.504	1.4540^{20}	
2015	2-Chloro-1,2-diphenylethanone		$C_{14}H_{11}ClO$	447-31-4	230.689	nd (al)	68.5	dec			s EtOH; sl chl; i alk
2016	Chlorodiphenylmethane		$C_{13}H_{11}Cl$	90-99-3	202.679		17.0(0.4)	140^3	1.140^{25}	1.5951^{20}	s chl
2017	1-Chlorododecane	Lauryl chloride	$C_{12}H_{25}Cl$	112-52-7	204.780	liq	-9.3	263(4)	0.8673^{20}	1.4434^{20}	i H_2O; vs EtOH; msc ace, ctc; s bz

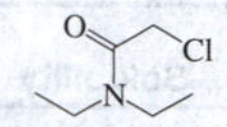
1977
2-Chloro-*N*,*N*-diethylacetamide

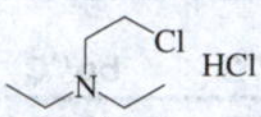
1978
2-Chloro-*N*,*N*-diethylethanamine, hydrochloride

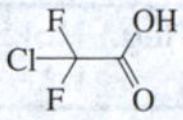
1979
Chlorodifluoroacetic acid

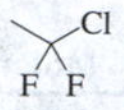
1980
1-Chloro-1,1-difluoroethane

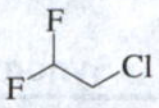
1981
1-Chloro-2,2-difluoroethane

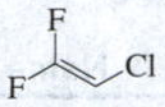
1982
1-Chloro-2,2-difluoroethene

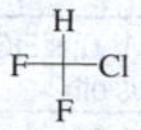
1983
Chlorodifluoromethane

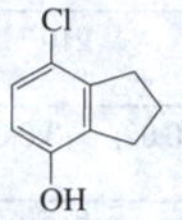
1984
7-Chloro-2,3-dihydro-1*H*-inden-4-ol

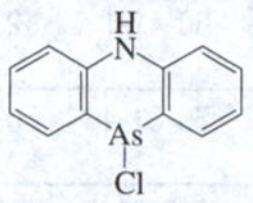
1985
10-Chloro-5,10-dihydrophenarsazine

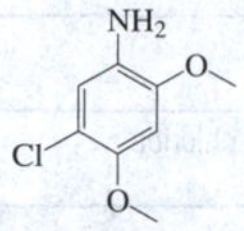
1986
5-Chloro-2,4-dimethoxyaniline

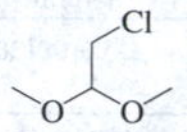
1987
2-Chloro-1,1-dimethoxyethane

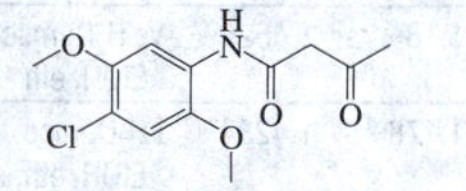
1988
N-(4-Chloro-2,5-dimethoxyphenyl)-3-oxobutanamide

1989
Chlorodimethylaluminum

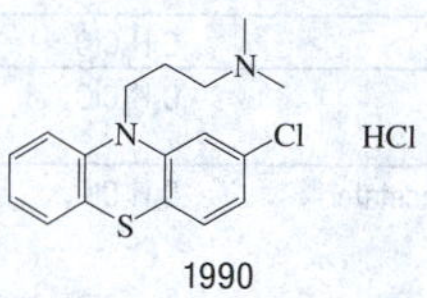
1990
2-Chloro-10-(3-dimethylaminopropyl)phenothiazine monohydrochloride

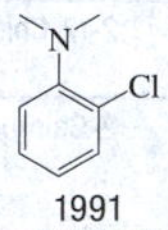
1991
2-Chloro-*N*,*N*-dimethylaniline

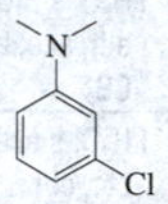
1992
3-Chloro-*N*,*N*-dimethylaniline

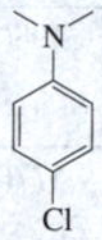
1993
4-Chloro-*N*,*N*-dimethylaniline

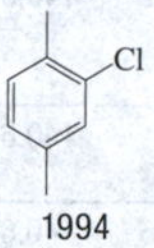
1994
2-Chloro-1,4-dimethylbenzene

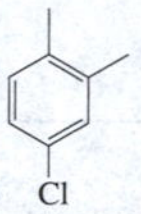
1995
4-Chloro-1,2-dimethylbenzene

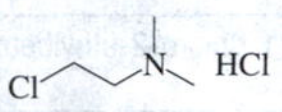
1996
2-Chloro-*N*,*N*-dimethylethanamine, hydrochloride

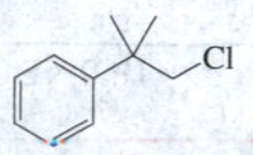
1997
(2-Chloro-1,1-dimethylethyl)benzene

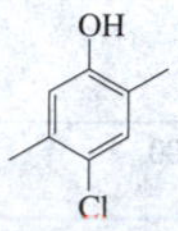
1998
4-Chloro-2,5-dimethylphenol

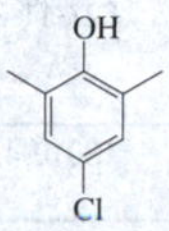
1999
4-Chloro-2,6-dimethylphenol

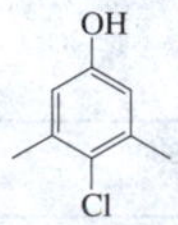
2000
4-Chloro-3,5-dimethylphenol

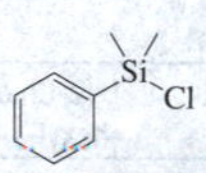
2001
Chlorodimethylphenylsilane

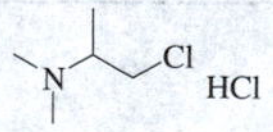
2002
1-Chloro-*N*,*N*-dimethyl-2-propanamine, hydrochloride

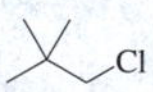
2003
1-Chloro-2,2-dimethylpropane

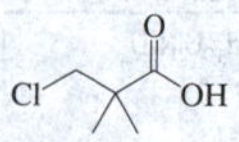
2004
3-Chloro-2,2-dimethylpropanoic acid

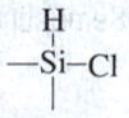
2005
Chlorodimethylsilane

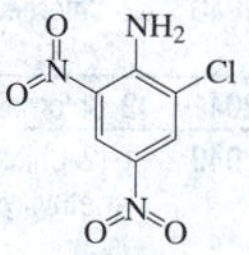
2006
2-Chloro-4,6-dinitroaniline

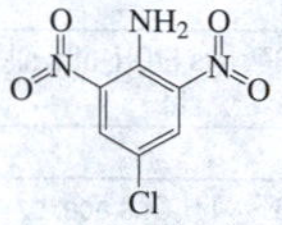
2007
4-Chloro-2,6-dinitroaniline

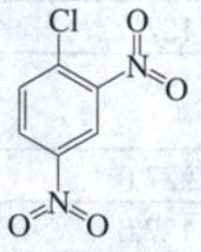
2008
1-Chloro-2,4-dinitrobenzene

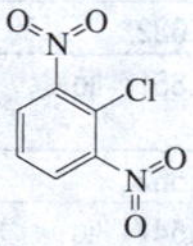
2009
2-Chloro-1,3-dinitrobenzene

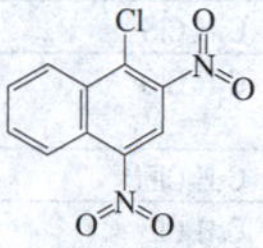
2010
1-Chloro-2,4-dinitronaphthalene

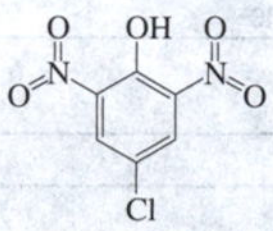
2011
4-Chloro-2,6-dinitrophenol

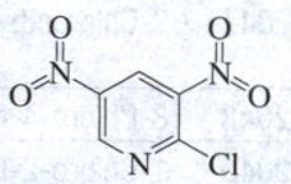
2012
2-Chloro-3,5-dinitropyridine

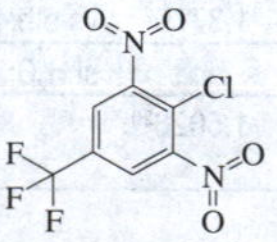
2013
2-Chloro-1,3-dinitro-5-(trifluoromethyl)benzene

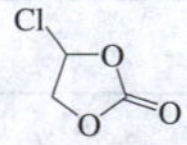
2014
4-Chloro-1,3-dioxolan-2-one

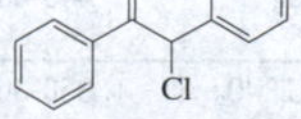
2015
2-Chloro-1,2-diphenylethanone

2016
Chlorodiphenylmethane

2017
1-Chlorododecane

Organic

No.	Name	Synonym	Mol. form.	CAS Reg. No.	Mol. wt.	Physical form	mp/°C	bp/°C	den/ g cm^{-3}	n_D	Solubility
2018	Chloroethane	Ethyl chloride	C_2H_5Cl	75-00-3	64.514	vol liq or gas	-138(2)	12.3(0.2)	0.9239^{0}	1.3676^{20}	sl H_2O, chl; vs EtOH; msc eth
2019	2-Chloroethanesulfonyl chloride		$C_2H_4Cl_2O_2S$	1622-32-8	163.023			201.5	1.555^{20}	1.4920^{20}	
2020	2-Chloroethanol	Ethylene chlorohydrin	C_2H_5ClO	107-07-3	80.513	liq	-68(2)	126(2)	1.2019^{20}	1.4419^{20}	msc H_2O, EtOH; sl eth; s chl
2021	2-Chloroethanol, 4-methylbenzenesulfonate		$C_9H_{11}ClO_3S$	80-41-1	234.699			210^{21}			i H_2O; s ctc
2022	Chloroethene	Vinyl chloride	C_2H_3Cl	75-01-4	62.498	col gas	-153.84(0.02)	-13.8(0.3)	0.9106^{20}	1.3700^{20}	sl H_2O; s EtOH; vs eth
2023	1-Chloro-4-ethoxybenzene		C_8H_9ClO	622-61-7	156.609		17.1(0.2)	212(1)	1.1254^{20}	1.5252^{20}	s EtOH, eth, HOAc; vs bz; sl ctc
2024	(2-Chloroethoxy)benzene		C_8H_9ClO	622-86-6	156.609		28	218.5			i H_2O; vs EtOH, eth, ace, bz; sl ctc
2025	1-Chloro-1-ethoxyethane		C_4H_9ClO	7081-78-9	108.566			93.5	0.9655^{20}	1.4053^{20}	
2026	2-(2-Chloroethoxy)ethanol		$C_4H_9ClO_2$	628-89-7	124.566			194(3)	1.18^{25}	1.4529^{20}	vs H_2O; msc EtOH, eth
2027	2-Chloroethyl acetate	β-Chloroethyl acetate	$C_4H_7ClO_2$	542-58-5	122.551			129(11)	1.178^{20}	1.4234^{20}	i H_2O; msc EtOH, eth; s ctc
2028	2-Chloroethyl acetoacetate		$C_6H_9ClO_3$	54527-68-3	164.586			198	1.2055^{21}	1.4430^{20}	vs bz, eth, EtOH
2029	2-Chloroethylamine hydrochloride	2-Chloroethanamine hydrochloride	$C_2H_7Cl_2N$	870-24-6	115.990		146.3				vs H_2O, ace, EtOH
2030	(1-Chloroethyl)benzene		C_8H_9Cl	672-65-1	140.610			105^{50}			
2031	(2-Chloroethyl)benzene		C_8H_9Cl	622-24-2	140.610			198(15)	1.069^{25}	1.5276^{20}	i H_2O; s EtOH, eth, ace, bz, CS_2
2032	1-Chloro-2-ethylbenzene		C_8H_9Cl	89-96-3	140.610	liq	-83.3(0.2)	177(3)	1.0569^{20}	1.5218^{20}	i H_2O; s ace, bz, ctc, chl
2033	1-Chloro-3-ethylbenzene		C_8H_9Cl	620-16-6	140.610	liq	-55.0(0.2)	180(3)	1.0529^{20}	1.5195^{20}	vs ace, bz, eth, EtOH
2034	1-Chloro-4-ethylbenzene		C_8H_9Cl	622-98-0	140.610	liq	-62.5(0.1)	184.4(0.9)	1.0455^{20}	1.5175^{20}	i H_2O; msc EtOH, eth, ace, peth; s HOAc
2035	2-Chloroethyl chloroformate		$C_3H_4Cl_2O_2$	627-11-2	142.969			155	1.3847^{20}	1.4483^{20}	i H_2O; s EtOH, eth, ace, bz; sl ctc
2036	1-(2-Chloroethyl)-3-cyclohexyl-1-nitrosourea	Lomustine	$C_9H_{16}ClN_3O_2$	13010-47-4	233.695	ye pow	90				i H_2O; s EtOH
2037	*N*-(2-Chloroethyl)-dibenzylamine	Dibenamine	$C_{16}H_{18}ClN$	51-50-3	259.774	oily liq		169^{3}			
2038	*N*-(2-Chloroethyl)-dibenzylamine hydrochloride	Dibenamine hydrochloride	$C_{16}H_{19}Cl_2N$	55-43-6	296.235	cry	194				i H_2O; s EtOH, dil acid
2039	Chloroethyldimethylsilane		$C_4H_{11}ClSi$	6917-76-6	122.669			89.5	0.8675^{20}	1.4105^{20}	
2040	2-Chloroethyl ethyl ether		C_4H_9ClO	628-34-2	108.566			98(4)	0.9895^{20}	1.4113^{20}	sl H_2O; msc eth; s chl
2041	2-Chloroethyl isocyanate		C_3H_4ClNO	1943-83-5	105.523			44^{17}			
2042	1-(2-Chloroethyl)-3-(4-methylcyclohexyl)-1-nitrosourea	Semustine	$C_{10}H_{18}ClN_3O_2$	13909-09-6	247.722	cry	64 dec				
2043	5-(2-Chloroethyl)-4-methylthiazole	Clomethiazole	C_6H_8ClNS	533-45-9	161.653	oil		92^{7}	1.233^{25}		
2044	*N*-(2-Chloroethyl)morpholine		$C_6H_{12}ClNO$	3240-94-6	149.618			42^{1}			
2045	4-(2-Chloroethyl)morpholine, hydrochloride		$C_6H_{13}Cl_2NO$	3647-69-6	186.079		185				
2046	1-Chloro-2-(ethylthio)ethane		C_4H_9ClS	693-07-2	124.632			157	1.0663^{25}		
2047	2-Chloroethyl vinyl ether		C_4H_7ClO	110-75-8	106.551	liq	-70	108	1.0495^{20}	1.4378^{20}	vs EtOH, eth; sl chl
2048	3-Chloro-4-fluoroaniline		C_6H_5ClFN	367-21-5	145.562		45.0	227.0			
2049	1-Chloro-2-fluorobenzene		C_6H_4ClF	348-51-6	130.547	liq	-43	137.6	1.2233^{30}	1.4918^{30}	i H_2O; s ace, bz
2050	1-Chloro-3-fluorobenzene		C_6H_4ClF	625-98-9	130.547			128(25)	1.221^{25}	1.4911	
2051	1-Chloro-4-fluorobenzene		C_6H_4ClF	352-33-0	130.547	liq	-26.8	130	1226^{15}	1.4990^{15}	i H_2O; s EtOH, eth, bz
2052	1-Chloro-1-fluoroethane		C_2H_4ClF	1615-75-4	82.504	vol liq or gas		16(5)			
2053	1-Chloro-2-fluoroethane		C_2H_4ClF	762-50-5	82.504			53.1(0.2)	1.1747^{20}	1.3775^{20}	vs eth, EtOH
2054	Chlorofluoromethane		CH_2ClF	593-70-4	68.478	col gas	-135.1	-9.1			sl H_2O; vs chl
2055	1-Chloro-3-fluoro-2-methylbenzene		C_7H_6ClF	443-83-4	144.574			154	1.191^{25}	1.5026^{20}	
2056	2-Chloro-1-fluoro-4-nitrobenzene	3-Chloro-4-fluoronitrobenzene	$C_6H_3ClFNO_2$	350-30-1	175.545		41.5	229.5			

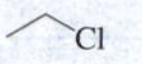
2018
Chloroethane

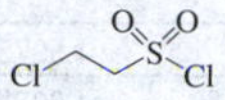
2019
2-Chloroethanesulfonyl chloride

2020
2-Chloroethanol

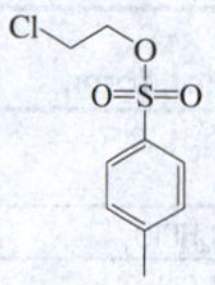
2021
2-Chloroethanol, 4-methylbenzenesulfonate

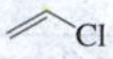
2022
Chloroethene

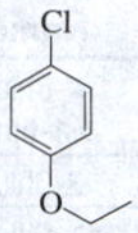
2023
1-Chloro-4-ethoxybenzene

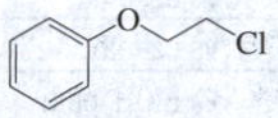
2024
(2-Chloroethoxy)benzene

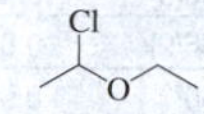
2025
1-Chloro-1-ethoxyethane

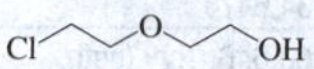
2026
2-(2-Chloroethoxy)ethanol

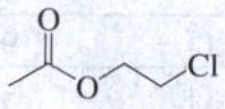
2027
2-Chloroethyl acetate

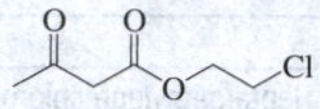
2028
2-Chloroethyl acetoacetate

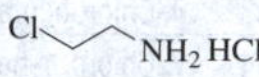
2029
2-Chloroethylamine hydrochloride

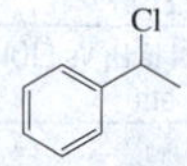
2030
(1-Chloroethyl)benzene

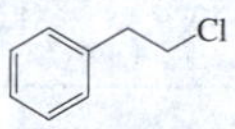
2031
(2-Chloroethyl)benzene

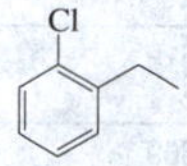
2032
1-Chloro-2-ethylbenzene

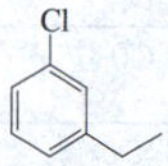
2033
1-Chloro-3-ethylbenzene

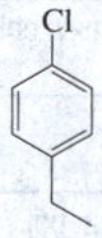
2034
1-Chloro-4-ethylbenzene

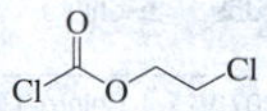
2035
2-Chloroethyl chloroformate

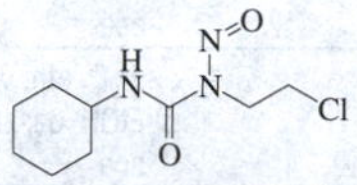
2036
1-(2-Chloroethyl)-3-cyclohexyl-1-nitrosourea

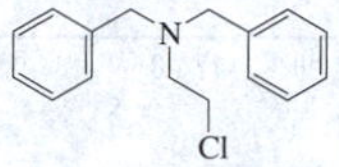
2037
N-(2-Chloroethyl)dibenzylamine

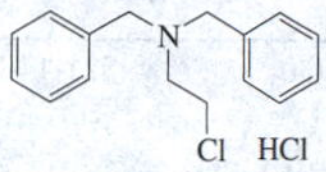
2038
N-(2-Chloroethyl)dibenzylamine hydrochloride

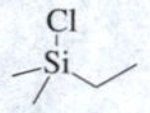
2039
Chloroethyldimethylsilane

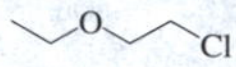
2040
2-Chloroethyl ethyl ether

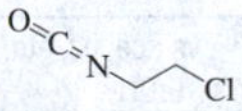
2041
2-Chloroethyl isocyanate

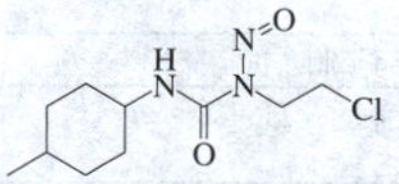
2042
1-(2-Chloroethyl)-3-(4-methylcyclohexyl)-1-nitrosourea

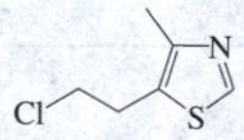
2043
5-(2-Chloroethyl)-4-methylthiazole

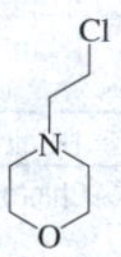
2044
N-(2-Chloroethyl)morpholine

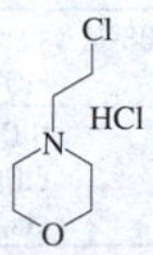
2045
4-(2-Chloroethyl)morpholine, hydrochloride

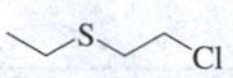
2046
1-Chloro-2-(ethylthio)ethane

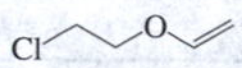
2047
2-Chloroethyl vinyl ether

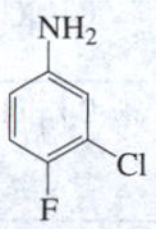
2048
3-Chloro-4-fluoroaniline

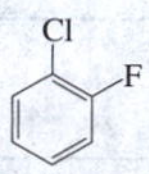
2049
1-Chloro-2-fluorobenzene

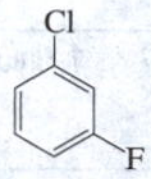
2050
1-Chloro-3-fluorobenzene

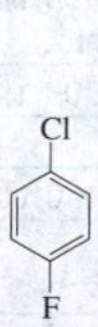
2051
1-Chloro-4-fluorobenzene

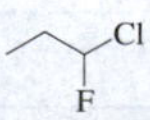
2052
1-Chloro-1-fluoroethane

2053
1-Chloro-2-fluoroethane

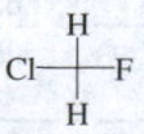
2054
Chlorofluoromethane

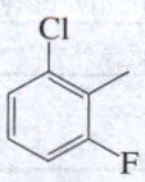
2055
1-Chloro-3-fluoro-2-methylbenzene

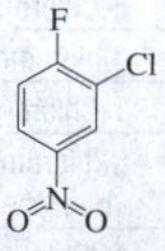
2056
2-Chloro-1-fluoro-4-nitrobenzene

No.	Name	Synonym	Mol. form.	CAS Reg. No.	Mol. wt.	Physical form	mp/°C	bp/°C	den/ g cm^{-3}	n_D	Solubility
2057	4-Chloro-1-(4-fluorophenyl)-1-butanone		$C_{10}H_{10}ClFO$	3874-54-2	200.636			136^{6}	1.22^{25}	1.5255^{20}	
2058	3-Chloro-2,5-furandione		C_4HClO_3	96-02-6	132.502		33	196	1.5375^{25}	1.4980^{20}	
2059	1-Chloro-1,2,2,3,3,4,4-heptafluorocyclobutane	Refrigerant C317	C_4ClF_7	377-41-3	216.485	liq or gas	-39.1	25	1.602^{15}		
2060	1-Chloroheptane	Heptyl chloride	$C_7H_{15}Cl$	629-06-1	134.647	liq	-69.4(0.4)	159(2)	0.8762^{20}	1.4264^{20}	i H_2O; msc EtOH, eth; sl ctc; s chl
2061	2-Chloroheptane		$C_7H_{15}Cl$	1001-89-4	134.647			61^{32}	0.8672^{20}	1.4221^{20}	i H_2O; vs eth; s bz, chl, HOAc
2062	3-Chloroheptane		$C_7H_{15}Cl$	999-52-0	134.647			156(9)	0.8690^{20}	1.4228^{20}	vs bz, eth
2063	4-Chloroheptane		$C_7H_{15}Cl$	998-95-8	134.647			156(9)	0.8710^{20}	1.4237^{20}	vs bz, eth
2064	7-Chloro-1-heptanol	Heptamethylene chlorohydrin	$C_7H_{15}ClO$	55944-70-2	150.646	cry (peth, bz)	11	150^{20}	0.9998^{15}	1.4537^{25}	vs EtOH, peth
2065	1-Chlorohexadecane		$C_{16}H_{33}Cl$	4860-03-1	260.886		17.9	326(4)	0.8635^{20}	1.4503^{20}	i H_2O
2066	1-Chlorohexane	Hexyl chloride	$C_6H_{13}Cl$	544-10-5	120.620	liq	-94(1)	135.0(0.5)	0.8738^{25}	1.4200^{20}	i H_2O; s EtOH, eth, ace, bz; vs chl; sl ctc
2067	2-Chlorohexane	2-Hexyl chloride	$C_6H_{13}Cl$	638-28-8	120.620			124(5)	0.8694^{21}	1.4142^{22}	vs ace, bz, eth, EtOH
2068	3-Chlorohexane	3-Hexyl chloride	$C_6H_{13}Cl$	2346-81-8	120.620			121(5)	0.8684^{20}	1.4163^{20}	vs ace, bz, eth, EtOH
2069	6-Chloro-1-hexanol		$C_6H_{13}ClO$	2009-83-8	136.619			107^{12}	1.0241^{20}	1.4550^{20}	sl H_2O; vs EtOH, eth
2070	4-Chloro-17-hydroxyandrost-4-en-3-one, (17β)	Clostebol	$C_{19}H_{27}ClO_2$	1093-58-9	322.869		189				
2071	5-Chloro-2-hydroxybenzaldehyde		$C_7H_5ClO_2$	635-93-8	156.567	pl (al)	100.3	105^{12}			i H_2O; vs EtOH; s eth, alk
2072	4-Chloro-α-hydroxybenzeneacetic acid		$C_8H_7ClO_3$	492-86-4	186.593		120.3				vs bz, EtOH
2073	3-Chloro-4-hydroxybenzoic acid		$C_7H_5ClO_3$	3964-58-7	172.566	nd (w)	171	sub			sl H_2O, bz, chl; vs EtOH, eth, ace
2074	5-Chloro-2-hydroxybenzoic acid		$C_7H_5ClO_3$	321-14-2	172.566	nd (w, al)	174.8				s H_2O, eth; vs EtOH, bz; sl ace
2075	2-Chloro-5-hydroxybenzophenone		$C_{13}H_9ClO_2$	85-19-8	232.662		95.3				i H_2O
2076	3-Chloro-4-hydroxy-5-methoxybenzaldehyde		$C_8H_7ClO_3$	19463-48-0	186.593	tetr	169.4(0.6)				i H_2O; s EtOH, HOAc
2077	1-Chloro-2-iodobenzene		C_6H_4ClI	615-41-8	238.453		0.7	234.5	1.9515^{25}	1.6331^{25}	i H_2O; s ace; sl ctc
2078	1-Chloro-3-iodobenzene		C_6H_4ClI	625-99-0	238.453			230	1.9255^{20}		i H_2O; s ace
2079	1-Chloro-4-iodobenzene		C_6H_4ClI	637-87-6	238.453	lf (ace, al)	53.57(0.05)	226(19)	1.886^{27}		i H_2O; s EtOH, $PhNO_2$; sl chl
2080	1-Chloro-4-iodobutane		C_4H_8ClI	10297-05-9	218.464	liq		116	1.785	1.5400^{20}	
2081	Chloroiodomethane		CH_2ClI	593-71-5	176.384			114(14)	2.422^{20}	1.5822^{20}	vs ace, bz, eth, EtOH
2082	1-Chloro-3-iodopropane		C_3H_6ClI	6940-76-7	204.437			163(6)	1.904^{20}	1.5472^{20}	i H_2O; s eth, bz, chl; sl ctc
2083	5-Chloro-7-iodo-8-quinolinol	Iodochlorhydroxyquin	C_9H_5ClINO	130-26-7	305.499	ye br nd (al)	178.5				sl EtOH; s HOAc
2084	1-Chloro-2-isocyanatobenzene		C_7H_4ClNO	3320-83-0	153.566		30.5	200.9(0.2)			sl ctc
2085	1-Chloro-3-isocyanatobenzene		C_7H_4ClNO	2909-38-8	153.566			113^{43}			sl chl
2086	1-Chloro-2-isopropylbenzene		$C_9H_{11}Cl$	2077-13-6	154.636	liq	-74.4	191(7)	1.0341^{20}	1.5168^{20}	vs ace, bz, eth, EtOH
2087	1-Chloro-4-isopropylbenzene		$C_9H_{11}Cl$	2621-46-7	154.636	liq	-12.3	193(6)	1.0208^{20}	1.5117^{20}	i H_2O; msc EtOH, eth, ace, ctc; vs bz
2088	1-Chloro-4-isothiocyanatobenzene		C_7H_4ClNS	2131-55-7	169.632	nd (al)	46	249.5			i H_2O; s EtOH
2089	Chloromethane	Methyl chloride	CH_3Cl	74-87-3	50.488	col gas	-97.6(0.2)	-24.1(0.3)	0.911^{25} (p>1 atm)	1.3389^{20}	sl H_2O; s EtOH; msc eth, ace, bz, chl
2090	4-Chloro-2-methoxyaniline	4-Chloro-2-anisidine	C_7H_8ClNO	93-50-5	157.598	nd or pr (dil al)	52	260			s EtOH, eth, bz, chl
2091	5-Chloro-2-methoxyaniline		C_7H_8ClNO	95-03-4	157.598	nd (dil al)	84				s EtOH; sl lig
2092	(Chloromethoxy)ethane	Chloromethyl ethyl ether	C_3H_7ClO	3188-13-4	94.540			83	1.0188^{15}	1.4040^{20}	
2093	1-Chloro-2-methoxyethane		C_3H_7ClO	627-42-9	94.540			79(13)	1.0345^{20}	1.4111^{20}	vs H_2O, eth
2094	[(Chloromethoxy)methyl]-benzene		C_8H_9ClO	3587-60-8	156.609			103^{13}	1.1350^{20}	1.5192^{20}	
2095	1-(Chloromethoxy)propane		C_4H_9ClO	3587-57-3	108.566			109	0.9884^{20}	1.4125^{20}	vs eth, EtOH

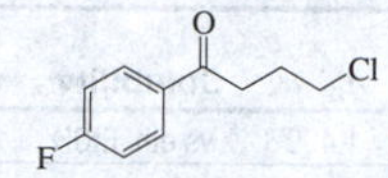
2057
4-Chloro-1-(4-fluorophenyl)-1-butanone

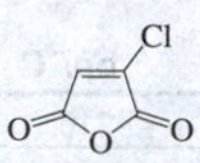
2058
3-Chloro-2,5-furandione

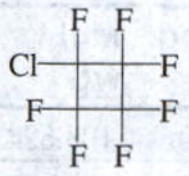
2059
1-Chloro-1,2,2,3,3,4,4-heptafluorocyclobutane

2060
1-Chloroheptane

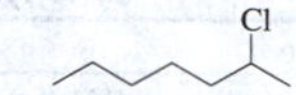
2061
2-Chloroheptane

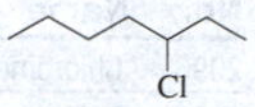
2062
3-Chloroheptane

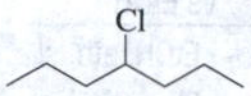
2063
4-Chloroheptane

2064
7-Chloro-1-heptanol

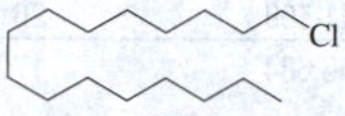
2065
1-Chlorohexadecane

2066
1-Chlorohexane

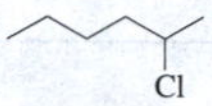
2067
2-Chlorohexane

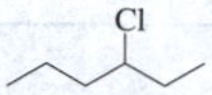
2068
3-Chlorohexane

2069
6-Chloro-1-hexanol

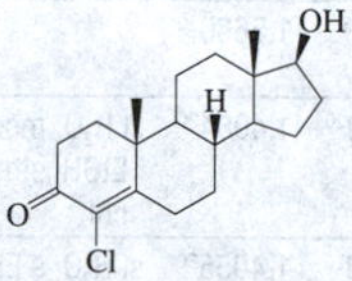
2070
4-Chloro-17-hydroxyandrost-4-en-3-one, (17β)

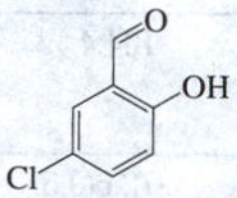
2071
5-Chloro-2-hydroxybenzaldehyde

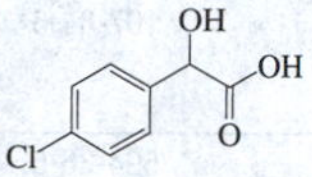
2072
4-Chloro-α-hydroxybenzeneacetic acid

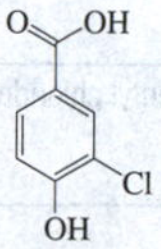
2073
3-Chloro-4-hydroxybenzoic acid

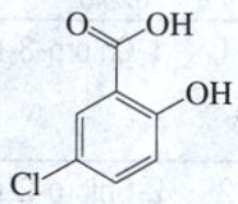
2074
5-Chloro-2-hydroxybenzoic acid

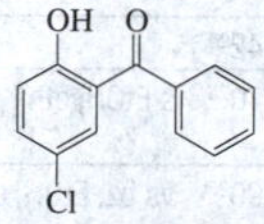
2075
2-Chloro-5-hydroxybenzophenone

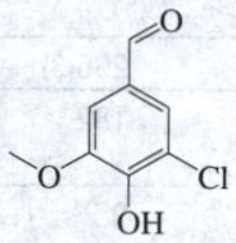
2076
3-Chloro-4-hydroxy-5-methoxybenzaldehyde

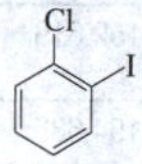
2077
1-Chloro-2-iodobenzene

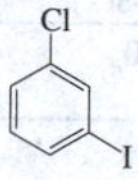
2078
1-Chloro-3-iodobenzene

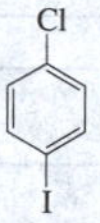
2079
1-Chloro-4-iodobenzene

2080
1-Chloro-4-iodobutane

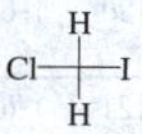
2081
Chloroiodomethane

2082
1-Chloro-3-iodopropane

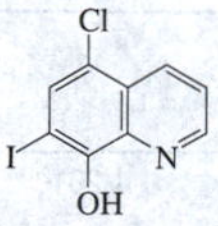
2083
5-Chloro-7-iodo-8-quinolinol

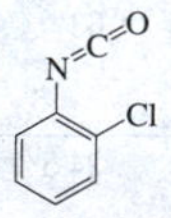
2084
1-Chloro-2-isocyanatobenzene

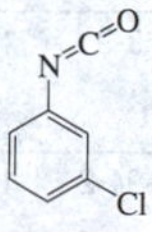
2085
1-Chloro-3-isocyanatobenzene

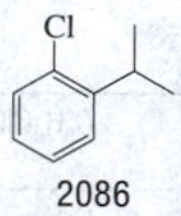
2086
1-Chloro-2-isopropylbenzene

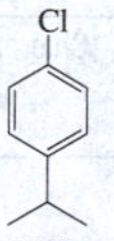
2087
1-Chloro-4-isopropylbenzene

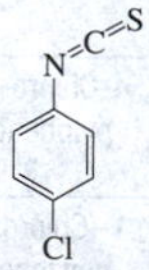
2088
1-Chloro-4-isothiocyanatobenzene

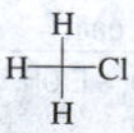
2089
Chloromethane

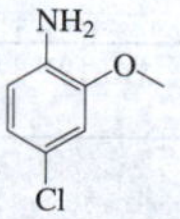
2090
4-Chloro-2-methoxyaniline

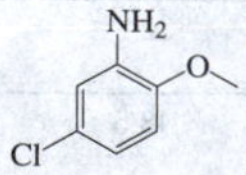
2091
5-Chloro-2-methoxyaniline

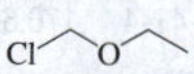
2092
(Chloromethoxy)ethane

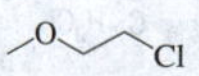
2093
1-Chloro-2-methoxyethane

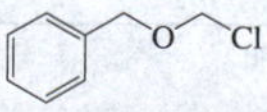
2094
[(Chloromethoxy)methyl]benzene

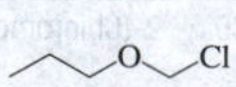
2095
1-(Chloromethoxy)propane

No.	Name	Synonym	Mol. form.	CAS Reg. No.	Mol. wt.	Physical form	mp/°C	bp/°C	den/ g cm^{-3}	n_D	Solubility
2096	Chloromethyl acetate		$C_3H_5ClO_2$	625-56-9	108.524			116	1.194^{20}	1.409^{20}	vs eth, EtOH
2097	5-Chloro-2-(methylamino)-benzophenone	*N*-Methyl-2-amino-5-chloro-benzophenone	$C_{14}H_{12}ClNO$	1022-13-5	245.704		92				
2098	4-Chloro-*N*-methylaniline		C_7H_8ClN	932-96-7	141.599			240	1.169^{11}	1.5835^{20}	s EtOH, ace, bz
2099	2-Chloro-4-methylaniline		C_7H_8ClN	615-65-6	141.599		7	220	1.151^{20}	1.5748^{22}	sl EtOH, bz
2100	2-Chloro-6-methylaniline		C_7H_8ClN	87-63-8	141.599			215			
2101	3-Chloro-2-methylaniline		C_7H_8ClN	87-60-5	141.599		1	245		1.5880^{20}	s H_2O, EtOH; i eth, bz
2102	3-Chloro-4-methylaniline		C_7H_8ClN	95-74-9	141.599		26	233(3)			s EtOH; sl ctc
2103	4-Chloro-2-methylaniline	*p*-Chloro-*o*-toluidine	C_7H_8ClN	95-69-2	141.599	lf (al)	30.3	244			s EtOH; sl ctc
2104	5-Chloro-2-methylaniline		C_7H_8ClN	95-79-4	141.599		26	239			vs EtOH
2105	1-Chloro-2-methyl-9,10-anthracenedione		$C_{15}H_9ClO_2$	129-35-1	256.684		170.5				i EtOH, eth; sl py
2106	(Chloromethyl)benzene	Benzyl chloride	C_7H_7Cl	100-44-7	126.584	liq	-39.4(0.6)	174(7)	1.1004^{20}	1.5391^{20}	i H_2O; msc EtOH, eth, chl; sl ctc
2107	3-Chloro-*N*-methylbenzene-methanamine		$C_8H_{10}ClN$	39191-07-6	155.625			88^{4}		1.5350^{25}	s chl
2108	α-(Chloromethyl)-benzenemethanol		C_8H_9ClO	1674-30-2	156.609			128^{17}	1.1926^{20}	1.5523^{20}	s EtOH; vs eth
2109	4-Chloro-α-methylbenzenemethanol		C_8H_9ClO	3391-10-4	156.609			121^{15}		1.5505^{20}	s ctc
2110	5-(Chloromethyl)-1,3-benzodi-oxole		$C_8H_7ClO_2$	20850-43-5	170.594		20.5	134^{14}	1.312^{25}	1.5660^{20}	
2111	1-Chloro-3-methylbutane	Isopentyl chloride	$C_5H_{11}Cl$	107-84-6	106.594	liq	-104.4	99(2)	0.8750^{20}	1.4084^{20}	sl H_2O; msc EtOH, eth; vs chl
2112	2-Chloro-2-methylbutane		$C_5H_{11}Cl$	594-36-5	106.594	liq	-72.6(0.5)	85(1)	0.8653^{20}	1.4055^{20}	sl H_2O; s EtOH, eth, ctc
2113	2-Chloro-3-methylbutane		$C_5H_{11}Cl$	631-65-2	106.594			92(2)	0.878^{20}		
2114	1-Chloro-3-methyl-2-butene		C_5H_9Cl	503-60-6	104.578			109	0.9273^{20}	1.4485^{20}	vs ace, eth, EtOH, chl
2115	3-Chloro-3-methyl-1-butyne		C_5H_7Cl	1111-97-3	102.563	liq	-61	76	0.9061^{20}		
2116	(Chloromethyl)cyclopropane		C_4H_7Cl	5911-08-0	90.552	liq	-90.9	87(6)	0.98^{25}	1.4350^{20}	
2117	1-(Chloromethyl)-2,4-dimeth-ylbenzene		$C_9H_{11}Cl$	824-55-5	154.636			233(18)	1.0580^{19}		vs bz, eth, EtOH
2118	(Chloromethyl)dimethylphenyl-silane		$C_9H_{13}ClSi$	1833-51-8	184.738			225	1.0240^{25}		s ctc, CS_2
2119	Chloromethyldiphenylsilane		$C_{13}H_{13}ClSi$	144-79-6	232.781			296(3)	1.1277^{20}	1.5742^{20}	
2120	1-Chloro-3-(1-methylethoxy)-2-propanol		$C_6H_{13}ClO_2$	4288-84-0	152.619			182	1.0910^{20}	1.4370^{25}	s EtOH, eth
2121	1-(Chloromethyl)-4-ethylben-zene		$C_9H_{11}Cl$	1467-05-6	154.636			95^{15}		1.5290^{25}	vs bz, EtOH, chl
2122	(1-Chloro-1-methylethyl)-benzene		$C_9H_{11}Cl$	934-53-2	154.636			98^{1}	1.192^{25}	1.5290^{25}	
2123	1-(Chloromethyl)-2-fluoroben-zene		C_7H_6ClF	345-35-7	144.574			172	1.216^{25}	1.5150^{20}	
2124	1-(Chloromethyl)-4-fluoroben-zene		C_7H_6ClF	352-11-4	144.574			82^{26}	1.2143^{20}	1.5130	
2125	2-(Chloromethyl)furan		C_5H_5ClO	617-88-9	116.546			49^{26}	1.1783^{20}	1.4941^{20}	vs bz, eth, EtOH
2126	3-(Chloromethyl)heptane	2-Ethylhexyl chloride	$C_8H_{17}Cl$	123-04-6	148.674			171(3)	0.8769^{20}	1.4319^{20}	i H_2O; s EtOH, eth, ace, bz; sl ctc
2127	4-Chloro-5-methyl-2-isopro-pylphenol	Chlorothymol	$C_{10}H_{13}ClO$	89-68-9	184.662		63	258.5			vs H_2O; s EtOH, eth, bz, ctc, peth, alk
2128	1-(Chloromethyl)-4-methoxy-benzene		C_8H_9ClO	824-94-2	156.609	nd	24.5	262.5	1.261^{20}	1.580^{20}	vs ace, bz, eth
2129	1-(Chloromethyl)-2-methyl-benzene		C_8H_9Cl	552-45-4	140.610			202(13)	1.063^{25}	1.5410^{25}	vs eth, EtOH
2130	1-(Chloromethyl)-3-methyl-benzene		C_8H_9Cl	620-19-9	140.610			195.5	1.064^{20}	1.5345^{20}	i H_2O; s EtOH, eth
2131	1-(Chloromethyl)-4-methyl-benzene		C_8H_9Cl	104-82-5	140.610			201	1.0512^{20}	1.5380	i H_2O; s EtOH; msc eth
2132	Chloromethyl methyl ether		C_2H_5ClO	107-30-2	80.513	liq	-103.5	59(3)	1.063^{10}	1.397^{20}	s EtOH, eth, ace, chl
2133	2-(Chloromethyl)-2-methyloxi-rane		C_4H_7ClO	598-09-4	106.551			122	1.1011^{20}	1.4310^{20}	vs H_2O, eth
2134	1-(Chloromethyl)naphthalene		$C_{11}H_9Cl$	86-52-2	176.642	pr	32	291(21)	1.1813^{20}	1.6380^{20}	i H_2O; s EtOH, ctc, peth
2135	2-(Chloromethyl)naphthalene		$C_{11}H_9Cl$	2506-41-4	176.642	lf (al)	48.5	169^{20}			i H_2O; s EtOH, peth

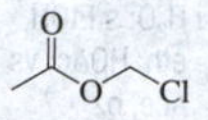

2096
Chloromethyl acetate

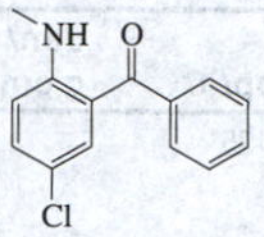

2097
5-Chloro-2-(methylamino)benzophenone

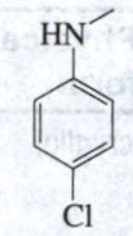

2098
4-Chloro-*N*-methylaniline

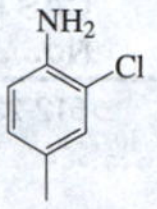

2099
2-Chloro-4-methylaniline

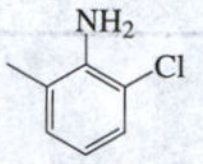

2100
2-Chloro-6-methylaniline

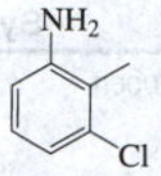

2101
3-Chloro-2-methylaniline

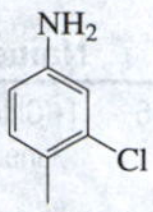

2102
3-Chloro-4-methylaniline

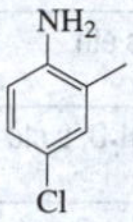

2103
4-Chloro-2-methylaniline

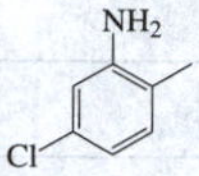

2104
5-Chloro-2-methylaniline

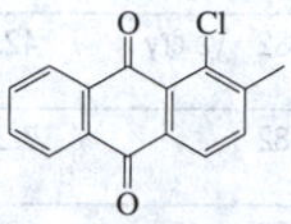

2105
1-Chloro-2-methyl-9,10-anthracenedione

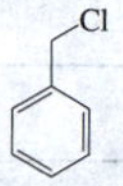

2106
(Chloromethyl)benzene

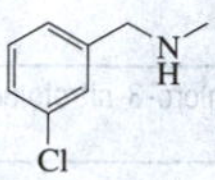

2107
3-Chloro-*N*-methylbenzenemethanamine

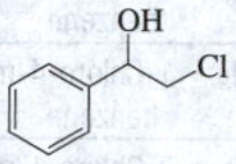

2108
α-(Chloromethyl)benzenemethanol

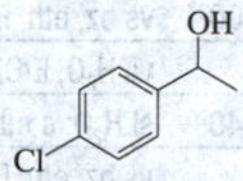

2109
4-Chloro-α-methylbenzenemethanol

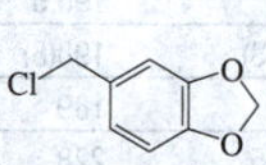

2110
5-(Chloromethyl)-1,3-benzodioxole

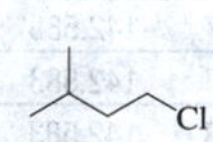

2111
1-Chloro-3-methylbutane

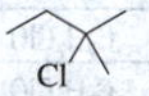

2112
2-Chloro-2-methylbutane

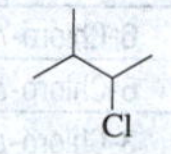

2113
2-Chloro-3-methylbutane

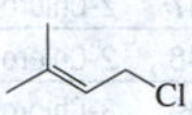

2114
1-Chloro-3-methyl-2-butene

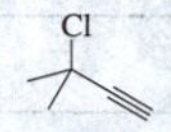

2115
3-Chloro-3-methyl-1-butyne

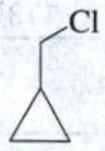

2116
(Chloromethyl)cyclopropane

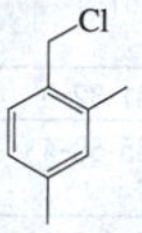

2117
1-(Chloromethyl)-2,4-dimethylbenzene

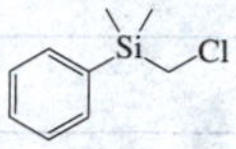

2118
(Chloromethyl)dimethylphenylsilane

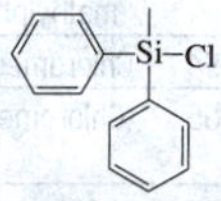

2119
Chloromethyldiphenylsilane

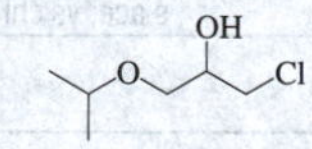

2120
1-Chloro-3-(1-methylethoxy)-2-propanol

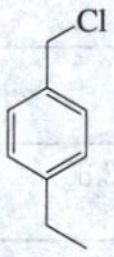

2121
1-(Chloromethyl)-4-ethylbenzene

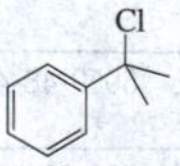

2122
(1-Chloro-1-methylethyl)benzene

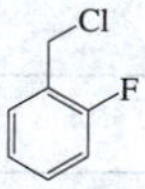

2123
1-(Chloromethyl)-2-fluorobenzene

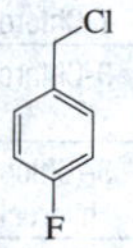

2124
1-(Chloromethyl)-4-fluorobenzene

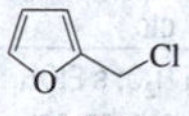

2125
2-(Chloromethyl)furan

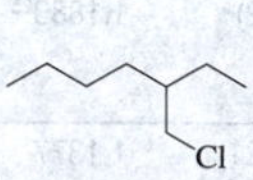

2126
3-(Chloromethyl)heptane

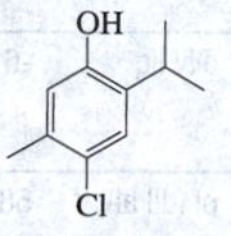

2127
4-Chloro-5-methyl-2-isopropylphenol

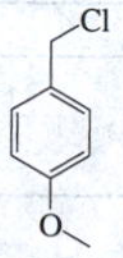

2128
1-(Chloromethyl)-4-methoxybenzene

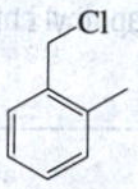

2129
1-(Chloromethyl)-2-methylbenzene

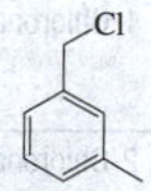

2130
1-(Chloromethyl)-3-methylbenzene

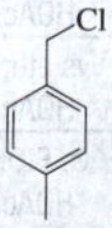

2131
1-(Chloromethyl)-4-methylbenzene

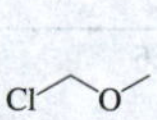

2132
Chloromethyl methyl ether

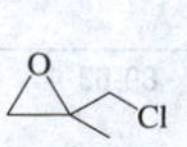

2133
2-(Chloromethyl)-2-methyloxirane

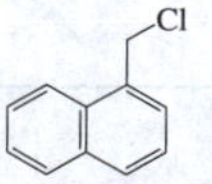

2134
1-(Chloromethyl)naphthalene

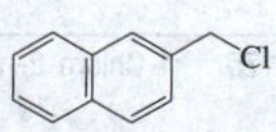

2135
2-(Chloromethyl)naphthalene

No.	Name	Synonym	Mol. form.	CAS Reg. No.	Mol. wt.	Physical form	mp/°C	bp/°C	den/ g cm^{-3}	n_D	Solubility
2136	1-(Chloromethyl)-2-nitrobenzene		$C_7H_6ClNO_2$	612-23-7	171.582	cry (lig)	47.7(0.2)	125^{4}		1.5557^{62}	i H_2O; s EtOH, eth, HOAc; vs ace, bz
2137	1-(Chloromethyl)-3-nitrobenzene		$C_7H_6ClNO_2$	619-23-8	171.582	pa ye nd (lig)	46	173^{34}		1.5577^{62}	vs ace, bz, eth, EtOH
2138	1-(Chloromethyl)-4-nitrobenzene	4-Nitrobenzyl chloride	$C_7H_6ClNO_2$	100-14-1	171.582	pl or nd (al)	71			1.5647^{62}	i H_2O; s EtOH, eth; vs ace, bz, AcOEt
2139	1-Chloro-2-methyl-3-nitrobenzene		$C_7H_6ClNO_2$	83-42-1	171.582	nd (dil al)	37.8	238		1.5377^{69}	i H_2O; s EtOH
2140	1-Chloro-2-methyl-4-nitrobenzene		$C_7H_6ClNO_2$	13290-74-9	171.582	ye cry	42.5	249			vs eth
2141	1-Chloro-4-methyl-2-nitrobenzene	4-Chloro-3-nitrotoluene	$C_7H_6ClNO_2$	89-60-1	171.582		7.2(0.2)	261		1.5572^{20}	i H_2O; s ctc
2142	2-Chloro-1-methyl-4-nitrobenzene		$C_7H_6ClNO_2$	121-86-8	171.582	nd (al)	66.5	260		1.5470^{69}	sl H_2O, chl; s EtOH, eth, HOAc
2143	4-Chloro-1-methyl-2-nitrobenzene		$C_7H_6ClNO_2$	89-59-8	171.582	mcl nd	36.5(1)	242	1.2559^{80}		i H_2O; s EtOH, eth; sl chl
2144	2-Chloro-4-methylpentane		$C_6H_{13}Cl$	25346-32-1	120.620			116(10)	0.8610^{20}	1.4113^{20}	vs eth
2145	3-(Chloromethyl)pentane		$C_6H_{13}Cl$	4737-41-1	120.620			126(5)	0.8914^{20}	1.4222^{20}	vs bz, eth, chl
2146	2-Chloro-4-methylphenol	2-Chloro-*p*-cresol	C_7H_7ClO	6640-27-3	142.583			195.5	1.1785^{27}	1.5200^{27}	vs bz, eth, EtOH
2147	2-Chloro-5-methylphenol	6-Chloro-*m*-cresol	C_7H_7ClO	615-74-7	142.583	pr (peth)	46(3)	198(8)	1.215^{15}		vs H_2O, EtOH
2148	2-Chloro-6-methylphenol	6-Chloro-*o*-cresol	C_7H_7ClO	87-64-9	142.583			189		1.5449^{20}	sl H_2O; s eth
2149	3-Chloro-4-methylphenol	3-Chloro-*p*-cresol	C_7H_7ClO	615-62-3	142.583	nd (al)	55.5	228			vs bz, eth, EtOH
2150	4-Chloro-2-methylphenol	4-Chloro-*o*-cresol	C_7H_7ClO	1570-64-5	142.583	nd (peth)	51	223			sl H_2O; s peth
2151	4-Chloro-3-methylphenol	4-Chloro-*m*-cresol	C_7H_7ClO	59-50-7	142.583	nd (peth)	55(2)	232(9)			sl H_2O, chl; s EtOH, eth, peth
2152	(4-Chloro-2-methylphenoxy)-acetic acid	MCPA	$C_9H_9ClO_3$	94-74-6	200.618	pl (bz, to)	119.7(0.9)				sl H_2O; vs EtOH, eth; s bz, ctc
2153	4-(4-Chloro-2-methylphenoxy)butanoic acid		$C_{11}H_{13}ClO_3$	94-81-5	228.672		100.4(0.5)				
2154	Chloromethylphenylsilane		C_7H_9ClSi	1631-82-9	156.685			113^{100}	1.043^{20}	1.5171^{20}	
2155	(Chloromethyl)phosphonic acid		CH_4ClO_3P	2565-58-4	130.468	nd (bz/$MeNO_2$)	90				
2156	*N*-Chloromethylphthalimide		$C_9H_6ClNO_2$	17564-64-6	195.603		133.5(0.5)				
2157	2-Chloro-2-methylpropanal		C_4H_7ClO	917-93-1	106.551			90	1.053^{15}	1.4160^{16}	vs eth, EtOH
2158	1-Chloro-2-methylpropane	Isobutyl chloride	C_4H_9Cl	513-36-0	92.567	liq	-130.3	69(1)	0.8773^{20}	1.3984^{20}	sl H_2O, ctc; s eth, ace, chl
2159	2-Chloro-2-methylpropane	*tert*-Butyl chloride	C_4H_9Cl	507-20-0	92.567	liq	-25.60(0.02)	50.9(0.5)	0.8420^{20}	1.3857^{20}	sl H_2O; msc EtOH, eth; s bz, ctc, chl
2160	1-Chloro-2-methylpropene	Dimethylvinyl chloride	C_4H_7Cl	513-37-1	90.552			68(4)	0.9186^{20}	1.4221^{20}	sl H_2O; s chl
2161	3-Chloro-2-methylpropene		C_4H_7Cl	563-47-3	90.552			72(2)	0.9165^{20}	1.4291^{20}	msc EtOH, eth; s ace; vs chl
2162	3-(Chloromethyl)pyridine, hydrochloride		$C_6H_7Cl_2N$	6959-48-4	164.033	hyg	143.8				
2163	Chloromethylsilane		CH_5ClSi	993-00-0	80.590	col gas	-135	7			
2164	1-Chloro-4-(methylsulfonyl)-benzene	4-Chlorobenzenethiol, *S*-methyl, *S,S*-dioxide	$C_7H_7ClO_2S$	98-57-7	190.648		98				
2165	1-Chloro-4-(methylthio)-benzene		C_7H_7ClS	123-09-1	158.649			105^{10}			
2166	1-Chloro-2-(methylthio)ethane		C_3H_7ClS	542-81-4	110.606			140	1.123^{20}	1.4902^{20}	s EtOH, eth, ace
2167	Chloro(methylthio)methane		C_2H_5ClS	2373-51-5	96.579			105	1.153^{25}	1.4963^{20}	
2168	(Chloromethyl)trimethylsilane		$C_4H_{11}ClSi$	2344-80-1	122.669			98.5	0.879^{25}	1.4175^{20}	
2169	1-Chloronaphthalene	1-Naphthyl chloride	$C_{10}H_7Cl$	90-13-1	162.616	oily liq	-6.0(0.2)	259(2)	1.1880^{25}	1.6326^{20}	i H_2O; s EtOH, eth, bz, CS_2; sl ctc
2170	2-Chloronaphthalene		$C_{10}H_7Cl$	91-58-7	162.616	pl (dil al), lf	58.02(0.05)	257(21)	1.1377^{71}	1.6079^{13}	i H_2O; s EtOH, eth, bz, chl, CS_2
2171	4-Chloro-1-naphthol		$C_{10}H_7ClO$	604-44-4	178.615	nd (chl, aq al)	120.5				s EtOH, eth, ace, bz, chl
2172	Chloroneb	Benzene, 1,4-dichloro-2,5-dimethoxy-	$C_8H_8Cl_2O_2$	2675-77-6	207.055		131.2(0.3)	268			
2173	2-Chloro-4-nitroaniline		$C_6H_5ClN_2O_2$	121-87-9	172.569	ye nd (w)	108.4(0.5)				vs eth, EtOH, HOAc
2174	2-Chloro-5-nitroaniline		$C_6H_5ClN_2O_2$	6283-25-6	172.569	ye nd (lig)	120.8(0.5)				vs eth, EtOH, HOAc
2175	4-Chloro-2-nitroaniline		$C_6H_5ClN_2O_2$	89-63-4	172.569	dk oran-ye pr (dil al)	116.5				vs EtOH, eth, HOAc; sl ace, lig

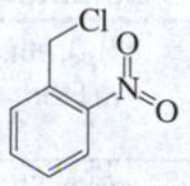
2136
1-(Chloromethyl)-2-nitrobenzene

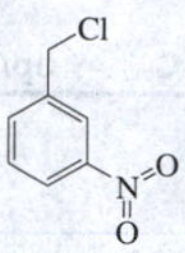
2137
1-(Chloromethyl)-3-nitrobenzene

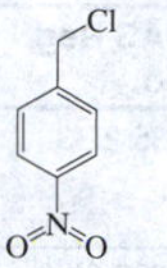
2138
1-(Chloromethyl)-4-nitrobenzene

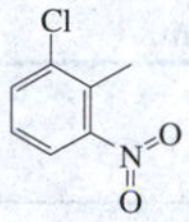
2139
1-Chloro-2-methyl-3-nitrobenzene

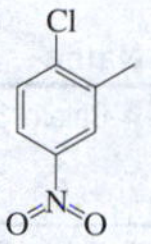
2140
1-Chloro-2-methyl-4-nitrobenzene

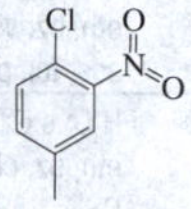
2141
1-Chloro-4-methyl-2-nitrobenzene

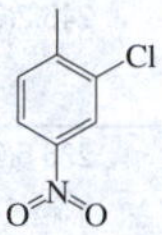
2142
2-Chloro-1-methyl-4-nitrobenzene

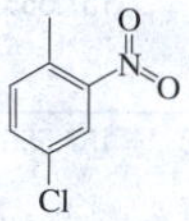
2143
4-Chloro-1-methyl-2-nitrobenzene

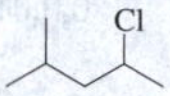
2144
2-Chloro-4-methylpentane

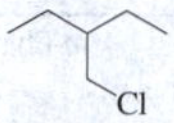
2145
3-(Chloromethyl)pentane

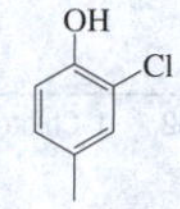
2146
2-Chloro-4-methylphenol

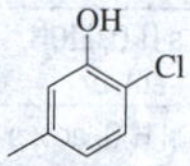
2147
2-Chloro-5-methylphenol

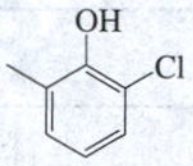
2148
2-Chloro-6-methylphenol

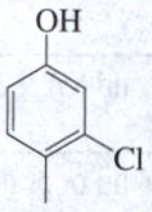
2149
3-Chloro-4-methylphenol

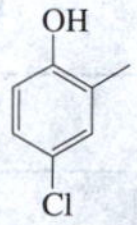
2150
4-Chloro-2-methylphenol

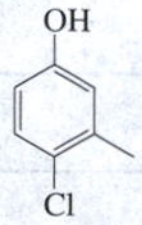
2151
4-Chloro-3-methylphenol

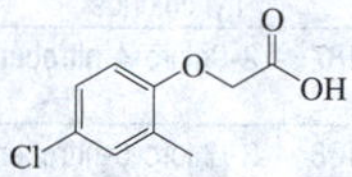
2152
(4-Chloro-2-methylphenoxy)acetic acid

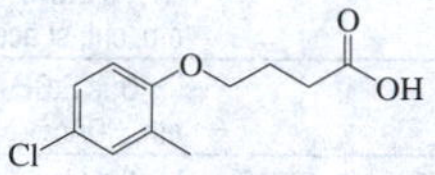
2153
4-(4-Chloro-2-methylphenoxy)butanoic acid

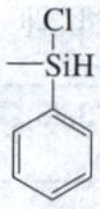
2154
Chloromethylphenylsilane

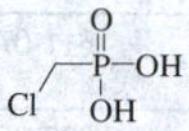
2155
(Chloromethyl)phosphonic acid

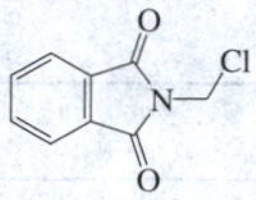
2156
N-Chloromethylphthalimide

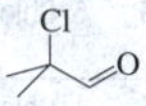
2157
2-Chloro-2-methylpropanal

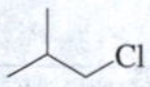
2158
1-Chloro-2-methylpropane

2159
2-Chloro-2-methylpropane

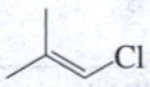
2160
1-Chloro-2-methylpropene

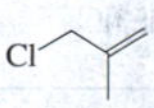
2161
3-Chloro-2-methylpropene

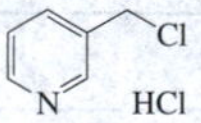
2162
3-(Chloromethyl)pyridine, hydrochloride

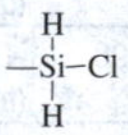
2163
Chloromethylsilane

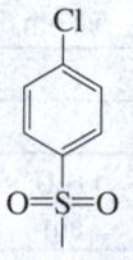
2164
1-Chloro-4-(methylsulfonyl)benzene

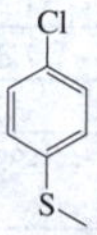
2165
1-Chloro-4-(methylthio)benzene

2166
1-Chloro-2-(methylthio)ethane

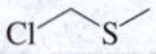
2167
Chloro(methylthio)methane

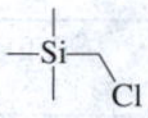
2168
(Chloromethyl)trimethylsilane

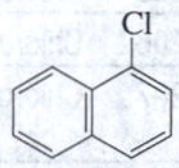
2169
1-Chloronaphthalene

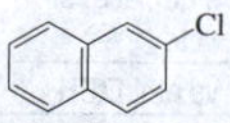
2170
2-Chloronaphthalene

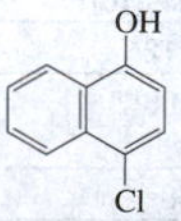
2171
4-Chloro-1-naphthol

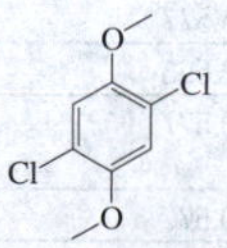
2172
Chloroneb

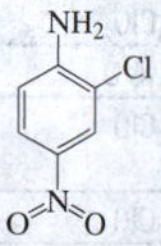
2173
2-Chloro-4-nitroaniline

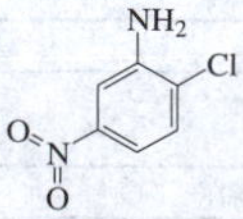
2174
2-Chloro-5-nitroaniline

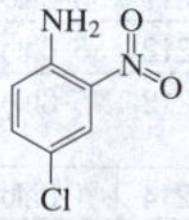
2175
4-Chloro-2-nitroaniline

No.	Name	Synonym	Mol. form.	CAS Reg. No.	Mol. wt.	Physical form	mp/°C	bp/°C	den/ g cm^{-3}	n_D	Solubility
2176	4-Chloro-3-nitroaniline		$C_6H_5ClN_2O_2$	635-22-3	172.569	ye nd or pr (w) nd (peth)	103				s H_2O, eth, chl; vs EtOH; sl lig
2177	5-Chloro-2-nitroaniline		$C_6H_5ClN_2O_2$	1635-61-6	172.569	ye nd (CS_2) ye lf (al, bz)	125.8(0.5)	sub			vs eth, EtOH
2178	1-Chloro-5-nitro-9,10-anthracenedione		$C_{14}H_6ClNO_4$	129-40-8	287.656		315.3				i H_2O, EtOH, eth, lig; sl bz; s py
2179	2-Chloro-5-nitrobenzaldehyde		$C_7H_4ClNO_3$	6361-21-3	185.565	cry (al)	81.3				vs EtOH, chl
2180	4-Chloro-3-nitrobenzaldehyde		$C_7H_4ClNO_3$	16588-34-4	185.565		64.5				sl H_2O; s chl
2181	1-Chloro-2-nitrobenzene	*o*-Chloronitrobenzene	$C_6H_4ClNO_2$	88-73-3	157.555	mcl nd	32.1(0.3)	246.2(0.7)	1.368^{242}		i H_2O; s EtOH, eth, bz; vs ace, tol, py
2182	1-Chloro-3-nitrobenzene	*m*-Chloronitrobenzene	$C_6H_4ClNO_2$	121-73-3	157.555	pa ye orth pr (al)	43.6(0.2)	236.5(0.6)	1.343^{50}	1.5374^{80}	i H_2O; s EtOH, eth, bz, chl, CS_2
2183	1-Chloro-4-nitrobenzene	*p*-Chloronitrobenzene	$C_6H_4ClNO_2$	100-00-5	157.555	mcl pr	82.2(0.7)	238(3)	1.2979^{90}	1.5376^{100}	i H_2O; sl EtOH; s eth, chl, CS_2
2184	5-Chloro-3-nitro-1,2-benzenediamine		$C_6H_6ClN_3O_2$	42389-30-0	187.584		167				
2185	4-Chloro-3-nitrobenzenesulfonamide		$C_6H_5ClN_2O_4S$	97-09-6	236.633	ye cry (EtOH)	175				
2186	4-Chloro-3-nitrobenzenesulfonyl chloride		$C_6H_3Cl_2NO_4S$	97-08-5	256.064		60.8				
2187	2-Chloro-4-nitrobenzoic acid		$C_7H_4ClNO_4$	99-60-5	201.565	nd (w)	141.8				s H_2O, EtOH, eth, bz
2188	2-Chloro-5-nitrobenzoic acid		$C_7H_4ClNO_4$	2516-96-3	201.565	nd or pr (w)	166.5		1.608^{18}		sl H_2O, ace; s EtOH, eth, bz
2189	4-Chloro-3-nitrobenzoic acid		$C_7H_4ClNO_4$	96-99-1	201.565	nd or pl (w)	182.8		1.645^{18}		i H_2O; sl EtOH, ace
2190	1-Chloro-1-nitroethane		$C_2H_4ClNO_2$	598-92-5	109.512			124.5	1.2837^{20}	1.4224^{20}	i H_2O; s EtOH, ctc, alk
2191	2-Chloro-4-nitrophenol		$C_6H_4ClNO_3$	619-08-9	173.554	wh nd (50% al)	107.5(0.4)				s H_2O, EtOH, eth, chl; sl bz
2192	4-Chloro-2-nitrophenol		$C_6H_4ClNO_3$	89-64-5	173.554	ye mcl pr (al)	87.2(0.4)				i H_2O; s EtOH, eth, chl; sl ace
2193	5-Chloro-2-nitrophenol		$C_6H_4ClNO_3$	611-07-4	173.554	ye pr or nd (w)	39(3)	sub			sl H_2O; s EtOH, eth, HOAc
2194	1-Chloro-1-nitropropane		$C_3H_6ClNO_2$	600-25-9	123.539			142	1.207^{20}	1.4251^{20}	sl H_2O, chl; s EtOH, eth, oils
2195	2-Chloro-2-nitropropane		$C_3H_6ClNO_2$	594-71-8	123.539		-21.5(0.2)	142(6)	1.2^{20}	1.4378^{19}	sl H_2O; s EtOH, eth, ctc, oils; i KOH
2196	2-Chloro-3-nitropyridine		$C_5H_3ClN_2O_2$	5470-18-8	158.543	nd (w)	104.0				
2197	1-Chloro-2-nitro-4-(trifluoromethyl)benzene		$C_7H_3ClF_3NO_2$	121-17-5	225.553	liq	-1.3	222	1.511^{25}	1.4893^{20}	
2198	1-Chloro-4-nitro-2-(trifluoromethyl)benzene		$C_7H_3ClF_3NO_2$	777-37-7	225.553		22	232	1.527^{25}	1.5083^{26}	
2199	1-Chlorononane		$C_9H_{19}Cl$	2473-01-0	162.700	liq	-39.4	204(3)	0.8674^{25}	1.4343^{20}	i H_2O; s eth, chl
2200	9-Chloro-1-nonanol		$C_9H_{19}ClO$	51308-99-7	178.699		28	147^{14}		1.4575^{20}	vs eth, EtOH
2201	1-Chlorooctadecane		$C_{18}H_{37}Cl$	3386-33-2	288.940		28.6	337(15)	0.8616^{20}	1.4524^{20}	i H_2O; sl ctc
2202	1-Chlorooctane	Octyl chloride	$C_8H_{17}Cl$	111-85-3	148.674	liq	-57.8	183(3)	0.8734^{20}	1.4309^{20}	i H_2O; vs EtOH, eth; sl ctc
2203	2-Chlorooctane		$C_8H_{17}Cl$	628-61-5	148.674			172(7)	0.8658^{17}	1.4273^{21}	i H_2O; vs EtOH, eth
2204	8-Chloro-1-octanol		$C_8H_{17}ClO$	23144-52-7	164.673			139^{19}		1.4563^{25}	vs eth, EtOH
2205	Chloropentafluoroacetone		C_3ClF_5O	79-53-8	182.476	col gas	-133	7.8(0.9)			
2206	Chloropentafluorobenzene		C_6ClF_5	344-07-0	202.509			117.96	1.568^{25}	1.4256^{20}	
2207	Chloropentafluoroethane	Refrigerant 115	C_2ClF_5	76-15-3	154.466	col gas	-99.4(0.1)	-39.2(0.2)	1.5678^{-42}	1.2678^{-42}	i H_2O; s EtOH, eth
2208	1-Chloropentane	Pentyl chloride	$C_5H_{11}Cl$	543-59-9	106.594	liq	-99.0	107.9(0.3)	0.8820^{20}	1.4126^{20}	i H_2O; msc EtOH, eth; s bz, ctc; vs chl
2209	2-Chloropentane, (+)	*sec*-Amyl chloride	$C_5H_{11}Cl$	29882-57-3	106.594	liq	-137	96.3(0.9)	0.8698^{20}	1.4069^{20}	i H_2O; s EtOH, eth, bz; vs chl
2210	3-Chloropentane		$C_5H_{11}Cl$	616-20-6	106.594	liq	-105(2)	95(3)	0.8731^{20}	1.4082^{20}	i H_2O; s EtOH, eth, bz; sl ace
2211	5-Chloropentanoic acid		$C_5H_9ClO_2$	1119-46-6	136.577		18	234(10)	1.3416^{25}	1.4555^{20}	vs eth, EtOH
2212	5-Chloro-1-pentanol		$C_5H_{11}ClO$	5259-98-3	122.593			112^{12}		1.4518^{20}	vs eth, EtOH
2213	5-Chloro-2-pentanone		C_5H_9ClO	5891-21-4	120.577			106^{110}	1.0523^{20}	1.4375^{20}	s eth, ace; sl ctc
2214	1-Chloro-3-pentanone		C_5H_9ClO	32830-97-0	120.577			68^{20}		1.4361^{20}	vs eth, EtOH
2215	5-Chloropentanoyl chloride		$C_5H_8Cl_2O$	1575-61-7	155.022			83^{12}	1.210^{18}	1.4639^{20}	vs eth

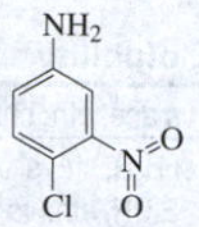
2176
4-Chloro-3-nitroaniline

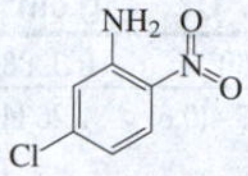
2177
5-Chloro-2-nitroaniline

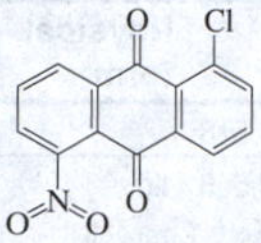
2178
1-Chloro-5-nitro-9,10-anthracenedione

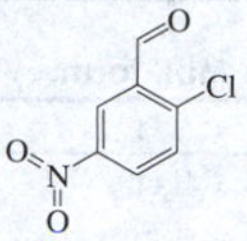
2179
2-Chloro-5-nitrobenzaldehyde

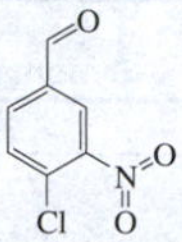
2180
4-Chloro-3-nitrobenzaldehyde

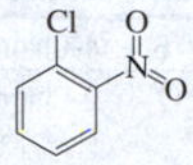
2181
1-Chloro-2-nitrobenzene

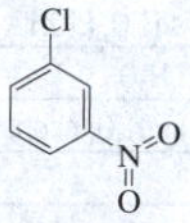
2182
1-Chloro-3-nitrobenzene

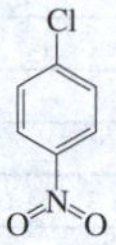
2183
1-Chloro-4-nitrobenzene

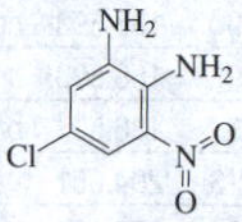
2184
5-Chloro-3-nitro-1,2-benzenediamine

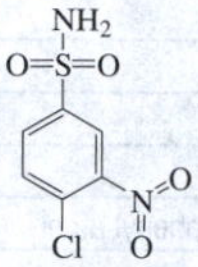
2185
4-Chloro-3-nitrobenzenesulfonamide

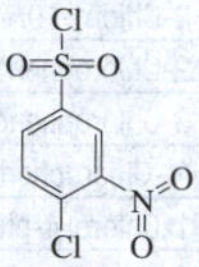
2186
4-Chloro-3-nitrobenzenesulfonyl chloride

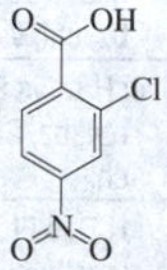
2187
2-Chloro-4-nitrobenzoic acid

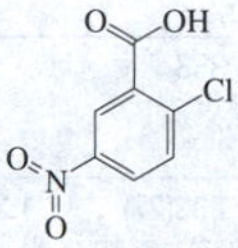
2188
2-Chloro-5-nitrobenzoic acid

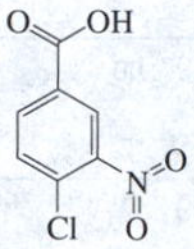
2189
4-Chloro-3-nitrobenzoic acid

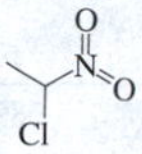
2190
1-Chloro-1-nitroethane

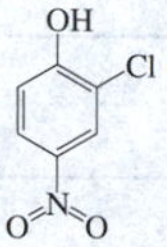
2191
2-Chloro-4-nitrophenol

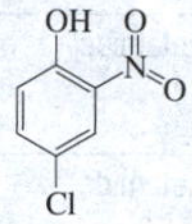
2192
4-Chloro-2-nitrophenol

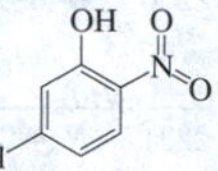
2193
5-Chloro-2-nitrophenol

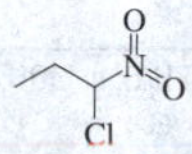
2194
1-Chloro-1-nitropropane

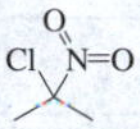
2195
2-Chloro-2-nitropropane

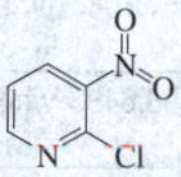
2196
2-Chloro-3-nitropyridine

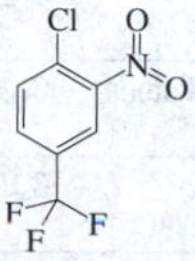
2197
1-Chloro-2-nitro-4-(trifluoromethyl)benzene

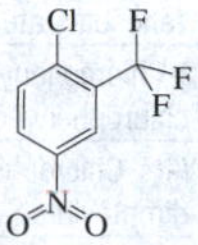
2198
1-Chloro-4-nitro-2-(trifluoromethyl)benzene

2199
1-Chlorononane

2200
9-Chloro-1-nonanol

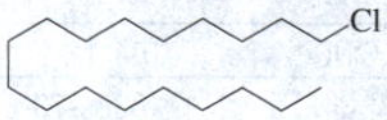
2201
1-Chlorooctadecane

2202
1-Chlorooctane

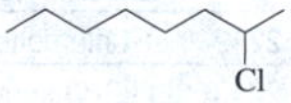
2203
2-Chlorooctane

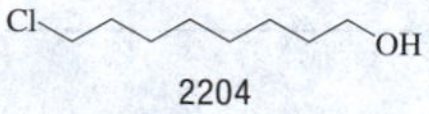
2204
8-Chloro-1-octanol

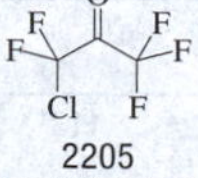
2205
Chloropentafluoroacetone

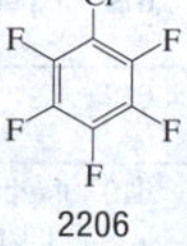
2206
Chloropentafluorobenzene

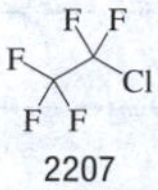
2207
Chloropentafluoroethane

2208
1-Chloropentane

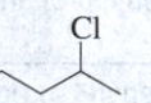
2209
2-Chloropentane, (+)

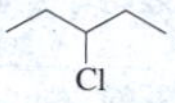
2210
3-Chloropentane

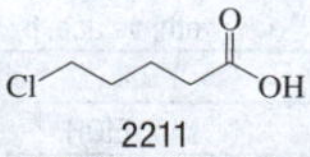
2211
5-Chloropentanoic acid

2212
5-Chloro-1-pentanol

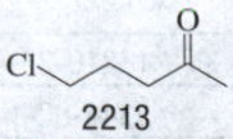
2213
5-Chloro-2-pentanone

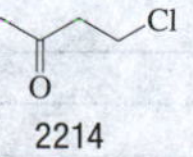
2214
1-Chloro-3-pentanone

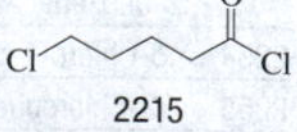
2215
5-Chloropentanoyl chloride

No.	Name	Synonym	Mol. form.	CAS Reg. No.	Mol. wt.	Physical form	mp/°C	bp/°C	den/ g cm^{-3}	n_D	Solubility
2216	4-Chloro-2-pentene		C_5H_9Cl	1458-99-7	104.578			97(5)	0.8988^{20}	1.4322^{20}	vs ace, eth, chl
2217	2-Chlorophenol		C_6H_5ClO	95-57-8	128.556	liq	8(1)	173.4(0.6)	1.2634^{20}	1.5524^{20}	sl H_2O, chl; s EtOH, eth; vs bz
2218	3-Chlorophenol		C_6H_5ClO	108-43-0	128.556		32.5(0.3)	210(3)	1.245^{45}	1.5565^{40}	sl H_2O, chl; s EtOH, eth; vs bz
2219	4-Chlorophenol		C_6H_5ClO	106-48-9	128.556		43.1(0.7)	219(4)	1.2651^{40}	1.5579^{40}	sl H_2O; vs EtOH, eth, bz; s alk
2220	Chlorophenol Red		$C_{19}H_{12}Cl_2O_5S$	4430-20-0	423.266	grn-br cry	261				sl H_2O; s EtOH
2221	2-Chloro-10*H*-phenothiazine		$C_{12}H_8ClNS$	92-39-7	233.717		198.5				
2222	2-Chlorophenoxyacetic acid		$C_8H_7ClO_3$	614-61-9	186.593	nd (w, al)	148.5				s H_2O, EtOH
2223	3-Chlorophenoxyacetic acid		$C_8H_7ClO_3$	588-32-9	186.593	cry (w)	110				i H_2O
2224	(4-Chlorophenoxy)acetic acid		$C_8H_7ClO_3$	122-88-3	186.593	pr or nd (w)	158.3(0.5)				vs H_2O; sl chl
2225	1-Chloro-4-phenoxybenzene	4-Chlorophenyl phenyl ether	$C_{12}H_9ClO$	7005-72-3	204.651			284.5	1.2026^{15}	1.599	
2226	3-(4-Chlorophenoxy)-1,2-propanediol	Chlorphenesin	$C_9H_{11}ClO_3$	104-29-0	202.634	cry	78	214^{19}			i H_2O; vs EtOH, eth; s bz, con sulf
2227	2-(3-Chlorophenoxy)propanoic acid	Cloprop	$C_9H_9ClO_3$	101-10-0	200.618	cry	113	100$^{1.5}$			
2228	2-Chloro-*N*-phenylacetamide		C_8H_8ClNO	587-65-5	169.609	nd (dil HOAc)		sub			vs bz, eth, EtOH
2229	*N*-(2-Chlorophenyl)acetamide		C_8H_8ClNO	533-17-5	169.609		86.7(0.5)				i H_2O; s EtOH, bz, chl; vs eth
2230	*N*-(3-Chlorophenyl)acetamide		C_8H_8ClNO	588-07-8	169.609	nd	76.6(0.5)	333			sl H_2O; vs EtOH, eth, bz, CS_2; s chl
2231	*N*-(4-Chlorophenyl)acetamide		C_8H_8ClNO	539-03-7	169.609		178.4(0.5)	332(27)	1.385^{22}		i H_2O; s EtOH; vs eth; sl ctc
2232	4-Chloro-α-phenylbenzenemethanol		$C_{13}H_{11}ClO$	119-56-2	218.678		59				sl chl
2233	4-Chlorophenyl benzenesulfonate		$C_{12}H_9ClO_3S$	80-38-6	268.715	col cry	59.3(0.5)		1.33		sl H_2O
2234	4-Chloro-1-phenyl-1-butanone		$C_{10}H_{11}ClO$	939-52-6	182.646		19.5	131^{4}	1.137^{25}	1.5459^{20}	
2235	4-Chlorophenyl 4-chlorobenzenesulfonate	Ovex	$C_{12}H_8Cl_2O_3S$	80-33-1	303.161		88(1)				i H_2O; sl EtOH; s ace
2236	(2-Chlorophenyl)(4-chlorophenyl)methanone	2,4'-Dichlorodiphenyl ketone	$C_{13}H_8Cl_2O$	85-29-0	251.108	pr (al)	65.7(0.5)	214^{22}	1.393^{14}		s EtOH; sl chl
2237	*N'*-(4-Chlorophenyl)-*N,N*-dimethylurea	Monuron	$C_9H_{11}ClN_2O$	150-68-5	198.648	wh pl (MeOH)	168.3(0.3)				i H_2O; sl EtOH, ace
2238	1-(3-Chlorophenyl)ethanone	*m*-Chloroacetophenone	C_8H_7ClO	99-02-5	154.594			244	1.2130^{40}	1.5494^{20}	s EtOH, eth, ace
2239	1-(4-Chlorophenyl)ethanone	*p*-Chloroacetophenone	C_8H_7ClO	99-91-2	154.594		18.4(0.2)	237(1)	1.1922^{20}	1.5550^{20}	i H_2O; msc EtOH, eth; s chl
2240	5-(4-Chlorophenyl)-6-ethyl-2,4-pyrimidinediamine	Pyrimethamine	$C_{12}H_{13}ClN_4$	58-14-0	248.711		233.5				
2241	2-(4-Chlorophenyl)-1*H*-indene-1,3(2*H*)-dione	Clorindione	$C_{15}H_9ClO_2$	1146-99-2	256.684	dk red nd (al)	145.5				vs bz, eth, EtOH
2242	4-Chlorophenyl isocyanate		C_7H_4ClNO	104-12-1	153.566		31.3	116^{45}			
2243	1-(2-Chlorophenyl)-2-methyl-2-propylamine	Clortermine	$C_{10}H_{14}ClN$	10389-73-8	183.678	liq		117^{16}			
2244	*N*-(2-Chlorophenyl)-3-oxo-butanamide		$C_{10}H_{10}ClNO_2$	93-70-9	211.645		106.5				s EtOH; i eth, lig
2245	(4-Chlorophenyl)-phenylmethanone		$C_{13}H_9ClO$	134-85-0	216.662	nd (al)	77.5	332			s EtOH, eth, ace; sl ctc
2246	3-(2-Chlorophenyl)propanoic acid		$C_9H_9ClO_2$	1643-28-3	184.619	nd or lf (w)	102				
2247	3-(3-Chlorophenyl)propanoic acid		$C_9H_9ClO_2$	21640-48-2	184.619	lf (peth)	77				
2248	3-(4-Chlorophenyl)propanoic acid		$C_9H_9ClO_2$	2019-34-3	184.619		126				
2249	3-Chloro-1-phenyl-1-propanone	2-Chloroethyl phenyl ketone	C_9H_9ClO	936-59-4	168.619	lf (eth), cry (al, peth)	49.5	113^{4}			
2250	1-(4-Chlorophenyl)-1-propanone		C_9H_9ClO	6285-05-8	168.619		37.3	135^{31}			i H_2O; s EtOH, CS_2; sl chl
2251	3-(3-Chlorophenyl)-2-propynoic acid		$C_9H_5ClO_2$	7396-28-3	180.588	cry (HOAc, bz-peth)	144.5				vs HOAc
2252	Chlorophenylsilane	Phenylchlorosilane	C_6H_7ClSi	4206-75-1	142.659			162.5	1.0683^{20}	1.5340^{20}	
2253	1-Chloro-4-(phenylsulfonyl)-benzene	Sulphenone	$C_{12}H_9ClO_2S$	80-00-2	252.716		94				i H_2O; sl EtOH; s eth; vs ace, bz
2254	5-Chloro-1-phenyltetrazole		$C_7H_5ClN_4$	14210-25-4	180.595		123				
2255	(2-Chlorophenyl)thiourea		$C_7H_7ClN_2S$	5344-82-1	186.662	nd or pl	143.3(0.5)				vs bz, EtOH

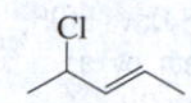
2216
4-Chloro-2-pentene

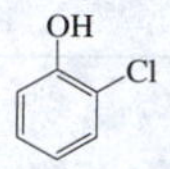
2217
2-Chlorophenol

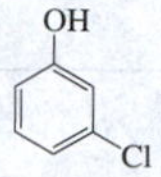
2218
3-Chlorophenol

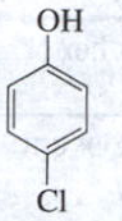
2219
4-Chlorophenol

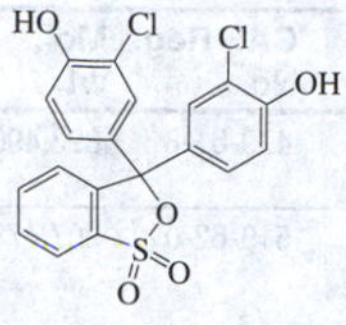
2220
Chlorophenol Red

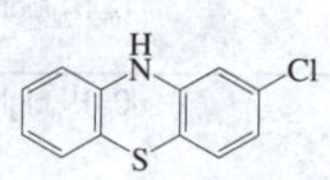
2221
2-Chloro-10*H*-phenothiazine

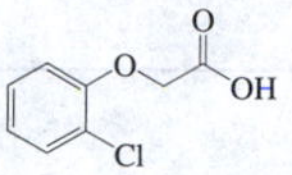
2222
2-Chlorophenoxyacetic acid

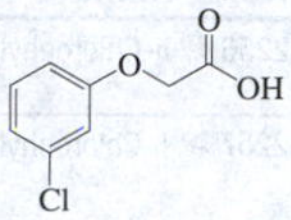
2223
3-Chlorophenoxyacetic acid

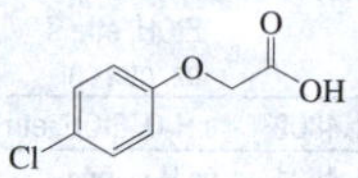
2224
(4-Chlorophenoxy)acetic acid

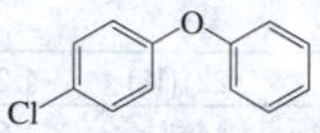
2225
1-Chloro-4-phenoxybenzene

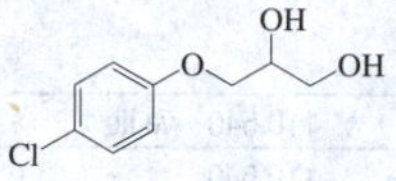
2226
3-(4-Chlorophenoxy)-1,2-propanediol

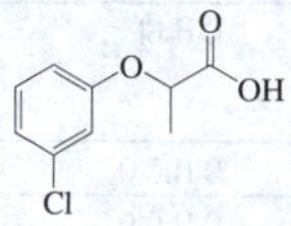
2227
2-(3-Chlorophenoxy)propanoic acid

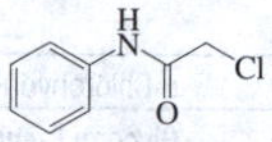
2228
2-Chloro-*N*-phenylacetamide

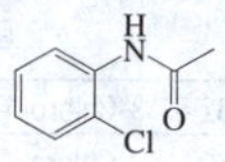
2229
N-(2-Chlorophenyl)acetamide

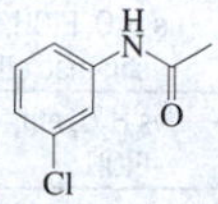
2230
N-(3-Chlorophenyl)acetamide

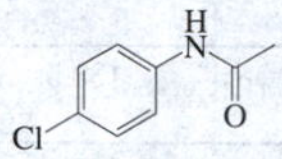
2231
N-(4-Chlorophenyl)acetamide

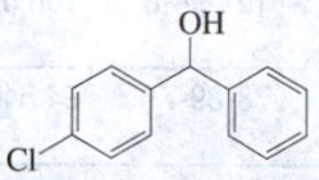
2232
4-Chloro-α-phenylbenzenemethanol

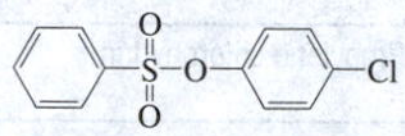
2233
4-Chlorophenyl benzenesulfonate

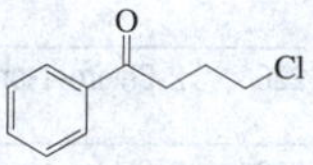
2234
4-Chloro-1-phenyl-1-butanone

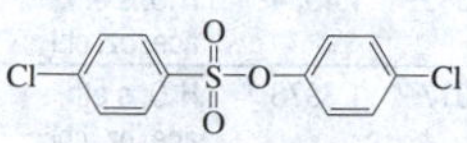
2235
4-Chlorophenyl 4-chlorobenzenesulfonate

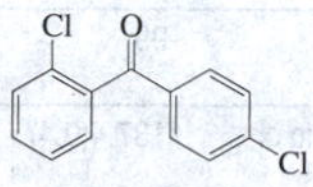
2236
(2-Chlorophenyl)(4-chlorophenyl)methanone

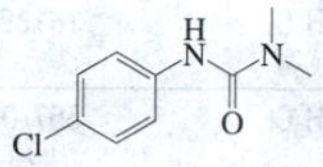
2237
N'-(4-Chlorophenyl)-*N*,*N*-dimethylurea

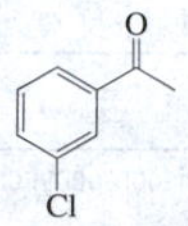
2238
1-(3-Chlorophenyl)ethanone

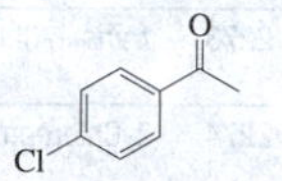
2239
1-(4-Chlorophenyl)ethanone

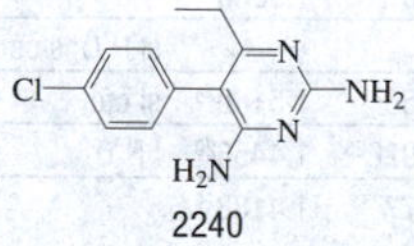
2240
5-(4-Chlorophenyl)-6-ethyl-2,4-pyrimidinediamine

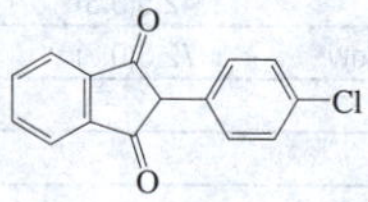
2241
2-(4-Chlorophenyl)-1*H*-indene-1,3(2*H*)-dione

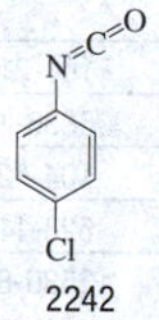
2242
4-Chlorophenyl isocyanate

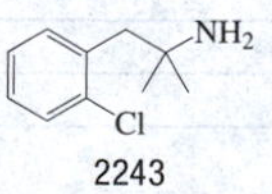
2243
1-(2-Chlorophenyl)-2-methyl-2-propylamine

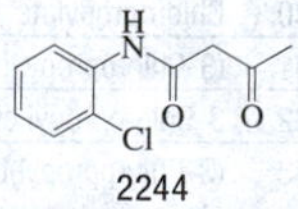
2244
N-(2-Chlorophenyl)-3-oxo-butanamide

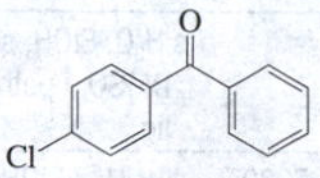
2245
(4-Chlorophenyl)phenylmethanone

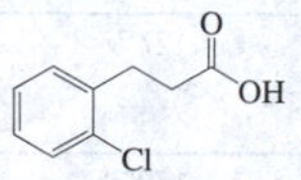
2246
3-(2-Chlorophenyl)propanoic acid

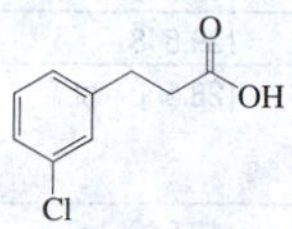
2247
3-(3-Chlorophenyl)propanoic acid

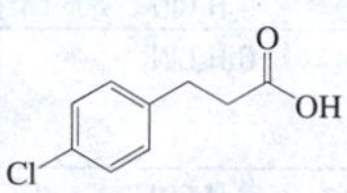
2248
3-(4-Chlorophenyl)propanoic acid

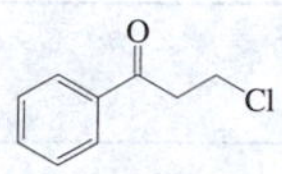
2249
3-Chloro-1-phenyl-1-propanone

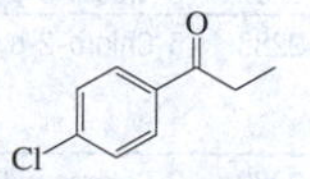
2250
1-(4-Chlorophenyl)-1-propanone

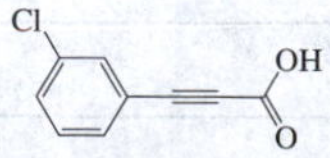
2251
3-(3-Chlorophenyl)-2-propynoic acid

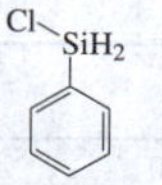
2252
Chlorophenylsilane

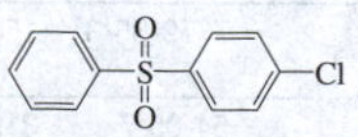
2253
1-Chloro-4-(phenylsulfonyl)benzene

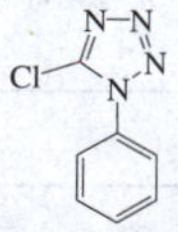
2254
5-Chloro-1-phenyltetrazole

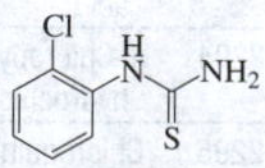
2255
(2-Chlorophenyl)thiourea

Organic

No.	Name	Synonym	Mol. form.	CAS Reg. No.	Mol. wt.	Physical form	mp/°C	bp/°C	den/ g cm^{-3}	n_D	Solubility
2256	α-Chlorophyll		$C_{55}H_{72}MgN_4O_5$	479-61-8	893.490	bl blk hex pl	152.3				i H_2O; vs EtOH, eth; s lig
2257	β-Chlorophyll		$C_{55}H_{70}MgN_4O_6$	519-62-0	907.473	bl-blk or grn pow	125				i H_2O; vs EtOH, eth, py; s MeOH
2258	Chloropropamide	4-Chloro-*N*-[(propylamino)-carbonyl]benzenesulfonamide	$C_{10}H_{13}ClN_2O_3S$	94-20-2	276.739	cry (EtOH)	128				i H_2O; s EtOH; sl eth, bz
2259	2-Chloropropanal		C_3H_5ClO	683-50-1	92.524			86	1.182^{15}	1.431^{17}	vs bz, eth
2260	1-Chloropropane	Propyl chloride	C_3H_7Cl	540-54-5	78.541	liq	-122.9(0.7)	46.2(0.5)	0.8899^{20}	1.3879^{20}	sl H_2O, ctc; msc EtOH, eth; s bz, chl
2261	2-Chloropropane	Isopropyl chloride	C_3H_7Cl	75-29-6	78.541	liq	-117.1(0.2)	35.0(0.6)	0.8617^{20}	1.3777^{20}	sl H_2O; msc EtOH, eth; s bz, ctc, chl
2262	3-Chloro-1,2-propanediol	α-Chlorohydrin	$C_3H_7ClO_2$	96-24-2	110.540	ye liq		221(18)	1.325^{18}	1.4809^{20}	s H_2O, EtOH, eth
2263	2-Chloro-1,3-propanediol	Glycerol β-chlorohydrin	$C_3H_7ClO_2$	497-04-1	110.540			146^{18}	1.3219^{20}	1.4831^{20}	vs H_2O, ace, EtOH
2264	3-Chloro-1,2-propanediol dinitrate	Clonitrate	$C_3H_5ClN_2O_6$	2612-33-1	200.534	sl ye liq		192.5	1.5112^{9}		vs ace, EtOH, chl
2265	3-Chloropropanenitrile	β-Chloropropionitrile	C_3H_4ClN	542-76-7	89.524	liq	-51.4(0.2)	175.5	1.1573^{20}	1.4360^{20}	sl ctc
2266	2-Chloropropanoic acid	2-Chloropropionic acid	$C_3H_5ClO_2$	598-78-7	108.524			185	1.2585^{20}	1.4380^{20}	msc H_2O, EtOH, eth; s ace
2267	3-Chloropropanoic acid	3-Chloropropionic acid	$C_3H_5ClO_2$	107-94-8	108.524	lf (w), hyg cry (lig)	41(4)	204 dec			s H_2O, EtOH, chl; msc eth
2268	2-Chloro-1-propanol	Propylene chlorohydrin	C_3H_7ClO	78-89-7	94.540			133.5	1.103^{20}	1.4390^{20}	vs H_2O, eth, EtOH
2269	3-Chloro-1-propanol		C_3H_7ClO	627-30-5	94.540			149(8)	1.1309^{20}	1.4459^{20}	vs H_2O; s EtOH, eth; sl ctc
2270	1-Chloro-2-propanol	*sec*-Propylene chlorohydrin	C_3H_7ClO	127-00-4	94.540			124.4(0.2)	1.113^{20}	1.4392^{20}	msc H_2O, EtOH, eth; sl ctc
2271	3-Chloropropanoyl chloride		$C_3H_4Cl_2O$	625-36-5	126.969			144	1.3307^{13}	1.4549^{20}	sl H_2O; vs EtOH, eth, chl
2272	*cis*-1-Chloropropene		C_3H_5Cl	16136-84-8	76.525	liq	-134.8	32(2)	0.9347^{20}	1.4055^{20}	i H_2O; s eth, ace, bz, chl
2273	*trans*-1-Chloropropene		C_3H_5Cl	16136-85-9	76.525	liq	-99	37(2)	0.9349^{20}	1.4054^{20}	i H_2O; s eth, ace, bz, chl
2274	2-Chloropropene	Isopropenyl chloride	C_3H_5Cl	557-98-2	76.525	vol liq or gas	-137.4(0.4)	23(2)	0.9017^{20}	1.3973^{20}	i H_2O; s eth, ace, bz, chl
2275	3-Chloropropene	Allyl chloride	C_3H_5Cl	107-05-1	76.525	liq	-136(2)	44.8(0.4)	0.9376^{20}	1.4157^{20}	i H_2O; msc EtOH, eth, ace, bz, lig; sl ctc
2276	2-Chloro-2-propenenitrile		C_3H_2ClN	920-37-6	87.508	liq	-65	88.5	1.096^{25}	1.4290^{20}	
2277	2-Chloropropenoic acid	2-Chloroacrylic acid	$C_3H_3ClO_2$	598-79-8	106.508		66	sub			
2278	*trans*-(3-Chloro-1-propenyl)-benzene		C_9H_9Cl	21087-29-6	152.620		8.5	106^{13}	1.0926^{20}	1.5851^{20}	vs ace, bz, eth, EtOH
2279	Chloropropham		$C_{10}H_{12}ClNO_2$	101-21-3	213.661		42.5(0.3)	149^{2}	1.18^{30}	1.5388^{20}	
2280	Chloropropylate		$C_{17}H_{16}Cl_2O_3$	5836-10-2	339.213	pow	72.3(0.3)				sl H_2O; s os
2281	(3-Chloropropyl)benzene		$C_9H_{11}Cl$	104-52-9	154.636			216(14)	1.056^{21}	1.5160^{25}	sl ctc
2282	3-Chloropropyl chloroformate		$C_4H_6Cl_2O_2$	628-11-5	156.996			177	1.2926^{25}	1.4456^{20}	i H_2O
2283	(3-Chloropropyl)trimethoxysilane		$C_6H_{15}ClO_3Si$	2530-87-2	198.720			91	1.077^{25}	1.4183^{25}	
2284	(3-Chloropropyl)trimethylsilane		$C_6H_{15}ClSi$	2344-83-4	150.722			151	0.8789^{20}	1.4319^{20}	
2285	3-Chloro-1-propyne	Propargyl chloride	C_3H_3Cl	624-65-7	74.509		-78	56(4)	1.030^{25}	1.4349^{20}	i H_2O; msc EtOH, eth, bz; s ctc
2286	6-Chloro-1*H*-purine	6-Chloropurine	$C_5H_3ClN_4$	87-42-3	154.558	nd (w)	176 dec				
2287	6-Chloro-3-pyridazinamine		$C_4H_4ClN_3$	5469-69-2	129.548		220				
2288	5-Chloro-2-pyridinamine		$C_5H_5ClN_2$	1072-98-6	128.560	pl	137	127^{11}			s H_2O, EtOH; sl DMSO; i peth, lig
2289	2-Chloropyridine		C_5H_4ClN	109-09-1	113.546	oil		170(2)	1.205^{15}	1.5320^{20}	sl H_2O; s EtOH, eth
2290	3-Chloropyridine		C_5H_4ClN	626-60-8	113.546			151(20)		1.5304^{20}	sl H_2O
2291	4-Chloropyridine		C_5H_4ClN	626-61-9	113.546	liq	-43.5	151(20)	1.2000^{25}		s H_2O; msc EtOH
2292	2-Chloro-3-pyridinecarboxylic acid		$C_6H_4ClNO_2$	2942-59-8	157.555		>175 dec				
2293	6-Chloro-3-pyridinecarboxylic acid		$C_6H_4ClNO_2$	5326-23-8	157.555		198 dec				
2294	4-Chloropyridine, hydrochloride		$C_5H_5Cl_2N$	7379-35-3	150.006			210 sub			
2295	Chloroquine		$C_{18}H_{26}ClN_3$	54-05-7	319.872		90				

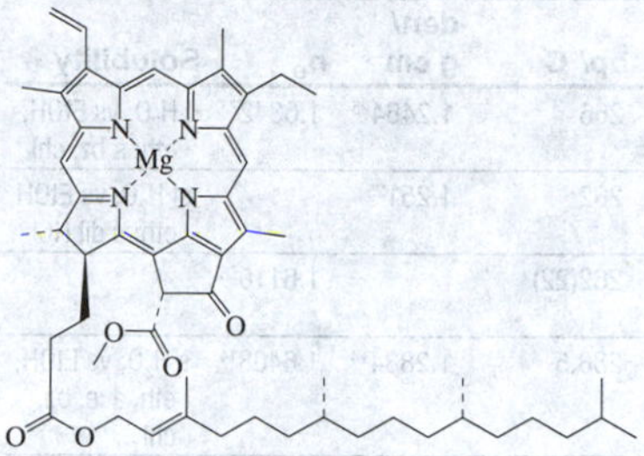
2256
α-Chlorophyll

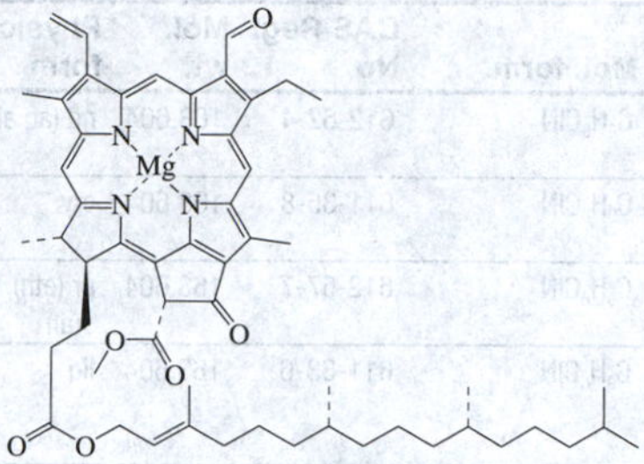
2257
β-Chlorophyll

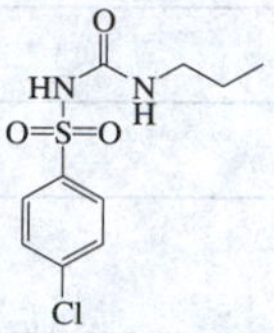
2258
Chloropropamide

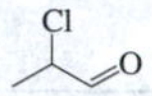
2259
2-Chloropropanal

2260
1-Chloropropane

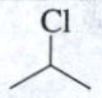
2261
2-Chloropropane

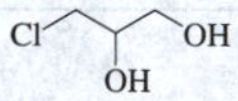
2262
3-Chloro-1,2-propanediol

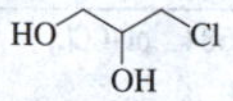
2263
2-Chloro-1,3-propanediol

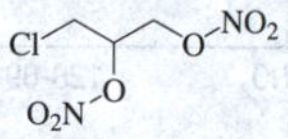
2264
3-Chloro-1,2-propanediol dinitrate

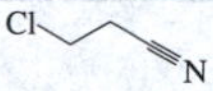
2265
3-Chloropropanenitrile

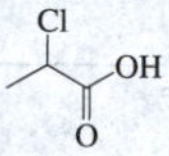
2266
2-Chloropropanoic acid

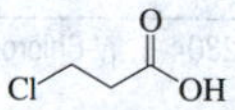
2267
3-Chloropropanoic acid

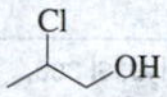
2268
2-Chloro-1-propanol

2269
3-Chloro-1-propanol

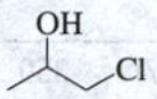
2270
1-Chloro-2-propanol

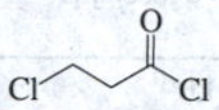
2271
3-Chloropropanoyl chloride

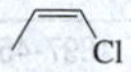
2272
cis-1-Chloropropene

2273
trans-1-Chloropropene

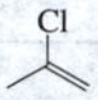
2274
2-Chloropropene

2275
3-Chloropropene

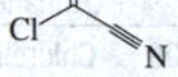
2276
2-Chloro-2-propenenitrile

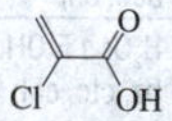
2277
2-Chloropropenoic acid

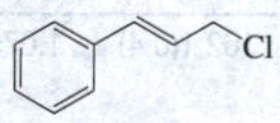
2278
trans-(3-Chloro-1-propenyl)benzene

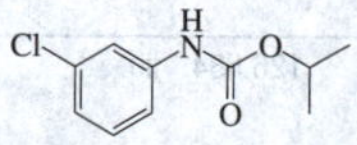
2279
Chloropropham

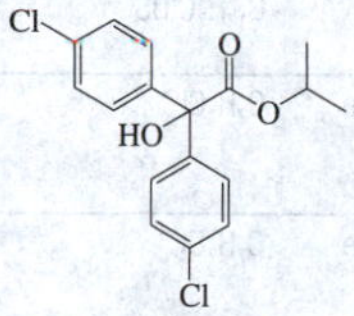
2280
Chloropropylate

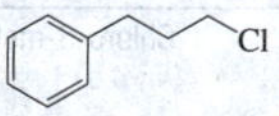
2281
(3-Chloropropyl)benzene

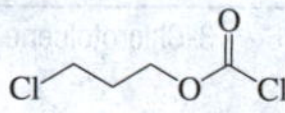
2282
3-Chloropropyl chloroformate

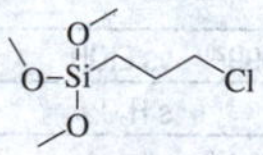
2283
(3-Chloropropyl)trimethoxysilane

2284
(3-Chloropropyl)trimethylsilane

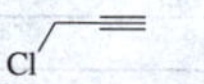
2285
3-Chloro-1-propyne

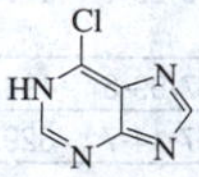
2286
6-Chloro-1*H*-purine

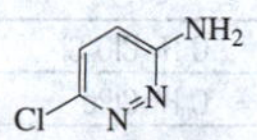
2287
6-Chloro-3-pyridazinamine

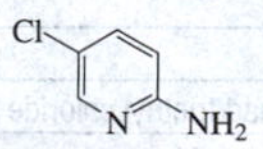
2288
5-Chloro-2-pyridinamine

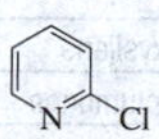
2289
2-Chloropyridine

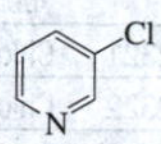
2290
3-Chloropyridine

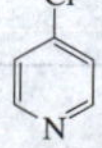
2291
4-Chloropyridine

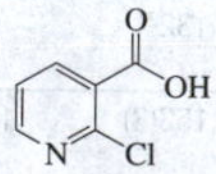
2292
2-Chloro-3-pyridinecarboxylic acid

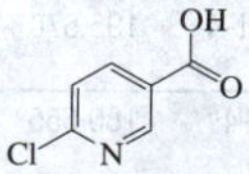
2293
6-Chloro-3-pyridinecarboxylic acid

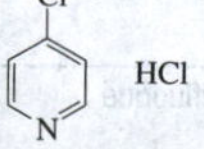
2294
4-Chloropyridine, hydrochloride

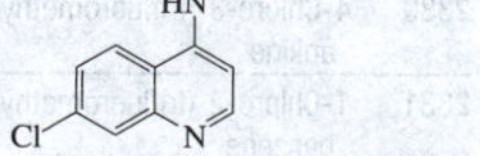
2295
Chloroquine

No.	Name	Synonym	Mol. form.	CAS Reg. No.	Mol. wt.	Physical form	mp/°C	bp/°C	den/ g cm^{-3}	n_D	Solubility
2296	2-Chloroquinoline		C_9H_6ClN	612-62-4	163.604	nd (aq al)	38.4(0.5)	266	1.2464^{25}	1.6342^{25}	i H_2O; vs EtOH, eth; s bz, chl
2297	4-Chloroquinoline		C_9H_6ClN	611-35-8	163.604	cry	34.9(0.5)	262	1.251^{25}		sl H_2O; vs EtOH, eth; s dil HCl
2298	6-Chloroquinoline		C_9H_6ClN	612-57-7	163.604	pr (eth), nd (al)	45.9(0.5)	262(22)		1.6110^{56}	
2299	8-Chloroquinoline		C_9H_6ClN	611-33-6	163.604	liq	-20	288.5	1.2834^{14}	1.6408^{14}	s H_2O; vs EtOH, eth, ace, bz, chl
2300	5-Chloro-8-quinolinol	Cloxyquin	C_9H_6ClNO	130-16-5	179.603	cry (al)	130				
2301	2-Chlorostyrene		C_8H_7Cl	2039-87-4	138.595	liq	-63.1	188.6(1)	1.1000^{20}	1.5649^{20}	s EtOH, eth, ace, ctc, HOAc; msc peth
2302	3-Chlorostyrene		C_8H_7Cl	2039-85-2	138.595			63^{6}	1.1033^{20}	1.5625^{20}	i H_2O; s EtOH, eth
2303	4-Chlorostyrene		C_8H_7Cl	1073-67-2	138.595		15.9	191(2)	1.0868^{20}	1.5660^{20}	i H_2O; s EtOH, eth; msc ace, bz, ctc
2304	*N*-Chlorosuccinimide		$C_4H_4ClNO_2$	128-09-6	133.534	pl (CCl_4)	150		1.65^{25}		sl H_2O, EtOH, bz, lig; s ace, HOAc
2305	1-Chlorotetradecane		$C_{14}H_{29}Cl$	2425-54-9	232.833		4.9	296(4)	0.8654^{20}	1.4474^{20}	i H_2O; s EtOH, chl; vs ace, bz; sl ctc
2306	6-Chloro-*N,N,N′,N′*-tetraethyl-1,3,5-triazine-2,4-diamine		$C_{11}H_{20}ClN_5$	580-48-3	257.764	oily liq	27	155^{9}	1.0956^{20}	1.5320^{20}	vs bz, chl, EtOH, lig
2307	1-Chloro-1,1,2,2-tetrafluoroethane		C_2HClF_4	354-25-6	136.476	col gas	-117	-13(6)			
2308	1-Chloro-1,2,2,2-tetrafluoroethane	HCFC-124	C_2HClF_4	2837-89-0	136.476	col gas	-199.15	-11.96(0.09)			
2309	Chlorothalonil		$C_8Cl_4N_2$	1897-45-6	265.911		253.1(0.7)	350	1.7^{25}		i H_2O; sl ace, cyhex
2310	Chlorothen	Chloromethapyrilene	$C_{14}H_{18}ClN_3S$	148-65-2	295.831			155^{10}	1.1751^{25}		
2311	Chlorothiazide		$C_7H_6ClN_3O_4S_2$	58-94-6	295.724		350 dec				
2312	2-Chlorothiophene	2-Thienyl chloride	C_4H_3ClS	96-43-5	118.585	liq	-71.85(0.05)	126(2)	1.2863^{20}	1.5487^{20}	i H_2O; msc EtOH, eth; sl chl
2313	5-Chloro-2-thiophenecarboxaldehyde		C_5H_3ClOS	7283-96-7	146.595			77.5^{5}		1.6036^{25}	sl chl
2314	2-Chloro-9*H*-thioxanthen-9-one		$C_{13}H_7ClOS$	86-39-5	246.712		153.5				
2315	2-Chlorotoluene	1-Chloro-2-methylbenzene	C_7H_7Cl	95-49-8	126.584	liq	-35.9(0.7)	158.8(0.4)	1.0825^{20}	1.5268^{20}	i H_2O; s EtOH, bz; msc eth, ace, chl
2316	3-Chlorotoluene	1-Chloro-3-methylbenzene	C_7H_7Cl	108-41-8	126.584	liq	-47.8	162.1(0.4)	1.075^{20}	1.5214^{19}	i H_2O; s EtOH, bz, ctc, chl; msc eth
2317	4-Chlorotoluene	1-Chloro-4-methylbenzene	C_7H_7Cl	106-43-4	126.584	liq	7.4(0.2)	161.8(0.2)	1.0697^{20}	1.5150^{20}	i H_2O; s EtOH, ctc, chl; msc eth
2318	6-Chloro-1,3,5-triazine-2,4-diamine		$C_3H_4ClN_5$	3397-62-4	145.551		>330				
2319	1-Chloro-2-(trichloromethyl)-benzene		$C_7H_4Cl_4$	2136-89-2	229.919		29.4(0.2)	263(3)	1.5187^{20}	1.5836^{20}	i H_2O; s eth, ace; sl ctc
2320	1-Chloro-4-(trichloromethyl)-benzene		$C_7H_4Cl_4$	5216-25-1	229.919			245	1.4463^{20}		vs ace, eth
2321	Chlorotriethoxysilane		$C_6H_{15}ClO_3Si$	4667-99-6	198.720	liq	-51	156	1.030^{20}	1.3999^{20}	vs EtOH
2322	Chlorotriethylplumbane	Lead triethyl chloride	$C_6H_{15}ClPb$	1067-14-7	329.8		123 dec				s H_2O
2323	Chlorotriethylsilane		$C_6H_{15}ClSi$	994-30-9	150.722			146(3)	0.8967^{20}	1.4314^{20}	
2324	1-Chloro-1,1,2-trifluoroethane		$C_2H_2ClF_3$	421-04-5	118.485	vol liq or gas		16(19)			
2325	1-Chloro-1,2,2-trifluoroethane		$C_2H_2ClF_3$	431-07-2	118.485	vol liq or gas		17.3			
2326	2-Chloro-1,1,1-trifluoroethane		$C_2H_2ClF_3$	75-88-7	118.485	col gas	-105.5	6.0(0.6)	1.389^{0}	1.3090^{0}	
2327	Chlorotrifluoroethene	Chlorotrifluoroethylene	C_2ClF_3	79-38-9	116.469	col gas	-158.14(0.05)	-28.3(0.3)	1.54^{-60}	1.38^{0}	s bz, chl
2328	Chlorotrifluoromethane	Refrigerant 13	$CClF_3$	75-72-9	104.459	col gas	-181.2	-81.37			i H_2O
2329	2-Chloro-5-(trifluoromethyl)-aniline		$C_7H_5ClF_3N$	121-50-6	195.570			103^{25}	1.428^{25}	1.4975^{20}	
2330	4-Chloro-3-(trifluoromethyl)-aniline		$C_7H_5ClF_3N$	320-51-4	195.570		36.5	132^{27}			
2331	1-Chloro-2-(trifluoromethyl)-benzene	*o*-Chlorobenzotrifluoride	$C_7H_4ClF_3$	88-16-4	180.555	liq	-6(2)	153(3)	1.2540^{30}	1.4513^{25}	s chl

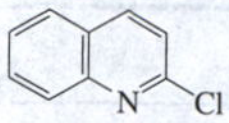
2296
2-Chloroquinoline

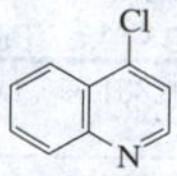
2297
4-Chloroquinoline

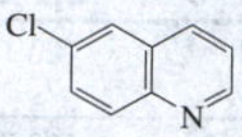
2298
6-Chloroquinoline

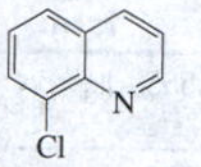
2299
8-Chloroquinoline

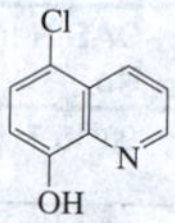
2300
5-Chloro-8-quinolinol

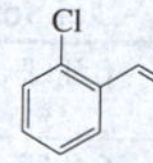
2301
2-Chlorostyrene

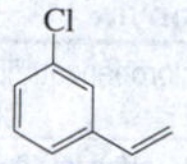
2302
3-Chlorostyrene

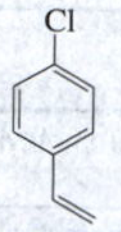
2303
4-Chlorostyrene

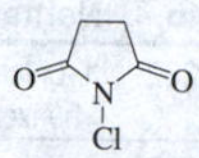
2304
N-Chlorosuccinimide

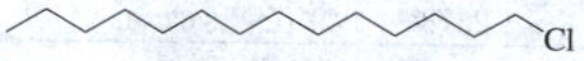
2305
1-Chlorotetradecane

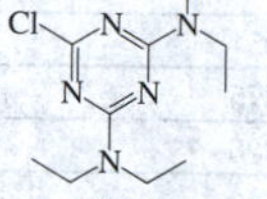
2306
6-Chloro-*N,N,N',N'*-tetraethyl-1,3,5-triazine-2,4-diamine

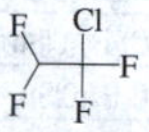
2307
1-Chloro-1,1,2,2-tetrafluoroethane

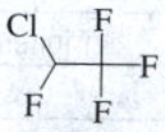
2308
1-Chloro-1,2,2,2-tetrafluoroethane

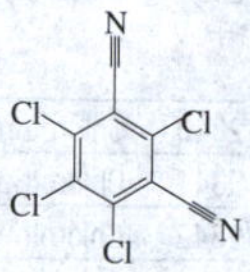
2309
Chlorothalonil

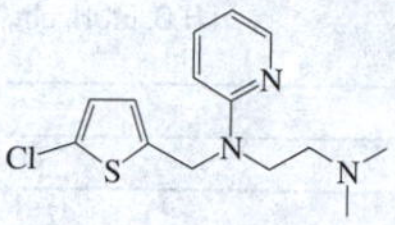
2310
Chlorothen

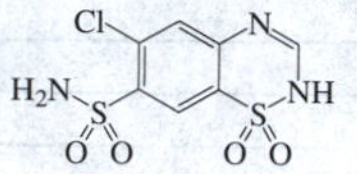
2311
Chlorothiazide

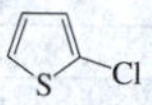
2312
2-Chlorothiophene

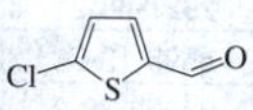
2313
5-Chloro-2-thiophenecarboxaldehyde

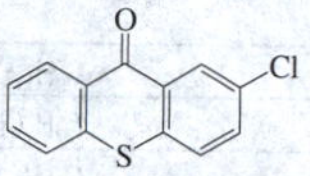
2314
2-Chloro-9*H*-thioxanthen-9-one

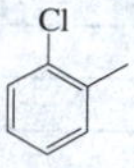
2315
2-Chlorotoluene

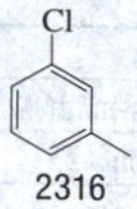
2316
3-Chlorotoluene

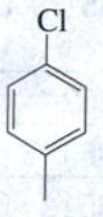
2317
4-Chlorotoluene

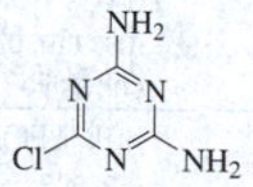
2318
6-Chloro-1,3,5-triazine-2,4-diamine

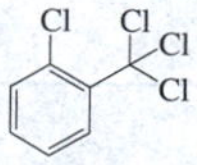
2319
1-Chloro-2-(trichloromethyl)benzene

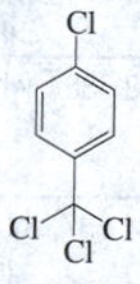
2320
1-Chloro-4-(trichloromethyl)benzene

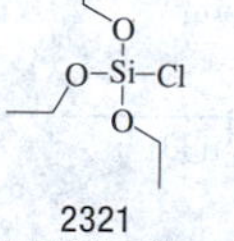
2321
Chlorotriethoxysilane

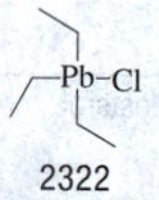
2322
Chlorotriethylplumbane

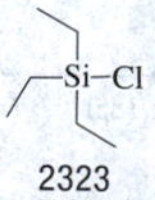
2323
Chlorotriethylsilane

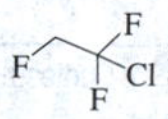
2324
1-Chloro-1,1,2-trifluoroethane

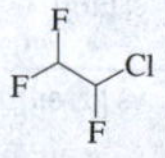
2325
1-Chloro-1,2,2-trifluoroethane

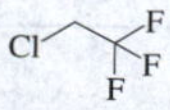
2326
2-Chloro-1,1,1-trifluoroethane

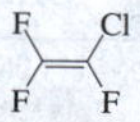
2327
Chlorotrifluoroethene

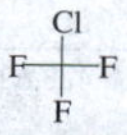
2328
Chlorotrifluoromethane

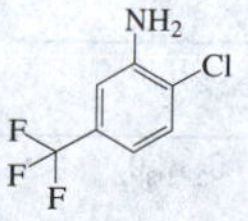
2329
2-Chloro-5-(trifluoromethyl)aniline

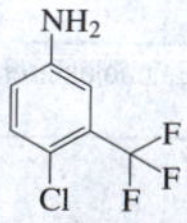
2330
4-Chloro-3-(trifluoromethyl)aniline

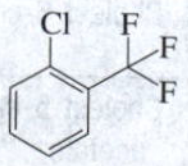
2331
1-Chloro-2-(trifluoromethyl)benzene

No.	Name	Synonym	Mol. form.	CAS Reg. No.	Mol. wt.	Physical form	mp/°C	bp/°C	den/g cm^{-3}	n_D	Solubility
2332	1-Chloro-3-(trifluoromethyl)-benzene	*m*-Chlorobenzotrifluoride	$C_7H_4ClF_3$	98-15-7	180.555	liq	-56	126(10)	1.3311^{25}	1.4438^{25}	
2333	1-Chloro-4-(trifluoromethyl)-benzene	*p*-Chlorobenzotrifluoride	$C_7H_4ClF_3$	98-56-6	180.555	liq	-33	138.5	1.3340^{25}	1.4431^{30}	
2334	3-Chloro-1,1,1-trifluoropropane		$C_3H_4ClF_3$	460-35-5	132.512	liq	-93.73(0.06)	45.1	1.3253^{20}	1.3350^{20}	i H_2O
2335	2-Chloro-2,4,4-trimethylpentane		$C_8H_{17}Cl$	6111-88-2	148.674		-26	150(8)	0.8746^{20}	1.4308^{20}	vs EtOH
2336	Chlorotrimethylstannane		C_3H_9ClSn	1066-45-1	199.266		38.5	148			s H_2O, chl, os
2337	2-Chloro-1,3,5-trinitrobenzene	Picryl chloride	$C_6H_2ClN_3O_6$	88-88-0	247.549	wh nd or pl (chl, al-lig)	82(1)		1.797^{20}		i H_2O; s EtOH, bz; sl eth; vs ace, tol
2338	Chlorotrinitromethane		$CClN_3O_6$	1943-16-4	185.480		5.8(0.2)	156(18)	1.6769^{20}	1.4500^{20}	vs eth, EtOH, chl
2339	Chlorotriphenylmethane		$C_{19}H_{15}Cl$	76-83-5	278.775	nd or pr (bz-peth)	109.2(0.5)	310			i H_2O; sl EtOH; vs eth, bz, chl; s ace
2340	Chlorotriphenylsilane		$C_{18}H_{15}ClSi$	76-86-8	294.851			241^{35}			
2341	Chlorotriphenylstannane	Triphenyltin chloride	$C_{18}H_{15}ClSn$	639-58-7	385.475		103.5				s chl
2342	Chlorotripropylstannane		$C_9H_{21}ClSn$	2279-76-7	283.426		-23.5	123^{13}	1.2678^{28}	1.49102^{28}	s ctc, os
2343	Chlorovinyldimethylsilane		C_4H_9ClSi	1719-58-0	120.653			82.3(0.4)	0.8744^{20}	1.4141^{20}	
2344	Chloroxuron		$C_{15}H_{15}ClN_2O_2$	1982-47-4	290.745		152.0(0.9)				
2345	Chlorozotocin		$C_9H_{16}ClN_3O_7$	54749-90-5	313.692	cry	147 dec				s H_2O
2346	Chlorphenesin carbamate		$C_{10}H_{12}ClNO_4$	886-74-8	245.660	cry (bz)	90				vs ace, EtOH, diox
2347	Chlorpheniramine		$C_{16}H_{19}ClN_2$	132-22-9	274.788	oily liq		142^{1}			
2348	Chlorpheniramine maleate	Chloroprophenpyridamine	$C_{20}H_{23}ClN_2O_4$	113-92-8	390.861		132.5				
2349	Chlorphentermine	2-(4-Chlorobenzyl)-2-propylamine	$C_{10}H_{14}ClN$	461-78-9	183.678	liq		231			
2350	Chlorpromazine	2-Chloro-*N,N*-dimethyl-10*H*-phenothiazine-10-propanamine	$C_{17}H_{19}ClN_2S$	50-53-3	318.864			$202^{0.8}$			i H_2O; vs EtOH, eth, bz, chl; s dil HCl
2351	Chlorprothixene		$C_{18}H_{18}ClNS$	113-59-7	315.861	pale ye cry	97.1(0.3)				i H_2O, EtOH, eth, chl
2352	Chlorpyrifos		$C_9H_{11}Cl_3NO_3PS$	2921-88-2	350.586		43(1)				
2353	Chlorpyrifos-methyl		$C_7H_7Cl_3NO_3PS$	5598-13-0	322.534		46.0(0.5)				
2354	Chlorsulfuron		$C_{12}H_{12}ClN_5O_4S$	64902-72-3	357.773		176				
2355	Chlortetracycline		$C_{22}H_{23}ClN_2O_8$	57-62-5	478.879	gold-ye	168.5				i H_2O, eth; sl EtOH, ace, bz; s diox
2356	Chlorthalidone		$C_{14}H_{11}ClN_2O_4S$	77-36-1	338.766	wh pow or cry	225 dec				s alk, EtOH; sl eth
2357	Chlorthion		$C_8H_9ClNO_5PS$	500-28-7	297.653	ye cry	21	$125^{0.1}$	1.437^{20}	1.5661^{20}	i H_2O; vs bz, eth, EtOH
2358	Chlorthiophos		$C_{11}H_{15}Cl_2O_3PS_2$	21923-23-9	361.245			$150^{0.001}$			
2359	Chlortoluron	*N'*-(3-Chloro-4-methylphenyl)-*N,N*-dimethylurea	$C_{10}H_{13}ClN_2O$	15545-48-9	212.675	cry	147				sl H_2O; s os
2360	Cholane		$C_{24}H_{42}$	548-98-1	330.590	pr (al)	90	$190^{0.001}$			
2361	Cholan-24-oic acid	Cholanic acid	$C_{24}H_{40}O_2$	25312-65-6	360.574	nd (al), cry (HOAc)	163.5				s EtOH, chl, HOAc
2362	Cholesta-3,5-diene		$C_{27}H_{44}$	747-90-0	368.638	wh nd (al)	80	260^{13}	0.925^{100}		i H_2O; s EtOH; msc eth, bz, chl; vs lig
2363	Cholesta-5,7-dien-3-ol, (3β)	7-Dehydrocholesterol	$C_{27}H_{44}O$	434-16-2	384.637	pl (+1w), (eth-MeOH)	150.5				i H_2O; sl EtOH; s eth, ace
2364	Cholesta-8,24-dien-3-ol, (3β,5α)		$C_{27}H_{44}O$	128-33-6	384.637	pl (MeOH),nd	110	$160^{0.001}$			s ace, chl, MeOH
2365	Cholestane, (5α)	28,29,30-Trinorlanostane	$C_{27}H_{48}$	481-21-0	372.670	sc or pl (eth-al, ace)	78.6(0.5)	250^{1}	0.9090^{88}	1.4887^{88}	i H_2O; sl EtOH; vs eth, bz, chl
2366	Cholestane, (5β)	Coprostane	$C_{27}H_{48}$	481-20-9	372.670	orth nd (al, ace)	72		0.9119^{87}	1.4884^{88}	vs eth, chl
2367	Cholestanol	Dihydrocholesterol	$C_{27}H_{48}O$	80-97-7	388.669	sc (al,+1w)	141.5				vs eth, chl
2368	Cholestan-3-ol, (3α,5α)	Epicholestanol	$C_{27}H_{48}O$	516-95-0	388.669	nd (al)	185.5				s chl
2369	Cholest-4-en-3-ol, (3β)	Allocholesterol	$C_{27}H_{46}O$	517-10-2	386.653	nd (eth-MeOH)	132				i H_2O; s EtOH; vs eth, ace, bz, chl
2370	Cholest-5-en-3-ol, (3α)	Epicholesterol	$C_{27}H_{46}O$	474-77-1	386.653	cry (al, chl-MeOH)	141.5				sl EtOH
2371	Cholest-5-en-3-ol (3β), acetate		$C_{29}H_{48}O_2$	604-35-3	428.690	wh nd (ace, al)	114.6(0.5)				vs bz, eth, chl

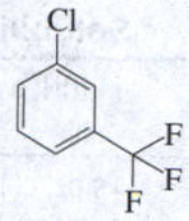
2332
1-Chloro-3-(trifluoromethyl)benzene

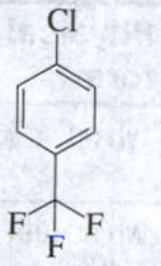
2333
1-Chloro-4-(trifluoromethyl)benzene

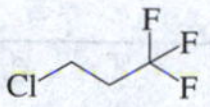
2334
3-Chloro-1,1,1-trifluoropropane

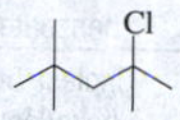
2335
2-Chloro-2,4,4-trimethylpentane

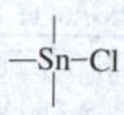
2336
Chlorotrimethylstannane

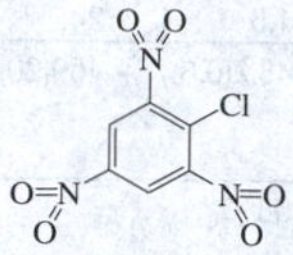
2337
2-Chloro-1,3,5-trinitrobenzene

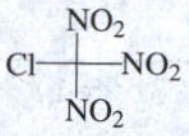
2338
Chlorotrinitromethane

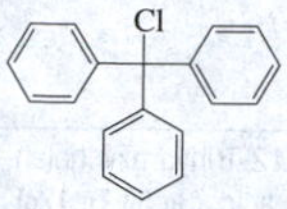
2339
Chlorotriphenylmethane

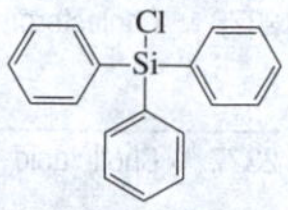
2340
Chlorotriphenylsilane

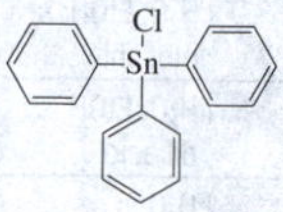
2341
Chlorotriphenylstannane

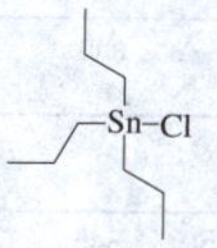
2342
Chlorotripropylstannane

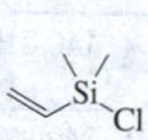
2343
Chlorovinyldimethylsilane

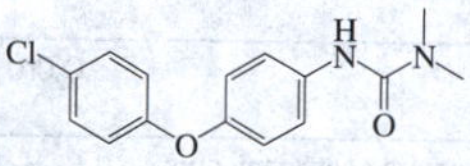
2344
Chloroxuron

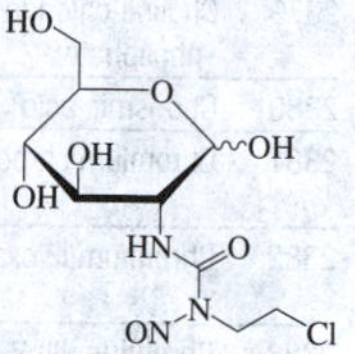
2345
Chlorozotocin

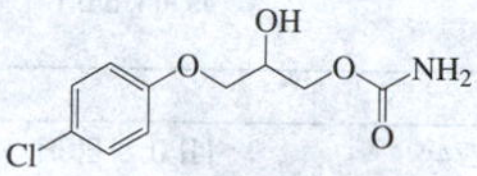
2346
Chlorphenesin carbamate

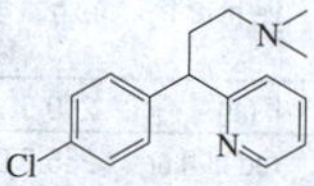
2347
Chlorpheniramine

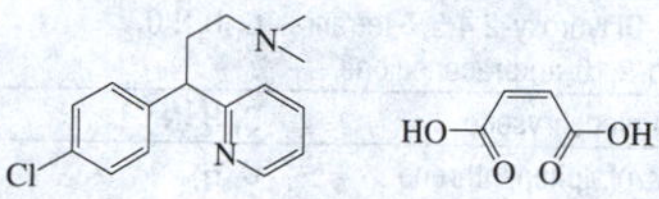
2348
Chlorpheniramine maleate

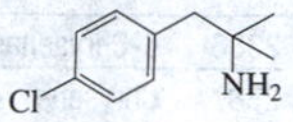
2349
Chlorphentermine

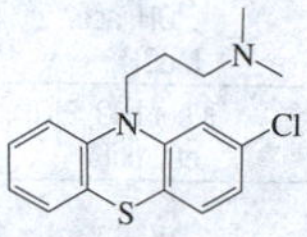
2350
Chlorpromazine

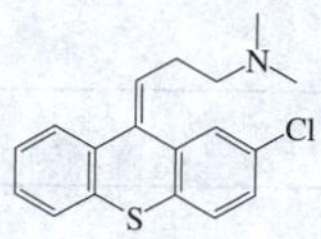
2351
Chlorprothixene

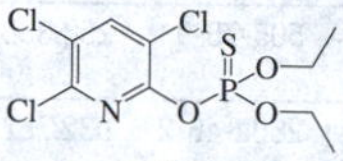
2352
Chlorpyrifos

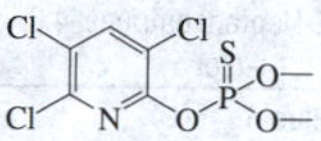
2353
Chlorpyrifos-methyl

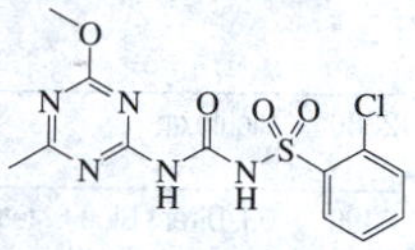
2354
Chlorsulfuron

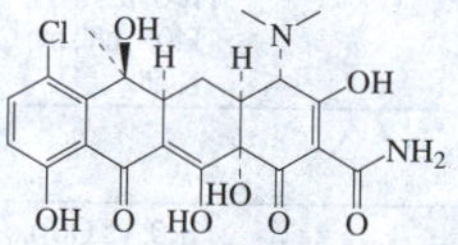
2355
Chlortetracycline

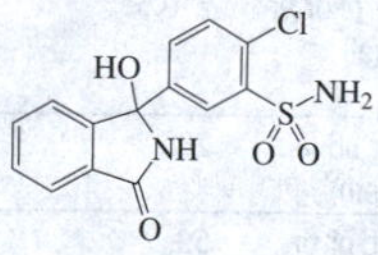
2356
Chlorthalidone

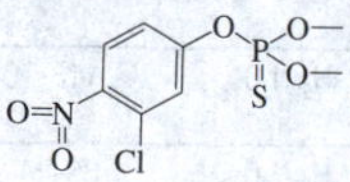
2357
Chlorthion

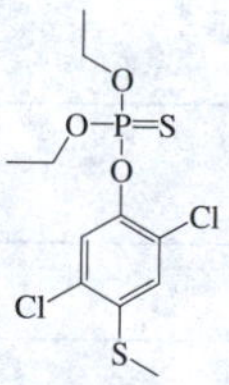
2358
Chlorthiophos

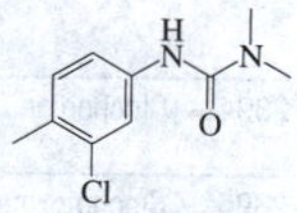
2359
Chlortoluron

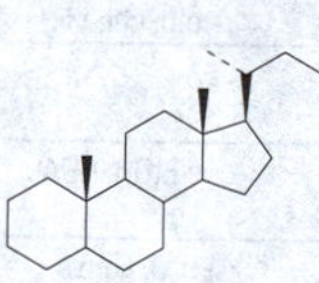
2360
Cholane

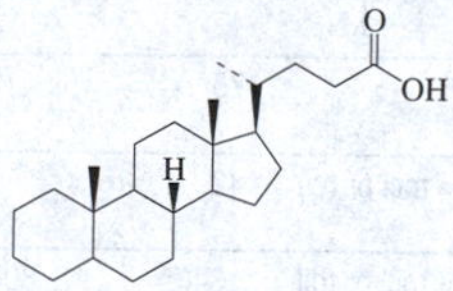
2361
Cholan-24-oic acid

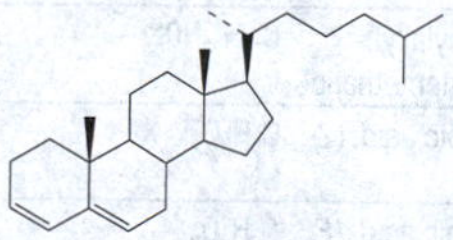
2362
Cholesta-3,5-diene

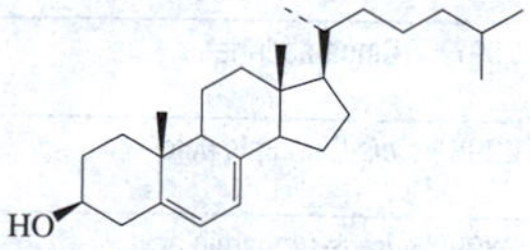
2363
Cholesta-5,7-dien-3-ol, (3β)

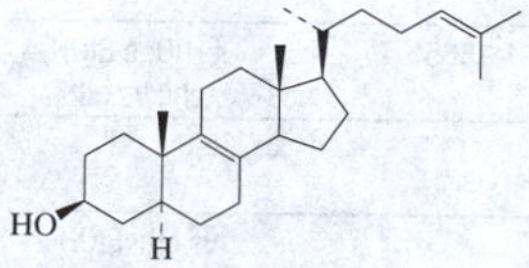
2364
Cholesta-8,24-dien-3-ol, (3β,5α)

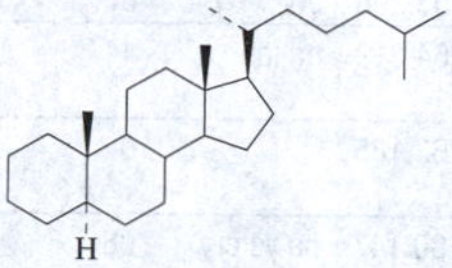
2365
Cholestane, (5α)

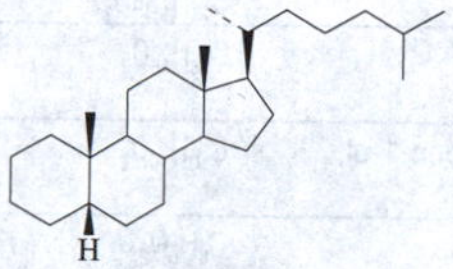
2366
Cholestane, (5β)

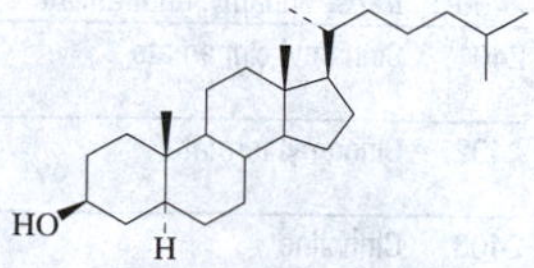
2367
Cholestanol

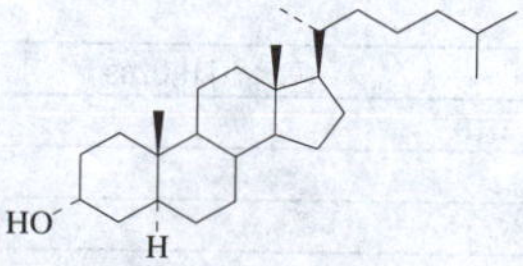
2368
Cholestan-3-ol, (3α,5α)

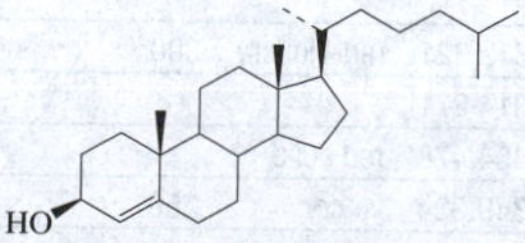
2369
Cholest-4-en-3-ol, (3β)

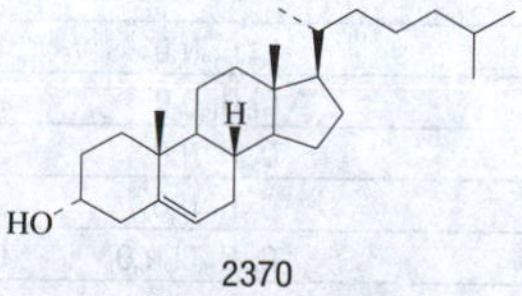
2370
Cholest-5-en-3-ol, (3α)

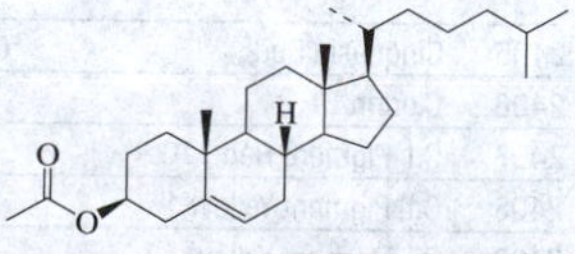
2371
Cholest-5-en-3-ol (3β), acetate

Organic

No.	Name	Synonym	Mol. form.	CAS Reg. No.	Mol. wt.	Physical form	mp/°C	bp/°C	den/g cm^{-3}	n_D	Solubility
2372	Cholest-5-en-3-ol (3β), benzoate		$C_{34}H_{50}O_2$	604-32-0	490.760	wh nd	150(1)		0.9413^{200}		i EtOH; s eth, chl
2373	Cholest-5-en-3-ol (3β)-, hexadecanoate		$C_{43}H_{76}O_2$	601-34-3	625.062	wh nd (eth al)	79.7(0.5)				vs bz, chl
2374	Cholest-5-en-3-ol (3β)-, *cis*-9-octadecenoate		$C_{45}H_{78}O_2$	303-43-5	651.100		47.9(0.5)				s chl
2375	Cholest-4-en-3-one		$C_{27}H_{44}O$	601-57-0	384.637	nd or pl (al)	81.5	245$^{0.03}$			
2376	Cholesterol		$C_{27}H_{46}O$	57-88-5	386.653	orth or tcl lf (al), nd (eth)	148.2(0.8)	459(20)	1.067^{20}		i H_2O; sl EtOH, ace; s bz, HOAc; vs diox
2377	Cholic acid	3,7,12-Trihydroxycholan-24-oic acid, (3α,5β,7α,12α)	$C_{24}H_{40}O_5$	81-25-4	408.572		198				sl H_2O; s EtOH, ace, alk; vs eth, chl
2378	Choline chloride		$C_5H_{14}ClNO$	67-48-1	139.624	hyg cry	305 dec				vs H_2O, EtOH
2379	Choline chloride dihydrogen phosphate	Phosphorylcholine	$C_5H_{15}ClNO_4P$	107-73-3	219.604	visc liq					
2380	Chorismic acid		$C_{10}H_{10}O_6$	617-12-9	226.182	cry	148				s H_2O
2381	Chromium carbonyl		C_6CrO_6	13007-92-6	220.056	col orth cry	dec 130	sub	1.77		i H_2O, EtOH; s eth, chl
2382	Chromium(II) oxalate		C_2CrO_4	814-90-4	140.015	ye-grn pow (hyd)					i H_2O, EtOH; s dil acid
2383	Chromium(III) 2,4-pentanedioate	Chromium acetylacetonate	$C_{15}H_{21}CrO_6$	21679-31-2	349.320	red mcl cry	208.7(0.5)	345	1.34		i H_2O; s bz
2384	Chromotrope 2B		$C_{16}H_9N_3Na_2O_{10}S_2$	548-80-1	513.366	red-br pow	300				s H_2O; i EtOH
2385	Chrysamminic acid	1,8-Dihydroxy-2,4,5,7-tetranitro-9,10-anthracenedione	$C_{14}H_4N_4O_{12}$	517-92-0	420.202	ye pl or lf	exp	dec			vs eth, EtOH
2386	6-Chrysenamine	6-Aminochrysene	$C_{18}H_{13}N$	2642-98-0	243.303	lf (al)	210.5				
2387	Chrysene	Benzo[*a*]phenanthrene	$C_{18}H_{12}$	218-01-9	228.288	red bl fl or orth pl (bz, HOAc)	255.0(0.1)	448	1.274^{20}		i H_2O; sl EtOH, eth, ace, bz, CS_2; s tol
2388	Ciafos		$C_9H_{10}NO_3PS$	2636-26-2	243.219	ye to red-ye liq	15	120$^{0.09}$ dec		1.5404^{32}	sl H_2O; vs chl, EtOH, ace, MeOH
2389	Cicutoxin	8,10,12-Heptadecatriene-4,6-diyne-1,14-diol	$C_{17}H_{22}O_2$	505-75-9	258.356	pr (eth/peth)	54				s hot H_2O, EtOH, eth, chl
2390	C.I. Direct Blue 6, tetrasodium salt	Direct Blue 6	$C_{32}H_{20}N_6Na_4O_{14}S_4$	2602-46-2	932.752	dk bronze pow					
2391	Cimetidine		$C_{10}H_{16}N_6S$	51481-61-9	252.339	cry	142				
2392	Cinchonamine		$C_{19}H_{24}N_2O$	482-28-0	296.406	orth nd (al) orth pr (MeOH)	186				i H_2O; vs EtOH, eth; s bz, chl
2393	Cinchonidine		$C_{19}H_{22}N_2O$	485-71-2	294.390	or pl or pr (al)	210.5	sub			i H_2O, bz; s EtOH, chl, py; sl eth
2394	Cinchonine		$C_{19}H_{22}N_2O$	118-10-5	294.390	pr nd (al, eth)	265				
2395	Cinchotoxine		$C_{19}H_{22}N_2O$	69-24-9	294.390	nd or pr (eth)	59				i H_2O; vs EtOH, eth, ace, bz, chl
2396	*trans*-Cinnamaldehyde	3-Phenyl-2-propenal, (*E*)-	C_9H_8O	14371-10-9	132.159	ye liq	-7.5	246	1.0497^{20}	1.6195^{20}	sl H_2O; s EtOH, eth, chl; i lig
2397	Cinnamedrine	α-[1-[Methyl(3-phenylallyl)-amino]ethyl]benzenemethanol	$C_{19}H_{23}NO$	90-86-8	281.392		75				
2398	*cis*-Cinnamic acid	3-Phenyl-2-propenoic acid, (*Z*)	$C_9H_8O_2$	102-94-3	148.159	mcl pr (w)	42				vs EtOH, HOAc, lig
2399	*trans*-Cinnamic acid	3-Phenyl-2-propenoic acid, (*E*)	$C_9H_8O_2$	140-10-3	148.159	mcl pr (dil al)	134(2)	300	1.2475^{4}		i H_2O, lig; vs EtOH; s eth, ace, bz
2400	*trans*-Cinnamyl anthranilate		$C_{16}H_{15}NO_2$	87-29-6	253.296	cry	64				
2401	Cinnamyl cinnamate		$C_{18}H_{16}O_2$	122-69-0	264.319	nd (al)	44		1.1565^{4}		i H_2O; s EtOH, chl; vs eth
2402	Cinnamyl formate	3-Phenyl-2-propen-1-ol, formate	$C_{10}H_{10}O_2$	104-65-4	162.185		0	252	1.086^{25}		
2403	Cinnoline	1,2-Benzodiazine	$C_8H_6N_2$	253-66-7	130.147	pa ye cry (lig)	38	114$^{0.3}$			vs eth, EtOH
2404	Cinoxate	3-(4-Methoxyphenyl)-2-propenoic acid, 2-ethoxyethyl ester	$C_{14}H_{18}O_4$	104-28-9	250.291	col liq	-25	185^{2}	1.102^{25}	1.567^{20}	i H_2O; msc EtOH
2405	Cinquasia Red	Quinacridone	$C_{20}H_{12}N_2O_2$	1047-16-1	312.321	red-viol cry	390				i H_2O, os
2406	Ciodrin		$C_{14}H_{19}O_6P$	7700-17-6	314.271			135$^{0.03}$	1.19^{25}		
2407	C.I. Pigment Red 170		$C_{26}H_{22}N_4O_4$	2786-76-7	454.478	red solid					
2408	C.I. Pigment Yellow 1		$C_{17}H_{16}N_4O_4$	2512-29-0	340.334	ye cry	256				
2409	C.I. Pigment Yellow 12		$C_{32}H_{26}Cl_2N_6O_4$	6358-85-6	629.492	ye cry	317				

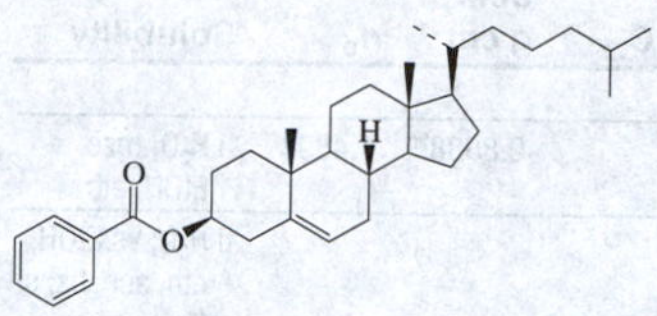
2372
Cholest-5-en-3-ol (3β), benzoate

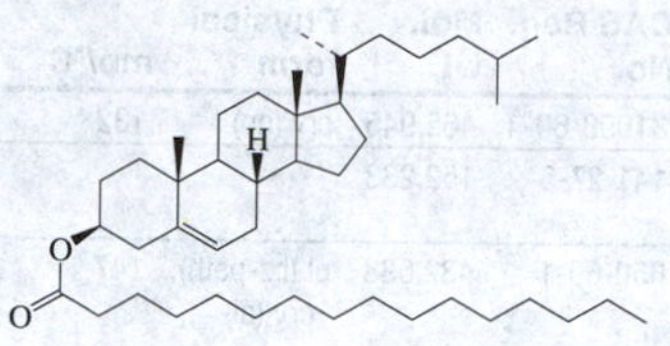
2373
Cholest-5-en-3-ol (3β)-, hexadecanoate

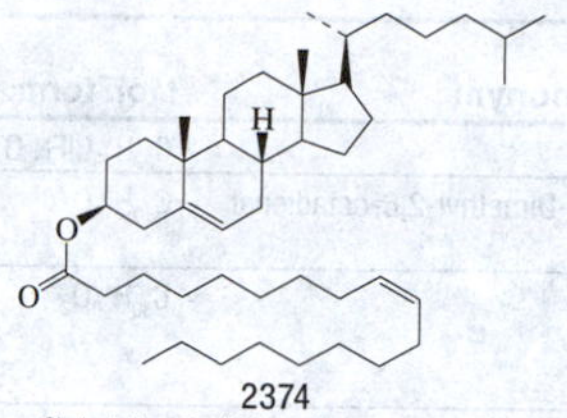
2374
Cholest-5-en-3-ol (3β)-, *cis*-9-octadecenoate

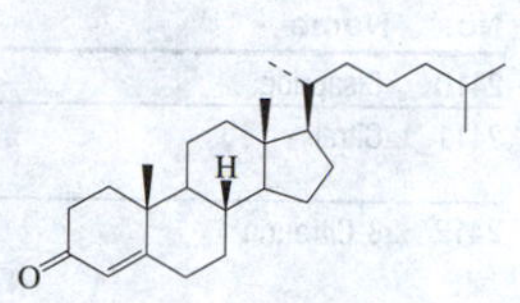
2375
Cholest-4-en-3-one

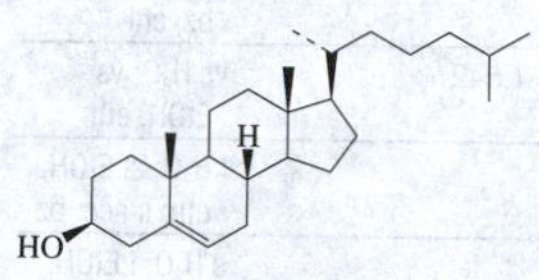
2376
Cholesterol

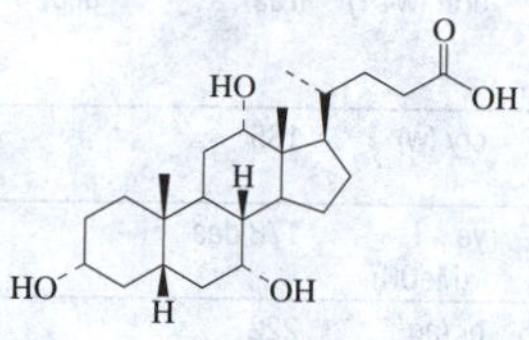
2377
Cholic acid

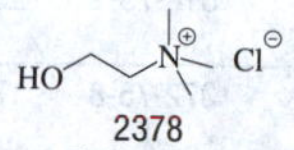
2378
Choline chloride

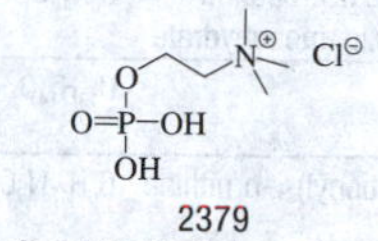
2379
Choline chloride dihydrogen phosphate

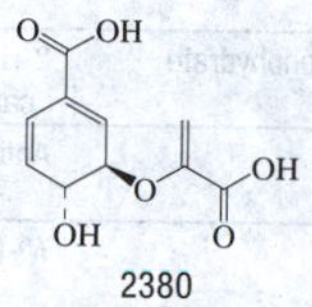
2380
Chorismic acid

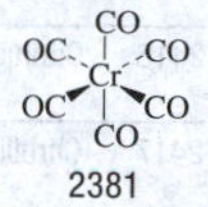
2381
Chromium carbonyl

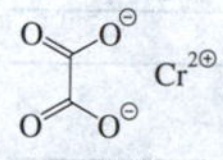
2382
Chromium(II) oxalate

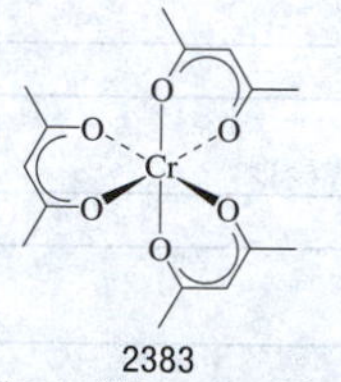
2383
Chromium(III) 2,4-pentanedioate

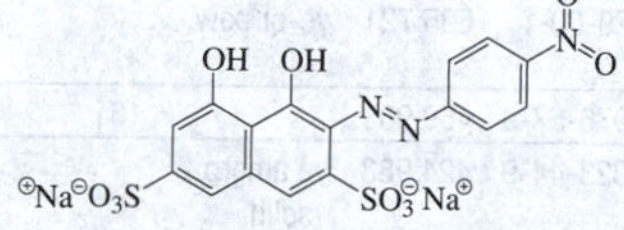
2384
Chromotrope 2B

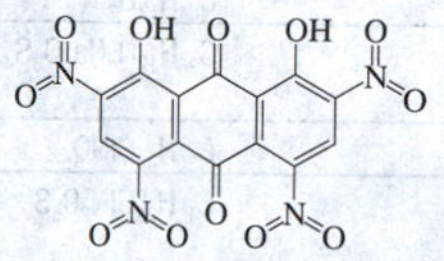
2385
Chrysamminic acid

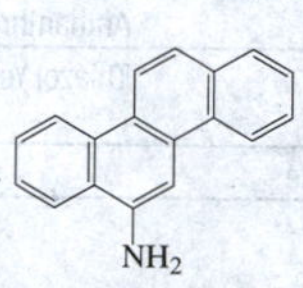
2386
6-Chrysenamine

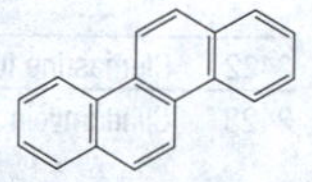
2387
Chrysene

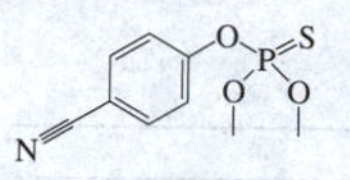
2388
Ciafos

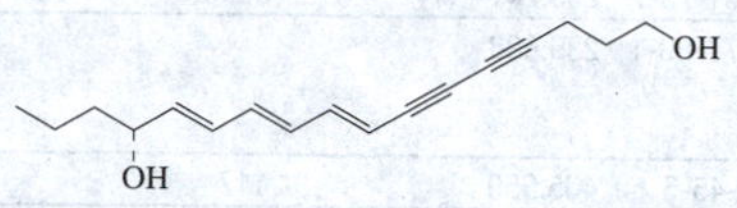
2389
Cicutoxin

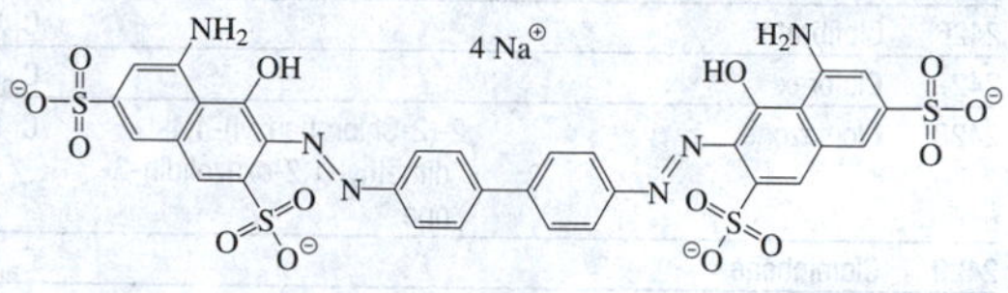
2390
C.I. Direct Blue 6, tetrasodium salt

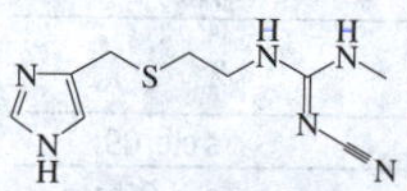
2391
Cimetidine

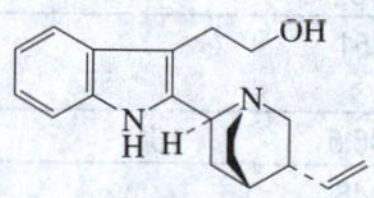
2392
Cinchonamine

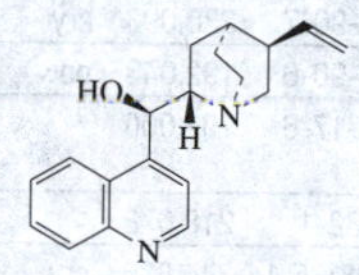
2393
Cinchonidine

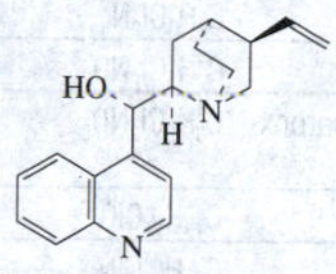
2394
Cinchonine

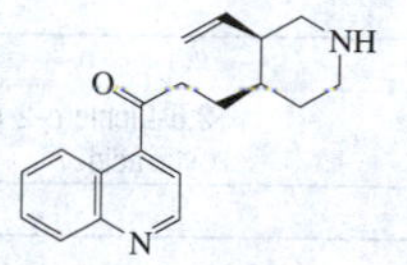
2395
Cinchotoxine

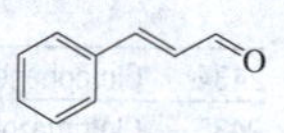
2396
trans-Cinnamaldehyde

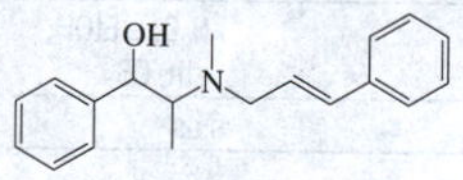
2397
Cinnamedrine

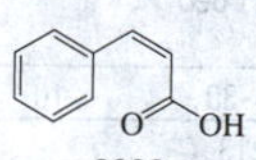
2398
cis-Cinnamic acid

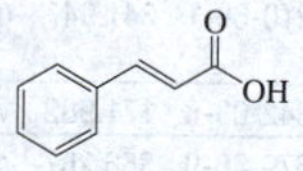
2399
trans-Cinnamic acid

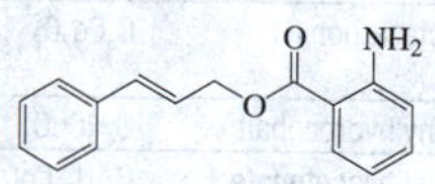
2400
trans-Cinnamyl anthranilate

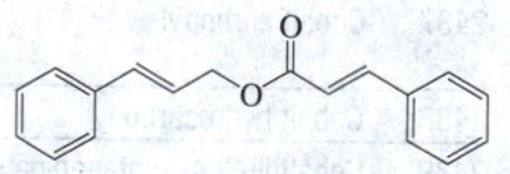
2401
Cinnamyl cinnamate

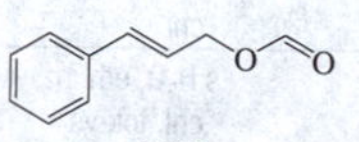
2402
Cinnamyl formate

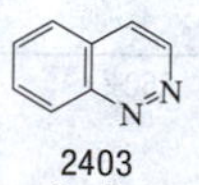
2403
Cinnoline

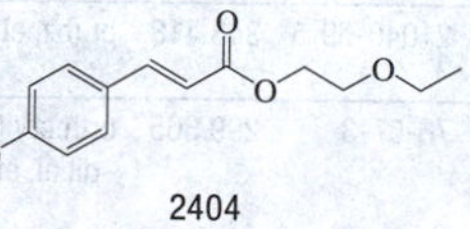
2404
Cinoxate

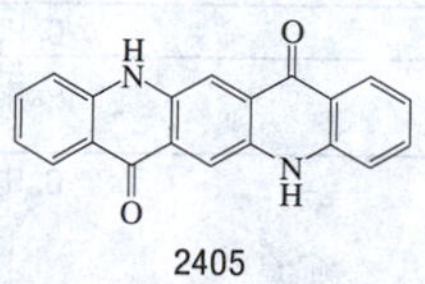
2405
Cinquasia Red

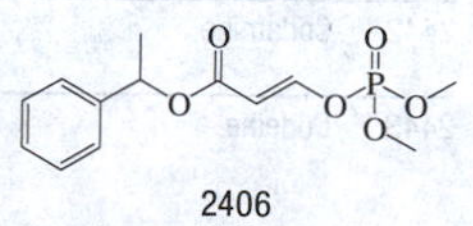
2406
Ciodrin

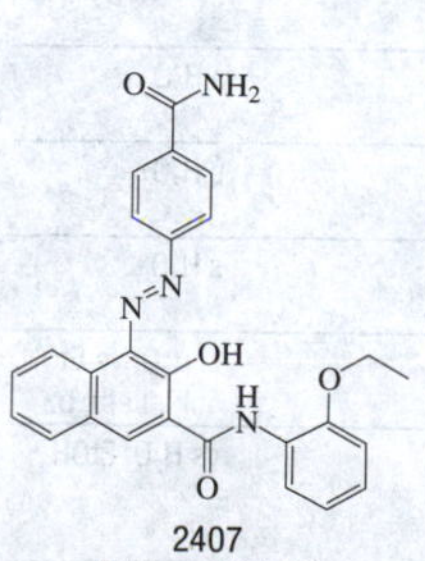
2407
C.I. Pigment Red 170

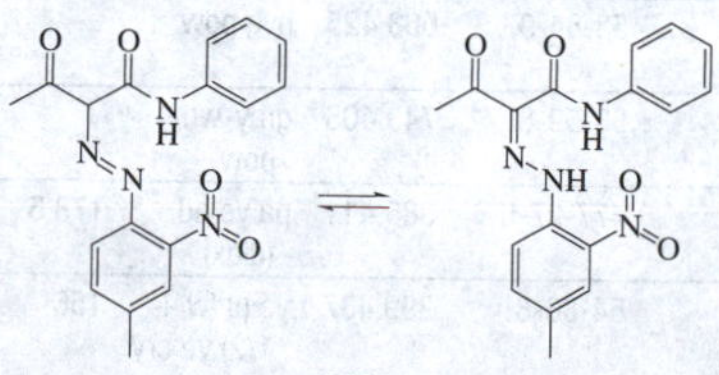
2408
C.I. Pigment Yellow 1

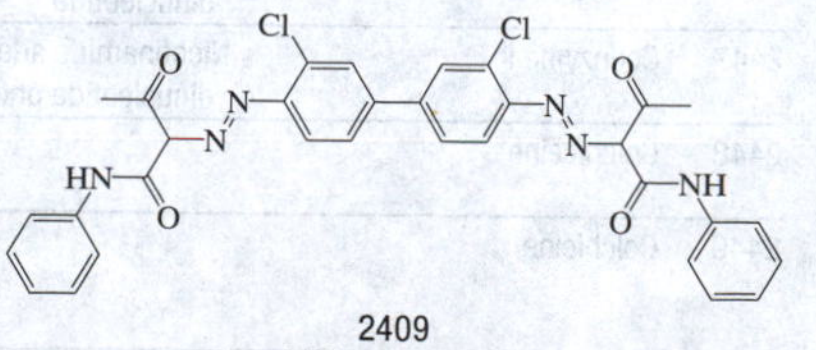
2409
C.I. Pigment Yellow 12

Organic

No.	Name	Synonym	Mol. form.	CAS Reg. No.	Mol. wt.	Physical form	mp/°C	bp/°C	den/ g cm⁻³	n_D	Solubility
2410	Cisapride		$C_{23}H_{29}ClFN_3O_4$	81098-60-4	465.945	cry (hp)	132				
2411	Citral	3,7-Dimethyl-2,6-octadienal	$C_{10}H_{16}O$	141-27-5	152.233			229	0.8888^{20}	1.4898^{20}	i H_2O; msc EtOH, eth
2412	β-Citraurin		$C_{30}H_{40}O_2$	650-69-1	432.638	pl (bz-peth), cry (al)	147				i H_2O; vs EtOH, eth, ace, bz; sl lig
2413	Citrazinic acid	1,2-Dihydro-6-hydroxy-2-oxo-4-pyridinecarboxylic acid	$C_6H_5NO_4$	99-11-6	155.109	ye pow	>300 dec				s H_2O, alk; sl HCl
2414	Citric acid	2-Hydroxy-1,2,3-propanetricarboxylic acid	$C_6H_8O_7$	77-92-9	192.124	orth (w+1)	153	dec	1.665^{20}		vs H_2O, EtOH; s eth, AcOEt; i bz, chl
2415	Citric acid monohydrate	2-Hydroxy-1,2,3-propanetricarboxylic acid, monohydrate	$C_6H_{10}O_8$	5949-29-1	210.138	cry (w)	135		1.542		vs H_2O; vs EtOH, eth
2416	Citrinin	Antimycin	$C_{13}H_{14}O_5$	518-75-2	250.247	ye nd (MeOH)	178 dec				i H_2O; sl EtOH, eth; s ace, bz
2417	Citrulline	N^5-(Aminocarbonyl)-L-ornithine	$C_6H_{13}N_3O_3$	372-75-8	175.185	pr (aq MeOH)	222				s H_2O; i EtOH, MeOH
2418	Citrus Red 2		$C_{18}H_{16}N_2O_3$	6358-53-8	308.331	cry	156				sl H_2O; s EtOH
2419	C.I. Vat Blue 6	7,16-Dichloro-6,15-dihydro-5,9,14,18-anthrazinetetrone	$C_{28}H_{12}Cl_2N_2O_4$	130-20-1	511.312	viol-bl pow					
2420	C.I. Vat Yellow 4	Anthanthrone	$C_{24}H_{12}O_2$	128-66-5	332.351	ye cry					
2421	Clayton Yellow	Thiazol Yellow G	$C_{28}H_{19}N_5Na_2O_6S_4$	1829-00-1	695.721	ye-br pow					s H_2O, EtOH, H_2SO_4
2422	Clemastine fumarate		$C_{25}H_{30}ClNO_5$	14976-57-9	459.963		181				
2423	Clindamycin		$C_{18}H_{33}ClN_2O_5S$	18323-44-9	424.983	ye amorp solid					
2424	Cloconazole		$C_{18}H_{15}ClN_2O$	77175-51-0	310.777		73				s EtOAc
2425	Clofentezine	3,6-Bis(2-chlorophenyl)-1,2,4,5-tetrazine	$C_{14}H_8Cl_2N_4$	74115-24-5	303.147		182				
2426	Clofibrate		$C_{12}H_{15}ClO_3$	637-07-0	242.698			149^{20}			
2427	Cloforex		$C_{13}H_{18}ClNO_2$	14261-75-7	255.741	cry	52.8	$89^{0.005}$			
2428	Clomazone	2-(2-Chlorobenzyl)-4,4-dimethyl-1,2-oxazolidin-3-one	$C_{12}H_{14}ClNO_2$	81777-89-1	239.698				1.192^{20}		
2429	Clomiphene		$C_{26}H_{28}ClNO$	911-45-5	405.959		117				
2430	Clonazepam		$C_{15}H_{10}ClN_3O_3$	1622-61-3	315.711	wh cry	237.5				i H_2O, bz; sl ace, MeOH, chl
2431	Clonidine		$C_9H_9Cl_2N_3$	4205-90-7	230.093	cry	137				
2432	Clopidol		$C_7H_7Cl_2NO$	2971-90-6	192.043	pow	>320				i H_2O
2433	Clopyralid	3,6-Dichloro-2-pyridinecarboxylic acid	$C_6H_3Cl_2NO_2$	1702-17-6	192.000		151				
2434	Clorophene		$C_{13}H_{11}ClO$	120-32-1	218.678		48.5	$161^{3.5}$	1.185^{58}		s ctc, CS_2
2435	Clotrimazole		$C_{22}H_{17}ClN_2$	23593-75-1	344.836	cry	148				sl H_2O, bz; s ace, chl, AcOEt, DMF
2436	Clozapine	Clozaril	$C_{18}H_{19}ClN_4$	5786-21-0	326.824	ye cry	183(1)				
2437	Cobalt carbonyl	Dicobalt octacarbonyl	$C_8Co_2O_8$	10210-68-1	341.947	oran cry	51 dec		1.78		i H_2O; s EtOH, eth, CS_2
2438	Cobalt hydrocarbonyl	Tetracarbonylhydrocobalt	C_4HCoO_4	16842-03-8	171.982	ye liq or gas	≈-30	10			s os
2439	Cobalt(III) 2,4-pentanedioate	Cobalt(III) acetylacetonate	$C_{15}H_{21}CoO_6$	21679-46-9	356.257	dark grn cry	213				s bz, ace
2440	Cocaine		$C_{17}H_{21}NO_4$	50-36-2	303.354	mcl pr (al)	98	$187^{0.1}$		1.5022^{98}	sl H_2O; vs EtOH, eth, bz, py; s CS_2
2441	Coclaurine		$C_{17}H_{19}NO_3$	486-39-5	285.338	pl (al)	220.5				
2442	Codamine		$C_{20}H_{25}NO_4$	21040-59-5	343.418	pr (bz, eth)	127				vs eth, EtOH, chl
2443	Codeine		$C_{18}H_{21}NO_3$	76-57-3	299.365	orth cry (w, dil al, eth)	157.5	250^{22}	1.32^{25}		s H_2O, eth, bz, chl, tol; vs EtOH; i peth
2444	Codeine phosphate		$C_{18}H_{24}NO_7P$	52-28-8	397.361	lf or pr (dil al)	227 dec				vs EtOH, chl
2445	Coenzyme A		$C_{21}H_{36}N_7O_{16}P_3S$	85-61-0	767.535	pow; unstab in air					s H_2O
2446	Coenzyme I	Nicotinamide adenine dinucleotide	$C_{21}H_{27}N_7O_{14}P_2$	53-84-9	663.425	hyg pow					s H_2O
2447	Coenzyme II	Nicotinamide adenine dinucleotide phosphate	$C_{21}H_{28}N_7O_{17}P_3$	53-59-8	743.405	gray-wh pow					s H_2O
2448	Colchiceine		$C_{21}H_{23}NO_6$	477-27-0	385.411	pa ye nd (diox)	178.5		1.24^{25}		sl H_2O; vs EtOH, chl; i eth, bz
2449	Colchicine		$C_{22}H_{25}NO_6$	64-86-8	399.437	ye pl (w + 1/2) ye cry (bz)	156				vs H_2O, EtOH

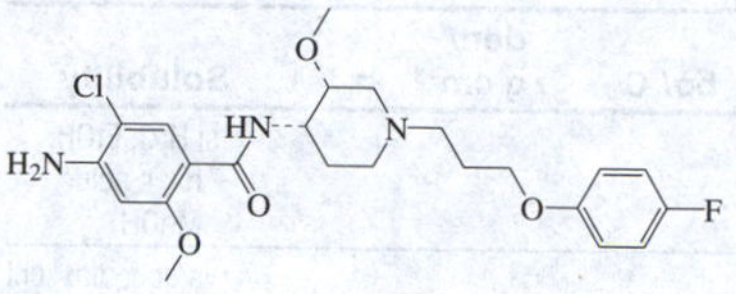
2410
Cisapride

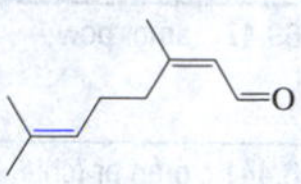
2411
Citral

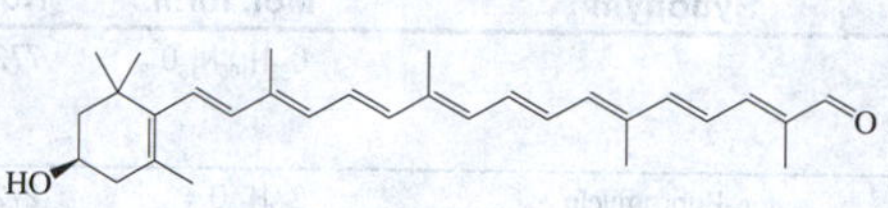
2412
β-Citraurin

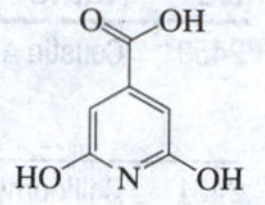
2413
Citrazinic acid

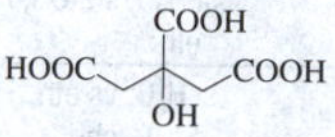
2414
Citric acid

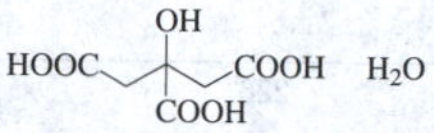
2415
Citric acid monohydrate

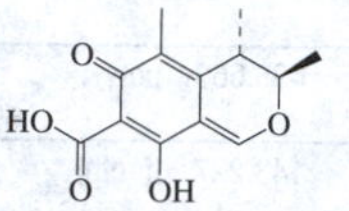
2416
Citrinin

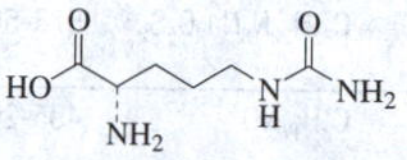
2417
Citrulline

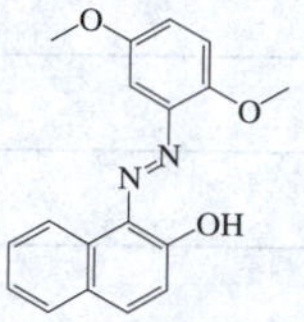
2418
Citrus Red 2

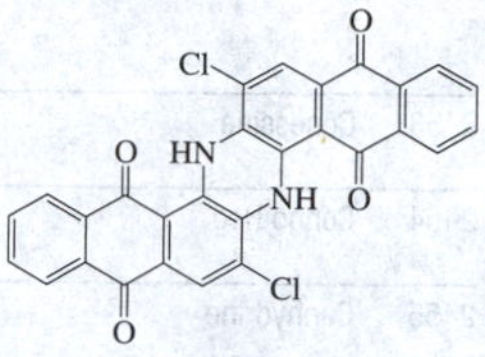
2419
C.I. Vat Blue 6

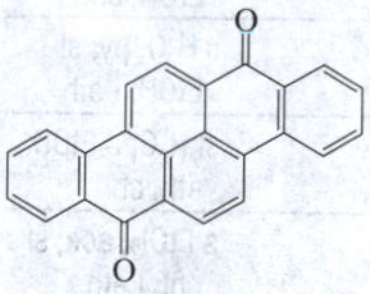
2420
C.I. Vat Yellow 4

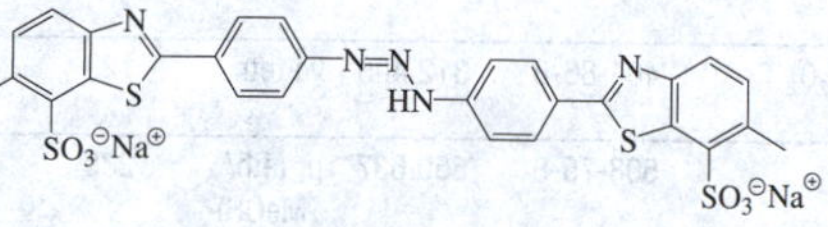
2421
Clayton Yellow

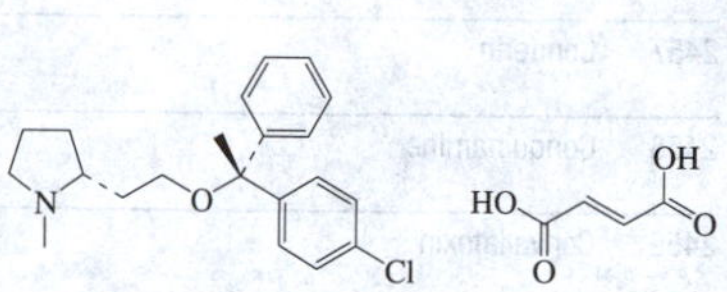
2422
Clemastine fumarate

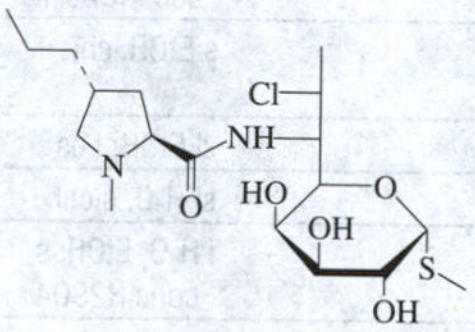
2423
Clindamycin

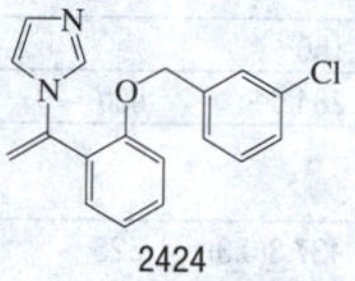
2424
Cloconazole

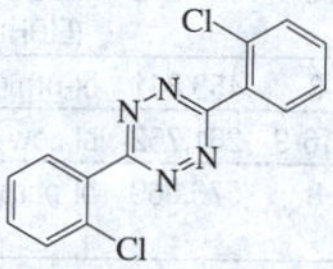
2425
Clofentezine

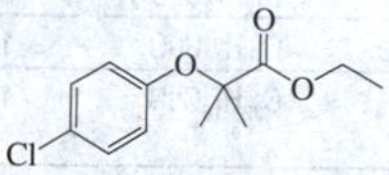
2426
Clofibrate

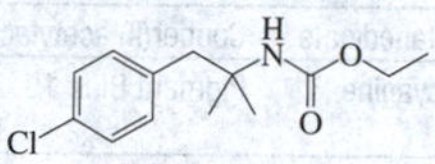
2427
Cloforex

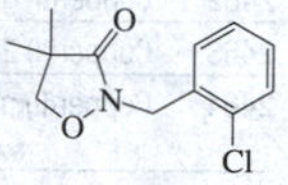
2428
Clomazone

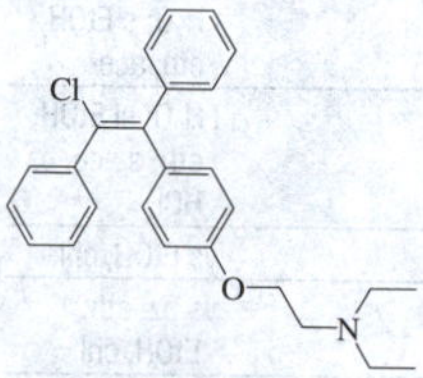
2429
Clomiphene

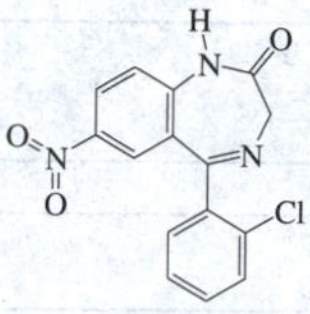
2430
Clonazepam

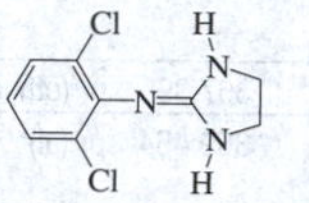
2431
Clonidine

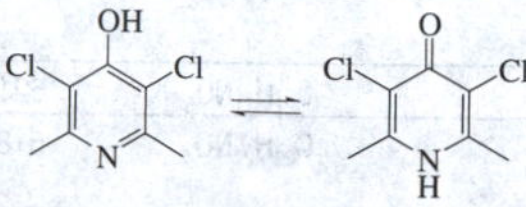
2432
Clopidol

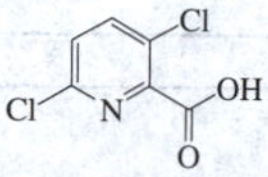
2433
Clopyralid

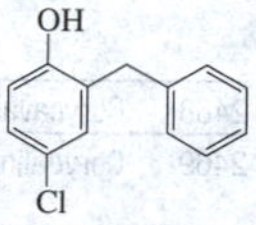
2434
Clorophene

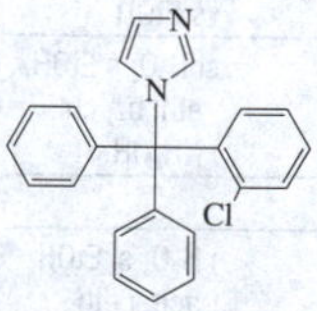
2435
Clotrimazole

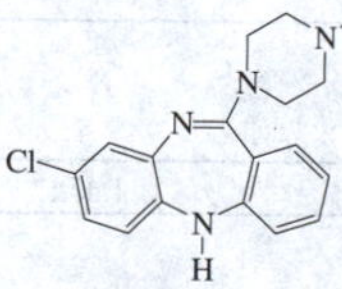
2436
Clozapine

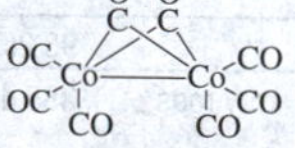
2437
Cobalt carbonyl

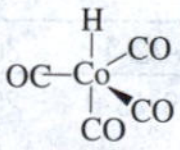
2438
Cobalt hydrocarbonyl

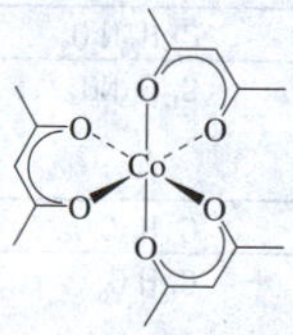
2439
Cobalt(III) 2,4-pentanedioate

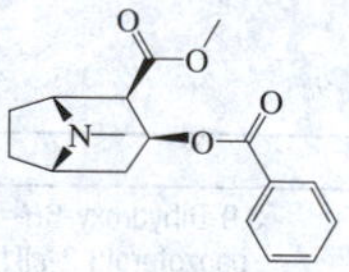
2440
Cocaine

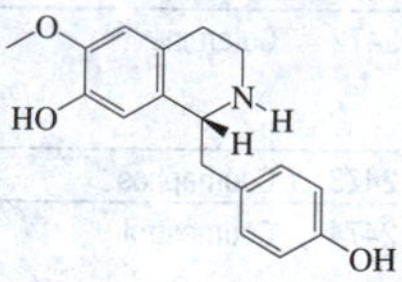
2441
Coclaurine

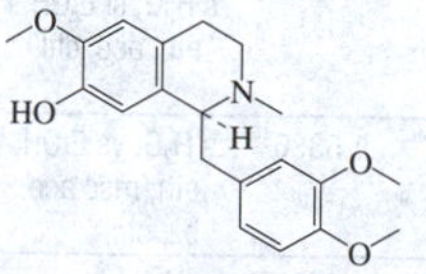
2442
Codamine

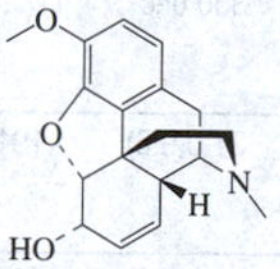
2443
Codeine

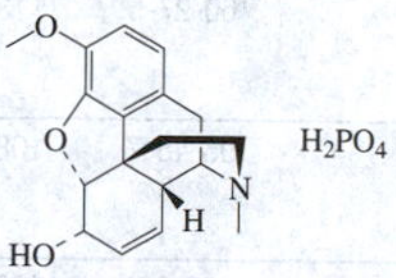
2444
Codeine phosphate

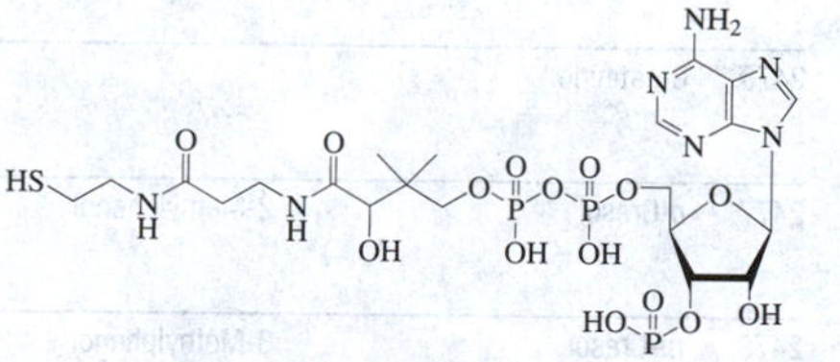
2445
Coenzyme A

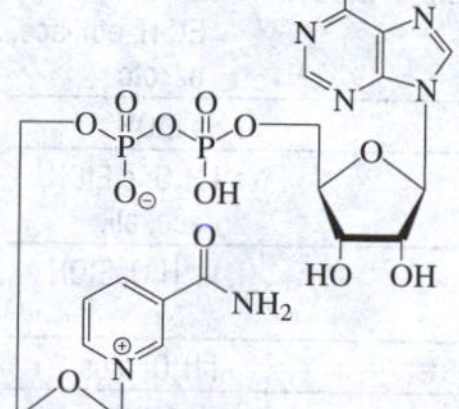
2446
Coenzyme I

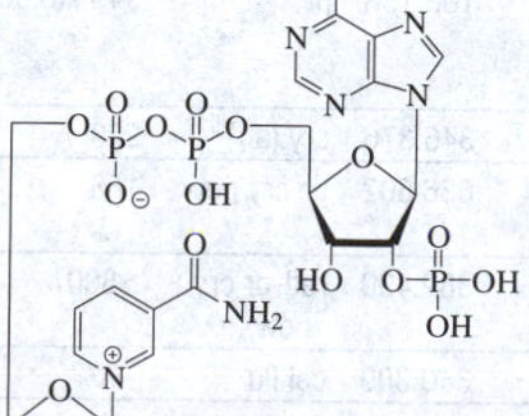
2447
Coenzyme II

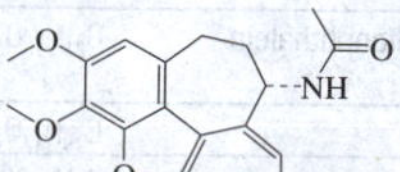
2448
Colchiceine

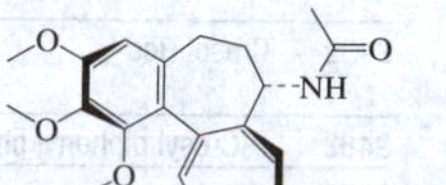
2449
Colchicine

Organic

No.	Name	Synonym	Mol. form.	CAS Reg. No.	Mol. wt.	Physical form	mp/°C	bp/°C	den/ g cm^{-3}	n_D	Solubility
2450	Colistin A		$C_{53}H_{100}N_{16}O_{13}$	7722-44-3	1169.47	amor pow					sl H_2O, EtOH, hx; s acids, MeOH
2451	Collinomycin	α-Rubromycin	$C_{27}H_{20}O_{12}$	27267-69-2	536.441	oran pr (chl-MeOH)	281				vs ace, diox, chl
2452	Columbin		$C_{20}H_{22}O_6$	546-97-4	358.385	nd (MeOH)	195.5				i H_2O; sl ace, AcOEt, MeOH; s chl
2453	Conessine		$C_{24}H_{40}N_2$	546-06-5	356.588	lf or pl (ace)	125.5	$166^{0.1}$			sl H_2O; s chl, HOAc
2454	Congo Red		$C_{32}H_{22}N_6Na_2O_6S_2$	573-58-0	696.663	pow	>360				sl H_2O; s EtOH; i eth
2455	Conhydrine		$C_8H_{17}NO$	495-20-5	143.227	lf (eth)	121	226			sl H_2O; vs eth, EtOH, chl
2456	Conhydrine, (+)	2-(α-Hydroxypropyl)piperidine	$C_8H_{17}NO$	495-20-5	143.227	lf (eth)	121	226			sl H_2O; vs eth, EtOH, chl
2457	Coniferin		$C_{16}H_{22}O_8$	531-29-3	342.341	nd (w+2)	186				s H_2O, py; sl EtOH; i eth
2458	Conquinamine		$C_{19}H_{24}N_2O_2$	464-86-8	312.406	ye tetr	123				sl H_2O; s EtOH, eth, chl
2459	Convallatoxin		$C_{29}H_{42}O_{10}$	508-75-8	550.637	pr (eth/ MeOH)	238				s EtOH, ace; sl chl; i eth
2460	Copaene		$C_{15}H_{24}$	3856-25-5	204.352			248.5	0.8996^{20}	1.4894^{20}	i H_2O; s eth, ace, HOAc, lig
2461	Copper(II) ethylacetoacetate	Bis(ethylacetoacetato)copper	$C_{12}H_{18}CuO_6$	14284-06-1	321.813	grn cry (EtOH)	192				s EtOH, chl
2462	Copper(II) gluconate	Cupric gluconate	$C_{12}H_{22}CuO_{14}$	527-09-3	453.841	bl-grn cry	156				sl EtOH; i os
2463	Copper(II) 2,4-pentanedioate	Copper(II) acetylacetonate	$C_{10}H_{14}CuO_4$	13395-16-9	261.762	bl pow	284 dec	sub			sl H_2O; s chl
2464	Copper(II) phthalocyanine	Pigment Blue 15	$C_{32}H_{16}CuN_8$	147-14-8	576.069	bl-purp cry					i H_2O, EtOH; s conc H2SO4
2465	Coronene		$C_{24}H_{12}$	191-07-1	300.352	ye nd (bz)	437.3(0.3)	525	1.371^{25}		i H_2O, con sulf; sl bz
2466	Corticosterone		$C_{21}H_{30}O_4$	50-22-6	346.461	nd (al, pl) (ace)	181				i H_2O; s EtOH, eth, ace
2467	Corybulbine		$C_{21}H_{25}NO_4$	518-77-4	355.429	nd (al)	237.5				i H_2O; sl EtOH, eth; s ace, bz, HCl
2468	Corycavamine		$C_{21}H_{21}NO_5$	521-85-7	367.396	pr (eth, al)	149				vs EtOH, chl
2469	Corydaline		$C_{22}H_{27}NO_4$	518-69-4	369.454	pr (al)	136				vs bz, eth, EtOH, chl
2470	Corydine		$C_{20}H_{23}NO_4$	476-69-7	341.402	tetr pr (eth)	149				vs eth, EtOH, chl
2471	Corynantheine		$C_{22}H_{26}N_2O_3$	18904-54-6	366.452		165.5				vs EtOH
2472	Cotarnine		$C_{12}H_{15}NO_4$	82-54-2	237.252	nd (bz), cry (eth)	132 dec				sl H_2O; s EtOH, eth, bz, chl, NH_4OH
2473	Coumaphos		$C_{14}H_{16}ClO_5PS$	56-72-4	362.766		95.2(0.2)		1.474		
2474	Coumestrol	3,9-Dihydroxy-6*H*-benzofuro[3,2-*c*][1] benzopyran-6-one	$C_{15}H_8O_5$	479-13-0	268.222	cry rods	385 dec				i H_2O; sl EtOH, ace; i eth
2475	Creatine		$C_4H_9N_3O_2$	57-00-1	131.133	mcl pr (w+1)	303 dec		1.33^{25}		s H_2O; sl EtOH; i eth
2476	Creatinine		$C_4H_7N_3O$	60-27-5	113.118	orth pr (w+2) lf (w)	300 dec				s H_2O; sl EtOH; i eth, ace, chl
2477	*o*-Cresol	2-Methylphenol	C_7H_8O	95-48-7	108.138		31.0(0.6)	191.0(0.1)	1.0327^{35}	1.5386^{35}	s H_2O; vs EtOH, eth; msc ace, bz, ctc
2478	*m*-Cresol	3-Methylphenol	C_7H_8O	108-39-4	108.138	liq	12.2(0.3)	202.2(0.1)	1.0339^{20}	1.5401^{20}	sl H_2O; msc EtOH, eth, ace, bz, ctc
2479	*p*-Cresol	4-Methylphenol	C_7H_8O	106-44-5	108.138	pr	34.77(0.05)	201.9(0.1)	1.0185^{40}	1.5312^{20}	sl H_2O; msc EtOH, eth, ace, bz, ctc
2480	*o*-Cresolphthalein		$C_{22}H_{18}O_4$	596-27-0	346.376	cry (al)	223				vs EtOH
2481	*o*-Cresolphthalein complexone	Metalphthalein	$C_{32}H_{32}N_2O_{12}$	2411-89-4	636.602	ye cry pow	186				i H_2O; s EtOH, ace, alk
2482	Cresol Red	*o*-Cresolsulfonphthalein	$C_{21}H_{18}O_5S$	1733-12-6	382.430	red-br cry pow	>300				vs H_2O, EtOH
2483	*p*-Cresyl diphenyl phosphate		$C_{19}H_{17}O_4P$	78-31-9	340.309	col liq	-40		1.208^{25}		i H_2O; s os
2484	Crimidine		$C_7H_{10}ClN_3$	535-89-7	171.627	br wax	87	143^4			vs EtOH
2485	Cromolyn	Cromoglicic acid	$C_{23}H_{16}O_{11}$	16110-51-3	468.366	col cry	241 dec				
2486	Crufomate		$C_{12}H_{19}ClNO_3P$	299-86-5	291.711		60.1(0.5)	$118^{0.01}$			

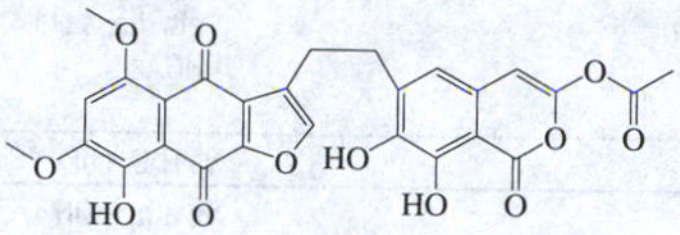
2451
Collinomycin

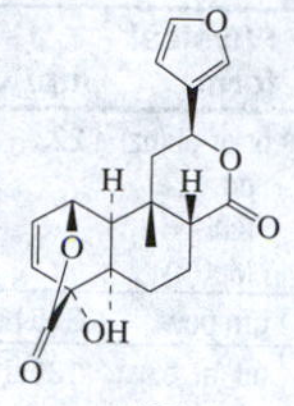
2452
Columbin

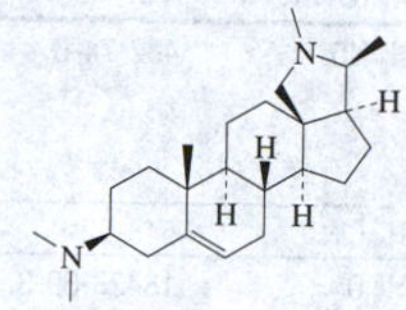
2453
Conessine

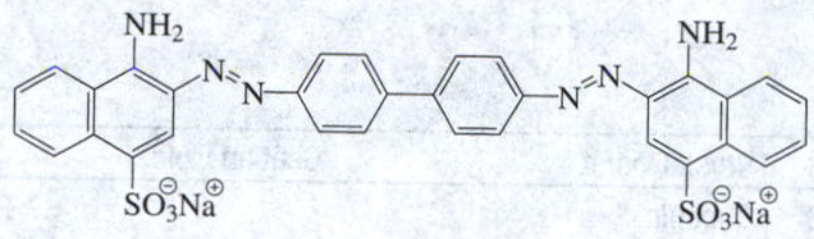
2454
Congo Red

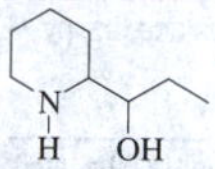
2455
Conhydrine

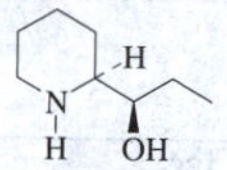
2456
Conhydrine, (+)

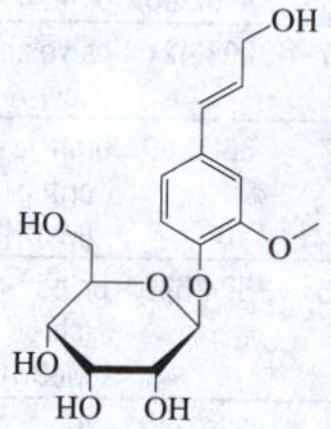
2457
Coniferin

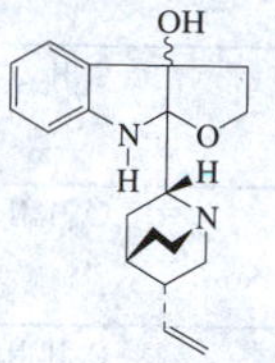
2458
Conquinamine

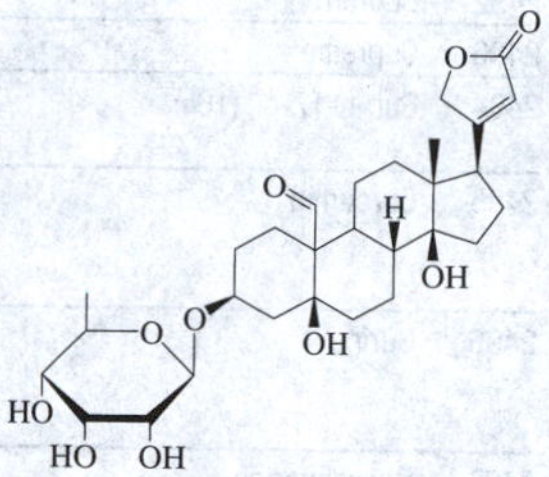
2459
Convallatoxin

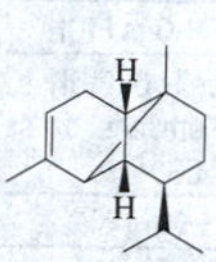
2460
Copaene

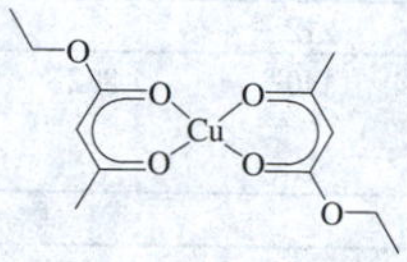
2461
Copper(II) ethylacetoacetate

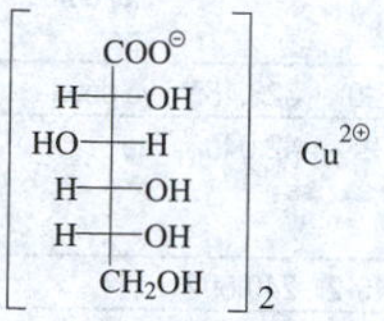
2462
Copper(II) gluconate

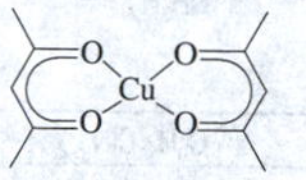
2463
Copper(II) 2,4-pentanedioate

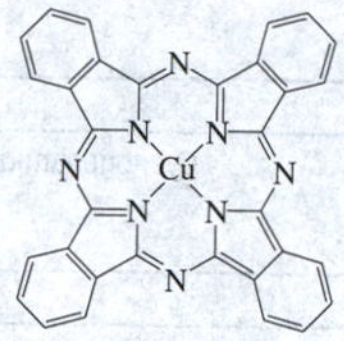
2464
Copper(II) phthalocyanine

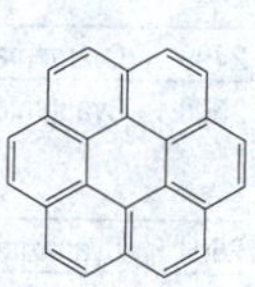
2465
Coronene

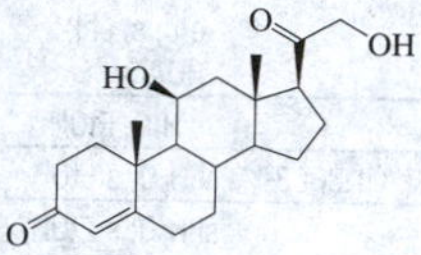
2466
Corticosterone

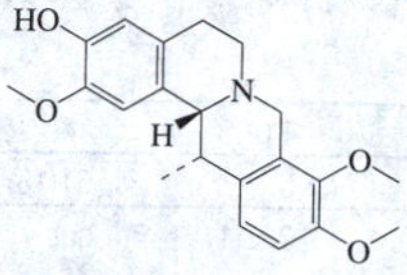
2467
Corybulbine

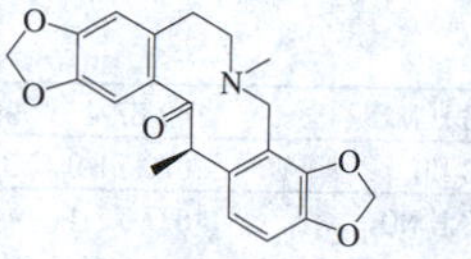
2468
Corycavamine

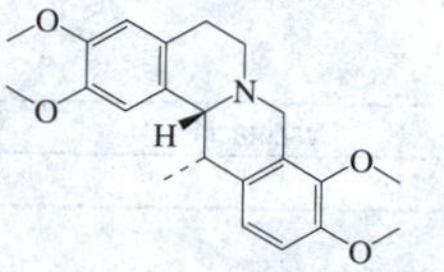
2469
Corydaline

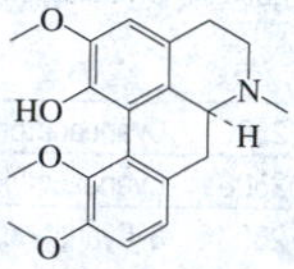
2470
Corydine

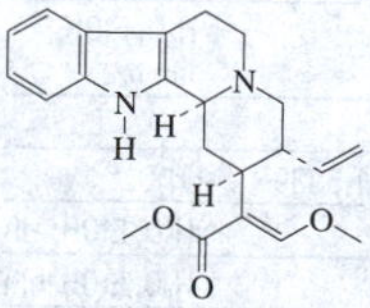
2471
Corynantheine

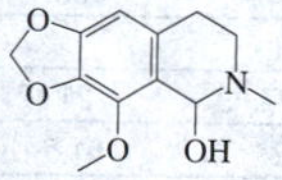
2472
Cotarnine

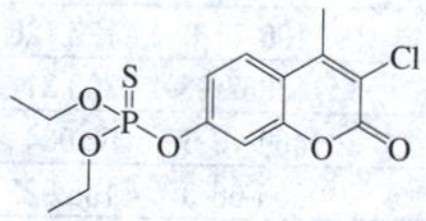
2473
Coumaphos

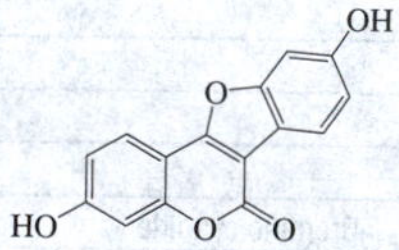
2474
Coumestrol

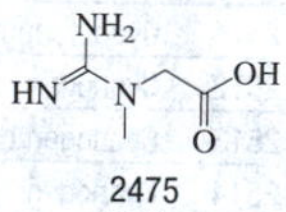
2475
Creatine

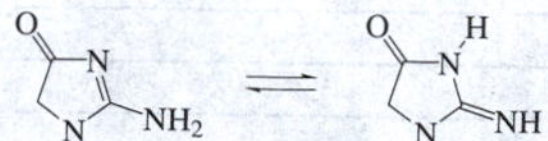
2476
Creatinine

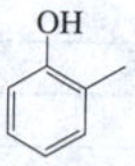
2477
o-Cresol

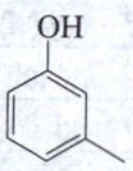
2478
m-Cresol

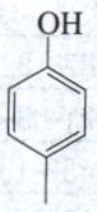
2479
p-Cresol

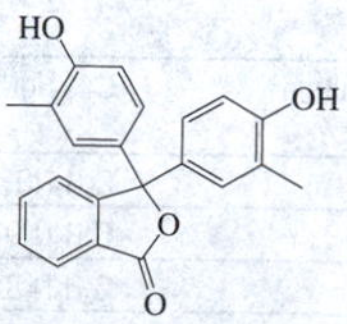
2480
o-Cresolphthalein

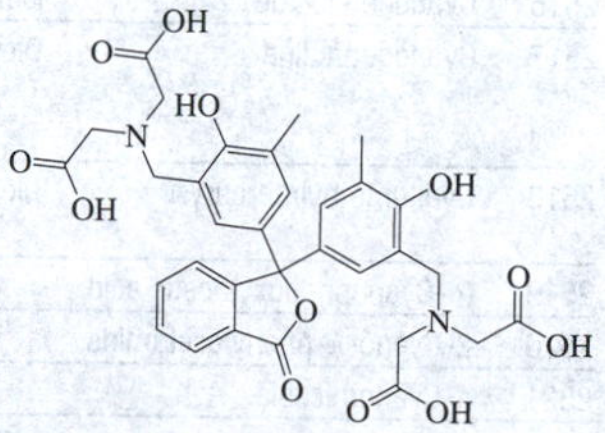
2481
o-Cresolphthalein complexone

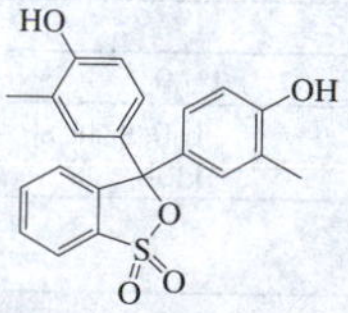
2482
Cresol Red

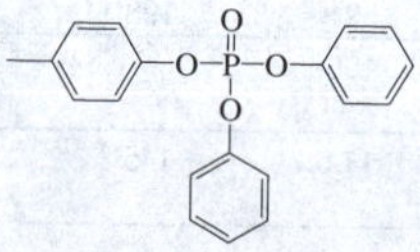
2483
p-Cresyl diphenyl phosphate

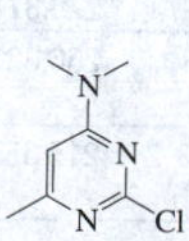
2484
Crimidine

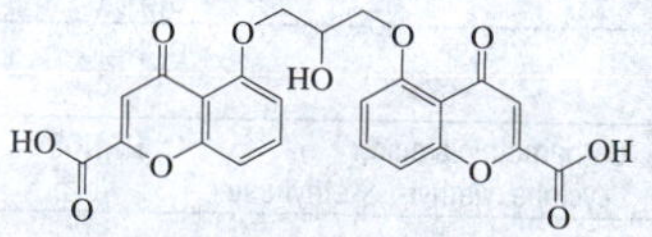
2485
Cromolyn

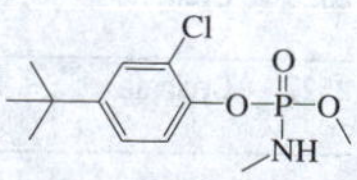
2486
Crufomate

Organic

No.	Name	Synonym	Mol. form.	CAS Reg. No.	Mol. wt.	Physical form	mp/°C	bp/°C	den/ g cm^{-3}	n_D	Solubility
2487	Cryptopine	Cryptocavine	$C_{21}H_{23}NO_5$	482-74-6	369.412	pr or pl (bz) nd (chl-MeOH)	223		1.315^{20}		i H_2O; sl EtOH, eth, bz; s chl, HOAc
2488	Crystal Violet	Gentian violet	$C_{25}H_{30}ClN_3$	548-62-9	407.979	grn pow	215 dec				vs H_2O, chl
2489	Cubebin		$C_{20}H_{20}O_6$	18423-69-3	356.369	nd (al, bz)	131.5				vs eth, EtOH, chl
2490	Cucurbitacin B		$C_{32}H_{46}O_8$	6199-67-3	558.702	cry (EtOH)	181				
2491	Cucurbitacin C		$C_{32}H_{48}O_8$	5988-76-1	560.718	cry (AcOEt)	207.5				
2492	Cupferron		$C_6H_9N_3O_2$	135-20-6	155.154		163.5				sl DMSO
2493	Cupreine		$C_{19}H_{22}N_2O_2$	524-63-0	310.390	pr (eth)	202				vs EtOH
2494	Curan-17-ol, (16α)	Geissoschizoline	$C_{19}H_{26}N_2O$	18397-07-4	298.421	pa ye amor pow	135 dec				i H_2O; vs EtOH, eth, chl
2495	Curcumin	Turmeric	$C_{21}H_{20}O_6$	458-37-7	368.380	oran ye pr, orth pr (MeOH)	183				vs EtOH, HOAc
2496	Curine		$C_{36}H_{38}N_2O_6$	436-05-5	594.696	pr, nd (chl-MeOH)	221				vs ace, bz, py
2497	Cuscohygrine		$C_{13}H_{24}N_2O$	454-14-8	224.342	oil		169^{23}	0.9733^{20}	1.4832^{20}	vs H_2O, bz, eth, EtOH
2498	Cusparine	2-[2-(1,3-Benzodioxol-5-yl)-ethyl]-4-methoxyquinoline	$C_{19}H_{17}NO_3$	529-92-0	307.343	(α) wh or ye nd (peth); (β) amber pr	92(α; 111(β)				i H_2O; vs ace, bz, eth, EtOH
2499	Cyamemazine		$C_{19}H_{21}N_3S$	3546-03-0	323.455	ye pow	92	$212^{0.25}$			i H_2O; s EtOH
2500	Cyanamide	Cyanogenamide	CH_2N_2	420-04-2	42.040	nd	45.55(0.04)	140^{19}	1.282^{20}	1.4418^{48}	vs H_2O, EtOH; s eth, ace, bz; sl CS_2
2501	Cyanazine		$C_9H_{13}ClN_6$	21725-46-2	240.692		165.6(0.5)				
2502	Cyanic acid	Hydrogen cyanate	CHNO	420-05-3	43.025	unstab liq or gas	-86	23	1.140^{20}		vs H_2O, bz, eth, chl
2503	2-Cyanoacetamide		$C_3H_4N_2O$	107-91-5	84.076	pl (w)	114.1(0.3)				vs H_2O
2504	Cyanoacetic acid		$C_3H_3NO_2$	372-09-8	85.062		66	160 dec			s H_2O, EtOH, eth; sl chl, HOAc
2505	Cyanoacetohydrazide	Cyacetacide	$C_3H_5N_3O$	140-87-4	99.091	pr (al)	114.5				vs H_2O, EtOH
2506	Cyanoacetylene		C_3HN	1070-71-9	51.047		5	61(18)	0.8167^{17}	1.3868^{25}	sl H_2O; s EtOH
2507	3-Cyanobenzoic acid		$C_8H_5NO_2$	1877-72-1	147.132	nd (w)	223(1)	sub			sl H_2O; s EtOH, eth
2508	4-Cyanobenzoic acid		$C_8H_5NO_2$	619-65-8	147.132		220(1)				s H_2O, EtOH, eth, HOAc; sl tfa
2509	4-Cyanobutanoic acid		$C_5H_7NO_2$	39201-33-7	113.116	hyg cry	45				s H_2O, EtOH, eth, bz
2510	2-Cyanoethyl acrylate		$C_6H_7NO_2$	106-71-8	125.126			108^{12}	1.062^{20}		
2511	Cyanofenphos		$C_{15}H_{14}NO_2PS$	13067-93-1	303.317		83			1.5839^{25}	sl H_2O
2512	Cyanogen		C_2N_2	460-19-5	52.034	col gas	-27.83	-21.1	0.787^{-21}		s H_2O, EtOH, eth
2513	Cyanogen bromide	Bromine cyanide	CBrN	506-68-3	105.922	nd	52	61.5	2.015^{20}		s H_2O, EtOH, eth
2514	Cyanogen chloride	Chlorine cyanide	CClN	506-77-4	61.471	col vol liq or gas	-6.55	13	1.186^{20}		s H_2O, EtOH; vs eth
2515	Cyanogen fluoride	Fluorine cyanide	CFN	1495-50-7	45.016	col gas	-82	-46			
2516	Cyanogen iodide	Iodine cyanide	CIN	506-78-5	152.922	nd (al, eth)	146.7	sub	2.84^{18}		vs eth, EtOH
2517	Cyanoguanidine	Dicyanodiamide	$C_2H_4N_4$	461-58-5	84.080		207(2)		1.404^{14}		s H_2O, EtOH, ace; i eth, bz, chl
2518	Cyanomethylmercury	Methylmercurynitrile	C_2H_3HgN	2597-97-9	241.64	cry (chl)	92				vs H_2O, EtOH, bz; s eth
2519	(4-Cyanophenoxy)acetic acid		$C_9H_7NO_3$	1878-82-6	177.157	cry (w)	178				
2520	2-Cyano-*N*-phenylacetamide		$C_9H_8N_2O$	621-03-4	160.172	nd (al)	199.5				
2521	4-Cyanothiazole		$C_4H_2N_2S$	1452-15-9	110.137	nd	58				
2522	Cyanuric acid	1,3,5-Triazine-2,4,6(1*H*,3*H*,5*H*)-trione	$C_3H_3N_3O_3$	108-80-5	129.074	wh cry	>330		1.75^{25}		sl hot H_2O, ace, bz, EtOH; s conc HCl
2523	Cyanuric fluoride	2,4,6-Trifluoro-1,3,5-triazine	$C_3F_3N_3$	675-14-9	135.047			72.8			
2524	Cycasin		$C_8H_{16}N_2O_7$	14901-08-7	252.222	nd (ace aq)	154 dec				
2525	Cyclandelate		$C_{17}H_{24}O_3$	456-59-7	276.371		52	193^{14}			i H_2O
2526	Cyclizine		$C_{18}H_{22}N_2$	82-92-8	266.381	cry (peth)	106				i H_2O; s chl; sl EtOH
2527	Cycloate	Carbamothioic acid, cyclohexylethyl-, *S*-ethyl ester	$C_{11}H_{21}NOS$	1134-23-2	215.356		11.5	145^{10}	1.0156^{30}		

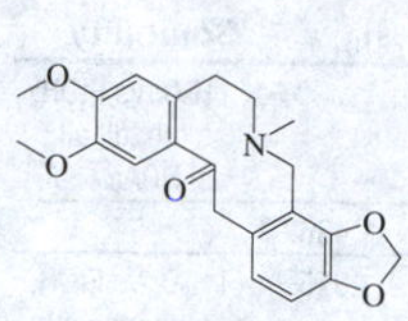
2487
Cryptopine

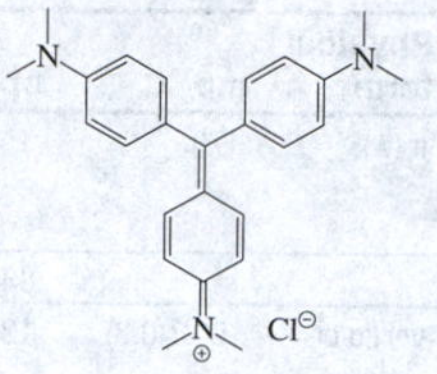
2488
Crystal Violet

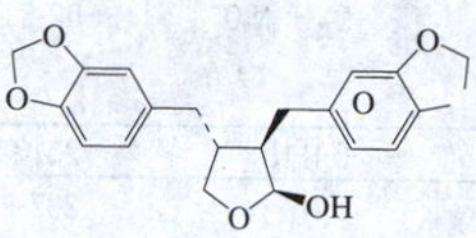
2489
Cubebin

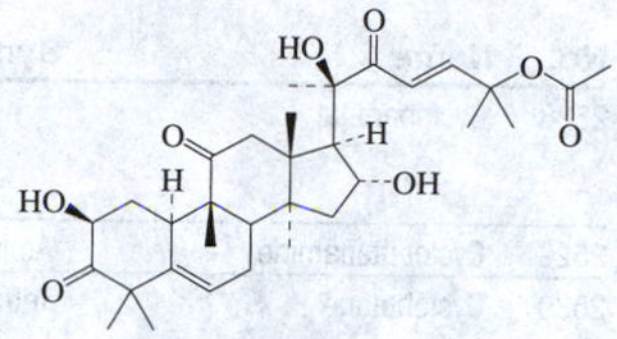
2490
Cucurbitacin B

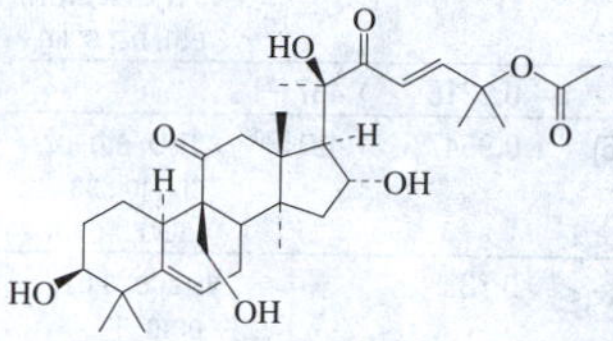
2491
Cucurbitacin C

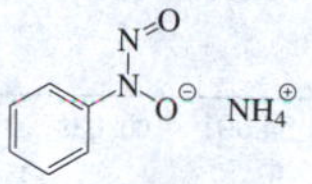
2492
Cupferron

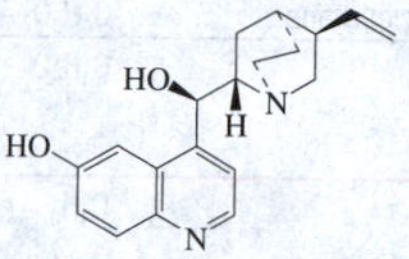
2493
Cupreine

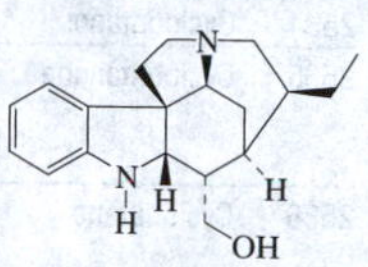
2494
Curan-17-ol, (16α)

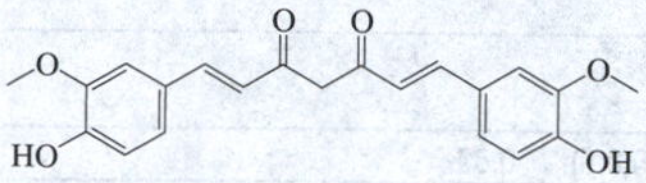
2495
Curcumin

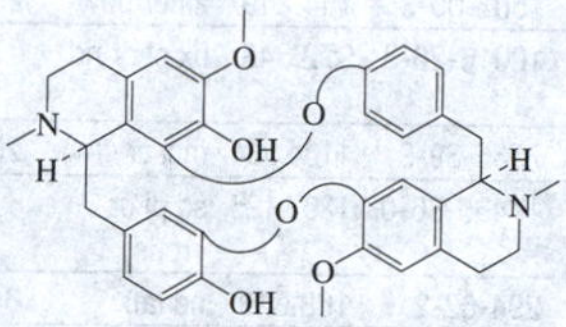
2496
Curine

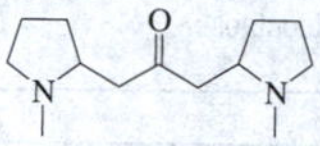
2497
Cuscohygrine

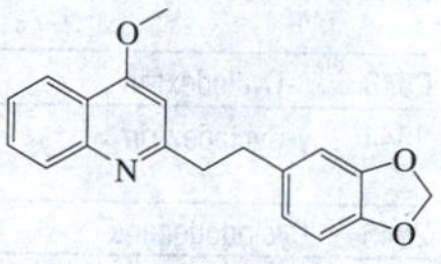
2498
Cusparine

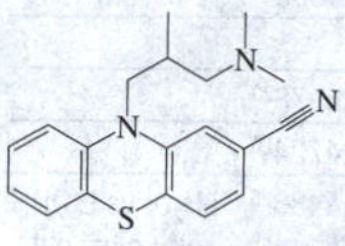
2499
Cyamemazine

$H_2N{-}C{\equiv}N$
2500
Cyanamide

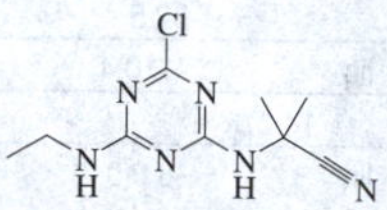
2501
Cyanazine

$HO{-}C{\equiv}N$
2502
Cyanic acid

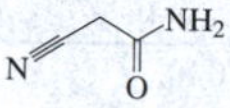
2503
2-Cyanoacetamide

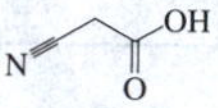
2504
Cyanoacetic acid

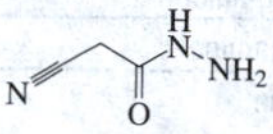
2505
Cyanoacetohydrazide

$HC{\equiv}C{-}C{\equiv}N$
2506
Cyanoacetylene

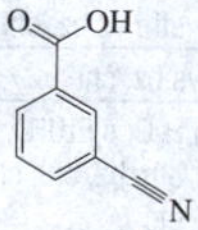
2507
3-Cyanobenzoic acid

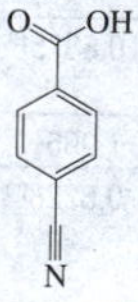
2508
4-Cyanobenzoic acid

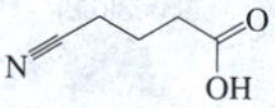
2509
4-Cyanobutanoic acid

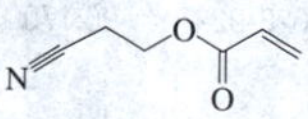
2510
2-Cyanoethyl acrylate

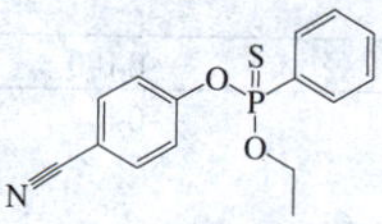
2511
Cyanofenphos

$N{\equiv}C{-}C{\equiv}N$
2512
Cyanogen

2513
Cyanogen bromide

$Cl{-}C{\equiv}N$
2514
Cyanogen chloride

$F{-}C{\equiv}N$
2515
Cyanogen fluoride

$I{-}C{\equiv}N$
2516
Cyanogen iodide

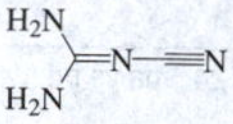
2517
Cyanoguanidine

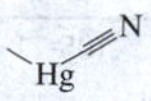
2518
Cyanomethylmercury

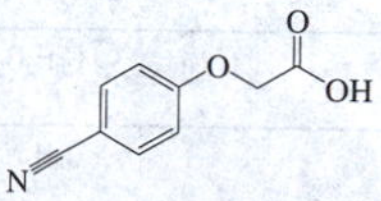
2519
(4-Cyanophenoxy)acetic acid

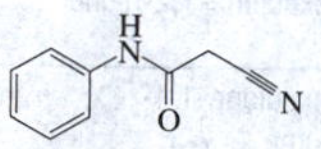
2520
2-Cyano-*N*-phenylacetamide

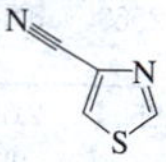
2521
4-Cyanothiazole

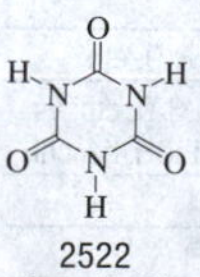
2522
Cyanuric acid

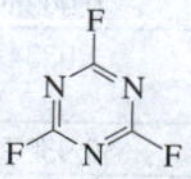
2523
Cyanuric fluoride

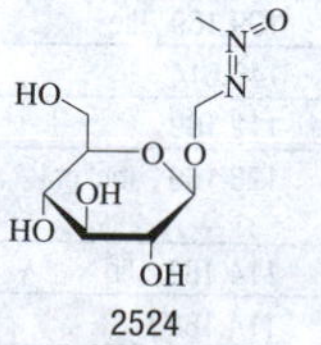
2524
Cycasin

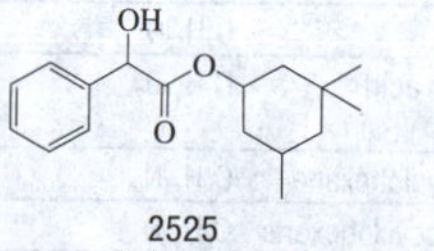
2525
Cyclandelate

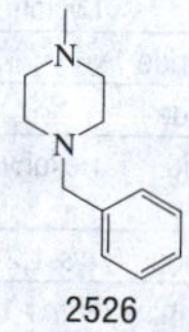
2526
Cyclizine

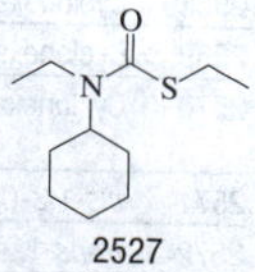
2527
Cycloate

No.	Name	Synonym	Mol. form.	CAS Reg. No.	Mol. wt.	Physical form	mp/°C	bp/°C	den/g cm⁻³	n_D	Solubility
2528	Cyclobarbital		$C_{12}H_{16}N_2O_3$	52-31-3	236.266	lf (w)	173				i H_2O; vs EtOH; s eth, dil alk; sl HOAc
2529	Cyclobutanamine	Aminocyclobutane	C_4H_9N	2516-34-9	71.121			83(20)	0.8328^{20}	1.4363^{19}	
2530	Cyclobutane	Tetramethylene	C_4H_8	287-23-0	56.107	vol liq or gas	-90.7(0.3)	12.5(0.2)	0.7038^{0}	1.375^{20}	i H_2O; vs EtOH, ace; msc eth; s bz
2531	Cyclobutanecarbonitrile	Cyanocyclobutane	C_5H_7N	4426-11-3	81.117			148(1)			
2532	Cyclobutanecarboxylic acid		$C_5H_8O_2$	3721-95-7	100.117	liq	-7(1)	192(4)	1.0599^{20}	1.4400^{20}	sl H_2O; msc EtOH, eth
2533	1,1-Cyclobutanedicarboxylic acid		$C_6H_8O_4$	5445-51-2	144.126	pr (w, eth)	158.0				vs H_2O; s EtOH, eth, bz; sl lig
2534	Cyclobutanol	Hydroxycyclobutane	C_4H_8O	2919-23-5	72.106			124(8)	0.9218^{15}	1.4371^{20}	
2535	Cyclobutanone		C_4H_6O	1191-95-3	70.090	liq	-50.9	98.8(0.6)	0.9547^{0}	1.4215^{20}	s H_2O, eth, bz, chl, tol; vs EtOH; i peth
2536	Cyclobutene		C_4H_6	822-35-5	54.091	col gas		2.5(0.4)	0.733^{0}		vs ace; s bz, peth
2537	Cyclochlorotine		$C_{24}H_{31}Cl_2N_5O_7$	12663-46-6	572.439	nd (MeOH)	255 dec				
2538	Cyclodecane		$C_{10}H_{20}$	293-96-9	140.266		10.4(0.9)	202.3(0.3)	0.8538^{25}	1.4716^{20}	
2539	1,2-Cyclodecanedione	Sebacil	$C_{10}H_{16}O_2$	96-01-5	168.233		41(3)	104^{10}			
2540	Cyclodecanol		$C_{10}H_{20}O$	1502-05-2	156.265		42(3)	125^{12}	0.9606^{20}	1.4926^{20}	s EtOH
2541	Cyclodecanone		$C_{10}H_{18}O$	1502-06-3	154.249	amor pow	23(2)	106^{13}	0.9654^{20}	1.4806^{20}	vs bz, eth, chl
2542	α-Cyclodextrin	Cyclomaltohexaose	$C_{36}H_{60}O_{30}$	10016-20-3	972.843	hx pl or nd					vs cold H_2O; i hot H_2O
2543	β-Cyclodextrin	Cyclomaltoheptaose	$C_{42}H_{70}O_{35}$	7585-39-9	1134.984	mcl cry (w)	260 dec				
2544	γ-Cyclodextrin	Cyclomaltooctaose	$C_{48}H_{80}O_{40}$	17465-86-0	1297.125	sq pl or rods					
2545	Cyclododecane		$C_{12}H_{24}$	294-62-2	168.319	nd (al)	60.8(0.4)	244.0(0.5)	0.82^{80}		
2546	Cyclododecanol		$C_{12}H_{24}O$	1724-39-6	184.318			286			
2547	Cyclododecanone		$C_{12}H_{22}O$	830-13-7	182.302		62.4(0.3)	127^{12}	0.9059^{66}	1.4571^{60}	
2548	1,5,9-Cyclododecatriene	CDT	$C_{12}H_{18}$	4904-61-4	162.271	liq	-17	240	0.84^{100}		
2549	*cis*-Cyclododecene		$C_{12}H_{22}$	1129-89-1	166.303			133^{35}		1.4840^{20}	vs bz, chl
2550	*trans*-Cyclododecene		$C_{12}H_{22}$	1486-75-5	166.303			113^{17}		1.4850^{20}	vs bz, chl
2551	*cis*-9-Cycloheptadecen-1-one	Civetone	$C_{17}H_{30}O$	542-46-1	250.419		32.5	343			
2552	1,3-Cycloheptadiene		C_7H_{10}	4054-38-0	94.154	liq	-110.4	120.5	0.868^{25}	1.4978^{20}	
2553	Cycloheptanamine		$C_7H_{15}N$	5452-35-7	113.201			54^{11}		1.4724^{20}	
2554	Cycloheptane		C_7H_{14}	291-64-5	98.186	liq	-8.0(0.2)	118.8(0.2)	0.8098^{20}	1.4436^{20}	i H_2O; vs EtOH, eth; s bz, chl
2555	1,2-Cycloheptanedione		$C_7H_{10}O_2$	3008-39-7	126.153		-40	108^{17}	1.0583^{22}	1.4689^{22}	s EtOH
2556	Cycloheptanol		$C_7H_{14}O$	502-41-0	114.185		7.15(0.05)	185	0.9554^{20}	1.40705^{20}	sl H_2O; vs EtOH, eth
2557	Cycloheptanone	Suberone	$C_7H_{12}O$	502-42-1	112.169			180.4(0.9)	0.9508^{20}	1.4608^{20}	i H_2O; vs EtOH, eth
2558	1,3,5-Cycloheptatriene	Tropilidene	C_7H_8	544-25-2	92.139	liq; cub cry (-80°C)	-75.18(0.06)	116.3(0.7)	0.8875^{19}	1.5343^{20}	i H_2O; s EtOH, eth; vs bz, chl
2559	2,4,6-Cycloheptatrien-1-one		C_7H_6O	539-80-0	106.122		-5(2)	113^{15}	1.095^{22}	1.6172^{22}	vs bz, chl
2560	Cycloheptene		C_7H_{12}	628-92-2	96.170	liq	-55.3(0.2)	115(3)	0.8228^{20}	1.4552^{20}	i H_2O; s EtOH, eth, bz, chl; sl ctc
2561	1,3-Cyclohexadiene		C_6H_8	592-57-4	80.128	liq	-89	80.3(0.3)	0.8405^{20}	1.4755^{20}	i H_2O; s EtOH, bz, chl, peth; vs eth
2562	1,4-Cyclohexadiene	1,4-Dihydrobenzene	C_6H_8	628-41-1	80.128	liq	-49(1)	89.5(0.2)	0.8471^{20}	1.4725^{20}	i H_2O; msc EtOH, eth; s bz, chl, peth
2563	3,5-Cyclohexadiene-1,2-dione		$C_6H_4O_2$	583-63-1	108.095	red pl or pr	≈65 dec				s eth, ace, bz; i peth
2564	2,5-Cyclohexadiene-1,4-dione, dioxime		$C_6H_6N_2O_2$	105-11-3	138.124	pa ye nd (w)	240 dec				s H_2O
2565	Cyclohexane	Hexahydrobenzene	C_6H_{12}	110-82-7	84.159	liq	6.7(0.2)	80.7(0.7)	0.7739^{25}	1.4235^{25}	i H_2O; msc EtOH, eth, ace, bz, lig, ctc
2566	Cyclohexaneacetic acid		$C_8H_{14}O_2$	5292-21-7	142.196	nd (HCO_2H)	28.9(1)	245	1.0423^{18}	1.4775^{20}	sl H_2O; s eth, ace
2567	Cyclohexanecarbonitrile	Cyclohexyl cyanide	$C_7H_{11}N$	766-05-2	109.169	liq	12.0(0.3)	188(3)	0.919	1.4505^{20}	
2568	Cyclohexanecarbonyl chloride		$C_7H_{11}ClO$	2719-27-9	146.614			183(6)	1.0962^{15}	1.4711^{29}	
2569	Cyclohexanecarboxaldehyde		$C_7H_{12}O$	2043-61-0	112.169			155(12)	0.9035^{20}	1.4496^{20}	s H_2O, eth
2570	Cyclohexanecarboxylic acid	Hexahydrobenzoic acid	$C_7H_{12}O_2$	98-89-5	128.169	mcl pr	28(3)	233(6)	1.0334^{22}	1.4530^{20}	sl H_2O, ctc; vs EtOH, bz, chl
2571	*cis*-1,2-Cyclohexanediamine	*cis*-1,2-Diaminocyclohexane	$C_6H_{14}N_2$	1436-59-5	114.188	liq		40^{2}	0.952^{20}	1.4951^{20}	
2572	*trans*-1,2-Cyclohexanediamine	*trans*-1,2-Diaminocyclohexane	$C_6H_{14}N_2$	1121-22-8	114.188		14.8	80^{15}	0.951^{20}		

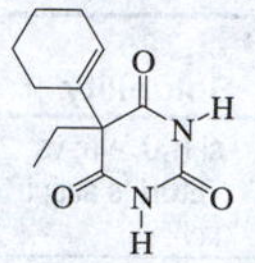
2528
Cyclobarbital

2529
Cyclobutanamine

2530
Cyclobutane

2531
Cyclobutanecarbonitrile

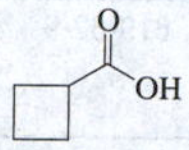
2532
Cyclobutanecarboxylic acid

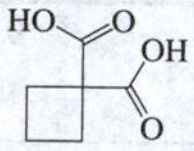
2533
1,1-Cyclobutanedicarboxylic acid

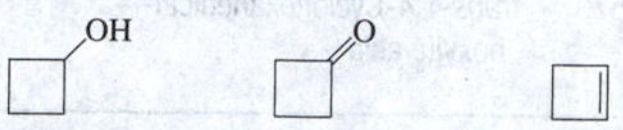
2534
Cyclobutanol

2535
Cyclobutanone

2536
Cyclobutene

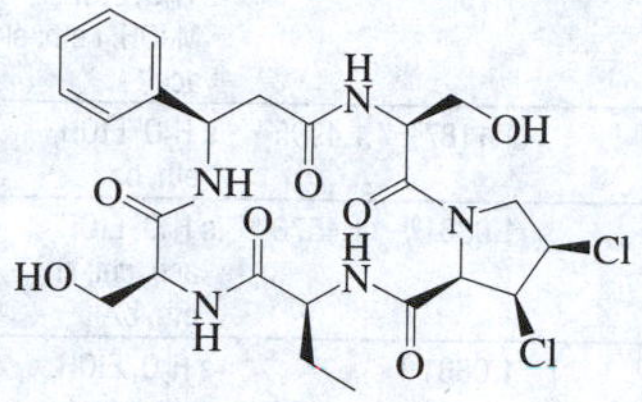
2537
Cyclochlorotine

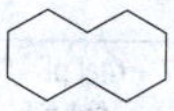
2538
Cyclodecane

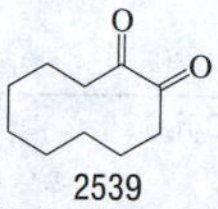
2539
1,2-Cyclodecanedione

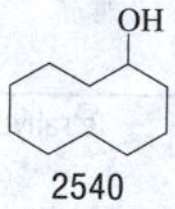
2540
Cyclodecanol

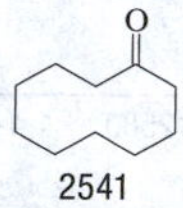
2541
Cyclodecanone

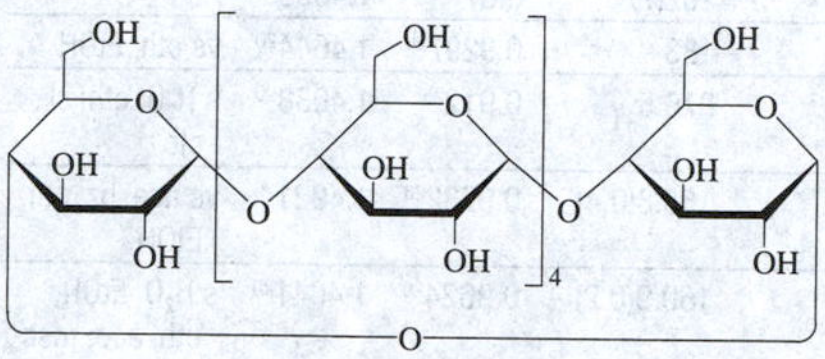
2542
α-Cyclodextrin

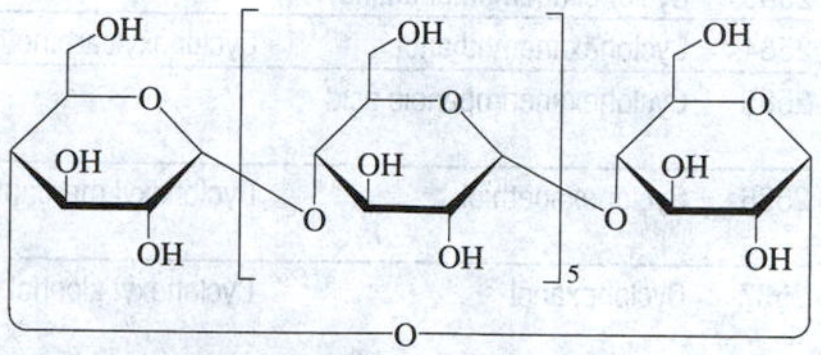
2543
β-Cyclodextrin

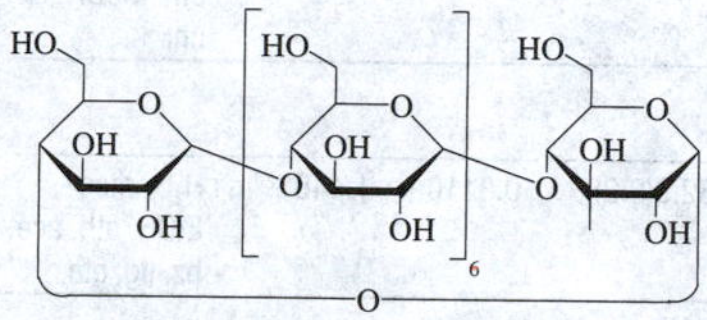
2544
γ-Cyclodextrin

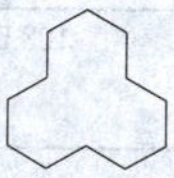
2545
Cyclododecane

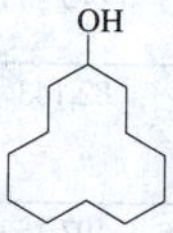
2546
Cyclododecanol

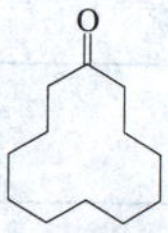
2547
Cyclododecanone

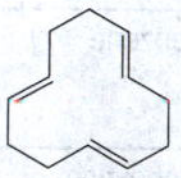
2548
1,5,9-Cyclododecatriene

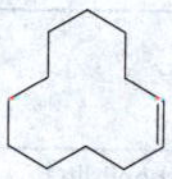
2549
cis-Cyclododecene

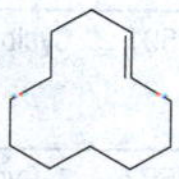
2550
trans-Cyclododecene

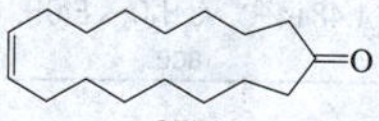
2551
cis-9-Cycloheptadecen-1-one

2552
1,3-Cycloheptadiene

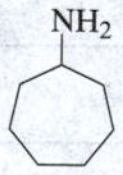
2553
Cycloheptanamine

2554
Cycloheptane

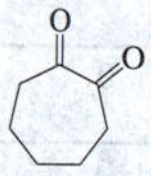
2555
1,2-Cycloheptanedione

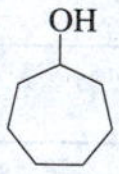
2556
Cycloheptanol

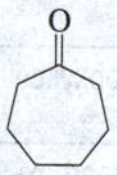
2557
Cycloheptanone

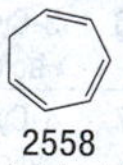
2558
1,3,5-Cycloheptatriene

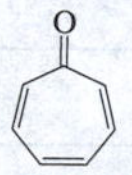
2559
2,4,6-Cycloheptatrien-1-one

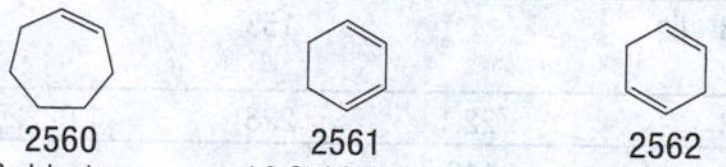
2560
Cycloheptene

2561
1,3-Cyclohexadiene

2562
1,4-Cyclohexadiene

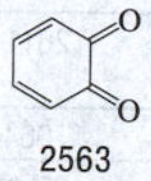
2563
3,5-Cyclohexadiene-1,2-dione

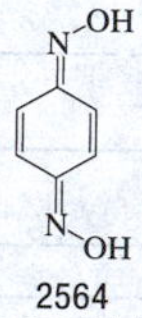
2564
2,5-Cyclohexadiene-1,4-dione, dioxime

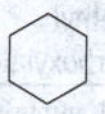
2565
Cyclohexane

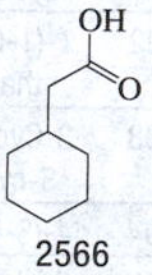
2566
Cyclohexaneacetic acid

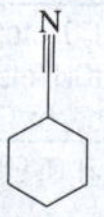
2567
Cyclohexanecarbonitrile

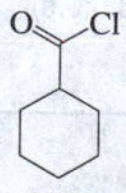
2568
Cyclohexanecarbonyl chloride

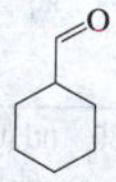
2569
Cyclohexanecarboxaldehyde

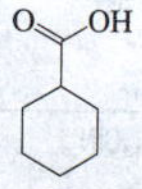
2570
Cyclohexanecarboxylic acid

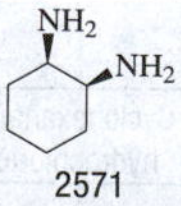
2571
cis-1,2-Cyclohexanediamine

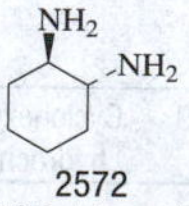
2572
trans-1,2-Cyclohexanediamine

No.	Name	Synonym	Mol. form.	CAS Reg. No.	Mol. wt.	Physical form	mp/°C	bp/°C	den/ g cm^{-3}	n_D	Solubility
2573	*trans*-1,4-Cyclohexanedicarboxylic acid		$C_8H_{12}O_4$	619-82-9	172.179	pr (w)	312.5	300 sub			sl H_2O, eth; vs EtOH; s ace; i chl
2574	1,3-Cyclohexanedimethanamine		$C_8H_{18}N_2$	2579-20-6	142.242		<-70	220	0.945^{20}		vs H_2O, eth, EtOH
2575	1,4-Cyclohexanedimethanol		$C_8H_{16}O_2$	105-08-8	144.212		43	283			
2576	*cis*-1,2-Cyclohexanediol		$C_6H_{12}O_2$	1792-81-0	116.158		95(4)	120^{15}	1.0297^{101}		s EtOH, ace, bz; sl chl
2577	*trans*-1,4-Cyclohexanediol		$C_6H_{12}O_2$	6995-79-5	116.158	mcl pr (ace)	143		1.18^{20}		s H_2O, EtOH, MeOH; i eth; sl ace
2578	1,2-Cyclohexanedione	1,2-Dioxocyclohexane	$C_6H_8O_2$	765-87-7	112.127	cry (peth)	40	194	1.1187^{21}	1.4995^{20}	s H_2O, EtOH, eth, bz
2579	1,3-Cyclohexanedione	Dihydroresorcinol	$C_6H_8O_2$	504-02-9	112.127	pr (bz)	105(2)		1.0861^{91}	1.4576^{102}	s H_2O, EtOH, ace, chl; sl eth, bz
2580	1,4-Cyclohexanedione	Tetrahydroquinone	$C_6H_8O_2$	637-88-7	112.127	mcl pl (w),nd (peth)	78.4(0.1)	132^{20}	1.0861^{91}		s H_2O, EtOH, eth, ace, bz, chl
2581	1,2-Cyclohexanedione dioxime	Nioxime	$C_6H_{10}N_2O_2$	492-99-9	142.155	nd (w, HOAc)	192				s H_2O, ace, chl; sl tfa
2582	Cyclohexaneethanol		$C_8H_{16}O$	4442-79-9	128.212			208(3)	0.9229^{20}	1.4641^{20}	s EtOH, eth, bz
2583	Cyclohexanemethanamine		$C_7H_{15}N$	3218-02-8	113.201			162(7)	0.87^{25}	1.4630^{20}	
2584	Cyclohexanemethanol	Cyclohexylcarbinol	$C_7H_{14}O$	100-49-2	114.185	liq	-43	183	0.9297^{20}	1.4644^{20}	vs eth, EtOH
2585	Cyclohexanepropanoic acid		$C_9H_{16}O_2$	701-97-3	156.222		16	276.5	0.912^{25}	1.4638^{20}	s H_2O, eth; sl ctc
2586	Cyclohexanethiol	Cyclohexyl mercaptan	$C_6H_{12}S$	1569-69-3	116.224			158.8(0.4)	0.9782^{20}	1.4921^{20}	vs ace, bz, eth, EtOH
2587	Cyclohexanol	Cyclohexyl alcohol	$C_6H_{12}O$	108-93-0	100.158	hyg nd	26(1)	160.9(0.2)	0.9624^{20}	1.4641^{20}	s H_2O, EtOH, eth, ace; msc bz; sl chl
2588	Cyclohexanone	Pimelic ketone	$C_6H_{10}O$	108-94-1	98.142	liq	-27.93(0.05)	155.4(0.1)	0.9478^{20}	1.4507^{20}	s H_2O, EtOH, eth, ace, bz, chl, ctc
2589	Cyclohexanone oxime		$C_6H_{11}NO$	100-64-1	113.157	hex pr (lig)	89.05(0.09)	208(2)			s H_2O, EtOH, eth, MeOH; sl chl
2590	Cyclohexanone peroxide		$C_{12}H_{22}O_5$	78-18-2	246.300	cry or long nd	79				
2591	Cyclohexene	Tetrahydrobenzene	C_6H_{10}	110-83-8	82.143	liq	-103.5(0.4)	82.9(0.2)	0.8110^{20}	1.4465^{20}	i H_2O; msc EtOH, eth, ace, bz, lig, ctc
2592	1-Cyclohexenecarbonitrile	1-Cyanocyclohexene	C_7H_9N	1855-63-6	107.153			81^{12}			
2593	1-Cyclohexene-1-carboxaldehyde		$C_7H_{10}O$	1192-88-7	110.153			69^{18}	0.9694^{20}	1.5005^{20}	s EtOH, eth
2594	3-Cyclohexene-1-carboxaldehyde		$C_7H_{10}O$	100-50-5	110.153		1.0	164(9)	0.9692^{20}	1.4745^{20}	s ace, MeOH; sl ctc
2595	1-Cyclohexene-1-carboxylic acid		$C_7H_{10}O_2$	636-82-8	126.153		38	241	1.109^{20}	1.4902^{20}	sl H_2O; s EtOH, ace
2596	3-Cyclohexene-1-carboxylic acid		$C_7H_{10}O_2$	4771-80-6	126.153		17	237(13)	1.0820^{20}	1.4814^{20}	vs H_2O; s EtOH, ace
2597	4-Cyclohexene-1,2-dicarboxylic acid		$C_8H_{10}O_4$	88-98-2	170.163	pr (w)	173.0				
2598	2-Cyclohexen-1-ol		$C_6H_{10}O$	822-67-3	98.142			176(10)	0.9923^{15}	1.4790^{25}	s EtOH, ace
2599	2-Cyclohexen-1-one		C_6H_8O	930-68-7	96.127	liq	-53	172.3(0.7)	0.9620^{25}	1.4883^{20}	vs EtOH; s ace
2600	1-Cyclohexen-1-ylbenzene		$C_{12}H_{14}$	771-98-2	158.239	liq	-11	268(15)	0.9939^{20}	1.5718^{20}	vs MeOH
2601	2-(1-Cyclohexen-1-yl)-cyclohexanone		$C_{12}H_{18}O$	1502-22-3	178.270			116^{3}		1.5070^{20}	
2602	1-(1-Cyclohexen-1-yl)-ethanone		$C_8H_{12}O$	932-66-1	124.180		73	202(12)	0.9655^{20}	1.4881^{20}	s EtOH, eth
2603	3-Cyclohexenylmethyl 3-cyclohexenecarboxylate		$C_{14}H_{20}O_2$	2611-00-9	220.308	liq		153^{7}			
2604	4-(3-Cyclohexen-1-yl)pyridine		$C_{11}H_{13}N$	70644-46-1	159.228		22.1	226	1.0222^{25}	1.5466^{25}	
2605	Cycloheximide		$C_{15}H_{23}NO_4$	66-81-9	281.349	pl (al)	119				vs EtOH
2606	Cyclohexyl acetate		$C_8H_{14}O_2$	622-45-7	142.196			174(4)	0.968^{20}	1.442^{20}	vs eth, EtOH
2607	Cyclohexyl acrylate		$C_9H_{14}O_2$	3066-71-5	154.206			183	1.0275^{20}	1.4673^{20}	i H_2O; msc EtOH, eth; s chl
2608	Cyclohexylamine	Cyclohexanamine	$C_6H_{13}N$	108-91-8	99.174	liq	-17.7(0.7)	133.6(0.5)	0.8191^{20}	1.4625^{15}	s H_2O, ctc; vs EtOH; msc eth, ace, bz
2609	Cyclohexylamine hydrochloride	Cyclohexanamine hydrochloride	$C_6H_{14}ClN$	4998-76-9	135.635	nd (w, al-eth)	206.5				vs H_2O, EtOH

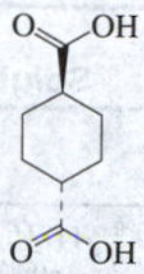
2573
trans-1,4-Cyclohexanedicarboxylic acid

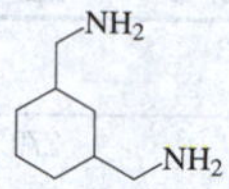
2574
1,3-Cyclohexanedimethanamine

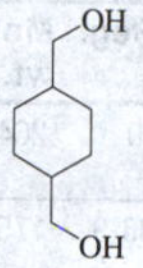
2575
1,4-Cyclohexanedimethanol

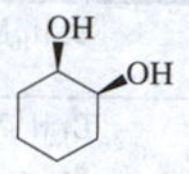
2576
cis-1,2-Cyclohexanediol

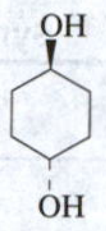
2577
trans-1,4-Cyclohexanediol

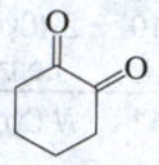
2578
1,2-Cyclohexanedione

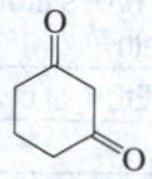
2579
1,3-Cyclohexanedione

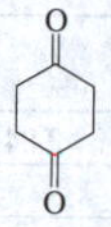
2580
1,4-Cyclohexanedione

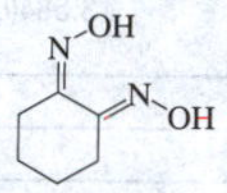
2581
1,2-Cyclohexanedione dioxime

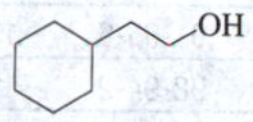
2582
Cyclohexaneethanol

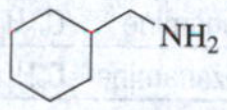
2583
Cyclohexanemethanamine

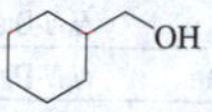
2584
Cyclohexanemethanol

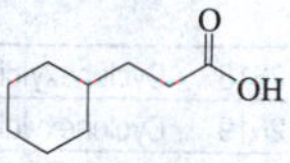
2585
Cyclohexanepropanoic acid

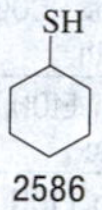
2586
Cyclohexanethiol

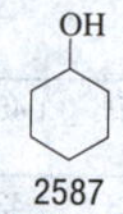
2587
Cyclohexanol

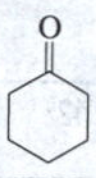
2588
Cyclohexanone

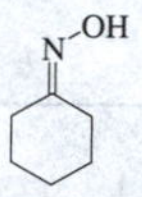
2589
Cyclohexanone oxime

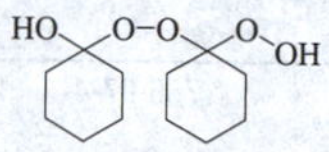
2590
Cyclohexanone peroxide

2591
Cyclohexene

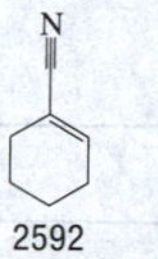
2592
1-Cyclohexenecarbonitrile

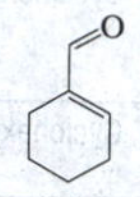
2593
1-Cyclohexene-1-carboxaldehyde

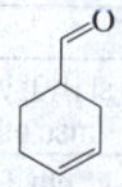
2594
3-Cyclohexene-1-carboxaldehyde

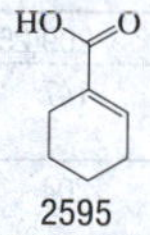
2595
1-Cyclohexene-1-carboxylic acid

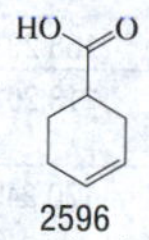
2596
3-Cyclohexene-1-carboxylic acid

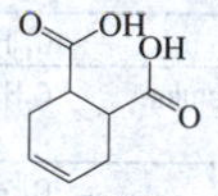
2597
4-Cyclohexene-1,2-dicarboxylic acid

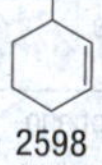
2598
2-Cyclohexen-1-ol

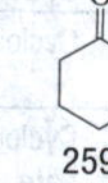
2599
2-Cyclohexen-1-one

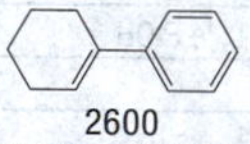
2600
1-Cyclohexen-1-ylbenzene

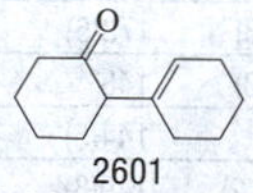
2601
2-(1-Cyclohexen-1-yl)cyclohexanone

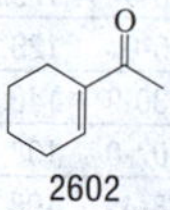
2602
1-(1-Cyclohexen-1-yl)ethanone

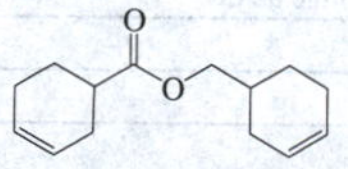
2603
3-Cyclohexenylmethyl 3-cyclohexenecarboxylate

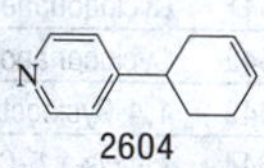
2604
4-(3-Cyclohexen-1-yl)pyridine

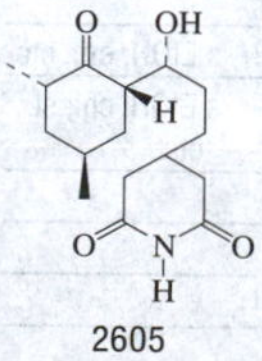
2605
Cycloheximide

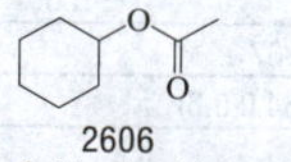
2606
Cyclohexyl acetate

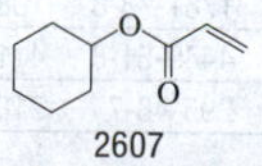
2607
Cyclohexyl acrylate

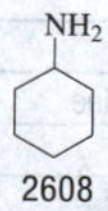
2608
Cyclohexylamine

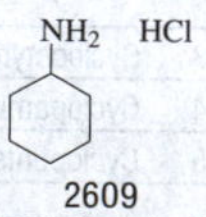
2609
Cyclohexylamine hydrochloride

No.	Name	Synonym	Mol. form.	CAS Reg. No.	Mol. wt.	Physical form	mp/°C	bp/°C	den/g cm^{-3}	n_D	Solubility
2610	2-(Cyclohexylaminothio)-benzothiazole		$C_{13}H_{16}N_2S_2$	95-33-0	264.409		103				
2611	*N*-Cyclohexylaniline		$C_{12}H_{17}N$	1821-36-9	175.270	mcl pr	16	279	1.0155^{20}	1.5610^{20}	i H_2O; s EtOH, eth, bz
2612	Cyclohexylbenzene	Phenylcyclohexane	$C_{12}H_{16}$	827-52-1	160.255	pl	7.02(0.1)	239(2)	0.9427^{20}	1.5329^{20}	i H_2O; vs EtOH; s eth; sl ctc
2613	Cyclohexyl benzoate		$C_{13}H_{16}O_2$	2412-73-9	204.265		<-10	285	1.0429^{20}	1.5200^{20}	i H_2O; s EtOH, eth
2614	Cyclohexyl butanoate		$C_{10}H_{18}O_2$	1551-44-6	170.249			213(8)	0.9572^{0}		i H_2O; s EtOH; sl ctc
2615	3-Cyclohexyl-2-butenoic acid	Cicrotoic acid	$C_{10}H_{16}O_2$	25229-42-9	168.233	pr (aq-MeOH)	85.5				
2616	Cyclohexyl chloroformate		$C_7H_{11}ClO_2$	13248-54-9	162.614			87.5^{27}			vs eth
2617	Cyclohexylcyclohexane		$C_{12}H_{22}$	92-51-3	166.303		3.684(0.002)	239(2)			sl H_2O; s EtOH, eth
2618	Cyclohexyldiethylamine	*N,N*-Diethylcyclohexanamine	$C_{10}H_{21}N$	91-65-6	155.281			192	0.8443^{25}		s EtOH; sl ctc
2619	Cyclohexyldimethylamine	*N,N*-Dimethylcyclohexanamine	$C_8H_{17}N$	98-94-2	127.228			161(21)			
2620	2-Cyclohexyl-4,6-dinitrophenol		$C_{12}H_{14}N_2O_5$	131-89-5	266.249	cry	105.5(0.3)				sl H_2O; s bz, DMF
2621	(1,2-Cyclohexylenedinitrilo)-tetraacetic acid monohydrate	CDTA	$C_{14}H_{24}N_2O_9$	13291-61-7	364.349	cry (w)	215				
2622	1-Cyclohexylethanone		$C_8H_{14}O$	823-76-7	126.196			180.5	0.9176^{20}	1.4565^{16}	i H_2O; s eth
2623	Cyclohexylethylamine	*N*-Ethylcyclohexanamine	$C_8H_{17}N$	5459-93-8	127.228			164	0.868^{0}		sl H_2O, ctc; msc EtOH, eth
2624	4-Cyclohexyl-3-ethyl-4*H*-1,2,4-triazole	Hexazole	$C_{10}H_{17}N_3$	4671-03-8	179.262	pr (eth)	89.5	227^{10}			vs H_2O, bz, chl
2625	Cyclohexyl formate		$C_7H_{12}O_2$	4351-54-6	128.169			162	1.0057^{0}	1.4430^{20}	i H_2O; s EtOH, HOAc, HCOOH; vs eth
2626	Cyclohexyl hydroperoxide		$C_6H_{12}O_2$	766-07-4	116.158		-20	$42^{0.1}$	1.019^{20}	1.4645^{25}	vs eth, EtOH, HOAc
2627	Cyclohexylideneacetonitrile		$C_8H_{11}N$	4435-18-1	121.180			107^{22}	0.9483^{15}	1.4382^{25}	vs eth, EtOH
2628	2-Cyclohexylidenecyclohexanone		$C_{12}H_{18}O$	1011-12-7	178.270	cry (MeOH aq)	56.5				
2629	Cyclohexyl isocyanate	Isocyanatocyclohexane	$C_7H_{11}NO$	3173-53-3	125.168			172	0.98^{25}	1.4551^{20}	
2630	Cyclohexylisopropylamine	*N*-Isopropylcyclohexanamine	$C_9H_{19}N$	1195-42-2	141.254			62^{12}	0.859^{25}	1.4480^{20}	
2631	Cyclohexyl isothiocyanate	Isothiocyanatocyclohexane	$C_7H_{11}NS$	1122-82-3	141.234			221	1.0339^{20}	1.5375^{20}	i H_2O; s EtOH, eth; sl ctc
2632	Cyclohexylmagnesium chloride		$C_6H_{11}ClMg$	931-51-1	142.909	hyg liq					s eth
2633	Cyclohexyl methacrylate		$C_{10}H_{16}O_2$	101-43-9	168.233			209(25)	0.9626^{20}	1.4578^{20}	
2634	Cyclohexylmethylamine	*N*-Methylcyclohexanamine	$C_7H_{15}N$	100-60-7	113.201			147	0.8660^{23}	1.4560^{20}	sl H_2O; vs EtOH; msc eth; s chl
2635	Cyclohexyl 2-methylpropanoate		$C_{10}H_{18}O_2$	1129-47-1	170.249			204	0.9489^{0}		vs eth, EtOH
2636	2-Cyclohexylphenol		$C_{12}H_{16}O$	119-42-6	176.254	nd (lig)	54.7(0.7)				vs EtOH, HOAc
2637	4-Cyclohexylphenol		$C_{12}H_{16}O$	1131-60-8	176.254	nd (bz)	130.6(0.5)	294(26)			i H_2O; vs EtOH, eth; s bz; sl lig
2638	α-Cyclohexyl-α-phenyl-1-piperidinepropanol	Trihexphenidyl	$C_{20}H_{31}NO$	144-11-6	301.466		114				
2639	Cyclohexyl propanoate		$C_9H_{16}O_2$	6222-35-1	156.222			194(6)	0.9359^{20}	1.4403^{20}	i H_2O; s EtOH, eth, ace, ctc
2640	Cyclohexylsulfamic acid	Cyclamic acid	$C_6H_{13}NO_3S$	100-88-9	179.237		169.5				vs alk
2641	Cyclononane		C_9H_{18}	293-55-0	126.239		10.7(0.3)	173(5)	0.8463^{25}	1.4666^{20}	
2642	Cyclononanone		$C_9H_{16}O$	3350-30-9	140.222		31.9(0.3)	148^{24}	0.9560^{20}	1.4729^{20}	s EtOH
2643	1,4-Cyclooctadiene		C_8H_{12}	1073-07-0	108.181	liq	-53(4)	144(5)	0.8754^{20}		
2644	*cis,cis*-1,5-Cyclooctadiene		C_8H_{12}	111-78-4	108.181	liq	-69.2(0.2)	149(3)	0.883^{20}	1.4905^{25}	vs bz
2645	Cyclooctanamine	Aminocyclooctane	$C_8H_{17}N$	5452-37-9	127.228	liq	-48	190	0.928^{25}	1.4804^{20}	
2646	Cyclooctane		C_8H_{16}	292-64-8	112.213		14.82(0.04)	151.1(0.1)	0.8349^{20}	1.4586^{20}	i H_2O; s bz, lig
2647	Cyclooctanol		$C_8H_{16}O$	696-71-9	128.212		25.1	99^{16}	0.9740^{20}	1.4871^{20}	s EtOH
2648	Cyclooctanone		$C_8H_{14}O$	502-49-8	126.196		44.2(0.3)	201.4(0.2)	0.9581^{20}	1.4694^{20}	i H_2O; s EtOH, ace, bz; sl ctc
2649	1,3,5,7-Cyclooctatetraene	[8]Annulene	C_8H_8	629-20-9	104.150	liq	-4.7(0.1)	140(3)	0.9206^{20}	1.5381^{20}	s EtOH, eth, ace, bz
2650	1,3,5-Cyclooctatriene		C_8H_{10}	1871-52-9	106.165	liq	-86(8)	144(11)	0.8971^{25}	1.5035^{25}	
2651	*cis*-Cyclooctene		C_8H_{14}	931-87-3	110.197	liq	-13(4)	145(3)	0.8472^{20}	1.4698^{20}	s EtOH, eth, ctc
2652	*trans*-Cyclooctene		C_8H_{14}	931-89-5	110.197	liq	-59(5)	146(5)	0.8483^{20}	1.4741^{25}	s EtOH, chl; sl ctc
2653	Cyclooctyne		C_8H_{12}	1781-78-8	108.181			158	0.868^{20}	1.4850^{20}	
2654	Cyclopamine	11-Deoxojervine	$C_{27}H_{41}NO_2$	4449-51-8	411.621	nd (EtOH)	237				
2655	Cyclopentadecane		$C_{15}H_{30}$	295-48-7	210.399	nd (MeOH)	64.0(0.6)		0.8364^{61}	1.4592^{61}	

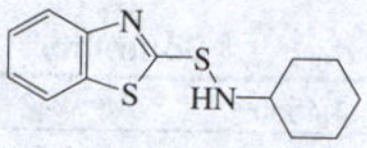
2610
2-(Cyclohexylaminothio)benzothiazole

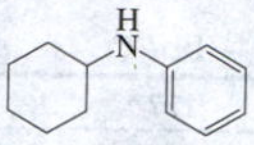
2611
N-Cyclohexylaniline

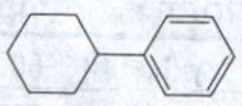
2612
Cyclohexylbenzene

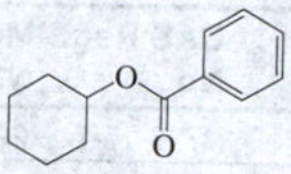
2613
Cyclohexyl benzoate

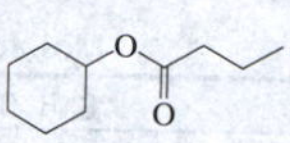
2614
Cyclohexyl butanoate

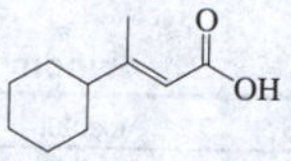
2615
3-Cyclohexyl-2-butenoic acid

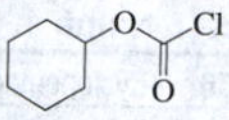
2616
Cyclohexyl chloroformate

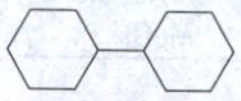
2617
Cyclohexylcyclohexane

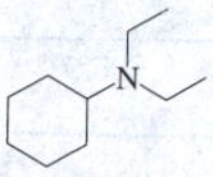
2618
Cyclohexyldiethylamine

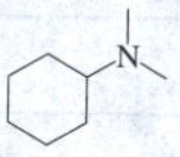
2619
Cyclohexyldimethylamine

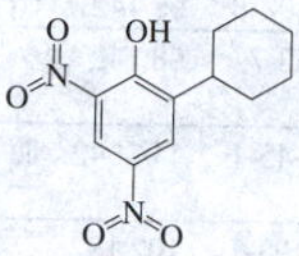
2620
2-Cyclohexyl-4,6-dinitrophenol

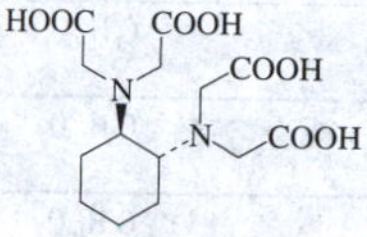
2621
(1,2-Cyclohexylenedinitrilo)tetraacetic acid monohydrate

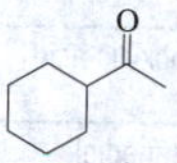
2622
1-Cyclohexylethanone

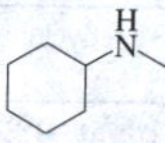
2623
Cyclohexylethylamine

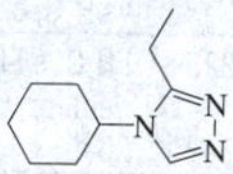
2624
4-Cyclohexyl-3-ethyl-4*H*-1,2,4-triazole

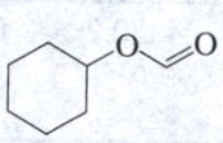
2625
Cyclohexyl formate

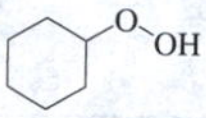
2626
Cyclohexyl hydroperoxide

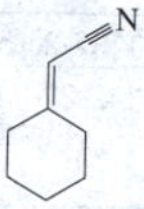
2627
Cyclohexylideneacetonitrile

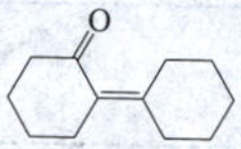
2628
2-Cyclohexylidenecyclohexanone

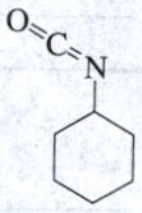
2629
Cyclohexyl isocyanate

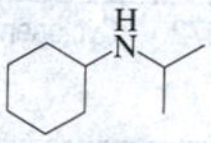
2630
Cyclohexylisopropylamine

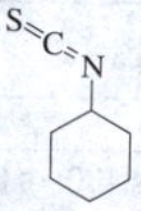
2631
Cyclohexyl isothiocyanate

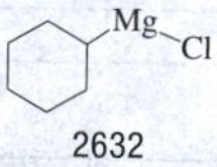
2632
Cyclohexylmagnesium chloride

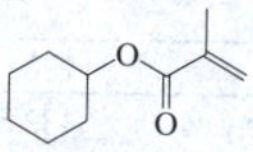
2633
Cyclohexyl methacrylate

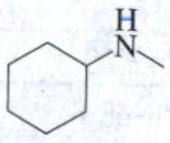
2634
Cyclohexylmethylamine

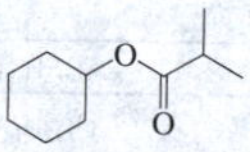
2635
Cyclohexyl 2-methylpropanoate

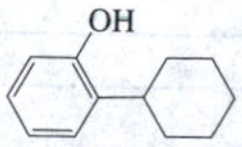
2636
2-Cyclohexylphenol

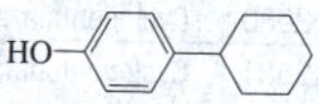
2637
4-Cyclohexylphenol

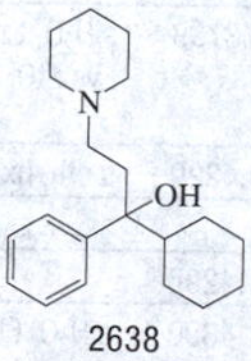
2638
α-Cyclohexyl-α-phenyl-1-piperidinepropanol

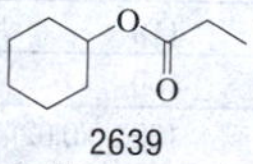
2639
Cyclohexyl propanoate

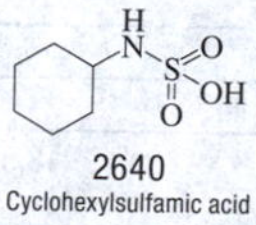
2640
Cyclohexylsulfamic acid

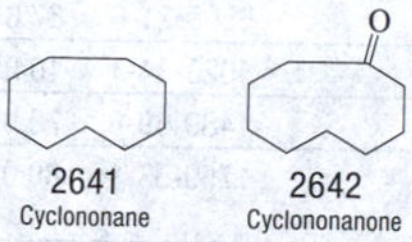
2641
Cyclononane

2642
Cyclononanone

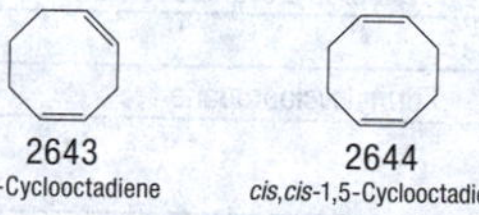
2643
1,4-Cyclooctadiene

2644
cis,cis-1,5-Cyclooctadiene

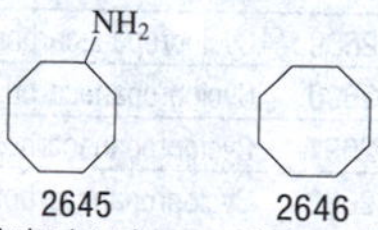
2645
Cyclooctanamine

2646
Cyclooctane

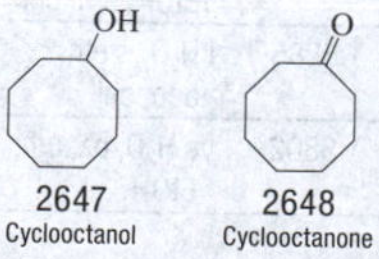
2647
Cyclooctanol

2648
Cyclooctanone

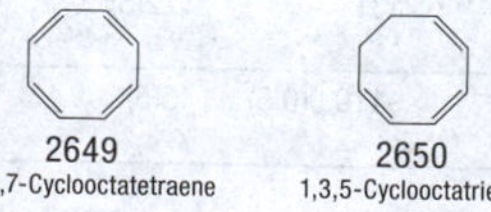
2649
1,3,5,7-Cyclooctatetraene

2650
1,3,5-Cyclooctatriene

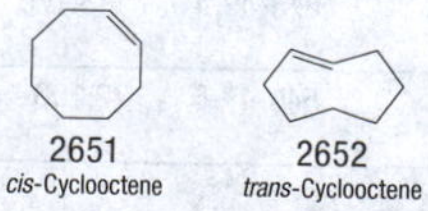
2651
cis-Cyclooctene

2652
trans-Cyclooctene

2653
Cyclooctyne

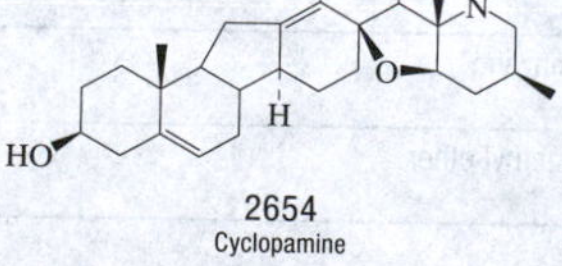
2654
Cyclopamine

2655
Cyclopentadecane

Organic

No.	Name	Synonym	Mol. form.	CAS Reg. No.	Mol. wt.	Physical form	mp/°C	bp/°C	den/g cm^{-3}	n_D	Solubility
2656	Cyclopentadecanol	Exaltol	$C_{15}H_{30}O$	4727-17-7	226.398	cry (MeOH)	80.5	177^{11}	0.930^{20}	1.4555^{98}	
2657	Cyclopentadecanone		$C_{15}H_{28}O$	502-72-7	224.382		65.2(0.5)	$120^{0.3}$	0.8895^{25}	1.4637^{60}	sl H_2O; s EtOH, ace
2658	1,3-Cyclopentadiene	Pyropentylene	C_5H_6	542-92-7	66.102	liq	-96.54(0.05)	41(1)	0.8021^{20}	1.4440^{20}	i H_2O; msc EtOH, eth, bz; s ace
2659	Cyclopentane	Pentamethylene	C_5H_{10}	287-92-3	70.133	liq	-93.4(0.3)	49.2(0.1)	0.7457^{20}	1.4065^{20}	i H_2O; msc EtOH, eth, ace, bz, peth, ctc
2660	Cyclopentaneacetic acid		$C_7H_{12}O_2$	1123-00-8	128.169	pl	13.5	228	1.0216^{18}	1.4523^{18}	
2661	Cyclopentanecarbonitrile	Cyanocyclopentane	C_6H_9N	4254-02-8	95.142	liq	-76	169(2)	0.912	1.4410^{20}	
2662	Cyclopentanecarboxaldehyde		$C_6H_{10}O$	872-53-7	98.142			133.5	0.9371^{20}	1.4432^{20}	vs H_2O, eth, EtOH
2663	Cyclopentanecarboxylic acid	Cyclopentanoic acid	$C_6H_{10}O_2$	3400-45-1	114.142	liq	-7	212	1.0527^{20}	1.4532^{20}	sl H_2O, ctc; s MeOH
2664	*cis*-1,2-Cyclopentanediol		$C_5H_{10}O_2$	5057-98-7	102.132		30	124^{29}			
2665	*trans*-1,2-Cyclopentanediol		$C_5H_{10}O_2$	5057-99-8	102.132		46(4)	226			
2666	Cyclopentanemethanol		$C_6H_{12}O$	3637-61-4	100.158			163	0.9332^{20}	1.4579^{20}	
2667	Cyclopentanepropanoic acid		$C_8H_{14}O_2$	140-77-2	142.196			158^{26}	1.0100^{17}	1.4570^{20}	
2668	Cyclopentanethiol	Cyclopentyl mercaptan	$C_5H_{10}S$	1679-07-8	102.198			132.2(0.6)	0.9550^{20}		
2669	Cyclopentanol	Cyclopentyl alcohol	$C_5H_{10}O$	96-41-3	86.132	liq	-17(2)	140.4(0.2)	0.9488^{20}	1.4530^{20}	sl H_2O, ctc; s EtOH, eth, ace
2670	Cyclopentanone	Adipic ketone	C_5H_8O	120-92-3	84.117	liq	-51.70(0.02)	130.5(0.2)	0.9487^{20}	1.4366^{20}	i H_2O; s EtOH, ace, ctc, hx; msc eth
2671	Cyclopentanone oxime		C_5H_9NO	1192-28-5	99.131		57.8	196			vs H_2O, bz
2672	Cyclopentene		C_5H_8	142-29-0	68.118	liq	-135.02(0.09)	44.2(0.2)	0.7720^{20}	1.4225^{20}	i H_2O; s EtOH, eth, bz, ctc, peth
2673	1-Cyclopentenecarbonitrile	1-Cyanocyclopentene	C_6H_7N	3047-38-9	93.127	liq		81^{30}			
2674	1-Cyclopentene-1-carboxaldehyde		C_6H_8O	6140-65-4	96.127	liq	-32	146	0.970^{21}	1.4872^{17}	
2675	2-Cyclopentene-1-tridecanoic acid, (*S*)-	Chaulmoogric acid	$C_{18}H_{32}O_2$	29106-32-9	280.446	pl or lf (al, HOAc)	68(2)	247^{20}			vs eth, chl
2676	2-Cyclopentene-1-undecanoic acid, (*R*)-	Hydnocarpic acid	$C_{16}H_{28}O_2$	459-67-6	252.392		60.5				vs EtOH, chl, peth
2677	2-Cyclopenten-1-one		C_5H_6O	930-30-3	82.101			136	0.989^{15}	1.4629^{15}	vs eth, EtOH
2678	3-Cyclopenten-1-one		C_5H_6O	14320-37-7	82.101	liq		28^{17}			
2679	*N*-(1-Cyclopenten-1-yl)-pyrrolidine	1-Pyrrolidinylcyclopentene	$C_9H_{15}N$	7148-07-4	137.222			105^{15}		1.5128^{20}	
2680	Cyclopenthiazide		$C_{13}H_{18}ClN_3O_4S_2$	742-20-1	379.883		238				
2681	Cyclopentobarbital		$C_{12}H_{14}N_2O_3$	76-68-6	234.250	cry (w, dil al)	139.5				sl H_2O; vs EtOH
2682	Cyclopentylamine	Cyclopentanamine	$C_5H_{11}N$	1003-03-8	85.148	liq	-82.69(0.05)	108.5(0.1)	0.8689^{20}	1.4728^{25}	s ace, bz, chl
2683	Cyclopentylbenzene		$C_{11}H_{14}$	700-88-9	146.229			223(4)	0.9462^{20}	1.5280^{20}	vs eth
2684	2-Cyclopentylidenecyclopentanone		$C_{10}H_{14}O$	825-25-2	150.217			135^{25}	1.0179^{18}	1.5215^{18}	
2685	Cyclopentyl methyl sulfide		$C_6H_{12}S$	7133-36-0	116.224			156.2(0.8)			
2686	Cyclophosphamide	Cyclophosphane	$C_7H_{15}Cl_2N_2O_2P$	50-18-0	261.086		50.4(0.5)				vs H_2O; sl bz, chl, diox, EtOH
2687	Cycloposine		$C_{33}H_{51}NO_7$	23185-94-6	573.761		268				
2688	Cyclopropane	Trimethylene	C_3H_6	75-19-4	42.080	col gas	-127.6(0.2)	-31(2)	0.617^{25} (p>1 atm)	1.3799^{-42}	s H_2O, bz, peth; vs EtOH, eth
2689	Cyclopropanecarbonitrile	Cyclopropyl cyanide	C_4H_5N	5500-21-0	67.090			135(1)	0.8946^{20}	1.4229^{20}	s eth, hx; sl ctc
2690	Cyclopropanecarbonyl chloride		C_4H_5ClO	4023-34-1	104.535			119	1.1516^{20}		
2691	Cyclopropanecarboxaldehyde	Formylcyclopropane	C_4H_6O	1489-69-6	70.090	liq		100	0.938	1.4298^{20}	
2692	Cyclopropanecarboxylic acid		$C_4H_6O_2$	1759-53-1	86.090		18.5	182.20(0.02)	1.0885^{20}	1.4390^{20}	s H_2O, EtOH, eth; sl ctc
2693	1,1-Cyclopropanedicarboxylic acid		$C_5H_6O_4$	598-10-7	130.100	pr or nd (chl) pr (w +1)	140.5				vs H_2O, eth
2694	Cyclopropanemethanol		C_4H_8O	2516-33-8	72.106			124	0.911^{25}		sl ctc
2695	Cyclopropanone		C_3H_4O	5009-27-8	56.063		stable only at low temp.				
2696	Cyclopropene		C_3H_4	2781-85-3	40.064	gas		-36(6)			
2697	Cyclopropylamine	Cyclopropanamine	C_3H_7N	765-30-0	57.095	liq	-35.38(0.05)	49.2(0.1)	0.8240^{20}	1.4210^{20}	msc H_2O; s EtOH, eth, chl
2698	Cyclopropylbenzene		C_9H_{10}	873-49-4	118.175	liq	-31	172(3)	0.9317^{20}	1.5285^{20}	i H_2O; s eth, ace, chl
2699	Cyclopropyl methyl ether		C_4H_8O	540-47-6	72.106	liq	-110.0(0.5)	45(5)	0.8100^{20}	1.3802^{20}	vs H_2O, bz, eth, EtOH

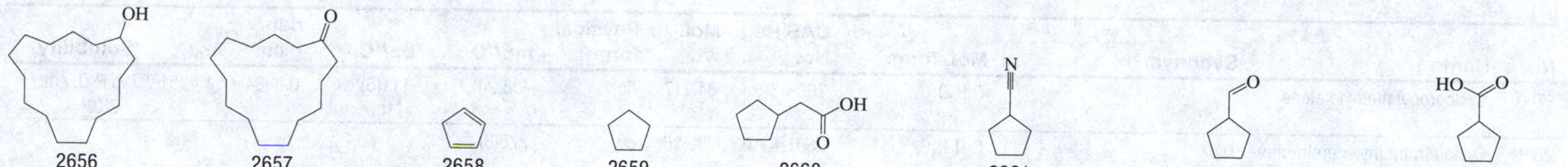

2656 Cyclopentadecanol
2657 Cyclopentadecanone
2658 1,3-Cyclopentadiene
2659 Cyclopentane
2660 Cyclopentaneacetic acid
2661 Cyclopentanecarbonitrile
2662 Cyclopentanecarboxaldehyde
2663 Cyclopentanecarboxylic acid

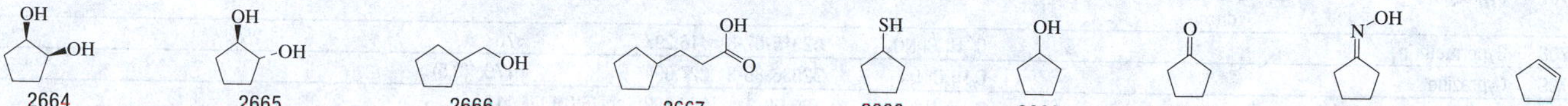

2664 *cis*-1,2-Cyclopentanediol
2665 *trans*-1,2-Cyclopentanediol
2666 Cyclopentanemethanol
2667 Cyclopentanepropanoic acid
2668 Cyclopentanethiol
2669 Cyclopentanol
2670 Cyclopentanone
2671 Cyclopentanone oxime
2672 Cyclopentene

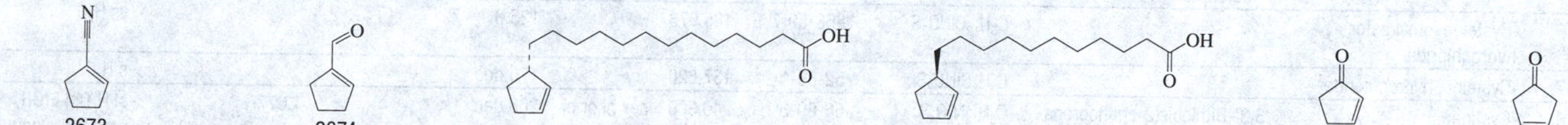

2673 1-Cyclopentenecarbonitrile
2674 1-Cyclopentene-1-carboxaldehyde
2675 2-Cyclopentene-1-tridecanoic acid, (*S*)-
2676 2-Cyclopentene-1-undecanoic acid, (*R*)-
2677 2-Cyclopenten-1-one
2678 3-Cyclopenten-1-one

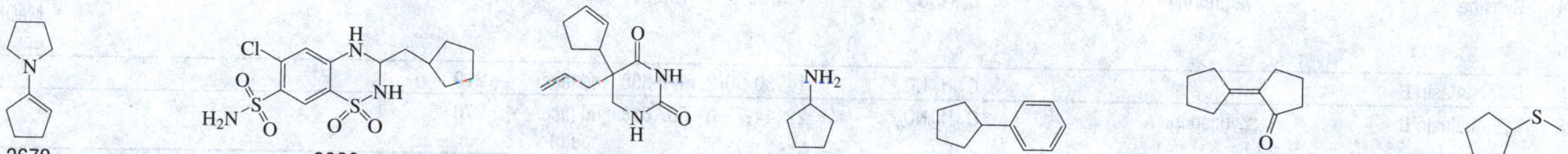

2679 *N*-(1-Cyclopenten-1-yl)pyrrolidine
2680 Cyclopenthiazide
2681 Cyclopentobarbital
2682 Cyclopentylamine
2683 Cyclopentylbenzene
2684 2-Cyclopentylidenecyclopentanone
2685 Cyclopentyl methyl sulfide

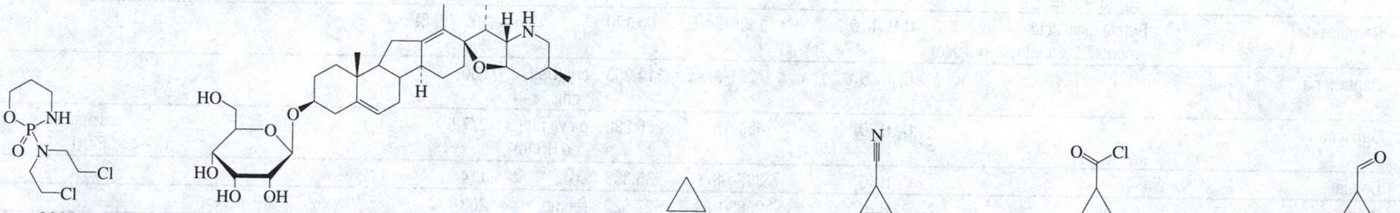

2686 Cyclophosphamide
2687 Cycloposine
2688 Cyclopropane
2689 Cyclopropanecarbonitrile
2690 Cyclopropanecarbonyl chloride
2691 Cyclopropanecarboxaldehyde

2692 Cyclopropanecarboxylic acid
2693 1,1-Cyclopropanedicarboxylic acid
2694 Cyclopropanemethanol
2695 Cyclopropanone
2696 Cyclopropene
2697 Cyclopropylamine
2698 Cyclopropylbenzene
2699 Cyclopropyl methyl ether

Organic

No.	Name	Synonym	Mol. form.	CAS Reg. No.	Mol. wt.	Physical form	mp/°C	bp/°C	den/ g cm^{-3}	n_D	Solubility
2700	Cyclopropyl methyl ketone		C_5H_8O	765-43-5	84.117	liq	-68.2(0.4)	111(5)	0.8984^{20}	1.4251^{20}	vs H_2O, eth, EtOH
2701	Cyclotetramethylenetetranitramine	HMX	$C_4H_8N_8O_8$	2691-41-0	296.156	cry	278(3)				
2702	Cyclothiazide		$C_{14}H_{16}ClN_3O_4S_2$	2259-96-3	389.878		234				
2703	Cycluron	*N*'-Cyclooctyl-*N*,*N*-dimethylurea	$C_{11}H_{22}N_2O$	2163-69-1	198.305	cry	138				sl H_2O; s bz, ace; vs MeOH
2704	Cyfluthrin		$C_{22}H_{18}Cl_2FNO_3$	68359-37-5	434.287		60				
2705	Cygon		$C_5H_{12}NO_3PS_2$	60-51-5	229.258		51.4(0.2)	$117^{0.1}$	1.277^{65}		
2706	Cyhalothrin	2,2-Dimethylcyclopropanecarboxylate	$C_{23}H_{19}ClF_3NO_3$	91465-08-6	449.850		49.2				
2707	Cyhexatin	Stannane, tricyclohexylhydroxy-	$C_{18}H_{34}OSn$	13121-70-5	385.172		196				
2708	Cypermethrin		$C_{22}H_{19}Cl_2NO_3$	52315-07-8	416.297		70		1.25^{20}		
2709	Cyprazine		$C_9H_{14}ClN_5$	22936-86-3	227.694		170.4(0.5)				
2710	Cyproheptadine		$C_{21}H_{21}N$	129-03-3	287.399	cry (EtOH aq)	113				
2711	Cyromazine	*N*-Cyclopropyl-1,3,5-triazine-2,4,6-triamine	$C_6H_{10}N_6$	66215-27-8	166.183	cry	220				
2712	Cystamine dihydrochloride		$C_4H_{14}Cl_2N_2S_2$	56-17-7	225.203	nd (MeOH)	218 dec				vs H_2O, EtOH
2713	Cysteamine		C_2H_7NS	60-23-1	77.149	cry (sub)	99.5	dec			vs H_2O, EtOH
2714	L-Cysteic acid		$C_3H_7NO_5S$	13100-82-8	169.157	cry	260 dec				s H_2O; i EtOH
2715	L-Cysteine	Propanoic acid, 2-amino-3-mercapto-, (*R*)-	$C_3H_7NO_2S$	52-90-4	121.159	cry (w)	240 dec				vs H_2O, ace, EtOH
2716	L-Cysteine, ethyl ester, hydrochloride		$C_5H_{12}ClNO_2S$	868-59-7	185.673		125.8				vs H_2O
2717	L-Cysteine, hydrochloride		$C_3H_8ClNO_2S$	52-89-1	157.620	cry	175 dec				s H_2O
2718	L-Cystine	3,3'-Dithiobis(2-aminopropanoic acid)	$C_6H_{12}N_2O_4S_2$	56-89-3	240.300	hex pl or pr (w)	260 dec		1.677^{25}		sl H_2O; i EtOH, eth, bz; s acid, alk
2719	Cytarabine	Cytosine arabinoside	$C_9H_{13}N_3O_5$	147-94-4	243.216	pr (EtOH aq)	212				s H_2O
2720	Cytidine	4-Amino-1-β-D-ribofuranosyl-2(1*H*)-pyrimidinone	$C_9H_{13}N_3O_5$	65-46-3	243.216	nd (dil al)	230 dec				vs H_2O; sl EtOH
2721	2'-Cytidylic acid	Cytidine 2'-monophosphate	$C_9H_{14}N_3O_8P$	85-94-9	323.196		239 dec				
2722	3'-Cytidylic acid	Cytidine 3'-monophosphate	$C_9H_{14}N_3O_8P$	84-52-6	323.196		233 dec				s H_2O, EtOH
2723	5'-Cytidylic acid	Cytidine 5'-monophosphate	$C_9H_{14}N_3O_8P$	63-37-6	323.196	orth nd	233 dec				vs H_2O, EtOH
2724	Cytisine	Sophorine	$C_{11}H_{14}N_2O$	485-35-8	190.241	pr	153	218^2			vs H_2O, EtOH, MeOH; s bz, ace
2725	Cytochalasin B		$C_{29}H_{37}NO_5$	14930-96-2	479.608	nd (ace)	219				
2726	Cytochalasin D	Zygosporin A	$C_{30}H_{37}NO_6$	22144-77-0	507.618	nd (ace/peth)	270				
2727	Cytochalasin E		$C_{28}H_{33}NO_7$	36011-19-5	495.565		207				
2728	Cytosine		$C_4H_5N_3O$	71-30-7	111.102	mcl or tcl pl (w+1)	265(1)				s H_2O; sl EtOH, chl; i eth
2729	Dacarbazine	5-(3,3-Dimethyl-1-triazenyl)-1*H*-imidazole-4-carboxamide	$C_6H_{10}N_6O$	4342-03-4	182.182	cry	205				
2730	Dactinomycin		$C_{62}H_{86}N_{12}O_{16}$	50-76-0	1255.416		245 dec				
2731	Daidzein	7-Hydroxy-3-(4-hydroxyphenyl)-4*H*-1-benzopyran-4-one	$C_{15}H_{10}O_4$	486-66-8	254.238	pa ye pr (50% al)	323 dec	sub			s EtOH, eth
2732	Daminozide	Butanedioic acid, mono(2,2-dimethylhydrazide)	$C_6H_{12}N_2O_3$	1596-84-5	160.170		152.7(0.3)				
2733	Dantrolene		$C_{14}H_{10}N_4O_5$	7261-97-4	314.253	cry (DMF aq)	280				
2734	Datiscetin		$C_{15}H_{10}O_6$	480-15-9	286.236	pa ye nd (al, aq HOAc)	277.5				vs ace, eth, EtOH
2735	Daucol		$C_{15}H_{26}O_2$	887-08-1	238.366	cry	114	128^2			
2736	Daunorubicin		$C_{27}H_{29}NO_{10}$	20830-81-3	527.520	red nd	208				
2737	Dazomet		$C_5H_{10}N_2S_2$	533-74-4	162.276	nd (bz)	106				reac H_2O; s EtOH
2738	Decabromobiphenyl ether	Bis(pentabromophenyl) ether	$C_{12}Br_{10}O$	1163-19-5	959.167	ye pr (tol)	305				i H_2O
2739	Decachlorobiphenyl		$C_{12}Cl_{10}$	2051-24-3	498.658	cry (bz)	306(1)				i H_2O
2740	1,3-Decadiene	1-Hexyl-1,3-butadiene	$C_{10}H_{18}$	2051-25-4	138.250			169	0.752^{30}		vs bz
2741	1,9-Decadiene		$C_{10}H_{18}$	1647-16-1	138.250			164(5)	0.75^{25}	1.4325^{20}	
2742	2,2',3,3',4,4',5,5',6,6'-Decafluoro-1,1'-biphenyl		$C_{12}F_{10}$	434-90-2	334.112		67.3(0.8)	207(2)	1.785^{20}		
2743	*cis*-Decahydronaphthalene	*cis*-Decalin	$C_{10}H_{18}$	493-01-6	138.250	liq	-42.9(0.3)	195.8(0.3)	0.8965^{20}	1.4810^{20}	i H_2O; msc EtOH; vs eth, ace, chl

2700
Cyclopropyl methyl ketone

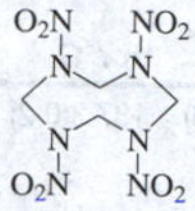
2701
Cyclotetramethylenetetranitramine

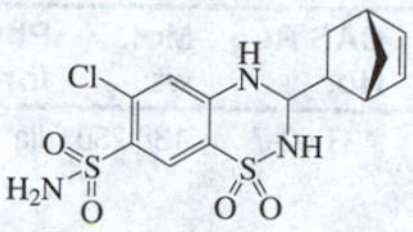
2702
Cyclothiazide

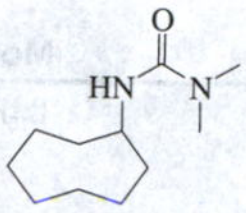
2703
Cycluron

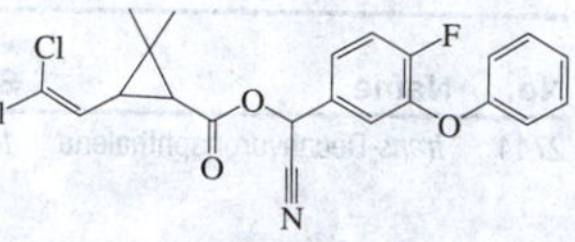
2704
Cyfluthrin

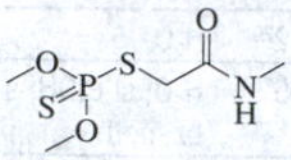
2705
Cygon

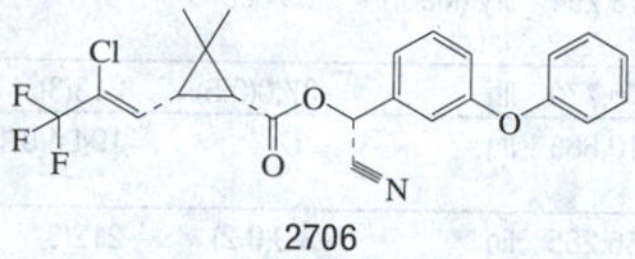
2706
Cyhalothrin

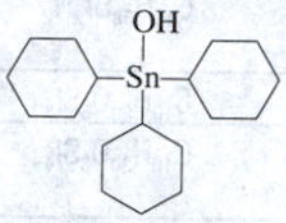
2707
Cyhexatin

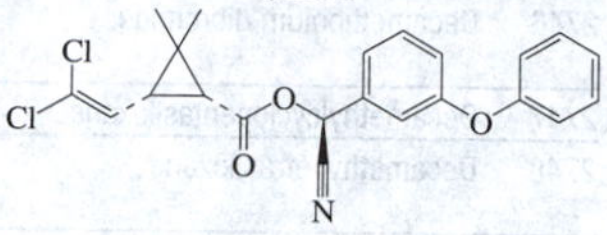
2708
Cypermethrin

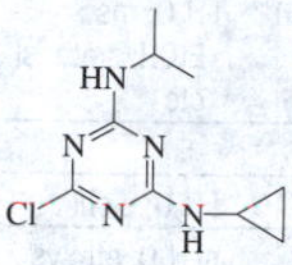
2709
Cyprazine

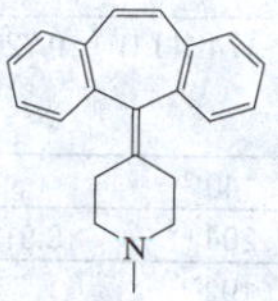
2710
Cyproheptadine

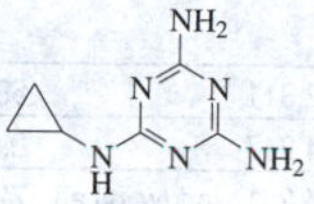
2711
Cyromazine

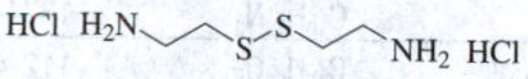
2712
Cystamine dihydrochloride

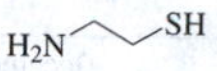
2713
Cysteamine

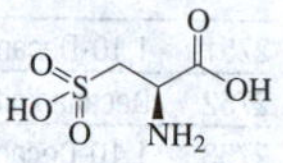
2714
L-Cysteic acid

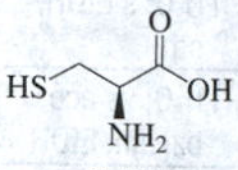
2715
L-Cysteine

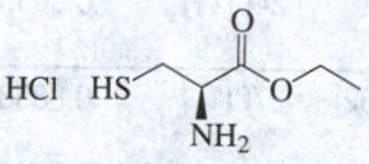
2716
L-Cysteine, ethyl ester, hydrochloride

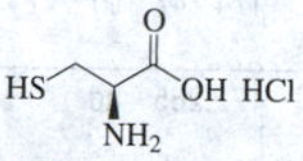
2717
L-Cysteine, hydrochloride

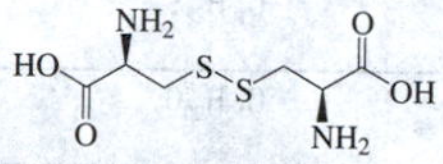
2718
L-Cystine

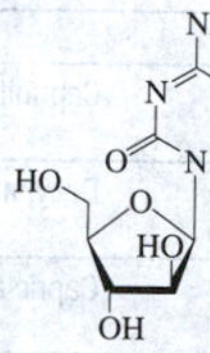
2719
Cytarabine

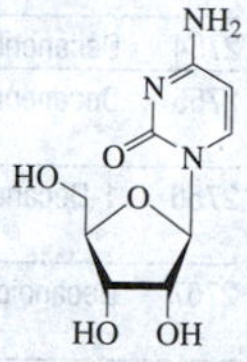
2720
Cytidine

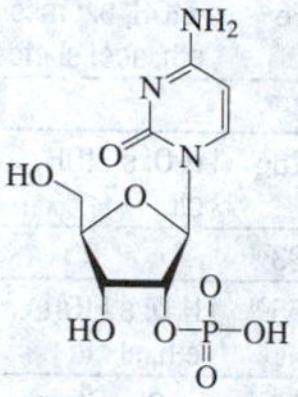
2721
2'-Cytidylic acid

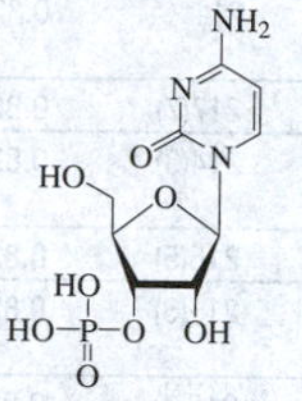
2722
3'-Cytidylic acid

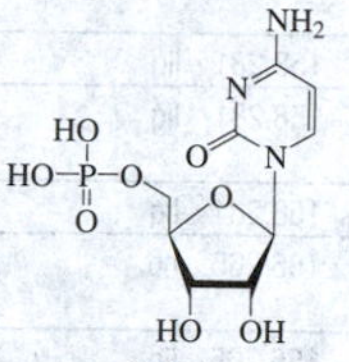
2723
5'-Cytidylic acid

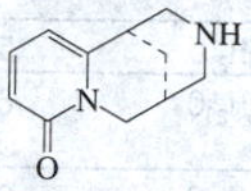
2724
Cytisine

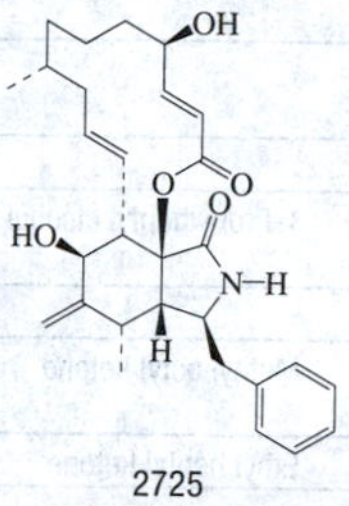
2725
Cytochalasin B

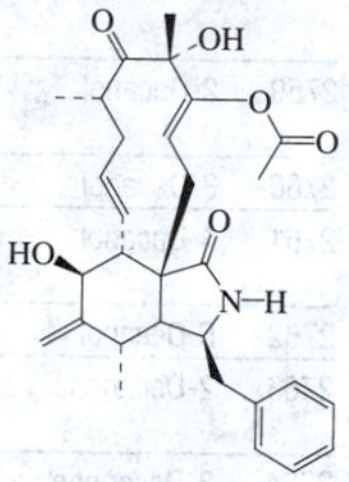
2726
Cytochalasin D

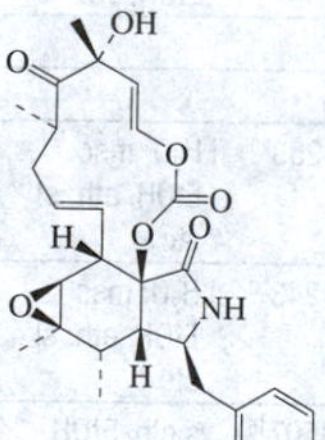
2727
Cytochalasin E

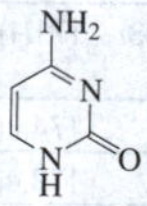
2728
Cytosine

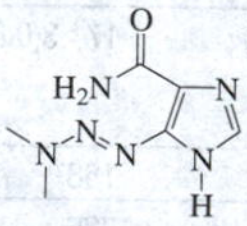
2729
Dacarbazine

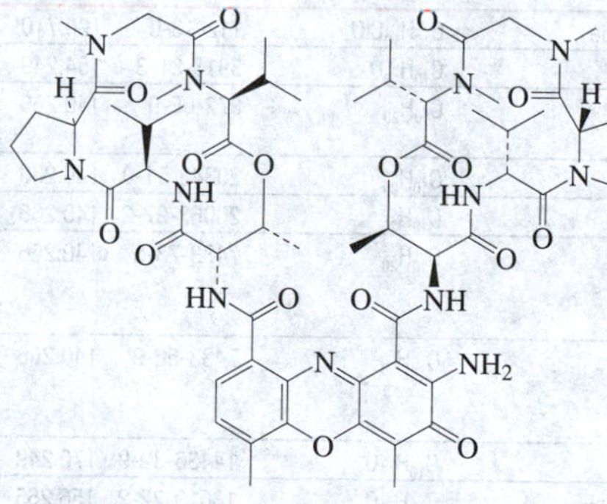
2730
Dactinomycin

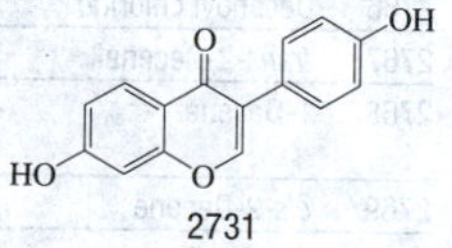
2731
Daidzein

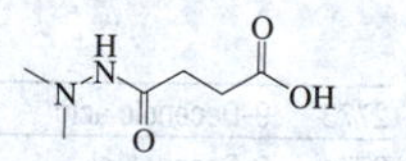
2732
Daminozide

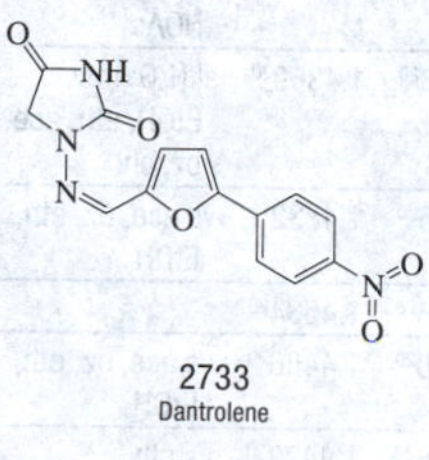
2733
Dantrolene

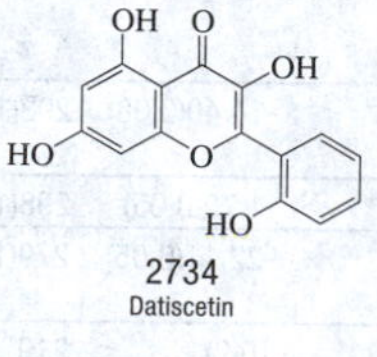
2734
Datiscetin

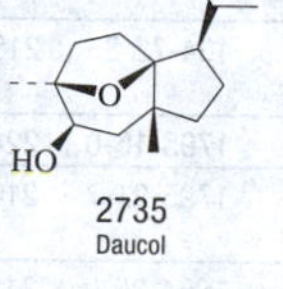
2735
Daucol

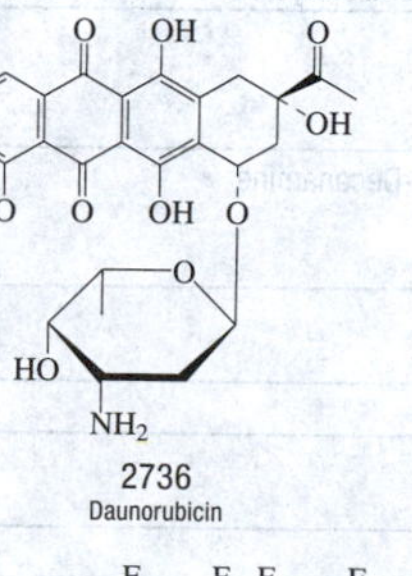
2736
Daunorubicin

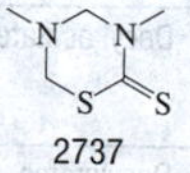
2737
Dazomet

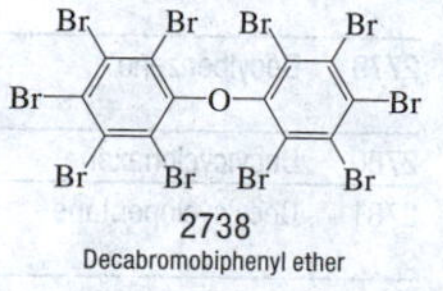
2738
Decabromobiphenyl ether

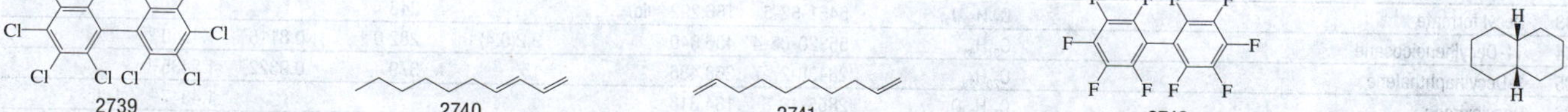
2739
Decachlorobiphenyl

2740
1,3-Decadiene

2741
1,9-Decadiene

2742
2,2',3,3',4,4',5,5',6,6'-Decafluoro-1,1'-biphenyl

2743
cis-Decahydronaphthalene

No.	Name	Synonym	Mol. form.	CAS Reg. No.	Mol. wt.	Physical form	mp/°C	bp/°C	den/ g cm^{-3}	n_D	Solubility
2744	*trans*-Decahydronaphthalene	*trans*-Decalin	$C_{10}H_{18}$	493-02-7	138.250	liq	-30.35(0.06)	187.3(0.2)	0.8659^{25}	1.4695^{20}	i H_2O; vs EtOH, eth, ace; msc bz; sl MeOH
2745	Decahydro-2-naphthol	Decahydro-β-naphthol	$C_{10}H_{18}O$	825-51-4	154.249			109^{14}	0.996^{25}	1.4992^{20}	
2746	Decamethonium dibromide		$C_{16}H_{38}Br_2N_2$	541-22-0	418.294	cry (MeOH/ ace)	269 dec				i eth
2747	Decamethylcyclopentasiloxane		$C_{10}H_{30}O_5Si_5$	541-02-6	370.770	liq	-37.0(0.5)	213(3)	0.9593^{20}	1.3982^{20}	i H_2O
2748	Decamethyltetrasiloxane		$C_{10}H_{30}O_3Si_4$	141-62-8	310.685	liq	-76	194.4(0.1)	0.8536^{25}	1.3895^{20}	i H_2O; sl EtOH; s bz, peth
2749	Decanal	Capraldehyde	$C_{10}H_{20}O$	112-31-2	156.265	liq	-3.9(0.2)	212(3)	0.830^{15}	1.4287^{20}	i H_2O; s EtOH, eth, ace; sl ctc
2750	Decane		$C_{10}H_{22}$	124-18-5	142.282	liq	-29.61(0.02)	174.1(0.1)	0.7266^{25}	1.4090^{25}	i H_2O; msc EtOH; s eth; sl ctc
2751	1,10-Decanediamine		$C_{10}H_{24}N_2$	646-25-3	172.311		59.7(0.4)	140^{12}			
2752	Decanedinitrile		$C_{10}H_{16}N_2$	1871-96-1	164.247		8.0(0.5)	204^{16}	0.913^{20}	1.4474^{20}	i H_2O; s chl
2753	1,10-Decanediol	Decamethylene glycol	$C_{10}H_{22}O_2$	112-47-0	174.281	nd (w, dil al)	72.4(0.2)	192^{20}			sl H_2O, eth; vs EtOH; s DMSO; i lig
2754	Decanedioyl dichloride		$C_{10}H_{16}Cl_2O_2$	111-19-3	239.139		-1.3	220^{75}	1.1212^{20}	1.4684^{18}	
2755	Decanenitrile	Caprinitrile	$C_{10}H_{19}N$	1975-78-6	153.265	liq	-14.4(0.4)	241(5)	0.8199^{20}	1.4296^{20}	vs ace, eth, EtOH, chl
2756	1-Decanethiol	Decyl mercaptan	$C_{10}H_{22}S$	143-10-2	174.347	liq	-25.290 (0.001)	240(11)	0.8443^{20}	1.4509^{20}	i H_2O; s EtOH, eth
2757	Decanoic acid	Capric acid	$C_{10}H_{20}O_2$	334-48-5	172.265	nd	31.39(0.02)	270(1)	0.8858^{40}	1.4288^{40}	i H_2O; vs ace, bz, eth, EtOH
2758	1-Decanol	Capric alcohol	$C_{10}H_{22}O$	112-30-1	158.281	oily liq	7(1)	229(3)	0.8297^{20}	1.4372^{20}	i H_2O; msc EtOH, eth, ace, bz, chl; s ctc
2759	2-Decanol		$C_{10}H_{22}O$	74742-10-2	158.281	liq	-2(1)	211	0.8250^{20}	1.4326^{25}	s EtOH, bz; msc eth, ace; sl ctc
2760	3-Decanol		$C_{10}H_{22}O$	1565-81-7	158.281	liq	-7.5	217(7)	0.827^{20}	1.434^{20}	
2761	4-Decanol	1-Propylheptyl alcohol	$C_{10}H_{22}O$	2051-31-2	158.281	liq	-11	214(3)	0.8261^{20}	1.4320^{20}	i H_2O; s EtOH, ctc
2762	5-Decanol		$C_{10}H_{22}O$	5205-34-5	158.281	liq	8.7	216(5)	0.824^{20}	1.4333^{20}	
2763	2-Decanone	Methyl octyl ketone	$C_{10}H_{20}O$	693-54-9	156.265	nd	14	211(3)	0.8248^{20}	1.4255^{20}	i H_2O; s EtOH, eth; sl ctc
2764	3-Decanone	Ethyl heptyl ketone	$C_{10}H_{20}O$	928-80-3	156.265	liq	2(4)	212(4)	0.8251^{20}	1.4252^{20}	s EtOH, eth, ctc
2765	4-Decanone	Hexyl propyl ketone	$C_{10}H_{20}O$	624-16-8	156.265	liq	-9(4)	206.5	0.824^{20}	1.4240^{21}	i H_2O; msc EtOH, eth
2766	Decanoyl chloride	Caprinoyl chloride	$C_{10}H_{19}ClO$	112-13-0	190.710	liq	-34.5	219(13)	0.919^{25}	1.4410^{20}	s eth, ctc
2767	*trans*-2-Decenal		$C_{10}H_{18}O$	3913-81-3	154.249			230			
2768	1-Decene		$C_{10}H_{20}$	872-05-9	140.266	liq	-66.21(0.03)	171(1)	0.7408^{20}	1.4215^{20}	i H_2O; msc EtOH, eth
2769	*cis*-2-Decene		$C_{10}H_{20}$	20348-51-0	140.266	col liq		174.2(0.7)			
2770	*trans*-2-Decene		$C_{10}H_{20}$	20063-97-2	140.266	col liq		173.4(0.5)			
2771	*cis*-5-Decene		$C_{10}H_{20}$	7433-78-5	140.266	col liq	-112(2)	170.4(0.8)	0.7445^{20}	1.4258^{20}	i H_2O; msc EtOH, eth; sl ctc
2772	*trans*-5-Decene		$C_{10}H_{20}$	7433-56-9	140.266	col liq	-73(1)	171.3(0.6)	0.7401^{20}	1.4243^{20}	i H_2O; msc EtOH, eth; sl ctc
2773	9-Decenoic acid	Caproleic acid	$C_{10}H_{18}O_2$	14436-32-9	170.249		26.5	158^{21}	0.9238^{15}	1.4507^{15}	vs eth, EtOH
2774	9-Decen-1-ol	Decylenic alcohol	$C_{10}H_{20}O$	13019-22-2	156.265			236	0.876^{25}	1.4480^{20}	
2775	3-Decen-2-one	Heptylidene acetone	$C_{10}H_{18}O$	10519-33-2	154.249			102$^{15.3}$	0.8473^{20}	1.4480^{20}	
2776	Declomycin	Demeclocycline	$C_{21}H_{21}ClN_2O_8$	127-33-3	464.853	cry	176 dec				
2777	Decyl acetate		$C_{12}H_{24}O_2$	112-17-4	200.318	liq	-15.0(0.4)	249(1)	0.8671^{20}	1.4273^{20}	i H_2O; s EtOH, eth, bz, ctc, HOAc
2778	Decylamine	1-Decanamine	$C_{10}H_{23}N$	2016-57-1	157.297		15(1)	217(2)	0.7936^{20}	1.4369^{20}	sl H_2O; msc EtOH, eth, ace, bz, chl
2779	Decylbenzene		$C_{16}H_{26}$	104-72-3	218.377	liq	-14.40(0.08)	298(1)	0.8555^{20}	1.4832^{20}	vs ace, bz, eth, EtOH
2780	Decylcyclohexane		$C_{16}H_{32}$	1795-16-0	224.425	liq	-1.72(0.05)	298(1)	0.8186^{20}	1.4534^{20}	
2781	Decylcyclopentane		$C_{15}H_{30}$	1795-21-7	210.399	liq	-22.11(0.05)	279(1)	0.8110^{20}	1.4486^{20}	vs ace, bz, eth, EtOH
2782	Decyl decanoate		$C_{20}H_{40}O_2$	1654-86-0	312.531		10(2)	219^{15}	0.8586^{20}	1.4423^{20}	vs eth
2783	Decyl formate		$C_{11}H_{22}O_2$	5451-52-5	186.292	liq		243			
2784	11-Decylheneicosane		$C_{31}H_{64}$	55320-06-4	436.840		9.2(0.4)	282.0^{10}	0.8116^{20}	1.4540^{20}	
2785	1-Decylnaphthalene		$C_{20}H_{28}$	26438-27-7	268.436		15	379	0.9322^{20}	1.5435^{20}	
2786	Decyloxirane		$C_{12}H_{24}O$	2855-19-8	184.318					1.4347^{25}	sl ctc

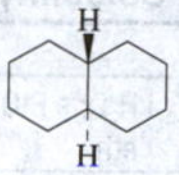
2744
trans-Decahydronaphthalene

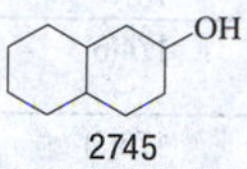
2745
Decahydro-2-naphthol

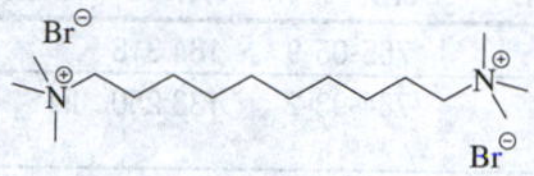
2746
Decamethonium dibromide

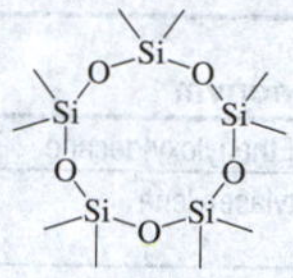
2747
Decamethylcyclopentasiloxane

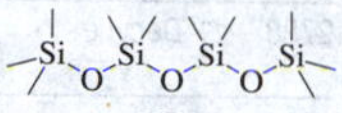
2748
Decamethyltetrasiloxane

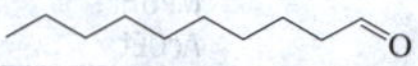
2749
Decanal

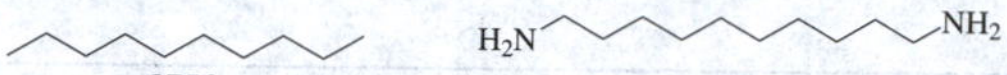
2750
Decane

2751
1,10-Decanediamine

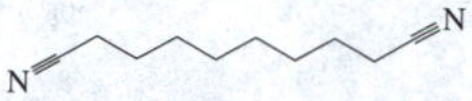
2752
Decanedinitrile

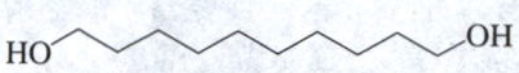
2753
1,10-Decanediol

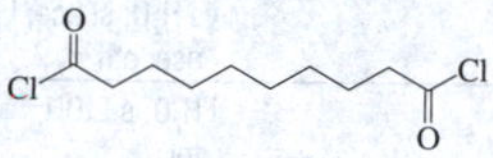
2754
Decanedioyl dichloride

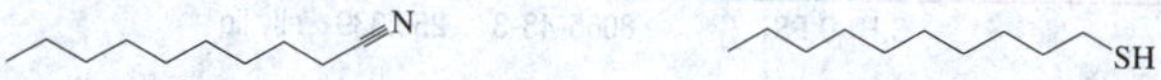
2755
Decanenitrile

2756
1-Decanethiol

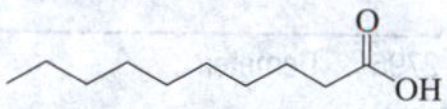
2757
Decanoic acid

2758
1-Decanol

2759
2-Decanol

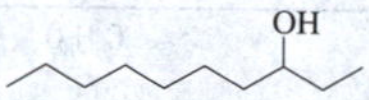
2760
3-Decanol

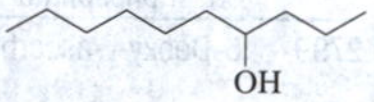
2761
4-Decanol

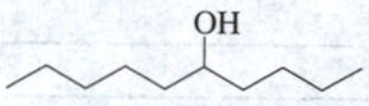
2762
5-Decanol

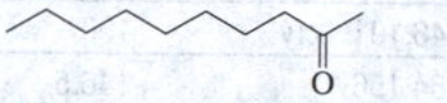
2763
2-Decanone

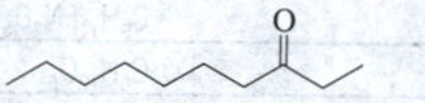
2764
3-Decanone

2765
4-Decanone

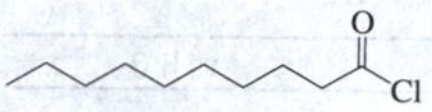
2766
Decanoyl chloride

2767
trans-2-Decenal

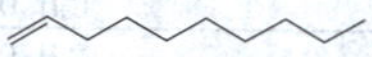
2768
1-Decene

2769
cis-2-Decene

2770
trans-2-Decene

2771
cis-5-Decene

2772
trans-5-Decene

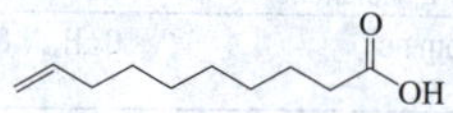
2773
9-Decenoic acid

2774
9-Decen-1-ol

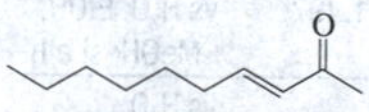
2775
3-Decen-2-one

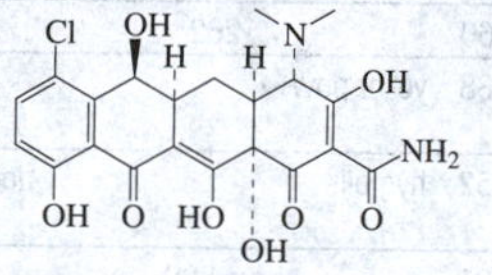
2776
Declomycin

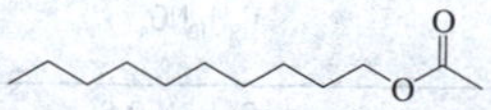
2777
Decyl acetate

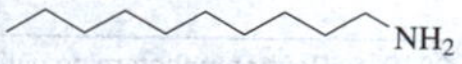
2778
Decylamine

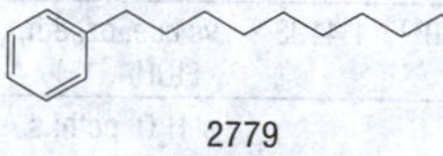
2779
Decylbenzene

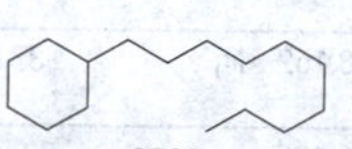
2780
Decylcyclohexane

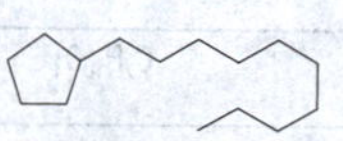
2781
Decylcyclopentane

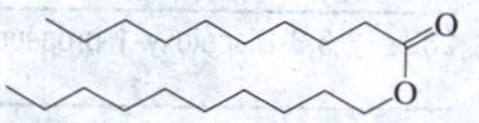
2782
Decyl decanoate

2783
Decyl formate

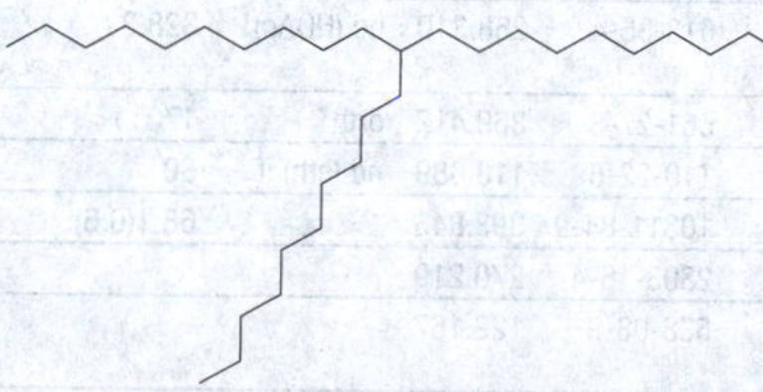
2784
11-Decylheneicosane

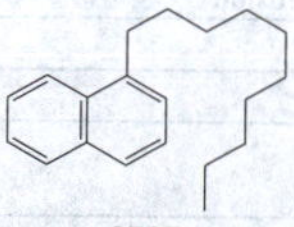
2785
1-Decylnaphthalene

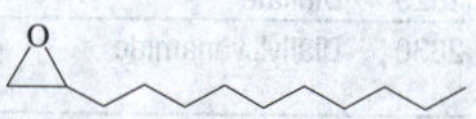
2786
Decyloxirane

No.	Name	Synonym	Mol. form.	CAS Reg. No.	Mol. wt.	Physical form	mp/°C	bp/°C	den/ g cm^{-3}	n_D	Solubility
2787	Decyl vinyl ether	1-(Ethenyloxy)decane	$C_{12}H_{24}O$	765-05-9	184.318		-41	101^{10}	0.812^{20}	1.4346^{20}	
2788	1-Decyne	Octylacetylene	$C_{10}H_{18}$	764-93-2	138.250	liq	-44	174(6)	0.7655^{20}	1.4265^{20}	i H_2O; s EtOH, eth
2789	5-Decyne	Dibutylacetylene	$C_{10}H_{18}$	1942-46-7	138.250	liq	-74(4)	178(3)	0.7690^{20}	1.4331^{20}	i H_2O; s EtOH, eth
2790	Dehydroabietic acid	8,11,13-Abietatrien-18-oic acid	$C_{20}H_{28}O_2$	1740-19-8	300.435	cry (EtOH aq)	172				
2791	Delphinidin		$C_{15}H_{11}ClO_7$	528-53-0	338.697		>350				vs H_2O, EtOH, MeOH; s AcOEt
2792	Delphinine		$C_{33}H_{45}NO_9$	561-07-9	599.712	orth (al)	199				i H_2O; s chl, ace, eth; vs EtOH
2793	Deltamethrin		$C_{22}H_{19}Br_2NO_3$	52918-63-5	505.199		100.4(0.5)				
2794	Demecarium bromide		$C_{32}H_{52}Br_2N_4O_4$	56-94-0	716.588	hyg pow	165 dec				vs H_2O; sl ace; i ace, eth
2795	Demeton	Systox	$C_8H_{19}O_3PS_2$	8065-48-3	258.339	oily liq		134^{2}			i H_2O; s EtOH, tol
2796	Demeton-*S*-methyl		$C_6H_{15}O_3PS_2$	919-86-8	230.285	ye liq		118^{1}	1.20^{20}	1.5063^{20}	i H_2O; s os
2797	2'-Deoxyadenosine		$C_{10}H_{13}N_5O_3$	958-09-8	251.242						sl H_2O
2798	2'-Deoxyadenosine 5'-triphosphate		$C_{10}H_{16}N_5O_{12}P_3$	1927-31-7	491.182	cry (EtOH aq)					
2799	6-Deoxy-L-ascorbic acid		$C_6H_8O_5$	528-81-4	160.125	pr (AcOEt)	168	160 sub			vs H_2O, ace, EtOH
2800	Deoxycholic acid	3,12-Dihydroxycholan-24-oic acid, (3α,5β,12α)	$C_{24}H_{40}O_4$	83-44-3	392.573	cry (al)	177				
2801	2'-Deoxycytidine 5'-monophosphate	2'-Deoxy-5'-cytidylic acid	$C_9H_{14}N_3O_7P$	1032-65-1	307.197	pow	183 dec				
2802	2'-Deoxy-5-fluorouridine	Floxuridine	$C_9H_{11}FN_2O_5$	50-91-9	246.191	cry	150				
2803	2-Deoxy-D-glucose		$C_6H_{12}O_5$	154-17-6	164.156		146.5				
2804	2'-Deoxyguanosine 5'-monophosphate	2'-Deoxy-5'-guanylic acid	$C_{10}H_{14}N_5O_7P$	902-04-5	347.222						s H_2O
2805	2-Deoxy-D-*chiro*-inositol	D-Quercitol	$C_6H_{12}O_5$	488-73-3	164.156	pr (w, dil al)	236		1.5845^{13}		vs H_2O
2806	1-Deoxy-1-(methylamino)-D-glucitol	*N*-Methylglucamine	$C_7H_{17}NO_5$	6284-40-8	195.214	cry (MeOH)	128.5				s H_2O
2807	6-Deoxy-3-*O*-methylgalactose	Digitalose	$C_7H_{14}O_5$	4481-08-7	178.183	nd (AcOEt)	119				vs H_2O
2808	D-2-Deoxyribose		$C_5H_{10}O_4$	533-67-5	134.131		90				
2809	Deserpidine		$C_{32}H_{38}N_2O_8$	131-01-1	578.652	nd or pr	230.5				i H_2O; s EtOH, chl
2810	Desethyl atrazine	6-Chloro-*N*-isopropyl-1,3,5-triazine-2,4-diamine	$C_6H_{10}ClN_5$	6190-65-4	187.630	cry	136				
2811	Desferrioxamine	Deferoxamine	$C_{25}H_{48}N_6O_8$	70-51-9	560.684	cry (EtOH aq)	139				
2812	Desipramine		$C_{18}H_{22}N_2$	50-47-5	266.381			$173^{0.02}$			
2813	Desmedipham		$C_{16}H_{16}N_2O_4$	13684-56-5	300.309		121.7(0.5)				
2814	Desmetryne		$C_8H_{15}N_5S$	1014-69-3	213.304	cry	85				
2815	Desthiobiotin		$C_{10}H_{18}N_2O_3$	533-48-2	214.261	lo nd (H_2O)	157				s H_2O
2816	Dexamethasone		$C_{22}H_{29}FO_5$	50-02-2	392.460		262				
2817	Dexon	Sodium dimethylaminobenzenediazosulfonate	$C_8H_{10}N_3NaO_3S$	140-56-7	251.238	ye-br pow					sl H_2O; s DMF
2818	Dexpanthenol		$C_9H_{19}NO_4$	81-13-0	205.252	hyg oil		dec	1.20^{20}	1.497^{20}	vs H_2O, EtOH, MeOH; sl eth
2819	Dextroamphetamine sulfate		$C_{18}H_{28}N_2O_4S$	51-63-8	368.491		>300		1.15^{25}		vs H_2O
2820	Dextromethorphan hydrobromide		$C_{18}H_{26}BrNO$	125-69-9	352.309	wh cry pow	123				s EtOH, chl; i eth
2821	Diacetone alcohol	4-Hydroxy-4-methyl-2-pentanone	$C_6H_{12}O_2$	123-42-2	116.158	liq	-47(2)	167.9	0.9387^{20}	1.4213^{20}	msc H_2O, EtOH, eth; s chl
2822	3,3-Diacetoxy-1-propene		$C_7H_{10}O_4$	869-29-4	158.152	liq	-37.6	176(18)	1.0760^{20}	1.4193^{20}	vs ace, bz, eth, EtOH
2823	1,3-Diacetylbenzene		$C_{10}H_{10}O_2$	6781-42-6	162.185		32	152^{15}			sl H_2O, peth; s EtOH, bz, chl, HOAc
2824	1,4-Diacetylbenzene	4-Acetylacetophenone	$C_{10}H_{10}O_2$	1009-61-6	162.185		113.0	128^{3}			vs EtOH; sl chl
2825	*N,N'*-Diacetyl-4,4'-diaminobiphenyl		$C_{16}H_{16}N_2O_2$	613-35-4	268.310	nd (HOAc)	328.3				
2826	Diacetylmorphine		$C_{21}H_{23}NO_5$	561-27-3	369.412	orth	173	273^{12}	1.56^{25}		vs bz, chl
2827	Diacetylperoxide	Acetyl peroxide	$C_4H_6O_4$	110-22-5	118.089	nd (eth) lf	30	63^{21}			vs eth, EtOH
2828	Dialifor		$C_{14}H_{17}ClNO_4PS_2$	10311-84-9	393.846		68.1(0.5)				
2829	Diallate		$C_{10}H_{17}Cl_2NOS$	2303-16-4	270.219			150^{9}			
2830	Diallylcyanamide		$C_7H_{10}N_2$	538-08-9	122.167			142^{90}			s EtOH; sl eth, ctc

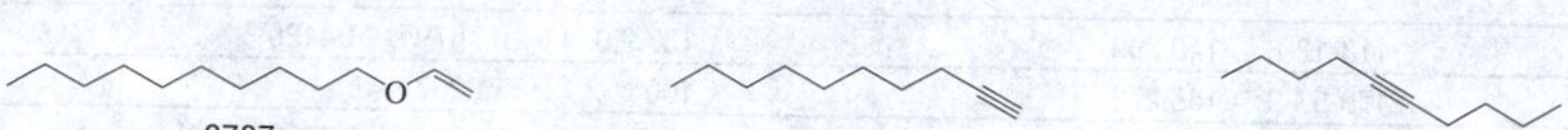

2787
Decyl vinyl ether

2788
1-Decyne

2789
5-Decyne

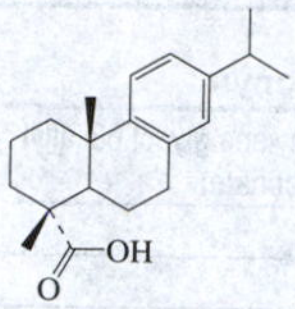

2790
Dehydroabietic acid

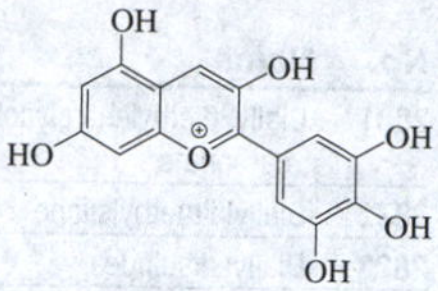

2791
Delphinidin

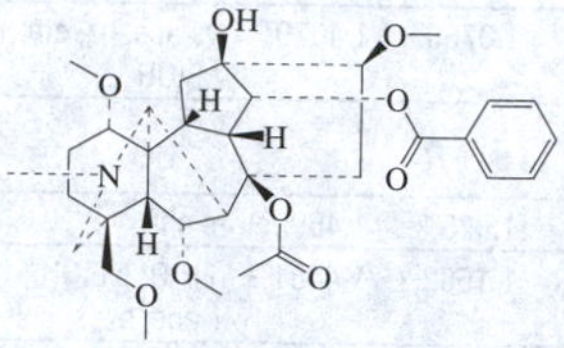

2792
Delphinine

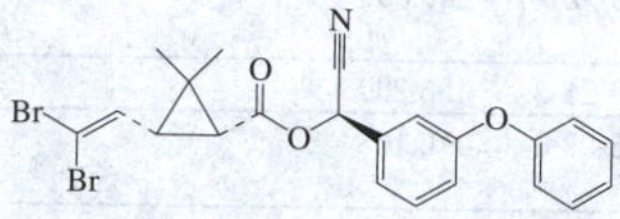

2793
Deltamethrin

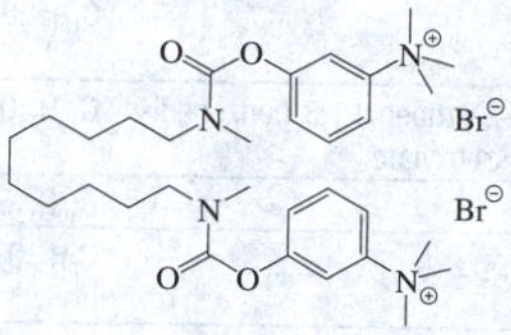

2794
Demecarium bromide

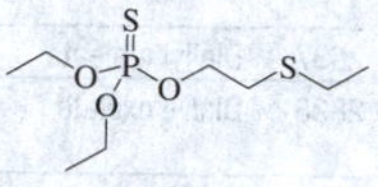

2795
Demeton

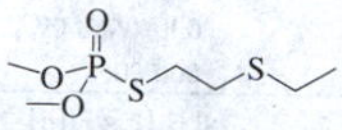

2796
Demeton-*S*-methyl

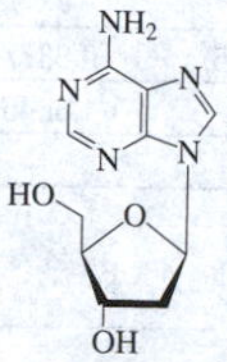

2797
2'-Deoxyadenosine

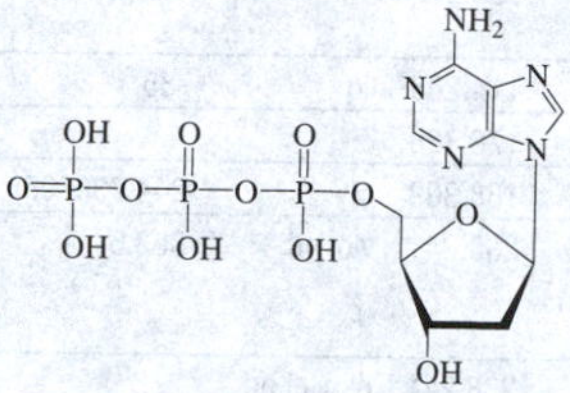

2798
2'-Deoxyadenosine 5'-triphosphate

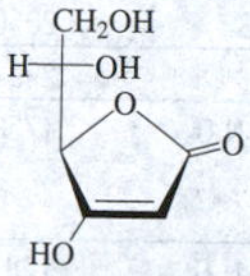

2799
6-Deoxy-L-ascorbic acid

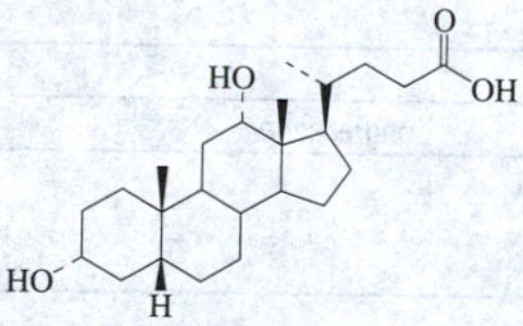

2800
Deoxycholic acid

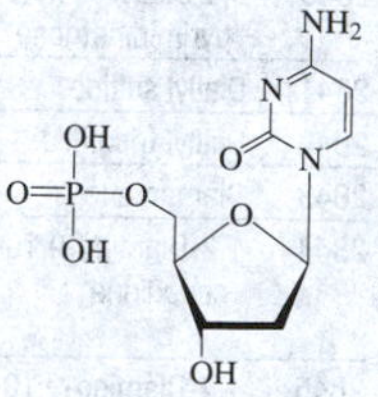

2801
2'-Deoxycytidine 5'-monophosphate

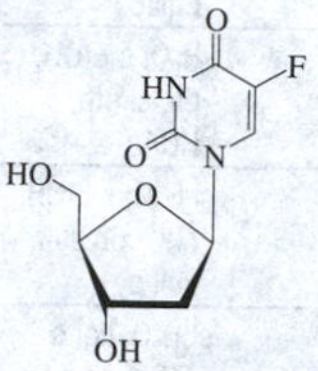

2802
2'-Deoxy-5-fluorouridine

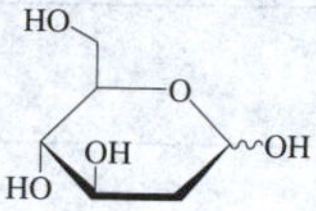

2803
2-Deoxy-D-glucose

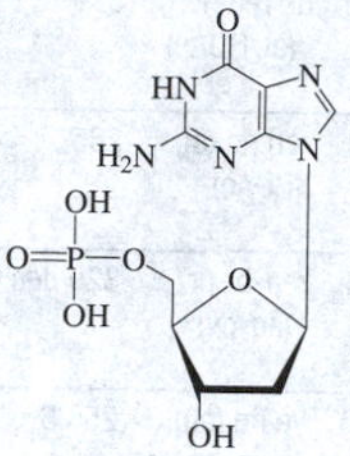

2804
2'-Deoxyguanosine 5'-monophosphate

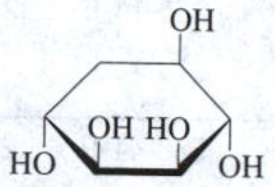

2805
2-Deoxy-D-*chiro*-inositol

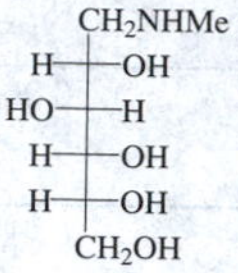

2806
1-Deoxy-1-(methylamino)-D-glucitol

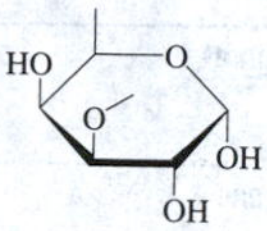

2807
6-Deoxy-3-*O*-methylgalactose

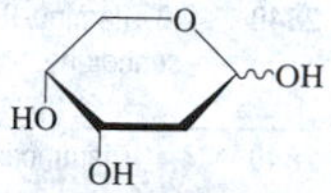

2808
D-2-Deoxyribose

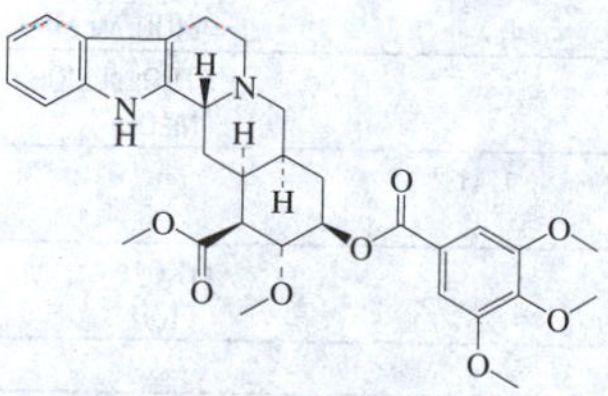

2809
Deserpidine

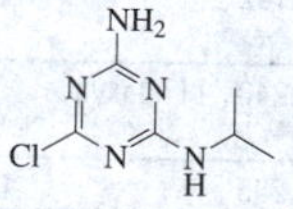

2810
Desethyl atrazine

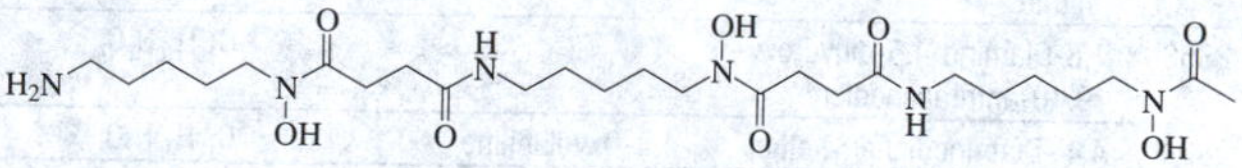

2811
Desferrioxamine

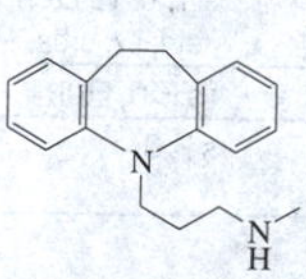

2812
Desipramine

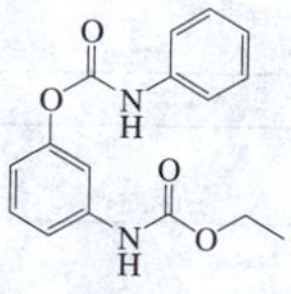

2813
Desmedipham

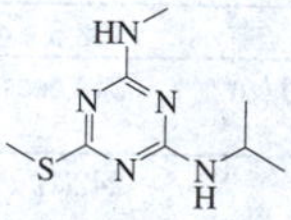

2814
Desmetryne

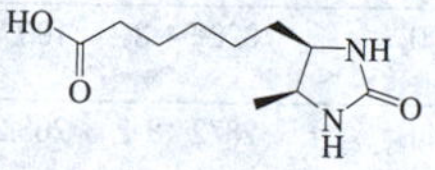

2815
Desthiobiotin

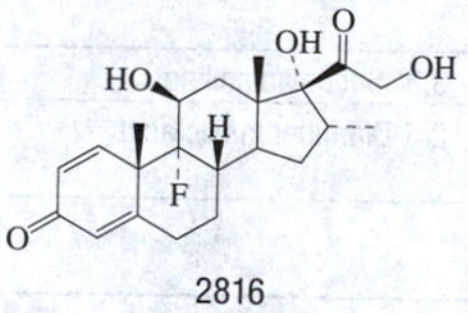

2816
Dexamethasone

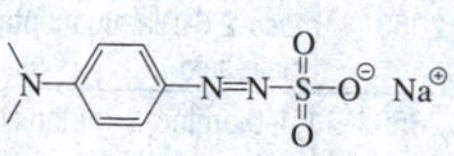

2817
Dexon

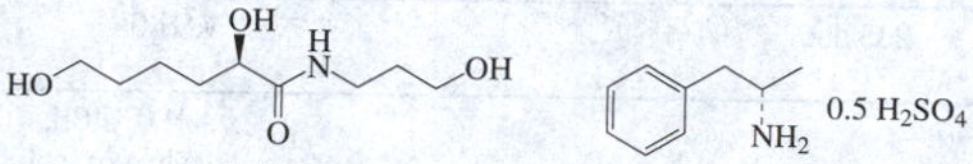

2818
Dexpanthenol

2819
Dextroamphetamine sulfate

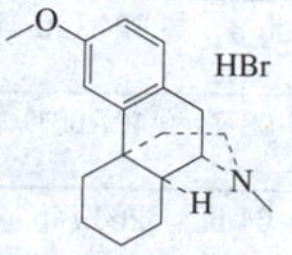

2820
Dextromethorphan hydrobromide

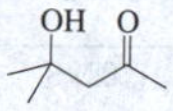

2821
Diacetone alcohol

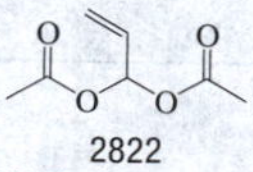

2822
3,3-Diacetoxy-1-propene

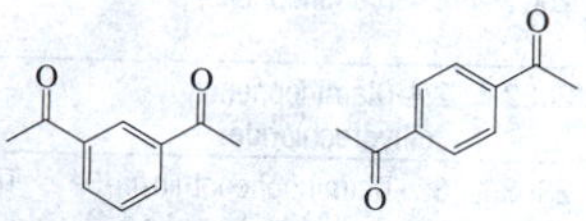

2823
1,3-Diacetylbenzene

2824
1,4-Diacetylbenzene

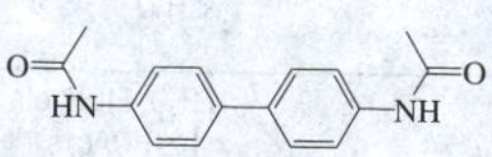

2825
N,N'-Diacetyl-4,4'-diaminobiphenyl

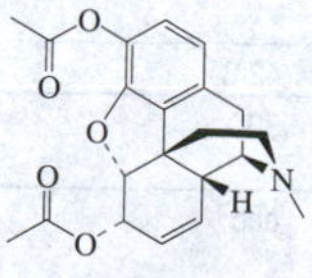

2826
Diacetylmorphine

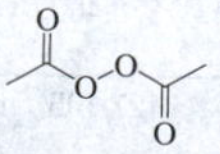

2827
Diacetylperoxide

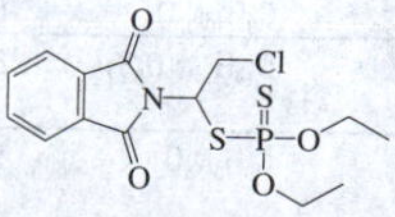

2828
Dialifor

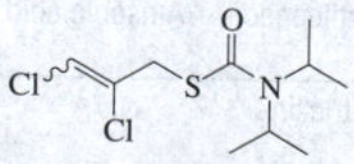

2829
Diallate

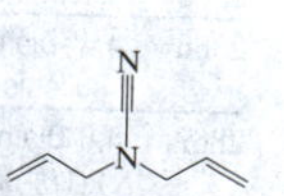

2830
Diallylcyanamide

No.	Name	Synonym	Mol. form.	CAS Reg. No.	Mol. wt.	Physical form	mp/°C	bp/°C	den/g cm⁻³	n_D	Solubility
2831	Diallyl diethylene glycol carbonate	Diethylene glycol bis(allyl carbonate)	$C_{12}H_{18}O_7$	142-22-3	274.267	col liq	-4	161^{2}	1.14^{20}		i H_2O; s os
2832	Diallyldimethylsilane		$C_8H_{16}Si$	1113-12-8	140.299			137.9(0.3)	0.7679^{20}	1.4420^{20}	
2833	Diallyl disulfide		$C_6H_{10}S_2$	2179-57-9	146.273			100^{48}	1.0237^{15}		
2834	Diallyl ether	Allyl ether	$C_6H_{10}O$	557-40-4	98.142	liq	-6	94(3)	0.8260^{20}	1.4163^{20}	i H_2O; msc EtOH, eth; vs ace; s chl
2835	Diallyl fumarate		$C_{10}H_{12}O_4$	2807-54-7	196.200			140^{3}	1.0768^{20}	1.4670^{25}	vs ace, bz, eth, EtOH
2836	Diallyl isophthalate	Di-2-propenyl 1,3-benzenedicarboxylate	$C_{14}H_{14}O_4$	1087-21-4	246.259			176^{5}			
2837	Diallyl maleate		$C_{10}H_{12}O_4$	999-21-3	196.200			129^{10}	1.075^{20}	1.4699^{20}	s chl
2838	Diallyl oxalate		$C_8H_{10}O_4$	615-99-6	170.163			217	1.1582^{20}	1.4481^{20}	i H_2O; s EtOH, ace, bz; sl chl
2839	*N,N*-Diallyl-2-propen-1-amine	Triallylamine	$C_9H_{15}N$	102-70-5	137.222		94	150(2)	0.809^{20}	1.4502^{20}	s EtOH, eth, ace, bz, acid
2840	5,5-Diallyl-2,4,6(1*H*,3*H*,5*H*)-pyrimidinetrione	Allobarbital	$C_{10}H_{12}N_2O_3$	52-43-7	208.213	lf	172(1)				sl H_2O, DMSO; s EtOH, eth, bz
2841	Diallyl sulfide		$C_6H_{10}S$	592-88-1	114.208	liq	-85	138(5)	0.8877^{27}	1.4870^{25}	vs eth, EtOH
2842	Diallyl trisulfide		$C_6H_{10}S_3$	2050-87-5	178.338			117^{16}	1.0845^{15}		vs eth
2843	Diamantane	Congressane	$C_{14}H_{20}$	2292-79-7	188.309	cry	244.73(0.05)				
2844	1,2-Diamino-9,10-anthracenedione		$C_{14}H_{10}N_2O_2$	1758-68-5	238.241	viol nd	303.5				sl EtOH, eth, chl, xyl; s py, con sulf
2845	1,4-Diamino-9,10-anthracenedione		$C_{14}H_{10}N_2O_2$	128-95-0	238.241	dk viol nd (py)	268				sl H_2O; s EtOH, bz, $PhNO_2$; vs py
2846	1,5-Diamino-9,10-anthracenedione		$C_{14}H_{10}N_2O_2$	129-44-2	238.241	dk red nd (al, HOAc)	319	sub			i H_2O; sl EtOH, eth, ace, bz; s $PhNO_2$
2847	1,8-Diamino-9,10-anthracenedione		$C_{14}H_{10}N_2O_2$	129-42-0	238.241	red nd (al, HOAc)	265				i H_2O; s EtOH, py; sl eth, HOAc
2848	2,6-Diamino-9,10-anthracenedione		$C_{14}H_{10}N_2O_2$	131-14-6	238.241	red-br pr (aq-py)	320 dec				sl H_2O; s EtOH, chl, con sulf, xyl, py
2849	4,4'-Diaminoazobenzene		$C_{12}H_{12}N_4$	538-41-0	212.250	ye nd (al), oran-ye pr (al)	250.5				sl H_2O, lig; s EtOH; vs bz, chl
2850	3,5-Diaminobenzoic acid		$C_7H_8N_2O_2$	535-87-5	152.151	nd (+1w)	228				sl H_2O, tfa; s EtOH; vs eth
2851	2,4-Diaminobutanoic acid		$C_4H_{10}N_2O_2$	305-62-4	118.134	hyg cry	118.1				s H_2O; sl EtOH, MeOH
2852	*cis*-2,3-Diamino-2-butenedinitrile		$C_4H_4N_4$	1187-42-4	108.102		178.5		1.41^{20}		
2853	1,8-Diamino-4,5-dihydroxy-9,10-anthracenedione		$C_{14}H_{10}N_2O_4$	128-94-9	270.240	bl nd (xyl)					i H_2O; s bz, xyl, EtOH
2854	4,4'-Diaminodiphenyl ether	4,4-Oxydianiline	$C_{12}H_{12}N_2O$	101-80-4	200.235		192.2(0.2)	>300			
2855	4,4'-Diaminodiphenylmethane	4,4'-Methylenedianiline	$C_{13}H_{14}N_2$	101-77-9	198.263	pl or nd (w) pl (bz)	90(1)	379(3)			sl H_2O; vs EtOH, eth, bz
2856	4,4'-Diaminodiphenyl sulfide	4,4'-Thiodianiline	$C_{12}H_{12}N_2S$	139-65-1	216.301	nd (w)	108.5				sl H_2O; vs EtOH, eth, bz; s tfa
2857	3,3'-Diaminodiphenyl sulfone	3,3'-Sulfonyldianiline	$C_{12}H_{12}N_2O_2S$	599-61-1	248.300		168.5				vs H_2O, EtOH
2858	*meso*-2,6-Diaminoheptanedioic acid	2,6-Diaminopimelic acid	$C_7H_{14}N_2O_4$	922-54-3	190.197	nd (w)	314 dec				s H_2O
2859	1,4-Diamino-2-methoxy-9,10-anthracenedione		$C_{15}H_{12}N_2O_3$	2872-48-2	268.267		242(1)				
2860	1,4-Diamino-5-nitro-9,10-anthracenedione		$C_{14}H_9N_3O_4$	82-33-7	283.239		278				
2861	2,4-Diaminophenol		$C_6H_8N_2O$	95-86-3	124.140	lf	79 dec				vs H_2O, ace, EtOH
2862	2,4-Diaminophenol, dihydrochloride		$C_6H_{10}Cl_2N_2O$	137-09-7	197.061	nd	235 dec				vs H_2O
2863	3,7-Diaminophenothiazin-5-ium chloride	Thionine	$C_{12}H_{10}ClN_3S$	581-64-6	263.745						sl H_2O, EtOH, eth; s bz, chl, acid
2864	4-[(2,4-Diaminophenyl)azo]-benzenesulfonamide	Prontosil	$C_{12}H_{14}ClN_5O_2S$	103-12-8	327.790		249.5				sl H_2O; s EtOH, ace, oils, fats
2865	1,3-Diamino-2-propanol		$C_3H_{10}N_2O$	616-29-5	90.123	cry	42.8				i eth, bz
2866	4,4'-Diamino-2,2'-stilbenedisulfonic acid	Amsonic acid	$C_{14}H_{14}N_2O_6S_2$	81-11-8	370.400	ye nd	300				sl H_2O
2867	4,6-Diamino-1,3,5-triazin-2(1*H*)-one		$C_3H_5N_5O$	645-92-1	127.105	nd (aq Na_2CO_3)	dec				i H_2O, EtOH, eth, bz, HOAc; s acid, alk

Organic

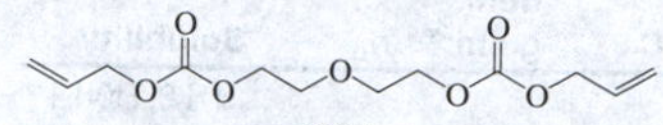
2831
Diallyl diethylene glycol carbonate

2832
Diallyldimethylsilane

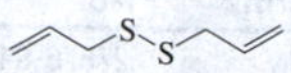
2833
Diallyl disulfide

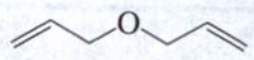
2834
Diallyl ether

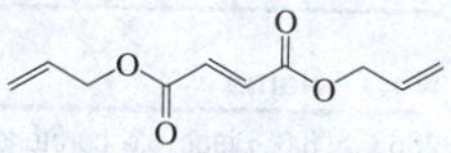
2835
Diallyl fumarate

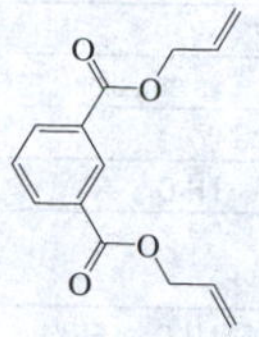
2836
Diallyl isophthalate

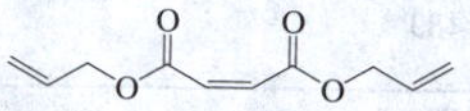
2837
Diallyl maleate

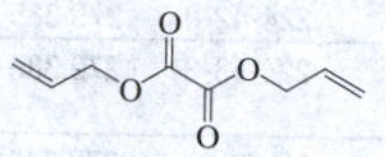
2838
Diallyl oxalate

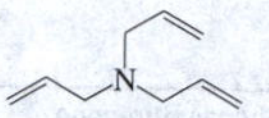
2839
N,N-Diallyl-2-propen-1-amine

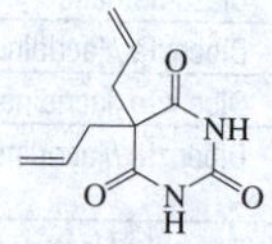
2840
5,5-Diallyl-2,4,6(1*H*,3*H*,5*H*)-pyrimidinetrione

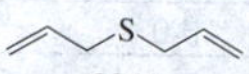
2841
Diallyl sulfide

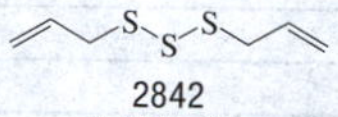
2842
Diallyl trisulfide

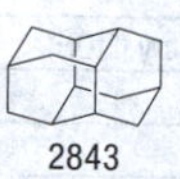
2843
Diamantane

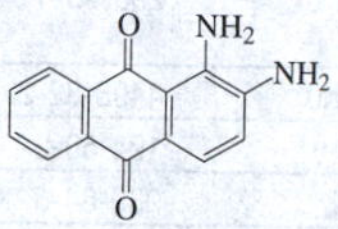
2844
1,2-Diamino-9,10-anthracenedione

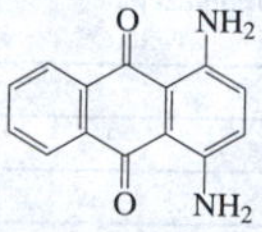
2845
1,4-Diamino-9,10-anthracenedione

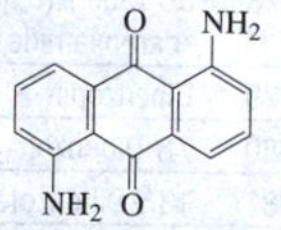
2846
1,5-Diamino-9,10-anthracenedione

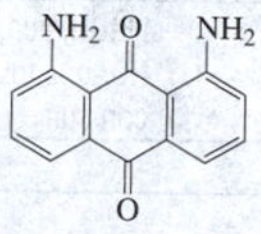
2847
1,8-Diamino-9,10-anthracenedione

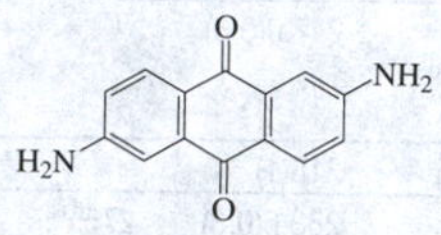
2848
2,6-Diamino-9,10-anthracenedione

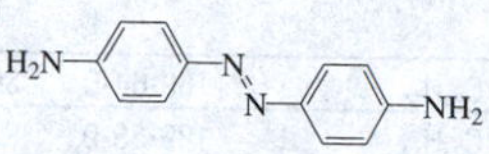
2849
4,4′-Diaminoazobenzene

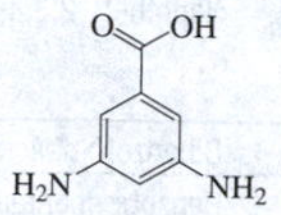
2850
3,5-Diaminobenzoic acid

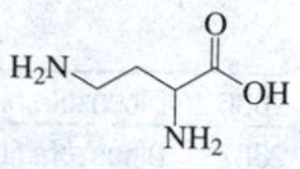
2851
2,4-Diaminobutanoic acid

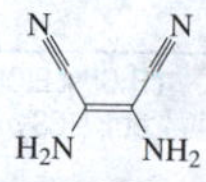
2852
cis-2,3-Diamino-2-butenedinitrile

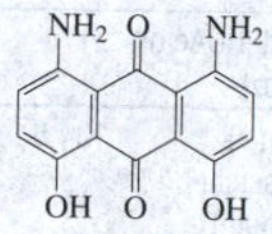
2853
1,8-Diamino-4,5-dihydroxy-9,10-anthracenedione

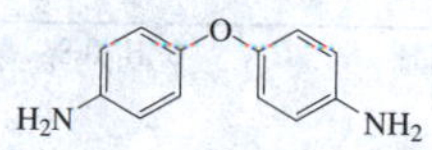
2854
4,4′-Diaminodiphenyl ether

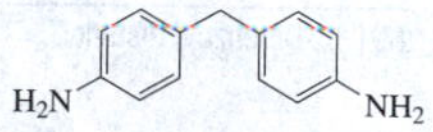
2855
4,4′-Diaminodiphenylmethane

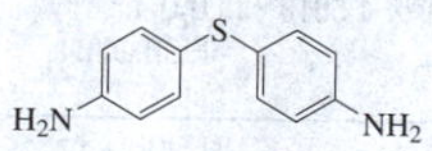
2856
4,4′-Diaminodiphenyl sulfide

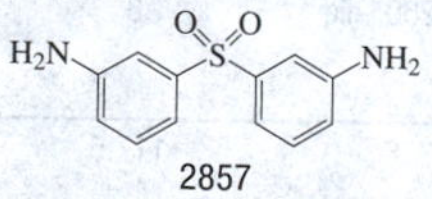
2857
3,3′-Diaminodiphenyl sulfone

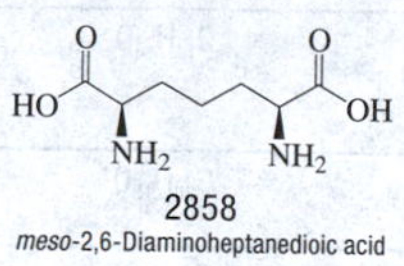
2858
meso-2,6-Diaminoheptanedioic acid

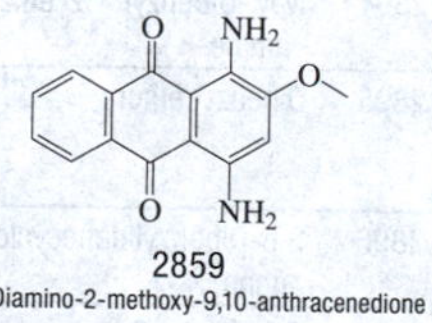
2859
1,4-Diamino-2-methoxy-9,10-anthracenedione

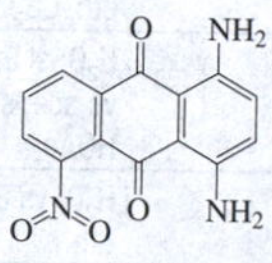
2860
1,4-Diamino-5-nitro-9,10-anthracenedione

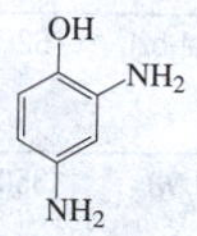
2861
2,4-Diaminophenol

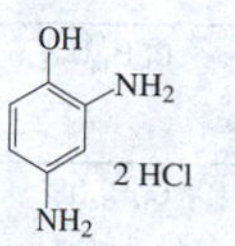
2862
2,4-Diaminophenol, dihydrochloride

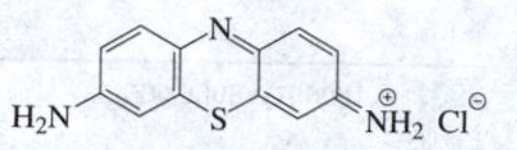
2863
3,7-Diaminophenothiazin-5-ium chloride

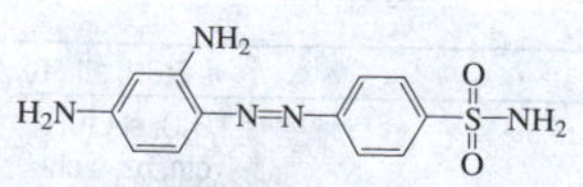
2864
4-[(2,4-Diaminophenyl)azo]benzenesulfonamide

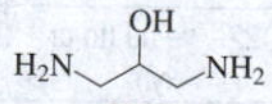
2865
1,3-Diamino-2-propanol

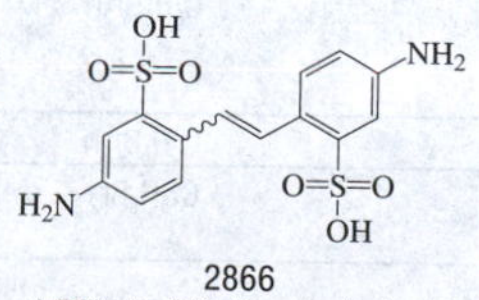
2866
4,4′-Diamino-2,2′-stilbenedisulfonic acid

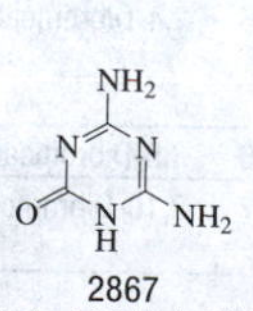
2867
4,6-Diamino-1,3,5-triazin-2(1*H*)-one

No.	Name	Synonym	Mol. form.	CAS Reg. No.	Mol. wt.	Physical form	mp/°C	bp/°C	den/ g cm^{-3}	n_D	Solubility
2868	8,8'-Diapo-ψ,ψ-carotenedioic acid	Crocetin	$C_{20}H_{24}O_4$	27876-94-4	328.403	brick red orth	286				sl H_2O, EtOH; i eth, bz; s py; vs NaOH
2869	Diatrizoic acid	*N,N*-Diacetyl-3,5-diamino-2,4,6-triiodobenzoic acid	$C_{11}H_9I_3N_2O_4$	117-96-4	613.913	cry (EtOH aq)	300				
2870	Diazenedicarboxamide	Azodicarbonamide	$C_2H_4N_4O_2$	123-77-3	116.079		225(1)				
2871	Diazinon		$C_{12}H_{21}N_2O_3PS$	333-41-5	304.345			$87^{0.05}$	1.1088^{20}	1.4922^{20}	
2872	Diazomethane		CH_2N_2	334-88-3	42.040	ye gas	-145	-23			vs eth, diox
2873	Dibenz[*a,h*]acridine		$C_{21}H_{13}N$	226-36-8	279.335	ye cry	226.6(0.7)				
2874	Dibenz[*a,j*]acridine	7-Azadibenz[*a,j*]anthracene	$C_{21}H_{13}N$	224-42-0	279.335		219.6(0.9)				i H_2O
2875	Dibenz[*c,h*]acridine		$C_{21}H_{13}N$	224-53-3	279.335	ye cry (EtOH)	189				
2876	Dibenz[*a,h*]anthracene	1,2:5,6-Dibenzanthracene	$C_{22}H_{14}$	53-70-3	278.346	pl (dil ace)	269(6)				i H_2O; sl EtOH; s ace, bz, CS_2
2877	Dibenz[*a,j*]anthracene		$C_{22}H_{14}$	224-41-9	278.346	oran lf or nd (bz)	198.2				i H_2O, HOAc; sl EtOH, eth, bz; s peth
2878	5*H*-Dibenz[*b,f*]azepine-5-carboxamide	Carbamazepine	$C_{15}H_{12}N_2O$	298-46-4	236.268		190.2				
2879	Dibenzepin		$C_{18}H_{21}N_3O$	4498-32-2	295.379		117	$185^{0.01}$			
2880	7*H*-Dibenzo[*c,g*]carbazole		$C_{20}H_{13}N$	194-59-2	267.324	cry (EtOH)	157(1)				
2881	13*H*-Dibenzo[*a,i*]carbazole		$C_{20}H_{13}N$	239-64-5	267.324		221.3				i H_2O
2882	Dibenzo[*b,k*]chrysene		$C_{26}H_{16}$	217-54-9	328.405		400				
2883	Dibenzo[*b,e*][1,4]dioxin	Diphenylene dioxide	$C_{12}H_8O_2$	262-12-4	184.191	nd (MeOH)	117.5(0.2)				
2884	Dibenzofuran	2,2'-Biphenylene oxide	$C_{12}H_8O$	132-64-9	168.191	lf or nd (al)	82.16(0.05)	285.2(0.3)	1.0886^{99}	1.6079^{99}	i H_2O; s EtOH, ace, bz; vs eth, HOAc
2885	Dibenzo[*a,e*]pyrene	Naphtho[1,2,3,4-*def*]chrysene	$C_{24}H_{14}$	192-65-4	302.368	pa ye nd(xyl)	247.0(0.6)				sl EtOH, ace, bz, HOAc; s tol, con sulf
2886	Dibenzo[*a,h*]pyrene	Dibenzo[*b,def*]chrysene	$C_{24}H_{14}$	189-64-0	302.368	oran pl	318(1)				
2887	Dibenzo[*a,i*]pyrene	Benzo[*rst*]pentaphene	$C_{24}H_{14}$	189-55-9	302.368		283.6(0.3)	$275^{0.05}$			
2888	Dibenzo[*a,l*]pyrene	Dibenzo[*def,p*]chrysene	$C_{24}H_{14}$	191-30-0	302.368	ye pl (bz/ EtOH)	164.5				
2889	Dibenzothiophene		$C_{12}H_8S$	132-65-0	184.257	nd (dil al, lig)	98.67(0.02)	331.6(0.4)			i H_2O; s chl, MeOH; vs EtOH, bz
2890	Dibenz[*c,e*]oxepin-5,7-dione		$C_{14}H_8O_3$	6050-13-1	224.212	nd (HOAc or bz)	217	sub			i H_2O; sl eth
2891	Dibenzoyl disulfide	Benzoyl disulfide	$C_{14}H_{10}O_2S_2$	644-32-6	274.358	pr(al), sc(chl-peth)	134.5	dec			i H_2O; sl EtOH, eth; s CS_2
2892	Dibenzylamine	*N*-Benzylbenzenemethanamine	$C_{14}H_{15}N$	103-49-1	197.276		-26(1)	300 dec	1.0256^{22}	1.5781^{20}	i H_2O; vs EtOH, eth; s ctc
2893	Dibenzyl disulfide		$C_{14}H_{14}S_2$	150-60-7	246.391	lf (al)	68.6(0.2)				sl H_2O; s EtOH, eth, bz, MeOH
2894	*N,N'*-Dibenzyl-1,2-ethanediamine	Benzathine	$C_{16}H_{20}N_2$	140-28-3	240.343	oily lig	26	195^{4}	1.024^{20}	1.5635^{20}	vs bz, eth, EtOH
2895	Dibenzyl ether	Benzyl ether	$C_{14}H_{14}O$	103-50-4	198.260	liq	1.8	298	1.0428^{20}	1.5618^{20}	i H_2O; msc EtOH, eth; s ctc
2896	2,6-Dibenzylidenecyclohexanone		$C_{20}H_{18}O$	897-78-9	274.356		117.5	190^{20}			sl EtOH; s bz, HOAc
2897	Dibenzyl malonate		$C_{17}H_{16}O_4$	15014-25-2	284.307			187^{2}	1.137^{25}	1.5447^{20}	
2898	Dibenzyl phosphite		$C_{14}H_{15}O_3P$	17176-77-1	262.241		-2.5	$162^{0.1}$		1.5521^{18}	
2899	Dibenzyl sulfide	Benzyl sulfide	$C_{14}H_{14}S$	538-74-9	214.326	pl (eth or chl)	48.3(0.5)	335(6)	1.0583^{50}		i H_2O; s EtOH, eth, CS_2
2900	Dibenzyl sulfone		$C_{14}H_{14}O_2S$	620-32-6	246.325	nd (al-bz)	152	290 dec			i H_2O; sl EtOH; vs ace; s bz, HOAc
2901	Dibenzyl sulfoxide		$C_{14}H_{14}OS$	621-08-9	230.325	lf (al, w)	135(3)	210 dec			i H_2O; vs EtOH, eth
2902	*N,N'*-Dibenzylurea		$C_{15}H_{16}N_2O$	1466-67-7	240.300	nd (al)	170(1)				vs EtOH, HOAc
2903	Dibromoacetic acid		$C_2H_2Br_2O_2$	631-64-1	217.844	hyg cry	49	195^{250}			vs H_2O; vs EtOH, eth
2904	Dibromoacetonitrile		C_2HBr_2N	3252-43-5	198.844			169	2.369^{20}	1.5393^{20}	
2905	2,4-Dibromoaniline		$C_6H_5Br_2N$	615-57-6	250.919	orth bipym (chl) nd or lf (al)	79.5	156^{74}	2.260^{20}		s EtOH, eth, chl, HOAc
2906	3,5-Dibromoaniline		$C_6H_5Br_2N$	626-40-4	250.919	nd (dil al)	57				vs EtOH, eth, bz
2907	9,10-Dibromoanthracene		$C_{14}H_8Br_2$	523-27-3	336.022	ye nd (to or xyl)	226	sub			i H_2O; sl EtOH, eth, bz; s chl

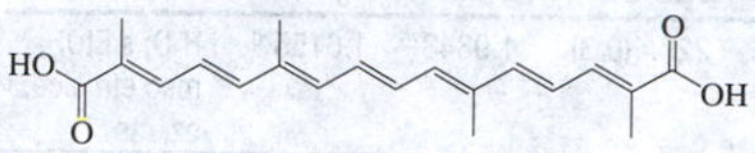
2868
8,8'-Diapo-ψ,ψ-carotenedioic acid

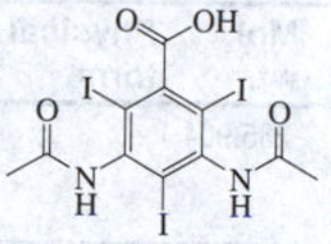
2869
Diatrizoic acid

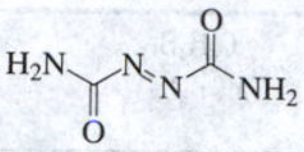
2870
Diazenedicarboxamide

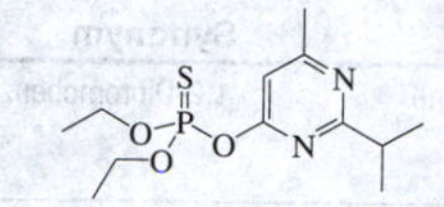
2871
Diazinon

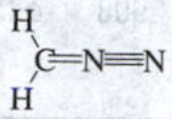
2872
Diazomethane

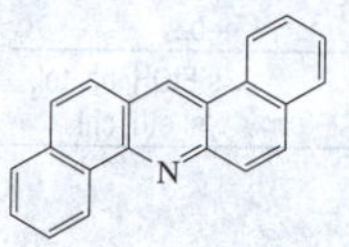
2873
Dibenz[a,h]acridine

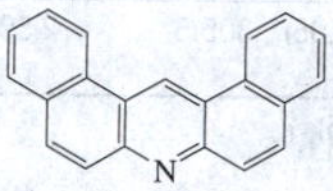
2874
Dibenz[a,j]acridine

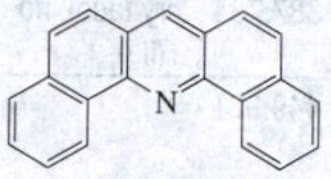
2875
Dibenz[c,h]acridine

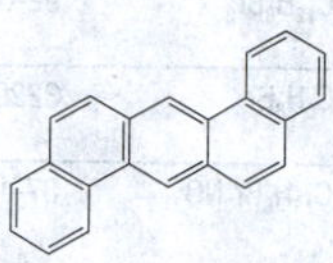
2876
Dibenz[a,h]anthracene

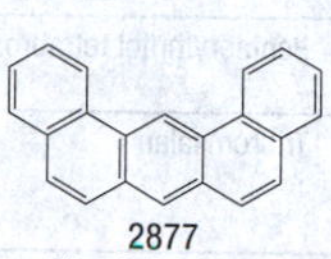
2877
Dibenz[a,j]anthracene

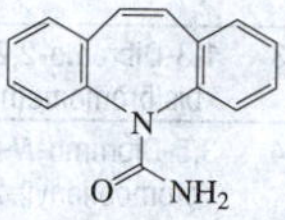
2878
5*H*-Dibenz[b,f]azepine-5-carboxamide

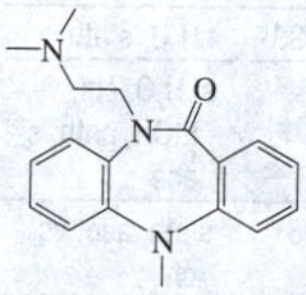
2879
Dibenzepin

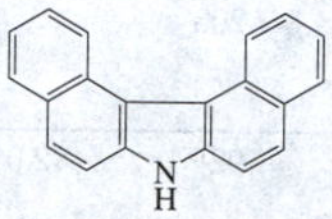
2880
7*H*-Dibenzo[c,g]carbazole

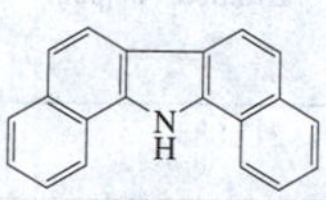
2881
13*H*-Dibenzo[a,i]carbazole

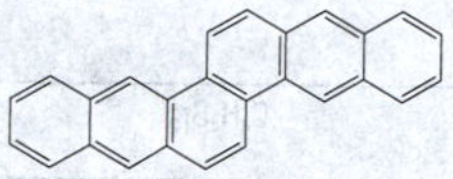
2882
Dibenzo[b,k]chrysene

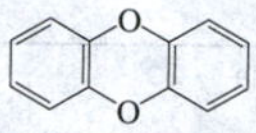
2883
Dibenzo[b,e][1,4]dioxin

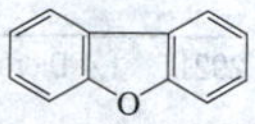
2884
Dibenzofuran

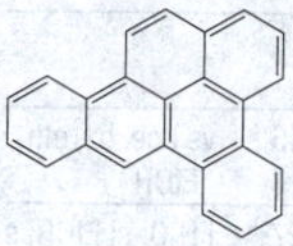
2885
Dibenzo[a,e]pyrene

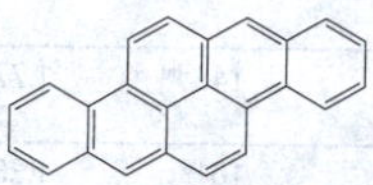
2886
Dibenzo[a,h]pyrene

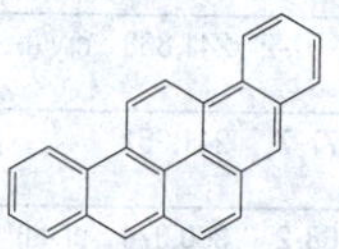
2887
Dibenzo[a,i]pyrene

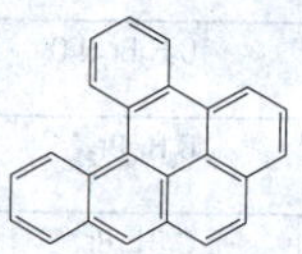
2888
Dibenzo[a,l]pyrene

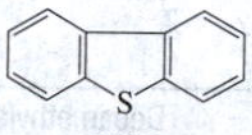
2889
Dibenzothiophene

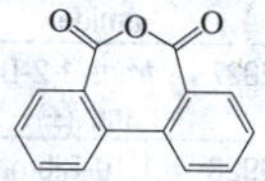
2890
Dibenz[c,e]oxepin-5,7-dione

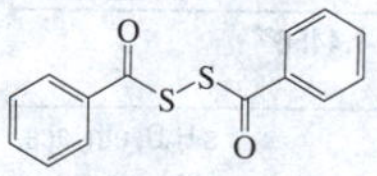
2891
Dibenzoyl disulfide

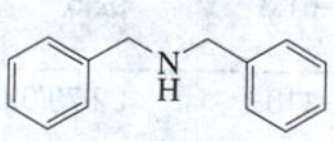
2892
Dibenzylamine

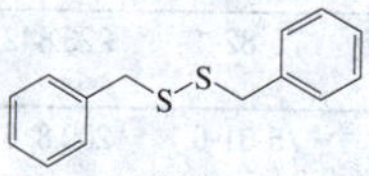
2893
Dibenzyl disulfide

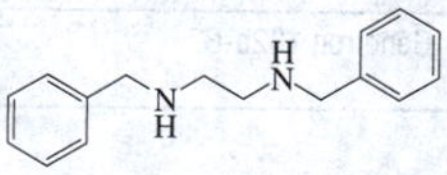
2894
N,N'-Dibenzyl-1,2-ethanediamine

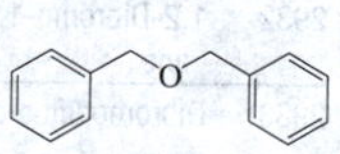
2895
Dibenzyl ether

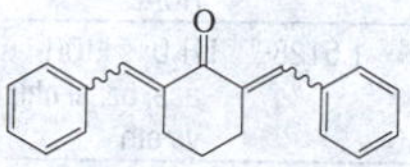
2896
2,6-Dibenzylidenecyclohexanone

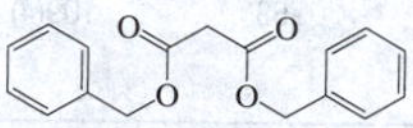
2897
Dibenzyl malonate

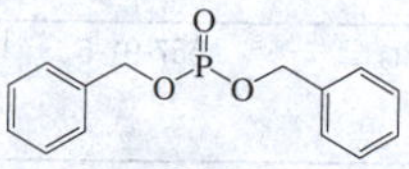
2898
Dibenzyl phosphite

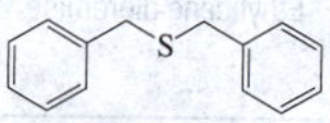
2899
Dibenzyl sulfide

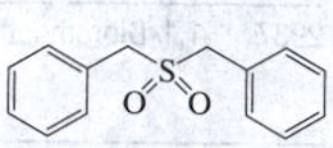
2900
Dibenzyl sulfone

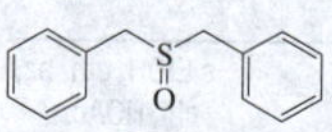
2901
Dibenzyl sulfoxide

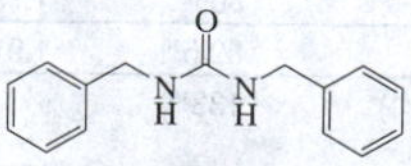
2902
N,N'-Dibenzylurea

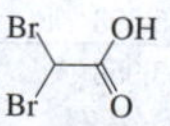
2903
Dibromoacetic acid

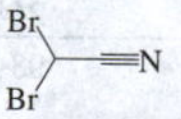
2904
Dibromoacetonitrile

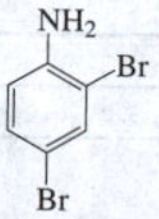
2905
2,4-Dibromoaniline

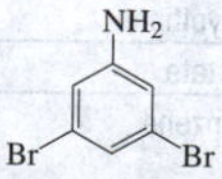
2906
3,5-Dibromoaniline

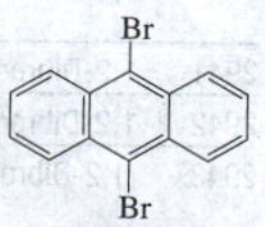
2907
9,10-Dibromoanthracene

No.	Name	Synonym	Mol. form.	CAS Reg. No.	Mol. wt.	Physical form	mp/°C	bp/°C	den/g cm^{-3}	n_D	Solubility
2908	*o*-Dibromobenzene	1,2-Dibromobenzene	$C_6H_4Br_2$	583-53-9	235.904		6(2)	220.4(0.3)	1.9843^{20}	1.6155^{20}	i H_2O; s EtOH; msc eth, ace, bz, ctc
2909	*m*-Dibromobenzene	1,3-Dibromobenzene	$C_6H_4Br_2$	108-36-1	235.904	liq	-6.9(0.5)	214(14)	1.9523^{20}	1.6083^{17}	i H_2O; s EtOH; msc eth
2910	*p*-Dibromobenzene	1,4-Dibromobenzene	$C_6H_4Br_2$	106-37-6	235.904	pl	87.3(0.1)	222(3)	2.261^{17}	1.5742	i H_2O; s EtOH, bz; vs eth, ace, CS_2
2911	4,4'-Dibromobenzophenone	Bis(4-bromophenyl) ketone	$C_{13}H_8Br_2O$	3988-03-2	340.010	pl (al)	177	394(24)			vs bz, HOAc, chl
2912	4,4'-Dibromo-1,1'-biphenyl		$C_{12}H_8Br_2$	92-86-4	312.000	mcl pr (MeOH)	164	357.5			i H_2O; sl EtOH; s bz
2913	1,3-Dibromo-2,2-bis(bromomethyl)propane	Pentaerythritol tetrabromide	$C_5H_8Br_4$	3229-00-3	387.734	cry (ace), nd (lig)	160.29(0.05)	305.5	2.596^{15}		s EtOH, bz, tol; sl eth, chl
2914	3,5-Dibromo-*N*-(4-bromophenyl)-2-hydroxybenzamide	Tribromsalan	$C_{13}H_8Br_3NO_2$	87-10-5	449.921		226.4(0.5)				
2915	1,1-Dibromobutane		$C_4H_8Br_2$	62168-25-6	215.915			158	1.784^{25}	1.4988^{25}	
2916	1,2-Dibromobutane	α-Butylene dibromide	$C_4H_8Br_2$	533-98-2	215.915	liq	-65.4(0.4)	161(4)	1.7915^{20}	1.4025^{20}	i H_2O; s eth, chl
2917	1,3-Dibromobutane		$C_4H_8Br_2$	107-80-2	215.915			176.4(0.4)	1.800^{20}	1.507^{20}	i H_2O; s eth, chl; sl ctc
2918	1,4-Dibromobutane		$C_4H_8Br_2$	110-52-1	215.915	liq	-21.1(0.5)	197(4)	1.8199^{25}	1.5167^{25}	i H_2O; sl ctc; s chl
2919	2,3-Dibromobutane		$C_4H_8Br_2$	5408-86-6	215.915	liq	-24	158(5)	1.7893^{22}	1.5133^{22}	i H_2O; s eth
2920	*trans*-1,4-Dibromo-2-butene		$C_4H_6Br_2$	821-06-7	213.899	pl (peth)	53.4	203			sl H_2O, chl; vs EtOH, peth; s ace
2921	1,4-Dibromo-2-butyne		$C_4H_4Br_2$	2219-66-1	211.883			92^{15}	2.014^{18}	1.588^{18}	s eth, ace; vs chl
2922	α,α'-Dibromo-*d*-camphor		$C_{10}H_{14}Br_2O$	514-12-5	310.025		61		1.854^{21}		i H_2O; vs EtOH, eth, bz, chl; s AcOEt
2923	Dibromochlorofluoromethane		CBr_2ClF	353-55-9	226.270			80.3	2.3173^{22}	1.4570^{20}	
2924	1,2-Dibromo-3-chloropropane		$C_3H_5Br_2Cl$	96-12-8	236.333			200(13)	2.093^{14}	1.553^{14}	i H_2O
2925	1,2-Dibromo-1-chloro-1,2,2-trifluoroethane		$C_2Br_2ClF_3$	354-51-8	276.277		50	92.8(0.2)			
2926	2,2-Dibromo-2-cyanoacetamide		$C_3H_2Br_2N_2O$	10222-01-2	241.868	cry (bz)	126				
2927	*trans*-1,2-Dibromocyclohexane, (±)-		$C_6H_{10}Br_2$	5183-77-7	241.951		-2.0	145^{100}	1.7759^{20}	1.5445^{19}	vs ace, bz, eth, EtOH
2928	1,10-Dibromodecane	Decamethylene dibromide	$C_{10}H_{20}Br_2$	4101-68-2	300.074	pl (al)	28	161^{9}	1.335^{30}	1.4927^{25}	i H_2O; sl EtOH; s eth
2929	1,2-Dibromo-1,1-dichloroethane		$C_2H_2Br_2Cl_2$	75-81-0	256.751	liq	-26	195	2.135^{20}	1.5662^{20}	vs ace, bz, eth, EtOH
2930	1,2-Dibromo-1,2-dichloroethane		$C_2H_2Br_2Cl_2$	683-68-1	256.751	liq	-26	195	2.135^{20}	1.5662^{20}	i H_2O; s EtOH, eth, ace, bz
2931	Dibromodichloromethane		CBr_2Cl_2	594-18-3	242.725		38	150.2	2.42^{25}		i H_2O; s EtOH, eth, ace, bz
2932	1,2-Dibromo-1,1-difluoroethane	Genetron 132b-B2	$C_2H_2Br_2F_2$	75-82-1	223.842	liq	-61.3	92.5	2.2238^{20}	1.4456^{20}	
2933	Dibromodifluoromethane		CBr_2F_2	75-61-6	209.816	vol liq or gas	-110.1	22.79(0.08)			s H_2O, eth, ace, bz
2934	1,3-Dibromo-5,5-dimethyl-2,4-imidazolidinedione	Dibromantine	$C_5H_6Br_2N_2O_2$	77-48-5	285.922		198 dec				
2935	1,3-Dibromo-2,2-dimethylpropane		$C_5H_{10}Br_2$	5434-27-5	229.941			185(11)	1.6775^{20}	1.5090	
2936	1,12-Dibromododecane		$C_{12}H_{24}Br_2$	3344-70-5	328.127	nd (al,HOAc)	41	215^{15}			i H_2O; vs EtOH, chl; s eth, HOAc
2937	1,1-Dibromoethane	Ethylidene dibromide	$C_2H_4Br_2$	557-91-5	187.861	liq	-63	109(4)	2.0555^{20}	1.5128^{20}	i H_2O; s EtOH, ace, bz; sl chl; vs eth
2938	1,2-Dibromoethane	Ethylene dibromide	$C_2H_4Br_2$	106-93-4	187.861	liq	9.8(0.1)	131.3(0.3)	2.1683^{25}	1.5356^{25}	vs ace, bz, eth, EtOH
2939	*cis*-1,2-Dibromoethene	*cis*-1,2-Dibromoethylene	$C_2H_2Br_2$	590-11-4	185.845	liq	-53	111(1)	2.2464^{20}	1.5428^{20}	i H_2O; vs EtOH, eth; s ace, bz, chl
2940	*trans*-1,2-Dibromoethene	*trans*-1,2-Dibromoethylene	$C_2H_2Br_2$	590-12-5	185.845	liq	-6.5	107(3)	2.2308^{20}	1.5505^{18}	i H_2O; vs EtOH, eth; s ace, bz, chl
2941	1,2-Dibromo-1-ethoxyethane		$C_4H_8Br_2O$	2983-26-8	231.914			80^{20}	1.7320^{20}	1.5044^{20}	vs EtOH, chl
2942	1,2-Dibromoethyl acetate		$C_4H_6Br_2O_2$	24442-57-7	245.898	liq		89.5^{16}	1.91^{20}		
2943	(1,2-Dibromoethyl)benzene		$C_8H_8Br_2$	93-52-7	263.958		73(1)	133^{19}			s EtOH, eth, bz, chl, HOAc, MeOH, lig

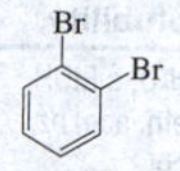
2908
o-Dibromobenzene

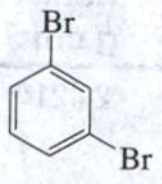
2909
m-Dibromobenzene

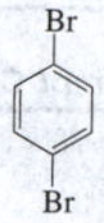
2910
p-Dibromobenzene

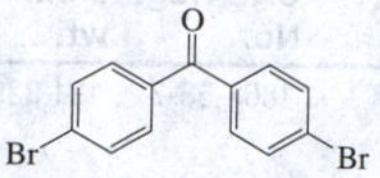
2911
4,4′-Dibromobenzophenone

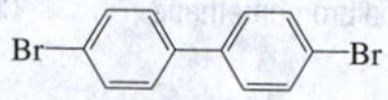
2912
4,4′-Dibromo-1,1′-biphenyl

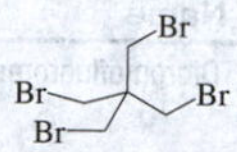
2913
1,3-Dibromo-2,2-bis(bromomethyl)propane

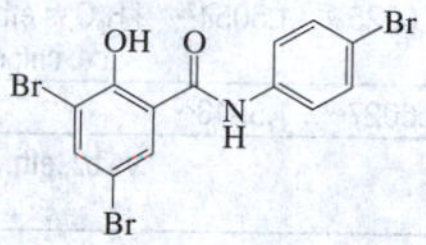
2914
3,5-Dibromo-*N*-(4-bromophenyl)-2-hydroxybenzamide

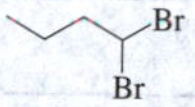
2915
1,1-Dibromobutane

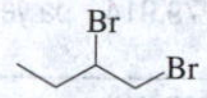
2916
1,2-Dibromobutane

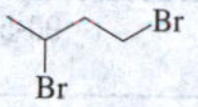
2917
1,3-Dibromobutane

2918
1,4-Dibromobutane

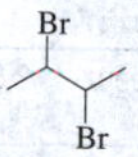
2919
2,3-Dibromobutane

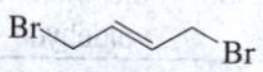
2920
trans-1,4-Dibromo-2-butene

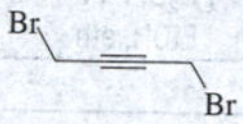
2921
1,4-Dibromo-2-butyne

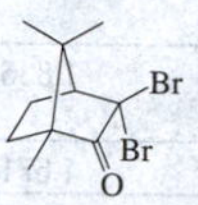
2922
α,α′-Dibromo-*d*-camphor

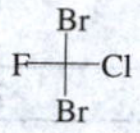
2923
Dibromochlorofluoromethane

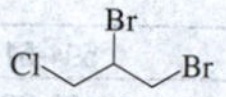
2924
1,2-Dibromo-3-chloropropane

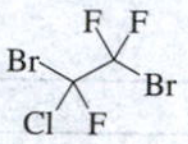
2925
1,2-Dibromo-1-chloro-1,2,2-trifluoroethane

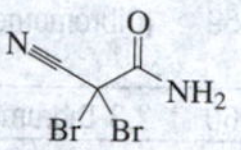
2926
2,2-Dibromo-2-cyanoacetamide

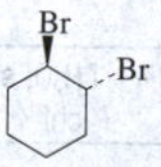
2927
trans-1,2-Dibromocyclohexane, (±)-

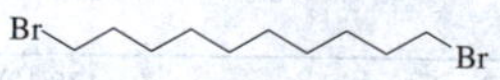
2928
1,10-Dibromodecane

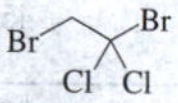
2929
1,2-Dibromo-1,1-dichloroethane

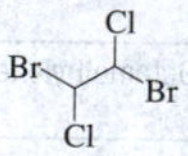
2930
1,2-Dibromo-1,2-dichloroethane

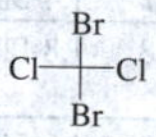
2931
Dibromodichloromethane

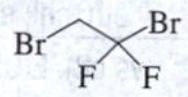
2932
1,2-Dibromo-1,1-difluoroethane

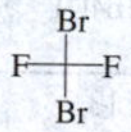
2933
Dibromodifluoromethane

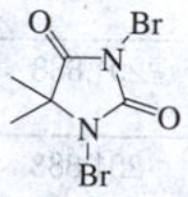
2934
1,3-Dibromo-5,5-dimethyl-2,4-imidazolidinedione

2935
1,3-Dibromo-2,2-dimethylpropane

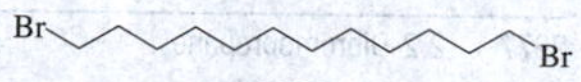
2936
1,12-Dibromododecane

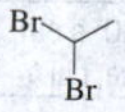
2937
1,1-Dibromoethane

2938
1,2-Dibromoethane

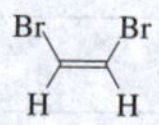
2939
cis-1,2-Dibromoethene

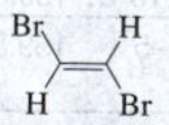
2940
trans-1,2-Dibromoethene

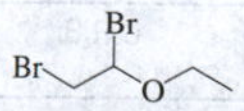
2941
1,2-Dibromo-1-ethoxyethane

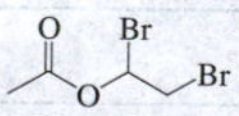
2942
1,2-Dibromoethyl acetate

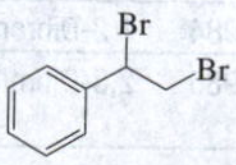
2943
(1,2-Dibromoethyl)benzene

No.	Name	Synonym	Mol. form.	CAS Reg. No.	Mol. wt.	Physical form	mp/°C	bp/°C	den/ g cm^{-3}	n_D	Solubility
2944	Dibromofluoromethane	Fluorodibromomethane	$CHBr_2F$	1868-53-7	191.825	liq	-78	64.9	2.421^{20}	1.4685^{20}	i H_2O; s EtOH, eth, ace, bz, chl
2945	1,2-Dibromoheptane		$C_7H_{14}Br_2$	42474-21-5	257.994			228	1.5086^{20}	1.4986^{20}	
2946	1,7-Dibromoheptane	Heptamethylene dibromide	$C_7H_{14}Br_2$	4549-31-9	257.994		41.7	247(12)	1.5306^{20}	1.5034^{20}	i H_2O; s eth, ace, bz, ctc, chl
2947	2,3-Dibromoheptane		$C_7H_{14}Br_2$	21266-88-6	257.994			101^{17}	1.5139^{20}	1.4992^{20}	
2948	3,4-Dibromoheptane		$C_7H_{14}Br_2$	21266-90-0	257.994			107^{24}	1.5182^{20}	1.5010^{20}	
2949	1,2-Dibromo-1,1,2,3,3,3-hexafluoropropane		$C_3Br_2F_6$	661-95-0	309.830			72.8	2.1630^{20}		i H_2O
2950	1,2-Dibromohexane		$C_6H_{12}Br_2$	624-20-4	243.967			103^{36}	1.5774^{20}	1.5024^{20}	vs bz, eth, chl
2951	1,6-Dibromohexane		$C_6H_{12}Br_2$	629-03-8	243.967	liq	-1.2	245.5	1.6025^{25}	1.5054^{25}	i H_2O; s eth, ace, chl; sl ctc
2952	3,4-Dibromohexane		$C_6H_{12}Br_2$	89583-12-0	243.967			80^{13}	1.6027^{20}	1.5043^{20}	
2953	3,5-Dibromo-2-hydroxybenzaldehyde	3,5-Dibromosalicylaldehyde	$C_7H_4Br_2O_2$	90-59-5	279.914	pa ye pr	86	sub			vs bz, eth, chl
2954	3,5-Dibromo-2-hydroxybenzoic acid	3,5-Dibromosalicylic acid	$C_7H_4Br_2O_3$	3147-55-5	295.913	nd	228				s ace
2955	3,5-Dibromo-4-hydroxybenzonitrile	Bromoxynil	$C_7H_3Br_2NO$	1689-84-5	276.913		190.5(0.8)				
2956	Dibromomethane	Methylene bromide	CH_2Br_2	74-95-3	173.835	liq	-52.1(0.7)	97.0(0.6)	2.4969^{20}	1.5420^{20}	sl H_2O; msc EtOH, eth, ace; s ctc
2957	1,4-Dibromo-2-methylbenzene	2,5-Dibromotoluene	$C_7H_6Br_2$	615-59-8	249.931		6(1)	236(11)	1.8127^{17}	1.5982^{18}	i H_2O
2958	2,4-Dibromo-1-methylbenzene		$C_7H_6Br_2$	31543-75-6	249.931		-10(1)	103^{11}	1.8176^{25}	1.5964^{25}	
2959	(Dibromomethyl)benzene		$C_7H_6Br_2$	618-31-5	249.931		1.0	156^{23}	1.8365^{28}	1.6147^{20}	i H_2O; msc EtOH, eth
2960	2,3-Dibromo-2-methylbutane		$C_5H_{10}Br_2$	594-51-4	229.941		7	62^{17}	1.6717^{20}	1.5729^{25}	
2961	2,4-Dibromo-6-methylphenol		$C_7H_6Br_2O$	609-22-3	265.930	nd (peth)	58	265 dec			s chl
2962	1,2-Dibromo-2-methylpropane		$C_4H_8Br_2$	594-34-3	215.915		10.5	139(2)	1.7827^{20}	1.5119^{20}	s EtOH, eth, chl
2963	1,4-Dibromonaphthalene		$C_{10}H_6Br_2$	83-53-4	285.963		83	310			i H_2O; s EtOH, eth; sl HOAc
2964	2,6-Dibromo-4-nitroaniline		$C_6H_4Br_2N_2O_2$	827-94-1	295.916	ye nd (al, HOAc)	207				sl H_2O; s HOAc
2965	2,6-Dibromo-4-nitrophenol		$C_6H_3Br_2NO_3$	99-28-5	296.901	pa ye pr or lf (al)	145 dec				i H_2O; vs EtOH, eth; sl ace, bz, HOAc
2966	1,9-Dibromononane		$C_9H_{18}Br_2$	4549-33-1	286.047	liq	-22.5	269(10)	1.4229^{20}		
2967	1,4-Dibromooctafluorobutane		$C_4Br_2F_8$	335-48-8	359.838			98(25)			
2968	1,8-Dibromooctane	Octamethylene dibromide	$C_8H_{16}Br_2$	4549-32-0	272.021		15.5	271	1.4594^{25}	1.4971^{25}	i H_2O; s eth, ctc, chl
2969	1,2-Dibromopentane		$C_5H_{10}Br_2$	3234-49-9	229.941			179(13)	1.668^{18}		
2970	1,4-Dibromopentane		$C_5H_{10}Br_2$	626-87-9	229.941		-34.4	146^{150}	1.6222^{20}	1.5086^{20}	
2971	1,5-Dibromopentane		$C_5H_{10}Br_2$	111-24-0	229.941	liq	-40.0(0.4)	222.3	1.6928^{25}	1.5102^{25}	i H_2O; s bz, chl; sl ctc
2972	2,4-Dibromopentane		$C_5H_{10}Br_2$	19398-53-9	229.941			75^{21}	1.6659^{20}	1.4987^{20}	
2973	2,4-Dibromophenol		$C_6H_4Br_2O$	615-58-7	251.903	nd (peth)	40(2)	238.5	2.0700^{20}		sl H_2O, ctc; vs EtOH, eth, bz
2974	2,6-Dibromophenol		$C_6H_4Br_2O$	608-33-3	251.903	nd (w)	56.5	255			s H_2O; vs EtOH, eth
2975	1,2-Dibromopropane	Propylene dibromide	$C_3H_6Br_2$	78-75-1	201.888	liq	-55.4(0.3)	140(1)	1.9324^{20}	1.5201^{20}	s EtOH, eth, chl; sl ctc
2976	1,3-Dibromopropane		$C_3H_6Br_2$	109-64-8	201.888	liq	-35(1)	164(1)	1.9701^{25}	1.5204^{25}	i H_2O; s EtOH, eth, chl; sl ctc
2977	2,2-Dibromopropane		$C_3H_6Br_2$	594-16-1	201.888			113	1.880^{20}		vs eth, EtOH, chl
2978	2,3-Dibromopropanoic acid		$C_3H_4Br_2O_2$	600-05-5	231.871		66.5	160^{20}			vs bz, eth, EtOH
2979	2,3-Dibromo-1-propanol	DBP	$C_3H_6Br_2O$	96-13-9	217.887			219	2.120^{20}		
2980	1,3-Dibromo-2-propanol		$C_3H_6Br_2O$	96-21-9	217.887	ye liq		219 dec	2.1364^{20}	1.5495^{25}	vs ace, eth, EtOH
2981	2,3-Dibromo-1-propanol, phosphate (3:1)	Tris(2,3-dibromopropyl) phosphate	$C_9H_{15}Br_6O_4P$	126-72-7	697.610						s chl
2982	1,3-Dibromo-2-propanone	1,3-Dibromoacetone	$C_3H_4Br_2O$	816-39-7	215.871	nd	26	97^{22}	2.1670^{18}		vs eth, CS_2
2983	1,1-Dibromo-1-propene		$C_3H_4Br_2$	13195-80-7	199.872			125	1.9767^{20}	1.5260^{20}	sl H_2O; s bz, ctc, chl
2984	1,2-Dibromo-1-propene		$C_3H_4Br_2$	26391-16-2	199.872			131.5	2.0076^{20}		
2985	2,3-Dibromo-1-propene		$C_3H_4Br_2$	513-31-5	199.872			142(18)	2.0345^{25}	1.5416^{25}	i H_2O; s eth, ace, chl

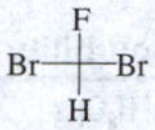
2944
Dibromofluoromethane

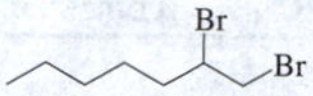
2945
1,2-Dibromoheptane

2946
1,7-Dibromoheptane

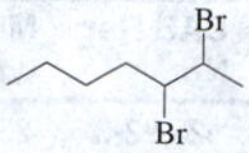
2947
2,3-Dibromoheptane

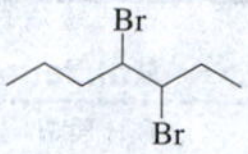
2948
3,4-Dibromoheptane

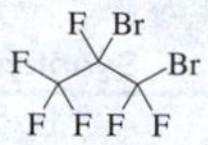
2949
1,2-Dibromo-1,1,2,3,3,3-hexafluoropropane

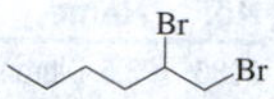
2950
1,2-Dibromohexane

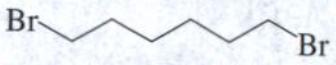
2951
1,6-Dibromohexane

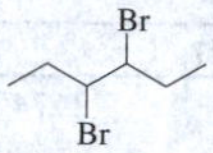
2952
3,4-Dibromohexane

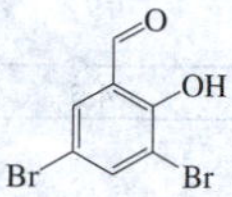
2953
3,5-Dibromo-2-hydroxybenzaldehyde

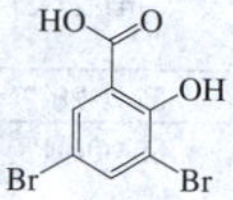
2954
3,5-Dibromo-2-hydroxybenzoic acid

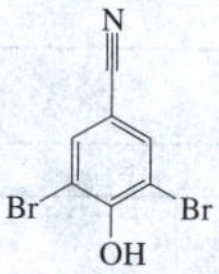
2955
3,5-Dibromo-4-hydroxybenzonitrile

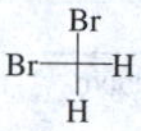
2956
Dibromomethane

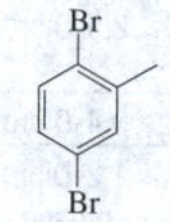
2957
1,4-Dibromo-2-methylbenzene

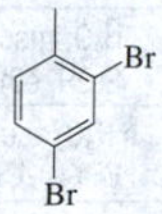
2958
2,4-Dibromo-1-methylbenzene

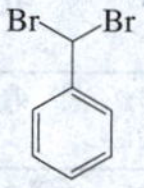
2959
(Dibromomethyl)benzene

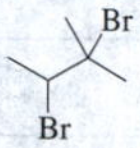
2960
2,3-Dibromo-2-methylbutane

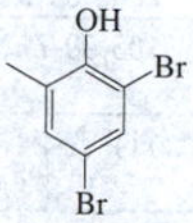
2961
2,4-Dibromo-6-methylphenol

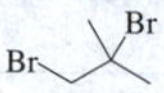
2962
1,2-Dibromo-2-methylpropane

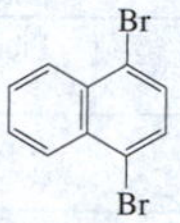
2963
1,4-Dibromonaphthalene

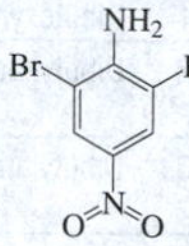
2964
2,6-Dibromo-4-nitroaniline

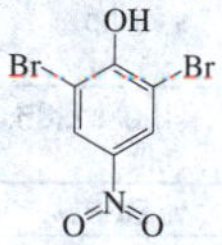
2965
2,6-Dibromo-4-nitrophenol

2966
1,9-Dibromononane

2967
1,4-Dibromooctafluorobutane

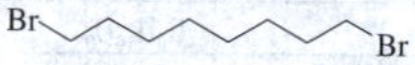
2968
1,8-Dibromooctane

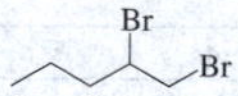
2969
1,2-Dibromopentane

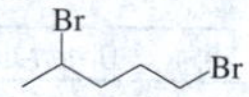
2970
1,4-Dibromopentane

2971
1,5-Dibromopentane

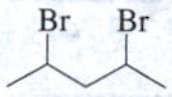
2972
2,4-Dibromopentane

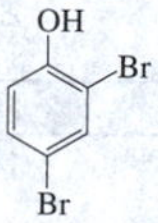
2973
2,4-Dibromophenol

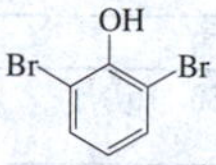
2974
2,6-Dibromophenol

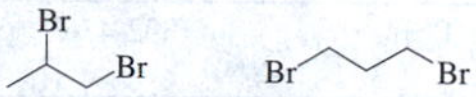
2975
1,2-Dibromopropane

2976
1,3-Dibromopropane

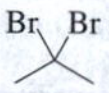
2977
2,2-Dibromopropane

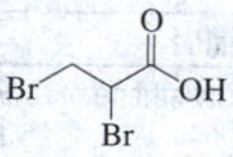
2978
2,3-Dibromopropanoic acid

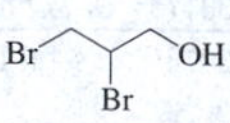
2979
2,3-Dibromo-1-propanol

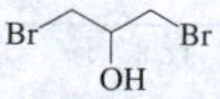
2980
1,3-Dibromo-2-propanol

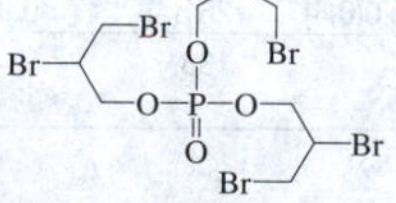
2981
2,3-Dibromo-1-propanol, phosphate (3:1)

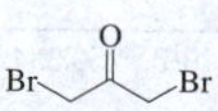
2982
1,3-Dibromo-2-propanone

2983
1,1-Dibromo-1-propene

2984
1,2-Dibromo-1-propene

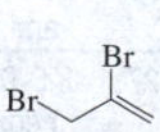
2985
2,3-Dibromo-1-propene

Organic

No.	Name	Synonym	Mol. form.	CAS Reg. No.	Mol. wt.	Physical form	mp/°C	bp/°C	den/ g cm⁻³	n_D	Solubility
2986	3,5-Dibromopyridine		$C_5H_3Br_2N$	625-92-3	236.893	nd (al)	112	222			sl H_2O; s EtOH, eth
2987	5,7-Dibromo-8-quinolinol	Broxyquinoline	$C_9H_5Br_2NO$	521-74-4	302.950	nd (al)	196	sub			i H_2O; s EtOH, ace, bz, chl, HOAc; sl eth
2988	2,6-Dibromoquinone-4-chlorimide	2,6-Dibromo-4-(chloroimino)-2,5-cyclohexadien-1-one	$C_6H_2Br_2ClNO$	537-45-1	299.347	ye pr (al or HOAc)	83				vs EtOH
2989	1,14-Dibromotetradecane	Tetradecamethylene dibromide	$C_{14}H_{28}Br_2$	37688-96-3	356.180	lf (al-eth) cry (al)	50.4	190^{8}			vs eth, EtOH, chl
2990	1,2-Dibromotetrafluoroethane	Refrigerant 114B2	$C_2Br_2F_4$	124-73-2	259.823	liq	-110(1)	47.1(0.2)	2.149^{25}	1.361^{25}	i H_2O
2991	2,3-Dibromothiophene		$C_4H_2Br_2S$	3140-93-0	241.932	liq	-17.5	218.5		1.6304^{22}	
2992	2,5-Dibromothiophene		$C_4H_2Br_2S$	3141-27-3	241.932	liq	-6	210.3	2.142^{23}	1.6288^{20}	i H_2O; vs EtOH, eth; s ctc
2993	3,4-Dibromothiophene		$C_4H_2Br_2S$	3141-26-2	241.932		4.5	217(17)			
2994	1,2-Dibromo-1,1,2-trifluoroethane	Halon 2302	$C_2HBr_2F_3$	354-04-1	241.832			76(4)	2.274^{27}	1.4191^{24}	
2995	2,6-Dibromo-3,4,5-trihydroxybenzoic acid	Dibromogallic acid	$C_7H_4Br_2O_5$	602-92-6	327.912	nd, pr or lf (w+1)	150				vs H_2O, eth, EtOH
2996	3,5-Dibromo-L-tyrosine		$C_9H_9Br_2NO_3$	300-38-9	338.980	nd or pl	245				sl H_2O, EtOH; i eth; s alk, acid
2997	Dibucaine	Cinchocaine	$C_{20}H_{29}N_3O_2$	85-79-0	343.463	hyg cry	64				
2998	Dibucaine hydrochloride		$C_{20}H_{30}ClN_3O_2$	61-12-1	379.924		94 dec				s chl
2999	1,4-Dibutoxybenzene		$C_{14}H_{22}O_2$	104-36-9	222.324		45.5	158^{15}			s ctc
3000	1,2-Dibutoxyethane	Ethylene glycol dibutyl ether	$C_{10}H_{22}O_2$	112-48-1	174.281	liq	-69.1	198(10)	0.8319^{25}	1.4112^{25}	
3001	Dibutoxymethane	Butylal	$C_9H_{20}O_2$	2568-90-3	160.254	liq	-59(1)	179.7(0.7)	0.8339^{20}	1.4072^{17}	
3002	Dibutyl adipate	Dibutyl hexanedioate	$C_{14}H_{26}O_4$	105-99-7	258.354		-32.4	165^{10}	0.9613^{20}	1.4369^{20}	i H_2O; msc EtOH, eth
3003	Dibutylamine	*N*-Butylbutanamine	$C_8H_{19}N$	111-92-2	129.244	liq	-61.8(0.5)	162(2)	0.7670^{20}	1.4177^{20}	s H_2O, ace, bz; vs EtOH, eth
3004	Di-*sec*-butylamine	*N*-*sec*-Butyl-2-butanamine	$C_8H_{19}N$	626-23-3	129.244			135(7)	0.7534^{20}	1.4162^{20}	vs H_2O; s EtOH
3005	2-Dibutylaminoethanol		$C_{10}H_{23}NO$	102-81-8	173.296			114^{18}			
3006	*N,N*-Dibutylaniline		$C_{14}H_{23}N$	613-29-6	205.340	liq	-32.2(0.2)	274.8	0.9037^{20}	1.5186^{20}	i H_2O; msc EtOH, eth; vs ace, bz; s ctc
3007	1,4-Di-*tert*-butylbenzene		$C_{14}H_{22}$	1012-72-2	190.325	nd (MeOH)	77.63(0.04)	237.3(0.5)	0.9850^{20}		i H_2O; s EtOH, eth
3008	2,5-Di-*tert*-butyl-1,4-benzenediol		$C_{14}H_{22}O_2$	88-58-4	222.324	cry (aq HOAc)	213.5				
3009	Dibutylbis(dodecylthio)-stannane	Dibutyltin bis(dodecyl sulfide)	$C_{32}H_{68}S_2Sn$	1185-81-5	635.722	col liq		$122^{0.3}$	1.05^{20}		s tol, hp
3010	Dibutyl carbonate		$C_9H_{18}O_3$	542-52-9	174.237			203(4)	0.9251^{20}	1.4117^{20}	i H_2O; s EtOH, eth
3011	Di-*tert*-butyl carbonate		$C_9H_{18}O_3$	34619-03-9	174.237	cry (al)	40	174(14)			vs EtOH
3012	2,5-Di-*tert*-butyl-2,5-cyclohexadiene-1,4-dione		$C_{14}H_{20}O_2$	2460-77-7	220.308	ye cry (al)	152.5				i H_2O; s EtOH, eth, bz, chl, HOAc
3013	2,6-Di-*tert*-butyl-2,5-cyclohexadiene-1,4-dione		$C_{14}H_{20}O_2$	719-22-2	220.308		69	$60^{0.01}$			
3014	2,6-Di-*tert*-butyl-4-(dimethylaminomethyl)phenol		$C_{17}H_{29}NO$	88-27-7	263.418	pl (EtOH)	94	179^{40}			
3015	2,2-Dibutyl-1,3,2-dioxastannepin-4,7-dione		$C_{12}H_{20}O_4Sn$	78-04-6	346.995	ye solid	110				
3016	Dibutyl disulfide		$C_8H_{18}S_2$	629-45-8	178.359	oil		236(4)	0.938^{20}	1.4923^{20}	i H_2O; msc EtOH, eth
3017	Di-*tert*-butyl disulfide		$C_8H_{18}S_2$	110-06-5	178.359		-2.5	88^{21}	0.9226^{20}	1.4899^{20}	
3018	*cis*-1,2-Di-*tert*-butylethene	*cis*-2,2,5,5-Tetramethyl-3-hexene	$C_{10}H_{20}$	692-47-7	140.266	liq		144(6)	0.744^{20}	1.4270^{20}	
3019	Dibutyl ether	Butyl ether	$C_8H_{18}O$	142-96-1	130.228	liq	-96(3)	141.6(0.3)	0.7684^{20}	1.3992^{20}	i H_2O; msc EtOH, eth; vs ace; sl ctc
3020	Di-*sec*-butyl ether		$C_8H_{18}O$	6863-58-7	130.228	liq		121.9(0.3)	0.756^{25}		
3021	Di-*tert*-butyl ether		$C_8H_{18}O$	6163-66-2	130.228	liq		107.1(0.7)	0.7658^{20}	1.3949^{20}	
3022	*N,N'*-Di-*tert*-butylethylenediamine	*N,N'*-Di-*tert*-butylethanediamine	$C_{10}H_{24}N_2$	4062-60-6	172.311	cry	53.3	189	0.69		
3023	2,6-Di-*tert*-butyl-4-ethylphenol		$C_{16}H_{26}O$	4130-42-1	234.376		44	272			i alk
3024	*N,N*-Dibutylformamide		$C_9H_{19}NO$	761-65-9	157.253						s ctc, CS_2
3025	Dibutyl fumarate		$C_{12}H_{20}O_4$	105-75-9	228.285	liq	-18.0(0.4)	281(4)	0.9775^{20}	1.4469^{20}	i H_2O; s ace, chl
3026	*N,N'*-Dibutyl-1,6-hexanediamine		$C_{14}H_{32}N_2$	4835-11-4	228.417			$138^{3.5}$		1.4470^{25}	
3027	3,5-Di-*tert*-butyl-2-hydroxybenzoic acid		$C_{15}H_{22}O_3$	19715-19-6	250.334		163.3				s chl

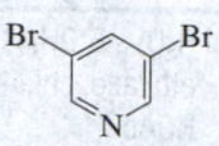
2986
3,5-Dibromopyridine

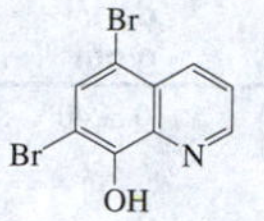
2987
5,7-Dibromo-8-quinolinol

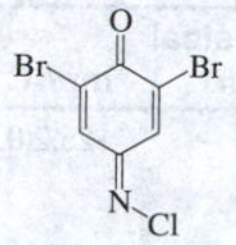
2988
2,6-Dibromoquinone-4-chlorimide

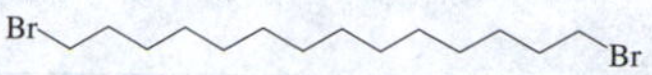
2989
1,14-Dibromotetradecane

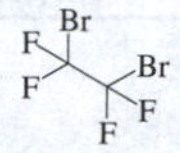
2990
1,2-Dibromotetrafluoroethane

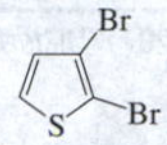
2991
2,3-Dibromothiophene

2992
2,5-Dibromothiophene

2993
3,4-Dibromothiophene

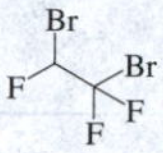
2994
1,2-Dibromo-1,1,2-trifluoroethane

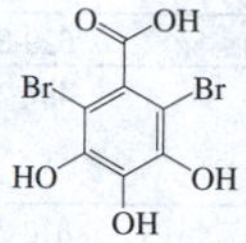
2995
2,6-Dibromo-3,4,5-trihydroxybenzoic acid

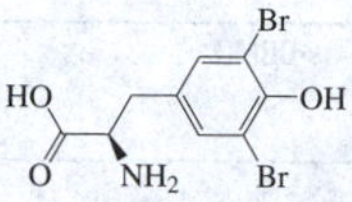
2996
3,5-Dibromo-L-tyrosine

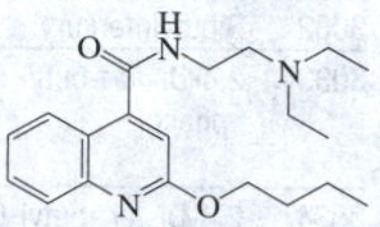
2997
Dibucaine

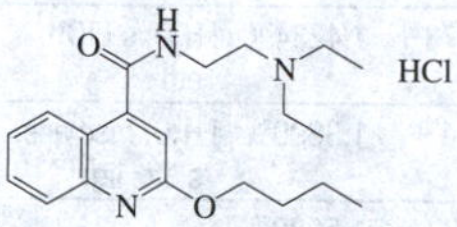
2998
Dibucaine hydrochloride

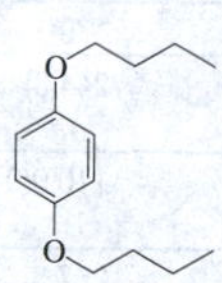
2999
1,4-Dibutoxybenzene

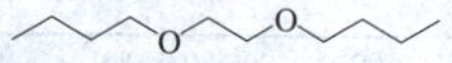
3000
1,2-Dibutoxyethane

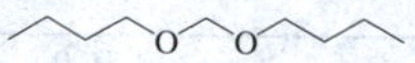
3001
Dibutoxymethane

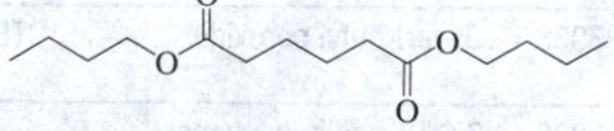
3002
Dibutyl adipate

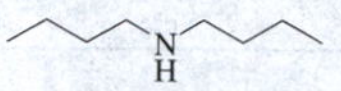
3003
Dibutylamine

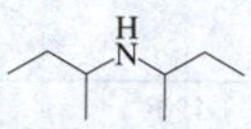
3004
Di-*sec*-butylamine

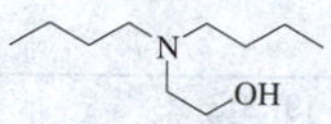
3005
2-Dibutylaminoethanol

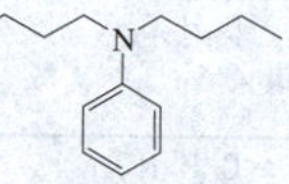
3006
N,N-Dibutylaniline

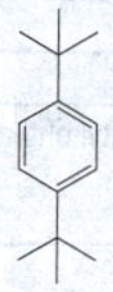
3007
1,4-Di-*tert*-butylbenzene

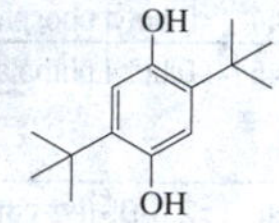
3008
2,5-Di-*tert*-butyl-1,4-benzenediol

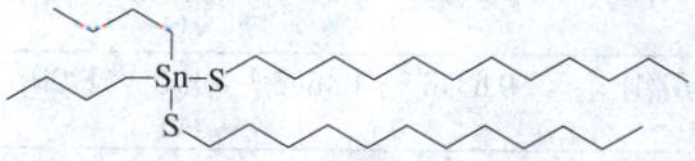
3009
Dibutylbis(dodecylthio)stannane

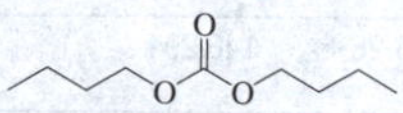
3010
Dibutyl carbonate

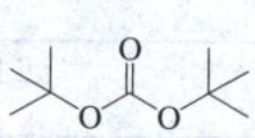
3011
Di-*tert*-butyl carbonate

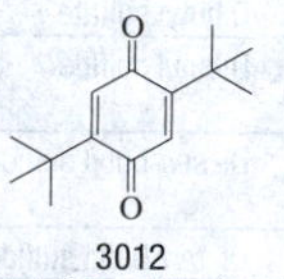
3012
2,5-Di-*tert*-butyl-2,5-cyclohexadiene-1,4-dione

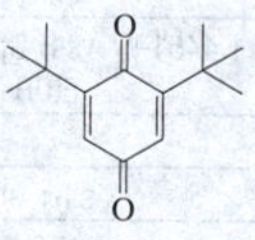
3013
2,6-Di-*tert*-butyl-2,5-cyclohexadiene-1,4-dione

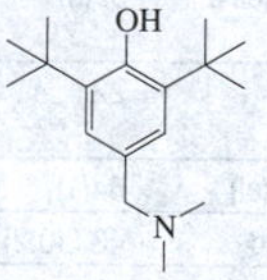
3014
2,6-Di-*tert*-butyl-4-(dimethylaminomethyl)phenol

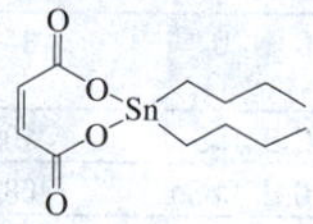
3015
2,2-Dibutyl-1,3,2-dioxastannepin-4,7-dione

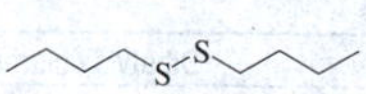
3016
Dibutyl disulfide

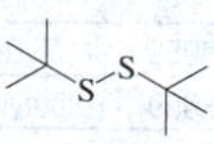
3017
Di-*tert*-butyl disulfide

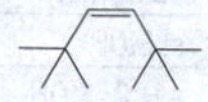
3018
cis-1,2-Di-*tert*-butylethene

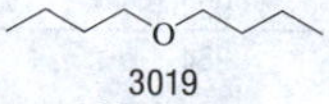
3019
Dibutyl ether

3020
Di-*sec*-butyl ether

3021
Di-*tert*-butyl ether

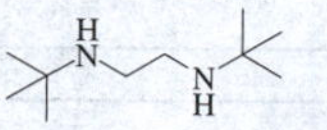
3022
N,N'-Di-*tert*-butylethylenediamine

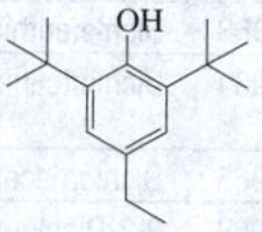
3023
2,6-Di-*tert*-butyl-4-ethylphenol

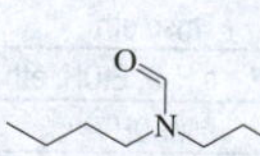
3024
N,N-Dibutylformamide

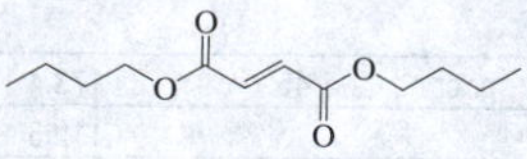
3025
Dibutyl fumarate

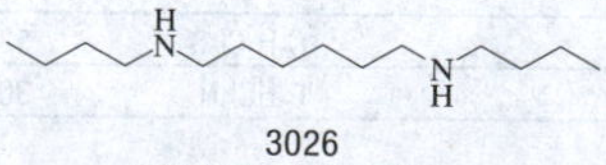
3026
N,N'-Dibutyl-1,6-hexanediamine

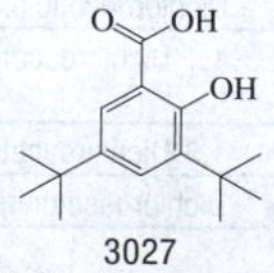
3027
3,5-Di-*tert*-butyl-2-hydroxybenzoic acid

No.	Name	Synonym	Mol. form.	CAS Reg. No.	Mol. wt.	Physical form	mp/°C	bp/°C	den/g cm⁻³	n_D	Solubility
3028	Di-*tert*-butyl ketone		$C_9H_{18}O$	815-24-7	142.238	liq	-25.2(0.2)	152(4)	0.8240^{18}	1.4194^{20}	i H_2O; s EtOH, eth, ace, chl, HOAc
3029	Dibutyl maleate		$C_{12}H_{20}O_4$	105-76-0	228.285		<-80	280			
3030	Dibutyl malonate		$C_{11}H_{20}O_4$	1190-39-2	216.275	liq	-95(3)	253(11)	0.9824^{20}	1.4262^{20}	i H_2O; s EtOH, eth, ace, bz, HOAc, ctc
3031	Di-*tert*-butyl malonate		$C_{11}H_{20}O_4$	541-16-2	216.275		-6	113^{31}	1.016^{20}	1.4184^{29}	s ace, chl
3032	Dibutylmercury		$C_8H_{18}Hg$	629-35-6	314.82			223	1.7779^{20}	1.5057^{20}	
3033	2,4-Di-*tert*-butyl-5-methylphenol	DBMC	$C_{15}H_{24}O$	497-39-2	220.351		62.1	282	0.912^{80}		i H_2O; s EtOH, eth, ace, bz, ctc
3034	2,4-Di-*tert*-butyl-6-methylphenol		$C_{15}H_{24}O$	616-55-7	220.351		51	269	0.891^{80}		i alk
3035	2,6-Di-*tert*-butyl-4-methylphenol		$C_{15}H_{24}O$	128-37-0	220.351		70.1(0.8)	265	0.8937^{75}	1.4859^{75}	i H_2O; s EtOH, ace, bz, peth; i alk
3036	Dibutyl nonanedioate		$C_{17}H_{32}O_4$	2917-73-9	300.434			170^{2}			sl chl
3037	Dibutyl oxalate		$C_{10}H_{18}O_4$	2050-60-4	202.248	liq	-30.5	244(3)	0.9873^{20}	1.4234^{20}	i H_2O; s EtOH, eth
3038	Di-*tert*-butyl peroxide	DTBP	$C_8H_{18}O_2$	110-05-4	146.228	liq	-40	110.0(0.2)	0.704^{20}	1.3890^{20}	i H_2O; msc ace; s ctc, lig
3039	2,6-Di-*sec*-butylphenol		$C_{14}H_{22}O$	5510-99-6	206.324	liq	-42	257.5		1.5080^{20}	
3040	2,4-Di-*tert*-butylphenol		$C_{14}H_{22}O$	96-76-4	206.324		58(1)	262(1)		1.5080^{20}	sl ctc; i alk
3041	2,6-Di-*tert*-butylphenol		$C_{14}H_{22}O$	128-39-2	206.324	pr (al)	37.5(0.3)	161^{50}		1.5001^{20}	sl EtOH; s ctc; i alk
3042	3,5-Di-*tert*-butylphenol		$C_{14}H_{22}O$	1138-52-9	206.324		88				
3043	Dibutyl phosphate		$C_8H_{19}O_4P$	107-66-4	210.208	oil		$136^{0.05}$	1.06^{20}		s ctc, BuOH
3044	Dibutyl phosphonate		$C_8H_{19}O_3P$	1809-19-4	194.209	oil		230	0.985^{25}	1.4220^{20}	
3045	Dibutyl phthalate	Butyl phthalate	$C_{16}H_{22}O_4$	84-74-2	278.344	liq	-35	338(9)	1.0465^{20}	1.4911^{20}	i H_2O; msc EtOH, eth, bz; s ctc
3046	2,6-Di-*tert*-butylpyridine		$C_{13}H_{21}N$	585-48-8	191.313			120^{20}			
3047	Dibutyl sebacate	Butyl sebacate	$C_{18}H_{34}O_4$	109-43-3	314.461	liq	-9.2(0.5)	356(9)	0.9405^{15}	1.4433^{15}	i H_2O; s eth, ctc
3048	Dibutyl succinate		$C_{12}H_{22}O_4$	141-03-7	230.301	liq	-29.2	269(3)	0.9752^{20}	1.4299^{20}	i H_2O; s EtOH, eth, bz, ctc
3049	Di-*tert*-butyl succinate		$C_{12}H_{22}O_4$	926-26-1	230.301		36.5	109^{9}			
3050	Dibutyl sulfate	Butyl sulfate	$C_8H_{18}O_4S$	625-22-9	210.292	liq		115^{6}			
3051	Dibutyl sulfide	Butyl sulfide	$C_8H_{18}S$	544-40-1	146.294	liq	-74.97(0.05)	168(4)	0.8386^{20}	1.4530^{20}	vs eth, EtOH, chl
3052	Di-*sec*-butyl sulfide		$C_8H_{18}S$	626-26-6	146.294			167(1)	0.8348^{20}	1.4506^{20}	i H_2O; vs EtOH, eth
3053	Di-*tert*-butyl sulfide		$C_8H_{18}S$	107-47-1	146.294	liq	-9.0	152.3(0.8)	0.815^{25}	1.4506^{20}	
3054	Dibutyl sulfite	Butyl sulfite	$C_8H_{18}O_3S$	626-85-7	194.292			230	0.9957^{20}	1.4310^{20}	s EtOH, eth
3055	Dibutyl sulfone		$C_8H_{18}O_2S$	598-04-9	178.293		44(1)	291	0.9885^{47}		i H_2O; s EtOH, eth
3056	Dibutyl sulfoxide		$C_8H_{18}OS$	2168-93-6	162.293	nd (dil al)	32(3)	290(13)	0.8317^{23}	1.4669^{20}	i H_2O; s EtOH, eth
3057	Dibutyl tartrate		$C_{12}H_{22}O_6$	87-92-3	262.299	pr	22	320	1.0909^{20}	1.4451^{20}	vs H_2O, ace, EtOH
3058	*N,N'*-Dibutylthiourea		$C_9H_{20}N_2S$	109-46-6	188.333	nd (al)	64(1)				
3059	Dibutyltin dichloride	Dibutyldichlorostannane	$C_8H_{18}Cl_2Sn$	683-18-1	303.845	solid	43.0(0.2)	135^{10}			s hx, eth, thf
3060	Dibutyltin dilaurate		$C_{32}H_{64}O_4Sn$	77-58-7	631.558	ye liq or cry	23				i H_2O, MeOH; s eth, bz, ctc
3061	Dicapthon		$C_8H_9ClNO_5PS$	2463-84-5	297.653	cry (MeOH)	50.5(0.9)				i H_2O; s ace, tol, xyl, AcOEt
3062	Dicentrine		$C_{20}H_{21}NO_4$	517-66-8	339.386						s chl
3063	Dichlofenthion		$C_{10}H_{13}Cl_2O_3PS$	97-17-6	315.153						s ctc, CS_2
3064	Dichlofluanid		$C_9H_{11}Cl_2FN_2O_2S_2$	1085-98-9	333.229	wh pow	105.3				i H_2O; s ace, MeOH, xyl
3065	Dichloroacetaldehyde		$C_2H_2Cl_2O$	79-02-7	112.942			90.5	1.436^{25}		sl EtOH
3066	2,2-Dichloroacetamide		$C_2H_3Cl_2NO$	683-72-7	127.957		99.4	234			s H_2O, EtOH, eth; sl ace
3067	Dichloroacetic acid		$C_2H_2Cl_2O_2$	79-43-6	128.942	liq	12(3)	193(3)	1.5634^{20}	1.4658^{20}	msc H_2O, EtOH, eth; s ace; sl ctc
3068	Dichloroacetic anhydride		$C_4H_2Cl_4O_3$	4124-30-5	239.869		18.0	215 dec	1.574^{24}		
3069	1,1-Dichloroacetone		$C_3H_4Cl_2O$	513-88-2	126.969			120	1.304^{18}		sl H_2O; s EtOH; msc eth
3070	1,3-Dichloroacetone		$C_3H_4Cl_2O$	534-07-6	126.969	pr or nd	45	173.4	1.3826^{46}	1.4716^{40}	s H_2O, EtOH, eth
3071	Dichloroacetonitrile		C_2HCl_2N	3018-12-0	109.942			112.5	1.369^{20}	1.4391^{25}	s MeOH

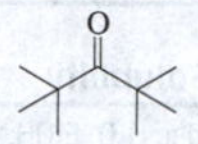
3028
Di-*tert*-butyl ketone

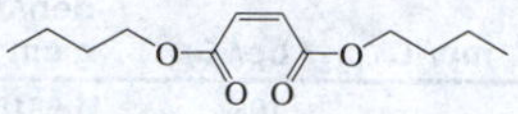
3029
Dibutyl maleate

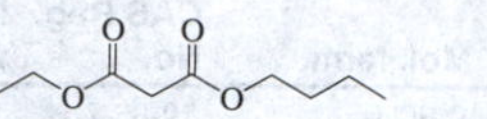
3030
Dibutyl malonate

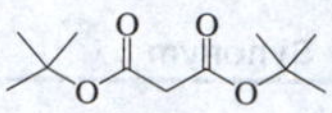
3031
Di-*tert*-butyl malonate

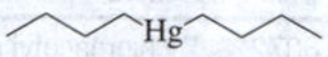
3032
Dibutylmercury

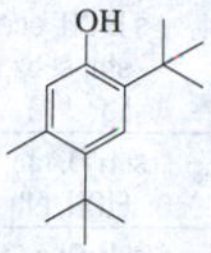
3033
2,4-Di-*tert*-butyl-5-methylphenol

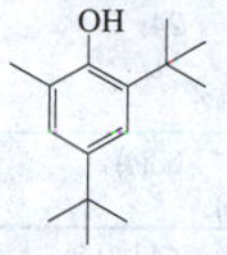
3034
2,4-Di-*tert*-butyl-6-methylphenol

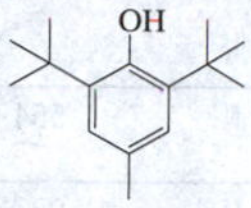
3035
2,6-Di-*tert*-butyl-4-methylphenol

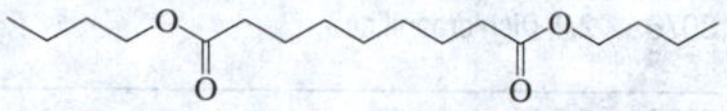
3036
Dibutyl nonanedioate

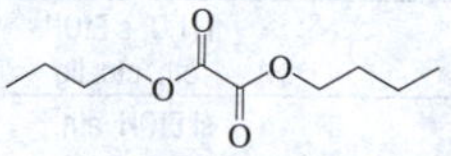
3037
Dibutyl oxalate

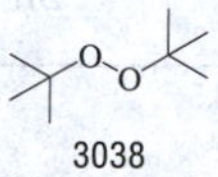
3038
Di-*tert*-butyl peroxide

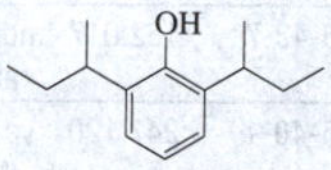
3039
2,6-Di-*sec*-butylphenol

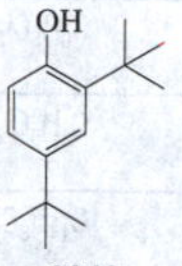
3040
2,4-Di-*tert*-butylphenol

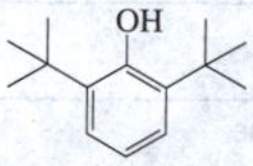
3041
2,6-Di-*tert*-butylphenol

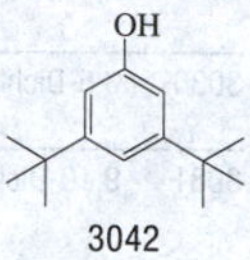
3042
3,5-Di-*tert*-butylphenol

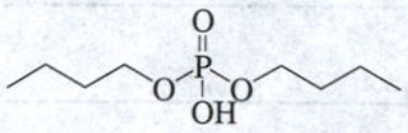
3043
Dibutyl phosphate

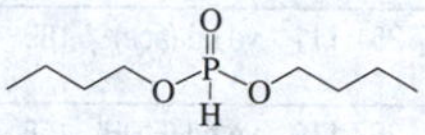
3044
Dibutyl phosphonate

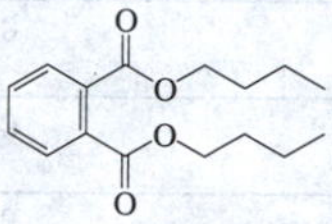
3045
Dibutyl phthalate

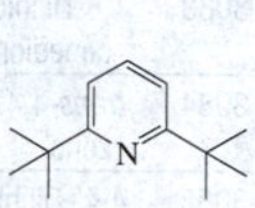
3046
2,6-Di-*tert*-butylpyridine

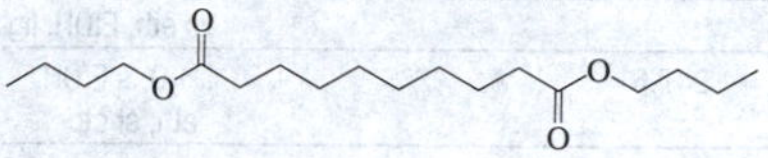
3047
Dibutyl sebacate

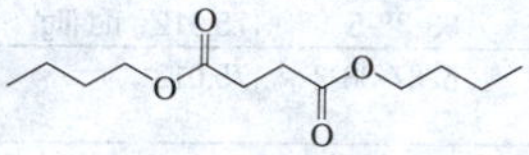
3048
Dibutyl succinate

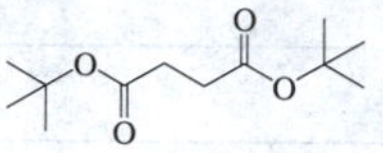
3049
Di-*tert*-butyl succinate

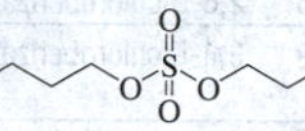
3050
Dibutyl sulfate

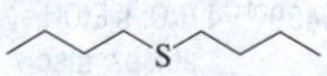
3051
Dibutyl sulfide

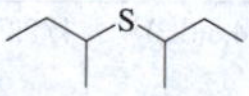
3052
Di-*sec*-butyl sulfide

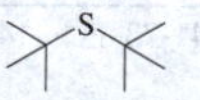
3053
Di-*tert*-butyl sulfide

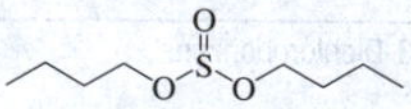
3054
Dibutyl sulfite

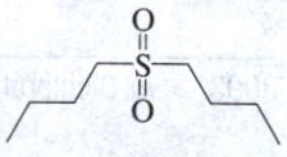
3055
Dibutyl sulfone

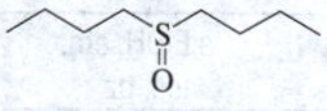
3056
Dibutyl sulfoxide

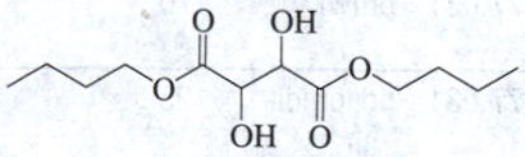
3057
Dibutyl tartrate

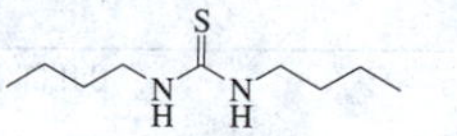
3058
N,N'-Dibutylthiourea

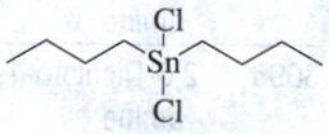
3059
Dibutyltin dichloride

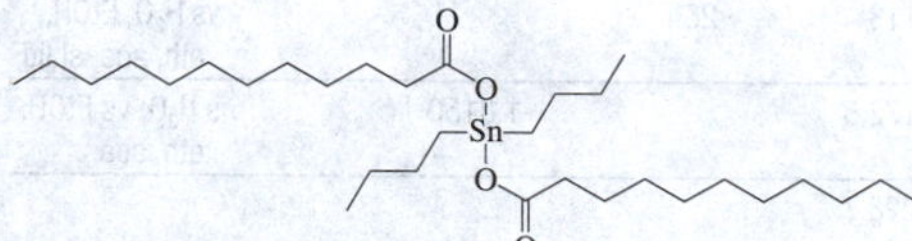
3060
Dibutyltin dilaurate

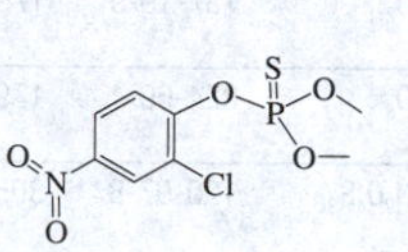
3061
Dicapthon

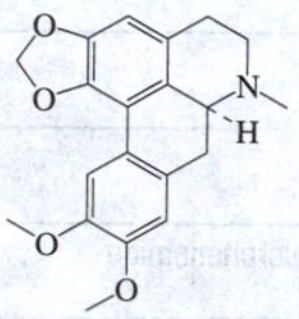
3062
Dicentrine

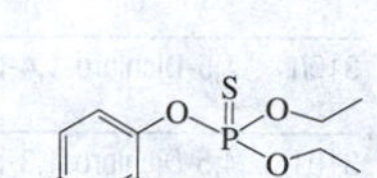
3063
Dichlofenthion

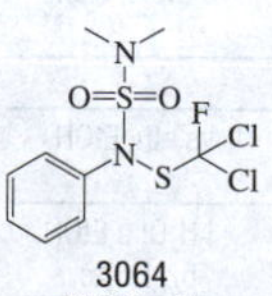
3064
Dichlofluanid

3065
Dichloroacetaldehyde

3066
2,2-Dichloroacetamide

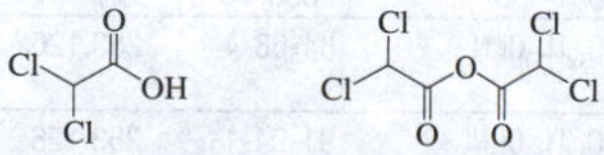
3067
Dichloroacetic acid

3068
Dichloroacetic anhydride

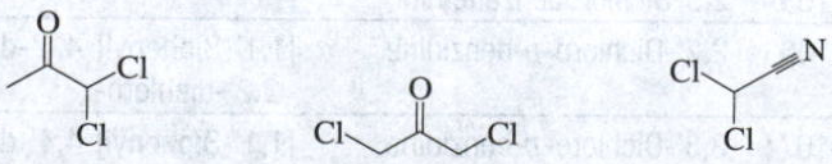
3069
1,1-Dichloroacetone

3070
1,3-Dichloroacetone

3071
Dichloroacetonitrile

No.	Name	Synonym	Mol. form.	CAS Reg. No.	Mol. wt.	Physical form	mp/°C	bp/°C	den/g cm^{-3}	n_D	Solubility
3072	Dichloroacetyl chloride		C_2HCl_3O	79-36-7	147.387			108	1.5315^{16}	1.4591^{20}	dec H_2O, EtOH; msc eth
3073	Dichloroacetylene		C_2Cl_2	7572-29-4	94.927	liq	-66	33	1.261^{20}	1.42790^{20}	s EtOH, eth, ace
3074	4-[(Dichloroamino)sulfonyl]-benzoic acid	Halazone	$C_7H_5Cl_2NO_4S$	80-13-7	270.091	pr (HOAc)	195 dec				sl H_2O, chl; vs HOAc; i peth
3075	2,3-Dichloroaniline		$C_6H_5Cl_2N$	608-27-5	162.017	nd (lig)	24	252			s EtOH, ace; vs eth; sl bz, ctc, lig
3076	2,4-Dichloroaniline		$C_6H_5Cl_2N$	554-00-7	162.017	pr (ace) nd (dil al) (lig)	63(4)	245	1.567^{20}		sl H_2O, chl; s EtOH, eth
3077	2,5-Dichloroaniline		$C_6H_5Cl_2N$	95-82-9	162.017	nd (lig)	44.9(0.2)	250(21)			sl H_2O; s EtOH, eth, bz, chl, CS_2
3078	2,6-Dichloroaniline		$C_6H_5Cl_2N$	608-31-1	162.017		39				sl H_2O; s EtOH, eth
3079	3,4-Dichloroaniline		$C_6H_5Cl_2N$	95-76-1	162.017	nd (lig)	72.0(0.5)	273.0(0.5)			s EtOH, eth; sl bz, chl
3080	3,5-Dichloroaniline		$C_6H_5Cl_2N$	626-43-7	162.017	nd (lig, dil al)	52	261			i H_2O; s EtOH, eth, ctc, lig
3081	9,10-Dichloroanthracene		$C_{14}H_8Cl_2$	605-48-1	247.120	ye nd (MeCOEt or CCl_4)	213.5				sl EtOH, eth, chl; s bz
3082	1,5-Dichloro-9,10-anthracenedione		$C_{14}H_6Cl_2O_2$	82-46-2	277.103	ye nd (to)	252				i H_2O; sl EtOH, ace; s bz, HOAc
3083	1,8-Dichloro-9,10-anthracenedione		$C_{14}H_6Cl_2O_2$	82-43-9	277.103	ye nd (HOAc)	202.5				i H_2O; sl EtOH; s bz, tol, $PhNO_2$
3084	*trans*-4,4'-Dichloroazobenzene		$C_{12}H_8Cl_2N_2$	1602-00-2	251.111	ye nd (ace)	189				
3085	4,4'-Dichloroazoxybenzene		$C_{12}H_8Cl_2N_2O$	614-26-6	267.110	ye nd (EtOH)	158				
3086	2,3-Dichlorobenzaldehyde		$C_7H_4Cl_2O$	6334-18-5	175.012	cry (dil al)	66				vs eth, EtOH
3087	2,4-Dichlorobenzaldehyde		$C_7H_4Cl_2O$	874-42-0	175.012	pr	74.1(0.3)	105^{15}			i H_2O; s EtOH, eth, bz, chl, HOAc
3088	2,6-Dichlorobenzaldehyde		$C_7H_4Cl_2O$	83-38-5	175.012	nd (lig)	69(3)				vs eth, EtOH, lig
3089	3,4-Dichlorobenzaldehyde		$C_7H_4Cl_2O$	6287-38-3	175.012		44	247.5			i H_2O; s EtOH, eth; sl ctc
3090	3,5-Dichlorobenzaldehyde		$C_7H_4Cl_2O$	10203-08-4	175.012	nd or lf (dil HOAc)	65	240			vs ace, bz, eth, EtOH
3091	2,6-Dichlorobenzamide		$C_7H_5Cl_2NO$	2008-58-4	190.027	cry	198				
3092	*o*-Dichlorobenzene	1,2-Dichlorobenzene	$C_6H_4Cl_2$	95-50-1	147.002	liq	-17.0(0.1)	180.2(0.3)	1.3059^{20}	1.5515^{20}	i H_2O; s EtOH, eth; msc ace, bz, ctc
3093	*m*-Dichlorobenzene	1,3-Dichlorobenzene	$C_6H_4Cl_2$	541-73-1	147.002	liq	-24.8(0.3)	172(2)	1.2884^{20}	1.5459^{20}	i H_2O; s EtOH, eth, bz; msc ace
3094	*p*-Dichlorobenzene	1,4-Dichlorobenzene	$C_6H_4Cl_2$	106-46-7	147.002	mcl pr, lf (ace)	53.1(0.2)	173.9(0.2)	1.2475^{55}	1.5285^{20}	i H_2O; msc EtOH, ace, bz; s eth, ctc
3095	2,5-Dichloro-1,4-benzenediamine		$C_6H_6Cl_2N_2$	20103-09-7	177.031	pr (w)	170				
3096	2,6-Dichloro-1,4-benzenediamine		$C_6H_6Cl_2N_2$	609-20-1	177.031	nd, pr (dil al)	125				s EtOH, eth, ace, bz
3097	3,5-Dichloro-1,2-benzenediol		$C_6H_4Cl_2O_2$	13673-92-2	179.001	pr	83.5(0.2)				sl H_2O; s EtOH; vs ace
3098	4,5-Dichloro-1,2-benzenediol		$C_6H_4Cl_2O_2$	3428-24-8	179.001	pr(chl-CS_2) nd(bz-peth)	116.5(0.2)				s H_2O; vs EtOH, bz
3099	4,6-Dichloro-1,3-benzenediol		$C_6H_4Cl_2O_2$	137-19-9	179.001		113	254			vs H_2O, EtOH, eth, ace; sl lig
3100	2,5-Dichloro-1,4-benzenediol		$C_6H_4Cl_2O_2$	824-69-1	179.001	nd or pr w, ace, bz)	172.5		1.8150^{24}		s H_2O; vs EtOH, eth, ace
3101	4,5-Dichloro-1,3-benzenedisulfonamide	Dichlorphenamide	$C_6H_6Cl_2N_2O_4S_2$	120-97-8	305.159		228.7				
3102	2,4-Dichlorobenzenemethanamine		$C_7H_7Cl_2N$	95-00-1	176.044			125^{13}		1.5762^{25}	s chl
3103	2,4-Dichlorobenzenemethanol	2,4-Dichlorobenzyl alcohol	$C_7H_6Cl_2O$	1777-82-8	177.028		59.5	150^{25}			s chl
3104	*N,N*-Dichlorobenzenesulfonamide		$C_6H_5Cl_2NO_2S$	473-29-0	226.081	ye mcl or pl	76				s EtOH; sl ctc
3105	2,5-Dichlorobenzenethiol		$C_6H_4Cl_2S$	5858-18-4	179.067			115^{50}			
3106	2,2'-Dichloro-*p*-benzidine	[1,1'-Biphenyl]-4,4'-diamine, 2,2'-dichloro-	$C_{12}H_{10}Cl_2N_2$	84-68-4	253.126	nd (w), pr (al)	165				vs eth, EtOH
3107	3,3'-Dichloro-*p*-benzidine	[1,1'-Biphenyl]-4,4'-diamine, 3,3'-dichloro-	$C_{12}H_{10}Cl_2N_2$	91-94-1	253.126	nd	132.5				i H_2O; s EtOH, bz, HOAc

Organic

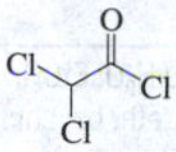
3072
Dichloroacetyl chloride

3073
Dichloroacetylene

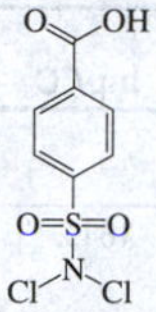
3074
4-[(Dichloroamino)sulfonyl]benzoic acid

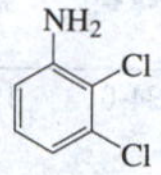
3075
2,3-Dichloroaniline

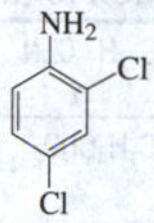
3076
2,4-Dichloroaniline

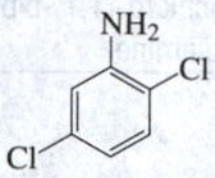
3077
2,5-Dichloroaniline

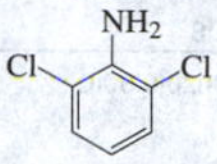
3078
2,6-Dichloroaniline

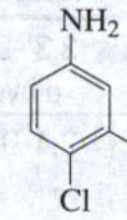
3079
3,4-Dichloroaniline

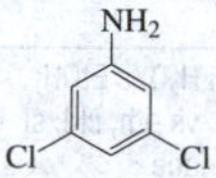
3080
3,5-Dichloroaniline

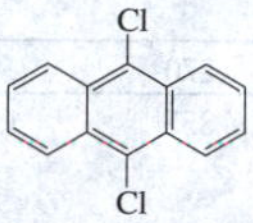
3081
9,10-Dichloroanthracene

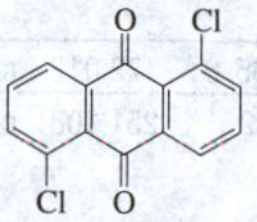
3082
1,5-Dichloro-9,10-anthracenedione

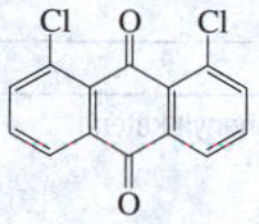
3083
1,8-Dichloro-9,10-anthracenedione

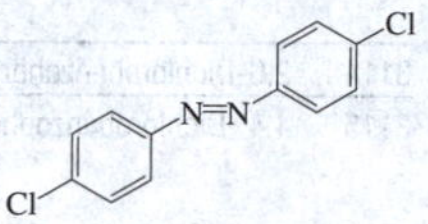
3084
trans-4,4′-Dichloroazobenzene

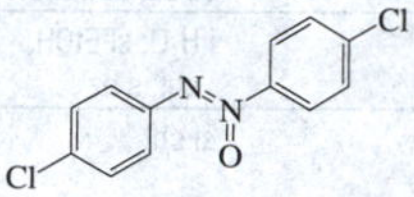
3085
4,4′-Dichloroazoxybenzene

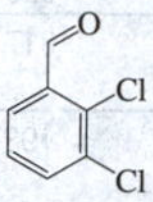
3086
2,3-Dichlorobenzaldehyde

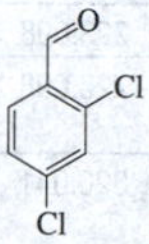
3087
2,4-Dichlorobenzaldehyde

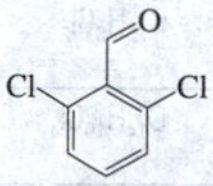
3088
2,6-Dichlorobenzaldehyde

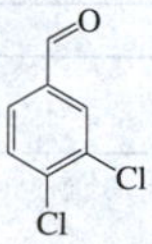
3089
3,4-Dichlorobenzaldehyde

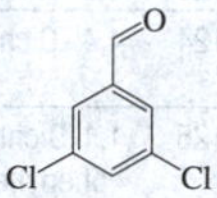
3090
3,5-Dichlorobenzaldehyde

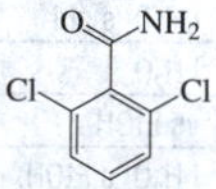
3091
2,6-Dichlorobenzamide

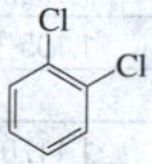
3092
o-Dichlorobenzene

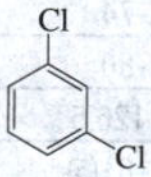
3093
m-Dichlorobenzene

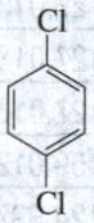
3094
p-Dichlorobenzene

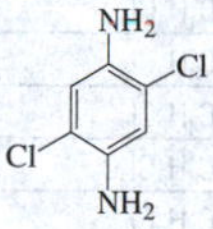
3095
2,5-Dichloro-1,4-benzenediamine

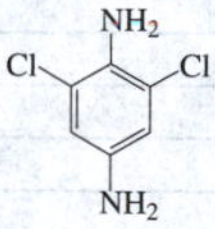
3096
2,6-Dichloro-1,4-benzenediamine

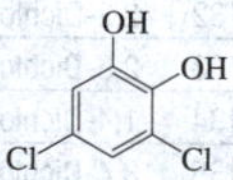
3097
3,5-Dichloro-1,2-benzenediol

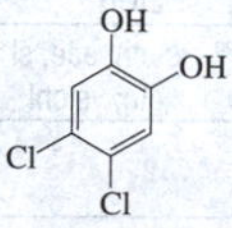
3098
4,5-Dichloro-1,2-benzenediol

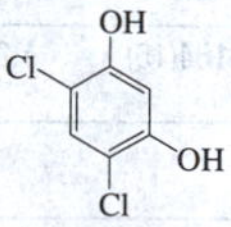
3099
4,6-Dichloro-1,3-benzenediol

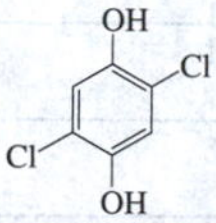
3100
2,5-Dichloro-1,4-benzenediol

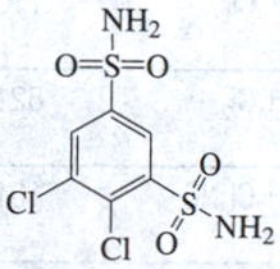
3101
4,5-Dichloro-1,3-benzenedisulfonamide

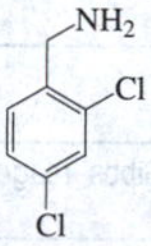
3102
2,4-Dichlorobenzenemethanamine

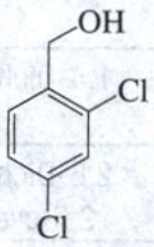
3103
2,4-Dichlorobenzenemethanol

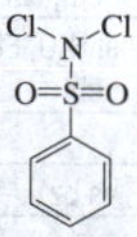
3104
N,N-Dichlorobenzenesulfonamide

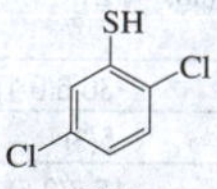
3105
2,5-Dichlorobenzenethiol

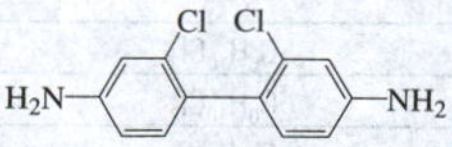
3106
2,2′-Dichloro-*p*-benzidine

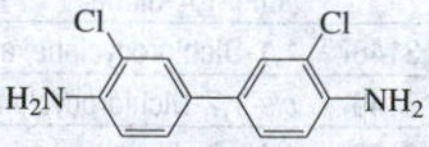
3107
3,3′-Dichloro-*p*-benzidine

No.	Name	Synonym	Mol. form.	CAS Reg. No.	Mol. wt.	Physical form	mp/°C	bp/°C	den/g cm⁻³	n_D	Solubility
3108	3,3'-Dichloro-*p*-benzidine dihydrochloride	3,3'-Dichloro-[1,1'-biphenyl]-4,4'-diamine	$C_{12}H_{12}Cl_4N_2$	612-83-9	326.048						i H_2O; vs EtOH
3109	2,4-Dichlorobenzoic acid		$C_7H_4Cl_2O_2$	50-84-0	191.012	nd (w or bz)	164.2	sub			s H_2O, EtOH, eth, bz, chl; sl ace
3110	2,5-Dichlorobenzoic acid		$C_7H_4Cl_2O_2$	50-79-3	191.012	nd (w)	154(1)	301			sl H_2O, DMSO; s EtOH, eth
3111	2,6-Dichlorobenzoic acid		$C_7H_4Cl_2O_2$	50-30-6	191.012	nd (al), pr (w)	142(2)	sub			s H_2O, EtOH, eth, bz, chl
3112	3,4-Dichlorobenzoic acid		$C_7H_4Cl_2O_2$	51-44-5	191.012	nd (w, al, bz)	204.5(0.4)				s H_2O, eth; vs EtOH; sl DMSO
3113	3,5-Dichlorobenzoic acid		$C_7H_4Cl_2O_2$	51-36-5	191.012	nd (al, w)	187.4(0.5)	sub			sl H_2O, lig, DMSO; s EtOH, eth
3114	2,6-Dichlorobenzonitrile	Dichlobenil	$C_7H_3Cl_2N$	1194-65-6	172.012	cry (peth)	143.6(0.7)	270			
3115	4,4'-Dichlorobenzophenone	Bis(4-chlorophenyl) ketone	$C_{13}H_8Cl_2O$	90-98-2	251.108	pl (al)	146.8(0.3)	353	1.4500^{20}		i H_2O; s EtOH; vs eth, chl; sl ace
3116	3,4-Dichlorobenzotrifluoride	1,2-Dichloro-4-(trifluoromethyl)benzene	$C_7H_3Cl_2F_3$	328-84-7	215.000	liq		175(3)	1.4729^{25}		
3117	2,3-Dichlorobenzoyl chloride		$C_7H_3Cl_3O$	2905-60-4	209.457	liq		140^{14}			
3118	2,4-Dichlorobenzoyl chloride		$C_7H_3Cl_3O$	89-75-8	209.457		16.5	150^{34}		1.5895^{20}	s ctc
3119	2,5-Dichlorobenzoyl chloride		$C_7H_3Cl_3O$	2905-61-5	209.457	liq		95.4^{1}			
3120	3,4-Dichlorobenzoyl chloride		$C_7H_3Cl_3O$	3024-72-4	209.457		25	242			sl ctc
3121	2,5-Dichlorobiphenyl		$C_{12}H_8Cl_2$	34883-39-1	223.098			182^{30}			i H_2O
3122	2,6-Dichlorobiphenyl		$C_{12}H_8Cl_2$	33146-45-1	223.098	cry	34.7(0.5)				i H_2O
3123	3,3'-Dichlorobiphenyl		$C_{12}H_8Cl_2$	2050-67-1	223.098	nd (dil al)	29	320			vs bz, eth, EtOH
3124	4,4'-Dichlorobiphenyl		$C_{12}H_8Cl_2$	2050-68-2	223.098	pr or nd (al, to-peth)	148(2)	317	1.4420^{0}		i H_2O; sl EtOH, chl; s bz
3125	1,1-Dichloro-2,2-bis(*p*-chlorophenyl)ethane		$C_{14}H_{10}Cl_4$	72-54-8	320.041		109(1)	193^{1}			sl chl
3126	2,2-Dichloro-1,1-bis(4-chlorophenyl)ethene		$C_{14}H_8Cl_4$	72-55-9	318.026		89.8(0.3)				
3127	2,3-Dichloro-1,3-butadiene		$C_4H_4Cl_2$	1653-19-6	122.981			101(15)	1.1829^{20}	1.4890^{20}	vs chl
3128	1,1-Dichlorobutane	Butylidene chloride	$C_4H_8Cl_2$	541-33-3	127.013			115(6)	1.0863^{20}	1.4355^{20}	i H_2O; s chl
3129	1,2-Dichlorobutane		$C_4H_8Cl_2$	616-21-7	127.013			123.9(0.8)	1.1116^{25}	1.4450^{20}	i H_2O; s eth, chl; sl ctc
3130	1,3-Dichlorobutane		$C_4H_8Cl_2$	1190-22-3	127.013			133.7(0.7)	1.1158^{20}	1.4445^{20}	i H_2O; s eth, chl; sl ctc
3131	1,4-Dichlorobutane		$C_4H_8Cl_2$	110-56-5	127.013	liq	-38.7(0.4)	155(3)	1.1331^{25}	1.4522^{25}	i H_2O; vs chl
3132	2,2-Dichlorobutane		$C_4H_8Cl_2$	4279-22-5	127.013	liq	-74	102(6)	1.1048^{25}	1.4295	i H_2O; s chl
3133	2,3-Dichlorobutane, (±)-		$C_4H_8Cl_2$	2211-67-8	127.013	liq	-80	119	1.105^{25}	1.4409^{25}	i H_2O
3134	1,4-Dichloro-2,3-butanediol		$C_4H_8Cl_2O_2$	2419-73-0	159.012		126.5	150^{30}			vs EtOH
3135	3,4-Dichloro-1-butene		$C_4H_6Cl_2$	760-23-6	124.997	liq	-61(5)	116	1.1170^{20}	1.4641^{20}	i H_2O; s EtOH, eth, ctc; vs chl, bz
3136	*cis*-1,3-Dichloro-2-butene		$C_4H_6Cl_2$	10075-38-4	124.997			127(16)	1.1605^{20}	1.4735^{20}	vs ace, bz, eth, EtOH
3137	*trans*-1,3-Dichloro-2-butene		$C_4H_6Cl_2$	7415-31-8	124.997			132	1.160^{20}	1.4719^{20}	vs ace, bz, eth, EtOH
3138	*cis*-1,4-Dichloro-2-butene		$C_4H_6Cl_2$	1476-11-5	124.997	liq	-42(2)	149(11)	1.188^{25}	1.4887^{25}	vs ace, bz, eth, EtOH
3139	*trans*-1,4-Dichloro-2-butene		$C_4H_6Cl_2$	110-57-6	124.997	col liq	3(2)	155.4	1.183^{25}	1.4871^{25}	vs ace, bz, eth, EtOH
3140	1,4-Dichloro-2-butyne		$C_4H_4Cl_2$	821-10-3	122.981			164(15)	1.258^{20}	1.5058^{20}	s eth, ace; sl ctc; vs chl
3141	2,6-Dichloro-4-(chloroimino)-2,5-cyclohexadien-1-one	Gibbs' reagent	$C_6H_2Cl_3NO$	101-38-2	210.445		66				
3142	1,2-Dichloro-4-(chloromethyl)-benzene		$C_7H_5Cl_3$	102-47-6	195.474		37.5	241			i H_2O; s EtOH, ctc
3143	2,4-Dichloro-1-(chloromethyl)-benzene		$C_7H_5Cl_3$	94-99-5	195.474			120^{13}			
3144	Dichloro(chloromethyl)-methylsilane		$C_2H_5Cl_3Si$	1558-33-4	163.506			121.5	1.2858^{20}	1.4500^{20}	
3145	Dichloro(2-chlorovinyl)arsine		$C_2H_2AsCl_3$	541-25-3	207.318	liq	0.1	190	1.888^{20}		
3146	2,5-Dichloro-2,5-cyclohexadi-ene-1,4-dione		$C_6H_2Cl_2O_2$	615-93-0	176.985	pa ye mcl pr (al)	162.3				i H_2O; sl EtOH; s eth, chl
3147	2,6-Dichloro-2,5-cyclohexadi-ene-1,4-dione		$C_6H_2Cl_2O_2$	697-91-6	176.985	ye orth (lig, bz)	121.8				sl H_2O, EtOH; s chl
3148	1,1-Dichlorocyclohexane		$C_6H_{10}Cl_2$	2108-92-1	153.049	liq	-36.6(0.1)	171	1.1559^{20}	1.4803^{20}	
3149	*cis*-1,2-Dichlorocyclohexane		$C_6H_{10}Cl_2$	10498-35-8	153.049	liq	-5(3)	191(10)	1.2021^{20}	1.4967^{20}	vs bz
3150	1,10-Dichlorodecane		$C_{10}H_{20}Cl_2$	2162-98-3	211.172		15.6(0.5)	167^{28}	0.9945^{25}	1.4586^{25}	

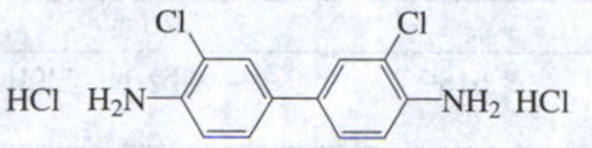
3108
3,3′-Dichloro-*p*-benzidine dihydrochloride

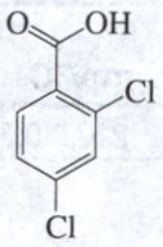
3109
2,4-Dichlorobenzoic acid

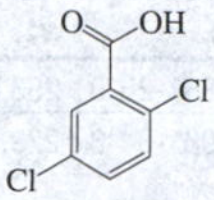
3110
2,5-Dichlorobenzoic acid

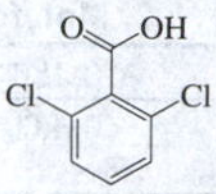
3111
2,6-Dichlorobenzoic acid

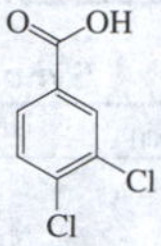
3112
3,4-Dichlorobenzoic acid

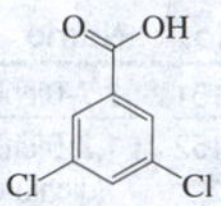
3113
3,5-Dichlorobenzoic acid

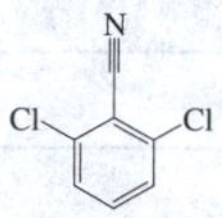
3114
2,6-Dichlorobenzonitrile

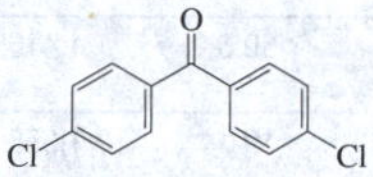
3115
4,4′-Dichlorobenzophenone

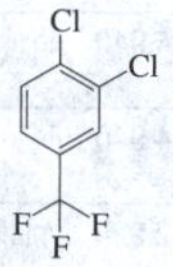
3116
3,4-Dichlorobenzotrifluoride

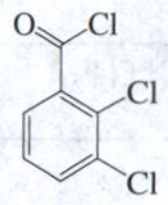
3117
2,3-Dichlorobenzoyl chloride

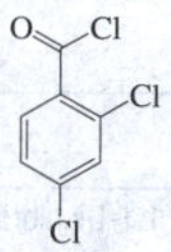
3118
2,4-Dichlorobenzoyl chloride

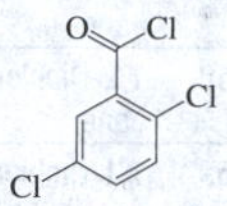
3119
2,5-Dichlorobenzoyl chloride

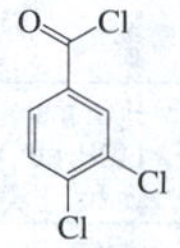
3120
3,4-Dichlorobenzoyl chloride

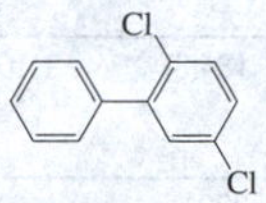
3121
2,5-Dichlorobiphenyl

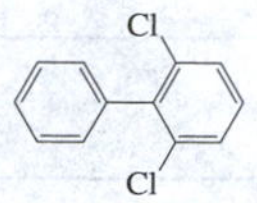
3122
2,6-Dichlorobiphenyl

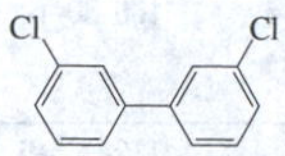
3123
3,3′-Dichlorobiphenyl

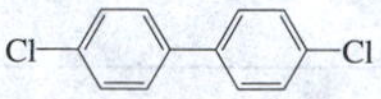
3124
4,4′-Dichlorobiphenyl

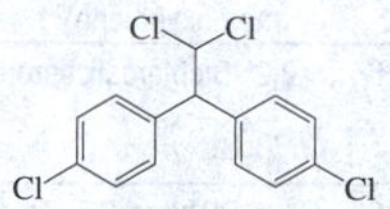
3125
1,1-Dichloro-2,2-bis(*p*-chlorophenyl)ethane

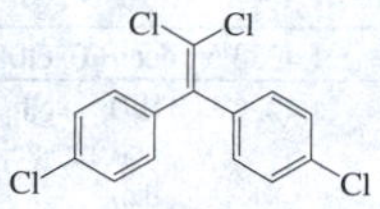
3126
2,2-Dichloro-1,1-bis(4-chlorophenyl)ethene

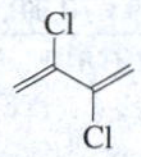
3127
2,3-Dichloro-1,3-butadiene

3128
1,1-Dichlorobutane

3129
1,2-Dichlorobutane

3130
1,3-Dichlorobutane

3131
1,4-Dichlorobutane

3132
2,2-Dichlorobutane

3133
2,3-Dichlorobutane, (±)-

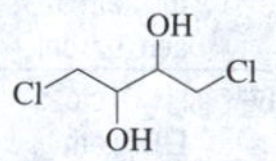
3134
1,4-Dichloro-2,3-butanediol

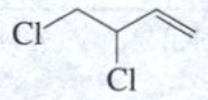
3135
3,4-Dichloro-1-butene

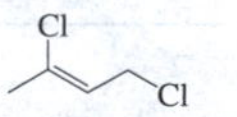
3136
cis-1,3-Dichloro-2-butene

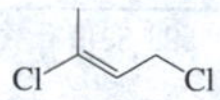
3137
trans-1,3-Dichloro-2-butene

3138
cis-1,4-Dichloro-2-butene

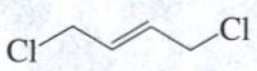
3139
trans-1,4-Dichloro-2-butene

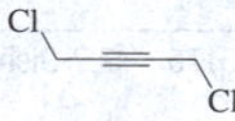
3140
1,4-Dichloro-2-butyne

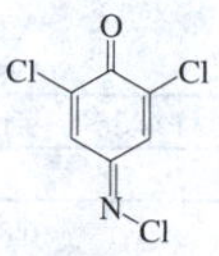
3141
2,6-Dichloro-4-(chloroimino)-2,5-cyclohexadien-1-one

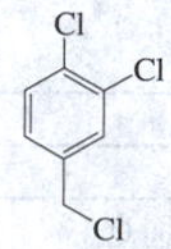
3142
1,2-Dichloro-4-(chloromethyl)benzene

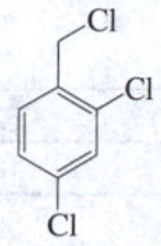
3143
2,4-Dichloro-1-(chloromethyl)benzene

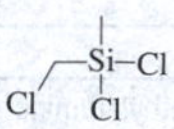
3144
Dichloro(chloromethyl)methylsilane

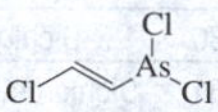
3145
Dichloro(2-chlorovinyl)arsine

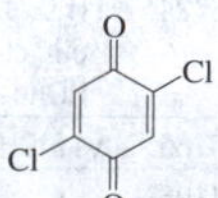
3146
2,5-Dichloro-2,5-cyclohexadiene-1,4-dione

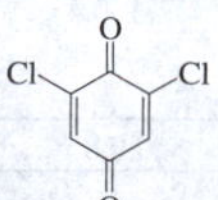
3147
2,6-Dichloro-2,5-cyclohexadiene-1,4-dione

3148
1,1-Dichlorocyclohexane

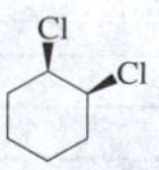
3149
cis-1,2-Dichlorocyclohexane

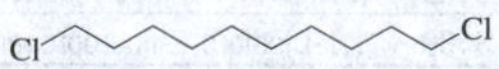
3150
1,10-Dichlorodecane

No.	Name	Synonym	Mol. form.	CAS Reg. No.	Mol. wt.	Physical form	mp/°C	bp/°C	den/ g cm⁻³	n_D	Solubility
3151	2,7-Dichlorodibenzo-*p*-dioxin		$C_{12}H_6Cl_2O_2$	33857-26-0	253.081	cry	210.0(0.5)				
3152	1,2-Dichloro-4-(dichloromethyl)benzene		$C_7H_4Cl_4$	56961-84-3	229.919			257	1.515^{22}		vs bz, eth, EtOH
3153	Dichloro(dichloromethyl)-methylsilane		$C_2H_4Cl_4Si$	1558-31-2	197.951			149	1.4116^{20}	1.4700^{20}	
3154	2,3-Dichloro-5,6-dicyanobenzoquinone		$C_8Cl_2N_2O_2$	84-58-2	227.004	ye-oran cry	214.5				vs bz, HOAc, diox
3155	Dichlorodiethylsilane		$C_4H_{10}Cl_2Si$	1719-53-5	157.114		-96.5	130(2)	1.0504^{20}	1.4309^{20}	
3156	1,1-Dichloro-1,2-difluoroethane		$C_2H_2Cl_2F_2$	25915-78-0	134.940	col liq		48.4			
3157	1,2-Dichloro-1,1-difluoroethane		$C_2H_2Cl_2F_2$	1649-08-7	134.940	liq	-101.2	47(1)	1.4163^{20}	1.36193^{20}	sl H_2O
3158	1,2-Dichloro-1,2-difluoroethane		$C_2H_2Cl_2F_2$	431-06-1	134.940	liq	-101.2	59.6	1.4163^{20}	1.3619^{20}	
3159	1,1-Dichloro-2,2-difluoroethene	1,1-Dichloro-2,2-difluoroethylene	$C_2Cl_2F_2$	79-35-6	132.924	vol liq or gas	-116	19	1.555^{-20}	1.383^{-20}	
3160	*cis*-1,2-Dichloro-1,2-difluoroethene	Refrigerant 1112	$C_2Cl_2F_2$	311-81-9	132.924	vol liq	-119.6	21.1	1.495^{0}		
3161	*trans*-1,2-Dichloro-1,2-difluoroethene		$C_2Cl_2F_2$	381-71-5	132.924	vol liq	-93.3	22	1.494^{0}		
3162	Dichlorodifluoromethane	Refrigerant 12	CCl_2F_2	75-71-8	120.914	col gas	-157.05(0.01)	-29.8(0.1)			sl H_2O; s EtOH, eth, HOAc
3163	2,2-Dichloro-1,1-difluoro-1-methoxyethane	Methoxyflurane	$C_3H_4Cl_2F_2O$	76-38-0	164.966	col liq	-35	105	1.43^{20}	1.3861^{20}	
3164	2,2′-Dichlorodiisopropyl ether		$C_6H_{12}Cl_2O$	108-60-1	171.064			184(3)	1.103^{20}	1.4505^{20}	i H_2O; msc EtOH, eth, ace; vs bz
3165	1,4-Dichloro-2,5-dimethylbenzene		$C_8H_8Cl_2$	1124-05-6	175.056		71	222			s chl
3166	2,5-Dichloro-2,5-dimethylhexane		$C_8H_{16}Cl_2$	6223-78-5	183.119	lf, nd	63(3)		0.9543^{20}		vs bz, eth, EtOH, chl
3167	1,3-Dichloro-5,5-dimethylhydantoin		$C_5H_6Cl_2N_2O_2$	118-52-5	197.019	pr	132		1.5^{20}		sl H_2O; s chl, ctc, bz
3168	2,4-Dichloro-3,5-dimethylphenol	Dichloroxylenol	$C_8H_8Cl_2O$	133-53-9	191.055		83				vs eth
3169	Dichlorodimethylsilane		$C_2H_6Cl_2Si$	75-78-5	129.061	liq	-16	70.5(0.5)	1.064^{25}	1.4038^{20}	dec H_2O, EtOH
3170	2,3-Dichloro-1,4-dioxane		$C_4H_6Cl_2O_2$	95-59-0	156.996		30	81^{10}	1.468^{20}	1.4928^{20}	i H_2O; vs eth, ace, bz, ctc, diox
3171	Dichlorodiphenylmethane		$C_{13}H_{10}Cl_2$	2051-90-3	237.124			299(17)	1.235^{18}		s eth, bz, ctc
3172	Dichlorodiphenylsilane		$C_{12}H_{10}Cl_2Si$	80-10-4	253.199			304(3)	1.204^{25}	1.5800^{20}	s EtOH, eth, ace, bz, ctc
3173	1,1-Dichloroethane	Ethylidene dichloride	$C_2H_4Cl_2$	75-34-3	98.959	liq	-96.93(0.06)	56.3(0.7)	1.1757^{20}	1.4164^{20}	sl H_2O; vs EtOH, eth; s ace, bz
3174	1,2-Dichloroethane	Ethylene dichloride	$C_2H_4Cl_2$	107-06-2	98.959	liq	-35.6(0.2)	83.4(0.1)	1.2454^{25}	1.4422^{25}	sl H_2O; vs EtOH; msc eth; s ace, bz, chl
3175	2,2-Dichloroethanol		$C_2H_4Cl_2O$	598-38-9	114.958			147(15)	1.4040^{25}	1.4626^{25}	sl H_2O, ctc; s EtOH, eth
3176	1,1-Dichloroethene	Vinylidene chloride	$C_2H_2Cl_2$	75-35-4	96.943	liq	-122.5(0.1)	31.6(0.3)	1.213^{20}	1.4249^{20}	i H_2O; s EtOH, ace, bz; vs eth, chl
3177	*cis*-1,2-Dichloroethene	*cis*-1,2-Dichloroethylene	$C_2H_2Cl_2$	156-59-2	96.943	liq	-80.0(0.2)	60(2)	1.2837^{20}	1.4490^{20}	sl H_2O; msc EtOH, eth, ace; vs bz, chl
3178	*trans*-1,2-Dichloroethene	*trans*-1,2-Dichloroethylene	$C_2H_2Cl_2$	156-60-5	96.943	liq	-49.8(0.2)	47.64(0.08)	1.2565^{20}	1.4454^{20}	sl H_2O; msc EtOH, eth, ace; vs bz, chl
3179	1,2-Dichloro-1-ethoxyethane		$C_4H_8Cl_2O$	623-46-1	143.012			144(4)	1.1370^{20}	1.4435^{20}	sl chl
3180	1,2-Dichloroethyl acetate		$C_4H_6Cl_2O_2$	10140-87-1	156.996	liq		79^{33}			
3181	Dichloroethylaluminum	Ethylaluminum chloride	$C_2H_5AlCl_2$	563-43-9	126.949	hyg solid or liq	32	115^{50}	1.207		reac H_2O
3182	Dichloroethylmethylsilane		$C_3H_8Cl_2Si$	4525-44-4	143.088			101	1.0047^{20}	1.4197^{20}	
3183	2′,7′-Dichlorofluorescein	2′,7′-Dichloro-3,6-fluorandiol	$C_{20}H_{10}Cl_2O_5$	76-54-0	401.196						sl DMSO
3184	1,1-Dichloro-1-fluoroethane	HCFC-141b	$C_2H_3Cl_2F$	1717-00-6	116.949	liq	-103.5(0.5)	32.05(0.09)	1.250^{10}	1.3600^{10}	i H_2O
3185	1,2-Dichloro-1-fluoroethane		$C_2H_3Cl_2F$	430-57-9	116.949	liq	-60	74(4)	1.3814^{20}	1.4132^{20}	
3186	1,1-Dichloro-2-fluoroethene	1,1-Dichloro-2-fluoroethylene	C_2HCl_2F	359-02-4	114.933	liq	-108.8	38(21)	1.3732^{16}	1.4031^{16}	
3187	Dichlorofluoromethane	Refrigerant 21	$CHCl_2F$	75-43-4	102.923	col gas	-130.4	8.9	1.405^{9}	1.3724^{9}	i H_2O; s EtOH, eth, ctc, chl, HOAc
3188	(Dichlorofluoromethyl)benzene		$C_7H_5Cl_2F$	498-67-9	179.019	liq	-26.8	179	1.3138^{11}	1.5180^{11}	vs EtOH
3189	1,1-Dichloro-2-fluoropropene		$C_3H_3Cl_2F$	430-95-5	128.960			78	1.3026^{25}	1.4196^{25}	
3190	1,7-Dichloroheptane		$C_7H_{14}Cl_2$	821-76-1	169.092			124^{35}	1.0408^{25}	1.4565^{25}	

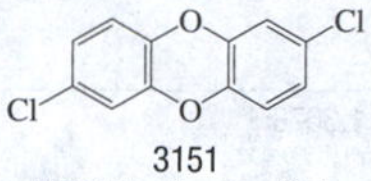
3151
2,7-Dichlorodibenzo-*p*-dioxin

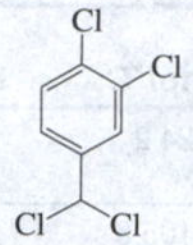
3152
1,2-Dichloro-4-(dichloromethyl)benzene

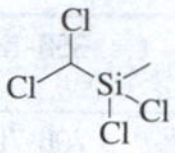
3153
Dichloro(dichloromethyl)methylsilane

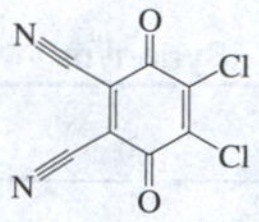
3154
2,3-Dichloro-5,6-dicyanobenzoquinone

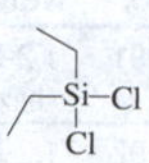
3155
Dichlorodiethylsilane

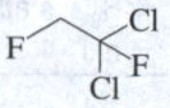
3156
1,1-Dichloro-1,2-difluoroethane

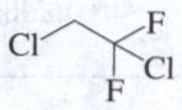
3157
1,2-Dichloro-1,1-difluoroethane

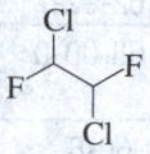
3158
1,2-Dichloro-1,2-difluoroethane

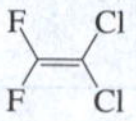
3159
1,1-Dichloro-2,2-difluoroethene

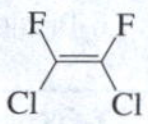
3160
cis-1,2-Dichloro-1,2-difluoroethene

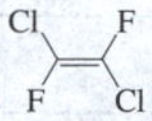
3161
trans-1,2-Dichloro-1,2-difluoroethene

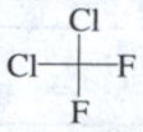
3162
Dichlorodifluoromethane

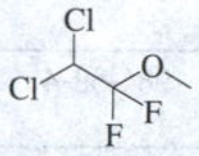
3163
2,2-Dichloro-1,1-difluoro-1-methoxyethane

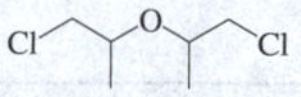
3164
2,2'-Dichlorodiisopropyl ether

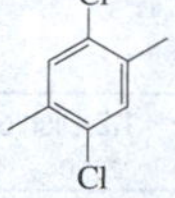
3165
1,4-Dichloro-2,5-dimethylbenzene

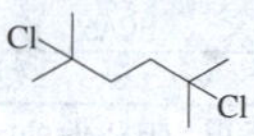
3166
2,5-Dichloro-2,5-dimethylhexane

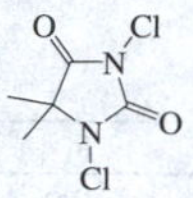
3167
1,3-Dichloro-5,5-dimethyl hydantoin

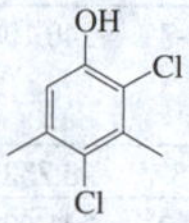
3168
2,4-Dichloro-3,5-dimethylphenol

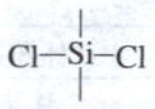
3169
Dichlorodimethylsilane

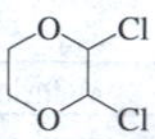
3170
2,3-Dichloro-1,4-dioxane

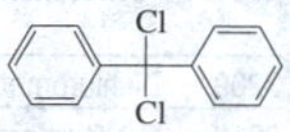
3171
Dichlorodiphenylmethane

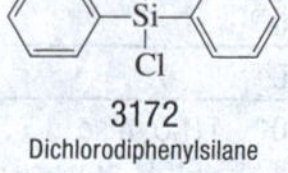
3172
Dichlorodiphenylsilane

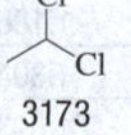
3173
1,1-Dichloroethane

3174
1,2-Dichloroethane

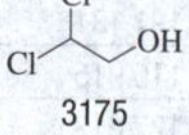
3175
2,2-Dichloroethanol

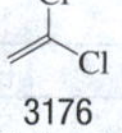
3176
1,1-Dichloroethene

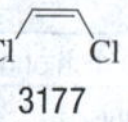
3177
cis-1,2-Dichloroethene

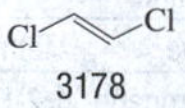
3178
trans-1,2-Dichloroethene

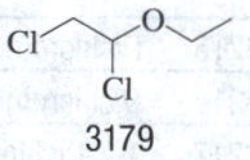
3179
1,2-Dichloro-1-ethoxyethane

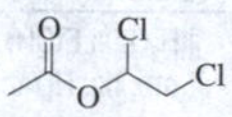
3180
1,2-Dichloroethyl acetate

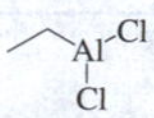
3181
Dichloroethylaluminum

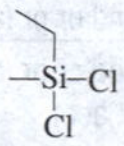
3182
Dichloroethylmethylsilane

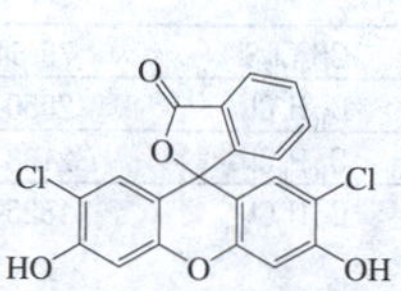
3183
2',7'-Dichlorofluorescein

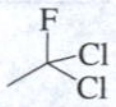
3184
1,1-Dichloro-1-fluoroethane

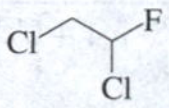
3185
1,2-Dichloro-1-fluoroethane

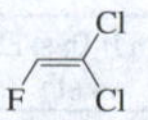
3186
1,1-Dichloro-2-fluoroethene

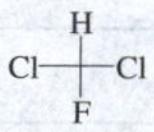
3187
Dichlorofluoromethane

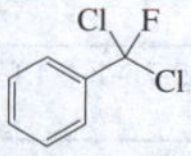
3188
(Dichlorofluoromethyl)benzene

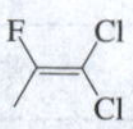
3189
1,1-Dichloro-2-fluoropropene

3190
1,7-Dichloroheptane

No.	Name	Synonym	Mol. form.	CAS Reg. No.	Mol. wt.	Physical form	mp/°C	bp/°C	den/ g cm^{-3}	n_D	Solubility
3191	1,2-Dichloro-1,2,3,3,4,4-hexafluorocyclobutane		$C_4Cl_2F_6$	356-18-3	232.939	liq	-24.2	59.5			
3192	1,2-Dichloro-3,3,4,4,5,5-hexafluorocyclopentene		$C_5Cl_2F_6$	706-79-6	244.949	liq	-105.8	90.7	1.6546^{20}	1.3676^{20}	
3193	1,2-Dichloro-1,1,2,3,3,3-hexafluoropropane		$C_3Cl_2F_6$	661-97-2	220.928			34(2)			i H_2O
3194	1,3-Dichloro-1,1,2,2,3,3-hexafluoropropane	Refrigerant 216	$C_3Cl_2F_6$	662-01-1	220.928	liq	-125.4	35.7	1.573^{20}	1.3030^{20}	
3195	1,5-Dichloro-1,1,3,3,5,5-hexamethyltrisiloxane		$C_6H_{18}Cl_2O_2Si_3$	3582-71-6	277.369	liq	-53	184	1.018^{20}		dec H_2O
3196	1,2-Dichlorohexane		$C_6H_{12}Cl_2$	2162-92-7	155.065			172.2(0.9)	1.085^{15}		vs eth, chl
3197	1,6-Dichlorohexane		$C_6H_{12}Cl_2$	2163-00-0	155.065			205(1)	1.0676^{25}	1.4555^{25}	i H_2O; s eth, ctc, chl
3198	3,5-Dichloro-2-hydroxybenzaldehyde		$C_7H_4Cl_2O_2$	90-60-8	191.012	ye orth (HOAc)	95				i H_2O
3199	3,5-Dichloro-2-hydroxybenzoic acid		$C_7H_4Cl_2O_3$	320-72-9	207.011	nd (dil al) orth pr	220.5	sub			sl H_2O; vs EtOH, eth
3200	2,6-Dichloroindophenol, sodium salt	Tillman's reagent	$C_{12}H_6Cl_2NNaO_2$	620-45-1	290.078	dk grn cry					s H_2O, EtOH, ace
3201	5,6-Dichloro-1,3-isobenzofurandione	4,5-Dichlorophthalic anhydride	$C_8H_2Cl_2O_3$	942-06-3	217.006	tab or pr (to)	188	313			vs eth, EtOH, tol
3202	Dichloromethane	Methylene chloride	CH_2Cl_2	75-09-2	84.933	liq	-95(2)	39.8(0.3)	1.3266^{20}	1.4242^{20}	sl H_2O; msc EtOH, eth; s ctc
3203	1,2-Dichloro-3-methoxybenzene		$C_7H_6Cl_2O$	1984-59-4	177.028		31.0(0.4)				
3204	1,3-Dichloro-2-methoxybenzene	2,6-Dichloroanisole	$C_7H_6Cl_2O$	1984-65-2	177.028	liq	10	105^{20}	1.291	1.5430^{20}	
3205	2,4-Dichloro-1-methoxybenzene		$C_7H_6Cl_2O$	553-82-2	177.028	pr	28.5	232			sl chl
3206	3,6-Dichloro-2-methoxybenzoic acid	Dicamba	$C_8H_6Cl_2O_3$	1918-00-9	221.038	cry (pent)	114.9(0.8)		1.57^{25}		
3207	(Dichloromethyl)benzene	Benzal chloride	$C_7H_6Cl_2$	98-87-3	161.029	liq	-17.0(0.5)	205	1.26^{25}	1.5502^{20}	i H_2O; vs eth, EtOH
3208	*N,N*-Dichloro-4-methylbenzenesulfonamide	Dichloramine-T	$C_7H_7Cl_2NO_2S$	473-34-7	240.108	pr(chl-peth)	83				i H_2O; s EtOH, eth, bz, ctc, HOAc
3209	Dichloromethylborane	Methyldichloroborane	CH_3BCl_2	7318-78-7	96.752	col gas		11			
3210	2,3-Dichloro-2-methylbutane	Amylene dichloride	$C_5H_{10}Cl_2$	507-45-9	141.038			129	1.0696^{15}	1.4450^{18}	i H_2O; vs eth, EtOH
3211	1,1-Dichloromethyl methyl ether	Methoxydichloromethane	$C_2H_4Cl_2O$	4885-02-3	114.958			86(9)	1.271^{25}	1.4300^{20}	
3212	2,4-Dichloro-3-methylphenol		$C_7H_6Cl_2O$	17788-00-0	177.028	pr (peth)	58	236			vs eth, chl
3213	2,4-Dichloro-6-methylphenol		$C_7H_6Cl_2O$	1570-65-6	177.028	nd (w, peth)	55				sl H_2O; vs EtOH, eth, chl, CS_2
3214	2,6-Dichloro-4-methylphenol		$C_7H_6Cl_2O$	2432-12-4	177.028	nd (lig)	39	231			i H_2O; vs eth, EtOH, HOAc
3215	Dichloromethylphenylsilane		$C_7H_8Cl_2Si$	149-74-6	191.131			206(3)	1.1866^{20}	1.5180^{20}	
3216	Dichloromethylphosphine	Methylphosphonous dichloride	CH_3Cl_2P	676-83-5	116.915			12^{50}	1.304^{20}	1.4940^{20}	
3217	1,2-Dichloro-2-methylpropane	1,2-Dichloroisobutane	$C_4H_8Cl_2$	594-37-6	127.013			108(2)	1.093^{20}	1.4370^{20}	i H_2O; msc EtOH, eth, ace, bz, ctc
3218	2,4-Dichloro-5-methylpyrimidine		$C_5H_4Cl_2N_2$	1780-31-0	163.004	pl (al)	26	235			sl H_2O; vs EtOH, eth, bz, chl
3219	2,4-Dichloro-6-methylpyrimidine		$C_5H_4Cl_2N_2$	5424-21-5	163.004	nd (lig)	46.5	219			vs bz, eth, EtOH, chl
3220	Dichloromethylsilane		CH_4Cl_2Si	75-54-7	115.035	liq	-93	40.9(0.1)	1.105^{25}		
3221	1,2-Dichloronaphthalene		$C_{10}H_6Cl_2$	2050-69-3	197.061	pl (al)	36	295(26)	1.3147^{49}	1.5338^{49}	s EtOH, eth
3222	1,3-Dichloronaphthalene		$C_{10}H_6Cl_2$	2198-75-6	197.061	nd or pr (al)	62.3	291			s EtOH
3223	1,4-Dichloronaphthalene		$C_{10}H_6Cl_2$	1825-31-6	197.061	nd or pr (al, ace)	67.5	288	1.2997^{76}	1.6228^{76}	i H_2O; sl EtOH; s eth, bz, HOAc; vs ace
3224	1,5-Dichloronaphthalene		$C_{10}H_6Cl_2$	1825-30-5	197.061	nd or lf (al) pr (sub)	107	sub	1.4900^{20}		i H_2O; sl EtOH; s eth
3225	1,6-Dichloronaphthalene		$C_{10}H_6Cl_2$	2050-72-8	197.061	nd or pr (al, peth)	49	sub			
3226	1,7-Dichloronaphthalene		$C_{10}H_6Cl_2$	2050-73-9	197.061	nd or pr (al, HOAc)	63.5	285.5	1.2611^{100}	1.6092^{100}	s EtOH, eth, bz, HOAc
3227	1,8-Dichloronaphthalene		$C_{10}H_6Cl_2$	2050-74-0	197.061	orth pl (hx) nd (al, sub)	89	sub	1.2924^{100}	1.6236^{100}	s EtOH, peth
3228	2,3-Dichloronaphthalene		$C_{10}H_6Cl_2$	2050-75-1	197.061	orth lf (al)	120				i H_2O; sl EtOH; vs eth

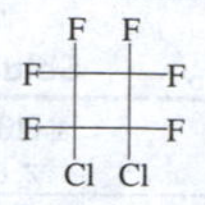
3191
1,2-Dichloro-1,2,3,3,4,4-hexafluorocyclobutane

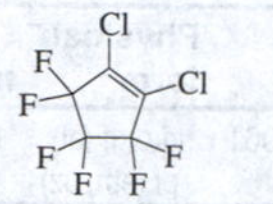
3192
1,2-Dichloro-3,3,4,4,5,5-hexafluorocyclopentene

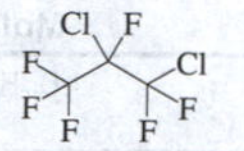
3193
1,2-Dichloro-1,1,2,3,3,3-hexafluoropropane

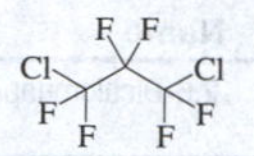
3194
1,3-Dichloro-1,1,2,2,3,3-hexafluoropropane

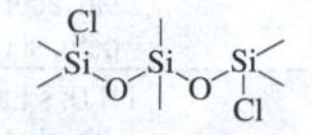
3195
1,5-Dichloro-1,1,3,3,5,5-hexamethyltrisiloxane

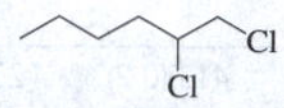
3196
1,2-Dichlorohexane

3197
1,6-Dichlorohexane

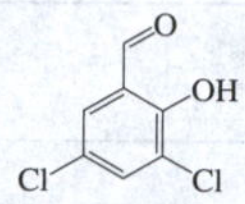
3198
3,5-Dichloro-2-hydroxybenzaldehyde

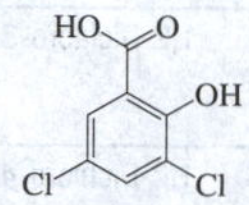
3199
3,5-Dichloro-2-hydroxybenzoic acid

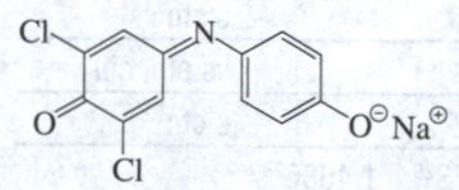
3200
2,6-Dichloroindophenol, sodium salt

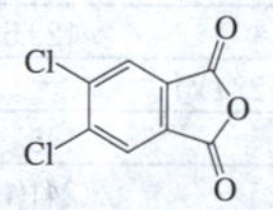
3201
5,6-Dichloro-1,3-isobenzofurandione

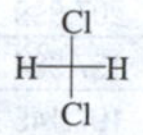
3202
Dichloromethane

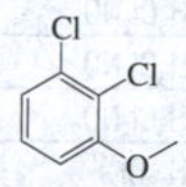
3203
1,2-Dichloro-3-methoxybenzene

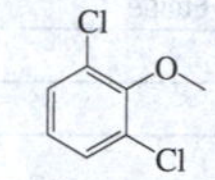
3204
1,3-Dichloro-2-methoxybenzene

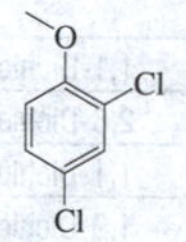
3205
2,4-Dichloro-1-methoxybenzene

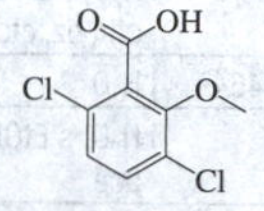
3206
3,6-Dichloro-2-methoxybenzoic acid

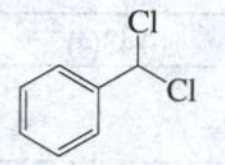
3207
(Dichloromethyl)benzene

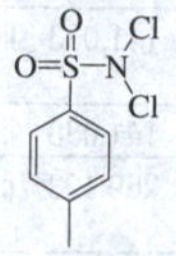
3208
N,N-Dichloro-4-methylbenzenesulfonamide

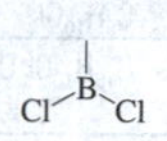
3209
Dichloromethylborane

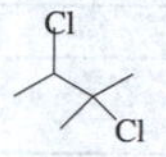
3210
2,3-Dichloro-2-methylbutane

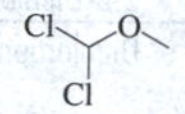
3211
1,1-Dichloromethyl methyl ether

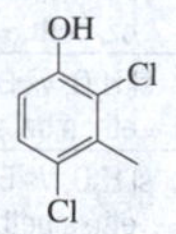
3212
2,4-Dichloro-3-methylphenol

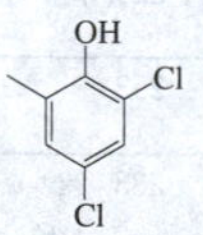
3213
2,4-Dichloro-6-methylphenol

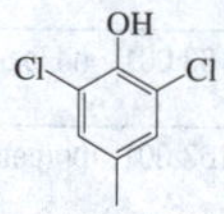
3214
2,6-Dichloro-4-methylphenol

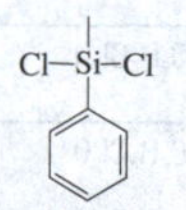
3215
Dichloromethylphenylsilane

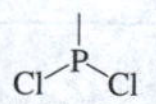
3216
Dichloromethylphosphine

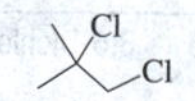
3217
1,2-Dichloro-2-methylpropane

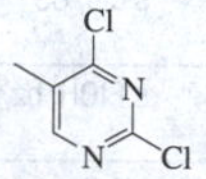
3218
2,4-Dichloro-5-methylpyrimidine

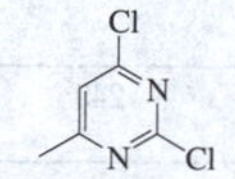
3219
2,4-Dichloro-6-methylpyrimidine

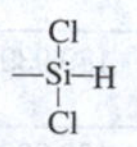
3220
Dichloromethylsilane

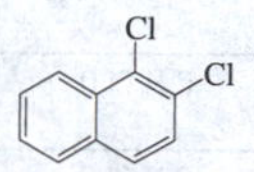
3221
1,2-Dichloronaphthalene

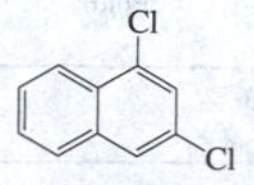
3222
1,3-Dichloronaphthalene

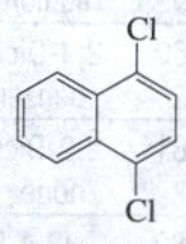
3223
1,4-Dichloronaphthalene

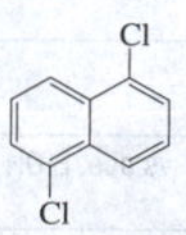
3224
1,5-Dichloronaphthalene

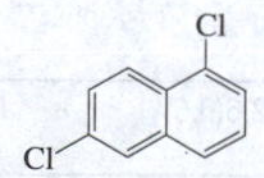
3225
1,6-Dichloronaphthalene

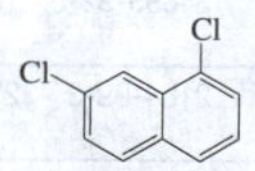
3226
1,7-Dichloronaphthalene

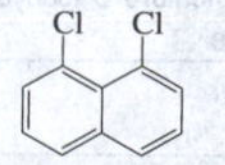
3227
1,8-Dichloronaphthalene

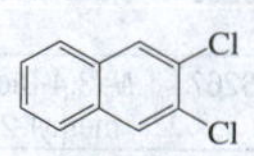
3228
2,3-Dichloronaphthalene

No.	Name	Synonym	Mol. form.	CAS Reg. No.	Mol. wt.	Physical form	mp/°C	bp/°C	den/g cm^{-3}	n_D	Solubility
3229	2,6-Dichloronaphthalene		$C_{10}H_6Cl_2$	2065-70-5	197.061	nd or lf (al) pl (eth, bz)	140.5	285			sl EtOH; s eth, bz, chl, HOAc
3230	2,7-Dichloronaphthalene		$C_{10}H_6Cl_2$	2198-77-8	197.061	pl or lf (al)	115.0				vs EtOH; s hx, HOAc
3231	2,3-Dichloro-1,4-naphthalenedione	Dichlone	$C_{10}H_4Cl_2O_2$	117-80-6	227.044	ye nd (al)	196.4(0.5)				i H_2O; sl EtOH, eth, bz; s chl
3232	2,4-Dichloro-1-naphthol	2,4-Dichloro-α-naphthol	$C_{10}H_6Cl_2O$	2050-76-2	213.060	nd (al, bz)	107.5	180			vs bz, eth, EtOH
3233	2,6-Dichloro-4-nitroaniline		$C_6H_4Cl_2N_2O_2$	99-30-9	207.014	ye nd (al, HOAc)	193.8(0.8)				s EtOH, acid; sl DMSO
3234	1,2-Dichloro-3-nitrobenzene		$C_6H_3Cl_2NO_2$	3209-22-1	192.000	mcl nd (peth, HOAc)	60.7(0.6)	257.5	1.721^{14}		i H_2O; s EtOH, eth, ace, bz, peth; sl chl
3235	1,2-Dichloro-4-nitrobenzene		$C_6H_3Cl_2NO_2$	99-54-7	192.000	nd (al)	41.0(0.2)	255.5	1.4558^{75}		i H_2O; s EtOH, eth; sl ctc
3236	1,3-Dichloro-5-nitrobenzene		$C_6H_3Cl_2NO_2$	618-62-2	192.000	mcl pr or lf (HOAc, al)	65(2)		1.4000^{100}		i H_2O; s EtOH, eth
3237	1,4-Dichloro-2-nitrobenzene		$C_6H_3Cl_2NO_2$	89-61-2	192.000	pl or pr (al) pl (AcOEt)	53.5(0.2)	267(10)	1.439^{75}	1.4390^{75}	i H_2O; s EtOH, eth, bz, CS_2; sl ctc
3238	2,4-Dichloro-1-nitrobenzene		$C_6H_3Cl_2NO_2$	611-06-3	192.000	nd (al)	33(2)	258.5	1.4790^{80}	1.5512^{70}	i H_2O; s EtOH, eth; sl chl
3239	1,1-Dichloro-1-nitroethane	Ethide	$C_2H_3Cl_2NO_2$	594-72-9	143.957			123.5			s ctc
3240	2,6-Dichloro-4-nitrophenol		$C_6H_3Cl_2NO_3$	618-80-4	207.999	br nd (w)	127 exp		1.822^{25}		vs eth, chl
3241	1,1-Dichloro-1-nitropropane		$C_3H_5Cl_2NO_2$	595-44-8	157.984			145	1.312^{20}		s ctc
3242	1,9-Dichlorononane		$C_9H_{18}Cl_2$	821-99-8	197.145			241(12)	1.0173^{25}	1.4586^{25}	
3243	1,8-Dichlorooctane		$C_8H_{16}Cl_2$	2162-99-4	183.119			246(1)	1.0248^{25}	1.4572^{25}	
3244	1,3-Dichloro-1,1,2,2,3-pentafluoropropane	HCFC-225cb	$C_3HCl_2F_5$	507-55-1	202.938	liq		55(2)	1.55^{25}		
3245	3,3-Dichloro-1,1,1,2,2-pentafluoropropane	Refrigerant 225ca	$C_3HCl_2F_5$	422-56-0	202.938	liq		50(2)	1.54^{25}		
3246	1,2-Dichloropentane		$C_5H_{10}Cl_2$	1674-33-5	141.038			148.2(0.7)	1.0872^{20}	1.4485^{20}	i H_2O; s EtOH; vs chl
3247	1,5-Dichloropentane		$C_5H_{10}Cl_2$	628-76-2	141.038	liq	-72.8	182.9(0.8)	1.0956^{25}	1.4545^{25}	i H_2O; s EtOH, eth, bz, ctc
3248	2,3-Dichloropentane		$C_5H_{10}Cl_2$	600-11-3	141.038	liq	-77.3	143(4)	1.0789^{20}	1.4464^{20}	i H_2O
3249	Dichlorophene		$C_{13}H_{10}Cl_2O_2$	97-23-4	269.123	cry (bz, peth)	177.5				i H_2O; s EtOH, ace
3250	2,3-Dichlorophenol		$C_6H_4Cl_2O$	576-24-9	163.001	cry (lig, bz)	56.8(0.3)				s EtOH, eth, bz, lig
3251	2,4-Dichlorophenol		$C_6H_4Cl_2O$	120-83-2	163.001	hex nd (bz)	43(2)	210			sl H_2O; s EtOH, eth, bz, chl
3252	2,5-Dichlorophenol		$C_6H_4Cl_2O$	583-78-8	163.001	pr (bz, peth)	57.8(0.3)	222(8)			sl H_2O; vs EtOH, eth; s bz, peth
3253	2,6-Dichlorophenol		$C_6H_4Cl_2O$	87-65-0	163.001	nd (peth)	66.6(0.4)	226(4)	1.653^{20}		vs EtOH, eth; s bz, peth
3254	3,4-Dichlorophenol		$C_6H_4Cl_2O$	95-77-2	163.001	nd (bz-peth)	67.8(0.3)	253			sl H_2O; vs EtOH, eth; s bz, peth
3255	3,5-Dichlorophenol		$C_6H_4Cl_2O$	591-35-5	163.001	pr (peth)	67.8(0.3)	233			sl H_2O; vs EtOH, eth; s peth
3256	(2,4-Dichlorophenoxy)acetic acid	2,4-D	$C_8H_6Cl_2O_3$	94-75-7	221.038	cry (bz)	140(2)	$160^{0.4}$			i H_2O; s EtOH; sl bz, DMSO
3257	4-(2,4-Dichlorophenoxy)-butanoic acid	Butyrac 118	$C_{10}H_{10}Cl_2O_3$	94-82-6	249.090		118(2)				
3258	2-(2,4-Dichlorophenoxy)-propanoic acid	Dichlorprop	$C_9H_8Cl_2O_3$	120-36-5	235.064		116.6(0.3)				sl H_2O, lig; s EtOH, eth
3259	Dichlorophenylarsine		$C_6H_5AsCl_2$	696-28-6	222.932	liq	-19	256(4)	1.6516^{20}	1.6386^{15}	vs bz, eth, EtOH
3260	2,4-Dichlorophenyl benzenesulfonate	Genite	$C_{12}H_8Cl_2O_3S$	97-16-5	303.161		45.5				s ctc, CS_2
3261	2,2-Dichloro-1-phenylethanone		$C_8H_6Cl_2O$	2648-61-5	189.039	amor	20.5	249	1.340^{16}	1.5686^{20}	s EtOH, bz, ctc
3262	1-(2,4-Dichlorophenyl)-ethanone		$C_8H_6Cl_2O$	2234-16-4	189.039		33.5			1.5640^{20}	i H_2O
3263	1-(2,5-Dichlorophenyl)-ethanone		$C_8H_6Cl_2O$	2476-37-1	189.039		12	118^{12}	1.321^{30}	1.5595^{30}	
3264	1-(3,4-Dichlorophenyl)-ethanone		$C_8H_6Cl_2O$	2642-63-9	189.039	nd (peth)	76	135^{12}			i H_2O; s ctc, lig
3265	3,4-Dichlorophenyl isocyanate	1,2-Dichloro-5-isocyanatobenzene	$C_7H_3Cl_2NO$	102-36-3	188.011	cry	42	112^{12}			
3266	3,5-Dichlorophenyl isocyanate	1,3-Dichloro-5-isocyanatobenzene	$C_7H_3Cl_2NO$	34893-92-0	188.011		33		1.380		
3267	*N*-(3,4-Dichlorophenyl)-2-methyl-2-propenamide	Dicryl	$C_{10}H_9Cl_2NO$	2164-09-2	230.090	cry (al-peth)	122.6(0.5)				vs ace, EtOH

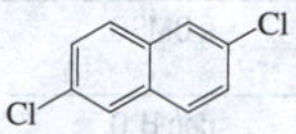
3229
2,6-Dichloronaphthalene

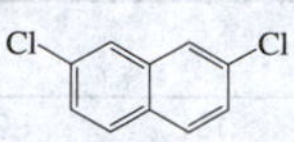
3230
2,7-Dichloronaphthalene

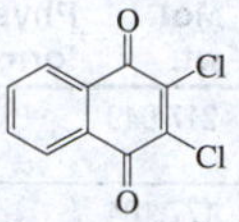
3231
2,3-Dichloro-1,4-naphthalenedione

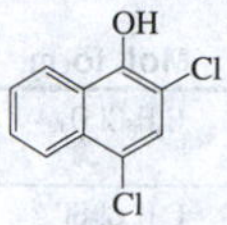
3232
2,4-Dichloro-1-naphthol

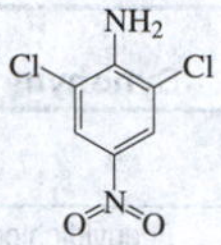
3233
2,6-Dichloro-4-nitroaniline

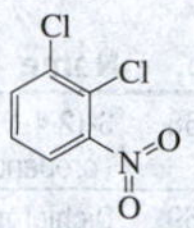
3234
1,2-Dichloro-3-nitrobenzene

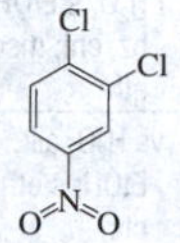
3235
1,2-Dichloro-4-nitrobenzene

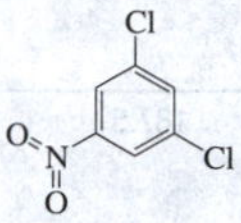
3236
1,3-Dichloro-5-nitrobenzene

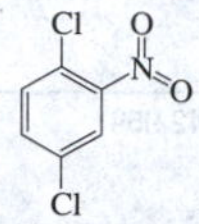
3237
1,4-Dichloro-2-nitrobenzene

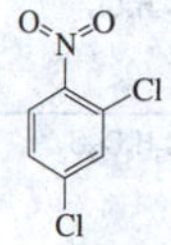
3238
2,4-Dichloro-1-nitrobenzene

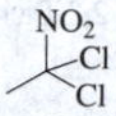
3239
1,1-Dichloro-1-nitroethane

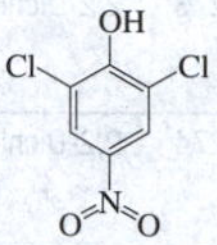
3240
2,6-Dichloro-4-nitrophenol

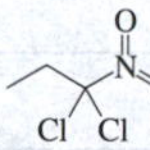
3241
1,1-Dichloro-1-nitropropane

3242
1,9-Dichlorononane

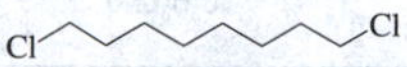
3243
1,8-Dichlorooctane

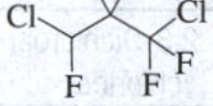
3244
1,3-Dichloro-1,1,2,2,3-pentafluoropropane

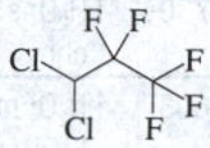
3245
3,3-Dichloro-1,1,1,2,2-pentafluoropropane

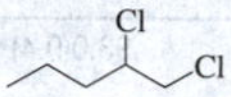
3246
1,2-Dichloropentane

3247
1,5-Dichloropentane

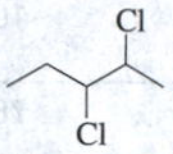
3248
2,3-Dichloropentane

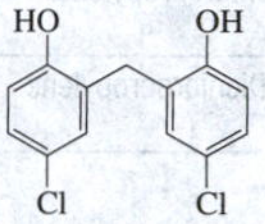
3249
Dichlorophene

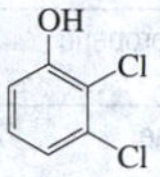
3250
2,3-Dichlorophenol

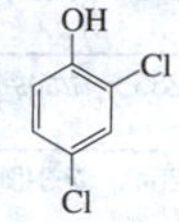
3251
2,4-Dichlorophenol

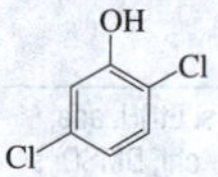
3252
2,5-Dichlorophenol

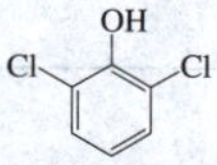
3253
2,6-Dichlorophenol

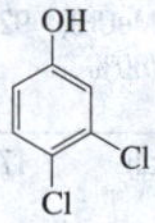
3254
3,4-Dichlorophenol

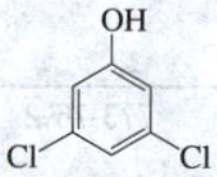
3255
3,5-Dichlorophenol

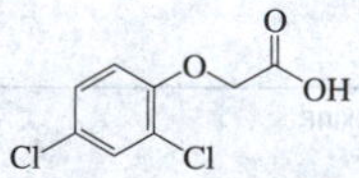
3256
(2,4-Dichlorophenoxy)acetic acid

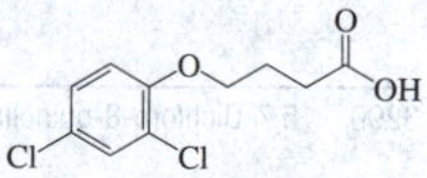
3257
4-(2,4-Dichlorophenoxy)butanoic acid

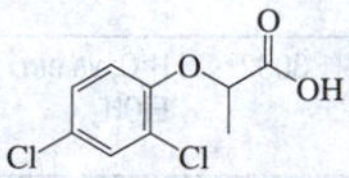
3258
2-(2,4-Dichlorophenoxy)propanoic acid

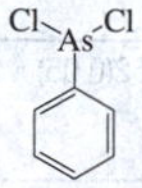
3259
Dichlorophenylarsine

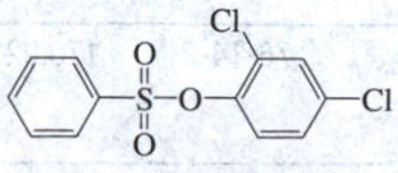
3260
2,4-Dichlorophenyl benzenesulfonate

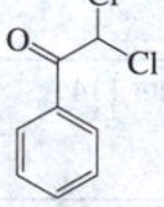
3261
2,2-Dichloro-1-phenylethanone

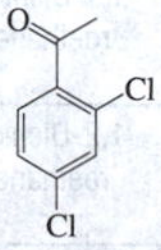
3262
1-(2,4-Dichlorophenyl)ethanone

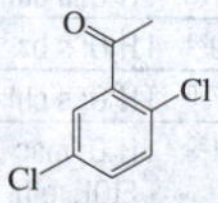
3263
1-(2,5-Dichlorophenyl)ethanone

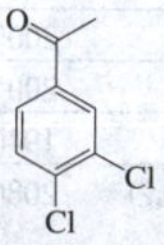
3264
1-(3,4-Dichlorophenyl)ethanone

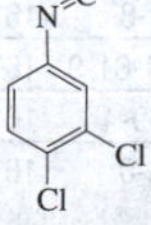
3265
3,4-Dichlorophenyl isocyanate

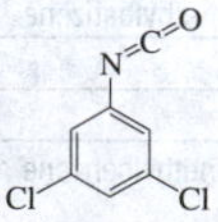
3266
3,5-Dichlorophenyl isocyanate

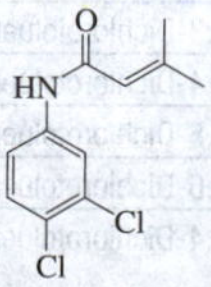
3267
N-(3,4-Dichlorophenyl)-2-methyl-2-propenamide

Organic

No.	Name	Synonym	Mol. form.	CAS Reg. No.	Mol. wt.	Physical form	mp/°C	bp/°C	den/ g cm^{-3}	n_D	Solubility
3268	3-(2,4-Dichlorophenyl)-2-propenoic acid		$C_9H_6Cl_2O_2$	1201-99-6	217.049		234				s DMSO
3269	Dichlorophenylsilane	Phenyldichlorosilane	$C_6H_6Cl_2Si$	1631-84-1	177.104			181	1.221^{25}		dec H_2O
3270	1,1-Dichloropropane	Propylidene chloride	$C_3H_6Cl_2$	78-99-9	112.986			88.4(0.5)	1.1321^{20}	1.4289^{20}	s EtOH, eth, bz, chl
3271	1,2-Dichloropropane, (±)-	Propylene dichloride	$C_3H_6Cl_2$	26198-63-0	112.986	liq	-100.53	96.4	1.1560^{20}	1.4394^{20}	sl H_2O; s EtOH, eth, bz, chl
3272	1,3-Dichloropropane		$C_3H_6Cl_2$	142-28-9	112.986	liq	-99.5	120.8(0.3)	1.1785^{25}	1.4455^{25}	sl H_2O; vs EtOH, eth; s bz, chl
3273	2,2-Dichloropropane		$C_3H_6Cl_2$	594-20-7	112.986	liq	-33.9(0.1)	69.51(0.1)	1.1136^{20}	1.4148^{20}	i H_2O; s EtOH, bz, chl; msc eth
3274	2,2-Dichloropropanoic acid	2,2-Dichloropropionic acid	$C_3H_4Cl_2O_2$	75-99-0	142.969			187.5	1.389^{12}		vs H_2O, alk, EtOH; s eth, ctc
3275	2,3-Dichloro-1-propanol		$C_3H_6Cl_2O$	616-23-9	128.985	visc		176(5)	1.3607^{20}	1.4819^{20}	sl H_2O, lig; msc EtOH, eth, ace, bz
3276	1,3-Dichloro-2-propanol		$C_3H_6Cl_2O$	96-23-1	128.985			171(4)	1.3506^{17}	1.4837^{20}	vs H_2O, EtOH; msc eth; s ace, chl
3277	2,3-Dichloro-1-propanol, phosphate (3:1)		$C_9H_{15}Cl_6O_4P$	78-43-3	430.904			$190^{0.1}$	1.517^{22}		
3278	2,3-Dichloropropanoyl chloride		$C_3H_3Cl_3O$	7623-13-4	161.414			53^{17}	1.4757^{20}	1.4764^{20}	
3279	1,1-Dichloropropene		$C_3H_4Cl_2$	563-58-6	110.970			76(5)	1.1864^{25}	1.4430^{25}	i H_2O; s eth, ace, chl
3280	*cis*-1,2-Dichloropropene		$C_3H_4Cl_2$	6923-20-2	110.970			105(5)		1.4549^{20}	i H_2O; s ace, bz, chl
3281	*trans*-1,2-Dichloropropene		$C_3H_4Cl_2$	7069-38-7	110.970			77	1.1818^{20}	1.4471^{20}	i H_2O; vs EtOH, ctc, MeOH
3282	*cis*-1,3-Dichloropropene	*cis*-1,3-Dichloropropylene	$C_3H_4Cl_2$	10061-01-5	110.970			104(1)	1.224^{20}	1.4682^{20}	i H_2O; s eth, bz, chl
3283	*trans*-1,3-Dichloropropene	*trans*-1,3-Dichloropropylene	$C_3H_4Cl_2$	10061-02-6	110.970			111(5)	1.217^{20}	1.4730^{20}	i H_2O; s eth, bz, chl
3284	2,3-Dichloropropene		$C_3H_4Cl_2$	78-88-6	110.970	liq	10	93.0(0.4)	1.211^{20}	1.4603^{20}	i H_2O; msc EtOH; s eth, bz, chl
3285	3,6-Dichloropyridazine		$C_4H_2Cl_2N_2$	141-30-0	148.978		68.8	$89^{0.2}$			s chl
3286	2,6-Dichloropyridine		$C_5H_3Cl_2N$	2402-78-0	147.990		87	211			
3287	4,6-Dichloro-2-pyrimidin-amine		$C_4H_3Cl_2N_3$	56-05-3	163.993		215				s DMSO
3288	2,4-Dichloropyrimidine		$C_4H_2Cl_2N_2$	3934-20-1	148.978		59	198			
3289	4,7-Dichloroquinoline		$C_9H_5Cl_2N$	86-98-6	198.049	cry (MeOH), nd (80% al)	92.4(0.5)	148^{10}			sl chl
3290	5,7-Dichloro-8-quinolinol	Chloroxine	$C_9H_5Cl_2NO$	773-76-2	214.048	cry (al)	179.5				sl EtOH, ace, chl, DMSO; s alk, bz, peth
3291	2,3-Dichloroquinoxaline		$C_8H_4Cl_2N_2$	2213-63-0	199.037	cry (al, bz)	151.2(0.4)				i H_2O; vs EtOH, bz, chl, HOAc
3292	2,5-Dichlorostyrene		$C_8H_6Cl_2$	1123-84-8	173.040		8.0	93^5	1.246^{20}	1.5798^{20}	
3293	1,2-Dichloro-3,4,5,6-tetrafluorobenzene		$C_6Cl_2F_4$	1198-59-0	218.964			157.7			
3294	1,1-Dichloro-1,2,2,2-tetrafluoroethane	Refrigerant 114a	$C_2Cl_2F_4$	374-07-2	170.921	col gas	-56.6	3(1)	1.455^{25} (p>1 atm)	1.3092^{0}	vs bz, eth, EtOH
3295	1,2-Dichloro-1,1,2,2-tetrafluoroethane	Refrigerant 114	$C_2Cl_2F_4$	76-14-2	170.921	col gas	-92.52(0.05)	3.6(0.5)	1.455^{25} (p>1 atm)	1.3092^{0}	i H_2O; vs eth, EtOH
3296	1,2-Dichloro-1,1,2,2-tetramethyldisilane		$C_4H_{12}Cl_2Si_2$	4342-61-4	187.215			148	1.010^{20}	1.4548^{20}	
3297	1,3-Dichloro-1,1,3,3-tetramethyldisiloxane		$C_4H_{12}Cl_2OSi_2$	2401-73-2	203.214	liq	-37.5	138	1.038^{20}		
3298	2,5-Dichlorothiophene		$C_4H_2Cl_2S$	3172-52-9	153.030	liq	-40.5	165(4)	1.4422^{20}	1.5626^{20}	i H_2O; msc EtOH, eth; s ctc
3299	2,3-Dichlorotoluene		$C_7H_6Cl_2$	32768-54-0	161.029		6	207.5	1.2458^{20}	1.5511^{20}	vs bz
3300	2,4-Dichlorotoluene	2,4-Dichloro-1-methylbenzene	$C_7H_6Cl_2$	95-73-8	161.029	liq	-13.5	200(8)	1.2476^{20}	1.5511^{20}	i H_2O; s ctc
3301	2,5-Dichlorotoluene		$C_7H_6Cl_2$	19398-61-9	161.029		2.5	200	1.2535^{20}	1.5449^{20}	i H_2O; s bz
3302	2,6-Dichlorotoluene		$C_7H_6Cl_2$	118-69-4	161.029		25.8	194(11)	1.2686^{20}	1.5507^{20}	i H_2O; s chl
3303	3,4-Dichlorotoluene	1,2-Dichloro-4-methylbenzene	$C_7H_6Cl_2$	95-75-0	161.029	liq	-15.2(0.2)	208(18)	1.2564^{20}	1.5471^{20}	i H_2O; msc EtOH, eth, ace, bz, lig, ctc

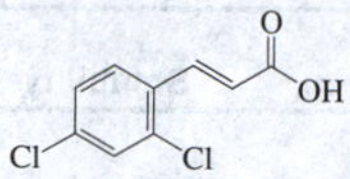
3268
3-(2,4-Dichlorophenyl)-2-propenoic acid

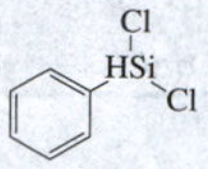
3269
Dichlorophenylsilane

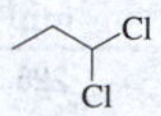
3270
1,1-Dichloropropane

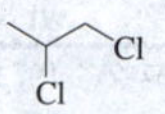
3271
1,2-Dichloropropane, (±)-

3272
1,3-Dichloropropane

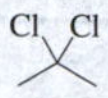
3273
2,2-Dichloropropane

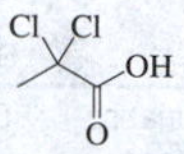
3274
2,2-Dichloropropanoic acid

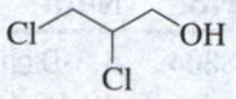
3275
2,3-Dichloro-1-propanol

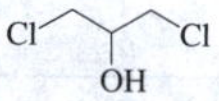
3276
1,3-Dichloro-2-propanol

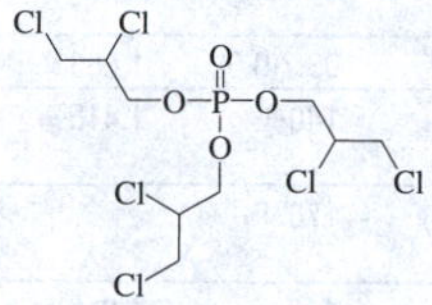
3277
2,3-Dichloro-1-propanol, phosphate (3:1)

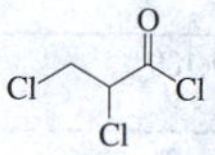
3278
2,3-Dichloropropanoyl chloride

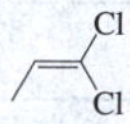
3279
1,1-Dichloropropene

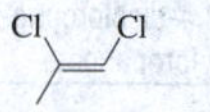
3280
cis-1,2-Dichloropropene

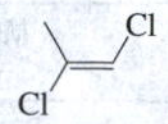
3281
trans-1,2-Dichloropropene

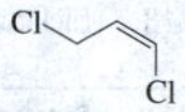
3282
cis-1,3-Dichloropropene

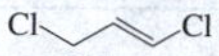
3283
trans-1,3-Dichloropropene

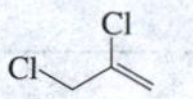
3284
2,3-Dichloropropene

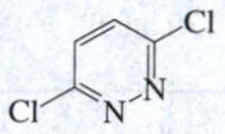
3285
3,6-Dichloropyridazine

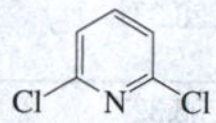
3286
2,6-Dichloropyridine

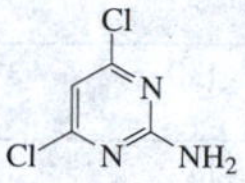
3287
4,6-Dichloro-2-pyrimidinamine

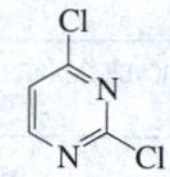
3288
2,4-Dichloropyrimidine

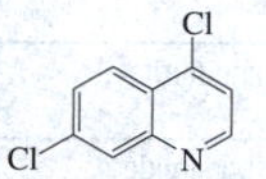
3289
4,7-Dichloroquinoline

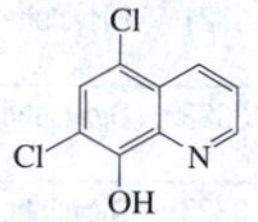
3290
5,7-Dichloro-8-quinolinol

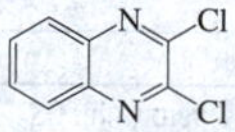
3291
2,3-Dichloroquinoxaline

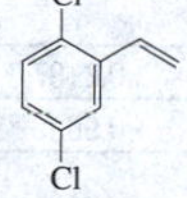
3292
2,5-Dichlorostyrene

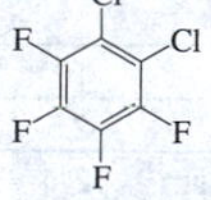
3293
1,2-Dichloro-3,4,5,6-tetrafluorobenzene

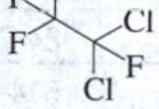
3294
1,1-Dichloro-1,2,2,2-tetrafluoroethane

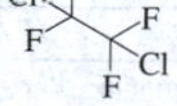
3295
1,2-Dichloro-1,1,2,2-tetrafluoroethane

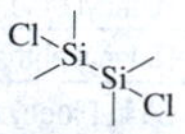
3296
1,2-Dichloro-1,1,2,2-tetramethyldisilane

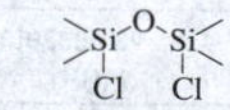
3297
1,3-Dichloro-1,1,3,3-tetramethyldisiloxane

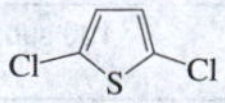
3298
2,5-Dichlorothiophene

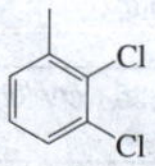
3299
2,3-Dichlorotoluene

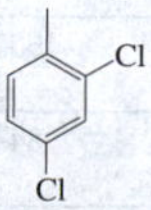
3300
2,4-Dichlorotoluene

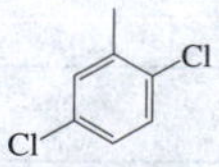
3301
2,5-Dichlorotoluene

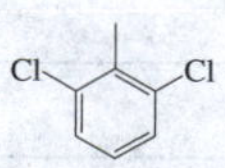
3302
2,6-Dichlorotoluene

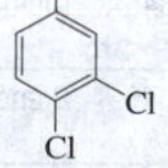
3303
3,4-Dichlorotoluene

No.	Name	Synonym	Mol. form.	CAS Reg. No.	Mol. wt.	Physical form	mp/°C	bp/°C	den/ g cm⁻³	n_D	Solubility
3304	1,3-Dichloro-1,3,5-triazine-2,4,6(1*H*,3*H*,5*H*)-trione	Dichlorocyanuric acid	$C_3HCl_2N_3O_3$	2782-57-2	197.964	cry	226.6				
3305	1,2-Dichloro-4-(trichloromethyl)benzene		$C_7H_3Cl_5$	13014-24-9	264.364		25.8	283.1	1.5913^{20}	1.5886^{20}	
3306	1,2-Dichloro-1,1,2-trifluoroethane	Refrigerant 123a	$C_2HCl_2F_3$	354-23-4	152.930	vol liq or gas	-78	30.0(0.1)	1.50^{25}		
3307	2,2-Dichloro-1,1,1-trifluoroethane	HCFC-123	$C_2HCl_2F_3$	306-83-2	152.930	vol liq or gas	-107	27.8(0.6)	1.4638^{25}		sl H_2O
3308	1,1-Dichloro-1,2,2-trifluoroethane	Refrigerant 123b	$C_2HCl_2F_3$	812-04-4	152.930			30.2			
3309	2,4-Dichloro-1-(trifluoromethyl)benzene	2,4-Dichlorobenzotrifluoride	$C_7H_3Cl_2F_3$	320-60-5	215.000					1.4802^{20}	
3310	4,5-Dichloro-2-(trifluoromethyl)-1*H*-benzimidazole	Chloroflurazole	$C_8H_3Cl_2F_3N_2$	3615-21-2	255.024		213.5				
3311	Dichlorovinylmethylsilane		$C_3H_6Cl_2Si$	124-70-9	141.072			93.7(0.5)	1.0868^{20}	1.4270^{20}	dec H_2O
3312	Dichlorvos	Phosphoric acid, 2,2-dichloroethenyl dimethyl ester	$C_4H_7Cl_2O_4P$	62-73-7	220.976			140^{20}	1.415^{25}		
3313	Diclofop-methyl	Methyl 2-[4-(2,4-dichlorophenoxy)phenoxy]propanoate	$C_{16}H_{14}Cl_2O_4$	51338-27-3	341.186		42.1(0.5)	$176^{0.1}$			
3314	Dicrotophos		$C_8H_{16}NO_5P$	141-66-2	237.191			400	1.216^{15}		
3315	Dicumarol		$C_{19}H_{12}O_6$	66-76-2	336.294	nd	290				
3316	Dicyanamide	Cyanocyanamide	C_2HN_3	504-66-5	67.049	aq soln only					
3317	*o*-Dicyanobenzene	*o*-Phthalodinitrile	$C_8H_4N_2$	91-15-6	128.131	nd (w, lig)	140.6(0.9)	150^{10}	1.1250^{25}		sl H_2O, lig; vs EtOH, bz; s eth, ace
3318	*m*-Dicyanobenzene	*m*-Phthalodinitrile	$C_8H_4N_2$	626-17-5	128.131	nd(al)	162	sub	0.992^{40}		sl H_2O; vs EtOH; s eth, bz, chl; i peth
3319	*p*-Dicyanobenzene	*p*-Phthalodinitrile	$C_8H_4N_2$	623-26-7	128.131	nd (w, MeOH)	224	sub			i H_2O; sl EtOH, eth; s bz; vs HOAc
3320	Dicyclohexyl adipate	Dicyclohexyl hexanedioate	$C_{18}H_{30}O_4$	849-99-0	310.429		35(1)				s chl
3321	Dicyclohexylamine	*N*-Cyclohexylcyclohexanamine	$C_{12}H_{23}N$	101-83-7	181.318	liq	-0.1	251(4)	0.9123^{20}	1.4842^{20}	sl H_2O, ctc; s EtOH, eth, bz
3322	Dicyclohexylamine nitrite	*N*-Cyclohexylcyclohexanamine, nitrite	$C_{12}H_{24}N_2O_2$	3129-91-7	228.331	cry	179.5(0.6)				
3323	Dicyclohexylcarbodiimide		$C_{13}H_{22}N_2$	538-75-0	206.327		34.5	123^{6}			
3324	Dicyclohexyl disulfide		$C_{12}H_{22}S_2$	2550-40-5	230.433	liq		195^{20}			
3325	Dicyclohexyl ether		$C_{12}H_{22}O$	4645-15-2	182.302	liq	-36	242.5	0.9227^{20}	1.4741^{20}	
3326	Dicyclohexylmethanone		$C_{13}H_{22}O$	119-60-8	194.313		57	159^{20}	0.986^{0}	1.4860^{20}	s eth, ace, ctc
3327	Dicyclohexylphosphine		$C_{12}H_{23}P$	829-84-5	198.285			281	0.904^{25}	1.5163^{20}	
3328	Dicyclohexyl phthalate		$C_{20}H_{26}O_4$	84-61-7	330.418	pr (al)	66	225^{4}	1.383^{20}	1.431^{20}	i H_2O; s EtOH, eth; sl chl
3329	*N,N'*-Dicyclohexylthiourea		$C_{13}H_{24}N_2S$	1212-29-9	240.408	cry (MeOH)	180				
3330	*N,N'*-Dicyclohexylurea		$C_{13}H_{24}N_2O$	2387-23-7	224.342		233.8				
3331	Dicyclomine hydrochloride	Dicycloverine hydrochloride	$C_{19}H_{36}ClNO_2$	67-92-5	345.948	cry	165				
3332	Dicyclopentadiene		$C_{10}H_{12}$	1755-01-7	132.202		32	170 dec	0.9302^{35}	1.5050^{35}	vs eth, EtOH
3333	Dicyclopentyl ether	Cyclopentyl ether	$C_{10}H_{18}O$	10137-73-2	154.249	liq		80^{13}			
3334	Dicyclopropyl ketone		$C_7H_{10}O$	1121-37-5	110.153			161	0.977^{25}	1.4670^{20}	
3335	Didecylamine	*N*-Decyl-1-decanamine	$C_{20}H_{43}N$	1120-49-6	297.562			359.0			
3336	Didecyl ether		$C_{20}H_{42}O$	2456-28-2	298.546		16	$196^{15.5}$	0.8187^{20}		
3337	Didecyl phthalate		$C_{28}H_{46}O_4$	84-77-5	446.663		2.5	240^{3}	0.9639^{20}		
3338	3',4'-Didehydro-β,ψ-caroten-16'-oic acid	Torularhodin	$C_{40}H_{52}O_2$	514-92-1	564.840	purp nd (MeOH-eth)	211				vs py, chl, CS_2
3339	2',3'-Dideoxyinosine	Didanosine	$C_{10}H_{12}N_4O_3$	69655-05-6	236.227	wh cry (EtOH aq)	162				
3340	2,6-Dideoxy-3-*O*-methyl-*ribo*-hexose	Cymarose	$C_7H_{14}O_4$	579-04-4	162.184	pr (eth-peth) nd (ace)	101				vs H_2O, ace, EtOH
3341	Didodecanoyl peroxide	Lauroyl peroxide	$C_{24}H_{46}O_4$	105-74-8	398.620	wh pl	49				i H_2O; s chl
3342	Didodecylamine	*N*-Dodecyl-1-dodecanamine	$C_{24}H_{51}N$	3007-31-6	353.669		46.9(0.5)	263^{27}			vs bz, eth, EtOH, chl
3343	Didodecyl phosphate		$C_{24}H_{51}O_4P$	7057-92-3	434.633	cry (MeOH)	59				
3344	Didodecyl phthalate	1,2-Benzenedicarboxylic acid, didodecyl ester	$C_{32}H_{54}O_4$	2432-90-8	502.769		22.0	256^{1}	0.9389^{20}		
3345	Dieldrin		$C_{12}H_8Cl_6O$	60-57-1	380.909		178.8(0.3)		1.75^{25}		i H_2O; sl EtOH; s ace, bz
3346	Dienestrol		$C_{18}H_{18}O_2$	84-17-3	266.335	cry (dil al)	227.5	130 sub			vs ace, eth, EtOH
3347	1,2:8,9-Diepoxy-*p*-menthane	Limonene diepoxide	$C_{10}H_{16}O_2$	96-08-2	168.233		242				

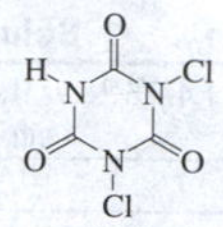
3304
1,3-Dichloro-1,3,5-triazine-2,4,6(1*H*,3*H*,5*H*)-trione

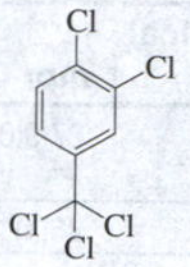
3305
1,2-Dichloro-4-(trichloromethyl)benzene

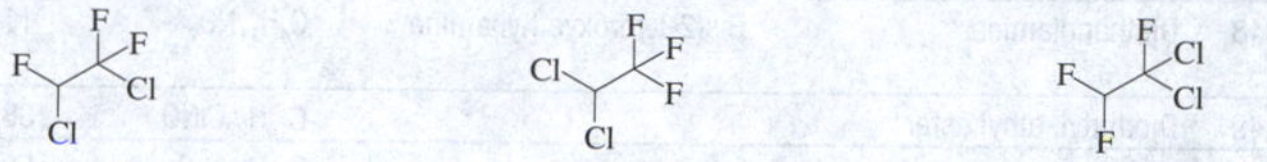
3306
1,2-Dichloro-1,1,2-trifluoroethane

3307
2,2-Dichloro-1,1,1-trifluoroethane

3308
1,1-Dichloro-1,2,2-trifluoroethane

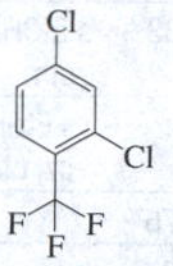
3309
2,4-Dichloro-1-(trifluoromethyl)benzene

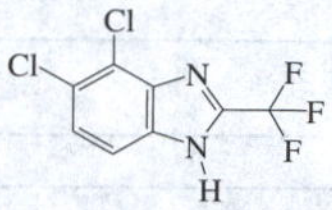
3310
4,5-Dichloro-2-(trifluoromethyl)-1*H*-benzimidazole

3311
Dichlorovinylmethylsilane

3312
Dichlorvos

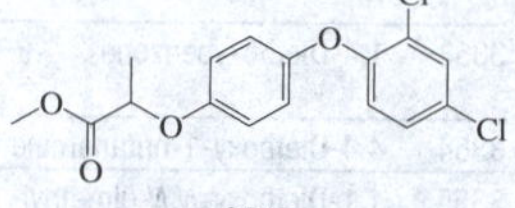
3313
Diclofop-methyl

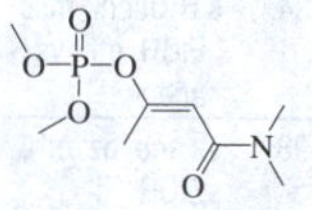
3314
Dicrotophos

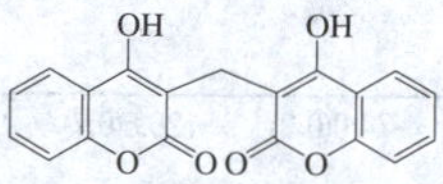
3315
Dicumarol

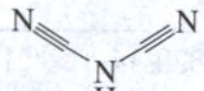
3316
Dicyanamide

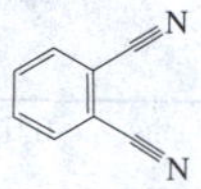
3317
o-Dicyanobenzene

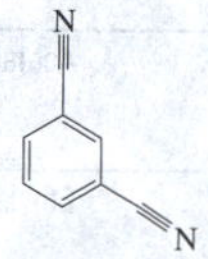
3318
m-Dicyanobenzene

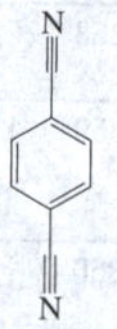
3319
p-Dicyanobenzene

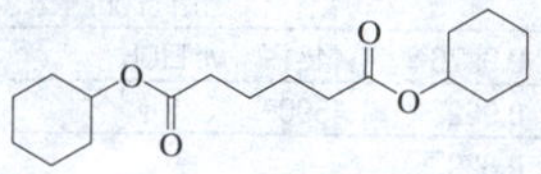
3320
Dicyclohexyl adipate

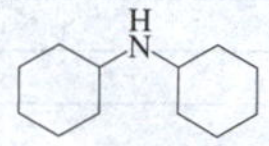
3321
Dicyclohexylamine

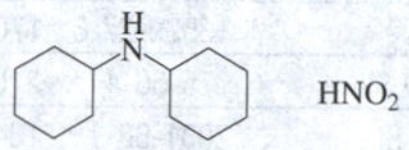
3322
Dicyclohexylamine nitrite

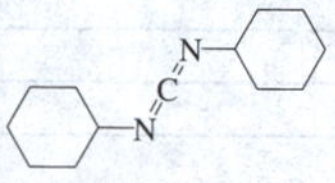
3323
Dicyclohexylcarbodiimide

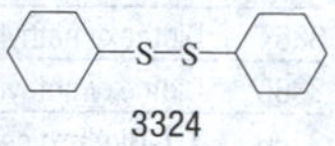
3324
Dicyclohexyl disulfide

3325
Dicyclohexyl ether

3326
Dicyclohexylmethanone

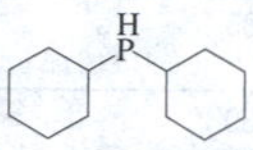
3327
Dicyclohexylphosphine

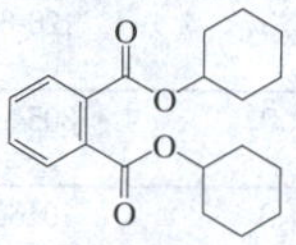
3328
Dicyclohexyl phthalate

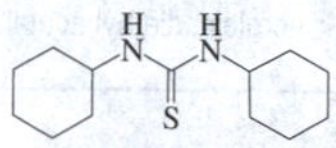
3329
N,N'-Dicyclohexylthiourea

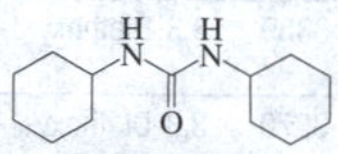
3330
N,N'-Dicyclohexylurea

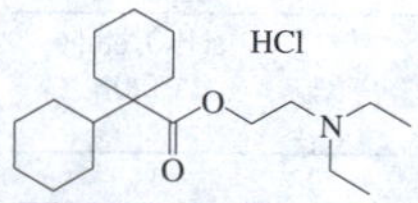
3331
Dicyclomine hydrochloride

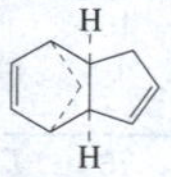
3332
Dicyclopentadiene

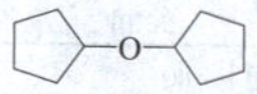
3333
Dicyclopentyl ether

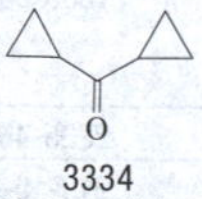
3334
Dicyclopropyl ketone

3335
Didecylamine

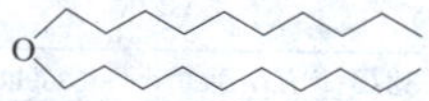
3336
Didecyl ether

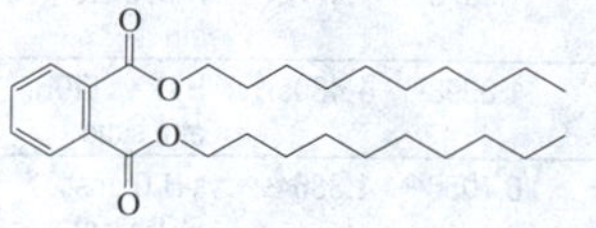
3337
Didecyl phthalate

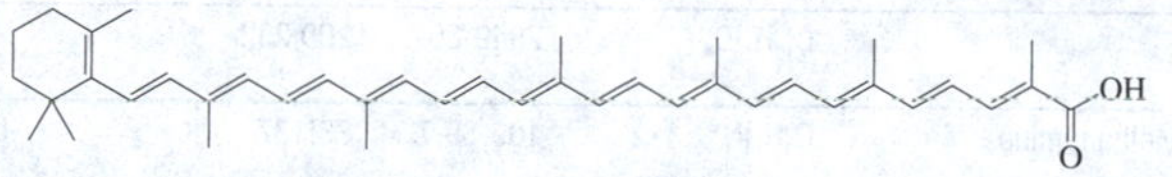
3338
3',4'-Didehydro-β,ψ-caroten-16'-oic acid

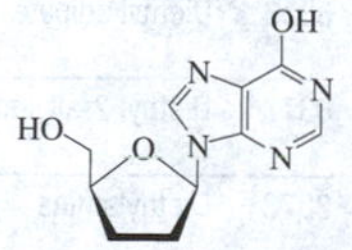
3339
2',3'-Dideoxyinosine

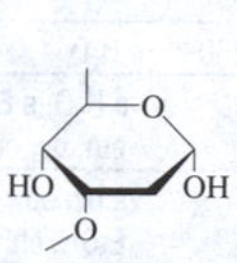
3340
2,6-Dideoxy-3-*O*-methyl-*ribo*-hexose

3341
Didodecanoyl peroxide

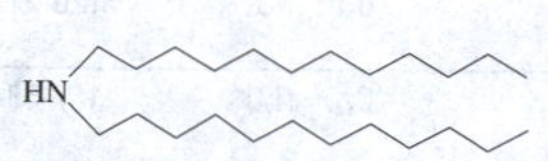
3342
Didodecylamine

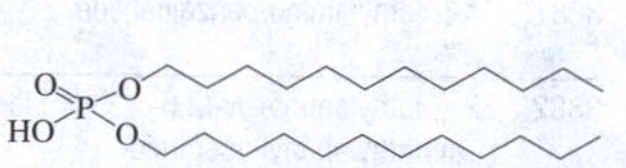
3343
Didodecyl phosphate

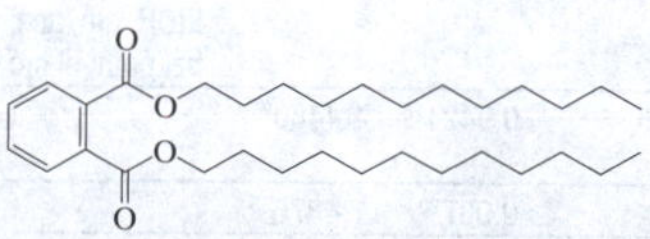
3344
Didodecyl phthalate

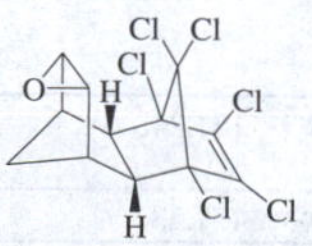
3345
Dieldrin

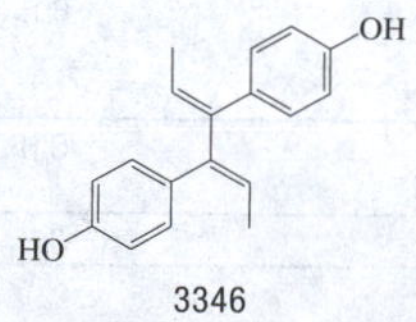
3346
Dienestrol

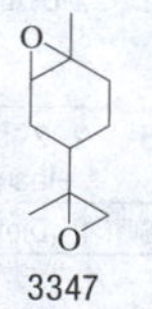
3347
1,2:8,9-Diepoxy-*p*-menthane

No.	Name	Synonym	Mol. form.	CAS Reg. No.	Mol. wt.	Physical form	mp/°C	bp/°C	den/ g cm^{-3}	n_D	Solubility
3348	Diethanolamine	Bis(2-hydroxyethyl)amine	$C_4H_{11}NO_2$	111-42-2	105.136		27.9(0.2)	271.2(0.7)	1.0966^{20}	1.4776^{20}	vs H_2O, EtOH; sl eth, bz
3349	Diethatyl, ethyl ester		$C_{16}H_{22}ClNO_3$	38727-55-8	311.804	cry	46(1)				
3350	4,4'-Diethoxyazobenzene		$C_{16}H_{18}N_2O_2$	588-52-3	270.326	ye lf (al)	162	dec			i H_2O; sl EtOH; s eth, bz, chl; vs HOAc
3351	3,4-Diethoxybenzaldehyde		$C_{11}H_{14}O_3$	2029-94-9	194.227		22	279	1.0100^{22}		vs EtOH
3352	1,2-Diethoxybenzene		$C_{10}H_{14}O_2$	2050-46-6	166.217	pr (peth, dil al)	44	219	1.0075^{20}	1.5083^{25}	s EtOH, ctc; vs eth
3353	1,4-Diethoxybenzene		$C_{10}H_{14}O_2$	122-95-2	166.217	pl (dil al)	72	246			vs EtOH; s eth, bz, ctc, chl
3354	4,4-Diethoxy-1-butanamine		$C_8H_{19}NO_2$	6346-09-4	161.243			196	0.933^{25}	1.4275^{20}	
3355	1,1-Diethoxy-*N,N*-dimethyl-methanamine		$C_7H_{17}NO_2$	1188-33-6	147.216			129	0.859^{25}	1.4007^{20}	
3356	Diethoxydimethylsilane	Dimethyldiethoxysilane	$C_6H_{16}O_2Si$	78-62-6	148.276	liq	-87	113(3)	0.865^{25}	1.3811^{20}	s ctc
3357	Diethoxydiphenylsilane		$C_{16}H_{20}O_2Si$	2553-19-7	272.415			296(3)	1.0329^{20}	1.5269^{20}	
3358	2,2-Diethoxyethanamine		$C_6H_{15}NO_2$	645-36-3	133.189	liq	-78	163	0.9159^{25}	1.4123^{25}	vs H_2O, eth, EtOH, chl
3359	1,1-Diethoxyethane	Acetal	$C_6H_{14}O_2$	105-57-7	118.174	liq	-106.1(0.6)	102(2)	0.8254^{20}	1.3834^{20}	s H_2O, chl; msc EtOH, eth; vs ace
3360	1,2-Diethoxyethane	Ethylene glycol diethyl ether	$C_6H_{14}O_2$	629-14-1	118.174	liq	-74.0(0.2)	120.6(0.7)	0.8351^{25}	1.3898^{25}	vs ace, bz, eth, EtOH
3361	1,1-Diethoxyethene		$C_6H_{12}O_2$	2678-54-8	116.158			68^{100}	0.7932^{20}	1.3643^{21}	
3362	Diethoxymethane		$C_5H_{12}O_2$	462-95-3	104.148	liq	-66(2)	86(2)	0.8319^{20}	1.3748^{18}	s H_2O; msc EtOH; vs ace, bz; sl chl
3363	2-(Diethoxymethyl)furan		$C_9H_{14}O_3$	13529-27-6	170.205			191.5	0.9976^{20}	1.4451^{20}	vs EtOH
3364	Diethoxymethylphenylsilane		$C_{11}H_{18}O_2Si$	775-56-4	210.346			217(3)	0.9627^{20}	1.4690^{20}	
3365	Diethoxymethylsilane		$C_5H_{14}O_2Si$	2031-62-1	134.250			98	0.829^{25}		
3366	1,1-Diethoxypentane		$C_9H_{20}O_2$	3658-79-5	160.254			59^{12}	0.829^{22}	1.4029^{22}	
3367	1,1-Diethoxypropane		$C_7H_{16}O_2$	4744-08-5	132.201			121(1)	0.825^{20}	1.3924^{19}	s H_2O, ace, bz; vs EtOH, eth
3368	2,2-Diethoxypropane		$C_7H_{16}O_2$	126-84-1	132.201			101(5)	0.8200^{21}	1.3891^{20}	s EtOH, ace, bz; vs eth; sl ctc
3369	3,3-Diethoxy-1-propene	Acrolein, diethyl acetal	$C_7H_{14}O_2$	3054-95-3	130.185			123.5	0.8543^{15}	1.4000^{20}	sl H_2O; msc EtOH, eth
3370	3,3-Diethoxy-1-propyne		$C_7H_{12}O_2$	10160-87-9	128.169			139	0.8942^{22}	1.4140^{20}	vs ace, eth, EtOH, chl
3371	*N,N*-Diethylacetamide		$C_6H_{13}NO$	685-91-6	115.173			199(10)	0.9130^{17}	1.4374^{17}	s H_2O, EtOH; msc eth, ace, bz; sl ctc
3372	Diethyl 2-acetamidomalonate		$C_9H_{15}NO_5$	1068-90-2	217.219	cry (al,bz-peth)	96.3	185^{20}			sl H_2O, eth; s tfa, EtOH
3373	*N,N*-Diethylacetoacetamide		$C_8H_{15}NO_2$	2235-46-3	157.211	liq		76^{13}			
3374	Diethyl acetylphosphonate		$C_6H_{13}O_4P$	919-19-7	180.138			114^{20}	1.1005^{20}	1.4200^{26}	
3375	Diethyl 2-acetylsuccinate		$C_{10}H_{16}O_5$	1115-30-6	216.231			255	1.081^{20}	1.4346^{20}	i H_2O; s EtOH, eth, bz; sl chl
3376	Diethyl adipate	Diethyl hexanedioate	$C_{10}H_{18}O_4$	141-28-6	202.248	liq	-20(2)	250(11)	1.0076^{20}	1.4272^{20}	i H_2O; s EtOH, eth
3377	Diethyl 2-allylmalonate		$C_{10}H_{16}O_4$	2049-80-1	200.232			222.5	1.0098^{20}	1.4305^{20}	i H_2O; vs EtOH, eth; s ctc
3378	Diethylamine	*N*-Ethylethanamine	$C_4H_{11}N$	109-89-7	73.137	liq	-50(2)	55.4(0.1)	0.7056^{20}	1.3864^{20}	vs H_2O; msc EtOH; s eth, ctc
3379	Diethylamine hydrochloride	*N*-Ethylethanamine hydrochloride	$C_4H_{12}ClN$	660-68-4	109.598	lf (al-eth)	228.5		1.0477^{22}		vs H_2O, EtOH
3380	(Diethylamino)acetonitrile		$C_6H_{12}N_2$	3010-02-4	112.172			169(6)	0.8660^{20}	1.4260^{20}	s H_2O
3381	4-(Diethylamino)benzaldehyde		$C_{11}H_{15}NO$	120-21-8	177.243	ye nd (w)	41	172^{10}			vs H_2O; s EtOH, eth, bz, ctc
3382	2-(Diethylamino)-*N*-(2,6-dimethylphenyl)acetamide	Lidocaine	$C_{14}H_{22}N_2O$	137-58-6	234.337	nd (bz, al)	68(1)	181^4			vs bz, eth, EtOH, chl
3383	2-(Diethylamino)-*N*-(2,6-dimethylphenyl)acetamide, monohydrochloride		$C_{14}H_{23}ClN_2O$	73-78-9	270.798		128				vs H_2O
3384	2-Diethylaminoethanol		$C_6H_{15}NO$	100-37-8	117.189	hyg		162.13(0.09)	0.8921^{20}	1.4412^{20}	msc H_2O; s EtOH, eth, ace, bz, peth; sl ctc
3385	2-[2-(Diethylamino)ethoxy]-ethanol		$C_8H_{19}NO_2$	140-82-9	161.243			221.5	0.9421^{25}	1.4480^{20}	
3386	2-(Diethylamino)ethyl acrylate		$C_9H_{17}NO_2$	2426-54-2	171.237		<-60	81^{10}	0.937^{20}	1.4376^{25}	

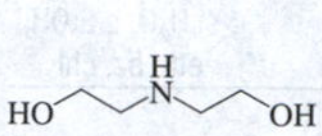
3348
Diethanolamine

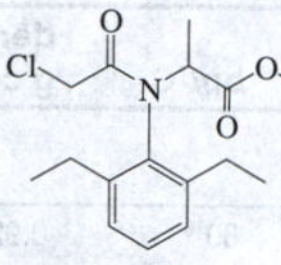
3349
Diethatyl, ethyl ester

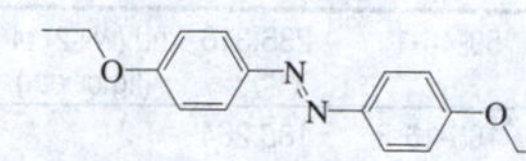
3350
4,4'-Diethoxyazobenzene

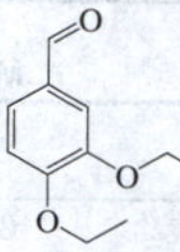
3351
3,4-Diethoxybenzaldehyde

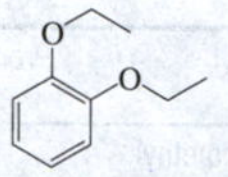
3352
1,2-Diethoxybenzene

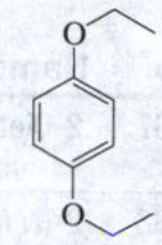
3353
1,4-Diethoxybenzene

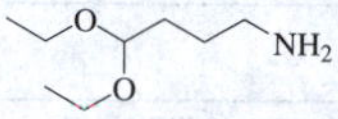
3354
4,4-Diethoxy-1-butanamine

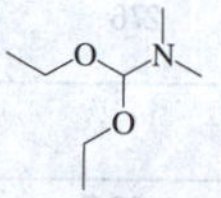
3355
1,1-Diethoxy-*N*,*N*-dimethylmethanamine

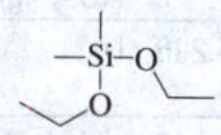
3356
Diethoxydimethylsilane

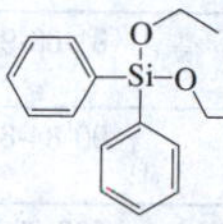
3357
Diethoxydiphenylsilane

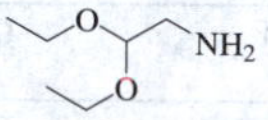
3358
2,2-Diethoxyethanamine

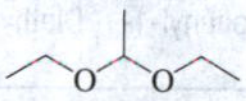
3359
1,1-Diethoxyethane

3360
1,2-Diethoxyethane

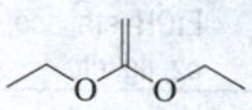
3361
1,1-Diethoxyethene

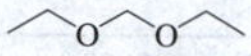
3362
Diethoxymethane

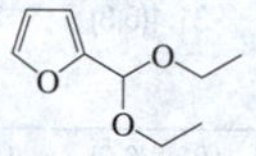
3363
2-(Diethoxymethyl)furan

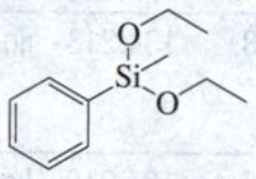
3364
Diethoxymethylphenylsilane

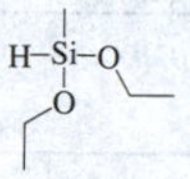
3365
Diethoxymethylsilane

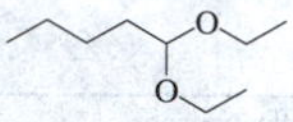
3366
1,1-Diethoxypentane

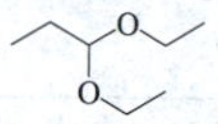
3367
1,1-Diethoxypropane

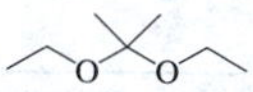
3368
2,2-Diethoxypropane

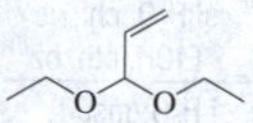
3369
3,3-Diethoxy-1-propene

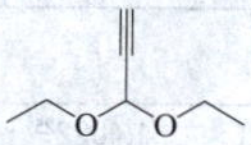
3370
3,3-Diethoxy-1-propyne

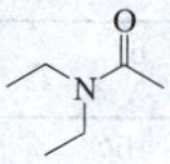
3371
N,*N*-Diethylacetamide

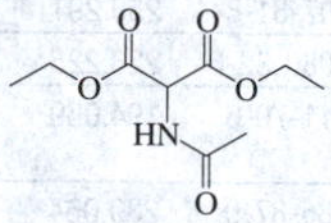
3372
Diethyl 2-acetamidomalonate

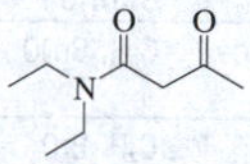
3373
N,*N*-Diethylacetoacetamide

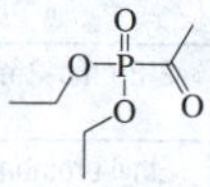
3374
Diethyl acetylphosphonate

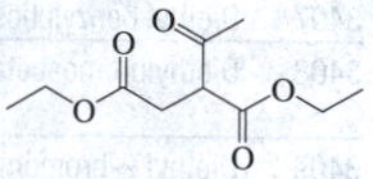
3375
Diethyl 2-acetylsuccinate

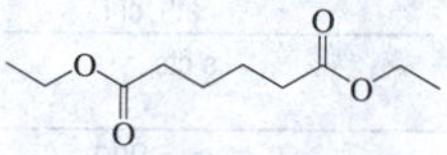
3376
Diethyl adipate

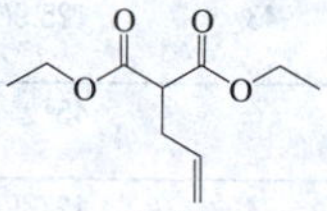
3377
Diethyl 2-allylmalonate

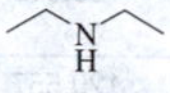
3378
Diethylamine

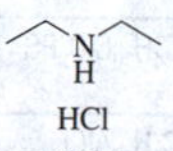
3379
Diethylamine hydrochloride

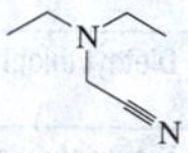
3380
(Diethylamino)acetonitrile

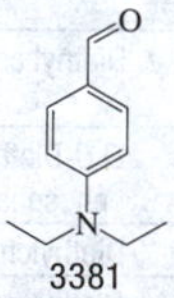
3381
4-(Diethylamino)benzaldehyde

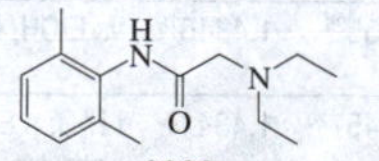
3382
2-(Diethylamino)-*N*-(2,6-dimethylphenyl)acetamide

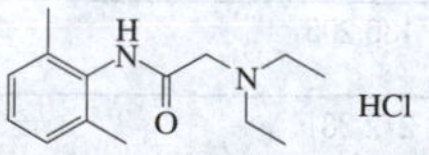
3383
2-(Diethylamino)-*N*-(2,6-dimethylphenyl)acetamide, monohydrochloride

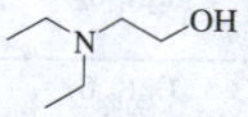
3384
2-Diethylaminoethanol

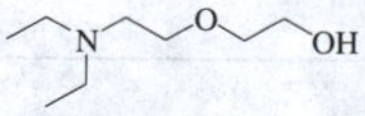
3385
2-[2-(Diethylamino)ethoxy]ethanol

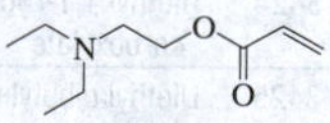
3386
2-(Diethylamino)ethyl acrylate

No.	Name	Synonym	Mol. form.	CAS Reg. No.	Mol. wt.	Physical form	mp/°C	bp/°C	den/g cm^{-3}	n_D	Solubility
3387	2-Diethylaminoethyl 4-aminobenzoate	Procaine	$C_{13}H_{20}N_2O_2$	59-46-1	236.310	nd (w+2) pl (lig or eth)	61				sl H_2O; s EtOH, eth, bz, chl
3388	2-(*N,N*-Diethylamino)ethyl methacrylate		$C_{10}H_{19}NO_2$	105-16-8	185.264			80^{10}	0.92^{30}		
3389	2-(Diethylamino)ethyl 2-phenylbutanoate	Butethamate	$C_{16}H_{25}NO_2$	14007-64-8	263.376			168^{11}		1.4909^{20}	
3390	4-(Diethylamino)-2-hydroxybenzaldehyde		$C_{11}H_{15}NO_2$	17754-90-4	193.243		65.0				
3391	Diethyl 2-aminomalonate		$C_7H_{13}NO_4$	6829-40-9	175.183			122^{16}	1.100^{16}	1.4353^{16}	vs H_2O, EtOH, eth; s ace, bz; i lig
3392	7-(Diethylamino)-4-methyl-2*H*-1-benzopyran-2-one		$C_{14}H_{17}NO_2$	91-44-1	231.291	cry (al, bz-lig)					sl H_2O; s EtOH, eth, ace
3393	3-(Diethylamino)phenol		$C_{10}H_{15}NO$	91-68-9	165.232	orth bipym (CS_2-lig)	78	276			s H_2O, EtOH, eth, CS_2; sl lig
3394	2-(Diethylamino)-1-phenyl-1-propanone	Diethylpropion	$C_{13}H_{19}NO$	90-84-6	205.296	liq		111^{14}			
3395	3-(Diethylamino)-1-propanol		$C_7H_{17}NO$	622-93-5	131.216			189.5	0.8600^{20}	1.4439^{20}	s EtOH; s eth, ace, bz; sl chl
3396	3-(Diethylamino)-1-propyne	*N,N*-Diethyl-2-propargylamine	$C_7H_{13}N$	4079-68-9	111.185	liq		120			
3397	2,6-Diethylaniline		$C_{10}H_{15}N$	579-66-8	149.233		1.5	251(8)	0.906^{25}	1.5452^{20}	
3398	*N,N*-Diethylaniline		$C_{10}H_{15}N$	91-66-7	149.233	ye oil	-21.3(0.2)	216(1)	0.9307^{20}	1.5409^{20}	sl H_2O; s EtOH, ace, ctc; vs eth, chl
3399	Diethylarsine		$C_4H_{11}As$	692-42-2	134.052			105	1.1338^{24}	1.4709	vs ace, bz, eth, EtOH
3400	*N,N*-Diethylbenzamide		$C_{11}H_{15}NO$	1696-17-9	177.243			132^{5}			
3401	*o*-Diethylbenzene	1,2-Diethylbenzene	$C_{10}H_{14}$	135-01-3	134.218	liq	-31.4(0.3)	183.4(0.4)	0.8800^{20}	1.5035^{20}	i H_2O; msc EtOH, eth, ace, bz, lig, ctc
3402	*m*-Diethylbenzene	1,3-Diethylbenzene	$C_{10}H_{14}$	141-93-5	134.218	liq	-83.9(0.2)	181.1(0.5)	0.8602^{20}	1.4955^{20}	i H_2O; msc EtOH, eth, ace, bz, lig, ctc
3403	*p*-Diethylbenzene	1,4-Diethylbenzene	$C_{10}H_{14}$	105-05-5	134.218	liq	-43.3(0.4)	184(1)	0.8620^{20}	1.4967^{20}	i H_2O; msc EtOH, eth, ace, bz, lig, ctc
3404	*N,N*-Diethyl-1,4-benzenediamine		$C_{10}H_{16}N_2$	93-05-0	164.247			261			vs bz
3405	Diethyl benzylidenemalonate	Diethyl benzalmalonate	$C_{14}H_{16}O_4$	5292-53-5	248.275		32	216^{30}	1.1045^{20}	1.5389^{20}	i H_2O; s EtOH, eth, ace, bz
3406	Diethyl benzylmalonate		$C_{14}H_{18}O_4$	607-81-8	250.291			300	1.076^{15}	1.4872^{20}	i H_2O; sl chl
3407	Diethyl benzylphosphonate		$C_{11}H_{17}O_3P$	1080-32-6	228.225			110^{2}		1.4930^{20}	s ctc
3408	Diethylbromoacetamide	2-Bromo-2-ethylbutanamide	$C_6H_{12}BrNO$	511-70-6	194.069		67				sl H_2O, chl; vs EtOH, eth, bz
3409	Diethyl 2-bromomalonate	Ethyl bromomalonate	$C_7H_{11}BrO_4$	685-87-0	239.064		-54	231(12)	1.4022^{25}	1.4521^{20}	i H_2O; msc EtOH, eth; s ace, ctc
3410	*N,N*-Diethylbutanamide		$C_8H_{17}NO$	1114-76-7	143.227			206	0.8884^{20}	1.4403^{25}	vs H_2O, EtOH
3411	Diethyl 2-butylmalonate	Pentane-1,1-dicarboxylic acid, diethyl ester	$C_{11}H_{20}O_4$	133-08-4	216.275			237(8)	0.9764^{20}	1.4250^{20}	vs EtOH, eth
3412	Diethyl 2-butynedioate		$C_8H_{10}O_4$	762-21-0	170.163		0.8	184^{200}	1.0075^{20}	1.4425^{20}	s EtOH, eth, ctc
3413	Diethylcarbamazine citrate		$C_{16}H_{29}N_3O_8$	1642-54-2	391.416	cry	138				
3414	Diethylcarbamic chloride		$C_5H_{10}ClNO$	88-10-8	135.592			186			
3415	*N,N'*-Diethylcarbanilide		$C_{17}H_{20}N_2O$	85-98-3	268.353	cry (al)	71(2)				i H_2O; vs EtOH; s chl
3416	Diethyl carbonate	Ethyl carbonate	$C_5H_{10}O_3$	105-58-8	118.131	liq	-43	125.9(0.9)	0.9692^{25}	1.3845^{20}	i H_2O; s EtOH, eth, chl
3417	*O,O*-Diethyl chloridothionophosphate	Diethyl thiophosphoryl chloride	$C_4H_{10}ClO_2PS$	2524-04-1	188.613			45^{3}			s ctc
3418	Diethylchloroaluminum	Diethylaluminum chloride	$C_4H_{10}AlCl$	96-10-6	120.557	col liq	-74	127^{50}	0.96		reac H2O
3419	Diethyl chloromalonate	Ethyl chloromalonate	$C_7H_{11}ClO_4$	14064-10-9	194.613			222	1.2040^{20}	1.4327^{20}	i H_2O; msc EtOH, eth, chl; s CS_2
3420	Diethyl chlorophosphonate	Diethoxyphosphoryl chloride	$C_4H_{10}ClO_3P$	814-49-3	172.547			93.5	1.205^{19}	1.4170^{20}	
3421	Diethylcyanamide		$C_5H_{10}N_2$	617-83-4	98.146	liq	-80.7(0.6)	187(15)	0.854^{20}	1.4126^{25}	i H_2O; s EtOH, eth
3422	Diethyl 1,1-cyclobutanedicarboxylate		$C_{10}H_{16}O_4$	3779-29-1	200.232			223(5)	1.0456^{20}	1.4330^{26}	vs EtOH; sl ctc
3423	1,1-Diethylcyclohexane		$C_{10}H_{20}$	78-01-3	140.266			178(6)			
3424	Diethyl 1,1-cyclopropanedicarboxylate		$C_9H_{14}O_4$	1559-02-0	186.205			214(7)	1.055^{25}	1.4345^{18}	vs EtOH, eth
3425	Diethyl dibutylmalonate		$C_{15}H_{28}O_4$	596-75-8	272.381			150^{12}	0.9457^{20}	1.4341^{20}	i H_2O; s EtOH, eth, ctc

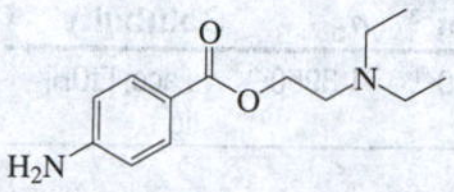
3387
2-Diethylaminoethyl 4-aminobenzoate

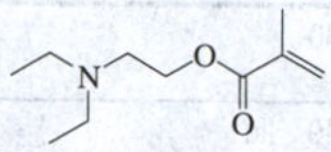
3388
2-(*N,N*-Diethylamino)ethyl methacrylate

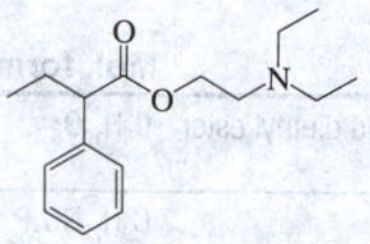
3389
2-(Diethylamino)ethyl 2-phenylbutanoate

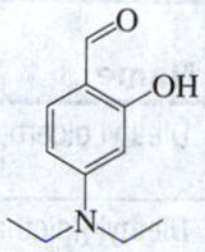
3390
4-(Diethylamino)-2-hydroxybenzaldehyde

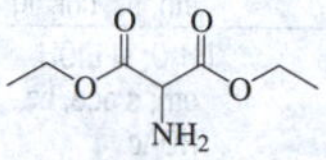
3391
Diethyl 2-aminomalonate

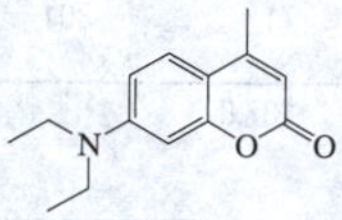
3392
7-(Diethylamino)-4-methyl-2*H*-1-benzopyran-2-one

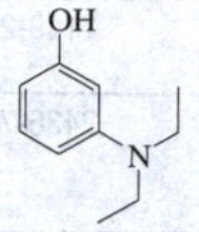
3393
3-(Diethylamino)phenol

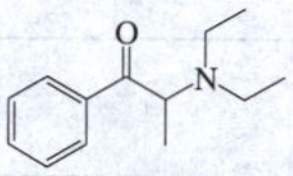
3394
2-(Diethylamino)-1-phenyl-1-propanone

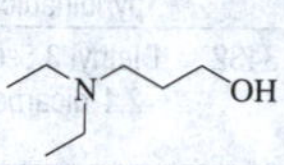
3395
3-(Diethylamino)-1-propanol

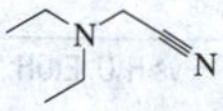
3396
3-(Diethylamino)-1-propyne

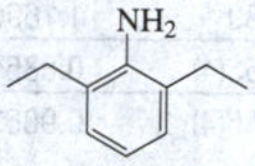
3397
2,6-Diethylaniline

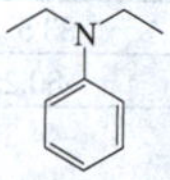
3398
N,N-Diethylaniline

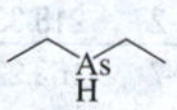
3399
Diethylarsine

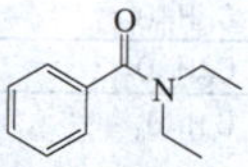
3400
N,N-Diethylbenzamide

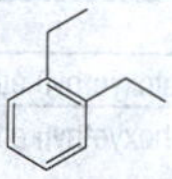
3401
o-Diethylbenzene

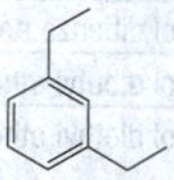
3402
m-Diethylbenzene

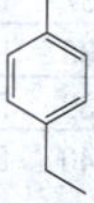
3403
p-Diethylbenzene

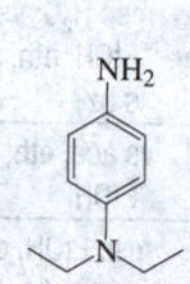
3404
N,N-Diethyl-1,4-benzenediamine

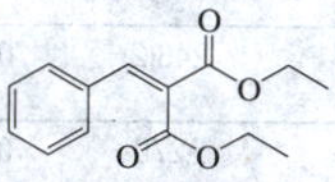
3405
Diethyl benzylidenemalonate

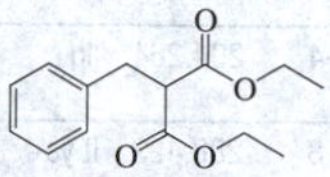
3406
Diethyl benzylmalonate

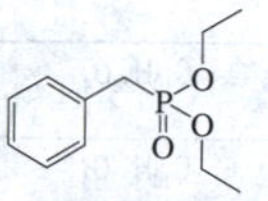
3407
Diethyl benzylphosphonate

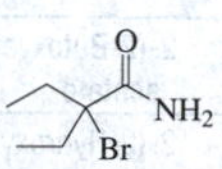
3408
Diethylbromoacetamide

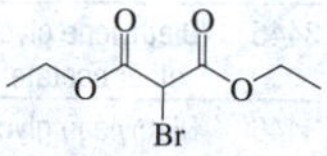
3409
Diethyl 2-bromomalonate

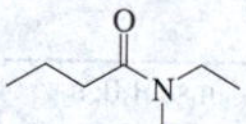
3410
N,N-Diethylbutanamide

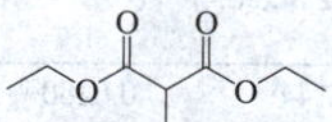
3411
Diethyl 2-butylmalonate

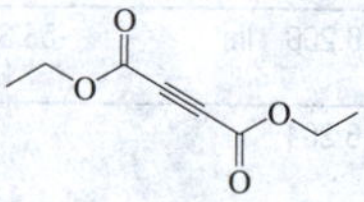
3412
Diethyl 2-butynedioate

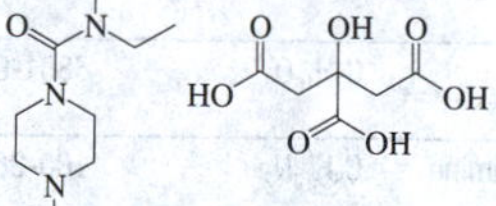
3413
Diethylcarbamazine citrate

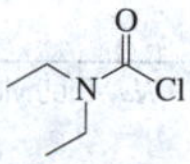
3414
Diethylcarbamic chloride

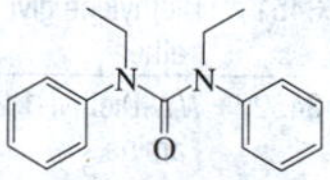
3415
N,N'-Diethylcarbanilide

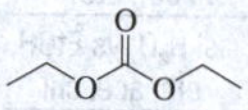
3416
Diethyl carbonate

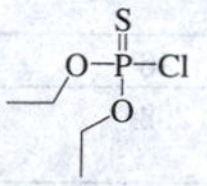
3417
O,O-Diethyl chloridothionophosphate

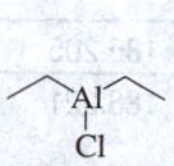
3418
Diethylchloroaluminum

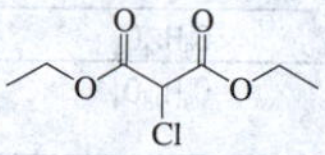
3419
Diethyl chloromalonate

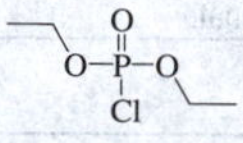
3420
Diethyl chlorophosphonate

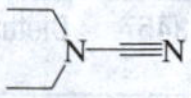
3421
Diethylcyanamide

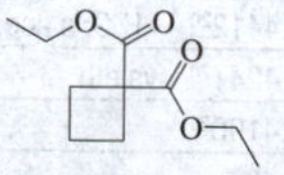
3422
Diethyl 1,1-cyclobutanedicarboxylate

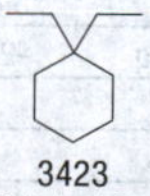
3423
1,1-Diethylcyclohexane

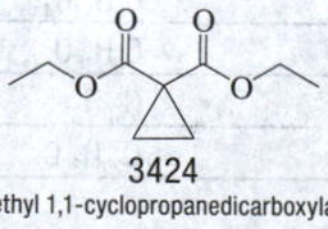
3424
Diethyl 1,1-cyclopropanedicarboxylate

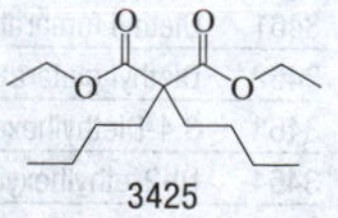
3425
Diethyl dibutylmalonate

No.	Name	Synonym	Mol. form.	CAS Reg. No.	Mol. wt.	Physical form	mp/°C	bp/°C	den/ g cm^{-3}	n_D	Solubility
3426	Diethyl dicarbonate	Pyrocarbonic acid diethyl ester	$C_6H_{10}O_5$	1609-47-8	162.140			93^{18}	1.120^{20}	1.3960^{20}	vs ace, EtOH, lig
3427	Diethyl [(diethanolamino)-methyl]phosphonate		$C_9H_{22}NO_5P$	2781-11-5	255.249	liq		$150^{0.01}$			
3428	5,5-Diethyldihydro-2*H*-1,3-oxazine-2,4(3*H*)-dione	Diethadione	$C_8H_{13}NO_3$	702-54-5	171.194	cry (eth)	97.5				
3429	Diethyl 1,4-dihydro-2,4,6-trimethyl-3,5-pyridinedicarboxylate	3,5-Diethoxycarbonyl-1,4-dihydrocollidine	$C_{14}H_{21}NO_4$	632-93-9	267.322	lt bl flr pl (al)	131				sl H_2O, EtOH, eth, CS_2; vs chl
3430	Diethyldimethyllead	Diethyldimethylplumbane	$C_6H_{16}Pb$	1762-27-2	295.4	col liq		51^{13}	1.79^{20}		
3431	Diethyl 2,6-dimethyl-3,5-pyridinedicarboxylate		$C_{13}H_{17}NO_4$	1149-24-2	251.279		71	301			i H_2O; s EtOH, eth, bz, chl, lig
3432	Diethyl 3,5-dimethylpyrrole-2,4-dicarboxylate		$C_{12}H_{17}NO_4$	2436-79-5	239.268	nd (dil al)	137.8				i H_2O; sl EtOH, eth; s ace, bz, HOAc
3433	Diethyl disulfide		$C_4H_{10}S_2$	110-81-6	122.252	liq	-101.5(0.1)	154.0(0.6)	0.9931^{20}	1.5073^{20}	sl H_2O; msc EtOH, eth
3434	*N,N*-Diethyldodecanamide		$C_{16}H_{33}NO$	3352-87-2	255.439			166^{2}	0.847^{25}	1.4545^{20}	s chl
3435	Diethylene glycol	Diglycol	$C_4H_{10}O_3$	111-46-6	106.120	liq	-10.3(0.3)	245.5(0.2)	1.1197^{15}	1.4472^{20}	s H_2O, EtOH, eth, chl
3436	Diethylene glycol, bischloroformate	Oxydi-2,1-ethanediyl carbonochloridate	$C_6H_8Cl_2O_5$	106-75-2	231.031	liq		126^{5}	1.39^{20}	1.4542^{20}	
3437	Diethylene glycol diacetate		$C_8H_{14}O_5$	628-68-2	190.194		18	200	1.1068^{15}	1.4348^{20}	vs EtOH
3438	Diethylene glycol dibenzoate		$C_{18}H_{18}O_5$	120-55-8	314.333		33.5	280^{24}	1.1690^{15}		vs H_2O, EtOH
3439	Diethylene glycol dibutyl ether	Bis(2-butoxyethyl) ether	$C_{12}H_{26}O_3$	112-73-2	218.332	liq	-60.2(0.2)	255(4)	0.885^{25}	1.4235^{20}	
3440	Diethylene glycol diethyl ether	Bis(2-ethoxyethyl) ether	$C_8H_{18}O_3$	112-36-7	162.227	liq	-44.3(0.2)	185(4)	0.9063^{20}	1.4115^{20}	vs H_2O, EtOH; s eth
3441	Diethylene glycol dimethacrylate	Oxydiethylene methacrylate	$C_{12}H_{18}O_5$	2358-84-1	242.268			>200	1.0821^{20}	1.4571^{25}	
3442	Diethylene glycol dimethyl ether	Diglyme	$C_6H_{14}O_3$	111-96-6	134.173	liq	-64.0(0.1)	162(2)	0.9434^{20}	1.4097^{20}	msc H_2O, EtOH, eth
3443	Diethylene glycol dinitrate	2,2'-Oxybisethanol, dinitrate	$C_4H_8N_2O_7$	693-21-0	196.116			$44^{0.01}$			
3444	Diethylene glycol monobutyl ether		$C_8H_{18}O_3$	112-34-5	162.227	liq	-68	232(4)	0.9553^{20}	1.4306^{20}	msc H_2O; vs EtOH, eth, ace; s bz
3445	Diethylene glycol monobutyl ether acetate	2-(2-Butoxyethoxy)ethyl acetate	$C_{10}H_{20}O_4$	124-17-4	204.264	liq	-32	248(2)	0.985^{20}	1.4262^{20}	vs ace, eth, EtOH
3446	Diethylene glycol monododecanoate	2-(2-Hydroxyethoxy)ethyl laurate	$C_{16}H_{32}O_4$	141-20-8	288.423	lt ye	17.5	>270	0.96^{25}		msc EtOH, eth, ace; s bz, tol
3447	Diethylene glycol monoethyl ether	Carbitol	$C_6H_{14}O_3$	111-90-0	134.173	hyg liq		202(3)	0.9885^{20}	1.4300^{20}	msc H_2O, EtOH, ace, bz; vs eth
3448	Diethylene glycol monoethyl ether acetate	Carbitol acetate	$C_8H_{16}O_4$	112-15-2	176.211	liq	-25	218(1)	1.0096^{20}	1.4213^{20}	vs H_2O, ace, eth, EtOH
3449	Diethylene glycol monohexyl ether	2-[2-(Hexyloxy)ethoxy]ethanol	$C_{10}H_{22}O_3$	112-59-4	190.280	col liq	-28	259(2)			
3450	Diethylene glycol monomethyl ether	2-(2-Methoxyethoxy)ethanol	$C_5H_{12}O_3$	111-77-3	120.147			194(2)	1.035^{20}	1.4264^{20}	msc H_2O, ace; vs EtOH, eth
3451	Diethylene glycol monopropyl ether		$C_7H_{16}O_3$	6881-94-3	148.200	liq	-53.3	214.8(0.4)			
3452	*N,N*-Diethyl-1,2-ethanediamine	*N,N*-Diethylethylenediamine	$C_6H_{16}N_2$	100-36-7	116.204			144	0.8280^{20}	1.4340^{20}	msc H_2O; s EtOH, eth, ctc, tol
3453	*N,N'*-Diethyl-1,2-ethanediamine		$C_6H_{16}N_2$	111-74-0	116.204			146	0.8280^{20}	1.4340^{20}	vs H_2O, eth, EtOH, tol
3454	Diethyl ether	Ethyl ether	$C_4H_{10}O$	60-29-7	74.121	liq	-116.22(0.04)	34.4(0.5)	0.7138^{20}	1.3526^{20}	sl H_2O; msc EtOH, bz, eth; vs ace
3455	Diethyl (ethoxymethylene)-malonate	2-Ethoxy-1,1-bis(ethoxycarbonyl)ethene	$C_{10}H_{16}O_5$	87-13-8	216.231			280 dec		1.4600^{20}	i H_2O; s EtOH, eth; sl chl
3456	Diethyl ethylidenemalonate		$C_9H_{14}O_4$	1462-12-0	186.205			116^{17}	1.0404^{20}	1.4308^{17}	vs eth, EtOH
3457	Diethyl ethylmalonate		$C_9H_{16}O_4$	133-13-1	188.221			208	1.006^{20}	1.4166^{20}	sl H_2O; vs EtOH, eth, ace, chl
3458	Diethyl ethylphenylmalonate		$C_{15}H_{20}O_4$	76-67-5	264.318			170^{19}	1.071^{20}	1.4896^{25}	i H_2O; s EtOH, eth; sl chl
3459	Diethyl ethylphosphonate		$C_6H_{15}O_3P$	78-38-6	166.155			198	1.0259^{20}	1.4163^{20}	sl H_2O; s EtOH, eth
3460	*N,N*-Diethylformamide		$C_5H_{11}NO$	617-84-5	101.147			175(10)	0.9080^{19}	1.4321^{25}	msc H_2O, ace, bz; vs EtOH, eth
3461	Diethyl fumarate		$C_8H_{12}O_4$	623-91-6	172.179		0.8	214(5)	1.0452^{20}	1.4412^{20}	i H_2O; s ace, chl
3462	Diethyl glutarate		$C_9H_{16}O_4$	818-38-2	188.221	syr liq	-24.1	237(3)	1.0220^{20}	1.4241^{20}	vs eth
3463	3,4-Diethylhexane		$C_{10}H_{22}$	19398-77-7	142.282			160(5)	0.7472^{25}	1.4190^{20}	
3464	Di-2-ethylhexyl maleate		$C_{20}H_{36}O_4$	142-16-5	340.498			156^{7}	0.94^{20}		

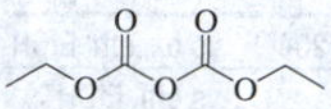
3426
Diethyl dicarbonate

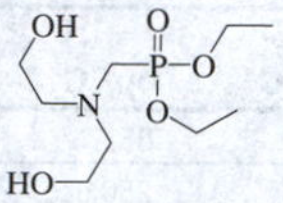
3427
Diethyl [(diethanolamino)methyl]phosphonate

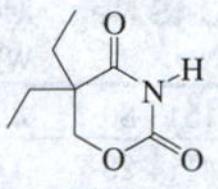
3428
5,5-Diethyldihydro-2*H*-1,3-oxazine-2,4(3*H*)-dione

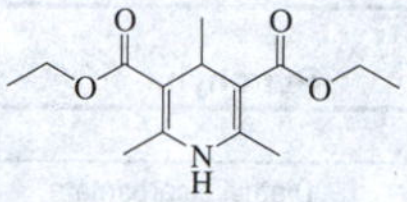
3429
Diethyl 1,4-dihydro-2,4,6-trimethyl-3,5-pyridinedicarboxylate

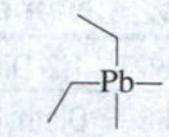
3430
Diethyldimethyllead

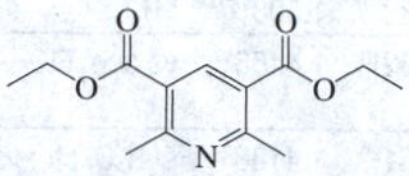
3431
Diethyl 2,6-dimethyl-3,5-pyridinedicarboxylate

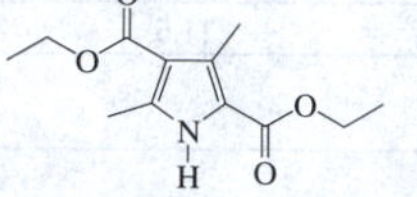
3432
Diethyl 3,5-dimethylpyrrole-2,4-dicarboxylate

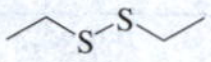
3433
Diethyl disulfide

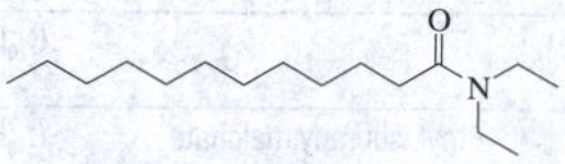
3434
N,N-Diethyldodecanamide

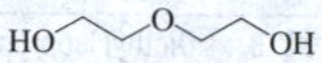
3435
Diethylene glycol

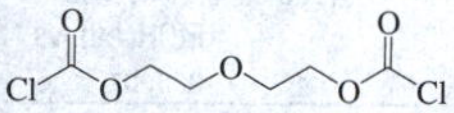
3436
Diethylene glycol, bischloroformate

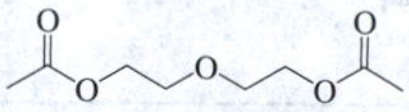
3437
Diethylene glycol diacetate

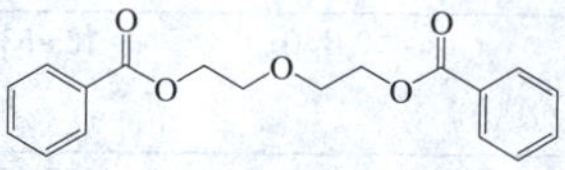
3438
Diethylene glycol dibenzoate

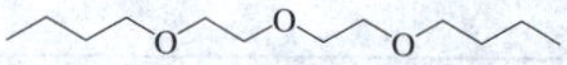
3439
Diethylene glycol dibutyl ether

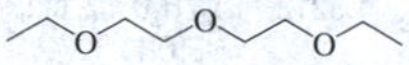
3440
Diethylene glycol diethyl ether

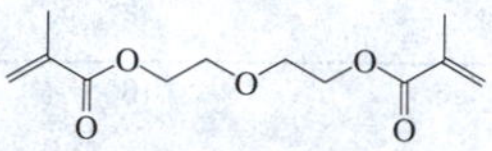
3441
Diethylene glycol dimethacrylate

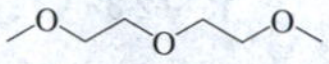
3442
Diethylene glycol dimethyl ether

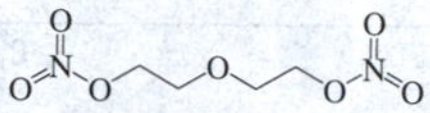
3443
Diethylene glycol dinitrate

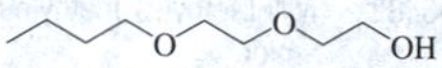
3444
Diethylene glycol monobutyl ether

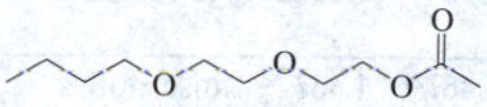
3445
Diethylene glycol monobutyl ether acetate

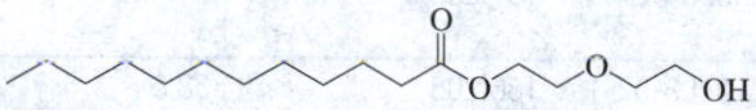
3446
Diethylene glycol monododecanoate

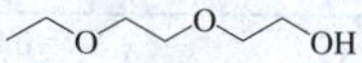
3447
Diethylene glycol monoethyl ether

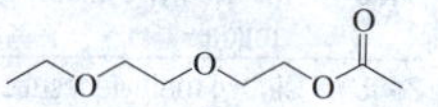
3448
Diethylene glycol monoethyl ether acetate

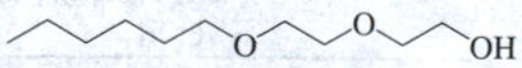
3449
Diethylene glycol monohexyl ether

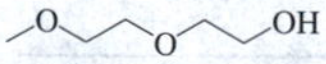
3450
Diethylene glycol monomethyl ether

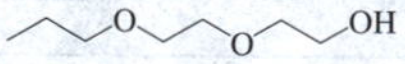
3451
Diethylene glycol monopropyl ether

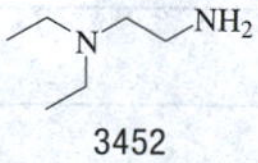
3452
N,N-Diethyl-1,2-ethanediamine

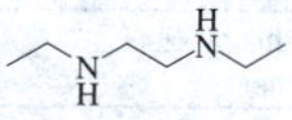
3453
N,N'-Diethyl-1,2-ethanediamine

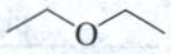
3454
Diethyl ether

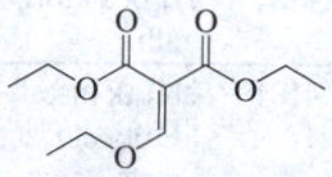
3455
Diethyl (ethoxymethylene)malonate

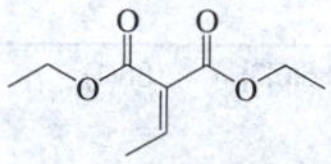
3456
Diethyl ethylidenemalonate

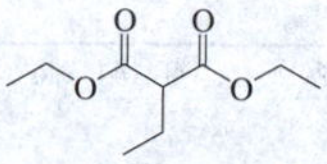
3457
Diethyl ethylmalonate

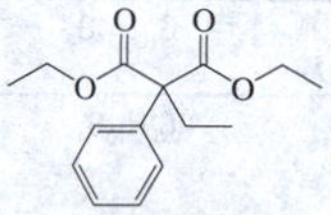
3458
Diethyl ethylphenylmalonate

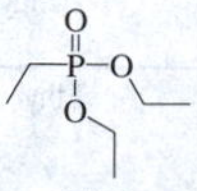
3459
Diethyl ethylphosphonate

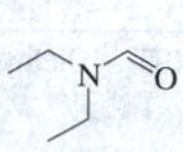
3460
N,N-Diethylformamide

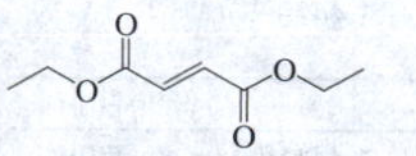
3461
Diethyl fumarate

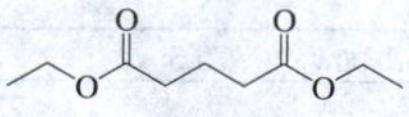
3462
Diethyl glutarate

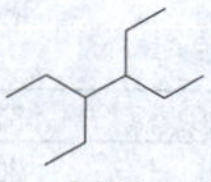
3463
3,4-Diethylhexane

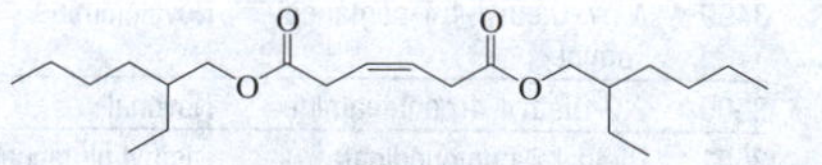
3464
Di-2-ethylhexyl maleate

No.	Name	Synonym	Mol. form.	CAS Reg. No.	Mol. wt.	Physical form	mp/°C	bp/°C	den/ g cm^{-3}	n_D	Solubility
3465	1,2-Diethylhydrazine		$C_4H_{12}N_2$	1615-80-1	88.151			85.5	0.797^{26}	1.4204^{20}	vs bz, eth, EtOH
3466	Diethyl 1,2-hydrazinedicarboxylate	Diethyl bicarbamate	$C_6H_{12}N_2O_4$	4114-28-7	176.170	nd (chl), pr (w)	135	250 dec	1.324^{8}		vs eth, EtOH
3467	Diethyl hydrogen phosphate	Diethyl phosphate	$C_4H_{11}O_4P$	598-02-7	154.101	syr		203 dec	1.1800^{20}	1.4170^{20}	vs eth
3468	*N,N*-Diethyl-4-hydroxy-3-methoxybenzamide	Ethamivan	$C_{12}H_{17}NO_3$	304-84-7	223.268		95				s chl
3469	Diethyl iminodiacetate		$C_8H_{15}NO_4$	6290-05-7	189.210	orth cry	247 dec				
3470	Diethyl isobutylmalonate		$C_{11}H_{20}O_4$	10203-58-4	216.275				0.9804^{20}	1.4236^{20}	i H_2O; vs EtOH, eth; s chl
3471	Diethyl isophthalate		$C_{12}H_{14}O_4$	636-53-3	222.237		11.5	298(6)	1.1239^{17}	1.508^{18}	i H_2O
3472	Diethyl isopropylidenemalonate		$C_{10}H_{16}O_4$	6802-75-1	200.232			176.5	1.0282^{18}	1.4486^{17}	vs ace, EtOH
3473	Diethyl isopropylmalonate	Ethyl isopropylmalonate	$C_{10}H_{18}O_4$	759-36-4	202.248			215	0.9961^{20}	1.4188^{21}	sl H_2O, ctc; vs EtOH, eth; s chl
3474	Diethyl ketomalonate	Ethyl mesoxalate	$C_7H_{10}O_5$	609-09-6	174.151	pa ye grn oil	-30	210	1.1419^{16}	1.4310^{22}	vs H_2O; s EtOH, eth, chl; i CS_2
3475	Diethyl malate	Diethyl hydroxybutanedioate	$C_8H_{14}O_5$	7554-12-3	190.194			253	1.1290^{20}		
3476	Diethyl maleate		$C_8H_{12}O_4$	141-05-9	172.179	liq	-8.8	222(8)	1.0662^{20}	1.4416^{20}	i H_2O; s EtOH, eth; sl chl
3477	Diethyl malonate	Ethyl malonate	$C_7H_{12}O_4$	105-53-3	160.168	liq	-50(2)	200(3)	1.0551^{20}	1.4139^{20}	sl H_2O; msc EtOH, eth; vs ace, bz
3478	Diethyl mercury		$C_4H_{10}Hg$	627-44-1	258.71			159	2.43^{20}		s eth; sl EtOH
3479	Diethylmethylamine	*N*-Ethyl-*N*-methylethanamine	$C_5H_{13}N$	616-39-7	87.164	liq	-196	65.9(0.3)	0.703^{25}	1.3879^{25}	vs H_2O, EtOH, eth
3480	*N,N*-Diethyl-2-methylaniline		$C_{11}H_{17}N$	606-46-2	163.260	liq	-60	209	0.9286^{20}	1.5153^{20}	sl H_2O; msc EtOH, eth; s ctc
3481	*N,N*-Diethyl-4-methylaniline		$C_{11}H_{17}N$	613-48-9	163.260			229	0.9242^{16}		sl H_2O; msc EtOH, eth
3482	*N,N*-Diethyl-3-methylbenzamide	DEET	$C_{12}H_{17}NO$	134-62-3	191.269			160^{19}	0.996^{20}	1.5212^{20}	vs H_2O, bz, eth, EtOH
3483	1,3-Diethyl-5-methylbenzene		$C_{11}H_{16}$	2050-24-0	148.245	liq	-74.1(0.3)	205	0.8748^{20}	1.5027^{20}	i H_2O; msc EtOH, eth, ace, bz, lig, ctc
3484	N^4,N^4-Diethyl-2-methyl-1,4-benzenediamine, monohydrochloride	4-*N,N*-Diethyl-1,4-diamino-2-methylbenzene, hydrochloride	$C_{11}H_{19}ClN_2$	2051-79-8	214.735	cry	250 dec				
3485	*N,N*-Diethyl-3-methylbutanamide	Isovaleryl diethylamide	$C_9H_{19}NO$	533-32-4	157.253			211	0.8764^{20}	1.4422^{20}	vs eth, EtOH
3486	Diethyl methylenesuccinate		$C_9H_{14}O_4$	2409-52-1	186.205		58.5	228	1.0467^{20}	1.4377^{20}	msc EtOH; s eth, bz; vs ace
3487	Diethyl methylmalonate		$C_8H_{14}O_4$	609-08-5	174.195			198.1(0.8)	1.0225^{20}	1.4126^{20}	sl H_2O; vs EtOH, eth, ace, chl
3488	Diethyl methylphosphonate		$C_5H_{13}O_3P$	683-08-9	152.129			181(6)	1.0406^{30}	1.4101^{30}	s H_2O, EtOH, eth; i bz
3489	*N,N*-Diethyl-4-methyl-1-piperazinecarboxamide	Diethylcarbamazine	$C_{10}H_{21}N_3O$	90-89-1	199.293		49.5(0.5)	110^{3}			
3490	3,3-Diethyl-5-methyl-2,4-piperidinedione		$C_{10}H_{17}NO_2$	125-64-4	183.248		75.5				s H_2O, bz, chl, EtOH
3491	*N,N*-Diethyl-1-naphthalenamine		$C_{14}H_{17}N$	84-95-7	199.292			285	1.013^{20}	1.5961^{20}	s EtOH, eth, bz; sl ctc
3492	*N,N*-Diethyl-4-nitroaniline		$C_{10}H_{14}N_2O_2$	2216-15-1	194.230	ye nd (lig) pl (al)	77.5		1.225^{25}		s EtOH; sl lig
3493	*N,N*-Diethyl-4-nitrosoaniline		$C_{10}H_{14}N_2O$	120-22-9	178.230	grn mcl pr (eth) grn lf (ace)	87.5		1.24^{15}		sl H_2O; s EtOH, eth, ace, chl
3494	Diethyl nonanedioate	Diethyl azelate	$C_{13}H_{24}O_4$	624-17-9	244.328	liq	-18.5	289(15)	0.9729^{20}	1.4351^{20}	i H_2O; s EtOH, eth
3495	Diethyl oxalate	Ethyl oxalate	$C_6H_{10}O_4$	95-92-1	146.141	liq	-40.6(0.3)	186(1)	1.0785^{20}	1.4101^{20}	sl H_2O; msc EtOH, eth, ace; s ctc
3496	Diethyl oxobutanedioate	Diethyl oxalacetate	$C_8H_{12}O_5$	108-56-5	188.178			131^{24}	1.131^{20}	1.4561^{17}	i H_2O; msc EtOH, eth, bz; vs ace
3497	Diethyl 3-oxo-1,5-pentanedioate	Diethyl 1,3-acetonedicarboxylate	$C_9H_{14}O_5$	105-50-0	202.204			250	1.113^{20}		sl H_2O; msc EtOH
3498	3,3-Diethylpentane	Tetraethylmethane	C_9H_{20}	1067-20-5	128.255	liq	-33.04(0.06)	146.2(0.3)	0.7536^{20}	1.4206^{20}	i H_2O; s eth, bz
3499	*N′,N′*-Diethyl-1,4-pentanediamine	Novoldiamine	$C_9H_{22}N_2$	140-80-7	158.284			201	0.814^{20}	1.4429^{20}	
3500	2,2-Diethyl-4-pentenamide	Novonal	$C_9H_{17}NO$	512-48-1	155.237	wh pow	75.5				vs eth, EtOH
3501	Diethyl 2-pentenedioate	Diethyl glutaconate	$C_9H_{14}O_4$	2049-67-4	186.205			237	1.0496^{20}	1.4411^{20}	vs eth, EtOH

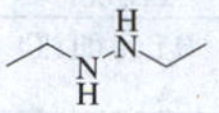
3465
1,2-Diethylhydrazine

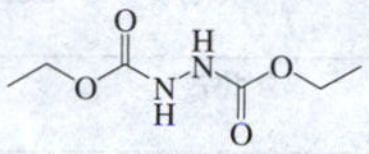
3466
Diethyl 1,2-hydrazinedicarboxylate

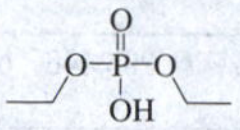
3467
Diethyl hydrogen phosphate

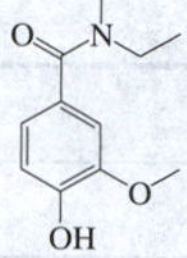
3468
N,N-Diethyl-4-hydroxy-3-methoxybenzamide

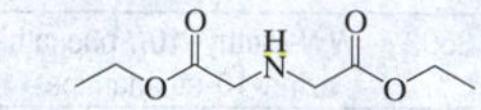
3469
Diethyl iminodiacetate

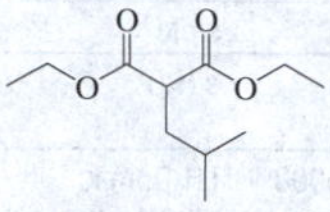
3470
Diethyl isobutylmalonate

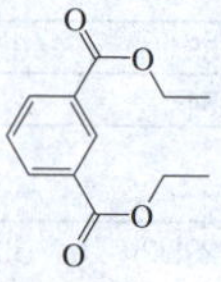
3471
Diethyl isophthalate

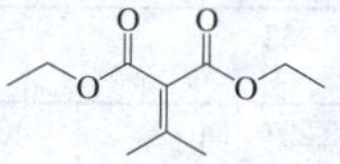
3472
Diethyl isopropylidenemalonate

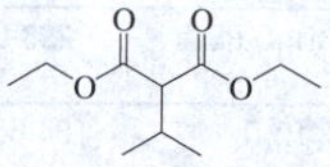
3473
Diethyl isopropylmalonate

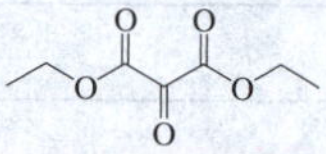
3474
Diethyl ketomalonate

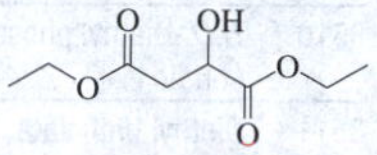
3475
Diethyl malate

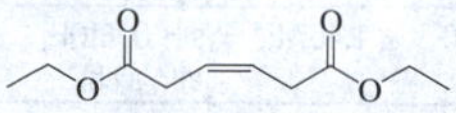
3476
Diethyl maleate

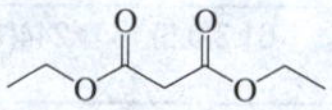
3477
Diethyl malonate

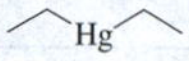
3478
Diethyl mercury

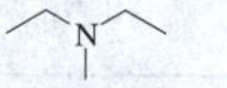
3479
Diethylmethylamine

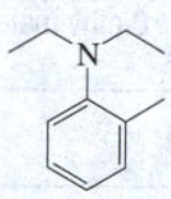
3480
N,N-Diethyl-2-methylaniline

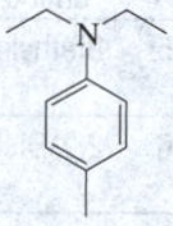
3481
N,N-Diethyl-4-methylaniline

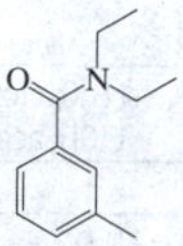
3482
N,N-Diethyl-3-methylbenzamide

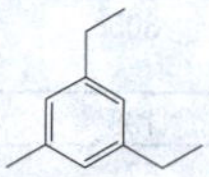
3483
1,3-Diethyl-5-methylbenzene

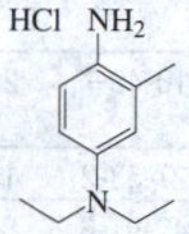
3484
*N*4,*N*4-Diethyl-2-methyl-1,4-benzenediamine, monohydrochloride

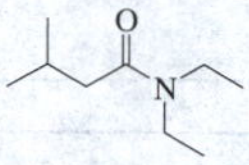
3485
N,N-Diethyl-3-methylbutanamide

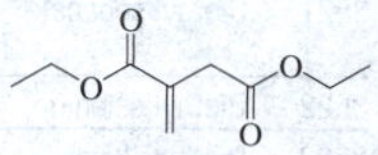
3486
Diethyl methylenesuccinate

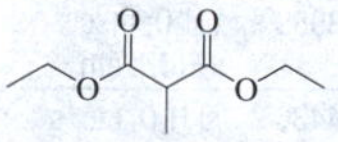
3487
Diethyl methylmalonate

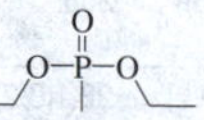
3488
Diethyl methylphosphonate

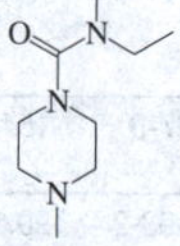
3489
N,N-Diethyl-4-methyl-1-piperazinecarboxamide

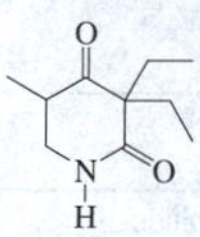
3490
3,3-Diethyl-5-methyl-2,4-piperidinedione

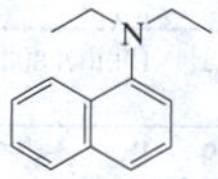
3491
N,N-Diethyl-1-naphthalenamine

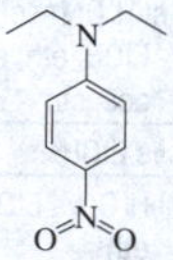
3492
N,N-Diethyl-4-nitroaniline

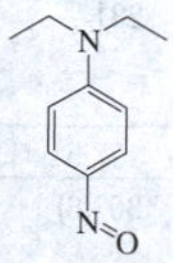
3493
N,N-Diethyl-4-nitrosoaniline

3494
Diethyl nonanedioate

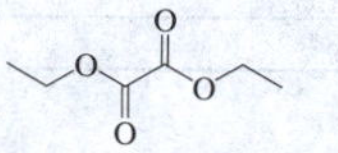
3495
Diethyl oxalate

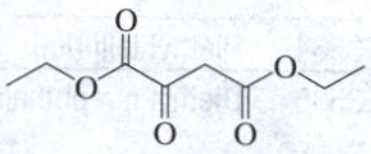
3496
Diethyl oxobutanedioate

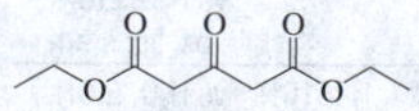
3497
Diethyl 3-oxo-1,5-pentanedioate

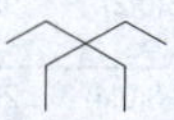
3498
3,3-Diethylpentane

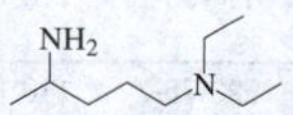
3499
N',*N'*-Diethyl-1,4-pentanediamine

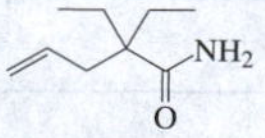
3500
2,2-Diethyl-4-pentenamide

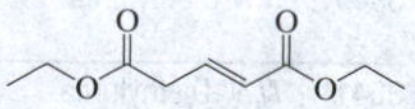
3501
Diethyl 2-pentenedioate

Organic

No.	Name	Synonym	Mol. form.	CAS Reg. No.	Mol. wt.	Physical form	mp/°C	bp/°C	den/ g cm⁻³	n_D	Solubility
3502	Diethylperoxide		$C_4H_{10}O_2$	628-37-5	90.121	liq	-70	46(9)	0.8240^{19}	1.3715^{17}	sl H_2O; msc EtOH, eth
3503	*N,N*-Diethyl-10*H*-phenothiazine-10-ethanamine	Diethazine	$C_{18}H_{22}N_2S$	60-91-3	298.446	oil		$167^{0.5}$			i H_2O; s dil HCl
3504	*N,N*-Diethyl-α-phenylbenzenemethanamine	*N,N*-Diethylbenzhydrylamine	$C_{17}H_{21}N$	519-72-2	239.356		58.5	170^{17}			
3505	Diethyl phenylmalonate		$C_{13}H_{16}O_4$	83-13-6	236.264		16.5	291(17)	1.0950^{20}	1.4977^{20}	vs ace, EtOH
3506	Diethyl phenylphosphonite		$C_{10}H_{15}O_2P$	1638-86-4	198.199			235	1.032^{16}		
3507	5,5-Diethyl-1-phenyl-2,4,6(1*H*,3*H*,5*H*)-pyrimidinetrione	Phenetharbital	$C_{14}H_{16}N_2O_3$	357-67-5	260.288		178				vs EtOH
3508	Diethylphosphine		$C_4H_{11}P$	627-49-6	90.104			85	0.786^{20}		
3509	Diethyl phosphonate		$C_4H_{11}O_3P$	762-04-9	138.102			54^{6}			s ctc
3510	*O,O'*-Diethyl phosphorodithioate		$C_4H_{11}O_2PS_2$	298-06-6	186.233						s H_2O
3511	Diethyl phthalate		$C_{12}H_{14}O_4$	84-66-2	222.237	liq	-40.5	298(2)	1.232^{14}	1.5000^{21}	i H_2O; msc EtOH, eth; s ace, bz, ctc
3512	3,3-Diethyl-2,4-piperidinedione	Piperidione	$C_9H_{15}NO_2$	77-03-2	169.221	nd (w)	104				vs H_2O, EtOH, chl, MeOH
3513	*N,N*-Diethylpropanamide		$C_7H_{15}NO$	1114-51-8	129.200			191	0.8972^{20}	1.4425^{20}	vs EtOH
3514	*N,N*-Diethyl-1,3-propanediamine		$C_7H_{18}N_2$	104-78-9	130.231			165(2)	0.822^{20}	1.443^{20}	
3515	Diethylpropanedioic acid	Diethylmalonic acid	$C_7H_{12}O_4$	510-20-3	160.168	pr (w,bz)	127 dec				vs H_2O, EtOH, eth; sl bz, chl
3516	2,2-Diethyl-1,3-propanediol		$C_7H_{16}O_2$	115-76-4	132.201		61.3(0.5)	244(12)	1.050^{20}	1.4574^{25}	vs H_2O, EtOH, eth; s chl
3517	Diethyl 2-propylmalonate		$C_{10}H_{18}O_4$	2163-48-6	202.248			221	0.989^{20}	1.4197^{20}	sl H_2O; vs EtOH, eth
3518	*N,N*-Diethyl-3-pyridinecarboxamide	Nikethamide	$C_{10}H_{14}N_2O$	59-26-7	178.230	ye solid or visc liq	25	280 dec	1.060^{25}	1.525^{20}	sl DMSO
3519	*N,N*-Diethyl-4-pyridinecarboxamide	Isonicotinic acid diethylamide	$C_{10}H_{14}N_2O$	530-40-5	178.230			119^{1}		1.525^{20}	vs H_2O, ace, eth, EtOH
3520	3,3-Diethyl-2,4(1*H*,3*H*)-pyridinedione	Pyrithyldione	$C_9H_{13}NO_2$	77-04-3	167.205		90.7				
3521	Diethyl sebacate		$C_{14}H_{26}O_4$	110-40-7	258.354		2.5	305	0.9646^{20}	1.4306^{20}	sl H_2O, ctc; s EtOH, ace; i bz
3522	Diethyl selenide		$C_4H_{10}Se$	627-53-2	137.08	pa ye	55	108	1.2300^{20}	1.4768^{20}	
3523	Diethylsilane		$C_4H_{12}Si$	542-91-6	88.224	liq	-134.3	57	0.6843^{20}	1.3921^{20}	i H_2O
3524	*trans*-Diethylstilbestrol		$C_{18}H_{20}O_2$	56-53-1	268.351	pl (bz)	172(3)				vs eth, EtOH, chl
3525	*trans*-Diethylstilbestrol dipropanoate	Clinestrol	$C_{24}H_{28}O_4$	130-80-3	380.477	pr (MeOH)	104				vs bz, eth, EtOH
3526	*trans*-Diethylstilbestrol monomethyl ether	Mestilbol	$C_{19}H_{22}O_2$	18839-90-2	282.377	nd (bz-peth)	117.5	$190^{0.3}$			vs ace, eth, EtOH
3527	Diethyl succinate	Ethyl succinate	$C_8H_{14}O_4$	123-25-1	174.195	liq	-21.6(0.8)	217(1)	1.0402^{20}	1.4201^{20}	i H_2O; msc EtOH, eth; s ace, chl
3528	Diethyl sulfate		$C_4H_{10}O_4S$	64-67-5	154.185	oil	-26.0(0.5)	208	1.172^{25}	1.3989^{20}	i H_2O; msc EtOH, eth
3529	Diethyl sulfide	Ethyl sulfide	$C_4H_{10}S$	352-93-2	90.187	liq	-103.9(0.1)	92.1(0.2)	0.8362^{20}	1.4430^{20}	sl H_2O, ctc; s EtOH, eth
3530	Diethyl sulfite	Ethyl sulfite	$C_4H_{10}O_3S$	623-81-4	138.185			158	1.1^{20}	1.4310^{20}	s EtOH, eth
3531	Diethyl sulfone	Ethyl sulfone	$C_4H_{10}O_2S$	597-35-3	122.186	orth pl	70(4)	264(14)	1.357^{20}		s H_2O, eth; vs bz; i peth
3532	Diethyl sulfoxide		$C_4H_{10}OS$	70-29-1	106.186	syr	14	104^{25}	1.0092^{22}		vs H_2O, eth, EtOH
3533	Diethyl DL-tartrate		$C_8H_{14}O_6$	57968-71-5	206.193		18.7	281	1.2046^{20}	1.4438^{20}	sl H_2O; msc EtOH, eth; s ace, ctc
3534	Diethyl telluride		$C_4H_{10}Te$	627-54-3	185.72	red-ye		137.5	1.599^{15}	1.5182^{15}	vs EtOH
3535	Diethyl terephthalate		$C_{12}H_{14}O_4$	636-09-9	222.237	mcl pr (al, peth)	43.2(0.6)	303(8)	1.0989^{45}		i H_2O; vs EtOH, eth
3536	Diethyl thiodipropionate		$C_{10}H_{18}O_4S$	673-79-0	234.313			174^{15}	1.1034^{20}	1.4655^{20}	
3537	*N,N'*-Diethylthiourea		$C_5H_{12}N_2S$	105-55-5	132.227		76.9(0.4)	287(17)			s H_2O, EtOH; vs eth; sl ctc
3538	*N,N*-Diethyl-1,1,1-trimethylsilanamine	(Diethylamino)trimethylsilane	$C_7H_{19}NSi$	996-50-9	145.319			126.3	0.7627^{20}	1.4112^{20}	
3539	Diethyltrisulfide		$C_4H_{10}S_3$	3600-24-6	154.317		-72.6	85^{26}	1.1082^{20}	1.5689^{13}	
3540	*N,N*-Diethylurea		$C_5H_{12}N_2O$	634-95-7	116.161	pl, nd (eth)	75.3(0.2)	$95^{0.02}$			vs H_2O, EtOH, bz, lig; s eth
3541	*N,N'*-Diethylurea		$C_5H_{12}N_2O$	623-76-7	116.161	tab (lig), hyg nd (al)	110.3(0.4)	263	1.0415^{25}	1.4616^{40}	vs H_2O, EtOH, eth

Organic

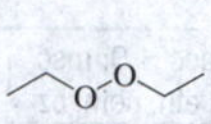
3502
Diethylperoxide

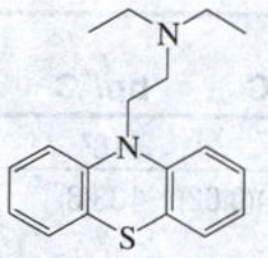
3503
N,N-Diethyl-10*H*-phenothiazine-10-ethanamine

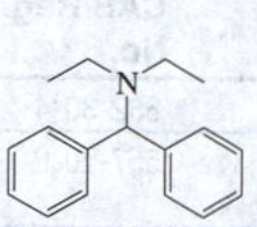
3504
N,N-Diethyl-α-phenylbenzenemethanamine

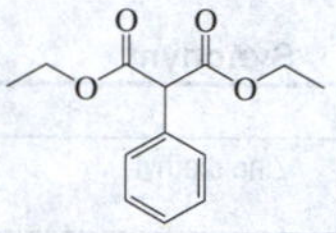
3505
Diethyl phenylmalonate

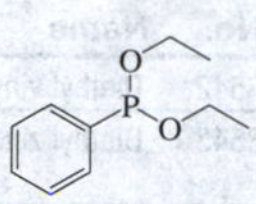
3506
Diethyl phenylphosphonite

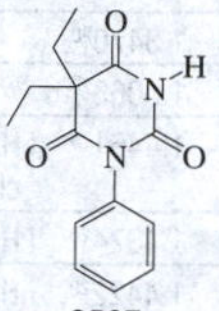
3507
5,5-Diethyl-1-phenyl-2,4,6(1*H*,3*H*,5*H*)-pyrimidinetrione

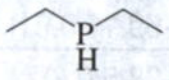
3508
Diethylphosphine

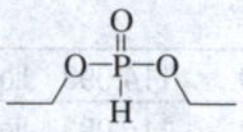
3509
Diethyl phosphonate

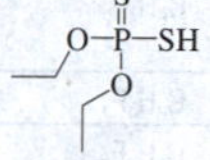
3510
O,O'-Diethyl phosphorodithioate

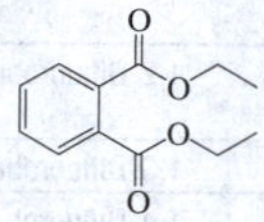
3511
Diethyl phthalate

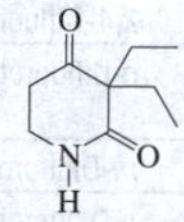
3512
3,3-Diethyl-2,4-piperidinedione

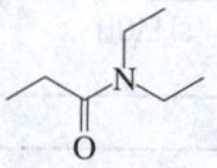
3513
N,N-Diethylpropanamide

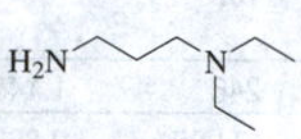
3514
N,N-Diethyl-1,3-propanediamine

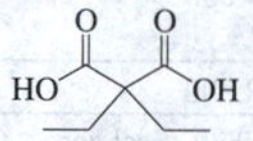
3515
Diethylpropanedioic acid

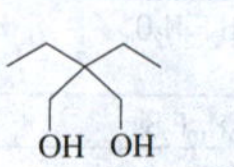
3516
2,2-Diethyl-1,3-propanediol

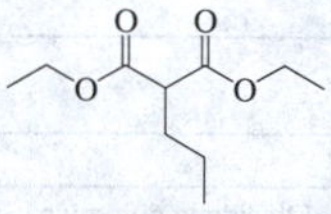
3517
Diethyl 2-propylmalonate

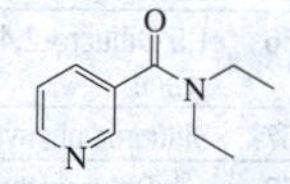
3518
N,N-Diethyl-3-pyridinecarboxamide

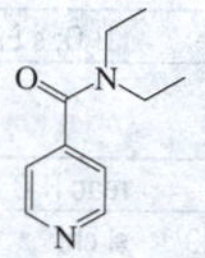
3519
N,N-Diethyl-4-pyridinecarboxamide

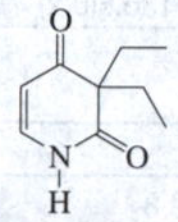
3520
3,3-Diethyl-2,4(1*H*,3*H*)-pyridinedione

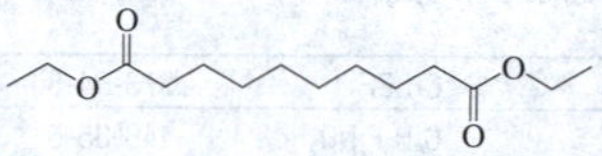
3521
Diethyl sebacate

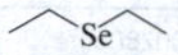
3522
Diethyl selenide

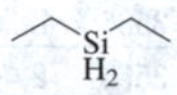
3523
Diethylsilane

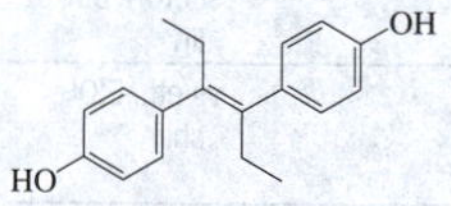
3524
trans-Diethylstilbestrol

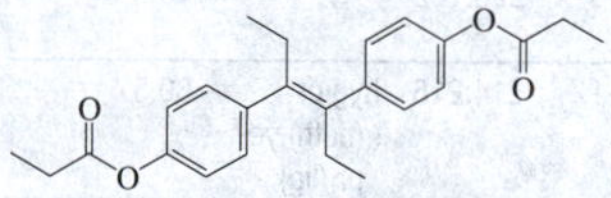
3525
trans-Diethylstilbestrol dipropanoate

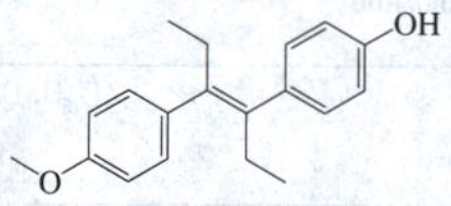
3526
trans-Diethylstilbestrol monomethyl ether

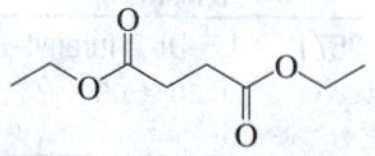
3527
Diethyl succinate

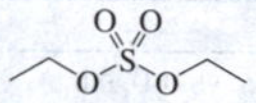
3528
Diethyl sulfate

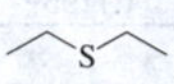
3529
Diethyl sulfide

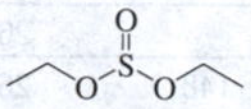
3530
Diethyl sulfite

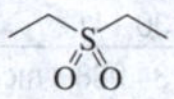
3531
Diethyl sulfone

3532
Diethyl sulfoxide

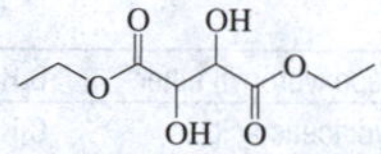
3533
Diethyl DL-tartrate

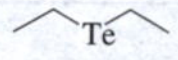
3534
Diethyl telluride

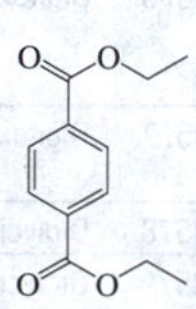
3535
Diethyl terephthalate

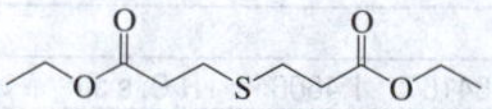
3536
Diethyl thiodipropionate

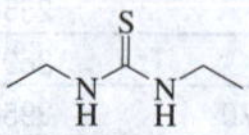
3537
N,N'-Diethylthiourea

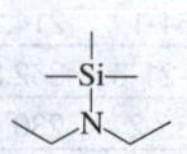
3538
N,N-Diethyl-1,1,1-trimethylsilanamine

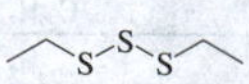
3539
Diethyltrisulfide

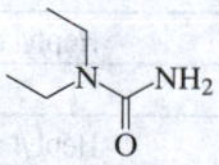
3540
N,N-Diethylurea

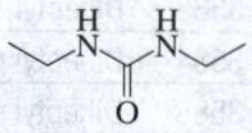
3541
N,N'-Diethylurea

No.	Name	Synonym	Mol. form.	CAS Reg. No.	Mol. wt.	Physical form	mp/°C	bp/°C	den/g cm⁻³	n_D	Solubility
3542	Diethyl vinylphosphonate		$C_6H_{13}O_3P$	682-30-4	164.139			110^{2}	1.068^{25}	1.4290^{20}	
3543	Diethyl zinc	Zinc diethyl	$C_4H_{10}Zn$	557-20-0	123.531	col liq	-33.34(0.02)	133(6)	1.2065^{20}	1.4936^{20}	dec H_2O; msc eth, peth, bz
3544	Difenoconazole		$C_{19}H_{17}Cl_2N_3O_3$	119446-68-3	406.262		76	$220^{0.03}$			
3545	Difenzoquat methyl sulfate	1*H*-Pyrazolium, 1,2-dimethyl-3,5-diphenyl-, methyl sulfate	$C_{18}H_{20}N_2O_4S$	43222-48-6	360.428		158.5(0.5)				
3546	Diflubenzuron	*N*-[[(4-Chlorophenyl)amino]carbonyl]-2,6-difluorobenzamide	$C_{14}H_9ClF_2N_2O_2$	35367-38-5	310.683		228(1)				
3547	Difluoroacetic acid		$C_2H_2F_2O_2$	381-73-7	96.033	liq	-1	133	1.526^{25}	1.3470^{20}	
3548	2,4-Difluoroaniline		$C_6H_5F_2N$	367-25-9	129.108	liq	-7.5	170	1.268^{25}	1.5063^{20}	
3549	*o*-Difluorobenzene	1,2-Difluorobenzene	$C_6H_4F_2$	367-11-3	114.093	liq	-47.1(0.1)	93.9(0.5)	1.1599^{18}	1.4451^{18}	i H_2O; s ace, bz, chl
3550	*m*-Difluorobenzene	1,3-Difluorobenzene	$C_6H_4F_2$	372-18-9	114.093	liq	-69.11(0.01)	83.0(0.5)	1.1572^{20}	1.4374^{20}	i H_2O; s ace, bz
3551	*p*-Difluorobenzene	1,4-Difluorobenzene	$C_6H_4F_2$	540-36-3	114.093	liq	-23.5(0.2)	88.9(0.3)	1.1701^{20}	1.4422^{20}	i H_2O; s ace, bz; sl ctc
3552	4,4'-Difluoro-1,1'-biphenyl	4,4'-Difluorodiphenyl	$C_{12}H_8F_2$	398-23-2	190.189	mcl pr (al) lf (w)	90(2)	254.5			i H_2O; vs EtOH, bz, chl; s eth, ace
3553	1,1-Difluorocyclohexane		$C_6H_{10}F_2$	371-90-4	120.140	liq		99.5			
3554	3,3-Difluorocyclopropene		$C_3H_2F_2$	56830-75-2	76.045	liq		34			
3555	Difluorodimethylsilane		$C_2H_6F_2Si$	353-66-2	96.152	col gas	-87.5	2.5			
3556	1,5-Difluoro-2,4-dinitrobenzene		$C_6H_2F_2N_2O_4$	327-92-4	204.088		75.5	132^{2}			sl EtOH
3557	Difluorodiphenylsilane		$C_{12}H_{10}F_2Si$	312-40-3	220.290			246	1.145^{17}	1.5221^{25}	
3558	1,1-Difluoroethane	Ethylidene difluoride	$C_2H_4F_2$	75-37-6	66.050	col gas	-118.6	-24.05	0.896^{25} (p>1 atm)	1.3011^{-72}	
3559	1,2-Difluoroethane	Ethylene difluoride	$C_2H_4F_2$	624-72-6	66.050	vol liq		26			vs bz, eth, chl
3560	1,1-Difluoroethene	Vinylidene fluoride	$C_2H_2F_2$	75-38-7	64.034	col gas	-144	-85.5(0.8)			vs eth, EtOH
3561	*cis*-1,2-Difluoroethene	*cis*-1,2-Difluoroethylene	$C_2H_2F_2$	1630-77-9	64.034	col gas		-45(19)			
3562	*trans*-1,2-Difluoroethene	*trans*-1,2-Difluoroethylene	$C_2H_2F_2$	1630-78-0	64.034	col gas		-53.1			
3563	Difluoromethane	Methylene fluoride	CH_2F_2	75-10-5	52.024	col gas	-136.8(0.2)	-51.65(0.07)	1.2139^{-52}		i H_2O; s EtOH
3564	2-(Difluoromethoxy)-1,1,1-trifluoroethane	Difluoromethyl 2,2,2-trifluoroethyl ether	$C_3H_3F_5O$	1885-48-9	150.047	col liq		29.2(0.2)			
3565	Difluoromethylborane		CH_3BF_2	373-64-8	63.843	gas		-78.5^{287}			reac H_2O
3566	2,4-Difluoro-1-nitrobenzene		$C_6H_3F_2NO_2$	446-35-5	159.091		9.8	207	1.4571^{14}	1.5149^{14}	sl chl
3567	2,2-Difluoropropane		$C_3H_6F_2$	420-45-1	80.077	col gas	-104.8	0(3)	0.9205^{20} (p>1 atm	1.2904^{20}	
3568	1,3-Difluoro-2-propanol		$C_3H_6F_2O$	453-13-4	96.076			127	1.24^{25}	1.3725^{20}	
3569	Di-2-furanylethanedione		$C_{10}H_6O_4$	492-94-4	190.153	ye nd (al), cry (bz)	166.3				sl H_2O; s EtOH, eth, bz, chl
3570	Di-2-furanylethanedione dioxime	α-Furildioxime	$C_{10}H_8N_2O_4$	522-27-0	220.182		167				sl EtOH, eth, bz, lig
3571	1,5-Di-2-furanyl-1,4-pentadien-3-one		$C_{13}H_{10}O_3$	886-77-1	214.216	hyg pr (peth) ye pr (lig)	60.5	181^{4}			vs eth, EtOH, chl
3572	Difurfuryl disulfide	Furfuryl disulfide	$C_{10}H_{10}O_2S_2$	4437-20-1	226.315		10	167^{13}			vs EtOH
3573	Difurfuryl ether	Furfuryl ether	$C_{10}H_{10}O_3$	4437-22-3	178.184			101^{2}	1.1405^{20}	1.5088^{20}	i H_2O
3574	Digitonin		$C_{56}H_{92}O_{29}$	11024-24-1	1229.312		237.5				
3575	Digitoxigenin		$C_{23}H_{34}O_4$	143-62-4	374.514		253				s EtOH; vs MeOH
3576	Digitoxin		$C_{41}H_{64}O_{13}$	71-63-6	764.939	pr (dil al)	255.5				sl H_2O; vs EtOH; s eth, chl, MeOH, py
3577	Digitoxose		$C_6H_{12}O_4$	527-52-6	148.157	cry (MeOH +eth)	112				vs H_2O, ace; s py, AcOEt
3578	Diglycidyl ether	Bis(2,3-epoxypropyl) ether	$C_6H_{10}O_3$	2238-07-5	130.141			260	1.1195^{20}		
3579	Diglycolic acid	2,2'-Oxydiacetic acid	$C_4H_6O_5$	110-99-6	134.088	mcl pr (w + 1)	148	269(18)			vs H_2O, eth, EtOH
3580	Digoxigenin		$C_{23}H_{34}O_5$	1672-46-4	390.513	pr (AcOEt)	222				vs EtOH, MeOH; sl chl
3581	Digoxin		$C_{41}H_{64}O_{14}$	20830-75-5	780.939	trc pl (dil al, py)	249 dec				vs EtOH
3582	Diheptylamine	*N*-Heptyl-1-heptanamine	$C_{14}H_{31}N$	2470-68-0	213.403	nd	31.5	266(8)	0.7956^{21}		sl H_2O; s EtOH; vs eth
3583	Diheptyl ether	Heptyl ether	$C_{14}H_{30}O$	629-64-1	214.387			258(4)	0.8008^{20}	1.4275^{20}	vs eth, EtOH
3584	Diheptyl phthalate		$C_{22}H_{34}O_4$	3648-21-3	362.503			360			
3585	Diheptyl sulfide	Heptyl sulfide	$C_{14}H_{30}S$	629-65-2	230.453		70	295(11)	0.8416^{20}	1.4606^{20}	i H_2O; s eth
3586	Dihexylamine	*N*-Hexyl-1-hexanamine	$C_{12}H_{27}N$	143-16-8	185.349	liq	-13.0(0.2)	236	0.7889^{20}	1.4339^{20}	s EtOH, eth

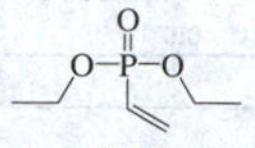
3542
Diethyl vinylphosphonate

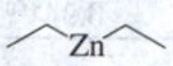
3543
Diethyl zinc

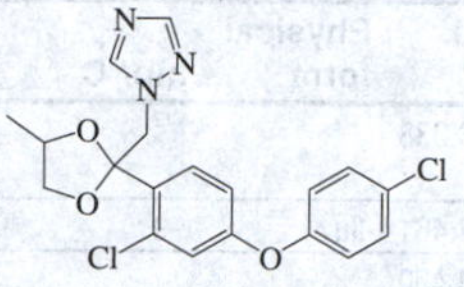
3544
Difenoconazole

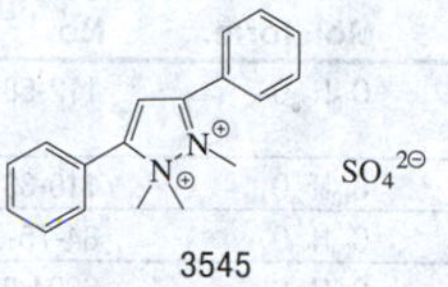
3545
Difenzoquat methyl sulfate

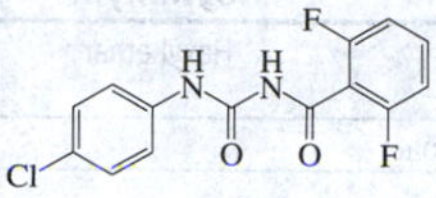
3546
Diflubenzuron

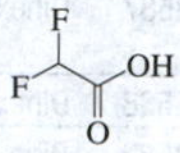
3547
Difluoroacetic acid

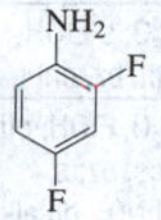
3548
2,4-Difluoroaniline

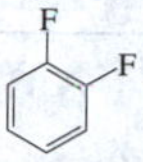
3549
o-Difluorobenzene

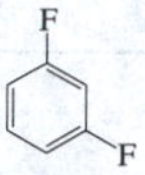
3550
m-Difluorobenzene

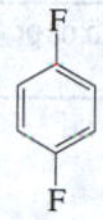
3551
p-Difluorobenzene

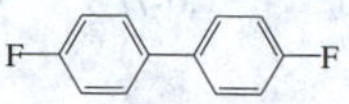
3552
4,4'-Difluoro-1,1'-biphenyl

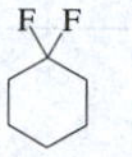
3553
1,1-Difluorocyclohexane

3554
3,3-Difluorocyclopropene

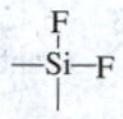
3555
Difluorodimethylsilane

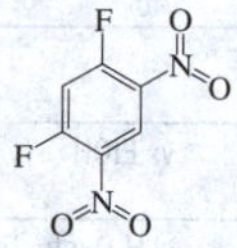
3556
1,5-Difluoro-2,4-dinitrobenzene

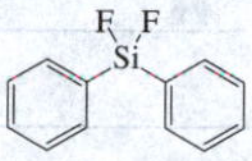
3557
Difluorodiphenylsilane

3558
1,1-Difluoroethane

3559
1,2-Difluoroethane

3560
1,1-Difluoroethene

3561
cis-1,2-Difluoroethene

3562
trans-1,2-Difluoroethene

3563
Difluoromethane

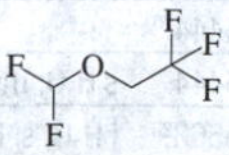
3564
2-(Difluoromethoxy)-1,1,1-trifluoroethane

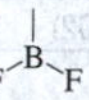
3565
Difluoromethylborane

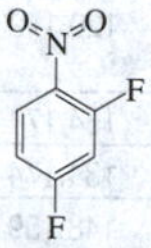
3566
2,4-Difluoro-1-nitrobenzene

3567
2,2-Difluoropropane

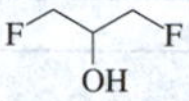
3568
1,3-Difluoro-2-propanol

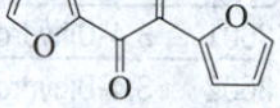
3569
Di-2-furanylethanedione

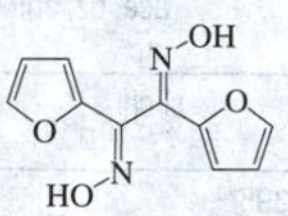
3570
Di-2-furanylethanedione dioxime

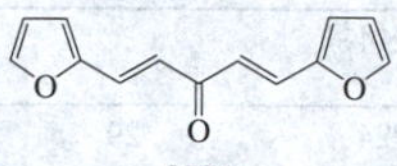
3571
1,5-Di-2-furanyl-1,4-pentadien-3-one

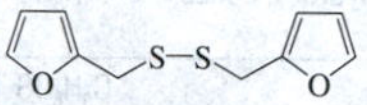
3572
Difurfuryl disulfide

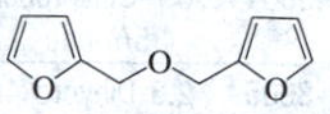
3573
Difurfuryl ether

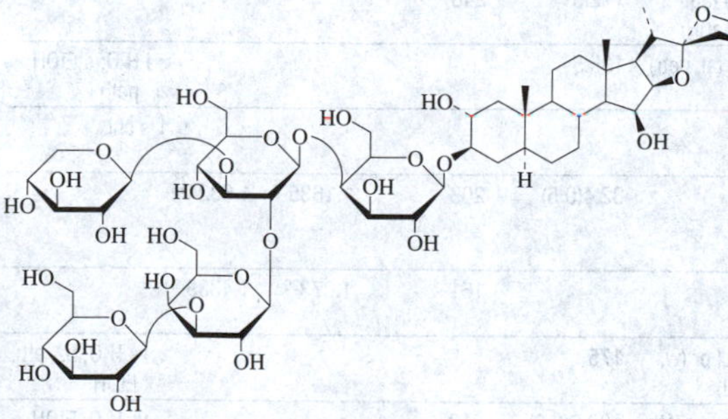
3574
Digitonin

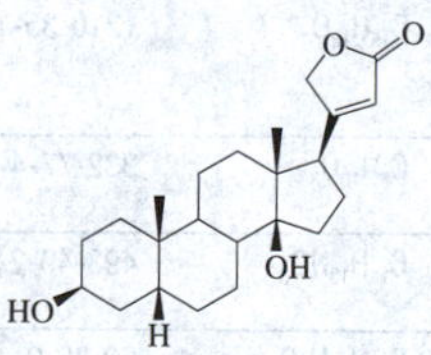
3575
Digitoxigenin

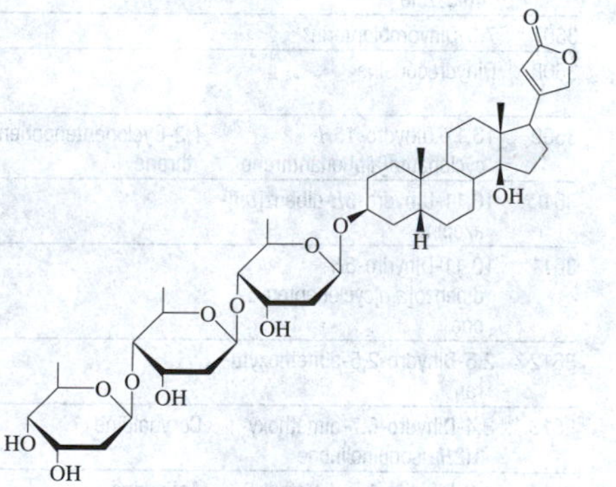
3576
Digitoxin

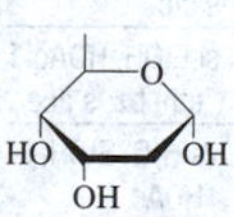
3577
Digitoxose

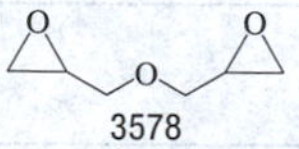
3578
Diglycidyl ether

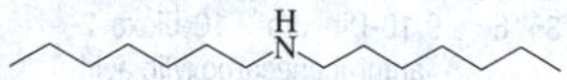
3582
Diheptylamine

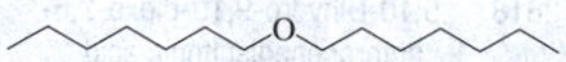
3583
Diheptyl ether

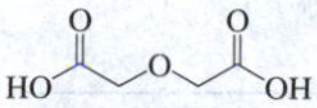
3579
Diglycolic acid

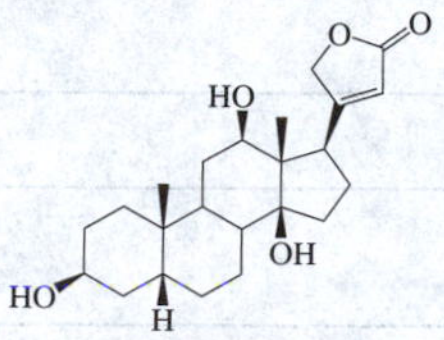
3580
Digoxigenin

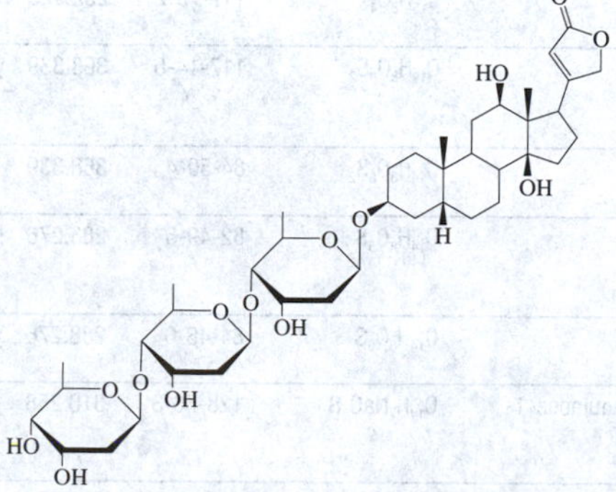
3581
Digoxin

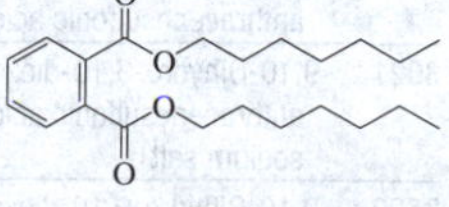
3584
Diheptyl phthalate

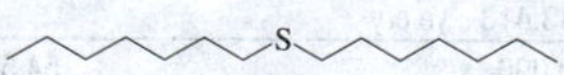
3585
Diheptyl sulfide

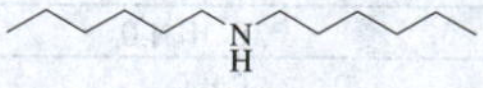
3586
Dihexylamine

No.	Name	Synonym	Mol. form.	CAS Reg. No.	Mol. wt.	Physical form	mp/°C	bp/°C	den/ g cm^{-3}	n_D	Solubility
3587	Dihexyl ether	Hexyl ether	$C_{12}H_{26}O$	112-58-3	186.333			220(4)	0.7936^{20}	1.4204^{20}	i H_2O; s eth; sl ctc
3588	Dihexyl hexanedioate		$C_{18}H_{34}O_4$	110-33-8	314.461	liq	-9	344(16)	0.941^{20}		
3589	Dihexyl phthalate		$C_{20}H_{30}O_4$	84-75-3	334.450			210^{5}			
3590	Dihexyl sulfide	Hexyl sulfide	$C_{12}H_{26}S$	6294-31-1	202.399			230	0.8411^{20}	1.4586^{20}	
3591	15,16-Dihydroaflatoxin G_1	Aflatoxin G_2	$C_{17}H_{14}O_7$	7241-98-7	330.289		239.3				
3592	9,10-Dihydroanthracene		$C_{14}H_{12}$	613-31-0	180.245	tab or pr	109.00(0.01)	305	1.215^{20}		i H_2O; s EtOH, eth, bz, chl
3593	6,15-Dihydro-5,9,14,18-anthrazinetetrone	Indanthrene	$C_{28}H_{14}N_2O_4$	81-77-6	442.422	bl nd	485 dec				i H_2O, EtOH, eth, ace, bz; s $PhNO_2$, dil alk
3594	1,2-Dihydrobenz[*j*]-aceanthrylene	Cholanthrene	$C_{20}H_{14}$	479-23-2	254.325	pa ye lf (bz-al)	170.4				i H_2O; s EtOH, bz, HOAc, lig, tol
3595	9,10-Dihydro-9,10[1',2']-benzenoanthracene	Triptycene	$C_{20}H_{14}$	477-75-8	254.325	cry (cyhex)	253.99(0.01)				
3596	1,3-Dihydro-2*H*-benzimid-azole-2-thione	2-Benzimidazolethiol	$C_7H_6N_2S$	583-39-1	150.201	pl (dil al or NH_3)	316.3(0.8)				vs EtOH
3597	1,3-Dihydro-2*H*-benzimidazol-2-one		$C_7H_6N_2O$	615-16-7	134.135	lf (w or al)	318 dec				sl H_2O, eth, bz; s ace; vs EtOH
3598	2,3-Dihydro-1,4-benzodioxin		$C_8H_8O_2$	493-09-4	136.149			213(1)	1.180^{20}	1.5485^{20}	
3599	2,3-Dihydrobenzofuran	Coumaran	C_8H_8O	496-16-2	120.149	liq	-21.5	187(19)	1.058^{25}	1.5497^{20}	vs eth, EtOH, chl
3600	3,4-Dihydro-1*H*-2-benzopyran	Isochroman	$C_9H_{10}O$	493-05-0	134.174		4.35(0.02)	110^{25}	1.067^{25}	1.5444^{20}	
3601	3,4-Dihydro-2*H*-1-benzopyran		$C_9H_{10}O$	493-08-3	134.174		-3.31(0.02)	215.6(0.8)	1.072^{20}	1.5444^{20}	s H_2O; msc os
3602	3,4-Dihydro-2*H*-1-benzopy-ran-2-one		$C_9H_8O_2$	119-84-6	148.159	lf	25	272	1.169^{18}	1.5563^{20}	i H_2O; sl EtOH, eth, ctc; s chl
3603	2,3-Dihydro-4*H*-1-benzopy-ran-4-one	4-Chromanone	$C_9H_8O_2$	491-37-2	148.159		39.1(0.4)	160^{50}	1.1291^{100}	1.5750	s EtOH; vs eth, ace, bz, chl; sl ctc
3604	6,7-Dihydrobenzo[*b*]thiophen-4(5*H*)-one	4,5,6,7-Tetrahydro-4-benzo-thiophenone	C_8H_8OS	13414-95-4	152.214						sl chl
3605	2,3-Dihydro-4*H*-1-benzothio-pyran-4-one		C_9H_8OS	3528-17-4	164.224		29	154^{12}	1.2487^{14}	1.6395^{20}	
3606	4,5-Dihydro-2-benzyl-1*H*-imidazole	Tolazoline	$C_{10}H_{12}N_2$	59-98-3	160.215	cry (peth)	67				
3607	7,8-Dihydrobiopterin		$C_9H_{13}N_5O_3$	6779-87-9	239.231	hyg nd (w)					s H_2O
3608	Dihydrocodeine		$C_{18}H_{23}NO_3$	125-28-0	301.381	cry (aq, MeOH)	112.5	248^{15}			
3609	16,17-Dihydro-15*H*-cyclopenta[*a*]phenanthrene	1,2-Cyclopentenophenan-threne	$C_{17}H_{14}$	482-66-6	218.293	nd (al, petr)	135.5				i H_2O; s EtOH, peth
3610	10,11-Dihydro-5*H*-dibenz[*b*,*f*]-azepine		$C_{14}H_{13}N$	494-19-9	195.260						s chl
3611	10,11-Dihydro-5*H*-dibenzo[*a*,*d*]cyclohepten-5-one		$C_{15}H_{12}O$	1210-35-1	208.255		32.4(0.5)	203^{7}	1.1635^{20}	1.6324^{20}	
3612	2,5-Dihydro-2,5-dimethoxyfu-ran		$C_6H_{10}O_3$	332-77-4	130.141			161	1.073^{25}	1.4339^{20}	
3613	3,4-Dihydro-6,7-dimethoxy-1(2*H*)-isoquinolinone	Corydaldine	$C_{11}H_{13}NO_3$	493-49-2	207.226	mcl pr (w, al)	175				vs H_2O, bz, eth, EtOH
3614	1,2-Dihydro-1,5-dimethyl-2-phenyl-3*H*-pyrazol-3-one	Antipyrine	$C_{11}H_{12}N_2O$	60-80-0	188.225	lf or sc (eth, bz)	108.0(0.2)	319			vs H_2O, EtOH
3615	2,3-Dihydro-1,4-dioxin		$C_4H_6O_2$	543-75-9	86.090			94.1	1.0836^{20}	1.4372^{20}	s ctc
3616	9,10-Dihydro-9,10-dioxo-2-anthracenecarboxylic acid		$C_{15}H_8O_4$	117-78-2	252.223	ye nd (HOAc)	291	sub			sl EtOH, HOAc; i eth, bz; s ace
3617	9,10-Dihydro-9,10-dioxo-1,5-anthracenedisulfonic acid		$C_{14}H_8O_8S_2$	117-14-6	368.339	ye nd (HCl +4w) pl (dil HOAc)	310 dec				vs H_2O, EtOH, HOAc
3618	9,10-Dihydro-9,10-dioxo-2,6-anthracenedisulfonic acid		$C_{14}H_8O_8S_2$	84-50-4	368.339						vs H_2O; s EtOH; i eth, bz
3619	9,10-Dihydro-9,10-dioxo-1-anthracenesulfonic acid		$C_{14}H_8O_5S$	82-49-5	288.276	lf (HOAc) ye lf (conc HCl, +3w)	216.0				vs H_2O, HOAc; s EtOH
3620	9,10-Dihydro-9,10-dioxo-2-anthracenesulfonic acid		$C_{14}H_8O_5S$	84-48-0	288.276	ye lf (+3w)					vs H_2O; s EtOH; i eth
3621	9,10-Dihydro-9,10-dioxo-1-anthracenesulfonic acid, sodium salt	Sodium anthraquinone-1-sulfonate	$C_{14}H_7NaO_5S$	128-56-3	310.258	ye lf (w)					sl H_2O
3622	9,10-Dihydro-9,10-dioxo-2-anthracenesulfonic acid, sodium salt		$C_{14}H_7NaO_5S$	131-08-8	310.258						sl DMSO
3623	7,8-Dihydrofolic acid		$C_{19}H_{21}N_7O_6$	4033-27-6	443.413	ye cry					
3624	2,3-Dihydrofuran		C_4H_6O	1191-99-7	70.090			54.5(0.2)	0.927^{25}	1.4239^{20}	

Organic

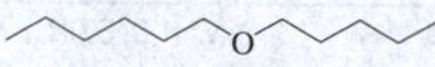
3587
Dihexyl ether

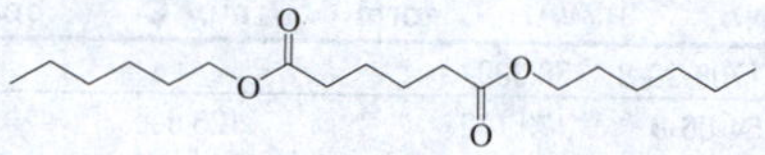
3588
Dihexyl hexanedioate

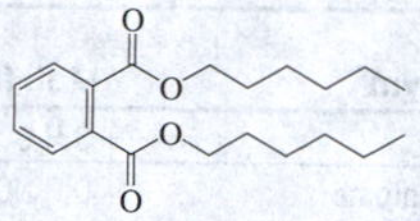
3589
Dihexyl phthalate

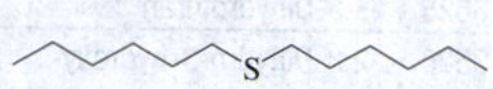
3590
Dihexyl sulfide

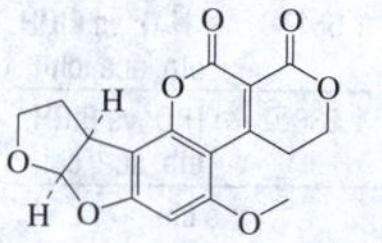
3591
15,16-Dihydroaflatoxin G_1

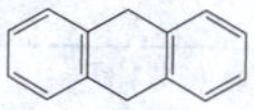
3592
9,10-Dihydroanthracene

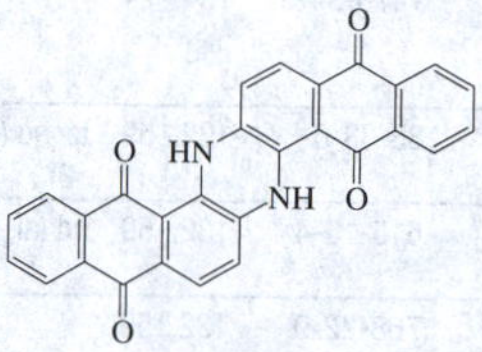
3593
6,15-Dihydro-5,9,14,18-anthrazinetetrone

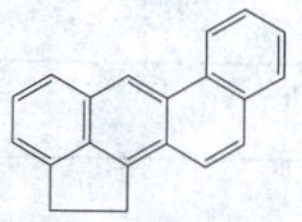
3594
1,2-Dihydrobenz[j]aceanthrylene

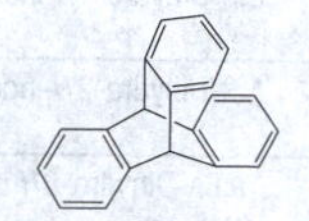
3595
9,10-Dihydro-9,10[1',2']-benzenoanthracene

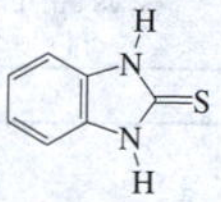
3596
1,3-Dihydro-2*H*-benzimidazole-2-thione

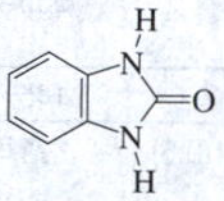
3597
1,3-Dihydro-2*H*-benzimidazol-2-one

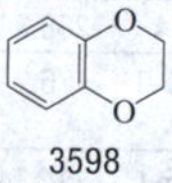
3598
2,3-Dihydro-1,4-benzodioxin

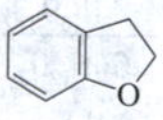
3599
2,3-Dihydrobenzofuran

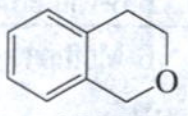
3600
3,4-Dihydro-1*H*-2-benzopyran

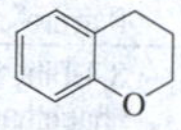
3601
3,4-Dihydro-2*H*-1-benzopyran

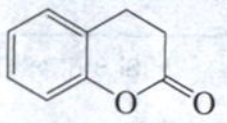
3602
3,4-Dihydro-2*H*-1-benzopyran-2-one

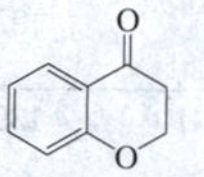
3603
2,3-Dihydro-4*H*-1-benzopyran-4-one

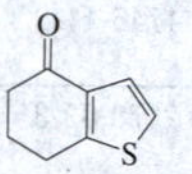
3604
6,7-Dihydrobenzo[b]thiophen-4(5*H*)-one

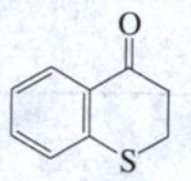
3605
2,3-Dihydro-4*H*-1-benzothiopyran-4-one

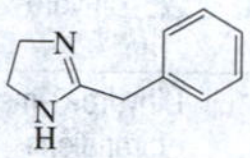
3606
4,5-Dihydro-2-benzyl-1*H*-imidazole

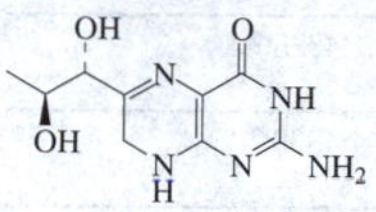
3607
7,8-Dihydrobiopterin

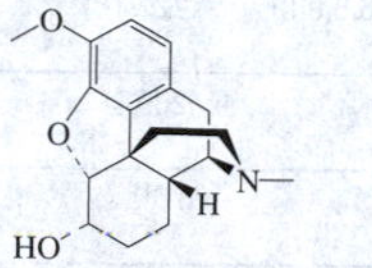
3608
Dihydrocodeine

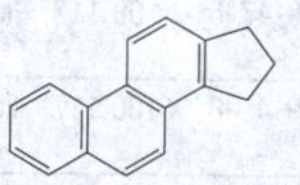
3609
16,17-Dihydro-15*H*-cyclopenta[a]phenanthrene

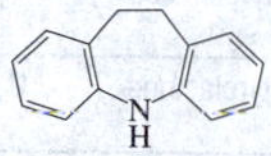
3610
10,11-Dihydro-5*H*-dibenz[b,f]azepine

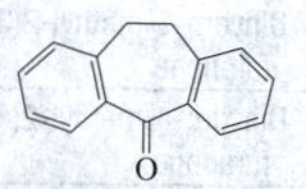
3611
10,11-Dihydro-5*H*-dibenzo[a,d]cyclohepten-5-one

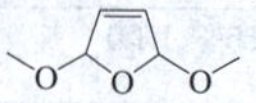
3612
2,5-Dihydro-2,5-dimethoxyfuran

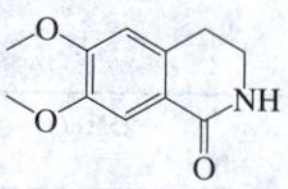
3613
3,4-Dihydro-6,7-dimethoxy-1(2*H*)-isoquinolinone

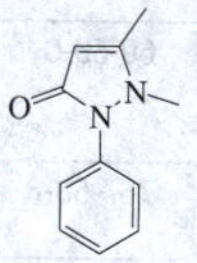
3614
1,2-Dihydro-1,5-dimethyl-2-phenyl-3*H*-pyrazol-3-one

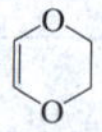
3615
2,3-Dihydro-1,4-dioxin

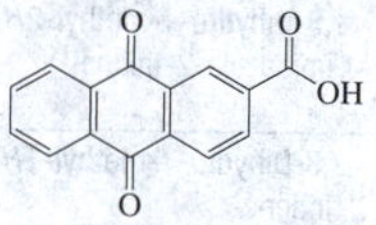
3616
9,10-Dihydro-9,10-dioxo-2-anthracenecarboxylic acid

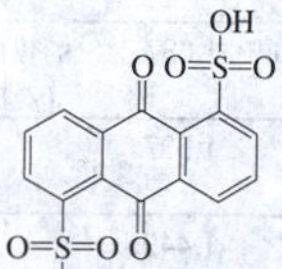
3617
9,10-Dihydro-9,10-dioxo-1,5-anthracenedisulfonic acid

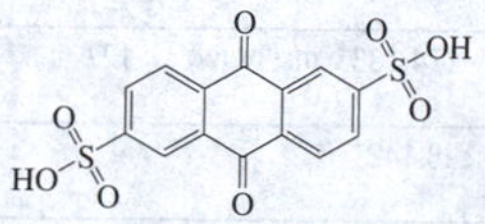
3618
9,10-Dihydro-9,10-dioxo-2,6-anthracenedisulfonic acid

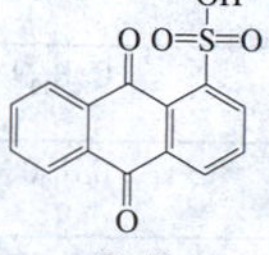
3619
9,10-Dihydro-9,10-dioxo-1-anthracenesulfonic acid

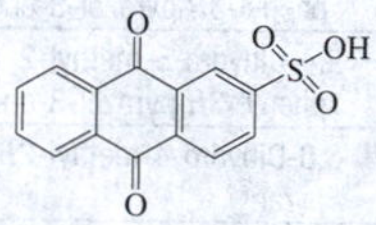
3620
9,10-Dihydro-9,10-dioxo-2-anthracenesulfonic acid

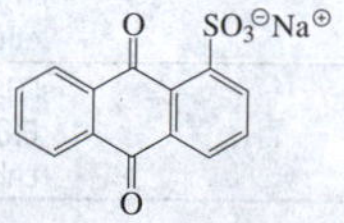
3621
9,10-Dihydro-9,10-dioxo-1-anthracenesulfonic acid, sodium salt

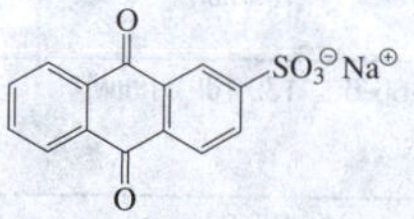
3622
9,10-Dihydro-9,10-dioxo-2-anthracenesulfonic acid, sodium salt

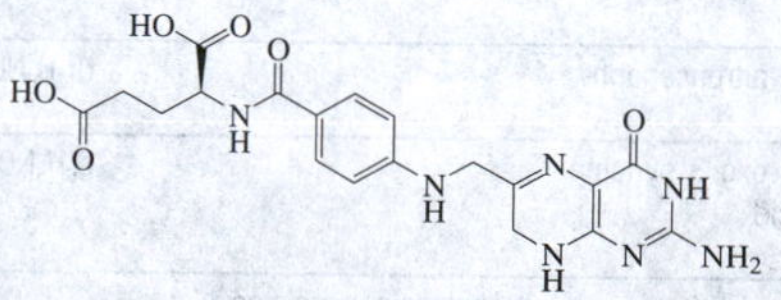
3623
7,8-Dihydrofolic acid

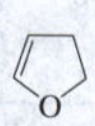
3624
2,3-Dihydrofuran

No.	Name	Synonym	Mol. form.	CAS Reg. No.	Mol. wt.	Physical form	mp/°C	bp/°C	den/g cm^{-3}	n_D	Solubility
3625	2,5-Dihydrofuran		C_4H_6O	1708-29-8	70.090					1.4311^{20}	
3626	2,3-Dihydro-3-hydroxy-1-methyl-1*H*-indole-5,6-dione	Adrenochrome	$C_9H_9NO_3$	54-06-8	179.172		125 dec				vs H_2O, EtOH; i eth, bz
3627	2,3-Dihydro-1*H*-inden-5-amine		$C_9H_{11}N$	24425-40-9	133.190	nd (peth)	37.5	248			sl H_2O, chl; s eth, ace, bz
3628	2,3-Dihydro-1*H*-inden-1-ol		$C_9H_{10}O$	6351-10-6	134.174	pl (peth)	55(2)	220			vs bz, EtOH, chl
3629	2,3-Dihydro-1*H*-inden-5-ol		$C_9H_{10}O$	1470-94-6	134.174		58	253			sl H_2O, peth; vs EtOH, eth; s sulf
3630	2,3-Dihydro-1*H*-inden-1-one		C_9H_8O	83-33-0	132.159	ta, nd (w + 3)	40(2)	243(8)	1.0943^{40}	1.561^{25}	sl H_2O; vs EtOH, eth, ace, chl
3631	1,3-Dihydro-2*H*-inden-2-one	2-Indanone	C_9H_8O	615-13-4	132.159	nd (al, eth)	58(3)	218 dec	1.0712^{69}	1.538^{67}	i H_2O; vs EtOH, eth, ace, chl
3632	1a,6a-Dihydro-6*H*-indeno[1,2-*b*]oxirene		C_9H_8O	768-22-9	132.159		24.5	113^{20}	1.1255^{24}		s chl
3633	2,3-Dihydro-1*H*-indole		C_8H_9N	496-15-1	119.164			230(13)	1.069^{20}	1.5923^{20}	sl H_2O; s eth, ace, bz
3634	1,3-Dihydro-2*H*-indol-2-one		C_8H_7NO	59-48-3	133.148	nd (w)	128	227^{23}			s H_2O, EtOH, eth
3635	2,3-Dihydro-1*H*-isoindol-1-one		C_8H_7NO	480-91-1	133.148	nd (w)	151	338			vs eth, EtOH, chl
3636	Dihydro-α-lipoic acid	6,8-Dimercaptooctanoic acid	$C_8H_{16}O_2S_2$	462-20-4	208.342	ye liq		$145^{0.2}$			
3637	3,4-Dihydro-6-methoxy-1(2*H*)-naphthalenone	6-Methoxy-α-tetralone	$C_{11}H_{12}O_2$	1078-19-9	176.212	cry (MeOH, lig)	78.1(0.5)	171^{11}			
3638	3,4-Dihydro-2-methoxy-2*H*-pyran		$C_6H_{10}O_2$	4454-05-1	114.142	liq		127(23)	1.006	1.4420^{20}	
3639	1,2-Dihydro-3-methylbenz[*j*]-aceanthrylene	3-Methylcholanthrene	$C_{21}H_{16}$	56-49-5	268.352	ye nd (bz)	178.0(0.2)	280^{80}	1.28^{20}		i H_2O
3640	2,3-Dihydro-2-methylbenzofuran		$C_9H_{10}O$	1746-11-8	134.174			197.5	1.061^{25}	1.5308	
3641	Dihydro-3-methylene-2,5-furandione		$C_5H_4O_3$	2170-03-8	112.084	orth bipym pr (eth, chl)	68(1)	139^{30}			sl eth; vs chl
3642	Dihydro-3-methylene-2(3*H*)-furanone	α-Methylene butyrolactone	$C_5H_6O_2$	547-65-9	98.101			85^{10}	1.1206^{20}	1.4650^{20}	s H_2O, eth, ace, bz; sl ctc; vs EtOH
3643	Dihydro-3-methyl-2,5-furandione		$C_5H_6O_3$	4100-80-5	114.100		36.8(0.8)	221(16)	1.22^{25}		
3644	Dihydro-3-methyl-2(3*H*)-furanone	2-Methyl-γ-butyrolactone	$C_5H_8O_2$	1679-47-6	100.117	liq		200	1.0570^{20}	1.4325^{20}	
3645	Dihydro-4-methyl-2(3*H*)-furanone	3-Methyl-γ-butyrolactone	$C_5H_8O_2$	1679-49-8	100.117	liq		76^{11}	1.058^{20}	1.4339^{20}	
3646	Dihydro-5-methyl-2(3*H*)-furanone, (±)-	(±)-γ-Valerolactone	$C_5H_8O_2$	57129-69-8	100.117	liq	-31	204(4)	1.0551^{20}	1.4328^{20}	msc H_2O; s EtOH, ace; sl ctc
3647	4,5-Dihydro-2-methyl-1*H*-imidazole	Lysidine	$C_4H_8N_2$	534-26-9	84.120	hyg	107	196.5			vs H_2O, EtOH; i eth; s chl
3648	1,3-Dihydro-1-methyl-2*H*-imidazole-2-thione	Methimazole	$C_4H_6N_2S$	60-56-0	114.169	lf (al)	146	280 dec			vs H_2O; s EtOH, chl; sl eth, bz, lig
3649	2,3-Dihydro-1-methyl-1*H*-indene		$C_{10}H_{12}$	767-58-8	132.202			191(7)	0.938^{25}	1.5266^{20}	i H_2O
3650	3,4-Dihydro-2-methyl-1(2*H*)-naphthalenone		$C_{11}H_{12}O$	1590-08-5	160.212		15	136^{16}	1.057^{25}	1.5535^{20}	
3651	4-(4,5-Dihydro-3-methyl-5-oxo-1*H*-pyrazol-1-yl)-benzenesulfonic acid		$C_{10}H_{10}N_2O_4S$	89-36-1	254.262	nd (w+1)	≈300 dec				
3652	1,2-Dihydro-5-methyl-2-phenyl-3*H*-pyrazol-3-one	5-Hydroxy-3-methyl-1-phenyl-pyrazole	$C_{10}H_{10}N_2O$	19735-89-8	174.198		128	287^{105}	1.2600^{20}	1.637	s H_2O, EtOH; sl bz; i peth
3653	2,4-Dihydro-5-methyl-2-phenyl-3*H*-pyrazol-3-one		$C_{10}H_{10}N_2O$	89-25-8	174.198	mcl pr (w)	127	287^{105}		1.637	
3654	3,6-Dihydro-4-methyl-2*H*-pyran		$C_6H_{10}O$	16302-35-5	98.142			118(6)	0.912^{25}	1.4495^{20}	
3655	4,5-Dihydro-2-methylthiazole		C_4H_7NS	2346-00-1	101.171	liq	-101	145	1.067^{25}	1.5200^{20}	
3656	1,2-Dihydronaphthalene		$C_{10}H_{10}$	447-53-0	130.186	liq	-8.77(0.05)	210.2(0.7)	0.9974^{20}	1.5814^{20}	
3657	1,4-Dihydronaphthalene	Δ 2-Dialin	$C_{10}H_{10}$	612-17-9	130.186	pl	24.6(0.9)	210(5)	0.9928^{33}	1.5577^{20}	
3658	3,4-Dihydro-2(1*H*)-naphthalenone		$C_{10}H_{10}O$	530-93-8	146.185		18	237	1.1055^{27}	1.5598^{20}	i H_2O; s eth, bz
3659	1,2-Dihydro-5-nitroacenaphthylene		$C_{12}H_9NO_2$	602-87-9	199.205		103				s H_2O, EtOH, eth, lig
3660	1,6-Dihydro-6-oxo-3-pyridinecarboxylic acid		$C_6H_5NO_3$	5006-66-6	139.109	nd(w)	310 dec	sub			sl H_2O, tfa; i EtOH, eth, bz, chl
3661	Dihydro-5-pentyl-2(3*H*)-furanone	4-Hydroxynonanoic acid lactone	$C_9H_{16}O_2$	104-61-0	156.222	oil		134^{12}			

Organic

3625
2,5-Dihydrofuran

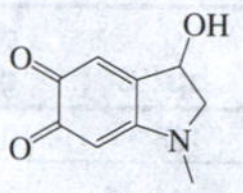
3626
2,3-Dihydro-3-hydroxy-1-methyl-1*H*-indole-5,6-dione

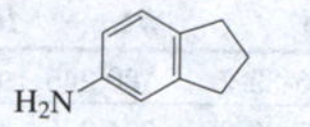
3627
2,3-Dihydro-1*H*-inden-5-amine

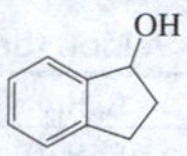
3628
2,3-Dihydro-1*H*-inden-1-ol

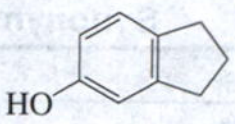
3629
2,3-Dihydro-1*H*-inden-5-ol

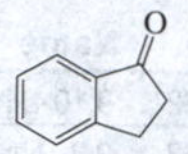
3630
2,3-Dihydro-1*H*-inden-1-one

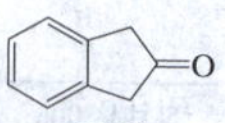
3631
1,3-Dihydro-2*H*-inden-2-one

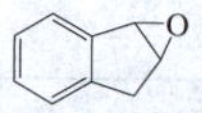
3632
1a,6a-Dihydro-6*H*-indeno[1,2-b]oxirene

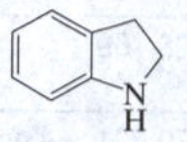
3633
2,3-Dihydro-1*H*-indole

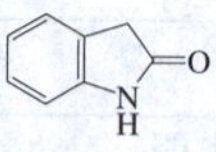
3634
1,3-Dihydro-2*H*-indol-2-one

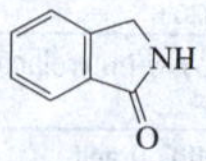
3635
2,3-Dihydro-1*H*-isoindol-1-one

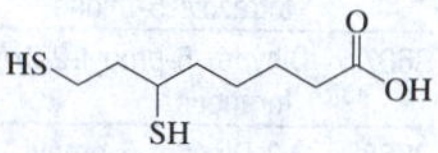
3636
Dihydro-α-lipoic acid

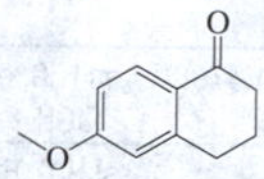
3637
3,4-Dihydro-6-methoxy-1(2*H*)-naphthalenone

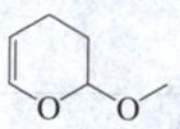
3638
3,4-Dihydro-2-methoxy-2*H*-pyran

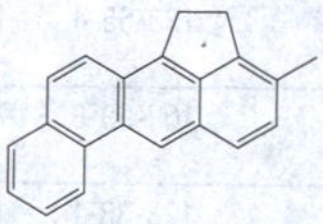
3639
1,2-Dihydro-3-methylbenz[j]aceanthrylene

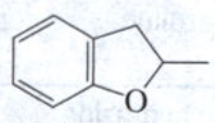
3640
2,3-Dihydro-2-methylbenzofuran

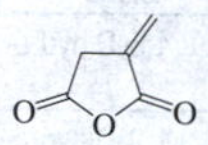
3641
Dihydro-3-methylene-2,5-furandione

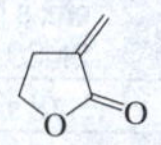
3642
Dihydro-3-methylene-2(3*H*)-furanone

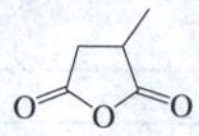
3643
Dihydro-3-methyl-2,5-furandione

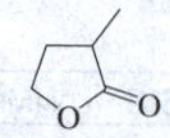
3644
Dihydro-3-methyl-2(3*H*)-furanone

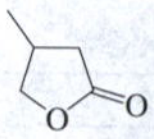
3645
Dihydro-4-methyl-2(3*H*)-furanone

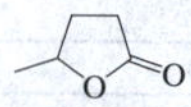
3646
Dihydro-5-methyl-2(3*H*)-furanone, (±)-

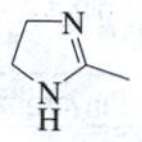
3647
4,5-Dihydro-2-methyl-1*H*-imidazole

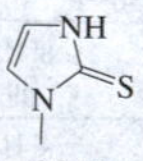
3648
1,3-Dihydro-1-methyl-2*H*-imidazole-2-thione

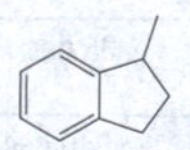
3649
2,3-Dihydro-1-methyl-1*H*-indene

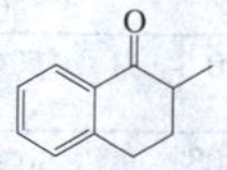
3650
3,4-Dihydro-2-methyl-1(2*H*)-naphthalenone

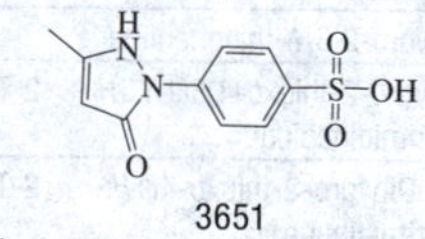
3651
4-(4,5-Dihydro-3-methyl-5-oxo-1*H*-pyrazol-1-yl)benzenesulfonic acid

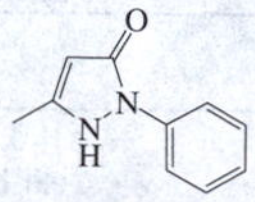
3652
1,2-Dihydro-5-methyl-2-phenyl-3*H*-pyrazol-3-one

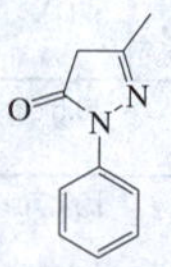
3653
2,4-Dihydro-5-methyl-2-phenyl-3*H*-pyrazol-3-one

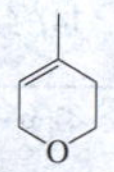
3654
3,6-Dihydro-4-methyl-2*H*-pyran

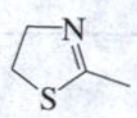
3655
4,5-Dihydro-2-methylthiazole

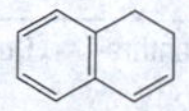
3656
1,2-Dihydronaphthalene

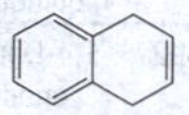
3657
1,4-Dihydronaphthalene

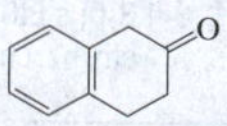
3658
3,4-Dihydro-2(1*H*)-naphthalenone

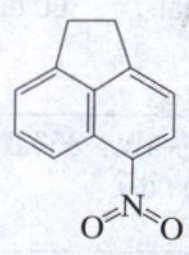
3659
1,2-Dihydro-5-nitroacenaphthylene

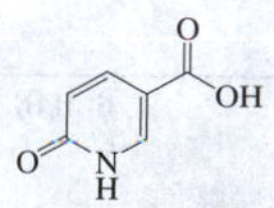
3660
1,6-Dihydro-6-oxo-3-pyridinecarboxylic acid

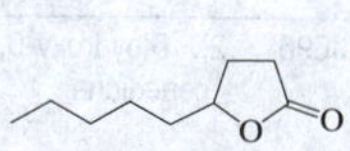
3661
Dihydro-5-pentyl-2(3*H*)-furanone

Organic

No.	Name	Synonym	Mol. form.	CAS Reg. No.	Mol. wt.	Physical form	mp/°C	bp/°C	den/ g cm⁻³	n_D	Solubility
3662	9,10-Dihydrophenanthrene		$C_{14}H_{12}$	776-35-2	180.245	nd (MeOH)	33.36(0.05)	168^{15}	1.0757^{40}	1.6415^{20}	s chl
3663	2,3-Dihydro-2-phenyl-4H-1-benzopyran-4-one		$C_{15}H_{12}O_2$	487-26-3	224.255	nd (lig)	76.3(0.5)				i H_2O; s ace, bz; sl ctc
3664	4,5-Dihydro-2-(phenylmethyl)-1H-imidazole, monohydrochloride		$C_{10}H_{13}ClN_2$	59-97-2	196.676		174				
3665	4,5-Dihydro-5-phenyl-2-oxazolamine	Aminorex	$C_9H_{10}N_2O$	2207-50-3	162.187	cry (bz)	137				
3666	1,4-Dihydro-1-phenyl-5H-tetrazole-5-thione	1-Phenyl-5-mercapto-1H-tetrazole	$C_7H_6N_4S$	86-93-1	178.215		145				
3667	Dihydro-5-propyl-2(3H)-furanone	γ-Propyl-γ-butyrolactone	$C_7H_{12}O_2$	105-21-5	128.169			84^{5}		1.4385^{25}	
3668	2,3-Dihydro-6-propyl-2-thioxo-4(1H)-pyrimidinone	Propylthiouracil	$C_7H_{10}N_2OS$	51-52-5	170.231	w pow (w)	219				sl H_2O, chl, DMSO, EtOH; i eth, bz
3669	1,7-Dihydro-6H-purine-6-thione	6-Mercaptopurine	$C_5H_4N_4S$	50-44-2	152.178	ye pr (w, + l w)	313 dec				i H_2O; s alk
3670	3,4-Dihydro-2H-pyran		C_5H_8O	110-87-2	84.117			85.5(0.2)	0.921^{19}	1.4402^{19}	s H_2O, EtOH; sl chl
3671	3,6-Dihydro-2H-pyran		C_5H_8O	3174-74-1	84.117	liq		95	0.94^{19}		
3672	Dihydro-2H-pyran-2,6(3H)-dione		$C_5H_6O_3$	108-55-4	114.100		56.3	158^{15}	1.4110^{20}		
3673	4,5-Dihydro-1H-pyrazole	2-Pyrazoline	$C_3H_6N_2$	109-98-8	70.093			144	1.0200^{17}	1.4796^{17}	vs H_2O, eth, EtOH
3674	1,2-Dihydro-3,6-pyridazinedione	Maleic hydrazide	$C_4H_4N_2O_2$	123-33-1	112.087	cry (w)	307				sl H_2O, EtOH, tfa
3675	Dihydro-2,4(1H,3H)-pyrimidinedione	5,6-Dihydrouracil	$C_4H_6N_2O_2$	504-07-4	114.103	nd (w)	275.5				vs H_2O; s EtOH, chl, MeOH
3676	2,5-Dihydro-1H-pyrrole	3-Pyrroline	C_4H_7N	109-96-6	69.106			90(5)	0.9097^{20}	1.4664^{20}	vs H_2O, ace, eth, EtOH
3677	3,4-Dihydro-2(1H)-quinolinone	Hydrocarbostyril	C_9H_9NO	553-03-7	147.173	pr (al, eth)	163.5	201^{45}			vs eth, EtOH
3678	1,4-Dihydro-2,3-quinoxalinedione	2,3-Quinoxalinediol	$C_8H_6N_2O_2$	15804-19-0	162.146	nd (w)	410				vs H_2O; sl EtOH, eth; s bz, DMSO, HOAc
3679	Dihydrotachysterol		$C_{28}H_{46}O$	67-96-9	398.664	cry (MeOH)	131				i H_2O; s os
3680	Dihydrothebaine		$C_{19}H_{23}NO_3$	561-25-1	313.391		162.5				i H_2O; s EtOH, bz, AcOEt
3681	4,5-Dihydro-2-thiazolamine		$C_3H_6N_2S$	1779-81-3	102.158	nd or lf (bz)	85.3	dec			vs H_2O, EtOH, bz, chl
3682	2,3-Dihydrothiophene		C_4H_6S	1120-59-8	86.156			112(13)			
3683	2,5-Dihydrothiophene		C_4H_6S	1708-32-3	86.156			122.4			
3684	2,5-Dihydrothiophene 1,1-dioxide	3-Sulfolene	$C_4H_6O_2S$	77-79-2	118.155		64.0(0.4)				s chl
3685	Dihydro-2(3H)-thiophenone		C_4H_6OS	1003-10-7	102.155			111^{52}	1.18^{25}	1.5230^{20}	
3686	Dihydro-2-thioxo-4,6(1H,5H)-pyrimidinedione	2-Thiobarbituric acid	$C_4H_4N_2O_2S$	504-17-6	144.152	pl (w)	235 dec				sl H_2O; s EtOH, dil alk, dil HCl
3687	2,3-Dihydro-2-thioxo-4(1H)-pyrimidinone	2-Thiouracil	$C_4H_4N_2OS$	141-90-2	128.152	pr (w, al)	>340 dec				sl H_2O, EtOH, DMSO; s anh HF
3688	1,2-Dihydro-3H-1,2,4-triazole-3-thione		$C_2H_3N_3S$	3179-31-5	101.130		222.5				s DMSO
3689	(1,3-Dihydro-1,3,3-trimethyl-2H-indol-2-ylidene) acetaldehyde		$C_{13}H_{15}NO$	84-83-3	201.264						s chl
3690	2,3-Dihydro-1,1,3-trimethyl-3-phenyl-1H-indene		$C_{18}H_{20}$	3910-35-8	236.352	tcl pr (al)	52(1)	308.5	1.0009^{20}	1.5681^{20}	i H_2O; s EtOH, bz, MeOH
3691	1,2-Dihydro-2,2,4-trimethylquinoline		$C_{12}H_{15}N$	147-47-7	173.254		26.5	260			
3692	1,4-Dihydroxy-9,10-anthracenedione	Quinizarin	$C_{14}H_8O_4$	81-64-1	240.212	ye red lf (eth) dk red nd	199.7(0.4)				s H_2O, EtOH, eth, bz, KOH, sulf
3693	1,5-Dihydroxy-9,10-anthracenedione	Anthrarufin	$C_{14}H_8O_4$	117-12-4	240.212	pa ye pl (gl HOAc)	280	sub			i H_2O; sl EtOH, eth, ace, CS_2; s bz
3694	1,8-Dihydroxy-9,10-anthracenedione	Danthron	$C_{14}H_8O_4$	117-10-2	240.212	red or red-ye nd or lf (al)	193	sub			i H_2O; sl EtOH, eth; s ace, HOAc, alk
3695	2,6-Dihydroxy-9,10-anthracenedione		$C_{14}H_8O_4$	84-60-6	240.212	ye nd (al)	360 dec				sl H_2O, EtOH; i eth, bz, chl; s alk
3696	2,7-Dihydroxy-9,10-anthracenedione		$C_{14}H_8O_4$	572-93-0	240.212	ye nd (+1w, dil al) nd (sub)	353.8	sub			i H_2O; s EtOH; sl eth, bz, chl

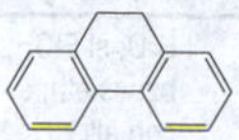
3662
9,10-Dihydrophenanthrene

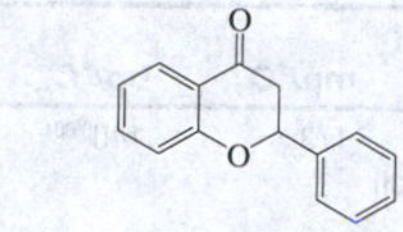
3663
2,3-Dihydro-2-phenyl-4*H*-1-benzopyran-4-one

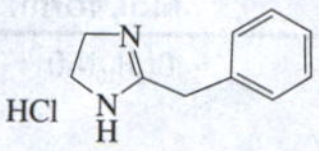

3664
4,5-Dihydro-2-(phenylmethyl)-1*H*-imidazole, monohydrochloride

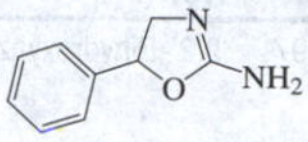
3665
4,5-Dihydro-5-phenyl-2-oxazolamine

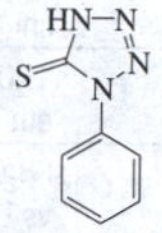
3666
1,4-Dihydro-1-phenyl-5*H*-tetrazole-5-thione

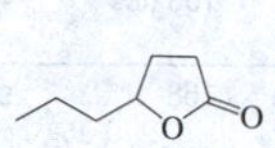
3667
Dihydro-5-propyl-2(3*H*)-furanone

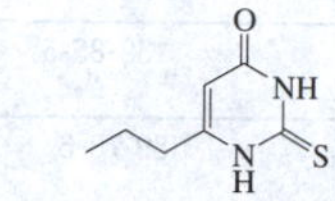
3668
2,3-Dihydro-6-propyl-2-thioxo-4(1*H*)-pyrimidinone

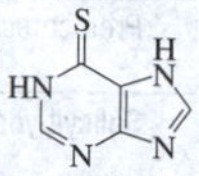
3669
1,7-Dihydro-6*H*-purine-6-thione

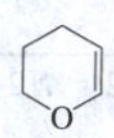
3670
3,4-Dihydro-2*H*-pyran

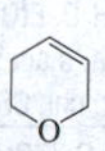
3671
3,6-Dihydro-2*H*-pyran

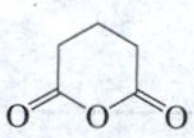
3672
Dihydro-2*H*-pyran-2,6(3*H*)-dione

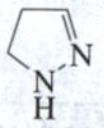
3673
4,5-Dihydro-1*H*-pyrazole

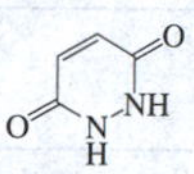
3674
1,2-Dihydro-3,6-pyridazinedione

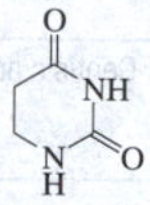
3675
Dihydro-2,4(1*H*,3*H*)-pyrimidinedione

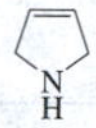
3676
2,5-Dihydro-1*H*-pyrrole

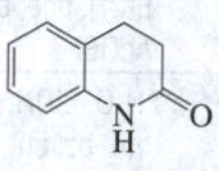
3677
3,4-Dihydro-2(1*H*)-quinolinone

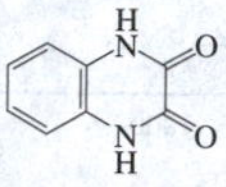
3678
1,4-Dihydro-2,3-quinoxalinedione

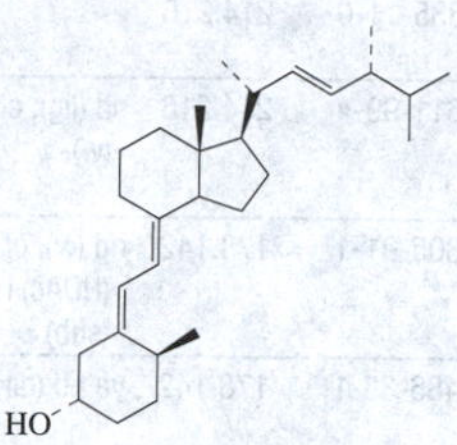
3679
Dihydrotachysterol

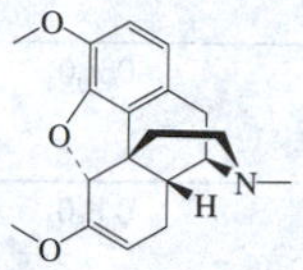
3680
Dihydrothebaine

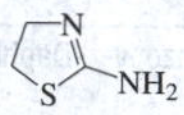
3681
4,5-Dihydro-2-thiazolamine

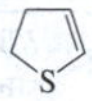
3682
2,3-Dihydrothiophene

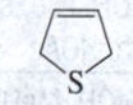
3683
2,5-Dihydrothiophene

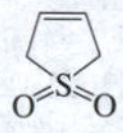
3684
2,5-Dihydrothiophene 1,1-dioxide

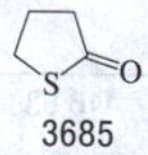
3685
Dihydro-2(3*H*)-thiophenone

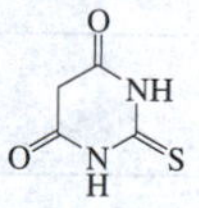
3686
Dihydro-2-thioxo-4,6(1*H*,5*H*)-pyrimidinedione

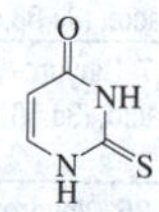
3687
2,3-Dihydro-2-thioxo-4(1*H*)-pyrimidinone

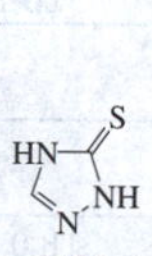
3688
1,2-Dihydro-3*H*-1,2,4-triazole-3-thione

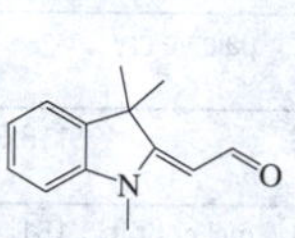
3689
(1,3-Dihydro-1,3,3-trimethyl-2*H*-indol-2-ylidene)acetaldehyde

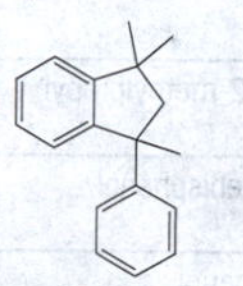
3690
2,3-Dihydro-1,1,3-trimethyl-3-phenyl-1*H*-indene

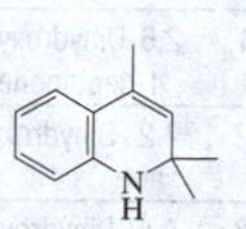
3691
1,2-Dihydro-2,2,4-trimethylquinoline

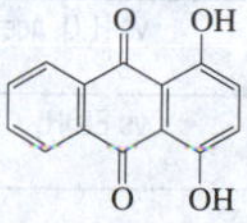
3692
1,4-Dihydroxy-9,10-anthracenedione

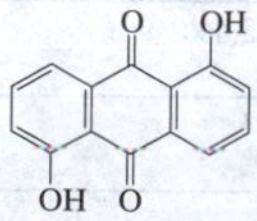
3693
1,5-Dihydroxy-9,10-anthracenedione

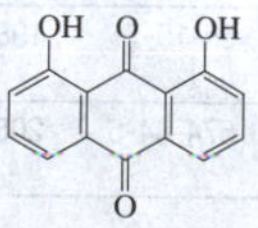
3694
1,8-Dihydroxy-9,10-anthracenedione

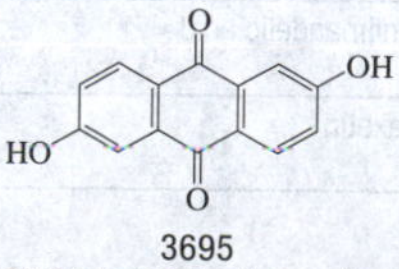
3695
2,6-Dihydroxy-9,10-anthracenedione

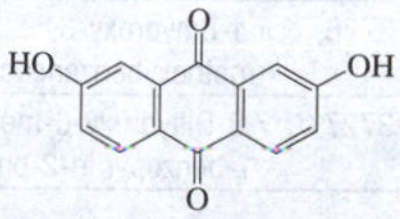
3696
2,7-Dihydroxy-9,10-anthracenedione

No.	Name	Synonym	Mol. form.	CAS Reg. No.	Mol. wt.	Physical form	mp/°C	bp/°C	den/g cm^{-3}	n_D	Solubility
3697	2,2'-Dihydroxyazobenzene		$C_{12}H_{10}N_2O_2$	2050-14-8	214.219	gold-ye lf (bz), nd (al)	173	$140^{0.001}$			i H_2O; sl EtOH, bz; vs eth; s con alk
3698	2,3-Dihydroxybenzaldehyde		$C_7H_6O_3$	24677-78-9	138.121	ye nd	108	235			vs ace, EtOH, HOAc
3699	2,4-Dihydroxybenzaldehyde	β-Resorcylaldehyde	$C_7H_6O_3$	95-01-2	138.121	nd (eth-lig)	135	226^{22}			s H_2O, HOAc; vs EtOH, eth, chl; sl bz
3700	2,5-Dihydroxybenzaldehyde		$C_7H_6O_3$	1194-98-5	138.121	ye nd (bz)	100.0				vs H_2O, EtOH, chl
3701	3,4-Dihydroxybenzaldehyde	Protocatechualdehyde	$C_7H_6O_3$	139-85-5	138.121	lf (w, to)	153 dec				s H_2O; vs EtOH, eth
3702	*N*,2-Dihydroxybenzamide	Salicylhydroxamic acid	$C_7H_7NO_3$	89-73-6	153.136	nd (HOAc)	168	sub			sl H_2O, DMSO; vs EtOH, eth; s HOAc
3703	2,5-Dihydroxybenzeneacetic acid	Homogentisic acid	$C_8H_8O_4$	451-13-8	168.148	pr (w+1), lf (al-chl)	153				vs H_2O, EtOH, eth; i bz, chl
3704	2,3-Dihydroxybenzoic acid		$C_7H_6O_4$	303-38-8	154.121	pr or nd (w+1)	205(4)		1.542^{20}		s H_2O, EtOH, eth; sl ace
3705	2,4-Dihydroxybenzoic acid	β-Resorcylic acid	$C_7H_6O_4$	89-86-1	154.121	cry (+w)	229(1)				s H_2O, EtOH, eth, bz; i CS_2
3706	2,5-Dihydroxybenzoic acid	Gentisic acid	$C_7H_6O_4$	490-79-9	154.121	nd or pr (w)	204(3)				vs H_2O, EtOH, eth; s ace; i bz, chl, CS_2
3707	2,6-Dihydroxybenzoic acid		$C_7H_6O_4$	303-07-1	154.121	nd (+w)	171(1)				s H_2O, EtOH, eth; i chl; sl tfa
3708	3,4-Dihydroxybenzoic acid	Protocatechuic acid	$C_7H_6O_4$	99-50-3	154.121	mcl nd (w+1)	202(1)		1.524^{4}		sl H_2O; vs EtOH; s eth; i bz
3709	3,5-Dihydroxybenzoic acid		$C_7H_6O_4$	99-10-5	154.121	pr or nd	235.3(0.8)				sl H_2O, ace; vs EtOH, eth
3710	2,2'-Dihydroxybenzophenone	Bis(2-hydroxyphenyl) ketone	$C_{13}H_{10}O_3$	835-11-0	214.216		59.5	333			i H_2O; s EtOH, eth, chl
3711	4,4'-Dihydroxybenzophenone	Bis(4-hydroxyphenyl) ketone	$C_{13}H_{10}O_3$	611-99-4	214.216	nd (lig), cry (w)	210		1.133^{131}		sl H_2O; s EtOH, eth, ace; i bz, CS_2
3712	6,7-Dihydroxy-2*H*-1-benzopyran-2-one	Esculetin	$C_9H_6O_4$	305-01-1	178.142	nd (w), pr (HOAc) lf (sub)	276	sub			sl H_2O, eth; s EtOH, ace, chl, AcOEt
3713	7,8-Dihydroxy-2*H*-1-benzopyran-2-one	Daphnetin	$C_9H_6O_4$	486-35-1	178.142	ye nd (dil al)	262	sub			s H_2O, EtOH; sl eth, bz, chl, CS_2
3714	2,4-Dihydroxybutanoic acid		$C_4H_8O_4$	1518-62-3	120.105	liq		96^{3}			
3715	3,6-Dihydroxycholan-24-oic acid, (3α,5β,6α)	Hyodeoxycholic acid	$C_{24}H_{40}O_4$	83-49-8	392.573	cry (AcOEt)	198.5				sl H_2O, eth, ace, bz; s EtOH, HOAc
3716	3,7-Dihydroxycholan-24-oic acid, (3α,5β,7β)	Ursodiol	$C_{24}H_{40}O_4$	128-13-2	392.573	pl (al)	203				vs EtOH; sl eth
3717	3,7-Dihydroxycholan-24-oic acid, (3α,5β,7α)	Chenodiol	$C_{24}H_{40}O_4$	474-25-9	392.573	nd (EtOAc +hep)	119				i H_2O, bz; vs EtOH, ace; s eth, HOAc
3718	1,25-Dihydroxycholecalciferol	Calcitriol	$C_{27}H_{44}O_3$	32222-06-3	416.636	wh cry pow	115				sl EtOH, MeOH, thf, AcOEt
3719	2,5-Dihydroxy-2,5-cyclohexadiene-1,4-dione		$C_6H_4O_4$	615-94-1	140.094	dk ye nd	211				sl H_2O, ace, DMSO; s EtOH, HOAc; i eth
3720	2,3-Dihydroxy-2-cyclopenten-1-one	Reductic acid	$C_5H_6O_3$	80-72-8	114.100		212				s H_2O, EtOH; sl eth, ace, AcOEt; i bz
3721	2,6-Dihydroxy-2,6-dimethyl-4-heptanone	Di(2-hydroxy-2-methylpropyl) ketone	$C_9H_{18}O_3$	3682-91-5	174.237	pale ye cry					
3722	2,2'-Dihydroxydiphenylmethane	2,2'-Methylenebisphenol	$C_{13}H_{12}O_2$	2467-02-9	200.233		118.3	363	1.280^{25}		
3723	4,4'-Dihydroxydiphenyl sulfide	4,4'-Thiobisphenol	$C_{12}H_{10}O_2S$	2664-63-3	218.271	mcl pr or lf (al)	151				sl H_2O, EtOH, eth, CS_2
3724	1,8-Dihydroxy-3-(hydroxymethyl)-9,10-anthracenedione	Aloe-emodol	$C_{15}H_{10}O_5$	481-72-1	270.237	oran ye nd (to, al)	223.5	sub			vs bz, eth, EtOH
3725	2,3-Dihydroxymaleic acid	Dihydroxymaleic acid	$C_4H_4O_6$	526-84-1	148.071	pl (w+2)	155 dec				sl H_2O, eth, MeOH; s EtOH
3726	α,4-Dihydroxy-3-methoxybenzeneacetic acid	Vanilmandelic acid	$C_9H_{10}O_5$	55-10-7	198.172	sc (bz-eth)	132 dec				vs H_2O, ace, eth
3727	7,8-Dihydroxy-6-methoxy-2*H*-1-benzopyran-2-one	Fraxetin	$C_{10}H_8O_5$	574-84-5	208.168	pl (dil al)	231				vs EtOH

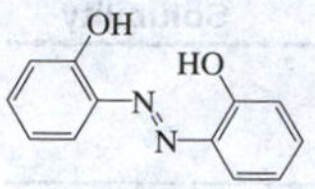
3697
2,2′-Dihydroxyazobenzene

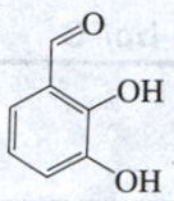
3698
2,3-Dihydroxybenzaldehyde

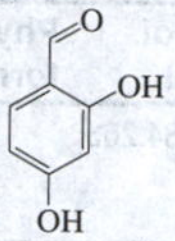
3699
2,4-Dihydroxybenzaldehyde

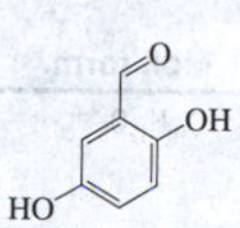
3700
2,5-Dihydroxybenzaldehyde

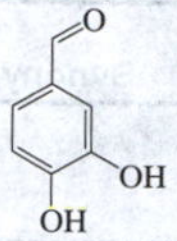
3701
3,4-Dihydroxybenzaldehyde

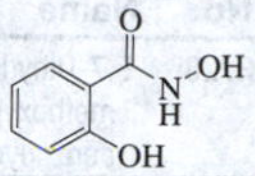
3702
N,2-Dihydroxybenzamide

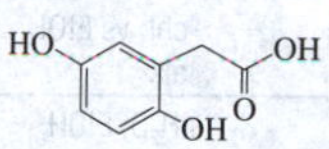
3703
2,5-Dihydroxybenzeneacetic acid

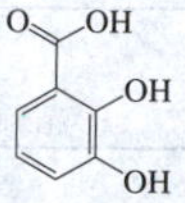
3704
2,3-Dihydroxybenzoic acid

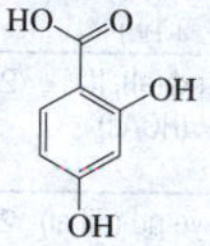
3705
2,4-Dihydroxybenzoic acid

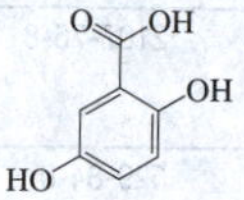
3706
2,5-Dihydroxybenzoic acid

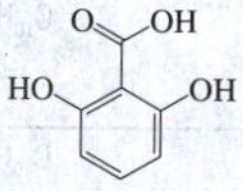
3707
2,6-Dihydroxybenzoic acid

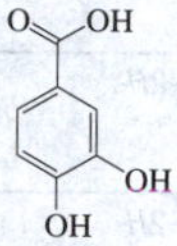
3708
3,4-Dihydroxybenzoic acid

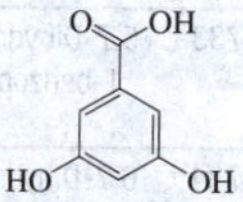
3709
3,5-Dihydroxybenzoic acid

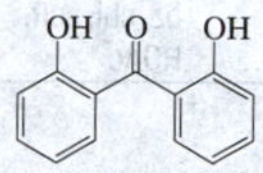
3710
2,2′-Dihydroxybenzophenone

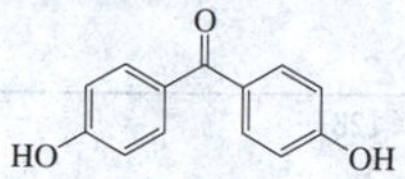
3711
4,4′-Dihydroxybenzophenone

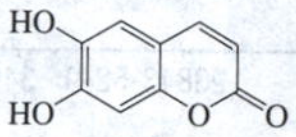
3712
6,7-Dihydroxy-2*H*-1-benzopyran-2-one

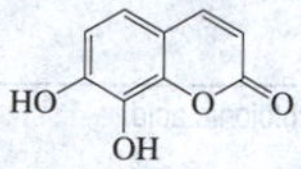
3713
7,8-Dihydroxy-2*H*-1-benzopyran-2-one

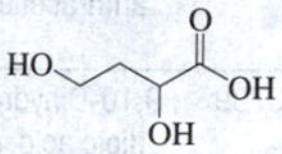
3714
2,4-Dihydroxybutanoic acid

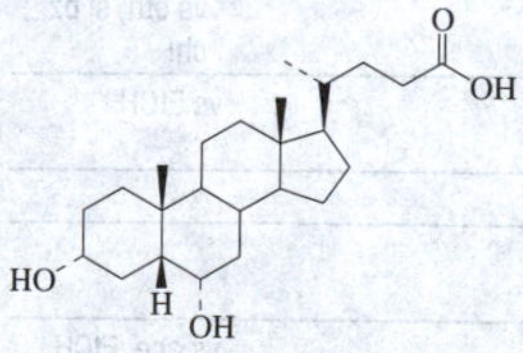
3715
3,6-Dihydroxycholan-24-oic acid, (3α,5β,6α)

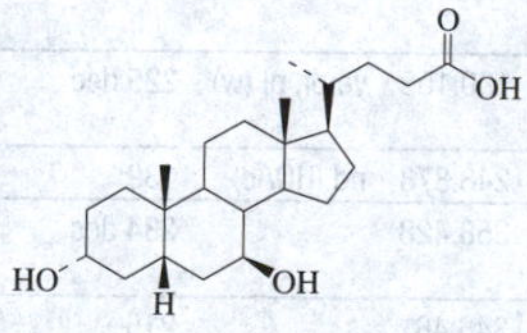
3716
3,7-Dihydroxycholan-24-oic acid, (3α,5β,7β)

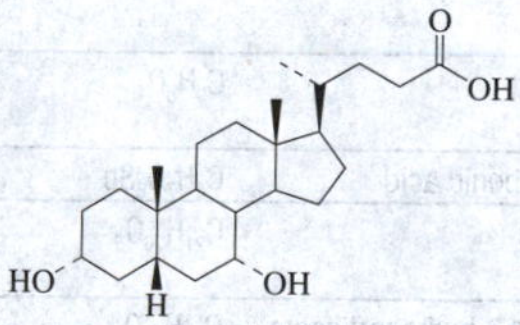
3717
3,7-Dihydroxycholan-24-oic acid, (3α,5β,7α)

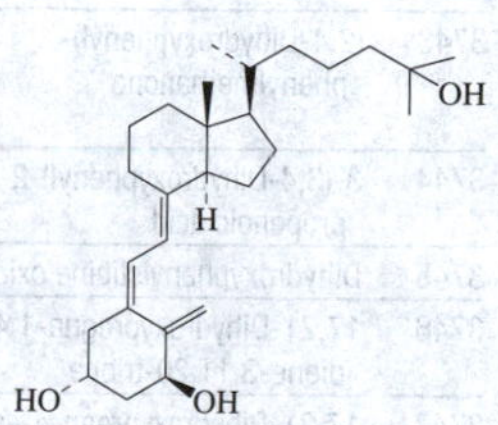
3718
1,25-Dihydroxycholecalciferol

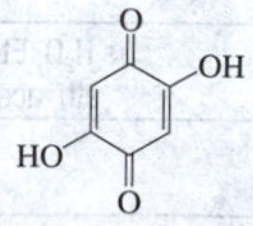
3719
2,5-Dihydroxy-2,5-cyclohexadiene-1,4-dione

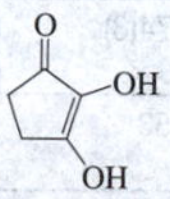
3720
2,3-Dihydroxy-2-cyclopenten-1-one

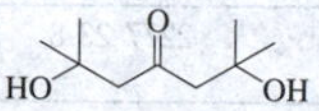
3721
2,6-Dihydroxy-2,6-dimethyl-4-heptanone

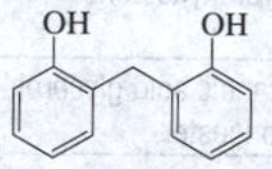
3722
2,2′-Dihydroxydiphenylmethane

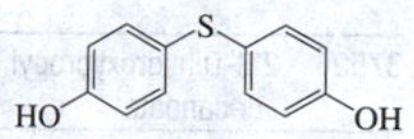
3723
4,4′-Dihydroxydiphenyl sulfide

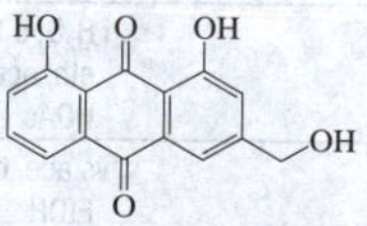
3724
1,8-Dihydroxy-3-(hydroxymethyl)-9,10-anthracenedione

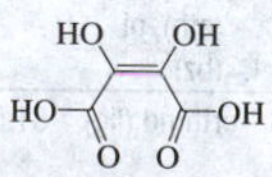
3725
2,3-Dihydroxymaleic acid

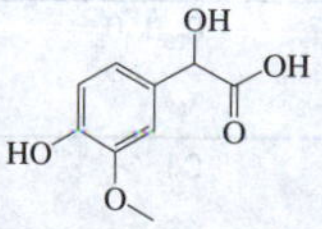
3726
α,4-Dihydroxy-3-methoxybenzeneacetic acid

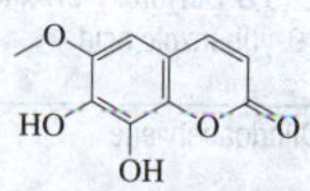
3727
7,8-Dihydroxy-6-methoxy-2*H*-1-benzopyran-2-one

No.	Name	Synonym	Mol. form.	CAS Reg. No.	Mol. wt.	Physical form	mp/°C	bp/°C	den/ g cm^{-3}	n_D	Solubility
3728	5,7-Dihydroxy-3-(4-methoxyphenyl)-4*H*-1-benzopyran-4-one		$C_{16}H_{12}O_5$	491-80-5	284.263		214.8				
3729	(2,6-Dihydroxy-4-methoxyphenyl)phenylmethanone	Cotoin	$C_{14}H_{12}O_4$	479-21-0	244.243	ye pr (chl) lf or nd (w)	130.5				vs ace, bz, eth, EtOH
3730	1,7-Dihydroxy-3-methoxy-9*H*-xanthen-9-one	Gentisin	$C_{14}H_{10}O_5$	437-50-3	258.226	ye orth	266.5				i H_2O; vs EtOH; i ace; s py
3731	1,8-Dihydroxy-3-methyl-9,10-anthracenedione	Chrysophanic acid	$C_{15}H_{10}O_4$	481-74-3	254.238	ye hex or mcl nd (sub)	196	sub	0.92^{25}		vs bz, HOAc
3732	2,4-Dihydroxy-6-methylbenzoic acid	*o*-Orsellinic acid	$C_8H_8O_4$	480-64-8	168.148	nd (dil HOAc, +1w)	176 dec				s EtOH, eth
3733	5,7-Dihydroxy-4-methyl-2*H*-1-benzopyran-2-one		$C_{10}H_8O_4$	2107-76-8	192.169	nd (al), lf (HOAc)	283				sl H_2O, eth, bz, chl; vs EtOH, alk
3734	6,7-Dihydroxy-4-methyl-2*H*-1-benzopyran-2-one		$C_{10}H_8O_4$	529-84-0	192.169	ye nd (dil al)	275				s H_2O, EtOH, HOAc
3735	5,8-Dihydroxy-1,4-naphthalenedione		$C_{10}H_6O_4$	475-38-7	190.153	dk red mcl pr (bz) red-br nd (al)	243(1)	sub			sl H_2O, EtOH, eth; s HOAc
3736	4,5-Dihydroxy-2,7-naphthalenedisulfonic acid	Chromotropic acid	$C_{10}H_8O_8S_2$	148-25-4	320.296	nd or lf (w+2)					s H_2O, alk; i EtOH, eth
3737	5,6-Dihydroxynaphtho[2,3-*f*]quinoline-7,12-dione	Alizarin Blue	$C_{17}H_9NO_4$	568-02-5	291.258	br-viol nd (bz)	269				vs bz, gl HOAc
3738	1,2-Dihydroxy-3-nitro-9,10-anthracenedione	Alizarin Orange	$C_{14}H_7NO_6$	568-93-4	285.209	oran nd or pl (HOAc)	244 dec	sub			sl H_2O; s EtOH, bz, chl, sulf, HOAc
3739	9,10-Dihydroxyoctadecanedioic acid, (*R**,*R**)-(±)-	Phloionic acid	$C_{18}H_{34}O_6$	23843-52-9	346.459	cry (al)	126				
3740	9,10-Dihydroxyoctadecanoic acid	9,10-Dihydroxystearic acid	$C_{18}H_{36}O_4$	120-87-6	316.477		90				i H_2O; sl EtOH, eth
3741	5,7-Dihydroxy-2-phenyl-4*H*-1-benzopyran-4-one	Chrysin	$C_{15}H_{10}O_4$	480-40-0	254.238	lt ye pr (MeOH)	285(2)				i H_2O; s EtOH, ace; sl eth, bz, CS_2
3742	1-(2,4-Dihydroxyphenyl)-ethanone	Resacetophenone	$C_8H_8O_3$	89-84-9	152.148	nd or lf	146		1.18^{141}		i H_2O, chl; s EtOH, py; sl eth, bz
3743	(2,4-Dihydroxyphenyl)-phenylmethanone	Benzoresorcinol	$C_{13}H_{10}O_3$	131-56-6	214.216	nd (w)	144				i H_2O; s EtOH; vs eth; sl bz, chl
3744	3-(3,4-Dihydroxyphenyl)-2-propenoic acid	Caffeic acid	$C_9H_8O_4$	331-39-5	180.158	ye pr, pl (w)	225 dec				vs EtOH
3745	Dihydroxyphenylstibine oxide	Benzenestibonic acid	$C_6H_7O_3Sb$	535-46-6	248.878	nd (HOAc)	139				
3746	17,21-Dihydroxypregna-1,4-diene-3,11,20-trione	Prednisone	$C_{21}H_{26}O_5$	53-03-2	358.428		234 dec				
3747	17,21-Dihydroxypregn-4-ene-3,20-dione	11-Deoxy-17-hydrocorticosterone	$C_{21}H_{30}O_4$	152-58-9	346.461		215				vs ace, EtOH, chl
3748	17,21-Dihydroxypregn-4-ene-3,11,20-trione	Cortisone	$C_{21}H_{28}O_5$	53-06-5	360.444		222				sl H_2O, eth, bz, chl; s EtOH, ace
3749	2,3-Dihydroxypropanal, (±)-		$C_3H_6O_3$	56-82-6	90.078	nd or pr (40% MeOH)	145	$145^{0.8}$	1.453^{18}		s H_2O; sl EtOH, eth; i bz, peth, lig
3750	2,3-Dihydroxypropanoic acid, (*R*)-	Glyceric acid	$C_3H_6O_4$	6000-40-4	106.078	thick gum					
3751	1,3-Dihydroxy-2-propanone	Dihydroxyacetone	$C_3H_6O_3$	96-26-4	90.078		74(3)				s H_2O, EtOH, eth, ace; i lig
3752	2,3-Dihydroxypropyl decanoate	Decanoic acid glycerol monoester	$C_{13}H_{26}O_4$	2277-23-8	246.343	pr (peth)	53				
3753	2,3-Dihydroxypropyl octanoate	Octanoic acid glycerol monoester	$C_{11}H_{22}O_4$	26402-26-6	218.291	cry (peth)	40				
3754	4,8-Dihydroxy-2-quinolinecarboxylic acid	Xanthurenic acid	$C_{10}H_7NO_4$	59-00-7	205.168	ye micry cry (w)	289				i H_2O; s EtOH, dil HCl; sl eth, bz
3755	Dihydroxytartaric acid		$C_4H_6O_8$	76-30-2	182.086		114.5				
3756	3,4-Dihydroxy-5-[(3,4,5-trihydroxybenzoyl)oxy]benzoic acid	Digallic acid	$C_{14}H_{10}O_9$	536-08-3	322.224	nf (dil al + 1w)	269 dec				vs ace, EtOH
3757	2-(3,6-Dihydroxy-9*H*-xanthen-9-yl)benzoic acid	Fluorescin	$C_{20}H_{14}O_5$	518-44-5	334.322	col or ye nd (eth), pl (bz)	126				i H_2O; s EtOH, eth, ace, bz, HOAc
3758	Diiodoacetylene		C_2I_2	624-74-8	277.830	orth nd (lig)	81.5	247(16)			vs ace, bz, eth, EtOH

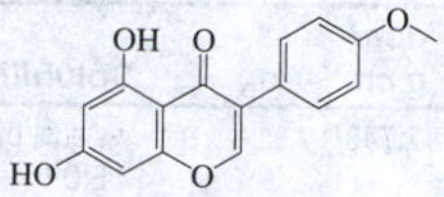
3728
5,7-Dihydroxy-3-(4-methoxyphenyl)-4*H*-1-benzopyran-4-one

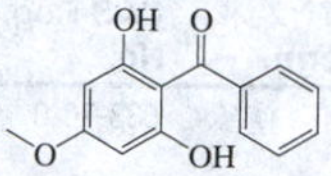
3729
(2,6-Dihydroxy-4-methoxyphenyl)phenylmethanone

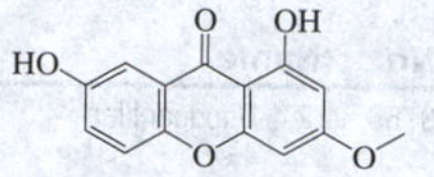
3730
1,7-Dihydroxy-3-methoxy-9*H*-xanthen-9-one

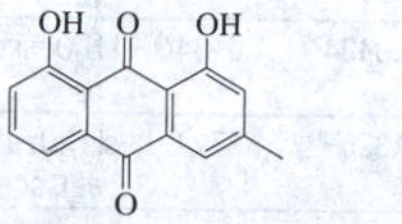
3731
1,8-Dihydroxy-3-methyl-9,10-anthracenedione

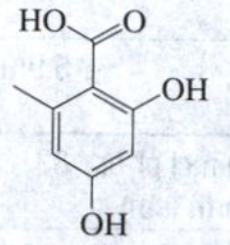
3732
2,4-Dihydroxy-6-methylbenzoic acid

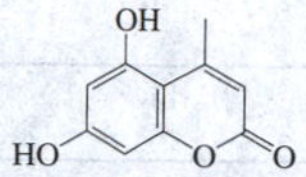
3733
5,7-Dihydroxy-4-methyl-2*H*-1-benzopyran-2-one

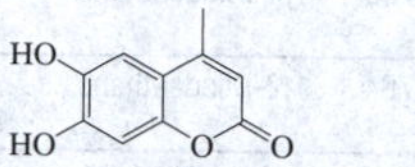
3734
6,7-Dihydroxy-4-methyl-2*H*-1-benzopyran-2-one

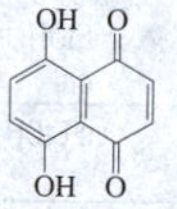
3735
5,8-Dihydroxy-1,4-naphthalenedione

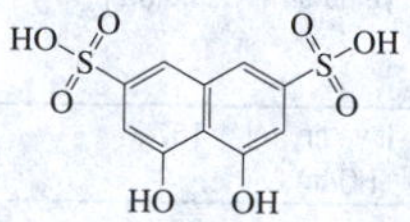
3736
4,5-Dihydroxy-2,7-naphthalenedisulfonic acid

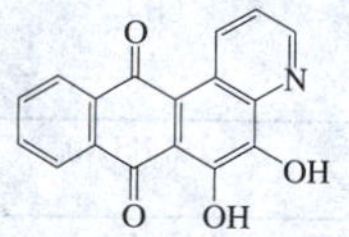
3737
5,6-Dihydroxynaphtho[2,3-f]quinoline-7,12-dione

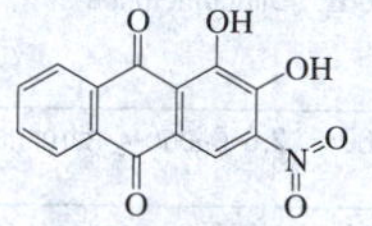
3738
1,2-Dihydroxy-3-nitro-9,10-anthracenedione

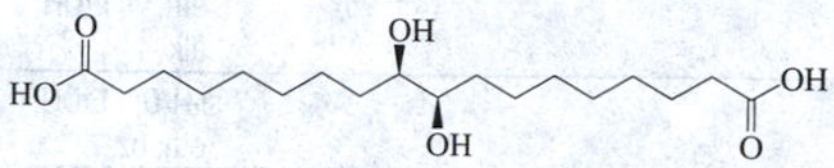
3739
9,10-Dihydroxyoctadecanedioic acid, (*R**,*R**)-(±)-

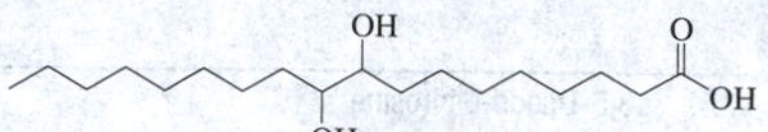
3740
9,10-Dihydroxyoctadecanoic acid

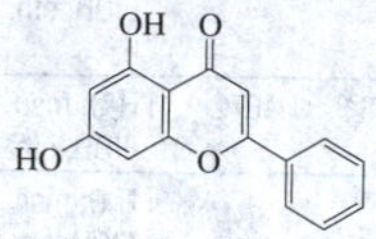
3741
5,7-Dihydroxy-2-phenyl-4*H*-1-benzopyran-4-one

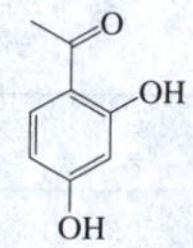
3742
1-(2,4-Dihydroxyphenyl)ethanone

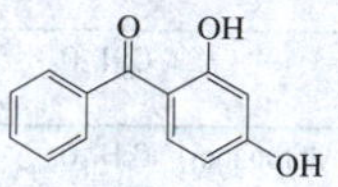
3743
(2,4-Dihydroxyphenyl)phenylmethanone

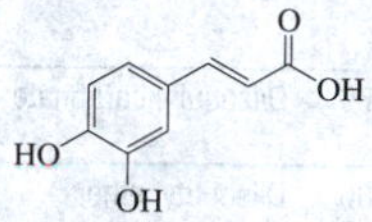
3744
3-(3,4-Dihydroxyphenyl)-2-propenoic acid

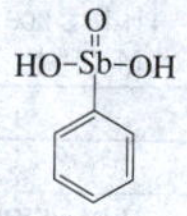
3745
Dihydroxyphenylstibine oxide

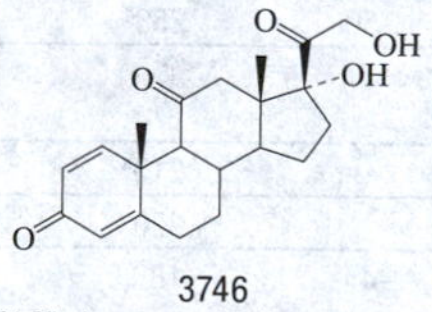
3746
17,21-Dihydroxypregna-1,4-diene-3,11,20-trione

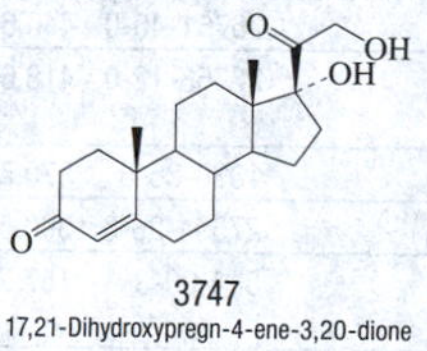
3747
17,21-Dihydroxypregn-4-ene-3,20-dione

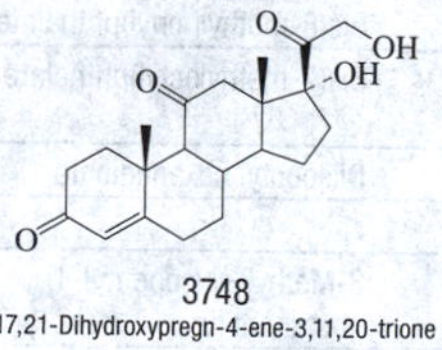
3748
17,21-Dihydroxypregn-4-ene-3,11,20-trione

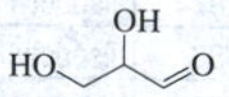
3749
2,3-Dihydroxypropanal, (±)-

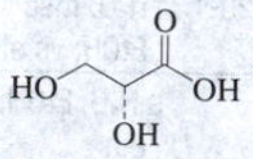
3750
2,3-Dihydroxypropanoic acid, (*R*)-

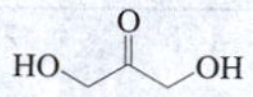
3751
1,3-Dihydroxy-2-propanone

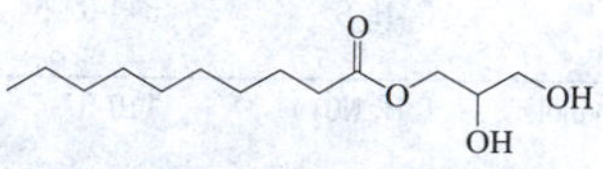
3752
2,3-Dihydroxypropyl decanoate

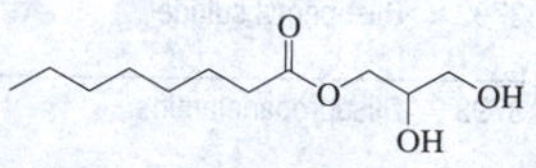
3753
2,3-Dihydroxypropyl octanoate

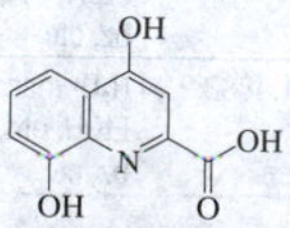
3754
4,8-Dihydroxy-2-quinolinecarboxylic acid

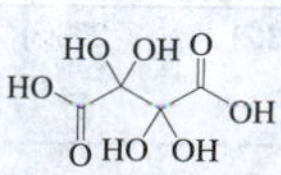
3755
Dihydroxytartaric acid

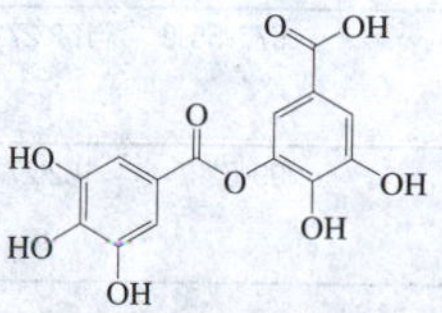
3756
3,4-Dihydroxy-5-[(3,4,5-trihydroxybenzoyl)oxy]benzoic acid

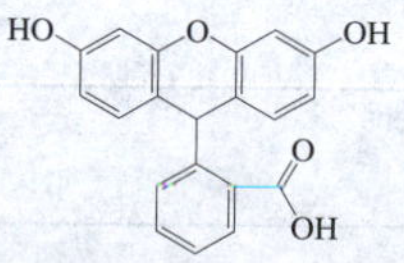
3757
2-(3,6-Dihydroxy-9*H*-xanthen-9-yl)benzoic acid

3758
Diiodoacetylene

No.	Name	Synonym	Mol. form.	CAS Reg. No.	Mol. wt.	Physical form	mp/°C	bp/°C	den/ g cm⁻³	n_D	Solubility
3759	2,4-Diiodoaniline		$C_6H_5I_2N$	533-70-0	344.920	br nd or orth cry (al)	95.5		2.748^{25}		vs ace, bz, eth, EtOH
3760	*o*-Diiodobenzene	1,2-Diiodobenzene	$C_6H_4I_2$	615-42-9	329.905	pl or pr (lig)	23.5(0.7)	287	2.54^{20}	1.7179^{20}	i H_2O; sl EtOH
3761	*m*-Diiodobenzene	1,3-Diiodobenzene	$C_6H_4I_2$	626-00-6	329.905	orth pl or pr (eth-al)	34.2(0.6)	285	2.47^{25}		i H_2O; vs eth, EtOH, chl
3762	*p*-Diiodobenzene	1,4-Diiodobenzene	$C_6H_4I_2$	624-38-4	329.905	orth lf (al)	129.25(0.05)	285			i H_2O; s EtOH; vs eth; sl chl
3763	1,4-Diiodobutane		$C_4H_8I_2$	628-21-7	309.916		5.9(0.4)	125^{15} dec	2.3494^{25}	1.6184^{25}	i H_2O; sl ctc; s os
3764	1,2-Diiodoethane		$C_2H_4I_2$	624-73-7	281.862	ye mcl pr or orth (eth)	83	200	3.325^{20}	1.871^{20}	sl H_2O; s EtOH, eth, ace, chl
3765	*cis*-1,2-Diiodoethene	*cis*-1,2-Diiodoethylene	$C_2H_2I_2$	590-26-1	279.846		-13(2)	72.5^{16}	3.0625^{20}		i H_2O; s eth, chl
3766	4,4'-Diiodofluorescein		$C_{20}H_{10}I_2O_5$	38577-97-8	584.099	oran-red pow					sl H_2O; s alk, EtOH
3767	1,6-Diiodohexane	Hexamethylene diiodide	$C_6H_{12}I_2$	629-09-4	337.968	nd	9.4(0.5)	163^{17}	2.0342^{25}	1.5837^{25}	i H_2O; vs EtOH, eth
3768	Diiodomethane	Methylene iodide	CH_2I_2	75-11-6	267.836	ye nd or lf	6.0(0.2)	182	3.3211^{20}	1.7411^{20}	sl H_2O, ctc; s EtOH, eth, bz, chl
3769	2,6-Diiodo-4-nitrophenol	Disophenol	$C_6H_3I_2NO_3$	305-85-1	390.902	lt ye cry (gl HOAc)	157				vs EtOH
3770	1,5-Diiodopentane	Pentamethylene diiodide	$C_5H_{10}I_2$	628-77-3	323.942		9	149^{20}	2.1692^{25}	1.5987^{25}	i H_2O; s eth, chl
3771	1,2-Diiodopropane		$C_3H_6I_2$	598-29-8	295.889				2.490^{18}		vs eth, EtOH
3772	1,3-Diiodopropane	Trimethylene diiodide	$C_3H_6I_2$	627-31-6	295.889		-20	222(12)	2.5612^{25}	1.6391^{25}	i H_2O; s eth, ctc, chl
3773	5,7-Diiodo-8-quinolinol	Iodoquinol	$C_9H_5I_2NO$	83-73-8	396.951	ye nd (HOAc, xyl)	210				sl H_2O, bz, chl, eth; vs EtOH; s alk
3774	3,5-Diiodo-L-tyrosine		$C_9H_9I_2NO_3$	300-39-0	432.981	ye nd (w, 70% al)	213				sl H_2O; i EtOH, eth, bz
3775	Diisobutyl adipate	Diisobutyl hexanedioate	$C_{14}H_{26}O_4$	141-04-8	258.354			288(16)	0.9543^{19}	1.4301^{20}	
3776	Diisobutylaluminum chloride	Diisobutyl aluminum chloride	$C_8H_{18}AlCl$	1779-25-5	176.664	hyg col liq	-40	152^{10}	0.905	1.4506^{20}	s eth, hx
3777	Diisobutylaluminum hydride		$C_8H_{19}Al$	1191-15-7	142.219	liq		140^{4}			s cyhex, eth, bz, tol
3778	Diisobutylamine	2-Methyl-*N*-(2-methylpropyl)-1-propanamine	$C_8H_{19}N$	110-96-3	129.244	liq	-73.5(0.4)	139.6		1.4090^{20}	sl H_2O, ctc; s EtOH, eth, ace, bz
3779	Diisobutyl carbonate		$C_9H_{18}O_3$	539-92-4	174.237			190(6)	0.9138^{20}	1.4072^{20}	i H_2O; msc EtOH, eth
3780	Diisobutyl ether	1,1'-Oxybis[2-methylpropane]	$C_8H_{18}O$	628-55-7	130.228			122.7(0.7)	0.761^{15}		i H_2O; msc EtOH, eth
3781	Diisobutyl phthalate		$C_{16}H_{22}O_4$	84-69-5	278.344			296.5	1.0490^{15}		s ctc
3782	Diisobutyl sulfide		$C_8H_{18}S$	592-65-4	146.294	liq	-105.5(0.5)	173(4)	0.8363^{10}		
3783	1,3-Diisocyanatobenzene		$C_8H_4N_2O_2$	123-61-5	160.130	cry	51	103^{8}			
3784	1,4-Diisocyanatobenzene		$C_8H_4N_2O_2$	104-49-4	160.130	cry	95	117^{14}			
3785	Diisodecyl phthalate	Bis(8-methylnonyl)phthalate	$C_{28}H_{46}O_4$	26761-40-0	446.663	liq	-50	253^{4}	0.966^{20}		i H_2O; s os
3786	Diisononyl phthalate	Bis(7-methyloctyl)phthalate	$C_{26}H_{42}O_4$	28553-12-0	418.609	col liq					i H_2O; s ace, MeOH; bz, eth
3787	Diisooctyl adipate	Diisooctyl hexanedioate	$C_{22}H_{42}O_4$	1330-86-5	370.566			210^{4}			
3788	Diisooctyl phthalate		$C_{24}H_{38}O_4$	27554-26-3	390.557			370			
3789	Diisopentylamine	3-Methyl-*N*-isopentyl-1-butanamine	$C_{10}H_{23}N$	544-00-3	157.297	liq	-44	187(5)	0.7672^{21}	1.4235^{20}	i H_2O; s EtOH; msc eth
3790	Diisopentyl ether	Diisoamyl ether	$C_{10}H_{22}O$	544-01-4	158.281	col liq		172(2)	0.7777^{20}	1.4085^{20}	i H_2O; vs ace, EtOH, chl
3791	Diisopentyl phthalate	Diisoamyl phthalate	$C_{18}H_{26}O_4$	605-50-5	306.397			367(16)	1.0209^{16}	1.4871^{20}	vs EtOH
3792	Diisopentyl sulfide	Isopentyl sulfide	$C_{10}H_{22}S$	544-02-5	174.347	liq	-74.6	211	0.8323^{20}	1.4520^{20}	i H_2O; msc EtOH; vs eth
3793	Diisopropanolamine	1,1'-Iminobis-2-propanol	$C_6H_{15}NO_2$	110-97-4	133.189	cry	44.5	250	0.989^{20}		s H_2O, EtOH; sl eth
3794	Diisopropyl adipate	Diisopropyl hexanedioate	$C_{12}H_{22}O_4$	6938-94-9	230.301		-0.6	$120^{6.5}$	0.9569^{20}	1.4247^{20}	vs ace, eth, EtOH
3795	Diisopropylamine	*N*-Isopropyl-2-propanamine	$C_6H_{15}N$	108-18-9	101.190	liq	-61	84(3)	0.7153^{20}	1.3924^{20}	vs ace, bz, eth, EtOH
3796	2,6-Diisopropylaniline		$C_{12}H_{19}N$	24544-04-5	177.286	liq	-45	272(9)	0.94^{25}	1.5332^{20}	
3797	1,2-Diisopropylbenzene		$C_{12}H_{18}$	577-55-9	162.271	liq	-57(4)	204.7(0.6)	0.8701^{20}	1.4960^{20}	i H_2O; msc EtOH, eth, ace, bz, ctc
3798	1,3-Diisopropylbenzene		$C_{12}H_{18}$	99-62-7	162.271	liq	-63(2)	203(3)	0.8559^{20}	1.4883^{20}	i H_2O; msc EtOH, eth, ace, bz, ctc

Organic

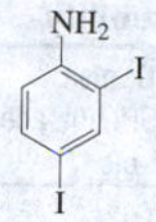
3759
2,4-Diiodoaniline

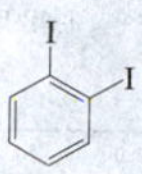
3760
o-Diiodobenzene

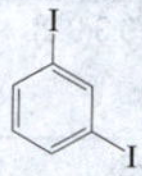
3761
m-Diiodobenzene

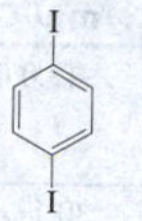
3762
p-Diiodobenzene

3763
1,4-Diiodobutane

3764
1,2-Diiodoethane

3765
cis-1,2-Diiodoethene

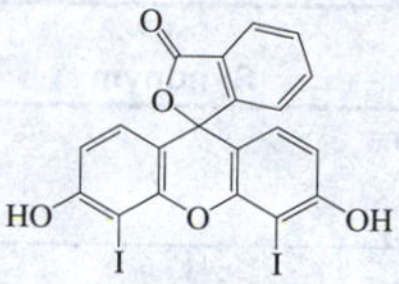
3766
4,4'-Diiodofluorescein

3767
1,6-Diiodohexane

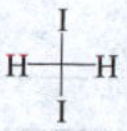
3768
Diiodomethane

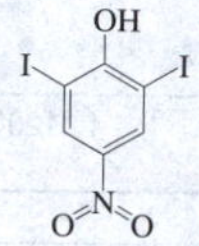
3769
2,6-Diiodo-4-nitrophenol

3770
1,5-Diiodopentane

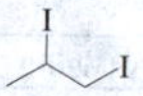
3771
1,2-Diiodopropane

3772
1,3-Diiodopropane

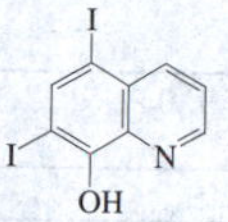
3773
5,7-Diiodo-8-quinolinol

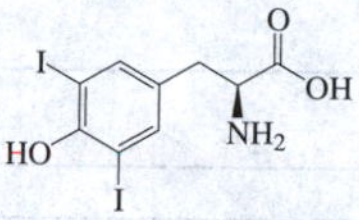
3774
3,5-Diiodo-L-tyrosine

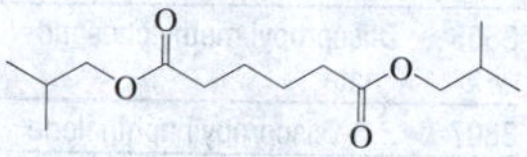
3775
Diisobutyl adipate

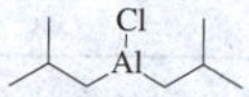
3776
Diisobutylaluminum chloride

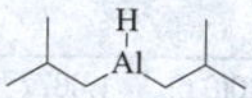
3777
Diisobutylaluminum hydride

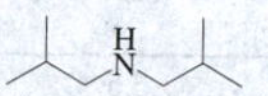
3778
Diisobutylamine

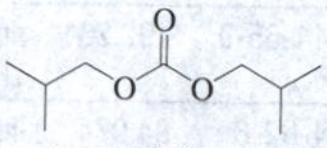
3779
Diisobutyl carbonate

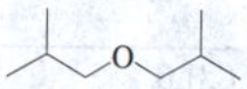
3780
Diisobutyl ether

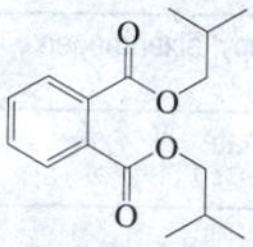
3781
Diisobutyl phthalate

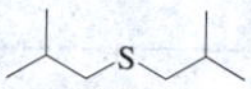
3782
Diisobutyl sulfide

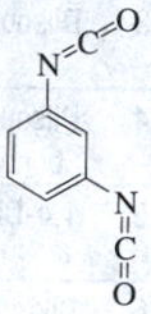
3783
1,3-Diisocyanatobenzene

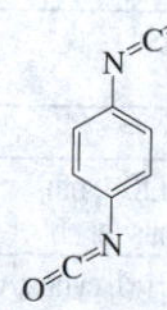
3784
1,4-Diisocyanatobenzene

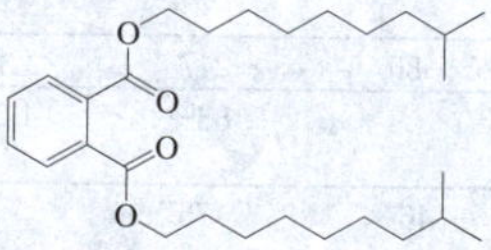
3785
Diisodecyl phthalate

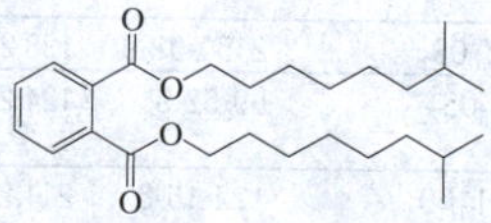
3786
Diisononyl phthalate

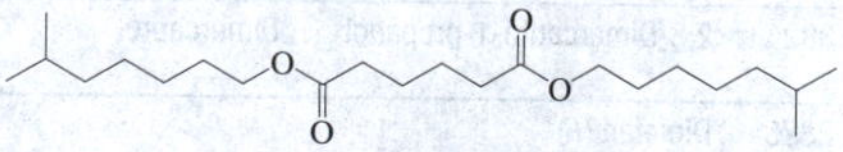
3787
Diisooctyl adipate

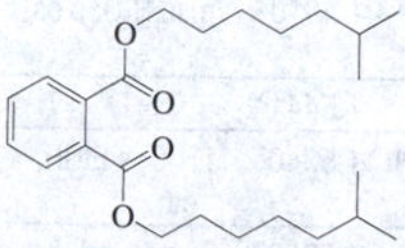
3788
Diisooctyl phthalate

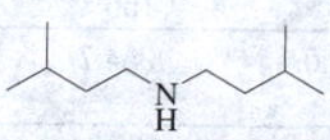
3789
Diisopentylamine

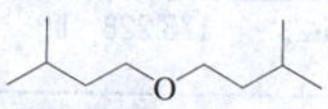
3790
Diisopentyl ether

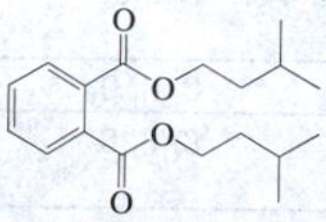
3791
Diisopentyl phthalate

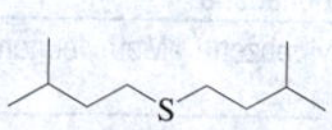
3792
Diisopentyl sulfide

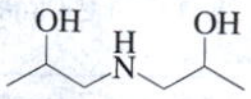
3793
Diisopropanolamine

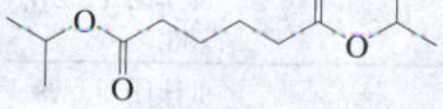
3794
Diisopropyl adipate

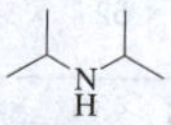
3795
Diisopropylamine

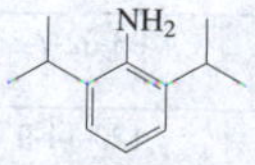
3796
2,6-Diisopropylaniline

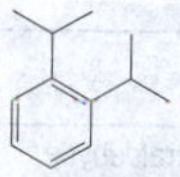
3797
1,2-Diisopropylbenzene

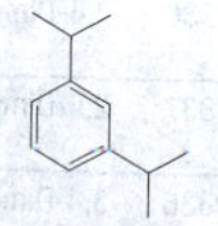
3798
1,3-Diisopropylbenzene

No.	Name	Synonym	Mol. form.	CAS Reg. No.	Mol. wt.	Physical form	mp/°C	bp/°C	den/ g cm⁻³	n_D	Solubility
3799	1,4-Diisopropylbenzene		$C_{12}H_{18}$	100-18-5	162.271	liq	-17.0(0.1)	210.3(0.2)	0.8568^{20}	1.4898^{20}	i H_2O; msc EtOH, eth, ace, bz, ctc
3800	*p*-Diisopropylbenzene hydroperoxide		$C_{12}H_{18}O_2$	98-49-7	194.270	waxy cry	30.1	123^{1}	0.9932^{20}		i H_2O
3801	*N,N*-Diisopropyl-2-benzothiazolesulfenamide		$C_{13}H_{18}N_2S_2$	95-29-4	266.425		59.0				
3802	*N,N'*-Diisopropylcarbodiimide		$C_7H_{14}N_2$	693-13-0	126.199			147	0.806^{25}	1.4320^{20}	
3803	Diisopropyl disulfide		$C_6H_{14}S_2$	4253-89-8	150.305	liq	-69.0(0.3)	177(1)	0.9435^{20}	1.4916^{20}	
3804	*N,N*-Diisopropylethanolamine	*N,N*-Diisopropyl-2-aminoethanol	$C_8H_{19}NO$	96-80-0	145.243			195(5)	0.826^{25}	1.4417^{20}	
3805	Diisopropyl ether	Isopropyl ether	$C_6H_{14}O$	108-20-3	102.174	liq	-85.37(0.05)	68.4(0.2)	0.7192^{25}	1.3658^{25}	sl H_2O; msc EtOH, eth; s ace, ctc
3806	Diisopropyl methylphosphonate		$C_7H_{17}O_3P$	1445-75-6	180.182			66^{3}		1.4120^{16}	
3807	2,6-Diisopropylnaphthalene		$C_{16}H_{20}$	24157-81-1	212.330	cry (MeOH)	70				
3808	Diisopropyl oxalate		$C_8H_{14}O_4$	615-81-6	174.195			189(8)	1.002^{20}	1.4100^{20}	vs eth, EtOH
3809	Diisopropyl phosphonate		$C_6H_{15}O_3P$	1809-20-7	166.155			97^{40}	0.9970^{18}		
3810	*O,O*-Diisopropyl phosphorodithioate		$C_6H_{15}O_2PS_2$	107-56-2	214.286	liq		71^{3}	1.09^{20}		s EtOH, bz, ace, ctc, chl
3811	Diisopropyl phthalate	1,2-Benzenedicarboxylic acid, diisopropyl ester	$C_{14}H_{18}O_4$	605-45-8	250.291			130^{12}	1.0615^{15}	1.4900^{20}	
3812	Diisopropyl sulfide		$C_6H_{14}S$	625-80-9	118.240	liq	-78.03(0.04)	120.0(0.3)	0.8142^{20}	1.4438^{20}	i H_2O; s EtOH, eth
3813	Diisopropyl tartrate, (±)-		$C_{10}H_{18}O_6$	58167-01-4	234.246		34	275	1.1166^{20}		vs ace, eth, EtOH
3814	Diisopropyl thioperoxydicarbonate	Diisopropyl dixanthogen	$C_8H_{14}O_2S_4$	105-65-7	270.456		52				s chl
3815	1,4-Diisothiocyanatobenzene	Bitoscanate	$C_8H_4N_2S_2$	4044-65-9	192.261	nd (ace, HOAc)	132				
3816	Diketene		$C_4H_4O_2$	674-82-8	84.074	liq	-6.5	127.0(0.6)	1.0877^{20}	1.4379^{20}	
3817	Dilactic acid	2,2'-Oxybispropanoic acid	$C_6H_{10}O_5$	19201-34-4	162.140	orth	112.5				vs H_2O, eth
3818	Dimefline		$C_{20}H_{21}NO_3$	1165-48-6	323.386		109.5				s chl
3819	Dimefox	Tetramethylphosphorodiamidic fluoride	$C_4H_{12}FN_2OP$	115-26-4	154.122	liq		86^{15}	1.1151^{20}	1.4267^{20}	vs H_2O, bz, eth
3820	Dimemorfan	3,17-Dimethylmorphinan, (9 α,13 α,14 α)-	$C_{18}H_{25}N$	36309-01-0	255.399	ye oil	92	$133^{0.3}$			
3821	2,3-Dimercaptobutanedioic acid, (*R**,*S**)	Succimer	$C_4H_6O_4S_2$	304-55-2	182.219	wh cry (MeOH)	193				
3822	1,4-Dimercapto-2,3-butanediol		$C_4H_{10}O_2S_2$	7634-42-6	154.251		42.5				s chl
3823	2,2'-Dimercaptodiethyl ether	2-Mercaptoethyl ether	$C_4H_{10}OS_2$	2150-02-9	138.251	liq	-80	217	1.114^{20}		
3824	2,3-Dimercapto-1-propanol	Dimercaprol	$C_3H_8OS_2$	59-52-9	124.225			$83^{0.8}$	1.2463^{20}	1.5749^{20}	s EtOH, eth, oils; sl chl
3825	Dimetan (*R*)-		$C_{11}H_{17}NO_3$	122-15-6	211.258	cry	46	175^{11}			s H_2O, cyhex; vs EtOH, eth, ace
3826	Dimethipin	2,3-Dihydro-5,6-dimethyl-1,4-dithiin, 1,1,4,4-tetraoxide	$C_6H_{10}O_4S_2$	55290-64-7	210.271		165				
3827	Dimethirimol	5-Butyl-2-(dimethylamino)-6-methylpyrimidin-4(1*H*)-one	$C_{11}H_{19}N_3O$	5221-53-4	209.288	nd	102				sl H_2O; vs chl, xyl; s EtOH, ace
3828	Dimethisoquin	2-[(3-Butyl-1-isoquinolinyl)-oxy]-*N,N*-dimethylethanamine	$C_{17}H_{24}N_2O$	86-80-6	272.385		146	156^{3}		1.5486^{20}	s H_2O, EtOH
3829	Dimethoxane	2,6-Dimethyl-1,3-dioxan-4-ol acetate	$C_8H_{14}O_4$	828-00-2	174.195	liq		86^{10}	1.0655^{20}	1.4310^{20}	msc H_2O; s os
3830	2',5'-Dimethoxyacetophenone		$C_{10}H_{12}O_3$	1201-38-3	180.200	cry	21	156^{14}	1.139	1.5441^{20}	
3831	1,2-Dimethoxy-4-allylbenzene	Methyleugenol	$C_{11}H_{14}O_2$	93-15-2	178.228	liq	-2.0	254.7	1.0396^{20}	1.5340^{20}	i H_2O; s EtOH, eth
3832	4,7-Dimethoxy-5-allyl-1,3-benzodioxole	Apiole	$C_{12}H_{14}O_4$	523-80-8	222.237	nd	29.2(0.2)	294	1.015^{20}	1.5360^{20}	vs ace, bz, EtOH, lig
3833	2,4-Dimethoxyaniline		$C_8H_{11}NO_2$	2735-04-8	153.179	pl (lig)	33.5	262.0			sl H_2O, chl; s EtOH, eth, bz, lig
3834	2,5-Dimethoxyaniline		$C_8H_{11}NO_2$	102-56-7	153.179		82.5	270(21)			s H_2O, EtOH, chl, lig
3835	3,4-Dimethoxyaniline		$C_8H_{11}NO_2$	6315-89-5	153.179	lf (eth)	87.5	159^{14}			s eth, chl
3836	2,4-Dimethoxybenzaldehyde		$C_9H_{10}O_3$	613-45-6	166.173	nd (al or lig)	72	290			i H_2O; s EtOH, eth, bz; sl chl
3837	2,5-Dimethoxybenzaldehyde		$C_9H_{10}O_3$	93-02-7	166.173		52	270			sl H_2O; s EtOH, eth
3838	3,4-Dimethoxybenzaldehyde	Veratraldehyde	$C_9H_{10}O_3$	120-14-9	166.173	nd (eth, lig, to)	43	281			sl H_2O, chl; vs EtOH, eth

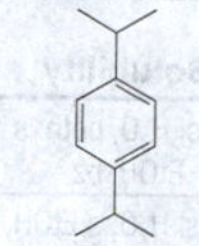
3799
1,4-Diisopropylbenzene

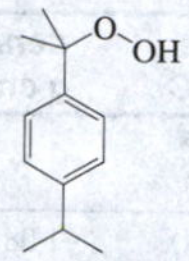
3800
p-Diisopropylbenzene hydroperoxide

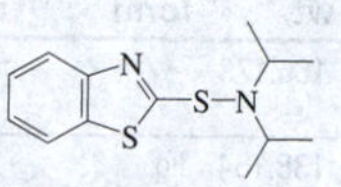
3801
N,N-Diisopropyl-2-benzothiazolesulfenamide

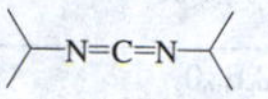
3802
N,N'-Diisopropylcarbodiimide

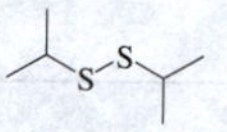
3803
Diisopropyl disulfide

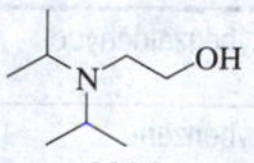
3804
N,N-Diisopropylethanolamine

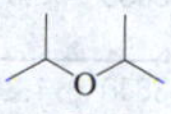
3805
Diisopropyl ether

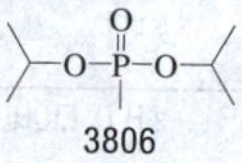
3806
Diisopropyl methylphosphonate

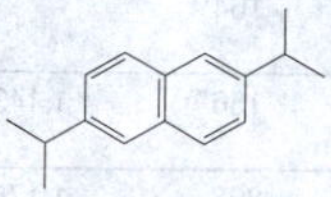
3807
2,6-Diisopropylnaphthalene

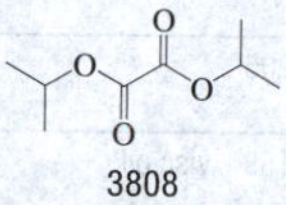
3808
Diisopropyl oxalate

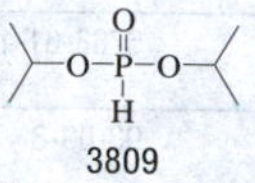
3809
Diisopropyl phosphonate

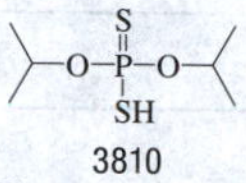
3810
O,O-Diisopropyl phosphorodithioate

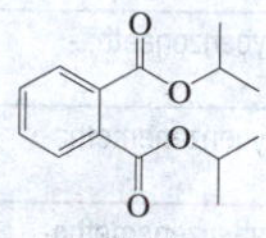
3811
Diisopropyl phthalate

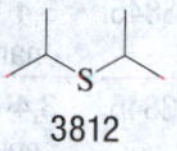
3812
Diisopropyl sulfide

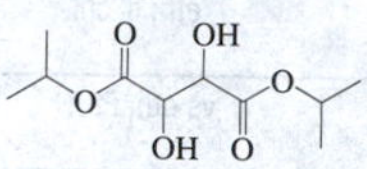
3813
Diisopropyl tartrate, (±)-

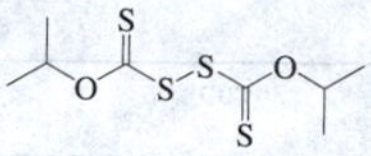
3814
Diisopropyl thioperoxydicarbonate

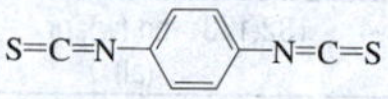
3815
1,4-Diisothiocyanatobenzene

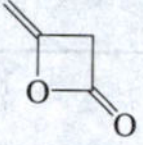
3816
Diketene

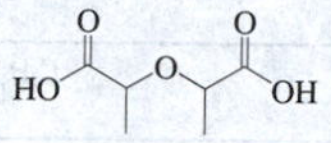
3817
Dilactic acid

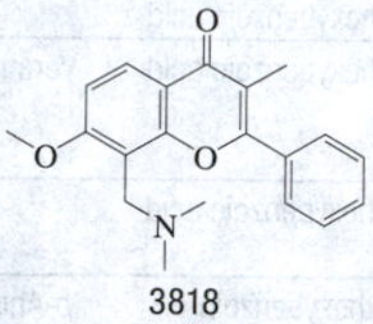
3818
Dimefline

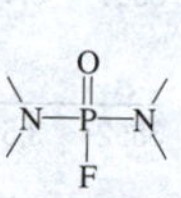
3819
Dimefox

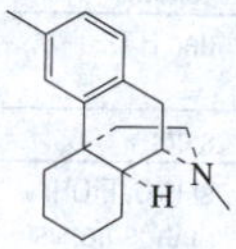
3820
Dimemorfan

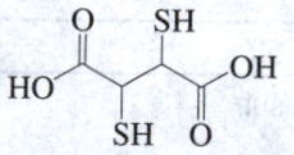
3821
2,3-Dimercaptobutanedioic acid, (*R**,*S**)

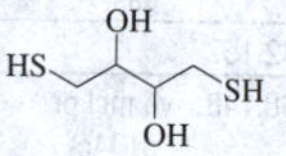
3822
1,4-Dimercapto-2,3-butanediol

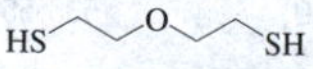
3823
2,2'-Dimercaptodiethyl ether

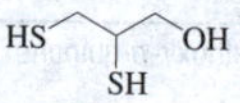
3824
2,3-Dimercapto-1-propanol

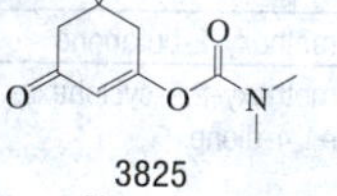
3825
Dimetan (*R*)-

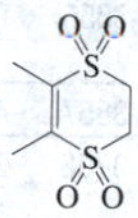
3826
Dimethipin

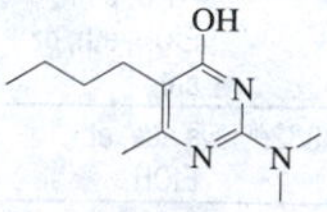
3827
Dimethirimol

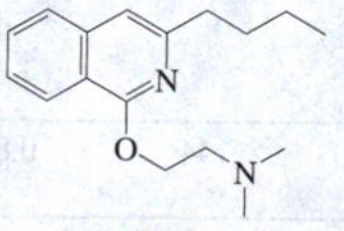
3828
Dimethisoquin

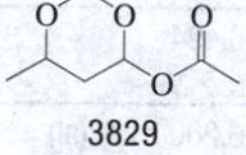
3829
Dimethoxane

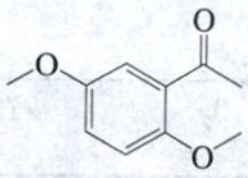
3830
2',5'-Dimethoxyacetophenone

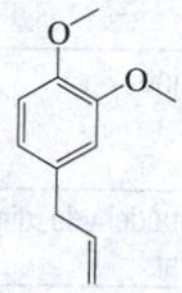
3831
1,2-Dimethoxy-4-allylbenzene

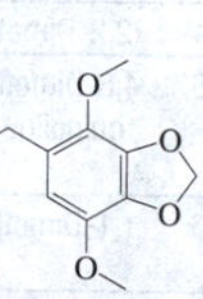
3832
4,7-Dimethoxy-5-allyl-1,3-benzodioxole

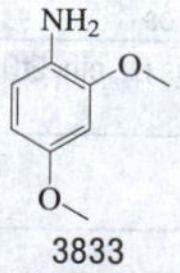
3833
2,4-Dimethoxyaniline

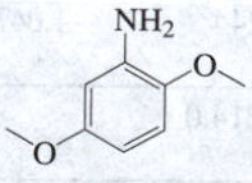
3834
2,5-Dimethoxyaniline

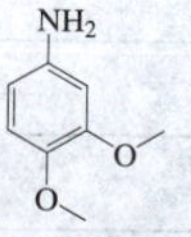
3835
3,4-Dimethoxyaniline

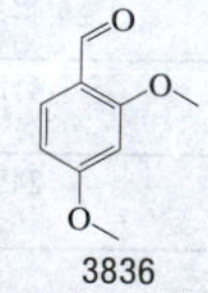
3836
2,4-Dimethoxybenzaldehyde

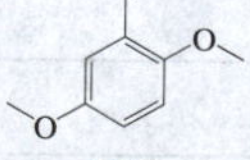
3837
2,5-Dimethoxybenzaldehyde

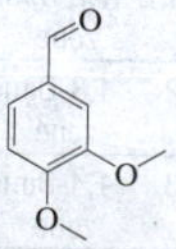
3838
3,4-Dimethoxybenzaldehyde

No.	Name	Synonym	Mol. form.	CAS Reg. No.	Mol. wt.	Physical form	mp/°C	bp/°C	den/ g cm^{-3}	n_D	Solubility
3839	3,5-Dimethoxybenzaldehyde		$C_9H_{10}O_3$	7311-34-4	166.173		46.3	151^{16}			sl H_2O, peth; s EtOH, bz
3840	1,2-Dimethoxybenzene	Veratrole	$C_8H_{10}O_2$	91-16-7	138.164	liq	22.5(0.2)	206(1)	1.0810^{25}	1.5827^{21}	sl H_2O; s EtOH, eth, ctc
3841	1,3-Dimethoxybenzene		$C_8H_{10}O_2$	151-10-0	138.164	liq	-35.3(0.4)	216(5)	1.0521^{25}	1.5231^{20}	sl H_2O; s EtOH, eth, bz, ctc, sulf
3842	1,4-Dimethoxybenzene		$C_8H_{10}O_2$	150-78-7	138.164	lf (w)	56.2(0.7)	213(5)	1.0375^{55}		sl H_2O; s EtOH, chl; vs eth, bz
3843	3,4-Dimethoxybenzeneacetic acid		$C_{10}H_{12}O_4$	93-40-3	196.200	cry (bz-peth) nd (w+1)	98				s H_2O, chl; vs EtOH, eth
3844	3,4-Dimethoxybenzeneethanamine		$C_{10}H_{15}NO_2$	120-20-7	181.232			164^{14}		1.5464^{20}	s ctc
3845	3,4-Dimethoxybenzenemethanamine		$C_9H_{13}NO_2$	5763-61-1	167.205			156^{12}	1.143^{25}		s chl
3846	3,4-Dimethoxybenzenemethanol		$C_9H_{12}O_3$	93-03-8	168.189	visc oil		298	1.178^{17}	1.555^{17}	s H_2O, EtOH
3847	3,3′-Dimethoxybenzidine	Dianisidine	$C_{14}H_{16}N_2O_2$	119-90-4	244.289	lf or nd (w)	137				i H_2O; s EtOH, eth, ace, bz, chl
3848	3,3′-Dimethoxybenzidine-4,4′-diisocyanate		$C_{16}H_{12}N_2O_4$	91-93-0	296.277	cry	112				
3849	2,4-Dimethoxybenzoic acid		$C_9H_{10}O_4$	91-52-1	182.173		107.2(0.4)				sl H_2O; s EtOH, eth, chl, HOAc
3850	2,6-Dimethoxybenzoic acid		$C_9H_{10}O_4$	1466-76-8	182.173		186 dec				
3851	3,4-Dimethoxybenzoic acid	Veratric acid	$C_9H_{10}O_4$	93-07-2	182.173	nd (w or HOAc) orth (sub)	181	sub			i H_2O; vs EtOH, eth; sl chl
3852	3,5-Dimethoxybenzoic acid		$C_9H_{10}O_4$	1132-21-4	182.173	nd (w), pr (al)	185.5	sub			vs eth, EtOH
3853	4,4′-Dimethoxybenzoin	*p*-Anisoin	$C_{16}H_{16}O_4$	119-52-8	272.296	pr (dil al)	109(3)				sl H_2O, chl, EtOH, eth; s ace
3854	5,7-Dimethoxy-2*H*-1-benzopyran-2-one	Limettin	$C_{11}H_{10}O_4$	487-06-9	206.195	pr or nd (al)	149	200 dec			sl H_2O; vs EtOH, ace, chl; i eth, lig
3855	4,4′-Dimethoxy-1,1′-biphenyl		$C_{14}H_{14}O_2$	2132-80-1	214.260	lf (bz)	175	sub			i H_2O, peth; vs EtOH, bz, chl; sl eth
3856	Dimethoxyborane		$C_2H_7BO_2$	4542-61-4	73.887	vol liq or gas	-130.6	25.9			dec H_2O
3857	4,4-Dimethoxy-2-butanone		$C_6H_{12}O_3$	5436-21-5	132.157			50^5			s ctc
3858	2,6-Dimethoxy-2,5-cyclohexadiene-1,4-dione	2,6-Dimethoxy-*p*-quinone	$C_8H_8O_4$	530-55-2	168.148	ye mcl pr (HOAc)	256	sub			sl H_2O, EtOH, eth; s tfa; vs alk, HOAc
3859	Dimethoxydimethylsilane		$C_4H_{12}O_2Si$	1112-39-6	120.223			82	0.8646^{20}	1.3708^{20}	dec H_2O
3860	Dimethoxydiphenylsilane		$C_{14}H_{16}O_2Si$	6843-66-9	244.362			286	1.0771^{20}	1.5447^{20}	
3861	1,1-Dimethoxydodecane	Lauraldehyde, dimethyl acetal	$C_{14}H_{30}O_2$	14620-52-1	230.387			133^5		1.4310^{25}	vs eth, EtOH
3862	2,2-Dimethoxyethanamine		$C_4H_{11}NO_2$	22483-09-6	105.136		-78	137^{95}	0.966^{25}	1.4170^{20}	
3863	1,2-Dimethoxyethane	Ethylene glycol dimethyl ether	$C_4H_{10}O_2$	110-71-4	90.121	liq	-69.0(0.2)	85.0(0.1)	0.8637^{25}	1.3770^{25}	s H_2O, EtOH, eth, ace, bz, chl, ctc
3864	(2,2-Dimethoxyethyl)benzene		$C_{10}H_{14}O_2$	101-48-4	166.217			193.5			
3865	4,8-Dimethoxyfuro[2,3-*b*]quinoline	Fagarine	$C_{13}H_{11}NO_3$	524-15-2	229.231	pr (al)	142				sl H_2O, peth; s EtOH, eth, bz, chl
3866	1,1-Dimethoxyhexadecane	Palmitaldehyde, dimethyl acetal	$C_{18}H_{38}O_2$	2791-29-9	286.494		10	144^2	0.8542^{20}	1.4382^{25}	vs ace, eth, EtOH
3867	2,4-Dimethoxy-6-hydroxyacetophenone	Xanthoxylin	$C_{10}H_{12}O_4$	90-24-4	196.200	cry (al)	82	185^{20}			vs eth, EtOH
3868	5,6-Dimethoxy-1-indanone		$C_{11}H_{12}O_3$	2107-69-9	192.211		119.5				sl ctc
3869	6,7-Dimethoxy-1(3*H*)-isobenzofuranone	Meconin	$C_{10}H_{10}O_4$	569-31-3	194.184	wh nd (w)	102.5				sl H_2O; s EtOH, eth, ace, bz, HOAc, chl
3870	Dimethoxymethane	Methylal	$C_3H_8O_2$	109-87-5	76.095	liq	-105.11(0.03)	42.3(0.2)	0.8593^{20}	1.3513^{20}	s H_2O; vs ace, bz, eth, EtOH
3871	1,2-Dimethoxy-4-methylbenzene		$C_9H_{12}O_2$	494-99-5	152.190	pr (eth)	24	221(4)	1.0509^{25}	1.5257^{25}	i H_2O; sl ctc; vs os
3872	1,3-Dimethoxy-5-methylbenzene		$C_9H_{12}O_2$	4179-19-5	152.190			244	1.0478^{15}	1.5234^{20}	vs bz, eth, EtOH
3873	1,4-Dimethoxy-2-methylbenzene		$C_9H_{12}O_2$	24599-58-4	152.190		21	214.0			

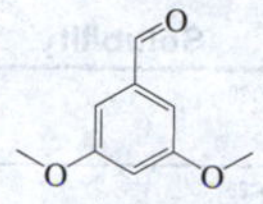
3839
3,5-Dimethoxybenzaldehyde

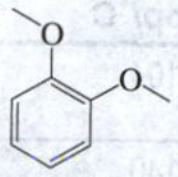
3840
1,2-Dimethoxybenzene

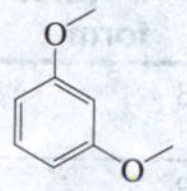
3841
1,3-Dimethoxybenzene

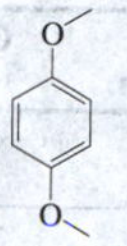
3842
1,4-Dimethoxybenzene

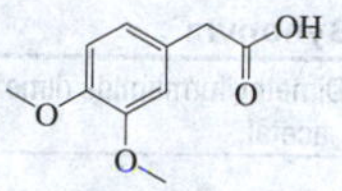
3843
3,4-Dimethoxybenzeneacetic acid

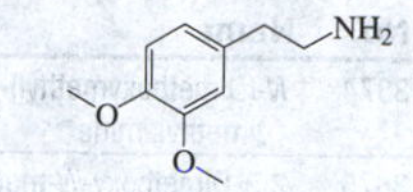
3844
3,4-Dimethoxybenzeneethanamine

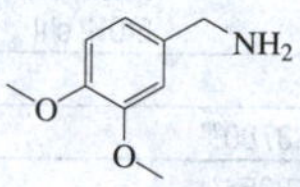
3845
3,4-Dimethoxybenzenemethanamine

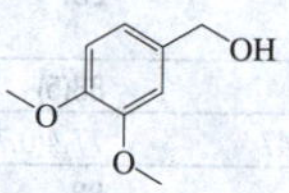
3846
3,4-Dimethoxybenzenemethanol

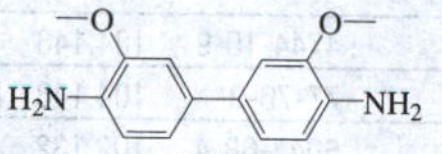
3847
3,3'-Dimethoxybenzidine

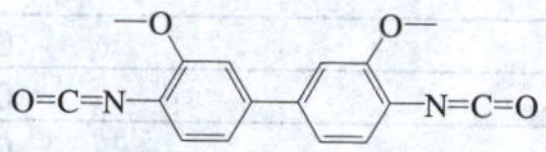
3848
3,3'-Dimethoxybenzidine-4,4'-diisocyanate

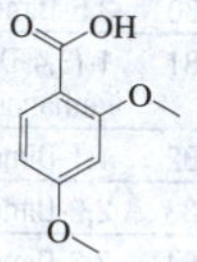
3849
2,4-Dimethoxybenzoic acid

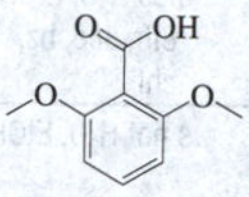
3850
2,6-Dimethoxybenzoic acid

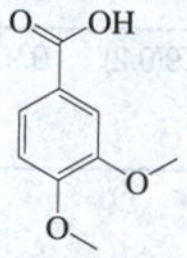
3851
3,4-Dimethoxybenzoic acid

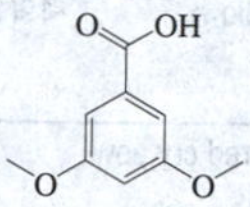
3852
3,5-Dimethoxybenzoic acid

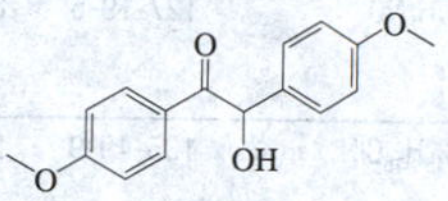
3853
4,4'-Dimethoxybenzoin

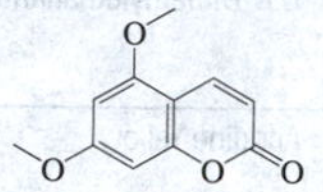
3854
5,7-Dimethoxy-2*H*-1-benzopyran-2-one

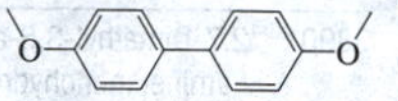
3855
4,4'-Dimethoxy-1,1'-biphenyl

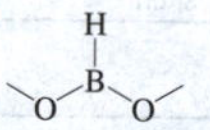
3856
Dimethoxyborane

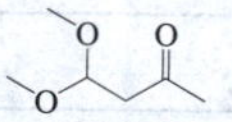
3857
4,4-Dimethoxy-2-butanone

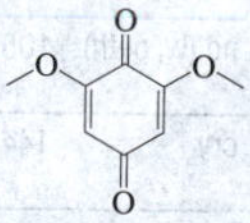
3858
2,6-Dimethoxy-2,5-cyclohexadiene-1,4-dione

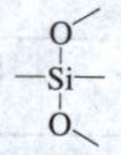
3859
Dimethoxydimethylsilane

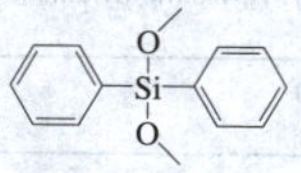
3860
Dimethoxydiphenylsilane

3861
1,1-Dimethoxydodecane

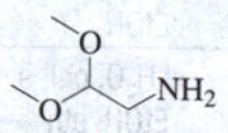
3862
2,2-Dimethoxyethanamine

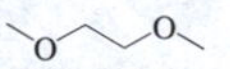
3863
1,2-Dimethoxyethane

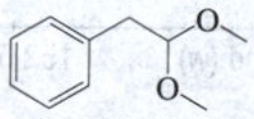
3864
(2,2-Dimethoxyethyl)benzene

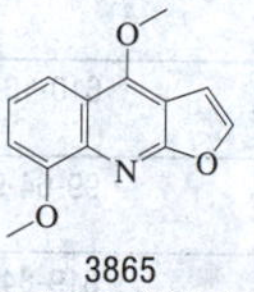
3865
4,8-Dimethoxyfuro[2,3-b]quinoline

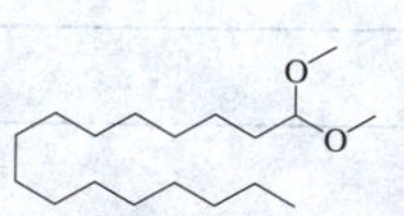
3866
1,1-Dimethoxyhexadecane

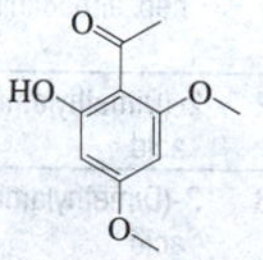
3867
2,4-Dimethoxy-6-hydroxyacetophenone

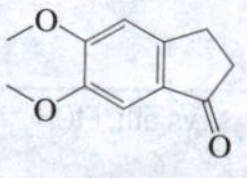
3868
5,6-Dimethoxy-1-indanone

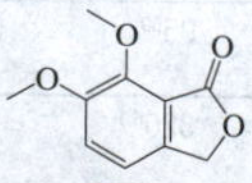
3869
6,7-Dimethoxy-1(3*H*)-isobenzofuranone

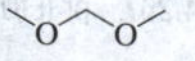
3870
Dimethoxymethane

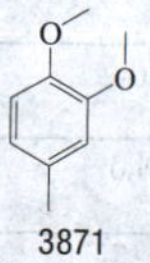
3871
1,2-Dimethoxy-4-methylbenzene

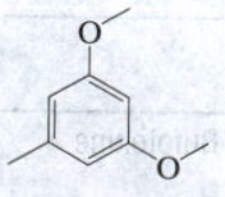
3872
1,3-Dimethoxy-5-methylbenzene

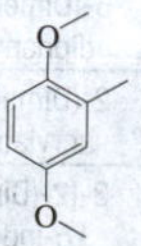
3873
1,4-Dimethoxy-2-methylbenzene

Organic

No.	Name	Synonym	Mol. form.	CAS Reg. No.	Mol. wt.	Physical form	mp/°C	bp/°C	den/ g cm^{-3}	n_D	Solubility
3874	*N*-(Dimethoxymethyl)-dimethylamine	Dimethylformamide dimethyl acetal	$C_5H_{13}NO_2$	4637-24-5	119.163			104	0.897^{25}	1.3972^{20}	
3875	2,2-Dimethoxy-*N*-methyletha-namine		$C_5H_{13}NO_2$	122-07-6	119.163			140	0.928^{25}	1.4115^{20}	
3876	Dimethoxymethylphenylsilane		$C_9H_{14}O_2Si$	3027-21-2	182.292			129^{79}		1.4795^{20}	
3877	1,2-Dimethoxy-4-nitroben-zene		$C_8H_9NO_4$	709-09-1	183.162	ye nd (al-w)	98	230^{15}	1.1888^{133}		i H_2O; vs EtOH, eth; s chl; sl lig
3878	1,4-Dimethoxy-2-nitroben-zene		$C_8H_9NO_4$	89-39-4	183.162	gold-ye nd (dil al)	72.5		1.1666^{132}		i H_2O; s EtOH, bz, chl, sulf
3879	2,6-Dimethoxyphenol		$C_8H_{10}O_3$	91-10-1	154.163	mcl pr (w)	56.5	261(22)			vs eth, EtOH
3880	3,5-Dimethoxyphenol		$C_8H_{10}O_3$	500-99-2	154.163		37	199^{35}			s eth, bz; sl lig
3881	1-(3,4-Dimethoxyphenyl)-ethanone		$C_{10}H_{12}O_3$	1131-62-0	180.200	pr (dil al)	51	287			vs H_2O, bz, EtOH, chl
3882	1,1-Dimethoxypropane		$C_5H_{12}O_2$	4744-10-9	104.148			88(5)	0.8648^{20}		
3883	2,2-Dimethoxypropane		$C_5H_{12}O_2$	77-76-9	104.148	liq	-47	77.4(0.7)	0.847^{25}	1.3780^{20}	
3884	3,3-Dimethoxy-1-propene		$C_5H_{10}O_2$	6044-68-4	102.132			88	0.862^{25}	1.3954^{20}	
3885	*trans*-1,2-Dimethoxy-4-(1-propenyl)benzene	Isoeugenyl methyl ether	$C_{11}H_{14}O_2$	6379-72-2	178.228		18	270.5	1.0521^{20}	1.5616^{20}	
3886	4,5-Dimethoxy-6-(2-propenyl)-1,3-benzodioxole	Apiole (Dill)	$C_{12}H_{14}O_4$	484-31-1	222.237	oil	29.5	285	1.1598^{15}	1.5305^{17}	
3887	1,2-Dimethoxy-4-vinylben-zene		$C_{10}H_{12}O_2$	6380-23-0	164.201					1.5711^{20}	s chl
3888	Dimethylacetal		$C_4H_{10}O_2$	534-15-6	90.121	liq	-113.2	63(2)	0.8501^{20}	1.3668^{20}	s H_2O, EtOH, eth, ctc, chl; vs ace
3889	*N,N*-Dimethylacetamide	*N,N*-Dimethylethanamide	C_4H_9NO	127-19-5	87.120	liq	-19(1)	165.9(0.2)	0.9372^{25}	1.4341^{25}	msc H_2O, EtOH, eth, ace, bz, chl
3890	2,7-Dimethyl-3,6-acridinedi-amine, monohydrochloride	Acridine Yellow	$C_{15}H_{16}ClN_3$	135-49-9	273.761	red cry pow					s hot H_2O, EtOH
3891	Dimethyl adipate	Dimethyl 1,6-hexanedioate	$C_8H_{14}O_4$	627-93-0	174.195	cry	10.3(0.5)	115^{13}	1.0600^{20}	1.4283^{20}	i H_2O; s EtOH, eth, ctc, HOAc
3892	3,3-Dimethylallyl diphosphate	3-Methyl-2-butenyl pyrophosphate	$C_5H_{12}O_7P_2$	358-72-5	246.092	cry (MeOH)					
3893	Dimethylamine	*N*-Methylmethanamine	C_2H_7N	124-40-3	45.084	col gas	-93(2)	7.3(0.4)	0.6804^{0}	1.350^{17}	vs H_2O; s EtOH, eth
3894	Dimethylamine hydrochloride	*N*-Methylmethanamine hydrochloride	C_2H_8ClN	506-59-2	81.545	orth nd (al)	171				vs H_2O, EtOH, chl
3895	(Dimethylamino)acetonitrile		$C_4H_8N_2$	926-64-7	84.120			136(6)	0.8649^{20}	1.4095^{20}	vs H_2O, EtOH
3896	4'-(Dimethylamino)-acetophenone	4-Acetyl-*N,N*-dimethylaniline	$C_{10}H_{13}NO$	2124-31-4	163.216	nd (w, peth)	105.5				vs H_2O, eth, lig; sl chl
3897	10-[(Dimethylamino)acetyl]-10*H*-phenothiazine	Ahistan	$C_{16}H_{16}N_2OS$	518-61-6	284.375	cry	144.5				
3898	4-(Dimethylamino)azobenzene		$C_{14}H_{15}N_3$	60-11-7	225.289	ye lf (al)	116(1)	dec			i H_2O; vs EtOH, py; s eth; sl chl, lig
3899	2',3-Dimethyl-4-aminoazo-benzene	4-*o*-Tolylazo-*o*-toluidine	$C_{14}H_{15}N_3$	97-56-3	225.289	ye lf (al)	102				vs eth, EtOH
3900	4-(Dimethylamino)-benzaldehyde	Ehrlich's reagent	$C_9H_{11}NO$	100-10-7	149.189	lf (w)	73.1(0.8)	176^{17}	1.0254^{100}		sl H_2O, chl; s EtOH, eth, ace, bz
3901	4-(Dimethylamino)-benzalrhodanine		$C_{12}H_{12}N_2OS_2$	536-17-4	264.365	dp red nd (xyl)	270 dec				i H_2O; sl EtOH, bz; vs eth, ctc; s ace
3902	2-(Dimethylamino)benzoic acid		$C_9H_{11}NO_2$	610-16-2	165.189	pr, nd (eth)	72	sub			vs H_2O, eth, EtOH
3903	3-(Dimethylamino)benzoic acid		$C_9H_{11}NO_2$	99-64-9	165.189	nd (w)	152.5				sl H_2O, chl; s EtOH, eth
3904	4-(Dimethylamino)benzoic acid		$C_9H_{11}NO_2$	619-84-1	165.189	nd (al)	242.5				s EtOH; sl eth
3905	4,4'-Dimethylaminobenzophe-nonimide	Brilliant Oil Yellow	$C_{17}H_{21}N_3$	492-80-8	267.369	ye or col pl (al)	136				i H_2O; s EtOH; sl eth
3906	(Dimethylamino)dimethylbo-rane		$C_4H_{12}BN$	1113-30-0	84.956	liq	-92	65			vs eth, ace
3907	6-(Dimethylamino)-4,4-diphenyl-3-heptanone		$C_{21}H_{27}NO$	76-99-3	309.445		99.5				vs EtOH
3908	6-(Dimethylamino)-4,4-diphenyl-3-hexanone	Normethadone	$C_{20}H_{25}NO$	467-85-6	295.419	oily liq		165^{3}			
3909	2-(Dimethylamino)ethyl acrylate		$C_7H_{13}NO_2$	2439-35-2	143.184		<-60	95^{50}	0.938^{20}		
3910	3-[2-(Dimethylamino)ethyl]-1*H*-indol-5-ol	Bufotenine	$C_{12}H_{16}N_2O$	487-93-4	204.267	pr (EtOAc)	146.5	$320^{0.1}$			vs eth, EtOH

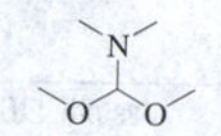
3874
N-(Dimethoxymethyl)dimethylamine

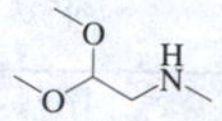
3875
2,2-Dimethoxy-*N*-methylethanamine

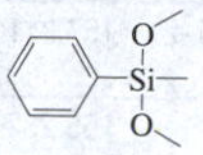
3876
Dimethoxymethylphenylsilane

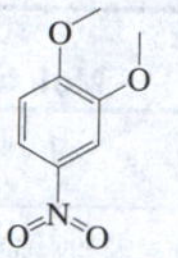
3877
1,2-Dimethoxy-4-nitrobenzene

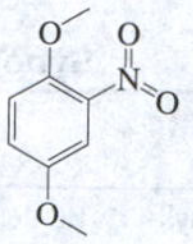
3878
1,4-Dimethoxy-2-nitrobenzene

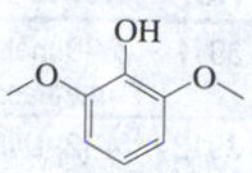
3879
2,6-Dimethoxyphenol

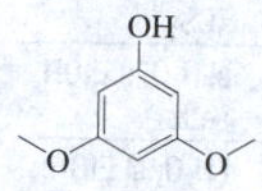
3880
3,5-Dimethoxyphenol

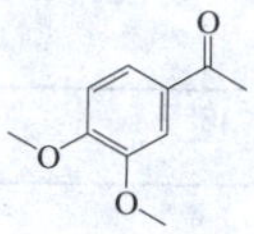
3881
1-(3,4-Dimethoxyphenyl)ethanone

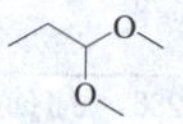
3882
1,1-Dimethoxypropane

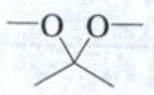
3883
2,2-Dimethoxypropane

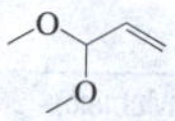
3884
3,3-Dimethoxy-1-propene

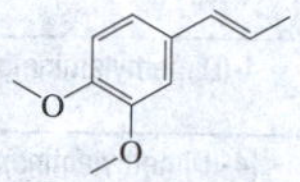
3885
trans-1,2-Dimethoxy-4-(1-propenyl)benzene

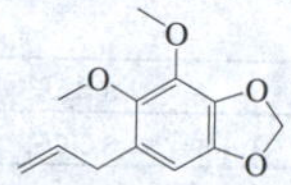
3886
4,5-Dimethoxy-6-(2-propenyl)-1,3-benzodioxole

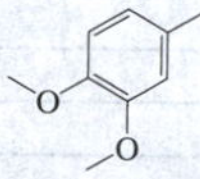
3887
1,2-Dimethoxy-4-vinylbenzene

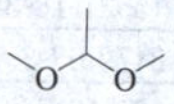
3888
Dimethylacetal

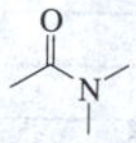
3889
N,N-Dimethylacetamide

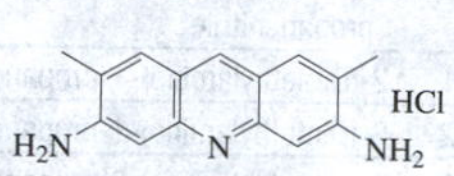
3890
2,7-Dimethyl-3,6-acridinediamine, monohydrochloride

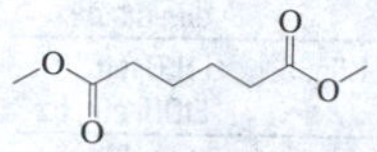
3891
Dimethyl adipate

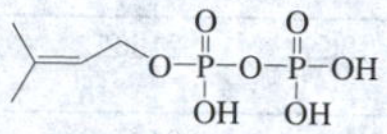
3892
3,3-Dimethylallyl diphosphate

3893
Dimethylamine

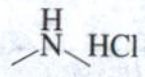
3894
Dimethylamine hydrochloride

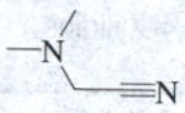
3895
(Dimethylamino)acetonitrile

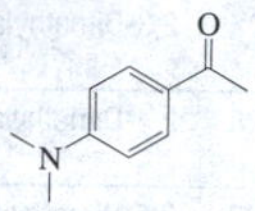
3896
4'-(Dimethylamino)acetophenone

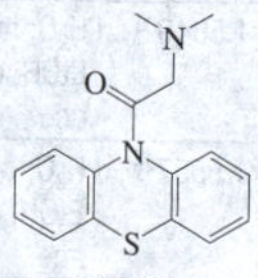
3897
10-[(Dimethylamino)acetyl]-10*H*-phenothiazine

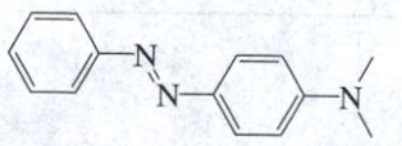
3898
4-(Dimethylamino)azobenzene

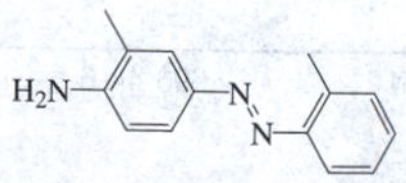
3899
2',3-Dimethyl-4-aminoazobenzene

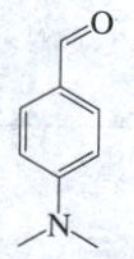
3900
4-(Dimethylamino)benzaldehyde

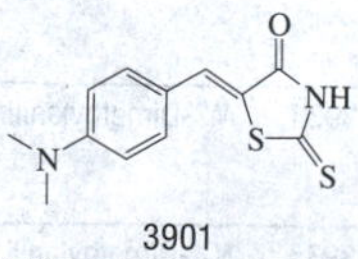
3901
4-(Dimethylamino)benzalrhodanine

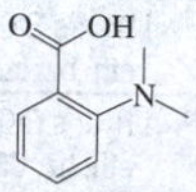
3902
2-(Dimethylamino)benzoic acid

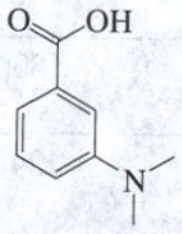
3903
3-(Dimethylamino)benzoic acid

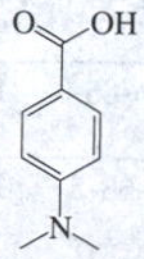
3904
4-(Dimethylamino)benzoic acid

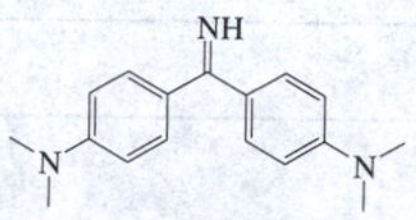
3905
4,4'-Dimethylaminobenzophenonimide

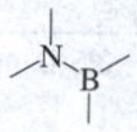
3906
(Dimethylamino)dimethylborane

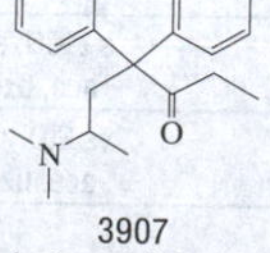
3907
6-(Dimethylamino)-4,4-diphenyl-3-heptanone

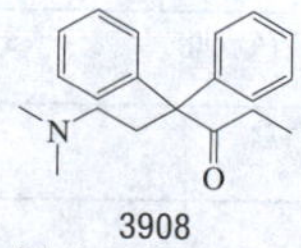
3908
6-(Dimethylamino)-4,4-diphenyl-3-hexanone

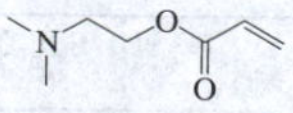
3909
2-(Dimethylamino)ethyl acrylate

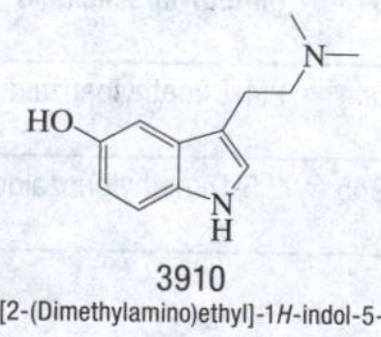
3910
3-[2-(Dimethylamino)ethyl]-1*H*-indol-5-ol

Organic

No.	Name	Synonym	Mol. form.	CAS Reg. No.	Mol. wt.	Physical form	mp/°C	bp/°C	den/ g cm⁻³	n_D	Solubility
3911	2-(Dimethylamino)ethyl methacrylate		$C_8H_{15}NO_2$	2867-47-2	157.211			63^{6}			
3912	4-[2-(Dimethylamino)ethyl]-phenol	Hordenine	$C_{10}H_{15}NO$	539-15-1	165.232	orth pr (al), nd (w)	117.5	173^{11}			vs eth, EtOH, chl
3913	*N*-[2-(Dimethylamino)ethyl]-*N,N′,N′*-trimethyl-1,2-ethanediamine		$C_9H_{23}N_3$	3030-47-5	173.299			84^{12}		1.4413^{25}	
3914	5-(Dimethylamino)-1-naphthalenesulfonyl chloride	Dansyl chloride	$C_{12}H_{12}ClNO_2S$	605-65-2	269.747		70				
3915	3-(Dimethylamino)phenol		$C_8H_{11}NO$	99-07-0	137.179	nd (lig)	86	266.5		1.5895^{26}	i H_2O; s EtOH, eth, ace, bz, CS_2
3916	4-(Dimethylamino)phenol		$C_8H_{11}NO$	619-60-3	137.179		77	165^{30}			sl H_2O; s EtOH, eth
3917	[4-(Dimethylamino)phenyl]-phenylmethanone	4-(Dimethylamino)-benzophenone	$C_{15}H_{15}NO$	530-44-9	225.286	ye lf (al) nd (peth)	92.5				i H_2O; sl EtOH; vs eth; s chl, peth
3918	3-(Dimethylamino)-1-phenyl-1-propanone, hydrochloride		$C_{11}H_{16}ClNO$	879-72-1	213.704		153.5				
3919	3-[4-(Dimethylamino)phenyl]-2-propenal	4-(Dimethylamino)-cinnamaldehyde	$C_{11}H_{13}NO$	6203-18-5	175.227		139.5				
3920	3-(Dimethylamino)-propanenitrile		$C_5H_{10}N_2$	1738-25-6	98.146			177(12)	0.8705^{20}		
3921	2-(Dimethylamino)-1-propanol		$C_5H_{13}NO$	15521-18-3	103.163			150.3	0.8820^{26}		s H_2O
3922	3-(Dimethylamino)-1-propanol		$C_5H_{13}NO$	3179-63-3	103.163			163.5	0.872^{25}	1.4360^{20}	s ctc
3923	1-(Dimethylamino)-2-propanol		$C_5H_{13}NO$	108-16-7	103.163			124.5	0.837^{25}	1.4193^{20}	s ctc
3924	3-(Dimethylamino)-1-propyne	*N,N*-Dimethyl-2-propargylamine	C_5H_9N	7223-38-3	83.132			81(3)	0.7792^{20}	1.4195^{20}	
3925	2-Dimethylaminopurine	*N,N*-Dimethyl-1*H*-purin-6-amine	$C_7H_9N_5$	938-55-6	163.180		263				
3926	2-(*p*-Dimethylaminostyryl)-benzothiazole		$C_{17}H_{16}N_2S$	1628-58-6	280.387	ye nd (MeOH)	207 dec				
3927	2,3-Dimethylaniline	2,3-Xylidine	$C_8H_{11}N$	87-59-2	121.180	liq	3(1)	223(17)	0.9931^{20}	1.5684^{20}	sl H_2O; vs EtOH, eth; s ctc
3928	2,4-Dimethylaniline	2,4-Xylidine	$C_8H_{11}N$	95-68-1	121.180	liq	-13(2)	215(2)	0.9723^{20}	1.5569^{20}	sl H_2O, ctc; s EtOH, eth, bz
3929	2,5-Dimethylaniline	2,5-Xylidine	$C_8H_{11}N$	95-78-3	121.180	ye lf (lig)	6(1)	214	0.9790^{21}	1.5591^{21}	sl H_2O; s eth, ctc
3930	2,6-Dimethylaniline	2,6-Xylidine	$C_8H_{11}N$	87-62-7	121.180	liq	11.0(0.4)	211(4)	0.9842^{20}	1.5610^{20}	vs eth, EtOH
3931	3,4-Dimethylaniline	3,4-Xylidine	$C_8H_{11}N$	95-64-7	121.180	pl or pr (lig)	49(1)	228	1.076^{18}		sl H_2O, chl; s eth; vs lig
3932	3,5-Dimethylaniline	3,5-Xylidine	$C_8H_{11}N$	108-69-0	121.180		10(1)	217(5)	0.9706^{20}	1.5581^{20}	sl H_2O; s eth, ctc
3933	*N*,2-Dimethylaniline		$C_8H_{11}N$	611-21-2	121.180			207.5	0.9709^{20}	1.5649^{20}	i H_2O; msc EtOH, eth; s ace
3934	*N*,3-Dimethylaniline		$C_8H_{11}N$	696-44-6	121.180			206.5	0.9660^{20}	1.5557^{25}	i H_2O; msc EtOH, eth; s ace
3935	*N*,4-Dimethylaniline		$C_8H_{11}N$	623-08-5	121.180			215(20)	0.9348^{55}	1.5568^{20}	i H_2O; msc EtOH, eth; s ace
3936	*N,N*-Dimethylaniline		$C_8H_{11}N$	121-69-7	121.180	pa ye	2.1(0.5)	193(1)	0.9557^{20}	1.5582^{20}	sl H_2O; s EtOH, eth, ace, bz; vs chl
3937	*N,N*-Dimethylaniline hydrochloride		$C_8H_{12}ClN$	5882-44-0	157.641	hyg pl (w, bz)	90		1.1156^{19}		vs H_2O, EtOH, chl
3938	2,6-Dimethylanisole		$C_9H_{12}O$	1004-66-6	136.190			183(11)	0.9619^{14}	1.5053^{14}	i H_2O; s EtOH, eth, bz, ctc
3939	3,5-Dimethylanisole		$C_9H_{12}O$	874-63-5	136.190			194	0.9627^{15}	1.5110^{20}	i H_2O; s EtOH, eth, bz, CS_2; sl ctc
3940	9,10-Dimethylanthracene		$C_{16}H_{14}$	781-43-1	206.282		186.4(0.5)	360.0			i H_2O
3941	1,4-Dimethyl-9,10-anthracenedione		$C_{16}H_{12}O_2$	1519-36-4	236.265	ye nd (al, sub)	140.5	sub			i H_2O; sl EtOH; s bz, xyl, HOAc
3942	Dimethylarsine		C_2H_7As	593-57-7	105.999	liq, ign in air	-136.1	36	1.208^{29}		vs ace, bz, eth, EtOH
3943	Dimethylarsinic acid	Cacodylic acid	$C_2H_7AsO_2$	75-60-5	137.998		199.5(0.6)	>200			vs H_2O; s EtOH; i eth
3944	2,4-Dimethylbenzaldehyde		$C_9H_{10}O$	15764-16-6	134.174	liq	-9	218			s EtOH; s eth, ace, bz; sl chl
3945	2,5-Dimethylbenzaldehyde	Isoxylaldehyde	$C_9H_{10}O$	5779-94-2	134.174			220	0.9500^{20}		vs EtOH; s eth, ace, bz, ctc

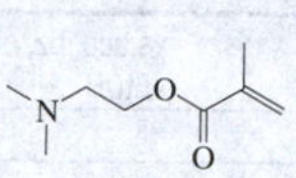
3911
2-(Dimethylamino)ethyl methacrylate

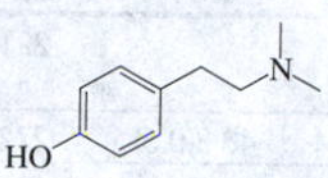
3912
4-[2-(Dimethylamino)ethyl]phenol

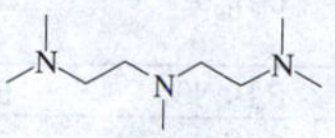
3913
N-[2-(Dimethylamino)ethyl]-*N,N',N'*-trimethyl-1,2-ethanediamine

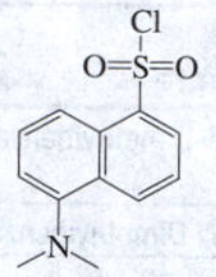
3914
5-(Dimethylamino)-1-naphthalenesulfonyl chloride

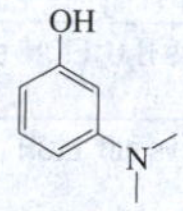
3915
3-(Dimethylamino)phenol

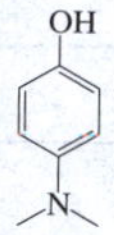
3916
4-(Dimethylamino)phenol

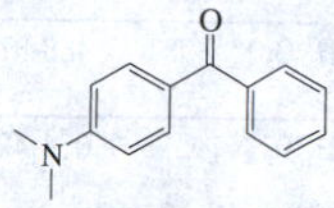
3917
[4-(Dimethylamino)phenyl]phenylmethanone

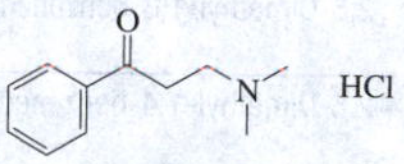
3918
3-(Dimethylamino)-1-phenyl-1-propanone, hydrochloride

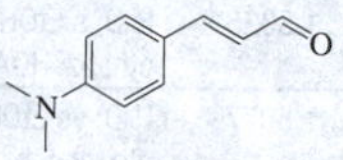
3919
3-[4-(Dimethylamino)phenyl]-2-propenal

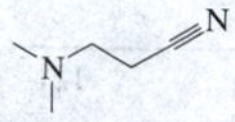
3920
3-(Dimethylamino)propanenitrile

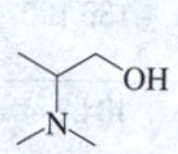
3921
2-(Dimethylamino)-1-propanol

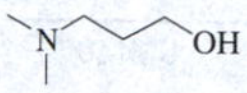
3922
3-(Dimethylamino)-1-propanol

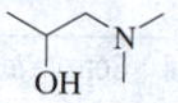
3923
1-(Dimethylamino)-2-propanol

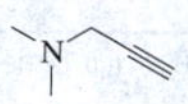
3924
3-(Dimethylamino)-1-propyne

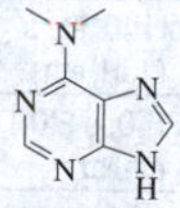
3925
2-Dimethylaminopurine

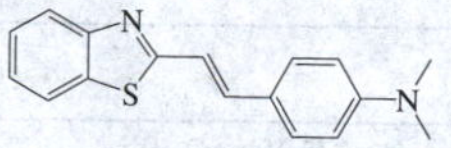
3926
2-(*p*-Dimethylaminostyryl)benzothiazole

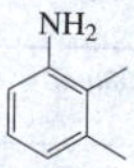
3927
2,3-Dimethylaniline

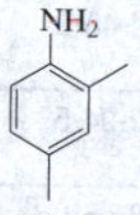
3928
2,4-Dimethylaniline

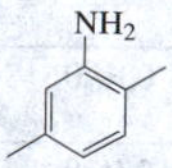
3929
2,5-Dimethylaniline

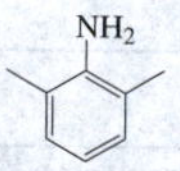
3930
2,6-Dimethylaniline

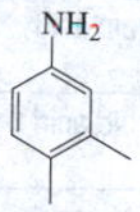
3931
3,4-Dimethylaniline

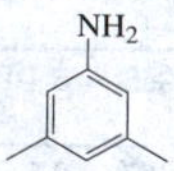
3932
3,5-Dimethylaniline

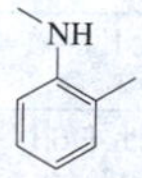
3933
N,2-Dimethylaniline

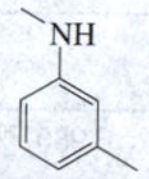
3934
N,3-Dimethylaniline

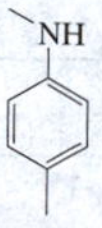
3935
N,4-Dimethylaniline

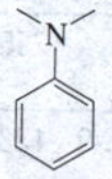
3936
N,N-Dimethylaniline

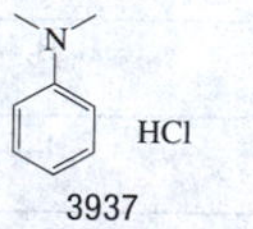
3937
N,N-Dimethylaniline hydrochloride

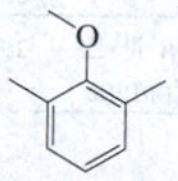
3938
2,6-Dimethylanisole

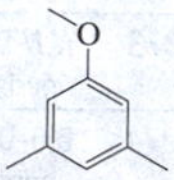
3939
3,5-Dimethylanisole

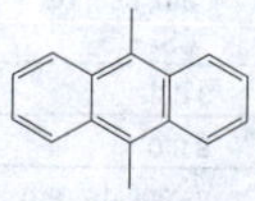
3940
9,10-Dimethylanthracene

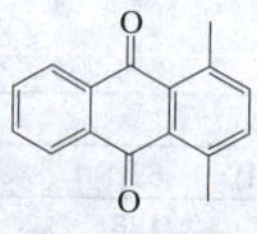
3941
1,4-Dimethyl-9,10-anthracenedione

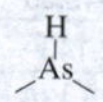
3942
Dimethylarsine

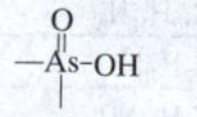
3943
Dimethylarsinic acid

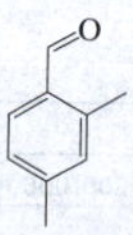
3944
2,4-Dimethylbenzaldehyde

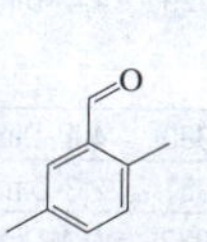
3945
2,5-Dimethylbenzaldehyde

No.	Name	Synonym	Mol. form.	CAS Reg. No.	Mol. wt.	Physical form	mp/°C	bp/°C	den/ g cm^{-3}	n_D	Solubility
3946	3,5-Dimethylbenzaldehyde		$C_9H_{10}O$	5779-95-3	134.174		9	221(12)			vs ace, bz, eth, EtOH
3947	*N,N*-Dimethylbenzamide		$C_9H_{11}NO$	611-74-5	149.189		43.8(0.5)	272.0			
3948	7,12-Dimethylbenz[*a*]anthracene	9,10-Dimethyl-1,2-benzanthracene	$C_{20}H_{16}$	57-97-6	256.341	pa ye pl (al, HOAc)	123(1)				vs ace, bz
3949	4,5-Dimethyl-1,2-benzenediamine		$C_8H_{12}N_2$	3171-45-7	136.194		128				
3950	*N,N*-Dimethyl-1,2-benzenediamine		$C_8H_{12}N_2$	2836-03-5	136.194	oil		218	0.995^{22}		sl H_2O; vs EtOH, eth, ace, bz
3951	*N,N*-Dimethyl-1,3-benzenediamine		$C_8H_{12}N_2$	2836-04-6	136.194		<-20	270	0.995^{25}		sl H_2O; vs EtOH, eth
3952	*N,N*-Dimethyl-1,4-benzenediamine	Dimethyl-*p*-phenylenediamine	$C_8H_{12}N_2$	99-98-9	136.194	nd (bz)	53	264(5)	1.036^{20}		s H_2O, chl; vs EtOH, eth, bz; sl lig
3953	2,5-Dimethyl-1,3-benzenediol		$C_8H_{10}O_2$	488-87-9	138.164	nd (bz), pr (w)	163	278.5			s H_2O, EtOH, eth
3954	2,6-Dimethyl-1,4-benzenediol		$C_8H_{10}O_2$	654-42-2	138.164	nd (xyl), cry (w)	152.3				vs eth, EtOH
3955	*N*,β-Dimethylbenzeneethanamine	Phenylpropylmethylamine	$C_{10}H_{15}N$	93-88-9	149.233			207.5	0.915^{25}		vs bz, eth, EtOH
3956	α,α-Dimethylbenzeneethanamine	Phentermine	$C_{10}H_{15}N$	122-09-8	149.233	oily liq		205			
3957	α,α-Dimethylbenzenemethanamine		$C_9H_{13}N$	585-32-0	135.206			196.5	0.9423^{20}	1.5181^{25}	
3958	α,4-Dimethylbenzenemethanol	1-(4-Methylphenyl)ethanol	$C_9H_{12}O$	536-50-5	136.190			215(11)	0.9668^{25}	1.5246^{20}	i H_2O; vs EtOH, eth
3959	α,α-Dimethylbenzenemethanol	α-Cumyl alcohol	$C_9H_{12}O$	617-94-7	136.190	pr	37(1)	202	0.9735^{20}	1.5325^{20}	i H_2O; s EtOH, eth, bz, HOAc
3960	α,α-Dimethylbenzenepropanol	Benzyl-*tert*-butanol	$C_{11}H_{16}O$	103-05-9	164.244	nd	24.5	121^{13}	0.9626^{21}	1.5077^{21}	i H_2O; vs EtOH, eth, ace, bz
3961	*N*,4-Dimethylbenzenesulfonamide		$C_8H_{11}NO_2S$	640-61-9	185.244	pl (dil al)	78.5		1.340^{25}		vs eth, EtOH
3962	5,6-Dimethyl-1*H*-benzimidazole	Dimedazole	$C_9H_{10}N_2$	582-60-5	146.188	cry (eth)	205.5	sub			s H_2O, EtOH, eth, chl, DMSO
3963	2,4-Dimethylbenzoic acid		$C_9H_{10}O_2$	611-01-8	150.174	mcl or tcl nd (w)	90	268			sl H_2O; s EtOH, ace, bz, chl, HOAc, tol
3964	2,5-Dimethylbenzoic acid		$C_9H_{10}O_2$	610-72-0	150.174	nd (al)	132(3)	sub	1.069^{21}		i H_2O; s EtOH, eth, ace, bz
3965	2,6-Dimethylbenzoic acid		$C_9H_{10}O_2$	632-46-2	150.174	nd (lig)	116	277(18)			sl H_2O, lig; s EtOH, eth
3966	3,4-Dimethylbenzoic acid		$C_9H_{10}O_2$	619-04-5	150.174	pr (al)	163(2)				i H_2O; s EtOH, eth, bz
3967	3,5-Dimethylbenzoic acid	Mesitylenic acid	$C_9H_{10}O_2$	499-06-9	150.174	nd (w, al)	171.1	275(19)			sl H_2O; vs EtOH, eth
3968	4,4'-Dimethylbenzophenone	Bis(4-methylphenyl) ketone	$C_{15}H_{14}O$	611-97-2	210.271	orth (al)	96.5	339(20)			vs ace, bz, eth, EtOH
3969	7,8-Dimethylbenzo[*g*]pteridine-2,4(1*H*,3*H*)-dione	Lumichrome	$C_{12}H_{10}N_4O_2$	1086-80-2	242.233	ye cry (chl)	300				sl H_2O, EtOH, chl
3970	2,5-Dimethylbenzoxazole		C_9H_9NO	5676-58-4	147.173			218.5	1.0880^{18}	1.5412^{20}	s ctc
3971	*N,N*-Dimethylbenzylamine	Dimethylbenzylamine	$C_9H_{13}N$	103-83-3	135.206			179(15)	0.915^{0}	1.5011^{20}	sl H_2O; msc EtOH, eth
3972	*N,N*-Dimethyl-*N'*-benzyl-1,2-ethanediamine	*N*-Benzyl-*N',N'*-dimethyl-1,2-ethanediamine	$C_{11}H_{18}N_2$	103-55-9	178.274			145^{30}	0.9343^{20}	1.5089^{20}	
3973	*N,N*-Dimethyl-*N'*-benzyl-*N'*-2-pyridinyl-1,2-ethanediamine	Tripelennamine	$C_{16}H_{21}N_3$	91-81-6	255.358	ye oil		$140^{0.1}$		1.576^{25}	misc H_2O
3974	6,6-Dimethylbicyclo[3.1.1]heptan-2-one, (1*R*)-		$C_9H_{14}O$	38651-65-9	138.206	liq	-1	209	0.9807^{20}	1.4787^{20}	vs eth, EtOH
3975	2,3-Dimethylbicyclo[2.2.1]hept-2-ene	2,3-Dimethyl-2-norbornene	C_9H_{14}	529-16-8	122.207			140.5	0.8698^{17}	1.4688^{17}	s eth, ace, bz
3976	6,6-Dimethylbicyclo[3.1.1]hept-2-ene-2-ethanol		$C_{11}H_{18}O$	128-50-7	166.260			235	0.973^{25}	1.4930^{20}	s chl
3977	2,2'-Dimethylbiphenyl		$C_{14}H_{14}$	605-39-0	182.261	cry (al)	19.94(0.01)	258(7)	0.9906^{20}	1.5752^{20}	i H_2O; vs EtOH, eth, bz; s ace
3978	3,3'-Dimethylbiphenyl		$C_{14}H_{14}$	612-75-9	182.261		8(4)	289.6(0.2)	0.9995^{20}	1.5946^{20}	i H_2O; vs EtOH, eth, bz; s ace
3979	4,4'-Dimethylbiphenyl		$C_{14}H_{14}$	613-33-2	182.261	mcl pr (eth)	120.9(0.4)	294(11)	0.917^{121}		i H_2O; sl EtOH; s eth, ace, bz, CS_2
3980	4,4'-Dimethyl-2,2'-bipyridine		$C_{12}H_{12}N_2$	1134-35-6	184.236		171.5				s chl
3981	2,3-Dimethyl-1,3-butadiene	Diisopropenyl	C_6H_{10}	513-81-5	82.143	liq	-76.0(0.1)	69(1)	0.7222^{25}	1.4394^{20}	s ctc
3982	*N,N*-Dimethylbutanamide		$C_6H_{13}NO$	760-79-2	115.173	liq	-40	186	0.9064^{25}	1.4391^{25}	vs ace, bz, eth, EtOH

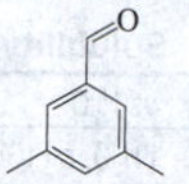
3946
3,5-Dimethylbenzaldehyde

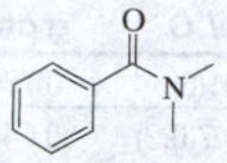
3947
N,N-Dimethylbenzamide

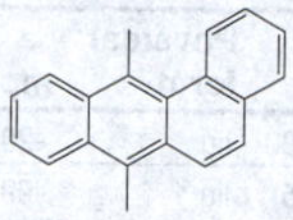
3948
7,12-Dimethylbenz[a]anthracene

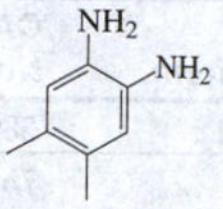
3949
4,5-Dimethyl-1,2-benzenediamine

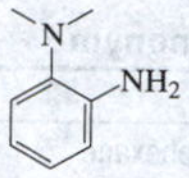
3950
N,N-Dimethyl-1,2-benzenediamine

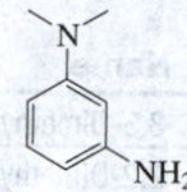
3951
N,N-Dimethyl-1,3-benzenediamine

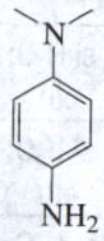
3952
N,N-Dimethyl-1,4-benzenediamine

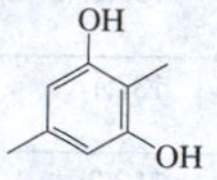
3953
2,5-Dimethyl-1,3-benzenediol

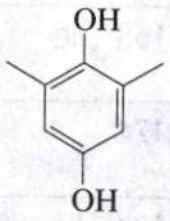
3954
2,6-Dimethyl-1,4-benzenediol

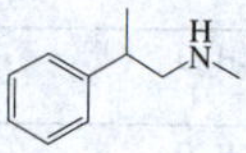
3955
N,β-Dimethylbenzeneethanamine

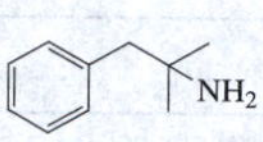
3956
α,α-Dimethylbenzeneethanamine

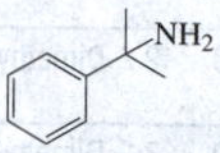
3957
α,α-Dimethylbenzenemethanamine

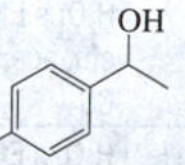
3958
α,4-Dimethylbenzenemethanol

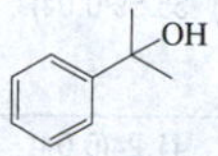
3959
α,α-Dimethylbenzenemethanol

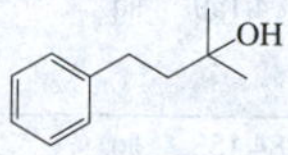
3960
α,α-Dimethylbenzenepropanol

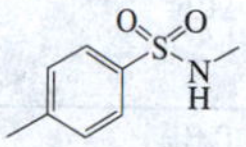
3961
N,4-Dimethylbenzenesulfonamide

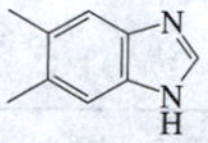
3962
5,6-Dimethyl-1*H*-benzimidazole

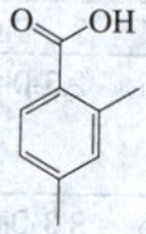
3963
2,4-Dimethylbenzoic acid

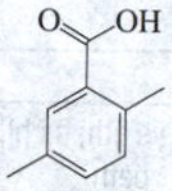
3964
2,5-Dimethylbenzoic acid

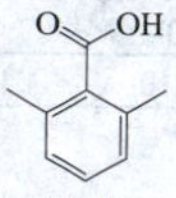
3965
2,6-Dimethylbenzoic acid

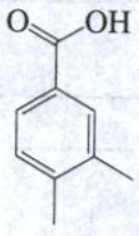
3966
3,4-Dimethylbenzoic acid

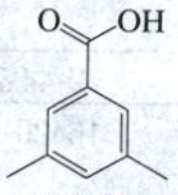
3967
3,5-Dimethylbenzoic acid

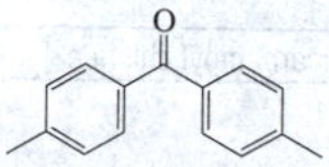
3968
4,4'-Dimethylbenzophenone

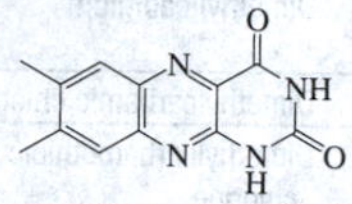
3969
7,8-Dimethylbenzo[g]pteridine-2,4(1*H*,3*H*)-dione

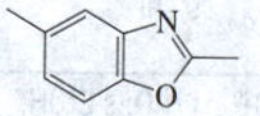
3970
2,5-Dimethylbenzoxazole

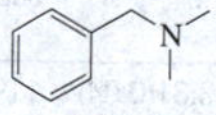
3971
N,N-Dimethylbenzylamine

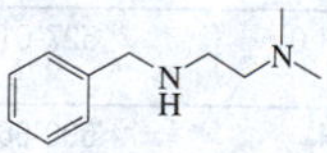
3972
N,N-Dimethyl-*N'*-benzyl-1,2-ethanediamine

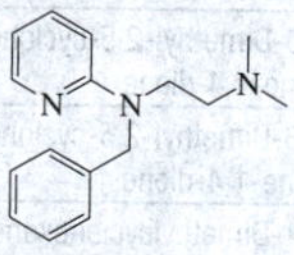
3973
N,N-Dimethyl-*N'*-benzyl-*N'*-2-pyridinyl-1,2-ethanediamine

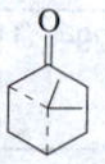
3974
6,6-Dimethylbicyclo[3.1.1]heptan-2-one, (1*R*)-

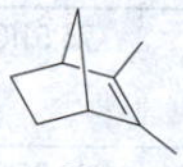
3975
2,3-Dimethylbicyclo[2.2.1]hept-2-ene

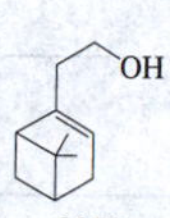
3976
6,6-Dimethylbicyclo[3.1.1]hept-2-ene-2-ethanol

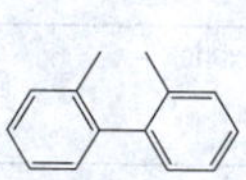
3977
2,2'-Dimethylbiphenyl

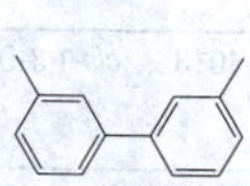
3978
3,3'-Dimethylbiphenyl

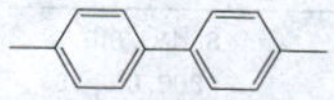
3979
4,4'-Dimethylbiphenyl

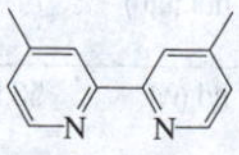
3980
4,4'-Dimethyl-2,2'-bipyridine

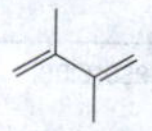
3981
2,3-Dimethyl-1,3-butadiene

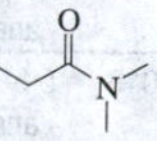
3982
N,N-Dimethylbutanamide

Organic

No.	Name	Synonym	Mol. form.	CAS Reg. No.	Mol. wt.	Physical form	mp/°C	bp/°C	den/g cm^{-3}	n_D	Solubility
3983	3,3-Dimethyl-2-butanamine		$C_6H_{15}N$	3850-30-4	101.190	liq	-20	101(6)	0.7668^{20}	1.4105^{25}	vs H_2O
3984	2,2-Dimethylbutane	Neohexane	C_6H_{14}	75-83-2	86.175	liq	-99.0(0.4)	49.7(0.2)	0.6444^{25}	1.3688^{20}	i H_2O; s EtOH, eth; vs ace, bz, peth, ctc
3985	2,3-Dimethylbutane	Diisopropyl	C_6H_{14}	79-29-8	86.175	liq	-128.1(0.2)	58.0(0.3)	0.6616^{20}	1.3750^{20}	i H_2O; s EtOH, eth; vs ace, bz, peth, ctc
3986	2,3-Dimethyl-2,3-butanediol	Pinacol	$C_6H_{14}O_2$	76-09-5	118.174	nd (al,eth)	43.3(0.2)	174(3)			sl H_2O, CS_2; vs EtOH, eth
3987	2,3-Dimethyl-2-butanethiol		$C_6H_{14}S$	1639-01-6	118.240	liq		126.1(0.7)			
3988	2,2-Dimethylbutanoic acid		$C_6H_{12}O_2$	595-37-9	116.158	liq	-15.0(0.5)	187.87(0.02)	0.9276^{20}	1.4145^{20}	sl H_2O; s EtOH, eth
3989	2,2-Dimethyl-1-butanol		$C_6H_{14}O$	1185-33-7	102.174		<-15	137(1)	0.8283^{20}	1.4208^{20}	sl H_2O; s EtOH, eth
3990	3,3-Dimethyl-1-butanol	Dimbunol	$C_6H_{14}O$	624-95-3	102.174	liq	-60	143(2)	0.844^{15}	1.4323^{15}	sl H_2O; s EtOH, eth, ace
3991	2,3-Dimethyl-2-butanol	Isopropyldimethylcarbinol	$C_6H_{14}O$	594-60-5	102.174	liq	-10.5(0.3)	118.7(0.5)	0.8236^{20}	1.4176^{20}	s H_2O; msc EtOH, eth
3992	3,3-Dimethyl-2-butanol, (±)-		$C_6H_{14}O$	20281-91-8	102.174		5.6	120.4	0.8122^{25}	1.4148^{20}	sl H_2O; vs EtOH, eth
3993	3,3-Dimethyl-2-butanone	Pinacolone	$C_6H_{12}O$	75-97-8	100.158	liq	-51.40(0.05)	106.1(0.2)	0.7229^{25}	1.3952^{20}	sl H_2O; s EtOH, eth, ace, ctc
3994	3,3-Dimethylbutanoyl chloride		$C_6H_{11}ClO$	7065-46-5	134.603			129(8)	0.969^{20}	1.4210^{20}	vs eth
3995	2,3-Dimethyl-1-butene		C_6H_{12}	563-78-0	84.159	liq	-157.27(0.09)	55.59(0.04)	0.6803^{20}	1.3995^{20}	i H_2O; s EtOH, eth, ace, ctc, CS_2
3996	3,3-Dimethyl-1-butene		C_6H_{12}	558-37-2	84.159	liq	-115.2(0.3)	41.24(0.04)	0.6529^{20}	1.3763^{20}	i H_2O; s EtOH, eth, ctc, chl
3997	2,3-Dimethyl-2-butene		C_6H_{12}	563-79-1	84.159	liq	-74.3(0.1)	73.19(0.06)	0.7080^{20}	1.4122^{20}	i H_2O; s EtOH, eth, ace, chl
3998	*N*-(1,3-Dimethylbutyl)-*N'*-phenyl-1,4-benzenediamine		$C_{18}H_{24}N_2$	793-24-8	268.397		46	164^{1}			
3999	3,3-Dimethyl-1-butyne	*tert*-Butylacetylene	C_6H_{10}	917-92-0	82.143	liq	-78.2(0.4)	38(2)	0.6623^{25}	1.3736^{20}	
4000	Dimethyl 2-butynedioate		$C_6H_6O_4$	762-42-5	142.110			199(9)	1.1564^{20}	1.4434^{20}	s EtOH, eth, ctc
4001	Dimethyl cadmium		C_2H_6Cd	506-82-1	142.480		-2.67(0.02)	105.5 (exp 150)	1.9846^{18}	1.5488	s peth
4002	Dimethylcarbamic chloride	Dimethylcarbamoyl chloride	C_3H_6ClNO	79-44-7	107.539	liq	-33	154(10)	1.168^{25}	1.4540^{20}	
4003	Dimethylcarbamothioic chloride		C_3H_6ClNS	16420-13-6	123.605	pr	42(2)	98^{10}			vs eth; s chl, peth
4004	Dimethyl carbate		$C_{11}H_{14}O_4$	39589-98-5	210.227	cry	38	$137^{12.5}$	1.164^{21}	1.4852^{20}	i H_2O
4005	Dimethyl carbonate	Methyl carbonate	$C_3H_6O_3$	616-38-6	90.078		-1(10)	90.11(0.09)	1.0636^{25}	1.3687^{20}	i H_2O; s EtOH, eth; sl ctc
4006	Dimethylcyanamide		$C_3H_6N_2$	1467-79-4	70.093			162(5)		1.4089^{19}	vs ace, eth, EtOH
4007	2,3-Dimethyl-2,5-cyclohexadiene-1,4-dione		$C_8H_8O_2$	526-86-3	136.149	ye nd	55	sub			sl H_2O; s EtOH, eth, chl
4008	2,5-Dimethyl-2,5-cyclohexadiene-1,4-dione		$C_8H_8O_2$	137-18-8	136.149	ye nd (al)	126.0				sl H_2O, EtOH; s eth, bz, chl
4009	2,6-Dimethyl-2,5-cyclohexadiene-1,4-dione		$C_8H_8O_2$	527-61-7	136.149	ye nd	72.5	sub	1.0479^{28}		s chl
4010	1,1-Dimethylcyclohexane		C_8H_{16}	590-66-9	112.213	liq	-33.31(0.04)	119.5(0.3)	0.7809^{20}	1.4290^{20}	i H_2O; s EtOH, eth, ace, bz; msc ctc
4011	*cis*-1,2-Dimethylcyclohexane		C_8H_{16}	2207-01-4	112.213	liq	-49.83(0.04)	129.7(0.6)	0.7963^{20}	1.4360^{20}	i H_2O; s EtOH, bz, ctc; msc eth, ace
4012	*trans*-1,2-Dimethylcyclohexane		C_8H_{16}	6876-23-9	112.213	liq	-88.12(0.02)	123.4(0.3)	0.7760^{20}	1.4270^{20}	i H_2O; s EtOH, eth; msc ace, bz; vs lig
4013	*cis*-1,3-Dimethylcyclohexane		C_8H_{16}	638-04-0	112.213	liq	-75.51(0.03)	124.4(0.6)	0.7660^{20}	1.4229^{20}	i H_2O; msc EtOH, eth, ace, bz, lig, ctc
4014	*trans*-1,3-Dimethylcyclohexane		C_8H_{16}	2207-03-6	112.213	liq	-90.05(0.03)	120.1(0.7)	0.79^{15}	1.4284^{25}	
4015	*cis*-1,4-Dimethylcyclohexane		C_8H_{16}	624-29-3	112.213	liq	-87.4(0.3)	124.3(0.7)	0.7829^{20}	1.4230^{20}	i H_2O; msc EtOH, eth, ace, bz, lig, ctc
4016	*trans*-1,4-Dimethylcyclohexane		C_8H_{16}	2207-04-7	112.213	liq	-36.9(0.2)	119.3(0.5)	0.77^{15}	1.4185^{25}	i H_2O
4017	Dimethyl *trans*-1,4-cyclohexanedicarboxylate		$C_{10}H_{16}O_4$	3399-22-2	200.232	ndl (eth)	71				s eth
4018	5,5-Dimethyl-1,3-cyclohexanedione	5,5-Dimethyldihydroresorcinol	$C_8H_{12}O_2$	126-81-8	140.180	nd (w)	52.5(0.5)				sl H_2O, eth; s ace, ctc; vs chl, HOAc

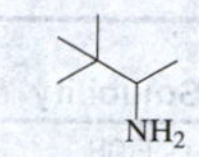
3983
3,3-Dimethyl-2-butanamine

3984
2,2-Dimethylbutane

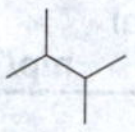
3985
2,3-Dimethylbutane

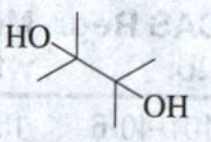
3986
2,3-Dimethyl-2,3-butanediol

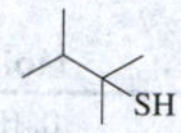
3987
2,3-Dimethyl-2-butanethiol

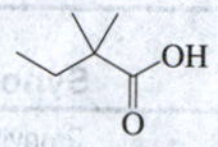
3988
2,2-Dimethylbutanoic acid

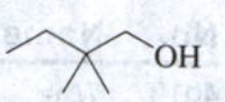
3989
2,2-Dimethyl-1-butanol

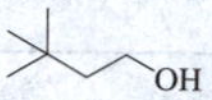
3990
3,3-Dimethyl-1-butanol

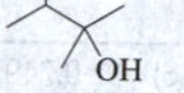
3991
2,3-Dimethyl-2-butanol

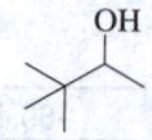
3992
3,3-Dimethyl-2-butanol, (±)-

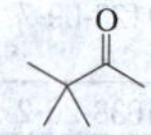
3993
3,3-Dimethyl-2-butanone

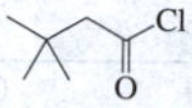
3994
3,3-Dimethylbutanoyl chloride

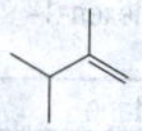
3995
2,3-Dimethyl-1-butene

3996
3,3-Dimethyl-1-butene

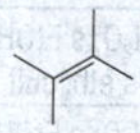
3997
2,3-Dimethyl-2-butene

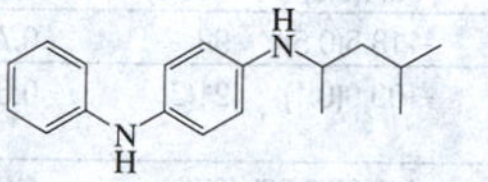
3998
N-(1,3-Dimethylbutyl)-*N'*-phenyl-1,4-benzenediamine

3999
3,3-Dimethyl-1-butyne

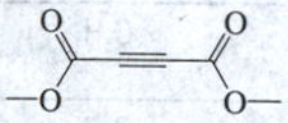
4000
Dimethyl 2-butynedioate

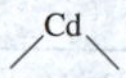
4001
Dimethyl cadmium

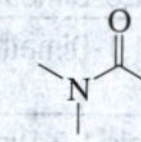
4002
Dimethylcarbamic chloride

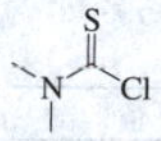
4003
Dimethylcarbamothioic chloride

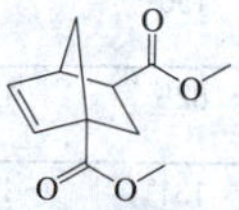
4004
Dimethyl carbate

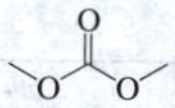
4005
Dimethyl carbonate

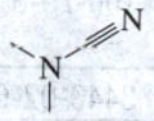
4006
Dimethylcyanamide

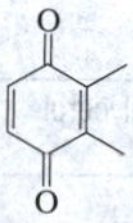
4007
2,3-Dimethyl-2,5-cyclohexadiene-1,4-dione

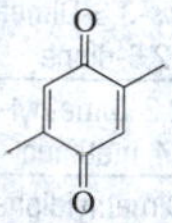
4008
2,5-Dimethyl-2,5-cyclohexadiene-1,4-dione

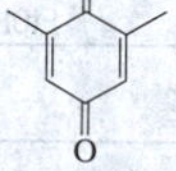
4009
2,6-Dimethyl-2,5-cyclohexadiene-1,4-dione

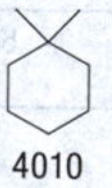
4010
1,1-Dimethylcyclohexane

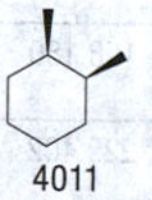
4011
cis-1,2-Dimethylcyclohexane

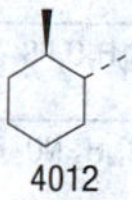
4012
trans-1,2-Dimethylcyclohexane

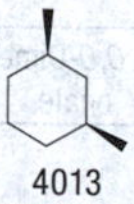
4013
cis-1,3-Dimethylcyclohexane

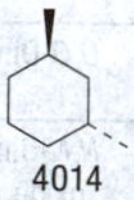
4014
trans-1,3-Dimethylcyclohexane

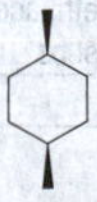
4015
cis-1,4-Dimethylcyclohexane

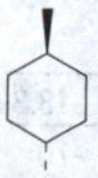
4016
trans-1,4-Dimethylcyclohexane

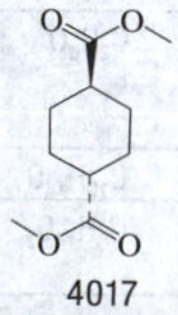
4017
Dimethyl *trans*-1,4-cyclohexanedicarboxylate

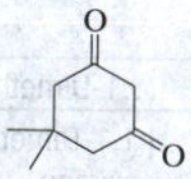
4018
5,5-Dimethyl-1,3-cyclohexanedione

Organic

No.	Name	Synonym	Mol. form.	CAS Reg. No.	Mol. wt.	Physical form	mp/°C	bp/°C	den/ g cm^{-3}	n_D	Solubility
4019	*N*,α-Dimethylcyclohexaneethanamine	Propylhexedrine	$C_{10}H_{21}N$	101-40-6	155.281			205	0.8501^{20}	1.4600^{20}	vs EtOH
4020	3,3-Dimethylcyclohexanol		$C_8H_{16}O$	767-12-4	128.212		11(2)	186(4)	0.9128^{14}	1.4606^{15}	
4021	2,2-Dimethylcyclohexanone		$C_8H_{14}O$	1193-47-1	126.196	liq	-20.5	172	0.9145^{20}	1.4486^{20}	
4022	2,6-Dimethylcyclohexanone		$C_8H_{14}O$	2816-57-1	126.196			175(15)	0.925^{25}	1.4460^{20}	
4023	3,3-Dimethylcyclohexanone		$C_8H_{14}O$	2979-19-3	126.196			180	0.909^{15}	1.4482^{17}	
4024	4,4-Dimethylcyclohexanone		$C_8H_{14}O$	4255-62-3	126.196		39	73^{14}	0.932^{20}	1.4537^{24}	
4025	1,2-Dimethylcyclohexene		C_8H_{14}	1674-10-8	110.197	liq	-84.1(0.2)	137(3)	0.8220^{25}	1.4620^{20}	
4026	1,3-Dimethylcyclohexene		C_8H_{14}	2808-76-6	110.197			126(4)	0.799^{25}	1.449^{20}	
4027	3,5-Dimethyl-2-cyclohexen-1-one		$C_8H_{12}O$	1123-09-7	124.180			208.5	0.9400^{20}	1.4812^{20}	s EtOH, eth
4028	1,1-Dimethylcyclopentane		C_7H_{14}	1638-26-2	98.186	liq	-69.43(0.01)	87.8(0.3)	0.7499^{25}	1.4136^{20}	
4029	*cis*-1,2-Dimethylcyclopentane		C_7H_{14}	1192-18-3	98.186	liq	-53.67(0.02)	99.5(0.3)	0.7680^{25}	1.4222^{20}	
4030	*trans*-1,2-Dimethylcyclopentane		C_7H_{14}	822-50-4	98.186	liq	-118(1)	91.9(0.4)	0.7468^{25}	1.4120^{20}	
4031	*cis*-1,3-Dimethylcyclopentane		C_7H_{14}	2532-58-3	98.186	liq	-133.67(0.03)	91.7(0.5)	0.7402^{25}	1.4089^{20}	
4032	*trans*-1,3-Dimethylcyclopentane		C_7H_{14}	1759-58-6	98.186	liq	-133.9(0.1)	90.7(0.6)	0.7443^{25}	1.4107^{20}	
4033	*N*,α-Dimethylcyclopentaneethanamine	Cyclopentamine	$C_9H_{19}N$	102-45-4	141.254			171		1.4500^{20}	
4034	1,2-Dimethylcyclopentene		C_7H_{12}	765-47-9	96.170	liq	-90.4(0.5)	105(3)	0.7928^{25}	1.4448^{20}	
4035	1,5-Dimethylcyclopentene		C_7H_{12}	16491-15-9	96.170	liq	-118.5(0.5)	99	0.780^{20}	1.4331^{20}	
4036	1,1-Dimethylcyclopropane		C_5H_{10}	1630-94-0	70.133	vol liq or gas	-109.0(0.1)	21(2)	0.6604^{20}	1.3668^{20}	i H_2O; s EtOH; vs eth, sulf
4037	*cis*-1,2-Dimethylcyclopropane		C_5H_{10}	930-18-7	70.133	liq	-140.92(0.09)	37(3)	0.6889^{25}	1.3829^{20}	i H_2O; s EtOH; vs eth; sl ctc
4038	*trans*-1,2-Dimethylcyclopropane		C_5H_{10}	2402-06-4	70.133	vol liq or gas	-149.7(0.1)	28(2)	0.6648^{25}	1.3713^{20}	vs eth, EtOH
4039	Dimethyldecylamine	*N,N*-Dimethyl-1-decanamine	$C_{12}H_{27}N$	1120-24-7	185.349			234(6)			
4040	Dimethyldiacetoxysilane	Bis(acetyloxy)dimethylsilane	$C_6H_{12}O_4Si$	2182-66-3	176.243	liq	-12.5	165	1.0540^{20}	1.4030^{20}	
4041	*trans*-Dimethyldiazene	Azomethane	$C_2H_6N_2$	4143-41-3	58.082	gas	-78	1.5	0.743^{0}	1.4199^{19}	vs ace, EtOH, eth; s ctc, hp
4042	2,2-Dimethyl-1,3-dioxane-4,6-dione	Meldrum's acid	$C_6H_8O_4$	2033-24-1	144.126		94				
4043	*cis*-3,6-Dimethyl-1,4-dioxane-2,5-dione		$C_6H_8O_4$	4511-42-6	144.126	orth (eth)	96.8	150^{25}			
4044	2,2-Dimethyl-1,3-dioxolane-4-methanol	Isopropylidene glycerol	$C_6H_{12}O_3$	100-79-8	132.157			82^{10}	1.064^{20}	1.4383^{20}	
4045	Dimethyldiphenoxysilane		$C_{14}H_{16}O_2Si$	3440-02-6	244.362		-23	131^{5}	1.0599^{25}	1.5330^{20}	
4046	2,3-Dimethyl-2,3-diphenylbutane	Dicumene	$C_{18}H_{22}$	1889-67-4	238.368	cry (MeOH)	119(1)				
4047	3,3'-Dimethyldiphenylmethane 4,4'-diisocyanate		$C_{17}H_{14}N_2O_2$	139-25-3	278.305						s chl
4048	2,9-Dimethyl-4,7-diphenyl-1,10-phenanthroline		$C_{26}H_{20}N_2$	4733-39-5	360.450		280 dec				
4049	Dimethyldiphenylsilane		$C_{14}H_{16}Si$	778-24-5	212.363			277	0.9867^{20}	1.5644^{20}	
4050	*N,N'*-Dimethyl-*N,N'*-diphenylurea		$C_{15}H_{16}N_2O$	611-92-7	240.300	pl (al)	121.6(0.7)	350			vs H_2O, EtOH, ace; sl eth, bz, CS_2
4051	Dimethyl disulfide	Methyl disulfide	$C_2H_6S_2$	624-92-0	94.199	liq	-84.67(0.1)	109.72(0.08)	1.0625^{20}	1.5289^{20}	i H_2O; msc EtOH, eth
4052	*O,O*-Dimethyl dithiophosphate	*O,O*-Dimethyl phosphorodithionate	$C_2H_7O_2PS_2$	756-80-9	158.180	liq		56^{4}	1.29^{20}		
4053	*N,N*-Dimethyldodecylamine oxide		$C_{14}H_{31}NO$	1643-20-5	229.402	hyg nd (tol)	130.5				
4054	1,2-Dimethylenecyclohexane		C_8H_{12}	2819-48-9	108.181			125(7)	0.8361^{20}	1.4718^{25}	i H_2O; s EtOH, eth, bz, chl; vs ace
4055	*N,N*-Dimethyl-1,2-ethanediamine		$C_4H_{12}N_2$	108-00-9	88.151			104	0.803^{25}	1.4260^{20}	
4056	*N,N'*-Dimethyl-1,2-ethanediamine		$C_4H_{12}N_2$	110-70-3	88.151			120	0.828^{15}		s EtOH, eth, dil HCl
4057	*N,N*-Dimethylethanolamine	Deanol	$C_4H_{11}NO$	108-01-0	89.136	liq	-65(1)	130.7(0.5)	0.8866^{20}	1.4300^{20}	msc H_2O, EtOH, eth; s chl
4058	Dimethyl ether	Methyl ether	C_2H_6O	115-10-6	46.068	col gas	-141.49(0.05)	-24.8(0.2)			s H_2O, EtOH, eth, ace, chl; sl bz
4059	(1,1-Dimethylethoxy)benzene		$C_{10}H_{14}O$	6669-13-2	150.217	liq	-18.3(0.4)	194(6)	0.9214^{20}		
4060	[(1,1-Dimethylethoxy)methyl]-oxirane		$C_7H_{14}O_2$	7665-72-7	130.185	liq	-70	152	0.898^{20}		

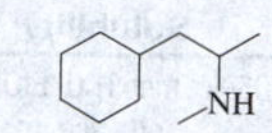
4019
N,α-Dimethylcyclohexaneethanamine

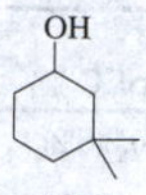
4020
3,3-Dimethylcyclohexanol

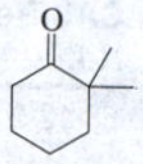
4021
2,2-Dimethylcyclohexanone

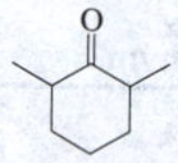
4022
2,6-Dimethylcyclohexanone

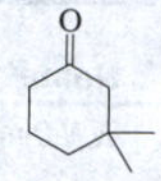
4023
3,3-Dimethylcyclohexanone

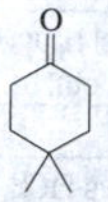
4024
4,4-Dimethylcyclohexanone

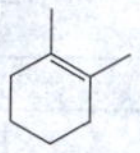
4025
1,2-Dimethylcyclohexene

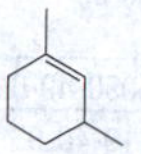
4026
1,3-Dimethylcyclohexene

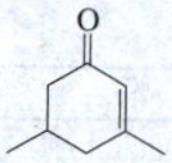
4027
3,5-Dimethyl-2-cyclohexen-1-one

4028
1,1-Dimethylcyclopentane

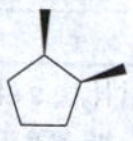
4029
cis-1,2-Dimethylcyclopentane

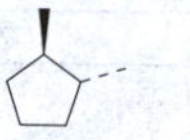
4030
trans-1,2-Dimethylcyclopentane

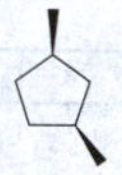
4031
cis-1,3-Dimethylcyclopentane

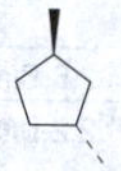
4032
trans-1,3-Dimethylcyclopentane

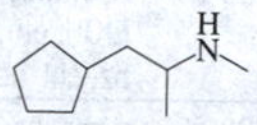
4033
N,α-Dimethylcyclopentaneethanamine

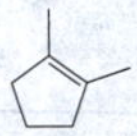
4034
1,2-Dimethylcyclopentene

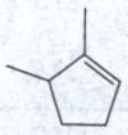
4035
1,5-Dimethylcyclopentene

4036
1,1-Dimethylcyclopropane

4037
cis-1,2-Dimethylcyclopropane

4038
trans-1,2-Dimethylcyclopropane

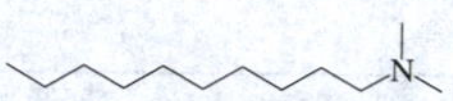
4039
Dimethyldecylamine

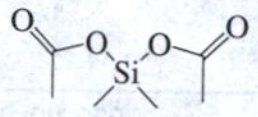
4040
Dimethyldiacetoxysilane

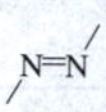
4041
trans-Dimethyldiazene

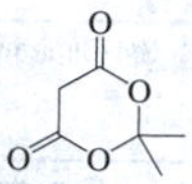
4042
2,2-Dimethyl-1,3-dioxane-4,6-dione

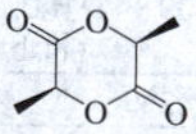
4043
cis-3,6-Dimethyl-1,4-dioxane-2,5-dione

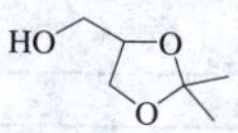
4044
2,2-Dimethyl-1,3-dioxolane-4-methanol

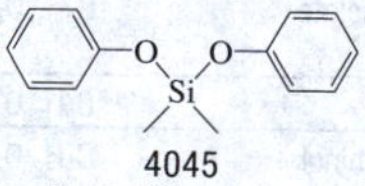
4045
Dimethyldiphenoxysilane

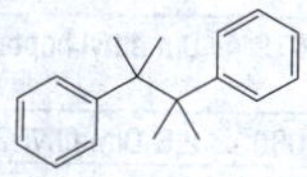
4046
2,3-Dimethyl-2,3-diphenylbutane

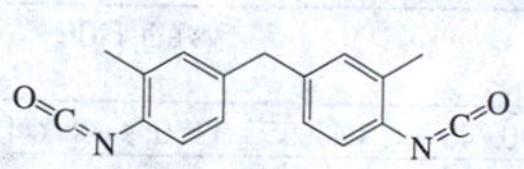
4047
3,3′-Dimethyldiphenylmethane 4,4′-diisocyanate

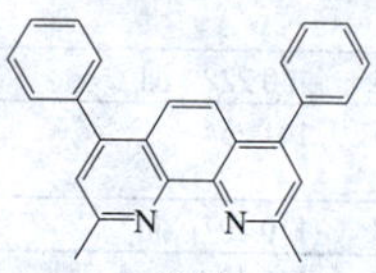
4048
2,9-Dimethyl-4,7-diphenyl-1,10-phenanthroline

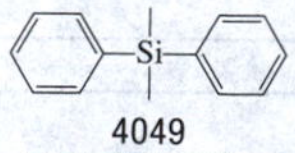
4049
Dimethyldiphenylsilane

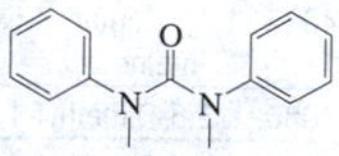
4050
N,*N*′-Dimethyl-*N*,*N*′-diphenylurea

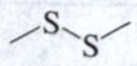
4051
Dimethyl disulfide

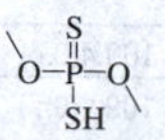
4052
O,*O*-Dimethyl dithiophosphate

4053
N,*N*-Dimethyldodecylamine oxide

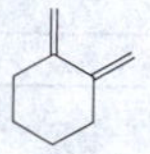
4054
1,2-Dimethylenecyclohexane

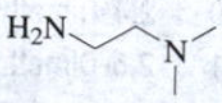
4055
N,*N*-Dimethyl-1,2-ethanediamine

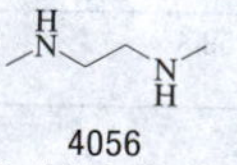
4056
N,*N*′-Dimethyl-1,2-ethanediamine

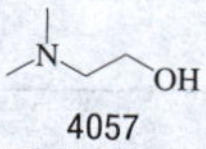
4057
N,*N*-Dimethylethanolamine

4058
Dimethyl ether

4059
(1,1-Dimethylethoxy)benzene

4060
[(1,1-Dimethylethoxy)methyl]oxirane

Organic

No.	Name	Synonym	Mol. form.	CAS Reg. No.	Mol. wt.	Physical form	mp/°C	bp/°C	den/ g cm^{-3}	n_D	Solubility
4061	*N,N*-Dimethylformamide	DMF	C_3H_7NO	68-12-2	73.094	liq	-60.3(0.2)	152.8(0.5)	0.9445^{25}	1.4305^{20}	msc H_2O, EtOH, eth, ace, bz; sl lig
4062	Dimethyl fumarate		$C_6H_8O_4$	624-49-7	144.126		101.7(0.7)	193	1.37^{20}	1.4062^{111}	i H_2O; s ace, chl
4063	2,5-Dimethylfuran		C_6H_8O	625-86-5	96.127	liq	-62.8	96(3)	0.8883^{20}	1.4363^{20}	i H_2O; s EtOH, eth, ace, bz, HOAc, chl
4064	3,4-Dimethyl-2,5-furandione		$C_6H_6O_3$	766-39-2	126.110	pl or lf (dil al)	96	223	1.107^{100}		sl H_2O; vs EtOH, eth, bz, chl
4065	Dimethyl germanium sulfide		C_2H_6GeS	16090-49-6	134.77	col cry	54.5	302			
4066	Dimethyl glutarate	Methyl glutarate	$C_7H_{12}O_4$	1119-40-0	160.168	liq	-42.5	216(4)	1.0876^{20}	1.4242^{20}	vs EtOH, eth; s chl
4067	*N,N*-Dimethylglycine		$C_4H_9NO_2$	1118-68-9	103.120	hyg nd (PrOH)	185.5				vs H_2O, MeOH; s EtOH, eth, ace
4068	Dimethylglyoxime		$C_4H_8N_2O_2$	95-45-4	116.119	nd (to or dil al)	245.5	234 sub			i H_2O; vs EtOH, eth; sl bz, tol
4069	2,6-Dimethyl-1,5-heptadiene		C_9H_{16}	6709-39-3	124.223	liq	-70	143(3)	0.7648^{25}		
4070	2,2-Dimethylheptane		C_9H_{20}	1071-26-7	128.255	liq	-113.05(0.1)	133(1)	0.7105^{20}	1.4016^{20}	i H_2O; s eth, ctc; vs ace, chl; msc bz
4071	2,3-Dimethylheptane		C_9H_{20}	3074-71-3	128.255	liq	-116	140.6(0.9)	0.7260^{20}	1.4088^{20}	i H_2O; msc EtOH, eth, ace, bz, peth, chl
4072	2,4-Dimethylheptane		C_9H_{20}	2213-23-2	128.255			132(2)	0.7115^{25}	1.4034^{20}	i H_2O; msc EtOH, eth, ace, bz, chl, peth
4073	2,5-Dimethylheptane		C_9H_{20}	2216-30-0	128.255			135(2)	0.7198^{20}	1.4033^{20}	vs ace, bz, eth, EtOH
4074	2,6-Dimethylheptane		C_9H_{20}	1072-05-5	128.255	liq	-103.1(0.5)	135(2)	0.7089^{20}	1.4011^{20}	sl chl
4075	3,3-Dimethylheptane		C_9H_{20}	4032-86-4	128.255			137(3)	0.7254^{20}	1.4087^{20}	i H_2O; msc EtOH; s eth; vs ace, bz
4076	3,4-Dimethylheptane		C_9H_{20}	922-28-1	128.255			140(3)	0.7314^{20}	1.4108^{20}	i H_2O; s eth, ctc; vs ace, chl; msc bz
4077	3,5-Dimethylheptane		C_9H_{20}	926-82-9	128.255			135(3)	0.7225^{20}	1.4083^{20}	i H_2O; s eth, ctc; vs ace, chl; msc bz
4078	4,4-Dimethylheptane		C_9H_{20}	1068-19-5	128.255			134(3)	0.7221^{20}	1.4076^{20}	i H_2O; s eth, ctc; vs ace, chl; msc bz
4079	Dimethyl heptanedioate	Dimethyl pimelate	$C_9H_{16}O_4$	1732-08-7	188.221		-21	120^{10}	1.0625^{20}	1.4309^{20}	sl H_2O; s EtOH, eth, bz
4080	2,6-Dimethyl-2-heptanol		$C_9H_{20}O$	13254-34-7	144.254			170(11)	0.8186^{20}	1.4242^{20}	
4081	2,6-Dimethyl-4-heptanol	Diisobutylcarbinol	$C_9H_{20}O$	108-82-7	144.254			193(6)	0.8114^{20}	1.4242^{20}	i H_2O; s EtOH, eth; sl ctc
4082	3,5-Dimethyl-4-heptanol		$C_9H_{20}O$	19549-79-2	144.254			179(5)	0.836^{18}	1.4283^{20}	sl H_2O
4083	2,6-Dimethyl-4-heptanone	Diisobutyl ketone	$C_9H_{18}O$	108-83-8	142.238	liq	-46.0(0.2)	157(3)	0.8062^{20}	1.412^{21}	i H_2O; msc EtOH, eth; s ctc
4084	2,6-Dimethyl-5-heptenal		$C_9H_{16}O$	106-72-9	140.222	oil		120^{100}			
4085	*N*,6-Dimethyl-5-hepten-2-amine	Isometheptene	$C_9H_{19}N$	503-01-5	141.254			177			vs eth, EtOH
4086	2,5-Dimethyl-1,5-hexadiene		C_8H_{14}	627-58-7	110.197	liq	-75.6(0.2)	114(2)	0.743^{20}	1.43995^{21}	i H_2O; s ace, chl
4087	2,5-Dimethyl-2,4-hexadiene		C_8H_{14}	764-13-6	110.197		13.8(0.3)	135.2(0.2)	0.7577^{25}	1.4785^{20}	i H_2O; s EtOH, eth, bz, chl
4088	2,2-Dimethylhexane		C_8H_{18}	590-73-8	114.229	liq	-121.19(0.07)	106.8(0.4)	0.6953^{20}	1.3935^{20}	vs ace, bz, eth, EtOH
4089	2,3-Dimethylhexane		C_8H_{18}	584-94-1	114.229			115.6(0.5)	0.6912^{25}	1.4011^{20}	vs ace, bz, EtOH, lig
4090	2,4-Dimethylhexane		C_8H_{18}	589-43-5	114.229			109.4(0.4)	0.6962^{25}	1.3929^{25}	
4091	2,5-Dimethylhexane	Biisobutyl	C_8H_{18}	592-13-2	114.229	liq	-91.14(0.02)	109.1(0.7)	0.6901^{25}	1.3925^{20}	i H_2O; msc EtOH, ace, bz; s eth
4092	3,3-Dimethylhexane		C_8H_{18}	563-16-6	114.229	liq	-126.2(0.1)	111.9(0.6)	0.7100^{20}	1.4001^{20}	i H_2O; msc EtOH; vs eth, ace, bz
4093	3,4-Dimethylhexane		C_8H_{18}	583-48-2	114.229			117.7(0.4)	0.7151^{25}	1.4041^{20}	i H_2O; s eth; msc EtOH, ace, bz
4094	2,5-Dimethyl-2,5-hexanediamine		$C_8H_{20}N_2$	23578-35-0	144.258			184	0.8485^{15}	1.4459^{20}	

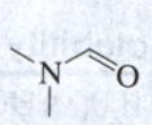
4061
N,N-Dimethylformamide

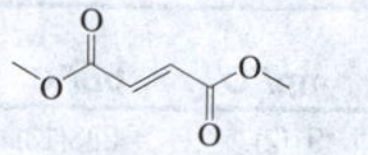
4062
Dimethyl fumarate

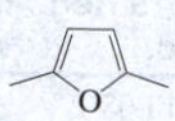
4063
2,5-Dimethylfuran

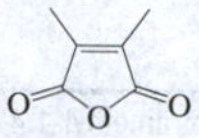
4064
3,4-Dimethyl-2,5-furandione

4065
Dimethyl germanium sulfide

4066
Dimethyl glutarate

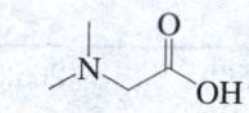
4067
N,N-Dimethylglycine

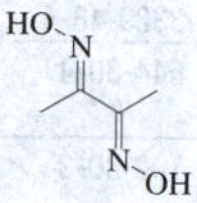
4068
Dimethylglyoxime

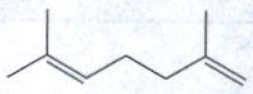
4069
2,6-Dimethyl-1,5-heptadiene

4070
2,2-Dimethylheptane

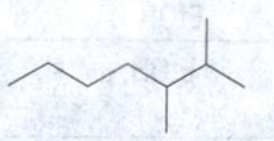
4071
2,3-Dimethylheptane

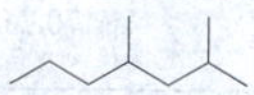
4072
2,4-Dimethylheptane

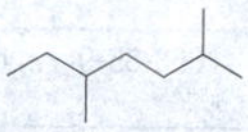
4073
2,5-Dimethylheptane

4074
2,6-Dimethylheptane

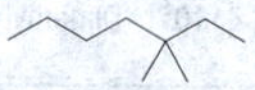
4075
3,3-Dimethylheptane

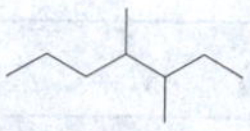
4076
3,4-Dimethylheptane

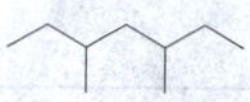
4077
3,5-Dimethylheptane

4078
4,4-Dimethylheptane

4079
Dimethyl heptanedioate

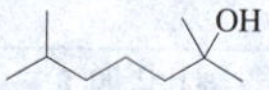
4080
2,6-Dimethyl-2-heptanol

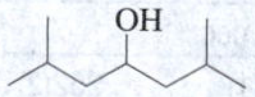
4081
2,6-Dimethyl-4-heptanol

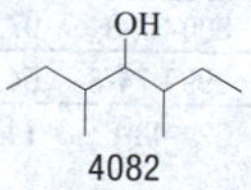
4082
3,5-Dimethyl-4-heptanol

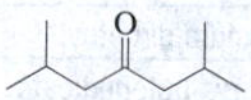
4083
2,6-Dimethyl-4-heptanone

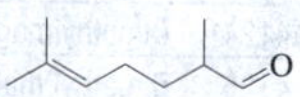
4084
2,6-Dimethyl-5-heptenal

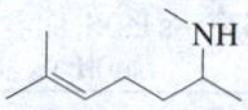
4085
N,6-Dimethyl-5-hepten-2-amine

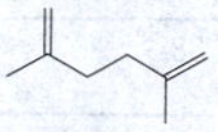
4086
2,5-Dimethyl-1,5-hexadiene

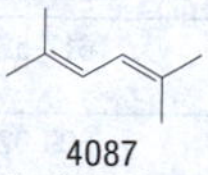
4087
2,5-Dimethyl-2,4-hexadiene

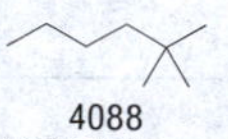
4088
2,2-Dimethylhexane

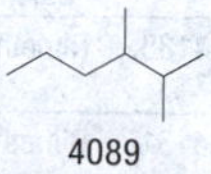
4089
2,3-Dimethylhexane

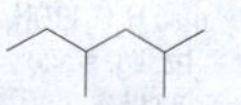
4090
2,4-Dimethylhexane

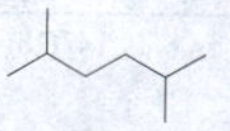
4091
2,5-Dimethylhexane

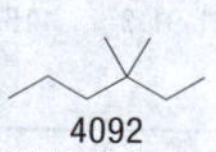
4092
3,3-Dimethylhexane

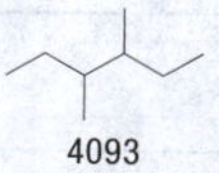
4093
3,4-Dimethylhexane

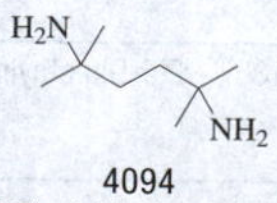
4094
2,5-Dimethyl-2,5-hexanediamine

No.	Name	Synonym	Mol. form.	CAS Reg. No.	Mol. wt.	Physical form	mp/°C	bp/°C	den/ g cm^{-3}	n_D	Solubility
4095	2,5-Dimethyl-2,5-hexanediol	1,1,4,4-Tetramethyl-1,4-butanediol	$C_8H_{18}O_2$	110-03-2	146.228	pr (AcOEt) fl (peth)	91(2)	233(13)	0.898^{20}		s H_2O; vs EtOH, bz, chl
4096	2,2-Dimethyl-1-hexanol		$C_8H_{18}O$	2370-13-0	130.228			95^{29}			
4097	2,3-Dimethyl-1-hexene		C_8H_{16}	16746-86-4	112.213			111(2)	0.7172^{25}	1.4113^{20}	
4098	5,5-Dimethyl-1-hexene		C_8H_{16}	7116-86-1	112.213			102(3)	0.705^{25}	1.4049^{20}	
4099	2,3-Dimethyl-2-hexene		C_8H_{16}	7145-20-2	112.213	liq	-115.0(0.1)	122(2)	0.7366^{25}	1.4268^{20}	
4100	2,5-Dimethyl-2-hexene		C_8H_{16}	3404-78-2	112.213			113(2)	0.7182^{20}	1.4140^{20}	
4101	*cis*-2,2-Dimethyl-3-hexene		C_8H_{16}	690-92-6	112.213	liq	-137.36(0.09)	105.4(0.8)	0.7086^{25}	1.4099^{20}	
4102	*trans*-2,2-Dimethyl-3-hexene		C_8H_{16}	690-93-7	112.213			100.9(0.2)	0.6995^{25}	1.4063^{20}	
4103	3,5-Dimethyl-1-hexen-3-ol		$C_8H_{16}O$	3329-48-4	128.212			146.5	0.8382^{20}	1.4342^{20}	
4104	1-(1,5-Dimethyl-4-hexenyl)-4-methylbenzene	α-Curcumene	$C_{15}H_{22}$	644-30-4	202.336			140^{19}	0.8805^{20}	1.4989^{20}	i H_2O; s bz
4105	2,5-Dimethyl-3-hexyne-2,5-diol		$C_8H_{14}O_2$	142-30-3	142.196		95(2)	206(9)	0.947^{20}		s H_2O, chl; vs EtOH, eth, ace, bz
4106	1,1-Dimethylhydrazine	UDMH	$C_2H_8N_2$	57-14-7	60.098	liq, fumes in air	-57.15(0.08)	62.4(0.8)	0.791^{22}	1.4075^{22}	vs H_2O, EtOH, eth, MeOH
4107	1,2-Dimethylhydrazine		$C_2H_8N_2$	540-73-8	60.098	fumes (air)	-8.86(0.09)	82(3)	0.8274^{20}	1.4209^{20}	msc H_2O, EtOH, eth
4108	1,2-Dimethylhydrazine dihydrochloride		$C_2H_{10}Cl_2N_2$	306-37-6	133.019	pr (w)	170 dec				vs H_2O, EtOH
4109	Dimethyl hydrogen phosphate	Dimethyl phosphate	$C_2H_7O_4P$	813-78-5	126.048			174 dec	1.3225^{20}	1.408^{25}	vs H_2O, ace, EtOH
4110	Dimethyl hydrogen phosphite		$C_2H_7O_3P$	868-85-9	110.049			170.2(0.5)	1.2002^{20}	1.4036^{20}	s EtOH, py; sl ctc
4111	1,2-Dimethyl-1*H*-imidazole		$C_5H_8N_2$	1739-84-0	96.131			206	1.0051^{11}		vs H_2O, eth, EtOH
4112	2,4-Dimethyl-1*H*-imidazole		$C_5H_8N_2$	930-62-1	96.131		92	267			
4113	5,5-Dimethyl-2,4-imidazolidinedione		$C_5H_8N_2O_2$	77-71-4	128.130	pr (dil al)	176.1(0.5)	sub			vs H_2O, EtOH, eth, ace, bz, chl; s DMSO
4114	1,1-Dimethylindan		$C_{11}H_{14}$	4912-92-9	146.229			194.0(0.8)	0.919^{20}	1.5135^{25}	
4115	1,3-Dimethyl-1*H*-indole		$C_{10}H_{11}N$	875-30-9	145.201	nd	142	258.5			s eth
4116	2,3-Dimethyl-1*H*-indole		$C_{10}H_{11}N$	91-55-4	145.201		107.5	284(20)			
4117	*N,N*-Dimethyl-1*H*-indole-3-ethanamine	*N,N*-Dimethyltryptamine	$C_{12}H_{16}N_2$	61-50-7	188.268		46				
4118	*N,N*-Dimethyl-1*H*-indole-3-methanamine	Gramine	$C_{11}H_{14}N_2$	87-52-5	174.242	nd or pl (ace)	138.5				i H_2O; s EtOH, eth, chl; i peth
4119	Dimethyl isophthalate		$C_{10}H_{10}O_4$	1459-93-4	194.184	nd(dil al)	68(1)	285.1(0.4)	1.194^{20}	1.5168^{20}	sl H_2O
4120	1,4-Dimethyl-7-isopropylazulene	Guaiazulene	$C_{15}H_{18}$	489-84-9	198.304	bl-viol pl (al)	31.5	167^{12}	0.973^{20}		s EtOH, eth, AcOEt
4121	1,6-Dimethyl-4-isopropylnaphthalene	Cadalene	$C_{15}H_{18}$	483-78-3	198.304			294	0.9667^{25}	1.5785^{25}	vs oils
4122	2,4-Dimethyl-3-isopropylpentane		$C_{10}H_{22}$	13475-79-1	142.282	liq	-81.7(0.2)	157(4)	0.7545^{25}	1.4246^{20}	
4123	3,5-Dimethylisoxazole		C_5H_7NO	300-87-8	97.116			142(7)	0.99^{25}	1.4421^{20}	
4124	Dimethylmagnesium	Magnesium dimethyl	C_2H_6Mg	2999-74-8	54.374	solid	220 dec				
4125	Dimethyl maleate	Methyl *cis*-butenedioate	$C_6H_8O_4$	624-48-6	144.126	liq	-19(1)	202(2)	1.1606^{20}	1.4416^{20}	sl H_2O, lig; s eth, ctc
4126	Dimethyl malonate	Methyl malonate	$C_5H_8O_4$	108-59-8	132.116	liq	-62(1)	181.1(0.6)	1.1595^{20}	1.4135^{20}	sl H_2O; msc EtOH; vs ace, bz; s chl
4127	Dimethylmalonic acid	Dimethylpropanedioc acid	$C_5H_8O_4$	595-46-0	132.116	pr (bz/peth)	191(2)				s hot H_2O
4128	Dimethyl mercury	Mercury dimethyl	C_2H_6Hg	593-74-8	230.66	liq		93	3.17^{25}	1.5452^{20}	i H_2O; vs EtOH, eth
4129	Dimethyl *cis*-2-methyl-2-butenedioate	Dimethyl citraconate	$C_7H_{10}O_4$	617-54-9	158.152			210.5	1.1153^{20}	1.4473^{20}	vs ace, eth, EtOH
4130	Dimethyl methylenesuccinate		$C_7H_{10}O_4$	617-52-7	158.152	hyg mcl (MeOH)	38	208	1.1241^{18}	1.4457^{20}	s EtOH, eth, MeOH; vs ace
4131	Dimethyl methylmalonate		$C_6H_{10}O_4$	609-02-9	146.141			177(15)	1.0977^{20}	1.4128^{20}	vs ace, eth, EtOH, chl
4132	Dimethyl methylphosphonate		$C_3H_9O_3P$	756-79-6	124.075			168.7(0.7)	1.1684^{20}	1.4099^{30}	s H_2O, EtOH, eth
4133	*trans*-2,2-Dimethyl-3-(2-methyl-1-propenyl) cyclopropanecarboxylic acid		$C_{10}H_{16}O_2$	4638-92-0	168.233	pr	20.0	245			vs eth, EtOH, chl
4134	Dimethyl 2-methylsuccinate		$C_7H_{12}O_4$	1604-11-1	160.168			196	1.076^{25}	1.4200^{20}	
4135	Dimethyl *p*-(methylthio)phenyl phosphate		$C_9H_{13}O_4PS$	3254-63-5	248.235	liq			1.273^{21}		sl H_2O; s ace, EtOH, diox, ctc, xyl
4136	2,6-Dimethylmorpholine		$C_6H_{13}NO$	141-91-3	115.173	liq	-88	147(18)	0.9329^{20}	1.4460^{20}	msc H_2O, EtOH, bz, lig; s ace; sl chl

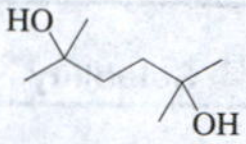

4095
2,5-Dimethyl-2,5-hexanediol

4096
2,2-Dimethyl-1-hexanol

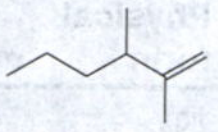

4097
2,3-Dimethyl-1-hexene

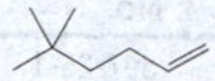

4098
5,5-Dimethyl-1-hexene

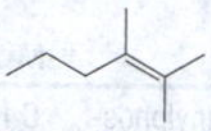

4099
2,3-Dimethyl-2-hexene

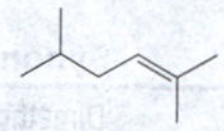

4100
2,5-Dimethyl-2-hexene

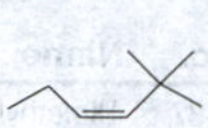

4101
cis-2,2-Dimethyl-3-hexene

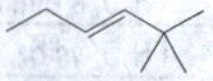

4102
trans-2,2-Dimethyl-3-hexene

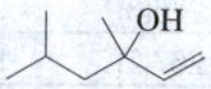

4103
3,5-Dimethyl-1-hexen-3-ol

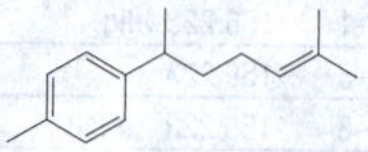

4104
1-(1,5-Dimethyl-4-hexenyl)-4-methylbenzene

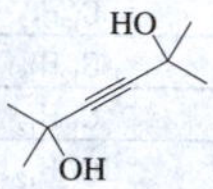

4105
2,5-Dimethyl-3-hexyne-2,5-diol

4106
1,1-Dimethylhydrazine

4107
1,2-Dimethylhydrazine

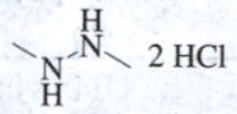

4108
1,2-Dimethylhydrazine dihydrochloride

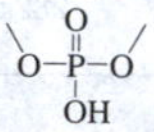

4109
Dimethyl hydrogen phosphate

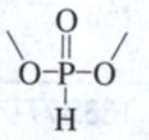

4110
Dimethyl hydrogen phosphite

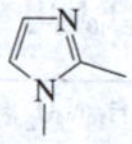

4111
1,2-Dimethyl-1*H*-imidazole

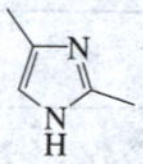

4112
2,4-Dimethyl-1*H*-imidazole

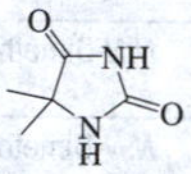

4113
5,5-Dimethyl-2,4-imidazolidinedione

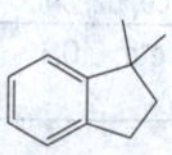

4114
1,1-Dimethylindan

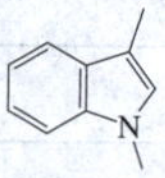

4115
1,3-Dimethyl-1*H*-indole

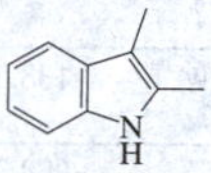

4116
2,3-Dimethyl-1*H*-indole

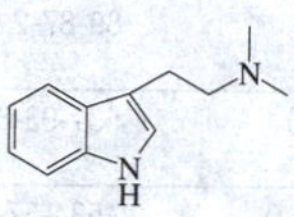

4117
N,N-Dimethyl-1*H*-indole-3-ethanamine

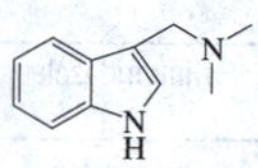

4118
N,N-Dimethyl-1*H*-indole-3-methanamine

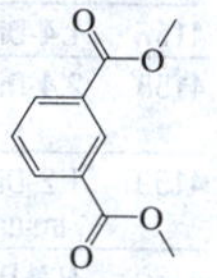

4119
Dimethyl isophthalate

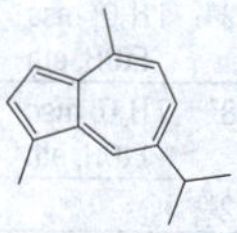

4120
1,4-Dimethyl-7-isopropylazulene

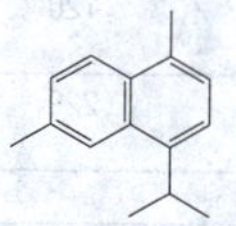

4121
1,6-Dimethyl-4-isopropylnaphthalene

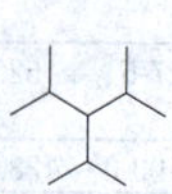

4122
2,4-Dimethyl-3-isopropylpentane

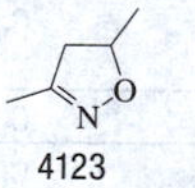

4123
3,5-Dimethylisoxazole

4124
Dimethylmagnesium

4125
Dimethyl maleate

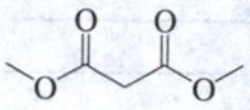

4126
Dimethyl malonate

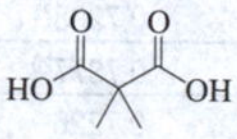

4127
Dimethylmalonic acid

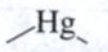

4128
Dimethyl mercury

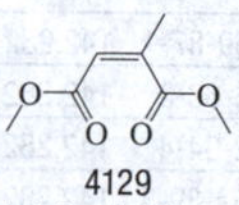

4129
Dimethyl *cis*-2-methyl-2-butenedioate

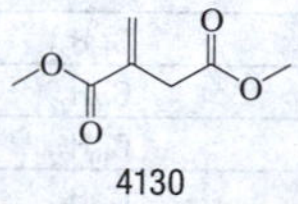

4130
Dimethyl methylenesuccinate

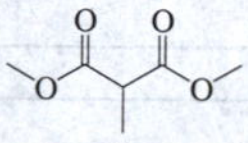

4131
Dimethyl methylmalonate

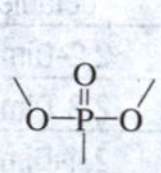

4132
Dimethyl methylphosphonate

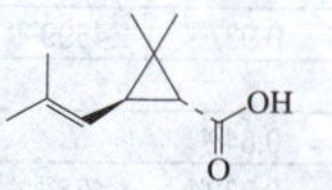

4133
trans-2,2-Dimethyl-3-(2-methyl-1-propenyl)cyclopropanecarboxylic acid

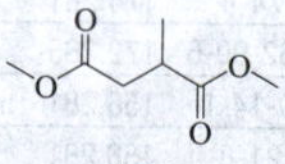

4134
Dimethyl 2-methylsuccinate

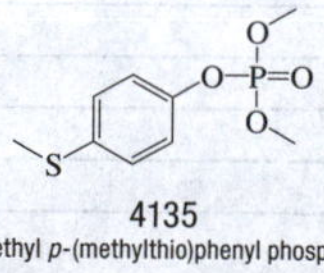

4135
Dimethyl *p*-(methylthio)phenyl phosphate

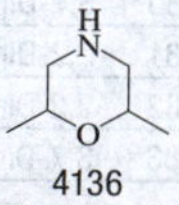

4136
2,6-Dimethylmorpholine

No.	Name	Synonym	Mol. form.	CAS Reg. No.	Mol. wt.	Physical form	mp/°C	bp/°C	den/ g cm^{-3}	n_D	Solubility
4137	Dimethyl morpholinophosphoramidate	Dimethyl 4-morpholinylphosphonate	$C_6H_{14}NO_4P$	597-25-1	195.153	liq		96^{1}			
4138	1,2-Dimethylnaphthalene		$C_{12}H_{12}$	573-98-8	156.223		-3.0(0.7)	267(5)	1.0179^{20}	1.6166^{20}	i H_2O; s eth, bz
4139	1,3-Dimethylnaphthalene		$C_{12}H_{12}$	575-41-7	156.223	liq	-6	265(5)	1.0144^{20}	1.6140^{20}	i H_2O; s eth, bz
4140	1,4-Dimethylnaphthalene		$C_{12}H_{12}$	571-58-4	156.223		7.6(0.4)	264(5)	1.0166^{20}	1.6127^{20}	i H_2O; vs EtOH; msc eth, ace, bz, ctc
4141	1,5-Dimethylnaphthalene		$C_{12}H_{12}$	571-61-9	156.223		81.5(0.9)	267(4)			i H_2O; vs bz, eth
4142	1,6-Dimethylnaphthalene		$C_{12}H_{12}$	575-43-9	156.223	liq	-16.2(0.5)	263(4)	1.0021^{20}	1.6166^{20}	i H_2O; s eth, bz
4143	1,7-Dimethylnaphthalene		$C_{12}H_{12}$	575-37-1	156.223	liq	-13.9	263(6)	1.0115^{20}	1.6083^{20}	i H_2O; s eth, bz
4144	1,8-Dimethylnaphthalene		$C_{12}H_{12}$	569-41-5	156.223		63.16(0.05)	276(3)	1.003^{20}		i H_2O; s eth, bz
4145	2,3-Dimethylnaphthalene	Guajen	$C_{12}H_{12}$	581-40-8	156.223	lf (al)	104.3(0.3)	267(2)	1.003^{20}	1.5060^{20}	i H_2O; vs bz, eth
4146	2,6-Dimethylnaphthalene		$C_{12}H_{12}$	581-42-0	156.223		110.1(0.2)	253(3)	1.003^{20}		i H_2O
4147	2,7-Dimethylnaphthalene		$C_{12}H_{12}$	582-16-1	156.223		96(1)	262.4(0.3)	1.003^{20}		
4148	*N,N*-Dimethyl-1-naphthylamine		$C_{12}H_{13}N$	86-56-6	171.238	viol flr cry		250	1.0423^{20}	1.624^{15}	i H_2O; s EtOH, eth, ctc
4149	*N,N*-Dimethyl-2-naphthylamine		$C_{12}H_{13}N$	2436-85-3	171.238	dk red nd	41.5(0.5)	304(15)	1.0279^{60}	1.6443^{53}	i H_2O; s EtOH, eth
4150	*N,N*-Dimethyl-2-nitroaniline		$C_8H_{10}N_2O_2$	610-17-3	166.177	ye-oran	-20	146^{20}	1.1794^{20}	1.6102^{20}	s H_2O, eth; vs EtOH, chl
4151	*N,N*-Dimethyl-3-nitroaniline		$C_8H_{10}N_2O_2$	619-31-8	166.177	red mcl pr (eth)	60(2)	282.5	1.313^{17}		i H_2O; s EtOH, eth
4152	*N,N*-Dimethyl-4-nitroaniline		$C_8H_{10}N_2O_2$	100-23-2	166.177	ye nd (al)	164(1)				i H_2O; s EtOH, eth, HOAc
4153	1,2-Dimethyl-3-nitrobenzene		$C_8H_9NO_2$	83-41-0	151.163	nd (al)	14(1)	240	1.1402^{20}	1.5441^{20}	i H_2O; s EtOH, ctc
4154	1,2-Dimethyl-4-nitrobenzene	4-Nitro-*o*-xylene	$C_8H_9NO_2$	99-51-4	151.163	ye pr (al)	28(1)	251	1.112^{15}	1.5202^{20}	i H_2O; msc EtOH
4155	1,3-Dimethyl-2-nitrobenzene		$C_8H_9NO_2$	81-20-9	151.163		15.2(0.7)	226	1.112^{15}	1.5202^{20}	i H_2O; vs EtOH; s ctc
4156	1,3-Dimethyl-5-nitrobenzene		$C_8H_9NO_2$	99-12-7	151.163	nd (al)	75	274			i H_2O; vs EtOH, eth
4157	1,4-Dimethyl-2-nitrobenzene		$C_8H_9NO_2$	89-58-7	151.163	pa ye liq	-25	240.5	1.132^{15}	1.5413^{20}	i H_2O; s EtOH
4158	2,4-Dimethyl-1-nitrobenzene		$C_8H_9NO_2$	89-87-2	151.163		9.6(0.2)	243(2)	1.135^{15}	1.5473^{25}	i H_2O; s eth, ace, bz, chl
4159	1,2-Dimethyl-5-nitro-1*H*-imidazole	Dimetridazole	$C_5H_7N_3O_2$	551-92-8	141.129	nd (w)	138.5				vs eth, EtOH
4160	*N,N*-Dimethyl-4-[2-(4-nitrophenyl)ethenyl]aniline		$C_{16}H_{16}N_2O_2$	4584-57-0	268.310		258.3				
4161	*N*,4-Dimethyl-*N*-nitrosobenzenesulfonamide	*p*-Tolylsulfonylmethylnitrosamide	$C_8H_{10}N_2O_3S$	80-11-5	214.241	cry	60				i H_2O; vs EtOH, eth
4162	Dimethyl nonanedioate	Methyl azelate	$C_{11}H_{20}O_4$	1732-10-1	216.275		-0.8	156^{20}	1.0082^{20}	1.4367^{20}	i H_2O; s EtOH, ace, bz, ctc
4163	6,6-Dimethyl-2-norpinene-2-carboxaldehyde	Myrtenal	$C_{10}H_{14}O$	564-94-3	150.217	unstab oil		99^{15}			
4164	*cis*-3,7-Dimethyl-2,6-octadienal		$C_{10}H_{16}O$	106-26-3	152.233			120^{20}	0.8869^{20}	1.4869^{20}	i H_2O; msc EtOH, eth
4165	*trans*-3,7-Dimethyl-2,6-octadienal	Citral	$C_{10}H_{16}O$	141-27-5	152.233			229	0.8888^{20}	1.4898^{20}	i H_2O; msc EtOH, eth
4166	3,7-Dimethyl-1,6-octadiene	Citronellene	$C_{10}H_{18}$	2436-90-0	138.250				0.7601^{20}	1.4362^{20}	
4167	3,7-Dimethyl-2,6-octadienoic acid	Geranic acid	$C_{10}H_{16}O_2$	459-80-3	168.233	oil					
4168	*cis*-3,7-Dimethyl-2,6-octadien-1-ol	Nerol	$C_{10}H_{18}O$	106-25-2	154.249		<-15	225	0.8756^{20}	1.4746^{20}	vs EtOH
4169	*cis*-3,7-Dimethyl-2,6-octadien-1-ol acetate		$C_{12}H_{20}O_2$	141-12-8	196.286			134^{25}	0.905^{15}	1.452^{20}	
4170	*trans*-3,7-Dimethyl-2,6-octadien-1-ol formate		$C_{11}H_{18}O_2$	105-86-2	182.260			229 dec	0.9086^{25}	1.4659^{20}	i H_2O; vs EtOH; s eth, ace
4171	2,2-Dimethyloctane		$C_{10}H_{22}$	15869-87-1	142.282			154(5)	0.7208^{25}	1.4082^{20}	
4172	2,3-Dimethyloctane		$C_{10}H_{22}$	7146-60-3	142.282			164(3)	0.7377^{20}	1.4146^{20}	
4173	2,4-Dimethyloctane		$C_{10}H_{22}$	4032-94-4	142.282			154(3)	0.7226^{25}	1.4091^{20}	
4174	2,5-Dimethyloctane		$C_{10}H_{22}$	15869-89-3	142.282			157(4)	0.7264^{25}	1.4112^{20}	
4175	2,6-Dimethyloctane		$C_{10}H_{22}$	2051-30-1	142.282			158(2)	0.7313^{20}	1.4097^{20}	
4176	2,7-Dimethyloctane		$C_{10}H_{22}$	1072-16-8	142.282	liq	-54(1)	160(1)	0.7202^{25}	1.4086^{20}	s eth, HOAc
4177	3,4-Dimethyloctane		$C_{10}H_{22}$	15869-92-8	142.282			162(7)	0.7410^{25}	1.4182^{20}	
4178	3,6-Dimethyloctane		$C_{10}H_{22}$	15869-94-0	142.282			160.8	0.7324^{25}	1.4139^{20}	
4179	Dimethyl octanedioate	Dimethyl suberate	$C_{10}H_{18}O_4$	1732-09-8	202.248	liq	6(1)	259(8)	1.0217^{20}	1.4341^{20}	i H_2O; s EtOH, eth, ace; sl ctc
4180	3,7-Dimethyl-1,7-octanediol		$C_{10}H_{22}O_2$	107-74-4	174.281			265	0.937^{20}	1.4599^{20}	sl bz, tol
4181	2,2-Dimethyloctanoic acid		$C_{10}H_{20}O_2$	29662-90-6	172.265			140^{13}			
4182	2,2-Dimethyl-1-octanol		$C_{10}H_{22}O$	2370-14-1	158.281	liq		211(6)	0.84^{20}		
4183	3,7-Dimethyl-1-octanol		$C_{10}H_{22}O$	106-21-8	158.281			224(11)	0.832^{25}	1.438^{25}	s eth

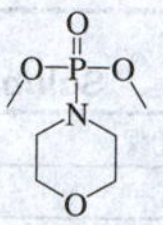
4137
Dimethyl morpholinophosphoramidate

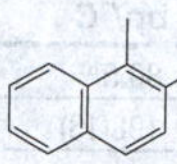
4138
1,2-Dimethylnaphthalene

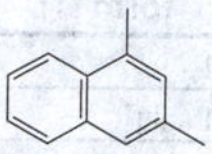
4139
1,3-Dimethylnaphthalene

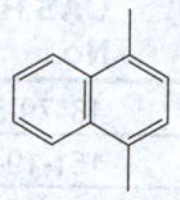
4140
1,4-Dimethylnaphthalene

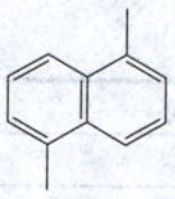
4141
1,5-Dimethylnaphthalene

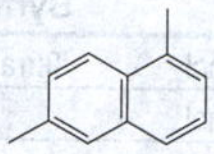
4142
1,6-Dimethylnaphthalene

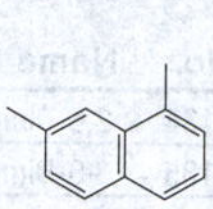
4143
1,7-Dimethylnaphthalene

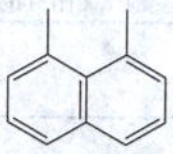
4144
1,8-Dimethylnaphthalene

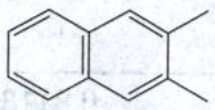
4145
2,3-Dimethylnaphthalene

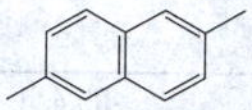
4146
2,6-Dimethylnaphthalene

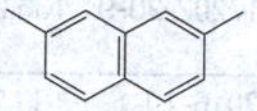
4147
2,7-Dimethylnaphthalene

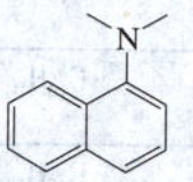
4148
N,*N*-Dimethyl-1-naphthylamine

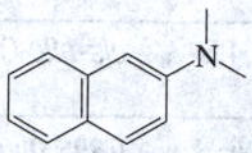
4149
N,*N*-Dimethyl-2-naphthylamine

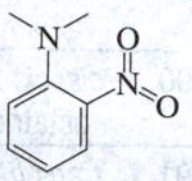
4150
N,*N*-Dimethyl-2-nitroaniline

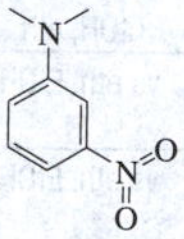
4151
N,*N*-Dimethyl-3-nitroaniline

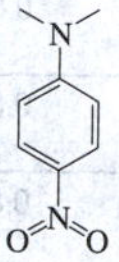
4152
N,*N*-Dimethyl-4-nitroaniline

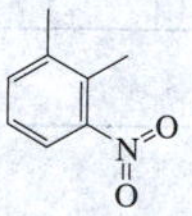
4153
1,2-Dimethyl-3-nitrobenzene

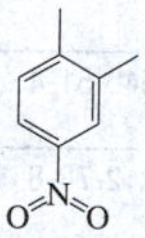
4154
1,2-Dimethyl-4-nitrobenzene

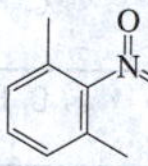
4155
1,3-Dimethyl-2-nitrobenzene

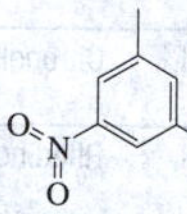
4156
1,3-Dimethyl-5-nitrobenzene

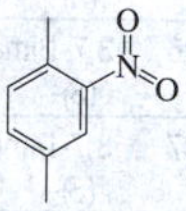
4157
1,4-Dimethyl-2-nitrobenzene

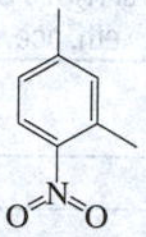
4158
2,4-Dimethyl-1-nitrobenzene

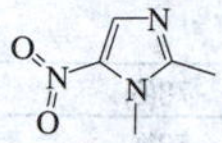
4159
1,2-Dimethyl-5-nitro-1*H*-imidazole

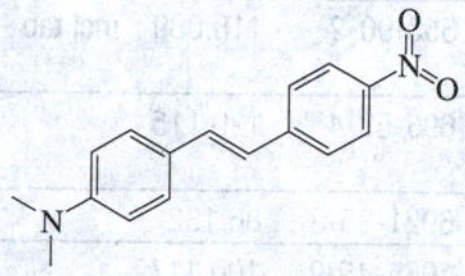
4160
N,*N*-Dimethyl-4-[2-(4-nitrophenyl)ethenyl]aniline

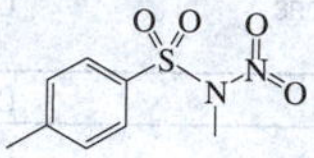
4161
N,4-Dimethyl-*N*-nitrosobenzenesulfonamide

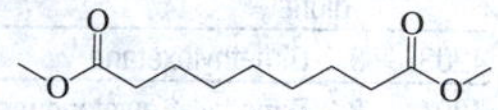
4162
Dimethyl nonanedioate

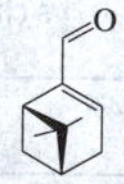
4163
6,6-Dimethyl-2-norpinene-2-carboxaldehyde

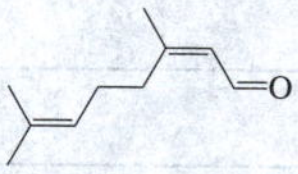
4164
cis-3,7-Dimethyl-2,6-octadienal

4165
trans-3,7-Dimethyl-2,6-octadienal

4166
3,7-Dimethyl-1,6-octadiene

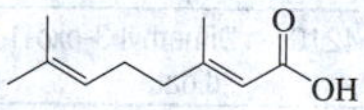
4167
3,7-Dimethyl-2,6-octadienoic acid

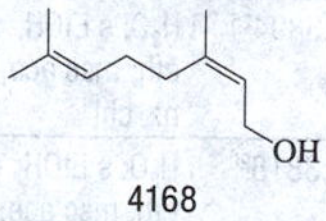
4168
cis-3,7-Dimethyl-2,6-octadien-1-ol

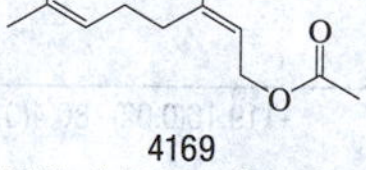
4169
cis-3,7-Dimethyl-2,6-octadien-1-ol acetate

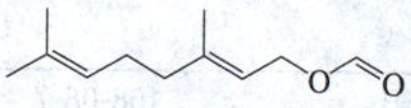
4170
trans-3,7-Dimethyl-2,6-octadien-1-ol formate

4171
2,2-Dimethyloctane

4172
2,3-Dimethyloctane

4173
2,4-Dimethyloctane

4174
2,5-Dimethyloctane

4175
2,6-Dimethyloctane

4176
2,7-Dimethyloctane

4177
3,4-Dimethyloctane

4178
3,6-Dimethyloctane

4179
Dimethyl octanedioate

4180
3,7-Dimethyl-1,7-octanediol

4181
2,2-Dimethyloctanoic acid

4182
2,2-Dimethyl-1-octanol

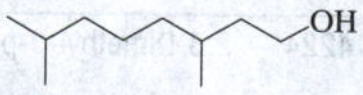
4183
3,7-Dimethyl-1-octanol

Organic

No.	Name	Synonym	Mol. form.	CAS Reg. No.	Mol. wt.	Physical form	mp/°C	bp/°C	den/ g cm⁻³	n_D	Solubility
4184	2,6-Dimethyl-2-octanol	Tetrahydromyrcenol	$C_{10}H_{22}O$	18479-57-7	158.281			80.5^{10}	0.8023^{25}	1.4220^{25}	
4185	3,6-Dimethyl-3-octanol		$C_{10}H_{22}O$	151-19-9	158.281	liq	-67.5	198(13)	0.8347^{22}	1.4370^{20}	
4186	3,7-Dimethyl-3-octanol		$C_{10}H_{22}O$	78-69-3	158.281			205.1	0.826^{25}	1.433^{25}	
4187	*cis*-3,7-Dimethyl-1,3,6-octatriene	*cis*-β-Ocimene	$C_{10}H_{16}$	3338-55-4	136.234				0.799^{20}		
4188	*trans*-3,7-Dimethyl-1,3,6-octatriene	*trans*-β-Ocimene	$C_{10}H_{16}$	3779-61-1	136.234				0.799^{20}		
4189	3,7-Dimethyl-1,3,7-octatriene	α-Ocimene	$C_{10}H_{16}$	502-99-8	136.234			177 dec	0.8000^{20}	1.4862^{20}	i H_2O; s EtOH, eth, chl, HOAc
4190	*cis, cis*-2,6-Dimethyl-2,4,6-octatriene	*cis-allo*-Ocimene	$C_{10}H_{16}$	17202-20-9	136.234	liq					
4191	*trans,trans*-2,6-Dimethyl-2,4,6-octatriene	*trans-allo*-Ocimene	$C_{10}H_{16}$	3016-19-1	136.234	liq	-35.4	188	0.8118^{20}	1.5446^{20}	
4192	3,7-Dimethyl-6-octenal	Citronellal	$C_{10}H_{18}O$	106-23-0	154.249	nd or orth cry		205(2)	0.853^{20}	1.4473^{20}	sl H_2O; s EtOH
4193	3,7-Dimethyl-1-octene		$C_{10}H_{20}$	4984-01-4	140.266	col liq		155(7)	0.7396^{20}	1.4212^{20}	
4194	3,7-Dimethyl-6-octenoic acid	Citronellic acid	$C_{10}H_{18}O_2$	502-47-6	170.249			257	0.9234^{21}		
4195	3,7-Dimethyl-6-octen-1-ol, (*R*)-	Citronellol, (+)	$C_{10}H_{20}O$	1117-61-9	156.265	oil		224	0.8550^{20}	1.4565^{20}	sl H_2O; msc EtOH, eth
4196	3,7-Dimethyl-6-octen-1-ol, (*S*)-	Citronellol, (-)	$C_{10}H_{20}O$	7540-51-4	156.265	oil		224	0.859^{18}	1.4576^{18}	vs eth, EtOH
4197	3,7-Dimethyl-7-octen-1-ol, (*S*)-	Rhodinol	$C_{10}H_{20}O$	6812-78-8	156.265			114^{12}	0.8549^{20}	1.4556^{20}	vs eth, EtOH
4198	3,7-Dimethyl-6-octen-3-ol		$C_{10}H_{20}O$	18479-51-1	156.265			94^{14}	0.8695^{15}	1.4569^{15}	
4199	3,7-Dimethyl-6-octen-1-ol, acetate	Citronellol acetate	$C_{12}H_{22}O_2$	150-84-5	198.302			115^{10}			
4200	Dimethyloldihydroxyethyleneurea	4,5-Dihydroxy-1,3-bis(hydroxymethyl)-2-imidazolidinone	$C_5H_{10}N_2O_5$	1854-26-8	178.143	hyg cry					
4201	Dimethyl oxalate		$C_4H_6O_4$	553-90-2	118.089	mcl tab	51(2)	163.4(0.5)	1.1716^{60}	1.379^{82}	sl H_2O; s EtOH, eth, ace, chl
4202	5,5-Dimethyl-2,4-oxazolidinedione	Dimethadione	$C_5H_7NO_3$	695-53-4	129.115		76.5				
4203	3,3-Dimethyloxetane		$C_5H_{10}O$	6921-35-3	86.132			78(6)	0.834^{25}	1.3965^{20}	
4204	3,3-Dimethyl-2-oxetanone		$C_5H_8O_2$	1955-45-9	100.117			58^{15}			
4205	2,2-Dimethyloxirane	2-Methyl-1,2-epoxypropane	C_4H_8O	558-30-5	72.106			51(2)	0.8112^{20}	1.3712^{22}	s EtOH, eth
4206	*cis*-2,3-Dimethyloxirane		C_4H_8O	1758-33-4	72.106	liq	-83.4(0.5)	60(4)	0.8226^{25}	1.3802^{20}	vs eth, ace, bz
4207	*trans*-2,3-Dimethyloxirane		C_4H_8O	6189-41-9	72.106	liq	-85	56(10)	0.8010^{25}	1.3736^{20}	vs eth, ace, bz
4208	3,3-Dimethyl-2-oxobutanoic acid		$C_6H_{10}O_3$	815-17-8	130.141		90.5	189			sl H_2O; s eth, bz, chl, CS_2
4209	*N*-(1,1-Dimethyl-3-oxobutyl)-2-propenamide	Diacetone acrylamide	$C_9H_{15}NO_2$	2873-97-4	169.221						s chl
4210	Dimethyl 3-oxo-1,5-pentanedioate	Dimethyl 1,3-acetonedicarboxylate	$C_7H_{10}O_5$	1830-54-2	174.151			150^{25}	1.185^{25}	1.4434^{20}	
4211	2,4-Dimethyl-1,3-pentadiene		C_7H_{12}	1000-86-8	96.170	liq	-116.0(0.3)	94(2)	0.7343^{23}	1.4390^{23}	
4212	*N,N*-Dimethylpentanamide		$C_7H_{15}NO$	6225-06-5	129.200		-51	141^{100}	0.8962^{25}	1.4419^{25}	vs H_2O, eth, EtOH
4213	2,2-Dimethylpentane		C_7H_{16}	590-35-2	100.202	liq	-123.71(0.04)	79.2(0.3)	0.6739^{20}	1.3822^{20}	i H_2O; s EtOH, eth; msc ace, bz, hp, chl
4214	2,3-Dimethylpentane		C_7H_{16}	565-59-3	100.202			89.8(0.6)	0.6908^{25}	1.3894^{25}	i H_2O; s EtOH, eth; msc ace, bz, chl
4215	2,4-Dimethylpentane		C_7H_{16}	108-08-7	100.202	liq	-119.16(0.02)	80.4(0.5)	0.6727^{20}	1.3815^{20}	i H_2O; s EtOH, eth; msc ace, bz, chl, hp
4216	3,3-Dimethylpentane		C_7H_{16}	562-49-2	100.202	liq	-134.4(0.4)	86.0(0.6)	0.6936^{20}	1.3909^{20}	i H_2O; s EtOH, eth; msc ace, bz, hp, chl
4217	3,3-Dimethylpentanedioic acid anhydride	Dihydro-4,4-dimethyl-2*H*-pyran-2,6(3*H*)-dione	$C_7H_{10}O_3$	4160-82-1	142.152		125.8	181^{25}			
4218	3,3-Dimethylpentanedioic acid		$C_7H_{12}O_4$	4839-46-7	160.168	mcl pl, nd (bz)	103.5	126^{415}	1.4278^{20}		vs H_2O, EtOH, eth; sl bz; i lig
4219	2,2-Dimethylpentanoic acid		$C_7H_{14}O_2$	1185-39-3	130.185	liq		98^{9}	0.9189^{20}		
4220	2,2-Dimethyl-1-pentanol		$C_7H_{16}O$	2370-12-9	116.201						s chl
4221	2,3-Dimethyl-2-pentanol		$C_7H_{16}O$	4911-70-0	116.201				0.804^{20}		sl H_2O
4222	2,4-Dimethyl-2-pentanol		$C_7H_{16}O$	625-06-9	116.201		<-20	133(3)	0.8103^{20}	1.4172^{20}	sl H_2O; s EtOH, eth, ctc
4223	2,2-Dimethyl-3-pentanol		$C_7H_{16}O$	3970-62-5	116.201	liq	9.3(0.5)	136(2)	0.8253^{20}	1.4223^{20}	i H_2O; s EtOH, eth
4224	2,3-Dimethyl-3-pentanol		$C_7H_{16}O$	595-41-5	116.201		<-30	140(4)	0.833^{20}	1.4287^{20}	sl H_2O, bz; s EtOH, eth

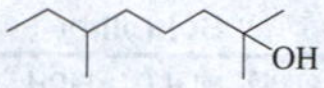
4184
2,6-Dimethyl-2-octanol

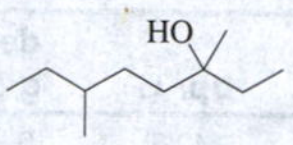
4185
3,6-Dimethyl-3-octanol

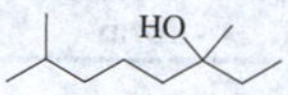
4186
3,7-Dimethyl-3-octanol

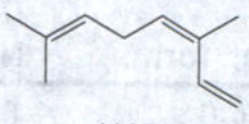
4187
cis-3,7-Dimethyl-1,3,6-octatriene

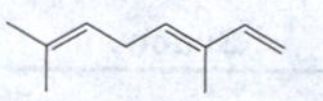
4188
trans-3,7-Dimethyl-1,3,6-octatriene

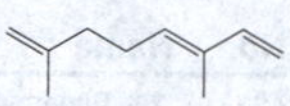
4189
3,7-Dimethyl-1,3,7-octatriene

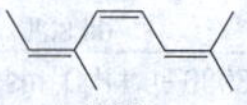
4190
cis, cis-2,6-Dimethyl-2,4,6-octatriene

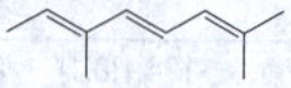
4191
trans,trans-2,6-Dimethyl-2,4,6-octatriene

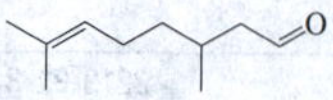
4192
3,7-Dimethyl-6-octenal

4193
3,7-Dimethyl-1-octene

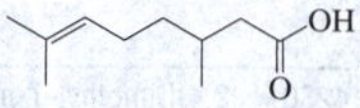
4194
3,7-Dimethyl-6-octenoic acid

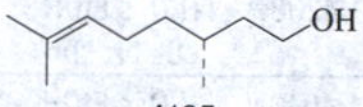
4195
3,7-Dimethyl-6-octen-1-ol, (*R*)-

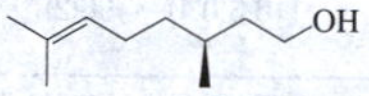
4196
3,7-Dimethyl-6-octen-1-ol, (*S*)-

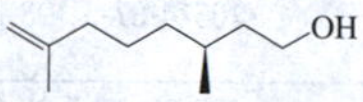
4197
3,7-Dimethyl-7-octen-1-ol, (*S*)-

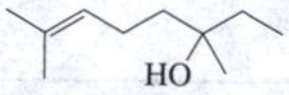
4198
3,7-Dimethyl-6-octen-3-ol

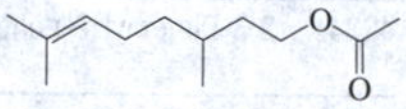
4199
3,7-Dimethyl-6-octen-1-ol, acetate

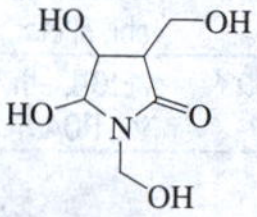
4200
Dimethyloldihydroxyethyleneurea

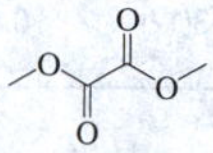
4201
Dimethyl oxalate

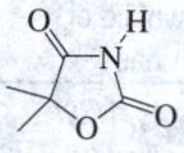
4202
5,5-Dimethyl-2,4-oxazolidinedione

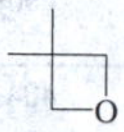
4203
3,3-Dimethyloxetane

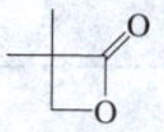
4204
3,3-Dimethyl-2-oxetanone

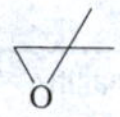
4205
2,2-Dimethyloxirane

4206
cis-2,3-Dimethyloxirane

4207
trans-2,3-Dimethyloxirane

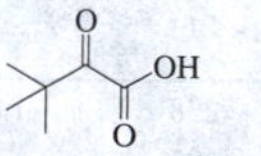
4208
3,3-Dimethyl-2-oxobutanoic acid

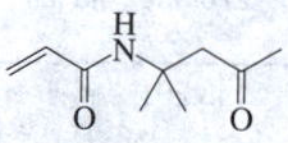
4209
N-(1,1-Dimethyl-3-oxobutyl)-2-propenamide

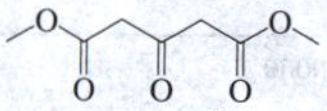
4210
Dimethyl 3-oxo-1,5-pentanedioate

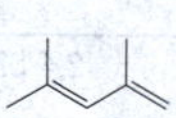
4211
2,4-Dimethyl-1,3-pentadiene

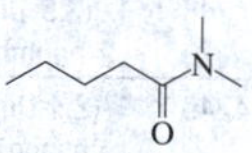
4212
N,N-Dimethylpentanamide

4213
2,2-Dimethylpentane

4214
2,3-Dimethylpentane

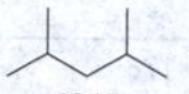
4215
2,4-Dimethylpentane

4216
3,3-Dimethylpentane

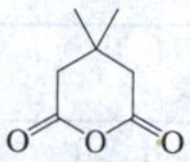
4217
3,3-Dimethylpentanedioic acid anhydride

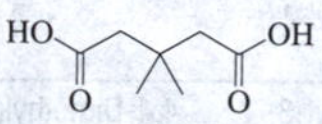
4218
3,3-Dimethylpentanedioic acid

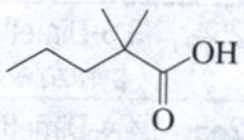
4219
2,2-Dimethylpentanoic acid

4220
2,2-Dimethyl-1-pentanol

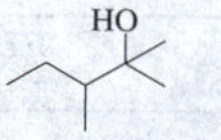
4221
2,3-Dimethyl-2-pentanol

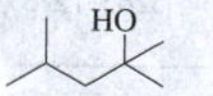
4222
2,4-Dimethyl-2-pentanol

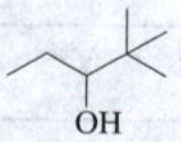
4223
2,2-Dimethyl-3-pentanol

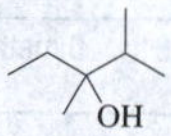
4224
2,3-Dimethyl-3-pentanol

No.	Name	Synonym	Mol. form.	CAS Reg. No.	Mol. wt.	Physical form	mp/°C	bp/°C	den/g cm⁻³	n_D	Solubility
4225	2,4-Dimethyl-3-pentanol		$C_7H_{16}O$	600-36-2	116.201		<-70	142(2)	0.8288^{20}	1.4250^{20}	sl H_2O; s EtOH, eth
4226	4,4-Dimethyl-2-pentanone		$C_7H_{14}O$	590-50-1	114.185	liq	-64	124(2)	0.809^{25}	1.4036^{20}	
4227	2,2-Dimethyl-3-pentanone		$C_7H_{14}O$	564-04-5	114.185	liq	-45	125(3)	0.8125^{20}	1.4065^{20}	sl H_2O; s EtOH, eth, ace, chl
4228	2,4-Dimethyl-3-pentanone	Diisopropyl ketone	$C_7H_{14}O$	565-80-0	114.185	liq	-68.4(0.6)	125.2(0.3)	0.8108^{20}	1.3999^{20}	sl H_2O; msc EtOH, eth; s bz; sl ctc
4229	2,3-Dimethyl-1-pentene		C_7H_{14}	3404-72-6	98.186	liq	-137(5)	84(1)	0.7051^{20}	1.4033^{20}	i H_2O; msc EtOH, eth; vs dil sulf
4230	2,4-Dimethyl-1-pentene		C_7H_{14}	2213-32-3	98.186	liq	-124.1(0.1)	81.6(0.9)	0.6943^{20}	1.3986^{20}	i H_2O; msc EtOH, eth; s bz, ctc, chl
4231	3,3-Dimethyl-1-pentene		C_7H_{14}	3404-73-7	98.186	liq	-134.4(0.1)	77(2)	0.6974^{20}	1.3984^{20}	i H_2O; msc EtOH, eth; s bz, chl
4232	3,4-Dimethyl-1-pentene		C_7H_{14}	7385-78-6	98.186			82(4)	0.6934^{25}	1.3992^{20}	
4233	4,4-Dimethyl-1-pentene		C_7H_{14}	762-62-9	98.186	liq	-136.6(0.1)	72.5(0.2)	0.6827^{20}	1.3818^{20}	i H_2O; msc EtOH, eth; s bz, ctc, chl
4234	2,3-Dimethyl-2-pentene		C_7H_{14}	10574-37-5	98.186	liq	-118.3(0.1)	96(2)	0.7277^{20}	1.4208^{20}	i H_2O; s EtOH, eth, bz, chl
4235	2,4-Dimethyl-2-pentene		C_7H_{14}	625-65-0	98.186	liq	-127.6(0.2)	83.3(0.7)	0.6954^{20}	1.4040^{20}	i H_2O; s EtOH, eth, bz, ctc
4236	*cis*-3,4-Dimethyl-2-pentene		C_7H_{14}	4914-91-4	98.186	liq	-124.2(0.1)	92(4)	0.7092^{25}	1.4104^{20}	
4237	*trans*-3,4-Dimethyl-2-pentene		C_7H_{14}	4914-92-5	98.186	liq	-113.4(0.1)	91(3)	0.7124^{25}	1.4128^{20}	
4238	*cis*-4,4-Dimethyl-2-pentene		C_7H_{14}	762-63-0	98.186	liq	-135.5(0.1)	80.4(0.9)	0.6951^{25}	1.4026^{20}	
4239	*trans*-4,4-Dimethyl-2-pentene		C_7H_{14}	690-08-4	98.186	liq	-115.2(0.1)	76.7(0.8)	0.6889^{20}	1.3982^{20}	i H_2O; s EtOH, eth, bz, chl
4240	4,4-Dimethyl-1-pentyne		C_7H_{12}	13361-63-2	96.170	liq	-75.7	75(3)	0.7142^{20}	1.3983^{20}	vs bz, eth, chl
4241	4,4-Dimethyl-2-pentyne		C_7H_{12}	999-78-0	96.170	liq	-82(2)	83(4)	0.7176^{20}	1.4071^{20}	i H_2O; s eth, bz, chl; sl ctc
4242	Dimethylperoxide		$C_2H_6O_2$	690-02-8	62.068	vol liq or gas	-100	-3(12)	0.8677^{0}	1.3503^{0}	sl EtOH, eth; s tol, HOAc
4243	2,9-Dimethyl-1,10-phenanthroline	Neocuproine	$C_{14}H_{12}N_2$	484-11-7	208.258	cry, 1/2w (w, lig)	163(1)				
4244	3,4-Dimethylphenol phosphate (3:1)		$C_{24}H_{27}O_4P$	3862-11-1	410.442		72	261^{7}			i H_2O; sl EtOH, chl, hx; s bz
4245	5-(2,5-Dimethylphenoxy)-2,2-dimethylpentanoic acid	Gemfibrozil	$C_{15}H_{22}O_3$	25812-30-0	250.334	cry	62.0(0.5)	$159^{0.02}$			
4246	*N*-(2,4-Dimethylphenyl)-acetamide		$C_{10}H_{13}NO$	2050-43-3	163.216	nd (al)	129.3	170^{10}			vs EtOH, chl
4247	1-[(2,4-Dimethylphenyl)azo]-2-naphthol	1-(2,4-Xylylazo)-2-naphthol	$C_{18}H_{16}N_2O$	3118-97-6	276.332	red nd (al)	166				vs eth, EtOH
4248	1-[(2,5-Dimethylphenyl)azo]-2-naphthol	1-(2,5-Xylylazo)-2-naphthol	$C_{18}H_{16}N_2O$	85-82-5	276.332	nd (al)	153				
4249	1-(2,4-Dimethylphenyl)-ethanone	2,4-Dimethylacetophenone	$C_{10}H_{12}O$	89-74-7	148.201			228	1.0121^{15}	1.5340^{20}	vs eth, EtOH
4250	1-(2,5-Dimethylphenyl)-ethanone	2,5-Dimethylacetophenone	$C_{10}H_{12}O$	2142-73-6	148.201	liq	-18.1(0.7)	232.5	0.9963^{19}	1.5291^{20}	i H_2O; vs EtOH, eth, bz, CS_2
4251	1-(3,4-Dimethylphenyl)-ethanone	3,4-Dimethylacetophenone	$C_{10}H_{12}O$	3637-01-2	148.201	liq	-5(1)	245(11)	1.0090^{14}	1.5413^{15}	i H_2O; vs EtOH, eth, bz; s ctc, HOAc
4252	4,4-Dimethyl-1-phenyl-1-penten-3-one		$C_{13}H_{16}O$	538-44-3	188.265		43	154^{25}	0.9508^{46}	1.5523^{25}	
4253	2,2-Dimethyl-1-phenyl-1-propanone		$C_{11}H_{14}O$	938-16-9	162.228			220	0.963^{26}	1.5086^{19}	s ace
4254	3,5-Dimethyl-1-phenyl-1*H*-pyrazole		$C_{11}H_{12}N_2$	1131-16-4	172.226			272	1.0566^{20}	1.5738^{19}	vs eth, EtOH, chl
4255	4,4-Dimethyl-1-phenyl-3-pyrazolidinone	4,4-Dimethylphenidone	$C_{11}H_{14}N_2O$	2654-58-2	190.241		176				
4256	*N*,*N*-Dimethyl-γ-phenyl-2-pyridinepropanamine	Pheniramine	$C_{16}H_{20}N_2$	86-21-5	240.343			181^{13}	1.0081^{25}	1.5519^{25}	vs bz, eth, EtOH, chl
4257	1,3-Dimethyl-3-phenyl-2,5-pyrrolidinedione	Methsuximide	$C_{12}H_{13}NO_2$	77-41-8	203.237		52.5	$121^{0.1}$			
4258	Dimethylphenylsilane		$C_8H_{12}Si$	766-77-8	136.267			159(3)	0.8891^{20}	1.4995^{20}	i H_2O
4259	*N*,*N*-Dimethyl-*N'*-phenylurea	Fenuron	$C_9H_{12}N_2O$	101-42-8	164.203	cry (hx)	132.9(0.5)				
4260	Dimethylphosphine		C_2H_7P	676-59-5	62.051	vol liq or gas		25			i H_2O; s EtOH, eth
4261	Dimethylphosphinic acid		$C_2H_7O_2P$	3283-12-3	94.050	cry (bz)	92	377			vs H_2O, EtOH, eth; s bz

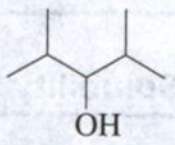
4225
2,4-Dimethyl-3-pentanol

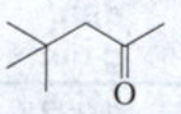
4226
4,4-Dimethyl-2-pentanone

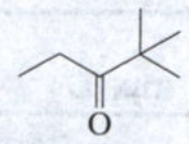
4227
2,2-Dimethyl-3-pentanone

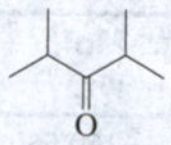
4228
2,4-Dimethyl-3-pentanone

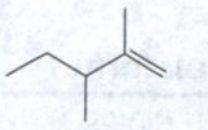
4229
2,3-Dimethyl-1-pentene

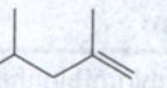
4230
2,4-Dimethyl-1-pentene

4231
3,3-Dimethyl-1-pentene

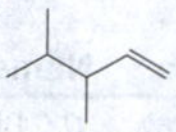
4232
3,4-Dimethyl-1-pentene

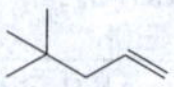
4233
4,4-Dimethyl-1-pentene

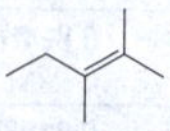
4234
2,3-Dimethyl-2-pentene

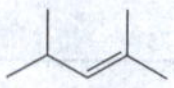
4235
2,4-Dimethyl-2-pentene

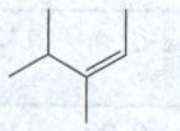
4236
cis-3,4-Dimethyl-2-pentene

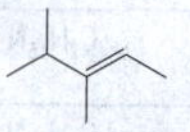
4237
trans-3,4-Dimethyl-2-pentene

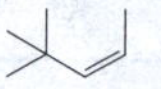
4238
cis-4,4-Dimethyl-2-pentene

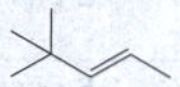
4239
trans-4,4-Dimethyl-2-pentene

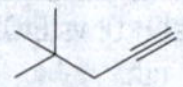
4240
4,4-Dimethyl-1-pentyne

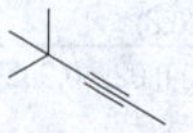
4241
4,4-Dimethyl-2-pentyne

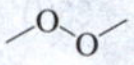
4242
Dimethylperoxide

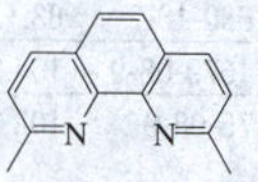
4243
2,9-Dimethyl-1,10-phenanthroline

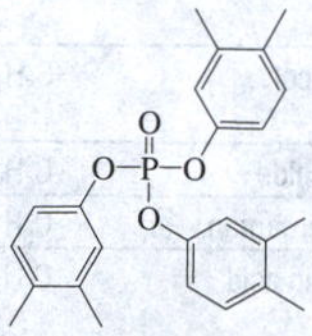
4244
3,4-Dimethylphenol phosphate (3:1)

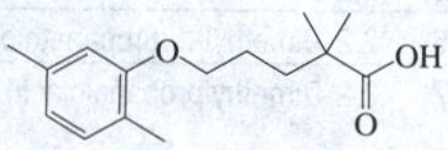
4245
5-(2,5-Dimethylphenoxy)-2,2-dimethylpentanoic acid

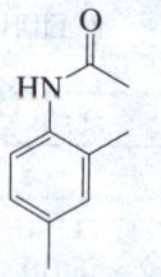
4246
N-(2,4-Dimethylphenyl)acetamide

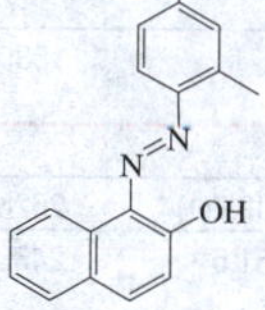
4247
1-[(2,4-Dimethylphenyl)azo]-2-naphthol

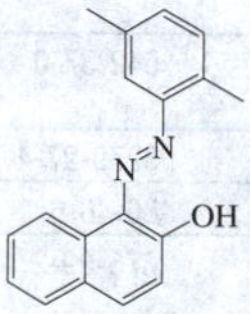
4248
1-[(2,5-Dimethylphenyl)azo]-2-naphthol

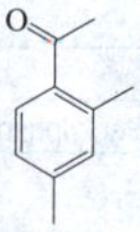
4249
1-(2,4-Dimethylphenyl)ethanone

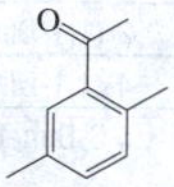
4250
1-(2,5-Dimethylphenyl)ethanone

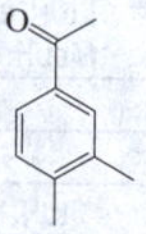
4251
1-(3,4-Dimethylphenyl)ethanone

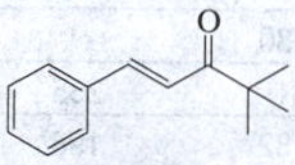
4252
4,4-Dimethyl-1-phenyl-1-penten-3-one

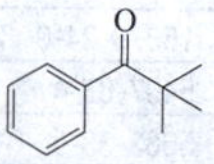
4253
2,2-Dimethyl-1-phenyl-1-propanone

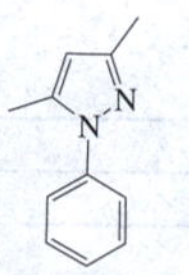
4254
3,5-Dimethyl-1-phenyl-1*H*-pyrazole

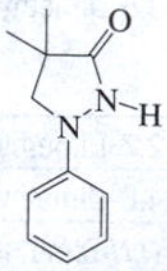
4255
4,4-Dimethyl-1-phenyl-3-pyrazolidinone

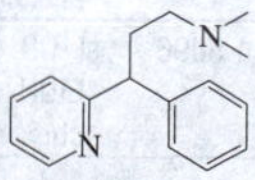
4256
N,*N*-Dimethyl-γ-phenyl-2-pyridinepropanamine

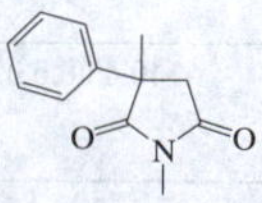
4257
1,3-Dimethyl-3-phenyl-2,5-pyrrolidinedione

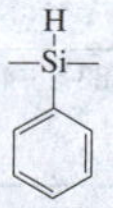
4258
Dimethylphenylsilane

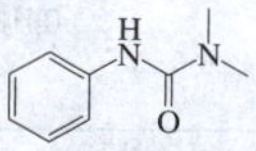
4259
N,*N*-Dimethyl-*N'*-phenylurea

4260
Dimethylphosphine

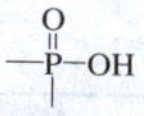
4261
Dimethylphosphinic acid

No.	Name	Synonym	Mol. form.	CAS Reg. No.	Mol. wt.	Physical form	mp/°C	bp/°C	den/ g cm^{-3}	n_D	Solubility
4262	*O,O*-Dimethyl phosphorochloridothioate	Dimethyl chlorothiophosphate	$C_2H_6ClO_2PS$	2524-03-0	160.560	hyg liq		68^{12}	1.322	1.4820^{20}	
4263	Dimethyl phthalate	Methyl phthalate	$C_{10}H_{10}O_4$	131-11-3	194.184	pa ye	1.03(0.02)	282.7(0.2)	1.1905^{20}	1.5138^{20}	i H_2O; msc EtOH, eth; s bz; sl ctc
4264	1,4-Dimethylpiperazine		$C_6H_{14}N_2$	106-58-1	114.188	liq	-1.0(0.3)	129.8(0.8)	0.8600^{20}	1.4474^{20}	vs H_2O, EtOH, eth
4265	*cis*-2,5-Dimethylpiperazine		$C_6H_{14}N_2$	6284-84-0	114.188	orth bipym nd or pr (chl)	114	164(19)		1.4720^{20}	vs H_2O, EtOH, chl; sl eth, bz
4266	1,2-Dimethylpiperidine, (±)-		$C_7H_{15}N$	2512-81-4	113.201			127.5	0.824^{15}	1.4395^{20}	vs H_2O, eth, EtOH
4267	2,6-Dimethylpiperidine		$C_7H_{15}N$	504-03-0	113.201			128(5)	0.8158^{25}	1.4377^{20}	msc H_2O, EtOH, eth; sl ctc; s acid
4268	3,5-Dimethylpiperidine	3,5-Lupetidine	$C_7H_{15}N$	35794-11-7	113.201			144(5)	0.853^{25}	1.4454^{20}	
4269	2,2-Dimethylpropanal	Pivaldehyde	$C_5H_{10}O$	630-19-3	86.132		1(2)	74(2)	0.7923^{17}	1.3791^{20}	s EtOH, eth
4270	2,2-Dimethylpropanamide		$C_5H_{11}NO$	754-10-9	101.147						s tfa
4271	*N,N*-Dimethylpropanamide		$C_5H_{11}NO$	758-96-3	101.147	liq	-45	176(7)	0.9269^{20}		
4272	*N,N*-Dimethyl-1-propanamine	Dimethylpropylamine	$C_5H_{13}N$	926-63-6	87.164			65(3)	0.7152^{20}	1.3860^{20}	vs bz, eth, EtOH
4273	*N,N*-Dimethyl-1,3-propanediamine		$C_5H_{14}N_2$	109-55-7	102.178			129(13)	0.8272^{20}		
4274	2,2-Dimethyl-1,3-propanediol	Neopentyl glycol	$C_5H_{12}O_2$	126-30-7	104.148	nd (bz)	129.3(0.9)	207(14)			s H_2O, bz, chl; vs EtOH, eth
4275	2,2-Dimethylpropanenitrile	*tert*-Butyl cyanide	C_5H_9N	630-18-2	83.132		18.97(0.05)	105.2(0.2)	0.7586^{25}	1.3774^{20}	
4276	2,2-Dimethyl-1-propanethiol	Neopentyl mercaptan	$C_5H_{12}S$	1679-08-9	104.214	liq		103.6(0.8)			
4277	2,2-Dimethylpropanoic acid	Trimethylacetic acid	$C_5H_{10}O_2$	75-98-9	102.132	nd	36(1)	164.20(0.02)	0.905^{50}	1.3931^{30}	sl H_2O; vs EtOH, eth
4278	2,2-Dimethyl-1-propanol	Neopentyl alcohol	$C_5H_{12}O$	75-84-3	88.148		55(3)	112(1)	0.812^{20}		sl H_2O; vs EtOH, eth; s ctc
4279	2,2-Dimethylpropanoyl chloride	Pivalic acid chloride	C_5H_9ClO	3282-30-2	120.577			104(4)	1.003^{20}	1.4139^{20}	vs eth
4280	*N,N*-Dimethyl-2-propenamide	*N,N*-Dimethylacrylamide	C_5H_9NO	2680-03-7	99.131	liq		81^{20}	0.962^{25}	1.4730^{20}	
4281	2,2-Dimethylpropylamine	2,2-Dimethyl-1-propanamine	$C_5H_{13}N$	5813-64-9	87.164			82(7)	0.7455^{20}	1.4023^{20}	vs eth
4282	(1,1-Dimethylpropyl)benzene		$C_{11}H_{16}$	2049-95-8	148.245			191(3)	0.8748^{20}	1.4958^{20}	
4283	(2,2-Dimethylpropyl)benzene		$C_{11}H_{16}$	1007-26-7	148.245			186(4)	0.8581^{18}	1.4884^{18}	
4284	4-(1,1-Dimethylpropyl)-cyclohexanone		$C_{11}H_{20}O$	16587-71-6	168.276		96	125^{16}	0.920^{25}	1.4677^{20}	
4285	1,1-Dimethylpropyl 3-methylbutanoate	*tert*-Pentyl isopentanoate	$C_{10}H_{20}O_2$	542-37-0	172.265			188(13)	0.8729^{0}		vs EtOH
4286	2-(1,1-Dimethylpropyl)phenol		$C_{11}H_{16}O$	3279-27-4	164.244						sl ctc
4287	4-(1,1-Dimethylpropyl)phenol	*p-tert*-Pentylphenol	$C_{11}H_{16}O$	80-46-6	164.244		92.7(1)	262(9)			
4288	4,6-Dimethyl-2*H*-pyran-2-one		$C_7H_8O_2$	675-09-2	124.138	lf (eth)	51.5	245			vs H_2O, eth, EtOH
4289	2,6-Dimethyl-4*H*-pyran-4-one		$C_7H_8O_2$	1004-36-0	124.138	pl, nd (sub)	132.1(0.2)	251	0.9953^{137}		s H_2O, EtOH, eth, ace
4290	2,3-Dimethylpyrazine		$C_6H_8N_2$	5910-89-4	108.141			161(16)	1.0281^{0}		s H_2O, EtOH, eth
4291	2,5-Dimethylpyrazine		$C_6H_8N_2$	123-32-0	108.141		15	152(4)	0.9887^{20}	1.4980^{20}	msc H_2O, EtOH, eth; s ace, chl
4292	2,6-Dimethylpyrazine		$C_6H_8N_2$	108-50-9	108.141	pr	47.5	155.6	0.9647^{50}		s H_2O, EtOH, eth; sl ctc
4293	1,3-Dimethyl-1*H*-pyrazole		$C_5H_8N_2$	694-48-4	96.131			137	0.9561^{17}	1.4734^{15}	vs H_2O
4294	3,5-Dimethyl-1*H*-pyrazole		$C_5H_8N_2$	67-51-6	96.131	cry (peth, al)	107.5	218	0.8839^{16}		s H_2O, ace; vs EtOH, eth, bz, MeOH
4295	2,7-Dimethylpyrene		$C_{18}H_{14}$	15679-24-0	230.304		230				
4296	4,6-Dimethyl-2-pyridinamine		$C_7H_{10}N_2$	5407-87-4	122.167		61	235			
4297	*N,N*-Dimethyl-2-pyridinamine		$C_7H_{10}N_2$	5683-33-0	122.167		182	196	1.0149^{14}	1.5663^{20}	s EtOH, eth, bz
4298	*N,N*-Dimethyl-4-pyridinamine		$C_7H_{10}N_2$	1122-58-3	122.167	pl (eth)	113.9(0.2)				vs H_2O, EtOH, bz, chl; s eth
4299	2,3-Dimethylpyridine	2,3-Lutidine	C_7H_9N	583-61-9	107.153			161.1(0.4)	0.9319^{25}	1.5057^{20}	s H_2O, EtOH, eth
4300	2,4-Dimethylpyridine	2,4-Lutidine	C_7H_9N	108-47-4	107.153	liq	-63.80(0.03)	158.4(0.3)	0.9309^{20}	1.5010^{20}	vs H_2O, EtOH, eth; s ace
4301	2,5-Dimethylpyridine	2,5-Lutidine	C_7H_9N	589-93-5	107.153	liq	-14.08(0.03)	157.00(0.05)	0.9297^{20}	1.5006^{20}	sl H_2O; vs EtOH; msc eth; s ace
4302	2,6-Dimethylpyridine	2,6-Lutidine	C_7H_9N	108-48-5	107.153	liq	-6.12(0.03)	144.0(0.1)	0.9226^{20}	1.4953^{20}	msc H_2O; sl EtOH; s eth, ace, chl
4303	3,4-Dimethylpyridine	3,4-Lutidine	C_7H_9N	583-58-4	107.153	liq	-10.45(0.03)	179.1(0.3)	0.9281^{20}	1.5096^{20}	sl H_2O, ctc; s EtOH, eth, ace, chl

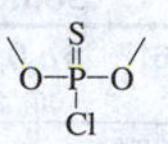
4262
O,O-Dimethyl phosphorochloridothioate

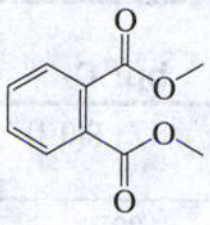
4263
Dimethyl phthalate

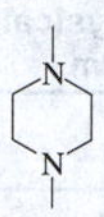
4264
1,4-Dimethylpiperazine

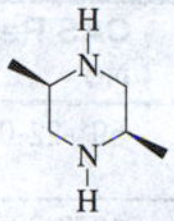
4265
cis-2,5-Dimethylpiperazine

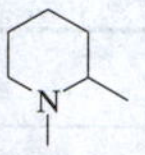
4266
1,2-Dimethylpiperidine, (±)-

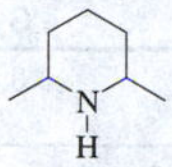
4267
2,6-Dimethylpiperidine

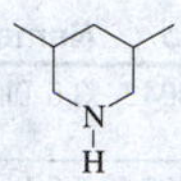
4268
3,5-Dimethylpiperidine

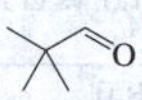
4269
2,2-Dimethylpropanal

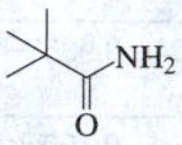
4270
2,2-Dimethylpropanamide

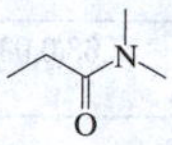
4271
N,N-Dimethylpropanamide

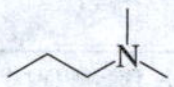
4272
N,N-Dimethyl-1-propanamine

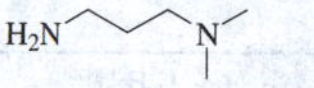
4273
N,N-Dimethyl-1,3-propanediamine

4274
2,2-Dimethyl-1,3-propanediol

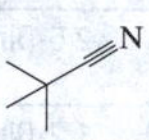
4275
2,2-Dimethylpropanenitrile

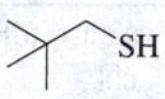
4276
2,2-Dimethyl-1-propanethiol

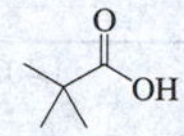
4277
2,2-Dimethylpropanoic acid

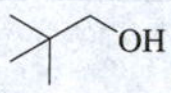
4278
2,2-Dimethyl-1-propanol

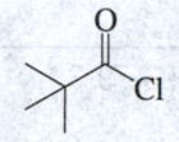
4279
2,2-Dimethylpropanoyl chloride

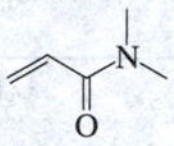
4280
N,N-Dimethyl-2-propenamide

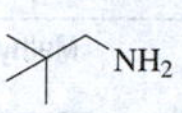
4281
2,2-Dimethylpropylamine

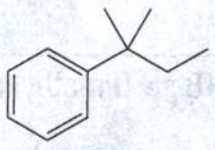
4282
(1,1-Dimethylpropyl)benzene

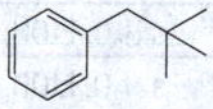
4283
(2,2-Dimethylpropyl)benzene

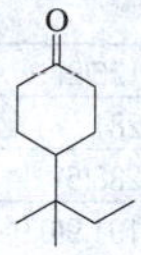
4284
4-(1,1-Dimethylpropyl)cyclohexanone

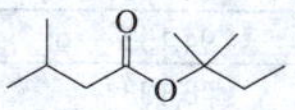
4285
1,1-Dimethylpropyl 3-methylbutanoate

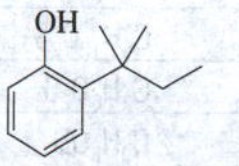
4286
2-(1,1-Dimethylpropyl)phenol

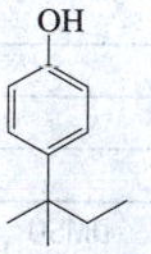
4287
4-(1,1-Dimethylpropyl)phenol

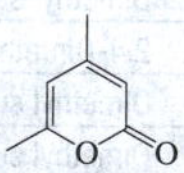
4288
4,6-Dimethyl-*2H*-pyran-2-one

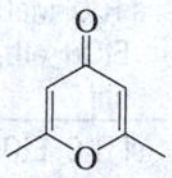
4289
2,6-Dimethyl-*4H*-pyran-4-one

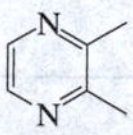
4290
2,3-Dimethylpyrazine

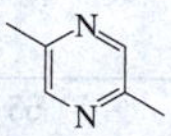
4291
2,5-Dimethylpyrazine

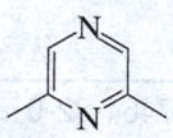
4292
2,6-Dimethylpyrazine

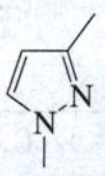
4293
1,3-Dimethyl-*1H*-pyrazole

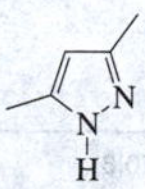
4294
3,5-Dimethyl-*1H*-pyrazole

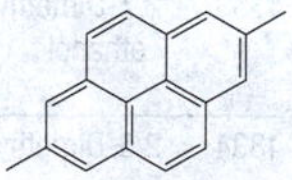
4295
2,7-Dimethylpyrene

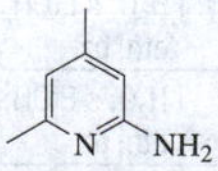
4296
4,6-Dimethyl-2-pyridinamine

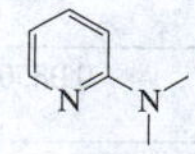
4297
N,N-Dimethyl-2-pyridinamine

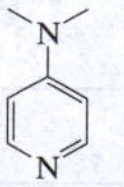
4298
N,N-Dimethyl-4-pyridinamine

4299
2,3-Dimethylpyridine

4300
2,4-Dimethylpyridine

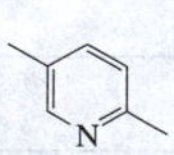
4301
2,5-Dimethylpyridine

4302
2,6-Dimethylpyridine

4303
3,4-Dimethylpyridine

Organic

No.	Name	Synonym	Mol. form.	CAS Reg. No.	Mol. wt.	Physical form	mp/°C	bp/°C	den/ g cm^{-3}	n_D	Solubility
4304	3,5-Dimethylpyridine	3,5-Lutidine	C_7H_9N	591-22-0	107.153	liq	-6.34(0.03)	171.9(0.1)	0.9419^{20}	1.5061^{20}	s H_2O, EtOH, eth, ace; sl ctc
4305	2,6-Dimethylpyridine-1-oxide		C_7H_9NO	1073-23-0	123.152	hyg	23.3(0.5)	133^{22}	1.073^{25}	1.5706^{20}	
4306	4,6-Dimethyl-2-pyrimidinamine		$C_6H_9N_3$	767-15-7	123.155		153.5				s H_2O, EtOH, ace, bz; i eth; vs chl
4307	2,6-Dimethyl-4-pyrimidinamine	Kyanmethin	$C_6H_9N_3$	461-98-3	123.155	nd (al), pl (bz)	183	sub			sl H_2O, EtOH, bz, chl
4308	4,6-Dimethylpyrimidine		$C_6H_8N_2$	1558-17-4	108.141		25	159(8)		1.4880^{20}	vs H_2O
4309	1,3-Dimethyl-2,4(1*H*,3*H*)-pyrimidinedione		$C_6H_8N_2O_2$	874-14-6	140.140		119.3(0.5)				sl EtOH; s eth, chl
4310	2,4-Dimethylpyrrole		C_6H_9N	625-82-1	95.142	pa bl flr cry		165(7)	0.9236^{20}	1.5048^{20}	sl H_2O; vs EtOH, eth, bz; s chl
4311	2,5-Dimethylpyrrole		C_6H_9N	625-84-3	95.142		7.68(0.08)	167.43(0.04)	0.9353^{20}	1.5036^{20}	i H_2O; vs EtOH, eth
4312	1,2-Dimethylpyrrolidine		$C_6H_{13}N$	765-48-0	99.174	oil		104(8)	0.799^{20}		s H_2O
4313	2,4-Dimethylquinoline	4-Methylquinaldine	$C_{11}H_{11}N$	1198-37-4	157.212	orth pr (eth)		267(2)	1.0611^{15}	1.6075^{20}	sl H_2O, chl; vs EtOH, eth
4314	2,6-Dimethylquinoline		$C_{11}H_{11}N$	877-43-0	157.212	orth pr (eth)	57.69(0.05)	268.1(0.4)			sl H_2O, EtOH, eth, chl; vs bz
4315	2,7-Dimethylquinoline	*m*-Toluquinaldine	$C_{11}H_{11}N$	93-37-8	157.212		61(2)	260(19)			sl H_2O; s EtOH, eth, chl
4316	2,3-Dimethylquinoxaline		$C_{10}H_{10}N_2$	2379-55-7	158.199	nd (w+3, ace)	106.3(0.4)				s EtOH, eth, ace, bz, chl, acid
4317	Dimethyl sebacate	Dimethyl decanedioate	$C_{12}H_{22}O_4$	106-79-6	230.301	lo pr	26.6(0.5)	175^{20}	0.9882^{28}	1.4355^{28}	i H_2O; s EtOH, eth, ace, ctc
4318	Dimethyl selenide	Methyl selenide	C_2H_6Se	593-79-3	109.03			58(3)	1.4077^{15}		vs eth, EtOH, chl
4319	Dimethylsilane	2-Silapropane	C_2H_8Si	1111-74-6	60.171	col gas	-150	-20	0.68^{-80}		
4320	Dimethylstearylamine	Dymanthine	$C_{20}H_{43}N$	124-28-7	297.562		22.9(0.2)				
4321	Dimethyl succinate	Methyl succinate	$C_6H_{10}O_4$	106-65-0	146.141		18.6(0.6)	197(1)	1.1198^{20}	1.4197^{20}	sl H_2O, ctc; s EtOH, ace; vs eth
4322	Dimethylsulfamoyl chloride	Dimethylaminosulfonyl chloride	$C_2H_6ClNO_2S$	13360-57-1	143.593			80^{16}			
4323	Dimethyl sulfate		$C_2H_6O_4S$	77-78-1	126.132	liq	-31.8(0.4)	186(3)	1.3322^{20}	1.3874^{20}	s H_2O, eth, bz, ctc; msc EtOH; i CS_2
4324	Dimethyl sulfide	2-Thiapropane	C_2H_6S	75-18-3	62.134	liq	-98.26(0.04)	37.32(0.05)	0.8483^{20}	1.4438^{20}	sl H_2O; s EtOH, eth
4325	Dimethyl sulfite		$C_2H_6O_3S$	616-42-2	110.132			126(2)	1.2129^{20}	1.4083^{20}	s H_2O, EtOH, eth
4326	2,4-Dimethylsulfolane		$C_6H_{12}O_2S$	1003-78-7	148.223	liq	-1.5	281	1.1362^{20}	1.4732^{20}	vs lig
4327	Dimethyl sulfone		$C_2H_6O_2S$	67-71-0	94.133	pr	108.83(0.05)	238(5)	1.1700^{110}	1.4226	s H_2O, EtOH, bz
4328	Dimethyl sulfoxide	DMSO	C_2H_6OS	67-68-5	78.133		18.52(0.05)	191.9(0.9)	1.1010^{25}	1.4793^{20}	s H_2O, EtOH, eth, ace, ctc, AcOEt
4329	Dimethyl L-tartrate	Dimethyl 2,3-dihydroxybutanedioate, [*R*-(*R**,*R**)]-	$C_6H_{10}O_6$	608-68-4	178.139	(i) cry (bz) (ii) cry (w)	50(form a); 61(form b)	280	1.306^{45}		vs H_2O, ace, eth, EtOH
4330	Dimethyl telluride		C_2H_6Te	593-80-6	157.67	pa ye		94			vs EtOH
4331	Dimethyl terephthalate		$C_{10}H_{10}O_4$	120-61-6	194.184		140.602 (0.004)	288	1.075^{141}		sl H_2O, EtOH, MeOH; s eth, chl
4332	Dimethyl tetrachloroterephthalate		$C_{10}H_6Cl_4O_4$	1861-32-1	331.965		158.2(0.7)				
4333	2,7-Dimethylthiachromine-8-ethanol		$C_{12}H_{14}N_4OS$	92-35-3	262.330	ye pr (chl)	228.8	sub			s H_2O, MeOH; sl EtOH, eth, ace, chl
4334	2,5-Dimethyl-1,3,4-thiadiazole		$C_4H_6N_2S$	27464-82-0	114.169		65	202.5			sl H_2O, EtOH, eth
4335	2,7-Dimethylthianthrene	Mesulphen	$C_{14}H_{12}S_2$	135-58-0	244.375	nd (HOAc,al)	123	184^{3}			vs ace, eth, peth, chl
4336	2,4-Dimethylthiazole		C_5H_7NS	541-58-2	113.182			145.60(0.09)	1.0562^{15}	1.5091^{20}	sl H_2O; s EtOH, eth, chl
4337	4,5-Dimethylthiazole		C_5H_7NS	3581-91-7	113.182		83.5	158	1.0699^{20}		vs eth, EtOH
4338	*N*,*N*-Dimethylthioacetamide		C_4H_9NS	631-67-4	103.186		74.5				
4339	Dimethyl thiodipropionate		$C_8H_{14}O_4S$	4131-74-2	206.260			162^{18}	1.1559^{20}	1.4740^{20}	
4340	2,3-Dimethylthiophene		C_6H_8S	632-16-6	112.193	liq	-49.0(0.6)	141(7)	1.0021^{20}	1.5192^{20}	i H_2O; vs EtOH, eth; s bz
4341	2,4-Dimethylthiophene		C_6H_8S	638-00-6	112.193			137(19)	0.9938^{20}	1.5104^{20}	i H_2O; s EtOH, eth, bz
4342	2,5-Dimethylthiophene		C_6H_8S	638-02-8	112.193	liq	-62.52(0.05)	139(2)	0.9850^{20}	1.5129^{20}	i H_2O; s EtOH, eth, bz

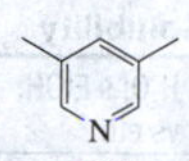
4304
3,5-Dimethylpyridine

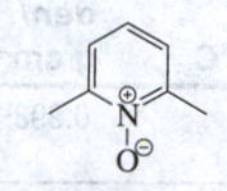
4305
2,6-Dimethylpyridine-1-oxide

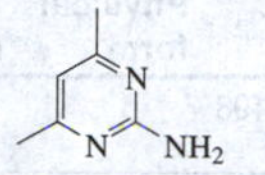
4306
4,6-Dimethyl-2-pyrimidinamine

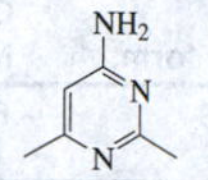
4307
2,6-Dimethyl-4-pyrimidinamine

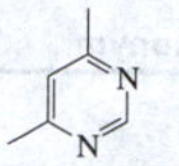
4308
4,6-Dimethylpyrimidine

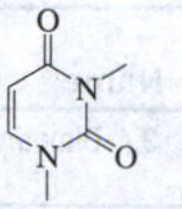
4309
1,3-Dimethyl-2,4(1*H*,3*H*)-pyrimidinedione

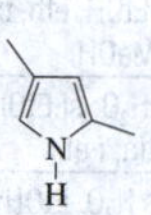
4310
2,4-Dimethylpyrrole

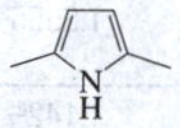
4311
2,5-Dimethylpyrrole

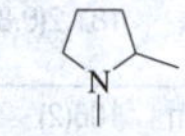
4312
1,2-Dimethylpyrrolidine

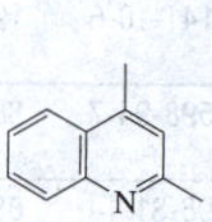
4313
2,4-Dimethylquinoline

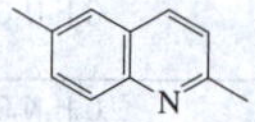
4314
2,6-Dimethylquinoline

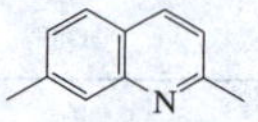
4315
2,7-Dimethylquinoline

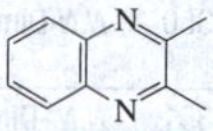
4316
2,3-Dimethylquinoxaline

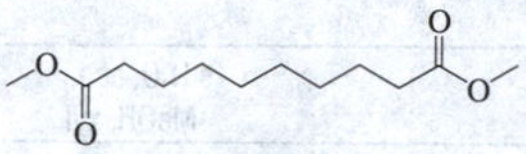
4317
Dimethyl sebacate

4318
Dimethyl selenide

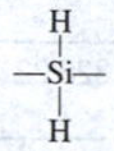
4319
Dimethylsilane

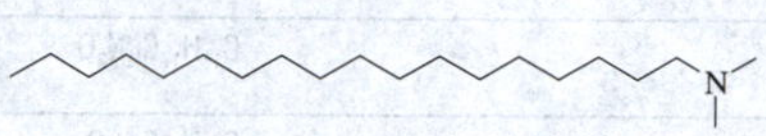
4320
Dimethylstearylamine

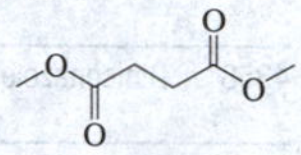
4321
Dimethyl succinate

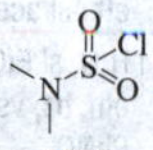
4322
Dimethylsulfamoyl chloride

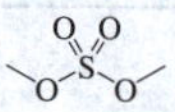
4323
Dimethyl sulfate

4324
Dimethyl sulfide

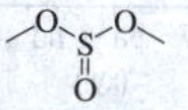
4325
Dimethyl sulfite

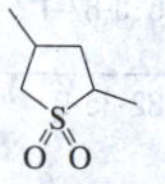
4326
2,4-Dimethylsulfolane

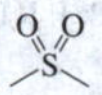
4327
Dimethyl sulfone

4328
Dimethyl sulfoxide

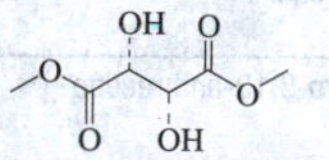
4329
Dimethyl L-tartrate

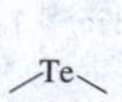
4330
Dimethyl telluride

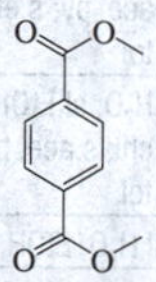
4331
Dimethyl terephthalate

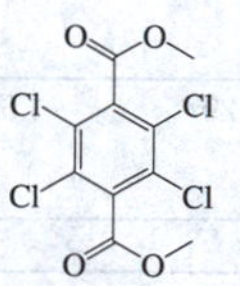
4332
Dimethyl tetrachloroterephthalate

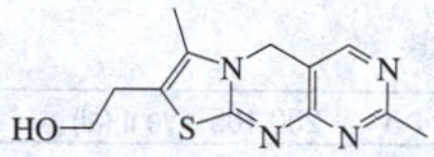
4333
2,7-Dimethylthiachromine-8-ethanol

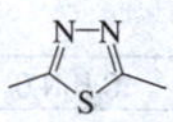
4334
2,5-Dimethyl-1,3,4-thiadiazole

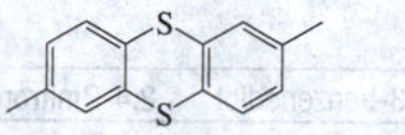
4335
2,7-Dimethylthianthrene

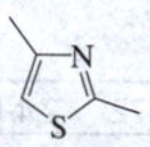
4336
2,4-Dimethylthiazole

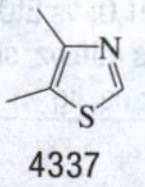
4337
4,5-Dimethylthiazole

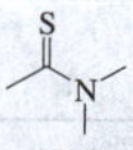
4338
N,*N*-Dimethylthioacetamide

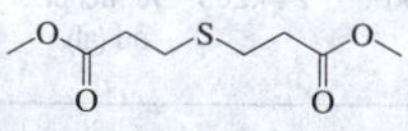
4339
Dimethyl thiodipropionate

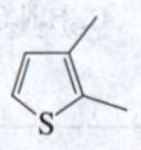
4340
2,3-Dimethylthiophene

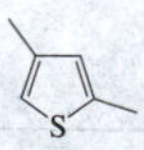
4341
2,4-Dimethylthiophene

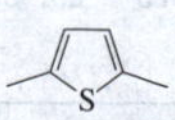
4342
2,5-Dimethylthiophene

No.	Name	Synonym	Mol. form.	CAS Reg. No.	Mol. wt.	Physical form	mp/°C	bp/°C	den/g cm^{-3}	n_D	Solubility
4343	3,4-Dimethylthiophene		C_6H_8S	632-15-5	112.193			145	0.993^{25}	1.5206^{20}	i H_2O; s EtOH; vs eth
4344	*N,N*-Dimethylthiourea		$C_3H_8N_2S$	6972-05-0	104.174	cry (w)	161.5				
4345	*N,N'*-Dimethylthiourea		$C_3H_8N_2S$	534-13-4	104.174	hyg pl	63.8(0.6)				vs H_2O, EtOH, ace; sl eth, bz; i CS_2
4346	2,6-Dimethyl-4-tridecylmorpholine	Tridemorph	$C_{19}H_{39}NO$	24602-86-6	297.519			$141^{1.3}$	0.86		
4347	*N,N*-Dimethyl-*N'*-[3-(trifluoromethyl)phenyl]urea	Fluometuron	$C_{10}H_{11}F_3N_2O$	2164-17-2	232.201		161(1)				vs ace, EtOH
4348	Dimethyl trisulfide		$C_2H_6S_3$	3658-80-8	126.264			41^{6}			
4349	6,10-Dimethyl-3,5,9-undecatrien-2-one	Pseudoionone	$C_{13}H_{20}O$	141-10-6	192.297	pa ye oil		144^{12}	0.8984^{20}	1.5335^{20}	s EtOH, eth, chl, MeOH
4350	*N,N*-Dimethylurea		$C_3H_8N_2O$	598-94-7	88.108	mcl pr (al, chl)	181.2(0.8)		1.2555^{25}		s H_2O; sl EtOH, tfa; i eth
4351	*N,N'*-Dimethylurea		$C_3H_8N_2O$	96-31-1	88.108	orth bipym (chl-eth)	106(2)	269	1.142^{25}		vs H_2O, EtOH; i eth; sl chl
4352	Dimethyl zinc		C_2H_6Zn	544-97-8	95.478	liq, ign in air	-43.01(0.02)	43(3)	1.386^{10}		s eth; msc peth
4353	Dimetilan		$C_{10}H_{16}N_4O_3$	644-64-4	240.259	col solid	69	205^{13}			s H_2O, chl, EtOH, ace, xyl
4354	Dimorpholamine		$C_{20}H_{38}N_4O_4$	119-48-2	398.541	cry (peth)	41.5	$229^{0.4}$			vs H_2O
4355	*N,N'*-Di-2-naphthyl-1,4-benzenediamine		$C_{26}H_{20}N_2$	93-46-9	360.450		235				i EtOH, eth, bz
4356	Di-2-naphthyl disulfide		$C_{20}H_{14}S_2$	5586-15-2	318.455	nd	139.5		1.144^{145}	1.4555^{20}	i H_2O; vs EtOH, eth; i lig
4357	*N,N'*-Di-1-naphthylurea		$C_{21}H_{16}N_2O$	607-56-7	312.364	nd (py, HOAc)	296	sub			vs py
4358	Diniconazole		$C_{15}H_{17}Cl_2N_3O$	83657-24-3	326.221	cry	149				s H_2O, ace, MeOH, xyl
4359	Dinitramine		$C_{11}H_{13}F_3N_4O_4$	29091-05-2	322.241		99.3(0.5)				
4360	2,3-Dinitroaniline		$C_6H_5N_3O_4$	602-03-9	183.122		128		1.646^{50}		i H_2O; s EtOH; sl eth
4361	2,4-Dinitroaniline		$C_6H_5N_3O_4$	97-02-9	183.122	ye nd (ace) grn ye tab (al)	180.1(0.5)		1.615^{14}		i H_2O; sl EtOH, ace, HCl
4362	2,5-Dinitroaniline		$C_6H_5N_3O_4$	619-18-1	183.122	oran nd (al)	138.0				vs EtOH
4363	2,6-Dinitroaniline		$C_6H_5N_3O_4$	606-22-4	183.122	gold lf (HOAc) ye nd (al)	138.0(0.5)				i H_2O, lig; sl EtOH; s eth, bz
4364	3,5-Dinitroaniline		$C_6H_5N_3O_4$	618-87-1	183.122	ye nd (dil al)	162(1)		1.601^{50}		i H_2O; s EtOH, eth; sl ace, bz
4365	1,5-Dinitro-9,10-anthracenedione		$C_{14}H_6N_2O_6$	82-35-9	298.207	pa ye nd (xyl)	385	sub			i H_2O; sl EtOH, eth, bz; vs $PhNO_2$
4366	1,8-Dinitro-9,10-anthracenedione		$C_{14}H_6N_2O_6$	129-39-5	298.207		312				
4367	2,4-Dinitrobenzaldehyde		$C_7H_4N_2O_5$	528-75-6	196.117	pa ye pr (al), pl (bz)	71.8(0.7)	200^{15}			sl H_2O, chl, lig; s EtOH, eth, bz
4368	3,5-Dinitrobenzamide	Nitromide	$C_7H_5N_3O_5$	121-81-3	211.132	lf (w)	184				vs H_2O
4369	1,2-Dinitrobenzene	*o*-Dinitrobenzene	$C_6H_4N_2O_4$	528-29-0	168.107	nd (bz), pl (al)	115.8(0.6)	319(3)	1.3119^{120}	1.565^{17}	i H_2O; s EtOH, bz, chl, AcOEt; sl DMSO
4370	1,3-Dinitrobenzene	*m*-Dinitrobenzene	$C_6H_4N_2O_4$	99-65-0	168.107	orth pl (al)	89.2(0.5)	296(2)	1.5751^{18}		sl H_2O; vs EtOH, ace, py; s eth, tol
4371	1,4-Dinitrobenzene	*p*-Dinitrobenzene	$C_6H_4N_2O_4$	100-25-4	168.107	nd (al)	171.1(0.9)	297	1.625^{18}		i H_2O; sl EtOH, chl; s ace, bz, tol
4372	2,4-Dinitro-1,3-benzenediol	2,4-Dinitroresorcinol	$C_6H_4N_2O_6$	519-44-8	200.105	ye lf (al)	147.3(0.4)				sl H_2O, EtOH
4373	2,4-Dinitrobenzenesulfenyl chloride		$C_6H_3ClN_2O_4S$	528-76-7	234.617	ye pr (bz-peth)	99				vs bz, chl, HOAc; sl peth
4374	2,4-Dinitrobenzenesulfonic acid		$C_6H_4N_2O_7S$	89-02-1	248.170	nd (w+3)	108				vs H_2O, EtOH; sl eth; i bz, peth
4375	2,4-Dinitrobenzoic acid		$C_7H_4N_2O_6$	610-30-0	212.116	nd (w)	182.7(0.5)		1.672^{20}		sl H_2O, EtOH, bz
4376	3,4-Dinitrobenzoic acid		$C_7H_4N_2O_6$	528-45-0	212.116	cry (dil al)	165.0(0.5)				sl H_2O; vs EtOH, eth
4377	3,5-Dinitrobenzoic acid		$C_7H_4N_2O_6$	99-34-3	212.116	mcl pr (al)	205				sl H_2O; vs EtOH, HOAc
4378	3,5-Dinitrobenzoyl chloride		$C_7H_3ClN_2O_5$	99-33-2	230.562	ye nd (bz)	74	196^{12}			s eth, chl
4379	2,2'-Dinitro-1,1'-biphenyl		$C_{12}H_8N_2O_4$	2436-96-6	244.203	ye mcl pr or nd (al)	127(1)	305	1.45^{25}		i H_2O; vs EtOH; s eth, bz; sl ace, lig

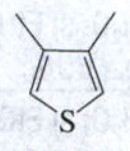
4343
3,4-Dimethylthiophene

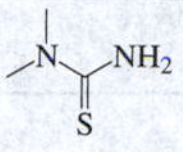
4344
N,N-Dimethylthiourea

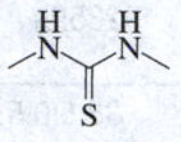
4345
N,N'-Dimethylthiourea

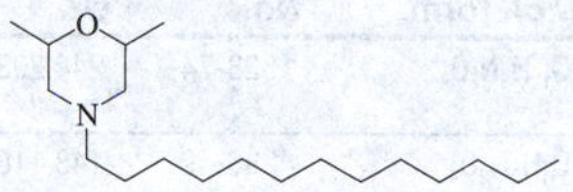
4346
2,6-Dimethyl-4-tridecylmorpholine

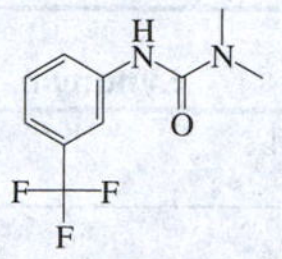
4347
N,N-Dimethyl-*N'*-[3-(trifluoromethyl)phenyl]urea

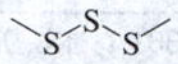
4348
Dimethyl trisulfide

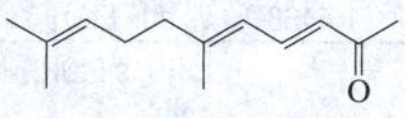
4349
6,10-Dimethyl-3,5,9-undecatrien-2-one

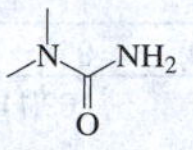
4350
N,N-Dimethylurea

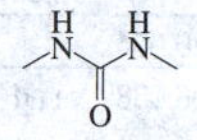
4351
N,N'-Dimethylurea

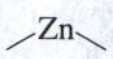
4352
Dimethyl zinc

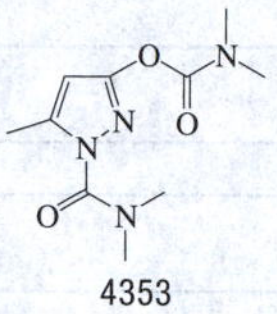
4353
Dimetilan

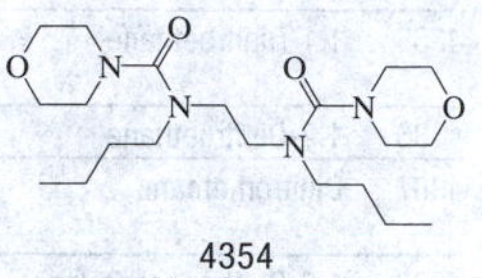
4354
Dimorpholamine

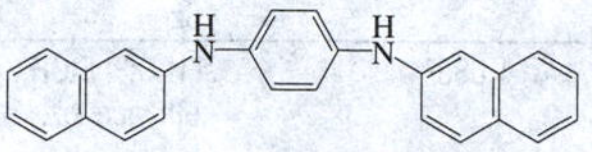
4355
N,N'-Di-2-naphthyl-1,4-benzenediamine

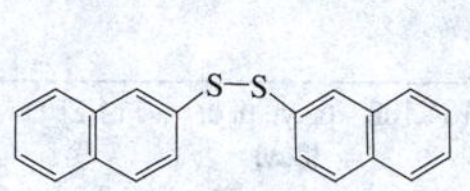
4356
Di-2-naphthyl disulfide

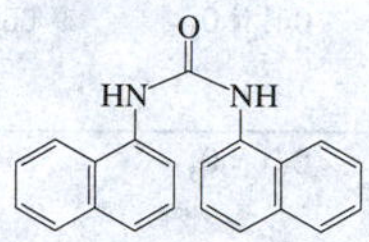
4357
N,N'-Di-1-naphthylurea

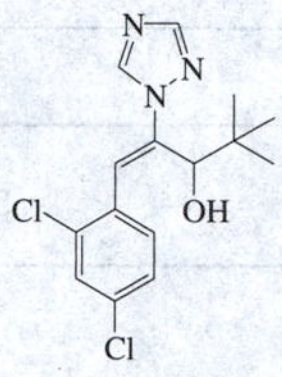
4358
Diniconazole

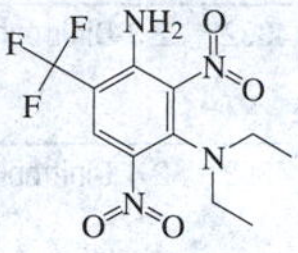
4359
Dinitramine

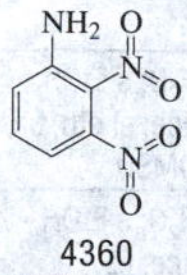
4360
2,3-Dinitroaniline

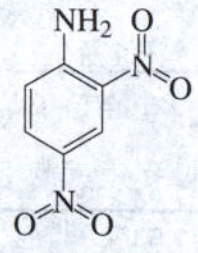
4361
2,4-Dinitroaniline

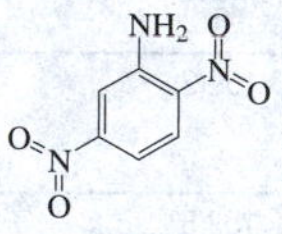
4362
2,5-Dinitroaniline

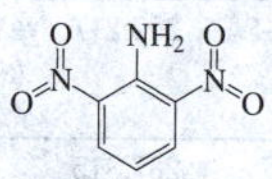
4363
2,6-Dinitroaniline

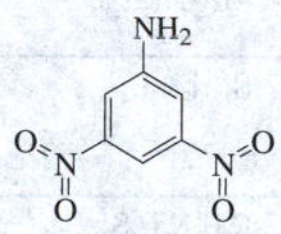
4364
3,5-Dinitroaniline

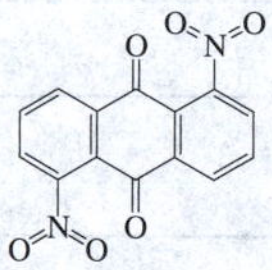
4365
1,5-Dinitro-9,10-anthracenedione

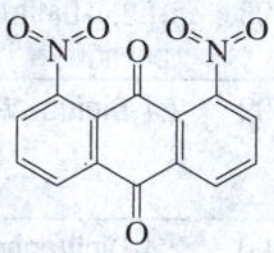
4366
1,8-Dinitro-9,10-anthracenedione

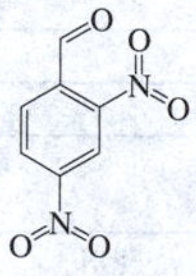
4367
2,4-Dinitrobenzaldehyde

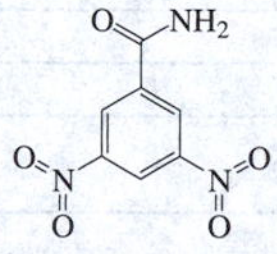
4368
3,5-Dinitrobenzamide

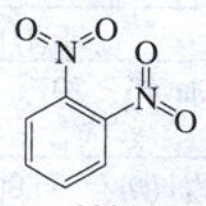
4369
1,2-Dinitrobenzene

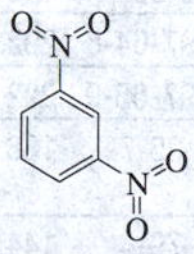
4370
1,3-Dinitrobenzene

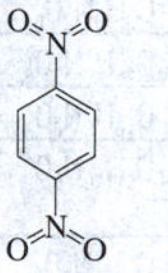
4371
1,4-Dinitrobenzene

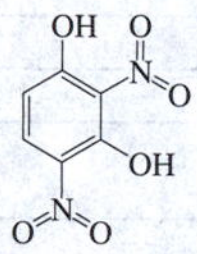
4372
2,4-Dinitro-1,3-benzenediol

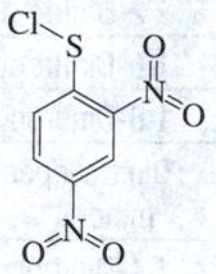
4373
2,4-Dinitrobenzenesulfenyl chloride

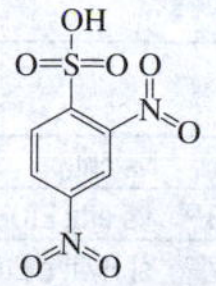
4374
2,4-Dinitrobenzenesulfonic acid

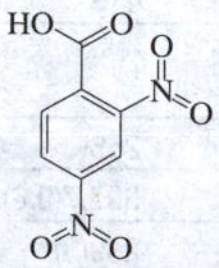
4375
2,4-Dinitrobenzoic acid

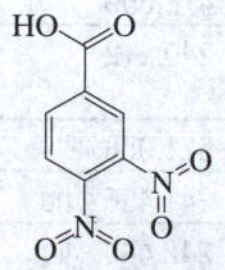
4376
3,4-Dinitrobenzoic acid

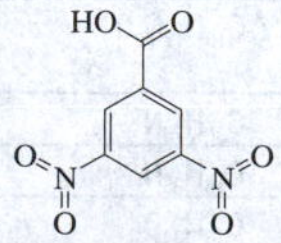
4377
3,5-Dinitrobenzoic acid

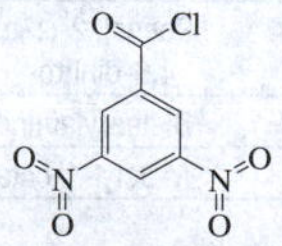
4378
3,5-Dinitrobenzoyl chloride

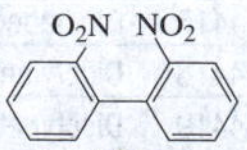
4379
2,2'-Dinitro-1,1'-biphenyl

No.	Name	Synonym	Mol. form.	CAS Reg. No.	Mol. wt.	Physical form	mp/°C	bp/°C	den/ g cm⁻³	n_D	Solubility
4380	4,4'-Dinitro-1,1'-biphenyl		$C_{12}H_8N_2O_4$	1528-74-1	244.203	nd (al)	235(6)				i H_2O; sl EtOH; s bz, HOAc
4381	1,4-Dinitrobutane		$C_4H_8N_2O_4$	4286-49-1	148.118	pl (al)	33.5(0.5)	176^{13}			i H_2O; sl EtOH; s eth, bz, MeOH
4382	4,4'-Dinitrodiphenylamine	4-Nitro-*N*-(4-nitrophenyl)-aniline	$C_{12}H_9N_3O_4$	1821-27-8	259.217	ye nd(al)	210(1)				i H_2O, tol; sl EtOH, bz; s ace, HOAc
4383	4,4'-Dinitrodiphenyl ether	Bis(4-nitrophenyl) ether	$C_{12}H_8N_2O_5$	101-63-3	260.202		143(1)				i H_2O; sl EtOH, eth; s bz, HOAc
4384	4,4'-Dinitrodiphenyl sulfide	Bis(4-nitrophenyl) sulfide	$C_{12}H_8N_2O_4S$	1223-31-0	276.268	oran pl (HOAc)	160.5				i H_2O; sl EtOH; s con sulf
4385	1,1-Dinitroethane		$C_2H_4N_2O_4$	600-40-8	120.064	ye mcl (bz, MeOH)		185.5	1.349^{24}		sl H_2O; s EtOH, eth
4386	1,2-Dinitroethane		$C_2H_4N_2O_4$	7570-26-5	120.064		38.2(0.4)	95^5	1.4597^{20}	1.4468^{20}	vs eth, EtOH
4387	Dinitromethane		$CH_2N_2O_4$	625-76-3	106.038	ye nd	<-15	169(11)			i H_2O; s EtOH, eth
4388	1,3-Dinitronaphthalene		$C_{10}H_6N_2O_4$	606-37-1	218.166	ye nd (bz, py-w)	148	sub			i H_2O; s EtOH, ace
4389	1,5-Dinitronaphthalene		$C_{10}H_6N_2O_4$	605-71-0	218.166	hex nd (ace, HOAc)	217(2)	371(3)	1.5860^{20}		i H_2O; sl EtOH, ace; s bz, py; vs eth
4390	1,8-Dinitronaphthalene		$C_{10}H_6N_2O_4$	602-38-0	218.166	ye orth pl (chl)	172(1)	445 dec			i H_2O; sl EtOH, bz; s ace, chl, py
4391	2,4-Dinitro-1-naphthol		$C_{10}H_6N_2O_5$	605-69-6	234.165	ye nd (al, chl)	138(1)				
4392	2,3-Dinitrophenol		$C_6H_4N_2O_5$	66-56-8	184.106	ye nd (w)	145.1(0.2)		1.681^{20}		sl H_2O, DMSO; vs EtOH, eth; s bz
4393	2,4-Dinitrophenol		$C_6H_4N_2O_5$	51-28-5	184.106	pa ye pl or lf (w)	114(2)	sub	1.683^{24}		sl H_2O; s EtOH, eth, ace, bz, tol, chl, py
4394	2,5-Dinitrophenol		$C_6H_4N_2O_5$	329-71-5	184.106	ye mcl pr or nd (w,lig)	105.6(0.2)				vs bz, eth
4395	2,6-Dinitrophenol		$C_6H_4N_2O_5$	573-56-8	184.106	pa ye orth nd or lf (dil al)	62.4(0.8)				i H_2O; vs EtOH, eth; s bz, chl; sl ctc
4396	3,4-Dinitrophenol		$C_6H_4N_2O_5$	577-71-9	184.106	tcl nd (w)	134.4(1)		1.672^{25}		vs bz, eth, EtOH
4397	2,4-Dinitrophenol, acetate		$C_8H_6N_2O_6$	4232-27-3	226.143	cry (MeOH)	72.5				
4398	4-[(2,4-Dinitrophenyl)amino]-phenol		$C_{12}H_9N_3O_5$	119-15-3	275.216	red lf	195.5				s alk
4399	2,4-Dinitro-*N*-phenylaniline		$C_{12}H_9N_3O_4$	961-68-2	259.217	ye red nd (al)	159(2)				i H_2O; s EtOH, ace; sl eth, bz, DMSO
4400	2,4-Dinitrophenyl dimethylcarbamodithioate		$C_9H_9N_3O_4S_2$	89-37-2	287.315		152.5		1.54^{20}		i H_2O; s EtOH, ace, bz
4401	(2,4-Dinitrophenyl)hydrazine		$C_6H_6N_4O_4$	119-26-6	198.137	blsh-red (al)	201.6(0.4)				i H_2O; s EtOH; sl eth, bz, chl, DMSO
4402	1,1-Dinitropropane		$C_3H_6N_2O_4$	601-76-3	134.091	liq	-42(1)	194(4)	1.2610^{25}	1.4339^{20}	s alk
4403	1,3-Dinitropropane		$C_3H_6N_2O_4$	6125-21-9	134.091		-21.4(0.2)	103^1	1.353^{26}	1.4654^{20}	i H_2O; s eth
4404	2,2-Dinitropropane		$C_3H_6N_2O_4$	595-49-3	134.091		52(2)	186.0(0.5)	1.30^{25}		sl H_2O
4405	2,2-Dinitro-1,3-propanediol		$C_3H_6N_2O_6$	2736-80-3	166.089	wh pl (bz)	142				
4406	1,6-Dinitropyrene		$C_{16}H_8N_2O_4$	42397-64-8	292.246		309(1)				
4407	1,8-Dinitropyrene		$C_{16}H_8N_2O_4$	42397-65-9	292.246		299(1)				
4408	Dinitrosopentamethylenetetra-mine		$C_5H_{10}N_6O_2$	101-25-7	186.172	cry (MeOH)	207				
4409	1,4-Dinitrosopiperazine		$C_4H_8N_4O_2$	140-79-4	144.133	pa ye pl (w)	155.9(0.7)				vs EtOH
4410	4,4'-Dinitro-2,2'-stilbenedisul-fonic acid		$C_{14}H_{10}N_2O_{10}S_2$	128-42-7	430.366	cry (AcOH)	266				
4411	Dinobuton	Dessin	$C_{14}H_{18}N_2O_7$	973-21-7	326.302	ye cry (EtOH)	60				
4412	Dinocap		$C_{18}H_{24}N_2O_6$	6119-92-2	364.393			$136^{0.01}$			
4413	Dinonyl adipate	Dinonyl hexanedioate	$C_{24}H_{46}O_4$	151-32-6	398.620			205^1			
4414	Dinonyl ether		$C_{18}H_{38}O$	2456-27-1	270.494	liq		318	0.81	1.4356^{20}	
4415	Dinonyl phthalate		$C_{26}H_{42}O_4$	84-76-4	418.609			413			
4416	Dinoseb	Phenol, 2-(1-methylpropyl)-4,6-dinitro-	$C_{10}H_{12}N_2O_5$	88-85-7	240.212		43.3(0.5)		1.265^{45}		
4417	Dioctadecylamine	Distearylamine	$C_{36}H_{75}N$	112-99-2	521.988		72.9	268^2			vs chl
4418	Dioctylamine	*N*-Octyl-1-octanamine	$C_{16}H_{35}N$	1120-48-5	241.456	nd	20(9)	307.7(0.6)	0.7963^{26}	1.4415^{26}	vs eth, EtOH
4419	Dioctyl ether		$C_{16}H_{34}O$	629-82-3	242.440	liq	-7.7(0.4)	289(3)	0.8063^{20}	1.4327^{20}	sl H_2O; s EtOH, eth, ctc

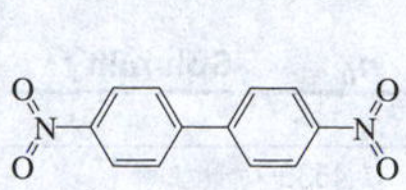
4380
4,4'-Dinitro-1,1'-biphenyl

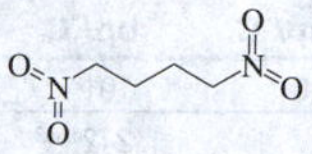
4381
1,4-Dinitrobutane

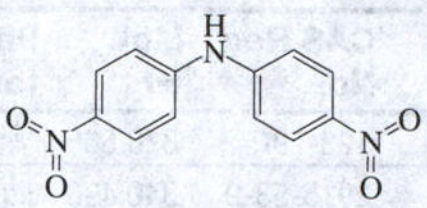
4382
4,4'-Dinitrodiphenylamine

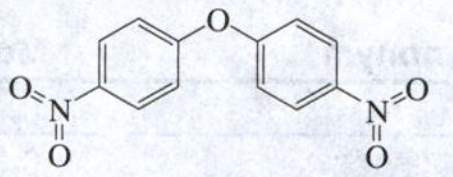
4383
4,4'-Dinitrodiphenyl ether

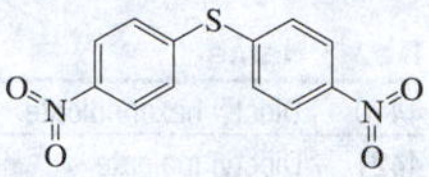
4384
4,4'-Dinitrodiphenyl sulfide

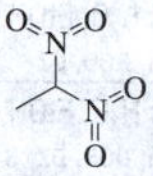
4385
1,1-Dinitroethane

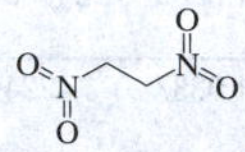
4386
1,2-Dinitroethane

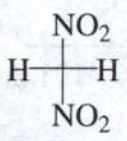
4387
Dinitromethane

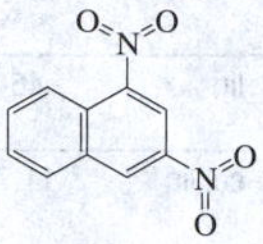
4388
1,3-Dinitronaphthalene

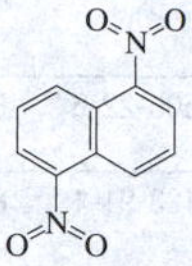
4389
1,5-Dinitronaphthalene

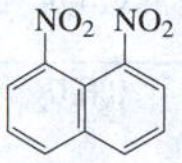
4390
1,8-Dinitronaphthalene

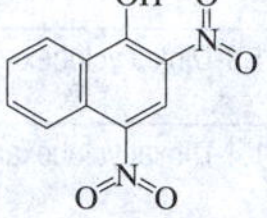
4391
2,4-Dinitro-1-naphthol

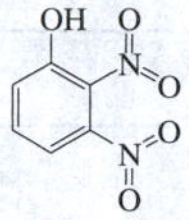
4392
2,3-Dinitrophenol

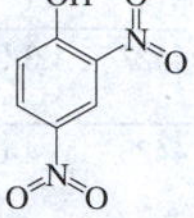
4393
2,4-Dinitrophenol

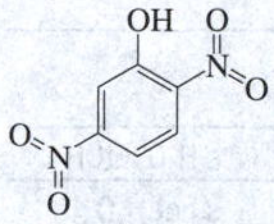
4394
2,5-Dinitrophenol

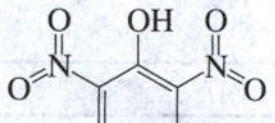
4395
2,6-Dinitrophenol

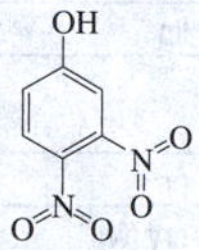
4396
3,4-Dinitrophenol

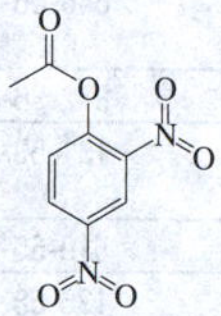
4397
2,4-Dinitrophenol, acetate

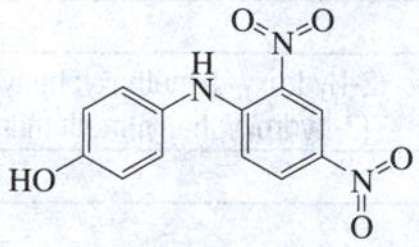
4398
4-[(2,4-Dinitrophenyl)amino]phenol

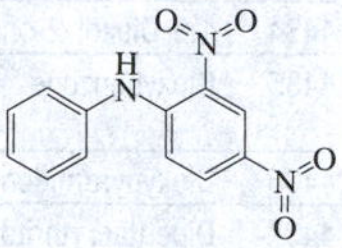
4399
2,4-Dinitro-*N*-phenylaniline

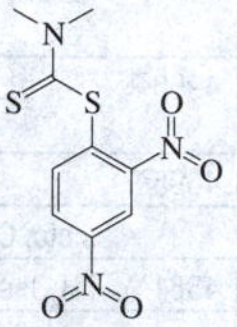
4400
2,4-Dinitrophenyl dimethylcarbamodithioate

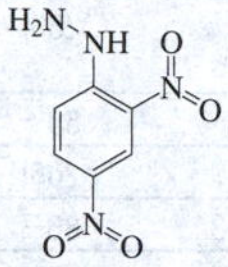
4401
(2,4-Dinitrophenyl)hydrazine

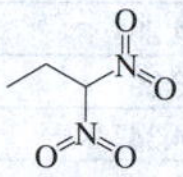
4402
1,1-Dinitropropane

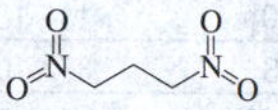
4403
1,3-Dinitropropane

4404
2,2-Dinitropropane

4405
2,2-Dinitro-1,3-propanediol

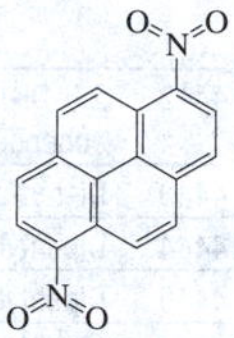
4406
1,6-Dinitropyrene

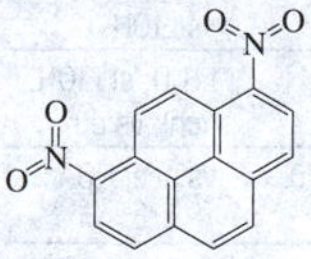
4407
1,8-Dinitropyrene

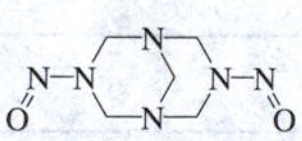
4408
Dinitrosopentamethylenetetramine

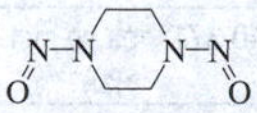
4409
1,4-Dinitrosopiperazine

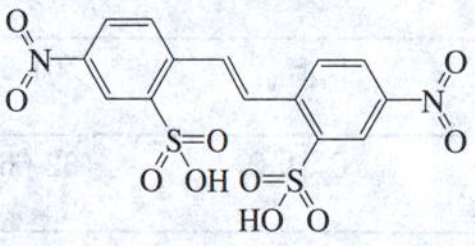
4410
4,4'-Dinitro-2,2'-stilbenedisulfonic acid

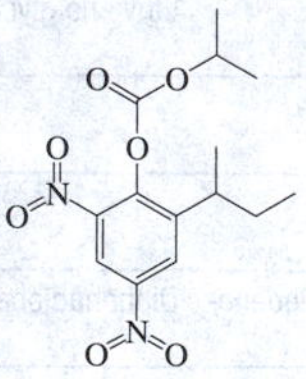
4411
Dinobuton

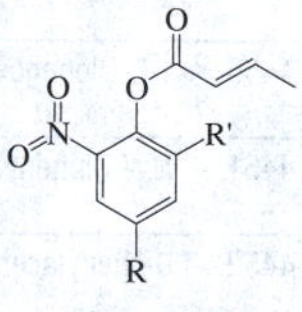
R = NO_2, R' = C_8H_{17}
and
R = C_8H_{17}, R' = NO_2
4412
Dinocap

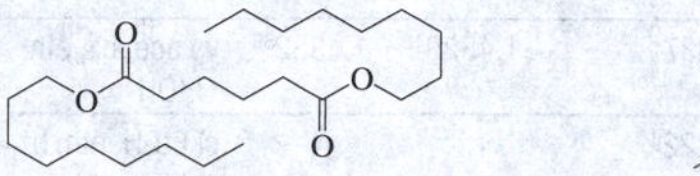
4413
Dinonyl adipate

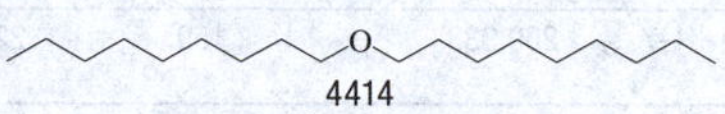
4414
Dinonyl ether

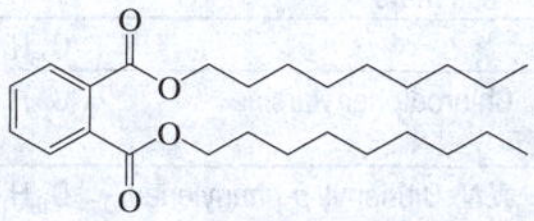
4415
Dinonyl phthalate

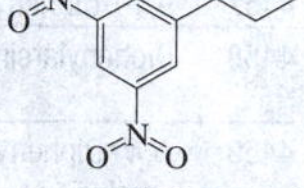
4416
Dinoseb

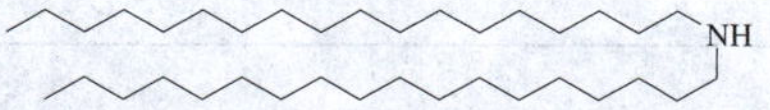
4417
Dioctadecylamine

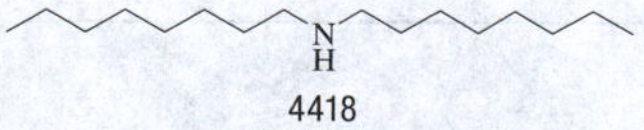
4418
Dioctylamine

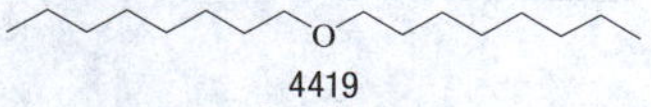
4419
Dioctyl ether

No.	Name	Synonym	Mol. form.	CAS Reg. No.	Mol. wt.	Physical form	mp/°C	bp/°C	den/g cm^{-3}	n_D	Solubility
4420	Dioctyl hexanedioate		$C_{22}H_{42}O_4$	123-79-5	370.566		9.6	191^{2}	0.922^{25}		
4421	Dioctyl maleate		$C_{20}H_{36}O_4$	2915-53-9	340.498	liq		$242^{0.002}$	0.94^{20}	1.4539^{20}	
4422	Dioctyl phthalate		$C_{24}H_{38}O_4$	117-84-0	390.557		25	220^{4}			
4423	Dioctyl sebacate	Dioctyl decanedioate	$C_{26}H_{50}O_4$	2432-87-3	426.673		18	$218^{0.5}$	0.9074^{25}		s ctc
4424	Dioctyl sulfide	Octyl sulfide	$C_{16}H_{34}S$	2690-08-6	258.506			202^{29}	0.842^{25}	1.4610^{20}	
4425	Dioctyl terephthalate		$C_{24}H_{38}O_4$	4654-26-6	390.557	cry		425	1.21^{62}		
4426	Dioscorine		$C_{13}H_{19}NO_2$	3329-91-7	221.296	grn-ye pr (eth)	34				s H_2O, ace, chl, EtOH; sl eth, bz
4427	1,3-Dioxane	1,3-Dioxacyclohexane	$C_4H_8O_2$	505-22-6	88.106	liq	-45	105(2)	1.0286^{25}	1.4165^{20}	msc H_2O, EtOH, eth, ace, bz
4428	1,4-Dioxane	1,4-Dioxacyclohexane	$C_4H_8O_2$	123-91-1	88.106	col liq	11.75(0.06)	101.2(0.3)	1.0337^{20}	1.4224^{20}	msc H_2O, EtOH, eth, ace, bz; s ctc
4429	1,4-Dioxane-2,5-dione		$C_4H_4O_4$	502-97-6	116.073	lf (al, al-chl)	83.0(0.1)				vs ace
4430	1,4-Dioxane-2,6-dione	Diglycollic anhydride	$C_4H_4O_4$	4480-83-5	116.073	cry (bz)	92.5	240.5			
4431	Dioxathion		$C_{12}H_{26}O_6P_2S_4$	78-34-2	456.538		-20		1.257^{26}		
4432	1,3-Dioxepane		$C_5H_{10}O_2$	505-65-7	102.132						s chl
4433	1,3-Dioxolane	1,3-Dioxacyclopentane	$C_3H_6O_2$	646-06-0	74.079	liq	-97.21(0.02)	75.3(0.1)	1.060^{20}	1.3974^{20}	msc H_2O; s EtOH, eth, ace
4434	1,3-Dioxol-2-one		$C_3H_2O_3$	872-36-6	86.046	liq	22	178(6)	1.35^{25}		
4435	Dioxybenzone	(2-Hydroxy-4-methoxyphenyl)-(2-hydroxyphenyl)methanone	$C_{14}H_{12}O_4$	131-53-3	244.243			172^{1}			
4436	Dioxypyramidon		$C_{13}H_{17}N_3O_3$	519-65-3	263.292	pr	105.5	197^{2}			s H_2O, EtOH
4437	Dipentaerythritol		$C_{10}H_{22}O_7$	126-58-9	254.278	cry (w)	221		1.366^{15}		s hot H_2O
4438	Dipentene	*p*-Menthadiene	$C_{10}H_{16}$	7705-14-8	136.234	liq	-95.5	176(5)	0.8402^{21}	1.4727^{20}	
4439	Dipentylamine	Diamylamine	$C_{10}H_{23}N$	2050-92-2	157.297			204(4)	0.7771^{20}	1.4272^{20}	sl H_2O; vs EtOH; msc eth; s ace
4440	Dipentyl *cis*-2-butenedioate	Dipentyl maleate	$C_{14}H_{24}O_4$	10099-71-5	256.339	liq		161^{10}	0.974^{20}		
4441	Dipentyl ether	Amyl ether	$C_{10}H_{22}O$	693-65-2	158.281	liq	-69.2(0.5)	187(2)	0.7833^{20}	1.4119^{20}	i H_2O; msc EtOH, eth; s chl
4442	2,6-Di-*tert*-pentyl-4-methylphenol	2,6-Bis(1,1-dimethylpropyl)-4-methylphenol	$C_{17}H_{28}O$	56103-67-4	248.403			283	0.931^{25}	1.4950^{20}	
4443	Di-*tert*-pentyl peroxide		$C_{10}H_{22}O_2$	10508-09-5	174.281		-55	58^{14}	0.808^{20}	1.4095^{20}	
4444	Dipentyl phthalate		$C_{18}H_{26}O_4$	131-18-0	306.397			205^{11}			s ctc, CS_2
4445	Dipentyl sulfide		$C_{10}H_{22}S$	872-10-6	174.347		-51.3	$86^{3.7}$	0.8407^{20}	1.4561^{20}	i H_2O; s eth
4446	Dipentyl sulfoxide		$C_{10}H_{22}OS$	1986-90-9	190.346		58	120^{1}			
4447	Diphenamid	Benzeneacetamide, *N*,*N*-dimethyl-α-phenyl-	$C_{16}H_{17}NO$	957-51-7	239.312		133(2)		$1.17^{23.3}$		
4448	Diphenidol	1,1-Diphenyl-4-piperidinyl-1-butanol	$C_{21}H_{27}NO$	972-02-1	309.445	nd (peth)	104.5				
4449	Diphenolic acid		$C_{17}H_{18}O_4$	126-00-1	286.323	cry (w)	171.5				vs H_2O, ace, EtOH
4450	1,2-Diphenoxyethane	Ethylene glycol diphenyl ether	$C_{14}H_{14}O_2$	104-66-5	214.260	lf (al)	98	182^{12}			i H_2O; sl EtOH; s eth, chl
4451	*N*,*N*-Diphenylacetamide		$C_{14}H_{13}NO$	519-87-9	211.259	wh cry pow	101.4(0.6)	sub			sl H_2O, eth, chl; s EtOH
4452	Diphenylacetylene		$C_{14}H_{10}$	501-65-5	178.229	mcl pr or pl (al)	62(2)	300	0.9657^{100}		i H_2O; sl EtOH, chl; vs eth
4453	2-(Diphenylacetyl)-1*H*-indene-1,3(2*H*)-dione	Diphenadione	$C_{23}H_{16}O_3$	82-66-6	340.371	pa ye mcl (al)	146.5			1.670	vs ace, HOAc
4454	Diphenylamine	*N*-Phenylbenzenamine	$C_{12}H_{11}N$	122-39-4	169.222	mcl lf(dil al)	53.2(0.3)	305.1(1)	1.158^{22}		i H_2O; vs EtOH, ace; s eth; sl chl
4455	Diphenylamine-2,2′-dicarboxylic acid		$C_{14}H_{11}NO_4$	579-92-0	257.242	ye cry (al)	296 dec				
4456	Diphenylamine-4-sulfonic acid, sodium salt	Sodium diphenylamine-4-sulfonate	$C_{12}H_{10}NNaO_3S$	6152-67-6	271.267	ye cry					
4457	9,10-Diphenylanthracene		$C_{26}H_{18}$	1499-10-1	330.421		237.0(0.2)				
4458	Diphenylarsinous chloride	Chlorodiphenylarsine	$C_{12}H_{10}AsCl$	712-48-1	264.582	orth pl (peth)	44	337	1.4820^{16}	1.6332^{56}	vs ace, bz, eth, EtOH
4459	*N*,*N*′-Diphenyl-1,4-benzenediamine	*N*,*N*′-Diphenyl-*p*-phenylenediamine	$C_{18}H_{16}N_2$	74-31-7	260.333		150	$222^{0.5}$			sl EtOH, eth, bz, chl; i acid
4460	α,α-Diphenylbenzeneethanol		$C_{20}H_{18}O$	4428-13-1	274.356	nd(bz-lig) pr (peth)	89.5	222^{11}			i H_2O; vs EtOH; sl eth, chl, peth
4461	α,α-Diphenylbenzenemethanethiol	Triphenylmethyl mercaptan	$C_{19}H_{16}S$	3695-77-0	276.395		105.8				

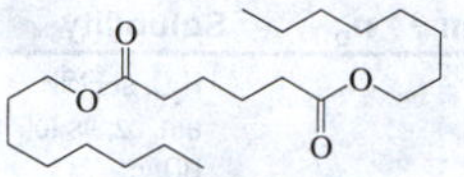
4420
Dioctyl hexanedioate

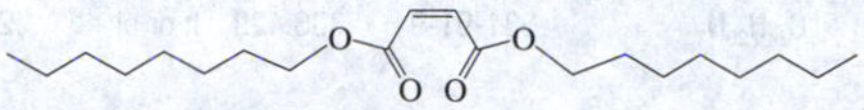
4421
Dioctyl maleate

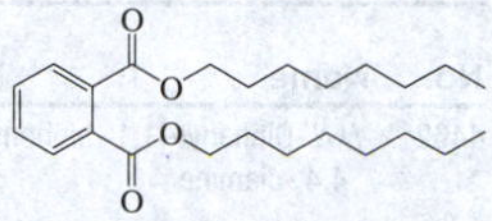
4422
Dioctyl phthalate

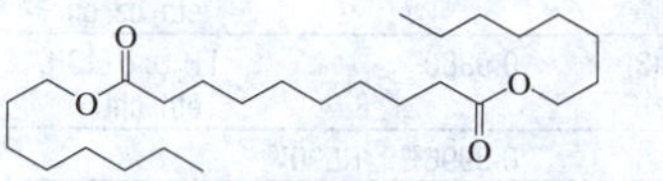
4423
Dioctyl sebacate

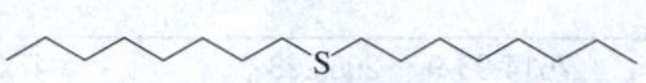
4424
Dioctyl sulfide

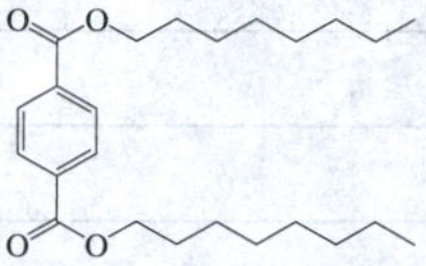
4425
Dioctyl terephthalate

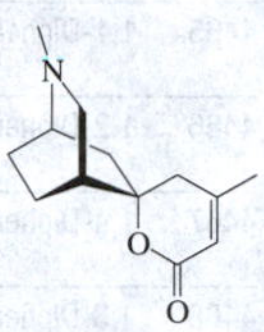
4426
Dioscorine

4427
1,3-Dioxane

4428
1,4-Dioxane

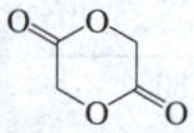
4429
1,4-Dioxane-2,5-dione

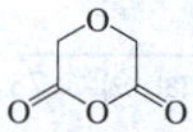
4430
1,4-Dioxane-2,6-dione

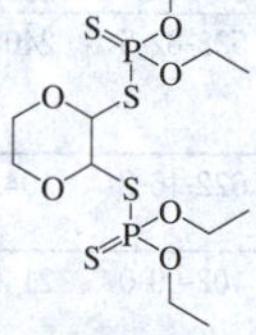
4431
Dioxathion

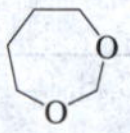
4432
1,3-Dioxepane

4433
1,3-Dioxolane

4434
1,3-Dioxol-2-one

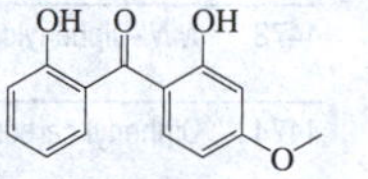
4435
Dioxybenzone

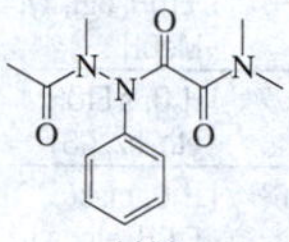
4436
Dioxypyramidon

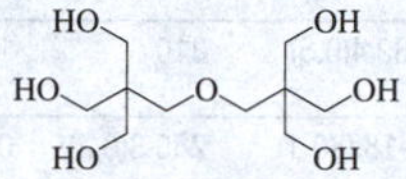
4437
Dipentaerythritol

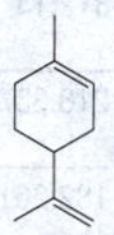
4438
Dipentene

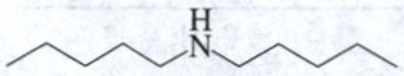
4439
Dipentylamine

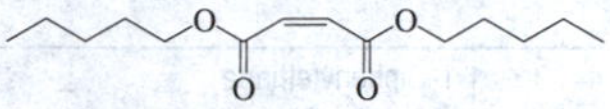
4440
Dipentyl *cis*-2-butenedioate

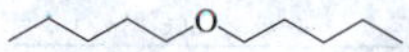
4441
Dipentyl ether

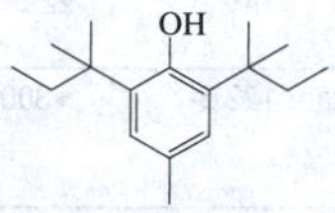
4442
2,6-Di-*tert*-pentyl-4-methylphenol

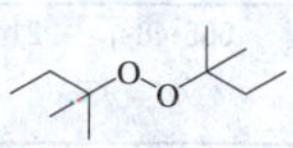
4443
Di-*tert*-pentyl peroxide

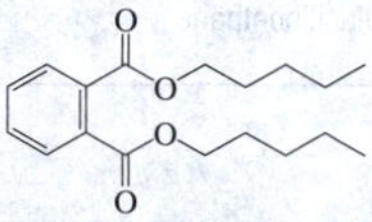
4444
Dipentyl phthalate

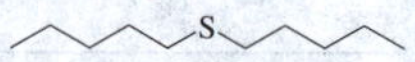
4445
Dipentyl sulfide

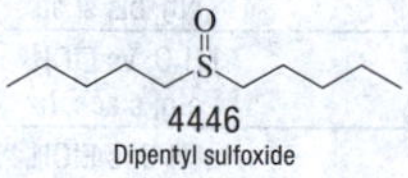
4446
Dipentyl sulfoxide

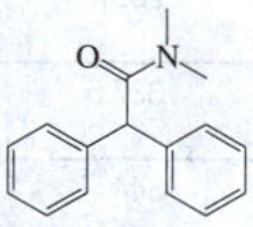
4447
Diphenamid

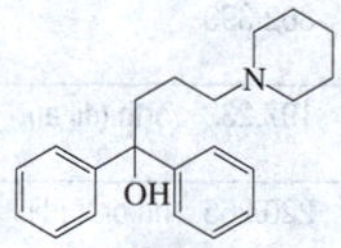
4448
Diphenidol

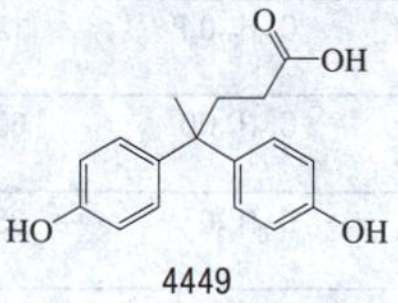
4449
Diphenolic acid

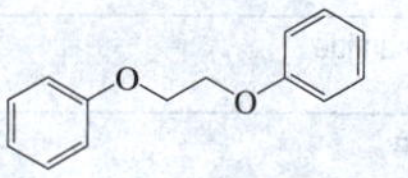
4450
1,2-Diphenoxyethane

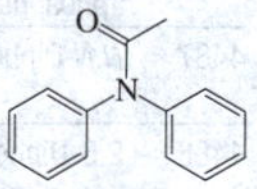
4451
N,N-Diphenylacetamide

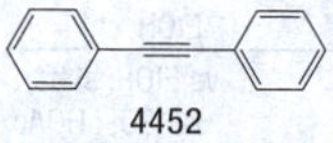
4452
Diphenylacetylene

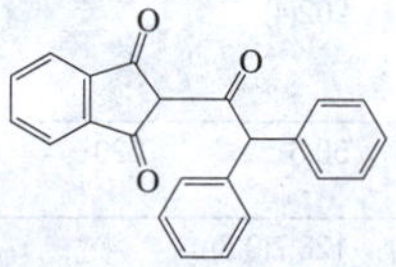
4453
2-(Diphenylacetyl)-1*H*-indene-1,3(2*H*)-dione

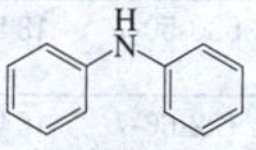
4454
Diphenylamine

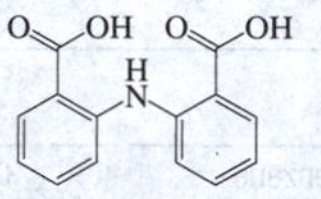
4455
Diphenylamine-2,2′-dicarboxylic acid

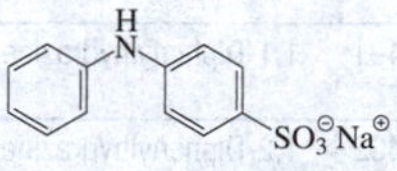
4456
Diphenylamine-4-sulfonic acid, sodium salt

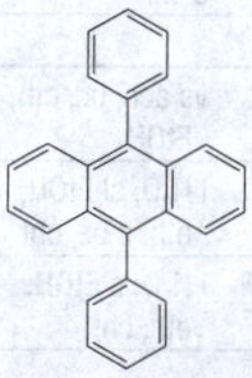
4457
9,10-Diphenylanthracene

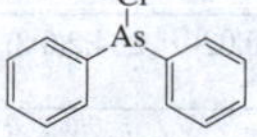
4458
Diphenylarsinous chloride

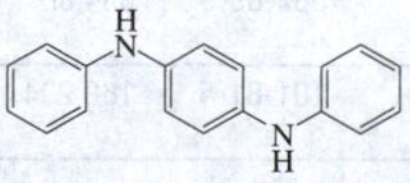
4459
N,N′-Diphenyl-1,4-benzenediamine

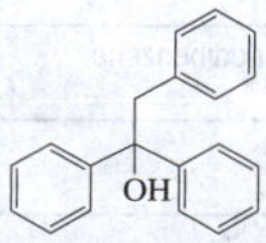
4460
α,α-Diphenylbenzeneethanol

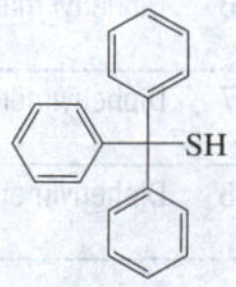
4461
α,α-Diphenylbenzenemethanethiol

No.	Name	Synonym	Mol. form.	CAS Reg. No.	Mol. wt.	Physical form	mp/°C	bp/°C	den/ g cm^{-3}	n_D	Solubility
4462	*N,N'*-Diphenyl-[1,1'-biphenyl]-4,4'-diamine	*N,N'*-Diphenylbenzidine	$C_{24}H_{20}N_2$	531-91-9	336.429	lf or pl	247				i H_2O; sl EtOH, eth, bz; vs tol, HOAc
4463	*trans,trans*-1,4-Diphenyl-1,3-butadiene		$C_{16}H_{14}$	538-81-8	206.282	lf (al, HOAc)	149(4)	352			vs bz, eth, EtOH, peth
4464	1,4-Diphenyl-1,3-butadiyne	Diphenyldiacetylene	$C_{16}H_{10}$	886-66-8	202.250		85.9(0.5)				
4465	1,1-Diphenylbutane		$C_{16}H_{18}$	719-79-9	210.314		27	287	0.9928^{20}	1.5664^{20}	i H_2O; s EtOH, eth, bz, chl
4466	1,2-Diphenylbutane		$C_{16}H_{18}$	5223-59-6	210.314			291(9)	0.9673^{20}	1.5554^{20}	i H_2O; s EtOH, eth, bz, chl
4467	1,4-Diphenylbutane		$C_{16}H_{18}$	1083-56-3	210.314		52.2(0.1)	316(3)	0.9880^{20}		i H_2O; s EtOH, eth, chl
4468	1,3-Diphenyl-1-butene		$C_{16}H_{16}$	7614-93-9	208.298		47.5	311	0.9996^{20}	1.590^{15}	
4469	*trans*-1,4-Diphenyl-2-butene-1,4-dione		$C_{16}H_{12}O_2$	959-28-4	236.265	ye nd (al, bz)	111				sl EtOH; s bz, HOAc; vs chl; i lig
4470	1,3-Diphenyl-2-buten-1-one	Dypnone	$C_{16}H_{14}O$	495-45-4	222.281			342.5	1.1080^{15}	1.6343^{20}	vs eth, EtOH
4471	Diphenylcarbamic chloride		$C_{13}H_{10}ClNO$	83-01-2	231.677	lf (al)	84.5				
4472	Diphenylcarbazone		$C_{13}H_{12}N_4O$	538-62-5	240.260	oran oran nd (bz) pr (al)	157 dec				i H_2O; vs EtOH, bz, chl
4473	*N,N'*-Diphenylcarbodiimide		$C_{13}H_{10}N_2$	622-16-2	194.231		169	331			sl H_2O, EtOH, eth; s bz
4474	Diphenyl carbonate	Phenyl carbonate	$C_{13}H_{10}O_3$	102-09-0	214.216	nd (al, bz)	78.9(0.3)	306	1.1215^{87}		i H_2O; s EtOH, eth, ctc, HOAc
4475	2,2'-Diphenylcarbonic dihydrazide	*sym*-Diphenylcarbazide	$C_{13}H_{14}N_4O$	140-22-7	242.276	cry (al + 1) cry (HOAc)	170	dec			sl H_2O, eth; s EtOH, ace, bz
4476	Diphenyl chlorophosphonate		$C_{12}H_{10}ClO_3P$	2524-64-3	268.632			314^{272}	1.296^{25}	1.5500^{20}	s tfa
4477	Diphenyl diselenide	Phenyl diselenide	$C_{12}H_{10}Se_2$	1666-13-3	312.13	ye nd	63.5	202^{11}	1.557^{80}	1.743^{20}	s EtOH, eth, xyl, MeOH
4478	Diphenyl disulfide	Phenyl disulfide	$C_{12}H_{10}S_2$	882-33-7	218.337	nd(al) or orth	60.4(0.5)	310	1.353^{20}		i H_2O; s EtOH, eth, bz, CS_2
4479	1,1-Diphenylethane		$C_{14}H_{14}$	612-00-0	182.261	liq	-18.0(0.1)	285.8(0.2)	0.9997^{20}	1.5756^{20}	i H_2O; msc EtOH, eth; s bz
4480	1,2-Diphenylethane	Dibenzyl	$C_{14}H_{14}$	103-29-7	182.261	mcl pr (MeOH)	51.18(0.06)	280(3)	0.9780^{25}	1.5476^{60}	i H_2O; s EtOH, eth, CS_2
4481	*N,N'*-Diphenylethanediamide		$C_{14}H_{12}N_2O_2$	620-81-5	240.257	lf (bz)	254	>360			vs bz
4482	*N,N'*-Diphenyl-1,2-ethanediamine	1,2-Dianilinoethane	$C_{14}H_{16}N_2$	150-61-8	212.290	cry (dil al)	74	229^{12}			i H_2O; s EtOH, eth; sl tfa
4483	1,2-Diphenyl-1,2-ethanediol, (*R**,*R**)-(±)-		$C_{14}H_{14}O_2$	655-48-1	214.260	nd (w,al),tab (eth)	122.5	>300			i H_2O, lig; vs EtOH, eth; s ace
4484	1,1-Diphenylethene		$C_{14}H_{12}$	530-48-3	180.245		8.2(0.4)	277(17)	1.0232^{20}	1.6085^{20}	i H_2O; s eth, chl
4485	Diphenyl ether	Oxybisbenzene	$C_{12}H_{10}O$	101-84-8	170.206		26.865 (0.003)	258.0(0.1)	1.0661^{30}	1.5787^{25}	i H_2O; s EtOH, eth, bz, HOAc; sl chl
4486	Diphenyl 2-ethylhexyl phosphate		$C_{20}H_{27}O_4P$	1241-94-7	362.399			232^{5}	1.090^{25}	1.510^{25}	
4487	*N,N*-Diphenylformamide		$C_{13}H_{11}NO$	607-00-1	197.232	orth (dil al)	73.5	337.5			i H_2O; s EtOH, eth, bz; sl ctc
4488	2,5-Diphenylfuran		$C_{16}H_{12}O$	955-83-9	220.265	nd or lf (dil al)	91	344			i H_2O; vs EtOH, eth; s ace, bz
4489	*N,N'*-Diphenylguanidine	1,3-Diphenylguanidine	$C_{13}H_{13}N_3$	102-06-7	211.262	mcl nd (al, to)	147(1)	170 dec	1.13^{20}		sl H_2O; s EtOH, ctc chl, tol; vs eth
4490	1,6-Diphenyl-1,3,5-hexatriene		$C_{18}H_{16}$	1720-32-7	232.320	lf (ace)	202(4)				i H_2O, EtOH, eth, HOAc; s ace; sl bz, chl
4491	1,1-Diphenylhydrazine		$C_{12}H_{12}N_2$	530-50-7	184.236	tab (lig)	50.5	220^{40}	1.190^{16}		vs bz, eth, EtOH, chl
4492	1,2-Diphenylhydrazine	Hydrazobenzene	$C_{12}H_{12}N_2$	122-66-7	184.236	tab (al-eth)	128.7(0.5)		1.158^{16}		vs EtOH; sl bz, DMSO; i HOAc
4493	5,5-Diphenyl-4-imidazolidinone	Doxenitoin	$C_{15}H_{14}N_2O$	3254-93-1	238.284	pl (MeOH)	183				
4494	Diphenyl isophthalate		$C_{20}H_{14}O_4$	744-45-6	318.323		138				s chl
4495	Diphenylketene	Diphenylethenone	$C_{14}H_{10}O$	525-06-4	194.228	red-ye liq		267.5	1.1107^{13}	1.615^{14}	
4496	Diphenyl maleate		$C_{16}H_{12}O_4$	7242-17-3	268.264	pl (lig)	73	226^{15}			vs ace, bz, eth, EtOH
4497	Diphenylmercury	Mercuriodibenzene	$C_{12}H_{10}Hg$	587-85-9	354.80			204^{10}	2.318^{25}		i H_2O; sl EtOH, eth; s bz, chl
4498	Diphenylmethane		$C_{13}H_{12}$	101-81-5	168.234	pr nd	25.22(0.02)	264.2(0.3)	1.001^{26}	1.5753^{20}	i H_2O; s EtOH, eth, chl

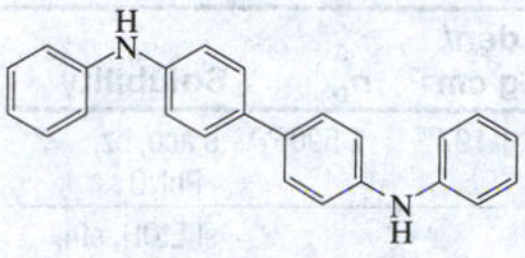
4462
N,N'-Diphenyl-[1,1'-biphenyl]-4,4'-diamine

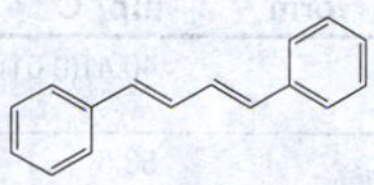
4463
trans,trans-1,4-Diphenyl-1,3-butadiene

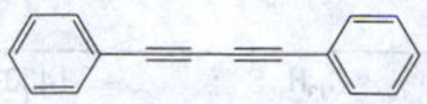
4464
1,4-Diphenyl-1,3-butadiyne

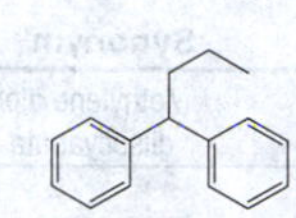
4465
1,1-Diphenylbutane

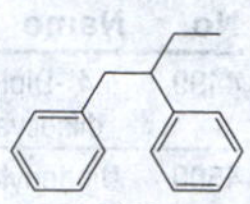
4466
1,2-Diphenylbutane

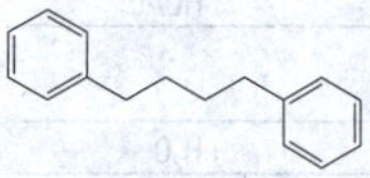
4467
1,4-Diphenylbutane

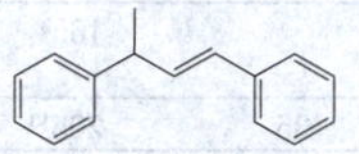
4468
1,3-Diphenyl-1-butene

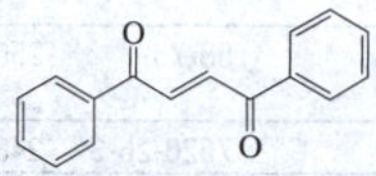
4469
trans-1,4-Diphenyl-2-butene-1,4-dione

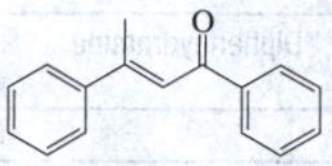
4470
1,3-Diphenyl-2-buten-1-one

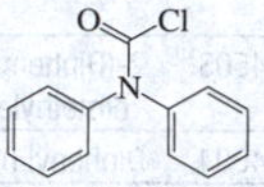
4471
Diphenylcarbamic chloride

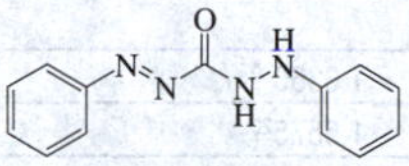
4472
Diphenylcarbazone

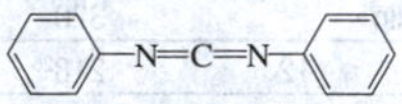
4473
N,N'-Diphenylcarbodiimide

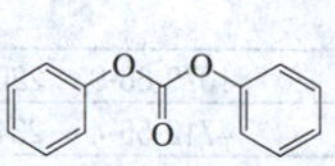
4474
Diphenyl carbonate

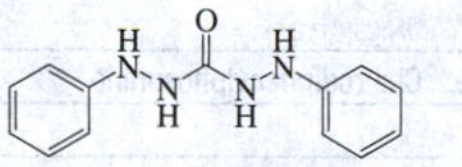
4475
2,2'-Diphenylcarbonic dihydrazide

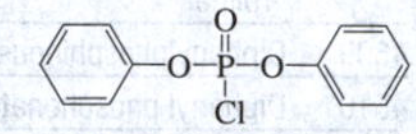
4476
Diphenyl chlorophosphonate

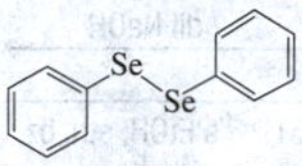
4477
Diphenyl diselenide

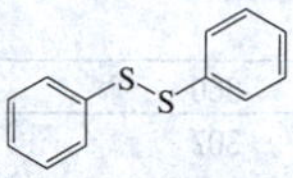
4478
Diphenyl disulfide

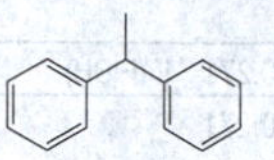
4479
1,1-Diphenylethane

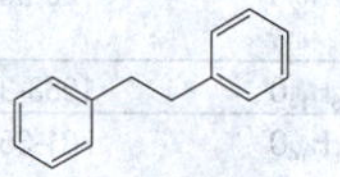
4480
1,2-Diphenylethane

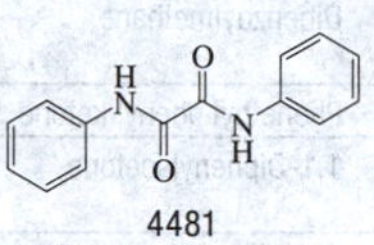
4481
N,N'-Diphenylethanediamide

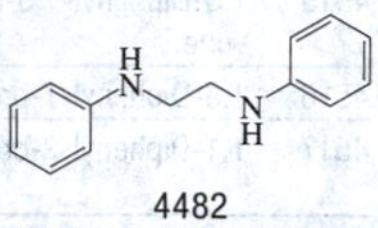
4482
N,N'-Diphenyl-1,2-ethanediamine

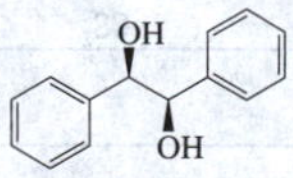
4483
1,2-Diphenyl-1,2-ethanediol, (*R**,*R**)-(±)-

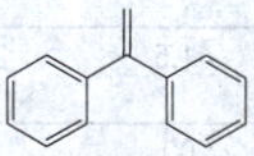
4484
1,1-Diphenylethene

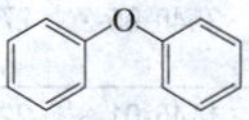
4485
Diphenyl ether

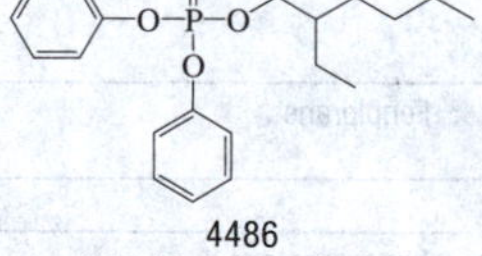
4486
Diphenyl 2-ethylhexyl phosphate

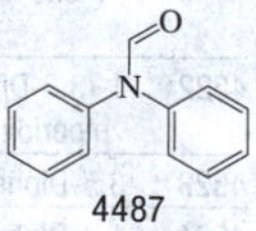
4487
N,N-Diphenylformamide

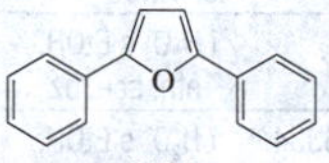
4488
2,5-Diphenylfuran

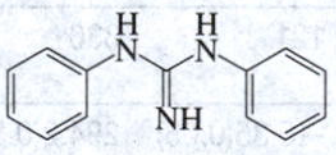
4489
N,N'-Diphenylguanidine

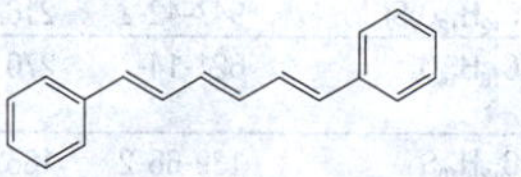
4490
1,6-Diphenyl-1,3,5-hexatriene

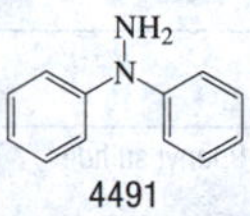
4491
1,1-Diphenylhydrazine

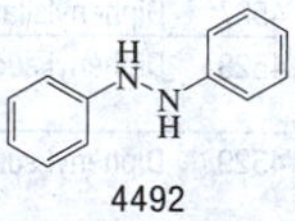
4492
1,2-Diphenylhydrazine

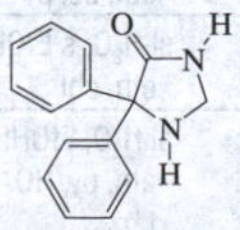
4493
5,5-Diphenyl-4-imidazolidinone

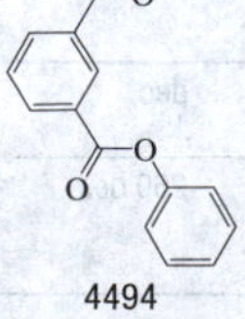
4494
Diphenyl isophthalate

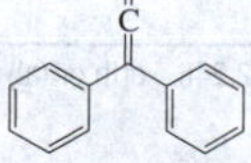
4495
Diphenylketene

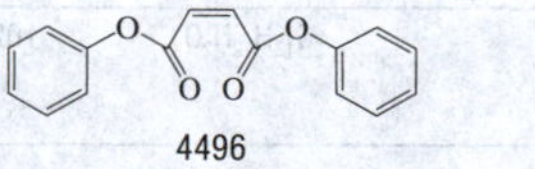
4496
Diphenyl maleate

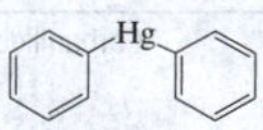
4497
Diphenylmercury

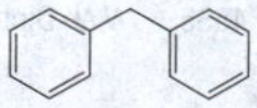
4498
Diphenylmethane

No.	Name	Synonym	Mol. form.	CAS Reg. No.	Mol. wt.	Physical form	mp/°C	bp/°C	den/ g cm^{-3}	n_D	Solubility
4499	4,4'-Diphenylmethane diisocyanate	Methylene diphenyl diisocyanate	$C_{15}H_{10}N_2O_2$	101-68-8	250.252		40.41(0.01)	196^{5}	1.197^{70}	1.5906^{50}	s ace, bz, $PhNO_2$
4500	Diphenylmethanethione		$C_{13}H_{10}S$	1450-31-3	198.283		53.5	174^{14}			sl EtOH, eth, peth; vs bz, chl
4501	*N,N'*-Diphenylmethanimidamide		$C_{13}H_{12}N_2$	622-15-1	196.247	nd (al)	142	>250			sl H_2O, peth; s EtOH, ace, bz; vs eth
4502	Diphenylmethanol	Benzohydrol	$C_{13}H_{12}O$	91-01-0	184.233	nd (lig)	65(1)	302(6)			sl H_2O; vs EtOH, eth, ctc, chl; s HOAc
4503	2-(Diphenylmethoxy)-*N,N*-dimethylethanamine	Diphenhydramine	$C_{17}H_{21}NO$	58-73-1	255.355	oil		165^{3}			
4504	Diphenyl methylphosphonate		$C_{13}H_{13}O_3P$	7526-26-3	248.214		35	205^{13}	1.2051^{20}		i H_2O
4505	2-(Diphenylmethyl)-1-piperidineethanol	Diphemethoxidine	$C_{20}H_{25}NO$	13862-07-2	295.419		106.5	$180^{0.1}$			
4506	2,5-Diphenyloxazole		$C_{15}H_{11}NO$	92-71-7	221.254	nd (lig)	71(3)	360	1.0940^{100}	1.6231^{100}	i H_2O; vs EtOH, eth; sl chl
4507	1,5-Diphenyl-1,4-pentadien-3-one	Dibenzalacetone	$C_{17}H_{14}O$	538-58-9	234.292	pl or lf (ace, AcOEt)	113 dec	dec			i H_2O; sl EtOH, eth; s ace, chl
4508	4,7-Diphenyl-1,10-phenanthroline		$C_{24}H_{16}N_2$	1662-01-7	332.397		220 dec				
4509	Diphenylphosphinous chloride	Chlorodiphenylphosphine	$C_{12}H_{10}ClP$	1079-66-9	220.634	hyg ye liq		320	1.229	1.6360^{20}	
4510	Diphenyl phosphonate		$C_{12}H_{11}O_3P$	4712-55-4	234.187		12	218^{26}	1.223^{25}	1.5575^{20}	
4511	Diphenyl phthalate	Phenyl phthalate	$C_{20}H_{14}O_4$	84-62-8	318.323	pr (al, lig)	73	253^{14}			i H_2O; sl EtOH, eth, ctc
4512	α,α-Diphenyl-2-piperidinemethanol	Pipradrol	$C_{18}H_{21}NO$	467-60-7	267.366	cry (hx)	97.5				
4513	1,3-Diphenylpropane		$C_{15}H_{16}$	1081-75-0	196.288	liq	6	300	1.007^{20}	1.5760^{20}	
4514	2,2-Diphenylpropane		$C_{15}H_{16}$	778-22-3	196.288		29.1(0.3)	281(6)	0.9980^{20}		
4515	1,3-Diphenyl-1,3-propanedione	Dibenzoylmethane	$C_{15}H_{12}O_2$	120-46-7	224.255		77(3)				s EtOH, eth, chl, dil NaOH
4516	1,3-Diphenyl-1-propanone	Phenethyl phenyl ketone	$C_{15}H_{14}O$	1083-30-3	210.271	lf (EtOH)	72.5	360			
4517	1,1-Diphenyl-2-propanone	1,1-Diphenylacetone	$C_{15}H_{14}O$	781-35-1	210.271		46	307		1.5361^{16}	s EtOH, eth, bz, chl, lig
4518	1,3-Diphenyl-2-propanone	Dibenzyl ketone	$C_{15}H_{14}O$	102-04-5	210.271	cry (al, peth)	34(1)	329(14)	1.195^{0}		i H_2O; s EtOH, eth, peth
4519	3,3-Diphenyl-2-propenal	β-Phenylcinnamaldehyde	$C_{15}H_{12}O$	1210-39-5	208.255	pa ye pr (lig)	44.8	205^{14}			
4520	1,1-Diphenyl-1-propene		$C_{15}H_{14}$	778-66-5	194.272		49(5)	289(6)	1.0250^{20}	1.5880^{20}	i H_2O; s EtOH, bz
4521	*trans*-1,3-Diphenyl-2-propen-1-one	Chalcone	$C_{15}H_{12}O$	614-47-1	208.255	pa ye lf, pr, nd (peth)	56(4)	346 dec	1.0712^{62}		i H_2O; sl EtOH; s eth, bz, chl, CS_2
4522	1-(3,3-Diphenylpropyl)-piperidine	Fenpiprane	$C_{20}H_{25}N$	3540-95-2	279.420		41.5	215^{8}			
4523	3,5-Diphenyl-1*H*-pyrazole		$C_{15}H_{12}N_2$	1145-01-3	220.269	cry (al)	200				
4524	1,4-Diphenyl-3,5-pyrazolidinedione	Phenopyrazone	$C_{15}H_{12}N_2O_2$	3426-01-5	252.268	cry (EtOAc, Diox)	233.5				
4525	Diphenyl selenide		$C_{12}H_{10}Se$	1132-39-4	233.17	ye nd (bz)	1.3	301.5	1.351^{20}	1.5500^{20}	i H_2O; msc EtOH, eth; s bz, xyl
4526	Diphenylsilane		$C_{12}H_{12}Si$	775-12-2	184.309			134^{16}	0.9969^{20}	1.5800^{20}	s ctc, CS_2
4527	Diphenylsilanediol		$C_{12}H_{12}O_2Si$	947-42-2	216.308						sl DMSO
4528	Diphenyl succinate		$C_{16}H_{14}O_4$	621-14-7	270.280	lf (al)	121	330			i H_2O; s EtOH, eth, ace, bz
4529	Diphenyl sulfide	Phenyl sulfide	$C_{12}H_{10}S$	139-66-2	186.272	liq	-15.35(0.05)	294.2(0.5)	1.1136^{20}	1.6334^{20}	i H_2O; s EtOH, ctc; msc eth, bz, CS_2
4530	Diphenyl sulfone		$C_{12}H_{10}O_2S$	127-63-9	218.271	mcl pr(bz), pl(al)	127(3)	379	1.252^{20}		i H_2O; s EtOH, eth, bz
4531	Diphenyl sulfoxide		$C_{12}H_{10}OS$	945-51-7	202.271	pr(lig)	70.2(0.2)	340^{16}			vs EtOH, eth, bz, HOAc; sl chl; i peth
4532	*N,N'*-Diphenylthiourea	*sym*-Diphenylthiourea	$C_{13}H_{12}N_2S$	102-08-9	228.312		153(1)		1.32^{25}		sl H_2O; vs EtOH, eth, chl, oils
4533	1,3-Diphenyl-1-triazene	Diazoaminobenzene	$C_{12}H_{11}N_3$	136-35-6	197.235	ye lf or pr (al)	98				i H_2O; vs EtOH, eth, bz, py
4534	*N,N*-Diphenylurea		$C_{13}H_{12}N_2O$	603-54-3	212.246	tab (al)	189	dec	1.276^{25}		sl H_2O; s EtOH, eth, chl
4535	*N,N'*-Diphenylurea	Carbanilide	$C_{13}H_{12}N_2O$	102-07-8	212.246	orth pr (al)	236(3)	260 dec	1.239^{25}		sl H_2O, EtOH; s eth, py, HOAc; i bz

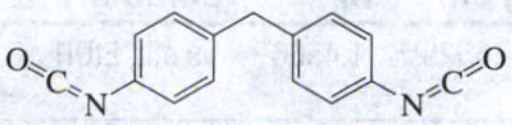
4499
4,4′-Diphenylmethane diisocyanate

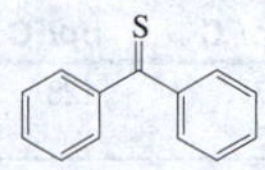
4500
Diphenylmethanethione

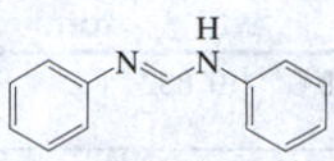
4501
N,N′-Diphenylmethanimidamide

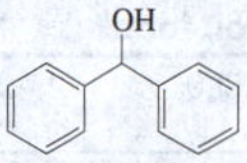
4502
Diphenylmethanol

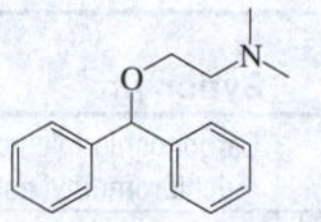
4503
2-(Diphenylmethoxy)-*N,N*-dimethylethanamine

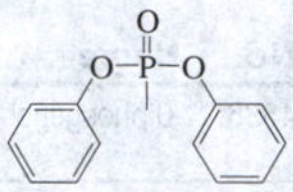
4504
Diphenyl methylphosphonate

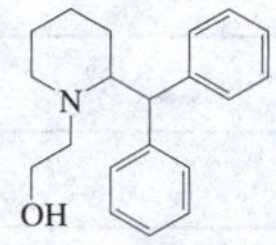
4505
2-(Diphenylmethyl)-1-piperidineethanol

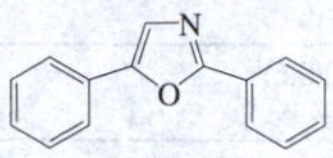
4506
2,5-Diphenyloxazole

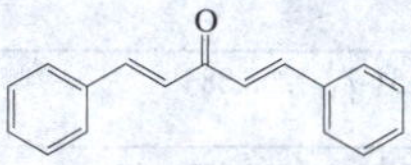
4507
1,5-Diphenyl-1,4-pentadien-3-one

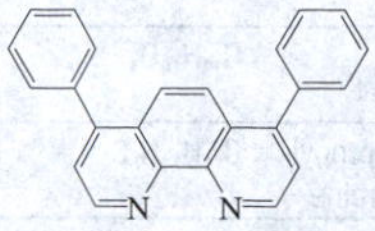
4508
4,7-Diphenyl-1,10-phenanthroline

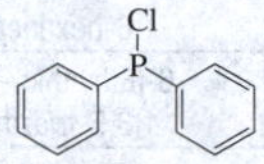
4509
Diphenylphosphinous chloride

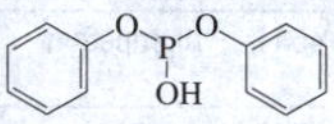
4510
Diphenyl phosphonate

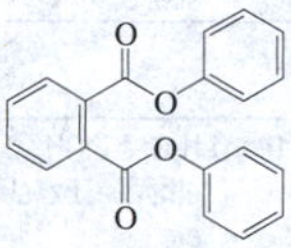
4511
Diphenyl phthalate

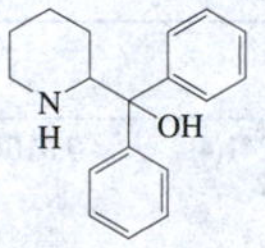
4512
α,α-Diphenyl-2-piperidinemethanol

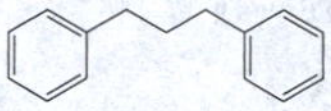
4513
1,3-Diphenylpropane

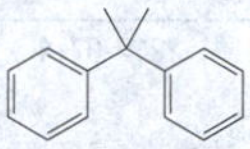
4514
2,2-Diphenylpropane

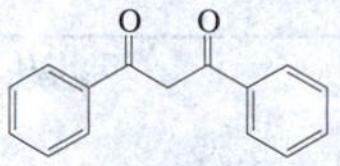
4515
1,3-Diphenyl-1,3-propanedione

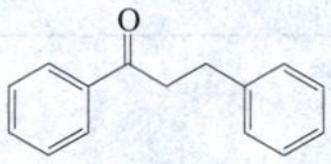
4516
1,3-Diphenyl-1-propanone

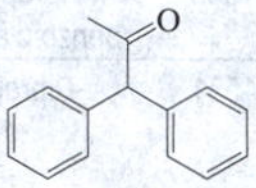
4517
1,1-Diphenyl-2-propanone

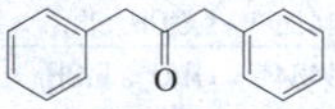
4518
1,3-Diphenyl-2-propanone

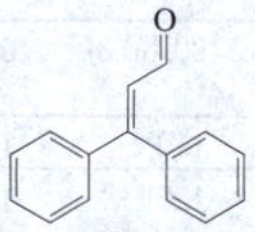
4519
3,3-Diphenyl-2-propenal

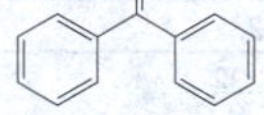
4520
1,1-Diphenyl-1-propene

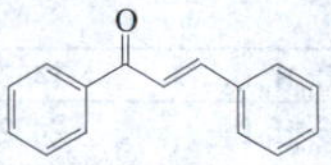
4521
trans-1,3-Diphenyl-2-propen-1-one

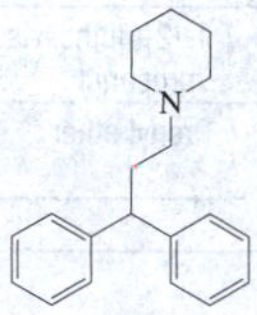
4522
1-(3,3-Diphenylpropyl)piperidine

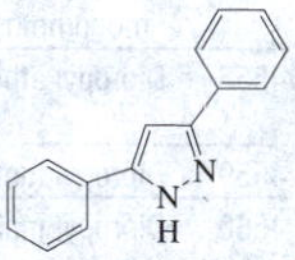
4523
3,5-Diphenyl-1*H*-pyrazole

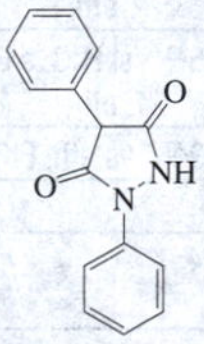
4524
1,4-Diphenyl-3,5-pyrazolidinedione

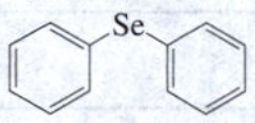
4525
Diphenyl selenide

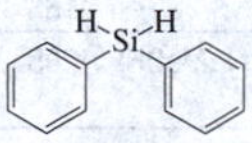
4526
Diphenylsilane

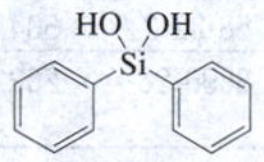
4527
Diphenylsilanediol

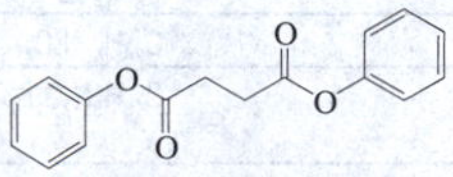
4528
Diphenyl succinate

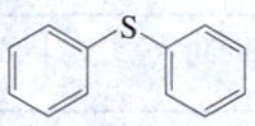
4529
Diphenyl sulfide

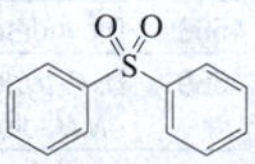
4530
Diphenyl sulfone

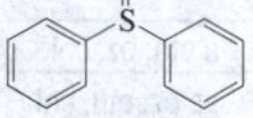
4531
Diphenyl sulfoxide

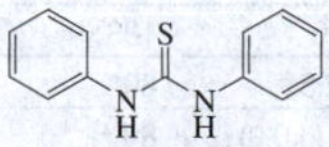
4532
N,N′-Diphenylthiourea

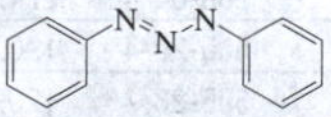
4533
1,3-Diphenyl-1-triazene

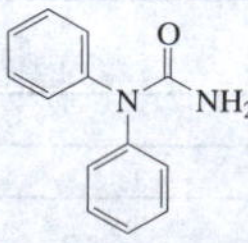
4534
N,N-Diphenylurea

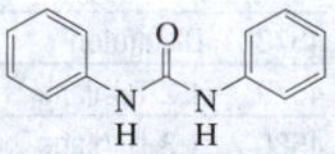
4535
N,N′-Diphenylurea

No.	Name	Synonym	Mol. form.	CAS Reg. No.	Mol. wt.	Physical form	mp/°C	bp/°C	den/ g cm^{-3}	n_D	Solubility
4536	Diphosgene	Carbonochloridic acid, trichloromethyl ester	$C_2Cl_4O_2$	503-38-8	197.832	liq	-57	128	1.6525^{14}	1.4566^{22}	vs eth, EtOH
4537	1,2-Dipiperidinoethane		$C_{12}H_{24}N_2$	1932-04-3	196.332	liq	-0.5	265	0.9160^{25}	1.4853^{25}	
4538	1,1'-Dipiperidinomethane	1,1'-Methylenedipiperidine	$C_{11}H_{22}N_2$	880-09-1	182.306			230	0.9269^{20}	1.4820^{20}	
4539	1,3-Di-4-piperidylpropane	4,4'-Trimethylenedipiperidine	$C_{13}H_{26}N_2$	16898-52-5	210.358		67.1	329			vs H_2O
4540	Diploicin		$C_{16}H_{10}Cl_4O_5$	527-93-5	424.059		232				
4541	Di-2-propenoyldiethyleneglycol		$C_{10}H_{14}O_5$	4074-88-8	214.215			200	1.1110^{25}	1.4595^{25}	
4542	Di-2-propenoyl-2,2-dimethyl-1,3-propanediol	2-Propenoic acid, 2,2-dimethyl-1,3-propanediyl ester	$C_{11}H_{16}O_4$	2223-82-7	212.243					1.4542^{25}	
4543	Di-2-propenoyl-1,6-hexanediol	2-Propenoic acid, 1,6-hexanediyl ester	$C_{12}H_{18}O_4$	13048-33-4	226.269				1.010^{25}		
4544	Dipropetryn	6-(Ethylthio)-*N,N'*-diisopropyl-1,3,5-triazine-2,4-diamine	$C_{11}H_{21}N_5S$	4147-51-7	255.384		105.0(0.5)				
4545	1,2-Dipropoxyethane		$C_8H_{18}O_2$	18854-56-3	146.228	liq		161(13)	0.8312^{25}	1.4013^{25}	
4546	Dipropoxymethane	Formaldehyde, dipropyl acetal	$C_7H_{16}O_2$	505-84-0	132.201	liq	-97.3	140.5	0.8345^{20}	1.3939^{19}	vs ace, bz, eth, EtOH
4547	*N,N*-Dipropylacetamide		$C_8H_{17}NO$	1116-24-1	143.227			227(13)	0.8992^{17}	1.4419^{17}	vs EtOH
4548	Dipropyl adipate	Dipropyl hexanedioate	$C_{12}H_{22}O_4$	106-19-4	230.301		-15.7	151^{11}	0.9790^{20}	1.4314^{20}	vs eth, EtOH, chl
4549	Dipropylamine	*N*-Propyl-1-propanamine	$C_6H_{15}N$	142-84-7	101.190	liq	-63	107.5(0.9)	0.7400^{20}	1.4050^{20}	s H_2O, EtOH; msc eth; vs ace, bz
4550	4-[(Dipropylamino)sulfonyl]-benzoic acid	Probenecid	$C_{13}H_{19}NO_4S$	57-66-9	285.360		195				
4551	*N,N*-Dipropylaniline		$C_{12}H_{19}N$	2217-07-4	177.286	ye lf		241(4)	0.9104^{20}	1.5271^{20}	i H_2O; s EtOH, eth, ace, bz; sl ctc
4552	Dipropylcarbamothioic acid, *S*-ethyl ester	EPTC	$C_9H_{19}NOS$	759-94-4	189.318			127^{20}	0.9546^{30}		
4553	Dipropyl carbonate		$C_7H_{14}O_3$	623-96-1	146.184			166(4)	0.9435^{20}	1.4008^{20}	sl H_2O; msc EtOH, eth
4554	Dipropyl disulfide		$C_6H_{14}S_2$	629-19-6	150.305	liq	-85.4(0.1)	196(1)	0.9599^{20}	1.4981^{20}	
4555	Dipropylene glycol		$C_6H_{14}O_3$	25265-71-8	134.173			231(2)	1.0206^{20}		msc H_2O; s EtOH
4556	Dipropylene glycol dibenzoate		$C_{20}H_{22}O_5$	27138-31-4	342.386			197^{1}			
4557	Dipropylene glycol monomethyl ether	1-(2-Methoxyisopropoxy)-2-propanol	$C_7H_{16}O_3$	34590-94-8	148.200	liq	-80	203(14)	0.95	1.4190^{20}	
4558	Dipropyl ether	Propyl ether	$C_6H_{14}O$	111-43-3	102.174	liq	-114.8(0.4)	90.1(0.3)	0.7466^{20}	1.3809^{20}	sl H_2O; vs eth, EtOH
4559	Dipropyl fumarate		$C_{10}H_{16}O_4$	14595-35-8	200.232			110^{5}	1.0129^{20}	1.4435^{20}	s EtOH, eth
4560	Dipropyl maleate		$C_{10}H_{16}O_4$	2432-63-5	200.232			126^{12}	1.0245^{20}	1.4434^{20}	i H_2O; s EtOH, eth, ace, bz
4561	Dipropyl oxalate		$C_8H_{14}O_4$	615-98-5	174.195	liq	-44.3	212(3)	1.0188^{20}	1.4158^{20}	sl H_2O; msc EtOH; s eth
4562	5,5-Dipropyl-2,4-oxazolidinedione		$C_9H_{15}NO_3$	512-12-9	185.220		42.5	149^{3}			
4563	Dipropyl succinate		$C_{10}H_{18}O_4$	925-15-5	202.248	liq	-5.9	248(7)	1.0020^{20}	1.4250^{20}	vs ace, bz, eth
4564	Dipropyl sulfate		$C_6H_{14}O_4S$	598-05-0	182.238			120^{20}	1.1064^{20}	1.4135^{20}	vs peth
4565	Dipropyl sulfide		$C_6H_{14}S$	111-47-7	118.240	liq	-102.68(0.04)	142.8(0.8)	0.814^{17}	1.4487^{20}	i H_2O; s EtOH, eth
4566	Dipropyl sulfone		$C_6H_{14}O_2S$	598-03-8	150.239	cry	28(2)		1.0278^{50}	1.4456^{30}	sl H_2O; s EtOH, eth
4567	Dipropyl sulfoxide		$C_6H_{14}OS$	4253-91-2	134.239	nd	25(1)	80^{2}	0.9654^{20}	1.4663^{20}	vs eth, EtOH
4568	Dipyridamole		$C_{24}H_{40}N_8O_4$	58-32-2	504.627		163				
4569	Di-2-pyridinyl disulfide, *N,N'*-dioxide	Dipyrithione	$C_{10}H_8N_2O_2S_2$	3696-28-4	252.313	cry (MeOH)	205				
4570	2,2'-Dipyrrolylmethane		$C_9H_{10}N_2$	21211-65-4	146.188	lf or nd (al)	73	164^{12}			vs bz, eth, EtOH
4571	Diquat		$C_{12}H_{12}N_2$	2764-72-9	184.236	Cation					
4572	Diquat dibromide		$C_{12}H_{12}Br_2N_2$	85-00-7	344.044		337		1.24^{20}		
4573	Disodium calcium EDTA	Edetate calcium disodium	$C_{10}H_{12}CaN_2Na_2O_8$	62-33-9	374.268	pow					s H_2O
4574	Disodium hydrogen citrate	Sodium acid citrate	$C_6H_6Na_2O_7$	144-33-2	236.088	wh pow (w)	149 dec				vs H_2O
4575	Disperse Blue No. 1	1,4,5,8-Tetraamino-9,10-anthracenedione	$C_{14}H_{12}N_4O_2$	2475-45-8	268.271	red-br nd	331				
4576	Distearyl thiodipropionate	Dioctadecyl thiobispropanoate	$C_{42}H_{82}O_4S$	693-36-7	683.163	cry	61				
4577	Disulfiram		$C_{10}H_{20}N_2S_4$	97-77-8	296.539		71.5	117^{17}			i H_2O; s EtOH; sl eth; vs chl
4578	Disulfoton		$C_8H_{19}O_2PS_3$	298-04-4	274.405		-25	$108^{0.01}$	1.144^{20}		
4579	1,2-Dithiane		$C_4H_8S_2$	505-20-4	120.237	nd	32.5	80^{14}		1.5981^{25}	s eth, bz, chl
4580	1,3-Dithiane		$C_4H_8S_2$	505-23-7	120.237		54.0(0.3)	89^{14}		1.5981^{25}	vs bz, eth, chl

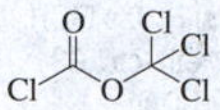
4536
Diphosgene

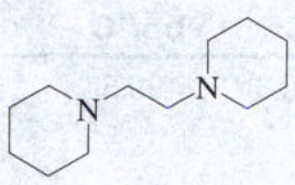
4537
1,2-Dipiperidinoethane

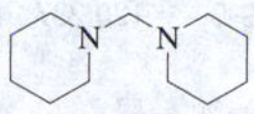
4538
1,1'-Dipiperidinomethane

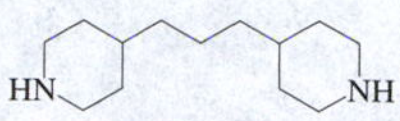
4539
1,3-Di-4-piperidylpropane

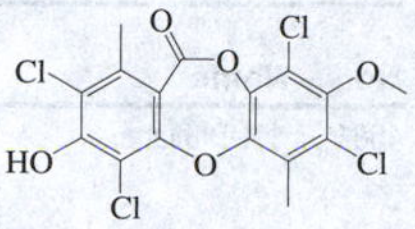
4540
Diploicin

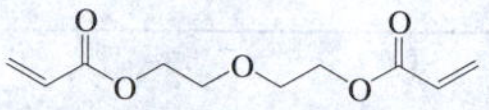
4541
Di-2-propenoyldiethyleneglycol

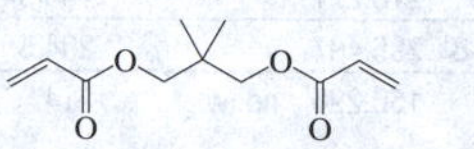
4542
Di-2-propenoyl-2,2-dimethyl-1,3-propanediol

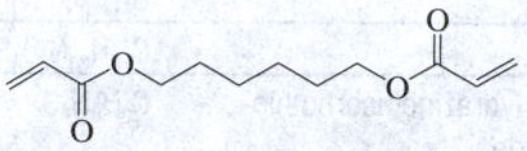
4543
Di-2-propenoyl-1,6-hexanediol

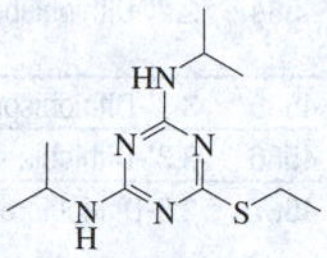
4544
Dipropetryn

4545
1,2-Dipropoxyethane

4546
Dipropoxymethane

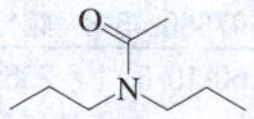
4547
N,N-Dipropylacetamide

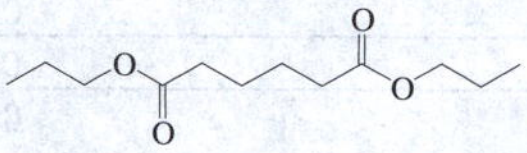
4548
Dipropyl adipate

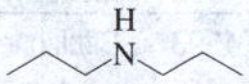
4549
Dipropylamine

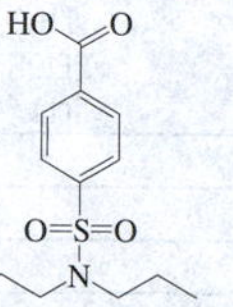
4550
4-[(Dipropylamino)sulfonyl]benzoic acid

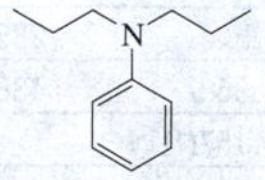
4551
N,N-Dipropylaniline

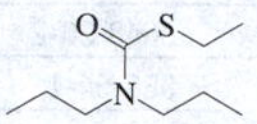
4552
Dipropylcarbamothioic acid, *S*-ethyl ester

4553
Dipropyl carbonate

4554
Dipropyl disulfide

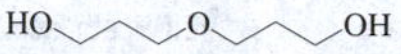
4555
Dipropylene glycol

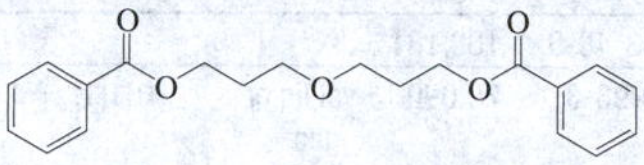
4556
Dipropylene glycol dibenzoate

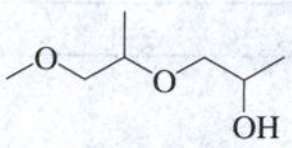
4557
Dipropylene glycol monomethyl ether

4558
Dipropyl ether

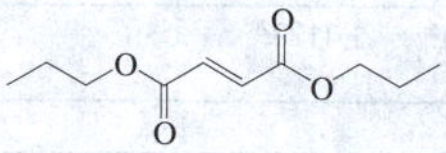
4559
Dipropyl fumarate

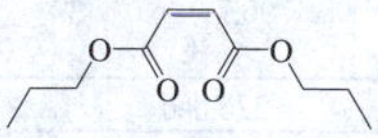
4560
Dipropyl maleate

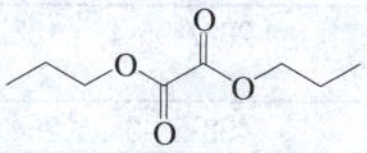
4561
Dipropyl oxalate

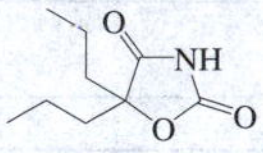
4562
5,5-Dipropyl-2,4-oxazolidinedione

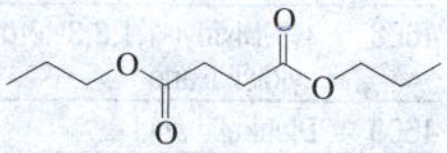
4563
Dipropyl succinate

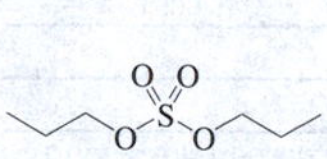
4564
Dipropyl sulfate

4565
Dipropyl sulfide

4566
Dipropyl sulfone

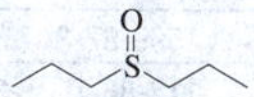
4567
Dipropyl sulfoxide

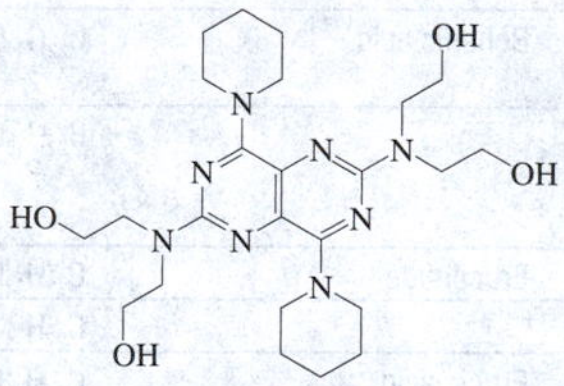
4568
Dipyridamole

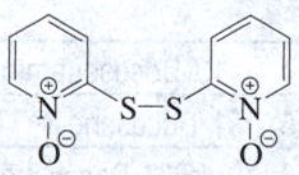
4569
Di-2-pyridinyl disulfide, *N,N'*-dioxide

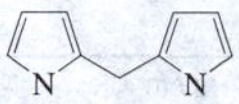
4570
2,2'-Dipyrrolylmethane

4571
Diquat

4572
Diquat dibromide

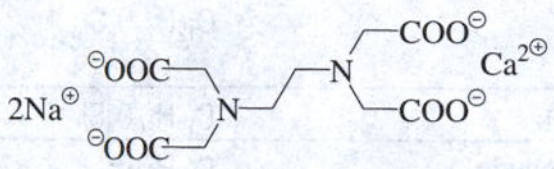
4573
Disodium calcium EDTA

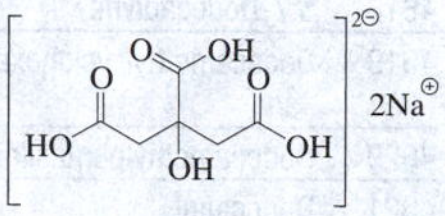
4574
Disodium hydrogen citrate

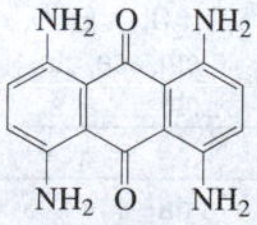
4575
Disperse Blue No. 1

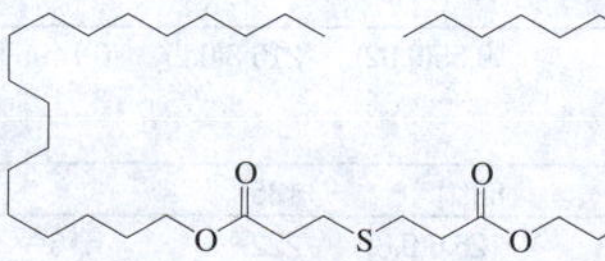
4576
Distearyl thiodipropionate

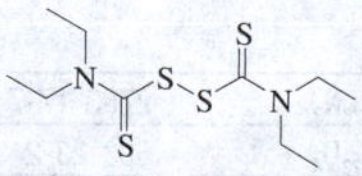
4577
Disulfiram

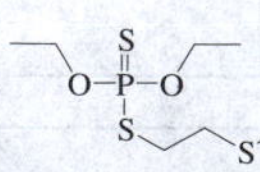
4578
Disulfoton

4579
1,2-Dithiane

4580
1,3-Dithiane

Organic

No.	Name	Synonym	Mol. form.	CAS Reg. No.	Mol. wt.	Physical form	mp/°C	bp/°C	den/g cm^{-3}	n_D	Solubility
4581	1,4-Dithiane		$C_4H_8S_2$	505-29-3	120.237	mcl pr	111.4(0.3)	199.5			sl H_2O; s EtOH, eth, ctc, CS_2, HOAc
4582	Dithianone		$C_{14}H_4N_2O_2S_2$	3347-22-6	296.324	nd (ace)	220				
4583	Dithiazanine iodide		$C_{23}H_{23}IN_2S_2$	514-73-8	518.476	grn nd (MeOH)	248 dec				i H_2O
4584	2,2'-Dithiobisbenzoic acid	Diphenyl disulfide-2,2'-dicarboxylic acid	$C_{14}H_{10}O_4S_2$	119-80-2	306.357		289.5				i H_2O; s EtOH, eth
4585	3,3'-Dithiobispropanoic acid		$C_6H_{10}O_4S_2$	1119-62-6	210.271		156(3)				
4586	3,3'-Dithiobis-D-valine		$C_{10}H_{20}N_2O_4S_2$	20902-45-8	296.407		204.5				
4587	2,5-Dithiobiurea	1,2-Hydrazinedicarbothioamide	$C_2H_6N_4S_2$	142-46-1	150.226	nd (w)	214				
4588	4,4'-Dithiodimorpholine		$C_8H_{16}N_2O_2S_2$	103-34-4	236.355		124.5				s chl
4589	1,2-Dithiolane		$C_3H_6S_2$	557-22-2	106.210		77	90^{27}			
4590	1,3-Dithiolane	1,3-Dithiacyclopentane	$C_3H_6S_2$	4829-04-3	106.210	liq	-50	175	1.259^{17}	1.5975^{15}	s EtOH, eth, xyl
4591	1,3-Dithiolane-2-thione		$C_3H_4S_3$	822-38-8	136.259		35	307			
4592	Dithiopyr		$C_{15}H_{16}F_5NO_2S_2$	97886-45-8	401.416		65				
4593	Dithizone		$C_{13}H_{12}N_4S$	60-10-6	256.326	bl-blk (chl-al)	167 dec				i H_2O; sl EtOH, eth; s chl, alk
4594	Di(*p*-tolyl)carbodiimide		$C_{15}H_{14}N_2$	726-42-1	222.285		58.5	221^{20}	1.1500^{20}		
4595	1,2-Di(*p*-tolyl)ethane	1,2-Bis(*p*-tolyl)ethane	$C_{16}H_{18}$	538-39-6	210.314	lf (al)	81.8(0.5)	178^{18}			i H_2O; sl EtOH; s bz, peth
4596	*N,N'*-Di(*o*-tolyl)guanidine		$C_{15}H_{17}N_3$	97-39-2	239.316	cry (dil al)	179		1.10^{20}		sl H_2O, tfa, EtOH; vs eth; s chl
4597	Ditridecyl phthalate		$C_{34}H_{58}O_4$	119-06-2	530.823	liq		$285^{3.5}$	0.952^{25}		
4598	Diundecyl phthalate		$C_{30}H_{50}O_4$	3648-20-2	474.716	cry (EtOH)	35.5	438(25)			
4599	Diuron		$C_9H_{10}Cl_2N_2O$	330-54-1	233.093		157(1)				
4600	*o*-Divinylbenzene	1,2-Divinylbenzene	$C_{10}H_{10}$	91-14-5	130.186			82^{14}	0.9325^{22}	1.5767^{20}	s ace, bz
4601	*m*-Divinylbenzene	1,3-Divinylbenzene	$C_{10}H_{10}$	108-57-6	130.186		-52.2(0.2)	121^{76}	0.9294^{20}	1.5760^{20}	s ace, bz
4602	*p*-Divinylbenzene	1,4-Divinylbenzene	$C_{10}H_{10}$	105-06-6	130.186		30(2)	95^{18}	0.913^{40}	1.5835^{25}	s ace, bz
4603	*cis*-1,2-Divinylcyclobutane		C_8H_{12}	16177-46-1	108.181			38^{38}	0.8010^{20}	1.4563^{20}	
4604	*trans*-1,2-Divinylcyclobutane		C_8H_{12}	6553-48-6	108.181			112.5	0.7817^{20}	1.4451^{20}	
4605	Divinyl ether		C_4H_6O	109-93-3	70.090	vol liq or gas	-101(1)	28(3)	0.773^{20}	1.3989^{20}	i H_2O; msc EtOH, eth, ace, chl
4606	Divinyl sulfide	Vinyl sulfide	C_4H_6S	627-51-0	86.156		20	84(4)	0.9174^{15}		sl H_2O; s ace; msc EtOH, eth
4607	Divinyl sulfone	Vinyl sulfone	$C_4H_6O_2S$	77-77-0	118.155	liq	-26	234.5	1.177^{25}	1.4765^{20}	
4608	1,3-Divinyl-1,1,3,3-tetramethyldisiloxane		$C_8H_{18}OSi_2$	2627-95-4	186.399	liq	-99.7	39	0.811^{20}	1.4123^{20}	
4609	Djenkolic acid		$C_7H_{14}N_2O_4S_2$	498-59-9	254.327	nd(w)	≈325 dec				
4610	DMPA		$C_{10}H_{14}Cl_2NO_2PS$	299-85-4	314.169	solid	51.8(0.3)	150^2			sl H_2O; vs bz, ctc, ace
4611	Docosane		$C_{22}H_{46}$	629-97-0	310.600	pl(to), cry (eth)	43.8(0.3)	369(5)	0.7944^{20}	1.4455^{20}	i H_2O; s EtOH, chl; vs eth
4612	Docosanoic acid	Behenic acid	$C_{22}H_{44}O_2$	112-85-6	340.583	nd	80.6(0.1)	306^{60}	0.8223^{90}	1.4270^{100}	sl H_2O, EtOH, eth
4613	1-Docosanol		$C_{22}H_{46}O$	661-19-8	326.599	cry (ace, chl)	71(1)	$180^{0.22}$			sl H_2O, eth; vs EtOH, MeOH; s chl
4614	13-Docosenamide	Erucamide	$C_{22}H_{43}NO$	112-84-5	337.582	cry	94				
4615	1-Docosene		$C_{22}H_{44}$	1599-67-3	308.584		38	367	0.794^{25}		
4616	*cis*-13-Docosenoic acid	Erucic acid	$C_{22}H_{42}O_2$	112-86-7	338.567	nd (al)	33.0(0.5)	265^{15}	0.860^{55}	1.4758^{20}	i H_2O; s EtOH, ctc; vs eth, MeOH
4617	*trans*-13-Docosenoic acid	Brassidic acid	$C_{22}H_{42}O_2$	506-33-2	338.567	pl (al)	61.9	282^{30}	0.8585^{57}	1.4347^{100}	
4618	5,7-Dodecadiyne	Dibutylbutadiyne	$C_{12}H_{18}$	1120-29-2	162.271			103^8			
4619	Dodecamethylcyclohexasiloxane		$C_{12}H_{36}O_6Si_6$	540-97-6	444.923	liq	-4.2(0.2)	245.0(0.2)	0.9672^{25}	1.4015^{20}	i H_2O
4620	Dodecamethylpentasiloxane		$C_{12}H_{36}O_4Si_5$	141-63-9	384.840	liq	-80	229.9(0.1)	0.8755^{20}	1.3925^{20}	s ctc, CS_2
4621	Dodecanal	Lauraldehyde	$C_{12}H_{24}O$	112-54-9	184.318	lf	44(4)	185^{100}	0.8352^{15}	1.435^{22}	i H_2O; sl EtOH; s eth
4622	Dodecanamide		$C_{12}H_{25}NO$	1120-16-7	199.333	nd	101(2)	199^{12}		1.4287^{110}	i H_2O; s EtOH, ace, ctc; sl eth, bz
4623	Dodecane		$C_{12}H_{26}$	112-40-3	170.334	liq	-9.55(0.02)	216.3(0.2)	0.7495^{20}	1.4210^{20}	i H_2O; vs EtOH, eth, ace, ctc, chl
4624	1,12-Dodecanediamine		$C_{12}H_{28}N_2$	2783-17-7	200.363		68(1)	135^3			
4625	Dodecanedioic acid		$C_{12}H_{22}O_4$	693-23-2	230.301		126.6(0.8)	222^{25}	1.15^{25}		s tfa

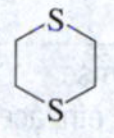
4581
1,4-Dithiane

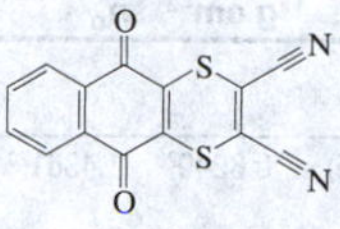
4582
Dithianone

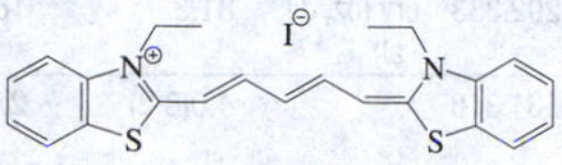
4583
Dithiazanine iodide

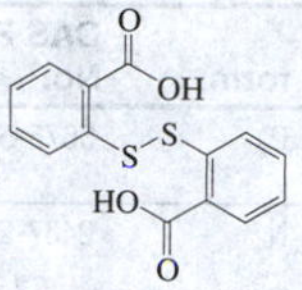
4584
2,2'-Dithiobisbenzoic acid

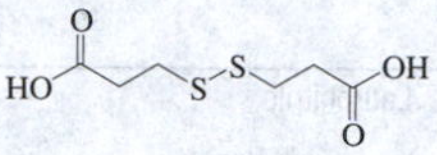
4585
3,3'-Dithiobispropanoic acid

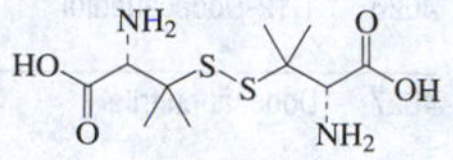
4586
3,3'-Dithiobis-D-valine

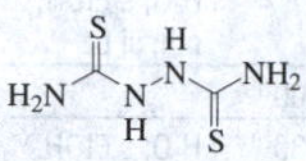
4587
2,5-Dithiobiurea

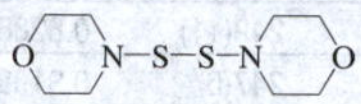
4588
4,4'-Dithiodimorpholine

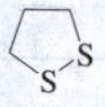
4589
1,2-Dithiolane

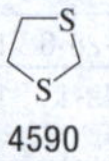
4590
1,3-Dithiolane

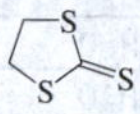
4591
1,3-Dithiolane-2-thione

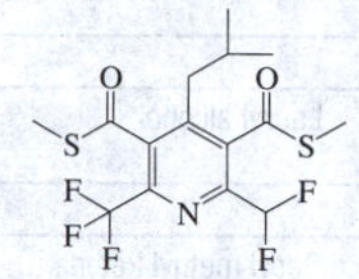
4592
Dithiopyr

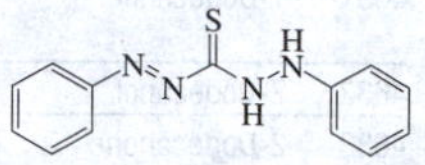
4593
Dithizone

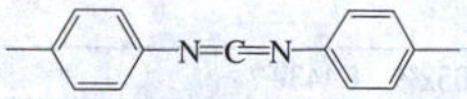
4594
Di(*p*-tolyl)carbodiimide

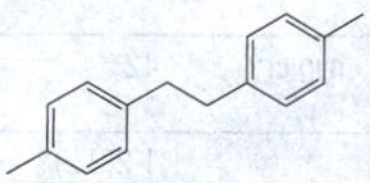
4595
1,2-Di(*p*-tolyl)ethane

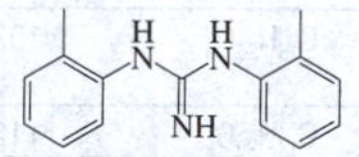
4596
N,N'-Di(*o*-tolyl)guanidine

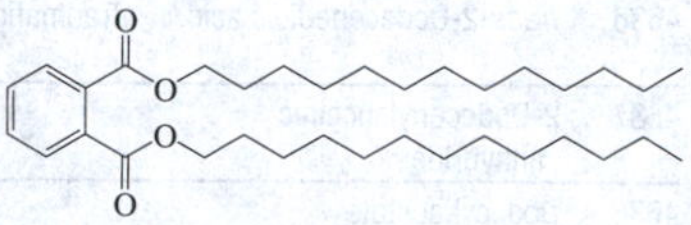
4597
Ditridecyl phthalate

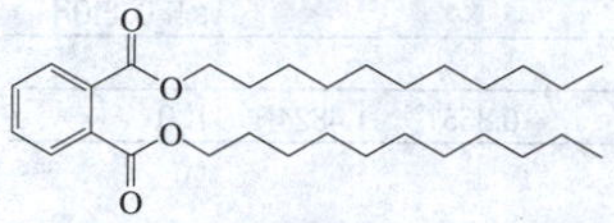
4598
Diundecyl phthalate

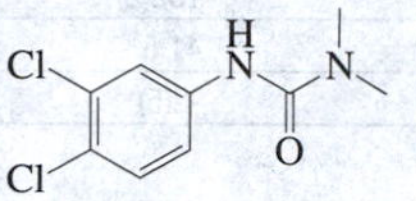
4599
Diuron

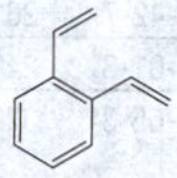
4600
o-Divinylbenzene

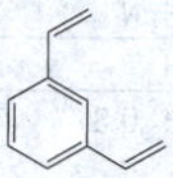
4601
m-Divinylbenzene

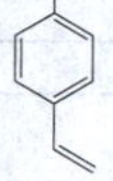
4602
p-Divinylbenzene

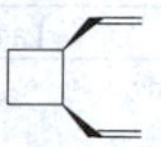
4603
cis-1,2-Divinylcyclobutane

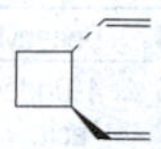
4604
trans-1,2-Divinylcyclobutane

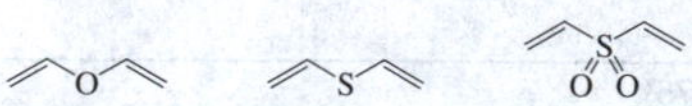
4605
Divinyl ether

4606
Divinyl sulfide

4607
Divinyl sulfone

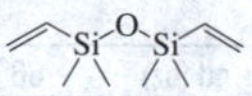
4608
1,3-Divinyl-1,1,3,3-tetramethyldisiloxane

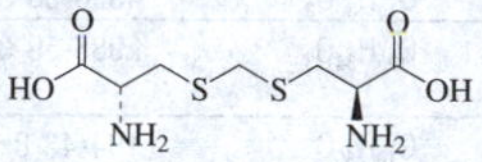
4609
Djenkolic acid

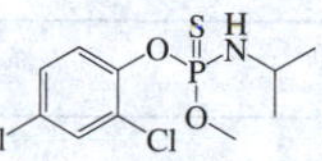
4610
DMPA

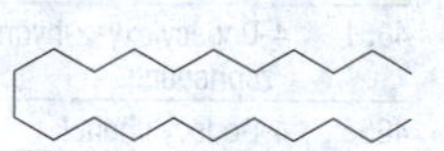
4611
Docosane

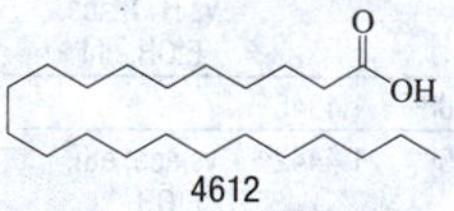
4612
Docosanoic acid

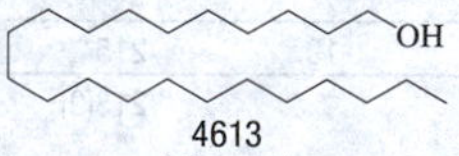
4613
1-Docosanol

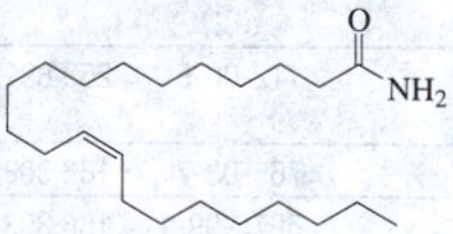
4614
13-Docosenamide

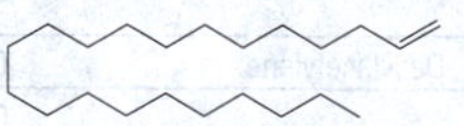
4615
1-Docosene

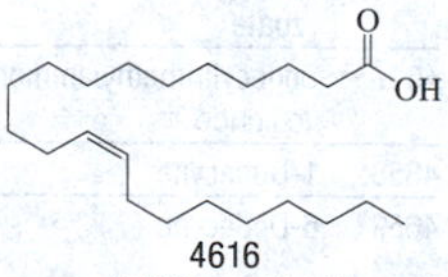
4616
cis-13-Docosenoic acid

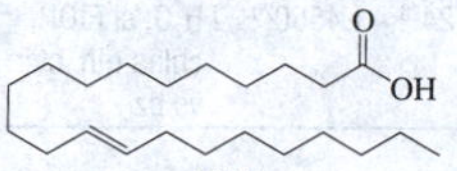
4617
trans-13-Docosenoic acid

4618
5,7-Dodecadiyne

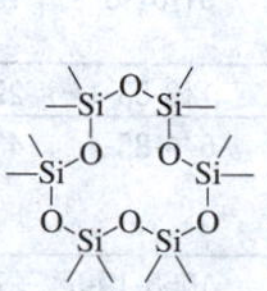
4619
Dodecamethylcyclohexasiloxane

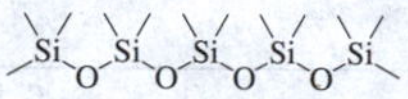
4620
Dodecamethylpentasiloxane

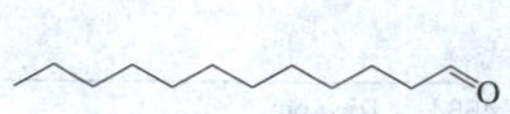
4621
Dodecanal

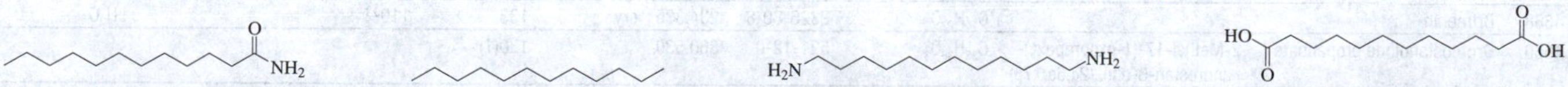
4622
Dodecanamide

4623
Dodecane

4624
1,12-Dodecanediamine

4625
Dodecanedioic acid

Organic

No.	Name	Synonym	Mol. form.	CAS Reg. No.	Mol. wt.	Physical form	mp/°C	bp/°C	den/ g cm^{-3}	n_D	Solubility
4626	1,12-Dodecanediol		$C_{12}H_{26}O_2$	5675-51-4	202.333	cry (bz, dil al)	81.3	189^{12}			s tfa
4627	Dodecanenitrile	Lauronitrile	$C_{12}H_{23}N$	2437-25-4	181.318		4.0(0.2)	276.1(0.5)	0.8240^{20}	1.4361^{20}	i H_2O; msc EtOH, eth, ace, bz, chl
4628	1-Dodecanethiol		$C_{12}H_{26}S$	112-55-0	202.399	liq	-6.7	277(3)	0.844^{20}	1.4589^{20}	i H_2O; s EtOH, eth, chl
4629	Dodecanoic acid	Lauric acid	$C_{12}H_{24}O_2$	143-07-7	200.318	nd (al)	43.82(0.02)	225^{100}	0.8679^{50}	1.4183^{82}	i H_2O; vs EtOH, eth; s ace; msc bz
4630	Dodecanoic anhydride		$C_{24}H_{46}O_3$	645-66-9	382.620	lf (al, eth)	41.8		0.8533^{70}	1.4292^{70}	vs EtOH
4631	1-Dodecanol	Lauryl alcohol	$C_{12}H_{26}O$	112-53-8	186.333	lf (dil al)	24.2(0.3)	264.1(0.3)	0.8309^{24}		i H_2O; s EtOH, eth; sl bz
4632	2-Dodecanol		$C_{12}H_{26}O$	10203-28-8	186.333		19	249(11)	0.8286^{20}	1.4400^{20}	
4633	2-Dodecanone	Decyl methyl ketone	$C_{12}H_{24}O$	6175-49-1	184.318		21	247(6)	0.8198^{20}	1.4330^{20}	i H_2O; s EtOH, eth, ace; sl ctc
4634	Dodecanoyl chloride		$C_{12}H_{23}ClO$	112-16-3	218.763		-17	145^{18}	0.9169^{25}	1.4458^{20}	vs eth
4635	1-Dodecene		$C_{12}H_{24}$	112-41-4	168.319	liq	-35.19(0.05)	213.4(0.9)	0.7584^{20}	1.4300^{20}	i H_2O; s EtOH, eth, ace, ctc, peth
4636	*trans*-2-Dodecenedioic acid	Traumatic acid	$C_{12}H_{20}O_4$	6402-36-4	228.285	cry (al,ace)	165.5				vs eth, EtOH, chl
4637	2-Dodecenylsuccinic anhydride		$C_{16}H_{26}O_3$	19780-11-1	266.375	hyg cry	42	181^{5}			
4638	Dodecyl acetate		$C_{14}H_{28}O_2$	112-66-3	228.371		1.2(0.2)	265	0.8652^{22}	1.4439^{20}	
4639	Dodecyl acrylate	Lauryl 2-propenoate	$C_{15}H_{28}O_2$	2156-97-0	240.382		4	$120^{0.8}$	0.8727^{20}		
4640	Dodecylamine	1-Dodecanamine	$C_{12}H_{27}N$	124-22-1	185.349		27.8(0.7)	255(2)	0.8015^{20}	1.4421^{20}	sl H_2O; msc EtOH, eth, bz, chl
4641	Dodecylamine, acetate	1-Dodecanamine, acetate	$C_{14}H_{31}NO_2$	2016-56-0	245.402		69.5				vs H_2O, EtOH
4642	Dodecylamine hydrochloride	Lauryl amine hydrochloride	$C_{12}H_{28}ClN$	929-73-7	221.810		182(2)				vs H_2O, EtOH
4643	4-Dodecylaniline		$C_{18}H_{31}N$	104-42-7	261.446		41.5	211^{10}			
4644	Dodecylbenzene	Laurylbenzene	$C_{18}H_{30}$	123-01-3	246.431		-2(5)	329(4)	0.8551^{20}	1.4824^{20}	i H_2O
4645	4-Dodecylbenzenesulfonic acid		$C_{18}H_{30}O_3S$	121-65-3	326.494			>205			
4646	Dodecylcyclohexane		$C_{18}H_{36}$	1795-17-1	252.479		12.6(0.4)	331	0.8223^{20}	1.4559^{20}	
4647	Dodecyl mercaptoacetate		$C_{14}H_{28}O_2S$	3746-39-2	260.436		1.5	171^{3}			
4648	Dodecyl methacrylate		$C_{16}H_{30}O_2$	142-90-5	254.408			142^{4}	0.866^{20}		
4649	Dodecyloxirane	1,2-Epoxytetradecane	$C_{14}H_{28}O$	3234-28-4	212.371	oil		$95^{0.4}$	0.845	1.4408^{20}	
4650	2-(Dodecyloxy)ethanol		$C_{14}H_{30}O_2$	4536-30-5	230.387			$143^{0.8}$			
4651	4-Dodecyloxy-2-hydroxybenzophenone		$C_{25}H_{34}O_3$	2985-59-3	382.536		43.5				
4652	4-Dodecylphenol		$C_{18}H_{30}O$	104-43-8	262.430	nd (bz)	66	175^{2}			
4653	1-Dodecylpiperidine		$C_{17}H_{35}N$	5917-47-5	253.467	pa ye		161^{5}	0.8378^{20}	1.4588^{20}	
4654	Dodecyl sulfate	Lauryl sulfate	$C_{12}H_{26}O_4S$	151-41-7	266.397	cry					s H_2O
4655	Dodecyltetraethylene glycol monoether	3,6,9,12-Tetraoxatetracosan-1-ol	$C_{20}H_{42}O_5$	5274-68-0	362.544			247^{10}			
4656	Dodecyl 3,4,5-trihydroxybenzoate		$C_{19}H_{30}O_5$	1166-52-5	338.438		96.5				s ace
4657	Dodecyltrimethylammonium chloride		$C_{15}H_{34}ClN$	112-00-5	263.891		246 dec				vs H_2O, ace, EtOH, chl
4658	1-Dodecyne	Decylacetylene	$C_{12}H_{22}$	765-03-7	166.303	liq	-19	215	0.7788^{20}	1.4340^{20}	
4659	6-Dodecyne		$C_{12}H_{22}$	6975-99-1	166.303			213(9)	0.785^{20}	1.4442^{20}	vs ace, eth, EtOH
4660	Dodine	Dodecylguanidine, monoacetate	$C_{15}H_{33}N_3O_2$	2439-10-3	287.442		137.5(0.5)				
4661	Dopamine	4(2-Aminoethyl)-1,2-benzenediol	$C_8H_{11}NO_2$	51-61-6	153.179	pr					
4662	Dothiepin		$C_{19}H_{21}NS$	113-53-1	295.442		56	$172^{0.05}$			
4663	Dotriacontane	Bicetyl	$C_{32}H_{66}$	544-85-4	450.866	pl (bz,chl, HOAc,eth)	69.7(0.3)	470(14)	0.8124^{20}	1.4550^{20}	i H_2O; sl EtOH, chl; s eth, ctc; vs bz
4664	Doxepin		$C_{19}H_{21}NO$	1668-19-5	279.376	oily liq		$155^{0.03}$			
4665	Doxorubicin	Adriamycin	$C_{27}H_{29}NO_{11}$	23214-92-8	543.519	cry	230				
4666	Doxorubicin hydrochloride	Adriamycin hydrochloride	$C_{27}H_{30}ClNO_{11}$	25316-40-9	579.980	oran-red nd	204 dec				s H_2O, MeOH; i ace, bz, chl, eth, peth
4667	Doxylamine		$C_{17}H_{22}N_2O$	469-21-6	270.369	liq		$139^{0.5}$			
4668	Drimenin		$C_{15}H_{22}O_2$	2326-89-8	234.335	cry	133	$110^{0.1}$			i H_2O
4669	Dromostanolone propanoate	2-Methyl-17-(1-oxopropoxy)-androstan-3-one, (2α,5α,17β)	$C_{23}H_{36}O_3$	521-12-0	360.530		130(1)				

4626
1,12-Dodecanediol

4627
Dodecanenitrile

4628
1-Dodecanethiol

4629
Dodecanoic acid

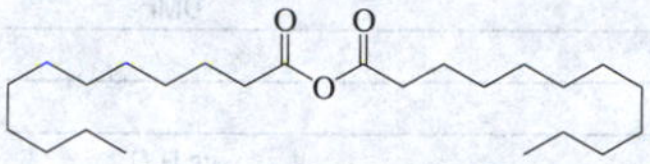

4630
Dodecanoic anhydride

4631
1-Dodecanol

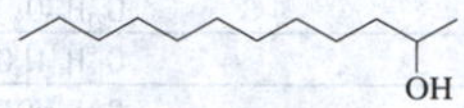

4632
2-Dodecanol

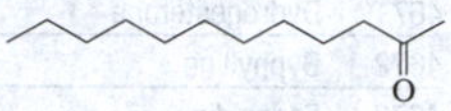

4633
2-Dodecanone

4634
Dodecanoyl chloride

4635
1-Dodecene

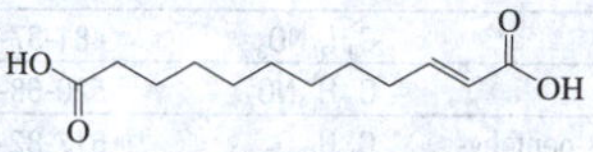

4636
trans-2-Dodecenedioic acid

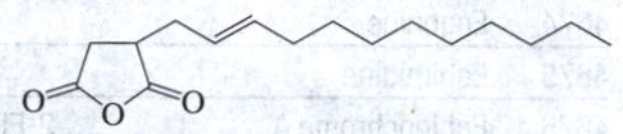

4637
2-Dodecenylsuccinic anhydride

4638
Dodecyl acetate

4639
Dodecyl acrylate

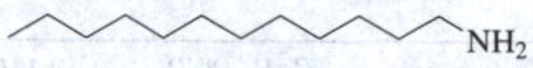

4640
Dodecylamine

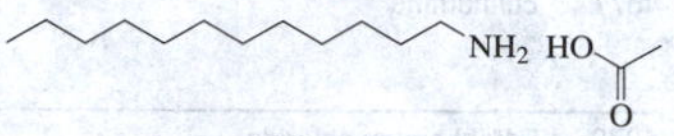

4641
Dodecylamine, acetate

4642
Dodecylamine hydrochloride

4643
4-Dodecylaniline

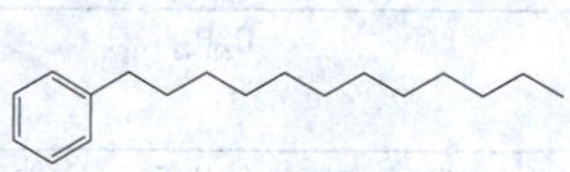

4644
Dodecylbenzene

4645
4-Dodecylbenzenesulfonic acid

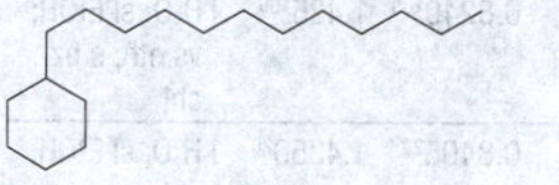

4646
Dodecylcyclohexane

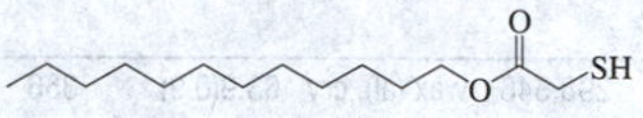

4647
Dodecyl mercaptoacetate

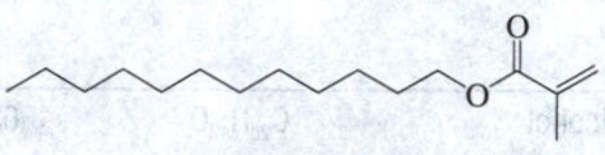

4648
Dodecyl methacrylate

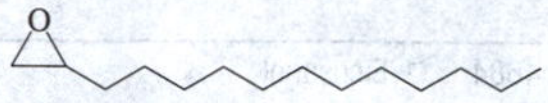

4649
Dodecyloxirane

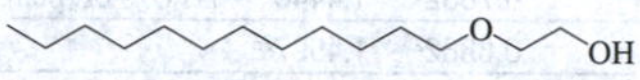

4650
2-(Dodecyloxy)ethanol

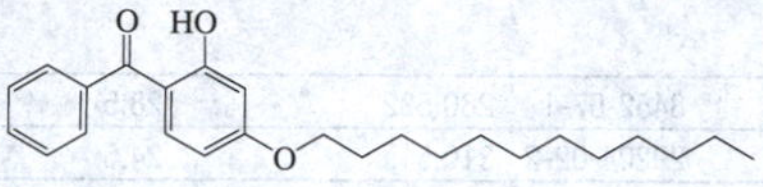

4651
4-Dodecyloxy-2-hydroxybenzophenone

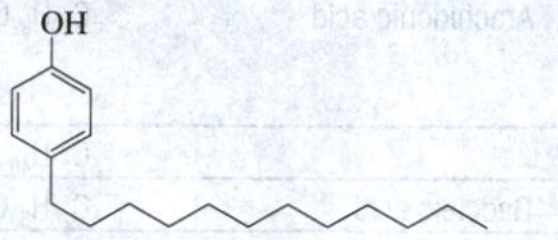

4652
4-Dodecylphenol

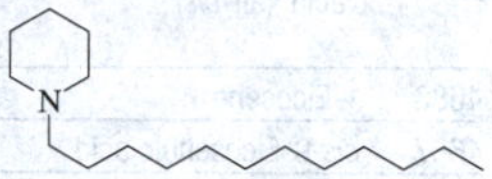

4653
1-Dodecylpiperidine

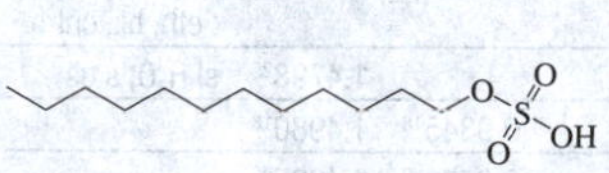

4654
Dodecyl sulfate

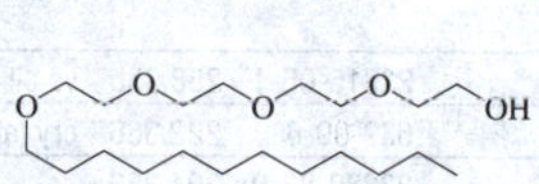

4655
Dodecyltetraethylene glycol monoether

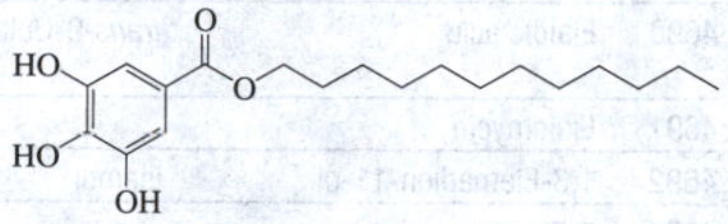

4656
Dodecyl 3,4,5-trihydroxybenzoate

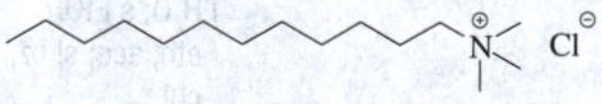

4657
Dodecyltrimethylammonium chloride

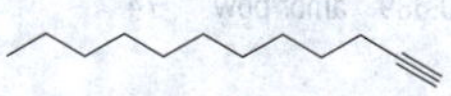

4658
1-Dodecyne

4659
6-Dodecyne

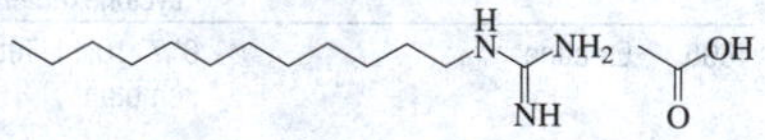

4660
Dodine

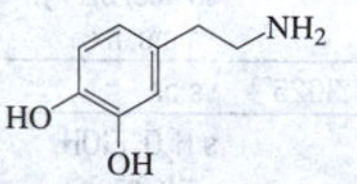

4661
Dopamine

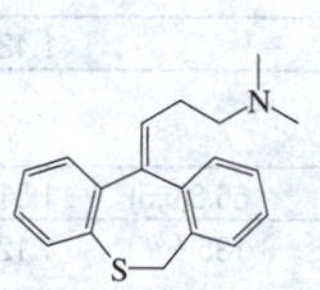

4662
Dothiepin

4663
Dotriacontane

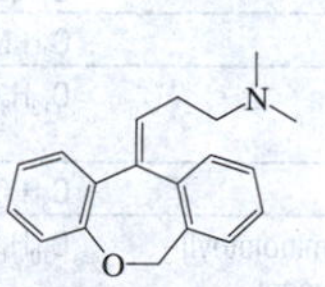

4664
Doxepin

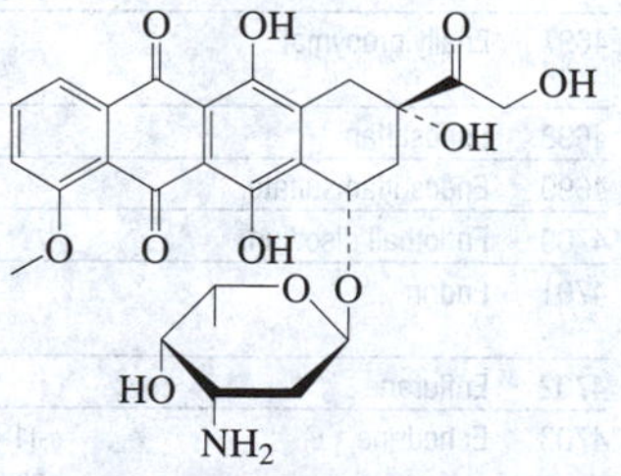

4665
Doxorubicin

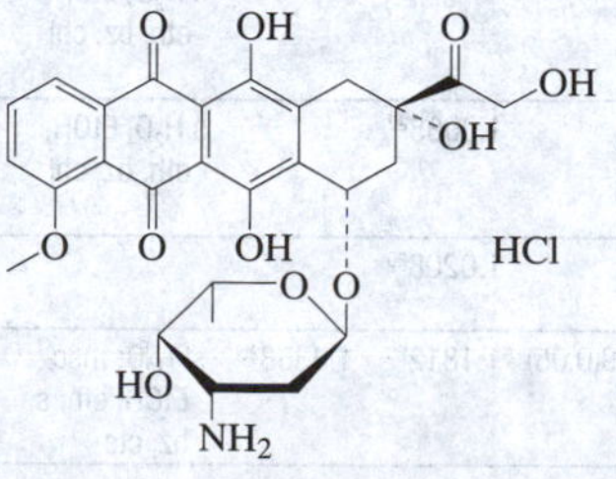

4666
Doxorubicin hydrochloride

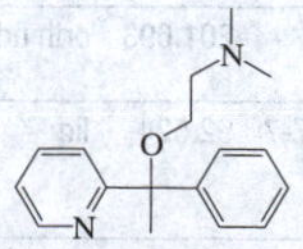

4667
Doxylamine

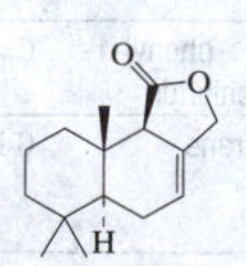

4668
Drimenin

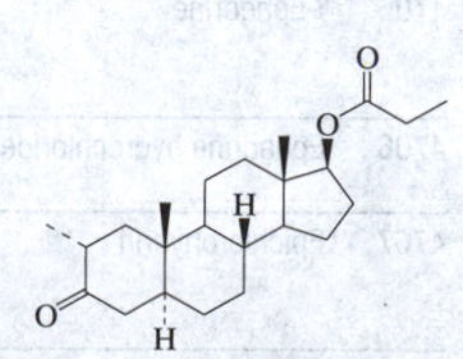

4669
Dromostanolone propanoate

No.	Name	Synonym	Mol. form.	CAS Reg. No.	Mol. wt.	Physical form	mp/°C	bp/°C	den/ g cm^{-3}	n_D	Solubility
4670	Droperidol	Dehydrobenzperidol	$C_{22}H_{22}FN_3O_2$	548-73-2	379.427	cry (w)	146 (hyd)				i H_2O; sl EtOH, eth, bz; s chl, DMF
4671	Dydrogesterone		$C_{21}H_{28}O_2$	152-62-5	312.446	cry (ace/hx)	172(1)				
4672	Dyphylline		$C_{10}H_{14}N_4O_4$	479-18-5	254.243		161.5				
4673	Ecgonidine		$C_9H_{13}NO_2$	484-93-5	167.205	cry (MeOH) (MeOH-eth)	228 dec				vs H_2O
4674	Ecgonine		$C_9H_{15}NO_3$	481-37-8	185.220	mcl pr	205				vs H_2O, EtOH
4675	Echimidine		$C_{20}H_{31}NO_7$	520-68-3	397.463	glass					
4676	Echinochrome A	2-Ethyl-3,5,6,7,8-pentahydroxy-1,4-naphthalenedione	$C_{12}H_{10}O_7$	517-82-8	266.203	red nd (Diox-w)	220 dec	120 sub			sl H_2O; s EtOH, ace; vs eth, bz
4677	Echitamine		$C_{22}H_{30}N_2O_5$	6871-44-9	402.483		206				s H_2O, EtOH, eth, chl, con sulf; i peth
4678	Edrophonium chloride		$C_{10}H_{16}ClNO$	116-38-1	201.693	cry	162				vs H_2O; s EtOH; i eth, chl
4679	Efloxate	Ethyl [(4-oxo-2-phenyl-4*H*-1-benzopyran-7-yl)oxy]acetate	$C_{19}H_{16}O_5$	119-41-5	324.327		123.7				s chl
4680	Eicosamethylnonasiloxane		$C_{20}H_{60}O_8Si_9$	2652-13-3	681.455			307.5	0.9173^{20}	1.3980^{20}	vs bz
4681	Eicosane	Icosane	$C_{20}H_{42}$	112-95-8	282.547	lf (al)	36.48(0.01)	344.1(0.9)	0.7886^{20}	1.4425^{20}	i H_2O; s eth, peth, bz; sl chl; vs ace
4682	Eicosanedioic acid	1,18-Octadecanedicarboxylic acid	$C_{20}H_{38}O_4$	2424-92-2	342.514	cry (bz,al)	125.5	233^2			s eth
4683	Eicosanoic acid	Arachidic acid	$C_{20}H_{40}O_2$	506-30-9	312.531	pl (al)	75.06(0.02)	400(6)	0.8240^{100}	1.425^{100}	i H_2O; sl EtOH; vs eth; s bz, chl
4684	1-Eicosanol	Arachic alcohol	$C_{20}H_{42}O$	629-96-9	298.546	wax (al), cry (chl)	63.9(0.9)	356	0.8405^{20}	1.4350^{20}	i H_2O; sl EtOH, chl; vs ace; s bz, peth
4685	5,8,11,14-Eicosatetraenoic acid, (all-*cis*)	Arachidonic acid	$C_{20}H_{32}O_2$	506-32-1	304.467		-38(2)	163^1	0.9082^{20}	1.4824^{20}	i H_2O; vs ace, eth, EtOH, peth
4686	1-Eicosene		$C_{20}H_{40}$	3452-07-1	280.532		28.5	341	0.7882^{30}	1.4440^{30}	i H_2O; s bz, peth
4687	*cis*-9-Eicosenoic acid	Gadoleic acid	$C_{20}H_{38}O_2$	29204-02-2	310.515		24.5	220^6	0.8882^{25}	1.4597^{25}	
4688	*trans*-9-Eicosenoic acid		$C_{20}H_{38}O_2$	506-31-0	310.515		54				
4689	*cis*-11-Eicosenoic acid	Gondoic acid	$C_{20}H_{38}O_2$	2462-94-4	310.515		24	267^{15}	0.8826^{25}		vs EtOH, MeOH
4690	Elaidic acid	*trans*-9-Octadecenoic acid	$C_{18}H_{34}O_2$	112-79-8	282.462	pl (al)	44.0(0.5)	288^{100}	0.8734^{45}	1.4499^{45}	i H_2O; s EtOH, eth, bz, chl
4691	Elaiomycin		$C_{13}H_{26}N_2O_3$	23315-05-1	258.356	ye oil				1.4798^{25}	sl H_2O; s os
4692	1,3-Elemadien-11-ol	Elemol	$C_{15}H_{26}O$	639-99-6	222.366	cry (al)	52.5	142^{12}	0.9345^{18}	1.4980^{18}	
4693	β-Elemene		$C_{15}H_{24}$	33880-83-0	204.352			120^{16}	0.8749^{20}	1.4935^{20}	
4694	Embelin	2,5-Dihydroxy-3-undecyl-2,5-cyclohexadiene-1,4-dione	$C_{17}H_{26}O_4$	550-24-3	294.386	oran pl (al)	142.5				vs bz, eth, EtOH
4695	Emetine	6',7',10,11-Tetramethoxyemetan	$C_{29}H_{40}N_2O_4$	483-18-1	480.639	amor pow	74				i H_2O; s EtOH, eth, ace; sl bz, chl
4696	Emylcamate	3-Methyl-3-pentanol, carbamate	$C_7H_{15}NO_2$	78-28-4	145.200	nd	57	35^1			sl H_2O; vs bz, eth, EtOH
4697	Enallylpropymal		$C_{11}H_{16}N_2O_3$	1861-21-8	224.256	cry (w, dil al)	56.5	177^{12}			vs bz, eth, EtOH, chl
4698	Endosulfan		$C_9H_6Cl_6O_3S$	115-29-7	406.925		106	$106^{0.7}$	1.745^{20}		
4699	Endosulfan sulfate		$C_9H_6Cl_6O_4S$	1031-07-8	422.925	cry (cyhex)	181				
4700	Endothall disodium		$C_8H_8Na_2O_5$	129-67-9	230.13		144		1.431^{20}		
4701	Endrin		$C_{12}H_8Cl_6O$	72-20-8	380.909	cry	dec 245				vs ace, bz, xyl; s ctc, hx
4702	Enflurane		$C_3H_2ClF_5O$	13838-16-9	184.492	liq		56.8(0.5)	1.5121^{25}	1.3025^{20}	vs os
4703	Ephedrine, (±)-	α-[1-(Methylamino)ethyl]-benzenemethanol, (*R**,*S**)-(±)-	$C_{10}H_{15}NO$	90-81-3	165.232	nd (eth, peth)	76.5	135^{12}	1.1220^{20}		s H_2O, EtOH, eth, bz, chl
4704	*d*-Ephedrine	α-[1-(Methylamino)ethyl]-benzenemethanol, [*S*-(*R**,*S**)]-	$C_{10}H_{15}NO$	321-98-2	165.232	pl (w)	40	225			s H_2O, EtOH, eth, bz, chl
4705	*l*-Ephedrine	α-[1-(Methylamino)ethyl]-benzenemethanol, [*R*-(*R**,*S**)]-	$C_{10}H_{15}NO$	299-42-3	165.232	pl (w + 1)	40	225	1.0085^{22}		s H_2O, EtOH, eth, bz, chl
4706	Ephedrine hydrochloride	2-(Methylamino)-1-phenyl-1-propanol, hydrochloride	$C_{10}H_{16}ClNO$	50-98-6	201.693	orth nd	219		1.0208^{20}		
4707	Epichlorohydrin	(Chloromethyl)oxirane	C_3H_5ClO	13403-37-7	92.524	liq	-26	111.99(0.05)	1.1812^{20}	1.4358^{25}	sl H_2O; msc EtOH, eth; s bz, ctc

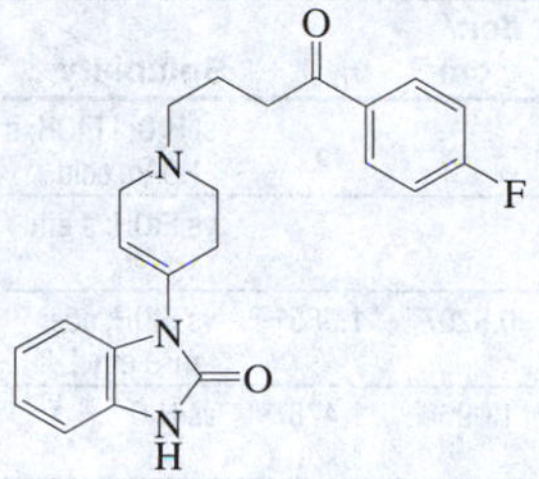
4670
Droperidol

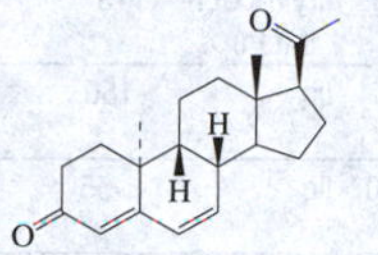
4671
Dydrogesterone

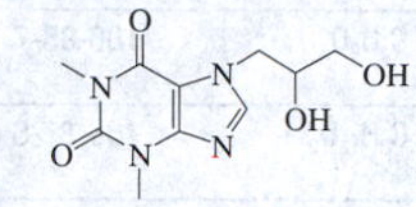
4672
Dyphylline

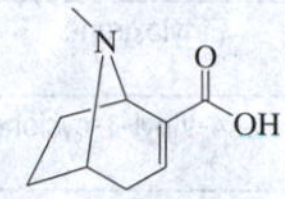
4673
Ecgonidine

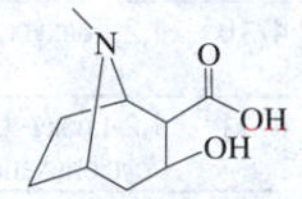
4674
Ecgonine

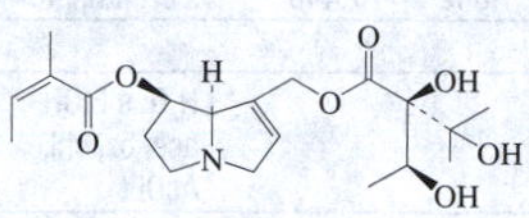
4675
Echimidine

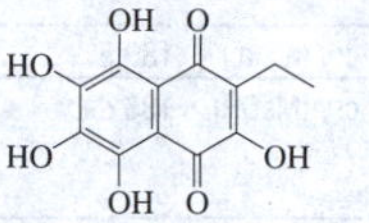
4676
Echinochrome A

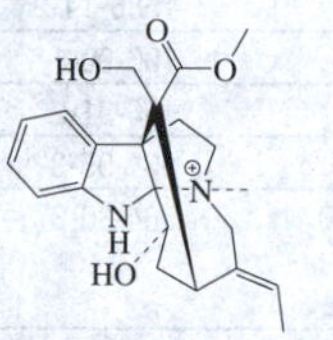
4677
Echitamine

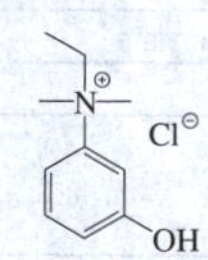
4678
Edrophonium chloride

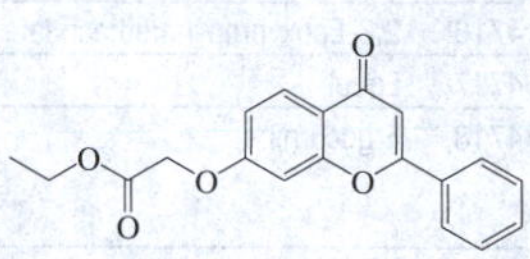
4679
Efloxate

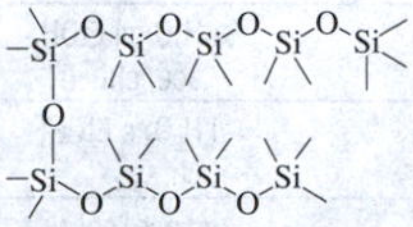
4680
Eicosamethylnonasiloxane

4681
Eicosane

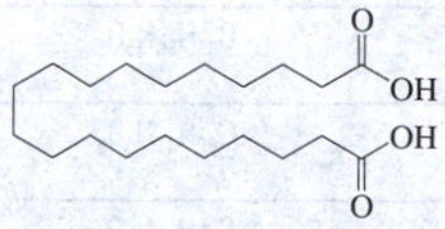
4682
Eicosanedioic acid

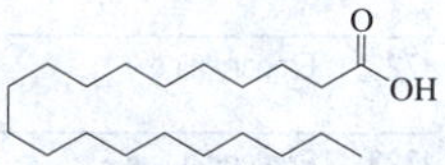
4683
Eicosanoic acid

4684
1-Eicosanol

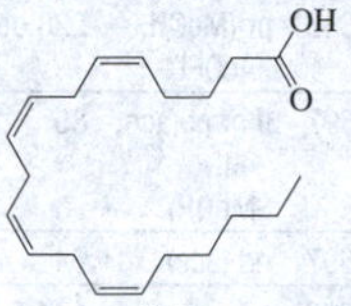
4685
5,8,11,14-Eicosatetraenoic acid, (all-*cis*)

4686
1-Eicosene

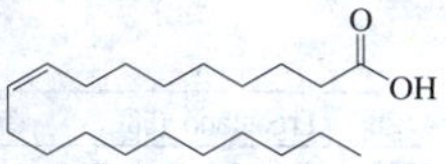
4687
cis-9-Eicosenoic acid

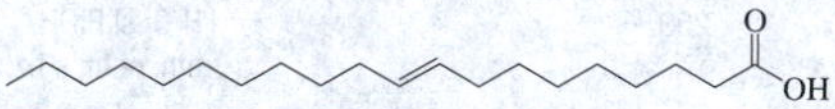
4688
trans-9-Eicosenoic acid

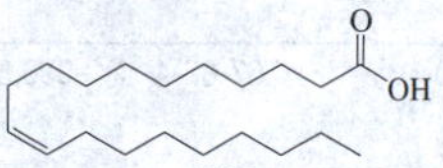
4689
cis-11-Eicosenoic acid

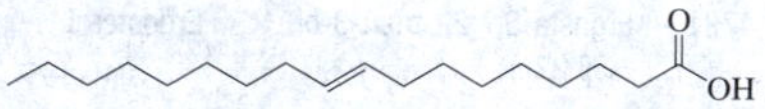
4690
Elaidic acid

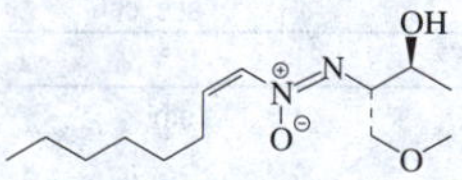
4691
Elaiomycin

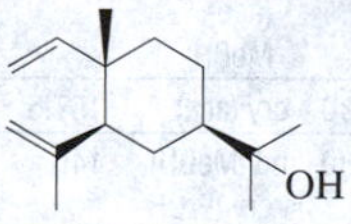
4692
1,3-Elemadien-11-ol

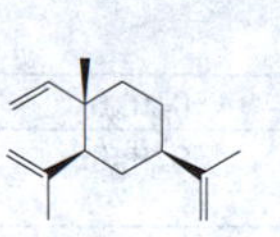
4693
β-Elemene

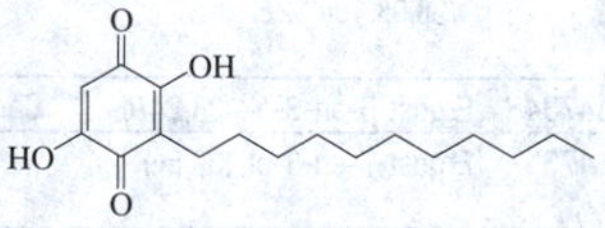
4694
Embelin

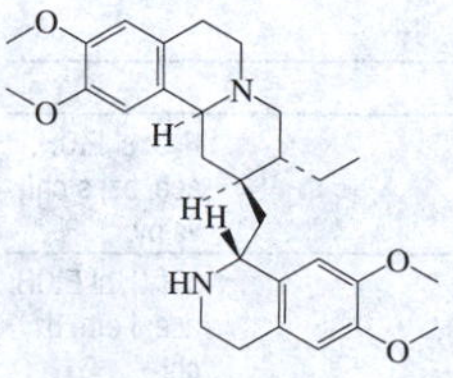
4695
Emetine

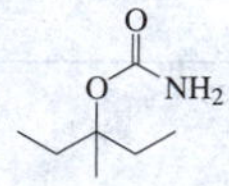
4696
Emylcamate

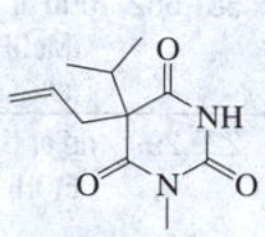
4697
Enallylpropymal

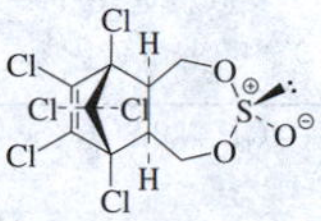
4698
Endosulfan

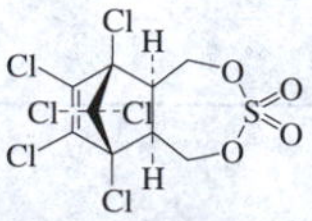
4699
Endosulfan sulfate

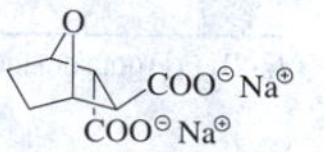
4700
Endothall disodium

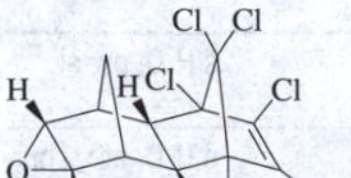
4701
Endrin

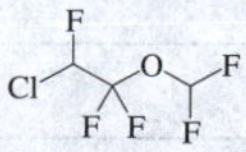
4702
Enflurane

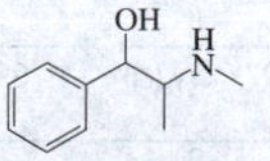
4703
Ephedrine, (±)-

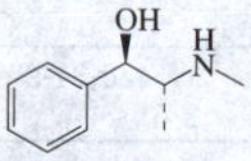
4704
d-Ephedrine

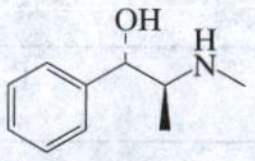
4705
l-Ephedrine

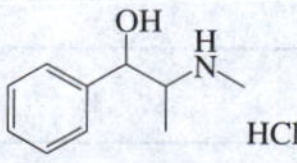
4706
Ephedrine hydrochloride

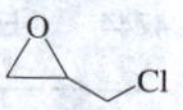
4707
Epichlorohydrin

No.	Name	Synonym	Mol. form.	CAS Reg. No.	Mol. wt.	Physical form	mp/°C	bp/°C	den/g cm^{-3}	n_D	Solubility
4708	Epinephrine	D-Adrenaline	$C_9H_{13}NO_3$	51-43-4	183.204	br (in air)	211.5				sl H_2O; i EtOH; s HOAc, acid
4709	Epiquinidine		$C_{20}H_{24}N_2O_2$	572-59-8	324.417	cry (AcOEt) lf (eth)	113				vs EtOH; s eth
4710	1,2-Epoxybutane	Ethyloxirane	C_4H_8O	106-88-7	72.106	liq	-150	63.4(0.2)	0.8297^{20}	1.3851^{20}	vs EtOH, ace; msc eth
4711	1,2-Epoxy-4-(epoxyethyl) cyclohexane	4-Vinyl-1-cyclohexene dioxide	$C_8H_{12}O_2$	106-87-6	140.180	liq	<-55	227	1.0966^{20}	1.4787^{20}	vs H_2O
4712	1,2-Epoxyhexadecane	Tetradecyloxirane	$C_{16}H_{32}O$	7320-37-8	240.424	hyg cry or liq	24.1	178^{12}	0.846	1.2240	
4713	1,2-Epoxyoctadecane	Hexadecyloxirane	$C_{18}H_{36}O$	7390-81-0	268.478	hyg cry	26.1	$137^{0.5}$			
4714	2,3-Epoxy-α-pinane	α-Pinene oxide	$C_{10}H_{16}O$	1686-14-2	152.233			85^{24}			
4715	2,3-Epoxypropyl acrylate	Glycidyl acrylate	$C_6H_8O_3$	106-90-1	128.126			53^{10}	1.1109^{20}	1.4490^{20}	vs bz
4716	2,3-Epoxypropyl methacrylate	Glycidol methacrylate	$C_7H_{10}O_3$	106-91-2	142.152			189	1.042^{20}	1.448^{25}	vs bz, eth, EtOH
4717	Equol		$C_{14}H_{14}O_3$	531-95-3	230.259	cry (aq, al)	189.5				
4718	Ergocornine		$C_{31}H_{39}N_5O_5$	564-36-3	561.673	cry (MeOH)	183 dec				i H_2O; s EtOH, ace, bz, chl, AcOEt
4719	Ergocorninine		$C_{31}H_{39}N_5O_5$	564-37-4	561.673	lo pr (al)	228 dec				vs ace, bz, EtOH, chl
4720	Ergocristine		$C_{35}H_{39}N_5O_5$	511-08-0	609.716	orth (bz)	175 dec				i H_2O; s EtOH, ace, chl
4721	Ergocristinine		$C_{35}H_{39}N_5O_5$	511-07-9	609.716	pr (al)	237 dec				i H_2O; sl EtOH, ace, chl
4722	Ergocryptine		$C_{32}H_{41}N_5O_5$	511-09-1	575.699	pr (al)	213 dec				i H_2O; s EtOH, chl
4723	Ergocryptinine		$C_{32}H_{41}N_5O_5$	511-10-4	575.70	lo pr (al)	245 dec				vs ace, chl
4724	Ergometrinine		$C_{19}H_{23}N_3O_2$	479-00-5	325.405	pr (ace)	196 dec				vs chl
4725	Ergonovine	Ergometrine	$C_{19}H_{23}N_3O_2$	60-79-7	325.405	pl or nd	162 dec				s H_2O, ace; vs EtOH; sl chl
4726	Ergosine		$C_{30}H_{37}N_5O_5$	561-94-4	547.646	pr (MeOH, AcOEt)	228 dec				s ace, chl; sl MeOH
4727	Ergostane, (5α)		$C_{28}H_{50}$	511-20-6	386.697	lf or pl (ace, eth-MeOH)	85				vs ace, eth, chl
4728	Ergostane, (5β)	Coproergostane	$C_{28}H_{50}$	511-21-7	386.697	nd (ace)	64				vs eth, chl
4729	Ergostan-3-ol, (3β,5α)	Ergostanol	$C_{28}H_{50}O$	6538-02-9	402.696	nd (MeOH-eth)	144.5				i H_2O; s eth, chl
4730	Ergosta-5,7,9(11),22-tetraen-3-ol, (3β,22*E*)-	Dehydroergosterol	$C_{28}H_{42}O$	516-85-8	394.632	lf (al) nd (eth),pl (al)	146	$230^{0.5}$			vs ace, bz, eth, EtOH
4731	Ergosta-5,7,22-trien-3-ol, (3β,22*E*)-	Ergosterol	$C_{28}H_{44}O$	57-87-4	396.648	pl (+w, al) nd (eth)	158(1)	$250^{0.01}$			i H_2O; sl EtOH, eth, peth; s bz, chl
4732	Ergosta-5,7,22-trien-3-ol, (3β,10α,22*E*)-	Pyrocalciferol	$C_{28}H_{44}O$	128-27-8	396.648	nd (MeOH)	94				i H_2O; s EtOH, chl, MeOH
4733	Ergosta-5,7,22-trien-3-ol, (3β,9β,10α,22*E*)-	Lumisterol	$C_{28}H_{44}O$	474-69-1	396.648	nd (ace-MeOH)	118				i H_2O; s EtOH, HOAc; vs eth, ace, chl
4734	Ergost-5-en-3-ol, (3β,24*R*)-	Campesterol	$C_{28}H_{48}O$	474-62-4	400.680	cry (ace)	157.5				
4735	Ergost-7-en-3-ol, (3β,5α)	γ-Ergostenol	$C_{28}H_{48}O$	516-78-9	400.680	nd (MeOH) cry (PrOH	146				s eth
4736	Ergost-8(14)-en-3-ol, (3β,5α)	α-Ergostenol	$C_{28}H_{48}O$	632-32-6	400.680	lf or nd (MeOH)	131				sl EtOH; s eth, bz, chl
4737	Ergotamine		$C_{33}H_{35}N_5O_5$	113-15-5	581.67	nd (al), pr (bz) pl (ace)	213 dec				vs bz, eth, chl
4738	Ergotamine tartrate (2:1)	Gynergen	$C_{35}H_{38}N_5O_8$	379-79-3	656.706		192 dec				
4739	Ergotaminine		$C_{33}H_{35}N_5O_5$	639-81-6	581.662	orth pl (MeOH) pl (al)	252 dec				i H_2O; sl EtOH, ace, bz; s chl; vs py
4740	Ergothioneine		$C_9H_{15}N_3O_2S$	497-30-3	229.299	nd or lf (dil EtOH)	290 dec				vs H_2O; sl EtOH, ace; i eth, bz, chl
4741	Eriochrome Black T		$C_{20}H_{12}N_3NaO_7S$	1787-61-7	461.380	br-blk pow					s H_2O, EtOH, MeOH
4742	Eriodictyol	3',4',5,7-Tetrahydroxyflavanone, (*S*)	$C_{15}H_{12}O_6$	552-58-9	288.252	pl or nd (EtOH)	267 dec				vs EtOH, HOAc
4743	Erythorbic acid	Isoascorbic acid	$C_6H_8O_6$	89-65-6	176.124	gran cry	168				s H_2O, py; sl ace
4744	β-Erythroidine		$C_{16}H_{19}NO_3$	466-81-9	273.327	cry (al)	99.5				s H_2O, eth, chl; vs EtOH, bz

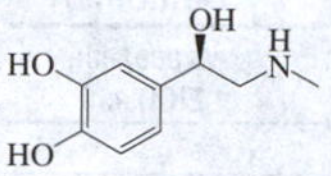
4708
Epinephrine

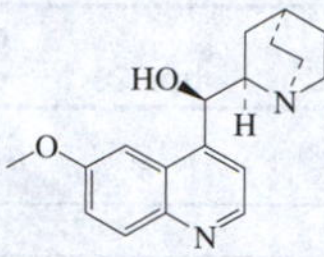
4709
Epiquinidine

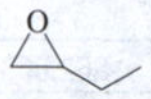
4710
1,2-Epoxybutane

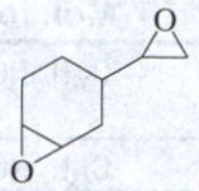
4711
1,2-Epoxy-4-(epoxyethyl)cyclohexane

4712
1,2-Epoxyhexadecane

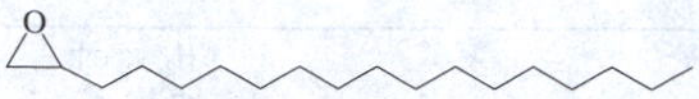
4713
1,2-Epoxyoctadecane

4714
2,3-Epoxy-α-pinane

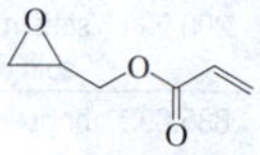
4715
2,3-Epoxypropyl acrylate

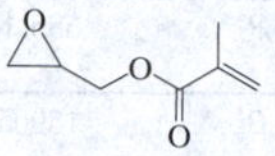
4716
2,3-Epoxypropyl methacrylate

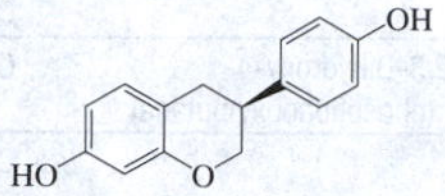
4717
Equol

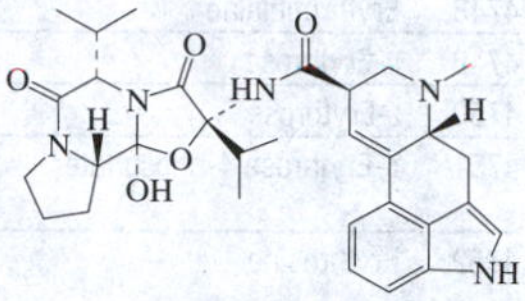
4718
Ergocornine

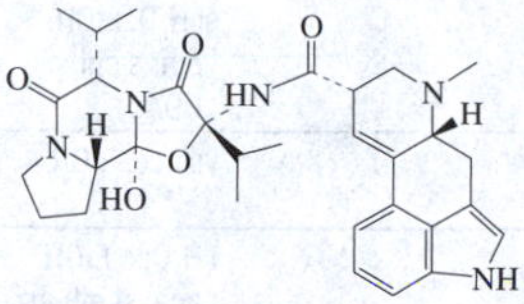
4719
Ergocorninine

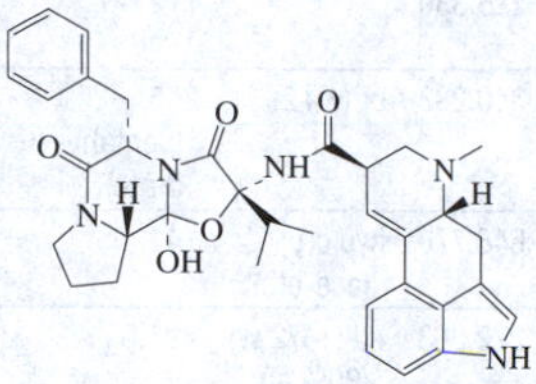
4720
Ergocristine

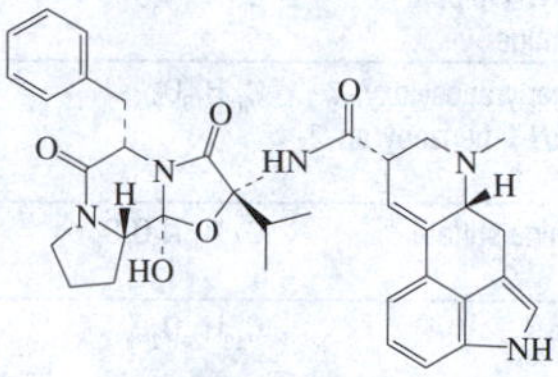
4721
Ergocristinine

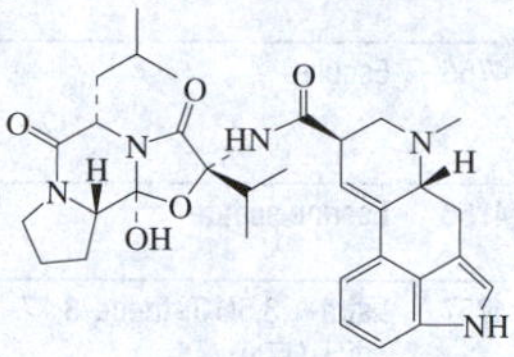
4722
Ergocryptine

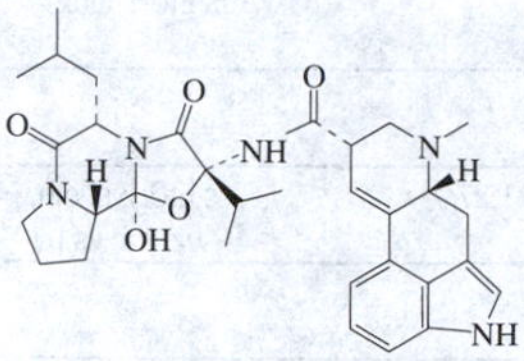
4723
Ergocryptinine

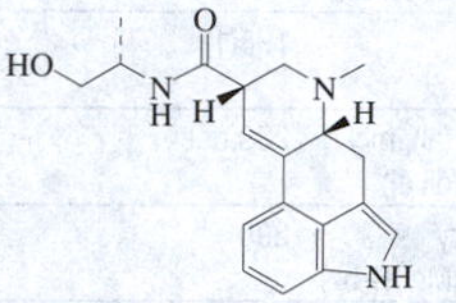
4724
Ergometrinine

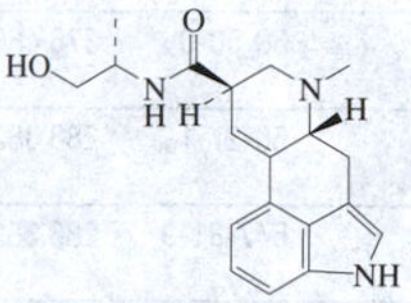
4725
Ergonovine

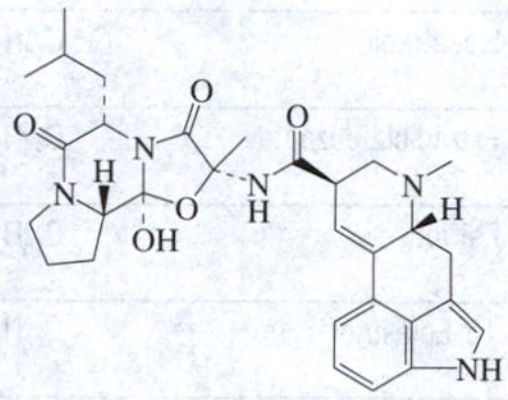
4726
Ergosine

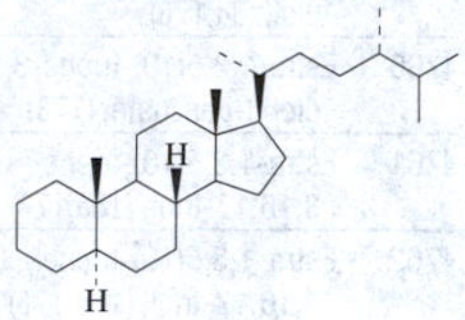
4727
Ergostane, (5α)

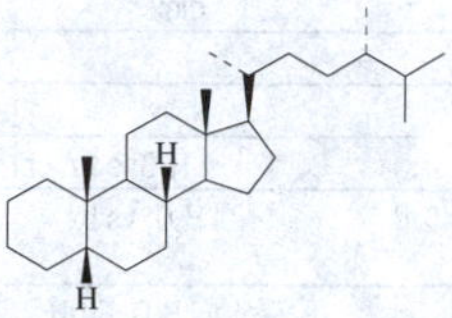
4728
Ergostane, (5β)

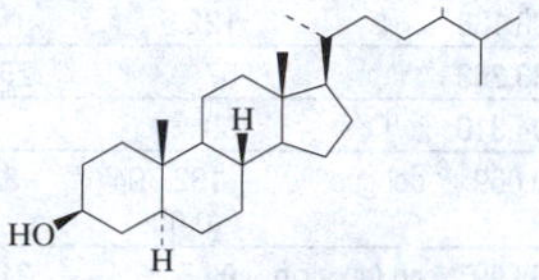
4729
Ergostan-3-ol, (3β,5α)

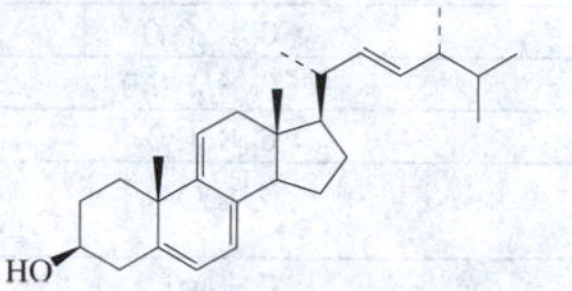
4730
Ergosta-5,7,9(11),22-tetraen-3-ol, (3β,22*E*)-

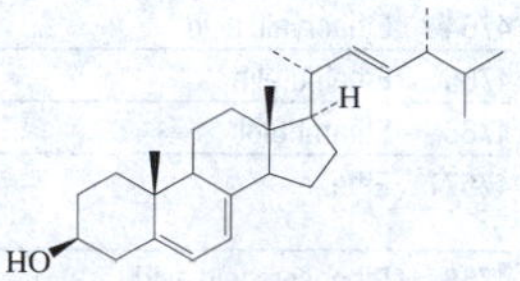
4731
Ergosta-5,7,22-trien-3-ol, (3β,22*E*)-

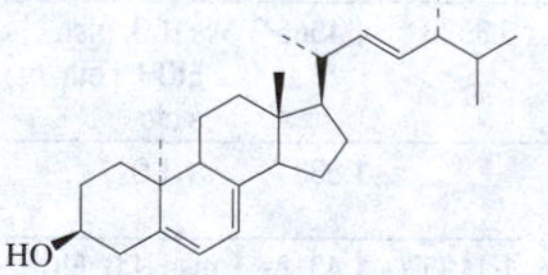
4732
Ergosta-5,7,22-trien-3-ol, (3β,10α,22*E*)-

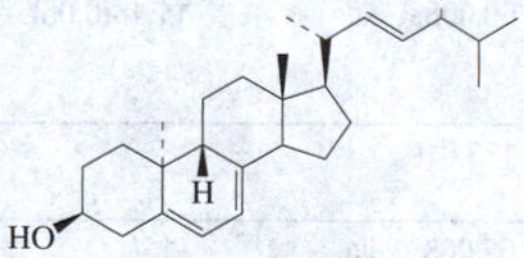
4733
Ergosta-5,7,22-trien-3-ol, (3β,9β,10α,22*E*)-

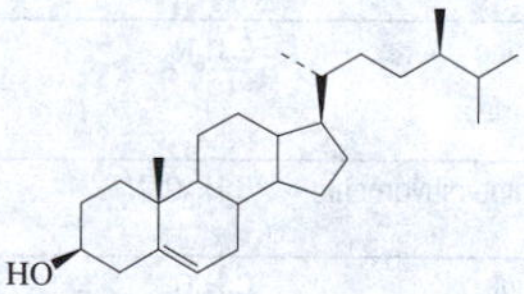
4734
Ergost-5-en-3-ol, (3β,24*R*)-

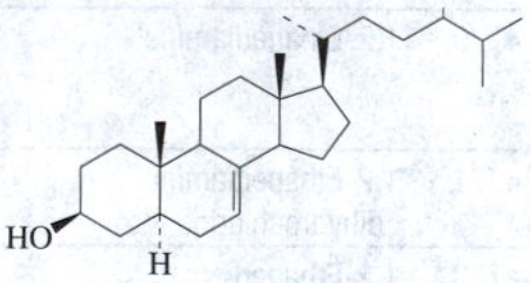
4735
Ergost-7-en-3-ol, (3β,5α)

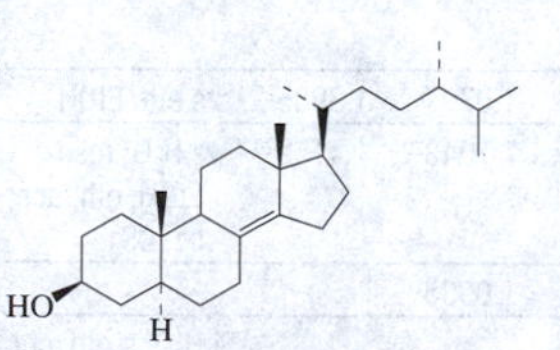
4736
Ergost-8(14)-en-3-ol, (3β,5α)

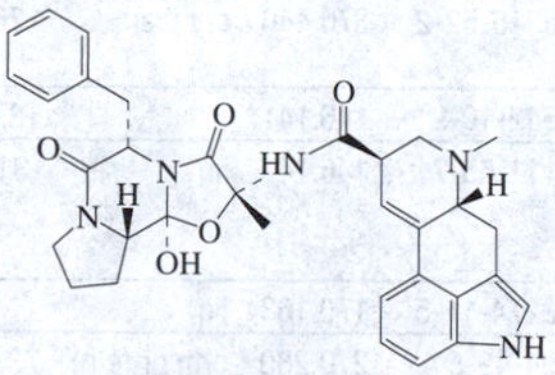
4737
Ergotamine

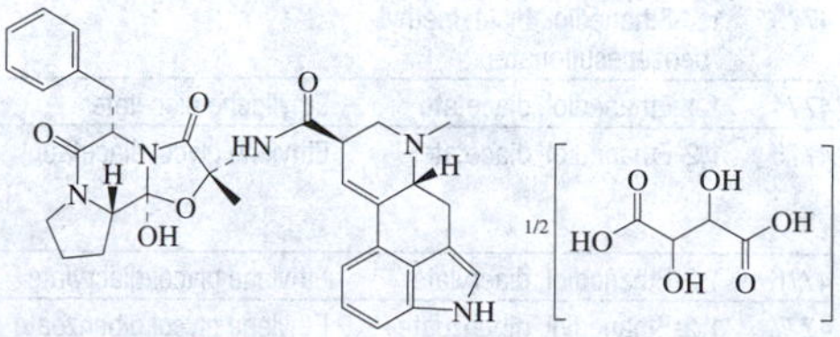
4738
Ergotamine tartrate (2:1)

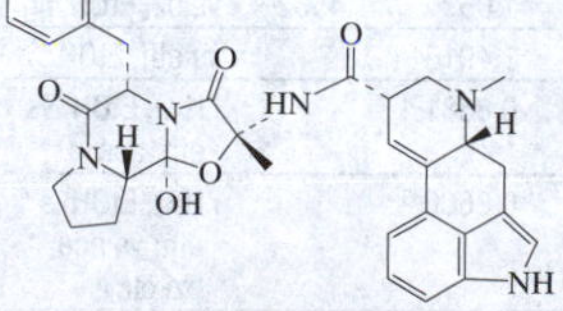
4739
Ergotaminine

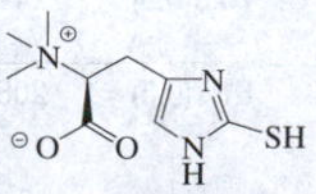
4740
Ergothioneine

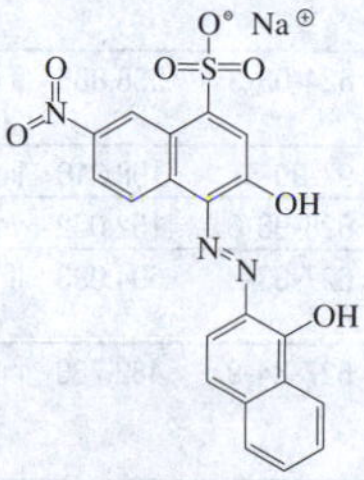
4741
Eriochrome Black T

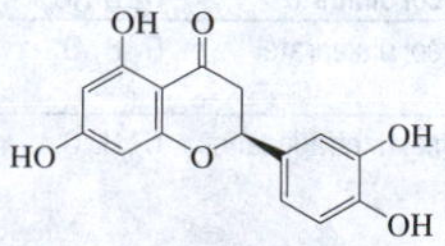
4742
Eriodictyol

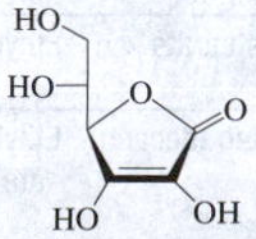
4743
Erythorbic acid

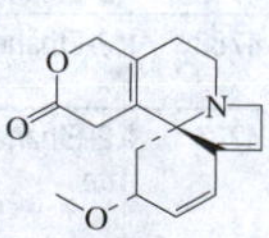
4744
β-Erythroidine

Organic

No.	Name	Synonym	Mol. form.	CAS Reg. No.	Mol. wt.	Physical form	mp/°C	bp/°C	den/g cm^{-3}	n_D	Solubility
4745	Erythromycin	Propiocine	$C_{37}H_{67}NO_{13}$	114-07-8	733.927	cry (w)	191				vs ace, eth, EtOH, chl
4746	Erythromycin ethyl succinate		$C_{43}H_{75}NO_{16}$	1264-62-6	862.053	cry (ace aq)	222				
4747	Erythromycin stearate		$C_{55}H_{103}NO_{15}$	643-22-1	1018.405	cry	92				i H_2O; sl EtOH, eth, chl
4748	Erythrophleine	Norcassamidine	$C_{24}H_{39}NO_5$	36150-73-9	421.571	glass	115				s H_2O, EtOH
4749	D-Erythrose		$C_4H_8O_4$	583-50-6	120.105	syr					s H_2O; vs EtOH
4750	L-Erythrose		$C_4H_8O_4$	533-49-3	120.105	syr					vs H_2O, EtOH
4751	D-Erythrose 4-phosphate	2,3-Dihydroxy-4-(phosphonooxy)butanal	$C_4H_9O_7P$	585-18-2	200.084	stab in aq soln only					s H_2O
4752	Erythrosine		$C_{20}H_8I_4O_5$	15905-32-5	835.893	br pow (Na salt)					s H_2; vs eth, EtOH
4753	L-Erythrulose		$C_4H_8O_4$	533-50-6	120.105	syr	dec	dec			vs H_2O, EtOH
4754	Esaprazole	*N*-Cyclohexyl-1-piperazineacetamide	$C_{12}H_{23}N_3O$	64204-55-3	225.330		112	$190^{0.5}$			
4755	Esculin	6-(β-D-Glucopyranosyloxy)-7-hydroxy-2*H*-1-benzopyran-2-one	$C_{15}H_{16}O_9$	531-75-9	340.283	pr (w+2)	205 (pentahydrate)				sl H_2O, EtOH, eth; s chl, py, HOAc
4756	Eserine sulfate	Physostigmine sulfate	$C_{30}H_{44}N_6O_8S$	64-47-1	648.770	hyg cry (ace-eth)	141				vs ace, EtOH
4757	Estra-1,3,5(10)-triene-3,17-diol, (17α)	α-Estradiol	$C_{18}H_{24}O_2$	57-91-0	272.383	nd (+1/2 w) (80% al)	221.5				i H_2O; s EtOH, ace; sl eth, bz
4758	Estra-1,3,5(10)-triene-3,17-diol (17β)	β-Estradiol	$C_{18}H_{24}O_2$	50-28-2	272.383	pr (80% al),	180(1)				vs ace, EtOH, Diox
4759	Estra-1,3,5(10)-triene-3,17-diol, (8α,17β)	Isoestradiol	$C_{18}H_{24}O_2$	517-04-4	272.383	cry (dil MeOH-chl)	181				s EtOH, diox
4760	Estra-1,3,5(10)-triene-3,17-diol 3-benzoate, (17β)	Estradiol benzoate	$C_{25}H_{28}O_3$	50-50-0	376.488		198(1)				
4761	Estra-1,3,5(10)-triene-3,16,17-triol, (16α,17β)	Estriol	$C_{18}H_{24}O_3$	50-27-1	288.382	lf (al), mcl (dil al)	288 dec		1.27^{25}		s EtOH; sl eth, bz, tfa; vs py
4762	Estra-1,3,5(10)-triene-3,16,17-triol, (16β,17β)	16-Epiestriol	$C_{18}H_{24}O_3$	547-81-9	288.382	cry (MeOH-bz)	290				
4763	Estrone		$C_{18}H_{22}O_2$	53-16-7	270.367	mcl, orth (al)	260.2		1.236^{25}		i H_2O; sl EtOH, eth, bz; s ace, diox
4764	Ethacrynic acid		$C_{13}H_{12}Cl_2O_4$	58-54-8	303.138		122.5				
4765	Ethalfluralin		$C_{13}H_{14}F_3N_3O_4$	55283-68-6	333.263		57	256 dec			
4766	Ethambutol		$C_{10}H_{24}N_2O_2$	74-55-5	204.310	cry	89				sl H_2O; s bz, chl
4767	Ethane		C_2H_6	74-84-0	30.069	col gas	-182.794 (0.005)	-88.6(0.4)	0.5446^{-89}		i H_2O; vs bz
4768	Ethanearsonic acid		$C_2H_7AsO_3$	507-32-4	153.997	nd (al), orth nd (w)	99.5	210^{12}			vs H_2O, EtOH
4769	Ethanedial dioxime		$C_2H_4N_2O_2$	557-30-2	88.065	orth pl (w)	178 dec	sub			vs H_2O, EtOH, eth
4770	1,2-Ethanediamine	Ethylenediamine	$C_2H_8N_2$	107-15-3	60.098	liq	11.14(0.05)	116.9(0.5)	0.8979^{20}	1.4565^{20}	vs H_2O; msc EtOH; i eth, bz; s ctc
4771	1,2-Ethanediamine, dihydrochloride	Ethylenediamine dihydrochloride	$C_2H_{10}Cl_2N_2$	333-18-6	133.019					1.633	vs H_2O
4772	1,2-Ethanediol	Ethylene glycol	$C_2H_6O_2$	107-21-1	62.068	liq	-13(1)	197.5(0.1)	1.1135^{20}	1.4318^{20}	msc H_2O, EtOH, ace; s eth, chl; sl bz
4773	1,2-Ethanediol, bis(4-methylbenzenesulfonate)		$C_{16}H_{18}O_6S_2$	6315-52-2	370.440	cry (bz)	126.5(0.5)				
4774	1,1-Ethanediol, diacetate	Ethylidene diacetate	$C_6H_{10}O_4$	542-10-9	146.141		18.9	168(3)	1.070^{25}	1.3985^{25}	vs eth, EtOH
4775	1,2-Ethanediol, diacetate	Ethylene glycol diacetate	$C_6H_{10}O_4$	111-55-7	146.141	liq	-31	184(4)	1.1043^{20}	1.4159^{20}	vs H_2O; msc EtOH, eth, ace, bz, CS_2
4776	1,2-Ethanediol, diacrylate	Ethylene glycol diacrylate	$C_8H_{10}O_4$	2274-11-5	170.163	liq		$55^{0.6}$	1.0935^{26}		
4777	1,2-Ethanediol, dibenzoate	Ethylene glycol dibenzoate	$C_{16}H_{14}O_4$	94-49-5	270.280	orth pr (eth)	73.5	360 dec			i H_2O; s eth, chl
4778	1,2-Ethanediol, didodecanoate	Ethylene glycol didodecanoate	$C_{26}H_{50}O_4$	624-04-4	426.673	pl (al)	56.6	188^{20}			vs eth, EtOH
4779	1,2-Ethanediol, diformate	Ethylene glycol diformate	$C_4H_6O_4$	629-15-2	118.089			174	1.193^{0}	1.3580	sl H_2O; s EtOH, eth
4780	1,2-Ethanediol, dihexadecanoate	Ethylene glycol dipalmitate	$C_{34}H_{66}O_4$	624-03-3	538.886	lf or nd (al-chl)	69.1(0.5)		0.8594^{78}		i H_2O, EtOH; s eth; vs ace
4781	1,2-Ethanediol, dimethacrylate	Ethylene glycol dimethacrylate	$C_{10}H_{14}O_4$	97-90-5	198.216	liq	-40	260	1.053^{20}	1.4532^{25}	vs bz, EtOH, lig
4782	1,2-Ethanediol, dinitrate	Ethylene glycol dinitrate	$C_2H_4N_2O_6$	628-96-6	152.062	ye liq	-22.5(0.6)	199(3)	1.4918^{20}		vs eth, EtOH
4783	1,2-Ethanediol, distearate	Ethylene glycol distearate	$C_{38}H_{74}O_4$	627-83-8	594.993	lf	75.3(0.5)	241^{20}	0.8581^{78}		i H_2O, EtOH; vs eth, ace
4784	1,2-Ethanediol, ditetradecanoate	Ethylene glycol ditetradecanoate	$C_{30}H_{58}O_4$	627-84-9	482.780	cry (eth, ace)	61.7(0.5)	208^{20}	0.8600^{80}		i H_2O, EtOH; s eth; vs ace, bz, ctc

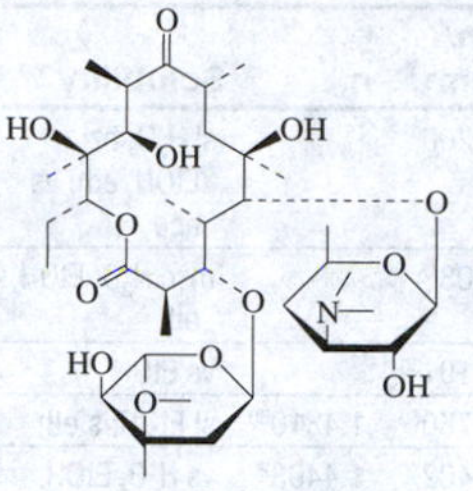
4745
Erythromycin

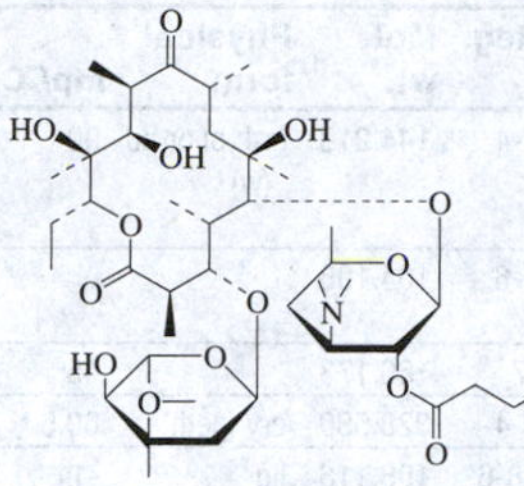
4746
Erythromycin ethyl succinate

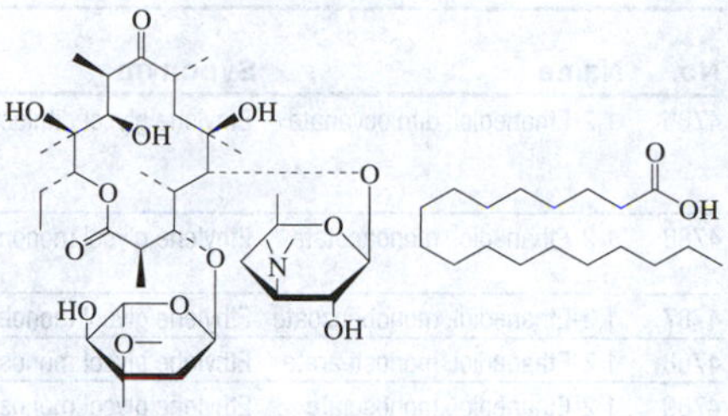
4747
Erythromycin stearate

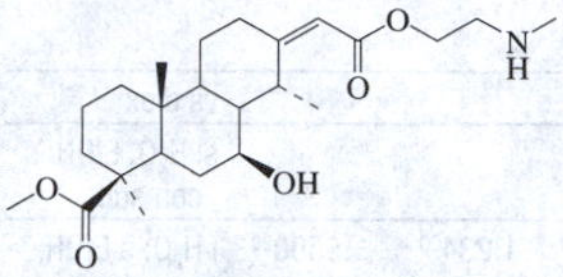
4748
Erythrophleine

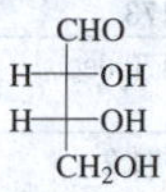
4749
D-Erythrose

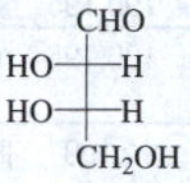
4750
L-Erythrose

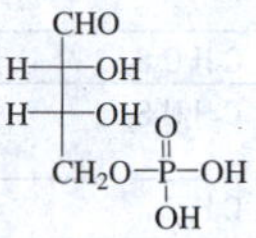
4751
D-Erythrose 4-phosphate

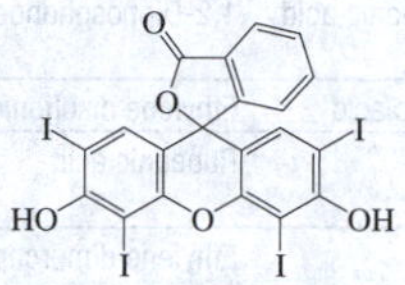
4752
Erythrosine

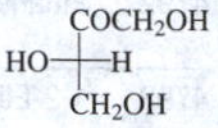
4753
L-Erythrulose

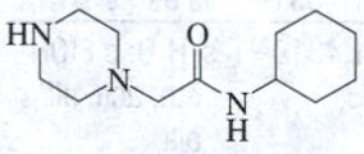
4754
Esaprazole

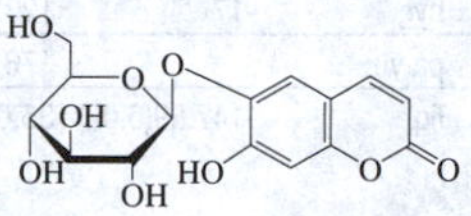
4755
Esculin

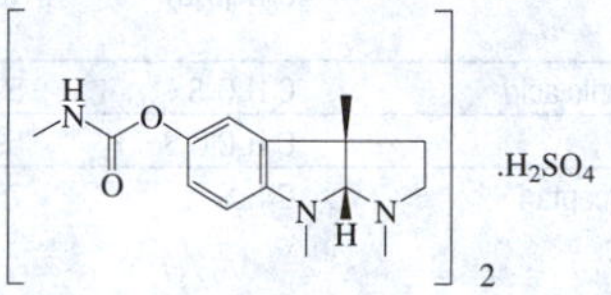
4756
Eserine sulfate

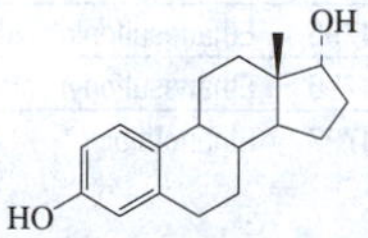
4757
Estra-1,3,5(10)-triene-3,17-diol, (17α)

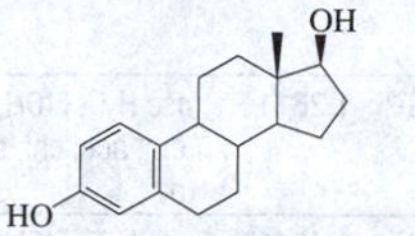
4758
Estra-1,3,5(10)-triene-3,17-diol (17β)

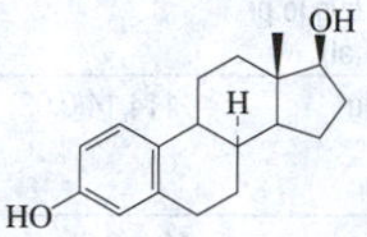
4759
Estra-1,3,5(10)-triene-3,17-diol, (8α,17β)

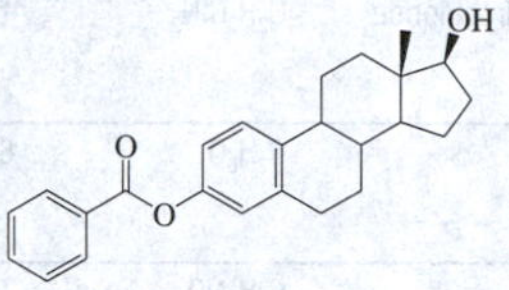
4760
Estra-1,3,5(10)-triene-3,17-diol 3-benzoate, (17β)

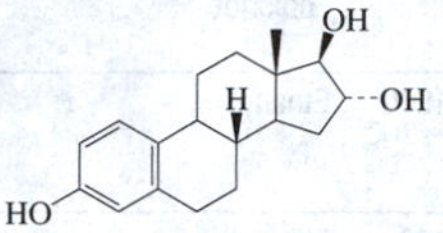
4761
Estra-1,3,5(10)-triene-3,16,17-triol, (16α,17β)

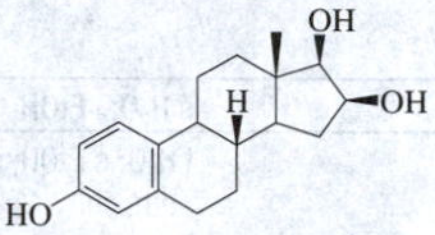
4762
Estra-1,3,5(10)-triene-3,16,17-triol, (16β,17β)

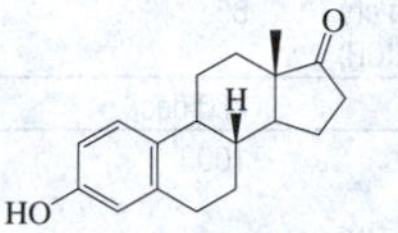
4763
Estrone

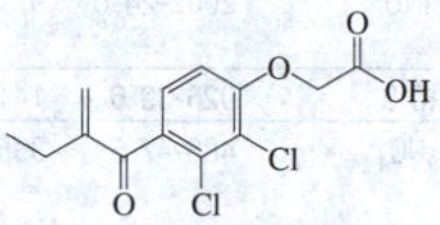
4764
Ethacrynic acid

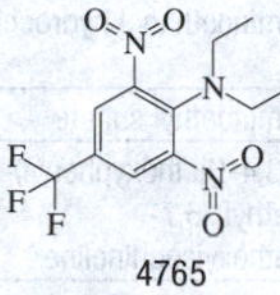
4765
Ethalfluralin

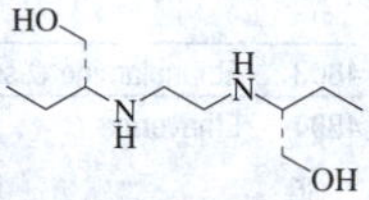
4766
Ethambutol

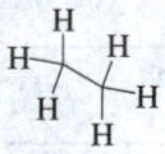
4767
Ethane

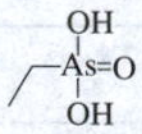
4768
Ethanearsonic acid

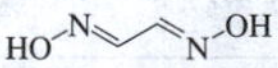
4769
Ethanedial dioxime

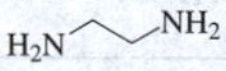
4770
1,2-Ethanediamine

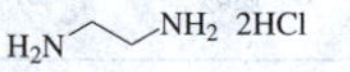
4771
1,2-Ethanediamine, dihydrochloride

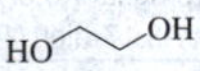
4772
1,2-Ethanediol

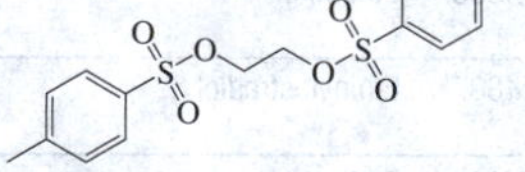
4773
1,2-Ethanediol, bis(4-methylbenzenesulfonate)

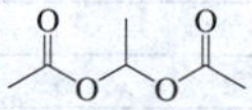
4774
1,1-Ethanediol, diacetate

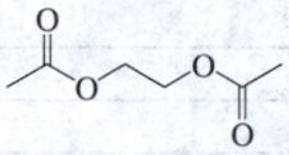
4775
1,2-Ethanediol, diacetate

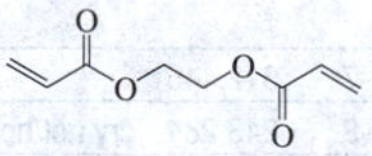
4776
1,2-Ethanediol, diacrylate

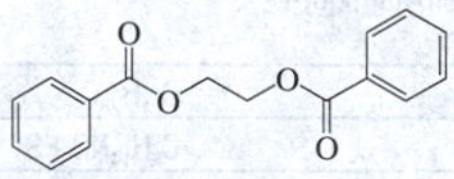
4777
1,2-Ethanediol, dibenzoate

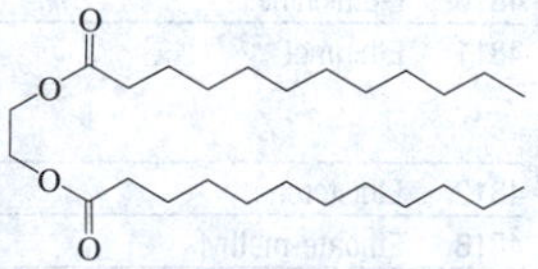
4778
1,2-Ethanediol, didodecanoate

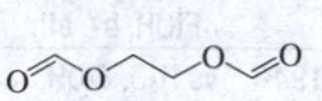
4779
1,2-Ethanediol, diformate

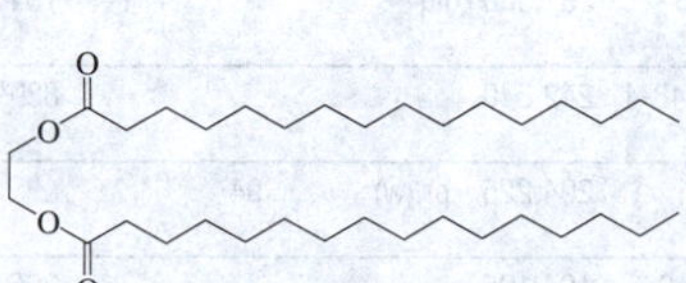
4780
1,2-Ethanediol, dihexadecanoate

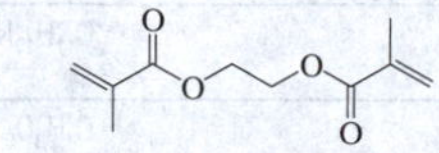
4781
1,2-Ethanediol, dimethacrylate

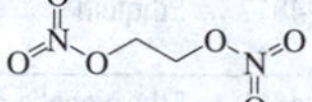
4782
1,2-Ethanediol, dinitrate

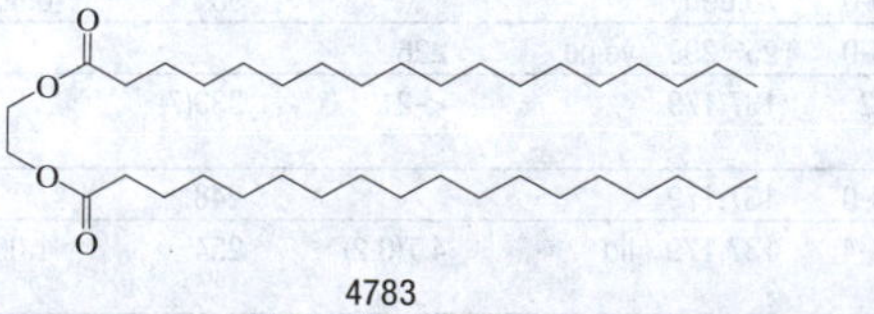
4783
1,2-Ethanediol, distearate

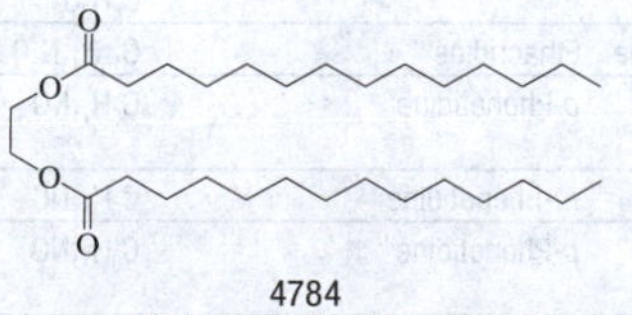
4784
1,2-Ethanediol, ditetradecanoate

No.	Name	Synonym	Mol. form.	CAS Reg. No.	Mol. wt.	Physical form	mp/°C	bp/°C	den/g cm^{-3}	n_D	Solubility
4785	1,2-Ethanediol, dithiocyanate	Ethylene glycol dithiocyanate	$C_4H_4N_2S_2$	629-17-4	144.218	orth pl or nd (w)	90	dec	1.4200^{0}		sl H_2O, bz; s EtOH, eth; vs ace
4786	1,2-Ethanediol, monoacetate	Ethylene glycol monoacetate	$C_4H_8O_3$	542-59-6	104.105			188(2)	1.108^{15}		msc H_2O, EtOH, eth
4787	1,2-Ethanediol, monobenzoate	Ethylene glycol monobenzoate	$C_9H_{10}O_3$	94-33-7	166.173		45	150^{10}	1.1101^{30}		vs EtOH
4788	1,2-Ethanediol, monostearate	Ethylene glycol monostearate	$C_{20}H_{40}O_3$	111-60-4	328.530	cry (peth)	60.5	190^{3}	0.8780^{60}	1.4310^{60}	sl EtOH; s eth
4789	1,2-Ethanediol, monosulfite	Ethylene glycol monosulfite	$C_2H_4O_3S$	3741-38-6	108.116	liq	-11	173	1.4402^{20}	1.4463^{20}	vs H_2O, EtOH, eth, ace, bz, AcOEt; sl chl
4790	1,2-Ethanediphosphonic acid	1,2-Diphosphonoethane	$C_2H_8O_6P_2$	6145-31-9	190.029	nd (EtOH/eth)	223				
4791	1,2-Ethanedisulfonic acid	Ethylene disulfonic acid	$C_2H_6O_6S_2$	110-04-3	190.195		173				vs diox
4792	Ethanedithioamide	Rubeanic acid	$C_2H_4N_2S_2$	79-40-3	120.196	red cry	170 dec				sl H_2O, EtOH; s con sulf
4793	1,2-Ethanedithiol	Ethylene dimercaptan	$C_2H_6S_2$	540-63-6	94.199	liq	-41.2	144(3)	1.234^{20}	1.5590^{20}	i H_2O; s EtOH, eth, ace, bz; vs alk
4794	1,2-Ethanediyl mercaptoacetate		$C_6H_{10}O_4S_2$	123-81-9	210.271			$138^{1.5}$			
4795	Ethanesulfonic acid	Ethylsulfonic acid	$C_2H_6O_3S$	594-45-6	110.132	hyg	-17	123^{1}	1.3341^{25}	1.4335^{20}	vs H_2O, EtOH
4796	Ethanesulfonyl chloride		$C_2H_5ClO_2S$	594-44-5	128.578	pa ye		176.5(1)	1.357^{22}	1.4531^{20}	vs eth; s CS_2
4797	Ethanethiol	Ethyl mercaptan	C_2H_6S	75-08-1	62.134	liq	-147.89(0.02)	35.0(0.1)	0.8315^{25}	1.4310^{20}	sl H_2O; s EtOH, eth, ace, dil alk
4798	Ethanimidamide		$C_2H_6N_2$	143-37-3	58.082		-35				sl H_2O; s EtOH, acid
4799	Ethanimidamide monohydrochloride	Acetamidine hydrochloride	$C_2H_7ClN_2$	124-42-5	94.543	nd or pr (al) hyg lo pr (al)	177.5				vs H_2O, EtOH
4800	Ethanol	Ethyl alcohol	C_2H_6O	64-17-5	46.068	liq	-114.14(0.03)	78.24(0.09)	0.7893^{20}	1.3611^{20}	msc H_2O, EtOH, eth, ace, chl; s bz
4801	Ethanolamine	Glycinol	C_2H_7NO	141-43-5	61.083	liq	10.4(0.2)	170.3(0.4)	1.0180^{20}	1.4541^{20}	msc H_2O, EtOH; sl eth, lig, bz; s chl
4802	Ethanolamine hydrochloride	2-Aminoethanol hydrochloride	C_2H_8ClNO	2002-24-6	97.544	hyg cry (EtOH)	85				
4803	Ethanolamine *O*-sulfate	2-Aminoethyl sulfate	$C_2H_7NO_4S$	926-39-6	141.147		230 dec				s H_2O; i EtOH
4804	Ethaverine	1-[(3,4-Diethoxyphenyl)-methyl]-6,7-diethoxyisoquinoline	$C_{24}H_{29}NO_4$	486-47-5	395.492		100				i H_2O; s EtOH; sl eth, chl
4805	Ethchlorvynol	1-Chloro-3-ethyl-1-penten-4-yn-ol	C_7H_9ClO	113-18-8	144.598	liq		181	1.07^{25}	1.474^{25}	i H_2O; s os
4806	Ethephon	Phosphonic acid, (2-chloroethyl)-	$C_2H_6ClO_3P$	16672-87-0	144.494		76.7(0.5)		1.2		
4807	Ethinylestradiol	19-Norpregna-1,3,5(10)-trien-20-yne-3,17-diol, (17α)-	$C_{20}H_{24}O_2$	57-63-6	296.404		186(1)				sl chl
4808	Ethion		$C_9H_{22}O_4P_2S_4$	563-12-2	384.476		-13	$165^{0.3}$	1.22^{20}		
4809	D-Ethionine	3-Ethylhomocysteine, (*R*)	$C_6H_{13}NO_2S$	535-32-0	163.238	cry (H_2O)	278 dec				
4810	L-Ethionine	3-Ethylhomocysteine, (*S*)	$C_6H_{13}NO_2S$	13073-35-3	163.238	cry (H_2O)	273 dec				
4811	Ethirimol	4(1*H*)-Pyrimidinone, 5-butyl-2-(ethylamino)-6-methyl-	$C_{11}H_{19}N_3O$	23947-60-6	209.288		161.2(0.5)		1.21^{25}		
4812	Ethisterone		$C_{21}H_{28}O_2$	434-03-7	312.446		272				
4813	Ethoate-methyl		$C_6H_{14}NO_3PS_2$	116-01-8	243.284	cry (tol/hp)	67				
4814	Ethofumesate		$C_{13}H_{18}O_5S$	26225-79-6	286.344		72.5(0.5)		1.14		
4815	Ethoheptazine	4-Carbethoxymethyl-4-phenyl-azacycloheptane	$C_{16}H_{23}NO_2$	77-15-6	261.360	liq		134^{1}	1.038^{26}	1.5210^{26}	
4816	Ethoprop	Phosphorodithioic acid, *O*-ethyl *S,S*-dipropyl ester	$C_8H_{19}O_2PS_2$	13194-48-4	242.340			$88^{0.2}$	1.094^{20}		
4817	Ethotoin		$C_{11}H_{12}N_2O_2$	86-35-1	204.225	pr (w)	94				s hot H_2O; vs EtOH, bz, eth
4818	Ethoxyacetic acid		$C_4H_8O_3$	627-03-2	104.105			206.5	1.1021^{20}	1.4194^{20}	vs H_2O, EtOH, eth; s chl
4819	4'-Ethoxyacetophenone		$C_{10}H_{12}O_2$	1676-63-7	164.201	pl (eth)	39	268			vs eth, EtOH
4820	Ethoxyacetylene		C_4H_6O	927-80-0	70.090			50	0.8000^{20}	1.3796^{20}	
4821	7-Ethoxy-3,9-acridinediamine	Ethacridine	$C_{15}H_{15}N_3O$	442-16-0	253.299	ye nd	226				
4822	2-Ethoxyaniline	*o*-Phenetidine	$C_8H_{11}NO$	94-70-2	137.179		<-21	233(7)		1.5560^{20}	sl H_2O, ctc; s EtOH, eth
4823	3-Ethoxyaniline	*m*-Phenetidine	$C_8H_{11}NO$	621-33-0	137.179			248			vs eth, EtOH
4824	4-Ethoxyaniline	*p*-Phenetidine	$C_8H_{11}NO$	156-43-4	137.179	liq	4.7(0.2)	254	1.0652^{16}	1.5528^{20}	sl H_2O; s EtOH, eth, chl

4785
1,2-Ethanediol, dithiocyanate

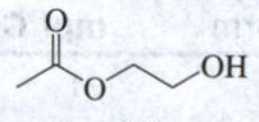
4786
1,2-Ethanediol, monoacetate

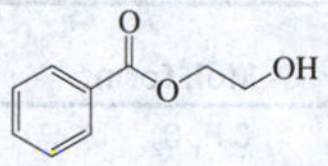
4787
1,2-Ethanediol, monobenzoate

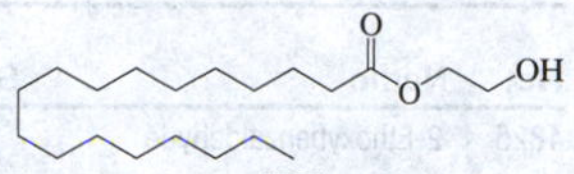
4788
1,2-Ethanediol, monostearate

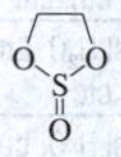
4789
1,2-Ethanediol, monosulfite

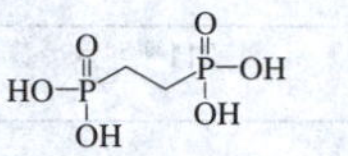
4790
1,2-Ethanediphosphonic acid

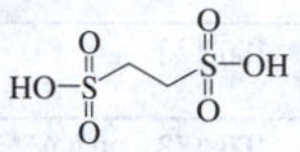
4791
1,2-Ethanedisulfonic acid

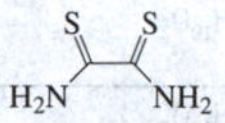
4792
Ethanedithioamide

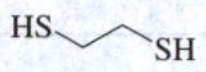
4793
1,2-Ethanedithiol

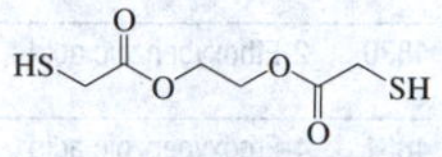
4794
1,2-Ethanediyl mercaptoacetate

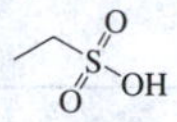
4795
Ethanesulfonic acid

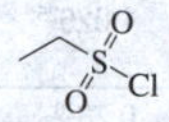
4796
Ethanesulfonyl chloride

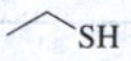
4797
Ethanethiol

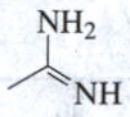
4798
Ethanimidamide

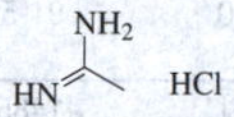
4799
Ethanimidamide monohydrochloride

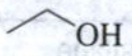
4800
Ethanol

4801
Ethanolamine

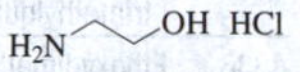
4802
Ethanolamine hydrochloride

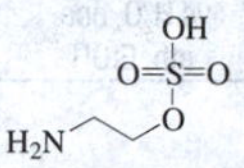
4803
Ethanolamine *O*-sulfate

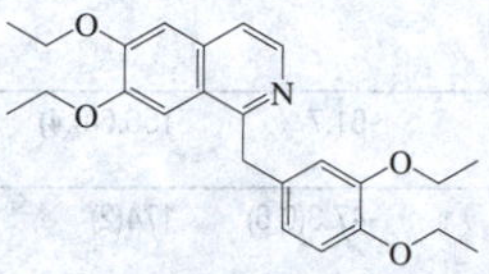
4804
Ethaverine

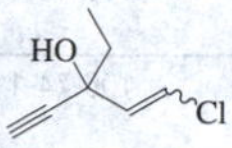
4805
Ethchlorvynol

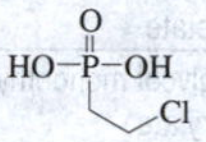
4806
Ethephon

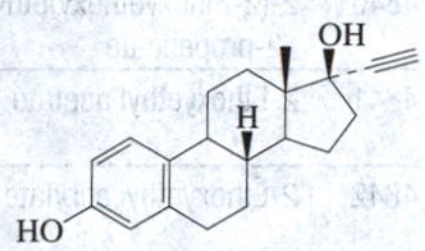
4807
Ethinylestradiol

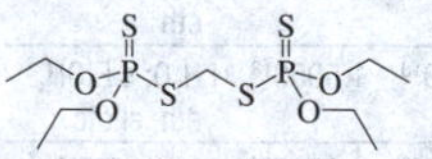
4808
Ethion

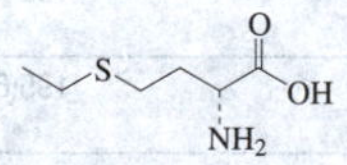
4809
D-Ethionine

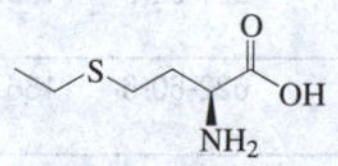
4810
L-Ethionine

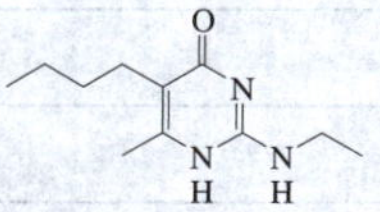
4811
Ethirimol

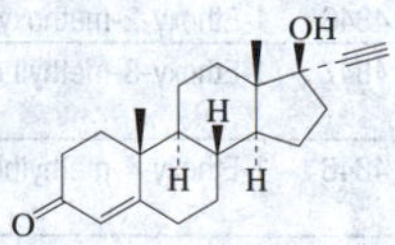
4812
Ethisterone

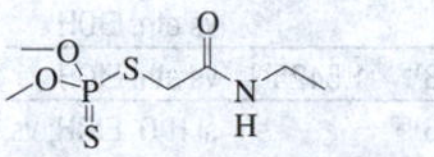
4813
Ethoate-methyl

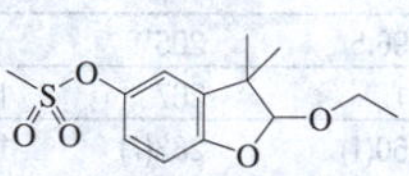
4814
Ethofumesate

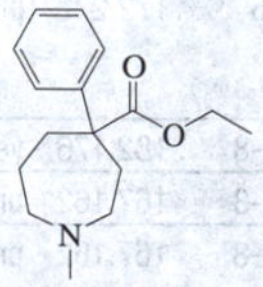
4815
Ethoheptazine

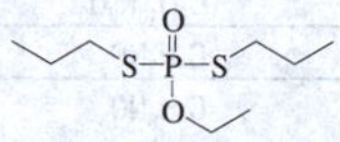
4816
Ethoprop

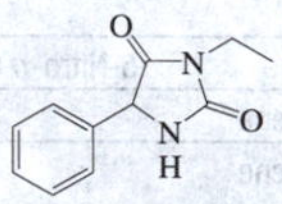
4817
Ethotoin

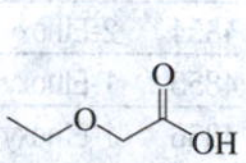
4818
Ethoxyacetic acid

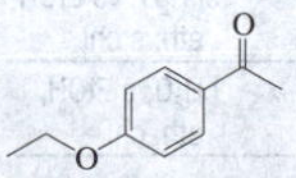
4819
4'-Ethoxyacetophenone

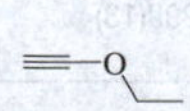
4820
Ethoxyacetylene

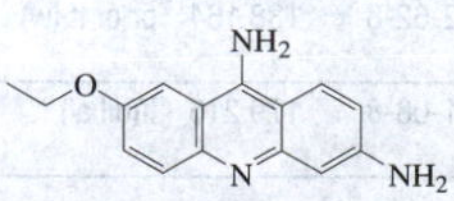
4821
7-Ethoxy-3,9-acridinediamine

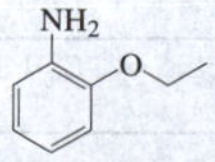
4822
2-Ethoxyaniline

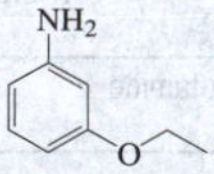
4823
3-Ethoxyaniline

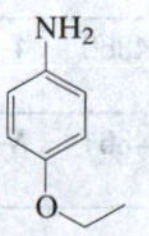
4824
4-Ethoxyaniline

No.	Name	Synonym	Mol. form.	CAS Reg. No.	Mol. wt.	Physical form	mp/°C	bp/°C	den/g cm^{-3}	n_D	Solubility
4825	2-Ethoxybenzaldehyde		$C_9H_{10}O_2$	613-69-4	150.174		21	248			msc EtOH, eth; sl chl
4826	4-Ethoxybenzaldehyde		$C_9H_{10}O_2$	10031-82-0	150.174		13.5	249	1.08^{21}		vs EtOH, eth, bz
4827	2-Ethoxybenzamide	Ethenzamide	$C_9H_{11}NO_2$	938-73-8	165.189	nd (w, al)	133				sl H_2O, chl; vs EtOH, eth
4828	Ethoxybenzene	Phenetole	$C_8H_{10}O$	103-73-1	122.164	liq	-29.6(0.4)	169.8(0.2)	0.9651^{20}	1.5076^{20}	i H_2O; s EtOH, eth, ctc
4829	4-Ethoxy-1,2-benzenediamine		$C_8H_{12}N_2O$	1197-37-1	152.193		71.5	295			vs H_2O; s EtOH, eth, chl
4830	2-Ethoxybenzoic acid		$C_9H_{10}O_3$	134-11-2	166.173		20.7	211^{39}			sl H_2O, EtOH, ctc
4831	4-Ethoxybenzoic acid		$C_9H_{10}O_3$	619-86-3	166.173	nd (w)	199(1)				sl H_2O, tfa; s EtOH, eth, bz
4832	6-Ethoxy-2-benzothiazolesulfonamide	Ethoxzolamide	$C_9H_{10}N_2O_3S_2$	452-35-7	258.316		191(1)				
4833	3-Ethoxy-*N,N*-diethylaniline		$C_{12}H_{19}NO$	1864-92-2	193.285			286		1.5325^{25}	s EtOH, bz, HOAc
4834	2-Ethoxy-3,4-dihydro-2*H*-pyran		$C_7H_{12}O_2$	103-75-3	128.169			132	0.9658^{25}	1.4394^{20}	
4835	6-Ethoxy-1,2-dihydro-2,2,4-trimethylquinoline	Ethoxyquin	$C_{14}H_{19}NO$	91-53-2	217.307			124^{2}	1.026^{25}	1.569^{25}	
4836	Ethoxydimethylsilane	Dimethylethoxysilane	$C_4H_{12}OSi$	14857-34-2	104.223	liq		54	0.76^{20}		
4837	2-Ethoxy-1,2-diphenylethanone		$C_{16}H_{16}O_2$	574-09-4	240.297	nd (lig)	62	194^{20}	1.1016^{17}	1.5727^{17}	vs bz, eth, EtOH, lig
4838	2-Ethoxyethanamine		$C_4H_{11}NO$	110-76-9	89.136			108(12)	0.8512^{20}	1.4101^{20}	msc H_2O, EtOH, eth; s ace, bz; sl chl
4839	2-Ethoxyethanol	Ethylene glycol monoethyl ether	$C_4H_{10}O_2$	110-80-5	90.121	liq	-70	134.7(0.2)	0.9253^{25}	1.4054^{25}	vs H_2O, ace, eth, EtOH
4840	2-(2-Ethoxyethoxy)ethyl 2-propenoate	Diethylene glycol ethyl ether acrylate	$C_9H_{16}O_4$	7328-17-8	188.221				1.13^{25}		
4841	2-Ethoxyethyl acetate	Ethylene glycol monoethyl ether acetate	$C_6H_{12}O_3$	111-15-9	132.157	liq	-61.7	156.6(0.4)	0.9740^{20}	1.4054^{20}	vs H_2O, ace, eth, EtOH
4842	2-Ethoxyethyl acrylate	Ethylene glycol monoethyl ether acrylate	$C_7H_{12}O_3$	106-74-1	144.168	liq	-47.3(0.6)	174(2)	0.983^{20}	1.4274^{20}	
4843	3-Ethoxy-2-hydroxybenzaldehyde		$C_9H_{10}O_3$	492-88-6	166.173		64.1(0.4)	264			
4844	3-Ethoxy-4-hydroxybenzaldehyde	Ethyl vanillin	$C_9H_{10}O_3$	121-32-4	166.173		76.7(0.4)	285			sl H_2O; s EtOH, eth, bz, chl
4845	4-Ethoxy-3-methoxybenzaldehyde		$C_{10}H_{12}O_3$	120-25-2	180.200	mcl pr	64.5	168^{13}			sl H_2O; s EtOH, eth, bz, chl, HOAc
4846	1-Ethoxy-2-methoxyethane		$C_5H_{12}O_2$	5137-45-1	104.148	liq		88(14)	0.8460^{25}	1.3843^{25}	
4847	1-Ethoxy-3-methylbenzene		$C_9H_{12}O$	621-32-9	136.190			192	0.949^{20}	1.513^{20}	i H_2O; s EtOH, eth
4848	1-Ethoxy-4-methylbenzene		$C_9H_{12}O$	622-60-6	136.190			195(6)	0.9509^{18}	1.5058^{18}	i H_2O; s EtOH, eth; sl ctc
4849	2-Ethoxy-2-methylbutane	Ethyl *tert*-pentyl ether	$C_7H_{16}O$	919-94-8	116.201			101.5(0.4)	0.7606^{25}	1.3886^{25}	vs eth, EtOH
4850	(Ethoxymethylene)-propanedinitrile		$C_6H_6N_2O$	123-06-8	122.124		66	160^{12}			s EtOH, eth; sl chl
4851	(Ethoxymethyl)oxirane	2,3-Epoxypropyl ethyl ether	$C_5H_{10}O_2$	4016-11-9	102.132			126(18)	0.9700^{20}	1.4320^{20}	s H_2O, EtOH, eth; sl ctc
4852	1-Ethoxynaphthalene		$C_{12}H_{12}O$	5328-01-8	172.222	nd	5.5	280(8)	1.060^{20}	1.5953^{25}	i H_2O; vs EtOH, eth
4853	2-Ethoxynaphthalene		$C_{12}H_{12}O$	93-18-5	172.222	pl (al)	37.5	282(8)	1.0640^{20}	1.5975^{36}	i H_2O; s EtOH, eth, tol, lig, CS_2
4854	2-Ethoxy-5-nitroaniline	5-Nitro-*o*-phenetidine	$C_8H_{10}N_2O_3$	136-79-8	182.176	ye nd (dil al)	96.5	205^{14}			vs eth, EtOH
4855	1-Ethoxy-2-nitrobenzene		$C_8H_9NO_3$	610-67-3	167.162	br ye	1.1	267	1.1903^{15}	1.5425^{20}	vs eth, EtOH
4856	1-Ethoxy-4-nitrobenzene		$C_8H_9NO_3$	100-29-8	167.162	pr (dil al, eth)	60(1)	283(7)	1.1176^{100}		sl H_2O, EtOH; vs eth; msc ace, bz; s peth
4857	*N*-(4-Ethoxy-3-nitrophenyl)-acetamide		$C_{10}H_{12}N_2O_4$	1777-84-0	224.213	nd (dil al)	124.0				vs ace, bz, EtOH
4858	2-Ethoxyphenol	Catechol monoethyl ether	$C_8H_{10}O_2$	94-71-3	138.164		29	217(6)	1.0903^{25}		sl H_2O, ctc; msc EtOH, eth
4859	3-Ethoxyphenol	Resorcinol monoethyl ether	$C_8H_{10}O_2$	621-34-1	138.164			251(19)	1.105^{15}		i H_2O; s EtOH, eth, bz; sl chl
4860	4-Ethoxyphenol	Hydroquinone monoethyl ether	$C_8H_{10}O_2$	622-62-8	138.164	pr or lf (w)	66.5	248(19)			sl H_2O; vs EtOH, eth; s chl
4861	*N*-(2-Ethoxyphenyl)acetamide		$C_{10}H_{13}NO_2$	581-08-8	179.216	lf(dil al)	79	299(13)			i H_2O; s EtOH, eth, chl

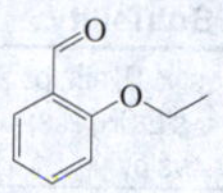
4825
2-Ethoxybenzaldehyde

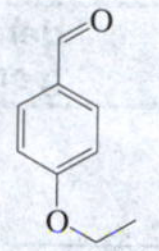
4826
4-Ethoxybenzaldehyde

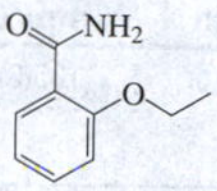
4827
2-Ethoxybenzamide

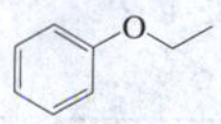
4828
Ethoxybenzene

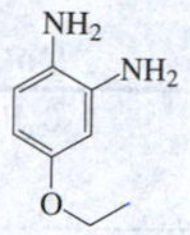
4829
4-Ethoxy-1,2-benzenediamine

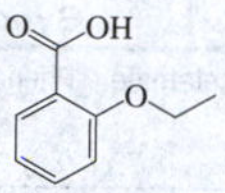
4830
2-Ethoxybenzoic acid

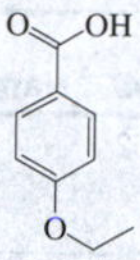
4831
4-Ethoxybenzoic acid

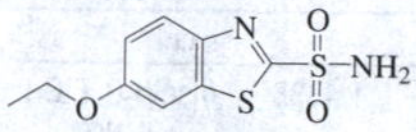
4832
6-Ethoxy-2-benzothiazolesulfonamide

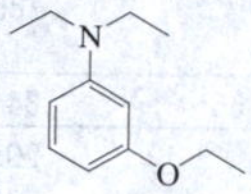
4833
3-Ethoxy-*N,N*-diethylaniline

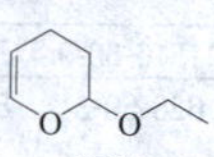
4834
2-Ethoxy-3,4-dihydro-2*H*-pyran

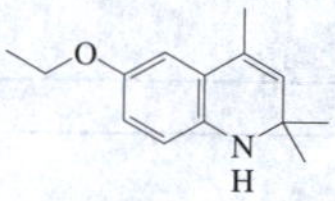
4835
6-Ethoxy-1,2-dihydro-2,2,4-trimethylquinoline

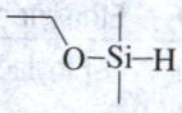
4836
Ethoxydimethylsilane

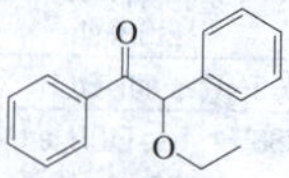
4837
2-Ethoxy-1,2-diphenylethanone

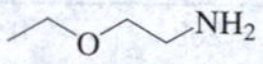
4838
2-Ethoxyethanamine

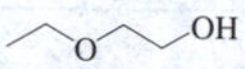
4839
2-Ethoxyethanol

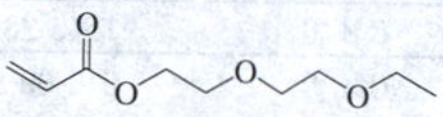
4840
2-(2-Ethoxyethoxy)ethyl 2-propenoate

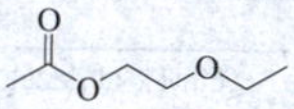
4841
2-Ethoxyethyl acetate

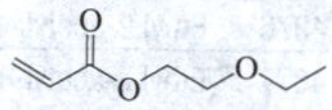
4842
2-Ethoxyethyl acrylate

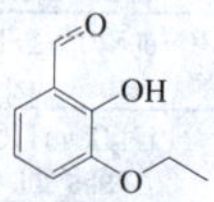
4843
3-Ethoxy-2-hydroxybenzaldehyde

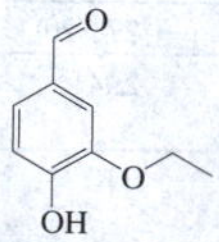
4844
3-Ethoxy-4-hydroxybenzaldehyde

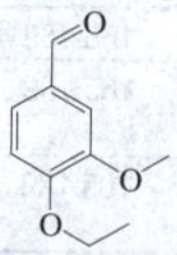
4845
4-Ethoxy-3-methoxybenzaldehyde

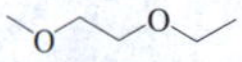
4846
1-Ethoxy-2-methoxyethane

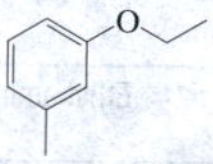
4847
1-Ethoxy-3-methylbenzene

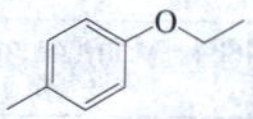
4848
1-Ethoxy-4-methylbenzene

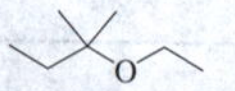
4849
2-Ethoxy-2-methylbutane

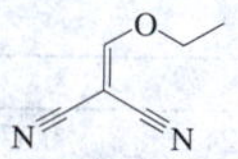
4850
(Ethoxymethylene)propanedinitrile

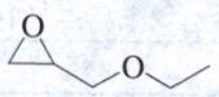
4851
(Ethoxymethyl)oxirane

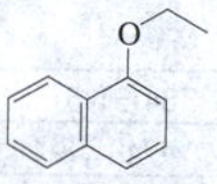
4852
1-Ethoxynaphthalene

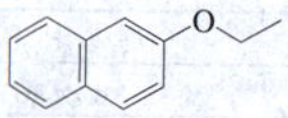
4853
2-Ethoxynaphthalene

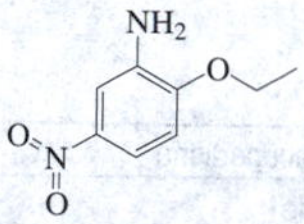
4854
2-Ethoxy-5-nitroaniline

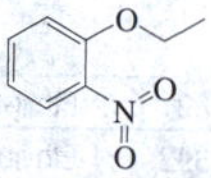
4855
1-Ethoxy-2-nitrobenzene

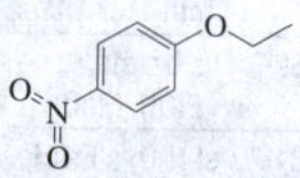
4856
1-Ethoxy-4-nitrobenzene

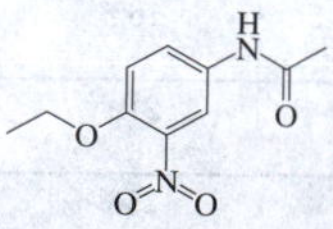
4857
N-(4-Ethoxy-3-nitrophenyl)acetamide

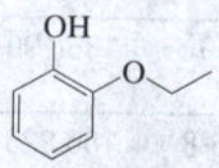
4858
2-Ethoxyphenol

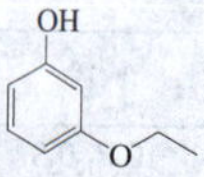
4859
3-Ethoxyphenol

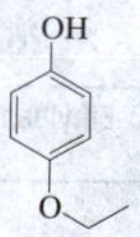
4860
4-Ethoxyphenol

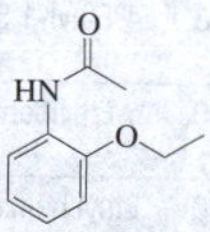
4861
N-(2-Ethoxyphenyl)acetamide

Organic

No.	Name	Synonym	Mol. form.	CAS Reg. No.	Mol. wt.	Physical form	mp/°C	bp/°C	den/ g cm^{-3}	n_D	Solubility
4862	*N*-(4-Ethoxyphenyl)acetamide	Phenacetin	$C_{10}H_{13}NO_2$	62-44-2	179.216	mcl pr	135(3)			1.571	sl H_2O, eth, bz; s EtOH, ace; vs py
4863	*N*-(4-Ethoxyphenyl)-2-hydroxypropanamide	*p*-Lactophenetide	$C_{11}H_{15}NO_3$	539-08-2	209.242		118				s H_2O; vs EtOH; sl eth, bz, chl, peth
4864	(4-Ethoxyphenyl)urea	Dulcin	$C_9H_{12}N_2O_2$	150-69-6	180.203	lf (dil al), pl (w)	173.5	dec			sl H_2O; s EtOH; vs AcOEt
4865	3-Ethoxypropanal		$C_5H_{10}O_2$	2806-85-1	102.132			135.2	0.9165^{20}		
4866	3-Ethoxypropanenitrile		C_5H_9NO	2141-62-0	99.131			171(3)	0.9285^{15}	1.4068^{20}	vs eth, EtOH
4867	8-Ethoxy-5-quinolinesulfonic acid	Actinoquinol	$C_{11}H_{11}NO_4S$	15301-40-3	253.275	br nd (w)	286 dec				s alk
4868	Ethoxytrimethylsilane		$C_5H_{14}OSi$	1825-62-3	118.250			76(3)	0.7573^{20}	1.3741^{20}	i H_2O; s EtOH, eth, ace
4869	Ethoxytriphenylsilane		$C_{20}H_{20}OSi$	1516-80-9	304.458		65	344(3)			s chl
4870	*N*-Ethylacetamide		C_4H_9NO	625-50-3	87.120			205(1)	0.942^{4}	1.4338^{20}	msc H_2O, EtOH; s chl, HOAc
4871	Ethyl acetate		$C_4H_8O_2$	141-78-6	88.106	liq	-83.8(0.3)	77.1(0.2)	0.9003^{20}	1.3723^{20}	s H_2O; msc EtOH, eth; vs ace, bz
4872	Ethyl acetoacetate	Ethyl 3-oxobutanoate	$C_6H_{10}O_3$	141-97-9	130.141	liq	-45	180(2)	1.0368^{10}	1.4171^{20}	s H_2O; msc EtOH, eth; s bz, chl
4873	4'-Ethylacetophenone	1-(4-Ethylphenyl)ethanone	$C_{10}H_{12}O$	937-30-4	148.201			114^{11}			
4874	Ethyl 2-acetylhexanoate		$C_{10}H_{18}O_3$	1540-29-0	186.248			221.5	0.9523^{20}	1.4301^{20}	vs ace, eth
4875	Ethyl 2-acetyl-3-methylbutanoate		$C_9H_{16}O_3$	1522-46-9	172.221			220(14)	0.9648^{18}	1.4256^{18}	i H_2O; msc EtOH, eth
4876	Ethyl 2-acetylpentanoate		$C_9H_{16}O_3$	1540-28-9	172.221			224	0.9661^{20}	1.4255^{20}	vs eth, EtOH
4877	Ethyl 2-acetyl-4-pentenoate	Ethyl 2-allylacetoacetate	$C_9H_{14}O_3$	610-89-9	170.205			208	0.9898^{20}	1.4388^{18}	msc EtOH, eth, bz
4878	Ethyl acrylate	Ethyl propenoate	$C_5H_8O_2$	140-88-5	100.117	liq	-71(2)	98.9(0.6)	0.9234^{20}	1.4068^{20}	sl H_2O, DMSO; msc EtOH, eth; s chl
4879	Ethylamine	Ethanamine	C_2H_7N	75-04-7	45.084	vol liq or gas	-81(2)	16.6(0.2)	0.689^{15}	1.3663^{20}	msc H_2O, EtOH, eth
4880	Ethylamine hydrochloride	Ethanamine hydrochloride	C_2H_8ClN	557-66-4	81.545	mcl pl (al)	109.5		1.2160^{20}		vs H_2O, EtOH
4881	Ethyl 2-aminoacetate	Glycine, ethyl ester	$C_4H_9NO_2$	459-73-4	103.120			149	1.0275^{10}	1.4242^{10}	msc H_2O, EtOH, eth, ace, bz; vs lig
4882	Ethyl 2-aminobenzoate		$C_9H_{11}NO_2$	87-25-2	165.189		14.3(0.2)	271(20)	1.1174^{20}	1.5646^{20}	vs eth, EtOH
4883	Ethyl 3-aminobenzoate		$C_9H_{11}NO_2$	582-33-2	165.189			294	1.171^{20}	1.5600^{22}	sl H_2O; vs EtOH, eth; s ctc
4884	Ethyl 4-aminobenzoate	Ethyl aminobenzoate	$C_9H_{11}NO_2$	94-09-7	165.189	nd (w), orth (eth)	89.6(0.3)	310			i H_2O; vs EtOH, eth; s chl, acid
4885	Ethyl (aminocarbonyl) carbamate		$C_4H_8N_2O_3$	626-36-8	132.118	nd (w, bz)	196.5	dec			i H_2O, eth; sl EtOH, bz, tfa
4886	2-(Ethylamino)ethanol		$C_4H_{11}NO$	110-73-6	89.136			178(8)	0.914^{20}	1.444^{20}	vs H_2O, EtOH, eth; s chl
4887	2-Ethylaniline		$C_8H_{11}N$	578-54-1	121.180	liq	-47(1)	213(4)	0.983^{22}	1.5584^{22}	sl H_2O, chl; vs EtOH, eth
4888	3-Ethylaniline		$C_8H_{11}N$	587-02-0	121.180	liq	-64	214(7)	0.9896^{25}		vs eth, EtOH
4889	4-Ethylaniline		$C_8H_{11}N$	589-16-2	121.180	liq	-5.1(0.3)	217.2(0.2)	0.9679^{20}	1.5554^{20}	sl H_2O, ctc; vs EtOH, eth
4890	*N*-Ethylaniline		$C_8H_{11}N$	103-69-5	121.180	liq	-63.4(0.4)	204(1)	0.9625^{20}	1.5559^{20}	i H_2O; msc EtOH, eth; vs ace, bz; s ctc
4891	2-Ethyl-9,10-anthracenedione		$C_{16}H_{12}O_2$	84-51-5	236.265		108.8				
4892	4-Ethylbenzaldehyde		$C_9H_{10}O$	4748-78-1	134.174			221	0.9790^{20}		
4893	*N*-Ethylbenzamide		$C_9H_{11}NO$	614-17-5	149.189	nd (w)	70.5				
4894	Ethylbenzene	Phenylethane	C_8H_{10}	100-41-4	106.165	liq	-94.95(0.02)	136.2(0.4)	0.8626^{25}	1.4930^{25}	i H_2O; msc EtOH, eth; sl chl
4895	α-Ethylbenzeneacetamide	α-Phenylbutyramide	$C_{10}H_{13}NO$	90-26-6	163.216	cry	86	185^{16}			s H_2O, ctc; sl ace
4896	α-Ethylbenzeneacetic acid		$C_{10}H_{12}O_2$	90-27-7	164.201	pl (eth)	47.5	271			s eth, bz, ctc
4897	α-Ethylbenzeneacetonitrile		$C_{10}H_{11}N$	769-68-6	145.201			241(8)	0.977^{14}		i H_2O; s EtOH, eth, bz
4898	4-Ethyl-1,3-benzenediol		$C_8H_{10}O_2$	2896-60-8	138.164	pr (chl, bz)	98.5	160^{24}			sl H_2O, EtOH, eth
4899	α-Ethylbenzenemethanol	α-Ethylbenzyl alcohol	$C_9H_{12}O$	93-54-9	136.190			219	0.9915^{25}	1.5169^{23}	vs bz, eth, EtOH, MeOH
4900	Ethyl benzenesulfonate		$C_8H_{10}O_3S$	515-46-8	186.228			156^{15}	1.2167^{20}	1.5081^{20}	sl H_2O; s EtOH; vs eth, chl

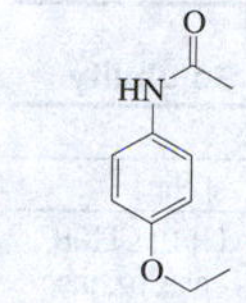
4862
N-(4-Ethoxyphenyl)acetamide

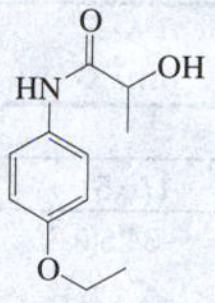
4863
N-(4-Ethoxyphenyl)-2-hydroxypropanamide

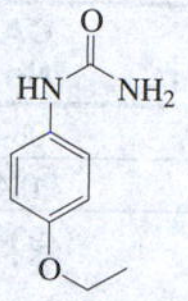
4864
(4-Ethoxyphenyl)urea

4865
3-Ethoxypropanal

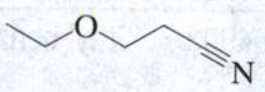
4866
3-Ethoxypropanenitrile

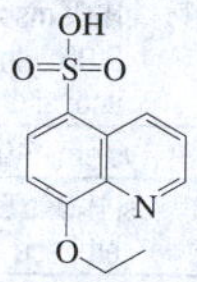
4867
8-Ethoxy-5-quinolinesulfonic acid

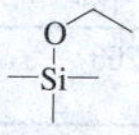
4868
Ethoxytrimethylsilane

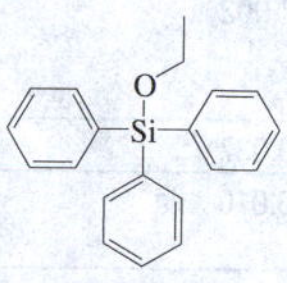
4869
Ethoxytriphenylsilane

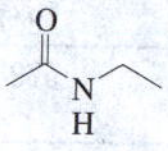
4870
N-Ethylacetamide

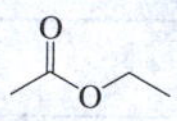
4871
Ethyl acetate

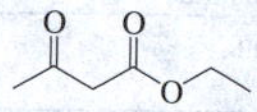
4872
Ethyl acetoacetate

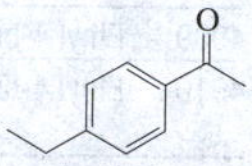
4873
4′-Ethylacetophenone

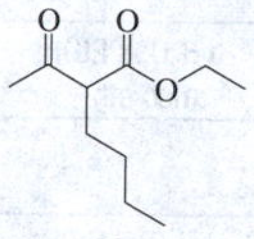
4874
Ethyl 2-acetylhexanoate

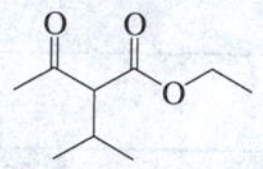
4875
Ethyl 2-acetyl-3-methylbutanoate

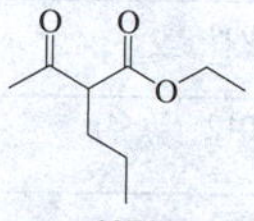
4876
Ethyl 2-acetylpentanoate

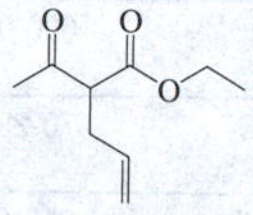
4877
Ethyl 2-acetyl-4-pentenoate

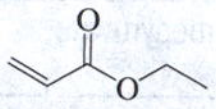
4878
Ethyl acrylate

4879
Ethylamine

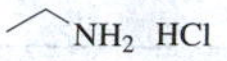
4880
Ethylamine hydrochloride

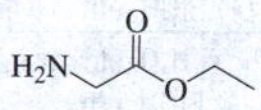
4881
Ethyl 2-aminoacetate

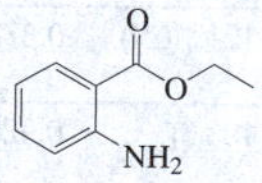
4882
Ethyl 2-aminobenzoate

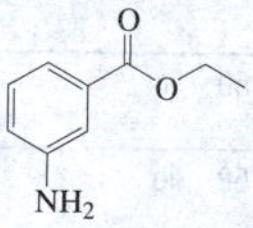
4883
Ethyl 3-aminobenzoate

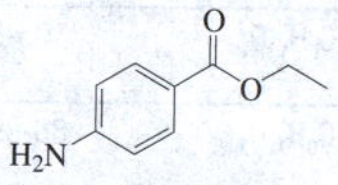
4884
Ethyl 4-aminobenzoate

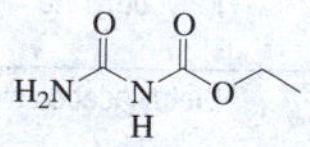
4885
Ethyl (aminocarbonyl)carbamate

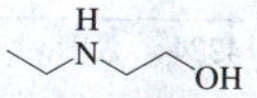
4886
2-(Ethylamino)ethanol

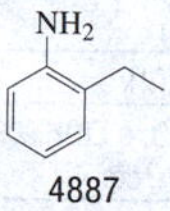
4887
2-Ethylaniline

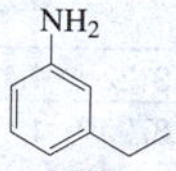
4888
3-Ethylaniline

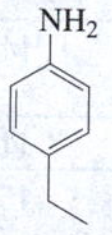
4889
4-Ethylaniline

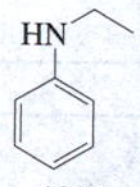
4890
N-Ethylaniline

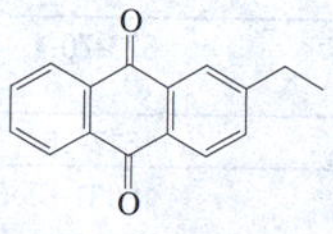
4891
2-Ethyl-9,10-anthracenedione

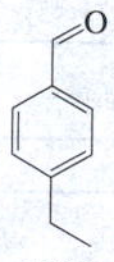
4892
4-Ethylbenzaldehyde

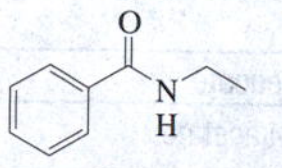
4893
N-Ethylbenzamide

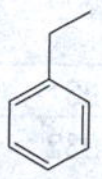
4894
Ethylbenzene

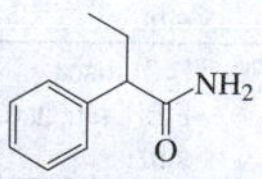
4895
α-Ethylbenzeneacetamide

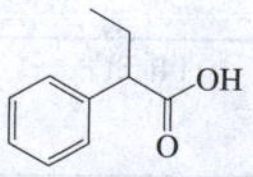
4896
α-Ethylbenzeneacetic acid

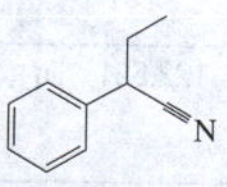
4897
α-Ethylbenzeneacetonitrile

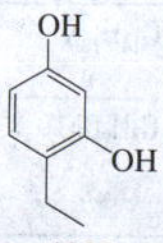
4898
4-Ethyl-1,3-benzenediol

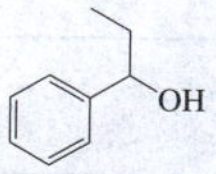
4899
α-Ethylbenzenemethanol

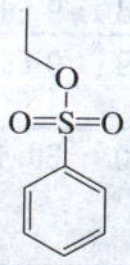
4900
Ethyl benzenesulfonate

No.	Name	Synonym	Mol. form.	CAS Reg. No.	Mol. wt.	Physical form	mp/°C	bp/°C	den/g cm⁻³	n_D	Solubility
4901	4-Ethylbenzenesulfonic acid		$C_8H_{10}O_3S$	98-69-1	186.228				1.23		
4902	2-Ethyl-1*H*-benzimidazole		$C_9H_{10}N_2$	1848-84-6	146.188		176.5				sl chl
4903	Ethyl benzoate	Ethyl benzenecarboxylate	$C_9H_{10}O_2$	93-89-0	150.174	liq	-34.5(0.3)	212.5(0.2)	1.0415^{25}	1.5007^{20}	i H_2O; s EtOH, ace, bz; msc eth; sl ctc
4904	Ethyl 1,3-benzodioxole-5-carboxylate		$C_{10}H_{10}O_4$	6951-08-2	194.184	pr	18.5	285.5			vs eth, EtOH, peth
4905	Ethyl benzoylacetate		$C_{11}H_{12}O_3$	94-02-0	192.211		<0	267 dec	1.1202^{15}	1.5317^{15}	sl H_2O; s EtOH, eth
4906	Ethyl *N*-benzylglycinate		$C_{11}H_{15}NO_2$	6436-90-4	193.243			177^{50}		1.5041^{20}	vs EtOH, eth, bz
4907	Ethyl 2-benzylideneacetoacetate		$C_{13}H_{14}O_3$	620-80-4	218.248	orth pl (dil al)	60.5	296			i H_2O; sl EtOH, eth, bz; vs chl
4908	Ethyl bromoacetate		$C_4H_7BrO_2$	105-36-2	167.002			159(4)	1.5032^{20}	1.4489^{20}	i H_2O; msc EtOH, eth; s ace; sl ctc
4909	Ethyl 4-bromoacetoacetate		$C_6H_9BrO_3$	13176-46-0	209.037			115^{14}	1.5278^{18}	1.5281^{20}	vs eth, EtOH
4910	Ethyl 4-bromobenzoate		$C_9H_9BrO_2$	5798-75-4	229.070	liq	-18	263	1.4332^{17}	1.5438^{17}	sl H_2O; s EtOH, eth, ace, bz
4911	Ethyl 2-bromobutanoate		$C_6H_{11}BrO_2$	533-68-6	195.054			177	1.3273^{20}	1.4475^{20}	i H_2O; msc EtOH, eth; s chl
4912	Ethyl 4-bromobutanoate		$C_6H_{11}BrO_2$	2969-81-5	195.054			192	1.3540^{20}	1.4559^{20}	
4913	Ethyl *trans*-4-bromo-2-butenoate		$C_6H_9BrO_2$	37746-78-4	193.038			100^{14}	1.402^{16}	1.4925^{20}	vs EtOH
4914	Ethyl 6-bromohexanoate	Ethyl 6-bromocaproate	$C_8H_{15}BrO_2$	25542-62-5	223.108	cry (peth)	33	126^{21}	1.238^{23}	1.4566^{21}	
4915	Ethyl 2-bromo-3-methylbutanoate		$C_7H_{13}BrO_2$	609-12-1	209.081			186	1.2760^{20}	1.4496^{20}	vs eth, EtOH
4916	Ethyl 2-bromo-2-methylpropanoate		$C_6H_{11}BrO_2$	600-00-0	195.054			178(5)	1.3263^{20}	1.4446^{20}	i H_2O; s EtOH; msc eth
4917	Ethyl 3-bromo-2-oxopropanoate	Ethyl 3-bromopyruvate	$C_5H_7BrO_3$	70-23-5	195.012			87^{9}			
4918	Ethyl 2-bromopentanoate		$C_7H_{13}BrO_2$	615-83-8	209.081			191	1.226^{18}	1.4496^{20}	i H_2O; s EtOH, eth
4919	Ethyl 5-bromopentanoate		$C_7H_{13}BrO_2$	14660-52-7	209.081			129^{35}	1.3085^{20}	1.4543^{20}	sl ctc
4920	Ethyl 2-bromopropanoate	Ethyl α-bromopropionate	$C_5H_9BrO_2$	535-11-5	181.028			152(13)	1.4135^{20}	1.4490^{20}	i H_2O; msc EtOH, eth; s chl
4921	Ethyl 3-bromopropanoate		$C_5H_9BrO_2$	539-74-2	181.028			163(13)	1.4123^{18}	1.4516^{20}	s EtOH, eth, ace; sl ctc
4922	2-Ethylbutanal	Diethylacetaldehyde	$C_6H_{12}O$	97-96-1	100.158			118^{160}	0.8110^{20}	1.4025^{20}	sl H_2O, ctc; msc EtOH, eth
4923	Ethyl butanoate	Ethyl butyrate	$C_6H_{12}O_2$	105-54-4	116.158	liq	-97(6)	121.1(0.4)	0.8735^{25}	1.3898^{25}	sl H_2O, ctc; s EtOH, eth
4924	2-Ethylbutanoic acid	Diethylacetic acid	$C_6H_{12}O_2$	88-09-5	116.158	liq	-31.8(0.5)	193(3)	0.9239^{20}	1.4132^{20}	sl H_2O, ctc; msc EtOH, eth
4925	2-Ethylbutanoic acid, triethyleneglycol diester		$C_{18}H_{34}O_6$	95-08-9	346.459			$181^{3.5}$			
4926	2-Ethyl-1-butanol	2-Ethylbutyl alcohol	$C_6H_{14}O$	97-95-0	102.174	liq	<-15	155(2)	0.8326^{20}	1.4220^{20}	sl H_2O; s EtOH, eth, chl
4927	2-Ethylbutanoyl chloride		$C_6H_{11}ClO$	2736-40-5	134.603			140(8)	0.9825^{20}	1.4234^{20}	vs eth
4928	2-Ethyl-1-butene		C_6H_{12}	760-21-4	84.159	liq	-132.0(0.2)	64.7(0.1)	0.6894^{20}	1.3969^{20}	i H_2O; s eth, ace, bz, chl
4929	Ethyl *cis*-2-butenoate	Ethyl isocrotonate	$C_6H_{10}O_2$	6776-19-8	114.142			136	0.9182^{20}	1.4242^{20}	vs ace, eth, EtOH
4930	Ethyl *trans*-2-butenoate	Ethyl crotonate	$C_6H_{10}O_2$	623-70-1	114.142			140(5)	0.9175^{20}	1.4243^{20}	i H_2O; s EtOH, eth
4931	Ethyl 3-butenoate		$C_6H_{10}O_2$	1617-18-1	114.142			119	0.9122^{20}	1.4105^{20}	s EtOH
4932	2-Ethylbutyl acetate		$C_8H_{16}O_2$	10031-87-5	144.212		<-100	161(9)	0.8790^{20}	1.4109^{20}	i H_2O; s EtOH, eth, ctc
4933	2-Ethylbutyl acrylate		$C_9H_{16}O_2$	3953-10-4	156.222	liq		80^{20}			
4934	2-Ethylbutylamine	2-Ethyl-1-butanamine	$C_6H_{15}N$	617-79-8	101.190	liq	125				
4935	Ethyl *N*-butylcarbamate		$C_7H_{15}NO_2$	591-62-8	145.200	liq	-22	221(12)	0.9434^{26}	1.4278^{26}	
4936	Ethyl 2-butynoate		$C_6H_8O_2$	4341-76-8	112.127			163	0.9641^{20}	1.4372^{20}	
4937	Ethyl carbamate	Urethane	$C_3H_7NO_2$	51-79-6	89.094	pr (bz, to)	48.25(0.04)	185	0.9862^{21}	1.4144^{51}	vs H_2O, EtOH, eth, bz, chl, py; sl lig
4938	9-Ethyl-9*H*-carbazol-3-amine		$C_{14}H_{14}N_2$	132-32-1	210.274		99				
4939	9-Ethyl-9*H*-carbazole		$C_{14}H_{13}N$	86-28-2	195.260	nd (al)	67(1)	190^{10}	1.059^{80}	1.6394^{80}	i H_2O; vs EtOH, eth
4940	Ethyl chloroacetate		$C_4H_7ClO_2$	105-39-5	122.551	liq	-21	144(3)	1.1585^{20}	1.4215^{20}	i H_2O; msc EtOH, eth, ace; s bz
4941	Ethyl 4-chloroacetoacetate		$C_6H_9ClO_3$	638-07-3	164.586		-8	220 dec	1.218^{25}	1.4520^{20}	

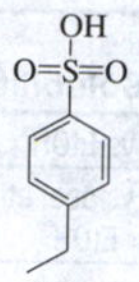
4901
4-Ethylbenzenesulfonic acid

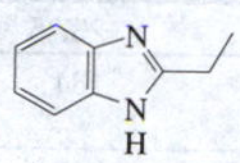
4902
2-Ethyl-1*H*-benzimidazole

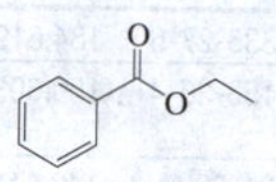
4903
Ethyl benzoate

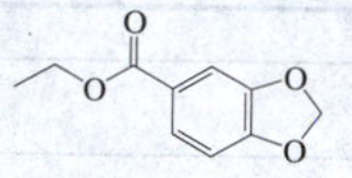
4904
Ethyl 1,3-benzodioxole-5-carboxylate

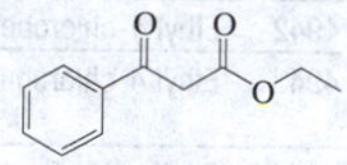
4905
Ethyl benzoylacetate

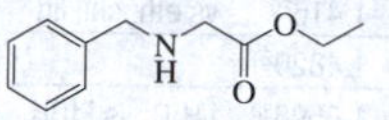
4906
Ethyl *N*-benzylglycinate

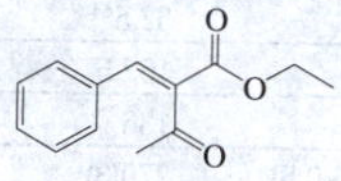
4907
Ethyl 2-benzylideneacetoacetate

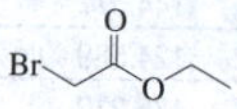
4908
Ethyl bromoacetate

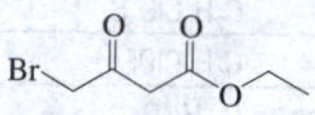
4909
Ethyl 4-bromoacetoacetate

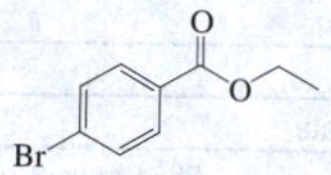
4910
Ethyl 4-bromobenzoate

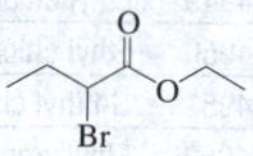
4911
Ethyl 2-bromobutanoate

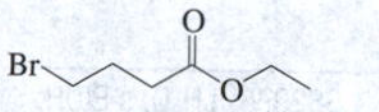
4912
Ethyl 4-bromobutanoate

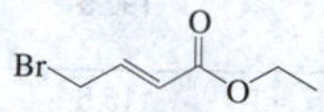
4913
Ethyl *trans*-4-bromo-2-butenoate

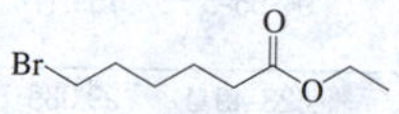
4914
Ethyl 6-bromohexanoate

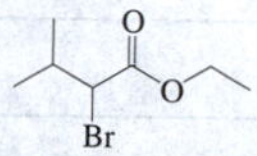
4915
Ethyl 2-bromo-3-methylbutanoate

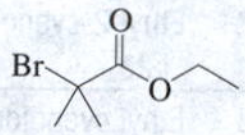
4916
Ethyl 2-bromo-2-methylpropanoate

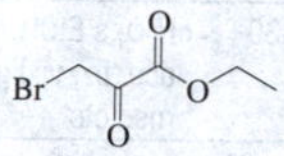
4917
Ethyl 3-bromo-2-oxopropanoate

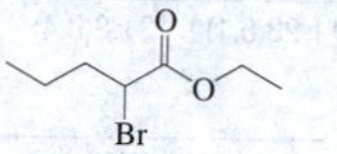
4918
Ethyl 2-bromopentanoate

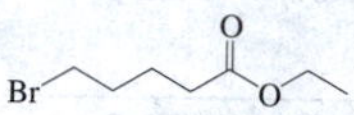
4919
Ethyl 5-bromopentanoate

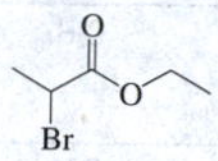
4920
Ethyl 2-bromopropanoate

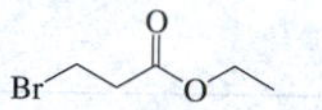
4921
Ethyl 3-bromopropanoate

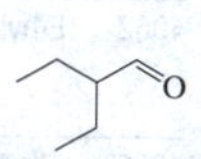
4922
2-Ethylbutanal

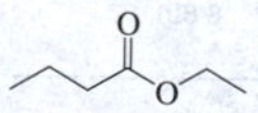
4923
Ethyl butanoate

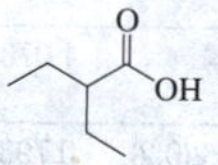
4924
2-Ethylbutanoic acid

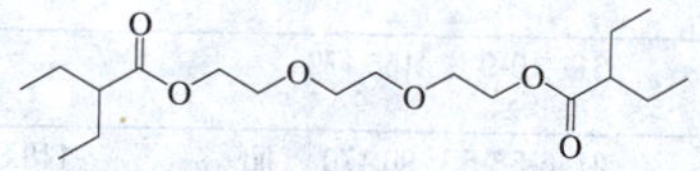
4925
2-Ethylbutanoic acid, triethyleneglycol diester

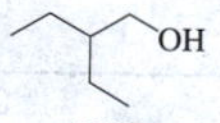
4926
2-Ethyl-1-butanol

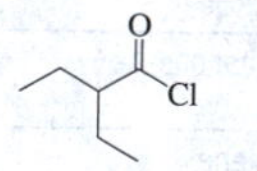
4927
2-Ethylbutanoyl chloride

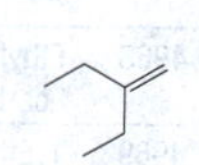
4928
2-Ethyl-1-butene

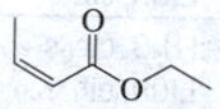
4929
Ethyl *cis*-2-butenoate

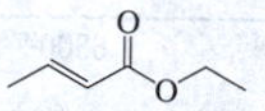
4930
Ethyl *trans*-2-butenoate

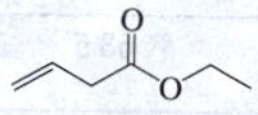
4931
Ethyl 3-butenoate

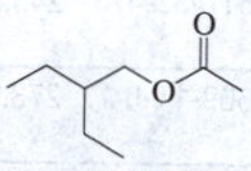
4932
2-Ethylbutyl acetate

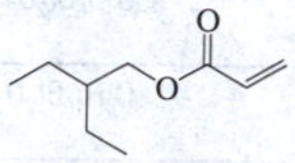
4933
2-Ethylbutyl acrylate

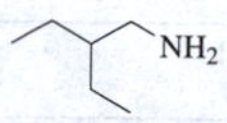
4934
2-Ethylbutylamine

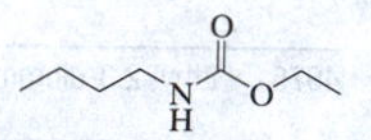
4935
Ethyl *N*-butylcarbamate

4936
Ethyl 2-butynoate

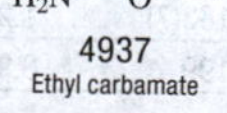
4937
Ethyl carbamate

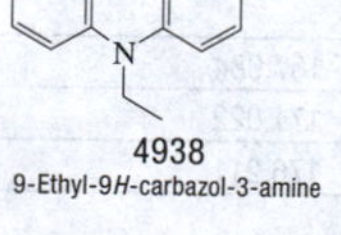
4938
9-Ethyl-9*H*-carbazol-3-amine

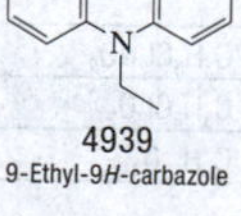
4939
9-Ethyl-9*H*-carbazole

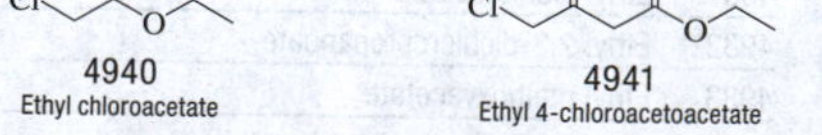
4940
Ethyl chloroacetate

4941
Ethyl 4-chloroacetoacetate

No.	Name	Synonym	Mol. form.	CAS Reg. No.	Mol. wt.	Physical form	mp/°C	bp/°C	den/ g cm^{-3}	n_D	Solubility
4942	Ethyl 4-chlorobenzoate		$C_9H_9ClO_2$	7335-27-5	184.619			237.5	1.1873^{14}		vs EtOH
4943	Ethyl 4-chlorobutanoate		$C_6H_{11}ClO_2$	3153-36-4	150.603			184	1.0756^{20}	1.4311^{20}	vs ace, eth, EtOH
4944	Ethyl chlorofluoroacetate		$C_4H_6ClFO_2$	401-56-9	140.541			129	1.225^{20}	1.3927^{20}	
4945	Ethyl chloroformate		$C_3H_5ClO_2$	541-41-3	108.524	liq	-80.6	91(2)	1.1352^{20}	1.3974^{20}	vs bz, eth, chl
4946	Ethyl 2-chloro-2-oxoacetate	Ethyl oxalyl chloride	$C_4H_5ClO_3$	4755-77-5	136.534	hyg		137	1.2226^{20}		vs bz, eth
4947	Ethyl 2-chloropropanoate	Ethyl α-chloropropionate	$C_5H_9ClO_2$	535-13-7	136.577			147(4)	1.0793^{20}	1.4178^{20}	i H_2O; msc EtOH, eth; sl ctc
4948	Ethyl 3-chloropropanoate		$C_5H_9ClO_2$	623-71-2	136.577			162	1.1086^{20}	1.4254^{20}	sl H_2O; msc EtOH, eth
4949	Ethyl chlorosulfinate		$C_2H_5ClO_2S$	6378-11-6	128.578			52.5^{44}	1.2837^{20}	1.4550^{25}	vs eth
4950	Ethyl chlorosulfonate		$C_2H_5ClO_3S$	625-01-4	144.577			152.5	1.3502^{25}	1.416^{20}	vs eth, chl, lig
4951	*S*-Ethyl chlorothioformate		C_3H_5ClOS	2941-64-2	124.589	liq		136	1.195^{20}	1.4820^{20}	
4952	Ethyl *trans*-cinnamate	Ethyl *trans*-3-phenyl-2-propenoate	$C_{11}H_{12}O_2$	4192-77-2	176.212		6.5(0.5)	270(3)	1.0491^{20}	1.5598^{20}	i H_2O; vs EtOH, eth, ace; s bz, ctc
4953	Ethyl cyanate		C_3H_5NO	627-48-5	71.078			162 dec	0.89^{20}	1.3788^{25}	vs eth, EtOH
4954	Ethyl cyanoacetate		$C_5H_7NO_2$	105-56-6	113.116	liq	-26.1(0.1)	216(4)	1.0654^{20}	1.4175^{20}	s H_2O; vs eth, EtOH
4955	Ethyl 2-cyanoacrylate	Ethyl 2-cyano-2-propenoate	$C_6H_7NO_2$	7085-85-0	125.126	liq		55^{3}			
4956	Ethyl 2-cyano-3,3-diphenyl-2-propenoate	Etocrilene	$C_{18}H_{15}NO_2$	5232-99-5	277.318		110.5	195^{3}			
4957	Ethyl 2-cyano-3-ethoxyacrylate		$C_8H_{11}NO_3$	94-05-3	169.178		52	190.5			
4958	Ethyl cyanoformate		$C_4H_5NO_2$	623-49-4	99.089			115.5	1.003^{25}	1.3820^{20}	i H_2O; s EtOH, eth, ctc
4959	Ethyl 2-cyano-2-phenylacetate		$C_{11}H_{11}NO_2$	4553-07-5	189.211	oil		275 dec	1.091^{20}	1.5012^{25}	vs ace, bz, eth, EtOH
4960	Ethyl 2-cyano-3-phenyl-2-propenoate	Ethyl 2-benzylidene-2-cyanoacetate	$C_{12}H_{11}NO_2$	2025-40-3	201.221	(i) nd (al) (ii) oil	51	188^{15}	1.1076^{25}	1.5033	vs ace, chl
4961	Ethylcyclobutane		C_6H_{12}	4806-61-5	84.159	liq	-142.9(0.2)	70(1)	0.7284^{20}	1.4020^{20}	i H_2O; msc EtOH, eth; s ace, bz, peth
4962	Ethylcyclohexane		C_8H_{16}	1678-91-7	112.213	liq	-111.28(0.1)	131.8(0.4)	0.7880^{20}	1.4330^{20}	i H_2O; s EtOH, ace, bz; vs lig; msc ctc
4963	Ethyl cyclohexanecarboxylate		$C_9H_{16}O_2$	3289-28-9	156.222			196(4)	0.9362^{20}	1.4501^{15}	vs ace, eth, EtOH, chl
4964	1-Ethylcyclohexene		C_8H_{14}	1453-24-3	110.197	liq	-109.9(0.1)	136(3)	0.8176^{25}	1.4567^{20}	
4965	Ethyl 3-cyclohexene-1-carboxylate		$C_9H_{14}O_2$	15111-56-5	154.206			194.5	0.9688^{20}	1.4578^{20}	
4966	Ethyl cyclohexylacetate		$C_{10}H_{18}O_2$	5452-75-5	170.249			211	0.9537^{14}	1.451^{14}	
4967	Ethylcyclopentane		C_7H_{14}	1640-89-7	98.186	liq	-138.42(0.02)	103.5(0.6)	0.7665^{20}	1.4198^{20}	i H_2O; msc EtOH, eth, ace; s bz, tol
4968	Ethyl 2-cyclopentanone-1-carboxylate		$C_8H_{12}O_3$	611-10-9	156.179			221	1.0781^{21}	1.4519^{20}	s eth, bz
4969	1-Ethylcyclopentene		C_7H_{12}	2146-38-5	96.170	liq	-118.8(0.4)	106.2(0.2)	0.7936^{25}	1.4412^{20}	
4970	Ethylcyclopropane		C_5H_{10}	1191-96-4	70.133	liq	-149.4(0.2)	36(2)	0.6790^{25}	1.3786^{20}	
4971	Ethyl cyclopropanecarboxylate		$C_6H_{10}O_2$	4606-07-9	114.142			128(5)	0.9608^{15}	1.4190^{20}	
4972	Ethyl decanoate	Ethyl caprate	$C_{12}H_{24}O_2$	110-38-3	200.318	liq	-20(2)	242(1)	0.8650^{20}	1.4256^{20}	i H_2O; vs eth, EtOH, chl
4973	Ethyl diazoacetate	Diazoacetic ester	$C_4H_6N_2O_2$	623-73-4	114.103	ye orth cry	-22	140 dec	1.0852^{18}	1.4605^{20}	sl H_2O; msc EtOH, eth, bz, lig
4974	Ethyl dibromoacetate		$C_4H_6Br_2O_2$	617-33-4	245.898			194	1.8991^{20}	1.5017^{13}	i H_2O; msc EtOH, eth
4975	Ethyl 2,3-dibromobutanoate		$C_6H_{10}Br_2O_2$	609-11-0	273.950	nd	58.5	113^{30}	1.6800^{20}		sl H_2O, ctc; s EtOH, eth
4976	Ethyl 2,4-dibromobutanoate		$C_6H_{10}Br_2O_2$	36847-51-5	273.950			149^{52}	1.6987^{20}	1.4960^{20}	i H_2O; s EtOH, eth
4977	Ethyl 2,3-dibromopropanoate		$C_5H_8Br_2O_2$	3674-13-3	259.925			214.5	1.7966^{20}	1.5007^{20}	s EtOH, eth
4978	Ethyl 3,6-di(*tert*-butyl)-1-naphthalenesulfonate	Ethyl dibunate	$C_{20}H_{28}O_3S$	5560-69-0	348.499						s chl
4979	Ethyl dichloroacetate		$C_4H_6Cl_2O_2$	535-15-9	156.996			155	1.2827^{20}	1.4386^{20}	sl H_2O; msc EtOH, eth; s ace, chl
4980	Ethyldichloroarsine	Dichloroethylarsine	$C_2H_5AsCl_2$	598-14-1	174.889			155.3	1.66^{20}		s H_2O; misc EtOH, bz
4981	Ethyl dichlorocarbamate		$C_3H_5Cl_2NO_2$	13698-16-3	157.984			66^{18}	1.304^{30}	1.4595^{20}	
4982	Ethyl 2,3-dichloropropanoate		$C_5H_8Cl_2O_2$	6628-21-3	171.022			183(7)	1.2401^{20}	1.4482^{20}	vs eth, EtOH
4983	Ethyl diethoxyacetate		$C_8H_{16}O_4$	6065-82-3	176.211			199	0.985^{25}	1.4100^{20}	

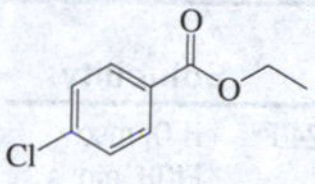
4942
Ethyl 4-chlorobenzoate

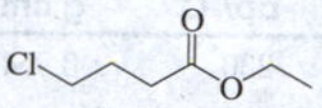
4943
Ethyl 4-chlorobutanoate

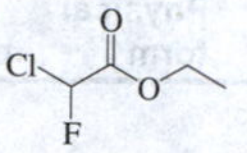
4944
Ethyl chlorofluoroacetate

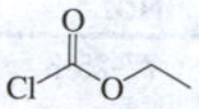
4945
Ethyl chloroformate

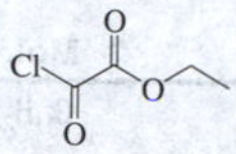
4946
Ethyl 2-chloro-2-oxoacetate

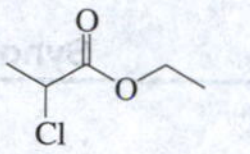
4947
Ethyl 2-chloropropanoate

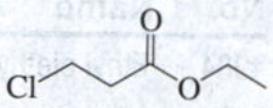
4948
Ethyl 3-chloropropanoate

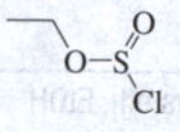
4949
Ethyl chlorosulfinate

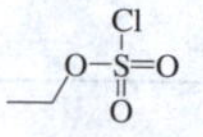
4950
Ethyl chlorosulfonate

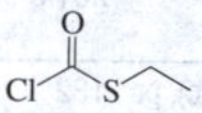
4951
S-Ethyl chlorothioformate

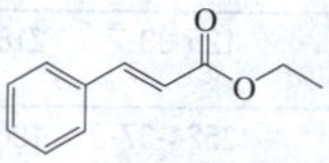
4952
Ethyl *trans*-cinnamate

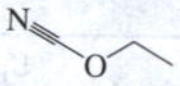
4953
Ethyl cyanate

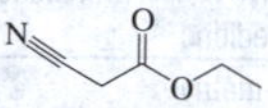
4954
Ethyl cyanoacetate

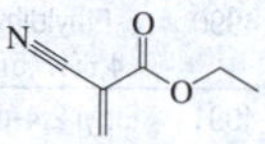
4955
Ethyl 2-cyanoacrylate

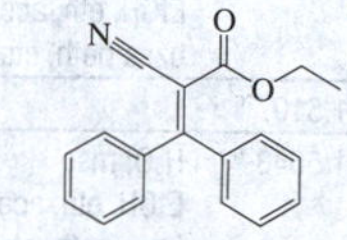
4956
Ethyl 2-cyano-3,3-diphenyl-2-propenoate

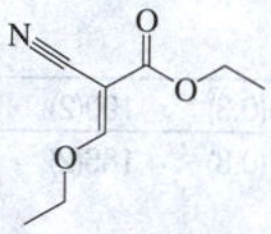
4957
Ethyl 2-cyano-3-ethoxyacrylate

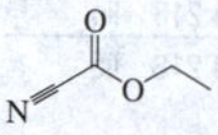
4958
Ethyl cyanoformate

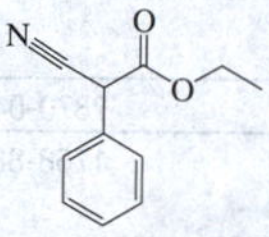
4959
Ethyl 2-cyano-2-phenylacetate

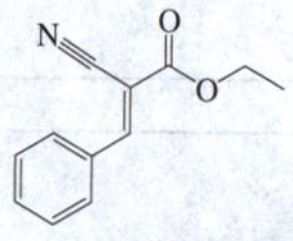
4960
Ethyl 2-cyano-3-phenyl-2-propenoate

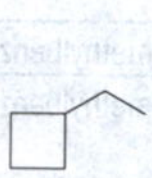
4961
Ethylcyclobutane

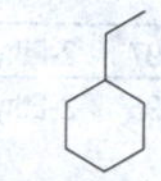
4962
Ethylcyclohexane

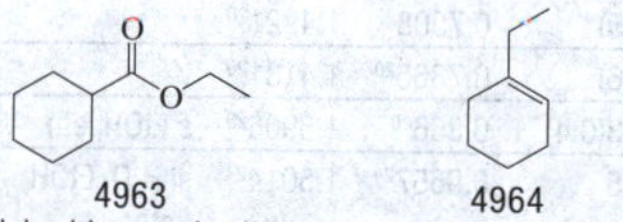
4963
Ethyl cyclohexanecarboxylate

4964
1-Ethylcyclohexene

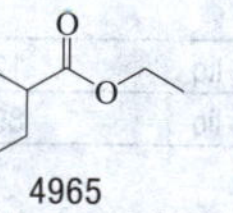
4965
Ethyl 3-cyclohexene-1-carboxylate

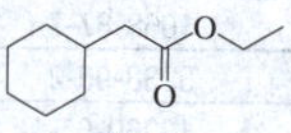
4966
Ethyl cyclohexylacetate

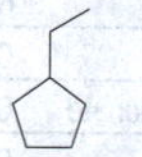
4967
Ethylcyclopentane

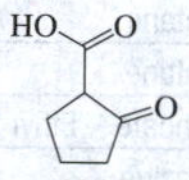
4968
Ethyl 2-cyclopentanone-1-carboxylate

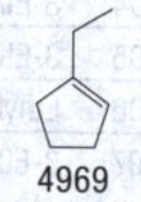
4969
1-Ethylcyclopentene

4970
Ethylcyclopropane

4971
Ethyl cyclopropanecarboxylate

4972
Ethyl decanoate

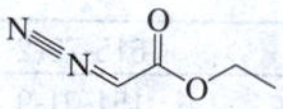
4973
Ethyl diazoacetate

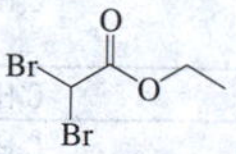
4974
Ethyl dibromoacetate

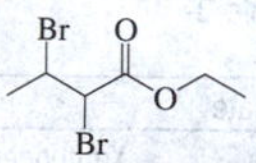
4975
Ethyl 2,3-dibromobutanoate

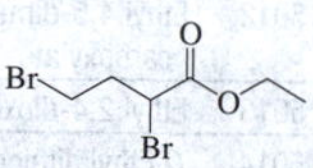
4976
Ethyl 2,4-dibromobutanoate

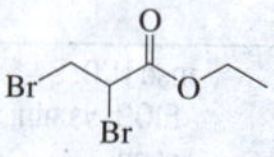
4977
Ethyl 2,3-dibromopropanoate

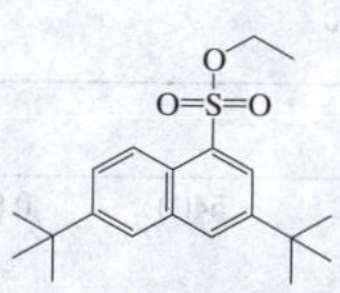
4978
Ethyl 3,6-di(*tert*-butyl)-1-naphthalenesulfonate

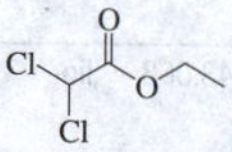
4979
Ethyl dichloroacetate

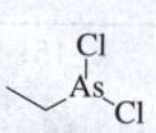
4980
Ethyldichloroarsine

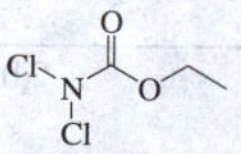
4981
Ethyl dichlorocarbamate

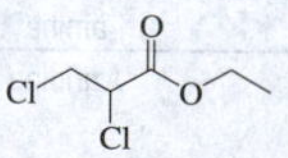
4982
Ethyl 2,3-dichloropropanoate

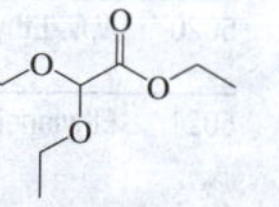
4983
Ethyl diethoxyacetate

No.	Name	Synonym	Mol. form.	CAS Reg. No.	Mol. wt.	Physical form	mp/°C	bp/°C	den/ g cm^{-3}	n_D	Solubility
4984	Ethyl diethylmalonate		$C_{11}H_{20}O_4$	77-25-8	216.275			230	0.9643^{30}	1.4240^{20}	i H_2O; msc EtOH, eth; s ctc
4985	Ethyl difluoroacetate		$C_4H_6F_2O_2$	454-31-9	124.087			100	1.1765^{20}		i H_2O
4986	Ethyldifluoroarsine		$C_2H_5AsF_2$	430-40-0	141.980	liq, fumes in air	-38.7	94.3	1.708^{17}		
4987	5-Ethyldihydro-5-*sec*-butyl-2-thioxo-4,6(1*H*,5*H*)-pyrimidinedione	Thiobutabarbital	$C_{10}H_{16}N_2O_2S$	2095-57-0	228.311		169				
4988	5-Ethyldihydro-2(3*H*)-furanone		$C_6H_{10}O_2$	695-06-7	114.142	liq	-18	218(4)	1.0261^{20}	1.4495^{20}	vs H_2O, EtOH
4989	Ethyl dihydrogen phosphate		$C_2H_7O_4P$	1623-14-9	126.048	hyg cry		dec	1.430^{25}	1.427	vs H_2O, ace, eth, EtOH
4990	5-Ethyldihydro-5-phenyl-4,6(1*H*,5*H*)-pyrimidinedione	Primidone	$C_{12}H_{14}N_2O_2$	125-33-7	218.251		281.5				
4991	Ethyl 2,4-dihydroxy-6-methylbenzoate		$C_{10}H_{12}O_4$	2524-37-0	196.200	lf (HOAc), pr (al)	132	sub			vs eth, EtOH
4992	*O*-Ethyl *S*-[2-(diisopropylamino)ethyl] methylphosphonothioate	VX Nerve agent	$C_{11}H_{26}NO_2PS$	50782-69-9	267.369	very toxic liq					
4993	Ethyldimethylamine	*N,N*-Dimethylethanamine	$C_4H_{11}N$	598-56-1	73.137	liq	-140(4)	36.5(0.6)	0.675^{20}	1.3705^{25}	
4994	Ethyl 4-(dimethylamino) benzoate		$C_{11}H_{15}NO_2$	10287-53-3	193.243		66.5	190^{14}	1.0099^{100}		
4995	1-Ethyl-2,4-dimethylbenzene		$C_{10}H_{14}$	874-41-9	134.218	liq	-62.9(0.2)	188(2)	0.8763^{20}	1.5038^{20}	vs ace, bz, eth, EtOH
4996	1-Ethyl-3,5-dimethylbenzene		$C_{10}H_{14}$	934-74-7	134.218	liq	-84.3(0.1)	184(2)	0.8608^{25}	1.4981^{20}	i H_2O; msc EtOH, eth, ace, bz; s peth, ctc
4997	2-Ethyl-1,3-dimethylbenzene		$C_{10}H_{14}$	2870-04-4	134.218	liq	-16.3(0.3)	190(2)	0.8864^{25}	1.5107^{20}	
4998	2-Ethyl-1,4-dimethylbenzene		$C_{10}H_{14}$	1758-88-9	134.218	liq	-53.7(0.3)	186(2)	0.8732^{25}	1.5043^{20}	i H_2O; msc EtOH, eth, ace, bz; s peth, ctc
4999	3-Ethyl-1,2-dimethylbenzene	3-Ethyl-*o*xylene	$C_{10}H_{14}$	933-98-2	134.218	liq	-49.5(0.3)	194(2)	0.8881^{25}	1.5117^{20}	
5000	4-Ethyl-1,2-dimethylbenzene		$C_{10}H_{14}$	934-80-5	134.218	liq	-66.9(0.2)	189(2)	0.8706^{25}	1.5031^{20}	i H_2O; msc EtOH, eth, ace, bz; s peth, ctc
5001	*N'*-Ethyl-*N,N*-dimethyl-1,2-ethanediamine		$C_6H_{16}N_2$	123-83-1	116.204			134.5	0.738^{25}	1.4222^{20}	
5002	Ethyl 4,4-dimethyl-3-oxopentanoate	Ethyl pivaloylacetate	$C_9H_{16}O_3$	17094-34-7	172.221	liq		83^{17}	0.97^{18}		
5003	3-Ethyl-2,2-dimethylpentane		C_9H_{20}	16747-32-3	128.255	liq	-99.4(0.1)	134(4)	0.7438^{20}	1.4123^{20}	
5004	3-Ethyl-2,3-dimethylpentane		C_9H_{20}	16747-33-4	128.255			142(5)	0.7508^{25}	1.4221^{20}	
5005	3-Ethyl-2,4-dimethylpentane		C_9H_{20}	1068-87-7	128.255	liq	-122.38(0.09)	123(6)	0.7365^{20}	1.4131^{20}	
5006	Ethyl 2,2-dimethylpropanoate	Ethyl 2,2-dimethylpropionate	$C_7H_{14}O_2$	3938-95-2	130.185	liq	-89.5(0.2)	118.3(0.4)	0.856^{20}	1.3906^{20}	s EtOH, eth
5007	3-Ethyl-2,5-dimethylpyrazine		$C_8H_{12}N_2$	13360-65-1	136.194			180.5	0.9657^{24}	1.5014^{24}	sl H_2O, EtOH, eth
5008	3-Ethyl-2,4-dimethyl-1*H*-pyrrole		$C_8H_{13}N$	517-22-6	123.196	pr	0	197(6)	0.913^{20}	1.4961^{20}	sl H_2O; s EtOH, eth, bz, chl
5009	Ethyl 3,5-dimethylpyrrole-2-carboxylate		$C_9H_{13}NO_2$	2199-44-2	167.205	cry (al)	125	135$^{10.5}$			s EtOH, ace
5010	Ethyl 2,4-dimethylpyrrole-3-carboxylate		$C_9H_{13}NO_2$	2199-51-1	167.205	cry (eth-lig, peth)	78.5	291			vs eth, EtOH
5011	Ethyl 2,5-dimethylpyrrole-3-carboxylate		$C_9H_{13}NO_2$	2199-52-2	167.205	orth (al)	117.5	291			vs EtOH
5012	Ethyl 4,5-dimethylpyrrole-3-carboxylate		$C_9H_{13}NO_2$	2199-53-3	167.205	cry (dil al)	111.3				vs eth, EtOH, chl
5013	Ethyl 2,4-dioxopentanoate		$C_7H_{10}O_4$	615-79-2	158.152		18	214	1.1251^{20}	1.4757^{17}	vs eth, EtOH
5014	*O*-Ethyl dithiocarbonate	Xanthogenic acid	$C_3H_6OS_2$	151-01-9	122.209	unstab liq	-53	25			
5015	Ethylene	Ethene	C_2H_4	74-85-1	28.053	col gas	-169.18(0.02)	-103.8(0.3)	0.5678^{-104}	1.363^{-100}	i H_2O; sl EtOH, bz, ace; s eth
5016	Ethylenebisdithiocarbamic acid		$C_4H_8N_2S_4$	111-54-6	212.380	unstab liq					
5017	Ethylene carbonate	Vinylene carbonate	$C_3H_4O_3$	96-49-1	88.062	mcl pl (al)	36.331 (0.004)	246(1)	1.3214^{39}	1.4148^{50}	msc H_2O, EtOH, eth, bz, chl, AcOEt
5018	Ethylenediaminetetraacetic acid	EDTA	$C_{10}H_{16}N_2O_8$	60-00-4	292.242	cry (w)	245 dec				
5019	Ethylenediaminetetraacetic acid, disodium salt, dihydrate	EDTA disodium	$C_{10}H_{18}N_2Na_2O_{10}$	6381-92-6	372.237		242 dec				
5020	*N,N'*-Ethylene distearylamide	*N,N'*-Dioctadecanoylethanediamine	$C_{38}H_{76}N_2O_2$	110-30-5	593.022	cry (EtOH)	149				
5021	Ethyleneimine	Aziridine	C_2H_5N	151-56-4	43.068	liq	-78.0(0.5)	54(1)	0.832^{25}		msc H_2O; s EtOH; vs eth; sl chl

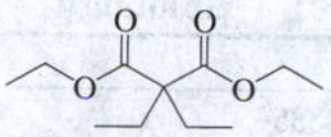
4984
Ethyl diethylmalonate

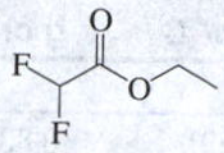
4985
Ethyl difluoroacetate

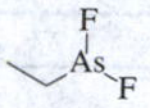
4986
Ethyldifluoroarsine

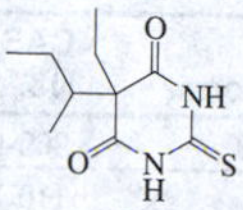
4987
5-Ethyldihydro-5-*sec*-butyl-2-thioxo-4,6(1*H*,5*H*)-pyrimidinedione

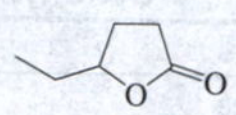
4988
5-Ethyldihydro-2(3*H*)-furanone

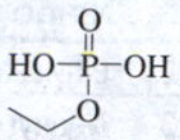
4989
Ethyl dihydrogen phosphate

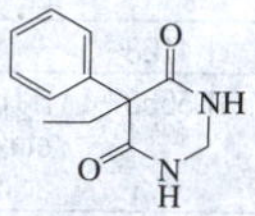
4990
5-Ethyldihydro-5-phenyl-4,6(1*H*,5*H*)-pyrimidinedione

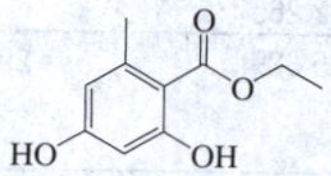
4991
Ethyl 2,4-dihydroxy-6-methylbenzoate

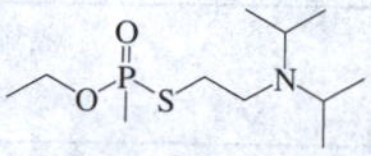
4992
O-Ethyl *S*-[2-(diisopropylamino)ethyl] methylphosphonothioate

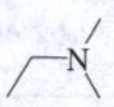
4993
Ethyldimethylamine

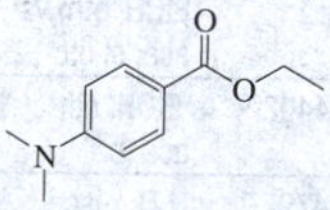
4994
Ethyl 4-(dimethylamino)benzoate

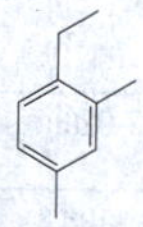
4995
1-Ethyl-2,4-dimethylbenzene

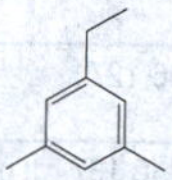
4996
1-Ethyl-3,5-dimethylbenzene

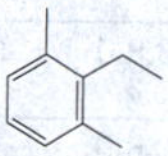
4997
2-Ethyl-1,3-dimethylbenzene

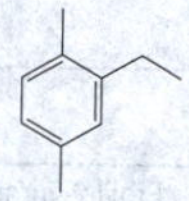
4998
2-Ethyl-1,4-dimethylbenzene

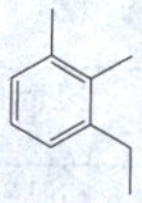
4999
3-Ethyl-1,2-dimethylbenzene

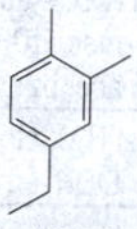
5000
4-Ethyl-1,2-dimethylbenzene

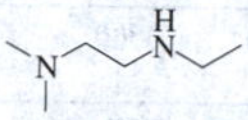
5001
N'-Ethyl-*N*,*N*-dimethyl-1,2-ethanediamine

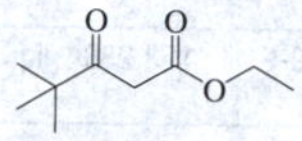
5002
Ethyl 4,4-dimethyl-3-oxopentanoate

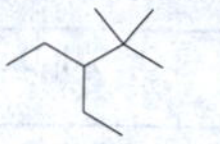
5003
3-Ethyl-2,2-dimethylpentane

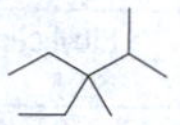
5004
3-Ethyl-2,3-dimethylpentane

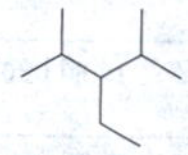
5005
3-Ethyl-2,4-dimethylpentane

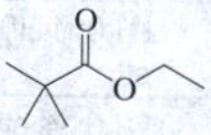
5006
Ethyl 2,2-dimethylpropanoate

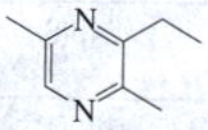
5007
3-Ethyl-2,5-dimethylpyrazine

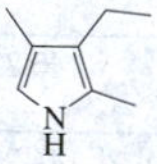
5008
3-Ethyl-2,4-dimethyl-1*H*-pyrrole

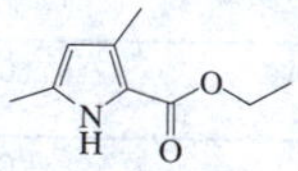
5009
Ethyl 3,5-dimethylpyrrole-2-carboxylate

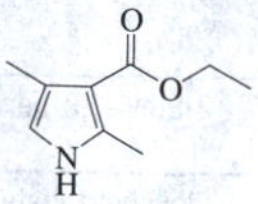
5010
Ethyl 2,4-dimethylpyrrole-3-carboxylate

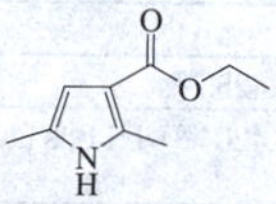
5011
Ethyl 2,5-dimethylpyrrole-3-carboxylate

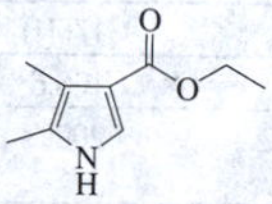
5012
Ethyl 4,5-dimethylpyrrole-3-carboxylate

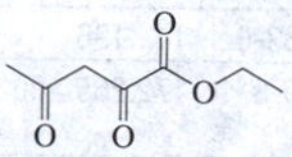
5013
Ethyl 2,4-dioxopentanoate

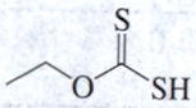
5014
O-Ethyl dithiocarbonate

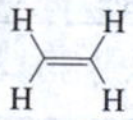
5015
Ethylene

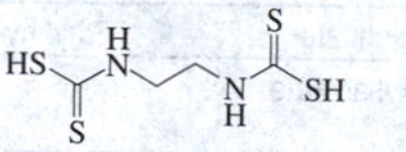
5016
Ethylenebisdithiocarbamic acid

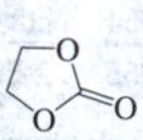
5017
Ethylene carbonate

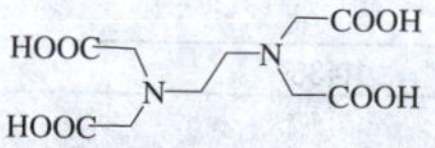
5018
Ethylenediaminetetraacetic acid

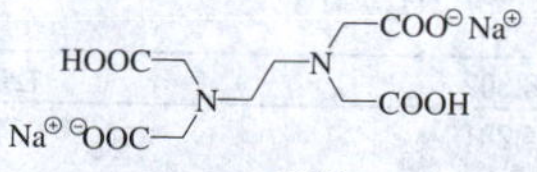
5019
Ethylenediaminetetraacetic acid, disodium salt, dihydrate

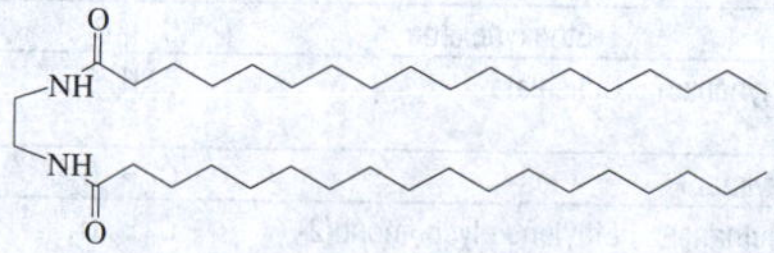
5020
N,*N'*-Ethylene distearylamide

5021
Ethyleneimine

Organic

No.	Name	Synonym	Mol. form.	CAS Reg. No.	Mol. wt.	Physical form	mp/°C	bp/°C	den/g cm^{-3}	n_D	Solubility
5022	Ethylestrenol		$C_{20}H_{32}O$	965-90-2	288.467	cry	77				
5023	*N*-Ethyl-1,2-ethanediamine		$C_4H_{12}N_2$	110-72-5	88.151			129	0.837^{25}	1.4385^{20}	
5024	Ethyl ethoxyacetate		$C_6H_{12}O_3$	817-95-8	132.157			158	0.9702^{20}	1.4039^{20}	s EtOH, eth, ace
5025	Ethyl 3-ethoxypropanoate		$C_7H_{14}O_3$	763-69-9	146.184			168(2)	0.9490^{20}	1.4065^{20}	
5026	Ethyl 2-ethoxy-1(2*H*)-quinolinecarboxylate	EEDQ	$C_{14}H_{17}NO_3$	16357-59-8	247.290		56.5	$126^{0.1}$			s chl
5027	Ethyl 2-ethylacetoacetate		$C_8H_{14}O_3$	607-97-6	158.195			197(4)	0.9847^{16}	1.4214^{25}	msc EtOH, eth
5028	Ethyl ethylcarbamate		$C_5H_{11}NO_2$	623-78-9	117.147			176	0.9813^{20}	1.4215^{20}	vs H_2O, eth, EtOH
5029	Ethyl 2-ethylhexanoate	Ethyl 2-ethylcaproate	$C_{10}H_{20}O_2$	2983-37-1	172.265			90^{28}	0.8586^{25}	1.4123^{25}	
5030	2-Ethyl-*N*-(2-ethylphenyl)-aniline		$C_{16}H_{19}N$	64653-59-4	225.329		29	336^{603}		1.5550^{25}	i H_2O; vs EtOH, eth; sl chl; s acid
5031	*O*-Ethyl ethylthiophosphonyl chloride		$C_4H_{10}ClOPS$	1497-68-3	172.613	liq		$35^{0.7}$	1.15^{20}		
5032	Ethyl fluoroacetate		$C_4H_7FO_2$	459-72-3	106.096			120	1.0912^{20}	1.3755^{20}	vs H_2O
5033	Ethyl 4-fluorobenzoate		$C_9H_9FO_2$	451-46-7	168.164	mcl pr (w)	26	210	1.146^{25}	1.4864^{20}	vs eth, EtOH
5034	*N*-Ethylformamide		C_3H_7NO	627-45-2	73.094			198	0.9552^{20}	1.4320^{20}	msc H_2O, EtOH, eth
5035	Ethyl formate		$C_3H_6O_2$	109-94-4	74.079	liq	-79.6(0.5)	54.09(0.1)	0.9208^{20}	1.3609^{20}	s H_2O; msc EtOH, eth; vs ace; sl ctc
5036	2-Ethylfuran		C_6H_8O	3208-16-0	96.127			92(3)	0.9018^{20}	1.4403^{20}	s EtOH, eth, ace, bz
5037	Ethyl 2-furancarboxylate	Ethyl 2-furanoate	$C_7H_8O_3$	614-99-3	140.137	lf or pr	34.5	196.8	1.1174^{21}	1.4797^{21}	i H_2O; msc EtOH, eth, ace; s bz
5038	γ-Ethyl L-glutamate		$C_7H_{13}NO_4$	1119-33-1	175.183		191				sl H_2O
5039	Ethyl heptafluorobutanoate		$C_6H_5F_7O_2$	356-27-4	242.092			95	1.394^{20}	1.3011^{20}	sl H_2O; s eth, ace
5040	3-Ethylheptane		C_9H_{20}	15869-80-4	128.255	liq	-114.9	143.1(0.7)	0.7225^{25}	1.4093^{20}	
5041	4-Ethylheptane		C_9H_{20}	2216-32-2	128.255			141(4)	0.7241^{25}	1.4096^{20}	i H_2O; s eth; msc EtOH, ace, bz
5042	Ethyl heptanoate	Ethyl oenanthate	$C_9H_{18}O_2$	106-30-9	158.238	liq	-66.2(0.5)	188(2)	0.8817^{20}	1.4100^{20}	sl H_2O, ctc; s EtOH, eth
5043	2-Ethylheptanoic acid		$C_9H_{18}O_2$	3274-29-1	158.238	liq		153^{31}		1.4255^{27}	
5044	4-Ethyl-4-heptanol		$C_9H_{20}O$	597-90-0	144.254			179(7)	0.8350^{20}	1.4332^{20}	vs eth, EtOH
5045	Ethyl *trans,trans*-2,4-hexadienoate	Ethyl sorbate	$C_8H_{12}O_2$	2396-84-1	140.180			195.5	0.9506^{20}	1.4951^{20}	vs eth, EtOH, chl
5046	2-Ethylhexanal		$C_8H_{16}O$	123-05-7	128.212	liq	<-100	161(7)	0.8540^{20}	1.4142^{20}	i H_2O; s EtOH, eth; sl ctc
5047	3-Ethylhexane		C_8H_{18}	619-99-8	114.229			118.5(0.5)	0.7136^{20}	1.4018^{20}	i H_2O; msc EtOH, eth, ace, bz, chl; s ctc
5048	2-Ethyl-1,3-hexanediol	Ethohexadiol	$C_8H_{18}O_2$	94-96-2	146.228	liq	-40	243(8)	0.9325^{22}	1.4497^{20}	sl H_2O; s EtOH, eth
5049	Ethyl hexanoate		$C_8H_{16}O_2$	123-66-0	144.212	liq	-67.6(0.4)	165(1)	0.873^{20}	1.4073^{20}	sl H_2O; vs eth, EtOH
5050	2-Ethylhexanoic acid		$C_8H_{16}O_2$	149-57-5	144.212			227.5(0.1)	0.9031^{25}	1.4241^{20}	s H_2O, eth, ctc; sl EtOH
5051	2-Ethyl-1-hexanol		$C_8H_{18}O$	104-76-7	130.228	liq	-70	186.2(0.2)	0.8319^{25}	1.4300^{20}	i H_2O; s EtOH, eth, ace, bz, chl
5052	2-Ethylhexanoyl chloride		$C_8H_{15}ClO$	760-67-8	162.657			101^{40}	0.939^{25}	1.4335^{20}	
5053	2-Ethyl-2-hexenal		$C_8H_{14}O$	645-62-5	126.196			175(1)	0.8554^{20}		
5054	Ethyl 3-hexenoate	Ethyl hydrosorbate	$C_8H_{14}O_2$	2396-83-0	142.196			166.5	0.8957^{20}	1.4255^{20}	
5055	2-Ethylhexyl acetate		$C_{10}H_{20}O_2$	103-09-3	172.265	liq	-80	200(1)	0.8718^{20}	1.4204^{20}	i H_2O; s EtOH, eth
5056	2-Ethylhexyl acrylate		$C_{11}H_{20}O_2$	103-11-7	184.276		-90	125^{60}	0.880^{25}	1.4332^{25}	
5057	2-Ethylhexylamine	2-Ethyl-1-hexanamine	$C_8H_{19}N$	104-75-6	129.244			172(12)			sl H_2O
5058	2-Ethylhexyl butyl phthalate	Butyl 2-ethylhexyl phthalate	$C_{20}H_{30}O_4$	85-69-8	334.450	col liq					sl H_2O
5059	2-Ethylhexyl dihydrogen phosphate	Mono(2-ethylhexyl) phosphate	$C_8H_{19}O_4P$	1070-03-7	210.208	liq					s H_2O, bz
5060	2-Ethylhexyl diphenyl phosphate	Forstab	$C_{20}H_{27}O_3P$	15647-08-2	346.400			$152^{0.15}$	1.054^{20}	1.5207^{27}	
5061	Ethyl hexyl ether	1-Ethoxyhexane	$C_8H_{18}O$	5756-43-4	130.228			142(4)	0.7722^{20}	1.4008^{20}	vs eth, EtOH
5062	2-Ethylhexyl 2-hydroxybenzoate	Octisalate	$C_{15}H_{22}O_3$	118-60-5	250.334	liq		190^{21}	1.01		
5063	2-Ethylhexyl methacrylate		$C_{12}H_{22}O_2$	688-84-6	198.302			120^{18}	0.880^{25}	1.436^{25}	
5064	2-[(2-Ethylhexyl)oxy]ethanol	Ethylene glycol mono(2-ethylhexyl) ether	$C_{10}H_{22}O_2$	1559-35-9	174.281			231(4)			

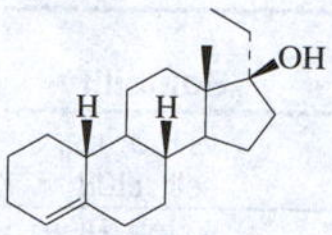
5022
Ethylestrenol

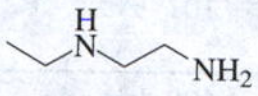
5023
N-Ethyl-1,2-ethanediamine

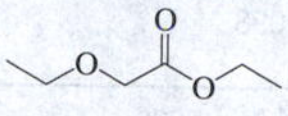
5024
Ethyl ethoxyacetate

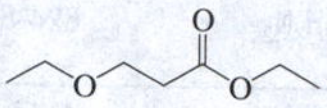
5025
Ethyl 3-ethoxypropanoate

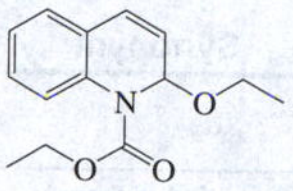
5026
Ethyl 2-ethoxy-1(2*H*)-quinolinecarboxylate

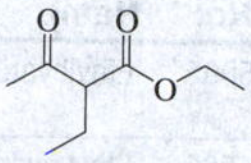
5027
Ethyl 2-ethylacetoacetate

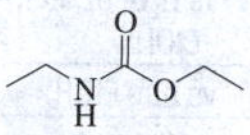
5028
Ethyl ethylcarbamate

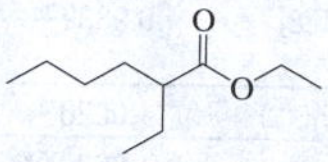
5029
Ethyl 2-ethylhexanoate

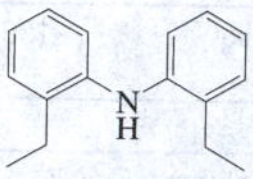
5030
2-Ethyl-*N*-(2-ethylphenyl)aniline

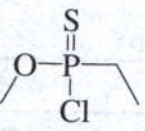
5031
O-Ethyl ethylthiophosphonyl chloride

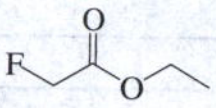
5032
Ethyl fluoroacetate

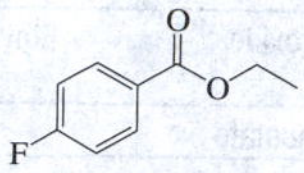
5033
Ethyl 4-fluorobenzoate

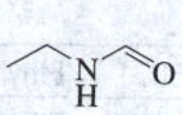
5034
N-Ethylformamide

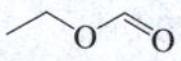
5035
Ethyl formate

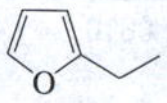
5036
2-Ethylfuran

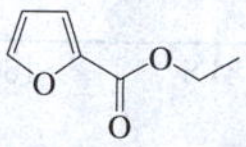
5037
Ethyl 2-furancarboxylate

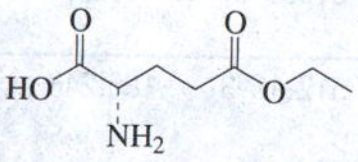
5038
γ-Ethyl L-glutamate

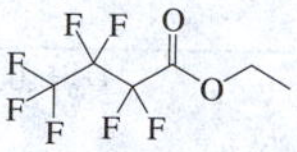
5039
Ethyl heptafluorobutanoate

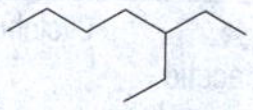
5040
3-Ethylheptane

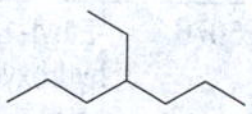
5041
4-Ethylheptane

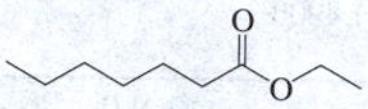
5042
Ethyl heptanoate

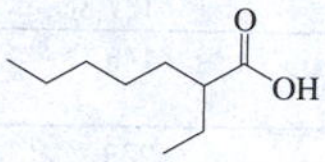
5043
2-Ethylheptanoic acid

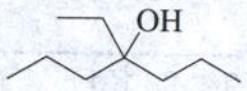
5044
4-Ethyl-4-heptanol

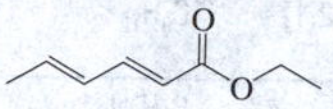
5045
Ethyl *trans,trans*-2,4-hexadienoate

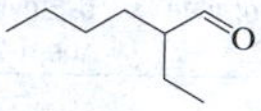
5046
2-Ethylhexanal

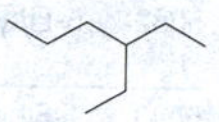
5047
3-Ethylhexane

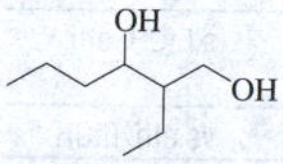
5048
2-Ethyl-1,3-hexanediol

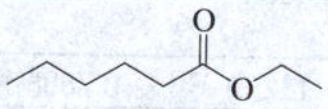
5049
Ethyl hexanoate

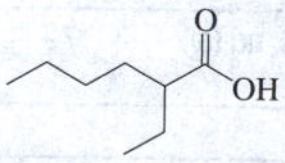
5050
2-Ethylhexanoic acid

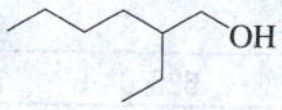
5051
2-Ethyl-1-hexanol

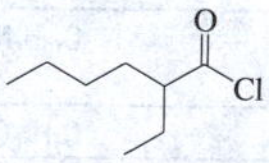
5052
2-Ethylhexanoyl chloride

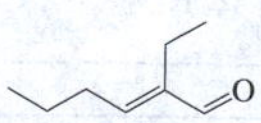
5053
2-Ethyl-2-hexenal

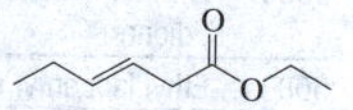
5054
Ethyl 3-hexenoate

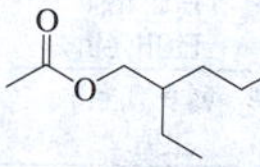
5055
2-Ethylhexyl acetate

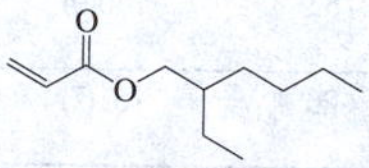
5056
2-Ethylhexyl acrylate

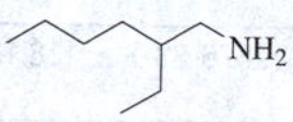
5057
2-Ethylhexylamine

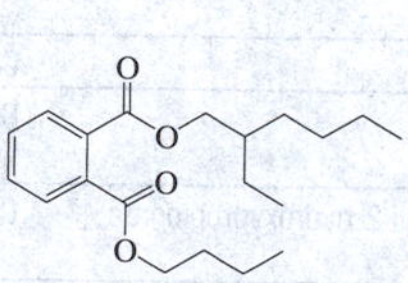
5058
2-Ethylhexyl butyl phthalate

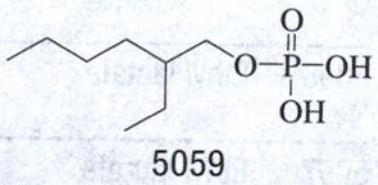
5059
2-Ethylhexyl dihydrogen phosphate

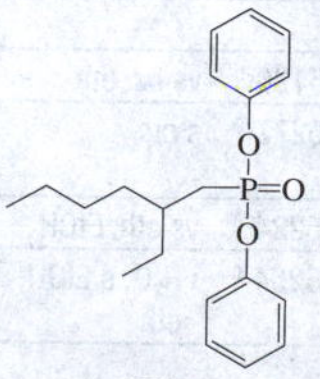
5060
2-Ethylhexyl diphenyl phosphite

5061
Ethyl hexyl ether

5062
2-Ethylhexyl 2-hydroxybenzoate

5063
2-Ethylhexyl methacrylate

5064
2-[(2-Ethylhexyl)oxy]ethanol

No.	Name	Synonym	Mol. form.	CAS Reg. No.	Mol. wt.	Physical form	mp/°C	bp/°C	den/g cm⁻³	n_D	Solubility
5065	Ethylhydrazine		$C_2H_8N_2$	624-80-6	60.098			101			vs H_2O, ace, eth, EtOH
5066	Ethyl hydrazinecarboxylate	Ethyl carbazate	$C_3H_8N_2O_2$	4114-31-2	104.108	cry	45.8(0.4)	198 dec			s EtOH, eth; sl chl
5067	Ethyl hydrogen adipate	Ethyl hydrogen hexanedioate	$C_8H_{14}O_4$	626-86-8	174.195	hyg cry (eth, peth)	29	285	0.9796^{20}	1.4311^{20}	s EtOH, eth, peth
5068	Ethyl hydrogen fumarate		$C_6H_8O_4$	2459-05-4	144.126		70	147^{16}	1.1109^{87}		s EtOH, ace; sl chl
5069	Ethyl hydrogen succinate	Butanedioic acid, monoethyl ester	$C_6H_{10}O_4$	1070-34-4	146.141	pr or nd	8	172^{42}	1.1466^{20}	1.4327^{20}	vs H_2O, eth, EtOH
5070	Ethyl hydroperoxide	Ethyl hydrogen peroxide	$C_2H_6O_2$	3031-74-1	62.068	liq	-100	95(8)	0.9332^{20}	1.3800^{20}	vs H_2O, bz, eth, EtOH
5071	Ethyl hydroxyacetate		$C_4H_8O_3$	623-50-7	104.105			156(3)	1.0826^{23}	1.4180^{20}	vs eth, EtOH
5072	Ethyl 3-hydroxybenzoate		$C_9H_{10}O_3$	7781-98-8	166.173	pl (bz)	74		1.0680^{131}		sl H_2O, chl; s EtOH, eth
5073	Ethyl 4-hydroxybenzoate	Ethylparaben	$C_9H_{10}O_3$	120-47-8	166.173	cry (dil al)	117	297.5			sl H_2O, chl, tfa; vs EtOH, eth; i CS_2
5074	Ethyl 3-hydroxybutanoate, (±)-		$C_6H_{12}O_3$	35608-64-1	132.157			185	1.017^{20}	1.4182^{20}	s H_2O, EtOH; sl ctc
5075	Ethyl 2-hydroxy-3-butenoate		$C_6H_{10}O_3$	91890-87-8	130.141			173 dec	1.0470^{15}	1.436^{13}	vs H_2O, eth, EtOH
5076	α-Ethyl-1-hydroxycyclohexaneacetic acid	Cyclobutyrol	$C_{10}H_{18}O_3$	512-16-3	186.248	cry (eth-peth)	81.5	164^{24}	1.0010^{18}	1.4680^{18}	vs ace, eth, EtOH, chl
5077	*N*-Ethyl-*N*-hydroxyethanamine	*N*,*N*-Diethylhydroxylamine	$C_4H_{11}NO$	3710-84-7	89.136		10	131.0(0.3)	0.8669^{20}	1.4195^{20}	
5078	2-Ethyl-3-hydroxyhexanal		$C_8H_{16}O_2$	496-03-7	144.212			138^{50}			
5079	Ethyl 4-hydroxy-3-methoxybenzoate		$C_{10}H_{12}O_4$	617-05-0	196.200	nd (dil al)	44	292			i H_2O; vs EtOH, eth; s chl
5080	Ethyl *cis*-12-hydroxy-9-octadecenoate, (*R*)-	Ethyl ricinoleate	$C_{20}H_{38}O_3$	55066-53-0	326.514			258^{13}	0.9180^{20}	1.4618^{22}	
5081	Ethylidenecyclohexane		C_8H_{14}	1003-64-1	110.197			137(3)	0.822^{25}	1.4618^{20}	
5082	5-Ethylidene-2-norbornene	5-Ethylidenebicyclo[2.2.1]hept-2-ene	C_9H_{12}	16219-75-3	120.191	liq		146	0.893	1.4900^{20}	
5083	1-Ethyl-1*H*-imidazole		$C_5H_8N_2$	7098-07-9	96.131			208	0.999^{25}		msc H_2O
5084	Ethyl iodoacetate		$C_4H_7IO_2$	623-48-3	214.002	oil		179(7)	1.8173^{13}	1.5079^{13}	s EtOH, eth
5085	Ethyl isobutylcarbamate	Isobutyl urethane	$C_7H_{15}NO_2$	539-89-9	145.200		<-65	110^{30}	0.9432^{20}	1.4288^{20}	vs eth, EtOH
5086	Ethyl isocyanate		C_3H_5NO	109-90-0	71.078			56(1)	0.9031^{20}	1.3808^{20}	i H_2O; msc EtOH, eth
5087	Ethyl isocyanide		C_3H_5N	624-79-3	55.079		<-66	79	0.7402^{20}	1.3622^{20}	vs H_2O; msc EtOH, eth; s ace
5088	*N*-Ethyl-1*H*-isoindole-1,3(2*H*)-dione		$C_{10}H_9NO_2$	5022-29-7	175.184	nd (al)	79	285.5			s EtOH, eth
5089	Ethyl isopentyl ether		$C_7H_{16}O$	628-04-6	116.201			112(5)	0.7688^{21}		vs eth, EtOH
5090	Ethylisopropylamine	*N*-Ethyl-2-propanamine	$C_5H_{13}N$	19961-27-4	87.164			68(7)		1.3872^{25}	
5091	1-Ethyl-2-isopropylbenzene		$C_{11}H_{16}$	18970-44-0	148.245			193(7)	0.888^{20}	1.508^{20}	vs ace, bz, eth, EtOH
5092	Ethyl isopropyl ether		$C_5H_{12}O$	625-54-7	88.148			54(3)	0.720^{25}	1.3698^{25}	s H_2O, ace, chl; msc EtOH, eth
5093	*N*-Ethyl-*N*-isopropyl-2-propanamine	Ethyldiisopropylamine	$C_8H_{19}N$	7087-68-5	129.244			114(10)	0.742^{25}	1.4138^{20}	s ctc
5094	Ethyl isopropyl sulfide		$C_5H_{12}S$	5145-99-3	104.214	liq	-122.2(0.2)	107.3(0.6)	0.8246^{20}		
5095	Ethyl isothiocyanate		C_3H_5NS	542-85-8	87.144	liq	-5.9	140(9)	0.9990^{20}	1.5130^{20}	i H_2O; msc EtOH, eth
5096	Ethyl lactate	Ethyl 2-hydroxypropionate	$C_5H_{10}O_3$	2676-33-7	118.131	liq	-26	151(7)	1.0328^{20}	1.4124^{20}	vs H_2O, eth, EtOH
5097	Ethyl laurate		$C_{14}H_{28}O_2$	106-33-2	228.371	liq	-1.8(0.4)	276(3)	0.8618^{20}	1.4311^{20}	i H_2O; vs EtOH; msc eth; sl ctc
5098	Ethyl levulinate		$C_7H_{12}O_3$	539-88-8	144.168			205.8	1.0111^{20}	1.4229^{20}	vs H_2O, EtOH
5099	Ethyl mercaptoacetate		$C_4H_8O_2S$	623-51-8	120.171			157	1.0964^{15}	1.4582^{20}	s EtOH, eth; sl ctc
5100	Ethyl methacrylate	Ethyl 2-methyl-2-propenoate	$C_6H_{10}O_2$	97-63-2	114.142			116(19)	0.9135^{20}	1.4147^{20}	sl H_2O, chl; msc EtOH, eth
5101	Ethyl methanesulfonate		$C_3H_8O_3S$	62-50-0	124.159			86^{10}			
5102	1-Ethyl-4-methoxybenzene		$C_9H_{12}O$	1515-95-3	136.190			196(8)	0.9624^{15}	1.5120^{20}	vs bz, eth
5103	α-Ethyl-4-methoxybenzenemethanol		$C_{10}H_{14}O_2$	5349-60-0	166.217			143^{20}		1.5277^{20}	s ctc
5104	Ethyl 2-methoxybenzoate		$C_{10}H_{12}O_3$	7335-26-4	180.200			262(7)	1.1124^{20}	1.5224^{20}	vs eth, EtOH
5105	Ethyl 4-methoxybenzoate		$C_{10}H_{12}O_3$	94-30-4	180.200		7.5	266(19)	1.1038^{20}	1.5254^{20}	i H_2O; s EtOH, eth
5106	4-Ethyl-2-methoxyphenol		$C_9H_{12}O_2$	2785-89-9	152.190	liq	-7	236.5	1.0931^{18}		

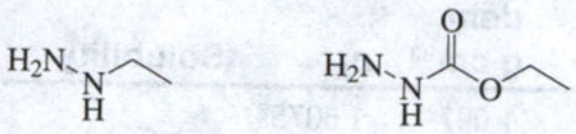

5065
Ethylhydrazine

5066
Ethyl hydrazinecarboxylate

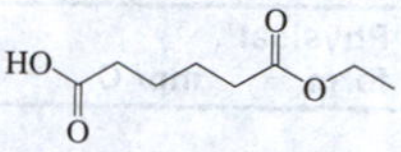

5067
Ethyl hydrogen adipate

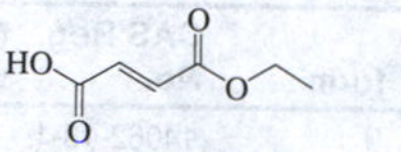

5068
Ethyl hydrogen fumarate

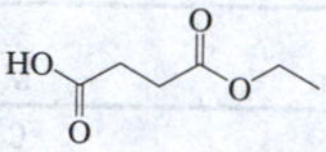

5069
Ethyl hydrogen succinate

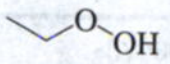

5070
Ethyl hydroperoxide

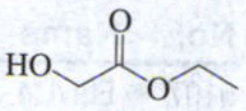

5071
Ethyl hydroxyacetate

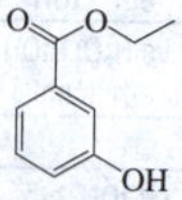

5072
Ethyl 3-hydroxybenzoate

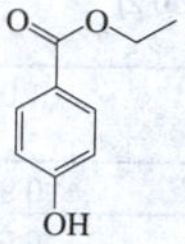

5073
Ethyl 4-hydroxybenzoate

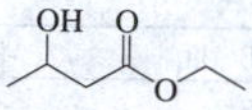

5074
Ethyl 3-hydroxybutanoate, (±)-

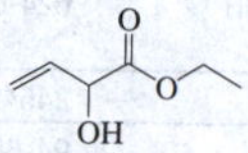

5075
Ethyl 2-hydroxy-3-butenoate

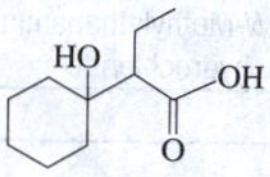

5076
α-Ethyl-1-hydroxycyclohexaneacetic acid

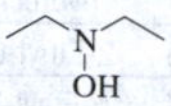

5077
N-Ethyl-*N*-hydroxyethanamine

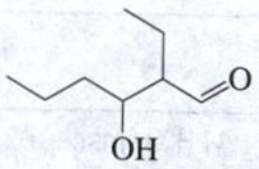

5078
2-Ethyl-3-hydroxyhexanal

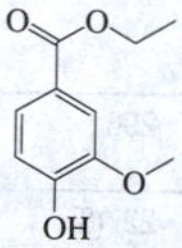

5079
Ethyl 4-hydroxy-3-methoxybenzoate

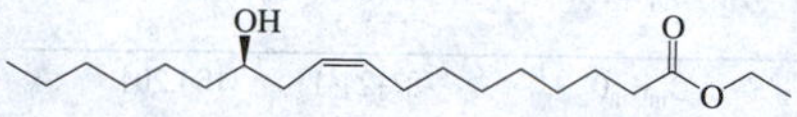

5080
Ethyl *cis*-12-hydroxy-9-octadecenoate, (*R*)-

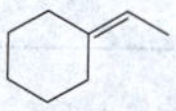

5081
Ethylidenecyclohexane

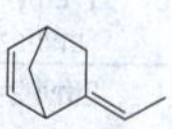

5082
5-Ethylidene-2-norbornene

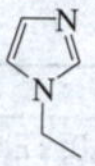

5083
1-Ethyl-1*H*-imidazole

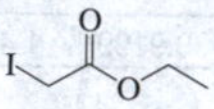

5084
Ethyl iodoacetate

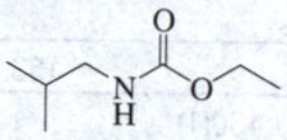

5085
Ethyl isobutylcarbamate

5086
Ethyl isocyanate

5087
Ethyl isocyanide

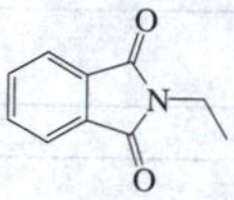

5088
N-Ethyl-1*H*-isoindole-1,3(2*H*)-dione

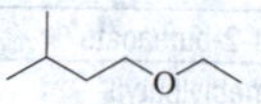

5089
Ethyl isopentyl ether

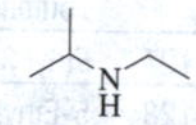

5090
Ethylisopropylamine

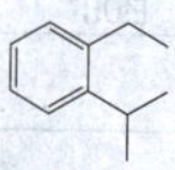

5091
1-Ethyl-2-isopropylbenzene

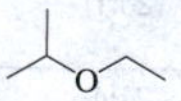

5092
Ethyl isopropyl ether

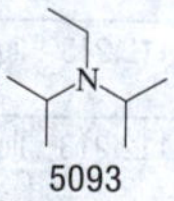

5093
N-Ethyl-*N*-isopropyl-2-propanamine

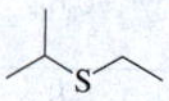

5094
Ethyl isopropyl sulfide

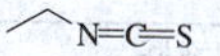

5095
Ethyl isothiocyanate

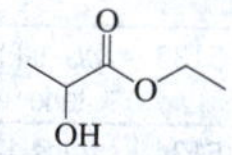

5096
Ethyl lactate

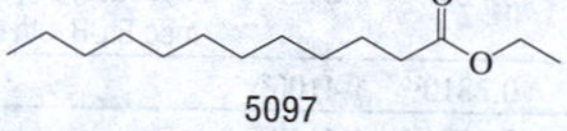

5097
Ethyl laurate

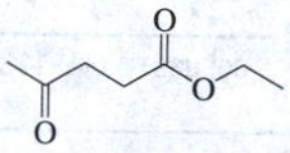

5098
Ethyl levulinate

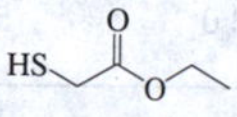

5099
Ethyl mercaptoacetate

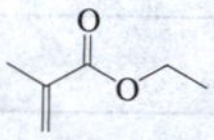

5100
Ethyl methacrylate

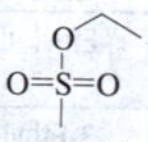

5101
Ethyl methanesulfonate

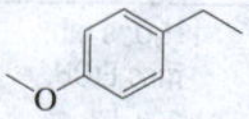

5102
1-Ethyl-4-methoxybenzene

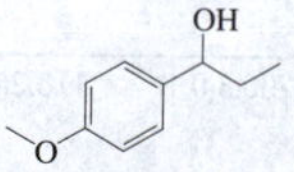

5103
α-Ethyl-4-methoxybenzenemethanol

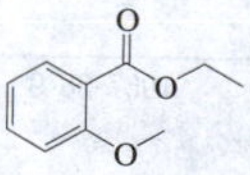

5104
Ethyl 2-methoxybenzoate

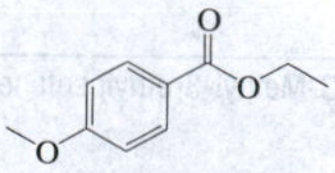

5105
Ethyl 4-methoxybenzoate

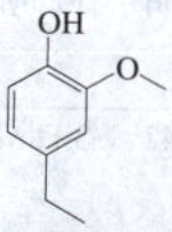

5106
4-Ethyl-2-methoxyphenol

No.	Name	Synonym	Mol. form.	CAS Reg. No.	Mol. wt.	Physical form	mp/°C	bp/°C	den/g cm^{-3}	n_D	Solubility
5107	Ethyl (4-methoxyphenyl)-acetate		$C_{11}H_{14}O_3$	14062-18-1	194.227			139^{70}	1.097^{25}	1.5075^{20}	
5108	Ethyl 2-methylacetoacetate		$C_7H_{12}O_3$	609-14-3	144.168			187	0.9941^{20}	1.4185^{20}	sl H_2O; s EtOH, eth; vs ace
5109	*N*-Ethyl-2-methylallylamine	*N*-Ethyl-2-methyl-2-propen-1-amine	$C_6H_{13}N$	18328-90-0	99.174	liq		104.7	0.753	1.4221^{20}	msc H_2O
5110	5-Ethyl-5-(2-methylallyl)-2-thiobarbituric acid	Methallatal	$C_{10}H_{14}N_2O_2S$	115-56-0	226.295		160.5				
5111	Ethylmethylamine	*N*-Methylethanamine	C_3H_9N	624-78-2	59.110			34.2(0.2)			vs H_2O, ace, eth, EtOH
5112	Ethylmethylamine hydrochloride	*N*-Methylethanamine hydrochloride	$C_3H_{10}ClN$	624-60-2	95.571	pl (al-eth)	128		1.0874^{20}		vs H_2O, EtOH; i eth; s chl
5113	2-Ethyl-6-methylaniline		$C_9H_{13}N$	24549-06-2	135.206	liq	-33	231	0.968^{25}	1.5525^{20}	
5114	*N*-Ethyl-2-methylaniline		$C_9H_{13}N$	94-68-8	135.206		<-15	216	0.948^{25}	1.5456^{20}	s EtOH, eth
5115	*N*-Ethyl-3-methylaniline		$C_9H_{13}N$	102-27-2	135.206			221	0.9263^{15}	1.5451^{20}	s EtOH, eth
5116	*N*-Ethyl-4-methylaniline	*N*-Ethyl-4-toluidine	$C_9H_{13}N$	622-57-1	135.206			217	0.9391^{16}		s EtOH, eth
5117	*N*-Ethyl-*N*-methylaniline		$C_9H_{13}N$	613-97-8	135.206			204(5)	0.92^{55}		i H_2O; msc EtOH, eth; s ctc
5118	*N*-Ethyl-α-methylbenzeneethanamine	*N*-Ethylamphetamine	$C_{11}H_{17}N$	457-87-4	163.260			105^{14}		1.4986^{25}	
5119	*N*-Ethyl-4-methylbenzenesulfonamide		$C_9H_{13}NO_2S$	80-39-7	199.270		64				s EtOH
5120	1-Ethyl-2-methyl-1*H*-benzimidazole		$C_{10}H_{12}N_2$	5805-76-5	160.215		51	296	1.073^{25}		
5121	Ethyl 2-methylbenzoate		$C_{10}H_{12}O_2$	87-24-1	164.201		<-10	227(8)	1.0325^{21}	1.507^{22}	i H_2O; msc EtOH, eth
5122	Ethyl 4-methylbenzoate		$C_{10}H_{12}O_2$	94-08-6	164.201			236(7)	1.0269^{18}	1.5089^{18}	i H_2O; msc EtOH, eth
5123	Ethyl 3-methylbutanoate	Ethyl isovalerate	$C_7H_{14}O_2$	108-64-5	130.185	liq	-99.3	135(3)	0.8656^{20}	1.3962^{20}	sl H_2O; vs EtOH, eth
5124	2-Ethyl-2-methylbutanoic acid		$C_7H_{14}O_2$	19889-37-3	130.185		<-20	207		1.4250^{20}	vs EtOH
5125	2-Ethyl-3-methyl-1-butene		C_7H_{14}	7357-93-9	98.186			86.3(0.5)	0.7150^{20}	1.410^{20}	i H_2O; s eth, ace, bz, chl
5126	Ethyl *trans*-2-methyl-2-butenoate		$C_7H_{12}O_2$	5837-78-5	128.169			156	0.9200^{20}	1.4340^{20}	
5127	Ethyl 3-methyl-2-butenoate		$C_7H_{12}O_2$	638-10-8	128.169			153.5	0.9199^{21}	1.4345^{20}	
5128	5-Ethyl-5-(1-methylbutyl)-2,4,6(1*H*,3*H*,5*H*)-pyrimidinetrione		$C_{11}H_{18}N_2O_3$	76-74-4	226.272		130(1)				sl H_2O; s EtOH, eth
5129	Ethyl *N*-methylcarbamate		$C_4H_9NO_2$	105-40-8	103.120			169(5)	1.0115^{20}	1.4183^{20}	vs H_2O, EtOH
5130	Ethyl methyl carbonate		$C_4H_8O_3$	623-53-0	104.105	liq	-14	107.5	1.012^{20}	1.3778^{20}	vs eth, EtOH
5131	*trans*-1-Ethyl-4-methylcyclohexane		C_9H_{18}	6236-88-0	126.239	liq	-80.8	147(6)	0.7798^{20}	1.4304^{20}	
5132	1-Ethyl-1-methylcyclopentane		C_8H_{16}	16747-50-5	112.213	liq	-143.8(0.2)	121.5(0.6)	0.7767^{25}	1.4272^{20}	vs ace, bz, eth, EtOH
5133	*cis*-1-Ethyl-2-methylcyclopentane		C_8H_{16}	930-89-2	112.213	liq	-105.9(0.1)	128(1)	0.7852^{20}	1.4293^{20}	
5134	*trans*-1-Ethyl-2-methylcyclopentane		C_8H_{16}	930-90-5	112.213	liq	-109(5)	121(4)	0.7649^{25}	1.4219^{20}	
5135	*cis*-1-Ethyl-3-methylcyclopentane		C_8H_{16}	2613-66-3	112.213			121(4)	0.7724^{20}	1.4203^{20}	vs ace, bz, eth, EtOH
5136	*trans*-1-Ethyl-3-methylcyclopentane		C_8H_{16}	2613-65-2	112.213	liq	-108	121(4)	0.7619^{20}	1.4186^{20}	
5137	1-Ethyl-1-methylcyclopropane		C_6H_{12}	53778-43-1	84.159	liq	-130.2(0.2)	57(4)	0.6968^{25}	1.3887^{20}	
5138	2-Ethyl-2-methyl-1,3-dioxolane		$C_6H_{12}O_2$	126-39-6	116.158			119(3)	0.9360^{20}		
5139	Ethyl methyl ether		C_3H_8O	540-67-0	60.095	col gas	-113	6(2)	0.7251^{0}	1.3420^{4}	s H_2O, ace, chl; msc EtOH, eth
5140	3-Ethyl-2-methylhexane		C_9H_{20}	16789-46-1	128.255			138(4)	0.7310^{20}	1.4106^{20}	
5141	3-Ethyl-3-methylhexane		C_9H_{20}	3074-76-8	128.255			140(3)	0.7371^{25}	1.4140^{20}	
5142	3-Ethyl-4-methylhexane	2,3-Diethylpentane	C_9H_{20}	3074-77-9	128.255			140(4)	0.7420^{20}	1.4134^{20}	
5143	4-Ethyl-2-methylhexane		C_9H_{20}	3074-75-7	128.255			134(3)	0.7195^{25}	1.4063^{20}	
5144	Ethyl 4-methylhexanoate	Ethyl 4-methylcaproate	$C_9H_{18}O_2$	1561-10-0	158.238			180	0.8708^{20}	1.4051^{20}	
5145	Ethyl 4-methyl-3-oxopentanoate		$C_8H_{14}O_3$	7152-15-0	158.195	liq	-9	173	0.98^{25}	1.250^{20}	
5146	3-Ethyl-2-methylpentane	2-Methyl-3-ethylpentane	C_8H_{18}	609-26-7	114.229	liq	-115.0(0.1)	115.6(0.6)	0.7193^{20}	1.4040^{20}	i H_2O; s eth; msc EtOH, ace, bz
5147	3-Ethyl-3-methylpentane	3-Methyl-3-ethylpentane	C_8H_{18}	1067-08-9	114.229	liq	-90.8(0.1)	118.2(0.9)	0.7274^{20}	1.4078^{20}	i H_2O; s eth; msc EtOH, ace, bz
5148	Ethyl 4-methylpentanoate		$C_8H_{16}O_2$	25415-67-2	144.212			163	0.8705^{20}	1.4050^{20}	

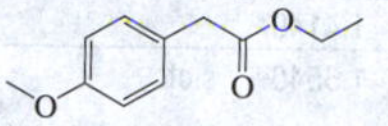
5107
Ethyl (4-methoxyphenyl)acetate

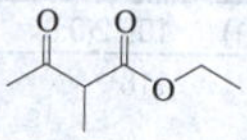
5108
Ethyl 2-methylacetoacetate

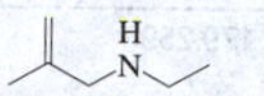
5109
N-Ethyl-2-methylallylamine

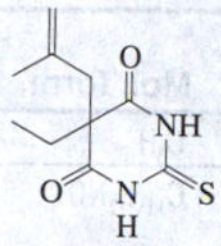
5110
5-Ethyl-5-(2-methylallyl)-2-thiobarbituric acid

5111
Ethylmethylamine

5112
Ethylmethylamine hydrochloride

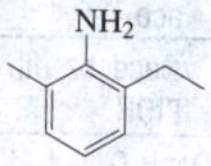
5113
2-Ethyl-6-methylaniline

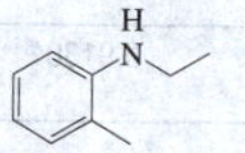
5114
N-Ethyl-2-methylaniline

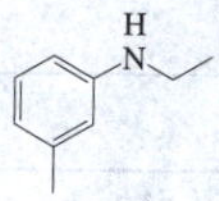
5115
N-Ethyl-3-methylaniline

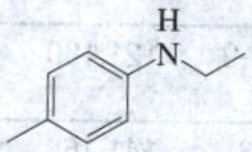
5116
N-Ethyl-4-methylaniline

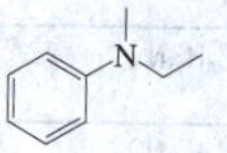
5117
N-Ethyl-*N*-methylaniline

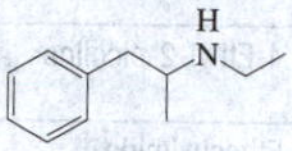
5118
N-Ethyl-α-methylbenzeneethanamine

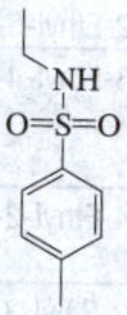
5119
N-Ethyl-4-methylbenzenesulfonamide

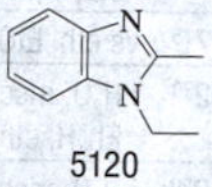
5120
1-Ethyl-2-methyl-1*H*-benzimidazole

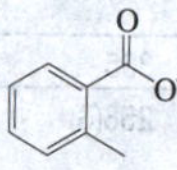
5121
Ethyl 2-methylbenzoate

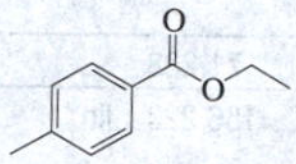
5122
Ethyl 4-methylbenzoate

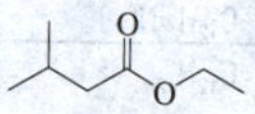
5123
Ethyl 3-methylbutanoate

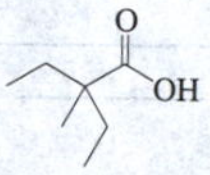
5124
2-Ethyl-2-methylbutanoic acid

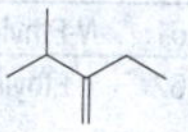
5125
2-Ethyl-3-methyl-1-butene

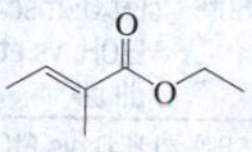
5126
Ethyl *trans*-2-methyl-2-butenoate

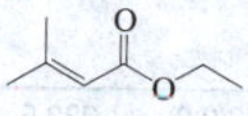
5127
Ethyl 3-methyl-2-butenoate

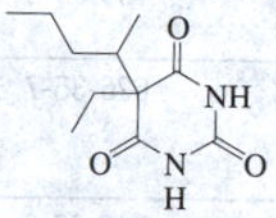
5128
5-Ethyl-5-(1-methylbutyl)-2,4,6(1*H*,3*H*,5*H*)-pyrimidinetrione

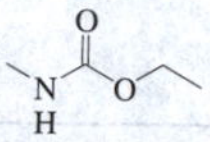
5129
Ethyl *N*-methylcarbamate

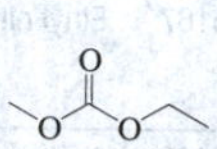
5130
Ethyl methyl carbonate

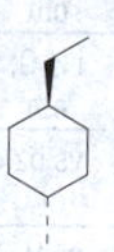
5131
trans-1-Ethyl-4-methylcyclohexane

5132
1-Ethyl-1-methylcyclopentane

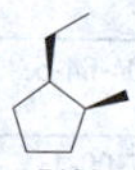
5133
cis-1-Ethyl-2-methylcyclopentane

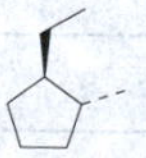
5134
trans-1-Ethyl-2-methylcyclopentane

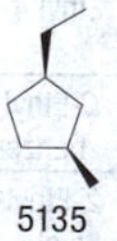
5135
cis-1-Ethyl-3-methylcyclopentane

5136
trans-1-Ethyl-3-methylcyclopentane

5137
1-Ethyl-1-methylcyclopropane

5138
2-Ethyl-2-methyl-1,3-dioxolane

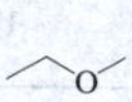
5139
Ethyl methyl ether

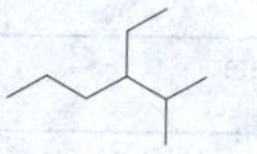
5140
3-Ethyl-2-methylhexane

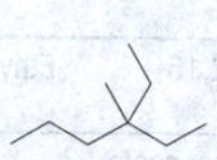
5141
3-Ethyl-3-methylhexane

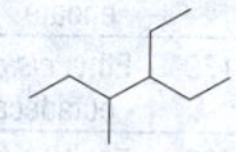
5142
3-Ethyl-4-methylhexane

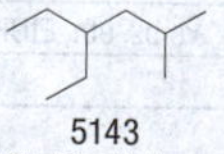
5143
4-Ethyl-2-methylhexane

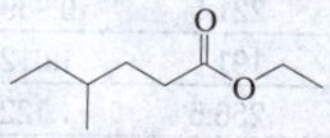
5144
Ethyl 4-methylhexanoate

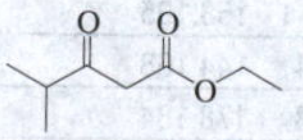
5145
Ethyl 4-methyl-3-oxopentanoate

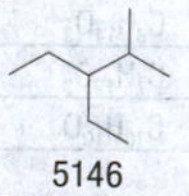
5146
3-Ethyl-2-methylpentane

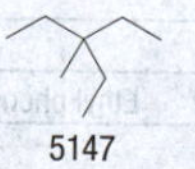
5147
3-Ethyl-3-methylpentane

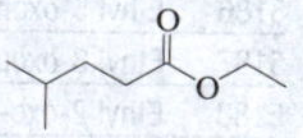
5148
Ethyl 4-methylpentanoate

Organic

No.	Name	Synonym	Mol. form.	CAS Reg. No.	Mol. wt.	Physical form	mp/°C	bp/°C	den/g cm⁻³	n_D	Solubility
5149	3-Ethyl-2-methyl-1-pentene		C_8H_{16}	19780-66-6	112.213	liq	-112.9(0.1)	109.2(0.9)	0.7262^{20}	1.4140^{20}	
5150	2-[Ethyl(3-methylphenyl)-amino]ethanol		$C_{11}H_{17}NO$	91-88-3	179.259			$118^{1.5}$		1.5540^{20}	s ctc
5151	Ethyl 3-methyl-3-phenyloxiranecarboxylate	Ethyl 3-methyl-3-phenylglycidate	$C_{12}H_{14}O_3$	77-83-8	206.237			273.5	1.044^{20}	1.5182^{20}	
5152	4-Ethyl-4-methyl-2,6-piperidinedione	Bemegride	$C_8H_{13}NO_2$	64-65-3	155.195	pl (w, ace-eth)	126.5	100 sub			s chl
5153	Ethyl 2-methylpropanoate	Ethyl isobutanoate	$C_6H_{12}O_2$	97-62-1	116.158	liq	-97.8(0.4)	111(2)	0.868^{20}	1.3869^{18}	sl H_2O, ctc; msc EtOH, eth; s ace
5154	2-Ethyl-5-methylpyrazine		$C_7H_{10}N_2$	13360-64-0	122.167	liq		79^{56}			
5155	3-Ethyl-4-methylpyridine	3-Ethyl-4-picoline	$C_8H_{11}N$	529-21-5	121.180			198	0.9286^{17}		sl H_2O; s EtOH, eth, chl; vs ace
5156	4-Ethyl-2-methylpyridine	4-Ethyl-2-picoline	$C_8H_{11}N$	536-88-9	121.180			180(6)	0.9130^{25}		vs ace, bz, eth, EtOH
5157	3-Ethyl-3-methyl-2,5-pyrrolidinedione	Ethosuximide	$C_7H_{11}NO_2$	77-67-8	141.168	cry (ace-eth)	64.5				vs H_2O
5158	Ethyl methyl sulfide		C_3H_8S	624-89-5	76.161	liq	-105.89(0.1)	66.6(0.3)	0.8422^{20}	1.4404^{20}	i H_2O; msc EtOH; s eth, chl
5159	*N*-Ethylmorpholine		$C_6H_{13}NO$	100-74-3	115.173			145(5)	0.8996^{20}	1.4400^{20}	msc H_2O, EtOH, eth; s ace, bz
5160	Ethyl myristate		$C_{16}H_{32}O_2$	124-06-1	256.424		12.3(0.8)	308(3)	0.8573^{25}	1.4362^{20}	i H_2O; s EtOH, ctc, lig; sl eth
5161	*N*-Ethyl-1-naphthalenamine		$C_{12}H_{13}N$	118-44-5	171.238			305	1.0652^{15}	1.6477^{15}	vs eth, EtOH
5162	1-Ethylnaphthalene		$C_{12}H_{12}$	1127-76-0	156.223	liq	-13.9(0.2)	258(3)	1.0082^{20}	1.6062^{20}	i H_2O; msc EtOH, eth
5163	2-Ethylnaphthalene		$C_{12}H_{12}$	939-27-5	156.223	liq	-7.4(0.9)	259(2)	0.9922^{20}	1.5999^{20}	i H_2O; msc EtOH, eth; sl chl
5164	Ethyl 1-naphthylacetate		$C_{14}H_{14}O_2$	2122-70-5	214.260	oil	88.5	222^{20}			s EtOH, eth
5165	Ethyl nitrate		$C_2H_5NO_3$	625-58-1	91.066	liq	-94.51(0.1)	89(2)	1.1084^{20}	1.3852^{20}	s H_2O; msc EtOH, eth
5166	Ethyl nitrite		$C_2H_5NO_2$	109-95-5	75.067	ye vol liq or gas		17.5(0.4)	0.899^{15}	1.3418^{10}	msc EtOH, eth
5167	Ethyl nitroacetate		$C_4H_7NO_4$	626-35-7	133.104			106^{25}	1.1953^{20}	1.4250^{20}	sl H_2O; msc EtOH; vs eth; s dil alk
5168	1-Ethyl-2-nitrobenzene		$C_8H_9NO_2$	612-22-6	151.163	liq	-12.2(0.2)	232.5	1.1207^{20}	1.5356^{20}	i H_2O; vs EtOH, eth; s ace; sl ctc
5169	1-Ethyl-4-nitrobenzene		$C_8H_9NO_2$	100-12-9	151.163	liq	-12.3	245.5	1.1192^{20}	1.5455^{20}	i H_2O; vs EtOH, eth; s ace; sl ctc
5170	Ethyl 3-nitrobenzoate		$C_9H_9NO_4$	618-98-4	195.172		47	297			i H_2O; vs EtOH, eth
5171	Ethyl 4-nitrobenzoate		$C_9H_9NO_4$	99-77-4	195.172		57	186.3			i H_2O; s EtOH, eth
5172	*O*-Ethyl *O*-*p*-nitrophenyl benzenethiophosphonate		$C_{14}H_{14}NO_4PS$	2104-64-5	323.304		39.2(0.3)		1.27^{25}	1.5978^{30}	vs bz, eth, EtOH
5173	2-Ethyl-2-nitro-1,3-propanediol		$C_5H_{11}NO_4$	597-09-1	149.146	nd (w)	57.5	dec			vs H_2O, eth, EtOH
5174	Ethyl 2-nitropropanoate		$C_5H_9NO_4$	2531-80-8	147.130			190.5		1.4210^{20}	vs bz, eth, EtOH
5175	*N*-Ethyl-*N*-nitrosourea	*N*-Nitroso-*N*-ethylurea	$C_3H_7N_3O_2$	759-73-9	117.107		100 dec				s chl
5176	Ethyl nonanoate		$C_{11}H_{22}O_2$	123-29-5	186.292	liq	-44.4(0.3)	224(5)	0.8657^{20}	1.4220^{20}	i H_2O; s EtOH, eth, ace, ctc
5177	5-Ethyl-2-norbornene		C_9H_{14}	15403-89-1	122.207	liq		143.6	0.86	1.4630^{20}	
5178	Ethyl *cis,cis*-9,12-octadecadienoate	Ethyl linoleate	$C_{20}H_{36}O_2$	544-35-4	308.499	ye or col	-55(5)	351(10)	0.8865^{20}		vs eth, EtOH
5179	Ethyl *cis,cis,cis*-9,12,15-octadecatrienoate	Ethyl linolenate	$C_{20}H_{34}O_2$	1191-41-9	306.483			357(10)	0.8919^{20}	1.4694^{20}	vs eth, EtOH
5180	Ethyl *trans*-9-octadecenoate		$C_{20}H_{38}O_2$	6114-18-7	310.515		5.1(0.3)	218^{15}	0.8664^{25}	1.4480^{25}	vs eth, EtOH
5181	3-Ethyloctane		$C_{10}H_{22}$	5881-17-4	142.282			165(6)	0.7359^{25}	1.4156^{20}	
5182	4-Ethyloctane		$C_{10}H_{22}$	15869-86-0	142.282			163(6)	0.7343^{25}	1.4151^{20}	
5183	Ethyl octanoate		$C_{10}H_{20}O_2$	106-32-1	172.265	liq	-44.7(0.3)	206(1)	0.866^{18}	1.4178^{20}	i H_2O; vs EtOH, eth; sl ctc
5184	Ethyl 1-octyl sulfide	1-(Ethylthio)octane	$C_{10}H_{22}S$	3698-94-0	174.347	liq		109^{14}			
5185	Ethyl oleate	Ethyl *cis*-9-octadecenoate	$C_{20}H_{38}O_2$	111-62-6	310.515		-21(2)	357(9)	0.8720^{20}	1.4515^{20}	vs eth, EtOH
5186	Ethyl 5-oxohexanoate		$C_8H_{14}O_3$	13984-57-1	158.195			221.5	0.989^{25}	1.4277^{20}	
5187	Ethyl 3-oxopentanoate		$C_7H_{12}O_3$	4949-44-4	144.168			191	1.0120^{20}	1.4230^{20}	vs bz, eth, EtOH
5188	Ethyl 2-oxo-2-phenylacetate	Ethyl phenylglyoxylate	$C_{10}H_{10}O_3$	1603-79-8	178.184			256.5	1.1222^{25}	1.5190^{25}	

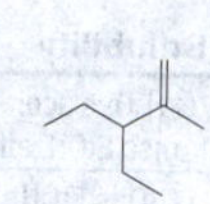

5149
3-Ethyl-2-methyl-1-pentene

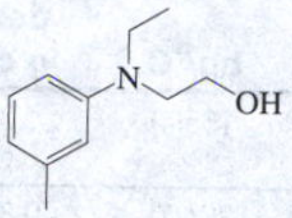

5150
2-[Ethyl(3-methylphenyl)amino]ethanol

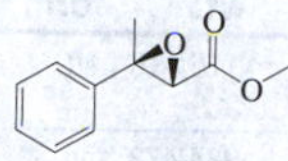

5151
Ethyl 3-methyl-3-phenyloxiranecarboxylate

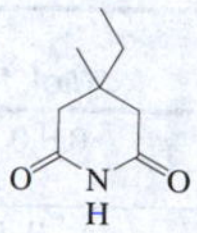

5152
4-Ethyl-4-methyl-2,6-piperidinedione

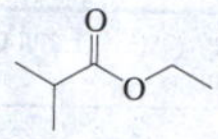

5153
Ethyl 2-methylpropanoate

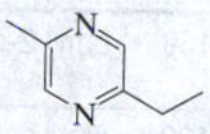

5154
2-Ethyl-5-methylpyrazine

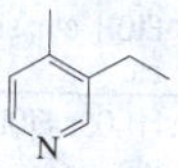

5155
3-Ethyl-4-methylpyridine

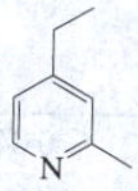

5156
4-Ethyl-2-methylpyridine

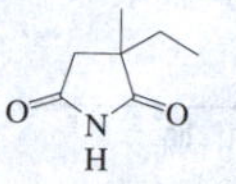

5157
3-Ethyl-3-methyl-2,5-pyrrolidinedione

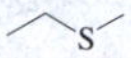

5158
Ethyl methyl sulfide

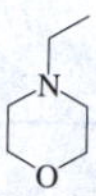

5159
N-Ethylmorpholine

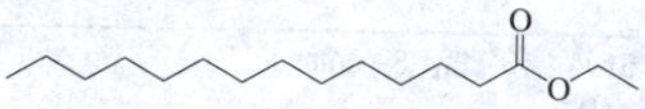

5160
Ethyl myristate

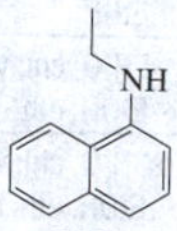

5161
N-Ethyl-1-naphthalenamine

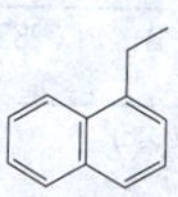

5162
1-Ethylnaphthalene

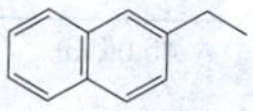

5163
2-Ethylnaphthalene

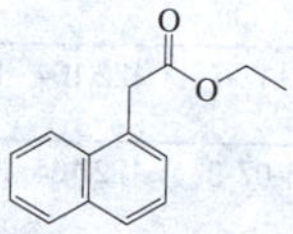

5164
Ethyl 1-naphthylacetate

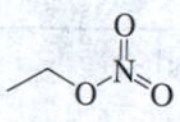

5165
Ethyl nitrate

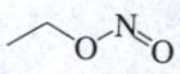

5166
Ethyl nitrite

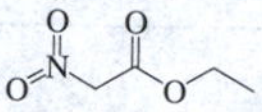

5167
Ethyl nitroacetate

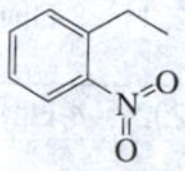

5168
1-Ethyl-2-nitrobenzene

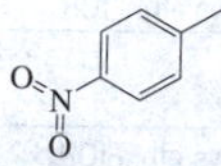

5169
1-Ethyl-4-nitrobenzene

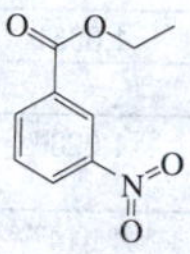

5170
Ethyl 3-nitrobenzoate

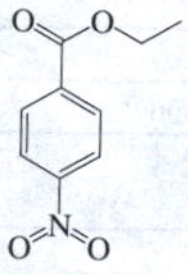

5171
Ethyl 4-nitrobenzoate

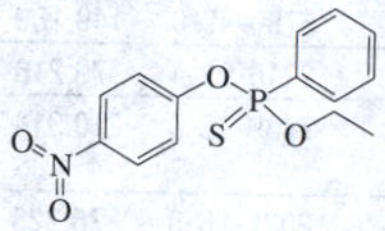

5172
O-Ethyl *O*-*p*-nitrophenyl benzenethiophosphonate

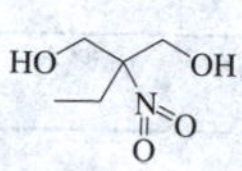

5173
2-Ethyl-2-nitro-1,3-propanediol

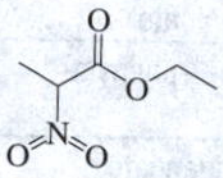

5174
Ethyl 2-nitropropanoate

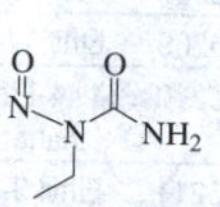

5175
N-Ethyl-*N*-nitrosourea

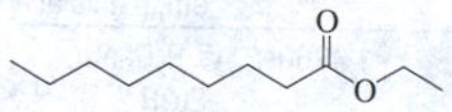

5176
Ethyl nonanoate

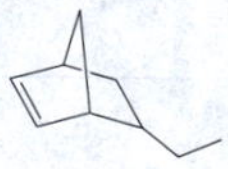

5177
5-Ethyl-2-norbornene

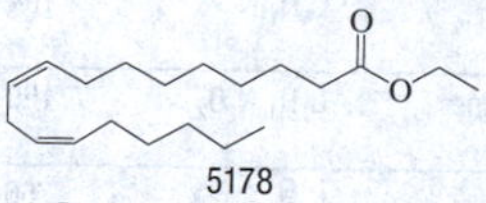

5178
Ethyl *cis,cis*-9,12-octadecadienoate

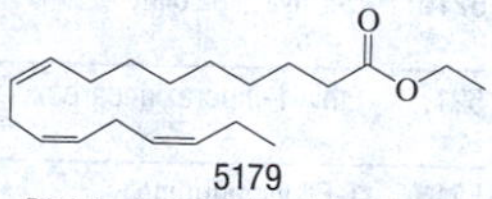

5179
Ethyl *cis,cis,cis*-9,12,15-octadecatrienoate

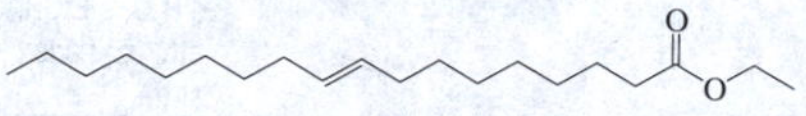

5180
Ethyl *trans*-9-octadecenoate

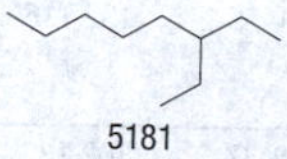

5181
3-Ethyloctane

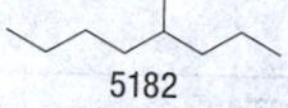

5182
4-Ethyloctane

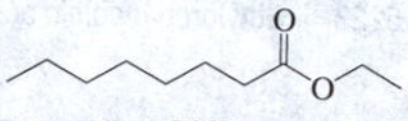

5183
Ethyl octanoate

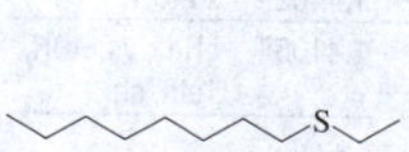

5184
Ethyl 1-octyl sulfide

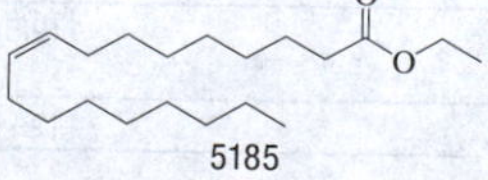

5185
Ethyl oleate

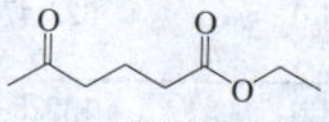

5186
Ethyl 5-oxohexanoate

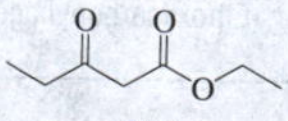

5187
Ethyl 3-oxopentanoate

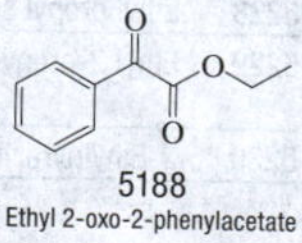

5188
Ethyl 2-oxo-2-phenylacetate

No.	Name	Synonym	Mol. form.	CAS Reg. No.	Mol. wt.	Physical form	mp/°C	bp/°C	den/g cm^{-3}	n_D	Solubility
5189	Ethyl 2-oxopropanoate	Ethyl pyruvate	$C_5H_8O_3$	617-35-6	116.116	liq	-50	155(4)	1.0596^{15}	1.4052^{20}	sl H_2O; s ace; msc EtOH, eth
5190	Ethyl palmitate		$C_{18}H_{36}O_2$	628-97-7	284.478	nd	24.2(0.4)	334(7)	0.8577^{25}	1.4347^{34}	i H_2O; s EtOH, eth, ace, bz, chl
5191	3-Ethylpentane		C_7H_{16}	617-78-7	100.202	liq	-118.55(0.01)	93.4(0.4)	0.6982^{20}	1.3934^{20}	i H_2O; s EtOH, eth; msc ace, bz, hp, chl
5192	3-Ethyl-2,4-pentanedione		$C_7H_{12}O_2$	1540-34-7	128.169			179(11)	0.9531^{19}	1.4408^{19}	vs eth, EtOH, chl
5193	Ethyl pentanoate	Ethyl valerate	$C_7H_{14}O_2$	539-82-2	130.185	liq	-91.2	142(3)	0.8770^{20}	1.4120^{20}	i H_2O; msc EtOH, eth; sl ctc
5194	3-Ethyl-3-pentanol		$C_7H_{16}O$	597-49-9	116.201	liq	-13(2)	142(3)	0.8407^{22}	1.4294^{20}	sl H_2O; s EtOH, eth
5195	2-Ethyl-1-pentene		C_7H_{14}	3404-71-5	98.186			94(2)	0.7079^{20}	1.405^{20}	vs bz, eth, EtOH
5196	3-Ethyl-1-pentene		C_7H_{14}	4038-04-4	98.186	liq	-127.51(0.09)	84(2)	0.6917^{25}	1.3982^{20}	
5197	3-Ethyl-2-pentene		C_7H_{14}	816-79-5	98.186			95(1)	0.7204^{20}	1.4148^{20}	i H_2O; s EtOH, eth, bz, chl
5198	Ethyl pentyl ether		$C_7H_{16}O$	17952-11-3	116.201			118(4)	0.7622^{20}	1.3927^{20}	vs eth, EtOH
5199	Ethyl 2-pentynoate		$C_7H_{10}O_2$	55314-57-3	126.153			67^{18}	0.962^{25}		
5200	2-Ethylphenol		$C_8H_{10}O$	90-00-6	122.164		18	204.5(0.1)	1.0146^{25}	1.5367^{20}	vs ace, bz, eth, EtOH
5201	3-Ethylphenol		$C_8H_{10}O$	620-17-7	122.164	liq	0(8)	218.4(0.1)	1.0283^{20}		sl H_2O, chl; vs EtOH, eth
5202	4-Ethylphenol		$C_8H_{10}O$	123-07-9	122.164	nd	45.0(0.3)	217.97(0.06)		1.5239^{25}	sl H_2O, chl; vs EtOH, eth, bz; s ace
5203	Ethyl phenoxyacetate		$C_{10}H_{12}O_3$	2555-49-9	180.200			247	1.0958^{30}	1.5080^{20}	
5204	*N*-Ethyl-*N*-phenylacetamide		$C_{10}H_{13}NO$	529-65-7	163.216		55	260	0.9938^{60}		s H_2O, eth, ctc
5205	Ethyl phenylacetate	Benzeneacetic acid, ethyl ester	$C_{10}H_{12}O_2$	101-97-3	164.201	liq	-29.4	228(2)	1.0333^{20}	1.4980^{20}	vs eth, EtOH
5206	2-(Ethylphenylamino)ethanol		$C_{10}H_{15}NO$	92-50-2	165.232						s chl
5207	Ethyl phenylcarbamate	Phenylurethane	$C_9H_{11}NO_2$	101-99-5	165.189	wh nd (w) pl (dil al)	53(1)	237 dec	1.1064^{30}	1.5376^{30}	i H_2O; vs EtOH, eth; s bz; sl ctc
5208	Ethyl *N*-phenylformimidate		$C_9H_{11}NO$	6780-49-0	149.189			214	1.0051^{20}	1.5279^{20}	s eth, bz
5209	Ethyl *N*-phenylglycinate		$C_{10}H_{13}NO_2$	2216-92-4	179.216	lf (dil al)	58	273.5			vs eth, EtOH
5210	1-(4-Ethylphenyl)-2-phenylethane		$C_{16}H_{18}$	7439-15-8	210.314	cry		294	1.028^{50}		
5211	Ethyl 3-phenylpropanoate		$C_{11}H_{14}O_2$	2021-28-5	178.228			247(3)	1.0147^{20}	1.4954^{20}	vs eth, EtOH
5212	Ethyl 3-phenylpropynoate	Ethyl phenylacetylenecarboxylate	$C_{11}H_{10}O_2$	2216-94-6	174.196			265	1.055^{25}	1.5520^{20}	s eth
5213	Ethyl phenyl sulfone		$C_8H_{10}O_2S$	599-70-2	170.229	lf (dil al)	42	160^{12}	1.1410^{20}		vs bz, eth, EtOH, chl
5214	Ethylphosphonic acid		$C_2H_7O_3P$	6779-09-5	110.049	hyg pl or nd	61.5	335^{8}			vs H_2O, eth, EtOH
5215	Ethyl phosphorodichloridate	Ethylphosphoric acid dichloride	$C_2H_5Cl_2O_2P$	1498-51-7	162.940			62^{10}		1.4338^{20}	
5216	5-Ethyl-2-picoline		$C_8H_{11}N$	104-90-5	121.180			178(3)	0.9202^{20}	1.4971^{20}	sl H_2O; s EtOH, eth, bz; vs ace
5217	Ethyl 1-piperazinecarboxylate	1-Carbethoxypiperazine	$C_7H_{14}N_2O_2$	120-43-4	158.198			237		1.4760^{25}	vs H_2O, eth, EtOH
5218	1-Ethylpiperidine		$C_7H_{15}N$	766-09-6	113.201			133(3)	0.8237^{20}	1.4480^{20}	
5219	Ethyl 4-piperidinecarboxylate		$C_8H_{15}NO_2$	1126-09-6	157.211	col oil		100^{10}		1.4591^{20}	vs H_2O, bz, eth, EtOH
5220	Ethyl 1-piperidinepropanoate		$C_{10}H_{19}NO_2$	19653-33-9	185.264			217	0.9627^{25}	1.4525^{25}	vs H_2O
5221	1-Ethyl-3-piperidinol		$C_7H_{15}NO$	13444-24-1	129.200			94^{15}		1.4777^{14}	
5222	*N*-Ethyl-1-propanamine		$C_5H_{13}N$	20193-20-8	87.164			85(8)	0.7204^{17}	1.3858^{25}	sl H_2O; vs ace, EtOH
5223	Ethylpropanedioic acid		$C_5H_8O_4$	601-75-2	132.116	pr (w+1)	112(2)	180$^{0.05}$			vs H_2O; s EtOH, eth, bz; i ace; sl tfa
5224	Ethyl propanoate	Ethyl propionate	$C_5H_{10}O_2$	105-37-3	102.132	liq	-73.6(0.5)	98.9(0.2)	0.8843^{25}	1.3839^{20}	sl H_2O, ctc; msc EtOH, eth; s ace
5225	Ethyl propyl ether		$C_5H_{12}O$	628-32-0	88.148	liq	-127.5(0.1)	63(3)	0.7386^{20}	1.3695^{20}	vs eth, EtOH, HOAc
5226	2-(1-Ethylpropyl)pyridine		$C_{10}H_{15}N$	7399-50-0	149.233			195.4	0.8981^{20}	1.4850^{25}	
5227	4-(1-Ethylpropyl)pyridine		$C_{10}H_{15}N$	35182-51-5	149.233		125.5	217	0.9085^{25}	1.40905^{25}	
5228	Ethyl propyl sulfide		$C_5H_{12}S$	4110-50-3	104.214	liq	-117.03(0.05)	118.5(0.8)	0.8370^{20}	1.4462^{20}	s EtOH
5229	Ethyl 2-propynoate	(Ethoxycarbonyl)acetylene	$C_5H_6O_2$	623-47-2	98.101			119(5)	0.9645^{16}	1.4105^{20}	i H_2O; vs EtOH, eth, chl
5230	2-Ethylpyrazine		$C_6H_8N_2$	13925-00-3	108.141			112^{200}			

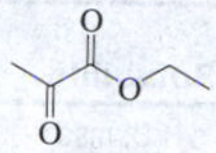
5189
Ethyl 2-oxopropanoate

5190
Ethyl palmitate

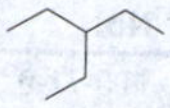
5191
3-Ethylpentane

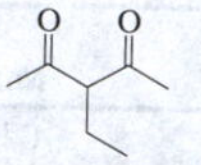
5192
3-Ethyl-2,4-pentanedione

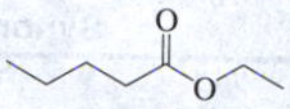
5193
Ethyl pentanoate

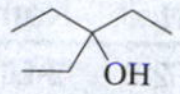
5194
3-Ethyl-3-pentanol

5195
2-Ethyl-1-pentene

5196
3-Ethyl-1-pentene

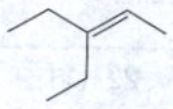
5197
3-Ethyl-2-pentene

5198
Ethyl pentyl ether

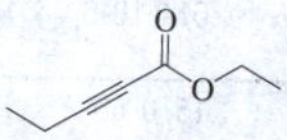
5199
Ethyl 2-pentynoate

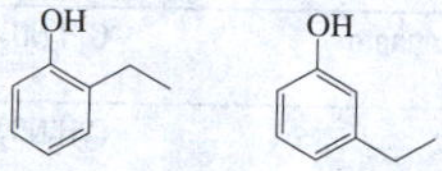
5200
2-Ethylphenol

5201
3-Ethylphenol

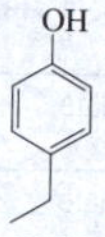
5202
4-Ethylphenol

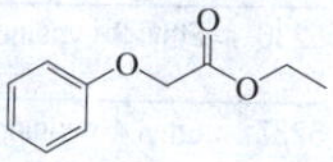
5203
Ethyl phenoxyacetate

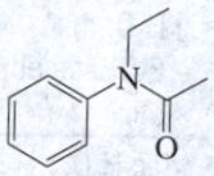
5204
N-Ethyl-*N*-phenylacetamide

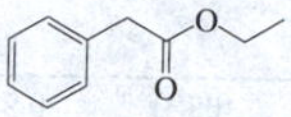
5205
Ethyl phenylacetate

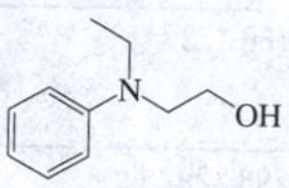
5206
2-(Ethylphenylamino)ethanol

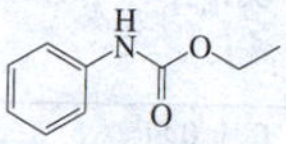
5207
Ethyl phenylcarbamate

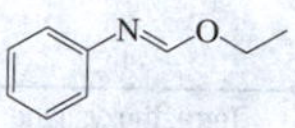
5208
Ethyl *N*-phenylformimidate

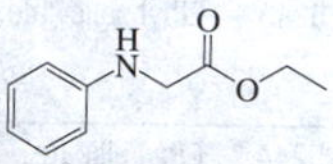
5209
Ethyl *N*-phenylglycinate

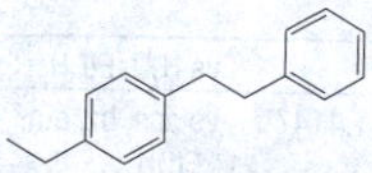
5210
1-(4-Ethylphenyl)-2-phenylethane

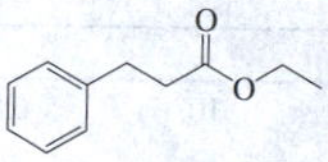
5211
Ethyl 3-phenylpropanoate

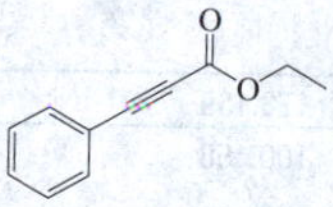
5212
Ethyl 3-phenylpropynoate

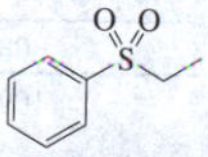
5213
Ethyl phenyl sulfone

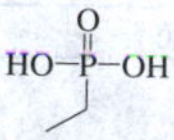
5214
Ethylphosphonic acid

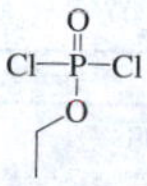
5215
Ethyl phosphorodichloridate

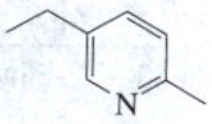
5216
5-Ethyl-2-picoline

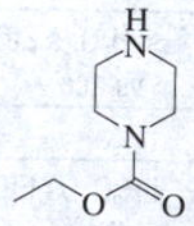
5217
Ethyl 1-piperazinecarboxylate

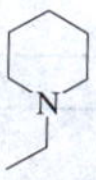
5218
1-Ethylpiperidine

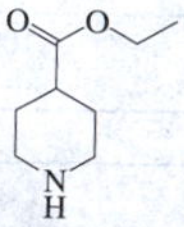
5219
Ethyl 4-piperidinecarboxylate

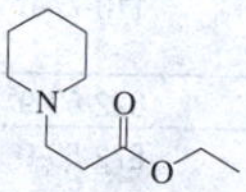
5220
Ethyl 1-piperidinepropanoate

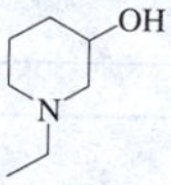
5221
1-Ethyl-3-piperidinol

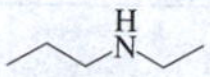
5222
N-Ethyl-1-propanamine

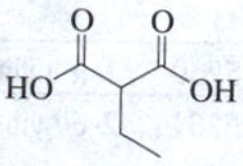
5223
Ethylpropanedioic acid

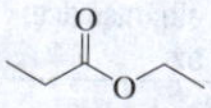
5224
Ethyl propanoate

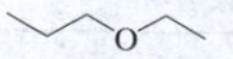
5225
Ethyl propyl ether

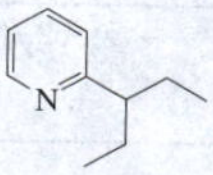
5226
2-(1-Ethylpropyl)pyridine

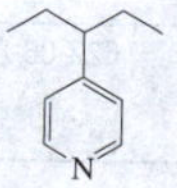
5227
4-(1-Ethylpropyl)pyridine

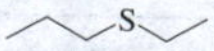
5228
Ethyl propyl sulfide

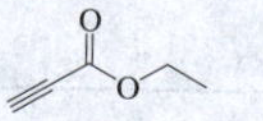
5229
Ethyl 2-propynoate

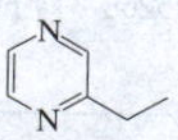
5230
2-Ethylpyrazine

No.	Name	Synonym	Mol. form.	CAS Reg. No.	Mol. wt.	Physical form	mp/°C	bp/°C	den/g cm^{-3}	n_D	Solubility
5231	2-Ethylpyridine		C_7H_9N	100-71-0	107.153	liq	-63.0(0.5)	149(1)	0.9502^{25}	1.4964^{20}	s H_2O; msc EtOH; vs eth, ace; sl ctc
5232	3-Ethylpyridine		C_7H_9N	536-78-7	107.153	liq	-76.9(0.5)	166(2)	0.9539^{25}	1.5021^{20}	s H_2O, EtOH, eth; vs ace; sl ctc
5233	4-Ethylpyridine		C_7H_9N	536-75-4	107.153	liq	-90.5(0.5)	168(2)	0.9417^{20}	1.5009^{20}	s H_2O, EtOH, eth; vs ace; sl ctc
5234	2-Ethyl-4-pyridinecarbothioamide	Ethionamide	$C_8H_{10}N_2S$	536-33-4	166.243		163				
5235	Ethyl 2-pyridinecarboxylate	Ethyl 2-picolinate	$C_8H_9NO_2$	2524-52-9	151.163	ye cry in air	1	243	1.1194^{20}	1.5104^{20}	vs H_2O, eth, EtOH
5236	Ethyl 3-pyridinecarboxylate	Ethyl nicotinate	$C_8H_9NO_2$	614-18-6	151.163		8.5	221(20)	1.1070^{20}	1.5024^{20}	vs H_2O, EtOH, eth, bz; sl ctc
5237	Ethyl 4-pyridinecarboxylate		$C_8H_9NO_2$	1570-45-2	151.163		23	221(11)	1.0091^{15}	1.5017^{20}	sl H_2O; s EtOH, bz; vs eth, chl
5238	*N*-Ethylpyridinium bromide		$C_7H_{10}BrN$	1906-79-2	188.065	cry (al)	111.5				s H_2O, EtOH; i eth
5239	1-Ethyl-1*H*-pyrrole		C_6H_9N	617-92-5	95.142			130(5)	0.9009^{20}	1.4841^{20}	vs EtOH
5240	1-Ethyl-1*H*-pyrrole-2,5-dione	*N*-Ethylmaleimide	$C_6H_7NO_2$	128-53-0	125.126	cry (bz)	45.5				sl H_2O; vs EtOH, eth; s chl
5241	1-Ethyl-2-pyrrolidinemethanamine		$C_7H_{16}N_2$	26116-12-1	128.215			59^{16}	0.887^{25}	1.4665^{20}	
5242	Ethyl Red	2-(4-Diethylaminophenylazo)-benzoic acid	$C_{17}H_{19}N_3O_2$	76058-33-8	297.352		135				
5243	Ethyl salicylate		$C_9H_{10}O_3$	118-61-6	166.173		45	150^{10}	1.1326^{20}	1.5296^{20}	i H_2O; msc EtOH; vs eth; s ctc
5244	Ethyl silicate	Tetraethoxysilane	$C_8H_{20}O_4Si$	78-10-4	208.329	liq	-82.2(0.1)	168(1)	0.9320^{20}	1.3928^{20}	dec H_2O
5245	Ethyl stearate	Ethyl octadecanoate	$C_{20}H_{40}O_2$	111-61-5	312.531		33.1(0.7)	356(6)	1.057^{20}	1.4349^{40}	i H_2O; s EtOH, eth, chl; vs ace
5246	2-Ethylstyrene		$C_{10}H_{12}$	7564-63-8	132.202	liq	-75.6(0.5)	189(1)	0.9017^{25}	1.5380^{20}	
5247	3-Ethylstyrene		$C_{10}H_{12}$	7525-62-4	132.202	liq	-101.3(0.6)	190(2)	0.8945^{20}	1.5351^{20}	
5248	4-Ethylstyrene		$C_{10}H_{12}$	3454-07-7	132.202	liq	-49.7(0.5)	192(9)	0.8884^{25}	1.5376^{20}	
5249	Ethyl sulfate		$C_2H_6O_4S$	540-82-9	126.132			280 dec	1.3657^{20}	1.4105^{20}	vs H_2O
5250	2-(Ethylsulfonyl)ethanol	Ethylsulfonylethyl alcohol	$C_4H_{10}O_3S$	513-12-2	138.185						sl chl
5251	2-Ethyl-5-(3-sulfophenyl)-isoxazolium hydroxide, inner salt	Woodward's Reagent K	$C_{11}H_{11}NO_4S$	4156-16-5	253.275		dec 207				
5252	Ethyl tartrate	Ethyl tartrate, acid	$C_6H_{10}O_6$	608-89-9	178.139		90				vs H_2O, EtOH
5253	2-Ethyltetrahydrofuran		$C_6H_{12}O$	1003-30-1	100.158			107(7)	0.8570^{19}	1.4147^{19}	vs ace, bz, eth, EtOH
5254	5-Ethyl-1,3,4-thiadiazol-2-amine		$C_4H_7N_3S$	14068-53-2	129.184		200.8				
5255	*S*-Ethyl thioacetate		C_4H_8OS	625-60-5	104.171			114(3)	0.9792^{20}	1.4583^{21}	i H_2O; vs EtOH, eth
5256	(Ethylthio)acetic acid		$C_4H_8O_2S$	627-04-3	120.171		-8.5	164^{83}	1.1497^{20}		vs H_2O, EtOH, eth
5257	(Ethylthio)benzene	Thiophenetole	$C_8H_{10}S$	622-38-8	138.230			207(1)	1.0211^{20}	1.5670^{20}	s EtOH
5258	Ethyl thiocyanate		C_3H_5NS	542-90-5	87.144	liq	-85.5(0.5)	144(7)	1.007^{23}	1.4684^{15}	i H_2O; msc EtOH, eth; s chl
5259	2-(Ethylthio)ethanol		$C_4H_{10}OS$	110-77-0	106.186	liq	-100	175(3)	1.0166^{20}	1.4867^{20}	sl H_2O; s EtOH; vs ace
5260	1-(Ethylthio)-4-methylbenzene		$C_9H_{12}S$	622-63-9	152.256			220	0.9996^{20}	1.555^{20}	
5261	2-Ethylthiophene		C_6H_8S	872-55-9	112.193			136(1)	0.9930^{20}	1.5122^{20}	i H_2O; vs EtOH, eth
5262	Ethyl thiophene-2-carboxylate		$C_7H_8O_2S$	2810-04-0	156.203			218	1.1623^{16}	1.5248^{20}	s EtOH, ace; sl ctc
5263	3-Ethyl-2-thioxo-4-thiazolidinone	3-Ethylrhodanine	$C_5H_7NOS_2$	7648-01-3	161.246		35.5				
5264	2-Ethyltoluene		C_9H_{12}	611-14-3	120.191	liq	-80.7(0.4)	165.1(0.4)	0.8807^{20}	1.5046^{20}	i H_2O; msc EtOH, eth, ace, bz, peth, ctc
5265	3-Ethyltoluene		C_9H_{12}	620-14-4	120.191	liq	-95.7(0.2)	161.3(0.5)	0.8645^{20}	1.4966^{20}	i H_2O; vs EtOH, eth; msc ace, bz
5266	4-Ethyltoluene		C_9H_{12}	622-96-8	120.191	liq	-62.7(0.5)	162.0(0.6)	0.8614^{20}	1.4959^{20}	i H_2O; vs EtOH, eth; msc ace, bz

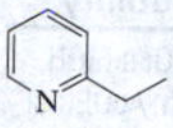
5231
2-Ethylpyridine

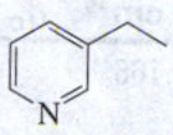
5232
3-Ethylpyridine

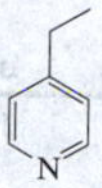
5233
4-Ethylpyridine

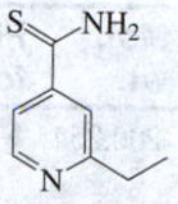
5234
2-Ethyl-4-pyridinecarbothioamide

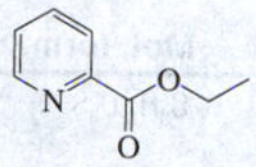
5235
Ethyl 2-pyridinecarboxylate

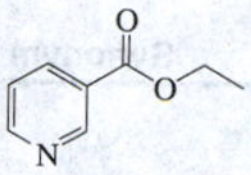
5236
Ethyl 3-pyridinecarboxylate

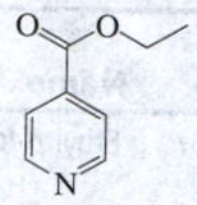
5237
Ethyl 4-pyridinecarboxylate

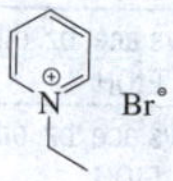
5238
N-Ethylpyridinium bromide

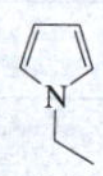
5239
1-Ethyl-1*H*-pyrrole

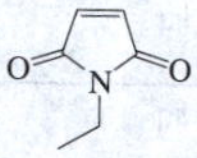
5240
1-Ethyl-1*H*-pyrrole-2,5-dione

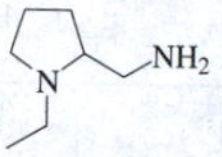
5241
1-Ethyl-2-pyrrolidinemethanamine

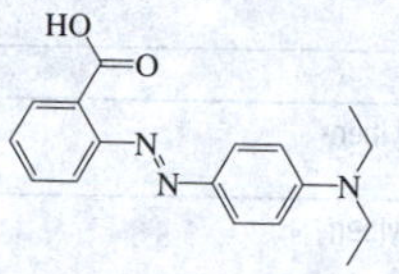
5242
Ethyl Red

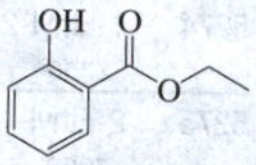
5243
Ethyl salicylate

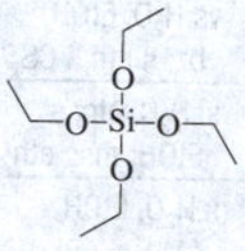
5244
Ethyl silicate

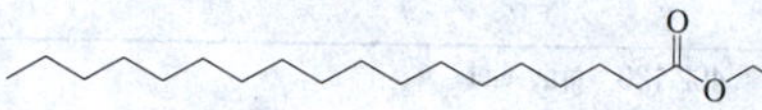
5245
Ethyl stearate

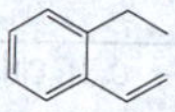
5246
2-Ethylstyrene

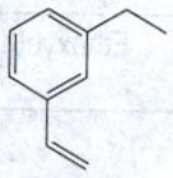
5247
3-Ethylstyrene

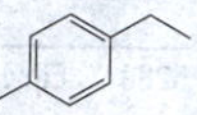
5248
4-Ethylstyrene

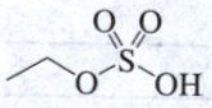
5249
Ethyl sulfate

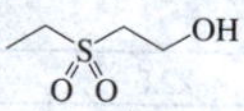
5250
2-(Ethylsulfonyl)ethanol

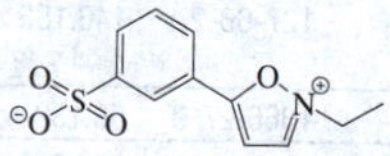
5251
2-Ethyl-5-(3-sulfophenyl)isoxazolium hydroxide, inner salt

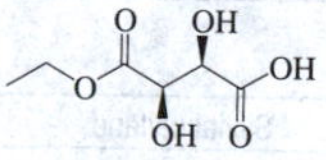
5252
Ethyl tartrate

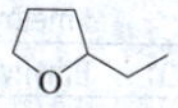
5253
2-Ethyltetrahydrofuran

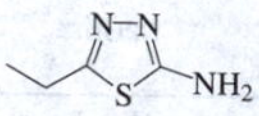
5254
5-Ethyl-1,3,4-thiadiazol-2-amine

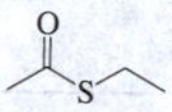
5255
S-Ethyl thioacetate

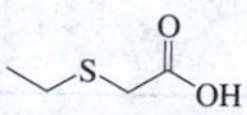
5256
(Ethylthio)acetic acid

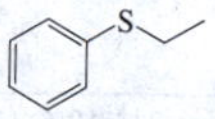
5257
(Ethylthio)benzene

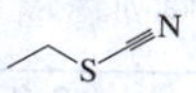
5258
Ethyl thiocyanate

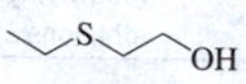
5259
2-(Ethylthio)ethanol

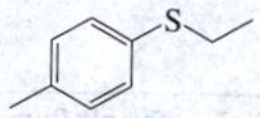
5260
1-(Ethylthio)-4-methylbenzene

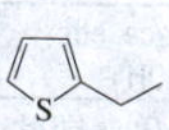
5261
2-Ethylthiophene

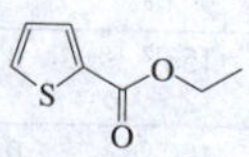
5262
Ethyl thiophene-2-carboxylate

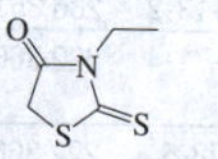
5263
3-Ethyl-2-thioxo-4-thiazolidinone

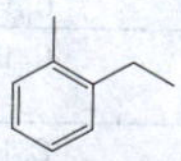
5264
2-Ethyltoluene

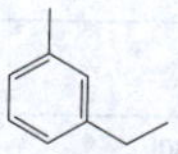
5265
3-Ethyltoluene

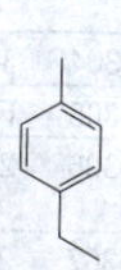
5266
4-Ethyltoluene

No.	Name	Synonym	Mol. form.	CAS Reg. No.	Mol. wt.	Physical form	mp/°C	bp/°C	den/ g cm⁻³	n_D	Solubility
5267	Ethyl *p*-toluenesulfonate		$C_9H_{12}O_3S$	80-40-0	200.254		33(4)	173^{15}	1.166^{48}		i H_2O; s EtOH, eth, AcOEt; sl ctc
5268	Ethyl trichloroacetate		$C_4H_5Cl_3O_2$	515-84-4	191.441			167(3)	1.3836^{20}	1.4505^{20}	i H_2O; s EtOH, eth, bz; sl chl
5269	Ethyl trifluoroacetate		$C_4H_5F_3O_2$	383-63-1	142.077			61	1.194^{20}	1.308^{20}	
5270	Ethyl 4,4,4-trifluoroacetoacetate		$C_6H_7F_3O_3$	372-31-6	184.113	liq	-39.1	132	1.2586^{15}	1.3783^{15}	s EtOH, eth
5271	Ethyl trifluoromethanesulfonate		$C_3H_5F_3O_3S$	425-75-2	178.130			115	1.3740^{0}		s eth
5272	Ethyl 3,4,5-trihydroxybenzoate		$C_9H_{10}O_5$	831-61-8	198.172	mcl pr (w+2 1/2) nd (chl)	163.0				sl H_2O, chl; s EtOH, eth, AcOEt
5273	Ethyltrimethoxysilane		$C_5H_{14}O_3Si$	5314-55-6	150.249			122.9(0.8)	0.9488^{20}	1.3838^{20}	vs EtOH
5274	1-Ethyl-2,4,5-trimethylbenzene		$C_{11}H_{16}$	17851-27-3	148.245	liq	-13.4(0.7)	212(1)	0.883^{20}	1.5075^{20}	vs ace, bz, eth, EtOH
5275	2-Ethyl-1,3,5-trimethylbenzene		$C_{11}H_{16}$	3982-67-0	148.245	liq	-15.5(0.3)	211(1)	0.883^{20}	1.5074^{20}	vs ace, bz, eth, EtOH
5276	Ethyltrimethyllead	Ethyltrimethylplumbane	$C_5H_{14}Pb$	1762-26-1	281.4	col liq		$27^{10.5}$	1.88^{20}		
5277	3-Ethyl-2,4,5-trimethylpyrrole		$C_9H_{15}N$	520-69-4	137.222	lf (eth)	66.5	214			
5278	4-Ethyl-2,6,7-trioxa-1-phosphabicyclo[2.2.2]octane	Trimethylolpropane phosphite	$C_6H_{11}O_3P$	824-11-3	162.123		53.7				s chl
5279	Ethyl undecanoate	Ethyl undecylate	$C_{13}H_{26}O_2$	627-90-7	214.344		-16(6)	131^{14}	0.8633^{20}	1.4285^{20}	i H_2O; s EtOH, eth, ace, bz
5280	Ethyl 10-undecenoate		$C_{13}H_{24}O_2$	692-86-4	212.329	liq	-38	264.5	0.8827^{15}	1.4449^{25}	i H_2O; s EtOH, eth, HOAc; sl ctc
5281	*N*-Ethylurea		$C_3H_8N_2O$	625-52-5	88.108	nd (bz, al-eth)	94(2)	dec	1.2130^{18}		vs H_2O, EtOH, bz; s eth; i CS_2
5282	Ethyl vinyl ether	Ethoxyethylene	C_4H_8O	109-92-2	72.106	liq	-115.8(0.4)	36(2)	0.7589^{20}	1.3767^{20}	sl H_2O, ctc; s EtOH; msc eth
5283	Ethyl Violet		$C_{31}H_{42}ClN_3$	2390-59-2	492.138	gray-viol cry					s H_2O, EtOH
5284	α-Ethynylbenzenemethanol	1-Phenylpropargyl alcohol	C_9H_8O	4187-87-5	132.159	pr	22	114^{12}	1.0655^{20}	1.5508^{20}	
5285	α-Ethynylbenzenemethanol carbamate	Carfimate	$C_{10}H_9NO_2$	3567-38-2	175.184	cry (al)	86.5				
5286	1-Ethynylcyclohexanamine		$C_8H_{13}N$	30389-18-5	123.196			65^{20}	0.913^{25}	1.4817^{20}	
5287	1-Ethynylcyclohexanol		$C_8H_{12}O$	78-27-3	124.180	cry (peth)	31(3)	174	0.9873^{20}	1.4822^{20}	i H_2O; s EtOH, bz, peth; sl chl
5288	1-Ethynylcyclohexanol, carbamate	Ethinamate	$C_9H_{13}NO_2$	126-52-3	167.205	nd	97	120^{3}		1.4441^{21}	sl H_2O; vs EtOH; s hx
5289	1-Ethynylcyclopentanol		$C_7H_{10}O$	17356-19-3	110.153		27	157.5	0.962^{25}	1.4751^{20}	
5290	α-Ethynyl-α-methylbenzenemethanol		$C_{10}H_{10}O$	127-66-2	146.185		52(3)	217.5	1.0314^{20}		
5291	Ethynylsilane	Silylacetylene	C_2H_4Si	1066-27-9	56.139	col gas		-22.5			
5292	Etioporphyrin		$C_{32}H_{38}N_4$	448-71-5	478.671		362				
5293	Etofylline		$C_9H_{12}N_4O_3$	519-37-9	224.216		158				vs H_2O; s EtOH; sl eth, bz
5294	Etoglucid	Oxirane, 2,2'-(2,5,8,11-tetraoxadodecane-1,12-diyl)bis-	$C_{12}H_{22}O_6$	1954-28-5	262.299	col liq	-13	196^{2}	1.1312^{20}	1.4622^{20}	
5295	Etoposide		$C_{29}H_{32}O_{13}$	33419-42-0	588.556	cry (MeOH)	≈243				s MeOH
5296	Etrimfos		$C_{10}H_{17}N_2O_4PS$	38260-54-7	292.291		-1.7		1.195^{20}		
5297	Eucalyptol	Cineole	$C_{10}H_{18}O$	470-82-6	154.249	oil	1.4(0.3)	176(4)	0.9267^{20}	1.4586^{20}	i H_2O; s EtOH, eth, chl; sl ctc
5298	Euparin	1-[6-Hydroxy-2-(1-methylvinyl)-5-benzofuranyl] ethanone	$C_{13}H_{12}O_3$	532-48-9	216.232		121.5				s eth, bz, chl; sl NaOH
5299	Evan's Blue		$C_{34}H_{24}N_6Na_4O_{14}S_4$	314-13-6	960.806						s H_2O, EtOH, acid
5300	Evodiamine		$C_{19}H_{17}N_3O$	518-17-2	303.357	ye lf (al)	28				
5301	Famotidine		$C_8H_{15}N_7O_2S_3$	76824-35-6	337.446	cry	170.6(0.6)				i EtOH, chl; vs DMF; s HOAc; sl MeOH
5302	Famphur		$C_{10}H_{16}NO_5PS_2$	52-85-7	325.342		57(2)				
5303	α-Farnesene		$C_{15}H_{24}$	502-61-4	204.352			130^{12}	0.8410^{20}	1.4836^{20}	i H_2O; s eth, ace; msc peth, lig
5304	β-Farnesene		$C_{15}H_{24}$	18794-84-8	204.352			121^{9}	0.8363^{20}	1.4899^{20}	vs ace, eth, chl
5305	Farnesic acid		$C_{15}H_{24}O_2$	7548-13-2	236.351	oil		204^{16}			
5306	2-*cis*,6-*trans*-Farnesol		$C_{15}H_{26}O$	3790-71-4	222.366	oil		156^{12}	0.8908^{20}	1.4877^{20}	vs ace, eth, EtOH
5307	2-*trans*,6-*trans*-Farnesol		$C_{15}H_{26}O$	106-28-5	222.366	oil		160^{10}	0.888^{20}	1.4877^{20}	i H_2O; vs EtOH; s eth, ace

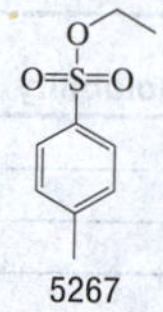
5267
Ethyl *p*-toluenesulfonate

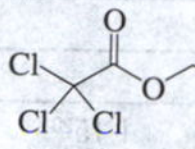
5268
Ethyl trichloroacetate

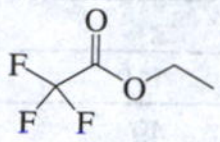
5269
Ethyl trifluoroacetate

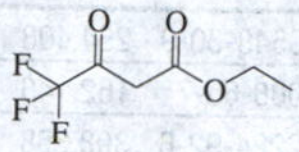
5270
Ethyl 4,4,4-trifluoroacetoacetate

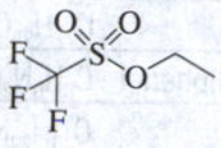
5271
Ethyl trifluoromethanesulfonate

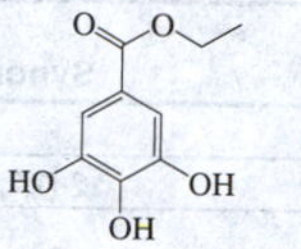
5272
Ethyl 3,4,5-trihydroxybenzoate

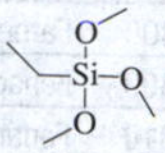
5273
Ethyltrimethoxysilane

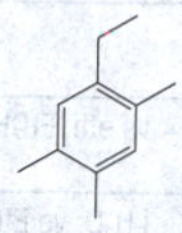
5274
1-Ethyl-2,4,5-trimethylbenzene

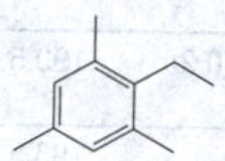
5275
2-Ethyl-1,3,5-trimethylbenzene

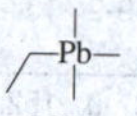
5276
Ethyltrimethyllead

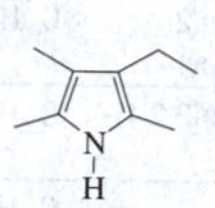
5277
3-Ethyl-2,4,5-trimethylpyrrole

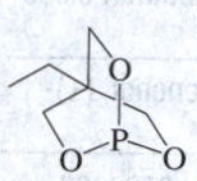
5278
4-Ethyl-2,6,7-trioxa-1-phosphabicyclo[2.2.2]octane

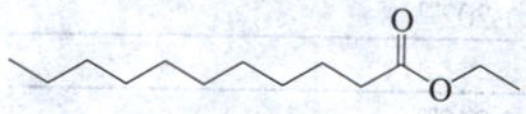
5279
Ethyl undecanoate

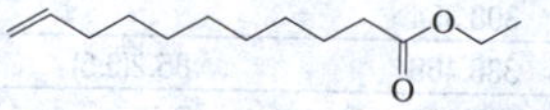
5280
Ethyl 10-undecenoate

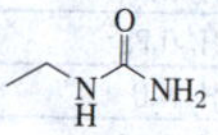
5281
N-Ethylurea

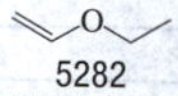
5282
Ethyl vinyl ether

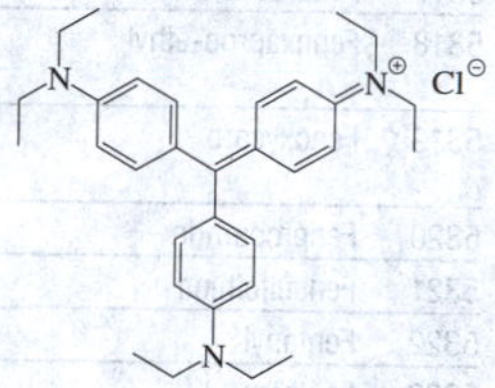
5283
Ethyl Violet

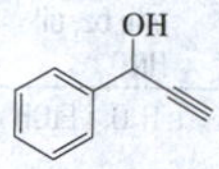
5284
α-Ethynylbenzenemethanol

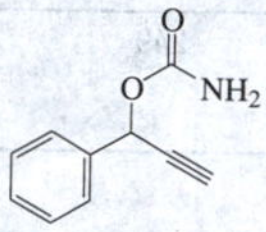
5285
α-Ethynylbenzenemethanol carbamate

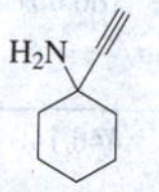
5286
1-Ethynylcyclohexanamine

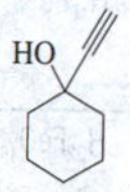
5287
1-Ethynylcyclohexanol

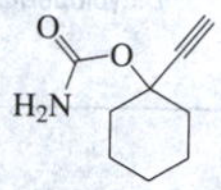
5288
1-Ethynylcyclohexanol, carbamate

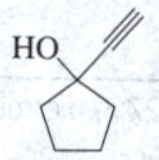
5289
1-Ethynylcyclopentanol

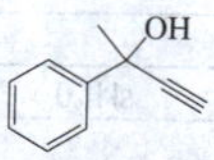
5290
α-Ethynyl-α-methylbenzenemethanol

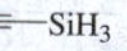
5291
Ethynylsilane

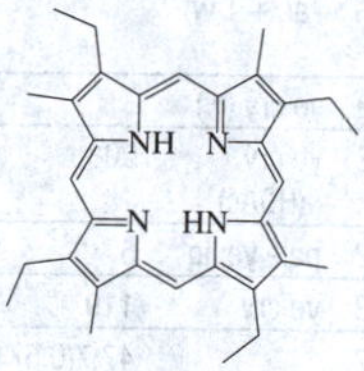
5292
Etioporphyrin

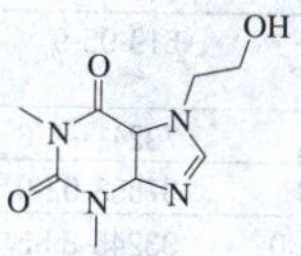
5293
Etofylline

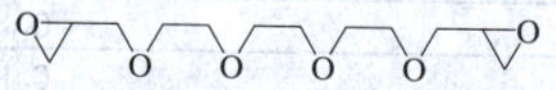
5294
Etoglucid

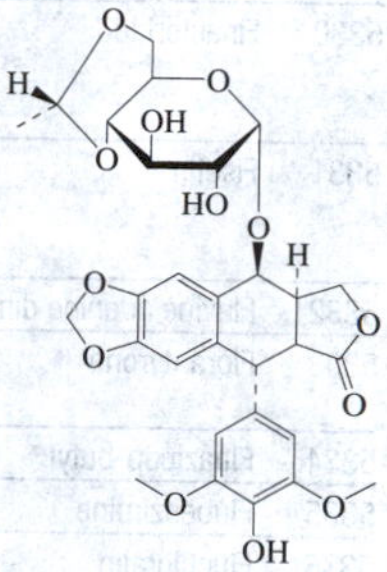
5295
Etoposide

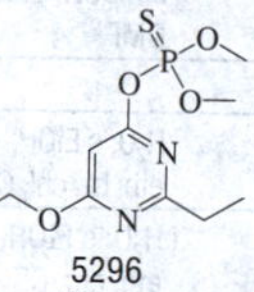
5296
Etrimfos

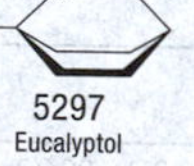
5297
Eucalyptol

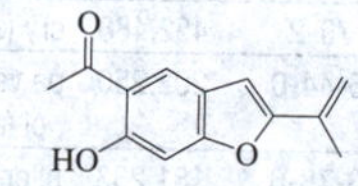
5298
Euparin

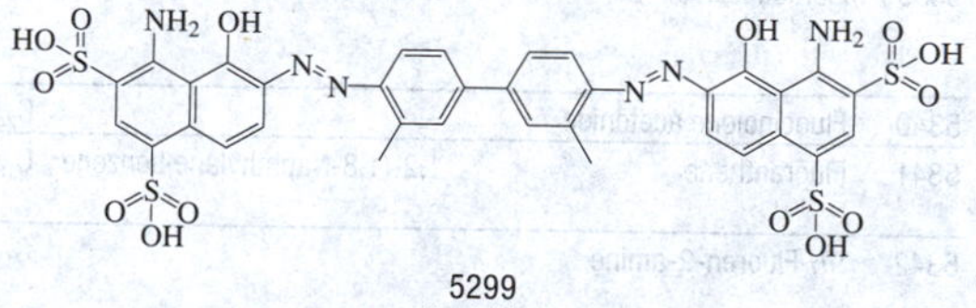
5299
Evan's Blue

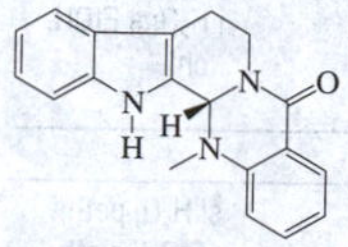
5300
Evodiamine

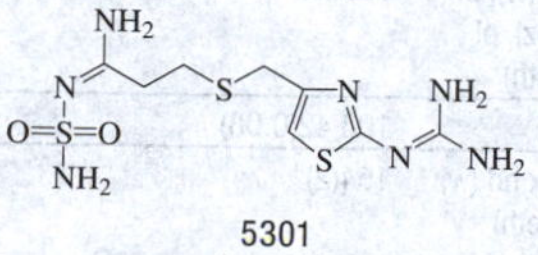
5301
Famotidine

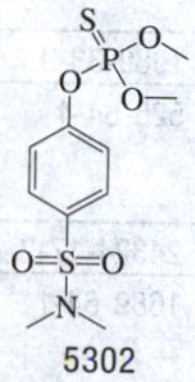
5302
Famphur

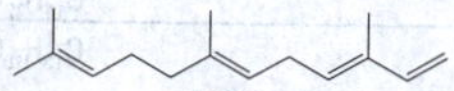
5303
α-Farnesene

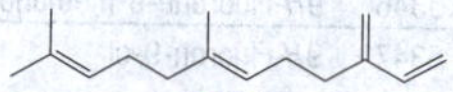
5304
β-Farnesene

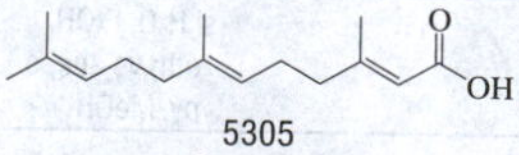
5305
Farnesic acid

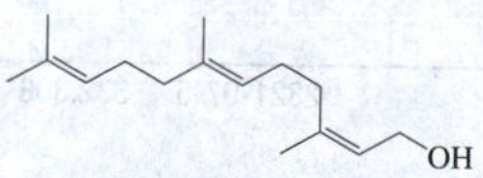
5306
2-*cis*,6-*trans*-Farnesol

5307
2-*trans*,6-*trans*-Farnesol

No.	Name	Synonym	Mol. form.	CAS Reg. No.	Mol. wt.	Physical form	mp/°C	bp/°C	den/ g cm^{-3}	n_D	Solubility
5308	Farnesol acetate		$C_{17}H_{28}O_2$	29548-30-9	264.403			168^{10}			
5309	Fenadiazole	2-(1,2,4-Oxadiazol-2-yl)phenol	$C_8H_6N_2O_2$	1008-65-7	162.146	cry	112	$180^{0.1}$			
5310	Fenamiphos		$C_{13}H_{22}NO_3PS$	22224-92-6	303.358		49		1.15^{20}		
5311	Fenarimol		$C_{17}H_{12}Cl_2N_2O$	60168-88-9	331.195		118				
5312	Fenbuconazole		$C_{20}H_{19}ClN_4$	114369-43-6	350.845		125				
5313	Fenbutatin oxide	Distannoxane, hexakis(2-methyl-2-phenylpropyl)-	$C_{60}H_{78}OSn_2$	13356-08-6	1052.68		145.8(0.5)				
5314	α-Fenchol, (±)-	1,3,3-Trimethylbicyclo[2.2.1]-heptan-2-ol, *endo*-(±)	$C_{10}H_{18}O$	36386-49-9	154.249		41.5(0.2)	199.5	0.9420^{40}		vs eth, EtOH
5315	(±)-Fenchone		$C_{10}H_{16}O$	18492-37-0	152.233	oily liq	6.1	193.5	0.9492^{15}	1.4702^{20}	i H_2O; vs EtOH; s eth, ace
5316	Fenfluramine		$C_{12}H_{16}F_3N$	458-24-2	231.257	cry (AcOEt)		110^{12}			
5317	Fenitrothion		$C_9H_{12}NO_5PS$	122-14-5	277.234			$118^{0.05}$	1.3227^{25}		
5318	Fenoxaprop-ethyl		$C_{18}H_{16}ClNO_5$	82110-72-3	361.777		85	$200^{0.001}$			sl H_2O, hx; s eth; vs ace, tol
5319	Fenoxycarb	Ethyl 2-(4-phenoxyphenoxy) ethylcarbamate	$C_{17}H_{19}NO_4$	79127-80-3	301.338		53				
5320	Fenpropathrin		$C_{22}H_{23}NO_3$	64257-84-7	349.423		49.3(0.4)		1.15^{25}		
5321	Fensulfothion		$C_{11}H_{17}O_4PS_2$	115-90-2	308.354			$140^{0.01}$	1.202^{20}		
5322	Fentanyl		$C_{22}H_{28}N_2O$	437-38-7	336.469		85.2(0.5)				
5323	Fenthion		$C_{10}H_{15}O_3PS_2$	55-38-9	278.328		7.5	$87^{0.01}$	1.246^{20}		
5324	Fenvalerate		$C_{25}H_{22}ClNO_3$	51630-58-1	419.901			dec	1.15^{25}		
5325	Ferbam	Iron, tris(dimethylcarba-modithioato-*S,S'*)-, (OC-6-11)-	$C_9H_{18}FeN_3S_6$	14484-64-1	416.494		180 dec				
5326	Ferrocene	Dicyclopentadienyl iron	$C_{10}H_{10}Fe$	102-54-5	186.031	oran needles	175(1)	249			i H_2O; s EtOH, eth, bz, dil HNO_3
5327	Ferrous gluconate		$C_{12}H_{22}FeO_{14}$	299-29-6	446.140	ye-gray pow (w)					s H_2O; i EtOH
5328	Ferrous lactate		$C_6H_{10}FeO_6$	5905-52-2	233.984	grn-wh pow (hyd)					s H_2O; i EtOH
5329	Fichtelite	18-Norabietane	$C_{19}H_{34}$	2221-95-6	262.473	cry	46	236^{43}	0.9380^{22}	1.5052^{20}	
5330	Finasteride	Proscar	$C_{23}H_{36}N_2O_2$	98319-26-7	372.544	wh cry	252(1)				sl H_2O; s chl, EtOH, MeOH, DMSO
5331	Fisetin		$C_{15}H_{10}O_6$	528-48-3	286.236	lt ye nd (dil al, + 1 w)	330				i H_2O; s EtOH, ace; sl eth, bz, peth
5332	Flavine adenine dinucleotide	FAD	$C_{27}H_{33}N_9O_{15}P_2$	146-14-5	785.550	ye cry (w)					
5333	Florantyrone		$C_{20}H_{14}O_3$	519-95-9	302.323	ye cry (HOAc)	208				s EtOH, MeOH
5334	Fluazipop-butyl		$C_{19}H_{20}F_3NO_4$	79241-46-6	383.362	pale ye liq	5				
5335	Flubenzimine		$C_{17}H_{10}F_6N_4S$	37893-02-0	416.343	ye cry	119				sl H_2O
5336	Fluchloralin		$C_{12}H_{13}ClF_3N_3O_4$	33245-39-5	355.697		47.7(0.5)				
5337	Flucythrinate	Cythrin	$C_{26}H_{23}F_2NO_4$	70124-77-5	451.463			$108^{0.35}$	1.189^{22}		
5338	Fludrocortisone		$C_{21}H_{29}FO_5$	127-31-1	380.450	cry (EtOH)	261 dec				
5339	Flumethiazide	Trifluoromethylthiazide	$C_8H_6F_3N_3O_4S_2$	148-56-1	329.277	cry	306				sl H_2O; i bz, tol; s MeOH, EtOH, DMF
5340	Fluocinolone acetonide		$C_{24}H_{30}F_2O_6$	67-73-2	452.488	cry (ace/hx)	266 dec				
5341	Fluoranthene	1,2-(1,8-Naphthylene)benzene	$C_{16}H_{10}$	206-44-0	202.250	pa ye nd or pl (al)	110.2(0.2)	380(5)	1.252^{0}		i H_2O; s EtOH, eth, bz, chl, CS_2
5342	9*H*-Fluoren-2-amine		$C_{13}H_{11}N$	153-78-6	181.233	pl or nd (dil al)	130.3				i H_2O; s EtOH, eth, ctc, CS_2
5343	9*H*-Fluorene	2,2'-Methylenebiphenyl	$C_{13}H_{10}$	86-73-7	166.218	lf (al)	114.76(0.03)	294(2)	1.203^{0}		i H_2O; sl EtOH; s eth, ace, bz, CS_2
5344	9*H*-Fluorene-9-carboxylic acid		$C_{14}H_{10}O_2$	1989-33-9	210.228		230(1)				
5345	9*H*-Fluorene-2,7-diamine	2,7-Diaminofluorene	$C_{13}H_{12}N_2$	525-64-4	196.247	nd (w), pr (bz), pl (eth)	166				i H_2O; s EtOH, chl
5346	9*H*-Fluorene-9-methanol		$C_{14}H_{12}O$	24324-17-2	196.244		103.42(0.08)				
5347	9*H*-Fluoren-9-ol		$C_{13}H_{10}O$	1689-64-1	182.217	hex nd (w, peth)	154(2)				sl H_2O, peth, EtOH; s eth, ace; vs bz
5348	9*H*-Fluoren-9-one	Fluorenone	$C_{13}H_8O$	486-25-9	180.202	ye orth bipym (al, bz-peth)	84(1)	343.1(0.6)	1.1300^{99}	1.6309^{99}	i H_2O; s EtOH, ace, bz; vs tol; sl ctc
5349	Fluorescein		$C_{20}H_{12}O_5$	2321-07-5	332.306	red orth pr	315 dec				sl H_2O, EtOH, eth; vs ace; s py, MeOH

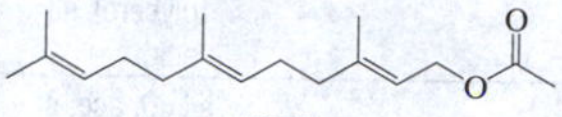
5308
Farnesol acetate

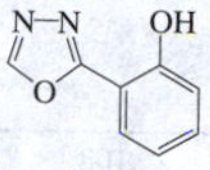
5309
Fenadiazole

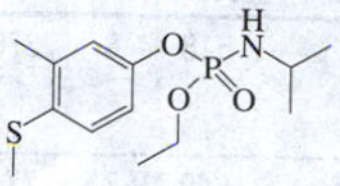
5310
Fenamiphos

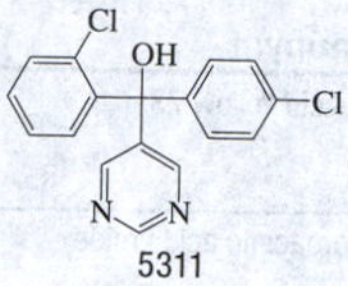
5311
Fenarimol

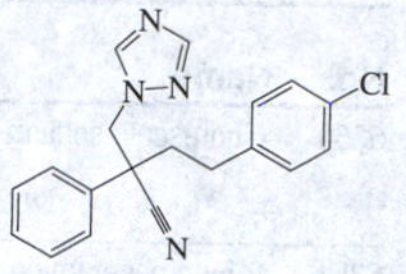
5312
Fenbuconazole

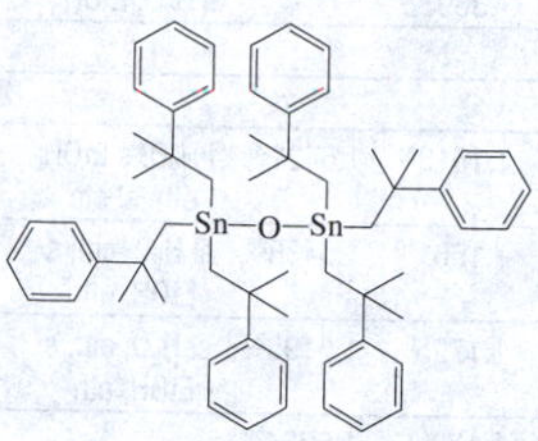
5313
Fenbutatin oxide

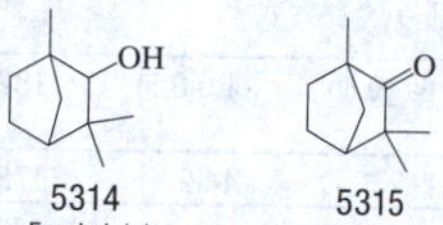
5314
α-Fenchol, (±)-

5315
(±)-Fenchone

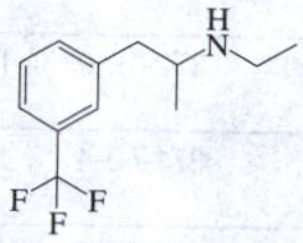
5316
Fenfluramine

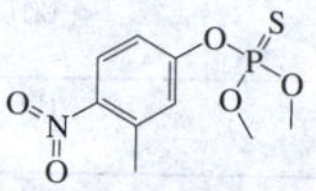
5317
Fenitrothion

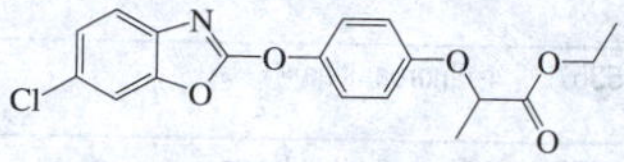
5318
Fenoxaprop-ethyl

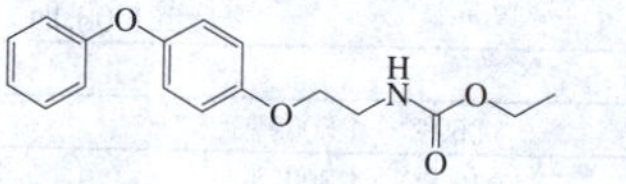
5319
Fenoxycarb

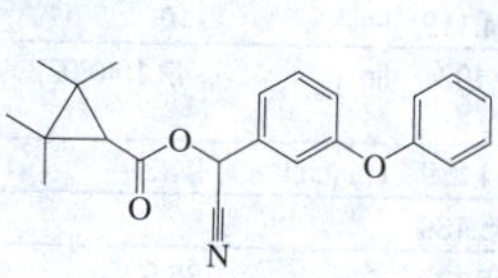
5320
Fenpropathrin

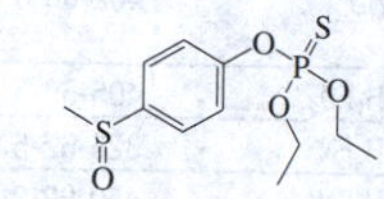
5321
Fensulfothion

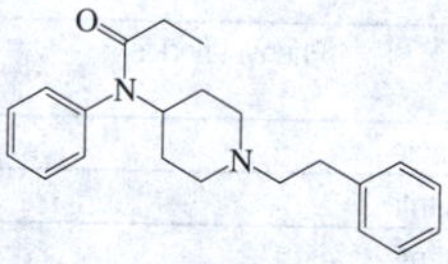
5322
Fentanyl

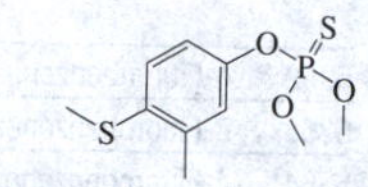
5323
Fenthion

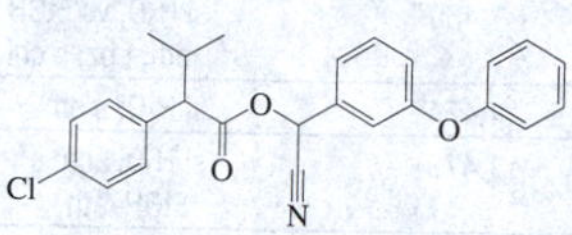
5324
Fenvalerate

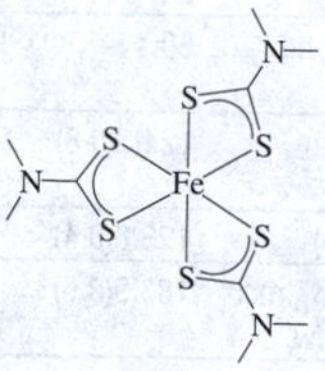
5325
Ferbam

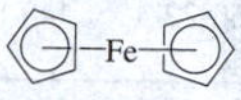
5326
Ferrocene

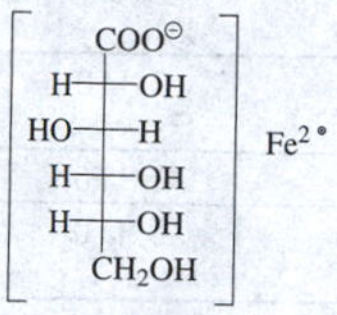
5327
Ferrous gluconate

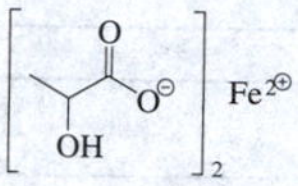
5328
Ferrous lactate

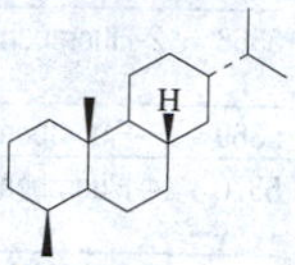
5329
Fichtelite

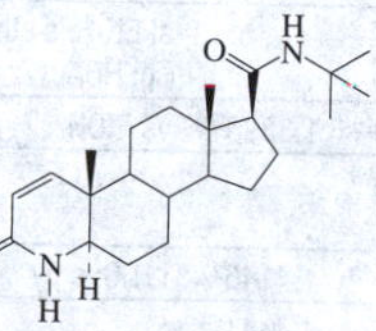
5330
Finasteride

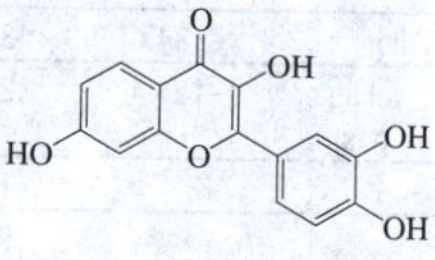
5331
Fisetin

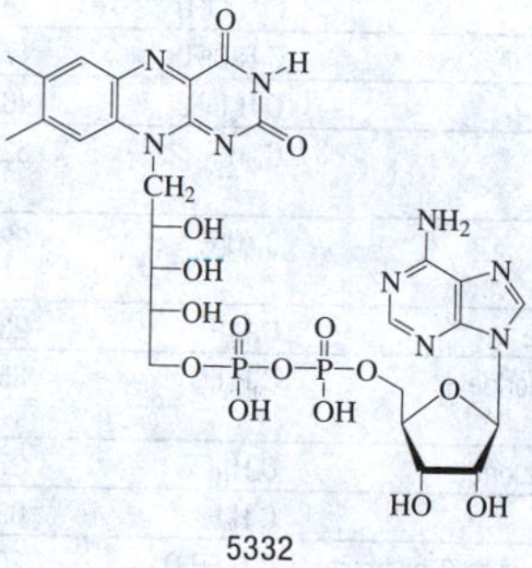
5332
Flavine adenine dinucleotide

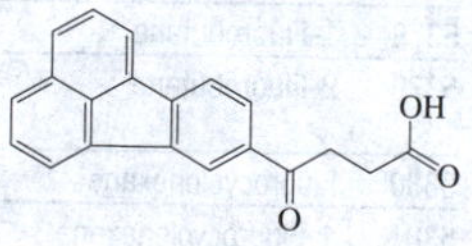
5333
Florantyrone

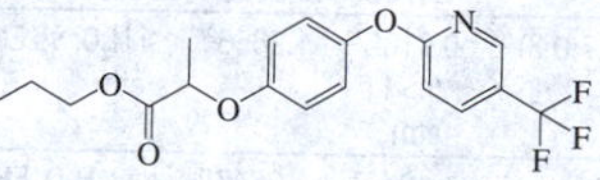
5334
Fluazipop-butyl

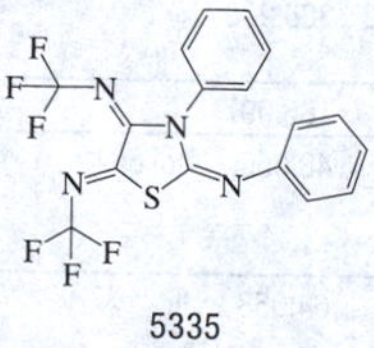
5335
Flubenzimine

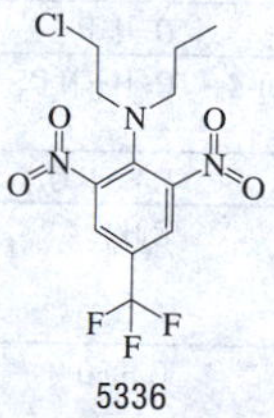
5336
Fluchloralin

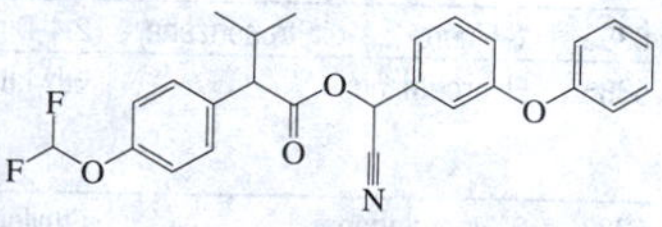
5337
Flucythrinate

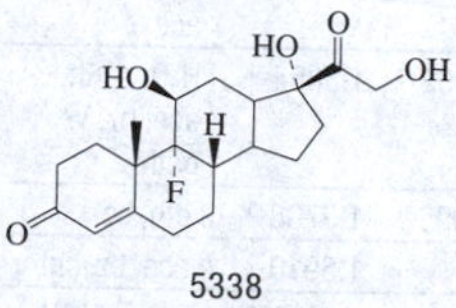
5338
Fludrocortisone

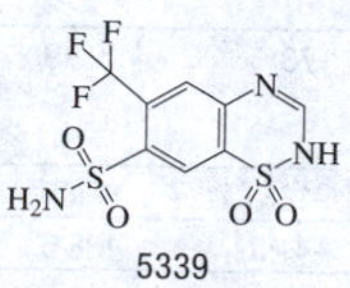
5339
Flumethiazide

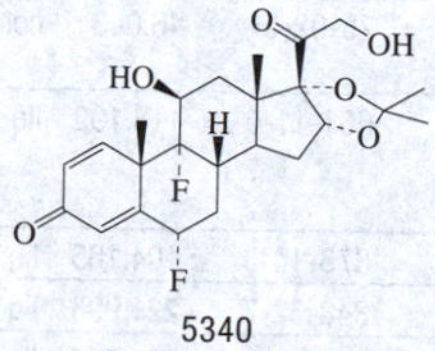
5340
Fluocinolone acetonide

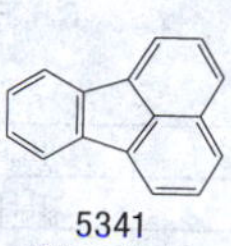
5341
Fluoranthene

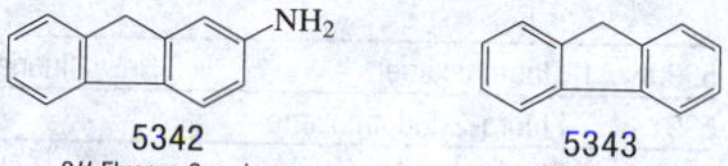
5342
9*H*-Fluoren-2-amine

5343
9*H*-Fluorene

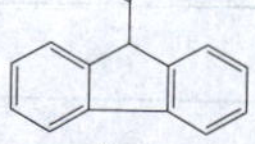
5344
9*H*-Fluorene-9-carboxylic acid

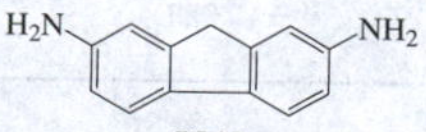
5345
9*H*-Fluorene-2,7-diamine

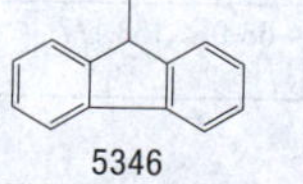
5346
9*H*-Fluorene-9-methanol

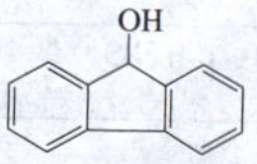
5347
9*H*-Fluoren-9-ol

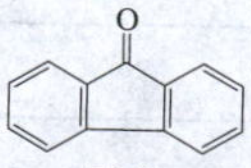
5348
9*H*-Fluoren-9-one

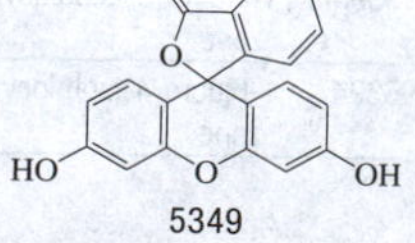
5349
Fluorescein

No.	Name	Synonym	Mol. form.	CAS Reg. No.	Mol. wt.	Physical form	mp/°C	bp/°C	den/ g cm⁻³	n_D	Solubility
5350	Fluorescein sodium	C.I. Acid Yellow 73	$C_{20}H_{10}Na_2O_5$	518-47-8	376.270	ye pow					s H_2O, EtOH, glycerol, dil acid
5351	2-Fluoroacetamide	Fluoroacetic acid amide	C_2H_4FNO	640-19-7	77.057		108	sub			s H_2O, ace; sl chl
5352	Fluoroacetic acid	Fluoroethanoic acid	$C_2H_3FO_2$	144-49-0	78.042	nd	35.2	168	1.3693^{36}		s H_2O, EtOH
5353	Fluoroacetyl chloride		C_2H_2ClFO	359-06-8	96.487	liq		72			
5354	Fluoroacetylene	Fluoroethyne	C_2HF	2713-09-9	44.027	gas	-196	-74(15)			
5355	2-Fluoroaniline		C_6H_6FN	348-54-9	111.117	pa ye liq	-29(1)	175	1.1513^{21}	1.5421^{20}	i H_2O; s EtOH, eth; sl ctc
5356	3-Fluoroaniline		C_6H_6FN	372-19-0	111.117			188	1.1561^{19}	1.5436^{20}	sl H_2O, chl; s EtOH, eth
5357	4-Fluoroaniline		C_6H_6FN	371-40-4	111.117	pa ye liq	-1.9(0.5)	182	1.1725^{20}	1.5195^{20}	sl H_2O, ctc; s EtOH, eth
5358	2-Fluorobenzaldehyde		C_7H_5FO	446-52-6	124.112	liq	-44.5	175	1.178^{25}	1.5234^{20}	
5359	3-Fluorobenzaldehyde		C_7H_5FO	456-48-4	124.112			173	1.17^{25}	1.5206^{20}	
5360	4-Fluorobenzaldehyde		C_7H_5FO	459-57-4	124.112	liq	-10	181.5	1.1810^{19}		
5361	Fluorobenzene	Phenyl fluoride	C_6H_5F	462-06-6	96.102	liq	-42.18(0.05)	84.7(0.3)	1.0225^{20}	1.4684^{30}	sl H_2O; vs bz, eth, EtOH, lig
5362	4-Fluorobenzeneacetic acid		$C_8H_7FO_2$	405-50-5	154.139	cry (chl)	94(3)	164^{2}			
5363	2-Fluorobenzeneacetonitrile		C_8H_6FN	326-62-5	135.139			232	1.059^{25}	1.5009^{20}	
5364	4-Fluorobenzeneacetonitrile		C_8H_6FN	459-22-3	135.139		86.0	228	1.1390^{20}	1.5002^{20}	
5365	4-Fluorobenzenemethanamine		C_7H_8FN	140-75-0	125.144			183		1.5139^{20}	
5366	4-Fluorobenzenemethanol		C_7H_7FO	459-56-3	126.128		23	210		1.5080^{20}	
5367	4-Fluorobenzenesulfonyl chloride		$C_6H_4ClFO_2S$	349-88-2	194.611	pl or nd	30	106^{9}			vs bz, eth, chl
5368	2-Fluorobenzoic acid		$C_7H_5FO_2$	445-29-4	140.112	nd (a)	124.2(0.8)		1.460^{25}		sl H_2O; vs EtOH, eth; i bz; s chl
5369	3-Fluorobenzoic acid		$C_7H_5FO_2$	455-38-9	140.112	lf (w)	123.6(0.4)		1.474^{25}		sl H_2O; s eth
5370	4-Fluorobenzoic acid		$C_7H_5FO_2$	456-22-4	140.112	pr (w), mcl pr (w)	183.9(0.6)		1.479^{25}		sl H_2O, ace; s EtOH, eth
5371	2-Fluorobenzonitrile		C_7H_4FN	394-47-8	121.112			93^{22}			
5372	4-Fluorobenzonitrile		C_7H_4FN	1194-02-1	121.112	nd (peth)	34.8	188.8	1.1070^{55}	1.4925^{55}	sl chl; s peth
5373	2-Fluorobenzoyl chloride		C_7H_4ClFO	393-52-2	158.557		2.0	91^{15}	1.328^{25}	1.5365^{20}	
5374	3-Fluorobenzoyl chloride		C_7H_4ClFO	1711-07-5	158.557	liq	-30	189	1.304^{25}	1.5285^{20}	
5375	4-Fluorobenzoyl chloride		C_7H_4ClFO	403-43-0	158.557		9	82^{20}	1.342^{25}	1.5296^{20}	
5376	2-Fluoro-1,1'-biphenyl		$C_{12}H_9F$	321-60-8	172.197		73.5	248	1.2452^{25}		s EtOH, eth, chl, peth; sl lig
5377	4-Fluoro-1,1'-biphenyl		$C_{12}H_9F$	324-74-3	172.197	pr	74.2	253	1.247^{25}		sl EtOH; s eth, gl HOAc
5378	1-Fluorobutane	Butyl fluoride	C_4H_9F	2366-52-1	76.112	liq	-134	31(8)	0.7789^{20}	1.3396^{20}	vs EtOH
5379	2-Fluorobutane	*sec*-Butyl fluoride	C_4H_9F	359-01-3	76.112	vol liq or gas	-121.4	23(9)	0.7559^{25}		
5380	Fluorocyclohexane	Cyclohexyl fluoride	$C_6H_{11}F$	372-46-3	102.149		13	103(5)	0.9279^{20}	1.4146^{20}	i H_2O; s py
5381	1-Fluorocyclohexene		C_6H_9F	694-51-9	100.133			96.5		1.4441^{25}	
5382	5-Fluorocytosine	4-Amino-5-fluoro-2-hydroxy-pyrimidine	$C_4H_4FN_3O$	2022-85-7	129.092	wh cry	296 dec				
5383	1-Fluorodecane	Decyl fluoride	$C_{10}H_{21}F$	334-56-5	160.272	liq	-35	189(8)	0.8194^{20}	1.4085	vs eth
5384	Fluorodifen	2-Nitro-1-(4-nitrophenoxy)-4-(trifluoromethyl)benzene	$C_{13}H_7F_3N_2O_5$	15457-05-3	328.200		91.7(0.5)				
5385	1-Fluoro-2,4-dinitrobenzene	2,4-Dinitrophenyl fluoride	$C_6H_3FN_2O_4$	70-34-8	186.097		25.8	296	1.4718^{54}	1.5690^{20}	s EtOH; sl chl
5386	Fluoroethane	Ethyl fluoride	C_2H_5F	353-36-6	48.059	col gas	-143.2	-37.7(0.3)	0.7182^{20} (p>1 atm	1.2656^{20}	sl H_2O; vs EtOH, eth
5387	2-Fluoroethanol	Ethylene fluorohydrin	C_2H_5FO	371-62-0	64.058	liq	-26.4	85(14)	1.1040^{20}	1.3647^{18}	msc H_2O, EtOH, eth; vs ace; sl chl
5388	Fluoroethene	Vinyl fluoride	C_2H_3F	75-02-5	46.043	col gas	-160.5	-72			i H_2O; s EtOH, ace
5389	1-Fluoroheptane		$C_7H_{15}F$	661-11-0	118.192	liq	-73	122(9)	0.8062^{20}	1.3854^{20}	i H_2O; s eth, ace, bz; vs peth
5390	1-Fluorohexane	Hexyl fluoride	$C_6H_{13}F$	373-14-8	104.165	liq	-103	88(8)	0.7995^{20}	1.3738^{20}	s eth, bz
5391	1-Fluoro-2-iodobenzene		C_6H_4FI	348-52-7	221.998	liq	-41.5	188.6		1.5910^{20}	s ace, bz, chl
5392	1-Fluoro-4-iodobenzene		C_6H_4FI	352-34-1	221.998	liq	-27	183	1.9523^{15}	1.5270^{22}	i H_2O; s EtOH, eth, ace
5393	1-Fluoro-3-isothiocyanatobenzene		C_7H_4FNS	404-72-8	153.177			227	1.27^{25}	1.6186^{20}	
5394	1-Fluoro-4-isothiocyanatobenzene		C_7H_4FNS	1544-68-9	153.177		27	228			

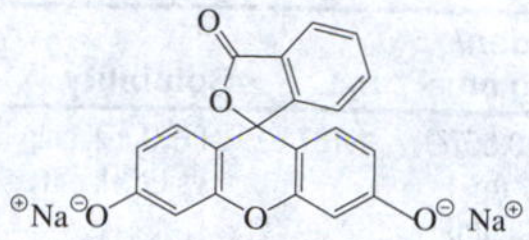
5350
Fluorescein sodium

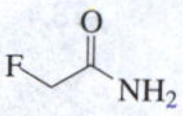
5351
2-Fluoroacetamide

5352
Fluoroacetic acid

5353
Fluoroacetyl chloride

5354
Fluoroacetylene

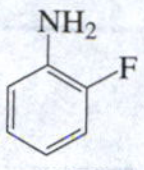
5355
2-Fluoroaniline

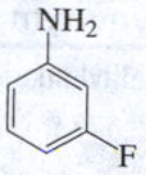
5356
3-Fluoroaniline

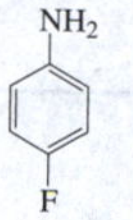
5357
4-Fluoroaniline

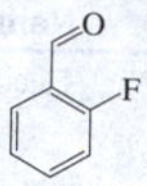
5358
2-Fluorobenzaldehyde

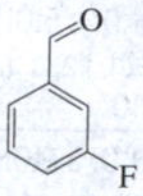
5359
3-Fluorobenzaldehyde

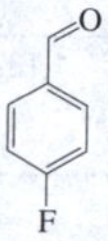
5360
4-Fluorobenzaldehyde

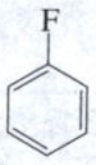
5361
Fluorobenzene

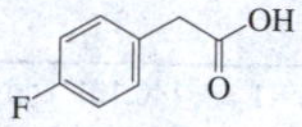
5362
4-Fluorobenzeneacetic acid

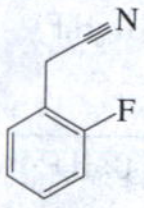
5363
2-Fluorobenzeneacetonitrile

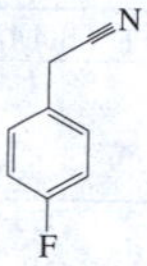
5364
4-Fluorobenzeneacetonitrile

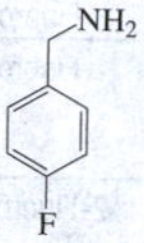
5365
4-Fluorobenzenemethanamine

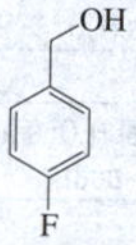
5366
4-Fluorobenzenemethanol

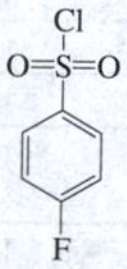
5367
4-Fluorobenzenesulfonyl chloride

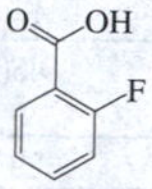
5368
2-Fluorobenzoic acid

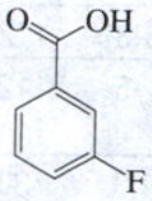
5369
3-Fluorobenzoic acid

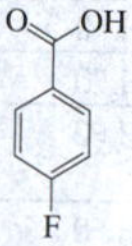
5370
4-Fluorobenzoic acid

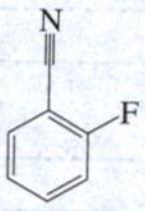
5371
2-Fluorobenzonitrile

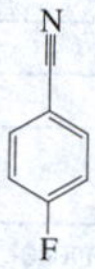
5372
4-Fluorobenzonitrile

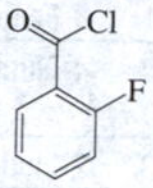
5373
2-Fluorobenzoyl chloride

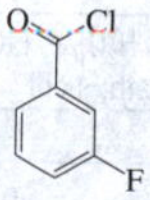
5374
3-Fluorobenzoyl chloride

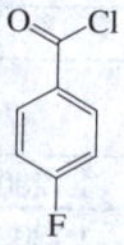
5375
4-Fluorobenzoyl chloride

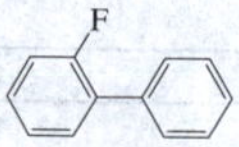
5376
2-Fluoro-1,1'-biphenyl

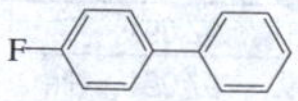
5377
4-Fluoro-1,1'-biphenyl

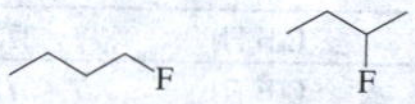
5378
1-Fluorobutane

5379
2-Fluorobutane

5380
Fluorocyclohexane

5381
1-Fluorocyclohexene

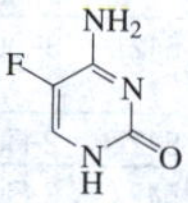
5382
5-Fluorocytosine

5383
1-Fluorodecane

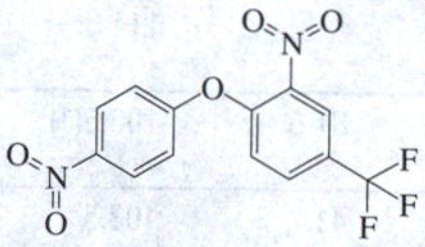
5384
Fluorodifen

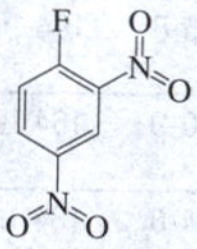
5385
1-Fluoro-2,4-dinitrobenzene

5386
Fluoroethane

5387
2-Fluoroethanol

5388
Fluoroethene

5389
1-Fluoroheptane

5390
1-Fluorohexane

5391
1-Fluoro-2-iodobenzene

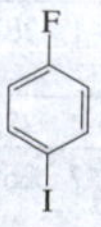
5392
1-Fluoro-4-iodobenzene

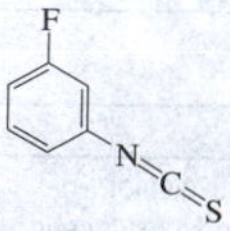
5393
1-Fluoro-3-isothiocyanatobenzene

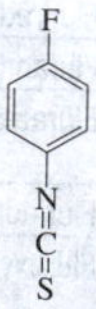
5394
1-Fluoro-4-isothiocyanatobenzene

Organic

No.	Name	Synonym	Mol. form.	CAS Reg. No.	Mol. wt.	Physical form	mp/°C	bp/°C	den/ g cm^{-3}	n_D	Solubility
5395	Fluoromethane	Methyl fluoride	CH_3F	593-53-3	34.033	col gas	-143.3	-78.4	0.557^{25} (p>1 atm)	1.1674^{25}	sl H_2O, bz, chl; vs EtOH, eth
5396	1-Fluoro-2-methoxybenzene		C_7H_7FO	321-28-8	126.128	liq	-39	154.5	1.1325^{17}	1.4969^{17}	i H_2O; s eth, ctc
5397	1-Fluoro-3-methoxybenzene		C_7H_7FO	456-49-5	126.128	liq	-35	159	1.104^{25}	1.4876^{20}	
5398	1-Fluoro-4-methoxybenzene		C_7H_7FO	459-60-9	126.128	liq	-45	157	1.1781^{18}	1.4886^{18}	s eth
5399	4-Fluoro-2-methylaniline		C_7H_8FN	452-71-1	125.144		14.2	94^{16}	1.1263^{18}	1.5363^{18}	s eth, ace, bz, ctc
5400	(Fluoromethyl)benzene		C_7H_7F	350-50-5	110.129	liq	-35	142(2)	1.0228^{25}	1.4892^{25}	s ctc
5401	2-Fluoro-4-methyl-1-nitrobenzene	3-Fluoro-4-nitrotoluene	$C_7H_6FNO_2$	446-34-4	155.127	nd (al)	53.2	97^{3}	1.4380^{25}		
5402	2-Fluoro-2-methylpropane	*tert*-Butyl fluoride	C_4H_9F	353-61-7	76.112	col gas		12.1			
5403	1-Fluoronaphthalene		$C_{10}H_7F$	321-38-0	146.161	liq	-14.0(0.2)	213(14)	1.1322^{20}	1.5939^{20}	i H_2O; s EtOH, eth, bz, chl, HOAc
5404	2-Fluoronaphthalene		$C_{10}H_7F$	323-09-1	146.161	nd (al)	58(1)	212			i H_2O; s EtOH, eth, bz, chl, HOAc
5405	1-Fluoro-2-nitrobenzene	*o*-Fluoronitrobenzene	$C_6H_4FNO_2$	1493-27-2	141.100	ye liq	-6	215 dec	1.3285^{18}	1.5489^{17}	vs eth, EtOH
5406	1-Fluoro-3-nitrobenzene	*m*-Fluoronitrobenzene	$C_6H_4FNO_2$	402-67-5	141.100	ye cry	41	197(9)	1.3254^{19}	1.5262^{15}	i H_2O; s EtOH, eth; sl bz
5407	1-Fluoro-4-nitrobenzene	*p*-Fluoronitrobenzene	$C_6H_4FNO_2$	350-46-9	141.100	ye nd	26.5(0.5)	205	1.3300^{20}	1.5316^{20}	i H_2O; s EtOH, eth; sl ctc
5408	1-Fluorooctane	Octyl fluoride	$C_8H_{17}F$	463-11-6	132.219	liq	-64	146(9)	0.8116^{20}	1.3946^{20}	
5409	1-Fluoropentane	Pentyl fluoride	$C_5H_{11}F$	592-50-7	90.139	liq	-120	63(3)	0.7907^{20}	1.3591^{2-}	vs eth, EtOH
5410	2-Fluorophenol		C_6H_5FO	367-12-4	112.101		16.1(0.5)	151(4)	1.120^{25}	1.5144^{20}	s H_2O
5411	3-Fluorophenol		C_6H_5FO	372-20-3	112.101		14(1)	178	1.238^{25}	1.5140^{20}	
5412	4-Fluorophenol		C_6H_5FO	371-41-5	112.101		48(1)	185.5	1.1889^{56}		sl H_2O; s ace, peth
5413	2-Fluoro-1-phenylethanone		C_8H_7FO	450-95-3	138.139	pl	29	90^{12}	1.152^{20}	1.5200^{20}	
5414	1-(4-Fluorophenyl)ethanone		C_8H_7FO	403-42-9	138.139	liq	-45	196	1.1382^{25}	1.5081^{25}	i H_2O; s bz, chl
5415	1-Fluoropropane	Propyl fluoride	C_3H_7F	460-13-9	62.086	col gas	-159	-3(2)	0.7596^{20} (p>1 atm	1.3115^{20}	sl H_2O; vs EtOH, eth
5416	2-Fluoropropane	Isopropyl fluoride	C_3H_7F	420-26-8	62.086	gas		-10(3)			sl H_2O
5417	1-Fluoro-2-propanone	Fluoroacetone	C_3H_5FO	430-51-3	76.069			77	1.0288^{20}	1.3700^{20}	
5418	*cis*-1-Fluoropropene		C_3H_5F	19184-10-2	60.070	col gas		-15(16)			
5419	*trans*-1-Fluoropropene		C_3H_5F	20327-65-5	60.070	col gas	≈-20				
5420	2-Fluoropropene		C_3H_5F	1184-60-7	60.070	col gas		-24			
5421	3-Fluoropropene		C_3H_5F	818-92-8	60.070	col gas		-7(15)			sl H_2O; vs EtOH, eth; s chl
5422	2-Fluoropyridine		C_5H_4FN	372-48-5	97.091			125	1.1280^{20}	1.4574^{20}	
5423	3-Fluoropyridine		C_5H_4FN	372-47-4	97.091	liq		107	1.130	1.4720^{20}	
5424	2-Fluorotoluene	*o*-Tolyl fluoride	C_7H_7F	95-52-3	110.129	liq	-62.5(0.5)	114(2)	1.0041^{13}	1.4704^{20}	i H_2O; vs EtOH, eth
5425	3-Fluorotoluene	*m*-Tolyl fluoride	C_7H_7F	352-70-5	110.129	liq	-89.2(0.4)	116(2)	0.9974^{20}	1.4691^{20}	i H_2O; vs EtOH, eth
5426	4-Fluorotoluene	*p*-Tolyl fluoride	C_7H_7F	352-32-9	110.129	liq	-56.6(0.1)	116.6(0.4)	0.9975^{20}	1.4699^{20}	i H_2O; vs EtOH, eth
5427	1-Fluoro-2-(trichloromethyl)-benzene		$C_7H_4Cl_3F$	488-98-2	213.464			95^{12}	1.453^{25}	1.5432^{20}	
5428	1-Fluoro-2-(trifluoromethyl)-benzene		$C_7H_4F_4$	392-85-8	164.101			114.5	1.293^{25}	1.4040^{25}	
5429	1-Fluoro-3-(trifluoromethyl)-benzene		$C_7H_4F_4$	401-80-9	164.101	liq	-81.5	100.6(1)	1.3021^{17}		
5430	1-Fluoro-4-(trifluoromethyl)-benzene		$C_7H_4F_4$	402-44-8	164.101	liq	-41.7	103.5	1.293^{25}	1.4025^{20}	
5431	Fluorotrimethylsilane	Trimethylsilyl fluoride	C_3H_9FSi	420-56-4	92.187	vol liq or gas		16.4			
5432	5-Fluorouracil	5-Fluoro-2,4(1*H*,3*H*)-Pyrimidinedione	$C_4H_3FN_2O_2$	51-21-8	130.077	cry (w, MeOH-eth)	284(4)	369(11)			
5433	Fluoxetine		$C_{17}H_{18}F_3NO$	54910-89-3	309.326	oil					
5434	Fluoxymesterone		$C_{20}H_{29}FO_3$	76-43-7	336.440		270				
5435	Fluphenazine		$C_{22}H_{26}F_3N_3OS$	69-23-8	437.520			$251^{0.3}$			
5436	Fluprednisolone		$C_{21}H_{27}FO_5$	53-34-9	378.434		210				
5437	Flurandrenolide	Fludroxycortide	$C_{24}H_{33}FO_6$	1524-88-5	436.513	cry (ace/hx)	251				
5438	Flurazepam		$C_{21}H_{23}ClFN_3O$	17617-23-1	387.878	wh rods (eth/peth)	80				
5439	Fluridone		$C_{19}H_{14}F_3NO$	59756-60-4	329.315		155				
5440	Fluroxypyr	[(4-Amino-3,5-dichloro-6-fluoro-2-pyridyl)oxy]acetic acid	$C_7H_5Cl_2FN_2O_3$	69377-81-7	255.030		232				

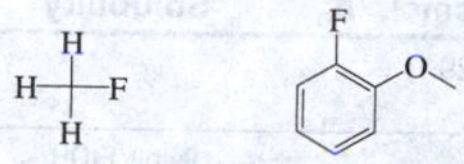
5395 Fluoromethane

5396 1-Fluoro-2-methoxybenzene

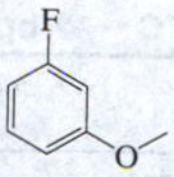
5397 1-Fluoro-3-methoxybenzene

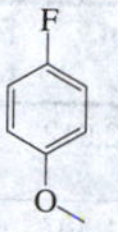
5398 1-Fluoro-4-methoxybenzene

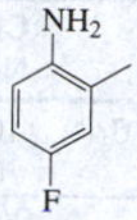
5399 4-Fluoro-2-methylaniline

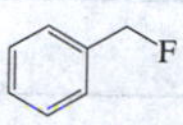
5400 (Fluoromethyl)benzene

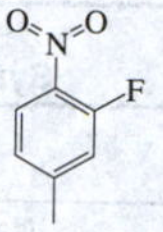
5401 2-Fluoro-4-methyl-1-nitrobenzene

5402 2-Fluoro-2-methylpropane

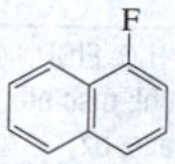
5403 1-Fluoronaphthalene

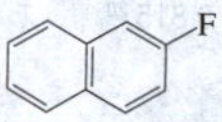
5404 2-Fluoronaphthalene

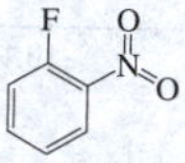
5405 1-Fluoro-2-nitrobenzene

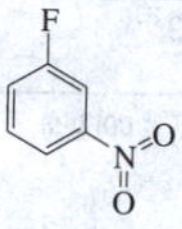
5406 1-Fluoro-3-nitrobenzene

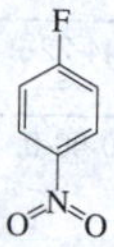
5407 1-Fluoro-4-nitrobenzene

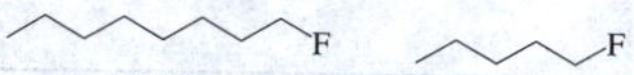
5408 1-Fluorooctane

5409 1-Fluoropentane

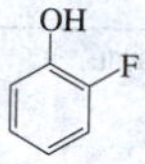
5410 2-Fluorophenol

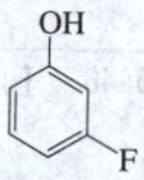
5411 3-Fluorophenol

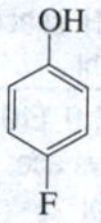
5412 4-Fluorophenol

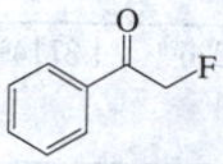
5413 2-Fluoro-1-phenylethanone

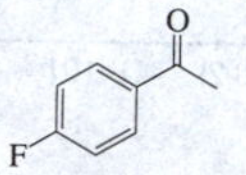
5414 1-(4-Fluorophenyl)ethanone

5415 1-Fluoropropane

5416 2-Fluoropropane

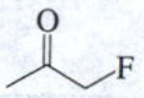
5417 1-Fluoro-2-propanone

5418 *cis*-1-Fluoropropene

5419 *trans*-1-Fluoropropene

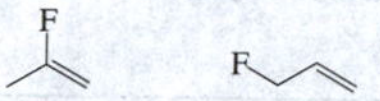
5420 2-Fluoropropene

5421 3-Fluoropropene

5422 2-Fluoropyridine

5423 3-Fluoropyridine

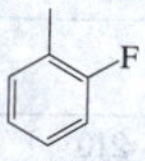
5424 2-Fluorotoluene

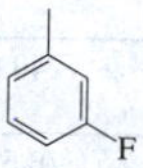
5425 3-Fluorotoluene

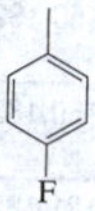
5426 4-Fluorotoluene

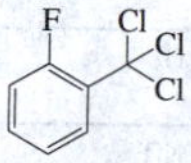
5427 1-Fluoro-2-(trichloromethyl)benzene

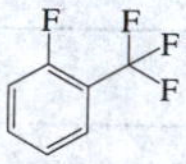
5428 1-Fluoro-2-(trifluoromethyl)benzene

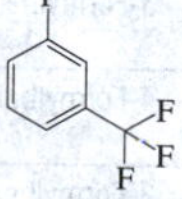
5429 1-Fluoro-3-(trifluoromethyl)benzene

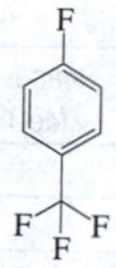
5430 1-Fluoro-4-(trifluoromethyl)benzene

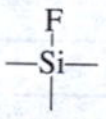
5431 Fluorotrimethylsilane

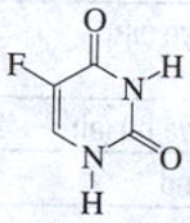
5432 5-Fluorouracil

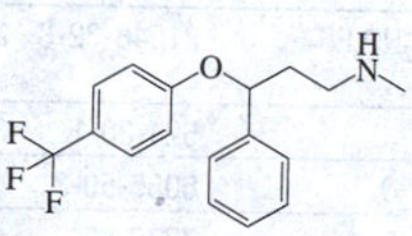
5433 Fluoxetine

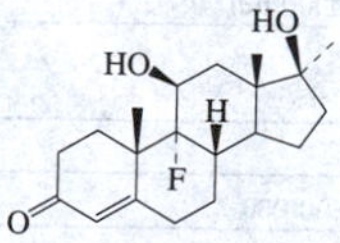
5434 Fluoxymesterone

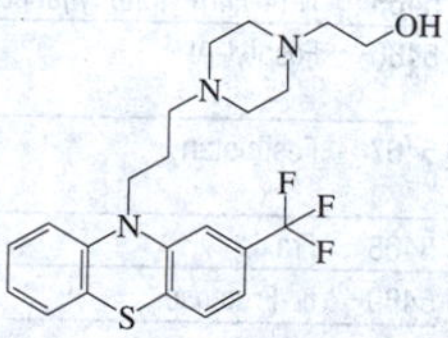
5435 Fluphenazine

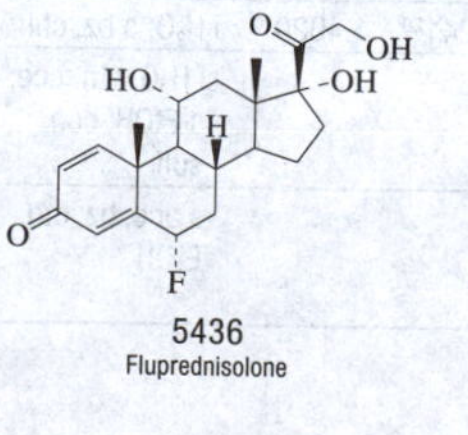
5436 Fluprednisolone

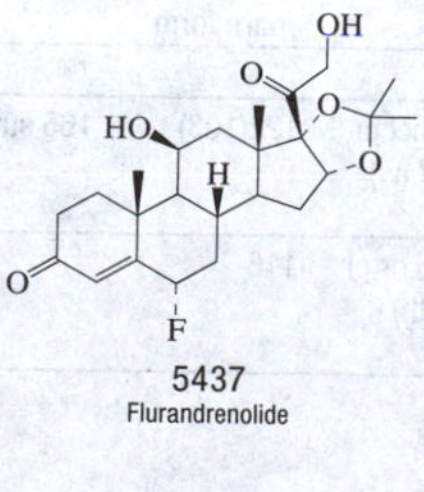
5437 Flurandrenolide

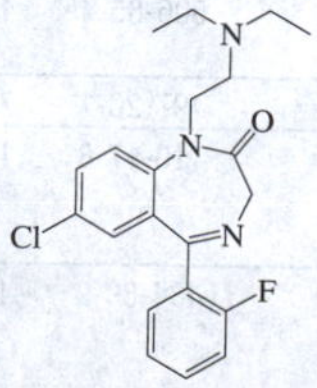
5438 Flurazepam

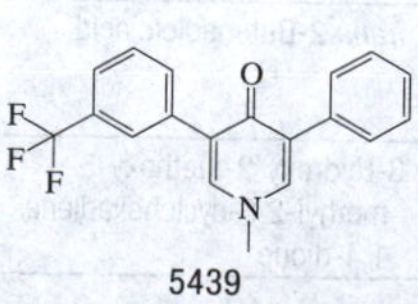
5439 Fluridone

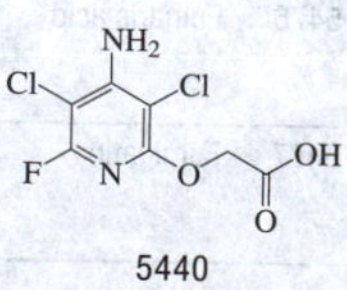
5440 Fluroxypyr

Organic

No.	Name	Synonym	Mol. form.	CAS Reg. No.	Mol. wt.	Physical form	mp/°C	bp/°C	den/ g cm^{-3}	n_D	Solubility
5441	Fluvalinate		$C_{26}H_{22}ClF_3N_2O_3$	102851-06-9	502.912			>450	1.29^{25}		
5442	Folic acid	Vitamin Bc	$C_{19}H_{19}N_7O_6$	59-30-3	441.397	ye-oran nd (w)	250 dec				vs py, EtOH, HOAc
5443	Folinic acid	5-Formyl-5,6,7,8-tetrahydrofolic acid	$C_{20}H_{23}N_7O_7$	58-05-9	473.440	cry (w + 3)	245 dec				sl H_2O
5444	Folpet	1*H*-Isoindole-1,3(2*H*)-dione, 2-[(trichloromethyl)thio]-	$C_9H_4Cl_3NO_2S$	133-07-3	296.558		181.0(0.6)				
5445	Fomesafen		$C_{15}H_{10}ClF_3N_2O_6S$	72178-02-0	438.762		220		1.28^{20}		
5446	Fomocaine	4-[3-[4-(Phenoxymethyl)phenyl]propyl]morpholine	$C_{20}H_{25}NO_2$	17692-39-6	311.419	col cry	53	$239^{1.1}$			
5447	Fonofos	Phosphonodithioic acid, ethyl-, *O*-ethyl *S*-phenyl ester	$C_{10}H_{15}OPS_2$	944-22-9	246.329			$130^{0.1}$	1.16^{25}		
5448	Formaldehyde	Methanal	CH_2O	50-00-0	30.026	col gas	-92	-19.1(0.5)	0.815^{-20}		s H_2O, EtOH, chl; msc eth, ace, bz
5449	Formaldehyde oxime	Formaldoxime	CH_3NO	75-17-2	45.041		1.3	109^{15}	1.133^{25}		s H_2O; vs EtOH, eth
5450	Formamide	Methanamide	CH_3NO	75-12-7	45.041	col liq	2.57(0.02)	217(3)	1.1334^{20}	1.4472^{20}	msc H_2O, EtOH; sl eth; s ace; i bz, chl
5451	Formamidinesulfinic acid	Aminoiminomethanesulfinic acid	$CH_4N_2O_2S$	1758-73-2	108.120	nd (al)	144 dec				vs H_2O; i eth, bz
5452	Formetanate hydrochloride		$C_{11}H_{16}ClN_3O_2$	23422-53-9	257.717	pow	201 dec				vs H_2O; s MeOH; sl ace, hx, chl
5453	Formic acid	Methanoic acid	CH_2O_2	64-18-6	46.026	col liq	8.3(0.2)	101	1.220^{20}	1.3714^{20}	msc H_2O, EtOH, eth; vs ace; s bz, tol
5454	*N*-Formimidoyl-L-glutamic acid	*N*-(Iminomethyl)-L-glutamic acid	$C_6H_{10}N_2O_4$	816-90-0	174.154		90				
5455	Formononetin	7-Hydroxy-3-(4-methoxyphenyl)-4*H*-1-benzopyran-4-one	$C_{16}H_{12}O_4$	485-72-3	268.264		256.5				
5456	Formothion		$C_6H_{12}NO_4PS_2$	2540-82-1	257.267	visc ye oil	25.5		1.361^{20}	1.5541^{20}	sl H_2O; misc os
5457	2-Formylbenzoic acid		$C_8H_6O_3$	119-67-5	150.132		100.5(0.2)		1.404^{25}		s H_2O; vs EtOH, eth
5458	3-Formylbenzoic acid		$C_8H_6O_3$	619-21-6	150.132	nd (w)	175.0(0.2)				vs H_2O, eth, EtOH
5459	4-Formylbenzoic acid		$C_8H_6O_3$	619-66-9	150.132		249.8(0.5)				sl H_2O; vs EtOH; s eth, chl
5460	3-Formylbenzonitrile		C_8H_5NO	24964-64-5	131.132		76.5	210			vs H_2O, EtOH, eth, chl
5461	4-Formylbenzonitrile		C_8H_5NO	105-07-7	131.132		100.5	133^{12}			s H_2O; vs EtOH, eth, chl
5462	6-Formyl-2,3-dimethoxybenzoic acid	Opianic acid	$C_{10}H_{10}O_5$	519-05-1	210.183	nd (w)	150				s EtOH, eth
5463	Formylferrocene		$C_{11}H_{10}FeO$	12093-10-6	214.041		118.5	$70^{0.1}$			
5464	Formyl fluoride	Fluoroformaldehyde	CHFO	1493-02-3	48.016	col gas	-142.2	-26.5	1.1950^{-30}		
5465	*N*-(4-Formylphenyl)acetamide		$C_9H_9NO_2$	122-85-0	163.173	pr (w)	158.0				vs H_2O, bz
5466	Fosetyl-Al	Aluminum tris(*O*-ethylphosphonate)	$C_6H_{18}AlO_9P_3$	39148-24-8	354.105		>300				
5467	Fosthietan		$C_6H_{12}NO_3PS_2$	21548-32-3	241.268	ye oil			1.3^{25}	1.5348^{25}	s ace, chl, MeOH, tol
5468	Fraxin		$C_{16}H_{18}O_{10}$	524-30-1	370.308	ye nd (al)	205				
5469	DL-Fructose	α-Acrose	$C_6H_{12}O_6$	6035-50-3	180.155	nd	130		1.665^{16}		
5470	L-Fructose		$C_6H_{12}O_6$	7776-48-9	180.155	wh cry	102				s H_2O
5471	β-D-Fructose	β-Levulose	$C_6H_{12}O_6$	53188-23-1	180.155	pr or nd (w) orth pr (al)	103 dec		1.60^{20}		vs H_2O, ace; s EtOH, MeOH, py
5472	D-Fructose 6-phosphate	Hexose monophosphate	$C_6H_{13}O_9P$	643-13-0	260.135						vs H_2O
5473	Fucoxanthin		$C_{42}H_{58}O_6$	3351-86-8	658.906	red pl (eth) hex pl (dil al)	168				vs eth, EtOH
5474	Fulminic acid	Carbyloxime	CHNO	506-85-4	43.025		unstable in pure form				s eth
5475	Fulvene		C_6H_6	497-20-1	78.112			7^{56}	0.8241^{20}	1.4920^{20}	i H_2O; s bz, chl
5476	Fumaric acid	*trans*-2-Butenedioic acid	$C_4H_4O_4$	110-17-8	116.073	nd, mcl pr or lf (w)	122(103)	165 sub	1.635^{20}		sl H_2O, eth, ace; s EtOH, con sulf
5477	Fumigatin	3-Hydroxy-2-methoxy-5-methyl-2,5-cyclohexadiene-1,4-dione	$C_8H_8O_4$	484-89-9	168.148	br nd or pl (peth)	116				vs ace, bz, eth, EtOH

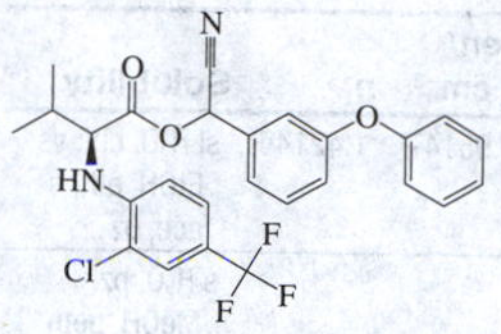
5441
Fluvalinate

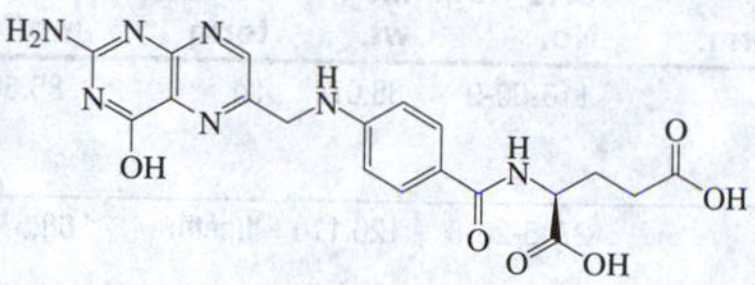
5442
Folic acid

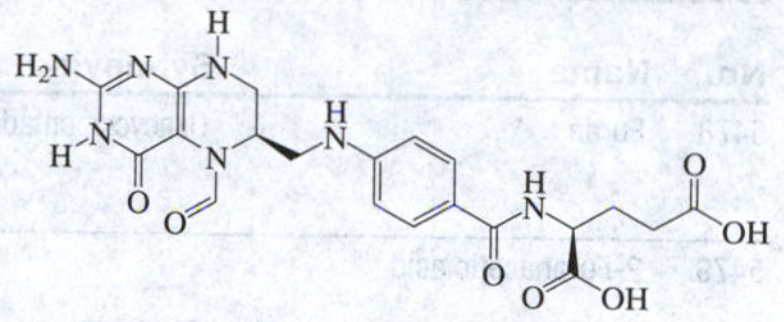
5443
Folinic acid

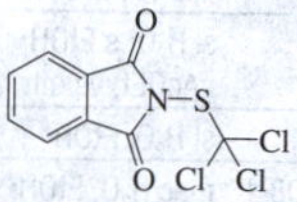
5444
Folpet

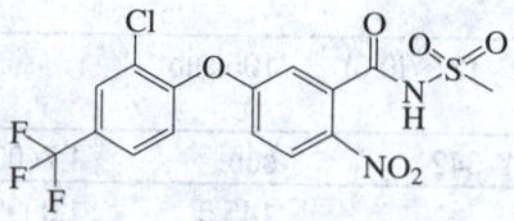
5445
Fomesafen

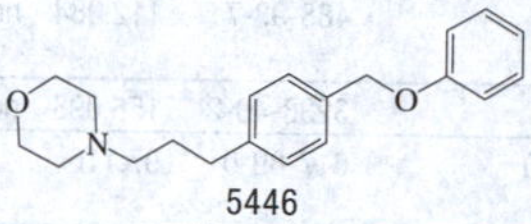
5446
Fomocaine

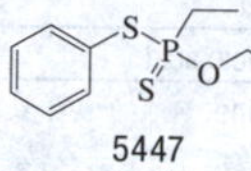
5447
Fonofos

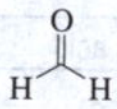
5448
Formaldehyde

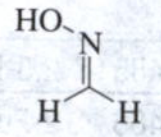
5449
Formaldehyde oxime

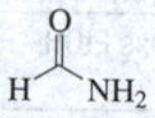
5450
Formamide

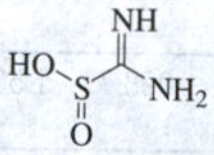
5451
Formamidinesulfinic acid

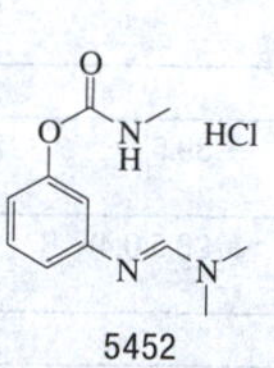
5452
Formetanate hydrochloride

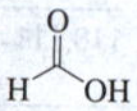
5453
Formic acid

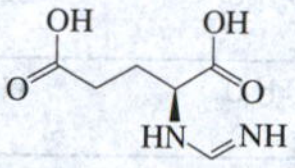
5454
N-Formimidoyl-L-glutamic acid

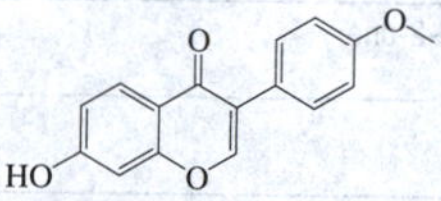
5455
Formononetin

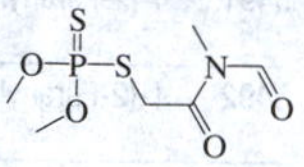
5456
Formothion

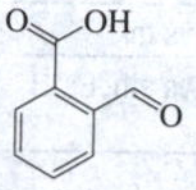
5457
2-Formylbenzoic acid

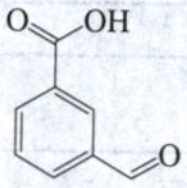
5458
3-Formylbenzoic acid

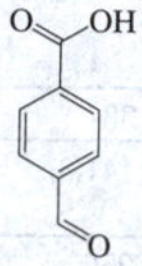
5459
4-Formylbenzoic acid

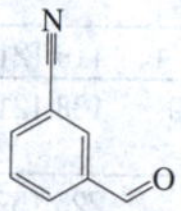
5460
3-Formylbenzonitrile

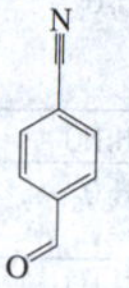
5461
4-Formylbenzonitrile

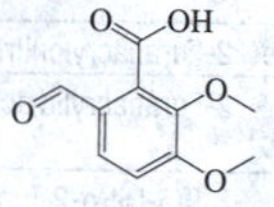
5462
6-Formyl-2,3-dimethoxybenzoic acid

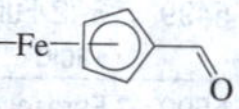
5463
Formylferrocene

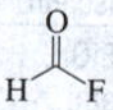
5464
Formyl fluoride

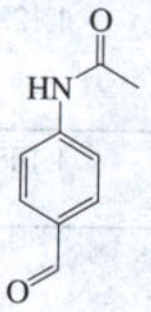
5465
N-(4-Formylphenyl)acetamide

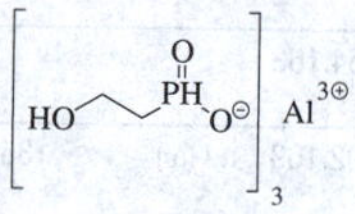
5466
Fosetyl-Al

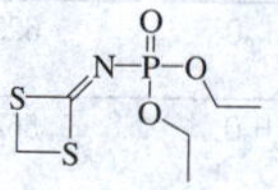
5467
Fosthietan

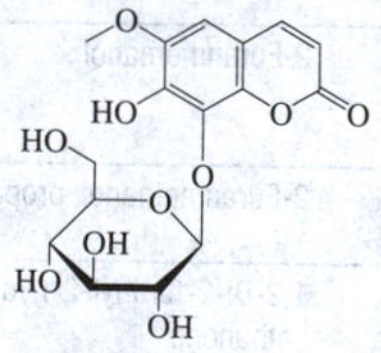
5468
Fraxin

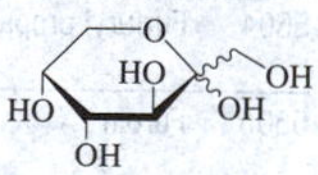
5469
DL-Fructose

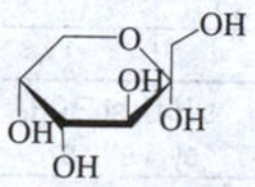
5470
L-Fructose

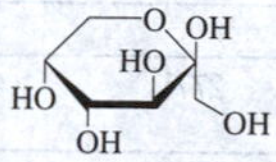
5471
β-D-Fructose

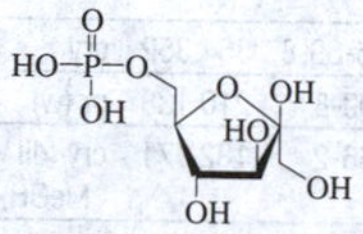
5472
D-Fructose 6-phosphate

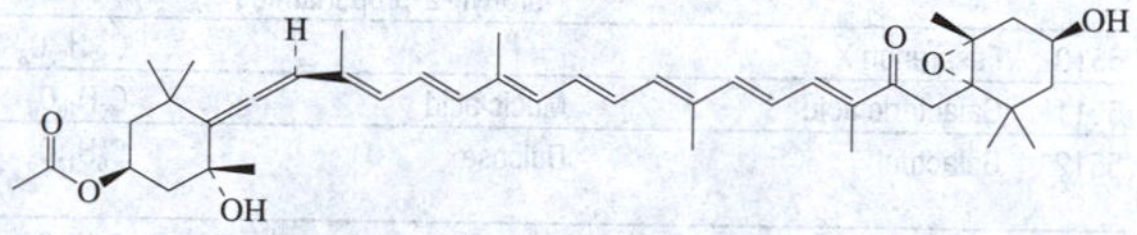
5473
Fucoxanthin

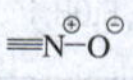
5474
Fulminic acid

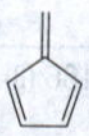
5475
Fulvene

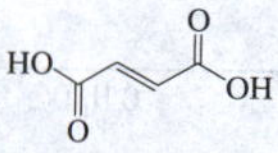
5476
Fumaric acid

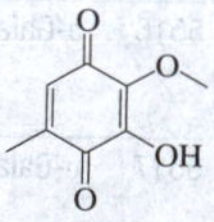
5477
Fumigatin

No.	Name	Synonym	Mol. form.	CAS Reg. No.	Mol. wt.	Physical form	mp/°C	bp/°C	den/ g cm^{-3}	n_D	Solubility
5478	Furan	Oxacyclopentadiene	C_4H_4O	110-00-9	68.074	liq	-85.58(0.05)	31.3(0.2)	0.9514^{20}	1.4214^{20}	sl H_2O, chl; vs EtOH, eth; s ace, bz
5479	2-Furanacetic acid		$C_6H_6O_3$	2745-26-8	126.110	lf(peth)	68.5	$102^{0.4}$			s H_2O, bz, MeOH, peth
5480	2-Furancarbonitrile		C_5H_3NO	617-90-3	93.084			147	1.0822^{20}	1.4798^{20}	s EtOH, eth
5481	2-Furancarbonyl chloride	2-Furoyl chloride	$C_5H_3ClO_2$	527-69-5	130.530	liq	-1.0	173	1.324^{25}	1.5310^{20}	i H_2O; s eth, chl; sl ctc
5482	3-Furancarboxaldehyde		$C_5H_4O_2$	498-60-2	96.085			145	1.110^{20}	1.4945^{20}	
5483	2-Furancarboxylic acid	2-Furoic acid	$C_5H_4O_3$	88-14-2	112.084	mcl nd or lf (w)	130(3)	231			s H_2O, EtOH; vs eth; sl ace
5484	3-Furancarboxylic acid		$C_5H_4O_3$	488-93-7	112.084	nd (w)	121.7(0.6)	105 sub			sl H_2O; s EtOH, AcOEt; vs eth
5485	2,5-Furandicarboxylic acid	Dehydromucic acid	$C_6H_4O_5$	3238-40-2	156.093	nd (w), lf (al)	342	sub	1.7400^{20}		sl H_2O, EtOH
5486	2-Furanmethanamine	Furfurylamine	C_5H_7NO	617-89-0	97.116			145.5	1.0995^{20}	1.4908^{20}	msc H_2O, EtOH; s eth, chl
5487	2-Furanmethanediol diacetate		$C_9H_{10}O_5$	613-75-2	198.172	nd or pl (eth-peth)	52(1)	220			vs bz, eth, EtOH
5488	2-Furanmethanethiol		C_5H_6OS	98-02-2	114.166			157	1.1319^{20}	1.5329^{20}	i H_2O; sl chl
5489	2-Furanmethanol acetate		$C_7H_8O_3$	623-17-6	140.137			182(6)	1.1175^{20}	1.4327^{20}	i H_2O; s EtOH, eth
5490	4-(2-Furanyl)-2-butanone		$C_8H_{10}O_2$	699-17-2	138.164	oil		203	1.0361^{19}	1.4696^{17}	
5491	4-(2-Furanyl)-3-buten-2-one		$C_8H_8O_2$	623-15-4	136.149		39.5	229 dec	1.0496^{57}	1.5788^{45}	i H_2O; vs EtOH, eth, chl; s peth
5492	1-(2-Furanyl)ethanone		$C_6H_6O_2$	1192-62-7	110.111	cry (lig)	28.5(0.4)	175	1.098^{20}	1.5017^{20}	i H_2O; s EtOH, eth
5493	2-Furanylmethyl pentanoate	Furfuryl valerate	$C_{10}H_{14}O_3$	36701-01-6	182.216			228	1.0284^{20}		vs eth, EtOH
5494	3-(2-Furanyl)-1-phenyl-2-propen-1-one		$C_{13}H_{10}O_2$	717-21-5	198.217		47	317	1.1140^{20}		s EtOH, eth
5495	1-(2-Furanyl)-1-propanone		$C_7H_8O_2$	3194-15-8	124.138	cry	28	88^{14}	1.0626^{28}	1.4922^{25}	s eth; sl ctc
5496	1-(2-Furanyl)-2-propanone	2-Furfuryl methyl ketone	$C_7H_8O_2$	6975-60-6	124.138		29	179.5	1.104^{20}	1.5035^{20}	
5497	3-(2-Furanyl)-2-propenal		$C_7H_6O_2$	623-30-3	122.122		52(1)	135^{14}			i H_2O; msc EtOH; s eth; sl chl
5498	3-(2-Furanyl)-2-propenenitrile	2-Furanacrylonitrile	C_7H_5NO	7187-01-1	119.121		38	96		1.5824^{25}	vs tol
5499	3-(2-Furanyl)-2-propenoic acid	2-Furanacrylic acid	$C_7H_6O_3$	539-47-9	138.121	nd (w)	141(3)	286			vs eth, EtOH
5500	Furazolidone	3-[[(5-Nitro-2-furanyl)-methylene]amino]-2-oxa-zolidinone	$C_8H_7N_3O_5$	67-45-8	225.159		255				
5501	Furethidine		$C_{21}H_{31}NO_4$	2385-81-1	361.476		28	$210^{0.5}$		1.5219^{20}	
5502	Furfural	2-Furaldehyde	$C_5H_4O_2$	98-01-1	96.085	liq	-38.3(0.8)	161.5(0.3)	1.1594^{20}	1.5261^{20}	s H_2O, bz, chl; vs EtOH, ace; msc eth
5503	Furfuryl alcohol	2-Furanmethanol	$C_5H_6O_2$	98-00-0	98.101	col-ye liq	-14.5(0.2)	168(2)	1.1296^{20}	1.4869^{20}	msc H_2O; vs EtOH, eth; s chl
5504	Furfuryl propanoate	2-Furanmethanol, propanoate	$C_8H_{10}O_3$	623-19-8	154.163			195	1.1085^{20}		sl H_2O; s EtOH, ace; msc eth
5505	Furoin	1,2-Di-2-furanyl-2-hydroxy-ethanone	$C_{10}H_8O_4$	552-86-3	192.169	nd (al)	135(3)				sl H_2O, EtOH, chl; s eth, MeOH
5506	Furonazide		$C_{12}H_{11}N_3O_2$	3460-67-1	229.234		202.3				
5507	Furosemide		$C_{12}H_{11}ClN_2O_5S$	54-31-9	330.743		204 dec				
5508	Fursultiamine		$C_{17}H_{26}N_4O_3S_2$	804-30-8	398.543	col pr	132 dec		1.29		sl H_2O
5509	Furylfuramide, (*E*)-	2-(2-Furanyl)-3-(5-nitro-2-furanyl)-2-propenamide	$C_{11}H_8N_2O_5$	18819-45-9	248.192	cry	154				
5510	Fusarenon X		$C_{17}H_{22}O_8$	23255-69-8	354.352	cry	182				
5511	Galactaric acid	Mucic acid	$C_6H_{10}O_8$	526-99-8	210.138	pr (w)	255 dec				
5512	Galactitol	Dulcose	$C_6H_{14}O_6$	608-66-2	182.171	cry (dil MeOH)	187.2(0.4)	277^{1}	1.47^{20}		s H_2O; sl EtOH, py; i eth, bz
5513	D-Galactonic acid, γ-lactone		$C_6H_{10}O_6$	2782-07-2	178.139	nd (w+1), nd (al)	112				vs H_2O
5514	α-D-Galactopyranose		$C_6H_{12}O_6$	3646-73-9	180.155		167				
5515	4-*O*-β-D-Galactopyranosyl-D-gluconic acid	Lactobionic acid	$C_{12}H_{22}O_{12}$	96-82-2	358.296	syr					vs H_2O; sl EtOH, MeOH, HOAc; i eth
5516	D-Galactose		$C_6H_{12}O_6$	59-23-4	180.155	pl or pr (al) pr or nd (w+1)	170				vs H_2O; sl EtOH; i eth, bz; s py
5517	D-Galacturonic acid		$C_6H_{10}O_7$	685-73-4	194.139	nd (w)	166 (β)				s H_2O, EtOH; i eth

5478
Furan

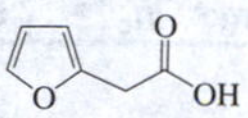
5479
2-Furanacetic acid

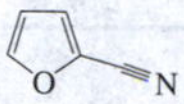
5480
2-Furancarbonitrile

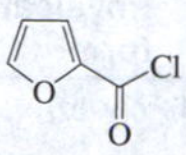
5481
2-Furancarbonyl chloride

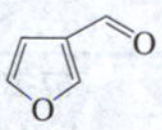
5482
3-Furancarboxaldehyde

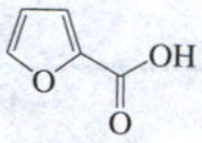
5483
2-Furancarboxylic acid

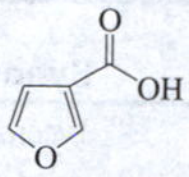
5484
3-Furancarboxylic acid

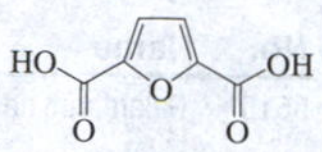
5485
2,5-Furandicarboxylic acid

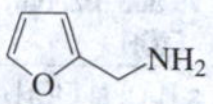
5486
2-Furanmethanamine

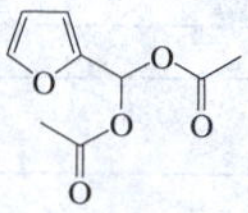
5487
2-Furanmethanediol diacetate

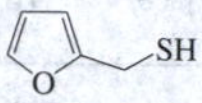
5488
2-Furanmethanethiol

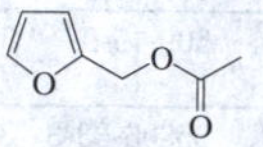
5489
2-Furanmethanol acetate

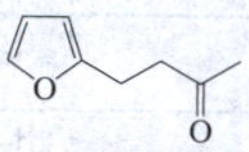
5490
4-(2-Furanyl)-2-butanone

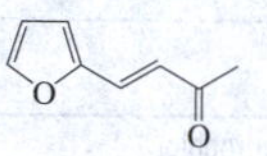
5491
4-(2-Furanyl)-3-buten-2-one

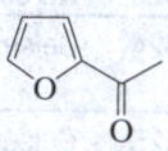
5492
1-(2-Furanyl)ethanone

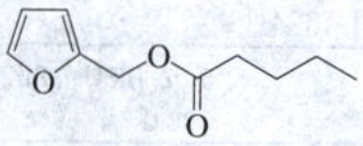
5493
2-Furanylmethyl pentanoate

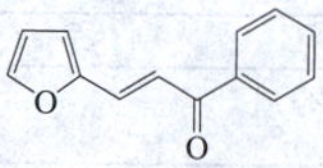
5494
3-(2-Furanyl)-1-phenyl-2-propen-1-one

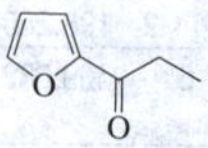
5495
1-(2-Furanyl)-1-propanone

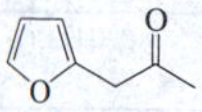
5496
1-(2-Furanyl)-2-propanone

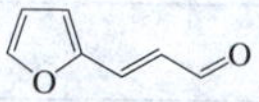
5497
3-(2-Furanyl)-2-propenal

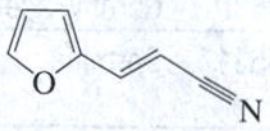
5498
3-(2-Furanyl)-2-propenenitrile

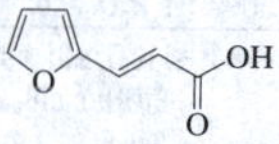
5499
3-(2-Furanyl)-2-propenoic acid

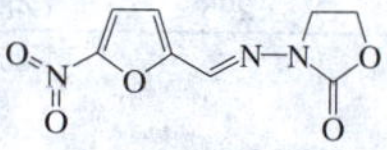
5500
Furazolidone

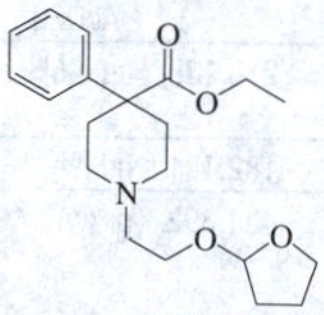
5501
Furethidine

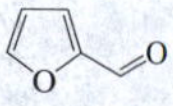
5502
Furfural

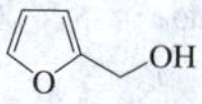
5503
Furfuryl alcohol

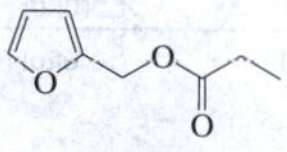
5504
Furfuryl propanoate

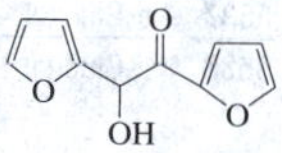
5505
Furoin

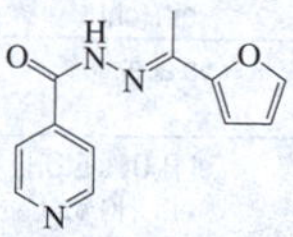
5506
Furonazide

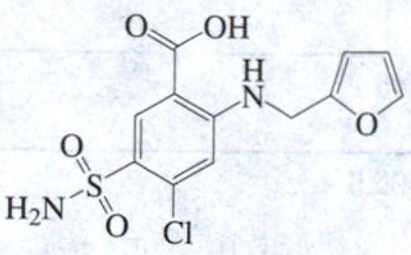
5507
Furosemide

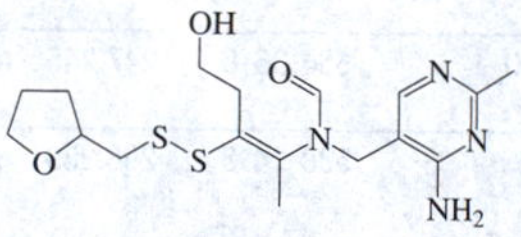
5508
Fursultiamine

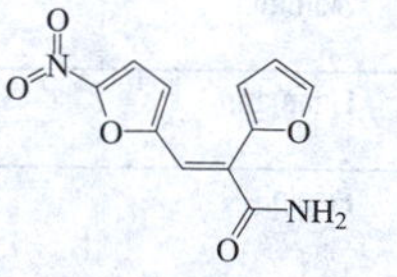
5509
Furylfuramide, *(E)*-

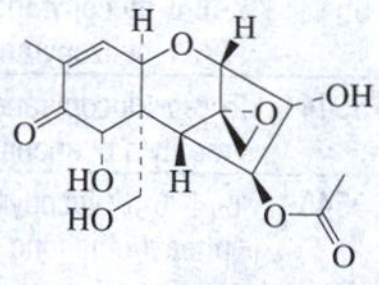
5510
Fusarenon X

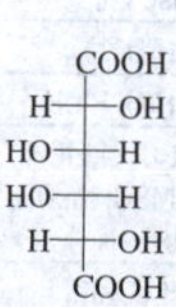
5511
Galactaric acid

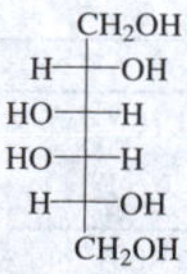
5512
Galactitol

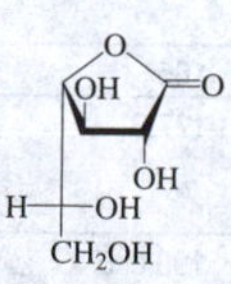
5513
D-Galactonic acid, γ-lactone

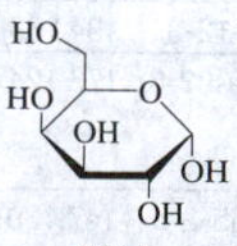
5514
α-D-Galactopyranose

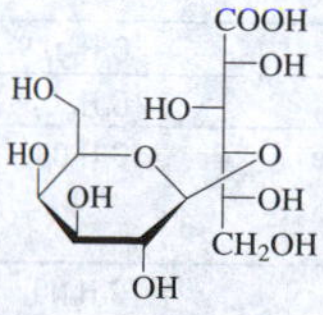
5515
4-*O*-β-D-Galactopyranosyl-D-gluconic acid

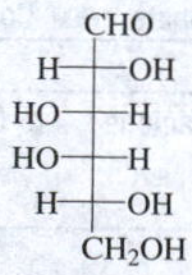
5516
D-Galactose

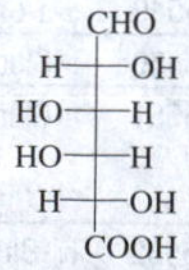
5517
D-Galacturonic acid

Organic

No.	Name	Synonym	Mol. form.	CAS Reg. No.	Mol. wt.	Physical form	mp/°C	bp/°C	den/ g cm⁻³	n_D	Solubility
5518	Galanthamine	Lycoremine	$C_{17}H_{21}NO_3$	357-70-0	287.354	cry (bz)	126.5				vs ace, EtOH, chl
5519	Galipine	2-[2-(3,4-Dimethoxyphenyl)-ethyl]-4-methoxyquinoline	$C_{20}H_{21}NO_3$	525-68-8	323.386	pr (al, eth) nd (peth)	115.5				vs ace, bz, eth, EtOH
5520	Gallamine triethiodide		$C_{30}H_{60}I_3N_3O_3$	65-29-2	891.528		147.5				vs H_2O, EtOH; sl eth, ace, bz, chl
5521	Gallein		$C_{20}H_{12}O_7$	2103-64-2	364.31	br-red pow (+1.5w) red (anh)	>300				vs ace, EtOH
5522	Ganciclovir		$C_9H_{13}N_5O_4$	82410-32-0	255.231	cry (MeOH)	250 dec				
5523	Gardol		$C_{15}H_{28}NNaO_3$	137-16-6	293.378						sl H_2O
5524	Gelsemine		$C_{20}H_{22}N_2O_2$	509-15-9	322.401	cry (ace)	178				vs ace, bz, eth, EtOH
5525	Gelsemine, monohydrochloride		$C_{20}H_{23}ClN_2O_2$	35306-33-3	358.862		326				s H_2O; sl EtOH
5526	Genistein	5,7-Dihydroxy-3-(4-hydroxyphenyl)-4*H*-1-benzopyran-4-one	$C_{15}H_{10}O_5$	446-72-0	270.237	nd(eth), pr(dil al)	301 dec				
5527	β-Gentiobiose	6-*O*-β-D-Glucopyranosyl-D-glucose	$C_{12}H_{22}O_{11}$	554-91-6	342.296	cry (EtOH)	192				s hot H_2O, hot MeOH
5528	Geraniol		$C_{10}H_{18}O$	106-24-1	154.249	oil	<-15	229(2)	0.8894^{20}	1.4766^{20}	i H_2O; s EtOH, eth, ace, chl
5529	Geranyl 2-methylpropanoate		$C_{14}H_{24}O_2$	2345-26-8	224.340			136^{13}	0.8997^{15}	1.4576^{20}	
5530	Geranyl acetate		$C_{12}H_{20}O_2$	16409-44-2	196.286			115^{12}	0.9163^{15}	1.4624^{20}	
5531	Germine		$C_{27}H_{43}NO_8$	508-65-6	509.632	pr or cry (MeOH)	220				s bz, MeOH, alk, acid
5532	Gibberellic acid		$C_{19}H_{22}O_6$	77-06-5	346.374	cry (EtOAc)	234				vs ace, EtOH, MeOH
5533	Gitoxigenin		$C_{23}H_{34}O_5$	545-26-6	390.513	pr (AcOEt) pr (+w, dil al)	234				i H_2O; sl eth; s chl
5534	Gitoxin		$C_{41}H_{64}O_{14}$	4562-36-1	780.939	pr (chl-MeOH)	285 dec				
5535	*d*-Glaucine		$C_{21}H_{25}NO_4$	475-81-0	355.429	pl, pr (eth, AcOEt)	120				vs ace, EtOH, chl
5536	D-Glucaric acid	D-Tetrahydroxyadipic acid	$C_6H_{10}O_8$	87-73-0	210.138	nd (45% al)	125.5				vs H_2O, EtOH; sl eth, chl
5537	D-Glucitol	Sorbitol	$C_6H_{14}O_6$	50-70-4	182.171	nd (w)	97(3)	$295^{3.5}$	1.489^{20}	1.3330^{20}	vs H_2O, ace
5538	D-Glucitol, hexaacetate	Sorbitol hexaacetate	$C_{18}H_{26}O_{12}$	7208-47-1	434.392	pr (w)	98(3)		1.30^{20}		sl H_2O, eth; vs EtOH; s chl, AcOEt
5539	D-Gluconic acid		$C_6H_{12}O_7$	526-95-4	196.155	nd (al-eth)	131(2)				s H_2O; sl EtOH; i eth, bz
5540	β-D-Glucopyranose		$C_6H_{12}O_6$	492-61-5	180.155	cry (hot EtOH)	149				
5541	6-*O*-α-D-Glucopyranosyl-D-fructose	Palatinose	$C_{12}H_{22}O_{11}$	13718-94-0	342.296						s H_2O
5542	2-(β-D-Glucopyranosyloxy)-benzaldehyde	Helicin	$C_{13}H_{16}O_7$	618-65-5	284.262	nd (w)	175				vs H_2O, EtOH
5543	7-(β-D-Glucopyranosyloxy)-2*H*-1-benzopyran-2-one	Skimmin	$C_{15}H_{16}O_8$	93-39-0	324.283	cry (w + 1)	220				s H_2O, EtOH; i eth, chl
5544	2-(β-D-Glucopyranosyloxy)-2-methylpropanenitrile	Linamarin	$C_{10}H_{17}NO_6$	554-35-8	247.245	nd (w, al)	145				vs ace
5545	1-[4-(β-D-Glucopyranosyloxy)-phenyl]ethanone	Picein	$C_{14}H_{18}O_7$	530-14-3	298.289	nd (w+1), nd (MeOH)	195.5				sl H_2O; s EtOH, eth, HOAc; i chl
5546	α-D-Glucose		$C_6H_{12}O_6$	26655-34-5	180.155		146 dec		1.5620^{18}		vs H_2O; sl EtOH; i ace, AcOEt; s py
5547	α-D-Glucose pentaacetate		$C_{16}H_{22}O_{11}$	604-68-2	390.339	pl or nd (al)	132.0(0.5)	sub			sl H_2O, EtOH, CS_2; s eth, chl, HOAc
5548	β-D-Glucose pentaacetate		$C_{16}H_{22}O_{11}$	604-69-3	390.339	nd (al)	134	sub	1.2740^{20}		i H_2O; sl EtOH, peth, eth; s bz; msc chl
5549	α-D-Glucose 1-phosphate	Cori ester	$C_6H_{13}O_9P$	59-56-3	260.135						vs H_2O
5550	D-Glucuronic acid		$C_6H_{10}O_7$	6556-12-3	194.139	nd (al)	165				vs H_2O, EtOH
5551	D-Glucuronic acid γ-lactone	D-Glucuronolactone	$C_6H_8O_6$	32449-92-6	176.124	mcl pl (w) cry (al)	177.5		1.76^{20}		s H_2O; sl EtOH, DMSO, MeOH; i bz
5552	DL-Glutamic acid		$C_5H_9NO_4$	617-65-2	147.130	orth (al,w)	199 dec		1.4601^{20}		sl H_2O, eth; i EtOH, CS_2, lig

Organic

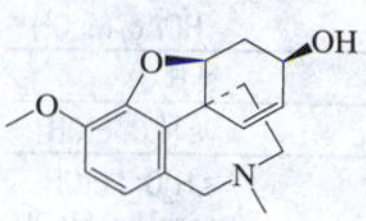

5518
Galanthamine

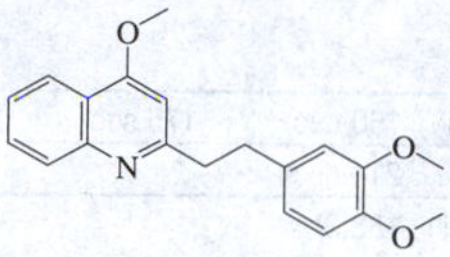

5519
Galipine

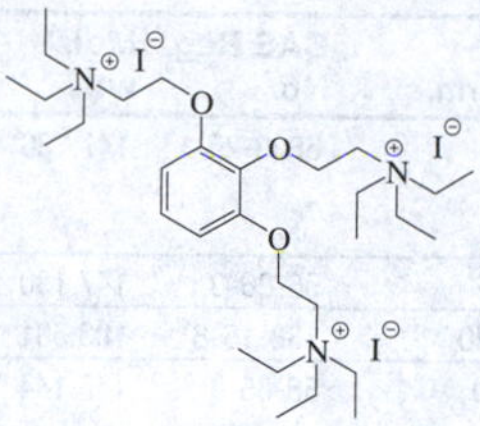

5520
Gallamine triethiodide

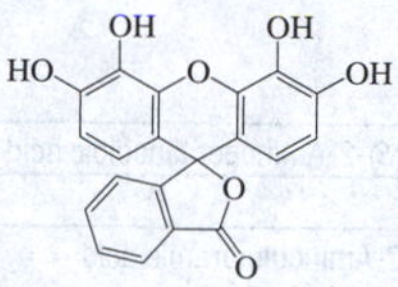

5521
Gallein

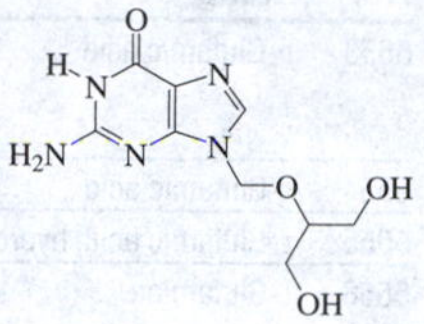

5522
Ganciclovir

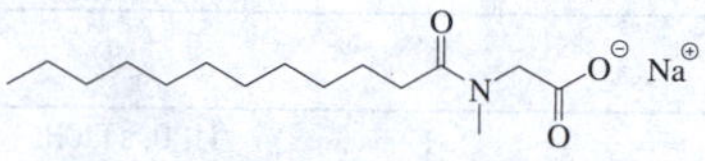

5523
Gardol

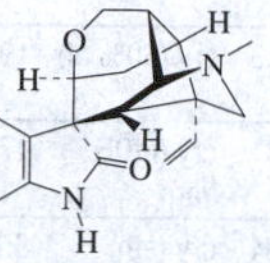

5524
Gelsemine

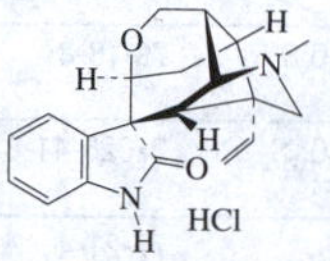

5525
Gelsemine, monohydrochloride

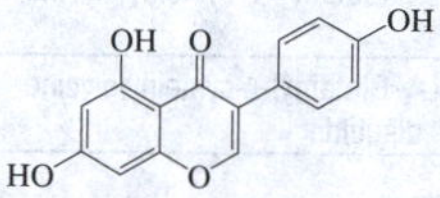

5526
Genistein

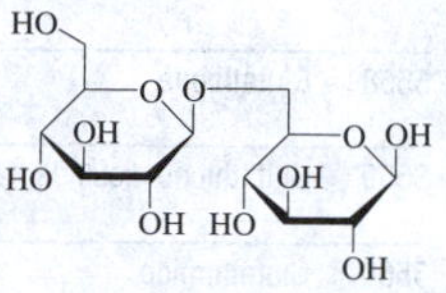

5527
β-Gentiobiose

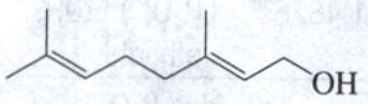

5528
Geraniol

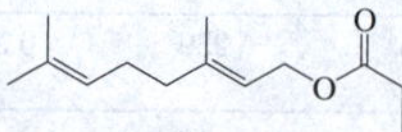

5529
Geranyl 2-methylpropanoate

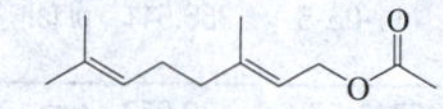

5530
Geranyl acetate

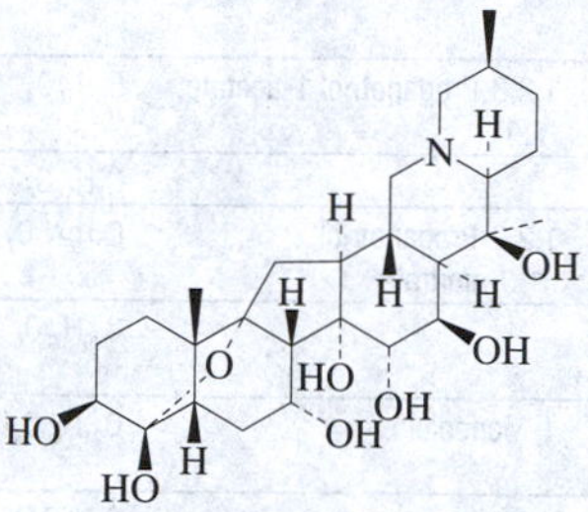

5531
Germine

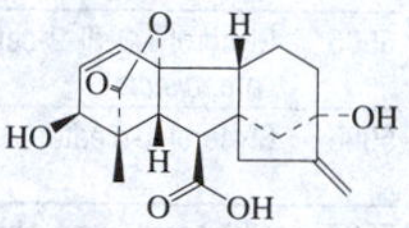

5532
Gibberellic acid

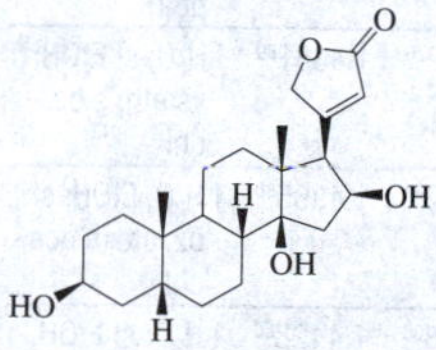

5533
Gitoxigenin

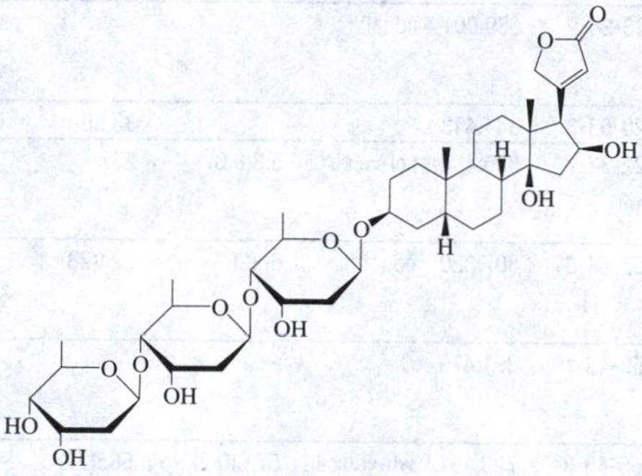

5534
Gitoxin

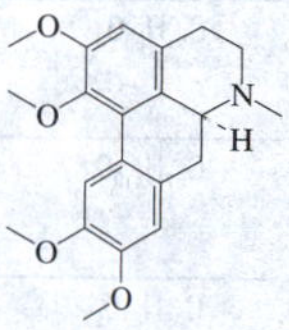

5535
d-Glaucine

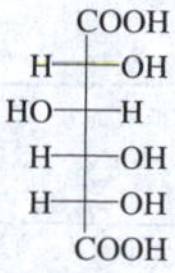

5536
D-Glucaric acid

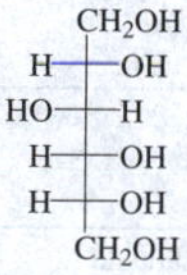

5537
D-Glucitol

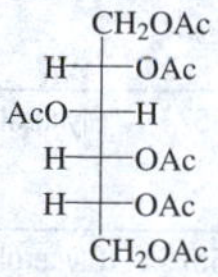

5538
D-Glucitol, hexaacetate

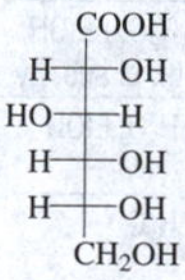

5539
D-Gluconic acid

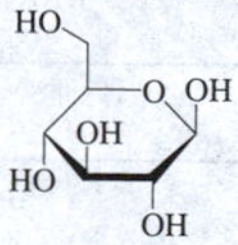

5540
β-D-Glucopyranose

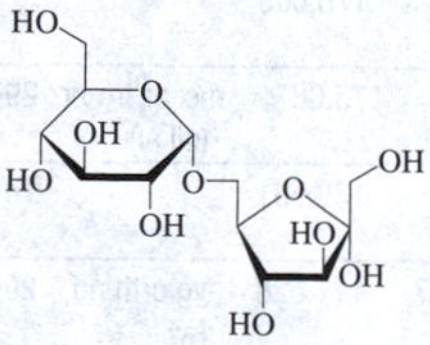

5541
6-*O*-α-D-Glucopyranosyl-D-fructose

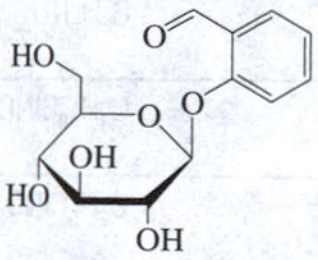

5542
2-(β-D-Glucopyranosyloxy)benzaldehyde

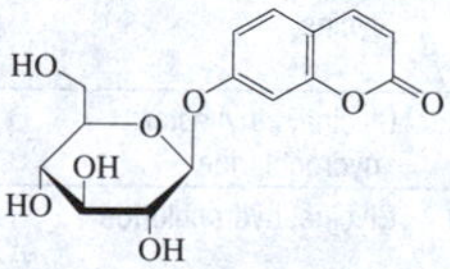

5543
7-(β-D-Glucopyranosyloxy)-2*H*-1-benzopyran-2-one

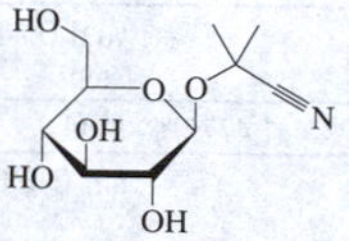

5544
2-(β-D-Glucopyranosyloxy)-2-methylpropanenitrile

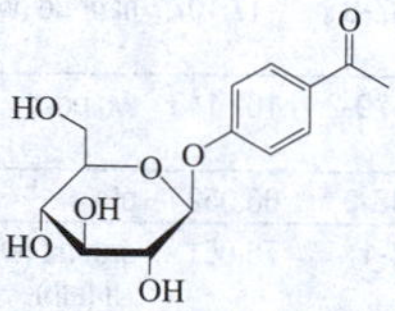

5545
1-[4-(β-D-Glucopyranosyloxy)phenyl]ethanone

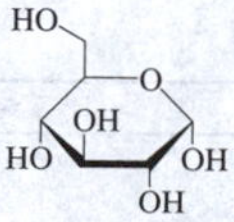

5546
α-D-Glucose

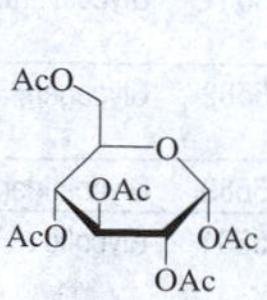

5547
α-D-Glucose pentaacetate

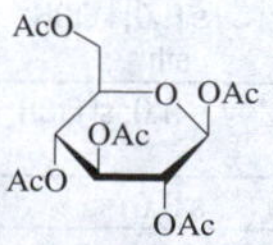

5548
β-D-Glucose pentaacetate

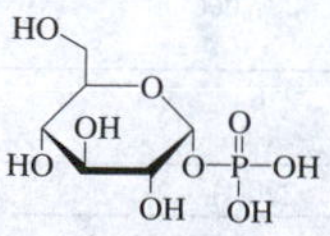

5549
α-D-Glucose 1-phosphate

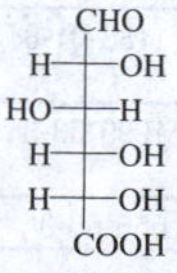

5550
D-Glucuronic acid

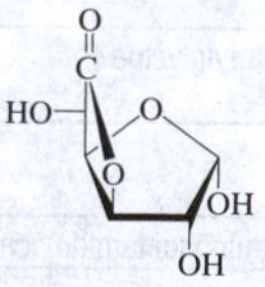

5551
D-Glucuronic acid γ-lactone

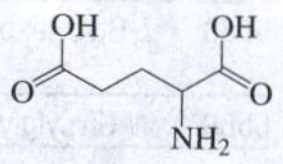

5552
DL-Glutamic acid

Organic

No.	Name	Synonym	Mol. form.	CAS Reg. No.	Mol. wt.	Physical form	mp/°C	bp/°C	den/g cm^{-3}	n_D	Solubility
5553	D-Glutamic acid		$C_5H_9NO_4$	6893-26-1	147.130	lf (w)	213 dec		1.538^{20}		sl H_2O; i EtOH, eth, ace, bz, HOAc, MeOH
5554	L-Glutamic acid	(*S*)-2-Aminopentanedioic acid	$C_5H_9NO_4$	56-86-0	147.130	orth (dil al)	160 dec	175 sub	1.538^{20}		sl H_2O
5555	L-Glutamic acid, hydrochloride		$C_5H_{10}ClNO_4$	138-15-8	183.591	orth pl (w)	214 dec				vs H_2O, EtOH
5556	L-Glutamine	2-Aminoglutaramic acid	$C_5H_{10}N_2O_3$	56-85-9	146.144	nd (w, dil al)	182(3)				s H_2O; i EtOH, eth, bz, MeOH
5557	Glutaric acid	Pentanedioic acid	$C_5H_8O_4$	110-94-1	132.116	nd (bz)	97.9(0.3)	273(10)	1.429^{15}	1.4188^{106}	vs H_2O, EtOH, eth; i bz; s chl, lig
5558	Glutathione	L-γ-Glutamyl-L-cysteinylglycine	$C_{10}H_{17}N_3O_6S$	70-18-8	307.323	cry (50% al)	195				vs H_2O; i EtOH, eth; s DMF
5559	Glutathione disulfide	L-γ-Glutamyl-L-cysteinylglycine disulfide	$C_{20}H_{32}N_6O_{12}S_2$	27025-41-8	612.631	cry (EtOH aq)	179				
5560	Glutethimide		$C_{13}H_{15}NO_2$	77-21-4	217.264	cry (eth)	85(1)				i H_2O; s EtOH; vs eth, ace
5561	Glycerol	1,2,3-Propanetriol	$C_3H_8O_3$	56-81-5	92.094	syr, orth pl	18.2(0.4)	289(3)	1.2613^{20}	1.4746^{20}	msc H_2O, EtOH; sl eth; i bz, ctc, chl
5562	Glycerol 1-acetate	1,2,3-Propanetriol 1-acetate, (±)	$C_5H_{10}O_4$	106-61-6	134.131			158^{165}	1.2060^{20}	1.4157^{20}	vs H_2O, EtOH
5563	Glycerol 1-butanoate		$C_7H_{14}O_4$	557-25-5	162.184			280	1.129^{18}	1.4531^{20}	vs H_2O, EtOH
5564	Glycerol 1,3-dinitrate	1,2,3-Propanetriol, 1,3-dinitrate	$C_3H_6N_2O_7$	623-87-0	182.089	pr (w), cry (eth)	26	148^{15}	1.523^{20}	1.4715^{20}	vs H_2O, eth, EtOH
5565	Glycerol 1,3-di-9-octadecenoate, *cis,cis*		$C_{39}H_{72}O_5$	2465-32-9	620.986	cry (eth/EtOH)	50.1				
5566	Glycerol 1-oleate	1-Monoolein	$C_{21}H_{40}O_4$	111-03-5	356.541	pl (al)	35	239^{3}	0.9420^{20}	1.4626^{20}	i H_2O; s EtOH, eth, chl
5567	L-Glycerol 1-phosphate	α-Glycerophosphoric acid	$C_3H_9O_6P$	5746-57-6	172.073	syr					dec H_2O
5568	Glycerol tridecanoate	Decanoic acid glycerol triester	$C_{33}H_{62}O_6$	621-71-6	554.841	cry (peth)	31(1)				
5569	Glycerol trielaidate	Trielaidin	$C_{57}H_{104}O_6$	537-39-3	885.432						vs bz, eth, chl
5570	Glycerol trilaurate	Trilaurin	$C_{39}H_{74}O_6$	538-24-9	639.001	nd (al)			0.8986^{55}	1.4404^{60}	i H_2O; s EtOH, eth, peth; vs ace, bz
5571	Glycerol tri-3-methylbutanoate	Triisovalerin	$C_{18}H_{32}O_6$	620-63-3	344.443			332.5	0.9984^{20}	1.4354^{20}	vs eth, EtOH
5572	Glycerol trioleate	Triolein	$C_{57}H_{104}O_6$	122-32-7	885.432	col-ye oil	5.3(0.6)	237^{18}	0.915^{15}	1.4676^{15}	i H_2O; sl EtOH; vs eth; s chl, peth
5573	Glycerol tripalmitate	Tripalmitin	$C_{51}H_{98}O_6$	555-44-2	807.320	nd (eth)	66(2)	624(28)	0.8752^{70}	1.4381^{80}	i H_2O; sl EtOH; vs eth; s bz, chl
5574	Glycerol tristearate	Tristearin	$C_{57}H_{110}O_6$	555-43-1	891.479				0.8559^{90}	1.4395^{80}	i H_2O, EtOH; sl bz, ctc; s ace, chl
5575	Glycerol tritetradecanoate	Trimyristin	$C_{45}H_{86}O_6$	555-45-3	723.161	wh-ye solid	57.0(0.2)	585(27)	0.8848^{60}	1.4428^{60}	i H_2O; sl EtOH, lig; s eth, ace, bz
5576	Glycerone phosphate	1-Hydroxy-3-(phosphonooxy)-2-propanone	$C_3H_7O_6P$	57-04-5	170.058						dec H_2O
5577	Glycine	Aminoacetic acid	$C_2H_5NO_2$	56-40-6	75.067	mcl or trg pr (dil al)	290 dec		1.161^{20}		vs H_2O; i EtOH, eth; sl ace, py
5578	Glycine, ethyl ester, hydrochloride	Ethyl aminoacetate hydrochloride	$C_4H_{10}ClNO_2$	623-33-6	139.581		144				vs H_2O, EtOH
5579	Glycine, hydrochloride		$C_2H_6ClNO_2$	6000-43-7	111.528	hyg orth nd (w)	200.5				vs H_2O
5580	Glycocholic acid		$C_{26}H_{43}NO_6$	475-31-0	465.622	nd (w)	166.5				sl H_2O, eth; vs EtOH
5581	Glycocyamine	Guanidinoacetic acid	$C_3H_7N_3O_2$	352-97-6	117.107	pl or nd (w)	282				sl H_2O, EtOH, eth
5582	Glycogen		$(C_6H_{10}O_5)_x$	9005-79-2	162.140	wh pow					vs H_2O; i EtOH, eth
5583	Glycolaldehyde		$C_2H_4O_2$	141-46-8	60.052	pl	97		1.366^{100}	1.4772^{19}	s chl
5584	Glycolic acid		$C_2H_4O_3$	79-14-1	76.051	orth nd (w) lf (eth)	79.5	100			s H_2O, EtOH, eth
5585	*N*-Glycolylneuraminic acid	*N*-(Hydroxyacetyl)neuraminic acid	$C_{11}H_{19}NO_{10}$	1113-83-3	325.270		186				
5586	Glycopyrrolate		$C_{19}H_{28}BrNO_3$	596-51-0	398.334		192.5				
5587	Glycylalanine	*N*-Alanylglycine	$C_5H_{10}N_2O_3$	1188-01-8	146.144		237 dec				s H_2O; i EtOH, eth
5588	L-Glycylasparagine		$C_6H_{11}N_3O_4$	1999-33-3	189.169	nd (EtOH aq)	216				s H_2O; sl EtOH
5589	*N*-Glycylglycine	2-(Aminoacetamido)acetic acid	$C_4H_8N_2O_3$	556-50-3	132.118		220(1)				s H_2O

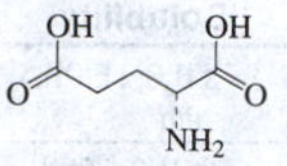
5553
D-Glutamic acid

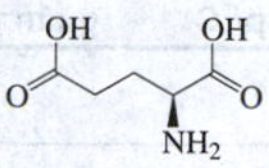
5554
L-Glutamic acid

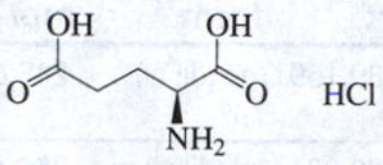
5555
L-Glutamic acid, hydrochloride

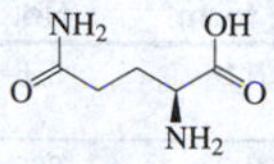
5556
L-Glutamine

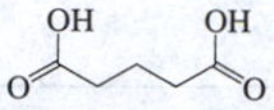
5557
Glutaric acid

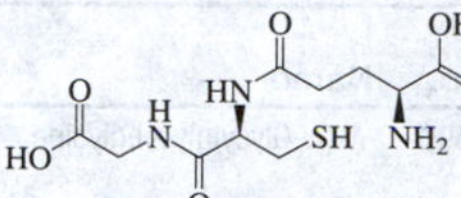
5558
Glutathione

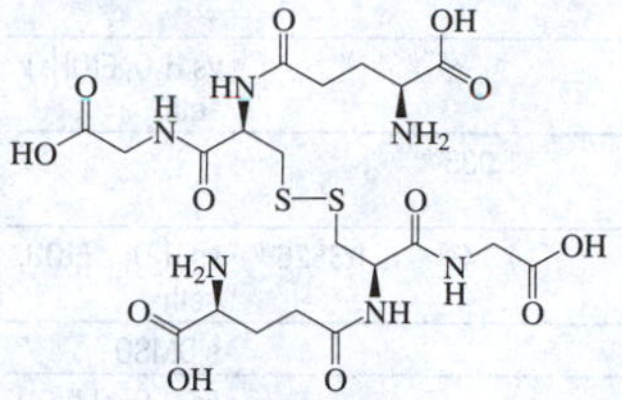
5559
Glutathione disulfide

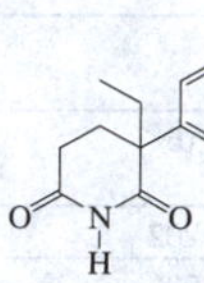
5560
Glutethimide

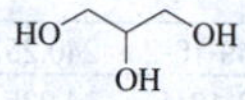
5561
Glycerol

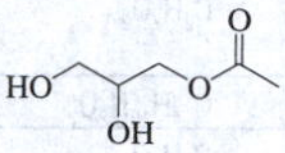
5562
Glycerol 1-acetate

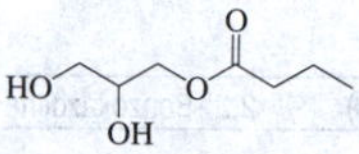
5563
Glycerol 1-butanoate

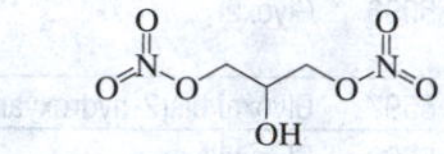
5564
Glycerol 1,3-dinitrate

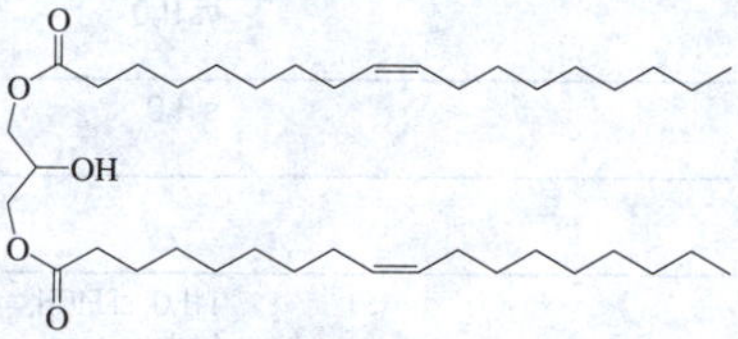
5565
Glycerol 1,3-di-9-octadecenoate, *cis,cis*

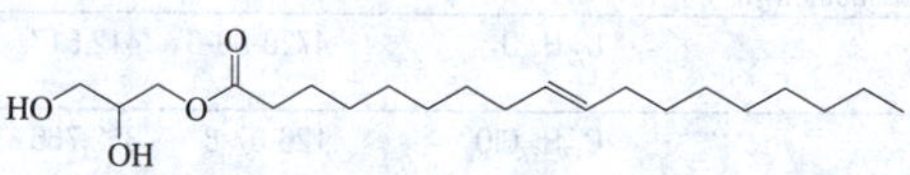
5566
Glycerol 1-oleate

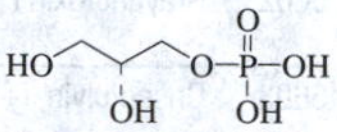
5567
L-Glycerol 1-phosphate

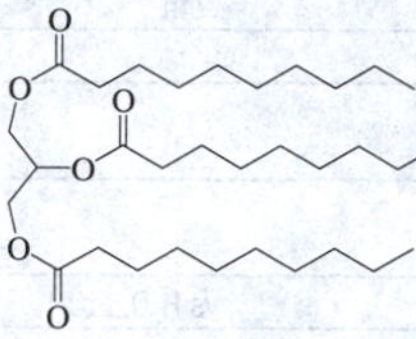
5568
Glycerol tridecanoate

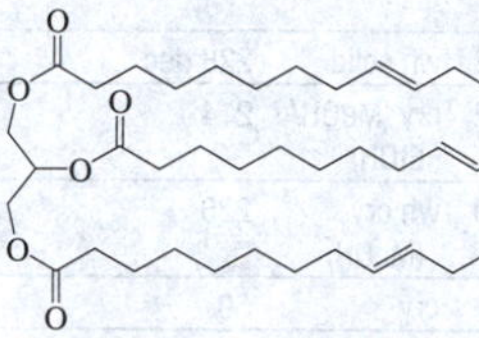
5569
Glycerol trielaidate

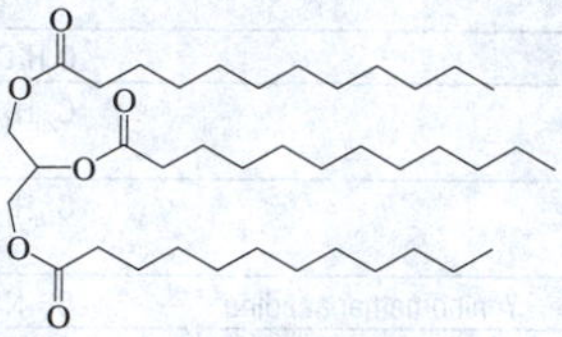
5570
Glycerol trilaurate

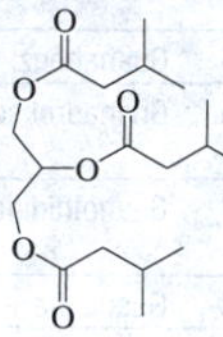
5571
Glycerol tri-3-methylbutanoate

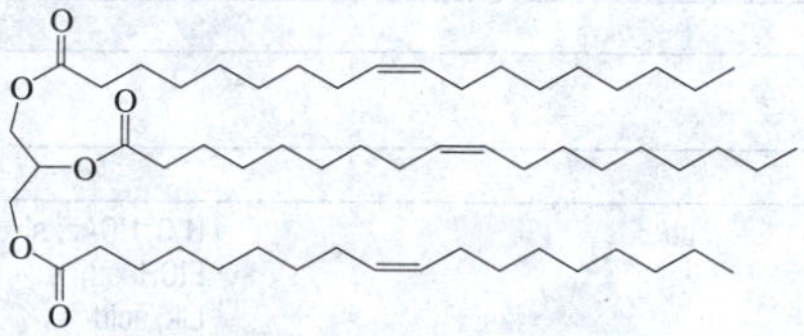
5572
Glycerol trioleate

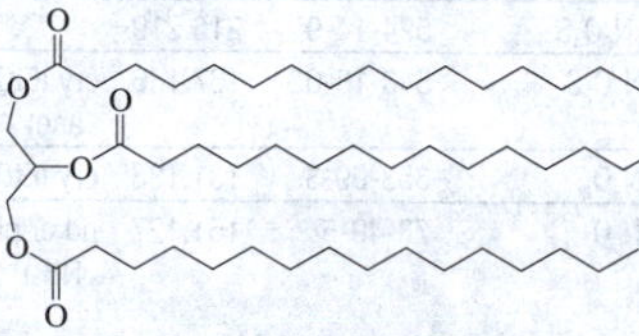
5573
Glycerol tripalmitate

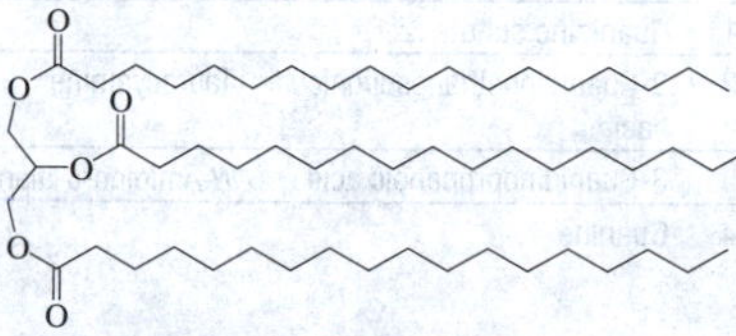
5574
Glycerol tristearate

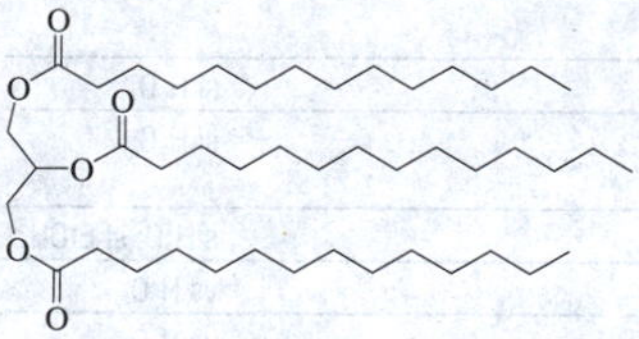
5575
Glycerol tritetradecanoate

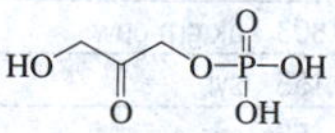
5576
Glycerone phosphate

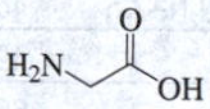
5577
Glycine

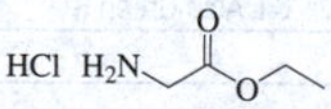
5578
Glycine, ethyl ester, hydrochloride

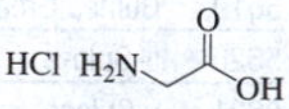
5579
Glycine, hydrochloride

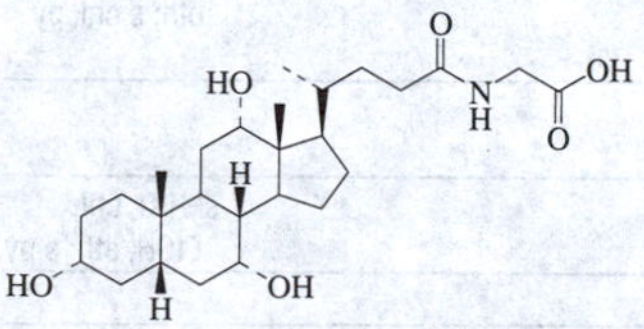
5580
Glycocholic acid

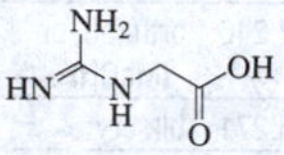
5581
Glycocyamine

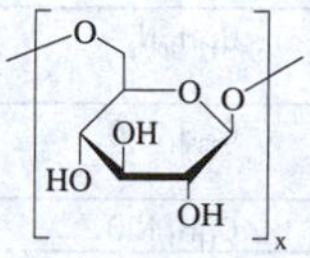
5582
Glycogen

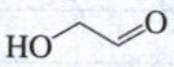
5583
Glycolaldehyde

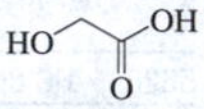
5584
Glycolic acid

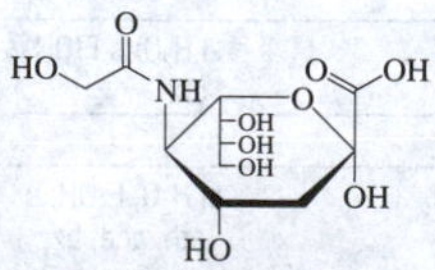
5585
N-Glycolylneuraminic acid

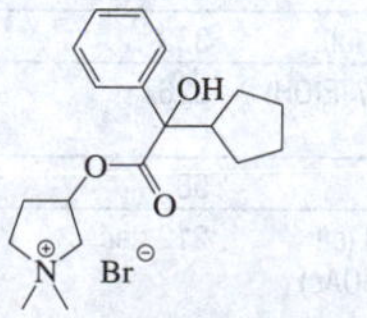
5586
Glycopyrrolate

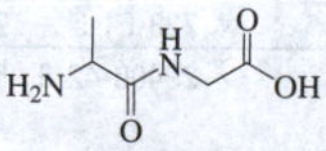
5587
Glycylalanine

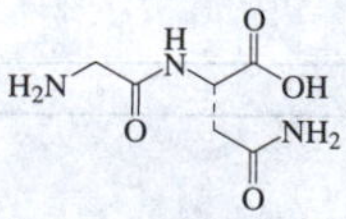
5588
L-Glycylasparagine

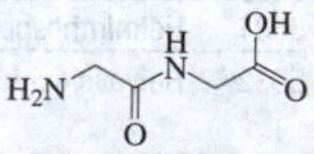
5589
N-Glycylglycine

Organic

No.	Name	Synonym	Mol. form.	CAS Reg. No.	Mol. wt.	Physical form	mp/°C	bp/°C	den/ g cm^{-3}	n_D	Solubility
5590	*N*-(*N*-Glycylglycyl)glycine		$C_6H_{11}N_3O_4$	556-33-2	189.169	nd (dil al)	246 dec				s H_2O; i EtOH, eth
5591	*N*-Glycyl-L-leucine		$C_8H_{16}N_2O_3$	869-19-2	188.224	pl (dil al) pl (dil al)	256 dec				vs H_2O; i EtOH
5592	*N*-Glycyl-L-phenylalanine		$C_{11}H_{14}N_2O_3$	3321-03-7	222.240		267.0(0.5)				s H_2O
5593	*N*-Glycylserine, (DL)-		$C_5H_{10}N_2O_4$	687-38-7	162.144		198 dec				
5594	Glycyrrhizic acid		$C_{42}H_{62}O_{16}$	1405-86-3	822.931	pl or pr (HOAc)	220 dec				vs H_2O, EtOH; i eth
5595	Glyodin	1*H*-Imidazole, 2-heptadecyl-4,5-dihydro-, monoacetate	$C_{22}H_{44}N_2O_2$	556-22-9	368.596				1.035^{20}		
5596	Glyoxal		$C_2H_2O_2$	107-22-2	58.036	ye pr	15	50.4	1.14^{20}	1.3826^{20}	vs H_2O; s EtOH, eth
5597	Glyoxal bis(2-hydroxyanil)	2,2'-Benzoxazoline	$C_{14}H_{12}N_2O_2$	1149-16-2	240.257		202				s DMSO
5598	Glyoxylic acid		$C_2H_2O_3$	298-12-4	74.035	orth pr (w+1/2)	98				vs H_2O; sl EtOH, eth, bz
5599	Glyphosate	Glycine, *N*-(phosphonomethyl)-	$C_3H_8NO_5P$	1071-83-6	169.074		230 dec				
5600	Glyphosate isopropylamine salt		$C_6H_{17}N_2O_5P$	38641-94-0	228.183	cry					vs H_2O
5601	Glyphosine	Glycine, *N,N*-bis(phosphonomethyl)-	$C_4H_{11}NO_8P_2$	2439-99-8	263.080	wh cry					s H_2O
5602	Grayanotoxin I		$C_{22}H_{36}O_7$	4720-09-6	412.517	cry (AcOEt/ C_5H_{12})	268				
5603	Griseofulvin, (+)		$C_{17}H_{17}ClO_6$	126-07-8	352.766	oct or orth cry (bz)	220				i H_2O; sl EtOH, eth, ace, bz, AcOEt, chl
5604	Guaiol		$C_{15}H_{26}O$	489-86-1	222.366	trg pr (al)	91	288 dec	0.9074^{100}	1.4716^{100}	i H_2O; s EtOH, eth
5605	Guanabenz		$C_8H_8Cl_2N_4$	5051-62-7	231.083	wh solid	228 dec				
5606	Guanadrel sulfate (2:1)		$C_{20}H_{40}N_6O_8S$	22195-34-2	524.632	cry (MeOH/ EtOH)	214				
5607	Guanethidine		$C_{10}H_{22}N_4$	55-65-2	198.309	wh cry (MeOH)	226				
5608	Guanidine	Aminomethanamidine	CH_5N_3	113-00-8	59.071	cry	50				vs H_2O, EtOH
5609	Guanidine monohydrochloride		CH_6ClN_3	50-01-1	95.532	orth bipym (al)	184(1)		1.354^{20}		vs H_2O, EtOH
5610	Guanidine mononitrate		$CH_6N_4O_3$	506-93-4	122.084	lf (w)	214.5(0.6)	dec			vs H_2O, EtOH
5611	Guanidine sulfate (2:1)		$C_2H_{12}N_6O_4S$	594-14-9	216.219		292 dec				
5612	2-Guanidinoethanesulfonic acid	Taurocyamine	$C_3H_9N_3O_3S$	543-18-0	167.186	cry (EtOH, ace)	227				
5613	3-Guanidinopropanoic acid	*N*-Amidino-β-alanine	$C_4H_9N_3O_2$	353-09-3	131.133	cry (EtOH)	210				
5614	Guanine		$C_5H_5N_5O$	73-40-5	151.127	nd or pl (aq NH_3)	360 dec	sub			i H_2O, HOAc; sl EtOH, eth; s alk, acid
5615	Guanosine	2-Amino-1,9-dihydro-9-β-D-ribofuranosyl-6*H*-purin-6-one	$C_{10}H_{13}N_5O_5$	118-00-3	283.241	nd (w)	239 dec				sl H_2O; i EtOH, eth; vs HOAc
5616	Guanosine 5'-diphosphate	Guanosine 5'-(trihydrogen diphosphate)	$C_{10}H_{15}N_5O_{11}P_2$	146-91-8	443.201	amorp solid					
5617	Guanosine 5'-monophosphate	5'-Guanylic acid	$C_{10}H_{14}N_5O_8P$	85-32-5	363.221	hyg cry	190 dec				sl H_2O
5618	Guanosine 5'-monophosphate, disodium salt	5'-Guanylic acid, disodium salt	$C_{10}H_{12}N_5Na_2O_8P$	5550-12-9	407.185		195 dec				sl H_2O
5619	Guinea Green B	C.I. Acid Green 3	$C_{37}H_{35}N_2NaO_6S_2$	4680-78-8	690.803	dk grn pow					s H_2O; sl EtOH
5620	D-Gulose		$C_6H_{12}O_6$	4205-23-6	180.155	syr		dec			vs H_2O
5621	L-Gulose		$C_6H_{12}O_6$	6027-89-0	180.155	syr		dec			vs H_2O
5622	Haloperidol		$C_{21}H_{23}ClFNO_2$	52-86-8	375.865		151.5				
5623	Harmaline	4,9-Dihydro-7-methoxy-1-methyl-3*H*-pyrido[3,4-*b*]-indole	$C_{13}H_{14}N_2O$	304-21-2	214.262	tab (MeOH) orth pr (al)	230				sl H_2O, EtOH, eth; s chl, py
5624	Harman	1-Methyl-9*H*-pyrido[3,4-*b*]-indole	$C_{12}H_{10}N_2$	486-84-0	182.220	bl flr orth cry (hp)	236.5				
5625	Harmine	7-Methoxy-1-methyl-9*H*-pyrido[3,4-*b*]indole	$C_{13}H_{12}N_2O$	442-51-3	212.246	orth (al), pr (MeOH)	273	sub			sl H_2O, chl, EtOH, eth; s py
5626	HC Blue No. 1		$C_{11}H_{17}N_3O_4$	2784-94-3	255.271	blk cry	100				
5627	HC Blue No. 2		$C_{12}H_{19}N_3O_5$	33229-34-4	285.296	dk bl-blk cry	110				
5628	Hectane		$C_{100}H_{202}$	6703-98-6	1404.67		115.2(0.4)				
5629	Hederagenin		$C_{30}H_{48}O_4$	465-99-6	472.700	pr (al)	333				
5630	Helenalin		$C_{15}H_{18}O_4$	6754-13-8	262.302	cry (EtOH)	226				sl H_2O; s EtOH, chl
5631	Helminthosporal		$C_{15}H_{22}O_2$	723-61-5	234.335		58	$117^{0.015}$			
5632	Helvolic acid		$C_{33}H_{44}O_8$	29400-42-8	568.697	nd (dil HOAc)	212 dec				sl H_2O, EtOH; s eth, ace, bz, diox

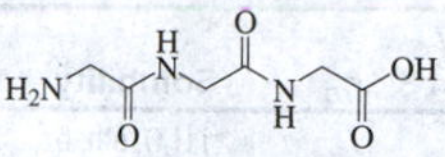
5590
N-(N-Glycylglycyl)glycine

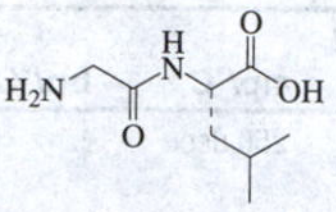
5591
N-Glycyl-L-leucine

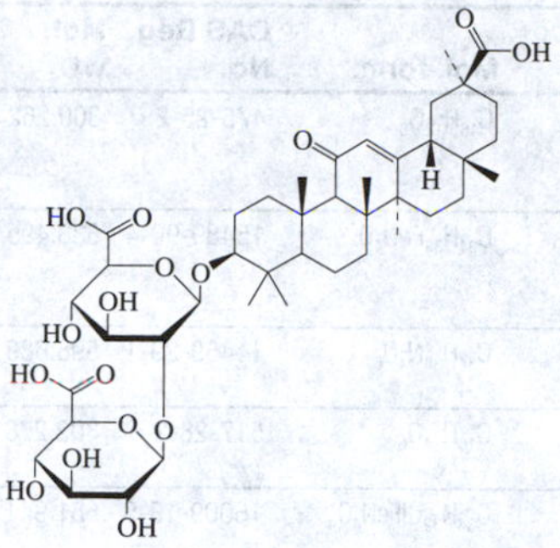
5594
Glycyrrhizic acid

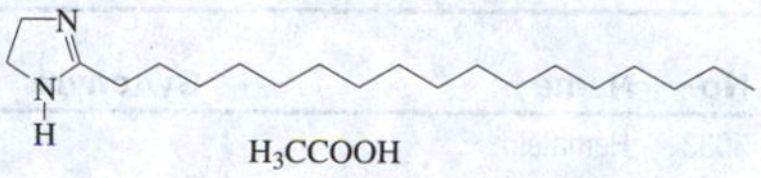
5595
Glyodin

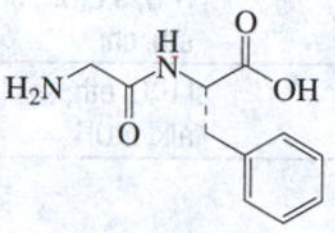
5592
N-Glycyl-L-phenylalanine

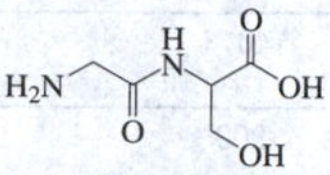
5593
N-Glycylserine, (DL)-

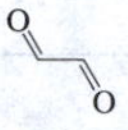
5596
Glyoxal

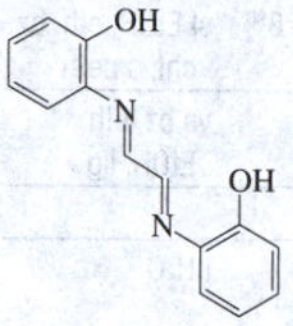
5597
Glyoxal bis(2-hydroxyanil)

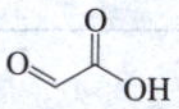
5598
Glyoxylic acid

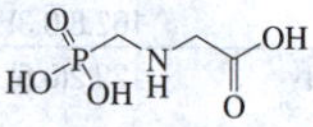
5599
Glyphosate

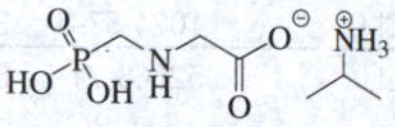
5600
Glyphosate isopropylamine salt

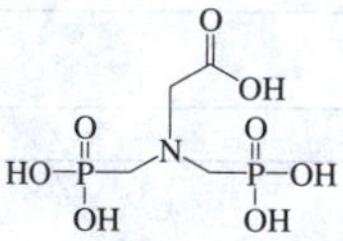
5601
Glyphosine

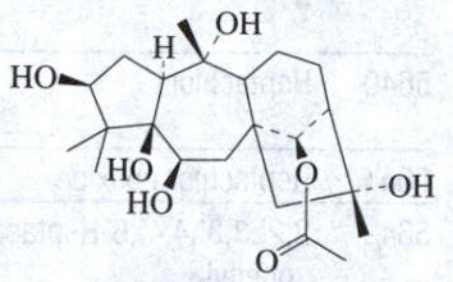
5602
Grayanotoxin I

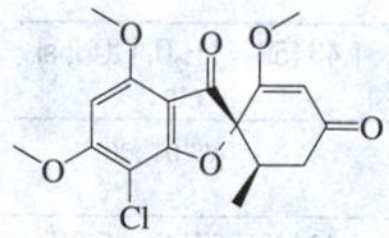
5603
Griseofulvin, (+)

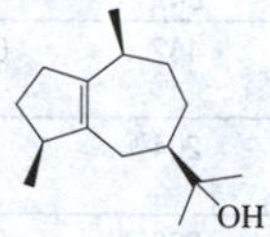
5604
Guaiol

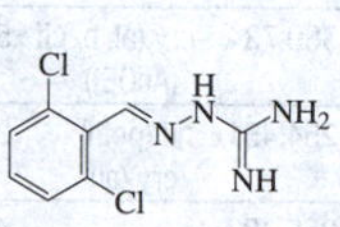
5605
Guanabenz

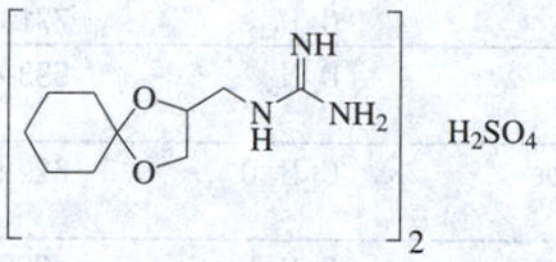
5606
Guanadrel sulfate (2:1)

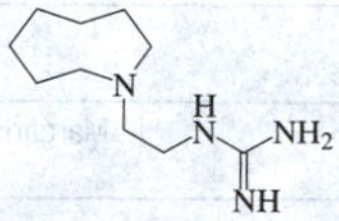
5607
Guanethidine

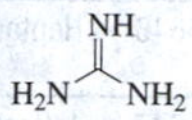
5608
Guanidine

5609
Guanidine monohydrochloride

5610
Guanidine mononitrate

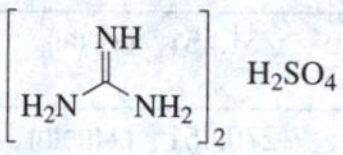
5611
Guanidine sulfate (2:1)

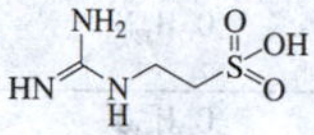
5612
2-Guanidinoethanesulfonic acid

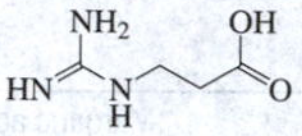
5613
3-Guanidinopropanoic acid

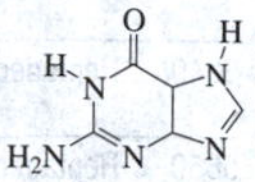
5614
Guanine

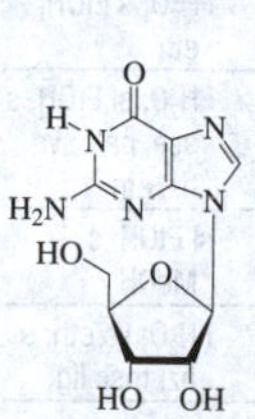
5615
Guanosine

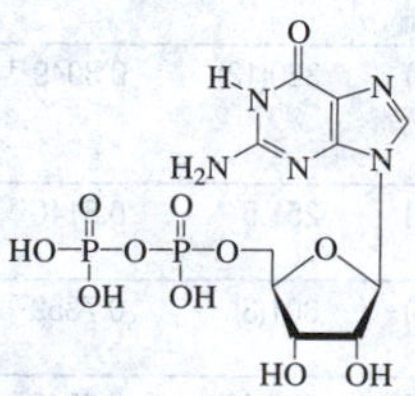
5616
Guanosine 5′-diphosphate

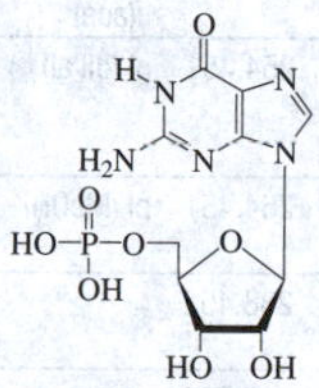
5617
Guanosine 5′-monophosphate

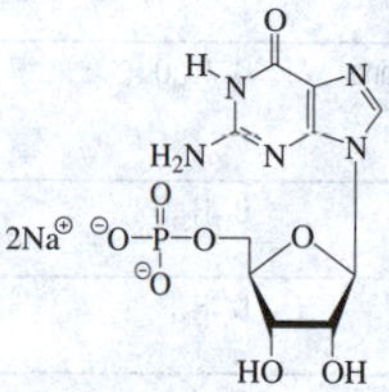
5618
Guanosine 5′-monophosphate, disodium salt

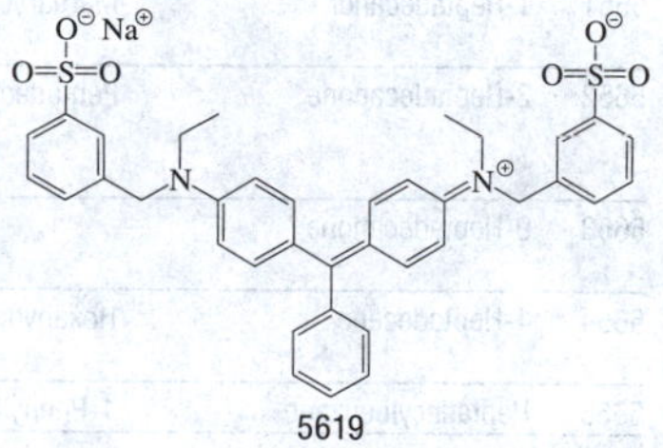
5619
Guinea Green B

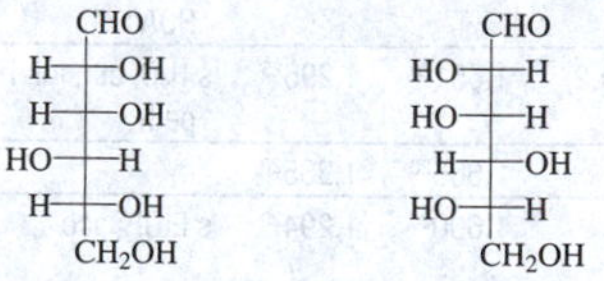
5620
D-Gulose

5621
L-Gulose

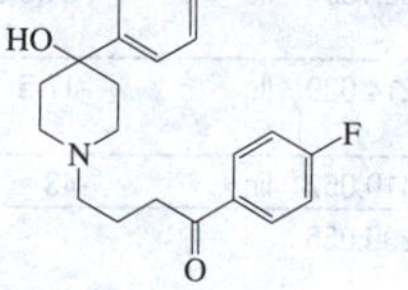
5622
Haloperidol

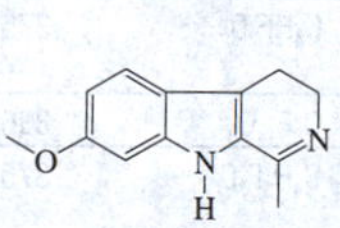
5623
Harmaline

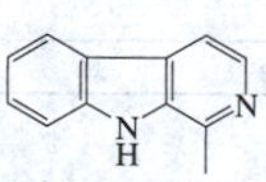
5624
Harman

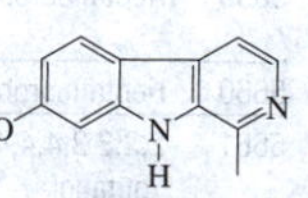
5625
Harmine

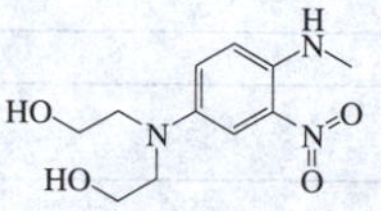
5626
HC Blue No. 1

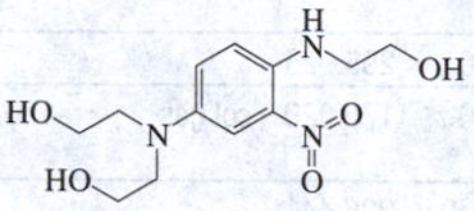
5627
HC Blue No. 2

$H_3C(CH_2)_{98}CH_3$
5628
Hectane

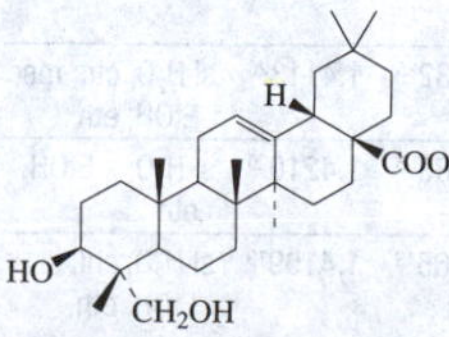
5629
Hederagenin

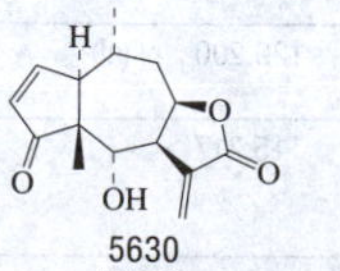
5630
Helenalin

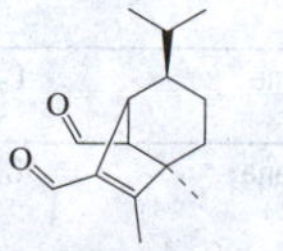
5631
Helminthosporal

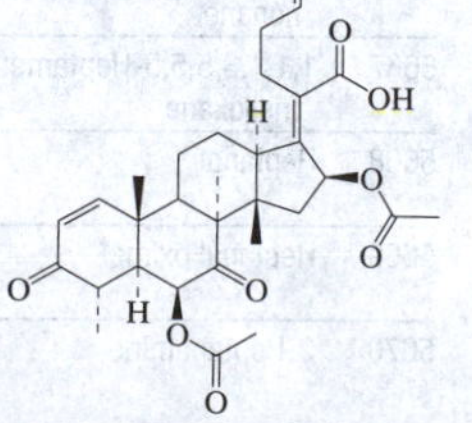
5632
Helvolic acid

No.	Name	Synonym	Mol. form.	CAS Reg. No.	Mol. wt.	Physical form	mp/°C	bp/°C	den/g cm^{-3}	n_D	Solubility
5633	Hematein		$C_{16}H_{12}O_6$	475-25-2	300.262	red-br cry	250 dec				i H_2O, eth, bz, chl; sl EtOH, HOAc
5634	Hematin		$C_{34}H_{33}FeN_4O_5$	15489-90-4	633.495	br pow (py)	>200				i H_2O, eth; s EtOH, alk; sl py, HOAc
5635	Hematoporphyrin		$C_{34}H_{38}N_4O_6$	14459-29-1	598.689	deep red cry	172.5				i H_2O; s EtOH; sl eth, chl
5636	Hematoxylin		$C_{16}H_{14}O_6$	517-28-2	302.278	ye cry	140				sl H_2O, eth; s alk, EtOH
5637	Hemin		$C_{34}H_{32}ClFeN_4O_4$	16009-13-5	651.941	long blades (gl HOAc)	>300				
5638	Heneicosane		$C_{21}H_{44}$	629-94-7	296.574	cry (w)	40.4(0.3)	359(6)	0.7919^{20}	1.4441^{20}	i H_2O; sl EtOH; s peth
5639	Hentriacontane	Untriacontane	$C_{31}H_{64}$	630-04-6	436.840	lf (AcOEt)	68(1)	458	0.781^{68}	1.4278^{90}	sl EtOH, eth, bz, chl; s peth
5640	Heptachlor		$C_{10}H_5Cl_7$	76-44-8	373.318	wh cry	95.8(0.3)		1.57^{9}		vs bz, eth, EtOH, lig
5641	Heptachlor epoxide		$C_{10}H_5Cl_7O$	1024-57-3	389.317		162.8(0.3)				
5642	2,2',3,3',4,4',6-Heptachlorobiphenyl		$C_{12}H_3Cl_7$	52663-71-5	395.323	cry	122.2(0.5)				i H_2O
5643	1,1,1,2,3,3,3-Heptachloropropane		C_3HCl_7	3849-33-0	285.211		11	249	1.7921^{34}	1.5427^{21}	vs chl
5644	Heptacontane		$C_{70}H_{142}$	7719-93-9	983.876		106(4)	647			
5645	Heptacosane		$C_{27}H_{56}$	593-49-7	380.734	cry (al, bz) lf (AcOEt)	58.8(0.1)	442	0.7796^{60}	1.4345^{65}	i H_2O, EtOH; sl eth
5646	Heptadecanal	Margaric aldehyde	$C_{17}H_{34}O$	629-90-3	254.451	nd (peth), cry (al)	36(3)	204^{26}			vs bz, eth
5647	1-Heptadecanamine		$C_{17}H_{37}N$	4200-95-7	255.483		49(1)	336(2)	0.8510^{20}	1.4510^{20}	i H_2O; s EtOH, eth
5648	Heptadecane		$C_{17}H_{36}$	629-78-7	240.468	hex lf	21.97(0.05)	303(2)	0.7780^{20}	1.4369^{20}	i H_2O; sl EtOH, ctc; s eth
5649	Heptadecanenitrile		$C_{17}H_{33}N$	5399-02-0	251.451	cry (al)	34.0(0.7)	349	0.8315^{20}	1.4467^{20}	i H_2O; sl EtOH, chl; vs eth
5650	Heptadecanoic acid	Margaric acid	$C_{17}H_{34}O_2$	506-12-7	270.451	pl (peth)	61.08(0.04)	227^{100}	0.8532^{60}	1.4342^{60}	i H_2O; sl EtOH; s eth, ace, bz, chl
5651	1-Heptadecanol	Margaryl alcohol	$C_{17}H_{36}O$	1454-85-9	256.467	lf (al), cry (ace)	53(1)	324	0.8475^{20}		i H_2O; s EtOH, eth
5652	2-Heptadecanone	Pentadecyl methyl ketone	$C_{17}H_{34}O$	2922-51-2	254.451	pl (dil al)	48.2(0.3)	320(13)	0.8049^{48}		i H_2O; sl EtOH; s ace, peth; vs bz, eth
5653	9-Heptadecanone		$C_{17}H_{34}O$	540-08-9	254.451	pl (MeOH)	50.7(0.2)	251.5	0.8140^{48}		sl EtOH; s MeOH
5654	1-Heptadecene	Hexahydroaplotaxene	$C_{17}H_{34}$	6765-39-5	238.452		10.7(0.6)	301(3)	0.7852^{20}	1.4432^{20}	i H_2O; vs eth; s bz; msc lig
5655	Heptadecylbenzene	1-Phenylheptadecane	$C_{23}H_{40}$	14752-75-1	316.564		19.7(0.4)	402(13)	0.8546^{20}	1.4810^{20}	
5656	*trans,trans*-2,4-Heptadienal		$C_7H_{10}O$	4313-03-5	110.153			84.5	0.881^{25}	1.5315^{20}	
5657	1,6-Heptadiene		C_7H_{12}	3070-53-9	96.170	liq		89(3)			
5658	1,6-Heptadiyne		C_7H_8	2396-63-6	92.139	liq	-84.8(0.6)	111(1)	0.8164^{17}	1.451^{17}	i H_2O; s bz, HOAc
5659	Heptafluorobutanoic acid		$C_4HF_7O_2$	375-22-4	214.039	liq	-17.5	121.5(0.2)	1.651^{20}	1.295^{25}	s H_2O, eth, tol; i peth
5660	Heptafluorobutanoic anhydride		$C_8F_{14}O_3$	336-59-4	410.062	liq	-43	106.5	1.665^{20}	1.285^{20}	
5661	2,2,3,3,4,4,4-Heptafluoro-1-butanol		$C_4H_3F_7O$	375-01-9	200.055			95(4)	1.600^{20}	1.294^{20}	s EtOH, ace
5662	Heptafluorobutanoyl chloride		C_4ClF_7O	375-16-6	232.484			39(10)	1.55^{20}	1.288^{20}	
5663	6,6,7,7,8,8,8-Heptafluoro-2,2-dimethyl-3,5-octanedione		$C_{10}H_{11}F_7O_2$	17587-22-3	296.182		38	46^{5}	1.273^{25}	1.3766^{20}	
5664	Heptafluoro-2-iodopropane	Perfluoroisopropyl iodide	C_3F_7I	677-69-0	295.925			38	1.3298^{20}		
5665	1,1,1,2,3,3,3-Heptafluoropropane	Refrigerant 227ea	C_3HF_7	431-89-0	170.029	col gas	-126.8	-16.34			
5666	2,2,4,4,6,8,8-Heptamethylnonane		$C_{16}H_{34}$	4390-04-9	226.441			246(1)			
5667	1,1,1,3,5,5,5-Heptamethyltrisiloxane		$C_7H_{22}O_2Si_3$	1873-88-7	222.506			143(1)	0.8194^{20}	1.3818^{20}	
5668	Heptanal	Heptaldehyde	$C_7H_{14}O$	111-71-7	114.185	liq	-43.94(0.02)	153(3)	0.8132^{25}	1.4113^{20}	sl H_2O, ctc; msc EtOH, eth
5669	Heptanal oxime	Enanthaldoxime	$C_7H_{15}NO$	629-31-2	129.200	pl (al)	57.5	195	0.8583^{55}	1.4210^{20}	sl H_2O; s EtOH, eth
5670	2-Heptanamine	Tuaminoheptane	$C_7H_{17}N$	123-82-0	115.217			141(4)	0.7665^{19}	1.4199^{19}	sl H_2O, chl; s EtOH, eth, peth

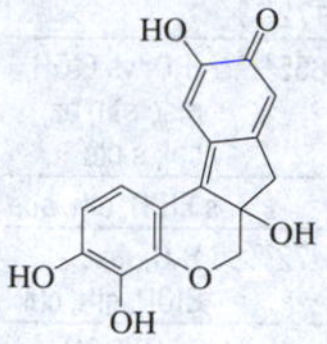
5633
Hematein

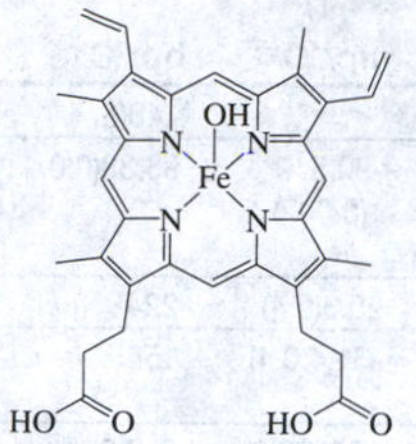
5634
Hematin

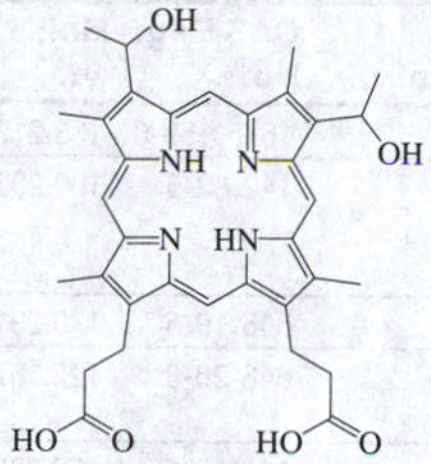
5635
Hematoporphyrin

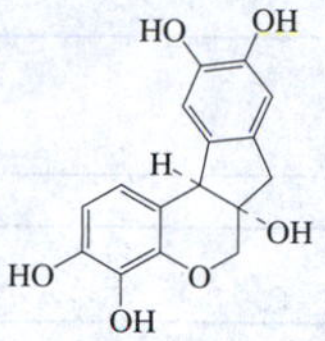
5636
Hematoxylin

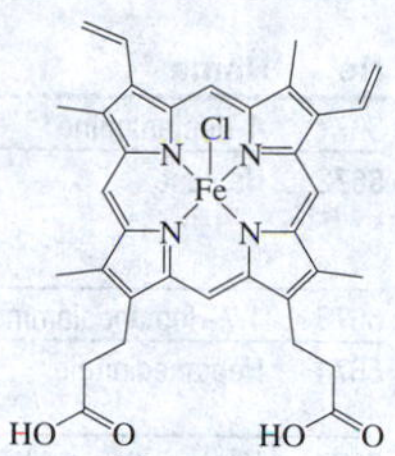
5637
Hemin

5638
Heneicosane

5639
Hentriacontane

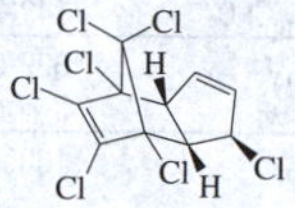
5640
Heptachlor

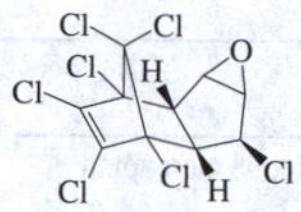
5641
Heptachlor epoxide

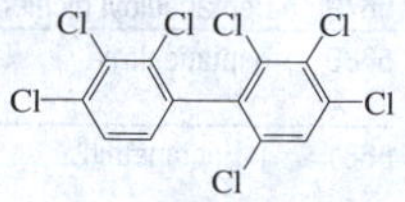
5642
2,2',3,3',4,4',6-Heptachlorobiphenyl

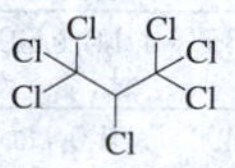
5643
1,1,1,2,3,3,3-Heptachloropropane

$H_3C(CH_2)_{68}CH_3$
5644
Heptacontane

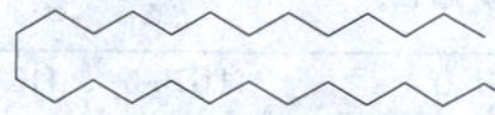
5645
Heptacosane

5646
Heptadecanal

5647
1-Heptadecanamine

5648
Heptadecane

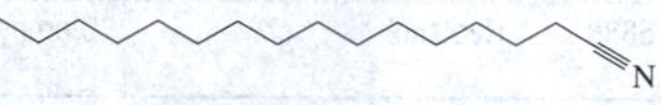
5649
Heptadecanenitrile

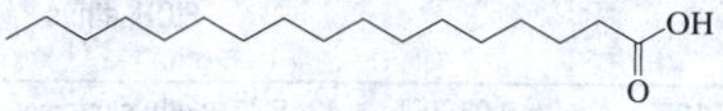
5650
Heptadecanoic acid

5651
1-Heptadecanol

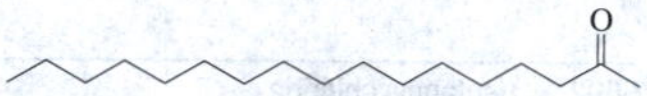
5652
2-Heptadecanone

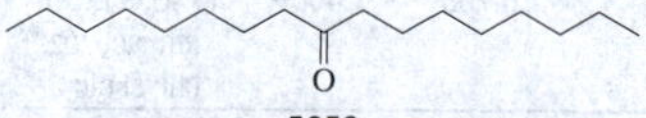
5653
9-Heptadecanone

5654
1-Heptadecene

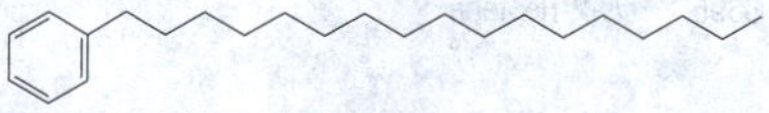
5655
Heptadecylbenzene

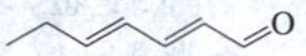
5656
trans,trans-2,4-Heptadienal

5657
1,6-Heptadiene

5658
1,6-Heptadiyne

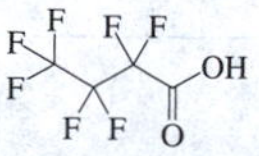
5659
Heptafluorobutanoic acid

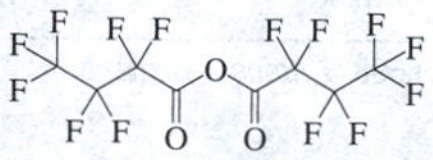
5660
Heptafluorobutanoic anhydride

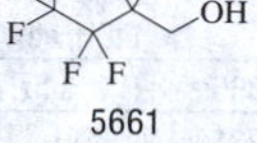
5661
2,2,3,3,4,4,4-Heptafluoro-1-butanol

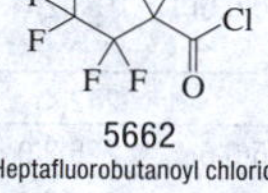
5662
Heptafluorobutanoyl chloride

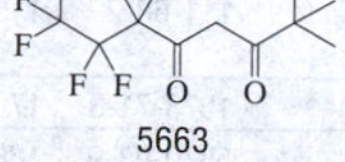
5663
6,6,7,7,8,8,8-Heptafluoro-2,2-dimethyl-3,5-octanedione

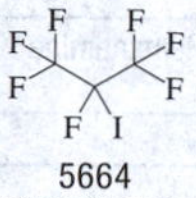
5664
Heptafluoro-2-iodopropane

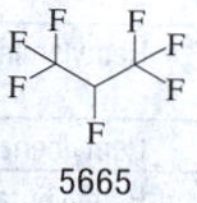
5665
1,1,1,2,3,3,3-Heptafluoropropane

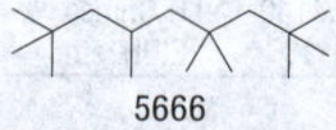
5666
2,2,4,4,6,8,8-Heptamethylnonane

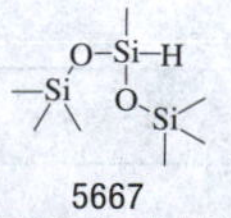
5667
1,1,1,3,5,5,5-Heptamethyltrisiloxane

5668
Heptanal

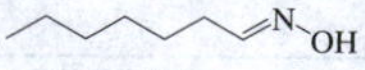
5669
Heptanal oxime

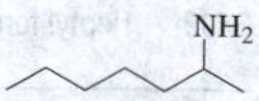
5670
2-Heptanamine

No.	Name	Synonym	Mol. form.	CAS Reg. No.	Mol. wt.	Physical form	mp/°C	bp/°C	den/ g cm^{-3}	n_D	Solubility
5671	4-Heptanamine		$C_7H_{17}N$	16751-59-0	115.217			140(6)	0.767^{20}	1.4172^{20}	
5672	Heptane		C_7H_{16}	142-82-5	100.202	liq	-90.549 (0.002)	98.38(0.07)	0.6795^{25}	1.3855^{25}	i H_2O; vs EtOH; msc eth, bz, chl; s ctc
5673	1,7-Heptanediamine		$C_7H_{18}N_2$	646-19-5	130.231		25.3(0.4)	224			s EtOH, eth, ace
5674	Heptanedinitrile		$C_7H_{10}N_2$	646-20-8	122.167		-31.4(0.4)	155^{14}	0.949^{18}	1.4472^{20}	i H_2O; msc EtOH, eth, chl
5675	Heptanedioic acid	Pimelic acid	$C_7H_{12}O_4$	111-16-0	160.168	pr (w)	104.4(0.3)	342.0	1.329^{15}		s H_2O, EtOH, eth; i bz
5676	1,7-Heptanediol		$C_7H_{16}O_2$	629-30-1	132.201		21.9(0.2)	262	0.9569^{25}	1.4520^{25}	vs eth, EtOH
5677	2,3-Heptanedione	Acetyl valeryl	$C_7H_{12}O_2$	96-04-8	128.169			144	0.919^{18}	1.4150^{18}	
5678	3,5-Heptanedione	Dipropionylmethane	$C_7H_{12}O_2$	7424-54-6	128.169			170(8)	0.945^{20}		
5679	Heptanedioyl dichloride		$C_7H_{10}Cl_2O_2$	142-79-0	197.059			137^{15}			
5680	Heptanenitrile		$C_7H_{13}N$	629-08-3	111.185	liq	-63.8(0.4)	183(2)	0.8106^{20}	1.4104^{30}	i H_2O; s eth, ace, bz, HOAc
5681	1-Heptanethiol	Heptyl mercaptan	$C_7H_{16}S$	1639-09-4	132.267	liq	-43.22(0.02)	176.9(0.7)	0.8427^{20}	1.4521^{20}	i H_2O; msc EtOH, eth; s chl
5682	2,4,6-Heptanetrione		$C_7H_{10}O_3$	626-53-9	142.152	lf	49	121^{10}	1.0599^{40}	1.4930^{20}	vs H_2O, eth, EtOH
5683	Heptanoic acid	Enanthic acid	$C_7H_{14}O_2$	111-14-8	130.185	liq	-7.17(0.05)	222(2)	0.9124^{25}	1.4170^{20}	sl H_2O, ctc; s EtOH, eth, ace
5684	Heptanoic anhydride		$C_{14}H_{26}O_3$	626-27-7	242.354	liq	-12.4(0.5)	269.5	0.9321^{20}	1.4335^{15}	i H_2O; s EtOH, eth
5685	1-Heptanol	Heptyl alcohol	$C_7H_{16}O$	111-70-6	116.201	liq	-33.2(0.1)	178(1)	0.8219^{20}	1.4249^{20}	sl H_2O, ctc; msc EtOH, eth
5686	2-Heptanol, (±)-		$C_7H_{16}O$	52390-72-4	116.201			159	0.8167^{20}	1.4210^{20}	sl H_2O, ctc; s EtOH, eth
5687	3-Heptanol, (*S*)-	Ethylbutylcarbinol	$C_7H_{16}O$	26549-25-7	116.201	liq	-70	163(2)	0.8227^{20}	1.4201^{20}	sl H_2O, ctc; s EtOH, eth
5688	4-Heptanol	Dipropylcarbinol	$C_7H_{16}O$	589-55-9	116.201	liq	-41.2	161(2)	0.8183^{20}	1.4205^{20}	sl H_2O; s EtOH, eth
5689	2-Heptanone	Methyl pentyl ketone	$C_7H_{14}O$	110-43-0	114.185	liq	-34.7(0.4)	151.0(0.3)	0.8111^{20}	1.4088^{20}	vs H_2O; s EtOH, eth
5690	3-Heptanone	Ethyl butyl ketone	$C_7H_{14}O$	106-35-4	114.185	liq	-37.2(0.4)	146(2)	0.8183^{20}	1.4057^{20}	sl H_2O, ctc; msc EtOH, eth
5691	4-Heptanone	Dipropyl ketone	$C_7H_{14}O$	123-19-3	114.185	liq	-32.1(0.8)	144(1)	0.8174^{20}	1.4069^{20}	i H_2O; msc EtOH, eth; s ctc
5692	Heptanoyl chloride		$C_7H_{13}ClO$	2528-61-2	148.630	liq	-83.8	125.2	0.9590^{20}	1.4345^{18}	s eth; sl ctc; vs lig
5693	2-Heptenal	Butylacrolein	$C_7H_{12}O$	2463-63-0	112.169			166	0.864^{17}	1.4468^{17}	
5694	1-Heptene		C_7H_{14}	592-76-7	98.186	liq	-118.83(0.06)	94(1)	0.6970^{20}	1.3998^{20}	i H_2O; s EtOH, eth; sl ctc
5695	*cis*-2-Heptene		C_7H_{14}	6443-92-1	98.186			97(2)	0.708^{20}	1.406^{20}	i H_2O; s EtOH, eth, ace, bz, chl; sl ctc
5696	*trans*-2-Heptene		C_7H_{14}	14686-13-6	98.186	liq	-109.47(0.09)	98(2)	0.7012^{20}	1.4045^{20}	i H_2O; s EtOH, eth, ace, bz, peth, chl
5697	*cis*-3-Heptene		C_7H_{14}	7642-10-6	98.186	liq	-136.6	96(2)	0.7030^{20}	1.4059^{20}	i H_2O; s EtOH, eth, ace, bz, peth, chl
5698	*trans*-3-Heptene		C_7H_{14}	14686-14-7	98.186	liq	-136.64(0.09)	96(2)	0.6981^{20}	1.4043^{20}	i H_2O; s EtOH, eth, ace, bz, chl; sl ctc
5699	6-Heptenoic acid		$C_7H_{12}O_2$	1119-60-4	128.169	liq	-6.5	226	0.9515^{14}	1.4404^{14}	
5700	1-Hepten-4-ol		$C_7H_{14}O$	3521-91-3	114.185			152.1	0.8384^{22}	1.4347^{20}	
5701	*trans*-2-Hepten-1-ol		$C_7H_{14}O$	33467-76-4	114.185			178	0.8516^{20}	1.4460^{20}	s EtOH, ace
5702	Heptyl acetate		$C_9H_{18}O_2$	112-06-1	158.238	liq	-50.3(0.6)	192.2(0.8)	0.8750^{15}	1.4150^{20}	i H_2O; s EtOH, eth, ctc
5703	Heptylamine	1-Heptanamine	$C_7H_{17}N$	111-68-2	115.217	liq	-23.0(0.6)	153(2)	0.7754^{20}	1.4251^{20}	sl H_2O, chl; msc EtOH, eth
5704	Heptylbenzene		$C_{13}H_{20}$	1078-71-3	176.298	liq	-48	242(4)	0.8567^{20}	1.4865^{20}	i H_2O; s bz, chl
5705	Heptyl butanoate		$C_{11}H_{22}O_2$	5870-93-9	186.292	liq	-57.5(0.5)	225.2(0.4)	0.8637^{20}	1.4231^{20}	vs EtOH
5706	Heptylcyclohexane		$C_{13}H_{26}$	5617-41-4	182.345	liq	-40.3(1)	244(2)	0.8109^{20}	1.4484^{20}	
5707	Heptylcyclopentane		$C_{12}H_{24}$	5617-42-5	168.319	liq	-61.0(0.2)	224(6)	0.8010^{20}	1.4421^{20}	vs ace, bz, eth, EtOH
5708	5-Heptyldihydro-2(3*H*)-furanone	4-Hydroxyundecanoic acid lactone	$C_{11}H_{20}O_2$	104-67-6	184.276			286	0.9494^{20}	1.4512^{20}	vs EtOH
5709	Heptyl formate		$C_8H_{16}O_2$	112-23-2	144.212			184(6)	0.8784^{20}	1.4140^{20}	i H_2O; msc EtOH, eth

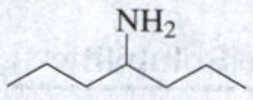
5671
4-Heptanamine

5672
Heptane

5673
1,7-Heptanediamine

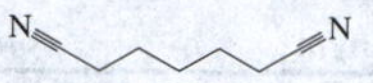
5674
Heptanedinitrile

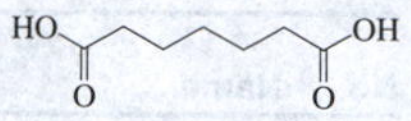
5675
Heptanedioic acid

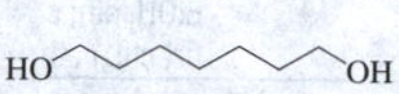
5676
1,7-Heptanediol

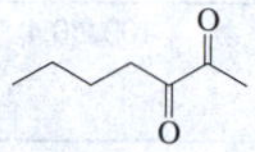
5677
2,3-Heptanedione

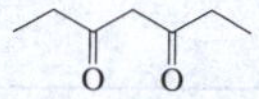
5678
3,5-Heptanedione

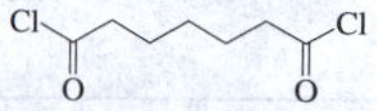
5679
Heptanedioyl dichloride

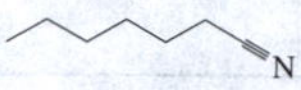
5680
Heptanenitrile

5681
1-Heptanethiol

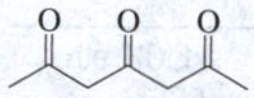
5682
2,4,6-Heptanetrione

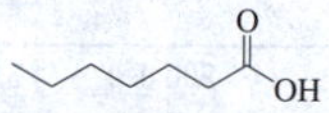
5683
Heptanoic acid

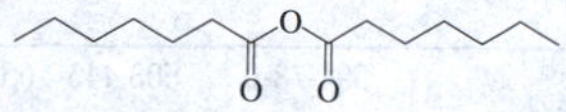
5684
Heptanoic anhydride

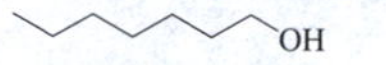
5685
1-Heptanol

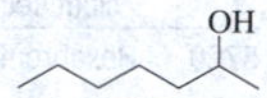
5686
2-Heptanol, (±)-

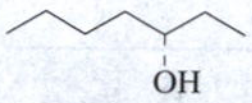
5687
3-Heptanol, (*S*)-

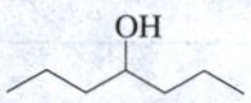
5688
4-Heptanol

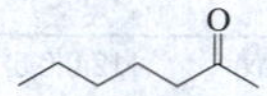
5689
2-Heptanone

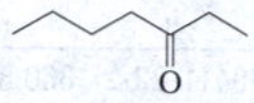
5690
3-Heptanone

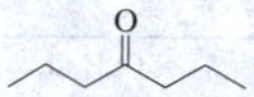
5691
4-Heptanone

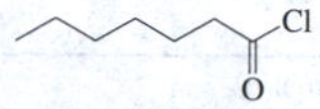
5692
Heptanoyl chloride

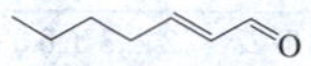
5693
2-Heptenal

5694
1-Heptene

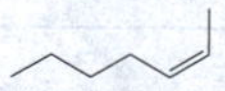
5695
cis-2-Heptene

5696
trans-2-Heptene

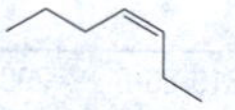
5697
cis-3-Heptene

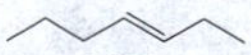
5698
trans-3-Heptene

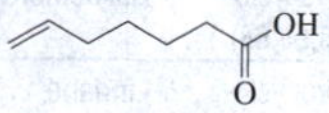
5699
6-Heptenoic acid

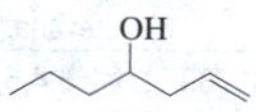
5700
1-Hepten-4-ol

5701
trans-2-Hepten-1-ol

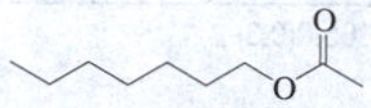
5702
Heptyl acetate

5703
Heptylamine

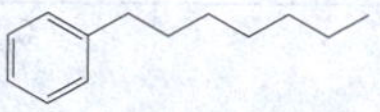
5704
Heptylbenzene

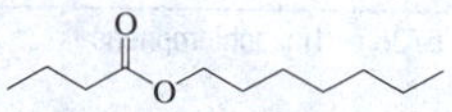
5705
Heptyl butanoate

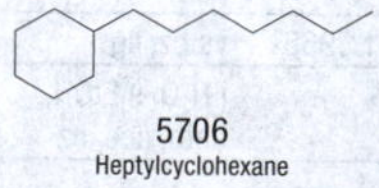
5706
Heptylcyclohexane

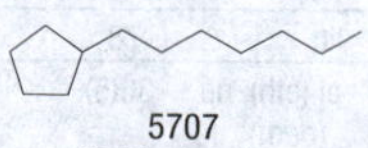
5707
Heptylcyclopentane

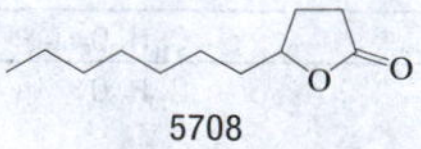
5708
5-Heptyldihydro-2(3*H*)-furanone

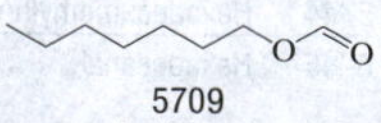
5709
Heptyl formate

No.	Name	Synonym	Mol. form.	CAS Reg. No.	Mol. wt.	Physical form	mp/°C	bp/°C	den/ g cm^{-3}	n_D	Solubility
5710	Heptyl pentanoate		$C_{12}H_{24}O_2$	5451-80-9	200.318	liq	-46.4(0.5)	241(4)	0.8623^{20}	1.4254^{15}	vs ace, eth, EtOH
5711	6-Heptyltetrahydro-2*H*-pyran-2-one	5-Dodecanolide	$C_{12}H_{22}O_2$	713-95-1	198.302	liq	-12	$101^{0.03}$			
5712	1-Heptyne		C_7H_{12}	628-71-7	96.170	liq	-80.9(0.2)	99.8(0.3)	0.7328^{20}	1.4087^{20}	sl H_2O; msc EtOH, eth; s bz, chl, peth
5713	2-Heptyne	1-Methyl-2-butylacetylene	C_7H_{12}	1119-65-9	96.170			113.6(0.3)	0.744^{25}	1.4230^{20}	i H_2O; msc EtOH, eth; s bz, chl, peth
5714	3-Heptyne	1-Ethyl-2-propylacetylene	C_7H_{12}	2586-89-2	96.170	liq	-130.5	109.2(0.4)	0.7336^{25}	1.4189^{20}	i H_2O; msc EtOH, eth; s bz, chl, peth
5715	Hesperetin		$C_{16}H_{14}O_6$	520-33-2	302.278	pl (dil al + 1/2 w)	226(2)	205 sub			vs eth, EtOH
5716	Hesperidin		$C_{28}H_{34}O_{15}$	520-26-3	610.561	wh nd (dil MeOH, HOAc)	262				vs py, EtOH, HOAc
5717	Hexabromobenzene		C_6Br_6	87-82-1	551.488	mcl nd (bz)	327				i H_2O; sl EtOH, eth; s bz, chl
5718	2,2',4,4',5,5'-Hexabromobiphenyl		$C_{12}H_4Br_6$	59080-40-9	627.584	cry (ctc)	160				
5719	1,2,5,6,9,10-Hexabromocyclododecane		$C_{12}H_{18}Br_6$	3194-55-6	641.695	cry	167				
5720	Hexabromoethane		C_2Br_6	594-73-0	503.445	orth pr (bz)		200 dec	3.823^{20}	1.863	sl EtOH, eth, CS_2
5721	Hexabutyldistannoxane	Bis(tributyltin) oxide	$C_{24}H_{54}OSn_2$	56-35-9	596.105	liq	-45	225^{10}	1.17^{20}	1.4870	
5722	Hexacene		$C_{26}H_{16}$	258-31-1	328.405	dk bl-grn cry (sub)	380	sub			i H_2O, EtOH
5723	Hexachloroacetone		C_3Cl_6O	116-16-5	264.749	liq	-1.0	203	1.7434^{12}	1.5112^{20}	sl H_2O; s ace
5724	Hexachlorobenzene	Perchlorobenzene	C_6Cl_6	118-74-1	284.782	nd (sub)	230(3)	325	2.044^{23}	1.5691^{23}	i H_2O; sl EtOH; s eth, chl; vs bz
5725	2,2',3,3',4,4'-Hexachlorobiphenyl		$C_{12}H_4Cl_6$	38380-07-3	360.878	cry	151.7(0.5)				i H_2O
5726	2,2',4,4',6,6'-Hexachlorobiphenyl		$C_{12}H_4Cl_6$	33979-03-2	360.878	cry	113.7(0.5)				i H_2O
5727	2,2',3,3',6,6'-Hexachlorobiphenyl		$C_{12}H_4Cl_6$	38411-22-2	360.878	cry (hx)	112.0(0.5)				i H_2O
5728	2,2',4,4',5,5'-Hexachlorobiphenyl		$C_{12}H_4Cl_6$	35065-27-1	360.878	cry	103.5				
5729	Hexachloro-1,3-butadiene		C_4Cl_6	87-68-3	260.761	liq	-21	216(2)	1.556^{25}	1.5542^{20}	i H_2O; s EtOH, eth
5730	1,2,3,4,5,6-Hexachlorocyclohexane, (1α,2α,3β,4α,5α,6β)	Lindane	$C_6H_6Cl_6$	58-89-9	290.830	nd (al)	115(2)	311(12)			vs ace, bz
5731	1,2,3,4,5,6-Hexachlorocyclohexane, (1α,2α,3β,4α,5β,6β)	α-Hexachlorocyclohexane	$C_6H_6Cl_6$	319-84-6	290.830	cry	157.4(0.7)				
5732	1,2,3,4,5,6-Hexachlorocyclohexane, (1α,2β,3α,4β,5α,6β)	β-Hexachlorocyclohexane	$C_6H_6Cl_6$	319-85-7	290.830	cry (bz, al, xyl)		$60^{0.50}$	1.89^{19}		i H_2O; sl EtOH, bz, chl, HOAc
5733	1,2,3,4,5,6-Hexachlorocyclohexane, (1α,2α,3α,4β,5α,6β)	δ-Lindane	$C_6H_6Cl_6$	319-86-8	290.830	pl	137.0(0.2)	$60^{0.36}$			
5734	Hexachloro-1,3-cyclopentadiene	Perchlorocyclopentadiene	C_5Cl_6	77-47-4	272.772	ye grn liq	-9	239	1.7019^{25}	1.5658^{20}	
5735	1,2,3,6,7,8-Hexachlorodibenzo-*p*-dioxin		$C_{12}H_2Cl_6O_2$	57653-85-7	390.861	cry	285				
5736	1,2,3,7,8,9-Hexachlorodibenzo-*p*-dioxin		$C_{12}H_2Cl_6O_2$	19408-74-3	390.861	cry	243				
5737	Hexachloroethane	Perchloroethane	C_2Cl_6	67-72-1	236.739	orth (al-eth)	186.8(0.2)	184.7 sp	2.091^{20}		i H_2O; vs EtOH, eth; s bz; sl liq HF
5738	Hexachlorophene		$C_{13}H_6Cl_6O_2$	70-30-4	406.904	nd (bz)	165.0(0.3)				i H_2O; s EtOH, eth, ace, chl, dil alk
5739	Hexachloropropene		C_3Cl_6	1888-71-7	248.750	liq	-72.9	214.2(0.8)	1.7632^{20}	1.5091^{20}	i H_2O; s ctc, chl
5740	Hexacontane		$C_{60}H_{122}$	7667-80-3	843.611		100(2)				
5741	Hexacosane		$C_{26}H_{54}$	630-01-3	366.707	mcl, tcl or orth (bz) cry (eth)	56.09(0.04)	415(11)	0.7783^{60}	1.4357^{60}	vs bz, lig, chl
5742	Hexacosanoic acid	Cerotic acid	$C_{26}H_{52}O_2$	506-46-7	396.690		87.7(0.5)		0.8198^{100}	1.4301^{100}	i H_2O; vs EtOH, eth
5743	1-Hexacosanol		$C_{26}H_{54}O$	506-52-5	382.706	orth pl (dil al)	80(1)	305^{20} dec			i H_2O; s EtOH, eth
5744	Hexadecamethylheptasiloxane		$C_{16}H_{48}O_6Si_7$	541-01-5	533.147	liq	-78	286.8(0.2)	0.9012^{20}	1.3965^{20}	vs bz, lig
5745	Hexadecanal		$C_{16}H_{32}O$	629-80-1	240.424	pl (eth), nd (peth)	36(5)	200^{29}			i H_2O; s EtOH, eth, ace, bz

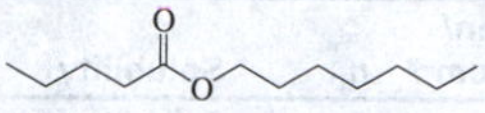
5710
Heptyl pentanoate

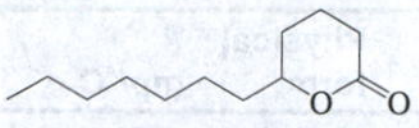
5711
6-Heptyltetrahydro-2*H*-pyran-2-one

5712
1-Heptyne

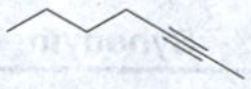
5713
2-Heptyne

5714
3-Heptyne

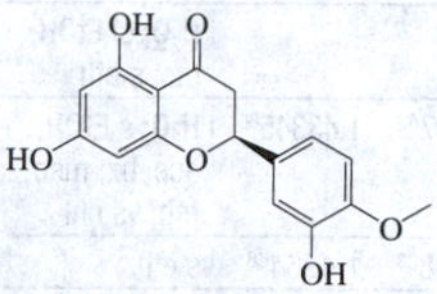
5715
Hesperetin

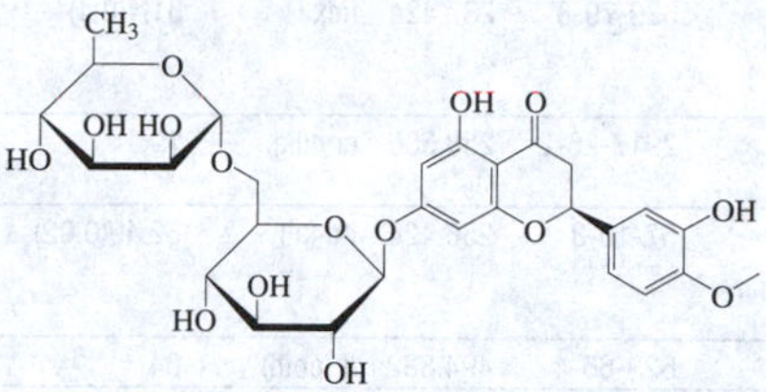
5716
Hesperidin

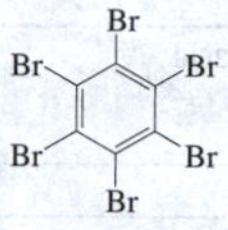
5717
Hexabromobenzene

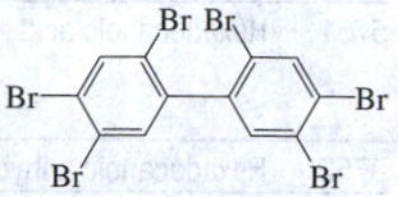
5718
2,2',4,4',5,5'-Hexabromobiphenyl

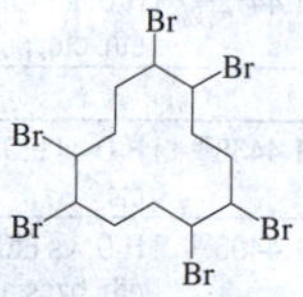
5719
1,2,5,6,9,10-Hexabromocyclododecane

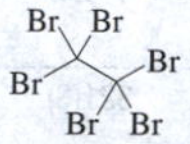
5720
Hexabromoethane

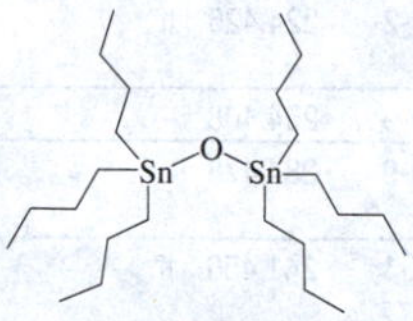
5721
Hexabutyldistannoxane

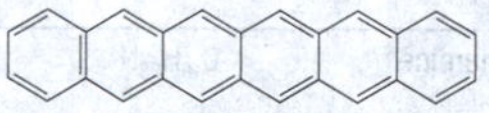
5722
Hexacene

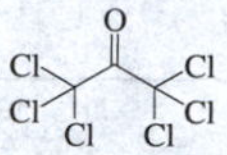
5723
Hexachloroacetone

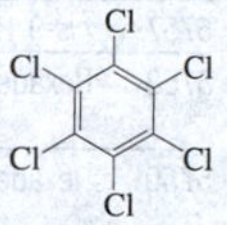
5724
Hexachlorobenzene

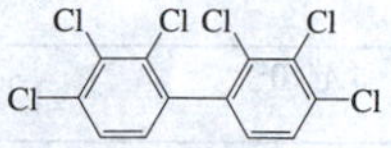
5725
2,2',3,3',4,4'-Hexachlorobiphenyl

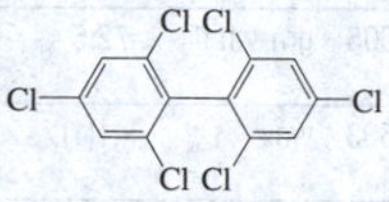
5726
2,2',4,4',6,6'-Hexachlorobiphenyl

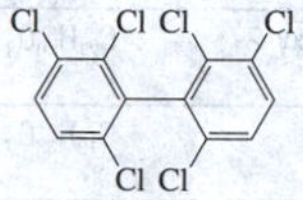
5727
2,2',3,3',6,6'-Hexachlorobiphenyl

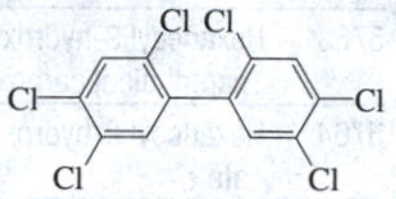
5728
2,2',4,4',5,5'-Hexachlorobiphenyl

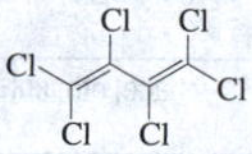
5729
Hexachloro-1,3-butadiene

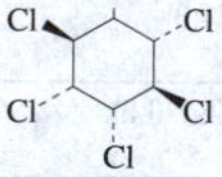
5730
1,2,3,4,5,6-Hexachlorocyclohexane, (1α,2α,3β,4α,5α,6β)

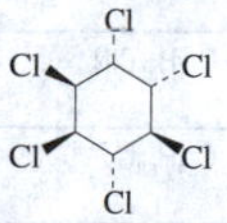
5731
1,2,3,4,5,6-Hexachlorocyclohexane, (1α,2α,3β,4α,5β,6β)

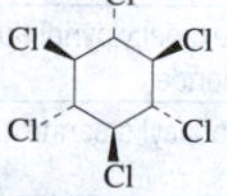
5732
1,2,3,4,5,6-Hexachlorocyclohexane, (1α,2β,3α,4β,5α,6β)

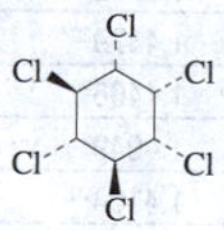
5733
1,2,3,4,5,6-Hexachlorocyclohexane, (1α,2α,3α,4β,5α,6β)

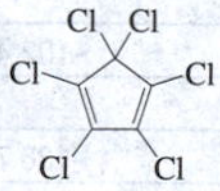
5734
Hexachloro-1,3-cyclopentadiene

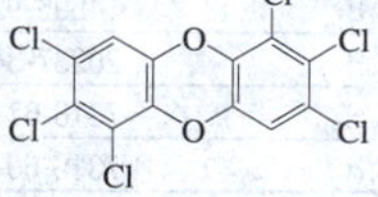
5735
1,2,3,6,7,8-Hexachlorodibenzo-*p*-dioxin

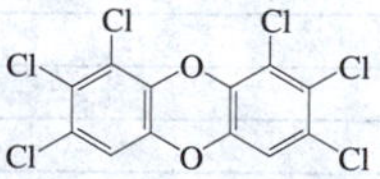
5736
1,2,3,7,8,9-Hexachlorodibenzo-*p*-dioxin

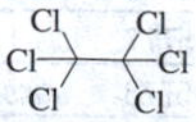
5737
Hexachloroethane

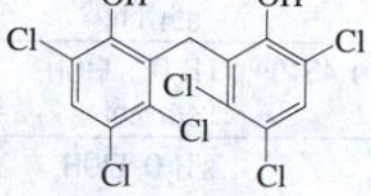
5738
Hexachlorophene

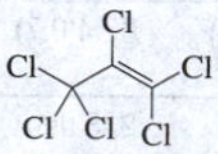
5739
Hexachloropropene

$H_3C(CH_2)_{58}CH_3$
5740
Hexacontane

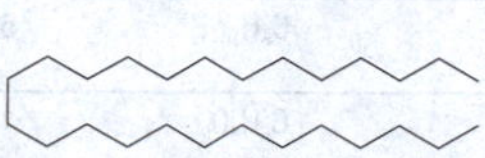
5741
Hexacosane

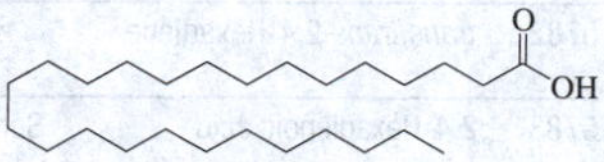
5742
Hexacosanoic acid

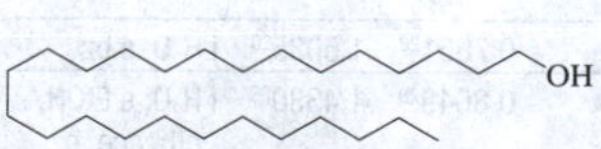
5743
1-Hexacosanol

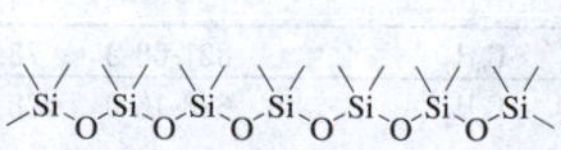
5744
Hexadecamethylheptasiloxane

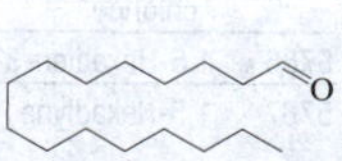
5745
Hexadecanal

Organic

No.	Name	Synonym	Mol. form.	CAS Reg. No.	Mol. wt.	Physical form	mp/°C	bp/°C	den/ g cm^{-3}	n_D	Solubility
5746	Hexadecanamide		$C_{16}H_{33}NO$	629-54-9	255.439	lf	105(3)	236^{12}	1.0000^{20}		i H_2O; sl EtOH, bz, ace, eth
5747	Hexadecane	Cetane	$C_{16}H_{34}$	544-76-3	226.441	lf (HOAc)	18.18(0.02)	286.9(0.7)	0.7701^{25}	1.4329^{25}	i H_2O; sl EtOH; msc eth; s ctc
5748	Hexadecanedioic acid		$C_{16}H_{30}O_4$	505-54-4	286.407	pl (al)	122(1)				vs ace, EtOH
5749	Hexadecanenitrile		$C_{16}H_{31}N$	629-79-8	237.424	hex	31.4(0.4)	333	0.8303^{20}	1.4450^{20}	i H_2O; vs EtOH, eth, ace, bz, chl
5750	1-Hexadecanethiol	Cetyl mercaptan	$C_{16}H_{34}S$	2917-26-2	258.506	cry (lig)	19	$125^{0.5}$			i H_2O; sl EtOH, ctc; s eth
5751	Hexadecanoic acid	Palmitic acid	$C_{16}H_{32}O_2$	57-10-3	256.424	nd (al)	62.49(0.02)	351(6)	0.8527^{62}	1.43345^{60}	i H_2O; s EtOH, ace, bz; msc eth; vs chl
5752	Hexadecanoic anhydride		$C_{32}H_{62}O_3$	623-65-4	494.832	lf (peth)	64		0.8388^{83}	1.4364^{68}	vs eth
5753	1-Hexadecanol	Cetyl alcohol	$C_{16}H_{34}O$	36653-82-4	242.440	fl (AcOEt)	49.30(0.01)	325(2)	0.8187^{50}	1.4283^{79}	i H_2O; sl EtOH; vs eth, bz, chl; s ace
5754	3-Hexadecanone		$C_{16}H_{32}O$	18787-64-9	240.424	lf (peth)	43	184^{17}			s chl
5755	Hexadecanoyl chloride		$C_{16}H_{31}ClO$	112-67-4	274.869		12	199^{20}	0.9016^{25}	1.4514^{20}	vs eth
5756	1-Hexadecene	1-Cetene	$C_{16}H_{32}$	629-73-2	224.425	lf	4.2(0.1)	285(1)	0.7811^{20}	1.4412^{20}	i H_2O; s EtOH, eth, ctc, peth
5757	*cis*-9-Hexadecenoic acid	Palmitoleic acid	$C_{16}H_{30}O_2$	373-49-9	254.408		2(2)	182^{1}			
5758	Hexadecyl acetate		$C_{18}H_{36}O_2$	629-70-9	284.478		-18.5	222^{205}	0.8574^{25}	1.4438^{20}	i H_2O; sl EtOH; s ctc
5759	Hexadecylamine	1-Hexadecanamine	$C_{16}H_{35}N$	143-27-1	241.456	lf	46(2)	321(3)	0.8129^{20}	1.4496^{20}	i H_2O; vs EtOH, eth, bz; s ace
5760	Hexadecylbenzene		$C_{22}H_{38}$	1459-09-2	302.537		26.4(0.4)	385	0.8547^{20}	1.4813^{20}	i H_2O; sl EtOH; vs eth, bz, CS_2
5761	Hexadecyldimethylamine	*N,N*-Dimethyl-1-hexadecanamine	$C_{18}H_{39}N$	112-69-6	269.510			330.0			
5762	Hexadecyl hexadecanoate	Cetyl palmitate	$C_{32}H_{64}O_2$	540-10-3	480.849	mcl lf	43.9(0.2)		0.989^{20}	1.4398^{70}	vs eth, EtOH
5763	Hexadecyl 3-hydroxy-2-naphthalenecarboxylate	Hexadecyl 3-hydroxy-2-naphthoate	$C_{27}H_{40}O_3$	531-84-0	412.605	grn-wh fl	72.5				vs bz, HOAc
5764	Hexadecyl 2-hydroxypropanoate	Cetyl lactate	$C_{19}H_{38}O_3$	35274-05-6	314.503	wax	41(1)	219^{10}		1.4410^{40}	
5765	Hexadecyl 2-methyl-2-propenoate		$C_{20}H_{38}O_2$	2495-27-4	310.515		24	183^{2}	0.87^{20}		
5766	3-(Hexadecyloxy)-1,2-propanediol, (*S*)-	Chimyl alcohol	$C_{19}H_{40}O_3$	506-03-6	316.519	lf (hx)	64	$120^{0.005}$			vs ace, peth, chl
5767	1-Hexadecylpyridinium bromide		$C_{21}H_{38}BrN$	140-72-7	384.438		61				
5768	1-Hexadecylpyridinium chloride	Cetylpyridinium chloride	$C_{21}H_{38}ClN$	123-03-5	339.987	wh pow	81(3)				vs H_2O, chl
5769	Hexadecyl stearate	Cetyl stearate	$C_{34}H_{68}O_2$	1190-63-2	508.903	lf or pl (eth, HOAc)	56.8(0.5)			1.4410^{70}	vs ace, eth, chl
5770	Hexadecyltrichlorosilane		$C_{16}H_{33}Cl_3Si$	5894-60-0	359.878			269			
5771	Hexadecyl vinyl ether	1-(Ethenyloxy)hexadecane	$C_{18}H_{36}O$	822-28-6	268.478		16	160^{2}	0.821^{27}	1.4444^{25}	
5772	1-Hexadecyne		$C_{16}H_{30}$	629-74-3	222.409		15(2)	284	0.7965^{20}	1.4440^{20}	vs bz
5773	*trans,trans*-2,4-Hexadienal	Sorbinaldehyde	C_6H_8O	142-83-6	96.127	liq	-16.5	174	0.898^{20}	1.5384^{20}	
5774	1,2-Hexadiene	Propylallene	C_6H_{10}	592-44-9	82.143			76(3)	0.7149^{20}	1.4282^{20}	vs eth, chl
5775	*cis*-1,3-Hexadiene		C_6H_{10}	14596-92-0	82.143			73(3)	0.7033^{25}	1.4379^{20}	
5776	*trans*-1,3-Hexadiene		C_6H_{10}	20237-34-7	82.143	liq	-102.4(0.6)	71.5(0.8)	0.6995^{25}	1.4406^{20}	
5777	*cis*-1,4-Hexadiene		C_6H_{10}	7318-67-4	82.143			70(13)	0.695^{25}	1.4049^{20}	vs eth
5778	*trans*-1,4-Hexadiene		C_6H_{10}	7319-00-8	82.143	liq	-138.7	65.0(0.5)	0.695^{25}	1.4104^{20}	
5779	1,5-Hexadiene	Biallyl	C_6H_{10}	592-42-7	82.143	liq	-140.7(0.1)	59.2(0.4)	0.6878^{25}	1.4042^{20}	i H_2O; s EtOH, eth, bz, chl; sl ctc
5780	*cis,cis*-2,4-Hexadiene		C_6H_{10}	6108-61-8	82.143	liq		85(3)	0.7298^{25}	1.4606^{20}	i H_2O; s EtOH, eth, chl
5781	*trans,cis*-2,4-Hexadiene		C_6H_{10}	5194-50-3	82.143	liq	-96.1(0.4)	83(3)	0.7185^{25}	1.4560^{20}	i H_2O; s EtOH, eth, chl
5782	*trans,trans*-2,4-Hexadiene		C_6H_{10}	5194-51-4	82.143	liq	-44.9(0.4)	82.4(0.7)	0.7101^{25}	1.4510^{20}	i H_2O; s EtOH, eth, chl
5783	2,4-Hexadienoic acid	Sorbic acid	$C_6H_8O_2$	110-44-1	112.127	nd (dil al) nd (w)	134.5	228 dec	1.204^{19}		s H_2O, EtOH, chl; vs eth
5784	2,4-Hexadien-1-ol	Sorbic alcohol	$C_6H_{10}O$	111-28-4	98.142	nd	29(2)	76^{12}	0.8967^{23}	1.4981^{20}	i H_2O; s EtOH, eth
5785	*trans,trans*-2,4-Hexadienoyl chloride		C_6H_7ClO	2614-88-2	130.572			82^{22}	1.0666^{19}	1.5545^{20}	vs ace
5786	1,5-Hexadien-3-yne	Divinylacetylene	C_6H_6	821-08-9	78.112	liq	-88	85(3)	0.7851^{20}	1.5035^{20}	i H_2O; s bz
5787	1,5-Hexadiyne	Bipropargyl	C_6H_6	628-16-0	78.112	liq	-5.5(0.2)	88(3)	0.8049^{20}	1.4380^{23}	i H_2O; s EtOH, eth, ace, bz

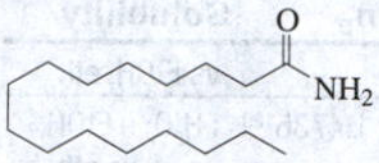
5746 Hexadecanamide

5747 Hexadecane

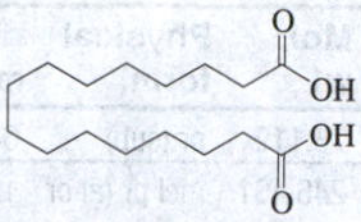
5748 Hexadecanedioic acid

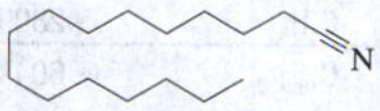
5749 Hexadecanenitrile

5750 1-Hexadecanethiol

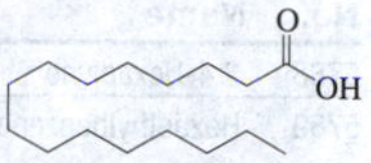
5751 Hexadecanoic acid

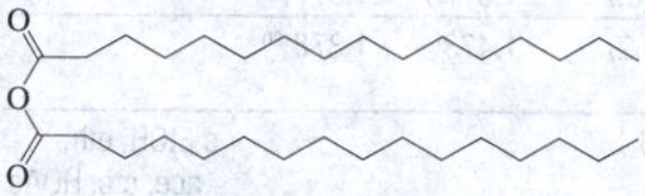
5752 Hexadecanoic anhydride

5753 1-Hexadecanol

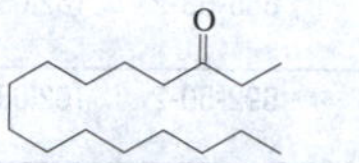
5754 3-Hexadecanone

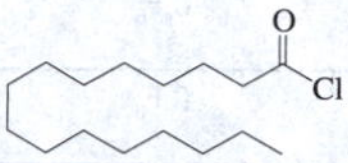
5755 Hexadecanoyl chloride

5756 1-Hexadecene

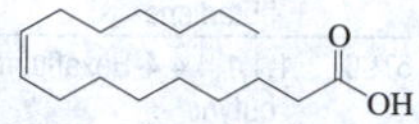
5757 *cis*-9-Hexadecenoic acid

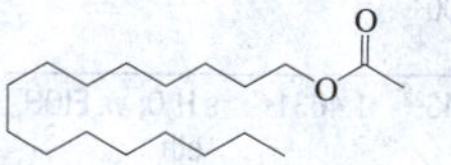
5758 Hexadecyl acetate

5759 Hexadecylamine

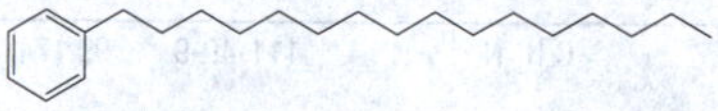
5760 Hexadecylbenzene

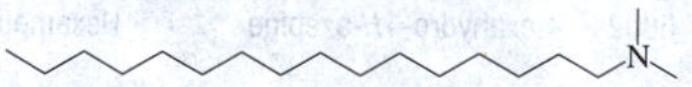
5761 Hexadecyldimethylamine

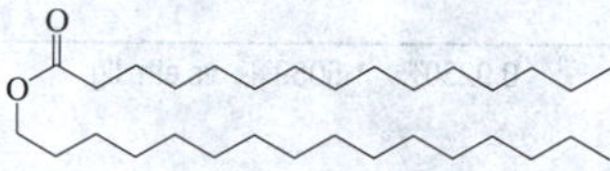
5762 Hexadecyl hexadecanoate

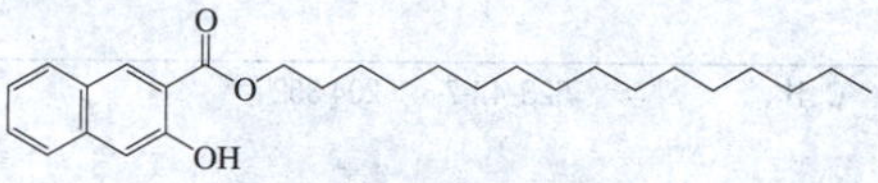
5763 Hexadecyl 3-hydroxy-2-naphthalenecarboxylate

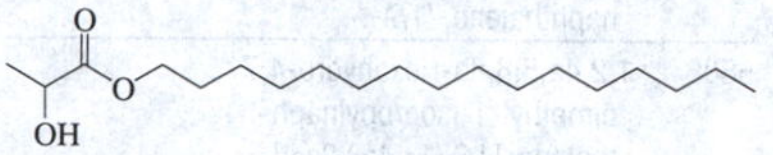
5764 Hexadecyl 2-hydroxypropanoate

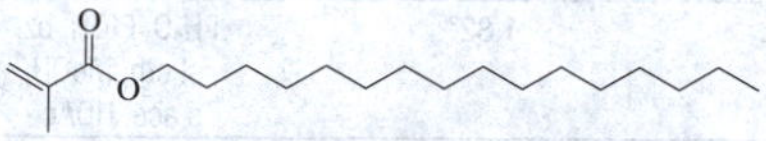
5765 Hexadecyl 2-methyl-2-propenoate

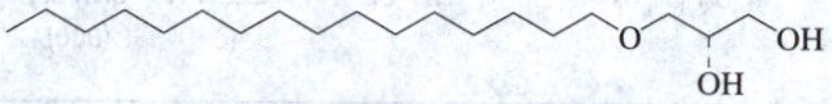
5766 3-(Hexadecyloxy)-1,2-propanediol, (*S*)-

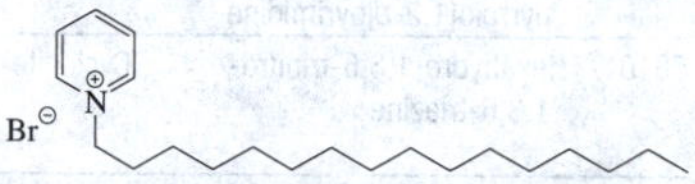
5767 1-Hexadecylpyridinium bromide

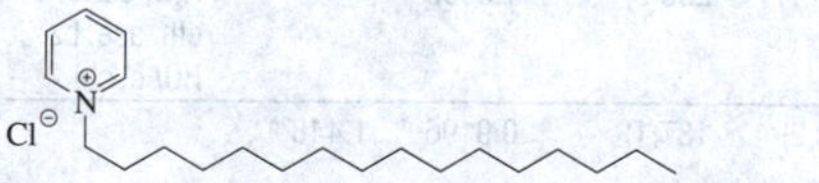
5768 1-Hexadecylpyridinium chloride

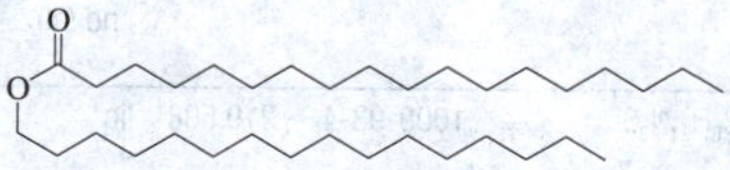
5769 Hexadecyl stearate

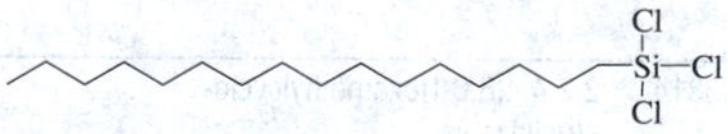
5770 Hexadecyltrichlorosilane

5771 Hexadecyl vinyl ether

5772 1-Hexadecyne

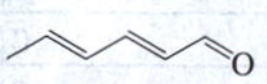
5773 *trans,trans*-2,4-Hexadienal

5774 1,2-Hexadiene

5775 *cis*-1,3-Hexadiene

5776 *trans*-1,3-Hexadiene

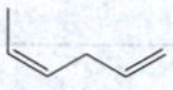
5777 *cis*-1,4-Hexadiene

5778 *trans*-1,4-Hexadiene

5779 1,5-Hexadiene

5780 *cis,cis*-2,4-Hexadiene

5781 *trans,cis*-2,4-Hexadiene

5782 *trans,trans*-2,4-Hexadiene

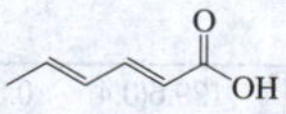
5783 2,4-Hexadienoic acid

5784 2,4-Hexadien-1-ol

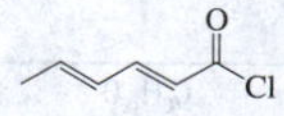
5785 *trans,trans*-2,4-Hexadienoyl chloride

5786 1,5-Hexadien-3-yne

5787 1,5-Hexadiyne

No.	Name	Synonym	Mol. form.	CAS Reg. No.	Mol. wt.	Physical form	mp/°C	bp/°C	den/ g cm⁻³	n_D	Solubility
5788	2,4-Hexadiyne	Dimethyldiacetylene	C_6H_6	2809-69-0	78.112	pr (sub)	64.67(0.01)	129.5(0.2)			vs EtOH, eth
5789	Hexaethylbenzene		$C_{18}H_{30}$	604-88-6	246.431	mcl pr (al or bz)	129(1)	298(3)	0.8305^{130}	1.4736^{130}	i H_2O; s EtOH, sulf; vs eth, bz
5790	Hexaethyldisiloxane		$C_{12}H_{30}OSi_2$	994-49-0	246.536			252(7)	0.8457^{20}	1.4340^{20}	
5791	Hexaethyl tetraphosphate	Ethyl tetraphosphate	$C_{12}H_{30}O_{13}P_4$	757-58-4	506.253	hyg	-40	150 dec	1.2917^{27}	1.4273^{27}	vs ace, bz, EtOH
5792	Hexafluorenium bromide		$C_{36}H_{42}Br_2N_2$	317-52-2	662.539	cry (PrOH)	188				
5793	Hexafluoroacetylacetone		$C_5H_2F_6O_2$	1522-22-1	208.059			69(2)	1.88^{20}	1.3333^{20}	
5794	Hexafluorobenzene	Perfluorobenzene	C_6F_6	392-56-3	186.054	liq	5.10(0.01)	80.2(0.2)	1.6175^{20}	1.3777^{20}	
5795	1,1,2,3,4,4-Hexafluoro-1,3-butadiene		C_4F_6	685-63-2	162.033	col gas	-132	5.4(0.2)	1.473^{5}	1.378^{-20}	
5796	1,1,1,4,4,4-Hexafluoro-2-butyne		C_4F_6	692-50-2	162.033	col gas	-117.4	-24.6			s EtOH, eth, ace, ctc, HOAc
5797	Hexafluorocyclobutene		C_4F_6	697-11-0	162.033	col gas	-60	2(5)	1.602^{-20}	1.298^{-20}	
5798	Hexafluoroethane	Perfluoroethane	C_2F_6	76-16-4	138.011	col gas	-100.015 (0.005)	-78.1(0.1)	1.590^{-78}		i H_2O; sl EtOH, eth
5799	1,1,1,2,3,3-Hexafluoropropane	Refrigerant 236ea	$C_3H_2F_6$	431-63-0	152.038	col gas		6.2	1.5026^{0}		
5800	1,1,1,3,3,3-Hexafluoropropane	Refrigerant 236fa	$C_3H_2F_6$	690-39-1	152.038	col gas	-93.6	-1.4(0.2)	1.4343^{0}		
5801	1,1,1,3,3,3-Hexafluoro-2-propanol		$C_3H_2F_6O$	920-66-1	168.037	liq	-2.0	59(3)	1.4600^{21}		
5802	Hexahydro-1*H*-azepine	Hexamethylenimine	$C_6H_{13}N$	111-49-9	99.174			136(2)	0.8643^{22}	1.4631^{20}	s H_2O; vs EtOH, eth
5803	Hexahydro-1*H*-1,4-diazepine		$C_5H_{12}N_2$	505-66-8	100.162	hyg	40.5	169			
5804	1,5a,6,9,9a,9b-Hexahydro-4a(4*H*)-dibenzofurancarboxaldehyde		$C_{13}H_{16}O_2$	126-15-8	204.265	liq	-80	307	1.10^{20}	1.5254^{20}	i H_2O
5805	*cis*-1,2,3,5,6,8a-Hexahydro-4,7-dimethyl-1-isopropylnaphthalene, (1*S*)-		$C_{15}H_{24}$	483-76-1	204.352			125^{12}	0.9160^{15}	1.5089^{15}	
5806	1,2,4a,5,8,8a-Hexahydro-4,7-dimethyl-1-isopropylnaphthalene, [1*S*-(1α,4aβ,8aα)]		$C_{15}H_{24}$	523-47-7	204.352			274	0.9230^{20}	1.5059^{20}	vs eth, lig
5807	Hexahydro-1,3-isobenzofurandione	Hexahydrophthalic anhydride	$C_8H_{10}O_3$	85-42-7	154.163		32	145^{18}			
5808	Hexahydro-1-methyl-1*H*-1,4-diazepine		$C_6H_{14}N_2$	4318-37-0	114.188			154	0.9111^{20}	1.4769^{20}	
5809	2,3,4,6,7,8-Hexahydropyrrolo[1,2-*a*]pyrimidine		$C_7H_{12}N_2$	3001-72-7	124.183			96^{7}	1.005^{25}	1.5190^{20}	
5810	Hexahydro-1,3,5-trinitro-1,3,5-triazine	Cyclonite	$C_3H_6N_6O_6$	121-82-4	222.116	orth cry (ace)	203.4(0.9)		1.82^{20}		i H_2O, EtOH, bz; sl eth, MeOH; s ace, HOAc
5811	Hexahydro-1,3,5-triphenyl-1,3,5-triazine		$C_{21}H_{21}N_3$	91-78-1	315.412		144	185			i H_2O; sl EtOH; s eth, ace, bz, tol
5812	1,2,3,5,6,7-Hexahydroxy-9,10-anthracenedione	Rufigallol	$C_{14}H_8O_8$	82-12-2	304.209	red rhom, red-ye nd (sub)		sub			i H_2O; sl EtOH, eth; s ace, alk
5813	Hexamethylbenzene	Mellitene	$C_{12}H_{18}$	87-85-4	162.271	orth pr or nd (al)	165.6(0.7)	268(3)	1.0630^{25}		i H_2O; s EtOH, eth, ace, bz, HOAc, chl
5814	2,2,4,4,6,6-Hexamethylcyclotrisilazane		$C_6H_{21}N_3Si_3$	1009-93-4	219.508	liq	-18.7(0.2)	187(1)	0.9196^{20}	1.448^{20}	
5815	Hexamethylcyclotrisiloxane	Dimethylsiloxane cyclic trimer	$C_6H_{18}O_3Si_3$	541-05-9	222.462		63.7(0.4)	135.1(0.4)	1.1200^{20}		i H_2O
5816	Hexamethyldisilane		$C_6H_{18}Si_2$	1450-14-2	146.378		14.4(0.6)	112.65(0.07)	0.7247^{22}	1.4229^{20}	i H_2O; s eth, ace, bz; dec alk
5817	Hexamethyldisilathiane		$C_6H_{18}SSi_2$	3385-94-2	178.443			162.5	0.851^{20}		
5818	Hexamethyldisilazane		$C_6H_{19}NSi_2$	999-97-3	161.393			125(1)	0.7741^{25}	1.4090^{20}	
5819	Hexamethyldisiloxane		$C_6H_{18}OSi_2$	107-46-0	162.377	liq	-68.2(0.1)	100.5(0.3)	0.7638^{20}	1.3774^{20}	i H_2O
5820	Hexamethylenediamine carbamate	(6-Aminohexyl)carbamic acid	$C_7H_{16}N_2O_2$	143-06-6	160.214	cry	150				
5821	Hexamethylene diisocyanate		$C_8H_{12}N_2O_2$	822-06-0	168.193			122^{10}	1.0528^{20}	1.4585^{20}	
5822	Hexamethylenetetramine	Methenamine	$C_6H_{12}N_4$	100-97-0	140.186	orth (al)	>250	sub	1.331^{-5}		vs H_2O; s EtOH, ace, chl; sl eth, bz
5823	Hexamethylolmelamine		$C_9H_{18}N_6O_6$	531-18-0	306.275		137				vs H_2O
5824	Hexamethylphosphoric triamide	Tris(dimethylamino)phosphine oxide	$C_6H_{18}N_3OP$	680-31-9	179.200	col liq	7.2	235(13)	1.03^{20}	1.4579^{20}	s EtOH, eth
5825	Hexamethylphosphorous triamide	Tris(dimethylamino)phosphine	$C_6H_{18}N_3P$	1608-26-0	163.201						s chl
5826	2,6,10,15,19,23-Hexamethyltetracosane	Squalane	$C_{30}H_{62}$	111-01-3	422.813	liq	-38	420(6)	0.8115^{15}	1.4530^{15}	i H_2O; sl EtOH, ace; s eth, chl; msc bz
5827	Hexanal	Caproaldehyde	$C_6H_{12}O$	66-25-1	100.158	liq	-58.2(0.2)	129.6(0.4)	0.8335^{20}	1.4039^{20}	sl H_2O; vs EtOH, eth; s ace, bz

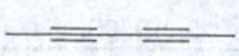

5788 2,4-Hexadiyne

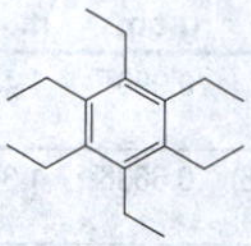
5789 Hexaethylbenzene

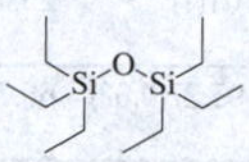
5790 Hexaethyldisiloxane

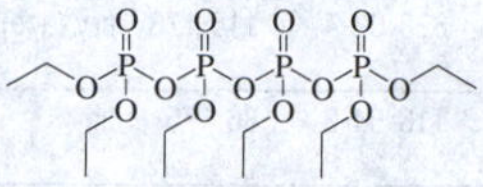
5791 Hexaethyl tetraphosphate

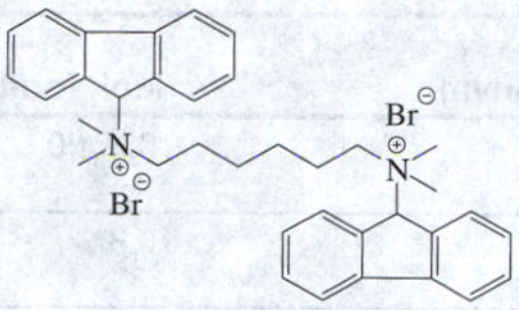
5792 Hexafluorenium bromide

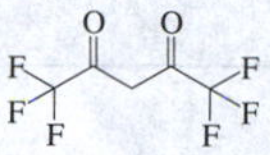
5793 Hexafluoroacetylacetone

5794 Hexafluorobenzene

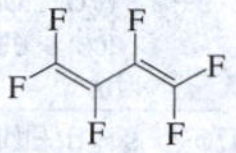
5795 1,1,2,3,4,4-Hexafluoro-1,3-butadiene

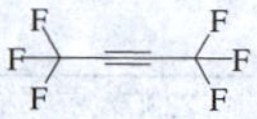
5796 1,1,1,4,4,4-Hexafluoro-2-butyne

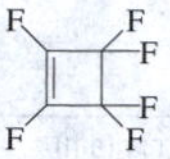
5797 Hexafluorocyclobutene

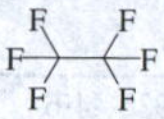
5798 Hexafluoroethane

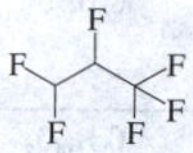
5799 1,1,1,2,3,3-Hexafluoropropane

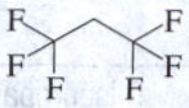
5800 1,1,1,3,3,3-Hexafluoropropane

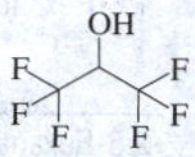
5801 1,1,1,3,3,3-Hexafluoro-2-propanol

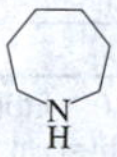
5802 Hexahydro-1*H*-azepine

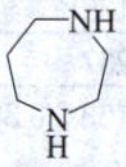
5803 Hexahydro-1*H*-1,4-diazepine

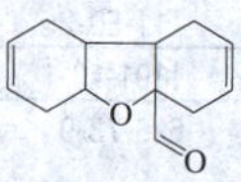
5804 1,5a,6,9,9a,9b-Hexahydro-4a(4*H*)-dibenzofurancarboxaldehyde

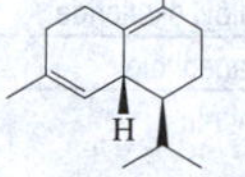
5805 *cis*-1,2,3,5,6,8a-Hexahydro-4,7-dimethyl-1-isopropylnaphthalene, (1*S*)-

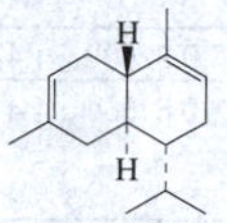
5806 1,2,4a,5,8,8a-Hexahydro-4,7-dimethyl-1-isopropylnaphthalene, [1*S*-(1α,4aβ,8aα)]

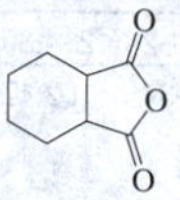
5807 Hexahydro-1,3-isobenzofurandione

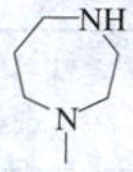
5808 Hexahydro-1-methyl-1*H*-1,4-diazepine

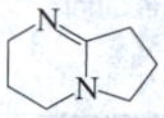
5809 2,3,4,6,7,8-Hexahydropyrrolo[1,2-a]pyrimidine

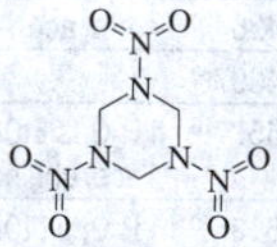
5810 Hexahydro-1,3,5-trinitro-1,3,5-triazine

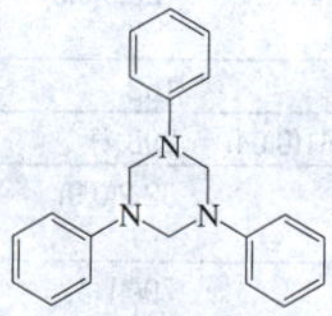
5811 Hexahydro-1,3,5-triphenyl-1,3,5-triazine

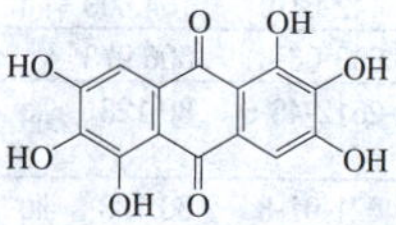
5812 1,2,3,5,6,7-Hexahydroxy-9,10-anthracenedione

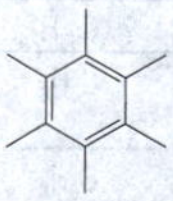
5813 Hexamethylbenzene

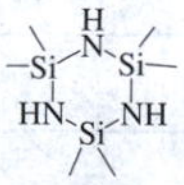
5814 2,2,4,4,6,6-Hexamethylcyclotrisilazane

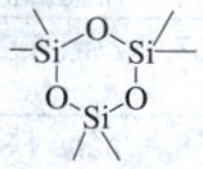
5815 Hexamethylcyclotrisiloxane

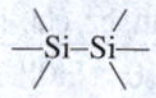
5816 Hexamethyldisilane

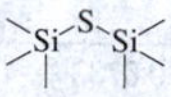
5817 Hexamethyldisilathiane

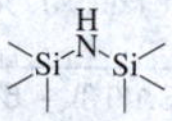
5818 Hexamethyldisilazane

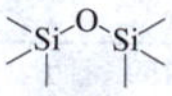
5819 Hexamethyldisiloxane

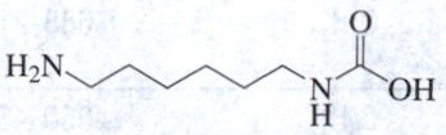
5820 Hexamethylenediamine carbamate

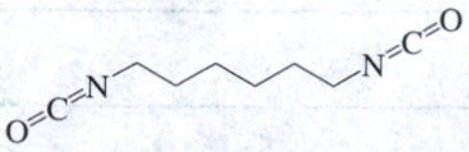
5821 Hexamethylene diisocyanate

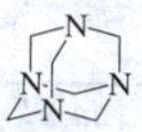
5822 Hexamethylenetetramine

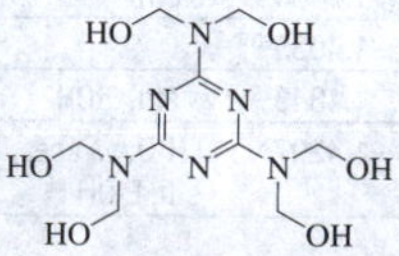
5823 Hexamethylolmelamine

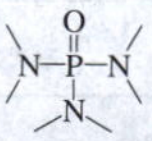
5824 Hexamethylphosphoric triamide

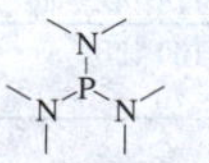
5825 Hexamethylphosphorous triamide

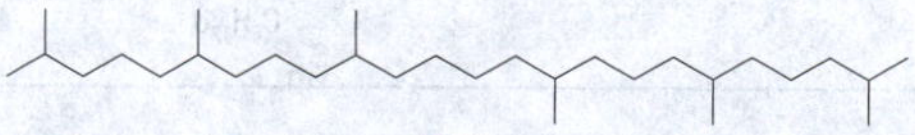
5826 2,6,10,15,19,23-Hexamethyltetracosane

5827 Hexanal

Organic

No.	Name	Synonym	Mol. form.	CAS Reg. No.	Mol. wt.	Physical form	mp/°C	bp/°C	den/ g cm⁻³	n_D	Solubility
5828	Hexanamide		$C_6H_{13}NO$	628-02-4	115.173	cry (ace)	101(1)	258(6)	0.999^{20}	1.4200^{110}	vs bz, eth, EtOH, chl
5829	Hexane		C_6H_{14}	110-54-3	86.175	liq	-95.27(0.02)	68.72(0.06)	0.6606^{25}	1.3727^{25}	i H_2O; vs EtOH; s eth, chl
5830	Hexanedial		$C_6H_{10}O_2$	1072-21-5	114.142		-8	93^{9}	1.003^{19}	1.4350^{20}	vs bz, eth, EtOH
5831	Hexanediamide		$C_6H_{12}N_2O_2$	628-94-4	144.171	pl	225.9(0.4)				vs EtOH
5832	1,6-Hexanediamine	Hexamethylenediamine	$C_6H_{16}N_2$	124-09-4	116.204	orth bipym pl	38.8(0.6)	197(2)			vs H_2O; s EtOH, bz
5833	Hexanedioic acid, dihydrazide		$C_6H_{14}N_4O_2$	1071-93-8	174.201		181.8				
5834	1,2-Hexanediol		$C_6H_{14}O_2$	6920-22-5	118.174		45	234(8)		1.4431^{20}	
5835	1,6-Hexanediol	Hexamethylene glycol	$C_6H_{14}O_2$	629-11-8	118.174		41.5(0.5)	208	1.079^{25}	1.4579^{25}	s H_2O, EtOH, ace; sl eth; i bz
5836	2,5-Hexanediol	Diisopropanol	$C_6H_{14}O_2$	2935-44-6	118.174	cry (eth)	43	229(7)	0.9610^{20}	1.4475^{20}	s H_2O, EtOH, eth; sl ctc
5837	1,6-Hexanediol dimethacrylate	Hexamethylene methacrylate	$C_{14}H_{22}O_4$	6606-59-3	254.323				0.998^{25}		
5838	2,3-Hexanedione	Acetylbutyryl	$C_6H_{10}O_2$	3848-24-6	114.142			128	0.934^{19}		
5839	2,4-Hexanedione	Propionylacetone	$C_6H_{10}O_2$	3002-24-2	114.142	oil		160	0.959^{20}	1.4516^{20}	
5840	2,5-Hexanedione	Acetonylacetone	$C_6H_{10}O_2$	110-13-4	114.142	liq	-5.5	194	0.7370^{20}	1.4232^{20}	vs H_2O, bz, eth, EtOH
5841	3,4-Hexanedione	Bipropionyl	$C_6H_{10}O_2$	4437-51-8	114.142	liq	-10	130	0.941^{21}	1.4130^{21}	
5842	Hexanedioyl dichloride		$C_6H_8Cl_2O_2$	111-50-2	183.033			126^{12}			sl chl
5843	1,6-Hexanedithiol		$C_6H_{14}S_2$	1191-43-1	150.305	liq	-21	237	0.9886^{25}	1.5110^{20}	
5844	Hexanenitrile	Capronitrile	$C_6H_{11}N$	628-73-9	97.158	liq	-80.3	163.5(0.3)	0.8051^{20}	1.4068^{20}	i H_2O; s EtOH, eth; sl chl
5845	1-Hexanethiol	Hexyl mercaptan	$C_6H_{14}S$	111-31-9	118.240	liq	-80.52(0.02)	152.7(0.6)	0.8424^{20}	1.4496^{20}	i H_2O; vs EtOH, eth
5846	2-Hexanethiol		$C_6H_{14}S$	1679-06-7	118.240	liq	-147.0(0.4)	139(1)	0.8345^{20}	1.4451^{20}	i H_2O; s EtOH, eth, bz
5847	1,2,6-Hexanetriol	1,2,6-Trihydroxyhexane	$C_6H_{14}O_3$	106-69-4	134.173			170^{3}	1.1049^{20}	1.58^{20}	
5848	Hexanoic acid	Caproic acid	$C_6H_{12}O_2$	142-62-1	116.158	liq	-4.1(0.7)	204.9(0.6)	0.9212^{25}	1.4163^{20}	sl H_2O; s EtOH, eth, chl
5849	Hexanoic anhydride		$C_{12}H_{22}O_3$	2051-49-2	214.301		-41	267(9)	0.9240^{15}	1.4297^{20}	vs eth, EtOH
5850	1-Hexanol	Caproyl alcohol	$C_6H_{14}O$	111-27-3	102.174	liq	-46.4(0.9)	156.9(0.7)	0.8136^{20}	1.4178^{20}	sl H_2O; s EtOH, ace, chl; msc eth, bz
5851	2-Hexanol		$C_6H_{14}O$	20281-86-1	102.174			138(6)	0.8159^{20}	1.4144^{20}	sl H_2O, ctc; s EtOH, eth
5852	3-Hexanol		$C_6H_{14}O$	17015-11-1	102.174			143(2)	0.8182^{20}	1.4167^{20}	sl H_2O; s EtOH, ace; msc eth
5853	2-Hexanone	Butyl methyl ketone	$C_6H_{12}O$	591-78-6	100.158	liq	-55.45(0.05)	127.6(0.1)	0.8113^{20}	1.4007^{20}	sl H_2O; s ace; msc EtOH, eth
5854	3-Hexanone	Ethyl propyl ketone	$C_6H_{12}O$	589-38-8	100.158	liq	-55.4(0.2)	123.5(0.3)	0.8118^{20}	1.4004^{20}	sl H_2O; s ace; msc EtOH, eth
5855	Hexanoyl chloride	Caproyl chloride	$C_6H_{11}ClO$	142-61-0	134.603	liq	-87	153	0.9784^{20}	1.4264^{20}	s eth, ace
5856	Hexatriacontane		$C_{36}H_{74}$	630-06-8	506.973		75.81(0.04)	298.4^{3}	0.7803^{80}	1.4397^{80}	
5857	*cis*-1,3,5-Hexatriene		C_6H_8	2612-46-6	80.128	liq	-12	82.2(0.9)	0.7175^{20}	1.4577^{20}	i H_2O; s EtOH, ace, chl, peth
5858	*trans*-1,3,5-Hexatriene		C_6H_8	821-07-8	80.128	liq	-12	79(4)	0.7369^{15}	1.5135^{20}	i H_2O; s EtOH, ace, chl, peth
5859	Hexazinone		$C_{12}H_{20}N_4O_2$	51235-04-2	252.313		117.2(0.5)	dec	1.25		
5860	*trans*-2-Hexenal		$C_6H_{10}O$	6728-26-3	98.142			146.5	0.8491^{20}	1.4480^{20}	
5861	*cis*-3-Hexenal		$C_6H_{10}O$	6789-80-6	98.142			121	0.8533^{22}	1.4300^{21}	
5862	1-Hexene		C_6H_{12}	592-41-6	84.159	liq	-139.76(0.05)	63.4(0.1)	0.6685^{25}	1.3852^{25}	i H_2O; vs bz, eth, EtOH, peth
5863	*cis*-2-Hexene		C_6H_{12}	7688-21-3	84.159	liq	-141.12(0.04)	68.9(0.5)	0.6824^{25}	1.3979^{20}	i H_2O; s EtOH, eth, bz, chl, lig
5864	*trans*-2-Hexene		C_6H_{12}	4050-45-7	84.159	liq	-133.1(0.3)	67.85(0.09)	0.6733^{25}	1.3936^{20}	i H_2O; s EtOH, eth, bz, chl, lig
5865	*cis*-3-Hexene		C_6H_{12}	7642-09-3	84.159	liq	-138.7(0.7)	66.4(0.5)	0.6778^{20}	1.3947^{20}	i H_2O; s EtOH, eth, bz, chl, lig
5866	*trans*-3-Hexene		C_6H_{12}	13269-52-8	84.159	liq	-113.7(0.5)	67.06(0.09)	0.6772^{20}	1.3943^{20}	i H_2O; s EtOH, eth, bz, chl, lig
5867	*trans*-3-Hexenedinitrile	*trans*-1,4-Dicyano-2-butene	$C_6H_6N_2$	1119-85-3	106.125	cry	76				
5868	2-Hexenoic acid		$C_6H_{10}O_2$	1191-04-4	114.142	nd (w, al)	36.5	216.5	0.965^{20}	1.4460^{40}	vs eth
5869	3-Hexenoic acid	Hydrosorbic acid	$C_6H_{10}O_2$	4219-24-3	114.142		12	208	0.9640^{23}	1.4935^{20}	
5870	5-Hexenoic acid	5-Hexanoic acid	$C_6H_{10}O_2$	1577-22-6	114.142	liq	-37	203	0.9610^{20}	1.4343^{20}	vs eth, EtOH
5871	1-Hexen-3-ol		$C_6H_{12}O$	4798-44-1	100.158			134(7)	0.834^{22}	1.4297^{18}	sl H_2O; vs ace, eth, EtOH

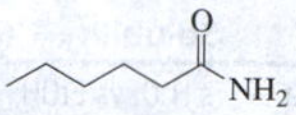

5828
Hexanamide

5829
Hexane

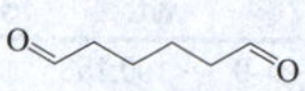

5830
Hexanedial

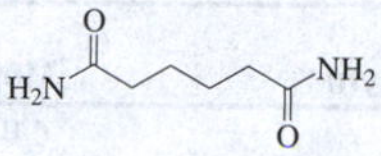

5831
Hexanediamide

5832
1,6-Hexanediamine

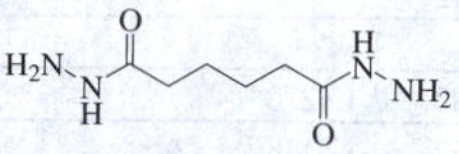

5833
Hexanedioic acid, dihydrazide

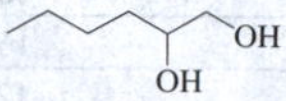

5834
1,2-Hexanediol

5835
1,6-Hexanediol

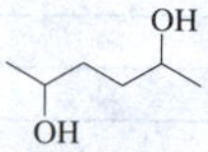

5836
2,5-Hexanediol

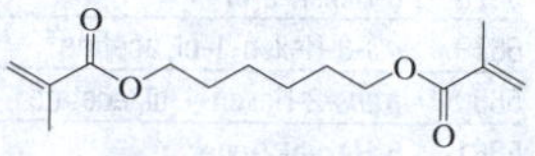

5837
1,6-Hexanediol dimethacrylate

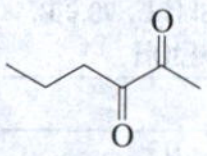

5838
2,3-Hexanedione

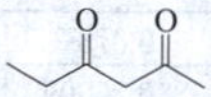

5839
2,4-Hexanedione

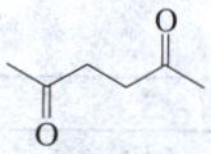

5840
2,5-Hexanedione

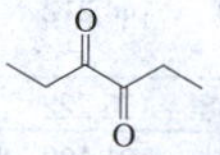

5841
3,4-Hexanedione

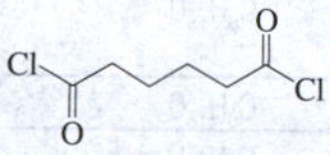

5842
Hexanedioyl dichloride

5843
1,6-Hexanedithiol

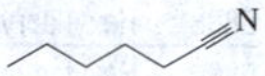

5844
Hexanenitrile

5845
1-Hexanethiol

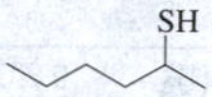

5846
2-Hexanethiol

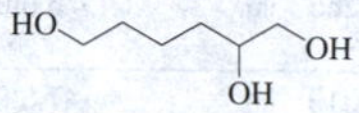

5847
1,2,6-Hexanetriol

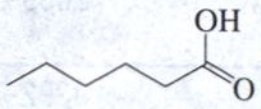

5848
Hexanoic acid

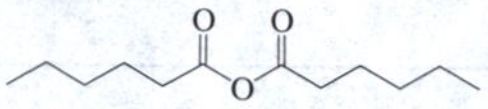

5849
Hexanoic anhydride

5850
1-Hexanol

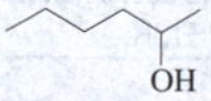

5851
2-Hexanol

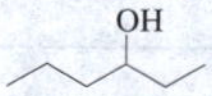

5852
3-Hexanol

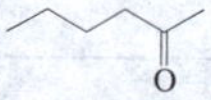

5853
2-Hexanone

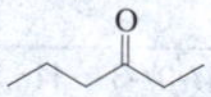

5854
3-Hexanone

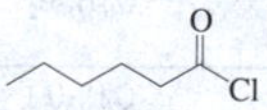

5855
Hexanoyl chloride

5856
Hexatriacontane

5857
cis-1,3,5-Hexatriene

5858
trans-1,3,5-Hexatriene

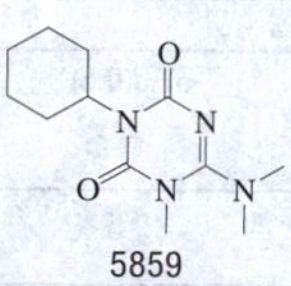

5859
Hexazinone

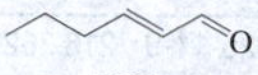

5860
trans-2-Hexenal

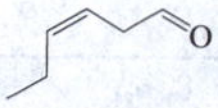

5861
cis-3-Hexenal

5862
1-Hexene

5863
cis-2-Hexene

5864
trans-2-Hexene

5865
cis-3-Hexene

5866
trans-3-Hexene

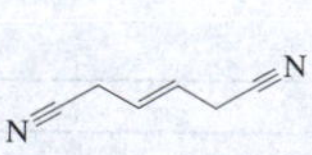

5867
trans-3-Hexenedinitrile

5868
2-Hexenoic acid

5869
3-Hexenoic acid

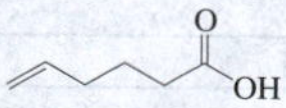

5870
5-Hexenoic acid

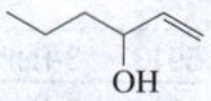

5871
1-Hexen-3-ol

No.	Name	Synonym	Mol. form.	CAS Reg. No.	Mol. wt.	Physical form	mp/°C	bp/°C	den/g cm⁻³	n_D	Solubility
5872	*cis*-2-Hexen-1-ol		$C_6H_{12}O$	928-94-9	100.158			157	0.8472^{20}	1.4397^{20}	s H_2O; vs EtOH; s eth, ace; sl ctc
5873	*trans*-2-Hexen-1-ol		$C_6H_{12}O$	928-95-0	100.158			172(8)	0.8490^{16}	1.4340^{20}	
5874	*cis*-3-Hexen-1-ol		$C_6H_{12}O$	928-96-1	100.158			157(9)	0.8478^{22}	1.4380^{20}	s H_2O; vs EtOH, eth
5875	*trans*-3-Hexen-1-ol		$C_6H_{12}O$	928-97-2	100.158			153(7)		1.4374^{20}	
5876	*trans*-4-Hexen-1-ol		$C_6H_{12}O$	928-92-7	100.158			159	0.8513^{20}	1.4402^{20}	
5877	4-Hexen-2-ol		$C_6H_{12}O$	52387-50-5	100.158			137.5	0.8405^{18}	1.4392^{20}	sl H_2O
5878	5-Hexen-2-ol		$C_6H_{12}O$	626-94-8	100.158			139(6)	0.842^{16}		sl H_2O
5879	*cis*-3-Hexen-1-ol, acetate		$C_8H_{14}O_2$	3681-71-8	142.196	liq		66^{12}			
5880	*trans*-2-Hexen-1-ol, acetate		$C_8H_{14}O_2$	2497-18-9	142.196	liq		165(11)	0.898	1.4270^{20}	
5881	5-Hexen-2-one		$C_6H_{10}O$	109-49-9	98.142			129.1(0.5)	0.833^{27}	1.4178^{27}	
5882	4-Hexen-3-one		$C_6H_{10}O$	2497-21-4	98.142			138(7)	0.8559^{20}	1.4388^{20}	s EtOH, eth; vs ace
5883	Hexestrol		$C_{18}H_{22}O_2$	84-16-2	270.367	nd (bz)	186.5				vs ace, eth, EtOH
5884	Hexobarbital		$C_{12}H_{16}N_2O_3$	56-29-1	236.266		145(1)				
5885	Hexocyclium methyl sulfate		$C_{21}H_{36}N_2O_5S$	115-63-9	428.586	cry	205				sl chl; i eth
5886	Hexyl acetate		$C_8H_{16}O_2$	142-92-7	144.212	liq	-61.0(0.2)	171.1(0.7)	0.8779^{15}	1.4092^{20}	i H_2O; vs eth, EtOH
5887	*sec*-Hexyl acetate	4-Methyl-2-pentyl acetate	$C_8H_{16}O_2$	108-84-9	144.212			147.5	0.8805^{25}	1.3980^{20}	sl H_2O; vs eth, EtOH
5888	Hexyl acrylate		$C_9H_{16}O_2$	2499-95-8	156.222		-45	40^{1}	0.878^{20}		
5889	Hexylamine	1-Hexanamine	$C_6H_{15}N$	111-26-2	101.190	liq	-21(1)	132(1)	0.7660^{20}	1.4180^{20}	sl H_2O; msc EtOH, eth; s chl
5890	Hexylbenzene		$C_{12}H_{18}$	1077-16-3	162.271	liq	-63.4(0.2)	226(2)	0.8575^{20}	1.4864^{20}	i H_2O; msc eth; s bz, peth
5891	4-Hexyl-1,3-benzenediol	4-Hexylresorcinol	$C_{12}H_{18}O_2$	136-77-6	194.270	nd (bz)	68.3(0.2)	334			vs ace, eth, EtOH, chl
5892	Hexyl benzoate		$C_{13}H_{18}O_2$	6789-88-4	206.281			272	0.9793^{20}		i H_2O; s EtOH, ace
5893	Hexyl butanoate		$C_{10}H_{20}O_2$	2639-63-6	172.265	liq	-78.0(0.5)	207(3)	0.8652^{20}	1.4160^{15}	i H_2O; s EtOH; sl chl
5894	Hexylcyclohexane		$C_{12}H_{24}$	4292-75-5	168.319	liq	-47.5(0.2)	225(1)	0.8076^{20}	1.4462^{20}	
5895	Hexylcyclopentane		$C_{11}H_{22}$	4457-00-5	154.293	liq	-73	206(10)	0.7965^{20}	1.4392^{20}	vs ace, bz, eth, EtOH
5896	2-Hexyldecanoic acid		$C_{16}H_{32}O_2$	25354-97-6	256.424	visc oil		$145^{0.02}$		1.4432^{24}	
5897	Hexyl formate		$C_7H_{14}O_2$	629-33-4	130.185	liq	-62.6(0.4)	154(5)	0.8813^{20}	1.4071^{20}	i H_2O; msc EtOH, eth
5898	Hexyl hexanoate	Hexyl caproate	$C_{12}H_{24}O_2$	6378-65-0	200.318	liq	-55.2(0.4)	241(4)	0.865^{18}	1.4264^{15}	vs ace, bz, eth, EtOH
5899	Hexyl isocyanate		$C_7H_{13}NO$	2525-62-4	127.184			44^{7}			
5900	Hexyl methacrylate		$C_{10}H_{18}O_2$	142-09-6	170.249			162	0.880^{25}	1.429^{25}	vs ace, bz, eth, EtOH
5901	Hexyl methyl ether		$C_7H_{16}O$	4747-07-3	116.201			125(4)			
5902	1-Hexylnaphthalene		$C_{16}H_{20}$	2876-53-1	212.330	liq	-18	322(9)	0.9566^{20}	1.5647^{20}	
5903	Hexyl octanoate		$C_{14}H_{28}O_2$	1117-55-1	228.371	liq	-31(1)	277(5)	0.8603^{20}	1.4323^{25}	i H_2O; s EtOH, eth, ace
5904	4-(Hexyloxy)benzoic acid		$C_{13}H_{18}O_3$	1142-39-8	222.280	cry	106				
5905	2-(Hexyloxy)ethanol	Ethylene glycol monohexyl ether	$C_8H_{18}O_2$	112-25-4	146.228	liq	-45.1	208(2)	0.8878^{20}	1.4291^{20}	sl H_2O; vs EtOH, eth
5906	Hexyl pentanoate		$C_{11}H_{22}O_2$	1117-59-5	186.292	liq	-63.0(0.5)	225(4)	0.8635^{20}	1.4228^{15}	vs ace, eth, EtOH
5907	4-Hexylphenol		$C_{12}H_{18}O$	2446-69-7	178.270			148^{9}			
5908	Hexyl propanoate		$C_9H_{18}O_2$	2445-76-3	158.238	liq	-57.5(0.5)	188(3)	0.8698^{20}	1.4162^{15}	i H_2O; s EtOH, eth, ace, AcOEt
5909	1-Hexyl-1,2,3,4-tetrahydro-naphthalene		$C_{16}H_{24}$	66325-11-9	216.362	liq		305	0.9176^{25}	1.5127^{25}	
5910	1-Hexyne	Butylacetylene	C_6H_{10}	693-02-7	82.143	liq	-132.1(0.4)	71.2(0.3)	0.7155^{25}	1.3989^{20}	i H_2O; s EtOH, eth, bz, chl; sl ctc
5911	2-Hexyne	1-Methyl-2-propylacetylene	C_6H_{10}	764-35-2	82.143	liq	-89.5(0.4)	84.3(0.5)	0.7315^{20}	1.4138^{20}	i H_2O; msc EtOH, eth; s bz, chl, peth
5912	3-Hexyne	Diethylacetylene	C_6H_{10}	928-49-4	82.143	liq	-104(3)	81.5(0.6)	0.7231^{20}	1.4115^{20}	i H_2O; s EtOH, eth, bz, chl, peth
5913	3-Hexyne-2,5-diol		$C_6H_{10}O_2$	3031-66-1	114.142			121^{15}	1.0180^{20}	1.4691^{20}	
5914	3-Hexyn-1-ol	3-Hexynol	$C_6H_{10}O$	1002-28-4	98.142			162	0.8982^{20}	1.4530^{20}	

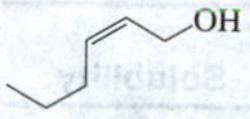
5872
cis-2-Hexen-1-ol

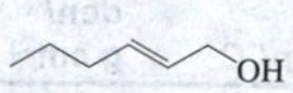
5873
trans-2-Hexen-1-ol

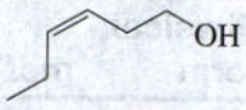
5874
cis-3-Hexen-1-ol

5875
trans-3-Hexen-1-ol

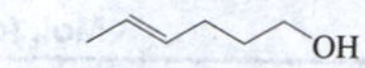
5876
trans-4-Hexen-1-ol

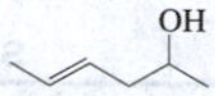
5877
4-Hexen-2-ol

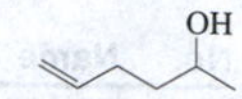
5878
5-Hexen-2-ol

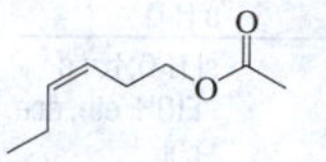
5879
cis-3-Hexen-1-ol, acetate

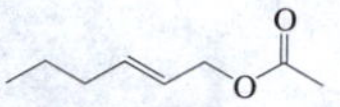
5880
trans-2-Hexen-1-ol, acetate

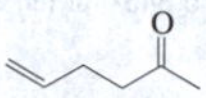
5881
5-Hexen-2-one

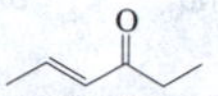
5882
4-Hexen-3-one

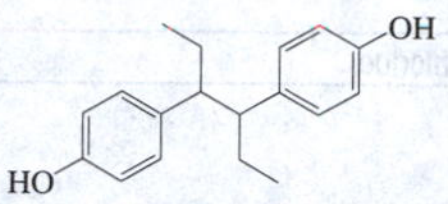
5883
Hexestrol

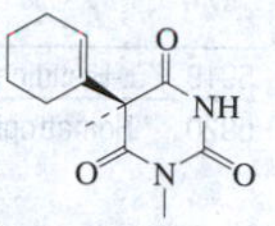
5884
Hexobarbital

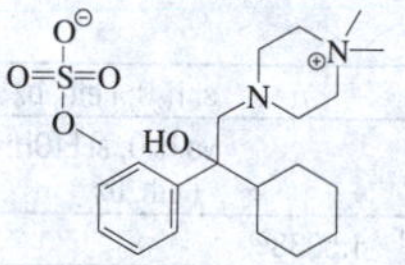
5885
Hexocyclium methyl sulfate

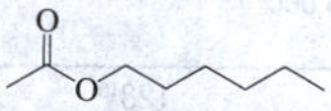
5886
Hexyl acetate

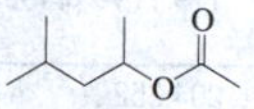
5887
sec-Hexyl acetate

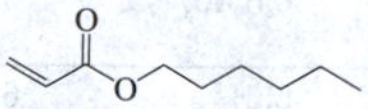
5888
Hexyl acrylate

5889
Hexylamine

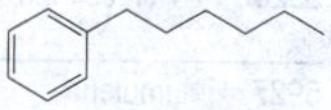
5890
Hexylbenzene

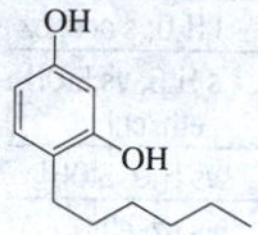
5891
4-Hexyl-1,3-benzenediol

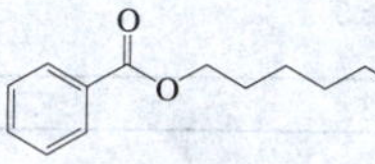
5892
Hexyl benzoate

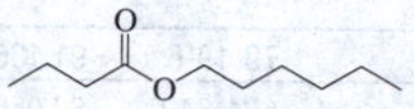
5893
Hexyl butanoate

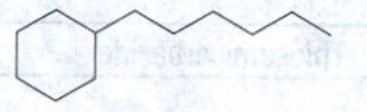
5894
Hexylcyclohexane

5895
Hexylcyclopentane

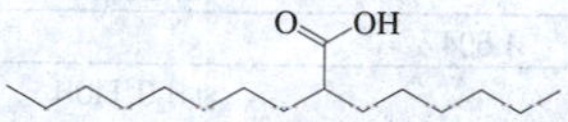
5896
2-Hexyldecanoic acid

5897
Hexyl formate

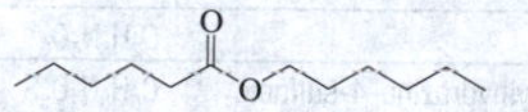
5898
Hexyl hexanoate

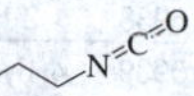
5899
Hexyl isocyanate

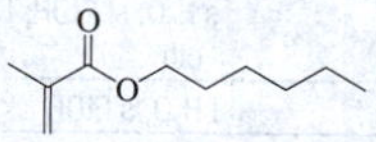
5900
Hexyl methacrylate

5901
Hexyl methyl ether

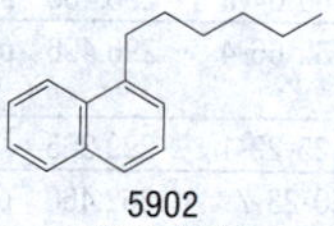
5902
1-Hexylnaphthalene

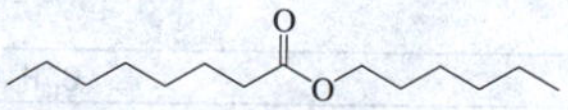
5903
Hexyl octanoate

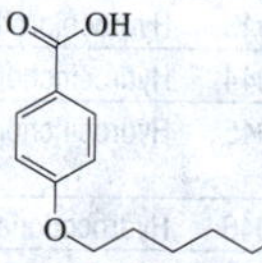
5904
4-(Hexyloxy)benzoic acid

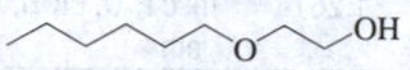
5905
2-(Hexyloxy)ethanol

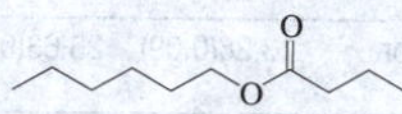
5906
Hexyl pentanoate

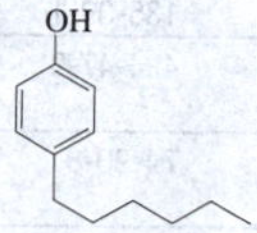
5907
4-Hexylphenol

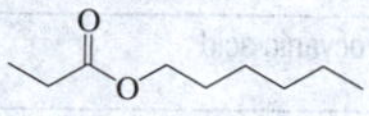
5908
Hexyl propanoate

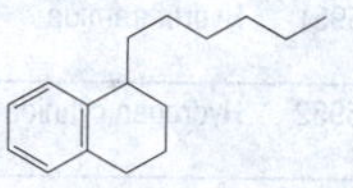
5909
1-Hexyl-1,2,3,4-tetrahydronaphthalene

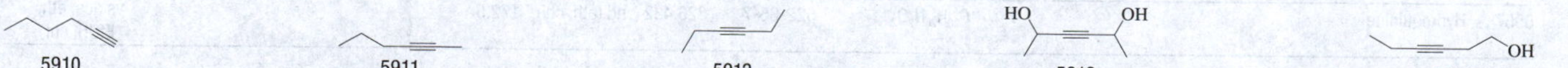
5910
1-Hexyne

5911
2-Hexyne

5912
3-Hexyne

5913
3-Hexyne-2,5-diol

5914
3-Hexyn-1-ol

No.	Name	Synonym	Mol. form.	CAS Reg. No.	Mol. wt.	Physical form	mp/°C	bp/°C	den/ g cm⁻³	n_D	Solubility
5915	1-Hexyn-3-ol		$C_6H_{10}O$	105-31-7	98.142	liq	-80	142	0.8704^{20}	1.4340^{25}	s ctc
5916	5-Hexyn-2-one		C_6H_8O	2550-28-9	96.127			149	0.9065^{20}	1.4366^{20}	
5917	Histamine		$C_5H_9N_3$	51-45-6	111.145	wh nd (chl)	83	209^{18}			s H_2O, EtOH, chl; sl eth
5918	L-Histidine	Glyoxaline-5-alanine	$C_6H_9N_3O_2$	71-00-1	155.154	nd or pl (dil al)	287 dec				s H_2O; sl EtOH; i eth, ace, bz, chl
5919	L-Histidine, monohydrochloride		$C_6H_{10}ClN_3O_2$	645-35-2	191.615		245 dec				s H_2O
5920	Homatropine		$C_{16}H_{21}NO_3$	87-00-3	275.343	pr (al, eth)	99.5				sl H_2O, bz; s EtOH, eth, ace, chl
5921	Homatropine hydrobromide	Tropanol mandelate	$C_{16}H_{22}BrNO_3$	51-56-9	356.255	orth pym or pl (w)	217 dec				vs H_2O, EtOH
5922	Homochlorocyclizine		$C_{19}H_{23}ClN_2$	848-53-3	314.852	oil		$177^{0.8}$			
5923	DL-Homocysteine	DL-2-Amino-4-mercaptobutanoic acid	$C_4H_9NO_2S$	454-29-5	135.185		272 dec				s H_2O; i eth, bz
5924	L-Homocysteine	L-2-Amino-4-mercaptobutanoic acid	$C_4H_9NO_2S$	6027-13-0	135.185	platelets	232				
5925	Homocystine		$C_8H_{16}N_2O_4S_2$	870-93-9	268.354		264				sl H_2O; i eth, bz
5926	L-Homoserine	2-Amino-4-hydroxybutanoic acid, (*S*)	$C_4H_9NO_3$	672-15-1	119.119	pr (90% al)	203 dec				vs H_2O; sl EtOH; i eth, bz
5927	Humulene		$C_{15}H_{24}$	6753-98-6	204.352			123^{10}	0.8905^{20}	1.5038^{20}	
5928	Humulon		$C_{21}H_{30}O_5$	26472-41-3	362.460	ye cry (eth)	66.5				sl H_2O; s EtOH, eth, ace, bz, alk
5929	Hydralazine	1-Hydrazinophthalazine	$C_8H_8N_4$	86-54-4	160.177	ye cry (MeOH)	172				s acid
5930	Hydramethylnon		$C_{25}H_{24}F_6N_4$	67485-29-4	494.476		193.3(0.5)				
5931	Hydrastine		$C_{21}H_{21}NO_6$	118-08-1	383.395	ye pr (al)	132				i H_2O; s ace, bz
5932	Hydrastinine		$C_{11}H_{13}NO_3$	6592-85-4	207.226	nd (lig), cry (eth)	116.5				s H_2O; vs EtOH, eth, chl
5933	Hydrazinecarbothioamide	Thiosemicarbazide	CH_5N_3S	79-19-6	91.136	lo nd (w)	183				vs H_2O, EtOH
5934	Hydrazinecarboxaldehyde		CH_4N_2O	624-84-0	60.055	ye lf or nd (al)	54				vs bz, eth, EtOH, chl
5935	Hydrazinecarboxamide		CH_5N_3O	57-56-7	75.070	pr (al)	96		1.484^{8}		vs H_2O; s EtOH; i eth, bz, chl
5936	Hydrazinecarboximidamide	Aminoguanidine	CH_6N_4	79-17-4	74.086	cry	dec				vs H_2O, EtOH
5937	1,2-Hydrazinedicarboxaldehyde		$C_2H_4N_2O_2$	628-36-4	88.065	pr (al)	161.0				vs H_2O; sl EtOH, DMSO; i eth
5938	1,2-Hydrazinedicarboxamide		$C_2H_6N_4O_2$	110-21-4	118.095	pl (w)	248(1)		1.604^{17}		
5939	4-Hydrazinobenzenesulfonic acid	Phenylhydrazine-4-sulfonic acid	$C_6H_8N_2O_3S$	98-71-5	188.204	nd, lf (w)	286				sl H_2O, EtOH
5940	4-Hydrazinobenzoic acid		$C_7H_8N_2O_2$	619-67-0	152.151	ye nd or pl (w)	221 dec				sl H_2O; i eth
5941	2-Hydrazinoethanol		$C_2H_8N_2O$	109-84-2	76.097	liq	-70	219	1.119^{25}		vs H_2O, EtOH, MeOH
5942	Hydrindantin		$C_{18}H_{10}O_6$	5103-42-4	322.268	pr (ace)	250 dec				
5943	Hydrochlorothiazide		$C_7H_8ClN_3O_4S_2$	58-93-5	297.740		274				
5944	Hydrocinchonidine		$C_{19}H_{24}N_2O$	485-64-3	296.406	lf (al)	229				vs EtOH
5945	Hydrocinchonine		$C_{19}H_{24}N_2O$	485-65-4	296.406	pr	268.5				s H_2O; sl EtOH; i eth
5946	Hydrocodone		$C_{18}H_{21}NO_3$	125-29-1	299.365		198				i H_2O; s EtOH
5947	Hydrocortisone		$C_{21}H_{30}O_5$	50-23-7	362.460	pl (al or i-PrOH)	220				sl H_2O; s EtOH, diox, HOAc
5948	Hydrocortisone 21-acetate	Cortisol acetate	$C_{23}H_{32}O_6$	50-03-3	404.496		223 dec		1.289^{20}		
5949	Hydrocotarnine		$C_{12}H_{15}NO_3$	550-10-7	221.252		56				i H_2O; s EtOH, eth, ace, bz, chl
5950	Hydroflumethiazide		$C_8H_8F_3N_3O_4S_2$	135-09-1	331.293		270.5				
5951	Hydrofuramide		$C_{15}H_{12}N_2O_3$	494-47-3	268.267	nd (al)	117				i H_2O; vs EtOH, eth
5952	Hydrogen cyanide	Hydrocyanic acid	CHN	74-90-8	27.026	vol liq or gas	-13.28(0.09)	25.63(0.04)	0.6876^{20}	1.2614^{20}	msc H_2O, EtOH, eth
5953	Hydrohydrastinine		$C_{11}H_{13}NO_2$	494-55-3	191.227	nd (lig), cry (peth)	66	303			vs ace, bz, eth, EtOH
5954	Hydromorphone	7,8-Dihydromorphin-6-one	$C_{17}H_{19}NO_3$	466-99-9	285.338	cry (EtOH)	266.5				
5955	Hydroprene		$C_{17}H_{30}O_2$	41096-46-2	266.419			174^{19}	0.8955^{20}		
5956	Hydroquinidine		$C_{20}H_{26}N_2O_2$	1435-55-8	326.432	nd (al)	168.5				s EtOH, eth, ace, chl
5957	Hydroquinine		$C_{20}H_{26}N_2O_2$	522-66-7	326.432	nd (eth, chl)	172.5				vs ace, eth, EtOH, chl

Organic

5915
1-Hexyn-3-ol

5916
5-Hexyn-2-one

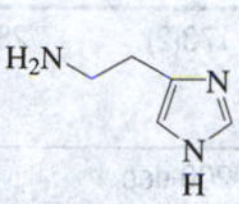

5917
Histamine

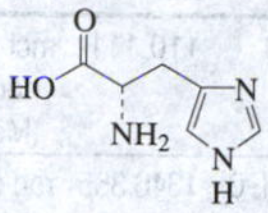

5918
L-Histidine

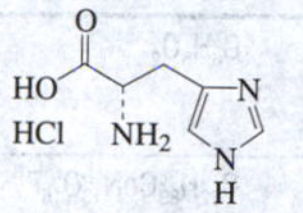

5919
L-Histidine, monohydrochloride

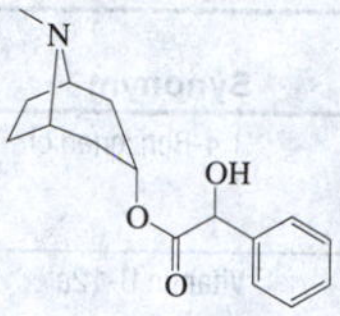

5920
Homatropine

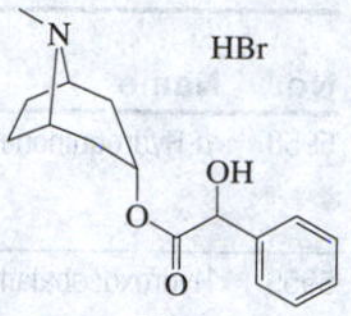

5921
Homatropine hydrobromide

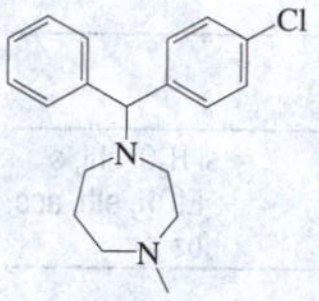

5922
Homochlorocyclizine

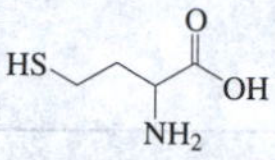

5923
DL-Homocysteine

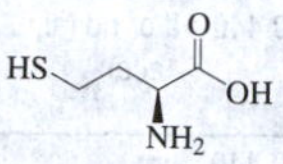

5924
L-Homocysteine

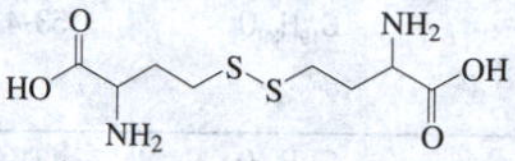

5925
Homocystine

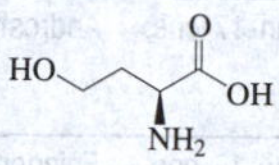

5926
L-Homoserine

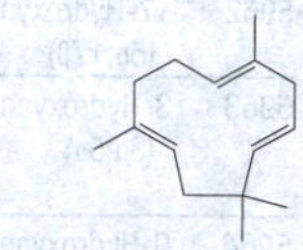

5927
Humulene

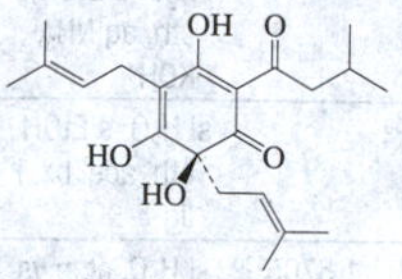

5928
Humulon

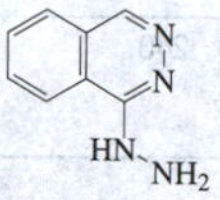

5929
Hydralazine

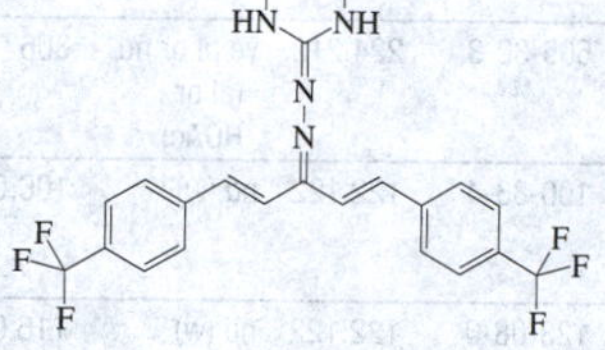

5930
Hydramethylnon

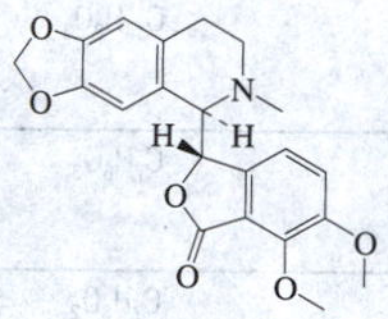

5931
Hydrastine

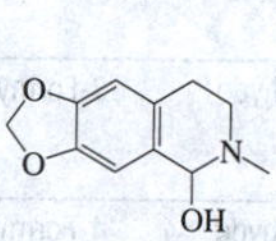

5932
Hydrastinine

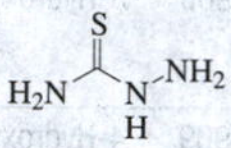

5933
Hydrazinecarbothioamide

5934
Hydrazinecarboxaldehyde

5935
Hydrazinecarboxamide

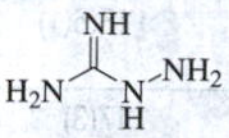

5936
Hydrazinecarboximidamide

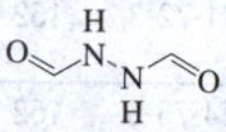

5937
1,2-Hydrazinedicarboxaldehyde

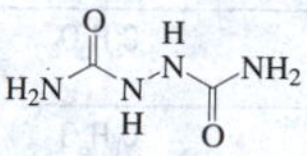

5938
1,2-Hydrazinedicarboxamide

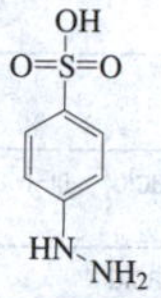

5939
4-Hydrazinobenzenesulfonic acid

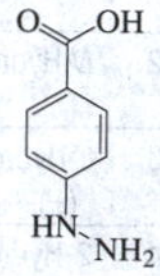

5940
4-Hydrazinobenzoic acid

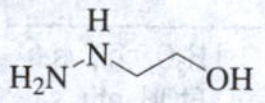

5941
2-Hydrazinoethanol

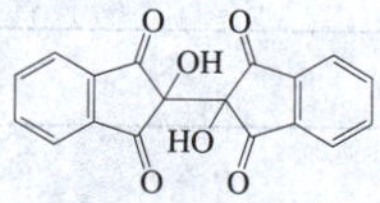

5942
Hydrindantin

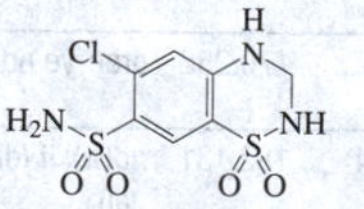

5943
Hydrochlorothiazide

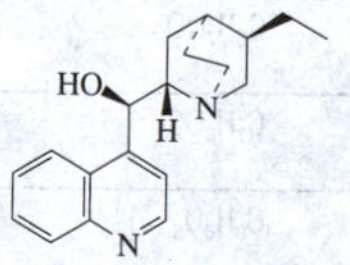

5944
Hydrocinchonidine

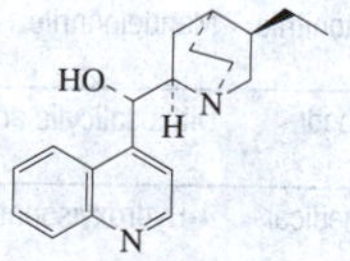

5945
Hydrocinchonine

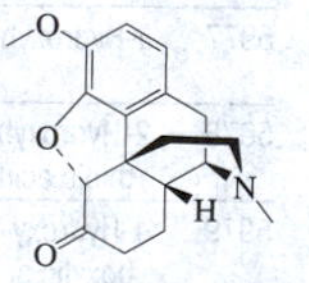

5946
Hydrocodone

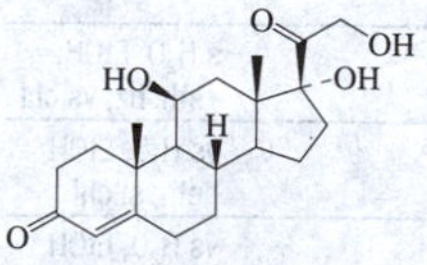

5947
Hydrocortisone

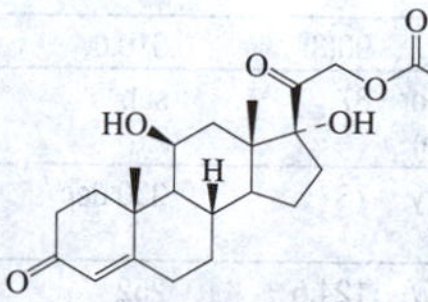

5948
Hydrocortisone 21-acetate

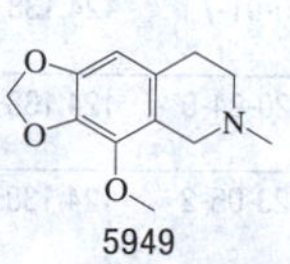

5949
Hydrocotarnine

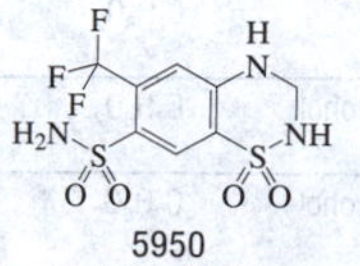

5950
Hydroflumethiazide

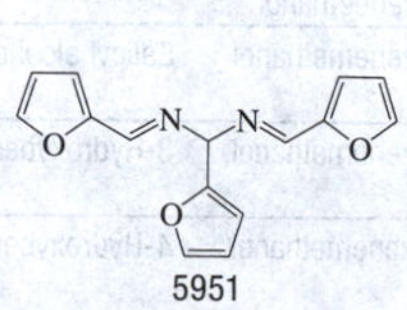

5951
Hydrofuramide

$HC \equiv N$

5952
Hydrogen cyanide

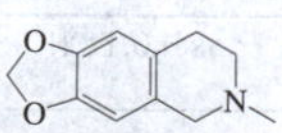

5953
Hydrohydrastinine

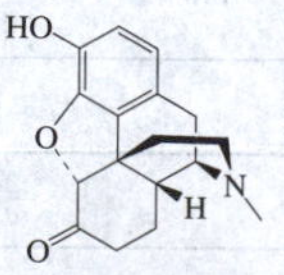

5954
Hydromorphone

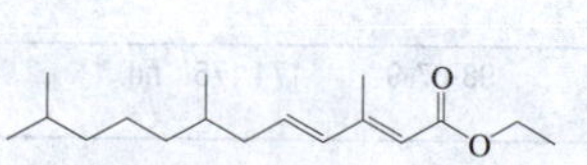

5955
Hydroprene

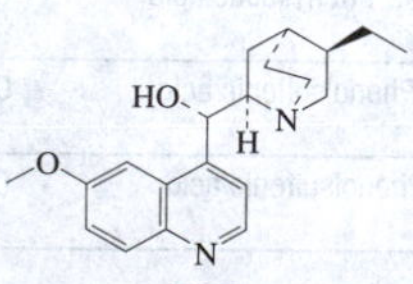

5956
Hydroquinidine

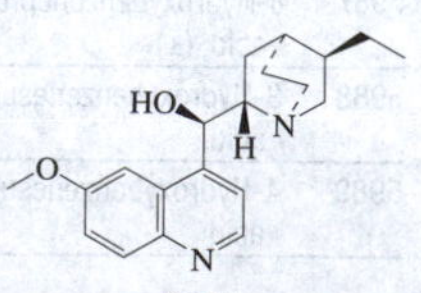

5957
Hydroquinine

No.	Name	Synonym	Mol. form.	CAS Reg. No.	Mol. wt.	Physical form	mp/°C	bp/°C	den/g cm^{-3}	n_D	Solubility
5958	*p*-Hydroquinone	1,4-Benzenediol	$C_6H_6O_2$	123-31-9	110.111	mcl pr (sub) nd(w) pr (MeOH)	173(2)	288(5)	1.330^{20}	1.632^{25}	s H_2O, eth; vs EtOH, ace; i bz
5959	Hydroxocobalamin	Vitamin B-12a	$C_{62}H_{89}CoN_{13}O_{15}P$	13422-51-0	1346.355	red cry (ace aq)	200 dec				s H_2O, EtOH; i ace, eth, bz
5960	Hydroxyacetonitrile	Glyconitrile	C_2H_3NO	107-16-4	57.051		<-72	183 dec		1.4117^{19}	vs H_2O, EtOH, eth; i bz, chl
5961	(Hydroxyacetyl)benzene		$C_8H_8O_2$	582-24-1	136.149	hex pl (al), pl (w or dil al)	90	125^{12}	1.0963^{99}		s H_2O, EtOH, eth, chl; sl lig
5962	17-Hydroxyandrostan-3-one, (5α,17β)	Stanolone	$C_{19}H_{30}O_2$	521-18-6	290.440		181	135 sub			
5963	3-Hydroxyandrostan-17-one, (3α,5α)	Androsterone	$C_{19}H_{30}O_2$	53-41-8	290.440	lf or nd (al, ace)	185				sl H_2O, chl; s EtOH, eth, ace, bz
5964	3-Hydroxyandrostan-17-one, (3β,5α)	Epiandrosterone	$C_{19}H_{30}O_2$	481-29-8	290.440	cry (bz-peth, ace)	178				
5965	17-Hydroxyandrost-4-en-3-one, (17β)	Testosterone	$C_{19}H_{28}O_2$	58-22-0	288.424	nd (dil ace)	151.0(0.3)				i H_2O; s EtOH, eth, ace
5966	1-Hydroxy-9,10-anthracenedione		$C_{14}H_8O_3$	129-43-1	224.212	red-oran nd (al)	193.8	sub			i H_2O; s EtOH, eth, bz; sl liq NH_3
5967	2-Hydroxy-9,10-anthracenedione		$C_{14}H_8O_3$	605-32-3	224.212	ye pl or nd (al or HOAc)	306	sub			i H_2O; s EtOH, eth, aq NH_3, KOH
5968	3-Hydroxybenzaldehyde	3-Formylphenol	$C_7H_6O_2$	100-83-4	122.122	nd (w)	106.0(0.2)	240	1.1179^{130}		sl H_2O; s EtOH, eth, ace, bz; i lig
5969	4-Hydroxybenzaldehyde	4-Formylphenol	$C_7H_6O_2$	123-08-0	122.122	nd (w)	116.0(0.2)		1.129^{130}	1.5705^{130}	sl H_2O, ace; vs EtOH, eth; s bz
5970	2-Hydroxybenzaldehyde, [(2-hydroxyphenyl)-methylene]hydrazone		$C_{14}H_{12}N_2O_2$	959-36-4	240.257		214				i H_2O; s EtOH, chl; vs bz, alk
5971	2-Hydroxybenzamide	Salicylamide	$C_7H_7NO_2$	65-45-2	137.137		140(2)	181.5^{14}	1.175^{140}		sl H_2O, eth, DMSO; s EtOH
5972	*N*-Hydroxybenzamide		$C_7H_7NO_2$	495-18-1	137.137	orth ta, lf (eth)	131 exp				s H_2O, EtOH; sl eth, bz
5973	α-Hydroxybenzeneacetic acid, (±)-	DL-Mandelic acid	$C_8H_8O_3$	611-72-3	152.148	orth pl	118.6(1)		1.2890^{20}		s H_2O, eth, EtOH, i-PrOH
5974	2-Hydroxybenzeneacetic acid		$C_8H_8O_3$	614-75-5	152.148		147(3)	240			sl H_2O, chl; s eth
5975	3-Hydroxybenzeneacetic acid		$C_8H_8O_3$	621-37-4	152.148	nd (bz-lig)	132	190^{11}			vs H_2O, EtOH, eth; s bz; sl lig
5976	4-Hydroxybenzeneacetic acid		$C_8H_8O_3$	156-38-7	152.148	nd (w)	150.2(0.8)	sub			sl H_2O; vs EtOH, eth
5977	α-Hydroxybenzeneacetonitrile	Mandelonitrile	C_8H_7NO	532-28-5	133.148	ye oily liq	-10		1.12		i H_2O; vs chl, eth, EtOH
5978	2-Hydroxybenzenecarbodithioic acid	Dithiosalicylic acid	$C_7H_6OS_2$	527-89-9	170.252	oran-ye nd	49				vs bz, eth, EtOH
5979	4-Hydroxy-1,3-benzenedicarboxylic acid	4-Hydroxyisophthalic acid	$C_8H_6O_5$	636-46-4	182.131	nd(w), lf (dil al)	310				i H_2O, chl; vs EtOH, eth; s HOAc
5980	5-Hydroxy-1,3-benzenedicarboxylic acid		$C_8H_6O_5$	618-83-7	182.131	nd(w+2) cr(aq-al)		sub			vs bz, eth, EtOH
5981	4-Hydroxy-1,3-benzenedisulfonic acid	Phenoldisulfonic acid	$C_6H_6O_7S_2$	96-77-5	254.238	nd (w)	>100 dec				vs H_2O, EtOH
5982	4-Hydroxybenzeneethanol		$C_8H_{10}O_2$	501-94-0	138.164		90(3)	310.0			
5983	2-Hydroxybenzenemethanol	Salicyl alcohol	$C_7H_8O_2$	90-01-7	124.138	lf (bz), nd or pl (w, eth)	87	sub	1.1613^{25}		s H_2O, EtOH, eth, bz; vs chl
5984	3-Hydroxybenzenemethanol	3-Hydroxybenzyl alcohol	$C_7H_8O_2$	620-24-6	124.138	nd (bz), cry (CCl_4)	73	300 dec	1.161^{25}		vs H_2O, EtOH, eth; sl chl
5985	4-Hydroxybenzenemethanol	4-Hydroxybenzyl alcohol	$C_7H_8O_2$	623-05-2	124.138	pr or nd (w)	124.5	252			vs H_2O, EtOH, bz, chl; s eth; sl DMSO
5986	4-Hydroxybenzenepropanoic acid	*p*-Hydroxyhydrocinnamic acid	$C_9H_{10}O_3$	501-97-3	166.173		129.3(0.5)	209^{14}			s H_2O, EtOH, eth, bz; i CS_2
5987	α-Hydroxybenzenepropanoic acid, (±)-	(±)-3-Phenyllactic acid	$C_9H_{10}O_3$	828-01-3	166.173	cry (chl, bz), pr (w)	98	149^{15}			vs H_2O, ace, eth, EtOH
5988	3-Hydroxybenzenesulfonic acid	*m*-Phenolsulfonic acid	$C_6H_6O_4S$	585-38-6	174.175	nd (w+2)					
5989	4-Hydroxybenzenesulfonic acid	*p*-Phenolsulfonic acid	$C_6H_6O_4S$	98-67-9	174.175	nd					vs H_2O, EtOH

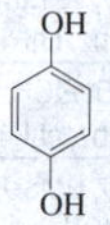
5958
p-Hydroquinone

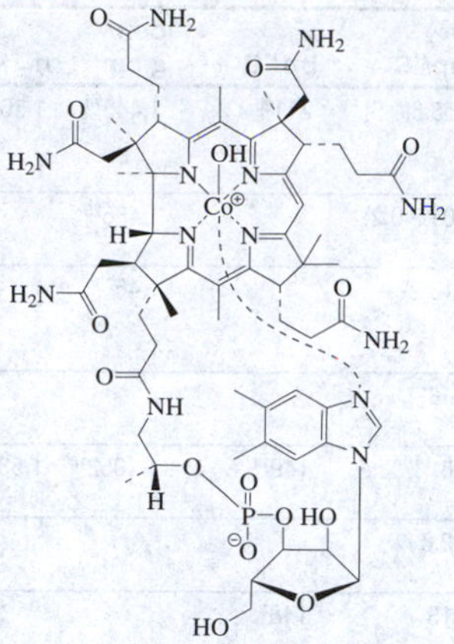
5959
Hydroxocobalamin

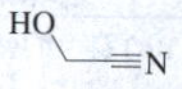
5960
Hydroxyacetonitrile

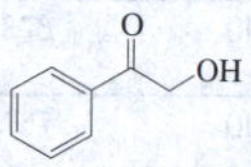
5961
(Hydroxyacetyl)benzene

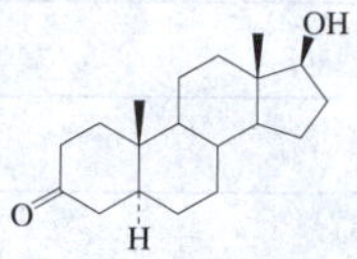
5962
17-Hydroxyandrostan-3-one, (5α,17β)

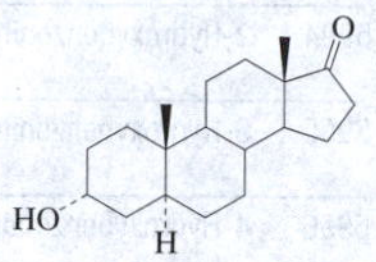
5963
3-Hydroxyandrostan-17-one, (3α,5α)

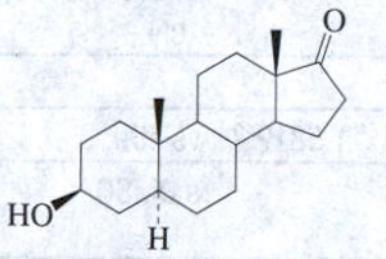
5964
3-Hydroxyandrostan-17-one, (3β,5α)

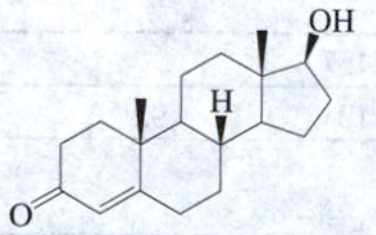
5965
17-Hydroxyandrost-4-en-3-one, (17β)

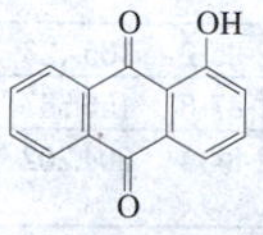
5966
1-Hydroxy-9,10-anthracenedione

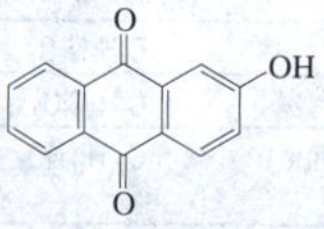
5967
2-Hydroxy-9,10-anthracenedione

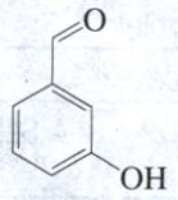
5968
3-Hydroxybenzaldehyde

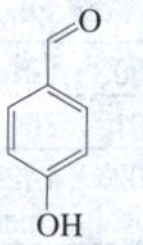
5969
4-Hydroxybenzaldehyde

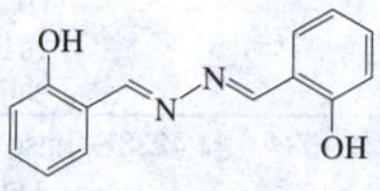
5970
2-Hydroxybenzaldehyde, [(2-hydroxyphenyl)methylene]hydrazone

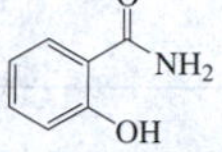
5971
2-Hydroxybenzamide

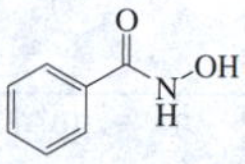
5972
N-Hydroxybenzamide

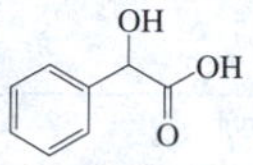
5973
α-Hydroxybenzeneacetic acid, (±)-

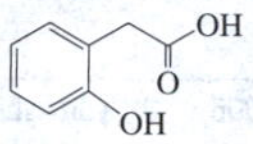
5974
2-Hydroxybenzeneacetic acid

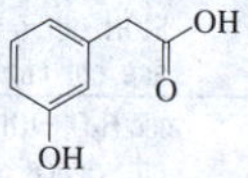
5975
3-Hydroxybenzeneacetic acid

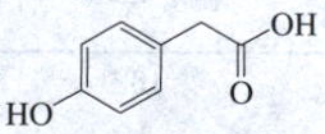
5976
4-Hydroxybenzeneacetic acid

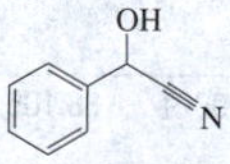
5977
α-Hydroxybenzeneacetonitrile

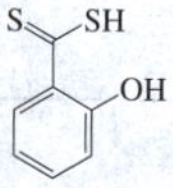
5978
2-Hydroxybenzenecarbodithioic acid

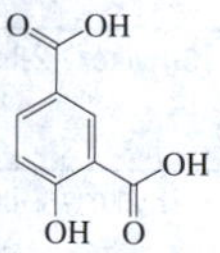
5979
4-Hydroxy-1,3-benzenedicarboxylic acid

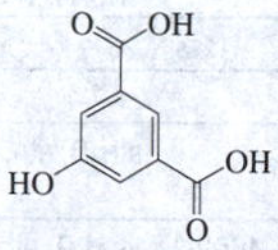
5980
5-Hydroxy-1,3-benzenedicarboxylic acid

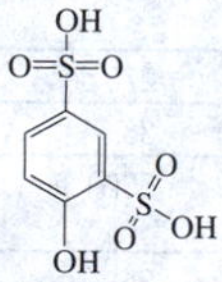
5981
4-Hydroxy-1,3-benzenedisulfonic acid

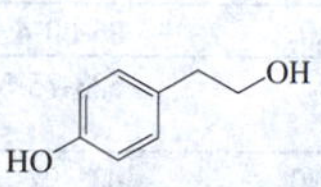
5982
4-Hydroxybenzeneethanol

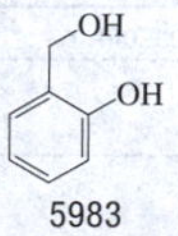
5983
2-Hydroxybenzenemethanol

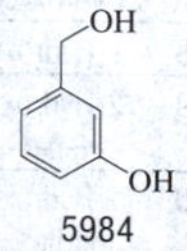
5984
3-Hydroxybenzenemethanol

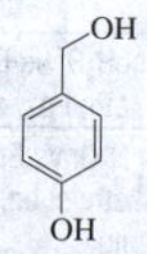
5985
4-Hydroxybenzenemethanol

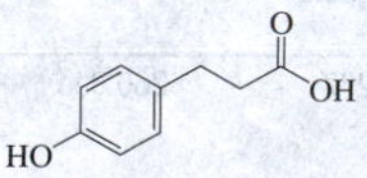
5986
4-Hydroxybenzenepropanoic acid

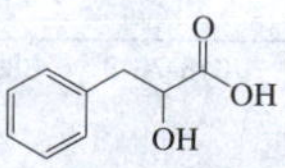
5987
α-Hydroxybenzenepropanoic acid, (±)-

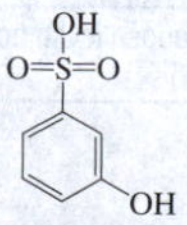
5988
3-Hydroxybenzenesulfonic acid

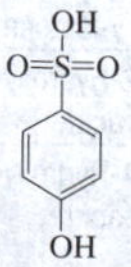
5989
4-Hydroxybenzenesulfonic acid

No.	Name	Synonym	Mol. form.	CAS Reg. No.	Mol. wt.	Physical form	mp/°C	bp/°C	den/g cm^{-3}	n_D	Solubility
5990	2-Hydroxybenzoic acid	Salicylic acid	$C_7H_6O_3$	69-72-7	138.121	nd (w), mcl pr (al)	158.6(0.5)	211^{20}	1.443^{20}	1.565	sl H_2O, bz, chl, ctc; vs EtOH, eth, ace
5991	3-Hydroxybenzoic acid		$C_7H_6O_3$	99-06-9	138.121	nd (w) pl, pr (al)	201.3(0.2)		1.485^{25}		sl H_2O; s EtOH, eth, ace; i bz
5992	4-Hydroxybenzoic acid		$C_7H_6O_3$	99-96-7	138.121	pr or pl (w, al) cry (ace)	213(2)		1.46^{25}		sl H_2O, bz; vs EtOH; s eth, ace
5993	2-Hydroxybenzoic acid, hydrazide		$C_7H_8N_2O_2$	936-02-7	152.151		148				vs bz, EtOH
5994	2-Hydroxybenzonitrile		C_7H_5NO	611-20-1	119.121		98	149^{14}	1.1052^{100}	1.5372^{100}	sl H_2O; vs EtOH, eth, bz, chl
5995	3-Hydroxybenzonitrile		C_7H_5NO	873-62-1	119.121	pr (al, eth) lf (w)	82.8				vs H_2O, EtOH, eth, bz, chl
5996	4-Hydroxybenzonitrile		C_7H_5NO	767-00-0	119.121	lf (w)	113	148^{1}			sl H_2O, DMSO; vs EtOH, eth, chl
5997	4-Hydroxybenzophenone	4-Hydroxyphenyl phenyl ketone	$C_{13}H_{10}O_2$	1137-42-4	198.217	nd (al), pr (dil al)	135		1.133^{172}		sl H_2O; vs EtOH, eth, HOAc
5998	4-Hydroxy-2*H*-1-benzopyran-2-one		$C_9H_6O_3$	1076-38-6	162.142	nd (w)	213.5				s H_2O, EtOH, eth; sl DMSO
5999	7-Hydroxy-2*H*-1-benzopyran-2-one	Umbelliferone	$C_9H_6O_3$	93-35-6	162.142	nd (w)	230.5	sub			vs EtOH, HOAc, chl
6000	1-Hydroxy-1*H*-benzotriazole		$C_6H_5N_3O$	2592-95-2	135.123		157.8				
6001	2-Hydroxybenzoyl chloride		$C_7H_5ClO_2$	1441-87-8	156.567		19	92^{15}	1.3112^{20}	1.5812^{20}	vs eth
6002	4-(2-Hydroxybenzoyl)-morpholine	4-Salicyloylmorpholine	$C_{11}H_{10}NO_3$	3202-84-4	204.202						s DMSO
6003	2-Hydroxybiphenyl	[1,1'-Biphenyl]-2-ol	$C_{12}H_{10}O$	90-43-7	170.206		57.6(0.7)	281(3)	1.213^{25}		i H_2O; s EtOH, ace, bz; vs eth, py
6004	3-Hydroxybiphenyl	[1,1'-Biphenyl]-3-ol	$C_{12}H_{10}O$	580-51-8	170.206		78	>300			sl H_2O; vs EtOH, eth, bz, py; s chl
6005	4-Hydroxybiphenyl	[1,1'-Biphenyl]-4-ol	$C_{12}H_{10}O$	92-69-3	170.206		170.0(0.5)	305			sl H_2O, DMSO; vs EtOH, eth, chl, py
6006	3-Hydroxybutanal	Aldol	$C_4H_8O_2$	107-89-1	88.106			83^{20}	1.103^{20}	1.4238^{20}	msc H_2O, EtOH; s eth; vs ace
6007	2-Hydroxybutanoic acid, (±)-		$C_4H_8O_3$	600-15-7	104.105		44.2	260 dec	1.125^{20}		s H_2O, EtOH, eth
6008	3-Hydroxybutanoic acid, (±)-		$C_4H_8O_3$	625-71-8	104.105		49	130^{12}		1.4424^{20}	vs H_2O, EtOH, eth; i bz
6009	4-Hydroxybutanoic acid		$C_4H_8O_3$	591-81-1	104.105		<-17	180 dec			
6010	1-Hydroxy-2-butanone		$C_4H_8O_2$	5077-67-8	88.106			160	1.0272^{20}	1.4189^{20}	vs H_2O, EtOH, eth
6011	3-Hydroxy-2-butanone, (±)-	Acetoin	$C_4H_8O_2$	52217-02-4	88.106		15	148	1.0044^{20}	1.4171^{20}	msc H_2O; sl EtOH, eth; s ace, chl; i lig
6012	4-Hydroxy-2-butanone		$C_4H_8O_2$	590-90-9	88.106			182	1.0233^{20}	1.4585^{14}	msc H_2O, EtOH, eth; vs ace
6013	2-Hydroxy-3-butenenitrile		C_4H_5NO	5809-59-6	83.089	liq		94^{17}			
6014	4-Hydroxybutyramide		$C_4H_9NO_2$	927-60-6	103.120		52				
6015	3-Hydroxycamphor	3-Hydroxy-1,7,7-trimethylbicyclo[2.2.1]heptan-2-one	$C_{10}H_{16}O_2$	10373-81-6	168.233	nd (bz-peth)	205.5				vs eth, EtOH, chl
6016	3-Hydroxycholan-24-oic acid, (3α,5β)	Lithocholic acid	$C_{24}H_{40}O_3$	434-13-9	376.573	hex lf (al) pr (dil al)	188(4)				i H_2O, lig; s EtOH, chl, HOAc; sl eth
6017	Hydroxycodeinone		$C_{18}H_{19}NO_4$	508-54-3	313.349		275 dec				
6018	2-Hydroxycyclodecanone	Sebacoin	$C_{10}H_{18}O_2$	96-00-4	170.249	cry (peth)	38.5	136^{14}			
6019	2-Hydroxy-2,4,6-cycloheptatrien-1-one		$C_7H_6O_2$	533-75-5	122.122	nd	50(1)	40 sub			s H_2O, eth, ace
6020	1-Hydroxycyclohexanecarbonitrile		$C_7H_{11}NO$	931-97-5	125.168		35	132^{20}	1.0172^{20}	1.4693^{20}	vs H_2O, eth
6021	2-Hydroxycyclohexanone		$C_6H_{10}O_2$	533-60-8	114.142	nd (al)				1.4785^{21}	vs H_2O, EtOH; i eth, bz, peth
6022	1-(1-Hydroxycyclohexyl)-ethanone		$C_8H_{14}O_2$	1123-27-9	142.196			125.5	1.0248^{25}	1.4670^{25}	vs eth, EtOH
6023	4-Hydroxydecanoic acid γ-lactone	5-Hexyldihydro-2(3*H*)-furanone	$C_{10}H_{18}O_2$	706-14-9	170.249	liq		301(8)			
6024	2-Hydroxy-3,5-diiodobenzoic acid	3,5-Diiodosalicylic acid	$C_7H_4I_2O_3$	133-91-5	389.914	nd (al)	235.5				sl H_2O; vs EtOH, eth; i bz, chl
6025	4-Hydroxy-3,5-diiodobenzoic acid		$C_7H_4I_2O_3$	618-76-8	389.914		237	260 dec			i H_2O; vs EtOH, eth; sl bz, chl, lig

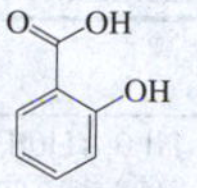
5990
2-Hydroxybenzoic acid

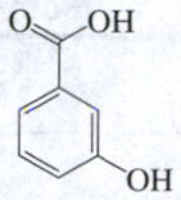
5991
3-Hydroxybenzoic acid

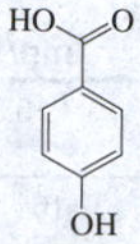
5992
4-Hydroxybenzoic acid

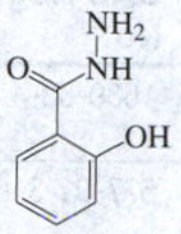
5993
2-Hydroxybenzoic acid, hydrazide

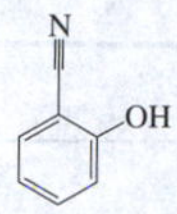
5994
2-Hydroxybenzonitrile

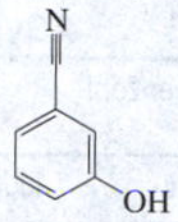
5995
3-Hydroxybenzonitrile

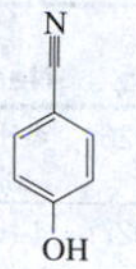
5996
4-Hydroxybenzonitrile

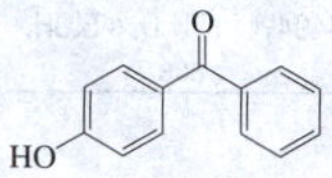
5997
4-Hydroxybenzophenone

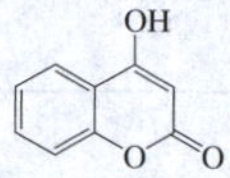
5998
4-Hydroxy-2*H*-1-benzopyran-2-one

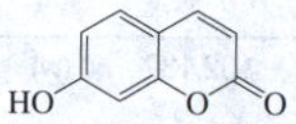
5999
7-Hydroxy-2*H*-1-benzopyran-2-one

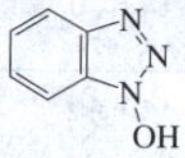
6000
1-Hydroxy-1*H*-benzotriazole

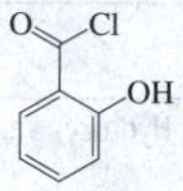
6001
2-Hydroxybenzoyl chloride

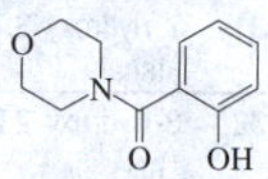
6002
4-(2-Hydroxybenzoyl)morpholine

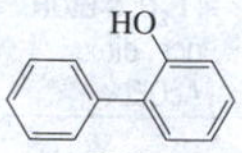
6003
2-Hydroxybiphenyl

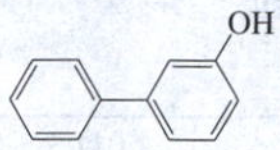
6004
3-Hydroxybiphenyl

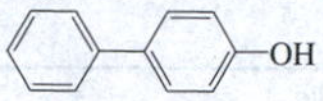
6005
4-Hydroxybiphenyl

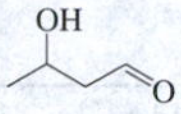
6006
3-Hydroxybutanal

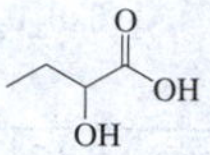
6007
2-Hydroxybutanoic acid, (±)-

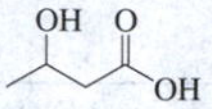
6008
3-Hydroxybutanoic acid, (±)-

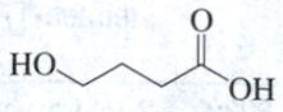
6009
4-Hydroxybutanoic acid

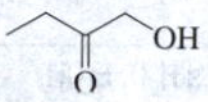
6010
1-Hydroxy-2-butanone

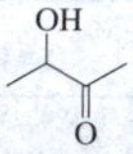
6011
3-Hydroxy-2-butanone, (±)-

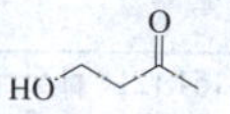
6012
4-Hydroxy-2-butanone

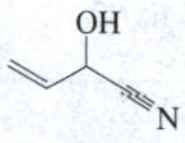
6013
2-Hydroxy-3-butenenitrile

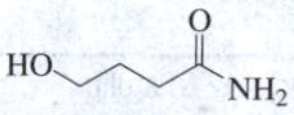
6014
4-Hydroxybutyramide

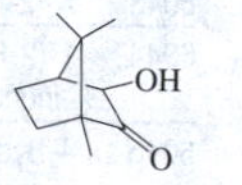
6015
3-Hydroxycamphor

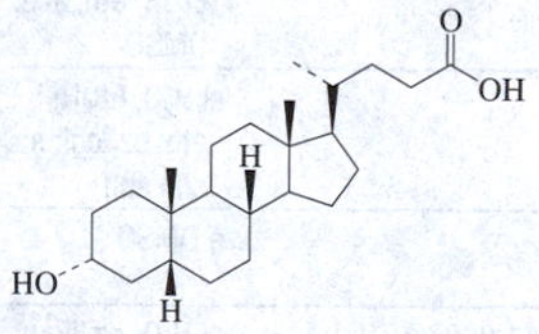
6016
3-Hydroxycholan-24-oic acid, (3α,5β)

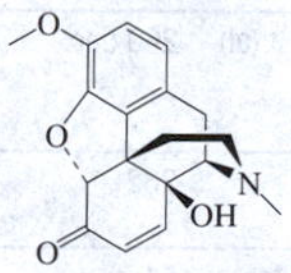
6017
Hydroxycodeinone

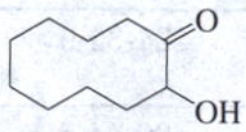
6018
2-Hydroxycyclodecanone

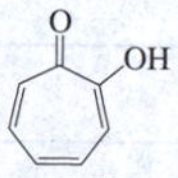
6019
2-Hydroxy-2,4,6-cycloheptatrien-1-one

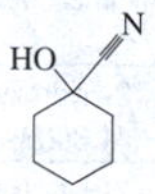
6020
1-Hydroxycyclohexanecarbonitrile

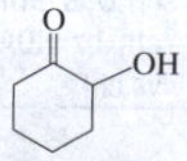
6021
2-Hydroxycyclohexanone

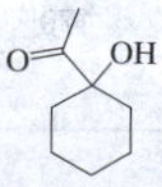
6022
1-(1-Hydroxycyclohexyl)ethanone

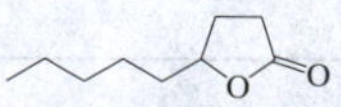
6023
4-Hydroxydecanoic acid γ-lactone

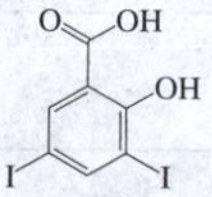
6024
2-Hydroxy-3,5-diiodobenzoic acid

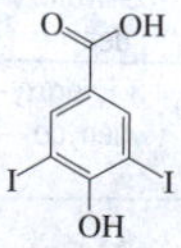
6025
4-Hydroxy-3,5-diiodobenzoic acid

No.	Name	Synonym	Mol. form.	CAS Reg. No.	Mol. wt.	Physical form	mp/°C	bp/°C	den/ g cm^{-3}	n_D	Solubility
6026	4-Hydroxy-3,5-diiodobenzonitrile		$C_7H_3I_2NO$	1689-83-4	370.914		215.8(0.5)				
6027	4-Hydroxy-3,5-diiodo-α-phenylbenzenepropanoic acid	Iodoalphionic acid	$C_{15}H_{12}I_2O_3$	577-91-3	494.063		164				i H_2O; s EtOH, eth; sl bz, chl
6028	2-Hydroxy-4,6-dimethoxybenzaldehyde		$C_9H_{10}O_4$	708-76-9	182.173		70	193^{25}			i H_2O; vs EtOH, eth, bz, chl, HOAc
6029	4-Hydroxy-3,5-dimethoxybenzaldehyde	Syringaldehyde	$C_9H_{10}O_4$	134-96-3	182.173	br nd (lig)	113	192^{14}			sl H_2O, lig; vs EtOH, eth, bz, chl
6030	4-Hydroxy-3,5-dimethoxybenzoic acid		$C_9H_{10}O_5$	530-57-4	198.172	nd (w)	204.5				sl H_2O; vs EtOH
6031	7-Hydroxy-3,7-dimethyloctanal		$C_{10}H_{20}O_2$	107-75-5	172.265			103^{3}	0.9220^{20}	1.4494^{20}	sl H_2O; s EtOH, ace
6032	3-Hydroxy-2,2-dimethylpropanal	Hydroxypivaldehyde	$C_5H_{10}O_2$	597-31-9	102.132	nd (w)	89.5	173			
6033	2-Hydroxy-3,5-dinitrobenzoic acid		$C_7H_4N_2O_7$	609-99-4	228.116	ye nd or pl (+1w)	182				s H_2O, EtOH, eth, bz
6034	11-Hydroxy-9,15-dioxoprosta-5,13-dien-1-oic acid, (5*Z*,11α,13*E*)-	15-Oxo-prostaglandin E2	$C_{20}H_{30}O_5$	26441-05-4	350.449	cry					
6035	1-Hydroxy-1,1-diphosphonoethane	Etidronic acid	$C_2H_8O_7P_2$	2809-21-4	206.028	cry (w)	105				s H_2O, EtOH, MeOH
6036	3-Hydroxyestra-1,3,5,7,9-pentaen-17-one	Equilenin	$C_{18}H_{18}O_2$	517-09-9	266.335		258.5	170 sub			sl EtOH, ace, chl
6037	3-Hydroxyestra-1,3,5(10),7-tetraen-17-one	Equilin	$C_{18}H_{20}O_2$	474-86-2	268.351	pl (AcOEt)	239	170 sub			sl H_2O; s EtOH, ace, diox, AcOEt
6038	2-Hydroxyethyl acrylate	2-Hydroxyethyl 2-propenoate	$C_5H_8O_3$	818-61-1	116.116	liq		191	1.011^{23}		
6039	*N*-(2-Hydroxyethyl)-dodecanamide		$C_{14}H_{29}NO_2$	142-78-9	243.386		88.5				
6040	*N*-(2-Hydroxyethyl)-ethylenediaminetriacetic acid		$C_{10}H_{18}N_2O_7$	150-39-0	278.259	cry	165 dec				
6041	2-Hydroxyethyl 2-hydroxybenzoate	Glycol salicylate	$C_9H_{10}O_4$	87-28-5	182.173		37	173^{15}	1.2526^{15}		sl H_2O; vs EtOH, eth, bz, chl
6042	2-Hydroxyethyl methacrylate	Ethylene glycol monomethacrylate	$C_6H_{10}O_3$	868-77-9	130.141			103^{13}	1.079^{20}	1.4515^{20}	
6043	*N*-(2-Hydroxyethyl)phthalimide		$C_{10}H_9NO_3$	3891-07-4	191.183	nd (al), lf (w)	130.3				sl H_2O
6044	1-(2-Hydroxyethyl)-2-pyrrolidinone		$C_6H_{11}NO_2$	3445-11-2	129.157		20	295	1.1435^{20}		
6045	4-Hydroxy-4*H*-furo[3,2-*c*]-pyran-2(6*H*)-one	Patulin	$C_7H_6O_4$	149-29-1	154.121	pl or pr (eth, chl)	111				s H_2O, EtOH, eth, ace, bz; i peth
6046	16-Hydroxyhexadecanoic acid	16-Hydroxypalmitic acid	$C_{16}H_{32}O_3$	506-13-8	272.423		96.5				i H_2O; s EtOH, ace; sl eth, bz
6047	2-Hydroxyhexanoic acid		$C_6H_{12}O_3$	6064-63-7	132.157	pr (eth)	60				vs H_2O
6048	6-Hydroxyhexanoic acid		$C_6H_{12}O_3$	1191-25-9	132.157	liq					
6049	3-Hydroxy-2-(hydroxymethyl)-2-methylpropanoic acid	Dimethylolpropionic acid	$C_5H_{10}O_4$	4767-03-7	134.131		195.5(0.5)				
6050	5-Hydroxy-2-(hydroxymethyl)-4*H*-pyran-4-one	Kojic acid	$C_6H_6O_4$	501-30-4	142.110	pr nd (ace)	153.5				sl H_2O, bz; s EtOH, eth, ace, DMSO
6051	8-Hydroxy-7-iodo-5-quinolinesulfonic acid	Ferron	$C_9H_6INO_4S$	547-91-1	351.118	ye pr, lf (al)	260 dec				sl H_2O, EtOH; i eth, bz, chl; s con sulf
6052	2-Hydroxy-1*H*-isoindole-1,3(2*H*)-dione		$C_8H_5NO_3$	524-38-9	163.131		232				s DMSO
6053	2-Hydroxy-4-isopropyl-2,4,6-cycloheptatrien-1-one		$C_{10}H_{12}O_2$	499-44-5	164.201	pa ye (peth)	51(2)	137^{10}	1.0606^{65}		sl H_2O, bz, lig; s ctc
6054	Hydroxylupanine		$C_{15}H_{24}N_2O_2$	15358-48-2	264.364	cry (ace)	169.5				vs H_2O, EtOH, chl
6055	*N*-Hydroxymethanamine	*N*-Methylhydroxylamine	CH_5NO	593-77-1	47.057	hyg nd	87.5	62.5^{15}	1.0003^{20}	1.4164^{20}	vs H_2O, EtOH
6056	2-Hydroxy-3-methoxybenzaldehyde		$C_8H_8O_3$	148-53-8	152.148	lt ye lf, grn nd (w, lig)	40.5(0.4)	265.5			sl H_2O, lig; vs EtOH, eth, ctc
6057	2-Hydroxy-4-methoxybenzaldehyde		$C_8H_8O_3$	673-22-3	152.148	nd (w), cry (al)	42.0				s EtOH, eth, bz, lig
6058	2-Hydroxy-5-methoxybenzaldehyde		$C_8H_8O_3$	672-13-9	152.148	ye liq (w)	4	247.5			vs eth, EtOH
6059	3-Hydroxy-4-methoxybenzaldehyde		$C_8H_8O_3$	621-59-0	152.148		117(3)	179^{15}	1.196^{25}		sl H_2O; s EtOH, eth, bz, HOAc; vs chl

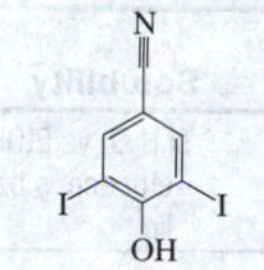

6026
4-Hydroxy-3,5-diiodobenzonitrile

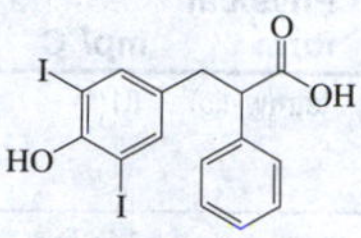

6027
4-Hydroxy-3,5-diiodo-α-phenylbenzenepropanoic acid

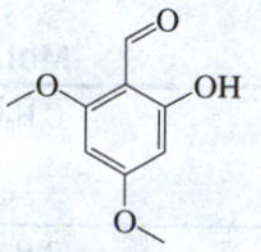

6028
2-Hydroxy-4,6-dimethoxybenzaldehyde

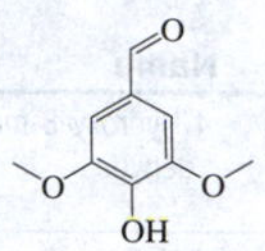

6029
4-Hydroxy-3,5-dimethoxybenzaldehyde

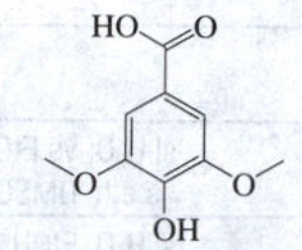

6030
4-Hydroxy-3,5-dimethoxybenzoic acid

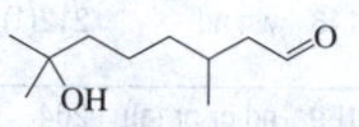

6031
7-Hydroxy-3,7-dimethyloctanal

6032
3-Hydroxy-2,2-dimethylpropanal

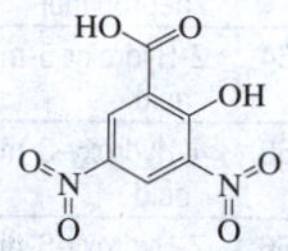

6033
2-Hydroxy-3,5-dinitrobenzoic acid

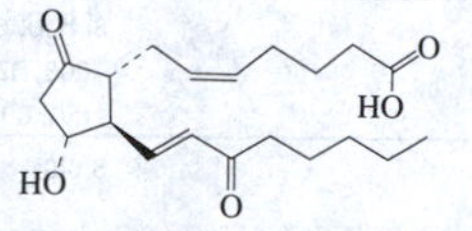

6034
11-Hydroxy-9,15-dioxoprosta-5,13-dien-1-oic acid, (5*Z*,11α,13*E*)-

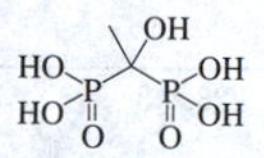

6035
1-Hydroxy-1,1-diphosphonoethane

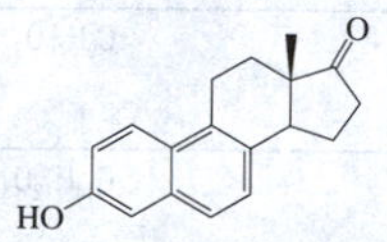

6036
3-Hydroxyestra-1,3,5,7,9-pentaen-17-one

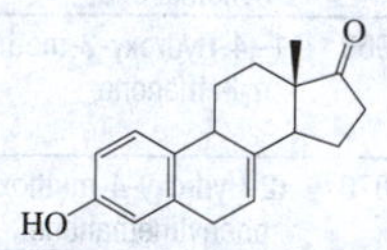

6037
3-Hydroxyestra-1,3,5(10),7-tetraen-17-one

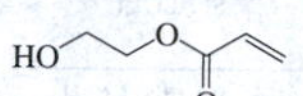

6038
2-Hydroxyethyl acrylate

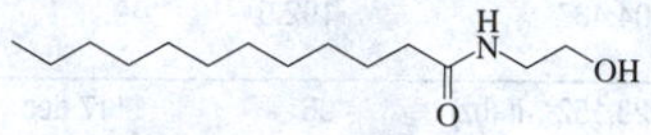

6039
N-(2-Hydroxyethyl)dodecanamide

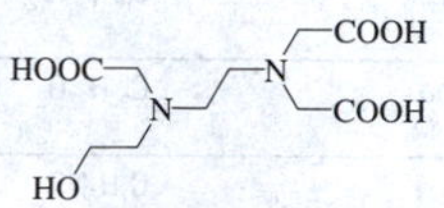

6040
N-(2-Hydroxyethyl)ethylenediaminetriacetic acid

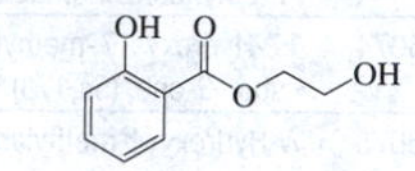

6041
2-Hydroxyethyl 2-hydroxybenzoate

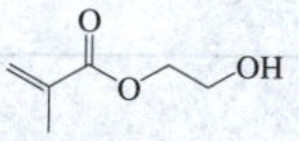

6042
2-Hydroxyethyl methacrylate

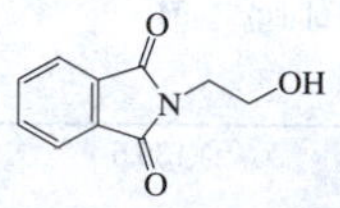

6043
N-(2-Hydroxyethyl)phthalimide

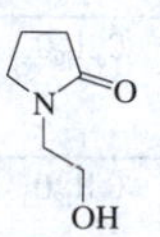

6044
1-(2-Hydroxyethyl)-2-pyrrolidinone

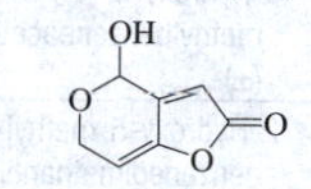

6045
4-Hydroxy-4*H*-furo[3,2-c]pyran-2(6*H*)-one

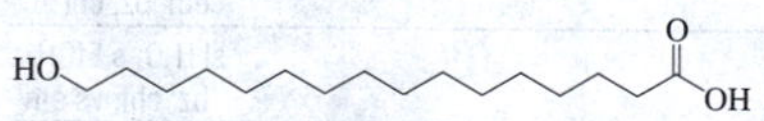

6046
16-Hydroxyhexadecanoic acid

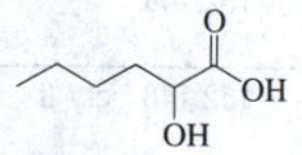

6047
2-Hydroxyhexanoic acid

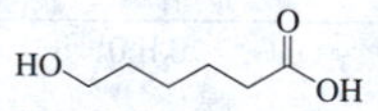

6048
6-Hydroxyhexanoic acid

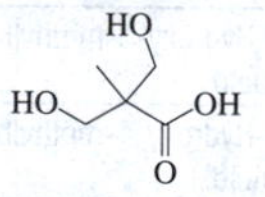

6049
3-Hydroxy-2-(hydroxymethyl)-2-methylpropanoic acid

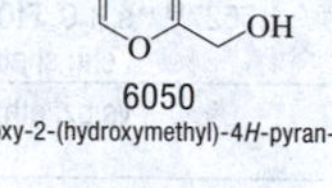

6050
5-Hydroxy-2-(hydroxymethyl)-4*H*-pyran-4-one

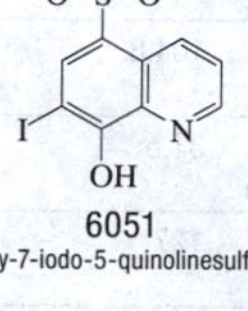

6051
8-Hydroxy-7-iodo-5-quinolinesulfonic acid

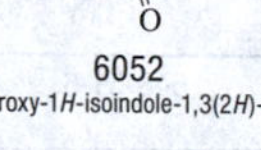

6052
2-Hydroxy-1*H*-isoindole-1,3(2*H*)-dione

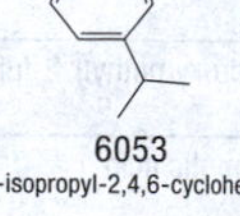

6053
2-Hydroxy-4-isopropyl-2,4,6-cycloheptatrien-1-one

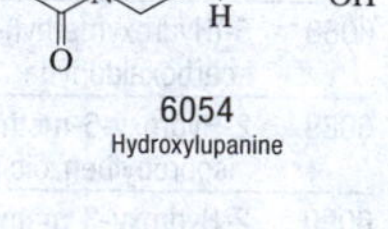

6054
Hydroxylupanine

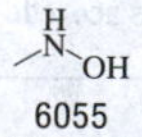

6055
N-Hydroxymethanamine

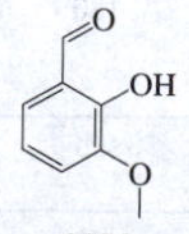

6056
2-Hydroxy-3-methoxybenzaldehyde

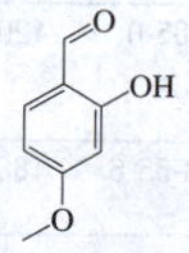

6057
2-Hydroxy-4-methoxybenzaldehyde

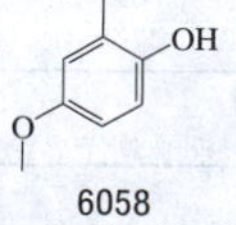

6058
2-Hydroxy-5-methoxybenzaldehyde

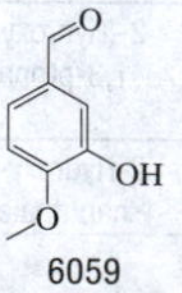

6059
3-Hydroxy-4-methoxybenzaldehyde

No.	Name	Synonym	Mol. form.	CAS Reg. No.	Mol. wt.	Physical form	mp/°C	bp/°C	den/g cm^{-3}	n_D	Solubility
6060	4-Hydroxy-3-methoxybenzaldehyde	Vanillin	$C_8H_8O_3$	121-33-5	152.148	tetr (w, lig)	81(1)	285	1.056^{25}		sl H_2O; vs EtOH, eth, ace; s bz, lig
6061	4-Hydroxy-3-methoxybenzeneacetic acid	Homovanillic acid	$C_9H_{10}O_4$	306-08-1	182.173		143.5				
6062	4-Hydroxy-3-methoxybenzenemethanol		$C_8H_{10}O_3$	498-00-0	154.163	pr (w), nd (bz)	115	dec			s H_2O, EtOH, eth, bz
6063	4-Hydroxy-3-methoxybenzenepropanol		$C_{10}H_{14}O_3$	2305-13-7	182.216		65	197^{15}		1.5545^{25}	vs eth, EtOH
6064	2-Hydroxy-5-methoxybenzoic acid		$C_8H_8O_4$	2612-02-4	168.148		142				
6065	4-Hydroxy-3-methoxybenzoic acid	Vanillic acid	$C_8H_8O_4$	121-34-6	168.148	wh nd	212(1)	sub			sl H_2O; vs EtOH; s eth, DMSO
6066	7-Hydroxy-6-methoxy-2*H*-1-benzopyran-2-one	Scopoletin	$C_{10}H_8O_4$	92-61-5	192.169	nd or pr (al)	204				sl H_2O, EtOH; s chl; i bz, CS_2
6067	4-(4-Hydroxy-3-methoxyphenyl)-2-butanone	Zingerone	$C_{11}H_{14}O_3$	122-48-5	194.227	cry (ace, eth)	40.5	187^{14}			vs eth
6068	1-(2-Hydroxy-4-methoxyphenyl)ethanone		$C_9H_{10}O_3$	552-41-0	166.173	nd (al)	52.5	158^{20}	1.3102^{81}	1.5452^{81}	vs bz, eth, EtOH, chl
6069	1-(4-Hydroxy-3-methoxyphenyl)ethanone	Apocynin	$C_9H_{10}O_3$	498-02-2	166.173	pr (w)	115	297			sl H_2O; s EtOH, ace, bz; vs eth, chl
6070	(2-Hydroxy-4-methoxyphenyl)phenylmethanone	Oxybenzone	$C_{14}H_{12}O_3$	131-57-7	228.243		65.5				s ctc
6071	3-(4-Hydroxy-3-methoxyphenyl)-2-propenal		$C_{10}H_{10}O_3$	458-36-6	178.184	cry (bz)	84		1.1562^{102}		vs bz, eth, EtOH
6072	*N*-Hydroxymethylamine hydrochloride	*N*-Methylhydroxylamine hydrochloride	CH_6ClNO	4229-44-1	83.518		83.5				
6073	4-Hydroxy-α-[(methylamino)methyl]benzenemethanol	Synephrine	$C_9H_{13}NO_2$	94-07-5	167.205		184.5				
6074	17-Hydroxy-17-methylandrostan-3-one, (5α,17β)	Mestanolone	$C_{20}H_{32}O_2$	521-11-9	304.467		192.5				sl AcOEt
6075	*N*-Hydroxy-4-methylaniline		C_7H_9NO	623-10-9	123.152	lf (bz)	96	117 dec			vs eth, EtOH, chl
6076	2-Hydroxy-5-methylbenzaldehyde		$C_8H_8O_2$	613-84-3	136.149	pl (aq, al)	55.1(0.2)	217.5	1.0913^{59}	1.547^{59}	vs eth, EtOH, chl
6077	α-(Hydroxymethyl)-benzeneacetic acid, (±)-	Tropic acid	$C_9H_{10}O_3$	552-63-6	166.173	nd, pl (al, bz, w)	118	dec			vs H_2O, eth, EtOH
6078	α-Hydroxy-α-methylbenzeneacetic acid, (±)-	Atrolactic acid	$C_9H_{10}O_3$	4607-38-9	166.173	nd, pl (lig)	94				vs ace, bz
6079	2-Hydroxy-5-methyl-1,3-benzenedimethanol		$C_9H_{12}O_3$	91-04-3	168.189		130.5				
6080	2-(Hydroxymethyl)-1,4-benzenediol	Gentisyl alcohol	$C_7H_8O_3$	495-08-9	140.137	nd (chl)	100	75 sub			vs H_2O, EtOH, chl
6081	2-Hydroxy-5-methylbenzoic acid	*p*-Cresotic acid	$C_8H_8O_3$	89-56-5	152.148		152.5(0.2)				sl H_2O; s EtOH, eth, bz, chl; i CS_2
6082	2-Hydroxy-3-methylbenzoic acid	*o*-Cresotic acid	$C_8H_8O_3$	83-40-9	152.148		167.0(0.2)				sl H_2O; s EtOH, eth, bz, chl
6083	2-Hydroxy-4-methylbenzoic acid	*m*-Cresotic acid	$C_8H_8O_3$	50-85-1	152.148	cry, lf	177.8(0.2)				sl H_2O; s EtOH, bz, chl; vs eth
6084	7-Hydroxy-4-methyl-2*H*-1-benzopyran-2-one	Hymecromone	$C_{10}H_8O_3$	90-33-5	176.169	nd (al)	194.5				sl H_2O, eth, chl; s EtOH, alk, HOAc
6085	3-Hydroxy-3-methylbutanoic acid		$C_5H_{10}O_3$	625-08-1	118.131		<-32	162^{12}	0.9384^{20}	1.5081^{20}	vs H_2O, eth, EtOH
6086	3-Hydroxy-3-methyl-2-butanone		$C_5H_{10}O_2$	115-22-0	102.132			148.4(0.5)	0.9526^{20}		s chl
6087	2-Hydroxy-3-methyl-2-cyclopenten-1-one		$C_6H_8O_2$	80-71-7	112.127		104.8				
6088	5-(Hydroxymethyl)-2-furancarboxaldehyde	5-(Hydroxymethyl)-2-furaldehyde	$C_6H_6O_3$	67-47-0	126.110	nd (eth-peth)	31.5	115^{1}	1.2062^{25}	1.5627^{18}	s H_2O, EtOH, bz, chl; sl eth, ctc
6089	2-Hydroxy-6-methyl-3-isopropylbenzoic acid	*o*-Thymotic acid	$C_{11}H_{14}O_3$	548-51-6	194.227	nd (w, bz, lig)	127	sub			vs bz, eth, EtOH
6090	2-Hydroxy-3-methyl-6-isopropyl-2-cyclohexen-1-one	Diosphenol	$C_{10}H_{16}O_2$	490-03-9	168.233		83	109^{10}			
6091	2-(Hydroxymethyl)-2-methyl-1,3-propanediol		$C_5H_{12}O_3$	77-85-0	120.147	wh pow or nd (al)	199(2)	136^{15}			msc H_2O, EtOH; i eth, bz; vs HOAc
6092	2-Hydroxy-3-methyl-1,4-naphthalenedione	Phthiocol	$C_{11}H_8O_3$	483-55-6	188.180	ye pr (eth-peth)	173.5	sub			vs ace, eth

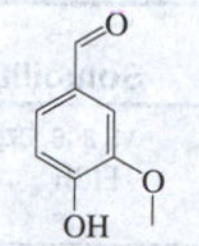
6060
4-Hydroxy-3-methoxybenzaldehyde

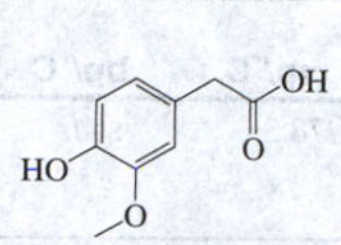
6061
4-Hydroxy-3-methoxybenzeneacetic acid

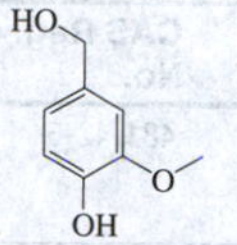
6062
4-Hydroxy-3-methoxybenzenemethanol

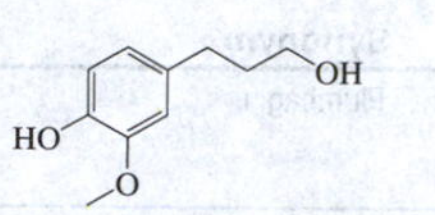
6063
4-Hydroxy-3-methoxybenzenepropanol

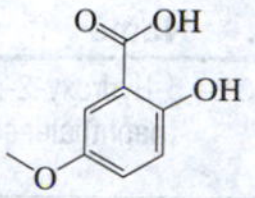
6064
2-Hydroxy-5-methoxybenzoic acid

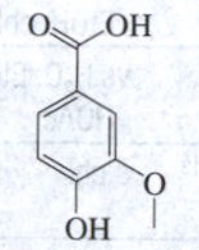
6065
4-Hydroxy-3-methoxybenzoic acid

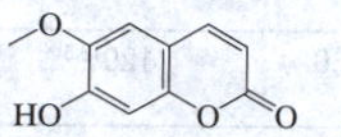
6066
7-Hydroxy-6-methoxy-2*H*-1-benzopyran-2-one

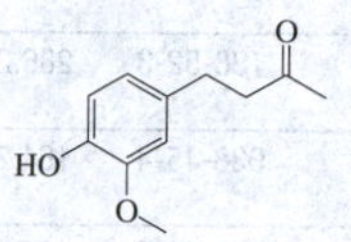
6067
4-(4-Hydroxy-3-methoxyphenyl)-2-butanone

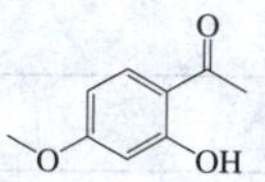
6068
1-(2-Hydroxy-4-methoxyphenyl)ethanone

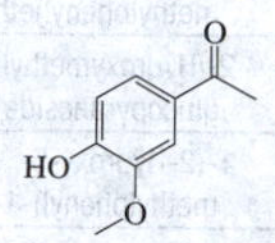
6069
1-(4-Hydroxy-3-methoxyphenyl)ethanone

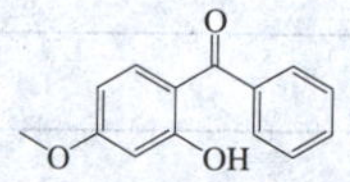
6070
(2-Hydroxy-4-methoxyphenyl)phenylmethanone

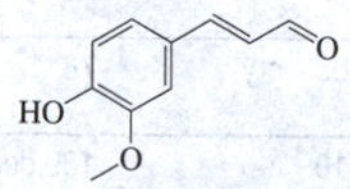
6071
3-(4-Hydroxy-3-methoxyphenyl)-2-propenal

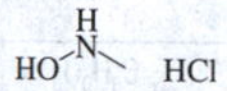
6072
N-Hydroxymethylamine hydrochloride

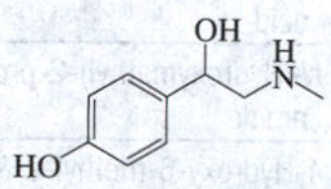
6073
4-Hydroxy-α-[(methylamino)methyl]benzenemethanol

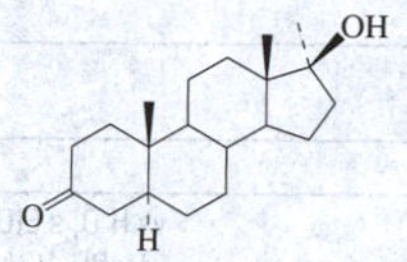
6074
17-Hydroxy-17-methylandrostan-3-one, (5α,17β)

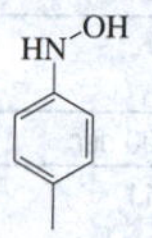
6075
N-Hydroxy-4-methylaniline

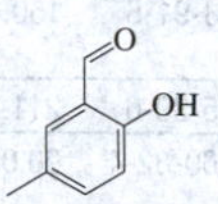
6076
2-Hydroxy-5-methylbenzaldehyde

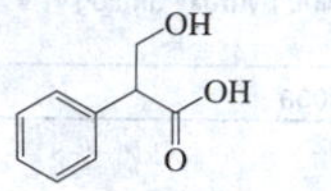
6077
α-(Hydroxymethyl)benzeneacetic acid, (±)-

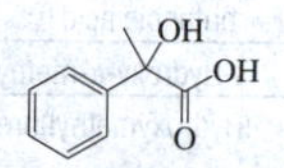
6078
α-Hydroxy-α-methylbenzeneacetic acid, (±)-

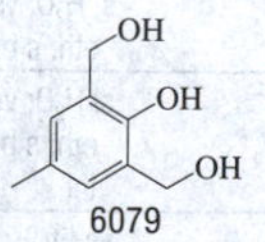
6079
2-Hydroxy-5-methyl-1,3-benzenedimethanol

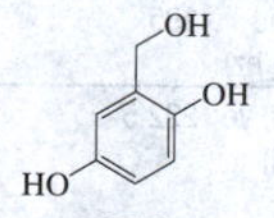
6080
2-(Hydroxymethyl)-1,4-benzenediol

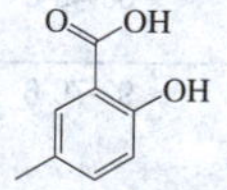
6081
2-Hydroxy-5-methylbenzoic acid

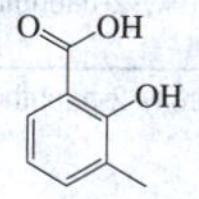
6082
2-Hydroxy-3-methylbenzoic acid

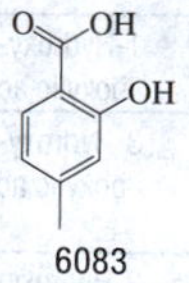
6083
2-Hydroxy-4-methylbenzoic acid

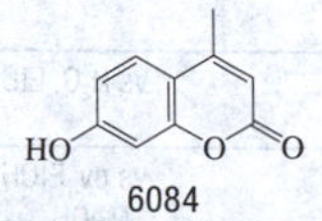
6084
7-Hydroxy-4-methyl-2*H*-1-benzopyran-2-one

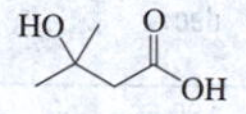
6085
3-Hydroxy-3-methylbutanoic acid

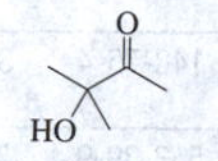
6086
3-Hydroxy-3-methyl-2-butanone

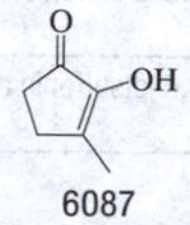
6087
2-Hydroxy-3-methyl-2-cyclopenten-1-one

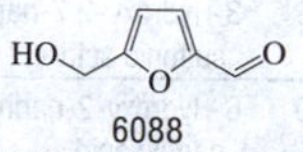
6088
5-(Hydroxymethyl)-2-furancarboxaldehyde

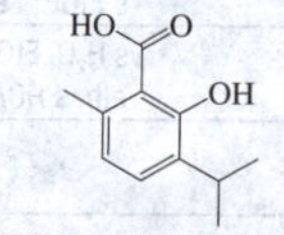
6089
2-Hydroxy-6-methyl-3-isopropylbenzoic acid

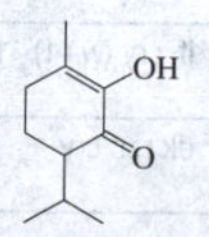
6090
2-Hydroxy-3-methyl-6-isopropyl-2-cyclohexen-1-one

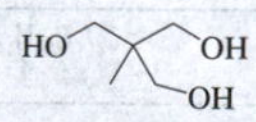
6091
2-(Hydroxymethyl)-2-methyl-1,3-propanediol

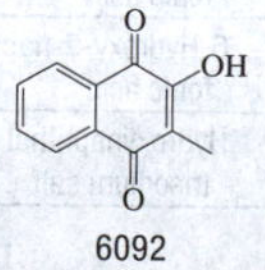
6092
2-Hydroxy-3-methyl-1,4-naphthalenedione

No.	Name	Synonym	Mol. form.	CAS Reg. No.	Mol. wt.	Physical form	mp/°C	bp/°C	den/ g cm⁻³	n_D	Solubility
6093	5-Hydroxy-2-methyl-1,4-naphthalenedione	Plumbagin	$C_{11}H_8O_3$	481-42-5	188.180	gold pr or oran-ye nd (dil al)	78.5	sub			vs ace, bz, eth, EtOH
6094	2-(Hydroxymethyl)-2-nitro-1,3-propanediol	Tris(hydroxymethyl)-nitromethane	$C_4H_9NO_5$	126-11-4	151.118	nd or pr	165	dec			vs H_2O, eth, EtOH
6095	2-Hydroxy-4-methylpentanoic acid, (*S*)-	L-Leucic acid	$C_6H_{12}O_3$	13748-90-8	132.157	orth (eth)	81.5				vs H_2O, eth, EtOH
6096	1-(2-Hydroxy-4-methylphenyl)ethanone		$C_9H_{10}O_2$	6921-64-8	150.174		21	245	1.1012^{10}	1.5527^{13}	
6097	1-(2-Hydroxy-5-methylphenyl)ethanone		$C_9H_{10}O_2$	1450-72-2	150.174	pr (lig)	50	210	1.0797^{53}		vs bz, eth, EtOH, chl
6098	2-(Hydroxymethyl)phenyl-β-D-glucopyranoside	Salicin	$C_{13}H_{18}O_7$	138-52-3	286.278	orth nd or lf (w)	207	240 dec	1.434^{20}		vs H_2O, EtOH, HOAc
6099	1-(2-Hydroxy-5-methylphenyl)-1-propanone		$C_{10}H_{12}O_2$	938-45-4	164.201		1.0	$129^{16.5}$	1.0841^{14}	1.549^{13}	s chl
6100	*N*-(Hydroxymethyl)phthalimide		$C_9H_7NO_3$	118-29-6	177.157	lf, pr (to)	141.5				i H_2O, eth, ctc; sl EtOH, bz; s tol
6101	3-Hydroxy-2-methylpropanal		$C_4H_8O_2$	38433-80-6	88.106	oil					
6102	2-Hydroxy-2-methylpropanoic acid		$C_4H_8O_3$	594-61-6	104.105	hyg pr (eth) nd (bz)	80(3)	212			vs H_2O, EtOH, eth; sl bz
6103	3-Hydroxy-2-methylpropanoic acid		$C_4H_8O_3$	2068-83-9	104.105	oil					
6104	*N*-(Hydroxymethyl)-2-propenamide	*N*-(Hydroxymethyl)acrylamide	$C_4H_7NO_2$	924-42-5	101.105	cry	76				
6105	4-Hydroxy-6-methyl-2*H*-pyran-2-one	Triacetic acid lactone	$C_6H_6O_3$	675-10-5	126.110		189 dec				
6106	3-Hydroxy-2-methyl-4*H*-pyran-4-one	Maltol	$C_6H_6O_3$	118-71-8	126.110	mcl pr (chl)	161.5	93 sub			sl H_2O, eth, bz; vs chl; s alk; peth
6107	5-Hydroxy-6-methyl-3,4-pyridinedimethanol	Pyridoxin	$C_8H_{11}NO_3$	65-23-6	169.178	nd (HOAc)	160	$140^{0.0001}$			
6108	4-Hydroxy-1-methyl-2-quinolinone	4-Hydroxy-*N*-methylcarbostyril	$C_{10}H_9NO_2$	1677-46-9	175.184		265				sl DMSO
6109	2-Hydroxy-4-(methylthio)-butanoic acid	Methionine hydroxy analog	$C_5H_{10}O_3S$	583-91-5	150.196	oil					
6110	3-Hydroxy-α-methyl-L-tyrosine	Methyldopa	$C_{10}H_{13}NO_4$	555-30-6	211.215	cry (MeOH)	300 dec				
6111	(Hydroxymethyl)urea		$C_2H_6N_2O_2$	1000-82-4	90.081	pr (al)	111				vs H_2O; s EtOH, MeOH, HOAc; i eth
6112	2-Hydroxy-1-naphthalenecarboxaldehyde		$C_{11}H_8O_2$	708-06-5	172.181	pr (al), nd (AcOEt)	83	192^{27}			i H_2O; s EtOH, eth, aq alk, sulf, peth
6113	2-Hydroxy-1-naphthalenecarboxylic acid	2-Hydroxy-1-naphthoic acid	$C_{11}H_8O_3$	2283-08-1	188.180		157.3				sl H_2O; vs EtOH; s eth, ace, bz, lig, chl
6114	1-Hydroxy-2-naphthalenecarboxylic acid	1-Hydroxy-2-naphthoic acid	$C_{11}H_8O_3$	86-48-6	188.180	cry (al) nd (al, eth, bz)	195				sl H_2O; vs EtOH, eth; s bz
6115	3-Hydroxy-2-naphthalenecarboxylic acid	3-Hydroxy-2-naphthoic acid	$C_{11}H_8O_3$	92-70-6	188.180	nd (dil al) ye lf (dil al)	222.5				sl H_2O; vs EtOH, eth; s bz, chl, tol
6116	2-Hydroxy-1,4-naphthalenedione	Lawsone	$C_{10}H_6O_3$	83-72-7	174.153	ye pr (HOAc)	195 dec				vs EtOH; i eth, bz, chl; s HOAc
6117	5-Hydroxy-1,4-naphthalenedione	Juglone	$C_{10}H_6O_3$	481-39-0	174.153	ye nd (bz) peth)	155	sub			i H_2O; s EtOH, eth, bz; vs chl; sl lig
6118	7-Hydroxy-1,3-naphthalenedisulfonic acid	2-Naphthol-6,8-disulfonic acid	$C_{10}H_8O_7S_2$	118-32-1	304.297						s H_2O
6119	3-Hydroxy-2,7-naphthalenedisulfonic acid	2-Naphthol-3,6-disulfonic acid	$C_{10}H_8O_7S_2$	148-75-4	304.297	hyg nd	dec				vs H_2O, EtOH
6120	6-Hydroxy-2-naphthalenepropanoic acid	Allenolic acid	$C_{13}H_{12}O_3$	553-39-9	216.232	cry (dil MeOH)	180.5				vs py, EtOH, MeOH
6121	4-Hydroxy-1-naphthalenesulfonic acid	1-Naphthol-4-sulfonic acid	$C_{10}H_8O_4S$	84-87-7	224.234	tab or pl (w)	170 dec				vs H_2O; i eth
6122	7-Hydroxy-1-naphthalenesulfonic acid	Croceic acid	$C_{10}H_8O_4S$	132-57-0	224.234						s H_2O
6123	1-Hydroxy-2-naphthalenesulfonic acid	1-Naphthol-2-sulfonic acid	$C_{10}H_8O_4S$	567-18-0	224.234	pl (w)	>250				sl H_2O, dil HCl; s EtOH; i eth
6124	6-Hydroxy-2-naphthalenesulfonic acid	2-Naphthol-6-sulfonic acid	$C_{10}H_8O_4S$	93-01-6	224.234	lf, cry (w+1)	125				vs H_2O, EtOH; i eth; s HOAc
6125	Hydroxynaphthol blue, trisodium salt		$C_{20}H_{14}N_2Na_3O_{11}S_3$	63451-35-4	620.471	dk red cry					

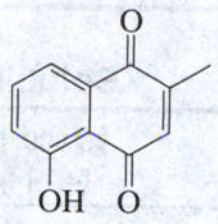
6093
5-Hydroxy-2-methyl-1,4-naphthalenedione

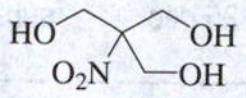
6094
2-(Hydroxymethyl)-2-nitro-1,3-propanediol

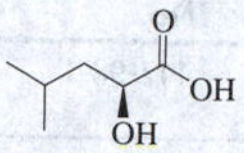
6095
2-Hydroxy-4-methylpentanoic acid, (*S*)-

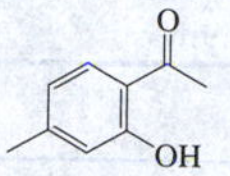
6096
1-(2-Hydroxy-4-methylphenyl)ethanone

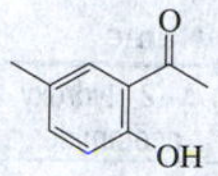
6097
1-(2-Hydroxy-5-methylphenyl)ethanone

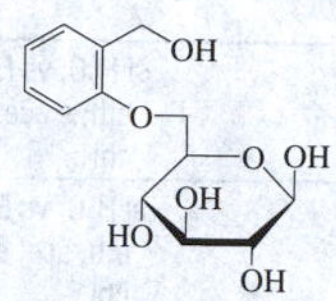
6098
2-(Hydroxymethyl)phenyl-β-D-glucopyranoside

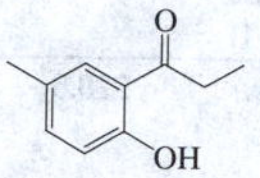
6099
1-(2-Hydroxy-5-methylphenyl)-1-propanone

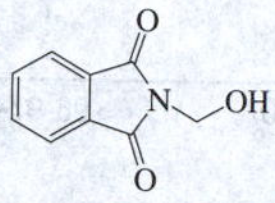
6100
N-(Hydroxymethyl)phthalimide

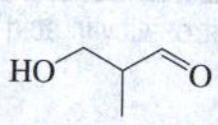
6101
3-Hydroxy-2-methylpropanal

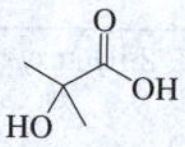
6102
2-Hydroxy-2-methylpropanoic acid

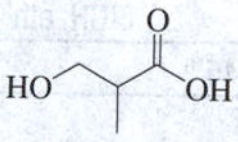
6103
3-Hydroxy-2-methylpropanoic acid

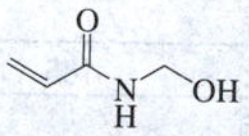
6104
N-(Hydroxymethyl)-2-propenamide

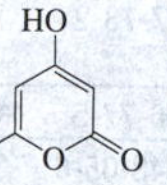
6105
4-Hydroxy-6-methyl-2*H*-pyran-2-one

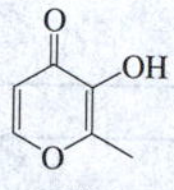
6106
3-Hydroxy-2-methyl-4*H*-pyran-4-one

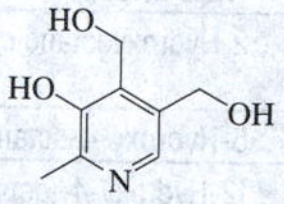
6107
5-Hydroxy-6-methyl-3,4-pyridinedimethanol

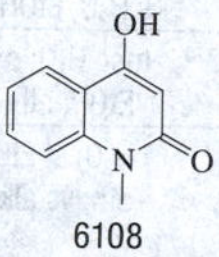
6108
4-Hydroxy-1-methyl-2-quinolinone

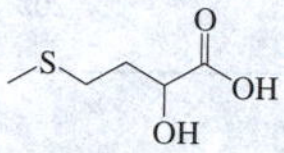
6109
2-Hydroxy-4-(methylthio)butanoic acid

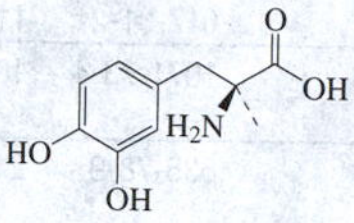
6110
3-Hydroxy-α-methyl-L-tyrosine

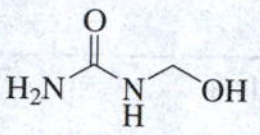
6111
(Hydroxymethyl)urea

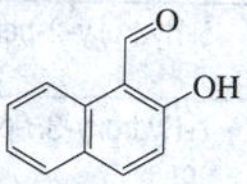
6112
2-Hydroxy-1-naphthalenecarboxaldehyde

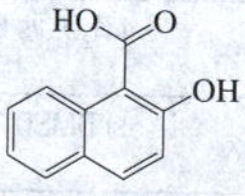
6113
2-Hydroxy-1-naphthalenecarboxylic acid

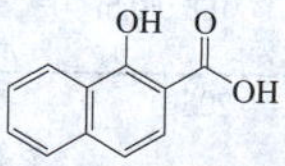
6114
1-Hydroxy-2-naphthalenecarboxylic acid

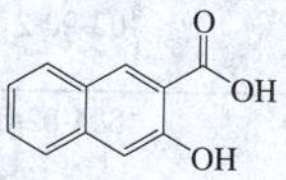
6115
3-Hydroxy-2-naphthalenecarboxylic acid

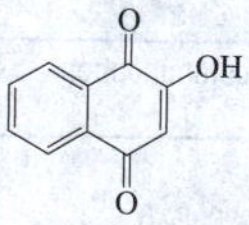
6116
2-Hydroxy-1,4-naphthalenedione

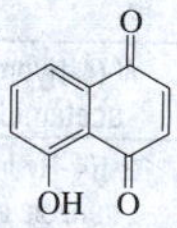
6117
5-Hydroxy-1,4-naphthalenedione

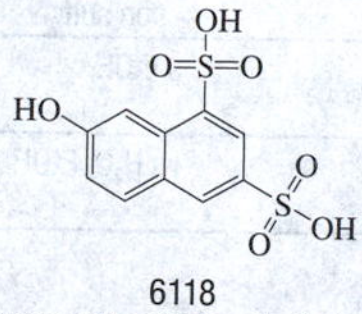
6118
7-Hydroxy-1,3-naphthalenedisulfonic acid

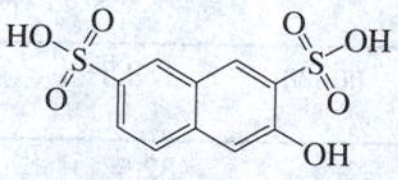
6119
3-Hydroxy-2,7-naphthalenedisulfonic acid

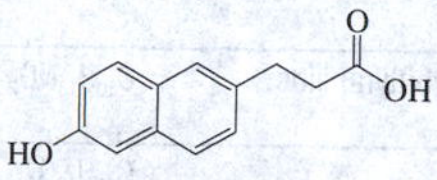
6120
6-Hydroxy-2-naphthalenepropanoic acid

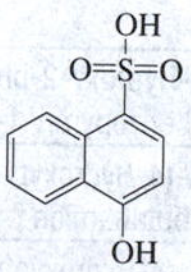
6121
4-Hydroxy-1-naphthalenesulfonic acid

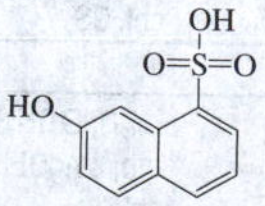
6122
7-Hydroxy-1-naphthalenesulfonic acid

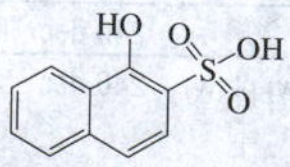
6123
1-Hydroxy-2-naphthalenesulfonic acid

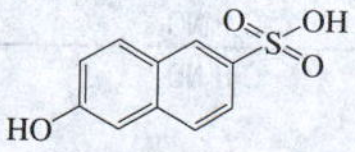
6124
6-Hydroxy-2-naphthalenesulfonic acid

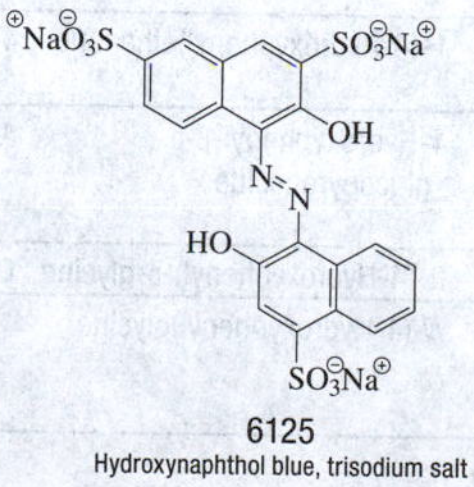
6125
Hydroxynaphthol blue, trisodium salt

No.	Name	Synonym	Mol. form.	CAS Reg. No.	Mol. wt.	Physical form	mp/°C	bp/°C	den/g cm^{-3}	n_D	Solubility
6126	*N*-(2-Hydroxy-1-naphthyl)-acetamide		$C_{12}H_{11}NO_2$	117-93-1	201.221	lf (w, dil al)	235 dec	sub			vs ace, bz, eth, EtOH
6127	1-(1-Hydroxy-2-naphthyl)-ethanone		$C_{12}H_{10}O_2$	711-79-5	186.206	pr (bz, lig) grn-ye nd (al)	98.6(0.2)	325 dec			vs bz, HOAc
6128	2-Hydroxy-3-nitrobenzalde-hyde		$C_7H_5NO_4$	5274-70-4	167.120	nd (HOAc)	109.5				vs bz, EtOH
6129	2-Hydroxy-5-nitrobenzalde-hyde		$C_7H_5NO_4$	97-51-8	167.120	cry (dil HOAc)	127.0				s ace
6130	2-Hydroxy-3-nitrobenzoic acid	3-Nitrosalicylic acid	$C_7H_5NO_5$	85-38-1	183.119	ye nd (HOAc, w+1)	148				sl H_2O; vs EtOH, eth; s ace, bz, chl
6131	2-Hydroxy-5-nitrobenzoic acid	5-Nitrosalicylic acid	$C_7H_5NO_5$	96-97-9	183.119	nd (w)	229.5		1.650^{20}		sl H_2O; vs EtOH, eth, ace, bz; s chl
6132	2-Hydroxy-1,2,3-nonadecane-tricarboxylic acid	Agaricic acid	$C_{22}H_{40}O_7$	666-99-9	416.549	cry pow	142 dec				s H_2O; sl EtOH, eth; i bz, chl
6133	12-Hydroxyoctadecanoic acid	12-Hydroxysteric acid	$C_{18}H_{36}O_3$	106-14-9	300.477	cry (al)	82				i H_2O; s EtOH, eth, chl
6134	*cis*-12-Hydroxy-9-octadece-noic acid, (*R*)-	Ricinoleic acid	$C_{18}H_{34}O_3$	141-22-0	298.461	visc liq	-8.28(0.02)	227^{10}	0.9450^{21}	1.4716^{21}	i H_2O; vs eth, EtOH
6135	2-Hydroxyoctanoic acid		$C_8H_{16}O_3$	617-73-2	160.211	pl	70	162^{10}			sl H_2O, chl; vs EtOH, eth
6136	5-Hydroxy-4-octanone	Butyroin	$C_8H_{16}O_2$	496-77-5	144.212	liq	-10	185	0.9107^{16}	1.4345^{16}	
6137	[2-Hydroxy-4-(octyloxy)-phenyl]phenylmethanone	Octabenzone	$C_{21}H_{26}O_3$	1843-05-6	326.429		48.5				
6138	3-Hydroxy-2-oxopropanoic acid	Hydroxypyruvic acid	$C_3H_4O_4$	1113-60-6	104.062		81 dec				
6139	3-Hydroxy-4-oxo-4*H*-pyran-2,6-dicarboxylic acid	Meconic acid	$C_7H_4O_7$	497-59-6	200.103	orth pl (w, dil HCl) (+3w)	120 dec				sl H_2O, MeOH, ace, eth; s EtOH, bz
6140	2-Hydroxypentanoic acid		$C_5H_{10}O_3$	617-31-2	118.131	hyg pl	34	sub			s H_2O, EtOH, eth
6141	5-Hydroxy-2-pentanone		$C_5H_{10}O_2$	1071-73-4	102.132			209	1.0071^{20}	1.4390^{20}	msc H_2O; s EtOH, eth
6142	7-Hydroxy-3*H*-phenoxazin-3-one	Resorufine	$C_{12}H_7NO_3$	635-78-9	213.189	br nd ($PhNO_2$) pr (HCl)					i H_2O; sl EtOH; i eth; vs alk
6143	*N*-(2-Hydroxyphenyl)-acetamide		$C_8H_9NO_2$	614-80-2	151.163	pl (dil al)	209				sl H_2O; vs EtOH, eth, bz; s DMSO
6144	*N*-(3-Hydroxyphenyl)-acetamide		$C_8H_9NO_2$	621-42-1	151.163	nd (w)	148.5				vs H_2O, EtOH; sl eth, bz, chl, DMSO
6145	*N*-(4-Hydroxyphenyl)-acetamide	Acetaminophen	$C_8H_9NO_2$	103-90-2	151.163	mcl pr (w)	168.0(0.5)		1.293^{21}		i H_2O; vs EtOH
6146	2-[(4-Hydroxyphenyl)azo]-benzoic acid		$C_{13}H_{10}N_2O_3$	1634-82-8	242.229		206				sl DMSO
6147	2-Hydroxy-*N*-phenylbenza-mide	Salicylanilide	$C_{13}H_{11}NO_2$	87-17-2	213.232	pr (w, al)	136.5				s H_2O; sl EtOH, eth, bz, chl
6148	*N*-Hydroxy-*N*-phenylbenza-mide		$C_{13}H_{11}NO_2$	304-88-1	213.232		121(1)				
6149	α-Hydroxy-α-phenylbenzeneacetic acid	Benzilic acid	$C_{14}H_{12}O_3$	76-93-7	228.243	mcl nd (w)	149(2)	180 dec			sl H_2O, ace; vs EtOH, eth; s con sulf
6150	3-Hydroxy-2-phenyl-4*H*-1-benzopyran-4-one		$C_{15}H_{10}O_3$	577-85-5	238.238	pa ye nd (al)	169.5				s EtOH
6151	*N*-(4-Hydroxyphenyl)-butanamide	4'-Hydroxybutyranilide	$C_{10}H_{13}NO_2$	101-91-7	179.216	nd (w)	139.5				vs H_2O, EtOH
6152	4-(4-Hydroxyphenyl)-2-butanone		$C_{10}H_{12}O_2$	5471-51-2	164.201		82.5				
6153	1-(2-Hydroxyphenyl)ethanone	2-Hydroxyacetophenone	$C_8H_8O_2$	118-93-4	136.149		2.5	218	1.1307^{20}	1.5584^{20}	vs eth, EtOH, HOAc
6154	1-(3-Hydroxyphenyl)ethanone	3-Hydroxyacetophenone	$C_8H_8O_2$	121-71-1	136.149	nd or lf	94(3)	296	1.0992^{109}	1.5348^{109}	sl H_2O; vs EtOH, eth, bz, chl; i lig
6155	1-(4-Hydroxyphenyl)ethanone	4-Hydroxyacetophenone	$C_8H_8O_2$	99-93-4	136.149	nd (eth, dil al)	108.2(0.5)	147^{3}	1.1090^{109}	1.5577^{109}	sl H_2O, DMSO; vs EtOH, eth
6156	4-Hydroxyphenyl-β-D-glucopyranoside	Arbutin	$C_{12}H_{16}O_7$	497-76-7	272.251	nd (w+1)	199.5				vs H_2O; s EtOH; sl eth; i bz, chl, CS_2
6157	2-(4-Hydroxyphenyl)-D-glycine	Oxfenicine	$C_8H_9NO_3$	22818-40-2	167.162	cry	240 dec				
6158	*N*-(4-Hydroxyphenyl)glycine		$C_8H_9NO_3$	122-87-2	167.162	lf (w) pl (w)	246 dec				sl H_2O, EtOH; i eth; s AcOEt, chl

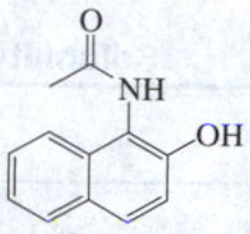
6126
N-(2-Hydroxy-1-naphthyl)acetamide

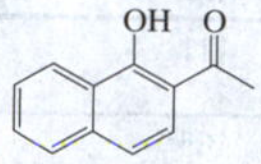
6127
1-(1-Hydroxy-2-naphthyl)ethanone

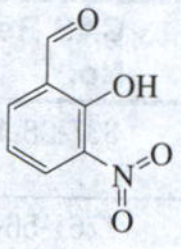
6128
2-Hydroxy-3-nitrobenzaldehyde

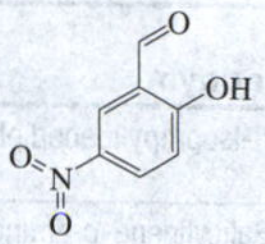
6129
2-Hydroxy-5-nitrobenzaldehyde

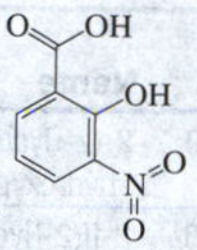
6130
2-Hydroxy-3-nitrobenzoic acid

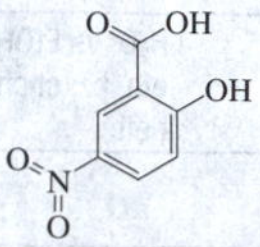
6131
2-Hydroxy-5-nitrobenzoic acid

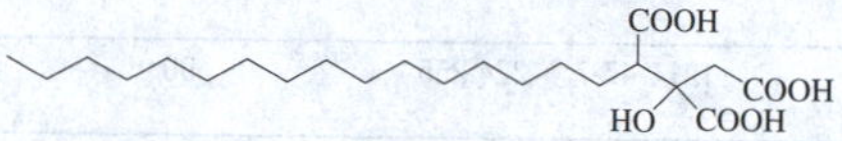
6132
2-Hydroxy-1,2,3-nonadecanetricarboxylic acid

6133
12-Hydroxyoctadecanoic acid

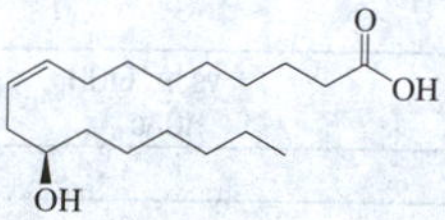
6134
cis-12-Hydroxy-9-octadecenoic acid, (*R*)-

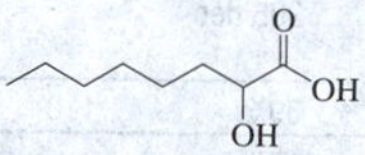
6135
2-Hydroxyoctanoic acid

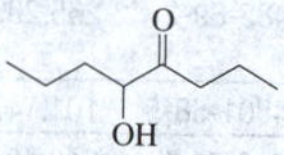
6136
5-Hydroxy-4-octanone

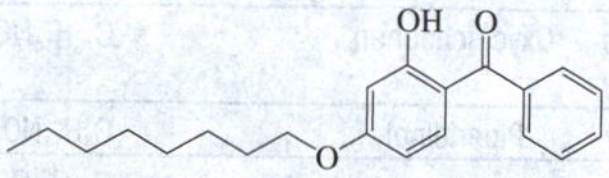
6137
[2-Hydroxy-4-(octyloxy)phenyl]phenylmethanone

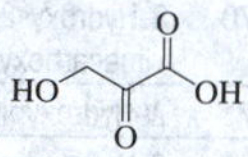
6138
3-Hydroxy-2-oxopropanoic acid

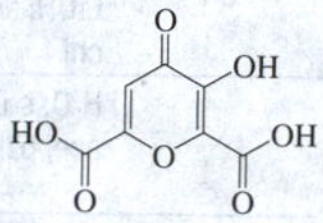
6139
3-Hydroxy-4-oxo-4*H*-pyran-2,6-dicarboxylic acid

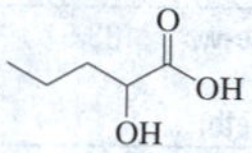
6140
2-Hydroxypentanoic acid

6141
5-Hydroxy-2-pentanone

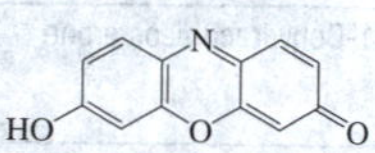
6142
7-Hydroxy-3*H*-phenoxazin-3-one

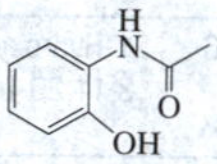
6143
N-(2-Hydroxyphenyl)acetamide

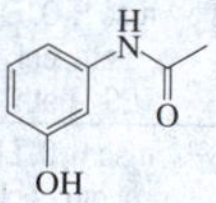
6144
N-(3-Hydroxyphenyl)acetamide

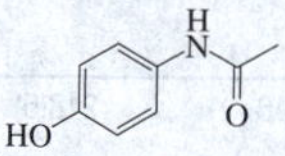
6145
N-(4-Hydroxyphenyl)acetamide

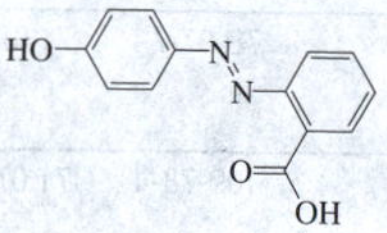
6146
2-[(4-Hydroxyphenyl)azo]benzoic acid

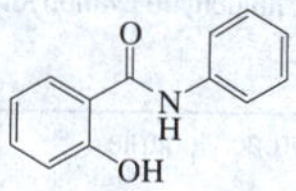
6147
2-Hydroxy-*N*-phenylbenzamide

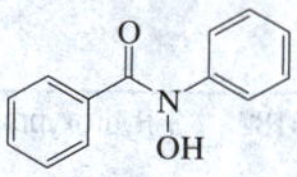
6148
N-Hydroxy-*N*-phenylbenzamide

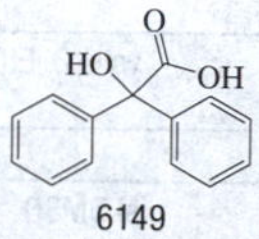
6149
α-Hydroxy-α-phenylbenzeneacetic acid

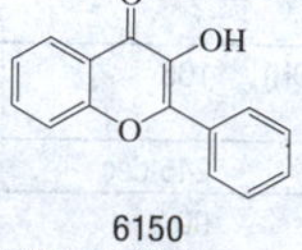
6150
3-Hydroxy-2-phenyl-4*H*-1-benzopyran-4-one

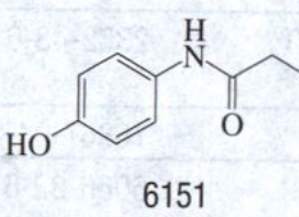
6151
N-(4-Hydroxyphenyl)butanamide

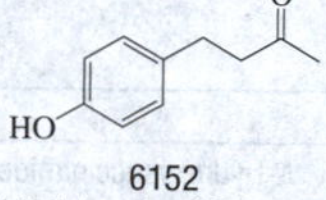
6152
4-(4-Hydroxyphenyl)-2-butanone

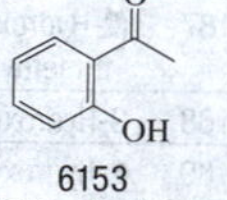
6153
1-(2-Hydroxyphenyl)ethanone

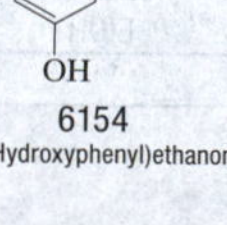
6154
1-(3-Hydroxyphenyl)ethanone

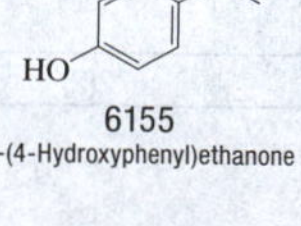
6155
1-(4-Hydroxyphenyl)ethanone

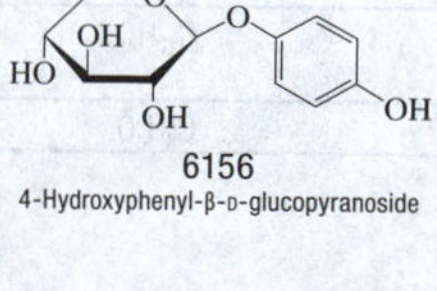
6156
4-Hydroxyphenyl-β-D-glucopyranoside

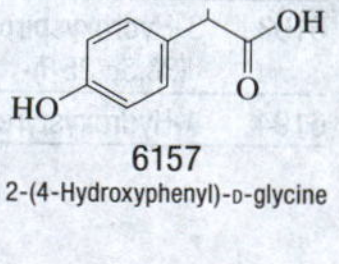
6157
2-(4-Hydroxyphenyl)-D-glycine

Organic

No.	Name	Synonym	Mol. form.	CAS Reg. No.	Mol. wt.	Physical form	mp/°C	bp/°C	den/g cm^{-3}	n_D	Solubility
6159	2-(2-Hydroxyphenyl)-2-(4-hydroxyphenyl)propane	2,4'-Isopropylidenediphenol	$C_{15}H_{16}O_2$	837-08-1	228.287	cry (bz)	111				
6160	2-[[(2-Hydroxyphenyl)imino]-methyl]phenol	*N*-Salicylidene-*o*-aminophenol	$C_{13}H_{11}NO_2$	1761-56-4	213.232		185				
6161	*N*-Hydroxy-*N*-(phenylmethyl)-benzenemethanamine		$C_{14}H_{15}NO$	621-07-8	213.275		122.5				s chl
6162	*N*-(4-Hydroxyphenyl)-octadecanamide		$C_{24}H_{41}NO_2$	103-99-1	375.589		133.8	239.5^{10}			i H_2O; sl eth, bz, chl; s ace
6163	3-(4-Hydroxyphenyl)-2-oxopropanoic acid	4-Hydroxy-α-oxobenzenepropanoic acid	$C_9H_8O_4$	156-39-8	180.158	cry (w)	220 dec				s H_2O; dec alk
6164	(2-Hydroxyphenyl)-phenylmethanone	2-Hydroxybenzophenone	$C_{13}H_{10}O_2$	117-99-7	198.217	pl (dil al)	35(3)	250^{560}			i H_2O; vs EtOH, eth, bz; sl chl, peth
6165	1-(2-Hydroxyphenyl)-3-phenyl-2-propen-1-one	2'-Hydroxychalcone	$C_{15}H_{12}O_2$	1214-47-7	224.255		90				
6166	2-Hydroxy-1-phenyl-1-propanone		$C_9H_{10}O_2$	5650-40-8	150.174	ye oil		251	1.1085^{18}	1.536^{23}	
6167	1-(2-Hydroxyphenyl)-1-propanone		$C_9H_{10}O_2$	610-99-1	150.174			150^{80}		1.5501^{20}	sl H_2O; s EtOH, eth, ctc, alk
6168	1-(4-Hydroxyphenyl)-1-propanone	Paroxypropione	$C_9H_{10}O_2$	70-70-2	150.174	wh nd or pl (w)	149				sl H_2O, ace; s EtOH, eth, alk
6169	3-(4-Hydroxyphenyl)-2-propenoic acid	*p*-Coumaric acid	$C_9H_8O_3$	7400-08-0	164.158	nd	212(11)				vs eth, EtOH
6170	3-Hydroxy-2-phenyl-4-quinolinecarboxylic acid	Oxycinchophen	$C_{16}H_{11}NO_3$	485-89-2	265.263	ye pr (al)	206 dec				vs bz, EtOH, HOAc
6171	*N*-Hydroxypiperidine	1-Piperidinol	$C_5H_{11}NO$	4801-58-5	101.147	hyg	39.3	110^{55}			
6172	3-Hydroxypregnan-20-one, (3α,5α)	Allopregnan-3α-ol-20-one	$C_{21}H_{34}O_2$	516-54-1	318.494	cry (al)	177				
6173	3-Hydroxypregnan-20-one, (3β,5α)	Allopregnan-3β-ol-20-one	$C_{21}H_{34}O_2$	516-55-2	318.494		170(15)				
6174	17-Hydroxypregn-4-ene-3,20-dione	17α-Hydroxyprogesterone	$C_{21}H_{30}O_3$	68-96-2	330.461						sl chl
6175	21-Hydroxypregn-4-ene-3,20-dione	Deoxycorticosterone	$C_{21}H_{30}O_3$	64-85-7	330.461	pl (eth)	141.5				sl H_2O, eth; vs EtOH, ace; s chl
6176	21-Hydroxypregn-4-ene-3,11,20-trione	11-Dehydrocorticosterone	$C_{21}H_{28}O_4$	72-23-1	344.445	pr (ace-w, al, ace-eth)	183.5				i H_2O; s EtOH, ace, bz
6177	*cis*-4-Hydroxy-L-proline		$C_5H_9NO_3$	618-27-9	131.130	nd (w+1)	239.5				vs H_2O
6178	*trans*-4-Hydroxy-L-proline		$C_5H_9NO_3$	51-35-4	131.130	lf (dil al) pr (w)	274				vs H_2O; sl EtOH
6179	3-Hydroxypropanal	Hydracrolein	$C_3H_6O_2$	2134-29-4	74.079			90^{18}			vs ace, eth, EtOH
6180	Hydroxypropanedioic acid	Tartronic acid	$C_3H_4O_5$	80-69-3	120.061	pr (w+1)	157	sub			s H_2O, EtOH; sl eth
6181	2-Hydroxypropanenitrile	Acetaldehyde cyanohydrin	C_3H_5NO	78-97-7	71.078	liq	-40.0(0.5)	184(3)	0.9877^{20}	1.4058^{18}	msc H_2O, EtOH; s eth, chl; i CS_2, peth
6182	3-Hydroxypropanenitrile	Hydracrylonitrile	C_3H_5NO	109-78-4	71.078	liq	-46	218(5)	1.0404^{25}	1.4248^{20}	msc H_2O, EtOH; sl eth; s chl; i CS_2
6183	3-Hydroxypropanoic acid	Hydracrylic acid	$C_3H_6O_3$	503-66-2	90.078	syr		dec		1.4489^{20}	vs H_2O; s EtOH; msc eth
6184	1-Hydroxy-2-propanone	Acetone alcohol	$C_3H_6O_2$	116-09-6	74.079	hyg liq	-17	145.5	1.0805^{20}	1.4295^{20}	vs H_2O, EtOH, eth
6185	4-(3-Hydroxy-1-propenyl)-2-methoxyphenol	Coniferyl alcohol	$C_{10}H_{12}O_3$	458-35-5	180.200	pr (eth-lig)	74	164^{3}			i H_2O; s EtOH, alk; vs eth
6186	2-Hydroxypropyl acrylate		$C_6H_{10}O_3$	999-61-1	130.141	liq		70^{2}			
6187	(2-Hydroxypropyl)-trimethylammonium chloride		$C_6H_{16}ClNO$	2382-43-6	153.650	pr (Bu OH)	165	dec			vs H_2O, EtOH
6188	3-Hydroxy-1*H*-pyridin-2-one		$C_5H_5NO_2$	16867-04-2	111.100		245 dec				
6189	1-Hydroxy-2,5-pyrrolidinedione	*N*-Hydroxysuccinimide	$C_4H_5NO_3$	6066-82-6	115.088	hyg	96.3				sl DMSO
6190	4-Hydroxy-2-quinolinecarboxylic acid	Kynurenic acid	$C_{10}H_7NO_3$	492-27-3	189.168	ye nd (+w, dil al)	282.5				sl H_2O; s EtOH; i eth; vs alk
6191	8-Hydroxy-5-quinolinesulfonic acid		$C_9H_7NO_4S$	84-88-8	225.222	ye lf, nd (+1w) (dil HCl)	322.5				sl H_2O
6192	4-Hydroxy-2-quinolinone	2,4-Quinolinediol	$C_9H_7NO_2$	86-95-3	161.158		353(4)				sl EtOH, $PhNO_2$, gl HOAc
6193	3-Hydroxyspirostan-12-one, (3β,5α,25*R*)-	Hecogenin	$C_{27}H_{42}O_4$	467-55-0	430.620	pl (eth)	266.5				vs ace, eth, EtOH
6194	4-Hydroxystyrene	4-Vinylphenol	C_8H_8O	2628-17-3	120.149		72(2)				

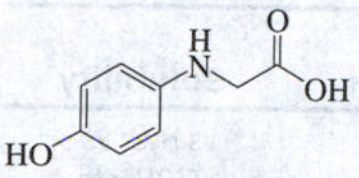
6158
N-(4-Hydroxyphenyl)glycine

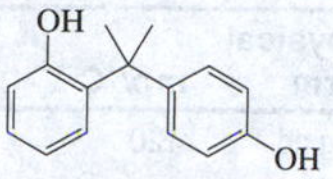
6159
2(2-Hydroxyphenyl)-2(4-hydroxyphenyl)propane

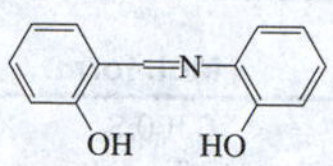
6160
2-[[(2-Hydroxyphenyl)imino]methyl]phenol

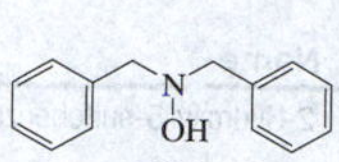
6161
N-Hydroxy-*N*-(phenylmethyl)benzenemethanamine

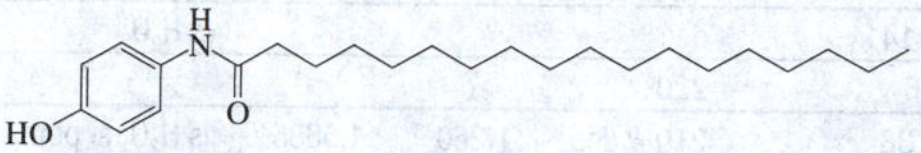
6162
N-(4-Hydroxyphenyl)octadecanamide

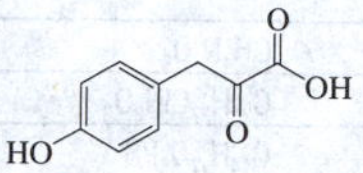
6163
3-(4-Hydroxyphenyl)-2-oxopropanoic acid

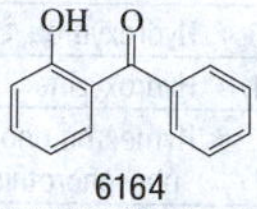
6164
(2-Hydroxyphenyl)phenylmethanone

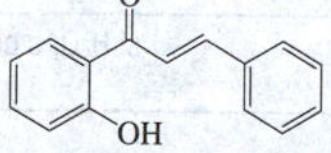
6165
1-(2-Hydroxyphenyl)-3-phenyl-2-propen-1-one

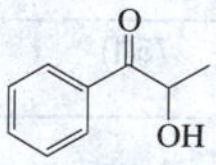
6166
2-Hydroxy-1-phenyl-1-propanone

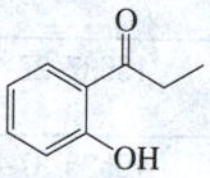
6167
1-(2-Hydroxyphenyl)-1-propanone

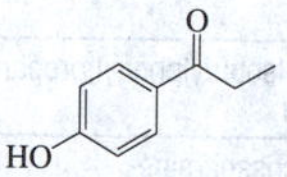
6168
1-(4-Hydroxyphenyl)-1-propanone

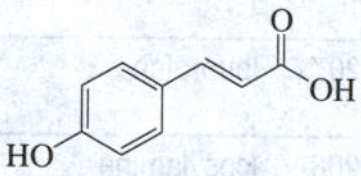
6169
3-(4-Hydroxyphenyl)-2-propenoic acid

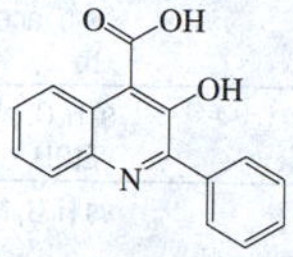
6170
3-Hydroxy-2-phenyl-4-quinolinecarboxylic acid

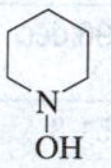
6171
N-Hydroxypiperidine

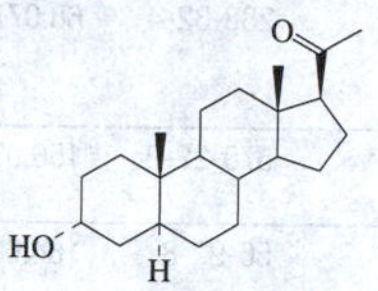
6172
3-Hydroxypregnan-20-one, (3α,5α)

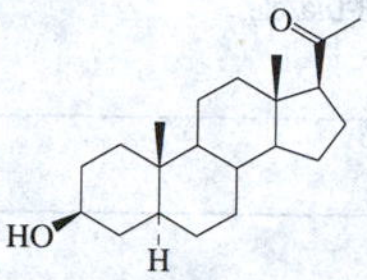
6173
3-Hydroxypregnan-20-one, (3β,5α)

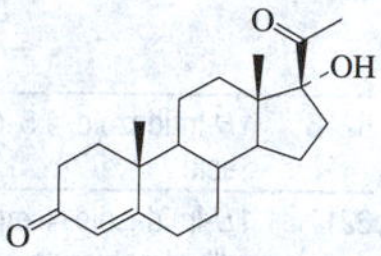
6174
17-Hydroxypregn-4-ene-3,20-dione

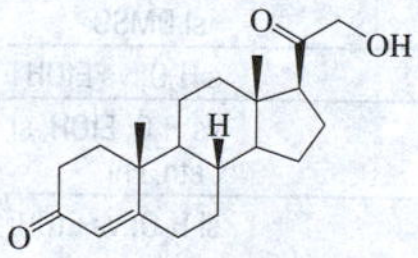
6175
21-Hydroxypregn-4-ene-3,20-dione

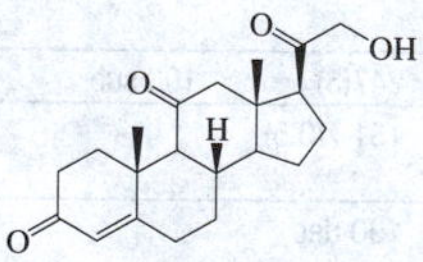
6176
21-Hydroxypregn-4-ene-3,11,20-trione

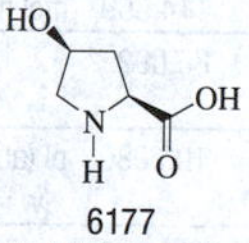
6177
cis-4-Hydroxy-L-proline

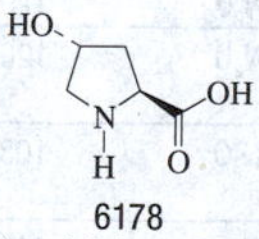
6178
trans-4-Hydroxy-L-proline

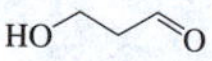
6179
3-Hydroxypropanal

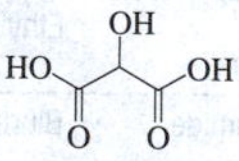
6180
Hydroxypropanedioic acid

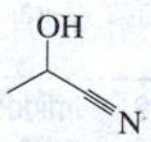
6181
2-Hydroxypropanenitrile

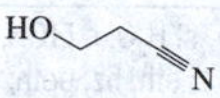
6182
3-Hydroxypropanenitrile

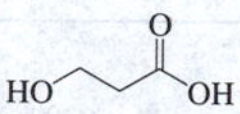
6183
3-Hydroxypropanoic acid

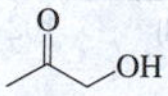
6184
1-Hydroxy-2-propanone

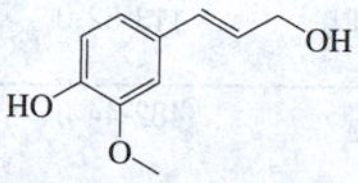
6185
4-(3-Hydroxy-1-propenyl)-2-methoxyphenol

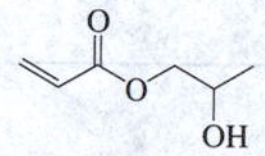
6186
2-Hydroxypropyl acrylate

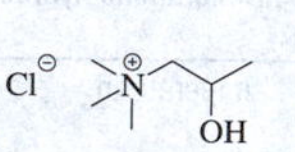
6187
(2-Hydroxypropyl)trimethylammonium chloride

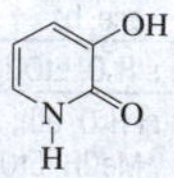
6188
3-Hydroxy-1*H*-pyridin-2-one

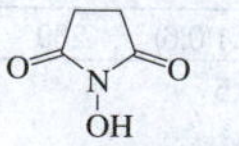
6189
1-Hydroxy-2,5-pyrrolidinedione

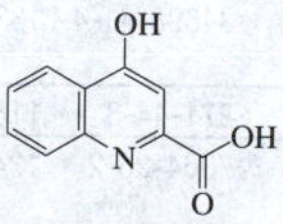
6190
4-Hydroxy-2-quinolinecarboxylic acid

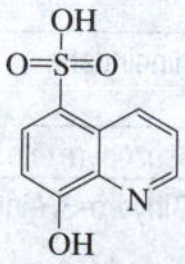
6191
8-Hydroxy-5-quinolinesulfonic acid

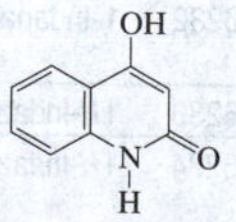
6192
4-Hydroxy-2-quinolinone

Organic

No.	Name	Synonym	Mol. form.	CAS Reg. No.	Mol. wt.	Physical form	mp/°C	bp/°C	den/ g cm^{-3}	n_D	Solubility
6195	2-Hydroxy-5-sulfobenzoic acid	5-Sulfosalicylic acid	$C_7H_6O_6S$	97-05-2	218.184	hyg nd	120				vs H_2O; vs EtOH, eth
6196	2-Hydroxy-5-sulfobenzoic acid dihydrate	5-Sulfosalicylic acid dihydrate	$C_7H_{10}O_8S$	5965-83-3	254.214	wh cry (w)					vs H_2O; vs EtOH, eth
6197	4-Hydroxy-2,2,6,6-tetramethylpiperidine	2,2,6,6-Tetramethyl-4-piperidinol	$C_9H_{19}NO$	2403-88-5	157.253		130	213.5			
6198	5-Hydroxytryptamine	3-(2-Aminoethyl)indol-5-ol	$C_{10}H_{12}N_2O$	50-67-9	176.214						s H_2O
6199	5-Hydroxy-DL-tryptophan		$C_{11}H_{12}N_2O_3$	114-03-4	220.224	rod or nd (al)	300 dec				
6200	Hydroxyurea		$CH_4N_2O_2$	127-07-1	76.055	nd (al)	141	dec			vs H_2O
6201	Hydroxyzine		$C_{21}H_{27}ClN_2O_2$	68-88-2	374.904	oil		$220^{0.5}$			
6202	Hymecromone *O,O*-diethyl phosphorothioate		$C_{14}H_{17}O_5PS$	299-45-6	328.321	nd	38	$210^{1.0}$ dec	1.260^{38}	1.5685^{37}	vs H_2O; sl peth
6203	Hymenoxone		$C_{15}H_{22}O_5$	57377-32-9	282.333	cry					
6204	Hyoscyamine	Tropine tropate	$C_{17}H_{23}NO_3$	101-31-5	289.370	tetr nd (dil al)	108.5				sl H_2O, eth, bz; vs EtOH, chl
6205	Hypoglycin A		$C_7H_{11}NO_2$	156-56-9	141.168	ye pl (Me aq)	282				
6206	Hypoxanthine		$C_5H_4N_4O$	68-94-0	136.112	oct nd (w)	150 dec				sl H_2O; s alk, dil acid
6207	Ibuprofen	2-(4-Isobutylphenyl)propanoic acid	$C_{13}H_{18}O_2$	15687-27-1	206.281	col cry	75(1)				sl H_2O; s os
6208	Icosylamine	1-Eicosanamine	$C_{20}H_{43}N$	10525-37-8	297.562			389(15)			
6209	D-Idose		$C_6H_{12}O_6$	5978-95-0	180.155	syr					vs H_2O
6210	L-Idose		$C_6H_{12}O_6$	5934-56-5	180.155	syr					vs H_2O
6211	Imazalil		$C_{14}H_{14}Cl_2N_2O$	35554-44-0	297.179		50.2(0.5)	dec	1.243^{23}		
6212	Imazapyr		$C_{13}H_{15}N_3O_3$	81334-34-1	261.276		171				
6213	Imazaquin		$C_{17}H_{17}N_3O_3$	81335-37-7	311.335		221				
6214	Imazethapyr		$C_{15}H_{19}N_3O_3$	81335-77-5	289.330		173				
6215	Imidazole	1,3-Diazole	$C_3H_4N_2$	288-32-4	68.077	mcl pr (bz)	89.52(0.04)	257	1.0303^{101}	1.4801^{101}	vs H_2O, EtOH; s eth, ace, py; sl bz
6216	1*H*-Imidazole-4,5-dicarboxylic acid		$C_5H_4N_2O_4$	570-22-9	156.097	pr	290 dec		1.749^{25}		sl H_2O, py; i EtOH, eth, bz
6217	1*H*-Imidazole-4-ethanamine, dihydrochloride		$C_5H_{11}Cl_2N_3$	56-92-8	184.066	pl (eth-HOAc), pr (w)	251.3		1.43^{20}		vs H_2O, MeOH
6218	2,4-Imidazolidinedione	Hydantoin	$C_3H_4N_2O_2$	461-72-3	100.076	nd (MeOH), lf (w)	220				s H_2O, EtOH, alk; sl eth; i peth
6219	2-Imidazolidinethione	Ethylene thiourea	$C_3H_6N_2S$	96-45-7	102.158	nd (al), pr (al)	203				vs H_2O; s EtOH; i eth, bz, chl; sl DMSO
6220	Imidazolidinetrione	Parabanic acid	$C_3H_2N_2O_3$	120-89-8	114.059	mcl nd (w)	247(3)	100 sub			s H_2O; vs EtOH
6221	2-Imidazolidinone	Ethylene urea	$C_3H_6N_2O$	120-93-4	86.092		131.7(0.5)				vs H_2O, EtOH; sl eth, chl
6222	Imidodicarbonic diamide	Biuret	$C_2H_5N_3O_2$	108-19-0	103.080	pl (al), nd (w+1)	190 dec				sl H_2O; vs EtOH; i eth
6223	3,3'-Iminobispropanenitrile	Bis(2-cyanoethyl)amine	$C_6H_9N_3$	111-94-4	123.155		-6(1)	162^5	1.0165^{20}		
6224	Iminodiacetic acid	Diglycine	$C_4H_7NO_4$	142-73-4	133.104	orth pr	247.5				sl H_2O; i EtOH, eth
6225	Iminodiacetic acid, dinitrile	2,2'-Iminobisacetonitrile	$C_4H_5N_3$	628-87-5	95.103		78				s H_2O, EtOH; sl eth, bz, chl
6226	Imipramine		$C_{19}H_{24}N_2$	50-49-7	280.407			$160^{0.1}$			
6227	Imipramine hydrochloride	Tofranil	$C_{19}H_{25}ClN_2$	113-52-0	316.868		174.5				vs H_2O; s EtOH; sl ace
6228	Imperatorin		$C_{16}H_{14}O_4$	482-44-0	270.280	cry (al)	102				sl H_2O; s EtOH, eth, bz, peth; vs chl
6229	Indaconitine		$C_{34}H_{47}NO_{10}$	4491-19-4	629.738	cry	202 dec				vs eth, EtOH, chl
6230	Indalone	Butopyronoxyl	$C_{12}H_{18}O_4$	532-34-3	226.269	ye-red liq		263	1.057^{20}	1.475^{25}	i H_2O; vs EtOH, eth, chl
6231	Indan		C_9H_{10}	496-11-7	118.175	liq	-51.34(0.02)	177.8(0.4)	0.9639^{20}	1.5378^{20}	i H_2O; msc EtOH, eth; sl chl
6232	1-Indanamine	1-Aminoindane	$C_9H_{11}N$	34698-41-4	133.190			221	1.038^{15}	1.5613^{20}	sl H_2O; s eth, ace, bz
6233	1*H*-Indazole	1*H*-Benzopyrazole	$C_7H_6N_2$	271-44-3	118.136	nd (al, w)	146.1(0.6)	269			s H_2O, EtOH, eth
6234	1*H*-Indazol-3-ol	1,2-Dihydro-3*H*-indazol-3-one	$C_7H_6N_2O$	7364-25-2	134.135	nd or lf (MeOH) pl or nd (al)	252.5				sl H_2O, eth; s MeOH, EtOH

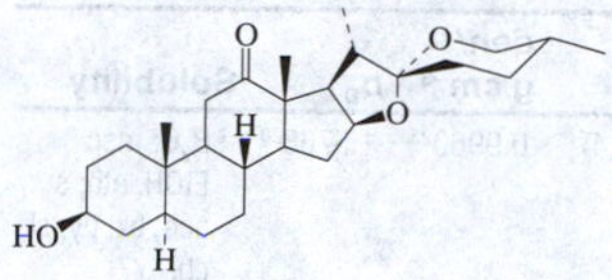

6193
3-Hydroxyspirostan-12-one, (3β,5α,25*R*)-

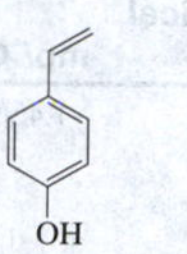

6194
4-Hydroxystyrene

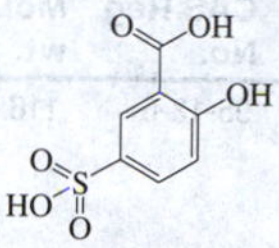

6195
2-Hydroxy-5-sulfobenzoic acid

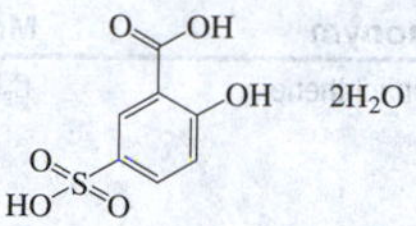

6196
2-Hydroxy-5-sulfobenzoic acid dihydrate

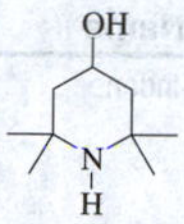

6197
4-Hydroxy-2,2,6,6-tetramethylpiperidine

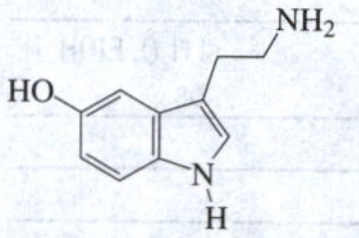

6198
5-Hydroxytryptamine

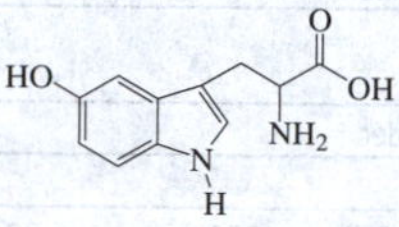

6199
5-Hydroxy-DL-tryptophan

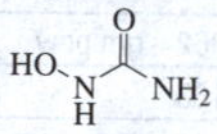

6200
Hydroxyurea

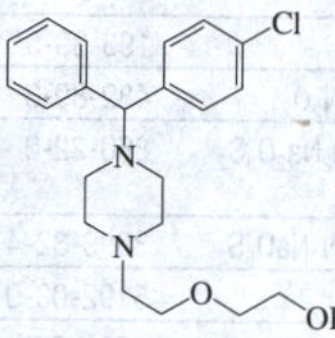

6201
Hydroxyzine

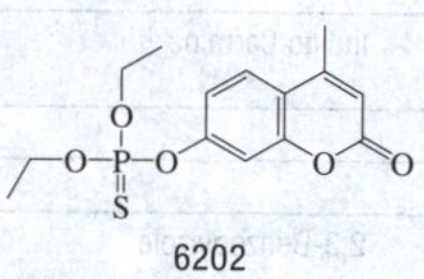

6202
Hymecromone *O,O*-diethyl phosphorothioate

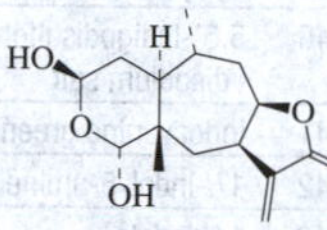

6203
Hymenoxone

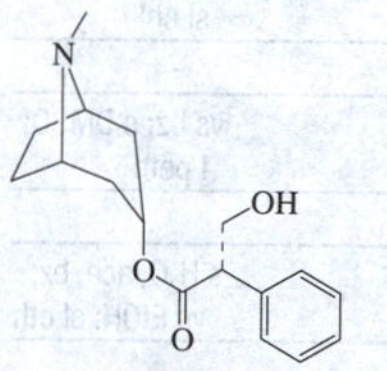

6204
Hyoscyamine

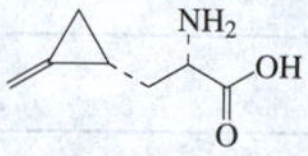

6205
Hypoglycin A

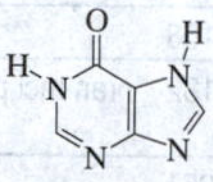

6206
Hypoxanthine

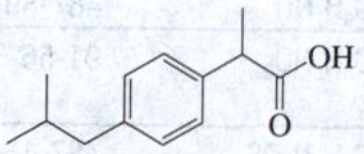

6207
Ibuprofen

6208
Icosylamine

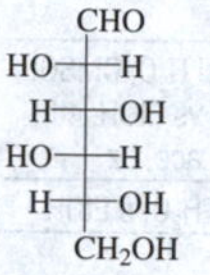

6209
D-Idose

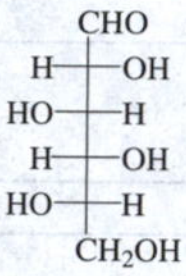

6210
L-Idose

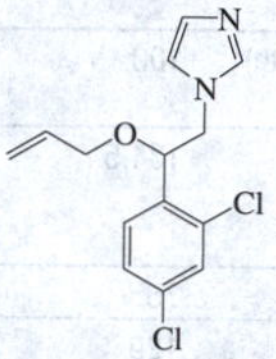

6211
Imazalil

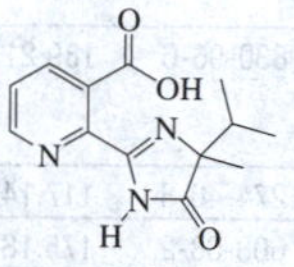

6212
Imazapyr

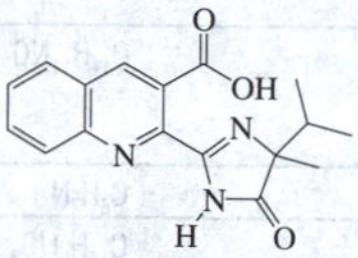

6213
Imazaquin

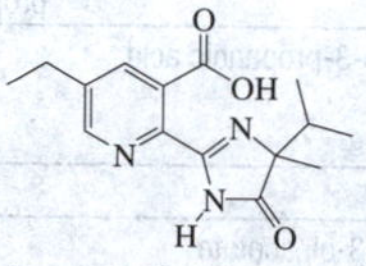

6214
Imazethapyr

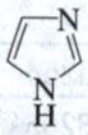

6215
Imidazole

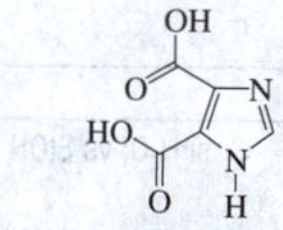

6216
1*H*-Imidazole-4,5-dicarboxylic acid

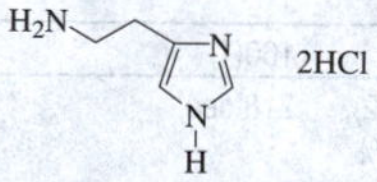

6217
1*H*-Imidazole-4-ethanamine, dihydrochloride

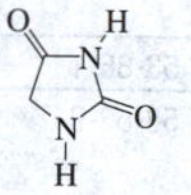

6218
2,4-Imidazolidinedione

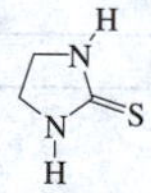

6219
2-Imidazolidinethione

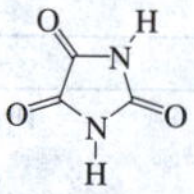

6220
Imidazolidinetrione

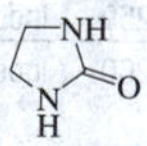

6221
2-Imidazolidinone

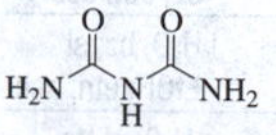

6222
Imidodicarbonic diamide

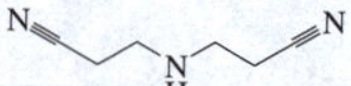

6223
3,3'-Iminobispropanenitrile

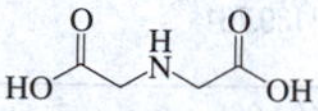

6224
Iminodiacetic acid

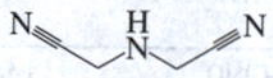

6225
Iminodiacetic acid, dinitrile

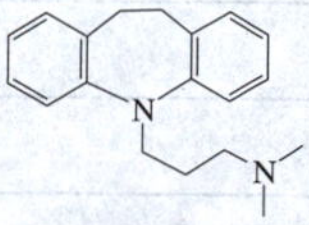

6226
Imipramine

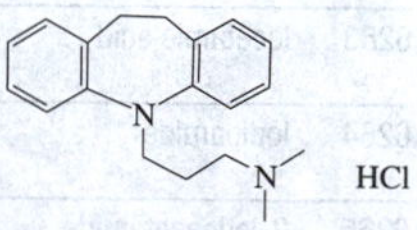

6227
Imipramine hydrochloride

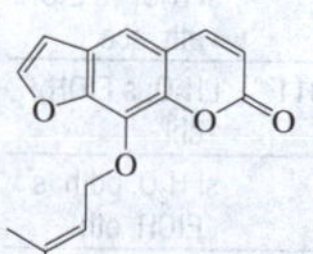

6228
Imperatorin

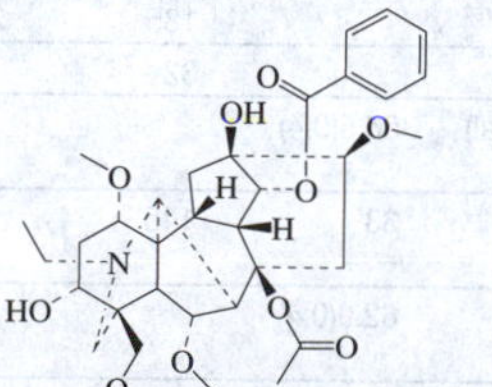

6229
Indaconitine

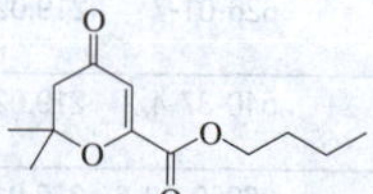

6230
Indalone

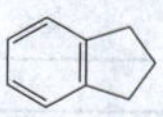

6231
Indan

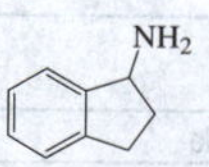

6232
1-Indanamine

6233
1*H*-Indazole

No.	Name	Synonym	Mol. form.	CAS Reg. No.	Mol. wt.	Physical form	mp/°C	bp/°C	den/ g cm^{-3}	n_D	Solubility
6235	Indene	Indonaphthene	C_9H_8	95-13-6	116.160	liq	-1.45(0.02)	182.5(0.4)	0.9960^{25}	1.5768^{20}	i H_2O; msc EtOH, eth; s ace, bz, py; sl chl
6236	1*H*-Indene-1,3(2*H*)-dione		$C_9H_6O_2$	606-23-5	146.143	nd (eth, lig)	129(3)		1.37^{21}		sl H_2O, ctc; vs EtOH; s eth, bz, alk
6237	1*H*-Indene-1,2,3-trione monohydrate	Ninhydrin	$C_9H_6O_4$	485-47-2	178.142	pa ye pr (w, al)	242 dec				vs H_2O; s EtOH, alk; sl eth
6238	Indeno[1,2,3-*cd*]pyrene	1,10-(1,2-Phenylene)pyrene	$C_{22}H_{12}$	193-39-5	276.330	ye cry (cy)	164(2)				
6239	Indigo		$C_{16}H_{10}N_2O_2$	482-89-3	262.262	dk bl pow	390 dec	300 sub			
6240	5,5'-Indigodisulfonic acid, disodium salt	Indigo Carmine	$C_{16}H_8N_2Na_2O_8S_2$	860-22-0	466.353	dk-bl pow					sl H_2O, EtOH; i os
6241	Indocyanine green		$C_{43}H_{47}N_2NaO_6S_2$	3599-32-4	774.962	grn pow	244 dec				
6242	1*H*-Indol-5-amine		$C_8H_8N_2$	5192-03-0	132.163		132				
6243	1*H*-Indole	2,3-Benzopyrrole	C_8H_7N	120-72-9	117.149	lf (w, peth) cry (eth)	52.3(0.6)	254(3)	1.22^{25}		s H_2O, bz; vs EtOH, eth, tol; sl ctc
6244	1*H*-Indole-3-acetic acid	Indoleacetic acid	$C_{10}H_9NO_2$	87-51-4	175.184	lf (bz), pl (chl)	168.5				i H_2O; vs EtOH; s eth, ace, bz; sl chl
6245	1*H*-Indole-3-acetonitrile		$C_{10}H_8N_2$	771-51-7	156.184		36	$160^{0.2}$			
6246	1*H*-Indole-3-butanoic acid	Indolebutyric acid	$C_{12}H_{13}NO_2$	133-32-4	203.237		124.5				vs bz; s DMSO; i peth
6247	1*H*-Indole-3-carboxaldehyde		C_9H_7NO	487-89-8	145.158		197.8				
6248	1*H*-Indole-2,3-dione	Isatin	$C_8H_5NO_2$	91-56-5	147.132	oran mcl pr	201(1)				s H_2O, ace, bz; vs EtOH; sl eth
6249	1*H*-Indole-2,3-dione, 3-thiosemicarbazone	Isatin, 3-thiosemicarbazone	$C_9H_8N_4OS$	487-16-1	220.251		283				
6250	1*H*-Indole-3-ethanamine, monohydrochloride	Tryptamine hydrochloride	$C_{10}H_{13}ClN_2$	343-94-2	196.676	nd (al-bz or lig)	255				vs ace, EtOH
6251	1*H*-Indole-3-ethanol	Tryptophol	$C_{10}H_{11}NO$	526-55-6	161.200	pr (bz-peth)	59	174^{2}			vs ace, eth, EtOH, chl
6252	1*H*-Indole-3-lactic acid, (*S*)-	α-Hydroxy-1*H*-indole-3-propanoic acid	$C_{11}H_{11}NO_3$	7417-65-4	205.210	cry (peth)	100				
6253	1*H*-Indole-3-propanoic acid		$C_{11}H_{11}NO_2$	830-96-6	189.211		134.5				sl H_2O, DMSO; vs EtOH, eth, ace, bz
6254	Indolizine		C_8H_7N	274-40-8	117.149	pl	75	205			i H_2O; s EtOH
6255	1*H*-Indol-3-ol, acetate		$C_{10}H_9NO_2$	608-08-2	175.184		129				
6256	1-(1*H*-Indol-3-yl)ethanone		$C_{10}H_9NO$	703-80-0	159.184	nd (bz)	192.3	144^{10}			vs EtOH
6257	1-(1*H*-Indol-3-yl)-2-propanone	3-Indolylacetone	$C_{11}H_{11}NO$	1201-26-9	173.211	br orth (bz), nd (aq MeOH)	116				
6258	3-(1*H*-Indol-3-yl)-2-propenoic acid	3-Indolylacrylic acid	$C_{11}H_9NO_2$	1204-06-4	187.195		185 dec				
6259	Indomethacin		$C_{19}H_{16}ClNO_4$	53-86-1	357.788		160(1)				
6260	Inosine	Hypoxanthine riboside	$C_{10}H_{12}N_4O_5$	58-63-9	268.226	pl (w + 2), nd (80% al)	218 dec				sl H_2O; vs EtOH
6261	Inosine 5'-monophosphate	5'-Inosinic acid	$C_{10}H_{13}N_4O_8P$	131-99-7	348.206	visc liq or glass					vs H_2O; sl EtOH, eth
6262	*myo*-Inositol	(1α,2α,3α,4β,5α,6β)-Cyclohexanehexol	$C_6H_{12}O_6$	87-89-8	180.155	cry (w)	223.8(0.4)		1.752		s H_2O
6263	Iocetamic acid		$C_{12}H_{13}I_3N_2O_3$	16034-77-8	613.955	wh-ye pow	225				i H_2O; sl EtOH, bz, eth, ace
6264	Iodipamide		$C_{20}H_{14}I_6N_2O_6$	606-17-7	1139.761		307 dec				i H_2O, bz; sl EtOH, eth, ace
6265	2-Iodoacetamide		C_2H_4INO	144-48-9	184.963		93.0				s H_2O; sl tfa
6266	Iodoacetic acid		$C_2H_3IO_2$	64-69-7	185.948		82.5	dec			s H_2O, EtOH, peth; sl eth, chl
6267	Iodoacetone		C_3H_5IO	3019-04-3	183.975			62^{12}	2.17^{15}		s EtOH
6268	Iodoacetonitrile		C_2H_2IN	624-75-9	166.948			185	2.307^{25}	1.5744^{20}	
6269	Iodoacetylene		C_2HI	14545-08-5	151.933			32			
6270	2-Iodoaniline		C_6H_6IN	615-43-0	219.023	nd (dil al)	56.5(0.4)				sl H_2O; vs EtOH, eth, ace
6271	3-Iodoaniline		C_6H_6IN	626-01-7	219.023	lf	33	145^{15}		1.6811^{20}	i H_2O; s EtOH, chl
6272	4-Iodoaniline		C_6H_6IN	540-37-4	219.023	nd (w)	62.9(0.4)				sl H_2O, peth; s EtOH, eth
6273	2-Iodobenzaldehyde		C_7H_5IO	26260-02-6	232.018		37	129^{14}			sl H_2O; s ace

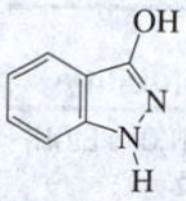
6234
1*H*-Indazol-3-ol

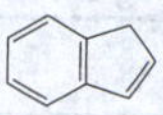
6235
Indene

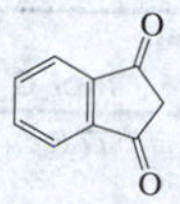
6236
1*H*-Indene-1,3(2*H*)-dione

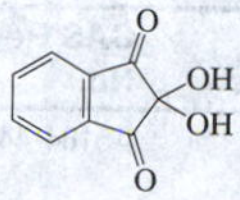
6237
1*H*-Indene-1,2,3-trione monohydrate

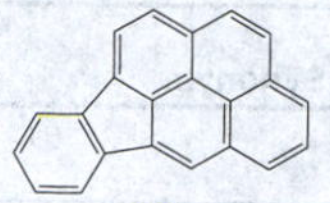
6238
Indeno[1,2,3-cd]pyrene

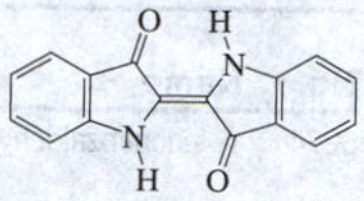
6239
Indigo

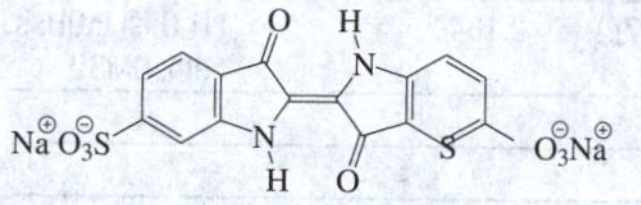
6240
5,5'-Indigodisulfonic acid, disodium salt

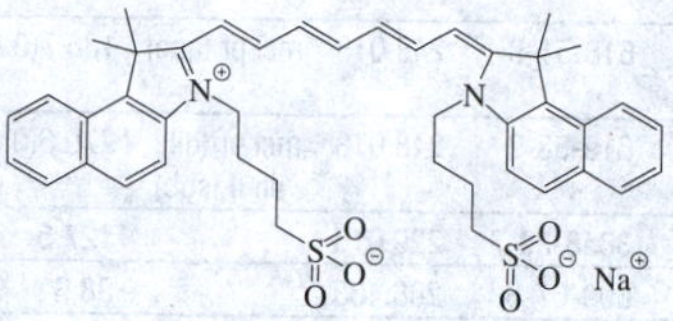
6241
Indocyanine green

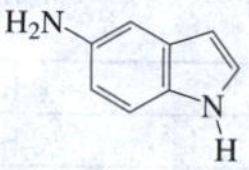
6242
1*H*-Indol-5-amine

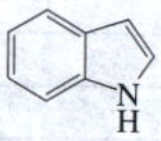
6243
1*H*-Indole

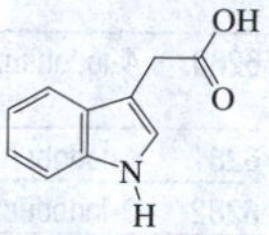
6244
1*H*-Indole-3-acetic acid

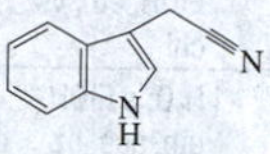
6245
1*H*-Indole-3-acetonitrile

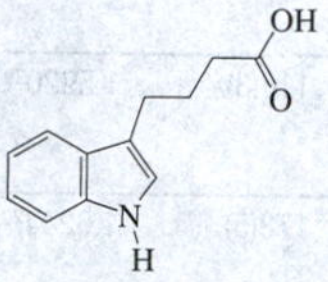
6246
1*H*-Indole-3-butanoic acid

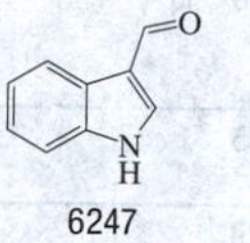
6247
1*H*-Indole-3-carboxaldehyde

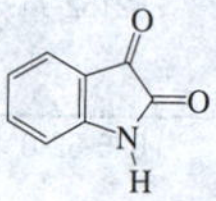
6248
1*H*-Indole-2,3-dione

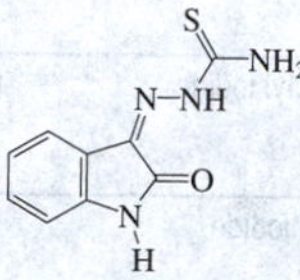
6249
1*H*-Indole-2,3-dione, 3-thiosemicarbazone

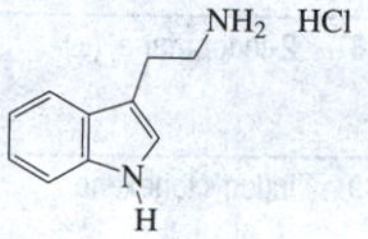
6250
1*H*-Indole-3-ethanamine, monohydrochloride

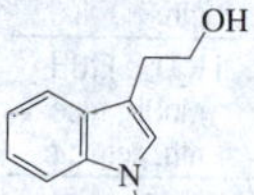
6251
1*H*-Indole-3-ethanol

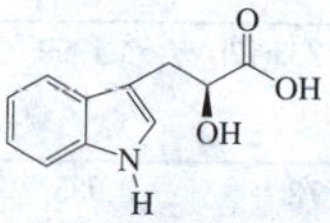
6252
1*H*-Indole-3-lactic acid, (*S*)-

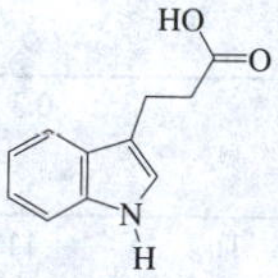
6253
1*H*-Indole-3-propanoic acid

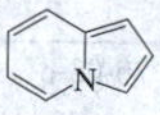
6254
Indolizine

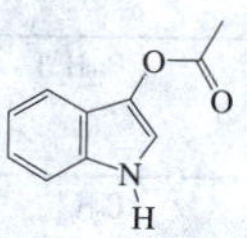
6255
1*H*-Indol-3-ol, acetate

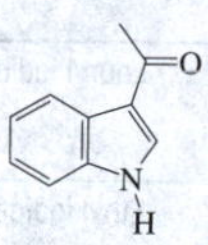
6256
1-(1*H*-Indol-3-yl)ethanone

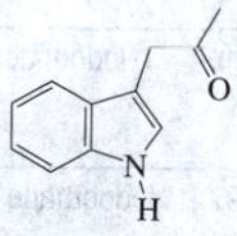
6257
1-(1*H*-Indol-3-yl)-2-propanone

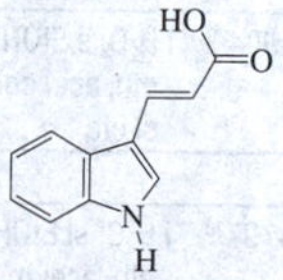
6258
3-(1*H*-Indol-3-yl)-2-propenoic acid

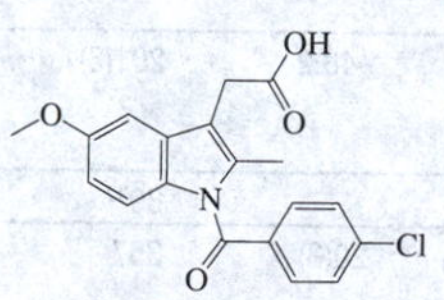
6259
Indomethacin

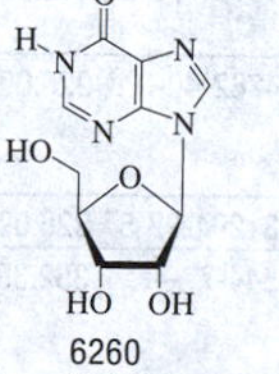
6260
Inosine

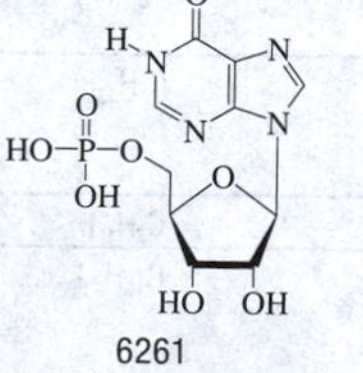
6261
Inosine 5'-monophosphate

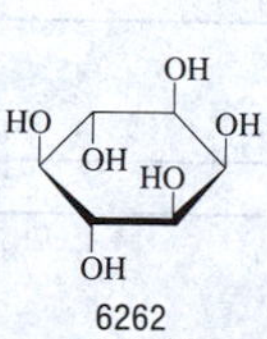
6262
myo-Inositol

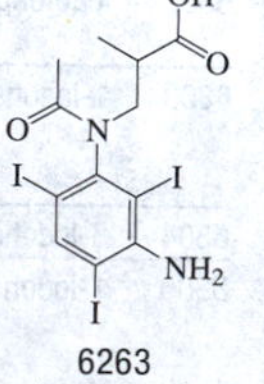
6263
Iocetamic acid

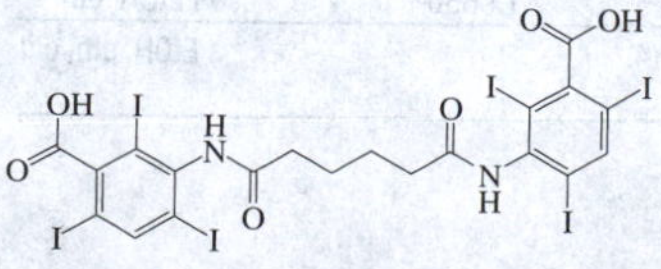
6264
Iodipamide

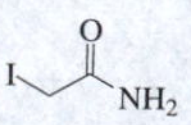
6265
2-Iodoacetamide

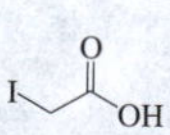
6266
Iodoacetic acid

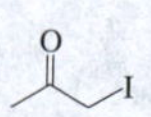
6267
Iodoacetone

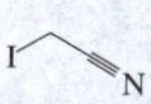
6268
Iodoacetonitrile

6269
Iodoacetylene

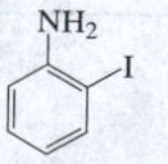
6270
2-Iodoaniline

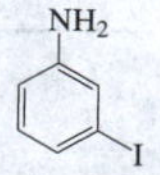
6271
3-Iodoaniline

No.	Name	Synonym	Mol. form.	CAS Reg. No.	Mol. wt.	Physical form	mp/°C	bp/°C	den/ g cm^{-3}	n_D	Solubility
6274	4-Iodobenzaldehyde		C_7H_5IO	15164-44-0	232.018		77.5	265			sl H_2O; s EtOH, bz
6275	Iodobenzene	Phenyl iodide	C_6H_5I	591-50-4	204.008	liq	-30.7(0.5)	188.5(0.6)	1.8308^{20}	1.6200^{20}	i H_2O; s EtOH; msc eth, ace, bz, ctc
6276	2-Iodobenzenemethanol		C_7H_7IO	5159-41-1	234.034		92	148^{32}		1.6349^{20}	
6277	4-Iodobenzenesulfonyl chloride	Pipsyl chloride	$C_6H_4ClIO_2S$	98-61-3	302.517		85				
6278	2-Iodobenzoic acid		$C_7H_5IO_2$	88-67-5	248.018	nd (w)	161.6(0.5)	exp	2.25^{25}		sl H_2O, ace; vs EtOH, eth
6279	3-Iodobenzoic acid		$C_7H_5IO_2$	618-51-9	248.018	mcl pr (ace)	186.7(0.8)	329(19)			sl H_2O, eth; vs EtOH
6280	4-Iodobenzoic acid		$C_7H_5IO_2$	619-58-9	248.018	mcl pr (dil al) lf (sub)	270.6(0.3)	316(17)	2.184^{20}		i H_2O; sl EtOH; s eth, DMSO
6281	4-Iodobenzonitrile		C_7H_4IN	3058-39-7	229.018		127.5				
6282	2-Iodobenzoyl chloride		C_7H_4ClIO	609-67-6	266.463		38.3	159^{27}			
6283	4-Iodobenzoyl chloride		C_7H_4ClIO	1711-02-0	266.463		64.5(0.8)	164^{32}			
6284	2-Iodo-1,1′-biphenyl		$C_{12}H_9I$	2113-51-1	280.103			190^{36}	1.5511^{25}	1.6620^{20}	i H_2O; s EtOH, eth, bz, HOAc
6285	3-Iodo-1,1′-biphenyl		$C_{12}H_9I$	20442-79-9	280.103		26.5	188^{16}	1.5967^{25}		
6286	4-Iodo-1,1′-biphenyl		$C_{12}H_9I$	1591-31-7	280.103	nd (al, HOAc)	113.5	320			i H_2O; s EtOH, eth, bz, HOAc
6287	1-Iodobutane	Butyl iodide	C_4H_9I	542-69-8	184.018	liq	-103.5(0.4)	130(2)	1.6154^{20}	1.5001^{20}	i H_2O; msc EtOH, eth; vs chl
6288	2-Iodobutane, (±)-	(±)-*sec*-Butyl iodide	C_4H_9I	52152-71-3	184.018	liq	-104.2(0.4)	118(3)	1.5920^{20}	1.4991^{20}	i H_2O; msc EtOH, eth; vs chl
6289	Iodocyclohexane	Cyclohexyl iodide	$C_6H_{11}I$	626-62-0	210.055			178(5)	1.6244^{20}	1.5477^{20}	i H_2O; s EtOH, eth, ace, bz
6290	Iodocyclopentane	Cyclopentyl iodide	C_5H_9I	1556-18-9	196.029			166.5	1.7096^{20}	1.5447^{20}	i H_2O; s eth, bz; sl ctc
6291	1-Iododecane		$C_{10}H_{21}I$	2050-77-3	268.178	liq	-16.3	263.7	1.2546^{20}	1.4858^{20}	i H_2O; s EtOH, eth, ctc
6292	1-Iodo-2,4-dimethylbenzene		C_8H_9I	4214-28-2	232.061			231(7)	1.6282^{16}	1.6008^{16}	i H_2O; s ace, bz
6293	2-Iodo-1,3-dimethylbenzene		C_8H_9I	608-28-6	232.061	oil	11.2(0.4)	228(3)	1.6158^{20}	1.6035^{20}	i H_2O; s ace, bz
6294	2-Iodo-1,4-dimethylbenzene		C_8H_9I	1122-42-5	232.061			232(11)	1.6168^{17}	1.5992^{17}	i H_2O; s ace, bz
6295	1-Iodo-2,2-dimethylpropane		$C_5H_{11}I$	15501-33-4	198.045			130(13)	1.4940^{20}	1.4890^{20}	i H_2O; s EtOH, eth
6296	1-Iodododecane	Lauryl iodide	$C_{12}H_{25}I$	4292-19-7	296.231		0.3	285(12)	1.1999^{20}	1.4840^{20}	i H_2O; s EtOH, MeOH; msc eth, ace, ctc
6297	Iodoethane	Ethyl iodide	C_2H_5I	75-03-6	155.965	liq	-111.0(0.4)	72(1)	1.9357^{20}	1.5133^{20}	sl H_2O; msc EtOH; s eth, chl
6298	2-Iodoethanol		C_2H_5IO	624-76-0	171.964			155(15)	2.1967^{20}	1.5713^{20}	vs H_2O, eth, EtOH
6299	Iodoethene	Vinyl iodide	C_2H_3I	593-66-8	153.949			57(4)	2.037^{20}	1.5385^{20}	vs eth, EtOH
6300	(2-Iodoethyl)benzene		C_8H_9I	17376-04-4	232.061	liq		122^{13}	1.603	1.6010^{20}	
6301	2-(1-Iodoethyl)-1,3-dioxolane-4-methanol	Iodinated glycerol	$C_6H_{11}IO_3$	5634-39-9	258.053	pale ye liq			1.797	1.547	s eth, chl, thf, AcOEt
6302	Iodofenphos		$C_8H_8Cl_2IO_3PS$	18181-70-9	412.997	wh cry	76				i H_2O; s ace, xyl; sl EtOH
6303	1-Iodoheptane		$C_7H_{15}I$	4282-40-0	226.098	liq	-48.2	201(6)	1.3719^{25}	1.4904^{20}	i H_2O; s EtOH, eth, ace, chl; sl ctc
6304	3-Iodoheptane		$C_7H_{15}I$	31294-92-5	226.098			89^{30}	1.3676^{20}		
6305	1-Iodohexadecane		$C_{16}H_{33}I$	544-77-4	352.337	pa ye liq	20(3)	357	1.1213^{25}	1.4797^{20}	i H_2O; sl EtOH; s eth, ace; msc bz; vs chl
6306	1-Iodohexane	Hexyl iodide	$C_6H_{13}I$	638-45-9	212.071	liq	-74.1(0.6)	182(3)	1.4305^{25}	1.4928^{20}	i H_2O
6307	Iodomethane	Methyl iodide	CH_3I	74-88-4	141.939	liq	-66(2)	42.4(0.2)	2.2789^{20}	1.5308^{20}	sl H_2O; s ace, bz, chl; msc EtOH, eth
6308	1-Iodo-2-methoxybenzene	*o*-Iodoanisole	C_7H_7IO	529-28-2	234.034			241	1.8^{20}		vs EtOH, eth, ace, bz, chl, lig
6309	1-Iodo-3-methoxybenzene	*m*-Iodoanisole	C_7H_7IO	766-85-8	234.034			244.5	1.9650^{20}		vs EtOH, eth
6310	1-Iodo-4-methoxybenzene	*p*-Iodoanisole	C_7H_7IO	696-62-8	234.034	lf (al), nd (MeOH)	53	247(4)			s EtOH, eth, chl

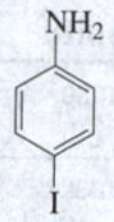
6272
4-Iodoaniline

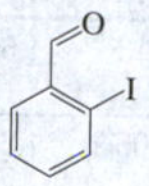
6273
2-Iodobenzaldehyde

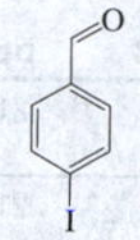
6274
4-Iodobenzaldehyde

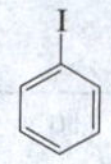
6275
Iodobenzene

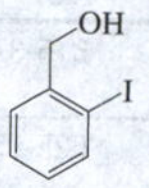
6276
2-Iodobenzenemethanol

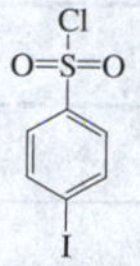
6277
4-Iodobenzenesulfonyl chloride

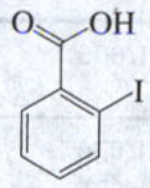
6278
2-Iodobenzoic acid

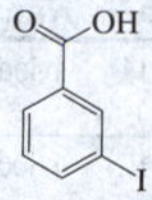
6279
3-Iodobenzoic acid

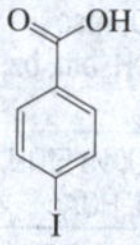
6280
4-Iodobenzoic acid

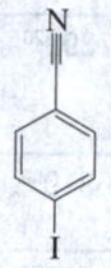
6281
4-Iodobenzonitrile

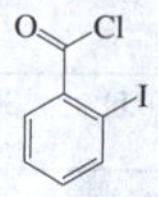
6282
2-Iodobenzoyl chloride

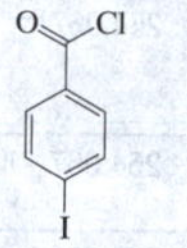
6283
4-Iodobenzoyl chloride

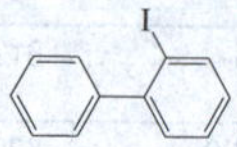
6284
2-Iodo-1,1'-biphenyl

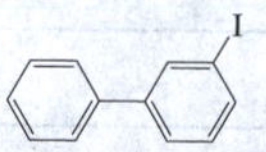
6285
3-Iodo-1,1'-biphenyl

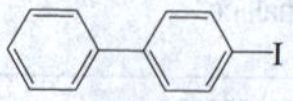
6286
4-Iodo-1,1'-biphenyl

6287
1-Iodobutane

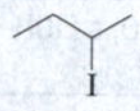
6288
2-Iodobutane, (±)-

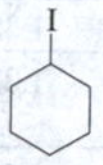
6289
Iodocyclohexane

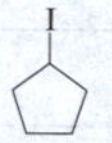
6290
Iodocyclopentane

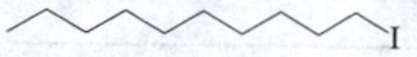
6291
1-Iododecane

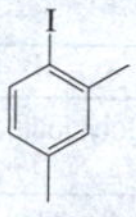
6292
1-Iodo-2,4-dimethylbenzene

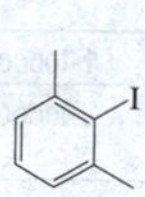
6293
2-Iodo-1,3-dimethylbenzene

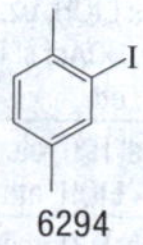
6294
2-Iodo-1,4-dimethylbenzene

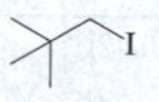
6295
1-Iodo-2,2-dimethylpropane

6296
1-Iodododecane

6297
Iodoethane

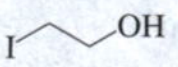
6298
2-Iodoethanol

6299
Iodoethene

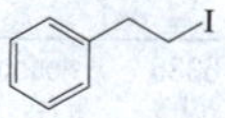
6300
(2-Iodoethyl)benzene

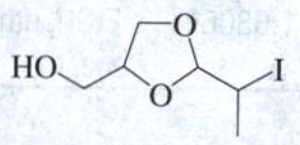
6301
2-(1-Iodoethyl)-1,3-dioxolane-4-methanol

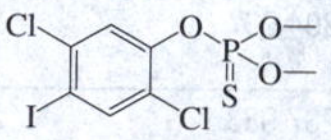
6302
Iodofenphos

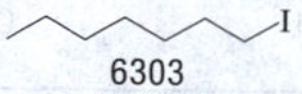
6303
1-Iodoheptane

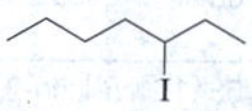
6304
3-Iodoheptane

6305
1-Iodohexadecane

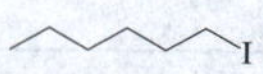
6306
1-Iodohexane

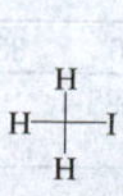
6307
Iodomethane

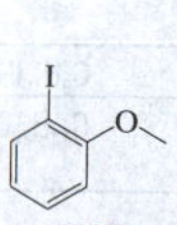
6308
1-Iodo-2-methoxybenzene

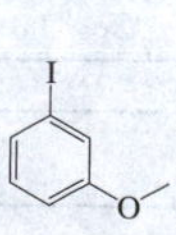
6309
1-Iodo-3-methoxybenzene

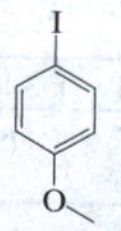
6310
1-Iodo-4-methoxybenzene

No.	Name	Synonym	Mol. form.	CAS Reg. No.	Mol. wt.	Physical form	mp/°C	bp/°C	den/g cm^{-3}	n_D	Solubility
6311	1-Iodo-2-methylbenzene		C_7H_7I	615-37-2	218.035			211.5	1.713^{20}	1.6079^{20}	i H_2O; msc EtOH, eth
6312	1-Iodo-3-methylbenzene		C_7H_7I	625-95-6	218.035	liq	-27.2	218(19)	1.705^{20}	1.6053^{20}	i H_2O; msc EtOH, eth
6313	(Iodomethyl)benzene		C_7H_7I	620-05-3	218.035	col or ye nd (MeOH)	26.3(0.2)	93^{10}	1.7335^{25}	1.6334^{25}	vs bz, eth, EtOH
6314	1-Iodo-3-methylbutane	Isopentyl iodide	$C_5H_{11}I$	541-28-6	198.045			145(2)	1.5118^{20}	1.4939^{20}	sl H_2O, ctc; msc EtOH, eth
6315	2-Iodo-2-methylbutane	*tert*-Pentyl iodide	$C_5H_{11}I$	594-38-7	198.045			124.5	1.4937^{20}	1.4981^{20}	i H_2O; msc EtOH, eth
6316	1-Iodo-2-methylpropane	Isobutyl iodide	C_4H_9I	513-38-2	184.018			121(1)	1.6035^{20}	1.4959^{20}	
6317	2-Iodo-2-methylpropane	*tert*-Butyl iodide	C_4H_9I	558-17-8	184.018	liq	-33.6(0.4)	98(6)	1.571^{25}	1.4918^{20}	msc EtOH, eth
6318	Iodomethylsilane		CH_5ISi	18089-64-0	172.041	col liq	-109.5	71.8			
6319	1-Iodonaphthalene		$C_{10}H_7I$	90-14-2	254.067		7(1)	303(77)	1.7399^{20}	1.7026^{20}	i H_2O; msc EtOH, eth, bz, CS_2
6320	2-Iodonaphthalene		$C_{10}H_7I$	612-55-5	254.067	lf (dil al)	54.3(0.5)	308	1.6319^{99}	1.6662^{99}	i H_2O; vs EtOH, eth, HOAc
6321	1-Iodo-2-nitrobenzene		$C_6H_4INO_2$	609-73-4	249.006	ye orth nd (al)	50(1)	290	1.9186^{75}		i H_2O; s EtOH, eth
6322	1-Iodo-3-nitrobenzene		$C_6H_4INO_2$	645-00-1	249.006	mcl pr	36(1)	280	1.9477^{50}		i H_2O; s EtOH, eth
6323	1-Iodo-4-nitrobenzene		$C_6H_4INO_2$	636-98-6	249.006	ye nd (al)	173(2)	288	1.8090^{155}		i H_2O; s EtOH, HOAc; sl DMSO
6324	1-Iodononane		$C_9H_{19}I$	4282-42-2	254.151	col liq	-20	245.0	1.2836^{25}	1.4848^{25}	
6325	1-Iodooctadecane		$C_{18}H_{37}I$	629-93-6	380.391	lf (lig), nd (ace, al-ace)	34(2)	379(20)	1.0994^{20}	1.4810^{20}	i H_2O; sl EtOH, eth
6326	1-Iodooctane		$C_8H_{17}I$	629-27-6	240.125	liq	-45.7(0.5)	226(4)	1.3298^{20}	1.4885^{20}	s EtOH, eth
6327	2-Iodooctane, (±)-	2-Octyl iodide, (±)	$C_8H_{17}I$	36049-78-2	240.125			212(12)	1.3251^{20}	1.4896^{20}	i H_2O; s EtOH, eth, lig
6328	1-Iodopentane	Amyl iodide	$C_5H_{11}I$	628-17-1	198.045	liq	-85.6	156(8)	1.5161^{20}	1.4959^{20}	s chl
6329	3-Iodopentane		$C_5H_{11}I$	1809-05-8	198.045			141(11)	1.5176^{20}	1.4974^{20}	vs ace, bz, eth
6330	2-Iodophenol		C_6H_5IO	533-58-4	220.007	nd	43	186^{160}	1.8757^{80}		s H_2O; vs EtOH, eth, CS_2
6331	3-Iodophenol		C_6H_5IO	626-02-8	220.007	nd (lig)	118	186^{100}			sl H_2O; s EtOH, eth
6332	4-Iodophenol		C_6H_5IO	540-38-5	220.007	nd (w or sub)	93.5	139^{5} dec	1.8573^{112}		sl H_2O; vs EtOH, eth
6333	1-(3-Iodophenyl)ethanone	3-Iodoacetophenone	C_8H_7IO	14452-30-3	246.045			129^{8}		1.622^{20}	s bz
6334	1-(4-Iodophenyl)ethanone	4-Iodoacetophenone	C_8H_7IO	13329-40-3	246.045		86	153^{18}			s EtOH, bz, CS_2, HOAc; sl lig, eth
6335	1-Iodopropane	Propyl iodide	C_3H_7I	107-08-4	169.992	liq	-101.4(0.4)	102(2)	1.7489^{20}	1.5058^{20}	sl H_2O, ctc; msc EtOH, eth
6336	2-Iodopropane	Isopropyl iodide	C_3H_7I	75-30-9	169.992	liq	-90.4(0.9)	89(3)	1.7042^{20}	1.5028^{20}	sl H_2O; msc EtOH, eth, bz, chl
6337	3-Iodopropanoic acid		$C_3H_5IO_2$	141-76-4	199.975	lf (w)	85				sl H_2O, chl; vs EtOH; s eth, ace
6338	3-Iodo-1-propanol		C_3H_7IO	627-32-7	185.991	visc oil		226	1.9976^{20}	1.5585^{20}	
6339	3-Iodopropene	Allyl iodide	C_3H_5I	556-56-9	167.976	ye liq	-98(2)	102(9)	1.848^{12}	1.5540^{21}	i H_2O; s EtOH, eth, chl
6340	2-Iodopyridine		C_5H_4IN	5029-67-4	204.997			100^{15}	1.928^{25}	1.6366^{20}	s EtOH, eth, ace, bz
6341	5-Iodo-2,4(1*H*,3*H*)-pyrimidinedione	5-Iodouracil	$C_4H_3IN_2O_2$	696-07-1	237.983		275 dec				
6342	1-Iodo-2,5-pyrrolidinedione	*N*-Iodosuccinimide	$C_4H_4INO_2$	516-12-1	224.985	cry (ace)	200.5		2.245^{25}		vs H_2O; s EtOH, ace; sl eth, DMSO
6343	Iodosylbenzene		C_6H_5IO	536-80-1	220.007	ye pow	210 exp				s H_2O, EtOH; i eth, ace, bz, peth
6344	2-Iodothiophene		C_4H_3IS	3437-95-4	210.036	liq	-40	167(8)	2.0595^{25}	1.6465^{25}	vs EtOH, eth; sl chl
6345	4-Iodotoluene		C_7H_7I	624-31-7	218.035	lf (al)	33.6(0.2)	214(4)	1.678^{20}		i H_2O; s EtOH, eth, CS_2; sl chl
6346	L-3-Iodotyrosine		$C_9H_{10}INO_3$	70-78-0	307.084	cry (w)	205 dec				
6347	*trans*-α-Ionone, (±)-		$C_{13}H_{20}O$	30685-95-1	192.297			146^{28}	0.9298^{21}	1.5041^{20}	vs ace, eth, EtOH

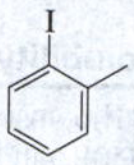
6311
1-Iodo-2-methylbenzene

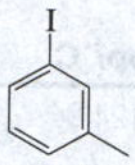
6312
1-Iodo-3-methylbenzene

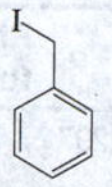
6313
(Iodomethyl)benzene

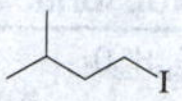
6314
1-Iodo-3-methylbutane

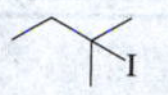
6315
2-Iodo-2-methylbutane

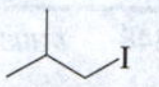
6316
1-Iodo-2-methylpropane

6317
2-Iodo-2-methylpropane

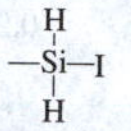
6318
Iodomethylsilane

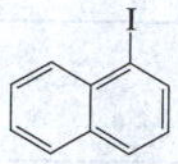
6319
1-Iodonaphthalene

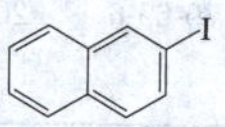
6320
2-Iodonaphthalene

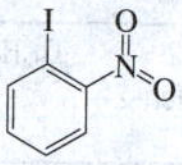
6321
1-Iodo-2-nitrobenzene

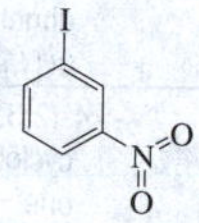
6322
1-Iodo-3-nitrobenzene

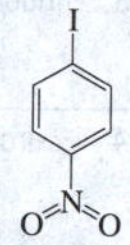
6323
1-Iodo-4-nitrobenzene

6324
1-Iodononane

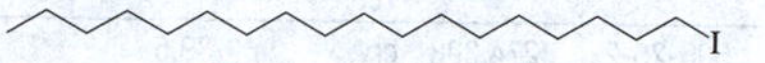
6325
1-Iodooctadecane

6326
1-Iodooctane

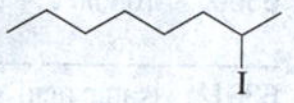
6327
2-Iodooctane, (±)-

6328
1-Iodopentane

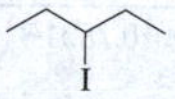
6329
3-Iodopentane

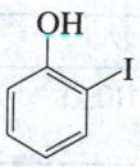
6330
2-Iodophenol

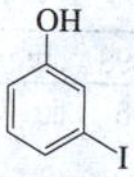
6331
3-Iodophenol

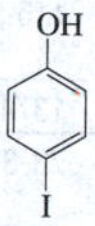
6332
4-Iodophenol

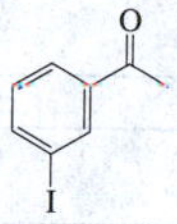
6333
1-(3-Iodophenyl)ethanone

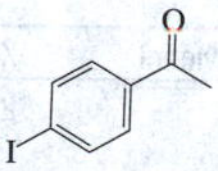
6334
1-(4-Iodophenyl)ethanone

6335
1-Iodopropane

6336
2-Iodopropane

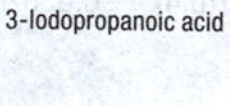
6337
3-Iodopropanoic acid

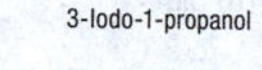
6338
3-Iodo-1-propanol

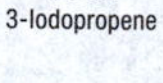
6339
3-Iodopropene

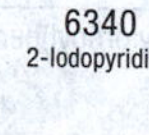
6340
2-Iodopyridine

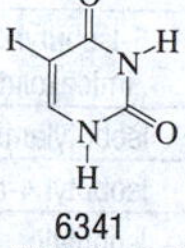
6341
5-Iodo-2,4(1*H*,3*H*)-pyrimidinedione

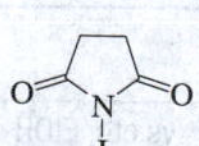
6342
1-Iodo-2,5-pyrrolidinedione

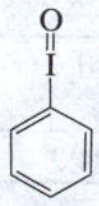
6343
Iodosylbenzene

6344
2-Iodothiophene

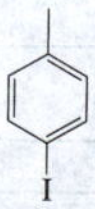
6345
4-Iodotoluene

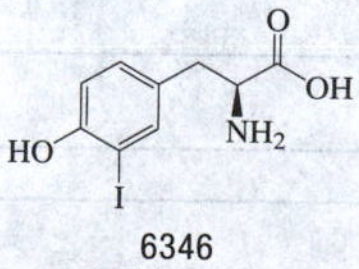
6346
L-3-Iodotyrosine

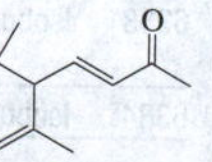
6347
trans-α-Ionone, (±)-

No.	Name	Synonym	Mol. form.	CAS Reg. No.	Mol. wt.	Physical form	mp/°C	bp/°C	den/g cm^{-3}	n_D	Solubility
6348	*trans*-β-Ionone		$C_{13}H_{20}O$	79-77-6	192.297			124^{10}	0.945^{20}	1.5198^{20}	sl H_2O; msc EtOH, eth; s chl
6349	Iopanoic acid		$C_{11}H_{12}I_3NO_2$	96-83-3	570.932	wh solid	156				i H_2O; s dil alk, EtOH
6350	Iophendylate	Ethyl 10-(4-iodophenyl)-undecanoate	$C_{19}H_{29}IO_2$	99-79-6	416.336	visc liq		197^{1}	1.25^{20}	1.525^{25}	sl H_2O; s EtOH, bz, chl
6351	Iopodic acid	Ipodate	$C_{12}H_{13}I_3N_2O_2$	5587-89-3	597.956	cry	168				i H_2O; vs EtOH, MeOH, chl, ace
6352	Iprodione		$C_{13}H_{13}Cl_2N_3O_3$	36734-19-7	330.166		136				
6353	Iridomyrmecin	Hexahydro-4,7-dimethylcyclopenta[*c*]pyran-3(1*H*)-one	$C_{10}H_{16}O_2$	485-43-8	168.233	pr	61	$106^{1.5}$		1.4607^{65}	sl H_2O; s eth
6354	α-Irone	4-(2,5,6,6-Tetramethyl-2-cyclohexen-1-yl)-3-buten-2-one	$C_{14}H_{22}O$	79-69-6	206.324			$90^{0.4}$	0.9362^{20}	1.5002^{20}	
6355	β-Irone	4-(2,5,6,6-Tetramethyl-1-cyclohexen-1-yl)-3-buten-2-one	$C_{14}H_{22}O$	79-70-9	206.324			125^{11}	0.9434^{21}	1.5162^{25}	sl H_2O; vs EtOH, eth, bz, chl
6356	Iron hydrocarbonyl	Hydrogen tetracarbonylferrate(II)	$C_4H_2FeO_4$	12002-28-7	169.902	col liq; unstab	-70	dec			s alk
6357	Iron nonacarbonyl	Diiron nonacarbonyl	$C_9Fe_2O_9$	15321-51-4	363.781	oran-ye cry	100 dec		2.85		
6358	Iron(III) NTA	Nitrilotriacetatoiron(III)	$C_6H_6FeNO_6$	16448-54-7	243.960	solid					s H_2O
6359	Iron pentacarbonyl		C_5FeO_5	13463-40-6	195.896	col to ye oily liq	-20	103	1.5^{20}	1.453^{22}	i H_2O; sl EtOH; s bz, ace, ctc
6360	Iron(III) 2,4-pentanedioate	Ferric acetylacetonate	$C_{15}H_{21}FeO_6$	14024-18-1	353.169	red-oran cry	188(1)		5.24		
6361	Isanic acid	17-Octadecene-9,11-diynoic acid	$C_{18}H_{26}O_2$	506-25-2	274.398	cry	39.5		0.9309^{45}	1.49148^{50}	s ace, EtOH, i-PrOH; sl peth
6362	Isatidine	Retrorsine *N*-oxide	$C_{18}H_{25}NO_7$	15503-86-3	367.395	cry	145				
6363	Isaxonine	*N*-Isopropyl-2-pyrimidinamine	$C_7H_{11}N_3$	4214-72-6	137.182		28	93^{12}			
6364	Isazophos		$C_9H_{17}ClN_3O_3PS$	67329-04-8	313.741			170	1.22^{20}		
6365	Isobenzan		$C_9H_4Cl_8O$	297-78-9	411.751	cry (hp)	122.2(0.3)				s eth, bz, xyl, tol
6366	1(3*H*)-Isobenzofuranone		$C_8H_6O_2$	87-41-2	134.133	nd or pl (w)	75	292(2)	1.1636^{99}	1.536^{99}	s H_2O; vs EtOH, eth; sl chl
6367	Isoborneol	1,7,7-Trimethylbicyclo[2.2.1]-heptan-2-ol, *exo*-(±)	$C_{10}H_{18}O$	24393-70-2	154.249	tab (peth)	212	sub	1.10^{20}		i H_2O; vs EtOH, eth, chl; sl bz
6368	Isobornyl thiocyanoacetate		$C_{13}H_{19}NO_2S$	115-31-1	253.361	ye oily liq		$95^{0.06}$	1.1465^{25}	1.512^{25}	i H_2O; vs EtOH, bz, chl, peth
6369	6-Isobornyl-3,4-xylenol	Xibornol	$C_{18}H_{26}O$	13741-18-9	258.398	cry	95	167^{3}	1.0240^{20}	1.5382^{20}	
6370	Isobutanal	2-Methyl-1-propanal	C_4H_8O	78-84-2	72.106	liq	-72.1(0.2)	64.1(0.2)	0.7891^{20}	1.3730^{20}	s H_2O, eth, ace, chl; sl ctc
6371	Isobutane	2-Methylpropane	C_4H_{10}	75-28-5	58.122	col gas	-159.59(0.02)	-11.7(0.5)	0.5510^{25} (p>1 atm)	1.3518^{-25}	sl H_2O; s EtOH, eth, chl
6372	Isobutene	2-Methyl-1-propene	C_4H_8	115-11-7	56.107	col gas	-140.7(0.2)	-7.0(0.2)	0.589^{25} (p>1 atm)	1.3926^{-25}	i H_2O; vs EtOH, eth; s bz, sulf
6373	Isobutyl acetate	2-Methylpropyl acetate	$C_6H_{12}O_2$	110-19-0	116.158	liq	-97.1(0.5)	116.9(0.6)	0.8712^{20}	1.3902^{20}	sl H_2O, ctc; msc EtOH, eth; s ace
6374	Isobutyl acrylate		$C_7H_{12}O_2$	106-63-8	128.169	liq	-61	137.0(0.8)	0.8896^{20}	1.4150^{20}	sl H_2O; s EtOH, eth, MeOH
6375	5-Isobutyl-3-allyl-2-thioxo-4-imidazolidinone	Albutoin	$C_{10}H_{16}N_2OS$	830-89-7	212.311		210.5				
6376	Isobutylamine	2-Methyl-1-propanamine	$C_4H_{11}N$	78-81-9	73.137	liq	-86(1)	68.8(0.1)	0.724^{25}	1.3988^{19}	
6377	Isobutyl 4-aminobenzoate	Isobutyl *p*-aminobenzoate	$C_{11}H_{15}NO_2$	94-14-4	193.243		64.5				
6378	Isobutylbenzene	(2-Methylpropyl)benzene	$C_{10}H_{14}$	538-93-2	134.218	liq	-51.6(0.2)	172.7(0.4)	0.8532^{20}	1.4866^{20}	i H_2O; msc EtOH, eth, ace, bz, peth, ctc
6379	Isobutyl benzoate		$C_{11}H_{14}O_2$	120-50-3	178.228			240(2)	0.9990^{20}		i H_2O; msc EtOH, eth; s ace, chl
6380	Isobutyl butanoate		$C_8H_{16}O_2$	539-90-2	144.212			157(1)	0.8364^{18}	1.4032^{20}	sl H_2O; msc EtOH, eth
6381	Isobutyl carbamate		$C_5H_{11}NO_2$	543-28-2	117.147	lf	67	207		1.4098^{76}	vs eth, EtOH
6382	Isobutyl chlorocarbonate		$C_5H_9ClO_2$	543-27-1	136.577			128.8	1.0426^{18}	1.4071^{18}	s EtOH, bz, chl; msc eth
6383	Isobutyl 2-chloropropanoate		$C_7H_{13}ClO_2$	114489-96-2	164.630			176	1.0312^{20}	1.4247^{20}	
6384	Isobutyl 3-chloropropanoate		$C_7H_{13}ClO_2$	62108-68-3	164.630			191.3	1.0323^{20}	1.4295^{20}	vs eth, EtOH

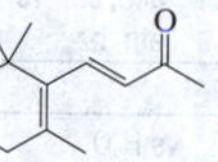
6348
trans-β-Ionone

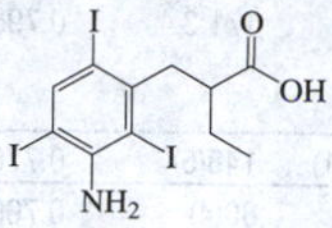
6349
Iopanoic acid

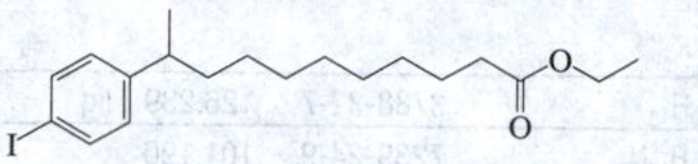
6350
Iophendylate

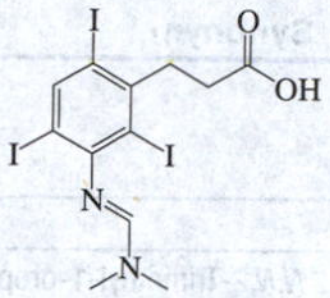
6351
Iopodic acid

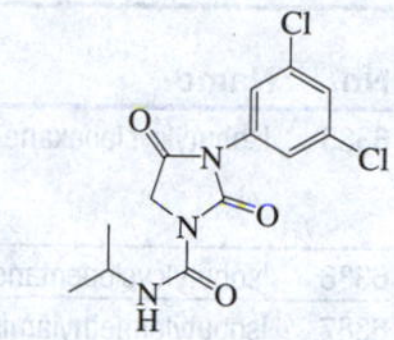
6352
Iprodione

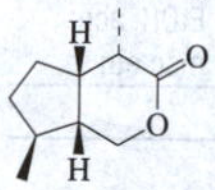
6353
Iridomyrmecin

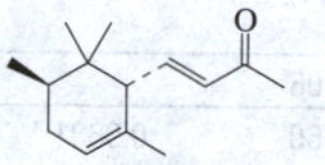
6354
α-Irone

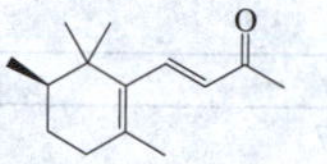
6355
β-Irone

$H_2Fe(CO)_4$
6356
Iron hydrocarbonyl

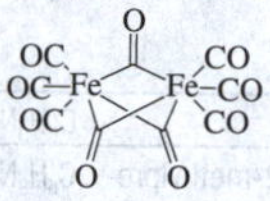
6357
Iron nonacarbonyl

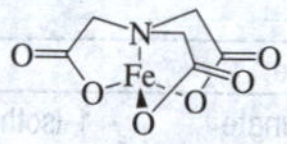
6358
Iron(III) NTA

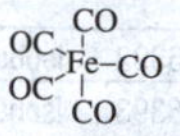
6359
Iron pentacarbonyl

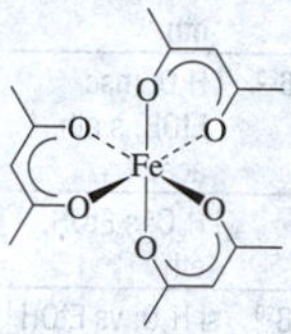
6360
Iron(III) 2,4-pentanedioate

6361
Isanic acid

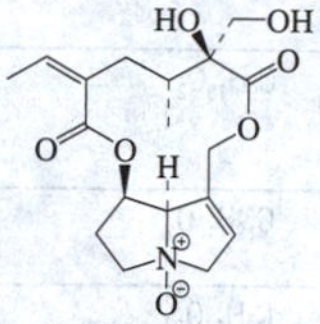
6362
Isatidine

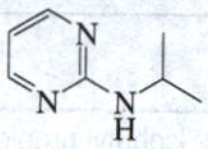
6363
Isaxonine

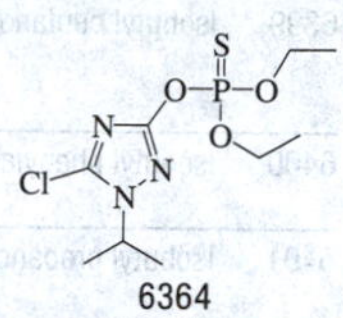
6364
Isazophos

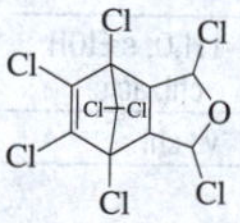
6365
Isobenzan

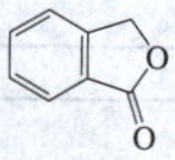
6366
1(3*H*)-Isobenzofuranone

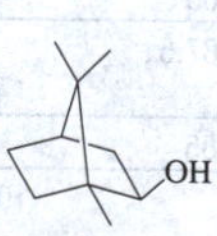
6367
Isoborneol

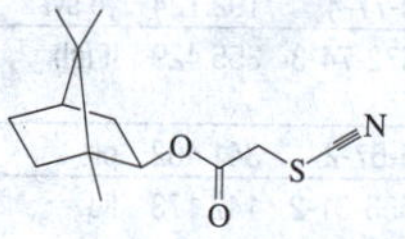
6368
Isobornyl thiocyanoacetate

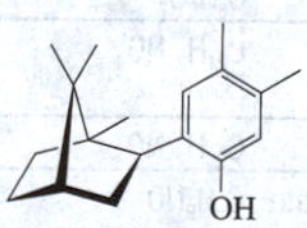
6369
6-Isobornyl-3,4-xylenol

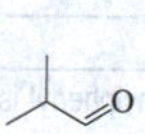
6370
Isobutanal

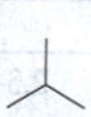
6371
Isobutane

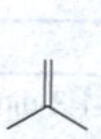
6372
Isobutene

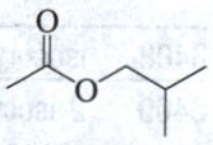
6373
Isobutyl acetate

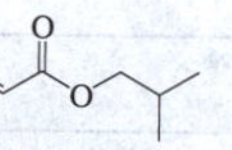
6374
Isobutyl acrylate

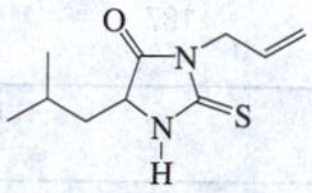
6375
5-Isobutyl-3-allyl-2-thioxo-4-imidazolidinone

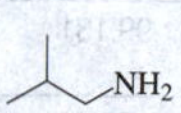
6376
Isobutylamine

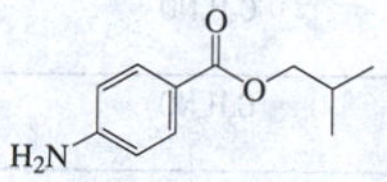
6377
Isobutyl 4-aminobenzoate

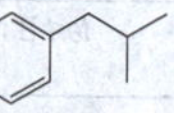
6378
Isobutylbenzene

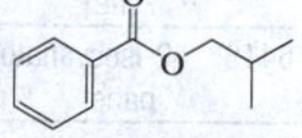
6379
Isobutyl benzoate

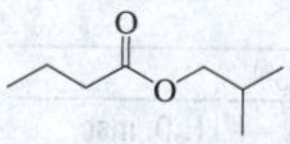
6380
Isobutyl butanoate

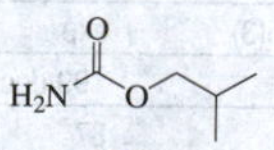
6381
Isobutyl carbamate

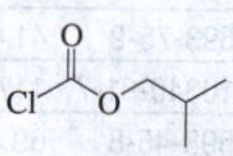
6382
Isobutyl chlorocarbonate

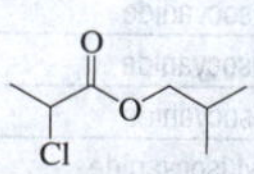
6383
Isobutyl 2-chloropropanoate

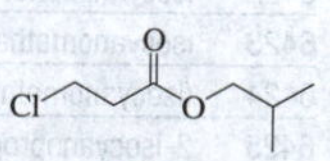
6384
Isobutyl 3-chloropropanoate

No.	Name	Synonym	Mol. form.	CAS Reg. No.	Mol. wt.	Physical form	mp/°C	bp/°C	den/ g cm^{-3}	n_D	Solubility
6385	Isobutylcyclohexane		$C_{10}H_{20}$	1678-98-4	140.266	liq	-95	171.3	0.7952^{20}	1.4386^{20}	i H_2O; s EtOH, ace, chl; vs eth, bz
6386	Isobutylcyclopentane		C_9H_{18}	3788-32-7	126.239	liq	-115.2(0.1)	148(5)	0.7769^{25}	1.4298^{20}	
6387	Isobutyldimethylamine	*N,N*,2-Trimethyl-1-propanamine	$C_6H_{15}N$	7239-24-9	101.190			80(4)	0.7097^{20}	1.3907^{20}	vs H_2O
6388	Isobutyl formate	2-Methylpropyl formate	$C_5H_{10}O_2$	542-55-2	102.132	liq	-95.5(0.7)	98.4(0.3)	0.8776^{20}	1.3857^{20}	sl H_2O, chl; msc EtOH, eth; vs ace
6389	Isobutyl heptanoate	Isobutyl enanthate	$C_{11}H_{22}O_2$	7779-80-8	186.292			208	0.8593^{20}		vs ace, bz, eth, EtOH
6390	Isobutyl 2-hydroxybenzoate	Isobutyl salicylate	$C_{11}H_{14}O_3$	87-19-4	194.227		5.9	261	1.0639^{20}	1.5087^{20}	i H_2O; s EtOH, eth, ctc
6391	Isobutyl isobutanoate		$C_8H_{16}O_2$	97-85-8	144.212	liq	-80.6(0.2)	148(3)	0.8542^{20}	1.3999^{20}	sl H_2O, ctc; s EtOH, ace; msc eth
6392	Isobutyl isocyanate		C_5H_9NO	1873-29-6	99.131			106			
6393	Isobutyl isothiocyanate	1-Isothiocyanato-2-methylpropane	C_5H_9NS	591-82-2	115.197			160	0.9631^{14}	1.5005^{14}	
6394	Isobutyl methacrylate		$C_8H_{14}O_2$	97-86-9	142.196			153(17)	0.8858^{20}	1.4199^{20}	i H_2O; msc EtOH, eth
6395	Isobutyl 3-methylbutanoate	Isobutyl isovalerate	$C_9H_{18}O_2$	589-59-3	158.238			169(3)	0.853^{20}	1.4057^{20}	i H_2O; msc EtOH, eth; vs ace; s chl
6396	Isobutyl methyl ether		$C_5H_{12}O$	625-44-5	88.148			58.6	0.7311^{20}		vs eth, EtOH
6397	Isobutyl nitrate		$C_4H_9NO_3$	543-29-3	119.119			123.4	1.0152^{20}	1.4028^{20}	
6398	Isobutyl nitrite		$C_4H_9NO_2$	542-56-3	103.120	col liq		68(3)	0.8699^{22}	1.3715^{22}	sl H_2O; s EtOH, eth
6399	Isobutyl pentanoate		$C_9H_{18}O_2$	10588-10-0	158.238			183(9)	0.8625^{25}	1.4046^{20}	i H_2O; msc EtOH; s eth, ace
6400	Isobutyl phenylacetate		$C_{12}H_{16}O_2$	102-13-6	192.254			247	0.999^{18}		i H_2O; s EtOH, eth
6401	Isobutyl propanoate	Isobutyl propionate	$C_7H_{14}O_2$	540-42-1	130.185	liq	-71.4(0.4)	136(2)	0.888^{0}	1.3973^{20}	sl H_2O; vs EtOH, eth; s ace, bz, chl, ctc
6402	Isobutyl stearate		$C_{22}H_{44}O_2$	646-13-9	340.583	wax	28.9	223^{15}	0.8498^{20}		vs eth
6403	Isobutyl thiocyanate		C_5H_9NS	591-84-4	115.197	liq	-58.9(0.5)	175(7)			vs eth, EtOH
6404	Isobutyl trichloroacetate		$C_6H_9Cl_3O_2$	33560-15-5	219.493			188	1.2636^{20}	1.4483^{20}	vs bz, eth, EtOH
6405	Isobutyl vinyl ether		$C_6H_{12}O$	109-53-5	100.158	liq	-112	83(3)	0.7645^{20}	1.3966^{20}	sl H_2O; vs EtOH, ace, bz; msc eth
6406	Isocitric acid		$C_6H_8O_7$	320-77-4	192.124	ye syr	105				
6407	Isocorybulbine		$C_{21}H_{25}NO_4$	22672-74-8	355.429	lf (al)	187.5		1.045^{20}		i H_2O; s EtOH, chl, acid
6408	Isocorydine		$C_{20}H_{23}NO_4$	475-67-2	341.402	pl	185				vs chl
6409	2-Isocyanato-1,3-dimethylbenzene	2,6-Dimethylphenyl isocyanate	C_9H_9NO	28556-81-2	147.173	liq		100^{13}			
6410	1-Isocyanato-2-methoxybenzene		$C_8H_7NO_2$	700-87-8	149.148			94^{17}			
6411	1-Isocyanato-3-methoxybenzene		$C_8H_7NO_2$	18908-07-1	149.148			102^{15}			
6412	1-Isocyanato-2-methylbenzene	2-Tolyl isocyanate	C_8H_7NO	614-68-6	133.148			185		1.5282^{20}	i H_2O; s eth
6413	1-Isocyanato-3-methylbenzene		C_8H_7NO	621-29-4	133.148			196.5	1.0330^{20}		vs bz, eth
6414	1-Isocyanato-4-methylbenzene		C_8H_7NO	622-58-2	133.148			187			vs bz, eth
6415	2-Isocyanato-2-methylpropane	*tert*-Butyl isocyanate	C_5H_9NO	1609-86-5	99.131		85.5		0.8670^{7}	1.4061^{20}	
6416	1-Isocyanatonaphthalene	1-Naphthyl isocyanate	$C_{11}H_7NO$	86-84-0	169.180			269	1.1774^{20}		s eth, bz
6417	1-Isocyanato-2-nitrobenzene	2-Nitrophenyl isocyanate	$C_7H_4N_2O_3$	3320-86-3	164.118	wh nd (peth)	41	137^{18}			vs bz, eth, chl
6418	1-Isocyanato-3-nitrobenzene	3-Nitrophenyl isocyanate	$C_7H_4N_2O_3$	3320-87-4	164.118	wh lf (lig)	51	130^{11}			vs bz, eth, chl
6419	1-Isocyanato-4-nitrobenzene	4-Nitrophenyl isocyanate	$C_7H_4N_2O_3$	100-28-7	164.118	pa ye nd	57	162^{20}			vs bz, eth, chl
6420	2-Isocyanatopropane	Isopropyl isocyanate	C_4H_7NO	1795-48-8	85.105			74.5	0.866^{25}	1.3825^{20}	
6421	1-Isocyanato-3-(trifluoromethyl)benzene	3-(Trifluoromethyl)phenyl isocyanate	$C_8H_4F_3NO$	329-01-1	187.119			54^{11}	1.3455^{20}	1.4690^{20}	
6422	Isocyanobenzene	Phenyl isocyanide	C_7H_5N	931-54-4	103.122	unstab liq		80^{40}	0.98^{15}		
6423	Isocyanomethane	Methyl isocyanide	C_2H_3N	593-75-9	41.052		-45(3)	exp	0.756^{4}		
6424	(Isocyanomethyl)benzene	Benzyl isocyanide	C_8H_7N	10340-91-7	117.149			199 dec	0.972^{15}	1.5193^{20}	
6425	2-Isocyanopropane	Isopropyl isocyanide	C_4H_7N	598-45-8	69.106			87	0.7596^{25}		i H_2O; msc EtOH, eth

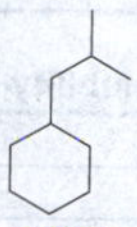
6385
Isobutylcyclohexane

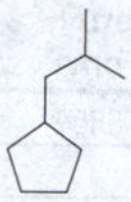
6386
Isobutylcyclopentane

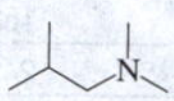
6387
Isobutyldimethylamine

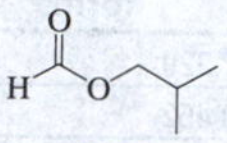
6388
Isobutyl formate

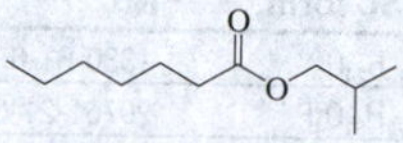
6389
Isobutyl heptanoate

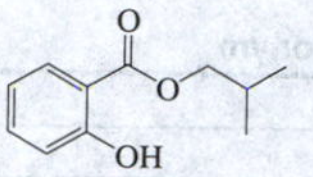
6390
Isobutyl 2-hydroxybenzoate

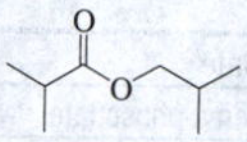
6391
Isobutyl isobutanoate

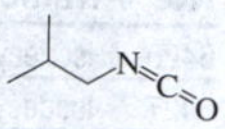
6392
Isobutyl isocyanate

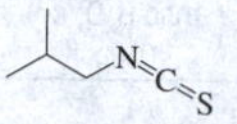
6393
Isobutyl isothiocyanate

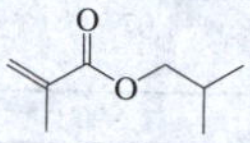
6394
Isobutyl methacrylate

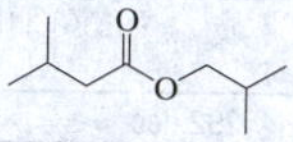
6395
Isobutyl 3-methylbutanoate

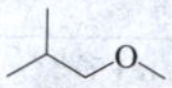
6396
Isobutyl methyl ether

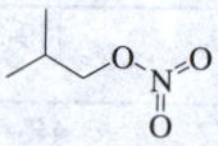
6397
Isobutyl nitrate

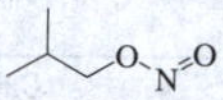
6398
Isobutyl nitrite

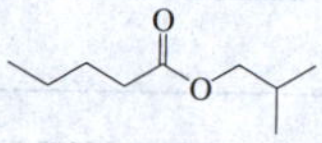
6399
Isobutyl pentanoate

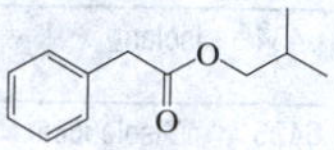
6400
Isobutyl phenylacetate

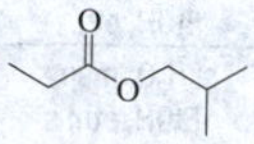
6401
Isobutyl propanoate

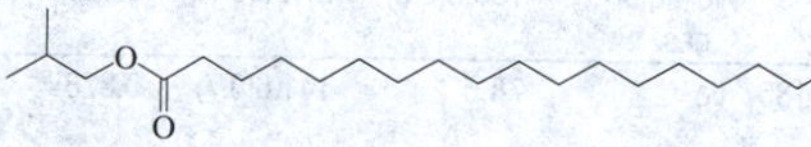
6402
Isobutyl stearate

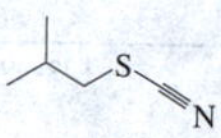
6403
Isobutyl thiocyanate

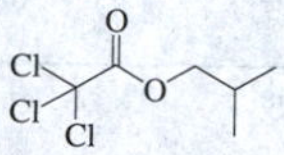
6404
Isobutyl trichloroacetate

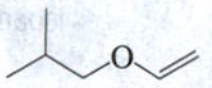
6405
Isobutyl vinyl ether

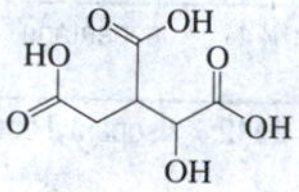
6406
Isocitric acid

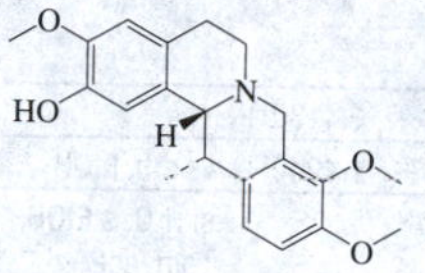
6407
Isocorybulbine

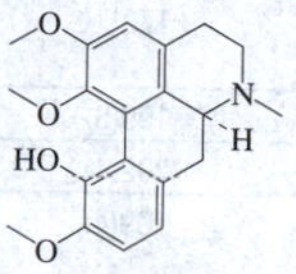
6408
Isocorydine

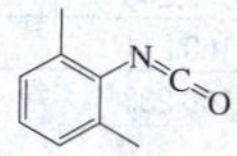
6409
2-Isocyanato-1,3-dimethylbenzene

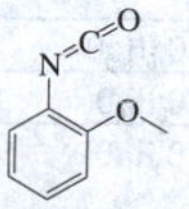
6410
1-Isocyanato-2-methoxybenzene

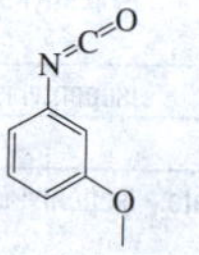
6411
1-Isocyanato-3-methoxybenzene

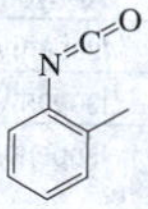
6412
1-Isocyanato-2-methylbenzene

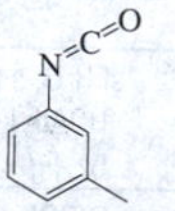
6413
1-Isocyanato-3-methylbenzene

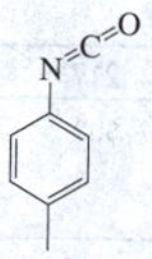
6414
1-Isocyanato-4-methylbenzene

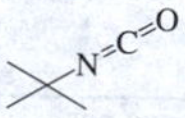
6415
2-Isocyanato-2-methylpropane

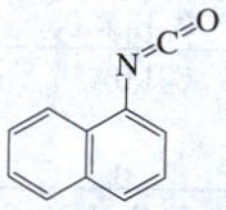
6416
1-Isocyanatonaphthalene

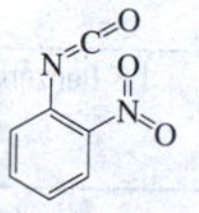
6417
1-Isocyanato-2-nitrobenzene

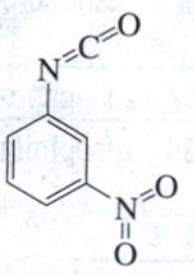
6418
1-Isocyanato-3-nitrobenzene

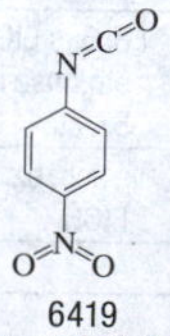
6419
1-Isocyanato-4-nitrobenzene

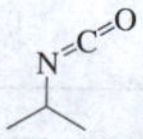
6420
2-Isocyanatopropane

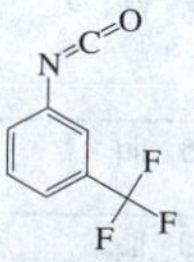
6421
1-Isocyanato-3-(trifluoromethyl)benzene

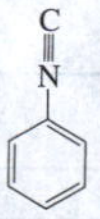
6422
Isocyanobenzene

—N≡C
6423
Isocyanomethane

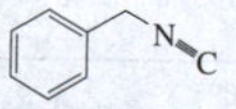
6424
(Isocyanomethyl)benzene

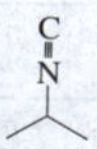
6425
2-Isocyanopropane

Organic

No.	Name	Synonym	Mol. form.	CAS Reg. No.	Mol. wt.	Physical form	mp/°C	bp/°C	den/ g cm⁻³	n_D	Solubility
6426	Isodecyl acrylate		$C_{13}H_{24}O_2$	1330-61-6	212.329		-100	158^{50}	0.885^{20}	1.4416^{20}	
6427	Isodecyl diphenyl phosphate		$C_{22}H_{31}O_4P$	29761-21-5	390.452			249^{10} dec			
6428	Isodecyl methacrylate		$C_{14}H_{26}O_2$	29964-84-9	226.355			126^{10}	0.876^{20}		
6429	8-Isoestrone		$C_{18}H_{22}O_2$	517-06-6	270.367	pr (MeOH)	254				vs eth, Diox
6430	Isoeugenol		$C_{10}H_{12}O_2$	97-54-1	164.201			264(6)	1.080^{25}	1.5739^{19}	vs eth, EtOH
6431	Isofenphos		$C_{15}H_{24}NO_4PS$	25311-71-1	345.395		<-12	$120^{0.01}$	1.134^{20}		
6432	Isoflurophate		$C_6H_{14}FO_3P$	55-91-4	184.145			62^{9}	1.055^{25}	1.3830^{25}	sl H_2O, lig; s eth; vs oils
6433	1*H*-Isoindole-1,3(2*H*)-dione	Phthalimide	$C_8H_5NO_2$	85-41-6	147.132	nd (w), pr (HOAc) lf (sub)	238(1)				vs bz
6434	Isolan		$C_{10}H_{17}N_3O_2$	119-38-0	211.261	col liq		$118^{2.5}$	1.07^{20}		msc H_2O; s EtOH, xyl
6435	DL-Isoleucine		$C_6H_{13}NO_2$	443-79-8	131.173		292 dec				
6436	L-Isoleucine	2-Amino-3-methylpentanoic acid	$C_6H_{13}NO_2$	73-32-5	131.173		284 dec				s H_2O; i EtOH
6437	Isolongifolene		$C_{15}H_{24}$	1135-66-6	204.352	liq		$82^{0.4}$			
6438	Isolysergic acid		$C_{16}H_{16}N_2O_2$	478-95-5	268.310	cry (w+2)	218 dec				sl H_2O, EtOH; s py
6439	α-Isomaltose	6-*O*-α-D-Glucopyranosyl-D-glucose	$C_{12}H_{22}O_{11}$	499-40-1	342.296		120				
6440	Isoniazid	4-Pyridinecarboxylic acid hydrazide	$C_6H_7N_3O$	54-85-3	137.139	cry (al)	171.4				vs H_2O, EtOH
6441	Isopentane	2-Methylbutane	C_5H_{12}	78-78-4	72.149	vol liq or gas	-159.8(0.2)	27.83(0.06)	0.6201^{20}	1.3537^{20}	i H_2O; msc EtOH, eth
6442	Isopentyl acetate	Isoamyl acetate	$C_7H_{14}O_2$	123-92-2	130.185	liq	-78.5	141.6(0.7)	0.876^{15}	1.4000^{20}	sl H_2O; msc EtOH, eth; s ace, chl
6443	Isopentylbenzene		$C_{11}H_{16}$	2049-94-7	148.245			196(7)	0.856^{20}	1.4867^{10}	i H_2O; s EtOH, eth; vs bz
6444	Isopentyl butanoate		$C_9H_{18}O_2$	106-27-4	158.238			184.8(0.3)	0.865^{19}	1.4110^{20}	i H_2O; vs EtOH, eth
6445	Isopentyl formate		$C_6H_{12}O_2$	110-45-2	116.158	liq	-93.5	124.0(0.8)	0.877^{20}	1.3967^{20}	sl H_2O, ctc; s EtOH; msc eth
6446	Isopentyl hexanoate	Isopentyl caproate	$C_{11}H_{22}O_2$	2198-61-0	186.292			225.5	0.861^{20}		i H_2O; s EtOH, eth
6447	Isopentyl α-hydroxybenzeneacetate	Isopentyl mandelate	$C_{13}H_{18}O_3$	5421-04-5	222.280	oily liq		172^{11}			
6448	Isopentyl isopentanoate	Isopentyl isovalerate	$C_{10}H_{20}O_2$	659-70-1	172.265			190(3)	0.8583^{19}	1.4130^{19}	
6449	Isopentyl lactate		$C_8H_{16}O_3$	19329-89-6	160.211			202.4	0.9589^{25}	1.4240^{25}	vs eth, EtOH
6450	Isopentyl 2-methylpropanoate	Isopentyl isobutyrate	$C_9H_{18}O_2$	2050-01-3	158.238			169(4)	0.8627^{20}		sl H_2O; s EtOH, eth, ace
6451	Isopentyl nitrite	Isoamyl nitrite	$C_5H_{11}NO_2$	110-46-3	117.147			99(3)	0.8828^{20}	1.3918^{20}	sl H_2O; msc EtOH, eth
6452	Isopentyl pentanoate		$C_{10}H_{20}O_2$	2050-09-1	172.265			193(6)			
6453	Isopentyl propanoate		$C_8H_{16}O_2$	105-68-0	144.212			173(4)	0.8697^{20}	1.4069^{20}	vs eth, EtOH
6454	Isopentyl salicylate		$C_{12}H_{16}O_3$	87-20-7	208.253			278	1.0535^{20}	1.5080^{20}	i H_2O; vs EtOH; s eth, chl; sl ctc
6455	Isopentyl trichloroacetate		$C_7H_{11}Cl_3O_2$	57392-55-9	233.520			217	1.2314^{20}	1.4521^{20}	vs eth, EtOH
6456	Isophorone	3,5,5-Trimethyl-2-cyclohexen-1-one	$C_9H_{14}O$	78-59-1	138.206	liq	-8.1	214.8(0.7)	0.9255^{20}	1.4766^{18}	
6457	Isophorone diisocyanate		$C_{12}H_{18}N_2O_2$	4098-71-9	222.283	liq	60	217^{100}	1.062^{20}		
6458	Isophthalic acid	1,3-Benzenedicarboxylic acid	$C_8H_6O_4$	121-91-5	166.132	nd (w, al)	348.0(0.2)	sub	1.538^{25}		sl H_2O; s EtOH, HOAc; i eth, bz, lig
6459	Isopilosine		$C_{16}H_{18}N_2O_3$	491-88-3	286.325	pl (al), pr (w, dil al)	187				vs EtOH
6460	Isopropalin	Benzenamine, 4-(1-methylethyl)-2,6-dinitro-*N*,*N*-dipropyl-	$C_{15}H_{23}N_3O_4$	33820-53-0	309.362	red-oran liq					i H_2O; s os
6461	Isopropamide iodide		$C_{23}H_{33}IN_2O$	71-81-8	480.424	cry or pow	190				s H_2O, EtOH, MeOH; i chl
6462	Isopropenyl acetate		$C_5H_8O_2$	108-22-5	100.117	liq	-92.9	97.4(0.9)	0.9090^{20}	1.4033^{20}	sl H_2O; s EtOH, chl, ace; vs eth
6463	Isopropenylbenzene	α-Methyl styrene	C_9H_{10}	98-83-9	118.175	liq	-22.36(0.05)	165.4(0.3)	0.9106^{20}	1.5386^{20}	i H_2O; s EtOH, eth; msc ace, bz, ctc
6464	*p*-Isopropenylisopropylbenzene		$C_{12}H_{16}$	2388-14-9	160.255	liq	-30.6	222(1)	0.8936^{20}	1.5238^{20}	vs ace, bz, eth, EtOH
6465	*p*-Isopropenylstyrene		$C_{11}H_{12}$	16262-48-9	144.213	liq		242	0.93	1.5684^{20}	
6466	4-Isopropoxydiphenylamine	4-Isopropoxy-*N*-phenylaniline	$C_{15}H_{17}NO$	101-73-5	227.302		83				

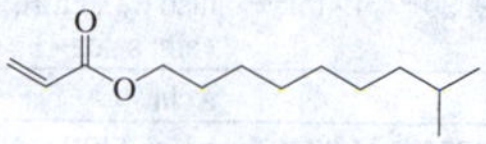
6426
Isodecyl acrylate

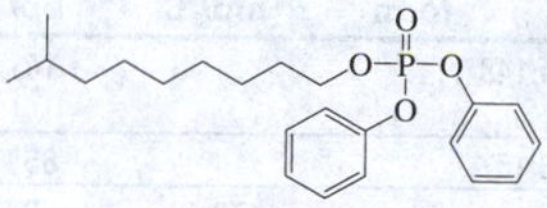
6427
Isodecyl diphenyl phosphate

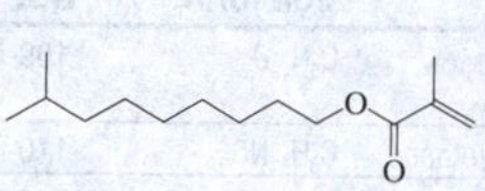
6428
Isodecyl methacrylate

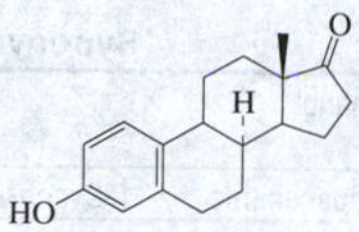
6429
8-Isoestrone

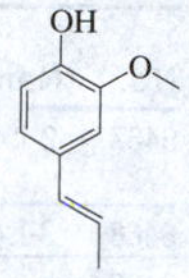
6430
Isoeugenol

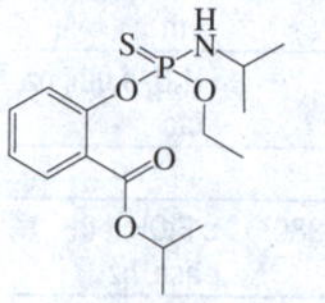
6431
Isofenphos

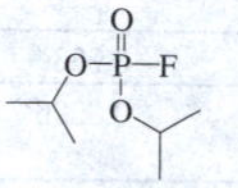
6432
Isoflurophate

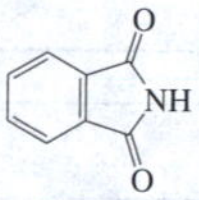
6433
1*H*-Isoindole-1,3(2*H*)-dione

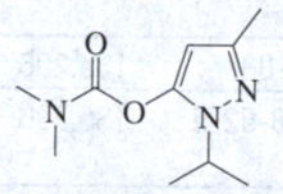
6434
Isolan

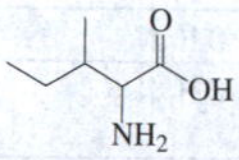
6435
DL-Isoleucine

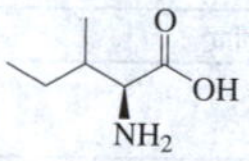
6436
L-Isoleucine

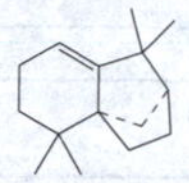
6437
Isolongifolene

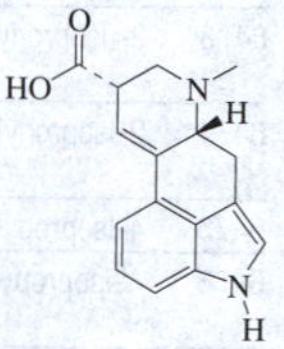
6438
Isolysergic acid

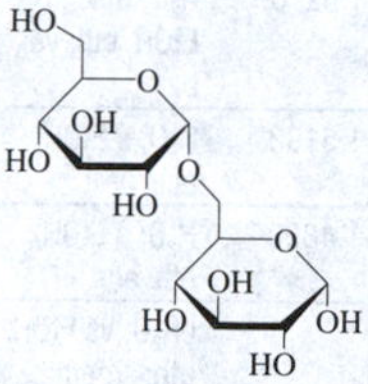
6439
α-Isomaltose

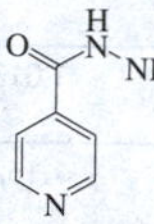
6440
Isoniazid

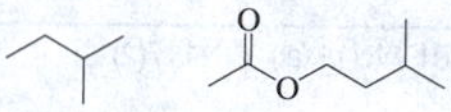
6441
Isopentane

6442
Isopentyl acetate

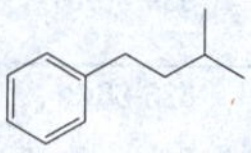
6443
Isopentylbenzene

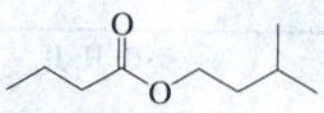
6444
Isopentyl butanoate

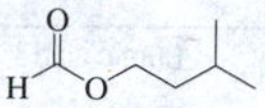
6445
Isopentyl formate

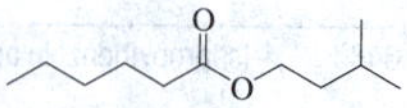
6446
Isopentyl hexanoate

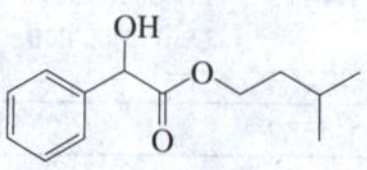
6447
Isopentyl α-hydroxybenzeneacetate

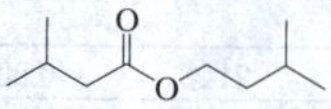
6448
Isopentyl isopentanoate

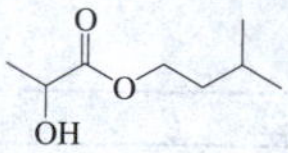
6449
Isopentyl lactate

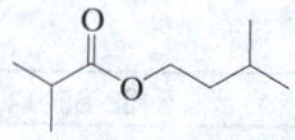
6450
Isopentyl 2-methylpropanoate

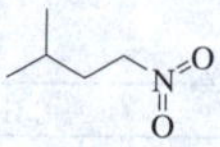
6451
Isopentyl nitrite

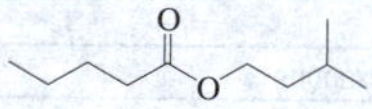
6452
Isopentyl pentanoate

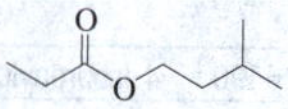
6453
Isopentyl propanoate

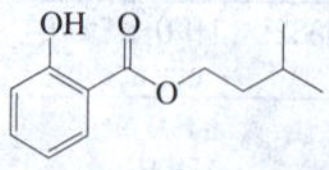
6454
Isopentyl salicylate

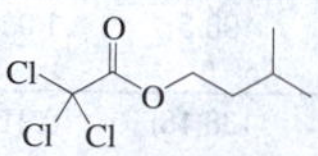
6455
Isopentyl trichloroacetate

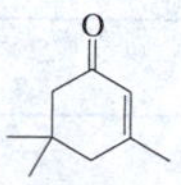
6456
Isophorone

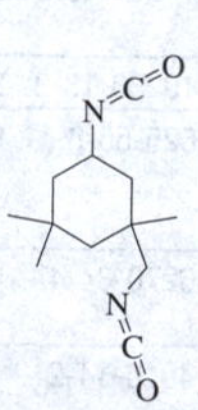
6457
Isophorone diisocyanate

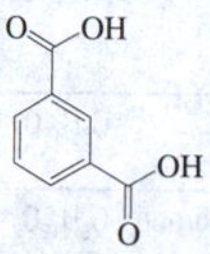
6458
Isophthalic acid

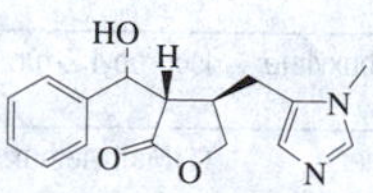
6459
Isopilosine

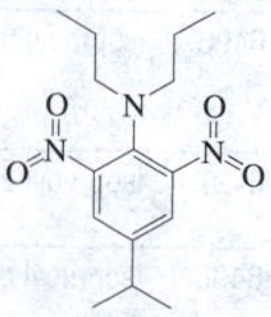
6460
Isopropalin

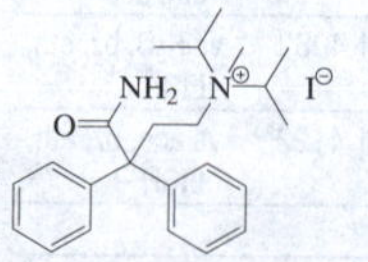
6461
Isopropamide iodide

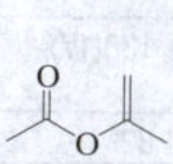
6462
Isopropenyl acetate

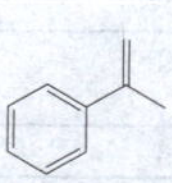
6463
Isopropenylbenzene

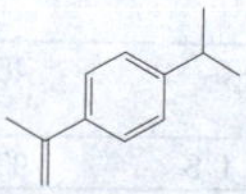
6464
p-Isopropenylisopropylbenzene

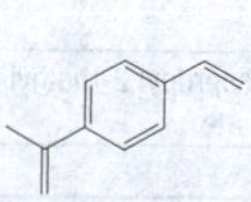
6465
p-Isopropenylstyrene

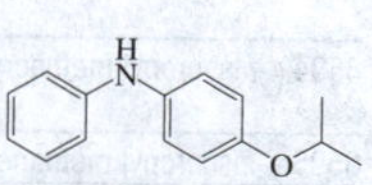
6466
4-Isopropoxydiphenylamine

Organic

No.	Name	Synonym	Mol. form.	CAS Reg. No.	Mol. wt.	Physical form	mp/°C	bp/°C	den/ g cm^{-3}	n_D	Solubility
6467	2-Isopropoxyethanol		$C_5H_{12}O_2$	109-59-1	104.148			141.6(0.4)	0.9030^{20}	1.4095^{20}	msc H_2O, EtOH, eth; s ace
6468	3-Isopropoxypropanenitrile	1-Cyano-2-isopropoxyethane	$C_6H_{11}NO$	110-47-4	113.157			65^{10}			s chl
6469	Isopropyl acetate	1-Methylethyl acetate	$C_5H_{10}O_2$	108-21-4	102.132	liq	-73.4	88.6(0.2)	0.8662^{25}	1.3746^{25}	s H_2O, EtOH, ace, chl; msc eth
6470	Isopropyl acrylate	Isopropyl 2-propenoate	$C_6H_{10}O_2$	689-12-3	114.142	liq		51^{103}			
6471	Isopropylamine	2-Propanamine	C_3H_9N	75-31-0	59.110	liq	-95.119 (0.001)	31.8(0.2)	0.6891^{20}	1.3742^{20}	msc H_2O, EtOH, eth; vs ace; s bz, chl
6472	Isopropylamine hydrochloride	2-Propanamine hydrochloride	$C_3H_{10}ClN$	15572-56-2	95.571		164				s DMSO
6473	2-(Isopropylamino)ethanol		$C_5H_{13}NO$	109-56-8	103.163		128.5	169(4)	0.8970^{20}	1.4395^{20}	msc H_2O, EtOH, eth
6474	2-Isopropylaniline		$C_9H_{13}N$	643-28-7	135.206			221	0.9760^{12}		i H_2O; s eth, bz, ctc
6475	4-Isopropylaniline	Cumidine	$C_9H_{13}N$	99-88-7	135.206			228(3)	0.953^{20}		
6476	*N*-Isopropylaniline		$C_9H_{13}N$	768-52-5	135.206			203	0.9526^{25}	1.5380^{20}	s EtOH, eth, ace, bz
6477	4-Isopropylbenzaldehyde	Cuminaldehyde	$C_{10}H_{12}O$	122-03-2	148.201			226(5)	0.9755^{20}	1.5301^{20}	i H_2O; s EtOH, eth; sl ctc
6478	Isopropylbenzene	Cumene	C_9H_{12}	98-82-8	120.191	liq	-96.01(0.05)	152.4(0.2)	0.8640^{25}	1.4915^{20}	i H_2O; msc EtOH, eth, ace, bz, peth, ctc
6479	Isopropylbenzene hydroperoxide	Cumene hydroperoxide	$C_9H_{12}O_2$	80-15-9	152.190	liq		153	1.03^{20}		
6480	4-Isopropylbenzenemethanol	Cumic alcohol	$C_{10}H_{14}O$	536-60-7	150.217		28	249	0.9818^{20}	1.5210^{20}	i H_2O; msc EtOH, eth; vs bz
6481	α-Isopropylbenzenemethanol	1-Phenyl-2-methylpropyl alcohol	$C_{10}H_{14}O$	611-69-8	150.217			223	0.9869^{14}	1.5193^{14}	i H_2O; s EtOH, ace
6482	Isopropyl benzoate		$C_{10}H_{12}O_2$	939-48-0	164.201			218(1)	1.0163^{15}	1.4890^{20}	i H_2O; s EtOH, eth, ace
6483	4-Isopropylbenzoic acid	Cumic acid	$C_{10}H_{12}O_2$	536-66-3	164.201	tcl pl (al)	117(2)	301(20)	1.162^{4}		sl H_2O; vs EtOH, eth; s peth
6484	Isopropyl butanoate		$C_7H_{14}O_2$	638-11-9	130.185			129(3)	0.8588^{20}	1.3936^{20}	i H_2O; s EtOH
6485	Isopropyl carbamate		$C_4H_9NO_2$	1746-77-6	103.120	nd	93	183	0.9951^{66}		
6486	Isopropyl chloroacetate		$C_5H_9ClO_2$	105-48-6	136.577			150(5)	1.0888^{20}	1.4382^{20}	vs eth
6487	Isopropyl chloroformate		$C_4H_7ClO_2$	108-23-6	122.551			105		1.4013^{20}	vs eth
6488	Isopropyl 2-chloropropanoate		$C_6H_{11}ClO_2$	40058-87-5	150.603			151.5	1.0315^{20}	1.4149^{20}	i H_2O; s EtOH, eth
6489	Isopropylcyclohexane		C_9H_{18}	696-29-7	126.239	liq	-89.9(0.5)	154.4(0.4)	0.8023^{20}	1.4410^{20}	i H_2O; vs EtOH, eth; msc ace, bz
6490	4-Isopropylcyclohexanone		$C_9H_{16}O$	5432-85-9	140.222			214	0.9099^{30}	1.4552^{25}	
6491	Isopropylcyclopentane		C_8H_{16}	3875-51-2	112.213	liq	-111.5(0.1)	126.4(0.8)	0.7765^{20}	1.4258^{20}	i H_2O; msc EtOH, ace, ctc; s eth, bz
6492	Isopropylcyclopropane		C_6H_{12}	3638-35-5	84.159	liq	-113.00(0.09)	58(2)	0.6936^{25}	1.3865^{20}	
6493	Isopropyl (2,4-dichlorophenoxy)acetate		$C_{11}H_{12}Cl_2O_3$	94-11-1	263.117		5	140^{1}	1.26^{25}	1.5209^{25}	
6494	*N*-Isopropyl-4,4-diphenylcyclohexanamine	Pramiverin	$C_{21}H_{27}N$	14334-40-8	293.446		70	$165^{0.05}$			
6495	Isopropyl dodecanoate	Isopropyl laurate	$C_{15}H_{30}O_2$	10233-13-3	242.398			196^{60}	0.8536^{20}	1.4280^{25}	vs eth, EtOH
6496	Isopropyl formate		$C_4H_8O_2$	625-55-8	88.106			68(2)	0.8728^{20}	1.3678^{20}	sl H_2O; msc EtOH, eth; vs ace; s chl
6497	Isopropyl 2-furancarboxylate	Isopropyl 2-furanoate	$C_8H_{10}O_3$	6270-34-4	154.163			198.5	1.0655^{24}	1.4682^{24}	i H_2O; s EtOH, eth, ace, bz
6498	Isopropyl glycidyl ether	(1-Methylethoxy)methyloxirane	$C_6H_{12}O_2$	4016-14-2	116.158			138(18)	0.9186^{20}		s H_2O, ace, EtOH
6499	4-Isopropylheptane		$C_{10}H_{22}$	52896-87-4	142.282			161(6)	0.7354^{25}	1.4153^{20}	
6500	Isopropylhydrazine		$C_3H_{10}N_2$	2257-52-5	74.124	liq		107			s H_2O, bz, EtOH; sl eth
6501	Isopropyl 2-hydroxybenzoate	Isopropyl salicylate	$C_{10}H_{12}O_3$	607-85-2	180.200			238	1.0729^{20}	1.5065^{20}	i H_2O; msc EtOH, eth
6502	Isopropyl isobutanoate	Isopropyl isobutyrate	$C_7H_{14}O_2$	617-50-5	130.185			123(9)	0.8471^{21}		i H_2O; s EtOH, eth, ace
6503	Isopropyl lactate		$C_6H_{12}O_3$	617-51-6	132.157			167	0.9980^{20}	1.4082^{25}	vs H_2O, bz, eth, EtOH
6504	Isopropyl methacrylate	Isopropyl 2-methyl-2-propenoate	$C_7H_{12}O_2$	4655-34-9	128.169			130(17)	0.8847^{20}	1.4122^{20}	vs ace, bz, eth, EtOH
6505	Isopropyl methanesulfonate		$C_4H_{10}O_3S$	926-06-7	138.185			82^{6}			
6506	Isopropylmethylamine	Methylisopropylamine	$C_4H_{11}N$	4747-21-1	73.137			53.3(0.3)			

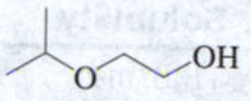
6467
2-Isopropoxyethanol

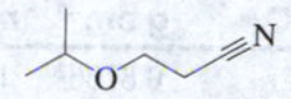
6468
3-Isopropoxypropanenitrile

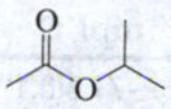
6469
Isopropyl acetate

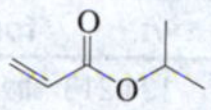
6470
Isopropyl acrylate

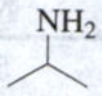
6471
Isopropylamine

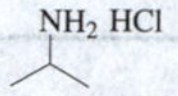
6472
Isopropylamine hydrochloride

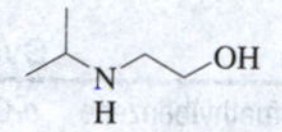
6473
2-(Isopropylamino)ethanol

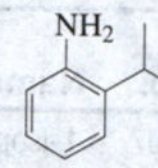
6474
2-Isopropylaniline

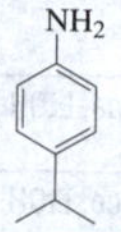
6475
4-Isopropylaniline

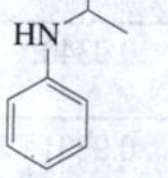
6476
N-Isopropylaniline

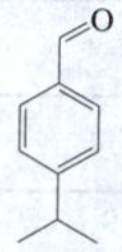
6477
4-Isopropylbenzaldehyde

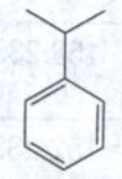
6478
Isopropylbenzene

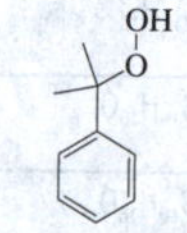
6479
Isopropylbenzene hydroperoxide

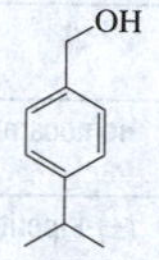
6480
4-Isopropylbenzenemethanol

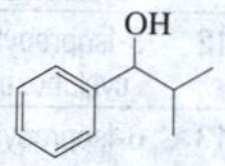
6481
α-Isopropylbenzenemethanol

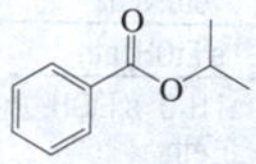
6482
Isopropyl benzoate

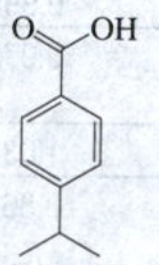
6483
4-Isopropylbenzoic acid

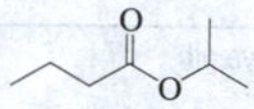
6484
Isopropyl butanoate

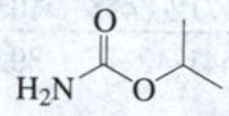
6485
Isopropyl carbamate

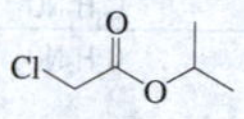
6486
Isopropyl chloroacetate

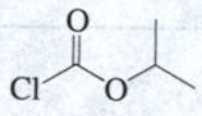
6487
Isopropyl chloroformate

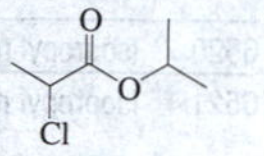
6488
Isopropyl 2-chloropropanoate

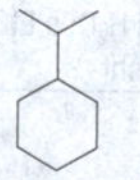
6489
Isopropylcyclohexane

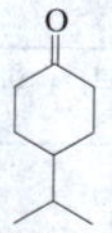
6490
4-Isopropylcyclohexanone

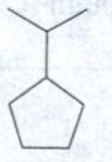
6491
Isopropylcyclopentane

6492
Isopropylcyclopropane

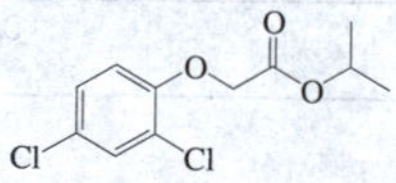
6493
Isopropyl (2,4-dichlorophenoxy)acetate

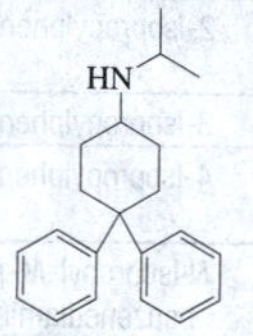
6494
N-Isopropyl-4,4-diphenylcyclohexanamine

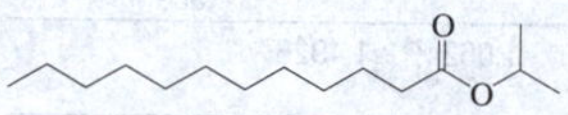
6495
Isopropyl dodecanoate

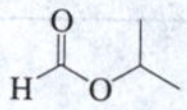
6496
Isopropyl formate

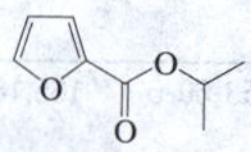
6497
Isopropyl 2-furancarboxylate

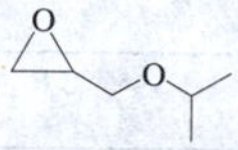
6498
Isopropyl glycidyl ether

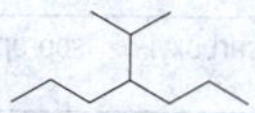
6499
4-Isopropylheptane

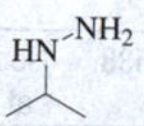
6500
Isopropylhydrazine

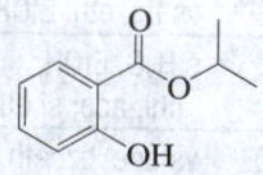
6501
Isopropyl 2-hydroxybenzoate

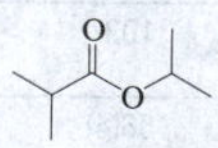
6502
Isopropyl isobutanoate

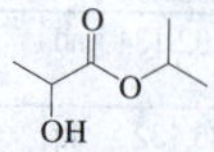
6503
Isopropyl lactate

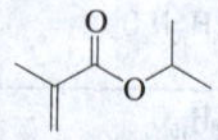
6504
Isopropyl methacrylate

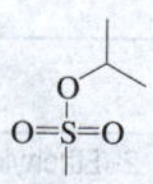
6505
Isopropyl methanesulfonate

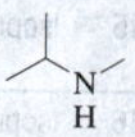
6506
Isopropylmethylamine

No.	Name	Synonym	Mol. form.	CAS Reg. No.	Mol. wt.	Physical form	mp/°C	bp/°C	den/g cm^{-3}	n_D	Solubility
6507	1-Isopropyl-2-methylbenzene	*o*-Cymene	$C_{10}H_{14}$	527-84-4	134.218	liq	-71.5(0.1)	178(2)	0.8766^{20}	1.5006^{20}	i H_2O; msc EtOH, eth, ace, bz, peth, ctc
6508	1-Isopropyl-3-methylbenzene	*m*-Cymene	$C_{10}H_{14}$	535-77-3	134.218	liq	-63.8(0.1)	175(4)	0.8610^{20}	1.4930^{20}	i H_2O; msc EtOH, eth, ace, bz, peth, ctc
6509	1-Isopropyl-4-methylbenzene	*p*-Cymene	$C_{10}H_{14}$	99-87-6	134.218	liq	-68.1(0.3)	177(2)	0.8573^{20}	1.4909^{20}	i H_2O; msc EtOH, eth, ace, bz, peth, ctc
6510	Isopropyl 3-methylbutanoate		$C_8H_{16}O_2$	32665-23-9	144.212			147(9)	0.8538^{17}	1.3960^{20}	vs ace, eth, EtOH
6511	5-Isopropyl-2-methyl-1,3-cyclohexadiene, (*R*)-		$C_{10}H_{16}$	4221-98-1	136.234			173	0.8421^{20}	1.4772^{19}	
6512	5-Isopropyl-3-methyl-2-cyclohexen-1-one, (±)-	Homocamfin	$C_{10}H_{16}O$	535-86-4	152.233	pa ye		244	0.9340^{21}	1.4865^{21}	vs ace, EtOH
6513	6-Isopropyl-3-methyl-2-cyclohexen-1-one, (±)-	(±)-Piperitone	$C_{10}H_{16}O$	6091-52-7	152.233	liq	-19	232.5	0.9331^{20}	1.4845^{20}	vs ace, EtOH
6514	Isopropyl methyl ether	2-Methoxypropane	$C_4H_{10}O$	598-53-8	74.121			30.8(0.5)	0.7237^{15}	1.3576^{20}	sl H_2O; msc EtOH, eth
6515	5-Isopropyl-2-methylphenol	Carvacrol	$C_{10}H_{14}O$	499-75-2	150.217	nd	2.5(0.2)	241.5(0.5)	0.9772^{20}	1.5230^{20}	sl H_2O; s EtOH, eth, ctc; vs ace
6516	2-Isopropyl-6-methyl-4-pyrimidinol		$C_8H_{12}N_2O$	2814-20-2	152.193	cry	173				
6517	Isopropyl methyl sulfide		$C_4H_{10}S$	1551-21-9	90.187	liq	-101.47(0.06)	84.7(0.5)	0.8291^{20}	1.4932^{20}	s EtOH, eth, ace
6518	1-Isopropylnaphthalene		$C_{13}H_{14}$	6158-45-8	170.250	liq	-15.6(0.2)	271(2)	0.9956^{20}	1.5952^{20}	
6519	2-Isopropylnaphthalene		$C_{13}H_{14}$	2027-17-0	170.250		15.0(0.4)	268(2)	0.9753^{20}	1.5848^{20}	i H_2O; vs EtOH, eth; s bz
6520	Isopropyl nitrate		$C_3H_7NO_3$	1712-64-7	105.093			101(2)	1.034^{19}	1.3912^{16}	s EtOH, eth
6521	Isopropyl nitrite		$C_3H_7NO_2$	541-42-4	89.094	pa ye oil		34(2)	0.8684^{15}		i H_2O; s EtOH, eth
6522	1-Isopropyl-4-nitrobenzene		$C_9H_{11}NO_2$	1817-47-6	165.189	pa ye oil		122^{9}	1.084^{20}	1.5367^{20}	i H_2O; s ace, bz, lig
6523	*N*-Isopropyl-*N*-nitroso-2-propanamine		$C_6H_{14}N_2O$	601-77-4	130.187	cry (eth,w)	48	194.5	0.9422^{20}		sl H_2O; s EtOH, eth, bz
6524	Isopropyl 3-oxobutanoate	Isopropyl acetoacetate	$C_7H_{12}O_3$	542-08-5	144.168	liq	-27.3	186	0.9835^{20}	1.4173^{20}	vs eth, EtOH, lig
6525	Isopropyl palmitate	Isopropyl hexadecanoate	$C_{19}H_{38}O_2$	142-91-6	298.504		12.8(0.5)	160^{2}	0.8404^{38}	1.4364^{25}	vs ace, bz, eth, EtOH
6526	Isopropyl pentanoate		$C_8H_{16}O_2$	18362-97-5	144.212				0.8579^{20}	1.4061^{20}	i H_2O; s EtOH, eth, ace
6527	2-Isopropylphenol		$C_9H_{12}O$	88-69-7	136.190		17(1)	213.84(0.08)	1.012^{20}	1.5315^{20}	sl H_2O; s EtOH, eth, bz, ctc
6528	3-Isopropylphenol		$C_9H_{12}O$	618-45-1	136.190		25.7(0.5)	230(3)		1.5261^{20}	vs eth
6529	4-Isopropylphenol		$C_9H_{12}O$	99-89-8	136.190	nd (peth)	62(2)	228.0(0.1)	0.990^{20}	1.5228^{20}	sl H_2O; s EtOH, chl
6530	*N*-Isopropyl-*N'*-phenyl-1,4-benzenediamine		$C_{15}H_{18}N_2$	101-72-4	226.317		72.5	148^{2}			
6531	Isopropyl phenylcarbamate	Propham	$C_{10}H_{13}NO_2$	122-42-9	179.216	wh nd (al)	85(2)		1.09^{20}	1.4989^{91}	vs bz, EtOH
6532	1-(4-Isopropylphenyl)ethanone		$C_{11}H_{14}O$	645-13-6	162.228			250(13)	0.9753^{15}	1.5235^{20}	
6533	Isopropyl propanoate		$C_6H_{12}O_2$	637-78-5	116.158			110(4)	0.8660^{20}	1.3872^{20}	sl H_2O; msc EtOH, eth
6534	*N*-Isopropyl-2-propenamide		$C_6H_{11}NO$	2210-25-5	113.157		64.5	110^{15}			
6535	Isopropylpropylamine	*N*-Propyl-2-propanamine	$C_6H_{15}N$	21968-17-2	101.190			96.2(0.3)			
6536	Isopropyl propyl sulfide		$C_6H_{14}S$	5008-73-1	118.240			132.0(0.8)	0.8269^{20}		
6537	4-Isopropylpyridine		$C_8H_{11}N$	696-30-0	121.180	liq	-54.9(0.5)	182(4)	0.9382^{25}	1.4962^{20}	sl H_2O; msc EtOH, eth; vs ace
6538	Isopropyl 3-pyridinecarboxylate	Isopropyl nicotinate	$C_9H_{11}NO_2$	553-60-6	165.189			126^{30}	1.0624^{20}	1.4926^{20}	
6539	Isopropyl silicate	Tetra(isopropoxy)silane	$C_{12}H_{28}O_4Si$	1992-48-9	264.434			184	0.8770^{20}		s ctc, CS_2
6540	Isopropyl stearate		$C_{21}H_{42}O_2$	112-10-7	326.557		24.1(0.5)	207^{6}	0.8403^{38}		vs ace, eth, EtOH, chl
6541	4-Isopropylstyrene		$C_{11}H_{14}$	2055-40-5	146.229	liq	-44.7	207(1)	0.8850^{20}	1.5289^{20}	vs ace, bz, eth, EtOH
6542	Isopropyl tetradecanoate	Isopropyl myristate	$C_{17}H_{34}O_2$	110-27-0	270.451			193^{20}	0.8532^{20}	1.4325^{25}	i H_2O; s EtOH, eth, chl; vs ace, bz
6543	(Isopropylthio)benzene		$C_9H_{12}S$	3019-20-3	152.256			208	0.9852^{20}	1.5464^{20}	
6544	Isopropyl trichloroacetate		$C_5H_7Cl_3O_2$	3974-99-0	205.468			174(7)	1.2911^{25}	1.4428^{20}	vs bz, eth, EtOH
6545	Isopropylurea		$C_4H_{10}N_2O$	691-60-1	102.134	nd		103$^{0.1}$			s H_2O, EtOH, chl, ace; sl eth
6546	Isopropyl vinyl ether	2-(Ethenyloxy)propane	$C_5H_{10}O$	926-65-8	86.132	liq	-140	56(4)	0.7534^{20}	1.3840^{20}	vs ace, bz, eth, EtOH

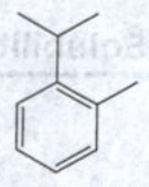
6507
1-Isopropyl-2-methylbenzene

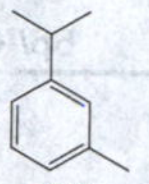
6508
1-Isopropyl-3-methylbenzene

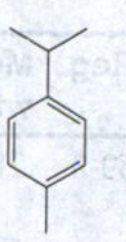
6509
1-Isopropyl-4-methylbenzene

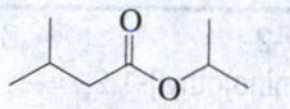
6510
Isopropyl 3-methylbutanoate

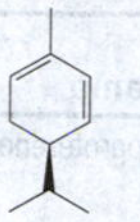
6511
5-Isopropyl-2-methyl-1,3-cyclohexadiene, (*R*)-

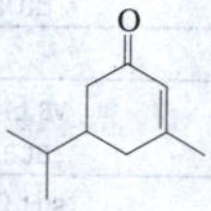
6512
5-Isopropyl-3-methyl-2-cyclohexen-1-one, (±)-

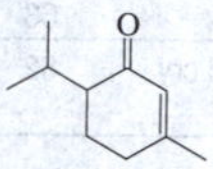
6513
6-Isopropyl-3-methyl-2-cyclohexen-1-one, (±)-

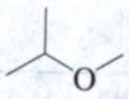
6514
Isopropyl methyl ether

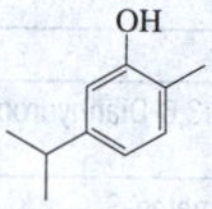
6515
5-Isopropyl-2-methylphenol

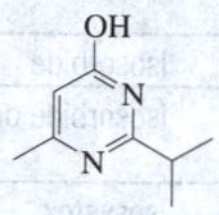
6516
2-Isopropyl-6-methyl-4-pyrimidinol

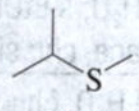
6517
Isopropyl methyl sulfide

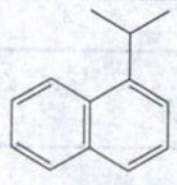
6518
1-Isopropylnaphthalene

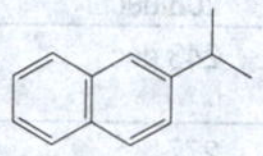
6519
2-Isopropylnaphthalene

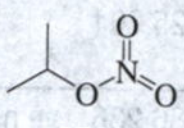
6520
Isopropyl nitrate

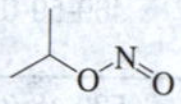
6521
Isopropyl nitrite

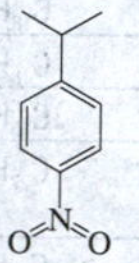
6522
1-Isopropyl-4-nitrobenzene

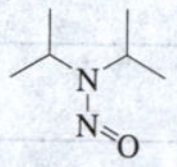
6523
N-Isopropyl-*N*-nitroso-2-propanamine

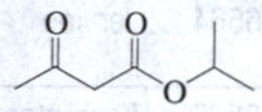
6524
Isopropyl 3-oxobutanoate

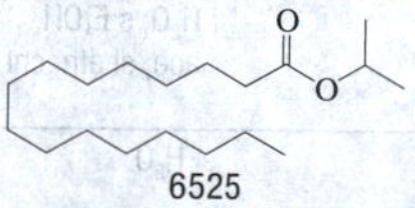
6525
Isopropyl palmitate

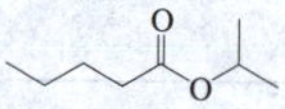
6526
Isopropyl pentanoate

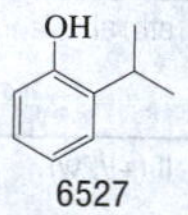
6527
2-Isopropylphenol

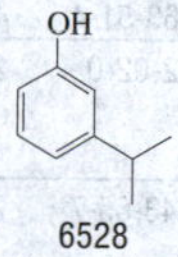
6528
3-Isopropylphenol

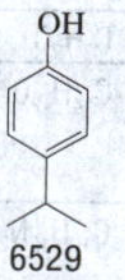
6529
4-Isopropylphenol

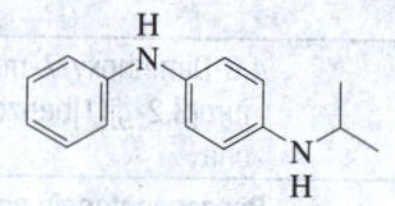
6530
N-Isopropyl-*N'*-phenyl-1,4-benzenediamine

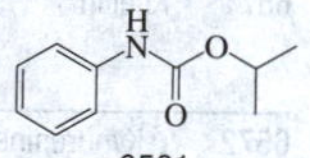
6531
Isopropyl phenylcarbamate

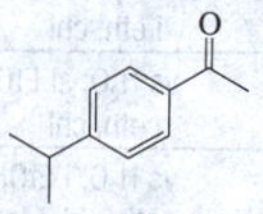
6532
1-(4-Isopropylphenyl)ethanone

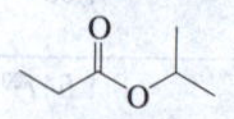
6533
Isopropyl propanoate

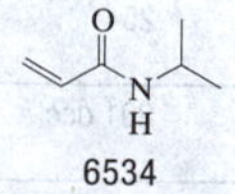
6534
N-Isopropyl-2-propenamide

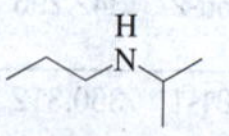
6535
Isopropylpropylamine

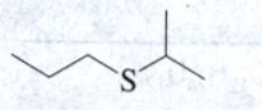
6536
Isopropyl propyl sulfide

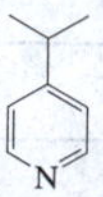
6537
4-Isopropylpyridine

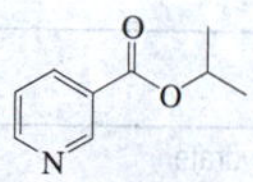
6538
Isopropyl 3-pyridinecarboxylate

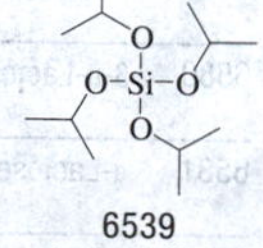
6539
Isopropyl silicate

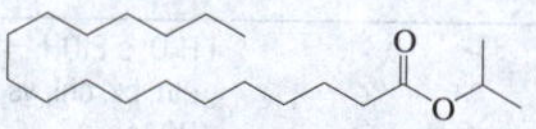
6540
Isopropyl stearate

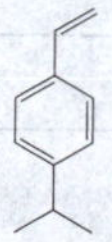
6541
4-Isopropylstyrene

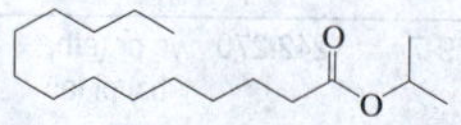
6542
Isopropyl tetradecanoate

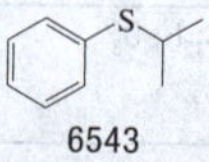
6543
(Isopropylthio)benzene

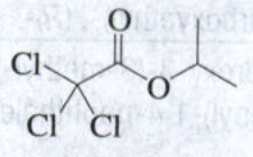
6544
Isopropyl trichloroacetate

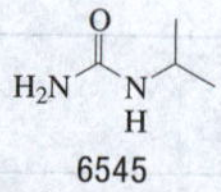
6545
Isopropylurea

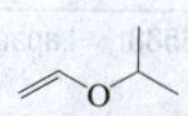
6546
Isopropyl vinyl ether

Organic

No.	Name	Synonym	Mol. form.	CAS Reg. No.	Mol. wt.	Physical form	mp/°C	bp/°C	den/ g cm−3	n_D	Solubility
6547	Isoproterenol	4-[1-Hydroxy-2-[isopropylamino]ethyl]-1,2-benzenediol	$C_{11}H_{17}NO_3$	7683-59-2	211.258		170.5				
6548	Isopsoralen		$C_{11}H_6O_3$	523-50-2	186.164		139				
6549	1-Isoquinolinamine		$C_9H_8N_2$	1532-84-9	144.173	pl(w)	123	164^{8}			sl H_2O, eth; vs EtOH
6550	3-Isoquinolinamine		$C_9H_8N_2$	25475-67-6	144.173		178.5				
6551	Isoquinoline	Benzo[c]pyridine	C_9H_7N	119-65-3	129.159	hyg pl	26.46(0.01)	243.2(0.6)	1.0910^{30}	1.6148^{20}	i H_2O; vs EtOH, chl; msc eth, bz
6552	7-Isoquinolinol		C_9H_7NO	7651-83-4	145.158		230				sl H_2O, eth; s EtOH
6553	Isosorbide		$C_6H_{10}O_4$	652-67-5	146.141		63	170^{2}			
6554	Isosorbide dinitrate	1,4:3,6-Dianhydroglucitol	$C_6H_8N_2O_8$	87-33-2	236.136	col cry	52				vs EtOH, eth, ace
6555	Isosystox	Demeton-S	$C_8H_{19}O_3PS_2$	126-75-0	258.339	liq		133^{2}	1.132^{21}		s H_2O
6556	Isothebaine		$C_{19}H_{21}NO_3$	568-21-8	311.375	orth cry (al)	203.5				i H_2O; msc EtOH, chl; sl eth; s MeOH
6557	Isothiocyanic acid		CHNS	3129-90-6	59.091	unstab gas					
6558	L-Isovaline	2-Amino-2-methylbutyric acid	$C_5H_{11}NO_2$	595-40-4	117.147	nd (w)	≈300				s EtOH; sl eth
6559	Isoxaben		$C_{18}H_{24}N_2O_4$	82558-50-7	332.395	wh cry	173				s EtOAc, MeCN, MeOH
6560	Isoxazole	1-Oxa-2-azacyclopentadiene	C_3H_3NO	288-14-2	69.062			95.4(0.7)	1.078^{20}	1.4298^{17}	s H_2O
6561	Isoxsuprine		$C_{18}H_{23}NO_3$	395-28-8	301.381	cry	103.0				
6562	Jacobine		$C_{18}H_{25}NO_6$	6870-67-3	351.395	pl (EtOH)	228				
6563	Javanicin		$C_{15}H_{14}O_6$	476-45-9	290.268	red cry (al)	208 dec				s alk
6564	Jervine		$C_{27}H_{39}NO_3$	469-59-0	425.604		243 dec				i H_2O; s EtOH, ace, chl; sl eth
6565	Kaempferol		$C_{15}H_{10}O_6$	520-18-3	286.236	ye nd (al, + 1 w)	277				sl H_2O, chl; vs EtOH, eth, ace; i bz
6566	Kainic acid		$C_{10}H_{15}NO_4$	487-79-6	213.231	cry (EtOH aq)	253 dec				s H_2O; i EtOH
6567	Kanamycin A		$C_{18}H_{36}N_4O_{11}$	59-01-8	484.499	cry (EtOH)					
6568	Kepone	Chlordecone	$C_{10}Cl_{10}O$	143-50-0	490.636		350 dec		1.61^{25}		
6569	Ketamine	2-(2-Chlorophenyl)-2-(methylamino)cyclohexanone, (±)	$C_{13}H_{16}ClNO$	6740-88-1	237.725	cry (eth-pentane)	92.5				
6570	Ketene		C_2H_2O	463-51-4	42.036	col gas	-151	-49.7(0.4)			sl eth, ace
6571	Khellin	4,9-Dimethoxy-7-methyl-5*H*-furo[3,2-*g*][1]benzopyran-5-one	$C_{14}H_{12}O_5$	82-02-0	260.242	eth, al	154 dec	$190^{0.05}$			i H_2O; s EtOH, ace; sl eth, chl
6572	L-Kynurenine	Benzenebutanoic acid, α,2-diamino-γ-oxo-	$C_{10}H_{12}N_2O_3$	343-65-7	208.213	lf (+l/2w)	194 dec				sl H_2O
6573	Labetalol		$C_{19}H_{24}N_2O_3$	36894-69-6	328.405	cry (MeOH)	164				
6574	DL-Lactic acid	2-Hydroxypropanoic acid, (±)	$C_3H_6O_3$	598-82-3	90.078	ye cry	16.9(0.2)	122^{15}	1.2060^{21}	1.4392^{20}	vs H_2O, EtOH; sl eth
6575	D-Lactic acid	D-Hydroxypropanoic acid	$C_3H_6O_3$	10326-41-7	90.078	pl (chl)	53	103^{2}			vs H_2O, EtOH
6576	L-Lactic acid	L-2-Hydroxypropanoic acid	$C_3H_6O_3$	79-33-4	90.078	hyg pr (eth)	53				vs H_2O, EtOH
6577	Lactofen		$C_{19}H_{15}ClF_3NO_7$	77501-63-4	461.773	ye pow (bz)	93				
6578	δ-Lactone-D-gluconic acid	δ-D-Gluconolactone	$C_6H_{10}O_6$	90-80-2	178.139	nd (al)					
6579	α-Lactose		$C_{12}H_{22}O_{11}$	14641-93-1	342.296	wh pow	222.8				vs H_2O; sl EtOH; i eth, chl
6580	β-D-Lactose		$C_{12}H_{22}O_{11}$	5965-66-2	342.296		254		1.59^{20}		vs H_2O; sl EtOH; i eth, chl
6581	α-Lactose monohydrate		$C_{12}H_{24}O_{12}$	5989-81-1	360.312	mcl (w)	201 dec		1.547^{20}		vs H_2O; i EtOH, eth, chl, MeOH
6582	Lactulose	4-*O*-β-D-Galactopyranosyl-D-fructose	$C_{12}H_{22}O_{11}$	4618-18-2	342.296	hx pl (MeOH)	169				vs H_2O
6583	Laminaribiose	3-*O*-β-D-Glucopyranosyl-D-glucose	$C_{12}H_{22}O_{11}$	34980-39-7	342.296		205				
6584	Lanosta-8,24-dien-3-ol, (3β)	Lanosterol	$C_{30}H_{50}O$	79-63-0	426.717	nd (eth), cry (MeOH-ace)	140.5				vs eth, EtOH, chl
6585	Lantadene A	Rehmannic acid	$C_{35}H_{52}O_5$	467-81-2	552.785	cry (MeOH)	297				
6586	Lantadene B		$C_{35}H_{52}O_5$	467-82-3	552.785	cry (EtOH)	302				
6587	L-Lanthionine	L-Cysteine, *S*-(2-amino-2-carboxyethyl)-, (*R*)-	$C_6H_{12}N_2O_4S$	922-55-4	208.235	hex pl	294 dec				sl H_2O
6588	Lapachol	2-Hydroxy-3-(3-methyl-2-butenyl)-1,4-naphthalenedione	$C_{15}H_{14}O_3$	84-79-7	242.270	ye pr (eth, bz) pl (al)	139.5				i H_2O; s EtOH, eth, bz, chl; vs HOAc

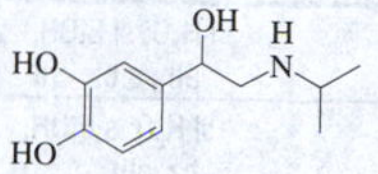
6547
Isoproterenol

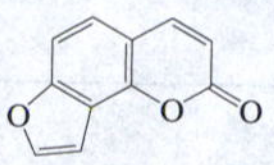
6548
Isopsoralen

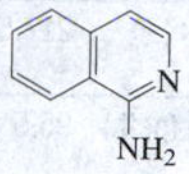
6549
1-Isoquinolinamine

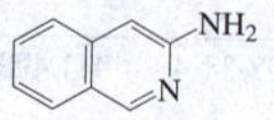
6550
3-Isoquinolinamine

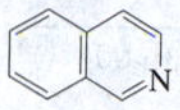
6551
Isoquinoline

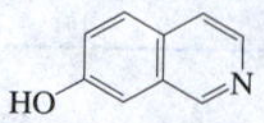
6552
7-Isoquinolinol

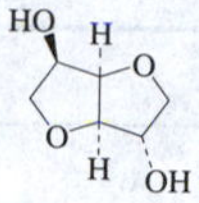
6553
Isosorbide

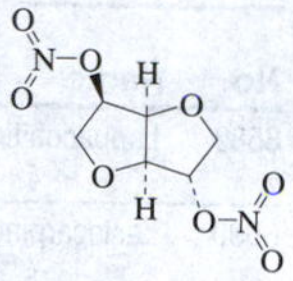
6554
Isosorbide dinitrate

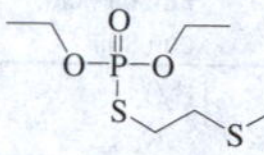
6555
Isosystox

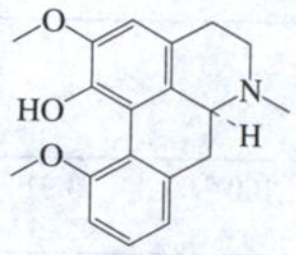
6556
Isothebaine

HN=C=S
6557
Isothiocyanic acid

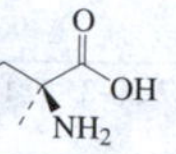
6558
L-Isovaline

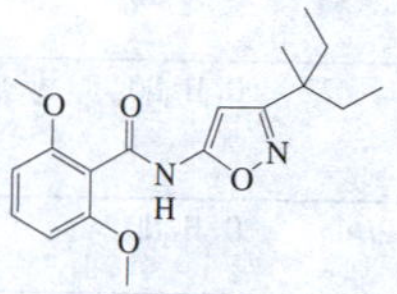
6559
Isoxaben

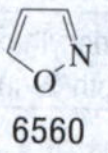
6560
Isoxazole

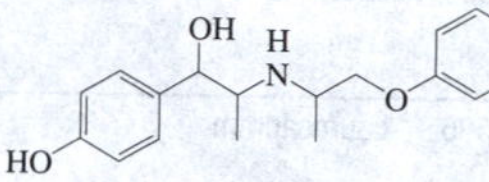
6561
Isoxsuprine

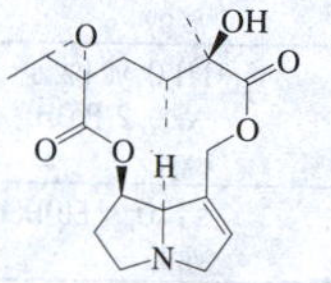
6562
Jacobine

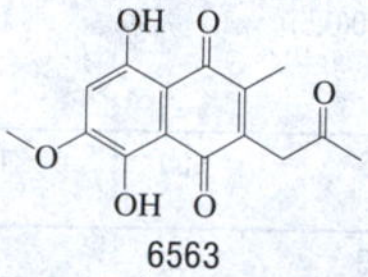
6563
Javanicin

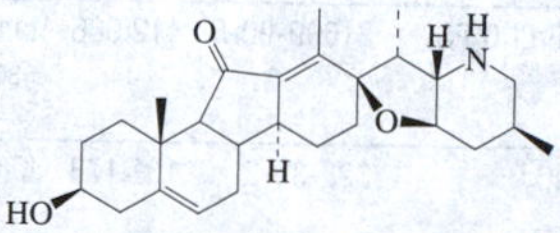
6564
Jervine

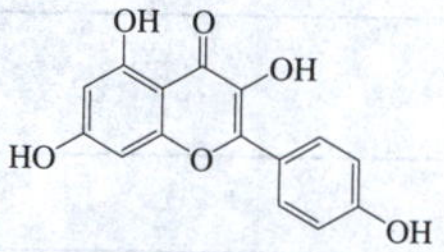
6565
Kaempferol

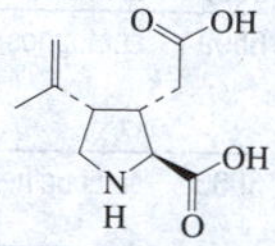
6566
Kainic acid

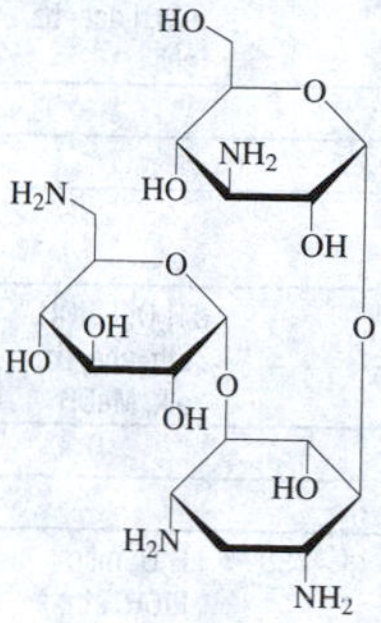
6567
Kanamycin A

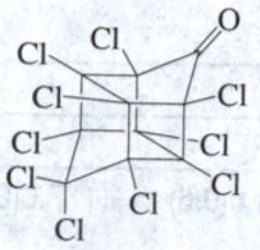
6568
Kepone

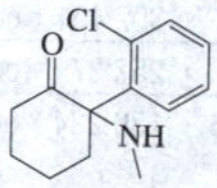
6569
Ketamine

$H_2C{=}C{=}O$
6570
Ketene

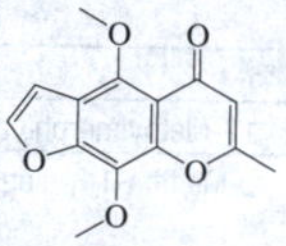
6571
Khellin

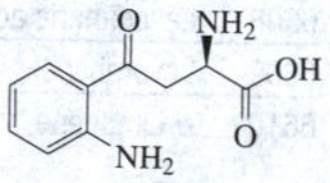
6572
L-Kynurenine

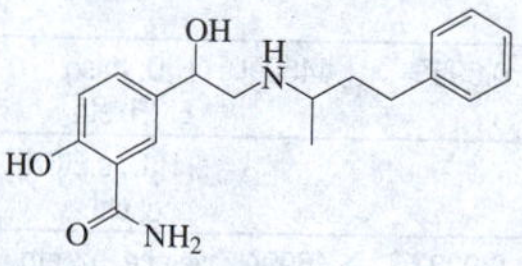
6573
Labetalol

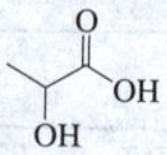
6574
DL-Lactic acid

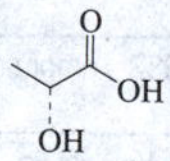
6575
D-Lactic acid

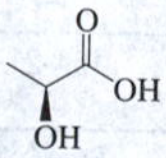
6576
L-Lactic acid

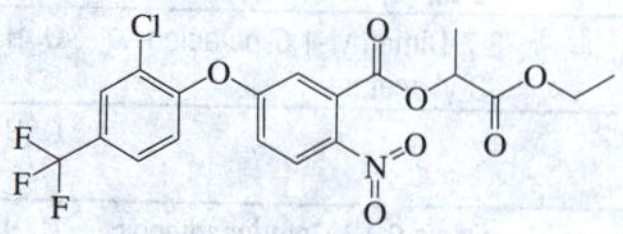
6577
Lactofen

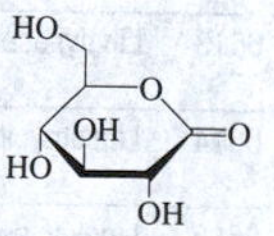
6578
δ-Lactone-D-gluconic acid

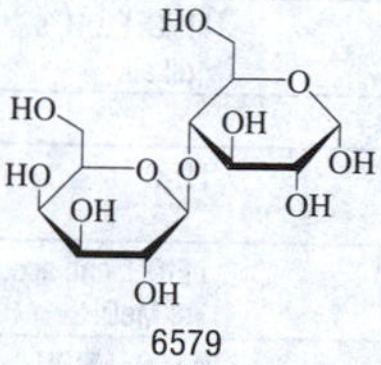
6579
α-Lactose

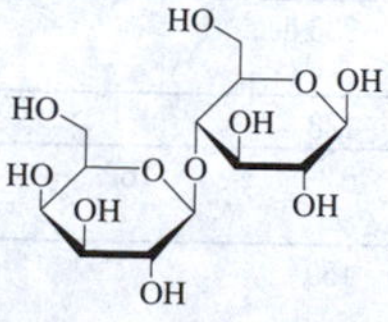
6580
β-D-Lactose

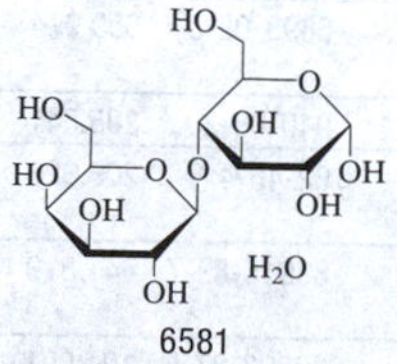
6581
α-Lactose monohydrate

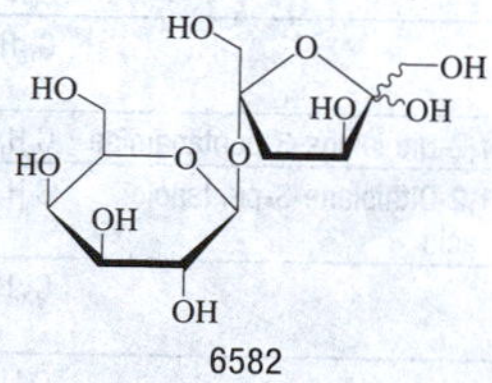
6582
Lactulose

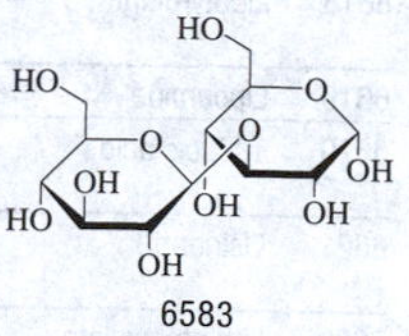
6583
Laminaribiose

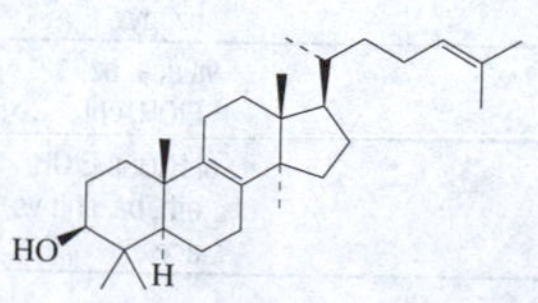
6584
Lanosta-8,24-dien-3-ol, (3β)

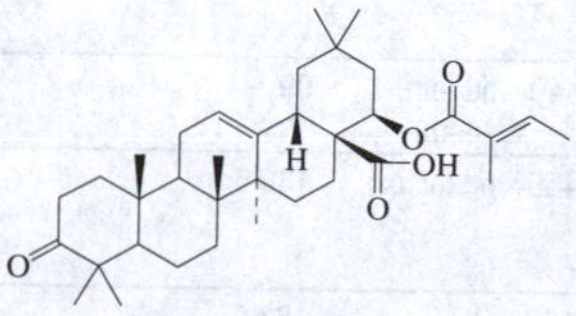
6585
Lantadene A

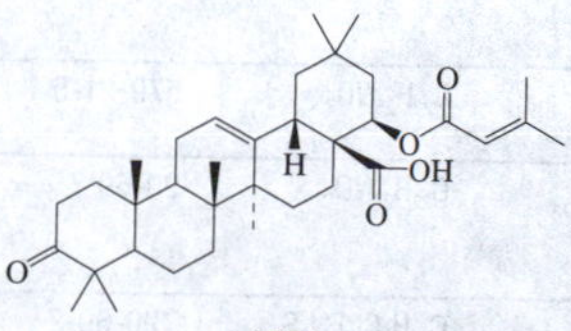
6586
Lantadene B

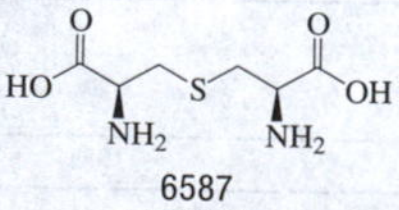
6587
L-Lanthionine

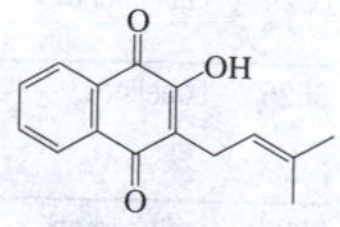
6588
Lapachol

No.	Name	Synonym	Mol. form.	CAS Reg. No.	Mol. wt.	Physical form	mp/°C	bp/°C	den/ g cm⁻³	n_D	Solubility
6589	Lappaconitine		$C_{32}H_{44}N_2O_8$	32854-75-4	584.699	hex pl (al)	217.5				i H_2O; sl EtOH, eth; s bz, chl
6590	Lasiocarpine		$C_{21}H_{33}NO_7$	303-34-4	411.490	col pl (peth)	95.5				sl H_2O; s EtOH, bz, eth
6591	Laudanidine		$C_{20}H_{25}NO_4$	301-21-3	343.418	hex pr (al)	184.5				vs H_2O, bz
6592	Laudanine		$C_{20}H_{25}NO_4$	85-64-3	343.418	ye wh pr (dil al, al-chl)	167		1.26^{20}		sl H_2O, EtOH, eth; s bz, chl
6593	Laudanosine		$C_{21}H_{27}NO_4$	2688-77-9	357.444	nd (peth), pr (al)	89				vs ace, eth, EtOH, chl
6594	Laureline		$C_{19}H_{19}NO_3$	81-38-9	309.359	tab (al) cubes (peth)	114				i H_2O; s EtOH, eth, dil acid, con sulf
6595	Laurocapram	1-Dodecylhexahydro-2*H*-azepin-2-one	$C_{18}H_{35}NO$	59227-89-3	281.477	col liq	-7	160^{50}	0.91	1.4701	i H_2O
6596	Lead bis(dimethyldithiocarbamate)		$C_6H_{12}N_2PbS_4$	19010-66-3	447.6	pale ye nd	258				
6597	Ledol		$C_{15}H_{26}O$	577-27-5	222.366	nd (al)	105	292	0.9078^{100}	1.4667^{110}	vs ace, eth, EtOH
6598	Lenacil		$C_{13}H_{18}N_2O_2$	2164-08-1	234.294		290		1.32^{25}		vs py
6599	Leptophos		$C_{13}H_{10}BrCl_2O_2PS$	21609-90-5	412.066	tan waxy solid	74.0(0.5)		1.53^{25}		i H_2O; vs bz; s ace, 2-PrOH, xyl
6600	DL-Leucine		$C_6H_{13}NO_2$	328-39-2	131.173	lf (w)	293	sub	1.293^{18}		s H_2O; sl EtOH; i eth
6601	D-Leucine		$C_6H_{13}NO_2$	328-38-1	131.173	pl (al)	293	sub			sl H_2O
6602	L-Leucine	2-Amino-4-methylpentanoic acid	$C_6H_{13}NO_2$	61-90-5	131.173	hex pl (dil al)	293	sub	1.293^{18}		sl H_2O; i EtOH, eth
6603	*N*-Leucylglycine		$C_8H_{16}N_2O_3$	686-50-0	188.224		248 dec				s H_2O; sl EtOH, eth; i ace, bz, chl
6604	Leuprolide		$C_{59}H_{84}N_{16}O_{12}$	53714-56-0	1209.398	fluffy solid					
6605	Leurosine		$C_{46}H_{56}N_4O_9$	23360-92-1	808.959	cry	203				
6606	Levallorphan	17-Allylmorphinan-3-ol	$C_{19}H_{25}NO$	152-02-3	283.408	cry (EtOH aq)	181				
6607	Levodopa	L-3,4-Dihydroxyphenylalanine	$C_9H_{11}NO_4$	59-92-7	197.188	pl (dil al) pr or nd ($w+SO_2$)	277(2)				s H_2O; i EtOH, eth, ace, bz; s alk, MeOH
6608	Levopimaric acid		$C_{20}H_{30}O_2$	79-54-9	302.451	orth cry	150				
6609	Levorphanol	17-Methylmorphinan-3-ol	$C_{17}H_{23}NO$	77-07-6	257.371	cry	198				
6610	*d*-Limonene	*p*-Mentha-1,8-diene, (*R*)	$C_{10}H_{16}$	5989-27-5	136.234	oil	-74.0(0.6)	177.6(0.5)	0.8411^{20}	1.4730^{20}	i H_2O; msc EtOH, eth; s ctc
6611	*l*-Limonene	*p*-Mentha-1,8-diene, (*S*)	$C_{10}H_{16}$	5989-54-8	136.234	oil		178(1)	0.843^{20}	1.4746^{20}	i H_2O; vs eth, EtOH
6612	Linalol	3,7-Dimethyl-1,6-octadien-3-ol, (±)-	$C_{10}H_{18}O$	22564-99-4	154.249			198	0.870^{15}	1.4627	
6613	Linalyl acetate	3,7-Dimethyl-1,6-octadien-3-yl acetate	$C_{12}H_{20}O_2$	115-95-7	196.286	liq		221(6)	0.895^{20}	1.4460^{20}	i H_2O; misc EtOH, eth
6614	Lincomycin		$C_{18}H_{34}N_2O_6S$	154-21-2	406.537	amor solid					sl H_2O; s EtOH, ace, chl
6615	Linoleic acid	*cis,cis*-9,12-Octadecadienoic acid	$C_{18}H_{32}O_2$	60-33-3	280.446	col liq	-6.9(0.7)	229^{16}	0.9022^{20}	1.4699^{20}	vs ace, bz, eth, EtOH
6616	Linolenic acid	*cis,cis,cis*-9,12,15-Octadecatrienoic acid	$C_{18}H_{30}O_2$	463-40-1	278.430		-10(2)	231^{17}	0.9164^{20}	1.4800^{20}	i H_2O; s EtOH, eth; sl bz
6617	Linuron		$C_9H_{10}Cl_2N_2O_2$	330-55-2	249.093		92.9(1)				
6618	Liothyronine		$C_{15}H_{12}I_3NO_4$	6893-02-3	650.974	cry	236 dec				i H_2O, EtOH; s dil alk
6619	Lipoamide	1,2-Dithiolane-3-pentanamide	$C_8H_{15}NOS_2$	940-69-2	205.341	cry	128				
6620	α-Lipoic acid	1,2-Dithiolane-3-pentanoic acid	$C_8H_{14}O_2S_2$	62-46-4	206.326	ye nd	61	162			i H_2O
6621	Lisinopril		$C_{21}H_{35}N_3O_7$	83915-83-7	441.519	wh cry pow	159				i EtOH, chl, ace; sl MeOH
6622	Lithium oxalate		$C_2Li_2O_4$	30903-87-8	101.901	col cry	dec		2.121^{17}		s H_2O; i EtOH, eth
6623	Lobelanidine		$C_{22}H_{29}NO_2$	552-72-7	339.471	sc (al, eth)	150				i H_2O; s EtOH; sl eth; vs ace, bz, py
6624	Lobelanine		$C_{22}H_{25}NO_2$	579-21-5	335.440	nd (eth, peth)	99				vs ace, bz, EtOH, chl
6625	Lobeline		$C_{22}H_{27}NO_2$	90-69-7	337.455	nd (al, bz)	130.5				sl H_2O; s EtOH, eth, bz, chl; vs ace
6626	Loflucarban		$C_{13}H_9Cl_2FN_2S$	790-69-2	315.192		163.5				

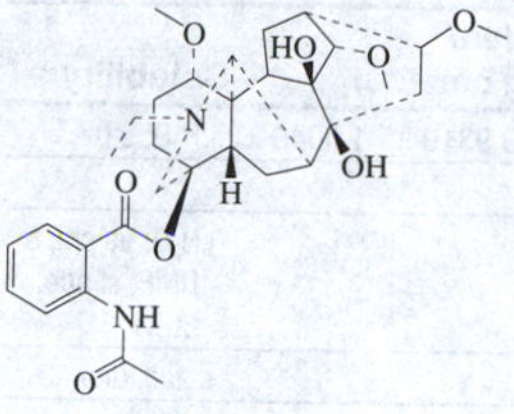
6589
Lappaconitine

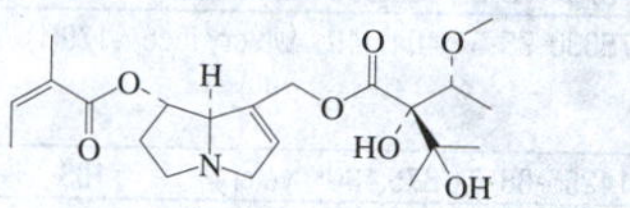
6590
Lasiocarpine

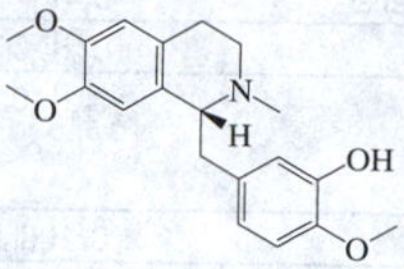
6591
Laudanidine

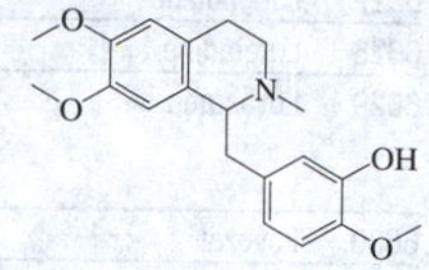
6592
Laudanine

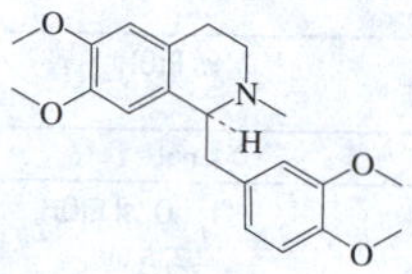
6593
Laudanosine

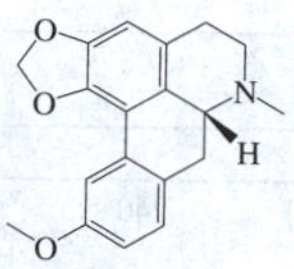
6594
Laureline

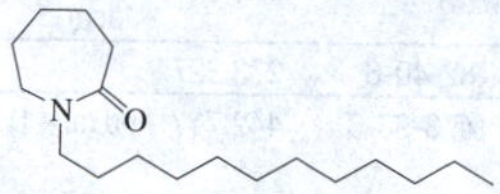
6595
Laurocapram

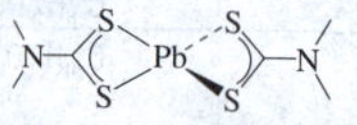
6596
Lead bis(dimethyldithiocarbamate)

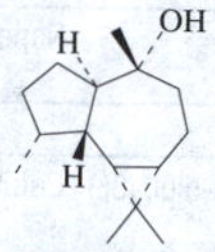
6597
Ledol

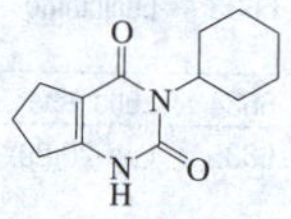
6598
Lenacil

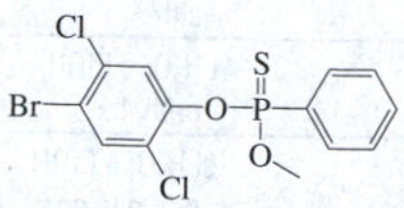
6599
Leptophos

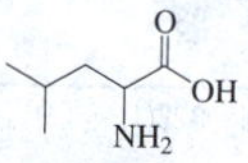
6600
DL-Leucine

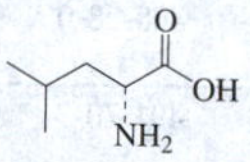
6601
D-Leucine

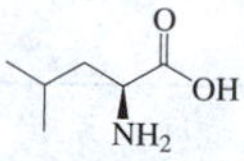
6602
L-Leucine

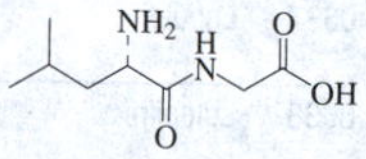
6603
N-Leucylglycine

H-5-oxoPro-His-Trp-Ser-Tyr-D-Leu-Leu-Arg-Pro-NHEt
6604
Leuprolide

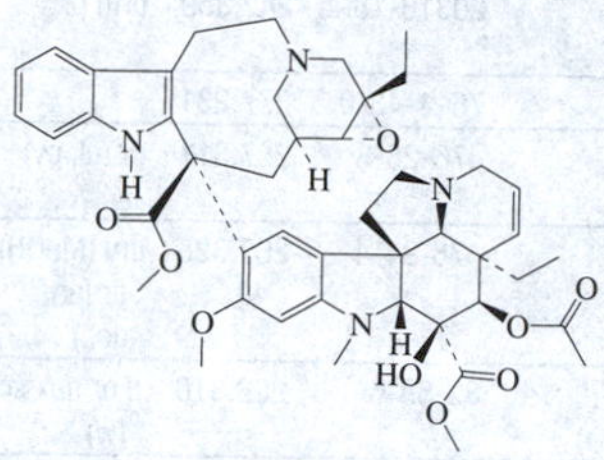
6605
Leurosine

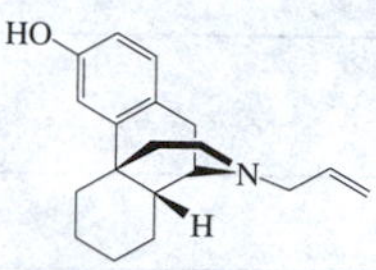
6606
Levallorphan

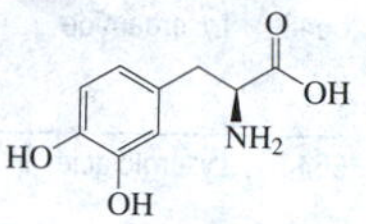
6607
Levodopa

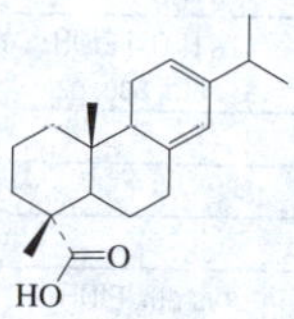
6608
Levopimaric acid

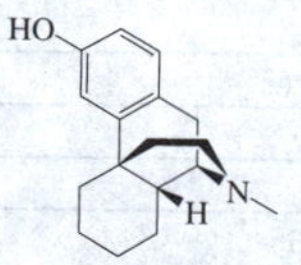
6609
Levorphanol

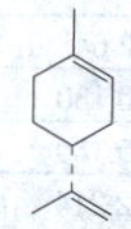
6610
d-Limonene

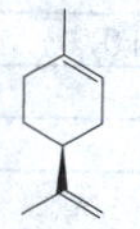
6611
l-Limonene

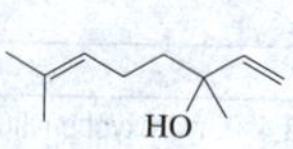
6612
Linalol

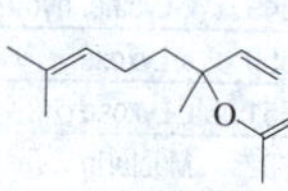
6613
Linalyl acetate

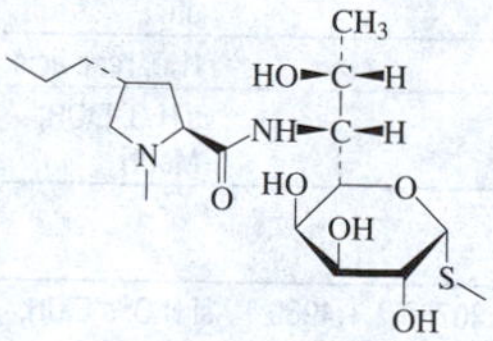
6614
Lincomycin

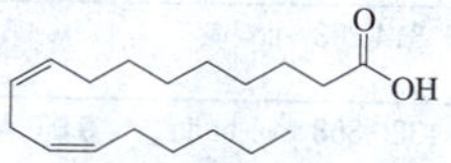
6615
Linoleic acid

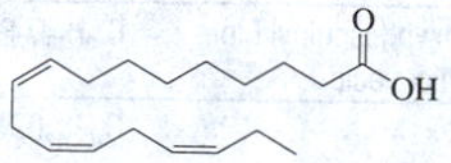
6616
Linolenic acid

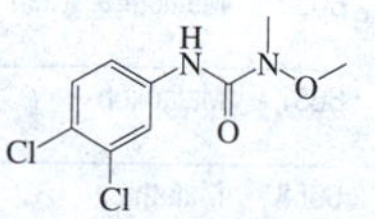
6617
Linuron

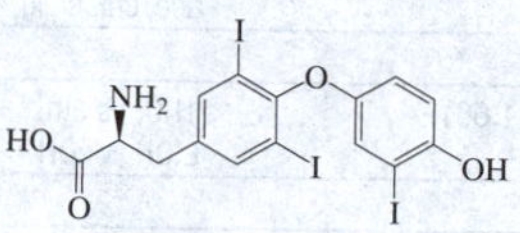
6618
Liothyronine

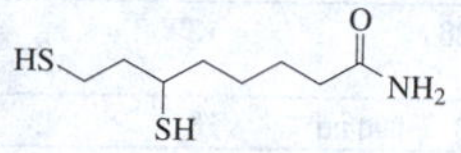
6619
Lipoamide

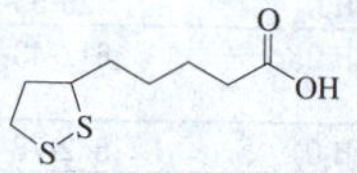
6620
α-Lipoic acid

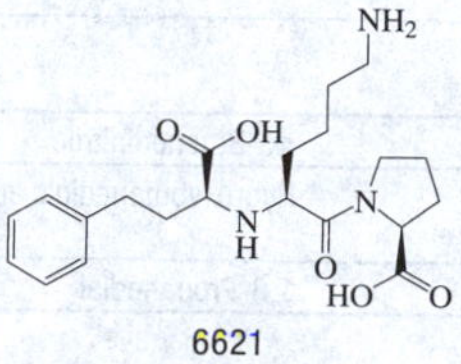
6621
Lisinopril

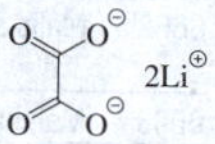
6622
Lithium oxalate

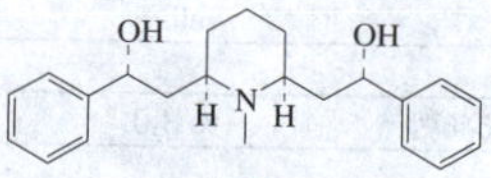
6623
Lobelanidine

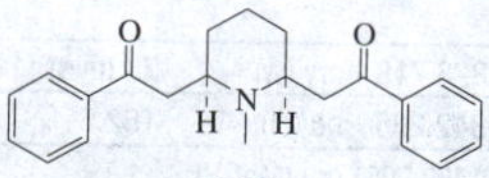
6624
Lobelanine

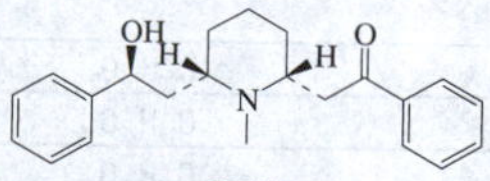
6625
Lobeline

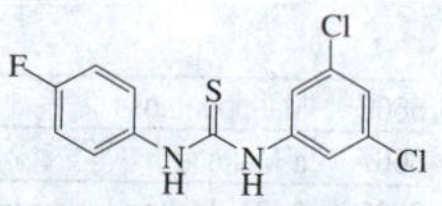
6626
Loflucarban

No.	Name	Synonym	Mol. form.	CAS Reg. No.	Mol. wt.	Physical form	mp/°C	bp/°C	den/ g cm^{-3}	n_D	Solubility
6627	Longifolene	Kuromatsuene	$C_{15}H_{24}$	475-20-7	204.352			258	0.9319^{18}	1.5040^{20}	i H_2O; s bz
6628	Loratadine	Claritin	$C_{22}H_{23}ClN_2O_2$	79794-75-5	382.883	cry (MeCN)	132				
6629	Lovastatin	Mevacor	$C_{24}H_{36}O_5$	75330-75-5	404.540	wh cry (ace aq)	179(1)				i H_2O; vs chl; s DMF; sl ace, EtOH
6630	Lovozal		$C_{15}H_7Cl_2F_3N_2O_2$	14255-88-0	375.130	ye cry	103				s ace, diox
6631	Loxapine		$C_{18}H_{18}ClN_3O$	1977-10-2	327.808	ye cry (peth)	109.5				
6632	Loxoprofen	α-Methyl-4-[(2-oxocyclopentyl)methyl] benzeneacetic acid	$C_{15}H_{18}O_3$	68767-14-6	246.302	col oil	110	192$^{0.3}$			
6633	Luciculine	Napelline	$C_{22}H_{35}NO_3$	5008-52-6	361.518	cry (+1w, ace)	149	165$^{0.02}$			vs EtOH
6634	Lunacrine		$C_{16}H_{19}NO_3$	82-40-6	273.327						s chl
6635	Lup-20(29)-ene-3,28-diol, (3β)	Betulin	$C_{30}H_{50}O_2$	473-98-3	442.717	nd (al +1)	256(2)	240 sub			i H_2O; sl EtOH, bz; s eth, AcOEt, lig
6636	Lup-20(29)-en-3-ol, (3β)	Lupeol	$C_{30}H_{50}O$	545-47-1	426.717	nd (al, ace)	216		0.9457^{218}	1.4910^{218}	i H_2O; vs EtOH, eth, ace, bz, chl
6637	Lupulon		$C_{26}H_{38}O_4$	468-28-0	414.578	pr (MeOH)	93				i H_2O; s EtOH, peth, hx
6638	Luteolin		$C_{15}H_{10}O_6$	491-70-3	286.236	ye nd (dil al, + 1 w)	328(1)				sl H_2O; s EtOH, eth, alk, con sulf
6639	Luteoskyrin	8,8'-Dihydroxyrugulosin	$C_{30}H_{22}O_{12}$	21884-44-6	574.489	ye nd (EtOH)	278 dec				
6640	Lycodine		$C_{16}H_{22}N_2$	20316-18-1	242.359	orth pr	99	190$^{1.0}$			s H_2O, chl, eth, EtOH; i peth
6641	Lycomarasmine		$C_9H_{15}N_3O_7$	7611-43-0	277.231		228 dec				
6642	Lycorine		$C_{16}H_{17}NO_4$	476-28-8	287.311	pr (al, py)	280	sub			i H_2O; sl EtOH, eth, chl
6643	Lysergamide		$C_{16}H_{17}N_3O$	478-94-4	267.325	cry (MeOH), pr (aq, ace)	137.5				sl EtOH, ace, os
6644	Lysergic acid		$C_{16}H_{16}N_2O_2$	82-58-6	268.310	lf or hex sc (w)	240 dec				sl H_2O, eth, bz; s EtOH, py
6645	Lysergide		$C_{20}H_{25}N_3O$	50-37-3	323.432		82				
6646	DL-Lysine	2,6-Diaminohexanoic acid, (±)	$C_6H_{14}N_2O_2$	70-54-2	146.187		224				sl H_2O
6647	D-Lysine	2,6-Diaminohexanoic acid, (D)	$C_6H_{14}N_2O_2$	923-27-3	146.187		218 dec				s H_2O
6648	L-Lysine	2,6-Diaminohexanoic acid, (L)	$C_6H_{14}N_2O_2$	56-87-1	146.187	nd (w, dil al)	224 dec				s H_2O; i EtOH, eth, ace, bz
6649	L-Lysine, hydrochloride		$C_6H_{15}ClN_2O_2$	10098-89-2	182.648		263 dec				
6650	D-Lyxose		$C_5H_{10}O_5$	1114-34-7	150.130		108		1.545^{20}		
6651	L-Lyxose		$C_5H_{10}O_5$	1949-78-6	150.130		110				
6652	Maclurin	(3,4-Dihydroxyphenyl)(2,4,6-trihydroxyphenyl)methanone	$C_{13}H_{10}O_6$	519-34-6	262.214	ye nd (al)	222.5				vs eth, EtOH
6653	Magenta base	Rosaniline	$C_{20}H_{19}N_3$	3248-93-9	301.385	br-red cry	186 dec				
6654	Magenta I	Rosaniline hydrochloride	$C_{20}H_{20}ClN_3$	632-99-5	337.846	grn cry	200 dec				sl H_2O, EtOH; i eth
6655	Magnesium stearate	Magnesium octadecanoate	$C_{36}H_{70}MgO_4$	557-04-0	591.244	wh pow	132				i H_2O; reac acid
6656	Malachite Green		$C_{23}H_{25}ClN_2$	569-64-2	364.911	grn cry					vs H_2O, EtOH, MeOH
6657	Malaoxon	(Dimethoxyphosphinylthio)-butanedioic acid	$C_{10}H_{19}O_7PS$	1634-78-2	314.293	liq		132$^{0.1}$			
6658	Malathion		$C_{10}H_{19}O_6PS_2$	121-75-5	330.358	ye-br liq	3.0(0.4)	156$^{0.7}$ dec	1.2076^{20}	1.4960^{20}	sl H_2O; s EtOH, eth, bz
6659	Maleic acid	*cis*-2-Butenedioic acid	$C_4H_4O_4$	110-16-7	116.073	mcl pr (w)	143.5(0.5)		1.590^{20}		vs H_2O, EtOH, ace; s eth; i bz, chl
6660	Maleic anhydride		$C_4H_2O_3$	108-31-6	98.057	nd (chl, eth)	52.56(0.04)	202	1.314^{60}		s H_2O; s eth, ace, chl; sl lig
6661	Maleonitrile	*cis*-Butenedinitrile	$C_4H_2N_2$	928-53-0	78.072	pr (EtOH)	31.5	111^{20}			
6662	Malic acid	Hydroxybutanedioic acid	$C_4H_6O_5$	617-48-1	134.088		132		1.601^{20}		s H_2O; vs eth, EtOH, MeOH
6663	Malonaldehyde	1,3-Propanedial	$C_3H_4O_2$	542-78-9	72.063	hyg nd	73				
6664	Malonic acid		$C_3H_4O_4$	141-82-2	104.062	tcl (al)	135(1)	sub	1.619^{10}		vs H_2O, py; s EtOH, eth; i bz
6665	Malononitrile		$C_3H_2N_2$	109-77-3	66.061		31.83(0.02)	219(3)	1.1910^{20}	1.4146^{34}	s H_2O, ace, bz, chl; vs EtOH, eth
6666	Maltopentaose		$C_{30}H_{52}O_{26}$	34620-76-3	828.718	cry (w)	78 (hyd)				
6667	α-Maltose		$C_{12}H_{22}O_{11}$	4482-75-1	342.296	nd (al)	162.5		1.546^{20}		vs H_2O
6668	6-*O*-α-Maltosyl-β-cyclodextrin		$C_{54}H_{90}O_{45}$	104723-60-6	1459.266	cry (MeOH)					

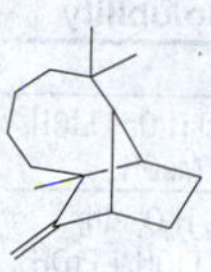
6627
Longifolene

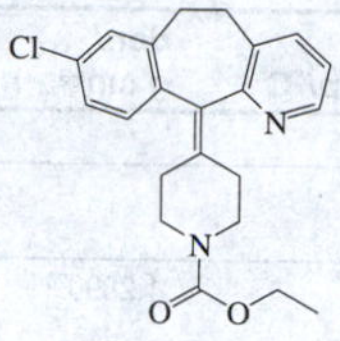
6628
Loratadine

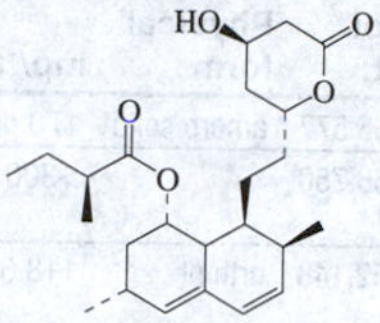
6629
Lovastatin

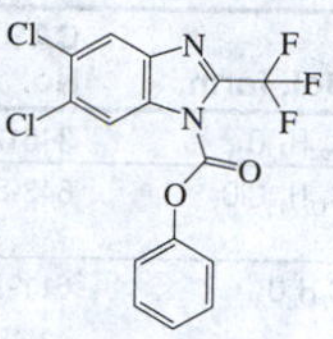
6630
Lovozal

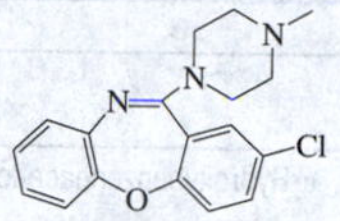
6631
Loxapine

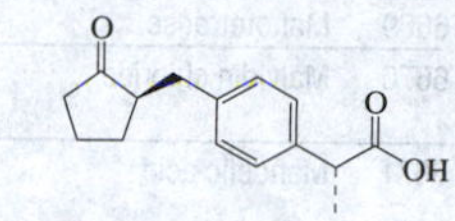
6632
Loxoprofen

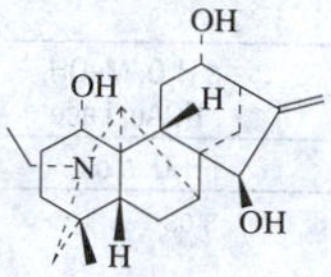
6633
Luciculine

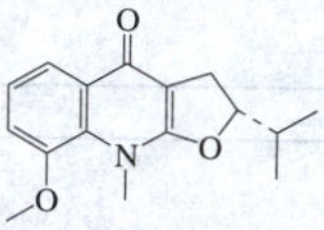
6634
Lunacrine

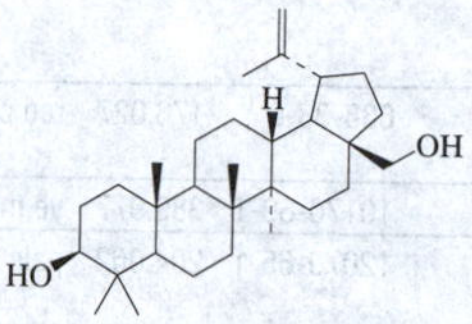
6635
Lup-20(29)-ene-3,28-diol, (3β)

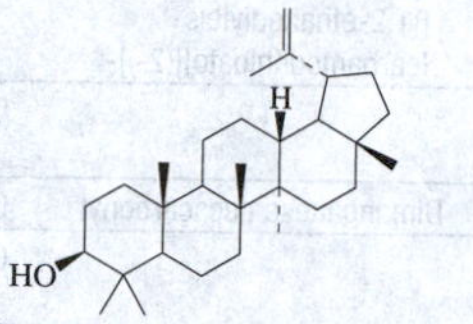
6636
Lup-20(29)-en-3-ol, (3β)

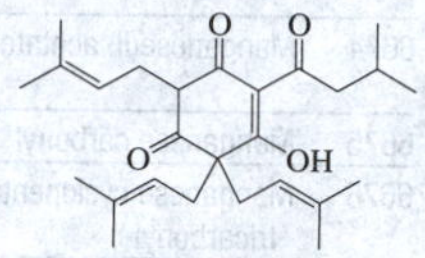
6637
Lupulon

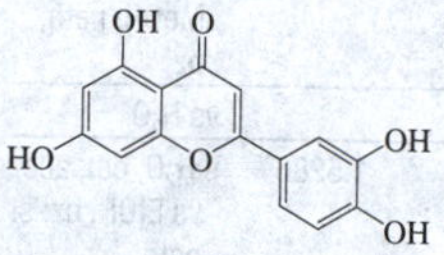
6638
Luteolin

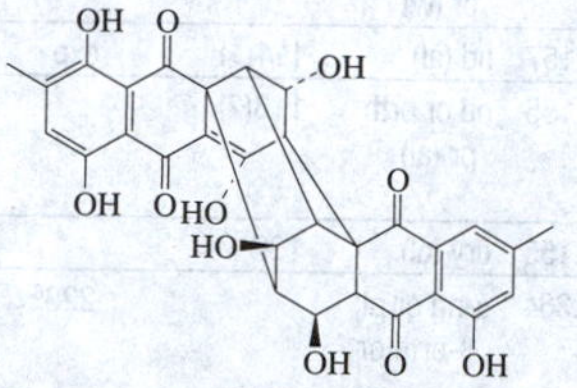
6639
Luteoskyrin

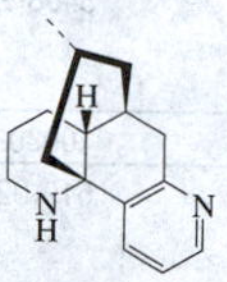
6640
Lycodine

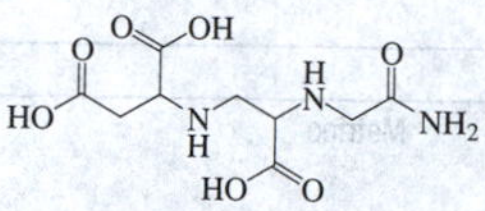
6641
Lycomarasmine

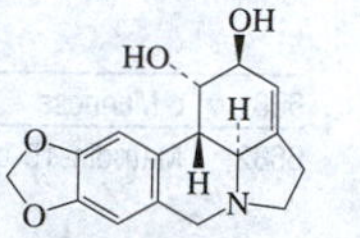
6642
Lycorine

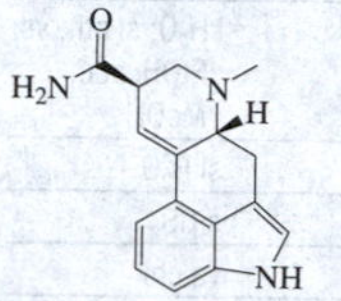
6643
Lysergamide

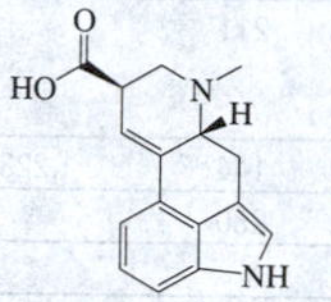
6644
Lysergic acid

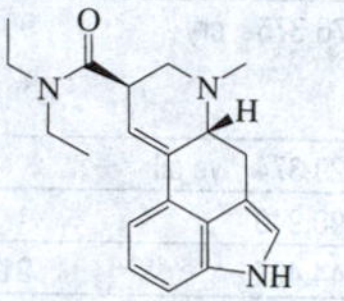
6645
Lysergide

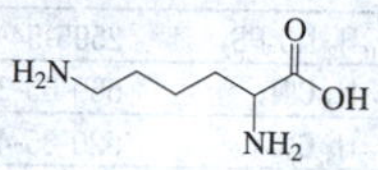
6646
DL-Lysine

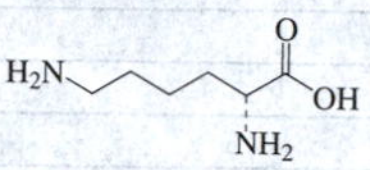
6647
D-Lysine

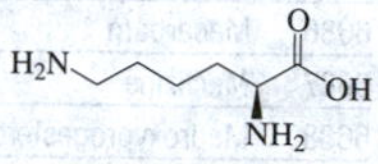
6648
L-Lysine

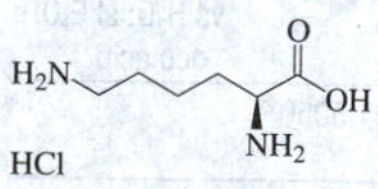
6649
L-Lysine, hydrochloride

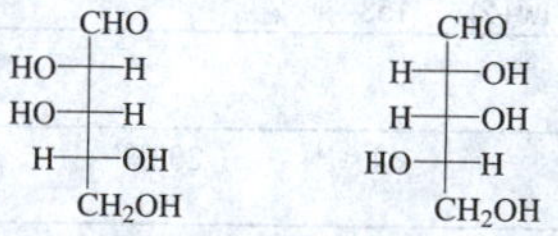
6650
D-Lyxose

6651
L-Lyxose

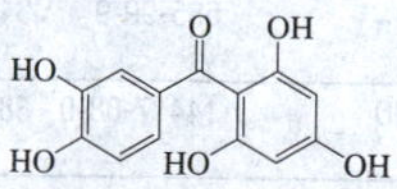
6652
Maclurin

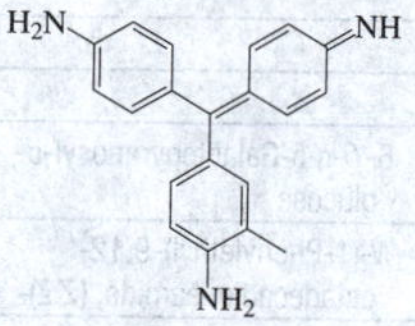
6653
Magenta base

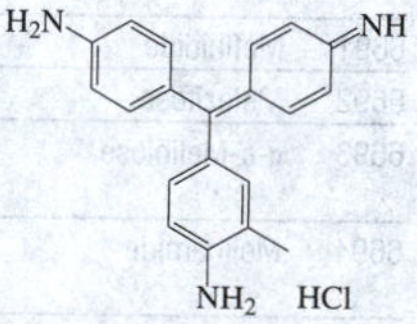
6654
Magenta I

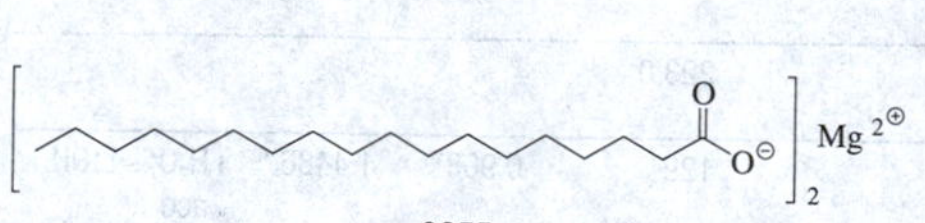
6655
Magnesium stearate

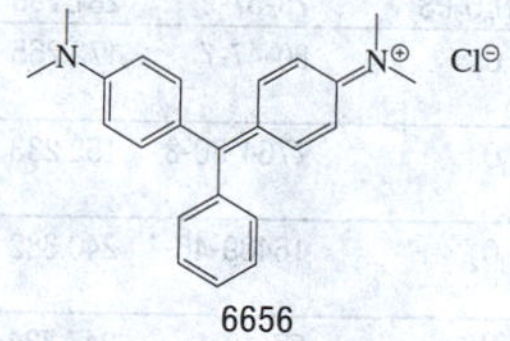
6656
Malachite Green

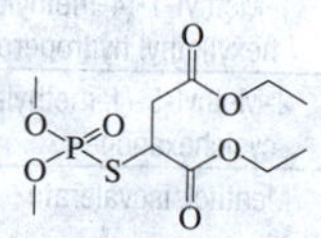
6657
Malaoxon

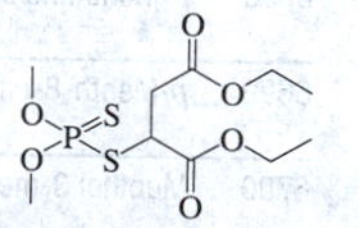
6658
Malathion

6659
Maleic acid

6660
Maleic anhydride

6661
Maleonitrile

6662
Malic acid

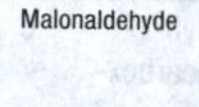
6663
Malonaldehyde

6664
Malonic acid

6665
Malononitrile

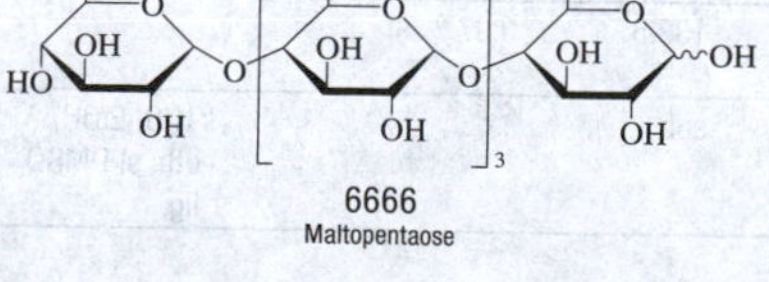
6666
Maltopentaose

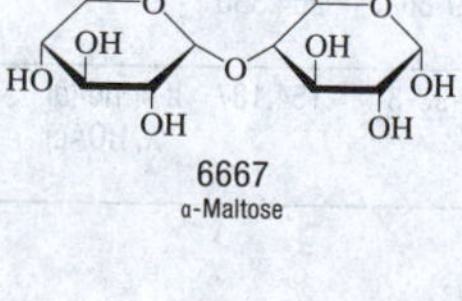
6667
α-Maltose

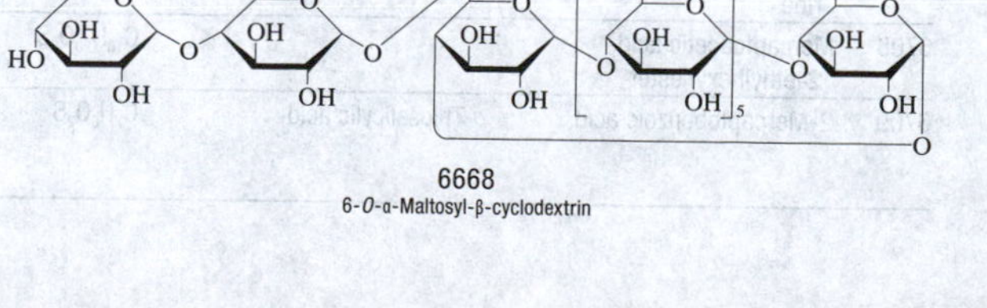
6668
6-*O*-α-Maltosyl-β-cyclodextrin

Organic

No.	Name	Synonym	Mol. form.	CAS Reg. No.	Mol. wt.	Physical form	mp/°C	bp/°C	den/ g cm^{-3}	n_D	Solubility
6669	Maltotetraose		$C_{24}H_{42}O_{21}$	34612-38-9	666.577	amorp solid	170 dec				
6670	Malvidin chloride		$C_{17}H_{15}ClO_7$	643-84-5	366.750		>300				sl H_2O; s EtOH, MeOH
6671	Mandelic acid	α-Hydroxybenzeneacetic acid	$C_8H_8O_3$	611-72-3	152.148	orth pl	118.6(1)		1.2890^{20}		s H_2O, eth, EtOH, i-PrOH
6672	Mandelonitrile glucoside		$C_{14}H_{17}NO_6$	138-53-4	295.288	wh nd or pl (al)	122				vs H_2O, EtOH
6673	Maneb	Manganese, [[1,2-ethanediylbis-[carbamodithioato]](2-)]-	$C_4H_6MnN_2S_4$	12427-38-2	265.302		dec 200				
6674	Manganese(II) acetate		$C_4H_6O_4Mn$	638-38-0	173.027	red cry (w)	210				s H_2O, MeOH, HOAc; i ace
6675	Manganese carbonyl	Dimanganese decacarbonyl	$C_{10}Mn_2O_{10}$	10170-69-1	389.977	ye mcl cry	154		1.75		i H_2O; s os
6676	Manganese cyclopentadienyl tricarbonyl		$C_8H_5MnO_3$	12079-65-1	204.062	pale ye cry	77(2)				s os
6677	Manganese 2-methylcyclo-pentadienyl tricarbonyl		$C_9H_7MnO_3$	12108-13-3	218.088	ye liq	1.5	233	1.388^{20}		i H_2O; misc bz
6678	D-Mannitol	Cordycepic acid	$C_6H_{14}O_6$	69-65-8	182.171	orth nd or pr (w)	164.1(0.1)	$295^{3.5}$	1.489^{20}	1.3330	vs H_2O; sl EtOH, py; i eth
6679	D-Mannitol hexanitrate		$C_6H_8N_6O_{18}$	15825-70-4	452.157	nd (al)	111(1)	exp	1.8^{20}		vs bz, eth, EtOH
6680	D-Mannose	Seminose	$C_6H_{12}O_6$	3458-28-4	180.155	nd or orth pr (al)	118(2)		1.539^{20}		vs H_2O; sl EtOH, MeOH; i eth, bz
6681	L-Mannose		$C_6H_{12}O_6$	10030-80-5	180.155	cry (al)	132				vs H_2O
6682	Matridin-15-one	Matrine	$C_{15}H_{24}N_2O$	519-02-8	248.364	α-nd or pl; β-orth pr		223^{6}		1.5286^{25}	s H_2O, eth, ace; vs EtOH, bz; sl peth
6683	Mazindol		$C_{16}H_{13}ClN_2O$	22232-71-9	284.739	cry (ace/hx)	198				i H_2O; s EtOH
6684	Mebendazole		$C_{16}H_{13}N_3O_3$	31431-39-7	295.292	cry (HOAc/ MeOH)	288.5				i H_2O, EtOH, eth, chl
6685	Mebhydroline		$C_{19}H_{20}N_2$	524-81-2	276.375	cry	95	211^{1}			i H_2O; sl eth; vs EtOH, ace, MeOH
6686	Mecarbam		$C_{10}H_{20}NO_5PS_2$	2595-54-2	329.374	ye oil		$144^{0.02}$	1.223^{20}		sl H_2O
6687	Meclizine		$C_{25}H_{27}ClN_2$	569-65-3	390.948			230			s CS_2
6688	Medroxyprogesterone		$C_{22}H_{32}O_3$	520-85-4	344.487		214.5				vs chl
6689	Mefenamic acid	2-[(2,3-Dimethylphenyl)-amino]benzoic acid	$C_{15}H_{15}NO_2$	61-68-7	241.286	hyg cry	230 dec				s alk; sl eth, chl
6690	Mefloquine		$C_{17}H_{16}F_6N_2O$	53230-10-7	378.311	cry (MeOH aq)	178.2				
6691	Mefluidide		$C_{11}H_{13}F_3N_2O_3S$	53780-34-0	310.292		184.3(0.5)				
6692	Melezitose		$C_{18}H_{32}O_{16}$	597-12-6	504.437	cry (w+2)	153		1.5565^{25}		vs H_2O
6693	α-D-Melibiose	6-*O*-α-D-Galactopyranosyl-D-glucose	$C_{12}H_{22}O_{11}$	585-99-9	342.296						vs H_2O; sl EtOH; dec acid
6694	Melinamide	*N*-(1-Phenylethyl)-9,12-octadecadieneamide, (*Z,Z*)-	$C_{26}H_{41}NO$	14417-88-0	383.610	oil	<4	$202^{0.07}$		1.5050^{23}	
6695	Melphalan	L-Phenylalanine, 4-[bis(2-chloroethyl)amino]-	$C_{13}H_{18}Cl_2N_2O_2$	148-82-3	305.200	nd	183 dec				i H_2O; s EtOH
6696	Menaquinone 7	Vitamin K_2(35)	$C_{46}H_{64}O_2$	2124-57-4	648.999	cry	54				
6697	Menazon		$C_6H_{12}N_5O_2PS_2$	78-57-9	281.296	cry (MeOH)	160				sl H_2O; s thf
6698	*p*-Menthane hydroperoxide	1-Methyl-1-(4-methylcyclo-hexyl)ethyl hydroperoxide	$C_{10}H_{20}O_2$	80-47-7	172.265			259	0.92		
6699	*p*-Menth-8-en-2-one	2-Methyl-5-(1-methylethenyl)-cyclohexanone	$C_{10}H_{16}O$	7764-50-3	152.233			223.0			
6700	Menthol 3-methylbutanoate	Menthol, isovalerate	$C_{15}H_{28}O_2$	16409-46-4	240.382			129^{9}	0.908^{15}	1.4486^{20}	i H_2O; s EtOH, ace
6701	Meperidine	Pethidine	$C_{15}H_{21}NO_2$	57-42-1	247.334		30	155^{5}			
6702	Mephenytoin		$C_{12}H_{14}N_2O_2$	50-12-4	218.251		136				
6703	Mephobarbital		$C_{13}H_{14}N_2O_3$	115-38-8	246.261	wh cry (w)	176				sl H_2O, eth, chl; vs EtOH
6704	Mephosfolan		$C_8H_{16}NO_3PS_2$	950-10-7	269.322	ye liq		$120^{0.001}$		1.5354^{26}	s ace, EtOH, bz
6705	Mepiquat chloride	Piperidinium, 1,1-dimethyl-, chloride	$C_7H_{16}ClN$	24307-26-4	149.662		223				
6706	Mepivacaine	*N*-(2,6-Dimethylphenyl)-1-methyl-2-piperidinecarbox-amide	$C_{15}H_{22}N_2O$	96-88-8	246.348	cry (eth)	150.5				s CS_2
6707	Mepivacaine monohydrochlo-ride	Carbocaine hydrochloride	$C_{15}H_{23}ClN_2O$	1722-62-9	282.809	cry	263				s H_2O
6708	Mercaptoacetic acid, 2-ethylhexyl ester		$C_{10}H_{20}O_2S$	7659-86-1	204.330			133.5	0.97^{20}		
6709	2-Mercaptobenzoic acid	*o*-Thiosalicylic acid	$C_7H_6O_2S$	147-93-3	154.187	lf or nd (al, w, HOAc)	168.5	sub			s H_2O, EtOH, eth; sl DMSO, lig

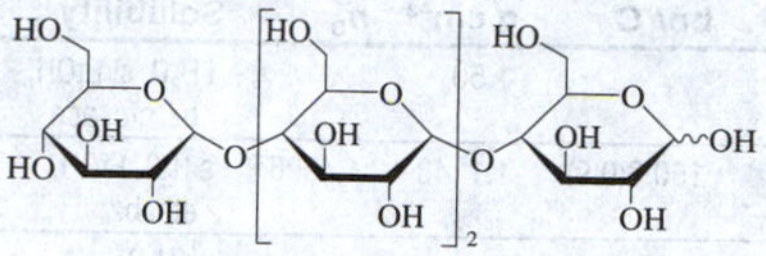
6669
Maltotetraose

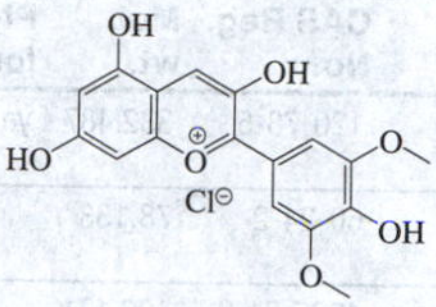
6670
Malvidin chloride

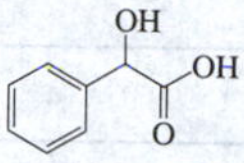
6671
Mandelic acid

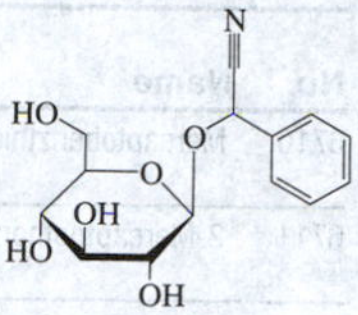
6672
Mandelonitrile glucoside

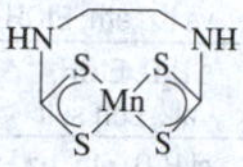
6673
Maneb

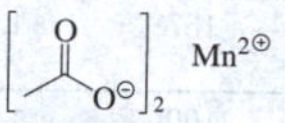
6674
Manganese(II) acetate

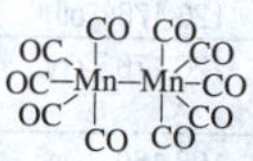
6675
Manganese carbonyl

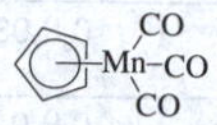
6676
Manganese cyclopentadienyl tricarbonyl

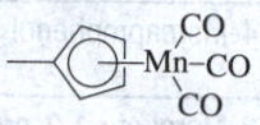
6677
Manganese 2-methylcyclopentadienyl tricarbonyl

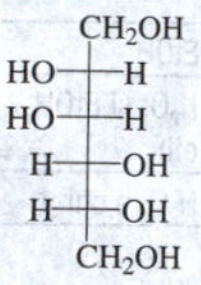
6678
D-Mannitol

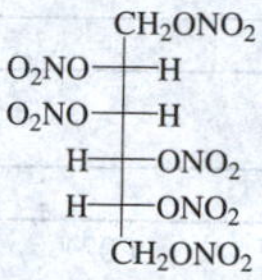
6679
D-Mannitol hexanitrate

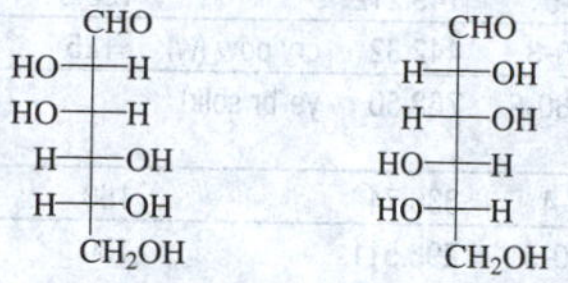
6680
D-Mannose

6681
L-Mannose

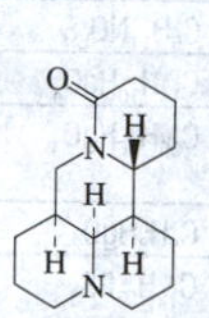
6682
Matridin-15-one

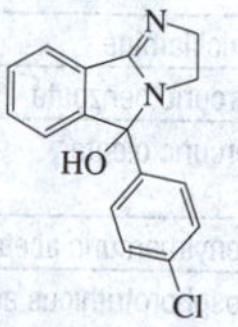
6683
Mazindol

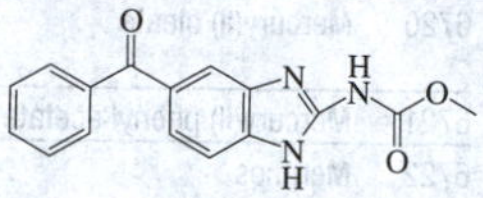
6684
Mebendazole

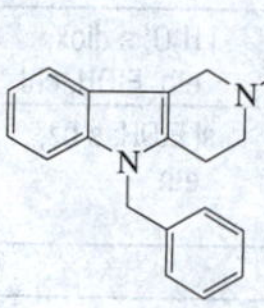
6685
Mebhydroline

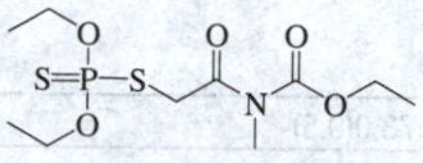
6686
Mecarbam

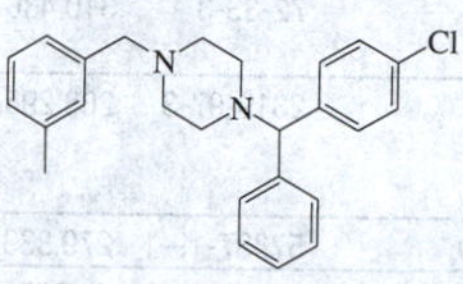
6687
Meclizine

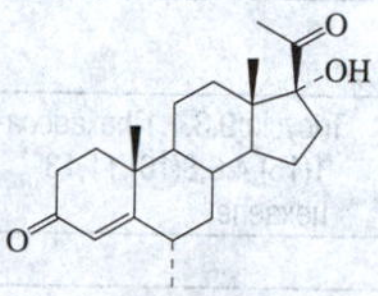
6688
Medroxyprogesterone

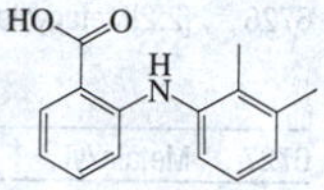
6689
Mefenamic acid

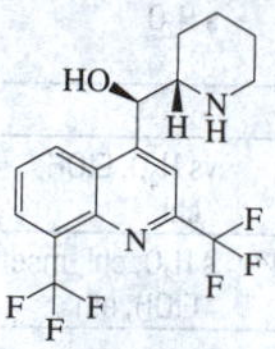
6690
Mefloquine

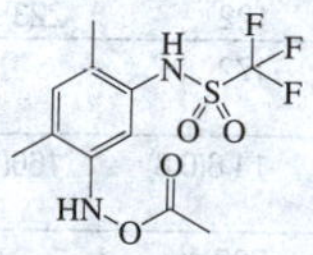
6691
Mefluidide

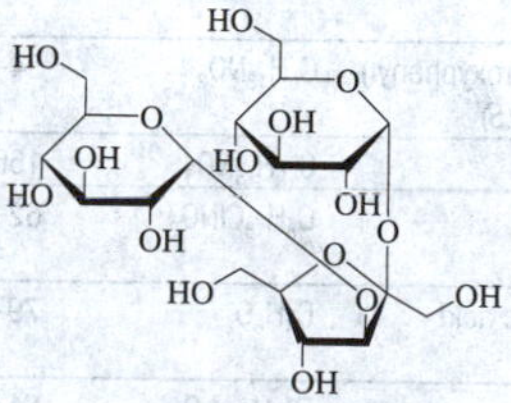
6692
Melezitose

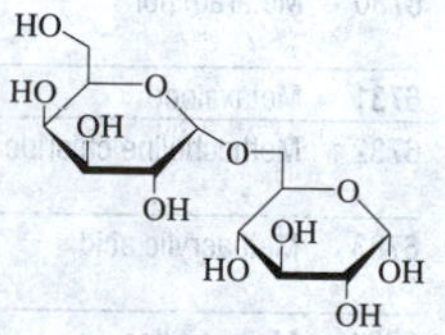
6693
α-D-Melibiose

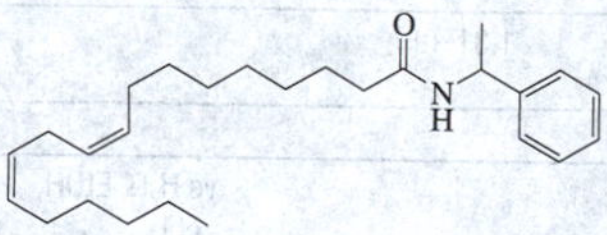
6694
Melinamide

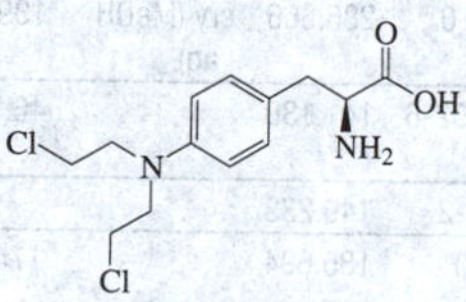
6695
Melphalan

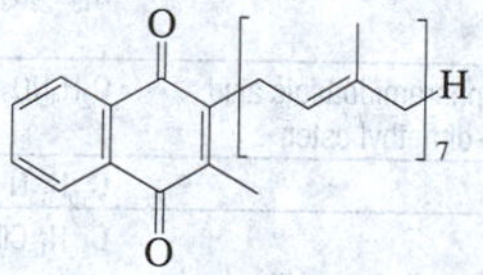
6696
Menaquinone 7

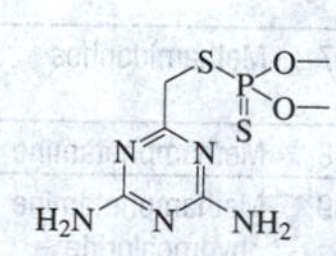
6697
Menazon

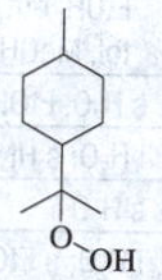
6698
p-Menthane hydroperoxide

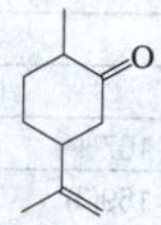
6699
p-Menth-8-en-2-one

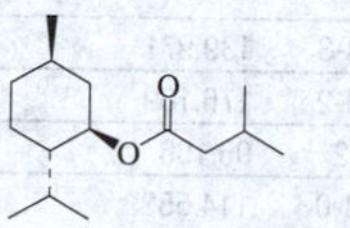
6700
Menthol 3-methylbutanoate

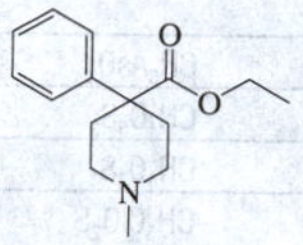
6701
Meperidine

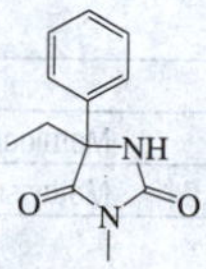
6702
Mephenytoin

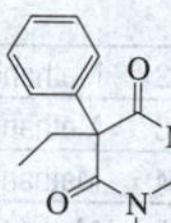
6703
Mephobarbital

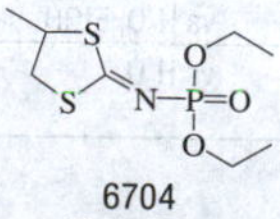
6704
Mephosfolan

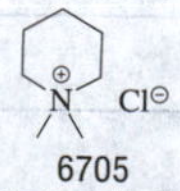
6705
Mepiquat chloride

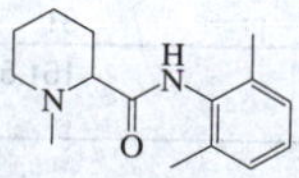
6706
Mepivacaine

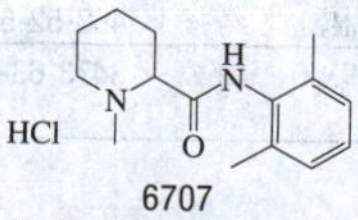
6707
Mepivacaine monohydrochloride

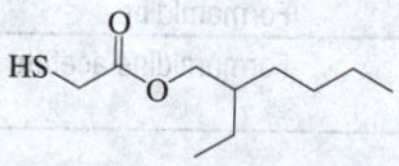
6708
Mercaptoacetic acid, 2-ethylhexyl ester

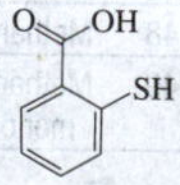
6709
2-Mercaptobenzoic acid

No.	Name	Synonym	Mol. form.	CAS Reg. No.	Mol. wt.	Physical form	mp/°C	bp/°C	den/ g cm⁻³	n_D	Solubility
6710	Mercaptobenzthiazyl ether	2,2'-Dithiobis[benzothiazole]	$C_{14}H_8N_2S_4$	120-78-5	332.487	ye nd	180		1.50		i H_2O; sl EtOH, bz, ctc, ace
6711	2-Mercaptoethanol		C_2H_6OS	60-24-2	78.133			150.0(0.9)	1.1143^{20}	1.4996^{20}	s H_2O, EtOH, eth, bz
6712	2-Mercapto-2-methylpropanoic acid		$C_4H_8O_2S$	4695-31-2	120.171		47	101^{15}			vs H_2O
6713	2-Mercapto-*N*-2-naphthylacetamide	Thionalide	$C_{12}H_{11}NOS$	93-42-5	217.286		111.5				i H_2O; vs EtOH, os
6714	2-Mercaptophenol		C_6H_6OS	1121-24-0	126.176	oil	5.5	217	1.2371^{0}		vs bz, eth, EtOH
6715	4-Mercaptophenol		C_6H_6OS	637-89-8	126.176	cry	29.5	167^{45}	1.1285^{25}	1.5101^{25}	s H_2O, EtOH, alk, con sulf
6716	3-Mercapto-1,2-propanediol	Thioglycerol	$C_3H_8O_2S$	96-27-5	108.160	visc		100^{1}	1.2455^{20}	1.5268^{20}	sl H_2O, eth, bz, chl; msc EtOH; vs ace
6717	3-Mercaptopropanoic acid		$C_3H_6O_2S$	107-96-0	106.144	amor	16.5(0.5)	111^{15}	1.218^{21}	1.494^{20}	s H_2O, EtOH, eth, ctc
6718	3-Mercapto-D-valine	Penicillamine	$C_5H_{11}NO_2S$	52-67-5	149.212		198.5				
6719	Mercury(II) benzoate	Mercuric benzoate	$C_{14}H_{10}HgO_4$	583-15-3	442.82	cry pow (w)	≈125				i EtOH
6720	Mercury(II) oleate	Mercuric oleate	$C_{36}H_{66}HgO_4$	1191-80-6	763.50	ye-br solid					i H_2O; sl EtOH, eth
6721	Mercury(II) phenyl acetate	Phenylmercuric acetate	$C_8H_8HgO_2$	62-38-4	336.74		153				i H_2O; s chl
6722	Merphos	Phosphorotrithious acid, *S,S,S*-tributyl ester	$C_{12}H_{27}PS_3$	150-50-5	298.511		100	$137^{0.7}$	1.02^{20}		
6723	Mesityl oxide	Isobutenyl methyl ketone	$C_6H_{10}O$	141-79-7	98.142	liq	-52.8(0.4)	129.7(0.4)	0.8653^{20}	1.4440^{20}	s H_2O, ace; msc EtOH, eth
6724	Mesoridazine		$C_{21}H_{26}N_2OS_2$	5588-33-0	386.573	oil					
6725	Mestranol		$C_{21}H_{26}O_2$	72-33-3	310.430	cry	154(1)				i H_2O; s diox, eth, EtOH, chl
6726	[2.2]Metacyclophane	Tricyclo[9.3.1.1]hexadeca-1(15),4,6,8(16),11,13-hexaene	$C_{16}H_{16}$	2319-97-3	208.298	orth pr	131(1)	290			sl EtOH; s bz, eth
6727	Metalaxyl		$C_{15}H_{21}NO_4$	57837-19-1	279.333		73.0(0.5)				
6728	Metaldehyde	Metacetaldehyde (polymer)	$(C_2H_4O)_x$	37273-91-9	44.052	tetr nd or pr (al)	246	115 sub			i H_2O, ace; sl EtOH, eth, bz, chl
6729	Metanil Yellow		$C_{18}H_{14}N_3NaO_3S$	587-98-4	375.377	br-ye pow					vs H_2O, EtOH; s bz, eth; sl ace
6730	Metaraminol	2-Amino-1-(3-hydroxyphenyl)-1-propanol, (1*R*,2*S*)	$C_9H_{13}NO_2$	54-49-9	167.205	hyg cry (HCl)					s H_2O
6731	Metaxalone		$C_{12}H_{15}NO_3$	1665-48-1	221.252	cry (AcOEt)	122	$223^{1.5}$			
6732	Methacholine chloride		$C_8H_{18}ClNO_2$	62-51-1	195.688	hyg cry	172				vs H_2O, EtOH, chl
6733	Methacrylic acid	2-Methylpropenoic acid	$C_4H_6O_2$	79-41-4	86.090	pr	14.6(0.8)	160(1)	1.0153^{20}	1.4314^{20}	s H_2O, chl; msc EtOH, eth
6734	Methacycline		$C_{22}H_{22}N_2O_8$	914-00-1	442.418	cry	205 dec				
6735	Methadone hydrochloride	6-(Dimethylamino)-4,4-diphenyl-3-heptanone hydrochloride	$C_{21}H_{28}ClNO$	1095-90-5	345.906	pl (al-eth)	235				vs H_2O, EtOH
6736	Methallenestril		$C_{18}H_{22}O_3$	517-18-0	286.366	cry (MeOH aq)	139				s eth
6737	Methamidophos	Phosphoramidothioic acid, *O,S*-dimethyl ester	$C_2H_8NO_2PS$	10265-92-6	141.130		46.7(0.5)		1.31^{20}		
6738	Methamphetamine		$C_{10}H_{15}N$	537-46-2	149.233			212			
6739	Methamphetamine hydrochloride	*N*,α-Dimethylbenzeneethanamine, hydrochloride, (*S*)-	$C_{10}H_{16}ClN$	51-57-0	185.694		174(1)				vs H_2O, EtOH, chl
6740	Methandrostenolone		$C_{20}H_{28}O_2$	72-63-9	300.435		166				
6741	Methane		CH_4	74-82-8	16.043	col gas	-182.4566 (0.0003)	-161.5(0.2)	0.4228^{-162}		sl H_2O, ace; s EtOH, eth, bz, tol, MeOH
6742	Methanearsonic acid		CH_5AsO_3	124-58-3	139.971		160.5				s H_2O, EtOH
6743	Methanedisulfonic acid	Methionic acid	$CH_4O_6S_2$	503-40-2	176.169		98				i H_2O; s HNO_3
6744	Methanesulfonic acid	Methylsulfonic acid	CH_4O_3S	75-75-2	96.106		20	167^{10}	1.4812^{18}	1.4317^{18}	s H_2O
6745	Methanesulfonyl chloride		CH_3ClO_2S	124-63-0	114.552			159(3)	1.4805^{18}	1.4573^{20}	i H_2O; s EtOH, eth
6746	Methanesulfonyl fluoride		CH_3FO_2S	558-25-8	98.097			123.5			
6747	Methanethiol	Methyl mercaptan	CH_4S	74-93-1	48.108	col gas	-122.98(0.09)	6.0(0.1)	0.8665^{20}		sl H_2O, chl; vs EtOH, eth
6748	Methanimidamide	Formamidine	CH_4N_2	463-52-5	44.056	pr	81	dec			vs H_2O, EtOH
6749	Methanimidamide, monoacetate	Formamidine acetate	$C_3H_8N_2O_2$	3473-63-0	104.108		161.5				vs H_2O

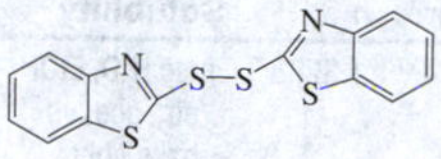
6710
Mercaptobenzthiazyl ether

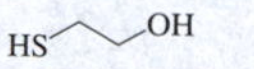
6711
2-Mercaptoethanol

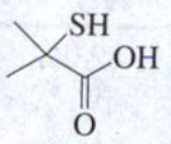
6712
2-Mercapto-2-methylpropanoic acid

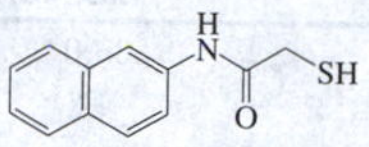
6713
2-Mercapto-*N*-2-naphthylacetamide

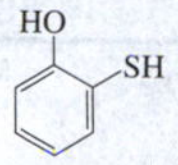
6714
2-Mercaptophenol

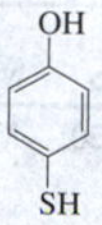
6715
4-Mercaptophenol

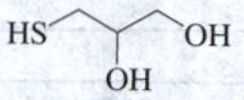
6716
3-Mercapto-1,2-propanediol

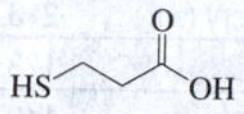
6717
3-Mercaptopropanoic acid

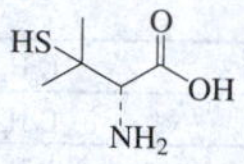
6718
3-Mercapto-D-valine

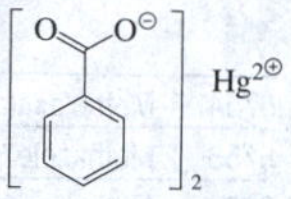
6719
Mercury(II) benzoate

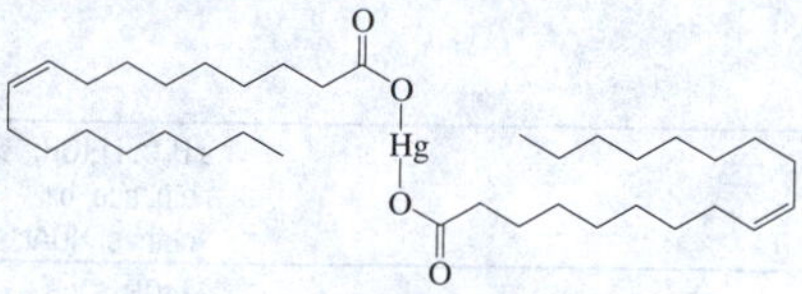
6720
Mercury(II) oleate

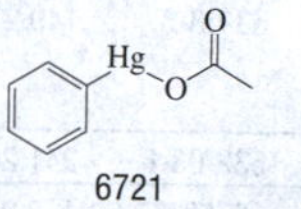
6721
Mercury(II) phenyl acetate

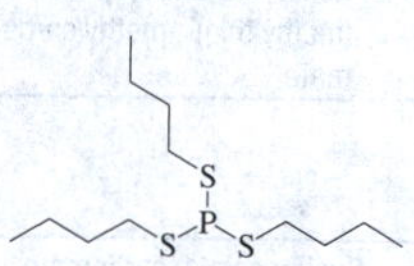
6722
Merphos

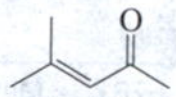
6723
Mesityl oxide

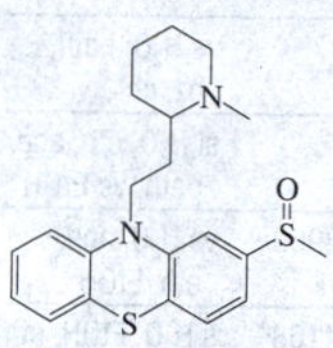
6724
Mesoridazine

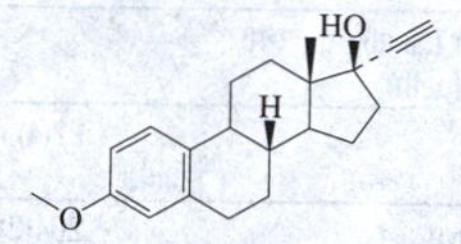
6725
Mestranol

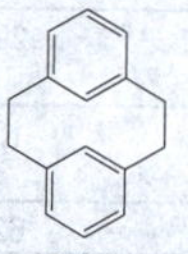
6726
[2.2]Metacyclophane

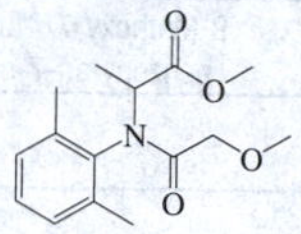
6727
Metalaxyl

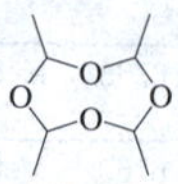
6728
Metaldehyde

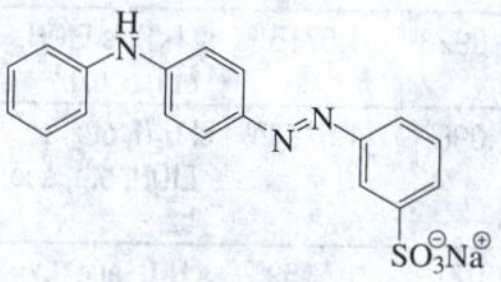
6729
Metanil Yellow

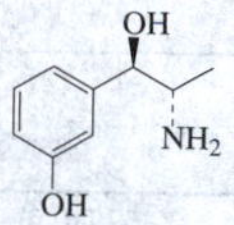
6730
Metaraminol

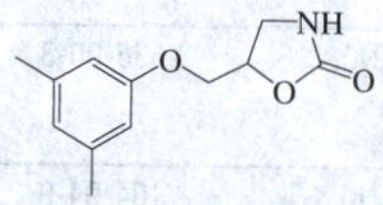
6731
Metaxalone

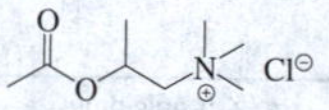
6732
Methacholine chloride

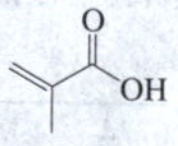
6733
Methacrylic acid

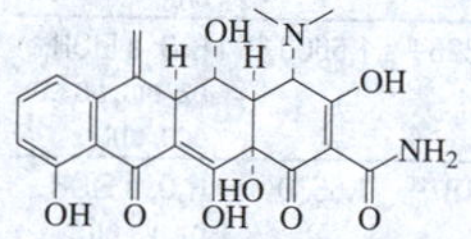
6734
Methacycline

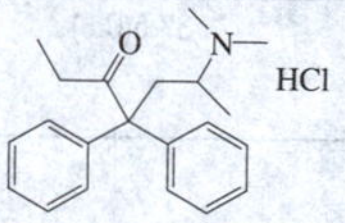
6735
Methadone hydrochloride

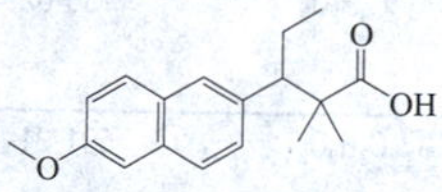
6736
Methallenestril

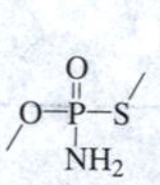
6737
Methamidophos

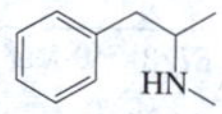
6738
Methamphetamine

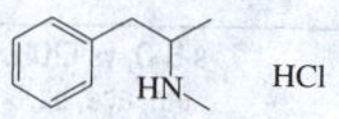
6739
Methamphetamine hydrochloride

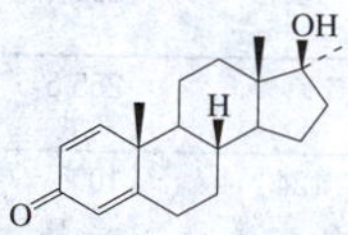
6740
Methandrostenolone

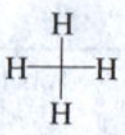
6741
Methane

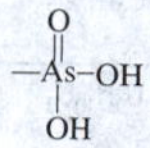
6742
Methanearsonic acid

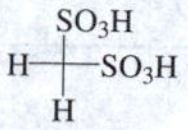
6743
Methanedisulfonic acid

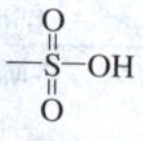
6744
Methanesulfonic acid

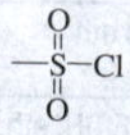
6745
Methanesulfonyl chloride

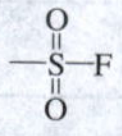
6746
Methanesulfonyl fluoride

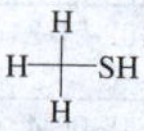
6747
Methanethiol

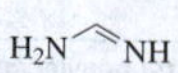
6748
Methanimidamide

6749
Methanimidamide, monoacetate

No.	Name	Synonym	Mol. form.	CAS Reg. No.	Mol. wt.	Physical form	mp/°C	bp/°C	den/ g cm^{-3}	n_D	Solubility
6750	Methanol	Methyl alcohol	CH_4O	67-56-1	32.042	liq	-97.5(0.1)	64.5(0.7)	0.7914^{20}	1.3288^{20}	msc H_2O, EtOH, eth, ace; vs bz; s chl
6751	Methantheline bromide		$C_{21}H_{26}BrNO_3$	53-46-3	420.340	cry (i-PrOH)	174.5				s H_2O, EtOH, chl; i eth
6752	Methapyrilene		$C_{14}H_{19}N_3S$	91-80-5	261.386			174^{3}		1.5915^{20}	
6753	Metharbital	5,5-Diethyl-1-methyl-2,4,6(1*H*,3*H*,5*H*)-pyrimidinetrione	$C_9H_{14}N_2O_3$	50-11-3	198.218	nd	150.5				s H_2O; sl chl
6754	Methazolamide		$C_5H_8N_4O_3S_2$	554-57-4	236.273	cry (w)	213 dec				
6755	Methazole		$C_9H_6Cl_2N_2O_3$	20354-26-1	261.061		123.9(0.5)		1.24^{25}		
6756	Methenamine allyl iodide	Allylhexamethylenetetramine iodide	$C_9H_{17}IN_4$	36895-62-2	308.162	cry	148 dec				vs H_2O; i chl, eth
6757	Methestrol		$C_{20}H_{26}O_2$	130-73-4	298.419	cry (dil HOAc)	145				
6758	Methidathion		$C_6H_{11}N_2O_4PS_3$	950-37-8	302.330		42.5(0.5)				
6759	Methiocarb	Phenol, 3,5-dimethyl-4-(methylthio)-, methylcarbamate	$C_{11}H_{15}NO_2S$	2032-65-7	225.308		121.5(0.4)				
6760	L-Methionine		$C_5H_{11}NO_2S$	63-68-3	149.212	hex pl (dil al)	281 dec				s H_2O; i EtOH, eth, ace, bz, peth; sl HOAc
6761	Methocarbamol	Guaifenesin-1-carbamate	$C_{11}H_{15}NO_5$	532-03-6	241.241	cry (bz)	93				s EtOH
6762	Methomyl		$C_5H_{10}N_2O_2S$	16752-77-5	162.210		79(2)		1.2946^{24}		
6763	Methoprene		$C_{19}H_{34}O_3$	40596-69-8	310.471			$100^{0.05}$	0.926^{20}		
6764	Methoprotryne		$C_{11}H_{21}N_5OS$	841-06-5	271.383	cry	69				sl H_2O; s os
6765	Methotrexate		$C_{20}H_{22}N_8O_5$	59-05-2	454.440	ye cry (w)	190 dec				
6766	Methoxamine hydrochloride		$C_{11}H_{18}ClNO_3$	61-16-5	247.719	cry	214				vs H_2O; i eth, bz, chl
6767	Methoxsalen	9-Methoxy-7*H*-furo[3,2-*g*][1]-benzopyran-7-one	$C_{12}H_8O_4$	298-81-7	216.190	pr (dil al) nd (peth)	148				sl H_2O, eth, ace, peth; vs EtOH
6768	Methoxyacetaldehyde		$C_3H_6O_2$	10312-83-1	74.079			92(4)	1.005^{25}	1.3950^{20}	vs H_2O, ace, eth, EtOH
6769	Methoxyacetic acid		$C_3H_6O_3$	625-45-6	90.078	hyg		204(3)	1.1768^{20}	1.4168^{20}	s H_2O, EtOH, eth
6770	Methoxyacetonitrile		C_3H_5NO	1738-36-9	71.078			121(4)	0.9492^{20}	1.3831^{20}	sl H_2O; s EtOH, eth, ace, chl, alk, acid
6771	Methoxyacetyl chloride		$C_3H_5ClO_2$	38870-89-2	108.524			112(7)	1.1871^{20}	1.4199^{20}	s eth, ace, ctc; vs chl
6772	2-Methoxyaniline	*o*-Anisidine	C_7H_9NO	90-04-0	123.152	ye liq	6.2	221(7)	1.0923^{20}	1.5715^{10}	sl H_2O; s EtOH, eth, ace, bz
6773	3-Methoxyaniline	*m*-Anisidine	C_7H_9NO	536-90-3	123.152	liq	-1	251	1.096^{20}	1.5794^{20}	sl H_2O, ctc; s EtOH, eth, ace, bz
6774	4-Methoxyaniline	*p*-Anisidine	C_7H_9NO	104-94-9	123.152	orth pl	57.8(0.9)	243(6)	1.071^{57}	1.5559^{60}	s H_2O, ace, bz; vs EtOH, eth
6775	2-Methoxyaniline hydrochloride	*o*-Anisidine hydrochloride	$C_7H_{10}ClNO$	134-29-2	159.613	nd	225				
6776	1-Methoxy-9,10-anthracenedione		$C_{15}H_{10}O_3$	82-39-3	238.238		170.3				sl EtOH; vs bz, chl
6777	2-Methoxybenzaldehyde		$C_8H_8O_2$	135-02-4	136.149	pr	37.5(0.5)	243(4)	1.1326^{20}	1.5600^{20}	i H_2O; s EtOH, bz, ctc; vs eth, ace, chl
6778	3-Methoxybenzaldehyde		$C_8H_8O_2$	591-31-1	136.149			220(5)	1.1187^{20}	1.5530^{20}	i H_2O; s EtOH, bz; vs eth, ace, chl
6779	4-Methoxybenzaldehyde	*p*-Anisaldehyde	$C_8H_8O_2$	123-11-5	136.149	liq	0	255(5)	1.119^{15}	1.5730^{20}	i H_2O; msc EtOH, eth; vs ace, chl; s bz
6780	4-Methoxybenzamide		$C_8H_9NO_2$	3424-93-9	151.163	nd or tab (w)	166.5	295			vs H_2O, EtOH
6781	4-Methoxybenzeneacetaldehyde		$C_9H_{10}O_2$	5703-26-4	150.174			255.5	1.096^{20}	1.5359^{20}	
6782	2-Methoxybenzeneacetic acid		$C_9H_{10}O_3$	93-25-4	166.173	nd (w)	124	100^{2}			s H_2O; vs EtOH, eth, ace, bz, chl
6783	4-Methoxybenzeneacetic acid		$C_9H_{10}O_3$	104-01-8	166.173	pl (w)	84.9(0.5)	138^{2}			i H_2O; vs EtOH; s eth, bz; sl chl, lig
6784	4-Methoxybenzeneacetonitrile		C_9H_9NO	104-47-2	147.173			286.5	1.0845^{20}	1.5309^{20}	s EtOH, eth, chl
6785	4-Methoxy-1,2-benzenediamine	4-Methoxy-*o*-phenylenediamine	$C_7H_{10}N_2O$	102-51-2	138.166	grn pl	51	200^{21}			vs eth
6786	4-Methoxy-1,3-benzenediamine	4-Methoxy-*m*-phenylenediamine	$C_7H_{10}N_2O$	615-05-4	138.166	nd (eth)	67.5				s EtOH, eth; sl DMSO

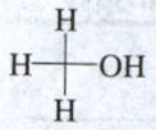
6750
Methanol

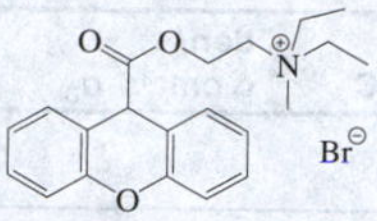
6751
Methantheline bromide

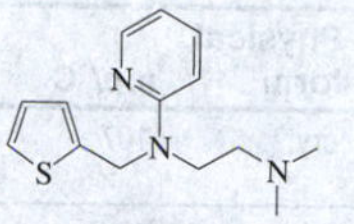
6752
Methapyrilene

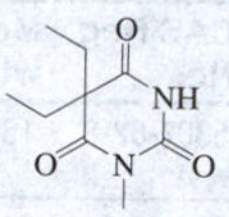
6753
Metharbital

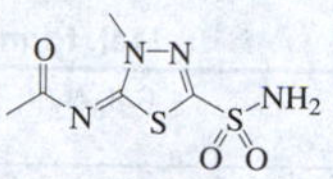
6754
Methazolamide

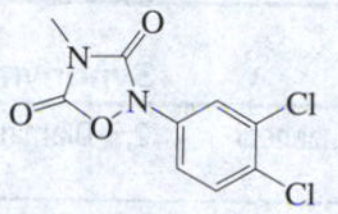
6755
Methazole

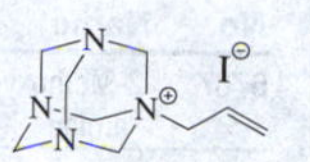
6756
Methenamine allyl iodide

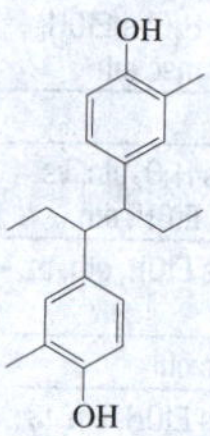
6757
Methestrol

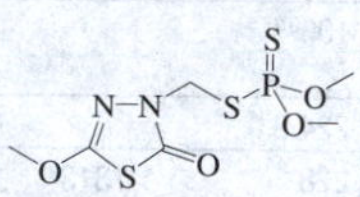
6758
Methidathion

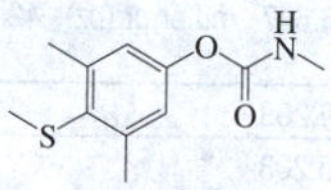
6759
Methiocarb

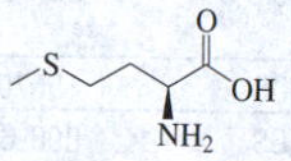
6760
L-Methionine

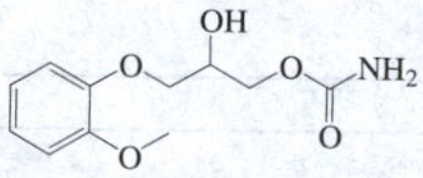
6761
Methocarbamol

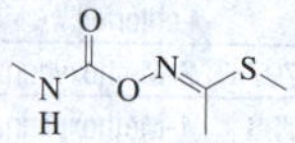
6762
Methomyl

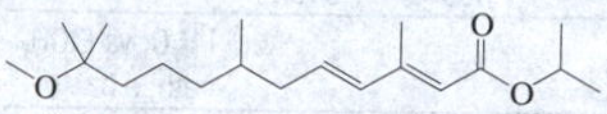
6763
Methoprene

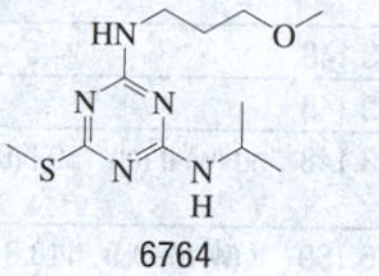
6764
Methoprotryne

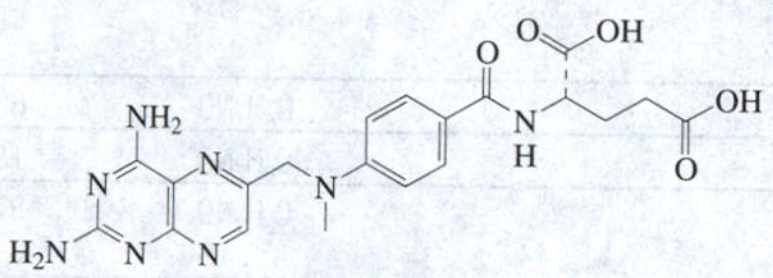
6765
Methotrexate

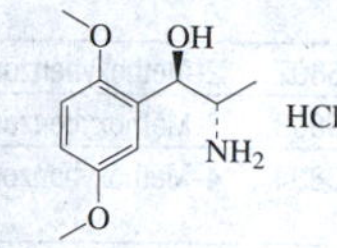
6766
Methoxamine hydrochloride

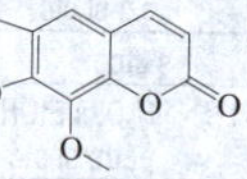
6767
Methoxsalen

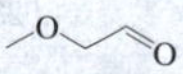
6768
Methoxyacetaldehyde

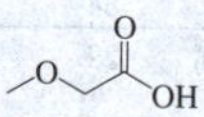
6769
Methoxyacetic acid

6770
Methoxyacetonitrile

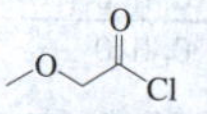
6771
Methoxyacetyl chloride

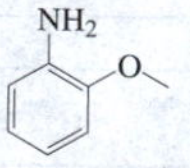
6772
2-Methoxyaniline

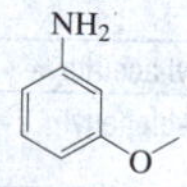
6773
3-Methoxyaniline

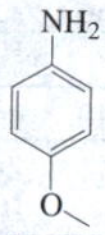
6774
4-Methoxyaniline

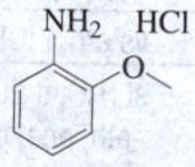
6775
2-Methoxyaniline hydrochloride

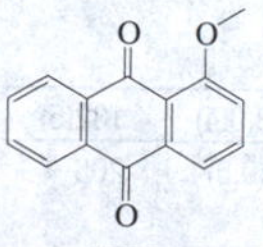
6776
1-Methoxy-9,10-anthracenedione

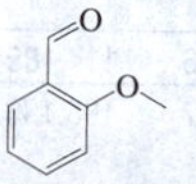
6777
2-Methoxybenzaldehyde

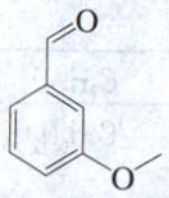
6778
3-Methoxybenzaldehyde

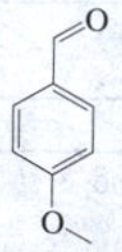
6779
4-Methoxybenzaldehyde

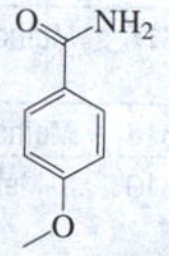
6780
4-Methoxybenzamide

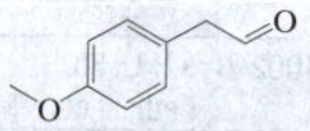
6781
4-Methoxybenzeneacetaldehyde

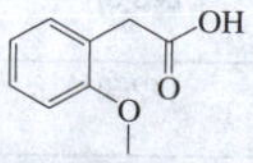
6782
2-Methoxybenzeneacetic acid

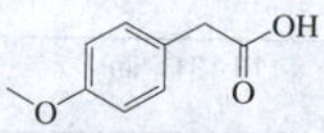
6783
4-Methoxybenzeneacetic acid

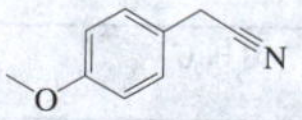
6784
4-Methoxybenzeneacetonitrile

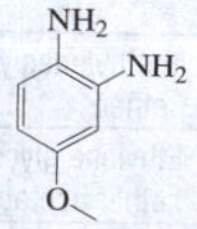
6785
4-Methoxy-1,2-benzenediamine

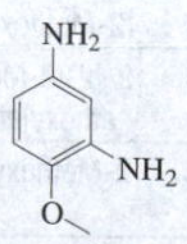
6786
4-Methoxy-1,3-benzenediamine

No.	Name	Synonym	Mol. form.	CAS Reg. No.	Mol. wt.	Physical form	mp/°C	bp/°C	den/ g cm⁻³	n_D	Solubility
6787	2-Methoxy-1,4-benzenediamine	2,5-Diaminoanisole	$C_7H_{10}N_2O$	5307-02-8	138.166	cry	107				
6788	3-Methoxy-1,2-benzenediol		$C_7H_8O_3$	934-00-9	140.137	nd	42.8	163^{48}			s chl
6789	4-Methoxybenzeneethanamine		$C_9H_{13}NO$	55-81-2	151.205			139^{20}		1.5379^{20}	
6790	4-Methoxybenzeneethanol		$C_9H_{12}O_2$	702-23-8	152.190		29	335			
6791	2-Methoxybenzenemethanamine		$C_8H_{11}NO$	6850-57-3	137.179			228	1.051^{25}	1.5475^{20}	
6792	4-Methoxybenzenemethanamine		$C_8H_{11}NO$	2393-23-9	137.179			236.5	1.050^{15}	1.5462^{20}	sl H_2O, EtOH, eth
6793	2-Methoxybenzenemethanol		$C_8H_{10}O_2$	612-16-8	138.164			249	1.0386^{25}	1.5455^{20}	i H_2O; s EtOH; msc eth
6794	3-Methoxybenzenemethanol		$C_8H_{10}O_2$	6971-51-3	138.164		30	252	1.112^{25}	1.5440^{20}	
6795	4-Methoxybenzenemethanol	Anise alcohol	$C_8H_{10}O_2$	105-13-5	138.164	nd	17(2)	254(3)	1.109^{26}	1.5420^{25}	s H_2O, ctc; vs EtOH, eth
6796	4-Methoxybenzenesulfonyl chloride		$C_7H_7ClO_3S$	98-68-0	206.647	nd or pr (bz)	42.5	$103^{0.25}$			s EtOH, eth, bz
6797	3-Methoxybenzenethiol		C_7H_8OS	15570-12-4	140.203			224.5		1.5874^{20}	s chl
6798	4-Methoxybenzenethiol		C_7H_8OS	696-63-9	140.203			228	1.1313^{25}	1.5801^{25}	s EtOH, eth, bz; sl chl
6799	2-Methoxybenzoic acid		$C_8H_8O_3$	579-75-9	152.148	pl (w)	100.9(0.5)	200			sl H_2O; vs EtOH, eth, chl; s bz, ctc
6800	3-Methoxybenzoic acid		$C_8H_8O_3$	586-38-9	152.148	nd (w)	107	170^{10}			sl H_2O, ctc; s EtOH, eth, bz; vs chl
6801	4-Methoxybenzoic acid	*p*-Anisic acid	$C_8H_8O_3$	100-09-4	152.148		184(2)	276.5			i H_2O; vs EtOH, MeOH, eth; s chl
6802	2-Methoxybenzonitrile		C_8H_7NO	6609-56-9	133.148		24.5	255.5	1.1063^{20}		s EtOH; vs eth
6803	3-Methoxybenzonitrile		C_8H_7NO	1527-89-5	133.148			140^{34}	1.089^{25}	1.5402^{20}	
6804	4-Methoxybenzonitrile		C_8H_7NO	874-90-8	133.148	nd (w) lf (al)	59.5(0.6)	256.5			i H_2O; vs EtOH, eth; s bz
6805	7-Methoxy-2*H*-1-benzopyran-2-one		$C_{10}H_8O_3$	531-59-9	176.169	lf (w, MeOH)	118.3				sl H_2O; s EtOH, eth, con sulf, alk
6806	6-Methoxy-2-benzothiazolamine		$C_8H_8N_2OS$	1747-60-0	180.227		166				
6807	2-(4-Methoxybenzoyl)benzoic acid	*o*-(*p*-Anisoyl)benzoic acid	$C_{15}H_{12}O_4$	1151-15-1	256.254	lf (w), cry (al, to)	146				vs eth, EtOH, tol
6808	2-Methoxybenzoyl chloride		$C_8H_7ClO_2$	21615-34-9	170.594			254			
6809	4-Methoxybenzoyl chloride	*p*-Anisoyl chloride	$C_8H_7ClO_2$	100-07-2	170.594	nd	21(3)	262.5	1.261^{20}	1.580^{20}	s eth, ace; vs bz; sl ctc
6810	4-Methoxybenzyl acetate		$C_{10}H_{12}O_3$	104-21-2	180.200		84	270	1.105^{25}		s ctc
6811	2-Methoxy-1,1'-biphenyl		$C_{13}H_{12}O$	86-26-0	184.233	pr (peth)	29	274	1.0233^{99}	1.5641^{99}	i H_2O; s EtOH, peth; sl ctc
6812	4-Methoxy-1,1'-biphenyl		$C_{13}H_{12}O$	613-37-6	184.233	pl (al)	90	157^{10}	1.0278^{100}	1.5744^{100}	i H_2O; s EtOH, eth
6813	1-Methoxy-1,3-butadiene		C_5H_8O	3036-66-6	84.117			92(2)	0.8296^{20}	1.4594^{20}	s H_2O, EtOH
6814	2-Methoxy-1,3-butadiene		C_5H_8O	3588-30-5	84.117			74(6)	0.8272^{20}	1.4442^{20}	vs ace, bz, eth, EtOH
6815	3-Methoxy-1-butanol		$C_5H_{12}O_2$	2517-43-3	104.148			159(7)	0.923^{23}	1.4148^{25}	vs EtOH, ace; s eth; sl chl
6816	1-Methoxy-1-buten-3-yne		C_5H_6O	2798-73-4	82.101			123 dec	0.906^{20}	1.4818^{20}	i H_2O; s chl
6817	Methoxychlor		$C_{16}H_{15}Cl_3O_2$	72-43-5	345.648	cry (dil al)	89(1)		1.41^{25}		i H_2O; s EtOH, ctc; vs eth, bz
6818	Methoxycyclohexane		$C_7H_{14}O$	931-56-6	114.185	liq	-74.3(0.4)	132(5)	0.8756^{20}	1.4355^{20}	vs eth, EtOH
6819	1-Methoxy-2,4-dinitrobenzene		$C_7H_6N_2O_5$	119-27-7	198.133	nd (al or w)	95.9(0.5)	206^{12}	1.3364^{131}	1.546^{15}	sl H_2O; s EtOH, eth, ace, bz; vs py
6820	1-Methoxy-3,5-dinitrobenzene	3,5-Dinitroanisole	$C_7H_6N_2O_5$	5327-44-6	198.133	nd (al)	105.3		1.558^{12}		vs ace, bz, MeOH
6821	2-Methoxy-1,2-diphenylethanone		$C_{15}H_{14}O_2$	3524-62-7	226.271	nd (lig)	49.5	188^{15}	1.1278^{14}		vs bz, eth, EtOH
6822	2-Methoxyethanol	Ethylene glycol monomethyl ether	$C_3H_8O_2$	109-86-4	76.095	liq	-85.1	124.3(0.1)	0.9647^{20}	1.4024^{20}	msc H_2O, eth, bz; vs EtOH; s ace; sl chl
6823	(2-Methoxyethoxy)ethene		$C_5H_{10}O_2$	1663-35-0	102.132			107(9)			
6824	2-[2-(2-Methoxyethoxy)-ethoxy]ethanol	Triethyleneglycol monomethyl ether	$C_7H_{16}O_4$	112-35-6	164.200			243(3)			
6825	2-Methoxyethyl acetate	Ethylene glycol monomethyl ether acetate	$C_5H_{10}O_3$	110-49-6	118.131	liq	-70	142(3)	1.0074^{19}	1.4002^{20}	s H_2O, EtOH, eth; sl ctc
6826	2-Methoxyethyl acrylate	2-Methoxyethyl 2-propenoate	$C_6H_{10}O_3$	3121-61-7	130.141			67^{16}	1.012^{20}		

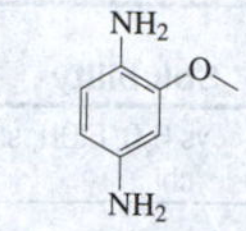
6787
2-Methoxy-1,4-benzenediamine

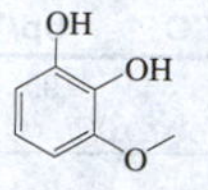
6788
3-Methoxy-1,2-benzenediol

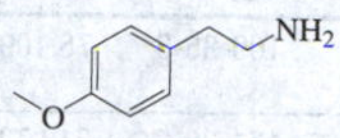
6789
4-Methoxybenzeneethanamine

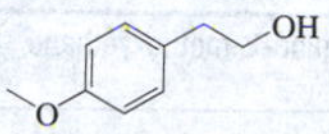
6790
4-Methoxybenzeneethanol

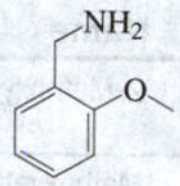
6791
2-Methoxybenzenemethanamine

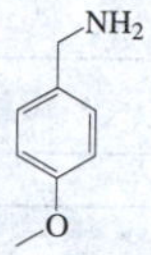
6792
4-Methoxybenzenemethanamine

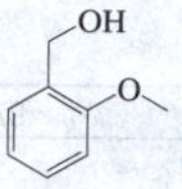
6793
2-Methoxybenzenemethanol

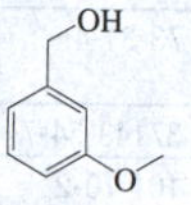
6794
3-Methoxybenzenemethanol

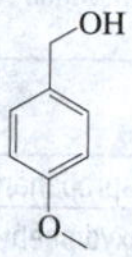
6795
4-Methoxybenzenemethanol

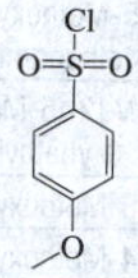
6796
4-Methoxybenzenesulfonyl chloride

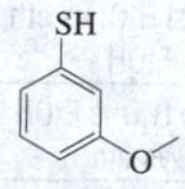
6797
3-Methoxybenzenethiol

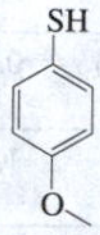
6798
4-Methoxybenzenethiol

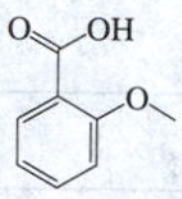
6799
2-Methoxybenzoic acid

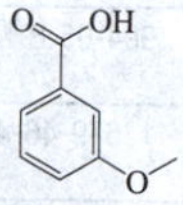
6800
3-Methoxybenzoic acid

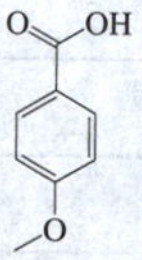
6801
4-Methoxybenzoic acid

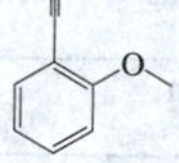
6802
2-Methoxybenzonitrile

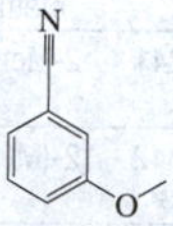
6803
3-Methoxybenzonitrile

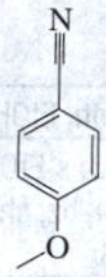
6804
4-Methoxybenzonitrile

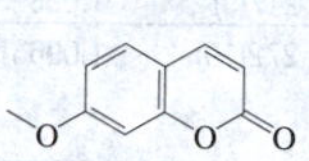
6805
7-Methoxy-2*H*-1-benzopyran-2-one

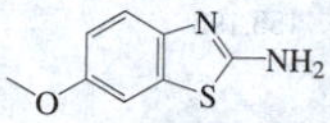
6806
6-Methoxy-2-benzothiazolamine

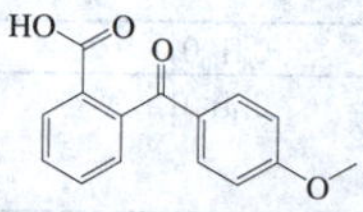
6807
2-(4-Methoxybenzoyl)benzoic acid

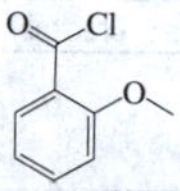
6808
2-Methoxybenzoyl chloride

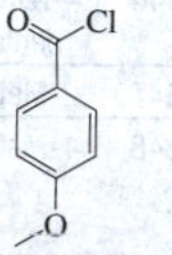
6809
4-Methoxybenzoyl chloride

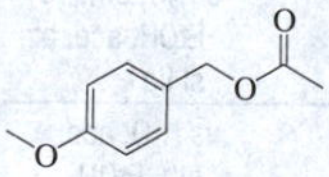
6810
4-Methoxybenzyl acetate

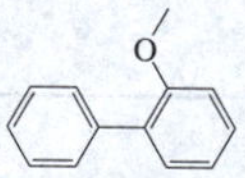
6811
2-Methoxy-1,1'-biphenyl

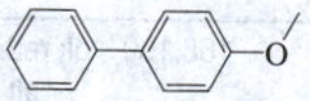
6812
4-Methoxy-1,1'-biphenyl

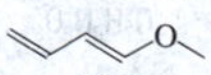
6813
1-Methoxy-1,3-butadiene

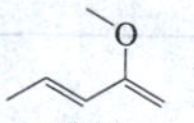
6814
2-Methoxy-1,3-butadiene

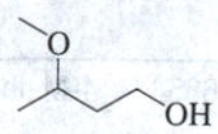
6815
3-Methoxy-1-butanol

6816
1-Methoxy-1-buten-3-yne

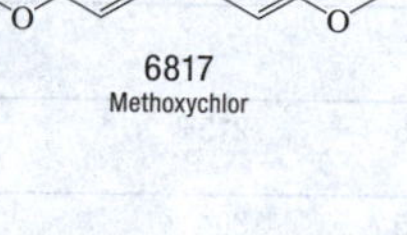
6817
Methoxychlor

6818
Methoxycyclohexane

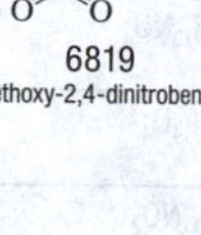
6819
1-Methoxy-2,4-dinitrobenzene

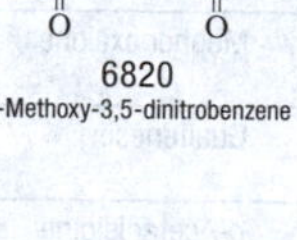
6820
1-Methoxy-3,5-dinitrobenzene

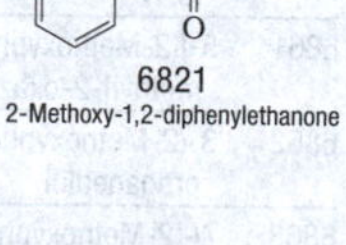
6821
2-Methoxy-1,2-diphenylethanone

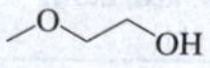
6822
2-Methoxyethanol

6823
(2-Methoxyethoxy)ethene

6824
2-[2-(2-Methoxyethoxy)ethoxy]ethanol

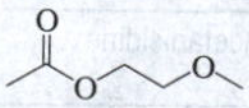
6825
2-Methoxyethyl acetate

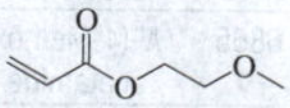
6826
2-Methoxyethyl acrylate

No.	Name	Synonym	Mol. form.	CAS Reg. No.	Mol. wt.	Physical form	mp/°C	bp/°C	den/g cm^{-3}	n_D	Solubility
6827	2-Methoxyethylamine	1-Amino-2-methoxyethane	C_3H_9NO	109-85-3	75.109			95			vs H_2O, EtOH; sl chl
6828	Methoxyethylmercuric acetate		$C_5H_{10}HgO_3$	151-38-2	318.72	nd (peth)	42				
6829	2-(2-Methoxyethyl)pyridine	Metyridine	$C_8H_{11}NO$	114-91-0	137.179			203	0.988^{20}	1.4975^{20}	vs H_2O, EtOH
6830	2-Methoxyfuran		$C_5H_6O_2$	25414-22-6	98.101			110.5	1.0646^{25}	1.4468^{25}	
6831	4-Methoxyfuro[2,3-*b*]quinoline	Dictamnine	$C_{12}H_9NO_2$	484-29-7	199.205	pr (al)	133.5				sl H_2O; vs EtOH; s eth, chl, AcOEt
6832	12-Methoxyibogamine	Ibogaine	$C_{20}H_{26}N_2O$	83-74-9	310.432		148				s chl
6833	5-Methoxy-1*H*-indole-3-ethanamine	5-Methoxytryptamine	$C_{11}H_{14}N_2O$	608-07-1	190.241	cry (al)	121.5				
6834	*N*-[2-(5-Methoxy-1*H*-indol-3-yl)ethyl]acetamide	Melatonin	$C_{13}H_{16}N_2O_2$	73-31-4	232.278	pa ye lf (bz)	117				
6835	3-Methoxyisopropylamine	1-Methoxy-2-propanamine	$C_4H_{11}NO$	37143-54-7	89.136			97		1.4031^{25}	
6836	4-Methoxy-*N*-(4-methoxyphenyl)aniline	4,4'-Dimethoxydiphenylamine	$C_{14}H_{15}NO_2$	101-70-2	229.275	lf (EtOH)	103				
6837	*N*-Methoxymethylamine	*N*-Methoxymethanamine	C_2H_7NO	1117-97-1	61.083	liq		42.4			
6838	2-Methoxy-5-methylaniline	5-Methyl-*o*-anisidine	$C_8H_{11}NO$	120-71-8	137.179		53	235			sl H_2O, chl; s EtOH, eth, bz, peth
6839	4-Methoxy-2-methylaniline		$C_8H_{11}NO$	102-50-1	137.179	cry (lig)	29.5	248.5	1.065^{25}	1.5647^{20}	vs EtOH
6840	4-Methoxy-α-methylbenzenemethanol		$C_9H_{12}O_2$	3319-15-1	152.190			257(19)	1.0794^{20}	1.5310^{25}	s ctc
6841	2-Methoxy-2-methylbutane	Methyl *tert*-pentyl ether	$C_6H_{14}O$	994-05-8	102.174			86.4(0.1)	0.7660^{25}	1.3862^{25}	sl H_2O; vs eth, EtOH
6842	2-(Methoxymethyl)furan		$C_6H_8O_2$	13679-46-4	112.127			132	1.0163^{20}	1.4570^{20}	i H_2O; s EtOH; vs eth
6843	2-(Methoxymethyl)-5-nitrofuran		$C_6H_7NO_4$	586-84-5	157.125			104^{3}	1.281^{20}	1.5325^{20}	vs EtOH
6844	(Methoxymethyl)oxirane		$C_4H_8O_2$	930-37-0	88.106			111(18)	0.9890^{20}	1.4320^{20}	vs H_2O, ace, eth, EtOH
6845	3-Methoxy-5-methyl-4-oxo-2,5-hexadienoic acid	Penicillic acid	$C_8H_{10}O_4$	90-65-3	170.163	orth or hex pl (+ 1w)	83				s H_2O, ace; vs EtOH, eth, bz; sl peth
6846	4-Methoxy-4-methyl-2-pentanone	Pentoxone	$C_7H_{14}O_2$	107-70-0	130.185			160	0.8980^{25}	1.418^{20}	
6847	2-Methoxy-4-methylphenol	Creosol	$C_8H_{10}O_2$	93-51-6	138.164	pr	5.5	217(9)	1.098^{20}	1.5353^{25}	vs eth, EtOH
6848	1-Methoxynaphthalene		$C_{11}H_{10}O$	2216-69-5	158.196		<-10	272(19)	1.0963^{14}	1.6940^{25}	i H_2O; s EtOH, eth, bz, chl; vs CS_2
6849	2-Methoxynaphthalene		$C_{11}H_{10}O$	93-04-9	158.196	lf (eth), pl (peth)	73.5	270(18)			vs bz, eth, chl
6850	2-Methoxy-1,4-naphthalenedione		$C_{11}H_8O_3$	2348-82-5	188.180		183.0				
6851	4-Methoxy-1-naphthol		$C_{11}H_{10}O_2$	84-85-5	174.196		129.8				
6852	2-Methoxy-4-nitroaniline		$C_7H_8N_2O_3$	97-52-9	168.150		141.0				s DMSO
6853	2-Methoxy-5-nitroaniline	5-Nitro-*o*-anisidine	$C_7H_8N_2O_3$	99-59-2	168.150		118(1)		1.2068^{15}		s H_2O, eth; vs EtOH, ace, bz; sl lig
6854	4-Methoxy-2-nitroaniline		$C_7H_8N_2O_3$	96-96-8	168.150	dk red pr (w or al)	129				vs H_2O, ace, eth, EtOH
6855	2-Methoxyphenol	Guaiacol	$C_7H_8O_2$	90-05-1	124.138	hex pr	28.3(0.6)	204(1)	1.1287^{21}	1.5429^{20}	sl H_2O; s EtOH, eth, ctc, chl
6856	3-Methoxyphenol		$C_7H_8O_2$	150-19-6	124.138		<-17	114^{5}	1.131^{25}	1.5510^{20}	sl H_2O, chl; msc EtOH, eth
6857	4-Methoxyphenol		$C_7H_8O_2$	150-76-5	124.138	pl	54(3)	253(2)			s H_2O, bz, ctc; vs EtOH, eth
6858	2-Methoxyphenol benzoate	Guaiacol benzoate	$C_{14}H_{12}O_3$	531-37-3	228.243		57.5				vs eth, chl
6859	2-Methoxyphenol carbonate (2:1)	Guaiacol carbonate	$C_{15}H_{14}O_5$	553-17-3	274.269	cry (al)	89				i H_2O; sl EtOH; s eth; vs chl
6860	2-Methoxyphenol phosphate (3:1)	Guaiacol phosphate	$C_{21}H_{21}O_7P$	563-03-1	416.362		91	277^{3}			vs ace, tol, chl
6861	5-[(2-Methoxyphenoxy)methyl]-2-oxazolidinone	Mephenoxalone	$C_{11}H_{13}NO_4$	70-07-5	223.226		144				
6862	3-(2-Methoxyphenoxy)-1,2-propanediol	Guaifenesin	$C_{10}H_{14}O_4$	93-14-1	198.216	orth pr (eth, eth-peth)	78.5	215^{19}			s H_2O, bz, chl; vs EtOH; i peth
6863	*N*-(2-Methoxyphenyl)acetamide	*o*-Acetanisidine	$C_9H_{11}NO_2$	93-26-5	165.189	nd (w)	87.5	304			vs H_2O, EtOH; s eth, ace, HOAc
6864	*N*-(3-Methoxyphenyl)acetamide	*m*-Acetanisidine	$C_9H_{11}NO_2$	588-16-9	165.189	nd or pl (w)	81				vs H_2O, EtOH; s eth, ace
6865	*N*-(4-Methoxyphenyl)acetamide	*p*-Acetanisidine	$C_9H_{11}NO_2$	51-66-1	165.189	pl (w)	127.2(0.5)				vs ace, EtOH, chl

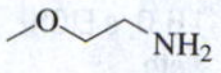
6827
2-Methoxyethylamine

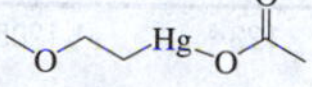
6828
Methoxyethylmercuric acetate

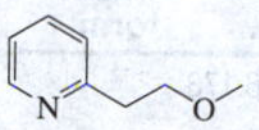
6829
2-(2-Methoxyethyl)pyridine

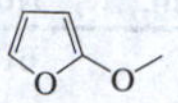
6830
2-Methoxyfuran

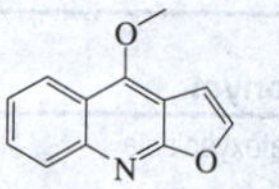
6831
4-Methoxyfuro[2,3-b]quinoline

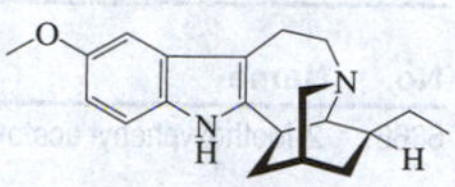
6832
12-Methoxyibogamine

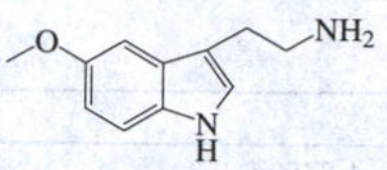
6833
5-Methoxy-1*H*-indole-3-ethanamine

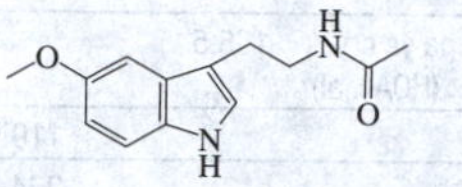
6834
N-[2-(5-Methoxy-1*H*-indol-3-yl)ethyl]acetamide

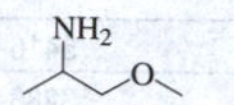
6835
3-Methoxyisopropylamine

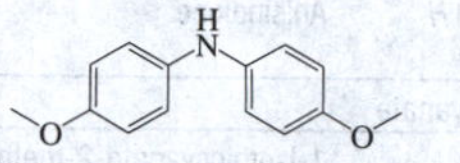
6836
4-Methoxy-*N*-(4-methoxyphenyl)aniline

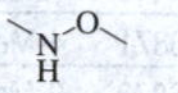
6837
N-Methoxymethylamine

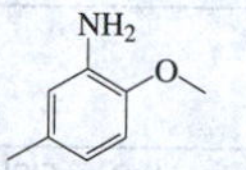
6838
2-Methoxy-5-methylaniline

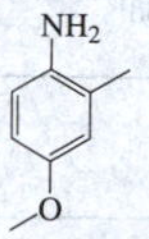
6839
4-Methoxy-2-methylaniline

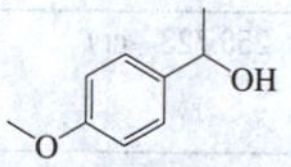
6840
4-Methoxy-α-methylbenzenemethanol

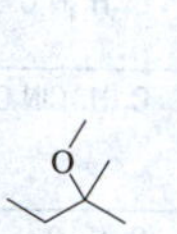
6841
2-Methoxy-2-methylbutane

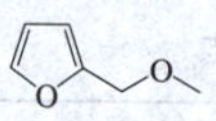
6842
2-(Methoxymethyl)furan

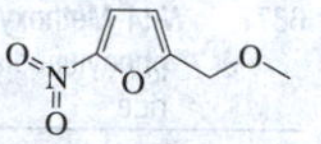
6843
2-(Methoxymethyl)-5-nitrofuran

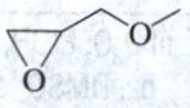
6844
(Methoxymethyl)oxirane

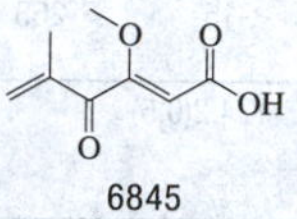
6845
3-Methoxy-5-methyl-4-oxo-2,5-hexadienoic acid

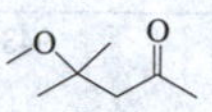
6846
4-Methoxy-4-methyl-2-pentanone

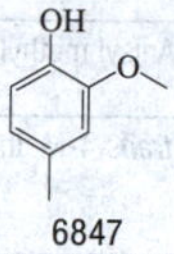
6847
2-Methoxy-4-methylphenol

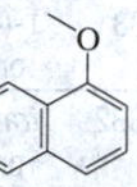
6848
1-Methoxynaphthalene

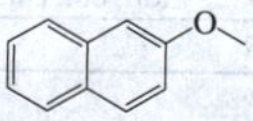
6849
2-Methoxynaphthalene

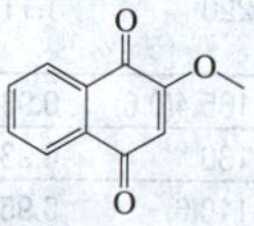
6850
2-Methoxy-1,4-naphthalenedione

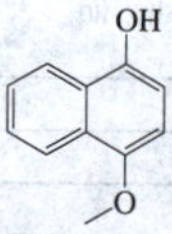
6851
4-Methoxy-1-naphthol

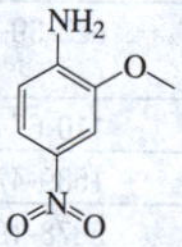
6852
2-Methoxy-4-nitroaniline

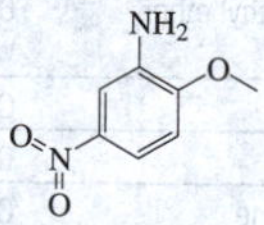
6853
2-Methoxy-5-nitroaniline

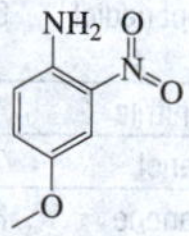
6854
4-Methoxy-2-nitroaniline

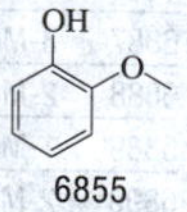
6855
2-Methoxyphenol

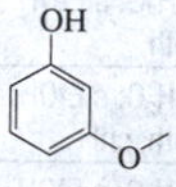
6856
3-Methoxyphenol

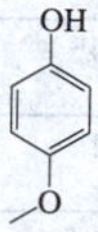
6857
4-Methoxyphenol

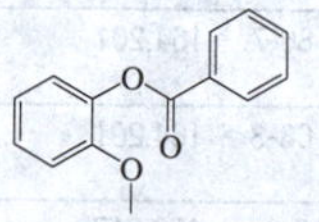
6858
2-Methoxyphenol benzoate

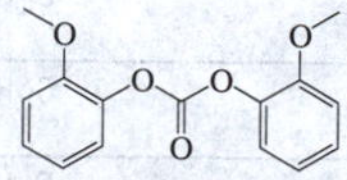
6859
2-Methoxyphenol carbonate (2:1)

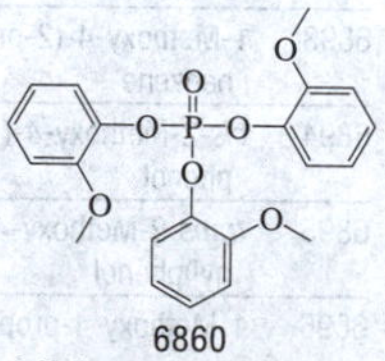
6860
2-Methoxyphenol phosphate (3:1)

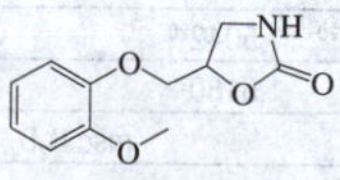
6861
5-[(2-Methoxyphenoxy)methyl]-2-oxazolidinone

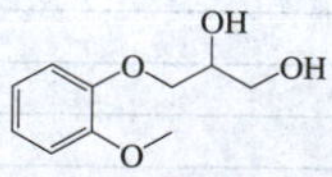
6862
3-(2-Methoxyphenoxy)-1,2-propanediol

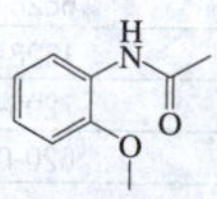
6863
N-(2-Methoxyphenyl)acetamide

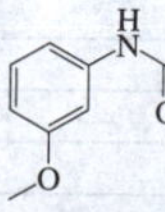
6864
N-(3-Methoxyphenyl)acetamide

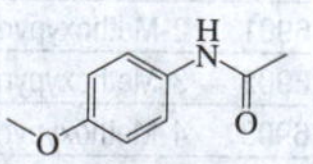
6865
N-(4-Methoxyphenyl)acetamide

No.	Name	Synonym	Mol. form.	CAS Reg. No.	Mol. wt.	Physical form	mp/°C	bp/°C	den/ g cm^{-3}	n_D	Solubility
6866	2-Methoxyphenyl acetate	2-Acetoxyanisole	$C_9H_{10}O_3$	613-70-7	166.173		31.5	123^{13}	1.1285^{25}	1.5101^{25}	i H_2O; s EtOH, eth
6867	4-(4-Methoxyphenyl)-3-buten-2-one		$C_{11}H_{12}O_2$	943-88-4	176.212	lf (al, eth, HOAc)	74.0	187.5^{19}			i H_2O; vs EtOH, eth; s bz, HOAc, sulf
6868	2-Methoxy-1-phenylethanone		$C_9H_{10}O_2$	4079-52-1	150.174	ye liq	8	241(19)	1.0897^{20}	1.5393^{20}	sl H_2O; s EtOH, ace
6869	1-(3-Methoxyphenyl)ethanone		$C_9H_{10}O_2$	586-37-8	150.174		95.5	240	1.0343^{19}	1.5410^{20}	s H_2O, EtOH, ace, ctc
6870	2-(4-Methoxyphenyl)-1*H*-indene-1,3(2*H*)-dione	Anisindione	$C_{16}H_{12}O_3$	117-37-3	252.264	pa ye cry (HOAc, al)	156.5				
6871	4-Methoxyphenyl isocyanate		$C_8H_7NO_2$	5416-93-3	149.148			110^{10}			
6872	2-Methoxyphenyl isothiocyanate	1-Isothiocyanato-2-methoxybenzene	C_8H_7NOS	3288-04-8	165.213			264	1.1878^{20}	1.6458^{20}	
6873	*N*-(4-Methoxyphenyl)-3-oxobutanamide		$C_{11}H_{13}NO_3$	5437-98-9	207.226		117.3				s EtOH, chl; sl eth
6874	2-Methoxyphenyl pentanoate	Guaiacol valerate	$C_{12}H_{16}O_3$	531-39-5	208.253			265	1.05^{25}		vs bz, eth, EtOH
6875	(4-Methoxyphenyl)-phenyldiazene		$C_{13}H_{12}N_2O$	2396-60-3	212.246	oran-red pl, lf (al, peth)	56	340	1.12^{75}		i H_2O; s EtOH, eth, ace
6876	*N*-(*p*-Methoxyphenyl)-*p*-phenylenediamine	*N*-(4-Methoxyphenyl)-1,4-benzenediamine	$C_{13}H_{14}N_2O$	101-64-4	214.262	nd	102	238^{12}			sl H_2O, peth; vs bz, eth, EtOH
6877	*N*-(4-Methoxyphenyl)-*p*-phenylenediamine hydrochloride		$C_{13}H_{15}ClN_2O$	3566-44-7	250.723	cry	245 dec				
6878	(4-Methoxyphenyl)-phenylmethanone		$C_{14}H_{12}O_2$	611-94-9	212.244	pr (eth)	61.5	355			i H_2O; vs EtOH, eth; s ace, bz, HOAc
6879	3-(4-Methoxyphenyl)-1-phenyl-2-propen-1-one		$C_{16}H_{14}O_2$	959-33-1	238.281	ye nd (al)	79	187^{19}			i H_2O; vs EtOH; s eth, ctc, chl, HOAc
6880	1-(4-Methoxyphenyl)-1-propanone	Ethyl 4-methoxyphenyl ketone	$C_{10}H_{12}O_2$	121-97-1	164.201		25.5	266	1.0798^{16}		s ctc
6881	1-(4-Methoxyphenyl)-2-propanone	Anisyl methyl ketone	$C_{10}H_{12}O_2$	122-84-9	164.201		<-15	268	1.0694^{17}	1.5253^{20}	vs eth, EtOH
6882	*trans*-3-(4-Methoxyphenyl)-2-propenoic acid	*trans*-4-Methoxycinnamic acid	$C_{10}H_{10}O_3$	943-89-5	178.184		174.2(0.5)				sl H_2O, EtOH, bz, DMSO; s ctc, HOAc
6883	*trans*-1-Methoxy-4-(2-phenylvinyl)benzene		$C_{15}H_{14}O$	1694-19-5	210.271		136.5	142.5^{15}			i H_2O; vs EtOH, eth, ace, bz; s peth
6884	1-Methoxy-1,2-propadiene	Methoxyallene	C_4H_6O	13169-00-1	70.090	oil		50(19)			
6885	3-Methoxy-1-propanamine		$C_4H_{11}NO$	5332-73-0	89.136			113(12)	0.8727^{20}	1.4391^{20}	s H_2O, ace, bz, ctc, chl, MeOH
6886	3-Methoxy-1,2-propanediol	Glycerol 3-methyl ether	$C_4H_{10}O_3$	623-39-2	106.120	hyg liq		220	1.114^{20}	1.442^{25}	vs H_2O, EtOH, ace; s eth
6887	3-Methoxypropanenitrile		C_4H_7NO	110-67-8	85.105			165.4(0.6)	0.9379^{20}	1.4043^{20}	s EtOH, eth, chl
6888	2-Methoxy-1-propanol		$C_4H_{10}O_2$	1589-47-5	90.121			130	0.938^{20}	1.4070^{20}	
6889	1-Methoxy-2-propanone	Methoxyacetone	$C_4H_8O_2$	5878-19-3	88.106			110(6)	0.957^{25}	1.3970^{20}	
6890	2-Methoxy-1-propene		C_4H_8O	116-11-0	72.106			35.7(0.3)	0.7372^{20}		
6891	3-Methoxy-1-propene	Allyl methyl ether	C_4H_8O	627-40-7	72.106			46(3)	0.77^{11}	1.3778^{20}	i H_2O; msc EtOH, eth; s ace
6892	*trans*-1-Methoxy-4-(1-propenyl)benzene	Anethole	$C_{10}H_{12}O$	4180-23-8	148.201	col oily liq	22.5	236(5)	0.9882^{20}	1.5615^{20}	sl H_2O; msc EtOH, eth; s ace; vs bz
6893	1-Methoxy-4-(2-propenyl)-benzene	Estragole	$C_{10}H_{12}O$	140-67-0	148.201			216(5)	0.965^{25}	1.5195^{20}	vs EtOH, chl
6894	*cis*-2-Methoxy-4-(1-propenyl)-phenol		$C_{10}H_{12}O_2$	5912-86-7	164.201			134^{13}	1.0837^{20}	1.5726^{20}	sl H_2O; s EtOH, eth
6895	*trans*-2-Methoxy-4-(1-propenyl)phenol		$C_{10}H_{12}O_2$	5932-68-3	164.201		33.5	141^{13}	1.0852^{20}	1.5784^{20}	sl H_2O; s EtOH, eth, chl
6896	1-Methoxy-4-propylbenzene		$C_{10}H_{14}O$	104-45-0	150.217			211.5	0.9472^{20}	1.5045^{20}	sl H_2O; s EtOH, ace, bz, chl; vs eth
6897	2-Methoxy-4-propylphenol		$C_{10}H_{14}O_2$	2785-87-7	166.217			121^{10}			
6898	3-Methoxy-1-propyne		C_4H_6O	627-41-8	70.090			63	0.83^{12}	1.5035^{20}	vs eth, EtOH
6899	5-Methoxypsoralen	Bergaptene	$C_{12}H_8O_4$	484-20-8	216.190	nd (EtOH)	188				i H_2O; sl EtOH, bz, chl
6900	6-Methoxy-3-pyridinamine		$C_6H_8N_2O$	6628-77-9	124.140		30	125^{10}		1.5745^{20}	
6901	2-Methoxypyridine		C_6H_7NO	1628-89-3	109.126			142(7)	1.0457^{20}	1.5042^{20}	
6902	3-Methoxypyridine		C_6H_7NO	7295-76-3	109.126	liq		178.5	1.083	1.5180^{20}	
6903	4-Methoxypyridine		C_6H_7NO	620-08-6	109.126			193(14)			msc H_2O

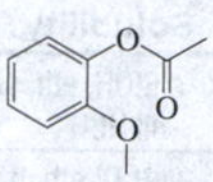
6866
2-Methoxyphenyl acetate

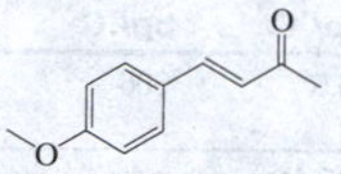
6867
4-(4-Methoxyphenyl)-3-buten-2-one

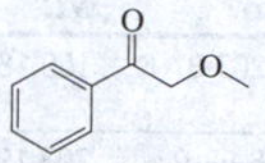
6868
2-Methoxy-1-phenylethanone

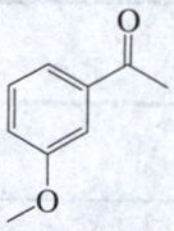
6869
1-(3-Methoxyphenyl)ethanone

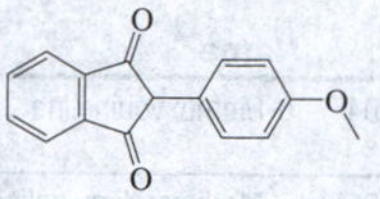
6870
2-(4-Methoxyphenyl)-1*H*-indene-1,3(2*H*)-dione

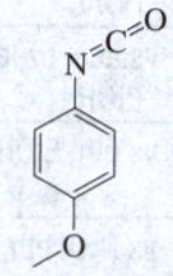
6871
4-Methoxyphenyl isocyanate

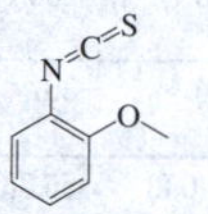
6872
2-Methoxyphenyl isothiocyanate

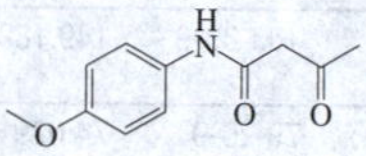
6873
N-(4-Methoxyphenyl)-3-oxobutanamide

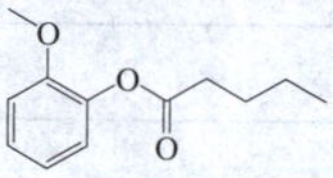
6874
2-Methoxyphenyl pentanoate

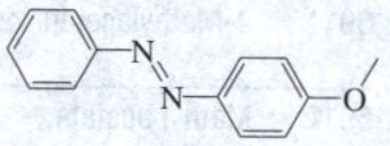
6875
(4-Methoxyphenyl)phenyldiazene

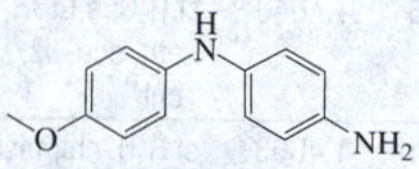
6876
N-(*p*-Methoxyphenyl)-*p*-phenylenediamine

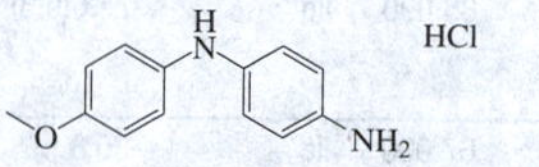
6877
N-(4-Methoxyphenyl)-*p*-phenylenediamine hydrochloride

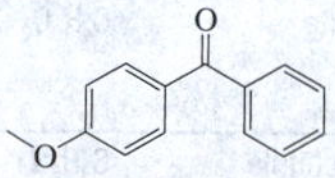
6878
(4-Methoxyphenyl)phenylmethanone

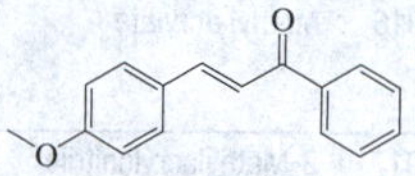
6879
3-(4-Methoxyphenyl)-1-phenyl-2-propen-1-one

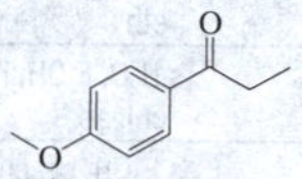
6880
1-(4-Methoxyphenyl)-1-propanone

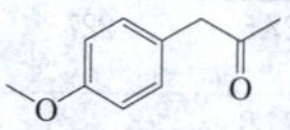
6881
1-(4-Methoxyphenyl)-2-propanone

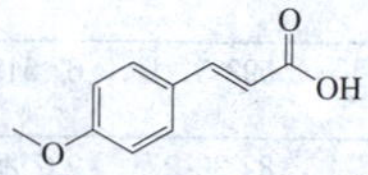
6882
trans-3-(4-Methoxyphenyl)-2-propenoic acid

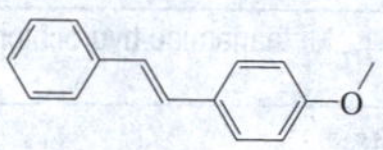
6883
trans-1-Methoxy-4-(2-phenylvinyl)benzene

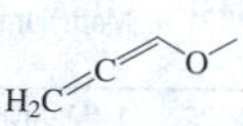
6884
1-Methoxy-1,2-propadiene

6885
3-Methoxy-1-propanamine

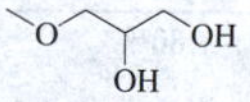
6886
3-Methoxy-1,2-propanediol

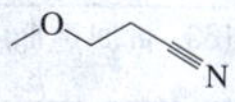
6887
3-Methoxypropanenitrile

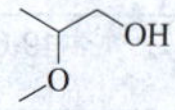
6888
2-Methoxy-1-propanol

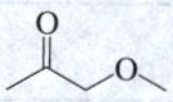
6889
1-Methoxy-2-propanone

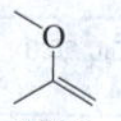
6890
2-Methoxy-1-propene

6891
3-Methoxy-1-propene

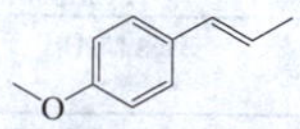
6892
trans-1-Methoxy-4-(1-propenyl)benzene

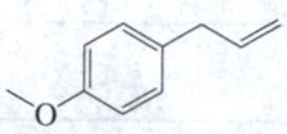
6893
1-Methoxy-4-(2-propenyl)benzene

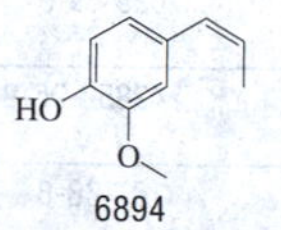
6894
cis-2-Methoxy-4-(1-propenyl)phenol

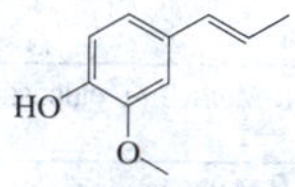
6895
trans-2-Methoxy-4-(1-propenyl)phenol

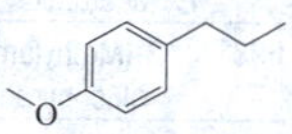
6896
1-Methoxy-4-propylbenzene

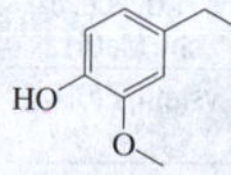
6897
2-Methoxy-4-propylphenol

6898
3-Methoxy-1-propyne

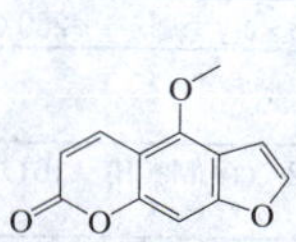
6899
5-Methoxypsoralen

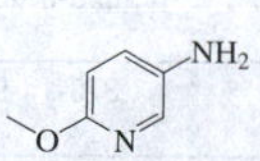
6900
6-Methoxy-3-pyridinamine

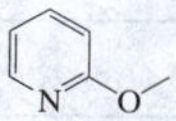
6901
2-Methoxypyridine

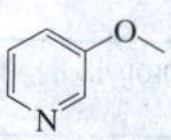
6902
3-Methoxypyridine

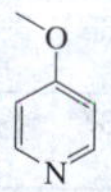
6903
4-Methoxypyridine

No.	Name	Synonym	Mol. form.	CAS Reg. No.	Mol. wt.	Physical form	mp/°C	bp/°C	den/ g cm⁻³	n_D	Solubility
6904	6-Methoxyquinoline		$C_{10}H_9NO$	5263-87-6	159.184	hyg lf	26.5	306	1.152^{20}		s EtOH, eth, chl, dil HCl
6905	6-Methoxy-4-quinolinecarboxylic acid	Quininic acid	$C_{11}H_9NO_3$	86-68-0	203.194	pa ye pr (dil al)	285 dec	sub			sl H_2O, eth, bz, tfa; i chl; s EtOH
6906	2-Methoxy-1,3,5-trinitrobenzene	Methyl picrate	$C_7H_5N_3O_7$	606-35-9	243.131	nd (dil MeOH)	67.6(0.3)		1.4947^{80}		i H_2O; vs EtOH, chl, bz; s eth
6907	(2-Methoxyvinyl)benzene		$C_9H_{10}O$	4747-15-3	134.174			211.5	0.9894^{23}	1.5620^{24}	
6908	Methscopolamine bromide	Scopolamine methobromide	$C_{18}H_{24}BrNO_4$	155-41-9	398.293	cry (EtOH)	215 dec				s H_2O; sl EtOH
6909	Methyl abietate		$C_{21}H_{32}O_2$	127-25-3	316.478	pa ye lf (liq)		225^{16}	1.049^{20}	1.5344	i H_2O; s EtOH, HOAc
6910	*N*-Methylacetamide		C_3H_7NO	79-16-3	73.094		30.6(0.1)	208(2)	0.9371^{25}	1.4301^{20}	vs ace, bz, eth, EtOH
6911	4-Methylacetanilide		$C_9H_{11}NO$	103-89-9	149.189	mcl cry or nd (dil al)	151(1)	307	1.2120^{15}		vs eth, EtOH
6912	Methyl acetate		$C_3H_6O_2$	79-20-9	74.079	liq	-98.2(0.2)	56.7(0.2)	0.9342^{20}	1.3614^{20}	vs H_2O, eth, EtOH
6913	Methyl acetoacetate	Methyl 3-oxobutanoate	$C_5H_8O_3$	105-45-3	116.116		27.5	168(3)	1.0762^{20}	1.4184^{20}	vs H_2O; msc EtOH, eth; s ctc
6914	4-Methylacetophenone		$C_9H_{10}O$	122-00-9	134.174	nd	28	225(8)	1.0051^{20}	1.5335^{20}	vs bz, eth, EtOH, chl
6915	Methyl 2-(acetyloxy)benzoate	Methyl *o*-acetylsalicylate	$C_{10}H_{10}O_4$	580-02-9	194.184	pl (peth)	51.5	135^9			vs eth, EtOH, chl
6916	Methyl acrylate	Methyl propenoate	$C_4H_6O_2$	96-33-3	86.090	liq	-75.6(0.3)	80.1(0.6)	0.9535^{20}	1.4040^{20}	sl H_2O; s EtOH, eth, ace, bz, chl
6917	2-Methylacrylonitrile	2-Methylpropenenitrile	C_4H_5N	126-98-7	67.090	liq	-35.8	90(3)	0.8001^{20}	1.4003^{20}	sl H_2O, chl; msc EtOH, eth, ace, tol
6918	2-Methylalanine	α-Aminoisobutyric acid	$C_4H_9NO_2$	62-57-7	103.120	mcl pr	335	280 sub			vs H_2O; sl EtOH; i eth
6919	5-Methyl-3-allyl-2,4-oxazolidinedione	Aloxidone	$C_7H_9NO_3$	526-35-2	155.151			138^{35}		1.4688^{25}	
6920	Methylamine	Methanamine	CH_5N	74-89-5	31.058	col gas	-93.42(0.09)	-6.4(0.3)	0.656^{25} (p>1 atm)		vs H_2O; s EtOH, ace, bz; msc eth
6921	Methylamine hydrochloride	Methanamine hydrochloride	CH_6ClN	593-51-1	67.519	hyg tetr tab (al)	227.5	227^{15}			s H_2O, EtOH; i chl, ace
6922	1-(Methylamino)-9,10-anthracenedione		$C_{15}H_{11}NO_2$	82-38-2	237.254	ye-red nd	171.0				s EtOH, bz, chl, HOAc
6923	Methyl 2-aminobenzoate	Methyl anthranilate	$C_8H_9NO_2$	134-20-3	151.163		24.4(0.2)	256	1.1682^{10}	1.5810	sl H_2O; vs EtOH, eth
6924	Methyl 3-aminobenzoate		$C_8H_9NO_2$	4518-10-9	151.163		39	152^{11}	1.232^{20}		vs EtOH, eth, bz, chl; s lig; sl peth
6925	Methyl 4-aminobenzoate		$C_8H_9NO_2$	619-45-4	151.163	lf or nd (aq MeOH)	111.8(0.7)				s chl
6926	2-(Methylamino)benzoic acid		$C_8H_9NO_2$	119-68-6	151.163	pl (al or lig)	180.5	$80^{0.01}$			sl H_2O; vs EtOH, eth, bz, chl
6927	3-(Methylamino)benzoic acid		$C_8H_9NO_2$	51524-84-6	151.163	pl (peth)	127				vs ace, bz, EtOH, chl
6928	4-(Methylamino)benzoic acid		$C_8H_9NO_2$	10541-83-0	151.163	nd (bz, w, dil al)	168				s H_2O, bz, AcOEt; vs EtOH, eth; sl tfa
6929	Methyl 3-amino-2-butenoate		$C_5H_9NO_2$	14205-39-1	115.131						s chl
6930	*N*-[(Methylamino)carbonyl]-acetamide		$C_4H_8N_2O_2$	623-59-6	116.119	tcl (w, al), pr (w)	180.5	311(17)			s H_2O, chl; sl EtOH, eth
6931	2-(Methylamino)-2-deoxy-α-L-glucopyranose	*N*-Methyl-α-L-glucosamine	$C_7H_{15}NO_5$	42852-95-9	193.198	glass					s MeOH
6932	2-(Methylamino)-ethanesulfonic acid	*N*-Methyltaurine	$C_3H_9NO_3S$	107-68-6	139.173		241.5				vs H_2O; i EtOH, eth
6933	4-[2-(Methylamino)ethyl]-1,2-benzenediol	Deoxyepinephrine	$C_9H_{13}NO_2$	501-15-5	167.205		188.5				
6934	Methyl 3-amino-4-hydroxy-benzoate	Orthocaine	$C_8H_9NO_3$	536-25-4	167.162	nd (bz or HOAc)	143				i H_2O; vs EtOH; s eth, alk; sl bz
6935	4-(Methylamino)phenol sulfate		$C_{14}H_{20}N_2O_6S$	1936-57-8	344.383	cry	260 dec				sl EtOH; i eth
6936	3-(Methylamino)propanenitrile		$C_4H_8N_2$	693-05-0	84.120			102^{49}	0.8992^{20}	1.4320^{20}	s H_2O, ace, bz, chl, MeOH
6937	4-[2-(Methylamino)propyl]-phenol	Pholedrine	$C_{10}H_{15}NO$	370-14-9	165.232	cry (MeOH)	161				vs eth, EtOH

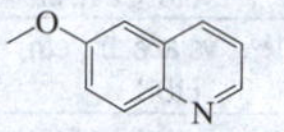
6904
6-Methoxyquinoline

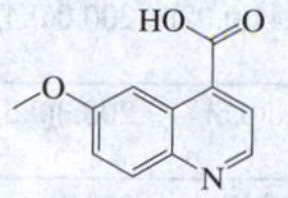
6905
6-Methoxy-4-quinolinecarboxylic acid

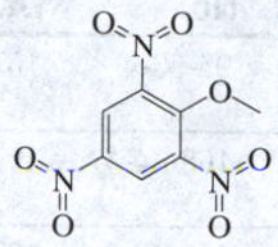
6906
2-Methoxy-1,3,5-trinitrobenzene

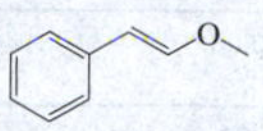
6907
(2-Methoxyvinyl)benzene

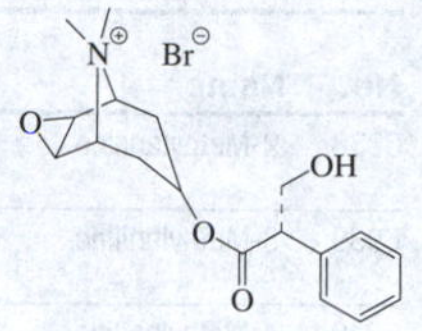
6908
Methscopolamine bromide

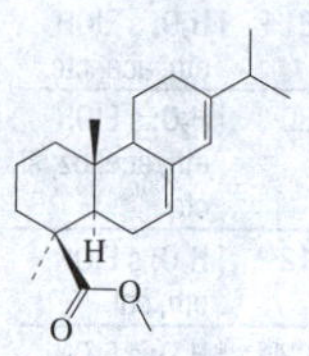
6909
Methyl abietate

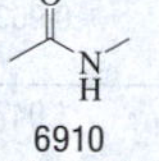
6910
N-Methylacetamide

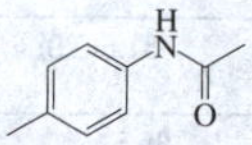
6911
4-Methylacetanilide

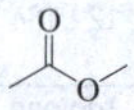
6912
Methyl acetate

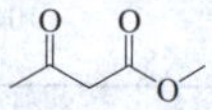
6913
Methyl acetoacetate

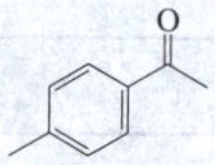
6914
4-Methylacetophenone

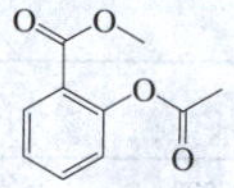
6915
Methyl 2-(acetyloxy)benzoate

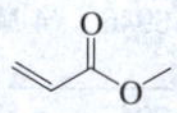
6916
Methyl acrylate

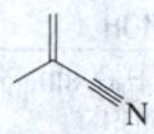
6917
2-Methylacrylonitrile

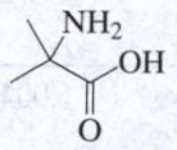
6918
2-Methylalanine

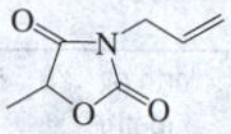
6919
5-Methyl-3-allyl-2,4-oxazolidinedione

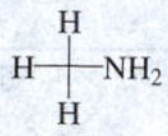
6920
Methylamine

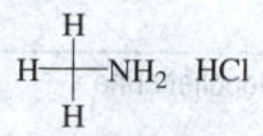
6921
Methylamine hydrochloride

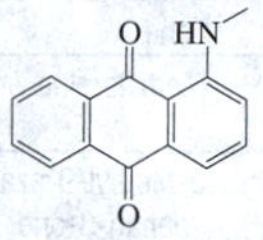
6922
1-(Methylamino)-9,10-anthracenedione

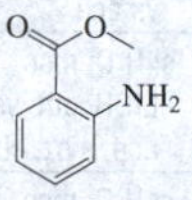
6923
Methyl 2-aminobenzoate

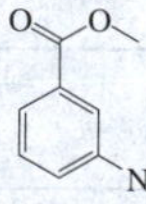
6924
Methyl 3-aminobenzoate

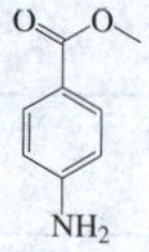
6925
Methyl 4-aminobenzoate

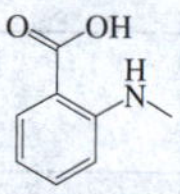
6926
2-(Methylamino)benzoic acid

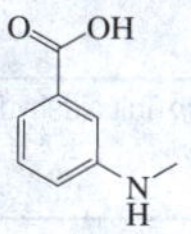
6927
3-(Methylamino)benzoic acid

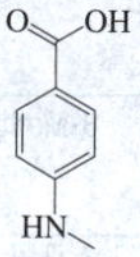
6928
4-(Methylamino)benzoic acid

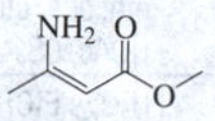
6929
Methyl 3-amino-2-butenoate

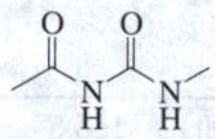
6930
N-[(Methylamino)carbonyl]acetamide

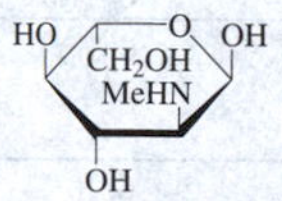
6931
2-(Methylamino)-2-deoxy-α-L-glucopyranose

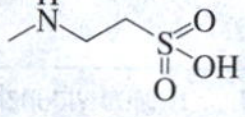
6932
2-(Methylamino)ethanesulfonic acid

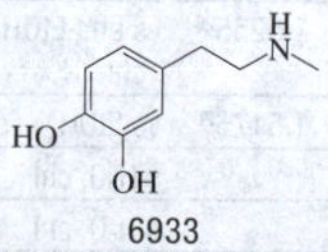
6933
4-[2-(Methylamino)ethyl]-1,2-benzenediol

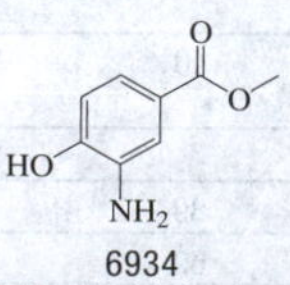
6934
Methyl 3-amino-4-hydroxybenzoate

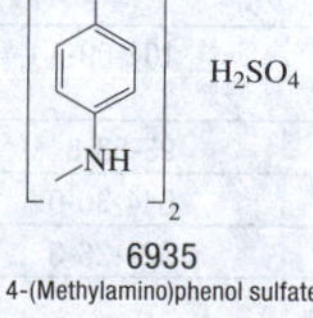
6935
4-(Methylamino)phenol sulfate

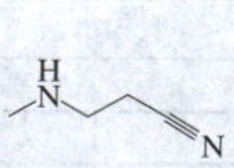
6936
3-(Methylamino)propanenitrile

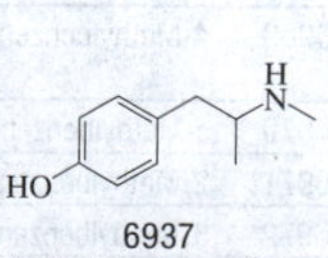
6937
4-[2-(Methylamino)propyl]phenol

Organic

No.	Name	Synonym	Mol. form.	CAS Reg. No.	Mol. wt.	Physical form	mp/°C	bp/°C	den/ g cm^{-3}	n_D	Solubility
6938	2-Methylaniline	*o*-Toluidine	C_7H_9N	95-53-4	107.153	liq	-14.41(0.02)	200.0(0.4)	0.9984^{20}	1.5725^{20}	sl H_2O; msc EtOH, eth, ctc
6939	3-Methylaniline	*m*-Toluidine	C_7H_9N	108-44-1	107.153	liq	-30.8(0.5)	203.3(0.5)	0.9889^{20}	1.5681^{20}	vs ace, bz, eth, EtOH
6940	4-Methylaniline	*p*-Toluidine	C_7H_9N	106-49-0	107.153	lf (w+1)	43.3(0.8)	201(1)	0.9619^{20}	1.5534^{45}	sl H_2O; vs EtOH, py; s eth, ace, ctc
6941	*N*-Methylaniline	*N*-Methylbenzenamine	C_7H_9N	100-61-8	107.153	liq	-57(1)	197(1)	0.9891^{20}	1.5684^{20}	i H_2O; s EtOH, eth, ctc, chl
6942	2-Methylaniline, hydrochloride	*o*-Toluidine, hydrochloride	$C_7H_{10}ClN$	636-21-5	143.614	mcl pr (w)	215				vs H_2O, EtOH
6943	4-Methylaniline, hydrochloride		$C_7H_{10}ClN$	540-23-8	143.614	mcl nd (eth-HOAc)	244.5	258	1.1930^{18}		vs H_2O, EtOH, HOAc
6944	2-Methylanisole		$C_8H_{10}O$	578-58-5	122.164	liq	-34.1(0.3)	173(2)	0.985^{25}	1.5161^{20}	i H_2O; s EtOH, eth, ace, ctc
6945	3-Methylanisole		$C_8H_{10}O$	100-84-5	122.164	liq	-55.5(0.8)	177(2)	0.969^{25}	1.5130^{20}	i H_2O; s EtOH, eth, ace, bz; sl ctc
6946	4-Methylanisole		$C_8H_{10}O$	104-93-8	122.164	liq	-31.6(0.9)	175(2)	0.969^{25}	1.5112^{20}	i H_2O; s EtOH, eth, chl
6947	1-Methylanthracene		$C_{15}H_{12}$	610-48-0	192.256	bl nd (MeOH) lf (al)	85.5	342(6)	1.0471^{99}	1.6802^{99}	i H_2O; s EtOH, eth, bz, chl, sulf
6948	2-Methylanthracene		$C_{15}H_{12}$	613-12-7	192.256	grn bl flr lf (sub)	209	340(6)	1.80^{0}		i H_2O, ace; sl EtOH, eth; s bz, CS_2
6949	9-Methylanthracene		$C_{15}H_{12}$	779-02-2	192.256	ye nd (dil al) pr (bz, al)	81.7(0.5)	196^{12}	1.065^{99}	1.6959^{99}	s EtOH, eth, ace, bz, chl
6950	2-Methyl-9,10-anthracenedione	2-Methylanthraquinone	$C_{15}H_{10}O_2$	84-54-8	222.239	ye nd (al, HOAc)	176.2(0.4)	sub			vs bz, EtOH, HOAc
6951	Methylarsine		CH_5As	593-52-2	91.973	col gas	-143	2			vs ace, eth, EtOH
6952	9-Methyl-9-azabicyclo[3.3.1]-nonan-3-one	Pseudopelletierine	$C_9H_{15}NO$	552-70-5	153.221	orth pr (peth)	54	246	1.001^{100}	1.4760^{100}	vs H_2O, eth, EtOH
6953	8-Methyl-8-azabicyclo[3.2.1]-octane	Tropane	$C_8H_{15}N$	529-17-9	125.212			166	0.9251^{15}		
6954	8-Methyl-8-azabicyclo[3.2.1]-octan-3-one		$C_8H_{13}NO$	532-24-1	139.195		43	227	1.9872^{100}	1.4598^{100}	s EtOH, eth, ace, bz, peth; sl chl
6955	Methyl azide		CH_3N_3	624-90-8	57.055			exp	0.869^{15}		
6956	Methylazoxymethanol acetate		$C_4H_8N_2O_3$	592-62-1	132.118			191			
6957	2-Methylbenzaldehyde	*o*-Tolualdehyde	C_8H_8O	529-20-4	120.149			195(4)	1.0328^{20}	1.5462^{20}	sl H_2O, ctc; s EtOH, eth, bz; vs ace
6958	3-Methylbenzaldehyde	*m*-Tolualdehyde	C_8H_8O	620-23-5	120.149			199	1.0189^{21}	1.5413^{21}	sl H_2O; msc EtOH, eth; vs ace; s bz, chl
6959	4-Methylbenzaldehyde	*p*-Tolualdehyde	C_8H_8O	104-87-0	120.149			202(7)	1.0194^{17}	1.5454^{20}	sl H_2O; msc EtOH, eth, ace; vs chl
6960	2-Methylbenzamide	*o*-Toluamide	C_8H_9NO	527-85-5	135.163		147				sl H_2O, eth, tfa, bz; vs EtOH
6961	4-Methylbenzamide	*p*-Toluamide	C_8H_9NO	619-55-6	135.163		162.5				sl H_2O, bz, chl; vs EtOH, eth; s tfa
6962	*N*-Methylbenzamide		C_8H_9NO	613-93-4	135.163		74.3(0.4)	294(5)			s EtOH, ace
6963	7-Methylbenz[*a*]anthracene		$C_{19}H_{14}$	2541-69-7	242.314	ye pl (al)	141				i H_2O; s EtOH, eth, ace, ctc, HOAc, CS_2
6964	8-Methylbenz[*a*]anthracene		$C_{19}H_{14}$	2381-31-9	242.314	pl (bz-al), nd (bz-lig)	156.5	272^{3}	1.2310^{0}		i H_2O; s EtOH, eth, bz, xyl
6965	9-Methylbenz[*a*]anthracene		$C_{19}H_{14}$	2381-16-0	242.314	nd (al)	152.5				i H_2O; s EtOH, eth, ctc, chl, CS_2, xyl
6966	10-Methylbenz[*a*]anthracene		$C_{19}H_{14}$	2381-15-9	242.314		184				i H_2O; s EtOH, HOAc
6967	12-Methylbenz[*a*]anthracene		$C_{19}H_{14}$	2422-79-9	242.314	pl (al)	150.5				i H_2O; s EtOH, CS_2, HOAc
6968	2-Methylbenzeneacetaldehyde		$C_9H_{10}O$	10166-08-2	134.174			221	1.0241^{10}		vs eth, EtOH, chl
6969	4-Methylbenzeneacetaldehyde		$C_9H_{10}O$	104-09-6	134.174		40	221.5	1.0052^{20}	1.5255^{20}	vs eth, EtOH, chl
6970	α-Methylbenzeneacetaldehyde		$C_9H_{10}O$	93-53-8	134.174			206.4(0.8)	1.0089^{20}	1.5176^{20}	vs EtOH
6971	2-Methylbenzeneacetic acid		$C_9H_{10}O_2$	644-36-0	150.174	nd (w)	89				s H_2O, chl
6972	3-Methylbenzeneacetic acid		$C_9H_{10}O_2$	621-36-3	150.174	nd (w)	62	121^{26}			s H_2O, chl

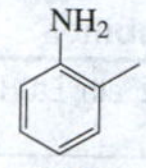
6938
2-Methylaniline

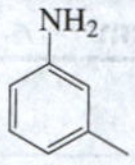
6939
3-Methylaniline

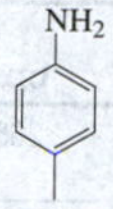
6940
4-Methylaniline

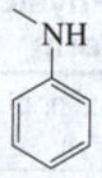
6941
N-Methylaniline

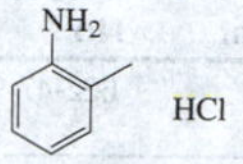
6942
2-Methylaniline, hydrochloride

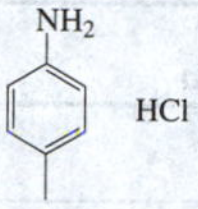
6943
4-Methylaniline, hydrochloride

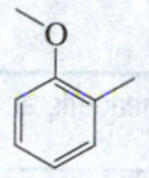
6944
2-Methylanisole

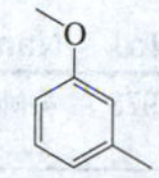
6945
3-Methylanisole

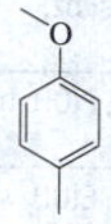
6946
4-Methylanisole

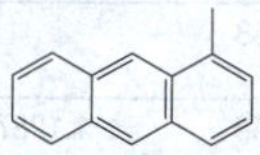
6947
1-Methylanthracene

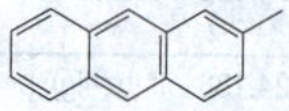
6948
2-Methylanthracene

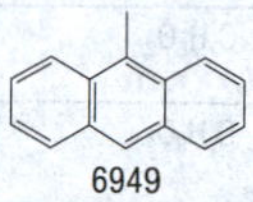
6949
9-Methylanthracene

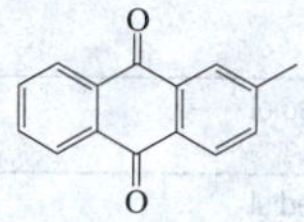
6950
2-Methyl-9,10-anthracenedione

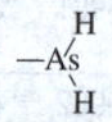
6951
Methylarsine

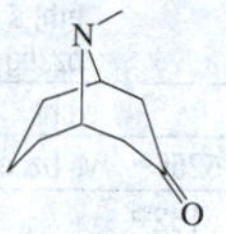
6952
9-Methyl-9-azabicyclo[3.3.1]nonan-3-one

6953
8-Methyl-8-azabicyclo[3.2.1]octane

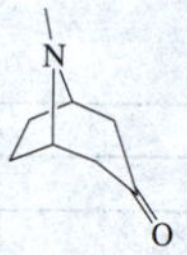
6954
8-Methyl-8-azabicyclo[3.2.1]octan-3-one

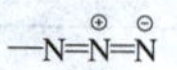
6955
Methyl azide

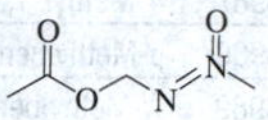
6956
Methylazoxymethanol acetate

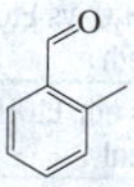
6957
2-Methylbenzaldehyde

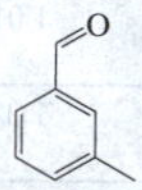
6958
3-Methylbenzaldehyde

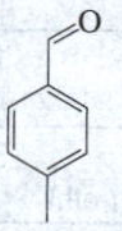
6959
4-Methylbenzaldehyde

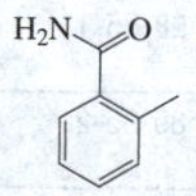
6960
2-Methylbenzamide

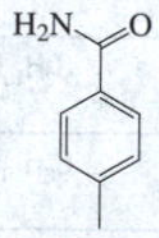
6961
4-Methylbenzamide

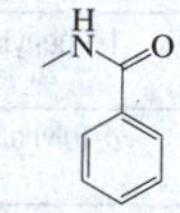
6962
N-Methylbenzamide

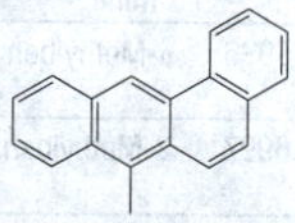
6963
7-Methylbenz[a]anthracene

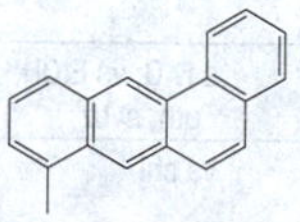
6964
8-Methylbenz[a]anthracene

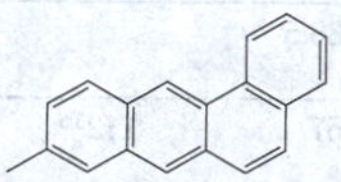
6965
9-Methylbenz[a]anthracene

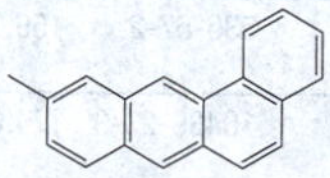
6966
10-Methylbenz[a]anthracene

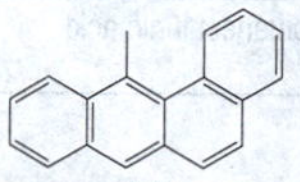
6967
12-Methylbenz[a]anthracene

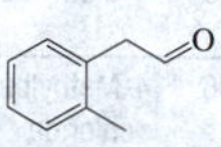
6968
2-Methylbenzeneacetaldehyde

6969
4-Methylbenzeneacetaldehyde

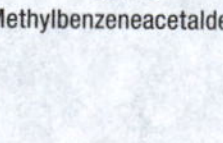
6970
α-Methylbenzeneacetaldehyde

6971
2-Methylbenzeneacetic acid

6972
3-Methylbenzeneacetic acid

No.	Name	Synonym	Mol. form.	CAS Reg. No.	Mol. wt.	Physical form	mp/°C	bp/°C	den/ g cm^{-3}	n_D	Solubility
6973	4-Methylbenzeneacetic acid		$C_9H_{10}O_2$	622-47-9	150.174	nd or pl (al, w)	93	265			vs bz, eth, EtOH
6974	α-Methylbenzeneacetic acid, (±)-		$C_9H_{10}O_2$	2328-24-7	150.174		<-20	263	1.1^{0}	1.5237^{20}	
6975	4-Methylbenzeneacetonitrile		C_9H_9N	2947-61-7	131.174		18	242.5	0.992^{25}	1.5190^{20}	i H_2O; s EtOH, eth, bz, ctc
6976	α-Methylbenzeneacetonitrile		C_9H_9N	1823-91-2	131.174			231	0.9854^{20}	1.5095^{25}	vs eth, EtOH
6977	3-Methyl-1,2-benzenediamine	Toluene-2,3-diamine	$C_7H_{10}N_2$	2687-25-4	122.167		63.5	255			vs ace, bz, EtOH
6978	4-Methyl-1,2-benzenediamine	Toluene-3,4-diamine	$C_7H_{10}N_2$	496-72-0	122.167	pl (lig)	89.5	265			vs H_2O; s lig
6979	2-Methyl-1,3-benzenediamine	Toluene-2,6-diamine	$C_7H_{10}N_2$	823-40-5	122.167	pr (bz, w)	106				s H_2O, EtOH, bz
6980	2-Methyl-1,4-benzenediamine	Toluene-2,5-diamine	$C_7H_{10}N_2$	95-70-5	122.167	pl (bz)	64	273.5			s H_2O, EtOH, eth; sl bz, HOAc
6981	3-Methyl-1,2-benzenediol		$C_7H_8O_2$	488-17-5	124.138	lf (bz)	68	248			s H_2O, EtOH, bz, chl
6982	4-Methyl-1,2-benzenediol		$C_7H_8O_2$	452-86-8	124.138	lf (bz-lig), pr (bz)	65	258	1.1287^{74}	1.5425^{74}	s H_2O, EtOH, eth, ace, chl; sl lig
6983	2-Methyl-1,3-benzenediol		$C_7H_8O_2$	608-25-3	124.138	pr (bz)	120	265			vs H_2O, bz, eth, EtOH
6984	4-Methyl-1,3-benzenediol		$C_7H_8O_2$	496-73-1	124.138	cry (bz-peth)	105	270			s H_2O, EtOH, eth; sl bz, peth
6985	5-Methyl-1,3-benzenediol	Orcinol	$C_7H_8O_2$	504-15-4	124.138	pr(w+1), lf(chl)	107	287	1.290^{4}		s H_2O, EtOH, eth, bz; sl lig, peth
6986	2-Methyl-1,4-benzenediol		$C_7H_8O_2$	95-71-6	124.138		125	283			vs H_2O, EtOH, eth; s ace; sl bz, lig
6987	4-Methyl-1,2-benzenedithiol	Toluene-3,4-dithiol	$C_7H_8S_2$	496-74-2	156.269		29				s chl
6988	β-Methylbenzeneethanamine		$C_9H_{13}N$	582-22-9	135.206			210	0.9433^{4}	1.5255^{20}	vs bz, eth, EtOH
6989	*N*-Methylbenzeneethanamine		$C_9H_{13}N$	589-08-2	135.206			206	0.93^{25}	1.5162^{20}	
6990	2-Methylbenzeneethanol		$C_9H_{12}O$	19819-98-8	136.190		1.0	243.5	1.016^{25}	1.5355^{20}	
6991	4-Methylbenzeneethanol		$C_9H_{12}O$	699-02-5	136.190			240(11)	1.0028^{20}	1.5267^{20}	
6992	2-Methylbenzenemethanamine		$C_8H_{11}N$	89-93-0	121.180	liq	-30	206	0.9766^{19}	1.5436^{19}	
6993	3-Methylbenzenemethanamine		$C_8H_{11}N$	100-81-2	121.180			203.5	0.966^{25}	1.5360^{20}	
6994	4-Methylbenzenemethanamine		$C_8H_{11}N$	104-84-7	121.180		12.5	195	0.952^{20}	1.5340^{20}	
6995	*N*-Methylbenzenemethanamine		$C_8H_{11}N$	103-67-3	121.180			180.5	0.9442^{18}		vs H_2O
6996	α-Methylbenzenemethanol	1-Phenylethanol	$C_8H_{10}O$	98-85-1	122.164		15(2)	205(4)	1.013^{25}	1.5265^{20}	i H_2O; vs EtOH, eth
6997	2-Methylbenzenemethanol	*o*-Tolyl alcohol	$C_8H_{10}O$	89-95-2	122.164	nd (peth-eth)	38	224	1.023^{40}		vs eth, EtOH, chl
6998	3-Methylbenzenemethanol	*m*-Tolyl alcohol	$C_8H_{10}O$	587-03-1	122.164		<-20	215.5	0.9157^{17}		sl H_2O; vs EtOH, eth; s chl
6999	4-Methylbenzenemethanol	*p*-Tolyl alcohol	$C_8H_{10}O$	589-18-4	122.164	nd (lig)	58.72(0.04)	217	0.978^{22}		vs eth, EtOH
7000	α-Methylbenzenemethanol, acetate		$C_{10}H_{12}O_2$	93-92-5	164.201	oil		109^{18}			
7001	4-Methylbenzenepropanal		$C_{10}H_{12}O$	5406-12-2	148.201			223	0.999^{14}	1.525^{14}	
7002	α-Methylbenzenepropanamine	1-Methyl-3-phenylpropylamine	$C_{10}H_{15}N$	22374-89-6	149.233		143	223	0.9289^{15}	1.5152^{20}	vs EtOH
7003	β-Methylbenzenepropanoic acid, (±)-		$C_{10}H_{12}O_2$	772-17-8	164.201		46.5	168^{14}	1.0701^{20}	1.5155^{20}	sl H_2O; s peth
7004	α-Methylbenzenepropanol		$C_{10}H_{14}O$	2344-70-9	150.217			239	0.9899^{16}	1.517^{16}	
7005	4-Methylbenzenesulfinic acid	*p*-Toluenesulfinic acid	$C_7H_8O_2S$	536-57-2	156.203	orth pl or nd (w)	86.5				s H_2O; vs EtOH, eth; sl bz
7006	4-Methylbenzenesulfinyl chloride		C_7H_7ClOS	10439-23-3	174.648	nd	57	$113^{3.5}$			vs chl
7007	2-Methylbenzenesulfonamide		$C_7H_9NO_2S$	88-19-7	171.217	oct cry (al), pr (w)	156.3(0.5)	214^{10}			sl H_2O, eth, DMSO; s EtOH
7008	4-Methylbenzenesulfonamide	*p*-Toluenesulfonamide	$C_7H_9NO_2S$	70-55-3	171.217	mcl pl (w+2)	138	214^{10}			sl H_2O, eth; s EtOH
7009	Methyl benzenesulfonate		$C_7H_8O_3S$	80-18-2	172.202		4.5	150^{15}	1.2730^{17}	1.5151^{20}	sl H_2O; vs EtOH, eth, chl
7010	2-Methylbenzenesulfonic acid		$C_7H_8O_3S$	88-20-0	172.202	hyg pl (w+2)	67.5	128.8^{25}			vs H_2O; s EtOH; i eth
7011	2-Methylbenzenesulfonyl chloride	*o*-Toluenesulfonyl chloride	$C_7H_7ClO_2S$	133-59-5	190.648		11(1)	154^{36}	1.3383^{20}	1.5565^{20}	i H_2O; s EtOH, eth, bz, ctc
7012	2-Methylbenzenethiol		C_7H_8S	137-06-4	124.204		15	192(20)	1.041^{20}	1.570^{20}	i H_2O; s EtOH; vs eth
7013	3-Methylbenzenethiol		C_7H_8S	108-40-7	124.204	liq	-20	196(20)	1.044^{20}	1.572^{20}	i H_2O; s EtOH; msc eth

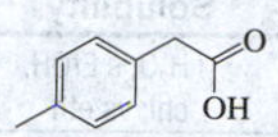
6973
4-Methylbenzeneacetic acid

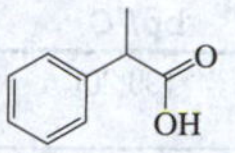
6974
α-Methylbenzeneacetic acid, (±)-

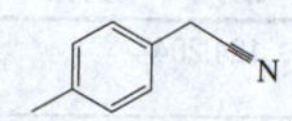
6975
4-Methylbenzeneacetonitrile

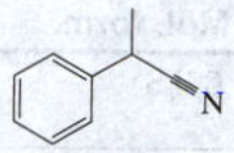
6976
α-Methylbenzeneacetonitrile

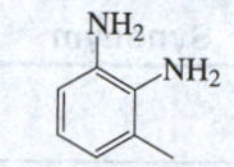
6977
3-Methyl-1,2-benzenediamine

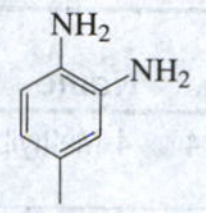
6978
4-Methyl-1,2-benzenediamine

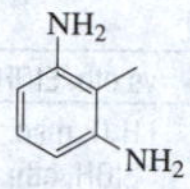
6979
2-Methyl-1,3-benzenediamine

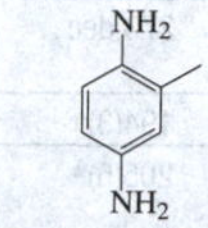
6980
2-Methyl-1,4-benzenediamine

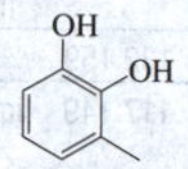
6981
3-Methyl-1,2-benzenediol

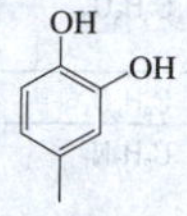
6982
4-Methyl-1,2-benzenediol

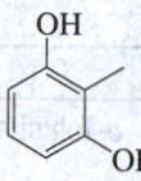
6983
2-Methyl-1,3-benzenediol

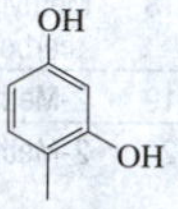
6984
4-Methyl-1,3-benzenediol

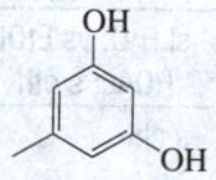
6985
5-Methyl-1,3-benzenediol

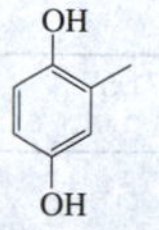
6986
2-Methyl-1,4-benzenediol

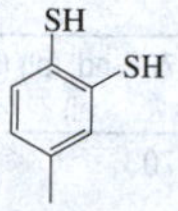
6987
4-Methyl-1,2-benzenedithiol

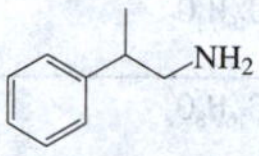
6988
β-Methylbenzeneethanamine

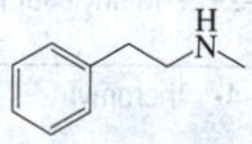
6989
N-Methylbenzeneethanamine

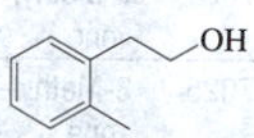
6990
2-Methylbenzeneethanol

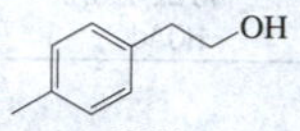
6991
4-Methylbenzeneethanol

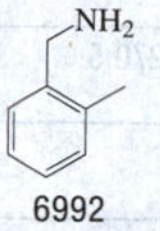
6992
2-Methylbenzenemethanamine

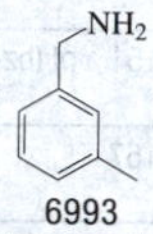
6993
3-Methylbenzenemethanamine

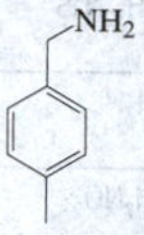
6994
4-Methylbenzenemethanamine

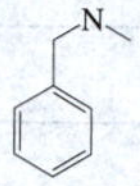
6995
N-Methylbenzenemethanamine

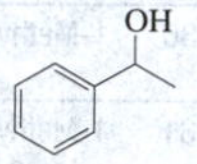
6996
α-Methylbenzenemethanol

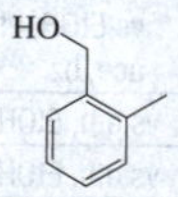
6997
2-Methylbenzenemethanol

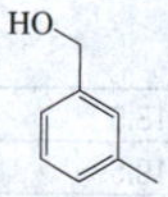
6998
3-Methylbenzenemethanol

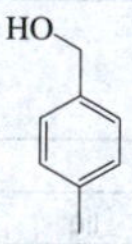
6999
4-Methylbenzenemethanol

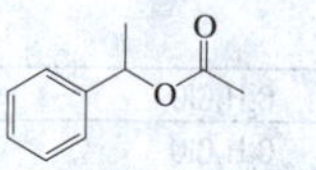
7000
α-Methylbenzenemethanol, acetate

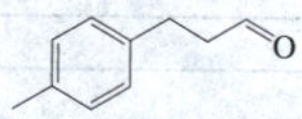
7001
4-Methylbenzenepropanal

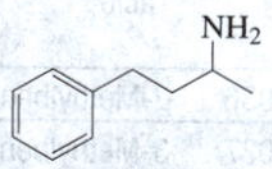
7002
α-Methylbenzenepropanamine

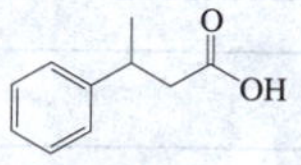
7003
β-Methylbenzenepropanoic acid, (±)-

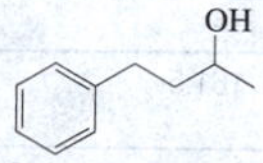
7004
α-Methylbenzenepropanol

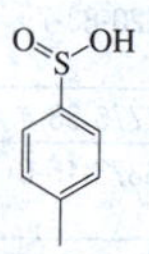
7005
4-Methylbenzenesulfinic acid

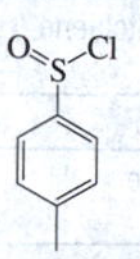
7006
4-Methylbenzenesulfinyl chloride

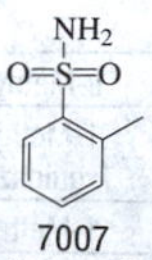
7007
2-Methylbenzenesulfonamide

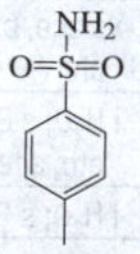
7008
4-Methylbenzenesulfonamide

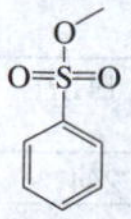
7009
Methyl benzenesulfonate

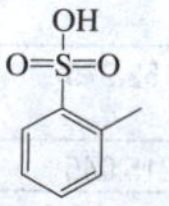
7010
2-Methylbenzenesulfonic acid

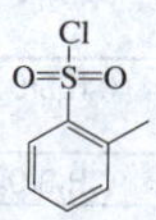
7011
2-Methylbenzenesulfonyl chloride

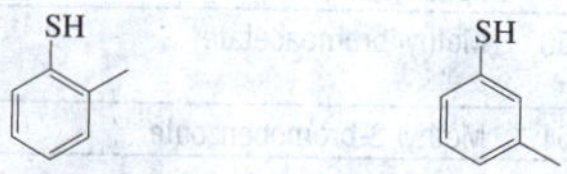
7012
2-Methylbenzenethiol

7013
3-Methylbenzenethiol

No.	Name	Synonym	Mol. form.	CAS Reg. No.	Mol. wt.	Physical form	mp/°C	bp/°C	den/g cm^{-3}	n_D	Solubility
7014	4-Methylbenzenethiol		C_7H_8S	106-45-6	124.204		43	199(20)	1.0220^{51}		i H_2O; s EtOH, chl; vs eth
7015	1-Methyl-1*H*-benzimidazole		$C_8H_8N_2$	1632-83-3	132.163	nd (peth), pl (al)	66	286	1.1254^{20}	1.6013^{7}	s peth
7016	2-Methyl-1*H*-benzimidazole		$C_8H_8N_2$	615-15-6	132.163	pr or nd (w)	178.3(0.3)				s H_2O; sl EtOH, eth; i bz
7017	Methyl benzoate	Methyl benzenecarboxylate	$C_8H_8O_2$	93-58-3	136.149	liq	-12.35(0.06)	199(2)	1.0837^{25}	1.5164^{20}	i H_2O; s EtOH, ctc, MeOH; msc eth
7018	Methyl 1,3-benzodioxole-5-carboxylate		$C_9H_8O_4$	326-56-7	180.158	nd or lf (peth)	53	273 dec			vs eth, EtOH
7019	2-Methylbenzofuran		C_9H_8O	4265-25-2	132.159			194(3)	1.0540^{20}	1.5495^{22}	vs eth, EtOH
7020	2-Methylbenzonitrile	*o*-Tolunitrile	C_8H_7N	529-19-1	117.149	liq	-10.5(0.3)	205(5)	0.9955^{20}	1.5279^{20}	i H_2O; msc EtOH, eth; sl ctc
7021	3-Methylbenzonitrile	*m*-Tolunitrile	C_8H_7N	620-22-4	117.149	liq	-23	215(7)	1.0316^{20}	1.5252^{20}	i H_2O; msc EtOH, eth; sl ctc
7022	4-Methylbenzonitrile	*p*-Tolunitrile	C_8H_7N	104-85-8	117.149		28(1)	218(3)	0.9762^{30}		i H_2O; vs EtOH, eth; sl ctc
7023	6-Methyl-2*H*-1-benzopyran-2-one		$C_{10}H_8O_2$	92-48-8	160.170		76.5	304			vs EtOH, eth, bz; sl chl, peth
7024	7-Methyl-2*H*-1-benzopyran-2-one	7-Methylcoumarin	$C_{10}H_8O_2$	2445-83-2	160.170	nd, (pl) (aq al)	128	171.5^{11}			sl H_2O; vs EtOH, HOAc; s eth
7025	3-Methyl-4*H*-1-benzopyran-4-one	Tricromyl	$C_{10}H_8O_2$	85-90-5	160.170						s chl
7026	6-Methyl-2-benzothiazolamine		$C_8H_8N_2S$	2536-91-6	164.228	nd (w) pr (dil al)	142				sl H_2O; s EtOH
7027	2-Methylbenzothiazole		C_8H_7NS	120-75-2	149.214		14	238	1.1763^{19}	1.6092^{19}	i H_2O; s EtOH, chl
7028	3-Methyl-2(3*H*)-benzothiazolethione		$C_8H_7NS_2$	2254-94-6	181.279	nd (al), pr (HOAc)	90	335			i H_2O; sl EtOH, eth; vs bz, chl
7029	4-(6-Methyl-2-benzothiazolyl)aniline		$C_{14}H_{12}N_2S$	92-36-4	240.323		194.8	434			sl EtOH, eth, bz, HOAc
7030	1-Methyl-1*H*-benzotriazole		$C_7H_7N_3$	13351-73-0	133.151	pl (bz-lig)	64.5	270.5			vs bz, EtOH, HOAc
7031	1-Methyl-2*H*-3,1-benzoxazine-2,4(1*H*)-dione		$C_9H_7NO_3$	10328-92-4	177.157		180				
7032	2-Methylbenzoxazole		C_8H_7NO	95-21-6	133.148		9.5	200.5	1.1211^{20}	1.5497^{20}	i H_2O; vs EtOH; msc eth
7033	Methyl benzoylacetate		$C_{10}H_{10}O_3$	614-27-7	178.184	pa ye		265 dec	1.158^{29}	1.537^{20}	vs ace, eth, EtOH
7034	Methyl 2-benzoylbenzoate		$C_{15}H_{12}O_3$	606-28-0	240.254	pl or mcl pr (dil al)	52	351	1.1903^{19}	1.591^{20}	i H_2O; vs EtOH, eth; s sulf
7035	2-(4-Methylbenzoyl)benzoic acid	2-(*p*-Toluoyl)benzoic acid	$C_{15}H_{12}O_3$	85-55-2	240.254		140.2(0.4)				sl H_2O, DMSO; vs EtOH, eth, ace, bz
7036	2-Methylbenzoyl chloride		C_8H_7ClO	933-88-0	154.594			213.5		1.5549^{20}	vs eth, EtOH
7037	3-Methylbenzoyl chloride		C_8H_7ClO	1711-06-4	154.594	liq	-23	219.5	1.0265^{21}	1.505^{22}	vs eth, EtOH
7038	4-Methylbenzoyl chloride		C_8H_7ClO	874-60-2	154.594	liq	-1.5	226	1.1686^{20}	1.5547^{20}	s ctc
7039	Methyl benzoylsalicylate	2-(Benzoyloxy)benzoic acid, methyl ester	$C_{15}H_{12}O_4$	610-60-6	256.254	cry	85	385			i H_2O; s bz, chl, eth, EtOH
7040	α-Methylbenzylamine, (±)-	1-Phenylethylamine	$C_8H_{11}N$	618-36-0	121.180		32	193(2)	0.9395^{15}	1.5238^{25}	s H_2O, chl; msc EtOH, eth
7041	1-Methyl-2-benzylbenzene	2-Benzyltoluene	$C_{14}H_{14}$	713-36-0	182.261		6.61(0.01)	281(7)	1.0020^{20}	1.5763^{20}	
7042	1-Methyl-4-benzylbenzene	4-Benzyltoluene	$C_{14}H_{14}$	620-83-7	182.261	liq	-30	282(6)	0.9976^{20}	1.5712^{20}	vs eth, bz, EtOH, chl
7043	α-Methylbenzyl formate		$C_9H_{10}O_2$	7775-38-4	150.174	liq					
7044	1-Methyl-2-benzyl-4(1*H*)-quinazolinone	Glycosine	$C_{16}H_{14}N_2O$	6873-15-0	250.294		161.5				
7045	1-Methylbicyclo[3.1.0]hexane		C_7H_{12}	4625-24-5	96.170			88(18)			
7046	2-Methylbiphenyl		$C_{13}H_{12}$	643-58-3	168.234	liq	-0.2(0.1)	258(4)	1.0113^{20}	1.5914^{20}	i H_2O; s EtOH, eth
7047	3-Methylbiphenyl		$C_{13}H_{12}$	643-93-6	168.234		4.5(0.2)	274(2)	1.0182^{17}	1.5972^{20}	i H_2O; s EtOH, eth, ctc
7048	4-Methylbiphenyl		$C_{13}H_{12}$	644-08-6	168.234	pl (lig, MeOH)	48.1(1)	273(6)	1.015^{27}		i H_2O; s EtOH, eth; sl ctc
7049	4-Methyl-*N*,*N*-bis(4-methylphenyl)aniline		$C_{21}H_{21}N$	1159-53-1	287.399	cry (HOAc)	115(1)				vs ace, bz, eth, chl
7050	Methyl bromoacetate		$C_3H_5BrO_2$	96-32-2	152.975			141(7)	1.6350^{20}	1.4520^{20}	i H_2O; s EtOH, eth, ace, bz
7051	Methyl 2-bromobenzoate		$C_8H_7BrO_2$	610-94-6	215.045			244			i H_2O; s EtOH

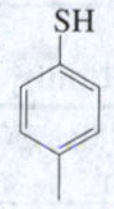
7014
4-Methylbenzenethiol

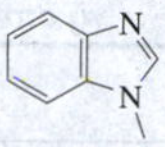
7015
1-Methyl-1*H*-benzimidazole

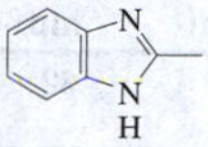
7016
2-Methyl-1*H*-benzimidazole

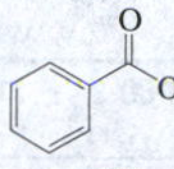
7017
Methyl benzoate

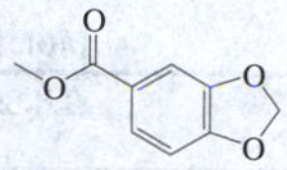
7018
Methyl 1,3-benzodioxole-5-carboxylate

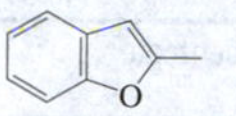
7019
2-Methylbenzofuran

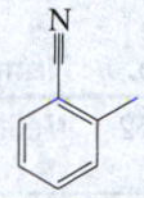
7020
2-Methylbenzonitrile

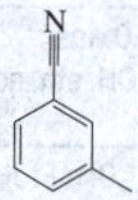
7021
3-Methylbenzonitrile

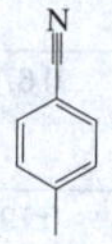
7022
4-Methylbenzonitrile

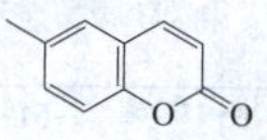
7023
6-Methyl-2*H*-1-benzopyran-2-one

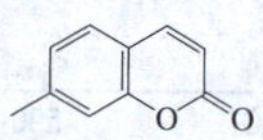
7024
7-Methyl-2*H*-1-benzopyran-2-one

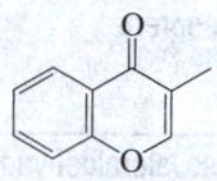
7025
3-Methyl-4*H*-1-benzopyran-4-one

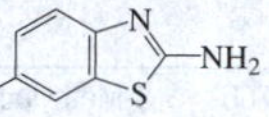
7026
6-Methyl-2-benzothiazolamine

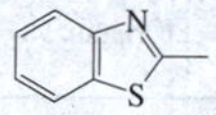
7027
2-Methylbenzothiazole

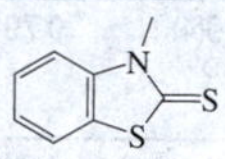
7028
3-Methyl-2(3*H*)-benzothiazolethione

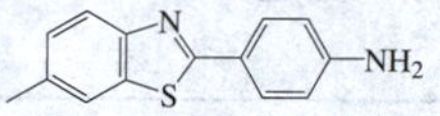
7029
4-(6-Methyl-2-benzothiazolyl)aniline

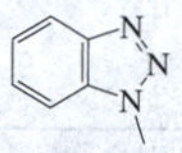
7030
1-Methyl-1*H*-benzotriazole

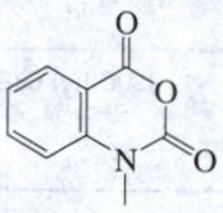
7031
1-Methyl-2*H*-3,1-benzoxazine-2,4(1*H*)-dione

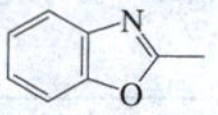
7032
2-Methylbenzoxazole

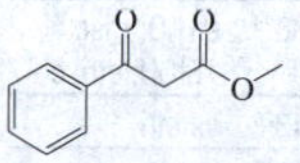
7033
Methyl benzoylacetate

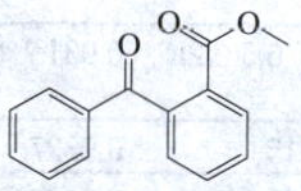
7034
Methyl 2-benzoylbenzoate

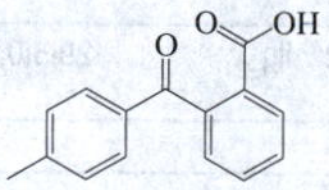
7035
2-(4-Methylbenzoyl)benzoic acid

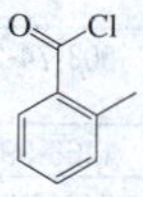
7036
2-Methylbenzoyl chloride

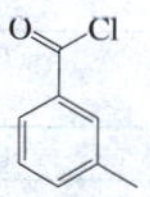
7037
3-Methylbenzoyl chloride

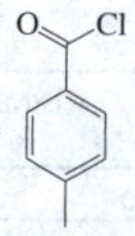
7038
4-Methylbenzoyl chloride

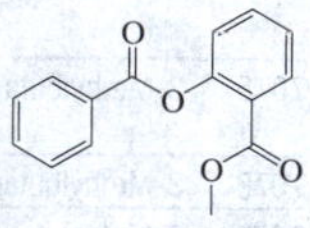
7039
Methyl benzoylsalicylate

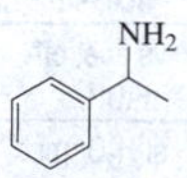
7040
α-Methylbenzylamine, (±)-

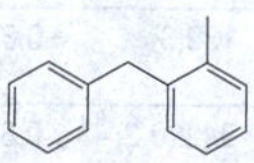
7041
1-Methyl-2-benzylbenzene

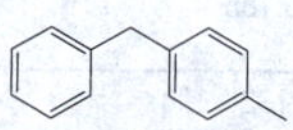
7042
1-Methyl-4-benzylbenzene

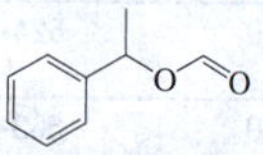
7043
α-Methylbenzyl formate

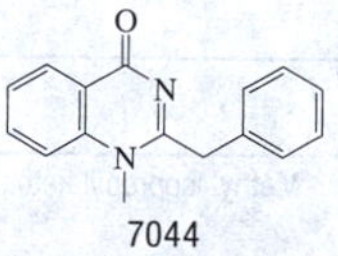
7044
1-Methyl-2-benzyl-4(1*H*)-quinazolinone

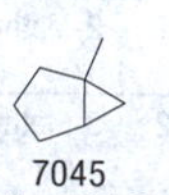
7045
1-Methylbicyclo[3,1,0]hexane

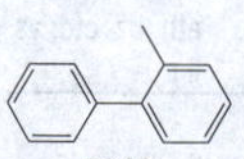
7046
2-Methylbiphenyl

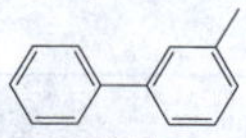
7047
3-Methylbiphenyl

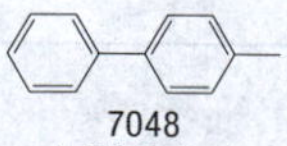
7048
4-Methylbiphenyl

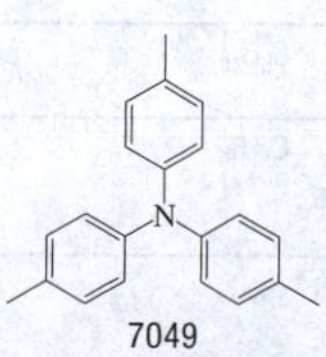
7049
4-Methyl-*N*,*N*-bis(4-methylphenyl)aniline

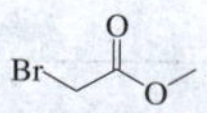
7050
Methyl bromoacetate

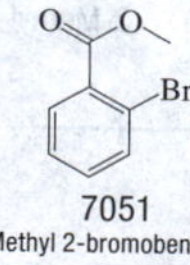
7051
Methyl 2-bromobenzoate

No.	Name	Synonym	Mol. form.	CAS Reg. No.	Mol. wt.	Physical form	mp/°C	bp/°C	den/ g cm^{-3}	n_D	Solubility
7052	Methyl 3-bromobenzoate		$C_8H_7BrO_2$	618-89-3	215.045	pl	32	125^{15}			sl H_2O; s EtOH, eth
7053	Methyl 4-bromobenzoate		$C_8H_7BrO_2$	619-42-1	215.045	lf (dil al), nd (eth)	81		1.689^{25}		s EtOH, eth, ace, peth; vs bz, chl
7054	Methyl 2-bromobutanoate		$C_5H_9BrO_2$	3196-15-4	181.028			168	1.4528^{20}	1.4029^{25}	vs EtOH
7055	Methyl 4-bromobutanoate		$C_5H_9BrO_2$	4897-84-1	181.028			186.5	1.4^{25}	1.4567^{25}	vs EtOH
7056	Methyl 4-bromo-2-butenoate		$C_5H_7BrO_2$	1117-71-1	179.013			84^{12}	1.490^{19}	1.498^{19}	
7057	Methyl 5-bromopentanoate		$C_6H_{11}BrO_2$	5454-83-1	195.054	liq		101^{14}	1.363	1.4630^{20}	
7058	Methyl 3-bromopropanoate		$C_4H_7BrO_2$	3395-91-3	167.002			105^{60}	1.4123^{18}	1.4542^{20}	s EtOH, eth, ace
7059	3-Methyl-1,2-butadiene		C_5H_8	598-25-4	68.118	liq	-113.61(0.08)	40.8(0.7)	0.6806^{25}	1.4203^{20}	vs ace, bz, eth, EtOH
7060	2-Methyl-1,3-butadiene	Isoprene	C_5H_8	78-79-5	68.118	liq	-146.1(0.5)	34.0(0.3)	0.679^{20}	1.4219^{20}	i H_2O; msc EtOH, eth, ace, bz
7061	3-Methylbutanal	Isovaleraldehyde	$C_5H_{10}O$	590-86-3	86.132	liq	-51	92.5(0.3)	0.7977^{20}	1.3902^{20}	sl H_2O; s EtOH, eth
7062	3-Methylbutanamide	Isovaleramide	$C_5H_{11}NO$	541-46-8	101.147	mcl lf (al)	137	226			s H_2O, EtOH, eth; vs peth
7063	3-Methyl-1-butanamine	Isopentylamine	$C_5H_{13}N$	107-85-7	87.164			96(4)	0.7505^{20}	1.4083^{20}	msc H_2O, EtOH, eth; s ace, chl
7064	2-Methyl-2-butanamine		$C_5H_{13}N$	594-39-8	87.164	liq	-105.0(0.6)	77(4)	0.731^{25}	1.3954^{25}	vs H_2O, ace, eth, EtOH
7065	3-Methyl-2-butanamine		$C_5H_{13}N$	598-74-3	87.164	liq	-50	81(6)	0.7574^{19}	1.4096^{18}	vs H_2O; s EtOH
7066	3-Methyl-1,3-butanediol		$C_5H_{12}O_2$	2568-33-4	104.148			208(5)	0.9448^{20}	1.4452^{20}	s H_2O, EtOH
7067	2-Methylbutanenitrile		C_5H_9N	18936-17-9	83.132			123(3)	0.7913^{15}	1.3933^{20}	vs eth, EtOH
7068	3-Methylbutanenitrile	Isobutyl cyanide	C_5H_9N	625-28-5	83.132	liq	-100.8(0.3)	129(4)	0.7914^{20}	1.3927^{20}	sl H_2O; msc EtOH, eth; vs ace
7069	2-Methyl-1-butanethiol, (+)		$C_5H_{12}S$	20089-07-0	104.214	liq		119.0(0.8)	0.8420^{20}	1.4440^{20}	
7070	3-Methyl-1-butanethiol	Isopentyl mercaptan	$C_5H_{12}S$	541-31-1	104.214	liq		118.3(0.8)	0.8350^{20}	1.4412^{20}	i H_2O; msc EtOH, eth; s ctc
7071	2-Methyl-2-butanethiol		$C_5H_{12}S$	1679-09-0	104.214	liq		99.1(0.5)	0.8120^{20}	1.4385^{20}	
7072	3-Methyl-2-butanethiol		$C_5H_{12}S$	2084-18-6	104.214	liq	-127.09(0.05)	109.8			
7073	Methyl butanoate		$C_5H_{10}O_2$	623-42-7	102.132	liq	-85.8	101.9(0.1)	0.8984^{20}	1.3878^{20}	sl H_2O, ctc; msc EtOH, eth
7074	2-Methylbutanoic acid	(±)-2-Methylbutyric acid	$C_5H_{10}O_2$	600-07-7	102.132		<-80	177	0.934^{20}	1.4051^{20}	sl H_2O; msc EtOH, eth; s chl
7075	3-Methylbutanoic acid	Isovaleric acid	$C_5H_{10}O_2$	503-74-2	102.132	liq	-29.6(0.7)	176.5(0.2)	0.931^{20}	1.4033^{20}	s H_2O; msc EtOH, eth, chl
7076	3-Methylbutanoic anhydride		$C_{10}H_{18}O_3$	1468-39-9	186.248			215	0.9327^{20}	1.4043^{20}	vs eth
7077	2-Methyl-1-butanol, (±)-		$C_5H_{12}O$	34713-94-5	88.148			129.0(0.4)	0.8152^{25}	1.4092^{20}	sl H_2O; msc EtOH, eth; vs ace
7078	3-Methyl-1-butanol	Isopentyl alcohol	$C_5H_{12}O$	123-51-3	88.148	liq	-117.2	130.8(0.3)	0.8104^{20}	1.4053^{20}	sl H_2O; vs ace, eth, EtOH
7079	2-Methyl-2-butanol	*tert*-Amyl alcohol	$C_5H_{12}O$	75-85-4	88.148	liq	-8.7(0.6)	102.4	0.8096^{20}	1.4052^{20}	s H_2O, bz, chl; msc EtOH, eth; vs ace
7080	3-Methyl-2-butanol, (±)-	Isopropylethanol	$C_5H_{12}O$	70116-68-6	88.148	liq		113.7(0.4)	0.8180^{20}	1.4089^{20}	sl H_2O; msc EtOH, eth; vs ace; s bz, ctc
7081	2-Methyl-1-butanol acetate		$C_7H_{14}O_2$	624-41-9	130.185			139(3)	0.8740^{20}	1.4040^{20}	vs ace, eth, EtOH
7082	3-Methyl-2-butanone	Methyl isopropyl ketone	$C_5H_{10}O$	563-80-4	86.132	liq	-93.13(0.05)	94.2(0.2)	0.8051^{20}	1.3880^{20}	sl H_2O; msc EtOH, eth; vs ace; s ctc
7083	2-Methylbutanoyl chloride, (±)-		C_5H_9ClO	57526-28-0	120.577			116	0.9917^{20}	1.4170^{20}	
7084	3-Methylbutanoyl chloride	Isovaleryl chloride	C_5H_9ClO	108-12-3	120.577			114	0.9844^{20}	1.4149^{20}	s eth
7085	*trans*-2-Methyl-2-butenal	Tiglic aldehyde	C_5H_8O	497-03-0	84.117	liq		114(3)	0.8710^{20}	1.4475^{20}	sl H_2O; vs EtOH
7086	3-Methyl-2-butenal	Senecialdehyde	C_5H_8O	107-86-8	84.117			134	0.8722^{20}	1.4528^{20}	s H_2O, EtOH, eth
7087	2-Methyl-1-butene		C_5H_{10}	563-46-2	70.133	liq	-137.53(0.02)	31.1(0.4)	0.6504^{20}	1.3778^{20}	i H_2O; s EtOH, eth, bz, ctc
7088	3-Methyl-1-butene		C_5H_{10}	563-45-1	70.133	vol liq or gas	-168.41(0.02)	20.1(0.2)	0.6213^{25}	1.3643^{20}	i H_2O; msc EtOH, eth; s bz
7089	2-Methyl-2-butene		C_5H_{10}	513-35-9	70.133	liq	-133.72(0.02)	38.5(0.4)	0.6623^{20}	1.3874^{20}	i H_2O; s EtOH, eth, bz, ctc; vs lig

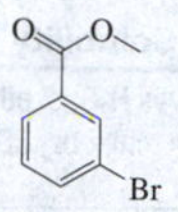

7052
Methyl 3-bromobenzoate

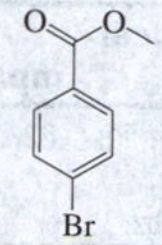

7053
Methyl 4-bromobenzoate

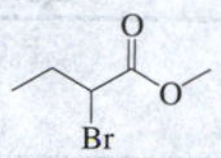

7054
Methyl 2-bromobutanoate

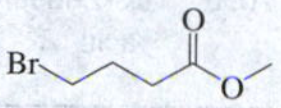

7055
Methyl 4-bromobutanoate

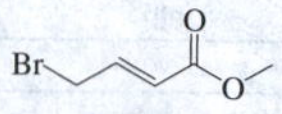

7056
Methyl 4-bromo-2-butenoate

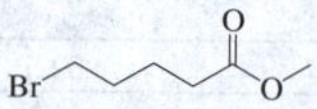

7057
Methyl 5-bromopentanoate

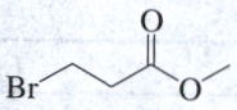

7058
Methyl 3-bromopropanoate

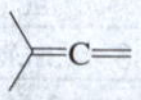

7059
3-Methyl-1,2-butadiene

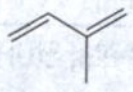

7060
2-Methyl-1,3-butadiene

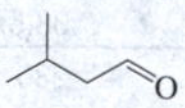

7061
3-Methylbutanal

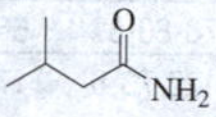

7062
3-Methylbutanamide

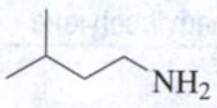

7063
3-Methyl-1-butanamine

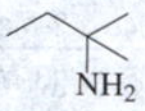

7064
2-Methyl-2-butanamine

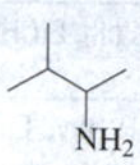

7065
3-Methyl-2-butanamine

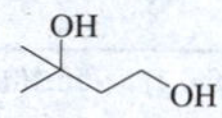

7066
3-Methyl-1,3-butanediol

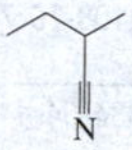

7067
2-Methylbutanenitrile

7068
3-Methylbutanenitrile

7069
2-Methyl-1-butanethiol, (+)

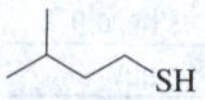

7070
3-Methyl-1-butanethiol

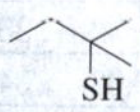

7071
2-Methyl-2-butanethiol

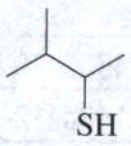

7072
3-Methyl-2-butanethiol

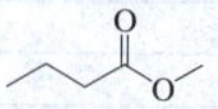

7073
Methyl butanoate

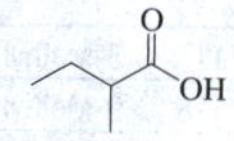

7074
2-Methylbutanoic acid

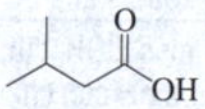

7075
3-Methylbutanoic acid

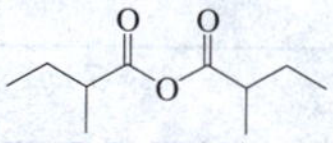

7076
3-Methylbutanoic anhydride

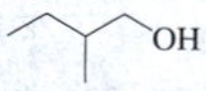

7077
2-Methyl-1-butanol, (±)-

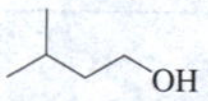

7078
3-Methyl-1-butanol

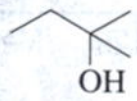

7079
2-Methyl-2-butanol

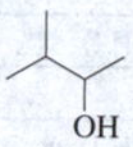

7080
3-Methyl-2-butanol, (±)-

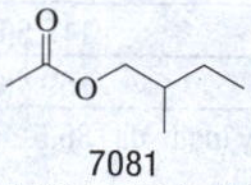

7081
2-Methyl-1-butanol acetate

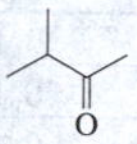

7082
3-Methyl-2-butanone

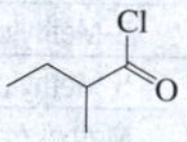

7083
2-Methylbutanoyl chloride, (±)-

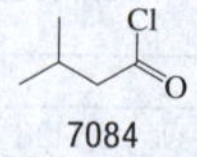

7084
3-Methylbutanoyl chloride

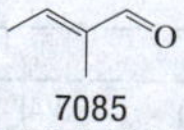

7085
trans-2-Methyl-2-butenal

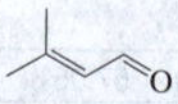

7086
3-Methyl-2-butenal

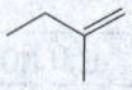

7087
2-Methyl-1-butene

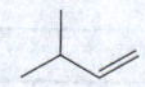

7088
3-Methyl-1-butene

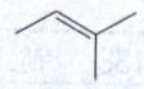

7089
2-Methyl-2-butene

No.	Name	Synonym	Mol. form.	CAS Reg. No.	Mol. wt.	Physical form	mp/°C	bp/°C	den/ g cm^{-3}	n_D	Solubility
7090	*cis*-2-Methyl-2-butenedioic acid	Citraconic acid	$C_5H_6O_4$	498-23-7	130.100	nd (eth-lig) tcl pr (eth-bz)	83.2(0.6)		1.617^{25}		vs H_2O; sl eth, chl; i bz, CS_2
7091	3-Methyl-2-butenenitrile		C_5H_7N	4786-24-7	81.117	liq		141			
7092	Methyl *cis*-2-butenoate	Methyl isocrotonate	$C_5H_8O_2$	4358-59-2	100.117			118		1.4175^{20}	
7093	Methyl *trans*-2-butenoate	Methyl crotonate	$C_5H_8O_2$	623-43-8	100.117	liq	-42	119(6)	0.9444^{20}	1.4242^{20}	i H_2O; vs EtOH, eth
7094	*cis*-2-Methyl-2-butenoic acid	Angelic acid	$C_5H_8O_2$	565-63-9	100.117	mcl pr or nd	45.5	185	0.9834^{49}	1.4434^{47}	sl H_2O; s EtOH; vs eth
7095	*trans*-2-Methyl-2-butenoic acid	Tiglic acid	$C_5H_8O_2$	80-59-1	100.117	tab (w)	63.5(0.5)	198.5	0.9641^{76}	1.4330^{76}	s H_2O; vs EtOH, eth
7096	3-Methyl-2-butenoic acid		$C_5H_8O_2$	541-47-9	100.117		69.5	197	1.0062^{24}		
7097	3-Methyl-2-buten-1-ol		$C_5H_{10}O$	556-82-1	86.132			144(6)	0.848^{25}	1.4412^{20}	
7098	3-Methyl-3-buten-1-ol		$C_5H_{10}O$	763-32-6	86.132			132(2)			
7099	2-Methyl-3-buten-2-ol		$C_5H_{10}O$	115-18-4	86.132	liq	-41(4)	99.3(0.8)	0.82^{20}		
7100	3-Methyl-3-buten-2-ol		$C_5H_{10}O$	10473-14-0	86.132			114	0.8531^{17}	1.4288^{17}	
7101	3-Methyl-3-buten-2-one	Isopropenyl methyl ketone	C_5H_8O	814-78-8	84.117	liq	-53.6(0.3)	97(3)	0.8527^{20}	1.4220^{20}	vs EtOH
7102	3-Methyl-2-butenoyl chloride		C_5H_7ClO	3350-78-5	118.562			146	1.065^{25}	1.4770^{20}	
7103	(3-Methyl-2-butenyl)guanidine	Galegine	$C_6H_{13}N_3$	543-83-9	127.187	hyg	62.5	dec			vs H_2O, EtOH
7104	2-Methyl-1-buten-3-yne	Isopropenylacetylene	C_5H_6	78-80-8	66.102	liq	-113	33(5)	0.6801^{11}	1.4140^{20}	s chl
7105	[(3-Methylbutoxy)methyl]-benzene		$C_{12}H_{18}O$	122-73-6	178.270			236	0.909^{20}	1.4792^{20}	vs eth, EtOH
7106	1-[2-(3-Methylbutoxy)-2-phenylethyl]pyrrolidine	Amixetrine	$C_{17}H_{27}NO$	24622-72-8	261.402			121^{2}		1.4978^{22}	
7107	2-Methylbutyl acrylate		$C_8H_{14}O_2$	44914-03-6	142.196			158(9)	0.8936^{20}	1.4240^{20}	vs eth, EtOH
7108	3-Methylbutyl benzoate	Isopentyl benzoate	$C_{12}H_{16}O_2$	94-46-2	192.254			259(4)	0.993^{15}		vs EtOH
7109	3-Methylbutyl 2-chloropropanoate		$C_8H_{15}ClO_2$	62108-69-4	178.657			208	1.0050^{20}	1.4289^{20}	
7110	3-Methylbutyl 3-chloropropanoate		$C_8H_{15}ClO_2$	62108-70-7	178.657			208	1.0171^{20}	1.4343^{20}	vs eth, EtOH
7111	Methyl *tert*-butyl ether	*tert*-Butyl methyl ether	$C_5H_{12}O$	1634-04-4	88.148	liq	-108.6(0.1)	55.1(0.1)	0.7353^{25}	1.3664^{25}	s H_2O; vs EtOH, eth
7112	3-Methylbutyl nitrate	Isopentyl nitrate	$C_5H_{11}NO_3$	543-87-3	133.146			148	0.996^{22}	1.4122^{21}	
7113	2-Methyl-3-butyn-2-amine		C_5H_9N	2978-58-7	83.132		18	79.5	0.79^{25}	1.4235^{20}	
7114	3-Methyl-1-butyne		C_5H_8	598-23-2	68.118	vol liq or gas	-89.7	28(4)	0.6660^{20}	1.3723^{20}	i H_2O; msc EtOH, eth
7115	2-Methyl-3-butyn-2-ol	1,1-Dimethylpropargyl alcohol	C_5H_8O	115-19-5	84.117		3.0(0.3)	105.7(0.5)	0.8618^{20}	1.4207^{20}	
7116	Methyl carbamate		$C_2H_5NO_2$	598-55-0	75.067	nd	55.4(0.6)	191(4)	1.1361^{56}	1.4125^{56}	vs H_2O; vs EtOH, eth
7117	3-Methyl-9*H*-carbazole		$C_{13}H_{11}N$	4630-20-0	181.233	pl (HOAc)	204(3)	365			vs bz, eth
7118	9-Methyl-9*H*-carbazole		$C_{13}H_{11}N$	1484-12-4	181.233	nd, lf (al)	89.32(0.03)	343.8(0.3)			vs eth
7119	Methyl chloroacetate		$C_3H_5ClO_2$	96-34-4	108.524	liq	-32.3(0.6)	130(2)	1.236^{20}	1.4218^{20}	vs ace, bz, eth, EtOH
7120	Methyl 2-chloroacrylate		$C_4H_5ClO_2$	80-63-7	120.535			52^{51}	1.189^{20}	1.4420^{20}	vs eth
7121	Methyl 2-chlorobenzoate		$C_8H_7ClO_2$	610-96-8	170.594			234			s EtOH
7122	Methyl 3-chlorobenzoate		$C_8H_7ClO_2$	2905-65-9	170.594		21	229			
7123	Methyl 4-chlorobenzoate		$C_8H_7ClO_2$	1126-46-1	170.594	nd or mcl pr	43.5		1.382^{20}		vs EtOH
7124	Methyl 4-chlorobutanoate		$C_5H_9ClO_2$	3153-37-5	136.577			161(11)	1.1293^{20}	1.4321^{20}	i H_2O; vs EtOH, eth; s ace
7125	Methyl chloroformate		$C_2H_3ClO_2$	79-22-1	94.497			70.5	1.2231^{20}	1.3868^{20}	msc EtOH, eth; s bz, ctc, chl
7126	Methyl 5-chloro-2-hydroxybenzoate		$C_8H_7ClO_3$	4068-78-4	186.593	nd (al)	50	249 dec			vs EtOH
7127	Methyl 5-chloro-2-nitrobenzoate		$C_8H_6ClNO_4$	51282-49-6	215.592	pl (MeOH)	48.5		1.453^{18}		vs MeOH
7128	Methyl chlorooxoacetate		$C_3H_3ClO_3$	5781-53-3	122.507			119	1.3316^{20}	1.4189^{20}	
7129	Methyl 2-chloropropanoate		$C_4H_7ClO_2$	17639-93-9	122.551			132.5	1.0750^{25}		
7130	3-Methylchrysene		$C_{19}H_{14}$	3351-31-3	242.314	lf (bz-peth)	171.9(0.7)				vs EtOH
7131	5-Methylchrysene		$C_{19}H_{14}$	3697-24-3	242.314		117.5(0.5)				i H_2O
7132	6-Methylchrysene		$C_{19}H_{14}$	1705-85-7	242.314		159.4(0.7)				
7133	Methyl *trans*-cinnamate	Methyl *trans*-3-phenyl-2-propenoate	$C_{10}H_{10}O_2$	1754-62-7	162.185	cry (peth, dil al)	36.5	261.9	1.042^{36}	1.5766^{22}	i H_2O; vs EtOH, eth; s ace, bz; sl chl
7134	*trans*-*o*-Methylcinnamic acid		$C_{10}H_{10}O_2$	2373-76-4	162.185	cry (EtOH)	175				
7135	*trans*-*m*-Methylcinnamic acid		$C_{10}H_{10}O_2$	3029-79-6	162.185	cry (w)	115				
7136	*trans*-*p*-Methylcinnamic acid		$C_{10}H_{10}O_2$	1866-39-3	162.185		198.5				
7137	Methyclothiazide		$C_9H_{11}Cl_2N_3O_4S_2$	135-07-9	360.237	cry (EtOH aq)	225				i H_2O, bz, chl; sl MeOH; vs ace, py
7138	Methyl cyanate		C_2H_3NO	1768-34-9	57.051	unstab gas	-30				
7139	Methyl cyanoacetate		$C_4H_5NO_2$	105-34-0	99.089	liq	-13.1(0.2)	203(4)	1.1225^{25}	1.4176^{20}	vs eth, EtOH

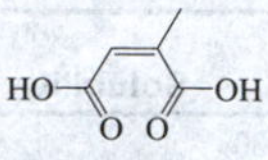
7090
cis-2-Methyl-2-butenedioic acid

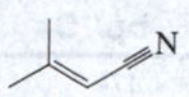
7091
3-Methyl-2-butenenitrile

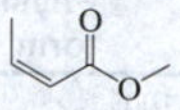
7092
Methyl *cis*-2-butenoate

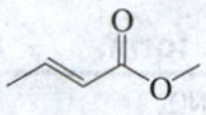
7093
Methyl *trans*-2-butenoate

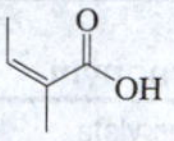
7094
cis-2-Methyl-2-butenoic acid

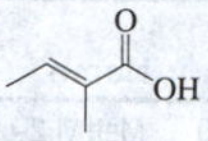
7095
trans-2-Methyl-2-butenoic acid

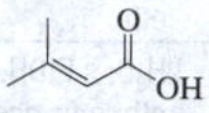
7096
3-Methyl-2-butenoic acid

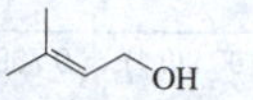
7097
3-Methyl-2-buten-1-ol

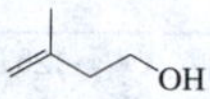
7098
3-Methyl-3-buten-1-ol

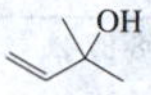
7099
2-Methyl-3-buten-2-ol

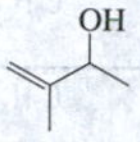
7100
3-Methyl-3-buten-2-ol

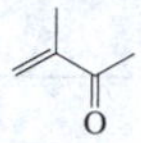
7101
3-Methyl-3-buten-2-one

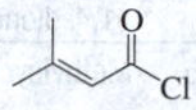
7102
3-Methyl-2-butenoyl chloride

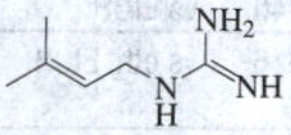
7103
(3-Methyl-2-butenyl)guanidine

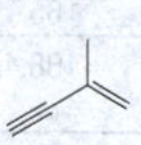
7104
2-Methyl-1-buten-3-yne

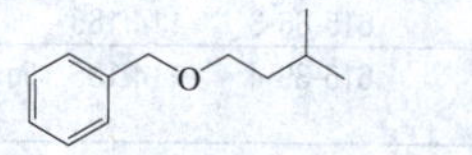
7105
[(3-Methylbutoxy)methyl]benzene

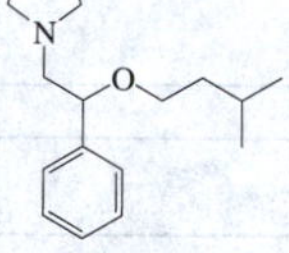
7106
1-[2-(3-Methylbutoxy)-2-phenylethyl]pyrrolidine

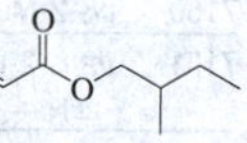
7107
2-Methylbutyl acrylate

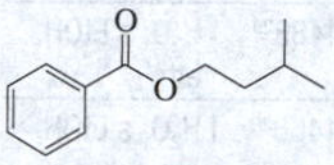
7108
3-Methylbutyl benzoate

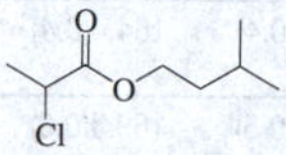
7109
3-Methylbutyl 2-chloropropanoate

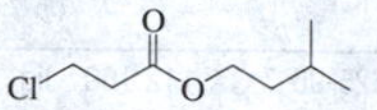
7110
3-Methylbutyl 3-chloropropanoate

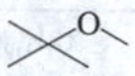
7111
Methyl *tert*-butyl ether

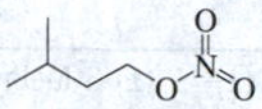
7112
3-Methylbutyl nitrate

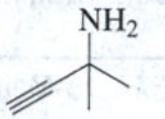
7113
2-Methyl-3-butyn-2-amine

7114
3-Methyl-1-butyne
7115
2-Methyl-3-butyn-2-ol

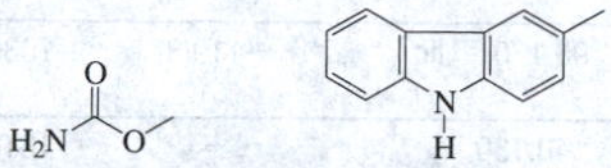
7116
Methyl carbamate
7117
3-Methyl-9*H*-carbazole

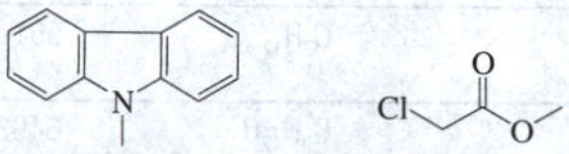
7118
9-Methyl-9*H*-carbazole
7119
Methyl chloroacetate

7120
Methyl 2-chloroacrylate
7121
Methyl 2-chlorobenzoate

7122
Methyl 3-chlorobenzoate
7123
Methyl 4-chlorobenzoate

7124
Methyl 4-chlorobutanoate
7125
Methyl chloroformate

7126
Methyl 5-chloro-2-hydroxybenzoate
7127
Methyl 5-chloro-2-nitrobenzoate

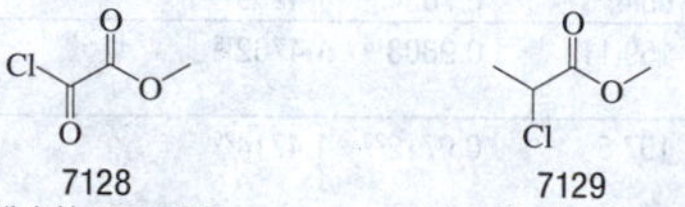
7128
Methyl chlorooxoacetate
7129
Methyl 2-chloropropanoate

7130
3-Methylchrysene
7131
5-Methylchrysene

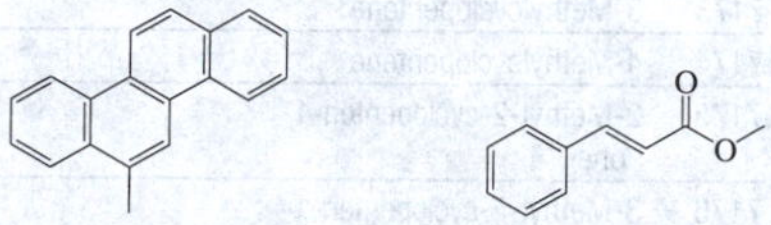
7132
6-Methylchrysene
7133
Methyl *trans*-cinnamate

7134
trans-*o*-Methylcinnamic acid
7135
trans-*m*-Methylcinnamic acid

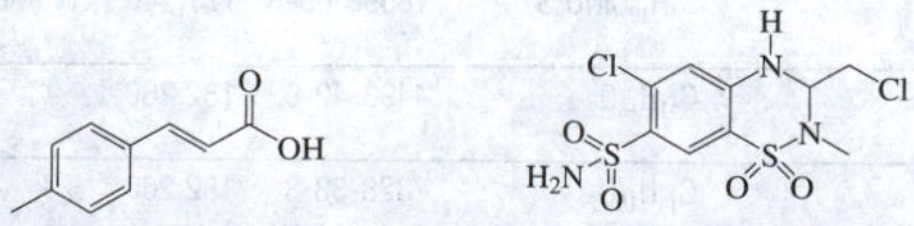
7136
trans-*p*-Methylcinnamic acid
7137
Methyclothiazide

7138
Methyl cyanate
7139
Methyl cyanoacetate

Organic

No.	Name	Synonym	Mol. form.	CAS Reg. No.	Mol. wt.	Physical form	mp/°C	bp/°C	den/ g cm^{-3}	n_D	Solubility
7140	Methyl 2-cyanoacrylate	Mecrylate	$C_5H_5NO_2$	137-05-3	111.100			47^{2}	1.1012^{20}	1.4430	
7141	Methylcyclobutane		C_5H_{10}	598-61-8	70.133	liq	-161.5	37(3)	0.6884^{20}	1.3866^{20}	i H_2O; msc EtOH, eth; s ace, bz, peth
7142	Methyl cyclobutanecarboxylate		$C_6H_{10}O_2$	765-85-5	114.142			135.5			
7143	2-Methyl-1,3-cyclohexadiene	4,5-Dihydrotoluene	C_7H_{10}	1489-57-2	94.154			110(19)	0.8260^{18}	1.4662^{18}	
7144	2-Methyl-2,5-cyclohexadiene-1,4-dione		$C_7H_6O_2$	553-97-9	122.122	ye pl or nd	68(2)	sub	1.08^{75}		sl H_2O; s EtOH, eth
7145	Methylcyclohexane		C_7H_{14}	108-87-2	98.186	liq	-126.6(0.4)	100.9(0.1)	0.7694^{20}	1.4231^{20}	i H_2O; s EtOH, eth; msc ace, bz, lig
7146	Methyl cyclohexanecarboxylate		$C_8H_{14}O_2$	4630-82-4	142.196			183(5)	0.9954^{15}	1.4433^{20}	i H_2O; s EtOH, eth, ace, chl
7147	α-Methylcyclohexanemethanol		$C_8H_{16}O$	1193-81-3	128.212			189	0.928^{25}	1.4656^{20}	vs EtOH, eth; sl ctc
7148	4-Methylcyclohexanemethanol		$C_8H_{16}O$	34885-03-5	128.212			$75^{2.5}$	0.9074^{20}	1.4617^{20}	
7149	1-Methylcyclohexanol		$C_7H_{14}O$	590-67-0	114.185		26.2(0.1)	155	0.9194^{20}	1.4595^{20}	i H_2O; s EtOH, bz, chl
7150	*cis*-2-Methylcyclohexanol		$C_7H_{14}O$	615-38-3	114.185		7	165	0.9360^{20}	1.4640^{20}	vs EtOH
7151	*trans*-2-Methylcyclohexanol, (±)-		$C_7H_{14}O$	615-39-4	114.185	liq	-2.0	168.4(0.2)	0.9247^{20}	1.4616^{20}	vs eth, EtOH
7152	*cis*-3-Methylcyclohexanol, (±)-		$C_7H_{14}O$	5454-79-5	114.185	liq	-5.5	168	0.9155^{20}	1.4752^{20}	vs eth, EtOH
7153	*trans*-3-Methylcyclohexanol, (±)-		$C_7H_{14}O$	7443-55-2	114.185	liq	-0.5	167	0.9214^{30}	1.4580^{20}	vs eth, EtOH
7154	*cis*-4-Methylcyclohexanol		$C_7H_{14}O$	7731-28-4	114.185	liq	-9.2	174(10)	0.9170^{20}	1.4614^{20}	vs eth, EtOH
7155	*trans*-4-Methylcyclohexanol		$C_7H_{14}O$	7731-29-5	114.185			175(3)	0.9118^{21}	1.4561^{20}	sl H_2O; msc EtOH; s eth
7156	2-Methylcyclohexanone, (±)-	(±)-2-Methylcyclohexanone	$C_7H_{12}O$	24965-84-2	112.169	liq	-13.9(0.4)	164.4(0.4)	0.9250^{20}	1.4483^{25}	i H_2O; s EtOH, eth
7157	3-Methylcyclohexanone, (±)-	(±)-3-Methylcyclohexanone	$C_7H_{12}O$	625-96-7	112.169	liq	-73.9(0.5)	169.3(0.6)	0.9136^{20}	1.4456^{20}	i H_2O; s EtOH, eth
7158	4-Methylcyclohexanone		$C_7H_{12}O$	589-92-4	112.169	liq	-40.6(0.5)	170.9(0.5)	0.9138^{20}	1.4451^{20}	i H_2O; s EtOH, eth; sl ctc
7159	1-Methylcyclohexene		C_7H_{12}	591-49-1	96.170	liq	-120.4(0.1)	110.3(0.4)	0.8102^{20}	1.4503^{20}	i H_2O; s eth, bz, ctc
7160	3-Methylcyclohexene, (±)-		C_7H_{12}	56688-75-6	96.170	liq	-123(2)	103(3)	0.7990^{20}	1.4414^{20}	vs bz, eth, chl, peth
7161	4-Methylcyclohexene		C_7H_{12}	591-47-9	96.170	liq	-119(3)	103(2)	0.7991^{20}	1.4414^{20}	i H_2O; s EtOH, eth
7162	Methyl 3-cyclohexene-1-carboxylate		$C_8H_{12}O_2$	6493-77-2	140.180			181(21)	1.0130^{20}	1.4610^{20}	
7163	2-Methyl-2-cyclohexen-1-one		$C_7H_{10}O$	1121-18-2	110.153			178.5	0.966^{20}	1.4833^{20}	s bz
7164	3-Methyl-2-cyclohexen-1-one		$C_7H_{10}O$	1193-18-6	110.153	liq	-21	201	0.9693^{20}	1.49475^{20}	msc H_2O; s bz
7165	3-Methylcyclopentadecanone	Muscone	$C_{16}H_{30}O$	541-91-3	238.408	oily liq		329	0.9221^{17}	1.4802^{17}	vs ace, eth, EtOH
7166	1-Methyl-1,3-cyclopentadiene		C_6H_8	96-39-9	80.128	liq		65(7)	0.81^{20}	1.4512^{20}	
7167	Methylcyclopentane		C_6H_{12}	96-37-7	84.159	liq	-142.419 (0.002)	71.8(0.2)	0.7486^{20}	1.4097^{20}	i H_2O; msc EtOH, eth, ace, bz, lig, ctc
7168	1-Methylcyclopentanol		$C_6H_{12}O$	1462-03-9	100.158	nd	35.4(0.3)	136(8)	0.9044^{23}	1.4429^{23}	
7169	*cis*-2-Methylcyclopentanol		$C_6H_{12}O$	25144-05-2	100.158			148.5	0.9379^{16}	1.4504^{16}	
7170	2-Methylcyclopentanone		$C_6H_{10}O$	1120-72-5	98.142	liq	-75(1)	140(3)	0.9139^{20}	1.4364^{20}	s H_2O; vs EtOH, eth, ace
7171	3-Methylcyclopentanone, (±)-		$C_6H_{10}O$	6195-92-2	98.142	liq	-58.4	144	0.913^{22}	1.4329^{20}	s H_2O; vs EtOH, eth, ace, HOAc
7172	1-Methylcyclopentene		C_6H_{10}	693-89-0	82.143	liq	-126.6(0.2)	75(1)	0.7748^{25}	1.4322^{20}	
7173	3-Methylcyclopentene		C_6H_{10}	1120-62-3	82.143			65(3)	0.7572^{25}	1.4216^{20}	
7174	4-Methylcyclopentene		C_6H_{10}	1759-81-5	82.143	liq	-160.8	68(4)	0.7634^{25}	1.4209^{20}	
7175	2-Methyl-2-cyclopenten-1-one		C_6H_8O	1120-73-6	96.127			159(11)	0.9808^{16}	1.4762^{15}	
7176	3-Methyl-2-cyclopenten-1-one		C_6H_8O	2758-18-1	96.127			157.5	0.9712^{20}	1.4714^{20}	
7177	Methylcyclopropane		C_4H_8	594-11-6	56.107	col gas	-177.2(0.1)	1(2)	0.6912^{-20}		vs eth, EtOH
7178	Methyl cyclopropanecarboxylate		$C_5H_8O_2$	2868-37-3	100.117			112(3)	0.9848^{20}	1.4144^{19}	s ace, chl
7179	α-Methylcyclopropanemethanol		$C_5H_{10}O$	765-42-4	86.132	liq	-32.1(0.6)	123(6)	0.8805^{20}	1.4316^{20}	
7180	Methyl L-cysteine hydrochloride		$C_4H_{10}ClNO_2S$	18598-63-5	171.646	cry (MeOH)	140.5				
7181	Methyl *trans*-2,*cis*-4-decadienoate		$C_{11}H_{18}O_2$	4493-42-9	182.260			$71^{0.15}$	0.9128^{22}	1.4874^{22}	
7182	Methyl *trans*-2,*trans*-4-decadienoate		$C_{11}H_{18}O_2$	7328-33-8	182.260			87^{13}	0.9082^{22}	1.4918^{22}	

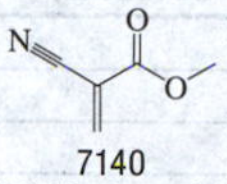
7140 Methyl 2-cyanoacrylate

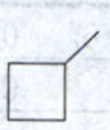
7141 Methylcyclobutane

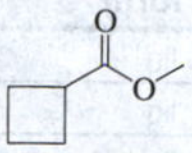
7142 Methyl cyclobutanecarboxylate

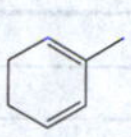
7143 2-Methyl-1,3-cyclohexadiene

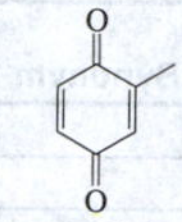
7144 2-Methyl-2,5-cyclohexadiene-1,4-dione

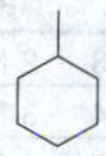
7145 Methylcyclohexane

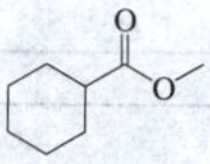
7146 Methyl cyclohexanecarboxylate

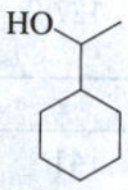
7147 α-Methylcyclohexanemethanol

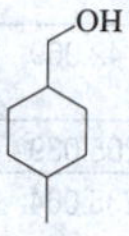
7148 4-Methylcyclohexanemethanol

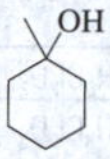
7149 1-Methylcyclohexanol

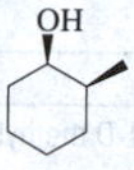
7150 *cis*-2-Methylcyclohexanol

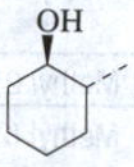
7151 *trans*-2-Methylcyclohexanol, (±)-

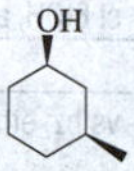
7152 *cis*-3-Methylcyclohexanol, (±)-

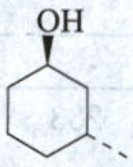
7153 *trans*-3-Methylcyclohexanol, (±)-

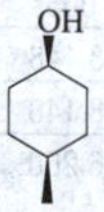
7154 *cis*-4-Methylcyclohexanol

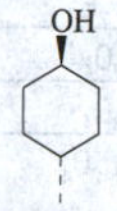
7155 *trans*-4-Methylcyclohexanol

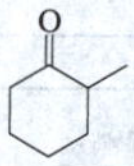
7156 2-Methylcyclohexanone, (±)-

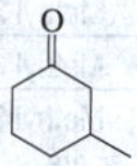
7157 3-Methylcyclohexanone, (±)-

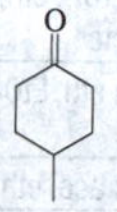
7158 4-Methylcyclohexanone

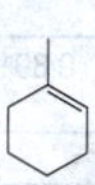
7159 1-Methylcyclohexene

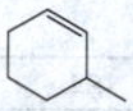
7160 3-Methylcyclohexene, (±)-

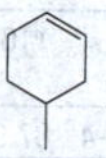
7161 4-Methylcyclohexene

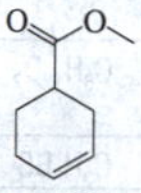
7162 Methyl 3-cyclohexene-1-carboxylate

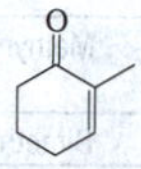
7163 2-Methyl-2-cyclohexen-1-one

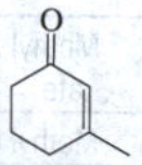
7164 3-Methyl-2-cyclohexen-1-one

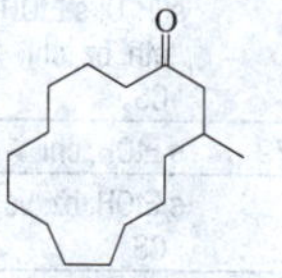
7165 3-Methylcyclopentadecanone

7166 1-Methyl-1,3-cyclopentadiene

7167 Methylcyclopentane

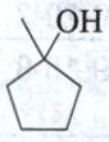
7168 1-Methylcyclopentanol

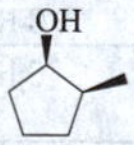
7169 *cis*-2-Methylcyclopentanol

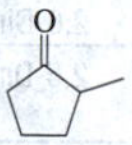
7170 2-Methylcyclopentanone

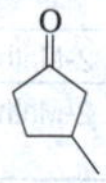
7171 3-Methylcyclopentanone, (±)-

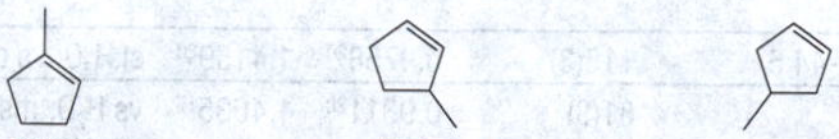
7172 1-Methylcyclopentene

7173 3-Methylcyclopentene

7174 4-Methylcyclopentene

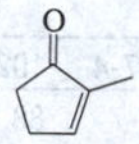
7175 2-Methyl-2-cyclopenten-1-one

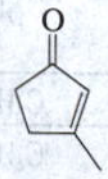
7176 3-Methyl-2-cyclopenten-1-one

7177 Methylcyclopropane

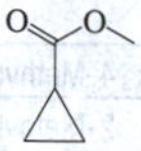
7178 Methyl cyclopropanecarboxylate

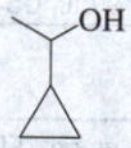
7179 α-Methylcyclopropanemethanol

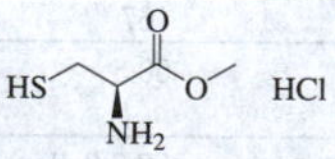
7180 Methyl L-cysteine hydrochloride

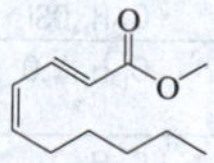
7181 Methyl *trans*-2,*cis*-4-decadienoate

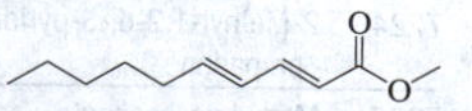
7182 Methyl *trans*-2,*trans*-4-decadienoate

No.	Name	Synonym	Mol. form.	CAS Reg. No.	Mol. wt.	Physical form	mp/°C	bp/°C	den/ g cm^{-3}	n_D	Solubility
7183	2-Methyldecane		$C_{11}H_{24}$	6975-98-0	156.309	liq	-48.83(0.03)	189.2(0.6)	0.7368^{20}	1.4154^{20}	
7184	3-Methyldecane		$C_{11}H_{24}$	13151-34-3	156.309	liq	-79.5(0.5)	192(1)	0.7422^{20}	1.4177^{20}	
7185	4-Methyldecane		$C_{11}H_{24}$	2847-72-5	156.309	liq	-92.8(0.5)	188(1)		1.4352^{20}	
7186	Methyl decanoate		$C_{11}H_{22}O_2$	110-42-9	186.292	liq	-12.8(0.2)	233(1)	0.8730^{20}	1.4259^{20}	i H_2O; vs EtOH, eth; sl ctc; msc chl
7187	Methyl demeton		$C_6H_{15}O_3PS_2$	8022-00-2	230.285	ye liq			1.20^{20}	1.5063^{20}	i H_2O; s os
7188	Methyldiborane(6)		CH_8B_2	23777-55-1	41.697	unstab gas					s eth
7189	Methyl 2,3-dibromopropanoate		$C_4H_6Br_2O_2$	1729-67-5	245.898			206	1.9333^{20}	1.5127^{20}	s EtOH
7190	Methyl dichloroacetate		$C_3H_4Cl_2O_2$	116-54-1	142.969	liq	-51.9(0.2)	127(9)	1.3774^{20}	1.4429^{20}	i H_2O; s EtOH, ctc
7191	Methyl 2,5-dichlorobenzoate		$C_8H_6Cl_2O_2$	2905-69-3	205.039	cry	38				
7192	Methyl (2,4-dichlorophenoxy)-acetate	2,4-D methyl ester	$C_9H_8Cl_2O_3$	1928-38-7	235.064		119	141^{18}			
7193	Methyl (3,4-dichlorophenyl)-carbamate	Swep	$C_8H_7Cl_2NO_2$	1918-18-9	220.054	nd	109.6(0.5)				
7194	Methyl 2,3-dichloropropanoate		$C_4H_6Cl_2O_2$	3674-09-7	156.996			92^{50}	1.3282^{20}		vs ace, eth, EtOH
7195	Methyldifluoroarsine		CH_3AsF_2	420-24-6	127.954	liq, fumes in air	-29.7	76.5	1.924^{18}		
7196	Methyldifluorophosphine	(Difluoro)methylphosphine	CH_3F_2P	753-59-3	84.006	gas	-110	-28			
7197	Methyl 2,4-dihydroxybenzoate		$C_8H_8O_4$	2150-47-2	168.148		116.5				sl EtOH, ace
7198	Methyl 3,5-dihydroxybenzoate		$C_8H_8O_4$	2150-44-9	168.148		165				
7199	Methyl 3,4-dimethoxybenzoate		$C_{10}H_{12}O_4$	2150-38-1	196.200	nd (dil al)	60.8	283			vs bz, eth, EtOH
7200	Methyldimethoxysilane		$C_3H_{10}O_2Si$	16881-77-9	106.196			61			
7201	3-Methyl-4'-(dimethylamino)-azobenzene		$C_{15}H_{17}N_3$	55-80-1	239.316	oran cry	122				
7202	2-Methyl-*N,N*-dimethylaniline	*N,N*-Dimethyl-*o*-toluidine	$C_9H_{13}N$	609-72-3	135.206	liq	-61(1)	185(2)	0.9286^{20}	1.5152^{20}	vs eth, EtOH
7203	3-Methyl-*N,N*-dimethylaniline	*N,N*-Dimethyl-*m*-toluidine	$C_9H_{13}N$	121-72-2	135.206			194(16)	0.9410^{20}	1.5492^{20}	msc EtOH, eth
7204	4-Methyl-*N,N*-dimethylaniline	*N,N*-Dimethyl-*p*-toluidine	$C_9H_{13}N$	99-97-8	135.206			211(4)	0.9366^{20}	1.5366^{20}	i H_2O; msc EtOH, eth; s ctc
7205	Methyl 2,2-dimethylpropanoate	Methyl 2,2-dimethylpropionate	$C_6H_{12}O_2$	598-98-1	116.158			102(1)	0.891^{0}	1.3905^{20}	vs eth, EtOH
7206	Methyl dimethylthioborane	Dimethyl(methylthio)borane	C_3H_9BS	19163-05-4	87.979	liq	-84	71			vs ace, eth
7207	2-Methyl-3,5-dinitrobenzamide	Dinitolmide	$C_8H_7N_3O_5$	148-01-6	225.159	cry	181				
7208	1-Methyl-2,3-dinitrobenzene	2,3-Dinitrotoluene	$C_7H_6N_2O_4$	602-01-7	182.134		56.6(0.2)				i H_2O; s EtOH, eth; sl chl
7209	1-Methyl-2,4-dinitrobenzene	2,4-Dinitrotoluene	$C_7H_6N_2O_4$	121-14-2	182.134	ye nd or mcl pr (CS_2)	69.6(0.9)	300 dec	1.3208^{71}	1.442	i H_2O; s EtOH, eth, chl, bz; vs ace, py
7210	1-Methyl-3,5-dinitrobenzene	3,5-Dinitrotoluene	$C_7H_6N_2O_4$	618-85-9	182.134	ye orth nd (HOAc)	82.7(0.9)	325(5)	1.2772^{111}		sl H_2O; s EtOH, eth, bz, chl, CS_2
7211	2-Methyl-1,3-dinitrobenzene	2,6-Dinitrotoluene	$C_7H_6N_2O_4$	606-20-2	182.134	orth nd (al)	63.8(0.9)	285	1.2833^{111}	1.479	s EtOH, chl
7212	2-Methyl-1,4-dinitrobenzene	2,5-Dinitrotoluene	$C_7H_6N_2O_4$	619-15-8	182.134	nd (al)	52.5		1.282^{111}		s EtOH, bz; vs CS_2
7213	4-Methyl-1,2-dinitrobenzene	3,4-Dinitrotoluene	$C_7H_6N_2O_4$	610-39-9	182.134	ye nd (CS_2)	58.3(0.9)		1.2594^{111}		i H_2O; s EtOH, CS_2; sl chl
7214	2-Methyl-4,6-dinitrophenol	4,6-Dinitro-*o*-cresol	$C_7H_6N_2O_5$	534-52-1	198.133	ye pr or nd (al)	86(1)				sl H_2O, peth; s EtOH, eth, ace, chl
7215	4-Methyl-2,6-dinitrophenol	2,6-Dinitro-*p*-cresol	$C_7H_6N_2O_5$	609-93-8	198.133	ye nd (eth, peth)	85				i H_2O; s EtOH, eth, bz
7216	Methyldioctylamine	*N*-Methyl-*N*-octyl-1-octanamine	$C_{17}H_{37}N$	4455-26-9	255.483		-30.1	158^{10}		1.4424^{20}	
7217	4-Methyl-1,3-dioxane		$C_5H_{10}O_2$	1120-97-4	102.132	liq	-44.5	115(3)	0.9758^{20}	1.4159^{20}	sl H_2O; vs os
7218	2-Methyl-1,3-dioxolane		$C_4H_8O_2$	497-26-7	88.106			81(3)	0.9811^{20}	1.4035^{17}	vs H_2O; msc EtOH, eth
7219	4-Methyl-1,3-dioxolane		$C_4H_8O_2$	1072-47-5	88.106	liq		84(19)	0.99^{20}	1.3980^{20}	
7220	Methyldiphenylamine	*N*-Methyl-*N*-phenylbenzenamine	$C_{13}H_{13}N$	552-82-9	183.249	liq	-7.6(0.3)	294(6)	1.0476^{20}	1.6193^{20}	i H_2O; sl EtOH, MeOH; s ctc
7221	4-Methyl-2,4-diphenyl-1-pentene		$C_{18}H_{20}$	6362-80-7	236.352	liq		172^{8}	0.99^{25}		
7222	Methyldiphenylsilane		$C_{13}H_{14}Si$	776-76-1	198.336			93.5^{1}	0.996^{20}	1.5694^{20}	s ctc
7223	Methyldiphenylsilanol		$C_{13}H_{14}OSi$	778-25-6	214.335		167	184^{24}	1.0840^{25}		s ctc, CS_2
7224	2-Methyl-1,2-di-3-pyridinyl-1-propanone	Metyrapone	$C_{14}H_{14}N_2O$	54-36-4	226.273		50.5				
7225	Methyl docosanoate	Methyl behenate	$C_{23}H_{46}O_2$	929-77-1	354.610	nd (ace)	53.3(0.4)	402(16)		1.4339^{60}	vs eth, EtOH

Organic

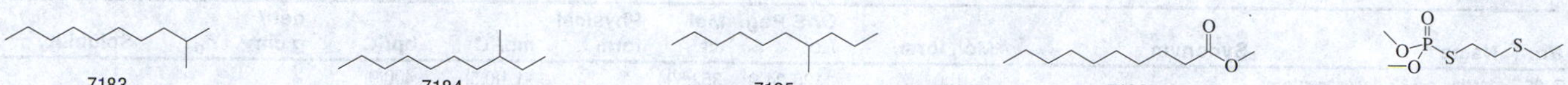

7183 2-Methyldecane | 7184 3-Methyldecane | 7185 4-Methyldecane | 7186 Methyl decanoate | 7187 Methyl demeton

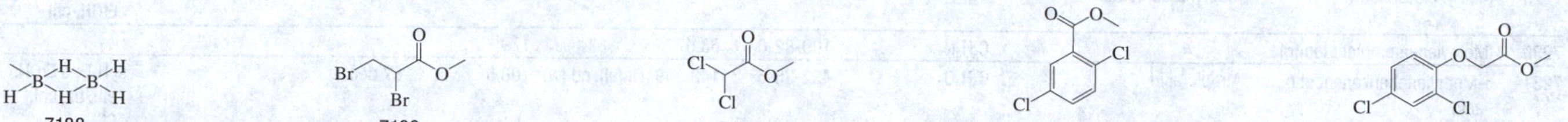

7188 Methyldiborane(6) | 7189 Methyl 2,3-dibromopropanoate | 7190 Methyl dichloroacetate | 7191 Methyl 2,5-dichlorobenzoate | 7192 Methyl (2,4-dichlorophenoxy)acetate

7193 Methyl (3,4-dichlorophenyl)carbamate | 7194 Methyl 2,3-dichloropropanoate | 7195 Methyldifluoroarsine | 7196 Methyldifluorophosphine | 7197 Methyl 2,4-dihydroxybenzoate | 7198 Methyl 3,5-dihydroxybenzoate

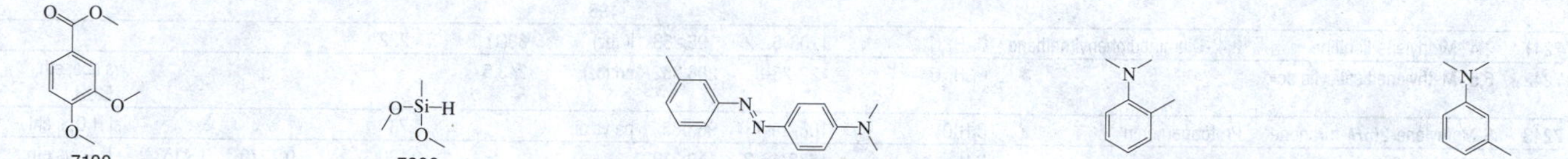

7199 Methyl 3,4-dimethoxybenzoate | 7200 Methyldimethoxysilane | 7201 3-Methyl-4'-(dimethylamino)azobenzene | 7202 2-Methyl-*N,N*-dimethylaniline | 7203 3-Methyl-*N,N*-dimethylaniline

7204 4-Methyl-*N,N*-dimethylaniline | 7205 Methyl 2,2-dimethylpropanoate | 7206 Methyl dimethylthioborane | 7207 2-Methyl-3,5-dinitrobenzamide | 7208 1-Methyl-2,3-dinitrobenzene | 7209 1-Methyl-2,4-dinitrobenzene

7210 1-Methyl-3,5-dinitrobenzene | 7211 2-Methyl-1,3-dinitrobenzene | 7212 2-Methyl-1,4-dinitrobenzene | 7213 4-Methyl-1,2-dinitrobenzene | 7214 2-Methyl-4,6-dinitrophenol | 7215 4-Methyl-2,6-dinitrophenol

7216 Methyldioctylamine | 7217 4-Methyl-1,3-dioxane | 7218 2-Methyl-1,3-dioxolane | 7219 4-Methyl-1,3-dioxolane | 7220 Methyldiphenylamine

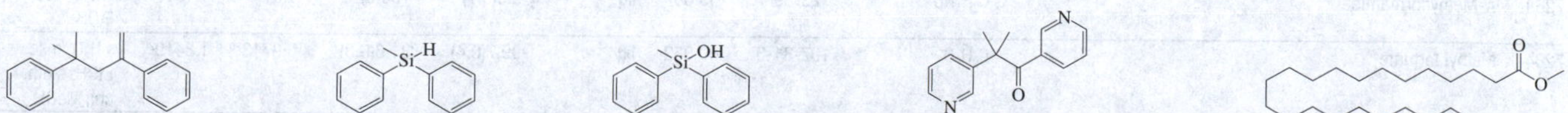

7221 4-Methyl-2,4-diphenyl-1-pentene | 7222 Methyldiphenylsilane | 7223 Methyldiphenylsilanol | 7224 2-Methyl-1,2-di-3-pyridinyl-1-propanone | 7225 Methyl docosanoate

Organic

No.	Name	Synonym	Mol. form.	CAS Reg. No.	Mol. wt.	Physical form	mp/°C	bp/°C	den/ g cm^{-3}	n_D	Solubility
7226	Methyl *cis*-13-docosenoate		$C_{23}H_{44}O_2$	1120-34-9	352.594		-1.1(0.5)	400(9)			
7227	Methyl dodecanoate	Methyl laurate	$C_{13}H_{26}O_2$	111-82-0	214.344		5.1(0.2)	268(2)	0.8702^{20}	1.4319^{20}	i H_2O; msc EtOH, eth, ace, bz; s chl, ctc
7228	2-Methyldodecanoic acid		$C_{13}H_{26}O_2$	2874-74-0	214.344	pl	22	153^{1}	0.890^{18}		
7229	Methyl eicosanoate	Methyl arachidate	$C_{21}H_{42}O_2$	1120-28-1	326.557	lf (MeOH)	46.4(0.3)	371(15)		1.4317^{60}	vs bz, eth, EtOH, chl
7230	(Methyleneamino)acetonitrile		$C_3H_4N_2$	109-82-0	68.077		129				
7231	α-Methylenebenzeneacetic acid	Atropic acid	$C_9H_8O_2$	492-38-6	148.159	lf (al), nd (w)	106.5	267 dec			sl H_2O; s EtOH, eth, bz, chl, CS_2
7232	Methylenebis(4-cyclohexyliso-cyanate)		$C_{15}H_{22}N_2O_2$	5124-30-1	262.348	liq			1.066	1.4970^{20}	
7233	4,4'-Methylenebis[2,6-di-*tert*-butylphenol]	Bis(3,5-di-*tert*-butyl-4-hydroxyphenyl)methane	$C_{29}H_{44}O_2$	118-82-1	424.658		154	289^{40}			
7234	4,4'-Methylenebis(*N*-methyl-aniline)	*N,N'*-Dimethyl-4,4'-diaminodi-phenylmethane	$C_{15}H_{18}N_2$	1807-55-2	226.317						s ctc, CS_2
7235	Methylene blue		$C_{16}H_{18}ClN_3S$	61-73-4	319.852	dk grn cry or pow (chl-eth)					s H_2O, EtOH, chl; i eth; sl py
7236	Methylenecyclobutane		C_5H_8	1120-56-5	68.118	liq	-134.52(0.07)	42(1)	0.7401^{20}	1.4210^{20}	
7237	Methylenecyclohexane		C_7H_{12}	1192-37-6	96.170	liq	-104.7(0.1)	103(3)	0.8074^{20}	1.4523^{20}	i H_2O; s eth, bz, chl
7238	2-Methylenecyclohexanol		$C_7H_{12}O$	4065-80-9	112.169			83^{13}	0.955^{20}	1.4843^{20}	
7239	Methylenecyclopentane		C_6H_{10}	1528-30-9	82.143			75(3)	0.7787^{20}	1.4355^{20}	s bz, chl
7240	Methylenecyclopropene		C_4H_4	4095-06-1	52.075	solid stab at -196					
7241	2,4'-Methylenedianiline	2,4'-Diaminodiphenylmethane	$C_{13}H_{14}N_2$	1208-52-2	198.263	lf (bz)	88(1)	222^{9}			
7242	5,5'-Methylenedisalicylic acid		$C_{15}H_{12}O_6$	122-25-8	288.252	nd (bz)	243.5				vs ace, eth, EtOH
7243	5-Methylene-2(5*H*)-furanone	Protoanemonin	$C_5H_4O_2$	108-28-1	96.085	pa ye oil		73^{11}			sl H_2O; s chl
7244	3-Methyleneheptane		C_8H_{16}	1632-16-2	112.213			120(3)	0.7270^{20}	1.4157^{20}	i H_2O; vs eth, bz, peth
7245	4-Methylene-1-isopropylbicy-clo[3.1.0]hexan-3-ol, [1*S*-(1α,3β,5α)]	4(10)-Thujene-3-ol	$C_{10}H_{16}O$	471-16-9	152.233			208	0.9488^{19}	1.4871^{25}	s eth
7246	4-Methylene-1-isopropylcy-clohexene		$C_{10}H_{16}$	99-84-3	136.234			173.5	0.838^{22}	1.4754^{22}	
7247	2-Methylenepentanedinitrile	2,4-Dicyano-1-butene	$C_6H_6N_2$	1572-52-7	106.125			103^{5}		1.4561^{20}	s chl
7248	Methylene thiocyanate	Dithiocyanatomethane	$C_3H_2N_2S_2$	6317-18-6	130.191	solid	102				
7249	2-Methylene-1,3,3-trimethyl-indoline	Fischer's base	$C_{12}H_{15}N$	118-12-7	173.254			244			sl H_2O; s EtOH, eth, bz, chl
7250	*N*-Methylephedrine, [*R*-(*R**,*S**)]	(1*R*,2*S*)-*N*-Methylephedrine	$C_{11}H_{17}NO$	552-79-4	179.259	nd or pl (al, eth)	87.5				i H_2O; s EtOH, eth, MeOH
7251	Methylergonovine	Methylergometrine	$C_{20}H_{25}N_3O_2$	113-42-8	339.432	pr (MeOH, ace)	172				i H_2O; s EtOH, ace
7252	*N*-Methyl-1,2-ethanediamine		$C_3H_{10}N_2$	109-81-9	74.124			115.5(0.6)	0.841^{25}	1.4395^{20}	
7253	*N*-Methyl-2-ethanolamine		C_3H_9NO	109-83-1	75.109			159.24(0.04)	0.937^{20}	1.4385^{20}	msc H_2O, EtOH, eth
7254	1-(1-Methylethoxy)butane	Butyl isopropyl ether	$C_7H_{16}O$	1860-27-1	116.201			107(9)	0.7594^{15}	1.3870^{15}	i H_2O; s EtOH, eth, ace, con sulf
7255	2-[2-(1-Methylethoxy)ethyl]-pyridine		$C_{10}H_{15}NO$	70715-19-4	165.232			133^{50}	0.9502^{25}	1.4820^{25}	vs H_2O
7256	1-(1-Methylethoxy)propane		$C_6H_{14}O$	627-08-7	102.174			82(8)	0.7370^{20}	1.376^{21}	sl H_2O; vs EtOH; s eth, ace
7257	1-(1-Methylethoxy)-2-propa-nol	1-Isopropoxy-2-propanol	$C_6H_{14}O_2$	3944-36-3	118.174			137.5	0.879^{20}	1.4070^{20}	
7258	Methyl 2-ethylacetoacetate		$C_7H_{12}O_3$	51756-08-2	144.168			182	0.995^{14}		vs ace, eth, EtOH
7259	5-(1-Methylethylidene)-1,3-cyclopentadiene		C_8H_{10}	2175-91-9	106.165		1(2)	155	0.881^{20}	1.5474^{20}	
7260	1-Methyl-9*H*-fluorene		$C_{14}H_{12}$	1730-37-6	180.245		87				
7261	9-Methyl-9*H*-fluorene		$C_{14}H_{12}$	2523-37-7	180.245	pr	45(1)	155^{15}	1.0263^{66}	1.610^{66}	i H_2O; s EtOH, eth, ace, bz, chl
7262	Methyl fluorosulfonate		CH_3FO_3S	421-20-5	114.096	col liq	-95	93	1.412	1.3326^{20}	
7263	*N*-Methylformamide		C_2H_5NO	123-39-7	59.067	liq	-2.5(0.7)	186(5)	1.011^{19}	1.4319^{20}	vs H_2O, ace, EtOH
7264	Methyl formate		$C_2H_4O_2$	107-31-3	60.052	liq	-99.7(0.4)	31.6(0.3)	0.9713^{20}	1.3419^{20}	vs H_2O; msc EtOH; s eth, chl, MeOH
7265	Methyl 4-formylbenzoate		$C_9H_8O_3$	1571-08-0	164.158	nd (w)	63	265			

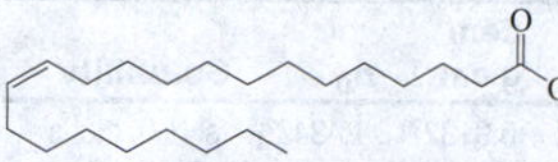
7226
Methyl *cis*-13-docosenoate

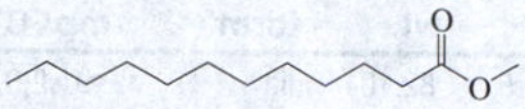
7227
Methyl dodecanoate

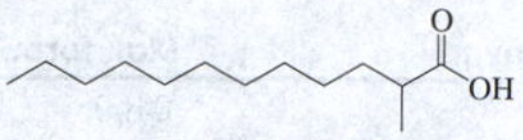
7228
2-Methyldodecanoic acid

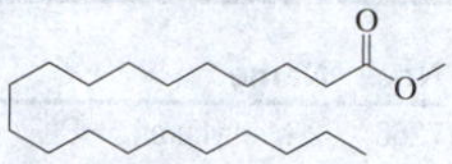
7229
Methyl eicosanoate

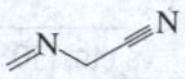
7230
(Methyleneamino)acetonitrile

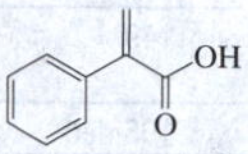
7231
α-Methylenebenzeneacetic acid

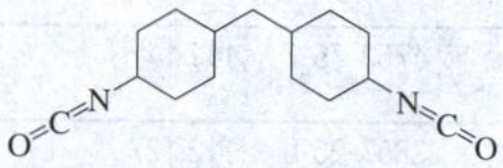
7232
Methylenebis(4-cyclohexylisocyanate)

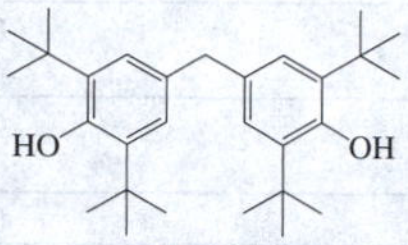
7233
4,4'-Methylenebis[2,6-di-*tert*-butylphenol]

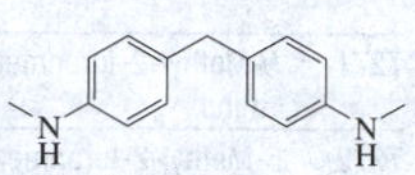
7234
4,4'-Methylenebis(*N*-methylaniline)

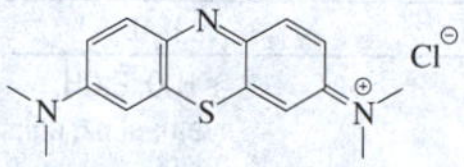
7235
Methylene blue

7236
Methylenecyclobutane

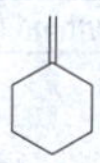
7237
Methylenecyclohexane

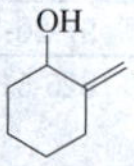
7238
2-Methylenecyclohexanol

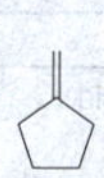
7239
Methylenecyclopentane

7240
Methylenecyclopropene

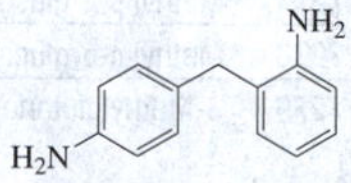
7241
2,4'-Methylenedianiline

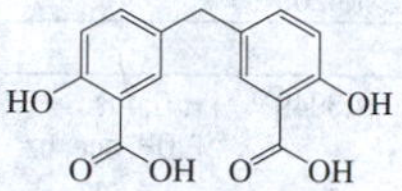
7242
5,5'-Methylenedisalicylic acid

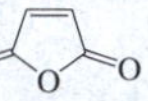
7243
5-Methylene-2(5*H*)-furanone

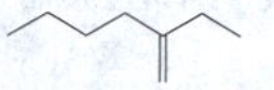
7244
3-Methyleneheptane

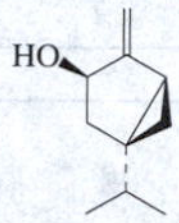
7245
4-Methylene-1-isopropylbicyclo[3.1.0]hexan-3-ol, [1*S*-(1α,3β,5α)]

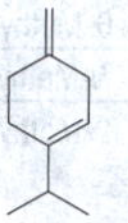
7246
4-Methylene-1-isopropylcyclohexene

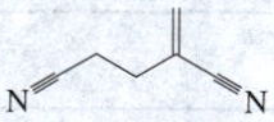
7247
2-Methylenepentanedinitrile

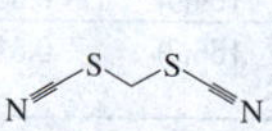
7248
Methylene thiocyanate

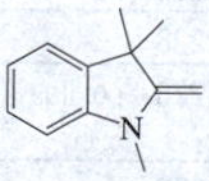
7249
2-Methylene-1,3,3-trimethylindoline

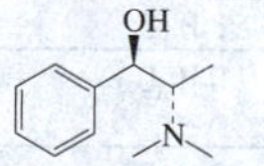
7250
N-Methylephedrine, [*R*-(*R**,*S**)]

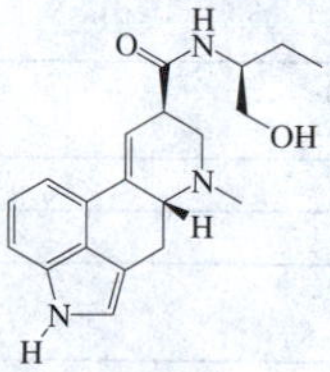
7251
Methylergonovine

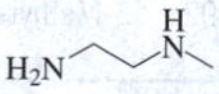
7252
N-Methyl-1,2-ethanediamine

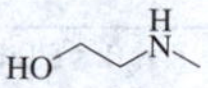
7253
N-Methyl-2-ethanolamine

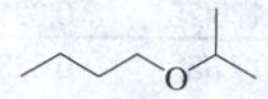
7254
1-(1-Methylethoxy)butane

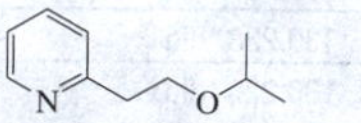
7255
2-[2-(1-Methylethoxy)ethyl]pyridine

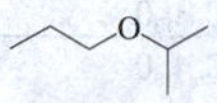
7256
1-(1-Methylethoxy)propane

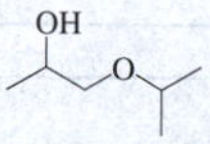
7257
1-(1-Methylethoxy)-2-propanol

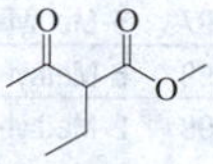
7258
Methyl 2-ethylacetoacetate

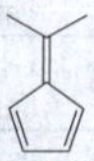
7259
5-(1-Methylethylidene)-1,3-cyclopentadiene

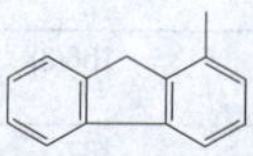
7260
1-Methyl-9*H*-fluorene

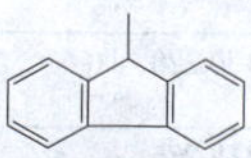
7261
9-Methyl-9*H*-fluorene

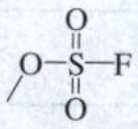
7262
Methyl fluorosulfonate

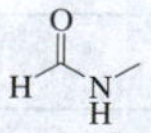
7263
N-Methylformamide

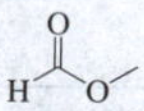
7264
Methyl formate

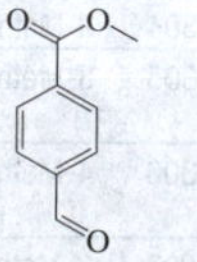
7265
Methyl 4-formylbenzoate

No.	Name	Synonym	Mol. form.	CAS Reg. No.	Mol. wt.	Physical form	mp/°C	bp/°C	den/ g cm⁻³	n_D	Solubility
7266	2-Methylfuran		C_5H_6O	534-22-5	82.101	liq	-91.2(0.9)	63.9(0.2)	0.9132^{20}	1.4342^{20}	sl H_2O, ctc; s EtOH, eth
7267	3-Methylfuran		C_5H_6O	930-27-8	82.101			66(5)	0.923^{18}	1.4330^{19}	i H_2O; s EtOH, eth
7268	5-Methyl-2-furancarboxaldehyde		$C_6H_6O_2$	620-02-0	110.111			186(1)	1.1072^{18}	1.5264^{20}	s H_2O; vs EtOH; msc eth; sl ctc
7269	Methyl 2-furancarboxylate	Methyl 2-furanoate	$C_6H_6O_3$	611-13-2	126.110			181.3	1.1786^{21}	1.4860^{20}	i H_2O; s EtOH, eth, bz, chl
7270	3-Methyl-2,5-furandione		$C_5H_4O_3$	616-02-4	112.084		6(2)	213.5	1.2469^{16}	1.4710^{21}	vs ace, eth, EtOH
7271	*N*-Methyl-2-furanmethanamine		C_6H_9NO	4753-75-7	111.141			149	0.989^{25}	1.4729^{20}	
7272	5-Methyl-2-furanmethanol		$C_6H_8O_2$	3857-25-8	112.127			195 dec	1.0769^{20}	1.4853^{20}	vs eth, EtOH
7273	α-Methyl-2-furanmethanol		$C_6H_8O_2$	4208-64-4	112.127			162.5	1.0739^{25}	1.4827^{15}	
7274	5-Methyl-2(3*H*)-furanone		$C_5H_6O_2$	591-12-8	98.101	nd	18	56^{12}	1.084^{20}	1.4476^{20}	s H_2O, EtOH, eth, CS_2; sl ctc
7275	5-Methyl-2(5*H*)-furanone		$C_5H_6O_2$	591-11-7	98.101		<-17	209	1.0810^{20}	1.4454^{20}	msc H_2O; s EtOH, eth
7276	Methylgermane		CH_6Ge	1449-65-6	90.70	col gas	-158	-23			
7277	Methyl β-D-glucopyranoside		$C_7H_{14}O_6$	709-50-2	194.182		109				s H_2O
7278	Methyl α-D-glucopyranoside	α-Methylglucoside	$C_7H_{14}O_6$	97-30-3	194.182	orth nd (al)	163.8(0.4)	200$^{0.2}$	1.46^{30}		vs H_2O
7279	3-Methylglutaric acid	3-Methylpentanedioic acid	$C_6H_{10}O_4$	626-51-7	146.141		87	166$^{0.5}$			s H_2O, EtOH, eth; sl bz, chl; i lig
7280	Methyl Green		$C_{27}H_{35}BrClN_3$	14855-76-6	516.944	grn pow (al)					vs H_2O
7281	Methyl heptadecanoate		$C_{18}H_{36}O_2$	1731-92-6	284.478	pl (al)	29.9(0.3)	185^{9}			i H_2O; s EtOH, ace, ctc; vs eth, bz
7282	Methyl heptafluorobutanoate		$C_5H_3F_7O_2$	356-24-1	228.066	liq	-81.7(0.3)	80	1.483^{20}	1.295^{20}	sl H_2O; s eth, ace
7283	6-Methyl-2-heptanamine, (±)-	Octodrine	$C_8H_{19}N$	5984-58-7	129.244	visc liq		156(12)	0.767^{25}	1.4209^{20}	
7284	*N*-Methyl-2-heptanamine		$C_8H_{19}N$	540-43-2	129.244			155			
7285	2-Methylheptane		C_8H_{18}	592-27-8	114.229	liq	-109(4)	117.6(0.9)	0.6980^{20}	1.3949^{20}	i H_2O; msc EtOH, ace, bz; s eth, ctc
7286	3-Methylheptane		C_8H_{18}	589-81-1	114.229	col liq	-120.48(0.09)	118.9(0.6)	0.7017^{25}	1.3961^{25}	i H_2O; s EtOH, eth; msc ace, bz, chl
7287	4-Methylheptane		C_8H_{18}	589-53-7	114.229	liq	-121.0(0.1)	117.7(0.5)	0.7046^{20}	1.3979^{20}	i H_2O; s eth; msc EtOH, ace, bz
7288	Methyl heptanoate		$C_8H_{16}O_2$	106-73-0	144.212	liq	-55.7(0.3)	169.7(0.4)	0.8815^{20}	1.4152^{20}	sl H_2O, ctc, ace; s EtOH, eth
7289	2-Methyl-1-heptanol, (±)-		$C_8H_{18}O$	111675-77-5	130.228	col liq	-112	175(2)	0.8022^{20}	1.424^{20}	
7290	3-Methyl-1-heptanol		$C_8H_{18}O$	1070-32-2	130.228	liq	-90	186(2)	0.824^{24}	1.4295^{25}	
7291	4-Methyl-1-heptanol		$C_8H_{18}O$	817-91-4	130.228			184(4)	0.8065^{25}	1.4253^{25}	vs EtOH
7292	5-Methyl-1-heptanol, (±)-		$C_8H_{18}O$	111767-95-4	130.228	col liq	-104	186(2)	0.8153^{25}	1.4272^{25}	
7293	6-Methyl-1-heptanol	Isooctyl alcohol	$C_8H_{18}O$	1653-40-3	130.228	liq	-106	191(2)	0.8176^{25}	1.4251^{25}	i H_2O; s EtOH, eth
7294	2-Methyl-2-heptanol		$C_8H_{18}O$	625-25-2	130.228	liq	-50.4	173(5)	0.8142^{20}	1.4250^{20}	i H_2O; s EtOH, eth
7295	3-Methyl-2-heptanol		$C_8H_{18}O$	31367-46-1	130.228			166(1)	0.8177^{25}	1.4199^{25}	i H_2O; s EtOH, eth, ctc
7296	4-Methyl-2-heptanol		$C_8H_{18}O$	56298-90-9	130.228	col liq	-102	171(1)	0.8027^{20}	1.424^{20}	
7297	5-Methyl-2-heptanol		$C_8H_{18}O$	54630-50-1	130.228	liq	-61	172(1)	0.8174^{21}		
7298	6-Methyl-2-heptanol		$C_8H_{18}O$	4730-22-7	130.228	liq	-105	173(2)	0.8218^{20}	1.4238^{10}	
7299	2-Methyl-3-heptanol, (±)-		$C_8H_{18}O$	100296-26-2	130.228	liq	-85	171(3)	0.8235^{20}	1.4265^{20}	sl H_2O; s EtOH, eth, ctc
7300	3-Methyl-3-heptanol	2-Ethyl-2-hexanol	$C_8H_{18}O$	5582-82-1	130.228	liq	-83	166(3)	0.8282^{20}	1.4279^{20}	i H_2O; s EtOH, eth, ctc
7301	4-Methyl-3-heptanol		$C_8H_{18}O$	14979-39-6	130.228	liq	-123	157(2)	0.827^{25}	1.4300^{20}	
7302	5-Methyl-3-heptanol		$C_8H_{18}O$	18720-65-5	130.228	liq	-91.2	154(2)	0.8425^{25}	1.433^{24}	
7303	6-Methyl-3-heptanol, (±)-		$C_8H_{18}O$	100295-85-0	130.228	col liq	-61	159(3)	0.8220^{20}	1.4254^{20}	
7304	2-Methyl-4-heptanol		$C_8H_{18}O$	21570-35-4	130.228	liq	-81	167(3)	0.8207^{20}	1.4203	vs eth, EtOH
7305	3-Methyl-4-heptanol		$C_8H_{18}O$	1838-73-9	130.228	liq		165(2)	0.8329^{25}	1.4211^{25}	sl H_2O; s EtOH, eth, ctc
7306	4-Methyl-4-heptanol		$C_8H_{18}O$	598-01-6	130.228	liq	-82	166(3)	0.8248^{20}	1.4258^{20}	i H_2O; s EtOH, eth, ctc
7307	6-Methyl-2-heptanol acetate		$C_{10}H_{20}O_2$	67952-57-2	172.265			187	0.8474^{20}	1.413^{20}	vs EtOH

7266
2-Methylfuran

7267
3-Methylfuran

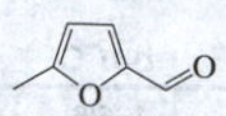

7268
5-Methyl-2-furancarboxaldehyde

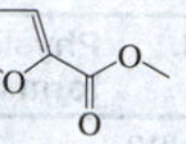

7269
Methyl 2-furancarboxylate

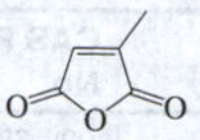

7270
3-Methyl-2,5-furandione

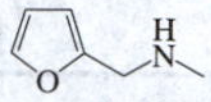

7271
N-Methyl-2-furanmethanamine

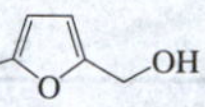

7272
5-Methyl-2-furanmethanol

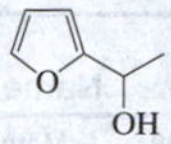

7273
α-Methyl-2-furanmethanol

7274
5-Methyl-2(3*H*)-furanone

7275
5-Methyl-2(5*H*)-furanone

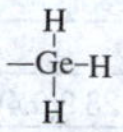

7276
Methylgermane

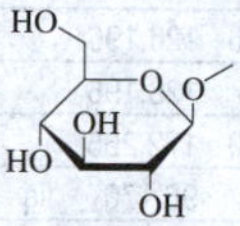

7277
Methyl β-D-glucopyranoside

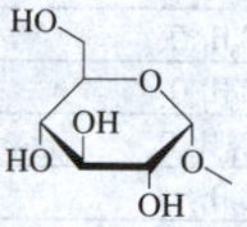

7278
Methyl α-D-glucopyranoside

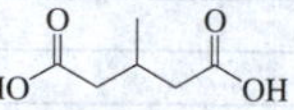

7279
3-Methylglutaric acid

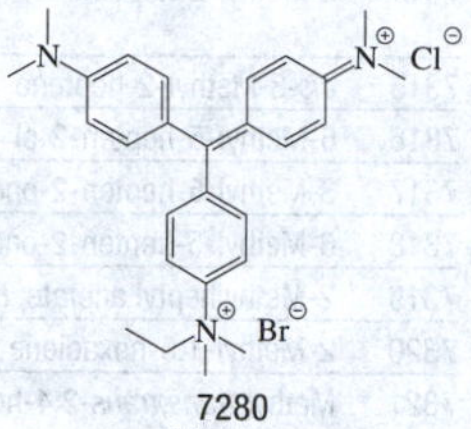

7280
Methyl Green

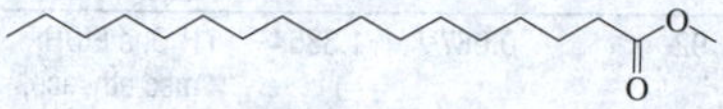

7281
Methyl heptadecanoate

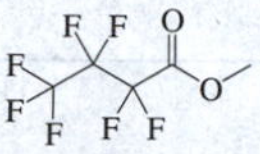

7282
Methyl heptafluorobutanoate

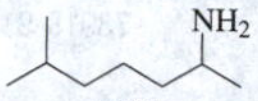

7283
6-Methyl-2-heptanamine, (±)-

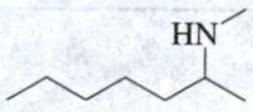

7284
N-Methyl-2-heptanamine

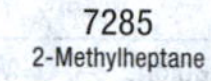

7285
2-Methylheptane

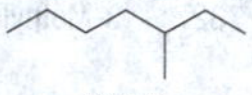

7286
3-Methylheptane

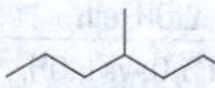

7287
4-Methylheptane

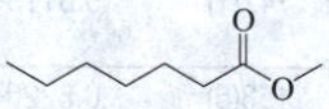

7288
Methyl heptanoate

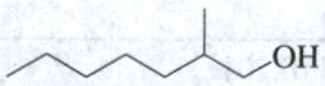

7289
2-Methyl-1-heptanol, (±)-

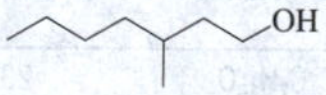

7290
3-Methyl-1-heptanol

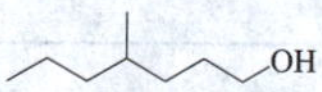

7291
4-Methyl-1-heptanol

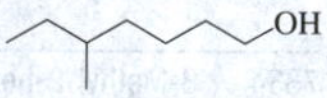

7292
5-Methyl-1-heptanol, (±)-

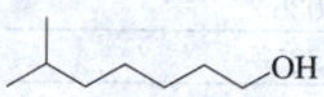

7293
6-Methyl-1-heptanol

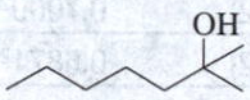

7294
2-Methyl-2-heptanol

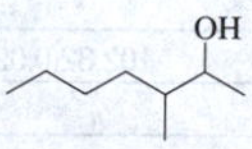

7295
3-Methyl-2-heptanol

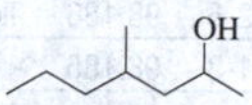

7296
4-Methyl-2-heptanol

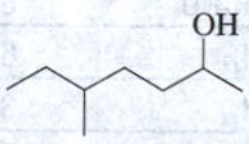

7297
5-Methyl-2-heptanol

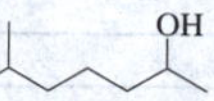

7298
6-Methyl-2-heptanol

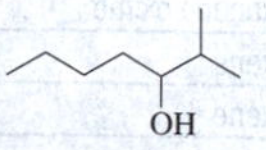

7299
2-Methyl-3-heptanol, (±)-

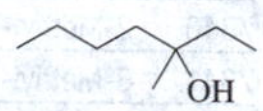

7300
3-Methyl-3-heptanol

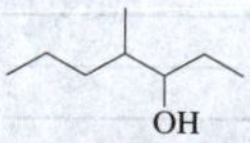

7301
4-Methyl-3-heptanol

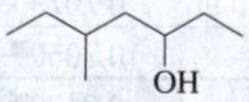

7302
5-Methyl-3-heptanol

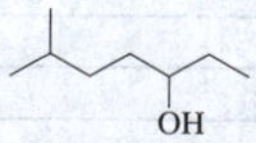

7303
6-Methyl-3-heptanol, (±)-

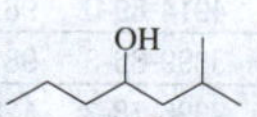

7304
2-Methyl-4-heptanol

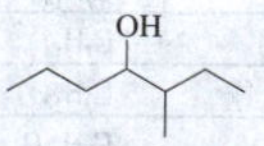

7305
3-Methyl-4-heptanol

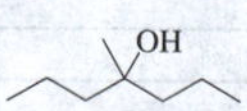

7306
4-Methyl-4-heptanol

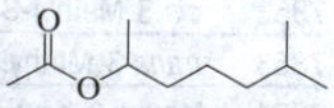

7307
6-Methyl-2-heptanol acetate

No.	Name	Synonym	Mol. form.	CAS Reg. No.	Mol. wt.	Physical form	mp/°C	bp/°C	den/ g cm^{-3}	n_D	Solubility
7308	6-Methyl-2-heptanone		$C_8H_{16}O$	928-68-7	128.212			166(6)	0.8151^{20}	1.4162^{20}	sl H_2O; vs EtOH, eth; msc ace, bz, chl
7309	5-Methyl-3-heptanone	Ethyl 2-methylbutyl ketone	$C_8H_{16}O$	541-85-5	128.212	liq		159(4)			
7310	6-Methyl-3-heptanone		$C_8H_{16}O$	624-42-0	128.212			167(10)	0.8304^{20}	1.4209^{20}	i H_2O; s EtOH, eth, bz, ctc
7311	2-Methyl-4-heptanone	Isobutyl propyl ketone	$C_8H_{16}O$	626-33-5	128.212			155(4)	0.813^{22}		i H_2O; s EtOH, eth
7312	2-Methyl-1-heptene		C_8H_{16}	15870-10-7	112.213	liq	-90.1(0.2)	119(2)	0.7104^{25}	1.4123^{20}	
7313	6-Methyl-1-heptene		C_8H_{16}	5026-76-6	112.213			113(1)	0.7079^{25}	1.4070^{20}	
7314	2-Methyl-2-heptene		C_8H_{16}	627-97-4	112.213			122(2)	0.7200^{25}	1.4170^{20}	i H_2O; s eth, bz, ctc, chl
7315	*cis*-3-Methyl-2-heptene		C_8H_{16}	22768-19-0	112.213			122	0.725^{25}	1.419^{20}	
7316	6-Methyl-5-hepten-2-ol		$C_8H_{16}O$	1569-60-4	128.212			186(5)	0.8545^{20}	1.4505^{20}	
7317	3-Methyl-5-hepten-2-one		$C_8H_{14}O$	38552-72-6	126.196			63^{20}	0.8463^{18}	1.4345^{18}	
7318	6-Methyl-5-hepten-2-one		$C_8H_{14}O$	110-93-0	126.196		173.5		0.8546^{16}	1.4445^{20}	vs eth, EtOH
7319	2-Methylheptyl acetate, (±)-		$C_{10}H_{20}O_2$	74112-36-0	172.265			195	0.8626^{14}	1.4146^{20}	vs eth, EtOH
7320	2-Methyl-1,5-hexadiene		C_7H_{12}	4049-81-4	96.170	liq	-128.9(0.4)	88(3)	0.7153^{25}	1.4183^{20}	
7321	Methyl *trans,trans*-2,4-hexadienoate	Methyl sorbate	$C_7H_{10}O_2$	689-89-4	126.153	lf	15	180	0.9777^{20}	1.5025^{22}	i H_2O; s EtOH, eth
7322	2-Methylhexanal		$C_7H_{14}O$	925-54-2	114.185	liq		141			
7323	3-Methylhexanal	3-Methylcaproaldehyde	$C_7H_{14}O$	19269-28-4	114.185			145(1)	0.8203^{20}	1.4122^{20}	i H_2O; s EtOH, eth
7324	3-Methyl-1-hexanamine		$C_7H_{17}N$	65530-93-0	115.217			149	0.772^{26}	1.4249^{25}	
7325	4-Methyl-2-hexanamine	Methylhexanamine	$C_7H_{17}N$	105-41-9	115.217			132.5	0.7655^{20}	1.4150^{25}	sl H_2O; vs EtOH, eth, chl, dil acid
7326	2-Methylhexane		C_7H_{16}	591-76-4	100.202	liq	-118.23(0.04)	90.0(0.8)	0.6787^{20}	1.3848^{20}	i H_2O; s EtOH; msc eth, ace, bz, lig, chl
7327	3-Methylhexane		C_7H_{16}	78918-91-9	100.202	liq	-119.4	92	0.687^{21}	1.3854^{25}	i H_2O; s EtOH; msc eth, ace, bz, lig, chl
7328	5-Methyl-2,3-hexanedione	2-Methylhexane-4,5-dione	$C_7H_{12}O_2$	13706-86-0	128.169			138	0.908^{22}	1.4119^{20}	
7329	Methyl hexanoate	Methyl caproate	$C_7H_{14}O_2$	106-70-7	130.185	liq	-70(2)	151(1)	0.8846^{20}	1.4049^{20}	i H_2O; vs EtOH, eth; s ace, bz, ctc
7330	2-Methylhexanoic acid		$C_7H_{14}O_2$	4536-23-6	130.185			214(9)	0.918^{20}	1.4193^{20}	vs ace, bz, eth, EtOH
7331	2-Methyl-1-hexanol, (±)-		$C_7H_{16}O$	111768-04-8	116.201			165(4)	0.826^{20}	1.4226^{20}	vs eth, EtOH
7332	5-Methyl-1-hexanol		$C_7H_{16}O$	627-98-5	116.201			170(3)	0.8192^{24}	1.4175^{20}	vs eth, EtOH
7333	2-Methyl-2-hexanol		$C_7H_{16}O$	625-23-0	116.201			142(1)	0.8119^{20}	1.4175^{20}	sl H_2O; msc EtOH, eth
7334	3-Methyl-2-hexanol		$C_7H_{16}O$	2313-65-7	116.201			153(5)	0.8220^{25}	1.4198^{18}	i H_2O; vs EtOH, eth; s ace
7335	5-Methyl-2-hexanol		$C_7H_{16}O$	627-59-8	116.201			151(5)	0.814^{20}	1.4180^{20}	sl H_2O; s EtOH, eth
7336	3-Methyl-3-hexanol		$C_7H_{16}O$	597-96-6	116.201			147(3)	0.8233^{20}	1.4231^{20}	sl H_2O; s EtOH, eth, ctc
7337	5-Methyl-2-hexanone	Methyl isopentyl ketone	$C_7H_{14}O$	110-12-3	114.185			139(2)	0.888^{20}	1.4062^{20}	sl H_2O; msc EtOH; vs ace, bz; s ctc
7338	2-Methyl-3-hexanone	Propyl isopropyl ketone	$C_7H_{14}O$	7379-12-6	114.185			134(3)	0.8091^{20}	1.4042^{20}	s EtOH, eth, chl; vs ace
7339	5-Methyl-2-hexanone oxime		$C_7H_{15}NO$	624-44-2	129.200			195.5	0.8881^{20}	1.4448^{20}	sl chl
7340	2-Methyl-1-hexene		C_7H_{14}	6094-02-6	98.186	liq	-102.82(0.09)	92(2)	0.7000^{20}	1.4035^{20}	
7341	3-Methyl-1-hexene		C_7H_{14}	3404-61-3	98.186			84(2)	0.6871^{25}	1.3965^{20}	
7342	4-Methyl-1-hexene		C_7H_{14}	3769-23-1	98.186	liq	-141.47(0.09)	87(2)	0.6942^{25}	1.4000^{20}	
7343	5-Methyl-1-hexene		C_7H_{14}	3524-73-0	98.186			85(1)	0.6877^{25}	1.3967^{20}	
7344	2-Methyl-2-hexene		C_7H_{14}	2738-19-4	98.186	liq	-130.36(0.09)	95(1)	0.7038^{25}	1.4106^{20}	
7345	*cis*-3-Methyl-2-hexene		C_7H_{14}	10574-36-4	98.186	liq	-118.5	97(3)	0.712^{20}	1.4126^{20}	
7346	*cis*-4-Methyl-2-hexene		C_7H_{14}	3683-19-0	98.186			86(2)	0.6952^{25}	1.4026^{20}	
7347	*trans*-4-Methyl-2-hexene		C_7H_{14}	3683-22-5	98.186	liq	-125.70(0.09)	88(2)	0.6925^{25}	1.4025^{20}	
7348	*cis*-5-Methyl-2-hexene		C_7H_{14}	13151-17-2	98.186			89(3)	0.697^{25}	1.404^{20}	
7349	*trans*-5-Methyl-2-hexene		C_7H_{14}	7385-82-2	98.186	liq	-124.3(0.2)	87(2)	0.6883^{25}	1.4006^{20}	
7350	*cis*-2-Methyl-3-hexene		C_7H_{14}	15840-60-5	98.186			86(3)	0.690^{25}	1.401^{20}	
7351	*trans*-2-Methyl-3-hexene		C_7H_{14}	692-24-0	98.186	liq	-141.6	88(3)	0.6853^{25}	1.4001^{20}	
7352	*cis*-3-Methyl-3-hexene		C_7H_{14}	4914-89-0	98.186			95.3(0.9)	0.7079^{25}	1.4126^{20}	
7353	*trans*-3-Methyl-3-hexene		C_7H_{14}	3899-36-3	98.186			94(1)	0.7050^{25}	1.4109^{20}	
7354	Methyl 3-hexenoate		$C_7H_{12}O_2$	2396-78-3	128.169			67^{34}	0.9132^{25}	1.4240^{23}	

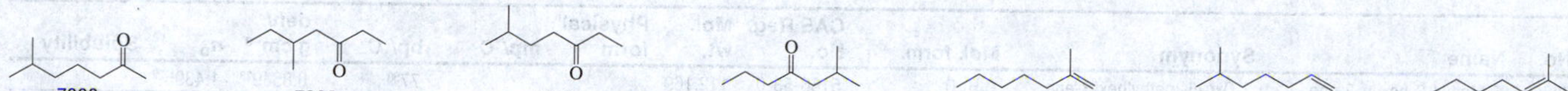

7308 6-Methyl-2-heptanone; 7309 5-Methyl-3-heptanone; 7310 6-Methyl-3-heptanone; 7311 2-Methyl-4-heptanone; 7312 2-Methyl-1-heptene; 7313 6-Methyl-1-heptene; 7314 2-Methyl-2-heptene

7315 *cis*-3-Methyl-2-heptene; 7316 6-Methyl-5-hepten-2-ol; 7317 3-Methyl-5-hepten-2-one; 7318 6-Methyl-5-hepten-2-one; 7319 2-Methylheptyl acetate, (±)-; 7320 2-Methyl-1,5-hexadiene

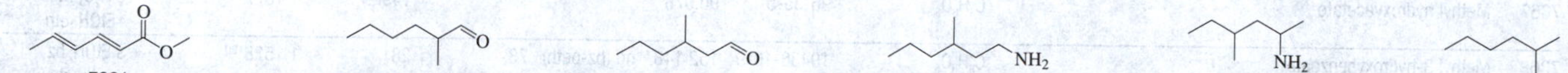

7321 Methyl *trans,trans*-2,4-hexadienoate; 7322 2-Methylhexanal; 7323 3-Methylhexanal; 7324 3-Methyl-1-hexanamine; 7325 4-Methyl-2-hexanamine; 7326 2-Methylhexane

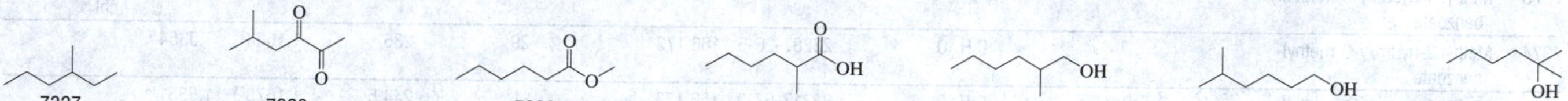

7327 3-Methylhexane; 7328 5-Methyl-2,3-hexanedione; 7329 Methyl hexanoate; 7330 2-Methylhexanoic acid; 7331 2-Methyl-1-hexanol, (±)-; 7332 5-Methyl-1-hexanol; 7333 2-Methyl-2-hexanol

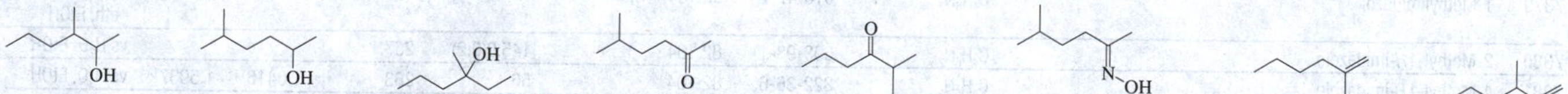

7334 3-Methyl-2-hexanol; 7335 5-Methyl-2-hexanol; 7336 3-Methyl-3-hexanol; 7337 5-Methyl-2-hexanone; 7338 2-Methyl-3-hexanone; 7339 5-Methyl-2-hexanone oxime; 7340 2-Methyl-1-hexene; 7341 3-Methyl-1-hexene

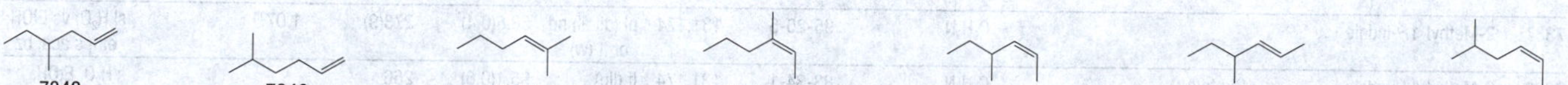

7342 4-Methyl-1-hexene; 7343 5-Methyl-1-hexene; 7344 2-Methyl-2-hexene; 7345 *cis*-3-Methyl-2-hexene; 7346 *cis*-4-Methyl-2-hexene; 7347 *trans*-4-Methyl-2-hexene; 7348 *cis*-5-Methyl-2-hexene

7349 *trans*-5-Methyl-2-hexene; 7350 *cis*-2-Methyl-3-hexene; 7351 *trans*-2-Methyl-3-hexene; 7352 *cis*-3-Methyl-3-hexene; 7353 *trans*-3-Methyl-3-hexene; 7354 Methyl 3-hexenoate

Organic

No.	Name	Synonym	Mol. form.	CAS Reg. No.	Mol. wt.	Physical form	mp/°C	bp/°C	den/g cm^{-3}	n_D	Solubility
7355	5-Methyl-3-hexen-2-one	2-Oxo-5-methylhex-3-ene	$C_7H_{12}O$	5166-53-0	112.169			77^{50}	0.8549^{28}	1.4395^{22}	
7356	5-Methyl-5-hexen-2-one		$C_7H_{12}O$	3240-09-3	112.169			150	0.8460^{20}	1.4348^{20}	vs ace, eth, EtOH
7357	5-Methyl-1-hexyne		C_7H_{12}	2203-80-7	96.170	liq	-124.6(0.5)	91(3)	0.7274^{20}	1.4059^{20}	i H_2O; s EtOH, eth, bz, chl, peth
7358	5-Methyl-2-hexyne		C_7H_{12}	53566-37-3	96.170	liq	-92.6(0.2)	102(3)	0.7378^{20}	1.4176^{20}	i H_2O; s eth, ace, bz, chl, peth
7359	2-Methyl-3-hexyne		C_7H_{12}	36566-80-0	96.170	liq	-116.7(0.4)	94(3)	0.7263^{20}	1.4120^{20}	vs bz, eth, chl, peth
7360	Methyl 2-hexynoate		$C_7H_{10}O_2$	18937-79-6	126.153			80^{23}	0.9648^{25}		
7361	L-1-Methylhistidine		$C_7H_{11}N_3O_2$	332-80-9	169.181	pl (DMF aq)	249				
7362	L-3-Methylhistidine		$C_7H_{11}N_3O_2$	368-16-1	169.181		250				
7363	Methylhydrazine		CH_6N_2	60-34-4	46.072	liq	-52.3(0.1)	83(3)		1.4325^{20}	s H_2O, eth, ctc; msc EtOH; i lig
7364	Methyl hydrazinecarboxylate	Methyl carbazate	$C_2H_6N_2O_2$	6294-89-9	90.081		73	108^{12}			s H_2O, EtOH; sl bz; i peth
7365	Methyl hydrogen succinate	Monomethyl succinate	$C_5H_8O_4$	3878-55-5	132.116		58	151^{20}			s H_2O
7366	Methyl hydroperoxide	Methyl hydrogen peroxide	CH_4O_2	3031-73-0	48.042	liq	-72	86	1.037^{15}	1.3641^{15}	vs H_2O, bz, eth, EtOH
7367	Methyl hydroxyacetate		$C_3H_6O_3$	96-35-5	90.078			149(2)	1.1677^{18}		s H_2O; msc EtOH, eth
7368	Methyl 3-hydroxybenzoate		$C_8H_8O_3$	19438-10-9	152.148	nd (bz-peth)	73	281	1.1528^{100}		s EtOH, bz, peth; sl chl
7369	Methyl 4-hydroxybenzoate	Methylparaben	$C_8H_8O_3$	99-76-3	152.148	nd (dil al)	125.2(0.5)	275 dec			sl H_2O; vs EtOH, eth, ace; s tfa
7370	Methyl α-hydroxydiphenylacetate	Methyl diphenylglycolate	$C_{15}H_{14}O_3$	76-89-1	242.270	mcl or tcl cry (al)	75.8	187^{13}			vs eth, EtOH
7371	*O*-Methylhydroxylamine	Methoxyamine	CH_5NO	67-62-9	47.057			49			
7372	*O*-Methylhydroxylamine hydrochloride	Methoxyamine hydrochloride	CH_6ClNO	593-56-6	83.518	pr	150.0				vs H_2O, EtOH
7373	Methyl 4-hydroxy-3-methoxybenzoate		$C_9H_{10}O_4$	3943-74-6	182.173	nd (dil al)	64	286			s EtOH, peth; sl chl
7374	Methyl 2-hydroxy-3-methylbenzoate		$C_9H_{10}O_3$	23287-26-5	166.173		29	235	1.1683^{25}	1.5354^{16}	
7375	Methyl 2-hydroxy-5-methylbenzoate		$C_9H_{10}O_3$	22717-57-3	166.173	liq	-1	244.5	1.1673^{25}	1.5351^{15}	
7376	Methyl 2-hydroxy-2-methylpropanoate	Methyl 2-methyllactate	$C_5H_{10}O_3$	2110-78-3	118.131			131(3)		1.4056^{20}	vs H_2O, EtOH
7377	Methyl 3-hydroxy-2-naphthalenecarboxylate	Methyl 3-hydroxy-2-naphthoate	$C_{12}H_{10}O_3$	883-99-8	202.205	pa ye orth nd (dil MeOH)	75.5	206			i H_2O; s EtOH
7378	Methyl α-hydroxyphenylacetate, (±)-	(±)-Methyl mandelate	$C_9H_{10}O_3$	4358-87-6	166.173	pl (bz-lig)	58	250 dec	1.1756^{20}		vs EtOH, chl
7379	1-Methylimidazol		$C_4H_6N_2$	616-47-7	82.104	liq	-6	195.5	1.0325^{20}	1.4970^{20}	vs H_2O, ace, eth, EtOH
7380	2-Methyl-1*H*-imidazole		$C_4H_6N_2$	693-98-1	82.104		145.9(0.2)	267			vs H_2O, EtOH
7381	4-Methyl-1*H*-imidazole		$C_4H_6N_2$	822-36-6	82.104		56	263	1.0416^{14}	1.5037^{14}	vs H_2O, EtOH
7382	*N*-Methyliminodiacetic acid	*N*-(Carboxymethyl)-*N*-methylglycine	$C_5H_9NO_4$	4408-64-4	147.130	cry (w)	226				s H_2O; i EtOH, eth
7383	1-Methyl-1*H*-indene		$C_{10}H_{10}$	767-59-9	130.186			199	0.970^{25}	1.5616^{20}	
7384	2-Methyl-1*H*-indene		$C_{10}H_{10}$	2177-47-1	130.186		80	209(18)	0.974^{25}	1.5652^{20}	i H_2O; s eth, ace, bz
7385	3-Methyl-1*H*-indene		$C_{10}H_{10}$	767-60-2	130.186			198	0.972^{25}	1.5621^{20}	i H_2O; s eth, ace, bz
7386	1-Methyl-1*H*-indole		C_9H_9N	603-76-9	131.174			241(20)	1.0707^{25}		i H_2O; s EtOH, eth, bz
7387	2-Methyl-1*H*-indole		C_9H_9N	95-20-5	131.174	pl (dil al) nd or lf (w)	58.6(0.4)	273(9)	1.07^{20}		sl H_2O; vs EtOH, eth; s ace, bz
7388	3-Methyl-1*H*-indole	Skatole	C_9H_9N	83-34-1	131.174	lf (lig)	95.1(0.6)	266			s H_2O, EtOH, eth, ace, bz, chl
7389	5-Methyl-1*H*-indole		C_9H_9N	614-96-0	131.174		61(2)	267	1.0202^{78}		s H_2O, EtOH, eth, bz, lig
7390	7-Methyl-1*H*-indole		C_9H_9N	933-67-5	131.174		83(2)	266	1.0202^{100}		
7391	Methyl 2-iodobenzoate		$C_8H_7IO_2$	610-97-9	262.045			280		1.6052^{20}	s EtOH
7392	Methyl 3-iodobenzoate		$C_8H_7IO_2$	618-91-7	262.045	nd (dil al)	54.5	277			i H_2O, lig; s EtOH; vs eth, ace
7393	Methyl 4-iodobenzoate		$C_8H_7IO_2$	619-44-3	262.045	nd (eth-al)	114.8	sub	2.0200^{10}		s EtOH, eth
7394	5-Methyl-1,3-isobenzofurandione		$C_9H_6O_3$	19438-61-0	162.142		93.0	295			

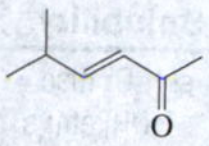
7355
5-Methyl-3-hexen-2-one

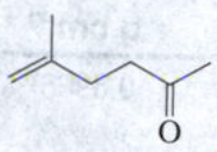
7356
5-Methyl-5-hexen-2-one

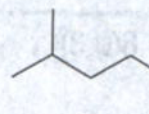
7357
5-Methyl-1-hexyne

7358
5-Methyl-2-hexyne

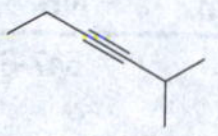
7359
2-Methyl-3-hexyne

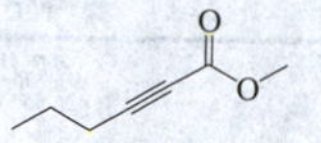
7360
Methyl 2-hexynoate

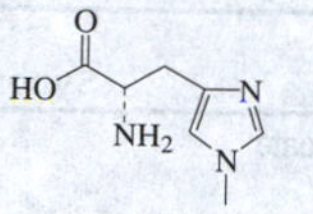
7361
L-1-Methylhistidine

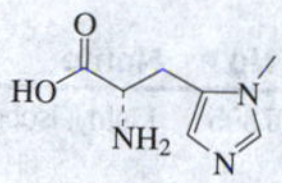
7362
L-3-Methylhistidine

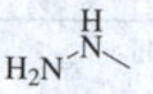
7363
Methylhydrazine

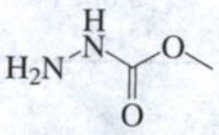
7364
Methyl hydrazinecarboxylate

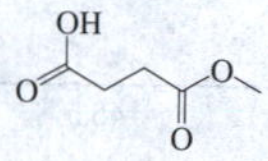
7365
Methyl hydrogen succinate

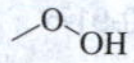
7366
Methyl hydroperoxide

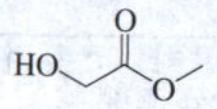
7367
Methyl hydroxyacetate

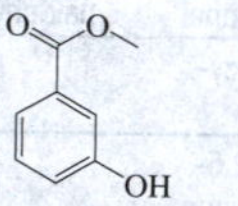
7368
Methyl 3-hydroxybenzoate

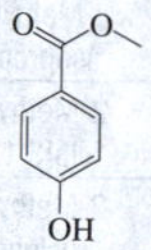
7369
Methyl 4-hydroxybenzoate

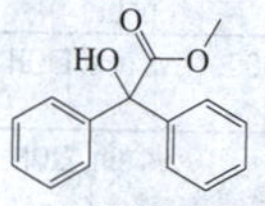
7370
Methyl α-hydroxydiphenylacetate

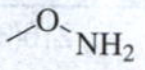
7371
O-Methylhydroxylamine

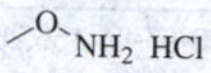
7372
O-Methylhydroxylamine hydrochloride

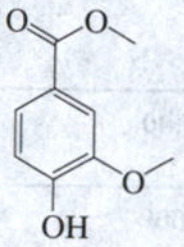
7373
Methyl 4-hydroxy-3-methoxybenzoate

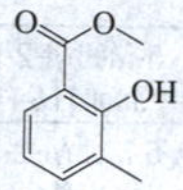
7374
Methyl 2-hydroxy-3-methylbenzoate

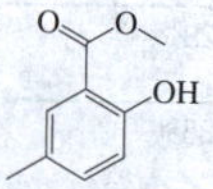
7375
Methyl 2-hydroxy-5-methylbenzoate

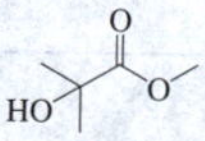
7376
Methyl 2-hydroxy-2-methylpropanoate

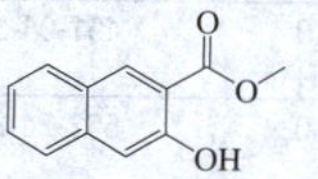
7377
Methyl 3-hydroxy-2-naphthalenecarboxylate

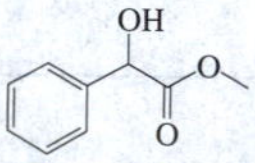
7378
Methyl α-hydroxyphenylacetate, (±)-

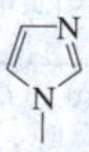
7379
1-Methylimidazol

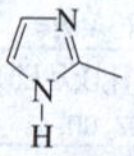
7380
2-Methyl-1*H*-imidazole

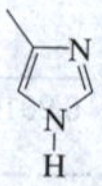
7381
4-Methyl-1*H*-imidazole

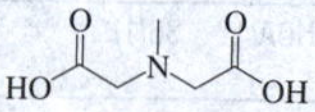
7382
N-Methyliminodiacetic acid

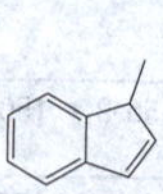
7383
1-Methyl-1*H*-indene

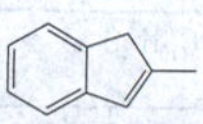
7384
2-Methyl-1*H*-indene

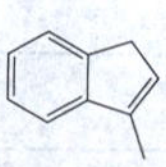
7385
3-Methyl-1*H*-indene

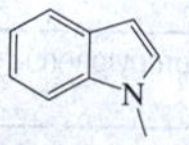
7386
1-Methyl-1*H*-indole

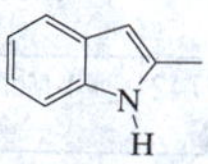
7387
2-Methyl-1*H*-indole

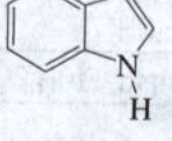
7388
3-Methyl-1*H*-indole

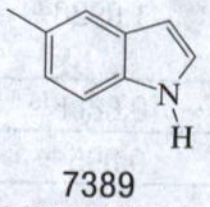
7389
5-Methyl-1*H*-indole

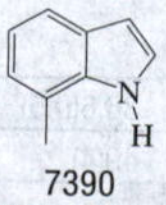
7390
7-Methyl-1*H*-indole

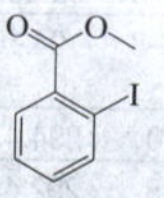
7391
Methyl 2-iodobenzoate

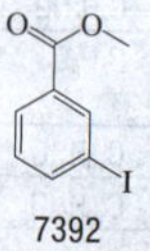
7392
Methyl 3-iodobenzoate

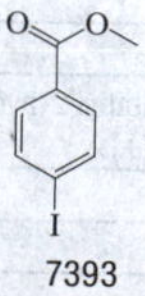
7393
Methyl 4-iodobenzoate

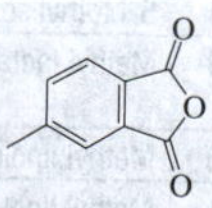
7394
5-Methyl-1,3-isobenzofurandione

No.	Name	Synonym	Mol. form.	CAS Reg. No.	Mol. wt.	Physical form	mp/°C	bp/°C	den/ g cm^{-3}	n_D	Solubility
7395	Methyl isobutanoate		$C_5H_{10}O_2$	547-63-7	102.132	liq	-84.6(0.3)	92(1)	0.8906^{20}	1.3840^{20}	sl H_2O; msc EtOH, eth; s ace, ctc
7396	Methyl isocyanate		C_2H_3NO	624-83-9	57.051	liq	-45	38.3(0.2)	0.9588^{20}	1.3694^{20}	vs H_2O
7397	2-Methyl-1*H*-isoindole-1,3(2*H*)-dione		$C_9H_7NO_2$	550-44-7	161.158	nd (al), lf (sub)	134	286			i H_2O; sl EtOH
7398	Methyl isopentanoate	Methyl isovalerate	$C_6H_{12}O_2$	556-24-1	116.158			116(2)	0.8808^{20}	1.3927^{20}	i H_2O; vs EtOH, eth, ace
7399	6-Methyl-*N*-isopentyl-2-heptanamine	Octamylamine	$C_{13}H_{29}N$	502-59-0	199.376			100^{7}			
7400	2-Methyl-5-isopropylaniline		$C_{10}H_{15}N$	2051-53-8	149.233	liq	-16	241	0.9942^{20}	1.5387^{20}	s ctc, CS_2
7401	α-Methyl-4-isopropylbenzenepropanal	3-*p*-Cumenyl-2-methylpropionaldehyde	$C_{13}H_{18}O$	103-95-7	190.281			270	0.9459^{20}	1.5068^{20}	vs bz, eth, EtOH
7402	2-Methyl-5-isopropylbicyclo[3.1.0]hex-2-ene		$C_{10}H_{16}$	2867-05-2	136.234			156(5)	0.8301^{20}	1.4515^{20}	
7403	2-Methyl-5-isopropyl-2,5-cyclohexadiene-1,4-dione		$C_{10}H_{12}O_2$	490-91-5	164.201		45.5	232			s chl
7404	*cis*-1-Methyl-4-isopropylcyclohexane		$C_{10}H_{20}$	6069-98-3	140.266	liq	-89.8(0.1)	172(6)	0.8039^{20}	1.4431^{20}	i H_2O; vs EtOH, eth; s bz, peth
7405	*trans*-1-Methyl-4-isopropylcyclohexane	*trans*-*p*-Menthane	$C_{10}H_{20}$	1678-82-6	140.266	oil	-86.3(0.1)	171(3)	0.7928^{20}	1.4366^{20}	vs bz, eth, EtOH, lig
7406	1-Methyl-4-isopropylcyclohexanol		$C_{10}H_{20}O$	21129-27-1	156.265			208.5	0.90^{20}	1.4619^{20}	
7407	5-Methyl-2-isopropylcyclohexanol, [1*S*-(1α,2β,5α)]-	(+)-Menthol	$C_{10}H_{20}O$	15356-60-2	156.265		42(2)	103^{9}			vs ace, bz, eth, EtOH
7408	5-Methyl-2-isopropylcyclohexanol, [1*R*-(1α,2β,5α)]-	(-)-Menthol	$C_{10}H_{20}O$	2216-51-5	156.265	nd (MeOH)	42.1(0.8)	214(12)	0.903^{15}	1.460^{22}	sl H_2O; vs EtOH, eth, ace, bz; s peth
7409	5-Methyl-2-isopropylcyclohexanol, [1*S*-(1α,2α,5β)]-	(+)-Neomenthol	$C_{10}H_{20}O$	2216-52-6	156.265	oil	-22	211.7	0.897^{22}	1.4600^{20}	vs ace, EtOH
7410	5-Methyl-2-isopropylcyclohexanol, [1*S*-(1α,2β,5β)]-	(+)-Isomenthol	$C_{10}H_{20}O$	23283-97-8	156.265	nd(dil al)	82.5	216(3)			vs eth, EtOH
7411	5-Methyl-2-isopropylcyclohexanol acetate, [1*R*-(1α,2α,5β)]		$C_{12}H_{22}O_2$	2623-23-6	198.302			223(16)	0.9244^{20}	1.4469^{20}	
7412	*cis*-5-Methyl-2-isopropylcyclohexanone	Menthone	$C_{10}H_{18}O$	491-07-6	154.249			205	0.8995^{20}	1.4527^{20}	
7413	*trans*-5-Methyl-2-isopropylcyclohexanone, (2*S*)-	*l*-Menthone	$C_{10}H_{18}O$	14073-97-3	154.249	liq	-6	207	0.8954^{20}	1.4505^{20}	sl H_2O; msc EtOH, eth, bz, CS_2; s ace
7414	1-Methyl-4-isopropylcyclohexene		$C_{10}H_{18}$	5502-88-5	138.250			170(9)	0.8457^{15}	1.4735^{20}	
7415	3-Methyl-6-isopropyl-2-cyclohexen-1-ol		$C_{10}H_{18}O$	491-04-3	154.249			$97^{15.5}$	0.9119^{25}	1.4729^{25}	
7416	4-Methyl-1-isopropyl-3-cyclohexen-1-ol		$C_{10}H_{18}O$	562-74-3	154.249			209	0.926^{20}	1.4785^{19}	
7417	5-Methyl-2-isopropylcyclohexyl ethoxyacetate, (1α,2β,5α)		$C_{14}H_{26}O_3$	579-94-2	242.354			155^{20}	0.9545^{20}		vs eth, EtOH, chl
7418	1-Methyl-4-isopropyl-2-nitrobenzene		$C_{10}H_{13}NO_2$	943-15-7	179.216			126^{10}	1.0744^{20}	1.5301^{20}	vs eth, EtOH
7419	1-Methyl-4-isopropyl-7-oxabicyclo[2.2.1]heptane		$C_{10}H_{18}O$	470-67-7	154.249		1	173.5	0.8997^{20}	1.4562^{20}	sl H_2O; msc EtOH, eth; s bz, lig
7420	1-Methyl-7-isopropylphenanthrene	Retene	$C_{18}H_{18}$	483-65-8	234.336		95.4(0.7)	391(24)	1.035^{25}		i H_2O; s EtOH, eth, bz, CS_2, HOAc
7421	4-Methyl-2-isopropylphenol		$C_{10}H_{14}O$	4427-56-9	150.217	nd (HOAc)	36(1)	231(6)	0.9910^{20}	1.5275^{20}	sl H_2O; s EtOH, bz, chl
7422	5-Methyl-2-isopropylphenyl acetate	Thymol, acetate	$C_{12}H_{16}O_2$	528-79-0	192.254			245	1.009^{9}		vs bz, eth, EtOH, chl
7423	1-Methylisoquinoline	Isoquinaldine	$C_{10}H_9N$	1721-93-3	143.185		10	252(9)	1.0777^{20}	1.6095^{20}	sl H_2O; s eth, ace, bz
7424	3-Methylisoquinoline		$C_{10}H_9N$	1125-80-0	143.185	cry (eth)	63.4(0.4)	253(1)			sl H_2O, chl; s eth, ace
7425	Methyl isothiocyanate		C_2H_3NS	556-61-6	73.117		36	119	1.0691^{37}	1.5258	sl H_2O; msc EtOH; vs eth
7426	5-Methyl-3-isoxazolamine		$C_4H_6N_2O$	1072-67-9	98.103		62				
7427	4-Methylisoxazole		C_4H_5NO	6454-84-8	83.089	liq		127			
7428	5-Methylisoxazole		C_4H_5NO	5765-44-6	83.089			123(7)	1.023^{20}	1.4386^{20}	s DMSO
7429	Methyl lactate, (±)-	Methyl 2-hydroxypropanoate, (±)	$C_4H_8O_3$	2155-30-8	104.105	oil		144.8	1.0928^{20}	1.4141^{20}	vs H_2O, eth, EtOH
7430	Methyl linoleate		$C_{19}H_{34}O_2$	112-63-0	294.472		-36.6(0.5)	347(6)	0.8886^{10}	1.4638^{20}	vs eth, EtOH
7431	Methyl linolenate		$C_{19}H_{32}O_2$	301-00-8	292.456		-49(4)	348(8)	0.895^{25}	1.4709^{20}	

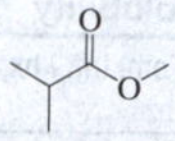
7395
Methyl isobutanoate

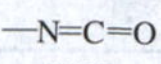

7396
Methyl isocyanate

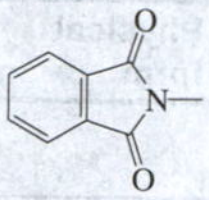
7397
2-Methyl-1*H*-isoindole-1,3(2*H*)-dione

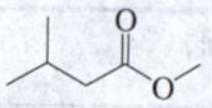
7398
Methyl isopentanoate

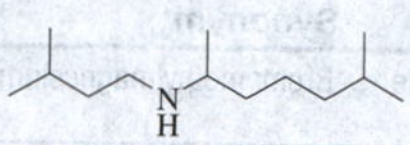
7399
6-Methyl-*N*-isopentyl-2-heptanamine

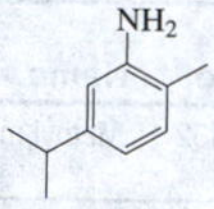
7400
2-Methyl-5-isopropylaniline

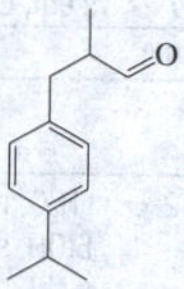
7401
α-Methyl-4-isopropylbenzenepropanal

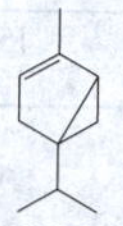
7402
2-Methyl-5-isopropylbicyclo[3.1.0]hex-2-ene

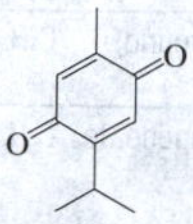
7403
2-Methyl-5-isopropyl-2,5-cyclohexadiene-1,4-dione

7404
cis-1-Methyl-4-isopropylcyclohexane

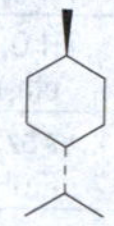
7405
trans-1-Methyl-4-isopropylcyclohexane

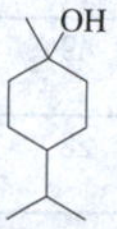
7406
1-Methyl-4-isopropylcyclohexanol

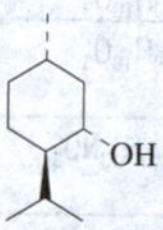
7407
5-Methyl-2-isopropylcyclohexanol, [1*S*-(1α,2β,5α)]-

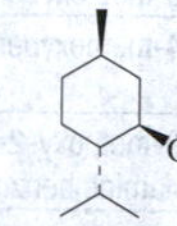
7408
5-Methyl-2-isopropylcyclohexanol, [1*R*-(1α,2β,5α)]-

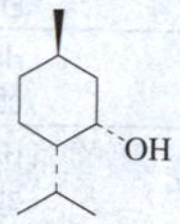
7409
5-Methyl-2-isopropylcyclohexanol, [1*S*-(1α,2α,5β)]-

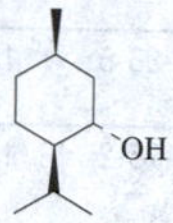
7410
5-Methyl-2-isopropylcyclohexanol, [1*S*-(1α,2β,5β)]-

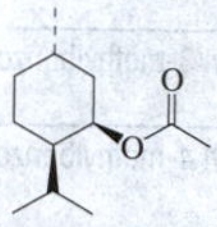
7411
5-Methyl-2-isopropylcyclohexanol acetate, [1*R*-(1α,2α,5β)]

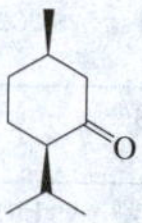
7412
cis-5-Methyl-2-isopropylcyclohexanone

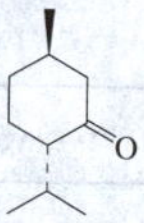
7413
trans-5-Methyl-2-isopropylcyclohexanone, (2*S*)-

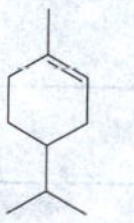
7414
1-Methyl-4-isopropylcyclohexene

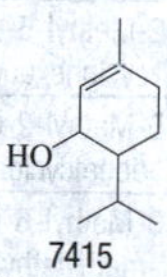
7415
3-Methyl-6-isopropyl-2-cyclohexen-1-ol

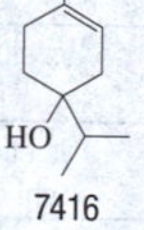
7416
4-Methyl-1-isopropyl-3-cyclohexen-1-ol

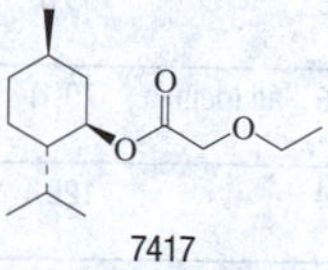
7417
5-Methyl-2-isopropylcyclohexyl ethoxyacetate, (1α,2β,5α)

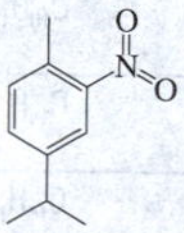
7418
1-Methyl-4-isopropyl-2-nitrobenzene

7419
1-Methyl-4-isopropyl-7-oxabicyclo[2.2.1]heptane

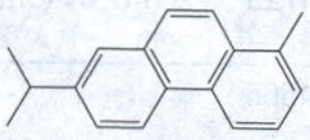
7420
1-Methyl-7-isopropylphenanthrene

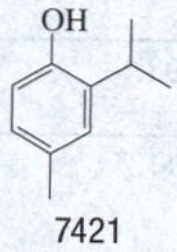
7421
4-Methyl-2-isopropylphenol

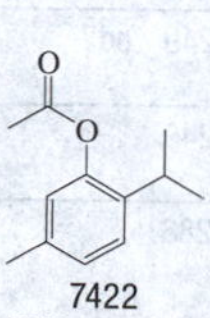
7422
5-Methyl-2-isopropylphenyl acetate

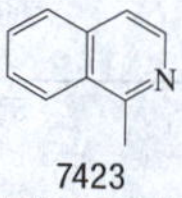
7423
1-Methylisoquinoline

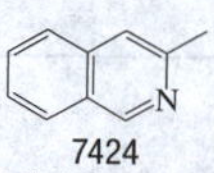
7424
3-Methylisoquinoline

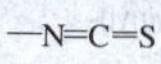

7425
Methyl isothiocyanate

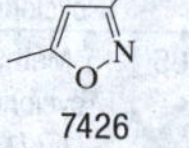
7426
5-Methyl-3-isoxazolamine

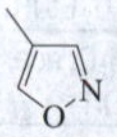
7427
4-Methylisoxazole

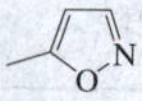
7428
5-Methylisoxazole

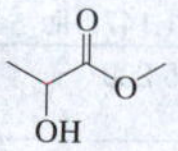
7429
Methyl lactate, (±)-

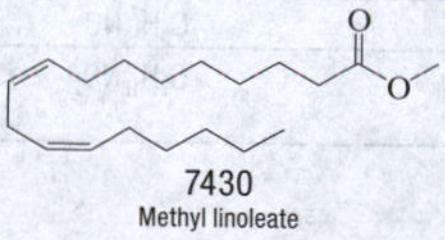
7430
Methyl linoleate

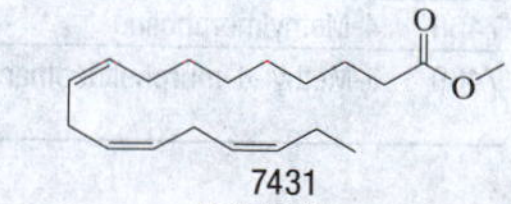
7431
Methyl linolenate

Organic

No.	Name	Synonym	Mol. form.	CAS Reg. No.	Mol. wt.	Physical form	mp/°C	bp/°C	den/ g cm^{-3}	n_D	Solubility
7432	Methyl magnesium bromide	Bromomethylmagnesium	CH_3BrMg	75-16-1	119.244						s eth, thf; i hx, bz
7433	Methylmagnesium chloride	Chloromethylmagnesium	CH_3ClMg	676-58-4	74.793	stab in thf soln					i peth, bz
7434	Methylmalonic acid	2-Methylpropanedioc acid	$C_4H_6O_4$	516-05-2	118.089	nd (bz-AcOEt) pr (eth-bz)	134(1)		1.455^{20}		vs H_2O, EtOH, eth; sl bz, tfa; s AcOEt
7435	Methyl mercaptoacetate		$C_3H_6O_2S$	2365-48-2	106.144			42^{10}		1.4657^{20}	vs eth, EtOH
7436	Methyl 3-mercaptopropanoate		$C_4H_8O_2S$	2935-90-2	120.171			54^{14}	1.085^{25}	1.4640^{20}	
7437	Methylmercuric dicyanamide	1-Cyano-3-(methylmercurio)-guanidine	$C_3H_6HgN_4$	502-39-6	298.70		157				
7438	Methyl methacrylate	Methyl 2-methyl-2-propenoate	$C_5H_8O_2$	80-62-6	100.117	liq	-47.55(0.02)	100.6(0.2)	0.9377^{25}	1.4142^{20}	sl H_2O; msc EtOH, eth, ace; s chl
7439	Methyl methanesulfonate		$C_2H_6O_3S$	66-27-3	110.132		20	185(9)	1.2943^{20}	1.4138^{20}	
7440	Methyl methoxyacetate		$C_4H_8O_3$	6290-49-9	104.105			131	1.0511^{20}	1.3962^{20}	sl H_2O; vs EtOH, eth, ace
7441	Methyl 2-methoxybenzoate		$C_9H_{10}O_3$	606-45-1	166.173			246.5	1.1571^{19}	1.534^{19}	i H_2O; s EtOH
7442	Methyl 3-methoxybenzoate		$C_9H_{10}O_3$	5368-81-0	166.173			248	1.1310^{20}	1.5224^{20}	i H_2O; s EtOH
7443	Methyl 4-methoxybenzoate		$C_9H_{10}O_3$	121-98-2	166.173	lf (al or eth)	49	259(2)			i H_2O; s EtOH, eth, chl
7444	Methyl 3-methoxy-2-(methylamino)benzoate	Damascenine	$C_{10}H_{13}NO_3$	483-64-7	195.215	pr (al)	28	271			vs bz, eth, EtOH, lig
7445	Methyl 3-methoxypropanoate		$C_5H_{10}O_3$	3852-09-3	118.131			141(2)	1.0139^{15}	1.4030^{20}	
7446	Methyl 2-methylacetoacetate		$C_6H_{10}O_3$	17094-21-2	130.141			177.4	1.0217^{25}	1.416^{24}	vs eth, EtOH
7447	Methyl 2-(methylamino)-benzoate		$C_9H_{11}NO_2$	85-91-6	165.189	cry (peth)	19	255	1.120^{15}	1.5839^{15}	i H_2O; s EtOH, eth
7448	Methyl 2-methylbenzoate		$C_9H_{10}O_2$	89-71-4	150.174		-44.38(0.04)	215	1.068^{20}		i H_2O; msc EtOH, eth
7449	Methyl 3-methylbenzoate		$C_9H_{10}O_2$	99-36-5	150.174			220(12)	1.061^{20}		i H_2O; s EtOH; sl ctc
7450	Methyl 4-methylbenzoate		$C_9H_{10}O_2$	99-75-2	150.174	cry (aq MeOH, peth)	33.35(0.04)	222(13)			i H_2O; vs EtOH, eth
7451	Methyl 2-methyl-2-butenoate, (*E*)-		$C_6H_{10}O_2$	6622-76-0	114.142			139	0.9349^{12}	1.4370^{20}	
7452	Methyl 3-methyl-2-butenoate		$C_6H_{10}O_2$	924-50-5	114.142		114	136.5	0.9337^{20}	1.432^{20}	
7453	3-Methyl-4-methylenehexane		C_8H_{16}	3404-67-9	112.213			113(3)	0.725^{25}	1.4142^{20}	
7454	2-Methyl-5-(1-methylethenyl)-cyclohexanone, (2*R*-*trans*)-		$C_{10}H_{16}O$	5524-05-0	152.233			221.5	0.928^{19}	1.4724	vs ace, eth
7455	5-Methyl-2-(1-methylethylidene)cyclohexanone		$C_{10}H_{16}O$	15932-80-6	152.233			93^{10}	0.9367^{20}	1.4869^{20}	
7456	3-Methyl-6-(1-methylethylidene)-2-cyclohexen-1-one	Piperitenone	$C_{10}H_{14}O$	491-09-8	150.217			120^{14}	0.9774^{20}	1.5294^{20}	vs EtOH, eth
7457	1-Methyl-4-(5-methyl-1-methylene-4-hexenyl)-cyclohexene, (*S*)-		$C_{15}H_{24}$	495-61-4	204.352			129^{10}	0.8673^{20}	1.4880^{20}	
7458	*N*-Methyl-*N*-(2-methylphenyl)-acetamide		$C_{10}H_{13}NO$	573-26-2	163.216		55.5	260			s EtOH, chl
7459	4-Methyl-*N*-(4-methylphenyl)-aniline		$C_{14}H_{15}N$	620-93-9	197.276	nd (peth)	79(1)	330.5			vs eth, peth
7460	2-Methyl-3-(2-methylphenyl)-4(3*H*)-quinazolinone	Methaqualone	$C_{16}H_{14}N_2O$	72-44-6	250.294		120				vs eth, EtOH, chl
7461	Methyl 3-(methylthio)-propanoate	2-Methoxycarbonylethyl methyl sulfide	$C_5H_{10}O_2S$	13532-18-8	134.197			75^{13}	1.077^{25}	1.4650^{20}	
7462	1-Methyl-4-(1-methylvinyl)-benzene		$C_{10}H_{12}$	1195-32-0	132.202	liq	-20	186(3)	0.8936^{23}	1.5283^{23}	
7463	1-Methyl-4-(1-methylvinyl)-cyclohexanol	β-Terpineol	$C_{10}H_{18}O$	138-87-4	154.249	nd	32.5	210	0.917^{20}	1.4747^{20}	
7464	5-Methyl-2-(1-methylvinyl)-cyclohexanol, [1*R*-(1α,2β,5α)]		$C_{10}H_{18}O$	89-79-2	154.249		78	93^{14}	0.911^{20}	1.4723^{20}	sl H_2O; s EtOH, eth
7465	5-Methyl-2-(1-methylvinyl)-cyclohexanol acetate, [1*R*-(1α,2β,5α)]		$C_{12}H_{20}O_2$	57576-09-7	196.286		85	113^{8}	0.925^{25}	1.4566^{20}	
7466	*trans*-5-Methyl-2-(1-methylvinyl)cyclohexanone		$C_{10}H_{16}O$	29606-79-9	152.233			100^{18}	0.9198^{20}	1.4675^{20}	
7467	2-Methyl-5-(1-methylvinyl)-2-cyclohexen-1-ol		$C_{10}H_{16}O$	99-48-9	152.233			228	0.9484^{25}	1.4942^{25}	
7468	4-Methylmorpholine		$C_5H_{11}NO$	109-02-4	101.147	liq	-64.40	121.4(1)	0.9051^{20}	1.4332^{20}	s H_2O, EtOH, eth
7469	α-Methyl-4-morpholineethanol		$C_7H_{15}NO_2$	2109-66-2	145.200			121^{18}	1.0174^{20}	1.4638^{20}	vs H_2O, ace, bz, EtOH

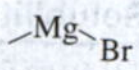
7432
Methyl magnesium bromide

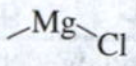
7433
Methylmagnesium chloride

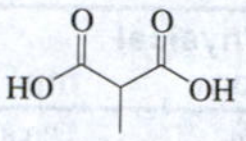
7434
Methylmalonic acid

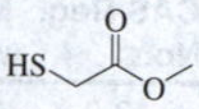
7435
Methyl mercaptoacetate

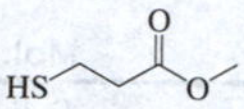
7436
Methyl 3-mercaptopropanoate

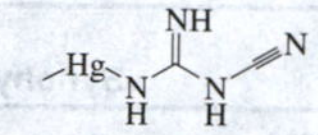
7437
Methylmercuric dicyanamide

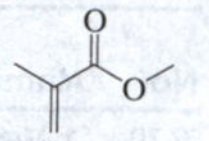
7438
Methyl methacrylate

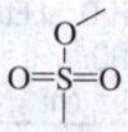
7439
Methyl methanesulfonate

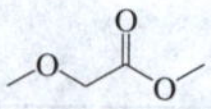
7440
Methyl methoxyacetate

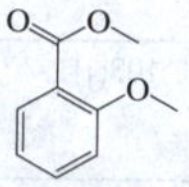
7441
Methyl 2-methoxybenzoate

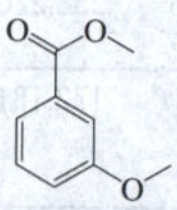
7442
Methyl 3-methoxybenzoate

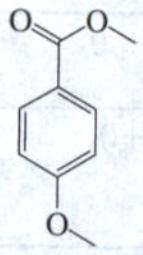
7443
Methyl 4-methoxybenzoate

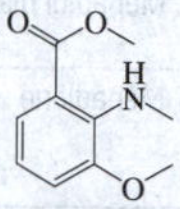
7444
Methyl 3-methoxy-2-(methylamino)benzoate

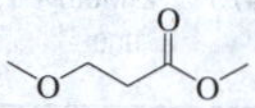
7445
Methyl 3-methoxypropanoate

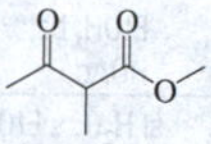
7446
Methyl 2-methylacetoacetate

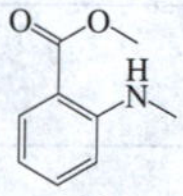
7447
Methyl 2-(methylamino)benzoate

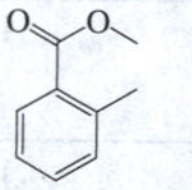
7448
Methyl 2-methylbenzoate

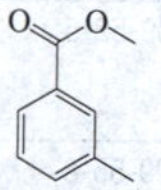
7449
Methyl 3-methylbenzoate

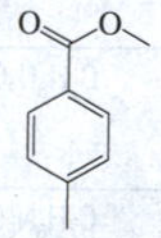
7450
Methyl 4-methylbenzoate

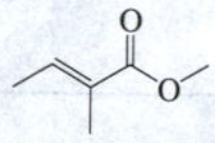
7451
Methyl 2-methyl-2-butenoate, (*E*)-

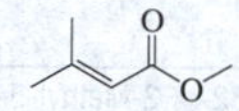
7452
Methyl 3-methyl-2-butenoate

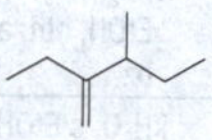
7453
3-Methyl-4-methylenehexane

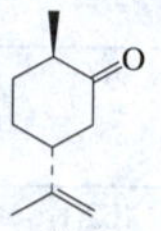
7454
2-Methyl-5-(1-methylethenyl)cyclohexanone, (2*R*-*trans*)-

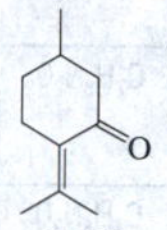
7455
5-Methyl-2-(1-methylethylidene)cyclohexanone

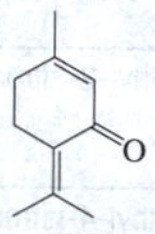
7456
3-Methyl-6-(1-methylethylidene)-2-cyclohexen-1-one

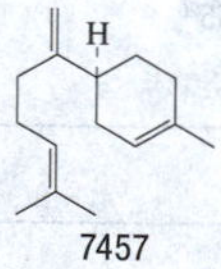
7457
1-Methyl-4-(5-methyl-1-methylene-4-hexenyl)cyclohexene, (*S*)-

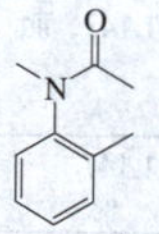
7458
N-Methyl-*N*-(2-methylphenyl)acetamide

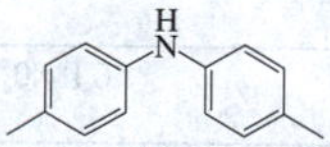
7459
4-Methyl-*N*-(4-methylphenyl)aniline

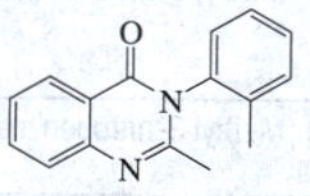
7460
2-Methyl-3-(2-methylphenyl)-4(3*H*)-quinazolinone

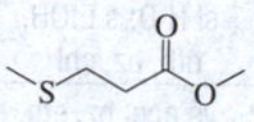
7461
Methyl 3-(methylthio)propanoate

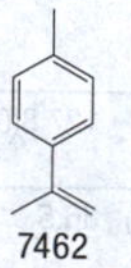
7462
1-Methyl-4-(1-methylvinyl)benzene

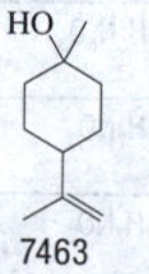
7463
1-Methyl-4-(1-methylvinyl)cyclohexanol

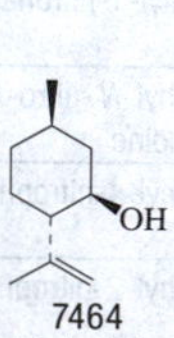
7464
5-Methyl-2-(1-methylvinyl)cyclohexanol, [1*R*-(1α,2β,5α)]

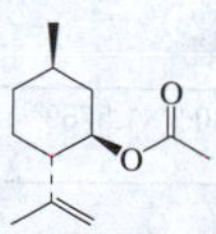
7465
5-Methyl-2-(1-methylvinyl)cyclohexanol acetate, [1*R*-(1α,2β,5α)]

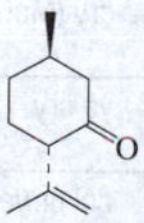
7466
trans-5-Methyl-2-(1-methylvinyl)cyclohexanone

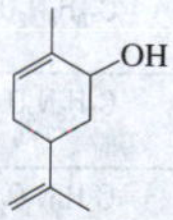
7467
2-Methyl-5-(1-methylvinyl)-2-cyclohexen-1-ol

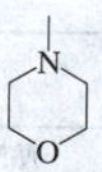
7468
4-Methylmorpholine

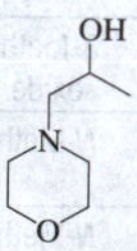
7469
α-Methyl-4-morpholineethanol

No.	Name	Synonym	Mol. form.	CAS Reg. No.	Mol. wt.	Physical form	mp/°C	bp/°C	den/ g cm⁻³	n_D	Solubility
7470	1-Methylnaphthalene		$C_{11}H_{10}$	90-12-0	142.197	liq	-30.43(0.07)	244.4(0.9)	1.0202^{20}	1.6170^{20}	i H_2O; vs EtOH, eth; s bz
7471	2-Methylnaphthalene		$C_{11}H_{10}$	91-57-6	142.197	mcl (al)	34.6(0.4)	241.1(0.3)	1.0058^{20}	1.6015^{40}	i H_2O; vs EtOH, eth; s bz, chl
7472	Methyl 1-naphthalenecarboxylate	Methyl 1-naphthoate	$C_{12}H_{10}O_2$	2459-24-7	186.206		59.5	168^{20}	1.1290^{20}	1.6086^{20}	vs bz, EtOH
7473	Methyl 2-naphthalenecarboxylate	Methyl 2-naphthoate	$C_{12}H_{10}O_2$	2459-25-8	186.206	lf (MeOH)	76.8(0.5)	290			vs bz, eth, EtOH, chl
7474	2-Methyl-1,4-naphthalenediol diacetate	Menadiol diacetate	$C_{15}H_{14}O_4$	573-20-6	258.270	pr (al)	113				vs EtOH
7475	2-Methyl-1,4-naphthalenedione	Menadione	$C_{11}H_8O_2$	58-27-5	172.181	ye nd (al, peth)	103(1)				i H_2O; sl EtOH, HOAc; s eth, bz, chl
7476	Methyl-1-naphthylamine	*N*-Methyl-1-naphthalenamine	$C_{11}H_{11}N$	2216-68-4	157.212	oil	174	294.5		1.6722^{20}	vs eth, EtOH
7477	Methyl nitrate		CH_3NO_3	598-58-3	77.040	exp gas	-82.9(0.5)	65(2)	1.2075^{20}	1.3748^{20}	sl H_2O; s EtOH, eth
7478	Methyl nitrite		CH_3NO_2	624-91-9	61.041	ye gas	-16(2)	-12	0.991^{15}		s EtOH, eth
7479	Methyl nitroacetate		$C_3H_5NO_4$	2483-57-0	119.077			107^{28}	1.320^{0}		
7480	2-Methyl-3-nitroaniline		$C_7H_8N_2O_2$	603-83-8	152.151	ye orth nd (w), ye lf (al)	92	305	1.3780^{15}		sl H_2O; s EtOH, eth, bz, chl
7481	2-Methyl-4-nitroaniline		$C_7H_8N_2O_2$	99-52-5	152.151		128(1)		1.1586^{140}		sl H_2O, DMSO; s EtOH, bz, HOAc
7482	2-Methyl-5-nitroaniline		$C_7H_8N_2O_2$	99-55-8	152.151		105.5				sl H_2O; s EtOH, eth, ace, bz, chl
7483	2-Methyl-6-nitroaniline		$C_7H_8N_2O_2$	570-24-1	152.151		96(1)		1.1900^{100}		sl H_2O; s EtOH, eth, bz, chl
7484	4-Methyl-2-nitroaniline		$C_7H_8N_2O_2$	89-62-3	152.151		117(1)		1.16^{121}		sl H_2O; s EtOH, chl
7485	4-Methyl-3-nitroaniline		$C_7H_8N_2O_2$	119-32-4	152.151		78(1)				sl H_2O, CS_2; s EtOH, eth, bz
7486	*N*-Methyl-2-nitroaniline		$C_7H_8N_2O_2$	612-28-2	152.151	red or oran nd (peth)	38	158^{18}			sl H_2O, lig; s EtOH, eth, ace, bz
7487	*N*-Methyl-4-nitroaniline		$C_7H_8N_2O_2$	100-15-2	152.151	br-ye pr (al) cry (eth)	152(1)	dec	1.201^{155}		i H_2O; s EtOH, bz, chl; sl eth, lig
7488	2-Methyl-1-nitro-9,10-anthracenedione		$C_{15}H_9NO_4$	129-15-7	267.237	pa ye nd (HOAc)	273.0				i H_2O, EtOH; sl eth, bz, chl; s $PhNO_2$
7489	2-Methyl-5-nitrobenzenesulfonic acid		$C_7H_7NO_5S$	121-03-9	217.200		135.8				vs H_2O, EtOH, eth, chl
7490	Methyl 2-nitrobenzoate		$C_8H_7NO_4$	606-27-9	181.147	liq	-13	275	1.2855^{20}		i H_2O; s EtOH, eth, bz, chl; i lig
7491	Methyl 3-nitrobenzoate		$C_8H_7NO_4$	618-95-1	181.147		78(1)	279^{60}			i H_2O; sl EtOH, eth, MeOH
7492	Methyl 4-nitrobenzoate		$C_8H_7NO_4$	619-50-1	181.147		96				i H_2O; s EtOH, eth, chl
7493	2-Methyl-4-nitro-1*H*-imidazole		$C_4H_5N_3O_2$	696-23-1	127.102		253				
7494	*N*-Methyl-*N*-nitromethanamine		$C_2H_6N_2O_2$	4164-28-7	90.081	nd(eth)	56(1)	187	1.1090^{72}	1.4462^{72}	vs H_2O, ace, eth, EtOH
7495	2-Methyl-1-nitronaphthalene		$C_{11}H_9NO_2$	881-03-8	187.195	ye pr or nd (al)	81.5	188^{20}			i H_2O; s EtOH; vs ace
7496	*N*-Methyl-*N'*-nitro-*N*-nitrosoguanidine		$C_2H_5N_5O_3$	70-25-7	147.093						s DMSO
7497	3-Methyl-4-nitrophenol		$C_7H_7NO_3$	2581-34-2	153.136	nd or pr (w)	127.8(0.2)				sl H_2O; s EtOH, eth, bz, chl
7498	4-Methyl-2-nitrophenol		$C_7H_7NO_3$	119-33-5	153.136	ye nd (al, w)	36.5	125^{22}	1.2399^{20}	1.5744^{40}	vs ace, bz, eth, EtOH
7499	1-Methyl-2-(4-nitrophenoxy)-benzene	2-Methylphenyl 4-nitrophenyl ether	$C_{13}H_{11}NO_3$	2444-29-3	229.231	ye cry (peth)		220^{27}			vs bz, eth, EtOH
7500	2-Methyl-2-nitro-1,3-propanediol		$C_4H_9NO_4$	77-49-6	135.119	mcl	150.7(0.5)	dec			vs H_2O, EtOH; sl DMSO
7501	2-Methyl-2-nitro-1-propanol		$C_4H_9NO_3$	76-39-1	119.119	nd or pl (MeOH)	88.5(0.5)	94^{10}			sl H_2O; vs EtOH, eth; s chl
7502	3-Methyl-4-nitroquinoline-*N*-oxide		$C_{10}H_8N_2O_3$	14073-00-8	204.182	cry (MeOH)	179				
7503	*N*-Methyl-*N*-nitrosoaniline		$C_7H_8N_2O$	614-00-6	136.151	ye cry	14.7	225 dec	1.1240^{20}	1.5769^{20}	i H_2O; s EtOH, eth
7504	*N*-Methyl-*N*-nitrosourea	*N*-Nitroso-*N*-methylurea	$C_2H_5N_3O_2$	684-93-5	103.080	col or ye pl (eth)	123 dec				sl H_2O, EtOH, eth

Organic

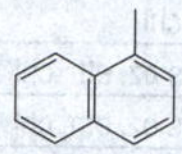
7470
1-Methylnaphthalene

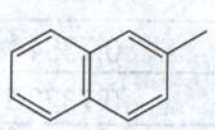
7471
2-Methylnaphthalene

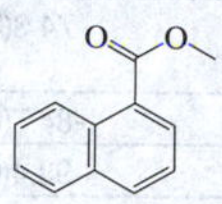
7472
Methyl 1-naphthalenecarboxylate

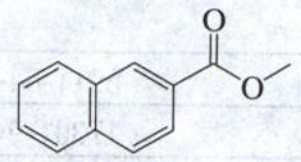
7473
Methyl 2-naphthalenecarboxylate

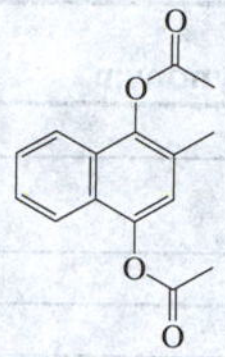
7474
2-Methyl-1,4-naphthalenediol diacetate

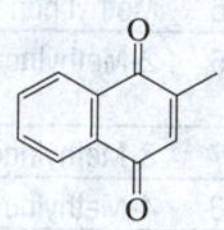
7475
2-Methyl-1,4-naphthalenedione

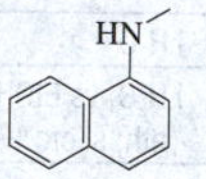
7476
Methyl-1-naphthylamine

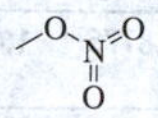
7477
Methyl nitrate

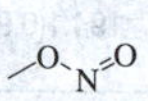
7478
Methyl nitrite

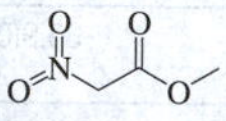
7479
Methyl nitroacetate

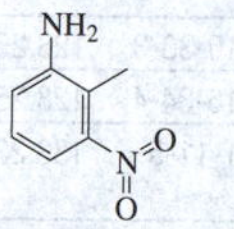
7480
2-Methyl-3-nitroaniline

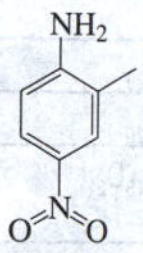
7481
2-Methyl-4-nitroaniline

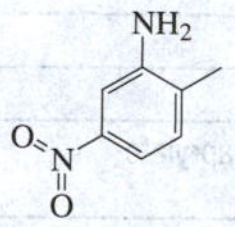
7482
2-Methyl-5-nitroaniline

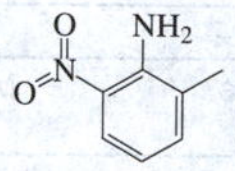
7483
2-Methyl-6-nitroaniline

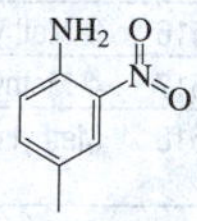
7484
4-Methyl-2-nitroaniline

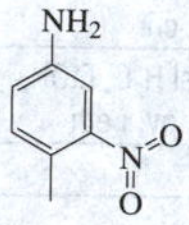
7485
4-Methyl-3-nitroaniline

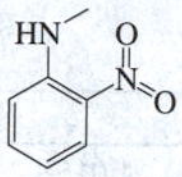
7486
N-Methyl-2-nitroaniline

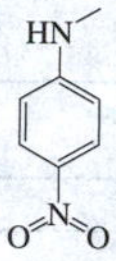
7487
N-Methyl-4-nitroaniline

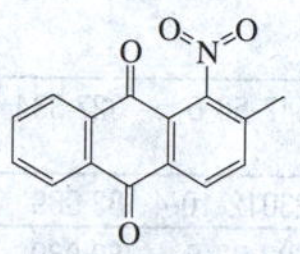
7488
2-Methyl-1-nitro-9,10-anthracenedione

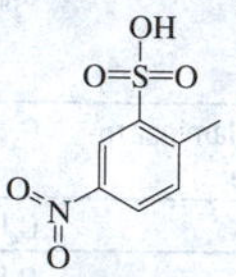
7489
2-Methyl-5-nitrobenzenesulfonic acid

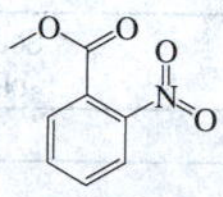
7490
Methyl 2-nitrobenzoate

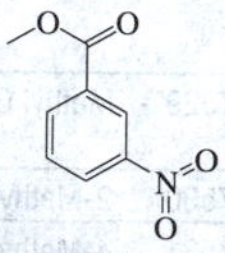
7491
Methyl 3-nitrobenzoate

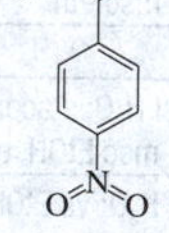
7492
Methyl 4-nitrobenzoate

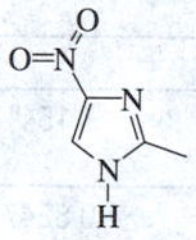
7493
2-Methyl-4-nitro-1*H*-imidazole

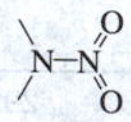
7494
N-Methyl-*N*-nitromethanamine

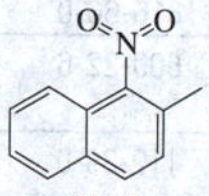
7495
2-Methyl-1-nitronaphthalene

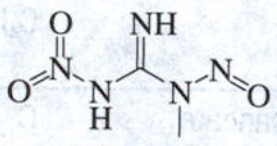
7496
N-Methyl-*N'*-nitro-*N*-nitrosoguanidine

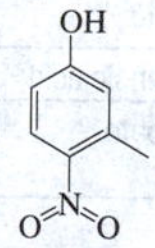
7497
3-Methyl-4-nitrophenol

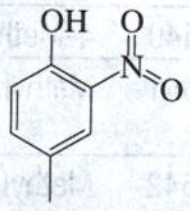
7498
4-Methyl-2-nitrophenol

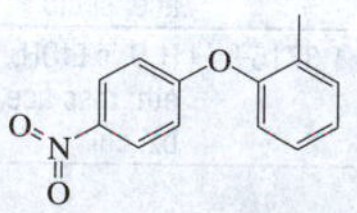
7499
1-Methyl-2-(4-nitrophenoxy)benzene

7500
2-Methyl-2-nitro-1,3-propanediol

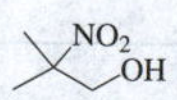
7501
2-Methyl-2-nitro-1-propanol

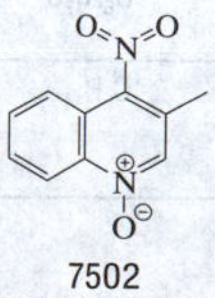
7502
3-Methyl-4-nitroquinoline-*N*-oxide

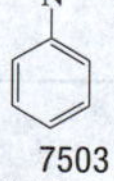
7503
N-Methyl-*N*-nitrosoaniline

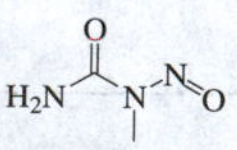
7504
N-Methyl-*N*-nitrosourea

Organic

No.	Name	Synonym	Mol. form.	CAS Reg. No.	Mol. wt.	Physical form	mp/°C	bp/°C	den/ g cm^{-3}	n_D	Solubility
7505	Methyl nonadecanoate		$C_{20}H_{40}O_2$	1731-94-8	312.531		38.3(0.6)	190^{4}			
7506	2-Methylnonane		$C_{10}H_{22}$	871-83-0	142.282	liq	-74.3(0.2)	167(4)	0.7281^{20}	1.4099^{20}	i H_2O; s eth, bz, chl
7507	3-Methylnonane		$C_{10}H_{22}$	5911-04-6	142.282	liq	-84.7(0.2)	168(2)	0.7354^{20}	1.4125^{20}	vs bz, eth, chl
7508	4-Methylnonane		$C_{10}H_{22}$	17301-94-9	142.282	liq	-98.4(0.2)	166(3)	0.7323^{20}	1.4123^{20}	vs bz, eth, chl
7509	5-Methylnonane		$C_{10}H_{22}$	15869-85-9	142.282	liq	-86.408(0.01)	164(5)	0.7326^{20}	1.4116^{20}	i H_2O; s eth, bz, chl
7510	Methyl nonanoate		$C_{10}H_{20}O_2$	1731-84-6	172.265			212(3)	0.8799^{15}	1.4214^{20}	i H_2O; s EtOH, eth; sl ctc
7511	8-Methyl-1-nonanol		$C_{10}H_{22}O$	55505-26-5	158.281			108^{10}			
7512	2-Methyl-1-nonene		$C_{10}H_{20}$	2980-71-4	140.266	liq	-64.2	167(2)	0.7412^{25}	1.4241^{20}	
7513	2-Methyl-2-norbornene	2-Methylbicyclo[2.2.1]hept-2-ene	C_8H_{12}	694-92-8	108.181	liq		118(7)			
7514	Methyl *trans*-9-octadecenoate		$C_{19}H_{36}O_2$	1937-62-8	296.488	liq	10.3(0.3)	218^{24}	0.8730^{20}	1.4513^{20}	vs eth, EtOH
7515	2-Methyloctane		C_9H_{20}	3221-61-2	128.255	liq	-80.3(0.3)	143(1)	0.7095^{25}	1.4031^{20}	i H_2O; s EtOH, eth; sl ctc; vs peth
7516	3-Methyloctane		C_9H_{20}	2216-33-3	128.255	liq	-108.0(0.2)	144(2)	0.717^{25}	1.4040^{25}	
7517	4-Methyloctane		C_9H_{20}	2216-34-4	128.255	liq	-116(5)	142(1)	0.716^{25}	1.4039^{25}	i H_2O
7518	Methyl octanoate	Methyl caprylate	$C_9H_{18}O_2$	111-11-5	158.238	liq	-36.9(0.4)	194.1(0.9)	0.8775^{20}	1.4170^{20}	i H_2O; vs EtOH, eth; sl ctc
7519	2-Methyloctanoic acid		$C_9H_{18}O_2$	3004-93-1	158.238			138^{14}		1.4281^{25}	
7520	2-Methyl-2-octanol		$C_9H_{20}O$	628-44-4	144.254			178	0.8210^{20}	1.4280^{20}	i H_2O; s EtOH, eth
7521	3-Methyl-3-octanol		$C_9H_{20}O$	5340-36-3	144.254			83^{18}	0.8108^{25}	1.4257^{25}	
7522	5-Methyl-2-octanone		$C_9H_{18}O$	58654-67-4	142.238			101^{50}			
7523	2-Methyl-1-octene		C_9H_{18}	4588-18-5	126.239	liq	-77.8(0.3)	144(3)	0.7343^{20}	1.4184^{20}	
7524	7-Methyl-1-octene		C_9H_{18}	13151-06-9	126.239	liq		139(3)			
7525	Methyloctylamine	*N*-Methyl-1-octanamine	$C_9H_{21}N$	2439-54-5	143.270			68^{8}			
7526	Methyl 2-octynoate		$C_9H_{14}O_2$	111-12-6	154.206			217	0.926^{20}	1.4464^{20}	
7527	3-Methyl-1-octyn-3-ol		$C_9H_{16}O$	23580-51-0	140.222			174	0.8547^{20}	1.443^{10}	
7528	Methyl oleate		$C_{19}H_{36}O_2$	112-62-9	296.488		-19.7(0.2)	347(5)	0.8739^{20}	1.4522^{20}	i H_2O; msc EtOH, eth; s chl
7529	Methyl Orange	Sodium *p*-dimethylaminoazobenzenesulfonate	$C_{14}H_{14}N_3NaO_3S$	547-58-0	327.334	oran, ye pl or sc (w)	dec				sl H_2O, EtOH, py; i eth
7530	2-Methyloxazole		C_4H_5NO	23012-10-4	83.089	liq		87.5			
7531	4-Methyloxazole		C_4H_5NO	693-93-6	83.089			88	1.015^{25}	1.4317^{20}	
7532	5-Methyloxazole		C_4H_5NO	66333-88-8	83.089	liq		88			
7533	2-Methyl-2-oxazoline		C_4H_7NO	1120-64-5	85.105			111	1.005^{25}	1.4340^{20}	
7534	2-Methyloxetane		C_4H_8O	2167-39-7	72.106	hyg		59	0.841^{25}	1.3885^{20}	
7535	4-Methyl-2-oxetanone	3-Hydroxybutyric acid lactone	$C_4H_6O_2$	3068-88-0	86.090			86^{50}	1.0555^{20}		
7536	Methyloxirane	1,2-Propylene oxide	C_3H_6O	16033-71-9	58.079	liq	-111.9	35	0.859^{0}	1.3660^{20}	vs H_2O, EtOH, eth; s chl
7537	3-Methyl-2-oxobutanoic acid		$C_5H_8O_3$	759-05-7	116.116		31.5	170.5	0.9968^{20}	1.3850^{16}	s H_2O, EtOH, eth
7538	*N*-Methyl-*N*-(1-oxododecyl)-glycine	*N*-Dodecanoylsarcosine	$C_{15}H_{29}NO_3$	97-78-9	271.396		44.5				s chl
7539	Methyl 4-oxopentanoate	Methyl levulinate	$C_6H_{10}O_3$	624-45-3	130.141			196	1.0511^{20}	1.4233^{20}	sl H_2O; s EtOH, ace, bz, ctc; msc eth
7540	4-Methyl-2-oxopentanoic acid		$C_6H_{10}O_3$	816-66-0	130.141	liq	10	84^{15}			
7541	Methyl 2-oxopropanoate	Methyl pyruvate	$C_4H_6O_3$	600-22-6	102.089			138(5)	1.154^{0}	1.4046^{25}	sl H_2O; s ace; msc EtOH, eth
7542	Methyl palmitate	Methyl hexadecanoate	$C_{17}H_{34}O_2$	112-39-0	270.451		29.6(0.5)	324(6)	0.8247^{75}		i H_2O; vs EtOH, ace, bz; s eth
7543	Methyl parathion		$C_8H_{10}NO_5PS$	298-00-0	263.208	cry	35.8(0.4)		1.358^{20}	1.5367^{25}	i H_2O; s os
7544	Methyl pentachlorophenyl sulfide	*S*-Methyl pentachlorobenzenethiol	$C_7H_3Cl_5S$	1825-19-0	296.429	cry (EtOH)	95.5				
7545	Methyl pentadecanoate		$C_{16}H_{32}O_2$	7132-64-1	256.424	nd (dil al)	19(2)	308(7)	0.8618^{25}	1.4390^{25}	s EtOH, eth
7546	*cis*-2-Methyl-1,3-pentadiene		C_6H_{10}	1501-60-6	82.143	liq	-117.6	80(16)	0.714^{25}	1.446^{20}	
7547	3-Methyl-1,3-pentadiene		C_6H_{10}	4549-74-0	82.143			77(5)	0.730^{25}	1.452^{20}	
7548	4-Methyl-1,3-pentadiene	1,1-Dimethyl-1,3-butadiene	C_6H_{10}	926-56-7	82.143			76(3)	0.7181^{20}	1.4532^{20}	
7549	Methyl pentafluoroethyl ether	1-Methoxyperfluoroethane	$C_3H_3F_5O$	22410-44-2	150.047	col gas		5.6(0.9)			
7550	Methyl pentafluoropropanoate		$C_4H_3F_5O_2$	378-75-6	178.058			59.5	1.390^{25}	1.2869^{25}	
7551	2-Methylpentanal	2-Methylvaleraldehyde	$C_6H_{12}O$	123-15-9	100.158			118(4)			s H_2O; s eth, ace; sl ctc
7552	2-Methylpentane	Isohexane	C_6H_{14}	107-83-5	86.175	liq	-153.60(0.09)	60.21(0.09)	0.650^{25}	1.3715^{20}	i H_2O; s EtOH, eth; msc ace, bz, chl

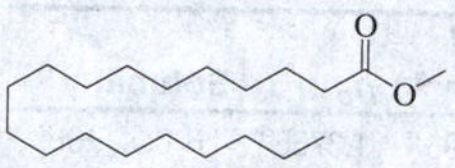
7505
Methyl nonadecanoate

7506
2-Methylnonane

7507
3-Methylnonane

7508
4-Methylnonane

7509
5-Methylnonane

7510
Methyl nonanoate

7511
8-Methyl-1-nonanol

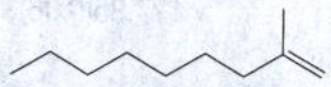
7512
2-Methyl-1-nonene

7513
2-Methyl-2-norbornene

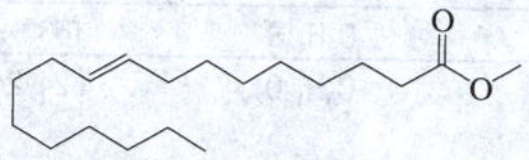
7514
Methyl *trans*-9-octadecenoate

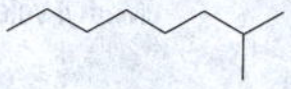
7515
2-Methyloctane

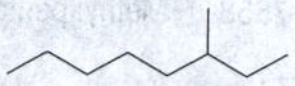
7516
3-Methyloctane

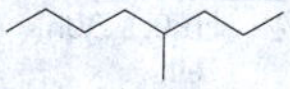
7517
4-Methyloctane

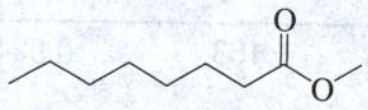
7518
Methyl octanoate

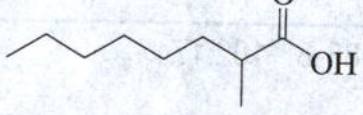
7519
2-Methyloctanoic acid

7520
2-Methyl-2-octanol

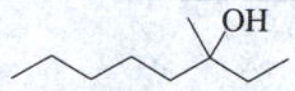
7521
3-Methyl-3-octanol

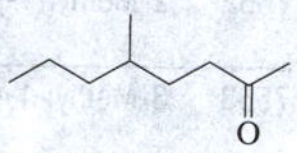
7522
5-Methyl-2-octanone

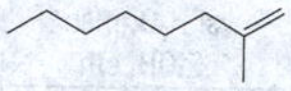
7523
2-Methyl-1-octene

7524
7-Methyl-1-octene

7525
Methyloctylamine

7526
Methyl 2-octynoate

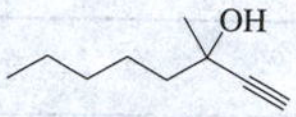
7527
3-Methyl-1-octyn-3-ol

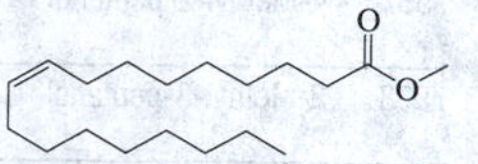
7528
Methyl oleate

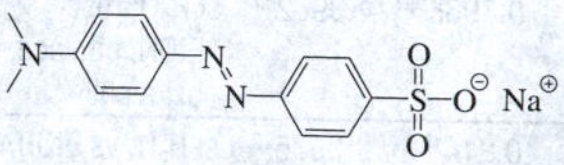
7529
Methyl Orange

7530
2-Methyloxazole

7531
4-Methyloxazole

7532
5-Methyloxazole

7533
2-Methyl-2-oxazoline

7534
2-Methyloxetane

7535
4-Methyl-2-oxetanone

7536
Methyloxirane

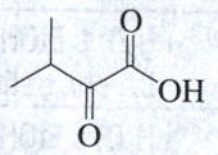
7537
3-Methyl-2-oxobutanoic acid

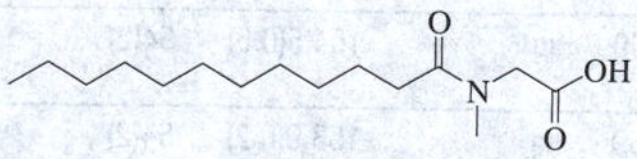
7538
N-Methyl-*N*-(1-oxododecyl)glycine

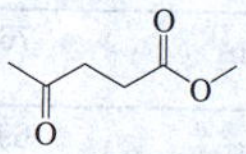
7539
Methyl 4-oxopentanoate

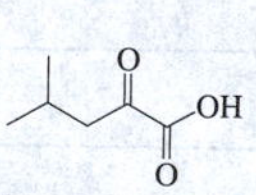
7540
4-Methyl-2-oxopentanoic acid

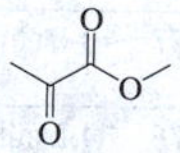
7541
Methyl 2-oxopropanoate

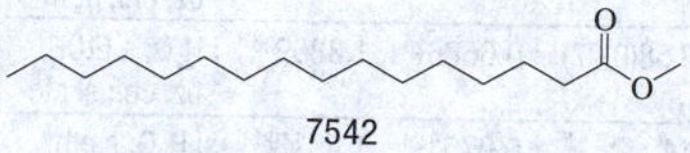
7542
Methyl palmitate

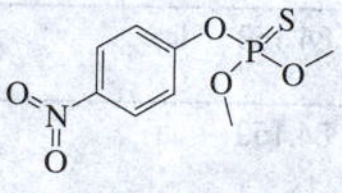
7543
Methyl parathion

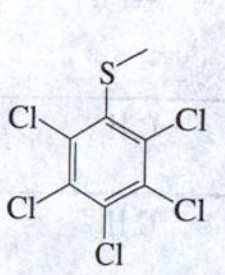
7544
Methyl pentachlorophenyl sulfide

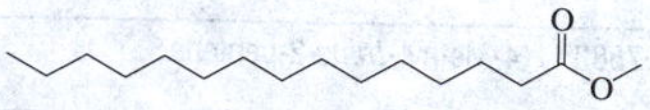
7545
Methyl pentadecanoate

7546
cis-2-Methyl-1,3-pentadiene

7547
3-Methyl-1,3-pentadiene

7548
4-Methyl-1,3-pentadiene

7549
Methyl pentafluoroethyl ether

7550
Methyl pentafluoropropanoate

7551
2-Methylpentanal

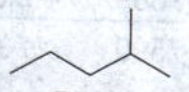
7552
2-Methylpentane

Organic

No.	Name	Synonym	Mol. form.	CAS Reg. No.	Mol. wt.	Physical form	mp/°C	bp/°C	den/ g cm^{-3}	n_D	Solubility
7553	3-Methylpentane		C_6H_{14}	96-14-0	86.175	liq	-162.89(0.05)	63.3(0.5)	0.6598^{25}	1.3765^{20}	i H_2O; s EtOH, ctc; msc eth, ace, bz, hp
7554	2-Methylpentanedinitrile	2-Methylglutaronitrile	$C_6H_8N_2$	4553-62-2	108.141	liq	-45	270	0.950	1.4340^{20}	s H_2O
7555	2-Methyl-2,4-pentanediol	Hexylene glycol	$C_6H_{14}O_2$	107-41-5	118.174	liq	-50	197.9(0.9)	0.923^{15}	1.4276^{20}	s H_2O, EtOH, eth; sl ctc
7556	4-Methylpentanenitrile	Isopentyl cyanide	$C_6H_{11}N$	542-54-1	97.158	liq	-51.1(0.4)	155(3)	0.8030^{20}	1.4059^{20}	i H_2O; s EtOH; msc eth; sl ctc
7557	2-Methyl-2-pentanethiol		$C_6H_{14}S$	1633-97-2	118.240	liq		125.0(0.5)			
7558	Methyl pentanoate	Methyl valerate	$C_6H_{12}O_2$	624-24-8	116.158			127.36(0.06)	0.8947^{20}	1.4003^{20}	sl H_2O, ctc; msc EtOH, eth; s ace
7559	2-Methylpentanoic acid, (±)-		$C_6H_{12}O_2$	22160-39-0	116.158			195(2)	0.9230^{20}	1.413^{20}	s H_2O, EtOH, eth; sl ctc
7560	3-Methylpentanoic acid, (±)-		$C_6H_{12}O_2$	22160-40-3	116.158	liq	-41.6(0.8)	197(1)	0.9262^{20}	1.4159^{20}	vs eth, EtOH
7561	4-Methylpentanoic acid		$C_6H_{12}O_2$	646-07-1	116.158	liq	-33	200(1)	0.9225^{20}	1.4144^{20}	sl H_2O; s EtOH, eth, chl
7562	2-Methyl-1-pentanol		$C_6H_{14}O$	105-30-6	102.174	liq		157(6)	0.8263^{20}	1.4182^{20}	sl H_2O; s EtOH, eth, ace, ctc
7563	3-Methyl-1-pentanol, (±)-		$C_6H_{14}O$	20281-83-8	102.174			153	0.8242^{20}	1.4112^{23}	i H_2O; s EtOH, eth
7564	4-Methyl-1-pentanol	Isohexyl alcohol	$C_6H_{14}O$	626-89-1	102.174			151(2)	0.8131^{20}	1.4134^{25}	i H_2O; s EtOH, eth
7565	2-Methyl-2-pentanol		$C_6H_{14}O$	590-36-3	102.174	liq	-103	121(1)	0.8350^{16}	1.4100^{20}	sl H_2O; s EtOH, eth
7566	3-Methyl-2-pentanol		$C_6H_{14}O$	565-60-6	102.174			142(2)	0.8307^{20}	1.4182^{20}	sl H_2O; s EtOH, eth
7567	4-Methyl-2-pentanol		$C_6H_{14}O$	108-11-2	102.174	liq	-90	132.0(0.5)	0.8075^{20}	1.4100^{20}	sl H_2O, ctc; s EtOH, eth
7568	2-Methyl-3-pentanol		$C_6H_{14}O$	565-67-3	102.174			127.9(0.2)	0.8243^{20}	1.4175^{20}	sl H_2O; msc EtOH, eth
7569	3-Methyl-3-pentanol		$C_6H_{14}O$	77-74-7	102.174	liq	-23.6(0.3)	129(4)	0.8286^{20}	1.4186^{20}	sl H_2O, ctc; msc EtOH, eth
7570	2-Methyl-1-pentanol acetate		$C_8H_{16}O_2$	7789-99-3	144.212			163	0.870^{25}		vs eth, EtOH
7571	3-Methyl-2-pentanone, (±)-	(±)-*sec*-Butyl methyl ketone	$C_6H_{12}O$	55156-16-6	100.158			117(2)	0.8130^{20}	1.4002^{20}	sl H_2O; msc EtOH, eth; s chl
7572	4-Methyl-2-pentanone	Isobutyl methyl ketone	$C_6H_{12}O$	108-10-1	100.158	liq	-85(2)	115.7(0.2)	0.7965^{25}	1.3962^{20}	sl H_2O; msc EtOH, eth, ace, bz; s chl
7573	2-Methyl-3-pentanone	Ethyl isopropyl ketone	$C_6H_{12}O$	565-69-5	100.158			118(1)	0.814^{18}	1.3975^{20}	sl H_2O; vs EtOH, bz; msc eth, ace; s chl
7574	4-Methylpentanoyl chloride		$C_6H_{11}ClO$	38136-29-7	134.603			143	0.9725^{20}		
7575	2-Methyl-2-pentenal		$C_6H_{10}O$	623-36-9	98.142			136(8)	0.8581^{20}	1.4488^{20}	i H_2O; s EtOH, eth, bz, MeOH
7576	2-Methyl-1-pentene		C_6H_{12}	763-29-1	84.159	liq	-135.7(0.4)	62.1(0.6)	0.6799^{20}	1.3920^{20}	i H_2O; s EtOH, bz, chl; sl ctc
7577	3-Methyl-1-pentene		C_6H_{12}	760-20-3	84.159	liq	-154.5(0.5)	54(2)	0.6675^{20}	1.3841^{20}	i H_2O; s EtOH, bz, chl, peth
7578	4-Methyl-1-pentene		C_6H_{12}	691-37-2	84.159	liq	-153.9(0.2)	54(2)	0.6642^{20}	1.3828^{20}	i H_2O; s EtOH, bz, chl, peth
7579	2-Methyl-2-pentene		C_6H_{12}	625-27-4	84.159	liq	-135.0(0.2)	67.3(0.5)	0.6863^{20}	1.4004^{20}	i H_2O; s EtOH, bz, ctc, chl
7580	3-Methyl-*cis*-2-pentene	*cis*-3-Methyl-2-pentene	C_6H_{12}	922-62-3	84.159	liq	-135.1(0.7)	70.4(0.1)	0.6886^{25}	1.4016^{20}	i H_2O; s EtOH, bz, chl, peth
7581	3-Methyl-*trans*-2-pentene	*trans*-3-Methyl-2-pentene	C_6H_{12}	616-12-6	84.159	liq	-138.5(0.3)	67.67(0.04)	0.6930^{25}	1.4045^{20}	i H_2O; s EtOH, bz, ctc, chl, peth
7582	4-Methyl-*cis*-2-pentene		C_6H_{12}	691-38-3	84.159	liq	-134.8	56.4(0.1)	0.6690^{20}	1.3800^{20}	i H_2O; s EtOH, bz, chl, peth
7583	4-Methyl-*trans*-2-pentene		C_6H_{12}	674-76-0	84.159	liq	-140.8	58.58(0.07)	0.6686^{20}	1.3889^{20}	i H_2O; s EtOH, bz, chl; sl ctc
7584	*trans*-2-Methyl-2-pentenoic acid		$C_6H_{10}O_2$	16957-70-3	114.142	pr	24.4	214	0.9751^{20}	1.4513^{20}	sl H_2O; s eth, chl, CS_2
7585	4-Methyl-2-pentenoic acid	4,4-Dimethyl-2-butenoic acid	$C_6H_{10}O_2$	10321-71-8	114.142		35	217	0.9529^{21}	1.4489^{21}	vs ace, eth, EtOH
7586	2-Methyl-3-pentenoic acid		$C_6H_{10}O_2$	37674-63-8	114.142			199	0.966^{15}	1.4402^{25}	
7587	4-Methyl-3-penten-2-ol		$C_6H_{12}O$	4325-82-0	100.158			134	0.840^{15}	1.9377^{15}	
7588	3-Methyl-2-penten-4-one		$C_6H_{10}O$	565-62-8	98.142			138		1.4508^{20}	
7589	4-Methyl-4-penten-2-one	Isomesityl oxide	$C_6H_{10}O$	3744-02-3	98.142	liq	-72.6(0.3)	124.2	0.8411^{20}		
7590	*cis*-3-Methyl-2-(2-pentenyl)-2-cyclopenten-1-one	Jasmone	$C_{11}H_{16}O$	488-10-8	164.244	ye oil		258	0.9437^{22}	1.4979^{22}	sl H_2O; s EtOH, eth, ctc, lig

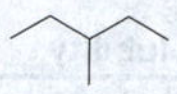
7553
3-Methylpentane

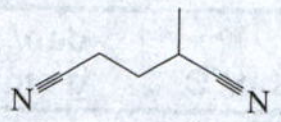
7554
2-Methylpentanedinitrile

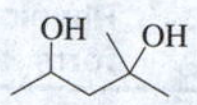
7555
2-Methyl-2,4-pentanediol

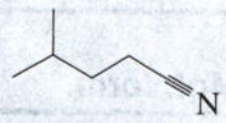
7556
4-Methylpentanenitrile

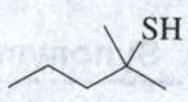
7557
2-Methyl-2-pentanethiol

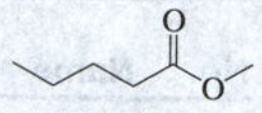
7558
Methyl pentanoate

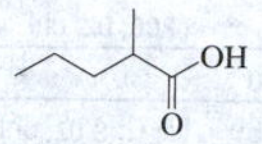
7559
2-Methylpentanoic acid, (±)-

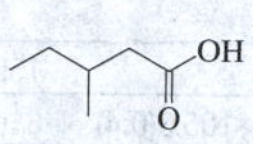
7560
3-Methylpentanoic acid, (±)-

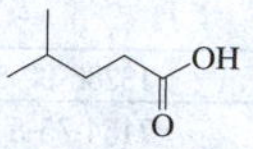
7561
4-Methylpentanoic acid

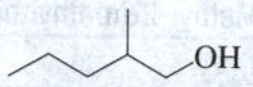
7562
2-Methyl-1-pentanol

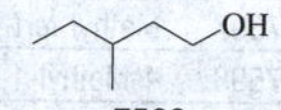
7563
3-Methyl-1-pentanol, (±)-

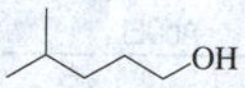
7564
4-Methyl-1-pentanol

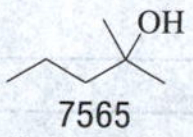
7565
2-Methyl-2-pentanol

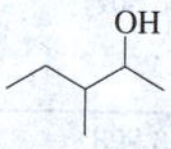
7566
3-Methyl-2-pentanol

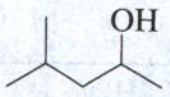
7567
4-Methyl-2-pentanol

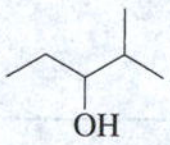
7568
2-Methyl-3-pentanol

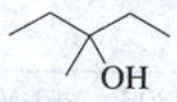
7569
3-Methyl-3-pentanol

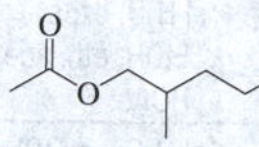
7570
2-Methyl-1-pentanol acetate

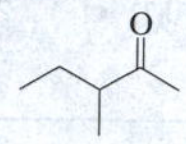
7571
3-Methyl-2-pentanone, (±)-

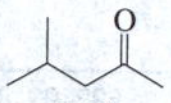
7572
4-Methyl-2-pentanone

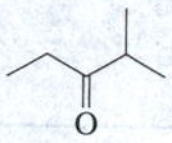
7573
2-Methyl-3-pentanone

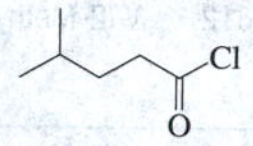
7574
4-Methylpentanoyl chloride

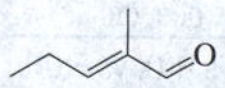
7575
2-Methyl-2-pentenal

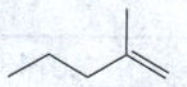
7576
2-Methyl-1-pentene

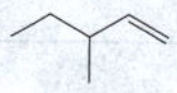
7577
3-Methyl-1-pentene

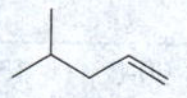
7578
4-Methyl-1-pentene

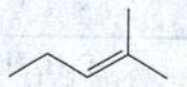
7579
2-Methyl-2-pentene

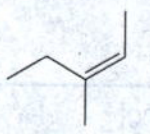
7580
3-Methyl-*cis*-2-pentene

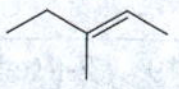
7581
3-Methyl-*trans*-2-pentene

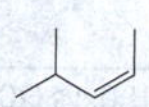
7582
4-Methyl-*cis*-2-pentene

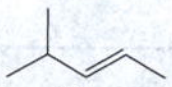
7583
4-Methyl-*trans*-2-pentene

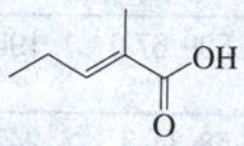
7584
trans-2-Methyl-2-pentenoic acid

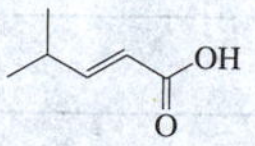
7585
4-Methyl-2-pentenoic acid

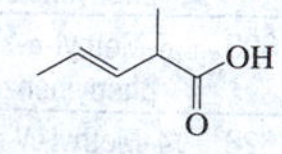
7586
2-Methyl-3-pentenoic acid

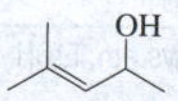
7587
4-Methyl-3-penten-2-ol

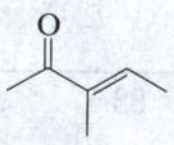
7588
3-Methyl-2-penten-4-one

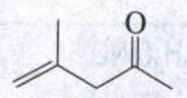
7589
4-Methyl-4-penten-2-one

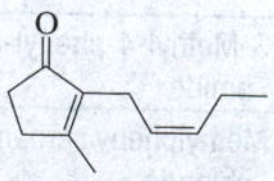
7590
cis-3-Methyl-2-(2-pentenyl)-2-cyclopenten-1-one

No.	Name	Synonym	Mol. form.	CAS Reg. No.	Mol. wt.	Physical form	mp/°C	bp/°C	den/ g cm^{-3}	n_D	Solubility
7591	3-(4-Methyl-3-pentenyl)furan		$C_{10}H_{14}O$	539-52-6	150.217			185.5	0.9017^{20}	1.4705^{21}	
7592	3-Methyl-3-penten-1-yne		C_6H_8	1574-33-0	80.128			69(7)	0.739^{20}	1.4332^{20}	s eth, bz
7593	3-Methyl-2-pentyl-2-cyclo-penten-1-one		$C_{11}H_{18}O$	1128-08-1	166.260			143^{22}	0.9165^{18}	1.4767^{20}	
7594	Methyl pentyl ether		$C_6H_{14}O$	628-80-8	102.174			99(3)	0.759^{22}	1.3862^{22}	vs ace, eth, EtOH
7595	5-Methyl-2-pentylphenol	6-*n*-Amyl-*m*-cresol	$C_{12}H_{18}O$	1300-94-3	178.270		24	138^{15}			vs ace, eth, EtOH
7596	Methyl pentyl sulfide		$C_6H_{14}S$	1741-83-9	118.240	liq	-94	148(4)	0.8431^{20}	1.4506^{20}	s EtOH, eth, ace, bz, chl
7597	Methyl *tert*-pentyl sulfide	2-Methyl-2-(methylthio)butane	$C_6H_{14}S$	13286-92-5	118.240	liq		150	0.84	1.4570^{20}	
7598	4-Methyl-1-pentyne		C_6H_{10}	7154-75-8	82.143	liq	-105.3(0.4)	61(2)	0.7000^{25}	1.3936^{20}	i H_2O; s bz, chl
7599	4-Methyl-2-pentyne		C_6H_{10}	21020-27-9	82.143	liq	-110.3(0.4)	73(2)	0.7112^{25}	1.4057^{20}	vs bz, chl
7600	3-Methyl-1-pentyn-3-ol	Meparfynol	$C_6H_{10}O$	77-75-8	98.142		30.5	120(22)	0.8688^{20}	1.4310^{20}	
7601	Methyl perfluorooctanoate		$C_9H_3F_{15}O_2$	376-27-2	428.095			158	1.684^{20}	1.304^{27}	
7602	1-Methylphenanthrene		$C_{15}H_{12}$	832-69-9	192.256	lf, pl (dil al)	123	344(6)			i H_2O; s EtOH
7603	3-Methylphenanthrene		$C_{15}H_{12}$	832-71-3	192.256	pr or nd (al)	65	350			i H_2O; s EtOH, ace; sl chl
7604	4-Methylphenanthrene		$C_{15}H_{12}$	832-64-4	192.256	pl (90% al)	51.76(0.05)	177^{10}			i H_2O; s EtOH, ctc
7605	Methylphenidate		$C_{14}H_{19}NO_2$	113-45-1	233.307			$136^{0.6}$			i H_2O, peth; s chl, EtOH, eth, AcOEt
7606	10-Methyl-10*H*-phenothiazine		$C_{13}H_{11}NS$	1207-72-3	213.298		101				
7607	10-Methyl-10*H*-phenothi-azine-2-acetic acid	Metiazinic acid	$C_{15}H_{13}NO_2S$	13993-65-2	271.335		144				s chl
7608	Methyl phenoxyacetate		$C_9H_{10}O_3$	2065-23-8	166.173			245	1.1493^{20}	1.5155^{20}	vs eth, EtOH
7609	1-Methyl-3-phenoxybenzene		$C_{13}H_{12}O$	3586-14-9	184.233			272	1.051^{25}	1.5727^{20}	
7610	[(2-Methylphenoxy)methyl]-oxirane		$C_{10}H_{12}O_2$	2210-79-9	164.201			123^{2}	1.0884^{20}		
7611	3-(2-Methylphenoxy)-1,2-propanediol	Mephenesin	$C_{10}H_{14}O_3$	59-47-2	182.216		70 dec				sl H_2O, eth; s EtOH
7612	*N*-(2-Methylphenyl)acetamide		$C_9H_{11}NO$	120-66-1	149.189	nd (al)	110(1)	296	1.168^{15}		sl H_2O, bz; s EtOH, eth, ace, HOAc
7613	*N*-(3-Methylphenyl)acetamide		$C_9H_{11}NO$	537-92-8	149.189	nd (w)	65.5	303	1.141^{15}		sl H_2O; vs EtOH, eth; s chl
7614	*N*-Methyl-*N*-phenylacetamide	*N*-Methylacetanilide	$C_9H_{11}NO$	579-10-2	149.189	nd (eth), pr (al)	103	256	1.0036^{105}	1.576	s H_2O, EtOH, eth, chl, lig
7615	2-Methylphenyl acetate	*o*-Cresyl acetate	$C_9H_{10}O_2$	533-18-6	150.174			208	1.0533^{15}	1.5002^{20}	vs eth, EtOH
7616	3-Methylphenyl acetate	*m*-Cresyl acetate	$C_9H_{10}O_2$	122-46-3	150.174		12	214(3)	1.043^{20}	1.4978^{20}	vs bz, eth, EtOH
7617	4-Methylphenyl acetate	*p*-Cresyl acetate	$C_9H_{10}O_2$	140-39-6	150.174			213(3)	1.0512^{17}	1.5163^{22}	sl H_2O, ctc; s EtOH, eth, chl
7618	Methyl 2-phenylacetate		$C_9H_{10}O_2$	101-41-7	150.174			215(9)	1.0622^{16}	1.5075^{20}	i H_2O; msc EtOH, eth; s ace, ctc
7619	2-(Methylphenylamino)ethanol		$C_9H_{13}NO$	93-90-3	151.205			218^{110}	1.0143^{0}		s H_2O; vs EtOH, eth, ace, bz
7620	2-[(2-Methylphenyl)amino]-ethanol		$C_9H_{13}NO$	136-80-1	151.205			285.5	1.0794^{20}	1.5675^{20}	vs eth, EtOH
7621	3-Methyl-*N*-phenylaniline		$C_{13}H_{13}N$	1205-64-7	183.249		30	316		1.6350^{20}	vs bz, eth, EtOH
7622	*N*-(4-Methylphenyl)benzamide		$C_{14}H_{13}NO$	582-78-5	211.259	orth nd (al)	158		1.202^{15}		vs eth, EtOH
7623	*N*-Methyl-*N*-phenylbenzene-methanamine		$C_{14}H_{15}N$	614-30-2	197.276						s ctc
7624	4-Methyl-α-phenylbenzenemethanol		$C_{14}H_{14}O$	1517-63-1	198.260		53(2)				
7625	α-Methyl-α-phenylbenzenemethanol		$C_{14}H_{14}O$	599-67-7	198.260			285	1.1059^{15}		
7626	4-Methyl-*N*-phenylbenzene-sulfonamide		$C_{13}H_{13}NO_2S$	68-34-8	247.313	(α) tcl, (β) mcl pr (al, bz)	103.5				i H_2O; vs EtOH; s bz, HOAc
7627	4-Methylphenyl benzoate		$C_{14}H_{12}O_2$	614-34-6	212.244	pl (eth-al)	71.5	316			vs eth, EtOH
7628	1-Methyl-*N*-phenyl-*N*-benzyl-4-piperidinamine	Bamipine	$C_{19}H_{24}N_2$	4945-47-5	280.407	cry (MeOH)	115				
7629	Methyl 2-phenylbutanoate		$C_{11}H_{14}O_2$	2294-71-5	178.228	nd (dil al)	77.5	228			vs eth, EtOH
7630	3-Methyl-1-phenyl-1-buta-none		$C_{11}H_{14}O$	582-62-7	162.228			236.5	0.9701^{16}	1.5139^{15}	i H_2O; msc EtOH, eth; vs ace
7631	3-Methyl-4-phenyl-3-buten-amide	β-Benzalbutyramide	$C_{11}H_{13}NO$	7236-47-7	175.227		133				
7632	Methylphenylcarbamic chloride		C_8H_8ClNO	4285-42-1	169.609	pl (al)	88.5	280			vs eth, EtOH

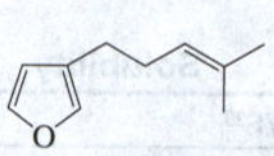
7591
3-(4-Methyl-3-pentenyl)furan

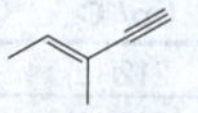
7592
3-Methyl-3-penten-1-yne

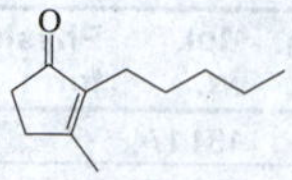
7593
3-Methyl-2-pentyl-2-cyclopenten-1-one

7594
Methyl pentyl ether

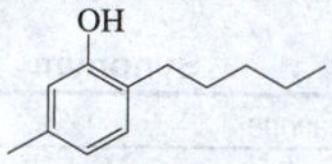
7595
5-Methyl-2-pentylphenol

7596
Methyl pentyl sulfide

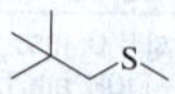
7597
Methyl *tert*-pentyl sulfide

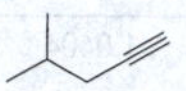
7598
4-Methyl-1-pentyne

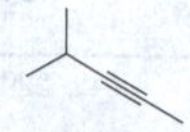
7599
4-Methyl-2-pentyne

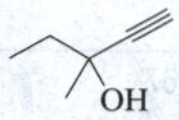
7600
3-Methyl-1-pentyn-3-ol

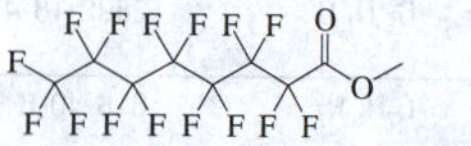
7601
Methyl perfluorooctanoate

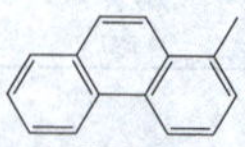
7602
1-Methylphenanthrene

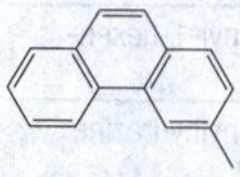
7603
3-Methylphenanthrene

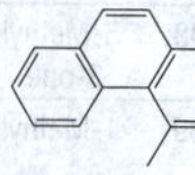
7604
4-Methylphenanthrene

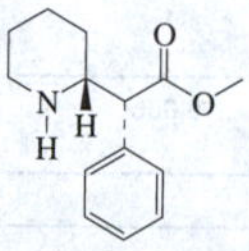
7605
Methylphenidate

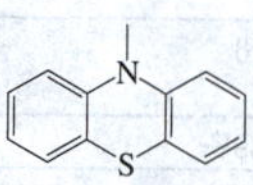
7606
10-Methyl-10*H*-phenothiazine

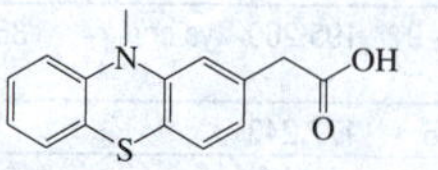
7607
10-Methyl-10*H*-phenothiazine-2-acetic acid

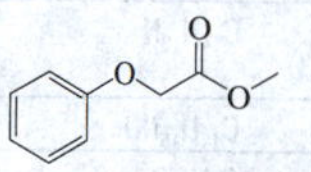
7608
Methyl phenoxyacetate

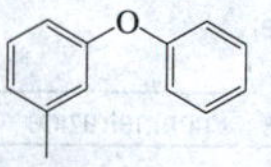
7609
1-Methyl-3-phenoxybenzene

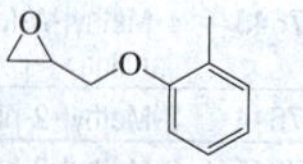
7610
[(2-Methylphenoxy)methyl]oxirane

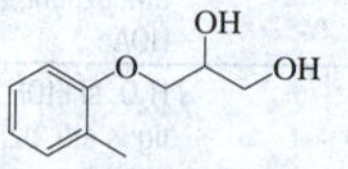
7611
3-(2-Methylphenoxy)-1,2-propanediol

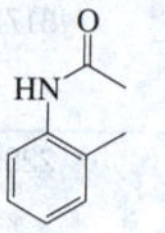
7612
N-(2-Methylphenyl)acetamide

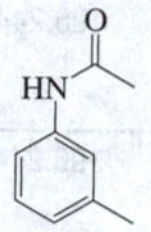
7613
N-(3-Methylphenyl)acetamide

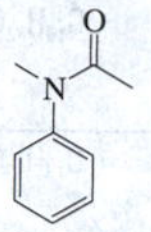
7614
N-Methyl-*N*-phenylacetamide

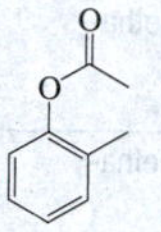
7615
2-Methylphenyl acetate

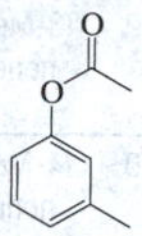
7616
3-Methylphenyl acetate

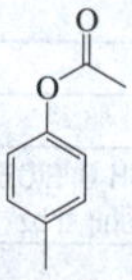
7617
4-Methylphenyl acetate

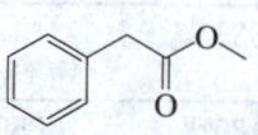
7618
Methyl 2-phenylacetate

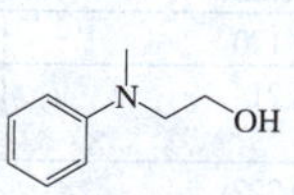
7619
2-(Methylphenylamino)ethanol

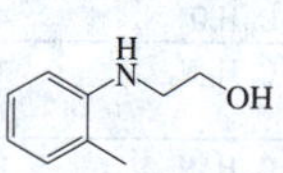
7620
2-[(2-Methylphenyl)amino]ethanol

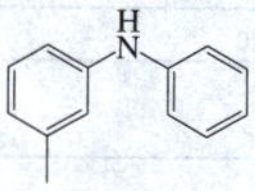
7621
3-Methyl-*N*-phenylaniline

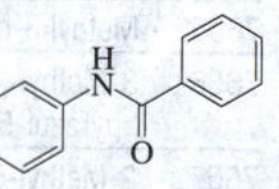
7622
N-(4-Methylphenyl)benzamide

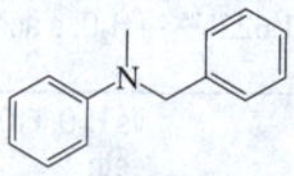
7623
N-Methyl-*N*-phenylbenzenemethanamine

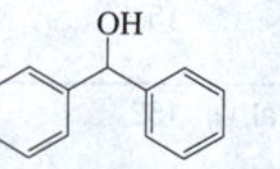
7624
4-Methyl-α-phenylbenzenemethanol

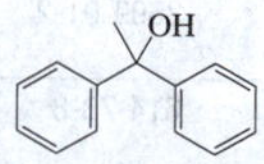
7625
α-Methyl-α-phenylbenzenemethanol

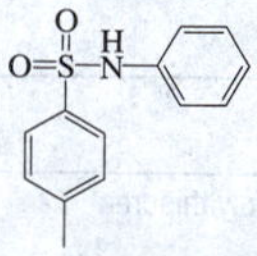
7626
4-Methyl-*N*-phenylbenzenesulfonamide

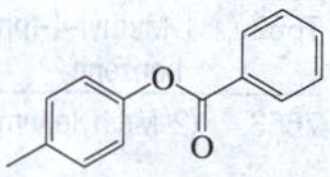
7627
4-Methylphenyl benzoate

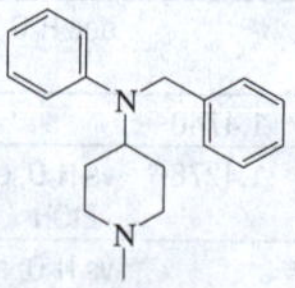
7628
1-Methyl-*N*-phenyl-*N*-benzyl-4-piperidinamine

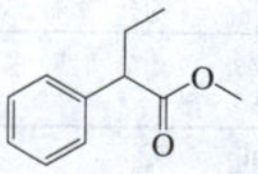
7629
Methyl 2-phenylbutanoate

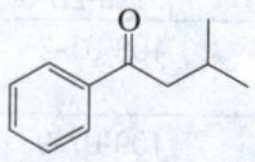
7630
3-Methyl-1-phenyl-1-butanone

7631
3-Methyl-4-phenyl-3-butenamide

7632
Methylphenylcarbamic chloride

No.	Name	Synonym	Mol. form.	CAS Reg. No.	Mol. wt.	Physical form	mp/°C	bp/°C	den/ g cm^{-3}	n_D	Solubility
7633	1-(2-Methylphenyl)ethanone		$C_9H_{10}O$	577-16-2	134.174			212(12)	1.026^{20}	1.5276^{20}	
7634	1-(3-Methylphenyl)ethanone		$C_9H_{10}O$	585-74-0	134.174			224(16)	1.0165^{0}	1.533^{15}	s EtOH, eth, ace; sl ctc
7635	4-(1-Methyl-1-phenylethyl)-phenol		$C_{15}H_{16}O$	599-64-4	212.287	pr (peth)	73.2(0.2)	341.2(0.7)			
7636	*N*-Methyl-*N*-phenylformamide		C_8H_9NO	93-61-8	135.163		14.5	243	1.0948^{20}	1.5589^{20}	sl H_2O, ctc; s EtOH, ace
7637	*N*-(2-Methylphenyl)formamide		C_8H_9NO	94-69-9	135.163	lf (al)	62	288	1.086^{55}		s H_2O; vs EtOH
7638	5-Methyl-1-phenyl-1-hexen-3-one		$C_{13}H_{16}O$	2892-18-4	188.265	cry	43	154^{25}	0.9509^{46}	1.5523^{25}	sl H_2O; s EtOH, bz, chl
7639	1-Methyl-1-phenylhydrazine		$C_7H_{10}N_2$	618-40-6	122.167			229.0(0.5)	1.0404^{20}	1.5691^{20}	sl H_2O; msc EtOH, eth, bz, chl
7640	3-Methyl-5-phenyl-2,4-imidazolidinedione	3-Methyl-5-phenylhydantoin	$C_{10}H_{10}N_2O_2$	6846-11-3	190.198		164.5				s chl
7641	1-Methyl-6-phenylimidazo[4,5-*b*]pyridin-2-amine	PhIP	$C_{13}H_{12}N_4$	105650-23-5	224.261	solid	327				
7642	2-[Methyl(phenylmethyl)-amino]ethanol		$C_{10}H_{15}NO$	101-98-4	165.232			134^{14}			
7643	4-Methyl-*N*-(phenylmethylene)aniline		$C_{14}H_{13}N$	2272-45-9	195.260	ye cry	35	318			vs ace
7644	3-Methyl-2-phenylmorpholine	Phenmetrazine	$C_{11}H_{15}NO$	134-49-6	177.243			139^{12}			
7645	2-Methyl-2-phenyloxirane		$C_9H_{10}O$	2085-88-3	134.174			84^{17}	1.0228^{20}	1.5232^{20}	
7646	*N*-(2-Methylphenyl)-3-oxo-butanamide		$C_{11}H_{13}NO_2$	93-68-5	191.227	pr (AcOEt)	107.5				vs bz, EtOH
7647	*N*-(4-Methylphenyl)-3-oxo-butanamide		$C_{11}H_{13}NO_2$	2415-85-2	191.227	pr (AcOEt)	95				sl H_2O, lig; s EtOH, bz
7648	(2-Methylphenyl)phenylmethanone		$C_{14}H_{12}O$	131-58-8	196.244		<-18	308	1.1098^{20}		i H_2O; vs EtOH
7649	(3-Methylphenyl)phenylmethanone		$C_{14}H_{12}O$	643-65-2	196.244	oil	2	317	1.095^{20}		i H_2O; s EtOH, eth, bz, chl, HOAc
7650	(4-Methylphenyl)phenylmethanone		$C_{14}H_{12}O$	134-84-9	196.244	mcl pr	54(2)	228^{70}	0.9926^{0}		i H_2O; sl EtOH, lig; s eth, bz, chl
7651	Methyl 3-phenylpropanoate	Methyl dihydrocinnamate	$C_{10}H_{12}O_2$	103-25-3	164.201			238.5	1.0455^{25}		i H_2O; s EtOH, eth, bz, AcOEt
7652	1-(4-Methylphenyl)-1-propanone		$C_{10}H_{12}O$	5337-93-9	148.201		7.2	236	0.9926^{20}	1.5278^{20}	i H_2O; s EtOH, eth, ace, bz, CS_2
7653	2-Methyl-1-phenyl-1-propanone		$C_{10}H_{12}O$	611-70-1	148.201	liq	-0.7	220	0.9863^{11}	1.5172^{20}	vs eth, EtOH
7654	2-Methyl-3-phenyl-2-propenal		$C_{10}H_{10}O$	101-39-3	146.185			251.6(1)	1.0407^{17}	1.6057^{17}	
7655	Methyl 3-phenyl-2-propynoate		$C_{10}H_8O_2$	4891-38-7	160.170		18(1)	158^{48}	1.0830^{25}	1.5618^{25}	
7656	3-Methyl-1-phenyl-1*H*-pyrazol-5-amine		$C_{10}H_{11}N_3$	1131-18-6	173.214		116	333			s H_2O, EtOH, chl; sl bz
7657	2-Methyl-5-phenylpyridine		$C_{12}H_{11}N$	3256-88-0	169.222			189^{50}	1.0590^{25}	1.6055^{25}	
7658	5-Methyl-5-phenyl-2,4,6(1*H*,3*H*,5*H*)-pyrimidinetrione	Phenylmethylbarbituric acid	$C_{11}H_{10}N_2O_3$	76-94-8	218.208	cry	220				i H_2O; s EtOH, eth, alk
7659	1-Methyl-3-phenyl-2,5-pyrrolidinedione	Phensuximide	$C_{11}H_{11}NO_2$	86-34-0	189.211	cry (hot al)	72				vs EtOH, MeOH
7660	Methylphenylsilane		$C_7H_{10}Si$	766-08-5	122.240			140	0.8895^{20}	1.5058^{20}	
7661	Methyl phenyl sulfone		$C_7H_8O_2S$	3112-85-4	156.203		88(3)				i H_2O; s EtOH, bz, chl; sl ctc
7662	1-Methyl-4-(phenylthio)-benzene		$C_{13}H_{12}S$	3699-01-2	200.299		15.7	317	1.0986^{25}	1.6225^{25}	i H_2O; s ace, bz
7663	(2-Methylphenyl)thiourea	*o*-Tolylthiourea	$C_8H_{10}N_2S$	614-78-8	166.243	nd (dil al, w)	162				vs H_2O, EtOH; sl eth
7664	*N*-Methyl-*N'*-phenylthiourea		$C_8H_{10}N_2S$	2724-69-8	166.243	ta, pl	112.5				vs EtOH
7665	Methyl phosphate	Methyl dihydrogen phosphate	CH_5O_4P	812-00-0	112.022	oil					
7666	Methylphosphine		CH_5P	593-54-4	48.025	col gas		-16			vs eth
7667	Methylphosphonic acid		CH_5O_3P	993-13-5	96.023	hyg pl	108.5	dec			vs H_2O, EtOH, eth; i bz, peth
7668	Methylphosphonic difluoride		CH_3F_2OP	676-99-3	100.005	liq		98	1.3314^{20}		
7669	Methylphosphonofluoridic acid, isopropyl ester	Sarin	$C_4H_{10}FO_2P$	107-44-8	140.093	liq	-57	147	1.10^{20}		dec H_2O
7670	Methyl phosphorodichloridite	Methyl dichlorophosphite	CH_3Cl_2OP	3279-26-3	132.914	hyg liq	-91	93	1.406	1.4740^{20}	
7671	1-Methylpiperazine		$C_5H_{12}N_2$	109-01-3	100.162			135(5)		1.4378^{20}	vs H_2O, eth, EtOH
7672	2-Methylpiperazine		$C_5H_{12}N_2$	109-07-9	100.162	hyg lf (al)	62	153			vs H_2O; s EtOH, eth, bz, chl

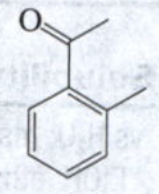
7633
1-(2-Methylphenyl)ethanone

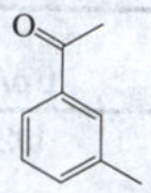
7634
1-(3-Methylphenyl)ethanone

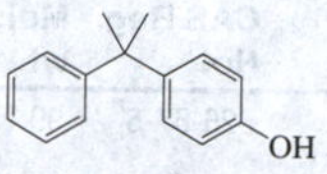
7635
4-(1-Methyl-1-phenylethyl)phenol

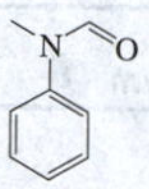
7636
N-Methyl-*N*-phenylformamide

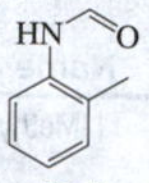
7637
N-(2-Methylphenyl)formamide

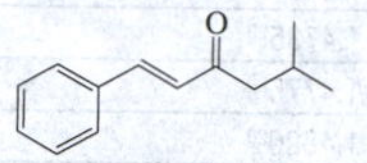
7638
5-Methyl-1-phenyl-1-hexen-3-one

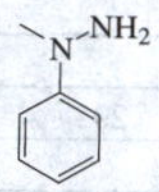
7639
1-Methyl-1-phenylhydrazine

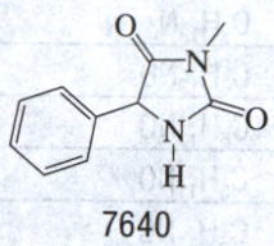
7640
3-Methyl-5-phenyl-2,4-imidazolidinedione

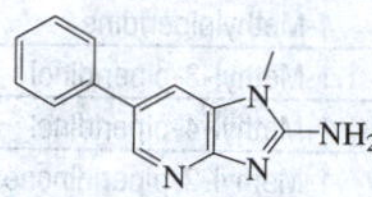
7641
1-Methyl-6-phenylimidazo[4,5-b]pyridin-2-amine

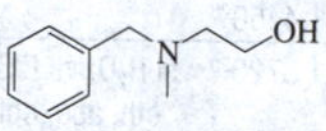
7642
2-[Methyl(phenylmethyl)amino]ethanol

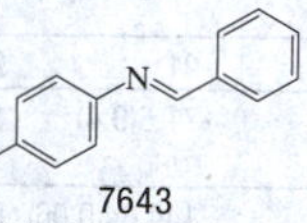
7643
4-Methyl-*N*-(phenylmethylene)aniline

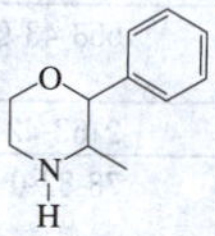
7644
3-Methyl-2-phenylmorpholine

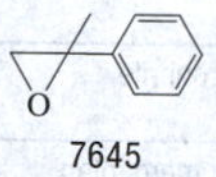
7645
2-Methyl-2-phenyloxirane

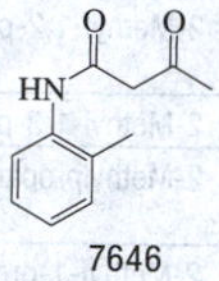
7646
N-(2-Methylphenyl)-3-oxo-butanamide

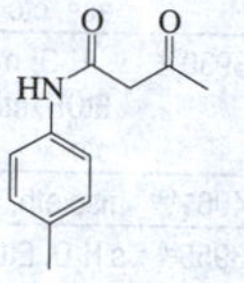
7647
N-(4-Methylphenyl)-3-oxo-butanamide

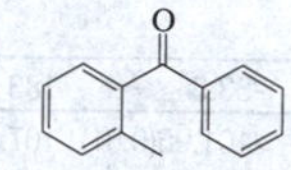
7648
(2-Methylphenyl)phenylmethanone

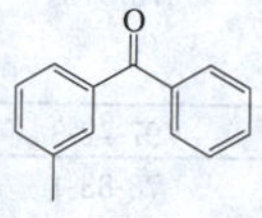
7649
(3-Methylphenyl)phenylmethanone

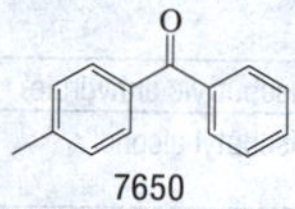
7650
(4-Methylphenyl)phenylmethanone

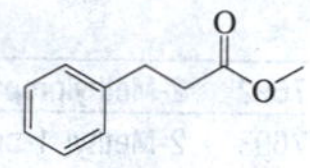
7651
Methyl 3-phenylpropanoate

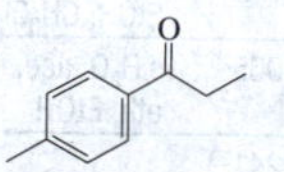
7652
1-(4-Methylphenyl)-1-propanone

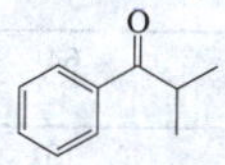
7653
2-Methyl-1-phenyl-1-propanone

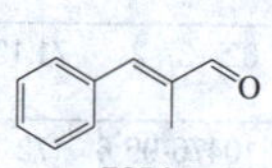
7654
2-Methyl-3-phenyl-2-propenal

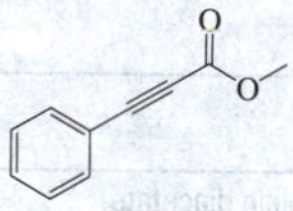
7655
Methyl 3-phenyl-2-propynoate

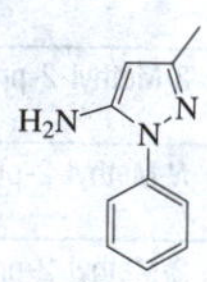
7656
3-Methyl-1-phenyl-1*H*-pyrazol-5-amine

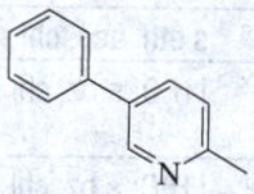
7657
2-Methyl-5-phenylpyridine

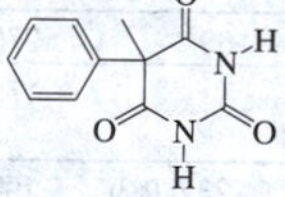
7658
5-Methyl-5-phenyl-2,4,6(1*H*,3*H*,5*H*)-pyrimidinetrione

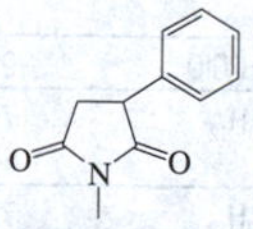
7659
1-Methyl-3-phenyl-2,5-pyrrolidinedione

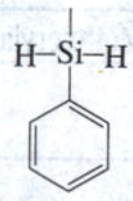
7660
Methylphenylsilane

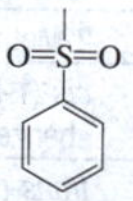
7661
Methyl phenyl sulfone

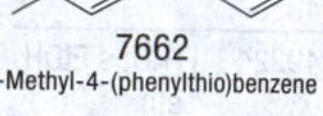
7662
1-Methyl-4-(phenylthio)benzene

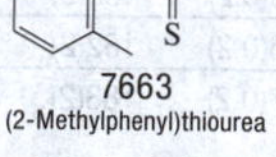
7663
(2-Methylphenyl)thiourea

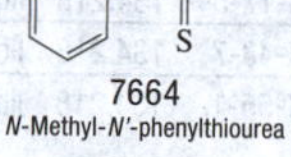
7664
N-Methyl-*N'*-phenylthiourea

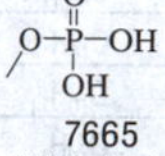
7665
Methyl phosphate

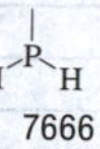
7666
Methylphosphine

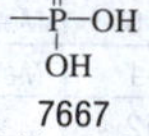
7667
Methylphosphonic acid

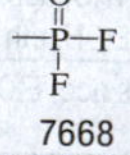
7668
Methylphosphonic difluoride

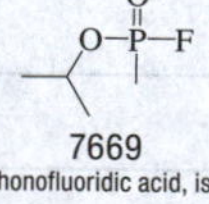
7669
Methylphosphonofluoridic acid, isopropyl ester

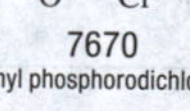
7670
Methyl phosphorodichloridite

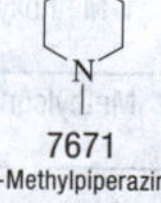
7671
1-Methylpiperazine

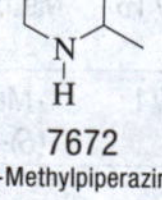
7672
2-Methylpiperazine

Organic

No.	Name	Synonym	Mol. form.	CAS Reg. No.	Mol. wt.	Physical form	mp/°C	bp/°C	den/ g cm^{-3}	n_D	Solubility
7673	1-Methylpiperidine		$C_6H_{13}N$	626-67-5	99.174	liq	-102.7	102.2(0.8)	0.8159^{20}	1.4355^{20}	vs H_2O; msc EtOH, eth; s ctc
7674	2-Methylpiperidine, (±)-		$C_6H_{13}N$	3000-79-1	99.174	liq	-2.5	118	0.8436^{24}	1.4459^{20}	vs H_2O; s EtOH, eth; sl chl; i dil KOH
7675	3-Methylpiperidine, (±)-		$C_6H_{13}N$	53152-98-0	99.174	liq	-27.2(0.3)	125(2)	0.8446^{26}	1.4470^{20}	vs H_2O; sl chl
7676	4-Methylpiperidine		$C_6H_{13}N$	626-58-4	99.174			126(6)	0.8674^{25}	1.4458^{20}	vs H_2O; sl chl
7677	1-Methyl-3-piperidinol		$C_6H_{13}NO$	3554-74-3	115.173			93^{26}	0.9635^{16}	1.4735^{20}	
7678	1-Methyl-4-piperidinol		$C_6H_{13}NO$	106-52-5	115.173		29	200		1.4775^{20}	
7679	1-Methyl-2-piperidinone		$C_6H_{11}NO$	931-20-4	113.157			221	1.0263^{25}	1.4820^{20}	
7680	1-Methyl-4-piperidinone		$C_6H_{11}NO$	1445-73-4	113.157			85^{45}	0.971^{25}	1.4580^{25}	
7681	Methylprednisolone		$C_{22}H_{30}O_5$	83-43-2	374.470	cry	232				
7682	2-Methylpropanamide		C_4H_9NO	563-83-7	87.120		127.1(0.4)	217	1.013^{20}		s chl
7683	*N*-Methylpropanamide	*N*-Methylpropionamide	C_4H_9NO	1187-58-2	87.120	liq	-30.9	207(3)	0.9305^{25}	1.4345^{25}	
7684	2-Methyl-1,2-propanediamine		$C_4H_{12}N_2$	811-93-8	88.151			131(3)	0.841^{25}	1.4410^{20}	s ctc
7685	2-Methyl-1,2-propanediol		$C_4H_{10}O_2$	558-43-0	90.121			177(5)	1.0024^{20}	1.4350^{20}	vs H_2O, eth, EtOH
7686	2-Methyl-1,3-propanediol		$C_4H_{10}O_2$	2163-42-0	90.121	liq	-91	221(4)	1.015^{20}	1.4450^{20}	
7687	2-Methylpropanenitrile	Isobutyronitrile	C_4H_7N	78-82-0	69.106	liq	-71.5(0.4)	102(2)	0.7704^{20}	1.3720^{20}	sl H_2O; vs EtOH, eth, ace, chl
7688	2-Methyl-1-propanethiol	Isobutyl mercaptan	$C_4H_{10}S$	513-44-0	90.187		-144.84(0.06)	88.5(0.6)	0.8357^{20}	1.4387^{20}	sl H_2O; vs EtOH, eth, ace; s ctc
7689	2-Methyl-2-propanethiol	*tert*-Butyl mercaptan	$C_4H_{10}S$	75-66-1	90.187	liq	1.27(0.06)	64.2(0.1)	0.7943^{25}	1.4232^{20}	i H_2O; s ctc, hp
7690	Methyl propanoate	Methyl propionate	$C_4H_8O_2$	554-12-1	88.106	liq	-87.5(0.5)	78.6(0.2)	0.9150^{20}	1.3775^{20}	sl H_2O; msc EtOH, eth; s ace, ctc
7691	2-Methylpropanoic acid	Isobutyric acid	$C_4H_8O_2$	79-31-2	88.106	liq	-46(1)	154.4(0.2)	0.9681^{20}	1.3930^{20}	vs H_2O; msc EtOH, eth; sl ctc
7692	2-Methylpropanoic anhydride	Isobutyric anhydride	$C_8H_{14}O_3$	97-72-3	158.195	liq	-53.5	183	0.9535^{20}	1.4061^{19}	msc eth; s chl
7693	2-Methyl-1-propanol	Isobutyl alcohol	$C_4H_{10}O$	78-83-1	74.121	liq	-101.96(0.01)	107.84(0.07)	0.8018^{20}	1.3955^{20}	s H_2O, EtOH, eth, ace, ctc
7694	2-Methyl-2-propanol	*tert*-Butyl alcohol	$C_4H_{10}O$	75-65-0	74.121		25.81(0.04)	82.3(0.1)	0.7887^{20}	1.3878^{20}	msc H_2O, EtOH, eth; s chl
7695	2-Methylpropanoyl chloride	Isobutyric acid chloride	C_4H_7ClO	79-30-1	106.551	liq	-90.0(0.7)	92(5)		1.4079^{20}	s eth
7696	2-Methylpropenal	Methacrolein	C_4H_6O	78-85-3	70.090			67(5)	0.840^{25}	1.4144^{20}	msc H_2O, EtOH, eth
7697	2-Methyl-2-propenamide		C_4H_7NO	79-39-0	85.105	cry (bz)	112.0(0.4)				sl eth, chl; s EtOH, CH_2Cl_2
7698	*N*-Methyl-2-propen-1-amine		C_4H_9N	627-37-2	71.121			64		1.4065^{20}	vs H_2O, ace, eth, EtOH
7699	2-Methyl-2-propene-1,1-diol diacetate	Methacrolein diacetate	$C_8H_{12}O_4$	10476-95-6	172.179			191		1.4241^{20}	
7700	2-Methyl-1-propene, tetramer		$C_{16}H_{32}$	15220-85-6	224.425	liq	-98	244	0.7944^{20}	1.4482^{20}	
7701	2-Methyl-2-propenoic anhydride	Methacrylic acid anhydride	$C_8H_{10}O_3$	760-93-0	154.163			89^{5}		1.4540^{20}	msc EtOH, eth
7702	2-Methyl-2-propenol	Methallyl alcohol	C_4H_8O	513-42-8	72.106			114.5	0.8515^{20}	1.4255^{20}	vs H_2O; msc EtOH, eth
7703	2-Methyl-2-propenoyl chloride	Methacrylic acid chloride	C_4H_5ClO	920-46-7	104.535	liq	-60	97(25)	1.0871^{20}	1.4435^{20}	s eth, ace, chl
7704	*cis*-(1-Methyl-1-propenyl)-benzene		$C_{10}H_{12}$	767-99-7	132.202			175(5)	0.9191^{25}	1.5402^{25}	i H_2O; s bz, chl
7705	*trans*-(1-Methyl-1-propenyl)-benzene		$C_{10}H_{12}$	768-00-3	132.202	liq	-23.5(0.3)	196(3)	0.9138^{25}	1.5425^{20}	i H_2O; s bz, chl
7706	(2-Methyl-1-propenyl)benzene		$C_{10}H_{12}$	768-49-0	132.202	liq	-51.0(0.2)	189(3)	0.900^{20}	1.5388^{20}	
7707	4-(2-Methylpropenyl)-morpholine	1-Morpholinoisobutene	$C_8H_{15}NO$	2403-55-6	141.211		120	89^{20}		1.4663^{20}	
7708	2-(2-Methylpropoxy)ethanol		$C_6H_{14}O_2$	4439-24-1	118.174			155(3)	0.8900^{20}	1.4143^{20}	
7709	Methylpropylamine	*N*-Methyl-1-propanamine	$C_4H_{11}N$	627-35-0	73.137			63(4)	0.7204^{17}		
7710	1-Methyl-2-propylbenzene		$C_{10}H_{14}$	1074-17-5	134.218	liq	-60.3(0.2)	185(2)	0.8697^{25}	1.4996^{20}	
7711	1-Methyl-3-propylbenzene		$C_{10}H_{14}$	1074-43-7	134.218	liq	-82.5(0.2)	182(2)	0.8569^{25}	1.4935^{20}	
7712	1-Methyl-4-propylbenzene		$C_{10}H_{14}$	1074-55-1	134.218	liq	-63.7(0.2)	183(2)	0.8544^{25}	1.4922^{20}	i H_2O; s EtOH, eth
7713	*cis*-1-Methyl-2-propylcyclo-pentane		C_9H_{18}	932-43-4	126.239	liq	-104.9(0.7)	152.6	0.7881^{25}	1.4343^{20}	
7714	*trans*-1-Methyl-2-propylcyclo-pentane		C_9H_{18}	932-44-5	126.239	liq	-123(2)	147(16)	0.7735^{25}	1.4274^{20}	
7715	Methyl propyl disulfide		$C_4H_{10}S_2$	2179-60-4	122.252	liq		70^{43}	0.980	1.5080^{20}	
7716	Methyl propyl ether	1-Methoxypropane	$C_4H_{10}O$	557-17-5	74.121			38.5(1)	0.7356^{13}	1.3579^{25}	s H_2O, ace; msc EtOH, eth
7717	1-Methyl-2-propylpiperidine, (*S*)-	Methylconiine	$C_9H_{19}N$	35305-13-6	141.254			174	0.8326^{22}	1.4538^{12}	vs ace, EtOH

7673
1-Methylpiperidine

7674
2-Methylpiperidine, (±)-

7675
3-Methylpiperidine, (±)-

7676
4-Methylpiperidine

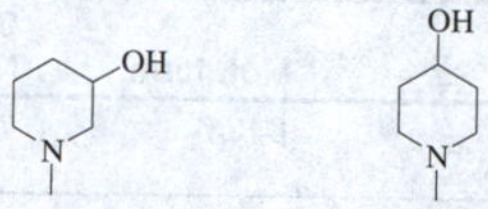

7677
1-Methyl-3-piperidinol

7678
1-Methyl-4-piperidinol

7679
1-Methyl-2-piperidinone

7680
1-Methyl-4-piperidinone

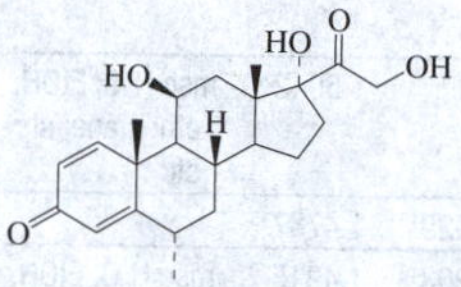

7681
Methylprednisolone

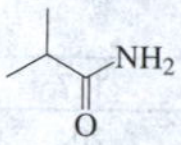

7682
2-Methylpropanamide

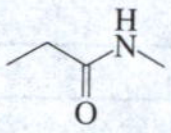

7683
N-Methylpropanamide

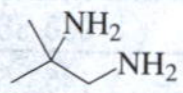

7684
2-Methyl-1,2-propanediamine

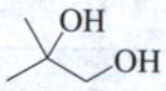

7685
2-Methyl-1,2-propanediol

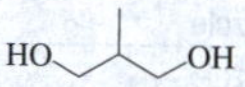

7686
2-Methyl-1,3-propanediol

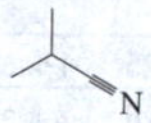

7687
2-Methylpropanenitrile

7688
2-Methyl-1-propanethiol

7689
2-Methyl-2-propanethiol

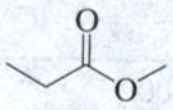

7690
Methyl propanoate

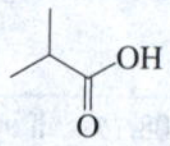

7691
2-Methylpropanoic acid

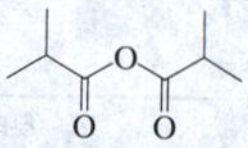

7692
2-Methylpropanoic anhydride

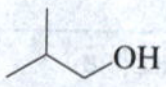

7693
2-Methyl-1-propanol

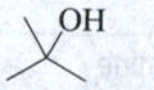

7694
2-Methyl-2-propanol

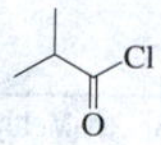

7695
2-Methylpropanoyl chloride

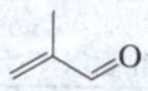

7696
2-Methylpropenal

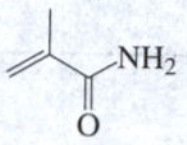

7697
2-Methyl-2-propenamide

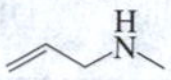

7698
N-Methyl-2-propen-1-amine

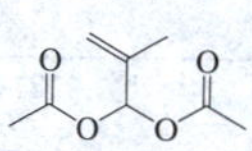

7699
2-Methyl-2-propene-1,1-diol diacetate

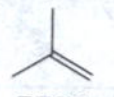

7700
2-Methyl-1-propene, tetramer

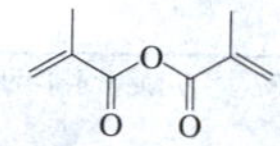

7701
2-Methyl-2-propenoic anhydride

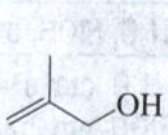

7702
2-Methyl-2-propenol

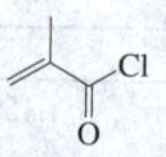

7703
2-Methyl-2-propenoyl chloride

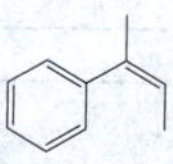

7704
cis-(1-Methyl-1-propenyl)benzene

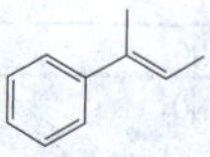

7705
trans-(1-Methyl-1-propenyl)benzene

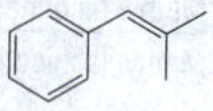

7706
(2-Methyl-1-propenyl)benzene

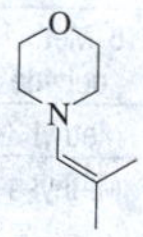

7707
4-(2-Methylpropenyl)morpholine

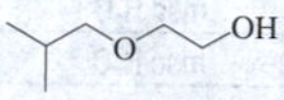

7708
2-(2-Methylpropoxy)ethanol

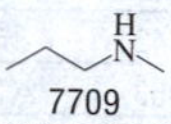

7709
Methylpropylamine

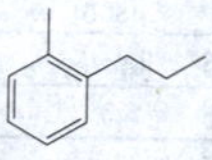

7710
1-Methyl-2-propylbenzene

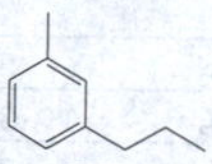

7711
1-Methyl-3-propylbenzene

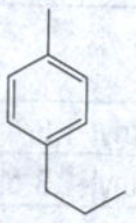

7712
1-Methyl-4-propylbenzene

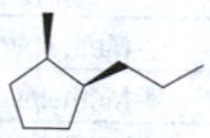

7713
cis-1-Methyl-2-propylcyclopentane

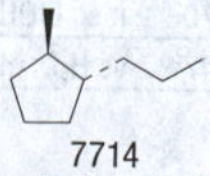

7714
trans-1-Methyl-2-propylcyclopentane

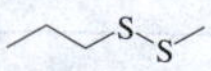

7715
Methyl propyl disulfide

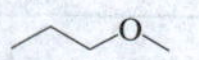

7716
Methyl propyl ether

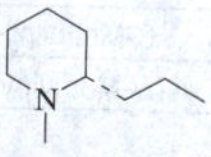

7717
1-Methyl-2-propylpiperidine, (*S*)-

No.	Name	Synonym	Mol. form.	CAS Reg. No.	Mol. wt.	Physical form	mp/°C	bp/°C	den/g cm^{-3}	n_D	Solubility
7718	2-Methyl-2-propyl-1,3-propanediol		$C_7H_{16}O_2$	78-26-2	132.201	cry (hx)	62.5	234			s H_2O, hx; sl chl
7719	2-Methyl-2-propyl-1,3-propanediol dicarbamate	Meprobamate	$C_9H_{18}N_2O_4$	57-53-4	218.250	cry (w)	105(3)				vs bz, eth, EtOH
7720	Methyl propyl sulfide		$C_4H_{10}S$	3877-15-4	90.187	liq	-112.96(0.06)	95.5(0.7)	0.8424^{20}	1.4442^{20}	s H_2O, EtOH, eth, ace
7721	*N*-Methyl-2-propyn-1-amine		C_4H_7N	35161-71-8	69.106			83	0.819^{25}	1.4332^{20}	
7722	*N*-Methyl-*N*-2-propynylbenzenemethanamine	Pargyline	$C_{11}H_{13}N$	555-57-7	159.228			96^{11}	0.944^{25}	1.5213^{20}	
7723	3-Methylpyrazinamine	2-Amino-3-methylpyrazine	$C_5H_7N_3$	19838-08-5	109.130	nd (hx/AcOEt)	174				
7724	2-Methylpyrazine		$C_5H_6N_2$	109-08-0	94.115	liq	-29	129(4)	1.03^{20}	1.5042^{20}	msc H_2O, EtOH, eth; s ace; sl ctc
7725	1-Methyl-1*H*-pyrazole		$C_4H_6N_2$	930-36-9	82.104			127	0.9929^{13}	1.4787^{13}	
7726	3-Methyl-1*H*-pyrazole		$C_4H_6N_2$	1453-58-3	82.104		36.5	204	1.0203^{16}	1.4915^{20}	msc H_2O, EtOH, eth
7727	4-Methyl-1*H*-pyrazole	Fomepizole	$C_4H_6N_2$	7554-65-6	82.104			206	1.015^{20}		
7728	3-Methyl-2-pyrazolin-5-one		$C_4H_6N_2O$	108-26-9	98.103		215				vs H_2O; sl EtOH
7729	1-Methylpyrene		$C_{17}H_{12}$	2381-21-7	216.277		74(2)	410			
7730	2-Methylpyrene		$C_{17}H_{12}$	3442-78-2	216.277	fl (EtOH)	143	409.8			
7731	3-Methylpyridazine	3-Methyl-1,2-diazine	$C_5H_6N_2$	1632-76-4	94.115		184	214	1.0450^{26}	1.5145^{20}	
7732	3-Methyl-2-pyridinamine	2-Amino-3-picoline	$C_6H_8N_2$	1603-40-3	108.141	hyg	33.5	222.8(0.5)			vs H_2O; s EtOH, eth, ace, bz, ctc; sl lig
7733	4-Methyl-2-pyridinamine	2-Amino-4-picoline	$C_6H_8N_2$	695-34-1	108.141	lf or pl (lig)	100	116^{11}			vs H_2O; s EtOH, eth, ace, bz; i lig; sl chl
7734	5-Methyl-2-pyridinamine		$C_6H_8N_2$	1603-41-4	108.141		77.3(0.2)	227			
7735	6-Methyl-2-pyridinamine	2-Amino-6-picoline	$C_6H_8N_2$	1824-81-3	108.141	hyg (lig)	41	208.5			vs H_2O; s EtOH, eth, ace, bz, lig
7736	*N*-Methyl-2-pyridinamine		$C_6H_8N_2$	4597-87-9	108.141		15	202(7)	1.048^{29}		s H_2O, bz; vs EtOH, eth, HOAc
7737	*N*-Methyl-4-pyridinamine		$C_6H_8N_2$	1121-58-0	108.141	pl (eth)	118.8				vs H_2O, ace, eth, EtOH
7738	2-Methylpyridine	2-Picoline	C_6H_7N	109-06-8	93.127	liq	-66.65(0.03)	129.4(0.2)	0.9443^{20}	1.4957^{20}	vs H_2O, ace; msc EtOH, eth; s ctc
7739	3-Methylpyridine	3-Picoline	C_6H_7N	108-99-6	93.127	liq	-18.1(0.3)	144.1(0.1)	0.9566^{20}	1.5040^{20}	msc H_2O, EtOH, eth; vs ace; s ctc
7740	4-Methylpyridine	4-Picoline	C_6H_7N	108-89-4	93.127	liq	3.68(0.02)	145.3(0.1)	0.9548^{20}	1.5037^{20}	msc H_2O, EtOH, eth; s ace, ctc
7741	6-Methyl-2-pyridinecarboxaldehyde		C_7H_7NO	1122-72-1	121.137		32	77^{12}			
7742	Methyl 3-pyridinecarboxylate	Methyl nicotinate	$C_7H_7NO_2$	93-60-7	137.137	cry	39.5(0.4)	204			s H_2O, EtOH, bz
7743	Methyl 4-pyridinecarboxylate	Methyl isonicotinate	$C_7H_7NO_2$	2459-09-8	137.137		10.2(0.9)	209(3)	1.1599^{20}	1.5135^{20}	sl H_2O, ctc; s EtOH, eth, bz
7744	2-Methylpyridine-1-oxide		C_6H_7NO	931-19-1	109.126		49.8(0.5)	260			
7745	3-Methylpyridine-1-oxide		C_6H_7NO	1003-73-2	109.126		41.0(0.5)	148^{15}			s chl
7746	4-Methylpyridine-1-oxide		C_6H_7NO	1003-67-4	109.126		185.8				
7747	1-Methyl-2(1*H*)-pyridinone		C_6H_7NO	694-85-9	109.126	nd	31	253(13)	1.1120^{20}		msc H_2O; sl peth, lig
7748	1-(6-Methyl-3-pyridinyl)ethanone		C_8H_9NO	36357-38-7	135.163		17.6	144^{50}	1.0168^{25}	1.5302^{25}	vs H_2O
7749	4-Methyl-2-pyrimidinamine		$C_5H_7N_3$	108-52-1	109.130	pl (w), nd (sub)	160.3	sub			s H_2O, EtOH; sl chl
7750	2-Methylpyrimidine	2-Methyl-1,3-diazine	$C_5H_6N_2$	5053-43-0	94.115	liq	-4	138			msc H_2O
7751	4-Methylpyrimidine	4-Methyl-1,3-diazine	$C_5H_6N_2$	3438-46-8	94.115		32	142	1.030^{16}	1.500^{20}	msc H_2O
7752	5-Methylpyrimidine	5-Methyl-1,3-diazine	$C_5H_6N_2$	2036-41-1	94.115		30.5	153			vs H_2O
7753	6-Methyl-2,4(1*H*,3*H*)-pyrimidinedione	6-Methyluracil	$C_5H_6N_2O_2$	626-48-2	126.114	oct pr or nd (w, al)	275 dec				s H_2O, EtOH; sl eth, tfa; vs NH_3
7754	1-Methylpyrrole		C_5H_7N	96-54-8	81.117	liq	-56.23(0.01)	112.7(0.2)	0.9145^{15}	1.4875^{20}	i H_2O; msc EtOH, eth
7755	2-Methylpyrrole		C_5H_7N	636-41-9	81.117	liq	-35.6(0.4)	148(4)	0.9446^{15}	1.5035^{16}	i H_2O; msc EtOH, eth
7756	3-Methylpyrrole		C_5H_7N	616-43-3	81.117	liq	-48.4	143(4)		1.4970^{20}	msc EtOH, eth
7757	*N*-Methylpyrrolidine		$C_5H_{11}N$	120-94-5	85.148			206.1(0.3)	0.8188^{20}	1.4247^{20}	vs H_2O, eth

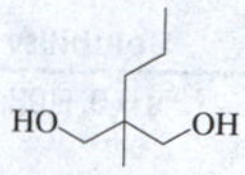
7718
2-Methyl-2-propyl-1,3-propanediol

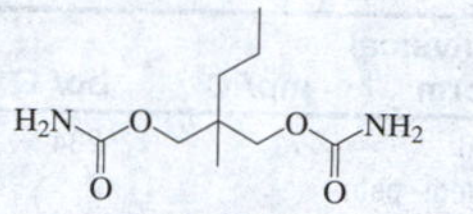
7719
2-Methyl-2-propyl-1,3-propanediol dicarbamate

7720
Methyl propyl sulfide

7721
N-Methyl-2-propyn-1-amine

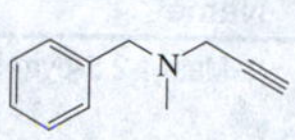
7722
N-Methyl-*N*-2-propynylbenzenemethanamine

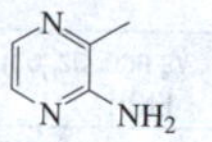
7723
3-Methylpyrazinamine

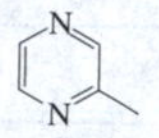
7724
2-Methylpyrazine

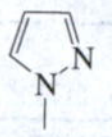
7725
1-Methyl-1*H*-pyrazole

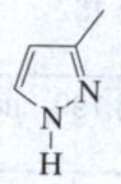
7726
3-Methyl-1*H*-pyrazole

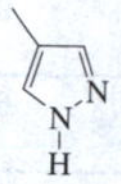
7727
4-Methyl-1*H*-pyrazole

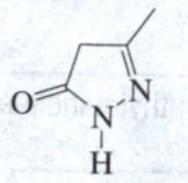
7728
3-Methyl-2-pyrazolin-5-one

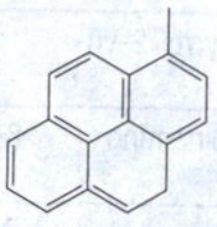
7729
1-Methylpyrene

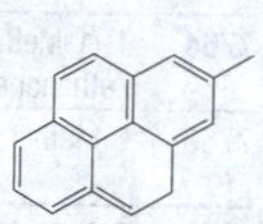
7730
2-Methylpyrene

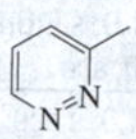
7731
3-Methylpyridazine

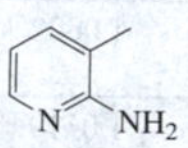
7732
3-Methyl-2-pyridinamine

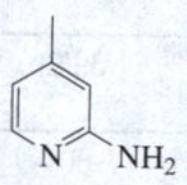
7733
4-Methyl-2-pyridinamine

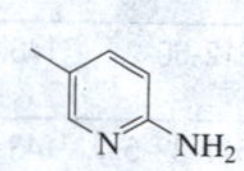
7734
5-Methyl-2-pyridinamine

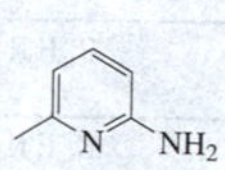
7735
6-Methyl-2-pyridinamine

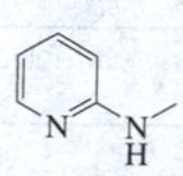
7736
N-Methyl-2-pyridinamine

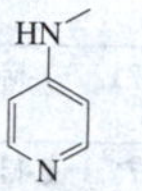
7737
N-Methyl-4-pyridinamine

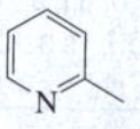
7738
2-Methylpyridine

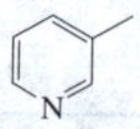
7739
3-Methylpyridine

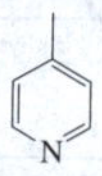
7740
4-Methylpyridine

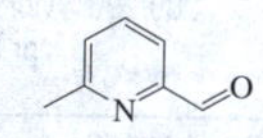
7741
6-Methyl-2-pyridinecarboxaldehyde

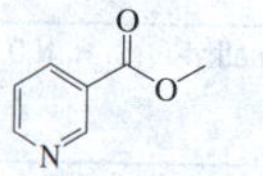
7742
Methyl 3-pyridinecarboxylate

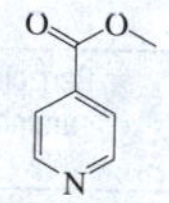
7743
Methyl 4-pyridinecarboxylate

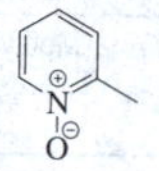
7744
2-Methylpyridine-1-oxide

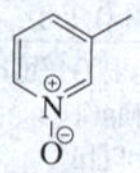
7745
3-Methylpyridine-1-oxide

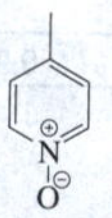
7746
4-Methylpyridine-1-oxide

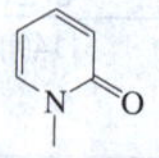
7747
1-Methyl-2(1*H*)-pyridinone

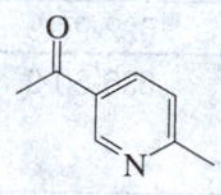
7748
1-(6-Methyl-3-pyridinyl)ethanone

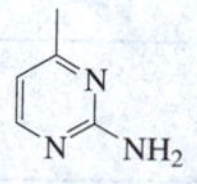
7749
4-Methyl-2-pyrimidinamine

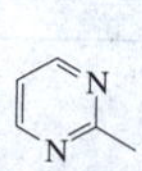
7750
2-Methylpyrimidine

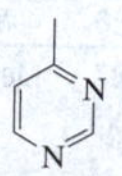
7751
4-Methylpyrimidine

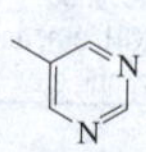
7752
5-Methylpyrimidine

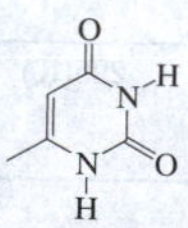
7753
6-Methyl-2,4(1*H*,3*H*)-pyrimidinedione

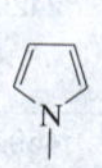
7754
1-Methylpyrrole

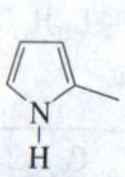
7755
2-Methylpyrrole

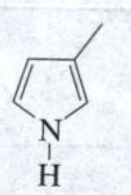
7756
3-Methylpyrrole

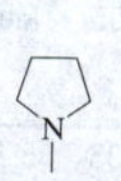
7757
N-Methylpyrrolidine

No.	Name	Synonym	Mol. form.	CAS Reg. No.	Mol. wt.	Physical form	mp/°C	bp/°C	den/ g cm⁻³	n_D	Solubility
7758	1-Methyl-2,5-pyrrolidinedione		$C_5H_7NO_2$	1121-07-9	113.116	nd (eth- peth, al, ace)	71	234			s H_2O, EtOH; vs eth
7759	*N*-Methyl-2-pyrrolidinethione		C_5H_9NS	10441-57-3	115.197	oil		$100^{0.08}$			
7760	5-Methyl-2-pyrrolidinone		C_5H_9NO	108-27-0	99.131		43	248	1.0458^{20}		
7761	1-(1-Methyl-2-pyrrolidinyl)-2-propanone, (*R*)-	Hygrine	$C_8H_{15}NO$	496-49-1	141.211			76.5^{11}		1.4555^{20}	vs EtOH, chl
7762	3-(1-Methyl-2-pyrrolidinyl)-pyridine, (±)-		$C_{10}H_{14}N_2$	22083-74-5	162.231			244	1.0082^{20}	1.5289^{20}	msc H_2O; vs EtOH, eth, chl; s lig
7763	*N*-Methyl-2-pyrrolidinone	1-Methyl-2-pyrrolidinone	C_5H_9NO	872-50-4	99.131	liq	-24.0(0.6)	204.2(0.3)	1.0230^{25}	1.4684^{20}	vs H_2O; s eth, ace, chl
7764	1-(1-Methyl-1*H*-pyrrol-2-yl)-ethanone		C_7H_9NO	932-16-1	123.152			201^{252}	1.0445^{15}	1.5403^{15}	s EtOH, bz, chl
7765	6-Methyl-8-quinolinamine	8-Amino-6-methylquinoline	$C_{10}H_{10}N_2$	68420-93-9	158.199	nd	73	sub			vs ace, bz, eth, EtOH
7766	2-Methylquinoline	Quinaldine	$C_{10}H_9N$	91-63-4	143.185	col oily liq	-2.67(0.05)	247.4(0.4)	1.06^{25}	1.6116^{20}	sl H_2O; s EtOH, eth, ace, ctc, chl
7767	3-Methylquinoline		$C_{10}H_9N$	612-58-8	143.185	pr	16.5	251(7)	1.0673^{20}	1.6171^{20}	vs ace, eth, EtOH
7768	4-Methylquinoline	Lepidine	$C_{10}H_9N$	491-35-0	143.185	col oily liq	9.5(0.3)	266(2)	1.083^{20}	1.6200^{20}	sl H_2O; s EtOH, eth, ace; i alk
7769	5-Methylquinoline		$C_{10}H_9N$	7661-55-4	143.185	col cry	19	257(5)	1.0832^{20}	1.6219^{20}	sl H_2O; s ace; msc EtOH, eth
7770	6-Methylquinoline		$C_{10}H_9N$	91-62-3	143.185	col oily liq	-22	265(1)	1.0654^{20}	1.6157^{20}	sl H_2O; s EtOH, eth, ace
7771	7-Methylquinoline	*m*-Toluquinoline	$C_{10}H_9N$	612-60-2	143.185	ye cry	39(2)	258(1)	1.0609^{20}	1.6150^{20}	sl H_2O; s EtOH, eth, ace
7772	8-Methylquinoline		$C_{10}H_9N$	611-32-5	143.185	col liq	-80	247.4(0.7)	1.0719^{20}	1.6164^{20}	sl H_2O; s ace; msc EtOH, eth
7773	2-Methyl-8-quinolinol		$C_{10}H_9NO$	826-81-3	159.184		73.8	267			i H_2O; s EtOH, eth, bz, ctc
7774	1-Methyl-2(1*H*)-quinolinone		$C_{10}H_9NO$	606-43-9	159.184	nd (lig)	74	325			sl H_2O, lig; s EtOH, eth, ace; vs bz
7775	1-Methyl-4(1*H*)-quinolinone	Echinopsine	$C_{10}H_9NO$	83-54-5	159.184	α-nd (bz); β-cry (al)	152				s H_2O; vs EtOH, bz, chl; sl eth
7776	2-Methylquinoxaline		$C_9H_8N_2$	7251-61-8	144.173	ye cry	180.5	244			msc H_2O, eth, ace, bz; vs EtOH; s ctc
7777	Methyl Red	Benzoic acid, 2-[[4-(dimethyl-amino)phenyl]azo]-	$C_{15}H_{15}N_3O_2$	493-52-7	269.299	viol or red pr (to, bz)	183				sl H_2O, lig; s EtOH; vs ace, bz, chl
7778	Methyl β-D-ribofuranoside		$C_6H_{12}O_5$	7473-45-2	164.156		80				
7779	Methyl salicylate	Methyl 2-hydroxybenzoate	$C_8H_8O_3$	119-36-8	152.148	liq	-8.5(0.5)	222.6(0.5)	1.181^{25}	1.535^{20}	sl H_2O; vs eth, EtOH, chl
7780	Methylsilane		CH_6Si	992-94-9	46.145	col gas	-156.5	-57.5			
7781	Methyl silyl ether		CH_6OSi	2171-96-2	62.144	col gas	-98.5	-21			
7782	Methylstannane		CH_6Sn	1631-78-3	136.769	col gas		1.4			dec H_2O
7783	Methyl stearate		$C_{19}H_{38}O_2$	112-61-8	298.504		38.7(0.6)	353(8)	0.8498^{40}	1.4367^{40}	vs eth, chl
7784	2-Methylstyrene		C_9H_{10}	611-15-4	118.175	liq	-68.5(0.3)	170(1)	0.9077^{25}	1.5437^{20}	i H_2O; s bz, chl
7785	3-Methylstyrene		C_9H_{10}	100-80-1	118.175	liq	-86.3(0.3)	170(1)	0.9076^{25}	1.5411^{20}	i H_2O; s EtOH, eth, bz
7786	4-Methylstyrene		C_9H_{10}	622-97-9	118.175	liq	-37.8(0.4)	172(1)	0.9173^{25}	1.5420	i H_2O; s bz
7787	Methylsuccinic acid		$C_5H_8O_4$	636-60-2	132.116	pr	114(4)	dec	1.4200^{0}	1.4303	vs H_2O, EtOH, MeOH; s eth; sl chl
7788	Methyl sulfate		CH_4O_4S	75-93-4	112.106		<-30	135 dec			vs H_2O, eth, EtOH
7789	(Methylsulfinyl)benzene		C_7H_8OS	1193-82-4	140.203		32.0	263.5		1.5885^{20}	
7790	1-(Methylsulfinyl)decane	Decyl methyl sulfoxide	$C_{11}H_{24}OS$	3079-28-5	204.373	cry	52.5				
7791	3-Methyl sulfolane		$C_5H_{10}O_2S$	872-93-5	134.197		1	277(13)	1.188^{25}	1.4772^{20}	
7792	(Methylsulfonyl)ethene		$C_3H_6O_2S$	3680-02-2	106.144			122^{24}	1.2117^{20}	1.4636^{20}	s eth, ace
7793	Methyl terephthalate	Methyl 1,4-benzenedicarboxyl-ate	$C_9H_8O_4$	1679-64-7	180.158	nd (w)	219.4(0.4)				
7794	17-Methyltestosterone	17-Hydroxy-17-methylandrost-4-en-3-one, (17β)	$C_{20}H_{30}O_2$	58-18-4	302.451		163.5				vs eth, EtOH
7795	Methyl tetradecanoate		$C_{15}H_{30}O_2$	124-10-7	242.398		19.0(0.5)	295(10)	0.8671^{20}	1.425^{45}	i H_2O; msc EtOH, eth, ace, bz, chl, ctc
7796	5-*N*-Methyl-5,6,7,8-tetrahy-drofolic acid		$C_{20}H_{25}N_7O_6$	134-35-0	459.456	cry (w)					

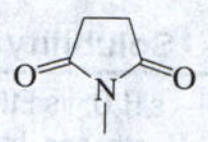
7758
1-Methyl-2,5-pyrrolidinedione

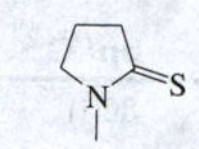
7759
N-Methyl-2-pyrrolidinethione

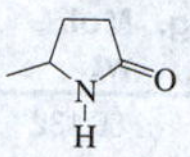
7760
5-Methyl-2-pyrrolidinone

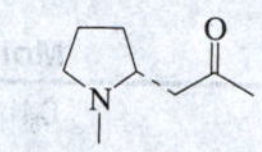
7761
1-(1-Methyl-2-pyrrolidinyl)-2-propanone, (*R*)-

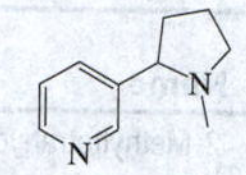
7762
3-(1-Methyl-2-pyrrolidinyl)pyridine, (±)-

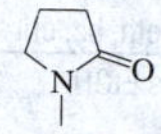
7763
N-Methyl-2-pyrrolidinone

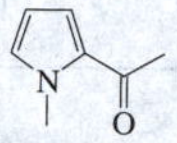
7764
1-(1-Methyl-1*H*-pyrrol-2-yl)ethanone

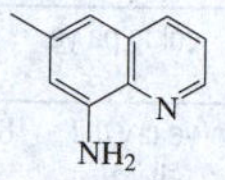
7765
6-Methyl-8-quinolinamine

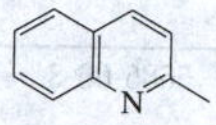
7766
2-Methylquinoline

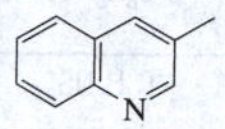
7767
3-Methylquinoline

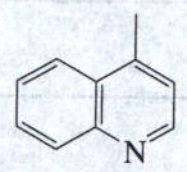
7768
4-Methylquinoline

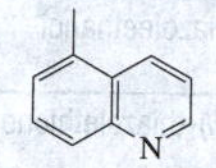
7769
5-Methylquinoline

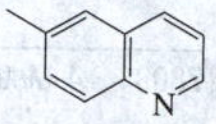
7770
6-Methylquinoline

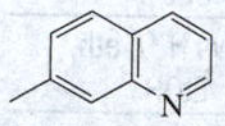
7771
7-Methylquinoline

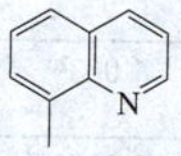
7772
8-Methylquinoline

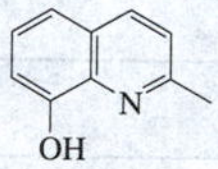
7773
2-Methyl-8-quinolinol

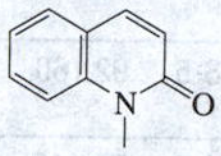
7774
1-Methyl-2(1*H*)-quinolinone

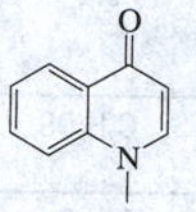
7775
1-Methyl-4(1*H*)-quinolinone

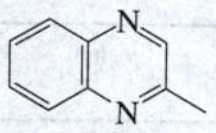
7776
2-Methylquinoxaline

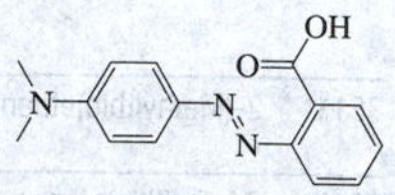
7777
Methyl Red

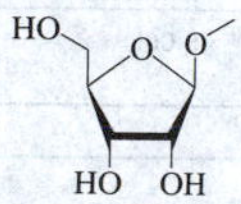
7778
Methyl β-D-ribofuranoside

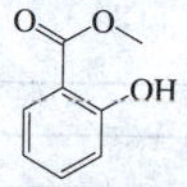
7779
Methyl salicylate

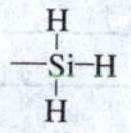
7780
Methylsilane

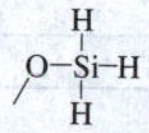
7781
Methyl silyl ether

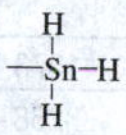
7782
Methylstannane

7783
Methyl stearate

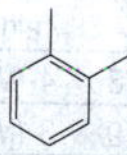
7784
2-Methylstyrene

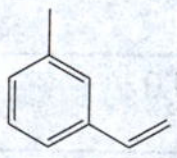
7785
3-Methylstyrene

7786
4-Methylstyrene

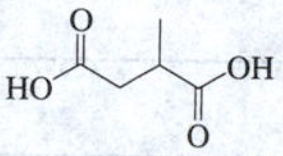
7787
Methylsuccinic acid

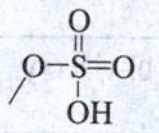
7788
Methyl sulfate

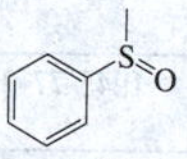
7789
(Methylsulfinyl)benzene

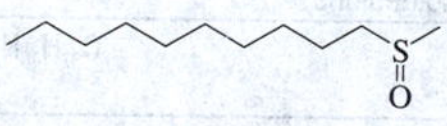
7790
1-(Methylsulfinyl)decane

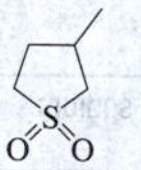
7791
3-Methyl sulfolane

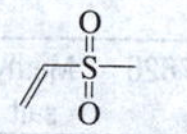
7792
(Methylsulfonyl)ethene

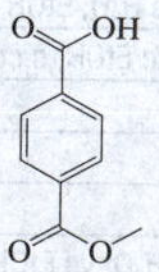
7793
Methyl terephthalate

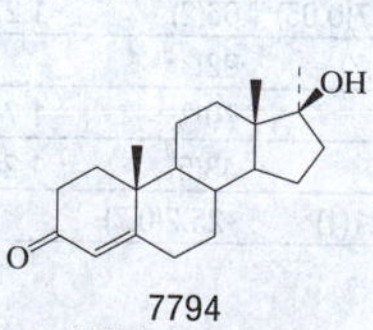
7794
17-Methyltestosterone

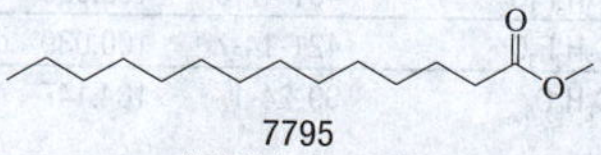
7795
Methyl tetradecanoate

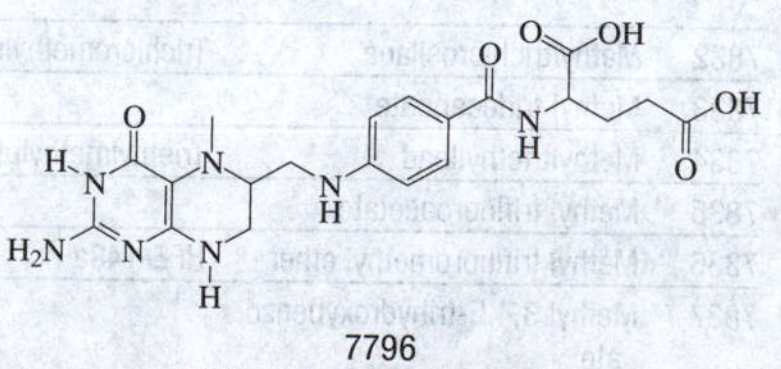
7796
5-*N*-Methyl-5,6,7,8-tetrahydrofolic acid

No.	Name	Synonym	Mol. form.	CAS Reg. No.	Mol. wt.	Physical form	mp/°C	bp/°C	den/g cm^{-3}	n_D	Solubility
7797	2-Methyltetrahydrofuran	2-Methyloxolane	$C_5H_{10}O$	96-47-9	86.132			80(1)	0.8552^{20}	1.4059^{21}	s H_2O; vs EtOH, eth, ace, bz; sl ctc
7798	*N*-Methyl-*N*,2,4,6-tetranitroaniline	Tetryl	$C_7H_5N_5O_8$	479-45-8	287.144	ye pr (al)	128(3)	exp	1.57^{10}		i H_2O; sl EtOH, eth, chl; s ace, bz, py
7799	4-Methyl-2-thiazolamine	2-Amino-4-methylthiazole	$C_4H_6N_2S$	1603-91-4	114.169		45.5	125^{20}			vs H_2O, EtOH, eth
7800	2-Methylthiazole		C_4H_5NS	3581-87-1	99.155			128.77(0.05)		1.510	msc H_2O; s EtOH, ace
7801	4-Methylthiazole		C_4H_5NS	693-95-8	99.155			133.2(0.2)	1.112^{25}		s H_2O, EtOH, eth
7802	4-Methyl-5-thiazoleethanol		C_6H_9NOS	137-00-8	143.206	col to pa ye		135^{7}	1.196^{24}		vs H_2O; s EtOH, eth, bz, chl
7803	4-Methyl-2(3*H*)-thiazolethione		$C_4H_5NS_2$	5685-06-3	131.220	ye cry (dil al)	89.3	188^{3}			vs EtOH
7804	Methylthiirane		C_3H_6S	1072-43-1	74.145	liq	-91	72.5	0.941^{20}	1.472^{20}	s chl
7805	(Methylthio)acetic acid		$C_3H_6O_2S$	2444-37-3	106.144		13.0	130^{27}	1.221^{20}	1.495^{20}	
7806	2-(Methylthio)aniline		C_7H_9NS	2987-53-3	139.218			234	1.111^{25}	1.6239^{20}	
7807	4-(Methylthio)aniline		C_7H_9NS	104-96-1	139.218			272.5	1.1379^{20}	1.6395^{20}	s EtOH, eth, ace, bz
7808	(Methylthio)benzene	Methyl phenyl sulfide	C_7H_8S	100-68-5	124.204			194.3(0.2)	1.0579^{20}	1.5868^{20}	i H_2O; s EtOH; vs ace
7809	2-(Methylthio)benzothiazole		$C_8H_7NS_2$	615-22-5	181.279	pr (dil al)	52	174^{22}			s EtOH, chl
7810	Methyl thiocyanate		C_2H_3NS	556-64-9	73.117	col liq	-2.5	132.9	1.0678^{25}	1.4669^{25}	sl H_2O; msc EtOH, eth; s ctc
7811	2-(Methylthio)ethanol		C_3H_8OS	5271-38-5	92.160			70^{20}	1.063^{20}	1.4861^{30}	vs H_2O, eth, EtOH
7812	(Methylthio)ethene	Methyl vinyl sulfide	C_3H_6S	1822-74-8	74.145			69(3)	0.9026^{20}	1.4837^{20}	s eth, ace, chl
7813	[(Methylthio)methyl]benzene		$C_8H_{10}S$	766-92-7	138.230	liq	-30	214(10)	1.0274^{20}	1.5620^{20}	
7814	4-(Methylthio)-2-oxobutanoic acid		$C_5H_8O_3S$	583-92-6	148.181	oil					
7815	2-Methylthiophene		C_5H_6S	554-14-3	98.167	liq	-63.3(0.2)	112.5(0.4)	1.0193^{20}	1.5203^{20}	i H_2O; msc EtOH, eth, ace, bz, hp, ctc
7816	3-Methylthiophene		C_5H_6S	616-44-4	98.167	liq	-68.91(0.06)	115.4(0.4)	1.0218^{20}	1.5204^{20}	i H_2O; msc EtOH, eth, ace, bz; vs chl
7817	5-Methyl-2-thiophenecarboxaldehyde		C_6H_6OS	13679-70-4	126.176			114^{25}		1.5825^{20}	s chl
7818	4-(Methylthio)phenol		C_7H_8OS	1073-72-9	140.203		84	154^{20}			
7819	3-(Methylthio)propanal		C_4H_8OS	3268-49-3	104.171			62^{11}			
7820	3-(Methylthio)propanoic acid	*S*-Methylpropiothetin	$C_4H_8O_2S$	646-01-5	120.171	ye oil or fl (hx)	21	132^{13}			
7821	3-(Methylthio)-1-propene		C_4H_8S	10152-76-8	88.172			94(18)	0.8767^{20}	1.4714^{20}	
7822	*N*-Methylthiosemicarbazide	*N*-Methylhydrazinecarbothioamide	$C_2H_7N_3S$	6610-29-3	105.162		136.5				s H_2O, EtOH, DMSO; i eth, bz, lig
7823	Methylthiouracil		$C_5H_6N_2OS$	56-04-2	142.179		330 dec	sub			i H_2O; sl EtOH, eth, MeOH, bz
7824	Methylthiourea		$C_2H_6N_2S$	598-52-7	90.147	pr (EtOH)	119.2(0.4)				vs H_2O, EtOH; sl eth; s ace
7825	1-Methylthymine	1,5-Dimethyl-2,4(1*H*,3*H*)-pyrimidinedione	$C_6H_8N_2O_2$	4160-72-9	140.140	nd (w)	295				s H_2O
7826	Methylthymol blue, sodium salt		$C_{37}H_{40}N_2O_{13}Na_4S$	1945-77-3	844.743	bl-viol cry					s H_2O
7827	Methyl 4-toluenesulfonate		$C_8H_{10}O_3S$	80-48-8	186.228		24.1(0.2)	292	1.2087^{40}		i H_2O; vs EtOH, bz; s eth, ctc; sl lig
7828	Methyltriacetoxysilane	Methylsilanetriol, triacetate	$C_7H_{12}O_6Si$	4253-34-3	220.252		40.5	111^{17}	1.1750^{20}	1.4083^{20}	
7829	6-Methyl-1,2,4-triazine-3,5(2*H*,4*H*)-dione	6-Azathymine	$C_4H_5N_3O_2$	932-53-6	127.102	cry (w)	211				s H_2O, EtOH, ace
7830	5-Methyl-[1,2,4]triazolo[1,5-*a*]pyrimidin-7-ol		$C_6H_6N_4O$	2503-56-2	150.138		>245				
7831	Methyl trichloroacetate		$C_3H_3Cl_3O_2$	598-99-2	177.414	liq	-17.5(0.4)	152(8)	1.4874^{20}	1.4572^{20}	i H_2O; vs EtOH, eth; s ctc
7832	Methyltrichlorosilane	Trichloromethylsilane	CH_3Cl_3Si	75-79-6	149.480	liq	-75.77(0.05)	66(2)	1.273^{20}	1.4106^{20}	dec H_2O, EtOH
7833	Methyl tridecanoate		$C_{14}H_{28}O_2$	1731-88-0	228.371		5.9(0.7)	92^{1}		1.4405^{20}	msc EtOH; s ctc
7834	Methyltriethyllead	Triethylmethylplumbane	$C_7H_{18}Pb$	1762-28-3	309.4	col liq		70^{16}	1.71^{20}		
7835	Methyl trifluoroacetate		$C_3H_3F_3O_2$	431-47-0	128.050			43(5)	1.28^{20}		
7836	Methyl trifluoromethyl ether	HFE-143a	$C_2H_3F_3O$	421-14-7	100.039	col gas	-149.1(1)	-25.2(0.7)			
7837	Methyl 3,4,5-trihydroxybenzoate		$C_8H_8O_5$	99-24-1	184.147	mcl pr (MeOH)	202				sl H_2O; vs EtOH, MeOH

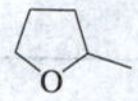
7797
2-Methyltetrahydrofuran

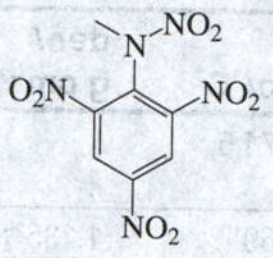
7798
N-Methyl-*N*,2,4,6-tetranitroaniline

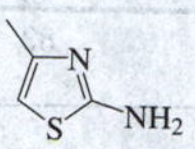
7799
4-Methyl-2-thiazolamine

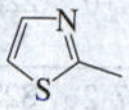
7800
2-Methylthiazole

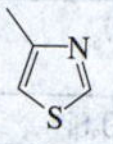
7801
4-Methylthiazole

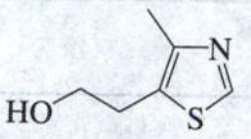
7802
4-Methyl-5-thiazoleethanol

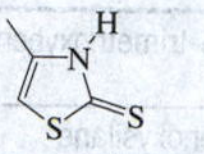
7803
4-Methyl-2(3*H*)-thiazolethione

7804
Methylthiirane

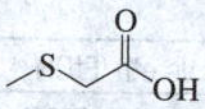
7805
(Methylthio)acetic acid

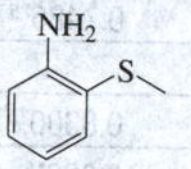
7806
2-(Methylthio)aniline

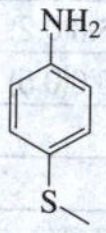
7807
4-(Methylthio)aniline

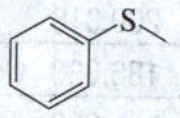
7808
(Methylthio)benzene

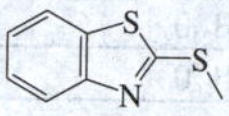
7809
2-(Methylthio)benzothiazole

7810
Methyl thiocyanate

7811
2-(Methylthio)ethanol

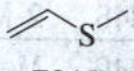
7812
(Methylthio)ethene

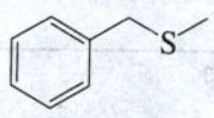
7813
[(Methylthio)methyl]benzene

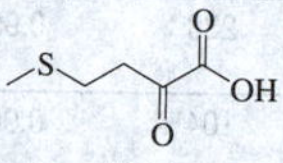
7814
4-(Methylthio)-2-oxobutanoic acid

7815
2-Methylthiophene

7816
3-Methylthiophene

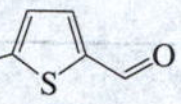
7817
5-Methyl-2-thiophenecarboxaldehyde

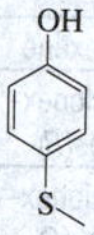
7818
4-(Methylthio)phenol

7819
3-(Methylthio)propanal

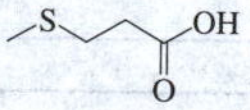
7820
3-(Methylthio)propanoic acid

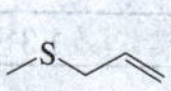
7821
3-(Methylthio)-1-propene

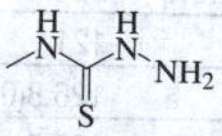
7822
N-Methylthiosemicarbazide

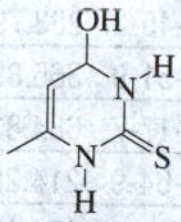
7823
Methylthiouracil

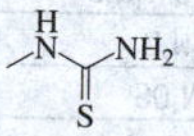
7824
Methylthiourea

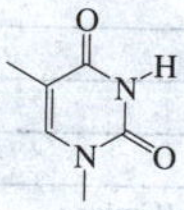
7825
1-Methylthymine

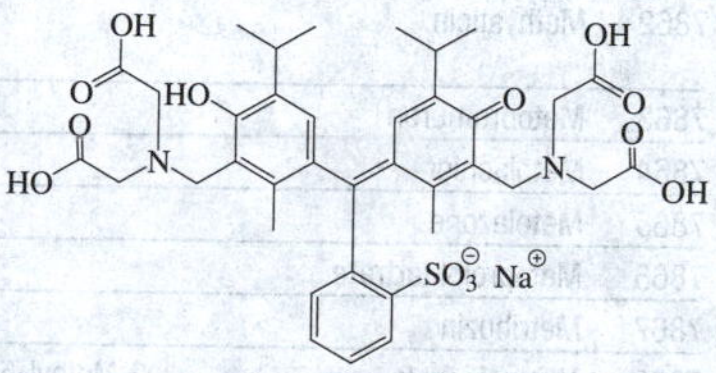
7826
Methylthymol blue, sodium salt

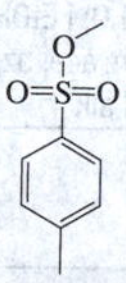
7827
Methyl 4-toluenesulfonate

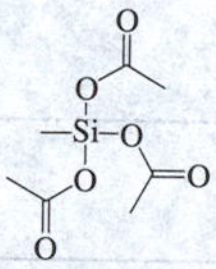
7828
Methyltriacetoxysilane

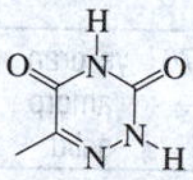
7829
6-Methyl-1,2,4-triazine-3,5(2*H*,4*H*)-dione

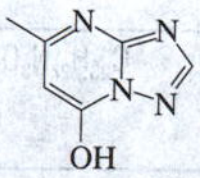
7830
5-Methyl-[1,2,4]triazolo[1,5-a]pyrimidin-7-ol

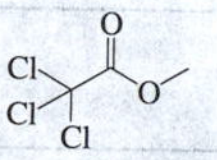
7831
Methyl trichloroacetate

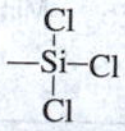
7832
Methyltrichlorosilane

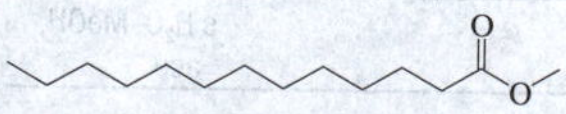
7833
Methyl tridecanoate

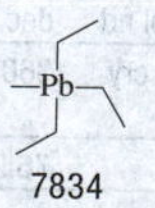
7834
Methyltriethyllead

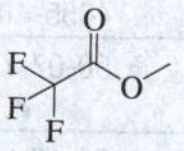
7835
Methyl trifluoroacetate

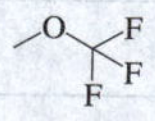
7836
Methyl trifluoromethyl ether

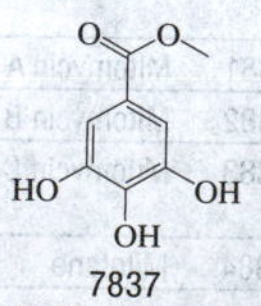
7837
Methyl 3,4,5-trihydroxybenzoate

No.	Name	Synonym	Mol. form.	CAS Reg. No.	Mol. wt.	Physical form	mp/°C	bp/°C	den/ g cm^{-3}	n_D	Solubility
7838	Methyl 3,4,5-trimethoxybenzoate		$C_{11}H_{14}O_5$	1916-07-0	226.226		83	274.5			
7839	Methyltriphenoxysilane		$C_{19}H_{18}O_3Si$	3439-97-2	322.430			269^{100}	1.135^{20}	1.5599^{20}	
7840	Methyl trithion		$C_9H_{12}ClO_2PS_3$	953-17-3	314.812	ye liq	-18				sl H_2O; misc os
7841	*N*-Methyl-L-tryptophan	L-Abrine	$C_{12}H_{14}N_2O_2$	526-31-8	218.251	pr (w)	295 dec				sl H_2O, MeOH; i eth; s alk
7842	*N*-Methyl-L-tyrosine	Surinamine	$C_{10}H_{13}NO_3$	537-49-5	195.215	nd	293				
7843	α-Methyl-DL-tyrosine, methyl ester, hydrochloride		$C_{11}H_{16}ClNO_3$	7361-31-1	245.703		190 dec				s H_2O
7844	2-Methylundecanal		$C_{12}H_{24}O$	110-41-8	184.318			119^{16}	0.832^{15}	1.4321^{20}	sl H_2O; s EtOH, eth
7845	2-Methylundecane		$C_{12}H_{26}$	7045-71-8	170.334	liq	-46.8(0.2)	211(1)		1.4191^{20}	
7846	3-Methylundecane		$C_{12}H_{26}$	1002-43-3	170.334	col liq	-58.0(0.5)	212(1)	0.7485^{25}	1.4208^{25}	
7847	Methyl undecanoate		$C_{12}H_{24}O_2$	1731-86-8	200.318			123^{10}			
7848	2-Methyl-1-undecanol		$C_{12}H_{26}O$	10522-26-6	186.333			129^{12}	0.8300^{15}	1.4382^{20}	vs eth, EtOH
7849	Methyl 10-undecenoate		$C_{12}H_{22}O_2$	111-81-9	198.302	liq	-27.5	246(11)	0.889^{15}	1.4393^{20}	i H_2O; s EtOH, eth, HOAc; sl ctc
7850	*N*-Methylurea		$C_2H_6N_2O$	598-50-5	74.081	orth pr (w, al)	101.3(0.8)	dec	1.2040^0		vs H_2O, EtOH; i eth, bz; s CS_2, lig
7851	5-Methyluridine	Thymine riboside	$C_{10}H_{14}N_2O_6$	1463-10-1	258.227	cry (EtOH)	184				
7852	3-Methyl-L-valine	L-*tert*-Leucine	$C_6H_{13}NO_2$	20859-02-3	131.173		248 dec				
7853	2-(1-Methylvinyl)aniline		$C_9H_{11}N$	52562-19-3	133.190			115^{20}	0.977^{25}	1.5722^{20}	
7854	1-Methyl-4-vinylcyclohexene		C_9H_{14}	17699-86-4	122.207	liq		152	0.85	1.4701^{20}	
7855	4-(1-Methylvinyl)-1-cyclohexene-1-carboxaldehyde, (*R*)-	*d*-Perillaldehyde	$C_{10}H_{14}O$	5503-12-8	150.217	oil		238	0.953^{20}	1.5058^{20}	s ctc
7856	4-(1-Methylvinyl)-1-cyclohexene-1-carboxaldehyde, (*S*)-	*l*-Perillaldehyde	$C_{10}H_{14}O$	18031-40-8	150.217	oil		104^{10}	0.9645^{20}	1.5072^{20}	
7857	4-(1-Methylvinyl)-1-cyclohexene-1-methanol		$C_{10}H_{16}O$	536-59-4	152.233			244	0.9690^{20}	1.5005^{20}	
7858	(1-Methylvinyl)cyclopropane		C_6H_{10}	4663-22-3	82.143	liq	-102.3(0.1)	70(2)	0.751^{20}	1.4252^{20}	
7859	Methyl vinyl ether		C_3H_6O	107-25-5	58.079	col gas	-122	6(14)	0.7725^0	1.3730^0	sl H_2O; vs EtOH, eth, ace, bz
7860	Methyl Violet	C.I. Basic Violet 1	$C_{24}H_{28}ClN_3$	8004-87-3	393.952	bl-viol pow	137 dec				s H_2O, EtOH
7861	Methysergide		$C_{21}H_{27}N_3O_2$	361-37-5	353.458	cry	195				
7862	Methysticin		$C_{15}H_{14}O_5$	495-85-2	274.269	nd (MeOH), pr (ace)	137				
7863	Metobromuron		$C_9H_{11}BrN_2O_2$	3060-89-7	259.099		95.9(0.4)		1.60^{20}		
7864	Metolachlor		$C_{15}H_{22}ClNO_2$	51218-45-2	283.795			$100^{0.001}$	1.12^{20}		
7865	Metolazone		$C_{16}H_{16}ClN_3O_3S$	17560-51-9	365.834	cry (EtOH)	254				
7866	Metoprolol tartrate		$C_{34}H_{56}N_2O_{12}$	56392-17-7	684.815	cry	121				
7867	Metribuzin		$C_8H_{14}N_4OS$	21087-64-9	214.288		126.8(0.5)		1.31^{20}		
7868	Metronidazole	2-Methyl-5-nitro-1*H*-imidazole-1-ethanol	$C_6H_9N_3O_3$	443-48-1	171.153		160.5				
7869	Metsulfuron-methyl		$C_{14}H_{15}N_5O_6S$	74223-64-6	381.364	wh cry	163				sl H_2O
7870	Mevinphos		$C_7H_{13}O_6P$	7786-34-7	224.148		21 (*E*), 6.9 (*Z*)	$101^{0.3}$			
7871	Mexacarbate	4-(Dimethylamino)-3,5-xylyl methylcarbamate	$C_{12}H_{18}N_2O_2$	315-18-4	222.283	cry	88.9(0.5)				vs EtOH, bz, ace
7872	MGK 264		$C_{17}H_{25}NO_2$	113-48-4	275.387		<-20	157	1.04		
7873	Mifepristone	RU-486	$C_{29}H_{35}NO_2$	84371-65-3	429.594	cry	150				
7874	Mimosine		$C_8H_{10}N_2O_4$	500-44-7	198.176	tab (w)	228 dec				sl H_2O; i EtOH, eth, ace, bz; s dil alk
7875	Minocycline		$C_{23}H_{27}N_3O_7$	10118-90-8	457.476	ye-oran amorp solid					
7876	Minoxidil		$C_9H_{15}N_5O$	38304-91-5	209.248	cry	248				i ace, bz, chl, sl; EtOH, MeOH
7877	Mipafox	Bis(isopropylamido) fluorophosphate	$C_6H_{16}FN_2OP$	371-86-8	182.175	cry (peth)	65	125^2			sl H_2O
7878	Mirex	Hexachloropentadiene dimer	$C_{10}Cl_{12}$	2385-85-5	545.543	cry (bz)	485 dec				vs bz, diox
7879	Misoprostol		$C_{22}H_{38}O_5$	59122-46-2	382.534	ye oil					s H_2O
7880	Mithramycin	Plicamycin	$C_{52}H_{76}O_{24}$	18378-89-7	1085.145	ye cry (ace)	182				s H_2O, EtOH, AcOEt; sl bz, eth
7881	Mitomycin A		$C_{16}H_{19}N_3O_6$	4055-39-4	349.338	purp nd	160 dec				
7882	Mitomycin B		$C_{16}H_{19}N_3O_6$	4055-40-7	349.338	purp-bl nd	dec				
7883	Mitomycin C		$C_{15}H_{18}N_4O_5$	50-07-7	334.328	bl-viol cry	360				s H_2O, MeOH, ace
7884	Mitotane		$C_{14}H_{10}Cl_4$	53-19-0	320.041		76.2(0.3)				

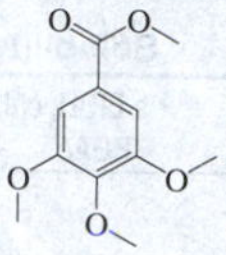
7838
Methyl 3,4,5-trimethoxybenzoate

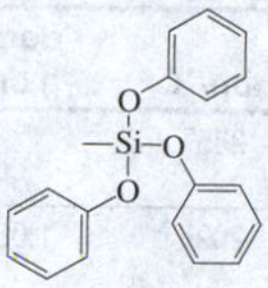
7839
Methyltriphenoxysilane

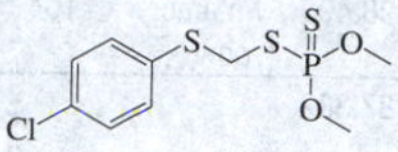
7840
Methyl trithion

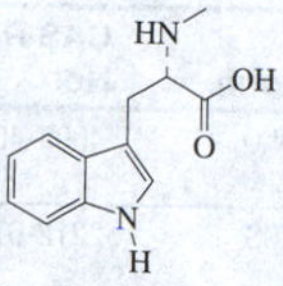
7841
N-Methyl-L-tryptophan

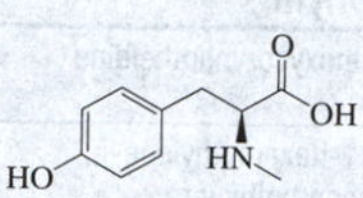
7842
N-Methyl-L-tyrosine

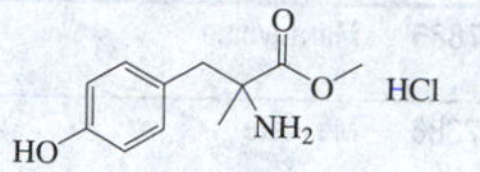
7843
α-Methyl-DL-tyrosine, methyl ester, hydrochloride

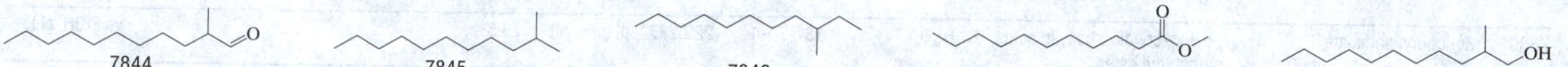
7844
2-Methylundecanal

7845
2-Methylundecane

7846
3-Methylundecane

7847
Methyl undecanoate

7848
2-Methyl-1-undecanol

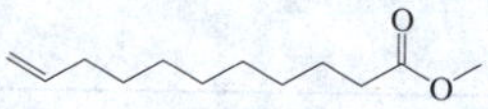
7849
Methyl 10-undecenoate

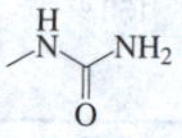
7850
N-Methylurea

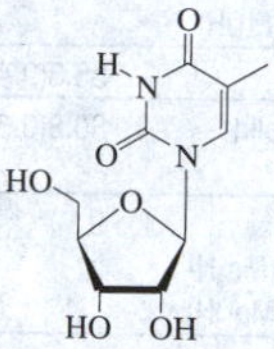
7851
5-Methyluridine

7852
3-Methyl-L-valine

7853
2-(1-Methylvinyl)aniline

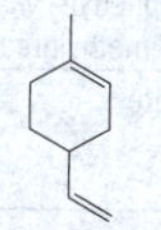
7854
1-Methyl-4-vinylcyclohexene

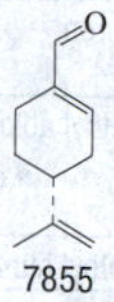
7855
4-(1-Methylvinyl)-1-cyclohexene-1-carboxaldehyde, (*R*)-

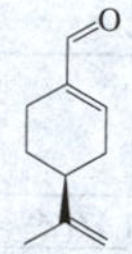
7856
4-(1-Methylvinyl)-1-cyclohexene-1-carboxaldehyde, (*S*)-

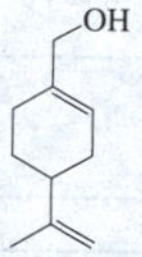
7857
4-(1-Methylvinyl)-1-cyclohexene-1-methanol

7858
(1-Methylvinyl)cyclopropane

7859
Methyl vinyl ether

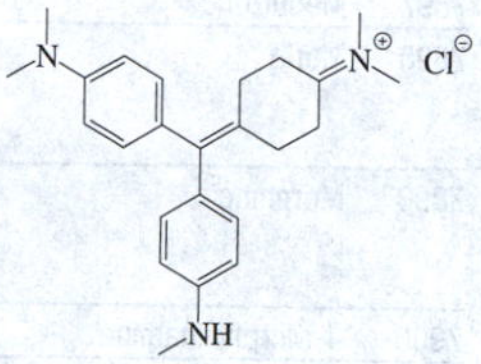
7860
Methyl Violet

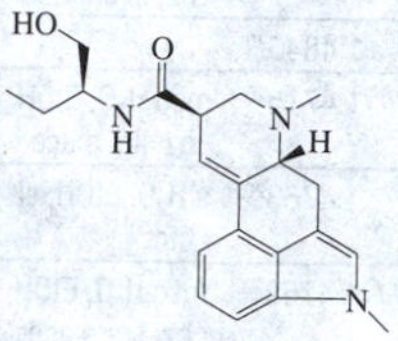
7861
Methysergide

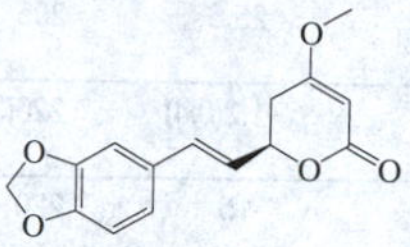
7862
Methysticin

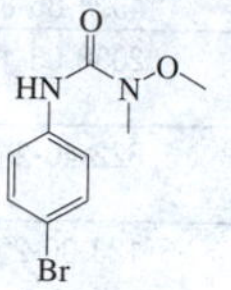
7863
Metobromuron

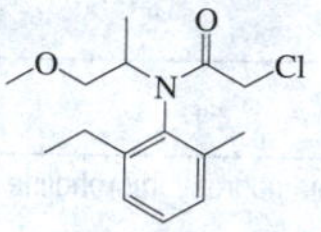
7864
Metolachlor

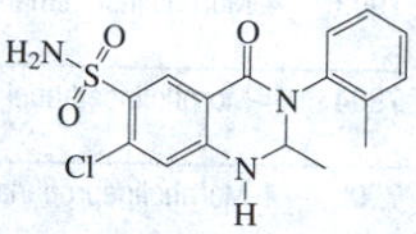
7865
Metolazone

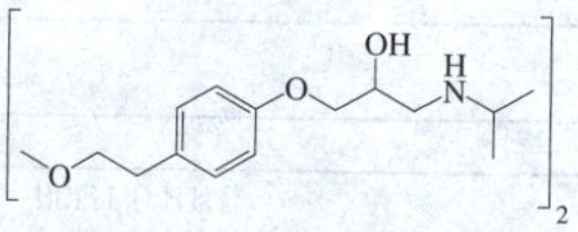
7866
Metoprolol tartrate

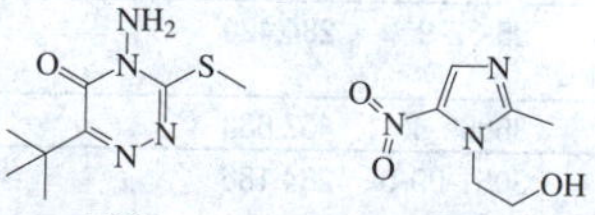
7867
Metribuzin

7868
Metronidazole

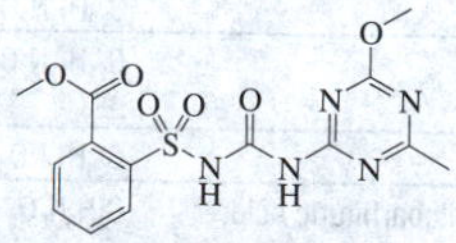
7869
Metsulfuron-methyl

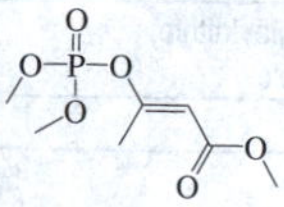
7870
Mevinphos

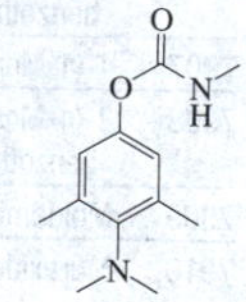
7871
Mexacarbate

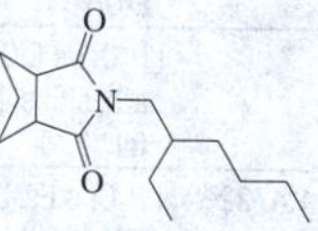
7872
MGK 264

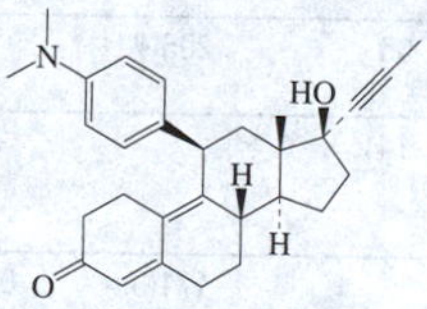
7873
Mifepristone

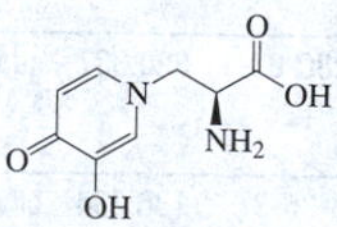
7874
Mimosine

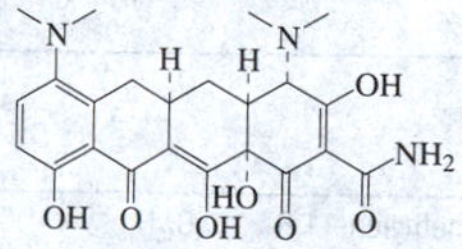
7875
Minocycline

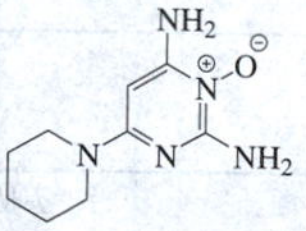
7876
Minoxidil

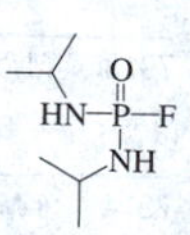
7877
Mipafox

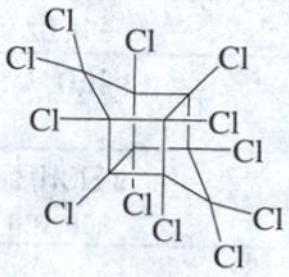
7878
Mirex

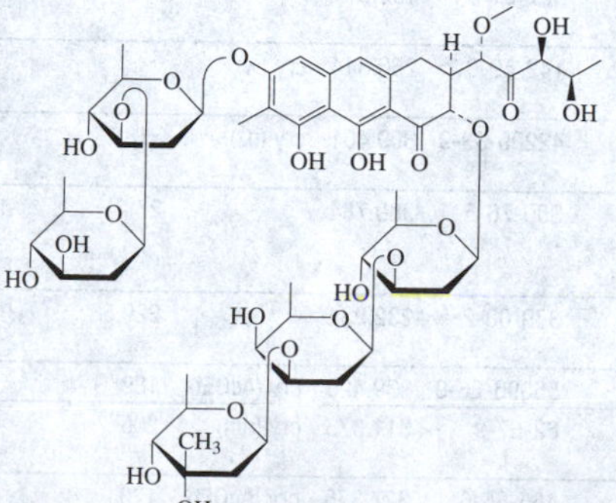
7880
Mithramycin

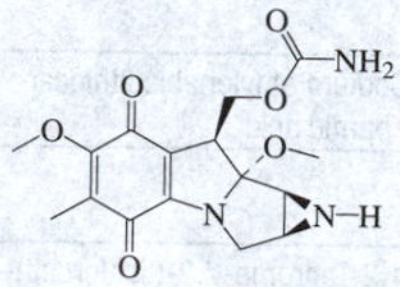
7881
Mitomycin A

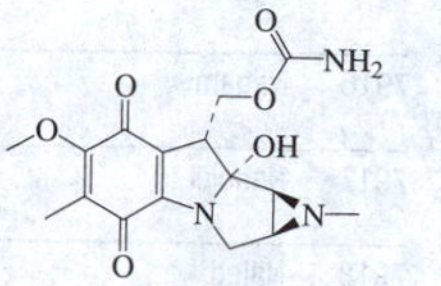
7882
Mitomycin B

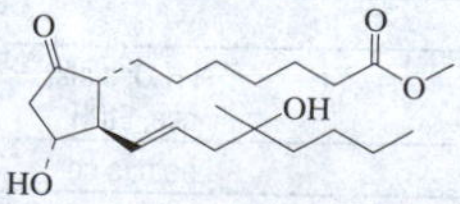
7879
Misoprostol

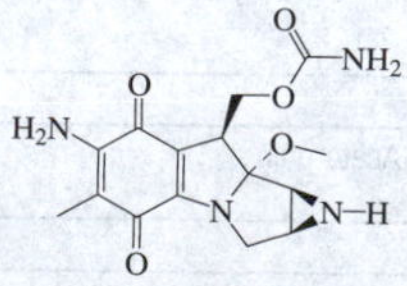
7883
Mitomycin C

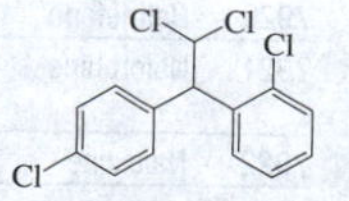
7884
Mitotane

Organic

No.	Name	Synonym	Mol. form.	CAS Reg. No.	Mol. wt.	Physical form	mp/°C	bp/°C	den/ g cm^{-3}	n_D	Solubility
7885	Mitragynine	9-Methoxycorynantheidine	$C_{23}H_{30}N_2O_4$	4098-40-2	398.495	wh amor pow	104	235^{5}			s EtOH, chl, HOAc
7886	Molinate	Ethyl 1-hexamethylene-iminecarbothiolate	$C_9H_{17}NOS$	2212-67-1	187.302			202^{10}	1.063^{20}		
7887	Molindone		$C_{16}H_{24}N_2O_2$	7416-34-4	276.374	cry	180				
7888	Molybdenum carbonyl	Molybdenum hexacarbonyl	C_6MoO_6	13939-06-5	264.00	wh cry	148	155 dec	1.96		i H_2O; s bz; sl eth
7889	Monobutyl phthalate	1,2-Benzenedicarboxylic acid, monobutyl ester	$C_{12}H_{14}O_4$	131-70-4	222.237	pl (ace, al)	73.5				vs EtOH, chl
7890	Monobutyltin trichloride		$C_4H_9Cl_3Sn$	1118-46-3	282.183	hyg liq	-63	93^{10}	0.85^{20}		s bz, CH_2Cl_2
7891	Monocrotaline		$C_{16}H_{23}NO_6$	315-22-0	325.357	wh pr (EtOH)	198 dec				
7892	Monocrotophos		$C_7H_{14}NO_5P$	6923-22-4	223.164		55.3(0.4)	$125^{0.0005}$	1.33^{20}		
7893	Monolinuron	*N*-(4-Chlorophenyl)-*N*-methoxy-*N*-methylurea	$C_9H_{11}ClN_2O_2$	1746-81-2	214.648	solid	80.8(0.5)				
7894	Monomethyl adipate	Methyl adipate	$C_7H_{12}O_4$	627-91-8	160.168	lf (Me_3N-MeOH)	9	158^{10}	1.0623^{20}	1.4283^{20}	s EtOH
7895	Monomethyl glutarate		$C_6H_{10}O_4$	1501-27-5	146.141			158^{27}	1.169^{25}	1.4381^{20}	
7896	Monosodium L-glutamate	Monosodium glutamate	$C_5H_8NNaO_4$	142-47-2	169.113						s H_2O
7897	Moquizone		$C_{20}H_{21}N_3O_3$	19395-58-5	351.399		136				s chl
7898	Morin		$C_{15}H_{10}O_7$	480-16-0	302.236	pa ye nd (+ 1 w, dil al)	303.5				sl H_2O, eth; vs EtOH; s bz, alk; i CS_2
7899	Morphine		$C_{17}H_{19}NO_3$	57-27-2	285.338	pr	255	190 sub			i H_2O, eth, ace; s MeOH, py; sl EtOH
7900	4-Morpholinamine		$C_4H_{10}N_2O$	4319-49-7	102.134			166	1.059^{25}	1.4772^{20}	
7901	Morpholine	Tetrahydro-1,4-oxazine	C_4H_9NO	110-91-8	87.120	hyg liq	-4.8(0.2)	128.2(0.2)	1.0005^{20}	1.4548^{20}	msc H_2O; s EtOH, eth, ace, bz; sl chl
7902	4-Morpholinecarboxaldehyde		$C_5H_9NO_2$	4394-85-8	115.131		21	238(1)	1.1520^{20}	1.4845^{20}	
7903	4-Morpholineethanamine		$C_6H_{14}N_2O$	2038-03-1	130.187		25.6	205	0.9897^{20}	1.4715^{20}	msc H_2O, EtOH, bz, lig; s ace
7904	4-Morpholineethanol		$C_6H_{13}NO_2$	622-40-2	131.173	liq	1.2(0.4)	220(3)	1.0710^{20}	1.4763^{20}	s H_2O, EtOH; sl ctc
7905	4-Morpholinepropanamine	4-(3-Aminopropyl)morpholine	$C_7H_{16}N_2O$	123-00-2	144.214	liq	-15	220	0.9854^{20}	1.4762^{20}	msc H_2O, EtOH, bz, lig; s ace; sl ctc
7906	2-(4-Morpholinothio)-benzothiazole	4-(2-Benzothiazolylthio)-morpholine	$C_{11}H_{12}N_2OS_2$	102-77-2	252.355	cry (EtOH)	86.7(0.7)				
7907	4-(4-Morpholinyl)aniline		$C_{10}H_{14}N_2O$	2524-67-6	178.230		131.6				
7908	2-(4-Morpholinyldithio)-benzothiazole		$C_{11}H_{12}N_2OS_3$	95-32-9	284.420		135				
7909	Muldamine		$C_{29}H_{47}NO_3$	36069-45-1	457.688		210				
7910	Murexide	5,5'-Nitrilobarbituric acid, ammonium salt	$C_8H_8N_6O_6$	3051-09-0	284.186						sl H_2O; i EtOH, eth; s alk
7911	Muscimol	5-(Aminomethyl)-3(2*H*)-isoxazolone	$C_4H_6N_2O_2$	2763-96-4	114.103	cry (EtOH)	175 dec				
7912	Myclobutanil		$C_{15}H_{17}ClN_4$	88671-89-0	288.776	ye cry	65	$205^{1.0}$			i H_2O, peth; s EtOH
7913	Mycophenolic acid		$C_{17}H_{20}O_6$	24280-93-1	320.337	nd (w)	141				i H_2O; vs EtOH, eth, chl; sl bz, tol
7914	β-Myrcene	7-Methyl-3-methylene-1,6-octadiene	$C_{10}H_{16}$	123-35-3	136.234	oil		171(3)	0.8013^{15}	1.4722^{20}	i H_2O; s EtOH, eth, bz, chl, HOAc
7915	Myristicin		$C_{11}H_{12}O_3$	607-91-0	192.211		<-20	276.5	1.1416^{20}	1.5403^{20}	i H_2O; sl EtOH; s eth, bz
7916	Nabam	Sodium ethylenebisdithiocar-bamic acid	$C_4H_6N_2Na_2S_4$	142-59-6	256.344	cry (w)					s H_2O
7917	Nadolol		$C_{17}H_{27}NO_4$	42200-33-9	309.401	cry (bz)	≈130				s EtOH; sl chl; i ace, eth, hx
7918	Naled	1,2-Dibromo-2,2-dichloroeth-ylphosphoric acid, dimethyl ester	$C_4H_7Br_2Cl_2O_4P$	300-76-5	380.784		27	$110^{0.5}$	1.96^{20}		
7919	Nalidixic acid		$C_{12}H_{12}N_2O_3$	389-08-2	232.234		229.5				sl EtOH, eth; s chl
7920	Nalmefene		$C_{21}H_{25}NO_3$	55096-26-9	339.429	cry (AcOEt)	189				
7921	Nalorphine	Acetorphin	$C_{19}H_{21}NO_3$	62-67-9	311.375	cry (eth)	208				sl H_2O; s alk, ace, EtOH
7922	Naloxone		$C_{19}H_{21}NO_4$	465-65-6	327.375	cry (AcOEt)	178				i peth; s chl
7923	Naltrexone		$C_{20}H_{23}NO_4$	16590-41-3	341.402	cry (ace)	169				

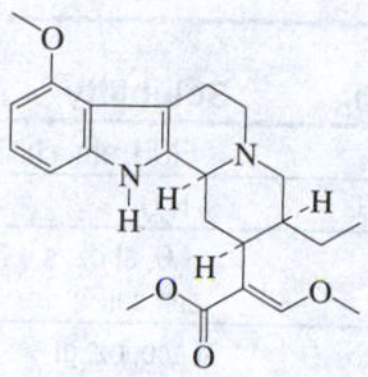

7885
Mitragynine

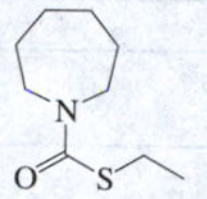

7886
Molinate

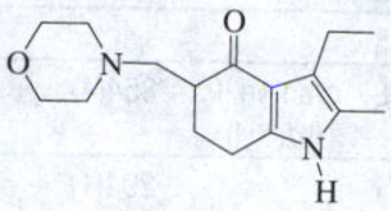

7887
Molindone

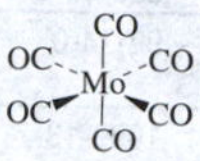

7888
Molybdenum carbonyl

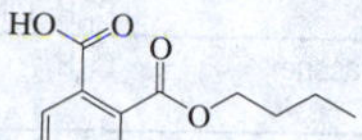

7889
Monobutyl phthalate

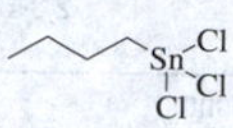

7890
Monobutyltin trichloride

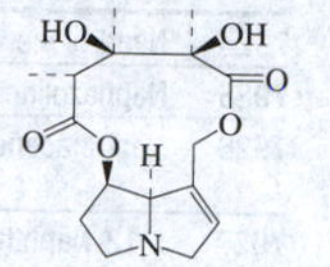

7891
Monocrotaline

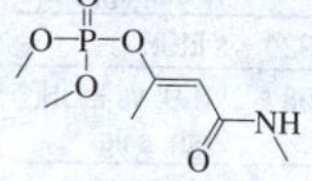

7892
Monocrotophos

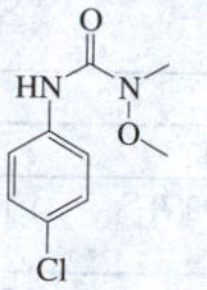

7893
Monolinuron

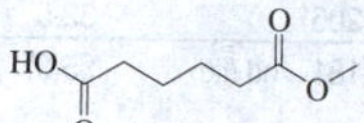

7894
Monomethyl adipate

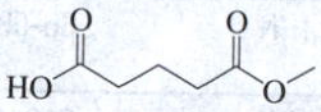

7895
Monomethyl glutarate

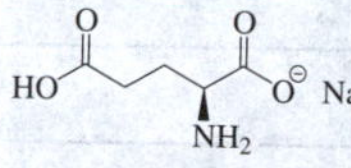

7896
Monosodium L-glutamate

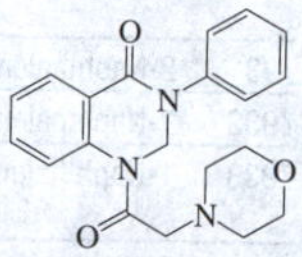

7897
Moquizone

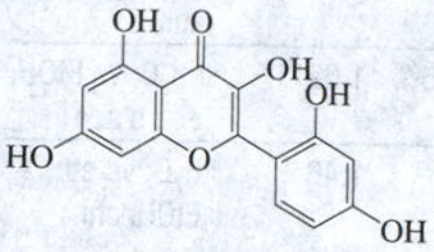

7898
Morin

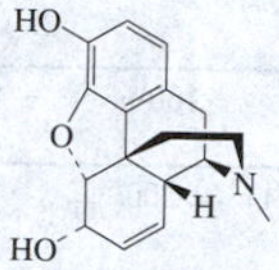

7899
Morphine

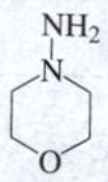

7900
4-Morpholinamine

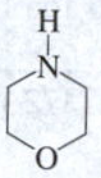

7901
Morpholine

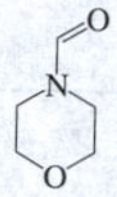

7902
4-Morpholinecarboxaldehyde

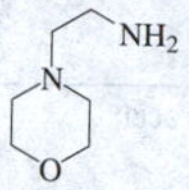

7903
4-Morpholineethanamine

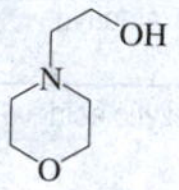

7904
4-Morpholineethanol

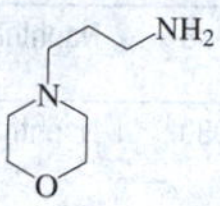

7905
4-Morpholinepropanamine

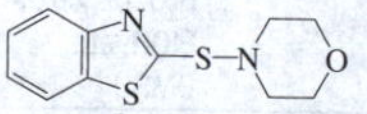

7906
2-(4-Morpholinothio)benzothiazole

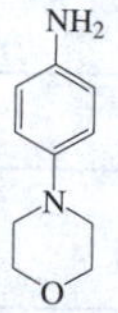

7907
4-(4-Morpholinyl)aniline

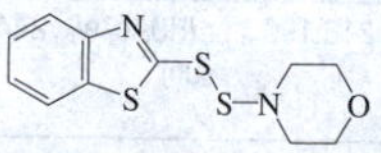

7908
2-(4-Morpholinyldithio)benzothiazole

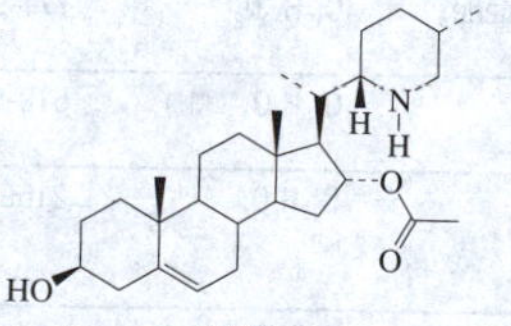

7909
Muldamine

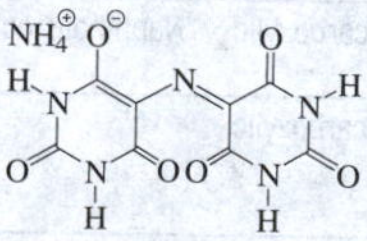

7910
Murexide

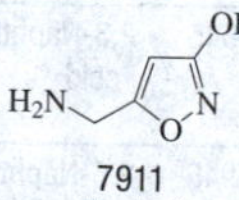

7911
Muscimol

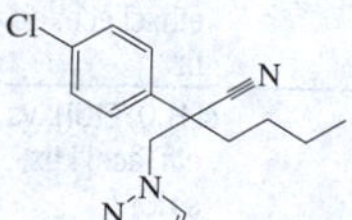

7912
Myclobutanil

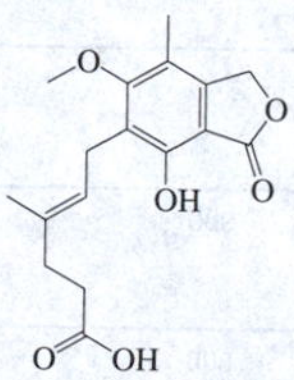

7913
Mycophenolic acid

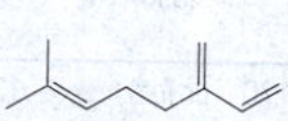

7914
β-Myrcene

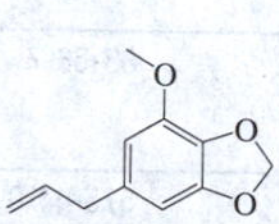

7915
Myristicin

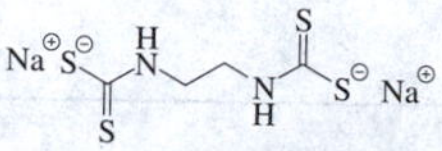

7916
Nabam

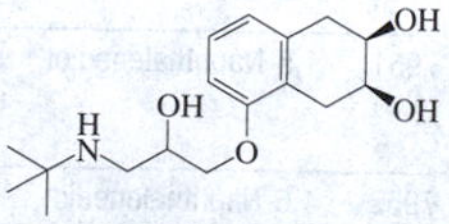

7917
Nadolol

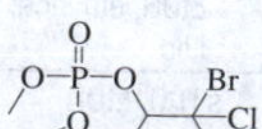

7918
Naled

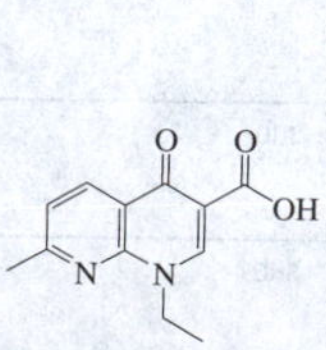

7919
Nalidixic acid

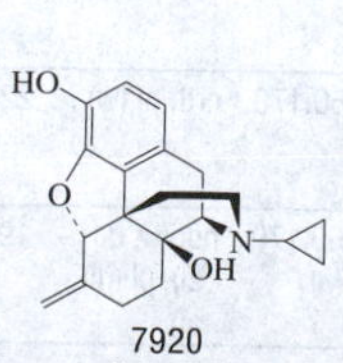

7920
Nalmefene

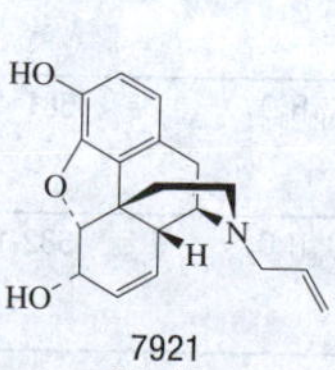

7921
Nalorphine

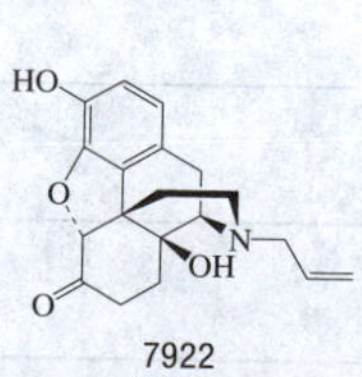

7922
Naloxone

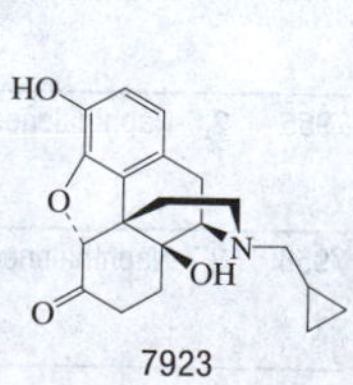

7923
Naltrexone

No.	Name	Synonym	Mol. form.	CAS Reg. No.	Mol. wt.	Physical form	mp/°C	bp/°C	den/ g cm^{-3}	n_D	Solubility
7924	Nandrolone	17-Hydroxyestr-4-en-3-one	$C_{18}H_{26}O_2$	434-22-0	274.398	cry	112				s EtOH, eth, chl
7925	Naphazoline hydrochloride		$C_{14}H_{15}ClN_2$	550-99-2	246.735						sl H_2O
7926	Naphthacene	2,3-Benzanthracene	$C_{18}H_{12}$	92-24-0	228.288	oran-ye lf (bz, xyl)	354(4)	sub			i H_2O; sl bz; s con sulf
7927	5,12-Naphthacenedione		$C_{18}H_{10}O_2$	1090-13-7	258.271		294(1)				sl ace, bz, gl HOAc
7928	Naphthalene		$C_{10}H_8$	91-20-3	128.171	mcl pl (al)	80.22(0.09)	218.0(0.1)	1.0253^{20}	1.5898^{25}	i H_2O; s EtOH; vs eth, ace, bz, CS_2
7929	1-Naphthaleneacetamide		$C_{12}H_{11}NO$	86-86-2	185.221	nd(w, al)		180 sub			i H_2O; s eth, bz, CS_2, HOAc
7930	1-Naphthaleneacetic acid	1-Naphthylacetic acid	$C_{12}H_{10}O_2$	86-87-3	186.206	nd (w)	132.8(0.5)	dec			sl H_2O, EtOH; vs eth, ace, chl; s bz
7931	2-Naphthaleneacetic acid	2-Naphthylacetic acid	$C_{12}H_{10}O_2$	581-96-4	186.206	lf(w) cry (bz)	142(1)				vs eth, lig, chl
7932	1-Naphthaleneacetonitrile		$C_{12}H_9N$	132-75-2	167.206		32.5	192^{18}		1.6192^{20}	s EtOH
7933	1-Naphthalenecarbonitrile		$C_{11}H_7N$	86-53-3	153.181	nd (lig)	35(2)	299(7)	1.1080^{25}	1.6298^{18}	i H_2O; vs EtOH, eth; s lig
7934	2-Naphthalenecarbonitrile		$C_{11}H_7N$	613-46-7	153.181	lf (lig)	59(2)	307(11)	1.0755^{60}		sl H_2O, chl; s EtOH, eth, lig
7935	1-Naphthalenecarbonyl chloride		$C_{11}H_7ClO$	879-18-5	190.626		20	297.5			
7936	2-Naphthalenecarbonyl chloride		$C_{11}H_7ClO$	2243-83-6	190.626	cry (peth)	51	305			vs bz, eth, chl
7937	1-Naphthalenecarboxaldehyde		$C_{11}H_8O$	66-77-3	156.181	pa ye	33.5	292	1.1503^{20}	1.6507^{20}	i H_2O; s EtOH, eth, ace, bz, sulf
7938	2-Naphthalenecarboxaldehyde		$C_{11}H_8O$	66-99-9	156.181	lf (w)	62	160^{19}	1.0775^{99}	1.6211^{99}	sl H_2O; vs EtOH, eth; s ace
7939	1-Naphthalenecarboxylic acid	1-Naphthoic acid	$C_{11}H_8O_2$	86-55-5	172.181	nd (HOAc-w, w, al)	160.8(0.8)	>300	1.398^{25}	1.46	i H_2O; vs eth, EtOH, chl
7940	2-Naphthalenecarboxylic acid	2-Naphthoic acid	$C_{11}H_8O_2$	93-09-4	172.181	nd (lig, chl, sub) pl (ace)	185(1)	>300	1.077^{100}		sl H_2O, DMSO, lig; s EtOH, eth, chl
7941	1,5-Naphthalenediamine	1,5-Diaminonaphthalene	$C_{10}H_{10}N_2$	2243-62-1	158.199	pr (eth, al, w)	189(2)	sub	1.4^{25}		s H_2O, EtOH, eth; vs chl
7942	1,8-Naphthalenediamine	1,8-Diaminonaphthalene	$C_{10}H_{10}N_2$	479-27-6	158.199		66.7(0.5)	205^{12}	1.1265^{90}	1.6828^{99}	vs eth, EtOH
7943	2,3-Naphthalenediamine	2,3-Diaminonaphthalene	$C_{10}H_{10}N_2$	771-97-1	158.199	lf (eth, w)	199		1.0968^{26}	1.6392^{26}	sl H_2O, DMSO; vs EtOH; s eth
7944	1,8-Naphthalenedicarboxylic acid	Naphthalic acid	$C_{12}H_8O_4$	518-05-8	216.190		260				i H_2O; sl EtOH, eth
7945	2,3-Naphthalenedicarboxylic acid		$C_{12}H_8O_4$	2169-87-1	216.190	pr (HOAc, w, sub)	244.5				i H_2O, bz, chl; sl EtOH, eth, DMSO
7946	2,6-Naphthalenedicarboxylic acid		$C_{12}H_8O_4$	1141-38-4	216.190	nd (al or sub)	>300 dec				vs EtOH
7947	2,6-Naphthalenedicarboxylic acid, dimethyl ester		$C_{14}H_{12}O_4$	840-65-3	244.243		191.3(0.4)				
7948	1,5-Naphthalene diisocyanate	1,5-Diisocyanatonaphthalene	$C_{12}H_6N_2O_2$	3173-72-6	210.188	cry	127	183^{10}			
7949	1,3-Naphthalenediol	Naphthoresorcinol	$C_{10}H_8O_2$	132-86-5	160.170	lf (w)	123.5				s H_2O, EtOH, eth; sl ace, bz, lig
7950	1,4-Naphthalenediol		$C_{10}H_8O_2$	571-60-8	160.170	mcl nd (bz, w)	192				s H_2O, EtOH, eth; sl ace; i bz
7951	1,5-Naphthalenediol		$C_{10}H_8O_2$	83-56-7	160.170	pr (w), nd (sub)	262 dec	sub			sl H_2O, EtOH; vs eth, ace; i bz; s HOAc
7952	1,6-Naphthalenediol		$C_{10}H_8O_2$	575-44-0	160.170	pr (bz)	138	sub			sl H_2O, EtOH; s eth, ace, bz, DMSO
7953	1,7-Naphthalenediol		$C_{10}H_8O_2$	575-38-2	160.170	nd (bz or sub)	180.5	sub			sl H_2O; vs EtOH, eth; s bz, HOAc
7954	2,3-Naphthalenediol		$C_{10}H_8O_2$	92-44-4	160.170	lf (w)	163.5				s H_2O, EtOH, eth, ace, bz, lig, HOAc
7955	2,6-Naphthalenediol		$C_{10}H_8O_2$	581-43-1	160.170	orth pl (w)	220	sub			sl H_2O, bz; s EtOH, eth, ace; i lig
7956	2,7-Naphthalenediol		$C_{10}H_8O_2$	582-17-2	160.170	nd, (w, dil al), pl (dil al)	193	sub			s H_2O, EtOH, eth, bz, chl; sl ace; i lig

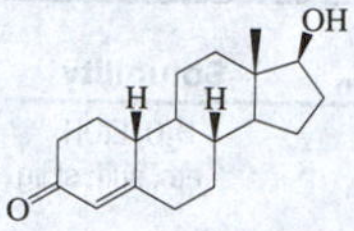
7924 Nandrolone

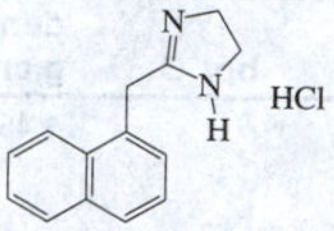
7925 Naphazoline hydrochloride

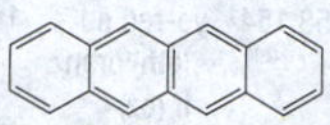
7926 Naphthacene

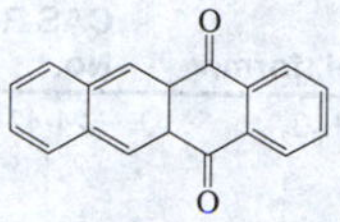
7927 5,12-Naphthacenedione

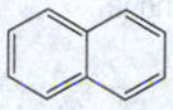
7928 Naphthalene

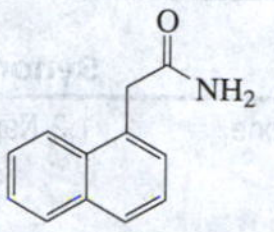
7929 1-Naphthaleneacetamide

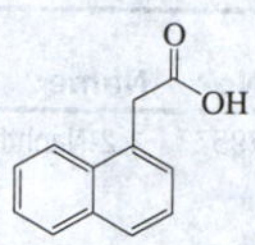
7930 1-Naphthaleneacetic acid

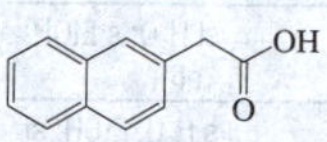
7931 2-Naphthaleneacetic acid

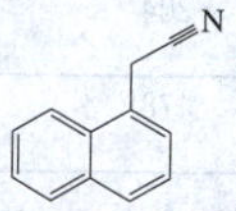
7932 1-Naphthaleneacetonitrile

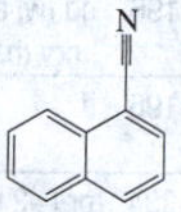
7933 1-Naphthalenecarbonitrile

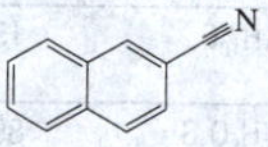
7934 2-Naphthalenecarbonitrile

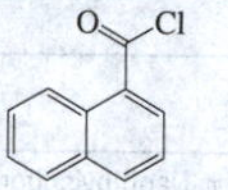
7935 1-Naphthalenecarbonyl chloride

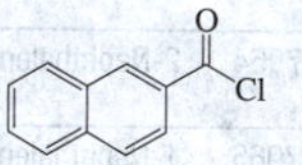
7936 2-Naphthalenecarbonyl chloride

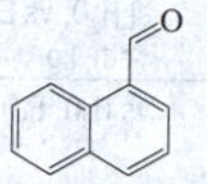
7937 1-Naphthalenecarboxaldehyde

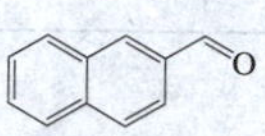
7938 2-Naphthalenecarboxaldehyde

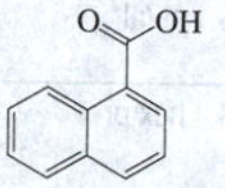
7939 1-Naphthalenecarboxylic acid

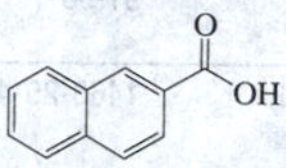
7940 2-Naphthalenecarboxylic acid

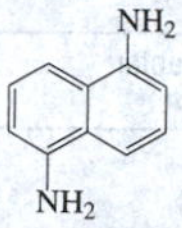
7941 1,5-Naphthalenediamine

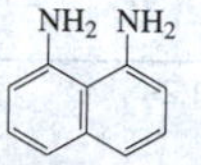
7942 1,8-Naphthalenediamine

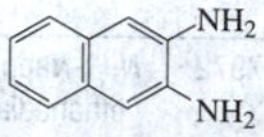
7943 2,3-Naphthalenediamine

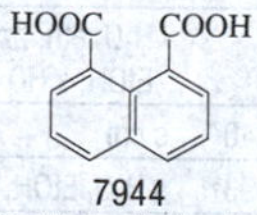
7944 1,8-Naphthalenedicarboxylic acid

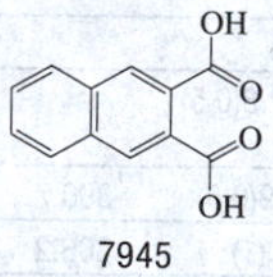
7945 2,3-Naphthalenedicarboxylic acid

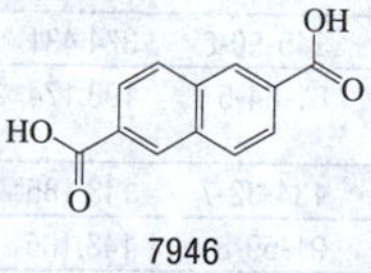
7946 2,6-Naphthalenedicarboxylic acid

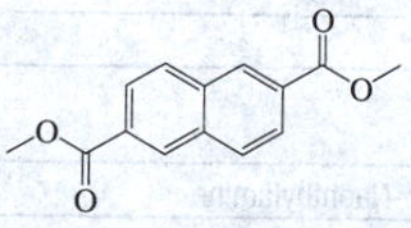
7947 2,6-Naphthalenedicarboxylic acid, dimethyl ester

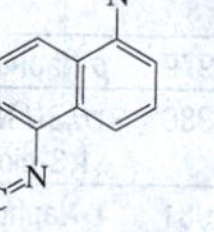
7948 1,5-Naphthalene diisocyanate

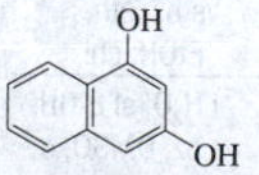
7949 1,3-Naphthalenediol

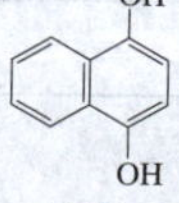
7950 1,4-Naphthalenediol

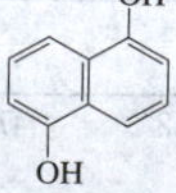
7951 1,5-Naphthalenediol

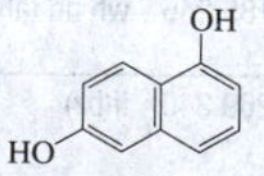
7952 1,6-Naphthalenediol

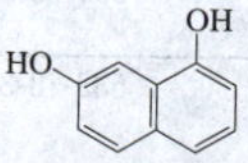
7953 1,7-Naphthalenediol

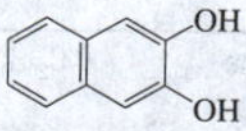
7954 2,3-Naphthalenediol

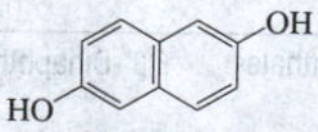
7955 2,6-Naphthalenediol

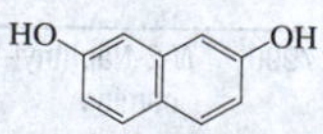
7956 2,7-Naphthalenediol

Organic

No.	Name	Synonym	Mol. form.	CAS Reg. No.	Mol. wt.	Physical form	mp/°C	bp/°C	den/g cm^{-3}	n_D	Solubility
7957	1,2-Naphthalenedione	1,2-Naphthoquinone	$C_{10}H_6O_2$	524-42-5	158.154	ye-red nd (eth) oran lf (bz)	146		1.450^{25}		s H_2O, EtOH, eth, sulf; sl lig
7958	1,4-Naphthalenedione	1,4-Naphthoquinone	$C_{10}H_6O_2$	130-15-4	158.154	bt ye nd (al, peth) ye (sub)	125.3(0.5)	sub			sl H_2O; vs EtOH; s eth, bz, chl, CS_2
7959	1,5-Naphthalenedisulfonic acid	Armstrong's acid	$C_{10}H_8O_6S_2$	81-04-9	288.297	pl (+4w, dil HOAc)	242 dec		1.493^{25}		vs H_2O; s EtOH; i eth
7960	1,6-Naphthalenedisulfonic acid	Naphthalene-1,6-disulfonic acid	$C_{10}H_8O_6S_2$	525-37-1	288.297	oran pr (+4w, HOAc or w)	125 dec				vs H_2O; s EtOH; i eth
7961	2,7-Naphthalenedisulfonic acid	Naphthalene-2,7-disulfonic acid	$C_{10}H_8O_6S_2$	92-41-1	288.297	hyg nd (conc HCl)	199				s H_2O; sl con HCl
7962	1-Naphthalenemethanamine		$C_{11}H_{11}N$	118-31-0	157.212			292	1.0958^{20}		s EtOH, eth, sulf, CS_2
7963	1-Naphthalenemethanol		$C_{11}H_{10}O$	4780-79-4	158.196	nd (w, al), cry (bz-lig)	64	304	1.1039^{80}		sl H_2O; vs EtOH, eth
7964	2-Naphthalenemethanol		$C_{11}H_{10}O$	1592-38-7	158.196	lf	81.3	178^{12}			sl H_2O; s EtOH, eth
7965	1-Naphthalenesulfonic acid	α-Naphthylsulfonic acid	$C_{10}H_8O_3S$	85-47-2	208.234	pr (+2 w, dil HCl)	140				s H_2O, EtOH; sl eth
7966	2-Naphthalenesulfonic acid	β-Naphthylsulfonic acid	$C_{10}H_8O_3S$	120-18-3	208.234	hyg pl (+1w), cry (+3w, HCl)	91	dec	1.441^{25}		vs H_2O, EtOH; s eth; sl bz
7967	1-Naphthalenesulfonyl chloride		$C_{10}H_7ClO_2S$	85-46-1	226.680	lf (eth)	68	209^{20}			vs bz, eth, EtOH
7968	2-Naphthalenesulfonyl chloride		$C_{10}H_7ClO_2S$	93-11-8	226.680	pow or lf (bz-peth)	81	201^{13}			i H_2O; s EtOH, bz, chl; sl peth; vs eth
7969	1,4,5,8-Naphthalenetetracarboxylic acid		$C_{14}H_8O_8$	128-97-2	304.209	lf or nd (w, dil HCl)	320				sl H_2O, bz, chl, EtOH; vs ace
7970	1-Naphthalenethiol	1-Naphthyl mercaptan	$C_{10}H_8S$	529-36-2	160.236			285 dec	1.1607^{20}	1.6802^{20}	sl H_2O, dil alk; vs EtOH, eth
7971	2-Naphthalenethiol	2-Naphthyl mercaptan	$C_{10}H_8S$	91-60-1	160.236	pl (al)	81	288	1.550^{25}		sl H_2O; vs EtOH, eth, lig
7972	*N*-(1-Naphthalenyl)-1,2-ethanediamine, dihydrochloride		$C_{12}H_{16}Cl_2N_2$	1465-25-4	259.174	hex pr	189				vs H_2O, EtOH
7973	1-Naphthalenylthiourea	ANTU	$C_{11}H_{10}N_2S$	86-88-4	202.275	pr (al)	198				i H_2O; sl EtOH, eth, ace
7974	Naphtho[2,3-*c*]furan-1,3-dione	2,3-Naphthalenedicarboxylic acid anhydride	$C_{12}H_6O_3$	716-39-2	198.174		246				sl EtOH, chl; s eth, bz
7975	1-Naphthol	1-Naphthalenol	$C_{10}H_8O$	90-15-3	144.170	ye nd (w)	95.1(0.6)	288(7)	1.0989^{99}	1.6224^{99}	i H_2O; vs EtOH, eth; s ace, bz; sl ctc
7976	2-Naphthol	2-Naphthalenol	$C_{10}H_8O$	135-19-3	144.170	mcl lf (w)	122(1)	286(1)	1.28^{20}		i H_2O; vs EtOH, eth; s bz, chl; sl lig
7977	1-Naphthol, acetate	1-Naphthyl acetate	$C_{12}H_{10}O_2$	830-81-9	186.206	nd or pl (al)	46.0(0.4)	114^{1}			i H_2O; s EtOH, eth
7978	2-Naphthol, acetate	2-Naphthyl acetate	$C_{12}H_{10}O_2$	1523-11-1	186.206	nd (al)	69.0(0.5)	132^{2}			i H_2O; s EtOH, eth, chl
7979	*p*-Naphtholbenzein		$C_{27}H_{18}O_2$	145-50-6	374.431		123				
7980	1*H*,3*H*-Naphtho[1,8-*cd*]pyran-1,3-dione		$C_{12}H_6O_3$	81-84-5	198.174		272.6(0.5)				i H_2O, eth, bz; sl EtOH; s HOAc
7981	1-Naphthylamine	α-Naphthylamine	$C_{10}H_9N$	134-32-7	143.185		48.9(0.5)	300.7	1.0228^{20}	1.6140^{20}	s chl
7982	2-Naphthylamine	β-Naphthylamine	$C_{10}H_9N$	91-59-8	143.185		110(1)	306.2	1.6414^{98}	1.6493^{98}	s H_2O, EtOH, eth
7983	2-[(1-Naphthylamino)-carbonyl]benzoic acid	Naptalam	$C_{18}H_{13}NO_3$	132-66-1	291.301		185		1.4^{20}		i H_2O; sl EtOH, ace, bz, tfa
7984	2-Naphthyl benzoate	2-Naphthalenol benzoate	$C_{17}H_{12}O_2$	93-44-7	248.276	nd or pr (al)	107.8(0.6)				i H_2O; s EtOH; sl eth, HOAc
7985	*N*-1-Naphthalenylacetamide		$C_{12}H_{11}NO$	575-36-0	185.221		160				s H_2O, EtOH; sl eth
7986	*N*-1-Naphthyl-1,2-ethanediamine	*N*-(1-Naphthyl)ethylenediamine	$C_{12}H_{14}N_2$	551-09-7	186.252	visc lig		204^{9}	1.114^{25}	1.6648^{25}	
7987	1-Naphthyl 2-hydroxybenzoate	1-Naphthyl salicylate	$C_{17}H_{12}O_3$	550-97-0	264.275		83				vs eth
7988	1-Naphthylhydroxylamine	*N*-Hydroxyl-1-naphthalenamine	$C_{10}H_9NO$	607-30-7	159.184		79				
7989	1-Naphthyl isothiocyanate	1-Isothiocyanatonaphthalene	$C_{11}H_7NS$	551-06-4	185.246	wh nd (al)	58				vs bz, eth, EtOH, chl
7990	*N*-2-Naphthyl-2-naphthalenamine	β,β'-Dinaphthylamine	$C_{20}H_{15}N$	532-18-3	269.340	lf(bz)	172.2	471			i H_2O; sl EtOH, bz, DMSO; s eth, HOAc

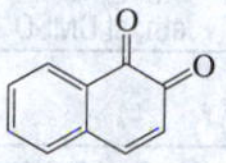
7957
1,2-Naphthalenedione

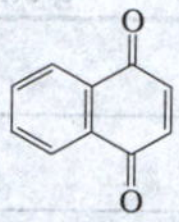
7958
1,4-Naphthalenedione

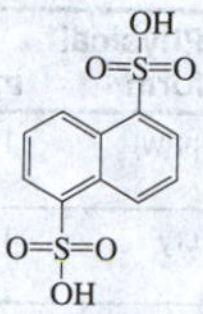
7959
1,5-Naphthalenedisulfonic acid

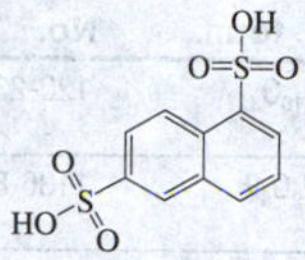
7960
1,6-Naphthalenedisulfonic acid

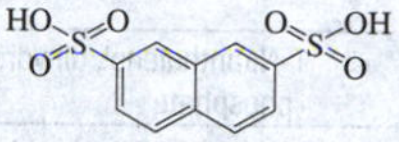
7961
2,7-Naphthalenedisulfonic acid

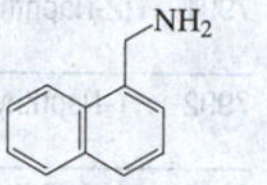
7962
1-Naphthalenemethanamine

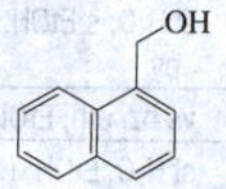
7963
1-Naphthalenemethanol

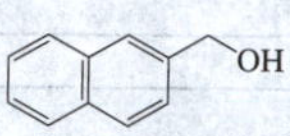
7964
2-Naphthalenemethanol

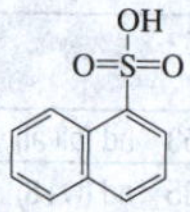
7965
1-Naphthalenesulfonic acid

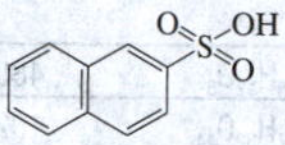
7966
2-Naphthalenesulfonic acid

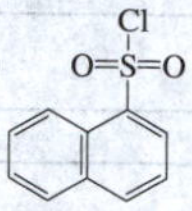
7967
1-Naphthalenesulfonyl chloride

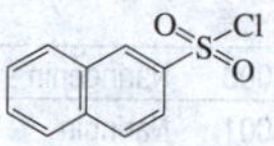
7968
2-Naphthalenesulfonyl chloride

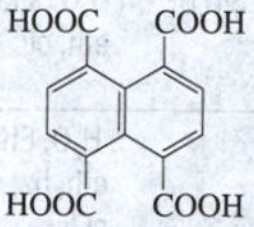
7969
1,4,5,8-Naphthalenetetracarboxylic acid

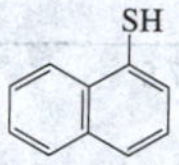
7970
1-Naphthalenethiol

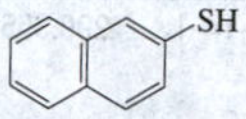
7971
2-Naphthalenethiol

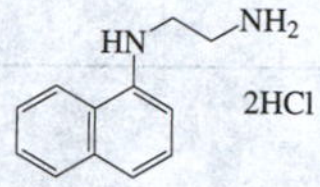
7972
N-(1-Naphthalenyl)-1,2-ethanediamine, dihydrochloride

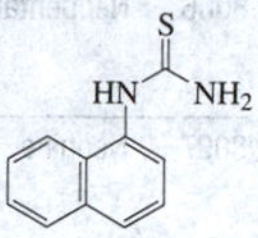
7973
1-Naphthalenylthiourea

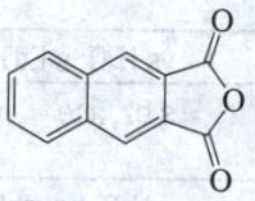
7974
Naphtho[2,3-c]furan-1,3-dione

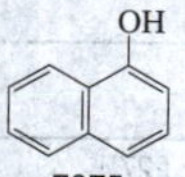
7975
1-Naphthol

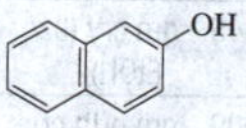
7976
2-Naphthol

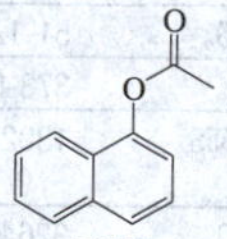
7977
1-Naphthol, acetate

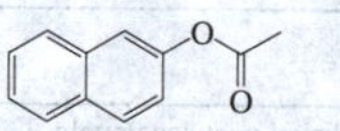
7978
2-Naphthol, acetate

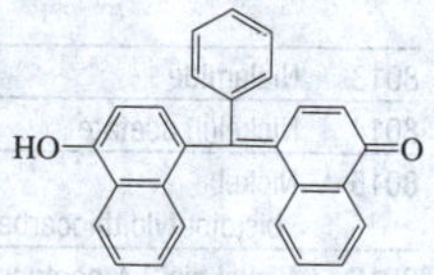
7979
p-Naphtholbenzein

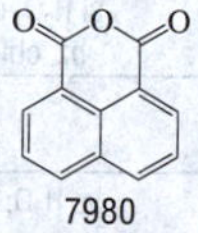
7980
1*H*,3*H*-Naphtho[1,8-cd]pyran-1,3-dione

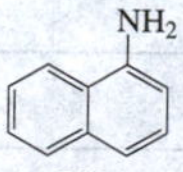
7981
1-Naphthylamine

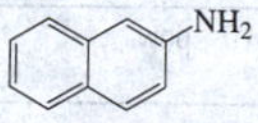
7982
2-Naphthylamine

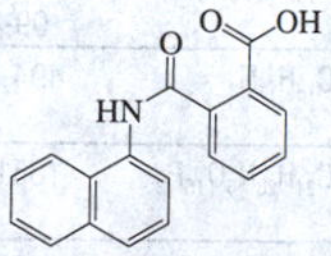
7983
2-[(1-Naphthylamino)carbonyl]benzoic acid

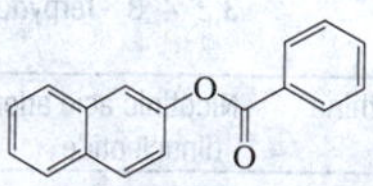
7984
2-Naphthyl benzoate

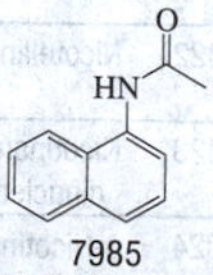
7985
N-1-Naphthalenylacetamide

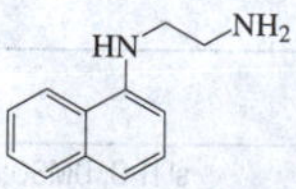
7986
N-1-Naphthyl-1,2-ethanediamine

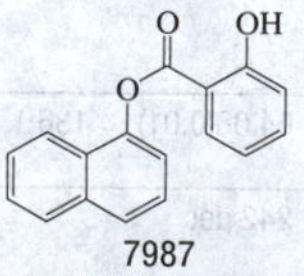
7987
1-Naphthyl 2-hydroxybenzoate

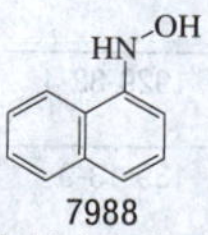
7988
1-Naphthylhydroxylamine

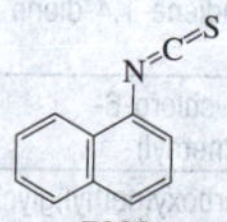
7989
1-Naphthyl isothiocyanate

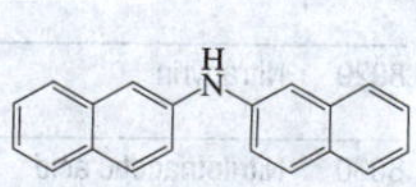
7990
N-2-Naphthyl-2-naphthalenamine

No.	Name	Synonym	Mol. form.	CAS Reg. No.	Mol. wt.	Physical form	mp/°C	bp/°C	den/ g cm⁻³	n_D	Solubility
7991	(2-Naphthyloxy)acetic acid	2-Naphthoxyacetic acid	$C_{12}H_{10}O_3$	120-23-0	202.205	pr(w)	156.2(0.3)				s H_2O, EtOH, eth; sl DMSO
7992	1-Naphthyl phosphate	1-Naphthalenol, dihydrogen phosphate	$C_{10}H_9O_4P$	1136-89-6	224.149	cry	160				
7993	2-Naphthyl salicylate	2-Naphthyl 2-hydroxybenzoate	$C_{17}H_{12}O_3$	613-78-5	264.275	cry (al)	95.5		1.11^{116}		i H_2O; sl EtOH; s eth, bz
7994	1,5-Naphthyridine	1,5-Diazanaphthalene	$C_8H_6N_2$	254-79-5	130.147	ye nd (peth)	75	112^{12}	1.2100^{20}		
7995	1,6-Naphthyridine		$C_8H_6N_2$	253-72-5	130.147		29.5				
7996	Napropamide	Propanamide, *N,N*-diethyl-2-(1-naphthalenyloxy)-	$C_{17}H_{21}NO_2$	15299-99-7	271.355		73.8(0.5)				
7997	Naproxen	6-Methoxy-α-methyl-2-naphthaleneacetic acid	$C_{14}H_{14}O_3$	22204-53-1	230.259	cry (ace/hx)	155(3)				i H_2O; sl eth; s MeOH, chl
7998	Narceine		$C_{23}H_{27}NO_8$	131-28-2	445.462		138				i H_2O
7999	Narcobarbital		$C_{11}H_{15}BrN_2O_3$	125-55-3	303.152		115				sl H_2O; s EtOH, py
8000	Naringenin		$C_{15}H_{12}O_5$	480-41-1	272.253	nd (dil al)	250(1)				vs bz, eth, EtOH
8001	Naringin		$C_{27}H_{32}O_{14}$	10236-47-2	580.535	nd (w+8)					sl H_2O, EtOH; i eth, bz, chl; s HOAc
8002	Nealbarbital		$C_{12}H_{18}N_2O_3$	561-83-1	238.282		156				vs ace, eth, EtOH
8003	Nellite	Diamidafos	$C_8H_{13}N_2O_2P$	1754-58-1	200.175	cry (ctc)	103.5				sl AcOEt, bz
8004	Neoabietic acid	8(14),13(15)-Abietadien-18-oic acid	$C_{20}H_{30}O_2$	471-77-2	302.451	cry (EtOH aq)	173				
8005	Neobornylamine		$C_{10}H_{19}N$	2223-67-8	153.265	pow	184				vs ace, eth
8006	Neopentane	2,2-Dimethylpropane	C_5H_{12}	463-82-1	72.149	col gas	-16.37(0.04)	9.50(0.06)	0.5852^{25} (p>1 atm	1.3476^{6}	i H_2O; s EtOH, eth, ctc
8007	Neopine		$C_{18}H_{21}NO_3$	467-14-1	299.365	nd (peth)	127.5				s H_2O, EtOH, eth, bz; vs chl; sl lig
8008	Neostigmine bromide		$C_{12}H_{19}BrN_2O_2$	114-80-7	303.195	cry (al-eth)	167 dec				vs H_2O; s EtOH
8009	Nepetalactone		$C_{10}H_{14}O_2$	490-10-8	166.217			$71^{0.05}$	1.0663^{25}	1.4859^{25}	
8010	*cis*-Nerolidol		$C_{15}H_{26}O$	142-50-7	222.366			276	0.8778^{20}	1.4898^{20}	vs EtOH; s eth, ace, HOAc
8011	Neurine		$C_5H_{13}NO$	463-88-7	103.163	syr					vs H_2O, eth, EtOH
8012	Neutral Red		$C_{15}H_{17}ClN_4$	553-24-2	288.776	grn pow					s H_2O, ethylene glycol, EtOH; i xyl
8013	Nialamide		$C_{16}H_{18}N_4O_2$	51-12-7	298.340		154(1)				
8014	Nickel(II) acetate		$C_4H_6NiO_4$	373-02-4	176.782						vs H_2O; s EtOH
8015	Nickel bis(dibutyldithiocarbamate)		$C_{18}H_{36}N_2NiS_4$	13927-77-0	467.445	grn cry (bz/EtOH)	91				s bz, ace
8016	Nickel bis(2,4-pentanedioate)	Nickel acetylacetonate	$C_{10}H_{14}NiO_4$	3264-82-2	256.909	grn orth cry	230	227^{11}			s H_2O, bz, chl, EtOH; i eth
8017	Nickel carbonyl	Nickel tetracarbonyl	C_4NiO_4	13463-39-3	170.734	col liq	-19.3	43 (exp 60)	1.31^{25}		i H_2O; s EtOH, bz, ace, ctc
8018	Nickelocene	Bis(η5-2,4-cyclopentadien-1-yl)nickel	$C_{10}H_{10}Ni$	1271-28-9	188.879	grn cry	173				
8019	Niclosamide		$C_{13}H_8Cl_2N_2O_4$	50-65-7	327.120		227				
8020	Nicofibrate		$C_{16}H_{16}ClNO_3$	31980-29-7	305.756		49	$180^{0.4}$			
8021	Nicosulfuron		$C_{15}H_{18}N_6O_6S$	111991-09-4	410.405		172				
8022	Nicotelline	3,2':4',3''-Terpyridine	$C_{15}H_{11}N_3$	494-04-2	233.268	prismatic nd	148	>300			sl H_2O, eth; s bz, chl, EtOH
8023	Nicotinamide hypoxanthine dinucleotide	Nicotinic acid adenine dinucleotide	$C_{21}H_{26}N_6O_{15}P_2$	1851-07-6	664.410	pow					
8024	β-Nicotinamide mononucleotide	NMN	$C_{11}H_{15}N_2O_8P$	1094-61-7	334.219	amor pow					vs H_2O; i ace
8025	L-Nicotine	3-(1-Methyl-2-pyrrolidinyl)pyridine, (*S*)-	$C_{10}H_{14}N_2$	54-11-5	162.231	hyg liq	-79	246(13)	1.0097^{20}	1.5282^{20}	msc H_2O; vs EtOH, eth, chl; s lig
8026	Nifurthiazole		$C_8H_6N_4O_4S$	3570-75-0	254.224	cry	215 dec				
8027	Nitralin	4-(Methylsulfonyl)-2,6-dinitro-*N,N*-dipropylaniline	$C_{13}H_{19}N_3O_6S$	4726-14-1	345.371		151.7(0.5)				
8028	Nitranilic acid	2,5-Dihydroxy-3,6-dinitro-2,5-cyclohexadiene-1,4-dione	$C_6H_2N_2O_8$	479-22-1	230.088	gold-ye pl (+w, dil HNO_3	170 dec				vs H_2O, EtOH; i eth
8029	Nitrapyrin	Pyridine, 2-chloro-6-(trichloromethyl)-	$C_6H_3Cl_4N$	1929-82-4	230.907		64.08(0.01)	136^{11}			
8030	Nitrilotriacetic acid	*N,N*-Bis(carboxymethyl)glycine	$C_6H_9NO_6$	139-13-9	191.138	pr cry (w)	242 dec				sl H_2O, DMSO; s EtOH

Organic

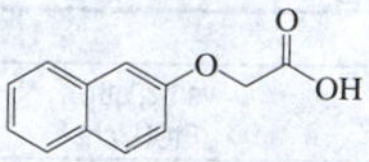

7991
(2-Naphthyloxy)acetic acid

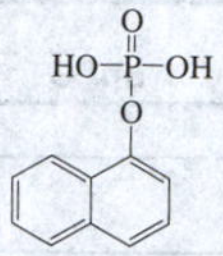

7992
1-Naphthyl phosphate

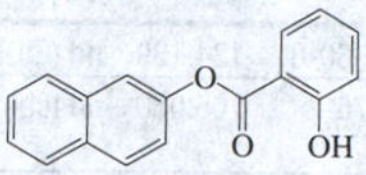

7993
2-Naphthyl salicylate

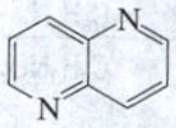

7994
1,5-Naphthyridine

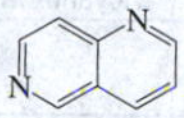

7995
1,6-Naphthyridine

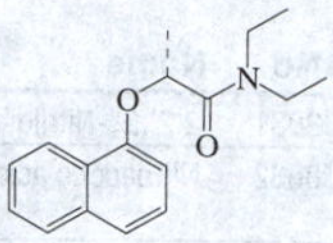

7996
Napropamide

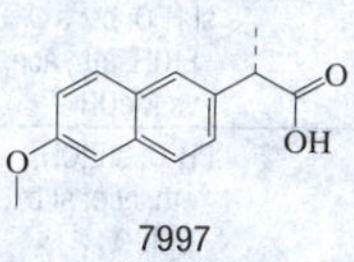

7997
Naproxen

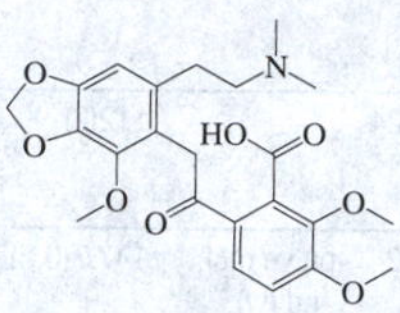

7998
Narceine

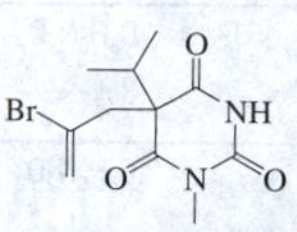

7999
Narcobarbital

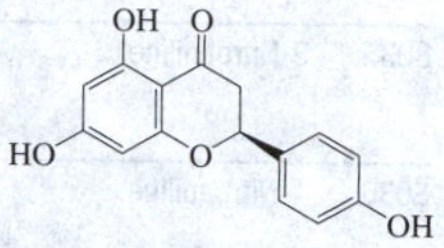

8000
Naringenin

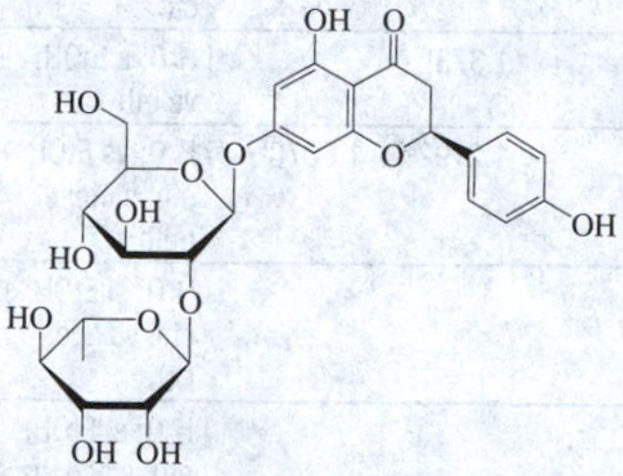

8001
Naringin

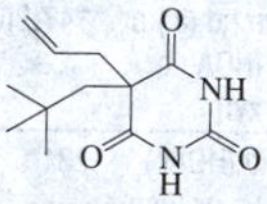

8002
Nealbarbital

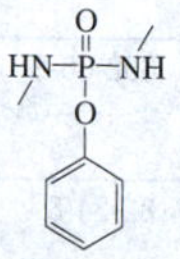

8003
Nellite

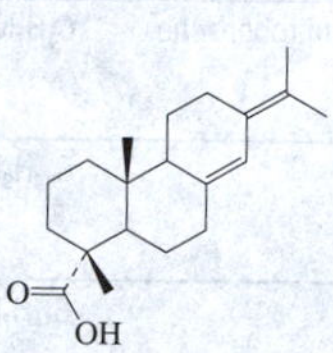

8004
Neoabietic acid

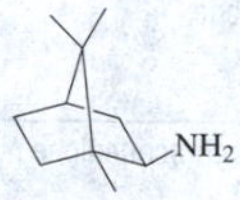

8005
Neobornylamine

8006
Neopentane

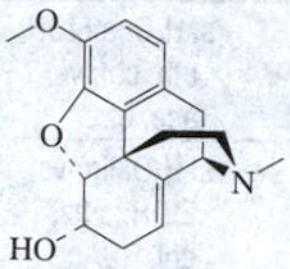

8007
Neopine

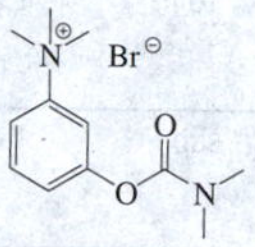

8008
Neostigmine bromide

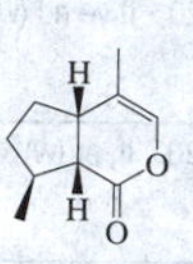

8009
Nepetalactone

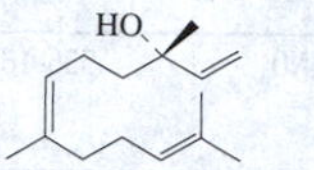

8010
cis-Nerolidol

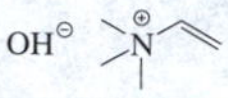

8011
Neurine

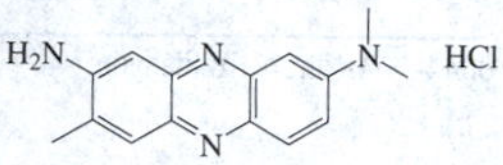

8012
Neutral Red

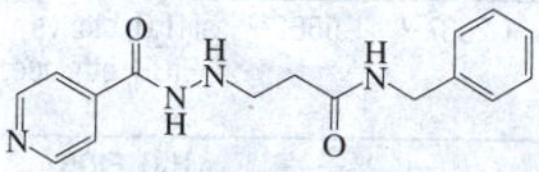

8013
Nialamide

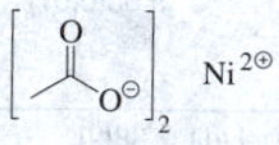

8014
Nickel(II) acetate

8015
Nickel bis(dibutyldithiocarbamate)

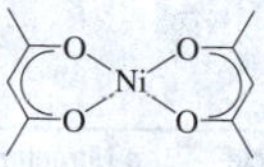

8016
Nickel bis(2,4-pentanedioate)

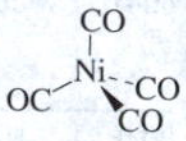

8017
Nickel carbonyl

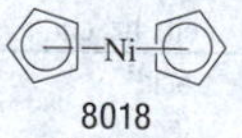

8018
Nickelocene

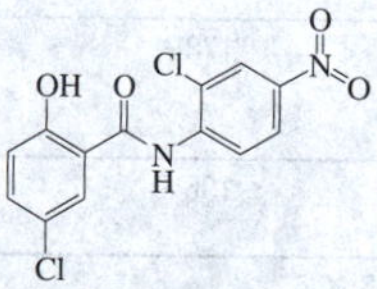

8019
Niclosamide

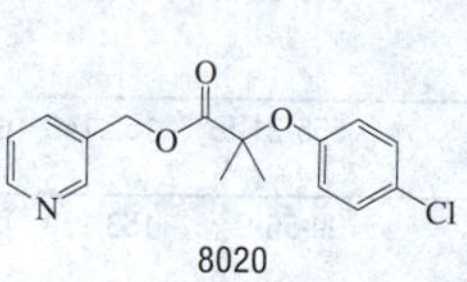

8020
Nicofibrate

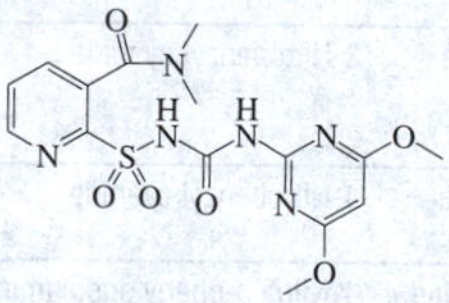

8021
Nicosulfuron

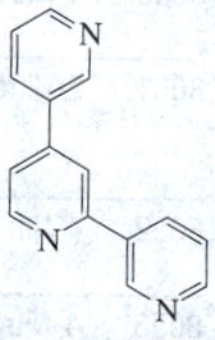

8022
Nicotelline

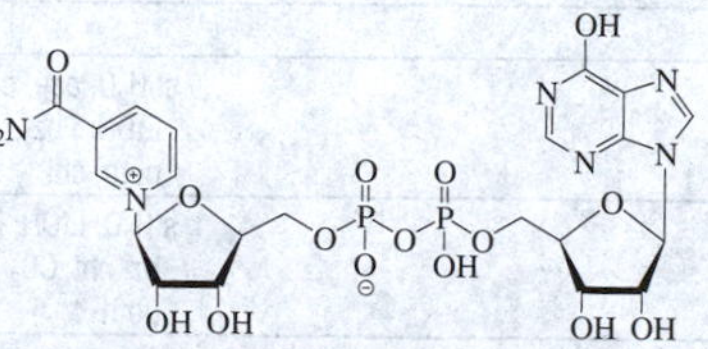

8023
Nicotinamide hypoxanthine dinucleotide

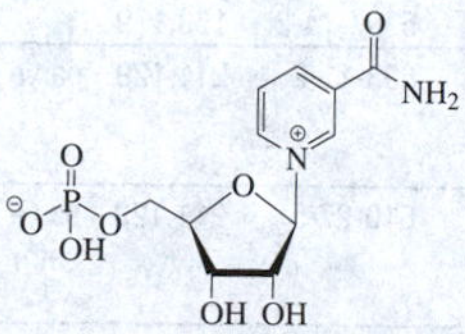

8024
β-Nicotinamide mononucleotide

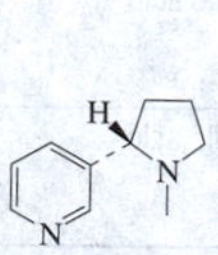

8025
L-Nicotine

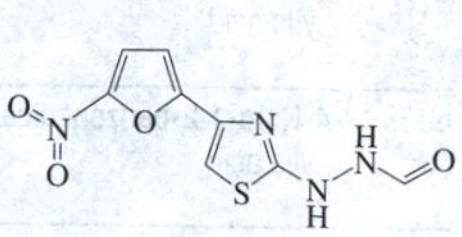

8026
Nifurthiazole

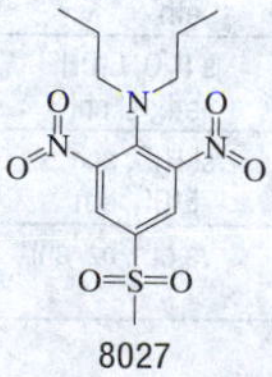

8027
Nitralin

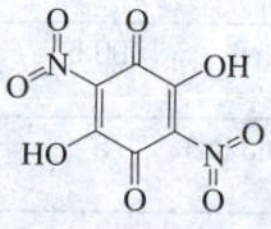

8028
Nitranilic acid

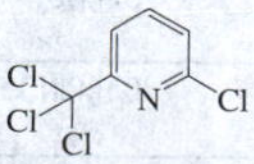

8029
Nitrapyrin

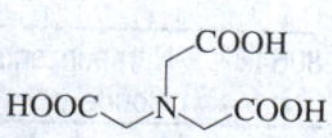

8030
Nitrilotriacetic acid

Organic

No.	Name	Synonym	Mol. form.	CAS Reg. No.	Mol. wt.	Physical form	mp/°C	bp/°C	den/ g cm^{-3}	n_D	Solubility
8031	2,2',2"-Nitrilotriacetonitrile	Tricyanotrimethylamine	$C_6H_6N_4$	7327-60-8	134.139	nd (EtOH)	125.5				
8032	Nitroacetic acid		$C_2H_3NO_4$	625-75-2	105.050	nd (chl)	92 dec				vs bz, eth, EtOH, chl
8033	Nitroacetone		$C_3H_5NO_3$	10230-68-9	103.077	pl, nd (eth, bz)	50(4)	103^{24}			vs bz, eth, EtOH
8034	2-Nitroaniline		$C_6H_6N_2O_2$	88-74-4	138.124		71(1)	285(2)	0.9015^{25}		sl H_2O; s EtOH; vs eth, ace, bz, chl
8035	3-Nitroaniline		$C_6H_6N_2O_2$	99-09-2	138.124		112(2)	312(3)	0.9011^{25}		sl H_2O, bz; s EtOH, eth, ace; vs MeOH
8036	4-Nitroaniline		$C_6H_6N_2O_2$	100-01-6	138.124	pa ye mcl nd (w)	147.7(0.5)	328(8)	1.424^{20}		i H_2O; s EtOH, eth, ace; sl bz, DMSO
8037	2-Nitroanisole	1-Methoxy-2-nitrobenzene	$C_7H_7NO_3$	91-23-6	153.136	oily liq	9.4(0.9)	272(7)	1.2540^{20}	1.5161^{20}	i H_2O; msc EtOH, eth; s ctc
8038	3-Nitroanisole	1-Methoxy-3-nitrobenzene	$C_7H_7NO_3$	555-03-3	153.136	nd (al), pl (bz-lig)	38(2)	258	1.373^{18}		i H_2O; s EtOH; vs eth
8039	4-Nitroanisole	1-Methoxy-4-nitrobenzene	$C_7H_7NO_3$	100-17-4	153.136	pr (al), nd (dil al)	54(1)	259(6)	1.2192^{60}	1.5070^{60}	i H_2O; vs EtOH, eth; s ctc; sl peth
8040	9-Nitroanthracene		$C_{14}H_9NO_2$	602-60-8	223.227	ye nd (al) pr (HOAc or xyl)	147.2(0.7)	275^{17}			i H_2O; sl EtOH, chl; vs ace, CS_2
8041	1-Nitro-9,10-anthracenedione		$C_{14}H_7NO_4$	82-34-8	253.211	nd (HOAc) ye pr (ace)	231.5	270^{7}			i H_2O; sl EtOH, eth; s ace, bz
8042	2-Nitrobenzaldehyde		$C_7H_5NO_3$	552-89-6	151.120	ye nd (w)	43.5(0.2)	153^{23}	1.2844^{20}		sl H_2O, chl; vs EtOH, eth, ace, bz
8043	3-Nitrobenzaldehyde		$C_7H_5NO_3$	99-61-6	151.120	lt ye nd (w)	58.0(0.2)	164^{23}	1.2792^{20}		sl H_2O; s EtOH, eth, chl; vs ace, bz
8044	4-Nitrobenzaldehyde		$C_7H_5NO_3$	555-16-8	151.120	lf, pr (w)	105(3)	sub	1.496^{25}		sl H_2O, lig; vs EtOH; s bz, chl, HOAc
8045	3-Nitrobenzamide		$C_7H_6N_2O_3$	645-09-0	166.134		142.7	312.5			s H_2O, EtOH, eth
8046	4-Nitrobenzamide		$C_7H_6N_2O_3$	619-80-7	166.134	nd (w)	190(2)				i H_2O; s EtOH, eth
8047	Nitrobenzene		$C_6H_5NO_2$	98-95-3	123.110	oily liq	5.65(0.07)	210.7(0.3)	1.2037^{20}	1.5562^{20}	sl H_2O, ctc; vs EtOH, eth, ace, bz
8048	2-Nitrobenzeneacetic acid	*o*-Nitrophenylacetic acid	$C_8H_7NO_4$	3740-52-1	181.147	nd (w, pl (dil al)	138(2)				s H_2O, EtOH
8049	3-Nitrobenzeneacetic acid		$C_8H_7NO_4$	1877-73-2	181.147	nd (w)	117(2)				vs EtOH
8050	4-Nitrobenzeneacetic acid	*p*-Nitrophenylacetic acid	$C_8H_7NO_4$	104-03-0	181.147	pa ye nd (w)	154				sl H_2O; s EtOH, eth, bz
8051	2-Nitrobenzeneacetonitrile	2-Nitrobenzyl cyanide	$C_8H_6N_2O_2$	610-66-2	162.146	nd (dil al), pr (HOAc, al)	84	178^{12}			vs ace, bz, eth, EtOH
8052	4-Nitrobenzeneacetonitrile	4-Nitrobenzyl cyanide	$C_8H_6N_2O_2$	555-21-5	162.146	pr (al)	117	196^{12}			sl H_2O; s EtOH, eth, bz, chl
8053	4-Nitro-1,2-benzenediamine	4-Nitro-*o*-phenylenediamine	$C_6H_7N_3O_2$	99-56-9	153.139	dk red nd (dil al)	199.5				s acid
8054	4-Nitro-1,3-benzenediamine		$C_6H_7N_3O_2$	5131-58-8	153.139	oran pr (w)	161				
8055	5-Nitro-1,3-benzenediamine	1,3-Diamino-5-nitrobenzene	$C_6H_7N_3O_2$	5042-55-7	153.139	red cry (w)	143				
8056	2-Nitro-1,4-benzenediamine		$C_6H_7N_3O_2$	5307-14-2	153.139		140.0				
8057	3-Nitro-1,2-benzenedicarboxylic acid	3-Nitrophthalic acid	$C_8H_5NO_6$	603-11-2	211.129	pa ye pr (w)	218				sl H_2O, ace; s EtOH; i bz, peth, chl
8058	4-Nitro-1,2-benzenedicarboxylic acid	4-Nitrophthalic acid	$C_8H_5NO_6$	610-27-5	211.129	pa ye nd (w, eth)	164.8				s H_2O, EtOH; i bz, chl, CS_2, peth
8059	2-Nitrobenzeneethanol		$C_8H_9NO_3$	15121-84-3	167.162		1.0	267	1.19^{25}	1.5637^{20}	
8060	4-Nitrobenzeneethanol		$C_8H_9NO_3$	100-27-6	167.162		63	148^{2}			
8061	2-Nitrobenzenemethanol	2-Nitrobenzyl alcohol	$C_7H_7NO_3$	612-25-9	153.136	nd (w)	74	270			sl H_2O; s EtOH, eth
8062	3-Nitrobenzenemethanol	3-Nitrobenzyl alcohol	$C_7H_7NO_3$	619-25-0	153.136	orth nd (w)	30.5	177^{3}	1.296^{19}		s H_2O, EtOH, eth; sl chl
8063	4-Nitrobenzenemethanol	4-Nitrobenzyl alcohol	$C_7H_7NO_3$	619-73-8	153.136	nd (w)	96.5	255 dec			sl H_2O, ace; s EtOH, eth
8064	2-Nitrobenzenesulfenyl chloride		$C_6H_4ClNO_2S$	7669-54-7	189.620	ye nd (bz)	75				vs eth, bz, chl

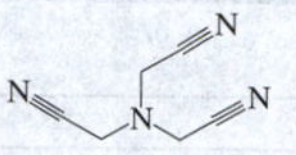
8031
2,2',2''-Nitrilotriacetonitrile

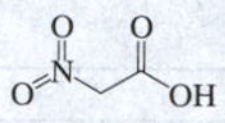
8032
Nitroacetic acid

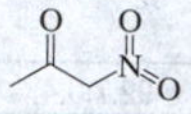
8033
Nitroacetone

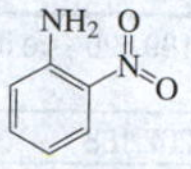
8034
2-Nitroaniline

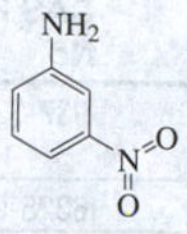
8035
3-Nitroaniline

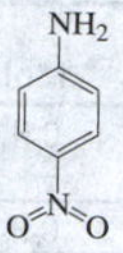
8036
4-Nitroaniline

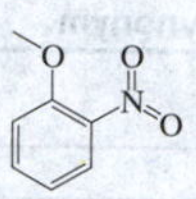
8037
2-Nitroanisole

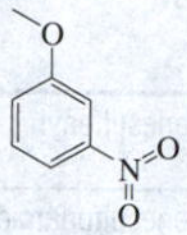
8038
3-Nitroanisole

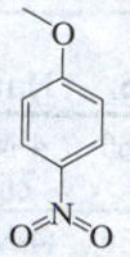
8039
4-Nitroanisole

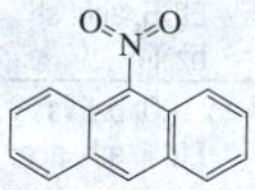
8040
9-Nitroanthracene

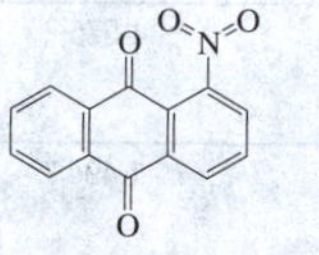
8041
1-Nitro-9,10-anthracenedione

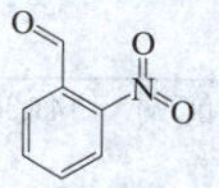
8042
2-Nitrobenzaldehyde

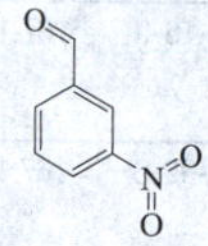
8043
3-Nitrobenzaldehyde

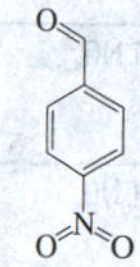
8044
4-Nitrobenzaldehyde

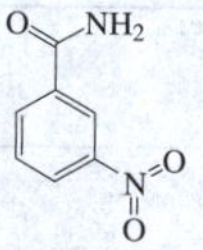
8045
3-Nitrobenzamide

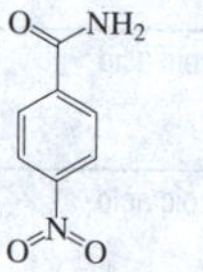
8046
4-Nitrobenzamide

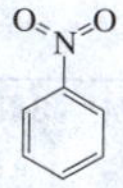
8047
Nitrobenzene

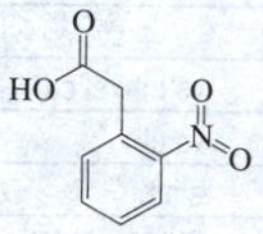
8048
2-Nitrobenzeneacetic acid

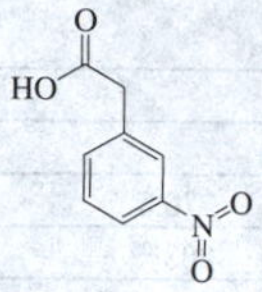
8049
3-Nitrobenzeneacetic acid

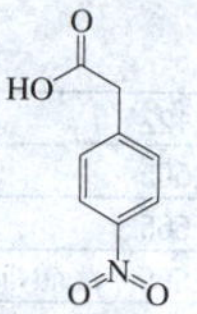
8050
4-Nitrobenzeneacetic acid

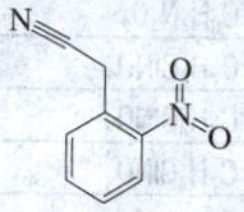
8051
2-Nitrobenzeneacetonitrile

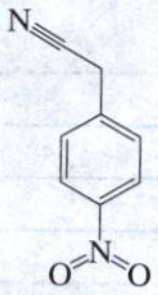
8052
4-Nitrobenzeneacetonitrile

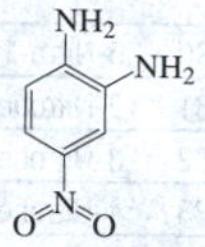
8053
4-Nitro-1,2-benzenediamine

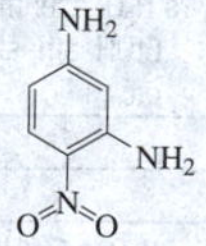
8054
4-Nitro-1,3-benzenediamine

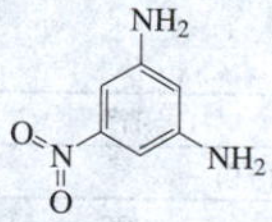
8055
5-Nitro-1,3-benzenediamine

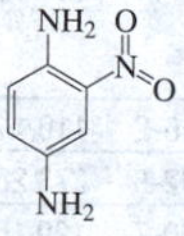
8056
2-Nitro-1,4-benzenediamine

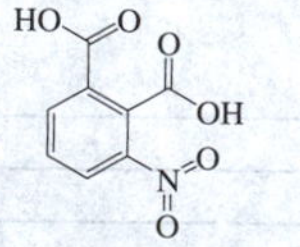
8057
3-Nitro-1,2-benzenedicarboxylic acid

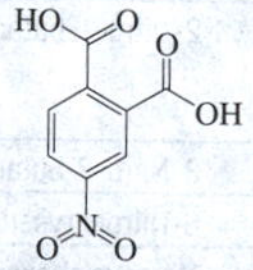
8058
4-Nitro-1,2-benzenedicarboxylic acid

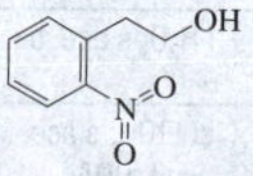
8059
2-Nitrobenzeneethanol

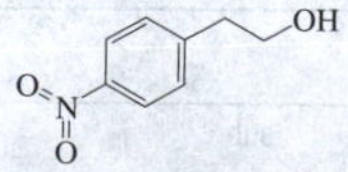
8060
4-Nitrobenzeneethanol

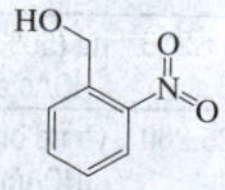
8061
2-Nitrobenzenemethanol

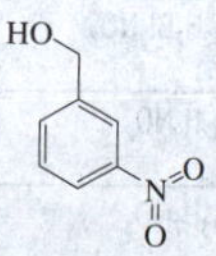
8062
3-Nitrobenzenemethanol

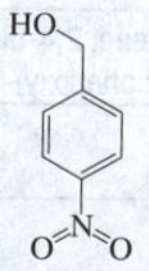
8063
4-Nitrobenzenemethanol

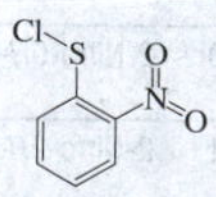
8064
2-Nitrobenzenesulfenyl chloride

No.	Name	Synonym	Mol. form.	CAS Reg. No.	Mol. wt.	Physical form	mp/°C	bp/°C	den/g cm⁻³	n_D	Solubility
8065	4-Nitrobenzenesulfenyl chloride		$C_6H_4ClNO_2S$	937-32-6	189.620	ye lf (peth)	52	$125^{0.1}$			vs bz
8066	4-Nitrobenzenesulfonamide		$C_6H_6N_2O_4S$	6325-93-5	202.188		-92.7(0.4)				
8067	3-Nitrobenzenesulfonic acid		$C_6H_5NO_5S$	98-47-5	203.173	pl	48				vs H_2O; s EtOH; i eth, bz
8068	4-Nitrobenzenesulfonic acid		$C_6H_5NO_5S$	138-42-1	203.173		95				vs H_2O
8069	2-Nitrobenzenesulfonyl chloride		$C_6H_4ClNO_4S$	1694-92-4	221.619	pr (lig, eth-peth)	68.5				s eth; sl peth
8070	3-Nitrobenzenesulfonyl chloride		$C_6H_4ClNO_4S$	121-51-7	221.619	mcl pr (eth) nd (lig)	64				i H_2O; s EtOH
8071	4-Nitrobenzenesulfonyl chloride		$C_6H_4ClNO_4S$	98-74-8	221.619	mcl pr (peth)	79.5	$143^{1.5}$			s peth
8072	5-Nitro-1*H*-benzimidazole		$C_7H_5N_3O_2$	94-52-0	163.134	nd (w)	207.8				i H_2O, eth, bz, chl; s acid; vs EtOH
8073	2-Nitrobenzoic acid		$C_7H_5NO_4$	552-16-9	167.120	tcl nd (w)	147(1)		1.575^{20}		s H_2O, eth; vs EtOH, ace; sl bz, lig
8074	3-Nitrobenzoic acid		$C_7H_5NO_4$	121-92-6	167.120	mcl pr (w)	141.3(0.4)		1.494^{20}		sl H_2O, bz; vs EtOH, eth, ace; s chl
8075	4-Nitrobenzoic acid		$C_7H_5NO_4$	62-23-7	167.120	mcl lf (w)	241(3)	320(14)	1.610^{20}		vs ace, eth, EtOH, chl, MeOH
8076	3-Nitrobenzoic acid, hydrazide		$C_7H_7N_3O_3$	618-94-0	181.149		153.5				sl H_2O, EtOH; i eth, bz, chl
8077	4-Nitrobenzoic acid, hydrazide		$C_7H_7N_3O_3$	636-97-5	181.149		215.5				sl H_2O, EtOH; i eth, bz, chl
8078	3-Nitrobenzonitrile		$C_7H_4N_2O_2$	619-24-9	148.119		116.6(0.6)	165^{16}			s H_2O, EtOH, bz; vs eth, ace; i peth
8079	4-Nitrobenzonitrile		$C_7H_4N_2O_2$	619-72-7	148.119		147.5(0.6)				sl H_2O, EtOH, eth; s chl, HOAc
8080	5-Nitro-1*H*-benzotriazole		$C_6H_4N_4O_2$	2338-12-7	164.122		217				
8081	2-Nitrobenzoyl chloride		$C_7H_4ClNO_3$	610-14-0	185.565		20				vs eth; sl ctc
8082	3-Nitrobenzoyl chloride		$C_7H_4ClNO_3$	121-90-4	185.565		36	276.5			vs eth
8083	4-Nitrobenzoyl chloride		$C_7H_4ClNO_3$	122-04-3	185.565	ye nd (lig)	71(1)	203^{105}			s eth
8084	2-Nitrobiphenyl	2-Nitro-1,1'-biphenyl	$C_{12}H_9NO_2$	86-00-0	199.205	pl (al, MeOH)	37(1)	320	1.44^{25}		i H_2O; s EtOH, eth, chl
8085	3-Nitrobiphenyl	3-Nitro-1,1'-biphenyl	$C_{12}H_9NO_2$	2113-58-8	199.205	ye pl or nd (dil al)	61(1)	227^{35}			i H_2O; s EtOH, eth, HOAc, lig
8086	4-Nitrobiphenyl	4-Nitro-1,1'-biphenyl	$C_{12}H_9NO_2$	92-93-3	199.205	ye nd (al)	112.9(0.7)	340			i H_2O; sl EtOH; s eth, bz, chl, HOAc
8087	2-Nitro-1,1-bis(*p*-chlorophenyl)propane		$C_{15}H_{13}Cl_2NO_2$	117-27-1	310.176	cry	82.0(0.5)	$180^{0.16}$			
8088	1-Nitrobutane		$C_4H_9NO_2$	627-05-4	103.120			152.8(0.4)	0.970^{25}	1.4303^{20}	sl H_2O; msc EtOH, eth; s alk
8089	2-Nitro-1-butanol		$C_4H_9NO_3$	609-31-4	119.119		-47	105^{10}	1.1332^{25}	1.4390^{20}	s H_2O, ace; msc EtOH, eth; sl ctc
8090	3-Nitro-2-butanol		$C_4H_9NO_3$	6270-16-2	119.119			91^{9}	1.1260^{20}	1.4414^{20}	
8091	6-Nitrochrysene		$C_{18}H_{11}NO_2$	7496-02-8	273.286	ye nd (bz)	213.5(0.9)				
8092	Nitrocyclohexane		$C_6H_{11}NO_2$	1122-60-7	129.157	liq	-34	205	1.0610^{20}	1.4612^{19}	i H_2O; s EtOH, lig
8093	1-Nitrodecane		$C_{10}H_{21}NO_2$	4609-87-4	187.280			86^{1}		1.4337^{20}	
8094	*N*-Nitrodiethylamine	*N*-Ethyl-*N*-nitroethanamine	$C_4H_{10}N_2O_2$	7119-92-8	118.134			206.5	1.057^{15}		vs eth, EtOH
8095	Nitroethane		$C_2H_5NO_2$	79-24-3	75.067	liq	-89.42(0.07)	114.1(0.2)	1.0448^{25}	1.3917^{20}	sl H_2O; msc EtOH, eth; s ace, chl
8096	2-Nitroethanol		$C_2H_5NO_3$	625-48-9	91.066	liq	-80	195(2)	1.270^{15}	1.4438^{19}	msc H_2O, EtOH, eth; i bz
8097	Nitroethene		$C_2H_3NO_2$	3638-64-0	73.051	liq	-55(2)	98.5	1.2212^{14}	1.4282^{20}	vs EtOH, eth, ace, bz, chl
8098	(2-Nitroethyl)benzene		$C_8H_9NO_2$	6125-24-2	151.163	liq	-23	250	1.126^{24}	1.5407^{19}	
8099	Nitrofen	Benzene, 2,4-dichloro-1-(4-nitrophenoxy)-	$C_{12}H_7Cl_2NO_3$	1836-75-5	284.095		71(2)				
8100	2-Nitro-9*H*-fluorene		$C_{13}H_9NO_2$	607-57-8	211.216	nd (50% HOAc ace)	155(1)				i H_2O; s ace, bz
8101	2-Nitro-9*H*-fluoren-9-one		$C_{13}H_7NO_3$	3096-52-4	225.200	ye nd or lf (HOAc)	224.3	sub			sl EtOH; s ace, sulf, HOAc

Organic

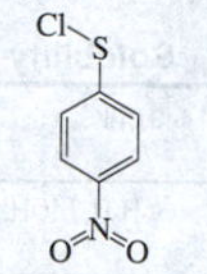
8065
4-Nitrobenzenesulfenyl chloride

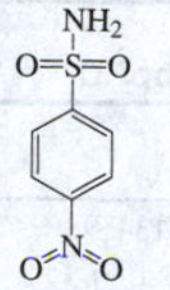
8066
4-Nitrobenzenesulfonamide

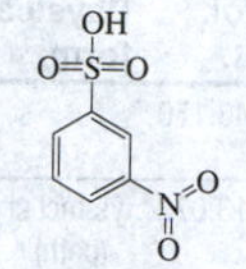
8067
3-Nitrobenzenesulfonic acid

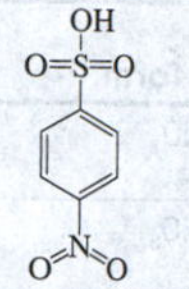
8068
4-Nitrobenzenesulfonic acid

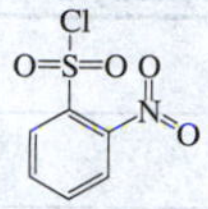
8069
2-Nitrobenzenesulfonyl chloride

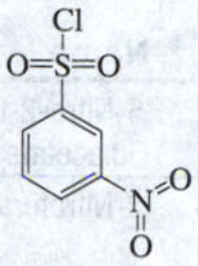
8070
3-Nitrobenzenesulfonyl chloride

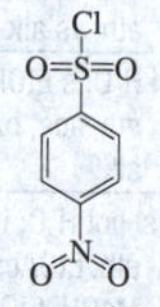
8071
4-Nitrobenzenesulfonyl chloride

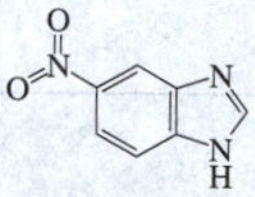
8072
5-Nitro-1*H*-benzimidazole

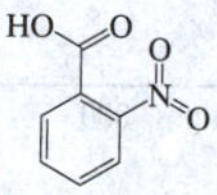
8073
2-Nitrobenzoic acid

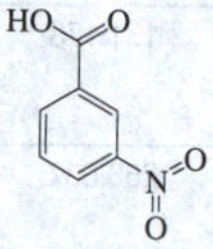
8074
3-Nitrobenzoic acid

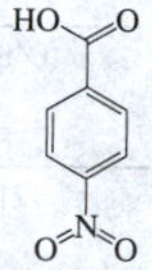
8075
4-Nitrobenzoic acid

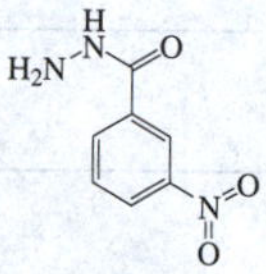
8076
3-Nitrobenzoic acid, hydrazide

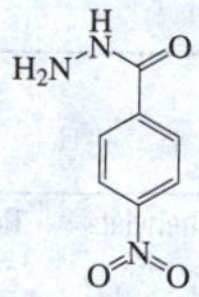
8077
4-Nitrobenzoic acid, hydrazide

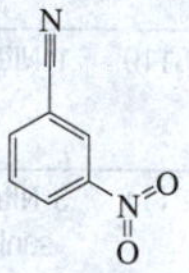
8078
3-Nitrobenzonitrile

Organic

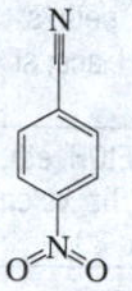
8079
4-Nitrobenzonitrile

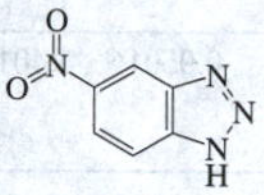
8080
5-Nitro-1*H*-benzotriazole

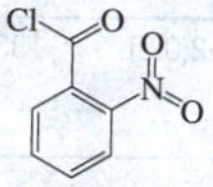
8081
2-Nitrobenzoyl chloride

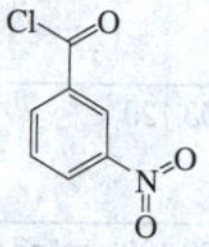
8082
3-Nitrobenzoyl chloride

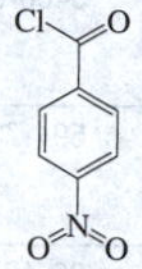
8083
4-Nitrobenzoyl chloride

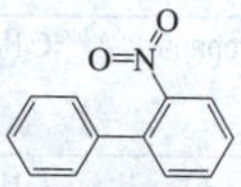
8084
2-Nitrobiphenyl

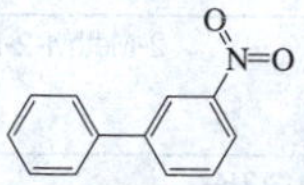
8085
3-Nitrobiphenyl

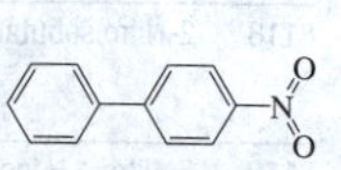
8086
4-Nitrobiphenyl

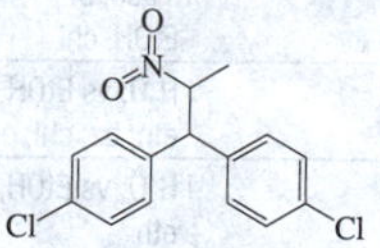
8087
2-Nitro-1,1-bis(*p*-chlorophenyl)propane

8088
1-Nitrobutane

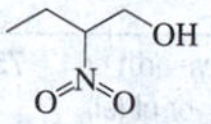
8089
2-Nitro-1-butanol

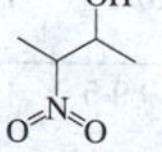
8090
3-Nitro-2-butanol

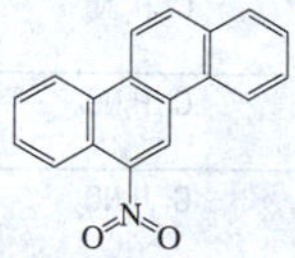
8091
6-Nitrochrysene

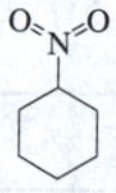
8092
Nitrocyclohexane

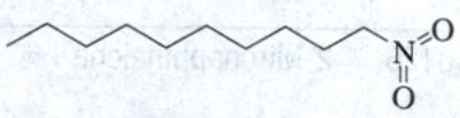
8093
1-Nitrodecane

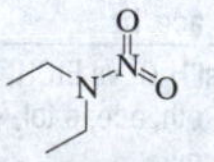
8094
N-Nitrodiethylamine

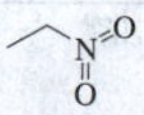
8095
Nitroethane

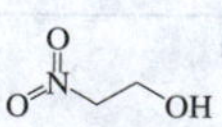
8096
2-Nitroethanol

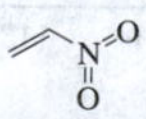
8097
Nitroethene

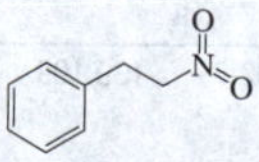
8098
(2-Nitroethyl)benzene

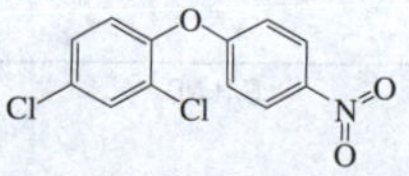
8099
Nitrofen

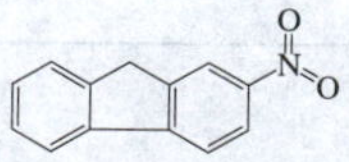
8100
2-Nitro-9*H*-fluorene

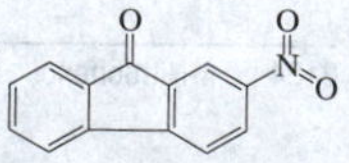
8101
2-Nitro-9*H*-fluoren-9-one

No.	Name	Synonym	Mol. form.	CAS Reg. No.	Mol. wt.	Physical form	mp/°C	bp/°C	den/ g cm⁻³	n_D	Solubility
8102	5-Nitro-2-furaldehyde diacetate		$C_9H_9NO_7$	92-55-7	243.170		92(1)				s chl
8103	2-Nitrofuran		$C_4H_3NO_3$	609-39-2	113.072	ye mcl cry (peth)	29(1)	134^{123}			s H_2O, EtOH, eth
8104	5-Nitro-2-furancarboxaldehyde		$C_5H_3NO_4$	698-63-5	141.083	pa ye (peth)	36(1)	130^{10}			sl H_2O; s peth
8105	5-Nitro-2-furancarboxylic acid		$C_5H_3NO_5$	645-12-5	157.082	pa ye pl (w)	186(1)	sub			s H_2O, EtOH, eth; sl ace, bz; i chl
8106	Nitrofurantoin		$C_8H_6N_4O_5$	67-20-9	238.158		263				
8107	Nitrofurazone	2-[(5-Nitro-2-furanyl)-methylene]hydrazinecarboxamide	$C_6H_6N_4O_4$	59-87-0	198.137	pa ye nd	237(1)				i H_2O, eth; sl EtOH, DMSO; s alk
8108	Nitrogen mustard *N*-oxide hydrochloride	Mechlorethamine oxide hydrochloride	$C_5H_{12}Cl_3NO$	302-70-5	208.514	pr (ace)	110				s H_2O
8109	Nitroguanidine		$CH_4N_4O_2$	556-88-7	104.069	nd or pr (w)	253(42)				sl H_2O, EtOH; i eth; vs alk
8110	1-Nitrohexane		$C_6H_{13}NO_2$	646-14-0	131.173			193	0.9396^{20}	1.4270^{20}	i H_2O; s EtOH, eth, ace, bz, alk
8111	3-Nitro-4-hydroxyphenylarsonic acid	Roxarsone	$C_6H_6AsNO_6$	121-19-7	263.037	ye nd or pl (w)	300				sl hot H_2O; i eth, EtOAc; vs MeOH, EtOH
8112	2-Nitro-1*H*-imidazole	Azomycin	$C_3H_3N_3O_2$	527-73-1	113.075	cry (MeOH)	287 dec				
8113	4-Nitro-1*H*-imidazole		$C_3H_3N_3O_2$	3034-38-6	113.075		303 dec				
8114	5-Nitro-1*H*-indazole		$C_7H_5N_3O_2$	5401-94-5	163.134	ye nd or col nd (al)	208				s EtOH, eth, bz; vs ace, HOAc; i lig
8115	6-Nitro-1*H*-indazole		$C_7H_5N_3O_2$	7597-18-4	163.134	nd (w, al, ace)	181 dec				s H_2O, EtOH, eth, bz; vs ace; i lig
8116	4-Nitro-1,3-isobenzofurandione		$C_8H_3NO_5$	641-70-3	193.114	nd (ace, al)	163.4(0.2)				i H_2O; s EtOH, ace, HOAc; sl bz
8117	5-Nitro-1,3-isobenzofurandione		$C_8H_3NO_5$	5466-84-2	193.114		115(1)	196^{8}			i H_2O, peth; s EtOH, ace; sl eth
8118	2-Nitroisobutane	2-Methyl-2-nitropropane	$C_4H_9NO_2$	594-70-7	103.120		26.2(0.2)	133.4(0.2)	0.9501^{28}	1.4015^{20}	msc EtOH, eth, ace, bz; vs chl; i alk
8119	5-Nitro-1*H*-isoindole-1,3(2*H*)-dione		$C_8H_4N_2O_4$	89-40-7	192.129	col nd (w), ye lf (al-ace)	202				vs ace
8120	Nitromersol		$C_7H_5HgNO_3$	133-58-4	351.71						i H_2O; sl ace, EtOH; s alk
8121	*N*-Nitromethanamine		$CH_4N_2O_2$	598-57-2	76.055		38	82^{10}	1.2433^{49}	1.4616^{49}	vs H_2O, EtOH, bz, chl; s eth; sl peth
8122	Nitromethane	Nitrocarbol	CH_3NO_2	75-52-5	61.041	liq	-28.7(0.8)	101.19(0.1)	1.1371^{20}	1.3817^{20}	s H_2O, EtOH, eth, ace, ctc, alk
8123	(Nitromethyl)benzene		$C_7H_7NO_2$	622-42-4	137.137	ye liq		236(6)	1.1596^{20}	1.5323^{20}	vs ace, eth
8124	Nitron		$C_{20}H_{16}N_4$	2218-94-2	312.368	ye lf (al), nd (chl)	189 dec				vs ace, bz, EtOH, chl
8125	1-Nitronaphthalene		$C_{10}H_7NO_2$	86-57-7	173.169	ye nd (al)	55.5(0.8)	180^{14}	1.332^{20}		i H_2O; vs EtOH, eth, bz, chl, py
8126	2-Nitronaphthalene		$C_{10}H_7NO_2$	581-89-5	173.169	ye orth nd or pl (al)	75(1)	314			i H_2O; vs EtOH, eth
8127	1-Nitro-2-naphthol		$C_{10}H_7NO_3$	550-60-7	189.168	ye nd, lf or pr (al)	104	$115^{0.05}$			s H_2O, EtOH; vs eth; sl chl
8128	1-Nitrooctane		$C_8H_{17}NO_2$	629-37-8	159.227		15	208.5	0.9346^{20}	1.4322^{20}	
8129	1-Nitropentane		$C_5H_{11}NO_2$	628-05-7	117.147			172.5	0.9525^{20}	1.4175^{20}	s EtOH, eth, bz
8130	3-Nitropentane		$C_5H_{11}NO_2$	551-88-2	117.147			154	0.957^{0}		vs ace, eth, EtOH
8131	5-Nitro-1,10-phenanthroline		$C_{12}H_7N_3O_2$	4199-88-6	225.203		202.3				
8132	2-Nitrophenol		$C_6H_5NO_3$	88-75-5	139.109	ye nd or pr (eth, al)	44.9(0.5)	216	1.2942^{40}	1.5723^{50}	sl H_2O; vs EtOH, eth, ace, bz, py
8133	3-Nitrophenol		$C_6H_5NO_3$	554-84-7	139.109	ye mcl (eth, aq HCl)	95(1)	194^{70}	1.2797^{100}		sl H_2O, DMSO; vs EtOH, eth, ace, bz
8134	4-Nitrophenol		$C_6H_5NO_3$	100-02-7	139.109	ye mcl pr (to)	113.8(0.2)		1.479^{20}		sl H_2O; vs EtOH, eth, ace; s tol, py

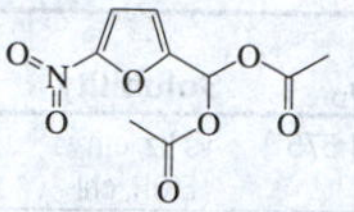
8102
5-Nitro-2-furaldehyde diacetate

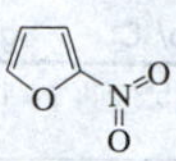
8103
2-Nitrofuran

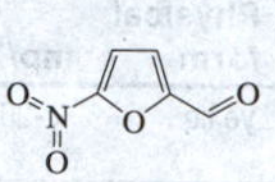
8104
5-Nitro-2-furancarboxaldehyde

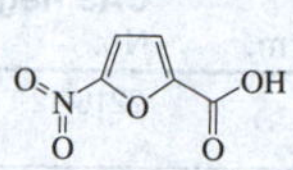
8105
5-Nitro-2-furancarboxylic acid

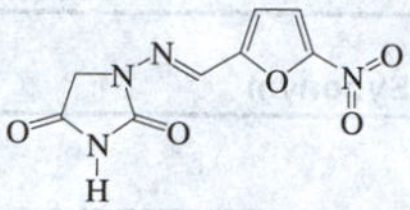
8106
Nitrofurantoin

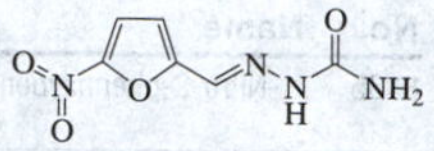
8107
Nitrofurazone

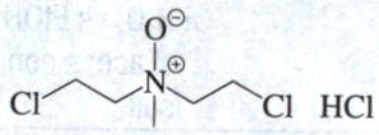
8108
Nitrogen mustard *N*-oxide hydrochloride

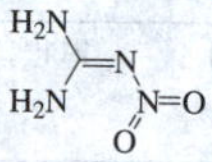
8109
Nitroguanidine

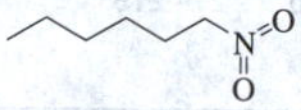
8110
1-Nitrohexane

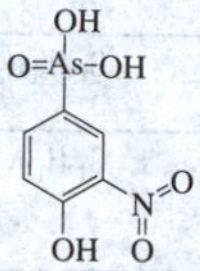
8111
3-Nitro-4-hydroxyphenylarsonic acid

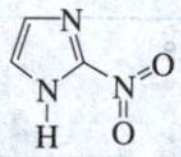
8112
2-Nitro-1*H*-imidazole

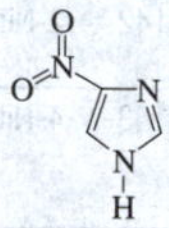
8113
4-Nitro-1*H*-imidazole

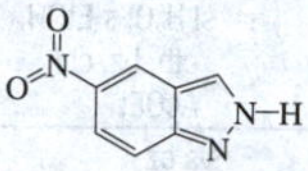
8114
5-Nitro-1*H*-indazole

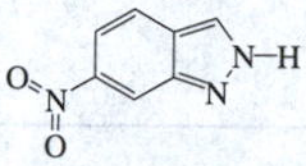
8115
6-Nitro-1*H*-indazole

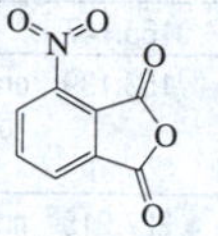
8116
4-Nitro-1,3-isobenzofurandione

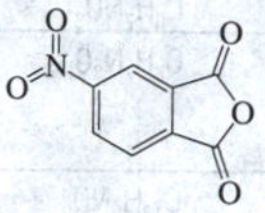
8117
5-Nitro-1,3-isobenzofurandione

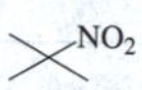
8118
2-Nitroisobutane

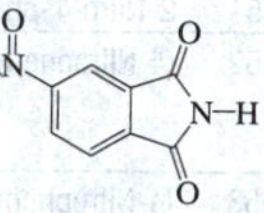
8119
5-Nitro-1*H*-isoindole-1,3(2*H*)-dione

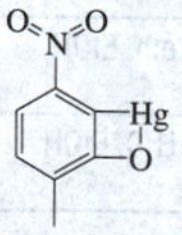
8120
Nitromersol

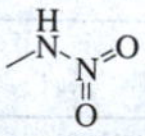
8121
N-Nitromethanamine

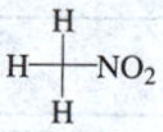
8122
Nitromethane

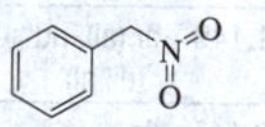
8123
(Nitromethyl)benzene

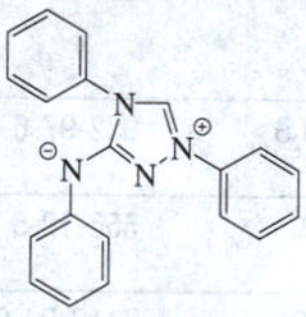
8124
Nitron

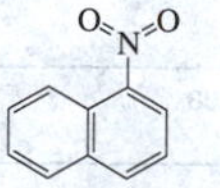
8125
1-Nitronaphthalene

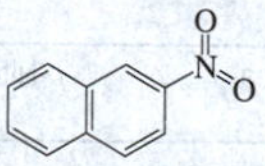
8126
2-Nitronaphthalene

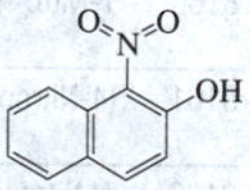
8127
1-Nitro-2-naphthol

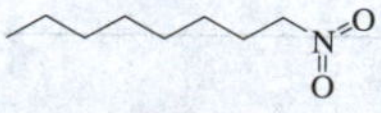
8128
1-Nitrooctane

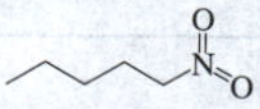
8129
1-Nitropentane

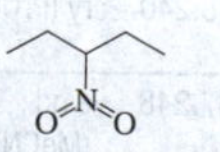
8130
3-Nitropentane

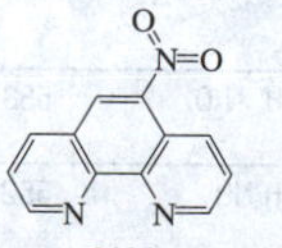
8131
5-Nitro-1,10-phenanthroline

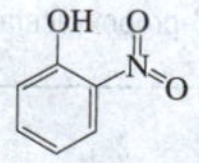
8132
2-Nitrophenol

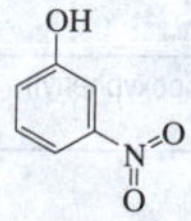
8133
3-Nitrophenol

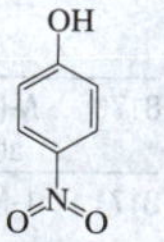
8134
4-Nitrophenol

No.	Name	Synonym	Mol. form.	CAS Reg. No.	Mol. wt.	Physical form	mp/°C	bp/°C	den/ g cm^{-3}	n_D	Solubility
8135	1-Nitro-2-phenoxybenzene		$C_{12}H_9NO_3$	2216-12-8	215.204	ye liq	<-20	235^{60}	1.2539^{22}	1.575^{20}	vs bz, eth, EtOH, chl
8136	1-Nitro-4-phenoxybenzene		$C_{12}H_9NO_3$	620-88-2	215.204	pl (peth), MeOH)	61	320			i H_2O; sl EtOH, ctc; s eth, bz
8137	*N*-(2-Nitrophenyl)acetamide		$C_8H_8N_2O_3$	552-32-9	180.161		93.0(0.5)	$100^{0.1}$	1.419^{15}		s H_2O, EtOH, bz, chl, lig; vs eth
8138	*N*-(3-Nitrophenyl)acetamide		$C_8H_8N_2O_3$	122-28-1	180.161	wh lf (al)	154.5(0.5)	$100^{0.0074}$			s H_2O, EtOH, chl; i eth; sl tfa
8139	*N*-(4-Nitrophenyl)acetamide		$C_8H_8N_2O_3$	104-04-1	180.161	ye pr (w)	216(1)	$100^{0.008}$			sl H_2O, eth, chl; s EtOH, tfa, alk
8140	2-Nitrophenyl acetate		$C_8H_7NO_4$	610-69-5	181.147	nd or pr (lig)	40.5	253 dec			s H_2O; vs EtOH, eth, ace, bz; sl lig
8141	4-Nitrophenyl acetate		$C_8H_7NO_4$	830-03-5	181.147	lf (dil al)	82.3				vs H_2O, bz; s EtOH, chl, lig
8142	2-Nitro-*N*-phenylaniline		$C_{12}H_{10}N_2O_2$	119-75-5	214.219		74.9(0.2)	215^{15}	1.3660^{20}		i H_2O; s EtOH; sl ctc
8143	4-Nitro-*N*-phenylaniline		$C_{12}H_{10}N_2O_2$	836-30-6	214.219		134(2)	211^{30}			i H_2O; vs EtOH; sl ace; s con sulf
8144	(4-Nitrophenyl)arsonic acid	Nitarsone	$C_6H_6AsNO_5$	98-72-6	247.038	lf or nd (w)	>310 dec				sl H_2O, EtOH, DMSO
8145	4-[(4-Nitrophenyl)azo]-1,3-benzenediol	Magneson	$C_{12}H_9N_3O_4$	74-39-5	259.217	red pow (al or MeOH)	200				i H_2O; sl EtOH, bz, HOAc, tol
8146	1-[(4-Nitrophenyl)azo]-2-naphthol		$C_{16}H_{11}N_3O_3$	6410-10-2	293.276	br-oran pl (to or bz)	257				vs bz, EtOH
8147	(3-Nitrophenyl)boronic acid		$C_6H_6BNO_4$	13331-27-6	166.928		274.5				
8148	1-(2-Nitrophenyl)ethanone	2-Nitroacetophenone	$C_8H_7NO_3$	577-59-3	165.147		28.5	178^{32}	1.2370^{25}	1.5468^{20}	i H_2O; vs EtOH, eth, chl
8149	1-(3-Nitrophenyl)ethanone	3-Nitroacetophenone	$C_8H_7NO_3$	121-89-1	165.147	nd (al)	81(1)	202			vs H_2O, eth; sl EtOH, chl
8150	1-(4-Nitrophenyl)ethanone	4-Nitroacetophenone	$C_8H_7NO_3$	100-19-6	165.147	ye pr (al)	80(2)	165^{5}			vs eth, EtOH
8151	2-Nitro-1-phenylethanone		$C_8H_7NO_3$	614-21-1	165.147		105(1)	158^{16}		1.5468^{30}	vs eth, EtOH
8152	(4-Nitrophenyl)hydrazine		$C_6H_7N_3O_2$	100-16-3	153.139	oran-red lf or nd (al)	158 dec				sl H_2O; s EtOH, eth, bz, chl, AcOEt
8153	(4-Nitrophenyl)phenylmethanone		$C_{13}H_9NO_3$	1144-74-7	227.215	nd or lf (al)	138		1.406^{9}		vs bz
8154	3-(4-Nitrophenyl)-1-phenyl-2-propen-1-one	Nitrochalcone	$C_{15}H_{11}NO_3$	1222-98-6	253.253	pa ye nd (al) pl (bz)	164				s EtOH, chl; i eth, lig
8155	4-Nitrophenyl phosphate	4-Nitrophenyl dihydrogen phosphate	$C_6H_6NO_6P$	330-13-2	219.089	ye-wh nd	155				i cold H_2O; s EtOH, chl, bz
8156	3-(2-Nitrophenyl)propanoic acid	2-Nitrobenzenepropanoic acid	$C_9H_9NO_4$	2001-32-3	195.172	ye cry	115				
8157	3-(4-Nitrophenyl)propanoic acid	4-Nitrobenzenepropanoic acid	$C_9H_9NO_4$	16642-79-8	195.172	nd (w)	163				
8158	3-(4-Nitrophenyl)-2-propenal	4-Nitrocinnamaldehyde	$C_9H_7NO_3$	1734-79-8	177.157	nd (w, al)	141.5				s H_2O, eth, ace, bz; vs EtOH
8159	3-(2-Nitrophenyl)-2-propynoic acid	*o*-Nitrophenylpropiolic acid	$C_9H_5NO_4$	530-85-8	191.141		≈157 dec; may explode				sl H_2O; vs EtOH, eth; i CS_2O
8160	1-Nitro-4-(phenylthio)benzene		$C_{12}H_9NO_2S$	952-97-6	231.270	pa ye mcl pr (lig)	56	288^{100}			vs eth, EtOH
8161	(4-Nitrophenyl)urea	*p*-Nitrophenylurea	$C_7H_7N_3O_3$	556-10-5	181.149	pr (al), nd (dil al)	238				vs H_2O, EtOH
8162	*N*-Nitropiperidine		$C_5H_{10}N_2O_2$	7119-94-0	130.145	liq	-5(1)	249(9)	1.1519^{26}	1.4954^{26}	
8163	1-Nitropropane		$C_3H_7NO_2$	108-03-2	89.094	liq	-104.3(0.6)	131.2(0.5)	0.9961^{25}	1.4018^{20}	sl H_2O; msc EtOH, eth; s chl
8164	2-Nitropropane	Isonitropropane	$C_3H_7NO_2$	79-46-9	89.094	liq	-91.3(0.2)	120.2(0.2)	0.9821^{25}	1.3944^{20}	sl H_2O; s chl
8165	3-Nitropropanoic acid		$C_3H_5NO_4$	504-88-1	119.077		62		1.59^{20}		vs H_2O, EtOH, eth; s chl; i lig
8166	2-Nitro-1-propanol		$C_3H_7NO_3$	2902-96-7	105.093			120^{32}	1.1841^{25}	1.4379^{20}	s H_2O, EtOH, eth; sl chl
8167	1-Nitro-1-propene		$C_3H_5NO_2$	3156-70-5	87.078			60^{34}	1.0661^{20}	1.4527^{20}	s eth, ace, chl
8168	2-Nitro-1-propene		$C_3H_5NO_2$	4749-28-4	87.078	ye-grn liq		52^{80}	1.0559^{25}	1.4358^{20}	s eth, ace, chl
8169	5-Nitro-2-propoxyaniline		$C_9H_{12}N_2O_3$	553-79-7	196.202	oran (PrOH-peth)	49				vs EtOH
8170	*N*-(5-Nitro-2-propoxyphenyl)-acetamide	5'-Nitro-2'-propoxyacetanilide	$C_{11}H_{14}N_2O_4$	553-20-8	238.240	cry (PrOH)	102.5				
8171	1-Nitropyrene		$C_{16}H_9NO_2$	5522-43-0	247.248	ye nd (MeCN)	152(1)				

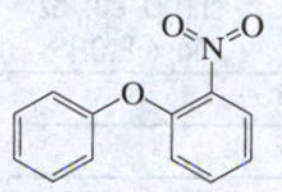
8135
1-Nitro-2-phenoxybenzene

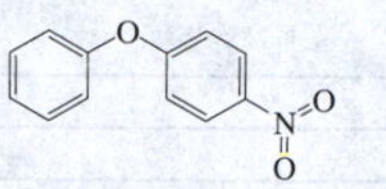
8136
1-Nitro-4-phenoxybenzene

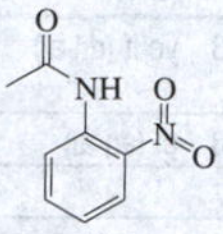
8137
N-(2-Nitrophenyl)acetamide

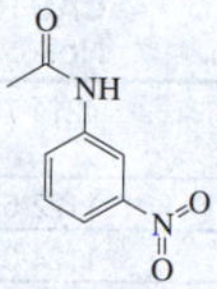
8138
N-(3-Nitrophenyl)acetamide

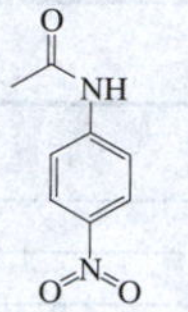
8139
N-(4-Nitrophenyl)acetamide

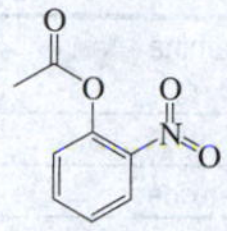
8140
2-Nitrophenyl acetate

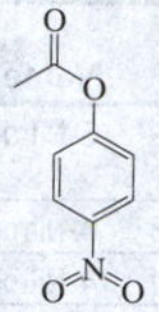
8141
4-Nitrophenyl acetate

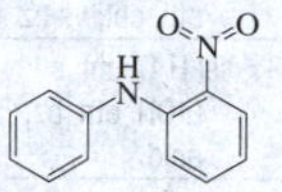
8142
2-Nitro-*N*-phenylaniline

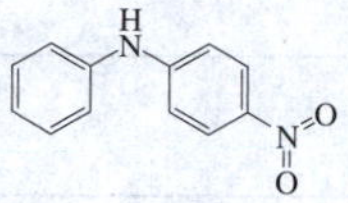
8143
4-Nitro-*N*-phenylaniline

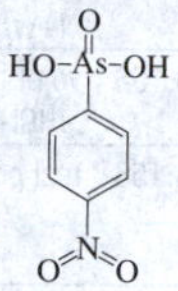
8144
(4-Nitrophenyl)arsonic acid

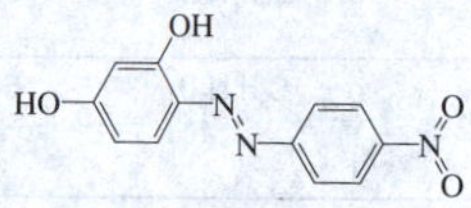
8145
4-[(4-Nitrophenyl)azo]-1,3-benzenediol

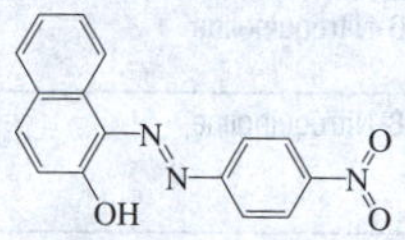
8146
1-[(4-Nitrophenyl)azo]-2-naphthol

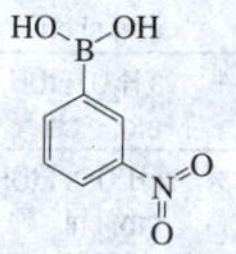
8147
(3-Nitrophenyl)boronic acid

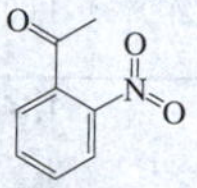
8148
1-(2-Nitrophenyl)ethanone

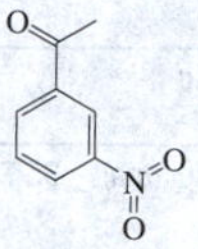
8149
1-(3-Nitrophenyl)ethanone

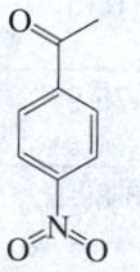
8150
1-(4-Nitrophenyl)ethanone

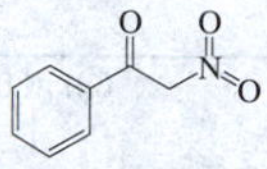
8151
2-Nitro-1-phenylethanone

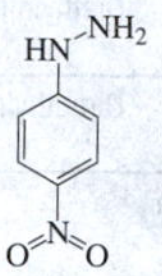
8152
(4-Nitrophenyl)hydrazine

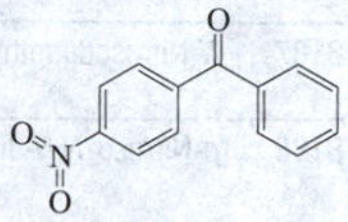
8153
(4-Nitrophenyl)phenylmethanone

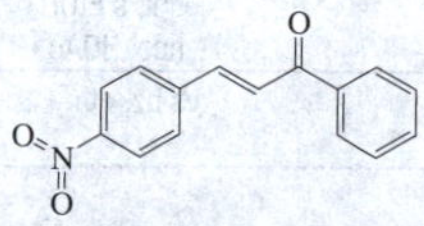
8154
3-(4-Nitrophenyl)-1-phenyl-2-propen-1-one

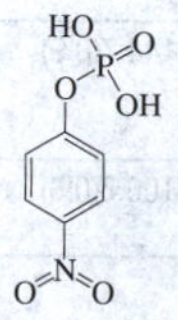
8155
4-Nitrophenyl phosphate

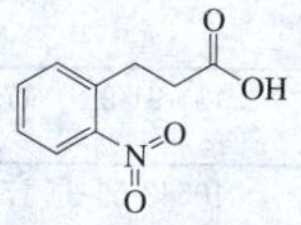
8156
3-(2-Nitrophenyl)propanoic acid

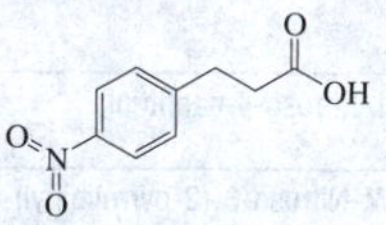
8157
3-(4-Nitrophenyl)propanoic acid

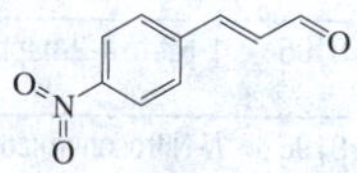
8158
3-(4-Nitrophenyl)-2-propenal

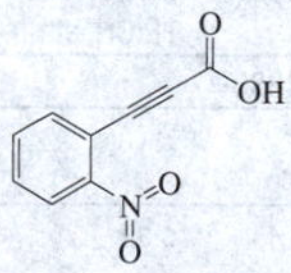
8159
3-(2-Nitrophenyl)-2-propynoic acid

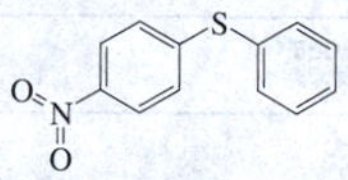
8160
1-Nitro-4-(phenylthio)benzene

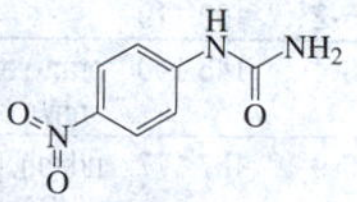
8161
(4-Nitrophenyl)urea

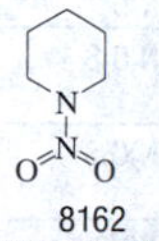
8162
N-Nitropiperidine

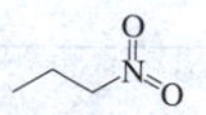
8163
1-Nitropropane

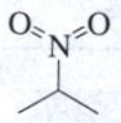
8164
2-Nitropropane

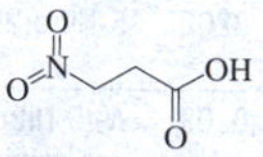
8165
3-Nitropropanoic acid

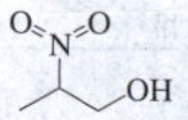
8166
2-Nitro-1-propanol

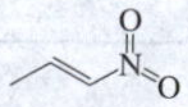
8167
1-Nitro-1-propene

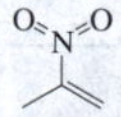
8168
2-Nitro-1-propene

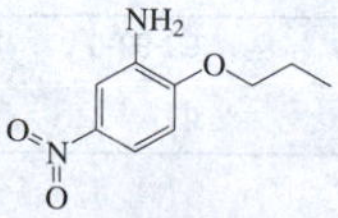
8169
5-Nitro-2-propoxyaniline

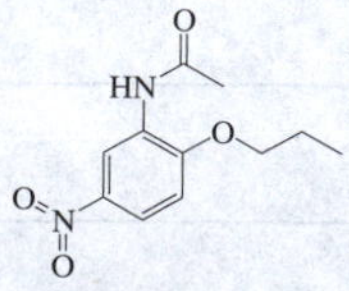
8170
N-(5-Nitro-2-propoxyphenyl)acetamide

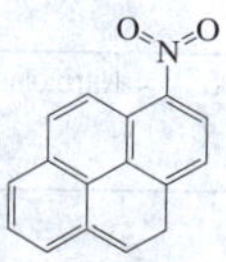
8171
1-Nitropyrene

Organic

No.	Name	Synonym	Mol. form.	CAS Reg. No.	Mol. wt.	Physical form	mp/°C	bp/°C	den/ g cm^{-3}	n_D	Solubility
8172	5-Nitro-2-pyridinamine		$C_5H_5N_3O_2$	4214-76-0	139.113	ye lf (dil al)	188				sl H_2O, eth, bz, lig; s EtOH
8173	4-Nitropyridine		$C_5H_4N_2O_2$	1122-61-8	124.098	pl (aq al)	50				
8174	4-Nitropyridine 1-oxide		$C_5H_4N_2O_3$	1124-33-0	140.097		160.5				
8175	5-Nitropyrimidinamine		$C_4H_4N_4O_2$	3073-77-6	140.101	nd (al)	236.5				sl H_2O, DMSO; s EtOH, ace; i eth, bz
8176	5-Nitro-2,4(1*H*,3*H*)-pyrimidinedione	5-Nitrouracil	$C_4H_3N_3O_4$	611-08-5	157.085	gold nd (al)	>300 exp				sl H_2O; s EtOH
8177	5-Nitro-2,4,6(1*H*,3*H*,5*H*)-pyrimidinetrione	5-Nitrobarbituric acid	$C_4H_3N_3O_5$	480-68-2	173.084	pr, lf (w+3)	180.5				s H_2O, EtOH; i eth
8178	5-Nitroquinoline		$C_9H_6N_2O_2$	607-34-1	174.156	pl (w, al) nd (+w)	88(1)	sub			sl H_2O, chl; s EtOH, bz
8179	6-Nitroquinoline		$C_9H_6N_2O_2$	613-50-3	174.156	ye pl (HCl-HOAc)	148(1)	$170^{0.2}$			s H_2O, EtOH; sl eth, chl; vs bz
8180	8-Nitroquinoline		$C_9H_6N_2O_2$	607-35-2	174.156	mcl pr (al)	88(1)				sl H_2O, chl; s EtOH, eth, bz, acid
8181	4-Nitroquinoline 1-oxide		$C_9H_6N_2O_3$	56-57-5	190.155	ye nd, pl (ace)	154				
8182	5-Nitro-8-quinolinol	Nitroxoline	$C_9H_6N_2O_3$	4008-48-4	190.155		180				
8183	Nitrosobenzene		C_6H_5NO	586-96-9	107.110	orth or mcl (al-eth)	67.8(1)	58^{18}			i H_2O; s EtOH, eth, bz, lig
8184	*N*-Nitrosodibutylamine	Dibutylnitrosamine	$C_8H_{18}N_2O$	924-16-3	158.241			105^{8}			
8185	*N*-Nitrosodiethanolamine	2,2'-(Nitrosoimino)ethanol	$C_4H_{10}N_2O_3$	1116-54-7	134.133	wh-ye oil		$125^{0.01}$		1.4849^{20}	
8186	*N*-Nitrosodiethylamine	Diethylnitrosamine	$C_4H_{10}N_2O$	55-18-5	102.134	ye oil		172(4)	0.9422^{20}	1.4386^{20}	s H_2O, EtOH, eth; sl chl
8187	*N*-Nitrosodimethylamine	Dimethylnitrosamine	$C_2H_6N_2O$	62-75-9	74.081	ye liq		146(2)	1.0048^{20}	1.4368^{20}	vs H_2O, EtOH, eth; s chl
8188	*p*-Nitroso-*N*,*N*-dimethylaniline		$C_8H_{10}N_2O$	138-89-6	150.177	grn pl (eth)	85.7(0.7)		1.145^{20}		sl H_2O; s EtOH, eth, chl, $HCONH_2$
8189	*N*-Nitrosodiphenylamine	*N*,*N*-Diphenylnitrosamine	$C_{12}H_{10}N_2O$	86-30-6	198.219	ye pl(lig)	66.7(0.6)				i H_2O; sl EtOH, chl; s bz
8190	4-(*N*-Nitrosomethylamino)-1-(3-pyridyl)-1-butanone	Ketone, 3-pyridyl-3-(*N*-methyl-*N*-nitrosamino)propyl	$C_{10}H_{13}N_3O_2$	64091-91-4	207.229		63				sl H_2O
8191	*N*-Nitrosomethylethylamine		$C_3H_8N_2O$	10595-95-6	88.108	ye liq		67^{40}			
8192	*N*-Nitroso-*N*-methylvinylamine	*N*-Methyl-*N*-nitrosoethenamine	$C_3H_6N_2O$	4549-40-0	86.092	ye liq		47			sl H_2O
8193	4-Nitrosomorpholine	*N*-Nitrosomorpholine	$C_4H_8N_2O_2$	59-89-2	116.119		29	225			s H_2O
8194	2-Nitroso-1-naphthol		$C_{10}H_7NO_2$	132-53-6	173.169		144.7(0.9)				sl H_2O, eth, bz, chl; s EtOH, ace, HOAc
8195	1-Nitroso-2-naphthol	1-Nitroso-β-naphthol	$C_{10}H_7NO_2$	131-91-9	173.169	ye-br nd (peth)	106.8(0.5)				vs bz, eth
8196	*N*-Nitrosonornicotine	*N*'-Nitroso-3-(2-pyrrolidinyl)-pyridine	$C_9H_{11}N_3O$	16543-55-8	177.202			$155^{0.2}$			
8197	4-Nitrosophenol		$C_6H_5NO_2$	104-91-6	123.110	pa ye orth nd (ace, bz)	144 dec				sl H_2O; s EtOH, eth, ace, bz, dil alk
8198	4-Nitroso-*N*-phenylaniline	*p*-Nitrosodiphenylamine	$C_{12}H_{10}N_2O$	156-10-5	198.219		143				sl H_2O, lig; vs EtOH, eth, bz
8199	*N*-Nitrosopiperidine	1-Nitrosopiperidine	$C_5H_{10}N_2O$	100-75-4	114.145	pa ye		211(7)	1.0631^{18}	1.4933^{18}	s H_2O, HCl
8200	*N*-Nitroso-*N*-propyl-1-propanamine	*N*-Nitrosodipropylamine	$C_6H_{14}N_2O$	621-64-7	130.187	gold		206	0.9163^{20}	1.4437^{20}	sl H_2O; msc EtOH, eth
8201	*N*-Nitrosopyrrolidine		$C_4H_8N_2O$	930-55-2	100.119			214	1.085^{25}	1.4880^{25}	
8202	5-Nitro-2-thiazolamine	2-Amino-5-nitrothiazole	$C_3H_3N_3O_2S$	121-66-4	145.140	oran-ye pow	202 dec				
8203	*N*-(5-Nitro-2-thiazolyl)-acetamide	Aminitrozole	$C_5H_5N_3O_3S$	140-40-9	187.177	nd (al), pl (HOAc)	264.5				s alk
8204	4-Nitrothioanisole		$C_7H_7NO_2S$	701-57-5	169.202		72	137^{2}	1.2391^{80}	1.6401^{20}	i H_2O; s ace, bz
8205	2-Nitrothiophene		$C_4H_3NO_2S$	609-40-5	129.138	lt ye mcl nd (peth)	46.5	224.5	1.3644^{43}		i H_2O; vs EtOH; s alk; sl peth
8206	2-Nitrotoluene		$C_7H_7NO_2$	88-72-2	137.137	liq	-3.6(0.8)	220.9(0.7)	1.1611^{19}	1.5450^{20}	i H_2O; msc EtOH, eth; s ctc
8207	3-Nitrotoluene		$C_7H_7NO_2$	99-08-1	137.137	pa ye	15.9(0.6)	232.1(0.4)	1.1581^{20}	1.5466^{20}	i H_2O; s EtOH, bz, ctc; msc eth
8208	4-Nitrotoluene		$C_7H_7NO_2$	99-99-0	137.137	orth cry (al, eth)	51.7(0.3)	238.66(0.09)	1.1038^{75}		i H_2O; s EtOH; vs eth, ace, bz, chl

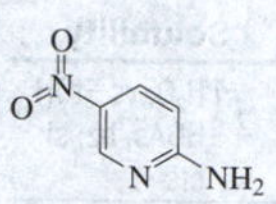
8172
5-Nitro-2-pyridinamine

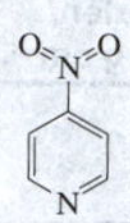
8173
4-Nitropyridine

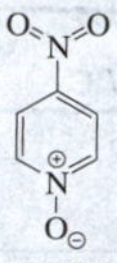
8174
4-Nitropyridine 1-oxide

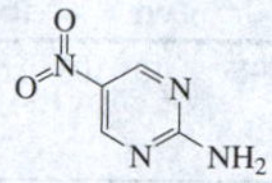
8175
5-Nitropyrimidinamine

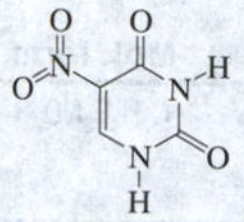
8176
5-Nitro-2,4(1*H*,3*H*)-pyrimidinedione

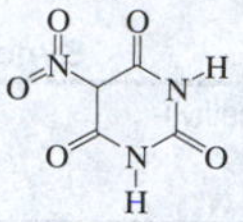
8177
5-Nitro-2,4,6(1*H*,3*H*,5*H*)-pyrimidinetrione

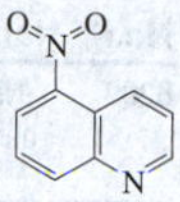
8178
5-Nitroquinoline

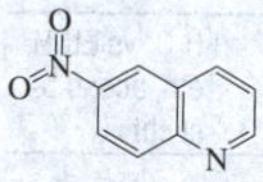
8179
6-Nitroquinoline

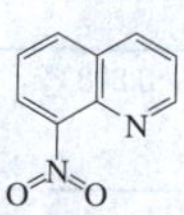
8180
8-Nitroquinoline

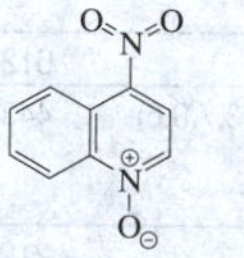
8181
4-Nitroquinoline 1-oxide

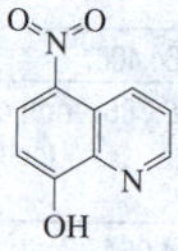
8182
5-Nitro-8-quinolinol

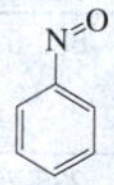
8183
Nitrosobenzene

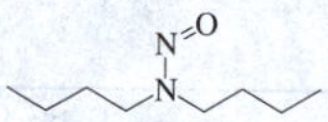
8184
N-Nitrosodibutylamine

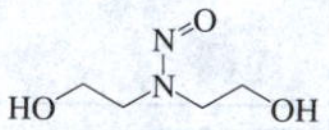
8185
N-Nitrosodiethanolamine

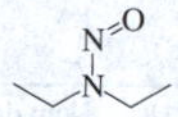
8186
N-Nitrosodiethylamine

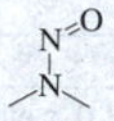
8187
N-Nitrosodimethylamine

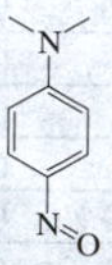
8188
p-Nitroso-*N*,*N*-dimethylaniline

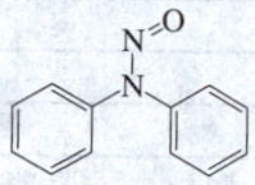
8189
N-Nitrosodiphenylamine

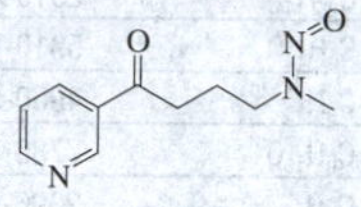
8190
4-(*N*-Nitrosomethylamino)-1-(3-pyridyl)-1-butanone

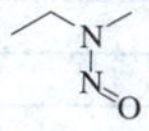
8191
N-Nitrosomethylethylamine

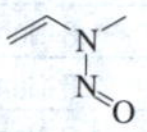
8192
N-Nitroso-*N*-methylvinylamine

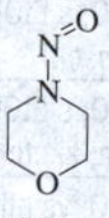
8193
4-Nitrosomorpholine

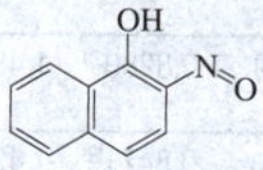
8194
2-Nitroso-1-naphthol

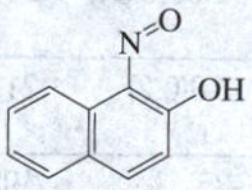
8195
1-Nitroso-2-naphthol

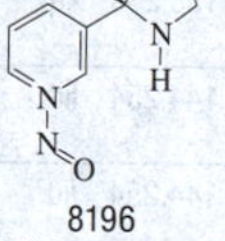
8196
N-Nitrosonornicotine

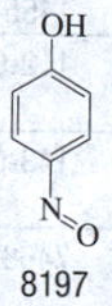
8197
4-Nitrosophenol

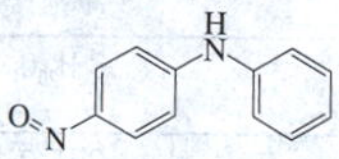
8198
4-Nitroso-*N*-phenylaniline

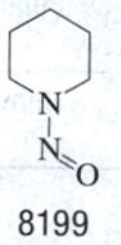
8199
N-Nitrosopiperidine

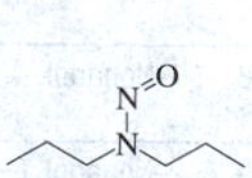
8200
N-Nitroso-*N*-propyl-1-propanamine

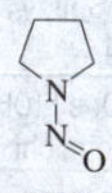
8201
N-Nitrosopyrrolidine

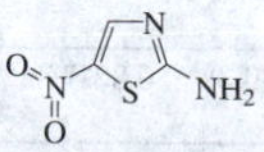
8202
5-Nitro-2-thiazolamine

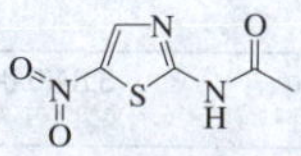
8203
N-(5-Nitro-2-thiazolyl)acetamide

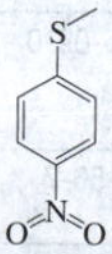
8204
4-Nitrothioanisole

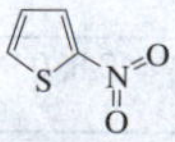
8205
2-Nitrothiophene

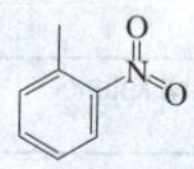
8206
2-Nitrotoluene

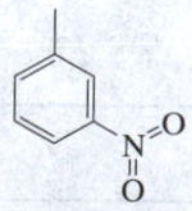
8207
3-Nitrotoluene

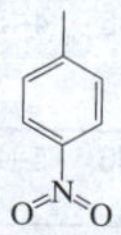
8208
4-Nitrotoluene

No.	Name	Synonym	Mol. form.	CAS Reg. No.	Mol. wt.	Physical form	mp/°C	bp/°C	den/ g cm^{-3}	n_D	Solubility
8209	1-Nitro-2-(trifluoromethyl)-benzene		$C_7H_4F_3NO_2$	384-22-5	191.108	cry (al)	32.5	217			i H_2O; vs EtOH, HOAc, bz; sl ctc
8210	1-Nitro-3-(trifluoromethyl)-benzene		$C_7H_4F_3NO_2$	98-46-4	191.108	liq	-2.4	202.8	1.4357^{15}	1.4719^{20}	i H_2O; s EtOH, eth; sl ctc
8211	Nitrourea		$CH_3N_3O_3$	556-89-8	105.053	pl (al-peth)	159(1)				vs ace, EtOH
8212	*trans*-(2-Nitrovinyl)benzene		$C_8H_7NO_2$	5153-67-3	149.148	ye pr (peth, al)	57(4)	255			i H_2O; s EtOH, ace; vs eth, chl, CS_2
8213	Nivalenol		$C_{15}H_{20}O_7$	23282-20-4	312.316	cry (MeOH)	224 dec				sl H_2O; s EtOH, MeOH
8214	Nizatidine		$C_{12}H_{21}N_5O_2S_2$	76963-41-2	331.458	cry (EtOH/ AcOEt)	131				sl H_2O; s MeOH; vs chl; i bz, eth
8215	2,2',3,3',4,5,5',6,6'-Nonachlorobiphenyl		$C_{12}HCl_9$	52663-77-1	464.213	cry	182.6(0.5)				i H_2O
8216	Nonacontane		$C_{90}H_{182}$	7667-51-8	1264.408			612^{200}			
8217	Nonacosane		$C_{29}H_{60}$	630-03-5	408.786	orth cry (peth)	63.7(0.6)	443(13)	0.8083^{20}	1.4529^{20}	i H_2O; vs EtOH, eth, ace; s bz; sl chl
8218	Nonadecafluorodecanoic acid		$C_{10}HF_{19}O_2$	335-76-2	514.084			219(2)			
8219	Nonadecane		$C_{19}H_{40}$	629-92-5	268.521	wax	31.5(0.4)	330(3)	0.7855^{20}	1.4409^{20}	i H_2O; sl EtOH; s eth, ace, ctc
8220	Nonadecanoic acid		$C_{19}H_{38}O_2$	646-30-0	298.504	lf (al)	68.06(0.04)	297^{100}	0.8468^{70}		i H_2O; vs EtOH, eth, bz, chl, lig
8221	1-Nonadecanol		$C_{19}H_{40}O$	1454-84-8	284.520	cry (ace)	61.4(0.5)	345		1.4328^{75}	s eth, ace
8222	2-Nonadecanone		$C_{19}H_{38}O$	629-66-3	282.504	pr (al)	55.1(0.9)	266^{110}	0.8108^{56}		i H_2O; sl EtOH; s ace, bz; vs eth, ctc
8223	10-Nonadecanone		$C_{19}H_{38}O$	504-57-4	282.504	lf(al)	56.8(0.3)	334(11)			i H_2O; sl EtOH; s eth, ace, lig; vs bz
8224	1-Nonadecene		$C_{19}H_{38}$	18435-45-5	266.505		23.4	331(17)	0.7886^{25}	1.4445^{25}	
8225	Nonadecylbenzene		$C_{25}H_{44}$	29136-19-4	344.617		29.5(0.3)	419	0.8545^{20}	1.4807^{20}	
8226	*trans,trans*-2,4-Nonadienal		$C_9H_{14}O$	5910-87-2	138.206			98^{10}	0.862^{25}	1.5207^{20}	
8227	1,8-Nonadiene		C_9H_{16}	4900-30-5	124.223			141(5)	0.7511^{20}	1.4302^{20}	
8228	2,6-Nonadien-1-ol		$C_9H_{16}O$	7786-44-9	140.222			108^{24}	0.8604^{25}	1.4598^{25}	
8229	1,8-Nonadiyne		C_9H_{12}	2396-65-8	120.191	liq	-27.3(0.2)	165(4)	0.8158^{20}	1.4490^{20}	i H_2O; s eth, ace
8230	Nonanal	Nonaldehyde	$C_9H_{18}O$	124-19-6	142.238		-19.3(0.2)	195(3)	0.8264^{22}	1.4273^{20}	s eth, chl
8231	Nonane		C_9H_{20}	111-84-2	128.255	liq	-53.47(0.03)	150.8(0.2)	0.7192^{20}	1.4058^{20}	i H_2O; vs EtOH, eth; msc ace, bz, hp
8232	Nonanedioic acid	Azelaic acid	$C_9H_{16}O_4$	123-99-9	188.221	lf or nd	106.5	357.1	1.225^{25}	1.4303^{111}	sl H_2O, eth, bz, DMSO; s EtOH
8233	1,9-Nonanediol		$C_9H_{20}O_2$	3937-56-2	160.254	cry (bz)	46.4(0.1)	173^{20}			sl H_2O; vs EtOH, eth; s bz; i lig
8234	Nonanedioyl dichloride		$C_9H_{14}Cl_2O_2$	123-98-8	225.112			166^{18}	1.143	1.4680^{20}	s eth; vs bz
8235	Nonanenitrile		$C_9H_{17}N$	2243-27-8	139.238	liq	-49.5(0.5)	225(5)	0.8178^{20}	1.4255^{20}	i H_2O; s EtOH, eth; sl ctc
8236	1-Nonanethiol	Nonyl mercaptan	$C_9H_{20}S$	1455-21-6	160.320	liq	-20.1(0.3)	220(1)	0.842^{25}	1.4548^{20}	
8237	Nonanoic acid	Pelargonic acid	$C_9H_{18}O_2$	112-05-0	158.238	liq	12.38(0.04)	256(1)	0.9052^{20}	1.4343^{19}	i H_2O; s EtOH, eth, chl
8238	1-Nonanol	Nonyl alcohol	$C_9H_{20}O$	143-08-8	144.254	liq	-5.0(0.2)	213.7(0.4)	0.8280^{20}	1.4333^{20}	i H_2O; s EtOH, eth; sl ctc
8239	2-Nonanol, (±)-		$C_9H_{20}O$	74683-66-2	144.254	liq	-35	193.5	0.8471^{20}	1.4353^{20}	i H_2O; vs eth, EtOH
8240	3-Nonanol, (±)-		$C_9H_{20}O$	74742-08-8	144.254		22	195	0.8250^{20}	1.4289^{20}	i H_2O; s EtOH, eth
8241	4-Nonanol		$C_9H_{20}O$	52708-03-9	144.254			192.5	0.8282^{20}	1.4197^{20}	i H_2O; s EtOH, eth
8242	5-Nonanol	Dibutylcarbinol	$C_9H_{20}O$	623-93-8	144.254		5.6(0.4)	195(2)	0.8220^{20}	1.4289^{20}	i H_2O; s EtOH
8243	2-Nonanone	Heptyl methyl ketone	$C_9H_{18}O$	821-55-6	142.238	liq	-7.4(0.2)	194(1)	0.8208^{20}	1.4210^{20}	i H_2O; s EtOH, eth, bz; vs ace, chl
8244	3-Nonanone	Ethyl hexyl ketone	$C_9H_{18}O$	925-78-0	142.238	liq	-8(4)	187(4)	0.8241^{20}	1.4208^{20}	i H_2O; s EtOH, eth, bz, chl; vs ace
8245	4-Nonanone	Pentyl propyl ketone	$C_9H_{18}O$	4485-09-0	142.238			188(4)	0.8190^{25}	1.4189^{20}	i H_2O; s EtOH, eth, chl; vs ace
8246	5-Nonanone	Dibutyl ketone	$C_9H_{18}O$	502-56-7	142.238	liq	-3.84(0.05)	188.4(0.3)	0.8217^{20}	1.4195^{20}	i H_2O; s EtOH; vs eth, chl
8247	Nonanoyl chloride		$C_9H_{17}ClO$	764-85-2	176.683	liq	-60.5	218(13)	0.9463^{15}		s eth, ace

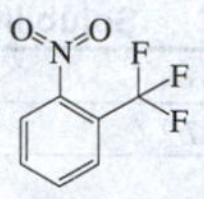
8209
1-Nitro-2-(trifluoromethyl)benzene

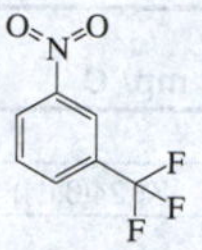
8210
1-Nitro-3-(trifluoromethyl)benzene

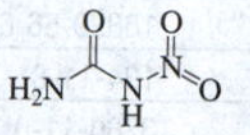
8211
Nitrourea

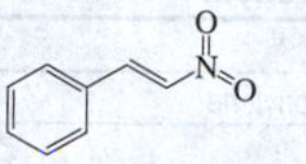
8212
trans-(2-Nitrovinyl)benzene

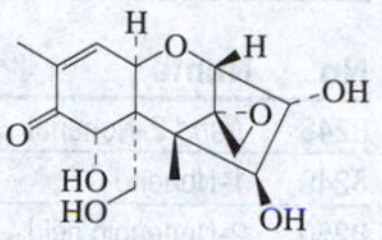
8213
Nivalenol

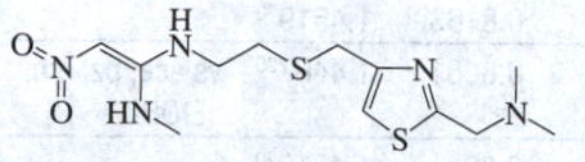
8214
Nizatidine

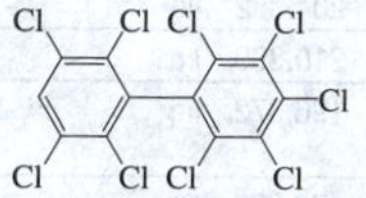
8215
2,2',3,3',4,5,5',6,6'-Nonachlorobiphenyl

$H_3C(CH_2)_{88}CH_3$
8216
Nonacontane

8217
Nonacosane

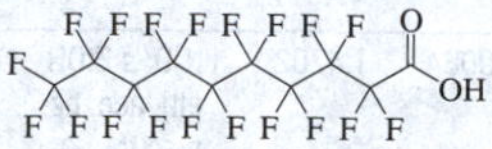
8218
Nonadecafluorodecanoic acid

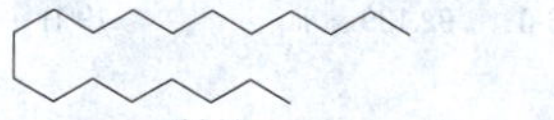
8219
Nonadecane

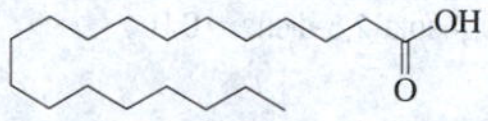
8220
Nonadecanoic acid

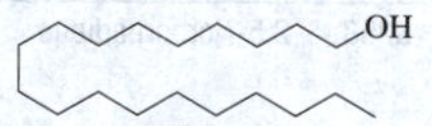
8221
1-Nonadecanol

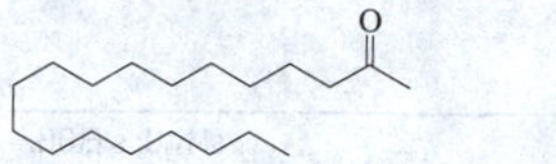
8222
2-Nonadecanone

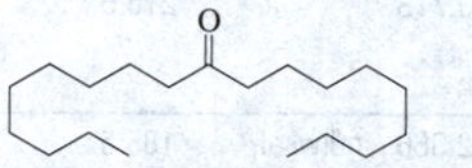
8223
10-Nonadecanone

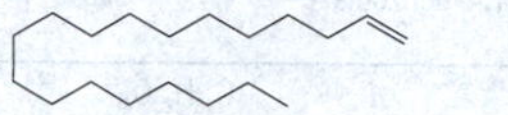
8224
1-Nonadecene

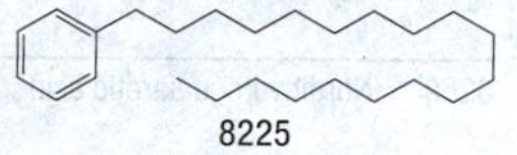
8225
Nonadecylbenzene

8226
trans,trans-2,4-Nonadienal

8227
1,8-Nonadiene

8228
2,6-Nonadien-1-ol

8229
1,8-Nonadiyne

8230
Nonanal

8231
Nonane

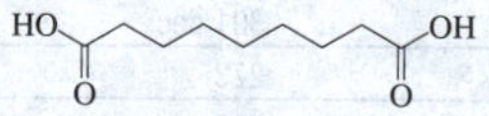
8232
Nonanedioic acid

8233
1,9-Nonanediol

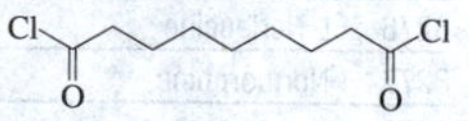
8234
Nonanedioyl dichloride

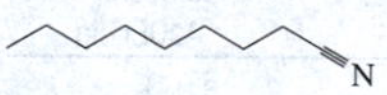
8235
Nonanenitrile

8236
1-Nonanethiol

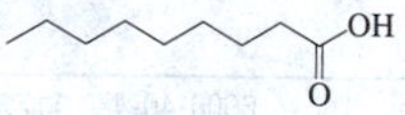
8237
Nonanoic acid

8238
1-Nonanol

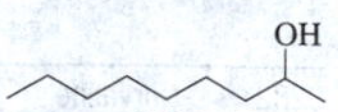
8239
2-Nonanol, (±)-

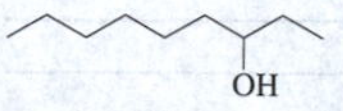
8240
3-Nonanol, (±)-

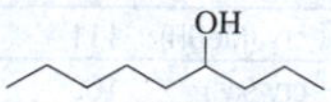
8241
4-Nonanol

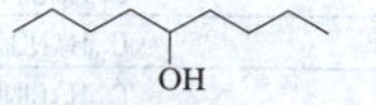
8242
5-Nonanol

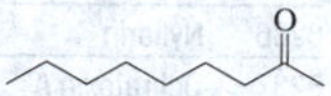
8243
2-Nonanone

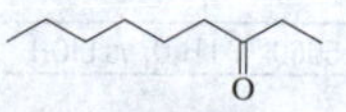
8244
3-Nonanone

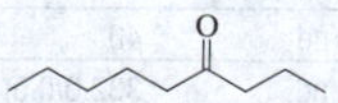
8245
4-Nonanone

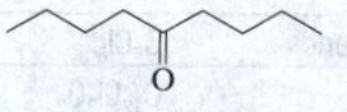
8246
5-Nonanone

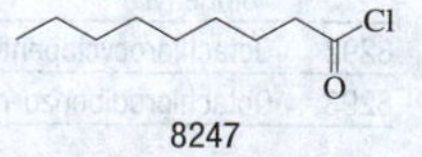
8247
Nonanoyl chloride

Organic

No.	Name	Synonym	Mol. form.	CAS Reg. No.	Mol. wt.	Physical form	mp/°C	bp/°C	den/ g cm⁻³	n_D	Solubility
8248	*trans*-2-Nonenal		$C_9H_{16}O$	18829-56-6	140.222	liq		101^{16}	0.846	1.4531^{20}	
8249	1-Nonene	1-Nonylene	C_9H_{18}	124-11-8	126.239	liq	-81.24(0.04)	146.9(0.6)	0.7253^{25}	1.4257^{20}	
8250	2-Nonenoic acid		$C_9H_{16}O_2$	3760-11-0	156.222			173^{20}			
8251	3-Nonenoic acid		$C_9H_{16}O_2$	4124-88-3	156.222		-4.4	156^{18}	0.9254^{20}	1.4454^{25}	
8252	1-Nonen-3-ol	1-Vinylheptanol	$C_9H_{18}O$	21964-44-3	142.238			193.5	0.824^{21}	1.4382^{15}	
8253	Nonyl acetate		$C_{11}H_{22}O_2$	143-13-5	186.292	liq	-26	225(8)	0.8785^{15}	1.426^{20}	
8254	Nonylamine	1-Nonanamine	$C_9H_{21}N$	112-20-9	143.270	liq	-1	198(3)	0.7886^{20}	1.4336^{20}	sl H_2O, chl; s EtOH, eth
8255	Nonylbenzene		$C_{15}H_{24}$	1081-77-2	204.352	liq	-24	280(3)	0.8584^{20}	1.4816^{20}	
8256	Nonylcyclohexane		$C_{15}H_{30}$	2883-02-5	210.399	liq	-10	281(3)	0.8163^{20}	1.4519^{20}	
8257	Nonylcyclopentane		$C_{14}H_{28}$	2882-98-6	196.372	liq	-29	262	0.8081^{20}	1.4467^{20}	vs ace, bz, eth, EtOH
8258	Nonyl formate		$C_{10}H_{20}O_2$	5451-92-3	172.265	liq	-33	216(13)	0.86	1.4216^{20}	
8259	1-Nonylnaphthalene		$C_{19}H_{26}$	26438-26-6	254.409		8	366	0.9371^{20}	1.5477^{20}	
8260	4-Nonylphenol		$C_{15}H_{24}O$	104-40-5	220.351	visc ye liq	42	317(19)	0.950^{20}	1.513^{20}	i H_2O; s bz, ctc, hp
8261	1-Nonyne	Heptylacetylene	C_9H_{16}	3452-09-3	124.223	liq	-50	150.8(0.4)	0.7658^{20}	1.4217^{20}	i H_2O; s eth, bz, ctc
8262	Norbormide		$C_{33}H_{25}N_3O_3$	991-42-4	511.570	cry (eth)	194				
8263	2,5-Norbornadiene	Bicyclo[2.2.1]hepta-2,5-diene	C_7H_8	121-46-0	92.139	liq	-19(1)	90(1)	0.9064^{20}	1.4702^{20}	i H_2O; s EtOH, eth, ace, bz; msc tol
8264	5-Norbornene-2,3-dicarboxylic acid anhydride		$C_9H_8O_3$	826-62-0	164.158		166				
8265	5-Norbornene-2-methylolacrylate		$C_{11}H_{14}O_2$	95-39-6	178.228	col liq		104	1.029^{25}		s os
8266	24-Norcholan-23-oic acid, (5β)	Norcholanic acid	$C_{23}H_{38}O_2$	511-18-2	346.547	nd(HOAc)	177				
8267	Nordazepam	7-Chloro-1,3-dihydro-5-phenyl-2*H*-1,4-benzodiazepin-2-one	$C_{15}H_{11}ClN_2O$	1088-11-5	270.713		216.5				
8268	Nordihydroguaiaretic acid		$C_{18}H_{22}O_4$	500-38-9	302.366	nd(w, al, HOAc)	185.5				sl H_2O; s EtOH, eth, ace, alk; i bz
8269	Norea		$C_{13}H_{22}N_2O$	18530-56-8	222.326		168.7(0.5)				
8270	Norepinephrine	Noradrenaline	$C_8H_{11}NO_3$	51-41-2	169.178		217 dec				sl H_2O, EtOH, eth; vs alk, dil HCl
8271	Norethisterone	19-Norpregn-4-en-20-yn-3-one, 17-hydroxy-, (17α)-	$C_{20}H_{26}O_2$	68-22-4	298.419	cry	207(2)				
8272	Norethynodrel		$C_{20}H_{26}O_2$	68-23-5	298.419	cry (MeOH)	170				
8273	Norflurazon		$C_{12}H_9ClF_3N_3O$	27314-13-2	303.666		178.0(0.6)				
8274	Norhyoscyamine		$C_{16}H_{21}NO_3$	537-29-1	275.343	nd	140.5				vs EtOH, chl
8275	DL-Norleucine	2-Aminohexanoic acid, (DL)	$C_6H_{13}NO_2$	616-06-8	131.173	lf(w)	327 dec		1.172^{25}		s H_2O; sl EtOH; i eth
8276	L-Norleucine	2-Aminohexanoic acid, (L)	$C_6H_{13}NO_2$	327-57-1	131.173		301 dec				sl H_2O
8277	Normorphine		$C_{16}H_{17}NO_3$	466-97-7	271.311		273				
8278	Norplant	Norgestrel, (-)	$C_{21}H_{28}O_2$	797-63-7	312.446	cry (MeOH)	206				
8279	19-Nortestosterone phenylpropionate	Nandrolone phenpropionate	$C_{27}H_{34}O_3$	62-90-8	406.557	cry	94(1)				
8280	Nortriptyline hydrochloride		$C_{19}H_{22}ClN$	894-71-3	299.838	cry (eth)	214				s H_2O, EtOH; i bz, eth, ace
8281	DL-Norvaline		$C_5H_{11}NO_2$	760-78-1	117.147	lf(al, w)	303	sub			s H_2O; i EtOH, eth, chl, AcOEt, lig
8282	L-Norvaline	2-Aminopentanoic acid, (*S*)	$C_5H_{11}NO_2$	6600-40-4	117.147	cry (dil al)	307				s H_2O
8283	Noscapine		$C_{22}H_{23}NO_7$	128-62-1	413.421	pr or nd (al)	176				i H_2O; s EtOH, bz, chl; sl eth; vs ace
8284	Novobiocin	Streptonivicin	$C_{31}H_{36}N_2O_{11}$	303-81-1	612.624	wh-ye orth cry	154		1.3448		i H_2O; s EtOH, EtOAc, ace, py
8285	Nuarimol		$C_{17}H_{12}ClFN_2O$	63284-71-9	314.740		126				
8286	Nylidrin	Buphenine	$C_{19}H_{25}NO_2$	447-41-6	299.408	cry (MeOH)	111				
8287	Ochratoxin A		$C_{20}H_{18}ClNO_6$	303-47-9	403.813	cry (xyl)	169				
8288	Ochratoxin B		$C_{20}H_{19}NO_6$	4825-86-9	369.368	cry (MeOH)	221				
8289	Ochratoxin C		$C_{22}H_{22}ClNO_6$	4865-85-4	431.866	amorp solid					
8290	Octacaine	3-(Diethylamino)-*N*-phenylbutanamide	$C_{14}H_{22}N_2O$	13912-77-1	234.337	cry	47	200^{1}			vs EtOH, bz, eth
8291	2,2',3,3',5,5',6,6'-Octachlorobiphenyl		$C_{12}H_2Cl_8$	2136-99-4	429.768	cry	160.6(0.5)				i H_2O
8292	Octachlorocyclopentene	Perchlorocyclopentene	C_5Cl_8	706-78-5	343.678	nd	40	283	1.8200^{50}	1.5660^{50}	i H_2O; vs EtOH
8293	Octachlorodibenzo-*p*-dioxin		$C_{12}Cl_8O_2$	3268-87-9	459.751	nd	332.0(0.5)				

8248
trans-2-Nonenal

8249
1-Nonene

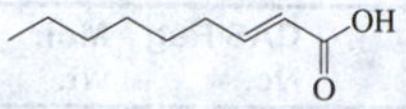

8250
2-Nonenoic acid

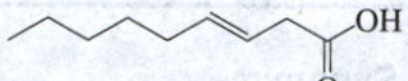

8251
3-Nonenoic acid

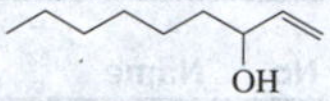

8252
1-Nonen-3-ol

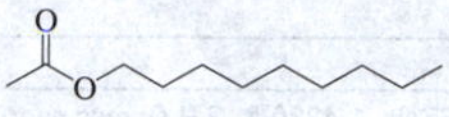

8253
Nonyl acetate

8254
Nonylamine

8255
Nonylbenzene

8256
Nonylcyclohexane

8257
Nonylcyclopentane

8258
Nonyl formate

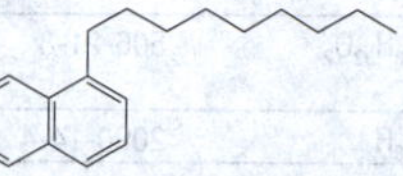

8259
1-Nonylnaphthalene

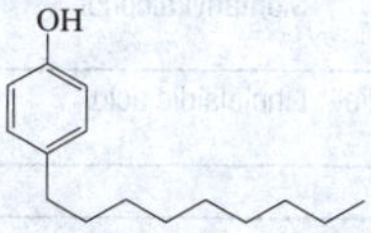

8260
4-Nonylphenol

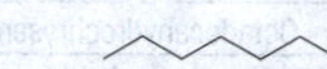

8261
1-Nonyne

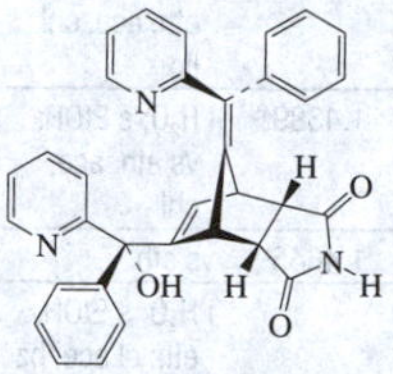

8262
Norbormide

8263
2,5-Norbornadiene

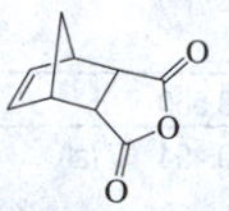

8264
5-Norbornene-2,3-dicarboxylic acid anhydride

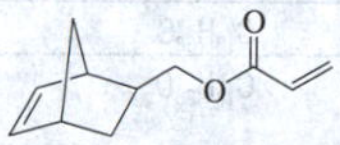

8265
5-Norbornene-2-methylolacrylate

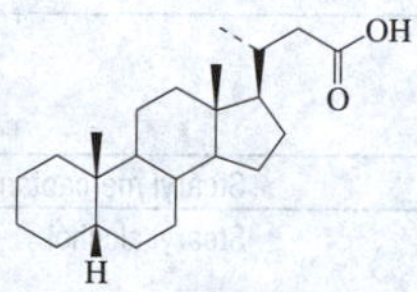

8266
24-Norcholan-23-oic acid, (5β)

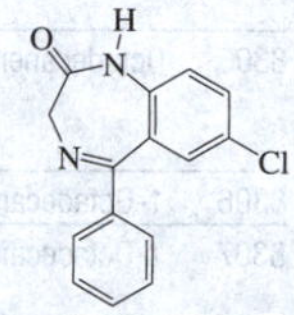

8267
Nordazepam

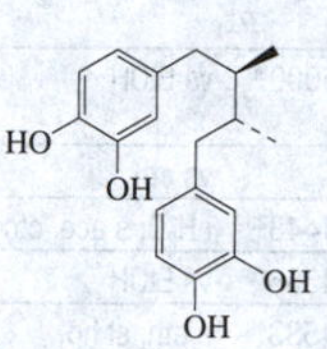

8268
Nordihydroguaiaretic acid

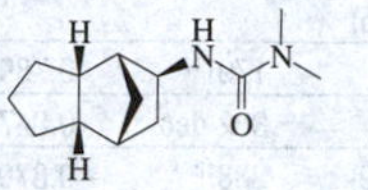

8269
Norea

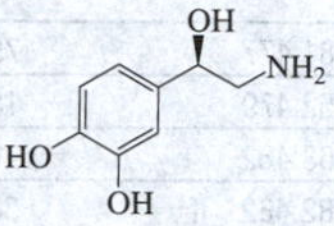

8270
Norepinephrine

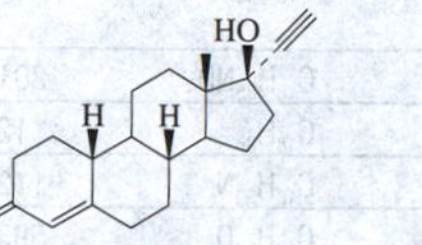

8271
Norethisterone

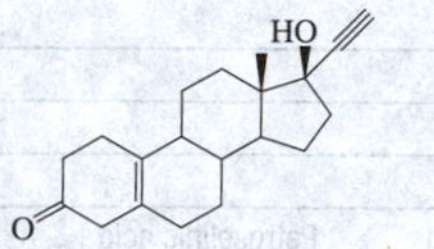

8272
Norethynodrel

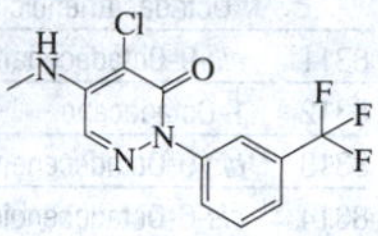

8273
Norflurazon

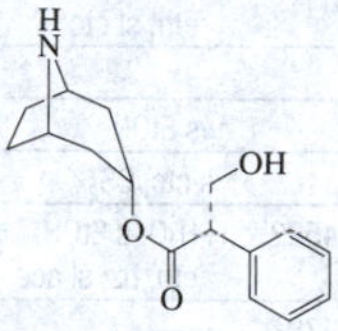

8274
Norhyoscyamine

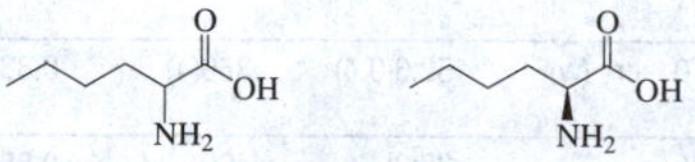

8275
DL-Norleucine

8276
L-Norleucine

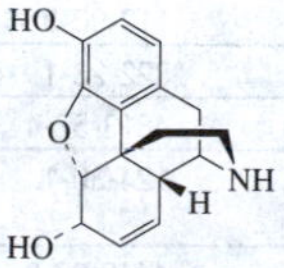

8277
Normorphine

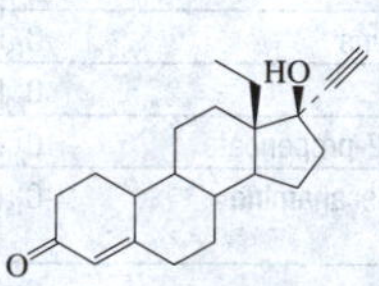

8278
Norplant

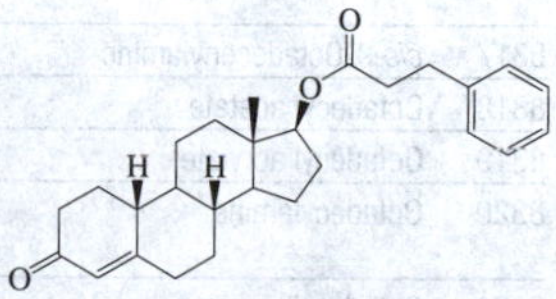

8279
19-Nortestosterone phenylpropionate

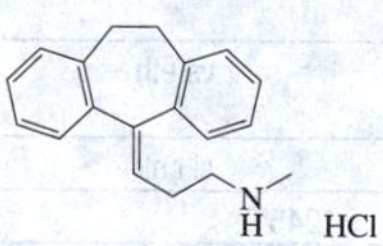

8280
Nortriptyline hydrochloride

8281
DL-Norvaline

8282
L-Norvaline

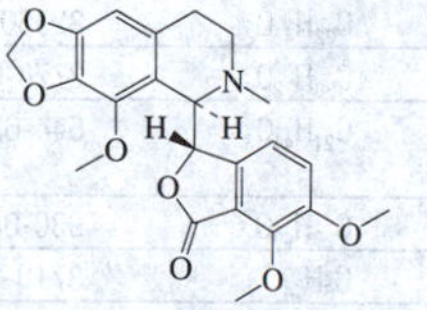

8283
Noscapine

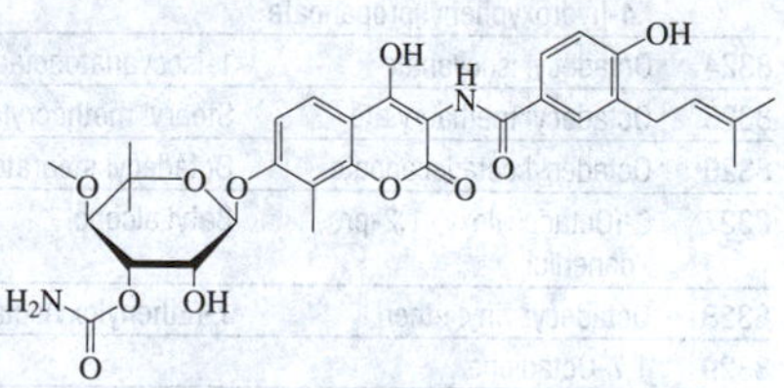

8284
Novobiocin

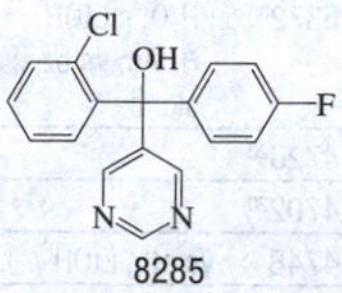

8285
Nuarimol

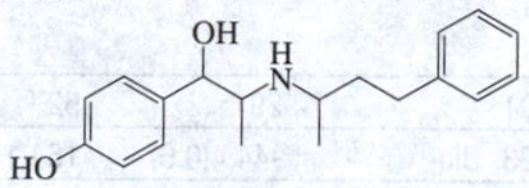

8286
Nylidrin

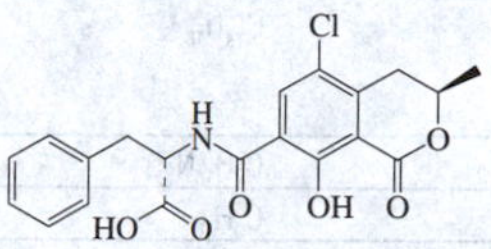

8287
Ochratoxin A

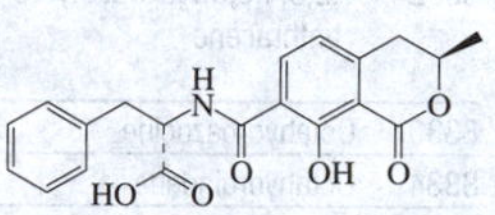

8288
Ochratoxin B

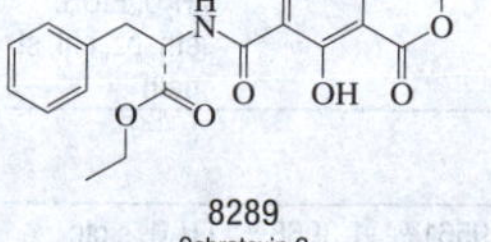

8289
Ochratoxin C

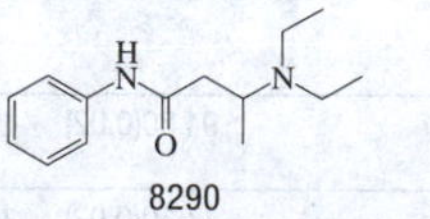

8290
Octacaine

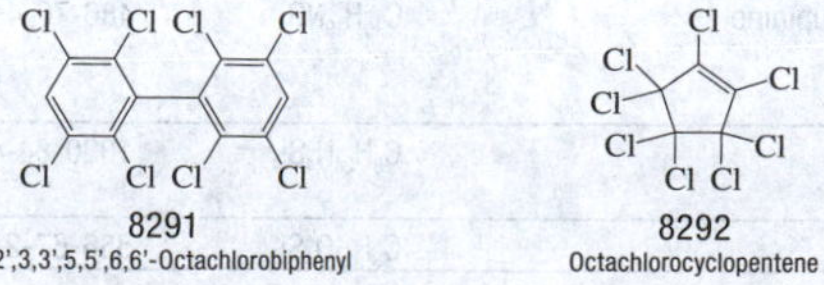

8291
2,2',3,3',5,5',6,6'-Octachlorobiphenyl

8292
Octachlorocyclopentene

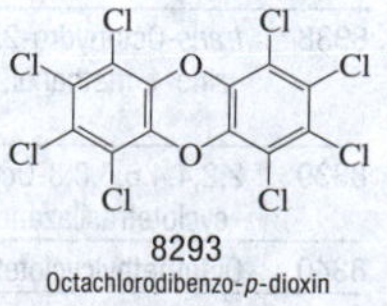

8293
Octachlorodibenzo-*p*-dioxin

No.	Name	Synonym	Mol. form.	CAS Reg. No.	Mol. wt.	Physical form	mp/°C	bp/°C	den/ g cm⁻³	n_D	Solubility
8294	Octachloronaphthalene	Perchloronaphthalene	$C_{10}Cl_8$	2234-13-1	403.731	nd (bz-CCl_4)	197.5	441^{7}			sl EtOH; vs bz, chl, lig
8295	Octachlorostyrene	Perchlorostyrene	C_8Cl_8	29082-74-4	379.710	cry (ace/ EtOH)	99				
8296	Octacontane		$C_{80}H_{162}$	7667-88-1	1124.142		112	672			
8297	Octacosane		$C_{28}H_{58}$	630-02-4	394.761	mcl or orth (bz-al)	61.3(0.1)	432(6)	0.8067^{20}	1.4330^{70}	i H_2O; msc ace; s bz, chl
8298	Octacosanoic acid	Montanic acid	$C_{28}H_{56}O_2$	506-48-9	424.744		90.9(0.5)		0.8191^{100}	1.4313^{100}	vs bz, chl
8299	1-Octacosanol	Montanyl alcohol	$C_{28}H_{58}O$	557-61-9	410.760	cry (ace, peth)	83.4	200^{1}			i H_2O; s CS_2
8300	*trans,trans*-9,12-Octadecadienoic acid	Linolelaidic acid	$C_{18}H_{32}O_2$	506-21-8	280.446	cry (MeOH)	28.5	$181^{0.8}$			sl H_2O; s ace, hx
8301	Octadecahydrochrysene		$C_{18}H_{30}$	2090-14-4	246.431		115	353			vs EtOH
8302	Octadecamethyloctasiloxane		$C_{18}H_{54}O_7Si_8$	556-69-4	607.302		-63	186^{20}	0.913^{25}	1.3970^{20}	vs bz, peth, lig
8303	Octadecanamide		$C_{18}H_{37}NO$	124-26-5	283.493	lf (al)	107(3)	250^{12}			vs eth, chl
8304	Octadecane		$C_{18}H_{38}$	593-45-3	254.495	nd (al, eth-MeOH)	28.17(0.05)	316(2)	0.7768^{28}	1.4390^{20}	i H_2O; sl EtOH; s eth, ace, chl, lig
8305	Octadecanenitrile		$C_{18}H_{35}N$	638-65-3	265.478		42(1)	362	0.8325^{20}	1.4389^{45}	i H_2O; s EtOH; vs eth, ace, chl
8306	1-Octadecanethiol	Stearyl mercaptan	$C_{18}H_{38}S$	2885-00-9	286.560		30	207^{11}	0.8475^{20}	1.4645^{20}	vs eth
8307	1-Octadecanol	Stearyl alcohol	$C_{18}H_{38}O$	112-92-5	270.494	lf (al)	58.0(0.2)	351(2)	0.8124^{59}		i H_2O; s EtOH, eth; sl ace, bz
8308	Octadecanoyl chloride		$C_{18}H_{35}ClO$	112-76-5	302.923		23	215^{15}	0.8969^{0}	1.4523^{24}	sl EtOH
8309	*cis,trans,trans*-9,11,13-Octadecatrienoic acid	*cis*-Eleostearic acid	$C_{18}H_{30}O_2$	506-23-0	278.430	nd (al)	49	235^{12} dec	0.9028^{50}	1.5112^{50}	vs eth, EtOH
8310	*trans,trans,trans*-9,11,13-Octadecatrienoic acid	*trans*-Eleostearic acid	$C_{18}H_{30}O_2$	544-73-0	278.430	lf (al)	71.5	188^{1}	0.8839^{80}	1.5000^{80}	vs EtOH
8311	*cis*-9-Octadecenamide		$C_{18}H_{35}NO$	301-02-0	281.477		75.9(0.5)				vs eth
8312	1-Octadecene		$C_{18}H_{36}$	112-88-9	252.479		18(3)	179^{15}	0.7891^{20}	1.4448^{20}	i H_2O; s ace, ctc
8313	*cis*-9-Octadecenenitrile		$C_{18}H_{33}N$	112-91-4	263.462		-1	332 dec	0.847^{17}	1.4566^{20}	vs EtOH
8314	*cis*-6-Octadecenoic acid	Petroselinic acid	$C_{18}H_{34}O_2$	593-39-5	282.462	lf	30.0(0.9)	238^{18}	0.8700^{40}	1.4533^{40}	s eth; sl hp, MeOH
8315	*trans*-11-Octadecenoic acid	Vaccenic acid	$C_{18}H_{34}O_2$	693-72-1	282.462		43(1)		0.8563^{70}	1.4499^{60}	s ace
8316	*cis*-9-Octadecen-1-ol	Oleyl alcohol	$C_{18}H_{36}O$	143-28-2	268.478		0(3)	207^{15}	0.8489^{20}	1.4606^{20}	i H_2O; s EtOH, eth; sl ctc
8317	*cis*-9-Octadecenylamine	Oleylamine	$C_{18}H_{37}N$	112-90-3	267.494	oil	25	147^{2}			
8318	Octadecyl acetate		$C_{20}H_{40}O_2$	822-23-1	312.531		32.2(0.6)	208^{9}	0.8510^{30}		vs EtOH
8319	Octadecyl acrylate	Stearyl 2-propenoate	$C_{21}H_{40}O_2$	4813-57-4	324.542						s ctc, CS_2
8320	Octadecylamine	1-Octadecanamine	$C_{18}H_{39}N$	124-30-1	269.510	cry (w)	52.6(0.4)	350(4)	0.8618^{20}	1.4522^{20}	i H_2O; s EtOH, eth, bz; sl ace
8321	Octadecylbenzene		$C_{24}H_{42}$	4445-07-2	330.590		35(2)	400	0.85^{36}	1.479^{36}	
8322	Octadecylcyclohexane		$C_{24}H_{48}$	4445-06-1	336.638		41.5(0.2)	409	0.8300^{20}	1.4610^{20}	
8323	Octadecyl 3-(3,5-di-*tert*-butyl-4-hydroxyphenyl)propanoate	Irganox 1076	$C_{35}H_{62}O_3$	2082-79-3	530.865	cry (MeOH/ AcOEt)	50				
8324	Octadecyl isocyanate	1-Isocyanatooctadecane	$C_{19}H_{37}NO$	112-96-9	295.503		15.5	172^{5}			
8325	Octadecyl methacrylate	Stearyl methacrylate	$C_{22}H_{42}O_2$	32360-05-7	338.567			195^{6}	0.880^{25}	1.429^{25}	
8326	Octadecyl octadecanoate	Octadecyl stearate	$C_{36}H_{72}O_2$	2778-96-3	536.956	cry (EtOH)	60				
8327	3-(Octadecyloxy)-1,2-propanediol	Batyl alcohol	$C_{21}H_{44}O_3$	544-62-7	344.572		70.5	217^{2}			vs eth
8328	Octadecyl vinyl ether	1-(Ethenyloxy)octadecane	$C_{20}H_{40}O$	930-02-9	296.531		30	182^{3}	0.8138^{40}		sl chl
8329	1,7-Octadiene		C_8H_{14}	3710-30-3	110.197			117.1(0.9)	0.734^{20}	1.4245^{20}	
8330	1,7-Octadiyne		C_8H_{10}	871-84-1	106.165			135.5	0.8169^{21}	1.4521^{18}	s eth
8331	2,2,3,3,4,4,5,5-Octafluoro-1-pentanol		$C_5H_4F_8O$	355-80-6	232.072			140(6)	1.6647^{20}	1.3178^{20}	
8332	1,2,3,4,5,6,7,8-Octahydroanthracene		$C_{14}H_{18}$	1079-71-6	186.293	pl (al)	72.22(0.05)	307(2)	0.9703^{80}	1.5372^{80}	i H_2O; s EtOH, HOAc; vs bz; sl ctc
8333	Octahydroazocine		$C_7H_{15}N$	1121-92-2	113.201		29	52^{15}	0.896^{25}	1.4720^{20}	
8334	Octahydroindene		C_9H_{16}	496-10-6	124.223	liq	-44.4(0.6)	164(15)	0.876^{25}	1.4702^{20}	
8335	Octahydroindolizine		$C_8H_{15}N$	13618-93-4	125.212			75^{43}	0.9074^{10}	1.4748	vs eth, EtOH
8336	*trans*-Octahydro-1(2*H*)-naphthalenone		$C_{10}H_{16}O$	21370-71-8	152.233		33	122^{20}	0.986^{20}	1.4849^{21}	
8337	1,2,3,4,5,6,7,8-Octahydrophenanthrene		$C_{14}H_{18}$	5325-97-3	186.293		16.7	295	1.026^{20}	1.5569^{17}	i H_2O; s ace, bz, CS_2, HOAc
8338	*trans*-Octahydro-2*H*-quinolizine-1-methanol, (1*R*)-	Lupinine	$C_{10}H_{19}NO$	486-70-4	169.264	orth (peth)	70	270			s H_2O, EtOH, eth, bz, chl; sl peth
8339	2,2,4,4,6,6,8,8-Octamethylcyclotetrasilazane		$C_8H_{28}N_4Si_4$	1020-84-4	292.677		94.50(0.02)				
8340	Octamethylcyclotetrasiloxane		$C_8H_{24}O_4Si_4$	556-67-2	296.617		17.10(0.02)	175.4(0.9)	0.9561^{20}	1.3968^{20}	i H_2O; s ctc

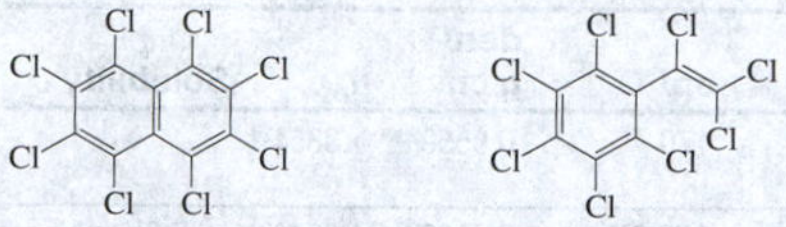

8294
Octachloronaphthalene

8295
Octachlorostyrene

$H_3C(CH_2)_{78}CH_3$

8296
Octacontane

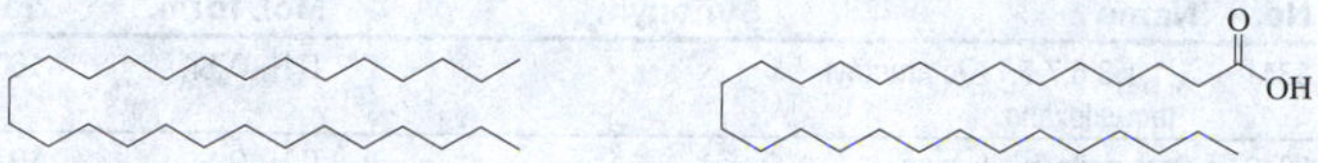

8297
Octacosane

8298
Octacosanoic acid

8299
1-Octacosanol

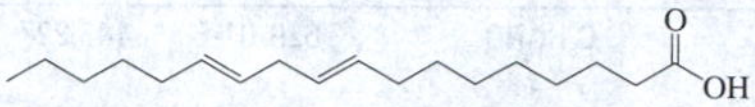

8300
trans,trans-9,12-Octadecadienoic acid

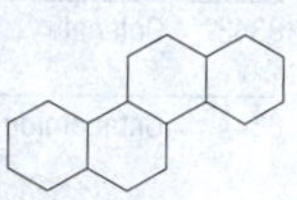

8301
Octadecahydrochrysene

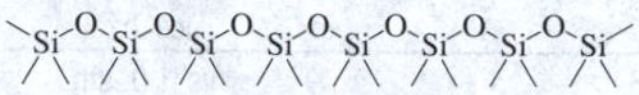

8302
Octadecamethyloctasiloxane

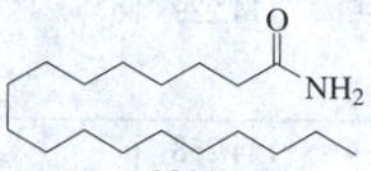

8303
Octadecanamide

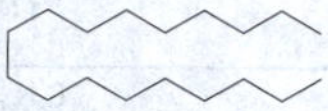

8304
Octadecane

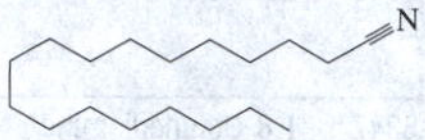

8305
Octadecanenitrile

8306
1-Octadecanethiol

8307
1-Octadecanol

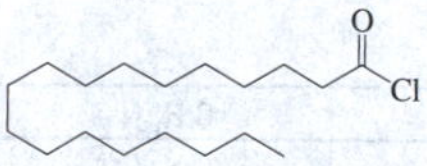

8308
Octadecanoyl chloride

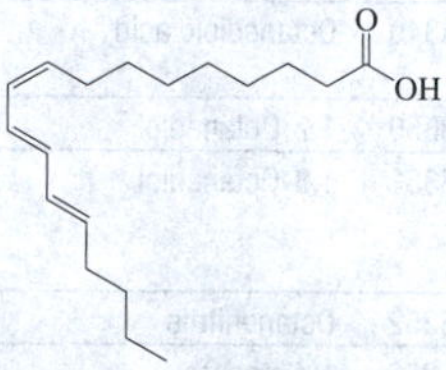

8309
cis,trans,trans-9,11,13-Octadecatrienoic acid

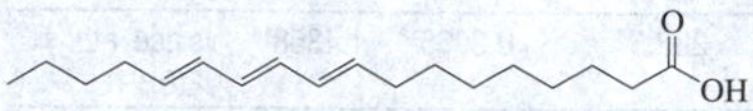

8310
trans,trans,trans-9,11,13-Octadecatrienoic acid

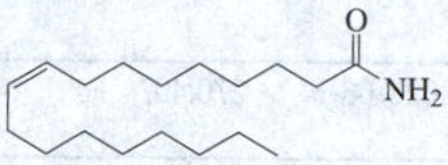

8311
cis-9-Octadecenamide

8312
1-Octadecene

8313
cis-9-Octadecenenitrile

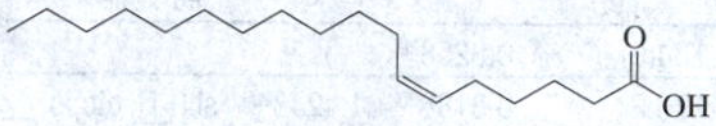

8314
cis-6-Octadecenoic acid

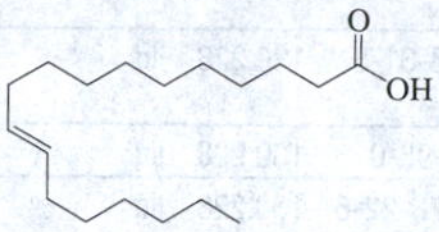

8315
trans-11-Octadecenoic acid

8316
cis-9-Octadecen-1-ol

8317
cis-9-Octadecenylamine

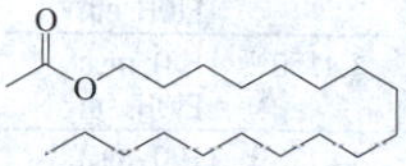

8318
Octadecyl acetate

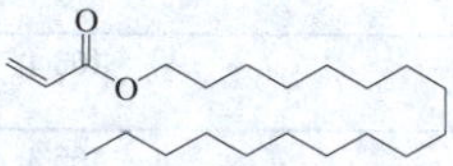

8319
Octadecyl acrylate

8320
Octadecylamine

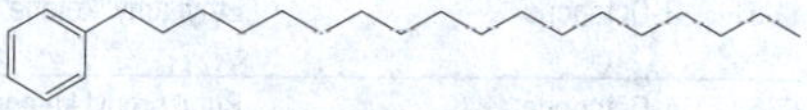

8321
Octadecylbenzene

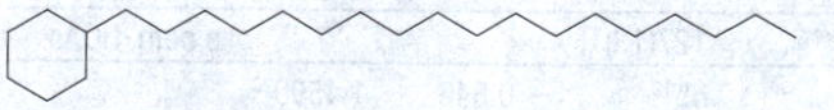

8322
Octadecylcyclohexane

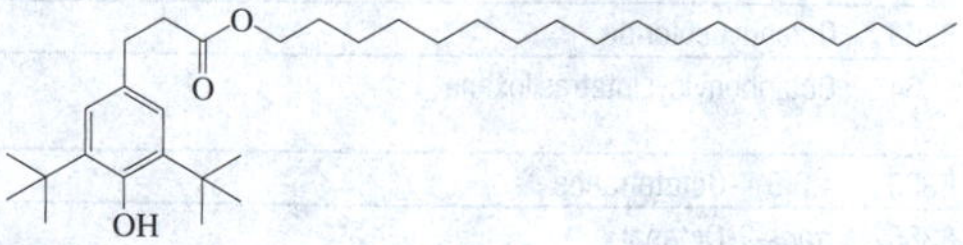

8323
Octadecyl 3-(3,5-di-*tert*-butyl-4-hydroxyphenyl)propanoate

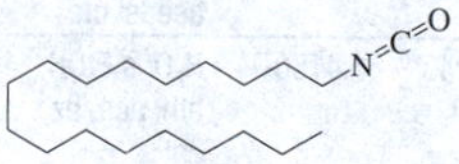

8324
Octadecyl isocyanate

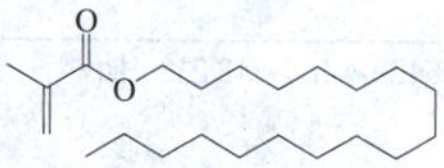

8325
Octadecyl methacrylate

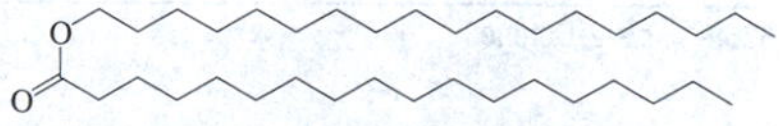

8326
Octadecyl octadecanoate

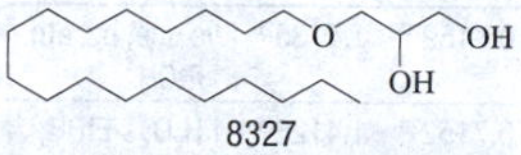

8327
3-(Octadecyloxy)-1,2-propanediol

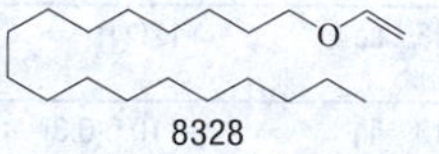

8328
Octadecyl vinyl ether

8329
1,7-Octadiene

8330
1,7-Octadiyne

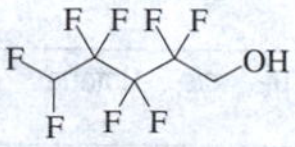

8331
2,2,3,3,4,4,5,5-Octafluoro-1-pentanol

8332
1,2,3,4,5,6,7,8-Octahydroanthracene

8333
Octahydroazocine

8334
Octahydroindene

8335
Octahydroindolizine

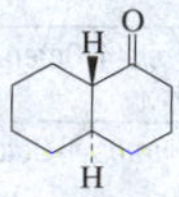

8336
trans-Octahydro-1(2*H*)-naphthalenone

8337
1,2,3,4,5,6,7,8-Octahydrophenanthrene

8338
trans-Octahydro-2*H*-quinolizine-1-methanol, (1*R*)-

8339
2,2,4,4,6,6,8,8-Octamethylcyclotetrasilazane

8340
Octamethylcyclotetrasiloxane

No.	Name	Synonym	Mol. form.	CAS Reg. No.	Mol. wt.	Physical form	mp/°C	bp/°C	den/ g cm⁻³	n_D	Solubility
8341	1,1,1,3,5,7,7,7-Octamethyltetrasiloxane		$C_8H_{26}O_3Si_4$	16066-09-4	282.632			170	0.8559^{20}	1.3854^{20}	
8342	Octamethyltrisiloxane		$C_8H_{24}O_2Si_3$	107-51-7	236.533	liq	-80	152.5(0.8)	0.8200^{20}	1.3840^{20}	sl EtOH; s bz, peth
8343	Octanal	Caprylic aldehyde	$C_8H_{16}O$	124-13-0	128.212			174(3)	0.8211^{20}	1.4217^{20}	vs ace, bz, eth, EtOH
8344	Octanamide		$C_8H_{17}NO$	629-01-6	143.227	lf, pl	105(1)	239	0.8450^{110}		sl H_2O, bz, chl; vs EtOH; s eth, ace
8345	2-Octanamine, (±)-		$C_8H_{19}N$	44855-57-4	129.244		97	163(4)	0.7744^{20}	1.4232^{25}	vs eth, EtOH
8346	Octane		C_8H_{18}	111-65-9	114.229	liq	-56.73(0.02)	125.62(0.1)	0.6986^{25}	1.3944^{25}	i H_2O; s eth; msc EtOH, ace, bz
8347	1,8-Octanediamine		$C_8H_{20}N_2$	373-44-4	144.258	pl	51.7(0.3)	225.6			vs H_2O, eth, EtOH
8348	Octanedinitrile	Suberonitrile	$C_8H_{12}N_2$	629-40-3	136.194		-4(1)	185^{15}	0.954^{25}	1.4436^{20}	
8349	Octanedioic acid	Suberic acid	$C_8H_{14}O_4$	505-48-6	174.195	lo nd or pl (w)	142.3(0.3)	345.5			i H_2O; msc eth, bz; sl DMSO
8350	1,2-Octanediol		$C_8H_{18}O_2$	1117-86-8	146.228		30.2(0.4)	131^{10}			
8351	1,8-Octanediol		$C_8H_{18}O_2$	629-41-4	146.228	nd (bz-lig), pr	59.6(0.1)	172^{20}			sl H_2O, eth, chl, lig; vs EtOH; s bz
8352	Octanenitrile	Caprylnitrile	$C_8H_{15}N$	124-12-9	125.212	liq	-53.5(0.5)	202(3)	0.8136^{20}	1.4203^{20}	vs eth
8353	1-Octanethiol	Octyl mercaptan	$C_8H_{18}S$	111-88-6	146.294	liq	-49.2(0.3)	199(1)	0.8433^{20}	1.4540^{20}	s EtOH; sl ctc
8354	Octanoic acid	Caprylic acid	$C_8H_{16}O_2$	124-07-2	144.212	liq	16.51(0.02)	240(1)	0.9073^{25}	1.4285^{20}	sl H_2O; msc EtOH, chl, CH_3CN
8355	Octanoic anhydride		$C_{16}H_{30}O_3$	623-66-5	270.407	liq	-1	282.5	0.9065^{18}	1.4358^{18}	vs ace, eth, EtOH
8356	1-Octanol	Capryl alcohol	$C_8H_{18}O$	111-87-5	130.228	liq	-14.7(0.4)	194.7(0.8)	0.8262^{25}	1.4295^{20}	i H_2O; msc EtOH, eth; s ctc
8357	2-Octanol	(±)-*sec*-Caprylic alcohol	$C_8H_{18}O$	4128-31-8	130.228	liq	-31.6	179(7)	0.8193^{20}	1.4203^{20}	sl H_2O; s EtOH, eth, ace
8358	3-Octanol		$C_8H_{18}O$	589-98-0	130.228	liq	-45	184(6)	0.8258^{20}		
8359	4-Octanol		$C_8H_{18}O$	74778-22-6	130.228	liq	-40.7	176.3	0.8186^{20}	1.4248^{20}	sl H_2O, ctc; s EtOH
8360	2-Octanone	Hexyl methyl ketone	$C_8H_{16}O$	111-13-7	128.212	liq	-20.31(0.08)	173(3)	0.820^{20}	1.4151^{20}	sl H_2O; msc EtOH, eth
8361	3-Octanone	Ethyl amyl ketone	$C_8H_{16}O$	106-68-3	128.212			166(4)	0.822^{25}	1.4150^{20}	i H_2O; msc EtOH, eth
8362	4-Octanone	Butyl propyl ketone	$C_8H_{16}O$	589-63-9	128.212			166(3)	0.8146^{25}	1.4173^{14}	i H_2O; msc EtOH, eth; s ctc
8363	Octanoyl chloride		$C_8H_{15}ClO$	111-64-8	162.657	liq	-63(1)	195(6)	0.9535^{15}	1.4335^{20}	s eth
8364	Octaphenylcyclotetrasiloxane		$C_{48}H_{40}O_4Si_4$	546-56-5	793.172	nd (bz-al, HOAc)	200.5	330^{1}			i H_2O; sl EtOH; s bz, chl, HOAc
8365	1,3,5,7-Octatetraene		C_8H_{10}	1482-91-3	106.165	cry (bz)	50	127(13)			s peth, HOAc
8366	*trans*-2-Octenal		$C_8H_{14}O$	2548-87-0	126.196	liq		85^{19}	0.846	1.4500^{20}	
8367	1-Octene	Caprylene	C_8H_{16}	111-66-0	112.213	liq	-101.66(0.05)	121.3(0.2)	0.7149^{20}	1.4087^{20}	i H_2O; msc EtOH; s eth, ace; sl ctc
8368	*cis*-2-Octene		C_8H_{16}	7642-04-8	112.213	liq	-101.3(0.8)	125.6(0.5)	0.7243^{20}	1.4150^{20}	i H_2O; s EtOH, eth, ace, bz, chl
8369	*trans*-2-Octene		C_8H_{16}	13389-42-9	112.213	liq	-88(2)	124.9(0.5)	0.7199^{20}	1.4132^{20}	i H_2O; s EtOH, eth, ace, bz; vs chl
8370	*cis*-3-Octene		C_8H_{16}	14850-22-7	112.213	liq	-126(1)	123(2)	0.7159^{20}	1.4135^{20}	vs ace, bz, eth, EtOH
8371	*trans*-3-Octene		C_8H_{16}	14919-01-8	112.213	liq	-110.0(0.3)	123.2(0.6)	0.7152^{20}	1.4126^{20}	i H_2O; s EtOH, eth, ace, bz, lig, ctc
8372	*cis*-4-Octene		C_8H_{16}	7642-15-1	112.213	liq	-119.1(0.4)	122.6(0.4)	0.7212^{20}	1.4148^{20}	vs ace, bz, eth, EtOH
8373	*trans*-4-Octene		C_8H_{16}	14850-23-8	112.213	liq	-93.8(0.1)	122.4(0.5)	0.7141^{20}	1.4114^{20}	i H_2O; s EtOH, eth, ace, bz, lig; sl ctc
8374	1-Octen-3-ol		$C_8H_{16}O$	3391-86-4	128.212			174	0.8395^{13}	1.4391^{12}	
8375	2-Octen-1-ol		$C_8H_{16}O$	22104-78-5	128.212			88^{11}	0.850^{20}	1.4470^{20}	
8376	1-Octen-3-yne		C_8H_{12}	17679-92-4	108.181			134(4)	0.7749^{20}	1.4592^{20}	vs eth
8377	Octhilinone	3(2*H*)-Isothiazolone, 2-octyl-	$C_{11}H_{19}NOS$	26530-20-1	213.340			$120^{0.01}$			

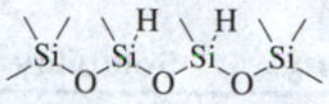
8341
1,1,1,3,5,7,7,7-Octamethyltetrasiloxane

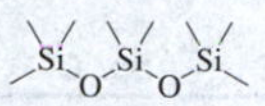
8342
Octamethyltrisiloxane

8343
Octanal

8344
Octanamide

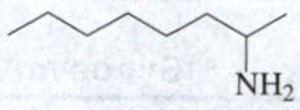
8345
2-Octanamine, (±)-

8346
Octane

8347
1,8-Octanediamine

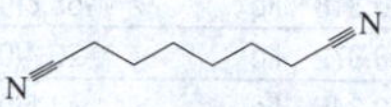
8348
Octanedinitrile

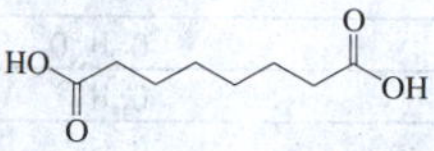
8349
Octanedioic acid

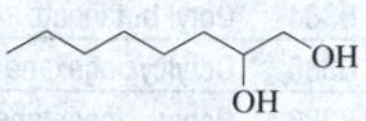
8350
1,2-Octanediol

8351
1,8-Octanediol

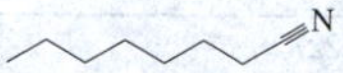
8352
Octanenitrile

8353
1-Octanethiol

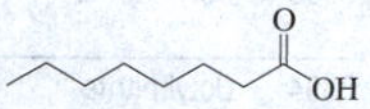
8354
Octanoic acid

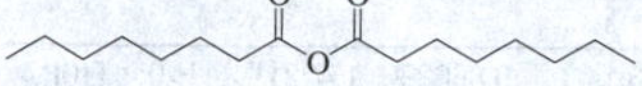
8355
Octanoic anhydride

8356
1-Octanol

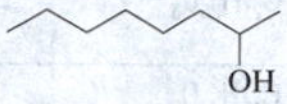
8357
2-Octanol

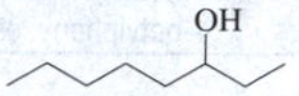
8358
3-Octanol

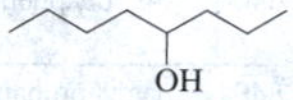
8359
4-Octanol

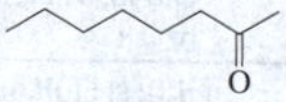
8360
2-Octanone

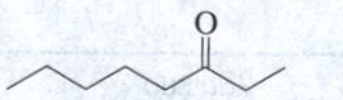
8361
3-Octanone

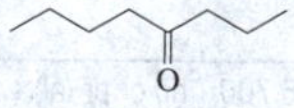
8362
4-Octanone

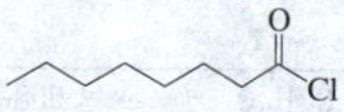
8363
Octanoyl chloride

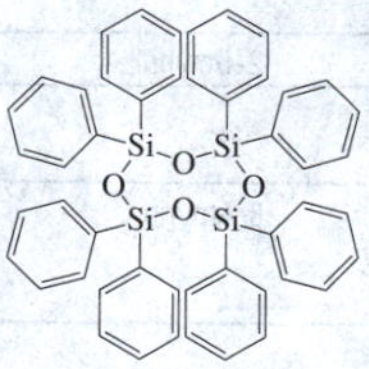
8364
Octaphenylcyclotetrasiloxane

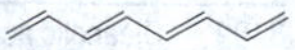
8365
1,3,5,7-Octatetraene

8366
trans-2-Octenal

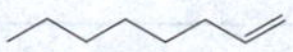
8367
1-Octene

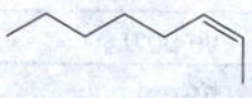
8368
cis-2-Octene

8369
trans-2-Octene

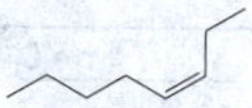
8370
cis-3-Octene

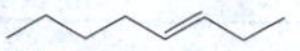
8371
trans-3-Octene

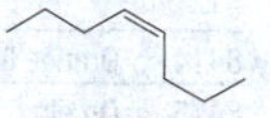
8372
cis-4-Octene

8373
trans-4-Octene

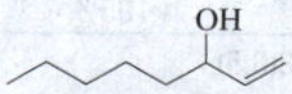
8374
1-Octen-3-ol

8375
2-Octen-1-ol

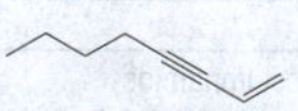
8376
1-Octen-3-yne

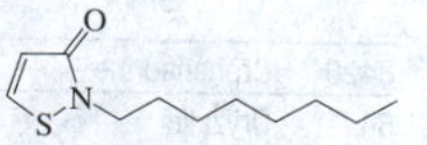
8377
Octhilinone

No.	Name	Synonym	Mol. form.	CAS Reg. No.	Mol. wt.	Physical form	mp/°C	bp/°C	den/g cm⁻³	n_D	Solubility
8378	Octyl acetate		$C_{10}H_{20}O_2$	112-14-1	172.265	liq	-38(2)	210(3)	0.8705^{20}	1.4150^{20}	i H_2O; s EtOH, eth; sl ctc
8379	Octyl acrylate	Octyl 2-propenoate	$C_{11}H_{20}O_2$	2499-59-4	184.276			227(5)	0.8810^{20}		
8380	Octylamine	1-Octanamine	$C_8H_{19}N$	111-86-4	129.244		0	178.6(0.2)	0.7826^{20}	1.4292^{20}	sl H_2O; vs EtOH, eth; s ctc
8381	Octylamine hydrochloride	1-Octanamine hydrochloride	$C_8H_{20}ClN$	142-95-0	165.705		196.5				s H_2O
8382	4-Octylaniline		$C_{14}H_{23}N$	16245-79-7	205.340		20	310	0.9128^{20}		vs eth
8383	Octylbenzene		$C_{14}H_{22}$	2189-60-8	190.325	liq	-36	263(2)	0.8562^{20}	1.4845^{20}	i H_2O; msc eth, bz
8384	Octyl butanoate		$C_{12}H_{24}O_2$	110-39-4	200.318	liq	-55.6(0.5)	239(4)	0.8629^{20}	1.4267^{15}	vs EtOH
8385	Octylcyclohexane		$C_{14}H_{28}$	1795-15-9	196.372	liq	-20.4(0.5)	263(2)	0.8138^{20}	1.4503^{20}	
8386	Octylcyclopentane		$C_{13}H_{26}$	1795-20-6	182.345	liq	-44.5(0.5)	242(15)	0.8048^{20}	1.4446^{20}	
8387	2-Octyldecanoic acid		$C_{18}H_{36}O_2$	619-39-6	284.478	nd or lf (al)	38.5	215^{13}	0.8447^{70}		vs eth, EtOH
8388	Octyldimethylamine	*N,N*-Dimethyl-1-octanamine	$C_{10}H_{23}N$	7378-99-6	157.297			192(6)			
8389	Octyl diphenyl phosphate		$C_{20}H_{27}O_4P$	115-88-8	362.399				1.09^{25}		
8390	Octyl formate		$C_9H_{18}O_2$	112-32-3	158.238	liq	-39.1(0.5)	198.8	0.8744^{20}	1.4208^{15}	i H_2O; s EtOH; msc eth; sl ctc
8391	Octyl isocyanate		$C_9H_{17}NO$	3158-26-7	155.237			78^{6}			
8392	Octyl methacrylate		$C_{12}H_{22}O_2$	2157-01-9	198.302			238(17)			
8393	Octyl nitrate		$C_8H_{17}NO_3$	629-39-0	175.226			110^{20}	0.975^{0}		sl H_2O; s EtOH, eth
8394	Octyl nitrite		$C_8H_{17}NO_2$	629-46-9	159.227			174.5	0.862^{17}	1.4127^{20}	sl H_2O; vs EtOH, eth
8395	Octyl octanoate		$C_{16}H_{32}O_2$	2306-88-9	256.424	liq	-18.1(0.3)	306.8	0.8554^{20}	1.4352^{20}	vs ace, eth, EtOH
8396	Octyloxirane		$C_{10}H_{20}O$	2404-44-6	156.265	liq		128^{95}			
8397	4-(Octyloxy)benzaldehyde		$C_{15}H_{22}O_2$	24083-13-4	234.335			$131^{0.5}$			
8398	4-Octylphenol		$C_{14}H_{22}O$	1806-26-4	206.324		43.0	169^{10}			
8399	Octyl phenyl ether	(Octyloxy)benzene	$C_{14}H_{22}O$	1818-07-1	206.324		8	284(7)	0.9131^{15}	1.4875^{20}	i H_2O; s EtOH, eth
8400	4-Octylphenyl salicylate	2-Hydroxybenzoic acid, 4-octylphenyl ester	$C_{21}H_{26}O_3$	2512-56-3	326.429	wh cry	73				
8401	Octyl propanoate		$C_{11}H_{22}O_2$	142-60-9	186.292	liq	-41.6(0.5)	226(3)	0.8663^{20}	1.4221^{15}	i H_2O; s EtOH, eth, bz; sl ctc
8402	1-Octyne	Hexylacetylene	C_8H_{14}	629-05-0	110.197	liq	-79.4(0.2)	126.2(0.3)	0.7461^{20}	1.4159^{20}	i H_2O; s EtOH, eth
8403	2-Octyne	Methylpentylacetylene	C_8H_{14}	2809-67-8	110.197	liq	-61.5(0.3)	138.0(0.2)	0.7596^{20}	1.4278^{20}	i H_2O; s EtOH, eth
8404	3-Octyne		C_8H_{14}	15232-76-5	110.197	liq	-103.9(0.4)	135.6(0.2)	0.7529^{20}	1.4250^{20}	i H_2O; s EtOH, eth
8405	4-Octyne	Dipropylacetylene	C_8H_{14}	1942-45-6	110.197	liq	-102.6(0.3)	133.5(0.4)	0.7509^{20}	1.4248^{20}	i H_2O; s EtOH, eth
8406	2-Octyn-1-ol	2-Octynol	$C_8H_{14}O$	20739-58-6	126.196		-18	98^{15}	0.8805^{20}	1.4556^{20}	vs eth
8407	Oleandrin		$C_{32}H_{48}O_9$	465-16-7	576.718	cry (EtOH)	250 dec				i H_2O; s EtOH, chl
8408	Olean-12-en-3-ol, (3β)	β-Amyrin	$C_{30}H_{50}O$	559-70-6	426.717	nd (lig or al)	197	260^{05}			i H_2O; sl EtOH, chl, lig; s eth, bz
8409	Oleanolic acid		$C_{30}H_{48}O_3$	508-02-1	456.700	nd or pr (al)	310 dec	280 sub			i H_2O; sl EtOH, eth, ace; vs py, HOAc
8410	Oleic acid	*cis*-9-Octadecenoic acid	$C_{18}H_{34}O_2$	112-80-1	282.462	liq	14(1)	360	0.8935^{20}	1.4582^{20}	i H_2O; msc EtOH, eth, ace, bz, chl, ctc
8411	Omeprazole		$C_{17}H_{19}N_3O_3S$	73590-58-6	345.416	cry (MeCN)	156				
8412	Omethoate		$C_5H_{12}NO_4PS$	1113-02-6	213.192	oil	≈135 dec		1.32^{20}	1.4987^{20}	msc H_2O; i hx
8413	Orange I	1-Naphthol Orange	$C_{16}H_{11}N_2NaO_4S$	523-44-4	350.324	red-br pow					s H_2O; sl EtOH; i bz
8414	Orange IV	Tropaeolin OO	$C_{18}H_{14}N_3NaO_3S$	554-73-4	375.377	ye pow					s H_2O
8415	Orcein			1400-62-0		br-red pow					
8416	L-Ornithine	2,5-Diaminopentanoic acid, (*S*)	$C_5H_{12}N_2O_2$	70-26-8	132.161	micro cry (al-eth)	140				vs H_2O, EtOH
8417	L-Ornithine, monohydrochloride		$C_5H_{13}ClN_2O_2$	3184-13-2	168.622	nd	215				vs H_2O
8418	Orotic acid	1,2,3,6-Tetrahydro-2,6-dioxo-4-pyrimidinecarboxylic acid	$C_5H_4N_2O_4$	65-86-1	156.097	cry (w)	345.5				sl H_2O; i os
8419	Oroxylin A	5,7-Dihydroxy-6-methoxy-2-phenyl-4*H*-1-benzopyran-4-one	$C_{16}H_{12}O_5$	480-11-5	284.263	ye nd (al)	231.5				vs ace, eth, EtOH
8420	Orphenadrine		$C_{18}H_{23}NO$	83-98-7	269.382			195^{12}			
8421	Oryzalin	Benzenesulfonamide, 4-(dipropylamino)-3,5-dinitro-	$C_{12}H_{18}N_4O_6S$	19044-88-3	346.359		142.3(0.5)				

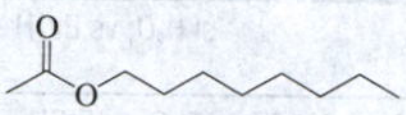
8378
Octyl acetate

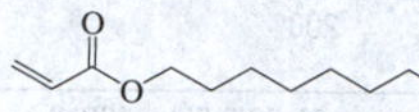
8379
Octyl acrylate

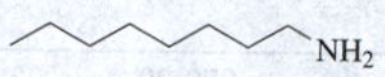
8380
Octylamine

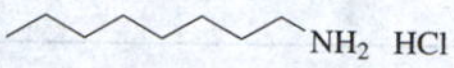
8381
Octylamine hydrochloride

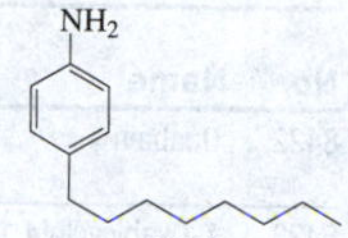
8382
4-Octylaniline

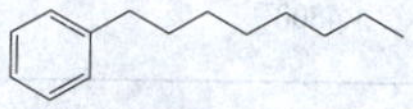
8383
Octylbenzene

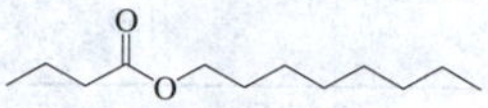
8384
Octyl butanoate

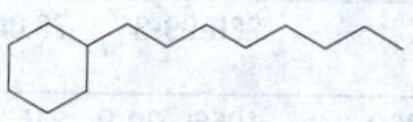
8385
Octylcyclohexane

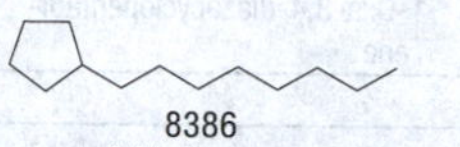
8386
Octylcyclopentane

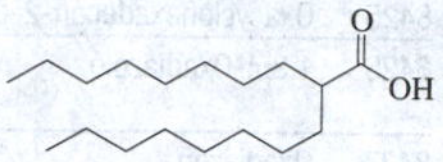
8387
2-Octyldecanoic acid

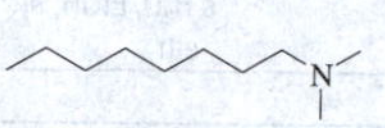
8388
Octyldimethylamine

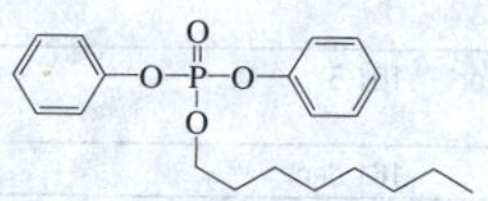
8389
Octyl diphenyl phosphate

8390
Octyl formate

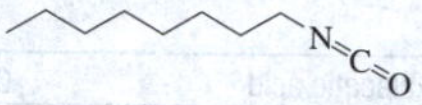
8391
Octyl isocyanate

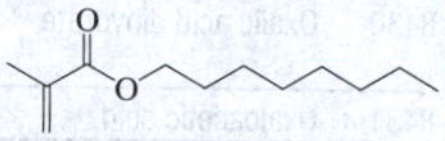
8392
Octyl methacrylate

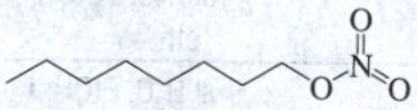
8393
Octyl nitrate

8394
Octyl nitrite

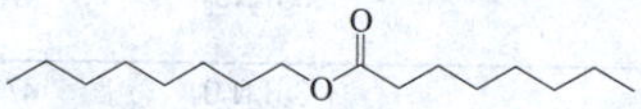
8395
Octyl octanoate

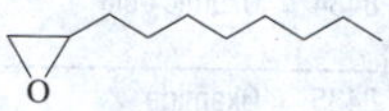
8396
Octyloxirane

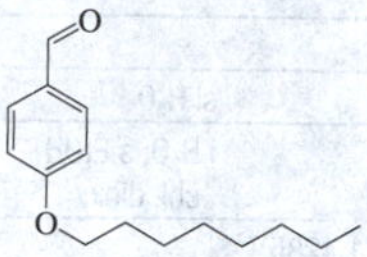
8397
4-(Octyloxy)benzaldehyde

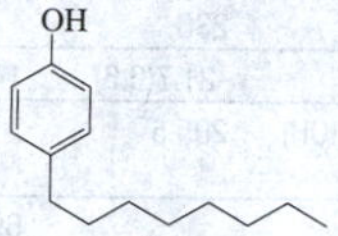
8398
4-Octylphenol

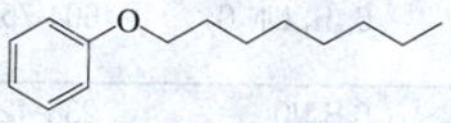
8399
Octyl phenyl ether

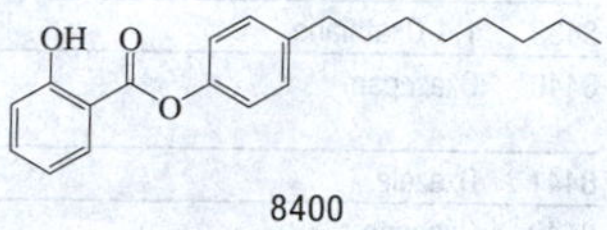
8400
4-Octylphenyl salicylate

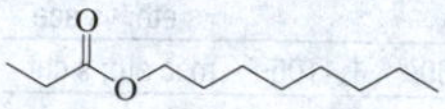
8401
Octyl propanoate

8402
1-Octyne

8403
2-Octyne

8404
3-Octyne

8405
4-Octyne

8406
2-Octyn-1-ol

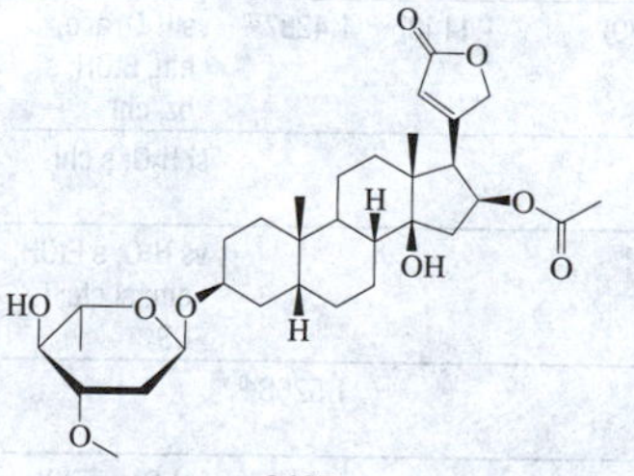
8407
Oleandrin

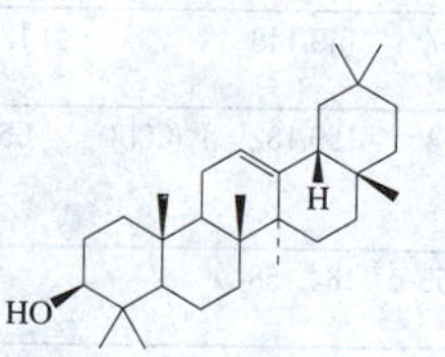
8408
Olean-12-en-3-ol, (3β)

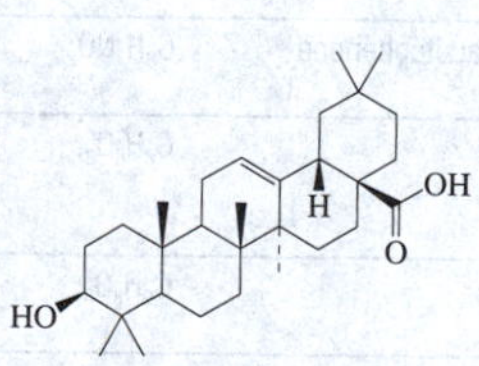
8409
Oleanolic acid

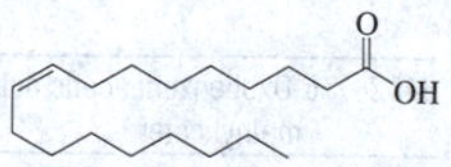
8410
Oleic acid

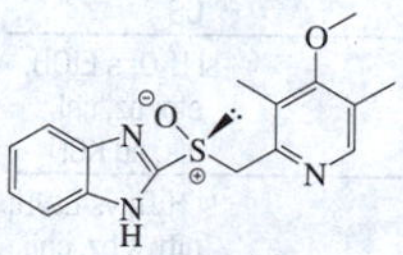
8411
Omeprazole

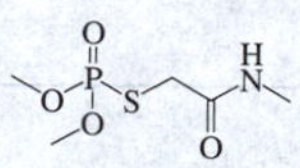
8412
Omethoate

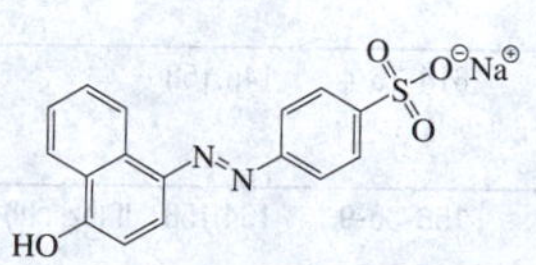
8413
Orange I

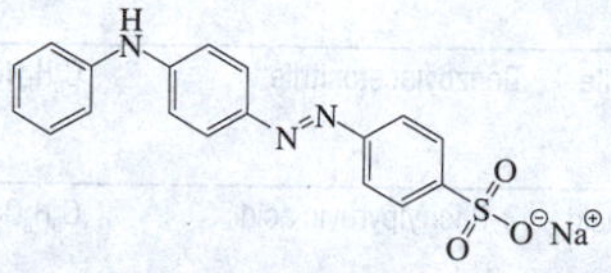
8414
Orange IV

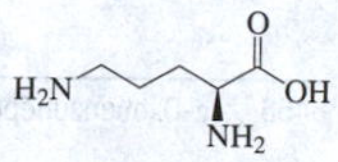
8416
L-Ornithine

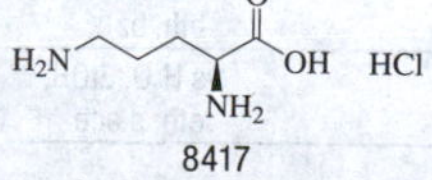
8417
L-Ornithine, monohydrochloride

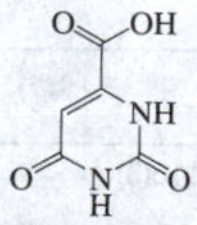
8418
Orotic acid

8419
Oroxylin A

8420
Orphenadrine

8421
Oryzalin

Organic

No.	Name	Synonym	Mol. form.	CAS Reg. No.	Mol. wt.	Physical form	mp/°C	bp/°C	den/ g cm^{-3}	n_D	Solubility
8422	Ouabain		$C_{29}H_{44}O_{12}$	630-60-4	584.652	hyg pl (+9w)	200				sl H_2O; vs EtOH
8423	7-Oxabicyclo[4.1.0]heptane		$C_6H_{10}O$	286-20-4	98.142		-35.00(0.05)	132(1)	0.9663^{20}	1.4519^{20}	i H_2O; vs EtOH, eth, ace, bz; s chl; sl ctc
8424	6-Oxabicyclo[3.1.0]hexane		C_5H_8O	285-67-6	84.117			101(6)	0.964^{25}	1.4336^{20}	
8425	Oxacyclohexadecan-2-one	Exaltolide	$C_{15}H_{28}O_2$	106-02-5	240.382	thick oil		176^{15}	0.9549^{20}	1.4708^{20}	
8426	1,3,4-Oxadiazole	1-Oxa-3,4-diazacyclopentadi-ene	$C_2H_2N_2O$	288-99-3	70.049			150		1.4300^{25}	
8427	Oxadiazon		$C_{15}H_{18}Cl_2N_2O_3$	19666-30-9	345.221		88.1(0.5)				
8428	Oxadixyl		$C_{14}H_{18}N_2O_4$	77732-09-3	278.304		104				
8429	Oxalic acid		$C_2H_2O_4$	144-62-7	90.035	orth pym or oct	189.5 dec	157 sub	1.900^{17}		s H_2O; vs EtOH; sl eth; i bz, chl, peth
8430	Oxalic acid dihydrate		$C_2H_6O_6$	6153-56-6	126.065	mcl tab or pr	101.5		1.653^{18}		s H_2O, EtOH; sl eth
8431	Oxaloacetic acid	Oxalacetic acid	$C_4H_4O_5$	328-42-7	132.072		161 dec				
8432	Oxalyl chloride	Oxalyl dichloride	$C_2Cl_2O_2$	79-37-8	126.926	liq	-16	63.5	1.4785^{20}	1.4316^{20}	s eth
8433	Oxalyl dihydrazide		$C_2H_6N_4O_2$	996-98-5	118.095	nd (w)	244.0		1.458^{22}		s H_2O; sl EtOH, eth, bz, chl
8434	Oxamic acid		$C_2H_3NO_3$	471-47-6	89.050	cry (w)	209(6)				sl H_2O; i EtOH, eth
8435	Oxamide		$C_2H_4N_2O_2$	471-46-5	88.065	nd (w)	299(3)		1.667^{20}		sl H_2O, EtOH; i eth
8436	Oxamniquine		$C_{14}H_{21}N_3O_3$	21738-42-1	279.335	ye-oran cry	149				s ace, chl, MeOH
8437	Oxamyl		$C_7H_{13}N_3O_3S$	23135-22-0	219.261		99.9(0.5)	dec	0.97^{25}		
8438	Oxandrolone		$C_{19}H_{30}O_3$	53-39-4	306.439		236				
8439	1,4-Oxathiane		C_4H_8OS	15980-15-1	104.171	liq	-21.7(0.6)	149(2)	1.1174^{20}		sl H_2O
8440	Oxazepam		$C_{15}H_{11}ClN_2O_2$	604-75-1	286.713	cry (EtOH)	205.5				i H_2O; s EtOH, chl, diox
8441	Oxazole		C_3H_3NO	288-42-6	69.062			69.5(0.2)		1.4285^{17}	
8442	Oxepane		$C_6H_{12}O$	592-90-5	100.158			118(3)	0.89^{25}	1.4400^{20}	
8443	2-Oxepanone	Caprolactone	$C_6H_{10}O_2$	502-44-3	114.142	liq	-1.02(0.04)	215	1.0761^{20}	1.4611^{20}	s EtOH, eth, ace
8444	Oxetane	Trimethylene oxide	C_3H_6O	503-30-0	58.079	liq	-97(1)	47.6(0.5)	0.8930^{25}	1.3961^{20}	msc H_2O, EtOH; s eth; vs ace
8445	2-Oxetanone	β-Propiolactone	$C_3H_4O_2$	57-57-8	72.063	liq	-33.283(0.01)	161(14)	1.1460^{20}	1.4105^{20}	msc eth; s chl
8446	3-Oxetanone		$C_3H_4O_2$	6704-31-0	72.063	unstab liq		106	1.137		
8447	Oxirane	Ethylene oxide	C_2H_4O	75-21-8	44.052	vol liq or gas	-112.46(0.05)	10.4(0.1)	0.8821^{10}	1.3597^{7}	s H_2O, EtOH, eth, ace, bz
8448	Oxiranecarboxaldehyde	Glycidaldehyde	$C_3H_4O_2$	765-34-4	72.063	liq	-62	112.5	1.1403^{20}	1.4265^{20}	
8449	Oxiranemethanol, (±)-	Glycidol	$C_3H_6O_2$	61915-27-3	74.079	liq	-45	156(9)	1.1143^{25}	1.4287^{20}	vs H_2O, ace, eth, EtOH; s bz, chl
8450	α-Oxobenzeneacetaldehyde aldoxime	Isonitrosoacetophenone	$C_8H_7NO_2$	532-54-7	149.148		129				sl H_2O; s chl
8451	α-Oxobenzeneacetic acid		$C_8H_6O_3$	611-73-4	150.132	pr (CCl_4)	65.8(0.5)	163^{15}			vs H_2O; s EtOH, eth; sl ctc; i CS_2
8452	α-Oxobenzeneacetic acid, methyl ester		$C_9H_8O_3$	15206-55-0	164.158			247		1.5268^{20}	
8453	α-Oxobenzeneacetonitrile		C_8H_5NO	613-90-1	131.132		32.5	206			i H_2O; vs EtOH, eth; sl chl
8454	γ-Oxobenzenebutanoic acid		$C_{10}H_{10}O_3$	2051-95-8	178.184	lf (dil al)	116.5				s H_2O, EtOH, eth, bz, chl, CS_2
8455	β-Oxobenzenepropanenitrile	Benzoylacetonitrile	C_9H_7NO	614-16-4	145.158		80.5	160^{10}			sl H_2O; s EtOH, eth, bz, chl, alk, aq KCN
8456	α-Oxobenzenepropanoic acid	3-Phenylpyruvic acid	$C_9H_8O_3$	156-06-9	164.158	lf (bz, chl)	157.5				sl H_2O; vs EtOH, eth; s bz, chl; i lig
8457	2-Oxo-2*H*-1-benzopyran-3-carboxylic acid	Coumarin-3-carboxylic acid	$C_{10}H_6O_4$	531-81-7	190.15	nd (w, bz)	190 dec				vs EtOH
8458	Oxobis(2,4-pentanedione)-vanadium	Vanadyl acetylacetonate	$C_{10}H_{14}O_5V$	3153-26-2	265.157	bl cry	258	$174^{0.2}$			i H_2O; s EtOH, MeOH, bz, chl
8459	2-Oxobutanoic acid		$C_4H_6O_3$	600-18-0	102.089		33	81^{16}	1.200^{17}	1.3972^{20}	vs H_2O, EtOH; sl eth
8460	4-Oxobutanoic acid		$C_4H_6O_3$	692-29-5	102.089	oil		135^{14}			s H_2O, EtOH, eth, bz
8461	2-Oxoglutaric acid	α-Ketoglutaric acid	$C_5H_6O_5$	328-50-7	146.099	cry (ace-bz)	116(3)				vs H_2O, EtOH, eth; s ace

Organic

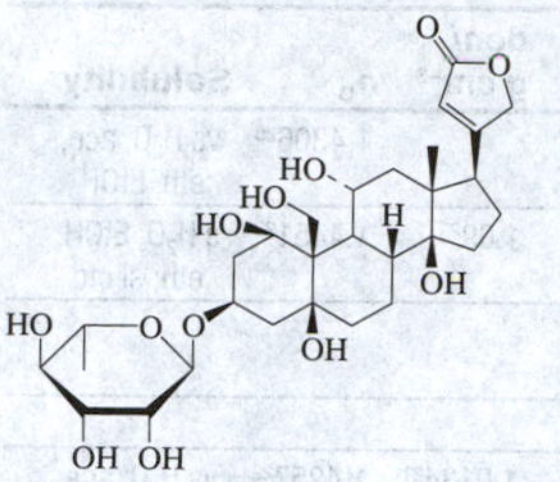
8422
Ouabain

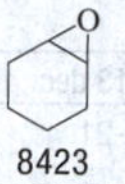
8423
7-Oxabicyclo[4.1.0]heptane

8424
6-Oxabicyclo[3.1.0]hexane

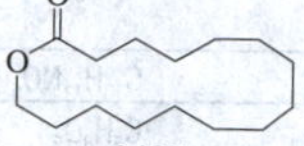
8425
Oxacyclohexadecan-2-one

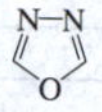
8426
1,3,4-Oxadiazole

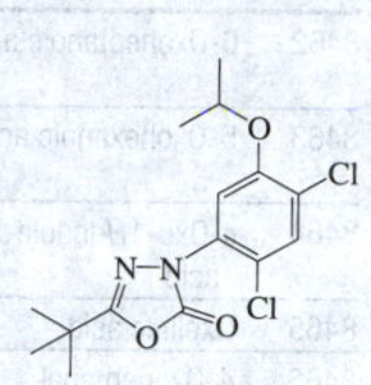
8427
Oxadiazon

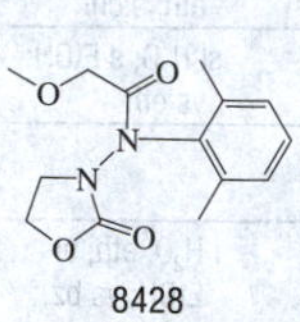
8428
Oxadixyl

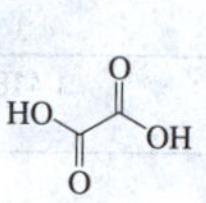
8429
Oxalic acid

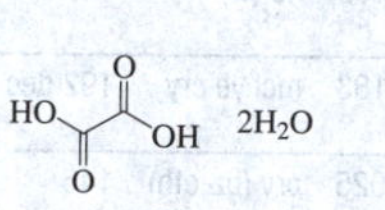
8430
Oxalic acid dihydrate

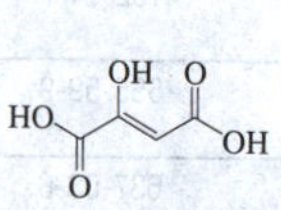
8431
Oxaloacetic acid

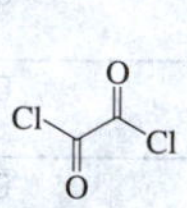
8432
Oxalyl chloride

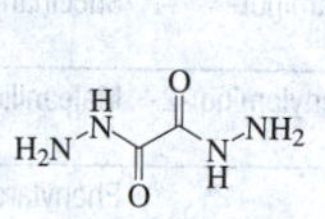
8433
Oxalyl dihydrazide

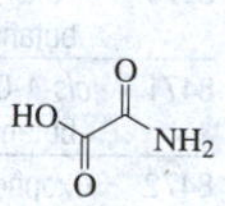
8434
Oxamic acid

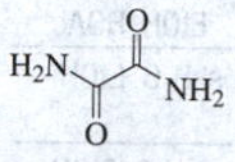
8435
Oxamide

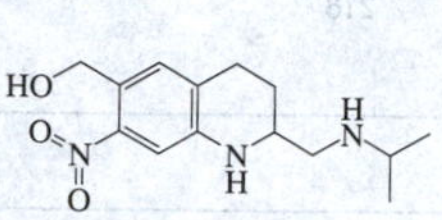
8436
Oxamniquine

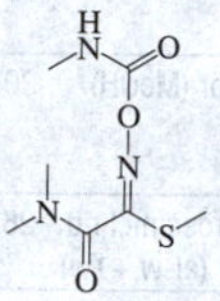
8437
Oxamyl

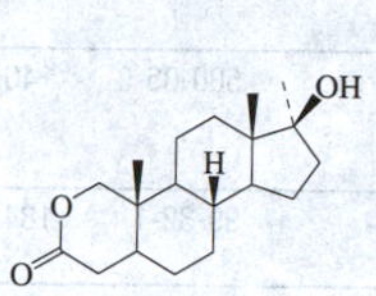
8438
Oxandrolone

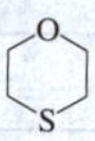
8439
1,4-Oxathiane

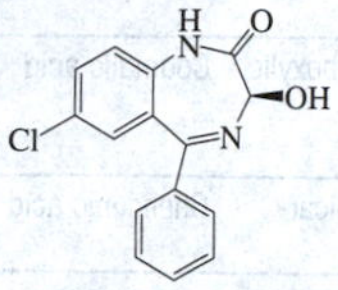
8440
Oxazepam

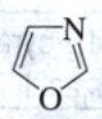
8441
Oxazole

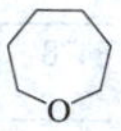
8442
Oxepane

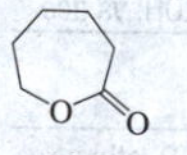
8443
2-Oxepanone

8444
Oxetane

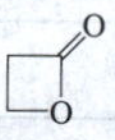
8445
2-Oxetanone

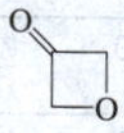
8446
3-Oxetanone

8447
Oxirane

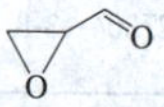
8448
Oxiranecarboxaldehyde

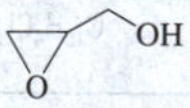
8449
Oxiranemethanol, (±)-

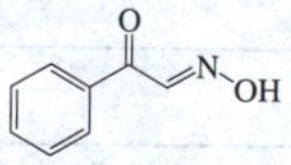
8450
α-Oxobenzeneacetaldehyde aldoxime

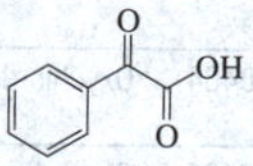
8451
α-Oxobenzeneacetic acid

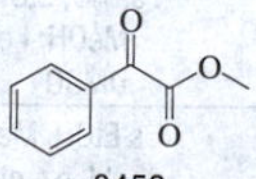
8452
α-Oxobenzeneacetic acid, methyl ester

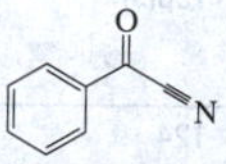
8453
α-Oxobenzeneacetonitrile

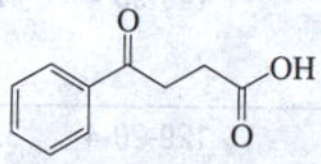
8454
γ-Oxobenzenebutanoic acid

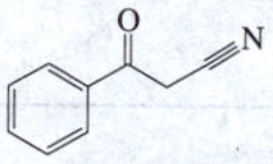
8455
β-Oxobenzenepropanenitrile

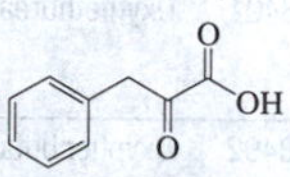
8456
α-Oxobenzenepropanoic acid

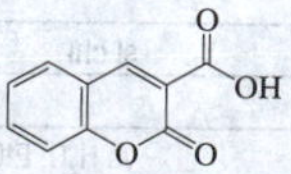
8457
2-Oxo-2*H*-1-benzopyran-3-carboxylic acid

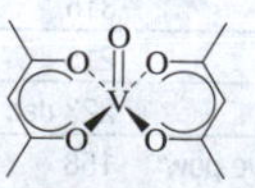
8458
Oxobis(2,4-pentanedione)vanadium

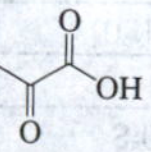
8459
2-Oxobutanoic acid

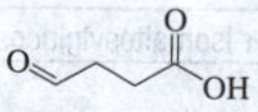
8460
4-Oxobutanoic acid

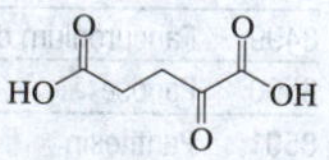
8461
2-Oxoglutaric acid

No.	Name	Synonym	Mol. form.	CAS Reg. No.	Mol. wt.	Physical form	mp/°C	bp/°C	den/g cm^{-3}	n_D	Solubility
8462	6-Oxoheptanoic acid		$C_7H_{12}O_3$	3128-07-2	144.168		40.2	251^{280}		1.4306^{25}	vs H_2O, ace, eth, EtOH
8463	5-Oxohexanoic acid		$C_6H_{10}O_3$	3128-06-1	130.141		13.5	274.5	1.09^{25}	1.4451^{20}	s H_2O, EtOH, eth; sl ctc
8464	α-Oxo-1*H*-indole-3-propanoic acid	Indole-3-pyruvic acid	$C_{11}H_9NO_3$	392-12-1	203.194	gray cry	211				
8465	Oxolinic acid		$C_{13}H_{11}NO_5$	14698-29-4	261.230	cry (DMF)	313 dec				
8466	4-Oxopentanal		$C_5H_8O_2$	626-96-0	100.117		<-21	187 dec	1.0134^{21}	1.4257^{22}	vs H_2O, ace, eth, EtOH
8467	3-Oxopentanedioic acid	Acetonedicarboxylic acid	$C_5H_6O_5$	542-05-2	146.099	nd (AcOEt)	138 dec				s H_2O, EtOH; sl eth; i bz, chl, lig
8468	2-Oxopentanoic acid		$C_5H_8O_3$	1821-02-9	116.116		6.5	179	1.0970^{14}		sl H_2O; s eth, bz, chl, lig, CS_2
8469	4-Oxopentanoic acid	Levulinic acid	$C_5H_8O_3$	123-76-2	116.116	lf or pl	33.0(0.7)	245 dec	1.1335^{20}	1.4396^{20}	vs H_2O, EtOH, eth; s chl
8470	4-Oxo-4-(phenylamino)-butanoic acid	Succinanilic acid	$C_{10}H_{11}NO_3$	102-14-7	193.199	nd (w)	148.5				sl H_2O; s EtOH; vs eth
8471	*cis*-4-Oxo-4-(phenylamino)-2-butenoic acid	Maleanilic acid	$C_{10}H_9NO_3$	555-59-9	191.183	mcl ye cry	192 dec		1.418^{30}		
8472	Oxophenylarsine	Phenylarsine oxide	C_6H_5AsO	637-03-6	168.025	cry (bz-eth) or (chl-eth)	145				i H_2O, eth; sl EtOH; vs bz, chl
8473	4-Oxo-4-phenyl-2-butenoic acid		$C_{10}H_8O_3$	583-06-2	176.169	nd or pr (tol)	99				sl H_2O, chl, lig; s EtOH, eth, tol
8474	2-Oxopropanal oxime	Isonitrosoacetone	$C_3H_5NO_2$	306-44-5	87.078	nd(CCl_4) lf (eth-peth)	69	sub	1.0744^{67}		s H_2O, eth; sl bz, ctc, chl
8475	2-Oxopropanenitrile	Acetyl cyanide	C_3H_3NO	631-57-2	69.062			92.3	0.9745^{20}	1.3764^{20}	s eth, ace, CH_3CN
8476	17-(1-Oxopropoxy)androst-4-en-3-one, (17β)	Testosterone-17-propionate	$C_{22}H_{32}O_3$	57-85-2	344.487		120				vs eth, py, EtOH
8477	2-Oxo-2*H*-pyran-5-carboxylic acid	Coumalic acid	$C_6H_4O_4$	500-05-0	140.094	pr (MeOH)	207 dec	218^{120}			sl H_2O, eth, ace; i bz, chl; s EtOH, HOAc
8478	4-Oxo-4*H*-pyran-2,6-dicarboxylic acid	Chelidonic acid	$C_7H_4O_6$	99-32-1	184.103	rose mcl nd (al-w,+1w)	262				sl H_2O, EtOH
8479	17-Oxosparteine		$C_{15}H_{24}N_2O$	489-72-5	248.364	ye to col hyg nd (peth)	84	209^{12}			vs H_2O, EtOH, eth; s chl
8480	4,4'-Oxybis(benzenesulfonyl chloride)	Diphenyl ether 4,4'-disulfonyl chloride	$C_{12}H_8Cl_2O_5S_2$	121-63-1	367.225	cry (peth)	128				
8481	4,4'-Oxybis(benzenesulfonyl hydrazide)		$C_{12}H_{14}N_4O_5S_2$	80-51-3	358.393	cry (H_2O)	164 dec				
8482	Oxybutynin		$C_{22}H_{31}NO_3$	5633-20-5	357.486	cry	114				
8483	Oxycarboxin	Carboxin *S,S*-dioxide	$C_{12}H_{13}NO_4S$	5259-88-1	267.301	pr (EtOH)	130.1(0.5)				sl H_2O; s bz, EtOH; vs ace
8484	Oxychlordane		$C_{10}H_4Cl_8O$	27304-13-8	423.762	cry (pentane)	100				
8485	Oxycodone	Dihydro-14-hydroxycodeinone	$C_{18}H_{21}NO_4$	76-42-6	315.365	rods (EtOH)	219				i H_2O, eth; s EtOH, chl
8486	Oxydemeton-methyl		$C_6H_{15}O_4PS_2$	301-12-2	246.284		<-20	$106^{0.01}$	1.289^{20}		
8487	10,10'-Oxydiphenoxarsine	10,10'-Oxybis[10*H*-phenoxarsine]	$C_{24}H_{16}As_2O_3$	58-36-6	502.225	col mcl cry	185		1.41		i H_2O; s EtOH, chl; i CH_2Cl_2
8488	Oxyfluorfen		$C_{15}H_{11}ClF_3NO_4$	42874-03-3	361.701		86.6(0.5)	358 dec	1.35^{73}		
8489	Oxymetazoline		$C_{16}H_{24}N_2O$	1491-59-4	260.374	cry (bz)	182				i eth, chl
8490	Oxymetholone		$C_{21}H_{32}O_3$	434-07-1	332.477	cry	179				
8491	Oxymethurea		$C_3H_8N_2O_3$	140-95-4	120.107	pr(al)	138(3)	149^{25}			s H_2O, EtOH, MeOH; i eth; sl DMSO
8492	Oxyphenbutazone		$C_{19}H_{20}N_2O_3$	129-20-4	324.373	cry (eth/peth)	124				s EtOH, MeOH, chl, bz, eth
8493	Oxyphenonium bromide		$C_{21}H_{34}BrNO_3$	50-10-2	428.404		191.5				vs H_2O; sl EtOH
8494	Oxytetracycline		$C_{22}H_{24}N_2O_9$	79-57-2	460.434		184.5		1.634^{20}		
8495	Oxytocin		$C_{43}H_{66}N_{12}O_{12}S_2$	50-56-6	1007.187	wh pow					s H_2O, BuOH
8496	Paclobutrazol		$C_{15}H_{20}ClN_3O$	76738-62-0	293.792	wh cry	166		1.22		i H_2O; vs ace, MeOH; s xyl, hx
8497	Palustric acid		$C_{20}H_{30}O_2$	1945-53-5	302.451	cry (MeOH)	164.5				
8498	Pamoic acid		$C_{23}H_{16}O_6$	130-85-8	388.369		315				
8499	Pancuronium dibromide		$C_{35}H_{60}Br_2N_2O_4$	15500-66-0	732.670	cry	215				sl chl
8500	Panose	4-α-Isomaltosylglucose	$C_{18}H_{32}O_{16}$	33401-87-5	504.437		223 dec				
8501	Panthesin		$C_{18}H_{32}N_2O_5S$	135-44-4	388.522	pa ye pow (al)	158				vs H_2O, EtOH

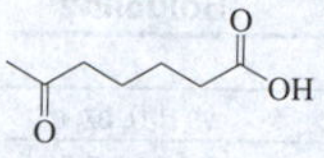
8462
6-Oxoheptanoic acid

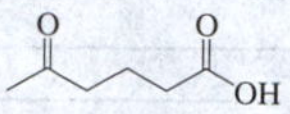
8463
5-Oxohexanoic acid

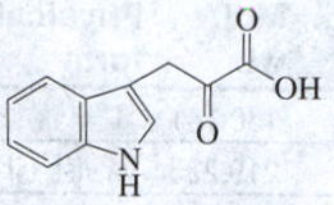
8464
α-Oxo-1*H*-indole-3-propanoic acid

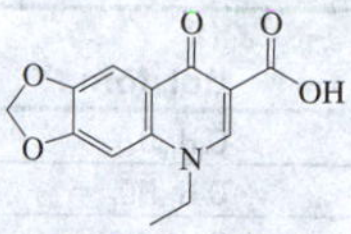
8465
Oxolinic acid

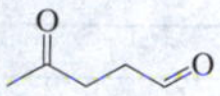
8466
4-Oxopentanal

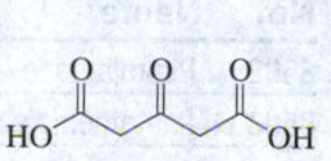
8467
3-Oxopentanedioic acid

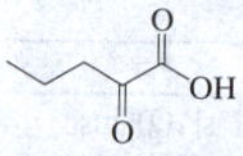
8468
2-Oxopentanoic acid

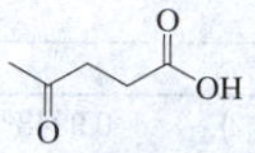
8469
4-Oxopentanoic acid

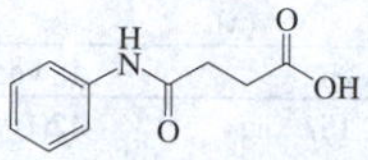
8470
4-Oxo-4-(phenylamino)butanoic acid

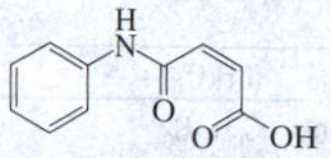
8471
cis-4-Oxo-4-(phenylamino)-2-butenoic acid

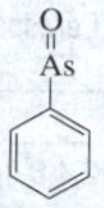
8472
Oxophenylarsine

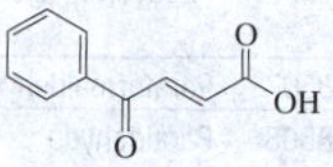
8473
4-Oxo-4-phenyl-2-butenoic acid

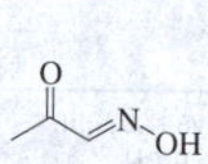
8474
2-Oxopropanal oxime

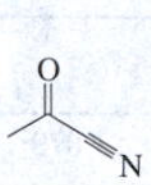
8475
2-Oxopropanenitrile

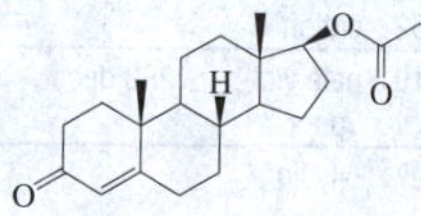
8476
17-(1-Oxopropoxy)-androst-4-en-3-one, (17β)

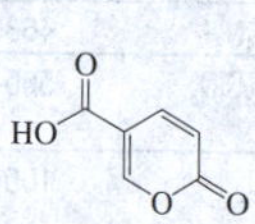
8477
2-Oxo-2*H*-pyran-5-carboxylic acid

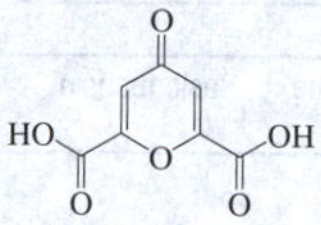
8478
4-Oxo-4*H*-pyran-2,6-dicarboxylic acid

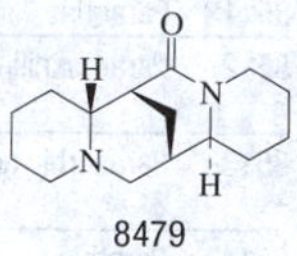
8479
17-Oxosparteine

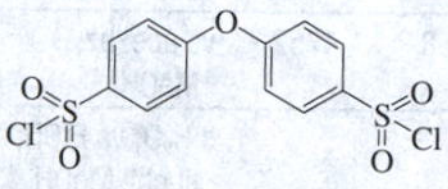
8480
4,4'-Oxybis(benzenesulfonyl chloride)

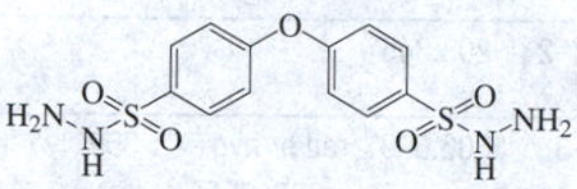
8481
4,4'-Oxybis(benzenesulfonyl hydrazide)

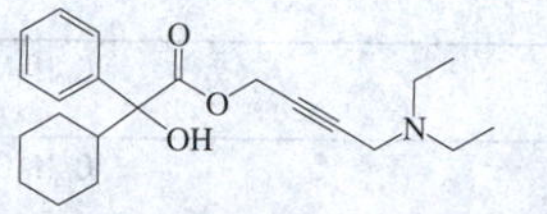
8482
Oxybutynin

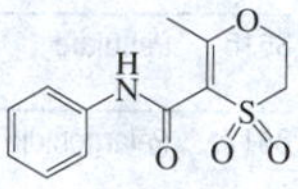
8483
Oxycarboxin

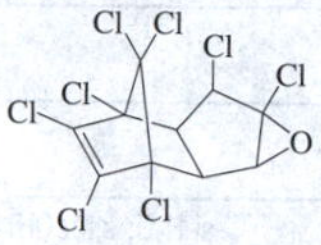
8484
Oxychlordane

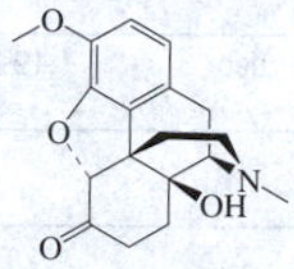
8485
Oxycodone

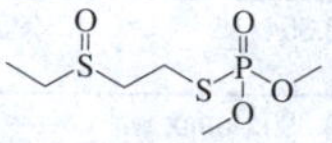
8486
Oxydemeton-methyl

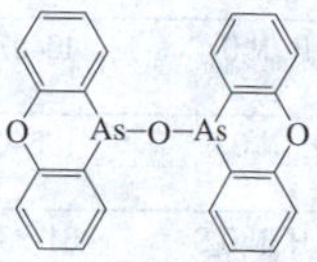
8487
10,10'-Oxydiphenoxarsine

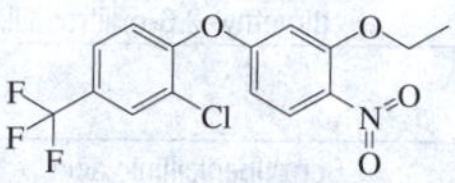
8488
Oxyfluorfen

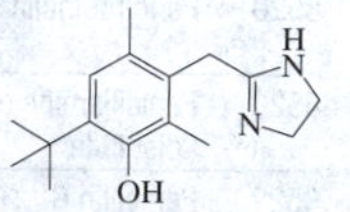
8489
Oxymetazoline

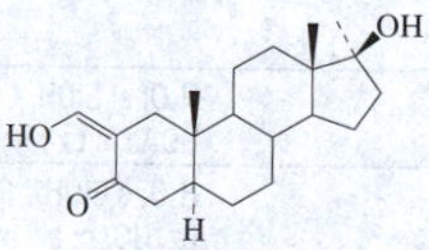
8490
Oxymetholone

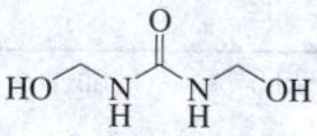
8491
Oxymethurea

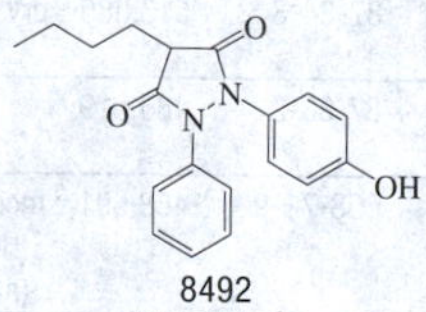
8492
Oxyphenbutazone

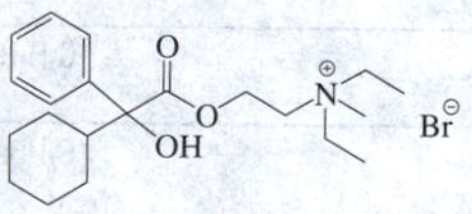
8493
Oxyphenonium bromide

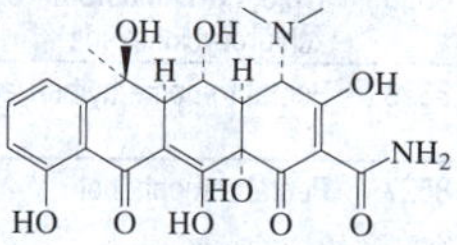
8494
Oxytetracycline

Cys-Tyr-Ile-Gln-Asn-Cys-Pro-Leu-Gly(NH_2)
8495
Oxytocin

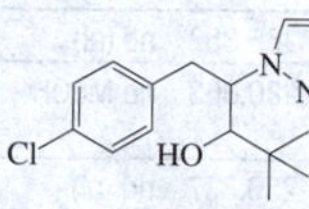
8496
Paclobutrazol

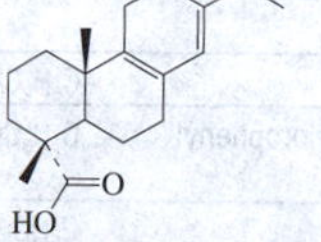
8497
Palustric acid

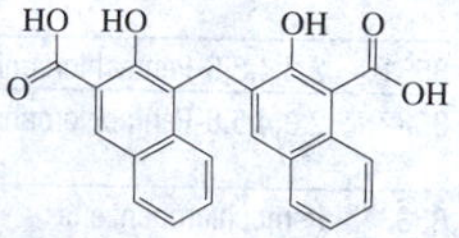
8498
Pamoic acid

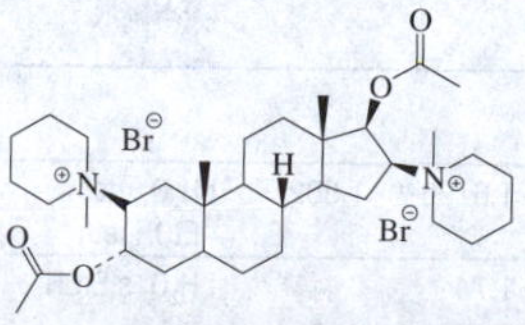
8499
Pancuronium dibromide

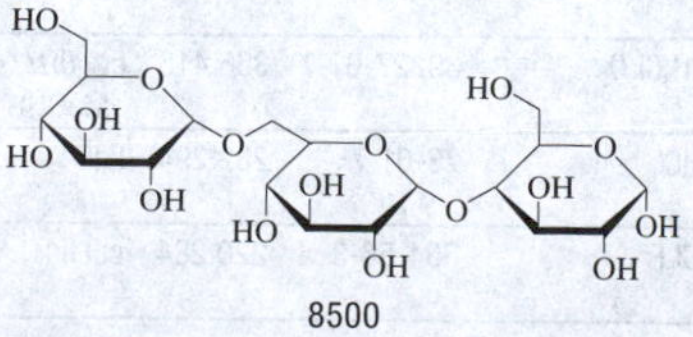
8500
Panose

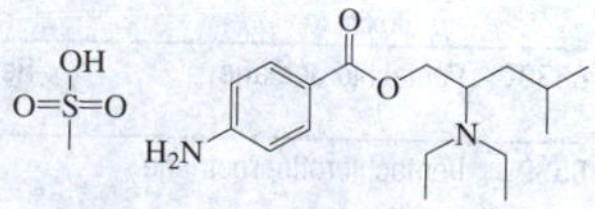
8501
Panthesin

Organic

No.	Name	Synonym	Mol. form.	CAS Reg. No.	Mol. wt.	Physical form	mp/°C	bp/°C	den/g cm^{-3}	n_D	Solubility
8502	Pantolactone		$C_6H_{10}O_3$	599-04-2	130.141		92				
8503	Pantothenic acid		$C_9H_{17}NO_5$	79-83-4	219.235	ye visc oil					vs H_2O, bz, eth
8504	Papaveraldine		$C_{20}H_{19}NO_5$	522-57-6	353.369	nd (al),cry (bz, peth)	210.5				i H_2O; sl EtOH, eth; s bz, chl
8505	Papaverine		$C_{20}H_{21}NO_4$	58-74-2	339.386	wh pr (al-eth), nd (chl-peth)	147.5	135 sub	1.337^{20}	1.625	sl H_2O; vs EtOH, chl; s ace, bz, py
8506	Papaverine hydrochloride	Cerespan	$C_{20}H_{22}ClNO_4$	61-25-6	375.847	wh mcl pr (w)	224.5				vs H_2O, EtOH
8507	Paraformaldehyde		$(CH_2O)_x$	30525-89-4	30.026		164 dec				
8508	Paraldehyde	2,4,6-Trimethyl-1,3,5-trioxane	$C_6H_{12}O_3$	123-63-7	132.157	liq	12(1)	124(2)	0.9943^{20}	1.4049^{20}	sl H_2O; msc EtOH, eth, chl
8509	Paramethadione		$C_7H_{11}NO_3$	115-67-3	157.167	liq			1.121^{25}	1.449^{25}	sl H_2O; s EtOH, chl, bz, eth
8510	Paraoxon	*O,O*-Diethyl *O*-(4-nitrophenyl) phosphate	$C_{10}H_{14}NO_6P$	311-45-5	275.195	oily liq		$161^{0.5}$	1.2683^{25}	1.5096	s eth
8511	Paraquat		$C_{12}H_{14}N_2$	4685-14-7	186.252	cation					
8512	Pararosaniline hydrochloride	Basic fuchsin	$C_{19}H_{18}ClN_3$	569-61-9	323.819	pale viol pow	269 dec				
8513	Parasorbic acid		$C_6H_8O_2$	10048-32-5	112.127	oily lig		100^{15}	1.079^{18}	1.4736^{20}	vs H_2O, eth, EtOH
8514	Parathion		$C_{10}H_{14}NO_5PS$	56-38-2	291.261	ye liq	6.1(0.9)	375	1.2681^{20}	1.5370^{25}	i H_2O; s eth, ace; sl ctc; vs EtOH, AcOEt
8515	Patchouli alcohol		$C_{15}H_{26}O$	5986-55-0	222.366		56		0.9906^{65}	1.5029^{65}	i H_2O; s EtOH, eth
8516	Pebulate		$C_{10}H_{21}NOS$	1114-71-2	203.345			142^{20}	0.9458^{20}	1.4752^{20}	vs ace, bz, MeOH
8517	Pelargonidin chloride		$C_{15}H_{11}ClO_5$	134-04-3	306.698	red br hyg (anh) pr or pl	>350				s H_2O; vs EtOH; sl chl, MeOH
8518	Pellotine		$C_{13}H_{19}NO_3$	83-14-7	237.295	pl (al, peth)	111.5				vs ace, eth, EtOH, peth
8519	Pemoline	2-Amino-5-phenyl-4(5*H*)-oxazolone	$C_9H_8N_2O_2$	2152-34-3	176.172	cry	256 dec				i H_2O, eth, ace; sl hot EtOH
8520	Pendimethalin	*N*-(1-Ethylpropyl)-3,4-dimethyl-2,6-dinitroaniline	$C_{13}H_{19}N_3O_4$	40487-42-1	281.308		54.7(0.5)	dec	1.19^{25}		
8521	Penicillamine cysteine disulfide		$C_8H_{16}N_2O_4S_2$	18840-45-4	268.354		195				
8522	Penicillin G	Benzylpenicillinic acid	$C_{16}H_{18}N_2O_4S$	61-33-6	334.390	amor wh pow					sl H_2O; s MeOH, EtOH, eth, chl, bz, ace
8523	Penicillin G procaine		$C_{29}H_{38}N_4O_6S$	54-35-3	570.700		108 dec		1.2555^{25}		s H_2O, EtOH, chl
8524	Penicillin V	Phenoxymethylpenicillin	$C_{16}H_{18}N_2O_5S$	87-08-1	350.389	cry	124 dec				sl H_2O; s os
8525	1,2,3,4,5-Pentabromo-6-chlorocyclohexane		$C_6H_6Br_5Cl$	87-84-3	513.085	cry	204				
8526	Pentabromomethylbenzene		$C_7H_3Br_5$	87-83-2	486.619		288		2.97^{17}		i H_2O; sl EtOH, HOAc; s bz
8527	Pentabromophenol		C_6HBr_5O	608-71-9	488.591	mcl pr (HOAc) nd (al)	229.5	sub			i H_2O; s EtOH, bz, HOAc; sl eth
8528	1,1,1,3,3-Pentabromo-2-propanone	Pentabromoacetone	C_3HBr_5O	79-49-2	452.559	nd (w, al) pr (eth)	79.5	sub			i H_2O; vs EtOH, eth, ace, chl
8529	Pentac	Dienochlor	$C_{10}Cl_{10}$	2227-17-0	474.637	tan cry (peth)	122				
8530	Pentacene	Benzo[*b*]naphthacene	$C_{22}H_{14}$	135-48-8	278.346	ye grn nd or lf (xyl)	271(2)				i H_2O; sl bz; s $PhNO_2$
8531	2,3,4,5,6-Pentachloroaniline		$C_6H_2Cl_5N$	527-20-8	265.352	nd (al)	233.9(0.5)				vs eth, EtOH, lig
8532	2,3,4,5,6-Pentachloroanisole	Methyl pentachlorophenyl ether	$C_7H_3Cl_5O$	1825-21-4	280.363	nd MeOH	108.5				
8533	Pentachlorobenzene		C_6HCl_5	608-93-5	250.337	nd (al)	84.2(0.6)	279(4)	1.8342^{16}		i H_2O, EtOH; sl eth, bz, chl, CS_2
8534	Pentachlorobenzenethiol	Pentachlorophenyl mercaptan	C_6HCl_5S	133-49-3	282.402		231.5				
8535	2,3,4,5,6-Pentachlorobiphenyl		$C_{12}H_5Cl_5$	18259-05-7	326.433	nd (peth)	124.4(0.5)				i H_2O
8536	2,2',4,5,5'-Pentachlorobiphenyl		$C_{12}H_5Cl_5$	37680-73-2	326.433	cry (EtOH)	76.9(0.5)				i H_2O
8537	1,2,3,4,7-Pentachlorodibenzo-*p*-dioxin		$C_{12}H_3Cl_5O_2$	39227-61-7	356.416	cry (bz/MeOH)	195				
8538	Pentachloroethane	Refrigerant 120	C_2HCl_5	76-01-7	202.294	liq	-29.0(0.2)	161(4)	1.6796^{20}	1.5025^{20}	i H_2O; msc EtOH, eth
8539	Pentachlorofluoroethane		C_2Cl_5F	354-56-3	220.284	col liq	101(4)	138	1.74^{25}		i H_2O; s EtOH, eth

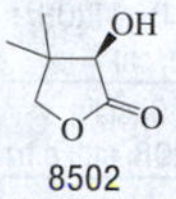
8502
Pantolactone

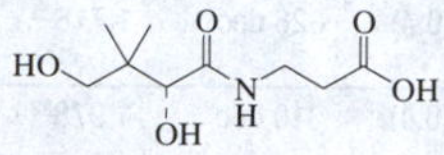
8503
Pantothenic acid

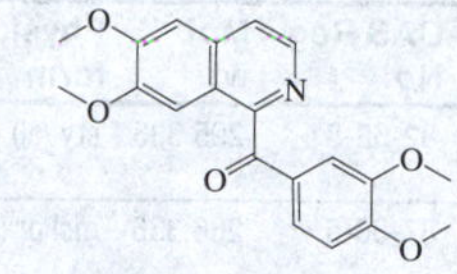
8504
Papaveraldine

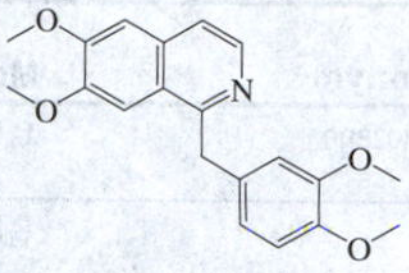
8505
Papaverine

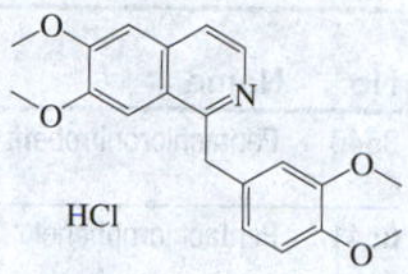
8506
Papaverine hydrochloride

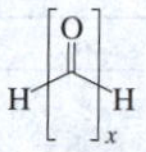
8507
Paraformaldehyde

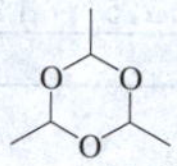
8508
Paraldehyde

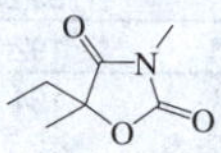
8509
Paramethadione

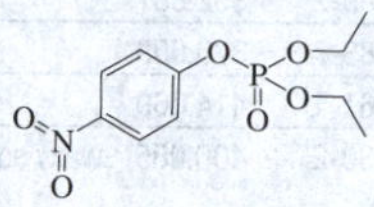
8510
Paraoxon

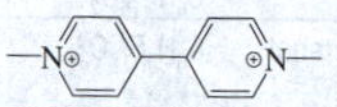
8511
Paraquat

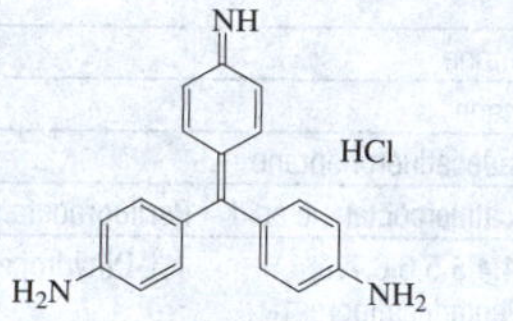
8512
Pararosaniline hydrochloride

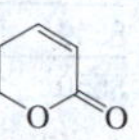
8513
Parasorbic acid

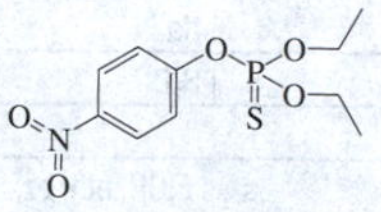
8514
Parathion

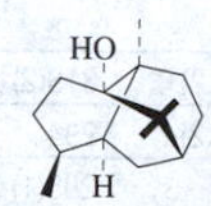
8515
Patchouli alcohol

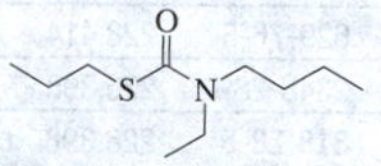
8516
Pebulate

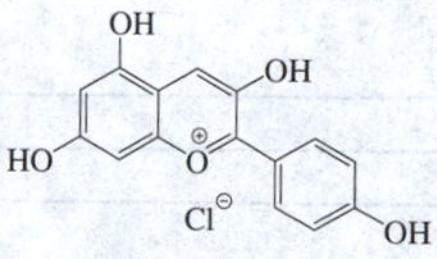
8517
Pelargonidin chloride

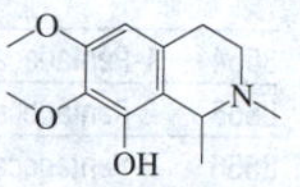
8518
Pellotine

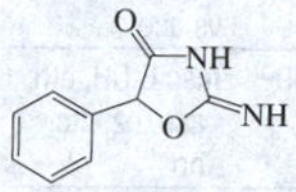
8519
Pemoline

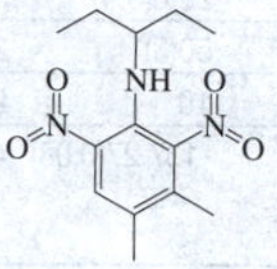
8520
Pendimethalin

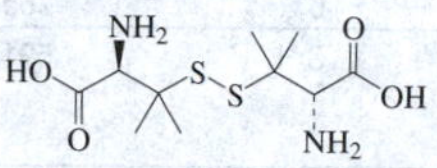
8521
Penicillamine cysteine disulfide

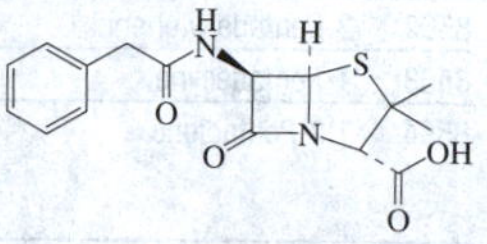
8522
Penicillin G

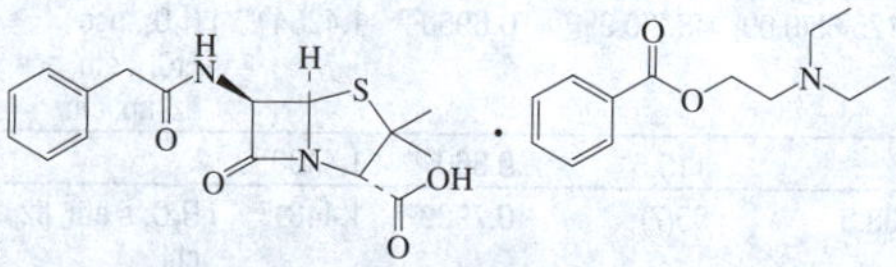
8523
Penicillin G procaine

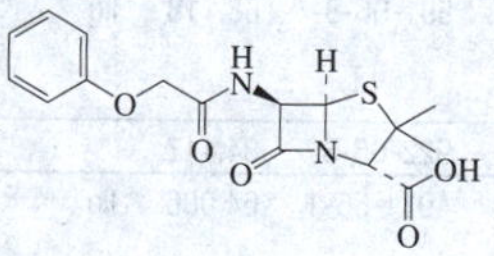
8524
Penicillin V

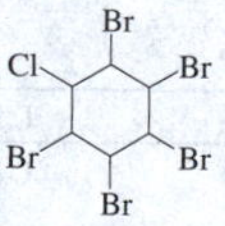
8525
1,2,3,4,5-Pentabromo-6-chlorocyclohexane

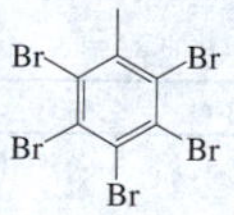
8526
Pentabromomethylbenzene

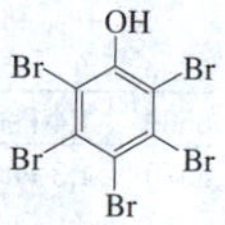
8527
Pentabromophenol

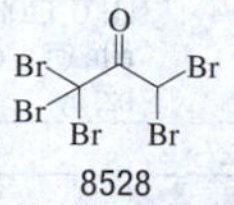
8528
1,1,1,3,3-Pentabromo-2-propanone

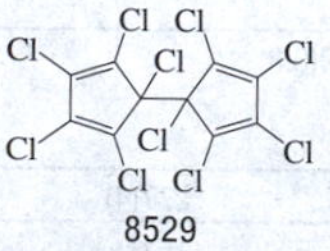
8529
Pentac

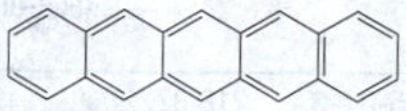
8530
Pentacene

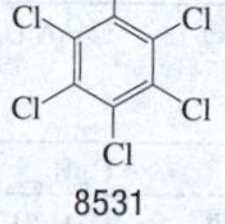
8531
2,3,4,5,6-Pentachloroaniline

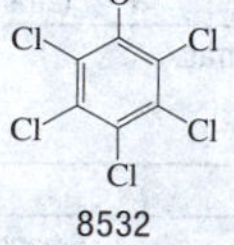
8532
2,3,4,5,6-Pentachloroanisole

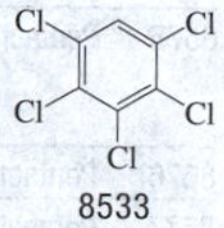
8533
Pentachlorobenzene

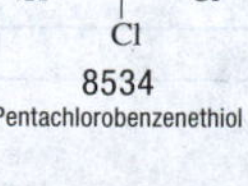
8534
Pentachlorobenzenethiol

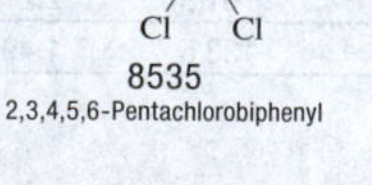
8535
2,3,4,5,6-Pentachlorobiphenyl

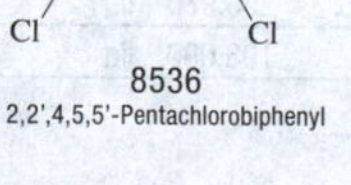
8536
2,2',4,5,5'-Pentachlorobiphenyl

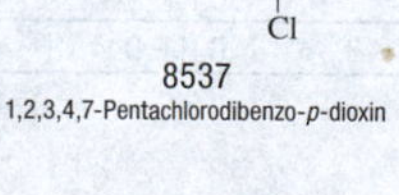
8537
1,2,3,4,7-Pentachlorodibenzo-*p*-dioxin

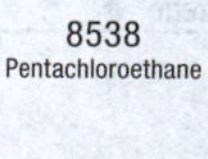
8538
Pentachloroethane

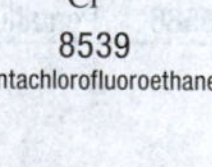
8539
Pentachlorofluoroethane

No.	Name	Synonym	Mol. form.	CAS Reg. No.	Mol. wt.	Physical form	mp/°C	bp/°C	den/ g cm^{-3}	n_D	Solubility
8540	Pentachloronitrobenzene	Quintozene	$C_6Cl_5NO_2$	82-68-8	295.335	cry (al)	144.7(0.4)	328 dec	1.718^{25}		i H_2O; sl EtOH; s bz, chl
8541	Pentachlorophenol		C_6HCl_5O	87-86-5	266.336	mcl pr (al + 1w) nd (bz)	189.5(0.4)	310 dec	1.978^{22}		i H_2O; sl lig; vs EtOH, eth; s bz
8542	1,1,2,2,3-Pentachloropropane		$C_3H_3Cl_5$	16714-68-4	216.321			181^{500}	1.633^{25}	1.5098^{25}	
8543	1,1,2,3,3-Pentachloro-1-propene		C_3HCl_5	1600-37-9	214.305			185	1.6317^{34}	1.5313^{20}	vs eth
8544	Pentachloropyridine		C_5Cl_5N	2176-62-7	251.326		124.0(0.5)	280			vs bz, EtOH, lig
8545	2,3,4,5,6-Pentachlorotoluene		$C_7H_3Cl_5$	877-11-2	264.364	nd (bz, peth)	224(2)	301			sl EtOH, eth, CS_2; s bz, tol, peth
8546	Pentacontane		$C_{50}H_{102}$	6596-40-3	703.345		91.7(0.6)	575.0			
8547	Pentacosane		$C_{25}H_{52}$	629-99-2	352.681		53.3(0.4)	401.9	0.8012^{20}	1.4491^{20}	s bz, chl
8548	1*H*-Pentadecafluoroheptane		C_7HF_{15}	375-83-7	370.059			95(2)	1.725^{25}	1.2690^{25}	
8549	Pentadecafluorooctanoic acid	Perfluorooctanoic acid	$C_8HF_{15}O_2$	335-67-1	414.069		54.3	192(1)			
8550	2,2,3,3,4,4,5,5,6,6,7,7,8,8,8-Pentadecafluoro-1-octanol	1,1-Dihydroperfluorooctanol	$C_8H_3F_{15}O$	307-30-2	400.085	waxy solid	47	164			
8551	Pentadecanal		$C_{15}H_{30}O$	2765-11-9	226.398	nd	24(2)	185^{25}			vs ace, eth, EtOH
8552	Pentadecane		$C_{15}H_{32}$	629-62-9	212.415		9.95(0.02)	270.6(0.4)	0.7685^{20}	1.4315^{20}	i H_2O; vs EtOH, eth
8553	Pentadecanoic acid	Pentadecylic acid	$C_{15}H_{30}O_2$	1002-84-2	242.398	pl (dil al, HOAc) cry (peth)	52.52(0.04)	257^{100}	0.8423^{80}	1.4254^{80}	i H_2O; vs EtOH, ace; s eth; sl tfa
8554	1-Pentadecanol		$C_{15}H_{32}O$	629-76-5	228.414		43.8(0.2)	318(2)	0.8347^{25}		i H_2O
8555	2-Pentadecanone		$C_{15}H_{30}O$	2345-28-0	226.398		39.0(0.2)	294	0.8182^{39}		
8556	8-Pentadecanone		$C_{15}H_{30}O$	818-23-5	226.398	cry (al)	42(1)	295(1)	0.8180^{39}		s EtOH, eth, bz, ctc, chl
8557	1-Pentadecene		$C_{15}H_{30}$	13360-61-7	210.399	liq	-3.8(0.1)	268.4(0.4)	0.7764^{20}	1.4389^{20}	i H_2O; s ace
8558	Pentadecylamine	Pentadecanamine	$C_{15}H_{33}N$	2570-26-5	227.430		36.5(0.4)	312(5)	0.8104^{20}	1.4480^{20}	vs eth, EtOH
8559	Pentadecylbenzene		$C_{21}H_{36}$	2131-18-2	288.511		12.0(0.4)	373	0.8548^{20}	1.4815^{20}	
8560	3-Pentadecyl-1,2-benzenediol	3-Pentadecylcatechol	$C_{21}H_{36}O_2$	492-89-7	320.510	nd (to, peth)	59.5				vs bz, eth, EtOH
8561	Pentadecylcyclohexane		$C_{21}H_{42}$	6006-95-7	294.558		25.0(0.4)	371(13)	0.8267^{20}	1.4588^{20}	
8562	3-Pentadecylphenol		$C_{21}H_{36}O$	501-24-6	304.510	nd (peth)	50(3)	230^{8}			vs ace, bz, EtOH
8563	1-Pentadecyne		$C_{15}H_{28}$	765-13-9	208.383		10	280(10)	0.7928^{20}	1.4419^{20}	vs ace
8564	1,2-Pentadiene	Ethylallene	C_5H_8	591-95-7	68.118	liq	-137.27(0.08)	44(2)	0.6926^{20}	1.4209^{20}	msc EtOH, eth, ace, bz, ctc, hp
8565	*cis*-1,3-Pentadiene	*cis*-Piperylene	C_5H_8	1574-41-0	68.118	liq	-140.81(0.09)	44.0(0.7)	0.6910^{20}	1.4363^{20}	msc EtOH, eth, ace, bz, ctc, hp
8566	*trans*-1,3-Pentadiene	*trans*-Piperylene	C_5H_8	2004-70-8	68.118	liq	-87.5(0.2)	42.0(0.3)	0.6710^{25}	1.4301^{20}	
8567	1,4-Pentadiene		C_5H_8	591-93-5	68.118	vol liq or gas	-148.3(0.3)	25.9(0.3)	0.6608^{20}	1.3888^{20}	i H_2O; vs EtOH, eth, ace, bz
8568	2,3-Pentadiene	1,3-Dimethylallene	C_5H_8	591-96-8	68.118	liq	-125.65(0.09)	48.2(0.9)	0.6950^{20}	1.4284^{20}	i H_2O; msc EtOH, eth, ace, bz, hp, ctc
8569	1,4-Pentadien-3-ol		C_5H_8O	922-65-6	84.117			115.5	0.860^{23}	1.4400^{17}	
8570	1,3-Pentadiyne	Methyldiacetylene	C_5H_4	4911-55-1	64.086	liq	-38.5	55(7)	0.7909^{20}	1.4431^{21}	i H_2O; s eth, bz, chl
8571	Pentaerythritol		$C_5H_{12}O_4$	115-77-5	136.147	cry (dil HCl)	258	sub		1.548	s H_2O; i eth, bz
8572	Pentaerythritol tetraacetate	2,2-Bis[(acetyloxy)methyl]-1,3-propanediol diacetate	$C_{13}H_{20}O_8$	597-71-7	304.293	tetr nd (w, bz)	83.5		1.273^{18}		s H_2O; vs EtOH, eth
8573	Pentaerythritol tetrakis(2-mercaptoacetate)		$C_{13}H_{20}O_8S_4$	10193-99-4	432.553	liq		250^{1}	1.385^{25}	1.5470^{20}	
8574	Pentaerythritol tetramethacrylate	Tetramethylolmethane tetramethacrylate	$C_{21}H_{28}O_8$	3253-41-6	408.442		53.5				
8575	Pentaerythritol tetranitrate		$C_5H_8N_4O_{12}$	78-11-5	316.138	tetr (ace) pr (ace-al)	140.9(0.8)		1.773^{20}		sl H_2O, EtOH, eth; vs ace; s bz, py
8576	Pentaethylbenzene		$C_{16}H_{26}$	605-01-6	218.377		<-20	277(4)	0.8971^{19}	1.5127^{20}	
8577	Pentaethyl tantalate	Ethanol, tantalum(5+) salt	$C_{10}H_{25}O_5Ta$	6074-84-6	406.251			151^{1}			
8578	2,3,4,5,6-Pentafluoroaniline		$C_6H_2F_5N$	771-60-8	183.079		34(1)	153.5			
8579	Pentafluorobenzaldehyde		C_7HF_5O	653-37-2	196.074		20	167		1.4506^{20}	
8580	Pentafluorobenzene		C_6HF_5	363-72-4	168.064	liq	-47.3(0.2)	85(3)	1.514^{25}	1.3905^{20}	
8581	Pentafluorobenzenethiol		C_6HF_5S	771-62-0	200.129	liq	-24	143	1.501^{25}	1.4645^{20}	
8582	Pentafluorobenzoic acid		$C_7HF_5O_2$	602-94-8	212.074		103.3(0.5)	220			
8583	Pentafluorobenzonitrile		C_7F_5N	773-82-0	193.074		3(2)	162(7)	1.563^{20}	1.4402^{25}	
8584	Pentafluoroethane	HFC-125	C_2HF_5	354-33-6	120.021	col gas	-100.6	-48.1			
8585	Pentafluoroiodobenzene		C_6F_5I	827-15-6	293.960	liq	-29	161(3)	2.212^{20}	1.4950^{25}	
8586	Pentafluoromethoxybenzene		$C_7H_3F_5O$	389-40-2	198.090	liq	-30(1)	138.5	1.493^{20}	1.4087^{20}	

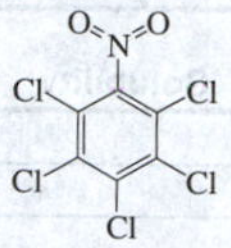
8540 Pentachloronitrobenzene

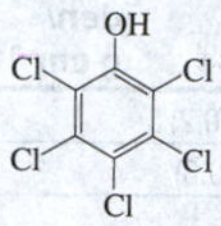
8541 Pentachlorophenol

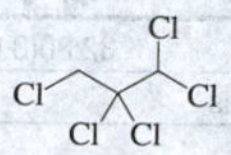
8542 1,1,2,2,3-Pentachloropropane

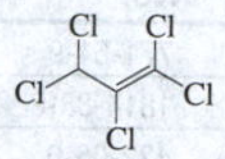
8543 1,1,2,3,3-Pentachloro-1-propene

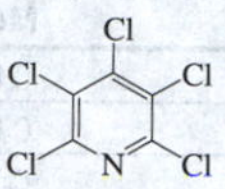
8544 Pentachloropyridine

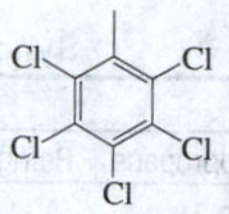
8545 2,3,4,5,6-Pentachlorotoluene

$H_3C(CH_2)_{48}CH_3$
8546 Pentacontane

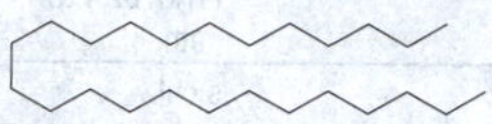
8547 Pentacosane

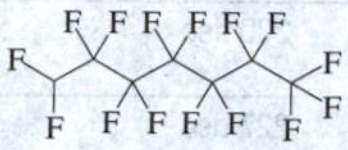
8548 1*H*-Pentadecafluoroheptane

8549 Pentadecafluorooctanoic acid

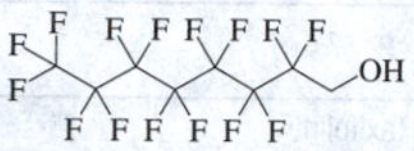
8550 2,2,3,3,4,4,5,5,6,6,7,7,8,8,8-Pentadecafluoro-1-octanol

8551 Pentadecanal

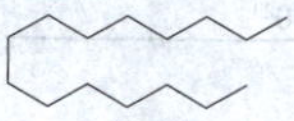
8552 Pentadecane

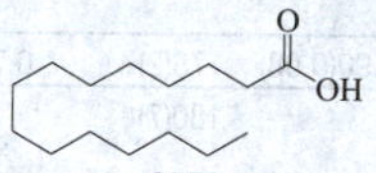
8553 Pentadecanoic acid

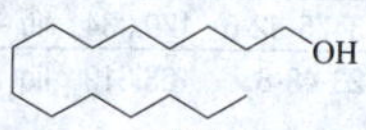
8554 1-Pentadecanol

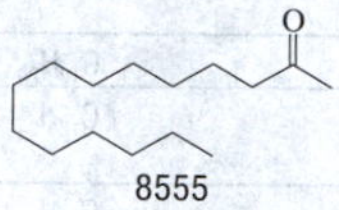
8555 2-Pentadecanone

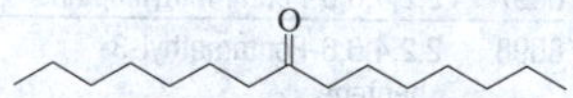
8556 8-Pentadecanone

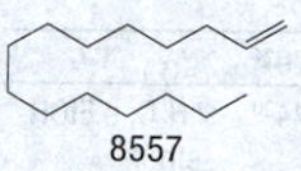
8557 1-Pentadecene

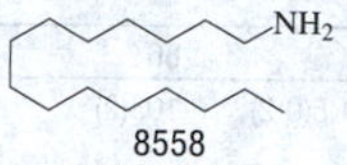
8558 Pentadecylamine

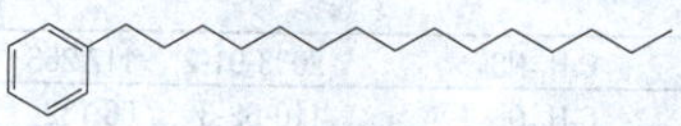
8559 Pentadecylbenzene

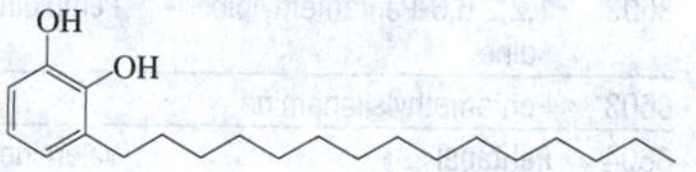
8560 3-Pentadecyl-1,2-benzenediol

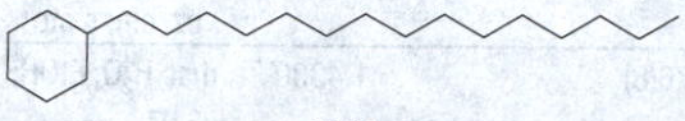
8561 Pentadecylcyclohexane

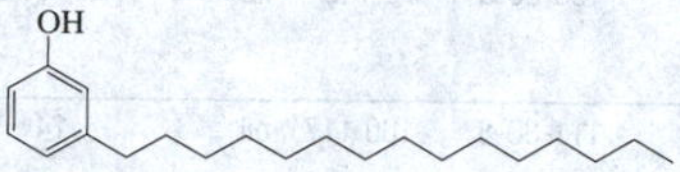
8562 3-Pentadecylphenol

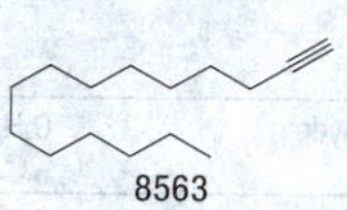
8563 1-Pentadecyne

8564 1,2-Pentadiene

8565 *cis*-1,3-Pentadiene

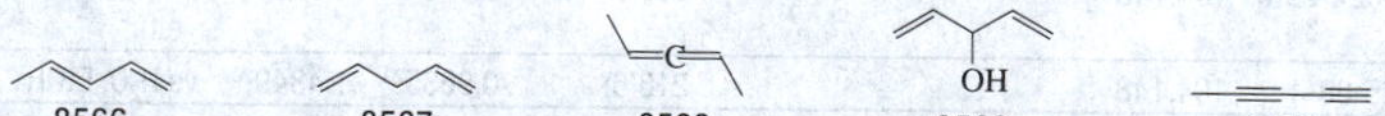
8566 *trans*-1,3-Pentadiene

8567 1,4-Pentadiene

8568 2,3-Pentadiene

8569 1,4-Pentadien-3-ol

8570 1,3-Pentadiyne

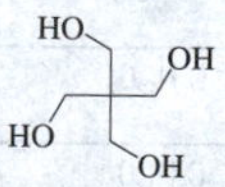
8571 Pentaerythritol

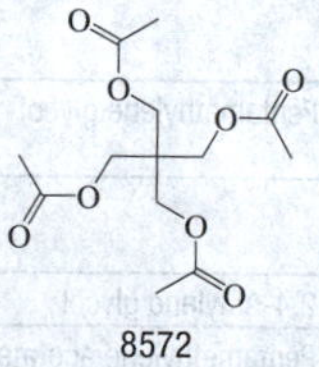
8572 Pentaerythritol tetraacetate

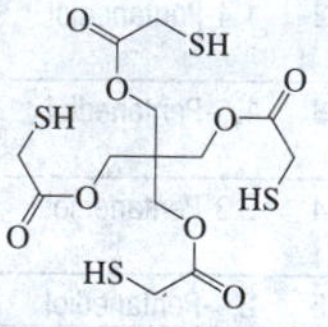
8573 Pentaerythritol tetrakis(2-mercaptoacetate)

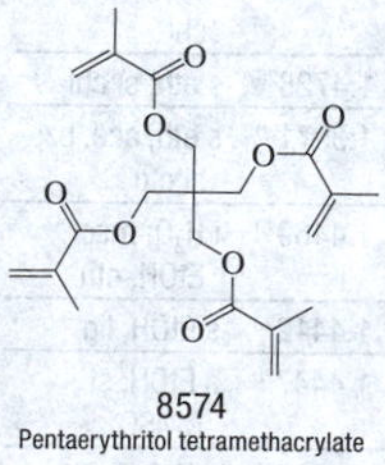
8574 Pentaerythritol tetramethacrylate

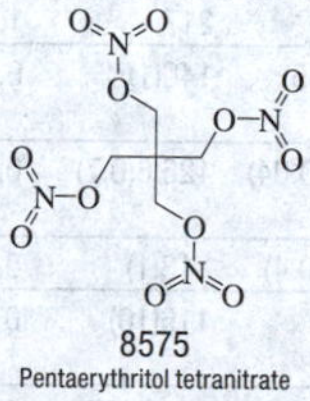
8575 Pentaerythritol tetranitrate

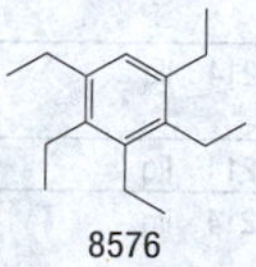
8576 Pentaethylbenzene

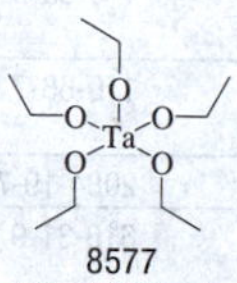
8577 Pentaethyl tantalate

8578 2,3,4,5,6-Pentafluoroaniline

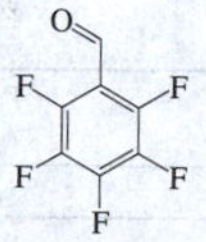
8579 Pentafluorobenzaldehyde

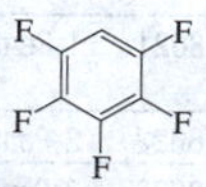
8580 Pentafluorobenzene

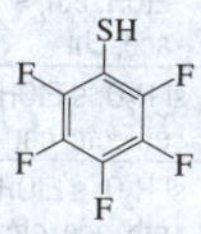
8581 Pentafluorobenzenethiol

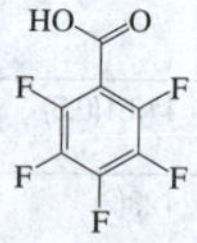
8582 Pentafluorobenzoic acid

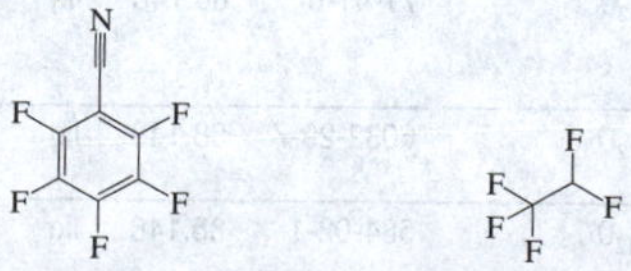
8583 Pentafluorobenzonitrile

8584 Pentafluoroethane

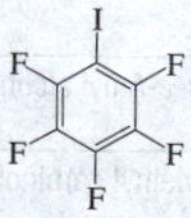
8585 Pentafluoroiodobenzene

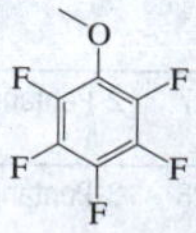
8586 Pentafluoromethoxybenzene

No.	Name	Synonym	Mol. form.	CAS Reg. No.	Mol. wt.	Physical form	mp/°C	bp/°C	den/ g cm^{-3}	n_D	Solubility
8587	Pentafluorophenol		C_6HF_5O	771-61-9	184.063		32.80(0.04)	145.1(0.2)		1.4263^{20}	
8588	1,1,1,2,2-Pentafluoropropane	Refrigerant 245cb	$C_3H_3F_5$	1814-88-6	134.048	col gas		-18.0(0.3)			
8589	2,2,3,3,3-Pentafluoro-1-propanol		$C_3H_3F_5O$	422-05-9	150.047			26^{50}			
8590	2,3,4,5,6-Pentafluorotoluene		$C_7H_3F_5$	771-56-2	182.091	liq	-29.79(0.05)	117(5)	1.440^{20}	1.4016^{25}	
8591	1,1,2,4,4-Pentafluoro-3-(trifluoromethyl)-1,3-butadiene		C_5F_8	384-04-3	212.041			39	1.527^{0}	1.3000^{0}	vs ace, bz, eth
8592	Pentagastrin		$C_{37}H_{49}N_7O_9S$	5534-95-2	767.892	col nd	230 dec				i H_2O, bz, EtOH, eth
8593	*trans*-3,3',4',5,7-Pentahydroxyflavanone, (±)-	Taxifolin	$C_{15}H_{12}O_7$	480-18-2	304.252		227 dec				s chl
8594	Pentamethonium bromide		$C_{11}H_{28}Br_2N_2$	541-20-8	348.161		301				sl H_2O
8595	Pentamethylbenzene		$C_{11}H_{16}$	700-12-9	148.245	pr (al)	54(2)	232	0.917^{20}	1.527^{20}	i H_2O; vs EtOH, bz; s chl
8596	2,4,6,8,10-Pentamethylcyclopentasiloxane		$C_5H_{20}O_5Si_5$	6166-86-5	300.638	liq	-108	169(6)	0.9985^{20}	1.3912^{20}	
8597	2,2,4,6,6-Pentamethylheptane		$C_{12}H_{26}$	13475-82-6	170.334	liq	-66.90(0.08)	178(2)	0.7463^{20}	1.4440^{20}	
8598	2,2,4,6,6-Pentamethyl-3-heptene		$C_{12}H_{24}$	123-48-8	168.319	liq		180(7)			
8599	2,2,3,3,4-Pentamethylpentane		$C_{10}H_{22}$	16747-44-7	142.282	liq	-36.5(0.2)	166(4)	0.7767^{25}	1.4361^{20}	
8600	2,2,3,4,4-Pentamethylpentane		$C_{10}H_{22}$	16747-45-8	142.282	liq	-38.8(0.2)	159(4)	0.7636^{25}	1.4307^{20}	
8601	Pentamethylphenol		$C_{11}H_{16}O$	2819-86-5	164.244	nd (al, peth, ace)	128	267			i H_2O; s EtOH
8602	1,2,2,6,6-Pentamethylpiperidine	Pempidine	$C_{10}H_{21}N$	79-55-0	155.281			147	0.8580^{0}	1.4550^{21}	
8603	Pentamethylsilanamine		$C_5H_{15}NSi$	2083-91-2	117.266			86	0.7400^{20}	1.4379^{24}	
8604	Pentanal	Valeraldehyde	$C_5H_{10}O$	110-62-3	86.132	liq	-81.5(0.2)	103(2)	0.8095^{20}	1.3944^{20}	sl H_2O; s EtOH, eth
8605	Pentanamide		$C_5H_{11}NO$	626-97-1	101.147	mcl pl (peth, al)	104(2)	225	0.8735^{110}	1.4183^{110}	vs H_2O, EtOH, eth; sl chl
8606	3-Pentanamine		$C_5H_{13}N$	616-24-0	87.164			89(5)	0.7487^{20}	1.4063^{20}	s EtOH; sl chl
8607	Pentane		C_5H_{12}	109-66-0	72.149	liq	-129.67(0.04)	36.06(0.07)	0.6262^{20}	1.3575^{20}	sl H_2O; msc EtOH, eth, ace, bz, chl; s ctc
8608	Pentanedial	Glutaraldehyde	$C_5H_8O_2$	111-30-8	100.117	oil	-14	176(8)		1.4330^{25}	msc H_2O, EtOH; s bz
8609	1,5-Pentanediamine	Cadaverine	$C_5H_{14}N_2$	462-94-2	102.178		11.8(0.4)	178(6)	0.873^{25}	1.463^{20}	s H_2O, EtOH; sl eth
8610	Pentanedinitrile	Glutaronitrile	$C_5H_6N_2$	544-13-8	94.115	liq	-28.9(0.1)	286	0.9911^{15}	1.4295^{20}	vs EtOH, chl
8611	1,2-Pentanediol, (±)-		$C_5H_{12}O_2$	91049-43-3	104.148			209	0.9723^{20}	1.4397^{19}	
8612	1,4-Pentanediol		$C_5H_{12}O_2$	626-95-9	104.148			202	0.9883^{20}	1.4452^{23}	vs H_2O, EtOH, chl
8613	1,5-Pentanediol	Pentamethylene glycol	$C_5H_{12}O_2$	111-29-5	104.148	liq	-20(7)	241(2)	0.9914^{20}	1.4494^{20}	s H_2O, EtOH; sl eth, bz
8614	2,3-Pentanediol		$C_5H_{12}O_2$	42027-23-6	104.148			187.5	0.9798^{19}	1.4412^{25}	s H_2O, EtOH; sl eth
8615	2,4-Pentanediol	2,4-Amylene glycol	$C_5H_{12}O_2$	625-69-4	104.148			218(6)	0.9635^{20}	1.4349^{20}	vs H_2O, EtOH
8616	1,5-Pentanediol diacetate	Pentamethylene acetate	$C_9H_{16}O_4$	6963-44-6	188.221		2	244(9)	1.0296^{20}	1.4261^{19}	
8617	2,3-Pentanedione	Acetylpropionyl	$C_5H_8O_2$	600-14-6	100.117	dk ye liq		109.9(0.4)	0.9565^{19}	1.4014^{19}	s H_2O; msc EtOH, eth, ace
8618	2,4-Pentanedione	Acetylacetone	$C_5H_8O_2$	123-54-6	100.117	liq	-18.3(0.2)	140.7(0.5)	0.9721^{25}	1.4494^{20}	vs H_2O; msc EtOH, eth, ace, chl
8619	Pentanedioyl dichloride		$C_5H_6Cl_2O_2$	2873-74-7	169.006			217	1.324^{20}	1.4728^{20}	s eth; sl chl
8620	Pentanenitrile	Valeronitrile	C_5H_9N	110-59-8	83.132	liq	-96.2(0.5)	140(1)	0.8008^{20}	1.3971^{20}	s eth, ace, bz; sl ctc
8621	1-Pentanethiol	Pentyl mercaptan	$C_5H_{12}S$	110-66-7	104.214	liq	-75.69(0.04)	126.6(0.7)	0.850^{20}	1.4469^{20}	i H_2O; msc EtOH, eth
8622	2-Pentanethiol	*sec*-Pentyl mercaptan	$C_5H_{12}S$	2084-19-7	104.214	liq	-169.0(0.4)	113(1)	0.8327^{20}	1.4412^{20}	s EtOH, lig
8623	3-Pentanethiol	3-Pentyl mercaptan	$C_5H_{12}S$	616-31-9	104.214	liq	-110.8	116(10)	0.8410^{20}	1.4447^{20}	s EtOH; sl DMSO
8624	Pentanoic acid	Valeric acid	$C_5H_{10}O_2$	109-52-4	102.132	liq	-33.63(0.05)	186.1(0.3)	0.9339^{25}	1.4085^{20}	s H_2O, EtOH, eth; sl ctc
8625	Pentanoic anhydride		$C_{10}H_{18}O_3$	2082-59-9	186.248	liq	-56.1(0.7)	244(17)	0.924^{20}	1.4171^{26}	vs eth, EtOH
8626	1-Pentanol	Amyl alcohol	$C_5H_{12}O$	71-41-0	88.148	liq	-77.58(0.04)	137.6(0.4)	0.8144^{20}	1.4101^{20}	sl H_2O; msc EtOH, eth; s ace, chl
8627	2-Pentanol	*sec*-Amyl alcohol	$C_5H_{12}O$	6032-29-7	88.148	liq	-73	119.1(0.5)	0.8094^{20}	1.4053^{20}	sl H_2O; s EtOH, eth, ctc, chl
8628	3-Pentanol	Diethyl carbinol	$C_5H_{12}O$	584-02-1	88.148	liq	-69.9(0.4)	123(2)	0.8203^{20}	1.4104^{20}	sl H_2O; s EtOH, eth, ace, ctc

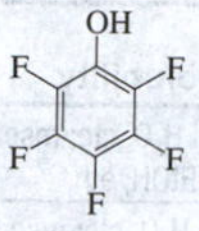
8587
Pentafluorophenol

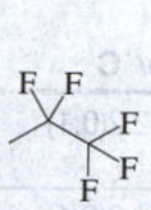
8588
1,1,1,2,2-Pentafluoropropane

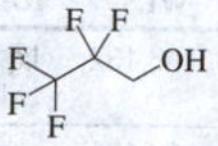
8589
2,2,3,3,3-Pentafluoro-1-propanol

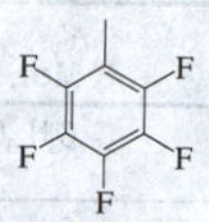
8590
2,3,4,5,6-Pentafluorotoluene

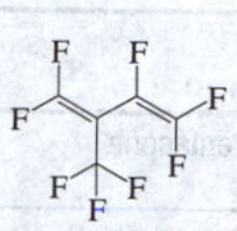
8591
1,1,2,4,4-Pentafluoro-3-(trifluoromethyl)-1,3-butadiene

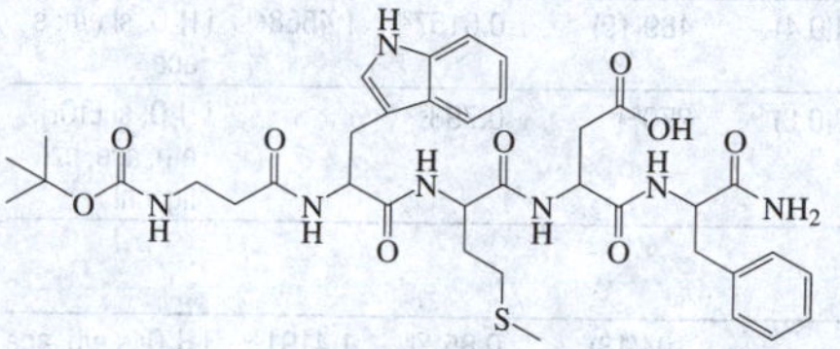
8592
Pentagastrin

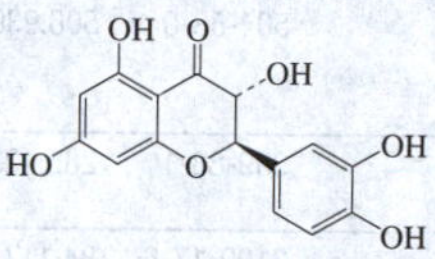
8593
trans-3,3',4',5,7-Pentahydroxyflavanone, (±)-

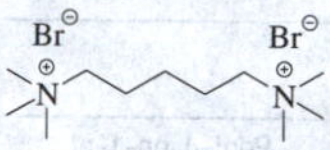
8594
Pentamethonium bromide

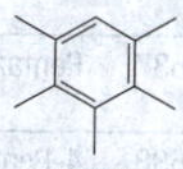
8595
Pentamethylbenzene

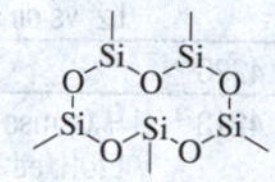
8596
2,4,6,8,10-Pentamethylcyclopentasiloxane

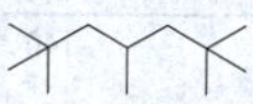
8597
2,2,4,6,6-Pentamethylheptane

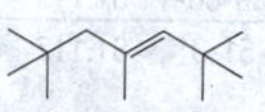
8598
2,2,4,6,6-Pentamethyl-3-heptene

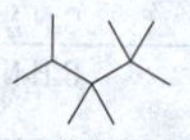
8599
2,2,3,3,4-Pentamethylpentane

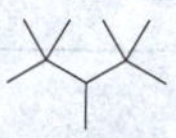
8600
2,2,3,4,4-Pentamethylpentane

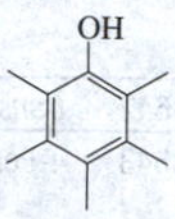
8601
Pentamethylphenol

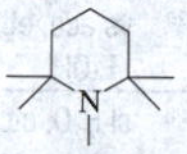
8602
1,2,2,6,6-Pentamethylpiperidine

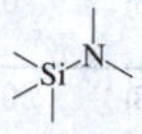
8603
Pentamethylsilanamine

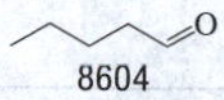
8604
Pentanal

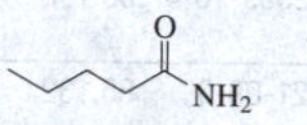
8605
Pentanamide

8606
3-Pentanamine

8607
Pentane

8608
Pentanedial

8609
1,5-Pentanediamine

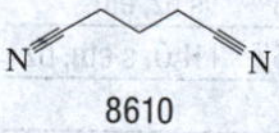
8610
Pentanedinitrile

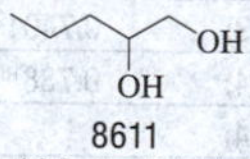
8611
1,2-Pentanediol, (±)-

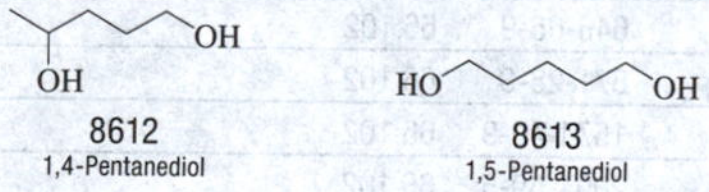
8612
1,4-Pentanediol

8613
1,5-Pentanediol

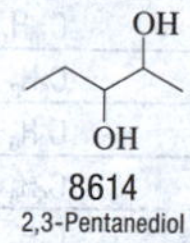
8614
2,3-Pentanediol

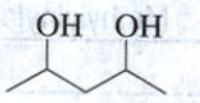
8615
2,4-Pentanediol

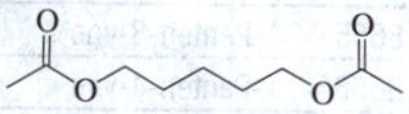
8616
1,5-Pentanediol diacetate

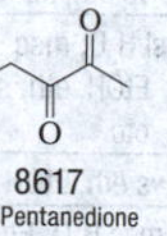
8617
2,3-Pentanedione

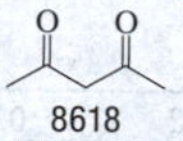
8618
2,4-Pentanedione

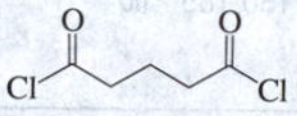
8619
Pentanedioyl dichloride

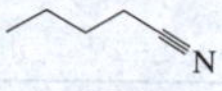
8620
Pentanenitrile

8621
1-Pentanethiol

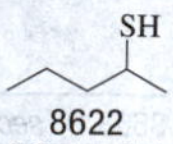
8622
2-Pentanethiol

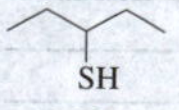
8623
3-Pentanethiol

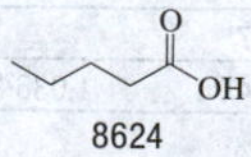
8624
Pentanoic acid

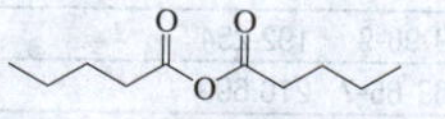
8625
Pentanoic anhydride

8626
1-Pentanol

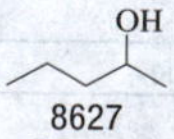
8627
2-Pentanol

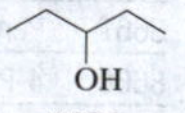
8628
3-Pentanol

No.	Name	Synonym	Mol. form.	CAS Reg. No.	Mol. wt.	Physical form	mp/°C	bp/°C	den/ g cm^{-3}	n_D	Solubility
8629	2-Pentanone	Methyl propyl ketone	$C_5H_{10}O$	107-87-9	86.132	liq	-76.83(0.05)	102.2(0.1)	0.809^{20}	1.3895^{20}	sl H_2O, ctc; msc EtOH, eth
8630	3-Pentanone	Diethyl ketone	$C_5H_{10}O$	96-22-0	86.132	liq	-38.98(0.05)	101.9(0.1)	0.8098^{25}	1.3905^{25}	s H_2O, ctc; msc EtOH, eth
8631	2-Pentanone oxime	Methyl propyl ketone oxime	$C_5H_{11}NO$	623-40-5	101.147			169(8)	0.9095^{20}	1.4450^{20}	vs H_2O, eth, EtOH
8632	Pentanoyl chloride	Valeroyl chloride	C_5H_9ClO	638-29-9	120.577	liq	-110	124(5)	1.0155^{15}	1.4200^{20}	
8633	Pentaphene	2,3:6,7-Dibenzphenanthrene	$C_{22}H_{14}$	222-93-5	278.346	ye grn lf(xyl)	257				i H_2O; sl EtOH, xyl, eth; s bz
8634	1,2,3,5,6-Pentathiepane	Lenthionine	$C_2H_4S_5$	292-46-6	188.378		60.5				
8635	Pentatriacontane		$C_{35}H_{72}$	630-07-9	492.947	cry (al)	74.4(0.4)	489(15)	0.8157^{20}	1.4568^{20}	i H_2O; sl eth; s ace
8636	18-Pentatriacontanone		$C_{35}H_{70}O$	504-53-0	506.930	lf (lig)	88.8(0.8)	$270^{0.1}$	0.793^{95}		i H_2O; sl EtOH, eth, ace, bz, lig, chl
8637	Pentazocine		$C_{19}H_{27}NO$	359-83-1	285.423	cry (MeOH aq)	147				
8638	4-Pentenal	Pent-4-en-1-al	C_5H_8O	2100-17-6	84.117			104(18)	0.852^{20}	1.4191^{20}	i H_2O; s eth, ace
8639	1-Pentene	α-Amylene	C_5H_{10}	109-67-1	70.133	vol liq or gas	-165.13(0.01)	30.0(0.3)	0.6405^{20}	1.3715^{20}	i H_2O; msc EtOH, eth; s bz; sl ctc
8640	*cis*-2-Pentene	*cis*-β-Amylene	C_5H_{10}	627-20-3	70.133	liq	-151.35(0.02)	36.9(0.2)	0.6556^{20}	1.3830^{20}	i H_2O; msc EtOH, eth; s bz, dil sulf
8641	*trans*-2-Pentene	*trans*-β-Amylene	C_5H_{10}	646-04-8	70.133	liq	-140.20(0.02)	36.3(0.6)	0.6431^{25}	1.3793^{20}	i H_2O; msc EtOH, eth; s bz; vs dil sulf
8642	*trans*-3-Pentenenitrile		C_5H_7N	16529-66-1	81.117	liq		144	0.837	1.4220^{20}	
8643	4-Pentenenitrile		C_5H_7N	592-51-8	81.117			140	0.8239^{24}	1.4213^{14}	i H_2O; msc EtOH, eth
8644	*trans*-3-Pentenoic acid		$C_5H_8O_2$	1617-32-9	100.117			193.2	0.989^{19}		
8645	4-Pentenoic acid	Allylacetic acid	$C_5H_8O_2$	591-80-0	100.117	liq	-22.5	188.5	0.9809^{20}	1.4281^{20}	sl H_2O; vs EtOH, eth
8646	1-Penten-3-ol		$C_5H_{10}O$	616-25-1	86.132			118(5)	0.839^{20}	1.4239^{20}	sl H_2O; msc EtOH, eth
8647	*cis*-2-Penten-1-ol		$C_5H_{10}O$	1576-95-0	86.132			138	0.8529^{20}	1.4354^{20}	s EtOH, eth, ace
8648	*trans*-2-Penten-1-ol		$C_5H_{10}O$	1576-96-1	86.132			138	0.8471^{20}	1.4341^{20}	s EtOH, eth, ace
8649	3-Penten-2-ol, (±)-		$C_5H_{10}O$	42569-16-4	86.132			123(9)	0.8328^{25}	1.4280^{20}	vs ace, eth, EtOH
8650	4-Penten-1-ol		$C_5H_{10}O$	821-09-0	86.132			136(8)	0.8457^{20}	1.4309^{20}	sl H_2O, ctc; s eth
8651	4-Penten-2-ol		$C_5H_{10}O$	625-31-0	86.132			116	0.8367^{20}	1.4225^{20}	vs H_2O; msc EtOH, eth
8652	1-Penten-3-one	Ethyl vinyl ketone	C_5H_8O	1629-58-9	84.117			103(6)	0.8468^{20}	1.4195^{20}	i H_2O; s EtOH, eth, ace, bz, chl
8653	*trans*-3-Penten-2-one		C_5H_8O	3102-33-8	84.117			122	0.8624^{20}	1.4350^{20}	s H_2O, eth, ace, ctc
8654	2-(3-Pentenyl)pyridine		$C_{10}H_{13}N$	2057-43-4	147.217			216	0.9234^{25}	1.5076^{25}	
8655	1-Penten-3-yne	Methylvinylacetylene	C_5H_6	646-05-9	66.102			59(3)	0.7401^{20}	1.4496^{20}	vs bz, eth
8656	1-Penten-4-yne		C_5H_6	871-28-3	66.102			42(4)	0.738^{16}	1.4125^{16}	i H_2O; s eth, bz
8657	*cis*-3-Penten-1-yne		C_5H_6	1574-40-9	66.102			44(4)			
8658	*trans*-3-Penten-1-yne		C_5H_6	2004-69-5	66.102			51(3)			
8659	Pentetic acid	Diethylenetriaminepentaacetic acid	$C_{14}H_{23}N_3O_{10}$	67-43-6	393.347	cry (w)	219				s H_2O, alk
8660	Pentostatin		$C_{11}H_{16}N_4O_4$	53910-25-1	268.270	wh cry (MeOH aq)	222				
8661	Pentryl	2-(*N*,2,4,6-Tetranitroanilino)-ethanol	$C_8H_6N_6O_{11}$	4481-55-4	362.167	wh-ye cry	129		1.82		i H_2O, ctc; s chl; vs eth, bz
8662	Pentyl acetate	Amyl acetate	$C_7H_{14}O_2$	628-63-7	130.185	liq	-70.9(0.7)	149.4(0.3)	0.8756^{20}	1.4023^{20}	sl H_2O; msc EtOH, eth; s ctc
8663	*sec*-Pentyl acetate, (*R*)-	*sec*-Amyl acetate (*R*)	$C_7H_{14}O_2$	54638-10-7	130.185			142	0.8803^{18}	1.4012^{20}	vs eth, EtOH
8664	Pentylamine	Amylamine	$C_5H_{13}N$	110-58-7	87.164	liq	-51(15)	104.7(0.2)	0.7544^{20}	1.448^{20}	msc H_2O, EtOH, eth; vs ace, bz; sl chl
8665	4-*tert*-Pentylaniline		$C_{11}H_{17}N$	2049-92-5	163.260			260.5			
8666	Pentylbenzene	Amylbenzene	$C_{11}H_{16}$	538-68-1	148.245	liq	-78.2(0.4)	203(3)	0.8585^{20}	1.4878^{20}	i H_2O; msc EtOH, eth, ace, bz, peth, ctc
8667	Pentyl benzoate		$C_{12}H_{16}O_2$	2049-96-9	192.254			137^{15}			
8668	4-Pentylbenzoyl chloride		$C_{12}H_{15}ClO$	49763-65-7	210.699			144^{10}	1.036^{25}	1.5300^{20}	

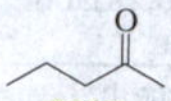
8629
2-Pentanone

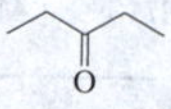
8630
3-Pentanone

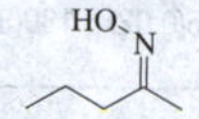
8631
2-Pentanone oxime

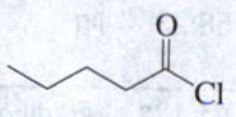
8632
Pentanoyl chloride

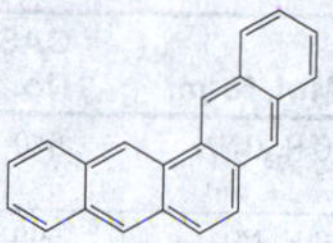
8633
Pentaphene

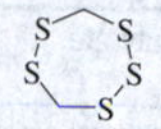
8634
1,2,3,5,6-Pentathiepane

8635
Pentatriacontane

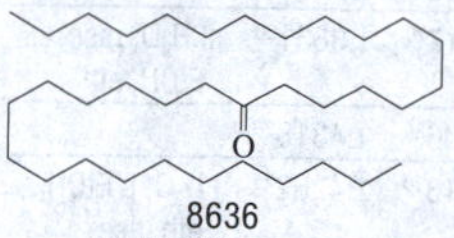
8636
18-Pentatriacontanone

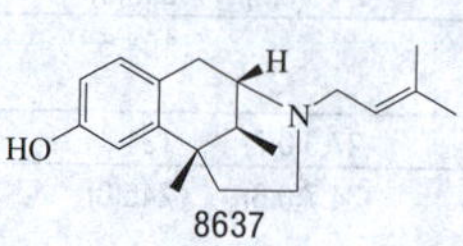
8637
Pentazocine

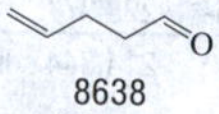
8638
4-Pentenal

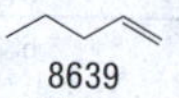
8639
1-Pentene

8640
cis-2-Pentene

8641
trans-2-Pentene

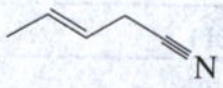
8642
trans-3-Pentenenitrile

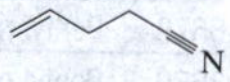
8643
4-Pentenenitrile

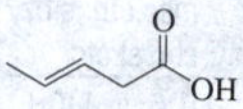
8644
trans-3-Pentenoic acid

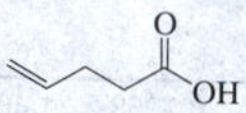
8645
4-Pentenoic acid

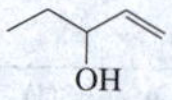
8646
1-Penten-3-ol

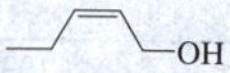
8647
cis-2-Penten-1-ol

8648
trans-2-Penten-1-ol

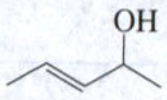
8649
3-Penten-2-ol, (±)-

8650
4-Penten-1-ol

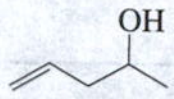
8651
4-Penten-2-ol

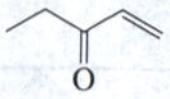
8652
1-Penten-3-one

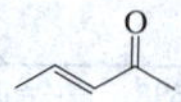
8653
trans-3-Penten-2-one

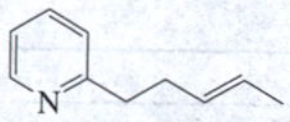
8654
2-(3-Pentenyl)pyridine

8655
1-Penten-3-yne

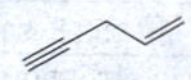
8656
1-Penten-4-yne

8657
cis-3-Penten-1-yne

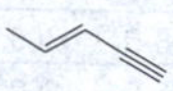
8658
trans-3-Penten-1-yne

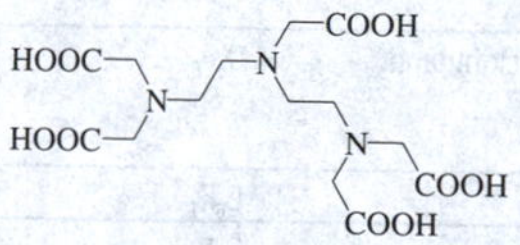
8659
Pentetic acid

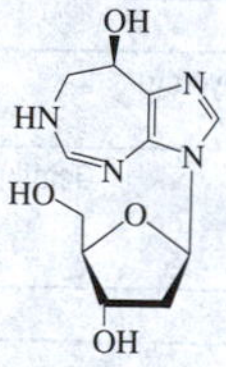
8660
Pentostatin

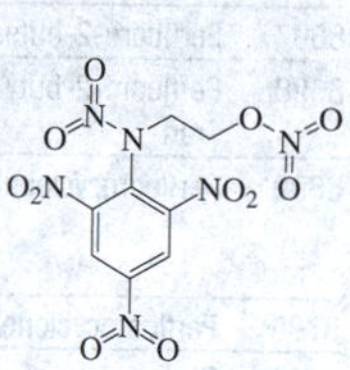
8661
Pentryl

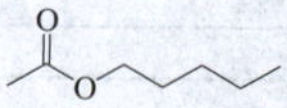
8662
Pentyl acetate

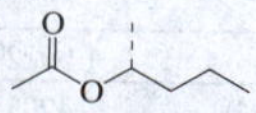
8663
sec-Pentyl acetate, (*R*)-

8664
Pentylamine

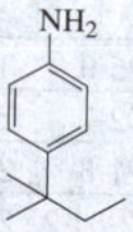
8665
4-*tert*-Pentylaniline

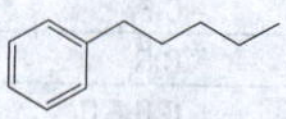
8666
Pentylbenzene

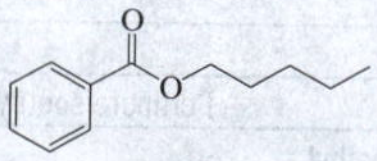
8667
Pentyl benzoate

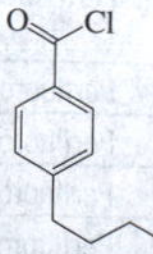
8668
4-Pentylbenzoyl chloride

Organic

No.	Name	Synonym	Mol. form.	CAS Reg. No.	Mol. wt.	Physical form	mp/°C	bp/°C	den/g cm^{-3}	n_D	Solubility
8669	Pentyl butanoate	Amyl butyrate	$C_9H_{18}O_2$	540-18-1	158.238	liq	-72.66(0.02)	180(3)	0.8713^{15}	1.4123^{20}	i H_2O; vs EtOH, eth
8670	*tert*-Pentyl carbamate	*tert*-Amyl carbamate	$C_6H_{13}NO_2$	590-60-3	131.173	nd (dil al)	86				vs ace, bz
8671	Pentyl chloroformate		$C_6H_{11}ClO_2$	638-41-5	150.603			61^{15}		1.4181^{18}	s eth
8672	Pentylcyclohexane		$C_{11}H_{22}$	4292-92-6	154.293	liq	-57.5	204(1)	0.8037^{20}	1.4437^{20}	vs ace, bz, eth, EtOH
8673	Pentylcyclopentane		$C_{10}H_{20}$	3741-00-2	140.266	liq	-83	180(7)	0.7912^{20}	1.4356^{20}	i H_2O; vs ace, bz, eth, EtOH
8674	Pentyl formate	Amyl formate	$C_6H_{12}O_2$	638-49-3	116.158	liq	-73(1)	126(3)	0.8853^{20}	1.3992^{20}	sl H_2O; msc EtOH, eth
8675	Pentyl heptanoate	Amyl enanthate	$C_{12}H_{24}O_2$	7493-82-5	200.318	liq	-49.3(0.7)	239(5)	0.8623^{20}	1.4263^{15}	vs ace, bz, eth, EtOH
8676	Pentyl hexanoate	Amyl caproate	$C_{11}H_{22}O_2$	540-07-8	186.292	liq	-49(3)	225(3)	0.8612^{25}	1.4202^{25}	s EtOH, eth, ace; sl ctc
8677	1-Pentylnaphthalene		$C_{15}H_{18}$	86-89-5	198.304	liq	-24.5(0.3)	305(5)	0.9656^{20}	1.5725^{20}	
8678	Pentyl nitrite	Amyl nitrite	$C_5H_{11}NO_2$	463-04-7	117.147			104.5	0.8817^{20}	1.3851^{20}	sl H_2O; msc EtOH, eth
8679	Pentyl nonanoate	Pentyl pelargonate	$C_{14}H_{28}O_2$	61531-45-1	228.371		-27.0(0.7)	131^{20}	0.8506^{25}	1.4318^{20}	
8680	Pentyl octanoate	Amyl octanoate	$C_{13}H_{26}O_2$	638-25-5	214.344	liq	-34.7(0.6)	245(8)	0.8613^{20}	1.4262^{25}	i H_2O; s EtOH, eth, ace
8681	4-(Pentyloxy)benzoyl chloride		$C_{12}H_{15}ClO_2$	36823-84-4	226.699			198^{30}	1.087^{25}	1.5434^{20}	
8682	Pentyl pentanoate		$C_{10}H_{20}O_2$	2173-56-0	172.265	liq	-78.8	204(6)	0.8638^{20}	1.4164^{20}	sl H_2O; msc EtOH, eth
8683	4-Pentylphenol		$C_{11}H_{16}O$	14938-35-3	164.244		18(1)	250.5	0.960^{20}	1.5272^{25}	vs eth, EtOH
8684	Pentyl propanoate		$C_8H_{16}O_2$	624-54-4	144.212	liq	-73.1	166(4)	0.8761^{25}	1.4096^{15}	i H_2O; msc EtOH, eth; s bz; sl ctc
8685	Pentyl salicylate		$C_{12}H_{16}O_3$	2050-08-0	208.253			270	1.064^{15}	1.506^{20}	sl H_2O; msc EtOH, eth
8686	Pentyl stearate		$C_{23}H_{46}O_2$	6382-13-4	354.610	pl	29.2(0.6)			1.4342^{50}	vs eth, EtOH
8687	1-Pentyne	Propylacetylene	C_5H_8	627-19-0	68.118	liq	-106.2(0.5)	39.9(1)	0.6901^{20}	1.3852^{20}	i H_2O; vs EtOH; msc eth; s bz, chl; sl ctc
8688	2-Pentyne		C_5H_8	627-21-4	68.118	liq	-109.3(0.2)	56(1)	0.7058^{25}	1.4039^{20}	i H_2O; vs EtOH; msc eth; s bz, chl
8689	4-Pentynoic acid	Propargylacetic acid	$C_5H_6O_2$	6089-09-4	98.101		55(3)	110^{30}			vs eth, EtOH
8690	2-Pentyn-1-ol		C_5H_8O	6261-22-9	84.117	liq	-49.7	150(9)	0.909^{20}	1.4518^{17}	
8691	3-Pentyn-1-ol		C_5H_8O	10229-10-4	84.117			154	0.9002^{20}	1.4454^{20}	
8692	4-Pentyn-1-ol		C_5H_8O	5390-04-5	84.117			150(11)	0.913^{20}	1.4414^{20}	
8693	Perazine	10-[3-(4-Methyl-1-piperazinyl)propyl]-10*H*-phenothiazine	$C_{20}H_{25}N_3S$	84-97-9	339.498	cry	52	$165^{0.001}$			
8694	Perfluidone		$C_{14}H_{12}F_3NO_4S_2$	37924-13-3	379.375		146.0(0.5)				
8695	Perfluoroacetone	Hexafluoroacetone	C_3F_6O	684-16-2	166.021	col gas	-125.45	-27.4(0.4)			
8696	Perfluorobutane	Decafluorobutane	C_4F_{10}	355-25-9	238.027	col gas	-129(1)	-2.1(0.8)	1.6484^{25}		s bz, chl
8697	Perfluoro-2-butene		C_4F_8	360-89-4	200.030	col gas	-129	-5.9(0.9)	1.5297^{25}		
8698	Perfluoro-2-butyltetrahydrofuran		$C_8F_{16}O$	335-36-4	416.059			105.2(0.6)			
8699	Perfluorocyclobutane	Octafluorocyclobutane	C_4F_8	115-25-3	200.030	col gas	-40.16(0.1)	-5.91	1.500^{25} (p>1 atm)		i H_2O; s eth
8700	Perfluorocyclohexane		C_6F_{12}	355-68-0	300.045		62.9(0.3)	52.8 sp			
8701	Perfluorocyclohexene		C_6F_{10}	355-75-9	262.048			51.7(0.1)	1.6650^{25}	1.293^{20}	
8702	Perfluorodecalin	PFC-9-1-18	$C_{10}F_{18}$	306-94-5	462.078	liq	-8.2(0.5)	143(1)	1.9305^{25}		
8703	Perfluorodecane		$C_{10}F_{22}$	307-45-9	538.072			135(3)			i H_2O
8704	Perfluorodimethoxymethane		$C_3F_8O_2$	53772-78-4	220.018	col gas	-161(2)	-10.0(0.7)			
8705	Perfluoro-2,3-dimethylbutane		C_6F_{14}	354-96-1	338.042	liq	-15.1(0.5)	59.8(0.3)			
8706	Perfluoroethyl ethyl ether		$C_4H_5F_5O$	22052-81-9	164.074	vol liq or gas		28(3)			
8707	Perfluoroethyl 2,2,2-trifluoroethyl ether		$C_4H_2F_8O$	156053-88-2	218.045	vol liq or gas		27.89			
8708	Perfluoroheptane		C_7F_{16}	335-57-9	388.049	liq	-51.3(0.1)	82.5(0.2)	1.7333^{20}	1.2618^{20}	i H_2O; vs ace, eth, EtOH, chl
8709	Perfluoro-1-heptene		C_7F_{14}	355-63-5	350.053			81(4)			
8710	Perfluorohexane	PFC-5-1-14	C_6F_{14}	355-42-0	338.042	liq	-86.1(0.3)	57.2(0.2)	1.6910^{20}	1.2515^{20}	i H_2O; s eth, bz, chl
8711	Perfluoro-1-hexene		C_6F_{12}	755-25-9	300.045			57.0			vs chl
8712	Perfluoroisobutane		C_4F_{10}	354-92-7	238.027	col gas		0			
8713	Perfluoroisobutene	Perfluoroisobutylene	C_4F_8	382-21-8	200.030	col gas	-130	7	1.5922^{0}		
8714	Perfluoroisopropyl methyl ether		$C_4H_3F_7O$	22052-84-2	200.055	vol liq or gas		29(1)	1.4205^{20}		

Organic

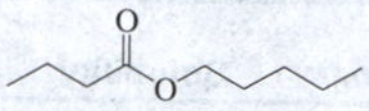
8669
Pentyl butanoate

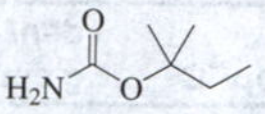
8670
tert-Pentyl carbamate

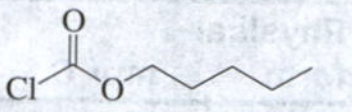
8671
Pentyl chloroformate

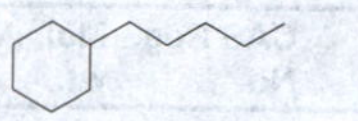
8672
Pentylcyclohexane

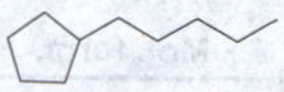
8673
Pentylcyclopentane

8674
Pentyl formate

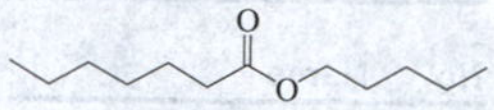
8675
Pentyl heptanoate

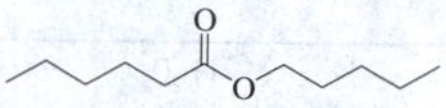
8676
Pentyl hexanoate

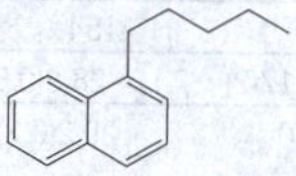
8677
1-Pentylnaphthalene

8678
Pentyl nitrite

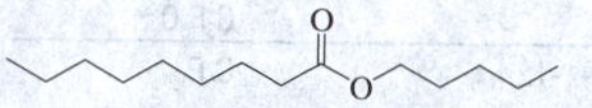
8679
Pentyl nonanoate

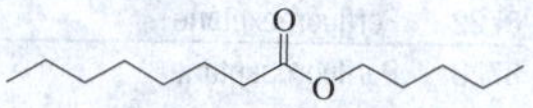
8680
Pentyl octanoate

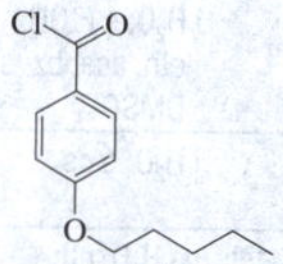
8681
4-(Pentyloxy)benzoyl chloride

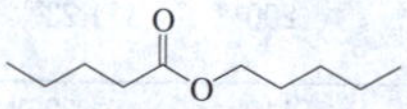
8682
Pentyl pentanoate

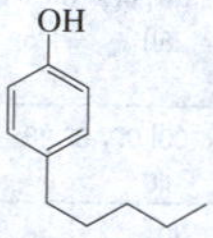
8683
4-Pentylphenol

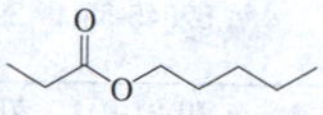
8684
Pentyl propanoate

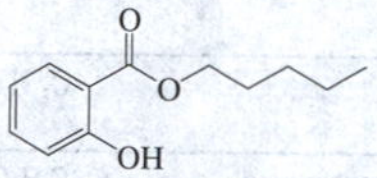
8685
Pentyl salicylate

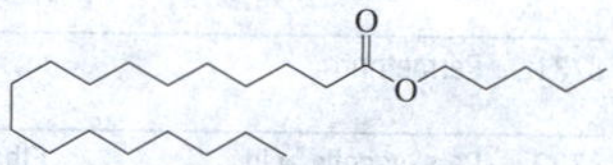
8686
Pentyl stearate

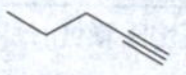
8687
1-Pentyne

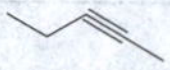
8688
2-Pentyne

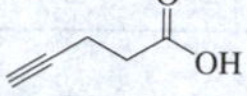
8689
4-Pentynoic acid

8690
2-Pentyn-1-ol

8691
3-Pentyn-1-ol

8692
4-Pentyn-1-ol

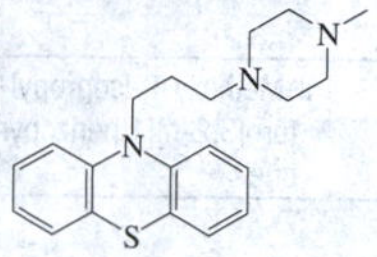
8693
Perazine

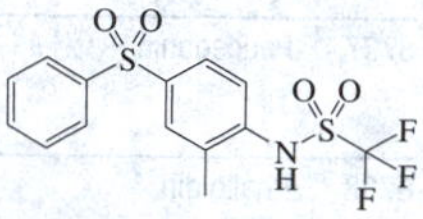
8694
Perfluidone

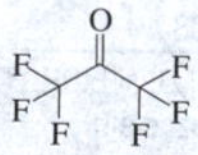
8695
Perfluoroacetone

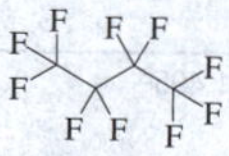
8696
Perfluorobutane

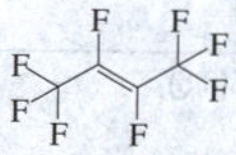
8697
Perfluoro-2-butene

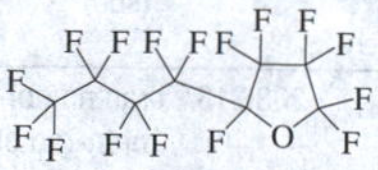
8698
Perfluoro-2-butyltetrahydrofuran

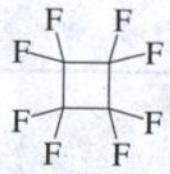
8699
Perfluorocyclobutane

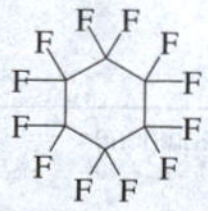
8700
Perfluorocyclohexane

8701
Perfluorocyclohexene

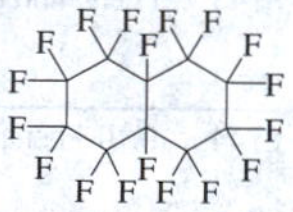
8702
Perfluorodecalin

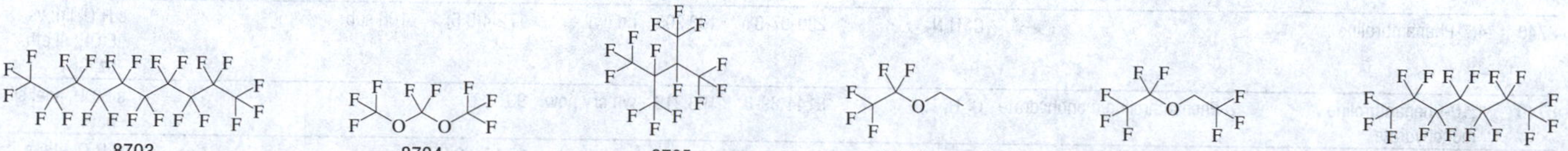
8703
Perfluorodecane

8704
Perfluorodimethoxymethane

8705
Perfluoro-2,3-dimethylbutane

8706
Perfluoroethyl ethyl ether

8707
Perfluoroethyl 2,2,2-trifluoroethyl ether

8708
Perfluoroheptane

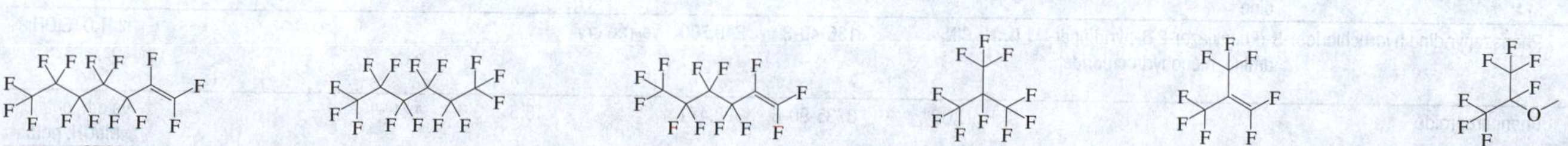
8709
Perfluoro-1-heptene

8710
Perfluorohexane

8711
Perfluoro-1-hexene

8712
Perfluoroisobutane

8713
Perfluoroisobutene

8714
Perfluoroisopropyl methyl ether

No.	Name	Synonym	Mol. form.	CAS Reg. No.	Mol. wt.	Physical form	mp/°C	bp/°C	den/ g cm^{-3}	n_D	Solubility
8715	Perfluoromethylcyclohexane		C_7F_{14}	355-02-2	350.053	liq	-39.0(0.2)	76.3(0.2)	1.7878^{25}	1.285^{17}	s ace, bz, ctc, tol, AcOEt
8716	Perfluoro-2-methylpentane		C_6F_{14}	355-04-4	338.042			57.6(0.3)	1.7326^{20}	1.2564^{22}	i H_2O; s bz
8717	Perfluoro-3-methylpentane		C_6F_{14}	865-71-4	338.042	liq	-115.4(0.5)	58(9)			s bz
8718	Perfluoronaphthalene		$C_{10}F_8$	313-72-4	272.094		79(1)	200(8)			
8719	Perfluorononane		C_9F_{20}	375-96-2	488.064			117(3)	1.8001^{20}		
8720	Perfluorooctane		C_8F_{18}	307-34-6	438.057			105(2)	1.73^{20}	1.282^{20}	i H_2O
8721	Perfluorooctylsulfonyl fluoride		$C_8F_{18}O_2S$	307-35-7	502.121	liq		154			
8722	Perfluorooxetane		C_3F_6O	425-82-1	166.021	col gas	-117(2)	-28.6(0.6)			
8723	Perfluoropentane	PFC-4-1-12	C_5F_{12}	678-26-2	288.035	vol liq or gas	-10	29.2(0.2)			i H_2O
8724	Perfluoropropane	PFC-218	C_3F_8	76-19-7	188.019	col gas	-147.7(0.1)	-36.8(0.3)			i H_2O
8725	Perfluoropropene	Hexafluoropropene	C_3F_6	116-15-4	150.022	col gas	-156.5	-30.2(0.5)		1.583^{-40}	i H_2O
8726	Perfluoropropyl methyl ether	HFE-347mcc3	$C_4H_3F_7O$	375-03-1	200.055			34(1)	1.4092^{20}		
8727	Perfluoropyridine	Pentafluoropyridine	C_5F_5N	700-16-3	169.053			83(26)			
8728	Perfluorotoluene		C_7F_8	434-64-0	236.062	liq	-65.48(0.06)	104.6(0.3)	1.6616^{25}	1.3670^{20}	
8729	Perfluorotripropylamine	Tris(perfluoropropyl)amine	$C_9F_{21}N$	338-83-0	521.069			129.7(0.5)	1.822^{4}	1.279^{25}	
8730	1*H*-Perimidine		$C_{11}H_8N_2$	204-02-4	168.195	grn cry (dil al)	223.0				i H_2O; s EtOH, eth, ace, bz; sl DMSO
8731	Permethrin		$C_{21}H_{20}Cl_2O_3$	52645-53-1	391.288	col cry or ye liq	34	$200^{0.01}$	1.23^{20}		i H_2O; s os
8732	Peroxyacetic acid	Ethaneperoxoic acid	$C_2H_4O_3$	79-21-0	76.051	liq	-0.2	110	1.226^{15}	1.3974^{20}	vs H_2O, eth, sulf; s EtOH
8733	Peroxypropanoic acid	Propaneperoxoic acid	$C_3H_6O_3$	4212-43-5	90.078			exp		1.4148^{15}	
8734	Perphenazine		$C_{21}H_{26}ClN_3OS$	58-39-9	403.968		97				
8735	Perthane	Ethane, 1,1-dichloro-2,2-bis(*p*-ethylphenyl)-	$C_{18}H_{20}Cl_2$	72-56-0	307.258		59.7(0.9)				
8736	Perylene	Dibenz[*de,kl*]anthracene	$C_{20}H_{12}$	198-55-0	252.309	gold-br, ye pl (bz, HOAc)	276(13)		1.35^{25}		i H_2O; sl EtOH, eth; vs ace, chl; s bz
8737	Peucedanin	3-Methoxy-2-isopropyl-7*H*-furo[3,2-*g*][1]benzopyran-7-one	$C_{15}H_{14}O_4$	133-26-6	258.270	pr or pl (bz-peth)	85	278^{17}			sl H_2O, bz; s EtOH, eth; vs chl, CS_2
8738	Phalloidin		$C_{35}H_{48}N_8O_{11}S$	17466-45-4	788.868	nd (w)	281 (hyd)				s EtOH, MeOH, py
8739	Phalloin		$C_{35}H_{48}N_8O_{10}S$	28227-92-1	772.869	cry (w)	250 dec				
8740	α-Phellandrene	2-Methyl-5-(1-methylethyl)-1,3-cyclohexadiene	$C_{10}H_{16}$	99-83-2	136.234		238	174.9	0.8410^{20}	1.471^{25}	i H_2O; s eth
8741	β-Phellandrene	*p*-Mentha-1(7),2-diene	$C_{10}H_{16}$	555-10-2	136.234			177(3)	0.8520^{20}	1.4788^{20}	i H_2O, EtOH; s eth
8742	9-Phenanthrenamine		$C_{14}H_{11}N$	947-73-9	193.244	lt ye cry (al)	138.3	sub			sl eth, bz, chl
8743	Phenanthrene		$C_{14}H_{10}$	85-01-8	178.229	mcl pl (al), lf (sub)	99(2)	338.4(0.9)	0.9800^{4}	1.5943	i H_2O; s EtOH, eth, ace, bz, CS_2
8744	9,10-Phenanthrenedione	Phenanthrenequinone	$C_{14}H_8O_2$	84-11-7	208.213	oran nd (to) oran-red pl (sub)	208.5(0.5)		1.405^{22}		i H_2O; sl EtOH, bz; s eth
8745	Phenanthridine		$C_{13}H_9N$	229-87-8	179.217	nd (dil al)	106.56(0.02)	350.3(0.6)			sl H_2O; vs EtOH, eth, bz, CS_2; s ace
8746	1,7-Phenanthroline		$C_{12}H_8N_2$	230-46-6	180.205	pl (anh), nd (w+2)	77.2(0.5)	360			s H_2O; vs EtOH; i eth, bz, lig
8747	1,10-Phenanthroline	*o*-Phenanthroline	$C_{12}H_8N_2$	66-71-7	180.205	wh nd (bz) cry (w+1)	118.56(0.05)	409.2(0.8)			vs H_2O; s EtOH, ace, bz; i peth
8748	4,7-Phenanthroline		$C_{12}H_8N_2$	230-07-9	180.205	nd (w)	172.4(0.5)	100 sub			s H_2O, lig; vs EtOH; sl eth, bz, CS_2
8749	1,10-Phenanthroline monohydrate	*o*-Phenanthroline monohydrate	$C_{12}H_{10}N_2O$	5144-89-8	198.219	wh cry pow	93				s EtOH, ace; sl bz
8750	Phenazine	Dibenzopyrazine	$C_{12}H_8N_2$	92-82-0	180.205	ye-red nd (HOAc)	174.76(0.05)				sl H_2O, eth; s bz, EtOH
8751	2,3-Phenazinediamine	2,3-Diaminophenazine	$C_{12}H_{10}N_4$	655-86-7	210.234	ye nd	264	sub			vs bz, EtOH
8752	1-Phenazinol	Hemipyocyanine	$C_{12}H_8N_2O$	528-71-2	196.204	ye nd (bz, dil MeOH)	158	sub			sl H_2O, EtOH; s bz, py, dil alk
8753	Phenazopyridine	2,6-Diamino-3-phenylazopyridine	$C_{11}H_{11}N_5$	94-78-0	213.239	red cry	139				
8754	Phenazopyridine hydrochloride	3-(Phenylazo)-2,6-pyridinediamine, monohydrochloride	$C_{11}H_{12}ClN_5$	136-40-3	249.700	ye-red cry					sl H_2O, EtOH; i bz, ace; s HOAc
8755	Phencarbamide		$C_{19}H_{24}N_2OS$	3735-90-8	328.471		48.5	$121^{0.01}$			vs eth, chl, MeOH, peth

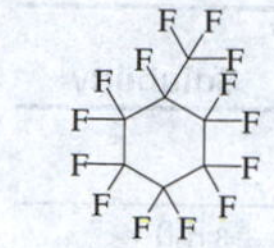
8715
Perfluoromethylcyclohexane

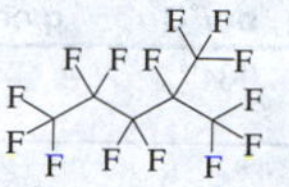
8716
Perfluoro-2-methylpentane

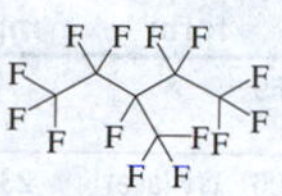
8717
Perfluoro-3-methylpentane

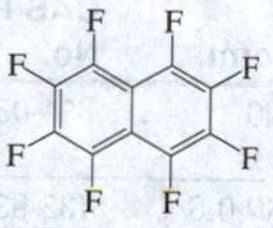
8718
Perfluoronaphthalene

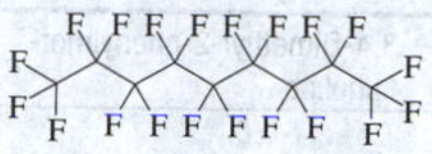
8719
Perfluorononane

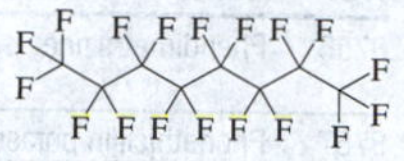
8720
Perfluorooctane

8721
Perfluorooctylsulfonyl fluoride

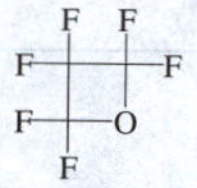
8722
Perfluorooxetane

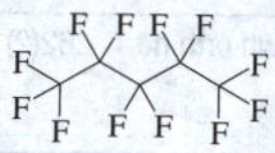
8723
Perfluoropentane

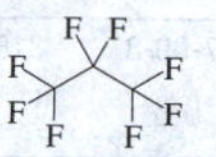
8724
Perfluoropropane

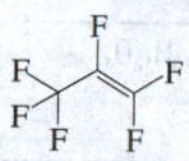
8725
Perfluoropropene

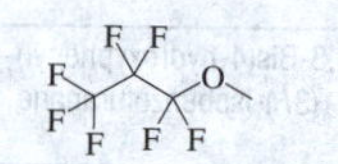
8726
Perfluoropropyl methyl ether

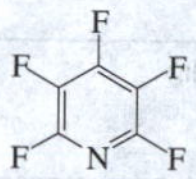
8727
Perfluoropyridine

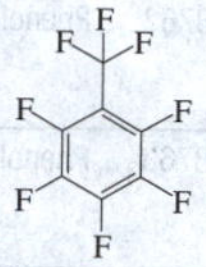
8728
Perfluorotoluene

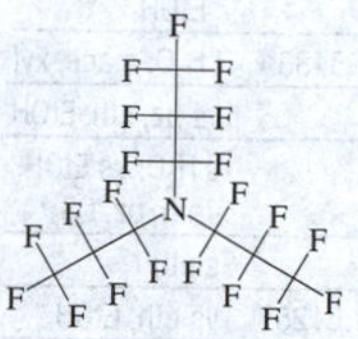
8729
Perfluorotripropylamine

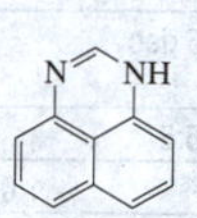
8730
1*H*-Perimidine

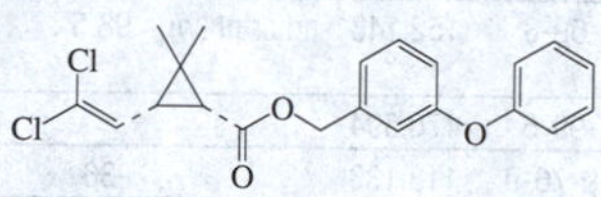
8731
Permethrin

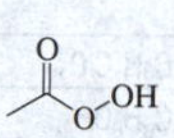
8732
Peroxyacetic acid

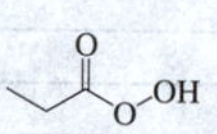
8733
Peroxypropanoic acid

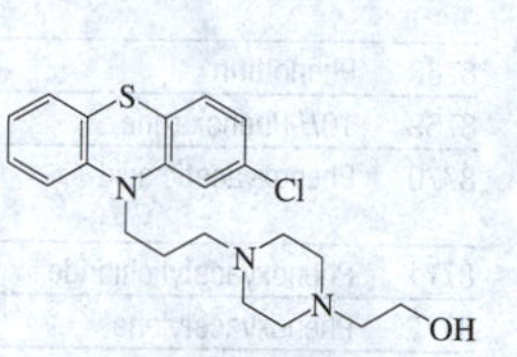
8734
Perphenazine

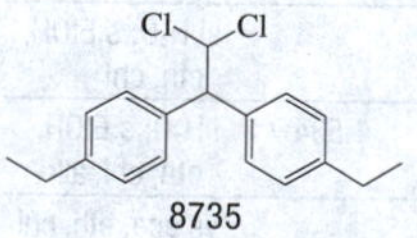
8735
Perthane

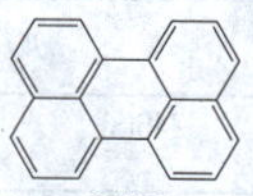
8736
Perylene

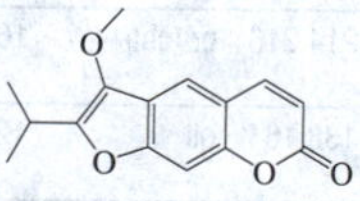
8737
Peucedanin

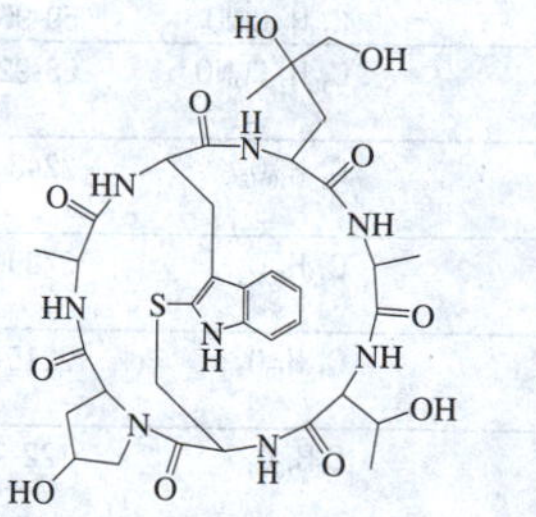
8738
Phalloidin

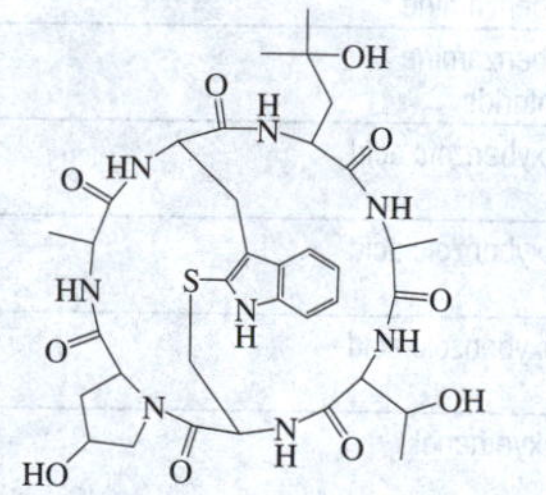
8739
Phalloin

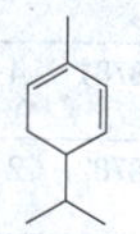
8740
α-Phellandrene

8741
β-Phellandrene

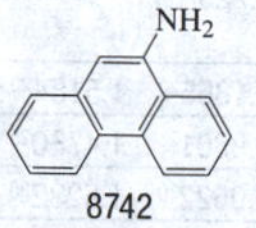
8742
9-Phenanthrenamine

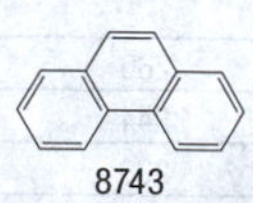
8743
Phenanthrene

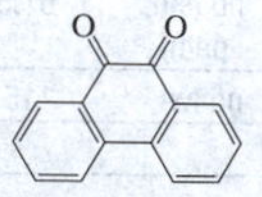
8744
9,10-Phenanthrenedione

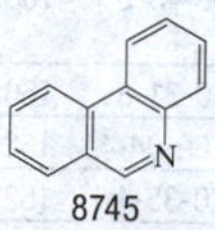
8745
Phenanthridine

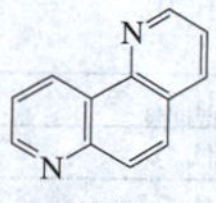
8746
1,7-Phenanthroline

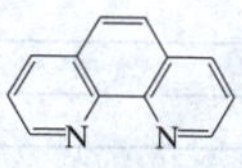
8747
1,10-Phenanthroline

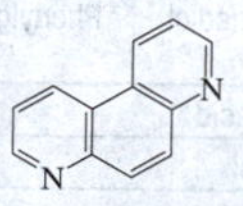
8748
4,7-Phenanthroline

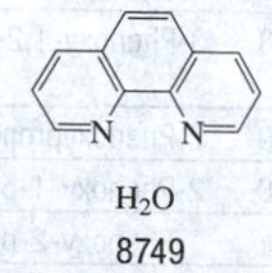
8749
1,10-Phenanthroline monohydrate

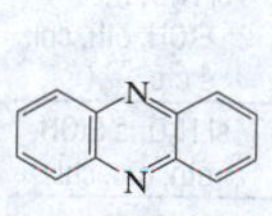
8750
Phenazine

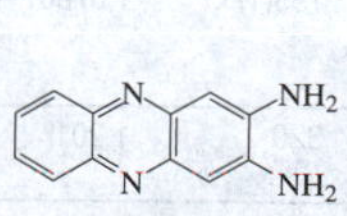
8751
2,3-Phenazinediamine

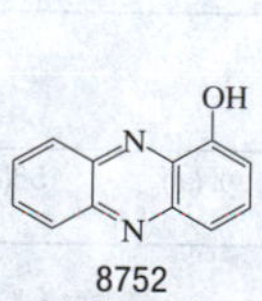
8752
1-Phenazinol

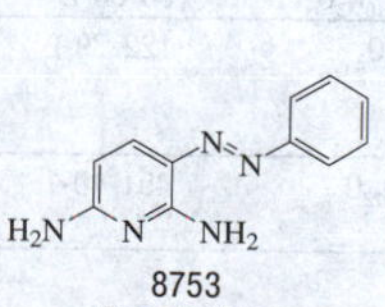
8753
Phenazopyridine

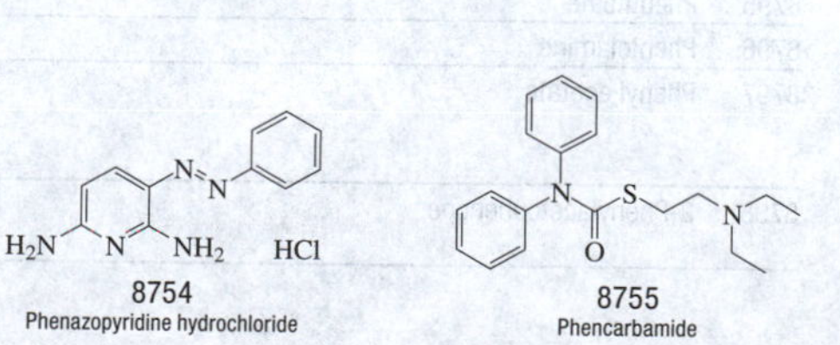
8754
Phenazopyridine hydrochloride

8755
Phencarbamide

No.	Name	Synonym	Mol. form.	CAS Reg. No.	Mol. wt.	Physical form	mp/°C	bp/°C	den/g cm^{-3}	n_D	Solubility
8756	Phendimetrazine	3,4-Dimethyl-2-phenylmorpholine	$C_{12}H_{17}NO$	634-03-7	191.269			134^{12}			
8757	Phenethicillin potassium		$C_{17}H_{19}KN_2O_5S$	132-93-4	402.506	cry (ace)	235				s H_2O
8758	Phenicin		$C_{14}H_{10}O_6$	128-68-7	274.225	ye-br (al)	230.5				sl H_2O; vs EtOH, chl, HOAc
8759	Phenindamine		$C_{19}H_{19}N$	82-88-2	261.361	cry	91		1.17		
8760	Phenmedipham		$C_{16}H_{16}N_2O_4$	13684-63-4	300.309		150.8(0.8)				
8761	Phenobarbital	5-Ethyl-5-phenyl-2,4,6(1*H*,3*H*,5*H*)-pyrimidinetrione	$C_{12}H_{12}N_2O_3$	50-06-6	232.234	pl (w)	176(1)				i H_2O, bz; s EtOH, eth; sl DMSO
8762	Phenol	Hydroxybenzene	C_6H_6O	108-95-2	94.111		40.89(0.01)	181.8(0.1)	1.0545^{45}	1.5408^{41}	s H_2O, EtOH; vs eth; msc ace, bz
8763	Phenolphthalein	3,3-Bis(4-hydroxyphenyl)-1(3*H*)-isobenzofuranone	$C_{20}H_{14}O_4$	77-09-8	318.323	wh orth nd	262(2)		1.277^{32}		i H_2O, bz; vs EtOH, ace; s eth, chl
8764	Phenolphthalin	2-[Bis(4-hydroxyphenyl)-methyl]benzoic acid	$C_{20}H_{16}O_4$	81-90-3	320.339	nd (w)	230.5				vs eth, EtOH
8765	Phenolphthalol		$C_{20}H_{18}O_3$	81-92-5	306.355	cry (dil al)	201.5				
8766	Phenol Red	Phenolsulfonphthalein	$C_{19}H_{14}O_5S$	143-74-8	354.376	dk red nd or pl	>300				sl H_2O, EtOH, ace, bz; i eth, chl
8767	10*H*-Phenothiazine	Thiodiphenylamine	$C_{12}H_9NS$	92-84-2	199.271	ye pr (al) ye lf or pl (tol)	184.9(0.6)	371			vs ace, bz, eth, EtOH
8768	Phenothrin		$C_{23}H_{26}O_3$	26002-80-2	350.450	col liq			1.061^{25}	1.5483^{25}	i H_2O; s ace, xyl
8769	10*H*-Phenoxazine		$C_{12}H_9NO$	135-67-1	183.205	lf (dil al, bz)	156.7(0.5)	dec			vs bz, eth, EtOH
8770	Phenoxyacetic acid		$C_8H_8O_3$	122-59-8	152.148	nd or pl (w)	98.5	285 dec			s H_2O; vs EtOH, eth, bz, CS_2
8771	Phenoxyacetyl chloride		$C_8H_7ClO_2$	701-99-5	170.594			225.5			s eth
8772	Phenoxyacetylene		C_8H_6O	4279-76-9	118.133		-36	61^{25}	1.0614^{20}	1.5125^{20}	vs eth, EtOH
8773	2-Phenoxyaniline		$C_{12}H_{11}NO$	2688-84-8	185.221	cry (lig)	45.8	308			s EtOH; s eth, ace, bz
8774	3-Phenoxyaniline		$C_{12}H_{11}NO$	3586-12-7	185.221	pr (lig)	37	315	1.1583^{25}		s EtOH, eth, ace, bz; sl lig
8775	4-Phenoxyaniline		$C_{12}H_{11}NO$	139-59-3	185.221	nd (w), cry (dil al)	85.5				s H_2O; vs EtOH, eth; sl lig
8776	3-Phenoxybenzaldehyde		$C_{13}H_{10}O_2$	39515-51-0	198.217		14.0	169^{11}	1.147^{25}	1.5954^{20}	
8777	Phenoxybenzamine		$C_{18}H_{22}ClNO$	59-96-1	303.827		39				s bz
8778	Phenoxybenzamine hydrochloride		$C_{18}H_{23}Cl_2NO$	63-92-3	340.288		139				sl H_2O; s EtOH
8779	2-Phenoxybenzoic acid		$C_{13}H_{10}O_3$	2243-42-7	214.216	lf (dil al)	113(2)	355	1.1553^{50}		i H_2O; vs EtOH, eth; s chl
8780	3-Phenoxybenzoic acid		$C_{13}H_{10}O_3$	3739-38-6	214.216	nd (aq al)	145.8				i H_2O; s EtOH, eth
8781	4-Phenoxybenzoic acid		$C_{13}H_{10}O_3$	2215-77-2	214.216	pr (chl)	161				sl H_2O; s EtOH, eth, chl
8782	2-Phenoxyethanol		$C_8H_{10}O_2$	122-99-6	138.164	oil	12(2)	246(3)	1.102^{22}	1.534^{20}	i H_2O; s EtOH, eth, chl, alk
8783	2-Phenoxyethyl acrylate	Phenyl Cellosolve acrylate	$C_{11}H_{12}O_3$	48145-04-6	192.211			110^{2}	1.090^{25}		vs ace, eth, chl
8784	2-Phenoxyethyl butanoate		$C_{12}H_{16}O_3$	23511-70-8	208.253			251	1.0388^{21}		vs ace, eth, EtOH
8785	3-Phenoxyphenol		$C_{12}H_{10}O_2$	713-68-8	186.206			175^{7}			
8786	4-Phenoxyphenol		$C_{12}H_{10}O_2$	831-82-3	186.206		84.0				
8787	2-(3-Phenoxyphenyl)propanoic acid, (±)-	Fenoprofen	$C_{15}H_{14}O_3$	31879-05-7	242.270	visc oil		$170^{0.11}$		1.5742^{25}	
8788	3-Phenoxy-1,2-propanediol	Phenylglyceryl ether	$C_9H_{12}O_3$	538-43-2	168.189	nd (eth, peth)	67.5	200^{22}	1.225^{20}		vs H_2O, bz, eth, EtOH
8789	2-Phenoxypropanoic acid		$C_9H_{10}O_3$	940-31-8	166.173	nd (w)	115.5	266	1.1865^{20}	1.5184^{20}	
8790	2-Phenoxy-1-propanol		$C_9H_{12}O_2$	4169-04-4	152.190			244	0.9801^{25}	1.4760^{25}	s EtOH, eth
8791	1-Phenoxy-2-propanol		$C_9H_{12}O_2$	770-35-4	152.190			233	1.0622^{20}	1.5232^{20}	
8792	1-Phenoxy-2-propanone	Phenoxyacetone	$C_9H_{10}O_2$	621-87-4	150.174			229.5	1.0903^{20}	1.5228^{20}	s eth, ace
8793	2-Phenoxypropanoyl chloride		$C_9H_9ClO_2$	122-35-0	184.619			147	1.1865^{20}	1.5178^{20}	s eth
8794	Phenprocoumon	3-(α-Ethylbenzyl)-4-hydroxycoumarin	$C_{18}H_{16}O_3$	435-97-2	280.318	pr (MeOH aq)	183(1)				
8795	Phenthoate		$C_{12}H_{17}O_4PS_2$	2597-03-7	320.364	ye oil		$123^{0.01}$			sl H_2O; s hx
8796	Phentolamine		$C_{17}H_{19}N_3O$	50-60-2	281.352		175				
8797	Phenyl acetate		$C_8H_8O_2$	122-79-2	136.149			195(1)	1.0780^{20}	1.5035^{20}	sl H_2O; msc EtOH, eth, chl; s ctc
8798	2-Phenylacetophenone		$C_{14}H_{12}O$	451-40-1	196.244	pl (al)	56(3)	320	1.201^{0}		sl H_2O; s EtOH, eth, ctc, chl

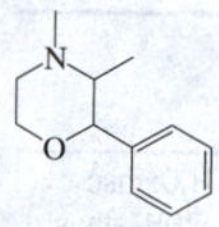
8756
Phendimetrazine

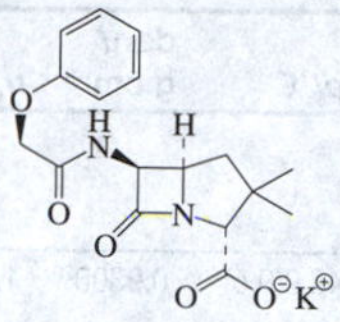
8757
Phenethicillin potassium

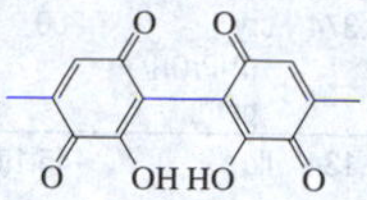
8758
Phenicin

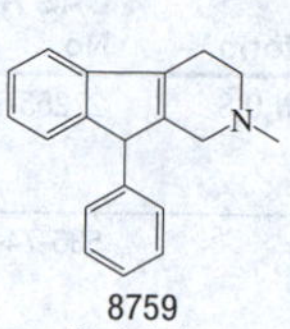
8759
Phenindamine

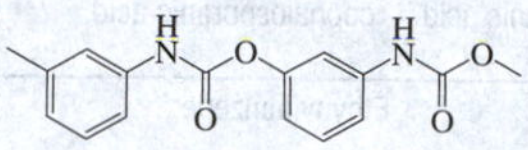
8760
Phenmedipham

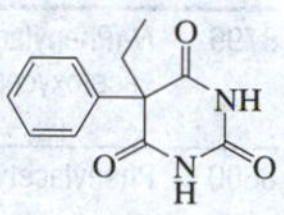
8761
Phenobarbital

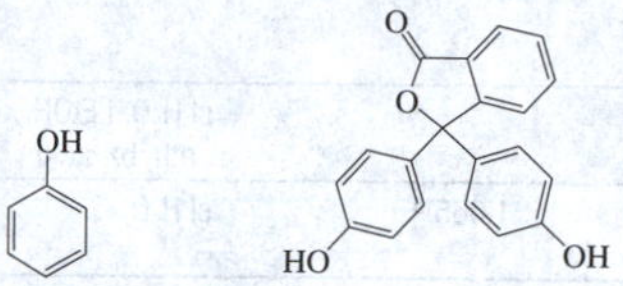
8762
Phenol

8763
Phenolphthalein

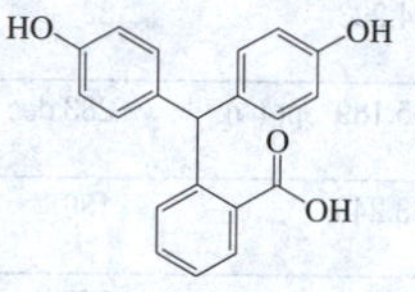
8764
Phenolphthalin

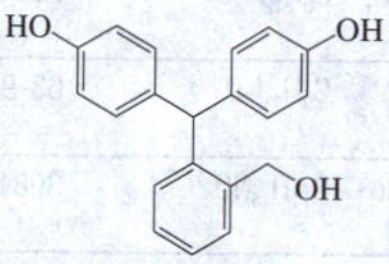
8765
Phenolphthalol

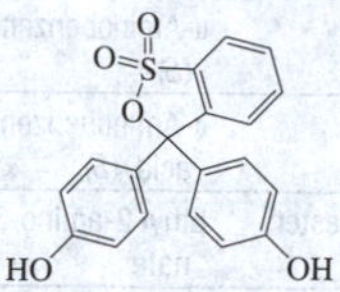
8766
Phenol Red

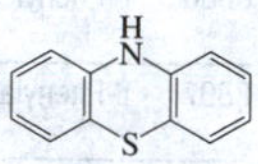
8767
10*H*-Phenothiazine

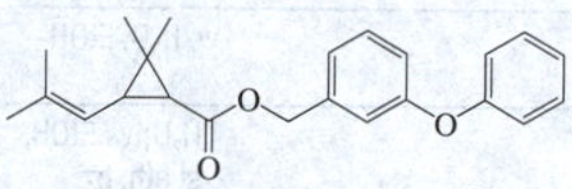
8768
Phenothrin

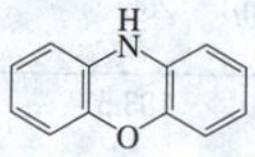
8769
10*H*-Phenoxazine

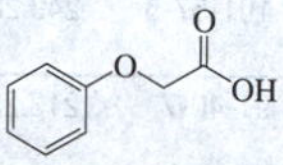
8770
Phenoxyacetic acid

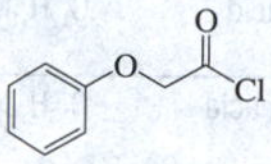
8771
Phenoxyacetyl chloride

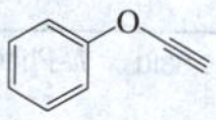
8772
Phenoxyacetylene

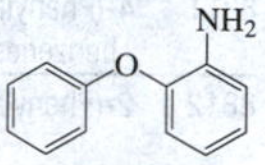
8773
2-Phenoxyaniline

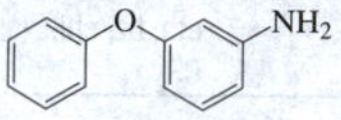
8774
3-Phenoxyaniline

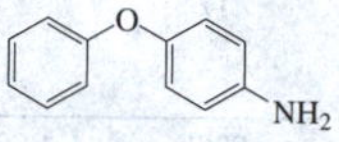
8775
4-Phenoxyaniline

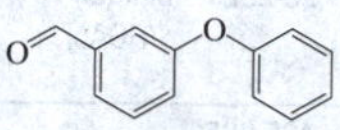
8776
3-Phenoxybenzaldehyde

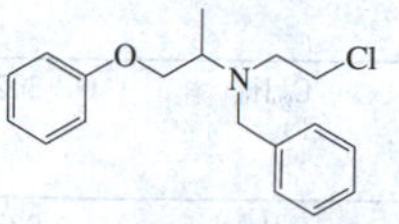
8777
Phenoxybenzamine

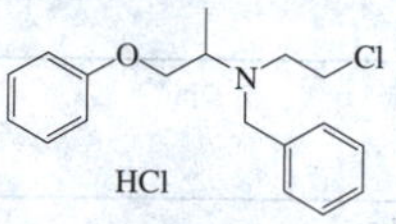
8778
Phenoxybenzamine hydrochloride

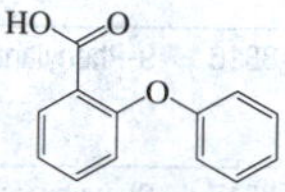
8779
2-Phenoxybenzoic acid

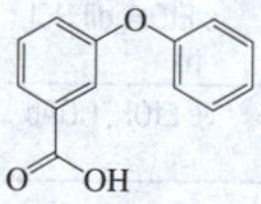
8780
3-Phenoxybenzoic acid

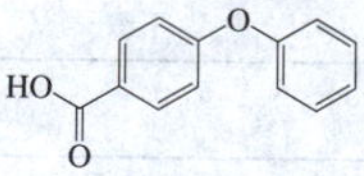
8781
4-Phenoxybenzoic acid

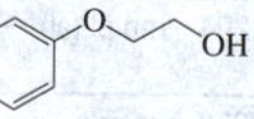
8782
2-Phenoxyethanol

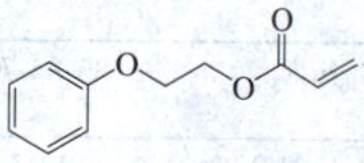
8783
2-Phenoxyethyl acrylate

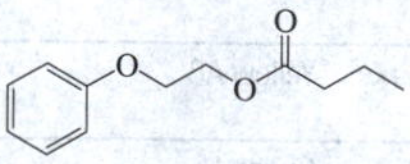
8784
2-Phenoxyethyl butanoate

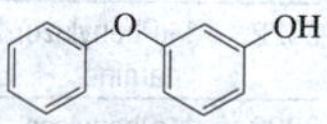
8785
3-Phenoxyphenol

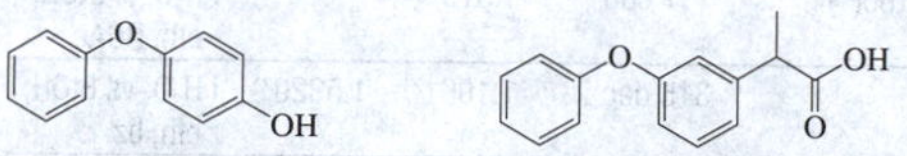
8786
4-Phenoxyphenol

8787
2-(3-Phenoxyphenyl)propanoic acid, (±)-

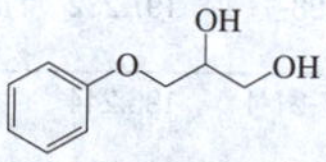
8788
3-Phenoxy-1,2-propanediol

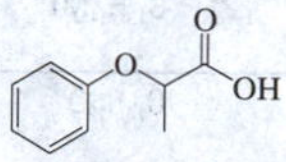
8789
2-Phenoxypropanoic acid

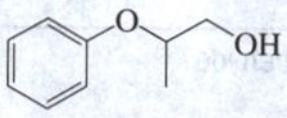
8790
2-Phenoxy-1-propanol

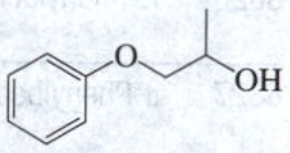
8791
1-Phenoxy-2-propanol

8792
1-Phenoxy-2-propanone

8793
2-Phenoxypropanoyl chloride

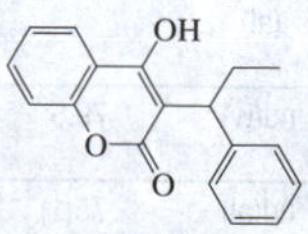
8794
Phenprocoumon

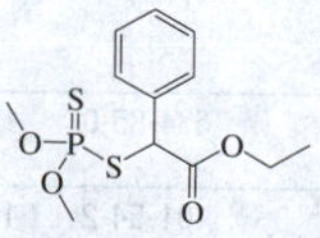
8795
Phenthoate

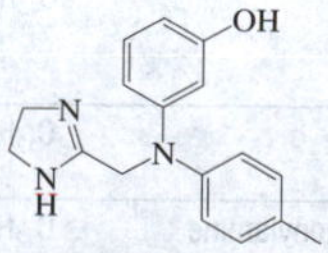
8796
Phentolamine

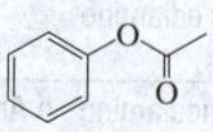
8797
Phenyl acetate

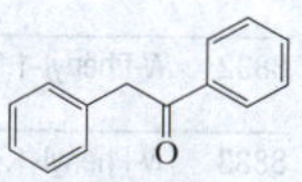
8798
2-Phenylacetophenone

No.	Name	Synonym	Mol. form.	CAS Reg. No.	Mol. wt.	Physical form	mp/°C	bp/°C	den/g cm^{-3}	n_D	Solubility
8799	*N*-(Phenylacetyl)-7-aminodeacetoxycephalosporanic acid	7-Phenylacetamidodeacetoxycephalosporanic acid	$C_{16}H_{16}N_2O_4S$	27255-72-7	332.374	cry (2-PrOH/peth)	200				
8800	Phenylacetylene	Ethynylbenzene	C_8H_6	536-74-3	102.134	liq	-45.1(0.4)	143.0(0.5)	0.9300^{20}	1.5470^{20}	i H_2O; msc EtOH, eth; s ace; sl chl
8801	(*N*-Phenylacetyl)glycine	Phenaceturic acid	$C_{10}H_{11}NO_3$	500-98-1	193.199	lf (EtOH)	143				
8802	Phenyl 2-(acetyloxy)benzoate	Phenyl acetylsalicylate	$C_{15}H_{12}O_4$	134-55-4	256.254		96				
8803	(Phenylacetyl)urea	Phenacemide	$C_9H_{10}N_2O_2$	63-98-9	178.187	cry (al)	215				vs bz, eth, EtOH
8804	9-Phenylacridine		$C_{19}H_{13}N$	602-56-2	255.313	ye nd, lf (al)	184	404			i H_2O; sl EtOH; s eth; vs bz
8805	L-Phenylalaninamide	α-Aminobenzenepropanamide, (*S*)-	$C_9H_{12}N_2O$	5241-58-7	164.203		82				
8806	L-Phenylalanine	α-Aminobenzenepropanoic acid, (*S*)	$C_9H_{11}NO_2$	63-91-2	165.189	pr (w)	283 dec				sl H_2O; i EtOH, eth, bz, acid
8807	L-Phenylalanine, ethyl ester	Ethyl 2-amino-3-phenylpropionate	$C_{11}H_{15}NO_2$	3081-24-1	193.243		136	148^{13}	1.065^{15}		sl H_2O
8808	L-Phenylalanylglycine		$C_{11}H_{14}N_2O_3$	721-90-4	222.240		262 dec				s H_2O
8809	3-Phenylallyl acetate		$C_{11}H_{12}O_2$	21040-45-9	176.212			265	1.0567^{20}	1.5425^{20}	i H_2O; s EtOH, eth, ace, bz, chl
8810	5-Phenyl-5-allyl-2,4,6(1*H*,3*H*,5*H*)-pyrimidinetrione	Phenallymal	$C_{13}H_{12}N_2O_3$	115-43-5	244.245		156.5				sl H_2O, bz, DMSO; vs EtOH, eth; i lig
8811	4-(Phenylamino)-benzenesulfonic acid	*N*-Phenylsulfanilic acid	$C_{12}H_{11}NO_3S$	101-57-5	249.285	pl (al-eth)	206				vs H_2O, EtOH
8812	2-(Phenylamino)benzoic acid	*N*-Phenylanthranilic acid	$C_{13}H_{11}NO_2$	91-40-7	213.232	lf (al)	183.5				i H_2O; vs EtOH; sl eth, bz
8813	Phenyl 4-amino-3-hydroxybenzoate	Phenyl *p*-aminosalicylate	$C_{13}H_{11}NO_3$	133-11-9	229.231		153				
8814	3-(Phenylamino)phenol		$C_{12}H_{11}NO$	101-18-8	185.221	lf (w)	81.5	340			sl H_2O; vs EtOH, eth, ace; s bz, acid
8815	4-(Phenylamino)phenol		$C_{12}H_{11}NO$	122-37-2	185.221	lf (w)	73	330			sl H_2O; vs EtOH, eth, bz, chl; s acid
8816	9-Phenylanthracene		$C_{20}H_{14}$	602-55-1	254.325	bl lf (al) (HOAc)	154(4)	417			i H_2O; s EtOH, eth, bz, chl, CS_2
8817	Phenylarsonous diiodide		$C_6H_5AsI_2$	6380-34-3	405.835		15	205^{14}	1.6264^{15}		
8818	4-(Phenylazo)-1,3-benzenediamine monohydrochloride	Chrysoidine hydrochloride	$C_{12}H_{13}ClN_4$	532-82-1	248.711	red-br cry pow	118.5				vs ace
8819	4-(Phenylazo)-1,3-benzenediol		$C_{12}H_{10}N_2O_2$	2051-85-6	214.219	dk red nd (dil al)	170				i H_2O; vs EtOH, eth, bz, HOAc
8820	4-Phenylazodiphenylamine	*N*-Phenyl-4-(phenylazo)-benzenamine	$C_{18}H_{15}N_3$	101-75-7	273.332	ye pl or pr	84.0				i H_2O; vs EtOH, eth, lig
8821	4-(Phenylazo)-1-naphthalenamine	α-Naphthyl Red	$C_{16}H_{13}N_3$	131-22-6	247.294	red-viol cry (EtOH)	123				s EtOH, dil HCl, bz
8822	1-(Phenylazo)-2-naphthalenamine	Yellow AB	$C_{16}H_{13}N_3$	85-84-7	247.294	red pl (al)	103				vs EtOH, HOAc
8823	1-(Phenylazo)-2-naphthol	Sudan I	$C_{16}H_{12}N_2O$	842-07-9	248.278	ye cry	132				
8824	4-(Phenylazo)phenol		$C_{12}H_{10}N_2O$	1689-82-3	198.219	ye lf (bz) oran pr (al)	151.5(0.8)	225^{20} dec			i H_2O; vs EtOH, eth; s bz, con sulf
8825	1-[[4-(Phenylazo)phenyl]azo]-2-naphthol	Sudan III	$C_{22}H_{16}N_4O$	85-86-9	352.388	br lf (grn lustre) (HOAc)	195				i H_2O; s EtOH, eth, ace, bz, xyl, chl
8826	*N*-Phenylbenzamide	Benzanilide	$C_{13}H_{11}NO$	93-98-1	197.232	lf (al)	160(1)	117 sub	1.315^{25}		i H_2O; sl EtOH, eth, HOAc
8827	α-Phenylbenzeneacetaldehyde		$C_{14}H_{12}O$	947-91-1	196.244			315 dec	1.1061^{21}	1.5920^{21}	i H_2O; vs EtOH, eth, bz
8828	α-Phenylbenzeneacetic acid	Diphenylacetic acid	$C_{14}H_{12}O_2$	117-34-0	212.244	nd (w), lf (al)	147.26(0.04)	194^{25}	1.257^{15}		sl H_2O; vs EtOH; s eth, chl
8829	α-Phenylbenzeneacetonitrile		$C_{14}H_{11}N$	86-29-3	193.244	pr (eth), lf (dil al)	74.3	184^{16}			s EtOH, chl; vs eth; sl lig
8830	α-Phenylbenzeneacetyl chloride		$C_{14}H_{11}ClO$	1871-76-7	230.689		56.5	170^{16}			s lig
8831	*N*-Phenylbenzenecarbothioamide		$C_{13}H_{11}NS$	636-04-4	213.298	ye pl or pr (al)	102	dec			i H_2O; vs EtOH; s eth, bz, chl; sl lig
8832	*N*-Phenyl-1,2-benzenediamine		$C_{12}H_{12}N_2$	534-85-0	184.236	nd(w)	79.5	313			sl H_2O, lig; s ace, bz, chl
8833	*N*-Phenyl-1,4-benzenediamine	*p*-Aminodiphenylamine	$C_{12}H_{12}N_2$	101-54-2	184.236	nd(al)	75(1)	354			sl H_2O, chl; vs EtOH; s eth, lig

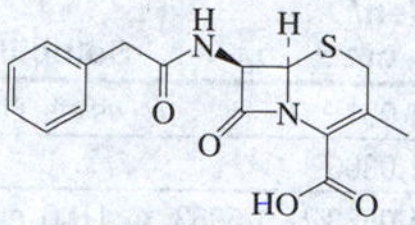
8799
N-(Phenylacetyl)-7-aminodeacetoxycephalosporanic acid

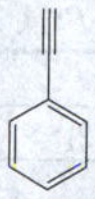
8800
Phenylacetylene

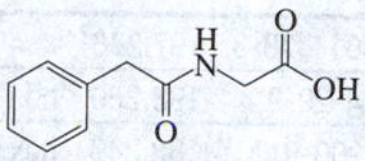
8801
(*N*-Phenylacetyl)glycine

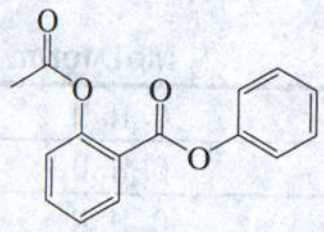
8802
Phenyl 2-(acetyloxy)benzoate

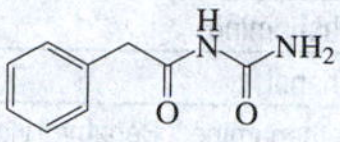
8803
(Phenylacetyl)urea

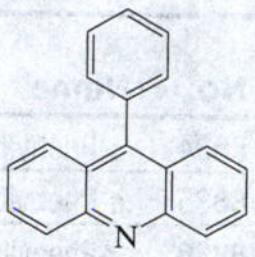
8804
9-Phenylacridine

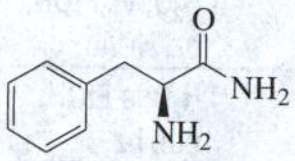
8805
L-Phenylalaninamide

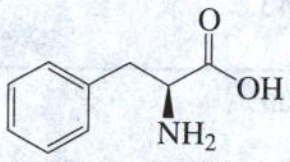
8806
L-Phenylalanine

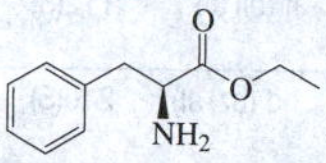
8807
L-Phenylalanine, ethyl ester

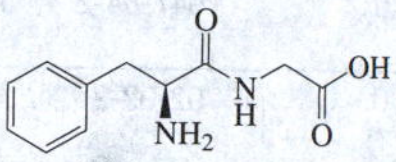
8808
L-Phenylalanylglycine

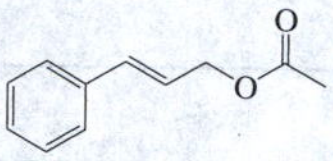
8809
3-Phenylallyl acetate

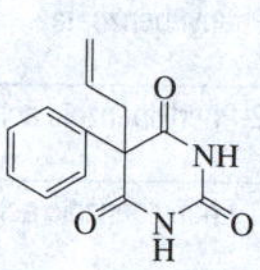
8810
5-Phenyl-5-allyl-2,4,6(1*H*,3*H*,5*H*)-pyrimidinetrione

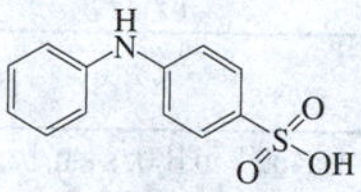
8811
4-(Phenylamino)benzenesulfonic acid

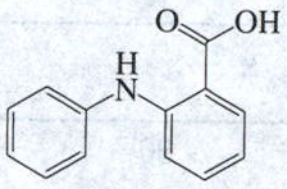
8812
2-(Phenylamino)benzoic acid

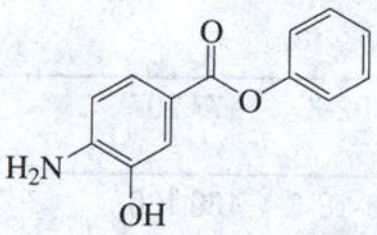
8813
Phenyl 4-amino-3-hydroxybenzoate

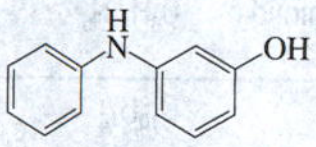
8814
3-(Phenylamino)phenol

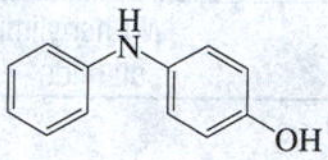
8815
4-(Phenylamino)phenol

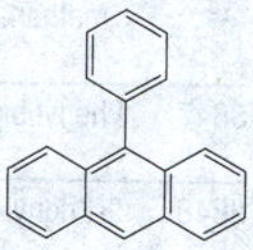
8816
9-Phenylanthracene

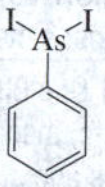
8817
Phenylarsonous diiodide

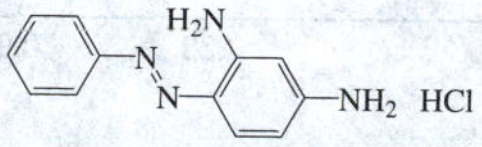
8818
4-(Phenylazo)-1,3-benzenediamine monohydrochloride

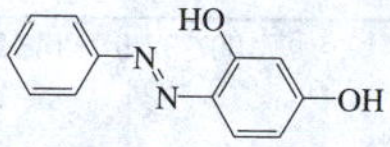
8819
4-(Phenylazo)-1,3-benzenediol

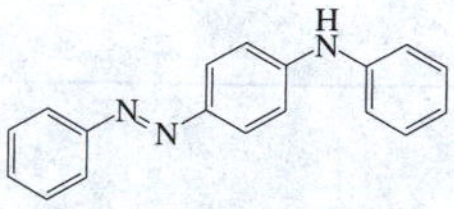
8820
4-Phenylazodiphenylamine

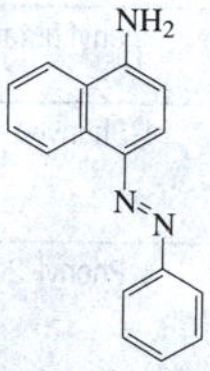
8821
4-(Phenylazo)-1-naphthalenamine

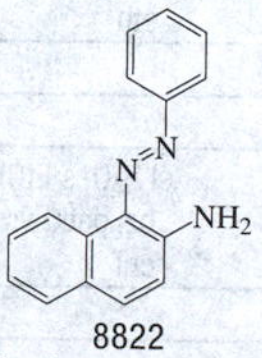
8822
1-(Phenylazo)-2-naphthalenamine

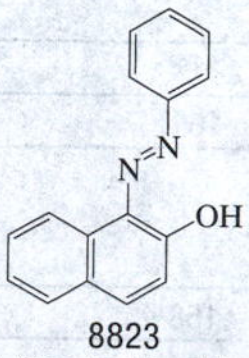
8823
1-(Phenylazo)-2-naphthol

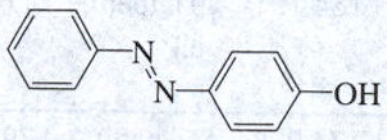
8824
4-(Phenylazo)phenol

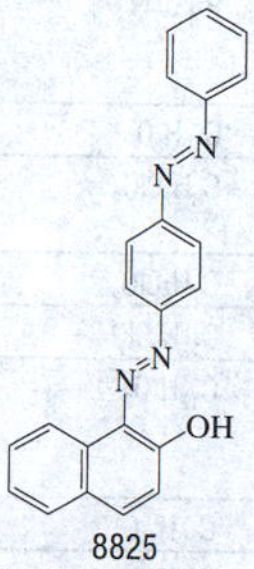
8825
1-[[4-(Phenylazo)phenyl]azo]-2-naphthol

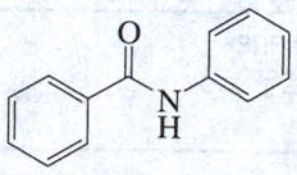
8826
N-Phenylbenzamide

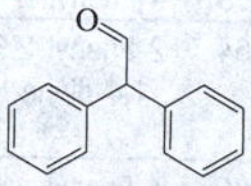
8827
α-Phenylbenzeneacetaldehyde

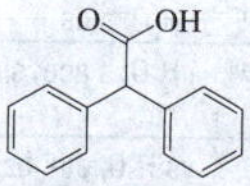
8828
α-Phenylbenzeneacetic acid

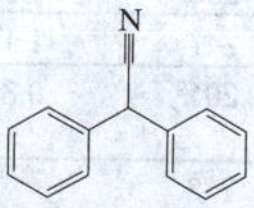
8829
α-Phenylbenzeneacetonitrile

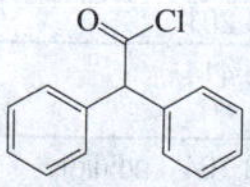
8830
α-Phenylbenzeneacetyl chloride

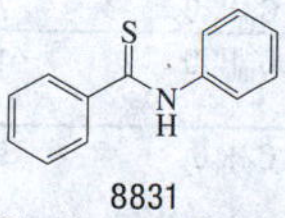
8831
N-Phenylbenzenecarbothioamide

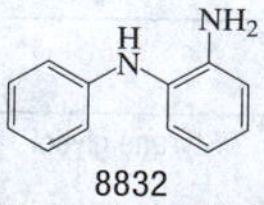
8832
N-Phenyl-1,2-benzenediamine

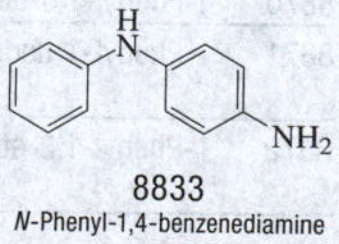
8833
N-Phenyl-1,4-benzenediamine

No.	Name	Synonym	Mol. form.	CAS Reg. No.	Mol. wt.	Physical form	mp/°C	bp/°C	den/ g cm⁻³	n_D	Solubility
8834	α-Phenylbenzeneethanamine		$C_{14}H_{15}N$	25611-78-3	197.276			311	1.031^{15}		vs eth, EtOH
8835	α-Phenylbenzeneethanol		$C_{14}H_{14}O$	614-29-9	198.260	nd (peth-bz)	53(2)	177^{15}	1.0360^{70}		
8836	α-Phenylbenzenemethanamine	Benzhydrylamine	$C_{13}H_{13}N$	91-00-9	183.249	hex pl	34	304	1.0633^{20}	1.5963	sl H_2O; s ace
8837	α-Phenylbenzenemethanimine		$C_{13}H_{11}N$	1013-88-3	181.233			282	1.0847^{19}	1.6191^{19}	vs eth
8838	β-Phenylbenzenepropanoic acid		$C_{15}H_{14}O_2$	606-83-7	226.271	nd (dil al)	156.0				sl H_2O; vs EtOH; s eth, ace
8839	2-Phenylbenzimidazole	Phenzidole	$C_{13}H_{10}N_2$	716-79-0	194.231	pl (HOAc) (al-w) nd (bz, w)	293				sl H_2O, bz; s EtOH, chl, HOAc
8840	Phenyl benzoate		$C_{13}H_{10}O_2$	93-99-2	198.217	mcl pr (eth-al)	63(8)	314	1.235^{20}		i H_2O; s EtOH, eth, chl
8841	2-Phenylbenzoic acid		$C_{13}H_{10}O_2$	947-84-2	198.217	lf (dil al)	112(3)	343.5			i H_2O; vs EtOH, bz, HOAc
8842	4-Phenylbenzoic acid		$C_{13}H_{10}O_2$	92-92-2	198.217	nd (bz, al)	210(5)	sub			i H_2O; s EtOH, eth, bz
8843	2-Phenyl-4*H*-1-benzopyran-4-one	Flavone	$C_{15}H_{10}O_2$	525-82-6	222.239	nd (lig), cry (30% al)	96.7(0.5)				i H_2O; s EtOH, eth, ace, bz
8844	3-Phenyl-4*H*-1-benzopyran-4-one	Isoflavone	$C_{15}H_{10}O_2$	574-12-9	222.239		148				
8845	2-Phenylbenzothiazole		$C_{13}H_9NS$	883-93-2	211.282	nd (dil al)	115	371			i H_2O; s EtOH, eth, CS_2
8846	*N*-Phenyl-*N*-benzylbenzene-methanamine		$C_{20}H_{19}N$	91-73-6	273.372		69	226^{10}	1.0444^{80}	1.6065^{80}	i H_2O; sl EtOH, HOAc; s eth, bz
8847	Phenyl biguanide	*N*-Phenylimidodicarbonimidic diamide	$C_8H_{11}N_5$	102-02-3	177.207		143				
8848	2-Phenyl-1,3-butadiene		$C_{10}H_{10}$	2288-18-8	130.186			60^{17}	0.925^{20}	1.5489^{20}	i H_2O; s eth, bz, chl
8849	*N*-Phenylbutanamide		$C_{10}H_{13}NO$	1129-50-6	163.216	mcl pr (al, bz, eth)	97	189^{15}	1.134^{25}		i H_2O; vs EtOH, eth; sl chl
8850	Phenylbutanedioic acid, (±)-		$C_{10}H_{10}O_4$	10424-29-0	194.184	lf or nd (w)	167(1)	dec			sl H_2O, chl; vs EtOH, eth, ace; i bz
8851	1-Phenyl-1,3-butanedione		$C_{10}H_{10}O_2$	93-91-4	162.185	pr	60(2)	257(9)	1.0599^{74}	1.5678^{78}	i H_2O; s eth; sl chl
8852	Phenyl butanoate	Phenyl butyrate	$C_{10}H_{12}O_2$	4346-18-3	164.201			228(6)	1.0382^{15}		i H_2O; s EtOH, eth
8853	1-Phenyl-1-butanone		$C_{10}H_{12}O$	495-40-9	148.201		12	228.5	0.988^{20}	1.5203^{20}	i H_2O; msc EtOH, eth; vs ace; s ctc
8854	1-Phenyl-2-butanone		$C_{10}H_{12}O$	1007-32-5	148.201			228	0.9877^{20}		i H_2O; s EtOH, ctc; msc eth; vs ace
8855	4-Phenyl-2-butanone		$C_{10}H_{12}O$	2550-26-7	148.201	liq	-13	235(7)	0.9849^{22}	1.511^{22}	i H_2O; s EtOH, eth, ctc; vs ace
8856	Phenylbutazone		$C_{19}H_{20}N_2O_2$	50-33-9	308.374		104(1)				
8857	2-Phenyl-1-butene	α-Ethylstyrene	$C_{10}H_{12}$	2039-93-2	132.202			190(6)	0.887^{25}	1.5288^{20}	
8858	1-Phenyl-2-buten-1-one		$C_{10}H_{10}O$	495-41-0	146.185		20.5	111^{9}	1.025^{15}	1.5626^{18}	
8859	*trans*-4-Phenyl-3-buten-2-one	Benzilideneacetone	$C_{10}H_{10}O$	1896-62-4	146.185	pl	41(3)	261	1.0097^{45}	1.5836^{45}	i H_2O; vs EtOH; s eth, ace, bz; sl peth
8860	4-Phenyl-3-butyn-2-one		$C_{10}H_8O$	1817-57-8	144.170		4.5	79^{2}	1.0215^{20}	1.5762^{20}	
8861	Phenyl chloroacetate		$C_8H_7ClO_2$	620-73-5	170.594	nd or pl (al)	44.5	232.5	1.2202^{44}	1.5146^{44}	i H_2O; vs EtOH, eth
8862	Phenyl chloroformate		$C_7H_5ClO_2$	1885-14-9	156.567			71^{9}			
8863	4-Phenyl-2-chlorophenol	3-Chloro-(1,1'-biphenyl)-4-ol	$C_{12}H_9ClO$	92-04-6	204.651	wh-ye cry	77	161^{7}			
8864	2-Phenyl-2,5-cyclohexadiene-1,4-dione		$C_{12}H_8O_2$	363-03-1	184.191	ye lf (peth, al)	114				sl H_2O; s EtOH, bz, peth; vs chl
8865	4-Phenylcyclohexanone		$C_{12}H_{14}O$	4894-75-1	174.238	cry (peth)	79	158^{12}			
8866	1-(1-Phenylcyclohexyl)-piperidine	Phencyclidine	$C_{17}H_{25}N$	77-10-1	243.388		46.5	$136^{1.0}$			
8867	3-Phenyl-2-cyclopenten-1-one		$C_{11}H_{10}O$	3810-26-2	158.196	liq	-23	234.2	0.9711^{20}	1.5440^{20}	s EtOH, ace, chl; sl eth
8868	*N*-Phenyl-*N*,*N*-diethanolamine		$C_{10}H_{15}NO_2$	120-07-0	181.232		57	200^{10}	1.201^{60}		vs ace, bz, eth, EtOH
8869	2-Phenyl-1,3-dioxane		$C_{10}H_{12}O_2$	772-01-0	164.201	nd (peth)	48(2)	253	1.6053^{60}		vs EtOH, eth
8870	4-Phenyl-1,3-dioxane		$C_{10}H_{12}O_2$	772-00-9	164.201			247	1.1038^{20}	1.5306^{18}	i H_2O; s os
8871	1-Phenyl-1-dodecanone		$C_{18}H_{28}O$	1674-38-0	260.414		44(1)	201^{9}	0.8794^{18}	1.4700^{18}	i H_2O; s ace; sl ctc
8872	1-Phenyl-1,2-ethanediol	Styrene glycol	$C_8H_{10}O_2$	93-56-1	138.164	nd (lig)	66(3)	273			vs H_2O, eth, bz, EtOH; sl lig

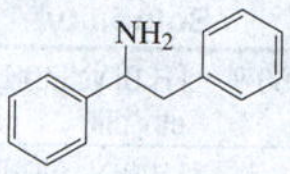
8834
α-Phenylbenzeneethanamine

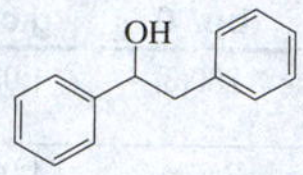
8835
α-Phenylbenzeneethanol

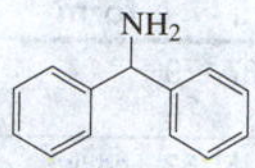
8836
α-Phenylbenzenemethanamine

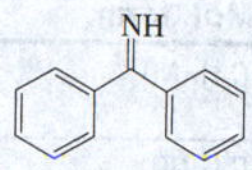
8837
α-Phenylbenzenemethanimine

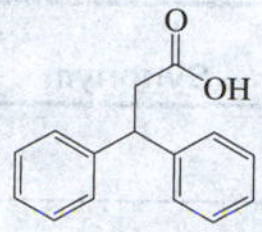
8838
β-Phenylbenzenepropanoic acid

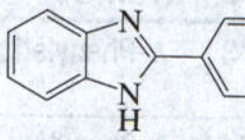
8839
2-Phenylbenzimidazole

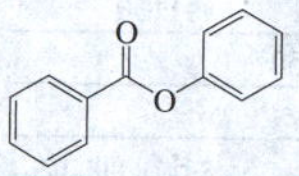
8840
Phenyl benzoate

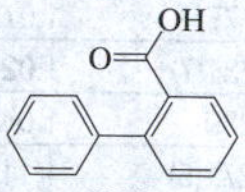
8841
2-Phenylbenzoic acid

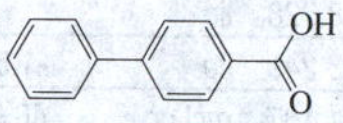
8842
4-Phenylbenzoic acid

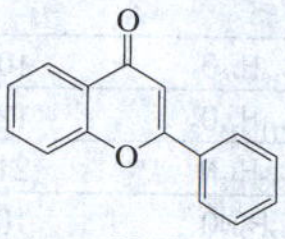
8843
2-Phenyl-4*H*-1-benzopyran-4-one

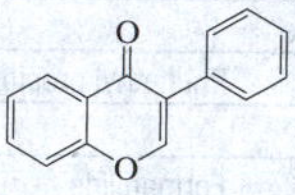
8844
3-Phenyl-4*H*-1-benzopyran-4-one

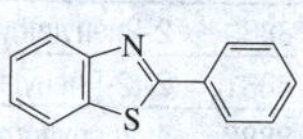
8845
2-Phenylbenzothiazole

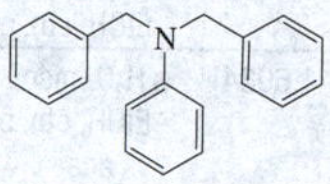
8846
N-Phenyl-*N*-benzylbenzenemethanamine

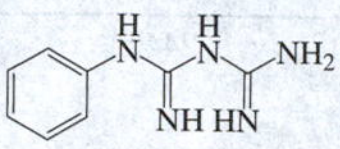
8847
Phenyl biguanide

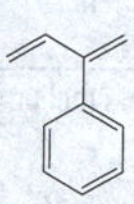
8848
2-Phenyl-1,3-butadiene

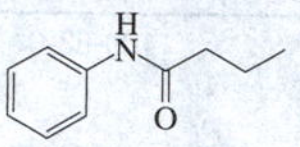
8849
N-Phenylbutanamide

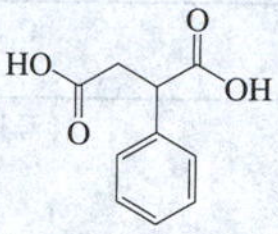
8850
Phenylbutanedioic acid, (±)-

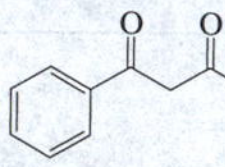
8851
1-Phenyl-1,3-butanedione

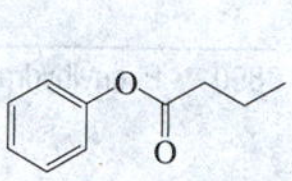
8852
Phenyl butanoate

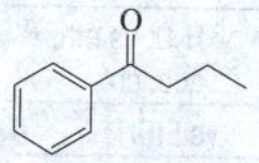
8853
1-Phenyl-1-butanone

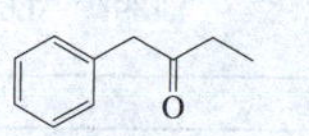
8854
1-Phenyl-2-butanone

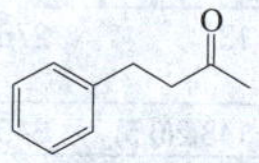
8855
4-Phenyl-2-butanone

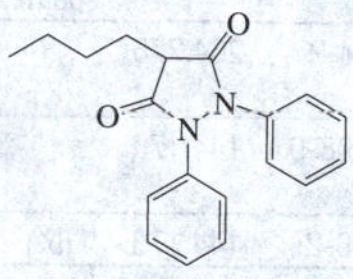
8856
Phenylbutazone

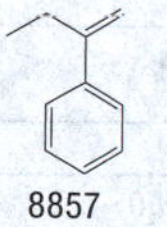
8857
2-Phenyl-1-butene

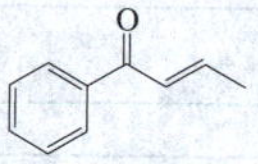
8858
1-Phenyl-2-buten-1-one

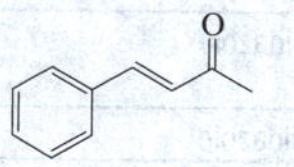
8859
trans-4-Phenyl-3-buten-2-one

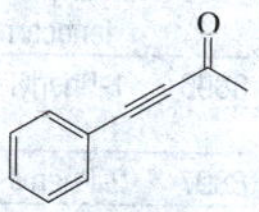
8860
4-Phenyl-3-butyn-2-one

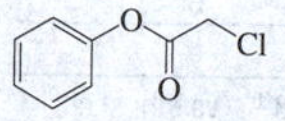
8861
Phenyl chloroacetate

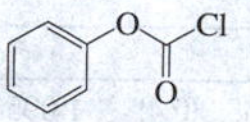
8862
Phenyl chloroformate

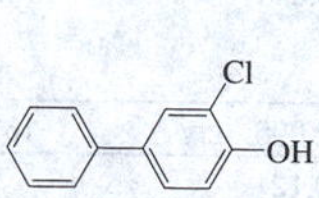
8863
4-Phenyl-2-chlorophenol

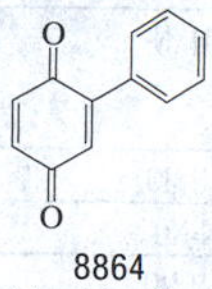
8864
2-Phenyl-2,5-cyclohexadiene-1,4-dione

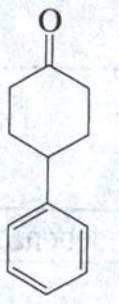
8865
4-Phenylcyclohexanone

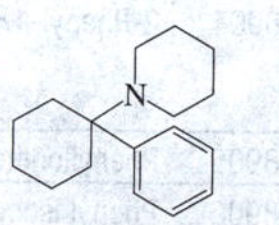
8866
1-(1-Phenylcyclohexyl)piperidine

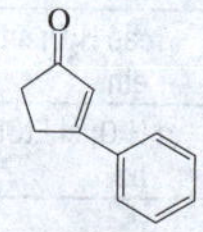
8867
3-Phenyl-2-cyclopenten-1-one

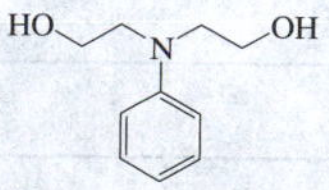
8868
N-Phenyl-*N,N*-diethanolamine

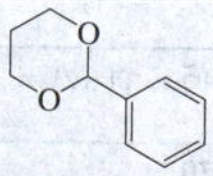
8869
2-Phenyl-1,3-dioxane

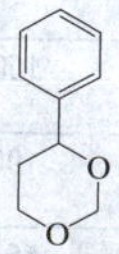
8870
4-Phenyl-1,3-dioxane

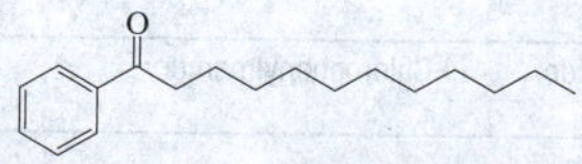
8871
1-Phenyl-1-dodecanone

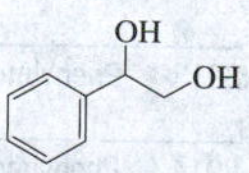
8872
1-Phenyl-1,2-ethanediol

Organic

No.	Name	Synonym	Mol. form.	CAS Reg. No.	Mol. wt.	Physical form	mp/°C	bp/°C	den/g cm^{-3}	n_D	Solubility
8873	*N*-Phenylethanolamine		$C_8H_{11}NO$	122-98-5	137.179			280(3)	1.0945^{20}	1.5760^{20}	sl H_2O; vs EtOH, eth, chl
8874	1-Phenylethanone oxime		C_8H_9NO	613-91-2	135.163	nd (w)	60	245	1.0515^{78}		sl H_2O; vs EtOH, eth, ace, bz; s ctc
8875	2-Phenylethyl acetate		$C_{10}H_{12}O_2$	103-45-7	164.201	liq	-31.1	233(1)	1.0883^{20}	1.5171^{20}	vs eth, EtOH
8876	1-Phenylethyl hydroperoxide		$C_8H_{10}O_2$	3071-32-7	138.164	liq		$50^{0.01}$			
8877	*N*-(2-Phenylethyl)-imidodicarbonimidic diamide, monohydrochloride	Phenformin hydrochloride	$C_{10}H_{16}ClN_5$	834-28-6	241.721	cry	178(1)				s H_2O
8878	2-Phenylethyl 2-methylpropanoate	Benzylcarbinol isobutyrate	$C_{12}H_{16}O_2$	103-48-0	192.254			250	0.9950^{15}	1.4871^{20}	
8879	2-Phenylethyl phenylacetate		$C_{16}H_{16}O_2$	102-20-5	240.297		26.5	$177^{4.5}$	1.077^{25}		vs EtOH
8880	2-Phenylethyl propanoate	Phenethyl propionate	$C_{11}H_{14}O_2$	122-70-3	178.228	liq		231(7)	1.02^{25}	1.4950^{20}	
8881	2-(2-Phenylethyl)pyridine		$C_{13}H_{13}N$	2116-62-3	183.249	liq	-1.5	289	1.0465^{0}		
8882	*N*-Phenylformamide	Formanilide	C_7H_7NO	103-70-8	121.137	mcl pr (lig-xyl)	46.5(0.6)	271(5)	1.1186^{50}		s H_2O, eth, bz; vs EtOH
8883	Phenyl formate		$C_7H_6O_2$	1864-94-4	122.122	liq		176(19)			
8884	2-Phenylfuran		$C_{10}H_8O$	17113-33-6	144.170			108^{18}	1.083^{20}	1.5920^{20}	vs ace, bz
8885	Phenyl α-D-glucopyranoside		$C_{12}H_{16}O_6$	4630-62-0	256.251		156(1)				
8886	Phenyl glycidyl ether		$C_9H_{10}O_2$	122-60-1	150.174			244.2(0.9)	1.1109^{21}	1.5307^{21}	
8887	*N*-Phenylglycine	Phenylaminoacetic acid	$C_8H_9NO_2$	103-01-5	151.163		127.5				vs H_2O, EtOH
8888	1-Phenyl-1-heptanone		$C_{13}H_{18}O$	1671-75-6	190.281	lf	17(3)	284(13)	0.9516^{20}	1.5060^{20}	vs ace, eth, EtOH
8889	1-Phenyl-1-hexanone		$C_{12}H_{16}O$	942-92-7	176.254	fl	27	265	0.9576^{20}	1.5027^{25}	sl H_2O, ctc; s EtOH, eth, ace
8890	Phenylhydrazine		$C_6H_8N_2$	100-63-0	108.141	mcl pr or pl	20(2)	244(5)	1.0986^{20}	1.6084^{10}	s H_2O; msc EtOH, eth, bz; vs ace
8891	2-Phenylhydrazinecarboxamide	Phenicarbazide	$C_7H_9N_3O$	103-03-7	151.165		172				sl H_2O, eth, bz, lig; s EtOH, ace
8892	*N*-Phenylhydrazinecarboxamide	4-Phenylsemicarbazide	$C_7H_9N_3O$	537-47-3	151.165	nd (bz), pl (w)	128				sl H_2O; vs EtOH, chl; i eth
8893	Phenylhydrazine monohydrochloride		$C_6H_9ClN_2$	59-88-1	144.601	lf (al)	244 dec	sub			vs H_2O, EtOH
8894	Phenylhydroxylamine	*N*-Hydroxybenzenamine	C_6H_7NO	100-65-2	109.126	nd (w, bz, peth)	79(2)				vs bz, eth, EtOH, chl
8895	Phenyl 1-hydroxy-2-naphthalenecarboxylate		$C_{17}H_{12}O_3$	132-54-7	264.275		96				vs bz, EtOH
8896	1-Phenyl-1*H*-imidazole		$C_9H_8N_2$	7164-98-9	144.173		13	276	1.1397^{15}	1.6025^{25}	i H_2O; vs eth, ace, chl
8897	2-Phenyl-1*H*-imidazole		$C_9H_8N_2$	670-96-2	144.173	lf (bz)	148.2(0.5)	340			vs EtOH
8898	5-Phenyl-2,4-imidazolidinedione	5-Phenylhydantoin	$C_9H_8N_2O_2$	89-24-7	176.172		184.5				
8899	Phenylimidocarbonyl chloride		$C_7H_5Cl_2N$	622-44-6	174.028	liq		210	1.28^{15}		
8900	2-[(Phenylimino)methyl]phenol		$C_{13}H_{11}NO$	779-84-0	197.232		49.5		1.087^{25}		i H_2O; s EtOH
8901	4-[(Phenylimino)methyl]phenol	*N*-(4-Hydroxybenzilidene)-aniline	$C_{13}H_{11}NO$	1689-73-2	197.232		196.0				i H_2O; s EtOH, eth; sl bz, chl
8902	1-Phenyl-1*H*-indene		$C_{15}H_{12}$	1961-96-2	192.256	oil		158^{7}			
8903	2-Phenyl-1*H*-indene-1,3(2*H*)-dione	Phenindione	$C_{15}H_{10}O_2$	83-12-5	222.239	lf (al, bz)	150				i H_2O; s EtOH, eth, ace, bz, MeOH, chl
8904	2-Phenyl-1*H*-indole		$C_{14}H_{11}N$	948-65-2	193.244		188(3)	250^{10}			sl H_2O; s eth, bz, chl, HOAc, CS_2
8905	Phenyliodine diacetate	Iodobenzene diacetate	$C_{10}H_{11}IO_4$	3240-34-4	322.096	cry	161				
8906	Phenyl isocyanate		C_7H_5NO	103-71-9	119.121			166.3(0.4)	1.0956^{20}	1.5368^{20}	vs eth; sl chl
8907	2-Phenyl-1*H*-isoindole-1,3(2*H*)-dione		$C_{14}H_9NO_2$	520-03-6	223.227	wh nd (al)	205(1)	sub			i H_2O; sl EtOH; msc chl
8908	Phenyl isopropyl ether	Isopropoxybenzene	$C_9H_{12}O$	2741-16-4	136.190	liq	-33.0(0.4)	180(4)	0.9408^{25}	1.4975^{20}	s H_2O, EtOH, ace, bz
8909	Phenyl isothiocyanate		C_7H_5NS	103-72-0	135.187	liq	-21	220(7)	1.1303^{20}	1.6492^{23}	i H_2O; s EtOH, eth, ctc
8910	3-Phenyl-2-isoxazolin-5-one		$C_9H_7NO_2$	1076-59-1	161.158		151				sl chl
8911	Phenyl laurate	Phenyl dodecanoate	$C_{18}H_{28}O_2$	4228-00-6	276.414	lf (al)	24.5	210^{15}	0.9354^{30}		vs ace, eth, EtOH
8912	Phenylmagnesium chloride	Chlorophenylmagnesium	C_6H_5ClMg	100-59-4	136.862	cry					reac H_2O; s thf, eth
8913	Phenylmercuric chloride	Chlorophenylmercury	C_6H_5ClHg	100-56-1	313.15	pl (bz)	251				i H_2O; sl EtOH, bz
8914	Phenylmercuric nitrate		$C_6H_5HgNO_3$	55-68-5	339.70		≈181				

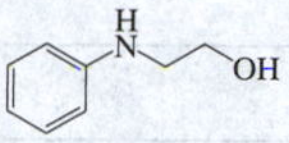
8873
N-Phenylethanolamine

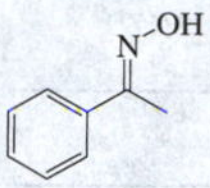
8874
1-Phenylethanone oxime

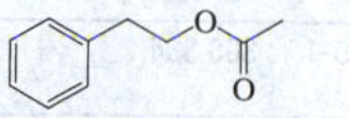
8875
2-Phenylethyl acetate

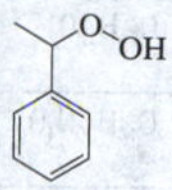
8876
1-Phenylethyl hydroperoxide

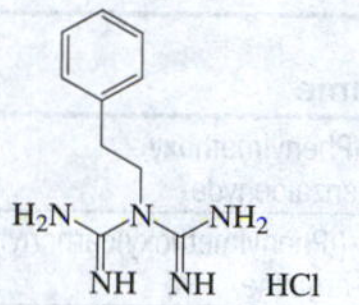
8877
N-(2-Phenylethyl)imidodicarbonimidic diamide, monohydrochloride

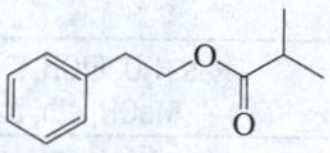
8878
2-Phenylethyl 2-methylpropanoate

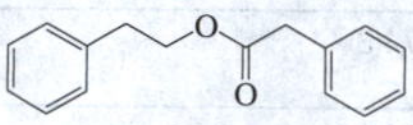
8879
2-Phenylethyl phenylacetate

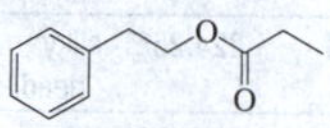
8880
2-Phenylethyl propanoate

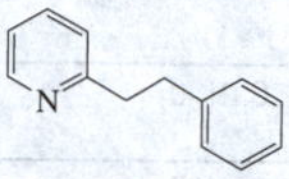
8881
2-(2-Phenylethyl)pyridine

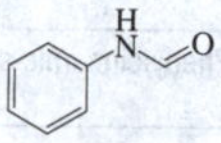
8882
N-Phenylformamide

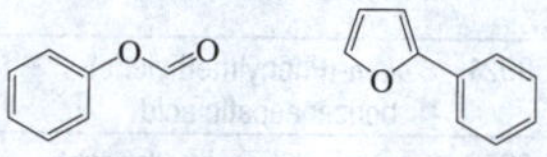
8883
Phenyl formate

8884
2-Phenylfuran

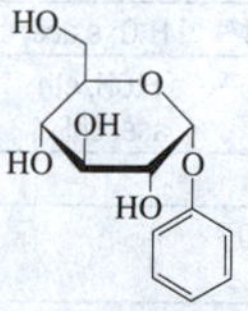
8885
Phenyl α-D-glucopyranoside

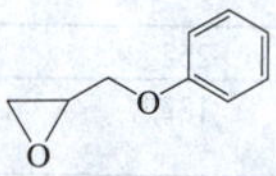
8886
Phenyl glycidyl ether

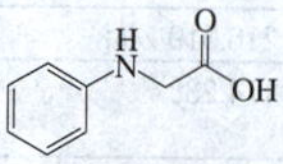
8887
N-Phenylglycine

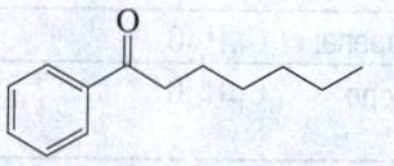
8888
1-Phenyl-1-heptanone

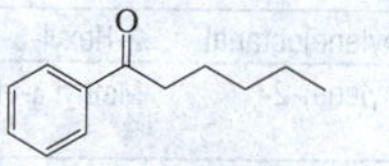
8889
1-Phenyl-1-hexanone

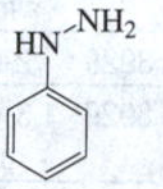
8890
Phenylhydrazine

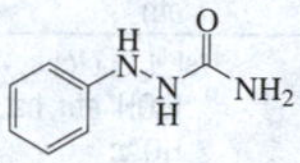
8891
2-Phenylhydrazinecarboxamide

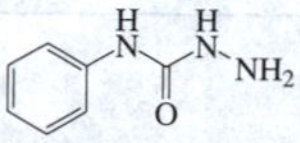
8892
N-Phenylhydrazinecarboxamide

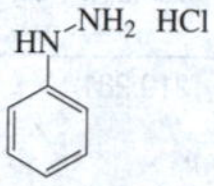
8893
Phenylhydrazine monohydrochloride

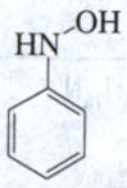
8894
Phenylhydroxylamine

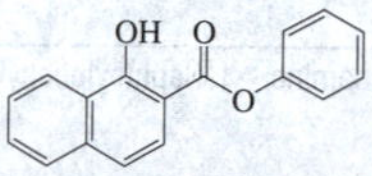
8895
Phenyl 1-hydroxy-2-naphthalenecarboxylate

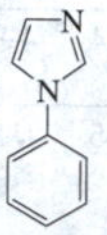
8896
1-Phenyl-1*H*-imidazole

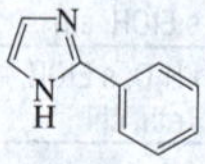
8897
2-Phenyl-1*H*-imidazole

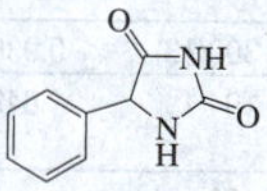
8898
5-Phenyl-2,4-imidazolidinedione

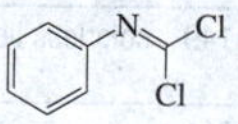
8899
Phenylimidocarbonyl chloride

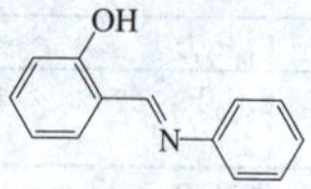
8900
2-[(Phenylimino)methyl]phenol

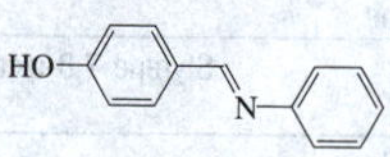
8901
4-[(Phenylimino)methyl]phenol

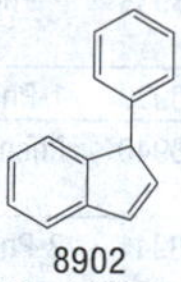
8902
1-Phenyl-1*H*-indene

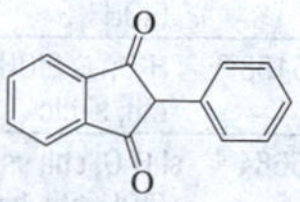
8903
2-Phenyl-1*H*-indene-1,3(2*H*)-dione

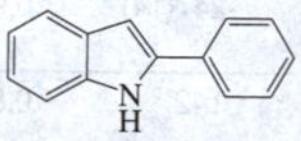
8904
2-Phenyl-1*H*-indole

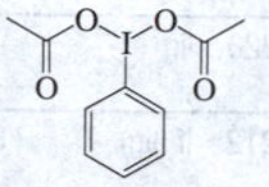
8905
Phenyliodine diacetate

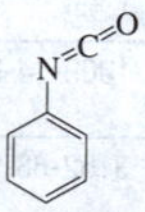
8906
Phenyl isocyanate

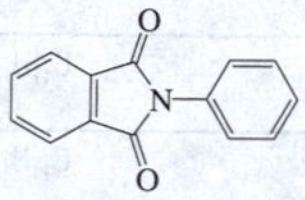
8907
2-Phenyl-1*H*-isoindole-1,3(2*H*)-dione

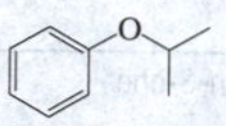
8908
Phenyl isopropyl ether

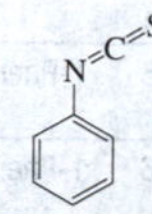
8909
Phenyl isothiocyanate

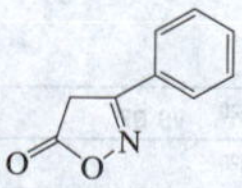
8910
3-Phenyl-2-isoxazolin-5-one

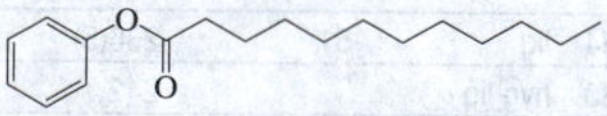
8911
Phenyl laurate

8912
Phenylmagnesium chloride

8913
Phenylmercuric chloride

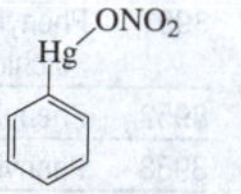
8914
Phenylmercuric nitrate

No.	Name	Synonym	Mol. form.	CAS Reg. No.	Mol. wt.	Physical form	mp/°C	bp/°C	den/ g cm⁻³	n_D	Solubility
8915	4-(Phenylmethoxy)-benzaldehyde		$C_{14}H_{12}O_2$	4397-53-9	212.244		73	217^{13}			
8916	N^2-[(Phenylmethoxy)carbonyl]-L-arginine		$C_{14}H_{20}N_4O_4$	1234-35-1	308.334		174				
8917	*N*-[(Phenylmethoxy)carbonyl]-L-aspartic acid		$C_{12}H_{13}NO_6$	1152-61-0	267.234		117.0				
8918	2-(Phenylmethoxy)phenol		$C_{13}H_{12}O_2$	6272-38-4	200.233			205^{20}	1.154^{22}	1.5906^{18}	vs eth, EtOH
8919	4-(Phenylmethoxy)phenol	Monobenzone	$C_{13}H_{12}O_2$	103-16-2	200.233	pl (w)	122				sl H_2O; vs EtOH, bz, eth; s ace
8920	*N*-(Phenylmethylene)aniline	Benzylideneaniline	$C_{13}H_{11}N$	538-51-2	181.233	pa ye nd (CS_2) pl (dil al)	56(5)	310	1.038^{55}	1.600^{100}	i H_2O; s EtOH, eth, NH_3; sl chl
8921	*cis*-α-(Phenylmethylene)-benzeneacetic acid	*cis*-α-Phenylcinnamic acid	$C_{15}H_{12}O_2$	91-47-4	224.255	silky needles	174(3)				s H_2O, EtOH, MeOH, eth, bz
8922	*trans*-α-(Phenylmethylene)-benzeneacetic acid	*trans*-α-Phenylcinnamic acid	$C_{15}H_{12}O_2$	91-48-5	224.255	prisms	138				vs H_2O; s EtOH, MeOH, eth, bz
8923	*N*-(Phenylmethylene)-benzenemethanamine		$C_{14}H_{13}N$	780-25-6	195.260			205^{20}			
8924	2-(Phenylmethylene)butanal		$C_{11}H_{12}O$	28467-92-7	160.212		18	243	1.0201^{22}	1.578^{20}	
8925	*N*-(Phenylmethylene)-ethanamine		$C_9H_{11}N$	6852-54-6	133.190			195	0.937^{20}	1.5378^{15}	i H_2O; s EtOH, eth
8926	2-(Phenylmethylene)heptanal		$C_{14}H_{18}O$	122-40-7	202.292	ye oil	80	174^{20}	0.9711^{20}	1.5381^{20}	i H_2O; s ace, ctc
8927	*N*-(Phenylmethylene)-methanamine	Benzylidenemethylamine	C_8H_9N	622-29-7	119.164			185	0.9671^{14}	1.5526^{20}	s EtOH, eth, ace, chl
8928	2-(Phenylmethylene)octanal	2-Hexyl-3-phenyl-2-propenal	$C_{15}H_{20}O$	101-86-0	216.319	liq	4	252			
8929	3-(Phenylmethylene)-2-pentanone	Methyl α-ethylstyryl ketone	$C_{12}H_{14}O$	3437-89-6	174.238			137^{12}	1.0005^{22}	1.5650^{22}	
8930	*N*-(Phenylmethyl)-1,2-ethane-diamine		$C_9H_{14}N_2$	4152-09-4	150.220			130^{11}			
8931	Phenylmethyl 4-hydroxyben-zoate		$C_{14}H_{12}O_3$	94-18-8	228.243						sl chl
8932	1-Phenyl-2-methyl-2-propanol		$C_{10}H_{14}O$	100-86-7	150.217	nd	24	215	0.9787^{16}	1.5173^{16}	
8933	*N*-(Phenylmethyl)-1*H*-purin-6-amine		$C_{12}H_{11}N_5$	1214-39-7	225.249		232.8				
8934	4-Phenylmorpholine		$C_{10}H_{13}NO$	92-53-5	163.216	cry (al-eth)	58.3				i H_2O, EtOH; vs eth
8935	*N*-Phenyl-1-naphthalenamine	1-Naphthylphenylamine	$C_{16}H_{13}N$	90-30-2	219.281		61				sl H_2O, ctc; s EtOH, eth, bz, HOAc
8936	*N*-Phenyl-2-naphthalenamine	*N*-Phenyl-β-naphthylamine	$C_{16}H_{13}N$	135-88-6	219.281		108(1)	395.5			i H_2O; s EtOH, eth, bz, HOAc; sl chl
8937	1-Phenylnaphthalene		$C_{16}H_{12}$	605-02-7	204.266	cry	45	334	1.096^{20}	1.6664^{20}	i H_2O; vs EtOH, eth, bz, HOAc; s ctc
8938	2-Phenylnaphthalene		$C_{16}H_{12}$	612-94-2	204.266	lf (al)	101(2)	345.5	1.2180^{20}		s EtOH, bz, chl, HOAc; vs eth
8939	1-Phenyl-1-octanone		$C_{14}H_{20}O$	1674-37-9	204.308		22.8	300(13)	0.9360^{30}		s EtOH, eth
8940	Phenyloxirane	Styrene-7,8-oxide	C_8H_8O	96-09-3	120.149	colorless liq	-35.6	194.1	1.0490^{25}	1.5342^{20}	i H_2O; s EtOH, eth, chl
8941	3-Phenyloxiranecarboxylic acid, ethyl ester		$C_{11}H_{12}O_3$	121-39-1	192.211			136^5			
8942	5-Phenyl-2,4-pentadienal		$C_{11}H_{10}O$	13466-40-5	158.196		42.5	160^3			i H_2O; msc EtOH, bz; vs eth
8943	1-Phenyl-1,4-pentanedione		$C_{11}H_{12}O_2$	583-05-1	176.212	ye oil		162^{12}		1.5250^{30}	vs ace
8944	1-Phenyl-1-pentanol		$C_{11}H_{16}O$	583-03-9	164.244			141^{25}	0.9655^{20}	1.4086^{25}	vs ace, eth, EtOH
8945	1-Phenyl-1-pentanone		$C_{11}H_{14}O$	1009-14-9	162.228	liq	-9.4	247(14)	0.986^{20}	1.5158^{20}	i H_2O; vs EtOH, eth; sl ctc
8946	1-Phenyl-1-penten-3-one		$C_{11}H_{12}O$	3152-68-9	160.212	lf (lig)	38.5	142^{12}	0.8697^{20}	1.5684^{20}	sl H_2O, chl; vs EtOH, eth, bz
8947	Phenylphosphine	Monophenylphosphine	C_6H_7P	638-21-1	110.094			160.5	1.001^{15}	1.5796^{20}	
8948	Phenylphosphinic acid	Benzenephosphinic acid	$C_6H_7O_2P$	1779-48-2	142.093		83.8				s H_2O; vs EtOH; sl eth, chl
8949	Phenylphosphonic acid	Benzenephosphonic acid	$C_6H_7O_3P$	1571-33-1	158.092	lf (w)	158.5(0.2)				vs H_2O; s EtOH, eth, ace; i bz
8950	Phenylphosphonic dichloride		$C_6H_5Cl_2OP$	824-72-6	194.983		1	258	1.3957^{25}	1.5581^{25}	sl DMSO
8951	Phenylphosphonothioic dichloride	Dichlorophenylphosphine sulfide	$C_6H_5Cl_2PS$	3497-00-5	211.049			205^{130}	1.376^{13}		
8952	Phenylphosphonous dichloride	Dichlorophenylphosphine	$C_6H_5Cl_2P$	644-97-3	178.984	liq	-51	236(5)	1.356^{20}	1.6030^{20}	vs bz
8953	Phenyl phosphorodichloridate	Phenyl dichlorophosphate	$C_6H_5Cl_2O_2P$	770-12-7	210.983	hyg liq		242	1.412^{20}	1.5230^{20}	

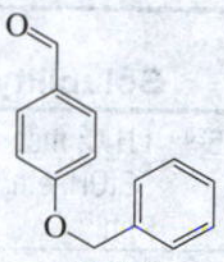
8915
4-(Phenylmethoxy)benzaldehyde

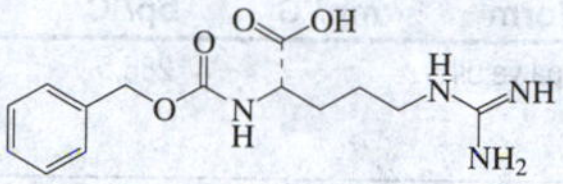
8916
*N*2-[(Phenylmethoxy)carbonyl]-L-arginine

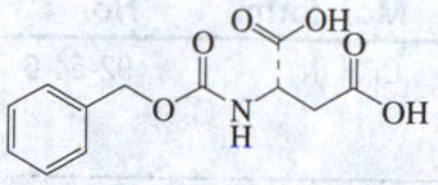
8917
N-[(Phenylmethoxy)carbonyl]-L-aspartic acid

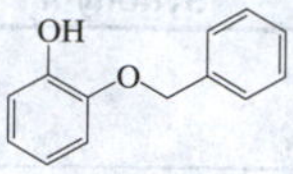
8918
2-(Phenylmethoxy)phenol

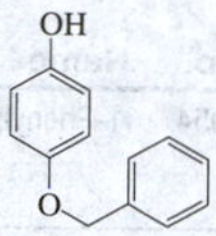
8919
4-(Phenylmethoxy)phenol

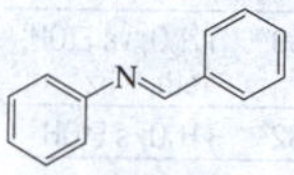
8920
N-(Phenylmethylene)aniline

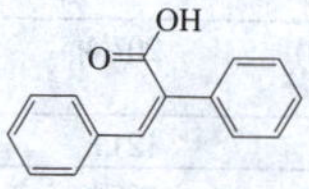
8921
cis-α-(Phenylmethylene)benzeneacetic acid

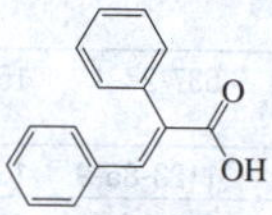
8922
trans-α-(Phenylmethylene)benzeneacetic acid

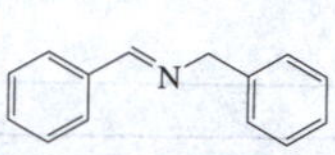
8923
N-(Phenylmethylene)benzenemethanamine

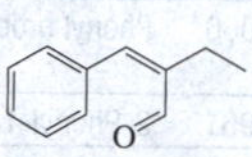
8924
2-(Phenylmethylene)butanal

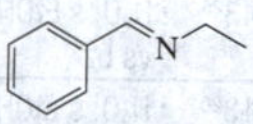
8925
N-(Phenylmethylene)ethanamine

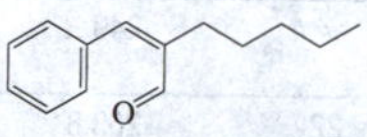
8926
2-(Phenylmethylene)heptanal

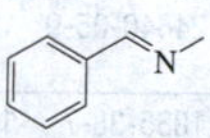
8927
N-(Phenylmethylene)methanamine

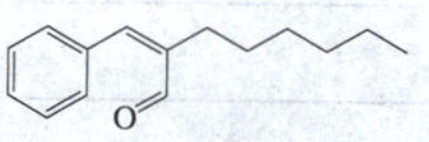
8928
2-(Phenylmethylene)octanal

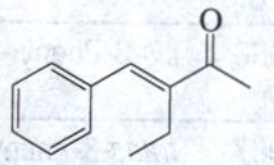
8929
3-(Phenylmethylene)-2-pentanone

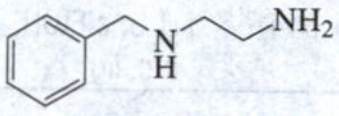
8930
N-(Phenylmethyl)-1,2-ethanediamine

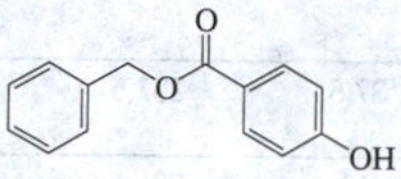
8931
Phenylmethyl 4-hydroxybenzoate

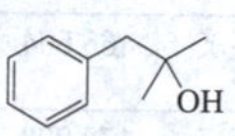
8932
1-Phenyl-2-methyl-2-propanol

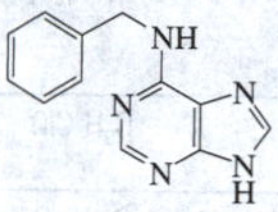
8933
N-(Phenylmethyl)-1*H*-purin-6-amine

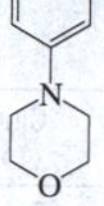
8934
4-Phenylmorpholine

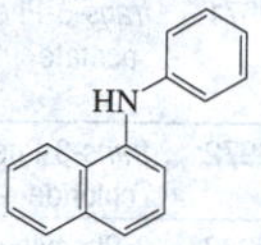
8935
N-Phenyl-1-naphthalenamine

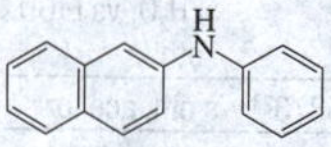
8936
N-Phenyl-2-naphthalenamine

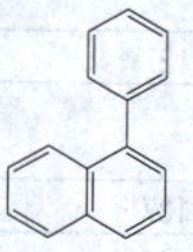
8937
1-Phenylnaphthalene

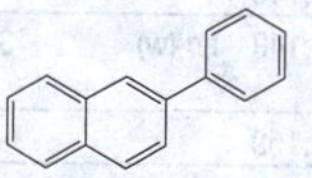
8938
2-Phenylnaphthalene

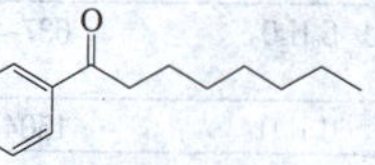
8939
1-Phenyl-1-octanone

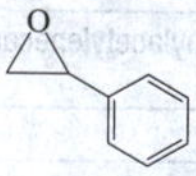
8940
Phenyloxirane

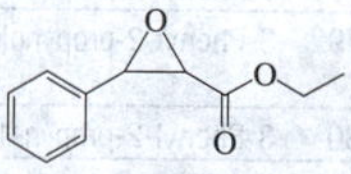
8941
3-Phenyloxiranecarboxylic acid, ethyl ester

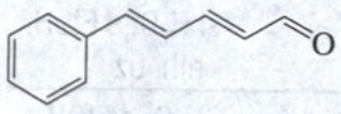
8942
5-Phenyl-2,4-pentadienal

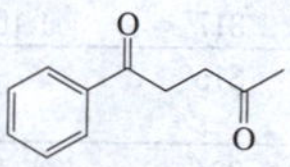
8943
1-Phenyl-1,4-pentanedione

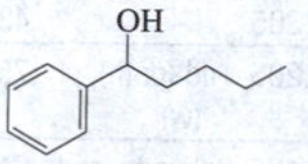
8944
1-Phenyl-1-pentanol

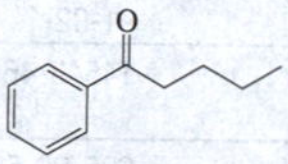
8945
1-Phenyl-1-pentanone

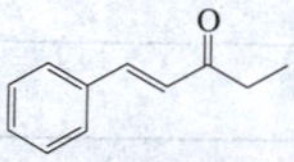
8946
1-Phenyl-1-penten-3-one

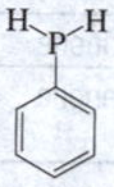
8947
Phenylphosphine

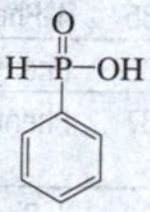
8948
Phenylphosphinic acid

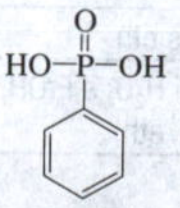
8949
Phenylphosphonic acid

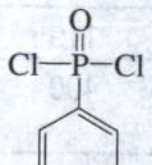
8950
Phenylphosphonic dichloride

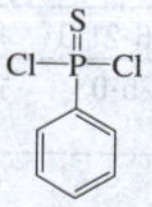
8951
Phenylphosphonothioic dichloride

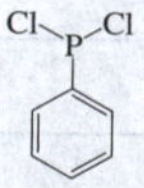
8952
Phenylphosphonous dichloride

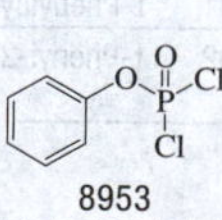
8953
Phenyl phosphorodichloridate

Organic

No.	Name	Synonym	Mol. form.	CAS Reg. No.	Mol. wt.	Physical form	mp/°C	bp/°C	den/ g cm⁻³	n_D	Solubility
8954	1-Phenylpiperazine		$C_{10}H_{14}N_2$	92-54-6	162.231	pa ye oil		286.5	1.0621^{20}	1.5875^{20}	i H_2O; msc EtOH, eth; s chl
8955	1-Phenylpiperidine		$C_{11}H_{15}N$	4096-20-2	161.244		4.7	260(5)	0.9944^{25}	1.5598^{25}	vs EtOH, eth, bz, chl
8956	4-Phenylpiperidine		$C_{11}H_{15}N$	771-99-3	161.244		60.5	257	0.9996^{16}		s chl
8957	*N*-Phenylpropanamide		$C_9H_{11}NO$	620-71-3	149.189	pl (eth, al, bz)	105.5	222.2	1.175^{25}		sl H_2O; vs EtOH, eth
8958	1-Phenyl-1,2-propanedione		$C_9H_8O_2$	579-07-7	148.159	ye oil	<20	222	1.1006^{20}	1.537^{10}	s H_2O, EtOH, eth
8959	1-Phenyl-1,2-propanedione, 2-oxime		$C_9H_9NO_2$	119-51-7	163.173	wh nd (w)	115				
8960	Phenyl propanoate		$C_9H_{10}O_2$	637-27-4	150.174	pr	19(1)	207(4)	1.0436^{25}	1.4980^{20}	i H_2O; vs EtOH, eth; s bz
8961	2-Phenyl-1-propanol		$C_9H_{12}O$	1123-85-9	136.190			121^{26}	0.975^{25}	1.5582^{2}	i H_2O; s EtOH
8962	1-Phenyl-2-propanol		$C_9H_{12}O$	698-87-3	136.190			125^{25}	0.991^{20}	1.5190^{20}	
8963	Phenylpropanolamine hydrochloride		$C_9H_{14}ClNO$	154-41-6	187.666		194				vs H_2O; s EtOH; i eth, bz, chl
8964	1-Phenyl-1-propanone	Propiophenone	$C_9H_{10}O$	93-55-0	134.174		18.6(0.2)	217.4(0.7)	1.0096^{20}	1.5269^{20}	i H_2O; s EtOH, eth, chl
8965	1-Phenyl-2-propanone	Phenylacetone	$C_9H_{10}O$	103-79-7	134.174	liq	-15.3(0.5)	214(17)	1.0157^{20}	1.5168^{20}	i H_2O; vs EtOH, eth; msc bz, xyl; s chl
8966	*cis*-3-Phenyl-2-propenenitrile		C_9H_7N	24840-05-9	129.159	liq	-4.4	249	1.0289^{20}	1.5843^{20}	i H_2O; s EtOH; vs bz
8967	*trans*-3-Phenyl-2-propenenitrile		C_9H_7N	1885-38-7	129.159		22	263.8	1.0304^{20}	1.6013^{20}	i H_2O; s EtOH, ace, ctc
8968	3-Phenyl-2-propenoic anhydride	Cinnamic anhydride	$C_{18}H_{14}O_3$	538-56-7	278.302	nd (bz or al) pt (al)	136				vs bz
8969	*cis*-3-Phenyl-2-propen-1-ol		$C_9H_{10}O$	4510-34-3	134.174	wh nd (eth-peth)	34	257.5	1.0440^{20}	1.5819^{20}	vs eth, EtOH
8970	*trans*-3-Phenyl-2-propen-1-ol		$C_9H_{10}O$	4407-36-7	134.174	wh nd (eth-peth)	34.8(0.8)	259(5)	1.0440^{20}	1.5819^{20}	sl H_2O, chl; vs EtOH, eth
8971	*trans*-3-Phenyl-2-propen-1-ol acetate	*trans*-Cinnamyl acetate	$C_{11}H_{12}O_2$	21040-45-9	176.212			265	1.0567^{20}	1.5425^{20}	i H_2O; s EtOH, eth, ace, bz, chl
8972	*trans*-3-Phenyl-2-propenoyl chloride	Cinnamoyl chloride	C_9H_7ClO	17082-09-6	166.604	ye cry	37.5	257.5	1.1617^{45}	1.614^{42}	i H_2O; s EtOH, ctc, lig
8973	3-Phenylpropyl acetate	Benzenepropanol, acetate	$C_{11}H_{14}O_2$	122-72-5	178.228	col liq	-40	69^{1}			
8974	1-Phenyl-2-propylamine, (±)-	Amphetamine	$C_9H_{13}N$	300-62-9	135.206	oil		198(12)	0.9306^{25}	1.518^{26}	sl H_2O, eth; s chl, EtOH
8975	1-Phenyl-2-propylamine, (*S*)-	Dexamphetamine	$C_9H_{13}N$	51-64-9	135.206	oil	27.5	203.5	0.949^{15}	1.4704^{20}	sl H_2O; s EtOH, eth
8976	Phenyl propyl ether	Propoxybenzene	$C_9H_{12}O$	622-85-5	136.190	liq	-28(1)	190.3(1)	0.9474^{20}	1.5014^{20}	s EtOH, eth
8977	4-(3-Phenylpropyl)pyridine		$C_{14}H_{15}N$	2057-49-0	197.276			322	1.024^{25}	1.5616^{25}	vs bz, eth, py, EtOH
8978	3-Phenyl-2-propynal		C_9H_6O	2579-22-8	130.143			127^{28}	1.0622^{20}	1.6079^{12}	
8979	3-Phenyl-2-propynoic acid	Phenylacetylenecarboxylic acid	$C_9H_6O_2$	637-44-5	146.143	nd (w)	137(2)		1.28^{20}		sl H_2O; vs EtOH, eth
8980	3-Phenyl-2-propyn-1-ol		C_9H_8O	1504-58-1	132.159			137^{15}	1.078^{20}	1.5873^{28}	s eth, ace, bz
8981	6-Phenyl-2,4,7-pteridinetriamine	Triamterene	$C_{12}H_{11}N_7$	396-01-0	253.262	ye pl (BuOH)	316				i eth; sl EtOH, chl
8982	1-Phenyl-3-pyrazolidinone		$C_9H_{10}N_2O$	92-43-3	162.187		126				i eth, lig
8983	2-Phenylpyridine		$C_{11}H_9N$	1008-89-5	155.196			271	1.0833^{25}	1.6210^{20}	sl H_2O; msc EtOH, eth
8984	3-Phenylpyridine		$C_{11}H_9N$	1008-88-4	155.196	pa ye oil	164	274(11)		1.6123^{25}	sl H_2O; s EtOH, eth
8985	4-Phenylpyridine		$C_{11}H_9N$	939-23-1	155.196	pl (w)	74(3)	281			s H_2O, EtOH, eth
8986	Phenyl-2-pyridinylmethanone		$C_{12}H_9NO$	91-02-1	183.205		43.3(0.2)	317	1.1556^{20}		s chl
8987	Phenyl-4-pyridinylmethanone		$C_{12}H_9NO$	14548-46-0	183.205	nd (peth), pl (w)	75(2)	315			sl H_2O; s EtOH, eth, bz
8988	1-Phenyl-1*H*-pyrrole		$C_{10}H_9N$	635-90-5	143.185	pl (sub), red in air	61(1)	234			i H_2O; s EtOH, eth, ace, bz; vs peth
8989	2-Phenyl-1*H*-pyrrole		$C_{10}H_9N$	3042-22-6	143.185	pl (al, sub)	130(2)	270(25)			i H_2O; vs EtOH, eth, bz, chl; sl lig
8990	1-Phenyl-1*H*-pyrrole-2,5-dione	*N*-Phenylmaleimide	$C_{10}H_7NO_2$	941-69-5	173.169	ye nd (bz-lig)	90.5	162^{12}			vs bz, eth, EtOH
8991	1-Phenylpyrrolidine		$C_{10}H_{13}N$	4096-21-3	147.217		11	119^{12}	1.018^{20}	1.5813^{20}	s eth
8992	1-Phenyl-2,5-pyrrolidinedione	Succinanil	$C_{10}H_9NO_2$	83-25-0	175.184	mcl pr or nd (w, al)	156	400	1.356^{25}		i H_2O; s EtOH, eth

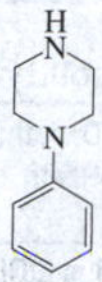
8954
1-Phenylpiperazine

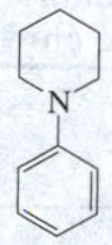
8955
1-Phenylpiperidine

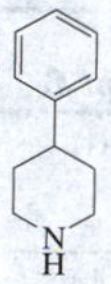
8956
4-Phenylpiperidine

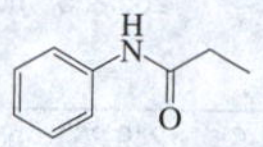
8957
N-Phenylpropanamide

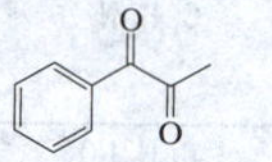
8958
1-Phenyl-1,2-propanedione

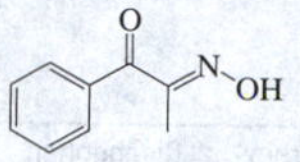
8959
1-Phenyl-1,2-propanedione, 2-oxime

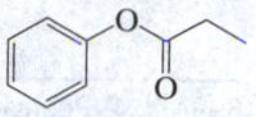
8960
Phenyl propanoate

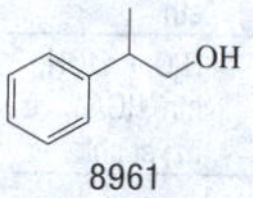
8961
2-Phenyl-1-propanol

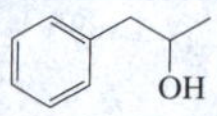
8962
1-Phenyl-2-propanol

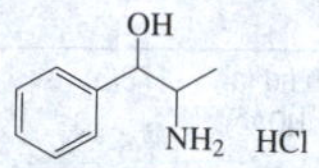
8963
Phenylpropanolamine hydrochloride

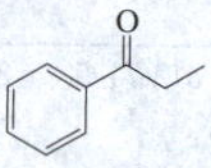
8964
1-Phenyl-1-propanone

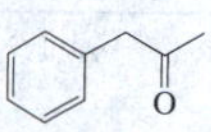
8965
1-Phenyl-2-propanone

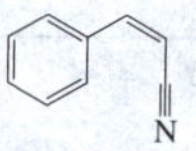
8966
cis-3-Phenyl-2-propenenitrile

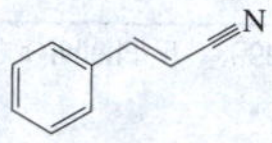
8967
trans-3-Phenyl-2-propenenitrile

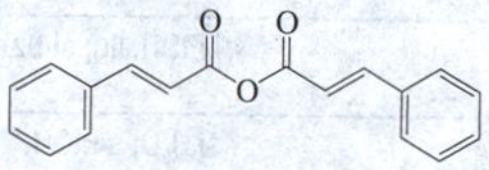
8968
3-Phenyl-2-propenoic anhydride

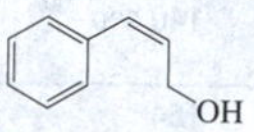
8969
cis-3-Phenyl-2-propen-1-ol

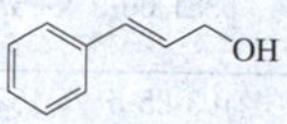
8970
trans-3-Phenyl-2-propen-1-ol

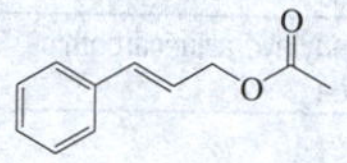
8971
trans-3-Phenyl-2-propen-1-ol acetate

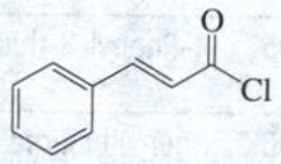
8972
trans-3-Phenyl-2-propenoyl chloride

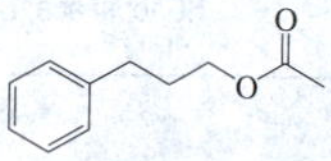
8973
3-Phenylpropyl acetate

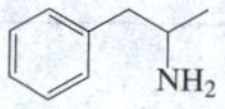
8974
1-Phenyl-2-propylamine, (±)-

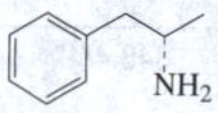
8975
1-Phenyl-2-propylamine, (*S*)-

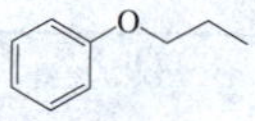
8976
Phenyl propyl ether

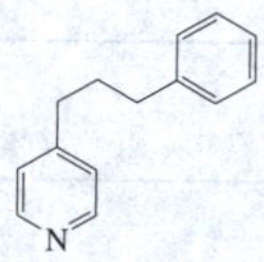
8977
4-(3-Phenylpropyl)pyridine

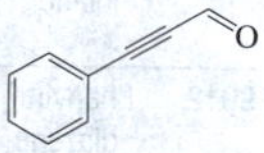
8978
3-Phenyl-2-propynal

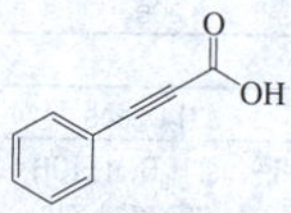
8979
3-Phenyl-2-propynoic acid

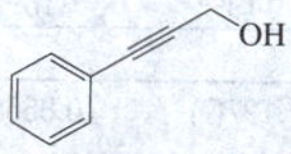
8980
3-Phenyl-2-propyn-1-ol

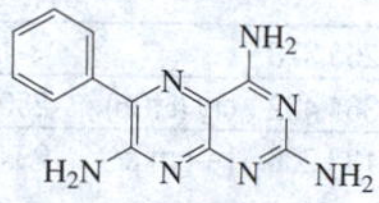
8981
6-Phenyl-2,4,7-pteridinetriamine

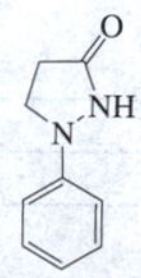
8982
1-Phenyl-3-pyrazolidinone

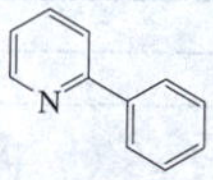
8983
2-Phenylpyridine

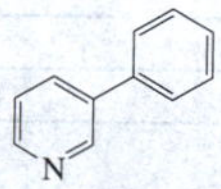
8984
3-Phenylpyridine

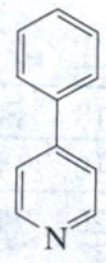
8985
4-Phenylpyridine

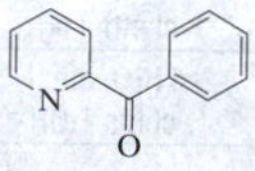
8986
Phenyl-2-pyridinylmethanone

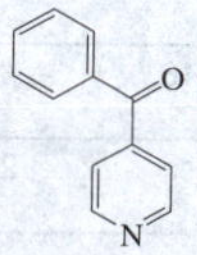
8987
Phenyl-4-pyridinylmethanone

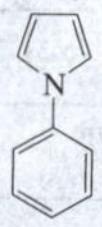
8988
1-Phenyl-1*H*-pyrrole

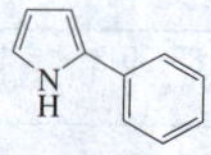
8989
2-Phenyl-1*H*-pyrrole

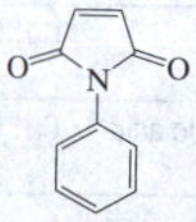
8990
1-Phenyl-1*H*-pyrrole-2,5-dione

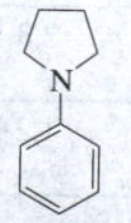
8991
1-Phenylpyrrolidine

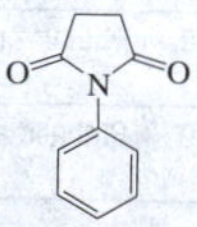
8992
1-Phenyl-2,5-pyrrolidinedione

Organic

No.	Name	Synonym	Mol. form.	CAS Reg. No.	Mol. wt.	Physical form	mp/°C	bp/°C	den/ g cm^{-3}	n_D	Solubility
8993	2-Phenylquinoline		$C_{15}H_{11}N$	612-96-4	205.255	nd (dil al)	83(1)	363			sl H_2O, peth; vs EtOH, eth, ace, bz
8994	2-Phenyl-4-quinolinecarboxylic acid	Cinchophen	$C_{16}H_{11}NO_2$	132-60-5	249.264	nd	214.5				i H_2O; s EtOH, eth, alk; sl ace, bz
8995	Phenyl salicylate		$C_{13}H_{10}O_3$	118-55-8	214.216		41.82(0.04)	173^{12}	1.2614^{30}		i H_2O; vs EtOH, ace, bz; s eth, HOAc
8996	Phenylsilane		C_6H_8Si	694-53-1	108.214			119	0.8681^{20}	1.5125^{20}	i H_2O
8997	1-Phenylsilatrane		$C_{12}H_{17}NO_3Si$	2097-19-0	251.354	pr or nd (ace)	209				
8998	Phenyl stearate		$C_{24}H_{40}O_2$	637-55-8	360.574		52	267^{15}			i H_2O; s EtOH, eth
8999	5'-Phenyl-1,1':3',1"-terphenyl		$C_{24}H_{18}$	612-71-5	306.400	orth nd (al or HOAc)	173(3)	462	1.199^{30}		i H_2O; s EtOH, eth, HOAc; vs bz; sl chl
9000	5-Phenyl-2,4-thiazolediamine	Amiphenazole	$C_9H_9N_3S$	490-55-1	191.252	fl (dil al) br in air	163 dec				
9001	Phenyl-2-thienylmethanone		$C_{11}H_8OS$	135-00-2	188.246	nd (dil al)	56.5	300	1.1890^{54}	1.6181^{54}	i H_2O; s EtOH, eth
9002	*N*-Phenylthioacetamide	Thioacetanilide	C_8H_9NS	637-53-6	151.229	nd (w)	75.5	dec			
9003	Phenyl thiocyanate		C_7H_5NS	5285-87-0	135.187			232.5	1.153^{18}		i H_2O; s EtOH, eth
9004	2-Phenylthiosemicarbazide	2-Phenylhydrazinecarbothioamide	$C_7H_9N_3S$	645-48-7	167.231	pr (al)	200 dec				
9005	4-Phenyl-3-thiosemicarbazide	*N*-Phenylhydrazinecarbothioamide	$C_7H_9N_3S$	5351-69-9	167.231	pl (al)	140 dec				i EtOH, lig; sl bz
9006	Phenylthiourea		$C_7H_8N_2S$	103-85-5	152.217	nd (w), pr (al)	154				sl H_2O; s EtOH, NaOH
9007	3-Phenyl-2-thioxo-4-thiazolidinone	3-Phenylrhodanine	$C_9H_7NOS_2$	1457-46-1	209.288	ye pr (HOAc) nd or pr (al)	194.5				i H_2O; sl EtOH, eth; s ace, chl, HOAc
9008	6-Phenyl-1,3,5-triazine-2,4-diamine	Benzoguanamine	$C_9H_9N_5$	91-76-9	187.201	nd, pl (al)	226.5				s EtOH, eth; sl tfa
9009	*N*-Phenyl-1,3,5-triazine-2,4-diamine	Amanozine	$C_9H_9N_5$	537-17-7	187.201	cry (diox, 50% al)	235.5				
9010	4-Phenyl-1,2,4-triazolidine-3,5-dione		$C_8H_7N_3O_2$	15988-11-1	177.161		205.5				
9011	Phenyltrimethylammonium iodide		$C_9H_{14}IN$	98-04-4	263.118	lf (al)	224				vs H_2O; s EtOH, HOAc; sl ace; i chl
9012	Phenyl(triphenylmethyl)diazene		$C_{25}H_{20}N_2$	981-18-0	348.440		111 dec				
9013	Phenylurea		$C_7H_8N_2O$	64-10-8	136.151	mcl pr (w, al)	147.4(0.5)	238	1.302^{25}		sl H_2O, eth, DMSO; s EtOH, AcOEt
9014	*trans*-5-(2-Phenylvinyl)-1,3-benzenediol	Pinosylvin	$C_{14}H_{12}O_2$	22139-77-1	212.244	nd (HOAc)	156				vs ace, bz, chl, HOAc
9015	Phenyl vinyl ether		C_8H_8O	766-94-9	120.149			155.5	0.9770^{20}	1.5224^{20}	i H_2O; vs eth
9016	Phenytoin	5,5-Diphenyl-2,4-imidazolidinedione	$C_{15}H_{12}N_2O_2$	57-41-0	252.268	nd (al)	286				i H_2O; s EtOH, ace; sl eth, bz
9017	Phloretin		$C_{15}H_{14}O_5$	60-82-2	274.269	nd (dil al), cry (ace)	263 dec				sl H_2O, chl; msc EtOH, bz; i eth; s ace
9018	Phorate		$C_7H_{17}O_2PS_3$	298-02-2	260.378		<-15	$119^{0.8}$	1.16^{25}		
9019	Phorbol		$C_{20}H_{28}O_6$	17673-25-5	364.432	cry (EtOH)	250 dec				s H_2O, ace
9020	Phorone		$C_9H_{14}O$	504-20-1	138.206	ye-grn pr	25.8(0.5)	197(7)	0.8850^{20}	1.4998^{20}	sl H_2O; s EtOH, eth, ace, ctc
9021	Phosalone		$C_{12}H_{15}ClNO_4PS_2$	2310-17-0	367.808		48.7(0.4)				
9022	Phosfolan		$C_7H_{14}NO_3PS_2$	947-02-4	255.295		36.5	$117^{0.001}$			vs H_2O, bz, ace; sl eth; s hx
9023	Phosmet		$C_{11}H_{12}NO_4PS_2$	732-11-6	317.321		72.0(0.2)	dec			
9024	Phosphamidon		$C_{10}H_{19}ClNO_5P$	13171-21-6	299.689	oil	-45	$162^{1.5}$	1.2132^{25}	1.4718^{25}	msc H_2O; s hx
9025	*N*-Phospho-L-arginine		$C_6H_{15}N_4O_5P$	1189-11-3	254.181	cry (ace aq)	177				
9026	*O*-Phosphorylethanolamine	Ethanolamine *O*-phosphate	$C_2H_8NO_4P$	1071-23-4	141.063	cry (EtOH aq)	242				
9027	*O*-Phosphoserine		$C_3H_8NO_6P$	407-41-0	185.073	cry	166 dec				
9028	Phthalazine	2,3-Benzodiazine	$C_8H_6N_2$	253-52-1	130.147		91.3(0.2)	316			s H_2O, EtOH, bz; sl eth; i lig
9029	Phthalic acid	1,2-Benzenedicarboxylic acid	$C_8H_6O_4$	88-99-3	166.132	pl (w)	207(3)	dec	2.18^{191}		sl H_2O, eth; i chl; s EtOH

Organic

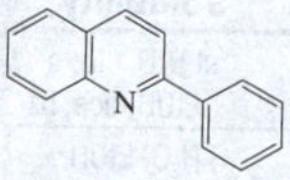
8993
2-Phenylquinoline

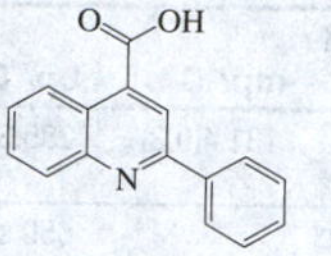
8994
2-Phenyl-4-quinolinecarboxylic acid

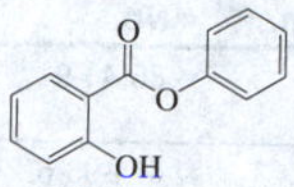
8995
Phenyl salicylate

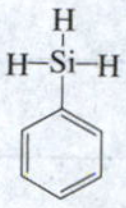
8996
Phenylsilane

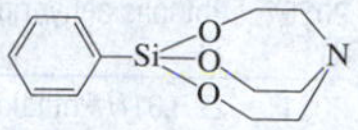
8997
1-Phenylsilatrane

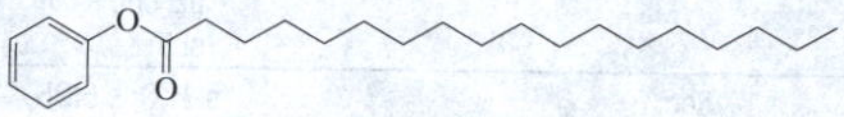
8998
Phenyl stearate

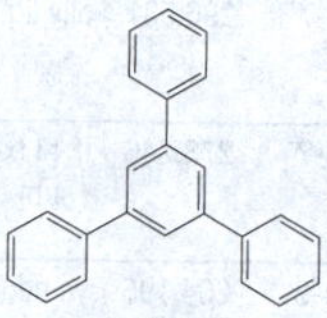
8999
5'-Phenyl-1,1':3',1''-terphenyl

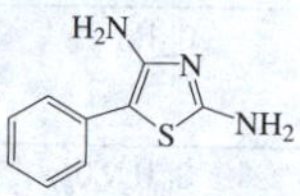
9000
5-Phenyl-2,4-thiazolediamine

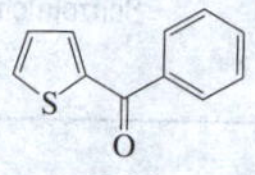
9001
Phenyl-2-thienylmethanone

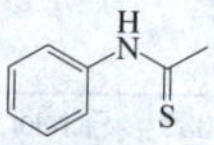
9002
N-Phenylthioacetamide

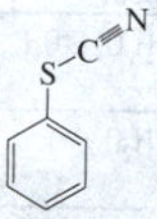
9003
Phenyl thiocyanate

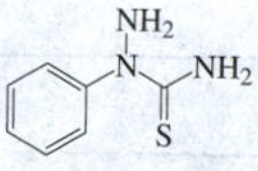
9004
2-Phenylthiosemicarbazide

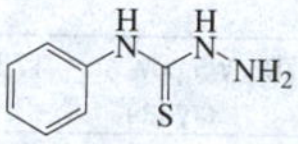
9005
4-Phenyl-3-thiosemicarbazide

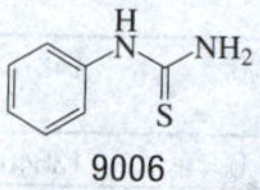
9006
Phenylthiourea

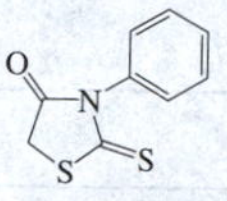
9007
3-Phenyl-2-thioxo-4-thiazolidinone

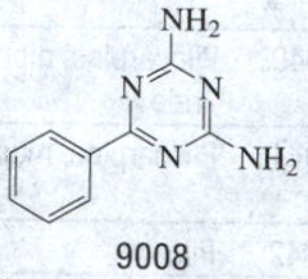
9008
6-Phenyl-1,3,5-triazine-2,4-diamine

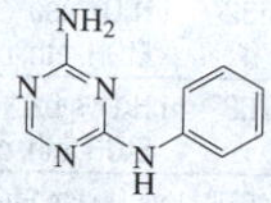
9009
N-Phenyl-1,3,5-triazine-2,4-diamine

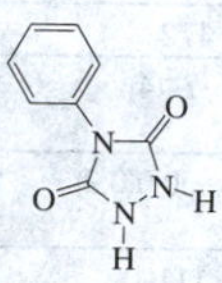
9010
4-Phenyl-1,2,4-triazolidine-3,5-dione

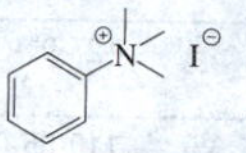
9011
Phenyltrimethylammonium iodide

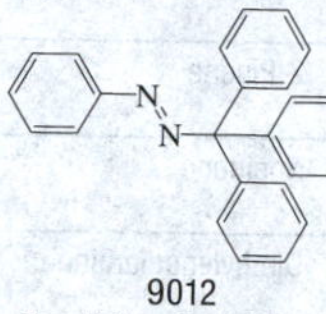
9012
Phenyl(triphenylmethyl)diazene

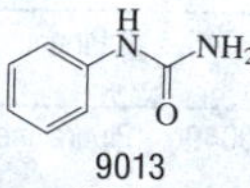
9013
Phenylurea

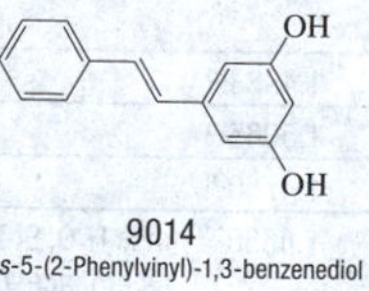
9014
trans-5-(2-Phenylvinyl)-1,3-benzenediol

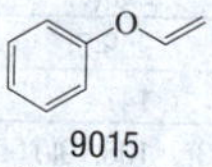
9015
Phenyl vinyl ether

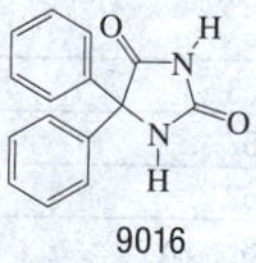
9016
Phenytoin

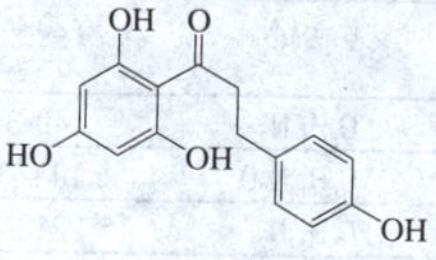
9017
Phloretin

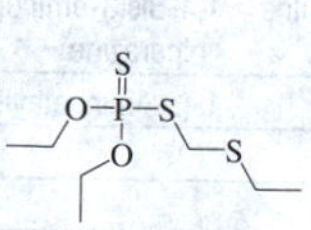
9018
Phorate

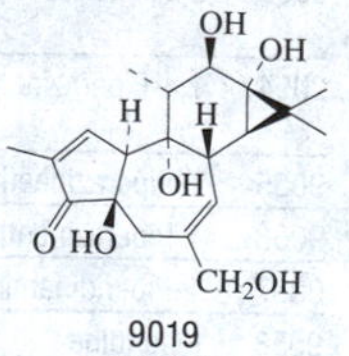
9019
Phorbol

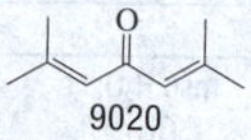
9020
Phorone

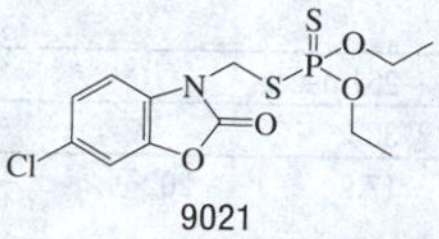
9021
Phosalone

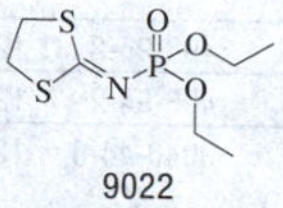
9022
Phosfolan

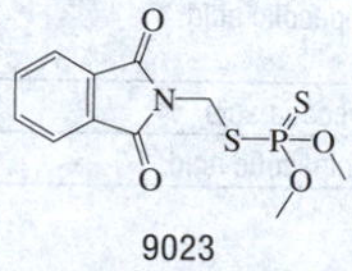
9023
Phosmet

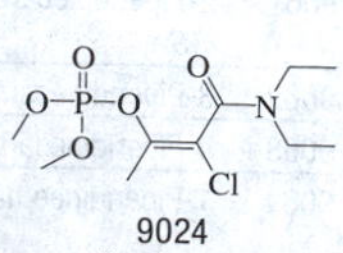
9024
Phosphamidon

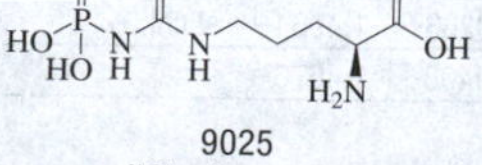
9025
N-Phospho-L-arginine

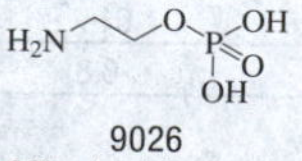
9026
O-Phosphorylethanolamine

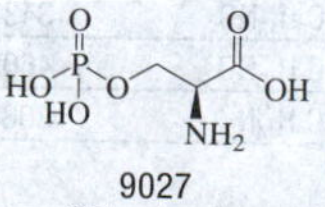
9027
O-Phosphoserine

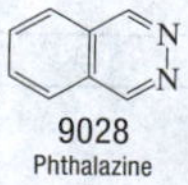
9028
Phthalazine

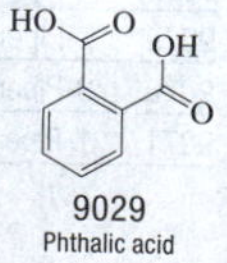
9029
Phthalic acid

No.	Name	Synonym	Mol. form.	CAS Reg. No.	Mol. wt.	Physical form	mp/°C	bp/°C	den/ g cm^{-3}	n_D	Solubility
9030	Phthalic anhydride		$C_8H_4O_3$	85-44-9	148.116	wh nd (al, bz)	131.4(0.3)	285.3(0.8)	1.527^{4}		sl H_2O, eth; s EtOH, ace, bz
9031	29*H*,31*H*-Phthalocyanine		$C_{32}H_{18}N_8$	574-93-6	514.539	grsh-bl mcl (quinoline)		550 sub			i H_2O, EtOH, eth; s $PhNH_2$
9032	Phthalylsulphathiazole		$C_{17}H_{13}N_3O_5S_2$	85-73-4	403.432		273				i H_2O, eth, chl; sl EtOH; s acid, alk
9033	Physostigmine		$C_{15}H_{21}N_3O_2$	57-47-6	275.347	orth pr (eth, bz)	105.5				sl H_2O; s EtOH, eth, bz, chl
9034	Phytol	3,7,11,15-Tetramethyl-2-hexadecen-1-ol, [*R*-[*R**,*R**-(*E*)]]	$C_{20}H_{40}O$	150-86-7	296.531	oily liq		203^{10}	0.8497^{25}	1.4595^{25}	
9035	Picene	Benzo[a]chrysene	$C_{22}H_{14}$	213-46-7	278.346	lf, pl (xyl, py, sub)	365(2)	519			i H_2O; sl EtOH, bz, chl; s con sulf
9036	Picrolonic acid		$C_{10}H_8N_4O_5$	550-74-3	264.195	ye nd (al)	116	dec			sl H_2O; s EtOH, eth, MeOH
9037	Picropodophyllin		$C_{22}H_{22}O_8$	477-47-4	414.405	col nd (al, bz)	228				vs ace, bz, eth, EtOH
9038	Picrotoxin		$C_{30}H_{34}O_{13}$	124-87-8	602.583	orth lf	203.5				vs py, EtOH
9039	Pilocarpine		$C_{11}H_{16}N_2O_2$	92-13-7	208.257	nd	34	260^{5}			s H_2O, EtOH; sl eth, bz; vs chl; i peth
9040	Pilocarpine, monohydrochloride		$C_{11}H_{17}ClN_2O_2$	54-71-7	244.718	hyg cry	204.5				vs H_2O, EtOH
9041	Pilocarpine, mononitrate		$C_{11}H_{17}N_3O_5$	148-72-1	271.270	wh pow or cry (al)	178				vs H_2O
9042	Pilosine		$C_{16}H_{18}N_2O_3$	13640-28-3	286.325	nd (al)	179				
9043	Pimaric acid	Dextropimaric acid	$C_{20}H_{30}O_2$	127-27-5	302.451	orth (ace) pr (al)	218.5	282^{18}			vs eth, py, EtOH
9044	Pinane	2,6,6-Trimethylbicyclo[3.1.1]-heptane	$C_{10}H_{18}$	473-55-2	138.250	oil	-53	167(5)	0.8467^{21}	1.4605^{21}	
9045	*trans*-2-Pinanol	Pinene hydrate	$C_{10}H_{18}O$	35408-04-9	154.249		60	81^{10}			
9046	Pindolol		$C_{14}H_{20}N_2O_2$	13523-86-9	248.321	cry (EtOH)	172				
9047	α-Pinene	2-Pinene	$C_{10}H_{16}$	80-56-8	136.234	liq	-74(4)	156.3(0.5)	0.8539^{25}	1.4632^{25}	i H_2O; msc EtOH, eth, chl
9048	β-Pinene	Nopinene	$C_{10}H_{16}$	127-91-3	136.234	liq	-50.0(0.2)	165.8(0.6)	0.860^{25}	1.4768^{25}	i H_2O; s bz, EtOH, eth, chl
9049	Piperazine	Diethylenediamine	$C_4H_{10}N_2$	110-85-0	86.135	hyg pl or lf (al)	111(1)	148.63(0.05)		1.446^{113}	vs H_2O; s EtOH, chl; i eth
9050	1-Piperazinecarboxaldehyde		$C_5H_{10}N_2O$	7755-92-2	114.145			$95^{0.5}$		1.5094^{20}	
9051	1,4-Piperazinediethanol		$C_8H_{18}N_2O_2$	122-96-3	174.241		132(2)	217^{30}			
9052	Piperazine dihydrochloride	Diethylenediamine dihydrochloride	$C_4H_{12}Cl_2N_2$	142-64-3	159.057						sl H_2O; i EtOH
9053	2,5-Piperazinedione		$C_4H_6N_2O_2$	106-57-0	114.103	tab or pl (w)	312 dec	260 sub			sl H_2O, EtOH; s HCl
9054	1,4-Piperazinedipropanamine	1,4-Bis(3-aminopropyl)-piperazine	$C_{10}H_{24}N_4$	7209-38-3	200.325		15	151^{2}	0.973^{25}	1.5015^{20}	
9055	1-Piperazineethanamine	1-(2-Aminoethyl)piperazine	$C_6H_{15}N_3$	140-31-8	129.203			225(2)	0.985^{25}	1.4983^{20}	
9056	1-Piperazineethanol		$C_6H_{14}N_2O$	103-76-4	130.187			259(6)	1.061^{25}	1.5065^{20}	
9057	1-Piperidinamine		$C_5H_{12}N_2$	2213-43-6	100.162			147	0.928^{25}	1.4750^{20}	
9058	Piperidine	Azacyclohexane	$C_5H_{11}N$	110-89-4	85.148	liq	-11.05(0.03)	106.19(0.09)	0.8606^{20}	1.4530^{20}	msc H_2O, EtOH; s eth, ace, bz, chl
9059	1-Piperidinecarboxaldehyde		$C_6H_{11}NO$	2591-86-8	113.157	liq	-30.8	221(14)	1.0158^{25}	1.4805^{25}	msc H_2O, EtOH, eth, bz, chl, lig
9060	4-Piperidinecarboxamide		$C_6H_{12}N_2O$	39546-32-2	128.171		138.5				
9061	2-Piperidinecarboxylic acid, (*S*)-	L-Pipecolic acid	$C_6H_{11}NO_2$	3105-95-1	129.157	nd (MeOH/eth)	260				
9062	3-Piperidinecarboxylic acid	Nipecotic acid	$C_6H_{11}NO_2$	498-95-3	129.157		261 dec				vs H_2O
9063	4-Piperidinecarboxylic acid	Isonipecotic acid	$C_6H_{11}NO_2$	498-94-2	129.157	nd	336				
9064	1-Piperidineethanol		$C_7H_{15}NO$	3040-44-6	129.200		17.9	202	0.9703^{25}	1.4749^{20}	msc H_2O; vs EtOH
9065	2-Piperidineethanol	2-(2-Hydroxyethyl)piperidine	$C_7H_{15}NO$	1484-84-0	129.200		69	204(13)	1.01^{27}		vs H_2O
9066	4-Piperidineethanol	4-(2-Hydroxyethyl)piperidine	$C_7H_{15}NO$	622-26-4	129.200	syr	132.5	227.5	1.0059^{15}	1.4907^{20}	vs H_2O, eth, EtOH
9067	Piperidine, hydrochloride	Piperidinium chloride	$C_5H_{12}ClN$	6091-44-7	121.609		142 dec				vs H_2O, chl
9068	4-Piperidinemethanamine	4-(Aminomethyl)piperidine	$C_6H_{14}N_2$	7144-05-0	114.188		25	200		1.4900^{20}	
9069	2-Piperidinemethanol		$C_6H_{13}NO$	3433-37-2	115.173		69(3)	104^{10}			sl chl
9070	3-Piperidinemethanol		$C_6H_{13}NO$	4606-65-9	115.173		61	$106^{3.5}$	1.0263^{20}	1.4964^{20}	sl chl
9071	1-Piperidinepropanenitrile		$C_8H_{14}N_2$	3088-41-3	138.210		-6.8	145^{50}	0.9403^{25}	1.4676^{25}	

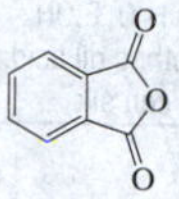
9030
Phthalic anhydride

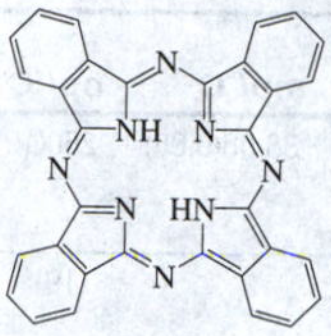
9031
29*H*,31*H*-Phthalocyanine

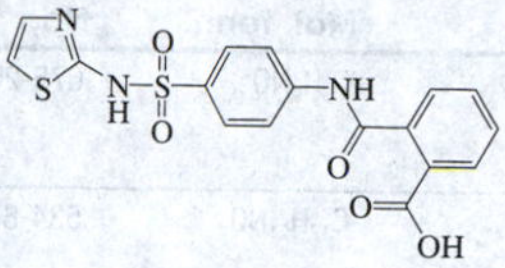
9032
Phthalylsulphathiazole

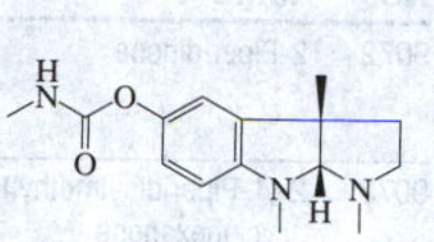
9033
Physostigmine

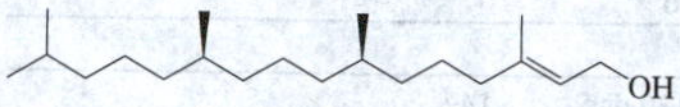
9034
Phytol

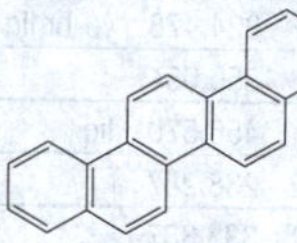
9035
Picene

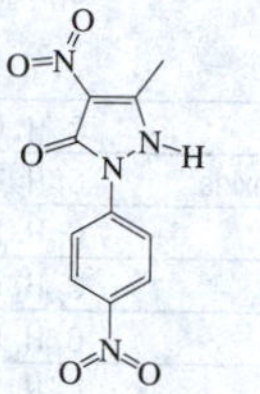
9036
Picrolonic acid

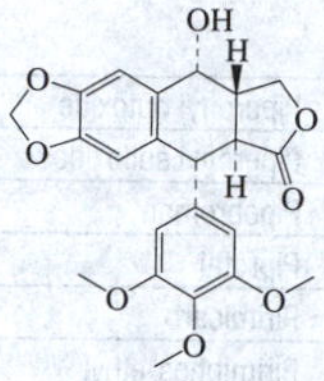
9037
Picropodophyllin

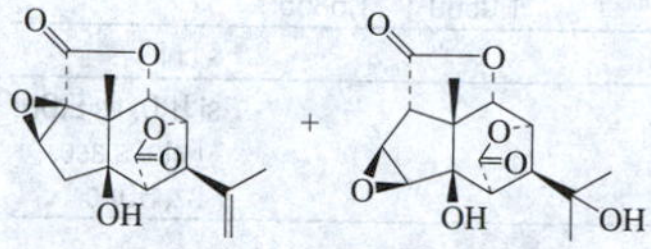
9038
Picrotoxin

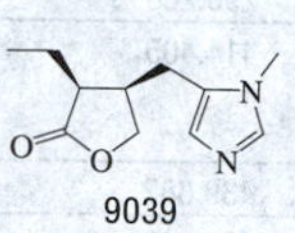
9039
Pilocarpine

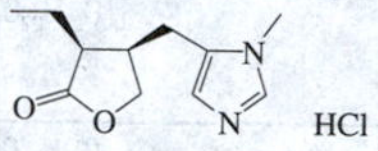
9040
Pilocarpine, monohydrochloride

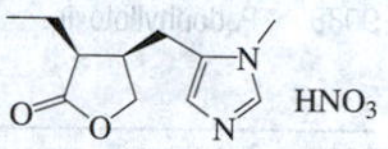
9041
Pilocarpine, mononitrate

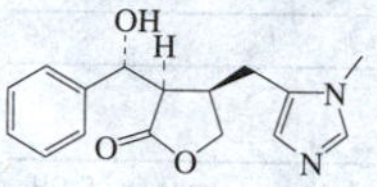
9042
Pilosine

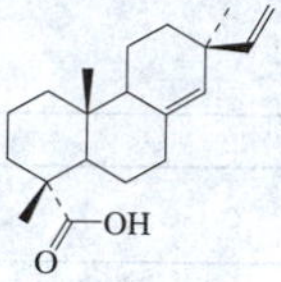
9043
Pimaric acid

9044
Pinane

9045
trans-2-Pinanol

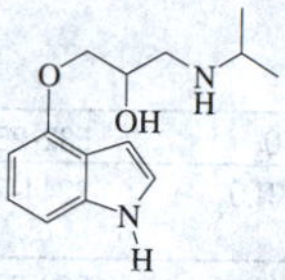
9046
Pindolol

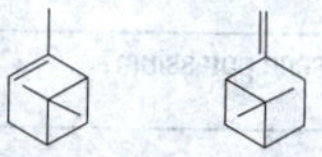
9047
α-Pinene

9048
β-Pinene

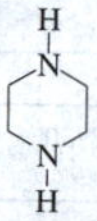
9049
Piperazine

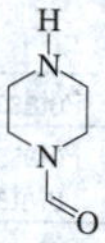
9050
1-Piperazinecarboxaldehyde

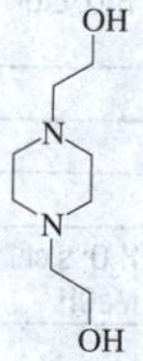
9051
1,4-Piperazinediethanol

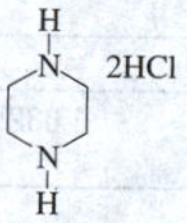
9052
Piperazine dihydrochloride

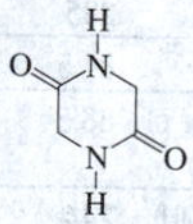
9053
2,5-Piperazinedione

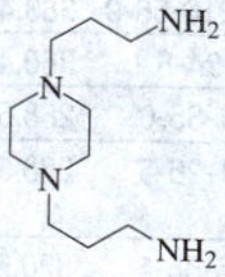
9054
1,4-Piperazinedipropanamine

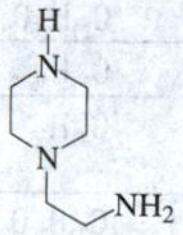
9055
1-Piperazineethanamine

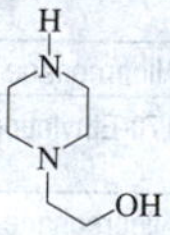
9056
1-Piperazineethanol

9057
1-Piperidinamine

9058
Piperidine

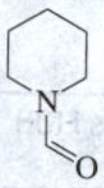
9059
1-Piperidinecarboxaldehyde

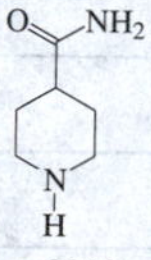
9060
4-Piperidinecarboxamide

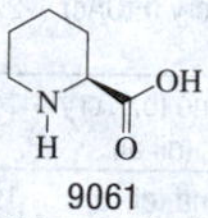
9061
2-Piperidinecarboxylic acid, (*S*)-

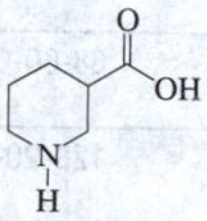
9062
3-Piperidinecarboxylic acid

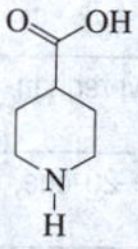
9063
4-Piperidinecarboxylic acid

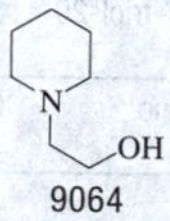
9064
1-Piperidineethanol

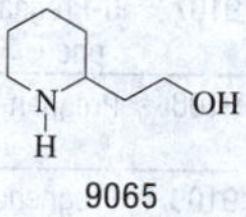
9065
2-Piperidineethanol

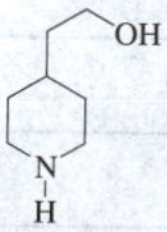
9066
4-Piperidineethanol

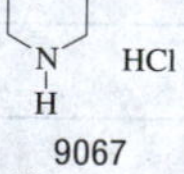
9067
Piperidine, hydrochloride

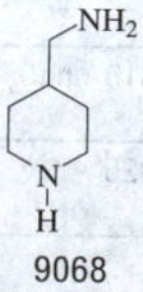
9068
4-Piperidinemethanamine

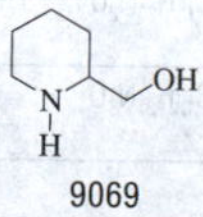
9069
2-Piperidinemethanol

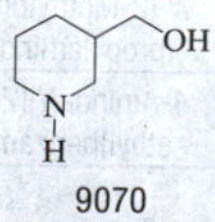
9070
3-Piperidinemethanol

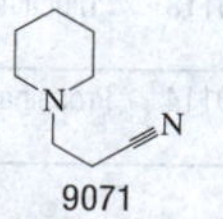
9071
1-Piperidinepropanenitrile

No.	Name	Synonym	Mol. form.	CAS Reg. No.	Mol. wt.	Physical form	mp/°C	bp/°C	den/g cm^{-3}	n_D	Solubility
9072	2-Piperidinone		C_5H_9NO	675-20-7	99.131	hyg	38.68(0.05)	260(19)			vs H_2O, EtOH, eth; s dil acid; i con alk
9073	2-(1-Piperidinylmethyl)-cyclohexanone	Pimeclone	$C_{12}H_{21}NO$	534-84-9	195.301			119^{14}			
9074	1-(2-Piperidinyl)-2-propanone, (±)-		$C_8H_{15}NO$	539-00-4	141.211	oil		91^{14}	0.9624^{20}	1.4683^{20}	vs EtOH, chl
9075	3-(2-Piperidinyl)pyridine, (*S*)-	Anabasine	$C_{10}H_{14}N_2$	494-52-0	162.231	liq	9	276(29)	1.0455^{20}	1.5430^{20}	msc H_2O; s EtOH, eth, bz
9076	Piperine		$C_{17}H_{19}NO_3$	94-62-2	285.338	pr (AcOEt) pl or mcl pr (al), cry	130(2)				i H_2O; s EtOH, bz, py; sl eth; vs chl
9077	Piperonyl butoxide		$C_{19}H_{30}O_5$	51-03-6	338.438			180^{1}	1.05^{25}		
9078	Piperonyl sulfoxide	Isosafrole octyl sulfoxide	$C_{18}H_{28}O_3S$	120-62-7	324.478	ye-br liq				1.530^{25}	sl H_2O; misc os
9079	Pipobroman		$C_{10}H_{16}Br_2N_2O_2$	54-91-1	356.054		106				
9080	Piprotal	Tropital	$C_{24}H_{40}O_8$	5281-13-0	456.570	liq		$215^{0.04}$			
9081	Pirimicarb		$C_{11}H_{18}N_4O_2$	23103-98-2	238.287		90.5				
9082	Pirimiphos-ethyl		$C_{13}H_{24}N_3O_3PS$	23505-41-1	333.387			>130 dec	1.14^{20}		
9083	Pirimiphos-methyl		$C_{11}H_{20}N_3O_3PS$	29232-93-7	305.334		15	dec	1.17^{20}		
9084	Pithecolobine		$C_{22}H_{46}N_4O_2$	22368-82-7	398.626	cry	68	$230^{0.007}$			s H_2O, chl, eth, EtOH, peth
9085	2-Pivaloyl-1,3-indandione	Pindone	$C_{14}H_{14}O_3$	83-26-1	230.259	ye cry	109.7(0.3)				
9086	Plasmocid		$C_{17}H_{25}N_3O$	551-01-9	287.400			$182^{1.0}$	1.0569^{24}	1.5855^{24}	
9087	Plumericin		$C_{15}H_{14}O_6$	77-16-7	290.268						s chl
9088	Podophyllotoxin		$C_{22}H_{22}O_8$	518-28-5	414.405		183				sl H_2O; vs EtOH; i eth; s ace, bz, HOAc
9089	Polythiazide		$C_{11}H_{13}ClF_3N_3O_4S_3$	346-18-9	439.882		214				
9090	Ponceau 3R	C.I. Food Red 6	$C_{19}H_{16}N_2Na_2O_7S_2$	3564-09-8	494.449	dk red pow					s H_2O; sl EtOH
9091	Populin		$C_{20}H_{22}O_8$	99-17-2	390.384	nd (w+2), pr (al)	180				
9092	21*H*,23*H*-Porphine		$C_{20}H_{14}N_4$	101-60-0	310.352	red or oran lf (chl-MeOH)	360	300 sub	1.336^{25}		i H_2O, eth, ace, bz; sl EtOH; s diox
9093	Potassium benzoate		$C_7H_5KO_2$	582-25-2	160.212	hyg cry					
9094	Potassium dichloroisocyan-urate	Troclosene potassium	$C_3Cl_2KN_3O_3$	2244-21-5	236.054	hyg cry	250 dec				
9095	Potassium D-gluconate		$C_6H_{11}KO_7$	299-27-4	234.245	ye-wh cry	183 dec				vs H_2O; i EtOH, eth, bz, chl
9096	Potassium *trans,trans*-2,4-hexadienoate	Potassium sorbate	$C_6H_7KO_2$	24634-61-5	150.217		>270 dec		1.361^{25}		vs H_2O; s EtOH
9097	Potassium hydrogen phthalate	Potassium biphthalate	$C_8H_5KO_4$	877-24-7	204.222				1.636^{25}		s H_2O; sl EtOH
9098	Potassium *cis*-9-octadeceno-ate	Potassium oleate	$C_{18}H_{33}KO_2$	143-18-0	320.552	ye-br solid					s H_2O, EtOH
9099	Prazosin		$C_{19}H_{21}N_5O_4$	19216-56-9	383.402	cry	279				
9100	Prednisolone		$C_{21}H_{28}O_5$	50-24-8	360.444		235				
9101	5α-Pregnane	Allopregnane	$C_{21}H_{36}$	641-85-0	288.511		84.5				
9102	5β-Pregnane	17β-Ethyletiocholane	$C_{21}H_{36}$	481-26-5	288.511	mcl sc or pl (MeOH)	83.5		1.032^{15}		i H_2O; s chl, MeOH
9103	5α-Pregnane-3α,20α-diol	Allopregnane-3α,20α-diol	$C_{21}H_{36}O_2$	566-58-5	320.510	cry (MeOH)	244				
9104	5β-Pregnane-3α,20*S*-diol	Pregnanediol	$C_{21}H_{36}O_2$	80-92-2	320.510	pl (ace)	243.5		1.15^{25}		sl EtOH, eth; s ace
9105	5α-Pregnane-3,20-dione	3,20-Allopregnanedione	$C_{21}H_{32}O_2$	566-65-4	316.478	cry	200				
9106	5β-Pregnane-3,20-dione		$C_{21}H_{32}O_2$	128-23-4	316.478	nd (dil al) cry (dil ace)	123				i H_2O; vs EtOH; s eth, ace
9107	5-Pregnane-3,17,21-triol-20-one	3,17,21-Trihydroxypregnan-20-one, (3α,5β)	$C_{21}H_{34}O_4$	68-60-0	350.493	cry (EtOAc)	226				
9108	Pregnan-3α-ol-20-one	3-Hydroxypregnan-20-one, (3α,5β)	$C_{21}H_{34}O_2$	128-20-1	318.494	nd (bz), cry (dil al)	149.5				vs EtOH
9109	Pregnenolone		$C_{21}H_{32}O_2$	145-13-1	316.478	nd (dil al)	192				
9110	Prenoxdiazine hydrochloride		$C_{23}H_{28}ClN_3O$	982-43-4	397.940		186.5				
9111	Prephenic acid		$C_{10}H_{10}O_6$	126-49-8	226.182	free acid unstab					
9112	Pridinol	1,1-Diphenyl-3-(1-piperidinyl)-1-propanol	$C_{20}H_{25}NO$	511-45-5	295.419	cry	120				s ace
9113	Prilocaine	*N*-(2-Methylphenyl)-2-(propylamino)propanamide	$C_{13}H_{20}N_2O$	721-50-6	220.310	nd	38	160^{1}		1.5299^{20}	
9114	Procainamide	4-Amino-*N*-[2-(diethylamino)-ethyl]benzamide	$C_{13}H_{21}N_3O$	51-06-9	235.325		47	212^{2}			

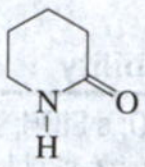
9072
2-Piperidinone

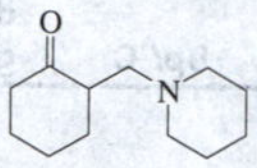
9073
2-(1-Piperidinylmethyl)cyclohexanone

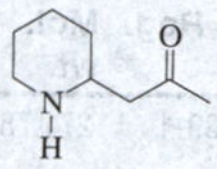
9074
1-(2-Piperidinyl)-2-propanone, (±)-

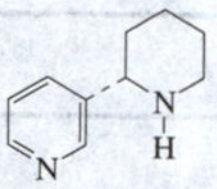
9075
3-(2-Piperidinyl)pyridine, (*S*)-

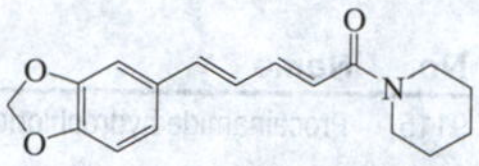
9076
Piperine

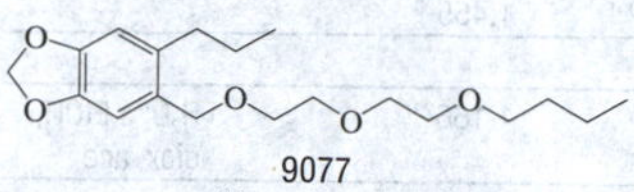
9077
Piperonyl butoxide

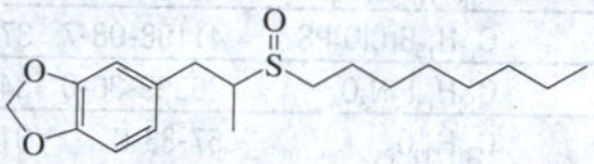
9078
Piperonyl sulfoxide

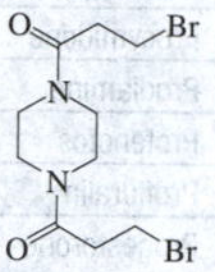
9079
Pipobroman

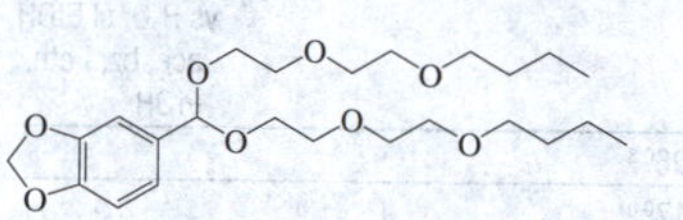
9080
Piprotal

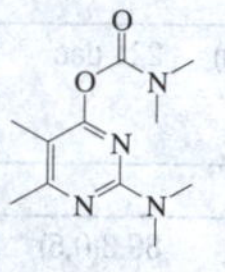
9081
Pirimicarb

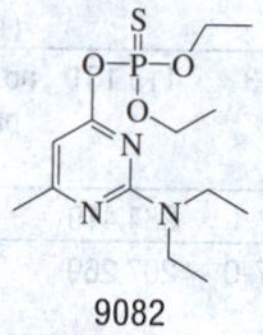
9082
Pirimiphos-ethyl

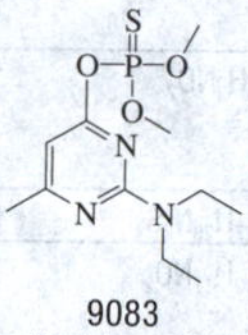
9083
Pirimiphos-methyl

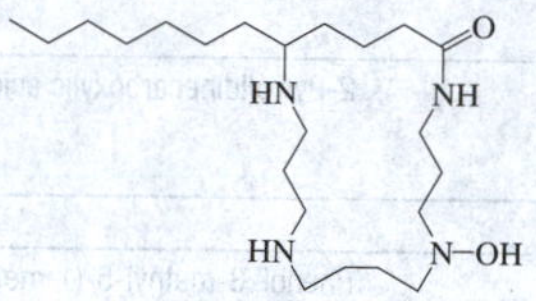
9084
Pithecolobine

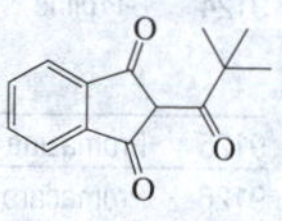
9085
2-Pivaloyl-1,3-indandione

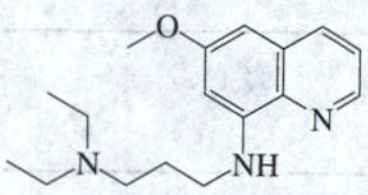
9086
Plasmocid

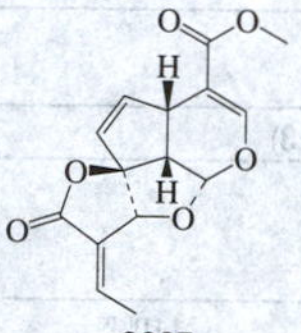
9087
Plumericin

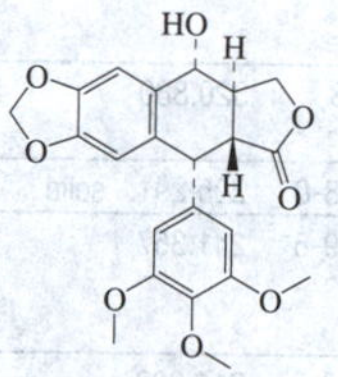
9088
Podophyllotoxin

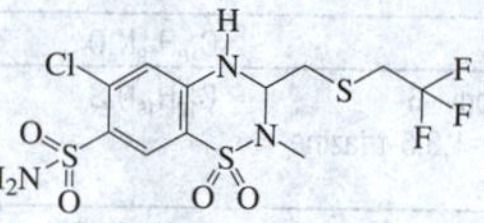
9089
Polythiazide

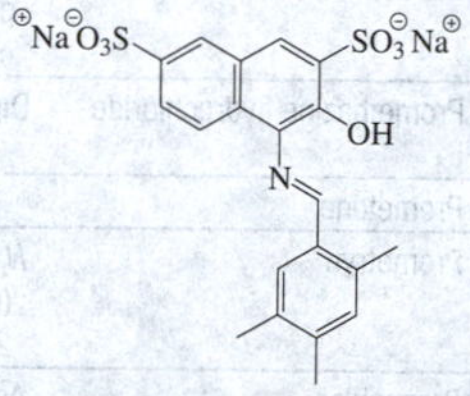
9090
Ponceau 3R

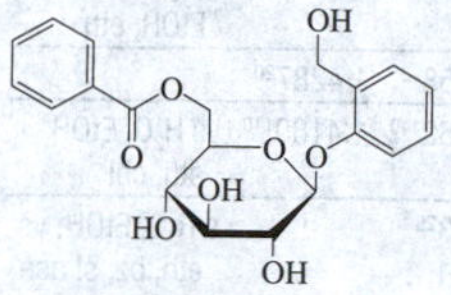
9091
Populin

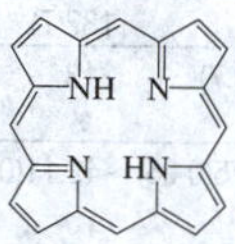
9092
21*H*,23*H*-Porphine

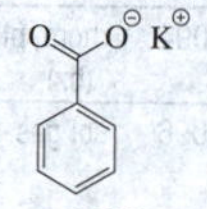
9093
Potassium benzoate

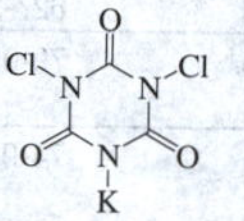
9094
Potassium dichloroisocyanurate

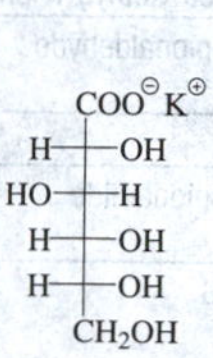
9095
Potassium D-gluconate

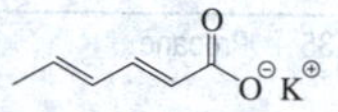
9096
Potassium *trans*,*trans*-2,4-hexadienoate

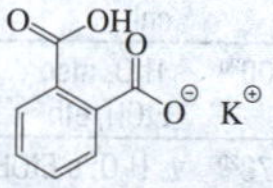
9097
Potassium hydrogen phthalate

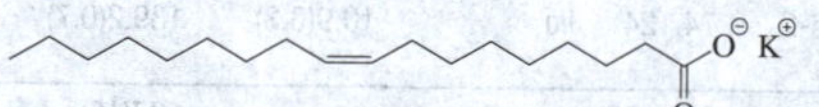
9098
Potassium *cis*-9-octadecenoate

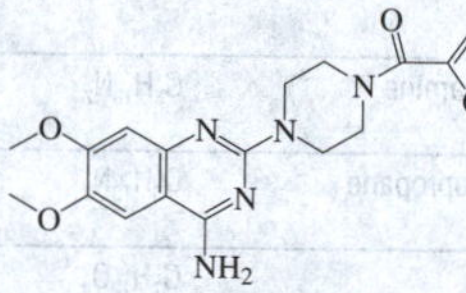
9099
Prazosin

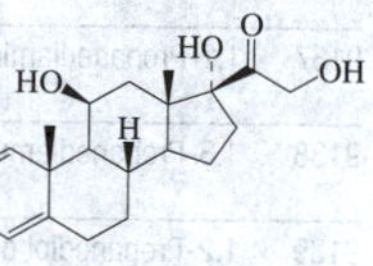
9100
Prednisolone

9101
5α-Pregnane

9102
5β-Pregnane

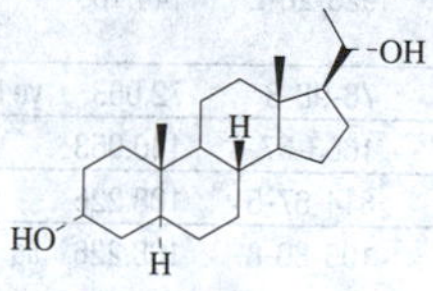
9103
5α-Pregnane-3α,20α-diol

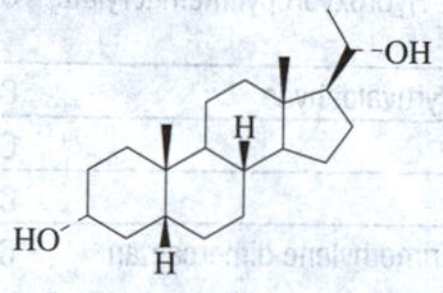
9104
5β-Pregnane-3α,20*S*-diol

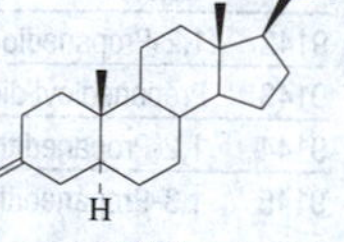
9105
5α-Pregnane-3,20-dione

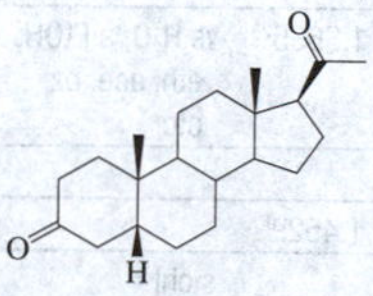
9106
5β-Pregnane-3,20-dione

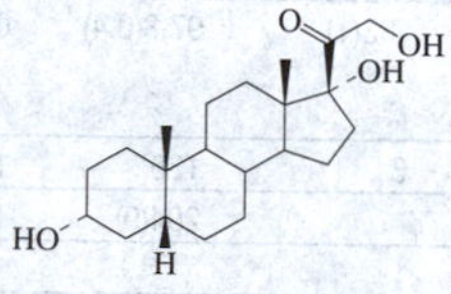
9107
5-Pregnane-3,17,21-triol-20-one

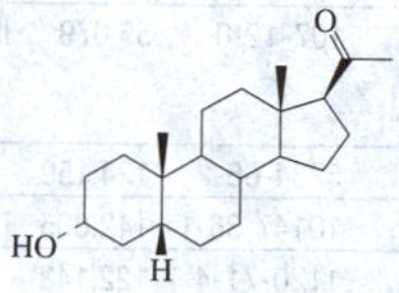
9108
Pregnan-3α-ol-20-one

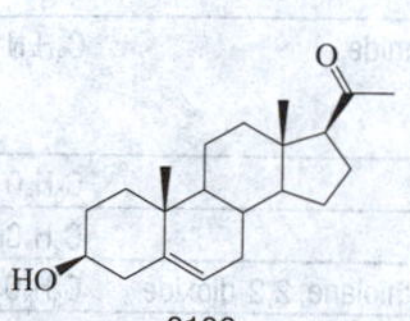
9109
Pregnenolone

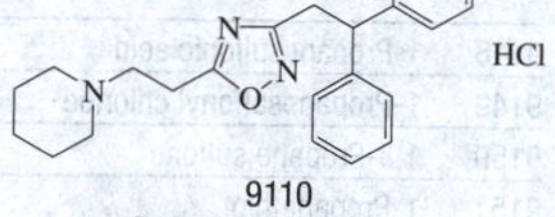
9110
Prenoxdiazine hydrochloride

9111
Prephenic acid

9112
Pridinol

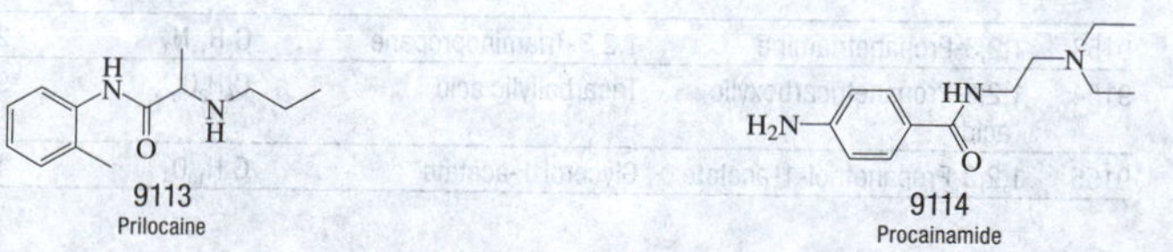
9113
Prilocaine

9114
Procainamide

Organic

No.	Name	Synonym	Mol. form.	CAS Reg. No.	Mol. wt.	Physical form	mp/°C	bp/°C	den/ g cm⁻³	n_D	Solubility
9115	Procainamide hydrochloride		$C_{13}H_{22}ClN_3O$	614-39-1	271.786		166				vs H_2O; s EtOH; i eth, bz; sl chl
9116	Procarbazine hydrochloride		$C_{12}H_{20}ClN_3O$	366-70-1	257.759	cry (MeOH)	225				
9117	Prochlorperazine		$C_{20}H_{24}ClN_3S$	58-38-8	373.943		228				
9118	Procymidone		$C_{13}H_{11}Cl_2NO_2$	32809-16-8	284.138		165.2(0.5)		1.452^{25}		
9119	Prodiamine		$C_{13}H_{17}F_3N_4O_4$	29091-21-2	350.294		124		1.47^{25}		
9120	Profenofos		$C_{11}H_{15}BrClO_3PS$	41198-08-7	373.631			$110^{0.001}$	1.455^{20}		
9121	Profluralin		$C_{14}H_{16}F_3N_3O_4$	26399-36-0	347.290		34.8(0.5)				
9122	Progesterone	Pregn-4-ene-3,20-dione	$C_{21}H_{30}O_2$	57-83-0	314.462	pr	131.0(0.3)		1.166^{23}		i H_2O; s EtOH, diox, ace
9123	DL-Proline		$C_5H_9NO_2$	609-36-9	115.131	hyg nd (al-eth) cry (+w)	205 dec				vs H_2O, EtOH
9124	L-Proline	2-Pyrrolidinecarboxylic acid	$C_5H_9NO_2$	147-85-3	115.131	nd (al-eth) pr (w)	221 dec				vs H_2O; sl EtOH, ace, bz; i eth, PrOH
9125	Promazine		$C_{17}H_{20}N_2S$	58-40-2	284.419			$206^{0.3}$			
9126	Promecarb	Phenol, 3-methyl-5-(1-methylethyl)-, methylcarbamate	$C_{12}H_{17}NO_2$	2631-37-0	207.269		89.3(0.5)	$117^{0.01}$			
9127	Promethazine	*N,N*,α-Trimethyl-10*H*-phenothiazine-10-ethanamine	$C_{17}H_{20}N_2S$	60-87-7	284.419		60	$191^{0.5}$			i H_2O; vs dil HCl
9128	Promethazine hydrochloride	Diprazin	$C_{17}H_{21}ClN_2S$	58-33-3	320.880		231				vs H_2O, EtOH, chl
9129	Prometone		$C_{10}H_{19}N_5O$	1610-18-0	225.291	solid	91.4(0.3)				
9130	Prometryn	*N,N'*-Diisopropyl-6-(methylthio)-1,3,5-triazine-2,4-diamine	$C_{10}H_{19}N_5S$	7287-19-6	241.357		121(1)		1.157^{20}		
9131	Propachlor	Acetamide, 2-chloro-*N*-(1-methylethyl)-*N*-phenyl-	$C_{11}H_{14}ClNO$	1918-16-7	211.688		78(1)	$110^{0.03}$	1.242^{25}		
9132	Propanal	Propionaldehyde	C_3H_6O	123-38-6	58.079	liq	-80	48.0(0.2)	0.8657^{25}	1.3636^{20}	s H_2O; msc EtOH, eth
9133	Propanal oxime		C_3H_7NO	627-39-4	73.094		40	132(7)	0.9258^{20}	1.4287^{20}	
9134	Propanamide	Propionamide	C_3H_7NO	79-05-0	73.094	rhom, pl (bz)	79.9(0.7)	213	0.9262^{110}	1.4180^{110}	vs H_2O, EtOH, eth, chl
9135	Propane	LPG	C_3H_8	74-98-6	44.096	col gas	-187.75(0.05)	-42.11(0.09)	0.493^{25} (p>1 atm)		s H_2O, EtOH; vs eth, bz; sl ace
9136	Propanediamide		$C_3H_6N_2O_2$	108-13-4	102.092	mcl pr(w)	170(1)				s H_2O; i EtOH, eth, bz; sl DMSO
9137	1,2-Propanediamine, (±)-	Propylenediamine	$C_3H_{10}N_2$	10424-38-1	74.124	hyg		118(1)	0.878^{15}	1.4460^{20}	vs H_2O; i eth; vs chl
9138	1,3-Propanediamine	1,3-Diaminopropane	$C_3H_{10}N_2$	109-76-2	74.124	liq	-10.9(0.3)	139.2(0.7)	0.884^{25}	1.4600^{20}	s H_2O; msc EtOH, eth
9139	1,2-Propanediol diacetate		$C_7H_{12}O_4$	623-84-7	160.168			187(15)	1.059^{20}	1.4173^{20}	vs H_2O; s EtOH, eth
9140	1,3-Propanediol diacetate		$C_7H_{12}O_4$	628-66-0	160.168			209.5	1.070^{14}	1.4192	vs H_2O; s EtOH
9141	1,2-Propanediol 1-methacrylate	2-Hydroxypropyl methacrylate	$C_7H_{12}O_3$	923-26-2	144.168			90^{9}	1.066^{25}	1.4458^{20}	
9142	1,2-Propanedione	Pyruvaldehyde	$C_3H_4O_2$	78-98-8	72.063	ye hyg liq		72(15)	1.0455^{20}	1.4002^{18}	s EtOH, eth, bz
9143	Propanedioyl dichloride		$C_3H_2Cl_2O_2$	1663-67-8	140.953			57^{28}	1.4509^{20}	1.4639^{20}	s eth, AcOEt
9144	1,2-Propanedithiol		$C_3H_8S_2$	814-67-5	108.226			164(10)	1.08^{20}	1.532^{20}	s chl
9145	1,3-Propanedithiol	Trimethylene dimercaptan	$C_3H_8S_2$	109-80-8	108.226	liq	-79	172.9	1.0772^{20}	1.5392^{20}	sl H_2O, ctc; msc EtOH, eth, bz
9146	2,2'-[1,3-Propanediylbis-(nitrilomethylidyne)]bisphenol	Disalicylidene-1,3-propanediamine	$C_{17}H_{18}N_2O_2$	120-70-7	282.337		54.3				
9147	Propanenitrile	Ethyl cyanide	C_3H_5N	107-12-0	55.079	liq	-93(1)	97.3(0.4)	0.7818^{20}	1.3655^{20}	vs H_2O; s EtOH, eth, ace, bz, ctc
9148	1-Propanesulfonic acid		$C_3H_8O_3S$	5284-66-2	124.159		8	136^{1}	1.2516^{25}		
9149	1-Propanesulfonyl chloride		$C_3H_7ClO_2S$	10147-36-1	142.605			204(9)	1.267^{20}	1.452^{20}	
9150	1,3-Propane sultone	1,2-Oxathiolane, 2,2-dioxide	$C_3H_6O_3S$	1120-71-4	122.143						s chl
9151	1-Propanethiol	Propyl mercaptan	C_3H_8S	107-03-9	76.161	liq	-113.12(0.07)	67.7(0.1)	0.8411^{20}	1.4380^{20}	sl H_2O; s EtOH, eth, ace, bz
9152	2-Propanethiol	Isopropyl mercaptan	C_3H_8S	75-33-2	76.161	liq	-130.50(0.05)	52.6(0.3)	0.8143^{20}	1.4255^{20}	sl H_2O; msc EtOH, eth; vs ace; s chl
9153	1,2,3-Propanetriamine	1,2,3-Triaminopropane	$C_3H_{11}N_3$	21291-99-6	89.139	visc oil		190			s H_2O
9154	1,2,3-Propanetricarboxylic acid	Tricarballylic acid	$C_6H_8O_6$	99-14-9	176.124	orth (eth)	166				vs H_2O, EtOH; sl eth
9155	1,2,3-Propanetriol-1-acetate	Glycerol 1-acetate	$C_5H_{10}O_4$	106-61-6	134.131			158^{165}	1.2060^{20}	1.4157^{20}	vs H_2O, EtOH

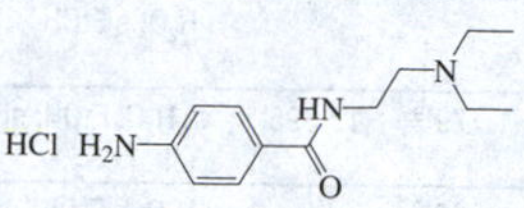
9115
Procainamide hydrochloride

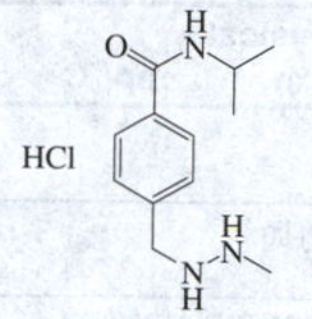
9116
Procarbazine hydrochloride

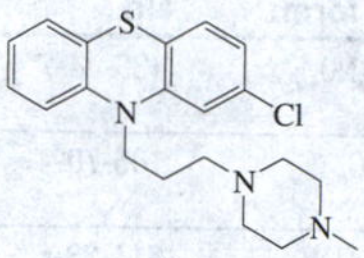
9117
Prochlorperazine

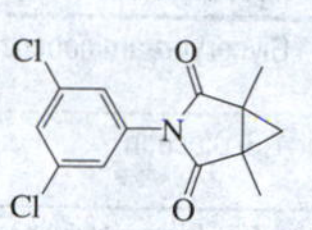
9118
Procymidone

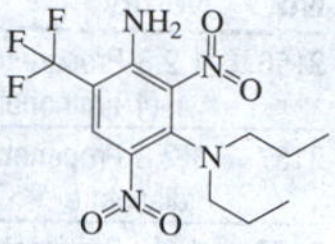
9119
Prodiamine

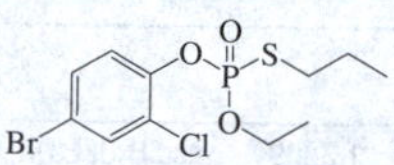
9120
Profenofos

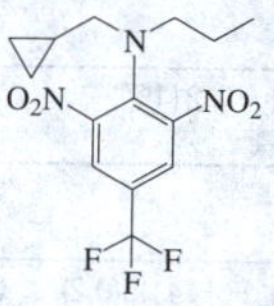
9121
Profluralin

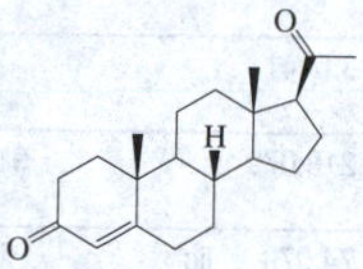
9122
Progesterone

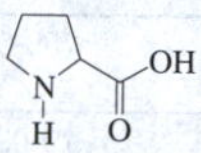
9123
DL-Proline

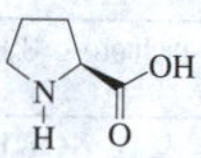
9124
L-Proline

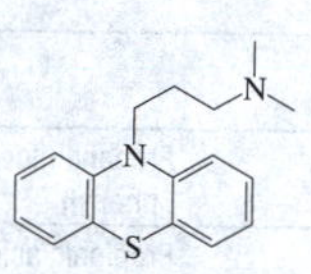
9125
Promazine

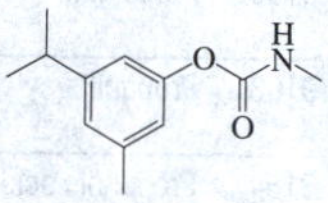
9126
Promecarb

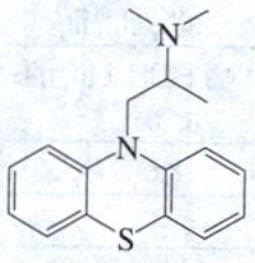
9127
Promethazine

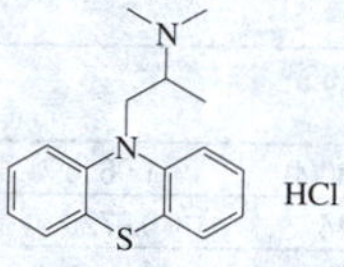
9128
Promethazine hydrochloride

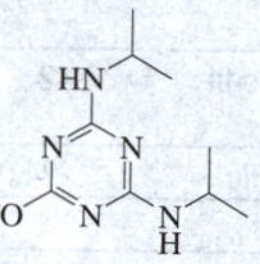
9129
Prometone

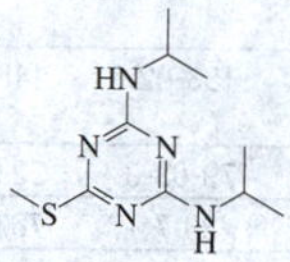
9130
Prometryn

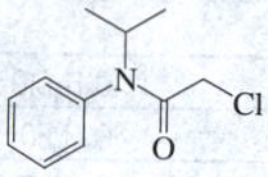
9131
Propachlor

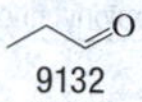
9132
Propanal

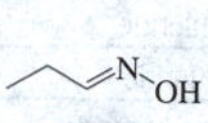
9133
Propanal oxime

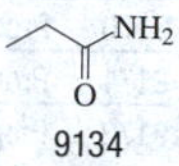
9134
Propanamide

9135
Propane

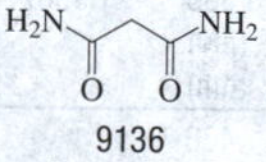
9136
Propanediamide

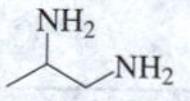
9137
1,2-Propanediamine, (±)-

9138
1,3-Propanediamine

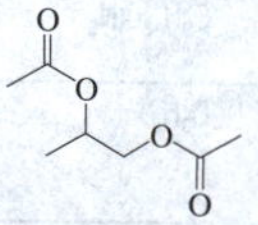
9139
1,2-Propanediol diacetate

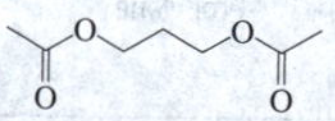
9140
1,3-Propanediol diacetate

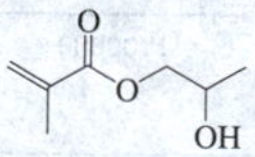
9141
1,2-Propanediol 1-methacrylate

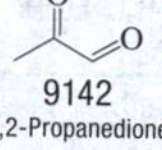
9142
1,2-Propanedione

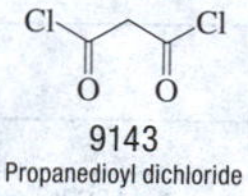
9143
Propanedioyl dichloride

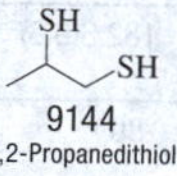
9144
1,2-Propanedithiol

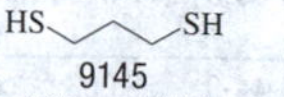
9145
1,3-Propanedithiol

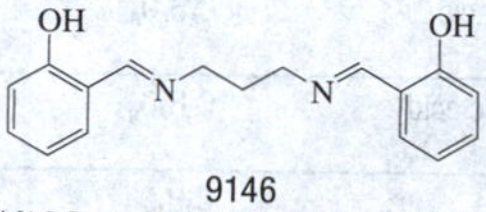
9146
2,2'-[1,3-Propanediylbis(nitrilomethylidyne)]bisphenol

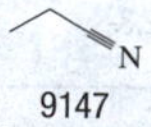
9147
Propanenitrile

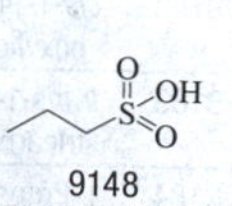
9148
1-Propanesulfonic acid

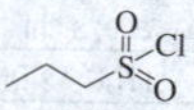
9149
1-Propanesulfonyl chloride

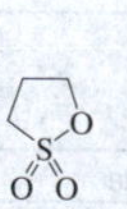
9150
1,3-Propane sultone

9151
1-Propanethiol

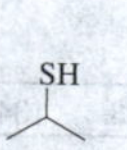
9152
2-Propanethiol

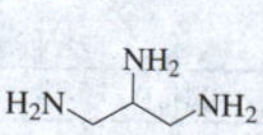
9153
1,2,3-Propanetriamine

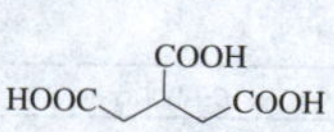
9154
1,2,3-Propanetricarboxylic acid

9155
1,2,3-Propanetriol-1-acetate

No.	Name	Synonym	Mol. form.	CAS Reg. No.	Mol. wt.	Physical form	mp/°C	bp/°C	den/ g cm^{-3}	n_D	Solubility
9156	1,2,3-Propanetriol 1-(4-aminobenzoate)	Glyceryl *p*-aminobenzoate	$C_{10}H_{13}NO_4$	136-44-7	211.215						i H_2O; s EtOH
9157	1,2,3-Propanetriol-1,3-diacetate	1,3-Diacetin	$C_7H_{12}O_5$	105-70-4	176.167	hyg liq		260(7)	1.179^{15}	1.4395^{20}	vs H_2O, EtOH; sl eth; i CS_2
9158	1,2,3-Propanetriol tribenzoate	1,2,3-Propanetriyl benzoate	$C_{24}H_{20}O_6$	614-33-5	404.412	nd (MeOH)	76(3)		1.228^{12}		i H_2O; s EtOH; vs eth, ace, bz, chl
9159	1,2,3-Propanetriol tripropanoate	1,2,3-Propanetriyl propanoate	$C_{12}H_{20}O_6$	139-45-7	260.283			175^{20}	1.108^{15}	1.4318^{19}	i H_2O; s EtOH, chl; vs eth
9160	1,2,3-Propanetriyl hexanoate	Glycerol trihexanoate	$C_{21}H_{38}O_6$	621-70-5	386.523		-60	373.1(0.9)	0.9867^{20}	1.4427^{20}	i H_2O; msc EtOH, eth, bz; vs ace
9161	1,2,3-Propanetriyl octanoate	Glycerol trioctanoate	$C_{27}H_{50}O_6$	538-23-8	470.682		9.6(0.6)	447(23)	0.9540^{20}	1.4482^{20}	i H_2O; msc EtOH; vs eth, bz, chl, lig
9162	Propanidid		$C_{18}H_{27}NO_5$	1421-14-3	337.411			211$^{0.7}$			i H_2O; s EtOH, chl
9163	Propanil	Propanamide, *N*-(3,4-dichloro-phenyl)-	$C_9H_9Cl_2NO$	709-98-8	218.079		91.3(0.3)		1.25^{25}		
9164	Propanoic acid	Propionic acid	$C_3H_6O_2$	79-09-4	74.079	liq	-20.5(0.5)	141.5(0.2)	0.9882^{25}	1.3809^{20}	msc H_2O, EtOH; s eth; sl chl
9165	Propanoic anhydride	Propionic anhydride	$C_6H_{10}O_3$	123-62-6	130.141	liq	-45.0(0.5)	168(1)	1.0110^{20}	1.4038^{20}	msc eth; sl ctc
9166	1-Propanol	Propyl alcohol	C_3H_8O	71-23-8	60.095	liq	-124.39(0.02)	97.04(0.09)	0.7997^{25}	1.3850^{20}	msc H_2O, EtOH, eth; s ace, chl; vs bz
9167	2-Propanol	Isopropyl alcohol	C_3H_8O	67-63-0	60.095	liq	-87.91(0.04)	82.21(0.09)	0.7809^{25}	1.3776^{20}	msc H_2O, EtOH, eth; s ace, chl; vs bz
9168	2-Propanone oxime	Acetoxime	C_3H_7NO	127-06-0	73.094	pr (al)	61	133(7)	0.9113^{62}	1.4156^{20}	s H_2O, EtOH, eth, chl, lig
9169	2-Propanone phenylhydrazone	Acetone, phenylhydrazone	$C_9H_{12}N_2$	103-02-6	148.204	orth	42	163^{50}			s EtOH, eth, dil acid
9170	Propanoyl chloride	Propionyl chloride	C_3H_5ClO	79-03-8	92.524	liq	-94.0(0.4)	80(4)	1.0646^{20}	1.4032^{20}	s eth
9171	Propanoyl fluoride	Propionyl fluoride	C_3H_5FO	430-71-7	76.069			44	0.972^{15}	1.329^{13}	
9172	Propantheline bromide		$C_{23}H_{30}BrNO_3$	50-34-0	448.393	cry	160				vs H_2O, EtOH, chl; i eth, bz
9173	Propargite		$C_{19}H_{26}O_4S$	2312-35-8	350.472				1.10^{25}		
9174	Propargyl acetate		$C_5H_6O_2$	627-09-8	98.101			124(5)	0.9982^{20}	1.4187^{20}	sl H_2O; s EtOH, eth
9175	Propargyl alcohol	3-Hydroxy-1-propyne	C_3H_4O	107-19-7	56.063	liq	-51.8(0.4)	113(3)	0.9478^{20}	1.4322^{20}	s H_2O, chl; msc EtOH, eth
9176	Propatyl nitrate	2-Ethyl-2-[(nitrooxy)methyl]-1,3-propanediol, dinitrate	$C_6H_{11}N_3O_9$	2921-92-8	269.166	wh pow	52		1.49		i H_2O; s EtOH, ace
9177	Propazine	6-Chloro-*N,N'*-diisopropyl-1,3,5-triazine-2,4-diamine	$C_9H_{16}ClN_5$	139-40-2	229.710		217(1)		1.162^{20}		
9178	Propene	Propylene	C_3H_6	115-07-1	42.080	col gas	-185.30(0.02)	-47.6(0.1)	0.505^{25} (p>1 atm)	1.3567^{-70}	sl H_2O; vs EtOH, HOAc
9179	*trans*-1-Propene-1,2-dicarboxylic acid	Mesaconic acid	$C_5H_6O_4$	498-24-8	130.100	orth nd or mcl pr (eth)	204.5	sub	1.466^{20}		sl H_2O, bz, CS_2; vs EtOH; s eth, tfa
9180	1-Propene-2,3-dicarboxylic acid	Itaconic acid	$C_5H_6O_4$	97-65-4	130.100	rhom (bz)	165.7(0.9)	dec	1.632^{25}		s H_2O, EtOH, ace; sl eth, bz, peth
9181	2-Propene-1-thiol		C_3H_6S	870-23-5	74.145			65	0.925^{23}	1.4832^{20}	i H_2O; msc EtOH, eth; s chl
9182	*cis*-1-Propene-1,2,3-tricarboxylic acid	*cis*-Aconitic acid	$C_6H_6O_6$	585-84-2	174.108	nd (w)	125				s H_2O; sl eth
9183	*trans*-1-Propene-1,2,3-tricarboxylic acid	*trans*-Aconitic acid	$C_6H_6O_6$	4023-65-8	174.108	lf (w) nd (w, eth)	182.6(0.3)				vs H_2O, EtOH
9184	1-Propen-1-one	Methylketene	C_3H_4O	6004-44-0	56.063	col gas	-80	-23			vs eth
9185	2-Propenoyl chloride	Acrylic acid chloride	C_3H_3ClO	814-68-6	90.508			75.5	1.1136^{20}	1.4343^{20}	vs chl
9186	*cis*-1-Propenylbenzene		C_9H_{10}	766-90-5	118.175	liq	-61.7(0.3)	169(4)	0.9088^{20}	1.5420^{20}	i H_2O; msc EtOH, eth, ace, bz, peth, ctc
9187	*trans*-1-Propenylbenzene		C_9H_{10}	873-66-5	118.175	liq	-29.6(0.7)	179(2)	0.9023^{25}	1.5506^{20}	i H_2O; msc EtOH, eth, ace, bz
9188	*trans*-5-(1-Propenyl)-1,3-benzodioxole		$C_{10}H_{10}O_2$	4043-71-4	162.185		6.8	253	1.1224^{20}	1.5782^{20}	i H_2O; msc EtOH, eth; vs ace; s chl
9189	4-(1-Propenyl)phenol	*p*-Anol	$C_9H_{10}O$	539-12-8	134.174	lf	94	250 dec			sl H_2O; vs DMF

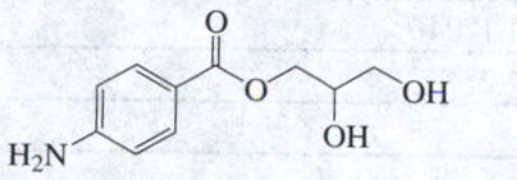
9156
1,2,3-Propanetriol 1-(4-aminobenzoate)

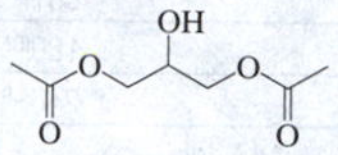
9157
1,2,3-Propanetriol-1,3-diacetate

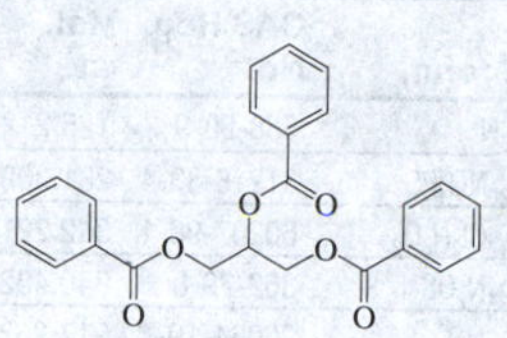
9158
1,2,3-Propanetriol tribenzoate

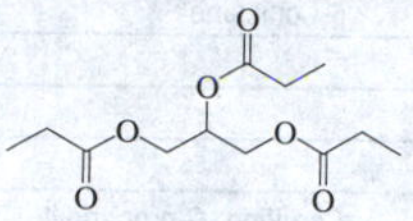
9159
1,2,3-Propanetriol tripropanoate

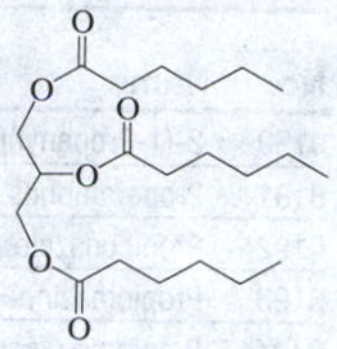
9160
1,2,3-Propanetriyl hexanoate

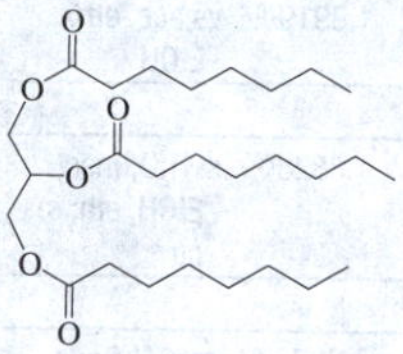
9161
1,2,3-Propanetriyl octanoate

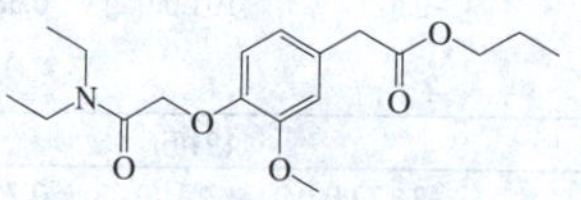
9162
Propanidid

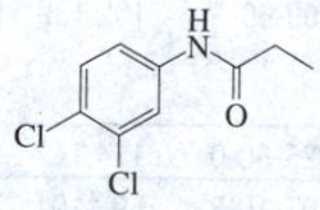
9163
Propanil

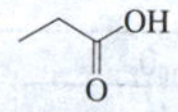
9164
Propanoic acid

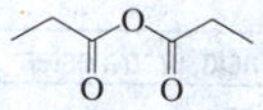
9165
Propanoic anhydride

9166
1-Propanol

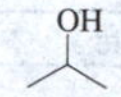
9167
2-Propanol

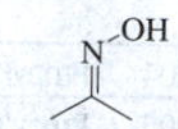
9168
2-Propanone oxime

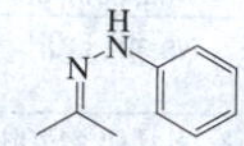
9169
2-Propanone phenylhydrazone

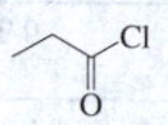
9170
Propanoyl chloride

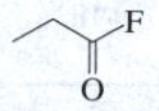
9171
Propanoyl fluoride

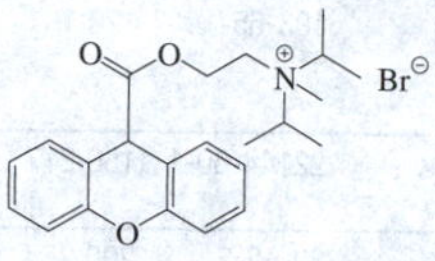
9172
Propantheline bromide

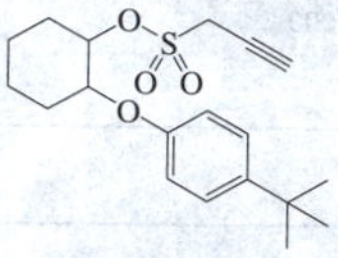
9173
Propargite

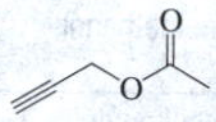
9174
Propargyl acetate

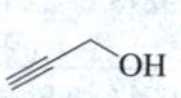
9175
Propargyl alcohol

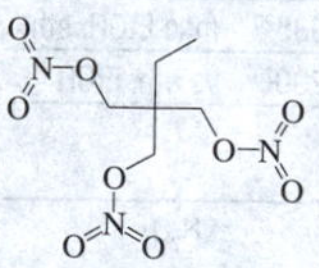
9176
Propatyl nitrate

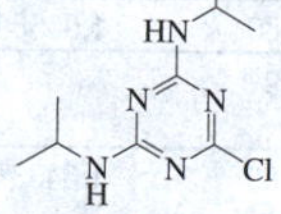
9177
Propazine

9178
Propene

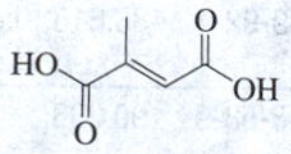
9179
trans-1-Propene-1,2-dicarboxylic acid

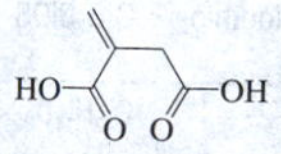
9180
1-Propene-2,3-dicarboxylic acid

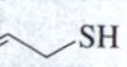
9181
2-Propene-1-thiol

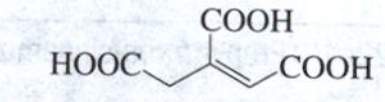
9182
cis-1-Propene-1,2,3-tricarboxylic acid

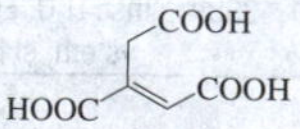
9183
trans-1-Propene-1,2,3-tricarboxylic acid

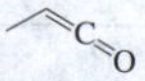
9184
1-Propen-1-one

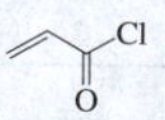
9185
2-Propenoyl chloride

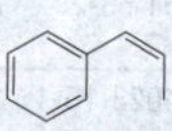
9186
cis-1-Propenylbenzene

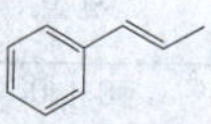
9187
trans-1-Propenylbenzene

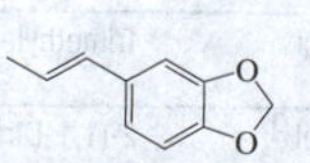
9188
trans-5-(1-Propenyl)-1,3-benzodioxole

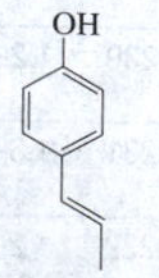
9189
4-(1-Propenyl)phenol

No.	Name	Synonym	Mol. form.	CAS Reg. No.	Mol. wt.	Physical form	mp/°C	bp/°C	den/g cm^{-3}	n_D	Solubility
9190	2-(1-Propenyl)piperidine	β-Coniceine	$C_8H_{15}N$	538-90-9	125.212		8	168	0.8716^{15}		
9191	Propetamphos		$C_{10}H_{20}NO_4PS$	31218-83-4	281.309			$88^{0.005}$	1.1294^{20}		
9192	Propiconazole		$C_{15}H_{17}Cl_2N_3O_2$	60207-90-1	342.221			$180^{0.1}$	1.27^{20}		
9193	Propiomazine		$C_{20}H_{24}N_2OS$	362-29-8	340.482			$240^{0.5}$			
9194	Propionyl-L-carnitine	Carnitine, *O*-propanoyl	$C_{10}H_{19}NO_4$	20064-19-1	217.263	hyg pr (2-PrOH)	147 dec				
9195	Propofol		$C_{12}H_{18}O$	2078-54-8	178.270		18(1)	247(8)	0.955^{20}	1.5140^{20}	
9196	Propoxur	Phenol, 2-(1-methylethoxy)-, methylcarbamate	$C_{11}H_{15}NO_3$	114-26-1	209.242		90.5(0.5)	dec	1.12^{20}		
9197	2-Propoxyethanol	Ethylene glycol monopropyl ether	$C_5H_{12}O_2$	2807-30-9	104.148			152(3)	0.9112^{20}	1.4133^{20}	s H_2O; vs EtOH, eth
9198	D-Propoxyphene	Dextropropoxyphene	$C_{22}H_{29}NO_2$	469-62-5	339.471	cry (peth)	75.5				
9199	L-Propoxyphene	Levopropoxyphene	$C_{22}H_{29}NO_2$	2338-37-6	339.471	cry (peth)	75.5				
9200	1-Propoxy-2-propanol	1,2-Propylene glycol 1-propyl ether	$C_6H_{14}O_2$	1569-01-3	118.174			150.2(0.7)	0.8886^{20}	1.4130^{20}	
9201	3-Propoxy-1-propene		$C_6H_{12}O$	1471-03-0	100.158			91	0.7764^{20}	1.3919^{20}	vs ace, eth, EtOH
9202	Propranolol		$C_{16}H_{21}NO_2$	525-66-6	259.344	cry (cyhex)	96				
9203	Propyl acetate		$C_5H_{10}O_2$	109-60-4	102.132	liq	-93(2)	101.0(0.2)	0.8820^{25}	1.3828^{25}	sl H_2O; msc EtOH, eth; s ctc
9204	Propyl acrylate	2-Propenoic acid, propyl ester	$C_6H_{10}O_2$	925-60-0	114.142			121(6)			
9205	Propylamine	1-Propanamine	C_3H_9N	107-10-8	59.110	liq	-84.78(0.04)	47.21(0.08)	0.7173^{20}	1.3870^{20}	msc H_2O; vs EtOH, ace; s bz, chl; sl ctc
9206	Propylamine hydrochloride	1-Propanamine hydrochloride	$C_3H_{10}ClN$	556-53-6	95.571		163.5				s DMSO
9207	Propyl 4-aminobenzoate	Risocaine	$C_{10}H_{13}NO_2$	94-12-2	179.216	pr	73.1(0.7)				vs bz, eth, EtOH, chl
9208	2-(Propylamino)ethanol		$C_5H_{13}NO$	16369-21-4	103.163			183(13)	0.9005^{20}	1.4428^{20}	
9209	4-Propylaniline		$C_9H_{13}N$	2696-84-6	135.206			229(14)			
9210	*N*-Propylaniline		$C_9H_{13}N$	622-80-0	135.206			220(3)	0.9443^{20}	1.5428^{20}	vs eth, EtOH
9211	Propylarsonic acid	1-Propanearsonic acid	$C_3H_9AsO_3$	107-34-6	168.023	nd (al), pl (w)	134.5				vs H_2O, EtOH; i eth
9212	Propylbenzene	Isocumene	C_9H_{12}	103-65-1	120.191	liq	-99.52(0.05)	159.2(0.5)	0.8593^{25}	1.4895^{25}	i H_2O; msc EtOH, eth, ace, bz, peth, ctc
9213	α-Propylbenzenemethanol, (*R*)-		$C_{10}H_{14}O$	22144-60-1	150.217		16	232	0.9740^{20}	1.5139^{20}	vs eth, EtOH
9214	Propyl benzenesulfonate		$C_9H_{12}O_3S$	80-42-2	200.254			162^{15}	1.1804^{17}	1.5035^{25}	sl H_2O; s EtOH; vs eth, chl
9215	Propyl benzoate	Propyl benzenecarboxylate	$C_{10}H_{12}O_2$	2315-68-6	164.201	liq	-51.6	231(2)	1.0230^{20}	1.5000^{20}	i H_2O; msc EtOH, eth
9216	5-Propyl-1,3-benzodioxole	Dihydrosafrole	$C_{10}H_{12}O_2$	94-58-6	164.201			228			s ctc
9217	Propyl butanoate		$C_7H_{14}O_2$	105-66-8	130.185	liq	-95.2(0.5)	144(2)	0.8730^{20}	1.4001^{20}	sl H_2O; msc EtOH, eth
9218	Propyl carbamate		$C_4H_9NO_2$	627-12-3	103.120	pr	60	196			vs ace, eth, EtOH
9219	Propyl chloroacetate		$C_5H_9ClO_2$	5396-24-7	136.577			160(6)	1.104^{20}	1.4261^{20}	vs eth
9220	Propyl 2-chlorobutanoate		$C_7H_{13}ClO_2$	62108-71-8	164.630			183	1.0252^{20}		
9221	Propyl chlorocarbonate		$C_4H_7ClO_2$	109-61-5	122.551			111(2)	1.0901^{20}	1.4035^{20}	msc EtOH, eth
9222	Propyl 3-chloropropanoate		$C_6H_{11}ClO_2$	62108-66-1	150.603			180	1.0656^{20}	1.4290^{20}	vs eth, EtOH
9223	*S*-Propyl chlorothioformate	*S*-Propyl carbonochloridothioate	C_4H_7ClOS	13889-92-4	138.616	liq		59^{26}			
9224	Propyl *trans*-cinnamate	Propyl *trans*-3-phenyl-2-propenoate	$C_{12}H_{14}O_2$	74513-58-9	190.238			285	1.0433^{0}		i H_2O
9225	Propylcyclohexane		C_9H_{18}	1678-92-8	126.239	liq	-94.86(0.05)	156.7(0.3)	0.7936^{20}	1.4370^{20}	i H_2O; msc EtOH, ace, ctc; s eth, bz
9226	2-Propylcyclohexanone		$C_9H_{16}O$	94-65-5	140.222			197	0.927^{20}	1.4538^{20}	i H_2O; s EtOH, ace; vs eth, bz
9227	Propylcyclopentane		C_8H_{16}	2040-96-2	112.213	liq	-117.34(0.06)	130.9(0.8)	0.7763^{20}	1.4266^{20}	i H_2O; msc EtOH, eth, ace; s bz; vs ctc
9228	1-Propylcyclopentanol		$C_8H_{16}O$	1604-02-0	128.212	liq	-37.5	171(9)	0.9040^{25}	1.4502^{25}	
9229	Propylene carbonate	4-Methyl-1,3-dioxolan-2-one	$C_4H_6O_3$	108-32-7	102.089	liq	-48.8	241.6(0.7)	1.2047^{20}	1.4189^{20}	vs H_2O, EtOH, eth, ace, bz
9230	1,2-Propylene glycol	Propylene glycol	$C_3H_8O_2$	57-55-6	76.095	liq	-60	187.3(0.2)	1.0361^{20}	1.4324^{20}	msc H_2O, EtOH; s eth, bz, chl
9231	1,3-Propylene glycol	Trimethylene glycol	$C_3H_8O_2$	504-63-2	76.095	liq	-27.6(0.1)	214.7(0.3)	1.0538^{20}	1.4398^{20}	msc H_2O, EtOH; vs eth; sl bz
9232	1,2-Propylene glycol 2-*tert*-butyl ether	2-(1,1-Dimethylethoxy)-1-propanol	$C_7H_{16}O_2$	94023-15-1	132.201	liq		152	0.87		

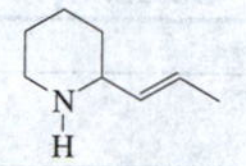
9190
2-(1-Propenyl)piperidine

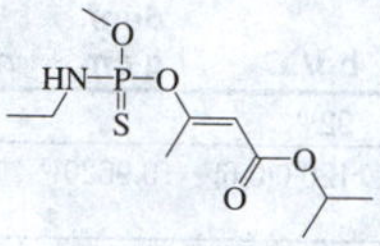
9191
Propetamphos

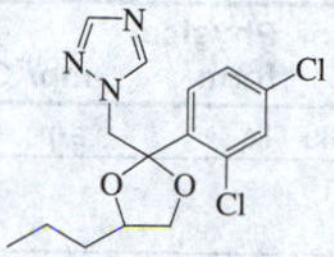
9192
Propiconazole

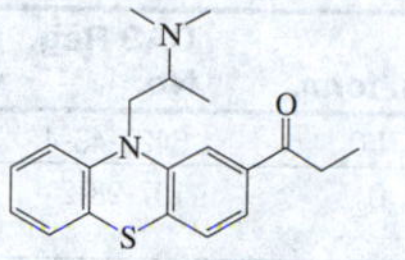
9193
Propiomazine

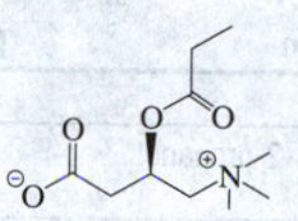
9194
Propionyl-L-carnitine

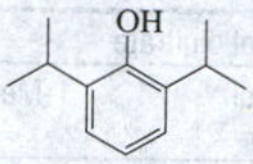
9195
Propofol

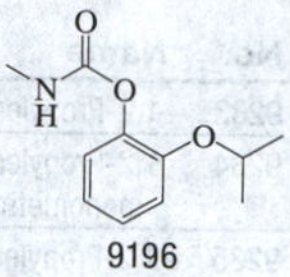
9196
Propoxur

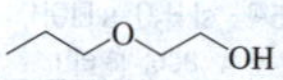
9197
2-Propoxyethanol

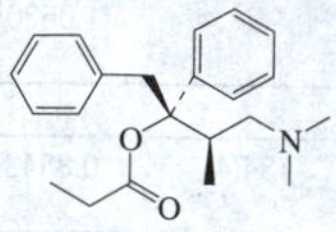
9198
D-Propoxyphene

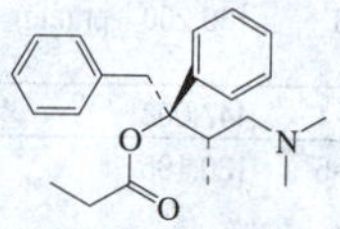
9199
L-Propoxyphene

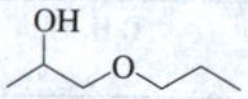
9200
1-Propoxy-2-propanol

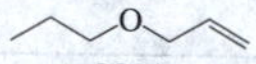
9201
3-Propoxy-1-propene

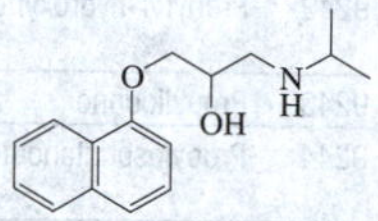
9202
Propranolol

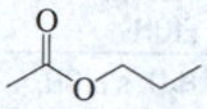
9203
Propyl acetate

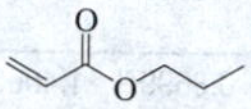
9204
Propyl acrylate

9205
Propylamine

9206
Propylamine hydrochloride

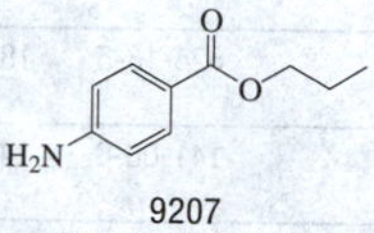
9207
Propyl 4-aminobenzoate

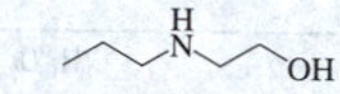
9208
2-(Propylamino)ethanol

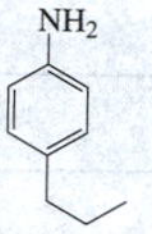
9209
4-Propylaniline

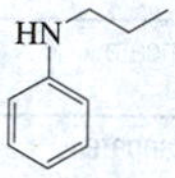
9210
N-Propylaniline

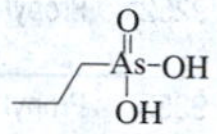
9211
Propylarsonic acid

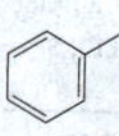
9212
Propylbenzene

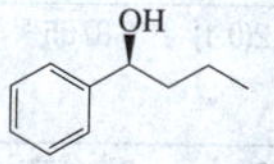
9213
α-Propylbenzenemethanol, (*R*)-

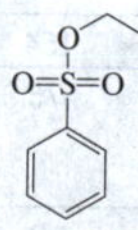
9214
Propyl benzenesulfonate

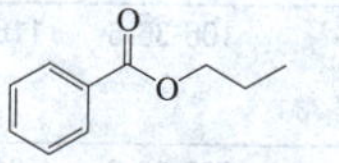
9215
Propyl benzoate

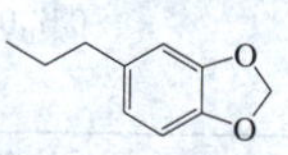
9216
5-Propyl-1,3-benzodioxole

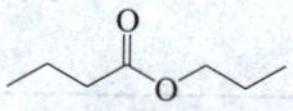
9217
Propyl butanoate

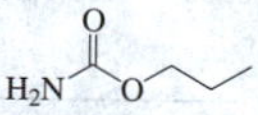
9218
Propyl carbamate

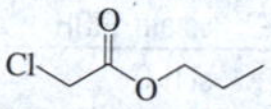
9219
Propyl chloroacetate

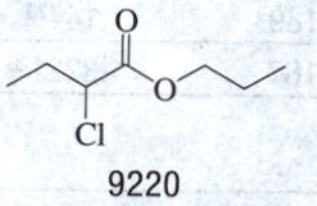
9220
Propyl 2-chlorobutanoate

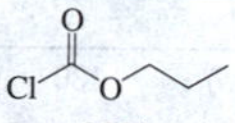
9221
Propyl chlorocarbonate

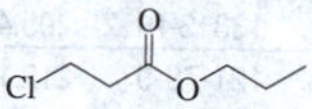
9222
Propyl 3-chloropropanoate

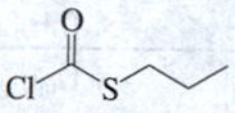
9223
S-Propyl chlorothioformate

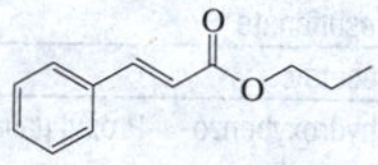
9224
Propyl *trans*-cinnamate

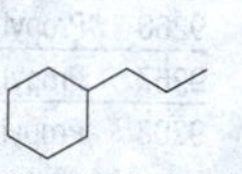
9225
Propylcyclohexane

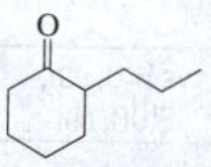
9226
2-Propylcyclohexanone

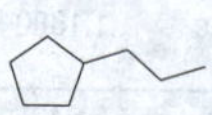
9227
Propylcyclopentane

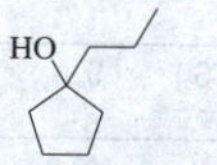
9228
1-Propylcyclopentanol

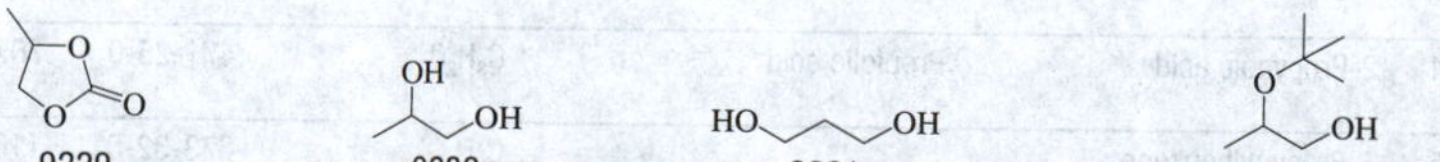
9229
Propylene carbonate

9230
1,2-Propylene glycol

9231
1,3-Propylene glycol

9232
1,2-Propylene glycol 2-*tert*-butyl ether

No.	Name	Synonym	Mol. form.	CAS Reg. No.	Mol. wt.	Physical form	mp/°C	bp/°C	den/ g cm^{-3}	n_D	Solubility
9233	1,2-Propylene glycol dinitrate		$C_3H_6N_2O_6$	6423-43-4	166.089	liq	exp	92^{10}			
9234	1,2-Propylene glycol monomethyl ether	1-Methoxy-2-propanol	$C_4H_{10}O_2$	107-98-2	90.121			120.0(0.6)	0.9620^{20}	1.4034^{20}	
9235	1,2-Propylene glycol monomethyl ether acetate	2-Acetoxy-1-methoxypropane	$C_6H_{12}O_3$	108-65-6	132.157	liq		146.0(0.4)			
9236	Propyleneimine	2-Methylaziridine	C_3H_7N	75-55-8	57.095			67	0.812^{16}		
9237	Propyl formate		$C_4H_8O_2$	110-74-7	88.106	liq	-92.9(0.4)	80.6(0.2)	0.9073^{20}	1.377^{20}	sl H_2O, ctc; msc EtOH, eth
9238	Propyl 3-(2-furyl)acrylate		$C_{10}H_{12}O_3$	623-22-3	180.200			113^{16}	1.0744^{20}	1.5392^{24}	vs bz, eth, EtOH
9239	4-Propylheptane		$C_{10}H_{22}$	3178-29-8	142.282			167(3)	0.7321^{25}	1.4135^{20}	
9240	Propyl hexanoate		$C_9H_{18}O_2$	626-77-7	158.238	liq	-74.0(0.5)	185(2)	0.8672^{20}	1.4170^{20}	vs eth, EtOH
9241	Propyl 2-hydroxybenzoate		$C_{10}H_{12}O_3$	607-90-9	180.200		97	239	1.0979^{20}	1.5161^{20}	s ctc, CS_2
9242	Propyl 4-hydroxybenzoate	Propylparaben	$C_{10}H_{12}O_3$	94-13-3	180.200	pr (eth)	96.1(0.5)		1.0630^{102}	1.5050^{102}	i H_2O; s EtOH, eth; sl chl
9243	Propyliodone		$C_{10}H_{11}I_2NO_3$	587-61-1	447.008		186				
9244	Propyl isobutanoate		$C_7H_{14}O_2$	644-49-5	130.185			134(4)	0.8843^{0}	1.3955^{20}	sl H_2O; s EtOH, ace; vs eth
9245	Propyl isocyanate	1-Isocyanatopropane	C_4H_7NO	110-78-1	85.105			85.4(0.4)	0.908^{25}	1.3970^{20}	
9246	Propyl isothiocyanate	1-Isothiocyanatopropane	C_4H_7NS	628-30-8	101.171			153	0.9781^{16}	1.5085^{16}	sl H_2O; msc EtOH, eth
9247	Propyl methacrylate		$C_7H_{12}O_2$	2210-28-8	128.169			140(19)	0.9022^{20}	1.4190^{20}	i H_2O; msc EtOH, eth
9248	Propyl 3-methylbutanoate	Propyl isopentanoate	$C_8H_{16}O_2$	557-00-6	144.212			155(3)	0.8617^{20}	1.4031^{20}	vs eth, EtOH
9249	1-Propylnaphthalene		$C_{13}H_{14}$	2765-18-6	170.250	liq	-8.6(0.3)	273(5)	0.9897^{20}	1.5923^{20}	
9250	Propyl nitrate		$C_3H_7NO_3$	627-13-4	105.093			110(1)	1.0538^{20}	1.3973^{20}	sl H_2O; s EtOH, eth, ctc
9251	Propyl nitrite		$C_3H_7NO_2$	543-67-9	89.094	liq		44(2)	0.886^{20}	1.3604^{20}	sl H_2O; s EtOH, eth
9252	Propyl octanoate		$C_{11}H_{22}O_2$	624-13-5	186.292	liq	-46(1)	222(3)	0.8659^{20}	1.4191^{25}	vs ace, eth, EtOH
9253	Propyl pentanoate		$C_8H_{16}O_2$	141-06-0	144.212	liq	-70.7	167(4)	0.8699^{20}	1.4065^{20}	i H_2O; s EtOH, eth, chl
9254	2-Propylpentanoic acid	Valproic acid	$C_8H_{16}O_2$	99-66-1	144.212	col liq		223(9)	0.904^{25}	1.425^{25}	sl H_2O
9255	2-Propylphenol		$C_9H_{12}O$	644-35-9	136.190		7(1)	221(4)	1.015^{20}		vs eth, EtOH
9256	4-Propylphenol		$C_9H_{12}O$	645-56-7	136.190		22(2)	232(4)	1.009^{20}	1.5379^{25}	sl H_2O, ctc; s EtOH
9257	2-Propylpiperidine, (*S*)-	Coniine	$C_8H_{17}N$	458-88-8	127.228	liq	-1.0	166.5	0.8440^{20}	1.4512^{22}	sl H_2O, chl; msc EtOH; vs eth; s bz
9258	*trans*-6-Propyl-3-piperidinol, (3*S*)-	Pseudoconhydrine	$C_8H_{17}NO$	140-55-6	143.227	hyg nd (eth)	106	236			vs H_2O, eth, EtOH
9259	*N*-Propylpropanamide		$C_6H_{13}NO$	3217-86-5	115.173		154	215	0.8985^{25}		sl H_2O, eth
9260	Propyl propanoate	Propyl propionate	$C_6H_{12}O_2$	106-36-5	116.158	liq	-75.9(0.4)	122.2(0.1)	0.8755^{25}	1.3909^{25}	sl H_2O, ctc; msc EtOH, eth; s ace
9261	2-Propylpyridine		$C_8H_{11}N$	622-39-9	121.180		1.0	170(2)	0.9119^{20}	1.4925^{20}	sl H_2O; msc EtOH, eth; vs ace
9262	4-Propylpyridine		$C_8H_{11}N$	1122-81-2	121.180			189(5)	0.9381^{15}	1.4966^{20}	vs eth, EtOH
9263	2-Propyl-4-pyridinecarbothioamide	Protionamide	$C_9H_{12}N_2S$	14222-60-7	180.269		136.7				
9264	Propyl Red	Benzoic acid, 2-[[4-(dipropylamino)phenyl]azo]-	$C_{19}H_{23}N_3O_2$	2641-01-2	325.405	viol-bl or purp red cry (al)					s EtOH, KOH
9265	(Propylthio)benzene		$C_9H_{12}S$	874-79-3	152.256	liq	-45	219(7)	0.9995^{20}	1.5571^{20}	
9266	Propyl 4-toluenesulfonate		$C_{10}H_{14}O_3S$	599-91-7	214.281		<-20	189^{9}	1.144^{20}	1.4998^{20}	
9267	Propyl trichloroacetate		$C_5H_7Cl_3O_2$	13313-91-2	205.468			187	1.3221^{20}	1.4501^{20}	vs eth, EtOH
9268	Propyl 3,4,5-trihydroxybenzoate	Propyl gallate	$C_{10}H_{12}O_5$	121-79-9	212.199	nd (w)	130				sl H_2O
9269	Propylurea		$C_4H_{10}N_2O$	627-06-5	102.134	pr (al)	107.8(0.4)				sl H_2O, DMSO; s EtOH
9270	Propyl vinyl ether	1-(Ethenyloxy)propane	$C_5H_{10}O$	764-47-6	86.132			64.7(0.4)	0.7674^{20}	1.3908^{20}	
9271	2-Propynal	Propargyl aldehyde	C_3H_2O	624-67-9	54.047			60	0.9152^{20}	1.4033^{25}	msc H_2O; s EtOH, eth, ace, bz, tol
9272	2-Propyn-1-amine		C_3H_5N	2450-71-7	55.079			84(3)	0.803^{25}	1.4480^{20}	
9273	Propyne	Methylacetylene	C_3H_4	74-99-7	40.064	col gas	-103.0(0.5)	-23.2	0.607^{25} (p>1 atm)	1.3863^{-40}	sl H_2O; vs EtOH; s bz, chl
9274	2-Propynoic acid	Propiolic acid	$C_3H_2O_2$	471-25-0	70.047	cry (CS_2)	9	144 dec	1.1380^{20}	1.4306^{20}	vs H_2O, eth, EtOH, chl
9275	1-Propynylbenzene		C_9H_8	673-32-5	116.160			186(10)	0.942^{15}	1.563^{15}	

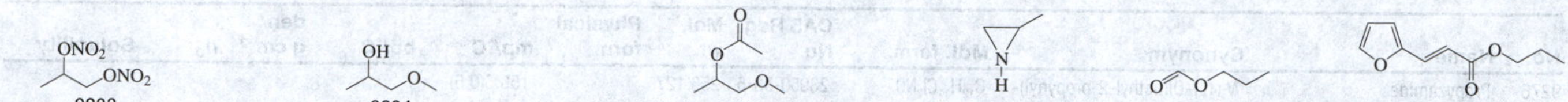

9233 1,2-Propylene glycol dinitrate
9234 1,2-Propylene glycol monomethyl ether
9235 1,2-Propylene glycol monomethyl ether acetate
9236 Propyleneimine
9237 Propyl formate
9238 Propyl 3-(2-furyl)acrylate

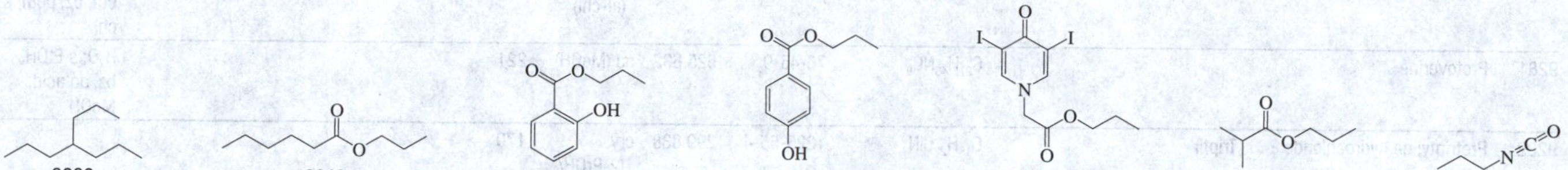

9239 4-Propylheptane
9240 Propyl hexanoate
9241 Propyl 2-hydroxybenzoate
9242 Propyl 4-hydroxybenzoate
9243 Propyliodone
9244 Propyl isobutanoate
9245 Propyl isocyanate

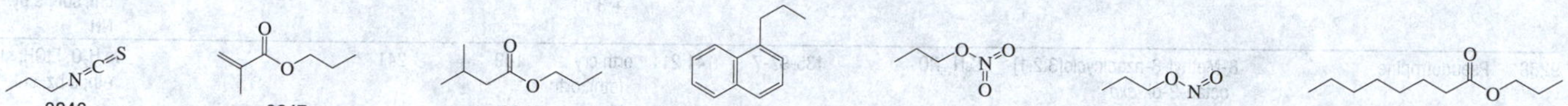

9246 Propyl isothiocyanate
9247 Propyl methacrylate
9248 Propyl 3-methylbutanoate
9249 1-Propylnaphthalene
9250 Propyl nitrate
9251 Propyl nitrite
9252 Propyl octanoate

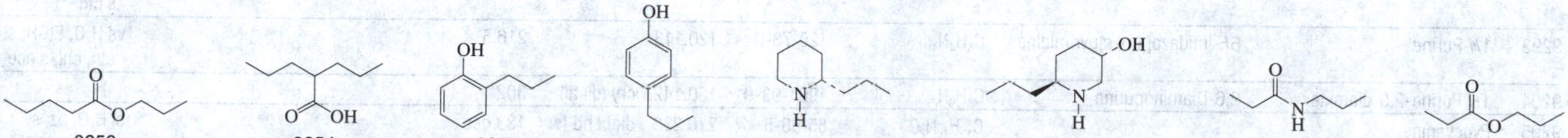

9253 Propyl pentanoate
9254 2-Propylpentanoic acid
9255 2-Propylphenol
9256 4-Propylphenol
9257 2-Propylpiperidine, (*S*)-
9258 *trans*-6-Propyl-3-piperidinol, (3*S*)-
9259 *N*-Propylpropanamide
9260 Propyl propanoate

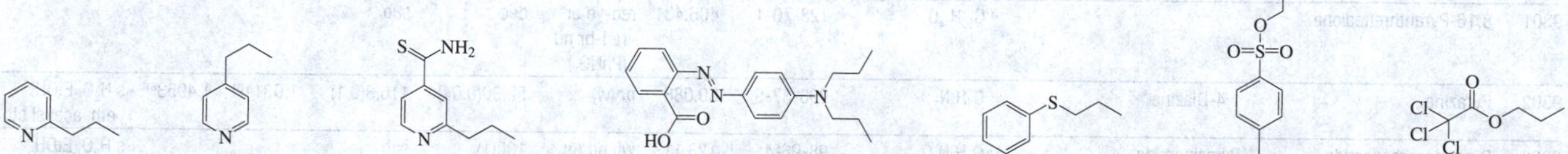

9261 2-Propylpyridine
9262 4-Propylpyridine
9263 2-Propyl-4-pyridinecarbothioamide
9264 Propyl Red
9265 (Propylthio)benzene
9266 Propyl 4-toluenesulfonate
9267 Propyl trichloroacetate

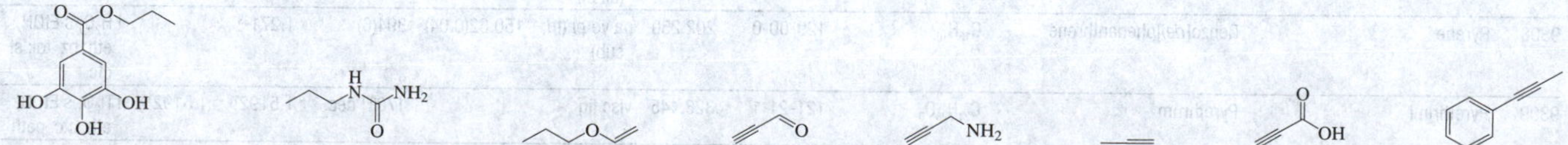

9268 Propyl 3,4,5-trihydroxybenzoate
9269 Propylurea
9270 Propyl vinyl ether
9271 2-Propynal
9272 2-Propyn-1-amine
9273 Propyne
9274 2-Propynoic acid
9275 1-Propynylbenzene

Organic

No.	Name	Synonym	Mol. form.	CAS Reg. No.	Mol. wt.	Physical form	mp/°C	bp/°C	den/ g cm^{-3}	n_D	Solubility
9276	Propyzamide	*N*-(1,1-Dimethyl-2-propynyl)-3,5-dichlorobenzamide	$C_{12}H_{11}Cl_2NO$	23950-58-5	256.127		156.5(0.5)				
9277	Prostaglandin E_1	11,15-Dihydroxy-9-oxo-13-prostenoic acid	$C_{20}H_{34}O_5$	745-65-3	354.481	cry (EtOAc)	115				s H_2O
9278	Prostaglandin E_2	11,15-Dihydroxy-9-oxo-5,13-prostadienoic acid	$C_{20}H_{32}O_5$	363-24-6	352.465	col cry	67				s H_2O, thf
9279	Prostaglandin $F_{2\alpha}$	9,11,15-Trihydroxyprosta-5,13-dienoic acid	$C_{20}H_{34}O_5$	551-11-1	354.481	oil or solid	≈30				sl H_2O; s EtOH, MeOH, chl, AcOEt
9280	Protopine	Fumarine	$C_{20}H_{19}NO_5$	130-86-9	353.369	mcl pr (al-chl)	208				i H_2O; sl EtOH, eth, bz, peth; s chl
9281	Protoverine		$C_{27}H_{43}NO_9$	76-45-9	525.632	nd (MeOH)	221				i H_2O; s EtOH, bz, aq acid, MeOH
9282	Protriptyline hydrochloride	Triptil	$C_{19}H_{22}ClN$	1225-55-4	299.838	cry (2-PrOH/ eth)	170				
9283	Prunetin		$C_{16}H_{12}O_5$	552-59-0	284.263		239.5				
9284	Pseudoaconitine		$C_{36}H_{51}NO_{12}$	127-29-7	689.790	tcl (MeOH)	214				vs eth, EtOH
9285	Pseudocodeine		$C_{18}H_{21}NO_3$	466-96-6	299.365	wh nd	181.5		1.290^{80}	1.574	
9286	Pseudojervine		$C_{33}H_{49}NO_8$	36069-05-3	587.744	wh nd or hex cry	304 dec				i H_2O, eth, bz, chl, tol, peth; s EtOH
9287	Pseudomorphine		$C_{34}H_{36}N_2O_6$	125-24-6	568.659	cry (aq NH_3, + 3 w)	282.5				i H_2O, EtOH, eth, chl, sulf; s py, NH_3
9288	Pseudotropine	8-Methyl-8-azabicyclo[3.2.1]-octan-3-ol, *exo*	$C_8H_{15}NO$	135-97-7	141.211	orth cry (eth), orth bipym (peth-bz)	109	241			vs H_2O, EtOH; sl eth; s bz, chl
9289	Psoralen		$C_{11}H_6O_3$	66-97-7	186.164	nd (w, EtOH)	171				
9290	Pteridine	Pyrazino[2,3-*d*]pyrimidine	$C_6H_4N_4$	91-18-9	132.123	ye pl (bz, sub)	139.5	125 sub			vs H_2O; s EtOH; sl eth, bz
9291	2,4(1*H*,3*H*)-Pteridinedione	Lumazine	$C_6H_4N_4O_2$	487-21-8	164.122	ye-oran nd (w)	348.5				vs HOAc
9292	Pulegone		$C_{10}H_{16}O$	89-82-7	152.233			224(7)	0.9346^{45}	1.4894^{20}	i H_2O; msc EtOH, eth, chl; s ctc
9293	1*H*-Purine	6*H*-Imidazo[4,5-*d*]pyrimidine	$C_5H_4N_4$	120-73-0	120.113		216.5				vs H_2O, EtOH; sl eth, chl; s ace
9294	1*H*-Purine-2,6-diamine	2,6-Diaminopurine	$C_5H_6N_6$	1904-98-9	150.142	cry (dil al)	302				
9295	Pyocyanine		$C_{13}H_{10}N_2O$	85-66-5	210.230	dk bl nd (w + 1) (chl-peth)	133 dec				sl H_2O, bz; s EtOH, ace; i eth; vs chl
9296	4*H*-Pyran	1,4-Pyran	C_5H_6O	289-65-6	82.101	unstab oil		80		1.4559^{20}	s EtOH, eth, bz
9297	2*H*-Pyran-2-one		$C_5H_4O_2$	504-31-4	96.085		8.5	207.5	1.200^{20}	1.5270^{25}	msc H_2O; vs ace
9298	4*H*-Pyran-4-one		$C_5H_4O_2$	108-97-4	96.085		32.5	212.5	1.190^{25}	1.5238	vs H_2O, chl, eth; s EtOH, bz; sl CS_2
9299	Pyrantel		$C_{11}H_{14}N_2S$	15686-83-6	206.307	cry (MeOH)	178				
9300	4*H*-Pyran-4-thione		C_5H_4OS	1120-93-0	112.150		49				s H_2O
9301	8,16-Pyranthrenedione		$C_{30}H_{14}O_2$	128-70-1	406.431	red-ye or red-br nd ($PhNO_2$)	dec	sub			
9302	Pyrazine	1,4-Diazine	$C_4H_4N_2$	290-37-9	80.088	pr (w)	52.30(0.04)	116.3(0.1)	1.0311^{61}	1.4953^{61}	s H_2O, EtOH, eth, ace; sl ctc
9303	Pyrazinecarboxamide	Pyrazinamide	$C_5H_5N_3O$	98-96-4	123.113	wh nd (w, al)	190(1)	sub			s H_2O, EtOH
9304	Pyrazinecarboxylic acid	Pyrazinoic acid	$C_5H_4N_2O_2$	98-97-5	124.098	wh nd (w)	225(2)	sub			
9305	2,3-Pyrazinedicarboxylic acid	2,3-Dicarboxypyrazine	$C_6H_4N_2O_4$	89-01-0	168.107	pr (w+2)	193 dec				vs H_2O; sl EtOH, eth, bz; s ace, MeOH
9306	1*H*-Pyrazole	1,2-Diazole	$C_3H_4N_2$	288-13-1	68.077	nd or pr (lig)	59.9(0.3)	187(5)		1.4203	s H_2O, EtOH, eth, bz; sl chl
9307	1-Pyrenamine		$C_{16}H_{11}N$	1606-67-3	217.265	ye nd (hx) lf (dil al)	117.5				s EtOH, ace, hx, acid; sl chl
9308	Pyrene	Benzo[*def*]phenanthrene	$C_{16}H_{10}$	129-00-0	202.250	pa ye pl (to, sub)	150.62(0.04)	394(6)	1.271^{23}		i H_2O; s EtOH, eth, bz, tol; sl ctc
9309	Pyrethrin I	Pyrethrum	$C_{21}H_{28}O_3$	121-21-1	328.445	visc liq		$170^{0.1}$ dec	1.5192^{18}	1.5192^{18}	i H_2O; s EtOH, eth, ctc, peth

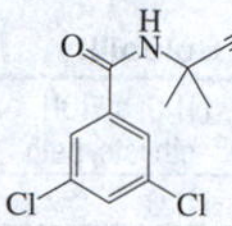
9276
Propyzamide

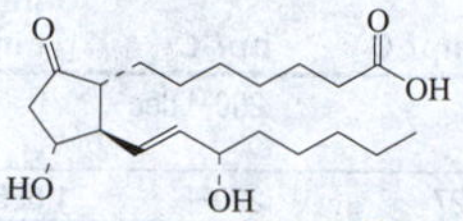
9277
Prostaglandin E_1

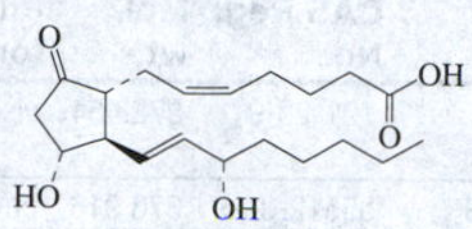
9278
Prostaglandin E_2

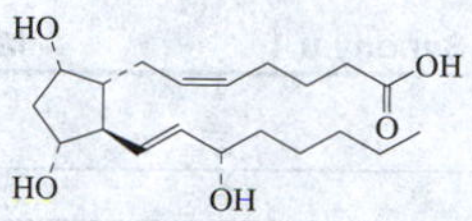
9279
Prostaglandin $F_{2\alpha}$

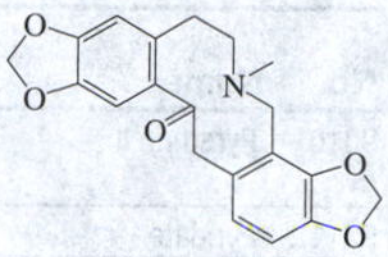
9280
Protopine

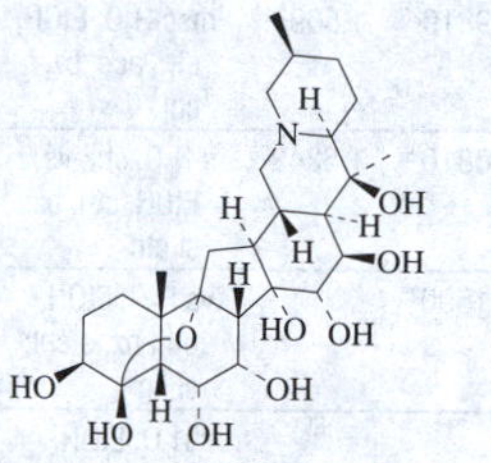
9281
Protoverine

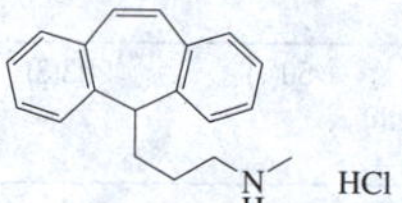
9282
Protriptyline hydrochloride

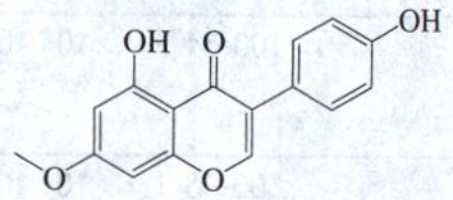
9283
Prunetin

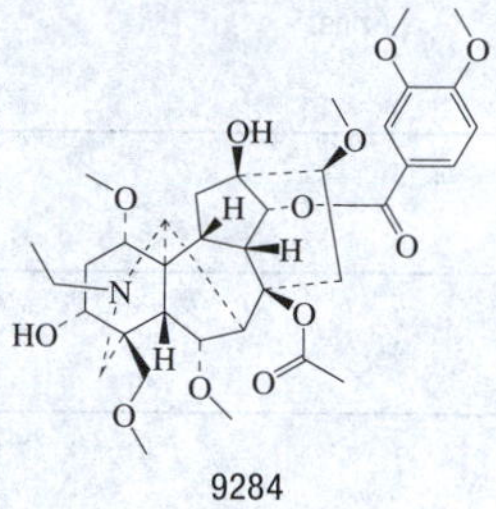
9284
Pseudoaconitine

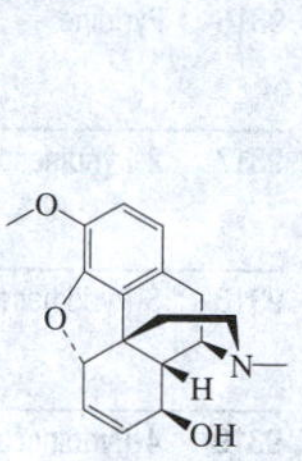
9285
Pseudocodeine

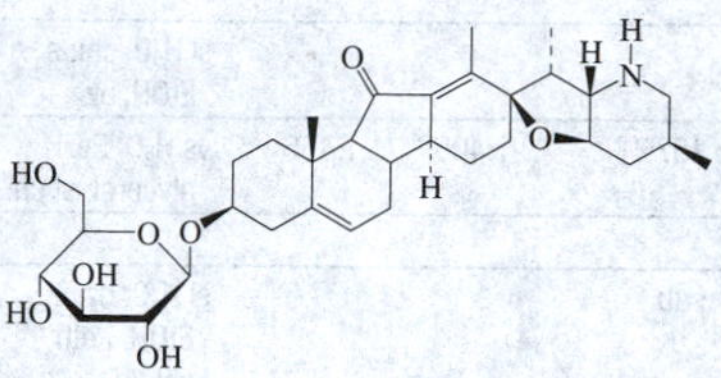
9286
Pseudojervine

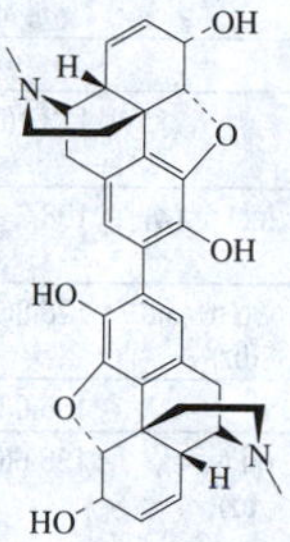
9287
Pseudomorphine

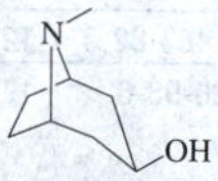
9288
Pseudotropine

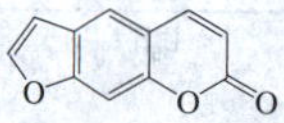
9289
Psoralen

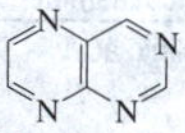
9290
Pteridine

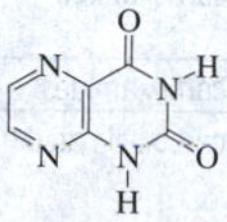
9291
2,4(1*H*,3*H*)-Pteridinedione

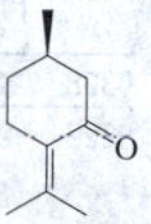
9292
Pulegone

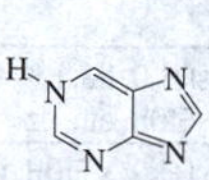
9293
1*H*-Purine

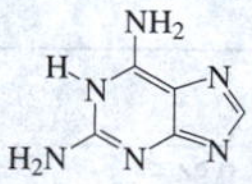
9294
1*H*-Purine-2,6-diamine

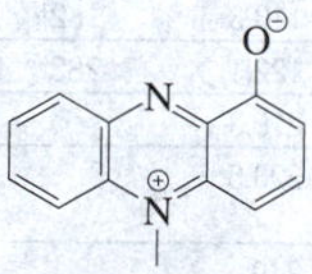
9295
Pyocyanine

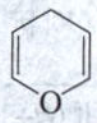
9296
4*H*-Pyran

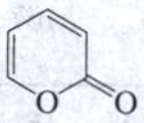
9297
2*H*-Pyran-2-one

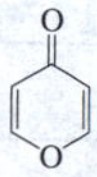
9298
4*H*-Pyran-4-one

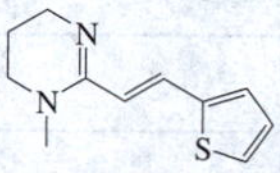
9299
Pyrantel

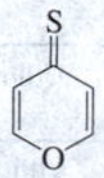
9300
4*H*-Pyran-4-thione

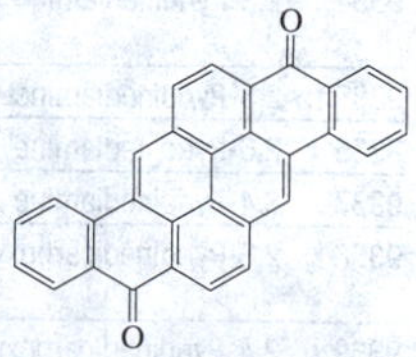
9301
8,16-Pyranthrenedione

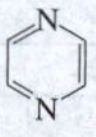
9302
Pyrazine

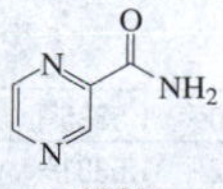
9303
Pyrazinecarboxamide

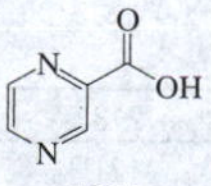
9304
Pyrazinecarboxylic acid

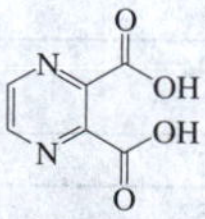
9305
2,3-Pyrazinedicarboxylic acid

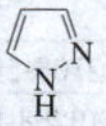
9306
1*H*-Pyrazole

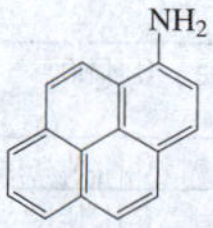
9307
1-Pyrenamine

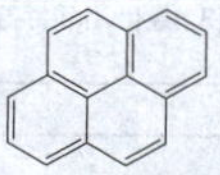
9308
Pyrene

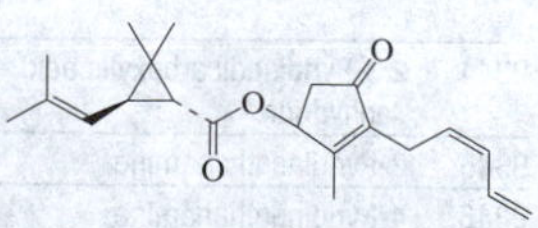
9309
Pyrethrin I

Organic

No.	Name	Synonym	Mol. form.	CAS Reg. No.	Mol. wt.	Physical form	mp/°C	bp/°C	den/ g cm⁻³	n_D	Solubility
9310	Pyrethrin II		$C_{22}H_{28}O_5$	121-29-9	372.454	visc liq		$200^{0.1}$ dec		1.5258^{20}	i H_2O; s EtOH, eth, ctc, peth
9311	Pyridate		$C_{19}H_{23}ClN_2O_2S$	55512-33-9	378.916	br oil	27	$220^{0.1}$	1.555^{20}	1.568^{20}	i H_2O
9312	Pyridazine	1,2-Diazabenzene	$C_4H_4N_2$	289-80-5	80.088	liq	-8	208	1.1035^{23}	1.5218^{20}	msc H_2O, EtOH; vs eth, ace, bz; i peth
9313	2-Pyridinamine	2-Aminopyridine	$C_5H_6N_2$	504-29-0	94.115	lf (lig)	58.2(0.9)	105^{20}			s EtOH, eth, ace, bz; sl chl
9314	3-Pyridinamine	3-Aminopyridine	$C_5H_6N_2$	462-08-8	94.115	lf (bz-lig)	64.5(0.3)	252			s H_2O, EtOH, eth; sl lig
9315	4-Pyridinamine	4-Aminopyridine	$C_5H_6N_2$	504-24-5	94.115	nd (bz)	159.0(0.5)	273			s H_2O, eth, bz; vs EtOH; sl lig
9316	Pyridine	Azine	C_5H_5N	110-86-1	79.101	liq	-41.63(0.03)	115.2(0.1)	0.9819^{20}	1.5095^{20}	msc H_2O, EtOH, eth, ace, bz, chl
9317	2-Pyridinecarbonitrile		$C_6H_4N_2$	100-70-9	104.109	nd or pr (eth)	29(2)	224.5	1.0810^{25}	1.5242^{25}	s H_2O, chl; vs EtOH, eth, bz; sl ctc
9318	3-Pyridinecarbonitrile		$C_6H_4N_2$	100-54-9	104.109	nd (lig), peth-eth)	50(1)	203(3)	1.1590^{25}		vs H_2O, EtOH, eth, bz; s chl; sl lig
9319	4-Pyridinecarbonitrile		$C_6H_4N_2$	100-48-1	104.109	nd(lig-eth)	79(1)	186			s H_2O, EtOH, eth, bz, chl; sl lig
9320	3-Pyridinecarbothioamide		$C_6H_6N_2S$	4621-66-3	138.190		192				
9321	4-Pyridinecarbothioamide		$C_6H_6N_2S$	2196-13-6	138.190		198 dec				
9322	2-Pyridinecarboxaldehyde		C_6H_5NO	1121-60-4	107.110			181(10)	1.1181^{25}	1.5389^{18}	s H_2O, EtOH, eth, AcOEt; sl ctc
9323	3-Pyridinecarboxaldehyde	Nicotinaldehyde	C_6H_5NO	500-22-1	107.110			92^{23}	1.1394^{25}		s H_2O, EtOH, ace, chl; sl eth, peth
9324	4-Pyridinecarboxaldehyde		C_6H_5NO	872-85-5	107.110			77^{12}		1.5423^{20}	s H_2O, eth, ctc
9325	2-Pyridinecarboxaldehyde oxime		$C_6H_6N_2O$	873-69-8	122.124		113.1(0.5)				
9326	2-Pyridinecarboxamide		$C_6H_6N_2O$	1452-77-3	122.124	mcl pr (w)	108.3				sl H_2O, chl; s EtOH, bz
9327	3-Pyridinecarboxamide	Niacinamide	$C_6H_6N_2O$	98-92-0	122.124	wh pw, nd (bz)	128.8(0.5)	$157^{0.0005}$	1.400^{25}	1.466	vs H_2O, EtOH, glycerol; sl chl
9328	4-Pyridinecarboxamide	Isonicotinamide	$C_6H_6N_2O$	1453-82-3	122.124		156.0(0.5)				
9329	2-Pyridinecarboxylic acid	Picolinic acid	$C_6H_5NO_2$	98-98-6	123.110	nd (w, al, bz)	136.6(0.8)	sub			sl H_2O, bz; s EtOH; i eth, chl, CS_2
9330	3-Pyridinecarboxylic acid	Nicotinic acid	$C_6H_5NO_2$	59-67-6	123.110	nd (al, w)	237(1)	sub	1.473^{25}		sl H_2O, EtOH, eth
9331	4-Pyridinecarboxylic acid	Isonicotinic acid	$C_6H_5NO_2$	55-22-1	123.110	nd(w)	318(1)	260 sub			sl H_2O, EtOH, eth, bz
9332	3-Pyridinecarboxylic acid 1-oxide	Oxiniacic acid	$C_6H_5NO_3$	2398-81-4	139.109	nd	254 dec				vs H_2O, MeOH
9333	4-Pyridinecarboxylic acid 1-oxide		$C_6H_5NO_3$	13602-12-5	139.109		273 dec				
9334	2,3-Pyridinediamine		$C_5H_7N_3$	452-58-4	109.130	lf or pl (dil al)	120.8	149^5			s H_2O, EtOH, bz
9335	2,5-Pyridinediamine	2,5-Diaminopyridine	$C_5H_7N_3$	4318-76-7	109.130	nd	110.3	182^{12}			vs H_2O, EtOH
9336	2,6-Pyridinediamine		$C_5H_7N_3$	141-86-6	109.130		122(1)	285			sl H_2O, ace
9337	3,4-Pyridinediamine		$C_5H_7N_3$	54-96-6	109.130	nd or lf	219.3				
9338	2,3-Pyridinedicarboxylic acid	Quinolinic acid	$C_7H_5NO_4$	89-00-9	167.120	mcl pr (w)	228.5				sl H_2O, tfa; i EtOH, eth, bz
9339	2,4-Pyridinedicarboxylic acid	Lutidinic acid	$C_7H_5NO_4$	499-80-9	167.120	lf (w+1)	249		0.942^{25}		sl H_2O; s EtOH; i eth, bz, CS_2
9340	2,5-Pyridinedicarboxylic acid	Isocinchomeronic acid	$C_7H_5NO_4$	100-26-5	167.120	lf or pr (dil HCl)	254				s H_2O, HCl; sl EtOH; i eth, bz
9341	2,6-Pyridinedicarboxylic acid	Dipicolinic acid	$C_7H_5NO_4$	499-83-2	167.120	nd (w+3/2)	252				sl H_2O, EtOH, HOAc
9342	3,4-Pyridinedicarboxylic acid	Cinchomeronic acid	$C_7H_5NO_4$	490-11-9	167.120	cry (w)	256	sub			sl H_2O, EtOH, bz; i eth, i chl
9343	3,5-Pyridinedicarboxylic acid	Dinicotinic acid	$C_7H_5NO_4$	499-81-0	167.120	cry (w)	324	sub			i H_2O; sl eth, HOAc; s DMSO, HCl
9344	2,3-Pyridinedicarboxylic acid anhydride	Furo[3,4-*b*]pyridine-5,7-dione	$C_7H_3NO_3$	699-98-9	149.104		138				
9345	2-Pyridineethanamine		$C_7H_{10}N_2$	2706-56-1	122.167			213	1.0220^{25}	1.5335^{25}	
9346	4-Pyridineethanamine		$C_7H_{10}N_2$	13258-63-4	122.167			121^{10}	1.0302^{25}	1.5381^{25}	vs H_2O

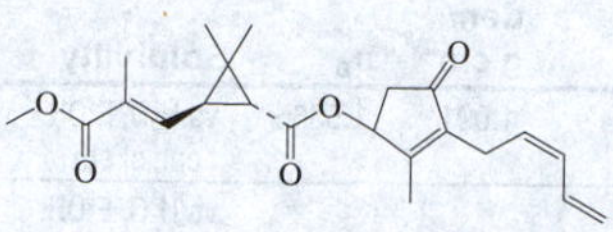
9310
Pyrethrin II

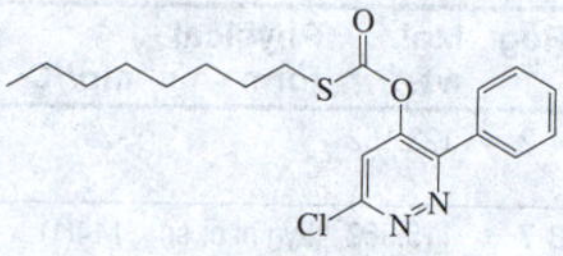
9311
Pyridate

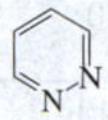
9312
Pyridazine

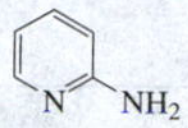
9313
2-Pyridinamine

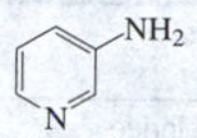
9314
3-Pyridinamine

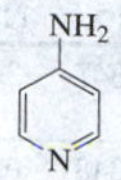
9315
4-Pyridinamine

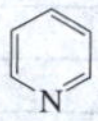
9316
Pyridine

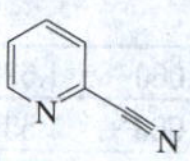
9317
2-Pyridinecarbonitrile

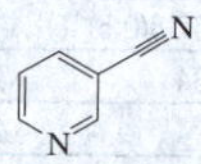
9318
3-Pyridinecarbonitrile

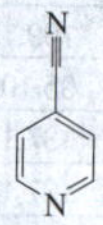
9319
4-Pyridinecarbonitrile

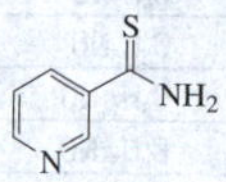
9320
3-Pyridinecarbothioamide

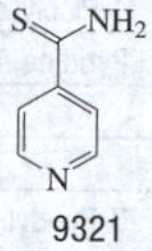
9321
4-Pyridinecarbothioamide

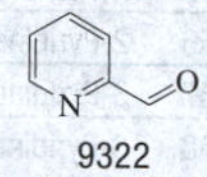
9322
2-Pyridinecarboxaldehyde

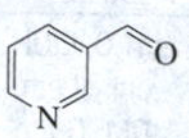
9323
3-Pyridinecarboxaldehyde

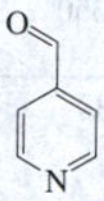
9324
4-Pyridinecarboxaldehyde

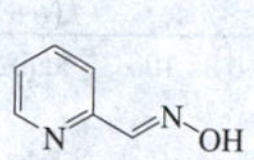
9325
2-Pyridinecarboxaldehyde oxime

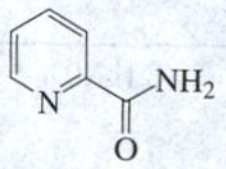
9326
2-Pyridinecarboxamide

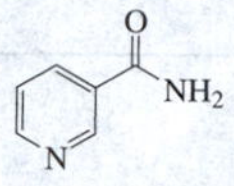
9327
3-Pyridinecarboxamide

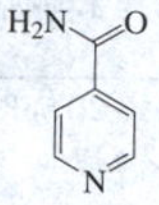
9328
4-Pyridinecarboxamide

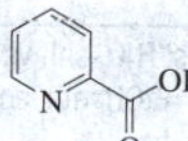
9329
2-Pyridinecarboxylic acid

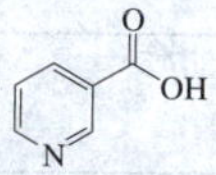
9330
3-Pyridinecarboxylic acid

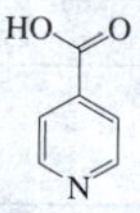
9331
4-Pyridinecarboxylic acid

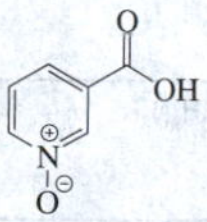
9332
3-Pyridinecarboxylic acid 1-oxide

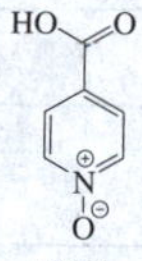
9333
4-Pyridinecarboxylic acid 1-oxide

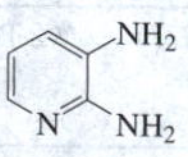
9334
2,3-Pyridinediamine

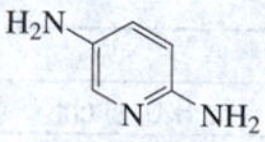
9335
2,5-Pyridinediamine

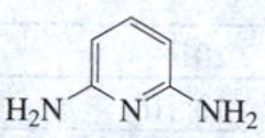
9336
2,6-Pyridinediamine

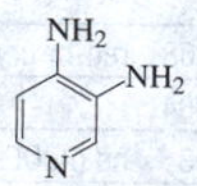
9337
3,4-Pyridinediamine

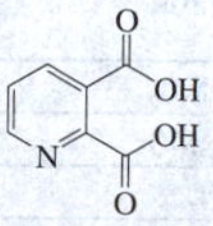
9338
2,3-Pyridinedicarboxylic acid

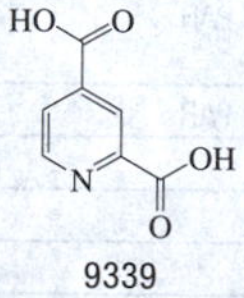
9339
2,4-Pyridinedicarboxylic acid

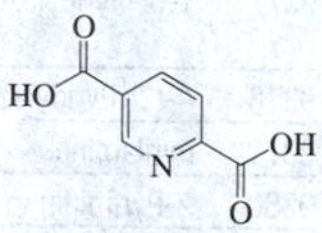
9340
2,5-Pyridinedicarboxylic acid

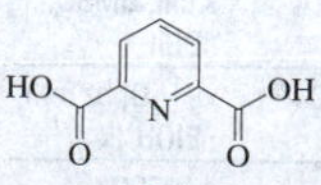
9341
2,6-Pyridinedicarboxylic acid

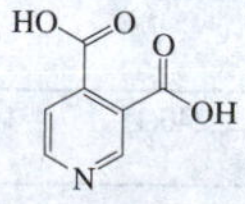
9342
3,4-Pyridinedicarboxylic acid

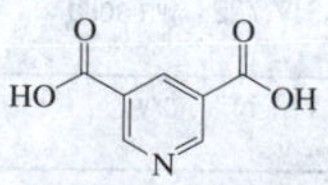
9343
3,5-Pyridinedicarboxylic acid

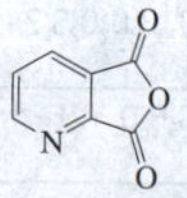
9344
2,3-Pyridinedicarboxylic acid anhydride

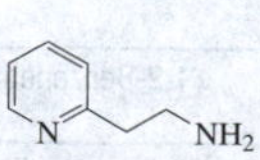
9345
2-Pyridineethanamine

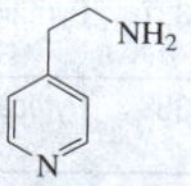
9346
4-Pyridineethanamine

No.	Name	Synonym	Mol. form.	CAS Reg. No.	Mol. wt.	Physical form	mp/°C	bp/°C	den/ g cm^{-3}	n_D	Solubility
9347	2-Pyridineethanol		C_7H_9NO	103-74-2	123.152		-7.8	190^{200}	1.091^{25}	1.5366^{20}	vs H_2O, EtOH, chl; sl eth
9348	Pyridine hydrochloride		C_5H_6ClN	628-13-7	115.562	hyg pl or sc (al)	144(1)	222			vs H_2O, EtOH, chl
9349	2-Pyridinemethanamine		$C_6H_8N_2$	3731-51-9	108.141			203	1.0525^{25}	1.5431^{25}	vs H_2O
9350	3-Pyridinemethanamine		$C_6H_8N_2$	3731-52-0	108.141	liq	-21.1	226	1.064^{20}	1.552^{20}	vs H_2O, eth, EtOH
9351	4-Pyridinemethanamine		$C_6H_8N_2$	3731-53-1	108.141	liq	-7.6	230	1.072^{20}	1.5495^{25}	vs H_2O
9352	2-Pyridinemethanol		C_6H_7NO	586-98-1	109.126			112^{16}	1.1317^{20}	1.5444^{20}	msc H_2O; vs EtOH, eth, ace, bz
9353	3-Pyridinemethanol	Nicotinyl alcohol	C_6H_7NO	100-55-0	109.126	liq	-6.5	266	1.131^{20}	1.5455^{20}	vs H_2O, eth
9354	4-Pyridinemethanol	4-Picolyl alcohol	C_6H_7NO	586-95-8	109.126		53	141^{12}			s chl
9355	Pyridine-1-oxide	Pyridine *N*-oxide	C_5H_5NO	694-59-7	95.100		67.8(0.5)	146^{13}			
9356	2-Pyridinepropanol		$C_8H_{11}NO$	2859-68-9	137.179		34	260.2	1.060^{25}	1.5298^{20}	vs H_2O
9357	3-Pyridinepropanol		$C_8H_{11}NO$	2859-67-8	137.179			284	1.063^{25}	1.5313^{20}	vs H_2O
9358	3-Pyridinesulfonic acid	3-Pyridylsulfonic acid	$C_5H_5NO_3S$	636-73-7	159.164	orth	357 dec		1.713^{25}		vs H_2O; sl EtOH; i eth
9359	2-Pyridinethiol, 1-oxide		C_5H_5NOS	1121-31-9	127.165		70.5				
9360	2(1*H*)-Pyridinethione		C_5H_5NS	2637-34-5	111.166		130.0				s H_2O, EtOH, bz, chl
9361	2-Pyridinol		C_5H_5NO	72762-00-6	95.100	nd (bz)	107.8		1.3910^{20}		vs H_2O, bz, EtOH
9362	3-Pyridinol		C_5H_5NO	109-00-2	95.100	nd (bz)	129				s H_2O, EtOH; sl eth, chl
9363	4-Pyridinol		C_5H_5NO	626-64-2	95.100	pr or nd (w+1)	148(1)	>350			s H_2O, EtOH; i eth, bz
9364	2(1*H*)-Pyridinone		C_5H_5NO	142-08-5	95.100	nd (bz)	106(3)	280	1.3910^{20}		s H_2O, EtOH, bz, chl; sl eth, DMSO
9365	2(1*H*)-Pyridinone hydrazone	2-Pyridinylhydrazine	$C_5H_7N_3$	4930-98-7	109.130		46.6	185^{140}			s chl
9366	α-[(2-Pyridinylamino)methyl]-benzenemethanol	Phenyramidol	$C_{13}H_{14}N_2O$	553-69-5	214.262	cry (dil MeOH)	83.5				
9367	1-(2-Pyridinyl)ethanone		C_7H_7NO	1122-62-9	121.137	ye in air		209(13)	1.077^{25}	1.5203^{20}	s EtOH, eth, HOAc; sl ctc
9368	1-(3-Pyridinyl)ethanone	Methyl pyridyl ketone	C_7H_7NO	350-03-8	121.137		13.5	225(11)		1.5341^{20}	s H_2O, EtOH, eth, acid
9369	1-(4-Pyridinyl)ethanone		C_7H_7NO	1122-54-9	121.137		16	224(12)	1.097^{25}	1.5282^{25}	sl EtOH, eth, acid
9370	*N*-(2-Pyridinylmethyl)-2-pyridinemethanamine		$C_{12}H_{13}N_3$	1539-42-0	199.251			200^{10}	1.1074^{25}	1.5757^{25}	
9371	*N*-2-Pyridinyl-2-pyridinamine		$C_{10}H_9N_3$	1202-34-2	171.198		90.5	307.5			sl H_2O, chl; vs EtOH, eth, ace, bz
9372	Pyridoxal hydrochloride	Vitamin B6	$C_8H_{10}ClNO_3$	65-22-5	203.623	orth	165 dec				vs H_2O; sl EtOH
9373	Pyridoxal 5-phosphate	Pyridoxal 5-(dihydrogen phosphate)	$C_8H_{10}NO_6P$	54-47-7	247.142	wh-ye pow or cry	141				
9374	Pyridoxamine	4-(Aminomethyl)-5-hydroxy-6-methyl-3-pyridinemethanol	$C_8H_{12}N_2O_2$	85-87-0	168.193	cry	198				s EtOH, acid
9375	Pyridoxamine dihydrochloride		$C_8H_{14}Cl_2N_2O_2$	524-36-7	241.115	pl (al)	226 dec				vs H_2O; sl EtOH
9376	Pyridoxine hydrochloride	5-Hydroxy-6-methyl-3,4-pyridinedimethanol hydrochloride	$C_8H_{12}ClNO_3$	58-56-0	205.639	pl (al, ace)	207	sub			vs H_2O
9377	1-(2-Pyridylazo)-2-naphthol	PAN	$C_{15}H_{11}N_3O$	85-85-8	249.267	red-br cry	130				i H_2O; s EtOH, eth, chl
9378	4-(2'-Pyridylazo)resorcinol	PAR	$C_{11}H_9N_3O_2$	1141-59-9	215.208	red-br cry	187 dec				
9379	Pyrilamine		$C_{17}H_{23}N_3O$	91-84-9	285.384			201^{5}			
9380	2-Pyrimidinamine		$C_4H_5N_3$	109-12-6	95.103	nd (AcOEt)	127.5	195(18)			s H_2O; sl chl
9381	4-Pyrimidinamine		$C_4H_5N_3$	591-54-8	95.103	pl (AcOEt)	151.5				vs H_2O, EtOH
9382	Pyrimidine	1,3-Diazine	$C_4H_4N_2$	289-95-2	80.088		22	123.8	1.0792^{25}	1.4998^{20}	msc H_2O; s EtOH
9383	2,4,5,6(1*H*,3*H*)-Pyrimidinetetrone	Alloxan	$C_4H_2N_2O_4$	50-71-5	142.070		256 dec	sub			vs H_2O; s EtOH, ace, bz, HOAc
9384	2,4,5,6(1*H*,3*H*)-Pyrimidinetetrone 5-oxime	Violuric acid	$C_4H_3N_3O_4$	87-39-8	157.085	pa ye orth	203 dec				sl H_2O; s EtOH
9385	2,4,6-Pyrimidinetriamine		$C_4H_7N_5$	1004-38-2	125.133		248 dec				
9386	Pyriminil		$C_{13}H_{12}N_4O_3$	53558-25-1	272.259	solid	224 dec				
9387	Pyrithione zinc		$C_{10}H_8N_2O_2S_2Zn$	13463-41-7	317.722	wh solid	262				s chl, DMSO, DMF
9388	Pyrocatechol	1,2-Benzenediol	$C_6H_6O_2$	120-80-9	110.111	cry	104.6(0.3)	246(1)	1.344^{20}	1.604^{25}	vs H_2O, bz, eth, EtOH
9389	L-Pyroglutamic acid	5-Oxo-L-proline	$C_5H_7NO_3$	98-79-3	129.115		162				s DMSO

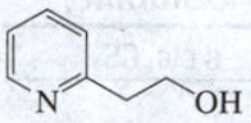
9347
2-Pyridineethanol

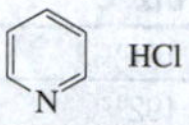
9348
Pyridine hydrochloride

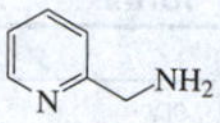
9349
2-Pyridinemethanamine

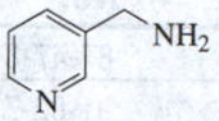
9350
3-Pyridinemethanamine

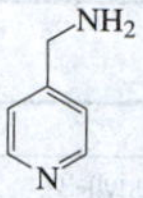
9351
4-Pyridinemethanamine

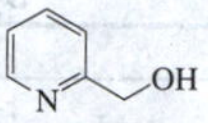
9352
2-Pyridinemethanol

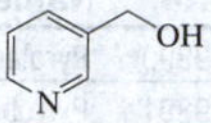
9353
3-Pyridinemethanol

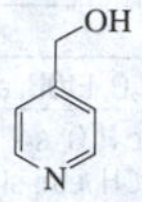
9354
4-Pyridinemethanol

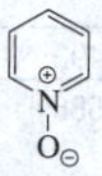
9355
Pyridine-1-oxide

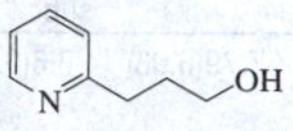
9356
2-Pyridinepropanol

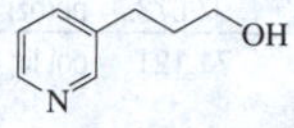
9357
3-Pyridinepropanol

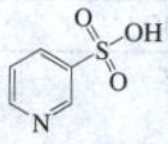
9358
3-Pyridinesulfonic acid

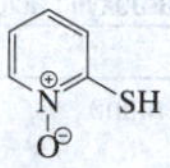
9359
2-Pyridinethiol, 1-oxide

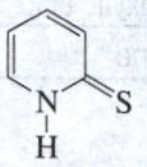
9360
2(1*H*)-Pyridinethione

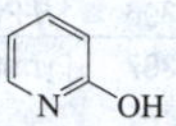
9361
2-Pyridinol

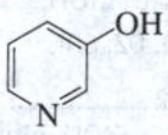
9362
3-Pyridinol

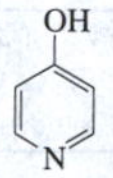
9363
4-Pyridinol

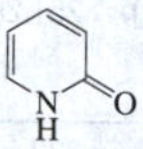
9364
2(1*H*)-Pyridinone

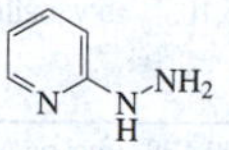
9365
2(1*H*)-Pyridinone hydrazone

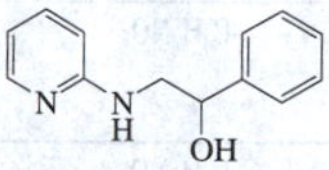
9366
α-[(2-Pyridinylamino)methyl]benzenemethanol

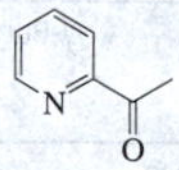
9367
1-(2-Pyridinyl)ethanone

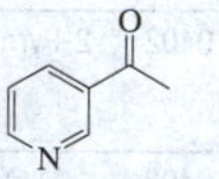
9368
1-(3-Pyridinyl)ethanone

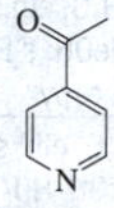
9369
1-(4-Pyridinyl)ethanone

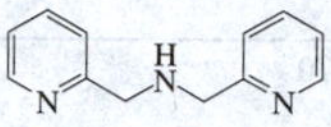
9370
N-(2-Pyridinylmethyl)-2-pyridinemethanamine

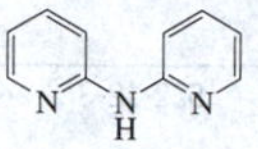
9371
N-2-Pyridinyl-2-pyridinamine

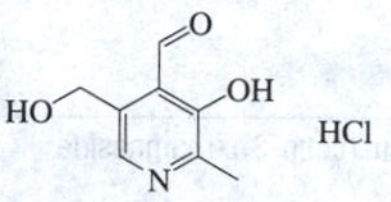
9372
Pyridoxal hydrochloride

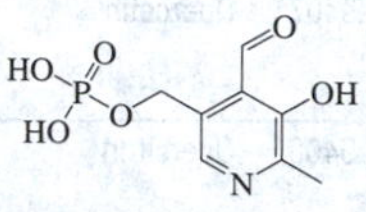
9373
Pyridoxal 5-phosphate

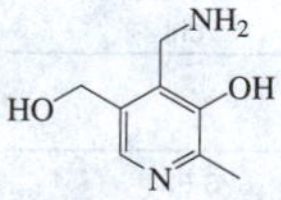
9374
Pyridoxamine

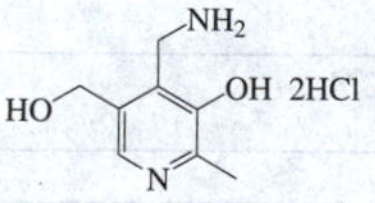
9375
Pyridoxamine dihydrochloride

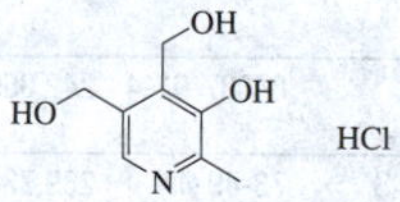
9376
Pyridoxine hydrochloride

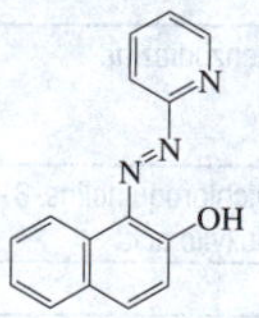
9377
1-(2-Pyridylazo)-2-naphthol

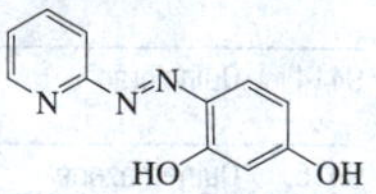
9378
4-(2'-Pyridylazo)resorcinol

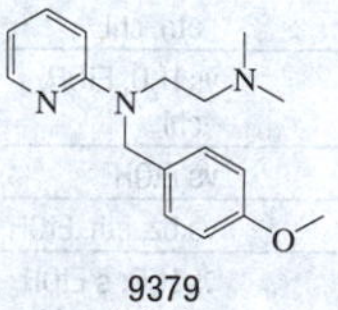
9379
Pyrilamine

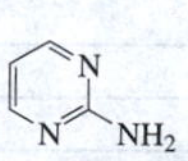
9380
2-Pyrimidinamine

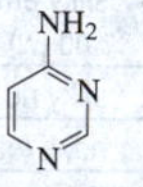
9381
4-Pyrimidinamine

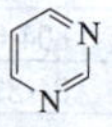
9382
Pyrimidine

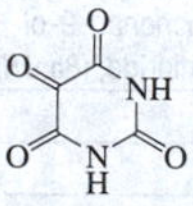
9383
2,4,5,6(1*H*,3*H*)-Pyrimidinetetrone

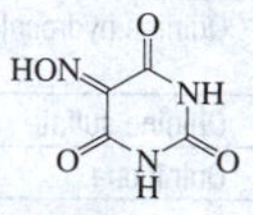
9384
2,4,5,6(1*H*,3*H*)-Pyrimidinetetrone 5-oxime

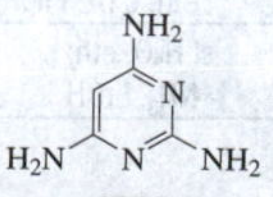
9385
2,4,6-Pyrimidinetriamine

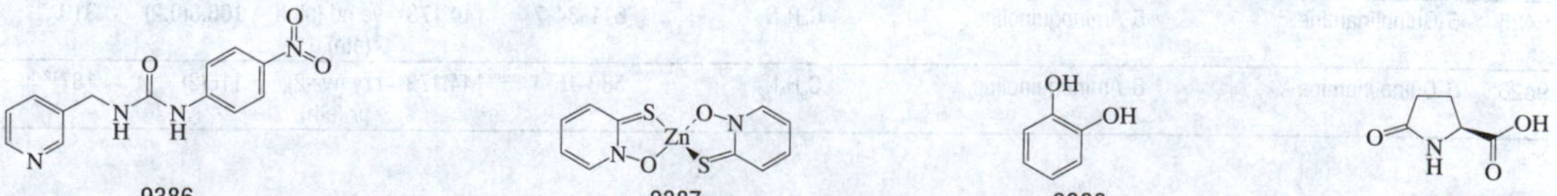
9386
Pyriminil

9387
Pyrithione zinc

9388
Pyrocatechol

9389
L-Pyroglutamic acid

Organic

No.	Name	Synonym	Mol. form.	CAS Reg. No.	Mol. wt.	Physical form	mp/°C	bp/°C	den/ g cm⁻³	n_D	Solubility
9390	Pyrolan		$C_{13}H_{15}N_3O_2$	87-47-8	245.277		52.0(0.5)	$161^{0.2}$			s ctc, CS_2
9391	Pyrrobutamine	1-[4-(4-Chlorophenyl)-3-phenyl-2-butenyl]pyrrolidine	$C_{20}H_{22}ClN$	91-82-7	311.849	cry	49	$192^{0.3}$			
9392	Pyrrole	Imidole	C_4H_5N	109-97-7	67.090	liq	-23.39(0.02)	129.74(0.04)	0.9698^{20}	1.5085^{20}	sl H_2O; s EtOH, eth, ace, bz, chl
9393	1*H*-Pyrrole-2-carboxaldehyde		C_5H_5NO	1003-29-8	95.100	orth pr (peth)	46.5	218		1.5939^{16}	sl chl, lig
9394	1*H*-Pyrrole-2-carboxylic acid		$C_5H_5NO_2$	634-97-9	111.100	lf (w)	191(5)				s H_2O, EtOH, eth
9395	1*H*-Pyrrole-3-carboxylic acid	3-Pyrrolecarboxylic acid	$C_5H_5NO_2$	931-03-3	111.100	nd (lig)	161.5				
9396	1*H*-Pyrrole-2,5-dione		$C_4H_3NO_2$	541-59-3	97.073	pl (bz)	94	sub	1.2493^{106}		s H_2O, EtOH, eth
9397	Pyrrolidine	Azacyclopentane	C_4H_9N	123-75-1	71.121	col liq	-57.79(0.03)	86.6(0.1)	0.8586^{20}	1.4431^{20}	msc H_2O; s EtOH, eth; sl bz, chl
9398	1-Pyrrolidineethanamine		$C_6H_{14}N_2$	7154-73-6	114.188			166	0.901^{25}	1.4687^{20}	
9399	1-Pyrrolidineethanol		$C_6H_{13}NO$	2955-88-6	115.173			187	0.9750^{20}	1.4713^{20}	
9400	1-[4-(1-Pyrrolidinyl)-2-butynyl]-2-pyrrolidinone	Oxotremorine	$C_{12}H_{18}N_2O$	70-22-4	206.283	pa ye liq		$124^{0.1}$	0.991^{25}	1.5160^{20}	
9401	3-(2-Pyrrolidinyl)pyridine, (*S*)-	Nornicotine	$C_9H_{12}N_2$	494-97-3	148.204	hyg		270	1.0737^{19}	1.5378^{18}	vs H_2O, ace, eth, EtOH
9402	2-Pyrrolidone	γ-Butyrolactam	C_4H_7NO	616-45-5	85.105	cry (peth)	25.92(0.01)	251.2(0.1)	1.120^{20}	1.4806^{30}	vs H_2O, EtOH, eth, bz, chl, CS_2
9403	1-(1*H*-Pyrrol-2-yl)ethanone		C_6H_7NO	1072-83-9	109.126	mcl nd (w)	89.8(0.4)	220			s H_2O, EtOH, eth
9404	Pyruvic acid		$C_3H_4O_3$	127-17-3	88.062		13.8	165 dec	1.2272^{20}	1.4280^{20}	msc H_2O, EtOH, eth; s ace
9405	Pyrvinium chloride		$C_{26}H_{28}ClN_3$	548-84-5	417.973	red pow (w)	250 dec				
9406	1,1':4',1":4",1'"-Quaterphenyl		$C_{24}H_{18}$	135-70-6	306.400		313.8(0.7)	428^{18}			i H_2O, EtOH, eth, chl; s bz, $PhNO_2$, HOAc
9407	Quercetin		$C_{15}H_{10}O_7$	117-39-5	302.236	ye nd (dil al, + 2 w)	322(1)	sub			sl H_2O, eth, MeOH; s EtOH, ace, py
9408	Quercitrin	Quercetin-3-L-rhamnoside	$C_{21}H_{20}O_{11}$	522-12-3	448.377	pa ye nd or pl (+2w, dil al)	170				i H_2O, eth; s EtOH, HOAc, MeOH, alk
9409	Quillaic acid		$C_{30}H_{46}O_5$	631-01-6	486.683	nd (dil al)	294				vs ace, eth, py, EtOH
9410	Quinacrine	Mepacrine	$C_{23}H_{30}ClN_3O$	83-89-6	399.956	ye oil	87				
9411	Quinaldine Red		$C_{21}H_{23}IN_2$	117-92-0	430.325	dk red pow					s H_2O; vs EtOH
9412	Quinamine		$C_{19}H_{24}N_2O_2$	464-85-7	312.406	pr (bz), nd (80% al)	185.5				i H_2O; vs EtOH, bz; s eth, ace
9413	Quinazoline	1,3-Benzodiazine	$C_8H_6N_2$	253-82-7	130.147	ye pl (peth)	47.7(0.2)	237(6)			vs H_2O; s EtOH, eth, ace, bz; sl chl
9414	Quinclorac	3,7-Dichloroquinoline-8-carboxylic acid	$C_{10}H_5Cl_2NO_2$	84087-01-4	242.059		274		1.75		
9415	Quinethazone		$C_{10}H_{12}ClN_3O_3S$	73-49-4	289.738						s tfa
9416	Quinic acid		$C_7H_{12}O_6$	77-95-2	192.166		162.5		1.64^{25}		vs H_2O, EtOH, HOAc
9417	Quinidine		$C_{20}H_{24}N_2O_2$	56-54-2	324.417	cry (+2.5w, dil al)	174				sl H_2O, eth; s EtOH, bz; vs chl; i peth
9418	Quinine	6'-Methoxycinchonan-9-ol, (8α,9*R*)	$C_{20}H_{24}N_2O_2$	130-95-0	324.417		57			1.625^{15}	sl H_2O, ace; vs EtOH, py; s eth, chl
9419	Quinine hydrochloride	6'-Methoxycinchonan-9-ol monohydrochloride, (8α,9*R*)	$C_{20}H_{25}ClN_2O_2$	130-89-2	360.878	silky efflor nd (w)	159				vs H_2O, EtOH, chl
9420	Quinine sulfate		$C_{40}H_{50}N_4O_8S$	804-63-7	746.912	silky nd (w)	235.2				vs EtOH
9421	Quininone		$C_{20}H_{22}N_2O_2$	84-31-1	322.401	nd, lf (eth)	108				vs bz, eth, EtOH
9422	2-Quinolinamine	2-Aminoquinoline	$C_9H_8N_2$	580-22-3	144.173	lf (w)	129(2)	sub			vs H_2O; s EtOH, eth, ace, chl; sl bz
9423	3-Quinolinamine	3-Aminoquinoline	$C_9H_8N_2$	580-17-6	144.173	orth (w, dil al)	94				vs eth, EtOH, chl
9424	4-Quinolinamine	4-Aminoquinoline	$C_9H_8N_2$	578-68-7	144.173	nd (bz, dil al)	154.8	180^{12}			s H_2O, bz, chl; vs EtOH, eth
9425	5-Quinolinamine	5-Aminoquinoline	$C_9H_8N_2$	611-34-7	144.173	ye nd (al) lf (eth)	106.0(0.2)	310			sl H_2O; vs EtOH, eth; s bz; i lig
9426	6-Quinolinamine	6-Aminoquinoline	$C_9H_8N_2$	580-15-4	144.173	cry (w+2), pr (eth)	116(2)	187^{12}			sl H_2O, eth; s NH_3, EtOH

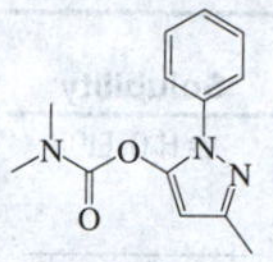
9390
Pyrolan

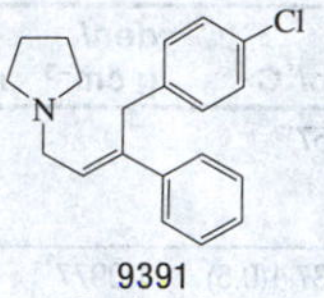
9391
Pyrrobutamine

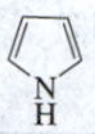
9392
Pyrrole

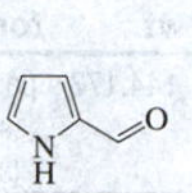
9393
1*H*-Pyrrole-2-carboxaldehyde

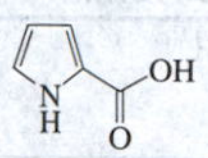
9394
1*H*-Pyrrole-2-carboxylic acid

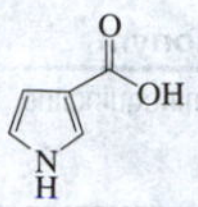
9395
1*H*-Pyrrole-3-carboxylic acid

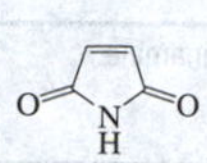
9396
1*H*-Pyrrole-2,5-dione

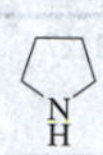
9397
Pyrrolidine

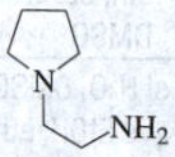
9398
1-Pyrrolidineethanamine

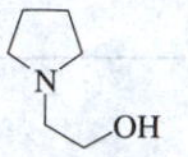
9399
1-Pyrrolidineethanol

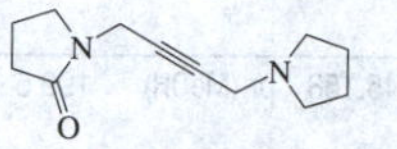
9400
1-[4-(1-Pyrrolidinyl)-2-butynyl]-2-pyrrolidinone

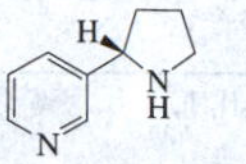
9401
3-(2-Pyrrolidinyl)pyridine, (*S*)-

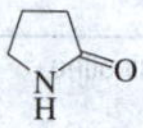
9402
2-Pyrrolidone

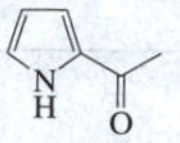
9403
1-(1*H*-Pyrrol-2-yl)ethanone

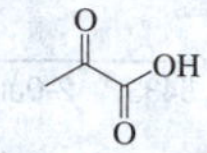
9404
Pyruvic acid

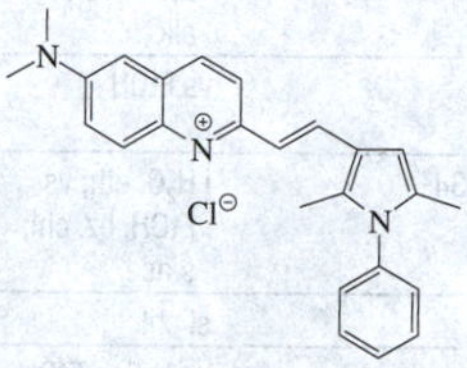
9405
Pyrvinium chloride

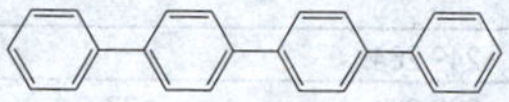
9406
1,1':4',1'':4'',1'''-Quaterphenyl

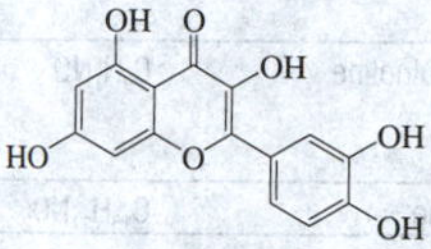
9407
Quercetin

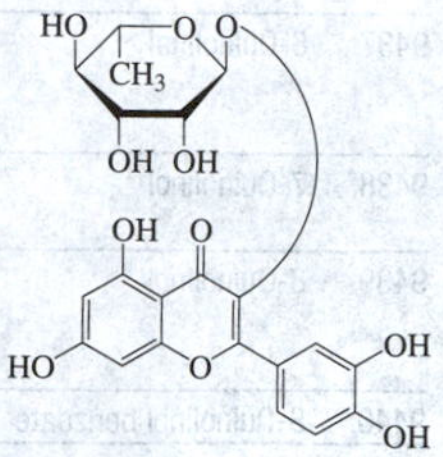
9408
Quercitrin

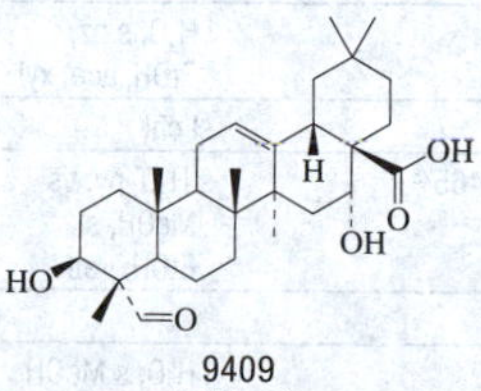
9409
Quillaic acid

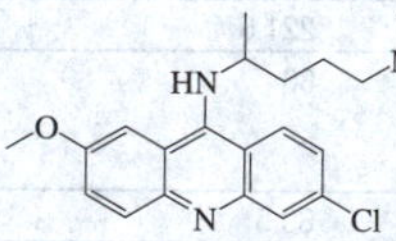
9410
Quinacrine

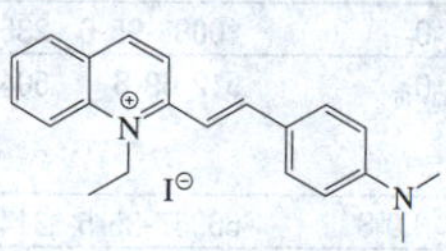
9411
Quinaldine Red

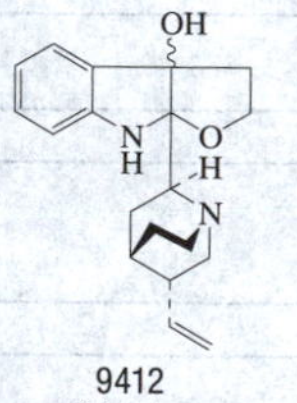
9412
Quinamine

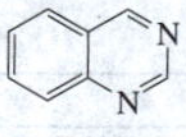
9413
Quinazoline

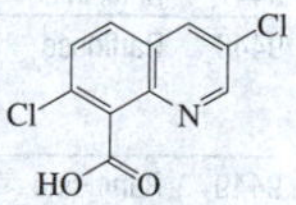
9414
Quinclorac

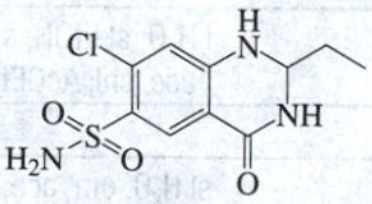
9415
Quinethazone

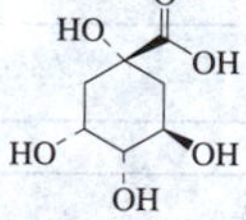
9416
Quinic acid

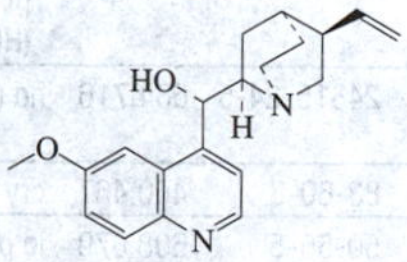
9417
Quinidine

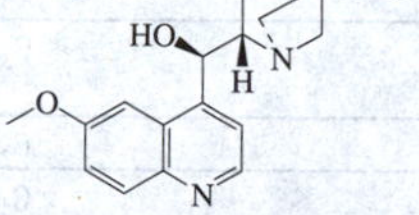
9418
Quinine

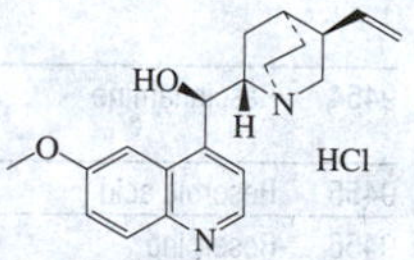
9419
Quinine hydrochloride

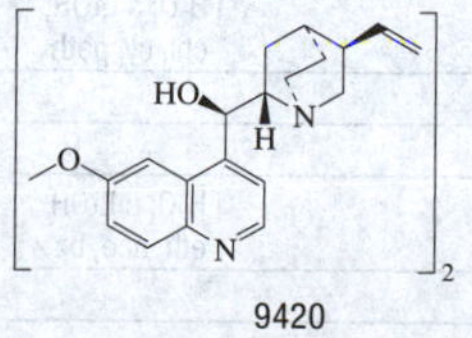
9420
Quinine sulfate

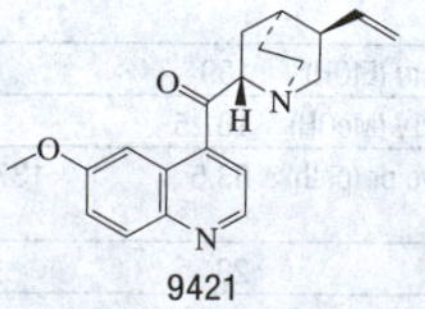
9421
Quininone

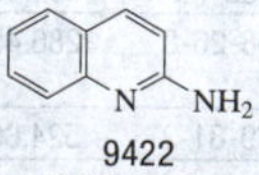
9422
2-Quinolinamine

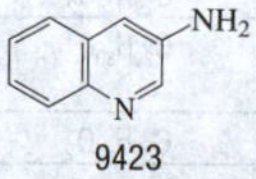
9423
3-Quinolinamine

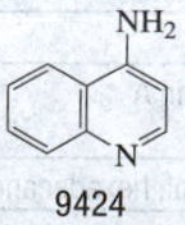
9424
4-Quinolinamine

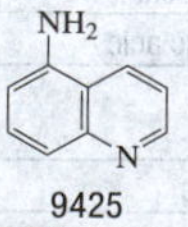
9425
5-Quinolinamine

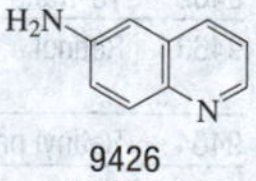
9426
6-Quinolinamine

No.	Name	Synonym	Mol. form.	CAS Reg. No.	Mol. wt.	Physical form	mp/°C	bp/°C	den/ g cm⁻³	n_D	Solubility
9427	8-Quinolinamine	8-Aminoquinoline	$C_9H_8N_2$	578-66-5	144.173	pa ye nd (sub) cry (al, lig)	64(2)	157^{19}			vs H_2O, EtOH
9428	Quinoline	1-Azanaphthalene	C_9H_7N	91-22-5	129.159	liq	-14.78(0.05)	237.1(0.5)	1.0977^{15}	1.6268^{20}	sl H_2O; msc EtOH, eth, ace, bz, CS_2; s ctc
9429	4-Quinolinecarboxaldehyde	Cinchoninaldehyde	$C_{10}H_7NO$	4363-93-3	157.169	nd (to-peth)	51	122^4			vs eth, tol
9430	2-Quinolinecarboxylic acid	Quinaldic acid	$C_{10}H_7NO_2$	93-10-7	173.169		156				s H_2O; vs bz
9431	8-Quinolinecarboxylic acid	8-Carboxyquinoline	$C_{10}H_7NO_2$	86-59-9	173.169	nd (w)	187	sub			vs EtOH
9432	2(1*H*)-Quinolinethione		C_9H_7NS	2637-37-8	161.224		187				i H_2O; vs EtOH, eth, bz; sl DMSO
9433	2-Quinolinol	2-Hydroxyquinoline	C_9H_7NO	59-31-4	145.158	pr (MeOH)	199.5	sub			sl H_2O, DMSO; vs EtOH, eth; s dil HCl
9434	3-Quinolinol	3-Hydroxyquinoline	C_9H_7NO	580-18-7	145.158	cry (bz, dil al)	201.3				i H_2O; s EtOH; sl eth, chl; vs bz
9435	4-Quinolinol	4-Hydroxyquinoline	C_9H_7NO	611-36-9	145.158	nd (w+3)	210				vs H_2O, EtOH; sl eth, bz, peth
9436	5-Quinolinol	5-Hydroxyquinoline	C_9H_7NO	578-67-6	145.158	nd (al), pl	224(3)	sub			s H_2O, bz, chl; sl EtOH; vs MeOH; i lig
9437	6-Quinolinol	6-Hydroxyquinoline	C_9H_7NO	580-16-5	145.158	pr (al, eth)	195	360			i H_2O, bz, chl; sl EtOH, eth; s alk
9438	7-Quinolinol	7-Hydroxyquinoline	C_9H_7NO	580-20-1	145.158	pr (al), nd (dil al-eth)	238(3)	sub			vs EtOH
9439	8-Quinolinol	8-Hydroxyquinoline	C_9H_7NO	148-24-3	145.158	nd (dil al)	74(2)	267	1.034^{20}		i H_2O, eth; vs EtOH, bz, chl; s ace
9440	8-Quinolinol benzoate	Benzoxiquine	$C_{16}H_{11}NO_2$	86-75-9	249.264						sl chl
9441	8-Quinolinol sulfate (2:1)	8-Hydroxyquinoline sulfate	$C_{18}H_{16}N_2O_6S$	134-31-6	388.394		177.5				vs H_2O; s EtOH; i eth
9442	Quinovic acid		$C_{30}H_{46}O_5$	465-74-7	486.683	pl or nd	298 dec				
9443	Quinovose		$C_6H_{12}O_5$	7658-08-4	164.156	cry (AcOEt)	136(1)				vs H_2O, EtOH
9444	Quinoxaline	1,4-Benzodiazine	$C_8H_6N_2$	91-19-0	130.147	cry (peth)	32.5(0.2)	227(7)	1.1334^{48}	1.6231^{48}	s H_2O; msc EtOH, eth, ace, bz; sl chl
9445	2(1*H*)-Quinoxalinone		$C_8H_6N_2O$	1196-57-2	146.146	lf (al)	269.4(0.4)	361(15)			
9446	Quizalofop-Ethyl		$C_{19}H_{17}ClN_2O_4$	76578-14-8	372.802	wh cry	93	$220^{0.2}$			i H_2O; s bz, EtOH, ace, xyl
9447	Radicinin		$C_{12}H_{12}O_5$	10088-95-6	236.220		221.5				sl chl
9448	Raffinose		$C_{18}H_{32}O_{16}$	512-69-6	504.437		80		1.465^{25}		s H_2O, py; vs MeOH; sl EtOH; i eth
9449	Ranitidine		$C_{13}H_{22}N_4O_3S$	66357-35-5	314.404	solid	69.5				
9450	Raubasine		$C_{21}H_{24}N_2O_3$	483-04-5	352.427		258 dec				i H_2O; s MeOH
9451	Raunescine		$C_{31}H_{36}N_2O_8$	117-73-7	564.626		165				i H_2O; s EtOH, chl, HOAc
9452	Reinecke salt	Ammonium tetrathiocyanodiamminechromate(III) monohydrate	$C_4H_{12}CrN_7OS_4$	13573-16-5	354.440	red cry (w)	270 dec				s H_2O, EtOH, ace; i bz
9453	Resazurin	7-Hydroxy-3*H*-phenoxazin-3-one, 10-oxide	$C_{12}H_7NO_4$	550-82-3	229.189	dk red to gr pr or pl (HOAc)		sub			i H_2O, eth; sl EtOH, HOAc; s alk
9454	Rescinnamine		$C_{35}H_{42}N_2O_9$	24815-24-5	634.716	nd (bz)	238.5				i H_2O; sl EtOH; s ace, chl, AcOEt
9455	Reserpic acid		$C_{22}H_{28}N_2O_5$	83-60-3	400.467	cry (MeOH)	242				
9456	Reserpine		$C_{33}H_{40}N_2O_9$	50-55-5	608.679	lo pr (dil ace)	264.5				sl H_2O, eth, ace; s EtOH, bz, AcOEt
9457	*cis*-Resmethrin, (-)		$C_{22}H_{26}O_3$	10453-86-8	338.439		57.5(0.5)				
9458	Resorcinol	1,3-Benzenediol	$C_6H_6O_2$	108-46-3	110.111	nd (bz), pl (w)	109.8(0.4)	280(2)	1.278^{20}	1.578^{25}	vs H_2O, ctc; s EtOH, eth; sl bz, chl
9459	11-*cis*-Retinal	Vitamin A_1 aldehyde	$C_{20}H_{28}O$	564-87-4	284.435	cry					
9460	Retinal (all *trans*)		$C_{20}H_{28}O$	116-31-4	284.435	oran cry	64				i H_2O; s EtOH, chl, cy, peth
9461	13-*cis*-Retinoic acid	Accutane	$C_{20}H_{28}O_2$	4759-48-2	300.435	cry (EtOH)	189				
9462	13-*trans*-Retinoic acid		$C_{20}H_{28}O_2$	302-79-4	300.435	cry (MeOH)	181.5				
9463	Retinol	Vitamin A	$C_{20}H_{30}O$	68-26-8	286.451	ye pr (peth)	63.5	$137^{0.000001}$			i H_2O; s EtOH, eth, ace, bz
9464	Retinyl palmitate	Retinol, hexadecanoate	$C_{36}H_{60}O_2$	79-81-2	524.860		28				

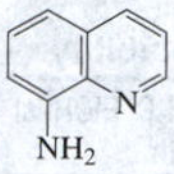
9427
8-Quinolinamine

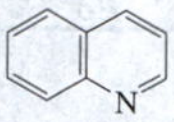
9428
Quinoline

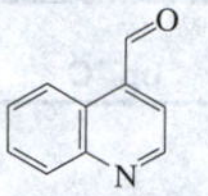
9429
4-Quinolinecarboxaldehyde

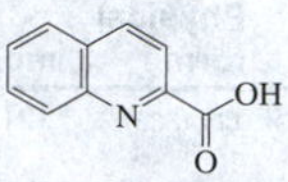
9430
2-Quinolinecarboxylic acid

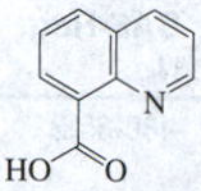
9431
8-Quinolinecarboxylic acid

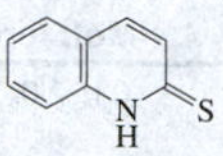
9432
2(1*H*)-Quinolinethione

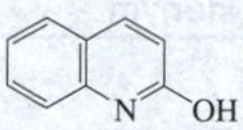
9433
2-Quinolinol

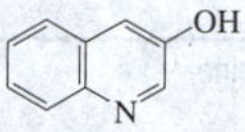
9434
3-Quinolinol

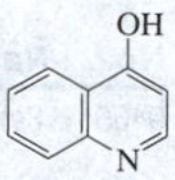
9435
4-Quinolinol

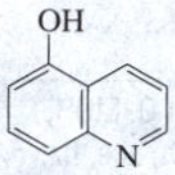
9436
5-Quinolinol

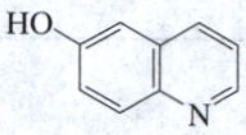
9437
6-Quinolinol

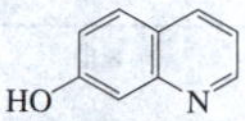
9438
7-Quinolinol

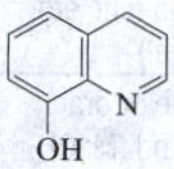
9439
8-Quinolinol

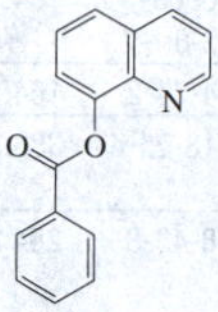
9440
8-Quinolinol benzoate

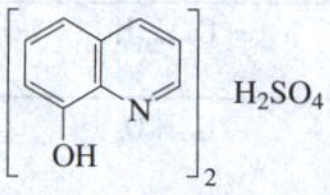
9441
8-Quinolinol sulfate (2:1)

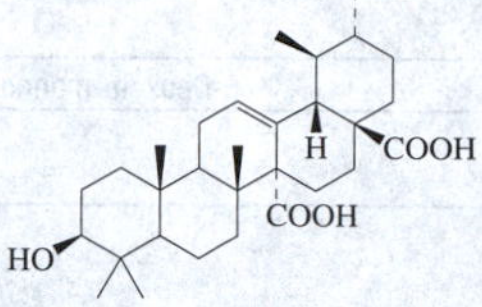
9442
Quinovic acid

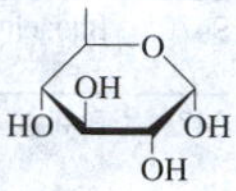
9443
Quinovose

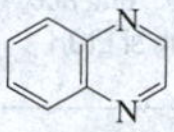
9444
Quinoxaline

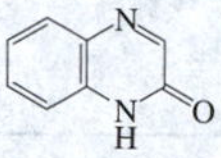
9445
2(1*H*)-Quinoxalinone

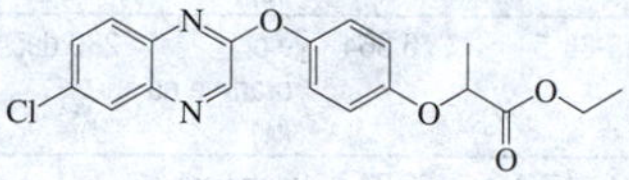
9446
Quizalofop-Ethyl

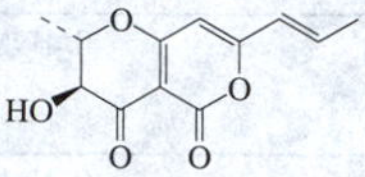
9447
Radicinin

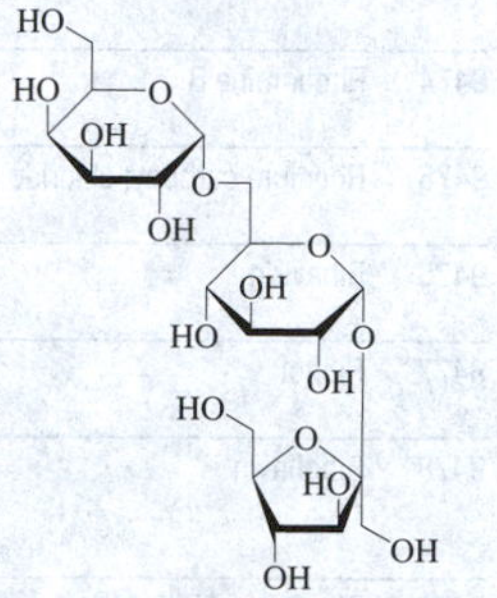
9448
Raffinose

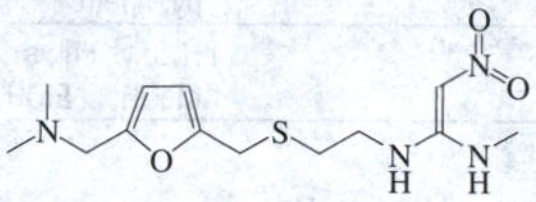
9449
Ranitidine

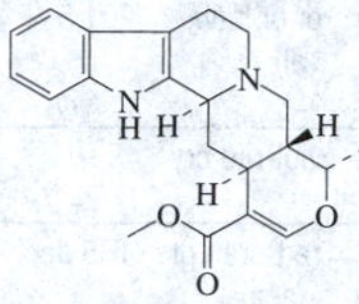
9450
Raubasine

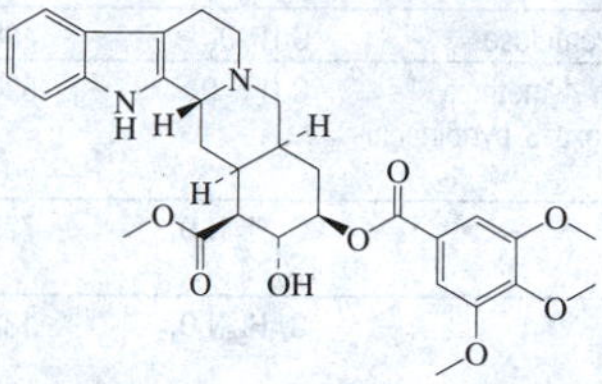
9451
Raunescine

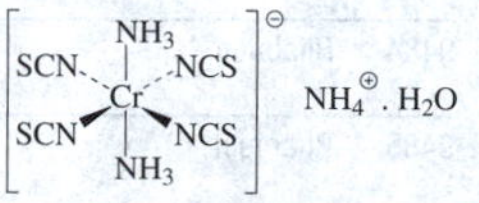
9452
Reinecke salt

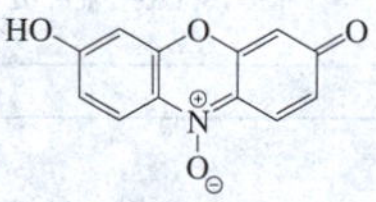
9453
Resazurin

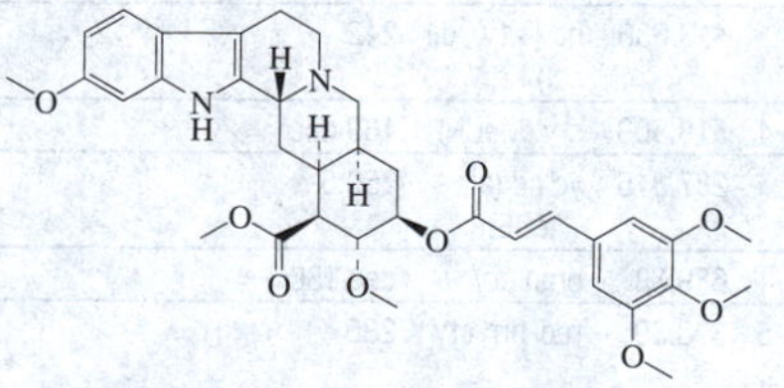
9454
Rescinnamine

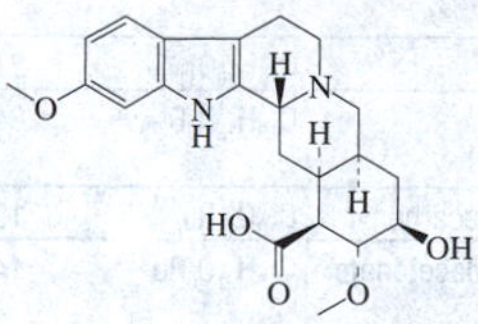
9455
Reserpic acid

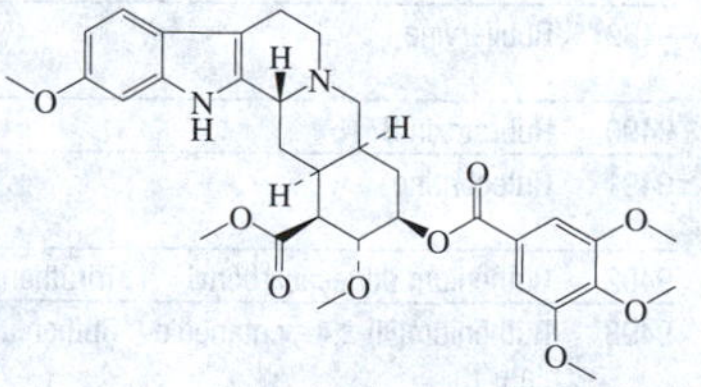
9456
Reserpine

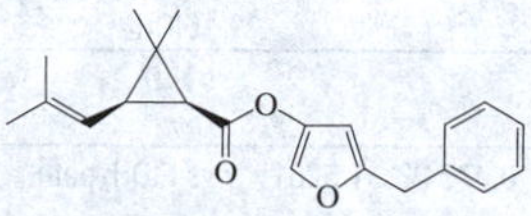
9457
cis-Resmethrin, (-)

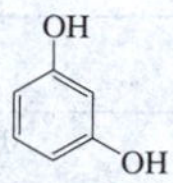
9458
Resorcinol

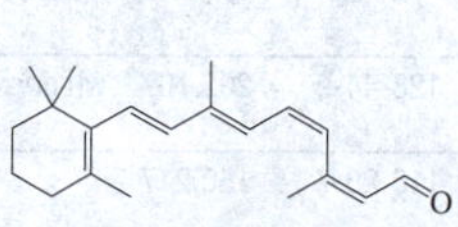
9459
11-*cis*-Retinal

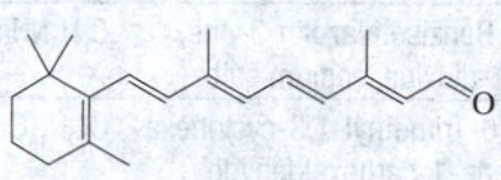
9460
Retinal (all *trans*)

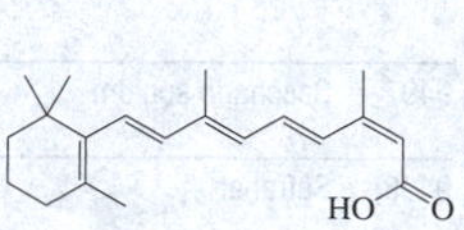
9461
13-*cis*-Retinoic acid

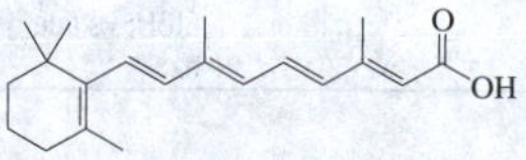
9462
13-*trans*-Retinoic acid

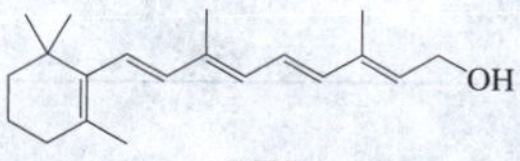
9463
Retinol

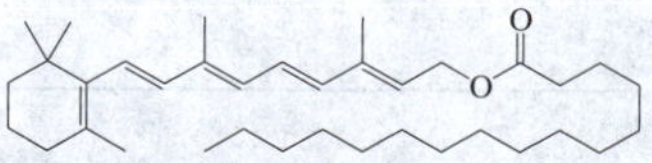
9464
Retinyl palmitate

No.	Name	Synonym	Mol. form.	CAS Reg. No.	Mol. wt.	Physical form	mp/°C	bp/°C	den/ g cm⁻³	n_D	Solubility
9465	Retronecine, (+)		$C_8H_{13}NO_2$	480-85-3	155.195	cry (ace)	121				s H_2O, EtOH; sl eth
9466	Retrorsine		$C_{18}H_{25}NO_6$	480-54-6	351.395	cry (AcOEt)	212				sl H_2O, ace; s EtOH, chl; i eth
9467	Rhamnetin		$C_{16}H_{12}O_7$	90-19-7	316.262	ye nd (al)	295				sl H_2O; s EtOH, ace, PhOH; vs dil alk
9468	DL-α-Rhamnose		$C_6H_{12}O_5$	116908-82-8	164.156	cry (w)	151				vs H_2O, EtOH
9469	D-Rhamnose	6-Deoxy-D-mannose	$C_6H_{12}O_5$	634-74-2	164.156						s H_2O
9470	Rheadine		$C_{21}H_{21}NO_6$	2718-25-4	383.395	nd (chl, eth, al)	257	sub			
9471	Rhein		$C_{15}H_8O_6$	478-43-3	284.221	ye or oran nd (MeOH, py)	321	sub			sl H_2O, EtOH, eth, ace, bz; vs py
9472	Rhenium carbonyl	Dirhenium decacarbonyl	$C_{10}O_{10}Re_2$	14285-68-8	652.515	ye-wh cry	170 dec		2.87		s os
9473	Rhizopterin		$C_{15}H_{12}N_6O_4$	119-20-0	340.294	lt ye pl (w)	>300				i H_2O, EtOH, eth; s aq alk, aq NH_3, py
9474	Rhodamine B		$C_{28}H_{32}ClN_2O_3$	81-88-9	480.018		165				s H_2O, EtOH, eth, bz, xyl
9475	Rhodium carbonyl chloride	Dirhodium tetracarbonyl dichloride	$C_4Cl_2O_4Rh_2$	14523-22-9	388.758	red-oran cry	124				s os
9476	Ribavirin	Tribavirin	$C_8H_{12}N_4O_5$	36791-04-5	244.205	col cry (EtOH)	175				s H_2O
9477	Ribitol	Adonitol	$C_5H_{12}O_5$	488-81-3	152.146	pr (w), nd (al)	96(1)				s H_2O, EtOH; i eth, lig
9478	Riboflavin		$C_{17}H_{20}N_4O_6$	83-88-5	376.364	ye or oran-ye nd (w)	280 dec				i H_2O, eth, ace, chl; sl EtOH
9479	Riboflavin-5′-phosphate		$C_{17}H_{21}N_4O_9P$	146-17-8	456.34	ye cry (w)					
9480	D-Ribose		$C_5H_{10}O_5$	50-69-1	150.130	pl (al)	88				s H_2O; sl EtOH
9481	L-Ribose		$C_5H_{10}O_5$	24259-59-4	150.130		81				
9482	D-Ribulose	*erythro*-2-Pentulose	$C_5H_{10}O_5$	488-84-6	150.130	syrup					vs H_2O
9483	Ricinine	1,2-Dihydro-4-methoxy-1-methyl-2-oxo-3-pyridinecarbonitrile	$C_8H_8N_2O_2$	524-40-3	164.162	pr or lf (w, al)	201.5	170 sub			s H_2O, chl; sl EtOH, bz; vs py; i peth
9484	Rifabutin		$C_{46}H_{62}N_4O_{11}$	72559-06-9	847.004	viol-red cry					i H_2O; vs chl; s MeOH; sl EtOH
9485	Rifampin		$C_{43}H_{58}N_4O_{12}$	13292-46-1	822.941	red-oran pl (ace)	185 dec				
9486	Rinderine	Echinatine-3′-epimer	$C_{15}H_{25}NO_5$	6029-84-1	299.364	cry (ace)	100.5				
9487	Ronnel		$C_8H_8Cl_3O_3PS$	299-84-3	321.546		42.6(0.9)	$152^{0.4}$	1.44^{32}	1.5335^{35}	
9488	Rotenone		$C_{23}H_{22}O_6$	83-79-4	394.417	nd or lf (al, aq-ace)	165.0(0.5)	$215^{0.5}$			i H_2O; s EtOH, ace, bz; sl eth; vs chl
9489	Rubijervine		$C_{27}H_{43}NO_2$	79-58-3	413.636	nd (+1w, dil al)	242				vs bz, EtOH, chl
9490	Rubratoxin B		$C_{26}H_{30}O_{11}$	21794-01-4	518.509	cry (MeCN)	169 dec				
9491	Rutecarpine		$C_{18}H_{13}N_3O$	84-26-4	287.315	ye nd (al, AcOEt)	259.5				sl EtOH, ace, bz
9492	Ruthenium dodecacarbonyl	Triruthenium dodecacarbonyl	$C_{12}O_{12}Ru_3$	15243-33-1	639.33	oran cry	dec 150				
9493	Ruthenium(III) 2,4-pentanedioate	Ruthenium(III) acetylacetonate	$C_{15}H_{21}O_6Ru$	14284-93-6	398.39	red-brn cry	230				
9494	Rutinose		$C_{12}H_{22}O_{10}$	90-74-4	326.297	hyg pow (al, eth)	190 dec				vs H_2O, EtOH
9495	Sabadine		$C_{29}H_{47}NO_8$	124-80-1	537.685	nd (eth)	258				vs ace, EtOH
9496	Saccharin		$C_7H_5NO_3S$	81-07-2	183.185	nd (ace) pr (al), lf (w)	227(2)	sub	0.828^{25}		sl H_2O, bz, eth, chl; s ace, EtOH
9497	Saccharin sodium	1,2-Benzisothiazolin-3-one, 1,1-dioxide, sodium salt	$C_7H_4NNaO_3S$	128-44-9	205.167	wh cry	229				s H_2O
9498	Safranal	2,6,6-Trimethyl-1,3-cyclohexadiene-1-carboxaldehyde	$C_{10}H_{14}O$	116-26-7	150.217			70^{1}	0.9734^{19}	1.5281^{19}	vs EtOH, peth
9499	Safrole	5-(2-Propenyl)-1,3-benzodioxole	$C_{10}H_{10}O_2$	94-59-7	162.185	mcl	11.2	235(3)	1.1000^{20}	1.5381^{20}	i H_2O; vs EtOH; msc eth, chl
9500	Salcomine	*N,N′*-Bis(salicylidene)-ethylenediaminocobalt(II)	$C_{16}H_{14}CoN_2O_2$	14167-18-1	325.227	red cry (DMF)					s bz, chl, py
9501	Salicylaldehyde	2-Hydroxybenzaldehyde	$C_7H_6O_2$	90-02-8	122.122	liq	-7(2)	208(5)	1.1674^{20}	1.5740^{20}	sl H_2O, chl; msc EtOH; vs ace, bz

Organic

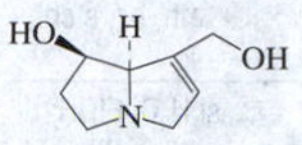
9465
Retronecine, (+)

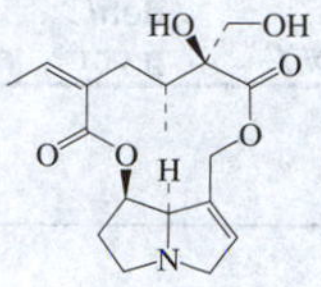
9466
Retrorsine

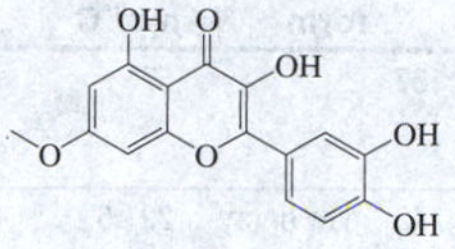
9467
Rhamnetin

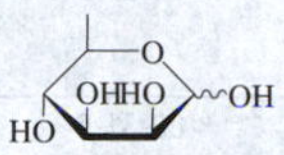
9468
DL-α-Rhamnose

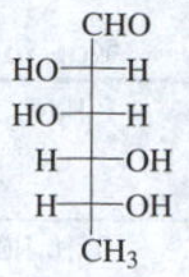
9469
D-Rhamnose

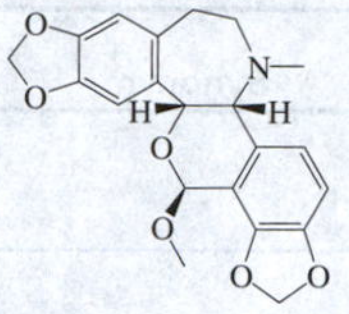
9470
Rheadine

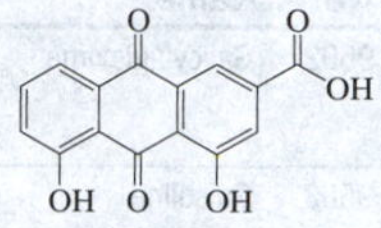
9471
Rhein

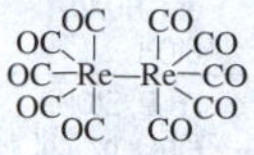
9472
Rhenium carbonyl

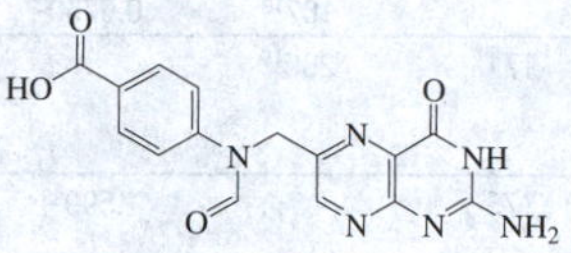
9473
Rhizopterin

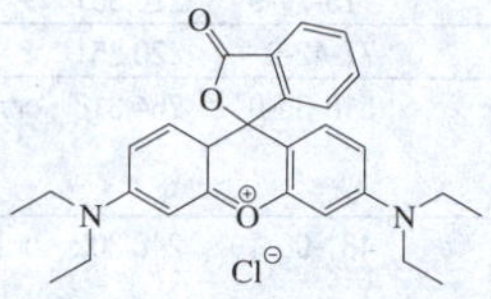
9474
Rhodamine B

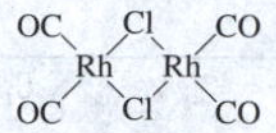
9475
Rhodium carbonyl chloride

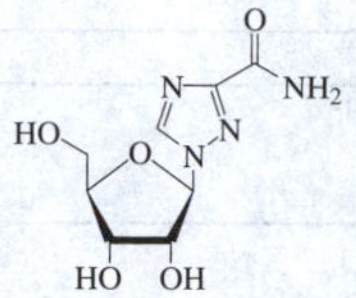
9476
Ribavirin

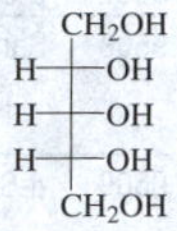
9477
Ribitol

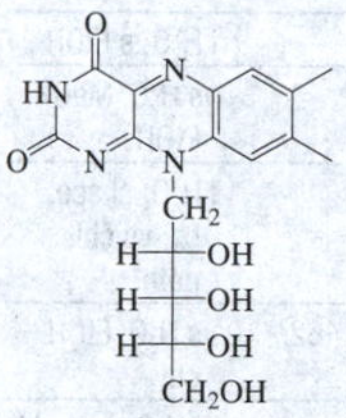
9478
Riboflavin

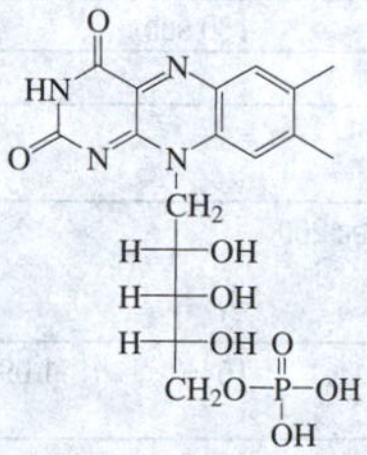
9479
Riboflavin-5'-phosphate

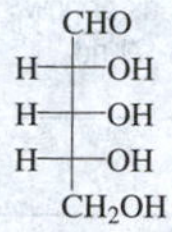
9480
D-Ribose

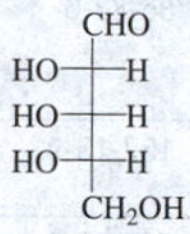
9481
L-Ribose

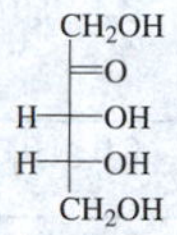
9482
D-Ribulose

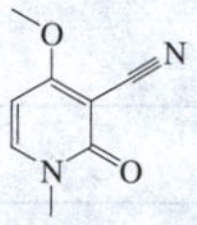
9483
Ricinine

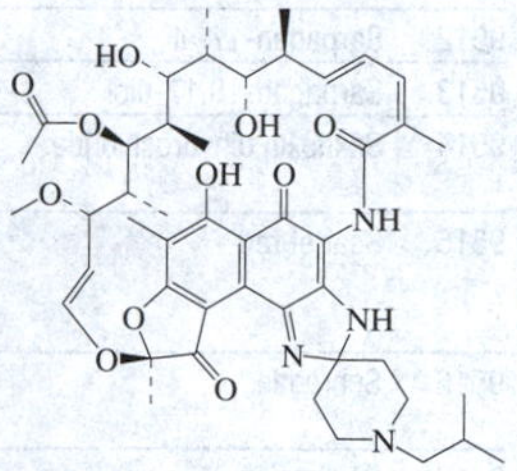
9484
Rifabutin

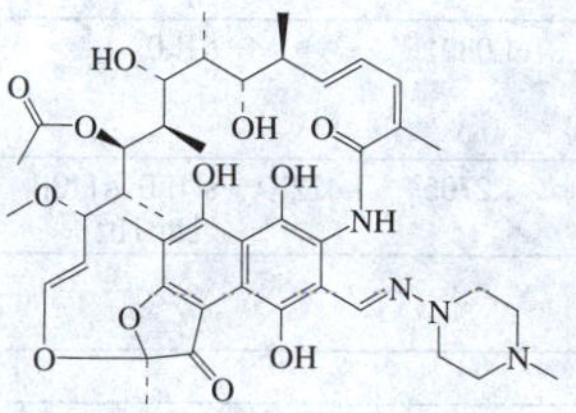
9485
Rifampin

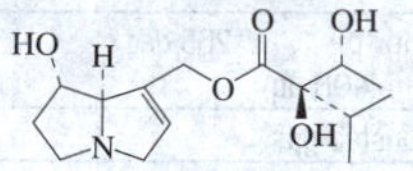
9486
Rinderine

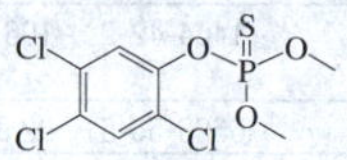
9487
Ronnel

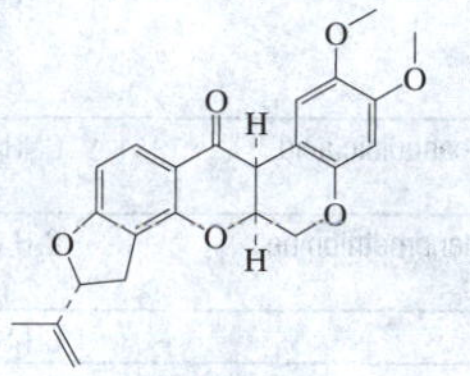
9488
Rotenone

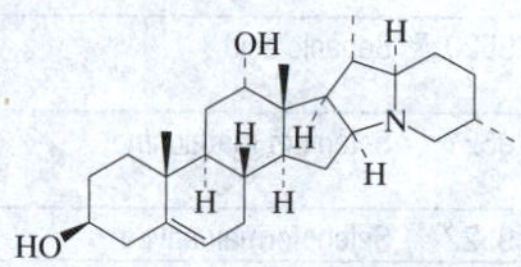
9489
Rubijervine

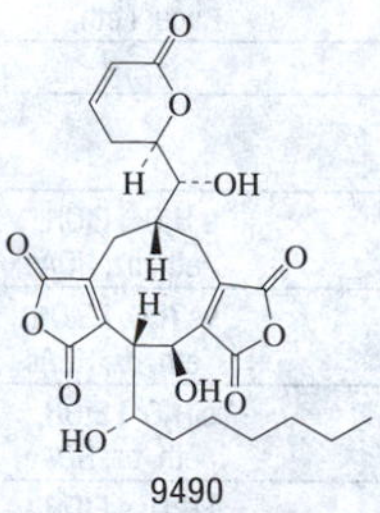
9490
Rubratoxin B

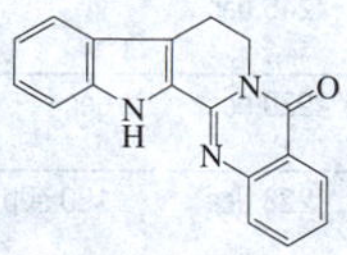
9491
Rutecarpine

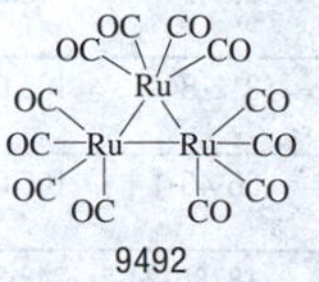
9492
Ruthenium dodecacarbonyl

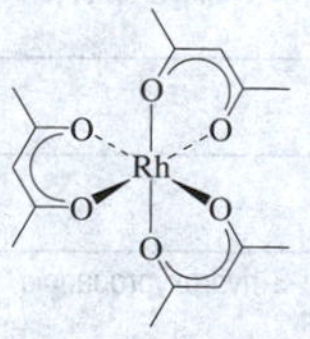
9493
Ruthenium(III) 2,4-pentanedioate

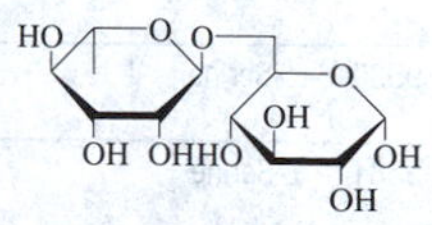
9494
Rutinose

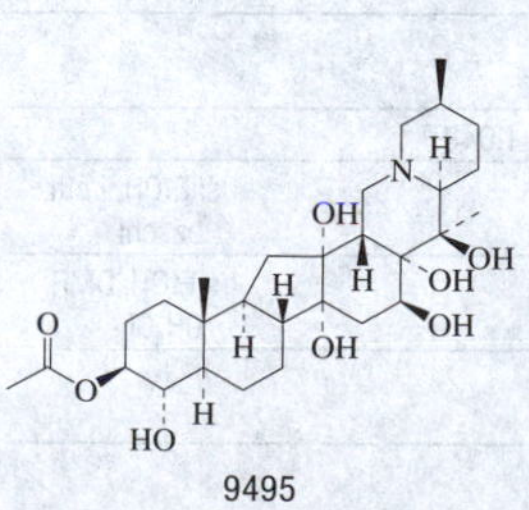
9495
Sabadine

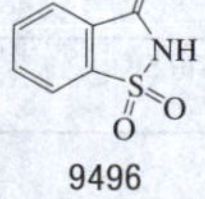
9496
Saccharin

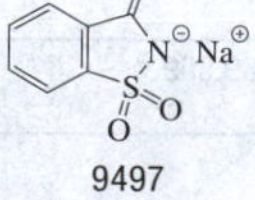
9497
Saccharin sodium

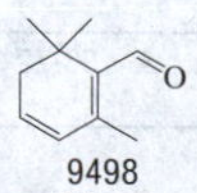
9498
Safranal

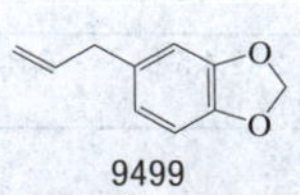
9499
Safrole

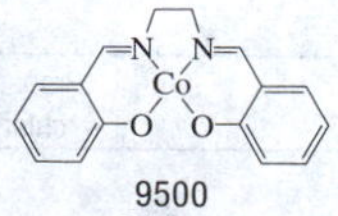
9500
Salcomine

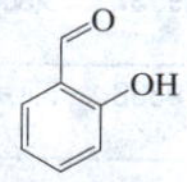
9501
Salicylaldehyde

No.	Name	Synonym	Mol. form.	CAS Reg. No.	Mol. wt.	Physical form	mp/°C	bp/°C	den/ g cm⁻³	n_D	Solubility
9502	Salicylaldoxime		$C_7H_7NO_2$	94-67-7	137.137		57				sl H_2O; vs EtOH, eth, bz; s chl; i lig
9503	Salsoline		$C_{11}H_{15}NO_2$	89-31-6	193.243	pow or cry (al)	221.5				sl H_2O, EtOH; i eth, peth; s chl, alk
9504	Salvarsan dihydrochloride	Arsphenamine	$C_{12}H_{14}As_2Cl_2N_2O_2$	139-93-5	439.000	ye hyg pow	190 dec				vs H_2O
9505	Sanguinarine		$C_{20}H_{15}NO_5$	2447-54-3	349.337	cry (eth, al)	266				vs ace, bz, eth, EtOH
9506	α-Santalol		$C_{15}H_{24}O$	115-71-9	220.351			301.5	0.9679^{20}	1.5023^{20}	i H_2O; s EtOH
9507	β-Santalol		$C_{15}H_{24}O$	77-42-9	220.351			167^{10}	0.9750^{20}	1.5115^{20}	
9508	Santonic acid		$C_{15}H_{20}O_4$	510-35-0	264.318	cry	171	285^{15}			sl H_2O; s chl, eth, HOAc, EtOH
9509	α-Santonin		$C_{15}H_{18}O_3$	481-06-1	246.302	orth (w, eth)	175		1.590^{25}		sl H_2O, EtOH, eth; s bz, chl; i peth
9510	Sarcosine	*N*-Methylglycine	$C_3H_7NO_2$	107-97-1	89.094	cry (al)	212 dec				s H_2O
9511	Sarmentogenin		$C_{23}H_{34}O_5$	76-28-8	390.513	pr (95% al, MeOH-eth)	280				i H_2O, eth, bz; s EtOH; sl ace, chl
9512	Sarpagan-17-al	Vellosimine	$C_{19}H_{20}N_2O$	6874-98-2	292.374	cry (MeOH)	305.5	180 sub			
9513	Sarpagan-10,17-diol	Sarpagine	$C_{19}H_{22}N_2O_2$	482-68-8	310.390	nd	320				i H_2O; s EtOH
9514	Saxitoxin dihydrochloride		$C_{10}H_{19}Cl_2N_7O_4$	35554-08-6	372.209	hyg wh solid					vs H_2O, MeOH, EtOH
9515	Scarlet red		$C_{24}H_{20}N_4O$	85-83-6	380.442	dk br pow or nd	185; dec 260				i H_2O; sl ace, bz; vs chl, peth
9516	Schradan		$C_8H_{24}N_4O_3P_2$	152-16-9	286.250		17	$154^{2.0}$	1.09^{25}	1.462^{25}	vs H_2O, EtOH, chl
9517	Scilliroside		$C_{32}H_{44}O_{12}$	507-60-8	620.684	lo pr (dil MeOH)	169	dec			sl H_2O, ace, chl; vs EtOH, diox; i eth
9518	Scopolamine		$C_{17}H_{21}NO_4$	51-34-3	303.354	visc liq					vs hot H_2O, EtOH, ace; sl bz
9519	Scopoline		$C_8H_{13}NO_2$	487-27-4	155.195	hyg nd (lig, eth, chl, peth)	108.5	248	1.0891^{134}		s H_2O
9520	Sebacic acid	Decanedioic acid	$C_{10}H_{18}O_4$	111-20-6	202.248	lf	131(1)	374(5)	1.2705^{20}	1.422^{133}	sl H_2O; s EtOH, eth; i bz
9521	Selenium methionine	Selenomethionine	$C_5H_{11}NO_2Se$	1464-42-2	196.11	hex pl (MeOH aq)	265 dec				
9522	Selenoformaldehyde		CH_2Se	6596-50-5	92.99	unstab gas					
9523	Selenourea	Carbamimidoselenoic acid	CH_4N_2Se	630-10-4	123.02	pr or nd (w)		200 dec			vs H_2O
9524	Semicarbazide hydrochloride		CH_6ClN_3O	563-41-7	111.531	pr (dil al)	176 dec				vs H_2O
9525	Senecionine		$C_{18}H_{25}NO_5$	130-01-8	335.396	pl	232				i H_2O; sl EtOH, eth; s chl
9526	Seneciphylline		$C_{18}H_{23}NO_5$	480-81-9	333.380	pl (AcOEt)	217 dec				s chl; sl EtOH, ace; i eth
9527	Senkirkin		$C_{19}H_{27}NO_6$	2318-18-5	365.420	pl (ace)	197				
9528	L-Sepiapterin	6-Lactoyl-7,8-dihydropterin	$C_9H_{11}N_5O_3$	17094-01-8	237.215	ye pow or cry					
9529	DL-Serine		$C_3H_7NO_3$	302-84-1	105.093	mcl pr or lf (w)	246 dec		1.603^{22}		s H_2O; i EtOH, eth, bz, HOAc
9530	D-Serine		$C_3H_7NO_3$	312-84-5	105.093	nd or hex pr (w)	229 dec	dec			vs H_2O; i EtOH, eth, bz, HOAc
9531	L-Serine	2-Amino-3-hydroxypropanoic acid, (*S*)	$C_3H_7NO_3$	56-45-1	105.093	hex pl or pr (w)	228 dec	150 sub	1.6^{22}		s H_2O; i EtOH, eth, bz, HOAc
9532	Serpentine alkaloid		$C_{21}H_{20}N_2O_3$	18786-24-8	348.395		175				i H_2O; s EtOH, eth, ace
9533	Sesin	2,4-Dichlorophenoxyethyl benzoate	$C_{15}H_{12}Cl_2O_3$	94-83-7	311.160	cry	66	$185^{1.5}$			
9534	Sesone	Sodium 2-(2,4-dichlorophenoxy)ethyl sulfate	$C_8H_7Cl_2NaO_5S$	136-78-7	309.100		245 dec				
9535	Sethoxydim		$C_{17}H_{29}NO_3S$	74051-80-2	327.482				1.043^{25}		
9536	Shikimic acid		$C_7H_{10}O_5$	138-59-0	174.151	nd	184				sl EtOH; i eth, bz, chl
9537	Siduron		$C_{14}H_{20}N_2O$	1982-49-6	232.321	cry solid	135				s EtOH, DMF, CH_2Cl_2
9538	Silvex	Propanoic acid, 2-(2,4,5-trichlorophenoxy)-	$C_9H_7Cl_3O_3$	93-72-1	269.509		180.7(0.5)				

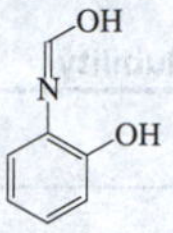
9502
Salicylaldoxime

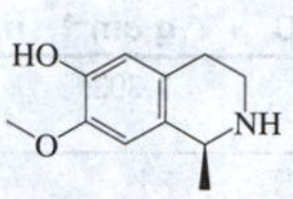
9503
Salsoline

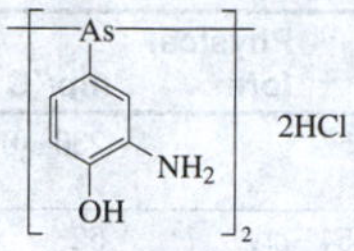
9504
Salvarsan dihydrochloride

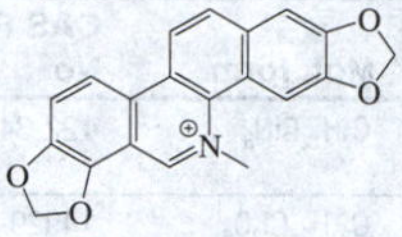
9505
Sanguinarine

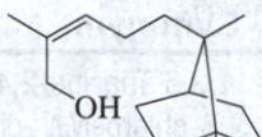
9506
α-Santalol

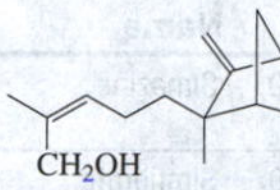
9507
β-Santalol

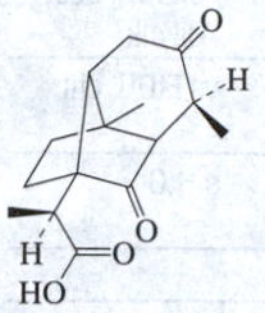
9508
Santonic acid

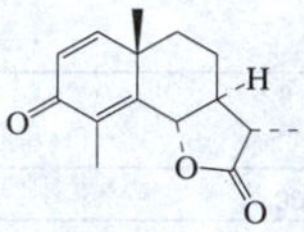
9509
α-Santonin

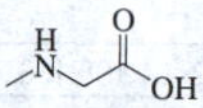
9510
Sarcosine

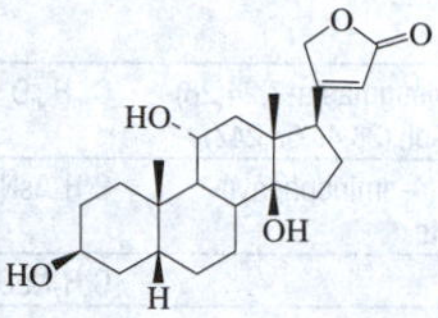
9511
Sarmentogenin

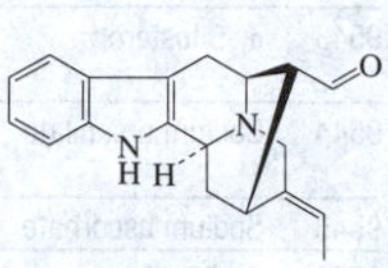
9512
Sarpagan-17-al

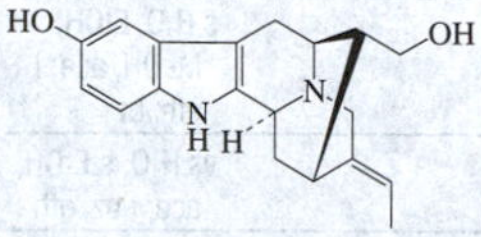
9513
Sarpagan-10,17-diol

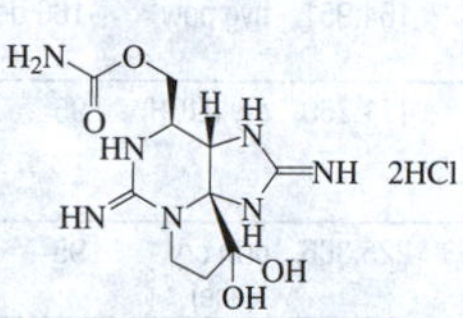
9514
Saxitoxin dihydrochloride

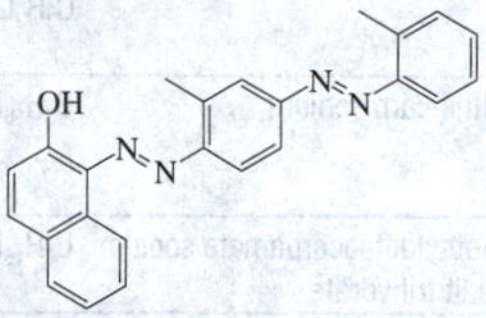
9515
Scarlet red

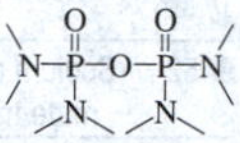
9516
Schradan

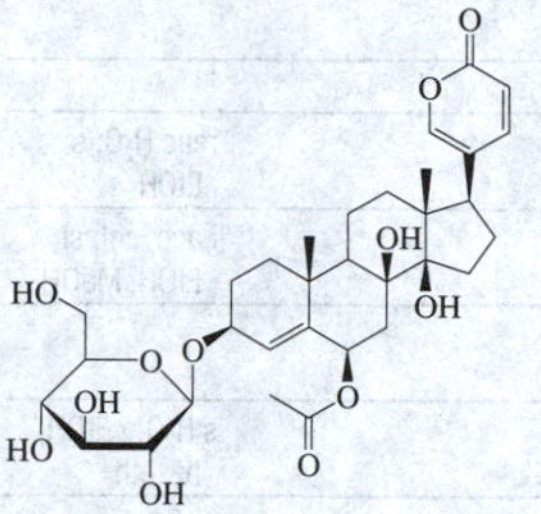
9517
Scilliroside

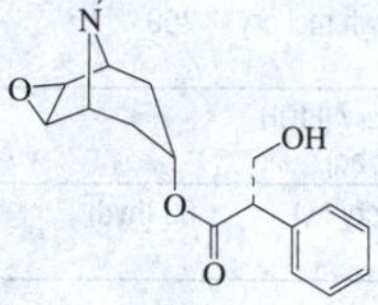
9518
Scopolamine

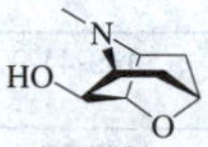
9519
Scopoline

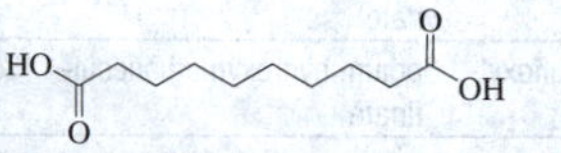
9520
Sebacic acid

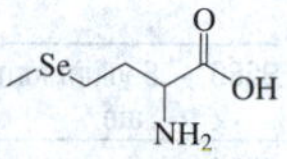
9521
Selenium methionine

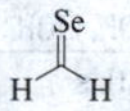
9522
Selenoformaldehyde

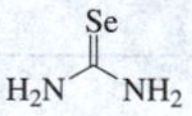
9523
Selenourea

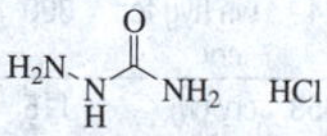
9524
Semicarbazide hydrochloride

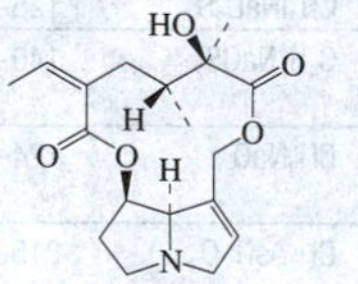
9525
Senecionine

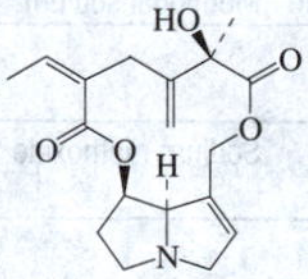
9526
Seneciphylline

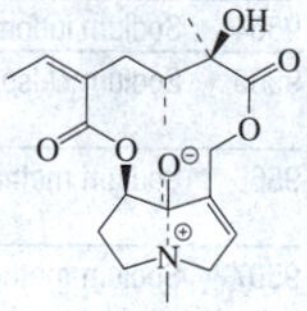
9527
Senkirkin

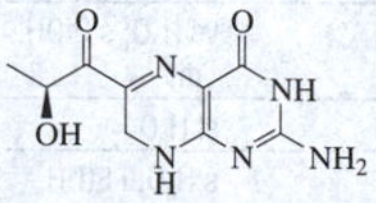
9528
L-Sepiapterin

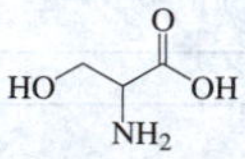
9529
DL-Serine

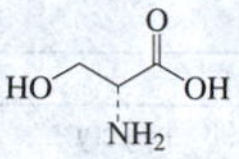
9530
D-Serine

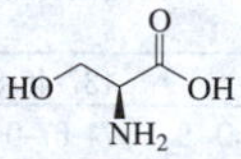
9531
L-Serine

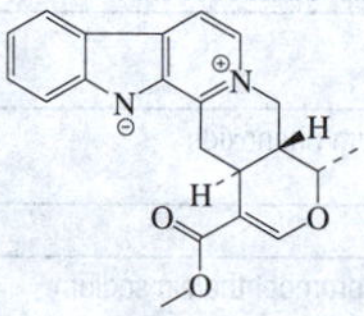
9532
Serpentine alkaloid

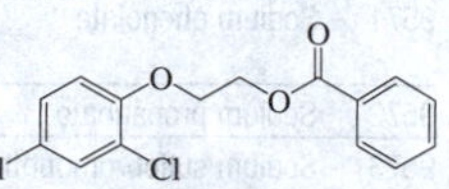
9533
Sesin

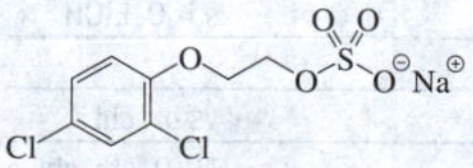
9534
Sesone

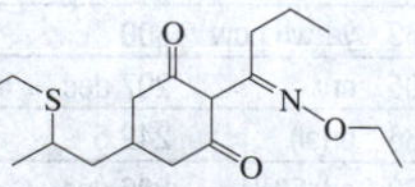
9535
Sethoxydim

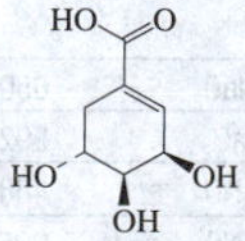
9536
Shikimic acid

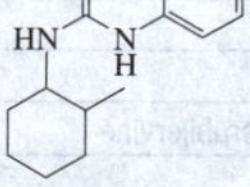
9537
Siduron

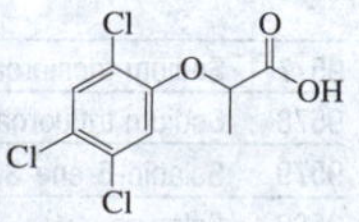
9538
Silvex

Organic

No.	Name	Synonym	Mol. form.	CAS Reg. No.	Mol. wt.	Physical form	mp/°C	bp/°C	den/ g cm^{-3}	n_D	Solubility
9539	Simazine	1,3,5-Triazine-2,4-diamine, 6-chloro-*N,N'*-diethyl-	$C_7H_{12}ClN_5$	122-34-9	201.657		230.4(0.9)		1.302^{20}		
9540	Simfibrate		$C_{23}H_{26}Cl_2O_6$	14929-11-4	469.354	cry	52	225$^{0.15}$			
9541	Sinapinic acid	3-(4-Hydroxy-3,5-dimethoxyphenyl)-2-propenoic acid	$C_{11}H_{12}O_5$	530-59-6	224.210	wh pow					i H_2O; s MeOH, ace
9542	Sinomenine		$C_{19}H_{23}NO_4$	115-53-7	329.391	nd (bz)	162				sl H_2O, eth, bz; s EtOH, ace, dil alk
9543	α_1-Sitosterol	4-Methylstigmasta-7,24(28)-dien-3-ol, (3β,4α,5α,24Z)	$C_{30}H_{50}O$	474-40-8	426.717	nd (al)	166				vs EtOH, chl
9544	Sodium arsanilate	Sodium (4-aminophenyl)-arsonate	$C_6H_7AsNNaO_3$	127-85-5	239.037	wh cry					s H_2O
9545	Sodium ascorbate		$C_6H_7NaO_6$	134-03-2	198.106	cry	218 dec				
9546	Sodium benzenesulfinate		$C_6H_5NaO_2S$	873-55-2	164.158	cry	300				
9547	Sodium benzenesulfonate	Monosodium benzenesulfonate	$C_6H_5NaO_3S$	515-42-4	180.157		>300				s H_2O; sl EtOH
9548	Sodium benzoate		$C_7H_5NaO_2$	532-32-1	144.104		>300				s H_2O
9549	Sodium cacodylate	Sodium dimethylarsonate	$C_2H_6AsNaO_2$	124-65-2	159.980	gran cry	60 (hyd)				vs H_2O; s EtOH
9550	Sodium 2,2-dichloropropanoate		$C_3H_3Cl_2NaO_2$	127-20-8	164.951	hyg pow	166 dec				
9551	Sodium diethyldithiocarbamate	Dithiocarb sodium	$C_5H_{10}NNaS_2$	148-18-5	171.260	cry (EtOH)	95				s H_2O, EtOH, MeOH, ace; i eth, bz
9552	Sodium diethyldithiocarbamate trihydrate	Diethyldithiocarbamate sodium salt trihydrate	$C_5H_{16}NNaO_3S_2$	20624-25-3	225.306	orth cry (ace)	95				vs H_2O; s EtOH, ace; i bz, eth
9553	Sodium 4,5-dihydroxy-2,7-naphthalenedisulfonic acid	Chromotropic acid disodium salt	$C_{10}H_6Na_2O_8S_2$	129-96-4	364.260	wh nd or lf (w)					vs H_2O
9554	Sodium dimethyldithiocarbamate		$C_3H_7NNaS_2$	128-04-1	144.215	col cry (w)	121 (hyd)				
9555	Sodium 4-dodecylbenzenesulfonate		$C_{18}H_{29}NaO_3S$	2211-98-5	348.476	cry	144				
9556	Sodium dodecyl sulfate	Sodium lauryl sulfate	$C_{12}H_{25}NaO_4S$	151-21-3	288.380	wh pow	205				
9557	Sodium ethanolate	Sodium ethoxide	C_2H_5NaO	141-52-6	68.050	hyg wh pow	260 dec				reac H_2O; s EtOH
9558	Sodium fluoroacetate		$C_2H_2FNaO_2$	62-74-8	100.024	wh mcl cry	200				i ace, chl; sl EtOH, MeOH
9559	Sodium formaldehyde bisulfite	Sodium hydroxymethanesulfonate	CH_3NaO_4S	870-72-4	134.088	cry (EtOH aq)					
9560	Sodium formaldehydesulfoxylate	Sodium hydroxymethanesulfinate	CH_3NaO_3S	149-44-0	118.088	cry (w)	63 (hyd)				s H_2O; i EtOH, bz, eth
9561	Sodium gluconate		$C_6H_{11}NaO_7$	527-07-1	218.137						s H_2O
9562	Sodium 2-hydroxyethanesulfonate	Monosodium 2-hydroxyethanesulfonate	$C_2H_5NaO_4S$	1562-00-1	148.114						s H_2O
9563	Sodium 2-hydroxy-2-propanesulfonate	Monosodium 2-hydroxy-2-propanesulfonate	$C_3H_7NaO_4S$	540-92-1	162.141	cry					s H_2O; sl EtOH
9564	Sodium iodomethanesulfonate	Methiodal sodium	CH_2INaO_3S	126-31-8	243.984	cry					sl EtOH, ace, bz
9565	Sodium *O*-isopropyl xanthate		$C_4H_7NaOS_2$	140-93-2	158.218	hyg wh-ye pow	150 dec				
9566	Sodium methanolate	Sodium methoxide	CH_3NaO	124-41-4	54.024	wh hyg tetr cry	300				reac H_2O; s MeOH, EtOH
9567	Sodium methylarsonate		CH_4AsNaO_3	2163-80-6	161.953	cry (w)	115				vs H_2O; s MeOH; i os
9568	Sodium methyldithiocarbamate	Metham sodium	$C_2H_4NNaS_2$	137-42-8	129.180	cry (w)					vs H_2O
9569	Sodium β-naphthoquinone-4-sulfonate	Sodium 3,4-dihydro-3,4-dioxo-1-naphthalenesulfonate	$C_{10}H_5NaO_5S$	521-24-4	260.199		287 dec				
9570	Sodium 2-oxopropanoate		$C_3H_3NaO_3$	113-24-6	110.044						s H_2O; sl abs EtOH
9571	Sodium phenolate	Sodium phenoxide	C_6H_5NaO	139-02-6	116.093	hyg cry	384				vs H_2O; s EtOH, thf
9572	Sodium propanoate		$C_3H_5NaO_2$	137-40-6	96.061						sl H_2O
9573	Sodium sulfobromophthalein	Sulfobromophthalein sodium	$C_{20}H_8Br_4Na_2O_{10}S_2$	71-67-0	837.998	hyg cry					s H_2O; i EtOH, ace
9574	Sodium tartrate		$C_4H_4Na_2O_6$	868-18-8	194.051						s H_2O
9575	Sodium tartrate dihydrate		$C_4H_8Na_2O_8$	6106-24-7	230.082				1.545^{25}		s H_2O; i EtOH
9576	Sodium tetraphenylborate		$C_{24}H_{20}BNa$	143-66-8	342.217	nd	300				s H_2O, EtOH, ace; sl eth, chl; i peth
9577	Sodium trichloroacetate		$C_2Cl_3NaO_2$	650-51-1	185.369	ye-wh pow	300				s H_2O, EtOH
9578	Sodium trifluoroacetate		$C_2F_3NaO_2$	2923-18-4	136.005	cry	207 dec				
9579	Solanid-5-ene-3,18-diol, (3β)	Isorubijervine	$C_{27}H_{43}NO_2$	468-45-1	413.636	pr(al)	242.5				vs bz, chl
9580	Solanine		$C_{45}H_{73}NO_{15}$	20562-02-1	868.060	nd (EtOH aq)	286 dec				i H_2O, eth, chl; s hot EtOH

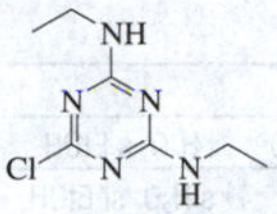
9539
Simazine

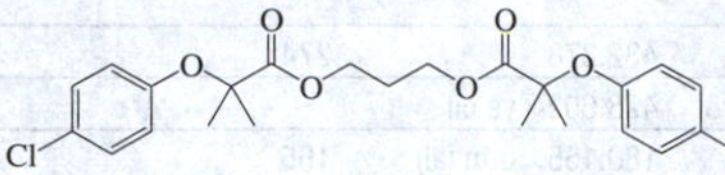
9540
Simfibrate

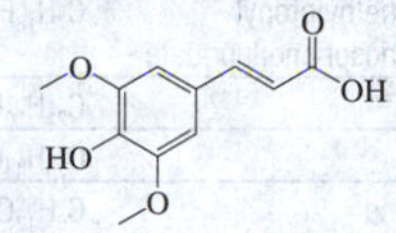
9541
Sinapinic acid

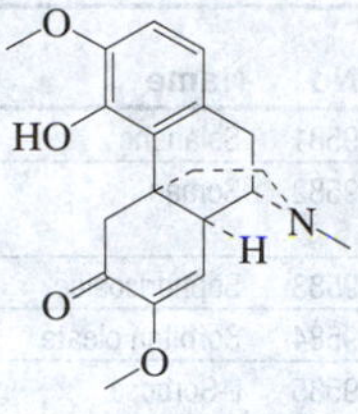
9542
Sinomenine

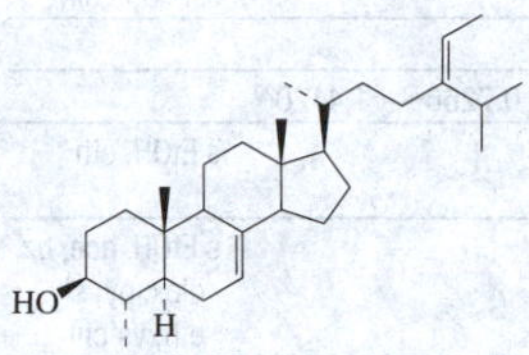
9543
α1-Sitosterol

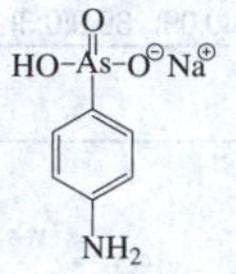
9544
Sodium arsanilate

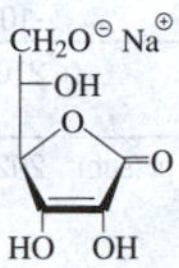
9545
Sodium ascorbate

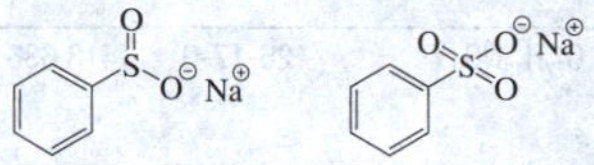
9546
Sodium benzenesulfinate

9547
Sodium benzenesulfonate

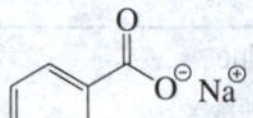
9548
Sodium benzoate

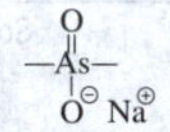
9549
Sodium cacodylate

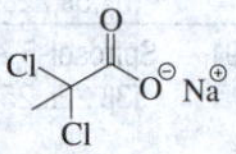
9550
Sodium 2,2-dichloropropanoate

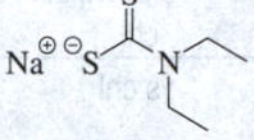
9551
Sodium diethyldithiocarbamate

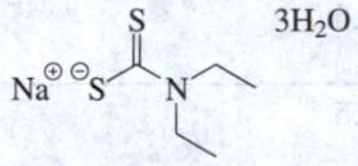
9552
Sodium diethyldithiocarbamate trihydrate

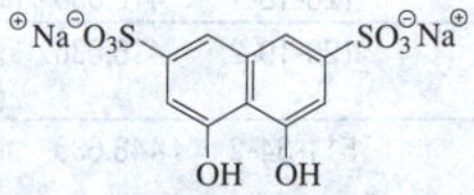
9553
Sodium 4,5-dihydroxy-2,7-naphthalenedisulfonic acid

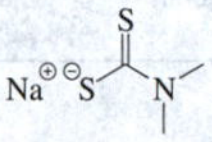
9554
Sodium dimethyldithiocarbamate

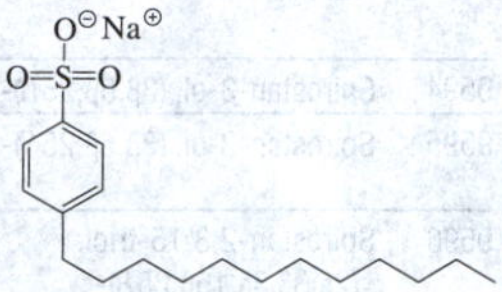
9555
Sodium 4-dodecylbenzenesulfonate

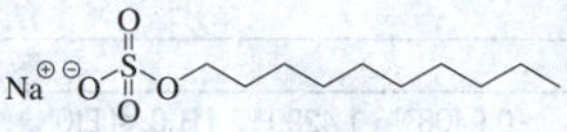
9556
Sodium dodecyl sulfate

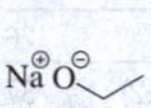
9557
Sodium ethanolate

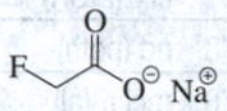
9558
Sodium fluoroacetate

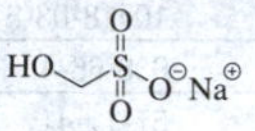
9559
Sodium formaldehyde bisulfite

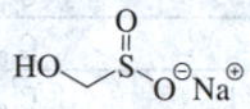
9560
Sodium formaldehydesulfoxylate

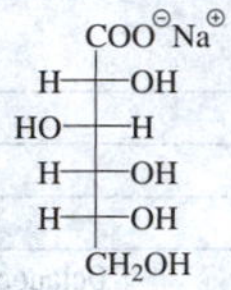
9561
Sodium gluconate

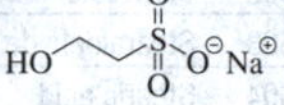
9562
Sodium 2-hydroxyethanesulfonate

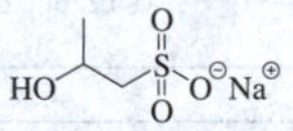
9563
Sodium 2-hydroxy-2-propanesulfonate

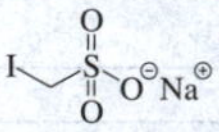
9564
Sodium iodomethanesulfonate

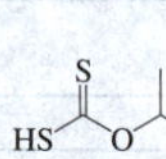
9565
Sodium *O*-isopropyl xanthate

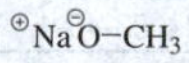
9566
Sodium methanolate

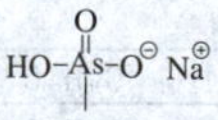
9567
Sodium methylarsonate

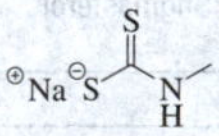
9568
Sodium methyldithiocarbamate

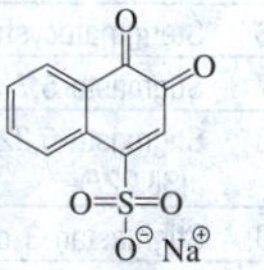
9569
Sodium β-naphthoquinone-4-sulfonate

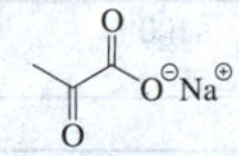
9570
Sodium 2-oxopropanoate

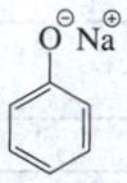
9571
Sodium phenolate

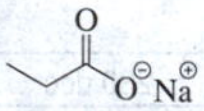
9572
Sodium propanoate

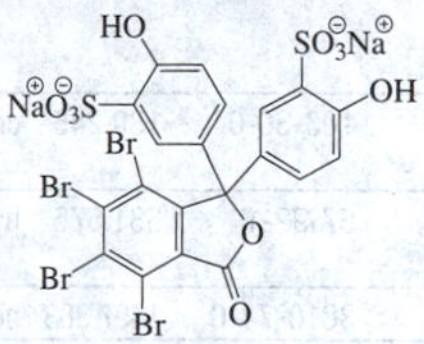
9573
Sodium sulfobromophthalein

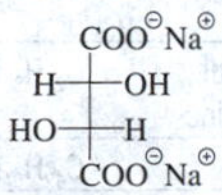
9574
Sodium tartrate

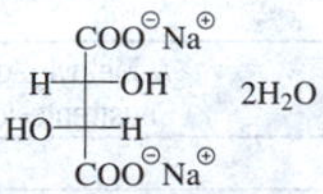
9575
Sodium tartrate dihydrate

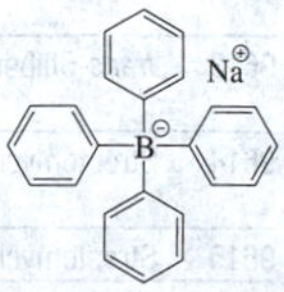
9576
Sodium tetraphenylborate

9577
Sodium trichloroacetate

9578
Sodium trifluoroacetate

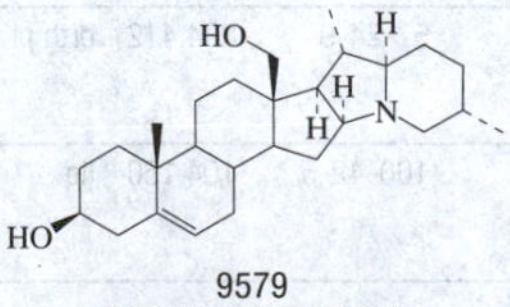
9579
Solanid-5-ene-3,18-diol, (3β)

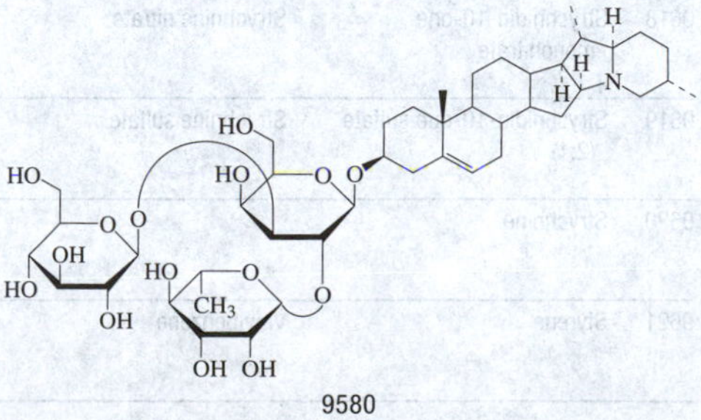
9580
Solanine

No.	Name	Synonym	Mol. form.	CAS Reg. No.	Mol. wt.	Physical form	mp/°C	bp/°C	den/ g cm^{-3}	n_D	Solubility
9581	Solanone		$C_{13}H_{22}O$	1937-54-8	194.313			60^{1}	0.870^{20}	1.4755^{20}	
9582	Soman	1,2,2-Trimethylpropyl methylphosphonofluoridate	$C_7H_{16}FO_2P$	96-64-0	182.173	liq					
9583	Sophoricoside		$C_{21}H_{20}O_{10}$	152-95-4	432.378		274				
9584	Sorbitan oleate		$C_{24}H_{44}O_6$	1338-43-8	428.602	ye oil			0.986	1.4800^{20}	i H_2O; s EtOH
9585	L-Sorbose	L-Sorbinose	$C_6H_{12}O_6$	87-79-6	180.155	orth (al)	165		1.612^{17}		s H_2O; sl EtOH, eth, MeOH
9586	Sparteine		$C_{15}H_{26}N_2$	90-39-1	234.380		30.5	325	1.0196^{20}	1.5312^{20}	vs eth, EtOH, chl
9587	Spinulosin		$C_8H_8O_5$	85-23-4	184.147	red-bl	202.5	120 sub			sl H_2O; s alk
9588	Spironolactone		$C_{24}H_{32}O_4S$	52-01-7	416.574			134			
9589	Spiro[2.2]pentane		C_5H_8	157-40-4	68.118	liq	-106.98(0.09)	39.0(0.3)	0.7266^{20}	1.4120^{20}	
9590	Spirosolan-3-ol, (3β,5α,22β,25*S*)-	Tomatidine	$C_{27}H_{45}NO_2$	77-59-8	415.652	pl	210.5				s EtOH, eth
9591	Spirosol-5-en-3-ol, (3β,22α,25*R*)-	Solasodine	$C_{27}H_{43}NO_2$	126-17-0	413.636	hex pl (sub)	202				s EtOH, ace, bz, diox, py; sl eth; vs chl
9592	Spirostan-2,3-diol, (2α,3β,5α,25*R*)-	Gitogenin	$C_{27}H_{44}O_4$	511-96-6	432.636	lf (bz), nd (eth)	271.5				i H_2O; s EtOH, chl; sl eth
9593	Spirostan-3-ol, (3β,5α,25*R*)-	Tigogenin	$C_{27}H_{44}O_3$	77-60-1	416.636	lf (al +1w) pr (ace)	205(5)				s EtOH, eth, ace, ctc, MeOH, peth
9594	Spirostan-3-ol, (3β,5β,25*R*)-	Smilagenin	$C_{27}H_{44}O_3$	126-18-1	416.636	nd (ace)	185				vs ace, bz, EtOH
9595	Spirostan-3-ol, (3β,5β,25*S*)-	Sarsasapogenin	$C_{27}H_{44}O_3$	126-19-2	416.636	lo pr, nd (ace)	200.5				s EtOH, ace, bz, chl
9596	Spirostan-2,3,15-triol, (2α,3β,5α,15β,25*R*)-	Digitogenin	$C_{27}H_{44}O_5$	511-34-2	448.635	nd (al)	281.5				vs chl
9597	Spirost-5-en-3-ol, (3β,25*R*)-	Diosgenin	$C_{27}H_{42}O_3$	512-04-9	414.620	cry (ace)	205.5				vs EtOH
9598	Spiro[5.5]undecane		$C_{11}H_{20}$	180-43-8	152.277			214(4)	0.8783^{20}	1.4731	
9599	S-Propyl thioacetate		$C_5H_{10}OS$	2307-10-0	118.197			138(7)	0.9535^{25}		
9600	Squalene		$C_{30}H_{50}$	111-02-4	410.718	oil	-4.8	421.3	0.8584^{20}	1.4990^{20}	i H_2O; sl EtOH; s eth, ace, ctc
9601	Stachydrine		$C_7H_{13}NO_2$	471-87-4	143.184	cry (w+1)	235				vs H_2O, EtOH
9602	Stanozolol		$C_{21}H_{32}N_2O$	10418-03-8	328.491	cry (EtOH)	≈236				
9603	Stearaldehyde		$C_{18}H_{36}O$	638-66-4	268.478	nd (peth)		261			
9604	Stearic acid	Octadecanoic acid	$C_{18}H_{36}O_2$	57-11-4	284.478	mcl lf (al)	69.3(0.2)	371(3)	0.9408^{20}	1.4299^{80}	i H_2O; sl EtOH, bz; s ace, chl, CS_2
9605	Stearic acid anhydride	Octadecanoic anhydride	$C_{36}H_{70}O_3$	638-08-4	550.939		72		0.8365^{82}	1.4362^{80}	i H_2O, EtOH; sl eth, bz
9606	Sterigmatocystin		$C_{18}H_{12}O_6$	10048-13-2	324.284	ye nd	246 dec				
9607	Stigmasta-5,7-dien-3-ol, (3β)	7-Dehydrositosterol	$C_{29}H_{48}O$	521-04-0	412.690		144.5				vs bz, eth, EtOH
9608	Stigmasta-5,22-dien-3-ol, (3β,22*E*)-	Stigmasterol	$C_{29}H_{48}O$	83-48-7	412.690		162(1)				vs bz, eth, EtOH
9609	Stigmastan-3-ol, (3β,5α)		$C_{29}H_{52}O$	83-45-4	416.722		144				
9610	Stigmast-5-en-3-ol, (3β,24*R*)-	β-Sitosterol	$C_{29}H_{50}O$	83-46-5	414.706	pl (al)	138(1)				s EtOH, eth, HOAc
9611	Stigmast-5-en-3-ol, (3β,24*S*)-	γ-Sitosterol	$C_{29}H_{50}O$	83-47-6	414.706	cry (EtOH)	148				s EtOH
9612	*cis*-Stilbene	*cis*-1,2-Diphenylethene	$C_{14}H_{12}$	645-49-8	180.245		-5	141^{12}	1.0143^{20}	1.6130^{20}	i H_2O; s EtOH, eth, ace, bz, peth, chl
9613	*trans*-Stilbene	*trans*-1,2-Diphenylethene	$C_{14}H_{12}$	103-30-0	180.245	cry (al)	124.82(0.02)	307	0.9707^{20}	1.6264^{17}	i H_2O; sl EtOH, chl; vs eth, bz
9614	Streptomycin	*N*-Methyl-L-glucosamidi-nostreptosidostreptidine	$C_{21}H_{39}N_7O_{12}$	57-92-1	581.575	hyg pow					s H_2O
9615	Streptomycin sulfate		$C_{42}H_{84}N_{14}O_{36}S_3$	3810-74-0	1457.383	pow	≈230 dec				
9616	Streptozotocin		$C_8H_{15}N_3O_7$	18883-66-4	265.221	pl	115 dec				s H_2O, EtOH
9617	Strophanthidin		$C_{23}H_{32}O_6$	66-28-4	404.496	orth tab (MeOH-w) lf (w+2)	173 dec				i H_2O, eth; s EtOH, ace, bz, HOAc, chl
9618	Strychnidin-10-one mononitrate	Strychnine nitrate	$C_{21}H_{23}N_3O_5$	66-32-0	397.425	nd (w)	295		1.627^{25}		vs H_2O, MeOH; sl bz; s chl, EtOH
9619	Strychnidin-10-one sulfate (2:1)	Strychnine sulfate	$C_{42}H_{46}N_4O_8S$	60-41-3	766.901		200 dec				s H_2O, EtOH, MeOH; i eth; sl chl
9620	Strychnine		$C_{21}H_{22}N_2O_2$	57-24-9	334.412	orth pr (al)	284(1)	270^{5}	1.36^{20}		sl H_2O, EtOH, ace, bz; i eth; s chl
9621	Styrene	Vinylbenzene	C_8H_8	100-42-5	104.150	liq	-30.65(0.06)	145.3(0.6)	0.9016^{25}	1.5440^{25}	i H_2O; s EtOH, eth, ace; msc bz; sl ctc

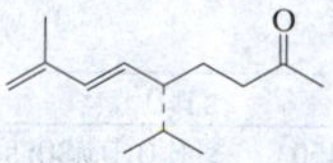
9581
Solanone

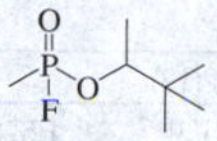
9582
Soman

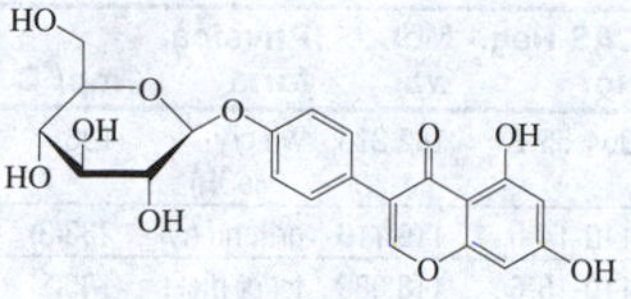
9583
Sophoricoside

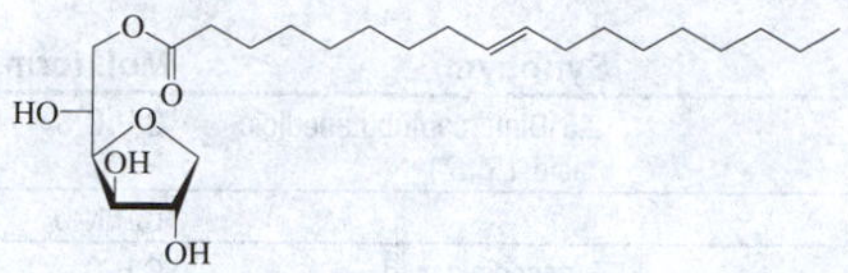
9584
Sorbitan oleate

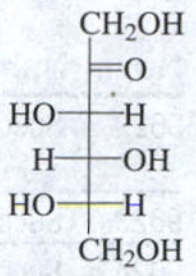
9585
L-Sorbose

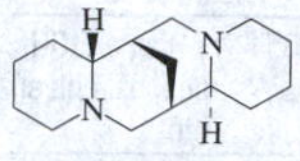
9586
Sparteine

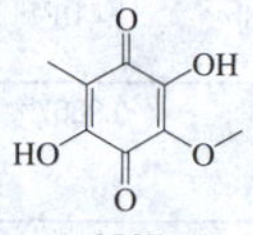
9587
Spinulosin

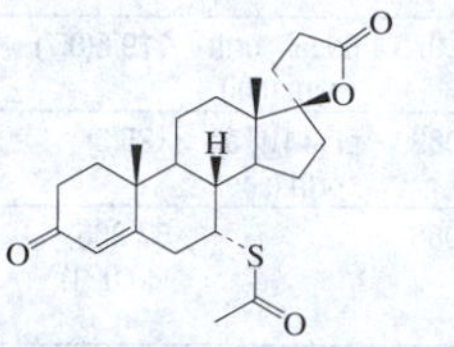
9588
Spironolactone

9589
Spiro[2.2]pentane

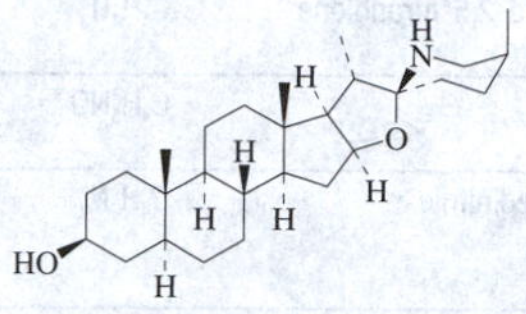
9590
Spirosolan-3-ol, (3β,5α,22β,25*S*)-

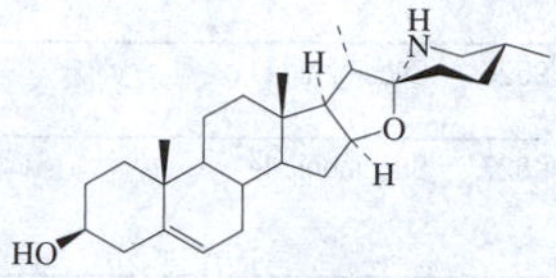
9591
Spirosol-5-en-3-ol, (3β,22α,25*R*)-

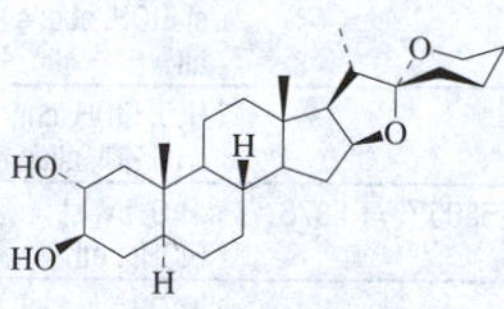
9592
Spirostan-2,3-diol, (2α,3β,5α,25*R*)-

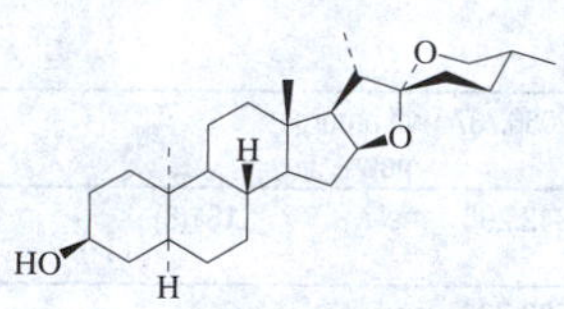
9593
Spirostan-3-ol, (3β,5α,25*R*)-

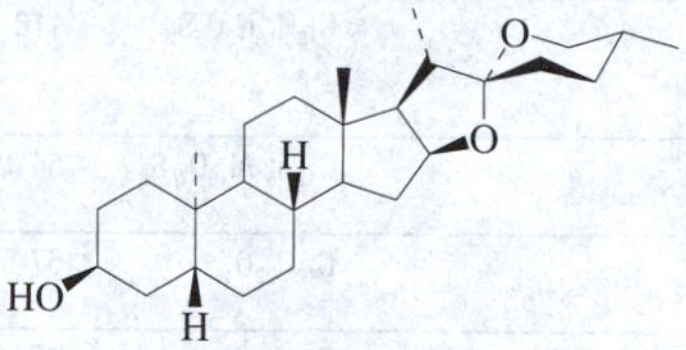
9594
Spirostan-3-ol, (3β,5β,25*R*)-

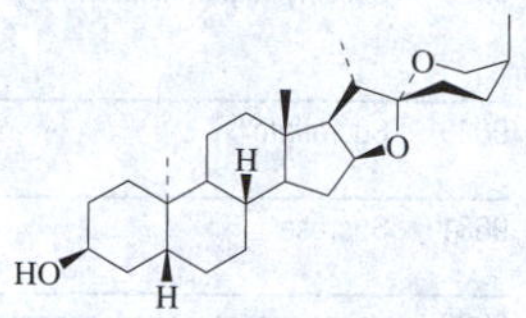
9595
Spirostan-3-ol, (3β,5β,25*S*)-

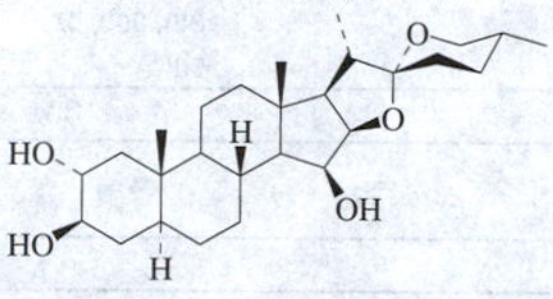
9596
Spirostan-2,3,15-triol, (2α,3β,5α,15β,25*R*)-

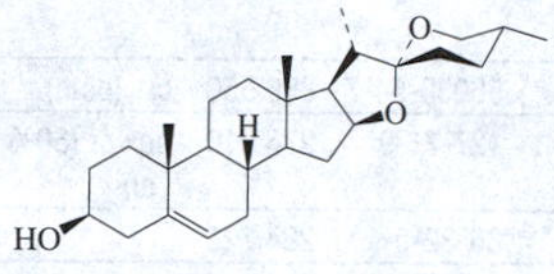
9597
Spirost-5-en-3-ol, (3β,25*R*)-

9598
Spiro[5.5]undecane

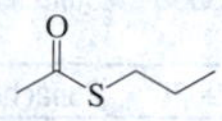
9599
S-Propyl thioacetate

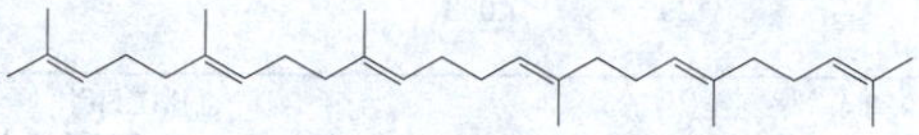
9600
Squalene

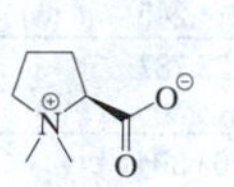
9601
Stachydrine

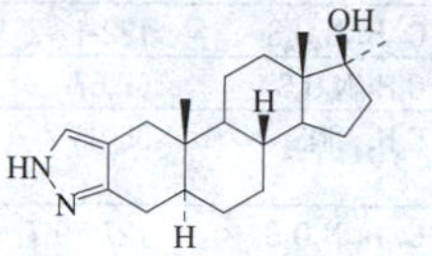
9602
Stanozolol

9603
Stearaldehyde

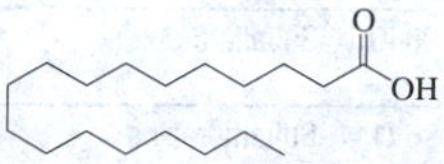
9604
Stearic acid

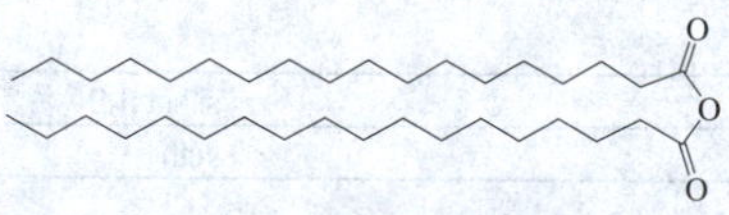
9605
Stearic acid anhydride

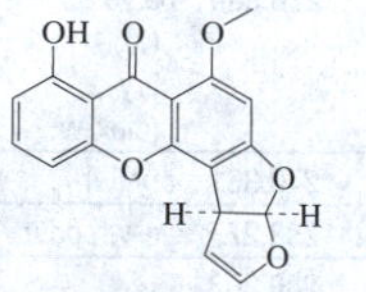
9606
Sterigmatocystin

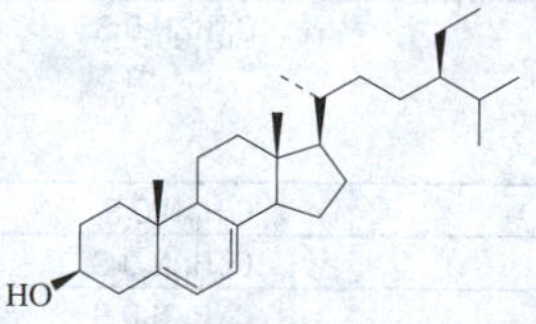
9607
Stigmasta-5,7-dien-3-ol, (3β)

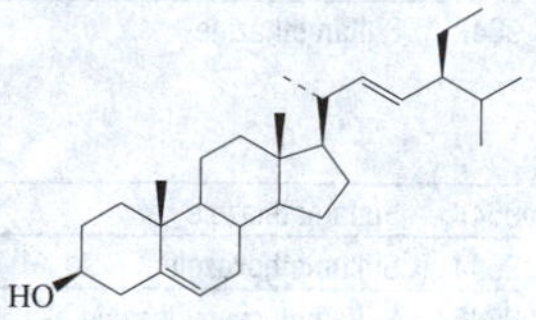
9608
Stigmasta-5,22-dien-3-ol, (3β,22*E*)-

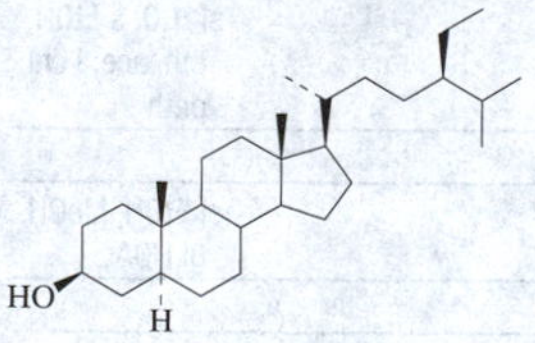
9609
Stigmastan-3-ol, (3β,5α)

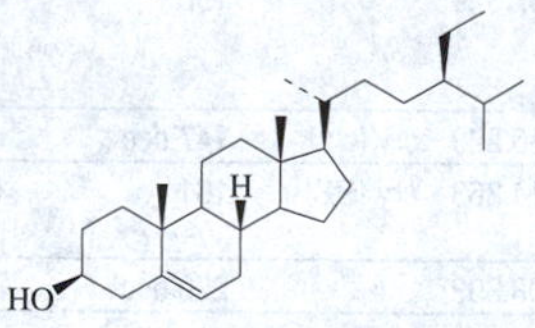
9610
Stigmast-5-en-3-ol, (3β,24*R*)-

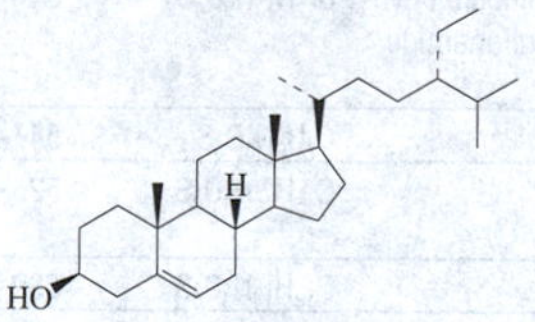
9611
Stigmast-5-en-3-ol, (3β,24*S*)-

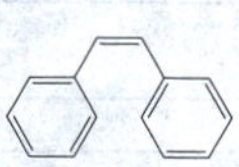
9612
cis-Stilbene

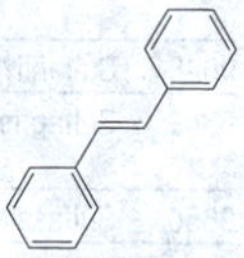
9613
trans-Stilbene

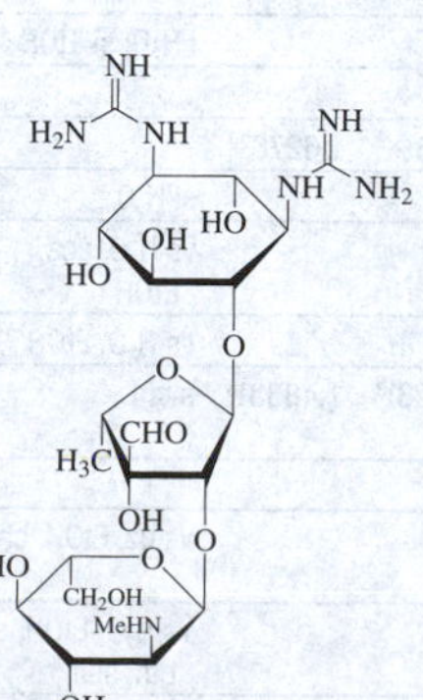
9614
Streptomycin

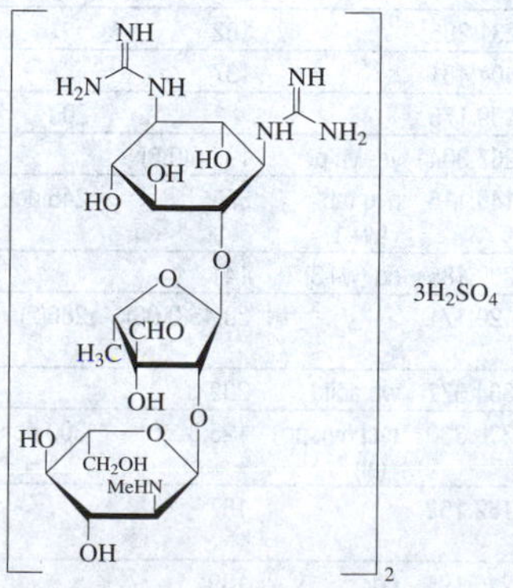
9615
Streptomycin sulfate

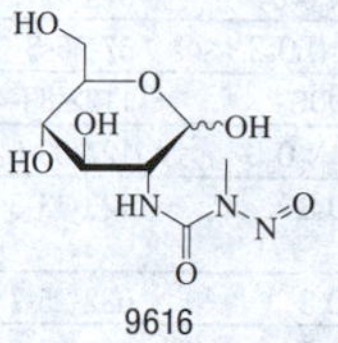
9616
Streptozotocin

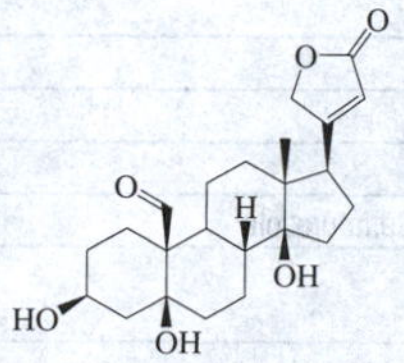
9617
Strophanthidin

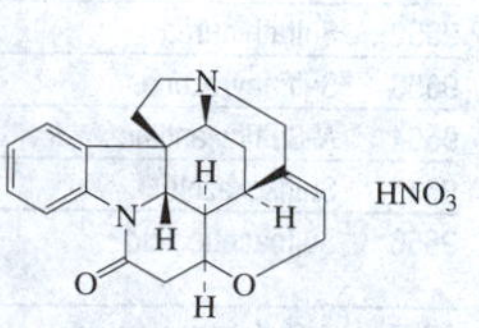
9618
Strychnidin-10-one mononitrate

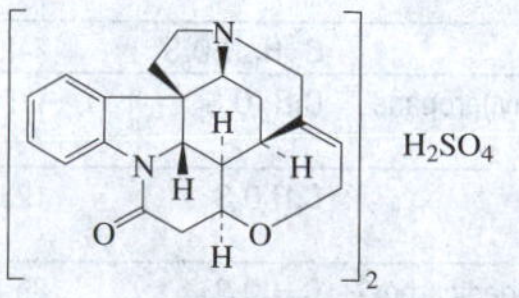
9619
Strychnidin-10-one sulfate (2:1)

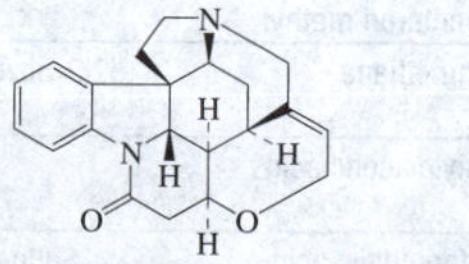
9620
Strychnine

9621
Styrene

Organic

No.	Name	Synonym	Mol. form.	CAS Reg. No.	Mol. wt.	Physical form	mp/°C	bp/°C	den/ g cm^{-3}	n_D	Solubility
9622	Succimer	2,3-Dimercaptobutanedioic acid, (*R**,*S**)	$C_4H_6O_4S_2$	304-55-2	182.219	wh cry (MeOH)	193				
9623	Succinamide		$C_4H_8N_2O_2$	110-14-5	116.119	orth nd (w)	263(3)	432(13)			s H_2O
9624	Succinic acid	Butanedioic acid	$C_4H_6O_4$	110-15-6	118.089	tcl or mcl pr	185(3)	234(3)	1.572^{25}	1.450	sl H_2O, DMSO; s EtOH, eth, ace; i bz
9625	Succinic anhydride	Dihydro-2,5-furandione	$C_4H_4O_3$	108-30-5	100.073	nd (al), orth pym (chl)	119.5(0.7)	261	1.2^{20}		i H_2O; s EtOH, chl; sl eth
9626	Succinimide		$C_4H_5NO_2$	123-56-8	99.089	pl (+1w, al) orth (ace)	125(3)	296(3)	1.418^{25}		s H_2O; sl EtOH, eth, ace
9627	Succinonitrile	Butanedinitrile	$C_4H_4N_2$	110-61-2	80.088		57.985 (0.001)	266	0.9867^{60}	1.4173^{60}	vs H_2O; s EtOH, ace, bz, chl; sl eth
9628	Succinylcholine chloride	Suxamethonium chloride	$C_{14}H_{30}Cl_2N_2O_4$	71-27-2	361.305	cry (w)	190				sl EtOH, bz, chl; i eth
9629	Succinylsulphathiazole		$C_{13}H_{13}N_3O_5S_2$	116-43-8	355.389	cry	193.5				i H_2O, eth, chl; sl EtOH, ace; s alk
9630	Sucralfate		$C_{12}H_{54}Al_{16}O_{75}S_8$	54182-58-0	2086.737	wh amorp pow					i H_2O, EtOH, chl; s dil HCl, alk
9631	Sucrose		$C_{12}H_{22}O_{11}$	57-50-1	342.296	mcl	181(8)		1.5805^{17}	1.5376	s H_2O, py; sl EtOH; i eth
9632	Sucrose monohexadecanoate	Sucrose palmitate	$C_{28}H_{52}O_{12}$	26446-38-8	580.706	cry	61				s H_2O
9633	Sucrose octaacetate		$C_{28}H_{38}O_{19}$	126-14-7	678.591	nd (al)	87(4)	250^{1}	1.27^{16}	1.4660	sl H_2O; s EtOH, eth, ace, bz, chl
9634	Sufentanil		$C_{22}H_{30}N_2O_2S$	56030-54-7	386.550	cry (peth)	96.6				
9635	Sulfabenzamide	*N*-[(4-Aminophenyl)sulfonyl]-benzamide	$C_{13}H_{12}N_2O_3S$	127-71-9	276.310	hex pr (60% al)	181.5				
9636	Sulfachlorpyridazine		$C_{10}H_9ClN_4O_2S$	80-32-0	284.722		187				
9637	Sulfacytine		$C_{12}H_{14}N_4O_3S$	17784-12-2	294.329	cry (MeOH/ BuOH)	167				i H_2O; s alk
9638	Sulfadimethoxine		$C_{12}H_{14}N_4O_4S$	122-11-2	310.329		203.5				
9639	Sulfaguanidine		$C_7H_{10}N_4O_2S$	57-67-0	214.245	nd (w)	191.5				
9640	Sulfallate	Carbamodithioic acid, diethyl-, 2-chloro-2-propenyl ester	$C_8H_{14}ClNS_2$	95-06-7	223.787			129^{1}	1.088		
9641	Sulfamerazine		$C_{11}H_{12}N_4O_2S$	127-79-7	264.304	cry	238(4)				sl H_2O, EtOH, ace, DMSO; i eth, chl
9642	Sulfamethazine		$C_{12}H_{14}N_4O_2S$	57-68-1	278.330	pa ye (w+1/2) cry (diox-w)	197(2)				s H_2O, acid, alk; sl DMSO
9643	Sulfamethiazole		$C_9H_{10}N_4O_2S_2$	144-82-1	270.331	cry (w)	210				sl hot H_2O
9644	Sulfamethoxazole		$C_{10}H_{11}N_3O_3S$	723-46-6	253.277	ye-wh pow	167(1)				i eth
9645	Sulfamethoxypyridazine		$C_{11}H_{12}N_4O_3S$	80-35-3	280.303		182.5				
9646	Sulfamethylthiazole		$C_{10}H_{11}N_3O_2S_2$	515-59-3	269.343		237				vs EtOH
9647	N^4-Sulfanilylsulfanilamide	4-Amino-*N*-[4-(aminosulfonyl)-phenyl]benzenesulfonamide	$C_{12}H_{13}N_3O_4S_2$	547-52-4	327.379		137				sl H_2O; s EtOH, eth, ace; i chl, peth
9648	Sulfanilylurea		$C_7H_9N_3O_3S$	547-44-4	215.229	cry (w)	147 dec				
9649	Sulfaphenazole		$C_{15}H_{14}N_4O_2S$	526-08-9	314.363	cry (EtOH)	181				sl EtOH, MeOH, gl HOAc
9650	Sulfasalazine		$C_{18}H_{14}N_4O_5S$	599-79-1	398.393		220 dec				
9651	Sulfathiazole	4-Amino-*N*-2-thiazolylbenzenesulfonamide	$C_9H_9N_3O_2S_2$	72-14-0	255.316	br pl, rods or pow (45% al)	200(1)				sl H_2O, EtOH, DMSO
9652	Sulfathiourea		$C_7H_9N_3O_2S_2$	515-49-1	231.295		182				i H_2O; sl EtOH
9653	Sulfinpyrazone		$C_{23}H_{20}N_2O_3S$	57-96-5	404.481		137				
9654	*N*-Sulfinylaniline		C_6H_5NOS	1122-83-4	139.175			200	1.236^{25}	1.6270^{20}	
9655	Sulfisoxazole	Sulfafurazole	$C_{11}H_{13}N_3O_3S$	127-69-5	267.304	ye-wh pr	195.0(0.5)				sl H_2O
9656	Sulfoacetic acid		$C_2H_4O_5S$	123-43-3	140.115	hyg tab (w+1)	85	245 dec			vs H_2O, ace, EtOH
9657	2-Sulfobenzoic acid		$C_7H_6O_5S$	632-25-7	202.185	nd (w+3)	141				vs H_2O, EtOH
9658	Sulfolane	Tetrahydrothiophene, 1-1-dioxide	$C_4H_8O_2S$	126-33-0	120.171		28.45(0.08)	286(2)	1.2723^{18}	1.4833^{18}	s chl
9659	Sulfometuron methyl		$C_{15}H_{16}N_4O_5S$	74222-97-2	364.377	wh solid	202				
9660	Sulfonmethane	2,2-Bis(ethylsulfonyl)propane	$C_7H_{16}O_4S_2$	115-24-2	228.330	mcl (w), pr (al)	125.8	300 dec			vs bz, EtOH, chl
9661	Sulfonyldiacetic acid		$C_4H_6O_6S$	123-45-5	182.152		187				vs H_2O, EtOH; s eth, sulf
9662	4-Sulfophthalic acid	4-Sulfo-1,2-benzenedicarboxylic acid	$C_8H_6O_7S$	89-08-7	246.195	cry	139				

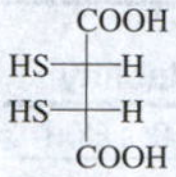
9622
Succimer

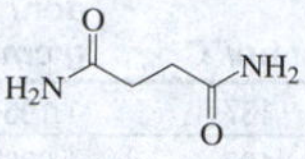
9623
Succinamide

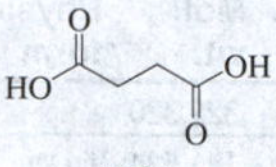
9624
Succinic acid

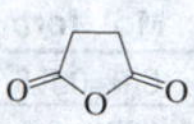
9625
Succinic anhydride

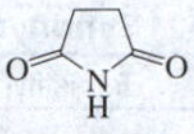
9626
Succinimide

9627
Succinonitrile

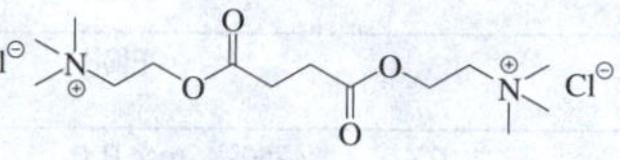
9628
Succinylcholine chloride

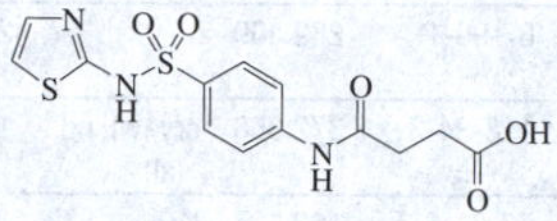
9629
Succinylsulphathiazole

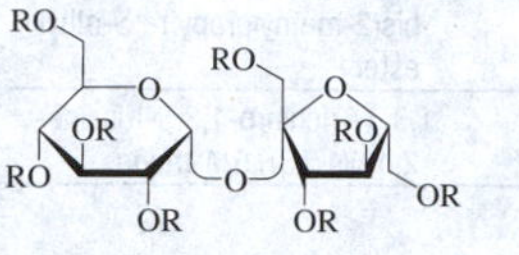
R = $SO_3[Al_2(OH)_5]$
9630
Sucralfate

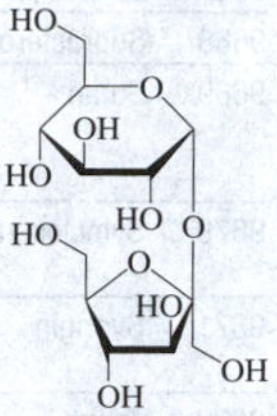
9631
Sucrose

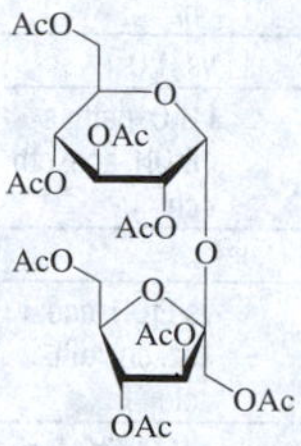
9633
Sucrose octaacetate

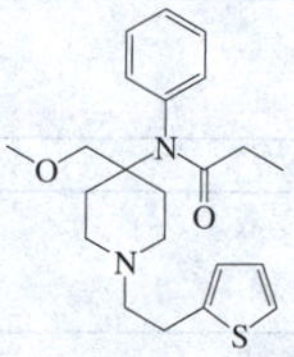
9634
Sufentanil

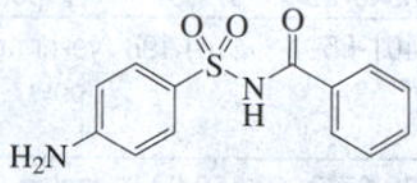
9635
Sulfabenzamide

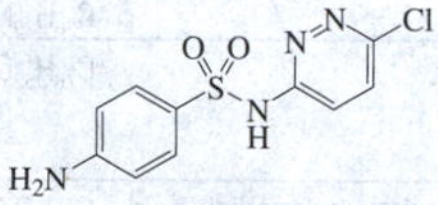
9636
Sulfachlorpyridazine

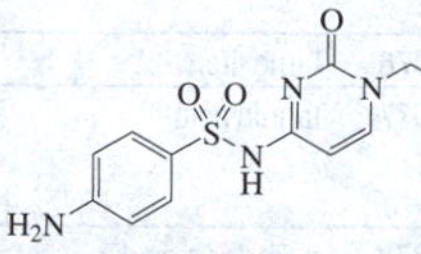
9637
Sulfacytine

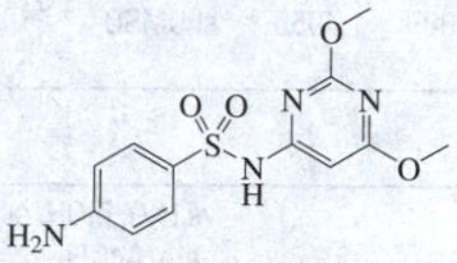
9638
Sulfadimethoxine

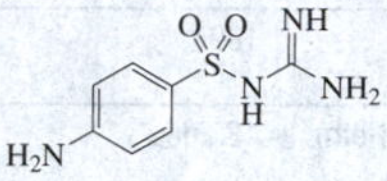
9639
Sulfaguanidine

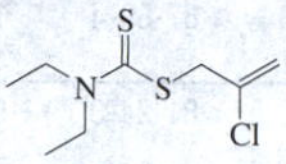
9640
Sulfallate

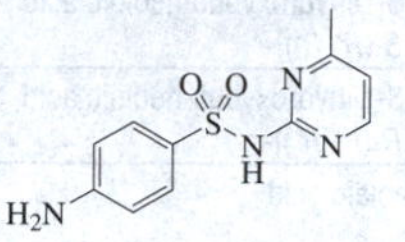
9641
Sulfamerazine

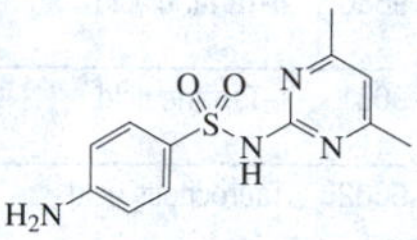
9642
Sulfamethazine

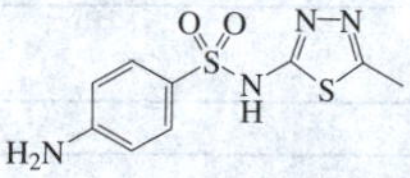
9643
Sulfamethiazole

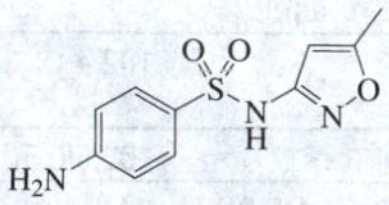
9644
Sulfamethoxazole

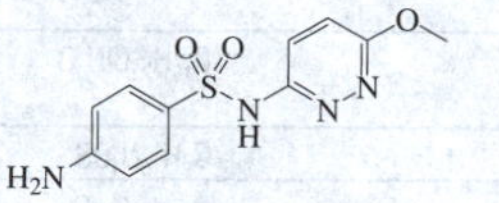
9645
Sulfamethoxypyridazine

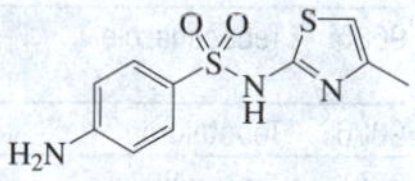
9646
Sulfamethylthiazole

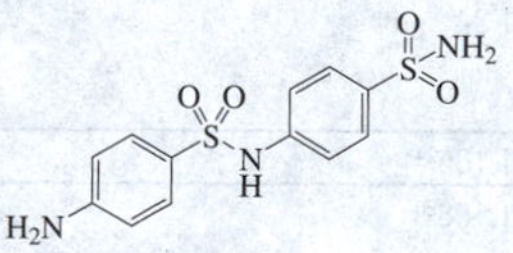
9647
N^4-Sulfanilylsulfanilamide

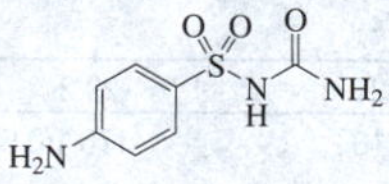
9648
Sulfanilylurea

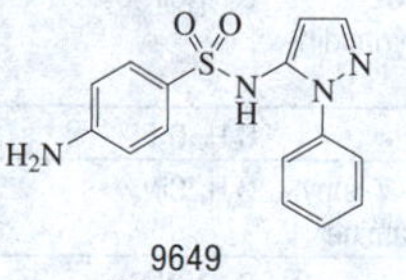
9649
Sulfaphenazole

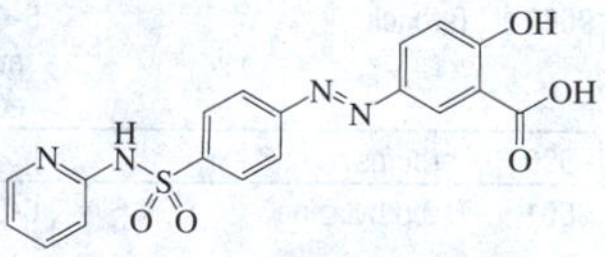
9650
Sulfasalazine

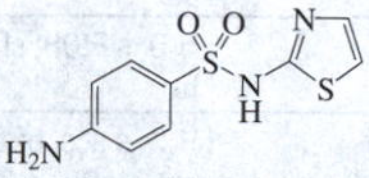
9651
Sulfathiazole

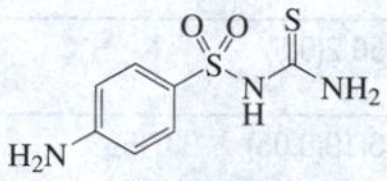
9652
Sulfathiourea

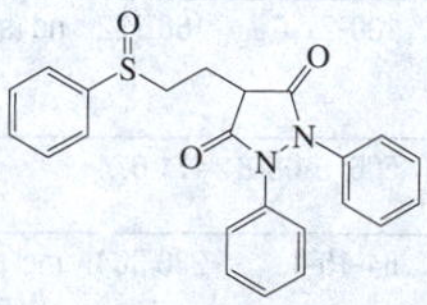
9653
Sulfinpyrazone

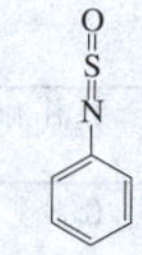
9654
N-Sulfinylaniline

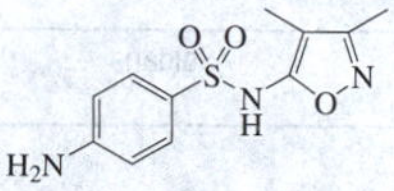
9655
Sulfisoxazole

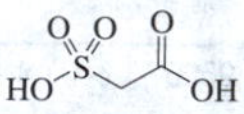
9656
Sulfoacetic acid

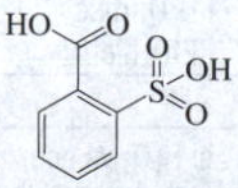
9657
2-Sulfobenzoic acid

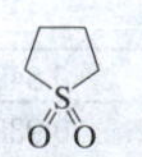
9658
Sulfolane

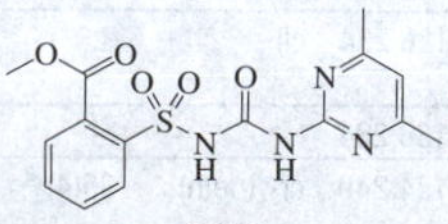
9659
Sulfometuron methyl

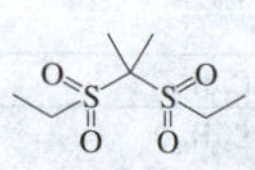
9660
Sulfonmethane

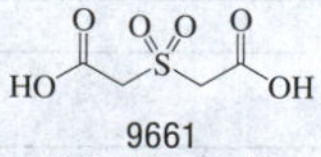
9661
Sulfonyldiacetic acid

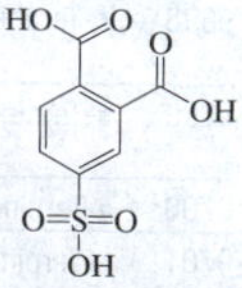
9662
4-Sulfophthalic acid

Organic

No.	Name	Synonym	Mol. form.	CAS Reg. No.	Mol. wt.	Physical form	mp/°C	bp/°C	den/ g cm⁻³	n_D	Solubility
9663	Sulfotep	Tetraethyl thiodiphosphate	$C_8H_{20}O_5P_2S_2$	3689-24-5	322.320			137^{2}	1.196^{25}	1.4753^{25}	i H_2O; s EtOH
9664	Sulfuryl chloride isocyanate		$CClNO_3S$	1189-71-5	141.534	liq	-44	107	1.626^{25}	1.4467^{20}	
9665	Sulphan Blue		$C_{27}H_{31}N_2NaO_6S_2$	129-17-9	566.664	viol pow					s EtOH
9666	Sulprofos		$C_{12}H_{19}O_2PS_3$	35400-43-2	322.447			$156^{0.1}$	1.20^{20}	1.5859	sl H_2O
9667	Sunset Yellow FCF	C.I. Food Yellow 3	$C_{16}H_{10}N_2Na_2O_7S_2$	2783-94-0	452.369	cry	>300				s H_2O; sl EtOH
9668	Suprasterol II		$C_{28}H_{44}O$	562-71-0	396.648	pr	110	$190^{0.005}$			s MeOH
9669	Sutan	Carbamothioic acid, bis(2-methylpropyl)-, *S*-ethyl ester	$C_{11}H_{23}NOS$	2008-41-5	217.372			138^{21}	0.9402^{25}		
9670	Symclosene	1,3,5-Trichloro-1,3,5-triazine-2,4,6(1*H*,3*H*,5*H*)-trione	$C_3Cl_3N_3O_3$	87-90-1	232.409		246.7 dec				
9671	Syringin		$C_{17}H_{24}O_9$	118-34-3	372.368	cry (w), nd (al)	192				vs EtOH
9672	Tabun	Dimethylphosphoroamidocyanidic acid, ethyl ester	$C_5H_{11}N_2O_2P$	77-81-6	162.127	liq	-50	240	1.077	1.4250^{20}	msc H_2O
9673	Tachysterol	9,10-Secoergosta-5(10),6,8,22-tetraen-3-ol, (3β,6*E*,22*E*)-	$C_{28}H_{44}O$	115-61-7	396.648						i H_2O, MeOH; s EtOH, eth, ace, bz
9674	D-Tagatose		$C_6H_{12}O_6$	87-81-0	180.155	cry (dil al)	134.5				vs H_2O
9675	Talbutal		$C_{11}H_{16}N_2O_3$	115-44-6	224.256	cry	109				i H_2O, peth; s EtOH, ace, eth, chl
9676	Tamoxifen		$C_{26}H_{29}NO$	10540-29-1	371.514	cry (peth)	97				
9677	Tannic acid	Tannin	$C_{76}H_{52}O_{46}$	1401-55-4	1701.198	ye-br amorp pow	≈210 dec				vs EtOH, ace; i bz, chl, eth, ctc
9678	DL-Tartaric acid	2,3-Dihydroxybutanedioic acid, (*R**,*R**)-(±)-	$C_4H_6O_6$	133-37-9	150.087	mcl pr (w, al +1w)	206		1.788^{25}		s H_2O, EtOH; sl eth; i bz
9679	*meso*-Tartaric acid	2,3-Dihydroxybutanedioic acid, (*R**,*R**)-	$C_4H_6O_6$	147-73-9	150.087	tcl pl (w)	147		1.666^{20}		vs H_2O, EtOH
9680	D-Tartaric acid	2,3-Dihydroxybutanedioic acid, [*S*-(*R**,*R**)]-	$C_4H_6O_6$	147-71-7	150.087	mcl, orth pr (w+1)	172.5		1.7598^{20}	1.4955^{20}	sl DMSO
9681	L-Tartaric acid	2,3-Dihydroxybutanedioic acid, [*R*-(*R**,*R**)]-	$C_4H_6O_6$	87-69-4	150.087		169				
9682	Taurocholic acid	Cholaic acid	$C_{26}H_{45}NO_7S$	81-24-3	515.703	pr (al-eth)	125 dec				vs H_2O, EtOH; sl eth, AcOEt
9683	Taxine A		$C_{35}H_{47}NO_{10}$	1361-49-5	641.749	cry (ace)	205				i H_2O; s EtOH, eth, chl
9684	Taxol	Paclitaxel	$C_{47}H_{51}NO_{14}$	33069-62-4	853.907	nd (MeOH aq)	215(4)				
9685	Tebuconazole		$C_{16}H_{22}ClN_3O$	107534-96-3	308.826		102.4				
9686	Tebuthiuron		$C_9H_{16}N_4OS$	34014-18-1	228.314		163.5(0.5)				
9687	Teniposide		$C_{32}H_{32}O_{13}S$	29767-20-2	656.653	cry (EtOH)	244				
9688	Tephrosin		$C_{23}H_{22}O_7$	76-80-2	410.417	pr (chl-MeOH)	198				vs ace, eth, chl
9689	Terbacil	5-Chloro-3-tert-butyl-6-methyl-2,4(1*H*,3*H*)-pyrimidinedione	$C_9H_{13}ClN_2O_2$	5902-51-2	216.664		176.3(0.5)	175 sub	1.34^{25}		
9690	Terbufos		$C_9H_{21}O_2PS_3$	13071-79-9	288.431		-29.2	$69^{0.01}$	1.105^{24}		
9691	Terbuthylazine	6-Chloro-*N*-*tert*-butyl-*N*'-ethyl-1,3,5-triazine-2,4-diamine	$C_9H_{16}ClN_5$	5915-41-3	229.710		176.1(0.5)		1.188^{20}		
9692	Terbutryn		$C_{10}H_{19}N_5S$	886-50-0	241.357		103.8(0.5)	$157^{0.06}$	1.115^{20}		
9693	Terebic acid	Tetrahydro-2,2-dimethyl-5-oxo-3-furancarboxylic acid	$C_7H_{10}O_4$	79-91-4	158.152	cry	175		0.815		sl H_2O; s EtOH
9694	Terephthalic acid	1,4-Benzenedicarboxylic acid	$C_8H_6O_4$	100-21-0	166.132	nd (sub)		300 sub	1.519^{25}		i H_2O, EtOH, eth, chl, HOAc; sl ctc
9695	Terfenadine	Seldane	$C_{32}H_{41}NO_2$	50679-08-8	471.674		150.2(0.5)				i H_2O; s EtOH; sl hx
9696	*o*-Terphenyl		$C_{18}H_{14}$	84-15-1	230.304	mcl pr (MeOH)	56.19(0.05)	337(5)	1.12^{25}		i H_2O; s ace, bz, chl, MeOH
9697	*m*-Terphenyl		$C_{18}H_{14}$	92-06-8	230.304	ye nd (al)	86.9(0.8)	375(1)	1.199^{20}		i H_2O; s EtOH, eth, bz, HOAc; sl chl
9698	*p*-Terphenyl		$C_{18}H_{14}$	92-94-4	230.304		213.8(0.7)	376	1.28^{25}		i H_2O; sl EtOH; s eth, bz, CS_2
9699	α-Terpinene	4-Isopropyl-1-methyl-1,3-cyclohexadiene	$C_{10}H_{16}$	99-86-5	136.234	oil		174(4)	0.8375^{19}	1.477^{19}	i H_2O; msc EtOH, eth
9700	γ-Terpinene		$C_{10}H_{16}$	99-85-4	136.234			183(5)	0.849^{20}	1.4765^{14}	
9701	α-Terpineol		$C_{10}H_{18}O$	2438-12-2	154.249	cry (peth)	35(4)	218(11)	0.9337^{20}	1.4831^{20}	sl H_2O; vs ace, bz, eth, EtOH

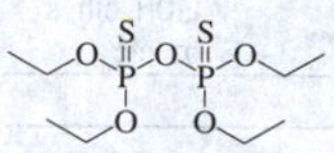
9663
Sulfotep

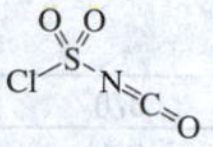
9664
Sulfuryl chloride isocyanate

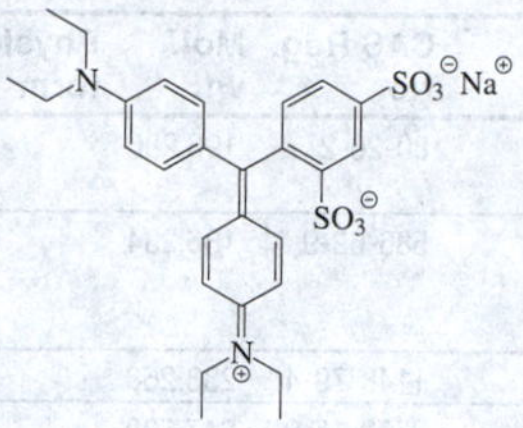
9665
Sulphan Blue

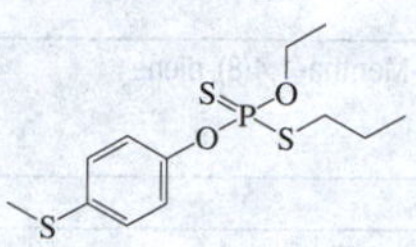
9666
Sulprofos

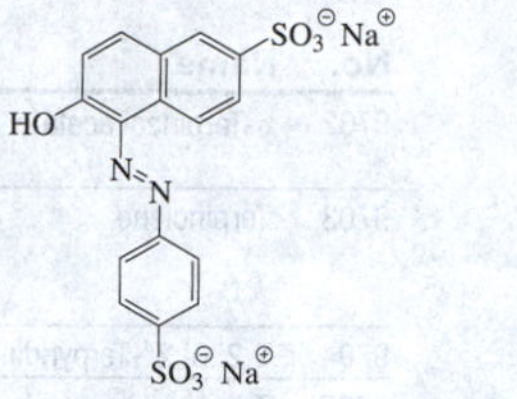
9667
Sunset Yellow FCF

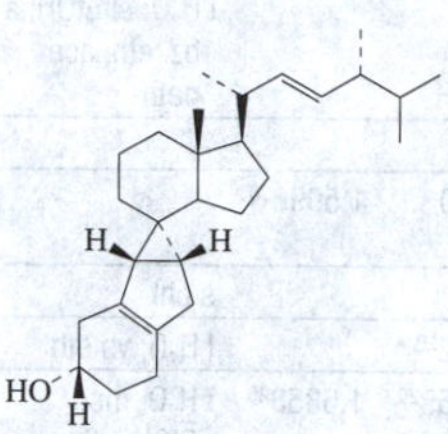
9668
Suprasterol II

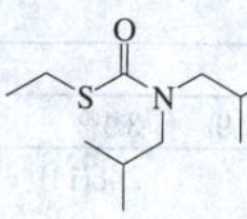
9669
Sutan

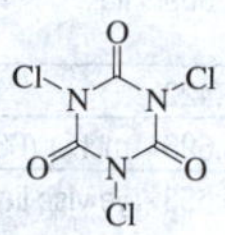
9670
Symclosene

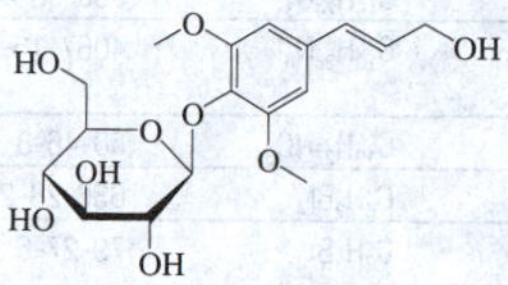
9671
Syringin

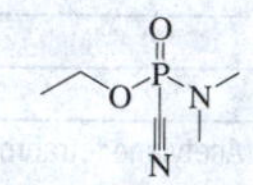
9672
Tabun

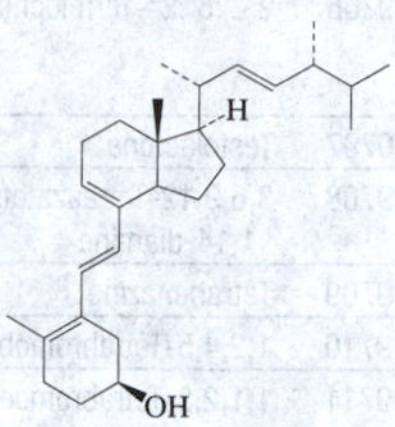
9673
Tachysterol

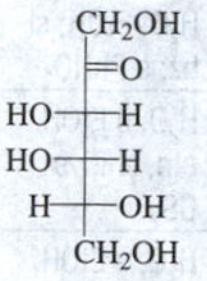
9674
D-Tagatose

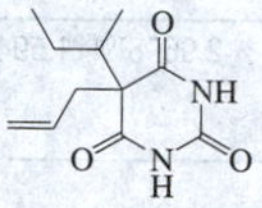
9675
Talbutal

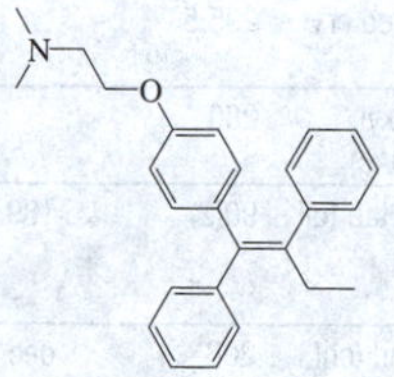
9676
Tamoxifen

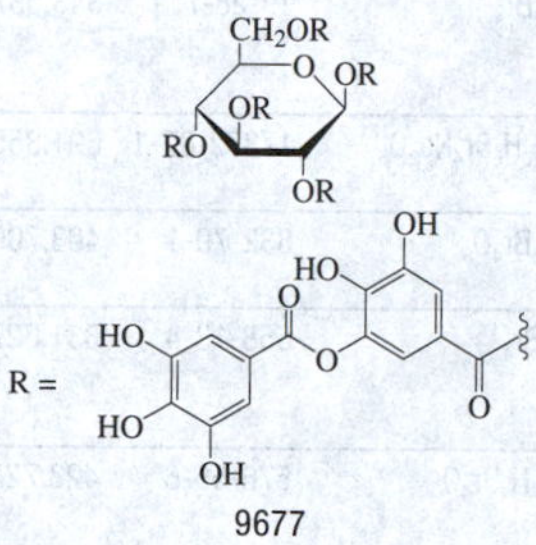
9677
Tannic acid

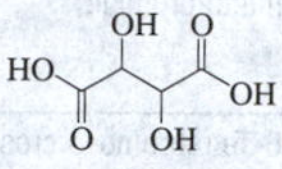
9678
DL-Tartaric acid

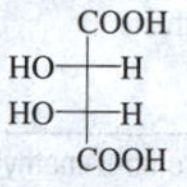
9679
meso-Tartaric acid

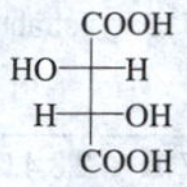
9680
D-Tartaric acid

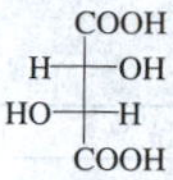
9681
L-Tartaric acid

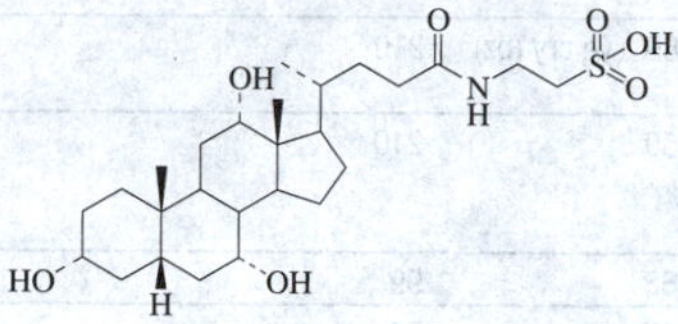
9682
Taurocholic acid

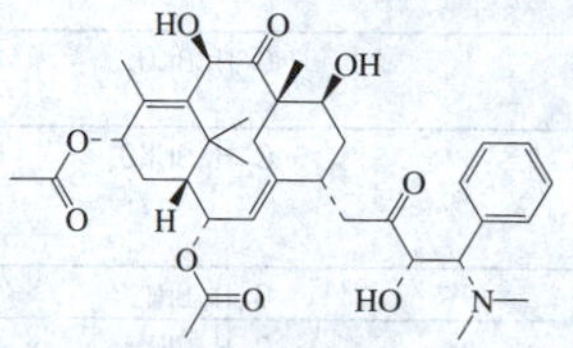
9683
Taxine A

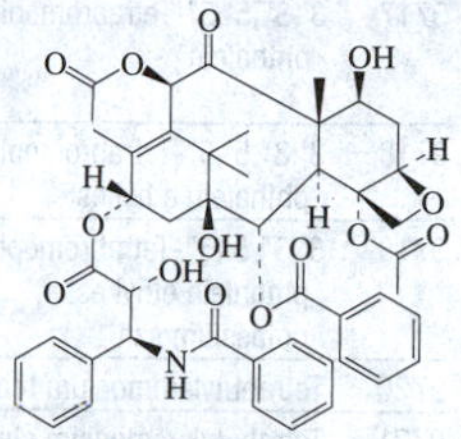
9684
Taxol

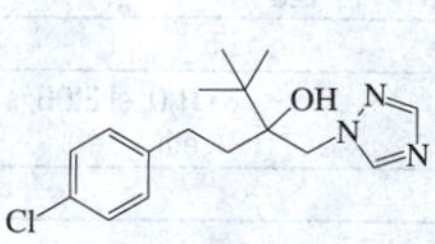
9685
Tebuconazole

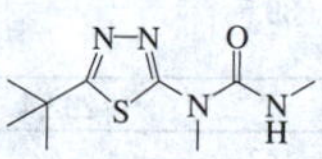
9686
Tebuthiuron

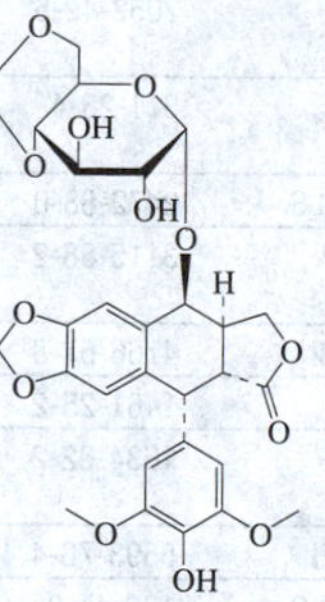
9687
Teniposide

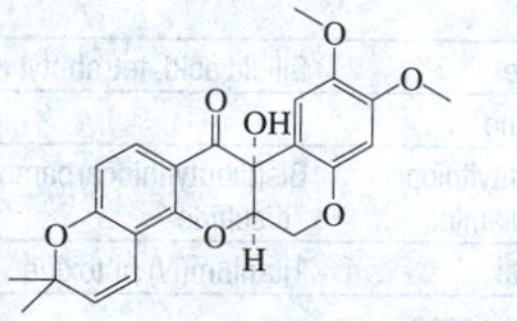
9688
Tephrosin

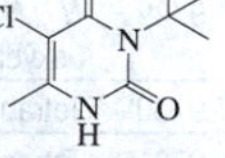
9689
Terbacil

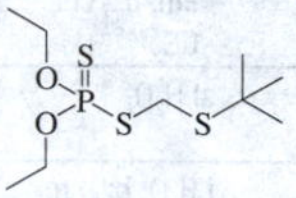
9690
Terbufos

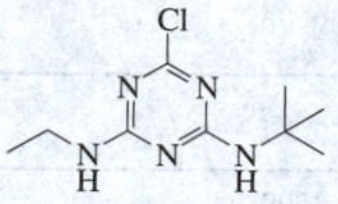
9691
Terbuthylazine

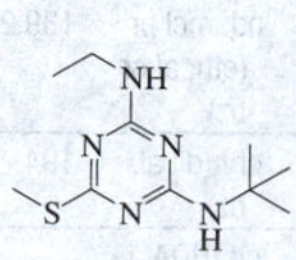
9692
Terbutryn

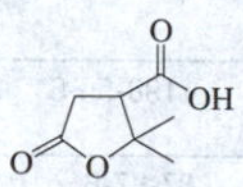
9693
Terebic acid

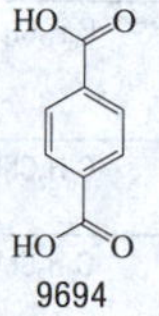
9694
Terephthalic acid

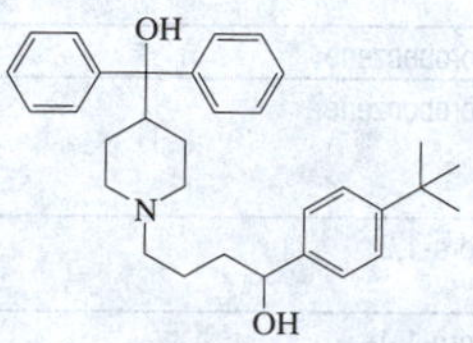
9695
Terfenadine

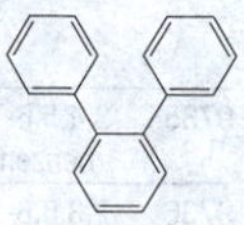
9696
o-Terphenyl

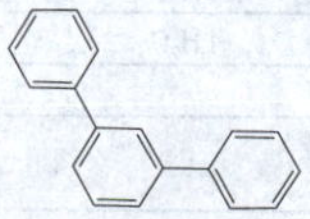
9697
m-Terphenyl

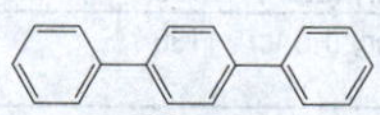
9698
p-Terphenyl

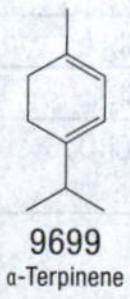
9699
α-Terpinene

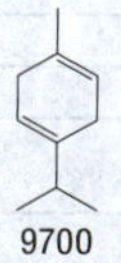
9700
γ-Terpinene

9701
α-Terpineol

No.	Name	Synonym	Mol. form.	CAS Reg. No.	Mol. wt.	Physical form	mp/°C	bp/°C	den/ g cm−3	n_D	Solubility
9702	α-Terpineol acetate		$C_{12}H_{20}O_2$	80-26-2	196.286			140^{40}	0.9659^{21}	1.4689^{21}	i H_2O; s EtOH, eth, bz
9703	Terpinolene	*p*-Mentha-1,4(8)-diene	$C_{10}H_{16}$	586-62-9	136.234			187(3)	0.8632^{15}	1.4883^{20}	i H_2O; msc EtOH, eth; s bz, ctc
9704	2,2':6',2"-Terpyridine		$C_{15}H_{11}N_3$	1148-79-4	233.268		88.0	370			
9705	Terrazole	1,2,4-Thiadiazole, 5-ethoxy-3-(trichloromethyl)-	$C_5H_5Cl_3N_2OS$	2593-15-9	247.530		19.9	95^1	1.503^{25}		
9706	2,2':5',2"-Terthiophene	α-Terthienyl	$C_{12}H_8S_3$	1081-34-1	248.387	ye-oran pl (MeOH)	93				i H_2O; sl EtOH; s bz, eth, ace, peth
9707	Testolactone		$C_{19}H_{24}O_3$	968-93-4	300.392	cry (ace)	218				
9708	3,6,9,12-Tetraazatetradecane-1,14-diamine	Pentaethylenehexamine	$C_{10}H_{28}N_6$	4067-16-7	232.369	liq			0.950	1.5096^{20}	
9709	Tetrabenazine		$C_{19}H_{27}NO_3$	58-46-8	317.422		128				s chl
9710	1,2,4,5-Tetrabromobenzene		$C_6H_2Br_4$	636-28-2	393.696	mcl pr (CS_2)	180.0(0.9)	3.1^{20}	3.072^{20}		i H_2O; vs eth
9711	1,1,2,2-Tetrabromoethane	Acetylene tetrabromide	$C_2H_2Br_4$	79-27-6	345.653	ye visc liq	0	248(11)	2.9655^{20}	1.6353^{20}	i H_2O; msc EtOH, eth; s ace, bz; sl ctc
9712	Tetrabromoethene	Tetrabromoethylene	C_2Br_4	79-28-7	343.637	pl (dil al), nd (al)	56.5	226			i H_2O; s EtOH, eth, ace; vs chl
9713	2',4',5',7'-Tetrabromofluorescein, disodium salt	Eosine YS	$C_{20}H_6Br_4Na_2O_5$	17372-87-1	691.855	ye-red cry	295.5				vs EtOH
9714	4,5,6,7-Tetrabromo-1,3-isobenzofurandione		$C_8Br_4O_3$	632-79-1	463.700	nd (xyl, HOAc)	280				i H_2O, EtOH; sl bz; s $PhNO_2$
9715	Tetrabromomethane	Carbon tetrabromide	CBr_4	558-13-4	331.627	mcl tab (dil al)	90(2)	189.5	2.9608^{100}	1.5942^{100}	i H_2O; s EtOH, eth, chl; vs CS_2
9716	2,3,4,5-Tetrabromo-6-methylphenol	3,4,5,6-Tetrabromo-*o*-cresol	$C_7H_4Br_4O$	576-55-6	423.722	ye nd (chl, HOAc)	208	dec			i H_2O; s EtOH, eth, bz, chl; sl lig, HOAc
9717	3',3",5',5"-Tetrabromophenolphthalein		$C_{20}H_{10}Br_4O_4$	76-62-0	633.907	nd (al, eth)	296				i H_2O; sl EtOH; vs eth; s alk, HOAc
9718	3',3",5',5"-Tetrabromophenolphthalein ethyl ester		$C_{22}H_{14}Br_4O_4$	1176-74-5	661.960	ye cry (bz)	210				
9719	3',3",5',5"-Tetrabromophenolphthalein ethyl ester, potassium salt		$C_{22}H_{13}Br_4KO_4$	62637-91-6	700.050		210				
9720	Tetrabutylammonium bromide	TMAB	$C_{16}H_{36}BrN$	1643-19-2	322.368		99				s chl
9721	Tetrabutylammonium chloride		$C_{16}H_{36}ClN$	1112-67-0	277.917	cry	74				
9722	Tetrabutylammonium fluoride		$C_{16}H_{36}FN$	429-41-4	261.462	cry (w)	37				
9723	Tetrabutylammonium hydroxide	TMAH	$C_{16}H_{37}NO$	2052-49-5	259.471	stab in soln					s H_2O, MeOH
9724	Tetrabutylammonium iodide		$C_{16}H_{36}IN$	311-28-4	369.368	lf (w, bz)	145(2)				sl H_2O, chl; vs EtOH
9725	Tetrabutylammonium sulfate		$C_{32}H_{72}N_2O_4S$	2472-88-0	580.990		170				sl chl
9726	Tetrabutylphosphonium bromide		$C_{16}H_{36}BrP$	3115-68-2	339.335	cry (ace/eth)	103.5(0.5)				
9727	Tetrabutyl silicate	Silicic acid, tetrabutyl ester	$C_{16}H_{36}O_4Si$	4766-57-8	320.541			256	0.8990^{20}	1.4128^{20}	
9728	Tetrabutylstannane		$C_{16}H_{36}Sn$	1461-25-2	347.167	liq	-97	145^{10}	1.06^{20}		
9729	*N,N,N',N'*-Tetrabutylthioperoxydicarbonic diamide	Bis(dibutylthiocarbamoyl) disulfide	$C_{18}H_{36}N_2S_4$	1634-02-2	408.752		39.5		1.03^{20}		i H_2O; sl EtOH; s eth
9730	Tetrabutyl titanate	Titanium(IV) butoxide	$C_{16}H_{36}O_4Ti$	5593-70-4	340.322			292.4			
9731	Tetracaine hydrochloride		$C_{15}H_{25}ClN_2O_2$	136-47-0	300.825		147				
9732	1,2,3,4-Tetrachlorobenzene		$C_6H_2Cl_4$	634-66-2	215.892	nd (al)	46.7(0.5)	254(3)			i H_2O; sl EtOH; vs eth, CS_2
9733	1,2,3,5-Tetrachlorobenzene		$C_6H_2Cl_4$	634-90-2	215.892	nd (al)	50.7(0.3)	245(3)			i H_2O
9734	1,2,4,5-Tetrachlorobenzene		$C_6H_2Cl_4$	95-94-3	215.892	nd, mcl pr (eth, al or bz)	139.2(0.4)	247(3)	1.858^{22}		i H_2O; sl EtOH; s eth, bz, chl, CS_2
9735	3,4,5,6-Tetrachloro-1,2-benzenediol		$C_6H_2Cl_4O_2$	1198-55-6	247.891	cry (dil al, bz)	194				sl H_2O
9736	2,3,5,6-Tetrachloro-1,4-benzenediol		$C_6H_2Cl_4O_2$	87-87-6	247.891	nd (HOAc)		sub			i H_2O, bz, ctc; vs EtOH, eth; sl HOAc
9737	2,2',4',5-Tetrachlorobiphenyl		$C_{12}H_6Cl_4$	41464-40-8	291.988	cry (MeOH)	65.9(0.5)				i H_2O
9738	2,3,4,5-Tetrachlorobiphenyl		$C_{12}H_6Cl_4$	33284-53-6	291.988	cry	90.7(0.5)				i H_2O
9739	3,3',4',4'-Tetrachlorobiphenyl		$C_{12}H_6Cl_4$	32598-13-3	291.988	cry (EtOH)	180				
9740	2,2',6,6'-Tetrachlorobisphenol A		$C_{15}H_{12}Cl_4O_2$	79-95-8	366.067	cry (HOAc)	136				

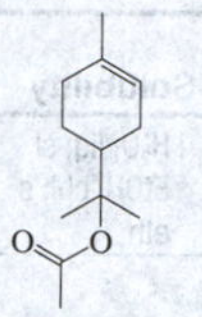
9702
α-Terpineol acetate

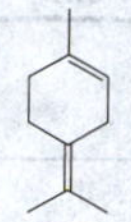
9703
Terpinolene

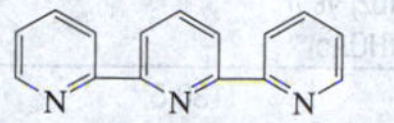
9704
2,2':6',2''-Terpyridine

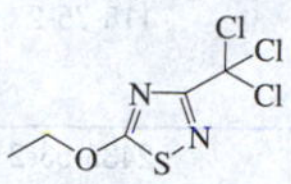
9705
Terrazole

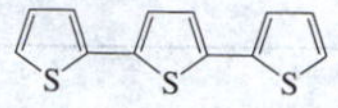
9706
2,2':5',2''-Terthiophene

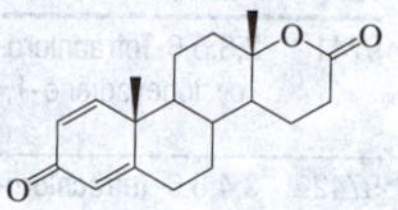
9707
Testolactone

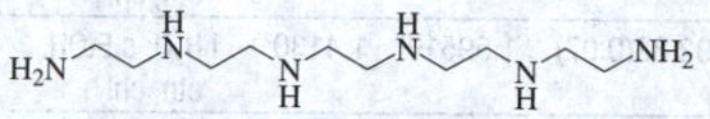
9708
3,6,9,12-Tetraazatetradecane-1,14-diamine

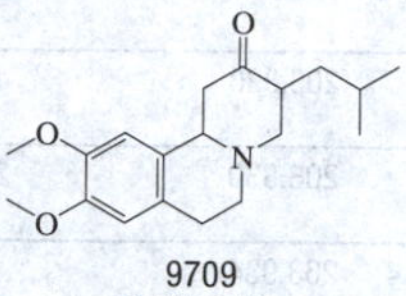
9709
Tetrabenazine

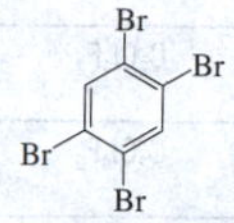
9710
1,2,4,5-Tetrabromobenzene

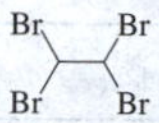
9711
1,1,2,2-Tetrabromoethane

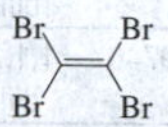
9712
Tetrabromoethene

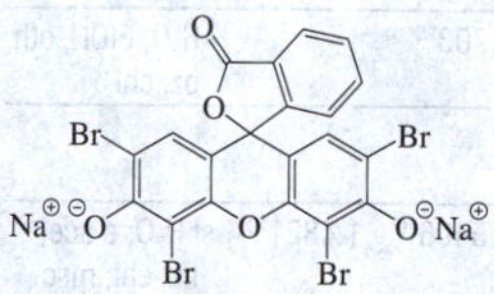
9713
2',4',5',7'-Tetrabromofluorescein, disodium salt

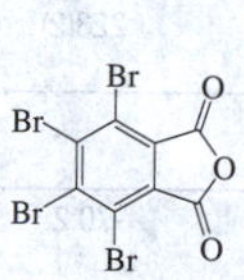
9714
4,5,6,7-Tetrabromo-1,3-isobenzofurandione

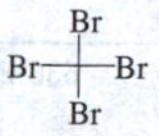
9715
Tetrabromomethane

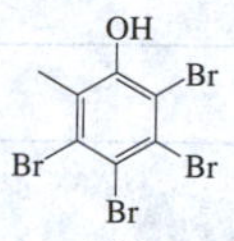
9716
2,3,4,5-Tetrabromo-6-methylphenol

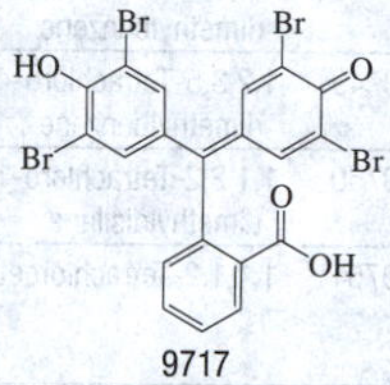
9717
3',3'',5',5''-Tetrabromophenolphthalein

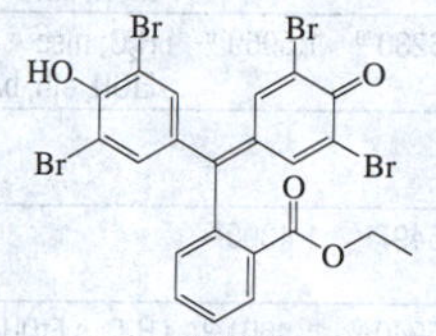
9718
3',3'',5',5''-Tetrabromophenolphthalein ethyl ester

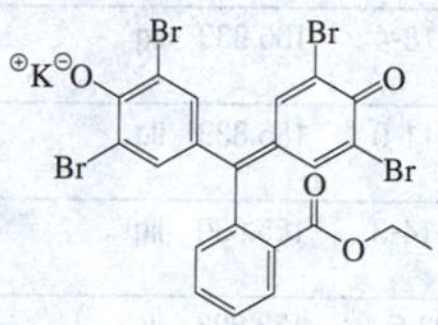
9719
3',3'',5',5''-Tetrabromophenolphthalein ethyl ester, potassium salt

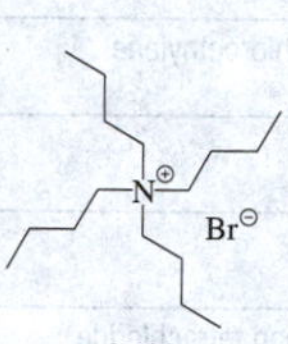
9720
Tetrabutylammonium bromide

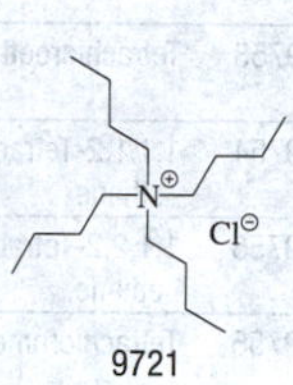
9721
Tetrabutylammonium chloride

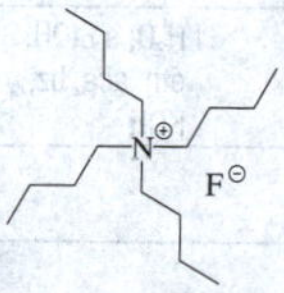
9722
Tetrabutylammonium fluoride

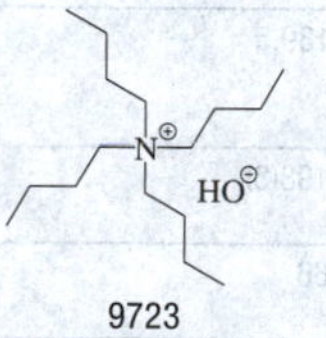
9723
Tetrabutylammonium hydroxide

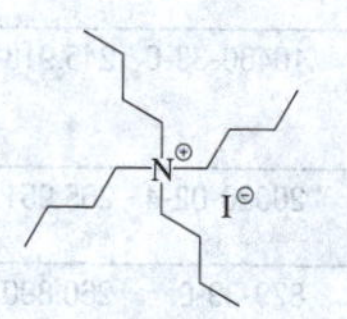
9724
Tetrabutylammonium iodide

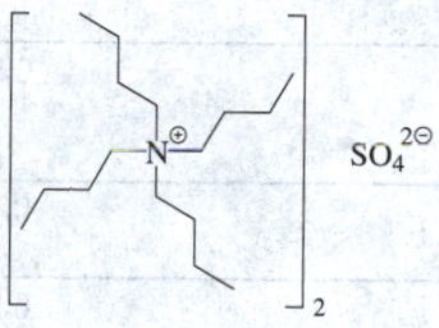
9725
Tetrabutylammonium sulfate

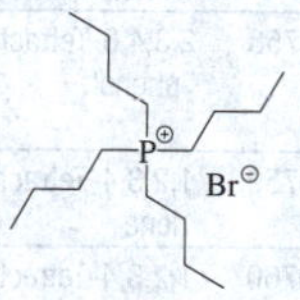
9726
Tetrabutylphosphonium bromide

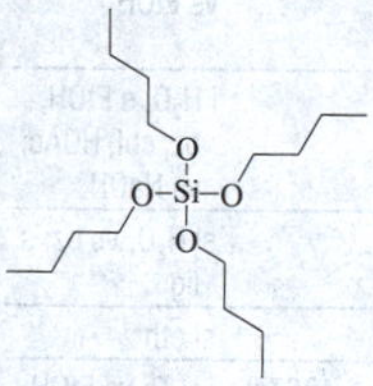
9727
Tetrabutyl silicate

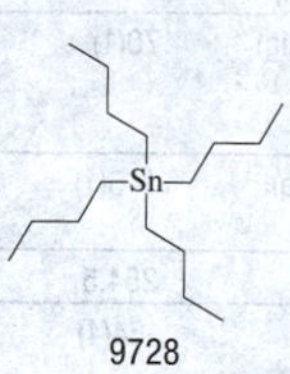
9728
Tetrabutylstannane

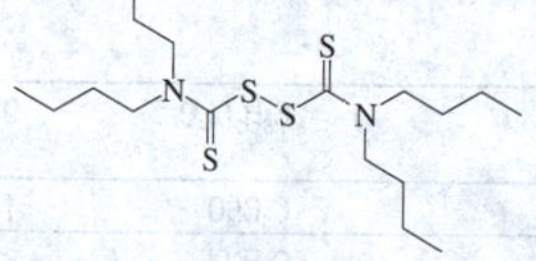
9729
N,N,N',N'-Tetrabutylthioperoxydicarbonic diamide

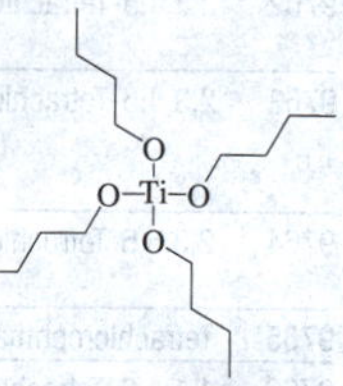
9730
Tetrabutyl titanate

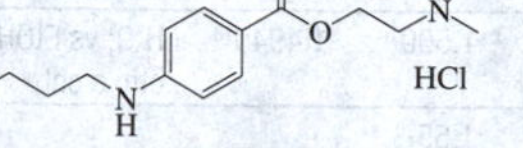
9731
Tetracaine hydrochloride

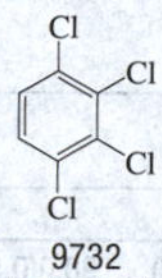
9732
1,2,3,4-Tetrachlorobenzene

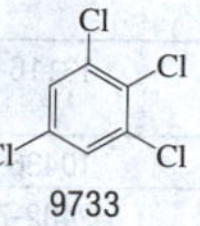
9733
1,2,3,5-Tetrachlorobenzene

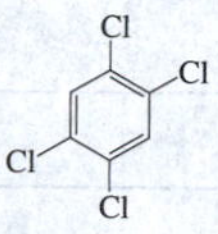
9734
1,2,4,5-Tetrachlorobenzene

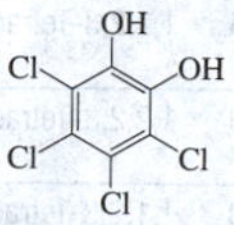
9735
3,4,5,6-Tetrachloro-1,2-benzenediol

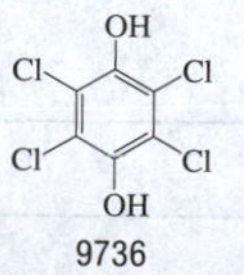
9736
2,3,5,6-Tetrachloro-1,4-benzenediol

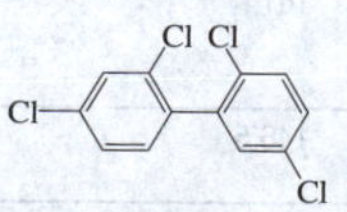
9737
2,2',4',5-Tetrachlorobiphenyl

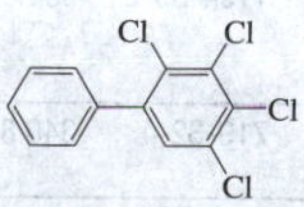
9738
2,3,4,5-Tetrachlorobiphenyl

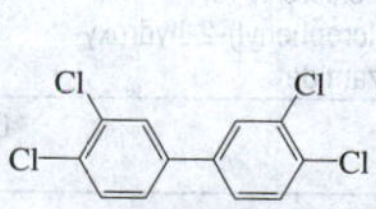
9739
3,3',4,4'-Tetrachlorobiphenyl

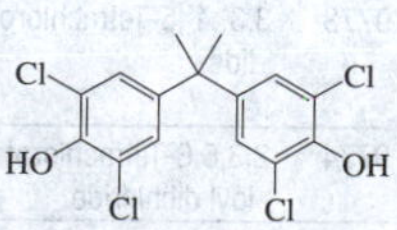
9740
2,2',6,6'-Tetrachlorobisphenol A

No.	Name	Synonym	Mol. form.	CAS Reg. No.	Mol. wt.	Physical form	mp/°C	bp/°C	den/ g cm^{-3}	n_D	Solubility
9741	2,3,5,6-Tetrachloro-2,5-cyclohexadiene-1,4-dione	Chloranil	$C_6Cl_4O_2$	118-75-2	245.875	ye mcl, pr (bz) ye lf (HOAc)	298.3(0.5)	sub			i H_2O, liq; sl EtOH, chl; s eth
9742	3,4,5,6-Tetrachloro-3,5-cyclohexadiene-1,2-dione		$C_6Cl_4O_2$	2435-53-2	245.875		130.5				
9743	2,3,7,8-Tetrachlorodibenzo-*p*-dioxin	Dioxin	$C_{12}H_4Cl_4O_2$	1746-01-6	321.971	nd	305(5)				
9744	2,3,7,8-Tetrachlorodibenzofuran		$C_{12}H_4Cl_4O$	51207-31-9	305.971	cry	227				
9745	1,1,1,2-Tetrachloro-2,2-difluoroethane		$C_2Cl_4F_2$	76-11-9	203.830		41.0(0.3)	96(3)	1.649^{25}		i H_2O; s EtOH, eth, chl
9746	1,1,2,2-Tetrachloro-1,2-difluoroethane		$C_2Cl_4F_2$	76-12-0	203.830		26.54(0.02)	92.83(0.07)	1.5951^{50}	1.4130^{25}	i H_2O; s EtOH, eth, chl
9747	1,2,3,4-Tetrachloro-5,5-dimethoxy-1,3-cyclopentadiene		$C_7H_6Cl_4O_2$	2207-27-4	263.934			109	1.501^{25}	1.5282^{20}	
9748	1,2,3,4-Tetrachloro-5,6-dimethylbenzene		$C_8H_6Cl_4$	877-08-7	243.946		228				i H_2O; s EtOH, eth, bz
9749	1,2,3,5-Tetrachloro-4,6-dimethylbenzene		$C_8H_6Cl_4$	877-09-8	243.946		223(2)		1.703^{25}		i H_2O, EtOH, eth, bz, chl
9750	1,1,2,2-Tetrachloro-1,2-dimethyldisilane		$C_2H_6Cl_4Si_2$	4518-98-3	228.052			154			
9751	1,1,1,2-Tetrachloroethane		$C_2H_2Cl_4$	630-20-6	167.849	liq	-70.2	130.2(0.2)	1.5406^{20}	1.4821^{20}	sl H_2O; s ace, bz, chl; msc EtOH, eth
9752	1,1,2,2-Tetrachloroethane	Acetylene tetrachloride	$C_2H_2Cl_4$	79-34-5	167.849	liq	-42.3(0.2)	146.0(0.3)	1.5953^{20}	1.4940^{20}	sl H_2O; s ace, bz, chl; msc EtOH, eth
9753	Tetrachloroethene	Perchloroethylene	C_2Cl_4	127-18-4	165.833	liq	-22.2(0.1)	121.2(0.3)	1.6230^{20}	1.5059^{20}	i H_2O; msc EtOH, eth, bz
9754	1,1,1,2-Tetrachloro-2-fluoroethane		C_2HCl_4F	354-11-0	185.839	liq	-95.3	117.1			
9755	1,1,2,2-Tetrachloro-1-fluoroethane		C_2HCl_4F	354-14-3	185.839	liq	-82.6	116.7	1.5497^{17}	1.4390^{20}	
9756	Tetrachloromethane	Carbon tetrachloride	CCl_4	56-23-5	153.823	liq	-22.8(0.1)	76.7(0.2)	1.5940^{20}	1.4601^{20}	i H_2O; s EtOH, ace; msc eth, bz, chl
9757	2,3,5,6-Tetrachloro-4-methoxyphenol	Drosophilin A	$C_7H_4Cl_4O_2$	484-67-3	261.918		116				
9758	2,3,4,6-Tetrachloro-5-methylphenol		$C_7H_4Cl_4O$	10460-33-0	245.918	nd (peth)	189.5				i H_2O; s EtOH, eth, ace, bz, KOH
9759	1,2,3,4-Tetrachloronaphthalene		$C_{10}H_4Cl_4$	20020-02-4	265.951		198(3)				
9760	1,2,3,4-Tetrachloro-5-nitrobenzene		$C_6HCl_4NO_2$	879-39-0	260.890		66				
9761	1,2,4,5-Tetrachloro-3-nitrobenzene		$C_6HCl_4NO_2$	117-18-0	260.890		100.4(0.5)	304	1.744^{25}		i H_2O; s EtOH, bz, chl
9762	2,3,4,5-Tetrachlorophenol		$C_6H_2Cl_4O$	4901-51-3	231.891	nd (peth, sub)	116(1)	sub			vs EtOH
9763	2,3,4,6-Tetrachlorophenol		$C_6H_2Cl_4O$	58-90-2	231.891	nd (lig)	70(1)	150^{15}			i H_2O; s EtOH, bz, chl, HOAc; vs NaOH
9764	2,3,5,6-Tetrachlorophenol		$C_6H_2Cl_4O$	935-95-5	231.891	lf (lig)	115(1)				sl H_2O; vs bz; s lig
9765	Tetrachlorophthalic anhydride		$C_8Cl_4O_3$	117-08-8	285.896		254.5	sub	1.49^{275}		sl eth
9766	1,1,1,2-Tetrachloropropane		$C_3H_4Cl_4$	812-03-3	181.876	liq	-64(4)	151(8)	1.473^{20}	1.4867^{20}	i H_2O; vs EtOH; s eth, chl
9767	1,1,1,3-Tetrachloropropane		$C_3H_4Cl_4$	1070-78-6	181.876			159(4)	1.4509^{20}	1.4825^{20}	i H_2O; vs EtOH, eth, bz, chl
9768	1,1,2,3-Tetrachloropropane		$C_3H_4Cl_4$	18495-30-2	181.876			181(4)	1.513^{17}	1.5037^{17}	i H_2O; s EtOH, chl; vs eth
9769	1,2,2,3-Tetrachloropropane		$C_3H_4Cl_4$	13116-53-5	181.876			165(4)	1.500^{18}	1.4940^{18}	i H_2O; vs EtOH, eth; s chl
9770	1,1,2,3-Tetrachloro-1-propene		$C_3H_2Cl_4$	10436-39-2	179.860	liq		167.2	1.55^{20}		
9771	2,3,5,6-Tetrachloropyridine		C_5HCl_4N	2402-79-1	216.881	cry (aq al)	90.0(0.5)	250.5			vs eth, EtOH, peth
9772	Tetrachloropyrimidine		$C_4Cl_4N_2$	1780-40-1	217.868		69.0				
9773	3,3′,4′,5-Tetrachlorosalicylanilide	3,5-Dichloro-*N*-(3,4-dichlorophenyl)-2-hydroxybenzamide	$C_{13}H_7Cl_4NO_2$	1154-59-2	351.013		161				
9774	2,3,5,6-Tetrachloroterphthaloyl dichloride		$C_8Cl_6O_2$	719-32-4	340.803	cry (ctc)	146.5				

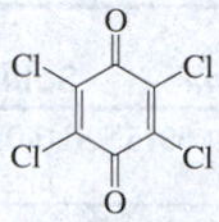
9741
2,3,5,6-Tetrachloro-2,5-cyclohexadiene-1,4-dione

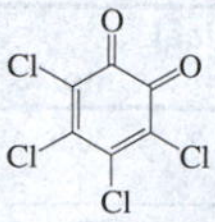
9742
3,4,5,6-Tetrachloro-3,5-cyclohexadiene-1,2-dione

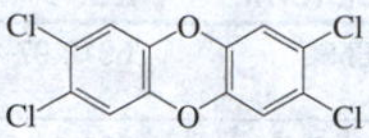
9743
2,3,7,8-Tetrachlorodibenzo-*p*-dioxin

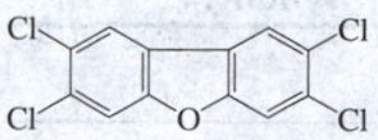
9744
2,3,7,8-Tetrachlorodibenzofuran

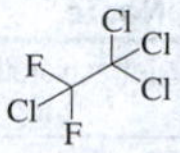
9745
1,1,1,2-Tetrachloro-2,2-difluoroethane

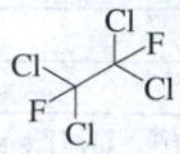
9746
1,1,2,2-Tetrachloro-1,2-difluoroethane

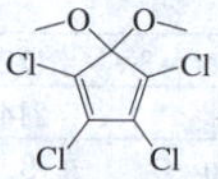
9747
1,2,3,4-Tetrachloro-5,5-dimethoxy-1,3-cyclopentadiene

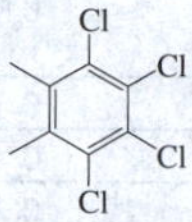
9748
1,2,3,4-Tetrachloro-5,6-dimethylbenzene

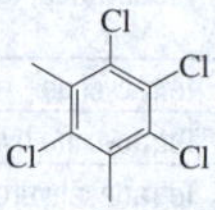
9749
1,2,3,5-Tetrachloro-4,6-dimethylbenzene

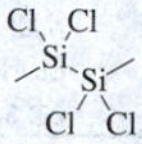
9750
1,1,2,2-Tetrachloro-1,2-dimethyldisilane

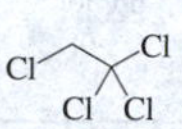
9751
1,1,1,2-Tetrachloroethane

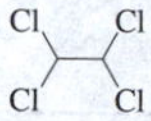
9752
1,1,2,2-Tetrachloroethane

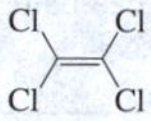
9753
Tetrachloroethene

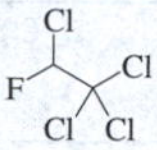
9754
1,1,1,2-Tetrachloro-2-fluoroethane

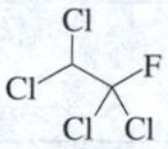
9755
1,1,2,2-Tetrachloro-1-fluoroethane

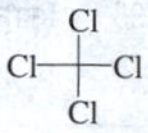
9756
Tetrachloromethane

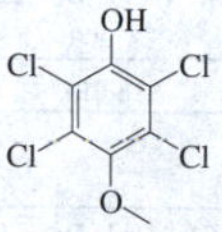
9757
2,3,5,6-Tetrachloro-4-methoxyphenol

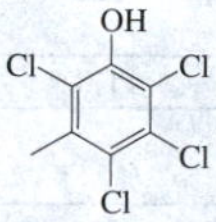
9758
2,3,4,6-Tetrachloro-5-methylphenol

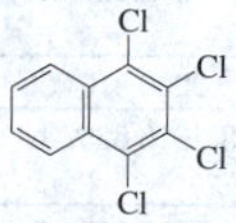
9759
1,2,3,4-Tetrachloronaphthalene

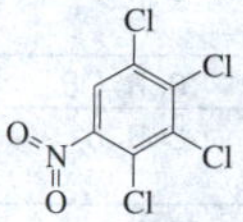
9760
1,2,3,4-Tetrachloro-5-nitrobenzene

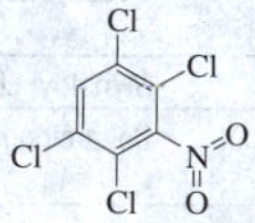
9761
1,2,4,5-Tetrachloro-3-nitrobenzene

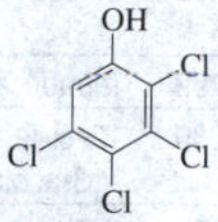
9762
2,3,4,5-Tetrachlorophenol

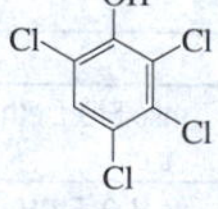
9763
2,3,4,6-Tetrachlorophenol

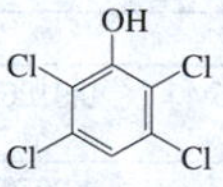
9764
2,3,5,6-Tetrachlorophenol

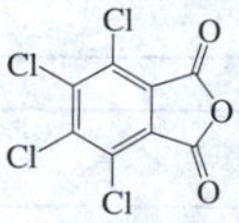
9765
Tetrachlorophthalic anhydride

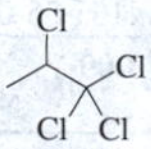
9766
1,1,1,2-Tetrachloropropane

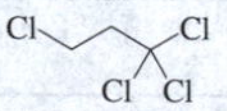
9767
1,1,1,3-Tetrachloropropane

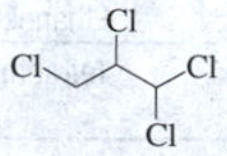
9768
1,1,2,3-Tetrachloropropane

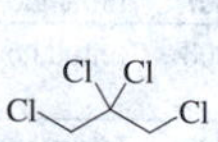
9769
1,2,2,3-Tetrachloropropane

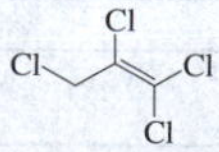
9770
1,1,2,3-Tetrachloro-1-propene

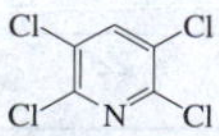
9771
2,3,5,6-Tetrachloropyridine

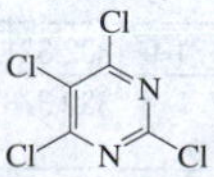
9772
Tetrachloropyrimidine

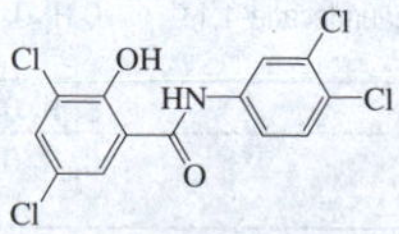
9773
3,3',4',5-Tetrachlorosalicylanilide

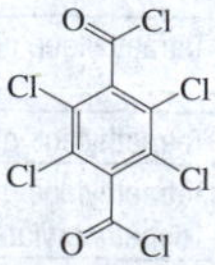
9774
2,3,5,6-Tetrachloroterphthaloyl dichloride

No.	Name	Synonym	Mol. form.	CAS Reg. No.	Mol. wt.	Physical form	mp/°C	bp/°C	den/g cm⁻³	n_D	Solubility
9775	Tetrachlorothiophene		C_4Cl_4S	6012-97-1	221.920	nd (dil al)	30.5	233.4	1.7036^{30}	1.5915^{30}	i H_2O; vs EtOH; msc eth
9776	Tetrachlorovinphos		$C_{10}H_9Cl_4O_4P$	961-11-5	365.961		97				
9777	Tetracontane		$C_{40}H_{82}$	4181-95-7	563.079		81.4(0.2)	522	0.8171^{25}	1.4572^{25}	
9778	Tetracosamethylundecasiloxane	Tetracosamethylhendecasiloxane	$C_{24}H_{72}O_{10}Si_{11}$	107-53-9	829.764			322.8	0.9247^{25}	1.3994^{20}	vs bz
9779	Tetracosane		$C_{24}H_{50}$	646-31-1	338.654	cry (eth)	50.3(0.7)	391(5)	0.7991^{20}	1.4283^{70}	i H_2O; sl EtOH; vs eth
9780	Tetracosanoic acid	Lignoceric acid	$C_{24}H_{48}O_2$	557-59-5	368.637		82(4)	272^{10}	0.8207^{100}	1.4287^{100}	vs bz, eth
9781	1-Tetracosanol		$C_{24}H_{50}O$	506-51-4	354.653		77	$210^{0.4}$			
9782	*cis*-15-Tetracosenoic acid	Nervonic acid	$C_{24}H_{46}O_2$	506-37-6	366.621		43				
9783	Tetracyanoethene	Tetracyanoethylene	C_6N_4	670-54-2	128.091		198.8(0.3)	223	1.348^{25}	1.560^{25}	sl eth, bz, ctc, chl; s ace
9784	Tetracycline		$C_{22}H_{24}N_2O_8$	60-54-8	444.434	cry (+3w)	172 dec				
9785	Tetracycline hydrochloride		$C_{22}H_{25}ClN_2O_8$	64-75-5	480.895		214				
9786	Tetradecahydrophenanthrene		$C_{14}H_{24}$	5743-97-5	192.341	liq	-3	284(6)	0.944^{20}	1.5011^{20}	i H_2O; s eth, ace, bz
9787	Tetradecamethylhexasiloxane		$C_{14}H_{42}O_5Si_6$	107-52-8	458.993	liq	-59	259.7(0.1)	0.8910^{20}	1.3948^{20}	vs bz
9788	Tetradecanal		$C_{14}H_{28}O$	124-25-4	212.371	lf	23(2)				i H_2O; s EtOH, eth, ace
9789	Tetradecanamide		$C_{14}H_{29}NO$	638-58-4	227.386	lf (ace)	104(1)	217^{12}			vs EtOH
9790	Tetradecane		$C_{14}H_{30}$	629-59-4	198.388	col liq	5.87(0.02)	253.5(0.4)	0.7596^{20}	1.4290^{20}	i H_2O; vs EtOH, eth; s ctc
9791	Tetradecanedioic acid		$C_{14}H_{26}O_4$	821-38-5	258.354		124.2(0.5)	366(10)			
9792	1,14-Tetradecanediol		$C_{14}H_{30}O_2$	19812-64-7	230.387	nd (bz)	85.8	200^{9}			vs eth, EtOH
9793	Tetradecanenitrile	Myristonitrile	$C_{14}H_{27}N$	629-63-0	209.371		19.1(0.3)	226^{100}	0.8281^{19}	1.4392^{23}	i H_2O; msc EtOH, eth, ace, bz; sl ctc
9794	1-Tetradecanethiol		$C_{14}H_{30}S$	2079-95-0	230.453		7	299(10)	0.8641^{20}	1.4597^{20}	i H_2O; s EtOH, eth, ctc
9795	Tetradecanoic acid	Myristic acid	$C_{14}H_{28}O_2$	544-63-8	228.371	lf (eth)	54.16(0.02)	250^{100}	0.8622^{54}	1.4723^{70}	i H_2O; s EtOH, ace, chl; sl eth; vs bz
9796	Tetradecanoic anhydride		$C_{28}H_{54}O_3$	626-29-9	438.727	lf (peth)	53.4		0.8502^{70}	1.4335^{70}	vs eth, EtOH
9797	1-Tetradecanol	Tetradecyl alcohol	$C_{14}H_{30}O$	112-72-1	214.387	lf	37.7(0.7)	295.8(0.4)	0.8236^{38}		i H_2O; vs EtOH, eth, ace, bz, chl
9798	2-Tetradecanone	Dodecylmethylketone	$C_{14}H_{28}O$	2345-27-9	212.371	cry (dil al)	33.5(0.2)	205^{100}			i H_2O; s EtOH, ace
9799	Tetradecanoyl chloride	Myristoyl chloride	$C_{14}H_{27}ClO$	112-64-1	246.816		-1	171^{16}	0.9078^{25}		s eth
9800	12-*O*-Tetradecanoylphorbol-13-acetate	Cocarcinogen A1	$C_{36}H_{56}O_8$	16561-29-8	616.825	oil					
9801	1-Tetradecene		$C_{14}H_{28}$	1120-36-1	196.372	liq	-12.88(0.1)	251.1(0.4)	0.7745^{25}	1.4351^{20}	i H_2O; vs EtOH, eth; s bz; sl ctc
9802	Tetradecyl acetate	1-Tetradecanol, acetate	$C_{16}H_{32}O_2$	638-59-5	256.424			173^{10}			
9803	Tetradecylamine	1-Tetradecanamine	$C_{14}H_{31}N$	2016-42-4	213.403		83.1	289(3)	0.8079^{20}	1.4463^{20}	i H_2O; vs EtOH, eth, bz, chl; s ace
9804	Tetradecylbenzene		$C_{20}H_{34}$	1459-10-5	274.484		9.7(0.2)	347(10)	0.8549^{20}	1.4818^{20}	
9805	Tetradecylcyclohexane		$C_{20}H_{40}$	1795-18-2	280.532		24(1)	358(15)	0.8254^{20}	1.4579^{20}	
9806	Tetradifon	1,2,4-Trichloro-5-[(4-chlorophenyl)sulfonyl]benzene	$C_{12}H_6Cl_4O_2S$	116-29-0	356.052		146.9(0.3)		1.151^{20}		
9807	Tetraethoxygermane	Ethanol, germanium(4+) salt	$C_8H_{20}GeO_4$	14165-55-0	252.88			139^{200}			
9808	Tetraethoxymethane	Tetraethyl orthocarbonate	$C_9H_{20}O_4$	78-09-1	192.253			159.5	0.9186^{20}	1.3905^{25}	msc EtOH, eth; s ctc
9809	Tetraethylammonium bromide		$C_8H_{20}BrN$	71-91-0	210.156	hyg (al)	174(2)		1.3970^{20}		vs H_2O, EtOH, chl, MeOH
9810	Tetraethylammonium chloride		$C_8H_{20}ClN$	56-34-8	165.705	hyg cry					vs H_2O, EtOH, ace, chl
9811	Tetraethylammonium iodide		$C_8H_{20}IN$	68-05-3	257.156	cry (w)	300 dec				s H_2O
9812	1,2,3,5-Tetraethylbenzene		$C_{14}H_{22}$	38842-05-6	190.325			250(3)			
9813	*N,N,N',N'*-Tetraethyl-1,2-benzenedicarboxamide	*N,N,N',N'*-Tetraethylphthalamide	$C_{16}H_{24}N_2O_2$	83-81-8	276.374		36	204^{16}			
9814	Tetra(2-ethylbutyl) silicate	Silicic acid, tetrakis(2-ethylbutyl) ester	$C_{24}H_{52}O_4Si$	78-13-7	432.754	liq			0.8920^{20}	1.4307^{20}	i H_2O; sl EtOH, ctc; s eth, bz
9815	Tetraethylene glycol	3,6,9-Trioxaundecane-1,11-diol	$C_8H_{18}O_5$	112-60-7	194.226	liq	-9.4(0.7)	315(7)	1.1285^{15}	1.4577^{20}	vs H_2O; s EtOH, eth, ctc, diox
9816	Tetraethylene glycol diacrylate		$C_{14}H_{22}O_7$	17831-71-9	302.321				1.125^{25}		
9817	Tetraethylene glycol dimethacrylate		$C_{16}H_{26}O_7$	109-17-1	330.373			220^{1}		1.4610^{25}	

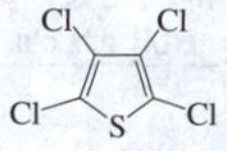
9775
Tetrachlorothiophene

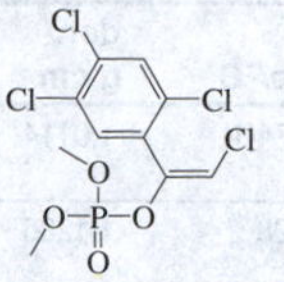
9776
Tetrachlorovinphos

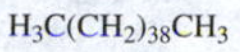
9777
Tetracontane

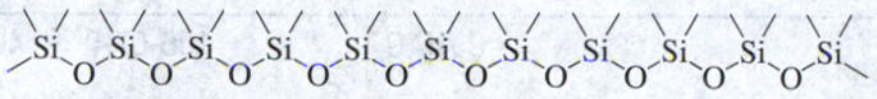
9778
Tetracosamethylundecasiloxane

9779
Tetracosane

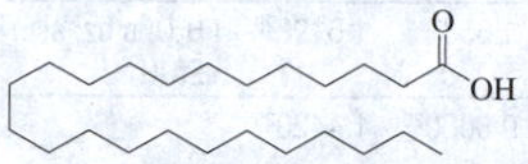
9780
Tetracosanoic acid

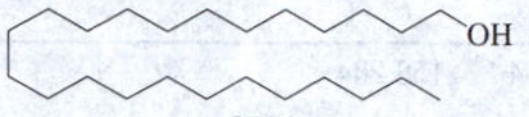
9781
1-Tetracosanol

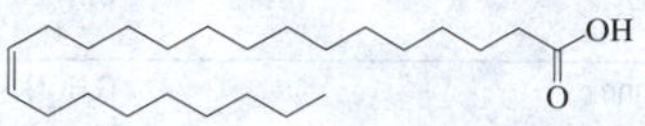
9782
cis-15-Tetracosenoic acid

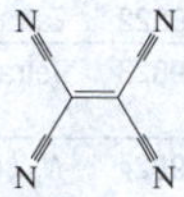
9783
Tetracyanoethene

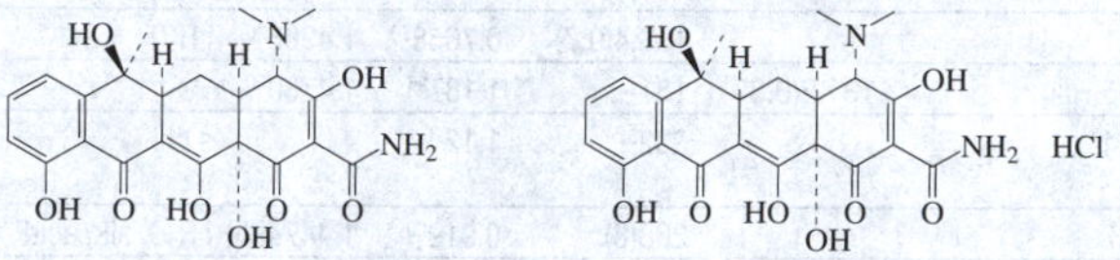
9784
Tetracycline
9785
Tetracycline hydrochloride

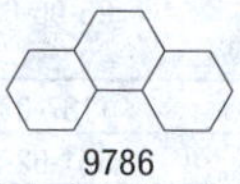
9786
Tetradecahydrophenanthrene

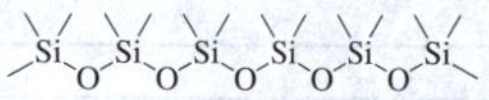
9787
Tetradecamethylhexasiloxane

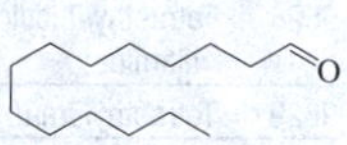
9788
Tetradecanal

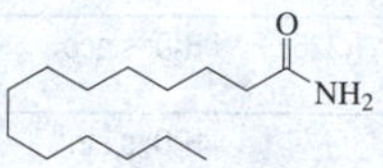
9789
Tetradecanamide

9790
Tetradecane

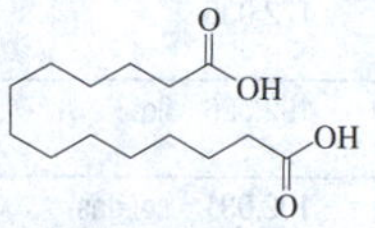
9791
Tetradecanedioic acid

9792
1,14-Tetradecanediol

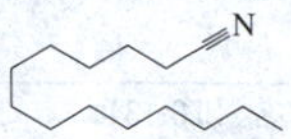
9793
Tetradecanenitrile

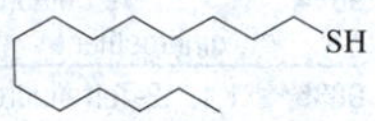
9794
1-Tetradecanethiol

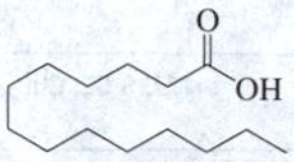
9795
Tetradecanoic acid

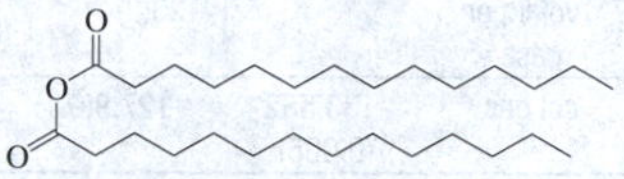
9796
Tetradecanoic anhydride

9797
1-Tetradecanol

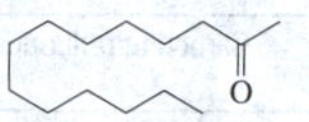
9798
2-Tetradecanone

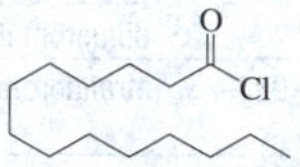
9799
Tetradecanoyl chloride

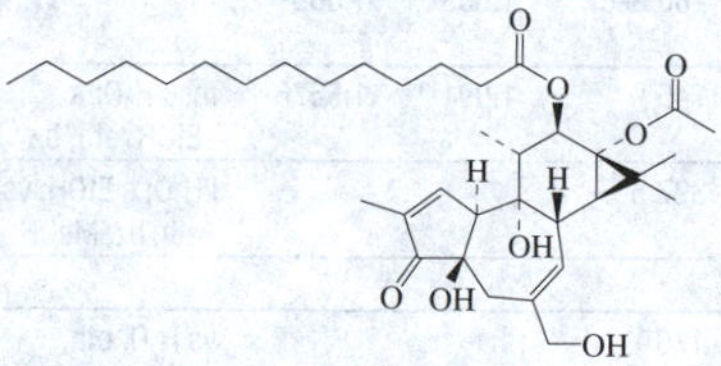
9800
12-*O*-Tetradecanoylphorbol-13-acetate

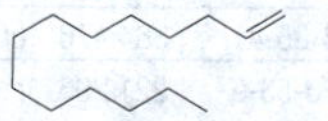
9801
1-Tetradecene

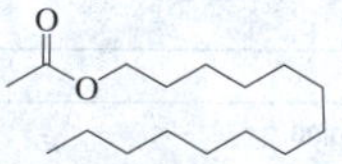
9802
Tetradecyl acetate

9803
Tetradecylamine

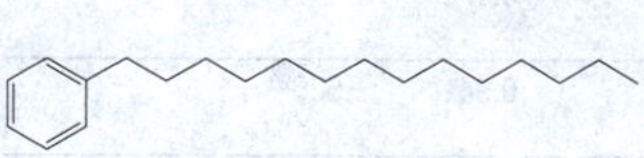
9804
Tetradecylbenzene

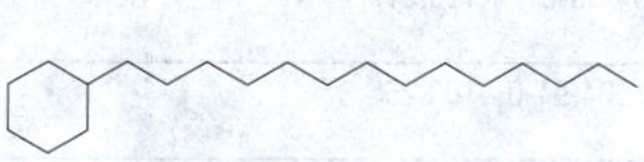
9805
Tetradecylcyclohexane

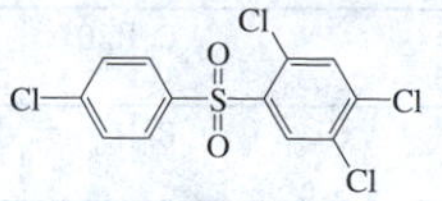
9806
Tetradifon

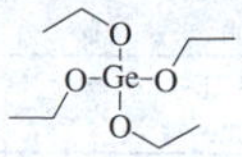
9807
Tetraethoxygermane

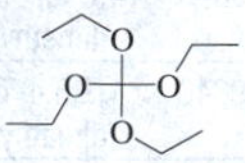
9808
Tetraethoxymethane

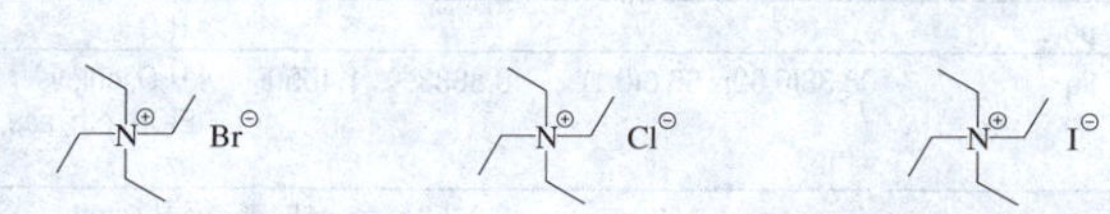
9809
Tetraethylammonium bromide
9810
Tetraethylammonium chloride
9811
Tetraethylammonium iodide

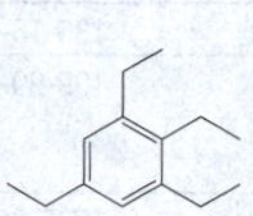
9812
1,2,3,5-Tetraethylbenzene

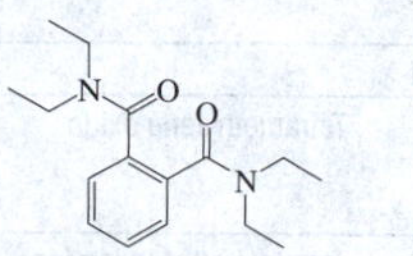
9813
N,N,N',N'-Tetraethyl-1,2-benzenedicarboxamide

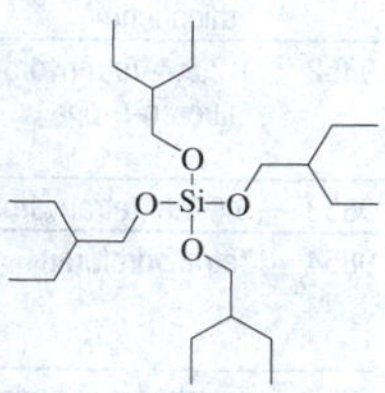
9814
Tetra(2-ethylbutyl) silicate

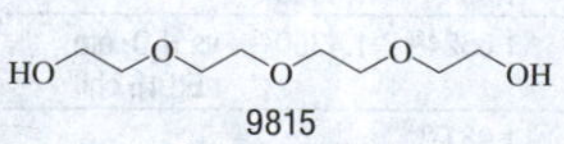
9815
Tetraethylene glycol

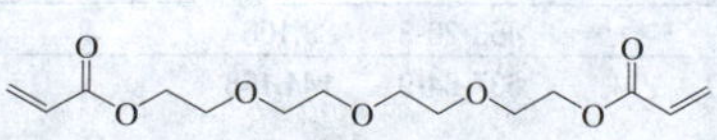
9816
Tetraethylene glycol diacrylate

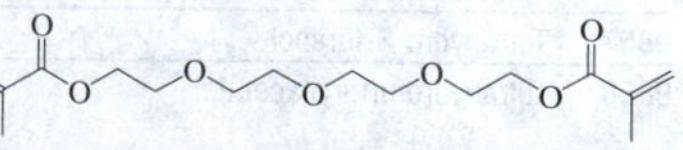
9817
Tetraethylene glycol dimethacrylate

No.	Name	Synonym	Mol. form.	CAS Reg. No.	Mol. wt.	Physical form	mp/°C	bp/°C	den/ g cm^{-3}	n_D	Solubility
9818	Tetraethylene glycol dimethyl ether		$C_{10}H_{22}O_5$	143-24-8	222.279			274(1)	1.0114^{20}		msc H_2O; s EtOH, eth, ctc
9819	Tetraethylene glycol monostearate		$C_{26}H_{52}O_6$	106-07-0	460.687		40	328	1.1285^{15}	1.4593^{20}	
9820	Tetraethylenepentamine		$C_8H_{23}N_5$	112-57-2	189.303			341.5		1.5042^{20}	s H_2O
9821	*N,N,N′,N′*-Tetraethyl-1,2-ethanediamine		$C_{10}H_{24}N_2$	150-77-6	172.311			192	0.808^{25}	1.4343^{20}	
9822	Tetraethylgermane		$C_8H_{20}Ge$	597-63-7	188.89			164.5	0.9914^{23}		
9823	Tetraethyl lead		$C_8H_{20}Pb$	78-00-2	323.4			183(5)	1.653^{20}	1.5198^{20}	i H_2O; s bz; sl EtOH
9824	*N,N,N′,N′*-Tetraethylmethanediamine		$C_9H_{22}N_2$	102-53-4	158.284			152(4)	0.8000^{20}	1.4420^{25}	
9825	Tetraethyl pyrophosphate		$C_8H_{20}O_7P_2$	107-49-3	290.188		170 dec	155^{3}	1.1847^{20}	1.4180^{20}	msc H_2O, EtOH, eth, ace, xyl, chl; sl ctc
9826	Tetraethylsilane		$C_8H_{20}Si$	631-36-7	144.331			153.4(0.7)	0.7658^{20}	1.4268^{20}	i H_2O
9827	Tetraethylstannane	Tin tetraethyl	$C_8H_{20}Sn$	597-64-8	234.955	liq	-131.2(0.3)	181	1.187^{25}	1.4730^{20}	
9828	Tetraethylthiodicarbonic diamide	Sulfiram	$C_{10}H_{20}N_2S_3$	95-05-6	264.474			232^{3}	1.12^{20}		s chl
9829	Tetraethylurea		$C_9H_{20}N_2O$	1187-03-7	172.267			208(8)	0.919^{20}	1.4474^{20}	i H_2O, alk, acid
9830	1,2,3,4-Tetrafluorobenzene		$C_6H_2F_4$	551-62-2	150.074			94.3(0.8)		1.4054^{20}	
9831	1,2,3,5-Tetrafluorobenzene		$C_6H_2F_4$	2367-82-0	150.074	liq	-46.2(0.1)	84.3(0.8)	1.319^{25}	1.4035^{20}	
9832	1,2,4,5-Tetrafluorobenzene		$C_6H_2F_4$	327-54-8	150.074		3.88(0.05)	90.2(0.3)	1.4255^{20}	1.4075^{20}	
9833	3,3,4,4-Tetrafluorodihydro-2,5-furandione	Tetrafluorosuccinic anhydride	$C_4F_4O_3$	699-30-9	172.035			54.5	1.6209^{20}	1.3240^{20}	
9834	1,1,2,2-Tetrafluoro-1,2-dinitroethane		$C_2F_4N_2O_4$	356-16-1	192.026	liq	-41.5	58.5	1.6024^{25}	1.3265^{25}	i H_2O; s ace
9835	1,1,1,2-Tetrafluoroethane	HFC-134a	$C_2H_2F_4$	811-97-2	102.031	col gas	-103.296 (0.008)	-26.1(0.1)	1.2072^{25}		i H_2O; s eth
9836	1,1,2,2-Tetrafluoroethane		$C_2H_2F_4$	359-35-3	102.031	col gas	-89	-20(1)			
9837	Tetrafluoroethene	Tetrafluoroethylene	C_2F_4	116-14-3	100.015	col gas	-131.14(0.02)	-76(1)	1.519^{-76}		i H_2O
9838	1,2,2,2-Tetrafluoroethyl difluoromethyl ether	Refrigerant 236me	$C_3H_2F_6O$	57041-67-5	168.037	vol liq or gas		23(2)	1.4540^{23}		
9839	Tetrafluoromethane	Carbon tetrafluoride	CF_4	75-73-0	88.005	col gas	-183.582 (0.005)	-127.9(0.1)	3.034^{25}		i H_2O; s bz, chl
9840	2,2,3,3-Tetrafluoro-1-propanol		$C_3H_4F_4O$	76-37-9	132.057	liq	-15	106(3)	1.4853^{20}	1.3197^{20}	s EtOH, ace, chl
9841	6,7,8,9-Tetrahydro-5*H*-benzocyclohepten-5-one		$C_{11}H_{12}O$	826-73-3	160.212			175^{40}	1.080^{20}	1.5698^{20}	s EtOH
9842	2,3,6,7-Tetrahydro-1*H*,5*H*-benzo[*ij*]quinolizine	Julolidine	$C_{12}H_{15}N$	479-59-4	173.254		40	280 dec	1.003^{20}	1.568^{25}	
9843	1,2,3,6-Tetrahydro-2,3′-bipyridine, (*S*)-	Anatabine	$C_{10}H_{12}N_2$	581-49-7	160.215			145^{10}	1.091^{19}	1.5676^{20}	msc H_2O; s EtOH, eth, bz
9844	2,3,4,9-Tetrahydro-1*H*-carbazole		$C_{12}H_{13}N$	942-01-8	171.238	lf (dil al)	120	327.5			i H_2O; s EtOH; vs eth, bz, MeOH
9845	Tetrahydrocortisone		$C_{21}H_{32}O_5$	53-05-4	364.476	cry (EtOAc)	190				
9846	1,2,3,4-Tetrahydro-6,7-dimethoxy-1,2-dimethylisoquinoline, (±)-	Carnegine	$C_{13}H_{19}NO_2$	490-53-9	221.296	pa br syr		170^{1}			vs H_2O, eth, EtOH
9847	Tetrahydro-2,5-dimethoxyfuran		$C_6H_{12}O_3$	696-59-3	132.157			145.7	1.02^{25}	1.4180^{20}	
9848	4,5,6,7-Tetrahydro-3,6-dimethylbenzofuran		$C_{10}H_{14}O$	494-90-6	150.217		86	80^{18}	0.972^{15}		
9849	1,2,3,4-Tetrahydro-1,5-dimethylnaphthalene		$C_{12}H_{16}$	21564-91-0	160.255			239	0.941^{20}	1.526^{20}	
9850	Tetrahydro-2,2-dimethyl-5-oxo-3-furanacetic acid	Terpenylic acid	$C_8H_{12}O_4$	26754-48-3	172.179	lf or pr (w+1)	90				vs H_2O
9851	*cis*-Tetrahydro-2,5-dimethylthiophene		$C_6H_{12}S$	5161-13-7	116.224	liq	-89	142.5(0.8)	0.9222^{20}	1.4799^{20}	vs ace, bz, eth, EtOH
9852	1,2,3,4-Tetrahydro-9*H*-fluoren-9-one	Phentydrone	$C_{13}H_{12}O$	634-19-5	184.233	lt ye nd or pr (pentane)	81.5	$139^{0.05}$			
9853	5,6,7,8-Tetrahydrofolic acid		$C_{19}H_{23}N_7O_6$	135-16-0	445.429	pow					s H_2O
9854	Tetrahydrofuran	Tetramethylene oxide	C_4H_8O	109-99-9	72.106	liq	-108.38(0.02)	66.0(0.1)	0.8833^{25}	1.4050^{25}	s H_2O, chl; vs EtOH, eth, ace, bz
9855	Tetrahydro-2-furanmethanamine	Tetrahydrofurfurylamine	$C_5H_{11}NO$	4795-29-3	101.147			153	0.9752^{20}	1.4551^{20}	vs H_2O, eth, EtOH
9856	Tetrahydro-2-furanmethanol propanoate		$C_8H_{14}O_3$	637-65-0	158.195			205.5	1.044^{20}		vs eth, EtOH, chl
9857	Tetrahydro-3-furanol		$C_4H_8O_2$	453-20-3	88.106			181	1.09^{25}	1.4500^{20}	
9858	Tetrahydrofurfuryl acetate		$C_7H_{12}O_3$	637-64-9	144.168			193	1.0624^{20}	1.4350^{25}	vs H_2O, eth, EtOH, chl
9859	Tetrahydrofurfuryl acrylate		$C_8H_{12}O_3$	2399-48-6	156.179		<-60	96^{6}	1.061^{20}		

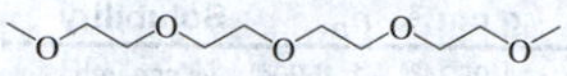
9818
Tetraethylene glycol dimethyl ether

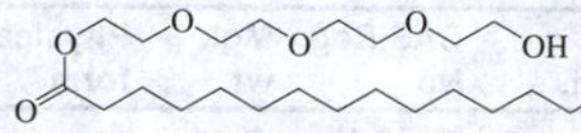
9819
Tetraethylene glycol monostearate

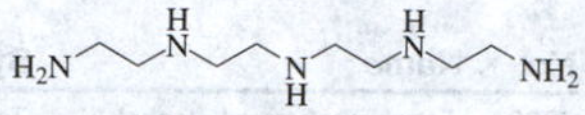
9820
Tetraethylenepentamine

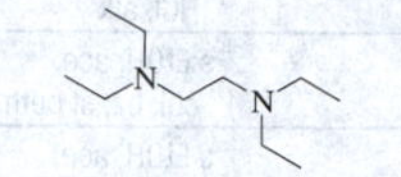
9821
N,N,N',N'-Tetraethyl-1,2-ethanediamine

9822
Tetraethylgermane

9823
Tetraethyl lead

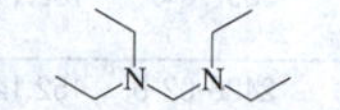
9824
N,N,N',N'-Tetraethylmethanediamine

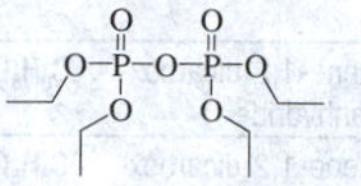
9825
Tetraethyl pyrophosphate

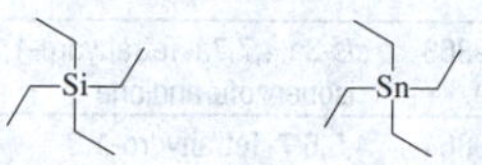
9826
Tetraethylsilane

9827
Tetraethylstannane

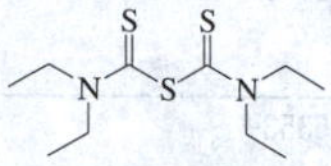
9828
Tetraethylthiodicarbonic diamide

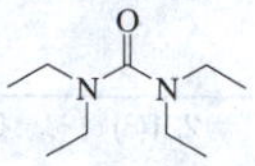
9829
Tetraethylurea

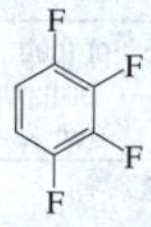
9830
1,2,3,4-Tetrafluorobenzene

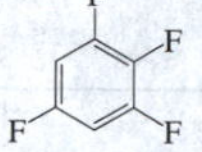
9831
1,2,3,5-Tetrafluorobenzene

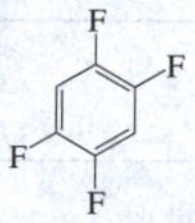
9832
1,2,4,5-Tetrafluorobenzene

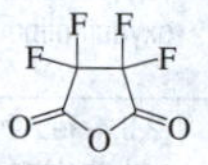
9833
3,3,4,4-Tetrafluorodihydro-2,5-furandione

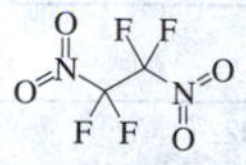
9834
1,1,2,2-Tetrafluoro-1,2-dinitroethane

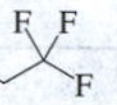
9835
1,1,1,2-Tetrafluoroethane

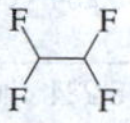
9836
1,1,2,2-Tetrafluoroethane

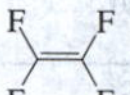
9837
Tetrafluoroethene

9838
1,2,2,2-Tetrafluoroethyl difluoromethyl ether

9839
Tetrafluoromethane

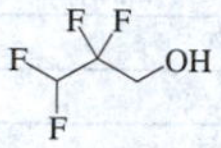
9840
2,2,3,3-Tetrafluoro-1-propanol

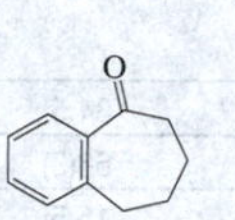
9841
6,7,8,9-Tetrahydro-5*H*-benzocyclohepten-5-one

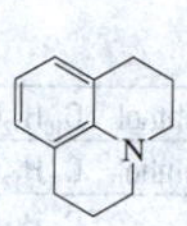
9842
2,3,6,7-Tetrahydro-1*H*,5*H*-benzo[ij]quinolizine

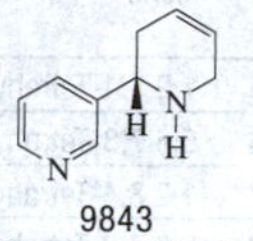
9843
1,2,3,6-Tetrahydro-2,3'-bipyridine, (*S*)-

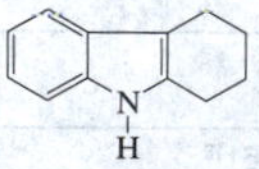
9844
2,3,4,9-Tetrahydro-1*H*-carbazole

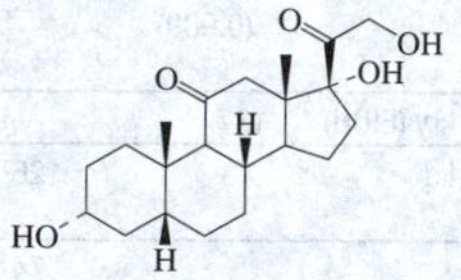
9845
Tetrahydrocortisone

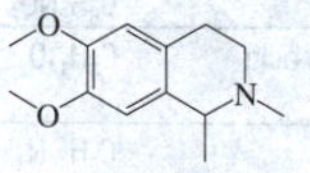
9846
1,2,3,4-Tetrahydro-6,7-dimethoxy-1,2-dimethylisoquinoline, (±)-

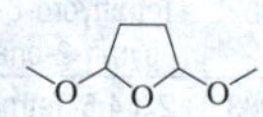
9847
Tetrahydro-2,5-dimethoxyfuran

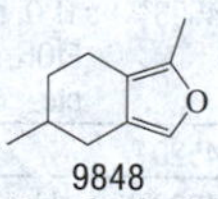
9848
4,5,6,7-Tetrahydro-3,6-dimethylbenzofuran

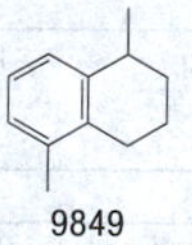
9849
1,2,3,4-Tetrahydro-1,5-dimethylnaphthalene

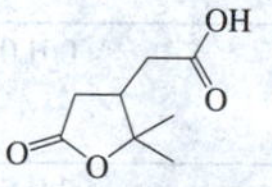
9850
Tetrahydro-2,2-dimethyl-5-oxo-3-furanacetic acid

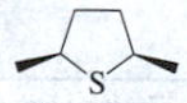
9851
cis-Tetrahydro-2,5-dimethylthiophene

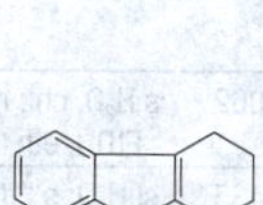
9852
1,2,3,4-Tetrahydro-9*H*-fluoren-9-one

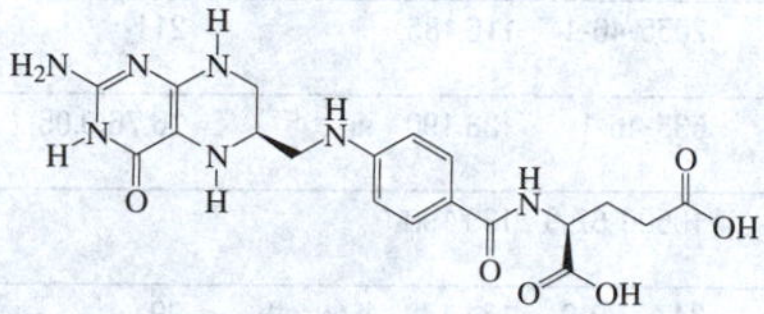
9853
5,6,7,8-Tetrahydrofolic acid

9854
Tetrahydrofuran

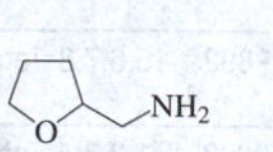
9855
Tetrahydro-2-furanmethanamine

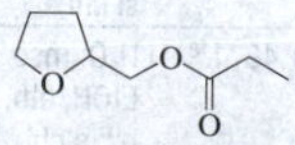
9856
Tetrahydro-2-furanmethanol propanoate

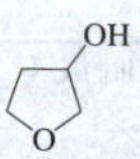
9857
Tetrahydro-3-furanol

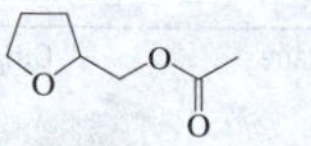
9858
Tetrahydrofurfuryl acetate

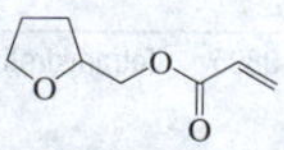
9859
Tetrahydrofurfuryl acrylate

No.	Name	Synonym	Mol. form.	CAS Reg. No.	Mol. wt.	Physical form	mp/°C	bp/°C	den/ g cm^{-3}	n_D	Solubility
9860	Tetrahydrofurfuryl alcohol	Tetrahydro-2-furancarbinol	$C_5H_{10}O_2$	97-99-4	102.132	liq	<-80	176.3(0.7)	1.0524^{20}	1.4520^{20}	vs ace, eth
9861	Tetrahydrofurfuryl methacrylate		$C_9H_{14}O_3$	2455-24-5	170.205			265	1.040^{25}	1.4554^{25}	
9862	Tetrahydroimidazo[4,5-*d*] imidazole-2,5(1*H*,3*H*)-dione	Acetyleneurea	$C_4H_6N_4O_2$	496-46-8	142.117	nd or pr (w)	297(2)				sl H_2O; i EtOH, HOAc; s eth, HCl, alk
9863	*cis*-3a,4,7,7a-Tetrahydro-1,3-isobenzofurandione	4-Cyclohexene-1,2-dicarboxylic acid, anhydride	$C_8H_8O_3$	935-79-5	152.148	cry (peth)	98(2)				s EtOH, ace, chl, bz; sl peth
9864	4,5,6,7-Tetrahydro-1,3-isobenzofurandione	1-Cyclohexene-1,2-dicarboxylic acid, anhydride	$C_8H_8O_3$	2426-02-0	152.148	pl (EtOH)	70.3(0.2)		1.2^{105}		s EtOH, ace, chl; vs eth
9865	1,2,3,4-Tetrahydroisoquinoline		$C_9H_{11}N$	91-21-4	133.190		<-15	232.5	1.0642^{24}	1.5668^{20}	i H_2O; s EtOH, bz, acid, xyl
9866	3,4,5,6-Tetrahydro-7-methoxy-2*H*-azepine		$C_7H_{13}NO$	2525-16-8	127.184	liq		49^{16}	0.887	1.4630^{20}	
9867	1,2,3,4-Tetrahydro-6-methoxyquinoline		$C_{10}H_{13}NO$	120-15-0	163.216	pr (peth, al) orth pym (w)	42.5	284		1.5718^{20}	s chl
9868	1,2,3,4-Tetrahydro-1-methylnaphthalene		$C_{11}H_{14}$	1559-81-5	146.229			221(6)	0.9583^{20}	1.5353^{20}	
9869	1,2,3,4-Tetrahydro-5-methylnaphthalene		$C_{11}H_{14}$	2809-64-5	146.229	liq	-23.0(0.3)	234(7)	0.9720^{20}	1.5439^{20}	
9870	1,2,3,4-Tetrahydro-6-methylnaphthalene		$C_{11}H_{14}$	1680-51-9	146.229	liq	-40	229	0.9537^{20}	1.5357^{20}	
9871	1,2,3,6-Tetrahydro-1-methyl-4-phenylpyridine	MPTP	$C_{12}H_{15}N$	28289-54-5	173.254	cry	41	87$^{0.8}$			
9872	Tetrahydro-3-methyl-2*H*-thiopyran		$C_6H_{12}S$	5258-50-4	116.224	liq	-60	158.1(0.4)	0.9473^{20}	1.4922^{20}	
9873	5,6,7,8-Tetrahydro-1-naphthalenamine		$C_{10}H_{13}N$	2217-41-6	147.217		38	278(14)	1.0625^{16}	1.5900^{20}	sl H_2O; s EtOH, eth, acid
9874	1,2,3,4-Tetrahydronaphthalene	Tetralin	$C_{10}H_{12}$	119-64-2	132.202	liq	-35.76(0.06)	207.2(0.3)	0.9645^{25}	1.5413^{20}	i H_2O; vs EtOH, eth; s chl, $PhNH_2$
9875	1,2,3,4-Tetrahydro-1-naphthol	1,2,3,4-Tetrahydro-α-naphthol	$C_{10}H_{12}O$	529-33-9	148.201		34.5	255	1.0996^{20}	1.5638^{20}	
9876	5,6,7,8-Tetrahydro-1-naphthol	5,6,7,8-Tetrahydro-α-naphthol	$C_{10}H_{12}O$	529-35-1	148.201		68(3)	266	1.0556^{75}		
9877	1,2,3,4-Tetrahydro-2-naphthol	Tetralol	$C_{10}H_{12}O$	530-91-6	148.201		15.5	140^{12}			
9878	5,6,7,8-Tetrahydro-2-naphthol	5,6,7,8-Tetrahydro-β-naphthol	$C_{10}H_{12}O$	1125-78-6	148.201		59(3)	275.5	1.0552^{65}		
9879	Tetrahydro-6-pentyl-2*H*-pyran-2-one	5-Hydroxydecanoic acid lactone	$C_{10}H_{18}O_2$	705-86-2	170.249	liq	-27	121^{3}			
9880	1,2,3,4-Tetrahydrophenanthrene		$C_{14}H_{14}$	1013-08-7	182.261	lf (MeOH)	29.404 (0.009)	173^{11}	1.0601^{40}		i H_2O; s EtOH, eth, ace, bz, HOAc, chl, lig
9881	1,2,3,6-Tetrahydrophthalimide		$C_8H_9NO_2$	85-40-5	151.163	cry (EtOH)	137				
9882	Tetrahydro-6-propyl-2*H*-pyran-2-one	5-Hydroxyoctanoic acid lactone	$C_8H_{14}O_2$	698-76-0	142.196	liq	-13	126^{15}			
9883	2,3,4,5-Tetrahydro-6-propylpyridine	γ-Coniceine	$C_8H_{15}N$	1604-01-9	125.212			174	0.8753^{15}	1.4661^{16}	
9884	Tetrahydropyran	Oxane	$C_5H_{10}O$	142-68-7	86.132	liq	-49.1(0.2)	88.0(0.4)	0.8814^{20}	1.4200^{20}	s EtOH, eth, bz, ctc
9885	Tetrahydro-2*H*-pyran-2-methanol		$C_6H_{12}O_2$	100-72-1	116.158			189(2)	1.027^{25}	1.458^{20}	
9886	Tetrahydro-2*H*-pyran-2-one		$C_5H_8O_2$	542-28-9	100.117	liq	-10.33(0.09)	228(1)	1.1082^{20}	1.4503^{20}	s H_2O; msc EtOH, eth; sl ctc
9887	Tetrahydro-4*H*-pyran-4-one		$C_5H_8O_2$	29943-42-8	100.117			166.5	1.084^{25}	1.4520^{20}	
9888	1,2,5,6-Tetrahydropyridine	Δ 3-Piperidine	C_5H_9N	694-05-3	83.132	liq	-48	117(4)	0.911^{25}	1.4800^{20}	s chl
9889	1,2,5,6-Tetrahydro-3-pyridinecarboxylic acid	Guvacine	$C_6H_9NO_2$	498-96-4	127.141	pr (w), rods (+1w dil al)	295 dec				vs H_2O
9890	3,4,5,6-Tetrahydro-2(1*H*)-pyrimidinethione	Hexahydropyrimidine-2-thione	$C_4H_8N_2S$	2055-46-1	116.185		211		1.33^{20}		
9891	1,2,3,4-Tetrahydroquinoline		$C_9H_{11}N$	635-46-1	133.190	nd	16.76(0.05)	250.5(0.2)	1.0588^{20}	1.6062^{19}	s H_2O, chl; msc EtOH, eth
9892	5,6,7,8-Tetrahydroquinoline	2,3-Cyclohexenopyridine	$C_9H_{11}N$	10500-57-9	133.190			223.5(0.9)	1.0304^{13}	1.5435^{20}	sl H_2O; s EtOH, eth, ace, bz
9893	1,2,3,4-Tetrahydroquinoxaline		$C_8H_{10}N_2$	3476-89-9	134.178	lf (w, eth, peth)	99	289			s H_2O, chl; vs EtOH, eth, bz; sl peth
9894	6,7,8,9-Tetrahydro-5*H*-tetrazolo[1,5-*a*]azepine	Pentylenetetrazole	$C_6H_{10}N_4$	54-95-5	138.170	cry (bz-lig)	59.5	194^{12}			vs H_2O, EtOH, ace; s eth, bz; sl chl
9895	Tetrahydrothiophene	Thiacyclopentane	C_4H_8S	110-01-0	88.172	liq	-96.13(0.05)	121.1(0.2)	0.9987^{20}	1.4871^{18}	i H_2O; msc EtOH, eth, ace, bz; s chl

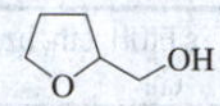
9860
Tetrahydrofurfuryl alcohol

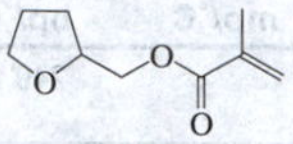
9861
Tetrahydrofurfuryl methacrylate

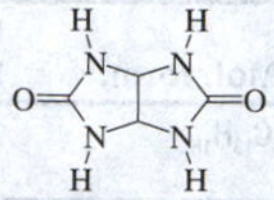
9862
Tetrahydroimidazo[4,5-d]imidazole-2,5(1*H*,3*H*)-dione

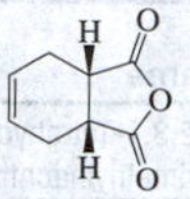
9863
cis-3a,4,7,7a-Tetrahydro-1,3-isobenzofurandione

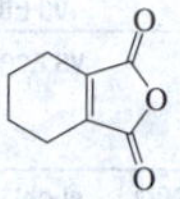
9864
4,5,6,7-Tetrahydro-1,3-isobenzofurandione

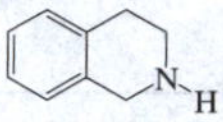
9865
1,2,3,4-Tetrahydroisoquinoline

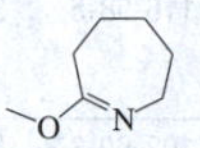
9866
3,4,5,6-Tetrahydro-7-methoxy-2*H*-azepine

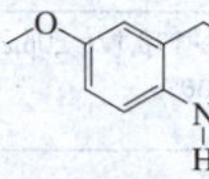
9867
1,2,3,4-Tetrahydro-6-methoxyquinoline

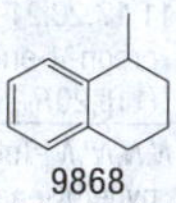
9868
1,2,3,4-Tetrahydro-1-methylnaphthalene

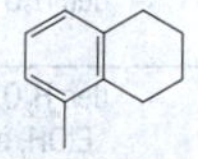
9869
1,2,3,4-Tetrahydro-5-methylnaphthalene

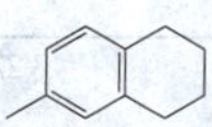
9870
1,2,3,4-Tetrahydro-6-methylnaphthalene

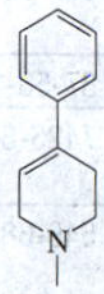
9871
1,2,3,6-Tetrahydro-1-methyl-4-phenylpyridine

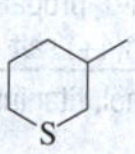
9872
Tetrahydro-3-methyl-2*H*-thiopyran

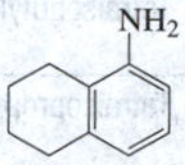
9873
5,6,7,8-Tetrahydro-1-naphthalenamine

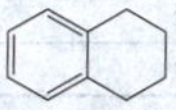
9874
1,2,3,4-Tetrahydronaphthalene

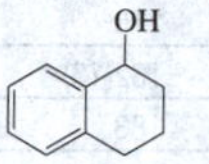
9875
1,2,3,4-Tetrahydro-1-naphthol

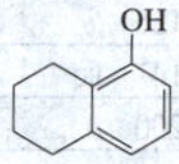
9876
5,6,7,8-Tetrahydro-1-naphthol

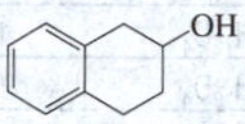
9877
1,2,3,4-Tetrahydro-2-naphthol

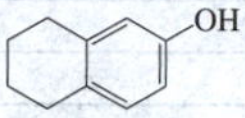
9878
5,6,7,8-Tetrahydro-2-naphthol

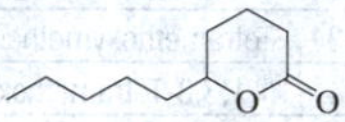
9879
Tetrahydro-6-pentyl-2*H*-pyran-2-one

9880
1,2,3,4-Tetrahydrophenanthrene

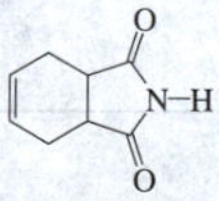
9881
1,2,3,6-Tetrahydrophthalimide

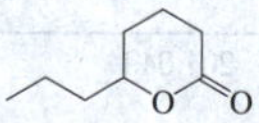
9882
Tetrahydro-6-propyl-2*H*-pyran-2-one

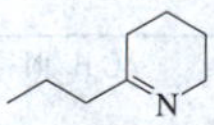
9883
2,3,4,5-Tetrahydro-6-propylpyridine

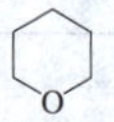
9884
Tetrahydropyran

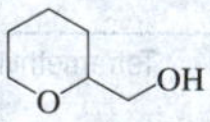
9885
Tetrahydro-2*H*-pyran-2-methanol

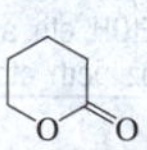
9886
Tetrahydro-2*H*-pyran-2-one

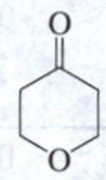
9887
Tetrahydro-4*H*-pyran-4-one

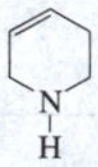
9888
1,2,5,6-Tetrahydropyridine

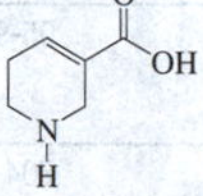
9889
1,2,5,6-Tetrahydro-3-pyridinecarboxylic acid

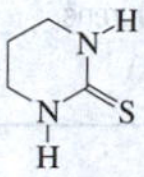
9890
3,4,5,6-Tetrahydro-2(1*H*)-pyrimidinethione

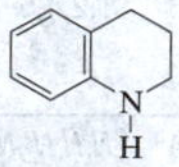
9891
1,2,3,4-Tetrahydroquinoline

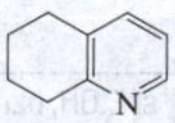
9892
5,6,7,8-Tetrahydroquinoline

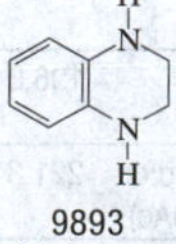
9893
1,2,3,4-Tetrahydroquinoxaline

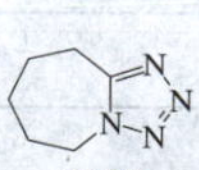
9894
6,7,8,9-Tetrahydro-5*H*-tetrazolo[1,5-a]azepine

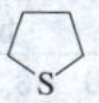
9895
Tetrahydrothiophene

Organic

No.	Name	Synonym	Mol. form.	CAS Reg. No.	Mol. wt.	Physical form	mp/°C	bp/°C	den/ g cm^{-3}	n_D	Solubility
9896	1,2,3,4-Tetrahydro-1,1,6-trimethylnaphthalene		$C_{13}H_{18}$	475-03-6	174.282			240	0.9303^{20}	1.5257^{20}	s EtOH, eth, bz, chl
9897	1,2,5,8-Tetrahydroxy-9,10-anthracenedione	Quinalizarin	$C_{14}H_8O_6$	81-61-8	272.210	oran nd	>275				sl H_2O, ace, bz, EtOH, eth
9898	2,3,4,6-Tetrahydroxy-5*H*-benzocyclohepten-5-one	Purpurogallin	$C_{11}H_8O_5$	569-77-7	220.179	red nd (gl HOAc)	274 dec				
9899	2,2',4,4'-Tetrahydroxybenzophenone		$C_{13}H_{10}O_5$	131-55-5	246.215	ye nd (w+1)	198.8(0.5)				vs H_2O, ace, eth, EtOH
9900	2,3,5,6-Tetrahydroxy-2,5-cyclohexadiene-1,4-dione	Tetroquinone	$C_6H_4O_6$	319-89-1	172.092	bl-blk cry					sl H_2O, eth, ctc; vs EtOH
9901	11,17,20,21-Tetrahydroxypregn-4-en-3-one, (11β,20*R*)-	4-Pregnene-11β,17α,20β,21-tetrol-3-one	$C_{21}H_{32}O_5$	116-58-5	364.476	cry (aq ace)	125 dec				vs ace, EtOH
9902	*N,N,N',N'*-Tetra(2-hydroxypropyl)ethylenediamine	ENTPROL	$C_{14}H_{32}N_2O_4$	102-60-3	292.415				1.030^{25}	1.478^{25}	sl chl
9903	Tetraiodoethene	Tetraiodoethylene	C_2I_4	513-92-8	531.639	ye lf, pr (eth)	187	sub	2.983^{20}		vs bz, chl
9904	4,5,6,7-Tetraiodo-1,3-isobenzofurandione		$C_8I_4O_3$	632-80-4	651.702	ye pr, nd (HOAc) nd (sub)	327.5	sub			i H_2O, EtOH, bz; sl HOAc
9905	Tetraiodomethane	Carbon tetraiodide	CI_4	507-25-5	519.629	red lf (bz, chl)	171	$135^{1.5}$	4.23^{20}		vs py, chl
9906	2,3,4,5-Tetraiodo-1*H*-pyrrole	Iodopyrrole	C_4HI_4N	87-58-1	570.676	ye nd (al)	155(6)				vs ace, eth, chl
9907	Tetraisobutyl titanate	2-Methyl-1-propanol, titanium(4+) salt	$C_{16}H_{36}O_4Ti$	7425-80-1	340.322			256^{500}	0.960^{50}		dec H_2O
9908	Tetraisopropyl titanate	2-Propanol, titanium(4+) salt	$C_{12}H_{28}O_4Ti$	546-68-9	284.215			227.5	0.9711^{20}		dec H_2O; s EtOH, eth, bz, chl
9909	*N,N,N',N'*-Tetrakis(2-hydroxyethyl)-1,2-ethanediamine		$C_{10}H_{24}N_2O_4$	140-07-8	236.309						sl H_2O, EtOH
9910	Tetrakis(hydroxymethyl)-phosphonium chloride		$C_4H_{12}ClO_4P$	124-64-1	190.562		152.5				s H_2O
9911	Tetrakis(methylthio)methane		$C_5H_{12}S_4$	6156-25-8	200.409						s chl
9912	1-Tetralone		$C_{10}H_{10}O$	529-34-0	146.185		8	115^{6}	1.0988^{16}	1.5672^{20}	
9913	Tetramethoxymethane		$C_5H_{12}O_4$	1850-14-2	136.147	liq	-2.5	132(13)	1.023^{25}	1.3845^{20}	
9914	1,1,3,3-Tetramethoxypropane		$C_7H_{16}O_4$	102-52-3	164.200			183	0.997^{25}	1.4081^{20}	
9915	Tetramethrin		$C_{19}H_{25}NO_4$	7696-12-0	331.407	wh cry	≈65-80		1.108^{20}	1.5175^{21}	
9916	*N,N,N',N'*-Tetramethyl-3,6-acridinediamine, monohydrochloride	Acridine Orange	$C_{17}H_{20}ClN_3$	65-61-2	301.814	oran-ye soln					s H_2O, EtOH
9917	Tetramethylammonium bromide		$C_4H_{12}BrN$	64-20-0	154.049	hyg bipym	230 dec		1.56^{25}		vs H_2O; sl EtOH; i eth, bz, chl; s MeOH
9918	Tetramethylammonium chloride		$C_4H_{12}ClN$	75-57-0	109.598	hyg bipym (dil al)	420 dec		1.169^{20}		s H_2O; sl EtOH; i eth, bz, chl; vs MeOH
9919	Tetramethylammonium iodide		$C_4H_{12}IN$	75-58-1	201.049		>230 dec		1.829^{25}		sl H_2O, alk, EtOH, ace; i eth, chl
9920	*N,N*,2,6-Tetramethylaniline		$C_{10}H_{15}N$	769-06-2	149.233	liq	-36	196	0.9147^{20}		
9921	1,2,3,4-Tetramethylbenzene		$C_{10}H_{14}$	488-23-3	134.218	liq	-6.7(1)	205(1)	0.9052^{20}	1.5203^{20}	i H_2O; msc EtOH, eth, ace, bz, peth, ctc
9922	1,2,3,5-Tetramethylbenzene	Isodurene	$C_{10}H_{14}$	527-53-7	134.218	liq	-23.8(0.2)	198(1)	0.8903^{20}	1.5130^{20}	i H_2O; msc EtOH, eth, ace, bz, peth, ctc
9923	1,2,4,5-Tetramethylbenzene	Durene	$C_{10}H_{14}$	95-93-2	134.218		79.2(0.2)	197(1)	0.8380^{81}	1.4790^{81}	i H_2O; msc EtOH, eth, ace, bz, peth, ctc
9924	*N,N,N',N'*-Tetramethyl-1,2-benzenediamine		$C_{10}H_{16}N_2$	704-01-8	164.247		8.9	215.5	0.9560^{20}		
9925	*N,N,N',N'*-Tetramethyl-1,4-benzenediamine	Tetramethyl-*p*-phenylenediamine	$C_{10}H_{16}N_2$	100-22-1	164.247	lf (dil al or lig)	51	260			sl H_2O; vs EtOH, eth, bz, chl
9926	2,3,5,6-Tetramethyl-1,4-benzenediol	Durohydroquinone	$C_{10}H_{14}O_2$	527-18-4	166.217	nd (al)	233				s EtOH; sl eth
9927	Tetramethyl 1,2,4,5-benzenetetracarboxylate		$C_{14}H_{14}O_8$	635-10-9	310.256	nd (al)	144.8(0.5)	sub			vs EtOH
9928	3,3',5,5'-Tetramethyl-[1,1'-biphenyl]-4,4'-diamine		$C_{16}H_{20}N_2$	54827-17-7	240.343		168.5				
9929	*N,N,N',N'*-Tetramethyl-[1,1'-biphenyl]-4,4'-diamine		$C_{16}H_{20}N_2$	366-29-0	240.343		196.0				
9930	3,3',5,5'-Tetramethyl-[1,1'-biphenyl]-4,4'-diol		$C_{16}H_{18}O_2$	2417-04-1	242.313	pa ye nd or pr (HOAc)	221.8	sub			sl EtOH, bz, gl HOAc, tol; i lig

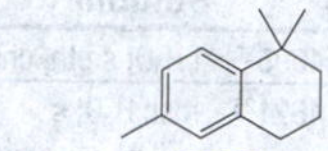
9896
1,2,3,4-Tetrahydro-1,1,6-trimethylnaphthalene

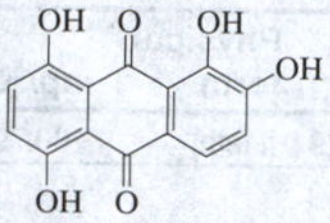
9897
1,2,5,8-Tetrahydroxy-9,10-anthracenedione

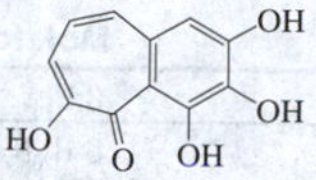
9898
2,3,4,6-Tetrahydroxy-5*H*-benzocyclohepten-5-one

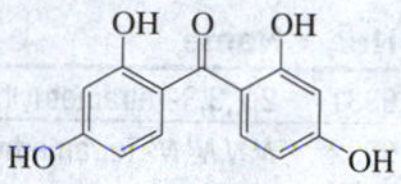
9899
2,2′,4,4′-Tetrahydroxybenzophenone

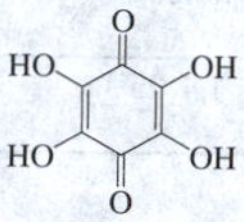
9900
2,3,5,6-Tetrahydroxy-2,5-cyclohexadiene-1,4-dione

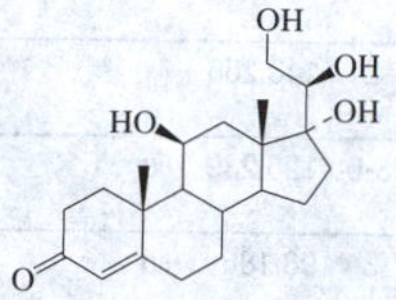
9901
11,17,20,21-Tetrahydroxypregn-4-en-3-one, (11β,20*R*)-

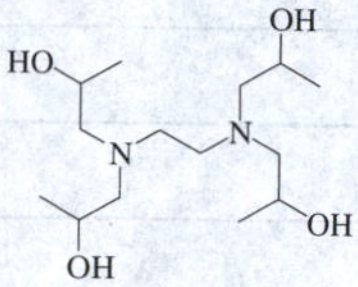
9902
N,N,N′,N′-Tetra(2-hydroxypropyl)ethylenediamine

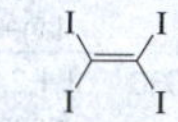
9903
Tetraiodoethene

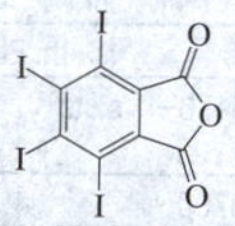
9904
4,5,6,7-Tetraiodo-1,3-isobenzofurandione

9905
Tetraiodomethane

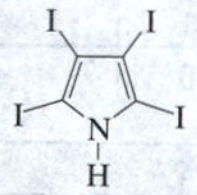
9906
2,3,4,5-Tetraiodo-1*H*-pyrrole

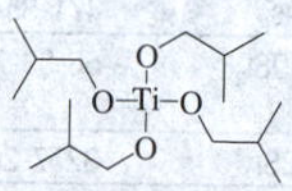
9907
Tetraisobutyl titanate

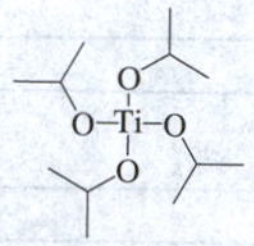
9908
Tetraisopropyl titanate

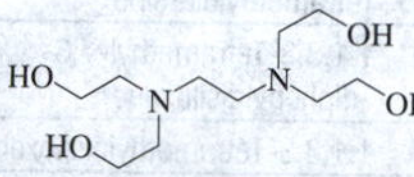
9909
N,N,N′,N′-Tetrakis(2-hydroxyethyl)-1,2-ethanediamine

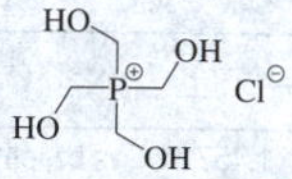
9910
Tetrakis(hydroxymethyl)phosphonium chloride

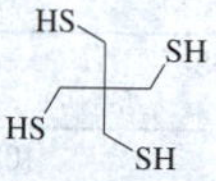
9911
Tetrakis(methylthio)methane

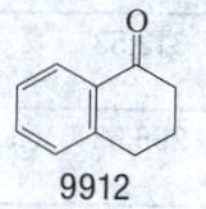
9912
1-Tetralone

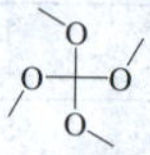
9913
Tetramethoxymethane

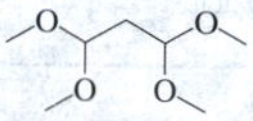
9914
1,1,3,3-Tetramethoxypropane

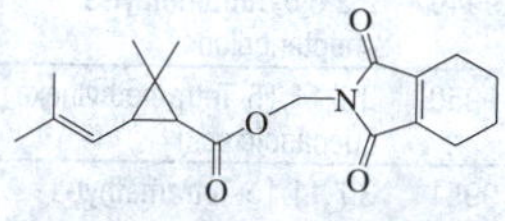
9915
Tetramethrin

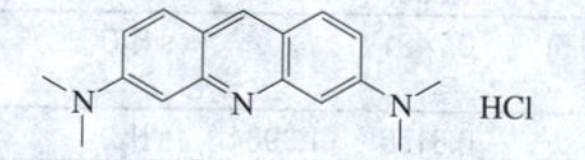
9916
N,N,N′,N′-Tetramethyl-3,6-acridinediamine, monohydrochloride

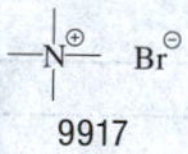
9917
Tetramethylammonium bromide

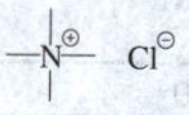
9918
Tetramethylammonium chloride

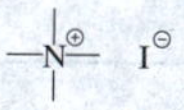
9919
Tetramethylammonium iodide

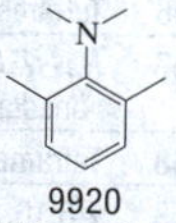
9920
N,N,2,6-Tetramethylaniline

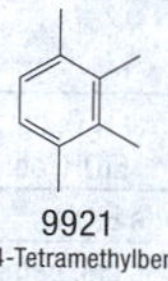
9921
1,2,3,4-Tetramethylbenzene

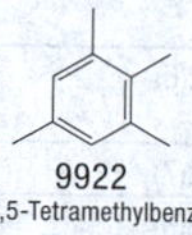
9922
1,2,3,5-Tetramethylbenzene

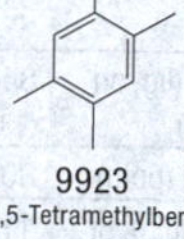
9923
1,2,4,5-Tetramethylbenzene

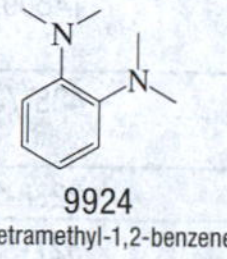
9924
N,N,N′,N′-Tetramethyl-1,2-benzenediamine

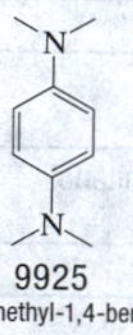
9925
N,N,N′,N′-Tetramethyl-1,4-benzenediamine

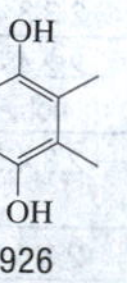
9926
2,3,5,6-Tetramethyl-1,4-benzenediol

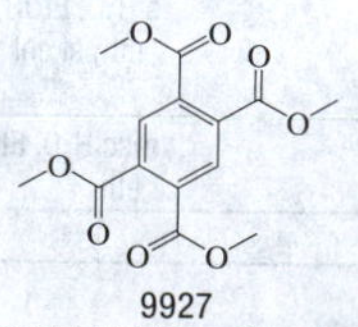
9927
Tetramethyl 1,2,4,5-benzenetetracarboxylate

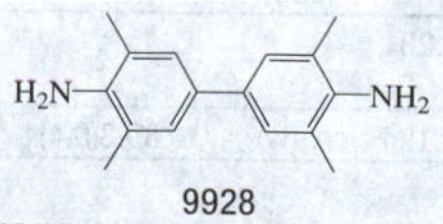
9928
3,3′,5,5′-Tetramethyl-[1,1′-biphenyl]-4,4′-diamine

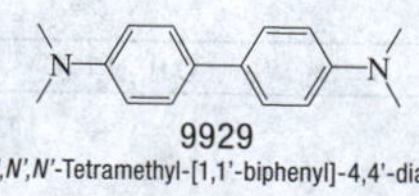
9929
N,N,N′,N′-Tetramethyl-[1,1′-biphenyl]-4,4′-diamine

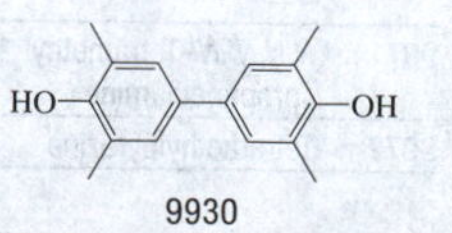
9930
3,3′,5,5′-Tetramethyl-[1,1′-biphenyl]-4,4′-diol

No.	Name	Synonym	Mol. form.	CAS Reg. No.	Mol. wt.	Physical form	mp/°C	bp/°C	den/ g cm^{-3}	n_D	Solubility
9931	2,2,3,3-Tetramethylbutane		C_8H_{18}	594-82-1	114.229	lf (eth)	100.79(0.05)	106.32(0.04)	0.8242^{20}	1.4695^{20}	i H_2O; s eth, chl
9932	*N,N,N',N'*-Tetramethyl-1,4-butanediamine		$C_8H_{20}N_2$	111-51-3	144.258			168	0.7942^{15}	1.4621^{25}	msc H_2O; s EtOH, eth
9933	4-(1,1,3,3-Tetramethylbutyl)-phenol		$C_{14}H_{22}O$	140-66-9	206.324		85.0(0.2)	279			
9934	2,2,4,4-Tetramethyl-1,3-cyclobutanedione		$C_8H_{12}O_2$	933-52-8	140.180						s chl
9935	2,3,5,6-Tetramethyl-2,5-cyclohexadiene-1,4-dione	Duroquinone	$C_{10}H_{12}O_2$	527-17-3	164.201	ye nd (al or lig)	111.69(0.09)				i H_2O; s EtOH, eth, ace, bz, sulf, chl
9936	1,2,3,4-Tetramethylcyclohexane		$C_{10}H_{20}$	3726-45-2	140.266				0.8219^{20}	1.4531^{20}	
9937	1,1,3,3-Tetramethylcyclopentane		C_9H_{18}	50876-33-0	126.239	liq	-88.3(0.2)	118(5)	0.7469^{25}	1.4125^{20}	
9938	1,1,2,2-Tetramethylcyclopropane		C_7H_{14}	4127-47-3	98.186	liq	-80.69(0.09)	76(3)			
9939	2,4,6,8-Tetramethylcyclotetrasiloxane		$C_4H_{16}O_4Si_4$	2370-88-9	240.510	liq	-65	134.5	0.9912^{20}	1.3870^{20}	i H_2O
9940	2,4,7,9-Tetramethyl-5-decyne-4,7-diol		$C_{14}H_{26}O_2$	126-86-3	226.355		47	165^{40}			
9941	*N,N,N',N'*-Tetramethyl-4,4'-diaminobenzophenone	Michler's ketone	$C_{17}H_{20}N_2O$	90-94-8	268.353	lf (al), nd (bz)	179	360 dec			i H_2O, eth; sl EtOH; vs bz; s chl
9942	Tetramethyldiarsine	Cacodyl	$C_4H_{12}As_2$	471-35-2	209.981	liq	-6	165	1.447^{15}		vs eth, EtOH
9943	1,1,3,3-Tetramethyl-1,3-diphenyldisiloxane		$C_{16}H_{22}OSi_2$	56-33-7	286.516	liq	-80	292	0.9763^{20}	1.5176^{20}	s ctc
9944	1,1,3,3-Tetramethyldisiloxane		$C_4H_{14}OSi_2$	3277-26-7	134.324			71	0.756^{20}	1.3700^{20}	
9945	1,1,3,3-Tetramethyl-1,3-disiloxanediol		$C_4H_{14}O_3Si_2$	1118-15-6	166.323		66		1.095^{25}		
9946	*N,N,N',N'*-Tetramethyl-1,2-ethanediamine	1,2-Dimethylaminoethane	$C_6H_{16}N_2$	110-18-9	116.204	liq	-58.0(0.3)	121(1)	0.77^{25}	1.4179^{20}	
9947	Tetramethylgermane	Germanium tetramethyl	$C_4H_{12}Ge$	865-52-1	132.78			32^{500}	1.006		
9948	1,1,3,3-Tetramethylguanidine		$C_5H_{13}N_3$	80-70-6	115.177						s ctc
9949	2,2,6,6-Tetramethyl-3,5-heptanedione	Dipivaloylmethane	$C_{11}H_{20}O_2$	1118-71-4	184.276			93^{35}	0.883^{25}	1.4589^{20}	sl ctc
9950	3,7,11,15-Tetramethylhexadecanoic acid	Phytanic acid	$C_{20}H_{40}O_2$	14721-66-5	312.531		-65				
9951	3,7,11,15-Tetramethyl-1-hexadecen-3-ol	Isophytol	$C_{20}H_{40}O$	505-32-8	296.531	oil		108$^{0.01}$	0.8519^{20}	1.4571^{20}	vs bz, eth, EtOH
9952	2,2,3,3-Tetramethylhexane		$C_{10}H_{22}$	13475-81-5	142.282	liq	-54.0(0.1)	160(2)	0.7609^{25}	1.4282^{20}	
9953	2,2,5,5-Tetramethylhexane		$C_{10}H_{22}$	1071-81-4	142.282	liq	-12.6(0.1)	137(2)	0.7148^{25}	1.4055^{20}	
9954	3,3,4,4-Tetramethylhexane		$C_{10}H_{22}$	5171-84-6	142.282			170(4)	0.7789^{25}	1.4368^{20}	
9955	*N,N,N',N'*-Tetramethyl-1,6-hexanediamine		$C_{10}H_{24}N_2$	111-18-2	172.311			209.5	0.806^{25}	1.4359^{20}	
9956	Tetramethyl lead		$C_4H_{12}Pb$	75-74-1	267.3	liq	-30.2(0.2)	110	1.995^{20}		
9957	*N,N,N',N'*-Tetramethylmethanediamine		$C_5H_{14}N_2$	51-80-9	102.178			83.5(0.5)	0.7491^{18}		s H_2O
9958	Tetramethyloxirane		$C_6H_{12}O$	5076-20-0	100.158			90.4	0.8156^{16}	1.3984^{16}	s H_2O
9959	2,6,10,14-Tetramethylpentadecane	Pristane	$C_{19}H_{40}$	1921-70-6	268.521			306(6)	0.7791^{25}	1.4370^{25}	vs bz, eth, chl, peth
9960	2,2,3,3-Tetramethylpentane		C_9H_{20}	7154-79-2	128.255	liq	-9.75(0.05)	140.2(0.4)	0.7530^{25}	1.4236^{20}	
9961	2,2,3,4-Tetramethylpentane		C_9H_{20}	1186-53-4	128.255	liq	-121.3(0.7)	133.0(0.8)	0.7389^{20}	1.4147^{20}	
9962	2,2,4,4-Tetramethylpentane	Di-*tert*-butylmethane	C_9H_{20}	1070-87-7	128.255	liq	-66.53(0.05)	122.2(1)	0.7195^{20}	1.4069^{20}	i H_2O; vs EtOH, bz
9963	2,3,3,4-Tetramethylpentane		C_9H_{20}	16747-38-9	128.255	liq	-102.1(0.1)	141.5(0.7)	0.7547^{20}	1.4222^{20}	
9964	2,2,4,4-Tetramethyl-3-pentanol		$C_9H_{20}O$	14609-79-1	144.254		49(3)	166(5)			
9965	2,3,4,5-Tetramethylphenol	Prehnitenol	$C_{10}H_{14}O$	488-70-0	150.217	nd (lig, aq al)	84(3)	262(8)			sl H_2O, lig; vs EtOH, eth
9966	2,3,4,6-Tetramethylphenol		$C_{10}H_{14}O$	3238-38-8	150.217	cry (peth)	80(2)	250(7)			s EtOH
9967	2,3,5,6-Tetramethylphenol		$C_{10}H_{14}O$	527-35-5	150.217	nd (lig), pr (al)	118(1)	248(6)			s chl, peth, HOAc
9968	2,2,6,6-Tetramethyl-4-piperidinamine		$C_9H_{20}N_2$	36768-62-4	156.268		17	188.5	0.912^{25}	1.4706^{20}	
9969	2,2,6,6-Tetramethylpiperidine	Norpempidine	$C_9H_{19}N$	768-66-1	141.254		28	153(8)	0.8367^{16}	1.4455^{20}	vs eth
9970	2,2,6,6-Tetramethyl-4-piperidinone		$C_9H_{17}NO$	826-36-8	155.237	orth pl (eth-w) nd (eth)	35.5(0.5)	205			s H_2O, EtOH, eth; sl chl
9971	*N,N,N',N'*-Tetramethyl-1,3-propanediamine		$C_7H_{18}N_2$	110-95-2	130.231			144	0.7837^{18}		msc H_2O, EtOH, eth
9972	Tetramethylpyrazine		$C_8H_{12}N_2$	1124-11-4	136.194	cry (w)	85.3(0.4)	193(16)			

9931
2,2,3,3-Tetramethylbutane

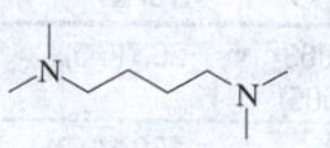
9932
N,N,N',N'-Tetramethyl-1,4-butanediamine

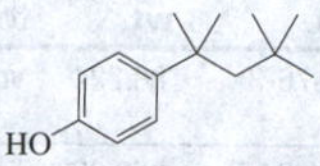
9933
4-(1,1,3,3-Tetramethylbutyl)phenol

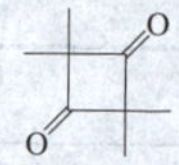
9934
2,2,4,4-Tetramethyl-1,3-cyclobutanedione

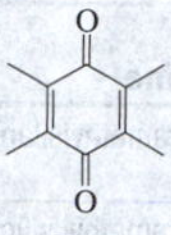
9935
2,3,5,6-Tetramethyl-2,5-cyclohexadiene-1,4-dione

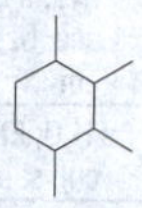
9936
1,2,3,4-Tetramethylcyclohexane

9937
1,1,3,3-Tetramethylcyclopentane

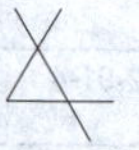
9938
1,1,2,2-Tetramethylcyclopropane

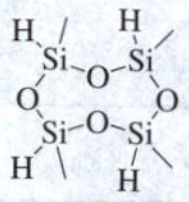
9939
2,4,6,8-Tetramethylcyclotetrasiloxane

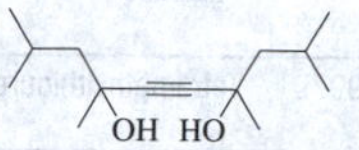
9940
2,4,7,9-Tetramethyl-5-decyne-4,7-diol

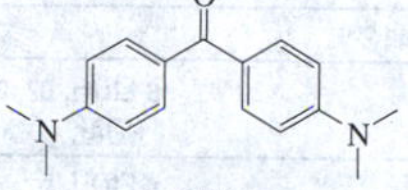
9941
N,N,N',N'-Tetramethyl-4,4'-diaminobenzophenone

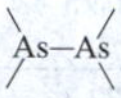
9942
Tetramethyldiarsine

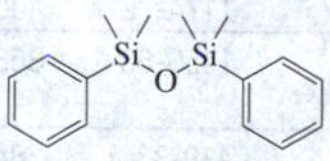
9943
1,1,3,3-Tetramethyl-1,3-diphenyldisiloxane

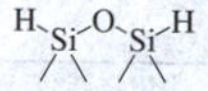
9944
1,1,3,3-Tetramethyldisiloxane

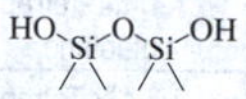
9945
1,1,3,3-Tetramethyl-1,3-disiloxanediol

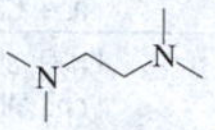
9946
N,N,N',N'-Tetramethyl-1,2-ethanediamine

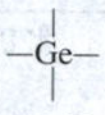
9947
Tetramethylgermane

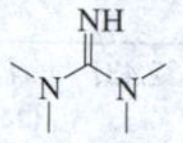
9948
1,1,3,3-Tetramethylguanidine

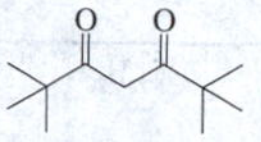
9949
2,2,6,6-Tetramethyl-3,5-heptanedione

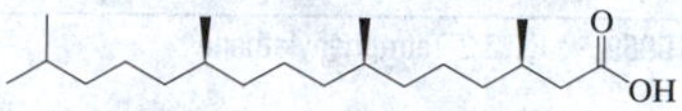
9950
3,7,11,15-Tetramethylhexadecanoic acid

9951
3,7,11,15-Tetramethyl-1-hexadecen-3-ol

9952
2,2,3,3-Tetramethylhexane

9953
2,2,5,5-Tetramethylhexane

9954
3,3,4,4-Tetramethylhexane

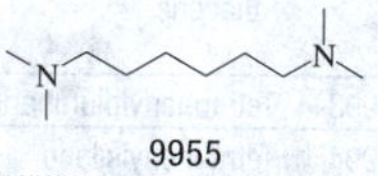
9955
N,N,N',N'-Tetramethyl-1,6-hexanediamine

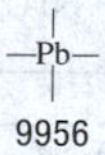
9956
Tetramethyl lead

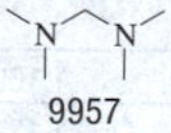
9957
N,N,N',N'-Tetramethylmethanediamine

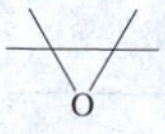
9958
Tetramethyloxirane

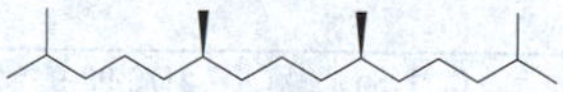
9959
2,6,10,14-Tetramethylpentadecane

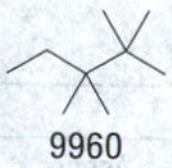
9960
2,2,3,3-Tetramethylpentane

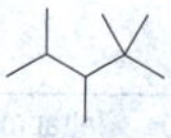
9961
2,2,3,4-Tetramethylpentane

9962
2,2,4,4-Tetramethylpentane

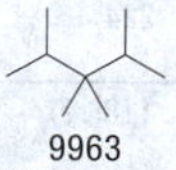
9963
2,3,3,4-Tetramethylpentane

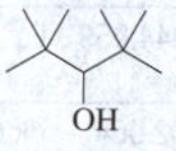
9964
2,2,4,4-Tetramethyl-3-pentanol

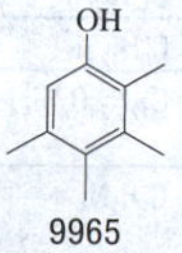
9965
2,3,4,5-Tetramethylphenol

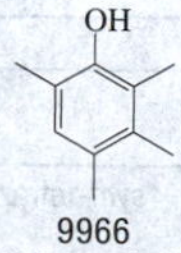
9966
2,3,4,6-Tetramethylphenol

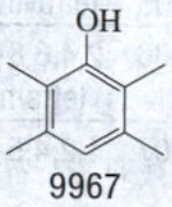
9967
2,3,5,6-Tetramethylphenol

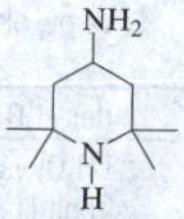
9968
2,2,6,6-Tetramethyl-4-piperidinamine

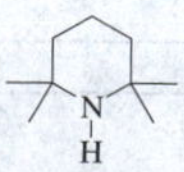
9969
2,2,6,6-Tetramethylpiperidine

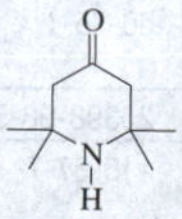
9970
2,2,6,6-Tetramethyl-4-piperidinone

9971
N,N,N',N'-Tetramethyl-1,3-propanediamine

9972
Tetramethylpyrazine

Organic

No.	Name	Synonym	Mol. form.	CAS Reg. No.	Mol. wt.	Physical form	mp/°C	bp/°C	den/ g cm⁻³	n_D	Solubility
9973	Tetramethylsilane	TMS	$C_4H_{12}Si$	75-76-3	88.224	vol liq or gas	-99.063 (0.005)	26.7(0.5)	0.648^{19}	1.3587^{20}	i H_2O; vs EtOH, eth; i sulf
9974	Tetramethyl silicate	Methyl silicate	$C_4H_{12}O_4Si$	681-84-5	152.222	liq	-1.0	120.1(0.7)	1.0232^{20}	1.3683^{20}	vs EtOH
9975	Tetramethylstannane		$C_4H_{12}Sn$	594-27-4	178.848	liq	-55.09(0.04)	76.8(1)	1.314^{25}	1.4386	i H_2O; s ctc, CS_2
9976	Tetramethylsuccinonitrile	Tetramethylbutanedinitrile	$C_8H_{12}N_2$	3333-52-6	136.194	mcl pl, lf, pr (dil al)	169.0(0.5)		1.070^{25}		s EtOH
9977	2,4,6,8-Tetramethyl-2,4,6,8-tetraphenylcyclotetrasiloxane		$C_{28}H_{32}O_4Si_4$	77-63-4	544.894	cry (HOAc)	100.2(0.1)	237^{115}	1.1183^{20}	1.5461^{20}	i H_2O; msc ace, hp
9978	Tetramethylthiodicarbonic diamide		$C_6H_{12}N_2S_3$	97-74-5	208.367		109.5		1.37^{25}		i H_2O; s EtOH, ace, bz, chl; sl eth
9979	Tetramethylthiourea		$C_5H_{12}N_2S$	2782-91-4	132.227		77.3(0.4)	245			s H_2O, EtOH, chl; sl eth
9980	Tetramethylurea		$C_5H_{12}N_2O$	632-22-4	116.161	liq	-2.67(0.02)	177.1(0.6)	0.9687^{20}	1.4496^{23}	sl EtOH, eth, ctc
9981	Tetranitromethane		CN_4O_8	509-14-8	196.033	liq	13.9(0.2)	125.6(0.4)	1.6380^{20}	1.4384^{20}	i H_2O; s EtOH, eth
9982	2,4,8,10-Tetraoxaspiro[5.5]-undecane		$C_7H_{12}O_4$	126-54-5	160.168		50.0(0.5)	147^{53}			vs H_2O, ace, eth, EtOH
9983	2,5,8,11-Tetraoxatridecan-13-ol		$C_9H_{20}O_5$	23783-42-8	208.252			164^{11}	0.987^{25}	1.4453^{20}	
9984	Tetraphenoxysilane	Phenyl silicate	$C_{24}H_{20}O_4Si$	1174-72-7	400.500		49	417	1.1412^{60}		
9985	1,1,4,4-Tetraphenyl-1,3-butadiene		$C_{28}H_{22}$	1450-63-1	358.475		203.5				s EtOH, bz, chl, HOAc
9986	2,3,4,5-Tetraphenyl-2,4-cyclopentadien-1-one		$C_{29}H_{20}O$	479-33-4	384.468	blk-viol lf (HOAc, xyl)	222.3				s EtOH, bz, xyl, HOAc
9987	1,1,2,2-Tetraphenylethane		$C_{26}H_{22}$	632-50-8	334.453	cry (bz), orth nd (chl)	211(2)	360			sl EtOH; s bz, HOAc
9988	1,1,2,2-Tetraphenyl-1,2-ethanediol	Benzopinacol	$C_{26}H_{22}O_2$	464-72-2	366.452	pr (bz), cry (ace)	182				i H_2O, peth; sl EtOH; s eth, ace, CS_2
9989	1,1,2,2-Tetraphenylethene		$C_{26}H_{20}$	632-51-9	332.437	mcl or orth (bz-eth or chl-al)	224.9(0.3)	420	1.155^{0}		i H_2O; sl EtOH, chl, eth; vs bz
9990	Tetraphenylgermane	Germanium tetraphenyl	$C_{24}H_{20}Ge$	1048-05-1	381.06		233(1)				
9991	Tetraphenylmethane		$C_{25}H_{20}$	630-76-2	320.427	orth nd (bz, sub)	281.4(0.4)	431			i H_2O, EtOH, eth, lig, HOAc; s bz, tol
9992	5,6,11,12-Tetraphenylnaphthacene	Rubrene	$C_{42}H_{28}$	517-51-1	532.671	oran-red (bz-lig)	332.5				i H_2O; sl EtOH, eth, ace, py; s bz
9993	Tetraphenylplumbane		$C_{24}H_{20}Pb$	595-89-1	515.6		228.3	126^{13}	1.5298^{20}		s chl
9994	Tetraphenylsilane		$C_{24}H_{20}Si$	1048-08-4	336.502		237.6(0.6)	228^{3}	1.078^{20}		s ctc, CS_2
9995	Tetraphenylstannane		$C_{24}H_{20}Sn$	595-90-4	427.126		227(2)	420			sl chl
9996	Tetrapropoxysilane	Propyl silicate	$C_{12}H_{28}O_4Si$	682-01-9	264.434			226	0.9158^{20}	1.4012^{20}	s ctc, CS_2
9997	Tetrapropylammonium bromide	*N,N,N*-Tripropyl-1-propanaminium bromide	$C_{12}H_{28}BrN$	1941-30-6	266.261		252				vs H_2O, chl
9998	Tetrapropylammonium iodide		$C_{12}H_{28}IN$	631-40-3	313.261	orth bipym	280 dec		1.3138^{25}		vs H_2O, chl; s EtOH, HOAc; sl eth
9999	Tetrapropylstannane		$C_{12}H_{28}Sn$	2176-98-9	291.060	liq	-109.1	228	1.1065^{20}	1.4745^{20}	
10000	Tetrapropyl thiodiphosphate	Aspon	$C_{12}H_{28}O_5P_2S_2$	3244-90-4	378.425	amber liq		$104^{0.1}$	1.12^{25}	1.4710^{21}	sl H_2O, peth
10001	Tetrapropyl titanate	1-Propanol, titanium(4+) salt	$C_{12}H_{28}O_4Ti$	3087-37-4	284.215			206^{100}			
10002	Tetrasodium EDTA	Edetate sodium	$C_{10}H_{12}N_2Na_4O_8$	64-02-8	380.169	amorp pow	300 (dihydrate)				sl EtOH
10003	Tetratetracontane		$C_{44}H_{90}$	7098-22-8	619.186		86.0(0.6)				
10004	Tetratriacontane		$C_{34}H_{70}$	14167-59-0	478.920	pl (eth)	72.8(0.3)	285.4^{3}	0.7728^{90}	1.4296^{90}	
10005	Tetravinylsilane		$C_8H_{12}Si$	1112-55-6	136.267			130.2	0.7999^{20}	1.4625^{20}	
10006	2,4,6,8-Tetravinyl-2,4,6,8-tetramethylcyclotetrasiloxane		$C_{12}H_{24}O_4Si_4$	2554-06-5	344.659	liq	-43.5	224	0.9875^{20}		s ctc, CS_2
10007	1,2,4,5-Tetrazine	*sym*-Tetrazine	$C_2H_2N_4$	290-96-0	82.064	dk red pr	99	118(17)			s H_2O, EtOH, eth, sulf
10008	1*H*-Tetrazol-5-amine		CH_3N_5	4418-61-5	85.069		198.8(0.9)				
10009	1*H*-Tetrazole		CH_2N_4	288-94-8	70.054	pl (al)	156.9(0.8)	sub	1.4060^{20}		sl H_2O
10010	Tetrodotoxin		$C_{11}H_{17}N_3O_8$	4368-28-9	319.268	cry	225 dec				sl H_2O,eth, EtOH; s dil HOAc
10011	Thalidomide	2-(2,6-Dioxo-3-piperidinyl)-1*H*-isoindole-1,3(2*H*)-dione	$C_{13}H_{10}N_2O_4$	50-35-1	258.229	nd	270				vs py, diox
10012	Thallium(I) ethanolate	Thallous ethoxide	C_2H_5OTl	20398-06-5	249.443	cloudy liq	-3	130 dec	3.49		dec H_2O
10013	Thebaine		$C_{19}H_{21}NO_3$	115-37-7	311.375	pl (eth), pr (dil al)	193	91 sub	1.305^{20}		i H_2O; vs EtOH, chl; sl eth; s bz

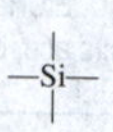
9973
Tetramethylsilane

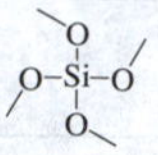
9974
Tetramethyl silicate

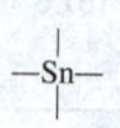
9975
Tetramethylstannane

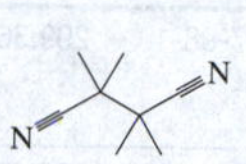
9976
Tetramethylsuccinonitrile

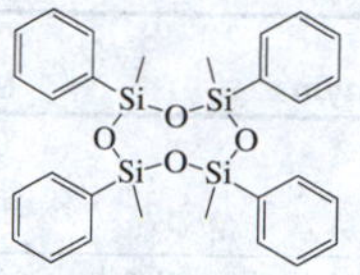
9977
2,4,6,8-Tetramethyl-2,4,6,8-tetraphenylcyclotetrasiloxane

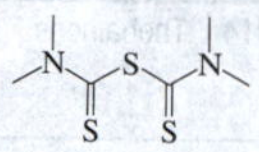
9978
Tetramethylthiodicarbonic diamide

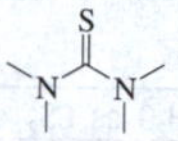
9979
Tetramethylthiourea

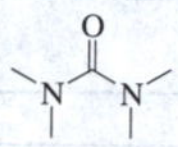
9980
Tetramethylurea

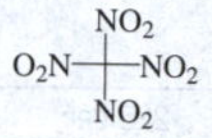
9981
Tetranitromethane

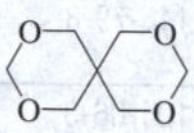
9982
2,4,8,10-Tetraoxaspiro[5.5]undecane

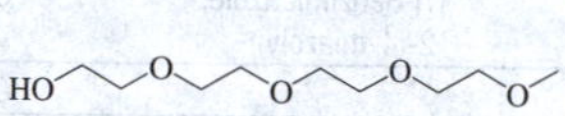
9983
2,5,8,11-Tetraoxatridecan-13-ol

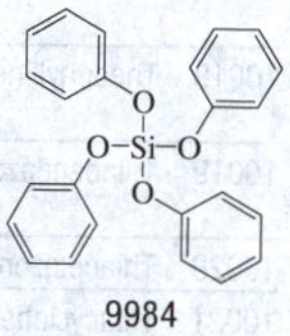
9984
Tetraphenoxysilane

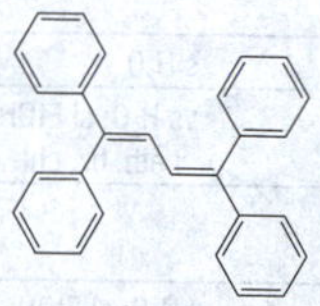
9985
1,1,4,4-Tetraphenyl-1,3-butadiene

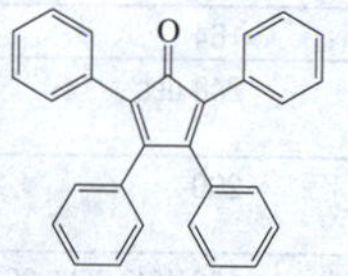
9986
2,3,4,5-Tetraphenyl-2,4-cyclopentadien-1-one

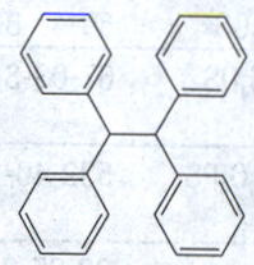
9987
1,1,2,2-Tetraphenylethane

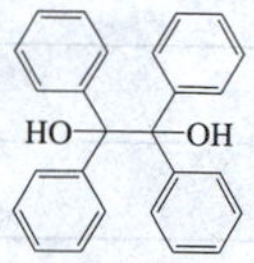
9988
1,1,2,2-Tetraphenyl-1,2-ethanediol

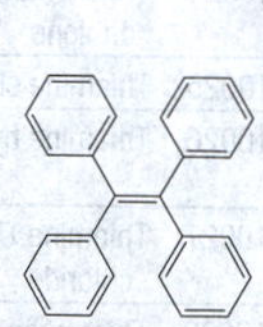
9989
1,1,2,2-Tetraphenylethene

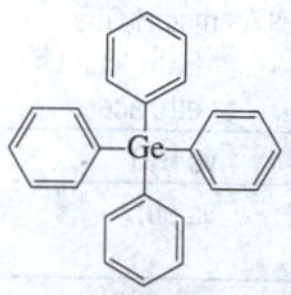
9990
Tetraphenylgermane

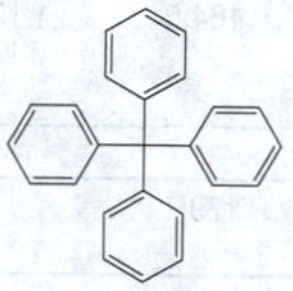
9991
Tetraphenylmethane

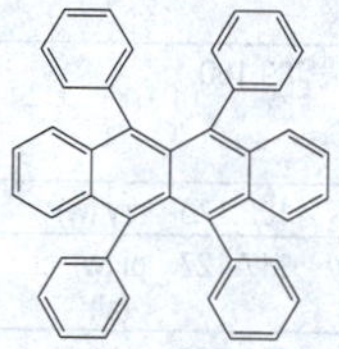
9992
5,6,11,12-Tetraphenylnaphthacene

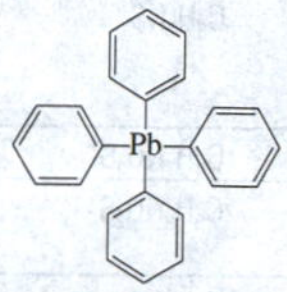
9993
Tetraphenylplumbane

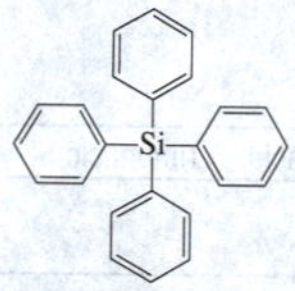
9994
Tetraphenylsilane

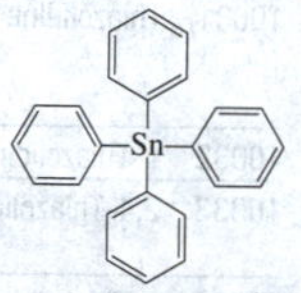
9995
Tetraphenylstannane

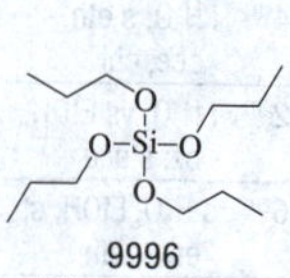
9996
Tetrapropoxysilane

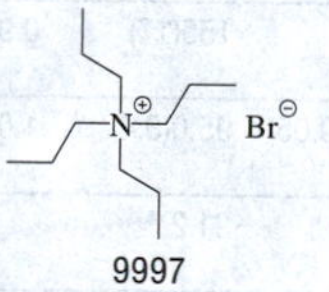
9997
Tetrapropylammonium bromide

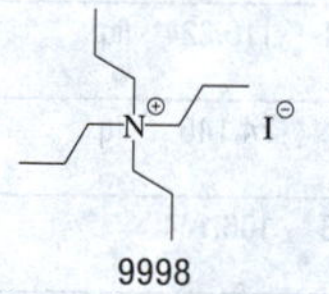
9998
Tetrapropylammonium iodide

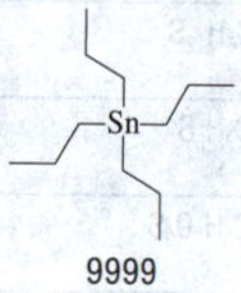
9999
Tetrapropylstannane

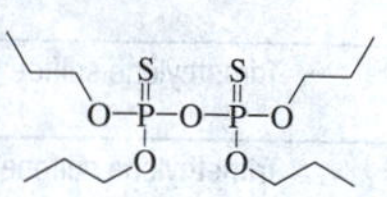
10000
Tetrapropyl thiodiphosphate

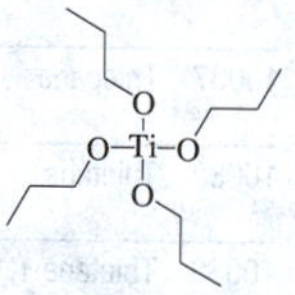
10001
Tetrapropyl titanate

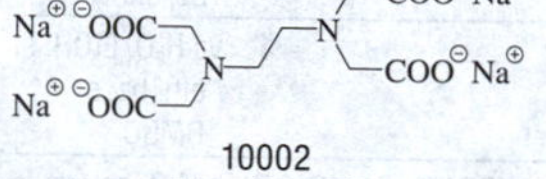
10002
Tetrasodium EDTA

$H_3C(CH_2)_{42}CH_3$
10003
Tetratetracontane

$H_3C(CH_2)_{32}CH_3$
10004
Tetratriacontane

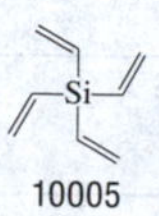
10005
Tetravinylsilane

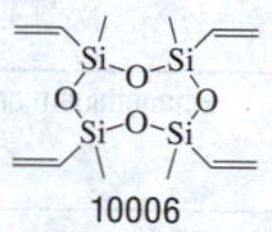
10006
2,4,6,8-Tetravinyl-2,4,6,8-tetramethylcyclotetrasiloxane

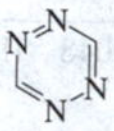
10007
1,2,4,5-Tetrazine

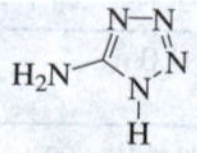
10008
1*H*-Tetrazol-5-amine

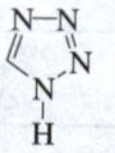
10009
1*H*-Tetrazole

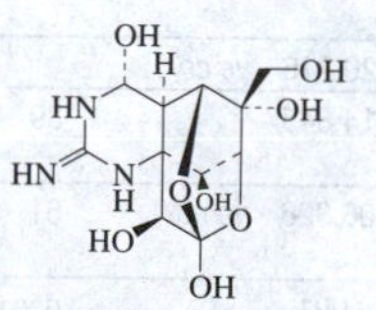
10010
Tetrodotoxin

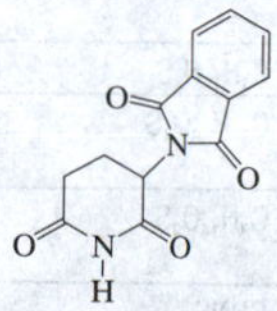
10011
Thalidomide

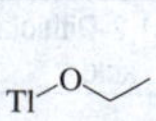
10012
Thallium(I) ethanolate

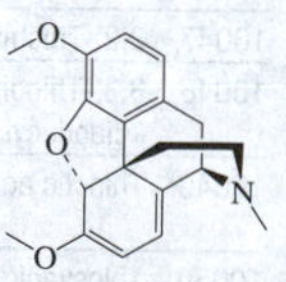
10013
Thebaine

Organic

No.	Name	Synonym	Mol. form.	CAS Reg. No.	Mol. wt.	Physical form	mp/°C	bp/°C	den/g cm^{-3}	n_D	Solubility
10014	Thebainone		$C_{18}H_{21}NO_3$	467-98-1	299.365	nd or pr (al)	151.5				sl H_2O, EtOH, eth; s ace, bz, AcOEt
10015	Thenaldine	1-Methyl-*N*-phenyl-*N*-(2-thienylmethyl)-4-piperidinamine	$C_{17}H_{22}N_2S$	86-12-4	286.435		96	$159^{0.02}$			
10016	Thenyldiamine		$C_{14}H_{19}N_3S$	91-79-2	261.386			170^{1}		1.5915^{20}	
10017	Theobromine		$C_7H_8N_4O_2$	83-67-0	180.165	orth or mcl nd (w)	357	290 sub			sl H_2O, EtOH; i eth, bz, ctc, lig, chl
10018	Theophylline	3,7-Dihydro-1,3-dimethyl-1*H*-purine-2,6-dione	$C_7H_8N_4O_2$	58-55-9	180.165	nd or pl (w+1)	274.7(0.5)				s H_2O; sl EtOH, eth, chl
10019	Thiabendazole	1*H*-Benzimidazole, 2-(4-thiazolyl)-	$C_{10}H_7N_3S$	148-79-8	201.248			305 sub			
10020	Thiacetazone		$C_{10}H_{12}N_4OS$	104-06-3	236.293		225 dec				i H_2O, os, CS_2
10021	Thiacyclohexane		$C_5H_{10}S$	1613-51-0	102.198		19.09(0.06)	141.73(0.04)	0.9861^{20}	1.5067^{20}	i H_2O; s EtOH, eth, ace, bz
10022	1,2,5-Thiadiazole	Piazthiole	$C_2H_2N_2S$	288-39-1	86.115	liq	-50.1	94	1.268^{25}	1.5150^{25}	
10023	1,3,4-Thiadiazole		$C_2H_2N_2S$	289-06-5	86.115	cry (sub)	42.5	204			
10024	1,3,4-Thiadiazolidine-2,5-dithione		$C_2H_2N_2S_3$	1072-71-5	150.245	ye cry (MeOH)	168				s H_2O
10025	Thiamine chloride		$C_{12}H_{17}ClN_4OS$	59-43-8	300.807	cry	164				s H_2O
10026	Thiamine hydrochloride		$C_{12}H_{18}Cl_2N_4OS$	67-03-8	337.268	mcl pl	248 dec				vs H_2O; sl EtOH; i eth, bz, chl
10027	Thiamine *O*-phosphate, chloride		$C_{12}H_{18}ClN_4O_4PS$	532-40-1	380.787		200				
10028	Thianthrene		$C_{12}H_8S_2$	92-85-3	216.322	mcl pr or pl (al)	156.43(0.05)	366.3(0.7)	1.4420^{20}		i H_2O; sl EtOH; s eth, bz, CS_2
10029	2-Thiazolamine	2-Aminothiazole	$C_3H_4N_2S$	96-50-4	100.142	ye pl (al)	93	140^{11}			sl H_2O, EtOH, eth, chl; vs dil HCl
10030	Thiazole		C_3H_3NS	288-47-1	85.128		-33.61(0.04)	118.2(0.1)	1.1998^{17}	1.5969^{20}	sl H_2O; s EtOH, eth, ace
10031	Thiazolidine		C_3H_7NS	504-78-9	89.160			164.5	1.131^{25}	1.551^{20}	msc H_2O; s EtOH, ctc; vs eth, ace
10032	4-Thiazolidinecarboxylic acid	Timonacic	$C_4H_7NO_2S$	444-27-9	133.170	cry (w)	196.5				vs H_2O
10033	2,4-Thiazolidinedione		$C_3H_3NO_2S$	2295-31-0	117.127	pl (w), pr (al)	128	179^{19}			vs eth
10034	2-Thiazolidinethione		$C_3H_5NS_2$	96-53-7	119.209	nd (w, MeOH)	107.3				s H_2O, bz, chl; sl EtOH; i eth, CS_2
10035	Thidiazuron		$C_9H_8N_4OS$	51707-55-2	220.251		211 dec				
10036	1-(2-Thienyl)ethanone		C_6H_6OS	88-15-3	126.176		10.4(0.2)	215(16)	1.1679^{20}	1.5667^{20}	sl H_2O; msc EtOH, eth; s ctc
10037	Thiepane	Hexamethylene sulfide	$C_6H_{12}S$	4753-80-4	116.224	liq	0.5	155(17)	0.991^{20}	1.5044^{18}	i H_2O; s eth, ace, chl
10038	Thietane	Trimethylene sulfide	C_3H_6S	287-27-4	74.145	liq	-73.19(0.05)	95.0(0.4)	1.0200^{20}	1.5102^{20}	i H_2O; vs EtOH, bz; s ace
10039	Thietane 1,1-dioxide	Trimethylene sulfone	$C_3H_6O_2S$	5687-92-3	106.144		75.5	91.2^{14}		1.5156^{20}	s H_2O, EtOH; sl eth, peth
10040	Thiethylperazine		$C_{22}H_{29}N_3S_2$	1420-55-9	399.615	cry	63	$227^{0.01}$			sl ace
10041	Thiirane	Ethylene sulfide	C_2H_4S	420-12-2	60.118		-109	54.9(0.5)	1.0130^{20}	1.4935^{20}	sl EtOH, eth; s ace, chl
10042	Thioacetaldehyde trimer	2,4,6-Trimethyl-1,3,5-trithiane	$C_6H_{12}S_3$	2765-04-0	180.354	α-mcl pl; β-nd (ace)	101	246.5			i H_2O; s EtOH, eth, ace; vs bz, chl
10043	Thioacetamide	Ethanethioamide	C_2H_5NS	62-55-5	75.133		113.3(0.5)				vs H_2O, EtOH; sl eth, bz; s DMSO
10044	Thioacetic acid		C_2H_4OS	507-09-5	76.117	ye fuming liq	<-17	93	1.064^{20}	1.4648^{20}	s H_2O, chl; vs EtOH, ace; msc eth
10045	Thiobencarb		$C_{12}H_{16}ClNOS$	28249-77-6	257.779		1.7	$127^{0.008}$	1.16^{20}		
10046	4,4'-Thiobis(6-*tert*-butyl-*m*-cresol)	Bis(5-*tert*-butyl-4-hydroxy-2-methylphenyl) sulfide	$C_{22}H_{30}O_2S$	96-69-5	358.537	cry	163				
10047	2,2'-Thiobisethanamine	Bis(2-aminoethyl) sulfide	$C_4H_{12}N_2S$	871-76-1	120.216	ye cry		232			
10048	3,3'-Thiobispropanoic acid, didodecyl ester	Didodecyl thiobispropanoate	$C_{30}H_{58}O_4S$	123-28-4	514.845		39				
10049	Thioctic acid	1,2-Dithiolane-3-pentanoic acid	$C_8H_{14}O_2S_2$	62-46-4	206.326	ye nd	61	162			i H_2O
10050	Thiocyanic acid		CHNS	463-56-9	59.091		dec 0				vs H_2O; s os

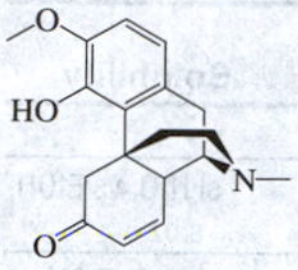
10014
Thebainone

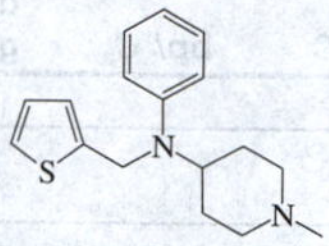
10015
Thenaldine

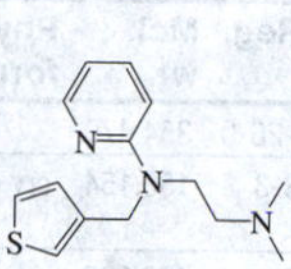
10016
Thenyldiamine

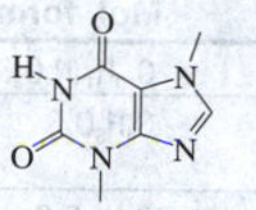
10017
Theobromine

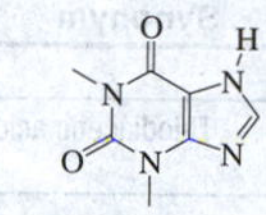
10018
Theophylline

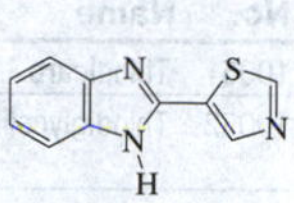
10019
Thiabendazole

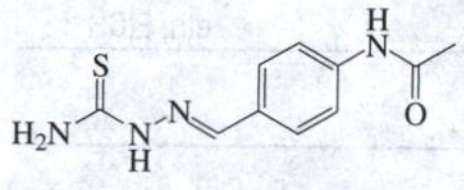
10020
Thiacetazone

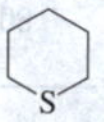
10021
Thiacyclohexane

10022
1,2,5-Thiadiazole

10023
1,3,4-Thiadiazole

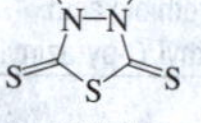
10024
1,3,4-Thiadiazolidine-2,5-dithione

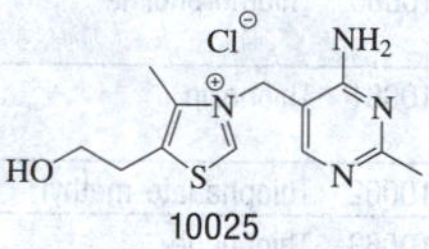
10025
Thiamine chloride

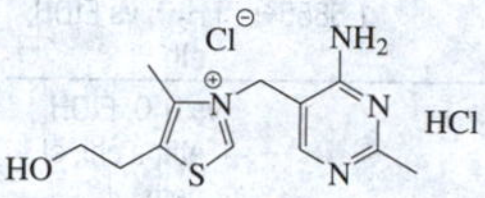
10026
Thiamine hydrochloride

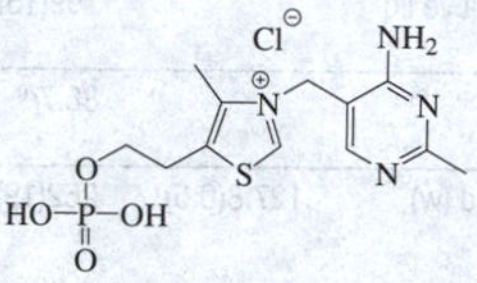
10027
Thiamine *O*-phosphate, chloride

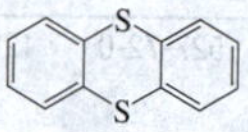
10028
Thianthrene

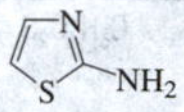
10029
2-Thiazolamine

10030
Thiazole

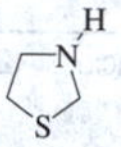
10031
Thiazolidine

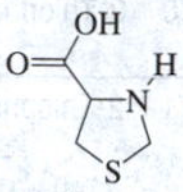
10032
4-Thiazolidinecarboxylic acid

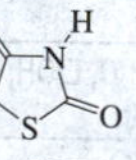
10033
2,4-Thiazolidinedione

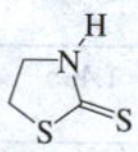
10034
2-Thiazolidinethione

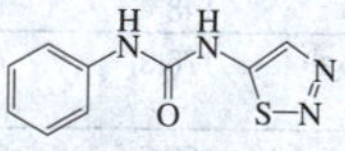
10035
Thidiazuron

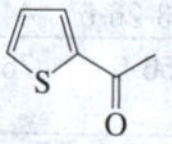
10036
1-(2-Thienyl)ethanone

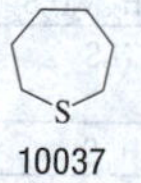
10037
Thiepane

10038
Thietane

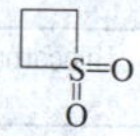
10039
Thietane 1,1-dioxide

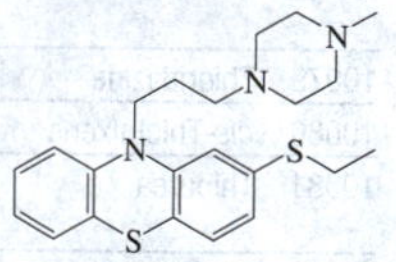
10040
Thiethylperazine

10041
Thiirane

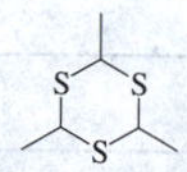
10042
Thioacetaldehyde trimer

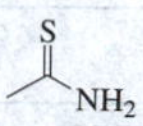
10043
Thioacetamide

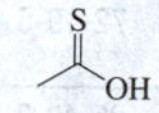
10044
Thioacetic acid

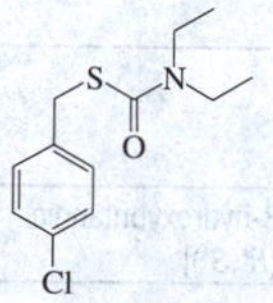
10045
Thiobencarb

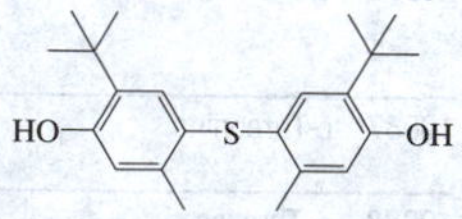
10046
4,4'-Thiobis(6-*tert*-butyl-*m*-cresol)

10047
2,2'-Thiobisethanamine

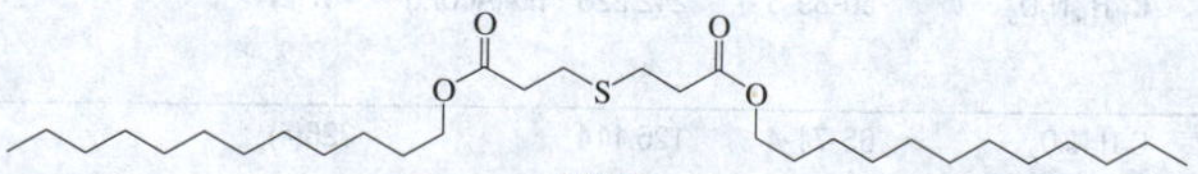
10048
3,3'-Thiobispropanoic acid, didodecyl ester

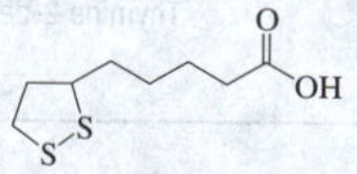
10049
Thioctic acid

10050
Thiocyanic acid

No.	Name	Synonym	Mol. form.	CAS Reg. No.	Mol. wt.	Physical form	mp/°C	bp/°C	den/ g cm^{-3}	n_D	Solubility
10051	Thiodicarb		$C_{10}H_{18}N_4O_4S_3$	59669-26-0	354.470		173		1.4^{20}		
10052	Thiodiglycolic acid	Thiodiacetic acid	$C_4H_6O_4S$	123-93-3	150.154	cry (w)	129				sl H_2O; vs EtOH; s bz
10053	3,3'-Thiodipropionic acid		$C_6H_{10}O_4S$	111-17-1	178.206	cry wh pow	129				vs H_2O, EtOH
10054	Thiofanox		$C_9H_{18}N_2O_2S$	39196-18-4	218.316		58.0(0.5)				
10055	Thioformaldehyde	Methanethial	CH_2S	865-36-1	46.092	unstab gas					
10056	Thioglycolic acid		$C_2H_4O_2S$	68-11-1	92.117		-16(2)	120^{20}	1.3253^{20}	1.5080^{20}	msc H_2O, EtOH, eth; sl chl
10057	Thioimidodicarbonic diamide	2,4-Dithiobiuret	$C_2H_5N_3S_2$	541-53-7	135.211	mcl cry	181 dec				vs ace
10058	Thiolactic acid		$C_3H_6O_2S$	71563-86-5	106.144		18.7(0.2)	106^{15}	1.1938^{20}	1.4810^{20}	s H_2O, EtOH, eth; sl chl
10059	Thiometon		$C_6H_{15}O_2PS_3$	640-15-3	246.351	oil		$110^{0.1}$	1.209^{20}		sl H_2O; s os
10060	Thiomorpholine	Thiamorpholine	C_4H_9NS	123-90-0	103.186			175	1.0882^{20}	1.5386^{20}	vs H_2O, ace, eth, EtOH
10061	Thionazin	Phosphorothioic acid, *O,O*-diethyl *O*-pyrazinyl ester	$C_8H_{13}N_2O_3PS$	297-97-2	248.239	liq	-0.9	80			
10062	Thiophanate-methyl		$C_{12}H_{14}N_4O_4S_2$	23564-05-8	342.394		172 dec				
10063	Thiophene	Thiofuran	C_4H_4S	110-02-1	84.140	liq	-38.12(0.05)	84.1(0.1)	1.0649^{20}	1.5289^{20}	msc EtOH, eth, ace, bz, ctc, diox, py; sl chl
10064	2-Thiopheneacetic acid		$C_6H_6O_2S$	1918-77-0	142.176	cry (w)	64.2(0.5)				vs H_2O, eth, EtOH
10065	2-Thiopheneacetonitrile		C_6H_5NS	20893-30-5	123.176			120^{23}	1.155^{25}	1.5425^{20}	
10066	2-Thiophenecarbonitrile	2-Cyanothiophene	C_5H_3NS	1003-31-2	109.150			193(23)	1.172^{25}	1.5629^{20}	s chl
10067	3-Thiophenecarbonitrile	3-Cyanothiophene	C_5H_3NS	1641-09-4	109.150	oil		204			
10068	2-Thiophenecarbonyl chloride		C_5H_3ClOS	5271-67-0	146.595			280			
10069	2-Thiophenecarboxaldehyde		C_5H_4OS	98-03-3	112.150	pa ye liq		199(15)	1.2127^{21}	1.5920^{20}	i H_2O; vs EtOH; s eth; sl chl
10070	3-Thiophenecarboxaldehyde	3-Formylthiophene	C_5H_4OS	498-62-4	112.150			86.7^{20}		1.5855^{20}	i H_2O; vs EtOH, eth
10071	2-Thiophenecarboxylic acid	2-Carboxythiophene	$C_5H_4O_2S$	527-72-0	128.150	nd (w)	127.8(0.5)	262(19)			vs H_2O, EtOH, eth; s chl; sl peth
10072	3-Thiophenecarboxylic acid	3-Thenoic acid	$C_5H_4O_2S$	88-13-1	128.150		139.8(0.5)				s H_2O
10073	2,5-Thiophenedicarboxylic acid	2,5-Dicarboxythiophene	$C_6H_4O_4S$	4282-31-9	172.159		359	409(10)			sl H_2O; s EtOH, eth
10074	2-Thiophenemethanol		C_5H_6OS	636-72-6	114.166			207	1.2053^{16}	1.5280^{20}	s EtOH, ace
10075	2-Thiophenesulfonyl chloride		$C_4H_3ClO_2S_2$	16629-19-9	182.649		28	100^{6}			s eth
10076	Thiopropazate		$C_{23}H_{28}ClN_3O_2S$	84-06-0	446.005			$216^{0.1}$			
10077	4*H*-Thiopyran-4-thione		$C_5H_4S_2$	1120-94-1	128.216		47				
10078	Thioquinox		$C_9H_4N_2S_3$	93-75-4	236.336	br-ye pow	180				i H_2O; sl ace, EtOH, peth
10079	Thioridazine		$C_{21}H_{26}N_2S_2$	50-52-2	370.58	cry	73	$230^{0.02}$			sl ace
10080	*cis*-Thiothixene		$C_{23}H_{29}N_3O_2S_2$	3313-26-6	443.625	cry	148				
10081	Thiourea	Thiocarbamide	CH_4N_2S	62-56-6	76.121	orth (al)	176(3)		1.405^{25}		s H_2O, EtOH; i eth
10082	9*H*-Thioxanthene	Dibenzothiapyran	$C_{13}H_{10}S$	261-31-4	198.283	nd (al-chl)	128.5	341			s chl
10083	9*H*-Thioxanthen-9-one	Thioxanthone	$C_{13}H_8OS$	492-22-8	212.267	ye nd (chl)	214.7(0.2)	373			i H_2O, peth; sl EtOH; s bz, chl, CS_2
10084	2-Thioxo-4-imidazolidinone	2-Thiohydantoin	$C_3H_4N_2OS$	503-87-7	116.141	wh nd (w)	230 dec				vs H_2O, EtOH; s eth, alk
10085	2-Thioxo-4-thiazolidinone	Rhodanine	$C_3H_3NOS_2$	141-84-4	133.192	lt ye pr (al, w)	170		0.868^{25}		sl H_2O, DMSO; vs EtOH, eth
10086	Thiram		$C_6H_{12}N_2S_4$	137-26-8	240.432	wh or ye mcl (chl-al)	155.6	129^{20}			vs chl
10087	L-Threonine	2-Amino-3-hydroxybutanoic acid, [*R*-(*R**,*S**)]	$C_4H_9NO_3$	72-19-5	119.119		256 dec				s H_2O; i EtOH, eth, chl
10088	D-Threose		$C_4H_8O_4$	95-43-2	120.105	hyg-syr or nd (w)	129				
10089	L-Threose		$C_4H_8O_4$	95-44-3	120.105						vs H_2O
10090	Thujic acid	5,5-Dimethyl-1,3,6-cycloheptatriene-1-carboxylic acid	$C_{10}H_{12}O_2$	499-89-8	164.201	cry (peth)	88.5				
10091	α-Thujone	4-Methyl-1-(1-methylethyl)-bicyclo[3.1.0]hexan-3-one, (*l*)	$C_{10}H_{16}O$	546-80-5	152.233			221(16)	0.9109^{25}	1.4490^{15}	i H_2O; s EtOH
10092	3-Thujopsene	Widdrene	$C_{15}H_{24}$	470-40-6	204.352	liq		122^{12}	0.932^{24}	1.5031^{25}	
10093	Thymidine	Thymine 2-desoxyriboside	$C_{10}H_{14}N_2O_5$	50-89-5	242.228	nd (AcOEt)	187(2)				s H_2O, EtOH, ace, py, HOAc; sl chl
10094	Thymine		$C_5H_6N_2O_2$	65-71-4	126.114		325(4)				sl H_2O, EtOH, eth, DMSO

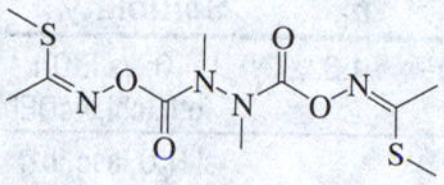
10051
Thiodicarb

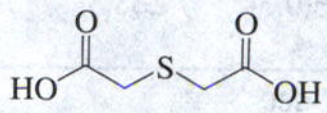
10052
Thiodiglycolic acid

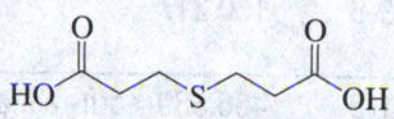
10053
3,3'-Thiodipropionic acid

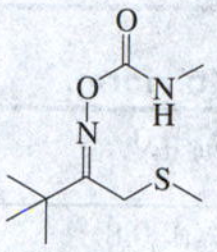
10054
Thiofanox

10055
Thioformaldehyde

10056
Thioglycolic acid

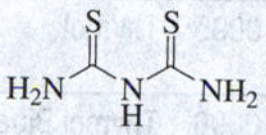
10057
Thioimidodicarbonic diamide

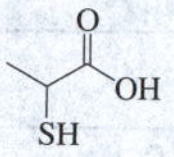
10058
Thiolactic acid

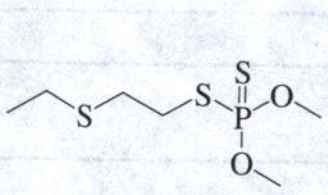
10059
Thiometon

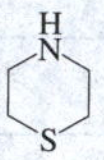
10060
Thiomorpholine

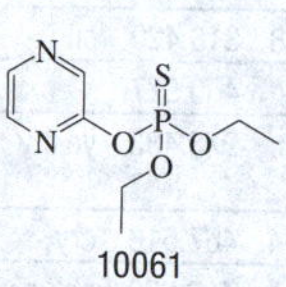
10061
Thionazin

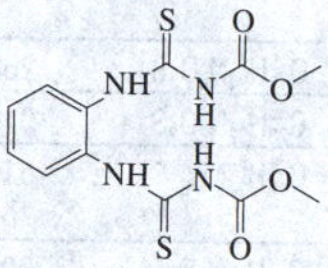
10062
Thiophanate-methyl

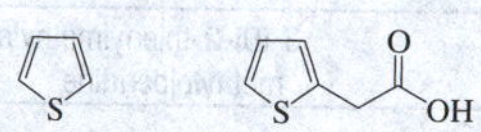
10063
Thiophene

10064
2-Thiopheneacetic acid

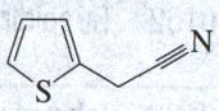
10065
2-Thiopheneacetonitrile

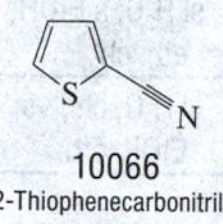
10066
2-Thiophenecarbonitrile

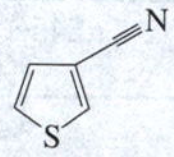
10067
3-Thiophenecarbonitrile

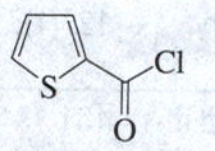
10068
2-Thiophenecarbonyl chloride

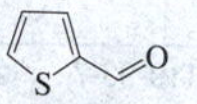
10069
2-Thiophenecarboxaldehyde

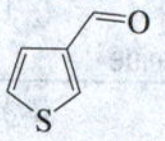
10070
3-Thiophenecarboxaldehyde

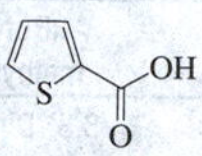
10071
2-Thiophenecarboxylic acid

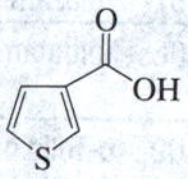
10072
3-Thiophenecarboxylic acid

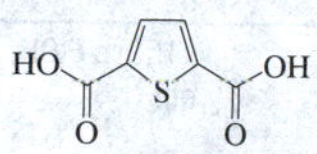
10073
2,5-Thiophenedicarboxylic acid

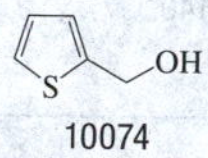
10074
2-Thiophenemethanol

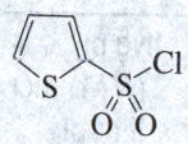
10075
2-Thiophenesulfonyl chloride

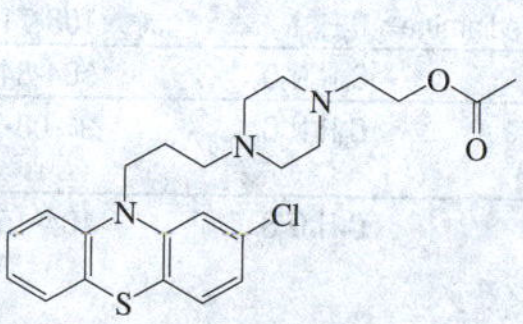
10076
Thiopropazate

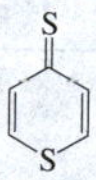
10077
4*H*-Thiopyran-4-thione

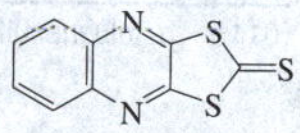
10078
Thioquinox

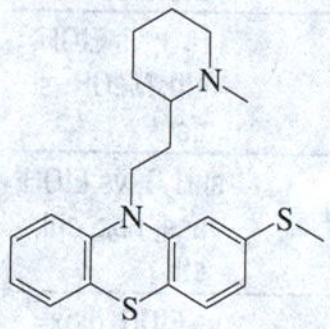
10079
Thioridazine

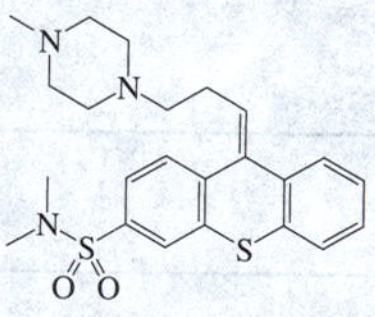
10080
cis-Thiothixene

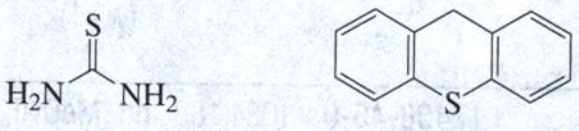
10081
Thiourea

10082
9*H*-Thioxanthene

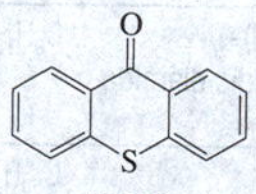
10083
9*H*-Thioxanthen-9-one

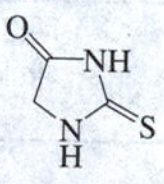
10084
2-Thioxo-4-imidazolidinone

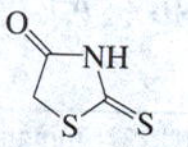
10085
2-Thioxo-4-thiazolidinone

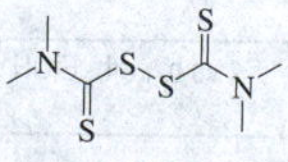
10086
Thiram

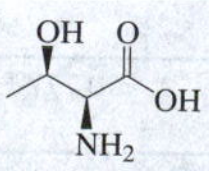
10087
L-Threonine

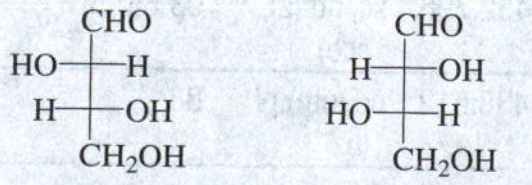
10088
D-Threose

10089
L-Threose

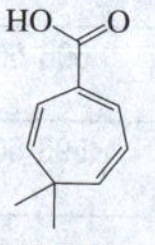
10090
Thujic acid

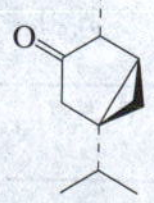
10091
α-Thujone

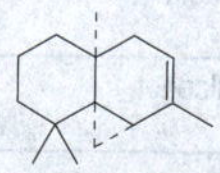
10092
3-Thujopsene

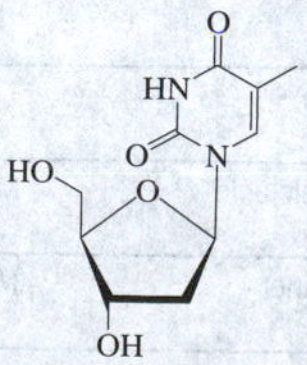
10093
Thymidine

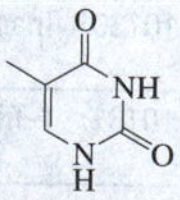
10094
Thymine

Organic

No.	Name	Synonym	Mol. form.	CAS Reg. No.	Mol. wt.	Physical form	mp/°C	bp/°C	den/ g cm^{-3}	n_D	Solubility
10095	Thymol	2-Isopropyl-5-methylphenol	$C_{10}H_{14}O$	89-83-8	150.217		49.6(0.3)	233(3)	0.970^{25}	1.5227^{20}	i H_2O; vs EtOH, eth, chl, AcOEt
10096	Thymol Blue		$C_{27}H_{30}O_5S$	76-61-9	466.589	grn-red (al, eth)	222 dec				sl H_2O, ace, bz; s EtOH, HOAc, $PhNH_2$
10097	Thymol iodide		$C_{20}H_{24}I_2O_2$	552-22-7	550.213	amorp					i H_2O; s eth; vs EtOH
10098	Thymolphthalein		$C_{28}H_{30}O_4$	125-20-2	430.536	pr or nd (al)	253				i H_2O; s EtOH, eth, ace; sl DMSO
10099	L-Thyroxine		$C_{15}H_{11}I_4NO_4$	51-48-9	776.871	nd	235				sl H_2O; i EtOH, bz
10100	Timolol		$C_{13}H_{24}N_4O_3S$	26839-75-8	316.420	oil					
10101	Tiocarlide		$C_{23}H_{32}N_2O_2S$	910-86-1	400.577		146				
10102	Tipepidine	3-(Di-2-thienylmethylene)-1-methylpiperidine	$C_{15}H_{17}NS_2$	5169-78-8	275.433	ye cry	65	$181^{4.5}$			
10103	Tobramycin		$C_{18}H_{37}N_5O_9$	32986-56-4	467.516	cry					s H_2O
10104	β-Tocopherol	5,8-Dimethyltocol	$C_{28}H_{48}O_2$	148-03-8	416.680	pa ye visc oil		$205^{0.1}$			vs ace, eth, EtOH, chl
10105	γ-Tocopherol	7,8-Dimethyltocol	$C_{28}H_{48}O_2$	7616-22-0	416.680	pa ye visc oil	-1.5	$205^{0.1}$			i H_2O; msc EtOH, eth, ace, chl
10106	δ-Tocopherol	8-Methyltocol	$C_{27}H_{46}O_2$	119-13-1	402.653	pa ye visc oil		$150^{0.001}$			i H_2O; vs EtOH, eth, ace, chl
10107	Tolazamide		$C_{14}H_{21}N_3O_3S$	1156-19-0	311.400	cry	172				
10108	Tolbutamide	*N*-[(Butylamino)carbonyl]-4-methylbenzenesulfonamide	$C_{12}H_{18}N_2O_3S$	64-77-7	270.347	orth cry	127(1)		1.245^{25}		sl H_2O; s EtOH, eth, chl
10109	*o*-Tolidine	3,3'-Dimethylbenzidine	$C_{14}H_{16}N_2$	119-93-7	212.290	wh-red lf (EtOH aq)	131				sl H_2O, chl; vs EtOH, eth
10110	Tolmetin		$C_{15}H_{15}NO_3$	26171-23-3	257.285	cry (MeCN)	156 dec				
10111	Toluene	Methylbenzene	C_7H_8	108-88-3	92.139	liq	-95.0(0.2)	110.60(0.07)	0.8623^{25}	1.4941^{25}	i H_2O; msc EtOH, eth; s ace, CS_2
10112	Toluene-2,4-diamine	4-Methyl-1,3-benzenediamine	$C_7H_{10}N_2$	95-80-7	122.167	nd (w), cry (al)	99	292			vs H_2O, EtOH, eth, bz; s chl
10113	Toluene-3,5-diamine	5-Methyl-1,3-benzenediamine	$C_7H_{10}N_2$	108-71-4	122.167	oil		284			
10114	Toluene-2,4-diisocyanate		$C_9H_6N_2O_2$	584-84-9	174.156		20.5	251	1.2244^{20}		vs ace, bz, eth
10115	Toluene-2,6-diisocyanate		$C_9H_6N_2O_2$	91-08-7	174.156		18.3				dec H_2O; s ace, bz
10116	*p*-Toluenesulfonic acid		$C_7H_8O_3S$	104-15-4	172.202	hyg pl (w+1) mcl lf or pl	104.5	140^{20}			vs H_2O; s EtOH, eth
10117	*p*-Toluenesulfonic acid monohydrate	4-Methylbenzenesulfonic acid, monohydrate	$C_7H_{10}O_4S$	6192-52-5	190.217		105.3				s H_2O
10118	*p*-Toluenesulfonyl chloride		$C_7H_7ClO_2S$	98-59-9	190.648	tcl (eth, peth)	71	145^{15}			i H_2O; s EtOH, eth, chl; vs bz
10119	*o*-Toluic acid	2-Methylbenzoic acid	$C_8H_8O_2$	118-90-1	136.149	pr or nd (w)	103.4(0.2)	259.5(0.6)	1.062^{115}	1.512^{115}	i H_2O; vs EtOH, eth; s chl
10120	*m*-Toluic acid	3-Methylbenzoic acid	$C_8H_8O_2$	99-04-7	136.149		109.3(0.7)		1.054^{112}	1.509	sl H_2O, chl; vs EtOH, eth
10121	*p*-Toluic acid	4-Methylbenzoic acid	$C_8H_8O_2$	99-94-5	136.149		180(1)				i H_2O; vs EtOH, eth, MeOH; sl tfa
10122	*N*-*o*-Tolylbiguanide	*N*-(2-Methylphenyl)-imidodicarbonimidic diamide	$C_9H_{13}N_5$	93-69-6	191.233	nd or pl (w+1)	145.0				sl H_2O; vs EtOH, ace; i bz, chl, eth
10123	Tomatine		$C_{50}H_{83}NO_{21}$	17406-45-0	1034.188	nd (MeOH)	270				vs EtOH, diox
10124	Tralomethrin		$C_{22}H_{19}Br_4NO_3$	66841-25-6	665.007	oran-ye solid					
10125	Tranylcypromine	2-Phenylcyclopropylamine	$C_9H_{11}N$	155-09-9	133.190	cry	44	127^{32}			
10126	Trehalose		$C_{12}H_{22}O_{11}$	99-20-7	342.296	orth cry	209.4(0.5)		1.58^{24}		vs H_2O; s EtOH; i eth, bz
10127	Triacetamide		$C_6H_9NO_3$	641-06-5	143.140	nd (eth)	79				vs eth
10128	Triacetin	Glycerol triacetate	$C_9H_{14}O_6$	102-76-1	218.203	col oily liq	-78	259(2)	1.1583^{20}	1.4301^{20}	sl H_2O; msc EtOH, eth, bz; vs ace
10129	Triacontane		$C_{30}H_{62}$	638-68-6	422.813	orth (eth, bz)	65.9(0.3)	451(7)	0.8097^{20}	1.4352^{70}	i H_2O; sl EtOH; s eth; vs bz
10130	Triacontanoic acid	Melissic acid	$C_{30}H_{60}O_2$	506-50-3	452.796	sc, nd (al, ace)	93.6(0.5)			1.4323^{100}	vs bz, CS_2, chl
10131	1-Triacontanol	Myricyl alcohol	$C_{30}H_{62}O$	593-50-0	438.812	nd (eth),pl (bz)	88		0.777^{95}		vs bz, eth, EtOH

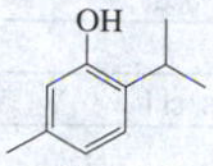
10095
Thymol

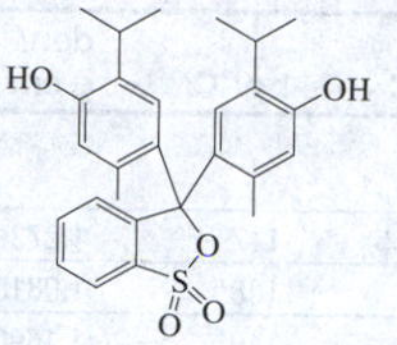
10096
Thymol Blue

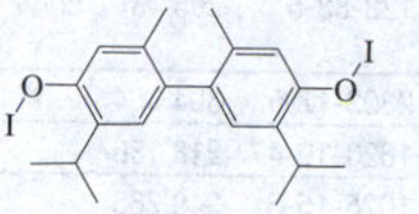
10097
Thymol iodide

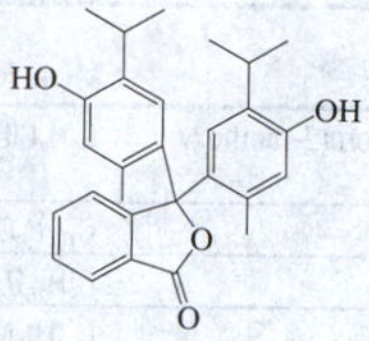
10098
Thymolphthalein

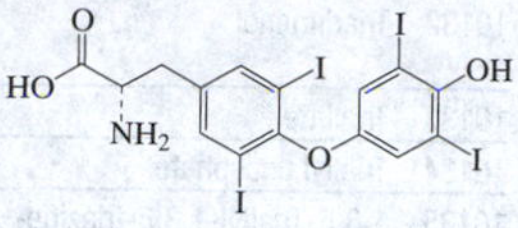
10099
L-Thyroxine

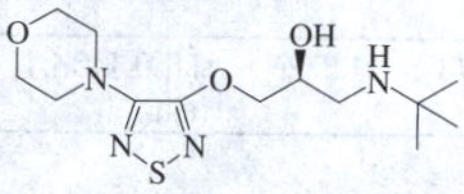
10100
Timolol

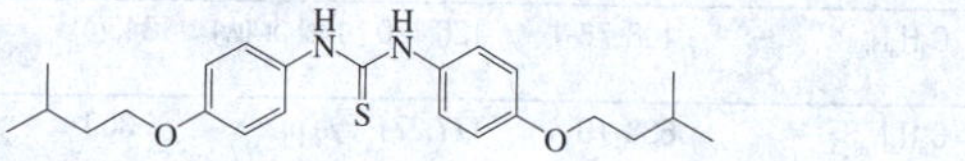
10101
Tiocarlide

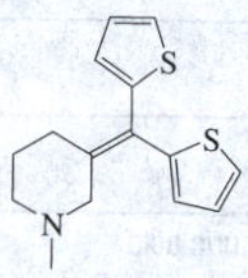
10102
Tipepidine

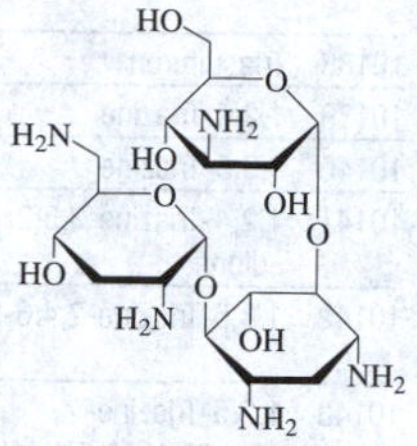
10103
Tobramycin

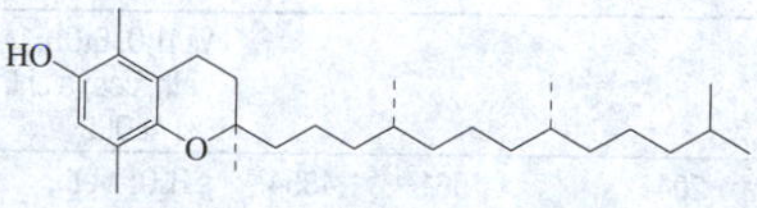
10104
β-Tocopherol

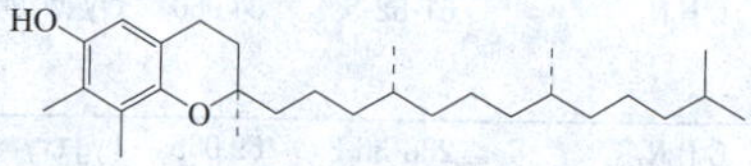
10105
γ-Tocopherol

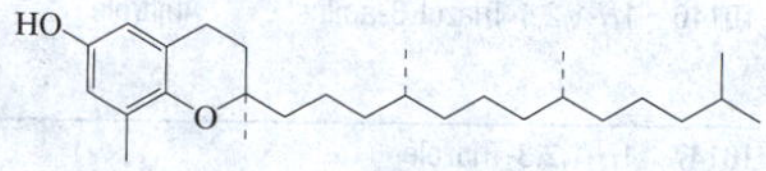
10106
δ-Tocopherol

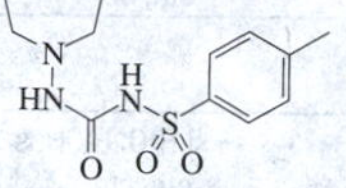
10107
Tolazamide

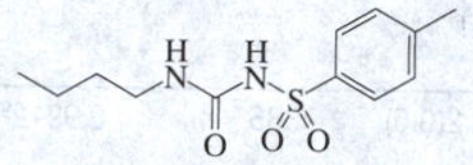
10108
Tolbutamide

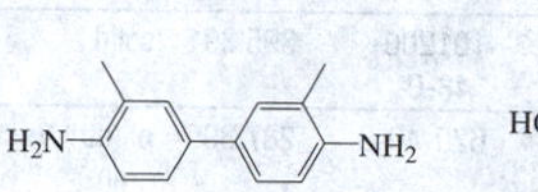
10109
o-Tolidine

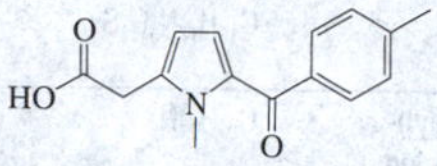
10110
Tolmetin

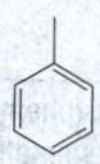
10111
Toluene

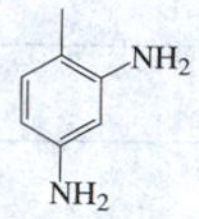
10112
Toluene-2,4-diamine

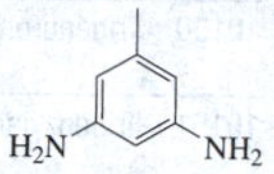
10113
Toluene-3,5-diamine

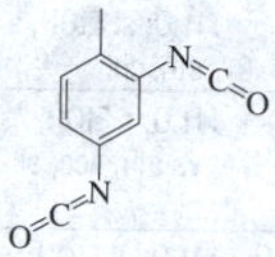
10114
Toluene-2,4-diisocyanate

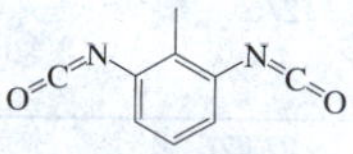
10115
Toluene-2,6-diisocyanate

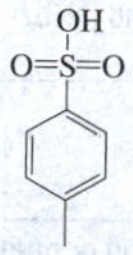
10116
p-Toluenesulfonic acid

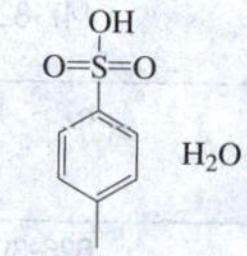
10117
p-Toluenesulfonic acid monohydrate

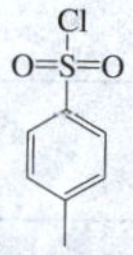
10118
p-Toluenesulfonyl chloride

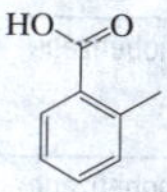
10119
o-Toluic acid

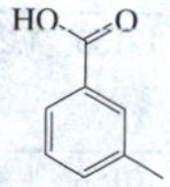
10120
m-Toluic acid

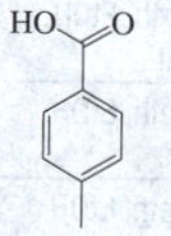
10121
p-Toluic acid

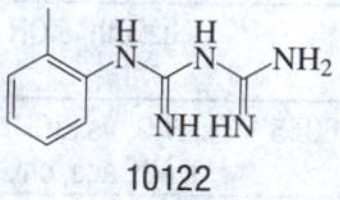
10122
N-*o*-Tolylbiguanide

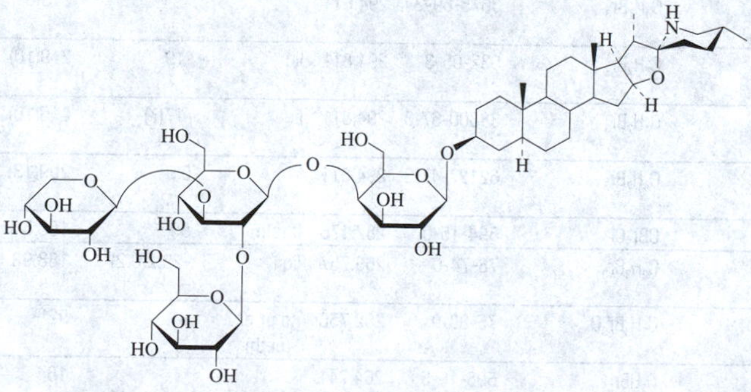
10123
Tomatine

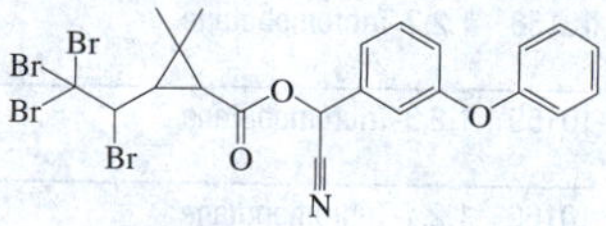
10124
Tralomethrin

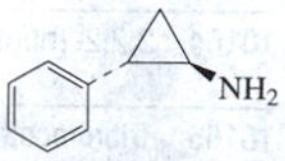
10125
Tranylcypromine

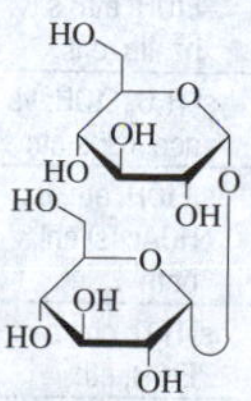
10126
Trehalose

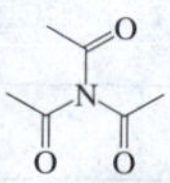
10127
Triacetamide

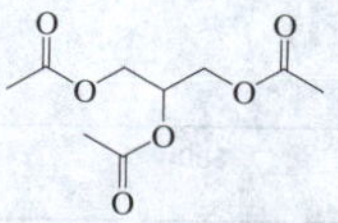
10128
Triacetin

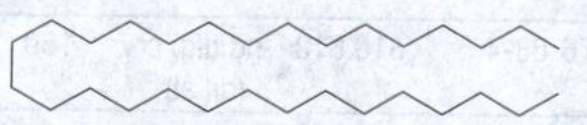
10129
Triacontane

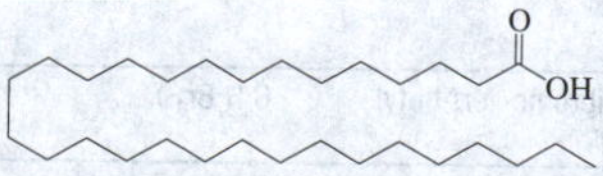
10130
Triacontanoic acid

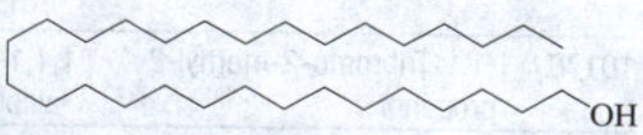
10131
1-Triacontanol

No.	Name	Synonym	Mol. form.	CAS Reg. No.	Mol. wt.	Physical form	mp/°C	bp/°C	den/ g cm^{-3}	n_D	Solubility
10132	Triadimenol	Mercury, chloro(2-methoxyethyl)-	C_3H_7ClHgO	123-88-6	295.13	cry	115				i H_2O; s EtOH, ace
10133	Triallate		$C_{10}H_{16}Cl_3NOS$	2303-17-5	304.664		34.0(0.5)	$117^{0.0003}$	1.273^{25}		
10134	Triallyl phosphate		$C_9H_{15}O_4P$	1623-19-4	218.186		-50	108^{7}	1.0815^{20}		sl chl
10135	1,3,5-Triallyl-1,3,5-triazine-2,4,6(1*H*,3*H*,5*H*)-trione		$C_{12}H_{15}N_3O_3$	1025-15-6	249.265		20.5	149^{4}	1.1590^{20}		
10136	Triamcinolone	Fluoxiprednisolone	$C_{21}H_{27}FO_6$	124-94-7	394.433	cry	270				
10137	Triamiphos		$C_{12}H_{19}N_6OP$	1031-47-6	294.292	cry (EtOH aq)	167				sl H_2O; s os
10138	Triasulfuron		$C_{14}H_{16}ClN_5O_5S$	82097-50-5	401.826		186				
10139	1,2,4-Triazine		$C_3H_3N_3$	290-38-0	81.076	pa ye oil	16.5	157		1.5149^{25}	
10140	1,3,5-Triazine		$C_3H_3N_3$	290-87-9	81.076		80.3(0.5)	113(5)	1.38^{25}		s EtOH, eth
10141	1,2,4-Triazine-3,5(2*H*,4*H*)-dione		$C_3H_3N_3O_2$	461-89-2	113.075		276.8				
10142	1,3,5-Triazine-2,4,6-triamine	Melamine	$C_3H_6N_6$	108-78-1	126.120	mcl pr (w)	343(4)	sub	1.573^{16}	1.872^{20}	sl H_2O, EtOH; i eth
10143	1,3,5-Triazine-2,4,6(1*H*,3*H*,5*H*)-trithione	Trithiocyanuric acid	$C_3H_3N_3S_3$	638-16-4	177.271	ye pr	>300	100^{22}			
10144	Triazofos		$C_{12}H_{16}N_3O_3PS$	24017-47-8	313.312	ye-br oil	5		1.2514^{20}		i H2O; s os
10145	Triazolam		$C_{17}H_{12}Cl_2N_4$	28911-01-5	343.210	tan cry (2-PrOH)	234				
10146	1*H*-1,2,4-Triazol-3-amine	Amitrole	$C_2H_4N_4$	61-82-5	84.080	cry (w, al)	155.8(0.6)				vs H_2O, EtOH; i eth, ace; s chl; sl AcOEt
10147	1*H*-1,2,3-Triazole		$C_2H_3N_3$	288-36-8	69.065	hyg cry	23	204	1.1861^{25}	1.4854^{25}	s H_2O; s eth, ace; i lig
10148	1*H*-1,2,4-Triazole	Pyrrodiazole	$C_2H_3N_3$	288-88-0	69.065	nd (bz/EtOH)	120(1)	204(6)			s H_2O, EtOH
10149	1*H*-1,2,4-Triazole-3,5-diamine		$C_2H_5N_5$	1455-77-2	99.095		211.5				s H_2O, EtOH; i eth, bz
10150	Tribenuron-methyl		$C_{15}H_{17}N_5O_6S$	101200-48-0	395.391	solid	141				
10151	Tribenzylamine	*N,N*-Bis(phenylmethyl)-benzenemethanamine	$C_{21}H_{21}N$	620-40-6	287.399	pl (eth), mcl (al)	92.2(0.5)	385	0.9912^{95}		sl H_2O, EtOH; s eth, ctc
10152	Tribromoacetaldehyde	Bromal	C_2HBr_3O	115-17-3	280.740			173(5)	2.6649^{25}	1.5939^{20}	vs ace, eth, EtOH
10153	Tribromoacetic acid		$C_2HBr_3O_2$	75-96-7	296.740	mcl	132	245 dec			s H_2O, EtOH, eth
10154	2,4,6-Tribromoaniline		$C_6H_4Br_3N$	147-82-0	329.815	nd (al, bz)	121(2)	300	2.35^{20}		i H_2O; sl EtOH; s eth, chl
10155	1,2,4-Tribromobenzene		$C_6H_3Br_3$	615-54-3	314.800		44.5	275			i H_2O; s EtOH; vs eth, ace; sl bz
10156	1,3,5-Tribromobenzene		$C_6H_3Br_3$	626-39-1	314.800	nd or pr (al)	121.8(0.2)	273(2)			i H_2O; sl EtOH; s eth, bz, chl
10157	1,1,2-Tribromobutane		$C_4H_7Br_3$	3675-68-1	294.811			219(10)	2.1835^{20}	1.5626^{17}	vs eth, EtOH, chl
10158	1,2,2-Tribromobutane		$C_4H_7Br_3$	3675-69-2	294.811			213.8	2.1692^{20}	1.568^{20}	vs eth, EtOH, chl
10159	1,2,3-Tribromobutane		$C_4H_7Br_3$	632-05-3	294.811	liq	-19	219(10)	2.1907^{20}	1.5680^{20}	vs eth, EtOH, chl
10160	1,2,4-Tribromobutane		$C_4H_7Br_3$	38300-67-3	294.811	liq	-17(7)	215(10)	2.170^{20}	1.5608^{20}	vs eth, EtOH, chl
10161	2,2,3-Tribromobutane		$C_4H_7Br_3$	62127-47-3	294.811		0.9	209(13)	2.1723^{20}	1.5602^{20}	i H_2O; s EtOH, eth, chl; sl ctc
10162	Tribromochloromethane		CBr_3Cl	594-15-0	287.176	lf (eth)	55	158.5	2.71^{15}		vs eth
10163	1,1,2-Tribromoethane		$C_2H_3Br_3$	78-74-0	266.757	liq	-29.2(0.2)	188.93	2.6210^{20}	1.5933^{20}	i H_2O; s EtOH, eth, bz, ctc
10164	2,2,2-Tribromoethanol		$C_2H_3Br_3O$	75-80-9	282.756	nd or pr (peth)	81	92^{10}			vs bz, eth, EtOH
10165	Tribromoethene		C_2HBr_3	598-16-3	264.741			164	2.708^{20}	1.6045^{16}	sl H_2O; vs EtOH; s eth, ace, chl
10166	Tribromofluoromethane		CBr_3F	353-54-8	270.721	liq	-73.6	108			i H_2O; s EtOH
10167	Tribromomethane	Bromoform	$CHBr_3$	75-25-2	252.731	liq	8.69(0.02)	149.2(0.5)	2.8788^{25}	1.5948^{25}	sl H_2O; msc EtOH, eth; s bz, lig, chl
10168	1,3,5-Tribromo-2-methoxybenzene		$C_7H_5Br_3O$	607-99-8	344.826	nd (al)	88	298	2.491^{25}		sl H_2O, EtOH; vs ace, bz; s ctc
10169	2,4,6-Tribromo-3-methylphenol	2,4,6-Tribromo-*m*-cresol	$C_7H_5Br_3O$	4619-74-3	344.826		84				s EtOH, eth, bz, HOAc; sl chl, peth
10170	1,1,1-Tribromo-2-methyl-2-propanol	1,1,1-Tribromo-*tert*-butyl alcohol	$C_4H_7Br_3O$	76-08-4	310.810	nd (lig) cry (dil al)	169	sub			sl H_2O, chl; s EtOH, eth

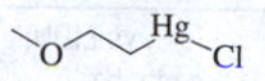
10132
Triadimenol

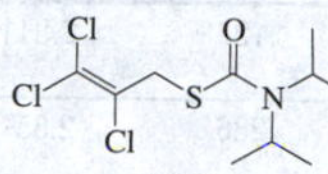
10133
Triallate

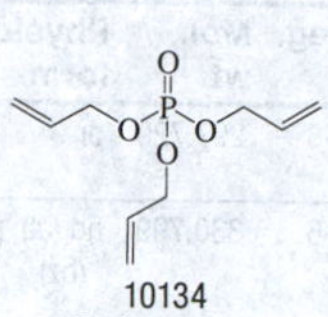
10134
Triallyl phosphate

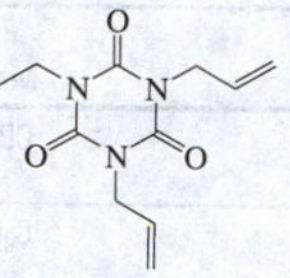
10135
1,3,5-Triallyl-1,3,5-triazine-2,4,6(1*H*,3*H*,5*H*)-trione

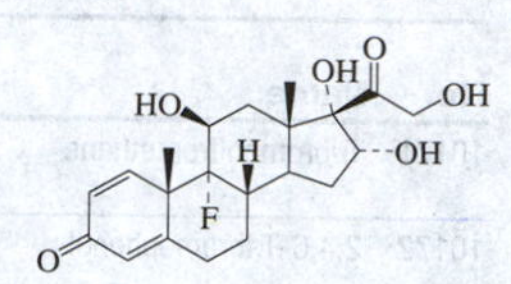
10136
Triamcinolone

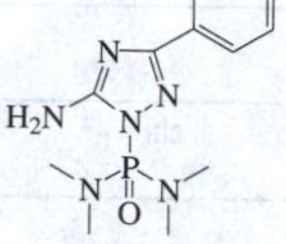
10137
Triamiphos

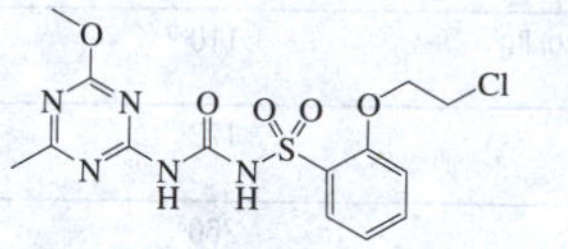
10138
Triasulfuron

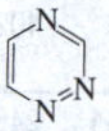
10139
1,2,4-Triazine

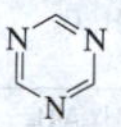
10140
1,3,5-Triazine

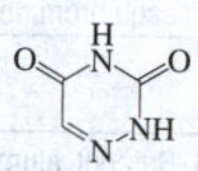
10141
1,2,4-Triazine-3,5(2*H*,4*H*)-dione

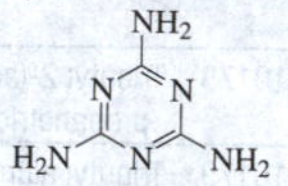
10142
1,3,5-Triazine-2,4,6-triamine

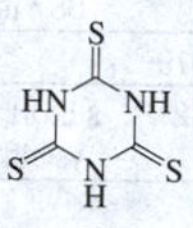
10143
1,3,5-Triazine-2,4,6(1*H*,3*H*,5*H*)-trithione

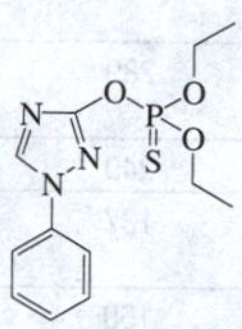
10144
Triazofos

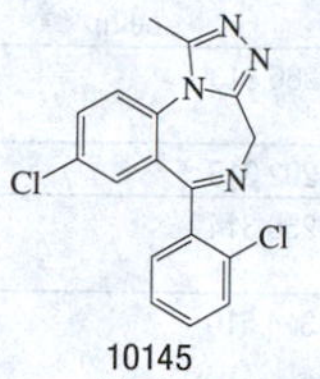
10145
Triazolam

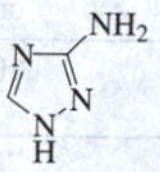
10146
1*H*-1,2,4-Triazol-3-amine

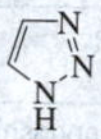
10147
1*H*-1,2,3-Triazole

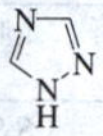
10148
1*H*-1,2,4-Triazole

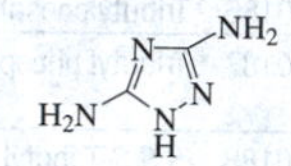
10149
1*H*-1,2,4-Triazole-3,5-diamine

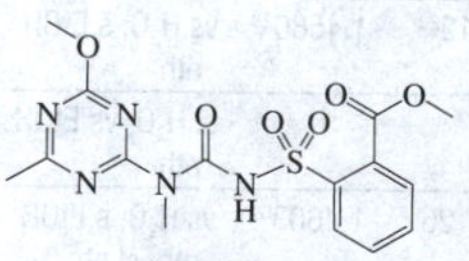
10150
Tribenuron-methyl

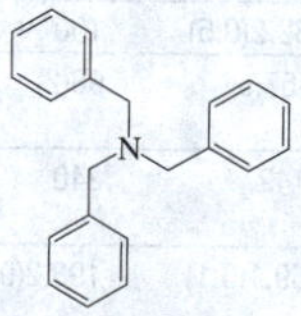
10151
Tribenzylamine

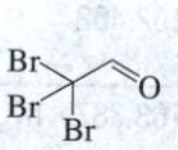
10152
Tribromoacetaldehyde

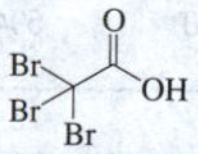
10153
Tribromoacetic acid

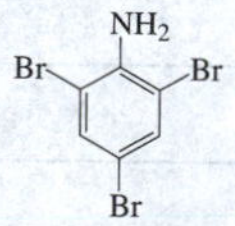
10154
2,4,6-Tribromoaniline

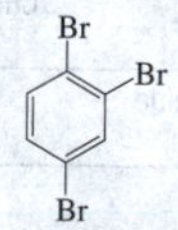
10155
1,2,4-Tribromobenzene

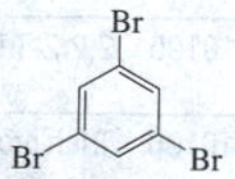
10156
1,3,5-Tribromobenzene

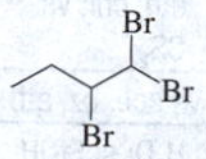
10157
1,1,2-Tribromobutane

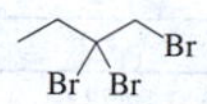
10158
1,2,2-Tribromobutane

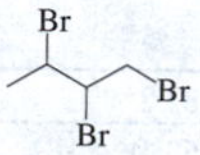
10159
1,2,3-Tribromobutane

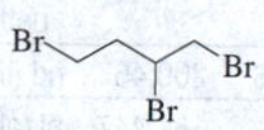
10160
1,2,4-Tribromobutane

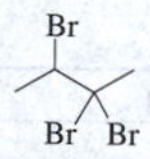
10161
2,2,3-Tribromobutane

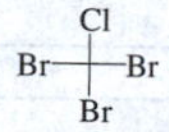
10162
Tribromochloromethane

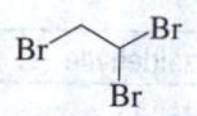
10163
1,1,2-Tribromoethane

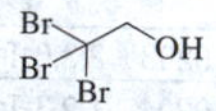
10164
2,2,2-Tribromoethanol

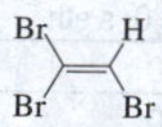
10165
Tribromoethene

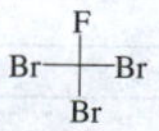
10166
Tribromofluoromethane

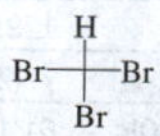
10167
Tribromomethane

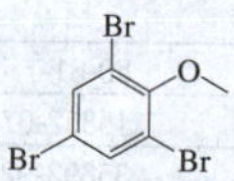
10168
1,3,5-Tribromo-2-methoxybenzene

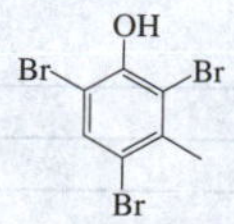
10169
2,4,6-Tribromo-3-methylphenol

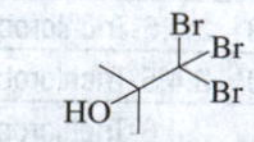
10170
1,1,1-Tribromo-2-methyl-2-propanol

Organic

No.	Name	Synonym	Mol. form.	CAS Reg. No.	Mol. wt.	Physical form	mp/°C	bp/°C	den/ g cm^{-3}	n_D	Solubility
10171	Tribromonitromethane		CBr_3NO_2	464-10-8	297.729	pr	10	127^{18}	2.811^{12}	1.5790^{20}	i H_2O; s EtOH, eth; vs ace, bz
10172	2,4,6-Tribromophenol		$C_6H_3Br_3O$	118-79-6	330.799	nd (al), pr (bz)	95(2)	286	2.55^{20}		i H_2O; vs EtOH; s eth, bz, HOAc, chl
10173	1,1,2-Tribromopropane		$C_3H_5Br_3$	14602-62-1	280.784			199(10)	2.3547^{20}	1.5790^{20}	i H_2O; s EtOH, chl, HOAc; vs eth
10174	1,2,2-Tribromopropane		$C_3H_5Br_3$	14476-30-3	280.784			190.5	2.2984^{20}	1.5670^{20}	vs eth, EtOH, chl
10175	1,2,3-Tribromopropane		$C_3H_5Br_3$	96-11-7	280.784		14(4)	222.1(0.7)	2.4208^{20}	1.5862^{20}	i H_2O; vs EtOH, eth; sl ctc
10176	2,3,5-Tribromothiophene		C_4HBr_3S	3141-24-0	320.828	nd (al)	29	260			s chl
10177	Tribromotrimethyldialuminum	Methyl aluminum sesquibromide	$C_3H_9Al_2Br_3$	12263-85-3	338.778	hyg col liq		110^{50}			
10178	Tributyl 2-(acetyloxy)-1,2,3-propanetricarboxylate		$C_{20}H_{34}O_8$	77-90-7	402.479			173^{1}			sl chl
10179	Tributyl aluminate	1-Butanol, aluminum salt	$C_{12}H_{27}AlO_3$	3085-30-1	246.322			260^{5}			
10180	Tributylaluminum		$C_{12}H_{27}Al$	1116-70-7	198.324			102^{2}			
10181	Tributylamine	*N,N*-Dibutyl-1-butanamine	$C_{12}H_{27}N$	102-82-9	185.349	liq	-70	207(2)	0.7770^{20}	1.4299^{20}	sl H_2O, ctc; vs EtOH, eth; s ace, bz
10182	Tributyl borate	Butyl borate	$C_{12}H_{27}BO_3$	688-74-4	230.151	oil	<-70	233.8(1)	0.8567^{20}	1.4106^{18}	s EtOH, bz; vs eth, MeOH
10183	Tributylfluorostannane	Tributyltin fluoride	$C_{12}H_{27}FSn$	1983-10-4	309.050	nd	≈260				
10184	2,4,6-Tri-*tert*-butylphenol		$C_{18}H_{30}O$	732-26-3	262.430	cry (al, peth)	132.5(0.3)	277(6)	0.864^{27}		i H_2O, alk; s EtOH, ace, ctc
10185	Tributyl phosphate	Butyl phosphate	$C_{12}H_{27}O_4P$	126-73-8	266.313			289	0.9727^{25}	1.4224^{25}	s H_2O, eth, bz, CS_2; msc EtOH
10186	Tributylphosphine		$C_{12}H_{27}P$	998-40-3	202.316			240	0.812^{25}	1.4619^{20}	
10187	Tributyl phosphite	Tributoxyphosphine	$C_{12}H_{27}O_3P$	102-85-2	250.314			137^{26}	0.9259^{20}	1.4321^{19}	s EtOH; sl ctc; vs eth
10188	*S,S,S*-Tributyl phosphorotrithioate	*S,S,S*-Tributyl trithiophosphate	$C_{12}H_{27}OPS_3$	78-48-8	314.510		<-25	$150^{0.3}$	1.057^{20}		
10189	Tributylsilane		$C_{12}H_{28}Si$	998-41-4	200.436			221	0.7794^{20}	1.4380^{20}	
10190	Tributylstannane	Tributyltin hydride	$C_{12}H_{28}Sn$	688-73-3	291.060	liq		113^{8}	1.103^{20}		
10191	Tributyrin	Butanoic acid, 1,2,3-propanetriyl ester	$C_{15}H_{26}O_6$	60-01-5	302.363	liq	-75	307.5	1.0350^{20}	1.4359^{20}	i H_2O; s EtOH, ace, bz; sl ctc; vs eth
10192	Tricalcium citrate	Calcium citrate	$C_{12}H_{10}Ca_3O_{14}$	813-94-5	498.433	cry (w)	≈100 dec (hyd)				sl H_2O; i EtOH
10193	Trichlorfon		$C_4H_8Cl_3O_4P$	52-68-6	257.437		82.2(0.5)	$100^{0.1}$	1.73^{20}		
10194	Trichloroacetaldehyde	Chloral	C_2HCl_3O	75-87-6	147.387	liq	-57.5	98(2)	1.512^{20}	1.4580^{20}	vs H_2O; s EtOH, eth
10195	2,2,2-Trichloroacetamide		$C_2H_2Cl_3NO$	594-65-0	162.402		142	240			sl H_2O; vs EtOH, eth
10196	Trichloroacetic acid		$C_2HCl_3O_2$	76-03-9	163.387	hyg cry	59.1(0.1)	198.2(0.1)	1.6126^{64}	1.4603^{61}	vs H_2O; s EtOH, eth; sl ctc
10197	Trichloroacetic anhydride		$C_4Cl_6O_3$	4124-31-6	308.759			223 dec	1.6908^{20}		vs eth, HOAc
10198	Trichloroacetonitrile		C_2Cl_3N	545-06-2	144.387	liq	-42	85.7	1.4403^{25}	1.4409^{20}	i H_2O
10199	Trichloroacetyl chloride		C_2Cl_4O	76-02-8	181.832			118.2(0.3)	1.6202^{20}	1.4695^{20}	msc eth
10200	2,3,4-Trichloroaniline		$C_6H_4Cl_3N$	634-67-3	196.462	nd (lig)	73	292			vs EtOH
10201	2,4,5-Trichloroaniline		$C_6H_4Cl_3N$	636-30-6	196.462	nd (lig)	96.5	270			s EtOH, eth; vs CS_2; sl lig
10202	2,4,6-Trichloroaniline		$C_6H_4Cl_3N$	634-93-5	196.462	cry (al), nd (lig or peth)	78.5	262			i H_2O; s EtOH, eth, chl; vs CS_2
10203	2,3,6-Trichlorobenzaldehyde		$C_7H_3Cl_3O$	4659-47-6	209.457	nd (lig)	87.3				vs ace, bz, eth
10204	1,2,3-Trichlorobenzene		$C_6H_3Cl_3$	87-61-6	181.447	pl (al)	53(1)	219(3)	1.4533^{25}		i H_2O; sl EtOH, chl; vs eth, bz
10205	1,2,4-Trichlorobenzene		$C_6H_3Cl_3$	120-82-1	181.447	orth	17.0(0.4)	213.5(0.3)	1.459^{25}	1.5717^{20}	i H_2O; sl EtOH, chl; vs eth
10206	1,3,5-Trichlorobenzene		$C_6H_3Cl_3$	108-70-3	181.447	nd	62.8(0.7)	209(1)			i H_2O; sl EtOH; vs eth, bz; s chl
10207	2,3,6-Trichlorobenzeneacetic acid	Chlorfenac	$C_8H_5Cl_3O_2$	85-34-7	239.484		160.4(0.5)				
10208	3,4,5-Trichloro-1,2-benzenediol		$C_6H_3Cl_3O_2$	56961-20-7	213.446	(i) pr (HOAc) (ii) pr (bz)	130(1)				sl H_2O; vs eth, EtOH, HOAc
10209	2,3,6-Trichlorobenzoic acid		$C_7H_3Cl_3O_2$	50-31-7	225.457		129.5(0.3)				sl H_2O; s eth
10210	2,4,5-Trichlorobiphenyl		$C_{12}H_7Cl_3$	15862-07-4	257.543	cry	77(2)				i H_2O
10211	2,4,6-Trichlorobiphenyl		$C_{12}H_7Cl_3$	35693-92-6	257.543	cry (EtOH aq)	61.1(0.5)	172^{15}			i H_2O

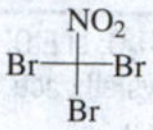
10171
Tribromonitromethane

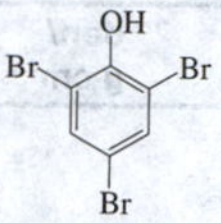
10172
2,4,6-Tribromophenol

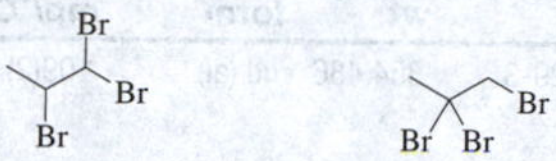
10173
1,1,2-Tribromopropane

10174
1,2,2-Tribromopropane

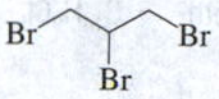
10175
1,2,3-Tribromopropane

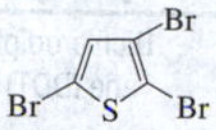
10176
2,3,5-Tribromothiophene

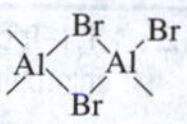
10177
Tribromotrimethyldialuminum

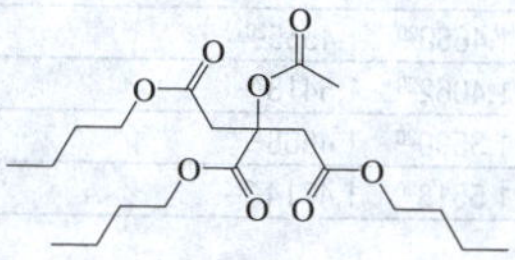
10178
Tributyl 2-(acetyloxy)-1,2,3-propanetricarboxylate

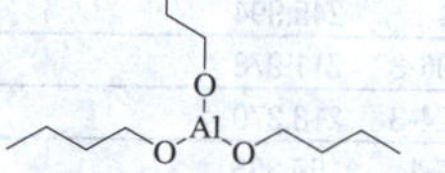
10179
Tributyl aluminate

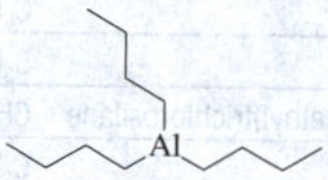
10180
Tributylaluminum

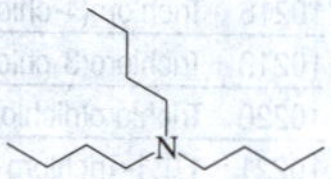
10181
Tributylamine

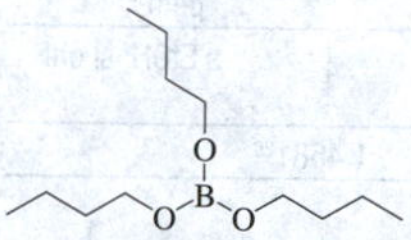
10182
Tributyl borate

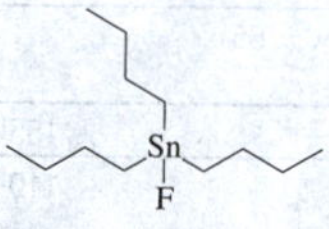
10183
Tributylfluorostannane

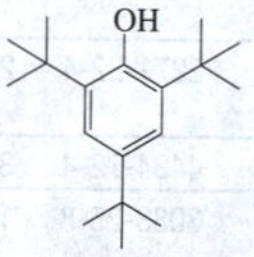
10184
2,4,6-Tri-*tert*-butylphenol

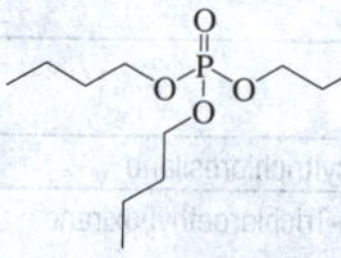
10185
Tributyl phosphate

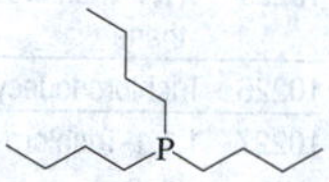
10186
Tributylphosphine

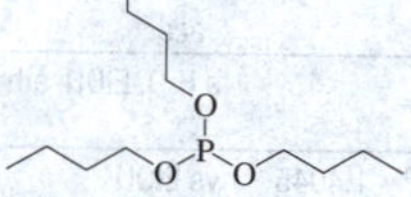
10187
Tributyl phosphite

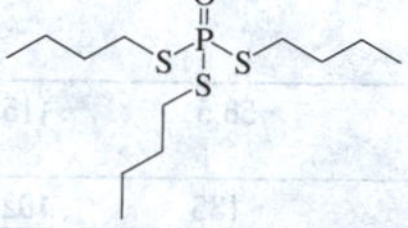
10188
S,S,S-Tributyl phosphorotrithioate

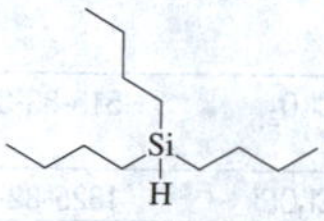
10189
Tributylsilane

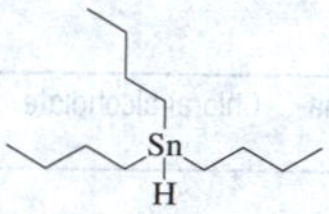
10190
Tributylstannane

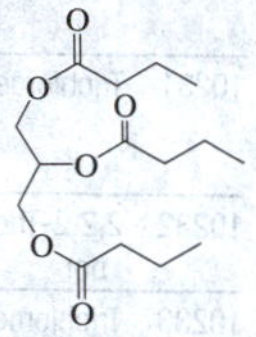
10191
Tributyrin

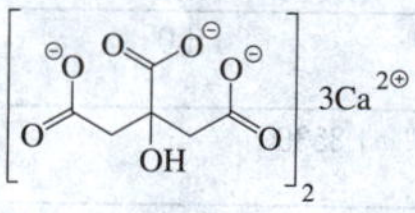
10192
Tricalcium citrate

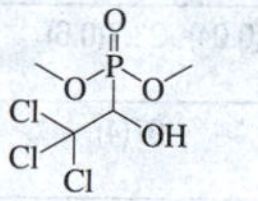
10193
Trichlorfon

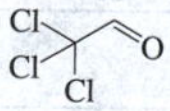
10194
Trichloroacetaldehyde

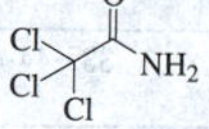
10195
2,2,2-Trichloroacetamide

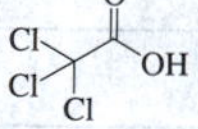
10196
Trichloroacetic acid

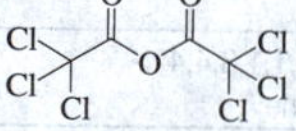
10197
Trichloroacetic anhydride

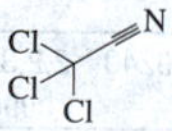
10198
Trichloroacetonitrile

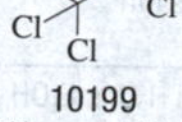
10199
Trichloroacetyl chloride

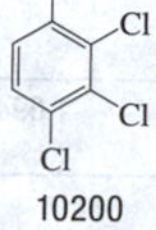
10200
2,3,4-Trichloroaniline

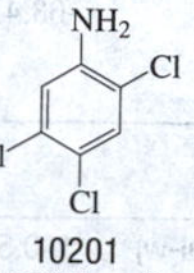
10201
2,4,5-Trichloroaniline

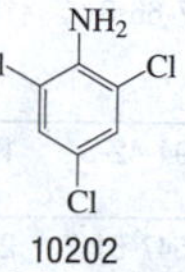
10202
2,4,6-Trichloroaniline

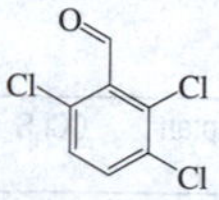
10203
2,3,6-Trichlorobenzaldehyde

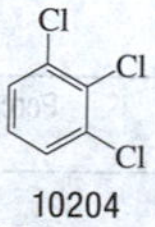
10204
1,2,3-Trichlorobenzene

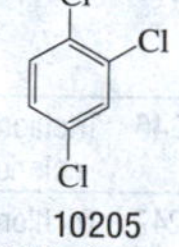
10205
1,2,4-Trichlorobenzene

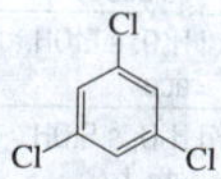
10206
1,3,5-Trichlorobenzene

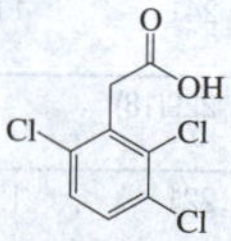
10207
2,3,6-Trichlorobenzeneacetic acid

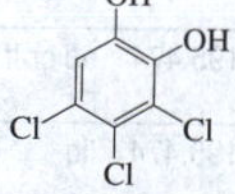
10208
3,4,5-Trichloro-1,2-benzenediol

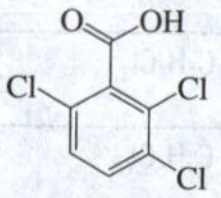
10209
2,3,6-Trichlorobenzoic acid

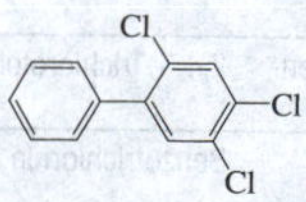
10210
2,4,5-Trichlorobiphenyl

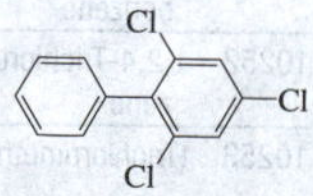
10211
2,4,6-Trichlorobiphenyl

No.	Name	Synonym	Mol. form.	CAS Reg. No.	Mol. wt.	Physical form	mp/°C	bp/°C	den/ g cm^{-3}	n_D	Solubility
10212	1,1,1-Trichloro-2,2-bis(4-chlorophenyl)ethane	Dichlorodiphenyltrichloroethane (DDT)	$C_{14}H_9Cl_5$	50-29-3	354.486	nd (al)	109(2)	260			i H_2O; sl EtOH; vs eth, ace, bz, py
10213	2,2,3-Trichlorobutanal	2,2,3-Trichlorobutyraldehyde	$C_4H_5Cl_3O$	76-36-8	175.441			166(7)	1.3956^{20}	1.4755^{20}	vs H_2O, eth, EtOH
10214	2,3,4-Trichloro-1-butene		$C_4H_5Cl_3$	2431-50-7	159.442			60^{20}	1.3430^{20}	1.4944^{20}	vs ace, chl
10215	3,4,4′-Trichlorocarbanilide	Triclocarban	$C_{13}H_9Cl_3N_2O$	101-20-2	315.581	fine pl	254.4(0.9)				
10216	1,2,4-Trichloro-5-(chloromethyl)benzene		$C_7H_4Cl_4$	3955-26-8	229.919			273	1.547^{20}		vs ace, eth, EtOH
10217	Trichloro(chloromethyl)silane	(Chloromethyl)trichlorosilane	CH_2Cl_4Si	1558-25-4	183.925			118	1.4650^{20}	1.4555^{20}	
10218	Trichloro(4-chlorophenyl)silane		$C_6H_4Cl_4Si$	825-94-5	245.994			233	1.4062^{20}	1.5418^{20}	
10219	Trichloro(3-chloropropyl)silane		$C_3H_6Cl_4Si$	2550-06-3	211.978			182.3(0.2)	1.3590^{20}	1.4668^{20}	
10220	Trichloro(dichloromethyl)silane	(Dichloromethyl)trichlorosilane	$CHCl_5Si$	1558-24-3	218.370			145	1.5518^{20}	1.4714^{20}	
10221	1,1,1-Trichloro-2,2-difluoroethane		$C_2HCl_3F_2$	354-12-1	169.385			75(16)			
10222	1,2,2-Trichloro-1,1-difluoroethane		$C_2HCl_3F_2$	354-21-2	169.385		-150.0(0.6)	71.9(0.4)	1.5447^{20}	1.3889^{20}	
10223	1,2,2-Trichloro-1,2-difluoroethane		$C_2HCl_3F_2$	354-15-4	169.385		-174	73.2(0.6)			
10224	2,4,6-Trichloro-3,5-dimethylphenol		$C_8H_7Cl_3O$	6972-47-0	225.500	ye nd (peth)	175				i H_2O; s chl; vs peth
10225	1,1,1-Trichloro-2,2-diphenylethane		$C_{14}H_{11}Cl_3$	2971-22-4	285.596		65				s EtOH; sl chl
10226	Trichlorododecylsilane	Dodecyltrichlorosilane	$C_{12}H_{25}Cl_3Si$	4484-72-4	303.772			155^{10}		1.4581^{20}	
10227	1,1,1-Trichloro-3,4-epoxybutane	(2,2,2-Trichloroethyl)oxirane	$C_4H_5Cl_3O$	3083-25-8	175.441	liq		110^{100}			
10228	1,1,1-Trichloroethane	Methyl chloroform	$C_2H_3Cl_3$	71-55-6	133.404	liq	-30(2)	74.02(0.08)	1.3390^{20}	1.4379^{20}	sl H_2O; s EtOH, chl; msc eth
10229	1,1,2-Trichloroethane	Vinyl trichloride	$C_2H_3Cl_3$	79-00-5	133.404	liq	-36.3(0.5)	113(1)	1.4397^{20}	1.4714^{20}	i H_2O; s EtOH, eth, chl
10230	2,2,2-Trichloroethanol		$C_2H_3Cl_3O$	115-20-8	149.403	hyg orth tab or pl	19	151(21)		1.4861^{20}	sl H_2O, ctc; msc EtOH, eth; s alk
10231	Trichloroethene	Trichloroethylene	C_2HCl_3	79-01-6	131.388	liq	-84.7(0.3)	86.8(0.1)	1.4642^{20}	1.4773^{20}	sl H_2O, ctc; msc EtOH, eth; s ace
10232	2,2,2-Trichloro-1-ethoxyethanol	Chloral alcoholate	$C_4H_7Cl_3O_2$	515-83-3	193.457		56.5	115.5	1.395^{40}		s H_2O, EtOH, eth
10233	Trichloroethoxysilane		$C_2H_5Cl_3OSi$	1825-82-7	179.505	liq	-135	102(3)	1.2274^{20}	1.4045^{20}	vs EtOH
10234	2,2,2-Trichloroethyl-β-D-glucopyranosiduronic acid	Urochloralic acid	$C_8H_{11}Cl_3O_7$	97-25-6	325.528	nd	142				vs H_2O, EtOH
10235	Trichloroethylsilane	Ethyltrichlorosilane	$C_2H_5Cl_3Si$	115-21-9	163.506	liq	-105.6	98.7(0.7)	1.2373^{20}	1.4256^{20}	s ctc
10236	1,1,1-Trichloro-2-fluoroethane	Refrigerant 131b	$C_2H_2Cl_3F$	2366-36-1	151.394	liq		86.5			
10237	1,1,2-Trichloro-1-fluoroethane	Refrigerant 131a	$C_2H_2Cl_3F$	811-95-0	151.394		-104.7	88.0	1.492^{20}		
10238	1,1,2-Trichloro-2-fluoroethane		$C_2H_2Cl_3F$	359-28-4	151.394			102.4	1.5393^{20}	1.4390^{20}	i H_2O
10239	Trichlorofluoromethane	Refrigerant 11	CCl_3F	75-69-4	137.368	vol liq or gas	-110.44(0.04)	23.7(0.6)	1.4879^{20}		i H_2O
10240	2,2,3-Trichloro-1,1,1,3,4,4,4-heptafluorobutane		$C_4Cl_3F_7$	335-44-4	287.391		2.0	97(4)	1.7484^{20}	1.3530^{20}	
10241	Trichlorohexylsilane	Hexyltrichlorosilane	$C_6H_{13}Cl_3Si$	928-65-4	219.612			190	1.1100^{20}		dec H_2O
10242	*N*-(2,2,2-Trichloro-1-hydroxyethyl)formamide	Chloral formamide	$C_3H_4Cl_3NO_2$	515-82-2	192.429	cry	120				vs ace, eth, EtOH
10243	3,3,3-Trichloro-2-hydroxypropanenitrile	Chlorocyanohydrin	$C_3H_2Cl_3NO$	513-96-2	174.413	pl (w)	61	217 dec			vs H_2O, eth, EtOH
10244	Trichloroisobutylsilane		$C_4H_9Cl_3Si$	18169-57-8	191.559			143.3	1.154^{20}		dec H_2O
10245	Trichloromethane	Chloroform	$CHCl_3$	67-66-3	119.378	liq	-63.47(0.07)	61.2(0.1)	1.4788^{25}	1.4459^{20}	sl H_2O; msc EtOH, eth, bz; s ace, ctc
10246	Trichloromethanesulfenyl chloride	Perchloromethyl mercaptan	CCl_4S	594-42-3	185.888	ye oil		149	1.6947^{20}	1.5484^{20}	s eth
10247	Trichloromethanesulfonyl chloride		CCl_4O_2S	2547-61-7	217.887	cry (al-w)	140.5	170			i H_2O; s EtOH, eth, CS_2
10248	Trichloromethanethiol	Trichloromethyl mercaptan	$CHCl_3S$	75-70-7	151.443	oran oil		125^{15}			
10249	Trichlormethiazide		$C_8H_8Cl_3N_3O_4S_2$	133-67-5	380.657		270 dec				sl H_2O; s EtOH
10250	1,2,4-Trichloro-5-methoxybenzene		$C_7H_5Cl_3O$	6130-75-2	211.473	nd (dil al)	77.5	254			vs EtOH, ace
10251	1,3,5-Trichloro-2-methoxybenzene	2,4,6-Trichloroanisole	$C_7H_5Cl_3O$	87-40-1	211.473	mcl nd (al)	61.5	241	1.640^{25}		s EtOH, bz, chl; vs ace
10252	1,2,4-Trichloro-5-methylbenzene	2,4,5-Trichlorotoluene	$C_7H_5Cl_3$	6639-30-1	195.474	nd or lf (al)	82.4	229(18)			i H_2O; s EtOH, ace
10253	(Trichloromethyl)benzene	Benzotrichloride	$C_7H_5Cl_3$	98-07-7	195.474	liq	-17.0(0.6)	221	1.3723^{20}	1.5580^{20}	i H_2O; s EtOH, eth, bz

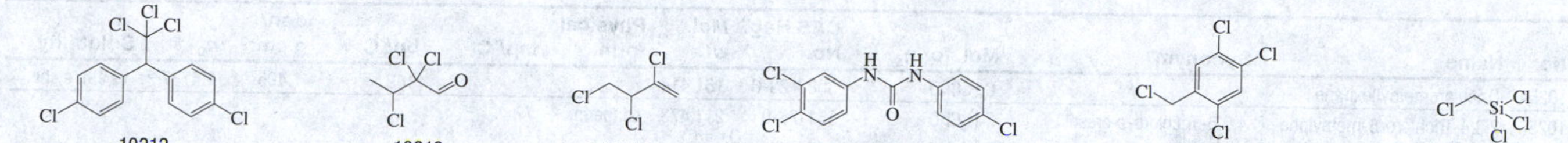

10212 1,1,1-Trichloro-2,2-bis(4-chlorophenyl)ethane

10213 2,2,3-Trichlorobutanal

10214 2,3,4-Trichloro-1-butene

10215 3,4,4'-Trichlorocarbanilide

10216 1,2,4-Trichloro-5-(chloromethyl)benzene

10217 Trichloro(chloromethyl)silane

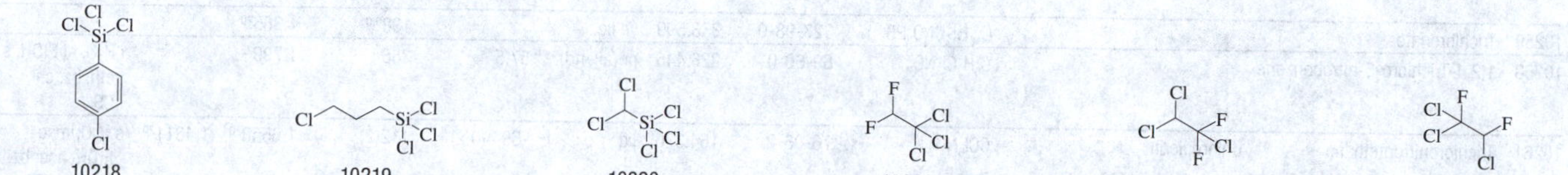

10218 Trichloro(4-chlorophenyl)silane

10219 Trichloro(3-chloropropyl)silane

10220 Trichloro(dichloromethyl)silane

10221 1,1,1-Trichloro-2,2-difluoroethane

10222 1,2,2-Trichloro-1,1-difluoroethane

10223 1,2,2-Trichloro-1,2-difluoroethane

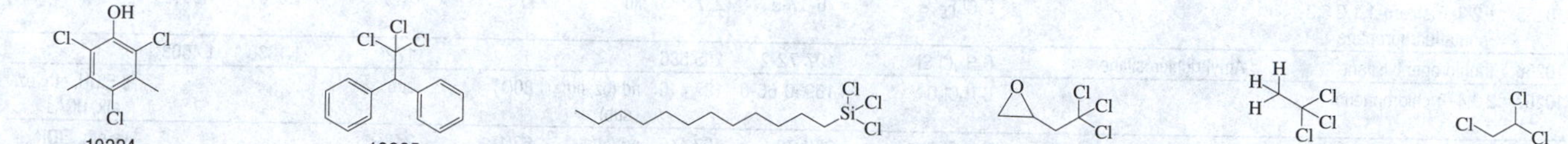

10224 2,4,6-Trichloro-3,5-dimethylphenol

10225 1,1,1-Trichloro-2,2-diphenylethane

10226 Trichlorododecylsilane

10227 1,1,1-Trichloro-3,4-epoxybutane

10228 1,1,1-Trichloroethane

10229 1,1,2-Trichloroethane

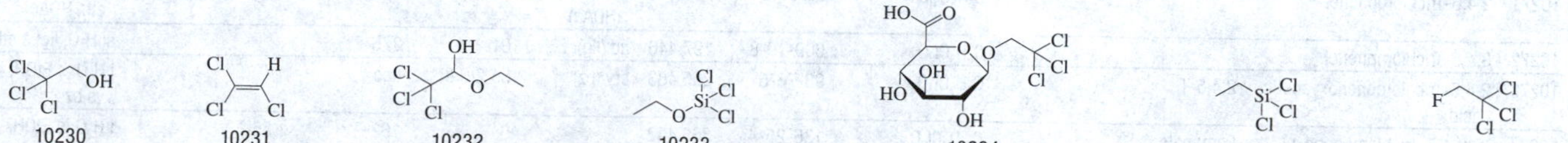

10230 2,2,2-Trichloroethanol

10231 Trichloroethene

10232 2,2,2-Trichloro-1-ethoxyethanol

10233 Trichloroethoxysilane

10234 2,2,2-Trichloroethyl-β-D-glucopyranosiduronic acid

10235 Trichloroethylsilane

10236 1,1,1-Trichloro-2-fluoroethane

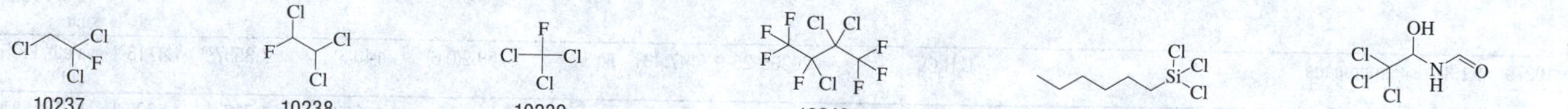

10237 1,1,2-Trichloro-1-fluoroethane

10238 1,1,2-Trichloro-2-fluoroethane

10239 Trichlorofluoromethane

10240 2,2,3-Trichloro-1,1,1,3,4,4,4-heptafluorobutane

10241 Trichlorohexylsilane

10242 *N*-(2,2,2-Trichloro-1-hydroxyethyl)formamide

10243 3,3,3-Trichloro-2-hydroxypropanenitrile

10244 Trichloroisobutylsilane

10245 Trichloromethane

10246 Trichloromethanesulfenyl chloride

10247 Trichloromethanesulfonyl chloride

10248 Trichloromethanethiol

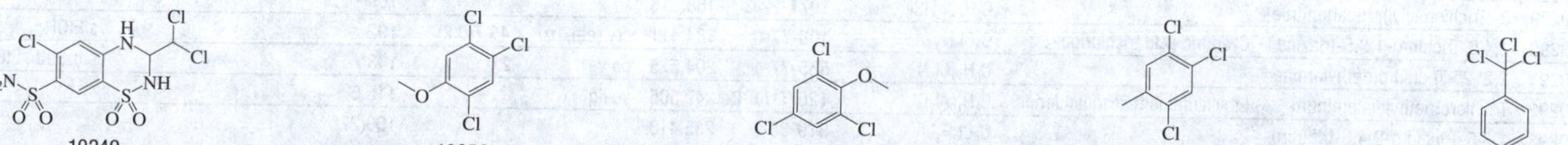

10249 Trichloromethiazide

10250 1,2,4-Trichloro-5-methoxybenzene

10251 1,3,5-Trichloro-2-methoxybenzene

10252 1,2,4-Trichloro-5-methylbenzene

10253 (Trichloromethyl)benzene

Organic

No.	Name	Synonym	Mol. form.	CAS Reg. No.	Mol. wt.	Physical form	mp/°C	bp/°C	den/g cm⁻³	n_D	Solubility
10254	(Trichloromethyl)oxirane		$C_3H_3Cl_3O$	3083-23-6	161.414			149	1.495^{20}	1.4737^{25}	vs eth; s chl
10255	2,3,4-Trichloro-6-methylphenol	4,5,6-Trichloro-*o*-cresol	$C_7H_5Cl_3O$	551-78-0	211.473	nd (peth)	77				
10256	2,3,6-Trichloro-4-methylphenol	2,3,6-Trichloro-*p*-cresol	$C_7H_5Cl_3O$	551-77-9	211.473	nd (HOAc, peth)	66.5				vs EtOH
10257	2,4,6-Trichloro-3-methylphenol	2,4,6-Trichloro-*m*-cresol	$C_7H_5Cl_3O$	551-76-8	211.473	nd or pl (w, peth)	46	265			i H_2O; vs EtOH, MeOH, chl
10258	1,1,1-Trichloro-2-methyl-2-propanol	1,1,1-Trichloro-*tert*-butyl alcohol	$C_4H_7Cl_3O$	57-15-8	177.457	hyg nd (w + 1)	99(2)	169(8)			i H_2O; s EtOH, eth, ace, bz, lig, chl
10259	Trichloronate		$C_{10}H_{12}Cl_3O_2PS$	327-98-0	333.599	ye liq		$108^{0.01}$	1.365^{20}		
10260	1,2,4-Trichloro-5-nitrobenzene		$C_6H_2Cl_3NO_2$	89-69-0	226.445	pr (al), nd (al)	57.5	288	1.790^{23}		i H_2O; sl EtOH; s eth, bz, chl, CS_2
10261	Trichloronitromethane	Chloropicrin	CCl_3NO_2	76-06-2	164.376	liq	-69.4(0.2)	112(2)	1.6558^{20}	1.4611^{20}	s H_2O; msc EtOH, ace, bz, MeOH, HOAc
10262	3,4,6-Trichloro-2-nitrophenol		$C_6H_2Cl_3NO_3$	82-62-2	242.444	pa ye cry (peth)	91(2)				
10263	Trichlorooctadecylsilane	Octadecyltrichlorosilane	$C_{18}H_{37}Cl_3Si$	112-04-9	387.932			223^{10}	0.984^{25}	1.4602^{20}	
10264	Trichlorooctylsilane	Octyltrichlorosilane	$C_8H_{17}Cl_3Si$	5283-66-9	247.666			232		1.4480^{20}	dec H_2O, EtOH; s ctc
10265	1,2,3-Trichloro-1,1,2,3,3-pentafluoropropane		$C_3Cl_3F_5$	76-17-5	237.383	liq	-72	73.7	1.6631^{20}	1.3512^{20}	
10266	Trichloropentylsilane	Amyltrichlorosilane	$C_5H_{11}Cl_3Si$	107-72-2	205.586			172	1.1330^{20}	1.4503^{20}	
10267	2,3,4-Trichlorophenol		$C_6H_3Cl_3O$	15950-66-0	197.446	nd (bz, lig, sub)	80(1)	sub			s EtOH, eth, bz, alk, HOAc
10268	2,3,5-Trichlorophenol		$C_6H_3Cl_3O$	933-78-8	197.446	nd (al)	62(1)	248^{250}			vs eth, EtOH
10269	2,3,6-Trichlorophenol		$C_6H_3Cl_3O$	933-75-5	197.446	nd (dil al, lig)	58(1)				sl H_2O; vs EtOH, eth, bz; s HOAc
10270	2,4,5-Trichlorophenol		$C_6H_3Cl_3O$	95-95-4	197.446	nd (al, peth)	68.4(0.5)	262(5)			sl H_2O; vs EtOH, eth, bz; s HOAc
10271	2,4,6-Trichlorophenol		$C_6H_3Cl_3O$	88-06-2	197.446	orth nd (HOAc)	69.5(0.2)	249(4)	1.4901^{75}		sl H_2O; s EtOH, eth, HOAc
10272	3,4,5-Trichlorophenol		$C_6H_3Cl_3O$	609-19-8	197.446	nd (lig)	101	275			sl H_2O, lig; s eth
10273	2,4,5-Trichlorophenoxyacetic acid	2,4,5-T	$C_8H_5Cl_3O_3$	93-76-5	255.483	cry (bz)	155.2(0.8)	dec			i H_2O; s EtOH; vs bz
10274	2-(2,4,5-Trichlorophenoxy)-ethyl 2,2-dichloropropanoate	Pentanate	$C_{11}H_9Cl_5O_3$	136-25-4	366.452		49	$162^{0.5}$	1.55^{50}		i H_2O; s EtOH, ace, xyl
10275	Trichloro(2-phenylethyl)silane		$C_8H_9Cl_3Si$	940-41-0	239.602			242	1.2397^{20}	1.5185^{20}	
10276	(2,4,6-Trichlorophenyl) hydrazine		$C_6H_5Cl_3N_2$	5329-12-4	211.476	cry (bz)	143				s H_2O, bz
10277	Trichlorophenylsilane		$C_6H_5Cl_3Si$	98-13-5	211.549			201(3)	1.321^{20}	1.5230^{20}	s ctc, chl, CS_2
10278	1,1,2-Trichloropropane		$C_3H_5Cl_3$	598-77-6	147.431			133(5)	1.372^{15}		i H_2O; s EtOH, chl; vs eth; sl ctc
10279	1,1,3-Trichloropropane		$C_3H_5Cl_3$	20395-25-9	147.431	liq	-58.9(0.6)	146(3)	1.3557^{20}	1.4718^{20}	vs eth, EtOH, chl
10280	1,2,2-Trichloropropane		$C_3H_5Cl_3$	3175-23-3	147.431			123(5)	1.318^{25}	1.4609^{20}	i H_2O; s EtOH, eth; vs chl
10281	1,2,3-Trichloropropane	Allyl trichloride	$C_3H_5Cl_3$	96-18-4	147.431	liq	-13.8(0.6)	158(2)	1.3889^{20}	1.4852^{20}	sl H_2O, ctc; s EtOH, eth; vs chl
10282	1,1,1-Trichloro-2-propanol		$C_3H_5Cl_3O$	76-00-6	163.430		47(5)	159(13)			vs ace, bz, eth, EtOH
10283	1,1,1-Trichloro-2-propanone	1,1,1-Trichloroacetone	$C_3H_3Cl_3O$	918-00-3	161.414			149	1.435^{20}	1.4635^{17}	i H_2O; vs EtOH, eth
10284	1,2,3-Trichloro-1-propene		$C_3H_3Cl_3$	96-19-5	145.415			142	1.412^{20}	1.5030^{20}	i H_2O; vs EtOH, eth; s bz, chl
10285	3,3,3-Trichloro-1-propene		$C_3H_3Cl_3$	2233-00-3	145.415	liq	-30	112(9)	1.367^{20}	1.4827^{20}	i H_2O; s EtOH, eth, bz, chl
10286	2,3,3-Trichloro-2-propenoyl chloride		C_3Cl_4O	815-58-7	193.843			158		1.5271^{18}	vs bz
10287	Trichloropropylsilane	Propyltrichlorosilane	$C_3H_7Cl_3Si$	141-57-1	177.533			123.5	1.195^{20}	1.4310^{20}	
10288	2,4,6-Trichloropyrimidine		$C_4HCl_3N_2$	3764-01-0	183.423		22.5	212.5		1.5700^{20}	
10289	3-(Trichlorosilyl)propanenitrile		$C_3H_4Cl_3NSi$	1071-22-3	188.516			109^{30}			
10290	2,4,6-Trichloro-1,3,5-triazine	Cyanuric acid trichloride	$C_3Cl_3N_3$	108-77-0	184.411	cry (eth, bz)	145.7(0.2)	192			vs EtOH
10291	2,2′,2″-Trichlorotriethylamine		$C_6H_{12}Cl_3N$	555-77-1	204.525	pa ye	-2.0	143^{15}			vs bz, eth, EtOH
10292	Trichlorotriethyldialuminum	Ethylaluminum sesquichloride	$C_6H_{15}Al_2Cl_3$	12075-68-2	247.505	ye liq		115.5^{50}			
10293	1,3,5-Trichloro-2,4,6-trifluorobenzene		$C_6Cl_3F_3$	319-88-0	235.418			199(27)			

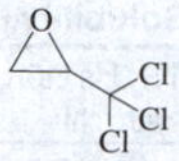
10254
(Trichloromethyl)oxirane

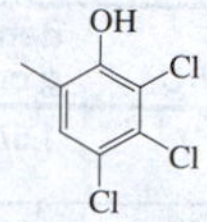
10255
2,3,4-Trichloro-6-methylphenol

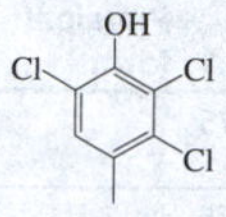
10256
2,3,6-Trichloro-4-methylphenol

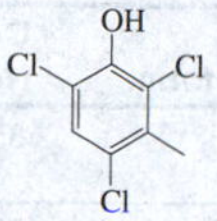
10257
2,4,6-Trichloro-3-methylphenol

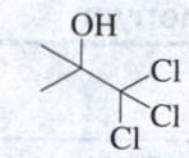
10258
1,1,1-Trichloro-2-methyl-2-propanol

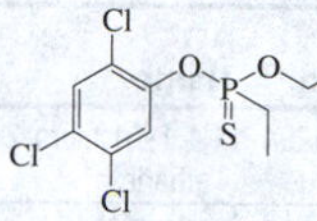
10259
Trichloronate

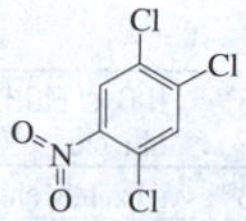
10260
1,2,4-Trichloro-5-nitrobenzene

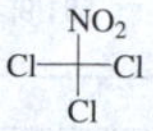
10261
Trichloronitromethane

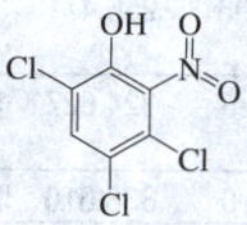
10262
3,4,6-Trichloro-2-nitrophenol

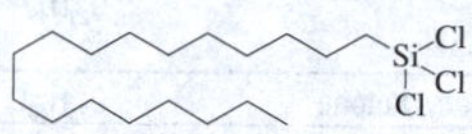
10263
Trichlorooctadecylsilane

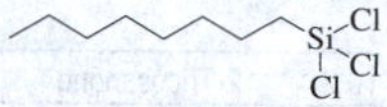
10264
Trichlorooctylsilane

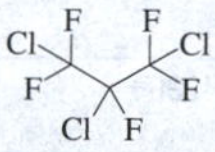
10265
1,2,3-Trichloro-1,1,2,3,3-pentafluoropropane

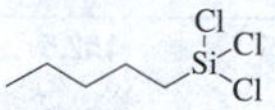
10266
Trichloropentylsilane

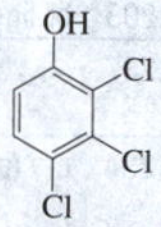
10267
2,3,4-Trichlorophenol

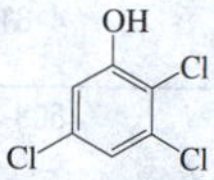
10268
2,3,5-Trichlorophenol

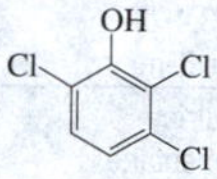
10269
2,3,6-Trichlorophenol

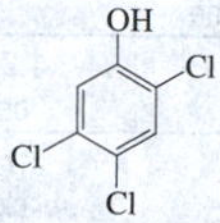
10270
2,4,5-Trichlorophenol

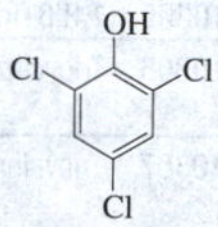
10271
2,4,6-Trichlorophenol

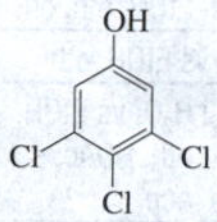
10272
3,4,5-Trichlorophenol

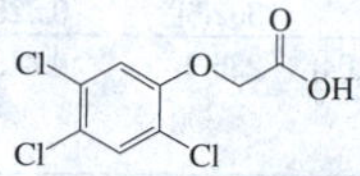
10273
2,4,5-Trichlorophenoxyacetic acid

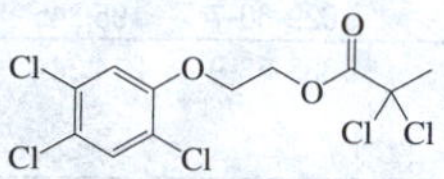
10274
2-(2,4,5-Trichlorophenoxy)ethyl 2,2-dichloropropanoate

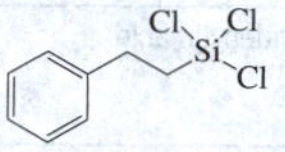
10275
Trichloro(2-phenylethyl)silane

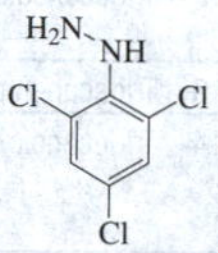
10276
(2,4,6-Trichlorophenyl)hydrazine

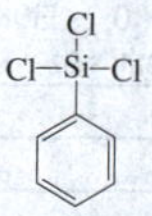
10277
Trichlorophenylsilane

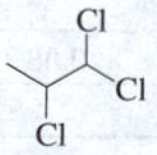
10278
1,1,2-Trichloropropane

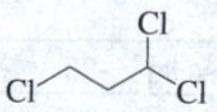
10279
1,1,3-Trichloropropane

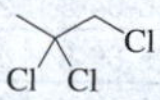
10280
1,2,2-Trichloropropane

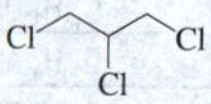
10281
1,2,3-Trichloropropane

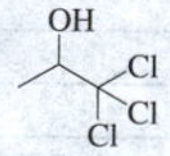
10282
1,1,1-Trichloro-2-propanol

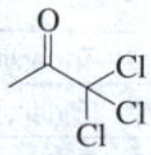
10283
1,1,1-Trichloro-2-propanone

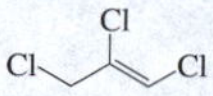
10284
1,2,3-Trichloro-1-propene

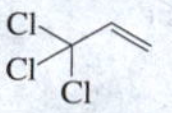
10285
3,3,3-Trichloro-1-propene

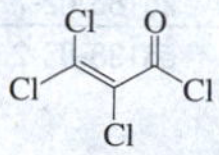
10286
2,3,3-Trichloro-2-propenoyl chloride

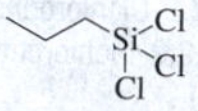
10287
Trichloropropylsilane

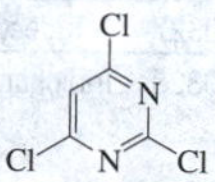
10288
2,4,6-Trichloropyrimidine

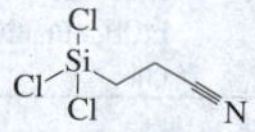
10289
3-(Trichlorosilyl)propanenitrile

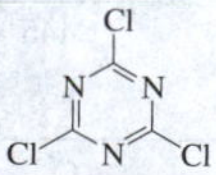
10290
2,4,6-Trichloro-1,3,5-triazine

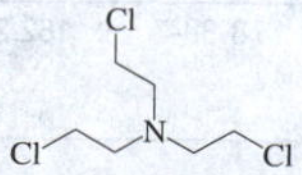
10291
2,2',2''-Trichlorotriethylamine

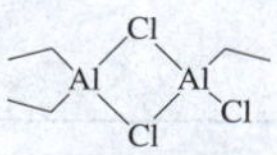
10292
Trichlorotriethyldialuminum

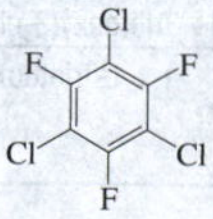
10293
1,3,5-Trichloro-2,4,6-trifluorobenzene

Organic

No.	Name	Synonym	Mol. form.	CAS Reg. No.	Mol. wt.	Physical form	mp/°C	bp/°C	den/ g cm⁻³	n_D	Solubility
10294	1,1,1-Trichloro-2,2,2-trifluoroethane		$C_2Cl_3F_3$	354-58-5	187.375		14.367(0.01)	46(1)	1.5790^{20}	1.3610^{35}	i H_2O; s EtOH, eth, chl
10295	1,1,2-Trichloro-1,2,2-trifluoroethane	1,1,2-Trichlorotrifluoroethane	$C_2Cl_3F_3$	76-13-1	187.375	liq	-36.2(0.9)	47.6(0.2)	1.5635^{25}	1.3557^{25}	i H_2O; s EtOH; msc eth, bz
10296	Trichlorovinylsilane	Vinyltrichlorosilane	$C_2H_3Cl_3Si$	75-94-5	161.490	liq	-95	90.9(0.4)	1.2426^{20}	1.4295^{20}	vs chl
10297	Trichodermin	12,13-Epoxytrichothec-9-en-4-ol acetate	$C_{17}H_{24}O_4$	4682-50-2	292.371	cry	59	$111^{0.05}$			sl H_2O; s EtOH, chl
10298	Triclofos	2,2,2-Trichloroethanol dihydrogen phosphate	$C_2H_4Cl_3O_4P$	306-52-5	229.383	cry (bz)	120.5				
10299	Triclopyr	Acetic acid, [(3,5,6-trichloro-2-pyridinyl)oxy]-	$C_7H_4Cl_3NO_3$	55335-06-3	256.471		150.7(0.5)	290 dec			
10300	Tricosane		$C_{23}H_{48}$	638-67-5	324.627	lf (eth-al)	47.4(0.2)	381(9)	0.7785^{48}	1.4468^{20}	i H_2O; sl EtOH; s eth, ctc
10301	12-Tricosanone	Diundecyl ketone	$C_{23}H_{46}O$	540-09-0	338.610	lf (al)	69.0(0.5)		0.8086^{69}	1.4283^{80}	vs bz, eth, chl
10302	Tri-*o*-cresyl phosphate	Tri-*o*-tolyl phosphate	$C_{21}H_{21}O_4P$	78-30-8	368.363	col or pa ye	11	410	1.1955^{20}	1.5575^{20}	i H_2O; vs EtOH, eth, ctc, tol; s HOAc
10303	Tri-*m*-cresyl phosphate	Tri-*m*-tolyl phosphate	$C_{21}H_{21}O_4P$	563-04-2	368.363	wax	25.5	260^{15}	1.150^{25}	1.5575^{20}	i H_2O; sl EtOH; s eth; vs ctc, tol
10304	Tri-*p*-cresyl phosphate	Tri-*p*-tolyl phosphate	$C_{21}H_{21}O_4P$	78-32-0	368.363	nd (al), tab (eth)	77.5	224^{35}	1.247^{25}		s EtOH, eth, bz, chl, HOAc
10305	1,3,6-Tricyanohexane		$C_9H_{11}N_3$	1772-25-4	161.203	br liq		257^{2}	1.040	1.4660^{20}	
10306	Tricyclazole	1,2,4-Triazolo[3,4-*b*]-benzothiazole, 5-methyl-	$C_9H_7N_3S$	41814-78-2	189.237		187.3(0.5)				
10307	Tricyclene	1,7,7-Trimethyltricyclo[2.2.1.0^{2,6}]heptane	$C_{10}H_{16}$	508-32-7	136.234	cry (al)	67.5	152.5	0.8668^{80}	1.4296^{80}	
10308	Tricyclo[3.3.1.1^{3,7}]decan-1-amine	Amantadine	$C_{10}H_{17}N$	768-94-5	151.249		180				sl H_2O
10309	Tricyclo[3.3.1.1^{3,7}]decane	Adamantane	$C_{10}H_{16}$	281-23-2	136.234	nd (sub)	270.1(0.8)	sub	1.07^{25}	1.568	s bz, ctc
10310	Tridecanal		$C_{13}H_{26}O$	10486-19-8	198.344		14(2)	156^{13}	0.8356^{18}	1.4384^{18}	i H_2O; s EtOH
10311	Tridecane		$C_{13}H_{28}$	629-50-5	184.361	liq	-5.35(0.02)	235.4(0.4)	0.7564^{20}	1.4256^{20}	i H_2O; vs EtOH, eth; s ctc
10312	Tridecanedioic acid		$C_{13}H_{24}O_4$	505-52-2	244.328		114.3(0.5)				sl H_2O, bz, tfa; s EtOH, eth, chl
10313	Tridecanenitrile		$C_{13}H_{25}N$	629-60-7	195.345		10(1)	302(6)	0.8257^{20}	1.4378^{20}	vs EtOH, eth
10314	Tridecanoic acid	Tridecylic acid	$C_{13}H_{26}O_2$	638-53-9	214.344	cry (peth ace)	41.85(0.04)	236^{100}	0.8458^{80}	1.4286^{60}	i H_2O; vs EtOH, eth, HOAc; s ace
10315	1-Tridecanol	Tridecyl alcohol	$C_{13}H_{28}O$	112-70-9	200.360	cry (al)	31(1)	287(8)	0.8223^{31}		i H_2O; s EtOH, eth
10316	2-Tridecanone	Methyl undecyl ketone	$C_{13}H_{26}O$	593-08-8	198.344		28(1)	268(1)	0.8217^{30}	1.4318^{20}	i H_2O; vs EtOH, eth, ace, bz, chl
10317	7-Tridecanone	Dihexyl ketone	$C_{13}H_{26}O$	462-18-0	198.344	lf (al)	33	266(9)	0.825^{30}		s EtOH, chl, lig; vs eth
10318	1-Tridecene		$C_{13}H_{26}$	2437-56-1	182.345	liq	-23.07(0.09)	232.8(0.7)	0.7658^{20}	1.4340^{20}	i H_2O; vs EtOH, eth; s bz
10319	Tridecyl acrylate		$C_{16}H_{30}O_2$	3076-04-8	254.408	liq		150^{10}	0.88^{20}		
10320	Tridecylaluminum		$C_{30}H_{63}Al$	1726-66-5	450.803	hyg visc liq	-38				
10321	Tridecylamine	*N,N*-Didecyl-1-decanamine	$C_{30}H_{63}N$	1070-01-5	437.828			406			
10322	(Tridecyl)amine	1-Tridecanamine	$C_{13}H_{29}N$	2869-34-3	199.376		27(2)	273(2)	0.8049^{20}	1.4443^{20}	sl H_2O; s EtOH, eth
10323	Tridecylbenzene	1-Phenyltridecane	$C_{19}H_{32}$	123-02-4	260.457		-0.3(0.4)	340(4)	0.8550^{20}	1.4821^{20}	
10324	Tridecylcyclohexane		$C_{19}H_{38}$	6006-33-3	266.505		14(1)	332(7)	0.8239^{20}	1.4570^{20}	
10325	Tridecyl methacrylate		$C_{17}H_{32}O_2$	2495-25-2	268.435			118^{1}	0.881^{20}	1.448^{25}	
10326	Tri(decyl) phosphite		$C_{30}H_{63}O_3P$	2929-86-4	502.793	liq		255^{3}			
10327	1-Tridecyne		$C_{13}H_{24}$	26186-02-7	180.330		1(3)	234	0.7842^{20}	1.4309^{20}	vs bz, eth
10328	Tridiphane	2-(3,5-Dichlorophenyl)-2-(2,2,2-trichloroethyl)oxirane, (±)	$C_{10}H_7Cl_5O$	58138-08-2	320.427		42.1(0.5)				
10329	Tridodecylamine	*N,N*-Didodecyl-1-dodecanamine	$C_{36}H_{75}N$	102-87-4	521.988		15.7(0.5)	$220^{0.03}$			
10330	Triethanolamine	Tris(2-hydroxyethyl)amine	$C_6H_{15}NO_3$	102-71-6	149.188	hyg cry	21.5(0.4)	350(5)	1.1242^{20}	1.4852^{20}	msc H_2O, EtOH; sl eth, bz; s chl
10331	1,3,5-Triethoxybenzene		$C_{12}H_{18}O_3$	2437-88-9	210.269	cry (al, dil al)	43.5	170^{24}			vs eth, EtOH
10332	Triethoxy(3-chloropropyl)silane	(3-Chloropropyl)triethoxysilane	$C_9H_{21}ClO_3Si$	5089-70-3	240.800	col gas		-149			
10333	1,1,1-Triethoxyethane		$C_8H_{18}O_3$	78-39-7	162.227			145	0.8847^{25}	1.3980^{20}	i H_2O; msc EtOH, eth, ctc, chl

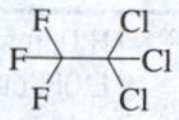
10294
1,1,1-Trichloro-2,2,2-trifluoroethane

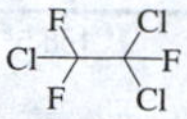
10295
1,1,2-Trichloro-1,2,2-trifluoroethane

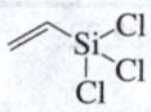
10296
Trichlorovinylsilane

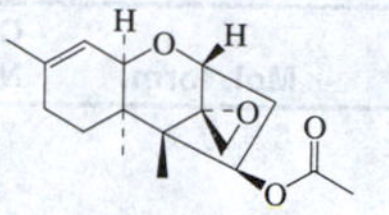
10297
Trichodermin

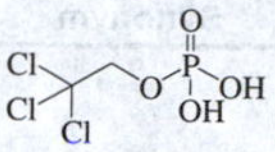
10298
Triclofos

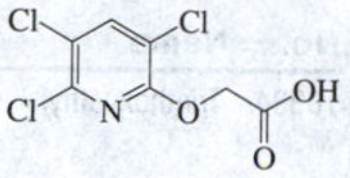
10299
Triclopyr

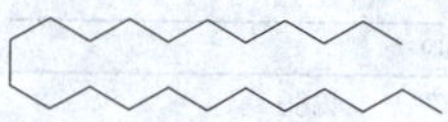
10300
Tricosane

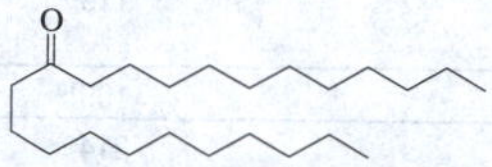
10301
12-Tricosanone

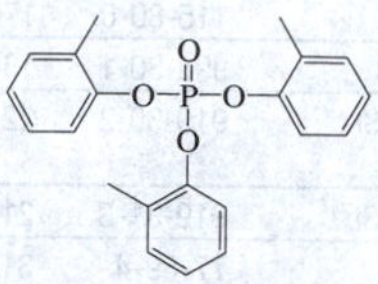
10302
Tri-*o*-cresyl phosphate

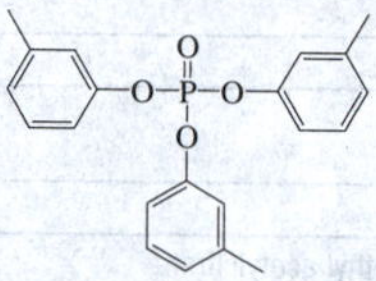
10303
Tri-*m*-cresyl phosphate

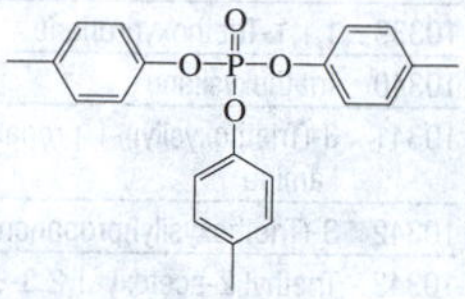
10304
Tri-*p*-cresyl phosphate

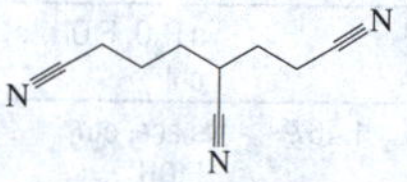
10305
1,3,6-Tricyanohexane

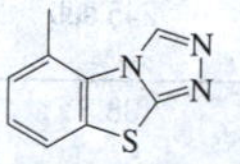
10306
Tricyclazole

10307
Tricyclene

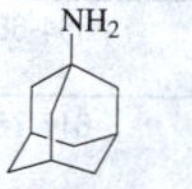
10308
Tricyclo[3.3.1.1^{3,7}]decan-1-amine

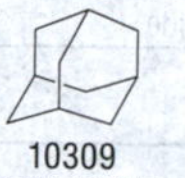
10309
Tricyclo[3.3.1.1^{3,7}]decane

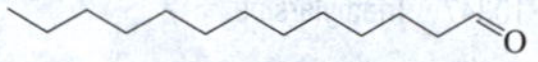
10310
Tridecanal

10311
Tridecane

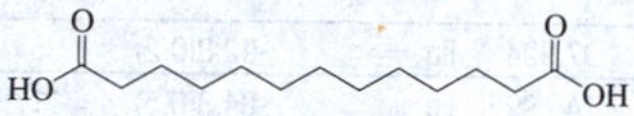
10312
Tridecanedioic acid

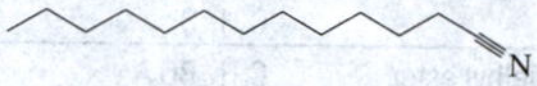
10313
Tridecanenitrile

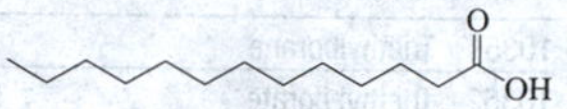
10314
Tridecanoic acid

10315
1-Tridecanol

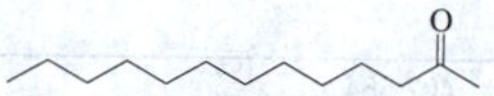
10316
2-Tridecanone

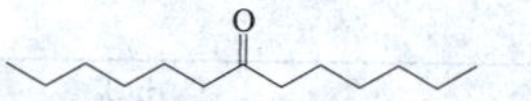
10317
7-Tridecanone

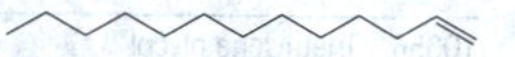
10318
1-Tridecene

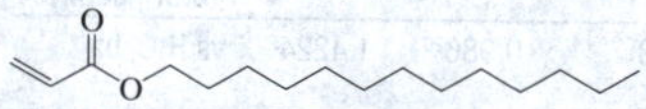
10319
Tridecyl acrylate

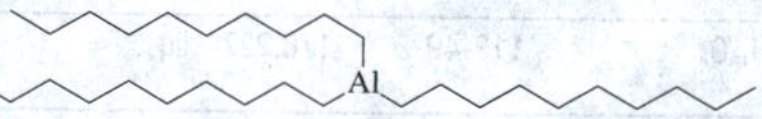
10320
Tridecylaluminum

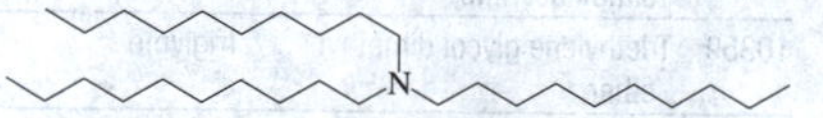
10321
Tridecylamine

10322
(Tridecyl)amine

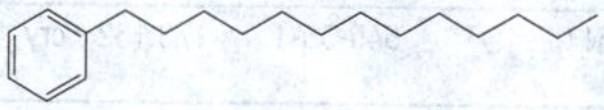
10323
Tridecylbenzene

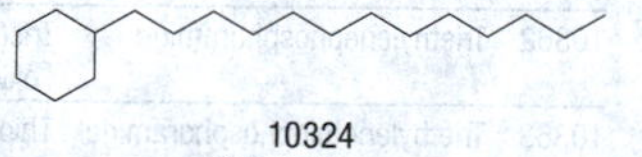
10324
Tridecylcyclohexane

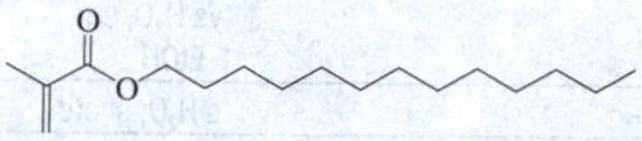
10325
Tridecyl methacrylate

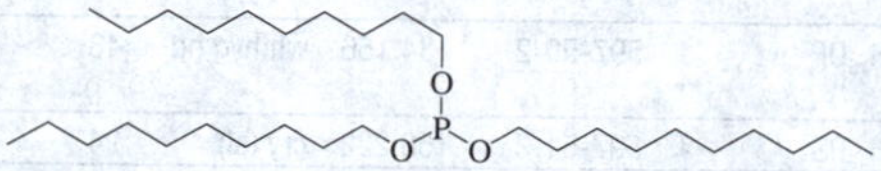
10326
Tri(decyl) phosphite

10327
1-Tridecyne

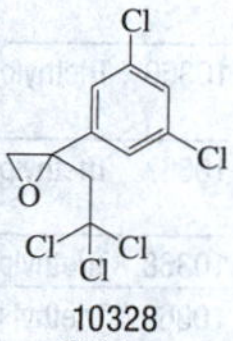
10328
Tridiphane

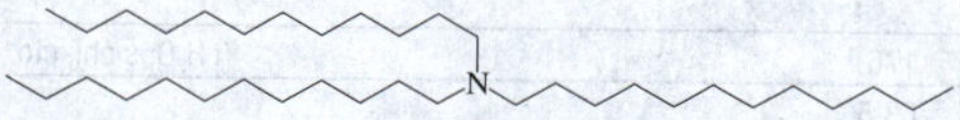
10329
Tridodecylamine

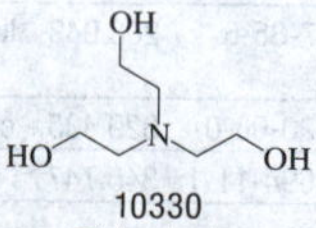
10330
Triethanolamine

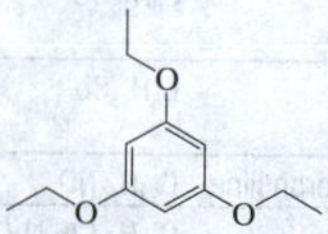
10331
1,3,5-Triethoxybenzene

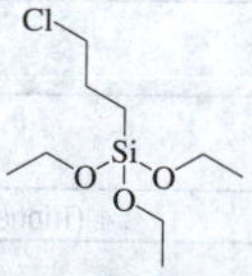
10332
Triethoxy(3-chloropropyl)silane

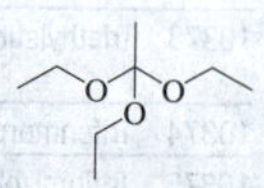
10333
1,1,1-Triethoxyethane

No.	Name	Synonym	Mol. form.	CAS Reg. No.	Mol. wt.	Physical form	mp/°C	bp/°C	den/ g cm^{-3}	n_D	Solubility
10334	Triethoxyethylsilane		$C_8H_{20}O_3Si$	78-07-9	192.329			158.5	0.8963^{20}	1.3955^{20}	i H_2O; msc EtOH, eth; s chl
10335	Triethoxymethane		$C_7H_{16}O_3$	122-51-0	148.200			145(3)	0.8909^{20}	1.3922^{20}	s EtOH, eth
10336	Triethoxymethylsilane		$C_7H_{18}O_3Si$	2031-67-6	178.302			142	0.8948^{25}	1.3832^{20}	
10337	Triethoxypentylsilane		$C_{11}H_{26}O_3Si$	2761-24-2	234.408			100^{30}	0.8862^{20}	1.4059^{20}	
10338	Triethoxyphenylsilane		$C_{12}H_{20}O_3Si$	780-69-8	240.371			233(3)	0.996^{25}	1.4604^{20}	
10339	1,1,1-Triethoxypropane		$C_9H_{20}O_3$	115-80-0	176.253			171		1.4000^{25}	vs eth, EtOH
10340	Triethoxysilane		$C_6H_{16}O_3Si$	998-30-1	164.275			133.5	0.8745^{20}		
10341	3-(Triethoxysilyl)-1-propanamine		$C_9H_{23}NO_3Si$	919-30-2	221.370			119^{29}	0.9506^{20}	1.4225^{20}	
10342	3-(Triethoxysilyl)propanenitrile		$C_9H_{19}NO_3Si$	919-31-3	217.338	liq		109^{10}	0.974^{20}		
10343	Triethyl 2-acetoxy-1,2,3-propanetricarboxylate	Triethyl acetylcitrate	$C_{14}H_{22}O_8$	77-89-4	318.320			214^{40}	1.135^{25}	1.4380	
10344	Triethylaluminum	Hexaethyldialuminum	$C_6H_{15}Al$	97-93-8	114.165	col hyg liq	-48.14(0.02)	193(1)	0.832^{25}		
10345	Triethylamine	*N,N*-Diethylethanamine	$C_6H_{15}N$	121-44-8	101.190	liq	-114.7(0.2)	88.8(0.2)	0.7275^{20}	1.4010^{20}	s H_2O, EtOH, eth, ctc; vs ace, bz, chl
10346	Triethylamine hydrochloride	*N,N*-Diethylethanamine hydrochloride	$C_6H_{16}ClN$	554-68-7	137.651	hex (al)	260 dec	245 sub	1.0689^{21}		vs H_2O, EtOH, chl
10347	Triethylarsine		$C_6H_{15}As$	617-75-4	162.105			138.5	1.150^{20}	1.467^{20}	vs ace, eth, EtOH
10348	1,2,3-Triethylbenzene		$C_{12}H_{18}$	42205-08-3	162.271	col liq	-26	172			
10349	1,2,4-Triethylbenzene		$C_{12}H_{18}$	877-44-1	162.271			217(3)	0.8738^{20}	1.5024^{20}	i H_2O; s EtOH, eth
10350	1,3,5-Triethylbenzene		$C_{12}H_{18}$	102-25-0	162.271	liq	-66.4(0.1)	215.8(0.9)	0.8631^{20}	1.4969^{20}	i H_2O; vs EtOH, eth
10351	Triethylborane		$C_6H_{15}B$	97-94-9	97.994	liq	-92.8(0.2)	95	0.70^{23}	1.3971	s EtOH, eth
10352	Triethyl borate	Boric acid, triethyl ester	$C_6H_{15}BO_3$	150-46-9	145.992	liq	-84.8(0.5)	117.9(0.7)	0.8546^{20}	1.3749^{20}	msc EtOH, eth
10353	Triethyl citrate		$C_{12}H_{20}O_7$	77-93-0	276.283			294	1.1369^{20}	1.4455^{20}	i H_2O; s EtOH, eth; sl ctc
10354	Triethylenediamine		$C_6H_{12}N_2$	280-57-9	112.172		159.83(0.02)				s chl
10355	Triethylene glycol	Triglycol	$C_6H_{14}O_4$	112-27-6	150.173	hyg liq	-9.4(0.5)	288.6(0.2)	1.1274^{15}	1.4531^{20}	msc H_2O, EtOH, bz; sl eth, chl; i peth
10356	Triethylene glycol bis(2-ethylhexanoate)		$C_{22}H_{42}O_6$	94-28-0	402.564						s chl
10357	Triethylene glycol diacetate		$C_{10}H_{18}O_6$	111-21-7	234.246	liq	-50	277(9)	1.1153^{20}		vs H_2O, eth, EtOH
10358	Triethylene glycol dimethacrylate		$C_{14}H_{22}O_6$	109-16-0	286.321			170^{5}	1.092^{20}	1.4595^{25}	vs ace, eth, EtOH, peth
10359	Triethylene glycol dimethyl ether	Triglyme	$C_8H_{18}O_4$	112-49-2	178.227	liq	-43.8(0.2)	218(3)	0.986^{20}	1.4224^{20}	vs H_2O, bz
10360	Triethylene glycol dinitrate	Ethanol, 2,2'-[1,2-ethanediylbis(oxy)]-bis-, dinitrate	$C_6H_{12}N_2O_8$	111-22-8	240.167			82$^{0.03}$			
10361	Triethylene glycol monoethyl ether	2-[2-(2-Ethoxyethoxy)ethoxy]-ethanol	$C_8H_{18}O_4$	112-50-5	178.227			256	1.0209^{20}		
10362	Triethylenephosphoramide	Tris(1-aziridinyl)phosphine, oxide	$C_6H_{12}N_3OP$	545-55-1	173.152	cry	41	91^{23}			vs H_2O, EtOH, eth, ace
10363	Triethylenethiophosphoramide	Thiotepa	$C_6H_{12}N_3PS$	52-24-4	189.218	cry	51.5				vs H_2O; s bz, chl, eth, EtOH
10364	1,3,5-Triethylhexahydro-1,3,5-triazine		$C_9H_{21}N_3$	7779-27-3	171.283			78^{6}		1.4580^{25}	
10365	Triethyl phosphate	Ethyl phosphate	$C_6H_{15}O_4P$	78-40-0	182.154	liq	-56.4	216(11)	1.0695^{20}	1.4053^{20}	s H_2O, eth, bz; vs EtOH; sl chl
10366	Triethylphosphine		$C_6H_{15}P$	554-70-1	118.157	liq	-88	129	0.8006^{19}	1.458^{15}	i H_2O; msc EtOH, eth
10367	Triethylphosphine oxide		$C_6H_{15}OP$	597-50-2	134.156	wh hyg nd	48	243			vs H_2O, eth, EtOH
10368	Triethylphosphine sulfide		$C_6H_{15}PS$	597-51-3	150.222	cry (al)	94				s H_2O; sl ctc
10369	Triethyl phosphite	Triethoxyphosphine	$C_6H_{15}O_3P$	122-52-1	166.155			157.9	0.9629^{20}	1.4127^{20}	i H_2O; vs EtOH, eth
10370	*O,O,O*-Triethyl phosphorothioate	*O,O,O*-Triethyl thiophosphate	$C_6H_{15}O_3PS$	126-68-1	198.220			217	1.0768^{20}	1.4480^{20}	
10371	Triethylsilane		$C_6H_{16}Si$	617-86-7	116.277	liq	-156.9	109	0.7302^{20}	1.447^{20}	i H_2O, sulf
10372	Triethylsilanol		$C_6H_{16}OSi$	597-52-4	132.276			154	0.8647^{20}	1.4329^{20}	i H_2O; msc EtOH, eth
10373	Triethylstibine		$C_6H_{15}Sb$	617-85-6	208.943	liq	-98	161.4	1.3224^{15}		i H_2O; s EtOH, eth
10374	Trifenmorph	4-(Triphenylmethyl)morpholine	$C_{23}H_{23}NO$	1420-06-0	329.435	cry (EtOH)	176				i H_2O; s chl, ctc
10375	Triflumizole		$C_{15}H_{15}ClF_3N_3O$	68694-11-1	345.747		63.5				

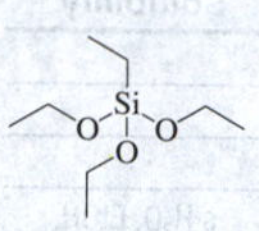
10334
Triethoxyethylsilane

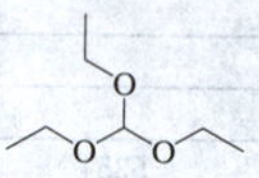
10335
Triethoxymethane

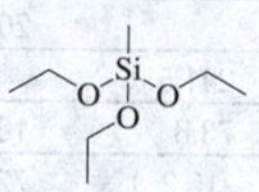
10336
Triethoxymethylsilane

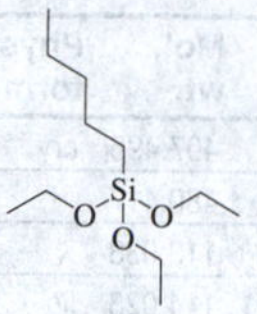
10337
Triethoxypentylsilane

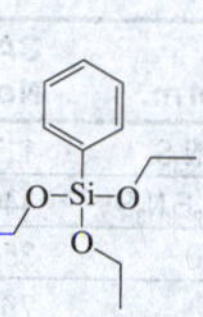
10338
Triethoxyphenylsilane

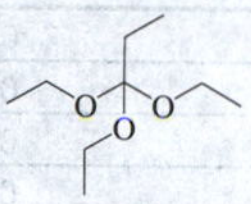
10339
1,1,1-Triethoxypropane

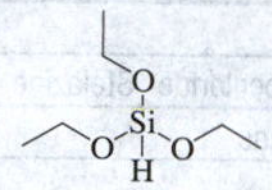
10340
Triethoxysilane

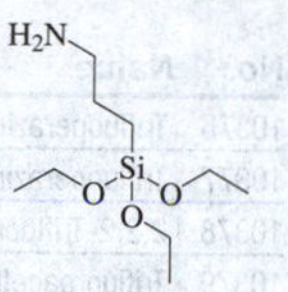
10341
3-(Triethoxysilyl)-1-propanamine

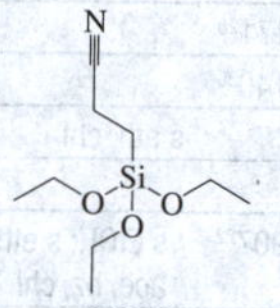
10342
3-(Triethoxysilyl)propanenitrile

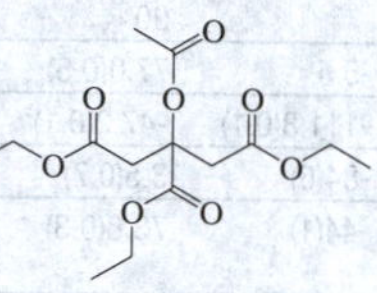
10343
Triethyl 2-acetoxy-1,2,3-propanetricarboxylate

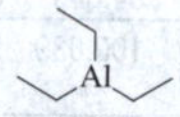
10344
Triethylaluminum

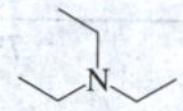
10345
Triethylamine

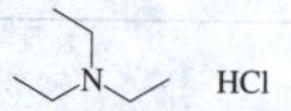
10346
Triethylamine hydrochloride

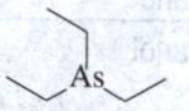
10347
Triethylarsine

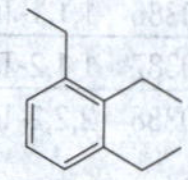
10348
1,2,3-Triethylbenzene

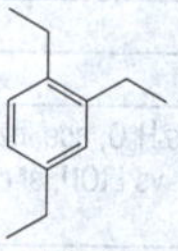
10349
1,2,4-Triethylbenzene

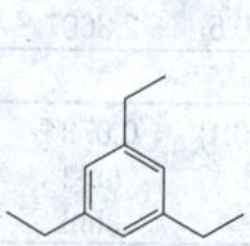
10350
1,3,5-Triethylbenzene

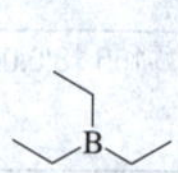
10351
Triethylborane

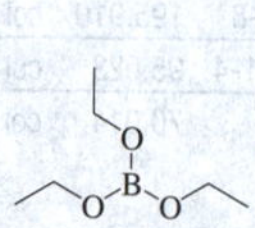
10352
Triethyl borate

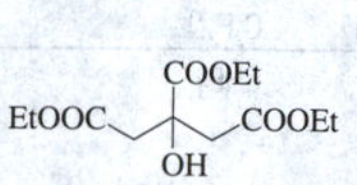
10353
Triethyl citrate

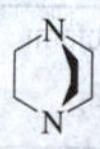
10354
Triethylenediamine

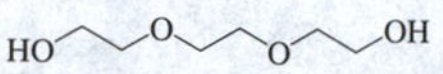
10355
Triethylene glycol

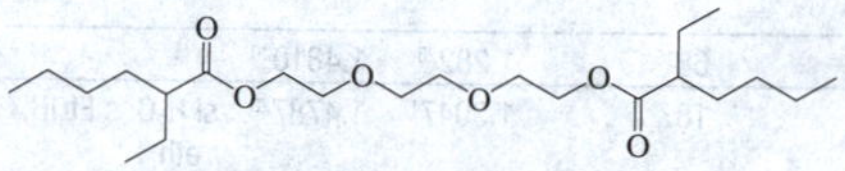
10356
Triethylene glycol bis(2-ethylhexanoate)

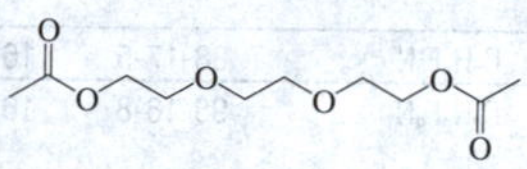
10357
Triethylene glycol diacetate

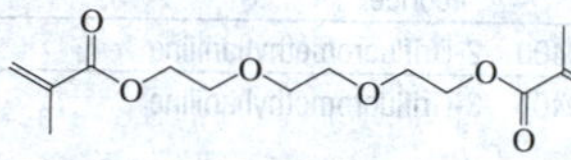
10358
Triethylene glycol dimethacrylate

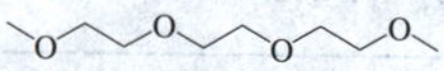
10359
Triethylene glycol dimethyl ether

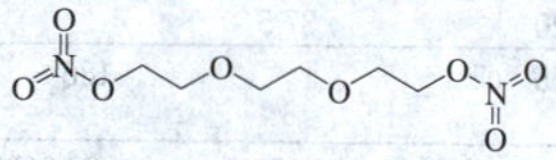
10360
Triethylene glycol dinitrate

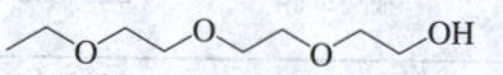
10361
Triethylene glycol monoethyl ether

10362
Triethylenephosphoramide

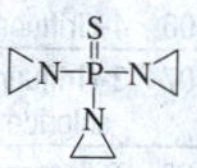
10363
Triethylenethiophosphoramide

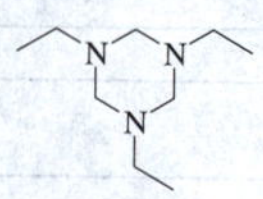
10364
1,3,5-Triethylhexahydro-1,3,5-triazine

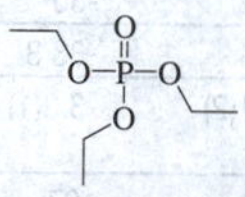
10365
Triethyl phosphate

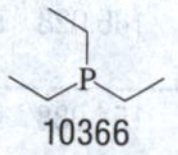
10366
Triethylphosphine

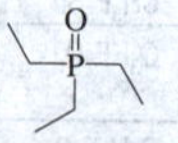
10367
Triethylphosphine oxide

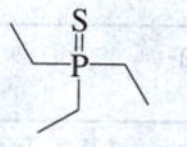
10368
Triethylphosphine sulfide

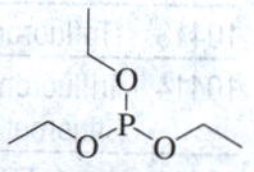
10369
Triethyl phosphite

10370
O,O,O-Triethyl phosphorothioate

10371
Triethylsilane

10372
Triethylsilanol

10373
Triethylstibine

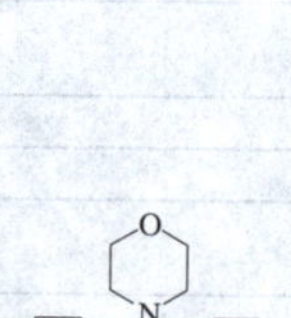
10374
Trifenmorph

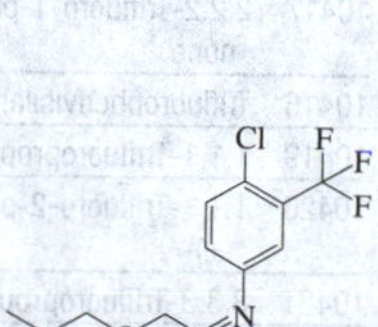
10375
Triflumizole

Organic

No.	Name	Synonym	Mol. form.	CAS Reg. No.	Mol. wt.	Physical form	mp/°C	bp/°C	den/ g cm^{-3}	n_D	Solubility
10376	Trifluoperazine		$C_{21}H_{24}F_3N_3S$	117-89-5	407.496	cry		$206^{0.7}$			
10377	Trifluoperazine dihydrochloride	Stelazine	$C_{21}H_{26}Cl_2F_3N_3S$	440-17-5	480.417		241.5				
10378	2,2,2-Trifluoroacetamide		$C_2H_2F_3NO$	354-38-1	113.038		73.8	162.5			
10379	Trifluoroacetic acid		$C_2HF_3O_2$	76-05-1	114.023	liq	-15.2	72(2)	1.5351^{25}		s H_2O, EtOH, eth, ace
10380	Trifluoroacetic acid anhydride		$C_4F_6O_3$	407-25-0	210.031	liq	-65	39.5	1.490^{25}	1.269^{25}	
10381	1,1,1-Trifluoroacetone	Methyl trifluoromethyl ketone	$C_3H_3F_3O$	421-50-1	112.050	vol liq or gas		22(5)	1.252^{25}		
10382	Trifluoroacetonitrile		C_2F_3N	353-85-5	95.023	col gas		-68.8(0.8)			
10383	Trifluoroacetyl chloride		C_2ClF_3O	354-32-5	132.468	col gas	-146(4)	-27(4)			
10384	1,2,4-Trifluorobenzene		$C_6H_3F_3$	367-23-7	132.083			90	1.264^{25}	1.4171^{20}	
10385	1,3,5-Trifluorobenzene		$C_6H_3F_3$	372-38-3	132.083	liq	-5.5	77.0(0.5)	1.277^{25}	1.4140^{20}	
10386	1,1,1-Trifluoroethane	Methyl fluoroform	$C_2H_3F_3$	420-46-2	84.040	col gas	-111.8(0.1)	-47.2(0.1)			s eth, chl
10387	1,1,2-Trifluoroethane		$C_2H_3F_3$	430-66-0	84.040	col gas	-84(5)	3.5(0.7)			
10388	2,2,2-Trifluoroethanol		$C_2H_3F_3O$	75-89-8	100.039	liq	-44(1)	73.8(0.3)	1.3842^{20}	1.2907^{22}	vs EtOH; s eth, ace, bz, chl
10389	Trifluoroethene	Trifluoroethylene	C_2HF_3	359-11-5	82.024	col gas		-53(6)	1.26^{-70}		i H_2O; sl EtOH; s eth
10390	2,2,2-Trifluoroethylamine	2,2,2-Trifluoroethanamine	$C_2H_4F_3N$	753-90-2	99.055			36	1.245^{25}		
10391	2,2,2-Trifluoroethyl methyl ether		$C_3H_5F_3O$	460-43-5	114.066			31.62			
10392	1,1,1-Trifluoro-2-iodoethane		$C_2H_2F_3I$	353-83-3	209.936			54.5	2.13^{25}	1.4009^{20}	
10393	Trifluoroiodomethane		CF_3I	2314-97-8	195.910	col gas		-21.8(0.6)	2.3607^{-32}	1.3790^{-32}	
10394	Trifluoroisocyanomethane	Trifluoromethyl isocyanide	C_2F_3N	19480-01-4	95.023	col gas		-80			
10395	Trifluoromethane	Fluoroform	CHF_3	75-46-7	70.014	col gas	-155.18(0.06)	-82.0(0.1)	0.673^{25} (p>1 atm)		s H_2O, ace, bz; vs EtOH; sl chl
10396	Trifluoromethanesulfenyl chloride		$CClF_3S$	421-17-0	136.524	col gas		-0.7			i H_2O
10397	Trifluoromethanesulfonic acid		CHF_3O_3S	1493-13-6	150.077	hyg liq	45	162			vs eth
10398	Trifluoromethanesulfonyl chloride		$CClF_3O_2S$	421-83-0	168.523			162		1.3344^{20}	i H_2O
10399	Trifluoromethanesulfonyl fluoride		CF_4O_2S	335-05-7	152.069	col gas		-21.7			
10400	2-(Trifluoromethyl)aniline		$C_7H_6F_3N$	88-17-5	161.125		35.5	68^{15}	1.282^{25}	1.4810^{20}	
10401	3-(Trifluoromethyl)aniline		$C_7H_6F_3N$	98-16-8	161.125		5.5	187	1.3047^{12}	1.4787^{20}	sl H_2O; s EtOH, eth
10402	4-(Trifluoromethyl)aniline		$C_7H_6F_3N$	455-14-1	161.125		38	117.5^{60}	1.283^{27}	1.4815^{25}	
10403	4-(Trifluoromethyl)-benzaldehyde		$C_8H_5F_3O$	455-19-6	174.120			80^{25}		1.4630^{20}	
10404	(Trifluoromethyl)benzene	Benzotrifluoride	$C_7H_5F_3$	98-08-8	146.110	liq	-28.99(0.06)	102.0(0.2)	1.1884^{20}	1.4146^{20}	msc EtOH, eth, ace, bz, ctc
10405	3-(Trifluoromethyl)benzonitrile		$C_8H_4F_3N$	368-77-4	171.120		14.5	189	1.2813^{20}	1.4508^{20}	
10406	4-(Trifluoromethyl)benzonitrile		$C_8H_4F_3N$	455-18-5	171.120		37.5				
10407	3-(Trifluoromethyl)benzoyl chloride		$C_8H_4ClF_3O$	2251-65-2	208.565	oil		186	1.383	1.4770^{20}	
10408	Trifluoromethyl difluoromethyl ether	HFE-125	C_2HF_5O	3822-68-2	136.020	col gas	-157(2)	-35.0(0.2)			
10409	2-(Trifluoromethyl)phenol		$C_7H_5F_3O$	444-30-4	162.109		45.5	147.5			
10410	3-(Trifluoromethyl)phenol		$C_7H_5F_3O$	98-17-9	162.109	liq	-0.9	178	1.3418^{25}		
10411	2-[[3-(Trifluoromethyl)phenyl]-amino]benzoic acid	Flufenamic acid	$C_{14}H_{10}F_3NO_2$	530-78-9	281.230		133.5				s DMSO
10412	Trifluoromethylsilane		CH_3F_3Si	373-74-0	100.116	col gas	-73	-30			
10413	(Trifluoromethyl)silane		CH_3F_3Si	10112-11-5	100.116	col gas	-124	-38.3			
10414	Trifluoromethyl 1,1,2,2-tetra-fluoroethyl ether		C_3HF_7O	2356-61-8	186.028	col gas	-141(2)	-3.3(1)			
10415	1,1,1-Trifluoro-2,4-pentanedi-one	1,1,1-Trifluoroacetylacetone	$C_5H_5F_3O_2$	367-57-7	154.088	liq		107			s os
10416	4,4,4-Trifluoro-1-phenyl-1,3-butanedione		$C_{10}H_7F_3O_2$	326-06-7	216.157	cry	39	224			i H_2O; s EtOH, ace
10417	2,2,2-Trifluoro-1-phenyletha-none		$C_8H_5F_3O$	434-45-7	174.120	liq	-40(6)	153(3)	1.279^{20}	1.4583^{20}	
10418	Trifluorophenylsilane		$C_6H_5F_3Si$	368-47-8	162.185	liq	-18	101.5	1.2169^{20}	1.4110^{20}	vs bz, EtOH
10419	1,1,1-Trifluoropropane		$C_3H_5F_3$	421-07-8	98.067	col gas		-13			
10420	1,1,1-Trifluoro-2-propanol, (±)-		$C_3H_5F_3O$	17556-48-8	114.066	liq	-52	76.2(0.7)	1.2632^{25}	1.3130^{25}	vs EtOH, eth; s ace, bz; sl ctc
10421	3,3,3-Trifluoropropene		$C_3H_3F_3$	677-21-4	96.051	col gas		-27(4)			
10422	3,3,3-Trifluoro-1-propyne	(Trifluoromethyl)acetylene	C_3HF_3	661-54-1	94.035	col gas		-48.3			
10423	4,4,4-Trifluoro-1-(2-thienyl)-1,3-butanedione	Thenoyltrifluoroacetone	$C_8H_5F_3O_2S$	326-91-0	222.185		42.8	97^8			

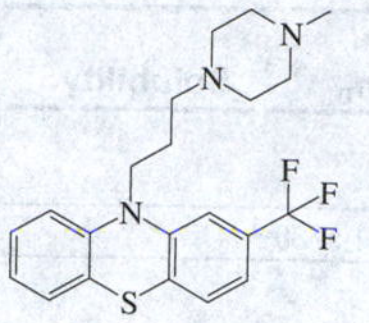
10376
Trifluoperazine

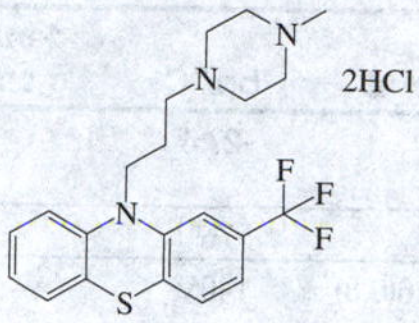
10377
Trifluoperazine dihydrochloride

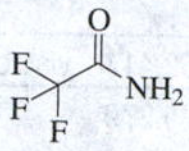
10378
2,2,2-Trifluoroacetamide

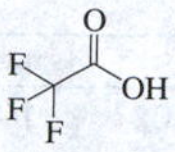
10379
Trifluoroacetic acid

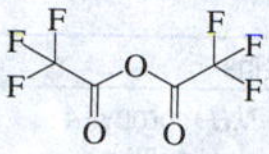
10380
Trifluoroacetic acid anhydride

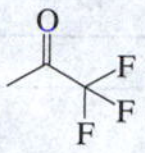
10381
1,1,1-Trifluoroacetone

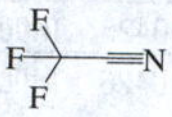
10382
Trifluoroacetonitrile

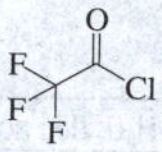
10383
Trifluoroacetyl chloride

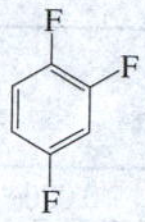
10384
1,2,4-Trifluorobenzene

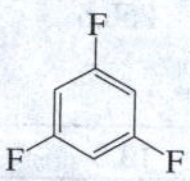
10385
1,3,5-Trifluorobenzene

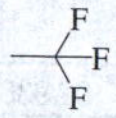
10386
1,1,1-Trifluoroethane

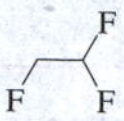
10387
1,1,2-Trifluoroethane

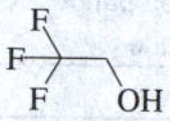
10388
2,2,2-Trifluoroethanol

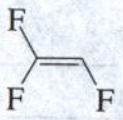
10389
Trifluoroethene

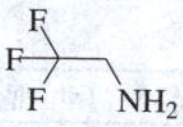
10390
2,2,2-Trifluoroethylamine

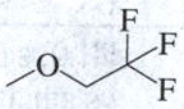
10391
2,2,2-Trifluoroethyl methyl ether

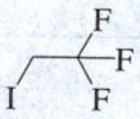
10392
1,1,1-Trifluoro-2-iodoethane

10393
Trifluoroiodomethane

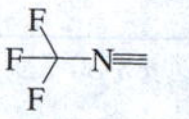
10394
Trifluoroisocyanomethane

10395
Trifluoromethane

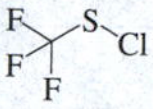
10396
Trifluoromethanesulfenyl chloride

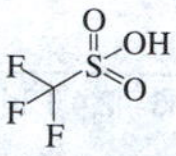
10397
Trifluoromethanesulfonic acid

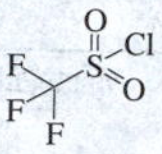
10398
Trifluoromethanesulfonyl chloride

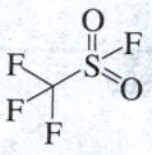
10399
Trifluoromethanesulfonyl fluoride

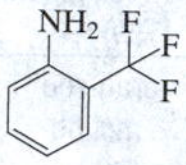
10400
2-(Trifluoromethyl)aniline

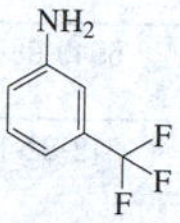
10401
3-(Trifluoromethyl)aniline

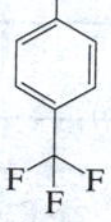
10402
4-(Trifluoromethyl)aniline

10403
4-(Trifluoromethyl)benzaldehyde

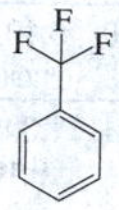
10404
(Trifluoromethyl)benzene

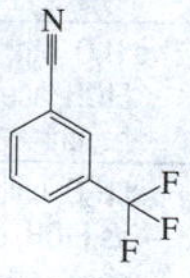
10405
3-(Trifluoromethyl)benzonitrile

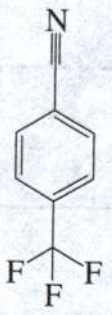
10406
4-(Trifluoromethyl)benzonitrile

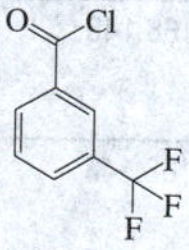
10407
3-(Trifluoromethyl)benzoyl chloride

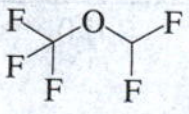
10408
Trifluoromethyl difluoromethyl ether

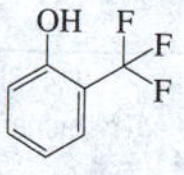
10409
2-(Trifluoromethyl)phenol

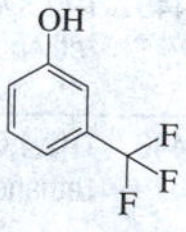
10410
3-(Trifluoromethyl)phenol

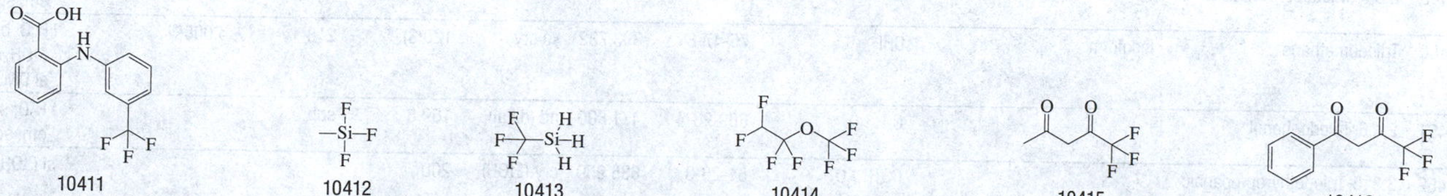
10411
2-[[3-(Trifluoromethyl)phenyl]amino]benzoic acid

10412
Trifluoromethylsilane

10413
(Trifluoromethyl)silane

10414
Trifluoromethyl 1,1,2,2-tetrafluoroethyl ether

10415
1,1,1-Trifluoro-2,4-pentanedione

10416
4,4,4-Trifluoro-1-phenyl-1,3-butanedione

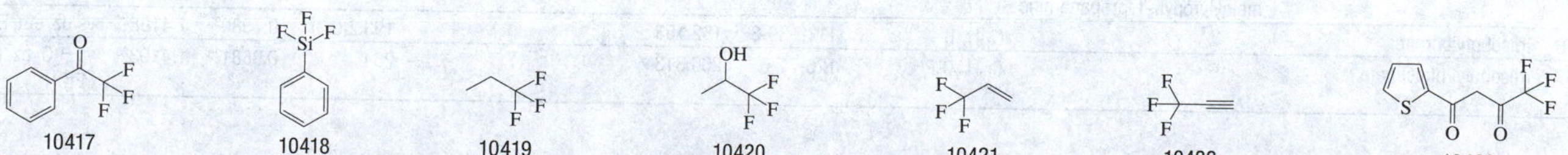
10417
2,2,2-Trifluoro-1-phenylethanone

10418
Trifluorophenylsilane

10419
1,1,1-Trifluoropropane

10420
1,1,1-Trifluoro-2-propanol, (±)-

10421
3,3,3-Trifluoropropene

10422
3,3,3-Trifluoro-1-propyne

10423
4,4,4-Trifluoro-1-(2-thienyl)-1,3-butanedione

No.	Name	Synonym	Mol. form.	CAS Reg. No.	Mol. wt.	Physical form	mp/°C	bp/°C	den/ g cm⁻³	n_D	Solubility
10424	Trifluoro(trifluoromethyl)-oxirane	Perfluoropropylene oxide	C_3F_6O	428-59-1	166.021	gas		-27.4			
10425	Triflupromazine	Fluopromazine	$C_{18}H_{19}F_3N_2S$	146-54-3	352.417	visc oil		$176^{0.7}$		1.5780^{23}	
10426	Trifluralin	2,6-Dinitro-*N,N*-dipropyl-4-(trifluoromethyl)aniline	$C_{13}H_{16}F_3N_3O_4$	1582-09-8	335.279		49.6(0.8)	$140^{4.2}$			
10427	Triforine		$C_{10}H_{14}Cl_6N_4O_2$	26644-46-2	434.962		155 dec				
10428	Trigonelline		$C_7H_7NO_2$	535-83-1	137.137	pr (aq, al, +1w)					vs H_2O
10429	Trihexylamine	*N,N*-Dihexyl-1-hexanamine	$C_{18}H_{39}N$	102-86-3	269.510			261.7	0.7976^{21}		i H_2O; vs EtOH, eth; s acid
10430	Trihexyl borate		$C_{18}H_{39}BO_3$	5337-36-0	314.312			143^{2}			sl ctc
10431	Trihexyphenidyl hydrochloride	α-Cyclohexyl-α-phenyl-1-piperidinepropanol hydrochloride	$C_{20}H_{32}ClNO$	52-49-3	337.927		258.5				
10432	Trihydro(pyridine)boron	Borane pyridine	C_5H_8BN	110-51-0	92.936		10.5		0.920^{20}	1.5280^{25}	i H_2O; dec acid
10433	1,2,3-Trihydroxy-9,10-anthracenedione	Anthragallol	$C_{14}H_8O_5$	602-64-2	256.211	ye nd (dil al)	313	290 sub			sl H_2O; s EtOH, eth, HOAc, CS_2
10434	1,2,4-Trihydroxy-9,10-anthracenedione	Purpurin	$C_{14}H_8O_5$	81-54-9	256.211	oran red or oran-ye nd (al)	259	sub			sl H_2O; vs EtOH, bz, HOAc; s eth
10435	2,3,4-Trihydroxybenzoic acid		$C_7H_6O_5$	610-02-6	170.120	nd (+w)	221	sub			sl H_2O; s EtOH, eth, ace; i bz, CS_2
10436	2,4,6-Trihydroxybenzoic acid		$C_7H_6O_5$	83-30-7	170.120	cry (w+1)	100 dec				sl H_2O; s EtOH; vs eth; i bz
10437	3,4,5-Trihydroxybenzoic acid	Gallic acid	$C_7H_6O_5$	149-91-7	170.120	pr (w+1)	262(1)		1.694^{6}		sl H_2O, eth; vs EtOH; s ace; i bz, chl
10438	2,3,4-Trihydroxybenzophenone	Alizarin Yellow A	$C_{13}H_{10}O_4$	1143-72-2	230.216	ye nd (dil al)	140.5				sl H_2O, bz; s EtOH, eth, ace, HOAc
10439	2',4,4'-Trihydroxychalcone	Isoliquiritigenin	$C_{15}H_{12}O_4$	961-29-5	256.254	ye nd (EtOH-w)	200				
10440	9,10,16-Trihydroxyhexadecanoic acid	Aleuritic acid	$C_{16}H_{32}O_5$	6949-98-0	304.422	lf (dil al), nd (w)	102				sl H_2O
10441	1,3,8-Trihydroxy-6-methyl-9,10-anthracenedione	Emodin	$C_{15}H_{10}O_5$	518-82-1	270.237	oran-red mcl nd (HOAc)	257	sub			vs eth, EtOH
10442	9,10,18-Trihydroxyoctadecanoic acid, (R^*,R^*)	Phloionolic acid	$C_{18}H_{36}O_5$	583-86-8	332.476	cry (dil al)	101.5				
10443	5,6,7-Trihydroxy-2-phenyl-4*H*-1-benzopyran-4-one	Baicalein	$C_{15}H_{10}O_5$	491-67-8	270.237	ye pr (al)	264 dec				sl H_2O, bz; s EtOH, eth, ace, HOAc
10444	1-(2,4,5-Trihydroxyphenyl)-1-butanone		$C_{10}H_{12}O_4$	1421-63-2	196.200		153.8				
10445	1-(2,3,4-Trihydroxyphenyl)-ethanone	Gallacetophenone	$C_8H_8O_4$	528-21-2	168.148		173				s H_2O, eth; vs EtOH, ace; sl bz, chl
10446	1-(2,4,6-Trihydroxyphenyl)-ethanone	2',4',6'-Trihydroxyacetophenone	$C_8H_8O_4$	480-66-0	168.148		221.0				sl H_2O, chl, bz; vs EtOH, eth, ace
10447	1-(2,4,6-Trihydroxyphenyl)-1-propanone	Flopropione	$C_9H_{10}O_4$	2295-58-1	182.173	nd (w, +1w)	175.5				vs eth, EtOH
10448	2,6,7-Trihydroxy-9-phenyl-3*H*-xanthen-3-one	Phenylfluorone	$C_{19}H_{12}O_5$	975-17-7	320.295	oran red (al-HCl)	>300				
10449	2,3,5-Triiodobenzoic acid		$C_7H_3I_3O_2$	88-82-4	499.811	pr (al)	230.6(0.3)				i H_2O; vs EtOH, eth; sl bz
10450	Triiodomethane	Iodoform	CHI_3	75-47-8	393.732	ye cry	120(3)	218	4.008^{25}		i H_2O, bz; s EtOH, eth, ace; sl DMSO
10451	2,4,6-Triiodophenol		$C_6H_3I_3O$	609-23-4	471.800	nd (dil al)	159.8	sub			i H_2O; sl EtOH; s eth, ace
10452	3,3',5-Triiodothyropropanoic acid		$C_{15}H_{11}I_3O_4$	51-26-3	635.959	cry (EtOH)	200				sl EtOH
10453	Triisobutyl aluminate	2-Methyl-1-propanol, aluminum salt	$C_{12}H_{27}AlO_3$	3453-79-0	246.322			275^{50}			
10454	Triisobutylaluminum		$C_{12}H_{27}Al$	100-99-2	198.324	liq	6	86^{10}			
10455	Triisobutylamine	2-Methyl-*N,N*-bis(2-methylpropyl)-1-propanamine	$C_{12}H_{27}N$	1116-40-1	185.349	liq	-21.8(0.5)	194(8)	0.7684^{20}	1.4252^{17}	vs eth, EtOH
10456	Triisobutylborane		$C_{12}H_{27}B$	1116-39-8	182.153			193.2(0.7)	0.7380^{25}	1.4188^{23}	vs bz, eth, EtOH
10457	Triisobutyl phosphate		$C_{12}H_{27}O_4P$	126-71-6	266.313			264	0.9681^{20}	1.4193^{20}	vs H_2O, bz, eth, EtOH

Organic

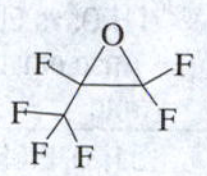
10424
Trifluoro(trifluoromethyl)oxirane

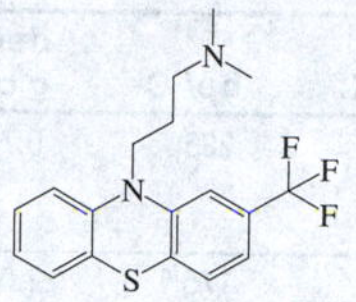
10425
Triflupromazine

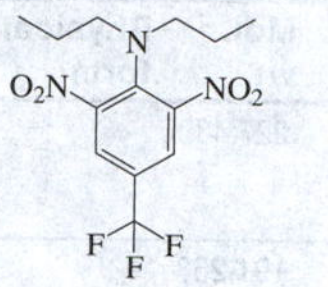
10426
Trifluralin

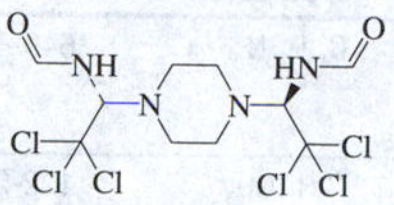
10427
Triforine

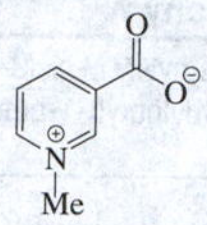
10428
Trigonelline

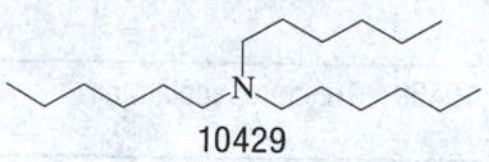
10429
Trihexylamine

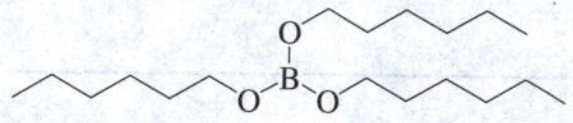
10430
Trihexyl borate

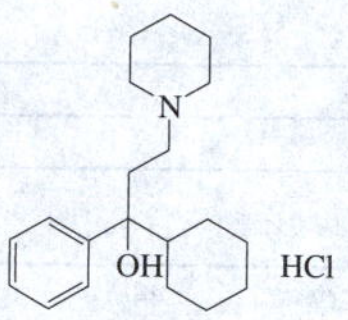
10431
Trihexyphenidyl hydrochloride

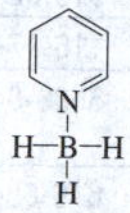
10432
Trihydro(pyridine)boron

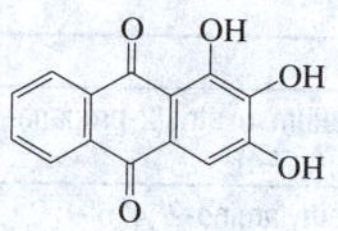
10433
1,2,3-Trihydroxy-9,10-anthracenedione

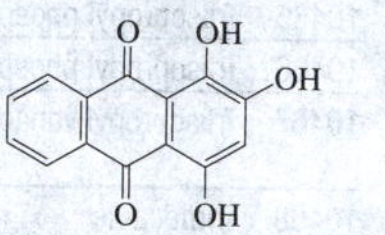
10434
1,2,4-Trihydroxy-9,10-anthracenedione

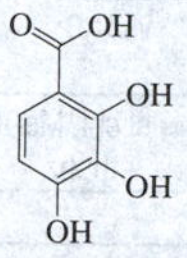
10435
2,3,4-Trihydroxybenzoic acid

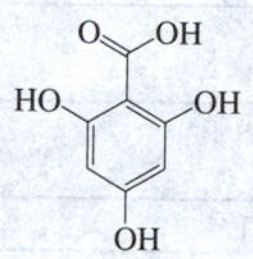
10436
2,4,6-Trihydroxybenzoic acid

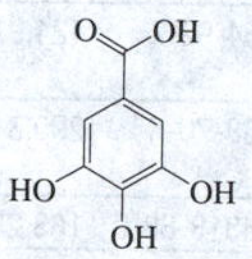
10437
3,4,5-Trihydroxybenzoic acid

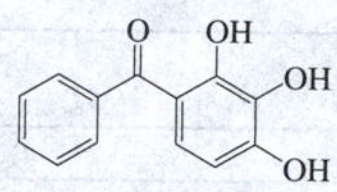
10438
2,3,4-Trihydroxybenzophenone

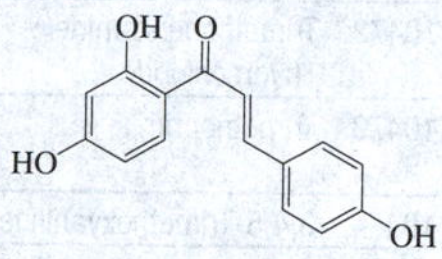
10439
2',4,4'-Trihydroxychalcone

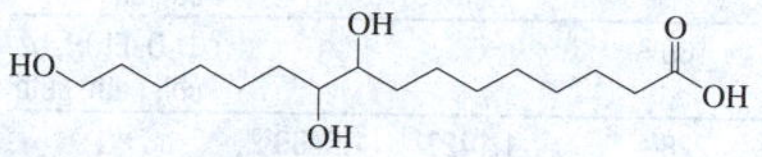
10440
9,10,16 Trihydroxyhexadecanoic acid

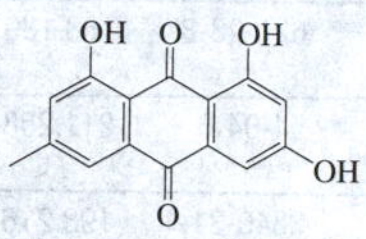
10441
1,3,8-Trihydroxy-6-methyl-9,10-anthracenedione

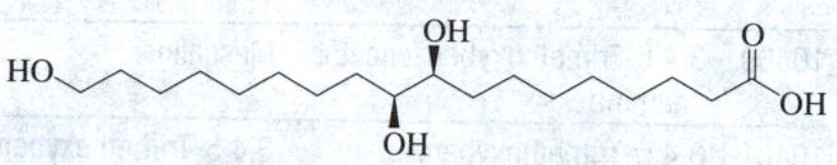
10442
9,10,18-Trihydroxyoctadecanoic acid, (*R**,*R**)

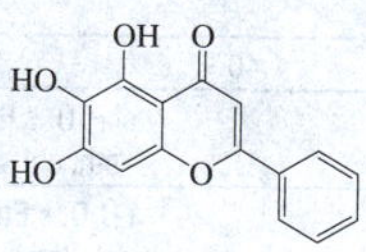
10443
5,6,7-Trihydroxy-2-phenyl-4*H*-1-benzopyran-4-one

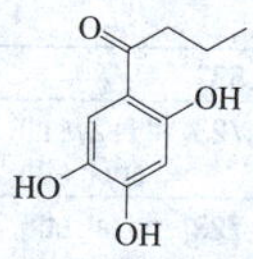
10444
1-(2,4,5-Trihydroxyphenyl)-1-butanone

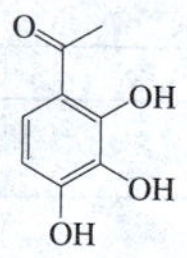
10445
1-(2,3,4-Trihydroxyphenyl)ethanone

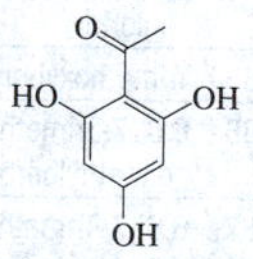
10446
1-(2,4,6-Trihydroxyphenyl)ethanone

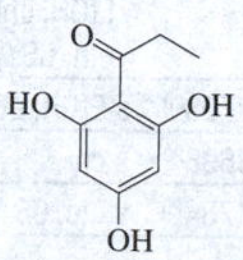
10447
1-(2,4,6-Trihydroxyphenyl)-1-propanone

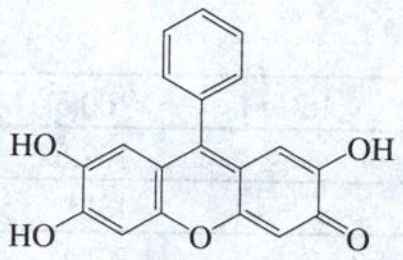
10448
2,6,7-Trihydroxy-9-phenyl-3*H*-xanthen-3-one

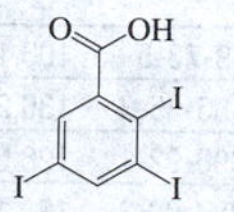
10449
2,3,5-Triiodobenzoic acid

10450
Triiodomethane

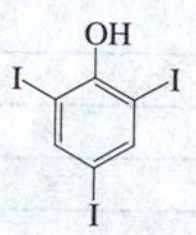
10451
2,4,6-Triiodophenol

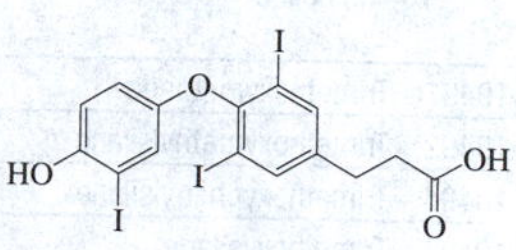
10452
3,3',5-Triiodothyropropanoic acid

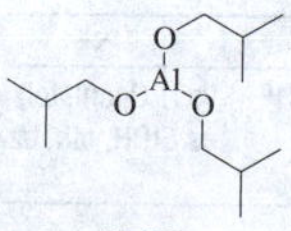
10453
Triisobutyl aluminate

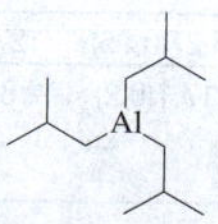
10454
Triisobutylaluminum

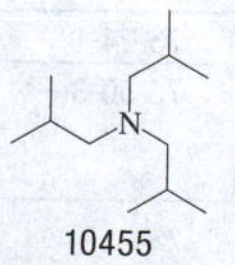
10455
Triisobutylamine

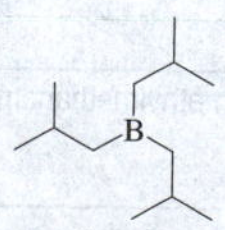
10456
Triisobutylborane

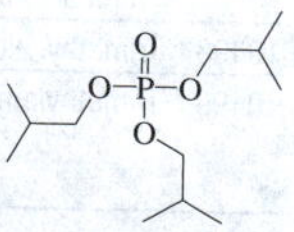
10457
Triisobutyl phosphate

No.	Name	Synonym	Mol. form.	CAS Reg. No.	Mol. wt.	Physical form	mp/°C	bp/°C	den/ g cm^{-3}	n_D	Solubility
10458	Triisopentylamine	3-Methyl-*N,N*-bis(3-methylbutyl)-1-butanamine	$C_{15}H_{33}N$	645-41-0	227.430			235	0.7848^{20}	1.4331^{20}	i H_2O; vs EtOH; msc eth, bz, ctc
10459	Triisopropanolamine		$C_9H_{21}NO_3$	122-20-3	191.268		45	175^{10}	1.0^{20}		s H_2O, EtOH; sl chl
10460	Triisopropoxymethane	Isopropyl orthoformate	$C_{10}H_{22}O_3$	4447-60-3	190.280			167	0.8621^{20}	1.4000^{20}	vs eth, EtOH
10461	Triisopropoxyvinylsilane		$C_{11}H_{24}O_3Si$	18023-33-1	232.393			179.5	0.8627^{25}	1.3981^{20}	s ctc
10462	1,2,4-Triisopropylbenzene		$C_{15}H_{24}$	948-32-3	204.352			244	0.8574^{25}	1.4896^{25}	
10463	1,3,5-Triisopropylbenzene		$C_{15}H_{24}$	717-74-8	204.352	liq	-9(2)	249(7)	0.8545^{20}	1.4882^{20}	s ace, bz, chl
10464	Triisopropyl borate		$C_9H_{21}BO_3$	5419-55-6	188.072			139.6(0.3)	0.8251^{20}	1.3777^{20}	vs EtOH, eth, bz, PrOH
10465	Triisopropyl phosphate		$C_9H_{21}O_4P$	513-02-0	224.234			219	0.9867^{20}	1.4057^{20}	vs EtOH
10466	Triisopropyl phosphite		$C_9H_{21}O_3P$	116-17-6	208.235			74^{20}	0.9063^{20}	1.4085^{25}	s EtOH, eth, chl
10467	Triisopropyl vanadate	Vanadium, oxotris(2-propanolato)-, (T-4)-	$C_9H_{21}O_4V$	5588-84-1	244.203			104^{10}			
10468	Trimecaine	2-Diethylamino-2',4',6'-trimethylacetanilide	$C_{15}H_{24}N_2O$	616-68-2	248.364	cry	44	187^{6}			
10469	Trimellitic anhydride		$C_9H_4O_5$	552-30-7	192.125		162	241^{14}			
10470	Trimeprazine	*N,N*,β-Trimethyl-10*H*-phenothiazine-10-propanamine	$C_{18}H_{22}N_2S$	84-96-8	298.446	cry	68	162$^{0.3}$			
10471	Trimethoate		$C_9H_{20}NO_3PS_2$	2275-18-5	285.364	solid	28.5	135$^{0.1}$			sl H_2O
10472	Trimethobenzamide hydrochloride		$C_{21}H_{29}ClN_2O_5$	554-92-7	424.918	cry	188				vs H_2O
10473	Trimethoprim		$C_{14}H_{18}N_4O_3$	738-70-5	290.318	ye cry	199				sl chl, MeOH; i eth, bz
10474	3,4,5-Trimethoxyaniline		$C_9H_{13}NO_3$	24313-88-0	183.204		112.8				
10475	2,3,4-Trimethoxybenzaldehyde		$C_{10}H_{12}O_4$	2103-57-3	196.200			122$^{0.5}$		1.5547^{20}	
10476	2,4,5-Trimethoxybenzaldehyde		$C_{10}H_{12}O_4$	4460-86-0	196.200		114				s H_2O, eth, chl, lig
10477	3,4,5-Trimethoxybenzaldehyde		$C_{10}H_{12}O_4$	86-81-7	196.200		72.5	148^{5}			s chl
10478	1,2,3-Trimethoxybenzene		$C_9H_{12}O_3$	634-36-6	168.189	orth nd (al)	48.5	235	1.1009^{45}		i H_2O; s EtOH, eth, bz
10479	1,3,5-Trimethoxybenzene		$C_9H_{12}O_3$	621-23-8	168.189	pr (al), lf (peth)	54.5	255.5			i H_2O; s EtOH, eth, bz
10480	3,4,5-Trimethoxybenzeneethanamine	Mescaline	$C_{11}H_{17}NO_3$	54-04-6	211.258	cry	35.5	180^{12}			s H_2O, EtOH, bz, chl; i eth, peth
10481	3,4,5-Trimethoxybenzenemethanol	3,4,5-Trimethoxybenzyl alcohol	$C_{10}H_{14}O_4$	3840-31-1	198.216		3	228^{25}	1.1427^{20}	1.5439^{20}	
10482	2,4,5-Trimethoxybenzoic acid		$C_{10}H_{12}O_5$	490-64-2	212.199	nd (al or bz-peth)	145(1)	300			vs H_2O, bz, EtOH, peth
10483	3,4,5-Trimethoxybenzoic acid		$C_{10}H_{12}O_5$	118-41-2	212.199	mcl nd (w)	171(1)	226^{10}			sl H_2O; vs EtOH, eth, chl
10484	3,4,5-Trimethoxybenzoyl chloride		$C_{10}H_{11}ClO_4$	4521-61-3	230.645		82	185^{18}			
10485	Trimethoxyboroxin		$C_3H_9B_3O_6$	102-24-9	173.532					1.40^{25}	
10486	6,6',7-Trimethoxy-2,2'-dimethylberbaman-12-ol	Berbamine	$C_{37}H_{40}N_2O_6$	478-61-5	608.723	lf (+2w, al) cry (peth)	198.5				sl H_2O; s EtOH, eth, chl, peth
10487	6,6',7-Trimethoxy-2,2'-dimethyloxyacanthan-12'-ol	Oxyacanthine	$C_{37}H_{40}N_2O_6$	548-40-3	608.723	nd (al, eth)	216.5				i H_2O; s EtOH, eth, bz, chl; i lig
10488	7',10,11-Trimethoxyemetan-6'-ol	Cephaeline	$C_{28}H_{38}N_2O_4$	483-17-0	466.613	nd (eth)	115.5				vs ace, EtOH, MeOH, chl
10489	1,1,1-Trimethoxyethane		$C_5H_{12}O_3$	1445-45-0	120.147			109(13)	0.9438^{25}	1.3859^{25}	vs eth, EtOH
10490	4,7,8-Trimethoxyfuro[2,3-*b*]-quinoline	Skimmianine	$C_{14}H_{13}NO_4$	83-95-4	259.258	pym (al)	177				i H_2O, peth; s EtOH, chl; sl eth, CS_2
10491	Trimethoxymethane		$C_4H_{10}O_3$	149-73-5	106.120		15	100(5)	0.9676^{20}	1.3793^{20}	s EtOH, eth
10492	Trimethoxymethylsilane		$C_4H_{12}O_3Si$	1185-55-3	136.222			102.5	0.9548^{20}	1.3696^{20}	s chl
10493	Trimethoxyphenylsilane		$C_9H_{14}O_3Si$	2996-92-1	198.291			130^{45}	1.064^{20}	1.4734^{20}	s ctc, CS_2
10494	Trimethoxysilane		$C_3H_{10}O_3Si$	2487-90-3	122.195			32^{100}			
10495	3-(Trimethoxysilyl)-1-propanethiol	(3-Mercaptopropyl)-trimethoxysilane	$C_6H_{16}O_3SSi$	4420-74-0	196.340			128^{50}	1.015^{25}	1.4420^{25}	
10496	*N*-[3-(Trimethoxysilyl)propyl]-1,2-ethanediamine		$C_8H_{22}N_2O_3Si$	1760-24-3	222.358			140.5^{15}	1.01^{25}	1.4416^{25}	
10497	3-(Trimethoxysilyl)propyl methacrylate		$C_{10}H_{20}O_5Si$	2530-85-0	248.349	liq		107^{5}			
10498	Trimethyl aluminum		C_3H_9Al	75-24-1	72.085		15.27(0.05)	127.11(0.05)	0.743^{20}		
10499	Trimethylamine	*N,N*-Dimethylmethanamine	C_3H_9N	75-50-3	59.110	col gas	-117.1(0.2)	2.8(0.2)	0.627^{25} (p>1 atm)	1.3631^{0}	vs H_2O, chl, tol; s EtOH, eth, bz

Organic

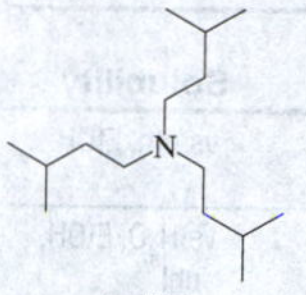
10458
Triisopentylamine

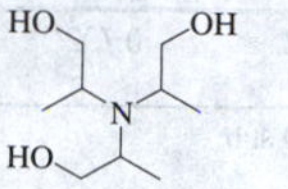
10459
Triisopropanolamine

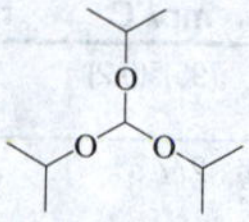
10460
Triisopropoxymethane

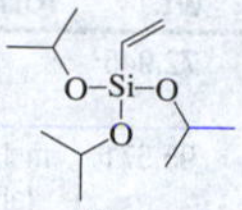
10461
Triisopropoxyvinylsilane

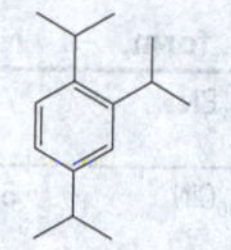
10462
1,2,4-Triisopropylbenzene

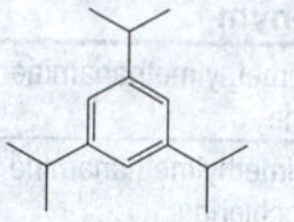
10463
1,3,5-Triisopropylbenzene

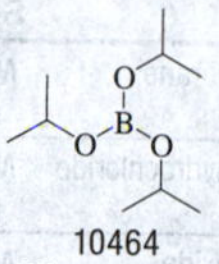
10464
Triisopropyl borate

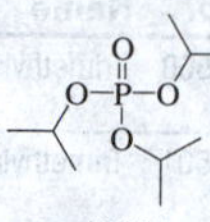
10465
Triisopropyl phosphate

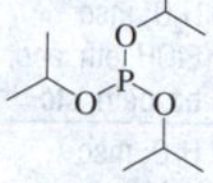
10466
Triisopropyl phosphite

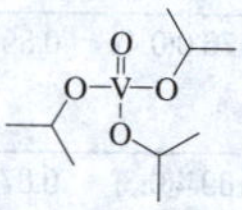
10467
Triisopropyl vanadate

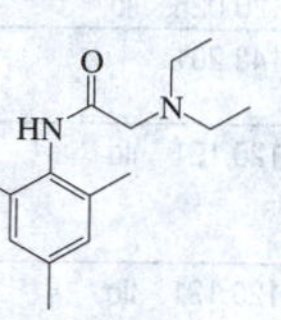
10468
Trimecaine

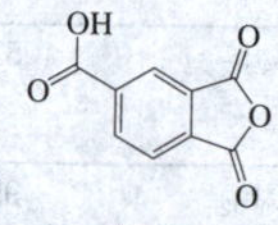
10469
Trimellitic anhydride

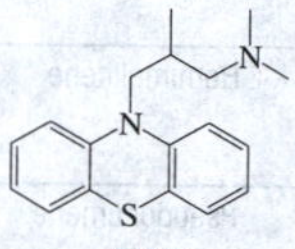
10470
Trimeprazine

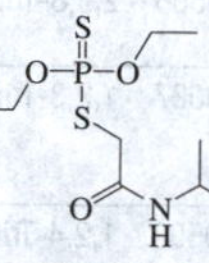
10471
Trimethoate

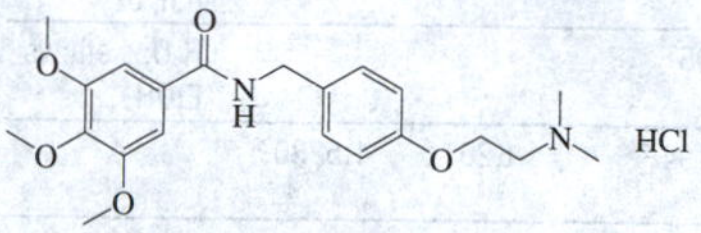
10472
Trimethobenzamide hydrochloride

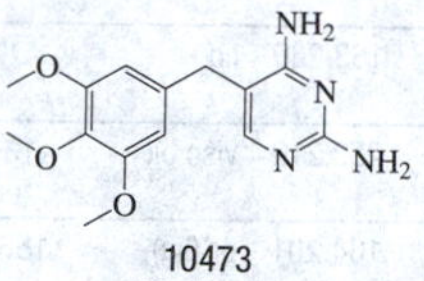
10473
Trimethoprim

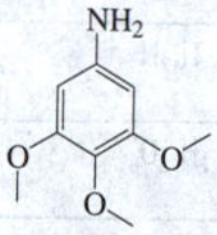
10474
3,4,5-Trimethoxyaniline

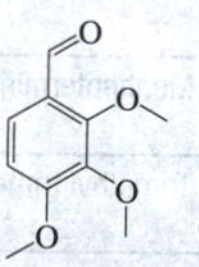
10475
2,3,4-Trimethoxybenzaldehyde

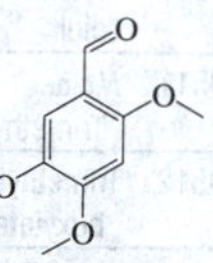
10476
2,4,5-Trimethoxybenzaldehyde

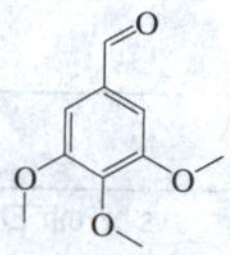
10477
3,4,5-Trimethoxybenzaldehyde

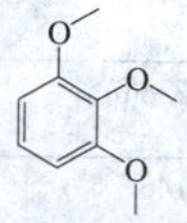
10478
1,2,3-Trimethoxybenzene

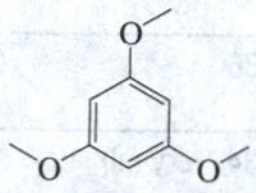
10479
1,3,5-Trimethoxybenzene

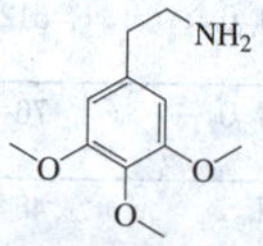
10480
3,4,5-Trimethoxybenzeneethanamine

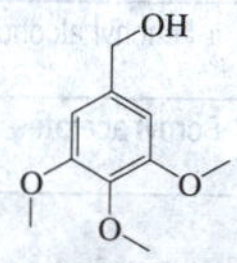
10481
3,4,5-Trimethoxybenzenemethanol

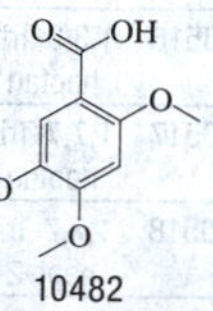
10482
2,4,5-Trimethoxybenzoic acid

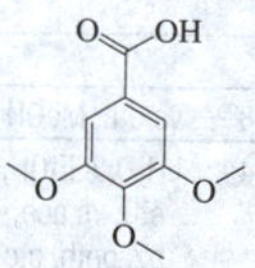
10483
3,4,5-Trimethoxybenzoic acid

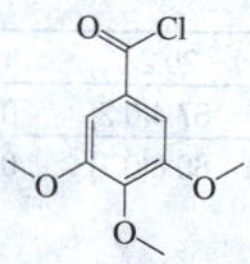
10484
3,4,5-Trimethoxybenzoyl chloride

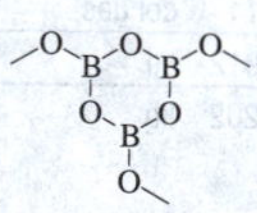
10485
Trimethoxyboroxin

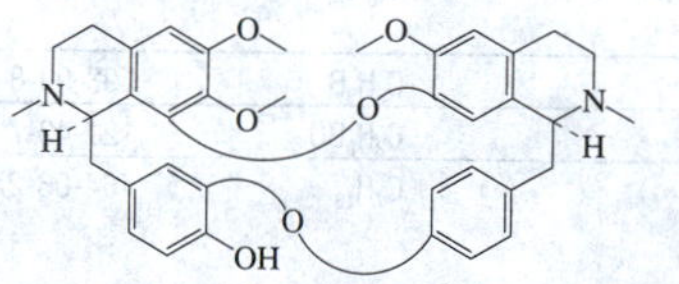
10486
6,6',7-Trimethoxy-2,2'-dimethylberbaman-12-ol

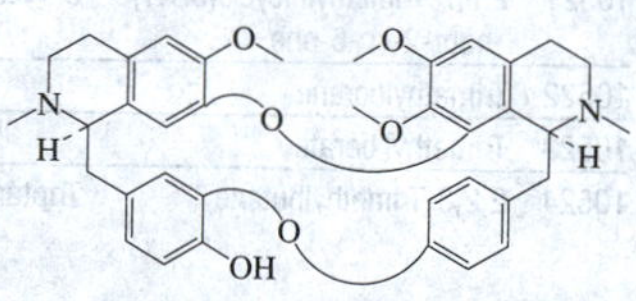
10487
6,6',7-Trimethoxy-2,2'-dimethyloxyacanthan-12'-ol

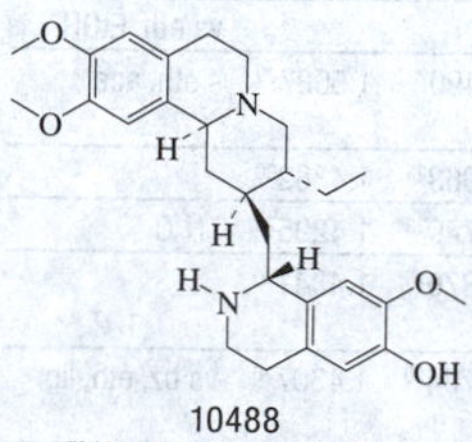
10488
7',10,11-Trimethoxyemetan-6'-ol

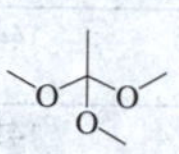
10489
1,1,1-Trimethoxyethane

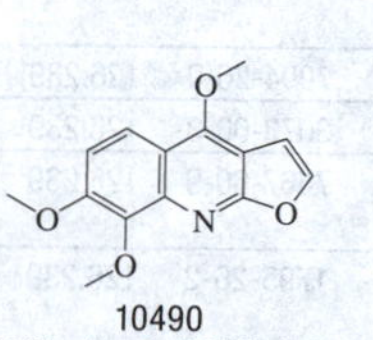
10490
4,7,8-Trimethoxyfuro[2,3-b]quinoline

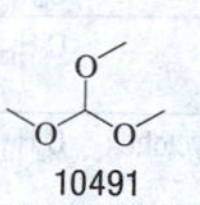
10491
Trimethoxymethane

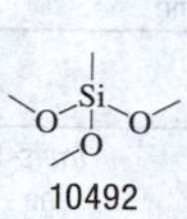
10492
Trimethoxymethylsilane

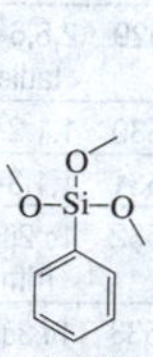
10493
Trimethoxyphenylsilane

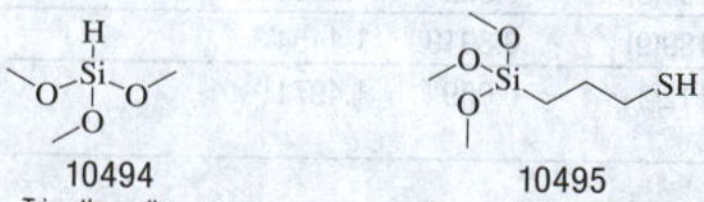
10494
Trimethoxysilane

10495
3-(Trimethoxysilyl)-1-propanethiol

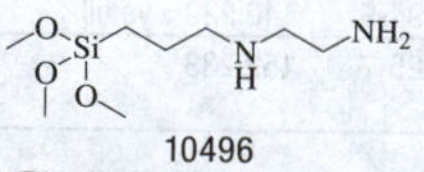
10496
N-[3-(Trimethoxysilyl)propyl]-1,2-ethanediamine

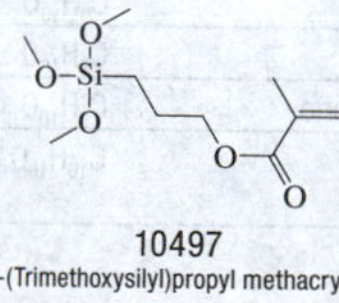
10497
3-(Trimethoxysilyl)propyl methacrylate

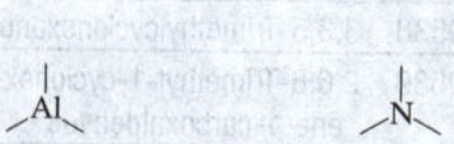
10498
Trimethyl aluminum

10499
Trimethylamine

Organic

No.	Name	Synonym	Mol. form.	CAS Reg. No.	Mol. wt.	Physical form	mp/°C	bp/°C	den/ g cm^{-3}	n_D	Solubility
10500	Trimethylamine borane	*N,N*-Dimethylmethanamine borane	$C_3H_{12}BN$	75-22-9	72.945		95.5(0.2)	172	0.792^{25}		vs eth, EtOH
10501	Trimethylamine hydrochloride	*N,N*-Dimethylmethanamine hydrochloride	$C_3H_{10}ClN$	593-81-7	95.571	mcl hyg nd (al)	277.5	200 sub			vs H_2O, EtOH, chl
10502	Trimethylamine oxide	*N,N*-Dimethylmethanamine oxide	C_3H_9NO	1184-78-7	75.109	hyg nd (w+2)	256				vs H_2O, EtOH
10503	2,4,5-Trimethylaniline		$C_9H_{13}N$	137-17-7	135.206	nd (w)	68	234.5	0.957^{25}		vs EtOH
10504	2,4,6-Trimethylaniline	Mesitylamine	$C_9H_{13}N$	88-05-1	135.206	liq	-4.9(0.4)	232.5	0.9633^{25}	1.5495^{20}	sl ctc
10505	Trimethylarsine		C_3H_9As	593-88-4	120.025	liq	-86.54(0.02)	52	1.144^{15}		vs bz, eth, EtOH
10506	2,4,6-Trimethylbenzaldehyde		$C_{10}H_{12}O$	487-68-3	148.201		14	238.5	1.0154^{25}		i H_2O; s EtOH, eth, ace, bz
10507	1,2,3-Trimethylbenzene	Hemimellitene	C_9H_{12}	526-73-8	120.191	liq	-25.32(0.04)	176.0(0.4)	0.8944^{20}	1.5139^{20}	i H_2O; msc EtOH, eth, ace, bz, peth, ctc
10508	1,2,4-Trimethylbenzene	Pseudocumene	C_9H_{12}	95-63-6	120.191	liq	-43.8(0.1)	169.4(0.3)	0.8758^{20}	1.5048^{20}	i H_2O; msc EtOH, eth, ace, bz, peth, ctc
10509	1,3,5-Trimethylbenzene	Mesitylene	C_9H_{12}	108-67-8	120.191	liq	-44.69(0.05)	164.7(0.3)	0.8615^{25}	1.4994^{20}	i H_2O; msc EtOH, eth, ace, bz, peth, ctc
10510	2,3,5-Trimethyl-1,4-benzenediol		$C_9H_{12}O_2$	700-13-0	152.190	nd (w)	170.0(0.5)				sl H_2O; vs EtOH, eth, bz
10511	*N*,α,α-Trimethylbenzeneethanamine	Mephentermine	$C_{11}H_{17}N$	100-92-5	163.260	liq		95^{9}			i H_2O; s eth; vs EtOH
10512	Trimethyl 1,2,4-benzenetricarboxylate	Trimethyl trimellitate	$C_{12}H_{12}O_6$	2459-10-1	252.219	visc oil	-13	194^{12}	1.261	1.5230^{20}	
10513	2,4,6-Trimethylbenzoic acid		$C_{10}H_{12}O_2$	480-63-7	164.201	pr (lig)	156.5				sl H_2O; s EtOH, eth, ace, chl
10514	Trimethylbenzylsilane		$C_{10}H_{16}Si$	770-09-2	164.320			190.5	0.8933^{20}	1.4941^{20}	
10515	1,7,7-Trimethylbicyclo[2.2.1]-heptane		$C_{10}H_{18}$	464-15-3	138.250	hex pl(al), pr(MeOH)		161			i H_2O; s EtOH, eth, AcOEt, MeOH
10516	1,3,3-Trimethylbicyclo[2.2.1]-heptan-2-ol, (1*S*-*endo*)	α-Fenchyl alcohol, (*l*)	$C_{10}H_{18}O$	512-13-0	154.249	pr	41.5(0.2)	94^{20}	0.9034^{84}		
10517	1,7,7-Trimethylbicyclo[2.2.1]-heptan-2-ol acetate, *endo*	Bornyl acetate	$C_{12}H_{20}O_2$	76-49-3	196.286		29	228(6)			
10518	1,7,7-Trimethylbicyclo[2.2.1]-hept-2-ene		$C_{10}H_{16}$	464-17-5	136.234	cry (al)	113	146			vs bz, eth, EtOH
10519	4,6,6-Trimethylbicyclo[3.1.1]-hept-3-en-2-ol, (1α,2α,5α)		$C_{10}H_{16}O$	1820-09-3	152.233		24	92^{10}	0.9657^{25}	1.4908^{25}	
10520	4,6,6-Trimethylbicyclo[3.1.1]-hept-3-en-2-ol, (1α,2β,5α)		$C_{10}H_{16}O$	1845-30-3	152.233		15.5	90^{10}	0.9684^{25}	1.4912^{25}	
10521	2,7,7-Trimethylbicyclo[3.1.1]-hept-2-en-6-one	Chrysanthenone	$C_{10}H_{14}O$	473-06-3	150.217			88^{12}		1.4720^{22}	vs EtOH
10522	Trimethylborane		C_3H_9B	593-90-8	55.914	col gas	-159.93(0.02)	-20.2			
10523	Trimethyl borate		$C_3H_9BO_3$	121-43-7	103.912	liq	-29.3(0.5)	67.4(0.2)	0.915^{25}	1.3568^{20}	vs eth, MeOH
10524	2,2,3-Trimethylbutane	Triptane	C_7H_{16}	464-06-2	100.202	liq	-24.56(0.05)	80.8(0.1)	0.6901^{20}	1.3864^{20}	i H_2O; s EtOH, eth; vs ace, bz, peth, ctc
10525	2,3,3-Trimethyl-2-butanol		$C_7H_{16}O$	594-83-2	116.201	cry (dil al +1/2w)	17(2)	128.3(0.8)	0.8380^{25}	1.4233^{22}	sl H_2O; vs ace, eth, EtOH
10526	2,3,3-Trimethyl-1-butene		C_7H_{14}	594-56-9	98.186	liq	-111(1)	77.8(0.6)	0.7050^{20}	1.4025^{20}	i H_2O; s eth, bz, chl, MeOH
10527	Trimethylchlorosilane		C_3H_9ClSi	75-77-4	108.642	liq	-55.17(0.02)	57.6(0.4)	0.856^{25}	1.3870^{20}	
10528	Trimethyl citrate		$C_9H_{14}O_7$	1587-20-8	234.203	tcl	79.3	285			vs eth, EtOH
10529	2,6,6-Trimethyl-2,4-cycloheptadien-1-one	Eucarvone	$C_{10}H_{14}O$	503-93-5	150.217			210	0.9490^{20}	1.5087^{20}	s eth, ace
10530	1,1,2-Trimethylcyclohexane		C_9H_{18}	7094-26-0	126.239	liq	-29.2(0.6)	146(3)	0.7963^{25}	1.4382^{20}	
10531	1,1,3-Trimethylcyclohexane		C_9H_{18}	3073-66-3	126.239	liq	-68.5(0.5)	136.6(0.4)	0.7749^{25}	1.4295^{20}	i H_2O
10532	1α,2β,4β-1,2,4-Trimethylcyclohexane		C_9H_{18}	7667-60-9	126.239	liq	-83.5	144(6)	0.7870^{25}	1.4341^{20}	
10533	1α,3α,5β-1,3,5-Trimethylcyclohexane	*trans*-1,3,5-Trimethylcyclohexane	C_9H_{18}	1795-26-2	126.239	liq	-107.4	141(2)	0.7794^{20}	1.4307^{20}	vs bz, eth, lig
10534	*cis*-3,3,5-Trimethylcyclohexanol		$C_9H_{18}O$	933-48-2	142.238		37.3	202	0.9006^{16}	1.4550^{16}	i H_2O; s EtOH, eth, chl
10535	*trans*-3,3,5-Trimethylcyclohexanol		$C_9H_{18}O$	767-54-4	142.238	cry (eth)	55.8	189.2	0.8631^{60}		i H_2O; s EtOH, eth, chl
10536	2,2,6-Trimethylcyclohexanone		$C_9H_{16}O$	2408-37-9	140.222	liq	-31.8	178.5	0.9043^{18}	1.4470^{20}	
10537	2,4,4-Trimethylcyclohexanone		$C_9H_{16}O$	2230-70-8	140.222			190(9)	0.902^{20}	1.4493^{20}	
10538	3,3,5-Trimethylcyclohexanone	Dihydroisophorone	$C_9H_{16}O$	873-94-9	140.222	ye oil		189(9)	0.8919^{19}	1.4454^{15}	
10539	2,6,6-Trimethyl-1-cyclohexene-1-carboxaldehyde	β-Cyclocitral	$C_{10}H_{16}O$	432-25-7	152.233			112^{29}	0.959^{15}	1.4971^{15}	

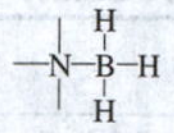
10500
Trimethylamine borane

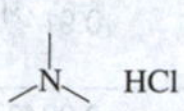
10501
Trimethylamine hydrochloride

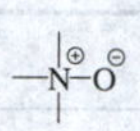
10502
Trimethylamine oxide

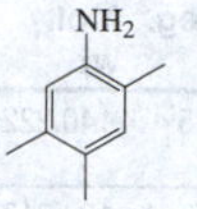
10503
2,4,5-Trimethylaniline

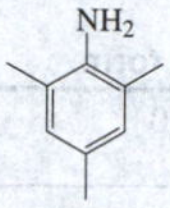
10504
2,4,6-Trimethylaniline

10505
Trimethylarsine

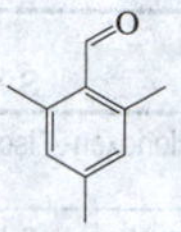
10506
2,4,6-Trimethylbenzaldehyde

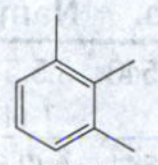
10507
1,2,3-Trimethylbenzene

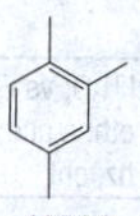
10508
1,2,4-Trimethylbenzene

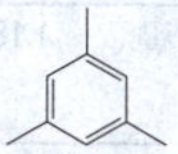
10509
1,3,5-Trimethylbenzene

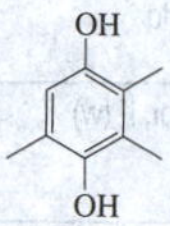
10510
2,3,5-Trimethyl-1,4-benzenediol

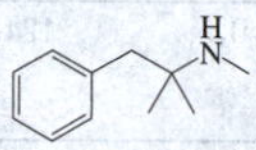
10511
N,α,α-Trimethylbenzeneethanamine

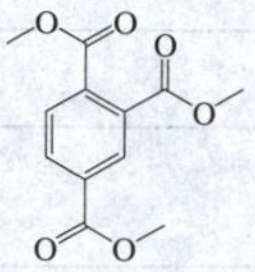
10512
Trimethyl 1,2,4-benzenetricarboxylate

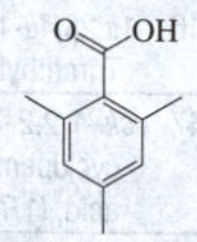
10513
2,4,6-Trimethylbenzoic acid

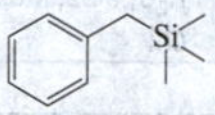
10514
Trimethylbenzylsilane

10515
1,7,7-Trimethylbicyclo[2.2.1]heptane

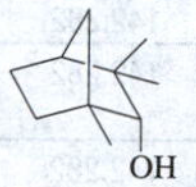
10516
1,3,3-Trimethylbicyclo[2.2.1]heptan-2-ol, (1*S*-*endo*)

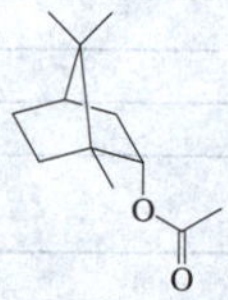
10517
1,7,7-Trimethylbicyclo[2.2.1]heptan-2-ol acetate, *endo*

10518
1,7,7-Trimethylbicyclo[2.2.1]hept-2-ene

10519
4,6,6-Trimethylbicyclo[3.1.1]hept-3-en-2-ol, (1α,2α,5α)

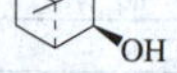
10520
4,6,6-Trimethylbicyclo[3.1.1]hept-3-en-2-ol, (1α,2β,5α)

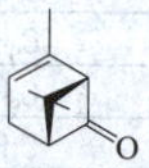
10521
2,7,7-Trimethylbicyclo[3.1.1]hept-2-en-6-one

10522
Trimethylborane

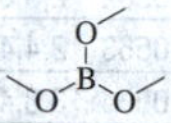
10523
Trimethyl borate

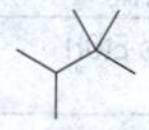
10524
2,2,3-Trimethylbutane

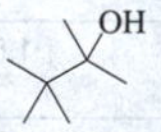
10525
2,3,3-Trimethyl-2-butanol

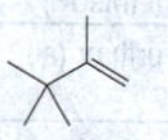
10526
2,3,3-Trimethyl-1-butene

10527
Trimethylchlorosilane

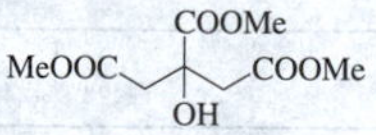
10528
Trimethyl citrate

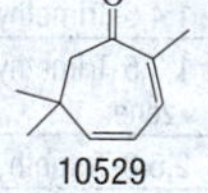
10529
2,6,6-Trimethyl-2,4-cycloheptadien-1-one

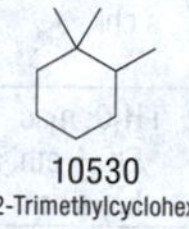
10530
1,1,2-Trimethylcyclohexane

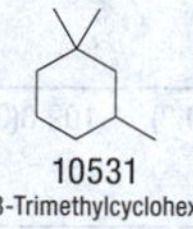
10531
1,1,3-Trimethylcyclohexane

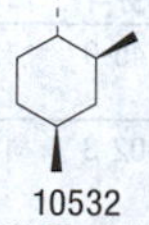
10532
1α,2β,4β-1,2,4-Trimethylcyclohexane

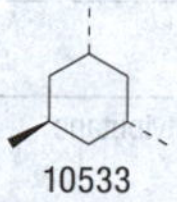
10533
1α,3α,5β-1,3,5-Trimethylcyclohexane

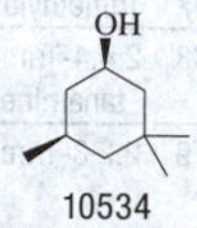
10534
cis-3,3,5-Trimethylcyclohexanol

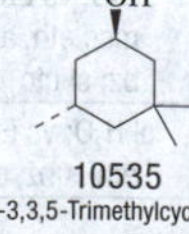
10535
trans-3,3,5-Trimethylcyclohexanol

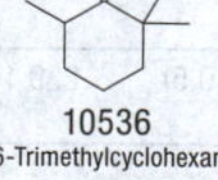
10536
2,2,6-Trimethylcyclohexanone

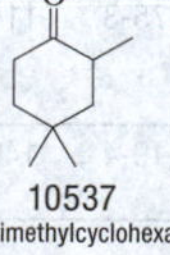
10537
2,4,4-Trimethylcyclohexanone

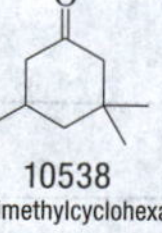
10538
3,3,5-Trimethylcyclohexanone

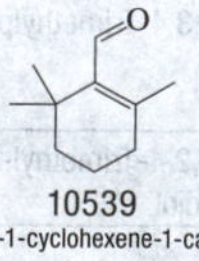
10539
2,6,6-Trimethyl-1-cyclohexene-1-carboxaldehyde

No.	Name	Synonym	Mol. form.	CAS Reg. No.	Mol. wt.	Physical form	mp/°C	bp/°C	den/ g cm^{-3}	n_D	Solubility
10540	3,5,5-Trimethyl-2-cyclohexen-1-ol	Isophorol	$C_9H_{16}O$	470-99-5	140.222			69^{5}	0.914^{20}	1.4717^{20}	
10541	4-(2,6,6-Trimethyl-1-cyclohexen-1-yl)-3-buten-2-ol	β-Ionol	$C_{13}H_{22}O$	22029-76-1	194.313			130^{14}	0.9243^{20}	1.4969^{20}	s EtOH, eth, ace
10542	4-(2,6,6-Trimethyl-2-cyclohexen-1-yl)-3-buten-2-ol	α-Ionol	$C_{13}H_{22}O$	25312-34-9	194.313	oil		127^{14}	0.9189^{20}	1.4735^{20}	
10543	1,1,2-Trimethylcyclopentane		C_8H_{16}	4259-00-1	112.213	liq	-21.7(0.2)	113.7(0.2)	0.7660^{20}	1.4199^{20}	
10544	1,1,3-Trimethylcyclopentane		C_8H_{16}	4516-69-2	112.213	liq	-142.5(0.2)	104.9(0.6)	0.7439^{25}	1.4112^{20}	i H_2O
10545	1α,2α,4β-1,2,4-Trimethylcyclopentane		C_8H_{16}	4850-28-6	112.213	liq	-131(5)	116.7(0.6)	0.7592^{25}	1.4186^{20}	
10546	1α,2β,4α-1,2,4-Trimethylcyclopentane		C_8H_{16}	16883-48-0	112.213	liq	-130.8(0.3)	110(3)	0.7430^{25}	1.4106^{20}	
10547	*cis*-1,2,2-Trimethyl-1,3-cyclopentanedicarboxylic acid, (1*R*)		$C_{10}H_{16}O_4$	124-83-4	200.232	pr, lf (w)	187		1.186^{20}		sl H_2O; vs EtOH, eth; s ace; i bz, chl
10548	2,2,4-Trimethylcyclopentanone		$C_8H_{14}O$	28056-54-4	126.196	liq	-40.6	158	0.877^{25}	1.4300^{20}	
10549	2,4,4-Trimethylcyclopentanone		$C_8H_{14}O$	4694-12-6	126.196	liq	-25.6	160.5	0.8785^{18}	1.433^{18}	
10550	1,1,2-Trimethylcyclopropane		C_6H_{12}	4127-45-1	84.159	liq	-138.2(0.1)	53(1)	0.6897^{25}	1.3864^{20}	
10551	3,7,11-Trimethyl-2,6,10-dodecatrienal		$C_{15}H_{24}O$	19317-11-4	220.351			172^{14}	0.893^{18}	1.4995	
10552	Trimethylgallium		C_3H_9Ga	1445-79-0	114.826			55.7			dec H_2O (exp)
10553	2,2,6-Trimethylheptane		$C_{10}H_{22}$	1190-83-6	142.282	liq	-105(1)	149(4)	0.7200^{25}	1.4078^{20}	
10554	2,5,5-Trimethylheptane		$C_{10}H_{22}$	1189-99-7	142.282			153(3)	0.7362^{25}	1.4149^{20}	
10555	3,3,5-Trimethylheptane		$C_{10}H_{22}$	7154-80-5	142.282			157(3)	0.7248^{20}	1.4170^{20}	i H_2O; s bz, ctc, chl
10556	3,4,5-Trimethylheptane		$C_{10}H_{22}$	20278-89-1	142.282			162.5	0.7519^{25}	1.4229^{20}	
10557	2,2,3-Trimethylhexane		C_9H_{20}	16747-25-4	128.255			134(2)	0.7257^{25}	1.4106^{20}	
10558	2,2,4-Trimethylhexane		C_9H_{20}	16747-26-5	128.255	liq	-122(4)	127(2)	0.711^{20}	1.4033^{20}	
10559	2,2,5-Trimethylhexane		C_9H_{20}	3522-94-9	128.255	liq	-105.9(0.1)	124(2)	0.7072^{20}	1.3997^{20}	i H_2O; vs EtOH, eth, ace, bz; s ctc
10560	2,3,3-Trimethylhexane		C_9H_{20}	16747-28-7	128.255	liq	-116.8(0.2)	137(3)	0.7345^{25}	1.4141^{20}	
10561	2,3,4-Trimethylhexane		C_9H_{20}	921-47-1	128.255			139(3)	0.7354^{25}	1.4144^{20}	
10562	2,3,5-Trimethylhexane		C_9H_{20}	1069-53-0	128.255	liq	-127.9(0.2)	131(2)	0.7218^{20}	1.4051^{20}	
10563	2,4,4-Trimethylhexane		C_9H_{20}	16747-30-1	128.255	liq	-113.4(0.1)	130.6(0.7)	0.7201^{25}	1.4074^{20}	
10564	3,3,4-Trimethylhexane		C_9H_{20}	16747-31-2	128.255	liq	-101.2(0.2)	139(4)	0.7414^{25}	1.4178^{20}	
10565	3,5,5-Trimethylhexanoic acid	Isononanoic acid	$C_9H_{18}O_2$	3302-10-1	158.238	liq		121^{10}			
10566	3,5,5-Trimethyl-1-hexanol		$C_9H_{20}O$	3452-97-9	144.254			193(5)	0.8236^{25}	1.4300^{25}	
10567	1,2,3-Trimethylindene		$C_{12}H_{14}$	4773-83-5	158.239	liq		100.5^{10}	0.9714^{20}	1.5521^{20}	
10568	Trimethylindium	Indium trimethyl	C_3H_9In	3385-78-2	159.921			135.7	1.568^{19}		
10569	2,3,3-Trimethyl-3*H*-indole		$C_{11}H_{13}N$	1640-39-7	159.228			107^{11}			
10570	Trimethyl(4-methylphenyl)-silane		$C_{10}H_{16}Si$	3728-43-6	164.320		38	192	0.8666^{20}	1.4900^{20}	
10571	1,4,5-Trimethylnaphthalene		$C_{13}H_{14}$	2131-41-1	170.250	lf (MeOH)	64.0(0.6)	145^{12}			i H_2O
10572	1,3,5-Trimethyl-2-nitrobenzene		$C_9H_{11}NO_2$	603-71-4	165.189	orth pr (al)	44(2)	255	1.51^{25}		vs EtOH
10573	2,6,8-Trimethyl-4-nonanol		$C_{12}H_{26}O$	123-17-1	186.333			225.4	0.8178^{20}		sl ctc
10574	2,4,7-Trimethyloctane		$C_{11}H_{24}$	62016-38-0	156.309			170(5)			
10575	Trimethylolpropane		$C_6H_{14}O_3$	77-99-6	134.173	wh pow or pl	60.2(0.2)	160^{5}			vs H_2O, EtOH
10576	3,5,5-Trimethyl-2,4-oxazolidinedione	Trimethadione	$C_6H_9NO_3$	127-48-0	143.140		46	79^{5}			s H_2O; vs EtOH, eth, ace, bz; i peth
10577	Trimethyloxonium fluoborate		$C_3H_9BF_4O$	420-37-1	147.907	hyg nd	148 dec				vs ace, chl
10578	2,4,4-Trimethyl-2-pentanamine		$C_8H_{19}N$	107-45-9	129.244						s chl
10579	2,2,3-Trimethylpentane	2-*tert*-Butylbutane	C_8H_{18}	564-02-3	114.229	liq	-112.4(0.3)	109.8(0.4)	0.7161^{20}	1.4030^{20}	i H_2O; msc EtOH, eth, ace, hp; s bz
10580	2,2,4-Trimethylpentane	Isooctane	C_8H_{18}	540-84-1	114.229	liq	-107.36(0.04)	99.2(0.2)	0.6878^{25}	1.3884^{25}	i H_2O; msc EtOH, ace, hp; s eth, ctc
10581	2,3,3-Trimethylpentane		C_8H_{18}	560-21-4	114.229	liq	-101.2(0.3)	114.7(0.3)	0.7262^{20}	1.4075^{20}	i H_2O; vs EtOH; msc eth, ace, bz, hp
10582	2,3,4-Trimethylpentane		C_8H_{18}	565-75-3	114.229	liq	-109.3(0.2)	113.4(0.3)	0.7191^{20}	1.4042^{20}	i H_2O; vs EtOH; msc eth, ace, bz; sl ctc
10583	2,2,4-Trimethyl-1,3-pentanediol		$C_8H_{18}O_2$	144-19-4	146.228	pl (bz)	55.2(0.5)	230.1(0.3)	0.936^{15}	1.4513^{15}	sl H_2O; vs EtOH, eth; s bz, chl

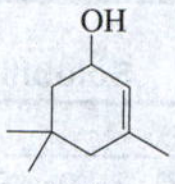
10540
3,5,5-Trimethyl-2-cyclohexen-1-ol

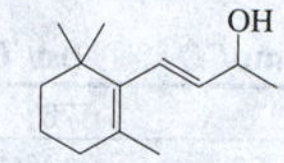
10541
4-(2,6,6-Trimethyl-1-cyclohexen-1-yl)-3-buten-2-ol

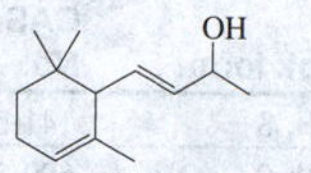
10542
4-(2,6,6-Trimethyl-2-cyclohexen-1-yl)-3-buten-2-ol

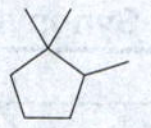
10543
1,1,2-Trimethylcyclopentane

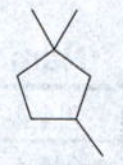
10544
1,1,3-Trimethylcyclopentane

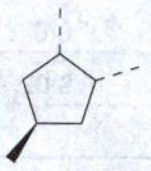
10545
1α,2α,4β-1,2,4-Trimethylcyclopentane

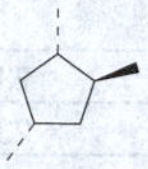
10546
1α,2β,4α-1,2,4-Trimethylcyclopentane

10547
cis-1,2,2-Trimethyl-1,3-cyclopentanedicarboxylic acid, (1*R*)

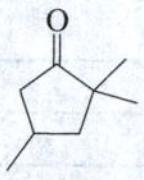
10548
2,2,4-Trimethylcyclopentanone

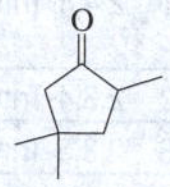
10549
2,4,4-Trimethylcyclopentanone

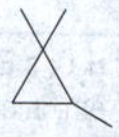
10550
1,1,2-Trimethylcyclopropane

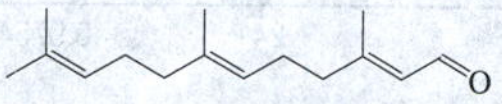
10551
3,7,11-Trimethyl-2,6,10-dodecatrienal

10552
Trimethylgallium

10553
2,2,6-Trimethylheptane

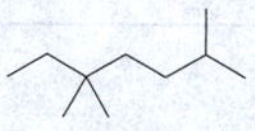
10554
2,5,5-Trimethylheptane

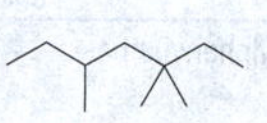
10555
3,3,5-Trimethylheptane

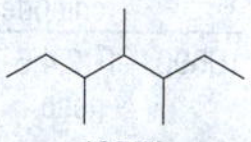
10556
3,4,5-Trimethylheptane

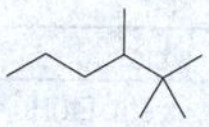
10557
2,2,3-Trimethylhexane

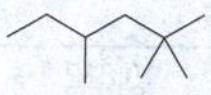
10558
2,2,4-Trimethylhexane

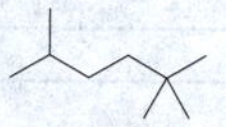
10559
2,2,5-Trimethylhexane

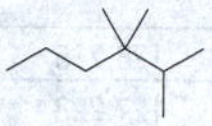
10560
2,3,3-Trimethylhexane

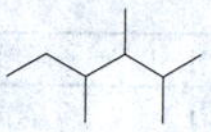
10561
2,3,4-Trimethylhexane

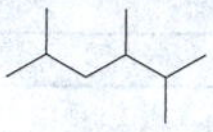
10562
2,3,5-Trimethylhexane

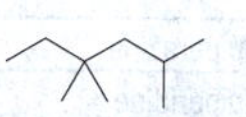
10563
2,4,4-Trimethylhexane

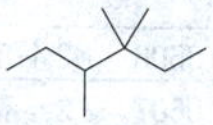
10564
3,3,4-Trimethylhexane

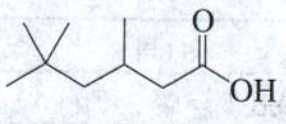
10565
3,5,5-Trimethylhexanoic acid

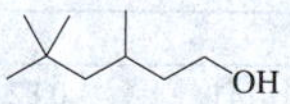
10566
3,5,5-Trimethyl-1-hexanol

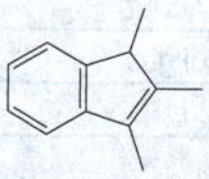
10567
1,2,3-Trimethylindene

10568
Trimethylindium

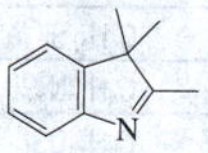
10569
2,3,3-Trimethyl-3*H*-indole

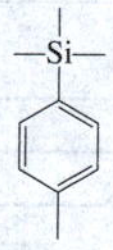
10570
Trimethyl(4-methylphenyl)silane

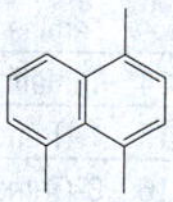
10571
1,4,5-Trimethylnaphthalene

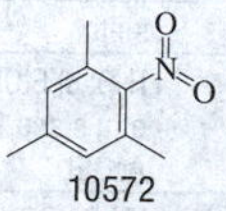
10572
1,3,5-Trimethyl-2-nitrobenzene

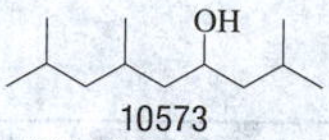
10573
2,6,8-Trimethyl-4-nonanol

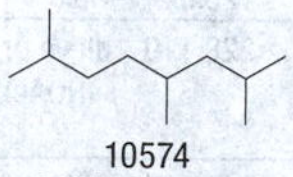
10574
2,4,7-Trimethyloctane

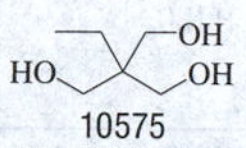
10575
Trimethylolpropane

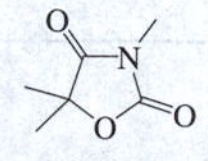
10576
3,5,5-Trimethyl-2,4-oxazolidinedione

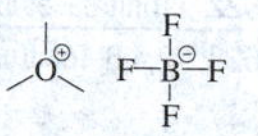
10577
Trimethyloxonium fluoborate

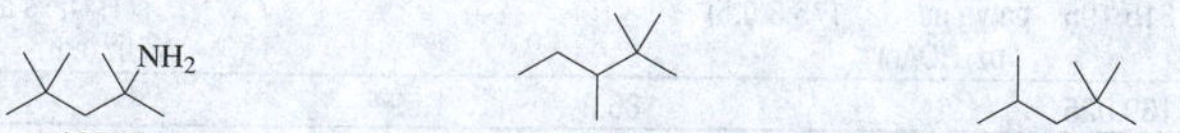
10578
2,4,4-Trimethyl-2-pentanamine

10579
2,2,3-Trimethylpentane

10580
2,2,4-Trimethylpentane

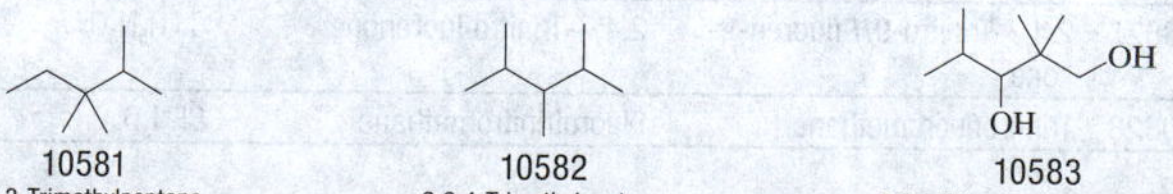
10581
2,3,3-Trimethylpentane

10582
2,3,4-Trimethylpentane

10583
2,2,4-Trimethyl-1,3-pentanediol

No.	Name	Synonym	Mol. form.	CAS Reg. No.	Mol. wt.	Physical form	mp/°C	bp/°C	den/g cm^{-3}	n_D	Solubility
10584	2,4,4-Trimethyl-2-pentanethiol		$C_8H_{18}S$	141-59-3	146.294	liq		76^{50}			
10585	2,4,4-Trimethyl-2-pentanol		$C_8H_{18}O$	690-37-9	130.228	liq	-20	146(4)	0.8225^{20}	1.4284^{20}	i H_2O; sl EtOH; s eth
10586	2,2,4-Trimethyl-3-pentanol		$C_8H_{18}O$	5162-48-1	130.228	liq	-13	151(3)	0.8297^{20}	1.4288^{20}	
10587	2,2,4-Trimethyl-3-pentanone	*tert*-Butyl isopropyl ketone	$C_8H_{16}O$	5857-36-3	128.212			146(3)	0.8065^{20}	1.4060	i H_2O; s eth, ace
10588	2,3,3-Trimethyl-1-pentene		C_8H_{16}	560-23-6	112.213	liq	-69(3)	108(2)	0.7308^{25}	1.4174^{20}	
10589	2,4,4-Trimethyl-1-pentene		C_8H_{16}	107-39-1	112.213	liq	-93.7(0.3)	101.3(0.4)	0.7150^{20}	1.4086^{20}	i H_2O; s eth, bz, ctc, chl, lig
10590	2,3,4-Trimethyl-2-pentene		C_8H_{16}	565-77-5	112.213	liq	-113.4(0.2)	116(3)	0.7434^{20}	1.4274^{20}	
10591	2,4,4-Trimethyl-2-pentene		C_8H_{16}	107-40-4	112.213	liq	-106.4(0.1)	104.9(0.7)	0.7218^{20}	1.4160^{20}	i H_2O; s eth, bz, ctc, chl; vs lig
10592	2,3,4-Trimethylphenol		$C_9H_{12}O$	526-85-2	136.190	nd (peth)	81(1)	236(6)			vs bz, eth, EtOH
10593	2,3,5-Trimethylphenol		$C_9H_{12}O$	697-82-5	136.190		95(1)	235(4)			
10594	2,3,6-Trimethylphenol		$C_9H_{12}O$	2416-94-6	136.190		63				
10595	2,4,5-Trimethylphenol		$C_9H_{12}O$	496-78-6	136.190	nd (lig)	71.6(1)	233(4)			i H_2O; vs EtOH, eth
10596	2,4,6-Trimethylphenol		$C_9H_{12}O$	527-60-6	136.190	nd (peth, MeOH)	72.1(0.4)	221(6)			vs eth, EtOH
10597	3,4,5-Trimethylphenol		$C_9H_{12}O$	527-54-8	136.190	nd (peth)	108(2)	249(4)			
10598	Trimethylphenoxysilane		$C_9H_{14}OSi$	1529-17-5	166.292	liq	-55	119	0.8681^{20}	1.5125^{20}	
10599	Trimethylphenylammonium chloride	Phenyltrimethylammonium chloride	$C_9H_{14}ClN$	138-24-9	171.667						vs H_2O, EtOH
10600	1-(2,4,6-Trimethylphenyl)ethanone		$C_{11}H_{14}O$	1667-01-2	162.228			241	0.9754^{20}	1.5175^{20}	i H_2O; s EtOH, eth, ace, bz, ctc
10601	1,1,1-Trimethyl-*N*-phenylsilanamine	Phenyl(trimethylsilyl)amine	$C_9H_{15}NSi$	3768-55-6	165.308			206	0.940^{20}		
10602	Trimethylphenylsilane		$C_9H_{14}Si$	768-32-1	150.293			169.5	0.8722^{20}	1.4907^{20}	s ctc, CS_2
10603	Trimethyl phosphate	Methyl phosphate	$C_3H_9O_4P$	512-56-1	140.074	liq	-46	197.2	1.2144^{20}	1.3967^{20}	vs H_2O; sl EtOH; s eth
10604	Trimethylphosphine		C_3H_9P	594-09-2	76.077	liq	-85	43(4)			i H_2O; s eth
10605	Trimethyl phosphite		$C_3H_9O_3P$	121-45-9	124.075			110(12)	1.0518^{20}	1.4095^{20}	vs EtOH, eth; sl ctc
10606	1,2,4-Trimethylpiperazine		$C_7H_{16}N_2$	120-85-4	128.215			149.5		1.4433^{20}	s ctc
10607	2,2,4-Trimethylpiperidine		$C_8H_{17}N$	101257-71-0	127.228			148	0.832^{15}	1.4458^{20}	vs eth, EtOH
10608	Trimethylpyrazine		$C_7H_{10}N_2$	14667-55-1	122.167			87^{35}			
10609	1,3,5-Trimethyl-1*H*-pyrazole		$C_6H_{10}N_2$	1072-91-9	110.156		37	170	0.9269^{40}	1.4589^{57}	
10610	2,3,6-Trimethylpyridine	2,3,6-Collidine	$C_8H_{11}N$	1462-84-6	121.180			170(15)	0.9220^{25}	1.5053^{20}	s H_2O, EtOH, eth, ace, bz
10611	2,4,6-Trimethylpyridine	2,4,6-Collidine	$C_8H_{11}N$	108-75-8	121.180	liq	-44.3(0.3)	170(1)	0.9166^{22}	1.4959^{25}	s H_2O, EtOH, eth, ace, ctc
10612	1,2,5-Trimethyl-1*H*-pyrrole		$C_7H_{11}N$	930-87-0	109.169			168(11)	0.807^{25}	1.4969^{20}	
10613	*N,N*,2-Trimethyl-6-quinolinamine		$C_{12}H_{14}N_2$	92-99-9	186.252	ye pr (HOAc, AcOEt)	101	319			s ctc, CS_2
10614	Trimethylsilane		$C_3H_{10}Si$	993-07-7	74.197	col gas	-135.9	6.7			
10615	1-(Trimethylsilyl)-1*H*-imidazole		$C_6H_{12}N_2Si$	18156-74-6	140.258						s chl
10616	3-(Trimethylsilyl)-1-propanol		$C_6H_{16}OSi$	2917-47-7	132.276			141	0.822^{25}	1.4298^{20}	
10617	Trimethylstibine		C_3H_9Sb	594-10-5	166.863	liq	-62	80.6	1.523^{15}	1.42^{15}	i H_2O; s EtOH, eth, CS_2
10618	Trimethylsulfonium iodide		C_3H_9IS	2181-42-2	204.072	cry (eth)	211 dec				
10619	Trimethylthiourea		$C_4H_{10}N_2S$	2489-77-2	118.200	pr (bz-lig)	87.5				vs bz, EtOH, chl
10620	2,4,6-Trimethyl-2,4,6-triphenylcyclotrisiloxane		$C_{21}H_{24}O_3Si_3$	546-45-2	408.671		100	$190^{1.5}$	1.1062^{20}	1.5397^{20}	
10621	Trimethylurea		$C_4H_{10}N_2O$	632-14-4	102.134	pr (eth)	74.0(0.2)	232.5	1.1900^{20}		s H_2O, EtOH; sl eth, bz
10622	Trinitroacetonitrile		$C_2N_4O_6$	630-72-8	176.044	wax	45(1)	exp			vs eth
10623	2,4,6-Trinitroaniline		$C_6H_4N_4O_6$	489-98-5	228.119	dk ye pr (HOAc)	190(1)	exp	1.762^{10}		i H_2O; sl EtOH, eth; s ace, bz, AcOEt
10624	1,3,5-Trinitrobenzene	*sym*-Trinitrobenzene	$C_6H_3N_3O_6$	99-35-4	213.104	orth pl (bz) lf (w)	121.3(0.4)	315	1.4775^{152}		sl H_2O, EtOH, eth; vs ace; s bz, py
10625	2,4,6-Trinitro-1,3-benzenediol	Styphnic acid	$C_6H_3N_3O_8$	82-71-3	245.103	hex ye cry (dil al)	178(1)	sub			vs eth, EtOH
10626	2,4,6-Trinitrobenzoic acid		$C_7H_3N_3O_8$	129-66-8	257.114	orth (w)	229(2)				sl H_2O, bz; vs EtOH; s eth, ace
10627	2,4,7-Trinitro-9*H*-fluoren-9-one	2,4,7-Trinitrofluorenone	$C_{13}H_5N_3O_7$	129-79-3	315.195	pa ye nd (bz, HOAc)	175.8(0.5)				sl H_2O; vs ace, bz, chl
10628	Trinitrofluoromethane	Fluorotrinitromethane	CFN_3O_6	1840-42-2	169.025			86.3	1.59^{20}		

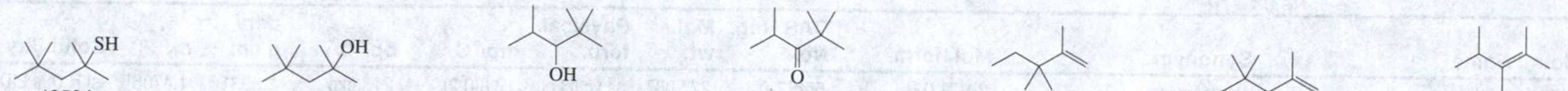

10584 2,4,4-Trimethyl-2-pentanethiol
10585 2,4,4-Trimethyl-2-pentanol
10586 2,2,4-Trimethyl-3-pentanol
10587 2,2,4-Trimethyl-3-pentanone
10588 2,3,3-Trimethyl-1-pentene
10589 2,4,4-Trimethyl-1-pentene
10590 2,3,4-Trimethyl-2-pentene

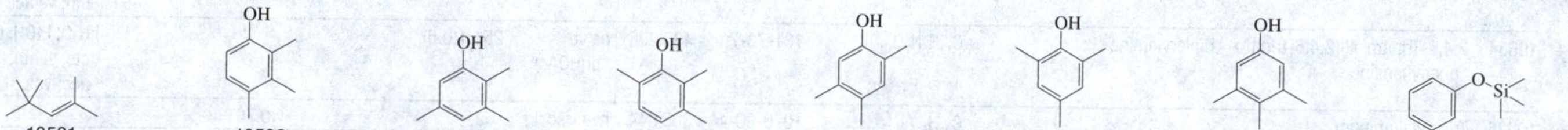

10591 2,4,4-Trimethyl-2-pentene
10592 2,3,4-Trimethylphenol
10593 2,3,5-Trimethylphenol
10594 2,3,6-Trimethylphenol
10595 2,4,5-Trimethylphenol
10596 2,4,6-Trimethylphenol
10597 3,4,5-Trimethylphenol
10598 Trimethylphenoxysilane

10599 Trimethylphenylammonium chloride
10600 1-(2,4,6-Trimethylphenyl)ethanone
10601 1,1,1-Trimethyl-*N*-phenylsilanamine
10602 Trimethylphenylsilane
10603 Trimethyl phosphate
10604 Trimethylphosphine
10605 Trimethyl phosphite

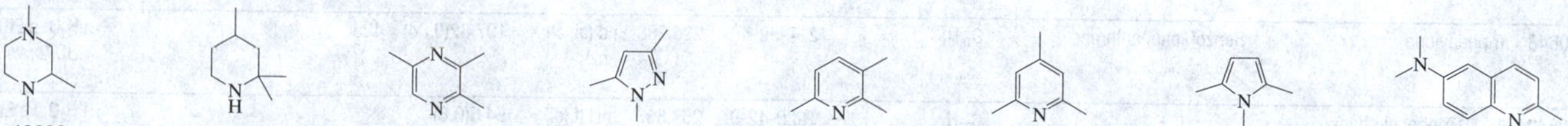

10606 1,2,4-Trimethylpiperazine
10607 2,2,4-Trimethylpiperidine
10608 Trimethylpyrazine
10609 1,3,5-Trimethyl-1*H*-pyrazole
10610 2,3,6-Trimethylpyridine
10611 2,4,6-Trimethylpyridine
10612 1,2,5-Trimethyl-1*H*-pyrrole
10613 *N,N*,2-Trimethyl-6-quinolinamine

10614 Trimethylsilane
10615 1-(Trimethylsilyl)-1*H*-imidazole
10616 3-(Trimethylsilyl)-1-propanol
10617 Trimethylstibine
10618 Trimethylsulfonium iodide
10619 Trimethylthiourea
10620 2,4,6-Trimethyl-2,4,6-triphenylcyclotrisiloxane
10621 Trimethylurea

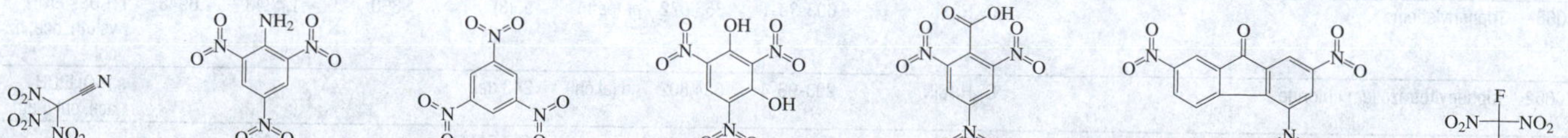

10622 Trinitroacetonitrile
10623 2,4,6-Trinitroaniline
10624 1,3,5-Trinitrobenzene
10625 2,4,6-Trinitro-1,3-benzenediol
10626 2,4,6-Trinitrobenzoic acid
10627 2,4,7-Trinitro-9*H*-fluoren-9-one
10628 Trinitrofluoromethane

Organic

No.	Name	Synonym	Mol. form.	CAS Reg. No.	Mol. wt.	Physical form	mp/°C	bp/°C	den/ g cm^{-3}	n_D	Solubility
10629	Trinitroglycerol	Nitroglycerin	$C_3H_5N_3O_9$	55-63-0	227.087	pa ye tcl or orth	12.8(0.2)	218 exp	1.5931^{20}	1.4786^{12}	sl H_2O; s EtOH, bz; msc eth; vs ace, chl
10630	Trinitromethane		CHN_3O_6	517-25-9	151.035		26(1)	exp	1.479^{20}	1.4451^{24}	vs ace, EtOH
10631	2,4,6-Trinitrophenol	Picric acid	$C_6H_3N_3O_7$	88-89-1	229.104	ye lf (w), pr (eth) pl (al)	121(2)	exp		1.763	sl H_2O; s EtOH, eth, bz, chl; vs ace
10632	2,4,6-Trinitrophenol, sodium salt	Sodium picrate	$C_6H_2N_3NaO_7$	3324-58-1	251.086	nd (w)	270.4				
10633	2,4,6-Trinitrotoluene	2-Methyl-1,3,5-trinitrobenzene	$C_7H_5N_3O_6$	118-96-7	227.131	orth (al)	80.9(0.8)	350(8)	1.654^{25}		i H_2O; sl EtOH; s eth; vs ace, bz
10634	2,4,6-Trinitro-*N*-(2,4,6-trinitrophenyl)aniline	Dipicrylamine	$C_{12}H_5N_7O_{12}$	131-73-7	439.208	pa ye pr(HOAc)	251.1(0.5)				i H_2O, EtOH, bz, ctc; sl eth, ace; vs py
10635	Trioctylaluminum		$C_{24}H_{51}Al$	1070-00-4	366.644	hyg visc liq	-62		0.701		
10636	Trioctylamine	*N,N*-Dioctyl-1-octanamine	$C_{24}H_{51}N$	1116-76-3	353.669	liq	-34.5(0.3)	379(4)	0.8110^{20}	1.4510^{19}	
10637	Trioctylphosphine oxide	TOPO	$C_{24}H_{51}OP$	78-50-2	386.635		53.5(0.5)	201^{2}			
10638	1,3,5-Trioxane	Formaldehyde, trimer	$C_3H_6O_3$	110-88-3	90.078	orth nd (eth)	60(2)	116(3)	1.17^{65}		vs H_2O; s EtOH, eth, bz, CS_2; i peth
10639	1,3,5-Trioxane-2,4,6-triimine	Cyamelide	$C_3H_3N_3O_3$	462-02-2	129.074	amor pow	dec	dec	1.127^{15}		vs eth, EtOH
10640	4,7,10-Trioxatridecane-1,13-diamine	Diethyleneglycol diaminopropyl ether	$C_{10}H_{24}N_2O_3$	4246-51-9	220.309	liq		147^{4}	1.005	1.4640^{20}	
10641	3,7,12-Trioxocholan-24-oic acid, (5β)	Dehydrocholic acid	$C_{24}H_{34}O_5$	81-23-2	402.524		237				i H_2O, eth; sl EtOH, bz; s ace, AcOEt
10642	Tripentylamine	*N,N*-Dipentyl-1-pentanamine	$C_{15}H_{33}N$	621-77-2	227.430			242.5	0.7907^{20}	1.4366^{20}	i H_2O; s EtOH, eth, acid
10643	Triphenylamine	*N,N*-Diphenylbenzenamine	$C_{18}H_{15}N$	603-34-9	245.319	mcl (MeOH, bz)	126.5(1)	362(10)	1.18^{25}		i H_2O; sl EtOH; s eth, bz, MeOH
10644	Triphenylarsine		$C_{18}H_{15}As$	603-32-7	306.234		60(2)	360	1.2634^{18}	1.6888^{21}	i H_2O; sl EtOH; vs eth, bz; s chl
10645	Triphenylarsine oxide		$C_{18}H_{15}AsO$	1153-05-5	322.233		192	324.0			
10646	Triphenylbismuthine		$C_{18}H_{15}Bi$	603-33-8	440.292		78(2)	242^{14}	1.715^{75}	1.7040^{75}	sl EtOH, chl; s eth, ace, bz, CS_2
10647	Triphenylborane		$C_{18}H_{15}B$	960-71-4	242.123	wh cry	142				i H_2O; sl eth; s bz, lig
10648	Triphenylene	Benzo[*l*]phenanthrene	$C_{18}H_{12}$	217-59-4	228.288	nd (al, chl, bz)	197.82(0.04)	425			i H_2O; s EtOH, HOAc; vs bz, chl
10649	1,1,2-Triphenylethane		$C_{20}H_{18}$	1520-42-9	258.357	mcl lf (dil al), nd (al)	54.5(0.8)				i H_2O; vs EtOH, eth, bz; sl MeOH
10650	1,1,2-Triphenylethene		$C_{20}H_{16}$	58-72-0	256.341	lf (al)	67.80(0.04)	220^{14}	1.0373^{78}	1.6292^{78}	i H_2O; s EtOH, chl, MeOH; vs eth
10651	*N,N',N''*-Triphenylguanidine		$C_{19}H_{17}N_3$	101-01-9	287.358	nd or pr (al)	146.5	dec	1.163^{20}		sl H_2O; s EtOH
10652	2,4,5-Triphenyl-1*H*-imidazole		$C_{21}H_{16}N_2$	484-47-9	296.365	nd (al)	275(2)	sub			i H_2O; s EtOH, eth
10653	Triphenylmethane		$C_{19}H_{16}$	519-73-3	244.330	orth (al)	92.0(0.7)	359	1.014^{99}	1.5839^{99}	i H_2O; sl EtOH; vs eth, py, chl; s bz
10654	Triphenylmethanol		$C_{19}H_{16}O$	76-84-6	260.329	pl (al), trg (bz)	162.3(0.4)	380	1.199^{0}		i H_2O, peth; vs EtOH, eth; s ace, bz
10655	Triphenyl phosphate		$C_{18}H_{15}O_4P$	115-86-6	326.283	cry (lig), pr (al), nd (eth)	49.39(0.04)	245^{11}	1.2055^{50}		i H_2O; s EtOH; vs eth, bz, ctc, chl
10656	Triphenylphosphine		$C_{18}H_{15}P$	603-35-0	262.286		80(2)	188^{1}	1.0749^{80}	1.6358^{80}	i H_2O; s EtOH, bz, chl; vs eth
10657	Triphenylphosphine oxide		$C_{18}H_{15}OP$	791-28-6	278.285	pr	158(3)	>360	1.2124^{23}		sl H_2O, eth, chl; vs EtOH, bz
10658	Triphenyl phosphite		$C_{18}H_{15}O_3P$	101-02-0	310.284		22(2)	360	1.1842^{20}	1.5900^{20}	i H_2O; vs EtOH
10659	Triphenylsilane		$C_{18}H_{16}Si$	789-25-3	260.406						s ctc, CS_2
10660	Triphenylsilanol		$C_{18}H_{16}OSi$	791-31-1	276.405		154.8		1.1777^{20}		s ctc, CS_2
10661	Triphenylstibine		$C_{18}H_{15}Sb$	603-36-1	353.072	pr (peth)	55(3)	>360	1.4343^{25}	1.6948^{42}	i H_2O; s EtOH; vs eth, ace, bz, chl
10662	Triphenyltetrazolium chloride		$C_{19}H_{15}ClN_4$	298-96-4	334.802	nd (al,chl)	243 dec				s H_2O, EtOH, ace, chl; i eth
10663	Triphenyltin hydroxide	Stannane, hydroxytriphenyl-	$C_{18}H_{16}OSn$	76-87-9	367.029		117.7(0.5)		1.54^{20}		
10664	2,4,6-Triphenyl-1,3,5-triazine		$C_{21}H_{15}N_3$	493-77-6	309.364		257				

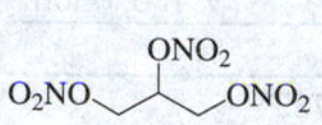
10629
Trinitroglycerol

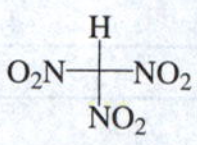
10630
Trinitromethane

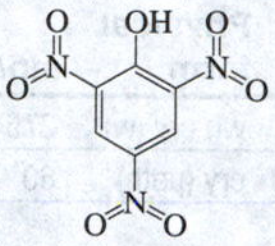
10631
2,4,6-Trinitrophenol

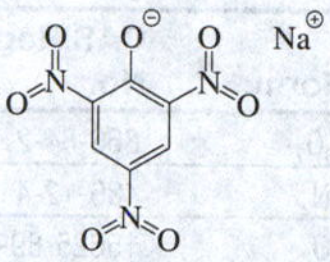
10632
2,4,6-Trinitrophenol, sodium salt

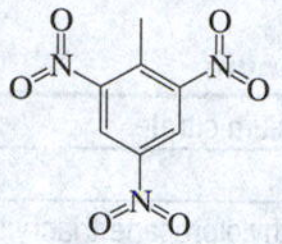
10633
2,4,6-Trinitrotoluene

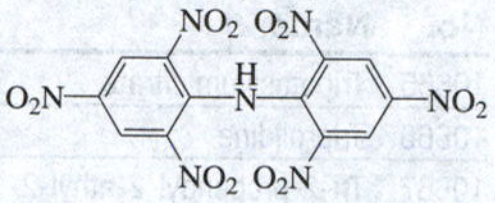
10634
2,4,6-Trinitro-*N*-(2,4,6-trinitrophenyl)aniline

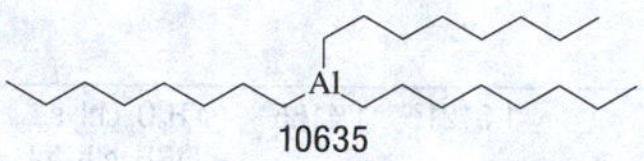
10635
Trioctylaluminum

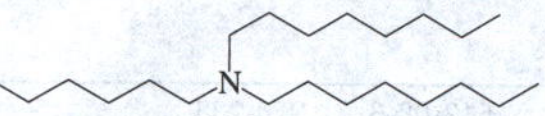
10636
Trioctylamine

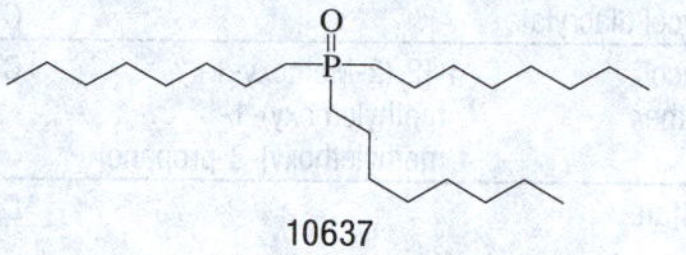
10637
Trioctylphosphine oxide

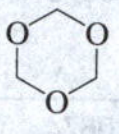
10638
1,3,5-Trioxane

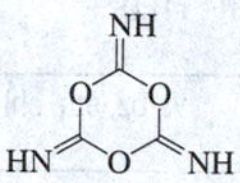
10639
1,3,5-Trioxane-2,4,6-triimine

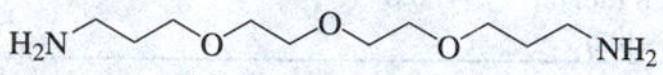
10640
4,7,10-Trioxatridecane-1,13-diamine

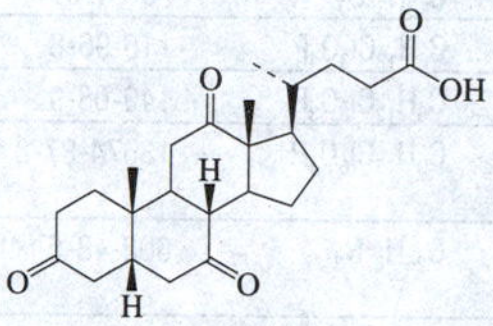
10641
3,7,12-Trioxocholan-24-oic acid, (5β)

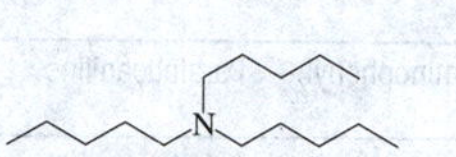
10642
Tripentylamine

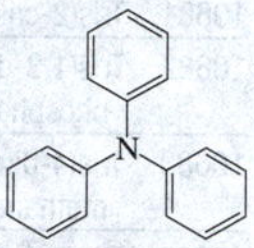
10643
Triphenylamine

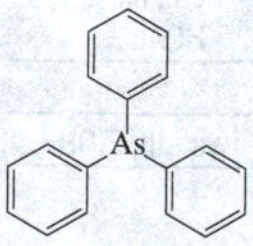
10644
Triphenylarsine

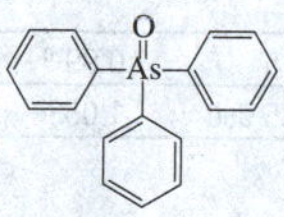
10645
Triphenylarsine oxide

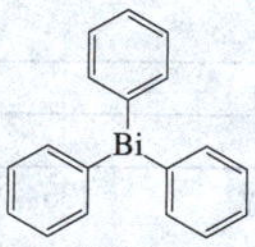
10646
Triphenylbismuthine

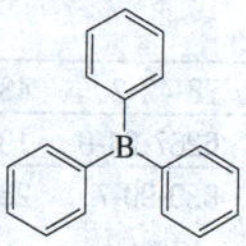
10647
Triphenylborane

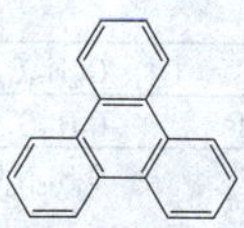
10648
Triphenylene

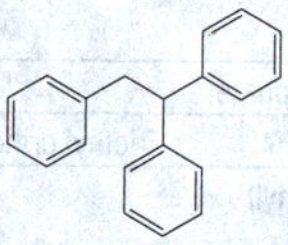
10649
1,1,2-Triphenylethane

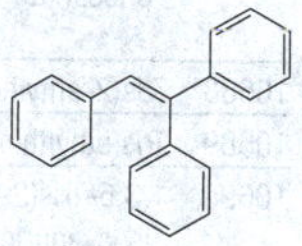
10650
1,1,2-Triphenylethene

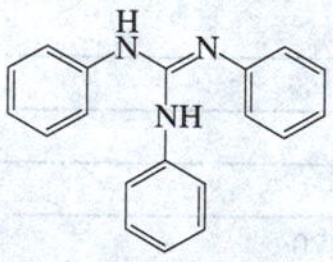
10651
N,N',N''-Triphenylguanidine

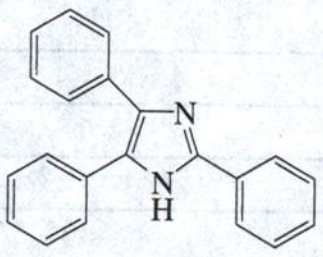
10652
2,4,5-Triphenyl-1*H*-imidazole

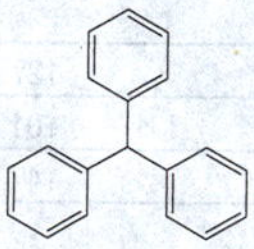
10653
Triphenylmethane

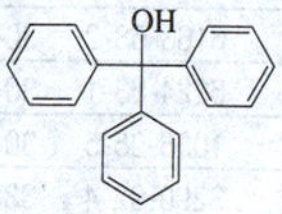
10654
Triphenylmethanol

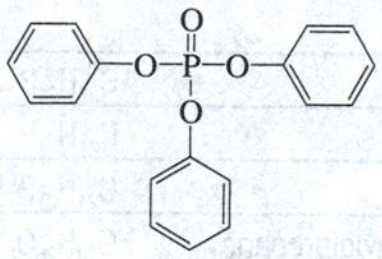
10655
Triphenyl phosphate

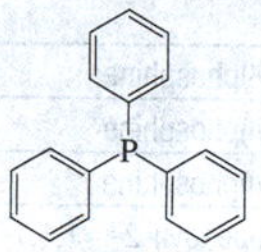
10656
Triphenylphosphine

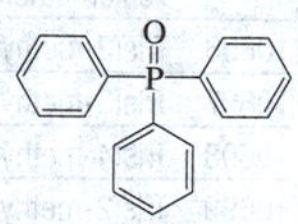
10657
Triphenylphosphine oxide

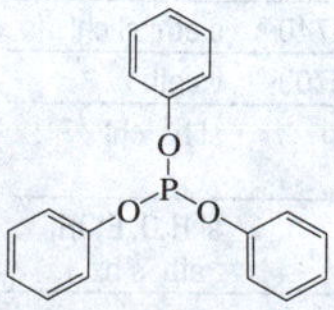
10658
Triphenyl phosphite

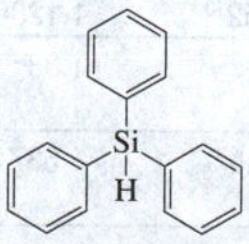
10659
Triphenylsilane

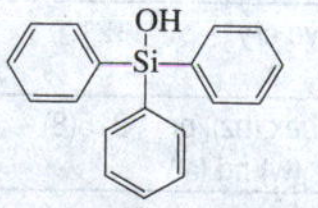
10660
Triphenylsilanol

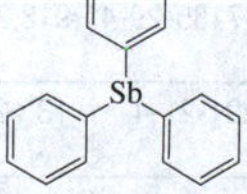
10661
Triphenylstibine

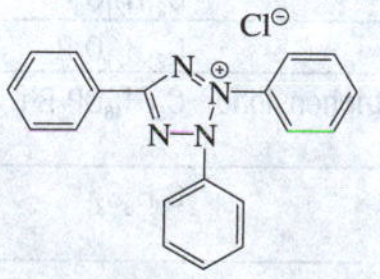
10662
Triphenyltetrazolium chloride

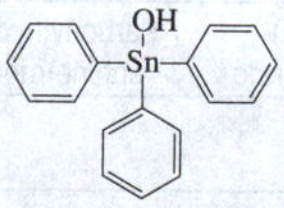
10663
Triphenyltin hydroxide

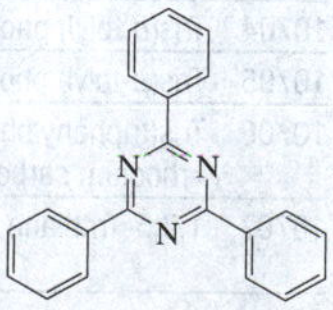
10664
2,4,6-Triphenyl-1,3,5-triazine

No.	Name	Synonym	Mol. form.	CAS Reg. No.	Mol. wt.	Physical form	mp/°C	bp/°C	den/ g cm^{-3}	n_D	Solubility
10665	Tripotassium citrate	Potassium citrate	$C_6H_5K_3O_7$	866-84-2	306.395	wh cry (w)	275 dec				vs H_2O; i EtOH
10666	Triprolidine		$C_{19}H_{22}N_2$	486-12-4	278.391	cry (peth)	60				
10667	Tri-2-propenoyl-2-ethyl-2-(hydroxymethyl)-1,3-propanediol	Trimethylolpropane triacrylate	$C_{15}H_{20}O_6$	15625-89-5	296.316					1.4735^{20}	
10668	Tripropylamine	*N,N*-Dipropyl-1-propanamine	$C_9H_{21}N$	102-69-2	143.270	liq	-100.5(0.5)	153(2)	0.7558^{20}	1.4181^{20}	vs eth, EtOH
10669	Tripropylborane		$C_9H_{21}B$	1116-61-6	140.074	liq	-56	159	0.7204^{25}	1.4135^{22}	
10670	Tripropyl borate	Boric acid, tripropyl ester	$C_9H_{21}BO_3$	688-71-1	188.072			178.7(0.4)	0.8576^{20}	1.3948^{20}	vs EtOH; msc eth; s PrOH
10671	Tripropylene glycol	[(1-Methyl-1,2-ethanediyl)-bis(oxy)]bispropanol	$C_9H_{20}O_4$	24800-44-0	192.253	liq		262(4)	1.02^{20}	1.4440^{20}	
10672	Tripropylene glycol diacrylate		$C_{15}H_{24}O_6$	42978-66-5	300.348						
10673	Tripropylene glycol monomethyl ether	1-[2-(2-Methoxy-1-methylethoxy)-1-methylethoxy]-2-propanol	$C_{10}H_{22}O_4$	20324-33-8	206.280			241.3			
10674	Tripropyl phosphate		$C_9H_{21}O_4P$	513-08-6	224.234			252	1.0121^{20}	1.4165^{20}	sl H_2O, chl; s EtOH, eth, tol, CS_2
10675	Tripropyl phosphite	Tripropoxyphosphine	$C_9H_{21}O_3P$	923-99-9	208.235			206.5	0.9417^{20}	1.4282^{20}	vs eth, EtOH
10676	Tripropylsilane		$C_9H_{22}Si$	998-29-8	158.357			172	0.7723^{0}	1.4280^{20}	i H_2O
10677	Tris(4-aminophenyl)methanol	C.I. Basic Red 9	$C_{19}H_{19}N_3O$	467-62-9	305.373	purp cry	205				
10678	2,4,6-Tris(1-aziridinyl)-1,3,5-triazine	Triethylenemelamine	$C_9H_{12}N_6$	51-18-3	204.231	cry pow	139 dec				s H_2O
10679	Tris(2-butoxyethyl) phosphate		$C_{18}H_{39}O_7P$	78-51-3	398.473	liq		255^{10}	1.02^{25}		i H_2O
10680	Tris(2-chloroethyl) phosphate		$C_6H_{12}Cl_3O_4P$	115-96-8	285.489			330	1.39^{25}	1.4721^{20}	s ctc
10681	Tris(2-chloroethyl) phosphite		$C_6H_{12}Cl_3O_3P$	140-08-9	269.490			120^{3}	1.3443^{26}	1.4868^{20}	
10682	Tris(1,3-dichloro-2-propyl) phosphate	Fyrol FR-2	$C_9H_{15}Cl_6O_4P$	13674-87-8	430.904	visc liq		236^{5}		1.5022^{20}	i H_2O
10683	Tris(4-dimethylaminophenyl)-methane	Paraleucaniline	$C_{25}H_{31}N_3$	603-48-5	373.534	lf (al), nd (bz)	176.5				vs bz, eth, chl
10684	Tris(2,4-dimethylphenyl) phosphate	2,4-Xylenol, phosphate (3:1)	$C_{24}H_{27}O_4P$	3862-12-2	410.442			233.5	1.142^{38}	1.5550^{20}	i H_2O; s bz, chl, hx
10685	Tris(2,5-dimethylphenyl) phosphate	2,5-Xylenol, phosphate (3:1)	$C_{24}H_{27}O_4P$	19074-59-0	410.442		79.8	262^{8}	1.197^{25}		i H_2O; sl EtOH, hx; s eth, bz, ctc
10686	Tris(2,6-dimethylphenyl) phosphate	2,6-Xylenol, phosphate (3:1)	$C_{24}H_{27}O_4P$	121-06-2	410.442	wax	137.8	263^{6}			i H_2O; sl EtOH, hx; s bz
10687	Tris(3,5-dimethylphenyl) phosphate		$C_{24}H_{27}O_4P$	25653-16-1	410.442		46.2	290^{10}			i H_2O; sl EtOH, chl, hx; s HOAc
10688	Tris(2-ethylhexyl) phosphate		$C_{24}H_{51}O_4P$	78-42-2	434.633	liq		215^{5}	0.99^{20}		
10689	Tris(ethylthio)methane	Triethyl orthothioformate	$C_7H_{16}S_3$	6267-24-9	196.397			235 dec	1.053^{20}	1.5410^{15}	vs eth, EtOH
10690	1,3,5-Tris(2-hydroxyethyl)-isocyanuric acid		$C_9H_{15}N_3O_6$	839-90-7	261.231	cry	136				
10691	1,1,1-Tris(hydroxymethyl)-ethane trinitrate	2-Methyl-2-[(nitrooxy)methyl]-1,3-propanediol, dinitrate	$C_5H_9N_3O_9$	3032-55-1	255.140			$83^{0.05}$			
10692	*N,N′,N″*-Tris(hydroxymethyl)-melamine	Trimethylolmelamine	$C_6H_{12}N_6O_3$	1017-56-7	216.197	cry	148				
10693	Tris(hydroxymethyl)-methylamine	2-Amino-2-(hydroxymethyl)-1,3-propanediol	$C_4H_{11}NO_3$	77-86-1	121.135		170.5(0.2)	219^{10}			vs H_2O; s MeOH
10694	Tris(methoxyethoxy)vinylsilane		$C_{11}H_{24}O_6Si$	1067-53-4	280.391						s ctc
10695	Tris(4-methoxyphenyl)-chloroethene	Chlorotrianisene	$C_{23}H_{21}ClO_3$	569-57-3	380.864		115				
10696	Tris(2-methylphenyl)phosphine		$C_{21}H_{21}P$	6163-58-2	304.366		127.0				
10697	Tris(3-methylphenyl)phosphine		$C_{21}H_{21}P$	6224-63-1	304.366		101.0				
10698	Tris(4-methylphenyl)phosphine		$C_{21}H_{21}P$	1038-95-5	304.366		147.0				
10699	Tris(2-methyl-2-propenoyl)-2-ethyl-2-hydroxymethyl-1,3-propanediol	1,1,1-Trimethylolpropane trimethacrylate	$C_{18}H_{26}O_6$	3290-92-4	338.395					1.470^{25}	
10700	Trisodium citrate	Sodium citrate	$C_6H_5Na_3O_7$	68-04-2	258.069	wh cry (w)	300				vs H_2O; i EtOH
10701	Trisodium *N*-hydroxyethylethylenediaminetriacetate	Versen-Ol	$C_{10}H_{15}N_2Na_3O_7$	139-89-9	344.204		288 (hyd)				
10702	Tris(perfluorobutyl)amine	Trinonafluorobutylamine	$C_{12}F_{27}N$	311-89-7	671.092			178(2)	1.884^{25}	1.291^{25}	s ace
10703	2,4,6-Tris(2-pyridinyl)-1,3,5-triazine	2,4,6-Tripyridyl-*s*-triazine	$C_{18}H_{12}N_6$	3682-35-7	312.328		210				
10704	Tris(*o*-tolyl) phosphite		$C_{21}H_{21}O_3P$	2622-08-4	352.364		11	238^{11}	1.1423^{20}	1.5740^{28}	s eth; sl chl
10705	Tris(*p*-tolyl) phosphite		$C_{21}H_{21}O_3P$	620-42-8	352.364	pa ye	52	252^{10}	1.1280^{25}	1.5703^{28}	vs eth
10706	Tris(triphenylphosphine)-rhodium carbonyl hydride	Carbonylhydrotris(triphenylphosphine)rhodium	$C_{55}H_{46}OP_3Rh$	17185-29-4	918.781	ye cry	121		1.33		sl bz, chl
10707	1,3,5-Trithiane		$C_3H_6S_3$	291-21-4	138.275	hex (bz), pr (w) nd (al)	215(8)	sub	1.6374^{24}		sl H_2O, EtOH, eth; s bz

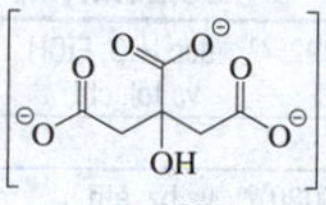
10665
Tripotassium citrate

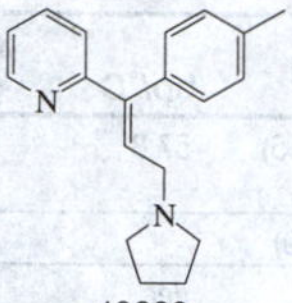
10666
Triprolidine

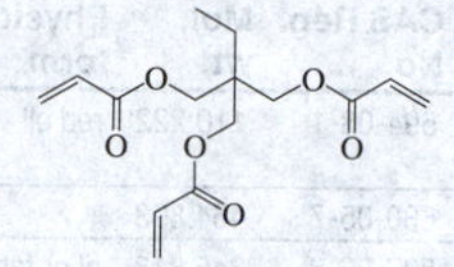
10667
Tri-2-propenoyl-2-ethyl-2-(hydroxymethyl)-1,3-propanediol

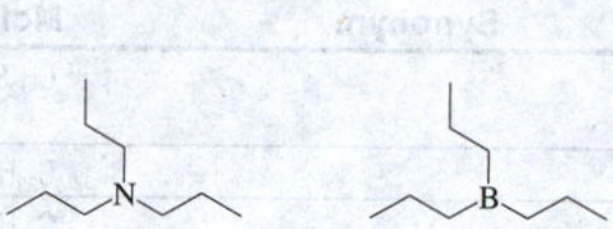
10668
Tripropylamine

10669
Tripropylborane

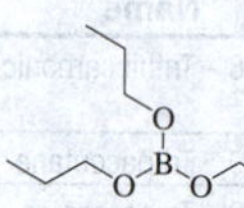
10670
Tripropyl borate

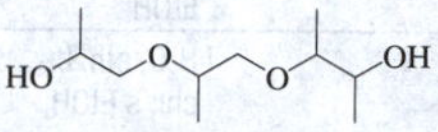
10671
Tripropylene glycol

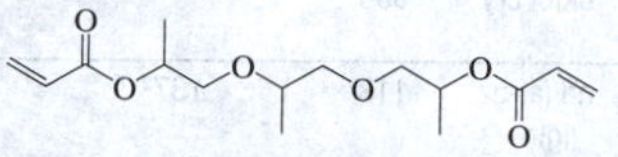
10672
Tripropylene glycol diacrylate

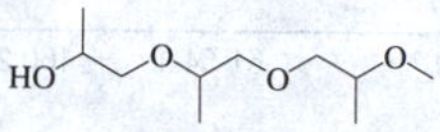
10673
Tripropylene glycol monomethyl ether

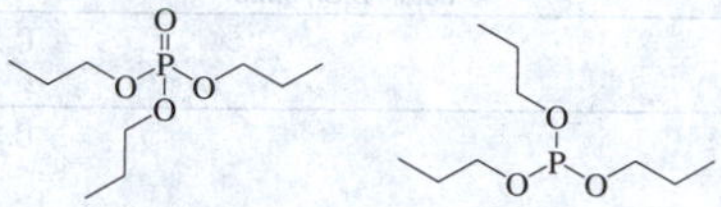
10674
Tripropyl phosphate

10675
Tripropyl phosphite

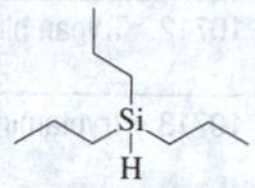
10676
Tripropylsilane

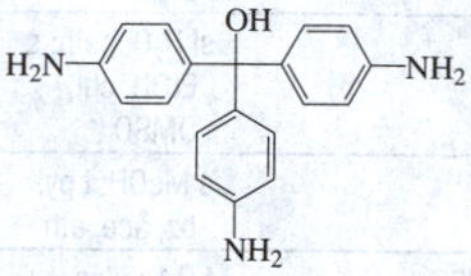
10677
Tris(4-aminophenyl)methanol

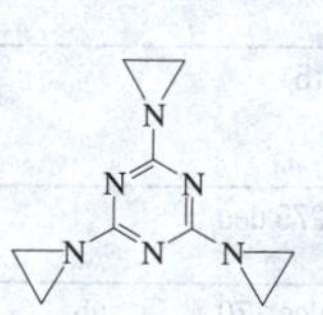
10678
2,4,6-Tris(1-aziridinyl)-1,3,5-triazine

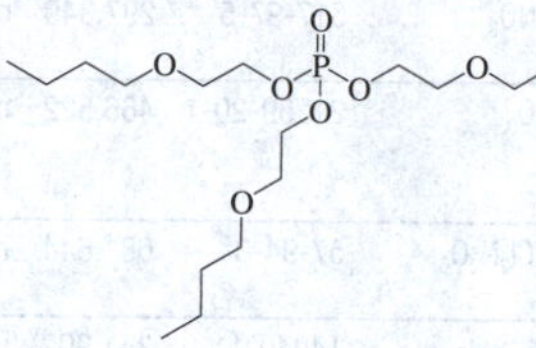
10679
Tris(2-butoxyethyl) phosphate

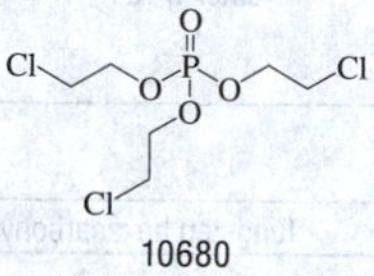
10680
Tris(2-chloroethyl) phosphate

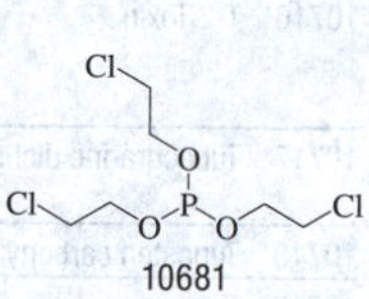
10681
Tris(2-chloroethyl) phosphite

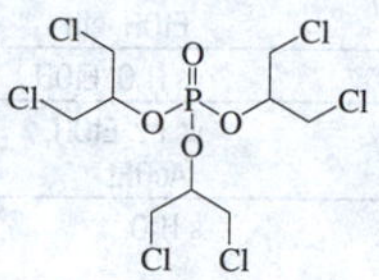
10682
Tris(1,3-dichloro-2-propyl) phosphate

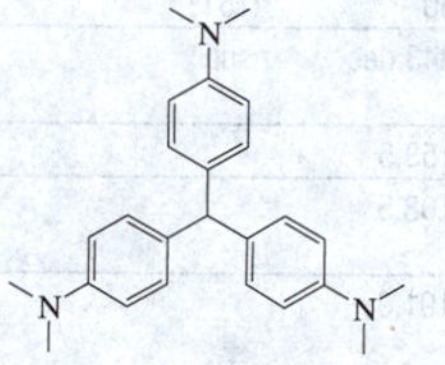
10683
Tris(4-dimethylaminophenyl)methane

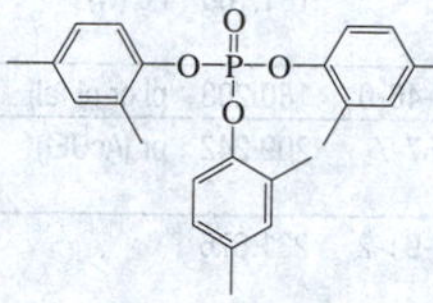
10684
Tris(2,4-dimethylphenyl) phosphate

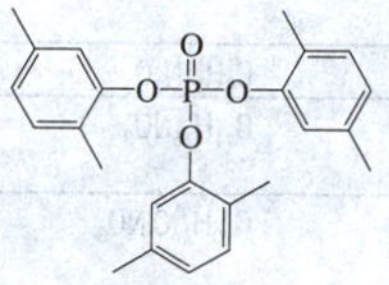
10685
Tris(2,5-dimethylphenyl) phosphate

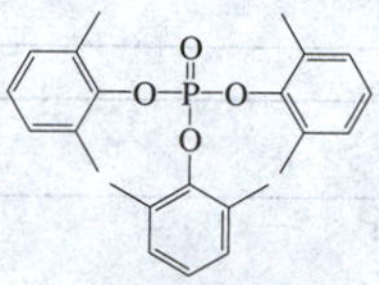
10686
Tris(2,6-dimethylphenyl) phosphate

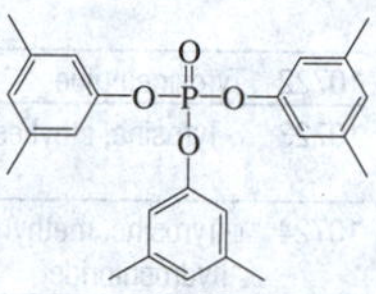
10687
Tris(3,5-dimethylphenyl) phosphate

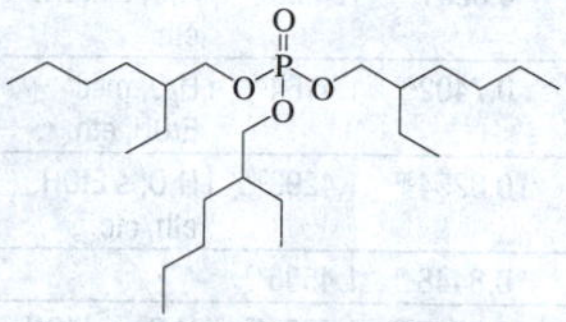
10688
Tris(2-ethylhexyl) phosphate

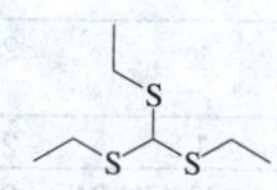
10689
Tris(ethylthio)methane

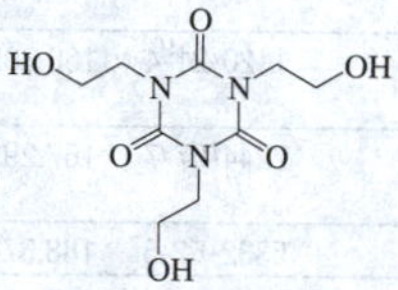
10690
1,3,5-Tris(2-hydroxyethyl) isocyanuric acid

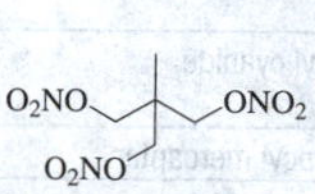
10691
1,1,1-Tris(hydroxymethyl)ethane trinitrate

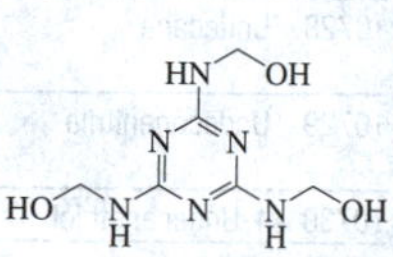
10692
N,N′,N′′-Tris(hydroxymethyl)melamine

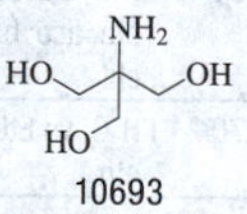
10693
Tris(hydroxymethyl)methylamine

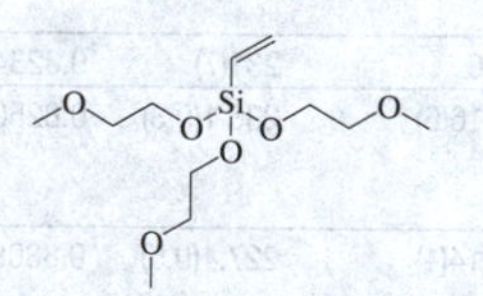
10694
Tris(methoxyethoxy)vinylsilane

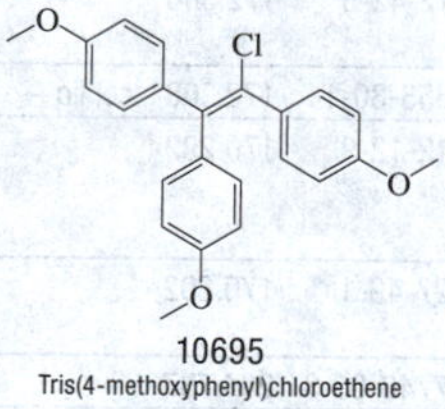
10695
Tris(4-methoxyphenyl)chloroethene

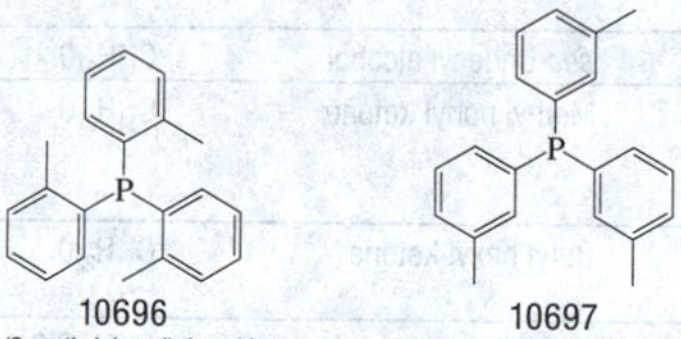
10696
Tris(2-methylphenyl)phosphine

10697
Tris(3-methylphenyl)phosphine

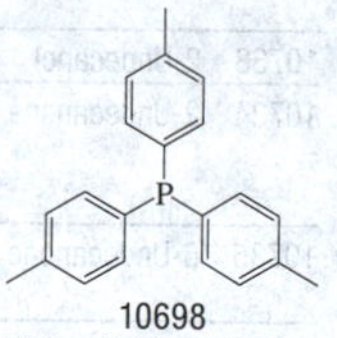
10698
Tris(4-methylphenyl)phosphine

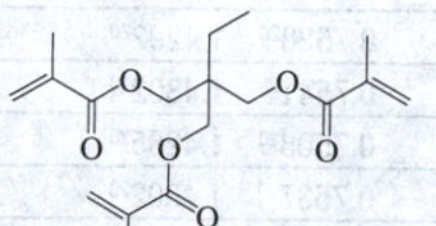
10699
Tris(2-methyl-2-propenoyl)-2-ethyl-2-hydroxymethyl-1,3-propanediol

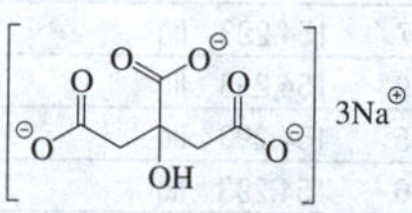
10700
Trisodium citrate

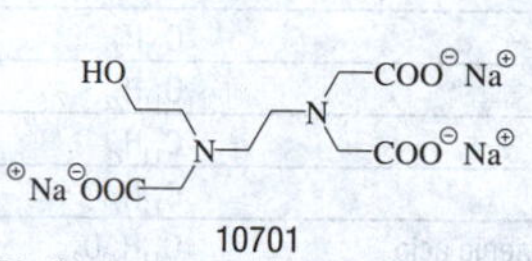
10701
Trisodium *N*-hydroxyethylethylenediaminetriacetate

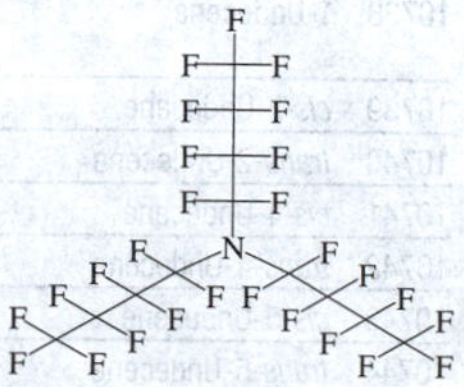
10702
Tris(perfluorobutyl)amine

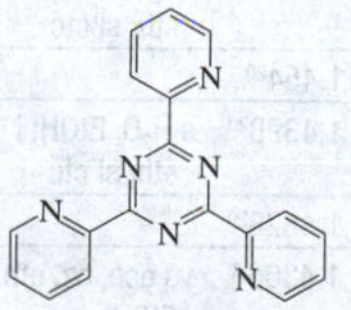
10703
2,4,6-Tris(2-pyridinyl)-1,3,5-triazine

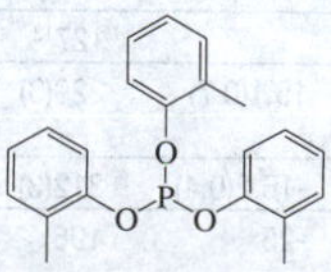
10704
Tris(*o*-tolyl) phosphite

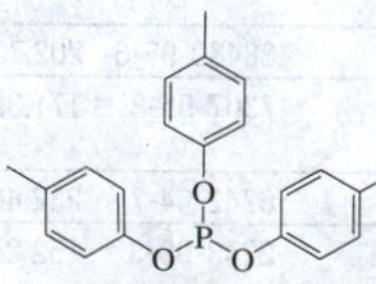
10705
Tris(*p*-tolyl) phosphite

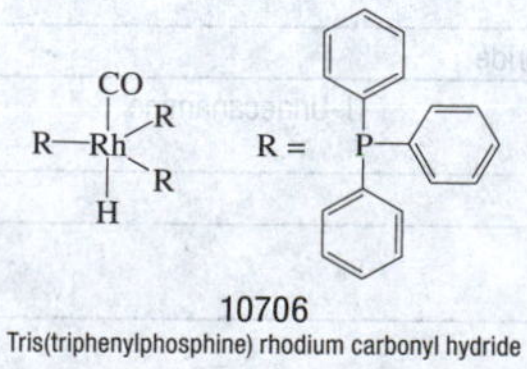
10706
Tris(triphenylphosphine) rhodium carbonyl hydride

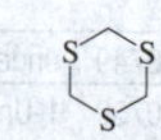
10707
1,3,5-Trithiane

Organic

No.	Name	Synonym	Mol. form.	CAS Reg. No.	Mol. wt.	Physical form	mp/°C	bp/°C	den/ g cm^{-3}	n_D	Solubility
10708	Trithiocarbonic acid		CH_2S_3	594-08-1	110.222	red oil	-26.8(0.5)	57.8	1.476^{25}	1.8225^{20}	dec H$_2$O, EtOH; vs tol, chl
10709	Tritriacontane		$C_{33}H_{68}$	630-05-7	464.893		71.2(0.9)				
10710	Tropacocaine		$C_{15}H_{19}NO_2$	537-26-8	245.318	pl or tab	49	dec	1.0426^{100}	1.5080^{100}	vs bz, eth, EtOH, peth
10711	Tropine	8-Methyl-8-azabicyclo[3.2.1]-octan-3-ol, *endo*	$C_8H_{15}NO$	120-29-6	141.211	hyg pl (eth)	63(2)	233	1.016^{100}	1.4811^{100}	vs H$_2$O, eth, EtOH
10712	Trypan blue		$C_{34}H_{24}N_6Na_4O_{14}S_4$	72-57-1	960.806	dk bl cry	300				s H$_2$O, acid; i EtOH
10713	Tryptamine		$C_{10}H_{12}N_2$	61-54-1	160.215	nd (al-bz, lig)	118	$137^{0.15}$			i H$_2$O, eth, bz, chl; s EtOH, ace
10714	L-Tryptophan	α-Aminoindole-3-propionic acid, (*l*)	$C_{11}H_{12}N_2O_2$	73-22-3	204.225	lf or pl (dil al)	289 dec				sl H$_2$O, HOAc; s EtOH; i eth, chl
10715	Tsuduranine		$C_{18}H_{19}NO_3$	517-97-5	297.349	nd (eth)	204				vs ace, eth, EtOH
10716	T-2 Toxin	Mycotoxin T2	$C_{24}H_{34}O_9$	21259-20-1	466.522	nd	151				sl H$_2$O, peth; s EtOH, chl, DMSO
10717	Tubocurarine dichloride		$C_{37}H_{42}Cl_2N_2O_6$	57-94-3	681.644	hyg cry	275 dec				s MeOH; i py, bz, ace, eth
10718	Tungsten carbonyl	Tungsten hexacarbonyl	C_6O_6W	14040-11-0	351.90	wh cry	dec 170	sub	2.65		i H$_2$O; s os
10719	Turanose		$C_{12}H_{22}O_{11}$	547-25-1	342.296	pr (w-al, MeOH)	168				vs H$_2$O; s EtOH, MeOH
10720	Tybamate		$C_{13}H_{26}N_2O_4$	4268-36-4	274.356	cry	50	$151^{0.06}$			
10721	L-Tyrosine	4-Hydroxy-L-phenylalanine	$C_9H_{11}NO_3$	60-18-4	181.188	nd (w)	343 dec	sub			sl H$_2$O, HOAc; i EtOH, eth
10722	Tyrosineamide		$C_9H_{12}N_2O_2$	4985-46-0	180.203	pl or pl (al)	153.5				vs H$_2$O, EtOH
10723	L-Tyrosine, ethyl ester		$C_{11}H_{15}NO_3$	949-67-7	209.242	pr (AcOEt)	108.5				vs bz, EtOH, AcOEt
10724	L-Tyrosine, methyl ester, hydrochloride		$C_{10}H_{14}ClNO_3$	3417-91-2	231.676		191.0				s H$_2$O
10725	1,10-Undecadiyne		$C_{11}H_{16}$	4117-15-1	148.245		-17(3)	83^{12}	0.8182^{21}	1.453^{21}	vs ace, bz
10726	Undecafluorocyclohexane		C_6HF_{11}	308-24-7	282.054			62.0			
10727	Undecanal		$C_{11}H_{22}O$	112-44-7	170.292		-4(3)	117^{18}	0.8251^{23}	1.4520^{20}	i H$_2$O; s EtOH, eth
10728	Undecane	Hendecane	$C_{11}H_{24}$	1120-21-4	156.309	liq	-25.54(0.05)	195.9(0.3)	0.7402^{20}	1.4164^{20}	i H$_2$O; msc EtOH, eth
10729	Undecanenitrile	Decyl cyanide	$C_{11}H_{21}N$	2244-07-7	167.292			265(6)	0.8254^{30}	1.4293^{30}	i H$_2$O; s EtOH, eth, ctc
10730	1-Undecanethiol	Undecyl mercaptan	$C_{11}H_{24}S$	5332-52-5	188.374	liq	-1.5	257.4	0.8448^{20}	1.4585^{20}	
10731	Undecanoic acid		$C_{11}H_{22}O_2$	112-37-8	186.292	cry (ace)	28.47(0.04)	280	0.8907^{20}	1.4294^{45}	i H$_2$O; vs EtOH, ace; s eth; msc bz
10732	1-Undecanol	Undecyl alcohol	$C_{11}H_{24}O$	112-42-5	172.308		16.6(0.2)	246(2)	0.8298^{20}	1.4392^{20}	i H$_2$O; s EtOH; vs eth
10733	2-Undecanol	*sec*-Undecyl alcohol	$C_{11}H_{24}O$	1653-30-1	172.308	col liq	0	231(7)	0.8234^{25}	1.4352^{25}	
10734	2-Undecanone	Methyl nonyl ketone	$C_{11}H_{22}O$	112-12-9	170.292		16(6)	233.1(0.3)	0.8250^{20}	1.4291^{20}	i H$_2$O; s EtOH, eth, ace, bz, chl
10735	6-Undecanone	Butyl hexyl ketone	$C_{11}H_{22}O$	927-49-1	170.292		14(1)	227.4(0.5)	0.8308^{20}	1.4270^{20}	i H$_2$O; vs EtOH, eth
10736	Undecanoyl chloride		$C_{11}H_{21}ClO$	17746-05-3	204.737						sl ctc
10737	10-Undecenal		$C_{11}H_{20}O$	112-45-8	168.276						sl ctc
10738	1-Undecene		$C_{11}H_{22}$	821-95-4	154.293	liq	-49.12(0.04)	192.7(0.5)	0.7503^{20}	1.4261^{20}	i H$_2$O; s eth, chl, lig
10739	*cis*-2-Undecene		$C_{11}H_{22}$	821-96-5	154.293	liq	-66.5	196.1	0.7576^{20}		
10740	*trans*-2-Undecene		$C_{11}H_{22}$	693-61-8	154.293	liq	-48.3	192.5	0.7528^{20}	1.4292^{20}	
10741	*cis*-4-Undecene		$C_{11}H_{22}$	821-98-7	154.293	liq	-97	192.6	0.7541^{20}	1.4302^{20}	
10742	*trans*-4-Undecene		$C_{11}H_{22}$	693-62-9	154.293	liq	-63.7	193	0.7508^{20}	1.4285^{20}	
10743	*cis*-5-Undecene		$C_{11}H_{22}$	764-96-5	154.293	liq	-106.5	192.3	0.7537^{20}	1.4302^{20}	
10744	*trans*-5-Undecene		$C_{11}H_{22}$	764-97-6	154.293	liq	-61.1	192	0.7497^{20}	1.4285^{20}	vs eth, chl, lig
10745	10-Undecenoic acid	Undecylenic acid	$C_{11}H_{20}O_2$	112-38-9	184.276	cry	24.4(0.3)	275	0.9072^{24}	1.4486^{24}	i H$_2$O; s EtOH, eth; sl ctc
10746	10-Undecen-1-ol		$C_{11}H_{22}O$	112-43-6	170.292	liq	-1.0	250	0.8495^{15}	1.4500^{20}	i H$_2$O; s EtOH, eth; sl ctc
10747	10-Undecenoyl chloride		$C_{11}H_{19}ClO$	38460-95-6	202.721			127^{13}	0.944^{20}	1.454^{20}	
10748	Undecylamine	1-Undecanamine	$C_{11}H_{25}N$	7307-55-3	171.324	cry (eth, al)	15.1(0.7)	229(3)	0.7979^{20}	1.4398^{20}	s H$_2$O, EtOH; i eth; sl ctc
10749	Undecylbenzene		$C_{17}H_{28}$	6742-54-7	232.404	liq	-16.2(0.4)	312(3)	0.8553^{20}	1.4828^{20}	
10750	1-Undecyne		$C_{11}H_{20}$	2243-98-3	152.277	liq	-25	196	0.7728^{20}	1.4306^{20}	vs ace, bz, eth, EtOH

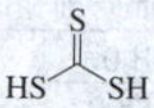
10708
Trithiocarbonic acid

10709
Tritriacontane

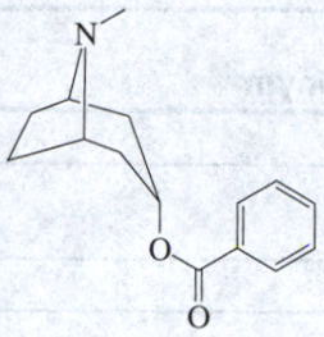
10710
Tropacocaine

10711
Tropine

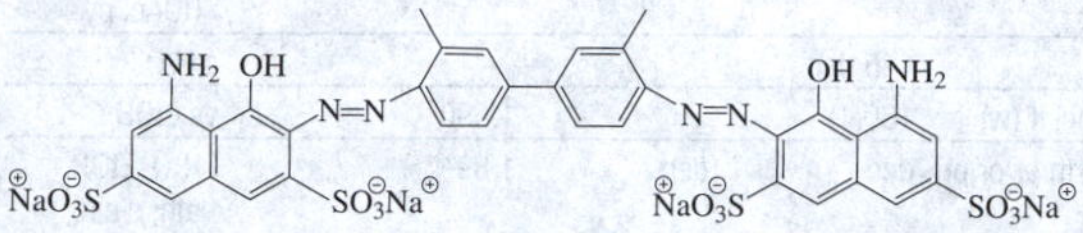
10712
Trypan blue

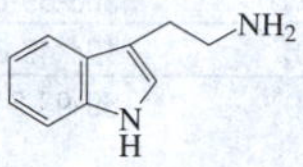
10713
Tryptamine

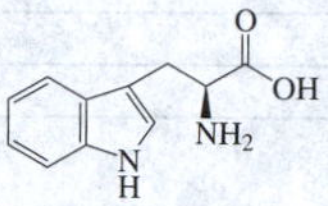
10714
L-Tryptophan

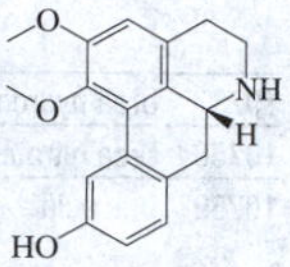
10715
Tsuduranine

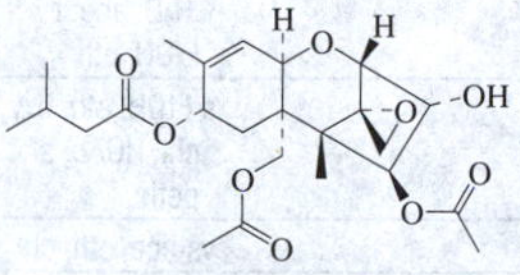
10716
T-2 Toxin

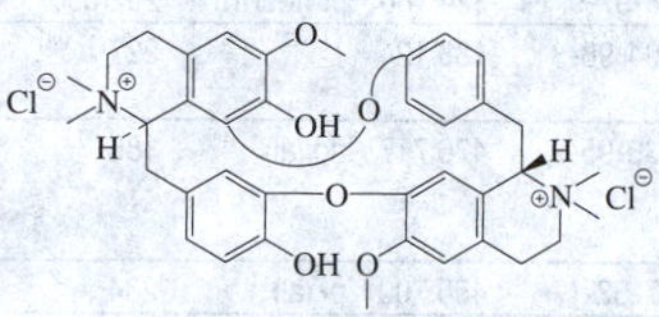
10717
Tubocurarine dichloride

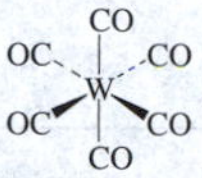
10718
Tungsten carbonyl

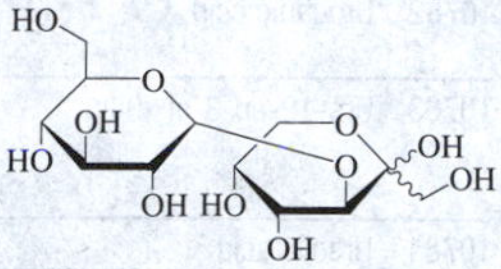
10719
Turanose

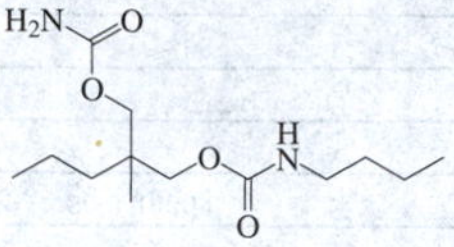
10720
Tybamate

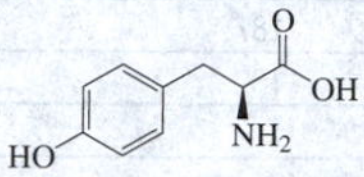
10721
L-Tyrosine

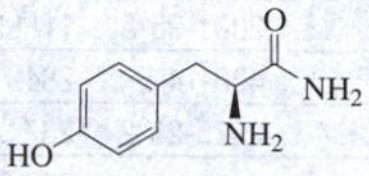
10722
Tyrosineamide

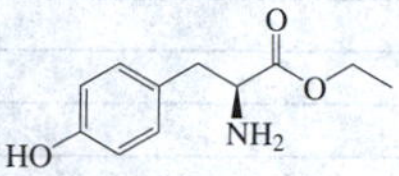
10723
L-Tyrosine, ethyl ester

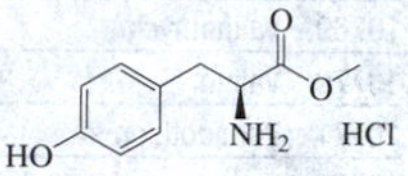
10724
L-Tyrosine, methyl ester, hydrochloride

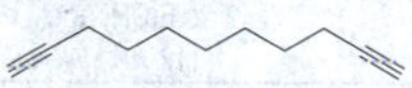
10725
1,10-Undecadiyne

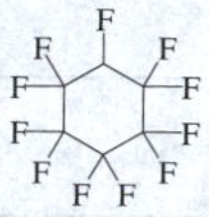
10726
Undecafluorocyclohexane

10727
Undecanal

10728
Undecane

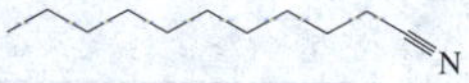
10729
Undecanenitrile

10730
1-Undecanethiol

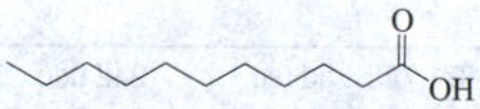
10731
Undecanoic acid

10732
1-Undecanol

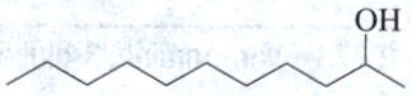
10733
2-Undecanol

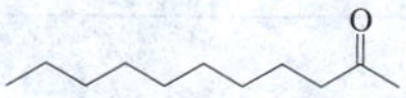
10734
2-Undecanone

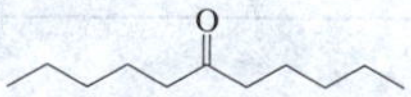
10735
6-Undecanone

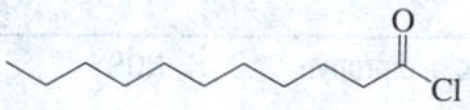
10736
Undecanoyl chloride

10737
10-Undecenal

10738
1-Undecene

10739
cis-2-Undecene

10740
trans-2-Undecene

10741
cis-4-Undecene

10742
trans-4-Undecene

10743
cis-5-Undecene

10744
trans-5-Undecene

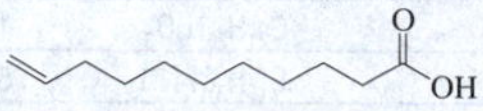
10745
10-Undecenoic acid

10746
10-Undecen-1-ol

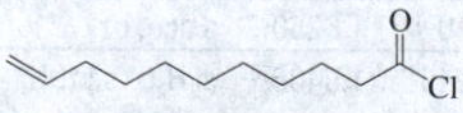
10747
10-Undecenoyl chloride

10748
Undecylamine

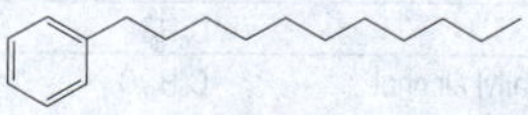
10749
Undecylbenzene

10750
1-Undecyne

No.	Name	Synonym	Mol. form.	CAS Reg. No.	Mol. wt.	Physical form	mp/°C	bp/°C	den/ g cm⁻³	n_D	Solubility
10751	2-Undecyne		$C_{11}H_{20}$	60212-29-5	152.277	liq	-30.1	206(4)	0.7827^{20}	1.4391^{20}	
10752	Uracil		$C_4H_4N_2O_2$	66-22-8	112.087	nd (w)	339(1)				sl H_2O; vs EtOH, eth; s dil NH_3
10753	Uracil mustard		$C_8H_{11}Cl_2N_3O_2$	66-75-1	252.098		206 dec				sl H_2O
10754	Uranyl acetate dihydrate		$C_4H_{10}O_8U$	6159-44-0	424.146	ye cry (HOAc)	80 dec		2.89		sl EtOH
10755	Urazole		$C_2H_3N_3O_2$	3232-84-6	101.064	lf (w)	249 dec				
10756	Urea	Carbamide	CH_4N_2O	57-13-6	60.055	tetr pr (al)	132.4(0.5)	dec	1.3230^{20}	1.484	vs H_2O, EtOH; i eth, bz; s HOAc, py
10757	Urea hydrochloride		CH_5ClN_2O	506-89-8	96.516		145 dec				s H_2O
10758	Urea nitrate		$CH_5N_3O_4$	124-47-0	123.069	mcl lf (w)	160(1)		1.690^{20}		vs EtOH
10759	Uric acid		$C_5H_4N_4O_3$	69-93-2	168.111	orth pr or pl	dec	dec	1.89^{25}		i H_2O, EtOH, eth; s alk, glycerol; sl acid
10760	Uridine	1-β-D-Ribofuranosyluracil	$C_9H_{12}N_2O_6$	58-96-8	244.200	nd (aq al)	165				s H_2O, EtOH, py
10761	5′-Uridylic acid	Uridine 5′-phosphoric acid	$C_9H_{13}N_2O_9P$	58-97-9	324.180	pr (MeOH)	202 dec				vs H_2O; s MeOH
10762	Urocanic acid	Imidazole-4-acrylic acid	$C_6H_6N_2O_2$	104-98-3	138.124		227				s H_2O, ace; i EtOH, eth
10763	Urs-12-en-3-ol, (3β)	α-Amyrin	$C_{30}H_{50}O$	638-95-9	426.717	nd (al)	186	$243^{0.5}$			s EtOH, eth, bz, chl, HOAc; sl peth
10764	Ursolic acid		$C_{30}H_{48}O_3$	77-52-1	456.700	pl (al)	284				vs ace, eth, chl
10765	Uzarin		$C_{35}H_{54}O_{14}$	20231-81-6	698.796	pr	269				
10766	Vacciniin	D-Glucose, 6-benzoate	$C_{13}H_{16}O_7$	14200-76-1	284.262	amor (aq ace, +1w)	122				vs H_2O, ace, EtOH, eth
10767	Validamycin A		$C_{20}H_{35}NO_{13}$	37248-47-8	497.491	amorp pow	95 dec				
10768	L-Valine	2-Aminoisovaleric acid	$C_5H_{11}NO_2$	72-18-4	117.147	lf (w-al)	315	sub	1.23^{25}		s H_2O
10769	Valinomycin		$C_{54}H_{90}N_6O_{18}$	2001-95-8	1111.322	cry	187				
10770	Valium		$C_{16}H_{13}ClN_2O$	439-14-5	284.739		132				
10771	Vamidothion		$C_8H_{18}NO_4PS_2$	2275-23-2	287.337	oil					i peth; s os
10772	Vanadium carbonyl	Vanadium hexacarbonyl	C_6O_6V	14024-00-1	219.002	bl-grn cry	dec 60	sub			
10773	Vanadium(III) 2,4-pentanedioate	Vanadium(III) acetylacetonate	$C_{15}H_{21}O_6V$	13476-99-8	348.266	brn cry	≈185	sub	≈1.0		s MeOH, ace, bz, chl
10774	DL-Vasicine	DL-Peganine	$C_{11}H_{12}N_2O$	6159-56-4	188.225	nd (al)	210.8				sl H_2O, eth, bz; s EtOH, ace, chl
10775	L-Vasicine	L-Peganine	$C_{11}H_{12}N_2O$	6159-55-3	188.225	nd (al)	211.5				sl H_2O, eth, bz; s EtOH, ace, chl
10776	Verapamil		$C_{27}H_{38}N_2O_4$	52-53-9	454.602	ye oil		$245^{0.01}$		1.5448^{25}	i H_2O; vs EtOH, ace; sl bz, hx
10777	Veratramine		$C_{27}H_{39}NO_2$	60-70-8	409.605	nd	206				s EtOH, bz, chl, dil acid; i dil alk
10778	Veratramine, 3-glucoside		$C_{33}H_{49}NO_7$	475-00-3	571.745	nd (aq. MeOH)	242 dec				
10779	Veratridine		$C_{36}H_{51}NO_{11}$	71-62-5	673.790	ye amorp pow	180				i H_2O; sl eth
10780	*d*-Verbenone		$C_{10}H_{14}O$	18309-32-5	150.217		9.8	227.5	0.9978^{20}	1.4993^{18}	s H_2O, EtOH, ace, bz
10781	Vernolate	Carbamothioic acid, dipropyl-, *S*-propyl ester	$C_{10}H_{21}NOS$	1929-77-7	203.345			150^{30}	0.952^{20}		
10782	Versalide		$C_{18}H_{26}O$	88-29-9	258.398	cry	46.5	130^{2}			s EtOH
10783	α-Vetivone	Isonootkatone	$C_{15}H_{22}O$	15764-04-2	218.335	cry (peth)	51.5	144^{2}	1.0035^{20}	1.5370^{20}	vs ace
10784	β-Vetivone		$C_{15}H_{22}O$	18444-79-6	218.335	cry (peth)	44.5	141^{2}	1.0001^{20}	1.5309^{20}	s ace
10785	Vicine	2,6-Diamino-5-(β-D-glucopyranosyloxy)-4(1*H*)-pyrimidinone	$C_{10}H_{16}N_4O_7$	152-93-2	304.257	nd (w, dil al, +1 w)	240 dec				sl H_2O, EtOH; vs acid, alk
10786	Vidarabine	β-D-9-Arabinofuranosyladenine	$C_{10}H_{15}N_5O_5$	5536-17-4	285.257	nd (w)	257				
10787	Vinblastine		$C_{46}H_{58}N_4O_9$	865-21-4	810.975	nd (MeOH)	216				i H_2O; s EtOH, ace, chl, AcOEt
10788	Vincamine		$C_{21}H_{26}N_2O_3$	1617-90-9	354.442		231.5				
10789	Vinclozolin		$C_{12}H_9Cl_2NO_3$	50471-44-8	286.110		110.5(0.5)	$131^{0.05}$	1.51		
10790	Vincristine		$C_{46}H_{56}N_4O_{10}$	57-22-7	824.958		219				
10791	Vinyl acetate		$C_4H_6O_2$	108-05-4	86.090	liq	-100.2(0.4)	72.6(0.3)	0.9256^{25}	1.3926^{25}	sl H_2O; msc EtOH; s eth, ace, bz, chl
10792	4-Vinylaniline		C_8H_9N	1520-21-4	119.164		23.5	116^{9}	1.010^{20}	1.6250^{22}	s ace, bz
10793	α-Vinylbenzenemethanol	1-Phenylallyl alcohol	$C_9H_{10}O$	4393-06-0	134.174				1.0249^{21}	1.5406^{20}	sl H_2O; s EtOH, eth, bz, chl

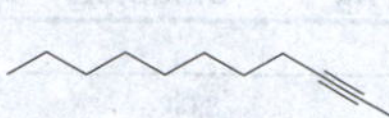
10751
2-Undecyne

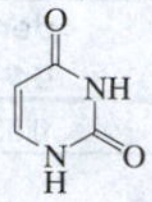
10752
Uracil

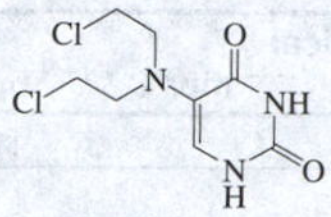
10753
Uracil mustard

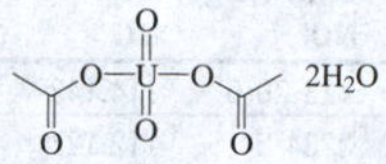
10754
Uranyl acetate dihydrate

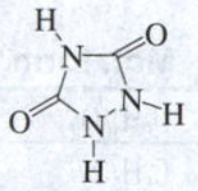
10755
Urazole

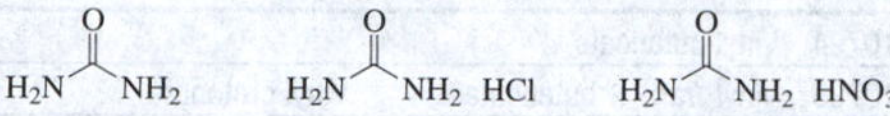
10756
Urea

10757
Urea hydrochloride

10758
Urea nitrate

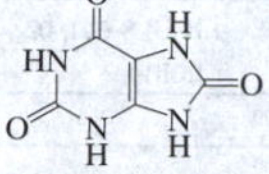
10759
Uric acid

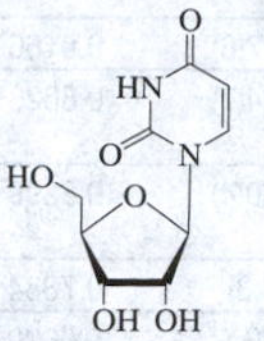
10760
Uridine

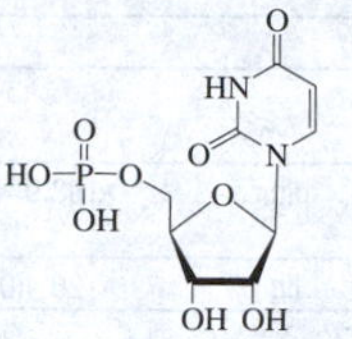
10761
5′-Uridylic acid

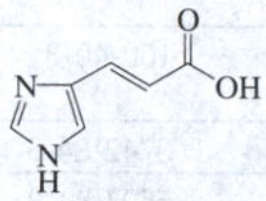
10762
Urocanic acid

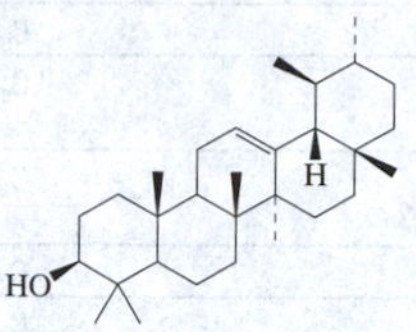
10763
Urs-12-en-3-ol, (3β)

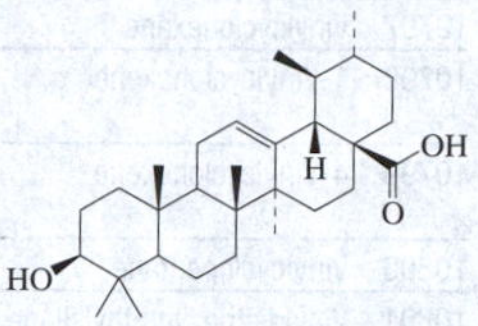
10764
Ursolic acid

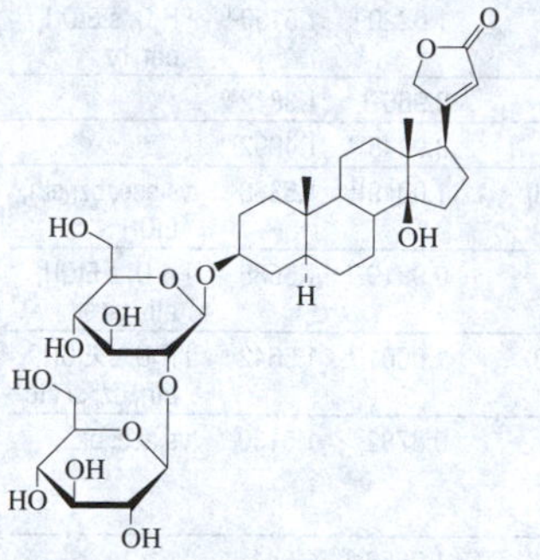
10765
Uzarin

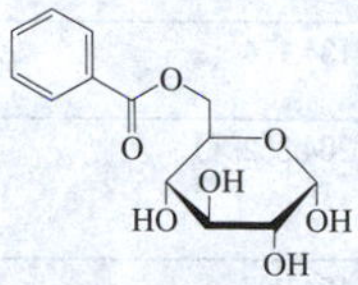
10766
Vacciniin

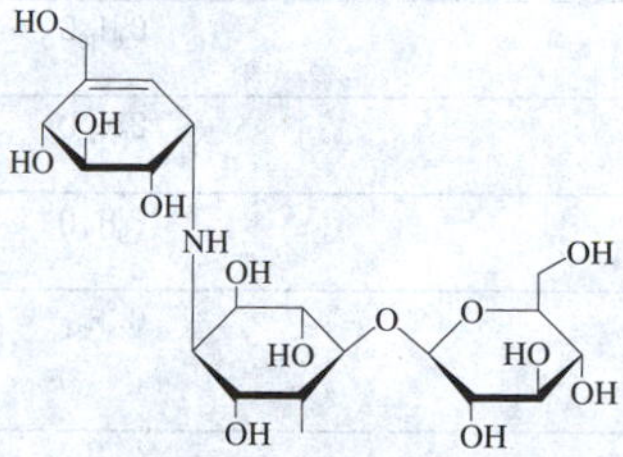
10767
Validamycin A

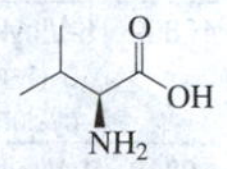
10768
L-Valine

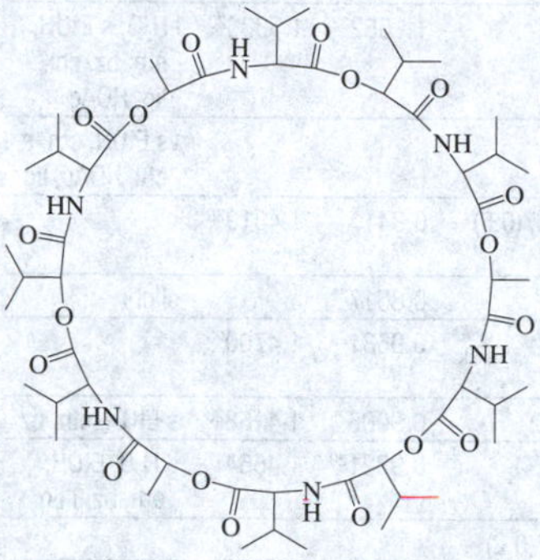
10769
Valinomycin

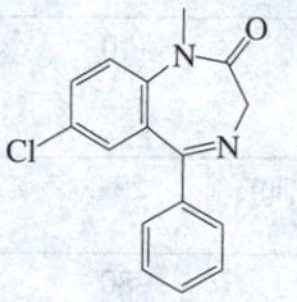
10770
Valium

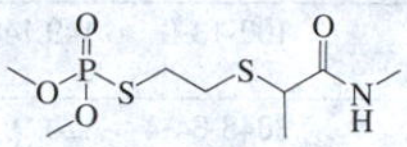
10771
Vamidothion

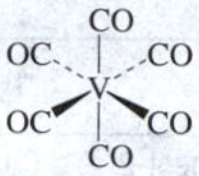
10772
Vanadium carbonyl

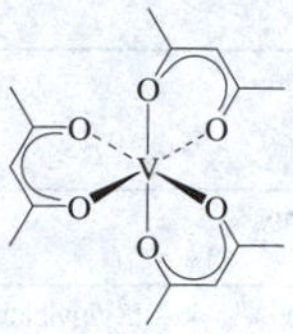
10773
Vanadium(III) 2,4-pentanedioate

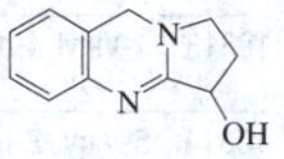
10774
DL-Vasicine

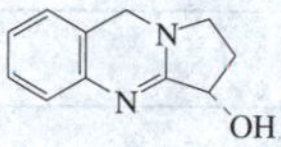
10775
L-Vasicine

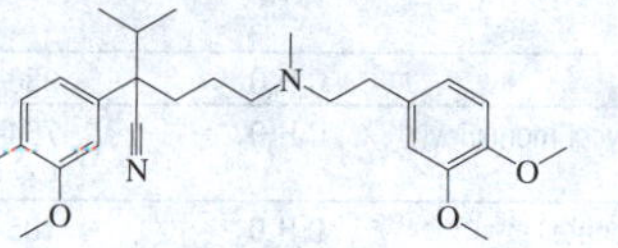
10776
Verapamil

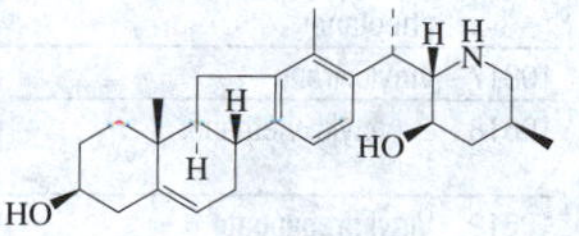
10777
Veratramine

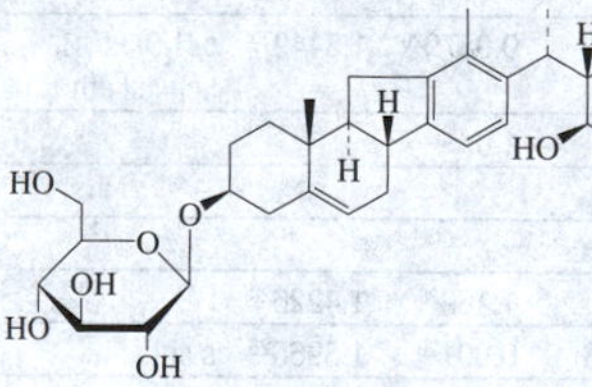
10778
Veratramine, 3-glucoside

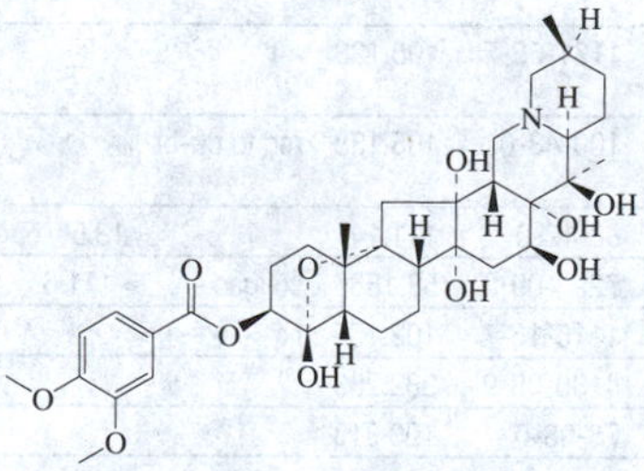
10779
Veratridine

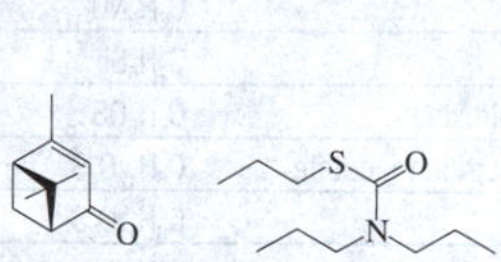
10780
d-Verbenone

10781
Vernolate

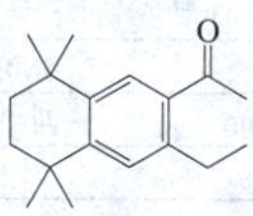
10782
Versalide

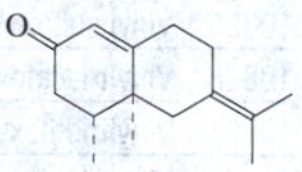
10783
α-Vetivone

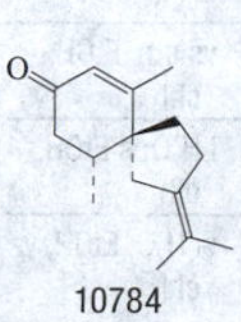
10784
β-Vetivone

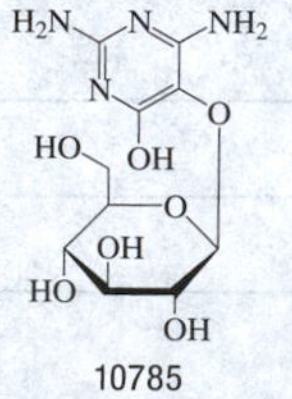
10785
Vicine

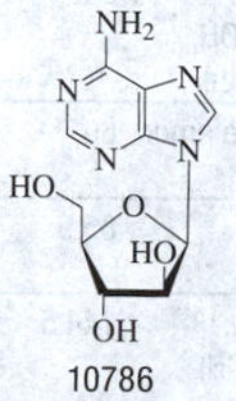
10786
Vidarabine

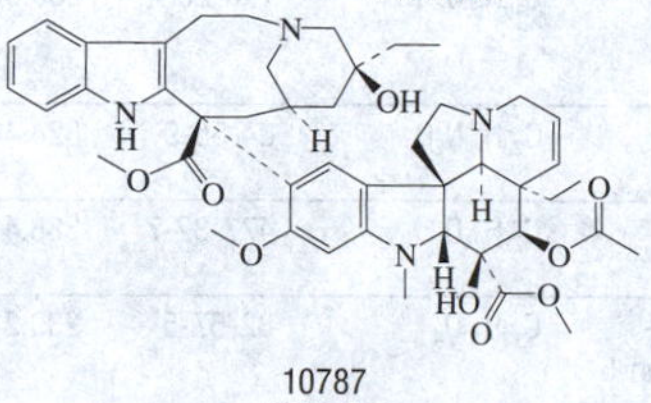
10787
Vinblastine

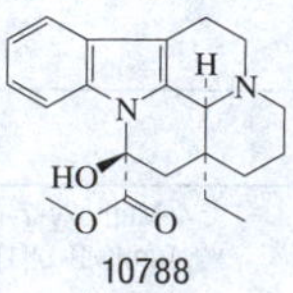
10788
Vincamine

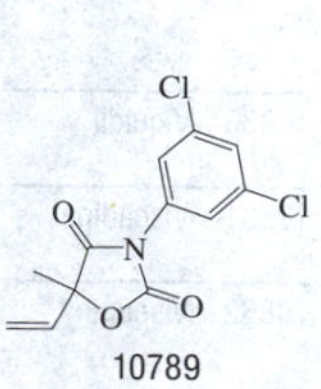
10789
Vinclozolin

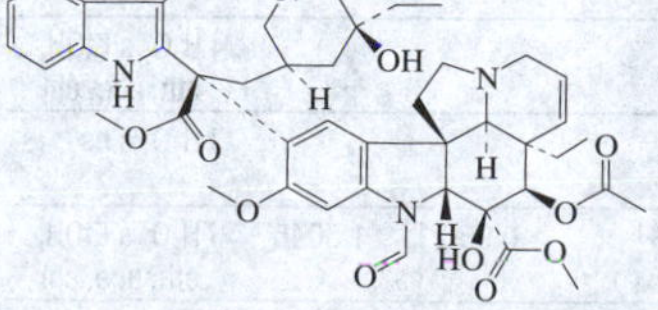
10790
Vincristine

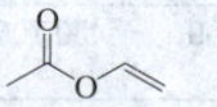
10791
Vinyl acetate

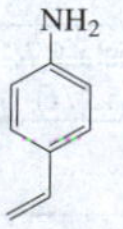
10792
4-Vinylaniline

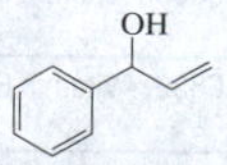
10793
α-Vinylbenzenemethanol

No.	Name	Synonym	Mol. form.	CAS Reg. No.	Mol. wt.	Physical form	mp/°C	bp/°C	den/ g cm^{-3}	n_D	Solubility
10794	Vinyl butanoate		$C_6H_{10}O_2$	123-20-6	114.142			116.7	0.9006^{20}		
10795	Vinyl *trans*-2-butenoate	Vinyl crotonate	$C_6H_8O_2$	3234-54-6	112.127						s ctc
10796	9-Vinyl-9*H*-carbazole		$C_{14}H_{11}N$	1484-13-5	193.244	cry (al)	66				i H_2O; sl EtOH; vs eth
10797	Vinylcyclohexane		C_8H_{14}	695-12-5	110.197			127(6)	0.8166^{19}	1.455^{19}	
10798	1-Vinylcyclohexene		C_8H_{12}	2622-21-1	108.181			144(7)	0.8623^{15}	1.4915^{20}	i H_2O; s eth, bz; vs MeOH
10799	4-Vinylcyclohexene		C_8H_{12}	100-40-3	108.181	liq	-108.9	130(4)	0.8299^{20}	1.4639^{20}	i H_2O; s eth, bz, peth
10800	Vinylcyclopentane		C_7H_{12}	3742-34-5	96.170	liq	-126.4(0.2)	99(3)	0.7834^{20}	1.4360^{20}	
10801	Vinyldiethoxymethylsilane		$C_7H_{16}O_2Si$	5507-44-8	160.287			133	0.8620^{20}	1.4001^{20}	
10802	Vinylethoxydimethylsilane		$C_6H_{14}OSi$	5356-83-2	130.260			99	0.790^{20}	1.3983^{20}	
10803	1-Vinyl-4-fluorobenzene		C_8H_7F	405-99-2	122.140		-34.5	67.4^{50}	1.0220^{20}	1.5150^{20}	i H_2O; s EtOH, eth, bz
10804	Vinyl formate		$C_3H_4O_2$	692-45-5	72.063	visc liq	-78	42(18)	0.965^{20}	1.3842^{20}	
10805	2-Vinylfuran		C_6H_6O	1487-18-9	94.111	liq	-94(4)	101(3)	0.9445^{19}	1.4992^{19}	
10806	1-Vinyl-2-methoxybenzene		$C_9H_{10}O$	612-15-7	134.174	nd	29	215(18)	1.0049^{17}	1.5388^{20}	vs ace, bz, eth, EtOH
10807	1-Vinyl-3-methoxybenzene		$C_9H_{10}O$	626-20-0	134.174			91^{15}	0.9919^{20}	1.5586^{23}	i H_2O; s EtOH, eth, bz
10808	1-Vinyl-4-methoxybenzene		$C_9H_{10}O$	637-69-4	134.174		2.0	208(19)	1.0001^{13}	1.5642^{13}	i H_2O; s EtOH, eth, bz; sl ctc
10809	6-Vinyl-6-methyl-1-isopropyl-3-(1-methylethylidene)-cyclohexene, (*S*)-		$C_{15}H_{24}$	5951-67-7	204.352			125^{8}	0.8782^{20}	1.5130^{26}	vs ace, bz
10810	1-Vinylnaphthalene		$C_{12}H_{10}$	826-74-4	154.207			124^{15}	1.0656^{20}	1.644^{20}	
10811	2-Vinylnaphthalene		$C_{12}H_{10}$	827-54-3	154.207		65(2)	135^{18}			i H_2O; s EtOH, ace, bz
10812	1-Vinyl-3-nitrobenzene		$C_8H_7NO_2$	586-39-0	149.148		-10	120^{11}	1.1552^{32}	1.5836^{20}	i H_2O; s EtOH, eth, bz, chl, lig, HOAc
10813	1-Vinyl-4-nitrobenzene		$C_8H_7NO_2$	100-13-0	149.148	pr (lig)	29	dec			vs EtOH, eth; s chl, HOAc, lig
10814	5-Vinyl-2-norbornene	5-Vinylbicyclo[2.2.1]hept-2-ene	C_9H_{12}	3048-64-4	120.191	liq	-80	140.7(0.5)	0.841	1.4810^{20}	
10815	Vinyl octadecanoate	Vinyl stearate	$C_{20}H_{38}O_2$	111-63-7	310.515		29	167^{2}	0.8517^{20}		sl chl
10816	3-Vinyl-7-oxabicyclo[4.1.0]-heptane		$C_8H_{12}O$	106-86-5	124.180		<-100	169	0.9581^{20}	1.4700^{20}	
10817	Vinyloxirane		C_4H_6O	930-22-3	70.090			68(2)	0.9006^{25}	1.4168^{20}	s EtOH, eth, bz
10818	2-(Vinyloxy)ethanol	Ethylene glycol monovinyl ether	$C_4H_8O_2$	764-48-7	88.106			139(4)	0.9821^{20}	1.4564^{17}	s H_2O, EtOH, eth, bz; i lig
10819	Vinyl propanoate	Vinyl propionate	$C_5H_8O_2$	105-38-4	100.117			94.8(0.2)			
10820	2-Vinylpyridine		C_7H_7N	100-69-6	105.138			159.5	0.9983^{20}	1.5495^{20}	sl H_2O; vs EtOH, eth, ace, chl
10821	3-Vinylpyridine		C_7H_7N	1121-55-7	105.138			162	0.9879^{20}	1.5530^{20}	sl H_2O; s EtOH, eth
10822	4-Vinylpyridine		C_7H_7N	100-43-6	105.138	red to dk-br		121^{150}	0.9879^{20}	1.5449^{20}	s H_2O, EtOH, chl; sl eth
10823	1-Vinyl-2-pyrrolidinone		C_6H_9NO	88-12-0	111.141		13.5	193^{400}	1.04^{20}		
10824	Vinylsilane		C_2H_6Si	7291-09-0	58.155	col gas	-171.6	-22.8			
10825	Vinyl sulfoxide	Divinyl sulfoxide	C_4H_6OS	1115-15-7	102.155	liq		86^{18}			
10826	Vinyltriacetoxysilane	Vinylsilanetriol, triacetate	$C_8H_{12}O_6Si$	4130-08-9	232.263			115^{10}	1.169^{20}	1.4226^{20}	
10827	Vinyltriethoxysilane		$C_8H_{18}O_3Si$	78-08-0	190.313			160.0(0.8)	0.901^{20}	1.3960^{25}	s chl
10828	Vinyltrimethylsilane		$C_5H_{12}Si$	754-05-2	100.235			55.3(0.2)	0.65^{20}	1.3914^{20}	i H_2O
10829	Violaxanthin		$C_{40}H_{56}O_4$	126-29-4	600.871	red pr (MeOH, al-eth)	208				s EtOH, eth, CS_2; i peth
10830	Viquidil		$C_{20}H_{24}N_2O_2$	84-55-9	324.417	red ye amor	60				vs eth, EtOH, chl
10831	Visnadine		$C_{21}H_{24}O_7$	477-32-7	388.412	nd	85.5				i H_2O; s EtOH, eth
10832	Visnagin	4-Methoxy-7-methyl-5*H*-furo[3,2-*g*][1]benzopyran-5-one	$C_{13}H_{10}O_4$	82-57-5	230.216	nd (w, MeOH)	144.5				sl H_2O, EtOH; vs chl
10833	Vitamin B12	Cyanocobalamin	$C_{63}H_{88}CoN_{14}O_{14}P$	68-19-9	1355.365		>300				
10834	Vitamin D2		$C_{28}H_{44}O$	50-14-6	396.648	pr (ace)	116.5	sub			i H_2O; s EtOH, eth, ace, chl
10835	Vitamin D3	9,10-Secocholesta-5,7,10(19)-trien-3-ol, (3β,5*Z*,7*E*)-	$C_{27}H_{44}O$	67-97-0	384.637		84.5				i H_2O; s os
10836	Vitamin E	α-Tocopherol	$C_{29}H_{50}O_2$	59-02-9	430.706	pale ye oil	3.0	$210^{0.1}$	0.950^{25}	1.5045^{25}	i H_2O; s EtOH, eth, ace, chl

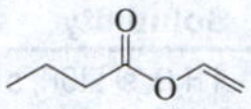
10794
Vinyl butanoate

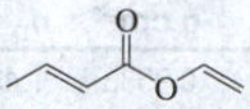
10795
Vinyl *trans*-2-butenoate

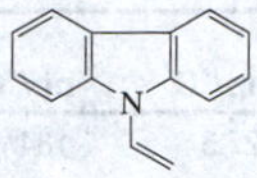
10796
9-Vinyl-9*H*-carbazole

10797
Vinylcyclohexane

10798
1-Vinylcyclohexene

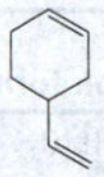
10799
4-Vinylcyclohexene

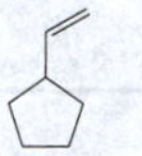
10800
Vinylcyclopentane

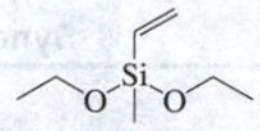
10801
Vinyldiethoxymethylsilane

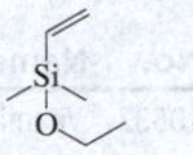
10802
Vinylethoxydimethylsilane

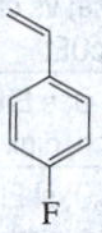
10803
1-Vinyl-4-fluorobenzene

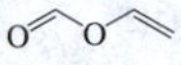
10804
Vinyl formate

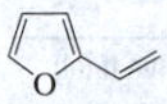
10805
2-Vinylfuran

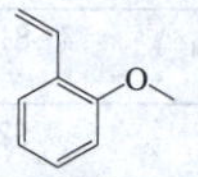
10806
1-Vinyl-2-methoxybenzene

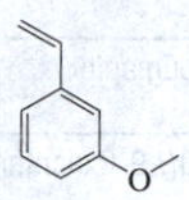
10807
1-Vinyl-3-methoxybenzene

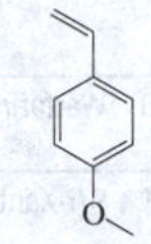
10808
1-Vinyl-4-methoxybenzene

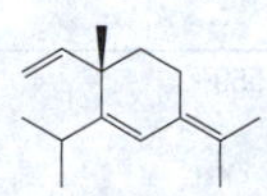
10809
6-Vinyl-6-methyl-1-isopropyl-3-(1-methylethylidene)cyclohexene, (*S*)-

10810
1-Vinylnaphthalene

10811
2-Vinylnaphthalene

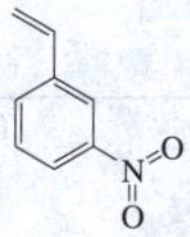
10812
1-Vinyl-3-nitrobenzene

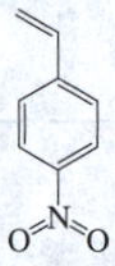
10813
1-Vinyl-4-nitrobenzene

10814
5-Vinyl-2-norbornene

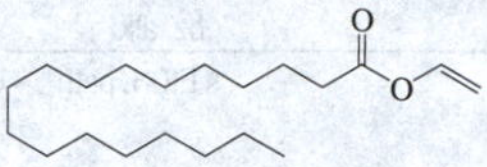
10815
Vinyl octadecanoate

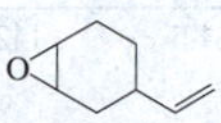
10816
3-Vinyl-7-oxabicyclo[4.1.0]heptane

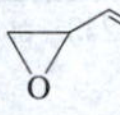
10817
Vinyloxirane

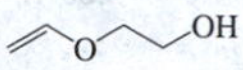
10818
2-(Vinyloxy)ethanol

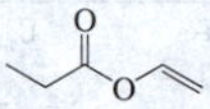
10819
Vinyl propanoate

10820
2-Vinylpyridine

10821
3-Vinylpyridine

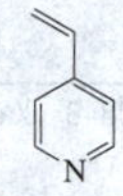
10822
4-Vinylpyridine

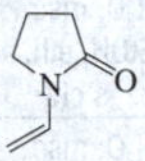
10823
1-Vinyl-2-pyrrolidinone

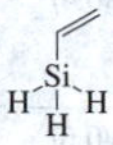
10824
Vinylsilane

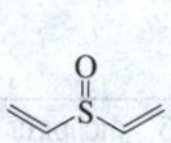
10825
Vinyl sulfoxide

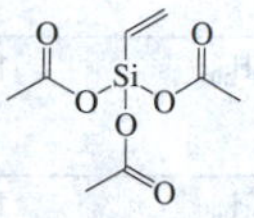
10826
Vinyltriacetoxysilane

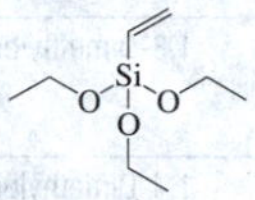
10827
Vinyltriethoxysilane

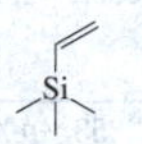
10828
Vinyltrimethylsilane

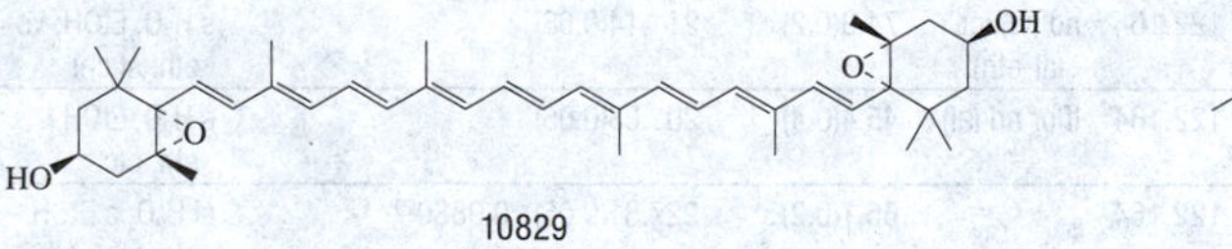
10829
Violaxanthin

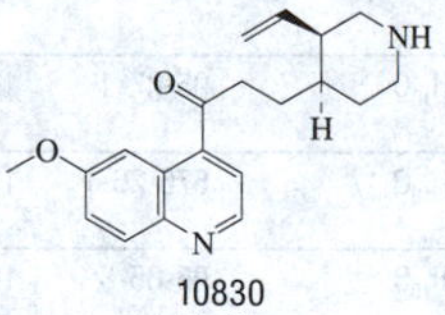
10830
Viquidil

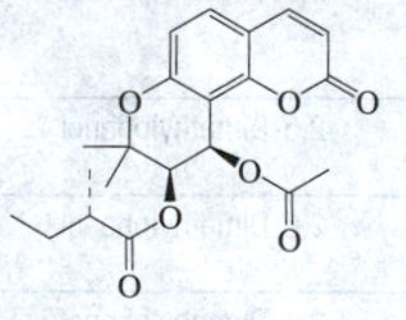
10831
Visnadine

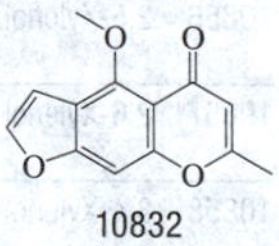
10832
Visnagin

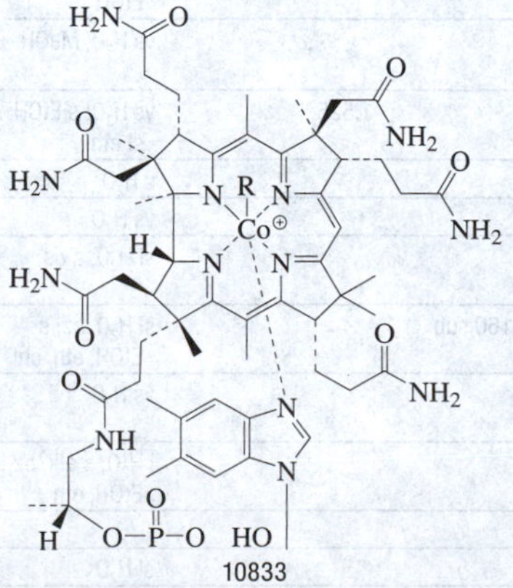
10833
Vitamin B12

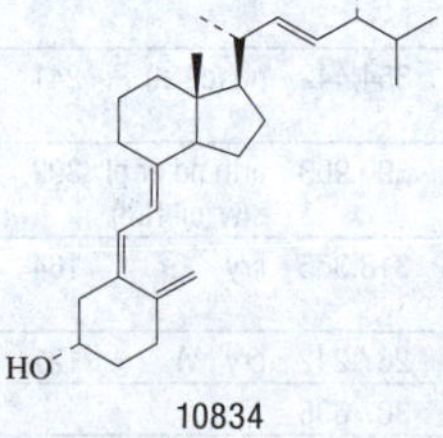
10834
Vitamin D2

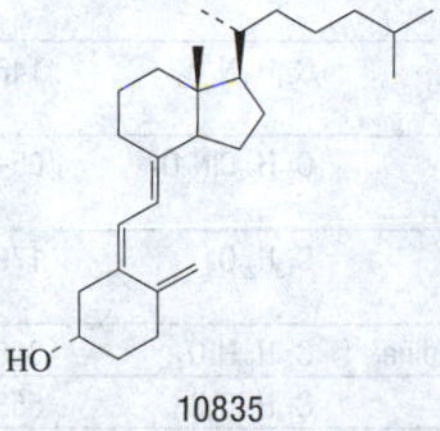
10835
Vitamin D3

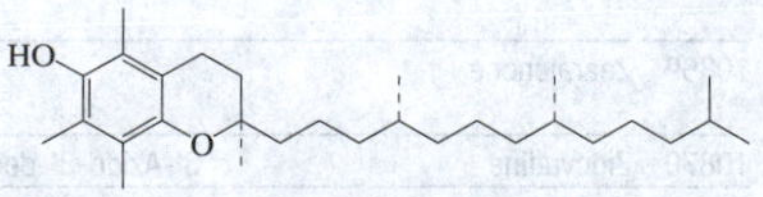
10836
Vitamin E

No.	Name	Synonym	Mol. form.	CAS Reg. No.	Mol. wt.	Physical form	mp/°C	bp/°C	den/ g cm^{-3}	n_D	Solubility
10837	Vitamin E acetate		$C_{31}H_{52}O_3$	58-95-7	472.743		-27.5	$184^{0.01}$	0.9533^{21}	1.497^{20}	i H_2O; sl EtOH; s eth, ace, chl
10838	Vitamin K1		$C_{31}H_{46}O_2$	84-80-0	450.696		-20	$142^{0.001}$	0.964^{25}	1.5250^{25}	i H_2O; s EtOH, eth, ace, bz, peth, chl
10839	Vomicine	4-Hydroxy-19-methyl-16,19-secostrychnidine-10,16-dione	$C_{22}H_{24}N_2O_4$	125-15-5	380.437	nd (80% al) pr (ace)	282				sl EtOH, eth, ace; vs chl; s AcOEt
10840	Warfarin	Coumadin	$C_{19}H_{16}O_4$	81-81-2	308.328	cry (al)	161				i H_2O; s EtOH, ace, diox
10841	9*H*-Xanthene	10*H*-9-Oxaanthracene	$C_{13}H_{10}O$	92-83-1	182.217	ye lf (al)	101(1)	311(2)			i H_2O; sl EtOH, ctc; s eth, bz, chl
10842	9*H*-Xanthen-9-ol		$C_{13}H_{10}O_2$	90-46-0	198.217	nd (aq al)	125				sl H_2O; s EtOH, eth, chl
10843	Xanthine		$C_5H_4N_4O_2$	69-89-6	152.112	ye pl (w)	dec	sub			i H_2O
10844	Xanthone		$C_{13}H_8O_2$	90-47-1	196.202	nd (al)	176.4(0.4)	351			i H_2O; s EtOH, eth, bz, chl; sl peth
10845	Xanthopterin		$C_6H_5N_5O_2$	119-44-8	179.137	hyg ye amor or oran pow (HOAc)	>410 dec	99^{18}	1.559^{25}		i H_2O; sl EtOH, eth; vs acid, alk
10846	Xanthosine		$C_{10}H_{12}N_4O_6$	146-80-5	284.225	pr cry (w)					sl cold H_2O; vs hot H_2O; dec acid
10847	Xanthoxyletin		$C_{15}H_{14}O_4$	84-99-1	258.270	pr (MeOH, peth)	133				i H_2O; s EtOH, ace; sl eth; vs bz, alk
10848	Xanthyletin	8,8-Dimethyl-2*H*,8*H*-benzo[1,2-*b*:5,4-*b*′]dipyran-2-one	$C_{14}H_{12}O_3$	553-19-5	228.243	pr (MeOH)	131.5	$142^{0.1}$			s EtOH, peth
10849	*p*-Xenylcarbimide	4-Isocyanato-1,1′-biphenyl	$C_{13}H_9NO$	92-95-5	195.216	nd	56	283 dec			vs eth
10850	Xibenolol		$C_{15}H_{25}NO_2$	81584-06-7	251.366	cry	57	$135^{0.7}$			s EtOH
10851	*o*-Xylene	1,2-Dimethylbenzene	C_8H_{10}	95-47-6	106.165	liq	-25.16(0.02)	144.4(0.4)	0.8755^{25}	1.5018^{25}	i H_2O; msc EtOH, eth, ace, bz, peth, ctc
10852	*m*-Xylene	1,3-Dimethylbenzene	C_8H_{10}	108-38-3	106.165	liq	-47.85(0.03)	139.1(0.4)	0.8598^{25}	1.4944^{25}	i H_2O; msc EtOH, eth, ace, bz; s chl
10853	*p*-Xylene	1,4-Dimethylbenzene	C_8H_{10}	106-42-3	106.165	mcl pr or liq	13.3(0.1)	138.3(0.5)	0.8565^{25}	1.4929^{25}	i H_2O; msc EtOH, eth, ace, bz; s chl
10854	2,3-Xylenol	2,3-Dimethylphenol	$C_8H_{10}O$	526-75-0	122.164	nd (w, dil al)	72.7(0.3)	216.88(0.05)		1.5420^{20}	sl H_2O; s EtOH, eth
10855	2,4-Xylenol	2,4-Dimethylphenol	$C_8H_{10}O$	105-67-9	122.164	nd (w)	25(1)	210.94(0.03)	0.9650^{20}	1.5420^{14}	sl H_2O; msc EtOH, eth; s ctc
10856	2,5-Xylenol	2,5-Dimethylphenol	$C_8H_{10}O$	95-87-4	122.164	nd (w), pr (al-eth)	74.9(0.2)	211.14(0.08)			s H_2O, EtOH; vs eth; sl chl
10857	2,6-Xylenol	2,6-Dimethylphenol	$C_8H_{10}O$	576-26-1	122.164	lf or nd (al)	45.4(0.4)	201.03(0.05)			s H_2O, EtOH, eth, ctc
10858	3,4-Xylenol	3,4-Dimethylphenol	$C_8H_{10}O$	95-65-8	122.164		65.1(0.2)	227.31(0.05)	0.9830^{20}		sl H_2O; s EtOH, ctc; msc eth
10859	3,5-Xylenol	3,5-Dimethylphenol	$C_8H_{10}O$	108-68-9	122.164	nd (w, peth)	63.4(0.3)	221.71(0.05)	0.9680^{20}		s H_2O, EtOH, ctc
10860	Xylenol orange		$C_{31}H_{32}N_2O_{13}S$	1611-35-4	672.656	dk red cry	286 dec				s H_2O
10861	Xylitol	Xylite	$C_5H_{12}O_5$	87-99-0	152.146	mcl (al)	95.9(0.9)	380(15)			vs H_2O, py, EtOH
10862	6-*O*-β-D-Xylopyranosyl-D-glucose	Primeverose	$C_{11}H_{20}O_{10}$	26531-85-1	312.271	cry (MeOH)	210				vs H_2O, MeOH
10863	D-Xylose		$C_5H_{10}O_5$	58-86-6	150.130	mcl nd	90.5		1.525^{20}		vs H_2O; s EtOH; sl eth
10864	D-Xylulose	D-*threo*-2-Pentulose	$C_5H_{10}O_5$	551-84-8	150.130	visc liq					s H_2O
10865	L-Xylulose	L-*threo*-2-Pentulose	$C_5H_{10}O_5$	527-50-4	150.130	syrup					vs H_2O
10866	3,5-Xylyl methylcarbamate	3,5-Dimethylphenyl methylcarbamate	$C_{10}H_{13}NO_2$	2655-14-3	179.216	cry	99				sl H_2O; s os
10867	Yohimbine		$C_{21}H_{26}N_2O_3$	146-48-5	354.442	nd (dil al)	241	160 sub			sl H_2O, bz; s EtOH, eth, chl
10868	Yohimbine hydrochloride	Tosanpin	$C_{21}H_{27}ClN_2O_3$	65-19-0	390.903	orth nd or pl (w, dil HCl)	302				vs H_2O
10869	Zearalenone		$C_{18}H_{22}O_5$	17924-92-4	318.365	cry	164				i H_2O; s alk, bz, EtOH, eth
10870	Zidovudine	3′-Azido-3′-deoxythymidine	$C_{10}H_{13}N_5O_4$	30516-87-1	267.242	cry (w)	121				
10871	Zinc benzoate		$C_{14}H_{10}O_4Zn$	553-72-0	307.636						sl H_2O

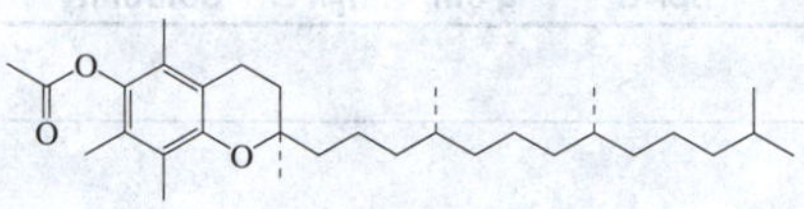
10837
Vitamin E acetate

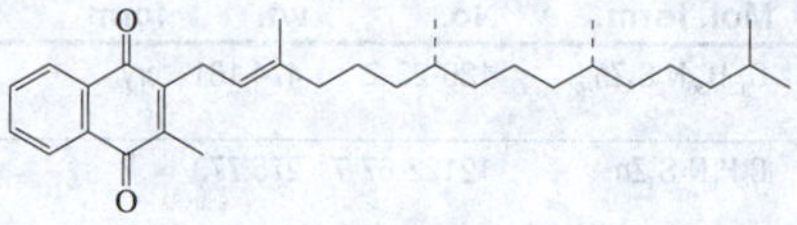
10838
Vitamin K1

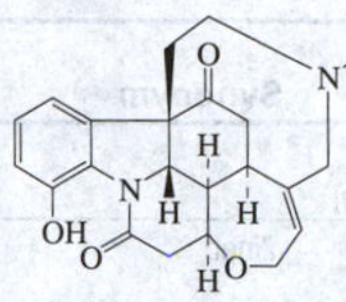
10839
Vomicine

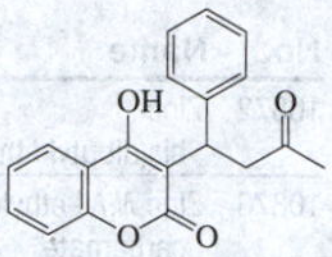
10840
Warfarin

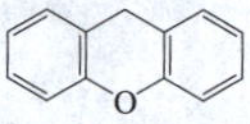
10841
9*H*-Xanthene

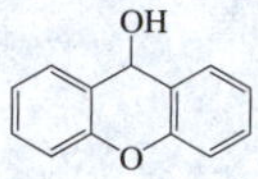
10842
9*H*-Xanthen-9-ol

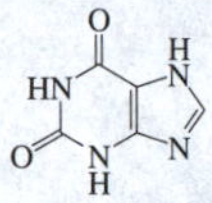
10843
Xanthine

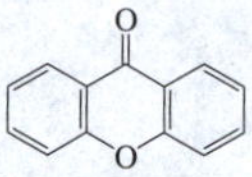
10844
Xanthone

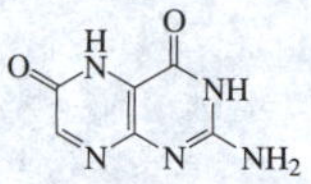
10845
Xanthopterin

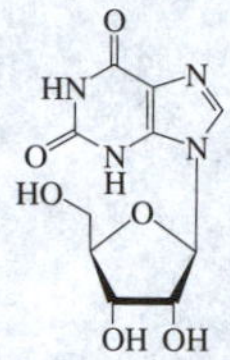
10846
Xanthosine

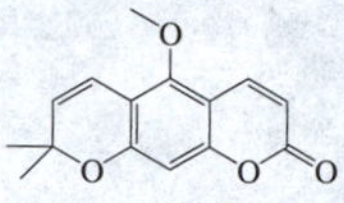
10847
Xanthoxyletin

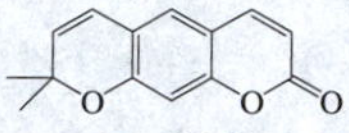
10848
Xanthyletin

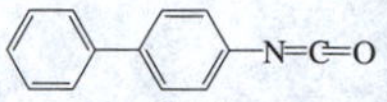
10849
p-Xenylcarbimide

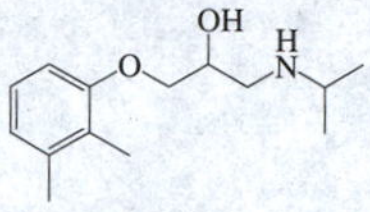
10850
Xibenolol

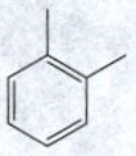
10851
o-Xylene

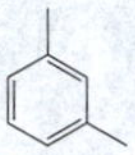
10852
m-Xylene

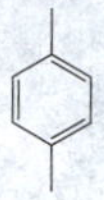
10853
p-Xylene

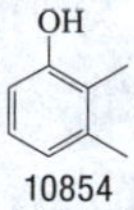
10854
2,3-Xylenol

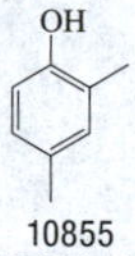
10855
2,4-Xylenol

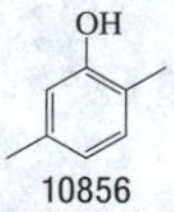
10856
2,5-Xylenol

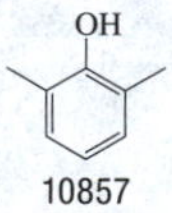
10857
2,6-Xylenol

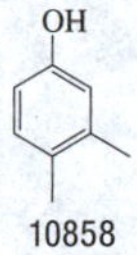
10858
3,4-Xylenol

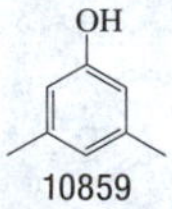
10859
3,5-Xylenol

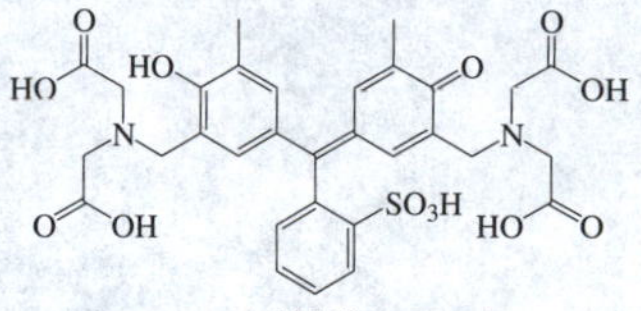
10860
Xylenol orange

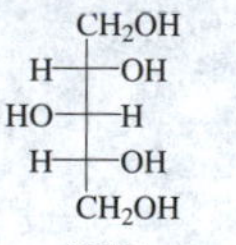
10861
Xylitol

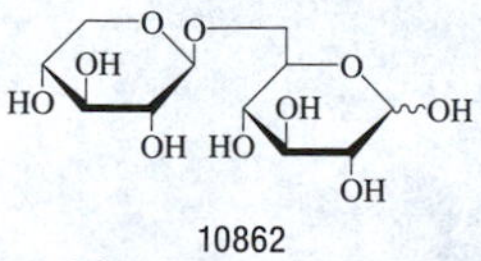
10862
6-*O*-β-D-Xylopyranosyl-D-glucose

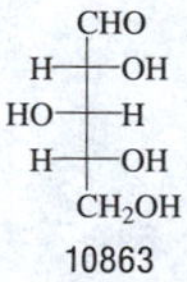
10863
D-Xylose

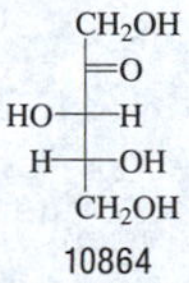
10864
D-Xylulose

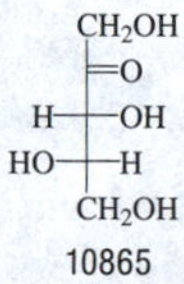
10865
L-Xylulose

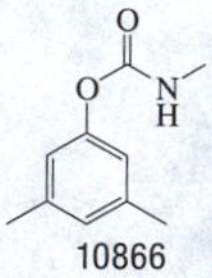
10866
3,5-Xylyl methylcarbamate

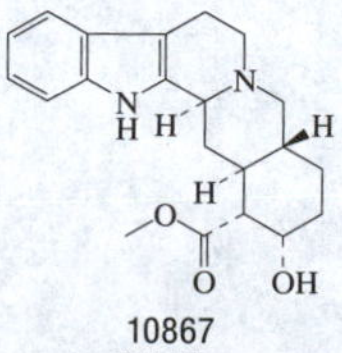
10867
Yohimbine

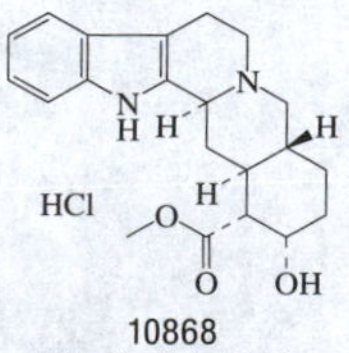
10868
Yohimbine hydrochloride

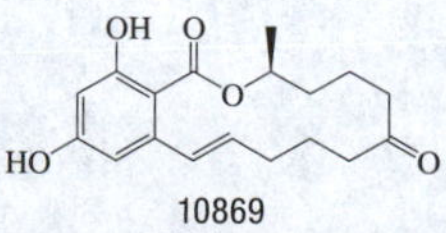
10869
Zearalenone

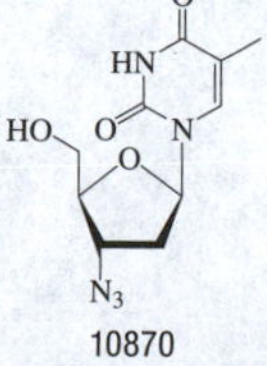
10870
Zidovudine

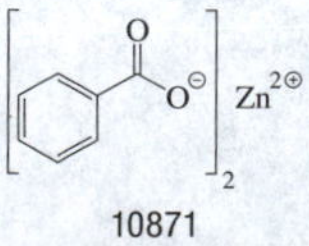
10871
Zinc benzoate

Organic

No.	Name	Synonym	Mol. form.	CAS Reg. No.	Mol. wt.	Physical form	mp/°C	bp/°C	den/ g cm⁻³	n_D	Solubility
10872	Zinc bis(dibutyldithiocarbamate)		$C_{18}H_{36}N_2S_4Zn$	136-23-2	474.161	cry	138				
10873	Zinc *N,N'*-ethylenebisdithiocarbamate	Zineb	$C_4H_6N_2S_4Zn$	12122-67-7	275.773		157 dec				
10874	Zinc gluconate		$C_{12}H_{22}O_{14}Zn$	4468-02-4	455.704	pow					
10875	Zinc 2,4-pentanedioate	Zinc acetylacetonate	$C_{10}H_{14}O_4Zn$	14024-63-6	263.625	cry	127.3(0.2)				sl H_2O; s EtOH, DMSO
10876	Zinc propanoate		$C_6H_{10}O_4Zn$	557-28-8	211.550	hyg pl or nd					sl EtOH
10877	Ziram	Zinc, bis(dimethylcarbamodithioato-*S,S'*)-, (T-4)-	$C_6H_{12}N_2S_4Zn$	137-30-4	305.841	cry	250		1.66^{25}		i H_2O; sl bz; s chl

Organic

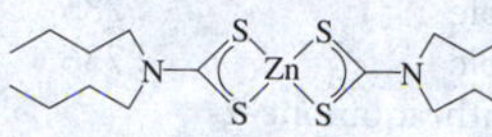

10872
Zinc bis(dibutyldithiocarbamate)

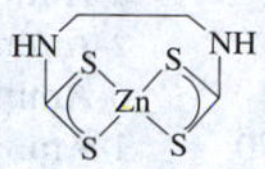

10873
Zinc *N,N'*-ethylenebisdithiocarbamate

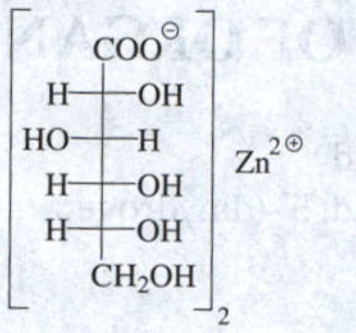

10874
Zinc gluconate

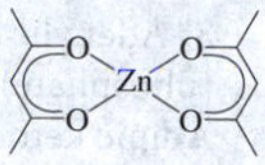

10875
Zinc 2,4-pentanedioate

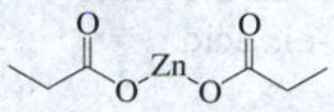

10876
Zinc propanoate

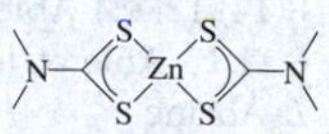

10877
Ziram

SYNONYM INDEX OF ORGANIC COMPOUNDS

Organic

1,2-Benzisothiazolin-3-one, 1,1-dioxide, sodium salt : 9497
Benzo[a]chrysene : 9035
1,2-Benzodiazine : 2403
1,3-Benzodiazine : 9413
1,4-Benzodiazine : 9444
2,3-Benzodiazine : 9028
1,3-Benzodioxol-4-ol, 2,2-dimethyl-, methylcarbamate : 578
2-[2-(1,3-Benzodioxol-5-yl)ethyl]-4-methoxyquinoline : 2498
Benzoguanamine : 9008
Benzohydrol : 4502
Benzoic acid amide : 591
Benzoic acid, bromide : 760
Benzoic acid, chloride : 761
Benzoic acid, 2-[[4-(dimethylamino)-phenyl]azo]- : 7777
Benzoic acid, 2-[[4-(dipropylamino)-phenyl]azo]- : 9264
Benzoic acid, fluoride : 765
Benzoic acid, hydrazide : 711
Benzoic acid, iodide : 767
Benzo[b]naphthacene : 8530
Benzo[rst]pentaphene : 2887
Benzo[a]phenanthrene : 2387
Benzo[l]phenanthrene : 10648
Benzo[def]phenanthrene : 9308
Benzopinacol : 9988
1*H*-Benzopyrazole : 6233
2,3-Benzopyrene : 729
Benzo[c]pyridine : 6551
2,3-Benzopyrrole : 6243
Benzoresorcinol : 3743
Benzosulfonazole : 737
(2-Benzothiazolylthio)methyl thiocyanate : 1390
4-(2-Benzothiazolylthio)morpholine : 7906
Benzotrichloride : 10253
Benzotrifluoride : 10404
Benzo[a]triphenylene : 687
2,2′-Benzoxazoline : 5597
Benzoxiquine : 9440
Benzoylacetonitrile : 8455
Benzoyl *tert*-butyl peroxide : 1638
2-Benzoyl-4-chloroaniline : 287
Benzoyl disulfide : 2891
2-(Benzoyloxy)benzoic acid, methyl ester : 7039
Benzoylpas : 754
1,12-Benzperylene : 716
1,2-Benzpyrene : 730
N-Benzylbenzenemethanamine : 2892
Benzyl bromide : 1247
Benzyl-*tert*-butanol : 3960
Benzyl carbazate : 1737
Benzylcarbinol isobutyrate : 8878
Benzyl chloride : 2106
Benzyl cyanide : 601
N-Benzyl-*N′*,*N′*-dimethyl-1,2-ethanediamine : 3972
Benzyl ether : 2895
Benzylideneaniline : 8920
Benzylidene hydrazine : 586
Benzylidenemethylamine : 8927
Benzyl isobutyrate : 810
Benzyl isocyanide : 6424
Benzyl laurate : 795
Benzyl nicotinate : 828
3-(Benzyloxy)-*L*-alanine : 832
N-(Benzyloxycarbonyl)-*L*-proline : 830
Benzylpenicillinic acid : 8522
o-Benzylphenol : 819
p-Benzylphenol : 820
Benzyl *trans*-3-phenyl-2-propenoate : 794
*N*4-Benzylsulfanilamide : 782
Benzyl sulfide : 2899
p-(Benzylsulfonamido)benzoic acid : 834
Berbamine : 10486
Bergaptene : 6899
Beryllium acetylacetonate : 844
Besylic acid : 662
Betulin : 6635
Biallyl : 5779
Bicetyl : 4663
Bicine : 1001
Bicyclo[5.3.0]decapentaene : 568
Bicyclo[2.2.1]hepta-2,5-diene : 8263
1,1′-Binaphthyl : 871
2,3,1′,8′-Binaphthylene : 702
p-Biphenylamine : 265
[1,1′-Biphenyl]-4,4′-diamine : 680
[1,1′-Biphenyl]-4,4′-diamine, 2,2′-dichloro- : 3106
[1,1′-Biphenyl]-4,4′-diamine, 3,3′-dichloro- : 3107
2,2′-Biphenylene oxide : 2884
[1,1′-Biphenyl]-2-ol : 6003
[1,1′-Biphenyl]-3-ol : 6004
[1,1′-Biphenyl]-4-ol : 6005
Bipropargyl : 5787
Bipropionyl : 5841
2,3′-Bipyridyl : 894
2,4′-Bipyridyl : 895
3,3′-Bipyridyl : 896
N,*N′*-Bis(acetoacetyl)-3,3′-dimethylbenzidine : 899
Bis(acetyloxy)dimethylsilane : 4040
2,2-Bis[(acetyloxy)methyl]-1,3-propanediol diacetate : 8572
Bis[(aminocarbonyl)amino]acetic acid : 156
Bis(2-aminoethyl) sulfide : 10047
Bis(3-aminopropyl)amine : 413
1,4-Bis(3-aminopropyl)piperazine : 9054
Bis(4-bromophenyl) ketone : 2911
Bis(2-butoxyethyl) ether : 3439
Bis(2-butoxyethyl) phthalate : 625
Bis(5-*tert*-butyl-4-hydroxy-2-methylphenyl) sulfide : 10046
N,*N*-Bis(carboxymethyl)glycine : 8030
Bis(4-chlorophenyl) ketone : 3115
3,6-Bis(2-chlorophenyl)-1,2,4,5-tetrazine : 2425
Bis(2-cyanoethyl)amine : 6223
Bis(η5-2,4-cyclopentadien-1-yl)nickel : 8018
Bis(3,5-di-*tert*-butyl-4-hydroxyphenyl)methane : 7233
Bis(dibutylthiocarbamoyl) disulfide : 9729
4,4′-Bis(dimethylamino)benzhydrol : 966
4,4′-Bis(dimethylamino)-thiobenzophenone : 965
2,6-Bis(1,1-dimethylpropyl)-4-methylphenol : 4442
1,4-Bis(2,3-epoxypropoxy)butane : 1411
Bis(2,3-epoxypropyl) ether : 3578
Bis(2-ethoxyethyl) ether : 3440
Bis(ethylacetoacetato)copper : 2461
Bis(2-ethylhexyl) phosphite : 984
2,2-Bis(ethylsulfonyl)propane : 9660
Bis(2-hydroxyethyl)amine : 3348
Bis(2-hydroxyethyl) 1,4-benzenedicarboxylate : 1006
Bis(hydroxymethyl)acetylene : 1680
3,3-Bis(4-hydroxyphenyl)-1(3*H*)-isobenzofuranone : 8763
Bis(2-hydroxyphenyl) ketone : 3710
Bis(4-hydroxyphenyl) ketone : 3711
2-[Bis(4-hydroxyphenyl)methyl]-benzoic acid : 8764
Bis(isopropylamido)fluorophosphate : 7877
Bis(2-methoxyethyl) phthalate : 626
Bis(8-methylnonyl)phthalate : 3785
Bis(7-methyloctyl)phthalate : 3786
Bis(4-methylphenyl) ketone : 3968
Bis(4-nitrophenyl) ether : 4383
Bis(4-nitrophenyl) sulfide : 4384
Bis(pentabromophenyl) ether : 2738
Bisphenol A : 1015
Bisphenol AD : 1014
Bisphenol A dimethacrylate : 1016
Bisphenol B : 1013
Bisphenol C : 1010
Bisphenol S : 1017
N,*N*-Bis(phenylmethyl)-benzenemethanamine : 10151
N,*N′*-Bis(salicylidene)-ethylenediaminocobalt(II) : 9500
1,2-Bis(*p*-tolyl)ethane : 4595
Bis(tributyltin) oxide : 5721
Bitoscanate : 3815
Biuret : 6222
Borane pyridine : 10432
Boric acid, triethyl ester : 10352
Boric acid, tripropyl ester : 10670
Bornyl acetate : 10517
d-Bornyl isovalerate : 1068
Brassidic acid : 4617
Brilliant Oil Yellow : 3905
Bromal : 10152
Bromcresol Green : 1156
Bromcresol Purple : 1157
Bromex : 913
Bromindione : 1322
Bromine cyanide : 2513
Bromisovalum : 279
p-Bromoacetanilide : 1318
ω-Bromoacetophenone : 1081
p-Bromoacetophenone : 1320
5-Bromoacetylsalicylic acid : 88
5-Bromoanthranilic acid : 266
p-Bromobenzyl bromide : 1117
p-Bromobenzyl chloride : 1148
α-Bromobenzyl cyanide : 1099
5-Bromo-3-*sec*-butyl-6-methyluracil : 1074
DL-α-Bromobutyric acid : 1125
α-Bromo-γ-butyrolactone : 1174
N′-(4-Bromo-3-chlorophenyl)-*N*-methoxy-*N*-methylurea : 1821
2-Bromoethanamine hydrobromide : 1200
2-Bromo-2-ethylbutanamide : 3408
2-Bromoethyl ethyl ether : 1198
Bromoform : 10167
α-Bromoisobutyric acid : 1281
β-Bromoisovaleric acid : 1254
p-Bromomandelic acid : 1231
2-(Bromomethyl)-1*H*-isoindole-1,3(2*H*)-dione : 1278
Bromomethylmagnesium : 7432
1-Bromo-β-naphthol : 1290
p-Bromophenacyl bromide : 1119

Organic

Organic

trans-1,2-Diaminocyclohexane : 2572
4,4'-Diaminodiphenylamine : 395
2,4'-Diaminodiphenylmethane : 7241
2,7-Diaminofluorene : 5345
2,6-Diamino-5-(β-*D*-glucopyranosyloxy)-4(1*H*)-pyrimidinone : 10785
2,6-Diaminohexanoic acid, (±) : 6646
2,6-Diaminohexanoic acid, (*D*) : 6647
2,6-Diaminohexanoic acid, (*L*) : 6648
1,5-Diaminonaphthalene : 7941
1,8-Diaminonaphthalene : 7942
2,3-Diaminonaphthalene : 7943
2,5-Diaminopentanoic acid, (*S*) : 8416
2,3-Diaminophenazine : 8751
2,6-Diamino-3-phenylazopyridine : 8753
2,6-Diaminopimelic acid : 2858
1,3-Diaminopropane : 9138
2,3-Diaminopropionic acid : 235
2,6-Diaminopurine : 9294
2,5-Diaminopyridine : 9335
1,3-Diamino-2-thiourea : 1751
Diamylamine : 4439
1,4:3,6-Dianhydroglucitol : 6554
1,2-Dianilinoethane : 4482
Dianisidine : 3847
1,2-Diazabenzene : 9312
1,5-Diazanaphthalene : 7994
1,3-Diazine : 9382
1,4-Diazine : 9302
Diazoacetic ester : 4973
Diazoaminobenzene : 4533
1,2-Diazole : 9306
1,3-Diazole : 6215
Dibenamine : 2037
Dibenamine hydrochloride : 2038
Dibenzalacetone : 4507
1,2:5,6-Dibenzanthracene : 2876
Dibenz[de,kl]anthracene : 8736
Dibenzo[b,def]chrysene : 2886
Dibenzo[def,p]chrysene : 2888
Dibenzo[a,jk]fluorene : 701
Dibenzopyrazine : 8750
Dibenzo[b,e]pyridine : 113
Dibenzopyrolle : 1730
Dibenzothiapyran : 10082
Dibenzoylmethane : 4515
2,3:6,7-Dibenzphenanthrene : 8633
Dibenzyl : 4480
Dibenzyl ketone : 4518
Dibromantine : 2934
1,3-Dibromoacetone : 2982
1,2-Dibromobenzene : 2908
1,3-Dibromobenzene : 2909
1,4-Dibromobenzene : 2910
4,4'-Dibromobenzilic acid isopropyl ester : 1343
2,6-Dibromo-4-(chloroimino)-2,5-cyclohexadien-1-one : 2988
1,2-Dibromo-2,2-dichloroethylphosphoric acid, dimethyl ester : 7918
1,2-Dibromo-2,4-dicyanobutane : 1118
cis-1,2-Dibromoethylene : 2939
trans-1,2-Dibromoethylene : 2940
Dibromogallic acid : 2995
3,5-Dibromosalicylaldehyde : 2953
3,5-Dibromosalicylic acid : 2954
2,5-Dibromotoluene : 2957
Dibutylacetylene : 2789
Dibutylbutadiyne : 4618
N,N-Dibutyl-1-butanamine : 10181
Dibutylcarbinol : 8242
Dibutyldichlorostannane : 3059
N,N'-Di-*tert*-butylethanediamine : 3022
Dibutyl ketone : 8246
Di-*tert*-butylmethane : 9962
Dibutylnitrosamine : 8184
Dibutyltin bis(dodecyl sulfide) : 3009
Dicamba : 3206
3,3'-Dicarboxybenzidine : 679
2,3-Dicarboxypyrazine : 9305
2,5-Dicarboxythiophene : 10073
Dichlobenil : 3114
Dichlone : 3231
Dichloramine-T : 3208
2,6-Dichloroanisole : 3204
1,2-Dichlorobenzene : 3092
1,3-Dichlorobenzene : 3093
1,4-Dichlorobenzene : 3094
2,4-Dichlorobenzotrifluoride : 3309
2,4-Dichlorobenzyl alcohol : 3103
3,3'-Dichloro-[1,1'-biphenyl]-4,4'-diamine : 3108
2,4-Dichloro-6-(*o*-chloroanilino)-*s*-triazine : 464
Dichlorocyanuric acid : 3304
3,5-Dichloro-*N*-(3,4-dichlorophenyl)-2-hydroxybenzamide : 9773
1,1-Dichloro-2,2-difluoroethylene : 3159
7,16-Dichloro-6,15-dihydro-5,9,14,18-anthrazinetetrone : 2419
2,5-Dichloro-3,6-dihydroxy-2,5-cyclohexadiene-1,4-dione : 1818
1,4-Dichloro-2,5-dimethoxybenzene : 2172
2,4'-Dichlorodiphenyl ketone : 2236
Dichlorodiphenyltrichloroethane (DDT) : 10212
Dichloroethylarsine : 4980
cis-1,2-Dichloroethylene : 3177
trans-1,2-Dichloroethylene : 3178
Dichloroethyl ether : 929
2',7'-Dichloro-3,6-fluorandiol : 3183
1,1-Dichloro-2-fluoroethylene : 3186
2,2'-Dichlorohydrazobenzene : 945
1,2-Dichloroisobutane : 3217
1,2-Dichloro-5-isocyanatobenzene : 3265
1,3-Dichloro-5-isocyanatobenzene : 3266
(Dichloromethyl)dimethylchlorosilane : 1973
(Dichloromethyl)trichlorosilane : 10220
2,4-Dichloro-α-naphthol : 3232
2,4-Dichloro-1-(4-nitrophenoxy)-benzene : 8099
2,4-Dichlorophenoxyethyl benzoate : 9533
Di(4-chlorophenoxy)methane : 941
N'-(3,4-Dichlorophenyl)-*N*-methoxy-*N*-methylurea : 6617
Dichlorophenylphosphine : 8952
Dichlorophenylphosphine sulfide : 8951
Di(*p*-chlorophenyl)thiourea : 948
2-(3,5-Dichlorophenyl)-2-(2,2,2-trichloroethyl)oxirane, (±) : 10328
4,5-Dichlorophthalic anhydride : 3201
2,2-Dichloropropionic acid : 3274
cis-1,3-Dichloropropylene : 3282
trans-1,3-Dichloropropylene : 3283
3,6-Dichloro-2-pyridinecarboxylic acid : 2433
3,7-Dichloroquinoline-8-carboxylic acid : 9414
1,2-Dichloro-4-(trifluoromethyl)-benzene : 3116
Dichloroxylenol : 3168
Dichlorphenamide : 3101
Dichlorprop : 3258
Dicobalt octacarbonyl : 2437
Dicryl : 3267
Dictamnine : 6831
Dicumene : 4046
Dicumyl peroxide : 1027
trans-1,4-Dicyano-2-butene : 5867
2,4-Dicyano-1-butene : 7247
Dicyanodiamide : 2517
Dicyclopentadienyl iron : 5326
Dicycloverine hydrochloride : 3331
Didanosine : 3339
N,N-Didecyl-1-decanamine : 10321
N,N-Didodecyl-1-dodecanamine : 10329
Didodecyl thiobispropanoate : 10048
Dienochlor : 8529
Diepoxybutane : 875
Diethadione : 3428
Diethanol-*m*-toluidine : 1003
Diethazine : 3503
3,5-Diethoxycarbonyl-1,4-dihydrocollidine : 3429
1-[(3,4-Diethoxyphenyl)methyl]-6,7-diethoxyisoquinoline : 4804
Diethoxyphosphoryl chloride : 3420
Diethylacetaldehyde : 4922
Diethylacetic acid : 4924
Diethyl 1,3-acetonedicarboxylate : 3497
Diethylacetylene : 5912
2-(Diethylamino)ethyl benzilate : 575
2-Diethylaminoethyl benzilate hydrochloride : 576
3-[2-(Diethylamino)ethyl]-3-phenyl-2(3*H*)-benzofuranone : 443
2-(4-Diethylaminophenylazo)benzoic acid : 5242
3-(Diethylamino)-*N*-phenylbutanamide : 8290
2-Diethylamino-2',4',6'-trimethylacetanilide : 10468
(Diethylamino)trimethylsilane : 3538
Diethyl azelate : 3494
5,5-Diethylbarbituric acid : 571
Diethyl benzalmalonate : 3405
1,2-Diethylbenzene : 3401
1,3-Diethylbenzene : 3402
1,4-Diethylbenzene : 3403
N,N-Diethylbenzhydrylamine : 3504
Diethyl bicarbamate : 3466
Diethylcarbamazine : 3489
Diethyl carbinol : 8628
N,N-Diethylcyclohexanamine : 2618
4-*N,N*-Diethyl-1,4-diamino-2-methylbenzene, hydrochloride : 3484
Diethyldiazine 1-oxide : 567
2,2'-Diethyldihexyl ether : 982
Diethyldimethylplumbane : 3430
Diethyldithiocarbamate sodium salt trihydrate : 9552
Diethylenediamine : 9049
Diethylenediamine dihydrochloride : 9052
Diethylene glycol bis(allyl carbonate) : 2831
Diethylene glycol diaminopropyl ether : 10640
Diethylene glycol divinyl ether : 1056
Diethylene glycol ethyl ether acrylate : 4840
Diethylenetriamine : 903

Organic

Organic

Organic

Organic

Organic

Organic

Organic

Organic

Organic

DIAMAGNETIC SUSCEPTIBILITY OF SELECTED ORGANIC COMPOUNDS

When a material is placed in a magnetic field $\boldsymbol{H}$, a magnetization $\boldsymbol{M}$ is induced in the material which is related to $\boldsymbol{H}$ by $\boldsymbol{M} = \kappa \boldsymbol{H}$, where κ is called the volume susceptibility. Since $\boldsymbol{H}$ and $\boldsymbol{M}$ have the same dimensions, κ is dimensionless. A more useful parameter is the molar susceptibility χ_m, defined by

$$\chi_m = \kappa V_m = \kappa M/\rho$$

where V_m is the molar volume of the substance, M the molar mass, and ρ the mass density. When the cgs system is used, the customary unit for χ_m is $cm^3\ mol^{-1}$; the corresponding SI base unit is $m^3\ mol^{-1}$. Substances with no unpaired electrons are called diamagnetic; they have negative values of χ_m.

This table gives values of the diamagnetic susceptibility for about 400 common organic compounds. Values are given for both the molar susceptibility χ_m and the volume susceptibility κ. Note that the tabulated numbers are the negatives of the susceptibility values. Most values of the volume susceptibility κ refer to solids or liquids at room temperature; an asterisk * following the value indicates a liquefied gas. A more extensive table may be found in Ref. 1.

In keeping with customary practice, the molar susceptibility is given here in units appropriate to the cgs system. These values should be multiplied by 4π to obtain values for use in SI equations (where the magnetic field strength $\boldsymbol{H}$ has units of $A\ m^{-1}$).

References

1. *Landolt-Börnstein, Numerical Data and Functional Relationships in Science and Technology, New Series*, II/16, *Diamagnetic Susceptibility*, Gupta, R. R., Ed., Springer-Verlag, Heidelberg, 1986.
2. Barter, C., Meisenheimer, R. G., and Stevenson, D. P., *J. Phys. Chem.* 64, 1312, 1960.
3. Broersma, S., *J. Chem. Phys.* 17, 873, 1949.

Molar and Volume Susceptibility of Organic Compounds

Name	Mol. form.	$-\chi_m/10^{-6}$ cm³ mol⁻¹	$-\kappa/10^{-6}$
Acenaphthene	$C_{12}H_{10}$	109.9	0.871
Acenaphthylene	$C_{12}H_8$	111.6	0.659
Acetaldehyde	C_2H_4O	22.2	0.395
Acetamide	C_2H_5NO	33.9	0.573
Acetic acid	$C_2H_4O_2$	31.8	0.553
Acetic anhydride	$C_4H_6O_3$	52.8	0.560
Acetone	C_3H_6O	33.8	0.457
Acetonitrile	C_2H_3N	27.8	0.532
Acetophenone	C_8H_8O	72.5	0.620
Acetyl chloride	C_2H_3ClO	39.3	0.553
Acetylene	C_2H_2	20.8	0.301*
Acridine	$C_{13}H_9N$	118.8	0.666
Allene	C_3H_4	25.3	0.369*
Allyl alcohol	C_3H_6O	36.7	0.540
Allylamine	C_3H_7N	40.1	0.532
Aniline	C_6H_7N	62.4	0.685
Anisole	C_7H_8O	72.2	0.664
Anthracene	$C_{14}H_{10}$	129.8	0.932
9,10-Anthracenedione	$C_{14}H_8O_2$	113.0	0.780
trans-Azobenzene	$C_{12}H_{10}N_2$	106.8	0.705
Azulene	$C_{10}H_8$	123.7	
Benzaldehyde	C_7H_6O	60.7	0.595
Benzamide	C_7H_7NO	72.0	0.641
Benzene	C_6H_6	54.8	0.615
Benzeneacetic acid	$C_8H_8O_2$	82.4	0.743
Benzeneacetonitrile	C_8H_7N	76.9	0.670
1,2-Benzenediamine	$C_6H_8N_2$	72.5	
1,3-Benzenediamine	$C_6H_8N_2$	70.4	0.657
1,4-Benzenediamine	$C_6H_8N_2$	70.7	
Benzil	$C_{14}H_{10}O_2$	106.8	0.551
Benzonitrile	C_7H_5N	65.2	0.638
Benzophenone	$C_{13}H_{10}O$	109.6	0.668
p-Benzoquinone	$C_6H_4O_2$	36	0.439
Benzyl acetate	$C_9H_{10}O_2$	93.2	0.655
Benzyl alcohol	C_7H_8O	71.8	0.692
Benzyl benzoate	$C_{14}H_{12}O_2$	132.2	0.693
Biphenyl	$C_{12}H_{10}$	103.3	0.697
Bromobenzene	C_6H_5Br	78.1	0.744
1-Bromobutane	C_4H_9Br	77.1	0.718
Bromochloromethane	CH_2BrCl	55.1	0.824
Bromodichloromethane	$CHBrCl_2$	66.3	0.801
Bromoethane	C_2H_5Br	54.7	0.733
Bromomethane	CH_3Br	42.8	0.755*
1-Bromo-2-methylpropane	C_4H_9Br	79.9	0.742
1-Bromonaphthalene	$C_{10}H_7Br$	115.9	0.828
1-Bromopropane	C_3H_7Br	65.6	0.722
2-Bromopropane	C_3H_7Br	65.1	0.696
3-Bromopropene	C_3H_5Br	58.6	0.677
4-Bromotoluene	C_7H_7Br	88.7	0.724
Bromotrichloromethane	$CBrCl_3$	73.2	0.743
1,2-Butadiene	C_4H_6	35.6	0.445
1,3-Butadiene	C_4H_6	32.1	0.365*
Butanal	C_4H_8O	45.9	0.510
Butane	C_4H_{10}	50.3	0.496*
1,3-Butanediol	$C_4H_{10}O_2$	61.8	0.689
1,4-Butanediol	$C_4H_{10}O_2$	61.8	0.697
Butanenitrile	C_4H_7N	50.4	0.579
1-Butanethiol	$C_4H_{10}S$	70.2	0.655
Butanoic acid	$C_4H_8O_2$	55.1	0.596
1-Butanol	$C_4H_{10}O$	55.9	0.611
2-Butanol	$C_4H_{10}O$	57.6	0.627
2-Butanone	C_4H_8O	45.6	0.506
1-Butene	C_4H_8	41.0	0.430*
cis-2-Butene	C_4H_8	42.6	0.468*
trans-2-Butene	C_4H_8	43.3	0.462*
Butylamine	$C_4H_{11}N$	58.9	0.597
Butylbenzene	$C_{10}H_{14}$	100.7	0.645
tert-Butylbenzene	$C_{10}H_{14}$	101.8	0.657
Butyl formate	$C_5H_{10}O_2$	65.8	0.577
4-*tert*-Butylphenol	$C_{10}H_{14}O$	108.0	0.653
Butyl propanoate	$C_7H_{14}O_2$	89.1	0.599
Camphor, (+)	$C_{10}H_{16}O$	103.0	0.670
Carbazole	$C_{12}H_9N$	119.9	0.930
Carbonyl chloride	CCl_2O	47.9	0.664*

Diamagnetic Susceptibility of Selected Organic Compounds

Name	Mol. form.	$-\chi_m/10^{-6}$ cm³ mol⁻¹	$-\kappa/10^{-6}$
2-Chloroaniline	C_6H_6ClN	79.5	0.755
3-Chloroaniline	C_6H_6ClN	76.6	0.730
4-Chloroaniline	C_6H_6ClN	76.7	0.859
Chlorobenzene	C_6H_5Cl	69.6	0.684
1-Chlorobutane	C_4H_9Cl	67.1	0.642
2-Chlorobutane	C_4H_9Cl	67.4	0.636
Chloroethene	C_2H_3Cl	35.9	0.523*
Chloromethane	CH_3Cl	32.0	0.577*
(Chloromethyl)benzene	C_7H_7Cl	81.2	0.706
1-Chloronaphthalene	$C_{10}H_7Cl$	107.6	0.786
1-Chloro-2-nitrobenzene	$C_6H_4ClNO_2$	75.5	0.656
1-Chloro-3-nitrobenzene	$C_6H_4ClNO_2$	77.2	0.658
1-Chloro-4-nitrobenzene	$C_6H_4ClNO_2$	74.7	0.615
1-Chlorooctane	$C_8H_{17}Cl$	114.9	0.675
2-Chlorophenol	C_6H_5ClO	77.3	0.760
3-Chlorophenol	C_6H_5ClO	77.6	0.752
4-Chlorophenol	C_6H_5ClO	77.7	0.765
1-Chloropropane	C_3H_7Cl	56.0	0.635
2-Chloropropene	C_3H_5Cl	47.8	0.563
3-Chloropropene	C_3H_5Cl	47.8	0.586
2-Chlorotoluene	C_7H_7Cl	82.4	0.705
3-Chlorotoluene	C_7H_7Cl	79.7	0.677
4-Chlorotoluene	C_7H_7Cl	80.3	0.679
Chlorotrifluoroethene	C_2ClF_3	49.1	0.649*
Chlorotrifluoromethane	$CClF_3$	45.3	
Chrysene	$C_{18}H_{12}$	148.0	0.826
o-Cresol	C_7H_8O	70.8	0.676
m-Cresol	C_7H_8O	71.9	0.687
p-Cresol	C_7H_8O	71.9	0.677
Cyanamide	CH_2N_2	24.8	0.756
Cyanogen	C_2N_2	21.6	0.396*
Cyanogen chloride	$CClN$	32.4	0.625
Cyclobutane	C_4H_8	40.0	0.502
Cycloheptane	C_7H_{14}	73.9	0.609
1,4-Cyclohexadiene	C_6H_8	48.7	0.515
Cyclohexane	C_6H_{12}	66.1	0.608
Cyclohexanol	$C_6H_{12}O$	73.4	0.705
Cyclohexanone	$C_6H_{10}O$	62.0	0.599
Cyclohexene	C_6H_{10}	58.0	0.573
Cyclooctane	C_8H_{16}	85.3	0.635
Cyclopentane	C_5H_{10}	59.2	0.629
Cyclopentanol	$C_5H_{10}O$	64.0	0.705
Cyclopentanone	C_5H_8O	51.6	0.582
Cyclopropane	C_3H_6	39.2	0.575*
cis-Decahydronaphthalene	$C_{10}H_{18}$	107.0	0.694
trans-Decahydronaphthalene	$C_{10}H_{18}$	107.6	0.674
Decane	$C_{10}H_{22}$	119.5	0.610
1,2-Dibromoethane	$C_2H_4Br_2$	78.8	0.910
Dibromomethane	CH_2Br_2	65.1	0.935
Dibutylamine	$C_8H_{19}N$	103.7	0.615
Dichloroacetyl chloride	C_2HCl_3O	69.0	0.717
o-Dichlorobenzene	$C_6H_4Cl_2$	84.4	0.750
m-Dichlorobenzene	$C_6H_4Cl_2$	84.1	0.737
p-Dichlorobenzene	$C_6H_4Cl_2$	81.7	0.693
Dichlorodifluoromethane	CCl_2F_2	52.2	
1,1-Dichloroethane	$C_2H_4Cl_2$	57.4	0.682
1,2-Dichloroethane	$C_2H_4Cl_2$	59.6	0.750
1,1-Dichloroethene	$C_2H_2Cl_2$	49.2	0.616
cis-1,2-Dichloroethene	$C_2H_2Cl_2$	51.0	0.675
trans-1,2-Dichloroethene	$C_2H_2Cl_2$	48.9	0.634
Dichloromethane	CH_2Cl_2	46.6	0.728
1,1-Diethoxyethane	$C_6H_{14}O_2$	81.4	0.569
Diethylamine	$C_4H_{11}N$	56.8	0.548
N,*N*-Diethylaniline	$C_{10}H_{15}N$	107.9	0.673
Diethyl carbonate	$C_5H_{10}O_3$	75.4	0.619
Diethyl ether	$C_4H_{10}O$	55.1	0.531
Diethyl malonate	$C_7H_{12}O_4$	91.8	0.605
Diethyl oxalate	$C_6H_{10}O_4$	81.7	0.603
Diethyl phthalate	$C_{12}H_{14}O_4$	127.5	0.707
Diethyl succinate	$C_8H_{14}O_4$	105.0	0.627
Diiodomethane	CH_2I_2	93.1	1.154
Dimethoxymethane	$C_3H_8O_2$	47.3	0.534
N,*N*-Dimethylaniline	$C_8H_{11}N$	85.1	0.671
2,2-Dimethylbutane	C_6H_{14}	76.2	0.570
2,3-Dimethylbutane	C_6H_{14}	76.2	0.585
2,3-Dimethyl-2-butene	C_6H_{12}	65.9	0.554
Dimethyl ether	C_2H_6O	26.3	
2,6-Dimethyl-4-heptanone	$C_9H_{18}O$	104.3	0.591
3,4-Dimethylhexane	C_8H_{18}	99.1	0.620
Dimethyl oxalate	$C_4H_6O_4$	55.7	0.553
2,2-Dimethylpentane	C_7H_{16}	87.0	0.585
2,3-Dimethylpentane	C_7H_{16}	87.5	0.603
2,4-Dimethylpentane	C_7H_{16}	87.5	0.587
3,3-Dimethylpentane	C_7H_{16}	89.5	0.620
2,4-Dimethyl-3-pentanone	$C_7H_{14}O$	81.1	0.576
2,4-Dimethylpyridine	C_7H_9N	71.3	0.619
2,6-Dimethylpyridine	C_7H_9N	72.5	0.624
Dimethyl sulfide	C_2H_6S	44.9	0.613
Dimethyl terephthalate	$C_{10}H_{10}O_4$	101.6	0.562
1,4-Dioxane	$C_4H_8O_2$	52.2	0.612
Diphenylacetylene	$C_{14}H_{10}$	116	0.629
Diphenylamine	$C_{12}H_{11}N$	108.4	0.742
1,2-Diphenylethane	$C_{14}H_{14}$	127.8	0.686
Diphenylmethane	$C_{13}H_{12}$	116.0	0.690
Dipropyl ether	$C_6H_{14}O$	79.4	0.580
Dodecanoic acid	$C_{12}H_{24}O_2$	113.0	0.490
1,2-Epoxybutane	C_4H_8O	54.8	0.631
Ethane	C_2H_6	26.8	0.485*
1,2-Ethanediamine	$C_2H_8N_2$	46.5	0.695
1,2-Ethanediol	$C_2H_6O_2$	38.8	0.696
Ethanethiol	C_2H_6S	47.0	0.629
Ethanol	C_2H_6O	33.6	0.576
Ethoxybenzene	$C_8H_{10}O$	84.5	0.668
Ethyl acetate	$C_4H_8O_2$	54.1	0.553
Ethyl acetoacetate	$C_6H_{10}O_3$	71.7	0.571
N-Ethylaniline	$C_8H_{11}N$	85.6	0.680
Ethylbenzene	C_8H_{10}	77.2	0.627
Ethyl benzoate	$C_9H_{10}O_2$	93.8	0.651
Ethyl carbamate	$C_3H_7NO_2$	57.0	0.631
Ethyl cyanoacetate	$C_5H_7NO_2$	67.3	0.634
Ethylene	C_2H_4	18.8	0.381*
Ethyl formate	$C_3H_6O_2$	43.0	0.534
3-Ethylhexane	C_8H_{18}	97.8	0.611
Ethyl 3-methylbutanoate	$C_7H_{14}O_2$	91.1	0.606
3-Ethylpentane	C_7H_{16}	86.2	0.601
Ethyl propanoate	$C_5H_{10}O_2$	66.3	0.574
Ethyl vinyl ether	C_4H_8O	47.9	0.504
Fluorobenzene	C_6H_5F	58.4	0.621
Fluoromethane	CH_3F	17.8	0.291*
Formaldehyde	CH_2O	18.6	0.505*

Name	Mol. form.	$-\chi_m/10^{-6}$ cm³ mol⁻¹	$-\kappa/10^{-6}$
Formamide	CH_3NO	23.0	0.579
Formic acid	CH_2O_2	19.9	0.527
Fumaric acid	$C_4H_4O_4$	49.1	0.692
Furan	C_4H_4O	43.1	0.602
Furfural	$C_5H_4O_2$	47.2	0.570
Furfuryl alcohol	$C_5H_6O_2$	61.0	0.702
D-Glucitol	$C_6H_{14}O_6$	107.8	0.881
Glycerol	$C_3H_8O_3$	57.1	0.782
Glycine	$C_2H_5NO_2$	39.6	0.612
Heptanal	$C_7H_{14}O$	81.0	0.577
Heptane	C_7H_{16}	85.2	0.578
Heptanoic acid	$C_7H_{14}O_2$	86.6	0.607
1-Heptanol	$C_7H_{16}O$	91.7	0.649
4-Heptanol	$C_7H_{16}O$	92.1	0.649
2-Heptanone	$C_7H_{14}O$	80.5	0.572
3-Heptanone	$C_7H_{14}O$	80.7	0.578
4-Heptanone	$C_7H_{14}O$	80.5	0.576
1-Heptene	C_7H_{14}	77.8	0.552
Hexachlorobenzene	C_6Cl_6	147.0	1.055
Hexachloroethane	C_2Cl_6	112.8	0.996
Hexadecane	$C_{16}H_{34}$	187.6	0.638
Hexadecanoic acid	$C_{16}H_{32}O_2$	198.6	0.660
1-Hexadecanol	$C_{16}H_{34}O$	183.5	0.620
1,5-Hexadiene	C_6H_{10}	55.1	0.461
Hexamethylbenzene	$C_{12}H_{18}$	122.5	0.802
Hexanal	$C_6H_{12}O$	69.4	0.578
Hexane	C_6H_{14}	74.1	0.568
1,6-Hexanediol	$C_6H_{14}O_2$	84.3	0.770
Hexanoic acid	$C_6H_{12}O_2$	78.1	0.619
1-Hexanol	$C_6H_{14}O$	79.2	0.631
2-Hexanone	$C_6H_{12}O$	69.2	0.561
3-Hexanone	$C_6H_{12}O$	69.0	0.559
1-Hexene	C_6H_{12}	66.4	0.527
Hexyl acetate	$C_8H_{16}O_2$	100.9	0.614
1-Hexyne	C_6H_{10}	64.5	0.562
p-Hydroquinone	$C_6H_6O_2$	64.7	0.781
2-Hydroxybenzoic acid	$C_7H_6O_3$	75	0.784
Indene	C_9H_8	83	0.712
1*H*-Indole	C_8H_7N	85.0	0.885
Iodobenzene	C_6H_5I	92.0	0.826
1-Iodobutane	C_4H_9I	93.6	0.822
Iodoethane	C_2H_5I	69.1	0.858
Iodomethane	CH_3I	57.2	0.918
1-Iodopropane	C_3H_7I	84.3	0.867
Isobutane	C_4H_{10}	50.5	0.479*
Isobutene	C_4H_8	40.8	0.428*
Isobutyl acetate	$C_6H_{12}O_2$	78.7	0.590
Isobutylamine	$C_4H_{11}N$	59.8	0.592
Isobutylbenzene	$C_{10}H_{14}$	101.7	0.646
Isobutyl formate	$C_5H_{10}O_2$	66.8	0.574
Isopentane	C_5H_{12}	64.4	0.553
Isopentyl acetate	$C_7H_{14}O_2$	89.4	0.602
Isopentyl formate	$C_6H_{12}O_2$	78.4	0.592
Isophthalic acid	$C_8H_6O_4$	84.6	0.783
Isopropenylbenzene	C_9H_{10}	80.0	0.616
Isopropyl acetate	$C_5H_{10}O_2$	67.0	0.568
Isopropylbenzene	C_9H_{12}	89.5	0.643
1-Isopropyl-4-methylbenzene	$C_{10}H_{14}$	102.8	0.657
Isoquinoline	C_9H_7N	83.9	0.709
d-Limonene	$C_{10}H_{16}$	98.0	0.605

Name	Mol. form.	$-\chi_m/10^{-6}$ cm³ mol⁻¹	$-\kappa/10^{-6}$
Maleic acid	$C_4H_4O_4$	49.6	0.679
Maleic anhydride	$C_4H_2O_3$	35.8	0.480
Methane	CH_4	17.4	0.459*
Methanol	CH_4O	21.4	0.529
2-Methoxyaniline	C_7H_9NO	79.1	0.702
Methylamine	CH_5N	27.0	0.570*
2-Methylaniline	C_7H_9N	74.9	0.698
3-Methylaniline	C_7H_9N	74.6	0.688
4-Methylaniline	C_7H_9N	72.5	0.651
N-Methylaniline	C_7H_9N	74.1	0.684
Methyl benzoate	$C_8H_8O_2$	81.6	0.650
2-Methyl-1,3-butadiene	C_5H_8	46.0	0.459
3-Methylbutanoic acid	$C_5H_{10}O_2$	67.7	0.617
2-Methyl-2-butene	C_5H_{10}	54.7	0.517
Methylcyclohexane	C_7H_{14}	78.9	0.618
Methylcyclopentane	C_6H_{12}	70.2	0.624
Methyl formate	$C_2H_4O_2$	31.1	0.503
4-Methylheptane	C_8H_{18}	97.3	0.600
Methyl methacrylate	$C_5H_8O_2$	57.3	0.537
1-Methylnaphthalene	$C_{11}H_{10}$	102.9	0.738
2-Methylnaphthalene	$C_{11}H_{10}$	102.7	0.726
Methyloxirane	C_3H_6O	42.5	0.629
2-Methylpentane	C_6H_{14}	75.3	0.568
3-Methylpentane	C_6H_{14}	75.5	0.578
4-Methyl-2-pentanol	$C_6H_{14}O$	80.4	0.635
4-Methyl-2-pentanone	$C_6H_{12}O$	69.7	0.554
Methyl propanoate	$C_4H_8O_2$	54.5	0.566
2-Methylpropanoic acid	$C_4H_8O_2$	56.1	0.616
2-Methyl-1-propanol	$C_4H_{10}O$	57.6	0.623
2-Methyl-2-propanol	$C_4H_{10}O$	56.6	0.602
4-Methylpyridine	C_6H_7N	59.8	0.613
Methyl salicylate	$C_8H_8O_3$	86.3	0.670
Morpholine	C_4H_9NO	55.0	0.632
Naphthalene	$C_{10}H_8$	91.6	0.733
1-Naphthol	$C_{10}H_8O$	96.2	0.733
2-Naphthol	$C_{10}H_8O$	96.8	0.859
1-Naphthylamine	$C_{10}H_9N$	92.5	0.661
2-Naphthylamine	$C_{10}H_9N$	98.0	1.123
Neopentane	C_5H_{12}	63.0	0.511*
2-Nitroaniline	$C_6H_6N_2O_2$	67.4	0.440
3-Nitroaniline	$C_6H_6N_2O_2$	69.7	0.455
4-Nitroaniline	$C_6H_6N_2O_2$	68.0	0.701
Nitrobenzene	$C_6H_5NO_2$	61.8	0.604
Nitroethane	$C_2H_5NO_2$	35.4	0.493
Nitromethane	CH_3NO_2	21.0	0.391
2-Nitrophenol	$C_6H_5NO_3$	68.9	0.641
3-Nitrophenol	$C_6H_5NO_3$	65.9	0.606
4-Nitrophenol	$C_6H_5NO_3$	66.9	0.711
1-Nitropropane	$C_3H_7NO_2$	45.0	0.503
2-Nitropropane	$C_3H_7NO_2$	45.4	0.500
2-Nitrotoluene	$C_7H_7NO_2$	72.2	0.611
3-Nitrotoluene	$C_7H_7NO_2$	72.7	0.614
4-Nitrotoluene	$C_7H_7NO_2$	73.3	0.590
Nonane	C_9H_{20}	108.1	0.606
1-Nonene	C_9H_{18}	100.1	0.575
cis-9-Octadecenoic acid	$C_{18}H_{34}O_2$	208.5	0.660
Octane	C_8H_{18}	96.6	0.591
Octanoic acid	$C_8H_{16}O_2$	99.5	0.626
1-Octanol	$C_8H_{18}O$	101.6	0.645
1-Octene	C_8H_{16}	88.8	0.566

Name	Mol. form.	$-\chi_m/10^{-6}$ cm^3 mol^{-1}	$-\kappa/10^{-6}$
Oxirane	C_2H_4O	30.5	0.611
Paraldehyde	$C_6H_{12}O_3$	86.1	0.648
Pentachloroethane	C_2HCl_5	99.1	0.823
Pentanal	$C_5H_{10}O$	57.5	0.540
Pentane	C_5H_{12}	63.1	0.548
1,5-Pentanediol	$C_5H_{12}O_2$	73.5	0.700
2,4-Pentanedione	$C_5H_8O_2$	54.9	0.533
Pentanoic acid	$C_5H_{10}O_2$	66.5	0.608
1-Pentanol	$C_5H_{12}O$	67.0	0.619
2-Pentanol	$C_5H_{12}O$	69.1	0.634
2-Pentanone	$C_5H_{10}O$	57.5	0.540
3-Pentanone	$C_5H_{10}O$	57.7	0.542
1-Pentene	C_5H_{10}	54.6	0.499
Pentyl acetate	$C_7H_{14}O_2$	88.9	0.598
Pentylamine	$C_5H_{13}N$	69.3	0.600
Perylene	$C_{20}H_{12}$	167.5	0.896
Phenanthrene	$C_{14}H_{10}$	127.6	0.702
Phenol	C_6H_6O	60.6	0.679
Phthalic acid	$C_8H_6O_4$	83.6	1.097
Phthalic anhydride	$C_8H_4O_3$	66.7	0.688
α-Pinene	$C_{10}H_{16}$	100.7	0.631
β-Pinene	$C_{10}H_{16}$	101.9	0.643
Piperidine	$C_5H_{11}N$	64.2	0.649
Propanal	C_3H_6O	34.2	0.510
Propane	C_3H_8	38.6	0.432*
1,3-Propanediol	$C_3H_8O_2$	50.2	0.695
Propanenitrile	C_3H_5N	38.6	0.548
Propanoic acid	$C_3H_6O_2$	43.5	0.580
1-Propanol	C_3H_8O	45.2	0.601
2-Propanol	C_3H_8O	45.7	0.594
Propene	C_3H_6	30.7	0.368*
Propyl acetate	$C_5H_{10}O_2$	65.9	0.569
Propylbenzene	C_9H_{12}	89.1	0.637
Propyl formate	$C_4H_8O_2$	55.0	0.566
Propyl propanoate	$C_6H_{12}O_2$	77.7	0.586
Pyrazine	$C_4H_4N_2$	37.8	0.487
Pyrene	$C_{16}H_{10}$	147	0.924
Pyridine	C_5H_5N	48.7	0.605
Pyrimidine	$C_4H_4N_2$	43.1	0.581
Pyrocatechol	$C_6H_6O_2$	68.2	0.832
Pyrrole	C_4H_5N	48.6	0.703
Pyrrolidine	C_4H_9N	54.8	0.662
Quinoline	C_9H_7N	86.1	0.732
Resorcinol	$C_6H_6O_2$	67.2	0.780
Safrole	$C_{10}H_{10}O_2$	97.5	0.661
Salicylaldehyde	$C_7H_6O_2$	66.8	0.639
Octadecanoic acid	$C_{18}H_{36}O_2$	220.8	0.730
Styrene	C_8H_8	68.2	0.590
Succinic acid	$C_4H_6O_4$	58.0	0.772
Succinic anhydride	$C_4H_4O_3$	47.5	0.570
Terephthalic acid	$C_8H_6O_4$	83.5	0.763
o-Terphenyl	$C_{18}H_{14}$	150.4	0.731
m-Terphenyl	$C_{18}H_{14}$	155.5	0.810
p-Terphenyl	$C_{18}H_{14}$	156.0	0.813
1,1,2,2-Tetrabromoethane	$C_2H_2Br_4$	123.4	1.059
Tetrabromomethane	CBr_4	93.7	0.837
1,1,2,2-Tetrachloroethane	$C_2H_2Cl_4$	89.8	0.853
Tetrachloroethene	C_2Cl_4	81.6	0.799
Tetrachloromethane	CCl_4	66.6	0.690
Tetradecane	$C_{14}H_{30}$	166.2	0.636
Tetradecanoic acid	$C_{14}H_{28}O_2$	176.0	0.664
Tetrahydrofurfuryl alcohol	$C_5H_{10}O_2$	69.4	0.715
Tetraiodomethane	CI_4	136	1.107
1,2,4,5-Tetramethylbenzene	$C_{10}H_{14}$	101.2	0.632
Tetranitromethane	CN_4O_8	43.0	0.359
Thiophene	C_4H_4S	57.3	0.725
Toluene	C_7H_8	66.1	0.619
o-Toluic acid	$C_8H_8O_2$	84.3	0.658
m-Toluic acid	$C_8H_8O_2$	83.0	0.643
p-Toluic acid	$C_8H_8O_2$	82.4	
Tribromomethane	$CHBr_3$	82.6	0.941
Trichloroacetaldehyde	C_2HCl_3O	73.0	0.749
Trichloroacetic acid	$C_2HCl_3O_2$	73.0	0.720
Trichloroethene	C_2HCl_3	65.8	0.733
Trichlorofluoromethane	CCl_3F	58.7	0.636
Trichloromethane	$CHCl_3$	59.3	0.735
Trichloronitromethane	CCl_3NO_2	75.3	0.759
Tridecane	$C_{13}H_{28}$	153.7	0.631
Triethylamine	$C_6H_{15}N$	83.3	0.599
Trifluoroacetic acid	$C_2HF_3O_2$	43.3	0.583
Triiodomethane	CHI_3	117.1	1.192
1,3,5-Trimethylbenzene	C_9H_{12}	92.3	0.662
2,2,4-Trimethylpentane	C_8H_{18}	98.3	0.592
2,3,4-Trimethylpentane	C_8H_{18}	99.8	0.628
2,4,6-Trimethylpyridine	$C_8H_{11}N$	83.1	0.629
Undecane	$C_{11}H_{24}$	131.8	0.624
Urea	CH_4N_2O	33.5	0.738
Vinyl acetate	$C_4H_6O_2$	46.4	0.499
Vinyl formate	$C_3H_4O_2$	34.7	0.465
o-Xylene	C_8H_{10}	77.8	0.642
m-Xylene	C_8H_{10}	76.6	0.620
p-Xylene	C_8H_{10}	76.8	0.620

* Value refers to a liquefied gas.

Section 4
Properties of the Elements and Inorganic Compounds

THE ELEMENTS

C. R. Hammond and David R. Lide

Each chemical element is characterized by the number of protons, *Z*, contained in the atomic nucleus. As of 2016, a total of 118 elements have been positively identified and assigned names and symbols. Elements can occur as two or more isotopes, differing in the number of neutrons, *N*, in the nucleus. An isotope is characterized by the sum of the protons and neutrons in its nucleus, which is called the mass number $A = Z + N$. Including those produced artificially, over 3000 isotopes have been identified, most of which are unstable to radioactive decay. A total of 288 stable isotopes of 84 elements exist on earth.[a]

The table below summarizes key properties of the 118 elements. More detailed information can be found in the tables of atomic weights and physical properties. Individual descriptions of the elements follow the table.

[a] These numbers include radioactive isotopes whose half-lives are comparable to or greater than the age of the earth, making it possible to define a terrestrial isotopic abundance. See Meija, J., et al., "Isotopic compositions of the elements 2013", *Pure Appl. Chem.* 2016; 88(3): 293–306.

Element	Symbol	Z	Atomic wt.	CAS Reg. No.	State at NTP	Primary valences	Mass no. of stable isotopes[a]	Melting point in °C	Boiling point in °C	Density in g/cm³
Actinium	Ac	89		7440-34-8	sol	+3		1050	≈3200	10
Aluminum	Al	13	26.9815385(7)	7429-90-5	sol	+3	27	660.323	2519	2.70
Americium	Am	95		7440-35-9	sol	+3, +4, +5, +6		1176	≈2011	12
Antimony	Sb	51	121.760(1)	7440-36-0	sol (gray Sb)	+3, +5	121, 123	630.628	1587	6.68
Argon	Ar	18	39.948(1)	7440-37-1	gas	0	36, 38, 40	-189.34	-185.848	1.395$^{-185.8}$ liq
Arsenic	As	33	74.921595(6)	7440-38-2	sol (gray As)	+3, +5, -3	75	817	616 sp	5.75
Astatine	At	85		7440-68-8	sol			302		
Barium	Ba	56	137.327(7)	7440-39-3	sol	+2	130, 132, 134, 135, 136, 137, 138	727	≈1845	3.62
Berkelium	Bk	97		7440-40-6	sol	+3, +4		986		13.25
Beryllium	Be	4	9.0121831(5)	7440-41-7	sol	+2	9	1287	2468	1.85
Bismuth	Bi	83	208.98040(1)	7440-69-9	sol	+3, +5	209	271.402	1564	9.79
Bohrium	Bh	107		54037-14-8	sol[b]	+7[b]				
Boron	B	5	[10.806, 10.821]	7440-42-8	sol	+3	10, 11	2077	4000	2.34
Bromine	Br	35	[79.901, 79.907]	7726-95-6	liq	+1, +5, -1	79, 81	-7.2	58.8	3.1028
Cadmium	Cd	48	112.414(4)	7440-43-9	sol	+2	106, 108, 110, 111, 112, 113, 114, 116	321.069	767	8.69
Calcium	Ca	20	40.078(4)	7440-70-2	sol	+2	40, 42, 43, 44, 46, 48	842	1484	1.54
Californium	Cf	98		7440-71-3	sol	+3		900		15.1
Carbon	C	6	[12.0096, 12.0106]	7440-44-0	sol (graphite)	+2, +4, -4	12, 13	4489 tp (10.3 MPa)	3825 sp	2.2
Carbon	C	6	[12.0096, 12.0106]	7440-44-0	sol (diamond)	+2, +4, -4	12, 13	4440 (12.4 GPa)		3.513
Cerium	Ce	58	140.116(1)	7440-45-1	sol	+3, +4	136, 138, 140, 142	799	3443	6.770
Cesium	Cs	55	132.90545196(6)	7440-46-2	sol	+1	133	28.5	671	1.873
Chlorine	Cl	17	[35.446, 35.457]	7782-50-5	gas	+1, +5, +7, -1	35,37	-101.5	-34.04	1.565$^{-34.0}$ liq
Chromium	Cr	24	51.9961(6)	7440-47-3	sol	+2, +3, +6	50, 52, 53, 54	1907	2671	7.15
Cobalt	Co	27	58.933194(4)	7440-48-4	sol	+2, +3	59	1495	2927	8.86
Copernicium	Cn	112		54084-26-3	liq/gas[b]	+2, +4[b]				
Copper	Cu	29	63.546(3)	7440-50-8	sol	+1, +2	63, 65	1084.62	2560	8.96
Curium	Cm	96		7440-51-9	sol	+3		1345		13.51
Darmstadtium	Ds	110		54083-77-1	sol[b]	+2, +4, +6[b]				
Dubnium	Db	105		53850-35-4	sol[b]	+5[b]				
Dysprosium	Dy	66	162.500(1)	7429-91-6	sol	+3	156, 158, 160, 161, 162, 163, 164	1412	2567	8.55
Einsteinium	Es	99		7429-92-7	sol	+3		860		
Erbium	Er	68	167.259(3)	7440-52-0	sol	+3	162, 164, 166, 167, 168, 170	1529	2868	9.07
Europium	Eu	63	151.964(1)	7440-53-1	sol	+2, +3	151, 153	822	1529	5.24
Fermium	Fm	100		7440-72-4	sol	+3		1527		
Flerovium	Fl	114		54085-16-4	liq/gas[b]	+2, +4[b]				

Element	Symbol	Z	Atomic wt.	CAS Reg. No.	State at NTP	Primary valences	Mass no. of stable isotopes[a]	Melting point in °C	Boiling point in °C	Density in g/cm³
Fluorine	F	9	18.998403163(6)	7782-41-4	gas	-1	19	-219.67	-188.11	$1.5127^{-188.1}$ liq
Francium	Fr	87		7440-73-5	sol/liq[b]	+1		≈21		
Gadolinium	Gd	64	157.25(3)	7440-54-2	sol	+3	152, 154, 155, 156, 157, 158, 160	1313	3273	7.90
Gallium	Ga	31	69.723(1)	7440-55-3	sol	+3	69, 71	29.7646	2229	5.91
Germanium	Ge	32	72.630(8)	7440-56-4	sol	+2, +4	70, 72, 73, 74, 76	938.25	2833	5.3234
Gold	Au	79	196.966569(5)	7440-57-5	sol	+1, +3	197	1064.18	2836	19.3
Hafnium	Hf	72	178.49(2)	7440-58-6	sol	+4	174, 176, 177, 178, 179, 180	2233	4600	13.3
Hassium	Hs	108		54037-57-9	sol[b]	+8[b]				
Helium	He	2	4.002602(2)	7440-59-7	gas	0	3, 4		-268.928	$0.1250^{-268.9}$ liq
Holmium	Ho	67	164.93033(2)	7440-60-0	sol	+3	165	1472	2700	8.80
Hydrogen	H	1	[1.00784, 1.00811]	1333-74-0	gas	+1	1, 2	-259.16	-252.879	$0.07083^{-252.9}$ liq
Indium	In	49	114.818(1)	7440-74-6	sol	+3	113, 115	156.5985	2027	7.31
Iodine	I	53	126.90447(3)	7553-56-2	sol	+1, +5, +7, -1	127	113.7	184.4	4.933
Iridium	Ir	77	192.217(3)	7439-88-5	sol	+3, +4	191, 193	2446	4428	22.562^{20}
Iron	Fe	26	55.845(2)	7439-89-6	sol	+2, +3	54, 56, 57, 58	1538	2861	7.87
Krypton	Kr	36	83.798(2)	7439-90-9	gas	0	78, 80, 82, 83, 84, 86	-157.37	-153.415	$2.417^{-153.4}$ liq
Lanthanum	La	57	138.90547(7)	7439-91-0	sol	+3	138, 139	920	3464	6.15
Lawrencium	Lr	103		22537-19-5	sol	+3		1627		
Lead	Pb	82	207.2(1)	7439-92-1	sol	+2, +4	204, 206, 207, 208	327.462	1749	11.3
Lithium	Li	3	[6.938, 6.997]	7439-93-2	sol	+1	6, 7	180.50	1342	0.534
Livermorium	Lv	116		54100-71-9	sol[b]	+2, +4[b]				
Lutetium	Lu	71	174.9668(1)	7439-94-3	sol	+3	175, 176	1663	3402	9.84
Magnesium	Mg	12	[24.304, 24.307]	7439-95-4	sol	+2	24, 25, 26	650	1090	1.74
Manganese	Mn	25	54.938044(3)	7439-96-5	sol	+2, +3, +4, +7	55	1246	2061	7.3
Meitnerium	Mt	109		54038-01-6	sol[b]	+1, +3, +6[b]				
Mendelevium	Md	101		7440-11-1	sol	+2, +3		827		
Mercury	Hg	80	200.592(3)	7439-97-6	liq	+1, +2	196, 198, 199, 200, 201, 202, 204	-38.8290	356.619	13.53359^{25} liq
Molybdenum	Mo	42	95.95(1)	7439-98-7	sol	+6	92, 94, 95, 96, 97, 98	2622	4639	10.2
Moscovium	Mc	115		54085-64-2	sol[b]	+1, +3[b]				
Neodymium	Nd	60	144.242(3)	7440-00-8	sol	+3	142, 143, 144, 145, 146, 148, 150	1016	3074	7.01
Neon	Ne	10	20.1797(6)	7440-01-9	gas	0	20, 21, 22	-248.59	-246.046	$1.207^{-246.0}$ liq
Neptunium	Np	93		7439-99-8	sol	+3, +4, +5, +6		644	≈3902	20.2
Nickel	Ni	28	58.6934(4)	7440-02-0	sol	+2, +3	58, 60, 61, 62, 64	1455	2913	8.90
Nihonium	Nh	113		54084-70-7	sol[b]	+1, +3[b]				
Niobium	Nb	41	92.90637(2)	7440-03-1	sol	+3, +5	93	2477	4741	8.57
Nitrogen	N	7	[14.00643, 14.00728]	7727-37-9	gas	+1, +2, +3, +4, +5, -1, -2, -3	14, 15	-210.0	-195.795	$0.8061^{-195.8}$ liq
Nobelium	No	102		10028-14-5	sol	+2, +3		827		
Oganesson	Og	118		54144-19-3	sol[b]	0, +2, +4[b]				
Osmium	Os	76	190.23(3)	7440-04-2	sol	+3, +4	184, 186, 187, 188, 189, 190, 192	3033	5008	22.587^{20}
Oxygen	O	8	[15.99903, 15.99977]	7782-44-7	gas	-2	16, 17, 18	-218.79	-182.962	$1.141^{-183.0}$ liq
Palladium	Pd	46	106.42(1)	7440-05-3	sol	+2, +3	102, 104, 105, 106, 108, 110	1554.8	2963	12.0
Phosphorus	P	15	30.973761998(5)	7723-14-1	sol (white P)	+3, +5, -3	31	44.15	280.5	1.823
Phosphorus	P	15	30.973761998(5)	7723-14-0	sol (red P)	+3, +5, -3	31	579.2	431 sp	2.16
Platinum	Pt	78	195.084(9)	7440-06-4	sol	+2, +4	190, 192, 194, 195, 196, 198	1768.2	3825	21.5

Element	Symbol	Z	Atomic wt.	CAS Reg. No.	State at NTP	Primary valences	Mass no. of stable isotopes[a]	Melting point in °C	Boiling point in °C	Density in g/cm³
Plutonium	Pu	94		7440-07-5	sol	+3, +4, +5, +6		640	3228	19.7
Polonium	Po	84		7440-08-6	sol	+2, +4		254	962	9.20
Potassium	K	19	39.0983(1)	7440-09-7	sol	+1	39, 40, 41	63.5	759	0.89
Praseodymium	Pr	59	140.90766(2)	7440-10-0	sol	+3	141	931	3520	6.77
Promethium	Pm	61		7440-12-2	sol	+3		1042	≈3000	7.26
Protactinium	Pa	91	231.03588(2)	7440-13-3	sol	+4, +5	231	1572		15.4
Radium	Ra	88		7440-14-4	sol	+2		696		5
Radon	Rn	86		10043-92-2	gas	0		-71	-61.7	0.009074^{25} gas
Rhenium	Re	75	186.207(1)	7440-15-5	sol	+4, +6, +7	185, 187	3185	5590	20.8
Rhodium	Rh	45	102.90550(2)	7440-16-6	sol	+3	103	1963	3695	12.4
Roentgenium	Rg	111		54386-24-2	sol[b]	+3, +5, -1[b]				
Rubidium	Rb	37	85.4678(3)	7440-17-7	sol	+1	85, 87	39.30	688	1.53
Ruthenium	Ru	44	101.07(2)	7440-18-8	sol	+3	100, 101, 102, 104	2333	4147	12.1
Rutherfordium	Rf	104		53850-36-5	sol[b]	+4				
Samarium	Sm	62	150.36(2)	7440-19-9	sol	+2, +3	144, 147, 148, 149, 150, 152, 154	1072	1794	7.52
Scandium	Sc	21	44.955908(5)	7440-20-2	sol	+3	45	1541	2836	2.99
Seaborgium	Sg	106		54038-81-2	sol[b]	+6[b]				
Selenium	Se	34	78.971(8)	7782-49-2	sol (gray Se)	+4, +6, -2	74, 76, 77, 78, 80, 82	220.8	685	4.809
Silicon	Si	14	[28.084, 28.086]	7440-21-3	sol	+2, +4, -4	28, 29	1414	3265	2.3296
Silver	Ag	47	107.8682(2)	7440-22-4	sol	+1	107, 109	961.78	2162	10.5
Sodium	Na	11	22.98976928(2)	7440-23-5	sol	+1	23	97.794	882.940	0.97
Strontium	Sr	38	87.62(1)	7440-24-6	sol	+2	84, 86, 87, 88	777	1377	2.64
Sulfur	S	16	[32.059, 32.076]	7704-34-9	sol (monoclinic S)	+4, +6, -2	32, 33, 34, 36	115.21	444.61	2.00
Sulfur	S	16	[32.059, 32.076]	7704-34-10	sol (rhombic S)	+4, +6, -2	32, 33, 34, 36	95.2 trans monocl	444.61	2.07
Tantalum	Ta	73	180.94788(2)	7440-25-7	sol	+5	180, 181	3017	5455	16.4
Technetium	Tc	43		7440-26-8	sol	+4, +6, +7		2157	4262	11
Tellurium	Te	52	127.60(3)	13494-80-9	sol	+4, +6, -2	120, 122, 123, 124, 125, 126, 128, 130	449.51	988	6.232
Tennessine	Ts	117		54101-14-3	sol[b]	+1, +3, +5[b]				
Terbium	Tb	65	158.92535(2)	7440-27-9	sol	+3	159	1359	3230	8.23
Thallium	Tl	81	[204.382, 204.385]	7440-28-0	sol	+1, +3	203, 205	304	1473	11.8
Thorium	Th	90	232.0377(4)	7440-29-1	sol	+4	232	1750	4785	11.7
Thulium	Tm	69	168.93422(2)	7440-30-4	sol	+3	169	1545	1950	9.32
Tin	Sn	50	118.710(7)	7440-31-6	sol (white Sn)	+2, +4	112, 114, 115, 116, 117, 118, 119, 120, 122, 124	231.928	2586	7.287
Tin	Sn	50	118.710(7)	7440-31-5	sol (gray Sn)	+2, +4	112, 114, 115, 116, 117, 118, 119, 120, 122, 124	13.2 trans white Sn	2586	5.769
Titanium	Ti	22	47.867(1)	7440-32-6	sol	+2, +3, +4	46, 47, 48, 49, 50	1670	3287	4.506
Tungsten	W	74	183.84(1)	7440-33-7	sol	+6	180, 182, 183, 184, 186	3414	5555	19.3
Uranium	U	92	238.02891(3)	7440-61-1	sol	+3, +4, +5, +6	234. 235, 238	1135	4131	19.1
Vanadium	V	23	50.9415(1)	7440-62-2	sol	+2, +3, +4, +5	50, 51	1910	3407	6.0
Xenon	Xe	54	131.293(6)	7440-63-3	gas	0	124, 126, 128, 129, 130, 131, 132, 134, 136	-111.75	-108.099	$2.942^{-108.1}$ liq
Ytterbium	Yb	70	173.045(10)	7440-64-4	sol	+2, +3	168, 170, 171, 172, 173, 174, 176	824	1196	6.90
Yttrium	Y	39	88.90584(2)	7440-65-5	sol	+3	89	1522	3345	4.47
Zinc	Zn	30	65.38(2)	7440-66-6	sol	+2	64, 66, 67, 68, 70	419.527	907	7.134
Zirconium	Zr	40	91.224(2)	7440-67-7	sol	+4	90, 91, 92, 94, 96	1854	4406	6.52

[a] Includes radioactive isotopes of very long lifetimes, whose terrestrial abundance is effectively stable.

[b] Predicted

Actinium — (Gr. *aktis, aktinos,* beam or ray), Ac. Discovered by Andre Debierne in 1899 and independently by F. Giesel in 1902. Occurs naturally in association with uranium minerals. Thirty-seven isotopes and isomers are now recognized. All are radioactive. Actinium-227, a decay product of uranium-235, is an alpha and beta emitter with a 21.77-year half-life. Its principal decay products are thorium-227 (18.72-day half-life), radium-223 (11.4-day half-life), and a number of short-lived products including radon, bismuth, polonium, and lead isotopes. In equilibrium with its decay products, it is a powerful source of alpha rays. Actinium metal has been prepared by the reduction of actinium fluoride with lithium vapor at about 1100 to 1300 °C. The chemical behavior of actinium is similar to that of the rare earths, particularly lanthanum. Purified actinium comes into equilibrium with its decay products at the end of 185 days, and then decays according to its 21.77-year half-life. It is about 150 times as active as radium, making it of value in the production of neutrons.

Aluminum — (L. *alumen, alum*), Al. The ancient Greeks and Romans used *alum* in medicine as an astringent, and as a mordant in dyeing. In 1761 de Morveau proposed the name *alumine* for the base in alum, and Lavoisier, in 1787, thought this to be the oxide of a still undiscovered metal. Wöhler is generally credited with having isolated the metal in 1827, although an impure form was prepared by Oersted two years earlier. In 1807, Davy proposed the name *alumium* for the metal, undiscovered at that time, and later agreed to change it to *aluminum*. Shortly thereafter, the name *aluminium* was adopted to conform with the "ium" ending of most elements, and this spelling is now in use elsewhere in the world. *Aluminium* was also the accepted spelling in the U.S. until 1925, at which time the American Chemical Society officially decided to use the name *aluminum* thereafter in their publications. The method of obtaining aluminum metal by the electrolysis of alumina dissolved in cryolite was discovered in 1886 by Hall in the U.S. and at about the same time by Heroult in France. Cryolite, a natural ore found in Greenland, is no longer widely used in commercial production, but has been replaced by an artificial mixture of sodium, aluminum, and calcium fluorides. Bauxite, an impure hydrated oxide ore, is found in large deposits in Jamaica, Australia, Suriname, Guyana, Russia, Arkansas, and elsewhere. The Bayer process is most commonly used today to refine bauxite so it can be accommodated in the Hall–Heroult refining process used to make most aluminum. Aluminum can now be produced from clay, but the process is not economically feasible at present. Aluminum is the most abundant metal to be found in the Earth's crust (8.1%), but is never found free in nature. In addition to the minerals mentioned above, it is found in feldspars, granite, and in many other common minerals. Natural aluminum is made of one isotope, ^{27}Al. Pure aluminum, a silvery-white metal, possesses many desirable characteristics. It is light, nontoxic, has a pleasing appearance, can easily be formed, machined, or cast, has a high thermal conductivity, and has excellent corrosion resistance. It is nonmagnetic and nonsparking, stands second among metals in the scale of malleability, and sixth in ductility. It is extensively used for kitchen utensils, outside building decoration, and in thousands of industrial applications where a strong, light, easily constructed material is needed. Although its electrical conductivity is only about 60% that of copper, it is used in electrical transmission lines because of its light weight. Pure aluminum is soft and lacks strength, but it can be alloyed with small amounts of copper, magnesium, silicon, manganese, and other elements to impart a variety of useful properties. These alloys are of vital importance in the construction of modern aircraft and rockets. Aluminum, evaporated in a vacuum, forms a highly reflective coating for both visible light and radiant heat. These coatings soon form a thin layer of the protective oxide and do not deteriorate as do silver coatings. They have found application in coatings for telescope mirrors, in making decorative paper, packages, toys, and in many other uses. The compounds of greatest importance are aluminum oxide, the sulfate, and the soluble sulfate with potassium (alum). The oxide, alumina, occurs naturally as ruby, sapphire, corundum, and emery, and is used in glassmaking and refractories. Synthetic ruby and sapphire have found application in the construction of lasers.

Americium — (the Americas), Am. Americium was the fourth transuranium element to be discovered; the isotope ^{241}Am was identified by Seaborg, James, Morgan, and Ghiorso late in 1944 at the wartime Metallurgical Laboratory of the University of Chicago as the result of successive neutron capture reactions by plutonium isotopes in a nuclear reactor. Since the isotope ^{241}Am can be prepared in relatively pure form by extraction as a decay product over a period of years from strongly neutron-bombarded plutonium, ^{241}Pu, this isotope is used for much of the chemical investigation of this element. Better suited is the isotope ^{243}Am due to its longer half-life (7370 years as compared to 432.2 years for ^{241}Am). A mixture of the isotopes ^{241}Am, ^{242}Am, and ^{243}Am can be prepared by intense neutron irradiation of ^{241}Am according to the reactions ^{241}Am (n, γ) → ^{242}Am (n, γ) → ^{243}Am. Nearly isotopically pure ^{243}Am can be prepared by a sequence of neutron bombardments and chemical separations starting with ^{242}Pu. Twenty-three radioactive isotopes and isomers are now recognized. Americium metal has been prepared by reducing the trifluoride with barium vapor at 1000 to 1200 °C or reducing the dioxide by lanthanum metal. The luster of freshly prepared americium metal is white and more silvery than plutonium or neptunium prepared in the same manner. It appears to be more malleable than uranium or neptunium and tarnishes slowly in dry air at room temperature. Americium is thought to exist in two forms: an alpha form which has a double hexagonal close-packed structure and a loose-packed cubic beta form. Americium must be handled with great care to avoid personal contamination. As little as 0.03 μCi of ^{241}Am is the maximum permissible total body burden. The alpha activity from ^{241}Am is about three times that of radium. When gram quantities of ^{241}Am are handled, the intense gamma activity makes exposure a serious problem. Americium dioxide, AmO_2, is the most important oxide. AmF_3, AmF_4, $AmCl_3$, $AmBr_3$, AmI_3, and other compounds have been prepared. The isotope ^{241}Am has been used as a portable source for gamma radiography. It has also been used as a radioactive glass thickness gage for the flat glass industry and as a source of ionization for smoke detectors.

Antimony — (Gr. *anti* plus *monos* - a metal not found alone), Sb. Antimony was recognized in compounds by the ancients and was known as a metal at the beginning of the 17th century and possibly much earlier. It is not abundant, but is found in over 100 mineral species. It is sometimes found native, but more frequently as the sulfide, *stibnite* (Sb_2S_3); it is also found as antimonides of the heavy metals, and as oxides. It is extracted from the sulfide by roasting to the oxide, which is reduced by salt and scrap iron; from its oxides it is also prepared by reduction with carbon. Two allotropic forms of antimony exist: the normal stable, metallic form (gray antimony), and the amorphous black form. The so-called explosive antimony is an ill-defined material always containing an appreciable amount of halogen; therefore, it no longer warrants consideration as a separate allotrope. The yellow form, obtained by oxidation of *stibine*, SbH_3, is probably impure, and is not a distinct form. Natural antimony is made of two stable isotopes, ^{121}Sb and ^{123}Sb. Metallic antimony is an extremely brittle metal of a flaky, crystalline texture. It is bluish white and has a metallic luster. It is not acted on by air at room temperature, but burns brilliantly when heated with the formation of white fumes of Sb_2O_3. It is a poor conductor of heat and electricity, and has a hardness of 3 to 3.5. Antimony, available commercially with a purity of 99.999 + %, is finding use in semiconductor technology for making infrared detectors, diodes, and Hall-effect devices. Commercial-grade antimony is widely used in alloys with percentages ranging from 1% to 20%. It greatly increases the hardness and mechanical strength of lead. Batteries, antifriction alloys, type metal, small arms and tracer bullets, cable sheathing, and minor products use about half the metal produced. Compounds taking up the other half are oxides, sulfides, sodium antimonate, and antimony trichloride. These are used in manufacturing flame-proofing compounds, paints, ceramic enamels, glass, and pottery. Tartar emetic (hydrated potassium antimonyl tartrate) has been used in medicine. Antimony and many of its compounds are toxic.

Argon — (Gr. *argos,* inactive), Ar. Its presence in air was suspected by Cavendish in 1785, discovered by Lord Rayleigh and Sir William Ramsay in 1894. The gas is prepared by fractionation of liquid air, the atmosphere containing 0.934% argon. The atmosphere of Mars contains 1.6% of ^{40}Ar and 5 ppm of ^{36}Ar. Argon is two and one half times as soluble in water as nitrogen, having about the same solubility as oxygen. It is recognized by the characteristic lines in the red end of the spectrum. It is used in electric light bulbs and in fluorescent tubes at a pressure of about 400 Pa, and in filling photo tubes, glow tubes, etc. Argon is also used as an inert gas shield for arc welding and cutting; as a blanket for the production of titanium and other reactive elements; and as a protective atmosphere for growing silicon and germanium crystals. Argon is colorless and odorless, both as a gas and liquid. It is available in high-purity form. Argon is considered to be more inert than the other rare gases, although a compound HArF has been prepared at extremely low temperatures. However, it does form a hydrate having a dissociation pressure of 105 atm at 0 °C. Ion molecules such as $(ArKr)^+$, $(ArXe)^+$, $(NeAr)^+$ have been observed spectroscopically. Argon also forms a clathrate with β-hydroquinone. This clathrate is stable and can be stored for a considerable time, but a true chemical bond does not exist.

Arsenic — (L. *arsenicum,* Gr. *arsenikon,* yellow orpiment, identified with *arsenikos,* male, from the belief that metals were different sexes; Arabic, *Az-zernikh,* the orpiment from Persian *zerni-zar,* gold), As. It is believed that Albertus Magnus obtained the element in 1250 A.D. In 1649 Schroeder published two methods of preparing the element. It is found native, in the sulfides *realgar* and *orpiment,* as arsenides and sulfarsenides of heavy metals, as the oxide, and as arsenates. *Mispickel,* arsenopyrite, (FeSAs) is the most common mineral, from which on heating the arsenic sublimes leaving ferrous sulfide. The most common form of the element, gray arsenic, is a steel gray, very brittle, crystalline, semimetallic solid; it tarnishes in air, and when heated is rapidly oxidized to arsenous oxide (As_2O_3) with the odor of garlic. Two other allotropes, black arsenic and yellow arsenic, can be made by condensing arsenic vapor. Arsenic and its compounds are extremely poisonous. Arsenic is also used in bronzing, pyrotechny, and for hardening and improving the sphericity of shot. The most important compounds are white arsenic (As_2O_3); the sulfide; Paris green, $3Cu(AsO_2)_2 \cdot Cu(C_2H_3O_2)_2$; calcium arsenate; and lead arsenate. The last three have been used as agricultural insecticides and poisons. Marsh's test for arsenic compounds makes use of the formation and ready decomposition of arsine (AsH_3); it has long been used in forensics. Arsenic is finding increasing uses as a doping agent in solid-state devices such as transistors. Gallium arsenide is used as a laser material and in various electronic devices.

Astatine — (Gr. *astatos,* unstable), At. Synthesized in 1940 by D.R. Corson, K. R. MacKenzie, and E. Segre at the University of California by bombarding bismuth with alpha particles. The longest-lived isotope, ^{210}At, has a half-life of only 8.1 hours. Over 50 other isotopes and isomers are now known. Minute quantities of ^{215}At, ^{218}At, and ^{219}At exist in equilibrium in nature with naturally occurring uranium and thorium isotopes, and traces of ^{217}At are in equilibrium with ^{233}U and ^{239}Np resulting from interaction of thorium and uranium with naturally produced neutrons. The total amount of astatine present in the Earth's crust, however, is probably less than 30 g. Only about 0.05 μg of astatine has been prepared to date. Mass spectrometers have been used to confirm that this highly radioactive halogen behaves chemically very much like other halogens, particularly iodine. The interhalogen compounds AtI, AtBr, and AtCl are known to form, but it is not yet known if astatine forms diatomic astatine molecules. HAt and CH_3At (methyl astatide) have been detected. Astatine is said to be more metallic that iodine, and, like iodine, it probably accumulates in the thyroid gland.

Barium — (Gr. *barys,* heavy), Ba. Baryta was distinguished from lime by Scheele in 1774; the element was discovered by Sir Humphrey Davy in 1808. It is found only in combination with other elements, chiefly in *barite* or *heavy spar* (sulfate) and *witherite* (carbonate) and is prepared by electrolysis of the chloride. Large deposits of barite are found in China, Germany, India, Morocco, and in the U.S. Barium is a metallic element, soft, and when pure is silvery white like lead; it belongs to the alkaline earth group, resembling calcium chemically. The metal oxidizes very easily and should be kept under petroleum or other suitable oxygen-free liquids to exclude air. It is decomposed by water or alcohol. The metal is used as a "getter" in vacuum tubes. The most important compounds are the peroxide (BaO_2), chloride, sulfate, carbonate, nitrate, and chlorate. Lithopone, a pigment containing barium sulfate and zinc sulfide, has good covering power, and does not darken in the presence of sulfides. The sulfate, as permanent white or *blanc fixe,* is also used in paint, in X-ray diagnostic work, and in glassmaking. *Barite* is extensively used as a weighting agent in oil well drilling fluids, and also in making rubber. The carbonate has been used as a rat poison, while the nitrate and chlorate give green colors in pyrotechny. The impure sulfide phosphoresces after exposure to the light. All barium compounds that are water or acid soluble are poisonous.

Berkelium — (*Berkeley,* home of the University of California), Bk. Berkelium, the eighth member of the actinide transition series, was discovered in December 1949 by Thompson, Ghiorso, and Seaborg, and was the fifth transuranium element synthesized. It was produced by cyclotron bombardment of milligram amounts of ^{241}Am with helium ions at Berkeley, California. The first isotope produced had a mass number of

243 and decayed with a half-life of 4.5 hours. Thirteen isotopes are now known and have been synthesized. The existence of ^{249}Bk, with a half-life of 320 days, makes it feasible to isolate berkelium in weighable amounts so that its properties can be investigated with macroscopic quantities. One of the first visible amounts of a pure berkelium compound, berkelium chloride, was produced in 1962. It weighed 3 ng. Elemental berkelium is expected to be a silvery metal, easily soluble in dilute mineral acids, and readily oxidized by air or oxygen at elevated temperatures to form the oxide. X-ray diffraction methods have been used to identify the following compounds: BkO_2, BkO_3, BkF_3, BkCl, and BkOCl. Berkelium most likely resembles terbium with respect to chemical properties. As with other actinide elements, berkelium tends to accumulate in the skeletal system. Because of its rarity, berkelium presently has no commercial or technological use.

Beryllium — (Gr. *beryllos, beryl;* also called Glucinium or Glucinum, Gr. *glykys,* sweet), Be. Discovered as the oxide by Vauquelin in beryl and in emeralds in 1798. The metal was isolated in 1828 by Wöhler and by Bussy independently by the action of potassium on beryllium chloride. Beryllium is found in some 30 mineral species, the most important of which are *bertrandite, beryl, chrysoberyl,* and *phenacite. Aquamarine* and *emerald* are precious forms of *beryl.* Beryllium minerals are found in the U.S., Brazil, Russia, Kazakhstan, and elsewhere. Colombia is known for its emeralds. *Beryl* ($3BeO \cdot Al_2O_3 \cdot 6SiO_2$) and *bertrandite* ($4BeO \cdot 2SiO_2 \cdot H_2O$) are the most important commercial sources of the element and its compounds. Most of the metal is now prepared by reducing beryllium fluoride with magnesium metal. Beryllium metal did not become readily available to industry until 1957. The metal, steel gray in color, has many desirable properties. It is one of the lightest of all metals, and has one of the highest melting points of the light metals. Its modulus of elasticity is about one third greater than that of steel. It resists attack by concentrated nitric acid, has excellent thermal conductivity, and is nonmagnetic. It has a high permeability to X-rays, and when bombarded by alpha particles, as from radium or polonium, neutrons are produced in the ratio of about 30 neutrons/million alpha particles. At ordinary temperatures beryllium resists oxidation in air, although its ability to scratch glass is probably due to the formation of a thin layer of the oxide. Beryllium is used as an alloying agent in producing beryllium copper, which is extensively used for springs, electrical contacts, spot-welding electrodes, and non-sparking tools. It has found application as a structural material for high-speed aircraft, missiles, spacecraft, and communication satellites. It was used in the windshield frame, brake discs, support beams, and other structural components of the space shuttle. Because beryllium is relatively transparent to X-rays, ultra-thin Be-foil is finding use in X-ray lithography for reproduction of microminiature integrated circuits. Natural beryllium is 100% ^{9}Be. Beryllium is used in nuclear reactors as a reflector or moderator for it has a low thermal neutron absorption cross section. It is used in gyroscopes, computer parts, and instruments where lightness, stiffness, and dimensional stability are required. The oxide has a very high melting point and is also used in nuclear technology and ceramic applications. Beryllium and its salts are toxic and should be handled with the greatest of care. They should not be tasted to verify the sweetish nature of beryllium (as did early experimenters). The metal, its alloys, and its salts can be handled safely if certain work codes are observed, but no attempt should be made to work with beryllium before becoming familiar with proper safeguards.

Bismuth — (Ger. *Weisse Masse,* white mass; later *Wisuth* and *Bisemutum*), Bi. In early times bismuth was confused with tin and lead. Claude Geoffroy the Younger showed it to be distinct from lead in 1753. It is a white crystalline, brittle metal with a pinkish tinge. It occurs native. The most important ores are *bismuthinite* or bismuth glance (Bi_2S_3) and *bismite* (Bi_2O_3). Peru, Japan, Mexico, Bolivia, and Canada are major bismuth producers. Much of the bismuth produced in the U.S. is obtained as a by-product in refining lead, copper, tin, silver, and gold ores. Bismuth is the most diamagnetic of all metals, and the thermal conductivity is lower than any metal, except mercury. It has a high electrical resistance, and has the highest Hall effect of any metal (i.e., greatest increase in electrical resistance when placed in a magnetic field). "Bismanol" is a permanent magnet of high coercive force, made of MnBi. Bismuth expands 3.32% on solidification. This property makes bismuth alloys particularly suited to the making of sharp castings of objects subject to damage by high temperatures. With other metals such as tin, cadmium, etc., bismuth forms low-melting alloys that are extensively used for safety devices in fire detection and extinguishing systems. Bismuth is used in producing malleable irons and is finding use as a catalyst for making acrylic fibers. When bismuth is heated in air it burns with a blue flame, forming yellow fumes of the oxide. The metal is also used as a thermocouple material, and has found application as a carrier for U^{235} or U^{233} fuel in atomic reactors. Its soluble salts are characterized by forming insoluble basic salts on the addition of water, a property sometimes used in detection work. Bismuth oxychloride is used extensively in cosmetics. Bismuth subnitrate and subcarbonate are used in medicine. Natural bismuth contains only one isotope ^{209}Bi.

Bohrium — (Named after Niels Bohr [1885–1962], Danish atomic and nuclear physicist), Bh. Bohrium is expected to have chemical properties similar to rhenium. This element was synthesized and unambiguously identified in 1981 using the Universal Linear Accelerator (UNILAC) at the Gesellschaft für Schwerionenforschung (G.S.I.) in Darmstadt, Germany. The discovery team was led by Armbruster and Münzenberg. The reaction producing the element was proposed and applied earlier by a Dubna Group led by Oganessian in 1976. A target of ^{209}Bi was bombarded by a beam of ^{54}Cr ions. In 1983 experiments at Dubna using the 157-inch cyclotron, produced ^{262}Bh by the reaction $^{209}Bi + {}^{54}Cr$. IUPAC adopted the name Bohrium with the symbol Bh for Element 107 in August 1997. Ten isotopes of bohrium are now recognized, two of which have lifetimes of around 1 minute.

Boron — (Ar. *Buraq,* Pers. *Burah*), B. Boron compounds have been known for thousands of years, but the element was not discovered until 1808 by Sir Humphry Davy and by Gay-Lussac and Thenard. The element is not found free in nature, but occurs as orthoboric acid, usually in certain volcanic spring waters, and as borates in *borax* and *colemanite. Ulexite,* another boron mineral, is interesting as it is nature's own version of fiber optics. Important sources of boron are the ores *rasorite (kernite)* and *tincal (borax ore).* Both of these ores are found in the Mojave Desert. *Tincal* is the most important source of boron from the Mojave. Extensive *borax* deposits are also found in Turkey. Boron exists naturally as 19.9% ^{10}B isotope and 80.1% ^{11}B isotope. High-purity crystalline boron may be prepared by the vapor phase reduction of boron trichloride or tribromide with hydrogen on electrically heated filaments. The impure, or amorphous, boron, a brownish-black powder, can be obtained by heating the trioxide with magnesium powder. Boron of 99.9999% purity has been produced and is available commercially. Elemental boron has an energy band gap of 1.50 to 1.56 eV, which is higher than that of either silicon or germanium. It has interesting optical characteristics, transmitting portions of the infrared, and is a poor conductor of electricity at room temperature, but a good conductor at high temperature. Amorphous boron is used in pyrotechnic flares to provide a distinctive green color, and in rockets as an igniter. By far the most commercially important boron compound in terms of dollar sales is $Na_2B_4O_7 \cdot 5H_2O$. This pentahydrate is used in very large quantities in the manufacture of insulation fiberglass and sodium perborate bleach. Boric acid is also an important boron compound with major markets in textile fiberglass and in cellulose insulation as a flame retardant. Next in order of importance is borax ($Na_2B_4O_7 \cdot 10H_2O$) which is used principally in laundry products. Use of borax as a mild antiseptic is minor in terms of dollars and tons. Boron compounds are also extensively used in the manufacture of borosilicate glasses. The isotope ^{10}B is used as a control for nuclear reactors, as a shield for nuclear radiation, and in instruments used for detecting neutrons. Boron nitride has remarkable properties and can be used to make a material as hard as diamond. The nitride also behaves like an electrical insulator but conducts heat like a metal. It also has lubricating properties similar to graphite. The hydrides are easily oxidized with considerable energy liberation, and have been studied for use as rocket fuels. Demand is increasing for boron filaments, a high-strength, lightweight material chiefly employed for advanced aerospace structures. Boron is similar to carbon in that it has a capacity to form stable covalently bonded molecular networks. Carboranes, metalloboranes, phosphacarboranes, and other families comprise thousands of compounds. Elemental boron and the borates are not considered to be toxic, and they do not require special care in handling. However, some of the more exotic boron hydrogen compounds are toxic and do require care.

Bromine — (Gr. *bromos*, stench), Br. Discovered by Balard in 1826, but not prepared in quantity until 1860. A member of the halogen group of elements, it is obtained from natural brines from wells in Michigan and Arkansas. Little bromine is extracted today from seawater, which contains only about 85 ppm. Bromine is the only liquid nonmetallic element. It is a heavy, mobile, reddish-brown liquid, volatilizing readily at room temperature to a red vapor with a strong disagreeable odor, resembling chlorine, and having a very irritating effect on the eyes and throat. It is readily soluble in water or carbon disulfide, forming a red solution; is less active than chlorine but more so than iodine. It unites readily with many elements and has a bleaching action; when spilled on the skin it produces painful sores. It presents a serious health hazard, and maximum safety precautions should be taken when handling it. Much of the bromine output in the U.S. was used in the production of ethylene dibromide, a lead scavenger used in making gasoline antiknock compounds. Since lead in gasoline was eliminated because of environmental considerations, production of bromine has decreased. Bromine is also used in making fumigants, flameproofing agents, water purification compounds, dyes, medicinals, sanitizers, etc. Organic bromides are also important.

Cadmium — (L. *cadmia;* Gr. *kadmeia* - ancient name for calamine, zinc carbonate), Cd. Discovered by Stromeyer in 1817 from an impurity in zinc carbonate. Cadmium most often occurs in small quantities associated with zinc ores, such as *sphalerite* (ZnS). *Greenockite* (CdS) is the only mineral of any consequence bearing cadmium. Almost all cadmium is obtained as a by-product in the treatment of zinc, copper, and lead ores. It is a soft, bluish-white metal which is easily cut with a knife. It is similar in many respects to zinc. It is a component of some of the lowest melting alloys; it is used in bearing alloys with low coefficients of friction and great resistance to fatigue; it is used extensively in electroplating, which accounts for about 60% of its use. It is also used in many types of solder, for standard E.M.F. cells, for Ni-Cd batteries, and as a barrier to control atomic fission. The market for Ni-Cd batteries has grown significantly. Cadmium compounds are used in blue and green phosphors for color TV tubes. It forms a number of salts, of which the sulfate is most common; the sulfide is used as a yellow pigment. Cadmium and solutions of its compounds are toxic. Failure to appreciate the toxic properties of cadmium may cause workers to be unwittingly exposed to dangerous fumes. Some silver solders, for example, contain cadmium and should be handled with care. Serious toxicity problems have been found from long-term exposure and work with cadmium plating baths. Cadmium is present in certain phosphate rocks. This has raised concerns that the long-term use of certain phosphate fertilizers might pose a health hazard from levels of cadmium that might enter the food chain. At one time the International Conference on Weights and Measures defined the meter in terms of the wavelength of the red cadmium spectral line.

Calcium — (L. *calx*, lime), Ca. Though lime was prepared by the Romans in the first century under the name calx, the metal was not discovered until 1808. After learning that Berzelius and Pontin prepared calcium amalgam by electrolyzing lime in mercury, Davy was able to isolate the impure metal. Calcium is a metallic element, fifth in abundance in the Earth's crust, of which it forms more than 3%. It is an essential constituent of leaves, bones, teeth, and shells. Never found in nature uncombined, it occurs abundantly as *limestone* ($CaCO_3$), *gypsum* ($CaSO_4 \cdot 2H_2O$), and *fluorite* (CaF_2); *apatite* is the fluorophosphate or chlorophosphate of calcium. The metal has a silvery color, is rather hard, and is prepared by electrolysis of the fused chloride to which calcium fluoride is added to lower the melting point. Chemically it is one of the alkaline earth elements; it readily forms a white coating of oxide in air, reacts with water, burns with a yellow-red flame, largely forming the oxide. The metal is used as a reducing agent in preparing other metals such as thorium, uranium, zirconium, etc., and is used as a deoxidizer, desulfurizer, and inclusion modifier for various ferrous and nonferrous alloys. It is also used as an alloying agent for aluminum, beryllium, copper, lead, and magnesium alloys, and serves as a "getter" for residual gases in vacuum tubes. Its natural and prepared compounds are widely used. Quicklime (CaO), made by heating limestone and changed into slaked lime by the careful addition of water, is the great cheap base of the chemical industry with countless uses. Mixed with sand it hardens as mortar and plaster by taking up carbon dioxide from the air. Calcium from limestone is an important element in Portland cement. The solubility of the carbonate in water containing carbon dioxide causes the formation of caves with stalactites and stalagmites and is responsible for hardness in water. Other important compounds are the carbide (CaC_2), chloride ($CaCl_2$), cyanamide ($CaCN_2$), hypochlorite ($Ca(OCl)_2$), nitrate ($Ca(NO_3)_2$), and sulfide (CaS). Calcium sulfide is phosphorescent after being exposed to light.

Californium — (State and University of California), Cf. Californium, the sixth transuranium element to be discovered, was produced by Thompson, Street, Ghioirso, and Seaborg in 1950 by bombarding microgram quantities of ^{242}Cm with 35 MeV helium ions in the Berkeley 60-inch cyclotron. The existence of the isotopes ^{249}Cf, ^{250}Cf, ^{251}Cf, and ^{252}Cf makes it feasible to isolate californium in weighable amounts so that its properties can be investigated. Californium-252 is a very strong neutron emitter. One microgram releases 170 million neutrons per minute, which presents biological hazards. Proper safeguards should be used in handling californium. Twenty isotopes of californium are now recognized. ^{249}Cf and ^{252}Cf have half-lives of 351 years and 900 years, respectively. Californium (III) is the only ion stable in aqueous solutions, all attempts to reduce or oxidize californium (III) having failed. In 1960 a few tenths of a microgram of californium trichloride, $CfCl_3$, californium oxychloride, CfOCl, and californium oxide, Cf_2O_3, were first prepared. Because californium is a very efficient source of neutrons, many new uses are expected for it. It has already found use in neutron moisture gages and in well-logging (the determination of water and oil-bearing layers). It is also being used as a portable neutron source for discovery of metals such as gold or silver by on-the-spot activation analysis.

Inorganic

Carbon — (L. *carbo,* charcoal), C. Carbon, an element of prehistoric discovery, is very widely distributed in nature. It is found in abundance in the sun, stars, comets, and atmospheres of most planets. Carbon in the form of microscopic diamonds is found in some meteorites. Natural diamonds are found in *kimberlite* or *lamporite* of ancient formations called "pipes," such as found in South Africa, Arkansas, and elsewhere. Diamonds are being recovered from the ocean floor off the Cape of Good Hope. About 30% of all industrial diamonds used in the U.S. are now made synthetically. The energy of the sun and stars can be attributed at least in part to the carbon-nitrogen cycle of nuclear reactions. Carbon is found free in nature in three allotropic forms: amorphous, graphite, and diamond. Graphite is one of the softest known materials while diamond is one of the hardest. Graphite exists in two forms: alpha and beta. These have identical physical properties, except for their crystal structure. Naturally occurring graphites are reported to contain as much as 30% of the rhombohedral (beta) form, whereas synthetic materials contain only the alpha form. The hexagonal alpha type can be converted to the beta by mechanical treatment, and the beta form reverts to the alpha on heating it above 1000 °C. Fullerenes or "buckyballs" have a number of unusual properties. These molecules, consisting of 60 or 70 carbon atoms linked together, are capable of withstanding great pressure and trapping foreign atoms inside their network of carbon. Fullerene films doped with alkali atoms remain superconductive at temperatures as high as 45 K. In combination, carbon is found as carbon dioxide in the atmosphere of the Earth and dissolved in all natural waters. On average, the earth's atmosphere contains about 0.04% CO_2, while the atmosphere of Mars contains 95.32% CO_2. It is a component of great rock masses in the form of carbonates of calcium (limestone), magnesium, and iron. Coal, petroleum, and natural gas are chiefly hydrocarbons. Carbon is unique among the elements in the vast number and variety of compounds it can form. With hydrogen, oxygen, nitrogen, and other elements, it forms a very large number of compounds, carbon atom often being linked to carbon atom. Many thousands of carbon compounds are vital to organic and life processes. Without carbon, the basis for life as we know it would be impossible. While it has been thought that silicon might take the place of carbon in forming a host of similar compounds, it is now not possible to form stable compounds with very long chains of silicon atoms. Some of the compounds of carbon of greatest industrial importance are carbon dioxide (CO_2), carbon monoxide (CO), carbon disulfide (CS_2), chloroform ($CHCl_3$), carbon tetrachloride (CCl_4), methane (CH_4), ethylene (C_2H_4), acetylene (C_2H_2), benzene (C_6H_6), ethyl alcohol (C_2H_5OH), acetic acid (CH_3COOH), and their derivatives. Natural carbon consists of 98.93% ^{12}C and 1.07% ^{13}C. In 1961 the International Union of Pure and Applied Chemistry adopted the isotope carbon-12 as the basis for atomic weights. Carbon-14, an isotope with a half-life of 5715 years, has been widely used to date such materials as wood, archeological specimens, etc. A brittle form of carbon, known as "glassy" or vitreous carbon, has been developed. It has a high resistance to corrosion, has good thermal stability, and is structurally impermeable to both gases and liquids. It has a randomized structure, making it useful in applications such as crystal growing, crucibles for high-temperature use, prosthetic devices, etc.

Cerium — (named for the asteroid *Ceres,* which was discovered in 1801 only 2 years before the element), Ce. Discovered in 1803 by Klaproth and by Berzelius and Hisinger; metal prepared by Hillebrand and Norton in 1875. Cerium is the most abundant of the metals of the so-called rare earths. It is found in a number of minerals including *allanite* (also known as *orthite*), *monazite, bastnasite, cerite,* and *samarskite.* Monazite and bastnasite are presently the two most important sources of cerium. Large deposits of monazite found on the beaches of Travancore, India, in river sands in Brazil, and deposits of *allanite* in the western United States, and *bastnasite* in Southern California will supply cerium, thorium, and the other rare-earth metals for many years to come. Metallic cerium is prepared by metallothermic reduction techniques, such as by reducing cerous fluoride with calcium, or by electrolysis of molten cerous chloride or other cerous halides. The metallothermic technique is used to produce high-purity cerium. Cerium is especially interesting because of its variable electronic structure. The energy of the inner 4f level is nearly the same as that of the outer or valence electrons, and only small amounts of energy are required to change the relative occupancy of these electronic levels. This gives rise to dual valency states. For example, a volume change of about 10% occurs when cerium is subjected to high pressures or low temperatures. It appears that the valence changes from about 3 to 4 when it is cooled or compressed. The low temperature behavior of cerium is complex. Four allotropic modifications are thought to exist: cerium at room temperature and at atmospheric pressure is known as γ cerium. Upon cooling to −16 °C, γ cerium changes to β cerium. The remaining γ cerium starts to change to α cerium when cooled to −172 °C, and the transformation is complete at −269 °C. α Cerium has a density of 8.16; δ cerium exists above 726 °C. At atmospheric pressure, liquid cerium is more dense than its solid form at the melting point. Cerium is an iron-gray lustrous metal. It is malleable, and oxidizes very readily at room temperature, especially in moist air. Except for europium, cerium is the most reactive of the "rare-earth" metals. It slowly decomposes in cold water, and rapidly in hot water. Alkali solutions and dilute and concentrated acids attack the metal rapidly. The pure metal is likely to ignite if scratched with a knife. Ceric salts are orange red or yellowish; cerous salts are usually white. Cerium is a component of misch metal, which is extensively used in the manufacture of pyrophoric alloys for cigarette lighters, etc. Natural cerium is stable and contains four isotopes. Thirty-two other radioactive isotopes and isomers are known. While cerium is not radioactive, the impure commercial grade may contain traces of thorium, which is radioactive. The oxide is an important constituent of incandescent gas mantles and it is emerging as a hydrocarbon catalyst in "self-cleaning" ovens. In this application it can be incorporated into oven walls to prevent the collection of cooking residues. As ceric sulfate it finds extensive use as a volumetric oxidizing agent in quantitative analysis. Cerium compounds are used in the manufacture of glass, both as a component and as a decolorizer. The oxide is finding increased use as a glass polishing agent instead of rouge, for it is much faster than rouge in polishing glass surfaces. Cerium compounds are finding use in automobile exhaust catalysts. Cerium is also finding use in making permanent magnets. Cerium, with other rare earths, is used in carbon-arc lighting, especially in the motion picture industry. It is also finding use as an important catalyst in petroleum refining and in metallurgical and nuclear applications. In small lots, cerium costs about $5/g (99.9%).

Cesium — (L. *caesius,* sky blue), Cs. Cesium was discovered spectroscopically by Bunsen and Kirchhoff in 1860 in mineral water from Durkheim. Cesium, an alkali metal, occurs in *lepidolite, pollucite* (a hydrated silicate of aluminum and cesium), and in other sources. One of the world's richest sources of cesium is located at Bernic Lake, Manitoba. The deposits are estimated to contain 300,000 tons of pollucite, averaging 20% cesium. It can be isolated by electrolysis of the fused cyanide and by a number of other methods. Very pure, gas-free cesium can be prepared by thermal decomposition of cesium azide. The metal is characterized by a spectrum containing two bright lines in the blue along with several others in the red, yellow, and green. It is silvery white, soft, and ductile. It is the most electropositive and most alkaline element. Cesium, gallium, and mercury are the only three metals that are liquid near room temperature. Cesium reacts explosively with cold water, and reacts with ice at temperatures above −116 °C. Cesium hydroxide, the strongest base known, attacks glass. Because of its great affinity for oxygen the metal is used as a "getter" in electron tubes. It is also used in photoelectric cells, as well as a catalyst in the hydrogenation of certain organic compounds. The metal has recently found application in ion propulsion systems. Cesium is used in atomic clocks; a second of time is now defined as being the duration of 9,192,631,770 periods of the radiation corresponding to the transition between the two hyperfine levels of the ground state of the cesium-133 atom. Its chief compounds are the chloride and the nitrate.

Chlorine — (Gr. *chloros,* greenish yellow), Cl. Discovered in 1774 by Scheele, who thought it contained oxygen; named in 1810 by Davy, who insisted it was an element. In nature it is found in the combined state only, chiefly with sodium and potassium as common salt (NaCl), *carnallite* ($KMgCl_3 \cdot 6H_2O$), and *sylvite* (KCl). It is a member of the halogen (salt-forming) group of elements and is obtained from chlorides by the action of oxidizing agents and more often by electrolysis; it is a greenish-yellow gas, combining directly with nearly all elements. At 10 °C one volume of water dissolves 3.10 volumes of chlorine, at 30 °C only 1.77 volumes. Chlorine is widely used in making many everyday products. It is used for producing safe drinking water the world over. Even the smallest water supplies are now usually chlorinated. It is also extensively used in the production of paper products, dyestuffs, textiles, petroleum products, medicines, antiseptics, insecticides, foodstuffs, solvents, paints, plastics, and many other consumer products. Most of the chlorine produced is used in the manufacture of chlorinated compounds for sanitation, pulp bleaching, disinfectants, and textile processing. Further use is in the manufacture of chlorates, chloroform, carbon tetrachloride, and in the extraction of bromine. Organic chemistry demands much from chlorine, both as an oxidizing agent and in substitution, since it often brings desired properties in an organic compound when substituted for hydrogen, as in one form of synthetic rubber. Chlorine is a respiratory irritant. The gas irritates the mucous membranes and the liquid burns the skin. As little as 3.5 ppm can be detected as an odor, and 1000 ppm is likely to be fatal after a few deep breaths. It was used as a war gas in 1915. Natural chlorine contains two isotopes, ^{35}Cl and ^{37}Cl.

Chromium — (Gr. *chroma,* color), Cr. Discovered in 1797 by Vauquelin, who prepared the metal the next year, chromium is a steel-gray, lustrous, hard metal that takes a high polish. The principal ore is *chromite* ($FeCr_2O_4$), which is found in Zimbabwe, Russia, South Africa, Turkey, Iran, Albania, Finland, Madagascar, the Philippines, and elsewhere. The U.S. has no appreciable chromite ore reserves. The metal is usually produced by reducing the oxide with aluminum. Chromium is used to harden steel, to manufacture stainless steel, and to form many useful alloys. Much is used in plating to produce a hard, beautiful surface and to prevent corrosion. Chromium is used to give glass an emerald green color. It finds wide use as a catalyst. All compounds of chromium are colored; the most important are the chromates of sodium and potassium (K_2CrO_4) and the dichromates ($K_2Cr_2O_7$) and the potassium and ammonium chrome alums, as $KCr(SO_4)_2 \cdot 12H_2O$. The dichromates are used as oxidizing agents in quantitative analysis, also in tanning leather. Other compounds are of industrial value; lead chromate is chrome yellow, a valued pigment. Chromium compounds are used in the textile industry as mordants, and by the aircraft and other industries for anodizing aluminum. The refractory industry has found chromite useful for forming bricks and shapes, as it has a high melting point, moderate thermal expansion, and stability of crystalline structure. Chromium is an essential trace element for human health. Many chromium compounds, however, are acutely or chronically toxic, and some are carcinogenic. They should be handled with proper safeguards.

Cobalt — (*Kobald,* from the German, goblin or evil spirit, *cobalos,* Greek, mine), Co. Discovered by Brandt about 1735. Cobalt occurs in the mineral *cobaltite, smaltite,* and *erythrite,* and is often associated with nickel, silver, lead, copper, and iron ores, from which it is most frequently obtained as a by-product. It is also present in meteorites. Important ore deposits are found in Congo-Kinshasa, Australia, Zambia, Russia, Canada, and elsewhere. The U.S. Geological Survey has announced that the bottom of the north central Pacific Ocean may have cobalt-rich deposits at relatively shallow depths in waters close to the Hawaiian Islands and other U.S. Pacific territories. Cobalt is a brittle, hard metal, closely resembling iron and nickel in appearance. It has a magnetic permeability of about two thirds that of iron. Cobalt tends to exist as a mixture of two allotropes over a wide temperature range; the β-form predominates below 400 °C, and the α above that temperature. The transformation is sluggish and accounts in part for the wide variation in reported data on physical properties of cobalt. It is alloyed with iron, nickel and other metals to make Alnico, an alloy of unusual magnetic strength with many important uses. Stellite alloys, containing cobalt, chromium, and tungsten, are used for high-speed, heavy-duty, high-temperature cutting tools, and for dies. Cobalt is also used in other magnet steels and stainless steels, and in alloys used in jet turbines and gas turbine generators. The metal is used in electroplating because of its appearance, hardness, and resistance to oxidation. The salts have been used for centuries for the production of brilliant and permanent blue colors in porcelain, glass, pottery, tiles, and enamels. It is the principal ingredient in Sevre's and Thenard's blue. A solution of the chloride ($CoCl_2 \cdot 6H_2O$) is used as sympathetic ink. The cobalt ammines are of interest; the oxide and the nitrate are important. Cobalt carefully used in the form of the chloride, sulfate, acetate, or nitrate has been found effective in correcting a certain mineral deficiency disease in animals. Soils should contain 0.13 to 0.30 ppm of cobalt for proper animal nutrition. Cobalt is found in Vitamin B-12, which is essential for human nutrition. Cobalt-60, an artificial isotope, is an important gamma ray source, and is extensively used as a tracer and a radiotherapeutic agent.

Copernicium — (After the astronomer Nicolaus Copernicus), Cn. In late February 1996, Siguard Hofmann and his collaborators at GSI Darmstadt announced their discovery of Element-112, having 112 protons and 165 neutrons, with an atomic mass of 277. This element was made by bombarding a lead target with high-energy zinc ions. A single nucleus of Element-112 was detected, which was identified by its decay pattern. Evidence indicates that nuclei with 162 neutrons are held together more strongly than nuclei with a smaller or larger number of neutrons. This suggests a narrow "peninsula" of relatively stable isotopes around Element-114. In 2010 IUPAC approved the name and symbol Copernicium, Cn, for Element-112. Theoretical calculations and limited experimental measurements suggest that copernecium will have properties similar to mercury, possibly existing as a gas at ambient temperature.

Copper — (L. *cuprum,* from the island of Cyprus), Cu. The discovery of copper dates from prehistoric times. It is said to have been mined for more than 5000 years. It is one of man's most important metals. Copper is reddish colored, takes on a bright metallic luster, and is malleable, ductile, and a good conductor of heat and electricity (second only to silver in electrical conductivity). The electrical industry is one of the greatest users of copper. Copper occasionally occurs native, and is found in many minerals such as *cuprite, malachite, azurite, chalcopyrite,* and *bornite.* Large copper ore deposits are found in the U.S., Chile, Zambia, Zaire, Peru, and Canada. The most important copper ores are the sulfides, oxides, and carbonates. From these, copper is obtained by smelting, leaching, and by electrolysis. Its alloys, brass and bronze, long used, are still very important; all American coins are now copper alloys; monel and gun metals also contain copper. The most important compounds are the oxide and the sulfate, blue vitriol; the latter has wide use as an agricultural poison and as an algicide in water purification. Copper compounds such as Fehling's solution are widely used in analytical chemistry in tests for sugar. High-purity copper (99.999 + %) is readily available commercially.

Curium — (Pierre and Marie Curie), Cm. Although curium follows americium in the periodic system, it was actually known before americium and was the third transuranium element to be discovered. It was identified by Seaborg, James, Morgan, and Ghiorso in 1944 at the wartime Metallurgical Laboratory in Chicago as a result of helium-ion bombardment of ^{239}Pu in the Berkeley, California, 60-inch cyclotron. Visible amounts (30 μg) of ^{242}Cm, in the form of the hydroxide, were first isolated by Werner and Perlman of the University of California in 1947. In 1950, Crane, Wallmann, and Cunningham found that the magnetic susceptibility of microgram samples of CmF_3 was of the same magnitude as that of GdF_3. This provided direct experimental evidence for assigning an electronic configuration to Cm^{+3}. In 1951, the same workers prepared curium in its elemental form for the first time. Sixteen isotopes of curium are now known. The most stable, ^{247}Cm, with a half-life of 16 million years, is so short compared to the Earth's age that any primordial curium must have disappeared long ago from the natural scene. Minute amounts of curium probably exist in natural deposits of uranium, as a result of a sequence of neutron captures and β decays sustained by the very low flux of neutrons naturally present in uranium ores. ^{242}Cm and ^{244}Cm are available in multigram quantities. ^{248}Cm has been produced only in milligram amounts. Curium is similar in some regards to gadolinium, its rare-earth homolog, but it has a more complex crystal structure. Curium is silver in color, is chemically reactive, and is more electropositive than aluminum. CmO_2, Cm_2O_3, CmF_3, CmF_4, $CmCl_3$, $CmBr_3$, and CmI_3 have been prepared. Most compounds of trivalent curium are faintly yellow in color. ^{242}Cm generates about three watts of thermal energy per gram. This compares to one-half watt per gram of ^{238}Pu. This suggests use for curium as a power source. Curium absorbed into the body accumulates in the bones, and is therefore very toxic as its radiation destroys the red-cell forming mechanism. The maximum permissible total body burden of ^{244}Cm (soluble) in a human being is 0.3 μCi.

Darmstadtium — (Darmstadt, city in Germany), Ds. In 1987 Oganessian et al., at Dubna, claimed discovery of this element. Their experiments indicated the spontaneous fissioning nuclide $^{272}110$ with a half-life of 10 ms. More recently a group led by Armbruster at G.S.I. in Darmstadt, Germany, reported evidence of $^{269}110$, which was produced by bombarding lead for many days with more than 10^{18} nickel atoms. A detector searched each collision for Element 110's distinct decay sequence. On November 9, 1994, evidence of 110 was detected. In 2003 IUPAC approved the name darmstadtium, symbol Ds, for Element 110.

Dubnium — (named after the Joint Institute of Nuclear Research in Dubna, Russia). Db. In 1967 G. N. Flerov reported that a Soviet team working at Dubna may have produced a few atoms of $^{260}105$ and $^{261}105$ by bombarding ^{243}Am with ^{22}Ne. Later, it was reported that Dubna scientists synthesized Element 105 and determined its decay pattern and chemical properties. In April 1970, it was announced that Ghiorso, Nurmia, Harris, K. A. Y. Eskola, and P. L. Eskola, working at the University of California at Berkeley, had positively identified Element 105. When a ^{15}N nucleus is absorbed by a ^{249}Cf nucleus, four neutrons are emitted and a new atom of $^{260}105$ with a half-life of 1.6 s is formed. Soon after the discovery the names Hahnium and Joliotium, named after Otto Hahn and Jean-Frederic Joliot and Mme. Joliot-Curie, were suggested as names for Element 105. The name Hahnium was used in the literature for several years, leading to considerable confusion. The IUPAC in August 1997 finally resolved the issue, naming Element 105 Dubnium with the symbol Db. Dubnium is thought to have properties similar to tantalum.

Dysprosium — (Gr. *dysprositos,* hard to get at), Dy. Dysprosium was discovered in 1886 by Lecoq de Boisbaudran, but not isolated. Neither the oxide nor the metal was available in relatively pure form until the development of ion-exchange separation and metallographic reduction techniques by Spedding and associates about 1950. Dysprosium occurs along with other so-called rare-earth or lanthanide elements in a variety of minerals such as *xenotime, fergusonite, gadolinite, euxenite, polycrase,* and *blomstrandine.* The most important sources, however, are from *monazite* and *bastnasite.* Dysprosium can be prepared by reduction of the trifluoride with calcium. The element has a metallic, bright silver luster. It is relatively stable in air at room temperature, and is readily attacked and dissolved, with the evolution of hydrogen, by dilute and concentrated mineral acids. The metal is soft enough to be cut with a knife and can be machined without sparking if overheating is avoided. Small amounts of impurities can greatly affect its physical properties. While dysprosium has not yet found many applications, its thermal neutron absorption cross-section and high melting point suggest metallurgical uses in nuclear control applications and for alloying with special stainless steels. A dysprosium oxide-nickel cermet has found use in cooling nuclear reactor rods. This cermet absorbs neutrons readily without swelling or contracting under prolonged neutron bombardment. In combination with vanadium, dysprosium has been used in making laser materials. Dysprosium-cadmium chalcogenides, as sources of infrared radiation, have been used for studying chemical reactions.

Einsteinium — (Albert Einstein [1879–1955]), Es. Einsteinium, the seventh transuranic element of the actinide series to be discovered, was identified by Ghiorso and co-workers at Berkeley in December 1952 in debris from the first large thermonuclear explosion, which took place in the Pacific in November 1952. The isotope produced was the 20-day ^{253}Es isotope. In 1961, a sufficient amount of einsteinium was produced to permit separation of a macroscopic amount of ^{253}Es. This sample weighed about 0.01 μg. A special magnetic-type balance was used in making this determination. ^{253}Es so produced was used to produce mendelevium. About 3 μg of einsteinium has been produced at Oak Ridge National Laboratories by irradiating for several years kilogram quantities of ^{239}Pu in a reactor to produce ^{242}Pu. This was then fabricated into pellets of plutonium oxide and aluminum powder, and loaded into target rods for an initial 1-year irradiation at the Savannah River Plant, followed by irradiation in a HFIR (High Flux Isotopic Reactor). After 4 months in the HFIR the targets were removed for chemical separation of the einsteinium from californium. Twenty-one isotopes and isomers of einsteinium are now recognized. ^{254}Es has the longest half-life (276 days). Tracer studies using ^{253}Es show that einsteinium has chemical properties typical of a heavy trivalent, actinide element. Einsteinium is extremely radioactive. Great care must be taken when handling it.

Erbium — (*Ytterby*, a town in Sweden), Er. Erbium, one of the rare-earth elements of the lanthanide series, is found in the minerals mentioned under dysprosium above. In 1842 Mosander separated "yttria," found in the mineral *gadolinite,* into three fractions which he called *yttria, erbia,* and *terbia.* The names *erbia* and *terbia* became confused in this early period. After 1860, Mosander's *terbia* was known as *erbia,* and after 1877, the earlier known *erbia* became *terbia*. The *erbia* of this period was later shown to consist of five oxides, now known as *erbia, scandia, holmia, thulia* and *ytterbia.* By 1905 Urbain and James independently succeeded in isolating fairly pure Er_2O_3. Klemm and Bommer first produced reasonably pure erbium metal in 1934 by reducing the anhydrous chloride with potassium vapor. The pure metal is soft and malleable and has a bright, silvery, metallic luster. As with other rare-earth metals, its properties depend to a certain extent on the impurities present. The metal is fairly stable in air and does not oxidize as rapidly as some of the other rare-earth metals. Recent production techniques, using ion-exchange reactions, have resulted in much lower prices of the rare-earth metals and their compounds. Erbium is finding nuclear and metallurgical uses. Added to vanadium, for example, erbium lowers the hardness and improves workability. Most of the rare-earth oxides have sharp absorption bands in the visible, ultraviolet, and near infrared. This property, associated with the electronic structure, gives beautiful pastel colors to many of the rare-earth salts. Erbium oxide gives a pink color and has been used as a colorant in glasses and porcelain enamel glazes.

Europium — (Europe), Eu. In 1890 Boisbaudran obtained basic fractions from samarium-gadolinium concentrates that had spark spectral lines not accounted for by samarium or gadolinium. These lines subsequently were shown to belong to europium. The discovery of europium is generally credited to Demarcay, who separated the rare earth in reasonably pure form in 1901. The pure metal was not isolated until much later. Europium is now prepared by mixing Eu_2O_3 with a 10% excess of lanthanum metal and heating the mixture in a tantalum crucible under high vacuum. The element is collected as a silvery-white metallic deposit on the walls of the crucible. As with other rare-earth metals, except for lanthanum, europium ignites in air at about 150 to 180 °C. Europium is about as hard as lead and is quite ductile. It is the most reactive of the rare-earth metals, quickly oxidizing in air. It resembles calcium in its reaction with water. *Bastnasite* and *monazite* are the principal ores containing europium. Europium has been identified spectroscopically in the sun and certain stars. Europium isotopes are good neutron absorbers and are being studied for use in nuclear control applications. Europium oxide is now widely used as a phosphor activator and europium-activated yttrium vanadate is in commercial use as the red phosphor in color TV tubes. Europium-doped plastic has been used as a laser material. Europium is one of the rarest and most costly of the rare-earth metals, but with the development of ion-exchange techniques and special processes, the cost of the metal has been greatly reduced in recent years

Fermium — (Enrico Fermi [1901–1954], nuclear physicist), Fm. Fermium, the eighth transuranium element of the actinide series to be discovered, was identified by Ghiorso and co-workers in 1952 in the debris from a thermonuclear explosion in the Pacific in work involving the University of California Radiation Laboratory, the Argonne National Laboratory, and the Los Alamos Scientific Laboratory. The isotope produced was the 20-hour ^{255}Fm. During 1953 and early 1954, while discovery of elements 99 and 100 was withheld from publication for security reasons, a group from the Nobel Institute of Physics in Stockholm bombarded ^{238}U with ^{16}O ions, and isolated a 30-min α-emitter, which they ascribed to ^{250}100, without claiming discovery of the element. This isotope has since been identified positively, and the 30-min half-life confirmed. The chemical properties of fermium have been studied solely with tracer amounts, and in normal aqueous media only the (III) oxidation state appears to exist. The isotope ^{254}Fm and heavier isotopes can be produced by intense neutron irradiation of lower elements such as plutonium by a process of successive neutron capture interspersed with beta decays until these mass numbers and atomic numbers are reached.

Twenty-two isotopes and isomers of fermium are known to exist. ^{257}Fm, with a half-life of 100.5 days, is the longest lived. ^{250}Fm, with a half-life of 28 min, has been shown to be a product of decay of ^{254}No. It was by chemical identification of ^{250}Fm that production of nobelium was confirmed. Fermium would probably have chemical properties resembling erbium.

Flerovium — (After the *Flerov* Laboratory of Nuclear Reactions, Russia), Fl. Although the synthesis of Element-114 was first reported in 1998 by a team at the Joint Institute for Nuclear Research (JINR) in Dubna, Russia, further experiments led to confusion, until the discovery was finally confirmed by groups in the U.S. and Germany in 2009-2011. IUPAC approved the name Flerovium in 2011. The most stable isotope so far discovered in ^{285}Fl, with a half-life of 0.13 s. Based on its electronic structure, flerovium is predicted to be a dense, volatile metal, possibly a gas at room temperature. It is calculated to be similar to lead in its chemistry, but probably less reactive.

Fluorine — (L. and F. *fluere,* flow, or flux), F. In 1529, Georgius Agricola described the use of fluorspar as a flux, and as early as 1670 Schwandhard found that glass was etched when exposed to fluorspar treated with acid. Scheele and many later investigators, including Davy, Gay-Lussac, Lavoisier, and Thenard, experimented with hydrofluoric acid, some experiments ending in tragedy. The element was finally isolated in 1886 by Moisson after nearly 74 years of continuous effort. Fluorine occurs chiefly in *fluorspar* (CaF_2) and *cryolite* (Na_2AlF_6), and is in *topaz* and other minerals. It is a member of the halogen family of elements, and is obtained by electrolyzing a solution of potassium hydrogen fluoride in anhydrous hydrogen fluoride in a vessel of metal or transparent fluorspar. Modern commercial production methods are essentially variations on the procedures first used by Moisson. Fluorine is the most electronegative and reactive of all elements. It is a pale yellow, corrosive gas, which reacts with practically all organic and inorganic substances. Finely divided metals, glass, ceramics, carbon, and even water burn in fluorine with a bright flame. Until World War II, there was no commercial production of elemental fluorine. The atomic bomb project and nuclear energy applications, however, made it necessary to produce large quantities. Safe handling techniques have now been developed and it is possible at present to transport liquid fluorine by the ton. Fluorine and its compounds are used in producing enriched uranium (from the hexafluoride) and more than 100 commercial fluorochemicals, including many well-known high-temperature plastics. Hydrofluoric acid is extensively used for etching the glass of light bulbs, etc. Fluorochlorohydrocarbons were extensively used in air conditioning and refrigeration systems, but these compounds have been replaced fluorohydrocarbons to reduce the depletion of stratospheric ozone. The presence of fluorine as a soluble fluoride in drinking water to the extent of 2 ppm may cause mottled enamel in teeth, when used by children acquiring permanent teeth; in smaller amounts, however, fluorides are beneficial and are used in water supplies to prevent dental cavities. Elemental fluorine is used as an oxidizer in rocket engines. A number of fluorides of the rare gases xenon, radon, and krypton have been prepared. Elemental fluorine and the fluoride ion are highly toxic. The free element has a characteristic pungent odor, detectable in concentrations as low as 20 ppb, which is below the safe working level.

Francium — (France), Fr. Discovered in 1939 by Mlle. Marguerite Perey of the Curie Institute, Paris. Francium, the heaviest known member of the alkali metal series, occurs from alpha disintegration of actinium. It can also be made artificially by bombarding thorium with protons. While it occurs naturally in uranium minerals, there is probably less than 30 g of francium at any time in the total crust of the earth. It has the highest equivalent weight of any element, and is the most unstable of the first 101 elements of the periodic system. Forty-six isotopes and isomers of francium are recognized. The longest lived, ^{223}Fr, a daughter of ^{227}Ac, has a half-life of 22 min. This is the only isotope of francium occurring in nature. Because all known isotopes are highly unstable, knowledge of the chemical properties of this element comes from radiochemical techniques. No weighable quantity of the element has been prepared or isolated. The chemical properties of francium should most closely resemble cesium, and francium salts are found to coprecipitate with cesium salts from aqueous solution. The predicted melting point is around 27 °C, so elemental francium might be either solid or liquid.

Gadolinium — (*gadolinite,* a mineral named for Gadolin, a Finnish chemist), Gd. Gadolinia, the oxide of gadolinium, was separated by Marignac in 1880, and Lecoq de Boisbaudran independently isolated the element from Mosander's "yttria" in 1886. The element was named for the mineral *gadolinite* from which this rare earth was originally obtained. Gadolinium is found in several other minerals, including *monazite* and *bastnasite,* which are of commercial importance. The element has been purified only in recent years. With the development of ion-exchange and solvent extraction techniques, the availability and price of gadolinium and the other rare-earth metals have greatly improved. The metal can be prepared by the reduction of the anhydrous fluoride with metallic calcium. As with other related rare-earth metals, it is silvery white, has a metallic luster, and is malleable and ductile. At room temperature, gadolinium crystallizes in the hexagonal, close-packed α form. Upon heating to 1235 °C, α-gadolinium transforms into the β form, which has a body-centered cubic structure. The metal is relatively stable in dry air, but in moist air it tarnishes with the formation of a loosely adhering oxide film which splits off and exposes more surface to oxidation. The metal reacts slowly with water and is soluble in dilute acid. Gadolinium has the highest thermal neutron capture cross-section of any known element (49,000 barns). Natural gadolinium is a mixture of seven isotopes. Two of these, ^{155}Gd and ^{157}Gd, have excellent capture characteristics, but they are present naturally in low concentrations. As a result, gadolinium has a very fast burnout rate and has limited use as a nuclear control rod material. It has been used in making gadolinium yttrium garnets, which have important electronic applications. Compounds of gadolinium are used in making phosphors for color TV tubes. The metal has unusual superconductive properties. As little as 1% gadolinium has been found to improve the workability and resistance of iron, chromium, and related alloys to high temperatures and oxidation. Gadolinium ethyl sulfate has extremely low noise characteristics and may find use in duplicating the performance of amplifiers, such as the maser. The metal is ferromagnetic. Gadolinium is unique for its high magnetic moment and for its special Curie temperature (above which ferromagnetism vanishes) lying just at room temperature. This suggests uses as a magnetic component that senses hot and cold.

Gallium — (L. *Gallia,* France), Ga. Predicted and described by Mendeleev as eka-aluminum, and discovered spectroscopically by Lecoq de Boisbaudran in 1875, who in the same year obtained the free metal by electrolysis of a solution of the hydroxide in KOH, Gallium is often found as a trace element

Inorganic

in *diaspore, sphalerite, germanite, bauxite,* and *coal.* Some flue dusts from burning coal have been shown to contain as much as 1.5% gallium. It is the only metal, except for mercury, cesium, and rubidium, which can be liquid near room temperature; this makes possible its use in high-temperature thermometers. It has one of the longest liquid ranges of any metal and has a low vapor pressure even at high temperatures. There is a strong tendency for gallium to supercool below its freezing point. Therefore, seeding may be necessary to initiate solidification. Ultra-pure gallium has a beautiful, silvery appearance, and the solid metal exhibits a conchoidal fracture similar to glass. The metal expands 3.1% on solidifying; therefore, it should not be stored in glass or metal containers, as they may break as the metal solidifies. Gallium wets glass or porcelain and forms a brilliant mirror when it is painted on glass. It is widely used in doping semiconductors and producing solid-state devices such as transistors. Gallium arsenide is capable of converting electricity directly into coherent light. High-purity gallium is attacked slowly only by mineral acids. Magnesium gallate containing divalent impurities such as Mn^{+2} is finding use in commercial ultraviolet activated powder phosphors. Gallium nitride has been used to produce blue light-emitting diodes such as those used in CD and DVD readers. Gallium was the key ingredient in the detector for the Gallex Experiment in the 1990s. In this experiment, 30.3 tons of gallium in the form of a $GaCl_3$-HCl solution was used to detect solar neutrinos. Gallium readily alloys with most metals, and has been used as a component in low melting alloys. Its toxicity appears to be of a low order, but it should be handled with care until more data are forthcoming.

Germanium — (L. *Germania,* Germany), Ge. Predicted by Mendeleev in 1871 as eka-silicon, and discovered by Winkler in 1886. The metal is found in *argyrodite,* a sulfide of germanium and silver; in *germanite,* which contains 8% of the element; in zinc ores; in coal; and in other minerals. The element is frequently obtained commercially from flue dusts of smelters processing zinc ores, and has been recovered from the by-products of combustion of certain coals. Its presence in coal insures a large reserve of the element in the years to come. Germanium can be separated from other metals by fractional distillation of its volatile tetrachloride. The tetrachloride may then be hydrolyzed to give GeO_2; the dioxide can be reduced with hydrogen to give the metal. The element is a gray-white metalloid, and in its pure state is crystalline and brittle, retaining its luster in air at room temperature. It is a very important semiconductor material. Zone-refining techniques have led to production of crystalline germanium for semiconductor use with an impurity of only one part in 10^{10}. Doped with arsenic, gallium, or other elements, it is used in thousands of electronic applications. Its application in fiber optics and infrared optical systems now provides the largest use for germanium. Germanium is also finding many other applications including use as an alloying agent, as a phosphor in fluorescent lamps, and as a catalyst. Germanium and germanium oxide are transparent to the infrared and are used in infrared spectrometers and other optical equipment, including extremely sensitive infrared detectors. Germanium oxide's high index of refraction and dispersion make it useful as a component of glasses used in wide-angle camera lenses and microscope objectives. The field of organogermanium chemistry is becoming increasingly important. Certain germanium compounds have a low mammalian toxicity, but a marked activity against certain bacteria, which makes them of interest as chemotherapeutic agents.

Gold — (Sanskrit *Jval;* Anglo-Saxon *gold*), Au (L. *aurum,* gold). Known and highly valued from earliest times, gold is found in nature as the free metal and in tellurides; it is very widely distributed and is almost always associated with quartz or pyrite. It occurs in veins and alluvial deposits, and is often separated from rocks and other minerals by sluicing and panning operations. About 25% of the world's gold output comes from South Africa, and about two thirds of the total U.S. production now comes from South Dakota and Nevada. The metal is recovered from its ores by cyaniding, amalgamating, and smelting processes. Refining is also frequently done by electrolysis. Gold occurs in sea water to the extent of 0.1 to 2 mg/ton, depending on the location where the sample is taken. As yet, no method has been found for recovering gold from sea water profitably. It is estimated that all the gold in the world, so far refined, could be placed in a single cube 60 ft on a side. Of all the elements, gold in its pure state is undoubtedly the most beautiful. It is metallic, having a yellow color when in a mass, but when finely divided it may be black, ruby, or purple. The Purple of Cassius is a delicate test for auric gold. It is the most malleable and ductile metal; 1 oz. of gold can be beaten out to 300 ft^2. It is a soft metal and is usually alloyed to give it more strength. The specific gravity of gold has been found to vary considerably depending on temperature, how the metal is precipitated, and how it is cold-worked. Gold is a good conductor of heat and electricity, and is unaffected by air and most reagents. It is used in coinage and is a standard for monetary systems in many countries. It is also extensively used for jewelry, decoration, dental work, and for plating. It is used for coating certain space satellites, as it is a good reflector of infrared radiation and is inert. Gold, like other precious metals, is measured in troy weight; when alloyed with other metals, the term *carat* is used to express the amount of gold present, 24 carats being pure gold. For many years the value of gold was set by the U.S. at $20.67/troy ounce; in 1934 this value was fixed by law at $35.00/troy ounce, 9/10th fine. On March 17, 1968, because of a financial crisis, a two-tiered pricing system was established whereby gold was still used to settle international accounts at the old $35.00/troy ounce price while the price of gold on the private market would be allowed to fluctuate. Since that time, the price of gold on the free market has fluctuated widely. The most common gold compounds are auric chloride ($AuCl_3$) and chlorauric acid ($HAuCl_4$), the latter being used in photography for toning the silver image. ^{198}Au, with a half-life of 2.7 days, is used for treating cancer and other diseases. Disodium aurothiomalate is administered intramuscularly as a treatment for arthritis. A mixture of one part nitric acid with three parts hydrochloric acid is called *aqua regia* (because it dissolved gold, the King of Metals). For many years the temperature assigned to the freezing point of gold was 1063.0 °C; this served as a calibration fixed point for the International Temperature Scales (ITS-27 and ITS-48) and the International Practical Temperature Scale (IPTS-48). In 1968, a new International Practical Temperature Scale (IPTS-68) was adopted, which required that the freezing point of gold be changed to 1064.43 °C. In 1990 a new International Temperature Scale (ITS-90) was adopted bringing the triple point of H_2O to 0.01 °C and the freezing point of gold to 1064.18 °C.

Hafnium — (*Hafnia,* Latin name for Copenhagen), Hf. Hafnium was thought to be present in various minerals and concentrations many years prior to its discovery in 1923, credited to D. Coster and G. von Hevesey. On the basis of the Bohr theory, the new element was expected to be associated with zirconium. It was finally identified in *zircon* from Norway, by means of X-ray spectroscopic analysis. It was named in honor of the city in which the discovery was made. Most zirconium minerals contain 1 to 5% hafnium. It was originally separated from zirconium by repeated recrystallization of the double ammonium or potassium fluorides by von Hevesey and Jantzen. Metallic hafnium was first prepared by van Arkel and deBoer by passing the vapor of the tetraiodide over a heated tungsten filament. Almost all hafnium metal now produced is made by reducing the tetrachloride with magnesium or with sodium (Kroll Process). Hafnium is a ductile metal with a brilliant silver luster. Its properties are considerably influenced by the impurities of zirconium present. Of all the elements, zirconium and hafnium are two of the most difficult to separate. Their chemistry is almost identical; however, the density of zirconium is about half that of hafnium. Very pure hafnium has been produced, with zirconium being the major impurity. Because hafnium has a good absorption cross section for thermal neutrons (almost 600 times that of zirconium), has excellent mechanical properties, and is extremely corrosion resistant, it is used for reactor control rods. Hafnium has been successfully alloyed with iron, titanium, niobium, tantalum, and other metals. Hafnium carbide is the most refractory binary composition known, and the nitride is the most refractory of all known metal nitrides (mp 3310 °C). Hafnium is used in gas-filled and incandescent lamps, and is an efficient "getter" for scavenging oxygen and nitrogen. Finely divided hafnium is pyrophoric and can ignite spontaneously in air. Care should be taken when machining the metal or when handling hot sponge hafnium. At 700 °C hafnium rapidly absorbs hydrogen to form the composition $HfH_{1.86}$. Hafnium is resistant to concentrated alkalis, but at elevated temperatures reacts with oxygen, nitrogen, carbon, boron, sulfur, and silicon. Halogens react directly to form tetrahalides.

Hassium — (named for the German state, Hesse) Hs. This element was first synthesized and identified in 1964 by the same G.S.I. Darmstadt Group who first identified *Bohrium* and *Meitnerium.* Presumably this element has chemical properties similar to osmium. Isotope ^{265}108 was produced using a beam of ^{58}Fe projectiles, produced by the Universal Linear Accelerator (UNILAC) to bombard a ^{208}Pb target. Discovery of *Bohrium* and *Meitnerium* was made using detection of isotopes with odd proton and neutron numbers. Elements having even atomic numbers have been thought to be less stable against spontaneous fusion than odd elements. The production of ^{265}108 in the same reaction as was used at G.S.I. was confirmed at Dubna with detection of the seventh member of the decay chain ^{253}Es. Isotopes of *Hassium* are believed to decay by spontaneous fission, explaining why 109 was produced before 108. IUPAC adopted the name *Hassium* after the German state of Hesse in September 1997. In June 2001 it was announced that hassium is the heaviest element to have its chemical properties analyzed. A research team at the UNILAC heavy-ion accelerator in Darmstadt, Germany built an instrument to detect and analyze hassium. Atoms of curium-248 were collided with atoms of magnesium-26, producing about 6 atoms of hassium with a half-life of 9 s. This was sufficiently long to obtain data showing that hassium atoms react with oxygen to form hassium oxide molecules. These condensed at a temperature consistent with the behavior of Group 8 elements. This experiment appears to confirm hassium's location under osmium in the periodic table.

Helium — (Gr. *helios,* the sun), He. Evidence of the existence of helium was first obtained by Janssen during the solar eclipse of 1868 when he detected a new line in the solar spectrum; Lockyer and Frankland suggested the name *helium* for the new element; in 1895, Ramsay discovered helium in the uranium mineral *cleveite,* and it was independently discovered in cleveite by the Swedish chemists Cleve and Langlet about the same time. Rutherford and Royds in 1907 demonstrated that α particles are helium nuclei. Except for hydrogen, helium is the most abundant element found throughout the universe. Helium is extracted from natural gas; all natural gas contains at least trace quantities of helium. It has been detected spectroscopically in great abundance, especially in the hotter stars, and it is an important component in both the proton–proton reaction and the carbon cycle, which account for the energy of the sun and stars. The fusion of hydrogen into helium provides the energy of the hydrogen bomb. The helium content of the atmosphere is about 1 part in 200,000. It is present in various radioactive minerals as a decay product. Much of the world's supply of helium is obtained from wells in Texas, Colorado, and Kansas. Helium has the lowest triple-point temperature of any element and has found wide use in cryogenic research, as its boiling point is near 4 K. Its use in the study of superconductivity is vital. Liquid helium (^{4}He) exists in two forms: He^{4}I and He^{4}II, with a sharp transition point at 2.174 K (3.83 cm Hg). ^{4}HeI (above this temperature) is a normal liquid, but ^{4}HeII (below it) is unlike any other known substance. It expands on cooling; its conductivity for heat is enormous; and neither its heat conduction nor viscosity obeys normal rules. It has other peculiar properties. Helium is the only liquid that cannot be solidified by lowering the temperature at atmospheric pressure. It remains liquid down to absolute zero at ordinary pressures, but it can readily be solidified by increasing the pressure. Solid ^{3}He and ^{4}He are unusual in that both can readily be changed in volume by more than 30% by application of pressure. The specific heat of helium gas is unusually high. The density of helium vapor at the normal boiling point is also very high, with the vapor expanding greatly when heated to room temperature. Containers filled with helium gas at 5 to 10 K should be treated as though they contained liquid helium due to the large increase in pressure resulting from warming the gas to room temperature. While helium normally has a 0 valence, it seems to have a weak tendency to combine with certain other elements. Means of preparing helium difluoride have been studied, and species such as HeNe and the molecular ions He^+ and He^{++} have been investigated. Helium is widely used as an inert gas shield for arc welding; as a protective gas in growing silicon and germanium crystals, and in titanium and zirconium production; as a cooling medium for nuclear reactors, and as a gas for supersonic wind tunnels. A mixture of helium and oxygen is used as an artificial atmosphere for divers and others working under pressure. Different ratios of He/O_2 are used for different depths at which the diver is operating. Helium is extensively used for filling balloons as it is a much safer gas than hydrogen. One of the recent largest uses for helium has been for pressurizing liquid fuel rockets. A Saturn booster such as used on the Apollo lunar missions required about 13 million ft^3 of helium for a firing, plus more for checkouts. Liquid helium's use in magnetic resonance imaging (MRI) continues to increase as the use of this diagnostic technique expands. Lifting gas applications are increasing. Various companies in addition to Goodyear, are now using "blimps" for advertising. The Navy and the Air Force are investigating the use of airships to provide early warning systems to detect low-flying cruise missiles. NASA is currently using helium-filled balloons to sample the atmosphere in Antarctica to study holes in the ozone layer. Research on and development of materials which become superconductive at temperatures well above the boiling point of helium could have a major impact on the demand for helium. Less costly refrigerants having boiling points considerably higher could replace the present need to cool such superconductive materials to the boiling point of helium.

Holmium — (L. *Holmia* , for Stockholm), Ho. The spectral absorption bands of holmium were noticed in 1878 by the Swiss chemists Delafontaine and Soret, who announced the existence of an "Element X." Cleve, of Sweden, later independently discovered the element while working on erbia earth. The element is named after Cleve's native city. Pure holmia, the yellow oxide, was prepared by Homberg in 1911. Holmium occurs in *gadolinite, monazite,* and in other rare-earth minerals. It is commercially obtained from monazite, occurring in that mineral to the extent of about 0.05%. It has been isolated by the reduction of its anhydrous chloride or fluoride with calcium metal. Pure holmium has a metallic to bright silver luster. It is relatively soft and malleable, and is stable in dry air at room temperature, but rapidly oxidizes in moist air and at elevated temperatures. Its high magnetic permeability leads to applications in high-field magnets. It is also used in near-infrared lasers.

Hydrogen — (Gr. *hydro*, water, and *genes,* forming), H. Hydrogen was prepared many years before it was recognized as a distinct substance by Cavendish in 1766. It was named by Lavoisier. Hydrogen is the most abundant of all elements in the universe, and it is thought that the heavier elements were, and still are, being built from hydrogen and helium. It has been estimated that hydrogen makes up more than 90% of all the atoms or three quarters of the mass of the universe. It is found in the sun and most stars, and plays an important part in the proton–proton reaction and carbon–nitrogen cycle, which accounts for the energy of the sun and stars. It is thought that hydrogen is a major component of the planet Jupiter and that at some depth in the planet's interior the pressure is so great that solid molecular hydrogen is converted into solid metallic hydrogen. On Earth, hydrogen occurs chiefly in combination with oxygen in water, but it is also present in organic matter such as living plants, petroleum, coal, etc. It is present as the free element in the atmosphere, but only to the extent of less than 1 ppm by volume. It is the lightest of all gases, and combines with other elements, sometimes explosively, to form compounds. Great quantities of hydrogen are required commercially for the fixation of nitrogen from the air in the Haber ammonia process and for the hydrogenation of fats and oils. It is also used in large quantities in methanol production, in hydrodealkylation, hydrocracking, and hydrodesulfurization. It is also used as a rocket fuel, for cutting and welding of metals, for production of hydrochloric acid, for the reduction of metallic ores, and for filling balloons. The lifting power of 1 ft^3 of hydrogen gas is about 0.076 lb at 0 °C, 760 mm pressure. Production of hydrogen in the U.S. alone amounts to about 3 billion cubic feet per year. It is prepared by the action of steam on heated carbon, by decomposition of certain hydrocarbons with heat, by the electrolysis of water, or by the displacement from acids by certain metals. It is also produced by the action of sodium or potassium hydroxide on aluminum. Liquid hydrogen is important in cryogenics and in the study of superconductivity, as its melting point is only about 20 K. Hydrogen consists of three isotopes, most of which is ^1H. In 1932, Urey announced the discovery of a stable isotope, deuterium (^2H or D). Deuterium is present in natural hydrogen to the extent of about 0.01%. Two years later an unstable isotope, tritium (^3H), was discovered. Tritium has a half-life of about 12.31 years. Tritium atoms are also present in natural hydrogen but in a much smaller proportion. Tritium is readily produced in nuclear reactors and is used in the production of the hydrogen bomb. It is also used as a radioactive agent in making luminous paints, and as a tracer. Heavy water, deuterium oxide (D_2O), is used as a moderator to slow down neutrons. Hydrogen gas under ordinary conditions is a mixture of two kinds of molecules, known as *ortho-* and *para-*hydrogen, which differ from one another by the spins of their electrons and nuclei. Normal hydrogen at room temperature contains 25% of the *para* form and 75% of the *ortho* form. The *ortho* form cannot be prepared in the pure state. Since the two forms differ in energy, the physical properties also differ. The melting and boiling points of *para*hydrogen are about 0.1 °C lower than those of normal hydrogen. Both stationary and mobile power sources using hydrogen combustion offer a major advantage for reducing atmospheric carbon dioxide and retarding global temperature rise. The "hydrogen economy" is expected to grow rapidly as steps are taken to reduce the combustion of fossil fuels.

Indium — (from the brilliant indigo line in its spectrum), In. Discovered spectroscopically in 1863 by Reich and Richter, who later isolated the metal. Indium is most frequently associated with zinc materials, and it is from these that most commercial indium is now obtained; however, it is also found in iron, lead, and copper ores. Until 1924, a gram or so constituted the world's supply of this element in isolated form. It is probably about as abundant as silver. About 4 million troy ounces of indium are now produced annually, much of it from Canada. It is available in ultrapure form. Indium is a very soft, silvery-white metal with a brilliant luster. The pure metal gives a high-pitched "cry" when bent. In liquid form it wets glass, as does gallium. It has found application in making low-melting alloys; an alloy of 24% indium–76% gallium is liquid at room temperature. Indium is used in making bearing alloys, and in the form of various semiconductors (InSb, InAs, etc.) it has many applications in transistors, rectifiers, thermistors, liquid crystal displays, batteries, solar cells, and photoconductors. It can be plated onto metal and evaporated onto glass, forming a mirror as good as that made with silver but with more resistance to atmospheric corrosion. Indium has a low order of toxicity.

Iodine — (Gr. *iodes,* violet), I. Discovered by Courtois in 1811. Iodine, a halogen, occurs sparingly in the form of iodides in sea water from which it is assimilated by seaweeds, in Chilean saltpeter and nitrate-bearing earth, known as *caliche* in brines from old sea deposits, and in brackish waters from oil and salt wells. Ultrapure iodine can be obtained from the reaction of potassium iodide with copper sulfate. Several other methods of isolating the element are known. Iodine is a bluish-black, lustrous solid, volatilizing at ordinary temperatures into a blue-violet gas with an irritating odor; it forms compounds with many elements, but is less active than the other halogens, which displace it from iodides. Iodine exhibits some metallic-like properties. It dissolves readily in chloroform, carbon tetrachloride, or carbon disulfide to form beautiful purple solutions. It is only slightly soluble in water. Iodine compounds are important in organic chemistry and very useful in medicine.

The artificial radioisotope ^{131}I, with a half-life of 8 days, has been used in treating the thyroid gland. The most common compounds are the iodides of sodium and potassium (Na I and KI) and the iodates (KIO_3). Lack of iodine is the cause of goiter. Iodides and thyroxin, which contains iodine, are used internally in medicine, and a solution of KI and iodine in alcohol is used for external wounds. Potassium iodide finds use in photography. The deep blue color with starch solution is characteristic of the free element. Care should be taken in handling and using iodine, as contact with the skin can cause lesions; iodine vapor is intensely irritating to the eyes and mucous membranes.

Iridium — (L. *iris*, rainbow), Ir. Discovered in 1803 by Tennant in the residue left when crude platinum is dissolved by aqua regia. The name iridium is appropriate, for its salts are highly colored. Iridium, a metal of the platinum family, is white, similar to platinum, but with a slight yellowish cast. It is very hard and brittle, making it hard to machine, form, or work. It is the most corrosion-resistant metal known, and was used in making the original standard meter bar, which is a 90% platinum–10% iridium alloy. Iridium is not attacked by any of the acids nor by aqua regia, but is attacked by molten salts, such as NaCl and NaCN. Iridium occurs uncombined in nature with platinum and other metals of this family in alluvial deposits. It is recovered as a by-product from the nickel mining industry. The largest reserves and production of the platinum group of metals, which includes iridium, is in South Africa, followed by Russia and Canada. Meteorites contain small amounts of iridium. The presence of iridium has recently been used in examining the Cretaceous-Tertiary (K-T) boundary. Because iridium is found widely distributed at the K-T boundary, it has been suggested that a large meteorite or asteroid collided with the Earth, killing the dinosaurs, and creating a large dust cloud and crater. Searches for such a crater point to one in the Yucatan, known as Chicxulub. Iridium has found use in making crucibles and apparatus for use at high temperatures. It is also used for electrical contacts. Its principal use is as a hardening agent for platinum. With osmium, it forms an alloy that is used for tipping pens and compass bearings. The density of iridium is only slightly lower than that of osmium, which has been generally credited as being the heaviest known element. Calculations of the densities of iridium and osmium from the space lattices give values of 22.65 and 22.61 g/cm^3, respectively. These values may be more reliable than actual physical measurements. At present, therefore, we know that either iridium or osmium is the densest known element, but the data do not yet allow selection between the two.

Iron — (Anglo-Saxon, *iron*), Fe (L. *ferrum*). The use of iron is prehistoric. Genesis mentions that Tubal-Cain, seven generations from Adam, was "an instructor of every artificer in brass and iron." A remarkable iron pillar, dating to about A.D. 400, remains standing today in Delhi, India. This solid shaft of wrought iron is about 7¼ m high by 40 cm in diameter. Corrosion to the pillar has been minimal although it has been exposed to the weather since its erection. Iron is a relatively abundant element in the universe. It is found in the sun and many types of stars in considerable quantity. It has been suggested that the iron we have here on Earth may have originated in a supernova. Iron is a very difficult element to produce in ordinary nuclear reactions, such as would take place in the sun. Iron is found native as a principal component of a class of iron–nickel meteorites known as *siderites*, and is a minor constituent of the other two classes of meteorites. The core of the Earth, 2150 miles in radius, is thought to be largely composed of iron with about 10% occluded hydrogen. The metal is the fourth most abundant element, by weight, making up the crust of the Earth. The most common ore is *hematite* (Fe_2O_3). Magnetite (Fe_3O_4) is frequently seen as *black sands* along beaches and banks of streams. *Lodestone* is another form of magnetite. *Taconite* is becoming increasingly important as a commercial ore. Iron is a vital constituent of plant and animal life, and appears in hemoglobin. It has a rich chemistry, both inorganic and organometallic. The pure metal is not often encountered in commerce, but is usually alloyed with carbon or other metals. The pure metal is very reactive chemically, and rapidly corrodes, especially in moist air or at elevated temperatures. It has four principal allotropic forms, known as α, γ, δ, and ε. The α-γ transition point is 912 °C; γ-δ is 1394 °C; and the ε phase appears at high pressures. The α form is magnetic, but the magnetism disappears above 771 °C, although the lattice remains unchanged. The relations of these forms are complex. Pig iron is an alloy containing about 3% carbon with varying amounts of S, Si, Mn, and P. It is hard, brittle, fairly fusible, and is used to produce other alloys, including steel. Wrought iron contains only a few tenths of a percent of carbon, is tough, malleable, less fusible, and usually has a "fibrous" structure. Carbon steel is an alloy of iron with carbon, with small amounts of Mn, S, P, and Si. Alloy steels are carbon steels with other additives such as nickel, chromium, vanadium, etc. Iron is the cheapest and most abundant, useful, and important of all metals.

Krypton — (Gr. *kryptos*, the hidden one), Kr. Discovered in 1898 by Ramsay and Travers in the residue left after liquid air had nearly boiled away, krypton is present in the air to the extent of about 1 ppm. The atmosphere of Mars has been found to contain 0.3 ppm of krypton. It is one of the noble gases. It is characterized by its brilliant green and orange spectral lines. The spectral lines of krypton are easily produced and some are very sharp; at one time the meter was defined in terms of the wavelength of a krypton line. Solid krypton is a white crystalline substance with a face-centered cubic structure that is common to all the rare gases. While krypton was first thought of as a noble gas that does not combine with other elements, the existence of several krypton compounds has been established. Krypton difluoride has been prepared in gram quantities and can be made by several methods. ^{85}Kr has found application in chemical analysis. By imbedding the isotope in various solids, *kryptonates* are formed. The activity of these kryptonates is sensitive to chemical reactions at the surface. Estimates of the concentration of reactants are therefore made possible. Krypton is used in certain photographic flash lamps for high-speed photography.

Lanthanum — (Gr. *lanthanein*, to lie hidden), La. Mosander in 1839 extracted a new earth *lanthana*, from impure cerium nitrate, and recognized the new element. Lanthanum is found in rare-earth minerals such as *cerite, monazite, allanite,* and *bastnasite.* Monazite and bastnasite are principal ores in which lanthanum occurs in percentages up to 25% and 38%, respectively. Misch metal, used in making lighter flints, contains about 25% lanthanum. Lanthanum was first isolated in relatively pure form in 1923. Ion-exchange and solvent extraction techniques have led to much easier isolation of the rare-earth elements, resulting in improved availability. The metal can be produced by reducing the anhydrous fluoride with calcium. Lanthanum is silvery white, malleable, ductile, and soft enough to be cut with a knife. It is one of the most reactive of the rare-earth metals. It oxidizes rapidly when exposed to air. Cold water attacks lanthanum slowly, and hot water attacks it much more rapidly. The metal reacts directly with elemental carbon, nitrogen, boron, selenium, silicon, phosphorus, sulfur, and with halogens. At 310 °C, lanthanum changes from a hexagonal to a face-centered cubic structure, and at 865 °C it again transforms into a body-centered cubic structure. Rare-earth compounds containing lanthanum are extensively used in carbon lighting applications, especially by the motion picture industry for studio lighting and projection. This application consumes about 25% of the rare-earth compounds produced. La_2O_3 improves the alkali resistance of glass, and is used in making special optical glasses. Small amounts of lanthanum, as an additive, can be used to produce nodular cast iron. Hydrogen sponge alloys containing lanthanum can take up to 400 times their own volume of hydrogen gas, and thus may be useful in energy storage systems. Lanthanum and its compounds have a low to moderate acute toxicity rating; therefore, care should be taken in handling them.

Lawrencium — (Ernest O. Lawrence [1901–1958], inventor of the cyclotron), Lr. This member of the 5f transition elements (actinide series) was discovered in March 1961 by A. Ghiorso, T. Sikkeland, A. E. Larsh, and R. M. Latimer. A 3-μg californium target, consisting of a mixture of isotopes of mass number 249, 250, 251, and 252, was bombarded with either ^{10}B or ^{11}B. The electrically charged transmutation nuclei recoiled with an atmosphere of helium and were collected on a thin copper conveyor tape which was then moved to place the collected atoms in front of a series of solid-state detectors. The isotope of element 103 produced in this way decayed by emitting an 8.6-MeV alpha particle with a half-life of 8 s. In 1967, Flerov and associates of the Dubna Laboratory reported their inability to detect an alpha emitter with a half-life of 8 s which was assigned by the Berkeley group to $^{257}103$. This assignment has been changed to ^{258}Lr or ^{259}Lr. In 1965, the Dubna workers found a longer-lived lawrencium isotope, ^{256}Lr, with a half-life of 35 s. In 1968, Ghiorso and associates at Berkeley were able to use a few atoms of this isotope to study the oxidation behavior of lawrencium. Using solvent extraction techniques and working very rapidly, they extracted lawrencium ions from a buffered aqueous solution into an organic solvent, completing each extraction in about 30 s. It was found that lawrencium behaves differently from dipositive nobelium and more like the tripositive elements earlier in the actinide series.

Lead — (Anglo-Saxon *lead*), Pb (L. *plumbum*). Long known, mentioned in Exodus. The alchemists believed lead to be the oldest metal and associated it with the planet Saturn. Native lead occurs in nature, but it is rare. Lead is obtained chiefly from *galena* (PbS) by a roasting process. *Anglesite* ($PbSO_4$), *cerussite* ($PbCO_3$), and *minim* (Pb_3O_4) are other common lead minerals. Lead is a bluish-white metal of bright luster, is very soft, highly malleable, ductile, and a poor conductor of electricity. It is very resistant to corrosion; lead pipes bearing the insignia of Roman emperors, used as drains from the baths, are still in service. It is used in containers for corrosive liquids (such as sulfuric acid) and may be toughened by the addition of a small percentage of antimony or other metals. Its alloys include solder, type metal, and various antifriction metals. Great quantities of lead, both as the metal and as the dioxide, are used in storage batteries. Lead is also used for cable covering, plumbing, and ammunition. The metal is very effective as a sound absorber, is used as a radiation shield around X-ray equipment and nuclear reactors, and is used to absorb vibration. Lead, alloyed with tin, is used in making organ pipes. White lead, the basic carbonate, sublimed white lead ($PbSO_4$), chrome yellow ($PbCrO_4$), red lead (Pb_3O_4), and other lead compounds are used extensively in paints, although in recent years the use of lead in paints has been drastically curtailed to eliminate or reduce health hazards. Lead oxide is used in producing fine "crystal glass" and "flint glass" of a high index of refraction for achromatic lenses. The nitrate and the acetate are soluble salts. Lead salts such as lead arsenate have been used as insecticides, but their use in recent years has been practically eliminated in favor of less harmful organic compounds. Care must be used in handling lead as it is a cumulative poison. Environmental concern with lead poisoning led to elimination of lead tetraethyl in gasoline. The U.S. Occupational Safety and Health Administration (OSHA) has recommended that industries limit airborne lead in the workplace to 50 μg/m^3.

Lithium — (Gr. *lithos*, stone), Li. Discovered by Arfvedson in 1817. Lithium is the lightest of all metals, with a density only about half that of water. It does not occur free in nature; combined it is found in small amounts in nearly all igneous rocks and in the waters of many mineral springs. *Lepidolite*, *spodumene*, *petalite*, and *amblygonite* are the more important minerals containing it. Lithium is presently being recovered from brines of Searles Lake, in California, and from Nevada, Chile, and Argentina. Large deposits of spodumene are found in North Carolina. The metal is produced electrolytically from the fused chloride. Lithium is silvery in appearance, much like Na, K, and other members of the alkali metal series. It reacts with water, but not as vigorously as sodium. Lithium imparts a beautiful crimson color to a flame, but when the metal burns strongly the flame is a dazzling white. In recent years, the production of lithium metal and its compounds has increased greatly. Because the metal has the highest specific heat of any solid element, it has found use in heat transfer applications; however, it is corrosive and requires special handling. The metal has been used as an alloying agent, is of interest in synthesis of organic compounds, has important pharmaceutical uses, and has nuclear applications. It ranks as a leading contender as a battery anode material because it has a high electrochemical potential. Lithium is also used in special glasses and ceramics. Lithium chloride is one of the most hygroscopic materials known, and it, as well as lithium bromide, is used in air conditioning and industrial drying systems. Lithium stearate is used as an all-purpose and high-temperature lubricant. The use of lithium-ion storage batteries in applications ranging from cell phones to automobiles is growing rapidly.

Livermorium — (Lawrence *Livermore* National Laboratory), Lv. A single atom of Element-116 was first synthesized in 2000 at Lawrence Livermore Laboratory in collaboration with workers at JINR, Dubna.The name Livermorium was adopted by IUPAC in 2012. Livermorium falls at the top of the chalcogen group: oxygen, sulfur, selenium, tellurium, and polonium. No information is available on its physical and chemical properties. The most stable isotope, ^{293}Lv, has a half-life of 61 ms.

Lutetium — (Lutetia, ancient name for Paris, sometimes called *cassiopeium* in Germany), Lu. In 1907, Urbain described a process by which Marignac's ytterbium (1879) could be separated into the two elements, ytterbium (neoytterbium) and lutetium. These elements were identical with "aldebaranium" and "cassiopeium," independently discovered by von Welsbach about the same time. Charles James of the University of New Hampshire also independently prepared the very pure oxide, *lutecia*, at this time. The spelling of the element was changed from *lutecium* to *lutetium* in 1949. Lutetium occurs in very small amounts in nearly all minerals containing yttrium, and is present in *monazite,* which is a commercial source, to the extent of about 0.003%. The pure metal was only isolated later and is one of the most difficult to prepare. It can be prepared by the reduction of anhydrous $LuCl_3$ or LuF_3 by an alkali or alkaline earth metal. The metal is silvery white and relatively stable in air. While new techniques, including ion-exchange reactions, have been developed to separate the various rare-earth elements, lutetium is still the most costly of all rare earths. Stable lutetium isotopes, which emit pure beta radiation after thermal neutron activation, can be used as catalysts in cracking, alkylation, hydrogenation, and polymerization. Virtually no other commercial uses have been found. Lutetium, like other rare-earth metals, is thought to have a low toxicity rating.

Magnesium — (*Magnesia,* district in Thessaly), Mg. Compounds of magnesium have long been known. Black recognized magnesium as an element in 1755. It was isolated by Davy in 1808, and prepared in coherent form by Bussy in 1831. Magnesium is the eighth most abundant element in the Earth's crust. It does not occur uncombined, but is found in large deposits in the form of *magnesite, dolomite,* and other minerals. The metal is now principally obtained in the U.S. by electrolysis of fused magnesium chloride derived from brines, wells, and sea water. Magnesium is a light, silvery-white, and fairly tough metal. It tarnishes slightly in air, and finely divided magnesium readily ignites upon heating in air and burns with a dazzling white flame. It has been used in flashlight photography, flares, and pyrotechnics, including incendiary bombs. It is one third lighter than aluminum, and in alloys is essential for airplane and missile construction. The metal improves the mechanical, fabrication, and welding characteristics of aluminum when used as an alloying agent. Magnesium is used in producing nodular graphite in cast iron, and is used as an additive to conventional propellants. It is also used as a reducing agent in the production of pure uranium and other metals from their salts. The hydroxide (*milk of magnesia*), chloride, sulfate (*Epsom salts*), and citrate are used in medicine. Dead-burned magnesite is employed for refractory purposes such as brick and liners in furnaces and converters. Calcined magnesia is also used for water treatment and in the manufacture of rubber, paper, etc. Organic magnesium compounds (Grignard's reagents) are important. Magnesium is an important element in both plant and animal life. Chlorophylls are magnesium-centered porphyrins. The adult daily requirement of magnesium is about 300 mg/day, but this is affected by various factors. Great care should be taken in handling magnesium metal, especially in the finely divided state, as serious fires can occur. Water should not be used on burning magnesium or on magnesium fires.

Manganese — (L. *magnes,* magnet, from magnetic properties of pyrolusite; It. *manganese,* corrupt form of *magnesia*), Mn. Recognized by Scheele, Bergman, and others as an element and isolated by Gahn in 1774 by reduction of the dioxide with carbon. Manganese minerals are widely distributed; oxides, silicates, and carbonates are the most common. The discovery of large quantities of manganese nodules on the floor of the oceans holds promise as a source of manganese. These nodules contain about 24% manganese together with many other elements in lesser abundance. Most manganese today is obtained from ores found in Ukraine, Brazil, Australia, Republic of So. Africa, Gabon, China, and India. *Pyrolusite* (MnO_2) and *rhodochrosite* ($MnCO_3$) are among the most common manganese minerals. The metal is obtained by reduction of the oxide with sodium, magnesium, aluminum, or by electrolysis. It is gray-white, resembling iron, but is harder and very brittle. The metal is reactive chemically, and decomposes in cold water slowly. Manganese is used to form many important alloys. In steel, manganese improves the rolling and forging qualities, strength, toughness, stiffness, wear resistance, hardness, and hardenability. With aluminum and antimony, especially with small amounts of copper, it forms highly ferromagnetic alloys. Manganese metal is ferromagnetic only after special treatment. The pure metal exists in four allotropic forms. The alpha form is stable up to 727 °C; gamma manganese, which is metastable at ordinary temperatures, is soft, easily cut, and capable of being bent. The dioxide (pyrolusite) is used as a depolarizer in dry cells, and is used to "decolorize" glass that is colored green by impurities of iron. Manganese by itself colors glass an amethyst color, and is responsible for the color of true amethyst. The dioxide is also used in the preparation of oxygen and chlorine, and in drying black paints. The permanganate is a powerful oxidizing agent and is used in quantitative analysis and in medicine. Manganese is widely distributed throughout the animal kingdom. It is an important trace element and may be essential for utilization of vitamin B_1.

Meitnerium — (Lise Meitner [1878–1968], Austrian–Swedish physicist and mathematician), Mt. In 1982, Element 109 was made and identified by physicists at the Heavy Ion Research Laboratory (G.S.I.), Darmstadt, Germany, by bombarding a target of ^{209}Bi with accelerated nuclei of ^{58}Fe. In August 1997 IUPAC adopted the name meitnerium for this element, honoring Lise Meitner. The production of Mt has been extremely small. It took a week of target bombardment (10^{11} nuclear encounters) to produce a single atom. Oganessian and his team at Dubna in 1984 repeated the Darmstadt experiment using a tenfold irradiation dose. One fission event from seven alpha decays of Mt was observed, thus indirectly confirming the existence of isotope ^{266}Mt.

Mendelevium — (Dmitri Mendeleev [1834–1907]), Md. Mendelevium, the ninth transuranium element of the actinide series to be discovered, was first identified by Ghiorso, Harvey, Choppin, Thompson, and Seaborg early in 1955 as a result of the bombardment of the isotope ^{253}Es with helium ions in the Berkeley 60-inch cyclotron. The isotope produced was ^{256}Md, which has a half-life of 78 min. This first identification was notable in that ^{256}Md was synthesized on a one-atom-at-a-time basis. ^{258}Md has a half-life of 51.5 days. It now appears possible that eventually enough ^{258}Md can be made so that some of its physical properties can be determined. ^{256}Md has been used to elucidate some of the chemical properties of mendelevium in aqueous solution. Experiments seem to show that the element possesses a moderately stable dipositive (II) oxidation state in addition to the tripositive (III) oxidation state, which is characteristic of actinide elements.

Mercury — (Planet *Mercury*), Hg (*hydrargyrum,* liquid silver). Known to ancient Chinese and Hindus; found in Egyptian tombs of 1500 B.C. Mercury is the only common metal liquid at ordinary temperatures. It only rarely occurs free in nature. The chief ore is *cinnabar* (HgS). Spain and China produce about 75% of the world's supply of the metal. The commercial unit for handling mercury is the "flask," which weighs 76 lb (34.46 kg). The metal is obtained by heating cinnabar in a current of air and by condensing the vapor. It is a heavy, silvery-white metal; a rather poor conductor of heat, as compared with other metals, and a fair conductor of electricity. It easily forms alloys with many metals, such as gold, silver, and tin, which are called *amalgams*. Its ease in amalgamating with gold is made use of in the recovery of gold from its ores. The metal is widely used in laboratory work for making thermometers, barometers, diffusion pumps, and many other instruments. It is used in making mercury-vapor lamps and advertising signs, etc. and is used in mercury switches and other electrical apparatus. Other uses are in making pesticides, mercury cells for caustic soda and chlorine production, antifouling paint, batteries, and catalysts. The previously heavy use in dental preparations is being phased out because of toxicity concerns. The most important salts are mercuric chloride $HgCl_2$ (corrosive sublimate — a violent poison), mercurous chloride Hg_2Cl_2 (calomel, occasionally still used in medicine), mercury fulminate ($Hg(ONC)_2$), a detonator widely used in explosives, and mercuric sulfide (HgS, vermillion, a high-grade paint pigment). Organic mercury compounds are important. It has been found that an electrical discharge causes mercury vapor to combine with neon, argon, krypton, and xenon. These products, held together with van der Waals' forces, correspond to HgNe, HgAr, HgKr, and HgXe. Mercury is a virulent poison and is readily absorbed through the respiratory tract, the gastrointestinal tract, or through unbroken skin. It acts as a cumulative poison and dangerous levels are readily attained in air. Air saturated with mercury vapor at 20 °C contains a concentration that exceeds the toxic limit many times. The danger increases at higher temperatures. *It is therefore important that mercury be handled with great care.* Containers of mercury should be securely covered and spillage should be avoided. If it is necessary to heat mercury or mercury compounds, it should be done in a well-ventilated hood. Methyl mercury is a dangerous pollutant and is now widely found in water and streams. The triple point of mercury, –38.8344 °C, is a fixed point on the International Temperature Scale (ITS-90).

Molybdenum — (Gr. *molybdos*, lead), Mo. Before Scheele recognized molybdenite as a distinct ore of a new element in 1778, it was confused with graphite and lead ore. The metal was prepared in an impure form in 1782 by Hjelm. Molybdenum does not occur native, but is obtained principally from *molybdenite* (MoS_2). *Wulfenite* ($PbMoO_4$) and *powellite* ($Ca(Mo,W)O_4$) are also minor commercial ores. Molybdenum is also recovered as a by-product of copper and tungsten mining operations. The U.S., Canada, Chile, and China produce most of the world's molybdenum ores. The metal is prepared from the powder made by the hydrogen reduction of purified molybdic trioxide or ammonium molybdate. The metal is silvery white, very hard, but is softer and more ductile than tungsten. It has a high elastic modulus, and only tungsten and tantalum, of the more readily available metals, have higher melting points. It is a valuable alloying agent, as it contributes to the hardenability and toughness of quenched and tempered steels. It also improves the strength of steel at high temperatures. It is used in certain nickel-based alloys, such as the Hastelloys® which are heat-resistant and corrosion-resistant to chemical solutions. Molybdenum oxidizes at elevated temperatures. The metal has found application as electrodes for electrically heated glass furnaces and forehearths. It is also used in nuclear energy applications and for missile and aircraft parts. Molybdenum is valuable as a catalyst in the refining of petroleum. It has found application as a filament material in electronic and electrical applications. Molybdenum is an essential trace element in plant nutrition. Some lands are barren for lack of this element in the soil. Molybdenum sulfide is useful as a lubricant, especially at high temperatures where oils would decompose. Almost all ultra-high strength steels with minimum yield points up to 300,000 lb/in^2 contain molybdenum in amounts from 0.25 to 8%.

Moscovium — (*Moscow* Oblast, location of Dubna), Mc. First detected in 2003 at JNR, Dubna, Russia. The initial claim could not be verified, but subsequent work in 2013-2015 in Germany, the U.S., and Russia provided the necessary confirmation. IUPAC approved the name Moscovium in 2016. Its most stable isotope is ^{289}Mc, with a half-life of 220 ms. Its position in the periodic table places it at the top of the group nitrogen, phosphorus, arsenic, antimony, and bismuth. However, no information is available yet of its actual chemical properties.

Neodymium — (Gr. *neos*, new, and *didymos*, twin), Nd. In 1841 Mosander extracted from *cerite* a new rose-colored oxide, which he believed contained a new element. He named the element *didymium*, as it was *an inseparable twin brother of lanthanum*. In 1885 von Welsbach separated didymium into two new elemental components, *neodymia* and *praseodymia*, by repeated fractionation of ammonium didymium nitrate. While the free metal is in *misch metal*, long known and used as a pyrophoric alloy for light flints, the element was not isolated in relatively pure form until 1925. Neodymium is present in misch metal to the extent of about 18%. It is present in the minerals *monazite* and *bastnasite*, which are principal sources of rare-earth metals. The element may be obtained by separating neodymium salts from other rare earths by ion-exchange or solvent extraction techniques, and by reducing anhydrous halides such as NdF_3 with calcium metal. Other separation techniques are possible. The metal has a bright silvery metallic luster. Neodymium is one of the more reactive rare-earth metals and quickly tarnishes in air, forming an oxide that splits off and exposes metal to oxidation. The metal, therefore, should be kept under light mineral oil or sealed in a plastic material. Neodymium exists in two allotropic forms, with a transformation from a double hexagonal to a body-centered cubic structure taking place at 855 °C. Didymium, of which neodymium is a component, is used for coloring glass to make welder's goggles. By itself, neodymium colors glass delicate shades ranging from pure violet through wine-red and warm gray. Light transmitted through such glass shows unusually sharp absorption bands. The glass has been used in astronomical work to produce sharp bands by which spectral lines may be calibrated. Glass containing neodymium can be used as a laser material. Neodymium salts are used as a colorant for enamels. The element is also being used with iron and boron to produce extremely strong magnets.

Neon — (Gr. *neos*, new), Ne. Discovered by Ramsay and Travers in 1898. Neon is a rare gaseous element present in the atmosphere to the extent of 1 part in 65,000 of air. It is obtained by liquefaction of air and separated from the other gases by fractional distillation. It is very inert element; no stable compounds of neon have been prepared. In a vacuum discharge tube, neon glows reddish orange. Of all the rare gases, the discharge of neon is the most intense at ordinary voltages and currents. The largest use of neon is in advertising signs. It is also used to make high-voltage indicators and lightning arrestors. Neon and helium were used in the first gas lasers. Liquid neon finds some use as an economical cryogenic refrigerant. It has over 40 times more refrigerating capacity per unit volume than liquid helium and more than three times that of liquid hydrogen.

Neptunium — (Planet *Neptune*), Np. Neptunium was the first synthetic transuranium element of the actinide series discovered; the isotope ^{239}Np was produced by McMillan and Abelson in 1940 at Berkeley, California, as the result of bombarding uranium with cyclotron-produced neutrons. The isotope ^{237}Np (half-life of 2.14×10^6 years) is currently obtained in gram quantities as a by-product from the production of plutonium in nuclear reactors. Trace quantities of the element are actually found in nature due to transmutation reactions in uranium ores produced by the neutrons which are present. Neptunium metal is prepared by the reduction of NpF_3 with barium or lithium vapor at about 1200 °C. It has a silvery appearance, is chemically reactive, and exists in at least three structural modifications: α-neptunium, orthorhombic; β-neptunium (above 280 °C), tetragonal; and γ-neptunium (above 576 °C), cubic. Neptunium has four ionic oxidation states in solution: Np^{+3} (pale purple), analogous to the rare earth ion Pm^{+3}; Np^{+4} (yellow green); NpO^{+} (green blue); and NpO^{++} (pale pink). These latter oxygenated species are in contrast to the rare earths that exhibit only simple ions of the (II), (III), and (IV) oxidation states in aqueous solution. The element forms tri- and tetrahalides such as NpF_3, NpF_4, $NpCl_4$, $NpBr_3$, NpI_3, and oxides of various compositions such as are found in the uranium-oxygen system, including Np_3O_8 and NpO_2.

Nickel — (Ger. *Nickel,* Satan or Old Nick's and from *kupfernickel,* Old Nick's copper), Ni. Discovered by Cronstedt in 1751 in kupfernickel (*niccolite*). Nickel is found as a constituent in most meteorites and often serves as one of the criteria for distinguishing a meteorite from other minerals. Iron meteorites, or *siderites,* may contain iron alloyed with from 5 to nearly 20% nickel. Nickel is obtained commercially from *pentlandite* and *pyrrhotite* of the Sudbury region of Ontario, a district that produces much of the world's nickel. It is now thought that the Sudbury deposit is the result of an ancient meteorite impact. Large deposits of nickel, cobalt, and copper have been developed at Voisey's Bay, Labrador. Other deposits of nickel are found in Russia, New Caledonia, Australia, Cuba, Indonesia, and elsewhere. Nickel is silvery white and takes on a high polish. It is hard, malleable, ductile, somewhat ferromagnetic, and a fair conductor of heat and electricity. It belongs to the iron-cobalt group of metals and is chiefly valuable for the alloys it forms. It is extensively used for making stainless steel and other corrosion-resistant alloys such as Invar®, Monel®, Inconel®, and the Hastelloys®. Tubing made of a copper-nickel alloy is used in desalination plants for converting sea water into fresh water. Nickel is also now used extensively in coinage and in making nickel steel for armor plate and burglar-proof vaults, and is a component in Nichrome®, Permalloy®, and constantan. Nickel added to glass gives a green color. Nickel plating is often used to provide a protective coating for other metals, and finely divided nickel is a catalyst for hydrogenating vegetable oils. It is also used in ceramics, in the manufacture of Alnico magnets, and in batteries. The sulfate and the oxides are important compounds. Nickel sulfide fume and dust, as well as other nickel compounds, are carcinogens.

Nihonium — (*Nihon*, a word for Japan), Nh. Priority for the discovery of Element 113 has been given to the RIKEN in Japan, which published evidence in 2004.Similar evidence was obtained at Dubna in Russia at about the same time. IUPAC confirmed the name nihonium in 2016. The most stable isotope, ^{286}Nh, has a half-life of 20 s. Nihonium falls in the boron, aluminum, etc. group., but here is no chemical evidence yet on the extent to which its properties differ from the lower members of the group.

Niobium — (*Niobe,* daughter of Tantalus); sometimes called Columbium (*Columbia,* name for America), Nb. Discovered in 1801 by Hatchett in an ore sent to England more than a century before by John Winthrop the Younger, first governor of Connecticut. The metal was first prepared in 1864 by Blomstrand, who reduced the chloride by heating it in a hydrogen atmosphere. The name *niobium* was adopted by the International Union of Pure and Applied Chemistry in 1950 after 100 years of controversy. Most leading chemical societies and government organizations refer to it by this name. Some metallurgists and commercial producers, however, still refer to the metal as "columbium." The element is found in *niobite* (or *columbite), niobite-tantalite, pyrochlore,* and *euxenite.* Large deposits of niobium have been found associated with carbonatites (carbon-silicate rocks), as a constituent of pyrochlore. Extensive ore reserves are found in Canada, Brazil, Congo-Kinshasa, Rwanda, and Australia. The metal can be isolated from tantalum, and prepared in several ways. It is a shiny, white, soft, and ductile metal, and takes on a bluish cast when exposed to air for a long time at room temperature. The metal starts to oxidize in air at 200 °C, and when processed at even moderate temperatures must be placed in a protective atmosphere. It is used in arc-welding rods for stabilized grades of stainless steel. Thousands of pounds of niobium have been used in advanced air frame systems such as were used in the Gemini space program. It has also found use in super-alloys for applications such as jet engine components, rocket subassemblies, and heat-resisting equipment. The element has superconductive properties; superconductive magnets made with Nb-Zr wire retain their superconductivity in strong magnetic fields.

Nitrogen — (L. *nitrum,* Fr. *nitre,* native soda; *-genes,* producing, N. Discovered by Daniel Rutherford in 1772, but Scheele, Cavendish, Priestley, and others about the same time studied "burnt or dephlogisticated air," as air without oxygen was then called. Nitrogen makes up 78% of the air, by volume. The atmosphere of Mars, by comparison, is 2.7% nitrogen. The estimated amount of this element in our atmosphere is more than 4000 trillion tons. From this inexhaustible source it can be obtained by liquefaction and fractional distillation. Nitrogen molecules give the orange-red, blue-green, blue-violet, and deep violet shades to the aurora. The element is so inert that Lavoisier named it *azote,* meaning without life, yet its compounds are so active as to be most important in foods, poisons, fertilizers, and explosives. Nitrogen can be also easily prepared by heating a water solution of ammonium nitrite. Nitrogen, as a gas, is colorless, odorless, and a generally inert element. As a liquid it is also colorless and odorless, and is similar in appearance to water. Two allotropic forms of solid nitrogen exist, with the transition from the α to the β form taking place at –237 °C. When nitrogen is heated, it combines directly with magnesium, lithium, or calcium; when mixed with oxygen and subjected to electric sparks, it forms first nitric oxide (NO) and then the dioxide (NO_2); when heated under pressure with a catalyst with hydrogen, ammonia is formed (Haber process). The ammonia thus formed is of the utmost importance as it is used in fertilizers, and it can be oxidized to nitric acid (Ostwald process). The ammonia industry is the largest consumer of nitrogen. Large amounts are also used by the electronics industry, which uses the gas as a blanketing medium during production of such components as transistors, diodes, etc. Large quantities of nitrogen are used in annealing stainless steel and other steel mill products. The drug industry also uses large quantities. Nitrogen is used as a refrigerant both for the immersion freezing of food products and for transportation of foods. Liquid nitrogen is also used in missile work as a purge for components, insulators for space chambers, etc., and by the oil industry to build up pressure in wells to force crude oil upward. Sodium and potassium nitrates are formed by the decomposition of organic matter with compounds of the metals present. In certain dry areas of the world these saltpeters are found in quantity. Ammonia, nitric acid, the nitrates, the five oxides (N_2O, NO, N_2O_3, NO_2, and N_2O_5), TNT, the cyanides, etc. are but a few of the important compounds.

Nobelium — (Alfred Nobel [1833–1896], inventor of dynamite), No. Several claims to the discovery of Element-102 were made in the 1957 to 1966 period, leading to a considerable amount of controversy and disputes over naming the element. The disputes were settled in 1966 with definitive evidence produced at the Joint Institute for Nuclear Research at Dubna. Although it had been used much earlier, the name Nobelium was finally ratified by IUPAC in 1994. ^{259}No has a half-life of 58 minutes. The chemical behavior is expected to be similar to ytterbium.

Oganesson — (Yuri *Oganessian*, Russian leader in heavy element research), Og. Element-118, the heaviest element so far found in the periodic table, was first detected in 2002 by Russian and American workers at JINR, Dubna. The name oganesson was approved by IUPAC in 2016. Only three or four atoms on ^{295}Og (half-life about 1 ms) have been produced. Considering its position in the periodic table, where it falls in the noble gases group, one might expect oganesson to be an unreactive gas at normal conditions. However, theoretical calculations suggest that it is a solid and possibly very reactive. Thus it might exhibit valences of +2 and +4, as well a zero.

Osmium — (Gr. *osme*, a smell), Os. Discovered in 1803 by Tennant in the residue left when crude platinum is dissolved by *aqua regia*. Osmium occurs in *iridosmine* and in platinum-bearing river sands of the Urals, North America, and South America. It is also found in the nickel-bearing ores of the Sudbury, Ontario, region along with other platinum metals. While the quantity of platinum metals in these ores is very small, the large tonnages of nickel ores processed make commercial recovery possible. The metal is lustrous, bluish white, extremely hard, and brittle even at high temperatures. It has the highest melting point and the lowest vapor pressure of the platinum group. The metal is very difficult to fabricate, but the powder can be sintered in a hydrogen atmosphere at a temperature of 2000 °C. The solid metal is not affected by air at room temperature, but the powdered or spongy metal slowly gives off osmium tetroxide, which is a powerful oxidizing agent and has a strong smell. The tetroxide is highly toxic, and boils at 130 °C. Concentrations in air as low as 10^{-7} g/m^3 can cause lung congestion, skin damage, or eye damage. The tetroxide has been used to detect fingerprints and to stain fatty tissue for microscope slides. The metal is almost entirely used with other metals of the platinum group to produce very hard alloys for fountain pen tips, instrument pivots, and electrical contacts. The measured densities of iridium and osmium seem to indicate that osmium is slightly more dense than iridium, so osmium is generally credited with being the heaviest known element.

Oxygen — (Gr. *oxys*, sharp, acid, and *-genes*, forming; acid former), O. For many centuries, workers occasionally realized air was composed of more than one component. The behavior of oxygen and nitrogen as components of air led to the advancement of the phlogiston theory of combustion, which captured the minds of chemists for a century. Oxygen was prepared by several workers, including Bayen and Borch, but they did not know how to collect it, did not study its properties, and did not recognize it as an elementary substance. Priestley is generally credited with its discovery, although Scheele also discovered it independently. Oxygen is the third most abundant element found in the sun, and it plays a part in the carbon–nitrogen cycle, one process thought to give the sun and stars their energy. Oxygen under excited conditions is responsible for the bright red and yellow-green colors of the aurora. Oxygen, as a gaseous element, forms 21% of the atmosphere by volume from which it can be obtained by liquefaction and fractional distillation. The atmosphere of Mars contains about 0.13% oxygen. The element and its compounds make up 49.2%, by weight, of the Earth's crust. About two thirds of the mass of the human body and nine tenths of water is oxygen. In the laboratory it can be prepared by the electrolysis of water or by heating potassium chlorate with manganese dioxide as a catalyst. The gas is colorless, odorless, and tasteless. The liquid and solid forms are a pale blue color and are strongly paramagnetic. Ozone (O_3), a highly active compound, is formed by the action of an electrical discharge or ultraviolet light on oxygen. Ozone's presence in the atmosphere (amounting to the equivalent of a layer 3 mm thick at ordinary pressures and temperatures) is of vital importance in preventing the harmful ultraviolet rays of the sun from reaching the Earth's surface. The use of certain halogenated compounds that have a detrimental effect on this ozone layer has been phased out by international agreement. Ozone is toxic and exposure should not exceed 0.1ppm. Undiluted ozone has a bluish color. Liquid ozone is bluish black, and solid ozone is violet-black. Oxygen is very reactive and capable of combining with most elements. It is a component of hundreds of thousands of organic compounds. It is essential for respiration of all plants and animals and for practically all combustion. In hospitals it is frequently used to aid respiration of patients. Its atomic weight was used as a standard until 1961 when the International Union of Pure and Applied Chemistry adopted carbon 12 as the new basis. Commercial oxygen consumption in the U.S. is estimated to be 20 million short tons per year and the demand is expected to increase. Oxygen enrichment of steel blast furnaces accounts for the greatest use of the gas. Large quantities are also used in making synthesis gas for ammonia and methanol, ethylene oxide, and for oxy-acetylene welding. Air separation plants produce about 99% of the gas, electrolysis plants about 1%.

Palladium — (named after the asteroid *Pallas*, discovered about the same time; Gr. *Pallas*, goddess of wisdom), Pd. Discovered in 1803 by Wollaston. Palladium is found along with platinum and other metals of the platinum group in deposits of Russia, South Africa, Canada (Ontario), and elsewhere. It is frequently found associated with the nickel-copper deposits such as those found in Ontario. Its separation from the platinum metals depends upon the type of ore in which it is found. It is a steel-white metal, does not tarnish in air, and is the least dense and lowest melting of the platinum group of metals. When annealed, it is soft and ductile; cold working greatly increases its strength and hardness. Palladium is attacked by nitric and sulfuric acid. At room temperatures the metal has the unusual property of absorbing up to 900 times its own volume of hydrogen. Hydrogen readily diffuses through heated palladium and this provides a means of purifying the gas. Finely divided palladium is a good catalyst for hydrogenation and dehydrogenation reactions. It is alloyed and used in jewelry trades. White gold is an alloy of gold decolorized by the addition of palladium. Like gold, palladium can be beaten into leaf as thin as 1/250,000 in. The metal is used in dentistry, watchmaking, and in making surgical instruments and electrical contacts. Palladium has been substituted for higher priced platinum in some catalytic converters. Palladium, however, is less resistant to poisoning by sulfur and lead than platinum.

Phosphorus — (Gr. *phosphoros*, light bearing; ancient name for the planet Venus when appearing before sunrise), P. Discovered in 1669 by Brand, who prepared it from urine. Phosphorus exists in several allotropic forms, the most common of which are white (or yellow), red, and black (or violet). White phosphorus has two modifications: α and β with a transition temperature at −190 °C. Red and black phosphorus have amorphous structures. Never found free in nature, it is widely distributed in combination with minerals. *Phosphate* rock, which contains the mineral *apatite*, an impure tricalcium phosphate, is an important source of the element. Large deposits are found in the Russia, China, Morocco, and in Florida, Tennessee, Utah, Idaho, and elsewhere. Phosphorus in an essential ingredient of all cell protoplasm, nervous tissue, and bones, and phosphate groups provide a key linkage in the DNA chain. Ordinary phosphorus is a waxy white solid; when pure it is colorless and transparent. It is insoluble in water, but soluble in carbon disulfide. It takes fire spontaneously in air, burning to the pentoxide. It is very poisonous, 50 mg constituting an approximate fatal dose. White phosphorus should be kept under water, as it is dangerously reactive in air, and it should be handled with forceps, as contact with the skin may cause severe burns. When exposed to sunlight or when heated in its own vapor to 250 °C, it is converted to the red variety, which does not phosphoresce in air as does the white form. Red phosphorus does not ignite spontaneously and it is not as dangerous as white phosphorus. It should, however, be handled with care as it does convert to the white form at some temperatures and it emits highly toxic fumes of the oxides of phosphorus when heated. The red modification is fairly stable, sublimes with a vapor pressure of 1 atm at 431 °C, and is used in the manufacture of safety matches, pyrotechnics, pesticides, incendiary shells, smoke bombs, tracer bullets, etc. White phosphorus may be made by several methods. By one process, tricalcium phosphate, the essential ingredient of phosphate rock, is heated in the presence of carbon and silica in an electric furnace or fuel-fired furnace. Elementary phosphorus is liberated as vapor and may be collected under water. If desired, the phosphorus vapor and carbon monoxide produced by the reaction can be oxidized at once in the presence of moisture to produce phosphoric acid, an important compound in making super-phosphate fertilizers. In recent years, concentrated phosphoric acids, which may contain as much as 70 to 75% P_2O_5, have become of great importance to agriculture and farm production. World-wide demand for fertilizers has caused record phosphate production. Phosphates are used in the production of special glasses, such as those used for sodium lamps. Bone-ash, calcium phosphate, is also used to produce fine chinaware and monocalcium phosphate used in baking powder. Phosphorus is also important in the production of steels, phosphor bronze, and many other products. Trisodium phosphate is important as a cleaning agent, as a water softener, and for preventing boiler scale and corrosion of pipes and boiler tubes. Organic compounds of phosphorus are important.

Platinum — (It. *platina*, silver), P. Discovered in South America by Ulloa in 1735 and by Wood in 1741. The metal was used by pre-Columbian Indians. Platinum occurs native, accompanied by small quantities of iridium, osmium, palladium, ruthenium, and rhodium, all belonging to the same group of metals. These are found in the alluvial deposits of the Ural Mountains and in Colombia. *Sperrylite* ($PtAs_2$), occurring with the nickel-bearing deposits of Sudbury, Ontario, is a source of a considerable amount of the metal. The large production of nickel offsets there being only one part of the platinum metals in two million parts of ore. The largest supplier of the platinum group of metals is South Africa, followed by Russia and Canada. Platinum is a beautiful silvery-white metal when pure and is malleable and ductile. It has a coefficient of expansion almost equal to that of soda–lime–silica glass, and is therefore used to make sealed electrodes in glass systems. The metal does not oxidize in air at any temperature, but is corroded by halogens, cyanides, sulfur, and caustic alkalis. It is insoluble in hydrochloric and nitric acid, but dissolves when they are mixed as *aqua regia*, forming chloroplatinic acid (H_2PtCl_6), an important compound. The metal is used extensively in jewelry, wire, and vessels for laboratory use, and in many valuable instruments including thermocouple elements. It is also used for electrical contacts, corrosion-resistant apparatus, and in dentistry. Platinum–cobalt alloys have magnetic properties. One such alloy made of 76.7% Pt and 23.3% Co, by weight, is a more powerful magnet than Alnico V. Platinum resistance wires are used for constructing high-temperature electric furnaces. The metal is used for coating missile nose cones, jet engine fuel nozzles, etc., which must perform reliably for long periods of time at high temperatures. The metal, like palladium, absorbs large volumes of hydrogen, retaining it at ordinary temperatures but giving it up at red heat. In the finely divided state platinum is an excellent catalyst, having long been used in the contact process for producing sulfuric acid. It is also used as a catalyst in cracking petroleum products. There is also much current interest in the use of platinum as a catalyst in fuel cells and in its use in catalytic convertors for automobiles. Platinum anodes are extensively used in cathodic protection systems for large ships and ocean-going vessels, pipelines, steel piers, etc. Hydrogen and oxygen explode in the presence of platinum. Pure platinum wire will glow red hot when placed in the vapor of methyl alcohol. It acts here as a catalyst, converting the alcohol to formaldehyde.

Plutonium — (planet *Pluto*), Pu. Plutonium was the second transuranium element of the actinide series to be discovered. The isotope ^{238}Pu was produced in 1940 by Seaborg, McMillan, Kennedy, and Wahl by deuteron bombardment of uranium in the 60-inch cyclotron at Berkeley, California. Plutonium also exists in trace quantities in naturally occurring uranium ores. It is formed in much the same manner as neptunium, by irradiation of natural uranium with the neutrons that are present. By far of greatest importance is the isotope Pu^{239}, with a half-life of 24,100 years, produced in extensive quantities in nuclear reactors from natural uranium. Plutonium has assumed the position of dominant importance among the transuranium elements because of its use as an explosive ingredient in nuclear weapons and the place it holds as a key material in the development of industrial sources of nuclear power. One kilogram is equivalent to about 22 million kilowatt hours of heat energy. The complete detonation of a kilogram of plutonium produces an explosion equal to about 20,000 tons of chemical explosive. Its importance depends on the nuclear property of being readily fissionable with neutrons and its availability in quantity. The various nuclear applications of plutonium are well known. ^{238}Pu was used in the Apollo lunar missions to power seismic and other equipment on the lunar surface. As with neptunium and uranium, plutonium metal can be prepared by reduction of the trifluoride with alkaline-earth metals. The metal has a silvery appearance and takes on a yellow tarnish when slightly oxidized. It is chemically reactive. A relatively large piece of plutonium is warm to the touch because of the energy given off in alpha decay. Larger pieces will produce enough heat to boil water. The metal readily dissolves in concentrated hydrochloric acid, hydroiodic acid, or perchloric acid with formation of the Pu^{+3} ion. The metal exhibits six allotropic modifications having various crystalline structures. The densities of these vary from 16.0 to 19.7 g/cm^3. Plutonium also exhibits four ionic valence states in aqueous solutions: Pu^{+3}(blue lavender), Pu^{+4} (yellow brown), PuO^+ (pink?), and PuO^{+2} (pink orange). The ion PuO^+ is unstable in aqueous solutions, disproportionating into Pu^{+4} and PuO^{+2}. The Pu^{+4} thus formed, however, oxidizes the PuO^+ into PuO^{+2}, itself being reduced to Pu^{+3}, giving finally Pu^{+3} and PuO^{+2}. Plutonium forms binary compounds with oxygen (PuO, PuO_2, and intermediate oxides of variable composition); with the halides (PuF_3, PuF_4, $PuCl_3$, $PuBr_3$, PuI_3); and with carbon, nitrogen, and silicon (PuC, PuN, $PuSi_2$). Oxyhalides are also well known: PuOCl, PuOBr, PuOI. Because of the high rate of emission of alpha particles and the element being specifically absorbed by bone marrow, plutonium, as well as all of the other transuranium elements except neptunium, are radiological poisons and must be handled with very special equipment and precautions. Plutonium is a very dangerous radiological hazard. Precautions must also be taken to prevent the unintentional formation of a critical mass. Plutonium in liquid solution is more likely to become critical than solid plutonium.

Polonium — (Poland, native country of Mme. Curie [1867–1934]), Po. Polonium was the first element discovered by Mme. Curie in 1898, while seeking the cause of radioactivity of pitchblende from Joachimsthal, Bohemia. Polonium is a very rare natural element. Uranium ores contain only about 100 µg of the element per ton. Its abundance is only about 0.2% of that of radium. In 1934, it was found that when natural bismuth (^{209}Bi) was bombarded by neutrons, ^{210}Bi, the parent of polonium, was obtained. Milligram amounts of polonium may now be prepared this way, by using the high neutron fluxes of nuclear reactors. Polonium is a low-melting, fairly volatile metal, 50% of which is vaporized in air in 45 hours at 55 °C. It is an alpha emitter with a half-life of 138.39 days. A milligram emits as many alpha particles as 5 g of radium. The energy released by its decay is so large (140 W/g) that a capsule containing about half a gram reaches a temperature above 500 °C. The capsule also presents a contact gamma-ray dose rate of 0.012 Gy/h. A few curies (1 curie = 3.7×10^{10} Bq) of polonium exhibit a blue glow, caused by excitation of the surrounding gas. Because almost all alpha radiation is stopped within the solid source and its container, giving up its energy, polonium has attracted attention for uses as a lightweight heat source for thermoelectric power in space satellites. Polonium-210 is the most readily available isotope. Metallic polonium has been prepared from polonium hydroxide and some other polonium compounds in the presence of concentrated aqueous or anhydrous liquid ammonia. Two allotropic modifications are known to exist. Polonium is readily dissolved in dilute acids, but is only slightly soluble in alkalis. Polonium salts of organic acids char rapidly; halide amines are reduced to the metal. Polonium can be mixed or alloyed with beryllium to provide a source of neutrons. It has been used in devices for eliminating static charges in textile mills, etc.; however, beta sources are more commonly used and are less dangerous. It is also used on brushes for removing dust from photographic films. The polonium for these is carefully sealed and controlled, minimizing hazards to the user. Polonium-210 is very dangerous to handle in even milligram or microgram amounts, and special equipment and strict control are necessary. Damage arises from the complete absorption of the energy of the alpha particles into tissue. The maximum permissible body burden for ingested polonium is only 0.03 µCi, which represents a particle weighing only 6.8×10^{-12} g. Weight for weight it is about 2.5×10^{11} times as toxic as hydrocyanic acid.

Potassium — (English, *potash* — pot ashes; L. *kalium*, Arab. *qali*, alkali), K. Discovered in 1807 by Davy, who obtained it from caustic potash (KOH); this was the first metal isolated by electrolysis. The metal is the seventh most abundant and makes up about 2.4% by weight of the Earth's crust. Most potassium minerals are insoluble and the metal is obtained from them only with great difficulty. Certain minerals, however, such as *sylvite, carnallite, langbeinite,* and *polyhalite* are found in ancient lake and sea beds and form rather extensive deposits from which potassium and its salts can readily be obtained. Potash is mined in Germany, New Mexico, California, Utah, and elsewhere. Large deposits of potash, found at a depth of some 1000 m in Saskatchewan, promise to be important in coming years. Potassium is also found in the ocean, but is present only in relatively small amounts compared to sodium. The greatest demand for potash has been in its use for fertilizers. Potassium is an essential constituent for plant growth and it is found in most soils. Potassium is never found free in nature, but is obtained by electrolysis of the hydroxide, much in the same manner as prepared by Davy. Thermal methods also are commonly used to produce potassium (such as by reduction of potassium compounds with CaC_2, C, Si, or Na). It is one of the most reactive and electropositive of metals. Except for lithium, it is the lightest known metal. It is soft, easily cut with a knife, and is silvery in appearance immediately after a fresh surface is exposed. It rapidly oxidizes in air and should be preserved in a mineral oil. As with other metals of the alkali group, it decomposes in water with the evolution of hydrogen. It catches fire spontaneously on water. Potassium and its salts impart a violet color to flames. An alloy of sodium and potassium (NaK) is used as a heat-transfer medium. Many potassium salts are of utmost importance, including the hydroxide, nitrate, carbonate, chloride, chlorate, bromide, iodide, cyanide, sulfate, chromate, and dichromate.

Praseodymium — (Gr. *prasios*, green, and *didymos*, twin), Pr. In 1841 Mosander extracted the rare earth *didymia* from *lanthana*; in 1879, Lecoq de Boisbaudran isolated a new earth, *samaria*, from didymia obtained from the mineral *samarskite*. Six years later, in 1885, von Welsbach separated didymia into two others, *praseodymia* and *neodymia*, which gave salts of different colors. As with other rare earths, compounds of these elements in solution have distinctive sharp spectral absorption bands or lines, some of which are only a few Ångströms wide. The element occurs along with other rare-earth elements in a variety of minerals. *Monazite* and *bastnasite* are the two principal commercial sources of the rare-earth metals. Ion-exchange and solvent extraction techniques have led to much easier isolation of the rare earths and the cost has dropped greatly. Praseodymium can be prepared by several methods, such as by calcium reduction of the anhydrous chloride or fluoride. Misch metal, used in making cigarette lighters, contains about 5% praseodymium metal. Praseodymium is soft, silvery, malleable, and ductile. It was first prepared in relatively pure form in 1931. It is somewhat more resistant to corrosion in air than europium, lanthanum, cerium, or neodymium, but it does develop a green oxide coating that splits off when exposed to air. As with other rare-earth metals it should be kept under a light mineral oil or sealed in plastic. The rare-earth oxides, including Pr_2O_3, are among the most refractory substances known. Salts of praseodymium are used to color glasses and enamels; when mixed with certain other materials, praseodymium produces an intense and unusually clean yellow color in glass. Didymium glass, of which praseodymium is a component, is a colorant for welder's goggles.

Promethium — (*Prometheus*, who, according to mythology, stole fire from heaven), Pm. In 1902 Branner predicted the existence of an element between neodymium and samarium, and this prediction was confirmed by Moseley in 1914. Unsuccessful searches were made for this predicted element over two decades, and various investigators proposed the names "illinium," "florentium," and "cyclonium" for this element. In 1941, workers at Ohio State University irradiated neodymium and praseodymium with neutrons, deuterons, and alpha particles, and produced several new radioactivities, which most likely were those of Element 61. Wu and Segre, and Bethe, in 1942, confirmed the formation; however, chemical proof of the production of Element 61 was lacking because of the difficulty in separating the rare earths from each other at that time. In 1945, Marinsky, Glendenin, and Coryell made the first chemical identification by using ion-exchange chromatography. Their work was done by fission of uranium and by neutron bombardment of neodymium. These investigators named the newly discovered element. Searches for the element on Earth have been fruitless, and it now appears that promethium is completely missing from the Earth's crust. Promethium, however, has been reported to be in the spectrum of the star HR^{465} in Andromeda. It must be formed near the star's surface, for no known isotope of promethium has a half-life longer than 17.7 years. That isotope, Promethium-145, is the most useful. Promethium-145 has a specific activity of 940 Ci/g. It is a soft beta emitter; although no gamma rays are emitted, X-radiation can be generated when beta particles impinge on elements of a high atomic number, and great care must be taken in handling it. Promethium salts luminesce in the dark with a pale blue or greenish glow, due to their high radioactivity. Ion-exchange methods led to the preparation of about 10 g of promethium from atomic reactor fuel processing wastes in early 1963. Little is yet known about the properties of metallic promethium. Two allotropic modifications exist. The element has applications as a beta source for thickness gages, and its radiation can be absorbed by a phosphor to produce light. Light produced in this manner can be used for signs or signals that require dependable operation; it can be used as a nuclear-powered battery by capturing light in photocells that convert it into electric current. Such a battery, using ^{147}Pm, would have a useful life of about 5 years. It is being used for fluorescent lighting starters and coatings for self-luminous watch dials. Promethium shows promise as a portable X-ray source, and it may become useful as a heat source to provide auxiliary power for space probes and satellites. More than 30 promethium compounds have been prepared. Most are colored.

Protactinium — (Gr. *protos*, first), Pa. The first isotope of Element 91 to be discovered was ^{234}Pa, also known as UX_2, a short-lived member of the naturally occurring ^{238}U decay series. It was identified by K. Fajans and O. H. Gohring in 1913 and they named the new element *brevium*. When the longer-lived isotope ^{231}Pa was identified by Hahn and Meitner in 1918, the name protoactinium was adopted as being more consistent with the characteristics of the most abundant isotope. Soddy, Cranson, and Fleck were also active in this work. The name *protoactinium* was shortened to *protactinium* in 1949. In 1927, Grosse prepared 2 mg of a white powder, which was shown to be Pa_2O_5. Later, in 1934, from 0.1 g of pure Pa_2O_5 he isolated the element by two methods, one of which was by converting the oxide to an iodide and "cracking" it in a high vacuum by an electrically heated filament by the reaction $2PaI_5 \rightarrow 2Pa + 5I_2$. Protactinium has a bright metallic luster that it retains for some time in air. The element occurs in *pitchblende* to the extent of about 1 part ^{231}Pa to 10 million parts of ore. Ores from Congo-Kinshasa have about 3 ppm. The most common isotope is ^{231}Pr with a half-life of 32,500 years. A number of protactinium compounds are known, some of which are colored. The element is superconductive below 1.4 K. The element is a dangerous toxic material and requires precautions similar to those used when handling plutonium. Protactinium is one of the rarest and most expensive naturally occurring elements.

Radium — (L. *radius*, ray), Ra. Radium was discovered in 1898 by Pierre and Marie Curie in the *pitchblende* or *uraninite* of North Bohemia (Czech Republic), where it occurs. There is about 1 g of radium in 7 tons of pitchblende. The element was isolated in 1911 by Mme. Curie and Debierne by the electrolysis of a solution of pure radium chloride, employing a mercury cathode; on distillation in an atmosphere of hydrogen this amalgam yielded the pure metal. Originally, radium was obtained from the rich pitchblende ore found at Joachimsthal, Bohemia. The *carnotite* sands of Colorado furnish some radium, but richer ores are found in the Democratic Republic of Congo and the Great Bear Lake region of Canada. Radium is present in all uranium minerals, and could be extracted, if desired, from the extensive wastes of uranium processing. Large uranium deposits are located in Ontario, New Mexico, Utah, Australia, and elsewhere. Radium is obtained commercially as the bromide or chloride; it is doubtful if any appreciable stock of the pure element now exists. The pure metal is brilliant white when freshly prepared, but blackens on exposure to air, probably due to formation of the nitride. It exhibits luminescence, as do its salts; it decomposes in water and is somewhat more volatile than barium. It is a member of the alkaline-earth group of metals. Radium imparts a carmine red color to a flame. Radium emits alpha, beta, and gamma rays and when mixed with beryllium produce neutrons. One gram of ^{226}Ra undergoes 3.7×10^{10} disintegrations per second. The *curie* (*Ci*) is defined as that amount of radioactivity which has the same disintegration rate as 1 g of ^{226}Ra. Radium-226, the most common isotope, has a half-life of 1599 years. One gram of radium produces about 0.0001 mL (stp) of emanation, or radon gas, per day. This is pumped from the radium and sealed in minute tubes, which are used in the treatment of cancer. One gram of radium yields about 4186 kJ of energy per year. Radium is used in producing neutron sources and in medicine for the treatment of cancer. Some of the more recently discovered radioisotopes, such as ^{60}Co, are now being used in place of radium. Some of these sources are much more powerful, and others are safer to use. Radium loses about 1% of its activity in 25 years, being transformed into elements of lower atomic weight. Lead is a final product of disintegration. Stored radium should be ventilated to prevent build-up of radon. Inhalation, injection, or body exposure to radium can cause cancer and other body disorders.

Radon — (from *radium*; called *niton* at first, L. *nitens*, shining), Rn. The element was discovered in 1900 by Dorn, who called it *radium emanation*. In 1908 Ramsay and Gray, who named it *niton*, isolated the element and determined its density, finding it to be the heaviest known gas. It is essentially inert and occupies the last place in the zero group of gases in the Periodic Table. Since 1923, it has been called radon. Radon-222, coming from radium, has a half-life of 3.823 days and is an alpha emitter; Radon-220, emanating naturally from thorium and called *thoron*, has a half-life of 55.6 s and is also an alpha emitter. Radon-219 emanates from actinium and is called *actinon*. It has a half-life of 3.96 s and is also an alpha emitter. It is estimated that every square mile of soil to a depth of 6 inches contains about 1 g of radium, which releases radon in tiny amounts to the atmosphere. Radon is present in some spring waters, such as those at Hot Springs, Arkansas. On the average, one part of radon is present to 1×10^{21} part of air. At ordinary temperatures radon is a colorless gas; when cooled below the freezing point, radon exhibits a brilliant phosphorescence which becomes yellow as the temperature is lowered and orange-red at the temperature of liquid air. It has been reported that fluorine reacts with radon, forming radon fluoride. Radon clathrates have also been reported. Radon is still produced for therapeutic use by a few hospitals by pumping it from a radium source and sealing it in minute tubes, called seeds or needles, for application to patients. This practice has now been largely discontinued as hospitals can order the seeds directly from suppliers, who make up the seeds with the desired activity for the day of use. Care must be taken in handling radon, as with other radioactive materials. The main hazard is from inhalation of the element and its solid daughters, which are collected on dust in the air. Good ventilation should be provided where radium, thorium, or actinium is stored to prevent build-up of this element. Radon build-up is a health consideration in uranium mines, and build-up in homes is a major concern in some areas. Many deaths from lung cancer are caused by radon exposure.

Rhenium — (L. *Rhenus*, Rhine), Re. Discovery of rhenium is generally attributed to Noddack, Tacke, and Berg, who announced in 1925 they had detected the element in platinum ores and *columbite*. They also found the element in *gadolinite* and *molybdenite*. By working up 660 kg of molybdenite they were able in 1928 to extract 1 g of rhenium. Rhenium does not occur free in nature or as a compound in a distinct mineral species. It is, however, widely spread throughout the Earth's crust to the extent of about 0.001 ppm. Commercial rhenium in the U.S. today is obtained from molybdenite roaster-flue dusts obtained from copper-sulfide ores mined in the vicinity of Miami, Arizona, and elsewhere in Arizona and Utah. Some molybdenites contain from 0.002 to 0.2% rhenium. It is estimated that in 1999 about 16,000 kg of rhenium was being produced. The total estimated world reserve of rhenium is 11,000,000 kg. Natural rhenium is a mixture of two isotopes, one of which is radioactive with a very long half-life. Rhenium metal is prepared by reducing ammonium perrhenate with hydrogen at elevated temperatures. The element is silvery white with a metallic luster; its density is exceeded by that of only platinum, iridium, and osmium, and its melting point is exceeded only by that of tungsten and carbon. It has other useful properties. The usual commercial form of the element is a powder, but it can be consolidated by pressing and resistance-sintering in a vacuum or hydrogen atmosphere. This produces a compact shape in excess of 90% of the density of the metal. Annealed rhenium is very ductile, and can be bent, coiled, or rolled. Rhenium is used as an additive to tungsten and molybdenum-based alloys to impart useful properties. It is widely used for filaments for mass spectrographs and ion gages. Rhenium-molybdenum alloys are superconductive at 10 K. Rhenium is also used as an electrical contact material as it has good wear resistance and withstands arc corrosion. Thermocouples made of Re-W are used for measuring temperatures up to 2200 °C, and rhenium wire has been used in photoflash lamps for photography. Rhenium catalysts are exceptionally resistant to poisoning from nitrogen, sulfur, and phosphorus, and are used for hydrogenation of fine chemicals, hydrocracking, reforming, and disproportionation of olefins. Rhenium has become especially important as a catalyst for petroleum refining and in making super-alloys for jet engines.

Rhodium — (Gr. *rhodon*, rose), Rh. Wollaston discovered rhodium in 1803-4 in crude platinum ore he presumably obtained from South America. Rhodium occurs native with other platinum metals in river sands of the Urals and in North and South America. It is also found with other platinum metals in the copper-nickel sulfide ores of the Sudbury, Ontario region. Although the quantity occurring here is very small, the large tonnages of nickel processed make the recovery commercially feasible. The annual world production of rhodium in 2000 was only about 9000 kg. The metal is silvery white and at red heat slowly changes in air to the sesquioxide. At higher temperatures it reverts to the element. Rhodium has a higher melting point and lower density than platinum. Its major use is as an alloying agent to harden platinum and palladium. Such alloys are used for furnace windings, thermocouple elements, bushings for glass fiber production, and laboratory crucibles. It is useful as an electrical contact material as it has a low electrical resistance, a low and stable contact resistance, and is highly resistant to corrosion. Plated rhodium, produced by electroplating or evaporation, is exceptionally hard and is used for optical instruments. It has a high reflectance and is hard and durable. Rhodium is also used for jewelry, for decoration, and as a catalyst.

Roentgenium — (Wilhelm Roentgen, discoverer of X-rays), Rg. On December 20, 1994, scientists at GSI Darmstadt, Germany announced they had detected three atoms of a new element with 111 protons and 161 neutrons. This element was made by bombarding ^{83}Bi with ^{28}Ni. Signals of Element 111 appeared for less than 0.002 s, then decayed into lighter elements including Element ^{268}109 and Element ^{264}107. These isotopes had not previously been observed. In 2004 IUPAC approved the name roentgenium for Element 111. Roentgenium is expected to have properties similar to gold.

Rubidium — (L. *rubidus*, deepest red), Rb. Discovered in 1861 by Bunsen and Kirchhoff in the mineral *lepidolite* by use of spectroscopy. The element is much more abundant than was thought at one time. It is now considered to be the 16th most abundant element in the Earth's crust. Rubidium occurs in *pollucite*, *carnallite*, *leucite*, and *zinnwaldite*, which contains traces up to 1%, in the form of the oxide. It is found in lepidolite to the extent of about 1.5%, and is recovered commercially from this source. Potassium minerals, such as those found at Searles Lake, California, and potassium chloride recovered from brines in Michigan also contain the element and are commercial sources. It is also found along with cesium in the extensive deposits of *pollucite* at Bernic Lake, Manitoba. Rubidium can be liquid at room temperature. It is a soft, silvery-white metallic element of the alkali group and is the second most electropositive and alkaline element. It ignites spontaneously in air and reacts violently in water, setting fire to the liberated hydrogen. As with other alkali metals, it forms amalgams with mercury and it alloys with gold, cesium, sodium, and potassium. It colors a flame yellowish violet. Rubidium metal can be prepared by reducing rubidium chloride with calcium, and by a number of other methods. It must be kept under a dry mineral oil or in a vacuum or inert atmosphere. Naturally occurring rubidium is made of two isotopes, ^{85}Rb and ^{87}Rb. Rubidium-87 is present to the extent of 27.83% in natural rubidium and is a beta emitter with a half-life of 4.9 $\times 10^{10}$ years. Ordinary rubidium is sufficiently radioactive to expose a photographic film in about 30 to 60 days. Rubidium forms four oxides: Rb_2O, Rb_2O_2, Rb_2O_3, Rb_2O_4. Because rubidium can be easily ionized, it has been considered for use in ion engines for space vehicles; however, cesium is somewhat more efficient for this purpose. It has also been proposed for use as a working fluid for vapor turbines and for use in a thermoelectric generator using magnetohydrodynamic principles. Rubidium has been used as a getter in vacuum tubes and as a photocell component. It has been used in making special glasses. $RbAg_4I_5$ has the highest room-temperature conductivity of any known ionic crystal. This suggests use in thin film batteries and other applications.

Ruthenium — (L. *Ruthenia*, Russia), Ru. Berzelius and Osann in 1827 examined the residues left after dissolving crude platinum from the Ural Mountains in *aqua regia*. While Berzelius found no unusual metals, Osann thought he found three new metals, one of which he named ruthenium. In 1844 Klaus, generally recognized as the discoverer, showed that Osann's ruthenium oxide was very impure and that it contained a new metal. Klaus obtained 6 g of ruthenium from the portion of crude platinum that is insoluble in *aqua regia*. A member of the platinum group, ruthenium occurs native with other members of the group of ores found in the Ural Mountains and in North and South America. It is also found along with other platinum metals in small but commercial quantities in *pentlandite* of the Sudbury, Ontario, nickel-mining region, and in *pyroxinite* deposits of South Africa. The metal is isolated commercially by a complex chemical process, the final stage of which is the hydrogen reduction of ammonium ruthenium chloride, which yields a powder. The powder is consolidated by powder metallurgy techniques or by argon-arc welding. Ruthenium is a hard, white metal and has four crystal modifications. It does not tarnish at room temperatures, but oxidizes in air at about 800 °C. The metal is not attacked by hot or cold acids or *aqua regia*, but when potassium chlorate is added to the solution, it oxidizes explosively. It is attacked by halogens and by hydroxides. Ruthenium can be plated by electrodeposition or by thermal decomposition methods. The metal is one of the most effective hardeners for platinum and palladium, and is alloyed with these metals to make electrical contacts for severe wear resistance. A ruthenium–molybdenum alloy is superconductive at 10.6 K. The corrosion resistance of titanium is improved a hundredfold by addition of 0.1% ruthenium. It is a versatile catalyst. Hydrogen sulfide can be split catalytically by light using an aqueous suspension of CdS particles loaded with ruthenium dioxide. This may have application to removal of H_2S in oil refining and other industrial processes. Compounds in at least eight oxidation states have been found, but of these, the +2. +3. and +4 states are the most common. Ruthenium compounds show a marked resemblance to those of osmium. Ruthenium tetroxide, like osmium tetroxide, is highly toxic and may explode.

Rutherfordium — (Ernest Rutherford [1871–1937], New Zealand, Canadian, and British physicist); Rf. An isotope of Element-104 was produced in 1964 at the Joint Nuclear Research Institute at Dubna by bombardment of plutonium with accelerated neon ions through the following reaction:

$$^{242}Pu + ^{22}Ne \rightarrow ^{260}104 + 4n$$

Other isotopes were later produced at Berkeley. A long-standing naming controversy was finally settled by IUPAC in 1997 through adoption of the name Rutherfordium.

Element 104, the first *transactinide* element, is expected to have chemical properties similar to those of hafnium. Experiments have shown that it forms volatile tetrahalides, $RfCl_4$ and $RfBr_4$, as well as the oxychloride, $RfOCl_2$. More complex compounds in which it shows a valence of 6 have also been produced. The pure element is expected to be a dense metal. The most stable isotope is probably ^{267}Rf, with a half-life of 1.3 minutes.

Samarium — (*Samarskite*, a mineral), Sm. Discovered spectroscopically by its sharp absorption lines in 1879 by Lecoq de Boisbaudran in the mineral *samarskite*, named in honor of a Russian mine official, Col. Samarski. Samarium is found along with other members of the rare-earth-elements in many minerals, including *monazite* and *bastnasite*, which are commercial sources. The largest producer of rare-earth minerals is now China, followed by the U.S., India, and Russia. It occurs in monazite to the extent of 2.8%. While *misch metal* containing about 1% of samarium metal has long been used, samarium was not isolated in relatively pure form until 1901. Samarium metal can be produced by reducing the oxide with barium or lanthanum. Samarium has a bright silver luster and is reasonably stable in air. Three crystal modifications of the metal exist, with transformations at 734 and 922 °C. The metal ignites in air at about 150 °C. Natural samarium is a mixture of seven isotopes, three of which are unstable but have long half-lives. The sulfide has excellent high-temperature stability and good thermoelectric efficiencies up to 1100 °C. $SmCo_5$ has been used in making a permanent magnet material with a very high resistance to demagnetization. Samarium oxide has been used in optical glass to absorb infrared radiation. Samarium is used to dope calcium fluoride crystals for use in lasers. Compounds of the metal act as sensitizers for phosphors excited in the infrared; the oxide exhibits catalytic properties in the dehydration and dehydrogenation of ethyl alcohol. It is used as a neutron absorber in nuclear reactors. Little is known of the toxicity of samarium; therefore, it should be handled carefully.

Scandium — (L. *Scandia*, Scandinavia), Sc. On the basis of the Periodic System, Mendeleev predicted the existence of *ekaboron*, which would have an atomic weight between 40 of calcium and 48 of titanium. The element was discovered by Nilson in 1878 in the minerals *euxenite* and *gadolinite*, which had not yet been found anywhere except in Scandinavia. By processing 10 kg of euxenite and other residues of rare-earth minerals, Nilson was able to prepare about 2 g of scandium oxide of high purity. Cleve later pointed out that Nilson's scandium was identical with Mendeleev's ekaboron. Scandium is apparently a much more abundant element in the sun and certain stars than here on Earth. It is about the 23rd most abundant element in the sun, compared to the 50th most abundant on Earth. It is widely distributed on Earth, occurring in very minute quantities in over 800 mineral species. The blue color of beryl (aquamarine variety) is said to be due to scandium. It occurs as a principal component in the rare mineral *thortveitite*, found in Scandinavia and Malagasy. It is also found in the residues remaining after the extraction of tungsten from Zinnwald *wolframite*, and in *wiikite* and *bazzite*. Most scandium is presently being recovered from *thortveitite* or is extracted as a by-product from uranium mill tailings. Metallic scandium was first prepared in 1937 by Fischer, Brunger, and Grieneisen, who electrolyzed a eutectic melt of potassium, lithium, and scandium chlorides at 700 to 800 °C. Tungsten wire and a pool of molten zinc served as the electrodes in a graphite crucible. Pure scandium is now produced by reducing scandium fluoride with calcium metal. The production of the first pound of 99% pure scandium metal was announced in 1960. Scandium is a silver-white metal that develops a slightly yellowish or pinkish cast upon exposure to air. It is relatively soft, and resembles yttrium and the rare-earth metals more than it resembles aluminum or titanium. It is a very light metal and has a much higher melting point than aluminum, making it of interest to designers of spacecraft. Scandium is not attacked by a 1:1 mixture of conc. HNO_3 and 48% HF. Scandium reacts rapidly with many acids. About 20 kg of scandium (as Sc_2O_3) are now being used yearly in the U.S. to produce high-intensity lights, and the radioactive isotope ^{46}Sc is used as a tracing agent in crude-oil refining. Scandium iodide added to mercury vapor lamps produces a highly efficient light source resembling sunlight. Little is yet known about the toxicity of scandium; therefore, it should be handled with care.

Seaborgium — (Glenn T. Seaborg [1912–1999], American chemist and nuclear physicist), Sg. The discovery of *Seaborgium*, Element 106, took place in 1974 almost simultaneously at the Lawrence-Berkeley Laboratory and at the Joint Institute for Nuclear Research at Dubna, Russia. The Berkeley Group, under direction of Ghiorso, used the Super-Heavy Ion Linear Accelerator (Super HILAC) as a source of heavy ^{18}O ions to bombard a 259 μg target of ^{249}Cf. This resulted in the production and positive identification of $^{263}106$, which decayed with a half-life of 0.8 s by the emission of alpha particles. The Dubna Team, directed by Flerov and Organessian, produced heavy ions of ^{54}Cr with their 310-cm heavy-ion cyclotron to bombard ^{207}Pb and ^{208}Pb and found a product that decayed with a half-life of 7 ms. They assigned $^{259}106$ to this isotope. Other isotopes of *Seaborgium* were later identified. Two of these, ^{269}Sg and ^{271}Sg, have half-lives of about 2 min. *Seaborgium* most likely would have properties resembling tungsten. IUPAC adopted the name *Seaborgium* in August 1997. Normally the naming of an element is not given until after the death of the person for which the element is named; however, in this case, it was named while Dr. Seaborg was still alive.

Selenium — (Gr. *Selene*, moon), Se. Discovered by Berzelius in 1817, who found it associated with tellurium, named for the Earth. Selenium is found in a few rare minerals, such as *crooksite* and *clausthalite*. In years past it has been obtained from flue dusts remaining from processing copper sulfide ores, but the anode muds from electrolytic copper refineries now provide the source of most of the world's selenium. Selenium is recovered by roasting the muds with soda or sulfuric acid, or by smelting them with soda and niter. Elemental selenium exists in several allotropic forms. Three are generally recognized, but as many as six have been claimed. Selenium can be prepared with either an amorphous or crystalline structure. The color of amorphous selenium is either red, in powder form, or black, in vitreous form. Crystalline monoclinic selenium is a deep red; crystalline hexagonal selenium, the most stable variety, is a metallic gray. The element is a member of the sulfur family and resembles sulfur both in its various forms and in its compounds. Selenium exhibits both photovoltaic action, where light is converted directly into electricity, and photoconductive action, where the electrical resistance decreases with increased illumination. These properties have made selenium useful in the production of photocells and exposure meters for photographic use, as well as solar cells. Selenium is also able to convert a.c. electricity to d.c., and is used in rectifiers. Below its melting point, selenium is a p-type semiconductor and has found uses in electronic and solid-state applications. It was formerly used in xerography for reproducing and copying documents, but it has been replaced in this application organic photoconductors. It is used by the glass industry to decolorize glass and to make ruby-colored glasses and enamels. It is also used as a photographic toner, and as an additive to stainless steel. Elemental selenium has low toxicity, and selenium is considered to be an essential trace element for humans; however, hydrogen selenide and other selenium compounds are extremely toxic, and resemble arsenic in their physiological reactions. Hydrogen selenide in a concentration of 1.5 ppm is intolerable to man. Selenium occurs in some soils in amounts sufficient to produce serious effects on animals feeding on plants, such as locoweed, grown in such soils.

Silicon — (L. *silex*, *silicis*, flint), Si. Davy in 1800 thought silica to be a compound and not an element; later in 1811, Gay Lussac and Thenard probably prepared impure amorphous silicon by heating potassium with silicon tetrafluoride. Berzelius, generally credited with the discovery, in 1824 succeeded in preparing amorphous silicon by the same general method as used earlier, but he purified the product by removing the fluosilicates by repeated washings. Deville in 1854 first prepared crystalline silicon, the second allotropic form of the element. Silicon is present in the sun and stars and is a principal component of a class of meteorites known as "aerolites." It is also a component of *tektites*, a natural glass of uncertain origin. Silicon makes up 25.7% of the Earth's crust, by weight, and is the second most abundant element, being exceeded only by oxygen. Silicon is not found free in nature, but occurs chiefly as the oxide and as silicates. *Sand, quartz, rock crystal, amethyst, agate, flint, jasper*, and *opal* are some of the forms in which the oxide appears. *Granite, hornblende, asbestos, feldspar, clay mica*, etc. are but a few of the numerous silicate minerals. Silicon is prepared commercially by heating silica and carbon in an electric furnace, using carbon electrodes. Several other methods can be used for preparing the element. Amorphous silicon can be prepared as a brown powder, which can be easily melted or vaporized. Crystalline silicon has a metallic luster and grayish color. The Czochralski process is commonly used to produce single crystals of silicon used for solid-state devices. Hyperpure silicon can be prepared by the thermal decomposition of ultra-pure trichlorosilane in a hydrogen atmosphere, and by a vacuum float zone process. This product can be doped with boron, gallium, phosphorus, or arsenic to produce semiconductors for use in transistors, solar cells, rectifiers, and other solid-state devices that are used extensively in the electronics industries.

Silicon is a relatively inert element, but it is attacked by halogens and dilute alkali. Most acids, except hydrofluoric, do not affect it. Hydrolysis and condensation of various substituted chlorosilanes can be used to produce a very great number of polymeric products, or silicones, ranging from liquids to hard, glasslike solids with many useful properties. Elemental silicon transmits more than 95% of all wavelengths of infrared light, from 1.3 to 6.7 μm. Silicon is one of man's most useful elements. In the form of sand and clay it is used to make concrete and brick; it is a useful refractory material for high-temperature work, and in the form of silicates it is used in making enamels, pottery, etc. Silica, as sand, is a principal ingredient of glass, one of the most inexpensive of materials with excellent mechanical, optical, thermal, and electrical properties. Glass can be made in a very great variety of shapes, and is used as containers, window glass, insulators, and thousands of other uses. Silicon is important in plant and animal life. Diatoms in both fresh and salt water extract silica from the water to build up their cell walls. Silica is present in ashes of plants and in the human skeleton. Silicon is an important ingredient in steel; silicon carbide is one of the most important abrasives. Miners, stonecutters, and other engaged in work where siliceous dust is breathed in large quantities often develop a serious lung disease known as *silicosis*.

Silver — (Anglo-Saxon, *Seolfor siolfur*), Ag (L. *argentum*). Silver has been known since ancient times. It is mentioned in Genesis. Slag dumps in Asia Minor and on islands in the Aegean Sea indicate that man learned to separate silver from lead as early as 3000 B.C. Silver occurs native and in ores such as *argentite* (Ag_2S) and *horn silver* ($AgCl$); lead, lead-zinc, copper, gold, and copper-nickel ores are principal sources. Mexico, Canada, Peru, and the U.S. are the principal silver producers in the western hemisphere. Silver is also recovered during electrolytic refining of copper. Commercial fine silver contains at least 99.9% silver. Purities of 99.999+% are available commercially. Pure silver has a brilliant white metallic luster. It is a little harder than gold and is very ductile and malleable, being exceeded only by gold and perhaps palladium. Pure silver has the highest electrical and thermal conductivity of all metals, and possesses the lowest contact resistance. It is stable in pure air and water, but tarnishes when exposed to ozone, hydrogen sulfide, or air containing sulfur. The alloys of silver are important. Sterling silver is used for jewelry, silverware, etc. where appearance is paramount. This alloy contains 92.5% silver, the remainder being copper or some other metal. Silver was once of great importance in photography, including production of X-ray images, but that application has diminished as digital photography became popular. It is used for dental alloys and in making solder and brazing alloys, electrical contacts, and high capacity silver–zinc and silver–cadmium batteries. Silver paints are used for making printed circuits. It is used in mirror production and may be deposited on glass or metals by chemical deposition, electrodeposition, or by evaporation. When freshly deposited, it is the best reflector of visible light known, but is rapidly tarnishes and loses much of its reflectance. It is a poor reflector of ultraviolet. Silver fulminate ($Ag_2C_2N_2O_2$), a powerful explosive, is sometimes formed during the silvering process. Silver nitrate, or *lunar caustic*, is an important silver compound. While silver itself is not considered to be toxic, most of its salts are poisonous. Silver compounds can be absorbed in the circulatory system and reduced silver deposited in the various tissues of the body. A condition, known as *argyria*, results with a greyish pigmentation of the skin and mucous membranes. Silver has germicidal effects and kills many lower organisms effectively without harm to higher animals. Silver has been used for coinage since ancient times. However, in the 20th century the U.S. and most other countries abandoned silver as a monetary standard and replaced silver coins by other metals, often alloys of copper, nickel, and zinc.

Sodium — (English, *soda*; Medieval Latin, *sodanum*, headache remedy), Na (L. *natrium*). Long recognized in compounds, sodium was first isolated by Davy in 1807 by electrolysis of caustic soda. Sodium is present in fair abundance in the sun and stars. The D lines of sodium are among the most prominent in the solar spectrum. Sodium is the sixth most abundant element on earth, comprising about 2.6% of the Earth's crust; it is the most abundant of the alkali group of metals of which it is a member. The most common compound is sodium chloride, but sodium occurs in many other minerals, such as *soda niter, cryolite, amphibole, zeolite, sodalite,* etc. It is a very reactive element and is never found free in nature. It is now obtained commercially by the electrolysis of absolutely dry fused sodium chloride. This method is much cheaper than that of electrolyzing sodium hydroxide, as was used in the past. Sodium is a soft, bright, silvery metal that floats on water, decomposing it with the evolution of hydrogen and the formation of the hydroxide. It may or may not ignite spontaneously on water, depending on the amount of oxide and metal exposed to the water. It normally does not ignite in air at temperatures below 115 °C. Sodium should be handled with respect, as it can be dangerous when improperly handled. Metallic sodium is vital in the manufacture of sodamide and esters, and in the preparation of organic compounds. The metal may be used to improve the structure of certain alloys, to descale metal, to purify molten metals, and as a heat transfer agent. An alloy of sodium with potassium, NaK, is also an important heat transfer agent. Sodium compounds are important to the paper, glass, soap, textile, petroleum, chemical, and metal industries. Soap is generally a sodium salt of certain fatty acids. The importance of common salt to animal nutrition has been recognized since prehistoric times. Among the many compounds that are of the greatest industrial importance are common salt ($NaCl$), soda ash (Na_2CO_3), baking soda ($NaHCO_3$), caustic soda ($NaOH$), Chile saltpeter ($NaNO_3$), di- and tri-sodium phosphates, sodium thiosulfate (hypo, $Na_2S_2O_3 \cdot 5H_2O$), and borax ($Na_2B_4O_7 \cdot 10H_2O$). Sodium metal should be handled with great care. It should be kept in an inert atmosphere and contact with water and other substances with which sodium reacts should be avoided.

Strontium — (*Strontian*, town in Scotland), Sr. Isolated by Davey by electrolysis in 1808; however, Adair Crawford in 1790 recognized a new mineral (*strontianite*) as differing from other barium minerals (*baryta*). Strontium is found chiefly as *celestite* ($SrSO_4$) and *strontianite* ($SrCO_3$). *Celestite* is found in Mexico, Turkey, Iran, Spain, Algeria, and in the U.K. The U.S. has no active *celestite* mines. The metal can be prepared by electrolysis of the fused chloride mixed with potassium chloride, or is made by reducing strontium oxide with aluminum in a vacuum at a temperature at which strontium distills off. Three allotropic forms of the metal exist, with transition points at 235 and 540 °C. Strontium is softer than calcium and decomposes water more vigorously. It does not absorb nitrogen below 380 °C. It should be kept under mineral oil to prevent oxidation. Freshly cut strontium has a silvery appearance, but rapidly turns a yellowish color with the formation of the oxide. The finely divided metal ignites spontaneously in air. Volatile strontium salts impart a beautiful crimson color to flames, and these salts are used in pyrotechnics and in the production of flares. The major use for strontium at present is in producing glass for color television picture tubes. All color TV and cathode ray tubes sold in the U.S. are required by law to contain strontium in the face plate glass to block X-ray emission. Strontium also improves the brilliance of the glass and the quality of the picture. It has also found use in producing ferrite magnets and in refining zinc. Strontium titanate is an interesting optical material as it has an extremely high refractive index and an optical dispersion greater than that of diamond. It has been used as a gemstone, but it is very soft. It does not occur naturally.

Sulfur — (Sanskrit, *sulvere*; L. *sulphurium*), S. Known to the ancients; referred to in Genesis as *brimstone*. Sulfur is found in meteorites. A dark area near the crater Aristarchus on the moon has been studied by R. W. Wood with ultraviolet light. This study suggests strongly that it is a sulfur deposit. Sulfur occurs native in the vicinity of volcanoes and hot springs. It is widely distributed in nature as *iron pyrites, galena, sphalerite, cinnabar, stibnite, gypsum, Epsom salts, celestite, barite,* etc. Sulfur is commercially recovered from wells sunk into the salt domes along the Gulf Coast of the U.S. It is obtained from these wells by the Frasch process, which forces heated water into the wells to melt the sulfur, which is then brought to the surface. Sulfur compounds also occur in natural gas and petroleum crudes and must be removed from these products. Formerly this was done chemically, which wasted the sulfur. New processes now permit recovery, and these sources promise to be very important. Large amounts of sulfur are being recovered from Alberta gas fields. Sulfur is a pale yellow, odorless, brittle solid that is insoluble in water but soluble in carbon disulfide. In every state, whether gas, liquid or solid, elemental sulfur occurs in more than one allotropic form or modification; these present a confusing multitude of forms whose relations are not yet fully understood. Amorphous or "plastic" sulfur is obtained by fast cooling of the crystalline form. X-ray studies indicate that amorphous sulfur may have a helical structure with eight atoms per spiral. Crystalline sulfur seems to be made of rings, each containing eight sulfur atoms that fit together to give a normal X-ray pattern. A finely divided form of sulfur, known as *flowers of sulfur*, is obtained by sublimation. Sulfur readily forms sulfides with many elements. Sulfur is a component of black gunpowder, and is used in the vulcanization of natural rubber and as a fungicide. It is also used extensively is making phosphate fertilizers. A large tonnage is used to produce sulfuric acid, the most important manufactured sulfur compound. It is used in making sulfite paper and other papers, as a fumigant, and in the bleaching of dried fruits. The element is a good electrical insulator. Organic compounds containing sulfur are very important. Calcium sulfate, ammonium sulfate, carbon disulfide, sulfur dioxide, and hydrogen sulfide are but a few of the many other important compounds of sulfur. Sulfur is essential to life. It is a minor constituent of fats, body fluids, and skeletal minerals. Carbon disulfide, hydrogen sulfide, and sulfur dioxide should be handled carefully. Hydrogen sulfide in small concentrations can be metabolized, but in higher concentrations it can quickly cause death by respiratory paralysis. It is insidious in that it quickly deadens the sense of smell. Sulfur dioxide is a dangerous component in atmospheric pollution.

Tantalum — (Gr. *Tantalos*, mythological character, father of *Niobe*), Ta. Discovered in 1802 by Ekeberg, but many chemists thought niobium and tantalum were identical elements until Rose, in 1844, and Marignac, in 1866, showed that niobic and tantalic acids were two different acids. The early investigators only isolated the impure metal. The first relatively pure ductile tantalum was produced by von Bolton in 1903. Tantalum occurs principally in the mineral *columbite-tantalite* (Fe,Mn) $(Nb,Ta)_2O_6$. Tantalum ores are found in Australia, Brazil, Rwanda, Zimbabwe, Congo-Kinshasa, Nigeria, and Canada. Separation of tantalum from niobium requires several complicated steps. Several methods are used to commercially produce the element, including electrolysis of molten potassium fluorotantalate, reduction of potassium fluorotantalate with sodium, or reacting tantalum carbide with tantalum oxide. Natural tantalum contains two isotopes, one of which is radioactive with a very long half-life. Tantalum is a gray, heavy, and very hard metal. When pure, it is ductile and can be drawn into fine wire, which is used as a filament for evaporating metals such as aluminum. Tantalum is almost completely immune to chemical attack at temperatures below 150 °C, and is attacked only by hydrofluoric acid, acidic solutions containing the fluoride ion, and free sulfur trioxide. Alkaline solutions attack it only slowly. At high temperatures, tantalum becomes much more reactive. The melting point of tantalum is exceeded only by tungsten and rhenium. Tantalum is used to make a variety of alloys with desirable properties such as high melting point, high strength, good ductility, etc. A tantalum carbide graphite composite material is one of the hardest materials ever made. Tantalum has good "gettering" ability at high temperatures, and tantalum oxide films are stable and have good rectifying and dielectric properties. Tantalum is used to make electrolytic capacitors and vacuum furnace parts, which account for about 60% of its production. The metal is also widely used to fabricate chemical process equipment, nuclear reactors, and aircraft and missile parts. Tantalum is completely immune to body liquids and is a nonirritating metal. It has, therefore, found wide use in making surgical appliances. Tantalum oxide is used to make special glass with a high index of refraction for camera lenses. The metal has many other uses.

Technetium — (Gr. *technetos*, artificial), Tc. Element 43 was predicted on the basis of the periodic table, and was erroneously reported as having been discovered in 1925, at which time it was named *masurium*. The element was actually discovered by Perrier and Segre in Italy in 1937. It was found in a sample of molybdenum that was bombarded by deuterons in the Berkeley cyclotron, and which E. Lawrence sent to these investigators. Technetium was the first element to be produced artificially. Since its discovery, searches for the element in terrestrial materials have been made without success. If it does exist, the concentration must be very small. Technetium has been found in the spectrum of S-, M-, and N-type stars, and its presence in stellar matter is leading to new theories of the production of heavy elements in the stars. ^{97}Tc has a half-life of 2.6×10^6 years. ^{98}Tc has a half-life of 4.2×10^6 years. The isomeric isotope ^{95m}Tc, with a half-life of 61 days, is useful for tracer work, as it produces energetic gamma rays. Technetium metal has been produced in kilogram quantities. The metal was first prepared by passing hydrogen gas at 1100 °C over Tc_2S_7. It is now conveniently prepared by the reduction of ammonium pertechnetate with hydrogen. Technetium is a silvery-gray metal that tarnishes slowly in moist air. The chemistry of technetium is similar to that of rhenium. Technetium dissolves in nitric acid, aqua regia, and concentrated sulfuric acid, but is not soluble in hydrochloric acid of any strength.

The element is a remarkable corrosion inhibitor for steel. It is reported that mild carbon steels may be effectively protected by as little as 55 ppm of $KTcO_4$ in aerated distilled water at temperatures up to 250 °C. This corrosion protection is limited to closed systems, since technetium is radioactive and must be confined. ^{99}Tc is a contamination hazard and should be handled in a glove box. The metal is an excellent superconductor at 11 K and below.

Tellurium — (L. *tellus*, earth), Te. Discovered by Muller von Reichenstein in 1782; named by Klaproth, who isolated it in 1798. Tellurium is occasionally found native, but is more often found as the telluride of gold (*calaverite*), and combined with other metals. It is recovered commercially from the anode muds produced during the electrolytic refining of blister copper. The U.S., Canada, Peru, and Japan are the largest producers of the element. Crystalline tellurium has a silvery-white appearance, and when pure exhibits a metallic luster. It is brittle and easily pulverized. Amorphous tellurium is formed by precipitating tellurium from a solution of telluric or tellurous acid. Whether this form is truly amorphous, or made of minute crystals, is open to question. Tellurium is a p-type semiconductor, and shows greater conductivity in certain directions, depending on alignment of the atoms. Its conductivity increases slightly with exposure to light. It can be doped with silver, copper, gold, tin, or other elements. In air, tellurium burns with a greenish-blue flame, forming the dioxide. Molten tellurium corrodes iron, copper, and stainless steel. Tellurium and its compounds are probably toxic and should be handled with care. Workmen exposed to as little as 0.01 mg/m^3 of air, or less, develop "tellurium breath," which has a garlic-like odor. Natural tellurium consists of eight isotopes, two of which are radioactive with very long half-lives. Tellurium improves the machinability of copper and stainless steel, and its addition to lead decreases the corrosive action of sulfuric acid on lead and improves its strength and hardness. Tellurium catalysts are used in the oxidation of organic compounds and are used in hydrogenation and halogenation reactions. Tellurium is also used in electronic and semiconductor devices and is a basic ingredient in blasting caps, and it is added to cast iron for chill control. Tellurium is used in ceramics. Bismuth telluride has been used in thermoelectric devices.

Tennessine — (*Tennessee*, location of Oak Ridge National Laboratory), Ts. A U.S.-Russian collaboration in 2010 detected Element-117 as a product of the fusion reaction of ^{48}Ca and ^{249}Bk. Confirming experimants were done in 2013. IUPAC approved the name Tennessine for Element-117 in 2016. In the periodic table the element falls in the halogen group, but relativistic effects are predicted to alter its chemical behavior beyond what would be expected from simple extrapolation from the lower halogens. The most stable isotope produced so far, ^{293}Ts, has a half-life of about 22 ms.

Terbium — (*Ytterby*, village in Sweden), Tb. Discovered by Mosander in 1843. Terbium is a member of the lanthanide or "rare earth" group of elements. It is found in *cerite, gadolinite*, and other minerals along with other rare earths. It is recovered commercially from *monazite* in which it is present to the extent of 0.03%, from *xenotime*, and from *euxenite*, a complex oxide containing 1% or more of terbia. As with other rare earths, elemental terbium can be produced by reducing the anhydrous chloride or fluoride with calcium metal in a tantalum crucible. Calcium and tantalum impurities can be removed by vacuum remelting. Other methods of isolation are possible. Terbium is reasonably stable in air. It is a silver-gray metal, and is malleable, ductile, and soft enough to be cut with a knife. Two crystal modifications exist, with a transformation temperature of 1289 °C. The oxide is a chocolate or dark maroon color. Sodium terbium borate is used as a laser material and emits coherent light at 0.546 μm. Terbium is used to dope calcium fluoride, calcium tungstate, and strontium molybdate, used in solid-state devices. The oxide has potential application as an activator for green phosphors. It can be used with ZrO_2 as a crystal stabilizer of fuel cells that operate at elevated temperature. Few other uses have been found. Little is known of the toxicity of terbium. It should be handled with care as with other lanthanide elements.

Thallium — (Gr. *thallos*, a green shoot or twig), Tl. Thallium was discovered spectroscopically in 1861 by Crookes. The element was named after the beautiful green spectral line, which identified the element. The metal was isolated both by Crookes and Lamy in 1862 about the same time. Thallium occurs in *crooksite, lorandite*, and *hutchinsonite*. It is also present in *pyrites* and is recovered from the roasting of this ore in connection with the production of sulfuric acid. It is also obtained from the smelting of lead and zinc ores. Extraction is somewhat complex and depends on the source of the thallium. Manganese nodules, found on the ocean floor, contain thallium. When freshly exposed to air, thallium exhibits a metallic luster, but soon develops a bluish-gray tinge, resembling lead in appearance. A heavy oxide builds up on thallium if left in air, and in the presence of water the hydroxide is formed. The metal is very soft and malleable. It can be cut with a knife. The element and its compounds are toxic and should be handled carefully. Contact of the metal with skin is dangerous, and when melting the metal adequate ventilation should be provided. Thallium is suspected of carcinogenic potential for man. Thallium sulfate has been widely employed as a rodenticide and ant killer. It is odorless and tasteless, giving no warning of its presence. Its use, however, has been prohibited in the U.S. since 1975 as a household insecticide and rodenticide. The electrical conductivity of thallium sulfide changes with exposure to infrared light, and this compound is used in photocells. Thallium bromide-iodide crystals have been used as infrared optical materials. Thallium has been used, with sulfur or selenium and arsenic, to produce low melting glasses which become fluid between 125 and 150 °C. These glasses have properties at room temperatures similar to ordinary glasses and are said to be durable and insoluble in water. Thallium oxide has been used to produce glasses with a high index of refraction. Thallium has been used in treating ringworm and other skin infections; however, its use has been limited because of the narrow margin between toxicity and therapeutic benefits. A mercury–thallium alloy, which forms a eutectic at 8.5% thallium, freezes at −60 °C, some 20° below the freezing point of mercury.

Thorium — (*Thor*, Scandinavian god of war), Th. Discovered by Berzelius in 1828. Thorium occurs in *thorite* ($ThSiO_4$) and in *thorianite* ($ThO_2 + UO_2$). Large deposits of thorium minerals have been reported in New England and elsewhere, but these have not yet been exploited. Thorium is now thought to be about three times as abundant as uranium and about as abundant as lead or molybdenum. The metal can a source of nuclear power. There is probably more energy available for use from thorium in the minerals of the Earth's crust than from both uranium and fossil fuels. However, any sizable demand for thorium as a nuclear fuel is still in the future. Work has been done in developing thorium cycle converter-reactor systems.

Several prototypes, including the HTGR (high-temperature gas-cooled reactor) and MSRE (molten salt converter reactor experiment), have operated. While the HTGR reactors are efficient, they are not expected to become important commercially for many years because of certain operating difficulties. Thorium is recovered commercially from the mineral *monazite*, which contains from 3 to 9% ThO_2 along with rare-earth minerals. Much of the internal heat the Earth produces has been attributed to thorium and uranium. Several methods are available for producing thorium metal: it can be obtained by reducing thorium oxide with calcium, by electrolysis of anhydrous thorium chloride in a fused mixture of sodium and potassium chlorides, by calcium reduction of thorium tetrachloride mixed with anhydrous zinc chloride, and by reduction of thorium tetrachloride with an alkali metal. Thorium was originally assigned a position in Group IV of the periodic table. Because of its atomic weight, valence, etc., it is now considered to be the second member of the *actinide* series of elements. When pure, thorium is a silvery-white metal which is air stable and retains its luster for several months. When contaminated with the oxide, thorium slowly tarnishes in air, becoming gray and finally black. The physical properties of thorium are greatly influenced by the degree of contamination with the oxide. The purest specimens often contain several tenths of a percent of the oxide. High-purity thorium has been made. Pure thorium is soft, very ductile, and can be cold-rolled, swaged, and drawn. Thorium is dimorphic, changing at 1360 °C from a cubic to a body-centered cubic structure. Thorium oxide has a melting point of 3300 °C, which is the highest of all oxides. Only a few elements, such as tungsten, and a few compounds, such as tantalum carbide, have higher melting points. Thorium is slowly attacked by water, but does not dissolve readily in most common acids, except hydrochloric. Powdered thorium metal is often pyrophoric and should be carefully handled. When heated in air, thorium turnings ignite and burn brilliantly with a white light. A former use of thorium was in the preparation of the Welsbach mantle, used for portable gas lights. These mantles, consisting of thorium oxide with about 1% cerium oxide and other ingredients, glow with a dazzling light when heated in a gas flame. Thorium is an important alloying element in magnesium, imparting high strength and creep resistance at elevated temperatures. Because thorium has a low work-function and high electron emission, it is used to coat tungsten wire used in electronic equipment. The oxide is also used to control the grain size of tungsten used for electric lamps; it is also used for high-temperature laboratory crucibles. Glasses containing thorium oxide have a high refractive index and low dispersion. Consequently, they find application in high quality lenses for cameras and scientific instruments. Thorium oxide has also found use as a catalyst in the conversion of ammonia to nitric acid, in petroleum cracking, and in producing sulfuric acid. Its radioactive nature and its handling and disposal problems limit other applications. ^{232}Th is sufficiently radioactive to expose a photographic plate in a few hours. Thorium disintegrates with the production of ^{220}Rn, which is an alpha emitter and presents a radiation hazard. Good ventilation of areas where thorium is stored or handled is therefore essential.

Thulium — (*Thule*, the earliest name for Scandinavia), Tm. Discovered in 1879 by Cleve. Thulium occurs in small quantities along with other rare earths in a number of minerals. It is obtained commercially from *monazite*, which contains about 0.007% of the element. Thulium is the least abundant of the rare-earth elements, but with new sources recently discovered, it is now considered to be about as rare as silver, gold, or cadmium. Ion-exchange and solvent extraction techniques have recently permitted much easier separation of the rare earths, with much lower costs. Thulium can be isolated by reduction of the oxide with lanthanum metal or by calcium reduction of the anhydrous fluoride. The pure metal has a bright, silvery luster. It is reasonably stable in air, but the metal should be protected from moisture in a closed container. The element is silver-gray, soft, malleable, and ductile, and can be cut with a knife. Because of the relatively high price of the metal, thulium has not yet found many practical applications. ^{169}Tm bombarded in a nuclear reactor can be used as a radiation source in portable X-ray equipment. ^{171}Tm is potentially useful as an energy source. Natural thulium also has possible use in *ferrites* (ceramic magnetic materials) used in microwave equipment. As with other lanthanides, thulium has a low-to-moderate acute toxicity rating. It should be handled with care.

Tin — (Anglo-Saxon, *tin*), Sn (L. *stannum*). Known to the ancients. Tin is found chiefly in *cassiterite* (SnO_2). Most of the world's supply comes from China, Indonesia, Peru, Brazil, and Bolivia. Cornwall was a major source in older periods. The U.S. produces almost none, although occurrences have been found in Alaska and Colorado. Tin is obtained by reducing the ore with coal in a reverberatory furnace. Ordinary tin is a silver-white metal, is malleable, somewhat ductile, and has a highly crystalline structure. Due to the breaking of these crystals, a "tin cry" is heard when a bar is bent. The element has two allotropic forms at normal pressure. On warming, gray, or α-tin, with a cubic structure, changes at 13.2 °C into white, or β-tin, the ordinary form of the metal. White tin has a tetragonal structure. When tin is cooled below 13.2 °C, it changes slowly from white to gray. This change is affected by impurities such as aluminum and zinc, and can be prevented by small additions of antimony or bismuth. This change from the α to β form is called the tin pest. Tin–lead alloys are used to make organ pipes. There are few if any uses for gray tin. Tin takes a high polish and is used to coat other metals to prevent corrosion or other chemical action. Such tin plate over steel is used in the so-called tin can for preserving food. Alloys of tin are very important. Soft solder, type metal, fusible metal, pewter, bronze, bell metal, Babbitt metal, white metal, die casting alloy, and phosphor bronze are some of the important alloys using tin. Tin resists distilled sea and soft tap water, but is attacked by strong acids, alkalis, and acid salts. Oxygen in solution accelerates the attack. When heated in air, tin forms SnO_2, which is feebly acid, forming stannate salts with basic oxides. The most important salt is the chloride ($SnCl_2 \cdot H_2O$), which is used as a reducing agent and as a mordant in calico printing. Tin salts sprayed onto glass are used to produce electrically conductive coatings. These have been used for panel lighting and for frost-free windshields. Most window glass is now made by floating molten glass on molten tin (float glass) to produce a flat surface (Pilkington process). Of recent interest is a crystalline tin–niobium alloy that is superconductive at very low temperatures. This promises to be important in the construction of superconductive magnets. Such magnets, made with tin–niobium wire, weigh but a few pounds and produce magnetic fields that are comparable to much larger electromagnets. The small amount of tin found in canned foods is quite harmless. The agreed limit of tin content in U.S. foods is 300 mg/kg. The trialkyl and triaryl tin compounds are used as biocides and must be handled carefully.

Titanium — (L. *Titans*, the first sons of the Earth, myth.), Ti. Discovered by Gregor in 1791; named by Klaproth in 1795. Impure titanium was prepared by Nilson and Pettersson in 1887; however, the pure metal (99.9%) was not made until 1910 by Hunter by heating $TiCl_4$ with sodium in a steel bomb. Titanium is present in meteorites and in the sun. Rocks obtained during the Apollo 17 lunar mission showed the presence of 12.1% TiO_2. Analyses of rocks obtained during earlier Apollo missions show lower percentages. Titanium oxide bands are prominent in the spectra of M-type stars. The element is the ninth most abundant in the crust of the Earth. Titanium is almost always present in igneous rocks and in the sediments derived from them. It occurs in the minerals *rutile, ilmenite,* and *sphene,* and is present in titanates and in many iron ores. Deposits of ilmenite and rutile are found in Florida, California, Tennessee, and New York. Australia, Norway, Malaysia, India, and China are also large suppliers of titanium minerals. Titanium is present in the ash of coal, in plants, and in the human body. The metal was a laboratory curiosity until Kroll, in 1946, showed that titanium could be produced commercially by reducing titanium tetrachloride with magnesium. This method is largely used for producing the metal today. The metal can be purified by decomposing the iodide. Titanium, when pure, is a lustrous, white metal. It has a low density, good strength, is easily fabricated, and has excellent corrosion resistance. It is ductile only when it is free of oxygen. The metal burns in air and is the only element that burns in nitrogen. Titanium is resistant to dilute sulfuric and hydrochloric acids, most organic acids, moist chlorine gas, and chloride solutions. The metal is dimorphic. The hexagonal α form changes to the cubic β form very slowly at 882 °C. The metal combines with oxygen at red heat, and with chlorine at 550 °C. Titanium is important as an alloying agent with aluminum, molybdenum, manganese, iron, and other metals. Alloys of titanium are principally used for aircraft and missiles where lightweight strength and ability to withstand extremes of temperature are important. Titanium is as strong as steel, but 45% lighter. It is 60% heavier than aluminum, but twice as strong. Titanium has potential use in desalination plants for converting sea water into fresh water. The metal has excellent resistance to sea water and is used for propeller shafts, rigging, and other parts of ships exposed to salt water. A titanium anode coated with platinum has been used to provide cathodic protection from corrosion by salt water. Titanium metal is considered to be physiologically inert; however, titanium powder may be a carcinogenic hazard. When pure, titanium dioxide is relatively clear and has an extremely high index of refraction with an optical dispersion higher than diamond. It is produced artificially for use as a gemstone, but it is relatively soft. Star sapphires and rubies exhibit their asterism as a result of the presence of TiO_2. Titanium dioxide is extensively used for both house paint and artist's paint, as it is permanent and has good covering power. Titanium oxide pigment accounts for the largest use of the element. Titanium paint is an excellent reflector of infrared, and is extensively used in solar observatories where heat causes poor seeing conditions. Titanium tetrachloride is used to iridize glass. This compound fumes strongly in air and has been used to produce smoke screens.

Tungsten — (Swedish, *tung sten*, heavy stone); also known as *wolfram* (from *wolframite*, said to be named from *wolf rahm* or *spumi lupi*, because the ore interfered with the smelting of tin and was supposed to devour the tin), W. In 1779 Peter Woulfe examined the mineral now known as *wolframite* and concluded it must contain a new substance. Scheele, in 1781, found that a new acid could be made from *tung sten* (a name first applied about 1758 to a mineral now known as *scheelite*). Scheele and Berman suggested the possibility of obtaining a new metal by reducing this acid. The de Elhuyar brothers found an acid in *wolframite* in 1783 that was identical to the acid of *tungsten* (tungstic acid) of Scheele, and in that year they succeeded in obtaining the element by reduction of this acid with charcoal. Tungsten occurs in *wolframite*, $(Fe,Mn)WO_4$; *scheelite*, $CaWO_4$; *huebnerite*, $MnWO_4$; and *ferberite*, $FeWO_4$. Important deposits of tungsten occur in California, Colorado, Bolivia, Russia, and Portugal. China is reported to have about 75% of the world's tungsten resources. The metal is obtained commercially by reducing tungsten oxide with hydrogen or carbon. Pure tungsten is a steel-gray to tin-white metal. Very pure tungsten can be cut with a hacksaw, and can be forged, spun, drawn, and extruded. The impure metal is brittle and can be worked only with difficulty. Tungsten has the highest melting point of all metals, and at temperatures over 1650 °C has the highest tensile strength. The metal oxidizes in air and must be protected at elevated temperatures. It has excellent corrosion resistance and is attacked only slightly by most mineral acids. The thermal expansion is about the same as borosilicate glass, which makes the metal useful for glass-to-metal seals. Tungsten and its alloys are used extensively for filaments for electric light bulbs and for metal evaporation work; for electrical contact points for automobile distributors; X-ray targets; windings and heating elements for electrical furnaces; and for numerous spacecraft and high-temperature applications. High-speed tool steels, Hastelloy®, Stellite®, and many other alloys contain tungsten. Tungsten carbide is of great importance to the metal-working, mining, and petroleum industries. Calcium and magnesium tungstates are widely used in fluorescent lighting; other salts of tungsten are used in the chemical and tanning industries. Tungsten disulfide is a dry, high-temperature lubricant, stable to 500 °C. Tungsten bronzes and other tungsten compounds are used in paints. Zirconium tungstate has found recent applications (see under Zirconium).

Uranium — (Planet *Uranus*), U. Yellow-colored glass, containing more than 1% uranium oxide and dating back to 79 A.D., has been found near Naples, Italy. Klaproth recognized an unknown element in *pitchblende* and attempted to isolate the metal in 1789. The metal apparently was first isolated in 1841 by Peligot, who reduced the anhydrous chloride with potassium. Uranium is not as rare as it was once thought. It is now considered to be more plentiful than mercury, antimony, silver, or cadmium, and is about as abundant as molybdenum or arsenic. It occurs in numerous minerals such as *pitchblende, uraninite, carnotite, autunite, uranophane, davidite,* and *tobernite*. It is also found in *phosphate rock, lignite, monazite sands,* and can be recovered commercially from these sources. Large deposits of uranium ore occur in Utah, Colorado, New Mexico, Canada, and elsewhere. Uranium can be made by reducing uranium halides with alkali or alkaline earth metals or by reducing uranium oxides by calcium, aluminum, or carbon at high temperatures. The metal can also be produced by electrolysis of KUF_5 or UF_4, dissolved in a molten mixture of $CaCl_2$ and NaCl. High-purity uranium can be prepared by the thermal decomposition of uranium halides on a hot filament. Uranium metal exhibits three crystallographic modifications α, β, and γ, with transitions at 669 °C and 776 °C. Uranium is a heavy, silvery-white metal that is pyrophoric when finely divided. It is a little softer than steel, and is attacked by cold water in a finely divided state. It is malleable, ductile, and slightly paramagnetic. In air, the metal becomes coated with a layer of oxide. Acids dissolve the metal, but it is unaffected by alkalis. Natural uranium is sufficiently radioactive to expose a photographic plate in an hour or so. Much of the internal heat of the Earth is thought to be attributable to the presence of uranium and thorium. ^{238}U, with a half-life of 4.46×10^9 years, has been used to estimate the age of igneous rocks. The origin of uranium, the highest member of the naturally occurring elements — except perhaps for traces of neptunium or plutonium — is not clearly understood, although it has been suggested that uranium might be a decay product of elements of higher atomic weight, which may have once been present on Earth or elsewhere in the universe. These original elements may have been formed as a result of the primordial big bang, in a supernova, or in some other stellar processes. The fact that recent studies show that most transuranic elements are extremely rare with very short half-lives indicates that it may be necessary to find some alternative explanation for the very large quantities of radioactive uranium we find on Earth. Studies of meteorites from other parts of the solar system show a relatively low radioactive content, compared to terrestrial rocks. Uranium is of great importance as a nuclear fuel. ^{238}U can be converted into fissionable plutonium in breeder reactors, where it is possible to produce more new fissionable material than the fissionable material used in maintaining the chain reaction. ^{235}U is of even greater importance, for it is the key to the utilization of uranium. ^{235}U, while occurring in natural uranium to the extent of only 0.72%, is so fissionable with slow neutrons that a self-sustaining fission chain reaction can be made to occur in a reactor constructed from natural uranium and a suitable moderator, such as heavy water or graphite, alone. ^{235}U can be concentrated by gaseous diffusion and other physical processes and used directly as a nuclear fuel, instead of natural uranium, or used as an explosive. Natural uranium, slightly enriched with ^{235}U by a small percentage, is used to fuel nuclear power reactors for the generation of electricity. One pound of completely fissioned uranium has the fuel value of over 1500 tons of coal. However, there are serious problems with nuclear waste disposal that have not been resolved. Depleted uranium has been used for inertial guidance devices, gyrocompasses, counterweights for aircraft control surfaces, ballast for missile reentry vehicles, and as a shielding material for tanks, etc. Concerns, however, have been raised over its low radioactive properties. Uranium metal is used for X-ray targets for production of high-energy X-rays. The nitrate has been used as photographic toner, and the acetate is used in analytical chemistry. Crystals of uranium nitrate are triboluminescent. Uranium salts have also been used for producing yellow "vaseline" glass and glazes. Uranium and its compounds are highly toxic, both from a chemical and radiological standpoint. Finely divided uranium metal, being pyrophoric, presents a fire hazard.

Vanadium — (Scandinavian goddess, *Vanadis*), V. Vanadium was first discovered by del Rio in 1801. Unfortunately, a French chemist incorrectly declared that del Rio's new element was only impure chromium; del Rio thought himself to be mistaken and accepted the French chemist's statement. The element was rediscovered in 1830 by Sefstrom, who named the element in honor of the Scandinavian goddess *Vanadis* because of its beautiful multicolored compounds. It was isolated in nearly pure form by Roscoe, in 1867, who reduced the chloride with hydrogen. Vanadium of 99.3 to 99.8% purity was not produced until 1927. Vanadium is found in about 65 different minerals among which *carnotite, roscoelite, vanadinite,* and *patronite* are important sources of the metal. Vanadium is also found in phosphate rock and certain iron ores, and is present in some crude oils in the form of organic complexes. It is also found in small percentages in meteorites. Commercial production from petroleum ash holds promise as an important source of the element. China, South Africa, and Russia supply much of the world's vanadium ores. High-purity ductile vanadium can be obtained by reduction of vanadium trichloride with magnesium or with magnesium–sodium mixtures. Much of the vanadium metal being produced is now made by calcium reduction of V_2O_5 in a pressure vessel, an adaptation of a process developed by McKechnie and Seybolt. Pure vanadium is a bright white metal, and is soft and ductile. It has good corrosion resistance to alkalis, sulfuric and hydrochloric acids, and salt water, but the metal oxidizes readily above 660 °C. The metal has good structural strength and a low-fission neutron cross section, making it useful in nuclear applications. Vanadium is used in producing rust-resistant spring and high-speed tool steels. It is an important carbide stabilizer in making steels. About 80% of the vanadium now produced is used as ferrovanadium or as a steel additive. Vanadium foil is used as a bonding agent in cladding titanium to steel. Vanadium pentoxide is used in ceramics and as a catalyst. Vanadium and its compounds are toxic and should be handled with care. Ductile vanadium is commercially available.

Xenon — (Gr. *xenon*, stranger), Xe. Discovered by Ramsay and Travers in 1898 in the residue left after evaporating liquid air components. Xenon is a member of the so-called noble or "inert" gases. It is present in the atmosphere at a concentration of 0.0114%. Xenon is present in the Martian atmosphere to the extent of 0.08 ppm. The element is found in the gases evolved from certain mineral springs, and is commercially obtained by extraction from liquid air. Before 1962, it had generally been assumed that xenon and other noble gases were unable to form compounds. However, it is now known that xenon, as well as other members of the zero valence elements, do form compounds. Among the compounds of xenon now reported are the difluoride, tetrafluoride, and hexafluoride; xenon hydrate, sodium perxenate, xenon deuterate, and $XePtF_6$ and $XeRhF_6$. Xenon trioxide, which is highly explosive, has been prepared. More than 80 xenon compounds have been made with xenon chemically bonded to fluorine and oxygen. Some xenon compounds are colored. Metallic xenon has also been produced, using several hundred kilobars of pressure. Xenon in a vacuum tube produces a beautiful blue glow when excited by an electrical discharge. The gas is used in making stroboscopic lamps, bactericidal lamps, and lamps used to excite ruby lasers. Xenon is used in the nuclear energy field in bubble chambers, probes, and other applications where its high molecular weight is of value. The perxenates are used in analytical chemistry as oxidizing agents. ^{133}Xe and ^{135}Xe are produced by neutron irradiation in air-cooled nuclear reactors. ^{133}Xe has useful applications as a radioisotope. Xenon is not toxic, but its compounds are highly toxic because of their strong oxidizing characteristics.

Ytterbium — (*Ytterby*, village in Sweden), Yb. Marignac in 1878 discovered a new component, which he called *ytterbia*, in the Earth then known as *erbia*. In 1907, Urbain separated ytterbia into two components, which he called *neoytterbia* and *lutecia*. The elements in these earths are now known as *ytterbium* and *lutetium*, respectively. These elements are identical with *aldebaranium* and *cassiopeium*, discovered independently and at about the same time by von Welsbach. Ytterbium occurs along with other rare earths in a number of rare minerals. It is commercially recovered principally from *monazite sand*, which contains about 0.03%. Ion-exchange and solvent extraction techniques developed in recent years have greatly simplified the separation of the rare earths from one another. The element was first prepared by Klemm and Bonner in 1937 by reducing ytterbium trichloride with potassium. Their metal was mixed, however, with KCl. Daane, Dennison, and Spedding prepared a much purer form in 1953 from which the chemical and physical properties of the element could be determined. Ytterbium has a bright silvery luster, is soft, malleable, and quite ductile. While the element is fairly stable, it should be kept in closed containers to protect it from air and moisture. Ytterbium is readily attacked and dissolved by dilute and concentrated mineral acids and reacts slowly with water. Ytterbium has three allotropic forms with transformation points at 3° and 795 °C. The beta form is a room-temperature, face-centered, cubic modification, while the high-temperature gamma form is a body-centered cubic form. Another body-centered cubic phase has recently been found to be stable at high pressures at room temperatures. The beta form ordinarily has metallic-type conductivity, but becomes a semiconductor when the pressure is increased above 16,000 atm. The electrical resistance increases tenfold as the pressure is increased to 39,000 atm and drops to about 80% of its standard temperature-pressure resistivity at a pressure of 40,000 atm. Ytterbium metal has possible use in improving the grain refinement, strength, and other mechanical properties of stainless steel. Few other uses have been found. Ytterbium has a low acute toxicity rating.

Yttrium — (*Ytterby*, village in Sweden near Vauxholm), Y. *Yttria*, which is an earth containing yttrium, was discovered by Gadolin in 1794. *Ytterby* is the site of a quarry which yielded many unusually minerals containing rare earths and other elements. This small town, near Stockholm, bears the honor of giving names to *erbium*, *terbium*, and *ytterbium* as well as *yttrium*. In 1843 Mosander showed that yttria could be resolved into the oxides (or earths) of three elements. The name yttria was reserved for the most basic one; the others were named *erbia* and *terbia*. Yttrium occurs in nearly all of the rare-earth minerals. Analysis of lunar rock samples obtained during the Apollo missions show a relatively high yttrium content. It is recovered commercially from *monazite sand*, which contains about 3%, and from *bastnasite*, which contains about 0.2%. Wohler obtained the impure element in 1828 by reduction of the anhydrous chloride with potassium. The metal is now produced commercially by reduction of the fluoride with calcium metal. It can also be prepared by other techniques. Yttrium has a silver-metallic luster and is relatively stable in air. Turnings of the metal, however, ignite in air if their temperature exceeds 400 °C, and finely divided yttrium is very unstable in air. Yttrium oxide is one of the most important compounds of yttrium and accounts for the largest use. At one time it was used in making YVO_4 europium, and Y_2O_3 europium phosphors to give the red color in color television tubes. Yttrium oxide also is used to produce yttrium iron garnets, which are very effective microwave filters. Yttrium iron, aluminum, and gadolinium garnets, with formulas such as $Y_3Fe_5O_{12}$ and $Y_3Al_5O_{12}$, have interesting magnetic properties. Yttrium iron garnet is also exceptionally efficient as both a transmitter and transducer of acoustic energy. These yttrium garnets have many uses in the electronics industry. Yttrium aluminum garnet, with a hardness of 8.5, is also finding use as a gemstone (simulated diamond). Small amounts of yttrium (0.1 to 0.2%) can be used to reduce the grain size in chromium, molybdenum, zirconium, and titanium, and to increase strength of aluminum and magnesium alloys. Alloys with other useful properties can be obtained by using yttrium as an additive. The metal can be used as a deoxidizer for vanadium and other nonferrous metals. Yttrium has been considered for use as a nodulizer for producing nodular cast iron, in which the graphite forms compact nodules instead of the usual flakes. Such iron has increased ductility. Yttrium is also finding application in laser systems and as a catalyst for ethylene polymerization. It also has potential use in ceramic and glass formulas, as the oxide has a high melting point and imparts shock resistance and low expansion characteristics to glass.

Zinc — (Ger. *Zink*, of obscure origin), Zn. Centuries before zinc was recognized as a distinct element, zinc ores were used for making brass. Tubal-Cain, seven generations from Adam, is mentioned as being an "instructor in every artificer in brass and iron." An alloy containing 87% zinc has been found in prehistoric ruins in Transylvania. Metallic zinc was produced in the 13th century A.D. in India by reducing calamine with organic substances such as wool. The metal was rediscovered in Europe by Marggraf in 1746, who showed that it could be obtained by reducing *calamine* with charcoal. The principal ores of zinc are *sphalerite* or *blende* (sulfide), *smithsonite* (carbonate), *calamine* (silicate), and *franklinite* (zinc, manganese, iron oxide). Canada, Japan, Belgium, Germany, and the Netherlands are suppliers of zinc ores. Zinc is also mined in Alaska, Tennessee, Missouri, and elsewhere in the U.S. Zinc can be obtained by roasting its ores to form the oxide and by reduction of the oxide with coal or carbon, with subsequent distillation of the metal. Other methods of extraction are possible. Zinc is a bluish-white, lustrous metal. It is brittle at ordinary temperatures but malleable at 100 to 150 °C. It is a fair conductor of electricity, and burns in air at high red heat with evolution of white clouds of the oxide. The metal is employed to form numerous alloys with other metals. Brass, nickel silver, typewriter metal, commercial bronze, spring brass, German silver, soft solder, and aluminum solder are some of the more important alloys. Large quantities of zinc are used to produce die castings, used extensively by the automotive, electrical, and hardware industries. An alloy called *Prestal*®, consisting of 78% zinc and 22% aluminum, is reported to be almost as strong as steel but as easy to mold as plastic. Zinc is also extensively used to galvanize other metals such as iron to prevent corrosion. Neither zinc nor zirconium is ferromagnetic; but $ZrZn_2$ exhibits ferromagnetism at temperatures below 35 K. Zinc oxide is a unique and very useful material to modern civilization. It is widely used in the manufacture of paints, rubber products, cosmetics, pharmaceuticals, floor coverings, plastics, printing inks, soap, storage batteries, textiles, electrical equipment, and other products. It has unusual electrical, thermal, optical, and solid-state properties that have not yet been fully investigated. Lithopone, a mixture of zinc sulfide and barium sulfate, is an important pigment. Zinc sulfide is used in making luminous dials and fluorescent lights. The chloride and chromate are also important compounds. Zinc is an essential element in the growth of human beings and animals. Tests show that zinc-deficient animals require 50% more food to gain the same weight as an animal supplied with sufficient zinc. Zinc is not considered to be toxic, but when freshly formed ZnO is inhaled a disorder known as the *oxide shakes* or *zinc chills* sometimes occurs. It is recommended that where zinc oxide is encountered good ventilation be provided.

Zirconium — (Syriac, *zargun*, color of gold), Zr. The name *zircon* may have originated from the Syriac word *zargono*, which describes the color of certain gemstones now known as *zircon, jargon, hyacinth, jacinth,* or *ligure*. This mineral, or its variations, is mentioned in biblical writings. These minerals were not known to contain this element until Klaproth, in 1789, analyzed a *jargon* from Sri Lanka and found a new earth, which Werner named zircon (*silex circonius*), and Klaproth called *Zirkonerde (zirconia)*. The impure metal was first isolated by Berzelius in 1824 by heating a mixture of potassium and potassium zirconium fluoride in a small iron tube. Pure zirconium was first prepared in 1914. Very pure zirconium was first produced in 1925 by van Arkel and de Boer by an iodide decomposition process they developed. Zirconium is found in abundance in S-type stars, and has been identified in the sun and meteorites. Analyses of lunar rock samples obtained during the various Apollo missions to the moon show a surprisingly high zirconium oxide content, compared with terrestrial rocks. *Zircon*, $ZrSiO_4$, the principal ore, is found in deposits in Florida, South Carolina, Australia, South Africa, and elsewhere. *Baddeleyite*, found in Brazil, is an important zirconium mineral. It is principally pure ZrO_2 in crystalline form having a hafnium content of about 1%. Zirconium also occurs in some 30 other recognized mineral species. Zirconium is produced commercially by reduction of the chloride with magnesium (the Kroll Process), and by other methods. It is a grayish-white lustrous metal. When finely divided, the metal may ignite spontaneously in air, especially at elevated temperatures. The solid metal is much more difficult to ignite. The inherent toxicity of zirconium compounds is low. Hafnium is invariably found in zirconium ores, and the separation is difficult. Commercial-grade zirconium contains from 1 to 3% hafnium. Zirconium has a low absorption cross section for neutrons, and is therefore used for nuclear energy applications, such as for cladding fuel elements. Commercial nuclear power generation now takes more than 90% of zirconium metal production. Reactor-grade zirconium is essentially free of hafnium. *Zircaloy*® is an important alloy developed specifically for nuclear applications. Zirconium is exceptionally resistant to corrosion by many common acids and alkalis, by sea water, and by other agents. It is used extensively by the chemical industry where corrosive agents are employed. Zirconium is used as a getter in vacuum tubes, as an alloying agent in steel, in surgical appliances, photoflash bulbs, explosive primers, rayon spinnerets, lamp filaments, etc. It is used in poison ivy lotions in the form of the carbonate as it combines with *urushiol*. With niobium, zirconium is superconductive at low temperatures and is used to make superconductive magnets. Alloyed with zinc, zirconium becomes magnetic at temperatures below 35 K. Zirconium oxide has a high index of refraction and is used as a gem material. The impure oxide, zirconia, is used for laboratory crucibles that will withstand heat shock, for linings of metallurgical furnaces, and by the glass and ceramic industries as a refractory material. Its use as a refractory material accounts for a large share of all zirconium consumed. Zirconium tungstate is an unusual material that shrinks, rather than expands, when heated. A few other compounds are known to possess this property, but they tend to shrink in one direction, while they stretch out in others in order to maintain an overall volume. Zirconium tungstate shrinks in all directions over a wide temperature range of from near absolute zero to +777 °C.

PHYSICAL CONSTANTS OF INORGANIC COMPOUNDS

The compounds in this table were selected on the basis of their laboratory and industrial importance, as well as their value in illustrating trends in the variation of physical properties with position in the periodic table. An effort has been made to include the most frequently encountered inorganic substances; a limited number of organometallics are also covered. Many, if not most, of the compounds that are solids at ambient temperature can exist in more than one crystalline modification. In the absence of other information, the data given here can be assumed to apply to the most stable or common crystalline form. In many cases, however, two or more forms are of practical importance, and separate entries will be found in the table.

Compounds are arranged primarily in alphabetical order by the most commonly used name; however, closely related compounds are often grouped together for convenience (e.g., silane, disilane, trisilane, etc.). The Formula Index following the table in the print version provides another means of locating a compound. An index to CAS Registry Numbers is available by request via e-mail (fiona.macdonald@taylorandfrancis.com).

The following data fields appear in the table:

Name: Systematic name for the substance. The valence state of a metallic element is indicated by a Roman numeral, e.g., copper in the +1 state is written as copper(I) rather than cuprous, iron in the +3 state is iron(III) rather than ferric.

Formula: The simplest descriptive formula is given, but this does not necessarily specify the actual structure of the compound. For example, aluminum chloride is designated as $AlCl_3$, even though a more accurate representation of the structure in the solid phase (and, under some conditions, in the gas phase) is Al_2Cl_6 A few exceptions are made, such as the use of Hg_2^{+2} for the mercury(I) ion.

CAS Reg. No.: Chemical Abstracts Service Registry Number. An asterisk* following the CAS Reg. No. for a hydrate indicates that the number refers to the anhydrous compound. In most cases the generic CAS Reg. No. for the compound is given rather than the number for a specific crystalline form or mineral.

Mol. wt.: Molecular weight (relative molar mass) as calculated with the 2005 IUPAC Recommended Atomic Weights. The number of decimal places corresponds to the number of places in the atomic weight of the least accurately known element (e.g., one place for lead compounds, two places for compounds of selenium, germanium, etc.); a maximum of three places is given. For compounds of radioactive elements for which IUPAC makes no recommendation, the mass number of the isotope with longest half-life is used.

Phys. form: The crystal system is given, when available, for compounds that are solid at room temperature, together with color and other descriptive features. Abbreviations are listed below.

mp: Normal melting point in °C. The notation tp indicates the temperature where solid, liquid, and gas are in equilibrium at a pressure greater than one atmosphere (i.e., the normal melting point does not exist). When available, the triple point pressure is listed.

bp: Normal boiling point in °C (referred to 101.325 kPa or 760 mmHg pressure). The notation sp following the number indicates the temperature where the pressure of the vapor in equilibrium with the solid reaches 101.325 kPa. See Ref. 8, p. 23, for further discussion of sublimation points and triple points. A notation "sublimes" without a temperature being given indicates that there is a perceptible sublimation pressure above the solid at ambient temperatures.

ρ: Density values for solids and liquids are always in units of grams per cubic centimeter and can be assumed to refer to temperatures near room temperature unless otherwise stated. Values for gases are the calculated ideal gas densities in grams per liter at 25 °C and 101.325 kPa; the unit is always specified for a gas value.

Sol.: Aqueous solubility is expressed as the number of grams of the compound (excluding any water of hydration) that will dissolve in 100 grams of water. The temperature in °C is given as a superscript. Solubility at other temperatures can be found for many compounds in the table "Aqueous Solubility of Inorganic Compounds at Various Temperatures" in Section 5.

Qualitative solubility: Qualitative information on the solubility in other solvents (and in water, if quantitative data are unavailable) is given here. The abbreviations are:

i = insoluble
sl = slightly soluble
s = soluble
vs = very soluble
reac = reacts with the solvent

Data were taken from a wide variety of reliable sources, including monographs, treatises, review articles, evaluated compilations and databases, and in many cases the primary literature. Some of the most useful references for the properties covered here are listed below.

List of Abbreviations

Ac: acetyl
ace: acetone
acid: acid solutions
alk: alkaline solutions
amorp: amorphous
anh: anhydrous
aq: aqueous
blk: black
brn: brown
bz: benzene
chl: chloroform
col: colorless
conc: concentrated
cry: crystals, crystalline
cub: cubic
cyhex: cyclohexane
dec: decomposes
dil: dilute
diox: dioxane
eth: ethyl ether
EtOH: ethanol
exp: explodes, explosive
extrap: extrapolated
flam: flammable
gl: glass, glassy
grn: green
hc: hydrocarbon solvents
hex: hexagonal, hexane
hp: heptane
HT: high temperature
hyd: hydrate
hyg: hygroscopic
i: insoluble in
liq: liquid
LT: low temperature
MeOH: methanol
monocl: monoclinic
octahed: octahedral
oran: orange
orth: orthorhombic
os: organic solvents
peth: petroleum ether
pow: powder
prec: precipitate
pur: purple
py: pyridine
reac: reacts with
refrac: refractory

rhom: rhombohedral
r.t.: room temperature
s: soluble in
silv: silvery
sl: slightly soluble in
soln: solution
sp: sublimation point
stab: stable
subl: sublimes
temp: temperature
tetr: tetragonal
thf: tetrahydrofuran
tol: toluene
tp: triple point
trans: transition, transformation
tricl: triclinic
trig: trigonal
unstab: unstable
viol: violet
visc: viscous
vs: very soluble in
wh: white
xyl: xylene
yel: yellow

References

1. Phillips, S. L., and Perry, D.L., *Handbook of Inorganic Compounds*, CRC Press, Boca Raton, FL, 1995; second edition by Perry, D. L., 2011.
2. Trotman-Dickenson, A. F., Executive Editor, *Comprehensive Inorganic Chemistry*, Vol. 1-5, Pergamon Press, Oxford, 1973.
3. Greenwood, N. N., and Earnshaw, A., *Chemistry of the Elements, Second Edition*, Butterworth-Heinemann, Oxford, 1997.
4. Wiberg, N., Wiberg, E., and Holleman, H. F., *Inorganic Chemistry, 34th Edition*, Academic Press, San Diego, 2001.
5. *GMELIN Handbook of Inorganic and Organometallic Chemistry*, Springer-Verlag, Heidelberg.
6. Chase, M.W., Davies, C.A., Downey, J.R., Frurip, D. J., McDonald, R.A., and Syverud, A.N.; *JANAF Thermochemical Tables, Third Edition, J. Phys. Chem. Ref. Data*, Vol. 14, Suppl. 1, 1985; Chase, M. W., *NISTJANAF Thermochemical Tables, Fourth Edition, J. Phys. Chem. Ref. Data*, Monograph No. 9, 1998.
7. *Landolt-Börnstein, Numerical Data and Functional Relationships in Science and Technology, New Series*, IV/19A, "Thermodynamic Properties of Inorganic Materials compiled by SGTE", Springer-Verlag, Heidelberg; Part 1, 1999; Part 2; 1999; Part 3, 2000; Part 4, 2001.
8. Lide, D. R., and Kehiaian, H.V., *CRC Handbook of Thermophysical and Thermochemical Data*, CRC Press, Boca Raton, FL, 1994.
9. *Kirk-Othmer Concise Encyclopedia of Chemical Technology*, Wiley-Interscience, New York, 1985.
10. *Dictionary of Inorganic Compounds*, Chapman & Hall, New York, 1992.
11. Massalski, T. B., Ed., *Binary Alloy Phase Diagrams, 2nd Edition*, ASM International, Metals Park, Ohio, 1990.
12. Dinsdale, A.T., "SGTE Data for Pure Elements", *CALPHAD*, 15, 317– 425, 1991.
13. Madelung, O., *Semiconductors: Group IV Elements and III-IV Compounds*, Springer-Verlag, Heidelberg, 1991.
14. Lidin, R. A., Andreeva, L. L., and Molochko, V. A., *Constants of Inorganic Substances*, Begell House, New York, 1995.
15. Gurvich, L. V., Veyts, I. V., and Alcock, C. B., *Thermodynamic Properties of Individual Substances, Fourth Edition*, Hemisphere Publishing Corp., New York, 1989.
16. *The Combined Chemical Dictionary on CDROM, Version 9:1*, Chapman & Hall/CRC, Boca Raton, FL, 2005.
17. Macdonald, F., Editor, *Chapman & Hall/CRC Combined Chemical Dictionary*, <ccd.chemnetbase.com/>.
18. Sangeeta, G., and LaGraff, J. R., *Inorganic Materials Chemistry, Second Edition*, CRC Press, Boca Raton, FL, 2005.
19. Stern, K. H., *High Temperature Properties and Thermal Decomposition of Inorganic Salts with Oxyanions*, CRC Press, Boca Raton, FL, 2001.
20. Donnay, J.D.H., and Ondik, H.M., *Crystal Data Determinative Tables, Third Edition, Volumes 2 and 4, Inorganic Compounds*, Joint Committee on Powder Diffraction Standards, Swarthmore, PA, 1973.
21. Robie, R., Bethke, P. M., and Beardsley, K. M., *Selected X-ray Crystallographic Data, Molar Volumes, and Densities of Minerals and Related Substances*, U.S. Geological Survey Bulletin 1248, 1967.
22. Carmichael, R. S., *Practical Handbook of Physical Properties of Rocks and Minerals*, CRC Press, Boca Raton, FL, 1989.
23. Deer, W. A., Howie, R.A., and Zussman, J., An *Introduction to the Rock-Forming Minerals, 2nd Edition*, Longman Scientific & Technical, Harlow, Essex, 1992.
24. Linstrom, P. J., and Mallard, W. G., Editors, NIST Chemistry WebBook, NIST Standard Reference Database No. 69, June 2005, National Institute of Standards and Technology, Gaithersburg, MD 20899, <webbook.nist.gov>.
25. *Phase Diagrams for Ceramists, Volumes 1–8; ACerS-NIST Phase Equilibrium Diagrams, Volumes 9–13*, American Ceramic Society, Westerville, Ohio, 1964–2001.

Physical Constants of Inorganic Compounds

No.	Name	Formula	CAS Reg. No.	Mol. wt.	Phys. form	mp/°C	bp/°C	ρ/g cm^{-3}	Sol. in g/100 g H_2O	Qualitative solubility
1	Actinium	Ac	7440-34-8	227	silv metal; cub	1050	≈3200	10		
2	Actinium bromide	$AcBr_3$	33689-81-5	467	wh hex cry		800 subl	5.85		s H_2O
3	Actinium chloride	$AcCl_3$	22986-54-5	333	wh hex cry		960 subl	4.81		
4	Actinium fluoride	AcF_3	33689-80-4	284	wh hex cry			7.88		i H_2O
5	Actinium iodide	AcI_3	33689-82-6	608	wh cry					s H_2O
6	Actinium oxide	Ac_2O_3	12002-61-8	502	wh hex cry	1977		9.19		i H_2O
7	Aluminum	Al	7429-90-5	26.982	silv-wh metal; cub cry	660.323	2519	2.70		i H_2O; s acid, alk
8	Aluminum acetate	$Al(C_2H_3O_2)_3$	139-12-8	204.113	wh hyg solid	dec				s H_2O; sl ace
9	Aluminum diacetate	$Al(OH)(C_2H_3O_2)_2$	142-03-0	162.078	wh amorp powder					i H_2O
10	Aluminum ammonium sulfate	$NH_4Al(SO_4)_2$	7784-25-0	237.146	wh powder					sl H_2O; i EtOH
11	Aluminum ammonium sulfate dodecahydrate	$NH_4Al(SO_4)_2 \cdot 12H_2O$	7784-26-1	453.329	col cry or powder	94.5	>280 dec	1.65		s H_2O; i EtOH
12	Aluminum antimonide	AlSb	25152-52-7	148.742	brn cub cry	1065		4.26		
13	Aluminum arsenide	AlAs	22831-42-1	101.903	oran cub cry; hyg	1740		3.76		
14	Aluminum borate	$2Al_2O_3 \cdot B_2O_3$	11121-16-7	273.543	needles	≈1050				i H_2O
15	Aluminum borohydride	$Al(BH_4)_3$	16962-07-5	71.510	flam liq	-64.5	44.5			reac H_2O
16	Aluminum bromate nonahydrate	$Al(BrO_3)_3 \cdot 9H_2O$	11126-81-1	572.826	wh hyg cry	62	>100 dec			s H_2O
17	Aluminum bromide	$AlBr_3$	7727-15-3	266.694	wh-yel monocl cry; hyg	97.5	255	3.2		reac H_2O; s bz, tol
18	Aluminum bromide hexahydrate	$AlBr_3 \cdot 6H_2O$	7784-11-4	374.785	col-yel hyg cry	93		2.54		s H_2O, EtOH, CS_2
19	Aluminum carbide	Al_4C_3	1299-86-1	143.958	yel hex cry	2100	>2200 dec	2.36		reac H_2O
20	Aluminum chlorate nonahydrate	$Al(ClO_3)_3 \cdot 9H_2O$	15477-33-5	439.473	hyg cry					vs H_2O; s EtOH
21	Aluminum chloride	$AlCl_3$	7446-70-0	133.341	wh hex cry or powder; hyg	192.6	180 sp	2.48	45.1^{25}	s bz, ctc, chl
22	Aluminum chloride hexahydrate	$AlCl_3 \cdot 6H_2O$	7784-13-6	241.432	col hyg cry	100 dec		2.398	45.1^{25}	s EtOH, eth
23	Dichloromethylaluminum	$AlCl_2CH_3$	917-65-7	112.923	cry	72.7				s bz, eth, hc

No.	Name	Formula	CAS Reg. No.	Mol. wt.	Phys. form	mp/°C	bp/°C	ρ/g cm^{-3}	Sol. in g/100 g H_2O	Qualitative solubility
24	Chlorodiethylaluminum	$AlCl(C_2H_5)_2$	96-10-6	120.557	col liq	-74		0.96		reac H_2O
25	Chlorodiisobutylaluminum	$AlCl(C_4H_9)_2$	1779-25-5	176.664	hyg col liq	-40		0.95^{20}		s eth, hx
26	Aluminum diboride	AlB_2	12041-50-8	48.604	powder	>920 dec		3.19		s dil HCl
27	Aluminum dodecaboride	AlB_{12}	12041-54-2	156.714	yel-brn prisms	2070		2.55		s hot HNO_3; i acid, alk
28	Aluminum ethanolate	$Al(C_2H_5O)_3$	555-75-9	162.163	liq, condenses to wh solid	140				reac H_2O; sl xyl
29	Aluminum fluoride	AlF_3	7784-18-1	83.977	wh hex cry	2250 tp (220 MPa)	1276 sp	3.10	0.50^{25}	
30	Aluminum fluoride monohydrate	$AlF_3{\cdot}H_2O$	32287-65-3	101.992	orth cry			2.17	0.50^{25}	
31	Aluminum fluoride trihydrate	$AlF_3{\cdot}3H_2O$	15098-87-0	138.023	wh hyg cry			1.914	0.50^{25}	
32	Aluminum hexafluorosilicate nonahydrate	$Al_2(SiF_6)_3{\cdot}9H_2O$	17099-70-6	642.329	hex prisms	>500 dec				s H_2O
33	Aluminum hydride	AlH_3	7784-21-6	30.006	col hex cry	>150 dec				reac H_2O
34	Aluminum hydroxide	$Al(OH)_3$	21645-51-2	78.004	wh amorp powder			2.42		i H_2O; s alk, acid
35	Aluminum hydroxychloride	$Al_2(OH)_5Cl{\cdot}2H_2O$	1327-41-9	210.483	gl solid					s H_2O
36	Aluminum iodide	AlI_3	7784-23-8	407.695	wh leaflets	188.28	382	3.98		reac H_2O
37	Aluminum iodide hexahydrate	$AlI_3{\cdot}6H_2O$	10090-53-6	515.786	yel hyg cry powder					vs H_2O; s EtOH, eth
38	Aluminum lactate	$Al(C_3H_5O_3)_3$	18917-91-4	294.192	powder					vs H_2O
39	Aluminum molybdate	$Al_2(MoO_4)_3$	15123-80-5	533.78	wh pow	≈950				
40	Aluminum nitrate	$Al(NO_3)_3$	13473-90-0	212.997	wh hyg solid	dec			68.9^{25}	vs EtOH; sl ace
41	Aluminum nitrate nonahydrate	$Al(NO_3)_3{\cdot}9H_2O$	7784-27-2	375.134	wh hyg monocl cry	73	135 dec	1.72	68.9^{25}	vs EtOH; i pyr
42	Aluminum nitride	AlN	24304-00-5	40.989	blue-wh hex cry	3000		3.255		reac H_2O
43	Aluminum oleate	$Al(C_{18}H_{33}O_2)_3$	688-37-9	871.342	yel solid					i H_2O; s EtOH, bz
44	Aluminum oxalate monohydrate	$Al_2(C_2O_4)_3{\cdot}H_2O$	814-87-9	336.035	wh pow					i H_2O, EtOH; s acid
45	Aluminum oxide (α)	Al_2O_3	1344-28-1	101.961	wh powder; hex	2053	2977	3.99		i H_2O, os; sl alk
46	Aluminum oxide (γ)	Al_2O_3	1344-28-1	101.961	soft wh pow	1200 trans corundum		3.97		i H_2O; s acid; sl alk
47	Aluminum oxyhydroxide	AlO(OH)	1318-23-6	59.989	wh orth cry	227 trans diasphore		3.07		i H_2O; s hot acid, alk
48	Aluminum oxyhydroxide	AlO(OH)	14457-84-2	59.989	orth cry	450 dec		3.38		i H_2O; s acid, alk
49	Aluminum palmitate	$Al(C_{15}H_{31}COO)_3$	555-35-1	793.230	wh-yel powder					i H_2O, EtOH; s peth
50	Aluminum 2,4-pentanedioate	$Al(CH_3COCHCOCH_3)_3$	13963-57-0	324.306	pale yel prisms	194.6	315	1.27		i H20; s bz, EtOH; sl hex
51	Aluminum perchlorate	$Al(ClO_4)_3$	14452-39-2	325.334	wh hyg cry				55^{0}	s H_2O, eth; i ctc
52	Aluminum perchlorate nonahydrate	$Al(ClO_4)_3{\cdot}9H_2O$	14452-39-2	487.471	wh hyg cry	82 dec		2.0	182.4^{0}	
53	Aluminum phosphate	$AlPO_4$	7784-30-7	121.953	wh rhomb plates	>1460		2.56		i H_2O; sl acid
54	Aluminum phosphate dihydrate	$AlPO_4{\cdot}2H_2O$	13477-75-3	157.984	wh rhom cry	1500 dec		2.54		i H_2O
55	Aluminum phosphate trihydroxide	$Al_2(OH)_3PO_4$	12004-29-4	199.957	wh or yel monocl cry			2.7		
56	Aluminum metaphosphate	$Al(PO_3)_3$	32823-06-6	263.898	col powder; tetr	≈1525		2.78		i H_2O
57	Aluminum hypophosphite	$Al(H_2PO_2)_3$	7784-22-7	221.948	cry powder	220 dec				i H_2O; s alk, acid
58	Aluminum phosphide	AlP	20859-73-8	57.956	grn or yel cub cry	2550		2.40		reac H_2O
59	Aluminum selenide	Al_2Se_3	1302-82-5	290.84	yel-brown powder	960		3.437		reac H_2O
60	Aluminum silicate	Al_2SiO_5	12183-80-1	162.046	gray-grn cry			3.145		
61	Aluminum silicate	Al_2SiO_5	1302-76-7	162.046	blue or gray tricl cry	1000 dec		3.68		
62	Aluminum silicate	$3Al_2O_3{\cdot}2SiO_2$	1302-93-8	426.052	col orth cry	1750		3.17		i H_2O, acid, HF
63	Aluminum silicate	Al_2SiO_5	12141-45-6	162.046	wh orth cry	1816		3.25		
64	Aluminum silicate dihydrate	$Al_2O_3{\cdot}2SiO_2{\cdot}2H_2O$	1332-58-7	258.161	wh-yel powder; tricl			2.59		i H_2O, acid, alk
65	Aluminum stearate	$Al(C_{18}H_{35}O_2)_3$	637-12-7	877.390	wh powder	115		1.070		i H_2O, EtOH, eth; s alk
66	Aluminum monostearate	$Al(OH)_2(C_{18}H_{35}O_2)$	7047-84-9	344.467	yel-wh pow	155		1.02		i H_2O
67	Aluminum distearate	$Al(OH)(C_{18}H_{35}O_2)_2$	300-92-5	610.928	wh pow	145				i H_2O
68	Aluminum sulfate	$Al_2(SO_4)_3$	10043-01-3	342.151	wh cry	1040 dec			38.5^{25}	i EtOH
69	Aluminum sulfate octadecahydrate	$Al_2(SO_4)_3{\cdot}18H_2O$	7784-31-8	666.426	col monocl cry	86 dec		1.69	38.5^{25}	
70	Aluminum sulfide	Al_2S_3	1302-81-4	150.158	yel-gray powder	1100		2.02		
71	Aluminum telluride	Al_2Te_3	12043-29-7	436.76	gray-blk hex cry	≈895		4.5		
72	Aluminum thiocyanate	$Al(SCN)_3$	538-17-0	201.229	yel powder					s H_2O; i EtOH, eth
73	Aluminum titanate	Al_2TiO_5	12004-39-6	181.827	refrac solid	1860				
74	Aluminum zirconium	Al_3Zr	12004-50-1	145.187	metallic solid	1645				
75	Americium	Am	7440-35-9	243	silv metal; hex or cub	1176	≈2011	12		s acid
76	Americium(III) oxide	Am_2O_3	12254-64-7	534	tan hex cry			11.77		s acid
77	Americium(III) bromide	$AmBr_3$	14933-38-1	483	wh orth cry			6.85		s H_2O
78	Americium(III) chloride	$AmCl_3$	13464-46-5	349	pink hex cry	500		5.87		
79	Americium(III) fluoride	AmF_3	13708-80-0	300	pink hex cry	1393		9.53		
80	Americium(III) iodide	AmI_3	13813-47-3	624	yel ortho cry	≈950		6.9		
81	Americium(IV) fluoride	AmF_4	15947-41-8	319	tan monocl cry			7.23		
82	Americium(IV) oxide	AmO_2	12005-67-3	275	blk cub cry	>1000 dec		11.68		s acid

No.	Name	Formula	CAS Reg. No.	Mol. wt.	Phys. form	mp/°C	bp/°C	ρ/g cm^{-3}	Sol. in g/100 g H_2O	Qualitative solubility
83	Ammonia	NH_3	7664-41-7	17.031	col gas	-77.65	-33.33	$0.7329^{-77.7}$ liq		vs H_2O; s EtOH, eth
84	Ammonium acetate	$NH_4C_2H_3O_2$	631-61-8	77.083	wh hyg cry	114		1.073	148^{4}	s EtOH; sl ace
85	Ammonium azide	NH_4N_3	12164-94-2	60.059	orth cry; flam	160	exp	1.346	20.2^{30}	
86	Ammonium benzoate	$NH_4C_7H_5O_2$	1863-63-4	139.152	wh cry or powder	198		1.26		s H_2O; sl EtOH
87	Ammonium bromate	NH_4BrO_3	13843-59-9	145.941	col hex cry	exp				vs H_2O
88	Ammonium bromide	NH_4Br	12124-97-9	97.943	wh hyg tetr cry	542 dec	396 sp	2.429	78.3^{25}	s EtOH, ace; sl eth
89	Ammonium caprylate	$NH_4C_8H_{15}O_2$	5972-76-9	161.243	hyg monocl cry	≈75				reac H_2O; s EtOH; i chl, bz
90	Ammonium carbamate	NH_2COONH_4	1111-78-0	78.071	cry powder					vs H_2O; s EtOH
91	Ammonium carbonate	$(NH_4)_2CO_3$	506-87-6	96.086	col cry powder	58 dec			100^{15}	
92	Ammonium chlorate	NH_4ClO_3	10192-29-7	101.490	wh cry	102 exp		1.80	28.7^{0}	
93	Ammonium chloride	NH_4Cl	12125-02-9	53.492	col cub cry	520.1 tp (dec)	338 sp	1.519	39.5^{25}	
94	Ammonium chromate	$(NH_4)_2CrO_4$	7788-98-9	152.071	yel cry	185 dec		1.90	37^{25}	sl ace, MeOH; i EtOH
95	Ammonium chromium sulfate dodecahydrate	$NH_4Cr(SO_4)_2 \cdot 12H_2O$	10022-47-6	478.343	blue-viol cry	94 dec		1.72		s H_2O; sl EtOH
96	Ammonium cobalt(II) phosphate	$CoNH_4PO_4$	14590-13-7	171.943	red-viol powder (hyd)					i H_2O; s acid
97	Ammonium cobalt(II) phosphate monohydrate	$CoNH_4PO_4 \cdot H_2O$	16827-96-6	189.959	red-purp orth plates	450 dec				s acid
98	Ammonium cobalt(II) sulfate hexahydrate	$(NH_4)_2Co(SO_4)_2 \cdot 6H_2O$	13586-38-4	395.227	red monocl prisms			1.90		s H_2O; i EtOH
99	Ammonium copper(II) chloride	$CuCl_2 \cdot 2NH_4Cl$	10060-13-6	241.435	yel hyg orth cry					s H_2O
100	Ammonium copper(II) chloride dihydrate	$CuCl_2 \cdot 2NH_4Cl \cdot 2H_2O$	10060-13-6	277.465	blue-grn tetr cry	110 dec		1.993		s H_2O, EtOH
101	Ammonium cyanide	NH_4CN	12211-52-8	44.056	col tetr cry	dec		1.10		vs H_2O
102	Ammonium dichromate	$(NH_4)_2Cr_2O_7$	7789-09-5	252.065	oran-red monocl cry; hyg	180 dec		2.155	35.6^{20}	
103	Ammonium dihydrogen arsenate	$NH_4H_2AsO_4$	13462-93-6	158.975	tetr cry	300 dec		2.311	52.7^{25}	
104	Ammonium dihydrogen phosphate	$NH_4H_2PO_4$	7722-76-1	115.026	wh tetr cry	190		1.80	40.4^{25}	sl EtOH; i ace
105	Ammonium *O,O*-diethyldithiophosphate	$(C_2H_5O)_2P(S)SNH_4$	1068-22-0	203.264	cry	165				
106	Ammonium dithiocarbamate	NH_4NH_2CSS	513-74-6	110.202	yel ortho cry	99 dec		1.45		s H_2O
107	Ammonium ferricyanide trihydrate	$(NH_4)_3Fe(CN)_6 \cdot 3H_2O$	14221-48-8	320.110	red cry					s H_2O; i EtOH
108	Ammonium ferrocyanide trihydrate	$(NH_4)_4Fe(CN)_6 \cdot 3H_2O$	14481-29-9	338.149	yel cry	dec				s H_2O; i EtOH
109	Ammonium fluoride	NH_4F	12125-01-8	37.037	wh hex cry; hyg	238		1.015	83.5^{25}	sl EtOH
110	Ammonium fluorosulfonate	NH_4SO_3F	13446-08-7	117.100	col needles	245				s H_2O, EtOH, MeOH
111	Ammonium formate	NH_4CHO_2	540-69-2	63.057	hyg cry	116		1.27	143^{20}	s EtOH
112	Ammonium heptafluorotantalate	$(NH_4)_2TaF_7$	12022-02-5	350.014	hyg cry					
113	Ammonium hexabromoosmate(IV)	$(NH_4)_2OsBr_6$	24598-62-7	705.73	small blk cubes					sl H_2O; s glycerol; i EtOH
114	Ammonium hexabromoplatinate(IV)	$(NH_4)_2PtBr_6$	17363-02-9	710.585	powder	145 dec			0.59^{20}	
115	Ammonium hexachloroiridate(III)	$(NH_4)_3IrCl_6$	15752-05-3	459.050	grn pow					
116	Ammonium hexachloroiridate(IV)	$(NH_4)_2IrCl_6$	16940-92-4	441.012	blk cry powder	dec		2.856	1.09^{25}	
117	Ammonium hexachloroosmate(IV)	$(NH_4)_2OsCl_6$	12125-08-5	439.03	red cry or powder		subl	2.93		s H_2O, EtOH
118	Ammonium hexachloropalladate(IV)	$(NH_4)_2PdCl_6$	19168-23-1	355.22	red-brn hyg cry	dec		2.418		
119	Ammonium hexachloroplatinate(IV)	$(NH_4)_2PtCl_6$	16919-58-7	443.879	red-oran cub cry	380 dec		3.065	0.5^{20}	i EtOH
120	Ammonium hexachlororuthenate(IV)	$(NH_4)_2RuCl_6$	18746-63-9	349.87	red cry					
121	Ammonium hexafluoroaluminate	$(NH_4)_3AlF_6$	7784-19-2	195.087	cub cry			1.78		s H_2O
122	Ammonium hexafluorogallate	$(NH_4)_3GaF_6$	14639-94-2	237.828	col cub cry	>200 dec		2.10		
123	Ammonium hexafluorogermanate	$(NH_4)_2GeF_6$	16962-47-3	222.71	wh cry	380	subl	2.564		s H_2O; i EtOH
124	Ammonium hexafluorophosphate	NH_4PF_6	16941-11-0	163.003	wh cub cry	58 dec		2.180		vs H_2O; s ace, EtOH, MeOH
125	Ammonium hexafluorosilicate	$(NH_4)_2SiF_6$	16919-19-0	178.153	wh cub or trig cry	dec		2.011	22.7^{25}	i EtOH, ace
126	Ammonium hexafluorotitanate	$(NH_4)_2TiF$	16962-40-6	197.934	wh solid					s H_2O
127	Ammonium hexafluorozirconate(IV)	$(NH_4)_2ZrF_6$	16919-31-6	241.291	wh hex cry			1.154		s H_2O
128	Ammonium hydrogen arsenate	$(NH_4)_2HAsO_4$	7784-44-3	176.004	wh powder			1.99		s H_2O
129	Ammonium hydrogen carbonate	NH_4HCO_3	1066-33-7	79.056	col or wh prisms	107 dec		1.586	24.8^{25}	i EtOH, bz
130	Ammonium hydrogen citrate	$(NH_4)_2HC_6H_5O_7$	3012-65-5	226.184	col cry			1.48		vs H_2O; sl EtOH
131	Ammonium hydrogen fluoride	NH_4HF_2	1341-49-7	57.044	wh orth cry	125	240 dec	1.50	60.2^{20}	
132	Ammonium hydrogen malate	$NH_4C_4H_5O_5$	5972-71-4	151.118	orth cry	160		1.15		s H_2O; sl EtOH
133	Ammonium hydrogen oxalate monohydrate	$NH_4HC_2O_4 \cdot H_2O$	5972-72-5	125.081	col rhomb cry	dec		1.56		sl H_2O, EtOH
134	Ammonium hydrogen phosphate	$(NH_4)_2HPO_4$	7783-28-0	132.055	wh cry	155 dec		1.619	69.5^{25}	i EtOH, ace
135	Ammonium hydrogen phosphite monohydrate	$(NH_4)_2HPO_3 \cdot H_2O$	51503-61-8	134.071	hyg cry					s H_2O
136	Ammonium hydrogen selenate	NH_4HSeO_4	10294-60-7	162.01	rhom cry	dec		2.162		
137	Ammonium hydrogen sulfate	NH_4HSO_4	7803-63-6	115.110	wh hyg cry	147		1.78	100^{20}	i EtOH, ace, py
138	Ammonium hydrogen sulfide	NH_4HS	12124-99-1	51.112	wh tetr or orth cry	dec		1.17	128^{0}	sl ace; i bz, eth

No.	Name	Formula	CAS Reg. No.	Mol. wt.	Phys. form	mp/°C	bp/°C	ρ/g cm^{-3}	Sol. in g/100 g H_2O	Qualitative solubility
139	Ammonium hydrogen sulfite	NH_4HSO_3	10192-30-0	99.110	col cry	dec		2.03	71.8^{0}	
140	Ammonium hydrogen tartrate	$NH_4HC_4H_4O_6$	3095-65-6	167.117	wh cry			1.68		sl H_2O; s alk; i EtOH
141	Ammonium hydroxide	NH_4OH	1336-21-6	35.046	exists only in soln					
142	Ammonium hypophosphite	$NH_4H_2PO_2$	7803-65-8	83.028	wh hyg cry	dec				vs H_2O; sl EtOH; i ace
143	Ammonium iodate	NH_4IO_3	13446-09-8	192.941	wh powder	150		3.3	3.84^{25}	
144	Ammonium iodide	NH_4I	12027-06-4	144.943	wh tetr cry; hyg	551 dec	405 sp	2.514	178^{25}	sl EtOH, MeOH
145	Ammonium iron(II) sulfate hexahydrate	$(NH_4)_2Fe(SO_4)_2 \cdot 6H_2O$	7783-85-9	392.139	blue-grn monocl cry	≈100 dec		1.86		s H_2O; i EtOH
146	Ammonium iron(III) chromate	$NH_4Fe(CrO_4)_2$	7789-08-4	305.871	red powder					i H_2O
147	Ammonium iron(III) oxalate trihydrate	$(NH_4)_3Fe(C_2O_4)_3 \cdot 3H_2O$	13268-42-3	428.063	grn monocl cry; hyg	≈160 dec		1.780		vs H_2O; i EtOH
148	Ammonium iron(III) sulfate dodecahydrate	$NH_4Fe(SO_4)_2 \cdot 12H_2O$	7783-83-7	482.192	col to viol cry	≈37		1.71		vs H_2O; i EtOH
149	Ammonium lactate	$NH_4C_3H_5O_3$	52003-58-4	107.108	col cry	92				s H_2O, EtOH; sl MeOH; i ace, eth
150	Ammonium magnesium chloride hexahydrate	$NH_4MgCl_3 \cdot 6H_2O$	39733-35-2	256.794	hyg cry	100 dec		1.46	17^{20}	s H_2O
151	Ammonium mercuric chloride dihydrate	$(NH_4)_2HgCl_4 \cdot 2H_2O$	33445-15-7	414.51	powder					s H_2O; sl EtOH
152	Ammonium metatungstate hexahydrate	$(NH_4)_6W_7O_{24} \cdot 6H_2O$	12028-48-7	1887.19	wh cry					s H_2O; i EtOH
153	Ammonium metavanadate	NH_4VO_3	7803-55-6	116.979	wh-yel cry	200 dec		2.326	4.8^{20}	
154	Ammonium molybdate(VI) tetrahydrate	$(NH_4)_6Mo_7O_{24} \cdot 4H_2O$	12054-85-2	1235.86	col or grn-yel cry	90 dec		2.498	43	i EtOH
155	Ammonium dimolybdate	$(NH_4)_2Mo_2O_7$	27546-07-2	339.95	cry					s H_2O
156	Ammonium molybdophosphate	$(NH_4)_3PO_4 \cdot 12MoO_3$	12026-66-3	1876.35	grn or yel cry	dec			0.02^{20}	sl H_2O; s alk
157	Ammonium nitrate	NH_4NO_3	6484-52-2	80.043	wh hyg cry; orth	169.7	200 -260 dec	1.72	213^{25}	sl MeOH
158	Ammonium nitrite	NH_4NO_2	13446-48-5	64.044	wh-yel cry	60 exp		1.69	221^{25}	i eth
159	Ammonium nitroferricyanide	$(NH_4)_2Fe(CN)_5NO$	14402-70-1	252.016	red-brn cry					s H_2O, EtOH
160	Ammonium oleate	$NH_4C_{18}H_{33}O_2$	544-60-5	299.493	yel-brn paste	21				s H_2O; sl ace
161	Ammonium oxalate	$(NH_4)_2C_2O_4$	1113-38-8	124.096	col sol			1.5	5.20^{25}	
162	Ammonium oxalate monohydrate	$(NH_4)_2C_2O_4 \cdot H_2O$	6009-70-7	142.110	wh orth cry	dec		1.50	5.20^{25}	sl EtOH
163	Ammonium palmitate	$NH_4C_{15}H_{31}CO_2$	593-26-0	273.455	yel-wh powder	22				s H_2O; sl bz, xyl; i ace, EtOH, ctc
164	Ammonium pentaborate tetrahydrate	$NH_4B_5O_8 \cdot 4H_2O$	12007-89-5	272.150	wh cry				7.03^{18}	
165	Ammonium pentachlororhodate(III) monohydrate	$(NH_4)_2RhCl_5 \cdot H_2O$	63771-33-5	334.262	red cry	210 dec				
166	Ammonium pentachlorozincate	$(NH_4)_3ZnCl_5$	14639-98-6	296.789	hyg orth cry			1.81		vs H_2O
167	Ammonium perchlorate	NH_4ClO_4	7790-98-9	117.490	wh orth cry	dec, exp		1.95	24.5^{25}	s MeOH; sl EtOH, ace; i eth
168	Ammonium permanganate	NH_4MnO_4	13446-10-1	136.975	purp rhomb cry	70 dec		2.22	7.9^{15}	
169	Ammonium peroxydisulfate	$(NH_4)_2S_2O_8$	7727-54-0	228.202	monocl cry or wh powder	dec		1.982	83.5^{25}	
170	Ammonium perrhenate	NH_4ReO_4	13598-65-7	268.244	col powder			3.97	6.23^{20}	
171	Ammonium phosphate trihydrate	$(NH_4)_3PO_4 \cdot 3H_2O$	10361-65-6	203.133	wh prisms				25.0^{25}	i ace
172	Ammonium phosphomolybdate monohydrate	$(NH_4)_3PO_4 \cdot 12MoO_3 \cdot H_2O$	54723-94-3	1894.36	yel cry or powder	dec			0.02	
173	Ammonium phosphotungstate dihydrate	$(NH_4)_3PO_4 \cdot 12WO_3 \cdot 2H_2O$	1311-90-6	2967.18	cry powder					sl H_2O
174	Ammonium picrate	$NH_4C_6H_2N_3O_7$	131-74-8	246.135	yel orth cry	exp		1.72		sl H_2O
175	Ammonium polysulfide	$(NH_4)_2S_x$	9080-17-5		yel unstab soln					reac acids
176	Ammonium salicylate	$NH_4C_7H_5O_3$	528-94-9	155.151	wh cry powder					vs H_2O; s EtOH
177	Ammonium selenate	$(NH_4)_2SeO_4$	7783-21-3	179.04	wh monocl cry	dec		2.194	117^{25}	i EtOH, ace
178	Ammonium selenite	$(NH_4)_2SeO_3$	7783-19-9	163.04	wh or red hyg cry	dec			121^{25}	
179	Ammonium stearate	$NH_4C_{18}H_{35}O_2$	1002-89-7	301.509	yel-wh powder	22		0.89		sl H_2O, bz; s EtOH, MeOH; i ace
180	Ammonium sulfamate	$NH_4NH_2SO_3$	7773-06-0	114.124	wh hyg cry	131	160 dec			vs H_2O; sl EtOH
181	Ammonium sulfate	$(NH_4)_2SO_4$	7783-20-2	132.140	wh or brn orth cry	280 dec		1.77	76.4^{25}	i EtOH, ace
182	Ammonium sulfide	$(NH_4)_2S$	12135-76-1	68.142	yel-oran cry	≈0 dec				s H_2O, EtOH, alk
183	Ammonium sulfite	$(NH_4)_2SO_3$	17026-44-7	116.140	wh hyg cry				64.2^{25}	
184	Ammonium sulfite monohydrate	$(NH_4)_2SO_3 \cdot H_2O$	7783-11-1	134.155	col cry	dec		1.41	64.2^{25}	i EtOH, ace
185	Ammonium tartrate	$(NH_4)_2C_4H_4O_6$	3164-29-2	184.147	wh cry	dec		1.601		s H_2O
186	Ammonium tellurate	$(NH_4)_2TeO_4$	13453-06-0	227.68	wh powder	dec		3.024		
187	Ammonium tetraborate tetrahydrate	$(NH_4)_2B_4O_7 \cdot 4H_2O$	12228-87-4	263.377	wh tetr cry	87 dec				vs H_2O; s HNO_3
188	Ammonium tetrachloroaluminate	NH_4AlCl_4	7784-14-7	186.833	wh hyg solid	304				s H_2O, eth
189	Ammonium tetrachloropalladate(II)	$(NH_4)_2PdCl_4$	13820-40-1	284.31	grn cry or red-brn pow					s H_2O
190	Ammonium tetrachloroplatinate(II)	$(NH_4)_2PtCl_4$	13820-41-2	372.973	red cry	dec		2.936		s H_2O; i EtOH

No.	Name	Formula	CAS Reg. No.	Mol. wt.	Phys. form	mp/°C	bp/°C	ρ/g cm^{-3}	Sol. in g/100 g H_2O	Qualitative solubility
191	Ammonium tetrachlorozincate	$(NH_4)_2ZnCl_4$	14639-97-5	243.298	wh orth plates; hyg	150 dec		1.879		vs H_2O
192	Ammonium tetrafluoroantimonate	NH_4SbF_4	14972-90-8	215.793	col cry					s H_2O
193	Ammonium tetrafluoroborate	NH_4BF_4	13826-83-0	104.844	wh powder; orth	487 dec		1.871	25^{20}	
194	Ammonium tetrathiocyanodiammin echromate(III) monohydrate	$NH_4[Cr(NH_3)_2(SCN)_4]\cdot H_2O$	13573-16-5	354.440	red cry	270 dec				s H_2O, EtOH, ace; i bz
195	Ammonium tetrathiomolybdate	$(NH_4)_2MoS_4$	15060-55-6	260.28	red cry	100 dec				vs H_2O
196	Ammonium tetrathiotungstate	$(NH_4)_2WS_4$	13862-78-7	348.18	oran cry	dec		2.71		s H_2O
197	Ammonium tetrathiovandate	$(NH_4)_3VS_4$	14693-56-2	233.317	dark viol cry					
198	Ammonium thiocyanate	NH_4SCN	1762-95-4	76.121	col hyg cry	≈149	dec	1.30	181^{25}	vs EtOH; s ace; i chl
199	Ammonium thiosulfate	$(NH_4)_2S_2O_3$	7783-18-8	148.205	wh cry	150 dec		1.678		vs H_2O; i EtOH, eth
200	Ammonium titanium oxalate monohydrate	$(NH_4)_2TiO(C_2O_4)_2\cdot H_2O$	10580-03-7	293.996	hyg cry					vs H_2O
201	Ammonium tungstate(VI)	$(NH_4)_{10}W_{12}O_{41}$	11120-25-5	3042.44	cry powder			2.3		s H_2O; i EtOH
202	Ammonium tungstate(VI) pentahydrate	$(NH4)_{10}W_{12}O_{41}\cdot 5H_2O$	1311-93-9	3132.52	cry pow or plates			2.3		vs H_2O; i EtOH
203	Ammonium uranate(VI)	$(NH_4)_2U_2O_7$	7783-22-4	624.131	red-yel amorp powder					i H_2O, alk; s acid
204	Ammonium uranium fluoride	$UO_2(NH_4)_3F_5$	18433-40-4	419.135	grn-yel monocl cry					s H_2O; i EtOH
205	Ammonium valerate	$NH_4C_4H_9CO_2$	42739-38-8	119.163	hyg cry	108				vs H_2O, EtOH; s eth
206	Antimony (gray)	Sb	7440-36-0	121.760	silv metal; hex	630.628	1587	6.68		i dil acid
207	Antimony (black)	Sb	7440-36-0	121.760	blk amorp solid	0 trans gray				
208	Stibine	SbH_3	7803-52-3	124.784	col gas; flam	-88	-17	0.005100^{25} gas		sl H_2O; s EtOH
209	Trimethylstibine	$Sb(CH_3)_3$	594-10-5	166.863	col flam liq	-62	80.6	1.523^{15}		
210	Pentamethylstibine	$Sb(CH_3)_5$	15120-50-0	196.933	col hyg liq	-19	127			
211	Tetramethyldistibine	$[Sb(CH_3)_2]_2$	41422-43-9	303.658	yel flam liq or red solid	17				reac H_2O
212	Antimony arsenide	SbAs	12322-34-8	196.682	hex cry	≈680		6.0		
213	Antimony potassium tartrate trihydrate	$K_2(SbC_4H_2O_6)_2\cdot 3H_2O$	28300-74-5	667.873	col cry			2.6		sl H_2O
214	Antimony(III) acetate	$Sb(C_2H_3O_2)_3$	3643-76-3	298.891	wh pow					
215	Antimony(III) bromide	$SbBr_3$	7789-61-9	361.472	yel orth cry; hyg	97	288	4.35		reac H_2O; s ace, bz, chl
216	Antimony(III) chloride	$SbCl_3$	10025-91-9	228.119	col orth cry; hyg	73.4	220.3	3.14	987^{25}	s acid, EtOH, bz, ace
217	Antimony(III) fluoride	SbF_3	7783-56-4	178.755	wh orth cry; hyg	287	376	4.38	492^{25}	
218	Antimony(III) iodide	SbI_3	7790-44-5	502.473	red rhomb cry	171	400	4.92		reac H_2O; s EtOH, ace; i ctc
219	Antimony(III) iodide sulfide	SbIS	13816-38-1	280.729	dark red prisms or needles	400				
220	Antimony(III) oxide (senarmontite)	Sb_2O_3	1309-64-4	291.518	col cub cry	570 trans	1425	5.58		sl H_2O; i os
221	Antimony(III) oxide (valentinite)	Sb_2O_3	1309-64-4	291.518	wh orth cry	655	1425	5.7		sl H_2O; i os
222	Antimony(III) oxychloride	SbOCl	7791-08-4	173.212	wh momo cry	170 dec				reac H_2O; i EtOH, eth
223	Antimony(III) phosphate	$SbPO_4$	12036-46-3	216.731	cry pow					reac H_2O
224	Antimony(III) potassium oxalate trihydrate	$K_3Sb(C_2O_4)_3\cdot 3H_2O$	5965-33-3	557.158	cry pow					s H_2O
225	Antimony(III) selenide	Sb_2Se_3	1315-05-5	480.40	grn orth cry	611		5.81		sl H_2O
226	Antimony(III) sulfate	$Sb_2(SO_4)_3$	7446-32-4	531.708	wh cry powder; hyg	dec		3.62		sl H_2O
227	Antimony(III) sulfide	Sb_2S_3	1345-04-6	339.715	gray-blk orth cry	550		4.562		i H_2O; s conc HCl
228	Antimony(III) telluride	Sb_2Te_3	1327-50-0	626.32	gray cry	620		6.5		
229	Antimony(III,V) oxide	Sb_2O_4	1332-81-6	307.518	yel orth cry			6.64		
230	Antimony(V) chloride	$SbCl_5$	7647-18-9	299.025	col or yel liq	4	140 dec	2.34		reac H_2O; s chl, ctc
231	Antimony(V) fluoride	SbF_5	7783-70-2	216.752	hyg visc liq	8.3	141	3.10		reac H_2O
232	Antimony(V) dichlorotrifluoride	$SbCl_2F_3$	7791-16-4	249.661	visc liq					reac H_2O
233	Antimony(V) oxide	Sb_2O_5	1314-60-9	323.517	yel powder; cub	dec		3.78	0.3^{20}	
234	Antimony(V) sulfide	Sb_2S_5	1315-04-4	403.845	oran-yel powder	75 dec		4.120		i H_2O; s acid, alk
235	Argon	Ar	7440-37-1	39.948	col gas	-189.34	-185.848	1.395$^{-185.8}$ liq		sl H_2O
236	Arsenic (gray)	As	7440-38-2	74.922	gray metal; rhomb	817	616 sp	5.75		i H_2O
237	Arsenic (black)	As	7440-38-2	74.922	blk amorp solid	270 trans gray As		4.9		
238	Arsenic (yellow)	As	7440-38-2	74.922	soft yel cub cry	358 trans gray As		1.97		s CS_2
239	Arsine	AsH_3	7784-42-1	77.946	col gas	-116	-62.5	0.003186^{25} gas		sl H_2O
240	Diarsine	As_2H_4	15942-63-9	153.875	unstab liq		≈100			
241	Arsenic acid	H_3AsO_4	7778-39-4	141.944	exists only in soln					
242	Arsenic acid hemihydrate	$H_3AsO_4\cdot 0.5H_2O$	7778-39-4	150.951	wh hyg cry	36.1		2.5		vs H_2O, EtOH
243	Arsenious acid	H_3AsO_3	13464-58-9	125.944	exists only in soln					
244	Arsenic diiodide	As_2I_4	13770-56-4	657.461	red cry	137				reac H_2O; s os

No.	Name	Formula	CAS Reg. No.	Mol. wt.	Phys. form	mp/°C	bp/°C	ρ/g cm^{-3}	Sol. in g/100 g H_2O	Qualitative solubility
245	Arsenic hemiselenide	As_2Se	1303-35-1	228.80	blk cry					i H_2O, os; dec acid, alk
246	Arsenic sulfide	As_4S_4	12279-90-2	427.946	red monocl cry	307	565	3.5		i H_2O; sl bz; s alk
247	Arsenic(III) bromide	$AsBr_3$	7784-33-0	314.634	col or yel orth cry; hyg	31.1	221	3.40		reac H_2O; s hc, ctc; vs eth, bz
248	Arsenic(III) chloride	$AsCl_3$	7784-34-1	181.281	col liq	-16	130	2.150		reac H_2O; vs chl, ctc, eth
249	Arsenic(III) ethoxide	$As(C_2H_5O)_3$	3141-12-6	210.103	liq		166	1.21		
250	Arsenic(III) fluoride	AsF_3	7784-35-2	131.917	col liq	-5.9	57.13	2.7		reac H_2O; s EtOH, eth, bz
251	Arsenic(III) iodide	AsI_3	7784-45-4	455.635	red hex cry	141	424	4.73		sl H_2O, EtOH, eth; s bz. tol
252	Arsenic(III) oxide (arsenolite)	As_2O_3	1327-53-3	197.841	wh cub cry	274	460	3.86	2.05^{25}	
253	Arsenic(III) oxide (claudetite)	As_2O_3	1327-53-3	197.841	wh monocl cry	314	460	3.74	2.05^{25}	s dil acid, alk; i EtOH
254	Arsenic(III) selenide	As_2Se_3	1303-36-2	386.72	brn-blk solid	377		4.75		i H_2O; s alk
255	Arsenic(III) sulfide	As_2S_3	1303-33-9	246.038	yel-oran monocl cry	312	707	3.46		i H_2O; s alk
256	Arsenic(III) telluride	As_2Te_3	12044-54-1	532.64	blk monocl cry	375		6.50		
257	Arsenic(V) chloride	$AsCl_5$	22441-45-8	252.187	stab at low temp	≈-50 dec				
258	Arsenic(V) fluoride	AsF_5	7784-36-3	169.914	col gas	-79.8	-52.8	0.006945^{25} gas		reac H_2O; s EtOH, bz, eth
259	Arsenic(V) oxide	As_2O_5	1303-28-2	229.840	wh amorp powder	730		4.32	65.8^{20}	vs EtOH
260	Arsenic(V) selenide	As_2Se_5	1303-37-3	544.64	blk solid	dec				i H_2O, EtOH, eth; s alk
261	Arsenic(V) sulfide	As_2S_5	1303-34-0	310.168	brn-yel amorp solid	dec				i H_2O; s alk
262	Astatine	At	7440-68-8	210	cry	302				s HNO_3, os
263	Barium	Ba	7440-39-3	137.327	silv-yel metal; cub	727	≈1845	3.62		reac H_2O; sl EtOH
264	Barium acetate	$Ba(C_2H_3O_2)_2$	543-80-6	255.416	wh powder			2.47	79.2^{25}	
265	Barium acetate monohydrate	$Ba(C_2H_3O_2)_2 \cdot H_2O$	5908-64-5	273.431	wh cry	110 dec		2.19	79.2^{25}	sl EtOH
266	Barium aluminate	$BaAl_2O_4$	12004-04-5	255.288	hex cry	1827				
267	Barium aluminide	$BaAl_4$	12672-79-6	245.253	metallic solid	1097				
268	Barium azide	$Ba(N_3)_2$	18810-58-7	221.367	monocl cry; exp	≈120 dec		2.936	17.3^{20}	sl EtOH; i eth
269	Barium bismuthate	$BaBiO_3$	12785-50-1	394.305	bronze cry	1040 dec				
270	Barium bromate	$Ba(BrO_3)_2$	13967-90-3	393.131	col monocl cry				0.79^{25}	s ace
271	Barium bromate monohydrate	$Ba(BrO_3)_2 \cdot H_2O$	10326-26-8	411.147	wh monocl cry	260 dec		3.99	0.831^{25}	i EtOH
272	Barium bromide	$BaBr_2$	10553-31-8	297.135	wh orth cry	857	1835	4.781	100^{25}	
273	Barium bromide dihydrate	$BaBr_2 \cdot 2H_2O$	7791-28-8	333.166	wh cry	75 dec		3.7	100^{25}	s MeOH; i EtOH, ace, diox
274	Barium calcium tungstate	Ba_2CaWO_6	15552-14-4	594.57	cub cry	1420				
275	Barium carbide	BaC_2	50813-65-5	161.348	gray tetr cry	dec		3.74		reac H_2O
276	Barium carbonate	$BaCO_3$	513-77-9	197.336	wh orth cry	1555		4.308	0.0014^{20}	s acid
277	Barium chlorate	$Ba(ClO_3)_2$	13477-00-4	304.229	wh cry	414			37.9^{25}	sl EtOH, ace
278	Barium chlorate monohydrate	$Ba(ClO_3)_2 \cdot H_2O$	10294-38-9	322.245	wh monocl cry	120 dec		3.179	37.9^{25}	s acid; sl EtOH, ace
279	Barium chloride	$BaCl_2$	10361-37-2	208.233	wh orth cry; hyg	961	1560	3.9	37.0^{25}	
280	Barium chloride dihydrate	$BaCl_2 \cdot 2H_2O$	10326-27-9	244.264	wh monocl cry	≈120 dec		3.097	37.0^{25}	i EtOH
281	Barium chloride fluoride	$BaClF$	13718-55-3	191.778	wh cry					
282	Barium chromate(V)	$Ba_3(CrO_4)_2$	12345-14-1	643.968	grn-blk hex cry			5.25		s H_2O
283	Barium chromate(VI)	$BaCrO_4$	10294-40-3	253.321	yel orth cry	1380		4.50	0.00026^{20}	reac acid
284	Barium citrate monohydrate	$Ba_3(C_6H_5O_7)_2 \cdot H_2O$	512-25-4	808.195	gray-wh cry					s H_2O, acid
285	Barium copper yttrium oxide	$BaCuY_2O_5$	82642-06-6	458.682	grn cry; not superconductor					
286	Barium copper yttrium oxide	$Ba_2Cu_3YO_7$	109064-29-1	666.194	blk solid; HT superconductor					
287	Barium copper yttrium oxide	$Ba_2Cu_4YO_8$	114104-80-2	745.739	HT superconductor					
288	Barium copper yttrium oxide	$Ba_4Cu_7Y_2O_{15}$	124365-83-9	1411.930	HT superconductor					
289	Barium cyanide	$Ba(CN)_2$	542-62-1	189.361	wh cry powder					vs H_2O; s EtOH
290	Barium dichromate dihydrate	$BaCr_2O_7 \cdot 2H_2O$	10031-16-0	389.346	brn-red needles	dec				reac H_2O
291	Barium disilicate	$BaSi_2O_5$	12650-28-1	273.495	wh orth cry	1420		3.70		
292	Barium dithionate dihydrate	$BaS_2O_6 \cdot 2H_2O$	13845-17-5	333.484	wh cry	140 dec		4.54	22.1^{20}	sl EtOH
293	Barium ferrite	$BaFe_{12}O_{19}$	11138-11-7	1111.460	magnetic solid					
294	Barium ferrocyanide hexahydrate	$Ba_2Fe(CN)_6 \cdot 6H_2O$	13821-06-2	594.694	yel monocl cry	80 dec				i H_2O, EtOH
295	Barium fluoride	BaF_2	7787-32-8	175.324	wh cub cry	1368	2260	4.893	0.161^{25}	
296	Barium formate	$Ba(COOH)_2$	541-43-5	227.362	cry			3.21		s H_2O; i EtOH
297	Barium hexaboride	BaB_6	12046-08-1	202.193	blk cub cry	2070		4.36		i H_2O; s acid; i EtOH
298	Barium hexafluorogermanate	$BaGeF_6$		323.96	wh cry	≈665		4.56		
299	Barium hexafluorosilicate	$BaSiF_6$	17125-80-3	279.403	wh orth needles	300 dec		4.29		i H_2O, EtOH; sl acid
300	Barium hydride	BaH_2	13477-09-3	139.343	gray orth cry	1200		4.16		reac H_2O
301	Barium hydrogen phosphate	$BaHPO_4$	10048-98-3	233.306	wh cry powder	400 dec		4.16	0.015^{20}	s dil acid
302	Barium hydrosulfide	$Ba(HS)_2$	25417-81-6	203.473	yel hyg cry					s H_2O
303	Barium hydrosulfide tetrahydrate	$Ba(HS)_2 \cdot 4H_2O$	12230-74-9	275.534	yel rhomb cry	50 dec				s H_2O
304	Barium hydroxide	$Ba(OH)_2$	17194-00-2	171.342	wh powder	408			4.91^{25}	
305	Barium hydroxide monohydrate	$Ba(OH)_2 \cdot H_2O$	22326-55-2	189.357	wh powder			3.743	4.91^{25}	s acid

No.	Name	Formula	CAS Reg. No.	Mol. wt.	Phys. form	mp/°C	bp/°C	ρ/g cm^{-3}	Sol. in g/100 g H_2O	Qualitative solubility
306	Barium hydroxide octahydrate	$Ba(OH)_2 \cdot 8H_2O$	12230-71-6	315.464	wh monocl cry	78 dec		2.18	4.91^{25}	
307	Barium hypophosphite monohydrate	$Ba(H_2PO_2)_2 \cdot H_2O$	14871-79-5	285.320	monocl plates			2.90		s H_2O; i EtOH
308	Barium iodate	$Ba(IO_3)_2$	10567-69-8	487.132	wh cry powder	476 dec		5.23	0.0396^{25}	
309	Barium iodate monohydrate	$Ba(IO_3)_2 \cdot H_2O$	7787-34-0	505.148	cry	130 dec		5.00	0.0396^{25}	s acid; i EtOH
310	Barium iodide	BaI_2	13718-50-8	391.136	wh orth cry	711		5.15	221^{25}	
311	Barium iodide dihydrate	$BaI_2 \cdot 2H_2O$	7787-33-9	427.167	col cry	740 dec		5.0	221^{25}	s EtOH, ace
312	Barium manganate(VI)	$BaMnO_4$	7787-35-1	256.263	grn-gray hyg cry			4.85	0.00041^{20}	
313	Barium metaborate monohydrate	$Ba(BO_2)_2 \cdot H_2O$	26124-86-7	240.962	wh powder	>900		3.3		sl H_2O
314	Barium metaborate dihydrate	$Ba(BO_2)_2 \cdot 2H_2O$	23436-05-7	258.977	wh prec	dec			1.3^{25}	sl H_2O
315	Barium metaphosphate	$Ba(PO_3)_2$	13466-20-1	295.271	wh powder	1560				i H_2O; sl acid
316	Barium metasilicate	$BaSiO_3$	13255-26-0	213.411	col rhomb powder	1605		4.40		i H_2O; s acid
317	Barium molybdate	$BaMoO_4$	7787-37-3	297.27	wh powder	1450		4.975	0.0021^{20}	
318	Barium niobate	$Ba(NbO_3)_2$	12009-14-2	419.136	yel orth cry	1455		5.44		i H_2O
319	Barium nitrate	$Ba(NO_3)_2$	10022-31-8	261.336	wh cub cry	590		3.24	10.3^{25}	sl EtOH, ace
320	Barium nitride	Ba_3N_2	12047-79-9	439.994	yel-brn cry	>500 dec		4.78		reac H_2O
321	Barium nitrite	$Ba(NO_2)_2$	13465-94-6	229.338	col hex cry	267		3.234	79.5^{25}	
322	Barium nitrite monohydrate	$Ba(NO_2)_2 \cdot H_2O$	7787-38-4	247.353	yel-wh hex cry	217 dec		3.18	79.5^{25}	i EtOH
323	Barium orthovanadate	$Ba_3(VO_4)_2$	39416-30-3	641.859	hex cry	707		5.14		
324	Barium oxalate	BaC_2O_4	516-02-9	225.346	wh powder	400 dec		2.658	0.0075	
325	Barium oxalate monohydrate	$BaC_2O_4 \cdot H_2O$	13463-22-1	243.361	wh cry powder			2.66	0.0075^{20}	s acid
326	Barium oxide	BaO	1304-28-5	153.326	wh-yel powder; cub and hex	1973		5.72 (cub)	1.5^{20}	s dil acid, EtOH; i ace
327	Barium 2,4-pentanedioate octahydrate	$Ba(CH_3COCHCOCH_3)_2 \cdot 8H_2O$	12084-29-6	479.665	col hyg cry	320 (anh)				
328	Barium perchlorate	$Ba(ClO_4)_2$	13465-95-7	336.228	col hex cry	505		3.20	312^{25}	vs EtOH
329	Barium perchlorate trihydrate	$Ba(ClO_4)_2 \cdot 3H_2O$	10294-39-0	390.274	col cry			2.74	312^{25}	s MeOH; sl EtOH, ace; i eth
330	Barium permanganate	$Ba(MnO_4)_2$	7787-36-2	375.198	brn-viol cry	200 dec		3.77	62.5^{20}	reac EtOH
331	Barium peroxide	BaO_2	1304-29-6	169.326	gray-wh tetr cry	450 dec		4.96	0.091^{20}	reac dil acid
332	Barium plumbate	$BaPbO_3$	12047-25-5	392.5	orth cry					
333	Barium potassium chromate	$BaK_2(CrO_4)_2$	27133-66-0	447.511	yel hex cry			3.63		vs H_2O
334	Barium pyrophosphate	$Ba_2P_2O_7$	13466-21-2	448.597	wh powder	1430		3.9	0.0088^{20}	s acid
335	Barium selenate	$BaSeO_4$	7787-41-9	280.29	wh rhomb cry	dec		4.75	0.015^{20}	
336	Barium selenide	BaSe	1304-39-8	216.29	cub cry powder	1780		5.02		reac H_2O
337	Barium selenite	$BaSeO_3$	13718-59-7	264.29	solid					i H_2O
338	Barium silicide	$BaSi_2$	1304-40-1	193.498	gray lumps	1180				reac H_2O
339	Barium sodium niobate	$Ba_2Na(NbO_3)_5$	12323-03-4	1002.170	wh orth cry	1437		5.40		i H_2O
340	Barium stannate	$BaSnO_3$	12009-18-6	304.035	cub cry			7.24		sl H_2O
341	Barium stannate trihydrate	$BaSnO_3 \cdot 3H_2O$	12009-18-6	358.081	wh cry powder					sl H_2O; s acid
342	Barium stearate	$Ba(C_{18}H_{35}O_2)_2$	6865-35-6	704.266	wh powder	160		1.145		i H_2O, EtOH
343	Barium strontium niobate	$BaSr(NbO_3)_4$	37185-09-4	788.57	pale yel solid					
344	Barium strontium tungstate	Ba_2SrWO_6	14871-56-8	642.11	hyg pow	1400				
345	Barium sulfate	$BaSO_4$	7727-43-7	233.390	wh orth cry	1580		4.49	0.00031^{20}	i EtOH
346	Barium sulfide	BaS	21109-95-5	169.392	col cub cry or gray powder	2227		4.3	8.94^{25}	
347	Barium sulfite	$BaSO_3$	7787-39-5	217.390	wh monocl cry	dec		4.44	0.0011^{25}	i EtOH
348	Barium tartrate	$BaC_4H_4O_6$	5908-81-6	285.398	wh cry			2.98		s H_2O; i EtOH
349	Barium tetracyanoplatinate(II) tetrahydrate	$BaPt(CN)_4 \cdot 4H_2O$	13755-32-3	508.543	yel powder or cry			2.076		sl H_2O; i EtOH
350	Barium tetraiodomercurate(II)	$BaHgI_4$	10048-99-4	845.54	yel-red hyg cry					vs H_2O, EtOH
351	Barium thiocyanate	$Ba(SCN)_2$	2092-17-3	253.491	hyg cry				167^{25}	s ace, MeOH, EtOH
352	Barium thiocyanate dihydrate	$Ba(SCN)_2 \cdot 2H_2O$	2092-17-3	289.522	hyg wh cry				167^{25}	s EtOH
353	Barium thiocyanate trihydrate	$Ba(SCN)_2 \cdot 3H_2O$	68016-36-4	307.537	wh needles; hyg			2.286	167^{25}	s EtOH
354	Barium thiosulfate	BaS_2O_3	35112-53-9	249.455	wh cry powder	220 dec			0.2^{20}	i EtOH
355	Barium thiosulfate monohydrate	$BaS_2O_3 \cdot H_2O$	7787-40-8	267.471	wh cry powder	dec		3.5	0.2	i EtOH
356	Barium titanate	$BaTiO_3$	12047-27-7	233.192	wh tetr cry	1625		6.02		i H_2O
357	Barium titanate [$BaTi_2O_5$]	$BaTi_2O_5$	12009-27-7	313.058	wh solid					
358	Barium titanate [$BaTi_4O_9$]	$BaTi_4O_9$	12009-31-3	472.790	wh solid					
359	Barium titanium silicate	$BaTi(SiO_3)_3$	15491-35-7	413.446	rhom blue-pur cry					
360	Barium tungstate	$BaWO_4$	7787-42-0	385.17	wh tetr cry	1475	1730	5.04	0.0016^{20}	
361	Barium uranium oxide	BaU_2O_7	10380-31-1	725.381	oran-yel powder					i H_2O; s acid
362	Barium yttrium tungsten oxide	$Ba_3Y_2WO_9$	37265-86-4	1006.53	cub cry	1470				
363	Barium zirconate	$BaZrO_3$	12009-21-1	276.549	gray-wh cub cry	2500		5.52		i H_2O, alk; sl acid
364	Barium zirconium silicate	$BaO \cdot ZrO_2 \cdot SiO_2$		336.634	wh pow					i H2O, alk; sl acid; s HF
365	Berkelium (α form)	Bk	7440-40-6	247	hex cry	930 trans β		14.78		
366	Berkelium (β form)	Bk	7440-40-6	247	cub cry	986		13.25		
367	Beryllium	Be	7440-41-7	9.012	hex cry	1287	2468	1.85		s acid, alk
368	Beryllium acetate	$Be(C_2H_3O_2)_2$	543-81-7	127.101	wh cry	60 dec				i H_2O, EtOH
369	Beryllium basic acetate	$Be_4O(C_2H_3O_2)_6$	1332-52-1	406.312	wh cry	285	330	1.25		i H_2O; s eth, os
370	Beryllium aluminate	$BeAl_2O_4$	12004-06-7	126.973	orth cry			3.65		
371	Beryllium aluminum metasilicate	$Be_3Al_2(SiO_3)_6$	1302-52-9	537.502	col or grn-yel cry; hex			2.64		
372	Beryllium boride [BeB_2]	BeB_2	12228-40-9	30.634	refrac solid	>1970				
373	Beryllium boride [BeB_6]	BeB_6	12429-94-6	73.878	red solid	2070				
374	Beryllium boride (Be_2B)	Be_2B	12536-51-5	28.835	pink cry	1520				
375	Beryllium boride (Be_4B)	Be_4B	12536-52-6	46.860	refrac solid	1160				
376	Beryllium borohydride	$Be(BH_4)_2$	17440-85-6	36.682	solid	125 dec	subl			reac H_2O

Inorganic

No.	Name	Formula	CAS Reg. No.	Mol. wt.	Phys. form	mp/°C	bp/°C	ρ/g cm^{-3}	Sol. in g/100 g H_2O	Qualitative solubility
377	Beryllium bromide	$BeBr_2$	7787-46-4	168.820	orth cry; hyg	508	473 sp	3.465		vs H_2O; s EtOH, pyr
378	Beryllium carbide	Be_2C	506-66-1	30.035	red cub cry	2127		1.90		reac H_2O
379	Beryllium carbonate tetrahydrate	$BeCO_3 \cdot 4H_2O$	60883-64-9	141.080	wh solid	100 dec			0.36^{0}	
380	Beryllium basic carbonate	$Be_3(OH)_2(CO_3)_2$	66104-24-3	181.069	wh powder					i H_2O; s acid, alk
381	Beryllium chloride	$BeCl_2$	7787-47-5	79.918	wh-yel orth cry; hyg	415	482	1.90	71.5^{25}	s EtOH, eth, py; i bz, tol
382	Beryllium fluoride	BeF_2	7787-49-7	47.009	tetr cry or gl; hyg	552	1283	2.1		vs H_2O; sl EtOH
383	Beryllium formate	$Be(CHO_2)_2$	1111-71-3	99.047	powder	>250 dec				reac H_2O; i os
384	Beryllium hydride	BeH_2	7787-52-2	11.028	wh amorp solid	250 dec		0.65		reac H_2O; i eth, tol
385	Beryllium hydrogen phosphate	$BeHPO_4$	13598-15-7	104.991	cry					i H_2O
386	Beryllium hydroxide (α)	$Be(OH)_2$	13327-32-7	43.027	wh powder or cry	≈200 dec		1.92		sl H_2O, alk; s acid
387	Beryllium hydroxide (β)	$Be(OH)_2$	13327-32-7	43.027	col tetr cry	138 dec				i H_2O; s, acid, alk
388	Beryllium iodide	BeI_2	7787-53-3	262.821	hyg needles	480	590	4.32		reac H_2O; s EtOH
389	Beryllium nitrate trihydrate	$Be(NO_3)_2 \cdot 3H_2O$	13597-99-4	187.068	yel-wh hyg cry	≈30	dec		107^{20}	s EtOH
390	Beryllium nitride	Be_3N_2	1304-54-7	55.050	gray refrac cry; cub	2200		2.71		reac acid, alk
391	Beryllium oxalate trihydrate	$BeC_2O_4 \cdot 3H_2O$	15771-43-4	151.077	rhom cry	320 dec				vs H_2O
392	Beryllium oxide	BeO	1304-56-9	25.011	wh hex cry	2578		3.01		i H_2O; sl acid, alk
393	Beryllium 2,4-pentanedioate	$Be(CH_3COCHCOCH_3)_2$	10210-64-7	207.228	monocl cry powder	108	270	1.168^{20}		i H_2O; vs EtOH, eth
394	Beryllium perchlorate tetrahydrate	$Be(ClO_4)_2 \cdot 4H_2O$	7787-48-6	279.975	hyg cry	250 dec			198^{25}	
395	Beryllium selenate tetrahydrate	$BeSeO_4 \cdot 4H_2O$	10039-31-3	224.03	orth cry	100 dec		2.03		vs H_2O
396	Beryllium sulfate	$BeSO_4$	13510-49-1	105.075	col tetr cry; hyg	1127		2.5	41.3^{25}	
397	Beryllium sulfate dihydrate	$BeSO_4 \cdot 2H_2O$	14215-00-0	141.105	col cry	92 dec				
398	Beryllium sulfate tetrahydrate	$BeSO_4 \cdot 4H_2O$	7787-56-6	177.136	col tetr cry	≈100 dec		1.71	41.3^{25}	i EtOH
399	Beryllium sulfide	BeS	13598-22-6	41.077	col cub cry	dec		2.36		reac hot H_2O
400	Bismuth	Bi	7440-69-9	208.980	gray-wh soft metal	271.402	1564	9.79		s acid
401	Bismuth acetate	$Bi(C_2H_3O_2)_3$	22306-37-2	386.111	col tablets	250				i H_2O
402	Bismuth subacetate	$BiOC_2H_3O_2$	5142-76-7	284.023	thin cry plates					i H_2O; s dil acid
403	Bismuth antimonide	$BiSb$	12323-19-2	330.740	cry	475				
404	Bismuth arsenate	$BiAsO_4$	13702-38-0	347.900	wh monocl cry			7.14		i H_2O; sl conc HNO_3
405	Bismuth basic carbonate	$(BiO)_2CO_3$	5892-10-4	509.969	wh powder			6.86		i H_2O; s acid
406	Bismuth basic dichromate	$Bi_2O_3 \cdot 2CrO_3$		665.948	red-oran amorp pow					i H_2O; s acid, alk
407	Bismuth citrate	$BiC_6H_5O_7$	813-93-4	398.080	wh powder			3.458		i H_2O; sl EtOH
408	Bismuth hydride	BiH_3	18288-22-7	212.004	col gas; unstab	-67	≈17	0.008665^{25} gas		
409	Bismuth hydroxide	$Bi(OH)_3$	10361-43-0	260.002	wh-yel amorp powder			4.962		i H_2O; s acid
410	Bismuth germanium oxide	$2Bi_2O_3 \cdot 3GeO_2$	12233-56-6	1245.84	wh pow	1044				
411	Bismuth hexafluoro-2,4-pentanedioate	$Bi(CF_3COCHCOCF_3)_3$	142617-56-9	830.132	powder	96				
412	Bismuth molybdate	Bi_2MoO_6	13565-96-3	609.90	yel solid			9.32		
413	Bismuth molybdate	$Bi_2(MoO_4)_3$	51898-99-8	897.77	monocl cry			5.95		
414	Bismuth nitrate pentahydrate	$Bi(NO_3)_3 \cdot 5H_2O$	10035-06-0	485.071	col tricl cry; hyg	≈75 dec		2.83		reac H_2O; s ace; i EtOH
415	Bismuth subnitrate	$Bi_5O(OH)_9(NO_3)_4$	1304-85-4	1461.990	hyg cry powder	260 dec		4.928		i H_2O, EtOH; s dil acid
416	Bismuth oleate	$Bi(C_{18}H_{33}O_2)_3$	52951-38-9	1053.340	soft yel-brn solid					i H_2O; s eth; sl bz
417	Bismuth oxalate	$Bi_2(C_2O_4)_3$	6591-55-5	682.018	wh powder					i H_2O, EtOH; s dil acid
418	Bismuth oxide	Bi_2O_3	1304-76-3	465.959	yel monocl cry or powder	825	1890	8.9		i H_2O; s acid
419	Bismuth tetroxide	Bi_2O_4	12048-50-9	481.959	red-oran powder	305		5.6		reac H_2O
420	Bismuth oxybromide	$BiOBr$	7787-57-7	304.883	col tetr cry	560 dec		8.08		i H_2O, EtOH; s acid
421	Bismuth oxychloride	$BiOCl$	7787-59-9	260.432	wh tetr cry	575 dec		7.72		i H_2O
422	Bismuth oxyiodide	$BiOI$	7787-63-5	351.883	red tetr cry	300 dec		7.92		i H_2O, EtOH, chl; s HCl
423	Bismuth oxynitrate	$BiONO_3$	10361-46-3	286.985	wh powder	260 dec		4.93		i H_2O, EtOH; s acid
424	Bismuth phosphate	$BiPO_4$	10049-01-1	303.951	monocl cry			6.32		sl H_2O, dil acid; i EtOH
425	Bismuth potassium iodide	BiK_4I_7	41944-01-8	1253.700	red cry					reac H_2O; s alk iodide soln
426	Bismuth selenide	Bi_2Se_3	12068-69-8	654.84	blk hex cry	710 dec		7.5		i H_2O
427	Bismuth stannate pentahydrate	$Bi_2(SnO_3)_3 \cdot 5H_2O$	12777-45-6	1008.160	wh cry					i H_2O
428	Bismuth sulfate	$Bi_2(SO_4)_3$	7787-68-0	706.149	wh needles or powder	405 dec		5.08		reac H_2O, EtOH
429	Bismuth sulfide	Bi_2S_3	1345-07-9	514.156	blk-brn orth cry	777		6.78		i H_2O; s acid
430	Bismuth telluride	Bi_2Te_3	1304-82-1	800.76	gray hex plates	580		7.74		i H_2O; s EtOH

No.	Name	Formula	CAS Reg. No.	Mol. wt.	Phys. form	mp/°C	bp/°C	ρ/g cm-3	Sol. in g/100 g H_2O	Qualitative solubility
431	Bismuth tribromide	$BiBr_3$	7787-58-8	448.692	yel cub cry	219	462	5.72		reac H_2O; s dil acid, ace; i EtOH
432	Bismuth trichloride	$BiCl_3$	7787-60-2	315.339	col or yel cub cry; hyg	234	441	4.75		reac H_2O; s acid, EtOH, ace
433	Bismuth trifluoride	BiF_3	7787-61-3	265.975	wh-gray cub cry	649	900	8.3		i H_2O
434	Bismuth pentafluoride	BiF_5	7787-62-4	303.972	wh tetr needles; hyg	151.4	230	5.55		reac H_2O
435	Bismuth triiodide	BiI_3	7787-64-6	589.693	blk-brn hex cry	408.6	542	5.778	0.00078^{20}	s EtOH
436	Bismuth trimethyl	$Bi(CH_3)_3$	593-91-9	254.083	col flam liq	-86	110	2.3		
437	Bismuth titanate	$Bi_4(TiO_4)_3$	12048-51-0	1171.520	wh orth cry			7.85		
438	Bismuth tungstate	$Bi_2(WO_4)_3$	13595-87-4	1161.47	wh powder					
439	Bismuth vanadate	$BiVO_4$	14059-33-7	323.920	orth cry	500 trans		6.25		i H_2O; s acid
440	Bismuth zirconate	$2Bi_2O_3 \cdot 3ZrO_2$	37306-42-6	1301.590	wh pow					
441	Boron	B	7440-42-8	10.81	blk rhomb cry	2077	4000	2.34		i H_2O
442	Diborane	B_2H_6	19287-45-7	27.670	col gas; flam	-164.85	-92.49	0.001131^{25} gas		reac H_2O
443	Tetraborane(10)	B_4H_{10}	18283-93-7	53.323	unstab col gas	-120	18	0.002180^{25} gas		reac H_2O
444	Pentaborane(9)	B_5H_9	19624-22-7	63.126	flam col liq	-46.74	60.10	0.60		reac hot H_2O
445	Pentaborane(11)	B_5H_{11}	18433-84-6	65.142	col liq; unstab	-122	65			reac H_2O
446	Hexaborane(10)	B_6H_{10}	23777-80-2	74.945	col liq	-62.3	108 dec	0.67		reac hot H_2O
447	Hexaborane(12)	B_6H_{12}	12008-19-4	76.961	col liq	-82.3	≈85			reac H_2O
448	Nonaborane(15)	B_9H_{15}	19465-30-6	112.418	col liq	2.7				
449	Decaborane(14)	$B_{10}H_{14}$	17702-41-9	122.221	wh orth cry	98.78	213	0.94		sl H_2O; s EtOH, bz, CS_2, ctc
450	Decaborane(16)	$B_{10}H_{16}$	71595-75-0	124.237	col cry	≈81	170 dec			
451	Dodecaborane(16)	$B_{12}H_{16}$	89711-39-7	145.859	col cry	65				s bz, hx
452	Tridecaborane(19)	$B_{13}H_{19}$	43093-20-5	159.694	yel cry	44				s hx, CH2Cl2
453	Tetradecaborane(18)	$B_{14}H_{18}$	55606-55-8	169.497	visc yel oil		100 dec			s cyhex, CS_2
454	Hexadecaborane(20)	$B_{16}H_{20}$	28265-11-4	193.135	col cry	≈110				s ctc, cyhex, thf
455	Octadecaborane(22)	$B_{18}H_{22}$	11071-61-7	216.773	yel cry	180				s os
456	Tetrabromodiborane	B_2Br_4	14355-29-4	341.238	col liq	≈1	20 dec			
457	Tetrachlorodiborane	B_2Cl_4	13701-67-2	163.434	col liq; flam	-92.6	66.5			reac H_2O
458	Tetrafluorodiborane	B_2F_4	13965-73-6	97.616	col gas; flam	-56	-34.0	0.003990^{25} gas		reac H_2O
459	Borane carbonyl	BH_3CO	13205-44-2	41.845	col gas	-137	-64	0.001710^{25} gas		reac H_2O
460	Borazine	$B_3N_3H_6$	6569-51-3	80.501	col liq	-58	53	0.824		reac H_2O
461	Boric acid	H_3BO_3	10043-35-3	61.833	col tricl cry	170.9		1.5	5.80^{25}	sl EtOH
462	Metaboric acid (α form)	HBO_2	13460-50-9	43.818	col orth cry; hyg	176		1.784		s H_2O
463	Metaboric acid (β form)	HBO_2	13460-50-9	43.818	col monocl cry; hyg	201		2.045		s H_2O
464	Metaboric acid (γ form)	HBO_2	13460-50-9	43.818	col cub cry	236		2.487		s H_2O
465	Tetrafluoroboric acid	HBF_4	16872-11-0	87.813	col liq		130 dec	≈1.8		vs H_2O, EtOH
466	Boron arsenide	BAs	12005-69-5	85.733	brn cub cry	1100 dec		5.22		
467	Boron carbide	B_4C	12069-32-8	55.255	hard blk cry	2350	>3500	2.50		i H_2O, acid
468	Boron nitride	BN	10043-11-5	24.818	wh powder; hex or cub cry	2967		2.18		i H_2O, acid
469	Boron oxide	B_2O_3	1303-86-2	69.620	col gl or hex cry; hyg	450		2.55	2.2^{20}	s EtOH
470	Boron phosphide	BP	20205-91-8	41.785	red cub cry or powder	1125 dec				reac H_2O, acid
471	Boron silicide	B_6Si	12008-29-6	92.952	blk cry	1980				
472	Boron sulfide	B_2S_3	12007-33-9	117.817	yel amorp solid	563		≈1.7		
473	Boron tribromide	BBr_3	10294-33-4	250.523	col liq; hyg	-46	91.3	2.6		reac H_2O, EtOH
474	Boron trichloride	BCl_3	10294-34-5	117.170	col liq or gas	-107.3	12.5	0.004789^{25} gas		reac H_2O, EtOH
475	Boron trifluoride	BF_3	7637-07-2	67.806	col gas	-126.8	-99.9	0.002772^{25} gas		s H_2O
476	Boron trifluoride etherate	$BF_3(C_2H_5)_2O$	109-63-7	141.927	liq	-60.4	125.5	1.125^{25}		reac H_2O; vs eth, EtOH
477	Boron triiodide	BI_3	13517-10-7	391.524	wh needles	49.7	209.5	3.35		i H_2O
478	Bromine	Br_2	7726-95-6	159.808	red liq	-7.2	58.8	3.1028		sl H_2O
479	Bromic acid	$HBrO_3$	7789-31-3	128.910	stab only in aq soln					s H_2O
480	Hypobromous acid	HOBr	13517-11-8	96.911	exists aq soln					s H_2O
481	Bromine dioxide	BrO_2	21255-83-4	111.903	unstab yel cry	≈0 dec				
482	Bromine monoxide	Br_2O	21308-80-5	175.807	unstab brn solid	-17.5 dec				
483	Dibromine trioxide	Br_2O_3	53809-75-9	207.806	oran needles (LT)	-40 dec				
484	Dibromine pentoxide	Br_2O_5	58572-43-3	239.805	col cry (low temp)	-20 dec				
485	Bromine azide	BrN_3	13973-87-0	121.924	red cry; exp	≈45	exp			
486	Bromine chloride	BrCl	13863-41-7	115.357	dark red liq (<5°C)	-66	5 dec			reac H_2O; s eth, CS_2
487	Bromine fluoride	BrF	13863-59-7	98.902	unstab red-brn gas	≈-33	≈20 dec	0.004043^{25} gas		
488	Bromine trifluoride	BrF_3	7787-71-5	136.899	col hyg liq	8.77	125.8	2.803		reac H_2O
489	Bromine pentafluoride	BrF_5	7789-30-2	174.896	col liq	-60.5	41.3	2.460		reac H_2O (exp)
490	Bromosyl trifluoride	$BrOF_3$	61519-37-7	152.898	col liq	-5	>20 dec			reac H_2O

No.	Name	Formula	CAS Reg. No.	Mol. wt.	Phys. form	mp/°C	bp/°C	ρ/g cm^{-3}	Sol. in g/100 g H_2O	Qualitative solubility
491	Bromyl fluoride	BrO_2F	22585-64-4	130.901	col liq	-9	55 dec			reac H_2O
492	Perbromyl fluoride	BrO_3F	37265-91-1	146.900	col gas	-110	20 dec			reac H_2O
493	Cadmium	Cd	7440-43-9	112.411	silv-wh metal	321.069	767	8.69		i H_2O; reac acid
494	Cadmium acetate	$Cd(C_2H_3O_2)_2$	543-90-8	230.500	col cry	255		2.34		s H_2O, EtOH
495	Cadmium acetate dihydrate	$Cd(C_2H_3O_2)_2 \cdot 2H_2O$	5743-04-4	266.529	wh cry	130 dec		2.01		vs H_2O; s EtOH
496	Cadmium antimonide	CdSb	12014-29-8	234.171	orth cry	456		6.92		
497	Cadmium arsenide	Cd_3As_2	12006-15-4	487.076	gray tetr cry	721		6.25		
498	Cadmium azide	$Cd(N_3)_2$	14215-29-3	196.451	yel-wh orth cry; exp	exp		3.24		
499	Cadmium borotungstate octadecahydrate	$Cd_5(BW_{12}O_{40}) \cdot 18H_2O$	1306-26-9	3743.20	yel cry					vs H_2O
500	Cadmium bromide	$CdBr_2$	7789-42-6	272.219	wh-yel hex cry; hyg	568	863	5.19	115^{25}	sl ace, eth
501	Cadmium bromide tetrahydrate	$CdBr_2 \cdot 4H_2O$	13464-92-1	344.281	wh-yel cry				115^{25}	s ace, EtOH
502	Cadmium carbonate	$CdCO_3$	513-78-0	172.420	wh hex cry	500 dec		5.026		i H_2O; s acid
503	Cadmium chlorate dihydrate	$Cd(ClO_3)_2 \cdot 2H_2O$	22750-54-5	315.344	col hyg cry	80 dec		2.28	2.64^{0}	
504	Cadmium chloride	$CdCl_2$	10108-64-2	183.317	rhom cry; hyg	568	964	4.08	120^{25}	s ace; sl EtOH; i eth
505	Cadmium chloride monohydrate	$CdCl_2 \cdot H_2O$	34330-64-8	201.332	wh cry				120^{25}	
506	Cadmium chloride hemipentahydrate	$CdCl_2 \cdot 2.5H_2O$	7790-78-5	228.354	wh rhomb leaflets			3.327	120^{25}	s ace
507	Cadmium chromate	$CdCrO_4$	14312-00-6	228.405	yel orth cry			4.5		i H_2O
508	Cadmium cyanide	$Cd(CN)_2$	542-83-6	164.445	wh cub cry			2.23	1.7^{15}	
509	Cadmium dichromate monohydrate	$CdCr_2O_7 \cdot H_2O$	69239-51-6	346.414	oran solid					s H_2O
510	Cadmium 2-ethylhexanoate	$Cd(C_8H_{15}O_2)_2$	2420-98-6	398.818	powder					
511	Cadmium fluoride	CdF_2	7790-79-6	150.408	cub cry	1075	1750	6.33	4.36^{25}	s acid; i EtOH
512	Cadmium hydroxide	$Cd(OH)_2$	21041-95-2	146.426	wh trig or hex cry	130 dec		4.79	0.00015^{20}	s dil acid
513	Cadmium iodate	$Cd(IO_3)_2$	7790-81-0	462.216	wh powder			6.48	0.091^{25}	s HNO_3
514	Cadmium iodide	CdI_2	7790-80-9	366.220	col hex flakes	388	744	5.64	86.2^{25}	vs H_2O; s EtOH, eth, ace
515	Cadmium metasilicate	$CdSiO_3$	13477-19-5	188.495	grn monocl cry	1252		5.10		
516	Cadmium molybdate	$CdMoO_4$	13972-68-4	272.35	col tetr cry	≈900 dec		5.4		i H_2O; s acid
517	Cadmium niobate	$Cd_2Nb_2O_7$	12187-14-3	522.631	cub cry	≈1410		6.28		i H_2O
518	Cadmium nitrate	$Cd(NO_3)_2$	10325-94-7	236.420	wh cub cry; hyg	360		3.6	156^{25}	s EtOH
519	Cadmium nitrate tetrahydrate	$Cd(NO_3)_2 \cdot 4H_2O$	10022-68-1	308.482	col orth cry; hyg	59.5		2.45	156^{25}	s EtOH, ace
520	Cadmium oxalate	CdC_2O_4	814-88-0	200.430	wh solid			3.32	0.0060^{25}	
521	Cadmium oxalate trihydrate	$CdC_2O_4 \cdot 3H_2O$	20712-42-9	254.476	wh amorp powder	340 dec			0.0060^{25}	i EtOH; s dil acid
522	Cadmium oxide	CdO	1306-19-0	128.410	brn cub cry		1559 sp	8.15		i H_2O; s dil acid
523	Cadmium 2,4-pentanedioate	$Cd(CH_3COCHCOCH_3)_2$	14689-45-3	310.627	wh solid or red cry	235				
524	Cadmium perchlorate hexahydrate	$Cd(ClO_4)_2 \cdot 6H_2O$	10326-28-0	419.404	wh hex cry			2.37	191.5^{25}	
525	Cadmium phosphate	$Cd_3(PO_4)_2$	13477-17-3	527.176	powder	≈1500				i H_2O
526	Cadmium phosphide	Cd_3P_2	12014-28-7	399.181	grn tetr needles	700		5.96		s dil HCl
527	Cadmium selenate dihydrate	$CdSeO_4 \cdot 2H_2O$	10060-09-0	291.40	orth cry	100 dec		3.62	70.5^{25}	
528	Cadmium selenide	CdSe	1306-24-7	191.37	wh cub cry	1240		5.81		i H_2O
529	Cadmium selenite	$CdSeO_3$	13814-59-0	239.37	col prisms					
530	Cadmium stearate	$Cd(C_{18}H_{35}O_2)_2$	2223-93-0	679.350	wh cry pow	134		1.21		
531	Cadmium succinate	$CdC_4H_4O_4$	141-00-4	228.484	wh pow or needles				0.37^{40}	sl H_2O; i EtOH
532	Cadmium sulfate	$CdSO_4$	10124-36-4	208.474	col orth cry	1000		4.69	76.7^{25}	i EtOH
533	Cadmium sulfate monohydrate	$CdSO_4 \cdot H_2O$	7790-84-3	226.489	monocl cry	105		3.79	76.7^{25}	
534	Cadmium sulfate octahydrate	$CdSO_4 \cdot 8H_2O$	15244-35-6	352.596	col monocl cry	40 dec		3.08	76.7^{25}	
535	Cadmium sulfide	CdS	1306-23-6	144.476	yel-oran hex cry	≈1480		4.826		i H_2O; s acid
536	Cadmium sulfite	$CdSO_3$	13477-23-1	192.474	col prisms	≈400 dec			0.05^{20}	sl H_2O
537	Cadmium telluride	CdTe	1306-25-8	240.01	brn-blk cub cry	1042		6.2		i H_2O, dil acid
538	Cadmium tellurite	$CdTeO_3$	15851-44-2	288.01	col monocl cry	695	1050 dec			
539	Cadmium tetrafluoroborate	$Cd(BF_4)_2$	14486-19-2	286.020	col hyg liq			1.6		vs H_2O, EtOH
540	Cadmium titanate	$CdTiO_3$	12014-14-1	208.276	orth cry			6.5		
541	Cadmium tungstate	$CdWO_4$	7790-85-4	360.25	wh monocl cry			8.0		i H_2O, acid; s NH_4OH
542	Calcium	Ca	7440-70-2	40.078	silv-wh metal	842	1484	1.54		reac H_2O; i bz
543	Calcium acetate	$Ca(C_2H_3O_2)_2$	62-54-4	158.167	wh hyg cry	160 dec		1.50		s H_2O; sl EtOH
544	Calcium acetate monohydrate	$Ca(C_2H_3O_2)_2 \cdot H_2O$	5743-26-0	176.182	wh needles or powder	≈150 dec				s H_2O; sl EtOH
545	Calcium acetate dihydrate	$Ca(C_2H_3O_2)_2 \cdot 2H_2O$	14977-17-4	194.196	long col needles					s H_2O
546	Calcium aluminate	$CaAl_2O_4$	12042-68-1	158.039	wh monocl cry	1605		2.98		reac H_2O
547	Calcium aluminate (β form)	$Ca_3Al_2O_6$	12042-78-3	270.193	wh cub cry; refr	1535		3.04		i H_2O
548	Calcium arsenate	$Ca_3(AsO_4)_2$	7778-44-1	398.072	wh powder	dec		3.6	0.0036^{20}	s dil acid
549	Calcium arsenite (1:1)	$CaAsO_3$	52740-16-6	162.998	wh powder					sl H_2O; s acid
550	Calcium borate hexahydrate	$CaB_4O_7 \cdot 6H_2O$	13701-64-9	303.409	wh cry pow	1162 (anh)				
551	Calcium boride	CaB_6	12007-99-7	104.944	refrac solid	2235		2.49		
552	Calcium bromate	$Ca(BrO_3)_2$	10102-75-7	295.882	wh pow	180				
553	Calcium bromate monohydrate	$Ca(BrO_3)_2 \cdot H_2O$	10102-75-7	313.898	wh monocl cry	180 dec		3.33		vs H_2O
554	Calcium bromide	$CaBr_2$	7789-41-5	199.886	rhom cry; hyg	742	1815	3.38	156^{25}	s EtOH, ace
555	Calcium bromide dihydrate	$CaBr_2 \cdot 2H_2O$	22208-73-7	235.917	wh cry pow					vs H_2O
556	Calcium bromide hexahydrate	$CaBr_2 \cdot 6H_2O$	13477-28-6	307.977	wh hyg powder	38 dec		2.29	156^{25}	
557	Calcium carbide	CaC_2	75-20-7	64.099	gray-blk orth cry	2300		2.22		reac H_2O
558	Calcium carbonate (aragonite)	$CaCO_3$	471-34-1	100.087	wh orth cry or powder	450 trans calcite		2.930	0.00066^{20}	s dil acid
559	Calcium carbonate (calcite)	$CaCO_3$	471-34-1	100.087	wh hex cry or powder	800		2.710	0.00066^{20}	s dil acid
560	Calcium carbonate (vaterite)	$CaCO_3$	471-34-1	100.087	col hex cry			2.653	0.0011^{25}	s dil acid

No.	Name	Formula	CAS Reg. No.	Mol. wt.	Phys. form	mp/°C	bp/°C	ρ/g cm⁻³	Sol. in g/100 g H_2O	Qualitative solubility
561	Calcium chlorate	$Ca(ClO_3)_2$	10137-74-3	206.980	wh cry	340			197^{25}	
562	Calcium chlorate dihydrate	$Ca(ClO_3)_2·2H_2O$	10035-05-9	243.011	wh monocl cry; hyg	100 dec		2.711	197^{25}	s EtOH
563	Calcium chloride	$CaCl_2$	10043-52-4	110.984	wh cub cry or powder; hyg	775	1935	2.15	81.3^{25}	vs EtOH
564	Calcium chloride monohydrate	$CaCl_2·H_2O$	13477-29-7	128.999	wh hyg cry	260 dec		2.24	81.3^{25}	s EtOH
565	Calcium chloride dihydrate	$CaCl_2·2H_2O$	10035-04-8	147.015	hyg flakes or powder	175 dec		1.85	81.3^{25}	vs EtOH
566	Calcium chloride tetrahydrate	$CaCl_2·4H_2O$	25094-02-4	183.046	col tricl cry			1.83		
567	Calcium chloride hexahydrate	$CaCl_2·6H_2O$	7774-34-7	219.075	wh hex cry; hyg	30 dec		1.71	81.3^{25}	
568	Calcium chlorite	$Ca(ClO_2)_2$	14674-72-7	174.982	wh cub cry			2.71		reac H_2O
569	Calcium chromate	$CaCrO_4$	13765-19-0	156.072	yel cry	1000 dec				sl H_2O; i EtOH, ace
570	Calcium chromate dihydrate	$CaCrO_4·2H_2O$	10060-08-9	192.102	yel orth cry	200 dec		2.50	13.2^{20}	s dil acids
571	Calcium citrate tetrahydrate	$Ca_3(C_6H_5O_7)_2·4H_2O$	5785-44-4	570.494	wh needles or pow	100 dec			0.096^{23}	i eth
572	Calcium cyanamide	$CaCN_2$	156-62-7	80.102	col hex cry	≈1340	subl	2.29		reac H_2O
573	Calcium cyanide	$Ca(CN)_2$	592-01-8	92.112	wh rhomb cry; hyg					s H_2O, EtOH
574	Calcium dichromate trihydrate	$CaCr_2O_7·3H_2O$	14307-33-6	310.112	red-oran cry	100 dec		2.37		vs H_2O; reac EtOH; i eth, ctc
575	Calcium dihydrogen phosphate monohydrate	$Ca(H_2PO_4)_2·H_2O$	10031-30-8	252.068	col tricl plates	100 dec		2.220		sl H_2O; s dil acid
576	Calcium 2-ethylhexanoate	$Ca(C_8H_{15}O_2)_2$	136-51-6	326.485	powder					
577	Calcium ferrocyanide dodecahydrate	$Ca_2Fe(CN)_6·12H_2O$		508.289	yel tricl cry	dec		1.68	87^{25}	vs H_2O; i EtOH
578	Calcium fluoride	CaF_2	7789-75-5	78.075	wh cub cry or powder	1418	2500	3.18	0.0016^{25}	sl acid
579	Calcium fluorophosphate	$Ca_5(PO_4)_3F$	12015-73-5	504.302	col hex cry	1650		3.201		i H_2O
580	Calcium fluorophosphate dihydrate	$CaPO_3F·2H_2O$	37809-19-1	174.079	col monocl cry				0.42^{27}	i os
581	Calcium formate	$Ca(CHO_2)_2$	544-17-2	130.113	orth cry	300 dec		2.02	16.6^{20}	i EtOH
582	Calcium hexaborate pentahydrate	$2CaO·3B_2O_3·5H_2O$	12291-65-5	411.091	col monocl cry	375 exp		2.42	1^{25}	sl acid
583	Calcium hexafluoro-2,4-pentanedioate	$Ca(CF_3COCHCOCF_3)_2$	121012-90-6	454.180	powder	135				
584	Calcium hexafluorosilicate dihydrate	$CaSiF_6·2H_2O$	16925-39-6	218.185	col tetr cry			2.25	0.52^{20}	i ace; reac hot H_2O
585	Calcium hydride	CaH_2	7789-78-8	42.094	gray orth cry or powder	1000		1.7		reac H_2O, EtOH
586	Calcium hydrogen phosphate	$CaHPO_4$	7757-93-9	136.057	wh tricl cry	dec		2.92	0.02^{25}	i EtOH
587	Calcium hydrogen phosphate dihydrate	$CaHPO_4·2H_2O$	7789-77-7	172.088	monocl cry	≈100 dec		2.31	0.02^{25}	i EtOH; s dil acid
588	Calcium hydrogen sulfite	$CaH_2(SO_3)_2$	13780-03-5	202.220				1.06		s H_2O
589	Calcium hydrosulfide hexahydrate	$Ca(HS)_2·6H_2O$		214.315	col cry	dec				s H_2O, EtOH
590	Calcium hydroxide	$Ca(OH)_2$	1305-62-0	74.093	soft hex cry			≈2.2	0.160^{20}	s acid
591	Calcium hydroxide phosphate	$Ca_5(OH)(PO_4)_3$	12167-74-7	502.311	col hex cry	>900 dec		3.155		i H_2O
592	Calcium hypochlorite	$Ca(OCl)_2$	7778-54-3	142.983	powder	100		2.350		
593	Calcium hypophosphite	$Ca(H_2PO_2)_2$	7789-79-9	170.055	wh monocl cry	300 dec				s H_2O; i EtOH
594	Calcium iodate	$Ca(IO_3)_2$	7789-80-2	389.883	wh monocl cry			4.52	0.306^{25}	s HNO_3; i EtOH
595	Calcium iodide	CaI_2	10102-68-8	293.887	hyg hex cry	783	1100	3.96	215^{25}	s MeOH, EtOH, ace; i eth
596	Calcium iodide hexahydrate	$CaI_2·6H_2O$	71626-98-7	401.978	wh hex needles or powder	42 dec		2.55	215^{25}	vs EtOH
597	Calcium metaborate	$Ca(BO_2)_2$	13701-64-9	125.698	powder				0.13^{20}	
598	Calcium metasilicate	$CaSiO_3$	1344-95-2	116.162	wh monocl cry	1540		2.92		i H_2O
599	Calcium molybdate	$CaMoO_4$	7789-82-4	200.02	wh tetr cry	1520		4.35	0.0011^{20}	i EtOH; s conc acid
600	Calcium nitrate	$Ca(NO_3)_2$	10124-37-5	164.087	wh cub cry; hyg	561		2.5	144^{25}	s EtOH, MeOH, ace
601	Calcium nitrate tetrahydrate	$Ca(NO_3)_2·4H_2O$	13477-34-4	236.149	wh cry	≈40 dec		1.82	144^{25}	s EtOH, ace
602	Calcium nitride	Ca_3N_2	12013-82-0	148.247	red-brn cub cry	1195		2.67		s H_2O, acid; i EtOH
603	Calcium nitrite	$Ca(NO_2)_2$	13780-06-8	132.089	wh-yel hex cry; hyg	392		2.23	94.6^{25}	sl EtOH
604	Calcium nitrite monohydrate	$Ca(NO_2)_2·H_2O$	10031-34-2	150.104	col or yel cry	100 dec				vs H_2O; sl EtOH
605	Calcium oleate	$Ca(C_{18}H_{33}O_2)_2$	142-17-6	602.985	pale yel solid	140 dec			0.04^{25}	sl H_2O; s bz; i EtOH, ace, eth
606	Calcium oxalate	CaC_2O_4	563-72-4	128.097	wh cry powder			2.2	0.00061^{20}	
607	Calcium oxalate monohydrate	$CaC_2O_4·H_2O$	5794-28-5	146.112	cub cry	200 dec		2.2	0.00061^{20}	s dil acid
608	Calcium oxide	CaO	1305-78-8	56.077	gray-wh cub cry	2613		3.34		reac H_2O; s acid
609	Calcium oxide silicate	Ca_3OSiO_4	12168-85-3	228.317	refrac solid	2150				
610	Calcium palmitate	$Ca(C_{16}H_{31}O_2)_2$	542-42-7	550.910	wh-yel pow	155 dec				i H_2O, EtOH, eth, ace; sl bz
611	Calcium perborate heptahydrate	$Ca(BO_3)_2·7H_2O$		283.803	gray-wh pow					s H_2O, acid
612	Calcium 2,4-pentanedioate	$Ca(CH_3COCHCOCH_3)_2$	19372-44-2	238.294	cry	dec				
613	Calcium perchlorate	$Ca(ClO_4)_2$	13477-36-6	238.979	wh cry	270 dec		2.65	188^{25}	s EtOH, MeOH
614	Calcium perchlorate tetrahydrate	$Ca(ClO_4)_2·4H_2O$	15627-86-8	311.041	wh cry					vs H_2O
615	Calcium permanganate	$Ca(MnO_4)_2$	10118-76-0	277.949	purp hyg cry			2.4	331^{20}	reac EtOH
616	Calcium peroxide	CaO_2	1305-79-9	72.077	wh-yel tetr cry; hyg	≈200 dec		2.9		sl H_2O; s acid
617	Calcium phosphate	$Ca_3(PO_4)_2$	7758-87-4	310.177	wh amorp powder	1670		3.14	0.00012^{20}	i EtOH; s dil acid

No.	Name	Formula	CAS Reg. No.	Mol. wt.	Phys. form	mp/°C	bp/°C	ρ/g cm^{-3}	Sol. in g/100 g H_2O	Qualitative solubility
618	Calcium phosphide	Ca_3P_2	1305-99-3	182.182	red-brn hyg cry	≈1600 dec		2.51		reac H_2O; i EtOH, eth
619	Calcium phosphonate monohydrate	$CaHPO_3 \cdot H_2O$	25232-60-4	138.073	col monocl cry	150 dec				sl H_2O; i EtOH
620	Calcium plumbate	Ca_2PbO_4	12013-69-3	351.4	oran-brn orth cry	dec		5.71		i H_2O; s acid
621	Calcium propanoate	$Ca(C_3H_5O_2)_2$	4075-81-4	186.219	monocl cry, powder					s H_2O; sl MeOH, EtOH; i ace, bz
622	Calcium pyrophosphate	$Ca_2P_2O_7$	7790-76-3	254.099	wh powder	1353		3.09		i H_2O; s dil acid
623	Calcium selenate dihydrate	$CaSeO_4 \cdot 2H_2O$	7790-74-1	219.07	wh monocl cry			2.75	8.3^{18}	
624	Calcium selenide	CaSe	1305-84-6	119.04	wh-brn cub cry	1400 dec		3.8		reac H_2O
625	Calcium silicide (CaSi)	CaSi	12013-55-7	68.164	orth cry	1324		2.39		
626	Calcium silicide [$CaSi_2$]	$CaSi_2$	12013-56-8	96.249	gray hex cry	1040		2.50		i cold H_2O; reac hot H_2O; s acid
627	Calcium stannate trihydrate	$CaSnO_3 \cdot 3H_2O$	12013-46-6	260.832	wh cry pow	≈350 dec				i H_2O
628	Calcium stearate	$Ca(C_{18}H_{35}O_2)_2$	1592-23-0	607.017	granular powder	180				i H_2O, EtOH
629	Calcium succinate trihydrate	$CaC_4H_4O_4 \cdot 3H_2O$	140-99-8	210.196	needles					sl H_2O; s dil acid; i EtOH
630	Calcium sulfate	$CaSO_4$	7778-18-9	136.141	orth cry	1460		2.96	0.205^{25}	
631	Calcium sulfate hemihydrate	$CaSO_4 \cdot 0.5H_2O$	10034-76-1	145.149	wh powder				0.205^{25}	
632	Calcium sulfate dihydrate	$CaSO_4 \cdot 2H_2O$	10101-41-4	172.171	monocl cry or powder	150 dec		2.32	0.205^{20}	i os
633	Calcium sulfide	CaS	20548-54-3	72.143	wh-yel cub cry; hyg	2524		2.59		sl H_2O; i EtOH
634	Calcium sulfite dihydrate	$CaSO_3 \cdot 2H_2O$	10257-55-3	156.172	wh powder				0.0070^{25}	sl EtOH; s acid
635	Calcium tartrate tetrahydrate	$CaC_4H_4O_6 \cdot 4H_2O$	3164-34-9	260.210	wh pow				0.04^{10}	s dil acid; sl EtOH
636	Calcium telluride	CaTe	12013-57-9	167.68	wh cub cry	1600 dec		4.87		
637	Calcium tetrahydroaluminate	$Ca(AlH_4)_2$	16941-10-9	102.105	gray powder; flam					reac H_2O; s thf; i eth, bz
638	Calcium thiocyanate tetrahydrate	$Ca(SCN)_2 \cdot 4H_2O$	2092-16-2	228.304	hyg cry	160 dec				vs H_2O; s EtOH, ace
639	Calcium thiosulfate hexahydrate	$CaS_2O_3 \cdot 6H_2O$	10124-41-1	260.298	tricl cry	45 dec		1.87		s H_2O; i EtOH
640	Calcium titanate	$CaTiO_3$	12049-50-2	135.943	cub cry	1980		3.98		
641	Calcium tungstate	$CaWO_4$	7790-75-2	287.92	wh tetr cry	1620		6.06	0.2^{18}	s hot acid
642	Calcium zirconate	$CaZrO_3$	12013-47-7	179.300	powder	2550				
643	Californium	Cf	7440-71-3	251	hex or cub metal	900		15.1		
644	Carbon (diamond)	C	7782-40-3	12.011	col cub cry	4440 (12.4 GPa)[a]		3.513		i H_2O
645	Carbon (graphite)	C	7782-42-5	12.011	soft blk hex cry	4489 tp (10.3 MPa)	3825 sp	2.2		i H_2O
646	Carbon black	C	1333-86-4	12.011	fine blk pow					i H_2O
647	Carbon [fullerene-C_{60}]	C_{60}	99685-96-8	720.642	yel needles or plates	>280				s os
648	Carbon [fullerene-C_{70}]	C_{70}	115383-22-7	840.749	red-brn solid	>280				s bz, tol
649	Fullerene fluoride	$C_{60}F_{60}$	134929-59-2	1860.550	col plates	287				vs ace; s thf; i chl
650	Carbon monoxide	CO	630-08-0	28.010	col gas	-205.1	-191.51	$0.8495^{-205.1}$ liq		sl H_2O; s chl, EtOH
651	Carbon dioxide	CO_2	124-38-9	44.010	col gas	-56.561 tp	-78.464 sp	1.56^{-79} solid		s H_2O
652	Carbon suboxide	C_3O_2	504-64-3	68.031	col gas	-112.5	6.8	1.114^{0}		reac H_2O
653	Carbon disulfide	CS_2	75-15-0	76.141	col or yel liq	-111.7	46.2	1.2632^{20}		i H_2O; vs EtOH, bz, os
654	Carbon subsulfide	C_3S_2	627-34-9	100.162	red liq	-1	90 dec	1.27		reac H_2O
655	Carbon diselenide	CSe_2	506-80-9	169.93	yel liq	-43.6	125.5	2.6823^{20}		i H_2O; vs ctc, tol
656	Carbon oxysulfide	OCS	463-58-1	60.075	col gas	-138.8	-50.2	0.002456^{25} gas		s H_2O, EtOH
657	Carbon oxyselenide	OCSe	1603-84-5	106.97	col gas; unstab	-124.4	-21.7	0.004372^{25} gas		reac H_2O
658	Carbon sulfide selenide	CSSe	5951-19-9	123.04	yel liq	-85	84.5	1.99		i H_2O
659	Carbon sulfide telluride	SCTe	10340-06-4	171.68	red-yel liq; unstab	-54	20 dec			reac H_2O
660	Carbonyl bromide	$COBr_2$	593-95-3	187.818	col liq		64	2.52^{15}		reac H_2O
661	Carbonyl chloride	$COCl_2$	75-44-5	98.916	col gas	-127.77	7.5	0.004043^{25} gas		sl H_2O; s bz, tol
662	Carbonyl fluoride	COF_2	353-50-4	66.007	col gas	-114	-84.5	0.002698^{25} gas		reac H_2O
663	Cyanogen	C_2N_2	460-19-5	52.034	col gas	-27.83	-21.1	0.787^{-21} liq		sl H_2O, eth; s EtOH
664	Cyanogen azide	N_3CN	764-05-6	68.038	col oily liq	exp				
665	Cyanogen bromide	BrCN	506-68-3	105.922	wh hyg needles	52	61.5	2.015^{20}		s H_2O, EtOH, eth
666	Cyanogen chloride	ClCN	506-77-4	61.471	col vol liq or gas	-6.55	13	1.186^{20}		s H_2O, EtOH, eth
667	Cyanogen fluoride	FCN	1495-50-7	45.016	col gas	-82	-46	0.001840^{25} gas		
668	Cyanogen iodide	ICN	506-78-5	152.922	col needles	146.7	subl	2.84^{18}		s H_2O, EtOH, eth
669	Cerium	Ce	7440-45-1	140.116	silv metal; cub or hex	799	3443	6.770		s dil acid

No.	Name	Formula	CAS Reg. No.	Mol. wt.	Phys. form	mp/°C	bp/°C	ρ/g cm⁻³	Sol. in g/100 g H_2O	Qualitative solubility
670	Cerium boride	CeB_6	12008-02-5	204.982	blue refrac solid; hex	2550		4.87		i H_2O, HCl
671	Cerium carbide	CeC_2	12012-32-7	164.137	red hex cry	2250		5.47		reac H_2O
672	Cerium carbide	Ce_2C_3	12115-63-8	316.264	yel-brn cub cry	1505		6.9		
673	Cerium nitride	CeN	25764-08-3	154.123	refrac cub cry	2557		7.89		
674	Cerium silicide	$CeSi_2$	12014-85-6	196.287	tetr cry	1420		5.31		i H_2O
675	Cerium(II) hydride	CeH_2	13569-50-1	142.132	cub cry			5.45		reac H_2O
676	Cerium(II) iodide	CeI_2	19139-47-0	393.925	bronze cry	808				
677	Cerium(II) sulfide	CeS	12014-82-3	172.181	yel cub cry	2445		5.9		
678	Cerium(III) acetate sesqihydrate	$Ce(C_2H_3O_2)_3 \cdot 1.5H_2O$	17829-82-2	344.271	col cry	115 dec			26^{15}	s H_2O
679	Cerium(III) ammonium nitrate tetrahydrate	$(NH_4)_2Ce(NO_3)_5 \cdot 4H_2O$	13083-04-0	558.279	col monocl cry	74				vs H_2O
680	Cerium(III) ammonium sulfate tetrahydrate	$NH_4Ce(SO_4)_2 \cdot 4H_2O$	21995-38-0	422.341	monocl cry					s H_2O
681	Cerium(III) bromide	$CeBr_3$	14457-87-5	379.828	wh hex cry; hyg	732	1457			s H_2O
682	Cerium(III) bromide heptahydrate	$CeBr_3 \cdot 7H_2O$	7789-56-2	505.935	col hyg needles	732				s H_2O, EtOH
683	Cerium(III) carbonate	$Ce_2(CO_3)_3$	537-01-9	460.259	wh pow	500 dec				i H_2O; s acid
684	Cerium(III) carbonate pentahydrate	$Ce_2(CO_3)_3 \cdot 5H_2O$	72520-94-6	550.335	wh powder					i H_2O; s dil acid
685	Cerium(III) chloride	$CeCl_3$	7790-86-5	246.475	wh hex cry	807		3.97		s H_2O, EtOH
686	Cerium(III) chloride heptahydrate	$CeCl_3 \cdot 7H_2O$	18618-55-8	372.582	yel orth cry; hyg	90 dec				vs H_2O, EtOH
687	Cerium(III) fluoride	CeF_3	7758-88-5	197.111	wh hex cry; hyg	1430	2180	6.157		i H_2O
688	Cerium(III) hydride	CeH_3	13864-02-3	143.140	blk pow or blue-blk cry	dec (flam)				reac H_2O
689	Cerium(III) hydroxide	$Ce(OH)_3$	15785-09-8	191.138	wh solid					i H_2O; s acid
690	Cerium(III) iodide	CeI_3	7790-87-6	520.829	yel orth cry; hyg	760				s H_2O
691	Cerium(III) iodide nonahydrate	$CeI_3 \cdot 9H_2O$	7790-87-6	682.967	wh-red cry					vs H_2O; s EtOH
692	Cerium(III) nitrate hexahydrate	$Ce(NO_3)_3 \cdot 6H_2O$	10108-73-3	434.222	col-red cry	150 dec			176^{25}	s ace
693	Cerium(III) oxalate nonahydrate	$Ce(C_2O_4)_3 \cdot 9H_2O$	13266-83-6	706.426	wh pow	dec				i H_2O, EtOH; s acid
694	Cerium(III) oxide	Ce_2O_3	1345-13-7	328.230	yel-grn cub cry	2250	3730	6.2		i H_2O; s acid
695	Cerium(III) 2,4-pentanedioate trihydrate	$Ce(CH_3COCHCOCH_3)_3 \cdot 3H_2O$	15653-01-7	491.486	yel hyg cry	≈150				vs EtOH
696	Cerium(III) perchlorate hexahydrate	$Ce(ClO_4)_3 \cdot 6H_2O$	36907-38-7	546.559	hyg col cry	200 dec				s H_2O, EtOH
697	Cerium(III) selenate	$Ce_2(SeO_4)_3$		709.11	rhom cry			4.46		s H_2O
698	Cerium(III) sulfate	$Ce_2(SO_4)_3$	13454-94-9	568.420	col hyg cry	920 dec				s H_2O
699	Cerium(III) sulfate octahydrate	$Ce_2(SO_4)_3 \cdot 8H_2O$	13454-94-9	712.542	wh orth cry	≈250 dec		2.87		s H_2O
700	Cerium(III) sulfide	Ce_2S_3	12014-93-6	376.427	red cub cry	2450		5.02		i H_2O
701	Cerium(III) tungstate	$Ce_2(WO_4)_3$	13454-74-5	1023.75	yel tetr cry	1089		6.77		i H_2O
702	Cerium(IV) ammonium nitrate	$(NH_4)_2Ce(NO_3)_6$	16774-21-3	548.223	red-oran cry					vs H_2O
703	Cerium(IV) ammonium sulfate dihydrate	$(NH_4)_4Ce(SO_4)_4 \cdot 2H_2O$	10378-47-9	632.551	cry pow	450 dec				
704	Cerium(IV) fluoride	CeF_4	10060-10-3	216.110	wh hyg powder	≈600 dec		4.77		i H_2O
705	Cerium(IV) hydroxide	$Ce(OH)_4$	12014-56-1	208.146	yel-wh pow					i H_2O; s conc acid
706	Cerium(IV) oxide	CeO_2	1306-38-3	172.115	wh-yel powder; cub	2480		7.216		i H_2O, dil acid
707	Cerium(IV) sulfate tetrahydrate	$Ce(SO_4)_2 \cdot 4H_2O$	10294-42-5	404.303	yel-oran orth cry	180 dec		3.91	9.66^{20}	
708	Cesium	Cs	7440-46-2	132.905	silv-wh metal	28.5	671	1.873		reac H_2O
709	Cesium acetate	$CsC_2H_3O_2$	3396-11-0	191.949	hyg lumps	194			10^{11}	
710	Cesium aluminum sulfate dodecahydrate	$CsAl(SO_4)_2 \cdot 12H_2O$	7784-17-0	568.196	col cub cry	117 dec		1.97		s H_2O; i EtOH
711	Cesium amide	$CsNH_2$	22205-57-8	148.928	wh tetr cry			3.70		
712	Cesium azide	CsN_3	22750-57-8	174.925	hyg tetr cry; exp	326		≈3.5	22^{40}	
713	Cesium bromate	$CsBrO_3$	13454-75-6	260.807	col hex cry			4.11	3.83^{25}	
714	Cesium bromide	CsBr	7787-69-1	212.809	wh cub cry; hyg	636	≈1300	4.43	123^{25}	s EtOH; i ace
715	Cesium carbonate	Cs_2CO_3	534-17-8	325.820	wh monocl cry; hyg	793		4.24	261^{15}	s EtOH, eth
716	Cesium chlorate	$CsClO_3$	13763-67-2	216.356	col hex cry	342		3.57	7.78^{25}	sl H_2O
717	Cesium chloride	CsCl	7647-17-8	168.358	wh cub cry; hyg	646	1297	3.988	191^{25}	s EtOH
718	Cesium chromate(IV)	Cs_4CrO_4	56320-90-2	647.616	yel hex cry	982		4.24		vs H_2O
719	Cesium cyanide	CsCN	21159-32-0	158.923	wh cub cry; hyg	350		3.34		vs H_2O
720	Cesium dibromoiodate	$CsIBr_2$	18278-82-5	419.617	dark oran cry	dec				s H_2O
721	Cesium fluoride	CsF	13400-13-0	151.903	wh cub cry; hyg	703		4.64	573^{25}	s MeOH; i diox, py
722	Cesium fluoroborate	$CsBF_4$	18909-69-8	219.710	wh orth cry	555 dec		3.2	1.6^{17}	sl H_2O
723	Cesium formate	$CsCHO_2$	3495-36-1	177.923	wh cry			1.017		vs H_2O
724	Cesium hexafluorogermanate	Cs_2GeF_6		452.44	wh cry	≈675		4.10		sl cold H_2O; s hot H_2O
725	Cesium hydride	CsH	58724-12-2	133.913	wh cub cry; flam	528		3.42		reac H_2O
726	Cesium hydrogen carbonate	$CsHCO_3$	15519-28-5	193.922	rhom cry	175 dec			209^{15}	s EtOH
727	Cesium hydrogen fluoride	$CsHF_2$	12280-52-3	171.910	tetr cry	170		3.86		
728	Cesium hydrogen sulfate	$CsHSO_4$	7789-16-4	229.976	col rhom prisms	dec		3.352		s H_2O
729	Cesium hydroxide	CsOH	21351-79-1	149.912	wh-yel hyg cry	342.3		3.68	300^{30}	s EtOH
730	Cesium iodate	$CsIO_3$	13454-81-4	307.807	wh monocl cry			4.85	2.6^{25}	
731	Cesium iodide	CsI	7789-17-5	259.809	col cub cry; hyg	632	≈1280	4.51	84.8^{25}	s EtOH, MeOH, ace
732	Cesium metaborate	$CsBO_2$	92141-86-1	175.715	cub cry	732		≈3.7		
733	Cesium molybdate	Cs_2MoO_4	13597-64-3	425.75	wh cry	956.3			67^{18}	s H_2O
734	Cesium nitrate	$CsNO_3$	7789-18-6	194.910	wh hex or cub cry	409		3.66	27.9^{25}	s ace; sl EtOH
735	Cesium nitrite	$CsNO_2$	13454-83-6	178.911	yel cry	406				s H_2O
736	Cesium oxide	Cs_2O	20281-00-9	281.810	yel-oran hex cry	495		4.65		vs H_2O
737	Cesium superoxide	CsO_2	12018-61-0	164.904	yel tetr cry	432		3.77		reac H_2O

No.	Name	Formula	CAS Reg. No.	Mol. wt.	Phys. form	mp/°C	bp/°C	ρ/g cm-3	Sol. in g/100 g H_2O	Qualitative solubility
738	Cesium trioxide	Cs_2O_3	12134-22-4	313.809	brn cry	≈400		4.25		reac H_2O
739	Cesium perchlorate	$CsClO_4$	13454-84-7	232.356	wh orth cry; hyg	≈600 dec		3.327	2.00^{25}	
740	Cesium periodate	$CsIO_4$	13478-04-1	323.807	wh rhom prisms			4.26	2.2^{15}	
741	Cesium sulfate	Cs_2SO_4	10294-54-9	361.874	wh orth cry or hex prisms; hyg	1005		4.24	182^{25}	i EtOH, ace, py
742	Cesium sulfide	Cs_2S	12214-16-3	297.876	yel orth hyg cry	520				vs H_2O
743	Cesium trifluoroacetate	$Cs(C_2F_3O_2)$	21907-50-6	245.920	hyg solid	115				vs H_2O
744	Chlorine	Cl_2	7782-50-5	70.90	grn-yel gas	-101.5	-34.04	$1.565^{-34.0}$ liq		sl H_2O
745	Hypochlorous acid	HOCl	7790-92-3	52.460	grn-yel; stable only in aq soln					s H_2O
746	Chloric acid	$HClO_3$	7790-93-4	84.459	exists only in aq soln					vs H_2O
747	Perchloric acid	$HClO_4$	7601-90-3	100.459	col hyg liq	-112	≈90 dec	1.77		s H_2O
748	Chlorine monoxide	Cl_2O	7791-21-1	86.905	yel-brn gas	-120.6	2.2	0.003552^{25} gas		vs H_2O
749	Chlorine dioxide	ClO_2	10049-04-4	67.452	oran-grn gas	-59	11	0.002757^{25} gas		sl H_2O
750	Dichlorine trioxide	Cl_2O_3	17496-59-2	118.904	dark brn solid	<25 exp				
751	Dichlorine hexoxide	Cl_2O_6	12442-63-6	166.902	red liq	3.5	≈200			reac H_2O
752	Dichlorine heptoxide	Cl_2O_7	10294-48-1	182.902	col oily liq; exp	-91.5	82	1.9		reac H_2O
753	Chlorine fluoride	ClF	7790-89-8	54.451	col gas	-155.6	-101.1	0.002226^{25} gas		reac H_2O
754	Chlorine trifluoride	ClF_3	7790-91-2	92.448	gas	-76.34	11.75	0.003779^{25} gas		reac H_2O
755	Chlorine pentafluoride	ClF_5	13637-63-3	130.445	col gas	-103	-13.1	0.005332^{25} gas		
756	Chlorosyl trifluoride	$ClOF_3$	30708-80-6	108.447	col liq	-42	27			reac H_2O
757	Chloryl fluoride	ClO_2F	13637-83-7	86.450	col gas	-115	-6	0.003534^{25} gas		reac H_2O
758	Chloryl trifluoride	ClO_2F_3	38680-84-1	124.447	col gas	-81.2	-21.6	0.005087^{25} gas		reac H_2O
759	Perchloryl fluoride	ClO_3F	7616-94-6	102.449	col gas	-147	-46.75	0.004188^{25} gas		
760	Chlorine perchlorate	$ClOClO_3$	27218-16-2	134.904	unstab yel liq	-117	≈45 dec	1.81^{0}		
761	Chromium	Cr	7440-47-3	51.996	blue-wh metal; cub	1907	2671	7.15		reac dil acid
762	Chromic acid	H_2CrO_4	7738-94-5	118.010	aq soln only					s H_2O
763	Chromium antimonide	CrSb	12053-12-2	173.756	hex cry	1110		7.11		
764	Chromium arsenide	Cr_2As	12254-85-2	178.914	tetr cry			7.04		
765	Chromium boride (CrB)	CrB	12006-79-0	62.807	refrac orth cry	2100		6.1		
766	Chromium boride [CrB_2]	CrB_2	12007-16-8	73.618	refrac solid; hex	2200		5.22		
767	Chromium boride (Cr_2B)	Cr_2B	12006-80-3	114.803	refrac solid	1875				
768	Chromium boride [Cr_5B_3]	Cr_5B_3	12007-38-4	292.414	tetr cry	1900		6.10		
769	Chromium carbide	Cr_3C_2	12012-35-0	180.009	gray orth cry	1895		6.68		
770	Chromium carbonyl	$Cr(CO)_6$	13007-92-6	220.056	col orth cry	130 dec	subl	1.77		i H_2O, EtOH; s eth, chl
771	Chromium nitride (CrN)	CrN	24094-93-7	66.003	gray cub cry	1080 dec		5.9		
772	Chromium nitride (Cr_2N)	Cr_2N	12053-27-9	117.999	hex cry	1650		6.8		
773	Chromium phosphide	CrP	26342-61-0	82.970	orth cry			5.25		
774	Chromium selenide	CrSe	12053-13-3	130.96	hex cry	≈1500		6.1		
775	Chromium silicide [$CrSi_2$]	$CrSi_2$	12018-09-6	108.167	gray hex cry	1490		4.91		
776	Chromium silicide (Cr_3Si)	Cr_3Si	12018-36-9	184.074	cub cry	1770		6.4		
777	Chromium(II) acetate monohydrate	$Cr(C_2H_3O_2)_2 \cdot H_2O$	628-52-4	188.100	red monocl cry			1.79		sl H_2O
778	Chromium(II) bromide	$CrBr_2$	10049-25-9	211.804	wh monocl cry; aq soln blue	842		4.236		s H_2O, EtOH
779	Chromium(II) chloride	$CrCl_2$	10049-05-5	122.902	wh hyg needles; aq soln blue	824	1120	2.88		s H_2O
780	Chromium(II) chloride tetrahydrate	$Cr(H_2O)_4Cl_2 \cdot 4H_2O$	13931-94-7	267.024	blue hyg cry	51 dec				s H_2O
781	Chromium(II) fluoride	CrF_2	10049-10-2	89.993	blue-grn monocl cry	894		3.79		sl H_2O; i EtOH
782	Chromium(II) formate monohydrate	$Cr(CHOO)_2 \cdot H_2O$	4493-37-2	160.046	red needles					s H_2O
783	Chromium(II) iodide	CrI_2	13478-28-9	305.805	red-brn cry; hyg	867		5.1		s H_2O
784	Chromium(II) oxalate monohydrate	$CrC_2O_4 \cdot H_2O$	814-90-4	158.030	yel-grn powder			2.468		sl H_2O
785	Chromium(II) sulfate pentahydrate	$CrSO_4 \cdot 5H_2O$	13825-86-0	238.135	blue cry				21^{0}	s dil acid; sl EtOH; i ace
786	Chromium(II,III) oxide	Cr_3O_4	12018-34-7	219.986	cub cry			6.1		
787	Chromium(III) acetate	$Cr(C_2H_3O_2)_3$	1066-30-4	229.127	blue-grn pwd					sl H_2O
788	Chromium(III) acetate monohydrate	$Cr(C_2H_3O_2)_3 \cdot H_2O$	25013-82-5	247.143	gray-grn pow					sl H_2O; i EtOH
789	Chromium(III) acetate hexahydrate	$Cr(C_2H_3O_2)_3 \cdot 6H_2O$	1066-30-4	337.220	blue needles					s H_2O
790	Chromium(III) acetate hydroxide	$Cr(C_2H_3O_2)_2(OH)$	39430-51-8	187.092	viol cry pow					vs H_2O
791	Chromium(III) bromide	$CrBr_3$	10031-25-1	291.708	dark grn hex cry	812		4.68		s hot H_2O, bz
792	Chromium(III) bromide hexahydrate (α)	$CrBr_3(H_2O)_4 \cdot 2H_2O$	18721-05-6	399.799	grn hyg cry					s H_2O, EtOH
793	Chromium(III) bromide hexahydrate (β)	$Cr(H_2O)_6Br_3$	10031-25-1	399.799	viol hyg cry					s H_2O; i EtOH, eth
794	Chromium(III) chloride	$CrCl_3$	10025-73-7	158.355	red-viol cry	827	1300 dec	2.76		sl H_2O
795	Chromium(III) chloride hexahydrate	$[CrCl_2(H_2O)_4]Cl \cdot 2H_2O$	10060-12-5	266.446	grn monocl cry; hyg					s H_2O, EtOH; sl ace; i eth
796	Chromium(III) fluoride	CrF_3	7788-97-8	108.991	grn needles	1425		3.8		i H_2O, EtOH
797	Chromium(III) fluoride trihydrate	$CrF_3 \cdot 3H_2O$	16671-27-5	163.037	grn hex cry			2.2		sl H_2O
798	Chromium(III) hydroxide sulfate	$Cr(OH)SO_4$	12336-95-7	165.066	grn cry					

No.	Name	Formula	CAS Reg. No.	Mol. wt.	Phys. form	mp/°C	bp/°C	ρ/g cm^{-3}	Sol. in g/100 g H_2O	Qualitative solubility
799	Chromium(III) fluoride nonahydrate	$Cr(H_2O)_6F_3 \cdot 3H_2O$	102430-09-1	271.129	rhom viol cry					sl H_2O
800	Chromium(III) hydroxide trihydrate	$Cr(OH)_3 \cdot 3H_2O$	1308-14-1	157.063	blue-grn powder					i H_2O; s acid
801	Chromium(III) iodide	CrI_3	13569-75-0	432.709	dark grn hex cry	857		5.32		sl H_2O
802	Chromium(III) nitrate	$Cr(NO_3)_3$	13548-38-4	238.011	grn hyg powder	>60 dec				vs H_2O
803	Chromium(III) nitrate nonahydrate	$Cr(NO_3)_3 \cdot 9H_2O$	7789-02-8	400.148	grn-blk monocl cry	66.3	>100 dec	1.80		vs H_2O
804	Chromium(III) oxide	Cr_2O_3	1308-38-9	151.990	grn hex cry	2432	≈3000	5.22		i H_2O, EtOH; sl acid, alk
805	Chromium(III) 2,4-pentanedioate	$Cr(CH_3COCHCOCH_3)_3$	21679-31-2	349.320	red monocl cry	208.7	345	1.34		i H_2O; s bz
806	Chromium(III) perchlorate	$Cr(ClO_4)_3$	27535-70-2	350.348	grn-blue cry				58^{25}	vs H_2O
807	Chromium(III) phosphate	$CrPO_4$	7789-04-0	146.967	blue orth cry	>1800		4.6		i H_2O, acid, aqua regia
808	Chromium(III) phosphate hemiheptahydrate	$CrPO_4 \cdot 3.5H_2O$	84359-31-9	210.021	blue-grn powder			2.15		i H_2O; s acid
809	Chromium(III) phosphate hexahydrate	$CrPO_4 \cdot 6H_2O$	84359-31-9	255.059	viol cry	>500 dec		2.121		i H_2O; s acid, alk
810	Chromium(III) potassium oxalate trihydrate	$K_3Cr(C_2O_4)_3 \cdot 3H_2O$	15275-09-9	487.394	blue-grn monocl cry					s H_2O
811	Chromium(III) potassium sulfate dodecahydrate	$CrK(SO_4)_2 \cdot 12H_2O$	7788-99-0	499.403	viol-blk cub cry	89 dec		1.83		s H_2O; i EtOH
812	Chromium(III) sulfate	$Cr_2(SO_4)_3$	10101-53-8	392.180	red pow	>700 dec		3.1	64^{25}	s H_2O; vs acid
813	Chromium(III) sulfate octadecahydrate	$Cr_2(SO_4)_3 \cdot 18H_2O$	10101-53-8	716.455	viol cry	115 dec		1.7		reac H_2O
814	Chromium(III) sulfide	Cr_2S_3	12018-22-3	200.187	brn-blk hex cry			3.8		
815	Chromium(III) telluride	Cr_2Te_3	12053-39-3	486.79	hex cry	≈1300		7.0		
816	Chromium(IV) chloride	$CrCl_4$	15597-88-3	193.808	gas, stable at HT		>600 dec	0.007922^{25} gas		
817	Chromium(IV) fluoride	CrF_4	10049-11-3	127.990	grn cry	277		2.89		reac H_2O
818	Chromium(IV) oxide	CrO_2	12018-01-8	83.995	brn-blk tetr powder	≈400 dec		4.89		i H_2O; s acid
819	Chromium(V) fluoride	CrF_5	14884-42-5	146.988	red orth cry	34	117			reac H_2O
820	Chromium(V) oxide	Cr_2O_5	12218-36-9	183.989	blk needles	200 dec				
821	Chromium(VI) fluoride	CrF_6	13843-28-2	165.986	yel solid; stable at low temp	-100 dec				
822	Chromium(VI) oxide	CrO_3	1333-82-0	99.994	red orth cry	197	≈250 dec	2.7	169^{25}	
823	Chromium(VI) tetrafluoride oxide	$CrOF_4$	23276-90-6	143.989	dark red solid	55				reac H_2O, ace, dmso
824	Chromium(VI) dichloride dioxide	CrO_2Cl_2	14977-61-8	154.901	red liq	-96.5	117	1.91		reac H_2O; s ctc, chl, bz
825	Chromium(VI) difluoride dioxide	CrO_2F_2	7788-96-7	121.992	red-viol cry	30	subl			reac H_2O
826	Cobalt	Co	7440-48-4	58.933	gray metal; hex or cub	1495	2927	8.86		s dil acid
827	Cobaltocene	$Co(C_5H_5)_2$	1277-43-6	189.119	blk-purp cry	173				
828	Cobalt antimonide	CoSb	12052-42-5	180.693	hex cry	1202		8.8		
829	Cobalt arsenic sulfide	CoAsS	12254-82-9	165.920	silv-wh solid			≈6.1		
830	Cobalt arsenide (CoAs)	CoAs	27016-73-5	133.855	orth cry	1180		8.22		
831	Cobalt arsenide [$CoAs_2$]	$CoAs_2$	12044-42-7	208.776	monocl cry			7.2		
832	Cobalt arsenide [$CoAs_3$]	$CoAs_3$	12256-04-1	283.698	cub cry	942		6.84		
833	Cobalt boride (CoB)	CoB	12006-77-8	69.744	refrac solid	1460		7.25		reac H_2O, HNO_3
834	Cobalt boride (Co_2B)	Co_2B	12045-01-1	128.677	refrac solid	1280		8.1		
835	Cobalt carbonyl	$Co_2(CO)_8$	10210-68-1	341.947	oran cry	51 dec		1.78		i H_2O; s EtOH, eth, CS_2
836	Cobalt disulfide	CoS_2	12013-10-4	123.063	cub cry			4.3		
837	Cobalt dodecacarbonyl	$Co_4(CO)_{12}$	17786-31-1	571.854	blk cry	60 dec		2.09		
838	Cobalt phosphide	Co_2P	12134-02-0	148.840	gray needles	1386		6.4		i H_2O; s HNO_3
839	Cobalt silicide	$CoSi_2$	12017-12-8	115.104	gray cub cry	1326		4.9		s hot HCl
840	Cobalt(II) acetate	$Co(C_2H_3O_2)_2$	71-48-7	177.022	pink cry					vs H_2O; s EtOH
841	Cobalt(II) acetate tetrahydrate	$Co(C_2H_3O_2)_2 \cdot 4H_2O$	6147-53-1	249.082	red monocl cry			1.705		s H_2O, EtOH, dil acid
842	Cobalt(II) aluminate	$CoAl_2O_4$	13820-62-7	176.894	blue cub cry			4.37		i H_2O
843	Cobalt(II) arsenate octahydrate	$Co_3(AsO_4)_2 \cdot 8H_2O$	24719-19-5	598.760	red monocl needles	400 dec	1000 dec	3.0		i H_2O; s dil acid
844	Cobalt(II) bromate hexahydrate	$Co(BrO_3)_2 \cdot 6H_2O$	13476-01-2	422.829	viol cry			≈2.5		vs H_2O
845	Cobalt(II) bromide	$CoBr_2$	7789-43-7	218.741	grn hex cry; hyg	678		4.91	113.2^{20}	s MeOH, EtOH, ace
846	Cobalt(II) bromide hexahydrate	$CoBr_2 \cdot 6H_2O$	13762-12-4	326.832	red hyg cry	47 dec	100 dec	2.46	113.2	
847	Cobalt(II) carbonate	$CoCO_3$	513-79-1	118.942	pink rhomb cry	280 dec		4.2	0.00014^{20}	i EtOH
848	Cobalt(II) basic carbonate	$2CoCO_3 \cdot 3Co(OH)_2 \cdot H_2O$	7542-09-8	534.743	red-viol cry	dec				i H_2O; s acid
849	Cobalt(II) chlorate hexahydrate	$Co(ClO_3)_2 \cdot 6H_2O$		333.927	dark red hyg cry	61 dec				s H_2O
850	Cobalt(II) chloride	$CoCl_2$	7646-79-9	129.839	blue hyg leaflets	737	1049	3.36	56.2^{25}	s EtOH, eth, ace, py
851	Cobalt(II) chloride dihydrate	$CoCl_2 \cdot 2H_2O$	16544-92-6	165.870	viol-blue cry			2.477	56.2^{25}	
852	Cobalt(II) chloride hexahydrate	$CoCl_2 \cdot 6H_2O$	7791-13-1	237.930	pink-red monocl cry	87 dec		1.924	56.2^{25}	s EtOH, ace, eth
853	Cobalt(II) chromate	$CoCrO_4$	24613-38-5	174.927	yel-brn orth cry			≈4.0		i H_2O; s acid
854	Cobalt(II) chromite	$CoCr_2O_4$	13455-25-9	226.923	blue-grn cub cry			5.14		i H_2O, conc acid
855	Cobalt(II) citrate dihydrate	$Co_3(C_6H_5O_7)_2 \cdot 2H_2O$	18727-04-3	591.028	rose red cry	150 dec			0.8^{15}	
856	Cobalt(II) cyanide	$Co(CN)_2$	542-84-7	110.967	blue hyg cry			1.872		i H_2O
857	Cobalt(II) cyanide dihydrate	$Co(CN)_2 \cdot 2H_2O$	20427-11-6	146.998	pink-brn needles					i H_2O, acid
858	Cobalt(II) diiron tetroxide	$CoFe_2O_4$	12052-28-7	234.621	blk solid					s hot HCl

Inorganic

No.	Name	Formula	CAS Reg. No.	Mol. wt.	Phys. form	mp/°C	bp/°C	ρ/g cm^{-3}	Sol. in g/100 g H_2O	Qualitative solubility
859	Cobalt(II) ferricyanide	$Co_3[Fe(CN)_6]_2$	14049-81-1	600.698	red needles					i H_2O, HCl; s NH_4OH
860	Cobalt(II) fluoride	CoF_2	10026-17-2	96.930	red tetr cry	1127	≈1400	4.46	1.4^{25}	s acid
861	Cobalt(II) fluoride tetrahydrate	$CoF_2 \cdot 4H_2O$	13817-37-3	168.992	red orth cry	dec		2.22	1.4^{25}	
862	Cobalt(II) formate dihydrate	$Co(CHO_2)_2 \cdot 2H_2O$	6424-20-0	184.998	red cry powder	140 dec		2.13	5.03^{20}	i EtOH
863	Cobalt(II) hexafluoro-2,4-pentanedioate	$Co(CF_3COCHCOCF_3)_2$	19648-83-0	473.035	powder	197				
864	Cobalt(II) hexafluorosilicate hexahydrate	$CoSiF_6 \cdot 6H_2O$	12021-68-0	309.100	pale red cry			2.087	76.8^{22}	
865	Cobalt(II) hydroxide	$Co(OH)_2$	21041-93-0	92.948	blue-grn cry	≈160 dec		3.60		sl H_2O; s acid
866	Cobalt(II) hydroxide monohydrate	$Co(OH)_2 \cdot H_2O$	35340-84-2	110.963	blue solid	136 dec				
867	Cobalt(II) iodate	$Co(IO_3)_2$	13455-28-2	408.738	blk-viol needles	200 dec		5.09	0.46^{20}	
868	Cobalt(II) iodide	CoI_2	15238-00-3	312.742	blk hex cry; hyg	520		5.60	203^{25}	
869	Cobalt(II) iodide dihydrate	$CoI_2 \cdot 2H_2O$	13455-29-3	348.773	hyg grn cry	100 dec				
870	Cobalt(II) iodide hexahydrate	$CoI_2 \cdot 6H_2O$	15238-00-3	420.833	red hex prisms	130 dec		2.90	203^{25}	s EtOH, eth, ace
871	Cobalt(II) molybdate	$CoMoO_4$	13762-14-6	218.87	blk monocl cry	1040		4.7		
872	Cobalt(II) molybdate monohydrate	$CoMoO_4 \cdot H_2O$	18601-87-1	236.89	blk pow					
873	Cobalt(II) nitrate	$Co(NO_3)_2$	10141-05-6	182.942	pale red powder	100 dec		2.49	103^{25}	
874	Cobalt(II) nitrate hexahydrate	$Co(NO_3)_2 \cdot 6H_2O$	10026-22-9	291.034	red monocl cry; hyg	≈55		1.88	103^{25}	s EtOH
875	Cobalt(II) nitrite	$Co(NO_2)_2$	18488-96-5	150.944					0.49^{25}	
876	Cobalt(II) oleate	$Co(C_{18}H_{33}O_2)_2$	14666-94-5	621.840	brn amorp pow					i H_2O; s EtOH, eth
877	Cobalt(II) orthosilicate	Co_2SiO_4	12017-08-2	209.950	red-viol orth cry	1345		4.63		i H_2O; s dil HCl
878	Cobalt(II) oxalate	CoC_2O_4	814-89-1	146.952	pink powder	250 dec		3.02	0.0037^{20}	s acid, NH_4OH
879	Cobalt(II) oxalate dihydrate	$CoC_2O_4 \cdot 2H_2O$	5965-38-8	182.982	pink needles	dec			0.0037	sl acid; s NH_4OH
880	Cobalt(II) oxide	CoO	1307-96-6	74.932	gray cub cry	1830		6.44		i H_2O; s acid
881	Cobalt(II) 2,4-pentanedioate	$Co(CH_3COCHCOCH_3)_2$	14024-48-7	257.149	bl-viol cry	167				
882	Cobalt(II) perchlorate	$Co(ClO_4)_2$	13455-31-7	257.834	red needles			3.33	113^{25}	i EtOH, ace
883	Cobalt(II) perchlorate hexahydrate	$Co(ClO_4)_2 \cdot 6H_2O$	13478-33-6	365.926	dark red cry	170 dec		3.33		vs H_2O
884	Cobalt(II) phosphate octahydrate	$Co_3(PO_4)_2 \cdot 8H_2O$	10294-50-5	510.865	pink amorp powder			2.77		i H_2O; s acid
885	Cobalt(II) potassium sulfate hexahydrate	$CoK_2(SO_4)_2 \cdot 6H_2O$	10026-20-7	437.347	red monocl cry	75 dec		2.22		vs H_2O
886	Cobalt(II) selenate pentahydrate	$CoSeO_4 \cdot 5H_2O$	14590-19-3	291.97	red tricl cry	dec		2.51	55^{15}	
887	Cobalt(II) selenide	CoSe	1307-99-9	137.89	yel hex cry	1055		7.65		i H_2O, alk; s aqua regia
888	Cobalt(II) selenite dihydrate	$CoSeO_3 \cdot 2H_2O$	19034-13-0	221.92	blue-red powder					i H_2O
889	Cobalt(II) stannate	Co_2SnO_4	12139-93-4	300.574	grn-blue cub cry			6.30		i H_2O; s alk
890	Cobalt(II) stearate	$Co(C_{18}H_{35}O_2)_2$	1002-88-6	625.872	purp solid	74		1.13		
891	Cobalt(II) sulfate	$CoSO_4$	10124-43-3	154.996	red orth cry	>700		3.71	38.3^{25}	
892	Cobalt(II) sulfate monohydrate	$CoSO_4 \cdot H_2O$	13455-34-0	173.011	red monocl cry			3.08	38.3^{25}	
893	Cobalt(II) sulfate heptahydrate	$CoSO_4 \cdot 7H_2O$	10026-24-1	281.102	pink monocl cry	41 dec		2.03	38.3^{25}	sl EtOH, MeOH
894	Cobalt(II) sulfide	CoS	1317-42-6	90.998	blk amorp powder	1117		5.45		i H_2O; s acid
895	Cobalt(II) telluride	CoTe	12017-13-9	186.53	hex cry			≈8.8		
896	Cobalt(II) thiocyanate	$Co(SCN)_2$	3017-60-5	175.097	yel-brn powder				103^{25}	s EtOH, MeOH, ace, eth
897	Cobalt(II) thiocyanate trihydrate	$Co(SCN)_2 \cdot 3H_2O$	97126-35-7	229.143	viol rhomb cry				103^{25}	s EtOH, eth, ace
898	Cobalt(II) titanate	$CoTiO_3$	12017-01-5	154.798	grn rhomb cry			5.0		
899	Cobalt(II) tungstate	$CoWO_4$	12640-47-0	306.77	blue monocl cry			≈7.8		i H_2O; s hot conc acid
900	Cobalt(II,III) oxide	Co_3O_4	1308-06-1	240.798	blk cub cry	900 dec		6.11		i H_2O; s acid, alk
901	Cobalt(III) acetate	$Co(C_2H_3O_2)_3$	917-69-1	236.064	grn hyg cry	100 dec				s H_2O, EtOH
902	Cobalt(III) ammonium tetranitrodiammine	$NH_4[Co(NH_3)_2(NO_2)_4]$	13600-89-0	295.054	red-brn orth cry			1.97		s H_2O
903	Cobalt(III) fluoride	CoF_3	10026-18-3	115.928	brn hex cry	927		3.88		reac H_2O; s EtOH, eth, bz
904	Cobalt(III) fluoride dihydrate	$Co_2F_6 \cdot 2H_2O$	54496-71-8	267.887	red rhomb cry			2.19		s H_2O; i EtOH
905	Cobalt(III) hexammine chloride	$Co(NH_3)_6Cl_3$	10534-89-1	267.475	red monocl cry			1.71		s H_2O; i EtOH
906	Cobalt(III) hydroxide	$Co(OH)_3$	1307-86-4	109.955	brn powder	dec		≈4		i H_2O; s acid
907	Cobalt(III) nitrate	$Co(NO_3)_3$	15520-84-0	244.948	grn cub cry; hyg			≈3.0		s H_2O; reac os
908	Cobalt(III) oxide	Co_2O_3	1308-04-9	165.864	gray-blk powder	895 dec		5.18		i H_2O; s conc acid
909	Cobalt(III) oxide monohydrate	$Co_2O_3 \cdot H_2O$	12016-80-7	183.880	brn-blk hex cry	150 dec				i H_2O; s acid
910	Cobalt(III) 2,4-pentanedioate	$Co(CH_3COCHCOCH_3)_3$	21679-46-9	356.257	dark grn cry	213				s bz, ace
911	Cobalt(III) potassium nitrite sesquihydrate	$CoK_3(NO_2)_6 \cdot 1.5H_2O$	13782-01-9	479.284	yel cub cry			2.6		sl H_2O; reac acid; i EtOH
912	Cobalt(III) sulfide	Co_2S_3	1332-71-4	214.061	blk cub cry			4.8		reac acid
913	Cobalt(III) titanate	Co_2TiO_4	12017-38-8	229.731	grn-blk cub cry			5.1		s conc HCl
914	Copper	Cu	7440-50-8	63.546	red metal; cub	1084.62	2560	8.96		sl dil acid
915	Copper arsenide	Cu_3As	12005-75-3	265.560	dark gray solid	827				
916	Copper nitride	Cu_3N	1308-80-1	204.645	cub cry	300 dec		5.84		
917	Copper phosphide	CuP_2	12019-11-3	125.494	monocl cry	≈900		4.20		
918	Copper silicide	Cu_5Si	12159-07-8	345.816	solid	825				
919	Copper(I) acetate	$CuC_2H_3O_2$	598-54-9	122.590	col cry	dec	subl			reac H_2O
920	Copper(I) acetylide	Cu_2C_2	1117-94-8	151.113	red amorp powder; exp					
921	Copper(I) azide	CuN_3	14336-80-2	105.566	tetr cry; exp					

No.	Name	Formula	CAS Reg. No.	Mol. wt.	Phys. form	mp/°C	bp/°C	ρ/g cm^{-3}	Sol. in g/100 g H_2O	Qualitative solubility
922	Copper(I) bromide	$CuBr$	7787-70-4	143.450	wh cub cry; hyg	483	1345	4.98	0.0012^{20}	i ace
923	Copper(I) chloride	$CuCl$	7758-89-6	98.999	wh cub cry	423	1490	4.14	0.0047^{20}	i EtOH, ace
924	Copper(I) cyanide	$CuCN$	544-92-3	89.564	wh powder or grn orth cry	474	dec	2.9		i H_2O, EtOH; s KCN soln
925	Copper(I) fluoride	CuF	13478-41-6	82.544	cub cry			7.1		
926	Copper(I) hydride	CuH	13517-00-5	64.554	red-brn solid	60 dec				
927	Copper(I) iodide	CuI	7681-65-4	190.450	wh cub cry	591	≈1290	5.67	0.000020^{20}	i dil acid
928	Copper(I) mercury iodide	Cu_2HgI_4	13876-85-2	835.30	red cry powder	≈60 trans brn				i H_2O, EtOH
929	Copper(I) oxide	Cu_2O	1317-39-1	143.091	red-brn cub cry	1244	1800 dec	6.0		i H_2O
930	Copper(I) selenide	Cu_2Se	20405-64-5	206.05	blue-blk tetr cry	1113		6.84		i H_2O; s acid
931	Copper(I) sulfide	Cu_2S	22205-45-4	159.157	blue-blk orth cry	1129		5.6		i H_2O; sl acid
932	Copper(I) sulfite hemihydrate	$Cu_2SO_3\cdot0.5H_2O$	13982-53-1	216.164	wh-yel hex cry					sl H_2O; s acid, alk; i EtOH, eth
933	Copper(I) sulfite monohydrate	$Cu_2SO_3\cdot H_2O$	35788-00-2	225.171	cry			3.83		sl H_2O; s HCl
934	Copper(I) telluride	Cu_2Te	12019-52-2	254.69	blue hex cry	1127		4.6		
935	Copper(I) thiocyanate	$CuSCN$	1111-67-7	121.629	wh-yel amorp powder	1084		2.85		i H_2O, dil acid, EtOH, ace; s eth
936	Copper(I,II) sulfite dihydrate	$Cu_2SO_3\cdot CuSO_3\cdot2H_2O$	13814-81-8	386.795	red prisms or powder					i H_2O, EtOH; s HCl
937	Copper(II) acetate	$Cu(C_2H_3O_2)_2$	142-71-2	181.635	blue-grn hyg powder					
938	Copper(II) acetate monohydrate	$Cu(C_2H_3O_2)_2\cdot H_2O$	6046-93-1	199.650	grn monocl cry	115	240 dec	1.88		s H_2O, EtOH; sl eth
939	Copper(II) acetate metaarsenite	$Cu(C_2H_3O_2)_2\cdot3Cu(AsO_2)_2$	12002-03-8	1013.800	grn cry powder					i H_2O; reac acid
940	Copper(II) basic acetate	$Cu(C_2H_3O_2)_2\cdot CuO\cdot6H_2O$	52503-64-7	369.271	blue-grn cry or powder					sl H_2O, EtOH; s dil acid, NH_4OH
941	Copper(II) acetylide	CuC_2	12540-13-5	87.567	brn-blk solid; exp	100 exp				
942	Copper(II) arsenate	$Cu_3(AsO_4)_2$	7778-41-8	468.476	blue-grn cry					i H_2O, EtOH; s dil acid
943	Copper(II) arsenite	$CuHAsO_3$	10290-12-7	187.474	yel-grn powder					i H_2O, EtOH; s acid
944	Copper(II) azide	$Cu(N_3)_2$	14215-30-6	147.586	brn orth cry; exp			≈2.6		
945	Copper(II) borate	$Cu(BO_2)_2$	39290-85-2	149.166	blue-grn powder			3.859		i H_2O; s acid
946	Copper(II) bromide	$CuBr_2$	7789-45-9	223.354	blk monocl cry; hyg	498	900	4.710	126^{25}	vs H_2O; s EtOH, ace; i bz, eth
947	Copper(II) butanoate monohydrate	$Cu(C_4H_7O_2)_2\cdot H_2O$	540-16-9	255.756	grn monocl plates					s H_2O, diox, bz; sl EtOH
948	Copper(II) carbonate	$CuCO_3$	1184-64-1	123.555	cry					i H_2O
949	Copper(II) carbonate hydroxide	$CuCO_3\cdot Cu(OH)_2$	12069-69-1	221.116	grn monocl cry	200 dec		4.0		i H_2O, EtOH; s dil acid
950	Copper(II) chlorate hexahydrate	$Cu(ClO_3)_2\cdot6H_2O$	14721-21-2	338.540	blue-grn hyg cry	65	100 dec		164^{18}	vs EtOH
951	Copper(II) chloride	$CuCl_2$	7447-39-4	134.452	yel-brn monocl cry; hyg	598	993	3.4	75.7^{25}	s EtOH, ace
952	Copper(II) chloride dihydrate	$CuCl_2\cdot2H_2O$	10125-13-0	170.483	grn-blue orth cry; hyg	100 dec		2.51	75.7^{20}	vs EtOH, MeOH; s ace; i eth
953	Copper(II) chloride hydroxide	$Cu_2(OH)_3Cl$	1332-65-6	213.567	pale grn cry					i H_2O; s acid
954	Copper(II) chromate	$CuCrO_4$	13548-42-0	179.540	red-brn cry					i H_2O; s EtOH
955	Copper(II) basic chromate	$CuCrO_4\cdot2Cu(OH)_2$	12433-14-6	374.661	brn pow	260 dec				i H_2O; s HNO_3
956	Copper(II) chromite	$CuCr_2O_4$	12018-10-9	231.536	gray-blk tetr cry			5.4		i H_2O, dil acid
957	Copper(II) citrate hemipentahydrate	$Cu_2C_6H_4O_7\cdot2.5H_2O$	10402-15-0	360.221	blue-grn cry	100 dec				sl H_2O; s dil acid
958	Copper(II) cyanide	$Cu(CN)_2$	14763-77-0	115.580	grn powder					i H_2O; s acid, alk
959	Copper(II) cyclohexanebutanoate	$Cu(C_{10}H_{17}O_2)_2$	2218-80-6	402.028	powder	126 dec				
960	Copper(II) dichromate dihydrate	$CuCr_2O_7\cdot2H_2O$	13675-47-3	315.565	red-brn tricl cry			2.286		vs H_2O
961	Copper(II) ethanolate	$Cu(C_2H_5O)_2$	2850-65-9	153.667	blue hyg solid	120 dec				i os
962	Copper(II) ethylacetoacetate	$Cu(C_2H_5CO_2CHCOCH_3)_2$	14284-06-1	321.813	grn cry	192				s EtOH, chl
963	Copper(II) 2-ethylhexanoate	$Cu(C_8H_{15}O_2)_2$	149-11-1	349.953	powder	252 dec				
964	Copper(II) ferrate	$CuFe_2O_4$	12018-79-0	239.234	blk cry					
965	Copper(II) ferrocyanide	$Cu_2Fe(CN)_6$	13601-13-3	339.041	red-br cub cry or powder			2.2		i H_2O, acid, os
966	Copper(II) ferrous sulfide	$CuFeS_2$	1308-56-1	183.521	yel tetr cry	950		4.2		i H_2O, HCl; s HNO_3
967	Copper(II) fluoride	CuF_2	7789-19-7	101.543	wh monocl cry	836	1676	4.23	0.075^{25}	
968	Copper(II) fluoride dihydrate	$CuF_2\cdot2H_2O$	13454-88-1	137.574	blue monocl cry	130 dec		2.934	0.075^{25}	
969	Copper(II) formate	$Cu(CHO_2)_2$	544-19-4	153.581	blue cry				12.5^{20}	i os
970	Copper(II) formate tetrahydrate	$Cu(CHO_2)_2\cdot4H_2O$	5893-61-8	225.641	blue monocl cry				12.5	sl EtOH; i os
971	Copper(II) gluconate	$CuC_{12}H_{22}O_{14}$	527-09-3	453.841	bl-grn cry	156				sl EtOH; i os
972	Copper(II) hexafluoro-2,4-pentanedioate	$Cu(CF_3COCHCOCF_3)_2$	14781-45-4	477.648	cry	98	220 dec			s MeOH, ace, tol
973	Copper(II) hexafluorosilicate tetrahydrate	$CuSiF_6\cdot4H_2O$	12062-24-7	277.684	blue monocl cry	dec		2.56	99.7^{17}	sl EtOH
974	Copper(II) hydroxide	$Cu(OH)_2$	20427-59-2	97.561	blue-grn powder			3.37		i H_2O; s acid, conc alk
975	Copper(II) iodate	$Cu(IO_3)_2$	13454-89-2	413.351	grn monocl cry	dec		5.241	0.15^{20}	s dil acid
976	Copper(II) iodate monohydrate	$Cu(IO_3)_2\cdot H_2O$	13454-90-5	431.367	blue tricl cry	248 dec		4.872	0.15^{20}	s dil H_2SO_4

No.	Name	Formula	CAS Reg. No.	Mol. wt.	Phys. form	mp/°C	bp/°C	ρ/g cm^{-3}	Sol. in g/100 g H_2O	Qualitative solubility
977	Copper(II) molybdate	$CuMoO_4$	13767-34-5	223.48	grn cry	≈500		3.4	0.038	
978	Copper(II) nitrate	$Cu(NO_3)_2$	3251-23-8	187.555	blue-grn orth cry; hyg	255	subl		145^{25}	s diox; reac eth
979	Copper(II) nitrate trihydrate	$Cu(NO_3)_2·3H_2O$	10031-43-3	241.602	blue rhomb cry	114	170 dec	2.32	145^{25}	vs EtOH
980	Copper(II) nitrate hexahydrate	$Cu(NO_3)_2·6H_2O$	13478-38-1	295.647	blue rhomb cry; hyg			2.07	145^{25}	s EtOH
981	Copper(II) oleate	$Cu(C_{18}H_{33}O_2)_2$	1120-44-1	626.453	blue-grn solid					i H_2O; sl EtOH; s eth
982	Copper(II) oxalate	CuC_2O_4	814-91-5	151.565	blue-wh powder	310 dec			0.0026^{20}	i EtOH, eth; s NH_4OH
983	Copper(II) oxalate hemihydrate	$CuC_2O_4·0.5H_2O$	814-91-5	160.572	blue-wh cry	200 dec			0.0026^{20}	s NH_4OH
984	Copper(II) oxide	CuO	1317-38-0	79.545	blk powder or monocl cry	1227		6.31		i H_2O, EtOH; s dil acid
985	Copper(II) 2,4-pentanedioate	$Cu(CH_3COCHCOCH_3)_2$	13395-16-9	261.762	blue powder	284 dec	subl			sl H_2O; s chl
986	Copper(II) oxychloride hemiheptahydrate	$CuCl_2·3CuO·3.5H_2O$	1332-40-7	436.142	blue-grn pow	140 dec				i H_2O; s acid, NH_4OH
987	Copper(II) perchlorate	$Cu(ClO_4)_2$	13770-18-8	262.447	grn hyg cry	130 dec			146^{30}	s eth, diox; i bz, ctc
988	Copper(II) perchlorate hexahydrate	$Cu(ClO_4)_2·6H_2O$	10294-46-9	370.539	blue monocl cry; hyg	82	120 dec	2.22	146^{30}	vs EtOH, HOAc, ace; sl eth
989	Copper(II) phosphate	$Cu_3(PO_4)_2$	7798-23-4	380.581	blue-grn tricl cry					i H_2O; s acid, NH_4OH
990	Copper(II) phosphate trihydrate	$Cu_3(PO_4)_2·3H_2O$	10031-48-8	434.627	blue-grn orth cry					i H_2O; s acid, NH_4OH
991	Copper(II) phthalocyanine	$CuC_{32}H_{16}N_8$	147-14-8	576.069	bl-purp cry					i H_2O, EtOH; s conc H_2SO_4
992	Copper(II) selenate pentahydrate	$CuSeO_4·5H_2O$	10031-45-5	296.58	blue tricl cry	80 dec		2.56	27.4^{25}	s acid, NH_4OH; sl ace; i EtOH
993	Copper(II) selenide	CuSe	1317-41-5	142.51	blue-blk needles or plates	550 dec		5.99		reac acid
994	Copper(II) selenite dihydrate	$CuSeO_3·2H_2O$	15168-20-4	226.54	blue orth cry			3.31		i H_2O; s acid, NH_4OH
995	Copper(II) silicate dihydrate	$CuSiO_3·2H_2O$	26318-99-0	175.661	grn-blue orth cry					
996	Copper(II) stannate	$CuSnO_3$	12019-07-7	230.254	blue pow					
997	Copper(II) stearate	$Cu(C_{18}H_{35}O_2)_2$	660-60-6	630.485	blue-grn amorp powder	≈250				i H_2O, EtOH, eth; s py
998	Copper(II) sulfate	$CuSO_4$	7758-98-7	159.609	wh-grn amorp powder or rhomb cry	560 dec		3.60	22.0^{25}	i EtOH
999	Copper(II) sulfate pentahydrate	$CuSO_4·5H_2O$	7758-99-8	249.685	blue tricl cry	110 dec		2.286	22.0^{25}	s MeOH; sl EtOH
1000	Copper(II) sulfate, basic	$Cu_3(OH)_4SO_4$	1332-14-5	354.730	grn rhomb cry			3.88		i H_2O
1001	Copper(II) sulfide	CuS	1317-40-4	95.611	blk hex cry	507 trans		4.76		i H_2O, EtOH, dil acid, alk
1002	Copper(II) tartrate trihydrate	$CuC_4H_4O_6·3H_2O$	815-82-7	265.663	blue-grn powder					sl H_2O; s acid, alk
1003	Copper(II) telluride	CuTe	12019-23-7	191.15	yel orth cry	≈400 trans		7.09		
1004	Copper(II) tellurite	$CuTeO_3$	13812-58-3	239.14	blk glassy solid					i H_2O
1005	Copper(II) tetrafluoroborate	$Cu(BF_4)_2$	14735-84-3	237.155	solid					s H_2O
1006	Copper(II) titanate	$CuTiO_3$	12019-08-8	159.411	gray pow					
1007	Copper(II) 1,1,1-trifluoro-2,4-pentanedioate	$Cu(CF_3COCHCOCH_3)_2$	14324-82-4	369.705	blue-purp cry	197	260 dec			s EtOH, tol
1008	Copper(II) tungstate	$CuWO_4$	13587-35-4	311.38	yel-brn powder			7.5		
1009	Copper(II) tungstate dihydrate	$CuWO_4·2H_2O$	13587-35-4	347.41	grn powder					i H_2O; sl HOAc; reac conc acid
1010	Copper(II) vanadate	$Cu(VO_3)_2$	12789-09-2	261.425	powder					
1011	Curium	Cm	7440-51-9	247	silv metal; hex or cub	1345		13.51		
1012	Dysprosium	Dy	7429-91-6	162.500	silv metal; hex	1412	2567	8.55		s dil acid
1013	Dysprosium boride	DyB_4	12310-43-9	205.744	tetr cry	2500		6.98		
1014	Dysprosium nitride	DyN	12019-88-4	176.507	cub cry			9.93		
1015	Dysprosium silicide	$DySi_2$	12133-07-2	218.671	orth cry	1550		5.2		
1016	Dysprosium(II) bromide	$DyBr_2$	83229-05-4	322.308	blk solid					
1017	Dysprosium(II) chloride	$DyCl_2$	13767-31-2	233.406	blk cry	721 dec				reac H_2O
1018	Dysprosium(II) iodide	DyI_2	36377-94-3	416.309	purp cry	659				reac H_2O
1019	Dysprosium(III) acetate tetrahydrate	$Dy(C_2H_3O_2)_3·4H_2O$	15280-55-4	411.693	yel needles	120 dec				s H_2O; sl EtOH
1020	Dysprosium(III) bromide	$DyBr_3$	14456-48-5	402.212	wh hyg cry	879				s H_2O
1021	Dysprosium(III) carbonate tetrahydrate	$Dy_2(CO_3)_3·4H_2O$	38245-35-1	577.088	wh cry pow					i H_2O
1022	Dysprosium(III) chloride	$DyCl_3$	10025-74-8	268.859	wh or yel cry	718	1530	3.67		s H_2O, MeOH
1023	Dysprosium(III) chloride hexahydrate	$DyCl_3·6H_2O$	15059-52-6	376.950	bright yel cry	162 dec				
1024	Dysprosium(III) fluoride	DyF_3	13569-80-7	219.495	grn cry	1157				
1025	Dysprosium(III) hydride	DyH_3	13537-09-2	165.524	hex cry			7.1		
1026	Dysprosium(III) hydroxide	$Dy(OH)_3$	1308-85-6	213.522	yel or wh needles	205 dec				i H_2O
1027	Dysprosium(III) iodide	DyI_3	15474-63-2	543.213	grn cry	978				
1028	Dysprosium(III) nitrate pentahydrate	$Dy(NO_3)_3·5H_2O$	10143-38-1	438.591	yel cry	88.6			208.4^{25}	
1029	Dysprosium(III) oxide	Dy_2O_3	1308-87-8	372.998	wh cub cry	2408	3900	7.81		s acid

No.	Name	Formula	CAS Reg. No.	Mol. wt.	Phys. form	mp/°C	bp/°C	ρ/g cm⁻³	Sol. in g/100 g H_2O	Qualitative solubility
1030	Dysprosium(III) sulfate octahydrate	$Dy_2(SO_4)_3 \cdot 8H_2O$	10031-50-2	757.310	pale yel cry	110				sl H_2O
1031	Dysprosium(III) sulfide	Dy_2S_3	12133-10-7	421.195	red-brn monocl cry			6.08		
1032	Dysprosium(III) telluride	Dy_2Te_3	12159-43-2	707.80	solid	≈1550				
1033	Einsteinium	Es	7429-92-7	252	metal; cub	860				
1034	Erbium	Er	7440-52-0	167.259	silv metal; hex	1529	2868	9.07		i H_2O; s acid
1035	Erbium boride	ErB_4	12310-44-0	210.503	tetr cry	2450		7.0		
1036	Erbium acetate tetrahydrate	$Er(C_2H_3O_2)_3 \cdot 4H_2O$	15280-57-6	416.452	pink or wh cry			2.11		s H_2O
1037	Erbium bromide	$ErBr_3$	13536-73-7	406.971	viol hyg cry	950	≈1460			s H_2O, thf
1038	Erbium bromide hexahydrate	$ErBr_3 \cdot 6H_2O$	14890-44-9	515.062	pink cry					s H_2O
1039	Erbium chloride	$ErCl_3$	10138-41-7	273.618	viol monocl cry; hyg	776		4.1		s H_2O
1040	Erbium chloride hexahydrate	$ErCl_3 \cdot 6H_2O$	10025-75-9	381.709	pink hyg cry	dec				s H_2O; sl EtOH
1041	Erbium fluoride	ErF_3	13760-83-3	224.254	pink orth cry	1146		7.8		i H_2O
1042	Erbium hydride	ErH_3	13550-53-3	170.283	hex cry			≈7.6		
1043	Erbium hydroxide	$Er(OH)_3$	14646-16-3	218.281	pink solid					i H_2O
1044	Erbium iodide	ErI_3	13813-42-8	547.972	viol hex cry; hyg	1014		≈5.5		s H_2O
1045	Erbium nitrate pentahydrate	$Er(NO_3)_3 \cdot 5H_2O$	10168-80-6	443.350	red cry	130 dec			240.8^{25}	s EtOH, ace
1046	Erbium nitride	ErN	12020-21-2	181.266	cub cry			10.6		
1047	Erbium oxide	Er_2O_3	12061-16-4	382.516	pink powder	2418	3920	8.64		i H_2O; s acid
1048	Erbium silicide	$ErSi_2$	12020-28-9	223.430	orth cry			7.26		
1049	Erbium sulfate	$Er_2(SO_4)_3$	13478-49-4	622.706	hyg powder	dec		3.68	13^{20}	
1050	Erbium sulfate octahydrate	$Er_2(SO_4)_3 \cdot 8H_2O$	10031-52-4	766.828	pink monocl cry	dec		3.20	13^{20}	
1051	Erbium sulfide	Er_2S_3	12159-66-9	430.713	red-brn monocl cry	1730		6.07		
1052	Erbium telluride	Er_2Te_3	12020-39-2	717.32	orth cry	1213		7.11		
1053	Europium	Eu	7440-53-1	151.964	soft silv metal; cub	822	1529	5.24		reac H_2O
1054	Europium boride	EuB_6	12008-05-8	216.830	cub cry	≈2600		4.91		
1055	Europium nitride	EuN	12020-58-5	165.971	cub cry			8.7		
1056	Europium silicide	$EuSi_2$	12434-24-1	208.135	tetr cry	1500		5.46		
1057	Europium(II) bromide	$EuBr_2$	13780-48-8	311.772	wh cry	683				s H_2O
1058	Europium(II) chloride	$EuCl_2$	13769-20-5	222.870	wh orth cry	731		4.9		s H_2O
1059	Europium(II) fluoride	EuF_2	14077-39-5	189.961	grn-yel cub cry	≈1380		6.5		
1060	Europium(II) iodide	EuI_2	22015-35-6	405.773	grn cry	580				s H_2O
1061	Europium(II) selenide	EuSe	12020-66-5	230.92	brn cub cry			6.45		
1062	Europium(II) sulfate	$EuSO_4$	10031-54-6	248.027	col orth cry			4.99		i H_2O
1063	Europium(II) sulfide	EuS	12020-65-4	184.029	cub cry			5.7		
1064	Europium(II) telluride	EuTe	12020-69-8	279.56	blk cub cry	1526		6.48		
1065	Europium(III) bromide	$EuBr_3$	13759-88-1	391.676	gray cry	dec				s H_2O
1066	Europium(III) chloride	$EuCl_3$	10025-76-0	258.323	grn-yel needles	623		4.89		
1067	Europium(III) chloride hexahydrate	$EuCl_3 \cdot 6H_2O$	13759-92-7	366.414	wh-yel hyg cry	850		4.89		s H_2O
1068	Europium(III) fluoride	EuF_3	13765-25-8	208.959	wh hyg cry	647				i H_2O
1069	Europium(III) iodide	EuI_3	13759-90-5	532.677	col cry; unstab	≈875				
1070	Europium(III) nitrate hexahydrate	$Eu(NO_3)_3 \cdot 6H_2O$	10031-53-5	446.070	wh-pink hyg cry	85 dec			193^{25}	
1071	Europium(III) oxalate	$Eu_2(C_2O_4)_3$	3269-12-3	567.985	wh solid					i H_2O; s acid
1072	Europium(III) oxide	Eu_2O_3	1308-96-9	351.926	pink powder	2350	3790	7.42		i H_2O; s acid
1073	Europium(III) perchlorate hexahydrate	$Eu(ClO_4)_3 \cdot 6H_2O$	36907-40-1	558.407	wh or pink cry					s H_2O, EtOH
1074	Europium(III) sulfate	$Eu_2(SO_4)_3$	13537-15-0	592.116	pale pink cry			4.99	2.1^{20}	
1075	Europium(III) sulfate octahydrate	$Eu_2(SO_4)_3 \cdot 8H_2O$	10031-52-4	736.238	pink cry	375 dec			2.1^{20}	
1076	Fermium	Fm	7440-72-4	257	metal	1527				
1077	Fluorine	F_2	7782-41-4	37.997	pale yel gas	-219.67	-188.11	$1.5127^{-188.1}$ liq		reac H_2O
1078	Fluorine monoxide	F_2O	7783-41-7	53.996	col gas	-223.8	-144.3	0.002207^{25} gas		sl H_2O
1079	Difluorine dioxide	F_2O_2	7783-44-0	69.996	red-oran solid, unstab gas	-163.5	-57 extrap	0.002861^{25} gas		
1080	Fluorine tetroxide	F_2O_4	107782-11-6	101.995	red-brn solid	-191	-185 dec			
1081	Fluorine nitrate	FNO_3	7789-26-6	81.003	col gas	-175	-46	0.003311^{25} gas		reac H_2O, EtOH, eth; s ace
1082	Fluorine perchlorate	$FOClO_3$	10049-03-3	118.449	col gas; exp	-167.3	-16	0.004841^{25} gas		reac H_2O
1083	Francium	Fr	7440-73-5	223	short-lived alkali metal	≈21				
1084	Gadolinium	Gd	7440-54-2	157.25	silv metal; hex	1313	3273	7.90		s dil acid
1085	Gadolinium boride	GdB_6	12008-06-9	222.12	blk-brn cub cry	2510		5.31		
1086	Gadolinium nitride	GdN	25764-15-2	171.26	cub cry			9.10		
1087	Gadolinium silicide	$GdSi_2$	12134-75-7	213.42	orth cry	1540		5.9		
1088	Gadolinium(II) iodide	GdI_2	13814-72-7	411.06	bronze cry	831				
1089	Gadolinium(II) selenide	GdSe	12024-81-6	236.21	cub cry	2170		8.1		
1090	Gadolinium(III) acetate tetrahydrate	$Gd(C_2H_3O_2)_2 \cdot 4H_2O$	15280-53-2	406.44	wh tricl cry	dec		1.61		s H_2O
1091	Gadolinium(III) bromide	$GdBr_3$	13818-75-2	396.96	wh monocl cry; hyg	785		4.56		
1092	Gadolinium(III) chloride	$GdCl_3$	10138-52-0	263.61	wh monocl cry; hyg	602		4.52		s H_2O
1093	Gadolinium(III) chloride hexahydrate	$GdCl_3 \cdot 6H_2O$	19423-81-5	371.70	col hyg cry			2.424		s H_2O
1094	Gadolinium(III) fluoride	GdF_3	13765-26-9	214.25	wh cry	1232				
1095	Gadolinium(III) iodide	GdI_3	13572-98-0	537.96	yel cry	930				

No.	Name	Formula	CAS Reg. No.	Mol. wt.	Phys. form	mp/°C	bp/°C	ρ/g cm⁻³	Sol. in g/100 g H_2O	Qualitative solubility
1096	Gadolinium(III) nitrate pentahydrate	$Gd(NO_3)_3 \cdot 5H_2O$	52788-53-1	433.34	wh cry	92 dec		2.41	190^{25}	
1097	Gadolinium(III) nitrate hexahydrate	$Gd(NO_3)_3 \cdot 6H_2O$	19598-90-4	451.36	hyg tricl cry	91 dec		2.33	190^{25}	s EtOH
1098	Gadolinium(III) oxalate decahydrate	$Gd_2(C_2O_4)_3 \cdot 10H_2O$	22992-15-0	758.71	wh monocl pow	110 dec				i H_2O; sl acid
1099	Gadolinium(III) oxide	Gd_2O_3	12064-62-9	362.50	wh hyg powder	2425	3900	7.41		i H_2O; s acid
1100	Gadolinium(III) sulfate	$Gd_2(SO_4)_3$	13628-54-1	602.69	col cry	500 dec		4.1	2.60^{20}	sl H_2O
1101	Gadolinium(III) sulfate octahydrate	$Gd_2(SO_4)_3 \cdot 8H_2O$	13450-87-8	746.81	col monocl cry	400 dec		4.14	2.3^{20}	sl H_2O
1102	Gadolinium(III) sulfide	Gd_2S_3	12134-77-9	410.70	yel cub cry			6.1		
1103	Gadolinium(III) telluride	Gd_2Te_3	12160-99-5	697.30	orth cry	1255		7.7		
1104	Gallium	Ga	7440-55-3	69.723	silv liq or gray orth cry	29.7646	2229	5.91		reac alk
1105	Gallium antimonide	GaSb	12064-03-8	191.483	brn cub cry	712		5.6137		
1106	Gallium arsenide	GaAs	1303-00-0	144.645	gray cub cry	1238		5.3176		
1107	Gallium nitride	GaN	25617-97-4	83.730	gray hex cry	>2500		6.1		
1108	Gallium phosphide	GaP	12063-98-8	100.697	yel cub cry	1457		4.138		
1109	Gallium suboxide	Ga_2O	12024-20-3	155.445	brn powder	>660	>800 dec	4.77		
1110	Gallium(II) chloride	$GaCl_2$	24597-12-4	140.629	wh orth cry	172.4	535	2.74		
1111	Gallium(II) selenide	GaSe	12024-11-2	148.68	hex cry	960		5.03		
1112	Gallium(II) sulfide	GaS	12024-10-1	101.788	hex cry	965		3.86		
1113	Gallium(II) telluride	GaTe	12024-14-5	197.32	monocl cry	824		5.44		
1114	Gallium(III) bromide	$GaBr_3$	13450-88-9	309.435	wh orth cry	123	279	3.69		
1115	Gallium(III) chloride	$GaCl_3$	13450-90-3	176.082	col needles or gl solid	77.9	201	2.47		
1116	Gallium(III) fluoride	GaF_3	7783-51-9	126.718	wh powder or col needles	>1000		4.47		i H_2O
1117	Gallium(III) fluoride trihydrate	$GaF_3 \cdot 3H_2O$	22886-66-4	180.764	wh cry	>140 dec				sl H_2O
1118	Gallium(III) hydride	GaH_3	13572-93-5	72.747	visc liq	-50	≈0 dec			
1119	Gallium(III) hydroxide	$Ga(OH)_3$	12023-99-3	120.745	unstab prec					
1120	Gallium(III) iodide	GaI_3	13450-91-4	450.436	monocl cry	212	340	4.5		
1121	Gallium(III) nitrate	$Ga(NO_3)_3$	13494-90-1	255.738	wh cry powder					s H_2O, EtOH, eth
1122	Gallium(III) oxide	Ga_2O_3	12024-21-4	187.444	wh cry	1807		≈6.0		s hot acid
1123	Gallium(III) oxide hydroxide	GaOOH	20665-52-5	102.730	orth cry			5.23		
1124	Gallium(III) 2,4-pentanedioate	$Ga(CH_3COCHCOCH_3)_3$	14405-43-7	367.047	wh powder	193	subl	1.42		
1125	Gallium(III) perchlorate hexahydrate	$Ga(ClO_4)_3 \cdot 6H_2O$	17835-81-3	476.166	cry	175 dec				
1126	Gallium(III) selenide	Ga_2Se_3	12024-24-7	376.33	cub cry	937		4.92		
1127	Gallium(III) sulfate	$Ga_2(SO_4)_3$	13494-91-2	427.634	hex cry					
1128	Gallium(III) sulfate octadecahydrate	$Ga_2(SO_4)_3 \cdot 18H_2O$	13780-42-2	751.909	octahed cry					s H_2O, EtOH
1129	Gallium(III) sulfide	Ga_2S_3	12024-22-5	235.641	monocl cry	1090		3.7		
1130	Gallium(III) telluride	Ga_2Te_3	12024-27-0	522.25	cub cry	790		5.57		
1131	Germanium	Ge	7440-56-4	72.63	gray-wh cub cry	938.25	2833	5.3234		i H_2O, dil acid, alk
1132	Germane	GeH_4	7782-65-2	76.67	col gas; flam	-165	-88.1	0.003134^{25} gas		i H_2O
1133	Digermane	Ge_2H_6	13818-89-8	151.33	col liq; flam	-109	29	1.98^{-109} liq		
1134	Trigermane	Ge_3H_8	14691-44-2	225.98	col liq	-105.6	110.5	2.20^{-105} liq		i H_2O
1135	Tetragermane	Ge_4H_{10}	14691-47-5	300.64	col liq		176.9			i H_2O
1136	Pentagermane	Ge_5H_{12}	15587-39-0	375.30	col liq		234			i H_2O
1137	Bromogermane	GeH_3Br	13569-43-2	155.57	col liq	-32	52	2.34		reac H_2O
1138	Chlorogermane	GeH_3Cl	13637-65-5	111.12	col liq	-52	28	1.75		reac H_2O
1139	Chlorotrifluorogermane	GeF_3Cl	14188-40-0	165.09	gas	-66.2	-20.3	0.006748^{25} gas		
1140	Dibromogermane	GeH_2Br_2	13769-36-3	234.46	col liq	-15	89	2.80		reac H_2O
1141	Dichlorogermane	GeH_2Cl_2	15230-48-5	145.56	col liq	-68	69.5	1.90		reac H_2O
1142	Dichlorodifluorogermane	GeF_2Cl_2	24422-21-7	181.54	col gas	-51.8	-2.8	0.007420^{25} gas		
1143	Dichlorodimethylgermane	$Ge(CH_3)_2Cl_2$	1529-48-2	173.62	liq	-22	124	1.49		
1144	Fluorogermane	GeH_3F	13537-30-9	94.66	col gas			0.003869^{25} gas		reac H_2O
1145	Iodogermane	GeH_3I	13573-02-9	202.57	liq	-15	≈90			reac H_2O
1146	Methylgermane	GeH_3CH_3	1449-65-6	90.70	col gas	-158	-23	0.003707^{25} gas		
1147	Tribromogermane	$GeHBr_3$	14779-70-5	313.36	col liq	-25	dec			reac H_2O
1148	Trichlorogermane	$GeHCl_3$	1184-65-2	180.01	liq	-71	75.3	1.93		reac H_2O
1149	Trichlorofluorogermane	$GeCl_3F$	24422-20-6	198.00	liq	-49.8	37.5			
1150	Germanium(II) bromide	$GeBr_2$	24415-00-7	232.45	yel monocl cry	122	150 dec			reac H_2O
1151	Germanium(II) chloride	$GeCl_2$	10060-11-4	143.55	wh-yel hyg powder	dec				reac H_2O; s eth, bz
1152	Germanium(II) fluoride	GeF_2	13940-63-1	110.64	wh orth cry; hyg	110	130 dec	3.64		reac H_2O
1153	Germanium(II) iodide	GeI_2	13573-08-5	326.45	oran-yel hex cry	428	550 dec	5.4		reac H_2O
1154	Germanium(II) oxide	GeO	20619-16-3	88.64	blk solid	700 dec				
1155	Germanium(II) selenide	GeSe	12065-10-0	151.60	gray orth cry or brn powder	675		5.6		
1156	Germanium(II) sulfide	GeS	12025-32-0	104.71	gray orth cry	658		4.1		
1157	Germanium(II) telluride	GeTe	12025-39-7	200.24	cub cry	724		6.16		i H_2O; s conc HNO_3
1158	Germanium(IV) bromide	$GeBr_4$	13450-92-5	392.26	wh cry	26.1	186.35	3.132		reac H_2O
1159	Germanium(IV) chloride	$GeCl_4$	10038-98-9	214.45	col liq	-51.50	86.55	1.88		reac H_2O; s bz, eth, EtOH, ctc

No.	Name	Formula	CAS Reg. No.	Mol. wt.	Phys. form	mp/°C	bp/°C	ρ/g cm^{-3}	Sol. in g/100 g H_2O	Qualitative solubility
1160	Germanium(IV) fluoride	GeF_4	7783-58-6	148.63	col gas	-15 tp	-36.5 sp	0.006075^{25} gas		reac H_2O
1161	Germanium(IV) iodide	GeI_4	13450-95-8	580.26	red-oran cub cry	146	348	4.322		reac H_2O
1162	Germanium(IV) nitride	Ge_3N_4	12065-36-0	273.95	orth cry	900 dec				i H_2O, acid, aqua regia
1163	Germanium(IV) oxide	GeO_2	1310-53-8	104.64	wh hex cry	1116		4.25		i H_2O
1164	Germanium(IV) selenide	$GeSe_2$	12065-11-1	230.56	yel-oran orth cry	707 dec		4.56		
1165	Germanium(IV) sulfide	GeS_2	12025-34-2	136.77	blk orth cry	840		3.01		
1166	Gold	Au	7440-57-5	196.967	soft yel metal	1064.18	2836	19.3		s aqua regia
1167	Bromoauric(III) acid pentahydrate	$HAuBr_4 \cdot 5H_2O$	17083-68-0	607.667	red-brn hyg cry	27				s H_2O, EtOH
1168	Chloroauric(III) acid tetrahydrate	$HAuCl_4 \cdot 4H_2O$	16903-35-8	411.848	yel monocl cry; hyg			≈3.9		vs H_2O, EtOH; s eth
1169	Gold(I) bromide	AuBr	10294-27-6	276.871	yel-gray tetr cry	165 dec		8.20		i H_2O
1170	Gold(I) chloride	AuCl	10294-29-8	232.420	yel orth cry	289 dec		7.6	0.000031^{20}	
1171	Gold(I) cyanide	AuCN	506-65-0	222.985	yel hex cry	dec		7.2		i H_2O, EtOH, eth, dil acid
1172	Gold(I) iodide	AuI	10294-31-2	323.871	yel-grn powder; tetr	120 dec		8.25		i H_2O; s CN soln
1173	Gold(I) sulfide	Au_2S	1303-60-2	425.998	brn-blk cub cry; unstab	240 dec		≈11		i H_2O, acid; s aqua regia
1174	Gold(III) bromide	$AuBr_3$	10294-28-7	436.679	red-br monocl cry	≈160 dec				s H_2O, EtOH
1175	Gold(III) chloride	$AuCl_3$	13453-07-1	303.326	red monocl cry	>160 dec		4.7	68^{20}	
1176	Gold(III) cyanide trihydrate	$Au(CN)_3 \cdot 3H_2O$	535-37-5	329.065	wh hyg cry	50 dec				vs H_2O; sl EtOH
1177	Gold(III) fluoride	AuF_3	14720-21-9	253.962	oran-yel hex cry	>300	subl	6.75		
1178	Gold(III) hydroxide	$Au(OH)_3$	1303-52-2	247.989	brn powder	≈100 dec				i H_2O; s acid
1179	Gold(III) iodide	AuI_3	31032-13-0	577.680	unstab grn powder	20 dec				
1180	Gold(III) oxide	Au_2O_3	1303-58-8	441.931	brn powder	≈150 dec				i H_2O; s acid
1181	Gold(III) selenate	$Au_2(SeO_4)_3$	10294-32-3	822.81	yel cry					i H_2O; s acid
1182	Gold(III) selenide	Au_2Se_3	1303-62-4	630.81	blk amorp solid	dec		4.65		s aqua regia
1183	Gold(III) sulfide	Au_2S_3	1303-61-3	490.128	unstab blk powder	200 dec				
1184	Hafnium	Hf	7440-58-6	178.49	gray metal; hex	2233	4600	13.3		s HF
1185	Hafnium boride	HfB_2	12007-23-7	200.11	gray hex cry	3100		10.5		
1186	Hafnium carbide	HfC	12069-85-1	190.50	refrac cub cry	≈3000		12.2		
1187	Hafnium hydride	HfH_2	12770-26-2	180.51	refrac tetr cry			11.4		
1188	Hafnium nitride	HfN	25817-87-2	192.50	yel-brn cub cry	3310		13.8		
1189	Hafnium phosphide	HfP	12325-59-6	209.46	hex cry			9.78		
1190	Hafnium silicide	$HfSi_2$	12401-56-8	234.66	gray orth cry	≈1700		7.6		
1191	Hafnocene dichloride	$Hf(C_5H_5)_2Cl_2$	12116-66-4	379.58	col hyg cry	235				s bz, chl; sl thf, eth; i hex
1192	Hafnium(II) bromide	$HfBr_2$	13782-95-1	338.30	blue-blk cry	400 dec				
1193	Hafnium(II) chloride	$HfCl_2$	13782-92-8	249.40	blk solid	400 dec				
1194	Hafnium(III) bromide	$HfBr_3$	13782-96-2	418.20	blue-blk cry	350 dec				
1195	Hafnium(III) chloride	$HfCl_3$	13782-93-9	284.85	blk solid	dec				
1196	Hafnium(III) iodide	HfI_3	13779-73-2	559.20	blk cry	dec				
1197	Hafnium(IV) bromide	$HfBr_4$	13777-22-5	498.11	wh cub cry	424 tp	323 sp	4.90		reac H_2O
1198	Hafnium(IV) chloride	$HfCl_4$	13499-05-3	320.30	wh monocl cry	432 tp	317 sp			reac H_2O
1199	Hafnium(IV) fluoride	HfF_4	13709-52-9	254.48	wh monocl cry	1025	970 sp	7.1		reac H_2O
1200	Hafnium(IV) iodide	HfI_4	13777-23-6	686.11	yel-oran cub cry	449 tp	394 sp	5.6		reac H_2O
1201	Hafnium(IV) oxide	HfO_2	12055-23-1	210.49	wh cub cry	2800	≈5400	9.68		i H_2O
1202	Hafnium(IV) oxychloride octahydrate	$HfOCl_2 \cdot 8H_2O$	14456-34-9	409.52	wh tetr cry	dec				s H_2O
1203	Hafnium(IV) selenide	$HfSe_2$	12162-21-9	336.41	brn hex cry			7.46		
1204	Hafnium(IV) silicate	$HfSiO_4$	13870-13-8	270.57	tetr cry	2758		7.0		
1205	Hafnium(IV) sulfate	$Hf(SO_4)_2$	15823-43-5	370.62	wh cry	>500 dec				
1206	Hafnium(IV) sulfide	HfS_2	18855-94-2	242.62	purp-brn hex cry			6.03		
1207	Hafnium(IV) titanate	$HfTiO_4$	12055-24-2	290.36	wh pow	1980 dec				
1208	Helium	He	7440-59-7	4.003	col gas		-268.928	$0.1250^{-268.9}$ liq		sl H_2O; i EtOH
1209	Holmium	Ho	7440-60-0	164.930	silv metal; hex	1472	2700	8.80		s dil acid
1210	Holmium acetate	$Ho(C_2H_3O_2)_3$	25519-09-9	342.062	yel cry	327 dec				s H_2O
1211	Holmium bromide	$HoBr_3$	13825-76-8	404.642	yel hyg cry	919	1470			
1212	Holmium chloride	$HoCl_3$	10138-62-2	271.289	yel monocl cry; hyg	720	1500	3.7		s H_2O
1213	Holmium chloride hexahydrate	$HoCl_3 \cdot 6H_2O$	14914-84-2	379.381	hyg yel cry	160 dec				s H_2O
1214	Holmium fluoride	HoF_3	13760-78-6	221.925	pink-yel orth cry; hyg	1143	>2200	7.664		s H_2O
1215	Holmium iodide	HoI_3	13813-41-7	545.643	yel hex cry	994		5.4		
1216	Holmium nitrate pentahydrate	$Ho(NO_3)_3 \cdot 5H_2O$	14483-18-2	441.022	hyg oran cry					s H_2O, EtOH, ace
1217	Holmium nitride	HoN	12029-81-1	178.937	cub cry			10.6		
1218	Holmium oxalate decahydrate	$Ho_2(C_2O_4)_3 \cdot 10H_2O$	28965-57-3	774.065	yel solid	40 dec				
1219	Holmium oxide	Ho_2O_3	12055-62-8	377.859	yel cub cry	2415	3900	8.41		s acid
1220	Holmium silicide	$HoSi_2$	12136-24-2	221.101	hex cry			7.1		
1221	Holmium sulfide	Ho_2S_3	12162-59-3	426.056	yel-oran monocl cry			5.92		
1222	Hydrazine	N_2H_4	302-01-2	32.045	col oily liq	1.54	113.55	1.0036		vs H_2O, EtOH, MeOH
1223	Hydrazine acetate	$N_2H_4 \cdot CH_3COOH$	13255-48-6	92.097	cry	100				
1224	Hydrazine azide	$N_2H_4 \cdot HN_3$	14662-04-5	75.074	hyg wh prism	75 exp				vs H_2O
1225	Hydrazine monohydrate	$N_2H_4 \cdot H_2O$	7803-57-8	50.060	fuming liq	-51.7	119	1.030		vs H_2O, EtOH; i chl, eth

No.	Name	Formula	CAS Reg. No.	Mol. wt.	Phys. form	mp/°C	bp/°C	ρ/g cm^{-3}	Sol. in g/100 g H_2O	Qualitative solubility
1226	Hydrazine hydrobromide	$N_2H_4 \cdot HBr$	13775-80-9	112.957	wh monocl cry flakes	84	≈190 dec	2.3		s H_2O, EtOH
1227	Hydrazine hydrochloride	$N_2H_4 \cdot HCl$	2644-70-4	68.506	wh orth cry	89	240	1.5		s H_2O; i os
1228	Hydrazine dihydrochloride	$N_2H_4 \cdot 2HCl$	5341-61-7	104.967	wh orth cry	198 dec		1.42		s H_2O; sl EtOH
1229	Hydrazine hydroiodide	$N_2H_4 \cdot HI$	10039-55-1	159.957	hyg cry	125				s H_2O
1230	Hydrazine nitrate	$N_2H_4 \cdot HNO_3$	13464-97-6	95.058	monocl cry; exp	70				vs H_2O
1231	Hydrazine dinitrate	$N_2H_4 \cdot 2HNO_3$	13464-98-7	158.071	needles	104 dec			20^{35}	s H_2O
1232	Hydrazine perchlorate hemihydrate	$N_2H_4 \cdot HClO_4 \cdot 0.5H_2O$	13762-65-7	141.511	solid	137	exp	1.94		reac H_2O, s EtOH; i eth, bz
1233	Hydrazine sulfate	$N_2H_4 \cdot H_2SO_4$	10034-93-2	130.124	col orth cry	254		1.378		sl H_2O; i EtOH
1234	Dihydrazine sulfate	$(N_2H_4)_2 \cdot H_2SO_4$	13464-80-7	162.169	hyg wh cry flakes	104	>180 dec		200^{25}	vs H_2O; i os
1235	Hydrazoic acid	HN_3	7782-79-8	43.028	col liq; exp	-80	35.7			s H_2O
1236	Hydroxylamine	H_2NOH	7803-49-8	33.030	wh orth flakes or needles	33.1	58	1.21		vs H_2O, MeOH
1237	Hydroxylamine hydrobromide	$H_2NOH \cdot HBr$	41591-55-3	113.942	monocl cry			2.35		s H_2O
1238	Hydroxylamine hydrochloride	$H_2NOH \cdot HCl$	5470-11-1	69.491	col monocl cry	159 dec	exp	1.68	94^{25}	vs H_2O
1239	Hydroxylamine perchlorate	$H_2NOH \cdot HClO_4$		133.489	orth cry	88	120 dec			
1240	Hydroxylamine sulfate	$2(H_2NOH) \cdot H_2SO_4$	10039-54-0	164.138	cry	170				vs H_2O
1241	Hydrogen	H_2	1333-74-0	2.016	col gas; flam	-259.16	-252.879	$0.07083^{-252.9}$ liq		sl H_2O
1242	Hydrogen-d_2	D_2	7782-39-0	4.028	col gas	-254.42	-249.48	0.000165^{25} gas		
1243	Hydrogen-t_2	T_2	10028-17-8	6.032	col gas	-252.53	-248.11	0.000247^{25} gas		
1244	Hydrogen-d_1	HD	13983-20-5	3.022	col gas	-256.55	-251.02	0.000124^{25} gas		
1245	Hydrogen-t_1	HT	14885-60-0	4.024	col gas	-254.7	-249.6			
1246	Hydrogen-d_1,t_1	DT	14885-61-1	5.030	col gas	-253.5	-238.9			
1247	Hydrogen bromide	HBr	10035-10-6	80.912	col gas	-86.80	-66.38	2.603^{-84} liq		vs H_2O; s EtOH
1248	Hydrogen bromide-d	DBr	13536-59-9	81.918	col gas	-87.54	-66.9			s H_2O
1249	Hydrogen chloride	HCl	7647-01-0	36.461	col gas	-114.17	-85	$1.187^{-114.1}$ liq		vs H_2O
1250	Hydrogen chloride dihydrate	$HCl \cdot 2H_2O$	13465-05-9	72.492	col liq	-17.7		1.46		
1251	Hydrogen chloride-d	DCl	7698-05-7	37.467	col gas	-114.72	-84.4			s H_2O
1252	Hydrogen cyanide	HCN	74-90-8	27.026	col liq or gas	-13.28	25.63	0.6876^{20}		vs H_2O, EtOH; sl eth
1253	Hydrogen fluoride	HF	7664-39-3	20.006	col gas	-83.36	20	1.002^{0} liq		vs H_2O, EtOH; sl eth
1254	Hydrogen iodide	HI	10034-85-2	127.912	col or yel gas	-50.76	-35.55	2.85^{-47} liq		vs H_2O; s os
1255	Hydrogen iodide-d	DI	14104-45-1	128.918	col gas	-51.93	-36.2			s H_2O
1256	Hydrogen peroxide	H_2O_2	7722-84-1	34.015	col liq	-0.43	150.2	1.44		vs H_2O
1257	Hydrogen selenide	H_2Se	7783-07-5	80.98	col gas; flam	-65.73	-41.25	0.003310^{25} gas		s H_2O
1258	Hydrogen sulfide	H_2S	7783-06-4	34.081	col gas; flam	-85.5	-59.55	$0.9923^{-85.5}$ liq		s H_2O
1259	Hydrogen disulfide	H_2S_2	13465-07-1	66.146	col liq		70.7	1.334		
1260	Hydrogen telluride	H_2Te	7783-09-7	129.62	col gas	-49	-2	0.005298^{25} gas		s H_2O, EtOH, alk
1261	Indium	In	7440-74-6	114.818	soft wh metal	156.5985	2027	7.31		s acid
1262	Indium antimonide	InSb	1312-41-0	236.578	blk cub cry	524		5.7747		
1263	Indium arsenide	InAs	1303-11-3	189.740	gray cub cry	942		5.67		i acid
1264	Indium nitride	InN	25617-98-5	128.825	brn hex cry	1100		6.88		
1265	Indium phosphide	InP	22398-80-7	145.792	blk cub cry	1062		4.81		sl acid
1266	Indium(I) bromide	InBr	14280-53-6	194.722	oran-red orth cry	285	656	4.96		reac H_2O
1267	Indium(I) chloride	InCl	13465-10-6	150.271	yel cub cry	225	608	4.19		reac H_2O
1268	Indium(I) iodide	InI	13966-94-4	241.722	orth cry	364.4	712	5.32		
1269	Indium(II) bromide	$InBr_2$	21264-43-7	274.626	orth cry			4.22		reac H_2O
1270	Indium(II) chloride	$InCl_2$	13465-11-7	185.724	col orth cry	235		3.64		reac H_2O
1271	Indium(II) sulfide	InS	12030-14-7	146.883	red-brn orth cry	692		5.2		
1272	Indium(III) bromide	$InBr_3$	13465-09-3	354.530	hyg yel-wh monocl cry	420		4.74	414^{20}	
1273	Indium(III) chloride	$InCl_3$	10025-82-8	221.177	yel monocl cry; hyg	583		4.0	195.1^{22}	s EtOH
1274	Indium(III) chloride tetrahydrate	$InCl_3 \cdot 4H_2O$	22519-64-8	293.239	wh cry					s H_2O
1275	Indium(III) fluoride	InF_3	7783-52-0	171.813	wh hex cry; hyg	1172	>1200	4.39		sl H_2O; s dil acid
1276	Indium(III) fluoride trihydrate	$InF_3 \cdot 3H_2O$	14166-78-0	225.859	wh cry	100 dec				s H_2O
1277	Indium(III) hydroxide	$In(OH)_3$	20661-21-6	165.840	cub cry			4.4		
1278	Indium(III) iodide	InI_3	13510-35-5	495.531	yel-red monocl cry; hyg	207		4.69	1308^{22}	
1279	Indium(III) nitrate trihydrate	$In(NO_3)_3 \cdot 3H_2O$	13770-61-1	354.879	col cry	100 dec				
1280	Indium(III) oxide	In_2O_3	1312-43-2	277.634	yel cub cry	1912		7.18		i H_2O; s hot acid
1281	Indium(III) perchlorate octahydrate	$In(ClO_4)_3 \cdot 8H_2O$	13465-15-1	557.292	wh cry	≈80	200 dec			
1282	Indium(III) phosphate	$InPO_4$	14693-82-4	209.789	wh orth cry			4.9		i H_2O
1283	Indium(III) selenide	In_2Se_3	1312-42-1	466.52	blk hex cry	660		5.8		
1284	Indium(III) sulfate	$In_2(SO_4)_3$	13464-82-9	517.824	hyg wh powder			3.44	117^{20}	
1285	Indium(III) sulfide	In_2S_3	12030-24-9	325.831	oran cub cry	1050		4.45		
1286	Indium(III) telluride	In_2Te_3	1312-45-4	612.44	blk cub cry	667		5.75		
1287	Iodine	I_2	7553-56-2	253.809	blue-blk plates	113.7	184.4	4.933	0.03^{20}	s bz, EtOH, eth, ctc, chl
1288	Iodic acid	HIO_3	7782-68-5	175.910	col orth cry	110 dec		4.63	308^{25}	i EtOH, eth

Inorganic

No.	Name	Formula	CAS Reg. No.	Mol. wt.	Phys. form	mp/°C	bp/°C	ρ/g cm⁻³	Sol. in g/100 g H_2O	Qualitative solubility
1289	Periodic acid dihydrate	$HIO_4·2H_2O$	10450-60-9	227.940	monocl hyg cry	122 dec				s H_2O, EtOH; sl eth
1290	Iodine tetroxide	I_2O_4	12399-08-5	317.807	yel cry	130	≈85 dec	4.2		sl H_2O
1291	Iodine pentoxide	I_2O_5	12029-98-0	333.806	hyg wh cry	≈300 dec		4.98	253.4^{20}	i EtOH, eth, CS_2
1292	Iodine hexoxide	I_2O_6	65355-99-9	349.805	yel solid	150 dec				reac H_2O
1293	Iodine nonaoxide	I_4O_9	73560-00-6	651.613	hyg yel powder	75 dec				
1294	Iodine bromide	IBr	7789-33-5	206.808	blk orth cry	40	116 dec	4.3		s H_2O, EtOH, eth
1295	Iodine chloride	ICl	7790-99-0	162.357	red cry or oily liq	27.38	97.0 dec	3.24		reac H_2O; s EtOH
1296	Iodine trichloride	ICl_3	865-44-1	233.263	yel tricl cry; hyg	101 tp (16 atm)	97 dec	3.2		reac H_2O; s EtOH, bz
1297	Iodine fluoride	IF	13873-84-2	145.902	wh pow (-78°C)	-14 dec				
1298	Iodine trifluoride	IF_3	22520-96-3	183.899	yel solid, stable at low temp	-28 dec				
1299	Iodine pentafluoride	IF_5	7783-66-6	221.896	yel liq	9.43	100.5	3.19		reac H_2O
1300	Iodine heptafluoride	IF_7	16921-96-3	259.893	col gas	6.5 tp	4.8 sp	0.010620^{25} gas		s H_2O
1301	Iodosyl trifluoride	IOF_3	19058-78-7	199.898	hyg col needles	>110 dec				reac H_2O
1302	Iodosyl pentafluoride	IOF_5	16056-61-4	237.895	col liq	4.5				
1303	Iodyl trifluoride	IO_2F_3	25402-50-0	215.898	yel solid	41	subl			
1304	Periodyl fluoride	IO_3F	30708-86-2	193.900	col cry	>100 dec				
1305	Iridium	Ir	7439-88-5	192.217	silv-wh metal; cub	2446	4428	22.562^{20}		s aqua regia
1306	Iridium carbonyl	$Ir_4(CO)_{12}$	11065-24-0	1104.990	yel cry	210 dec				
1307	Iridium(III) bromide	$IrBr_3$	10049-24-8	431.929	red-brn monocl cry			6.82		i H_2O, acid, alk
1308	Iridium(III) bromide tetrahydrate	$IrBr_3·4H_2O$	10049-24-8	503.991	grn-brn cry					s H_2O; i EtOH
1309	Iridium(III) chloride	$IrCl_3$	10025-83-9	298.576	brn monocl cry	763 dec		5.30		i H_2O, acid, alk
1310	Iridium(III) fluoride	IrF_3	23370-59-4	249.212	blk hex cry	250 dec		≈8.0		i H_2O, dil acid
1311	Iridium(III) iodide	IrI_3	7790-41-2	572.930	dark brn monocl cry			≈7.4		i H_2O, acid, bz, chl; s alk
1312	Iridium(III) oxide	Ir_2O_3	1312-46-5	432.432	blue-blk cry	1000 dec				i H_2O; sl hot HCl
1313	Iridium(III) 2,4-pentanedioate	$Ir(CH_3COCHCOCH_3)_3$	15635-87-7	489.541	oran-yel cry	270	subl			sl H_2O; s tol, chl, ace, MeOH
1314	Iridium(III) sulfide	Ir_2S_3	12136-42-4	480.629	orth cry			10.2		
1315	Iridium(IV) chloride	$IrCl_4$	10025-97-5	334.029	brn hyg solid	≈700 dec				s H_2O, EtOH
1316	Iridium(IV) oxide	IrO_2	12030-49-8	224.216	brn tetr cry	1100 dec		11.7		
1317	Iridium(IV) sulfide	IrS_2	12030-51-2	256.347	orth cry			9.3		
1318	Iridium(VI) fluoride	IrF_6	7783-75-7	306.207	yel cub cry; hyg	44	53.6	4.8		reac H_2O
1319	Iron	Fe	7439-89-6	55.845	silv-wh or gray met	1538	2861	7.87		s dil acid
1320	Ferrocene	$Fe(C_5H_5)_2$	102-54-5	186.031	oran needles	175	249			i H_2O; s EtOH, eth, bz, dil HNO_3
1321	Tetracarbonyldihydroiron	$Fe(CO)_4H_2$	12002-28-7	169.902	col liq, stab low temp	-70	-20 dec			s alk
1322	Iron pentacarbonyl	$Fe(CO)_5$	13463-40-6	195.896	yel oily liq; flam	-20	103	1.5^{20}		i H_2O; s eth, bz, ace
1323	Iron nonacarbonyl	$Fe_2(CO)_9$	15321-51-4	363.781	oran-yel cry	100 dec		2.85		
1324	Iron dodecacarbonyl	$Fe_3(CO)_{12}$	12088-65-2	503.656	dark grn cry	140		2.00		
1325	Iron arsenide	FeAs	12044-16-5	130.767	gray orth cry	1030		7.85		
1326	Iron boride (FeB)	FeB	12006-84-7	66.656	refrac solid; orth	1658		≈7		
1327	Iron boride (Fe_2B)	Fe_2B	12006-86-9	122.501	refrac solid; tetr	1389		7.3		
1328	Iron carbide	Fe_3C	12011-67-5	179.546	gray cub cry	1227		7.694		
1329	Iron phosphide (FeP)	FeP	26508-33-8	86.819	rhom cry			6.07		
1330	Iron phosphide (Fe_2P)	Fe_2P	1310-43-6	142.664	gray hex needles	1370		6.8		i H_2O, dil acid, alk
1331	Iron phosphide (Fe_3P)	Fe_3P	12023-53-9	198.509	gray solid	1100		6.74		i H_2O
1332	Iron disulfide	FeS_2	1317-66-4	119.975	blk cub cry	>600 dec		5.02		i H_2O
1333	Iron silicide	FeSi	12022-95-6	83.931	gray cub cry	1410		6.1		
1334	Iron disilicide	$FeSi_2$	12022-99-0	112.016	gray tetr cry	1220		4.74		
1335	Iron(II) acetate	$Fe(C_2H_3O_2)_2$	3094-87-9	173.934	wh cry	190 dec				s H_2O
1336	Iron(II) acetate tetrahydrate	$Fe(C_2H_3O_2)_2·4H_2O$	3094-87-9	245.994	grn cry	dec				s H_2O, EtOH
1337	Iron(II) aluminate	$Fe(AlO_2)_2$	12068-49-4	173.806	blk cub cry			4.3		
1338	Iron(II) arsenate	$Fe_3(AsO_4)_2$	10102-50-8	445.373	grn powder					i H_2O
1339	Iron(II) arsenate hexahydrate	$Fe_3(AsO_4)_2·6H_2O$	10102-50-8	553.465	grn amorp powder	dec				i H_2O; s acid
1340	Iron(II) bromide	$FeBr_2$	7789-46-0	215.653	yel-brn hex cry; hyg	691	dec	4.636	120^{25}	vs EtOH
1341	Iron(II) bromide hexahydrate	$FeBr_2·6H_2O$	13463-12-2	323.744	grn hyg cry	27 dec		4.64	120^{25}	s EtOH
1342	Iron(II) carbonate	$FeCO_3$	563-71-3	115.854	gray-brn hex cry			3.944	0.000062^{20}	
1343	Iron(II) chloride	$FeCl_2$	7758-94-3	126.751	wh hex cry; hyg	677	1023	3.16	65.0^{25}	vs EtOH, ace; sl bz
1344	Iron(II) chloride dihydrate	$FeCl_2·2H_2O$	16399-77-2	162.782	wh-grn monocl cry	120 dec		2.39	65.0^{25}	
1345	Iron(II) chloride tetrahydrate	$FeCl_2·4H_2O$	13478-10-9	198.813	grn monocl cry	105 dec		1.93	65.0^{25}	s EtOH
1346	Iron(II) chromite	$FeCr_2O_4$	1308-31-2	223.835	blk cub cry			5.0		

No.	Name	Formula	CAS Reg. No.	Mol. wt.	Phys. form	mp/°C	bp/°C	ρ/g cm⁻³	Sol. in g/100 g H_2O	Qualitative solubility
1347	Iron(II) fluoride	FeF_2	7789-28-8	93.842	wh tetr cry	1100		4.09		sl H_2O; s dil HF; i EtOH, eth
1348	Iron(II) fluoride tetrahydrate	$FeF_2 \cdot 4H_2O$	13940-89-1	165.904	col hex cry			2.20		
1349	Iron(II) hydroxide	$Fe(OH)_2$	18624-44-7	89.860	wh-grn hex cry			3.4	0.000052^{20}	
1350	Iron(II) iodide	FeI_2	7783-86-0	309.654	red-viol hex cry; hyg	594		5.3		s H_2O, EtOH, eth
1351	Iron(II) iodide tetrahydrate	$FeI_2 \cdot 4H_2O$	7783-86-0	381.716	blk hyg leaflets	90 dec		2.87		s H_2O, EtOH
1352	Iron(II) molybdate	$FeMoO_4$	13718-70-2	215.78	brn-yel monocl cry	1115		5.6		i H_2O
1353	Iron(II) nitrate	$Fe(NO_3)_2$	14013-86-6	179.854	grn solid				87.5^{25}	
1354	Iron(II) nitrate hexahydrate	$Fe(NO_3)_2 \cdot 6H_2O$	14013-86-6	287.946	grn solid	60 dec			87.5^{25}	
1355	Iron(II) orthosilicate	Fe_2SiO_4	10179-73-4	203.774	brn orth cry			4.30		
1356	Iron(II) oxalate dihydrate	$FeC_2O_4 \cdot 2H_2O$	6047-25-2	179.894	yel cry	150 dec		2.28	0.078^{25}	s acid
1357	Iron(II) oxide	FeO	1345-25-1	71.844	blk cub cry	1377		6.0		i H_2O, alk; s acid
1358	Iron(II) 2,4-pentanedioate	$Fe(CH_3COCHCOCH_3)_2$	14024-17-0	254.061	oran-brn cry	170	subl			sl bz, tol
1359	Iron(II) perchlorate	$Fe(ClO_4)_2$	13933-23-8	254.746	grn-wh hyg needles	>100 dec			210^{25}	
1360	Iron(II) phosphate octahydrate	$Fe_3(PO_4)_2 \cdot 8H_2O$	14940-41-1	501.600	gray-blue monocl cry; hyg			2.58		i H_2O; s acid
1361	Iron(II) selenide	FeSe	1310-32-3	134.81	blk hex cry			6.7		i H_2O
1362	Iron(II) sulfate	$FeSO_4$	7720-78-7	151.908	wh orth cry; hyg			3.65	29.5^{25}	
1363	Iron(II) sulfate monohydrate	$FeSO_4 \cdot H_2O$	17375-41-6	169.923	wh-yel monocl cry	300 dec		3.0	29.5^{25}	
1364	Iron(II) sulfate heptahydrate	$FeSO_4 \cdot 7H_2O$	7782-63-0	278.014	blue-grn monocl cry	≈60 dec		1.895	29.5^{25}	i EtOH
1365	Iron(II) sulfide	FeS	1317-37-9	87.910	col hex or tetr cry; hyg	1188	dec	4.7		i H_2O; reac acid
1366	Iron(II) tantalate	$Fe(TaO_3)_2$	12140-41-9	513.737	brn tetr cry			7.33		
1367	Iron(II) tartrate	$FeC_4H_4O_6$		203.916	wh cry				0.88	vs acid; s NH_4OH
1368	Iron(II) telluride	FeTe	12125-63-2	183.45	tetr cry	914		6.8		
1369	Iron(II) thiocyanate trihydrate	$Fe(SCN)_2 \cdot 3H_2O$	6010-09-9	226.055	grn monocl cry					s H_2O, EtOH, eth
1370	Iron(II) titanate	$FeTiO_3$	12168-52-4	151.710	blk rhomb cry	≈1470		4.72		
1371	Iron(II) tungstate	$FeWO_4$	13870-24-1	303.68	monocl cry			7.51		
1372	Iron(II,III) oxide	Fe_3O_4	1317-61-9	231.533	blk cub cry or amorp powder	1597		5.17		i H_2O; s acid
1373	Iron(III) acetate, basic	$FeOH(C_2H_3O_2)_2$	10450-55-2	190.941	brn-red amorp powder					i H_2O; s EtOH, acid
1374	Iron(III) ammonium citrate	$Fe(NH_4)_3(C_6H_5O_7)_2$	1185-57-5	488.160	red or brn pow; hyg					s H_2O; i EtOH
1375	Iron(III) arsenate dihydrate	$FeAsO_4 \cdot 2H_2O$	10102-49-5	230.795	grn-brn powder	dec		3.18		i H_2O; s dil acid
1376	Iron(III) bromide	$FeBr_3$	10031-26-2	295.557	dark red hex cry; hyg	dec		4.5	455^{25}	s EtOH, eth
1377	Iron(III) chloride	$FeCl_3$	7705-08-0	162.204	grn hex cry; hyg	307.6	≈316	2.90	91.2^{25}	s EtOH, eth, ace
1378	Iron(III) chloride hexahydrate	$FeCl_3 \cdot 6H_2O$	10025-77-1	270.295	yel-oran monocl cry; hyg	37 dec		1.82	91.2^{25}	s EtOH, eth, ace
1379	Iron(III) chromate	$Fe_2(CrO_4)_3$	10294-52-7	459.671	yel powder					i H_2O, EtOH; s acid
1380	Iron(III) citrate pentahydrate	$FeC_6H_5O_7 \cdot 5H_2O$	3522-50-7	335.021	red-brn cry					s H_2O; i EtOH
1381	Iron(III) dichromate	$Fe_2(Cr_2O_7)_3$	10294-53-8	759.654	red-brn solid					s H_2O, acid
1382	Iron(III) ferrocyanide	$Fe_4[Fe(CN)_6]_3$	14038-43-8	859.229	dark blue powder			1.80		i H_2O, dil acid, os
1383	Iron(III) fluoride	FeF_3	7783-50-8	112.840	grn hex cry	367		3.87	5.92^{25}	i EtOH, eth, bz
1384	Iron(III) fluoride trihydrate	$FeF_3 \cdot 3H_2O$	15469-38-2	166.886	yel-brn tetr cry			2.3	5.92^{25}	
1385	Iron(III) formate	$Fe(CHO_2)_3$	555-76-0	190.897	red-yel cry powder					s H_2O; sl EtOH
1386	Iron(III) hydroxide	$Fe(OH)_3$	1309-33-7	106.867	yel monocl cry			3.12		
1387	Iron(III) hydroxide oxide	$FeO(OH)$	20344-49-4	88.852	red-brn orth cry			4.26		i H_2O; s acid
1388	Iron(III) metavanadate	$Fe(VO_3)_3$	65842-03-7	352.665	gray-brn powder					i H_2O, EtOH; s acid
1389	Iron(III) nitrate	$Fe(NO_3)_3$	10421-48-4	241.860	cry				82.5^{20}	
1390	Iron(III) nitrate hexahydrate	$Fe(NO_3)_3 \cdot 6H_2O$	13476-08-9	349.951	viol cub cry	35 dec			82.5^{20}	
1391	Iron(III) nitrate nonahydrate	$Fe(NO_3)_3 \cdot 9H_2O$	7782-61-8	403.997	viol-gray hyg cry	47 dec		1.68	82.5^{20}	vs EtOH, ace
1392	Iron(III) oxalate	$Fe_2(C_2O_4)_3$	19469-07-9	375.747	yel amorp powder	100 dec				s H_2O, acid; i alk
1393	Iron(III) oxide	Fe_2O_3	1309-37-1	159.688	red-brn hex cry	1539		5.25		i H_2O; s acid
1394	Iron(III) 2,4-pentanedioate	$Fe(CH_3COCHCOCH_3)_3$	14024-18-1	353.169	red-oran cry	188		5.24		sl H_2O; s os
1395	Iron(III) perchlorate hexahydrate	$Fe(ClO_4)_3 \cdot 6H_2O$	32963-81-8	462.288	viol cry					
1396	Iron(III) phosphate dihydrate	$FePO_4 \cdot 2H_2O$	10045-86-0	186.847	gray-wh orth cry			2.87		i H_2O; s HCl
1397	Iron(III) pyrophosphate nonahydrate	$Fe_4(P_2O_7)_3 \cdot 9H_2O$	10058-44-3	907.348	yel powder					i H_2O; s acid
1398	Iron(III) hypophosphite	$Fe(H_2PO_2)_3$	7783-84-8	250.811	wh-gray powder					i H_2O
1399	Iron(III) sodium pyrophosphate	$FeNaP_2O_7$	10045-87-1	252.778	wh powder			1.5		i H_2O; s HCl
1400	Iron(III) sulfate	$Fe_2(SO_4)_3$	10028-22-5	399.878	gray-wh rhomb cry; hyg			3.10	440^{20}	sl EtOH; i ace
1401	Iron(III) sulfate nonahydrate	$Fe_2(SO_4)_3 \cdot 9H_2O$	13520-56-4	562.015	yel hex cry	400 dec		2.1	440^{20}	
1402	Iron(III) thiocyanate	$Fe(SCN)_3$	4119-52-2	230.092	red-viol hyg cry	dec				s H_2O, EtOH, ace; i tol, chl
1403	Krypton	Kr	7439-90-9	83.798	col gas	-157.37	-153.415	$2.417^{-153.4}$ liq		sl H_2O

No.	Name	Formula	CAS Reg. No.	Mol. wt.	Phys. form	mp/°C	bp/°C	ρ/g cm^{-3}	Sol. in g/100 g H_2O	Qualitative solubility
1404	Krypton difluoride	KrF_2	13773-81-4	121.795	col tetr cry	≈25 dec		3.24		reac H_2O
1405	Krypton fluoride hexafluoroantimonate	$KrFSb_2F_{11}$	39578-36-4	555.299	wh solid	45 dec				
1406	Lanthanum	La	7439-91-0	138.905	silv metal; hex	920	3464	6.15		s dil acid
1407	Lanthanum aluminum oxide	$LaAlO_3$	12003-65-5	213.885	wh rhom cry	500 trans cubic				
1408	Lanthanum boride	LaB_6	12008-21-8	203.771	blk cub cry; refrac	2715		4.76		
1409	Lanthanum bromate nonahydrate	$LaBrO_3 \cdot 9H_2O$		684.749	hex cry	100 dec		5.06		vs H_2O
1410	Lanthanum bromide	$LaBr_3$	13536-79-3	378.617	wh hex cry; hyg	788		5.1		s H_2O
1411	Lanthanum carbide	LaC_2	12071-15-7	162.926	tetr cry	2360		5.29		
1412	Lanthanum carbonate octahydrate	$La_2(CO_3)_3 \cdot 8H_2O$	6487-39-4	601.960	wh cry powder			2.6		i H_2O; s dil acid
1413	Lanthanum chloride	$LaCl_3$	10099-58-8	245.264	wh hex cry; hyg	858		3.84	95.7^{25}	
1414	Lanthanum chloride heptahydrate	$LaCl_3 \cdot 7H_2O$	20211-76-1	371.371	wh tricl cry; hyg	91 dec			95.7^{25}	s EtOH
1415	Lanthanum fluoride	LaF_3	13709-38-1	195.900	wh hex cry; hyg	1493		5.9		i H_2O, acid
1416	Lanthanum hydride	LaH_3	13864-01-2	141.929	blk cub cry			5.36		
1417	Lanthanum hydroxide	$La(OH)_3$	14507-19-8	189.927	wh amorp solid	dec			0.000020^{20}	
1418	Lanthanum iodate	$La(IO_3)_3$	13870-19-4	663.614	col cry				1.7	
1419	Lanthanum iodide	LaI_3	13813-22-4	519.619	wh orth cry; hyg	778		5.6		s H_2O
1420	Lanthanum nitrate hexahydrate	$La(NO_3)_3 \cdot 6H_2O$	10277-43-7	433.011	wh hyg tricl cry	≈40 dec			200^{25}	vs EtOH; s ace
1421	Lanthanum nitride	LaN	25764-10-7	152.912	cub cry			6.73		
1422	Lanthanum oxide	La_2O_3	1312-81-8	325.809	wh amorp powder	2304	3620	6.51		i H_2O; s dil acid
1423	Lanthanum perchlorate hexahydrate	$La(ClO_4)_3 \cdot 6H_2O$	36907-37-6	475.021	hyg col cry	100 dec				vs H_2O; s EtOH
1424	Lanthanum silicide	$LaSi_2$	12056-90-5	195.076	gray tetr cry	1520		5.0		
1425	Lanthanum sulfate	$La_2(SO_4)_3$	10099-60-2	565.999	hyg wh pow	1150				sl H_2O
1426	Lanthanum sulfate octahydrate	$La_2(SO_4)_3 \cdot 8H_2O$	57804-25-8	710.10	col cry	dec		2.82		sl H_2O
1427	Lanthanum sulfate nonahydrate	$La_2(SO_4)_3 \cdot 9H_2O$	10294-62-9	728.136	hex cry			2.82	2.7^{20}	i EtOH
1428	Lanthanum monosulfide	LaS	12031-30-0	170.970	yel cub cry	2300		5.61		
1429	Lanthanum sulfide	La_2S_3	12031-49-1	374.006	red cub cry	2110		4.9		
1430	Lawrencium	Lr	22537-19-5	262	metal	1627				
1431	Lead	Pb	7439-92-1	207.2	soft silv-gray metal; cub	327.462	1749	11.3		s conc acid
1432	Plumbane	PbH_4	15875-18-0	211.2	unstab col gas		-13			
1433	Lead(II) acetate	$Pb(C_2H_3O_2)_2$	301-04-2	325.3	wh cry	280	dec	3.25	44.3^{20}	
1434	Lead(II) acetate trihydrate	$Pb(C_2H_3O_2)_2 \cdot 3H_2O$	6080-56-4	379.3	col cry	75 dec		2.55		vs H_2O; sl EtOH
1435	Lead(II) acetate, basic	$Pb(C_2H_3O_2)_2 \cdot 2Pb(OH)_2$	1335-32-6	807.7	wh powder	dec			6.3^{0}	
1436	Lead(II) antimonate	$Pb_3(SbO_4)_2$	13510-89-9	993.1	oran-yel powder			6.58		i H_2O, dil acid
1437	Lead(II) arsenate	$Pb_3(AsO_4)_2$	3687-31-8	899.4	wh cry	1042 dec		5.8		i H_2O; s HNO_3
1438	Lead(II) arsenite	$Pb(AsO_2)_2$	10031-13-7	421.0	wh powder			5.85		i H_2O; s dil HNO_3
1439	Lead(II) azide	$Pb(N_3)_2$	13424-46-9	291.2	col orth needles; exp	≈350 exp		4.7	0.023^{18}	vs HOAc
1440	Lead(II) borate monohydrate	$Pb(BO_2)_2 \cdot H_2O$	10214-39-8	310.8	wh powder	500 dec		5.6		i H_2O; s dil HNO_3
1441	Lead(II) bromate monohydrate	$Pb(BrO_3)_2 \cdot H_2O$	10031-21-7	481.0	col cry	≈180 dec		5.53	1.33^{20}	
1442	Lead(II) bromide	$PbBr_2$	10031-22-8	367.0	wh orth cry	371	892	6.69	0.975^{25}	i EtOH
1443	Lead(II) butanoate	$Pb(C_4H_7O_2)_2$	819-73-8	381.4	col solid	≈90				i H_2O; s dil HNO_3
1444	Lead(II) carbonate	$PbCO_3$	598-63-0	267.2	col orth cry	≈315 dec		6.582		i H_2O
1445	Lead(II) carbonate, basic	$Pb(OH)_2 \cdot 2PbCO_3$	1319-46-6	775.6	wh hex cry	400 dec		≈6.5		i H_2O, EtOH; s acid
1446	Lead(II) chlorate	$Pb(ClO_3)_2$	10294-47-0	374.1	col hyg cry	230 dec		3.9	144^{18}	vs EtOH
1447	Lead(II) chloride	$PbCl_2$	7758-95-4	278.1	wh orth needles or powder	501	951	5.98	1.08^{25}	s alk
1448	Lead(II) chloride fluoride	PbClF	13847-57-9	261.7	tetr cry			7.05	0.035^{20}	
1449	Lead(II) chlorite	$Pb(ClO_2)_2$	13453-57-1	342.1	yel monocl cry	126 dec			0.2^{25}	sl H_2O; s alk
1450	Lead(II) chromate	$PbCrO_4$	7758-97-6	323.2	yel-oran monocl cry	844		6.12	0.000017^{20}	s alk, dil acid
1451	Lead(II) chromate(VI) oxide	$PbCrO_4 \cdot PbO$	18454-12-1	546.4	red powder					i H_2O
1452	Lead(II) citrate trihydrate	$Pb_3(C_6H_5O_7)_2 \cdot 3H_2O$	512-26-5	1053.8	wh cry powder					s H_2O; sl EtOH
1453	Lead(II) cyanide	$Pb(CN)_2$	592-05-2	259.2	wh-yel powder					sl H_2O; reac acid
1454	Lead(II) 2-ethylhexanoate	$Pb(C_7H_{15}CO_2)_2$	301-08-6	493.6	visc liq			1.56		
1455	Lead(II) fluoride	PbF_2	7783-46-2	245.2	wh orth cry	830	1293	8.44	0.0670^{25}	
1456	Lead(II) fluoroborate	$Pb(BF_4)_2$	13814-96-5	380.8	stab only in aq soln					s H_2O
1457	Lead(II) formate	$Pb(CHO_2)_2$	811-54-1	297.2	wh prisms or needles	190 dec		4.63	1.6^{16}	i EtOH
1458	Lead(II) hexafluoro-2,4-pentanedioate	$Pb(CF_3COCHCOCF_3)_2$	19648-88-5	621.3	cry	155	210			
1459	Lead(II) hydrogen arsenate	$PbHAsO_4$	7784-40-9	347.1	wh monocl cry	280 dec		5.943		i H_2O; s HNO_3, alk
1460	Lead(II) hydrogen phosphate	$PbHPO_4$	15845-52-0	303.2	wh monocl cry	dec		5.66		
1461	Lead(II) hydroxide	$Pb(OH)_2$	19783-14-3	241.2	wh powder	145 dec		5.69	0.00012^{20}	s acid
1462	Lead(II) iodate	$Pb(IO_3)_2$	25659-31-8	557.0	wh orth cry			6.50	0.0025^{25}	
1463	Lead(II) iodide	PbI_2	10101-63-0	461.0	yel hex cry or powder	410	872 dec	6.16	0.076^{25}	i EtOH
1464	Lead(II) lactate	$Pb(C_3H_5O_3)_2$	18917-82-3	385.3	wh cry powder					s H_2O, hot EtOH
1465	Lead(II) molybdate	$PbMoO_4$	10190-55-3	367.1	yel tetr cry	≈1060		6.7		i H_2O; s HNO_3, NaOH

Inorganic

No.	Name	Formula	CAS Reg. No.	Mol. wt.	Phys. form	mp/°C	bp/°C	ρ/g cm^{-3}	Sol. in g/100 g H_2O	Qualitative solubility
1466	Lead(II) niobate	$Pb(NbO_3)_2$	12034-88-7	489.0	rhom or tetr cry	1343		6.6		i H_2O
1467	Lead(II) nitrate	$Pb(NO_3)_2$	10099-74-8	331.2	col cub cry	470		4.53	59.7^{25}	sl EtOH
1468	Lead(II) oleate	$Pb(C_{18}H_{33}O_2)_2$	1120-46-3	770.1	wax-like solid					i H_2O; s EtOH, bz, eth
1469	Lead(II) oxalate	PbC_2O_4	814-93-7	295.2	wh powder	300 dec		5.28	0.00025^{20}	s dil HNO_3
1470	Lead(II) oxide (litharge)	PbO	1317-36-8	223.2	red tetr cry	489 trans massicot		9.35		i H_2O, EtOH; s dil HNO_3
1471	Lead(II) oxide (massicot)	PbO	1317-36-8	223.2	yel orth cry	887		9.64		i H_2O, EtOH; s dil HNO_3
1472	Lead(II) oxide hydrate	$3PbO{\cdot}H_2O$	1311-11-1	687.6	wh powder			7.41		i H_2O; s dil acid
1473	Lead(II) 2,4-pentanedioate	$Pb(CH_3COCHCOCH_3)_2$	15282-88-9	405.4	cry	143				
1474	Lead(II) perchlorate	$Pb(ClO_4)_2$	13453-62-8	406.1	wh cry				441^{25}	
1475	Lead(II) perchlorate trihydrate	$Pb(ClO_4)_2{\cdot}3H_2O$	13637-76-8	460.1	wh cry	100 dec		2.6	441^{25}	s EtOH
1476	Lead(II) phosphate	$Pb_3(PO_4)_2$	7446-27-7	811.5	wh hex cry	1014		7.01		i H_2O, EtOH
1477	Lead(II) hypophosphite	$Pb(H_2PO_2)_2$	10294-58-3	337.2	hyg cry powder	dec				sl H_2O; i EtOH
1478	Lead(II) metasilicate	$PbSiO_3$	10099-76-0	283.3	wh monocl cry powder	764		6.49		i H_2O, os
1479	Lead(II) orthosilicate	Pb_2SiO_4	13566-17-1	506.5	monocl cry	743		7.60		
1480	Lead(II) hexafluorosilicate dihydrate	$PbSiF_6{\cdot}2H_2O$	1310-03-8	385.3	col cry	dec				vs H_2O
1481	Lead(II) selenate	$PbSeO_4$	7446-15-3	350.2	orth cry			6.37	0.013^{25}	s conc acid
1482	Lead(II) selenide	PbSe	12069-00-0	286.2	gray cub cry	1078		8.1		i H_2O; s HNO_3
1483	Lead(II) selenite	$PbSeO_3$	7488-51-9	334.2	wh monocl cry	≈500		7.0		i H_2O
1484	Lead(II) sodium thiosulfate	$Na_4Pb(S_2O_3)_3$	10101-94-7	635.5	wh cry					sl H_2O
1485	Lead(II) stearate	$Pb(C_{18}H_{35}O_2)_2$	1072-35-1	774.1	wh powder	≈100		1.4		i H_2O; s hot EtOH
1486	Lead(II) sulfate	$PbSO_4$	7446-14-2	303.3	orth cry	1087		6.29	0.0044^{25}	i acid; sl alk
1487	Lead(II) sulfide	PbS	1314-87-0	239.3	blk powder or silv cub cry	1113		7.60		i H_2O; s acid
1488	Lead(II) sulfite	$PbSO_3$	7446-10-8	287.3	wh powder	dec				i H_2O; s HNO_3
1489	Lead(II) tantalate	$Pb(TaO_3)_2$	12065-68-8	665.1	orth cry			7.9		i H_2O
1490	Lead(II) telluride	PbTe	1314-91-6	334.8	gray cub cry	924		8.164		i H_2O, acid
1491	Lead(II) thiocyanate	$Pb(SCN)_2$	592-87-0	323.4	wh-yel powder			3.82	0.05^{20}	
1492	Lead(II) thiosulfate	PbS_2O_3	13478-50-7	319.3	wh cry	dec		5.18		i H_2O; s acid
1493	Lead(II) titanate	$PbTiO_3$	12060-00-3	303.1	yel tetr cry			7.9		i H_2O; reac HCl
1494	Lead(II) tungstate (stolzite)	$PbWO_4$	7759-01-5	455.0	yel tetr cry	1130		8.24	0.03^{20}	s alk
1495	Lead(II) tungstate (raspite)	$PbWO_4$	7759-01-5	455.0	monocl cry	400 trans		8.46	0.03^{20}	s alk
1496	Lead(II) metavanadate	$Pb(VO_3)_2$	10099-79-3	405.1	yel powder					i H_2O; reac HNO_3
1497	Lead(II) zirconate	$PbZrO_3$	12060-01-4	346.4	col orth cry			≈8		i H_2O, alk; s acid
1498	Lead(II,IV) oxide	Pb_2O_3	1314-27-8	462.4	blk monocl cry or red amorp powder	530 dec		10.05		i H_2O; s alk; reac conc HCl
1499	Lead(II,II,IV) oxide	Pb_3O_4	1314-41-6	685.6	red tetr cry	830		8.92		i H_2O, EtOH; s hot HCl
1500	Lead(IV) acetate	$Pb(C_2H_3O_2)_4$	546-67-8	443.4	col monocl cry	≈175		2.23		reac H_2O, EtOH; s bz, chl
1501	Lead(IV) bromide	$PbBr_4$	13701-91-2	526.8	unstab liq					
1502	Lead(IV) chloride	$PbCl_4$	13463-30-4	349.0	yel oily liq	-15	≈50 dec			
1503	Lead(IV) fluoride	PbF_4	7783-59-7	283.2	wh tetr cry; hyg	≈600		6.7		
1504	Lead(IV) oxide	PbO_2	1309-60-0	239.2	red tetr cry or brn powder	290 dec		9.64		
1505	Lithium	Li	7439-93-2	6.94	soft silv-wh metal	180.50	1342	0.534		reac H_2O
1506	Lithium acetate	$LiC_2H_3O_2$	546-89-4	65.985	cry	286			45.0^{25}	vs EtOH
1507	Lithium acetate dihydrate	$LiC_2H_3O_2{\cdot}2H_2O$	6108-17-4	102.016	wh rhomb cry	58 dec		1.3	45.0^{25}	s EtOH
1508	Lithium aluminum hydride	$LiAlH_4$	16853-85-3	37.955	gray-wh monocl cry	>125 dec		0.917		reac H_2O, EtOH; s eth, thf
1509	Lithium aluminum silicate	$LiAlSi_2O_6$	12068-40-5	186.090	wh monocl cry	1430		3.188		
1510	Lithium amide	$LiNH_2$	7782-89-0	22.964	tetr cry	380		1.18		reac H_2O
1511	Lithium arsenate	Li_3AsO_4	13478-14-3	159.743	col orth cry			3.07		sl H_2O; s HOAc
1512	Lithium azide	LiN_3	19597-69-4	48.961	hyg monocl cry; exp			1.83		vs H_2O
1513	Lithium borohydride	$LiBH_4$	16949-15-8	21.784	wh-gray orth cry or powder	268	380 dec	0.66		s alk, eth, thf
1514	Lithium bromate	$LiBrO_3$	13550-28-2	134.843	hyg col orth cry	260			65.4^{25}	vs H_2O
1515	Lithium bromide	LiBr	7550-35-8	86.845	wh cub cry; hyg	550	≈1300	3.464	181^{25}	s EtOH, eth
1516	Lithium bromide monohydrate	$LiBr{\cdot}H_2O$	23303-71-1	104.860	wh orth cry	33 trans cubic		3.46	145^{4}	vs H_2O
1517	Lithium carbide	Li_2C_2	1070-75-3	37.903	wh hyg cry			1.65		reac H_2O; i os
1518	Lithium carbonate	Li_2CO_3	554-13-2	73.891	wh monocl cry	732	1300 dec	2.11	1.30^{25}	s acid; i EtOH
1519	Lithium chlorate	$LiClO_3$	13453-71-9	90.392	col hyg rhom needles	127.6	300 dec	1.119	459^{25}	vs EtOH; sl ace
1520	Lithium chloride	LiCl	7447-41-8	42.394	wh cub cry or powder; hyg	610	1383	2.07	84.5^{25}	s EtOH, ace, py
1521	Lithium chloride monohydrate	$LiCl{\cdot}H_2O$	16712-20-2	60.409	hyg wh tetr cry	98 dec		1.78		vs H_2O
1522	Lithium chromate dihydrate	$Li_2CrO_4{\cdot}2H_2O$	7789-01-7	165.906	yel orth cry; hyg	75 dec		2.15		vs H_2O; s EtOH
1523	Lithium citrate tetrahydrate	$Li_3C_6H_5O_7{\cdot}4H_2O$	6680-58-6	281.983	wh cry	210 (anh)			75^{25}	vs H_2O; sl EtOH
1524	Lithium cobaltite	$LiCoO_2$	12190-79-3	97.873	dark gray pow					i H_2O
1525	Lithium cyanide	LiCN	2408-36-8	32.959	wh orth cry	160				

No.	Name	Formula	CAS Reg. No.	Mol. wt.	Phys. form	mp/°C	bp/°C	ρ/g cm^{-3}	Sol. in g/100 g H_2O	Qualitative solubility
1526	Lithium hydride-*d*	LiD	13587-16-1	8.955	hyg wh cry	694		0.82		reac H_2O
1527	Lithium dichromate dihydrate	$Li_2Cr_2O_7 \cdot 2H_2O$	10022-48-7	265.901	yel-red hyg cry	130 dec		2.34		vs H_2O
1528	Lithium dihydrogen phosphate	LiH_2PO_4	13453-80-0	103.928	col hyg cry	>100		2.461	126^{0}	
1529	Lithium diisopropylamide	$LiN(C_3H_7)_2$	4111-54-0	107.123	hyg col cry	dec				s eth; i hc
1530	Lithium ferrosilicon	LiFeSi	64082-35-5	90.872	dark brittle cry					reac H_2O
1531	Lithium fluoride	LiF	7789-24-4	25.939	wh cub cry or powder	848.2	1673	2.640	0.134^{25}	s acid
1532	Lithium formate monohydrate	$Li(CHO_2) \cdot H_2O$	6108-23-2	69.974	col-wh cry			1.46		s H_2O
1533	Lithium hexafluoroantimonate	$LiSbF_6$	18424-17-4	242.691	hyg pow	dec				
1534	Lithium hexafluoroarsenate	$LiAsF_6$	29935-35-1	310.672	rhom wh cry; hyg					
1535	Lithium hexafluorophosphate	$LiPF_6$	21324-40-3	151.905	wh pow	194		2.84		
1536	Lithium hexafluorosilicate	Li_2SiF_6	17347-95-4	155.958	col hex cry	350 dec				sl ace
1537	Lithium hexafluorostannate	Li_2SnF_6	17029-16-2	246.582	wh pow					
1538	Lithium hydride	LiH	7580-67-8	7.949	gray cub cry or powder; hyg	692		0.78		reac H_2O, EtOH
1539	Lithium hydrogen carbonate	$LiHCO_3$	5006-97-3	67.958	wh pow					sl H_2O
1540	Lithium hydroxide	LiOH	1310-65-2	23.948	col tetr cry	473	1626	1.45	12.5^{25}	sl EtOH
1541	Lithium hydroxide monohydrate	$LiOH \cdot H_2O$	1310-66-3	41.964	wh monocl cry or powder			1.51	12.5^{25}	sl EtOH
1542	Lithium hypochlorite	LiOCl	13840-33-0	58.393	wh pow					vs H_2O
1543	Lithium iodate	$LiIO_3$	13765-03-2	181.843	wh hyg hex cry	450		4.502	77.9^{25}	i EtOH
1544	Lithium iodide	LiI	10377-51-2	133.845	wh cub cry; hyg	469	1171	4.06	165^{25}	
1545	Lithium iodide trihydrate	$LiI \cdot 3H_2O$	7790-22-9	187.891	wh hyg cry	73		2.38	165^{25}	vs EtOH, ace
1546	Lithium manganate	Li_2MnO_3	12163-00-7	116.818	red-brn monocl cry			3.90		i H_2O
1547	Lithium metaborate	$LiBO_2$	13453-69-5	49.751	wh monocl cry; hyg	844		2.18	2.6^{20}	sl H_2O; s EtOH
1548	Lithium metaborate dihydrate	$LiBO_2 \cdot 2H_2O$	15293-74-0	85.782	wh cry pow			1.8		s H_2O
1549	Lithium metaphosphate	$LiPO_3$	13762-75-9	85.913	wh cry or gl solid			1.8		i H_2O
1550	Lithium metasilicate	Li_2SiO_3	10102-24-6	89.966	wh orth needles	1201		2.52		i cold H_2O; reac dil acid
1551	Lithium molybdate	Li_2MoO_4	13568-40-6	173.82	hyg wh cry	702		2.66	44.8^{25}	s H_2O
1552	Lithium niobate	$LiNbO_3$	12031-63-9	147.845	wh hex cry	1240		4.30		
1553	Lithium nitrate	$LiNO_3$	7790-69-4	68.946	col hex cry; hyg	253		2.38	102^{25}	s EtOH
1554	Lithium nitride	Li_3N	26134-62-3	34.830	red hex cry	813		1.27		reac H_2O
1555	Lithium nitrite	$LiNO_2$	13568-33-7	52.947	wh hyg cry	222				vs H_2O
1556	Lithium nitrite monohydrate	$LiNO_2 \cdot H_2O$	13568-33-7	70.962	col needles	>100		1.615	139.5^{25}	vs H_2O, EtOH
1557	Lithium orthosilicate	$LiSiO_4$	13453-84-4	99.025	wh rhom cry	1256		2.39		
1558	Lithium oxalate	$Li_2C_2O_4$	30903-87-8	101.901	col cry	dec		2.121^{17}		s H_2O; i EtOH, eth
1559	Lithium phosphate	Li_3PO_4	10377-52-3	115.794	wh orth cry	1205		2.46	0.027^{25}	
1560	Lithium oxide	Li_2O	12057-24-8	29.881	wh cub cry	1438		2.013		
1561	Lithium perchlorate	$LiClO_4$	7791-03-9	106.392	wh orth cry or powder	236	430 dec	2.428	58.7^{25}	s EtOH, ace, eth
1562	Lithium perchlorate trihydrate	$LiClO_4 \cdot 3H_2O$	13453-78-6	160.438	wh hex cry	95 dec		1.84		vs H_2O, EtOH, ace; i eth
1563	Lithium peroxide	Li_2O_2	12031-80-0	45.881	wh hex cry			2.31		s H_2O; i EtOH
1564	Lithium selenate monohydrate	$Li_2SeO_4 \cdot H_2O$	7790-71-8	174.86	monocl cry			2.56		vs H_2O
1565	Lithium selenite monohydrate	$Li_2SeO_3 \cdot H_2O$	15593-51-8	158.86	hyg cry					
1566	Lithium stearate	$LiC_{18}H_{35}O_2$	4485-12-5	290.411	cry	≈220				
1567	Lithium sulfate	Li_2SO_4	10377-48-7	109.945	wh monocl cry; hyg	860		2.21	34.2^{25}	
1568	Lithium sulfate monohydrate	$Li_2SO_4 \cdot H_2O$	10102-25-7	127.960	col cry	130 dec		2.06	34.2^{25}	sl EtOH
1569	Lithium sulfide	Li_2S	12136-58-2	45.947	wh cub cry; hyg	1372		1.64		
1570	Lithium tantalate	$LiTaO_3$	12031-66-2	235.887	wh pow	1650		7.45		
1571	Lithium tetraborate	$Li_2B_4O_7$	12007-60-2	169.122	wh tetr cry	917			2.9^{20}	sl H_2O
1572	Lithium tetraborate pentahydrate	$Li_2B_4O_7 \cdot 5H_2O$	1303-94-2	259.198	wh cry pow	200 dec				vs H_2O; i EtOH
1573	Lithium tetracyanoplatinate pentahydrate	$Li_2Pt(CN)_4 \cdot 5H_2O$	14402-73-4	403.112	grn-yel cry					sl H_2O
1574	Lithium tetrafluoroborate	$LiBF_4$	14283-07-9	93.746	hyg wh pow	dec				vs H_2O
1575	Lithium thiocyanate	LiSCN	556-65-0	65.024	wh hyg cry				120^{25}	
1576	Lithium titanate	Li_2TiO_3	12031-82-2	109.747	wh pow	1325				i H_2O
1577	Lithium tungstate	Li_2WO_4	13568-45-1	261.72	wh trig pow	740		3.71		s H_2O
1578	Lithium vanadate	$LiVO_3$	15060-59-0	105.881	yel pow		1400 subl			
1579	Lithium zirconate	Li_2ZrO_3	12031-83-3	153.104	wh solid					
1580	Lutetium	Lu	7439-94-3	174.967	silv metal; hex	1663	3402	9.84		s dil acid
1581	Lutetium boride	LuB_4	12688-52-7	218.211	tetr cry	2600		≈7.0		
1582	Lutetium bromide	$LuBr_3$	14456-53-2	414.679	wh hyg cry	1025				vs H_2O
1583	Lutetium chloride	$LuCl_3$	10099-66-8	281.326	wh monocl cry; hyg	925		3.98		s H_2O
1584	Lutetium chloride hexahydrate	$LuCl_3 \cdot 6H_2O$	15230-79-2	389.417	col cry	150 dec				s H_2O, EtOH
1585	Lutetium fluoride	LuF_3	13760-81-1	231.962	orth cry	1182	2200	8.3		i H_2O
1586	Lutetium iodide	LuI_3	13813-45-1	555.680	brn hex cry; hyg	1050		≈5.6		vs H_2O
1587	Lutetium iron oxide	$Lu_3Fe_5O_{12}$	12023-71-1	996.119	cry					
1588	Lutetium nitrate	$Lu(NO_3)_3$	10099-67-9	360.982	hyg col solid					s H_2O, EtOH
1589	Lutetium nitride	LuN	12125-25-6	188.974	cub cry			11.6		
1590	Lutetium oxide	Lu_2O_3	12032-20-1	397.932	wh cub cry or powder	2490	3980	9.41		
1591	Lutetium perchlorate hexahydrate	$Lu(ClO_4)_3 \cdot 6H_2O$	14646-29-8	581.410	col cry	350 (anh) dec				s H_2O, MeOH
1592	Lutetium sulfate	$Lu_2(SO_4)_3$	14986-89-1	638.122	wh pow	>850 dec				vs H_2O
1593	Lutetium sulfate octahydrate	$Lu_2(SO_4)_3 \cdot 8H_2O$	13473-77-3	782.244	col cry				42.3^{20}	s H_2O

No.	Name	Formula	CAS Reg. No.	Mol. wt.	Phys. form	mp/°C	bp/°C	ρ/g cm⁻³	Sol. in g/100 g H_2O	Qualitative solubility
1594	Lutetium sulfide	Lu_2S_3	12163-20-1	446.129	gray rhomb cry	1750 dec		6.26		
1595	Lutetium telluride	Lu_2Te_3	12163-22-3	732.73	orth cry			7.8		
1596	Magnesium	Mg	7439-95-4	24.305	silv-wh metal	650	1090	1.74		s dil acid
1597	Magnesium acetate	$Mg(C_2H_3O_2)_2$	142-72-3	142.394	wh orth/monocl cry	323 dec		1.50	65.6^{25}	
1598	Magnesium acetate monohydrate	$Mg(C_2H_3O_2)_2·H_2O$	60582-92-5	160.409	orth cry			1.55		
1599	Magnesium acetate tetrahydrate	$Mg(C_2H_3O_2)_2·4H_2O$	16674-78-5	214.454	col monocl cry; hyg	80 dec		1.45	65.6^{25}	vs EtOH
1600	Magnesium aluminate	$Mg(AlO_2)_2$	12068-51-8	142.266	col cub cry	2105		3.55		i H_2O
1601	Magnesium aluminum silicate	$Mg_2Al_3(AlSi_5O_{18})$	1302-88-1	584.953	blue cry			2.6		
1602	Magnesium amide	$Mg(NH_2)_2$	7803-54-5	56.350	wh powder; flam	dec		1.39		reac H_2O
1603	Magnesium ammonium phosphate hexahydrate	$MgNH_4PO_4·6H_2O$	13478-16-5	245.407	wh pow	dec		1.71		i H_2O, EtOH; s acid
1604	Magnesium antimonide	Mg_3Sb_2	12057-75-9	316.435	hex cry	1245		3.99		
1605	Magnesium arsenide	Mg_3As_2	12044-49-4	222.758	solid	≈1200		3.15		i H_2O
1606	Magnesium diboride	MgB_2	12007-25-9	45.927	hex cry	800 dec		2.57		
1607	Magnesium hexaboride	MgB_6	12008-22-9	89.171	refrac solid	1100 dec				i H_2O
1608	Magnesium dodecaboride	MgB_{12}	12230-32-9	154.037	refrac solid	1300 dec				
1609	Magnesium bromate hexahydrate	$Mg(BrO_3)_2·6H_2O$	7789-36-8	388.201	col cub cry	200 dec		2.29	98^{25}	
1610	Magnesium bromide	$MgBr_2$	7789-48-2	184.113	wh hex cry; hyg	711		3.72	102^{25}	
1611	Magnesium bromide hexahydrate	$MgBr_2·6H_2O$	13446-53-2	292.204	col monocl cry	165 dec		2.0	102^{25}	s EtOH
1612	Magnesium carbonate	$MgCO_3$	546-93-0	84.314	wh hex cry	990		3.010	0.18^{20}	i EtOH; s acid
1613	Magnesium carbonate dihydrate	$MgCO_3·2H_2O$	5145-48-2	120.345	col tricl cry			2.8		i H_2O, ace, NH_4OH
1614	Magnesium carbonate trihydrate	$MgCO_3·3H_2O$	14457-83-1	138.360	col monocl cry	165		1.8	0.18^{16}	
1615	Magnesium carbonate pentahydrate	$MgCO_3·5H_2O$	61042-72-6	174.390	wh monocl cry	>400 dec		3.04	0.38^{16}	
1616	Magnesium carbonate hydroxide tetrahydrate	$4MgCO_3·Mg(OH)_2·4H_2O$	39409-82-0	467.636	wh monocl cry			2.3		
1617	Magnesium carbonate hydroxide pentahydrate	$4MgCO_3·Mg(OH)_2·5H_2O$	56378-72-4	485.652	wh pow	700 dec				i H_2O; s dil acid; i EtOH
1618	Magnesium carbonate dihydroxide trihydrate	$MgCO_3·Mg(OH)_2·3H_2O$	12143-96-3	196.680	wh monocl cry	dec		2.04		
1619	Magnesium chlorate hexahydrate	$Mg(ClO_3)_2·6H_2O$	13446-19-0	299.299	wh hyg cry	≈35 dec		1.80	142^{25}	sl EtOH
1620	Magnesium chloride	$MgCl_2$	7786-30-3	95.211	wh hex leaflets; hyg	714	1412	2.325	56.0^{25}	
1621	Magnesium chloride hexahydrate	$MgCl_2·6H_2O$	7791-18-6	203.302	wh hyg cry	≈100 dec		1.56	56.0^{25}	s EtOH
1622	Magnesium chromate heptahydrate	$MgCrO_4·7H_2O$	13423-61-5	266.405	yel rhom cry			1.695	54.8^{25}	
1623	Magnesium chromite	$MgCr_2O_4$	12053-26-8	192.295	deep grn cry	2390		4.4		
1624	Magnesium citrate	$Mg_3(C_6H_5O_7)_2$	3344-18-1	451.114	wh cry					sl H_2O
1625	Magnesium citrate tetradecahydrate	$Mg_3(C_6H_5O_7)_2·14H_2O$	3344-18-1	703.328	wh cry pow					sl H_2O; s acid
1626	Magnesium fluoride	MgF_2	7783-40-6	62.302	wh tetr cry	1263	2227	3.148	0.013^{25}	
1627	Magnesium formate dihydrate	$Mg(CHO_2)_2·2H_2O$	6150-82-9	150.370	wh cry	dec				s H_2O; i EtOH
1628	Magnesium germanate	Mg_2GeO_4	12025-13-7	185.25	wh prec					i H_2O
1629	Magnesium germanide	Mg_2Ge	1310-52-7	121.25	cub cry	1117		3.09		
1630	Magnesium hydride	MgH_2	7693-27-8	26.321	wh tetr cry	327		1.45		reac H_2O
1631	Magnesium hydrogen phosphate trihydrate	$MgHPO_4·3H_2O$	7757-86-0	174.331	wh powder	550 dec		2.13		sl H_2O; s dil acid
1632	Magnesium hydroxide	$Mg(OH)_2$	1309-42-8	58.320	wh hex cry	350		2.37	0.00069^{20}	s dil acid
1633	Magnesium iodate tetrahydrate	$Mg(IO_3)_2·4H_2O$	7790-32-1	446.172	col monocl cry	210 dec		3.3	11.1^{25}	
1634	Magnesium iodide	MgI_2	10377-58-9	278.114	wh hex cry; hyg	634		4.43	146^{25}	
1635	Magnesium iodide hexahydrate	$MgI_2·6H_2O$	66778-21-0	386.205	wh monocl cry			2.35		
1636	Magnesium iodide octahydrate	$MgI_2·8H_2O$	7790-31-0	422.236	wh orth cry; hyg	41 dec		2.10	146^{25}	s EtOH
1637	Magnesium metaborate octahydrate	$Mg(BO_2)_2·8H_2O$	13703-82-7	254.047	wh pow	988 (anh)				sl H_2O
1638	Magnesium metasilicate	$MgSiO_3$	13776-74-4	100.389	wh monocl cry	≈1550 dec		3.19		i H_2O; sl HF
1639	Magnesium metatitanate	$MgTiO_3$	12032-30-3	120.170	col hex cry	1565		3.85		
1640	Magnesium molybdate	$MgMoO_4$	12013-21-7	184.24	wh pow	≈1060		2.2	15.9^{25}	s H_2O
1641	Magnesium nitrate	$Mg(NO_3)_2$	10377-60-3	148.314	wh cub cry			≈2.3	71.2^{25}	
1642	Magnesium nitrate dihydrate	$Mg(NO_3)_2·2H_2O$	15750-45-5	184.345	wh cry	≈100 dec		1.45	71.2^{25}	s EtOH
1643	Magnesium nitrate hexahydrate	$Mg(NO_3)_2·6H_2O$	13446-18-9	256.406	col monocl cry; hyg	≈95 dec		1.46	71.2^{25}	s EtOH
1644	Magnesium nitride	Mg_3N_2	12057-71-5	100.928	yel cub cry	≈1500 dec		2.71		
1645	Magnesium nitrite trihydrate	$Mg(NO_2)_2·3H_2O$	15070-34-5	170.362	wh hyg prisms	100 dec			129.9^{25}	s EtOH
1646	Magnesium orthosilicate	Mg_2SiO_4	26686-77-1	140.694	wh orth cry	1897		3.21		i H_2O
1647	Magnesium orthotitanate	Mg_2TiO_4	12032-52-9	160.475	wh cub cry	1840		3.53		
1648	Magnesium oxalate	MgC_2O_4	547-66-0	112.324	wh powder				0.038^{25}	
1649	Magnesium oxalate dihydrate	$MgC_2O_4·2H_2O$	6150-88-5	148.354	wh powder				0.038^{25}	i EtOH; s dil acid
1650	Magnesium oxide	MgO	1309-48-4	40.304	wh cub cry	2825	3600	3.6		sl H_2O; i EtOH
1651	Magnesium perborate heptahydrate	$Mg(BO_3)_2·7H_2O$	14635-87-1	268.030	wh pow					sl H_2O
1652	Magnesium perchlorate	$Mg(ClO_4)_2$	10034-81-8	223.206	wh hyg powder	250 dec		2.2	100^{25}	
1653	Magnesium perchlorate hexahydrate	$Mg(ClO_4)_2·6H_2O$	13446-19-0	331.298	wh hyg cry	190 dec		1.98	100^{25}	s EtOH
1654	Magnesium permanganate hexahydrate	$Mg(MnO_4)_2·6H_2O$	10377-62-5	370.268	blue-blk cry	dec		2.18		s H_2O
1655	Magnesium peroxide	MgO_2	1335-26-8	56.304	wh cub cry	100 dec		≈3.0		i H_2O; s dil acid
1656	Magnesium phosphate pentahydrate	$Mg_3(PO_4)_2·5H_2O$	7757-87-1	352.934	wh cry	400 dec			0.00009^{20}	s dil acid
1657	Magnesium phosphate octahydrate	$Mg_3(PO_4)_2·8H_2O$	13446-23-6	406.980	wh monocl cry			2.17	0.00009^{20}	s acid

No.	Name	Formula	CAS Reg. No.	Mol. wt.	Phys. form	mp/°C	bp/°C	ρ/g cm⁻³	Sol. in g/100 g H_2O	Qualitative solubility
1658	Magnesium pyrophosphate	$Mg_2P_2O_7$	13446-24-7	222.553	col monocl plates	1395		2.56		
1659	Magnesium pyrophosphate trihydrate	$Mg_2P_2O_7 \cdot 3H_2O$	10102-34-8	276.600	wh powder	100 dec		2.56		i H_2O; s acid
1660	Magnesium phosphide	Mg_3P_2	12057-74-8	134.863	yel cub cry			2.06		reac H_2O
1661	Magnesium selenate hexahydrate	$MgSeO_4 \cdot 6H_2O$	13446-28-1	275.35	wh monocl cry			1.928	55.5^{25}	
1662	Magnesium selenide	MgSe	1313-04-8	103.27	brn cub cry			4.2		reac H_2O
1663	Magnesium selenite hexahydrate	$MgSeO_3 \cdot 6H_2O$	15593-61-0	259.36	col hex cry			2.09		i H_2O; s dil acid
1664	Magnesium hexafluorosilicate hexahydrate	$MgSiF_6 \cdot 6H_2O$	60950-56-3	274.472	wh cry	120 dec		1.79	39.3^{18}	i EtOH
1665	Magnesium silicide	Mg_2Si	22831-39-6	76.696	gray cub cry	1102		1.99		reac H_2O
1666	Magnesium stannide	Mg_2Sn	1313-08-2	167.320	blue cub cry	771		3.60		s H_2O, dil HCl
1667	Magnesium sulfate	$MgSO_4$	7487-88-9	120.368	col orth cry	1137		2.66	35.7^{25}	
1668	Magnesium sulfate monohydrate	$MgSO_4 \cdot H_2O$	14168-73-1	138.383	col monocl cry	150 dec		2.57	35.7^{25}	
1669	Magnesium sulfate heptahydrate	$MgSO_4 \cdot 7H_2O$	10034-99-8	246.474	col orth cry	150 dec		1.67	35.7^{25}	sl EtOH
1670	Magnesium sulfide	MgS	12032-36-9	56.370	red-brn cub cry	2226		2.68		reac H_2O
1671	Magnesium sulfite trihydrate	$MgSO_3 \cdot 3H_2O$	19086-20-5	158.414	col orth cry			2.12	0.79^{25}	
1672	Magnesium sulfite hexahydrate	$MgSO_3 \cdot 6H_2O$	13446-29-2	212.460	wh hex cry	200 dec		1.72	0.79^{25}	i EtOH
1673	Magnesium tetrahydrogen phosphate dihydrate	$Mg(H_2PO_4)_2 \cdot 2H_2O$	15609-80-0	254.311	wh hyg cry	90 dec				s H_2O; i EtOH
1674	Magnesium thiocyanate tetrahydrate	$Mg(SCN)_2 \cdot 4H_2O$	306-61-6	212.531	wh hyg cry					vs H_2O, EtOH
1675	Magnesium thiosulfate hexahydrate	$MgS_2O_3 \cdot 6H_2O$	13446-30-5	244.525	col cry	170 dec		1.82	93^{25}	i EtOH
1676	Magnesium trisilicate	$Mg_2Si_3O_8$	14987-04-3	260.862	wh powder					i H_2O, EtOH
1677	Magnesium tungstate	$MgWO_4$	13573-11-0	272.14	wh monocl cry			6.89	0.016^{20}	i EtOH
1678	Magnesium vanadate	$Mg_2V_2O_7$	13568-63-3	262.489	tricl cry			3.1		
1679	Magnesium zirconate	$MgZrO_3$	12032-31-4	163.527	col cry	2060		4.23		
1680	Magnesium zirconium silicate	$MgO \cdot ZrO_2 \cdot SiO_2$	52110-05-1	223.612	wh solid					i H_2O, alk; sl acid
1681	Manganese	Mn	7439-96-5	54.938	hard gray metal	1246	2061	7.3		s dil acids
1682	Manganocene	$Mn(C_5H_5)_2$	1271-27-8	185.124	yel-brn cry	173				s py, thf; sl bz
1683	Manganese antimonide (MnSb)	MnSb	12032-82-5	176.698	hex cry	840		6.9		
1684	Manganese antimonide (Mn_2Sb)	Mn_2Sb	12032-97-2	231.636	tetr cry	948		7.0		
1685	Manganese boride (MnB)	MnB	12045-15-7	65.749	orth cry	1890		6.45		
1686	Manganese boride [MnB_2]	MnB_2	12228-50-1	76.560	hex cry	1827		5.3		
1687	Manganese boride (Mn_2B)	Mn_2B	12045-16-8	120.687	red-brn tetr cry	1580		7.20		
1688	Manganese carbide	Mn_3C	12266-65-8	176.825	refrac solid	1520		6.89		
1689	Manganese carbonyl	$Mn_2(CO)_{10}$	10170-69-1	389.977	yel monocl cry	154		1.75		i H_2O; s os
1690	Manganese pentacarbonyl bromide	$Mn(CO)_5Br$	14516-54-2	274.893	oran-yel cry					s os
1691	Manganese phosphide (MnP)	MnP	12032-78-9	85.912	orth cry	1147		5.49		
1692	Manganese phosphide (Mn_2P)	Mn_2P	12333-54-9	140.850	hex cry	1327		6.0		
1693	Manganese silicide [$MnSi_2$]	$MnSi_2$	12032-86-9	111.109	gray solid	1152 dec				
1694	Manganese(II) acetate tetrahydrate	$Mn(C_2H_3O_2)_2 \cdot 4H_2O$	6156-78-1	245.087	red monocl cry	80		1.59		s H_2O, EtOH
1695	Manganese(II) bromide	$MnBr_2$	13446-03-2	214.746	pink hex cry	698		4.385	151^{25}	
1696	Manganese(II) bromide tetrahydrate	$MnBr_2 \cdot 4H_2O$	10031-20-6	286.808	red hyg cry	64 dec			151^{25}	
1697	Manganese(II) carbonate	$MnCO_3$	598-62-9	114.947	pink hex cry	>200 dec		3.70	0.00008^{20}	s dil acid
1698	Manganese(II) chloride	$MnCl_2$	7773-01-5	125.844	pink trig cry; hyg	650	1190	2.977	77.3^{25}	s py, EtOH; i eth
1699	Manganese(II) chloride tetrahydrate	$MnCl_2 \cdot 4H_2O$	13446-34-9	197.906	red monocl cry; hyg	87.5		1.913	77.3^{25}	s EtOH; i eth
1700	Manganese(II) dihydrogen phosphate dihydrate	$Mn(H_2PO_4)_2 \cdot 2H_2O$	18718-07-5	284.944	col hyg cry					s H_2O; i EtOH
1701	Manganese(II) fluoride	MnF_2	7782-64-1	92.935	red tetr cry	900		3.98	1.02^{25}	i EtOH
1702	Manganese(II) hydroxide	$Mn(OH)_2$	18933-05-6	88.953	pink hex cry	dec		3.26	0.00034^{20}	
1703	Manganese(II) hypophosphite monohydrate	$Mn(H_2PO_2)_2 \cdot H_2O$	10043-84-2	202.931	pink cry	>250			15^{20}	s H_2O
1704	Manganese(II) iodide	MnI_2	7790-33-2	308.747	wh hex cry; hyg	638		5.04		s H_2O, EtOH
1705	Manganese(II) iodide tetrahydrate	$MnI_2 \cdot 4H_2O$	7790-33-2	380.809	red cry					vs H_2O; s EtOH
1706	Manganese(II) metasilicate	$MnSiO_3$	7759-00-4	131.022	red orth cry	1291		3.48		i H_2O
1707	Manganese(II) molybdate	$MnMoO_4$	14013-15-1	214.88	yel monocl cry			4.05		
1708	Manganese(II) nitrate	$Mn(NO_3)_2$	10377-93-2	178.947	col orth cry; hyg			2.2	161^{25}	s diox, thf
1709	Manganese(II) nitrate tetrahydrate	$Mn(NO_3)_2 \cdot 4H_2O$	20694-39-7	251.009	pink hyg cry	37.1 dec		2.13	161^{25}	s EtOH
1710	Manganese(II) nitrate hexahydrate	$Mn(NO_3)_2 \cdot 6H_2O$	10377-66-9	287.039	rose monocl cry	28 dec		1.8	161^{25}	vs EtOH
1711	Manganese(II) orthosilicate	Mn_2SiO_4	13568-32-6	201.960	orth cry			4.11		i H_2O
1712	Manganese(II) oxalate dihydrate	$MnC_2O_4 \cdot 2H_2O$	6556-16-7	178.987	wh cry powder	150 dec		2.45	0.032^{20}	s acid
1713	Manganese(II) oxide	MnO	1344-43-0	70.937	grn cub cry or powder	1842		5.37		i H_2O; s acid
1714	Manganese(II) perchlorate hexahydrate	$Mn(ClO_4)_2 \cdot 6H_2O$	15364-94-0	361.931	pink hex cry			2.10		
1715	Manganese(II) pyrophosphate	$Mn_2P_2O_7$	53731-35-4	283.819	wh monocl cry	1196		3.71		i H_2O
1716	Manganese(II) selenide	MnSe	1313-22-0	133.90	gray cub cry	1460		5.45		i H_2O
1717	Manganese(II) sulfate	$MnSO_4$	7785-87-7	151.001	wh orth cry	700	850 dec	3.25	63.7^{25}	
1718	Manganese(II) sulfate monohydrate	$MnSO_4 \cdot H_2O$	10034-96-5	169.016	red monocl cry			2.95	63.7^{25}	i EtOH
1719	Manganese(II) sulfate tetrahydrate	$MnSO_4 \cdot 4H_2O$	10101-68-5	223.062	red monocl cry	38 dec		2.26	63.7^{25}	i EtOH
1720	Manganese(II) sulfide (α form)	MnS	18820-29-6	87.003	grn cub cry	1530		4.0		i H_2O; s dil acid
1721	Manganese(II) sulfide (β form)	MnS	18820-29-6	87.003	red cub cry			3.3		i H_2O; s dil acid
1722	Manganese(II) sulfide (γ form)	MnS	18820-29-6	87.003	red hex cry			≈3.3		i H_2O; s dil acid
1723	Manganese(II) telluride	MnTe	12032-88-1	182.54	hex cry	≈1150		6.0		
1724	Manganese(II) tetraborate octahydrate	$MnB_4O_7 \cdot 8H_2O$	12228-91-0	354.300	red solid					i H_2O, EtOH; s dil acid
1725	Manganese(II) titanate	$MnTiO_3$	12032-74-5	150.803	red hex cry	1360		4.55		

No.	Name	Formula	CAS Reg. No.	Mol. wt.	Phys. form	mp/°C	bp/°C	ρ/g cm^{-3}	Sol. in g/100 g H_2O	Qualitative solubility
1726	Manganese(II) tungstate	$MnWO_4$	13918-22-4	302.78	wh monocl cry			7.2	0.0054^{20}	
1727	Manganese(II,III) oxide	Mn_3O_4	1317-35-7	228.812	brn tetr cry	1567		4.84		i H_2O; s HCl
1728	Manganese(III) acetate dihydrate	$Mn(C_2H_3O_2)_3 \cdot 2H_2O$	19513-05-4	268.100	brn cry					s eth, HOAc
1729	Manganese(III) fluoride	MnF_3	7783-53-1	111.933	red monocl cry; hyg	>600 dec		3.54		reac H_2O
1730	Manganese(III) hydroxide	MnO(OH)	1332-63-4	87.945	blk monocl cry	250 dec		≈4.3		i H_2O
1731	Manganese(III) oxide	Mn_2O_3	1317-34-6	157.874	blk cub cry	1080 dec		≈5.0		i H_2O
1732	Manganese(IV) oxide	MnO_2	1313-13-9	86.937	blk tetr cry	535 dec		5.08		i H_2O, HNO_3
1733	Manganese(VII) oxide	Mn_2O_7	12057-92-0	221.872	grn oil; exp	5.9	95 exp	2.40		vs H_2O
1734	Mendelevium	Md	7440-11-1	258	metal	827				
1735	Mercury	Hg	7439-97-6	200.59	heavy silv liq	-38.8290	356.619	13.53359^{25} liq		i H_2O
1736	Dimethyl mercury	$Hg(CH_3)_2$	593-74-8	230.66	liq		93	3.17^{25}		i H_2O; vs EtOH, eth
1737	Mercury(I) acetate	$Hg_2(C_2H_3O_2)_2$	631-60-7	519.27	col scales	dec				sl H_2O; i EtOH, eth
1738	Mercury(I) bromate	$Hg_2(BrO_3)_2$	13465-33-3	656.98	col cry	dec				i H_2O; sl acid
1739	Mercury(I) bromide	Hg_2Br_2	15385-58-7	560.99	wh tetr cry or powder	345 dec		7.307		i H_2O, EtOH, eth
1740	Mercury(I) carbonate	Hg_2CO_3	6824-78-8	461.19	yel-brn cry	130 dec			0.0000045	i EtOH
1741	Mercury(I) chlorate	$Hg_2(ClO_3)_2$	10294-44-7	568.08	wh rhom cry	≈250 dec		6.409		sl H_2O; s EtOH
1742	Mercury(I) chloride	Hg_2Cl_2	10112-91-1	472.09	wh tetr cry	525 tp	383 sp	7.16	0.0004^{25}	i EtOH, eth
1743	Mercury(I) chromate	Hg_2CrO_4	13465-34-4	517.17	brn-red solid					i H_2O EtOH; s conc HNO_3
1744	Mercury(I) fluoride	Hg_2F_2	13967-25-4	439.18	yel cub cry	570 dec	subl	8.73		reac H_2O
1745	Mercury(I) iodate	$Hg_2(IO_3)_2$	13465-35-5	750.99	yel-wh pow	175 dec			0.0032^{20}	
1746	Mercury(I) iodide	Hg_2I_2	15385-57-6	654.99	yel amorp powder	290		7.70		i H_2O, EtOH, eth
1747	Mercury(I) nitrate	$Hg_2(NO_3)_2$	10415-75-5	525.19	cry					sl H_2O
1748	Mercury(I) nitrate dihydrate	$Hg_2(NO_3)_2 \cdot 2H_2O$	14836-60-3	561.22	col cry	70 dec		4.8		sl H_2O
1749	Mercury(I) nitrite	$Hg_2(NO_2)_2$	13492-25-6	493.19	yel cry	100 dec		7.3		reac H_2O
1750	Mercury(I) oxalate	$Hg_2C_2O_4$	2949-11-3	489.20	cry					i H_2O; sl HNO_3
1751	Mercury(I) oxide	Hg_2O	15829-53-5	417.18	prob mixture of HgO+Hg	100 dec		9.8		i H_2O; s HNO_3
1752	Mercury(I) perchlorate tetrahydrate	$Hg_2(ClO_4)_2 \cdot 4H_2O$	65202-12-2	672.14	cry	64			442^{25}	
1753	Mercury(I) sulfate	Hg_2SO_4	7783-36-0	497.24	wh-yel cry powder			7.56	0.051^{25}	s dil HNO_3
1754	Mercury(I) sulfide	Hg_2S	51595-71-2	433.25	unstab blk pow	dec				i H_2O
1755	Mercury(I) thiocyanate	$Hg_2(SCN)_2$	13465-37-7	517.34	col powder	dec			0.03^{25}	s HCl, KCNS
1756	Mercury(I) tungstate	Hg_2WO_4	38705-19-0	649.02	yel amorp solid	dec				i H_2O, EtOH
1757	Mercury(II) acetate	$Hg(C_2H_3O_2)_2$	1600-27-7	318.68	wh-yel cry or powder	179 dec		3.28	25^{10}	s EtOH
1758	Mercury(II) amide chloride	$Hg(NH_2)Cl$	10124-48-8	252.07	wh solid		subl	5.38		i H_2O, EtOH; s warm acid
1759	Mercury(II) benzoate monohydrate	$Hg(C_7H_5O_2)_2 \cdot H_2O$	32839-04-6	460.83	wh cry	165			1.2^{15}	sl EtOH
1760	Mercury(II) bromate	$Hg(BrO_3)_2$	26522-91-8	456.39	cry	130 dec			0.15	s acid
1761	Mercury(II) bromide	$HgBr_2$	7789-47-1	360.40	wh rhomb cry or powder	241	318	6.05	0.61^{25}	sl chl; s EtOH, MeOH
1762	Mercury(II) chlorate	$Hg(ClO_3)_2$	13465-30-0	367.49	wh needles	dec		4.998	25	
1763	Mercury(II) chloride	$HgCl_2$	7487-94-7	271.50	wh orth cry	277	304	5.6	7.31^{25}	sl bz; s EtOH, MeOH, ace, eth
1764	Mercury(II) chromate	$HgCrO_4$	13444-75-2	316.58	red monocl cry			6.06		sl H_2O
1765	Mercury(II) cyanide	$Hg(CN)_2$	592-04-1	252.62	col tetr cry	320 dec		4.00	11.4^{25}	s EtOH; sl eth
1766	Mercury(II) dichromate	$HgCr_2O_7$	7789-10-8	416.58	red cry powder					i H_2O; s acid
1767	Mercury(II) fluoride	HgF_2	7783-39-3	238.59	wh cub cry; hyg	645 dec		8.95		reac H_2O
1768	Mercury(II) fulminate	$Hg(CNO)_2$	628-86-4	284.62	gray cry	exp		4.42		sl H_2O; s EtOH, NH_4OH
1769	Mercury(II) hydrogen arsenate	$HgHAsO_4$	7784-37-4	340.52	yel powder					i H_2O; s acid
1770	Mercury(II) iodate	$Hg(IO_3)_2$	7783-32-6	550.40	wh powder	175 dec				i H_2O
1771	Mercury(II) iodide (yellow)	HgI_2	7774-29-0	454.40	yel tetr cry or powder	256	351	6.28	0.0055^{25}	sl EtOH, ace, eth
1772	Mercury(II) iodide (red)	HgI_2	7774-29-0	454.40	red pow	127 trans yel			0.006^{25}	sl EtOH, ace, eth, chl
1773	Mercury(II) nitrate	$Hg(NO_3)_2$	10045-94-0	324.60	col hyg cry	79		4.3		s H_2O; i EtOH
1774	Mercury(II) nitrate monohydrate	$Hg(NO_3)_2 \cdot H_2O$	7783-34-8	342.62	wh-yel hyg cry			4.3		s H_2O, dil acid
1775	Mercury(II) nitrate dihydrate	$Hg(NO_3)_2 \cdot 2H_2O$	22852-67-1	360.63	monocl cry			4.78		s H_2O
1776	Mercury(II) oxalate	HgC_2O_4	3444-13-1	288.61	powder	165 dec				i H_2O
1777	Mercury(II) oxide	HgO	21908-53-2	216.59	red or yel orth cry	500 dec		11.14		i H_2O, EtOH; s dil acid
1778	Mercury(II) oxide sulfate	$(Hg_3O_2)SO_4$	1312-03-4	729.83	yel powder					i H_2O; s acid
1779	Mercury(II) oxycyanide	$Hg(CN)_2 \cdot HgO$	1335-31-5	469.21	wh orth cry	exp		4.44	11.4^{25}	
1780	Mercury(II) perchlorate trihydrate	$Hg(ClO_4)_2 \cdot 3H_2O$	7616-83-3	453.54	cry					
1781	Mercury(II) phosphate	$Hg_3(PO_4)_2$	7782-66-3	791.71	wh-yel powder					i H_2O, EtOH; s acid
1782	Mercury(II) selenide	HgSe	20601-83-6	279.55	gray cub cry		subl	8.21		i H_2O
1783	Mercury(II) sulfate	$HgSO_4$	7783-35-9	296.65	wh monocl cry			6.47		reac H_2O
1784	Mercury(II) sulfide (black)	HgS	1344-48-5	232.66	blk cub cry or powder	820		7.70		i H_2O; s acid, EtOH
1785	Mercury(II) sulfide (red)	HgS	1344-48-5	232.66	red hex cry	344 trans blk HgS		8.17		i H_2O, acid; s aqua regia
1786	Mercury(II) telluride	HgTe	12068-90-5	328.19	gray cub cry	673		8.63		
1787	Mercury(II) thiocyanate	$Hg(SCN)_2$	592-85-8	316.75	monocl cry	≈165 dec		3.71	0.070^{25}	s dil HCl

No.	Name	Formula	CAS Reg. No.	Mol. wt.	Phys. form	mp/°C	bp/°C	ρ/g cm⁻³	Sol. in g/100 g H_2O	Qualitative solubility
1788	Mercury(II) tungstate	$HgWO_4$	37913-38-5	448.43	yel cry	dec				i H_2O, EtOH
1789	Molybdenum	Mo	7439-98-7	95.96	gray-blk metal; cub	2622	4639	10.2		i H_2O, dil acid, alk
1790	Molybdophosphoric acid	$H_3P(Mo_3O_{10})_4$	51429-74-4	1825.25	bright yel cry					
1791	Molybdenum boride (Mo_2B)	Mo_2B	12006-99-4	202.69	refrac tetr cry	2000		9.2		
1792	Molybdenum boride [Mo_2B_5]	Mo_2B_5	12007-97-5	245.94	refrac hex cry	2210		≈7.2		
1793	Molybdenum carbide (MoC)	MoC	12011-97-1	107.95	refrac solid; cub	2577				
1794	Molybdenum carbide (Mo_2C)	Mo_2C	12069-89-5	203.89	gray orth cry	2687		9.18		
1795	Molybdenum carbonyl	$Mo(CO)_6$	13939-06-5	264.00	wh orth cry	148	155 dec	1.96		i H_2O; s bz; sl eth
1796	Molybdenum nitride (MoN)	MoN	12033-19-1	109.95	hex cry	1750		9.20		
1797	Molybdenum nitride (Mo_2N)	Mo_2N	12033-31-7	205.89	gray cub cry	790 dec		9.46		
1798	Molybdenum phosphide	MoP	12163-69-8	126.91	blk hex cry			7.34		
1799	Molybdenum silicide [$MoSi_2$]	$MoSi_2$	12136-78-6	152.11	gray tetr cry	≈1900		6.2		i H_2O; s HF
1800	Molybdenum(II) bromide	$MoBr_2$	13446-56-5	255.75	yel-red cry	700 dec		4.88		i H_2O, EtOH
1801	Molybdenum(II) chloride	$MoCl_2$	13478-17-6	166.85	yel cry	500 dec		3.71		i H_2O
1802	Molybdenum(II) iodide	MoI_2	14055-74-4	349.75	blk hyg cry	700		5.28		i H_2O
1803	Molybdenum(III) bromide	$MoBr_3$	13446-57-6	335.65	grn hex cry	500 dec		4.89		i H_2O, EtOH
1804	Molybdenum(III) chloride	$MoCl_3$	13478-18-7	202.30	dark red monocl cry	400 dec		3.74		i H_2O, os
1805	Molybdenum(III) fluoride	MoF_3	20193-58-2	152.94	yel-brn hex cry	>600		4.64		i H_2O
1806	Molybdenum(III) iodide	MoI_3	14055-75-5	476.65	blk solid	927				i H_2O
1807	Molybdenum(III) oxide	Mo_2O_3	1313-29-7	239.88	gray-blk powder					i H_2O; sl acid
1808	Molybdenum(IV) bromide	$MoBr_4$	13520-59-7	415.56	blk cry	110 dec				reac H_2O
1809	Molybdenum(IV) chloride	$MoCl_4$	13320-71-3	237.75	blk cry	317				reac H_2O, sl chl; i eth, bz
1810	Molybdenum(IV) fluoride	MoF_4	23412-45-5	171.93	grn cry	dec				reac H_2O
1811	Molybdenum(IV) iodide	MoI_4	14055-76-6	603.56	blk cry	100 dec				i H_2O
1812	Molybdenum(IV) oxide	MoO_2	18868-43-4	127.94	brn-viol tetr cry	≈1800 dec		6.47		i H_2O, acid, alk
1813	Molybdenum(IV) selenide	$MoSe_2$	12058-18-3	253.86	gray hex cry	>1200		6.90		
1814	Molybdenum(IV) sulfide	MoS_2	1317-33-5	160.07	blk powder or hex cry	1750		5.06		i H_2O; s conc acid
1815	Molybdenum(IV) telluride	$MoTe_2$	12058-20-7	351.14	gray hex cry			7.7		
1816	Molybdenum(V) chloride	$MoCl_5$	10241-05-1	273.21	grn-blk monocl cry; hyg	194	268	2.93		reac H_2O; s EtOH, eth
1817	Molybdenum(V) fluoride	MoF_5	13819-84-6	190.93	yel monocl cry	45.67	213.6	3.5		reac H_2O
1818	Molybdenum(V) oxytrichloride	$MoOCl_3$	13814-74-9	218.30	blk monocl cry	310	subl			reac H_2O
1819	Molybdenum(VI) acid monohydrate	$H_2MoO_4 \cdot H_2O$	7782-91-4	179.97	wh powder			3.1		sl H_2O; s alk
1820	Molybdenum(VI) dioxydichloride	MoO_2Cl_2	13637-68-8	198.85	yel-oran solid	176	250	3.31		reac H_2O
1821	Molybdenum(VI) dioxydifluoride	MoO_2F_2	13824-57-2	165.94	pale lilac cry	270		3.5		i MeCN, chl; sl HF
1822	Molybdenum(VI) fluoride	MoF_6	7783-77-9	209.93	wh cub cry or col liq; hyg	17.5	34.0	2.54		reac H_2O; vs hex, ctc
1823	Molybdenum(VI) metaphosphate	$Mo(PO_3)_6$	133863-98-6	569.77	yel powder			3.28		i H_2O, acid
1824	Molybdenum(VI) oxide	MoO_3	1313-27-5	143.94	wh-yel rhomb cry	802	1155	4.70	0.14^{20}	sl H_2O; s alk, acid
1825	Molybdenum(VI) oxytetrachloride	$MoOCl_4$	13814-75-0	253.75	grn hyg powder	105	159			
1826	Molybdenum(VI) oxytetrafluoride	$MoOF_4$	14459-59-7	187.93	volatile solid	97.2	186.0			
1827	Molybdenum(VI) sulfide	MoS_3	12033-29-3	192.14	blk solid	350 dec				i H_2O, os
1828	Neodymium	Nd	7440-00-8	144.242	silv metal; hex	1016	3074	7.01		
1829	Neodymium boride	NdB_6	12008-23-0	209.108	blk cub cry	2610		4.93		
1830	Neodymium nitride	NdN	25764-11-8	158.249	blk cub cry			7.69		
1831	Neodymium(II) acetate	$Nd(C_2H_3O_2)_2$	6192-13-8	321.373	red-purp cry					s H_2O
1832	Neodymium(II) chloride	$NdCl_2$	25469-93-6	215.148	grn hyg solid	841				
1833	Neodymium(III) bromate nonahydrate	$Nd(BrO_3)_3 \cdot 9H_2O$	15162-92-2	690.086	red hex cry	66 dec				
1834	Neodymium(III) bromide	$NdBr_3$	13536-80-6	383.954	viol orth cry; hyg	682	1540	5.3		s H_2O
1835	Neodymium(III) chloride	$NdCl_3$	10024-93-8	250.601	viol hex cry	759	1600	4.13	100^{25}	vs EtOH; i eth, chl
1836	Neodymium(III) chloride hexahydrate	$NdCl_3 \cdot 6H_2O$	13477-89-9	358.692	purp cry	124 dec		2.3	100^{25}	s EtOH
1837	Neodymium(III) fluoride	NdF_3	13709-42-7	201.237	viol hex cry; hyg	1377	2300	6.51		i H_2O
1838	Neodymium(III) hydroxide	$Nd(OH)_3$	16469-17-3	195.264	blue solid	210 dec				i H_2O
1839	Neodymium(III) iodide	NdI_3	13813-24-6	524.955	grn orth cry; hyg	787		5.85		s H_2O
1840	Neodymium(III) nitrate	$Nd(NO_3)_3$	10045-95-1	330.257	viol hyg. cry				152^{25}	s EtOH
1841	Neodymium(III) nitrate hexahydrate	$Nd(NO_3)_3 \cdot 6H_2O$	14517-29-4	438.348	purp hyg cry				152^{25}	s EtOH, ace
1842	Neodymium(III) oxide	Nd_2O_3	1313-97-9	336.482	blue hex cry; hyg	2233	3760	7.24		i H_2O; s dil acid
1843	Neodymium(III) sulfate	$Nd_2(SO_4)_3$	13477-91-3	576.672	pink needles	≈700 dec			7.1^{20}	
1844	Neodymium(III) sulfate octahydrate	$Nd_2(SO_4)_3 \cdot 8H_2O$	13477-91-3	720.794	red cry	350 dec		2.85		sl H_2O
1845	Neodymium(III) sulfide	Nd_2S_3	12035-32-4	384.679	orth cry	2207		5.46		
1846	Neodymium(III) telluride	Nd_2Te_3	12035-35-7	671.28	gray orth cry	1377		7.0		
1847	Neodymium(III) tris(cyclopentadienyl)	$Nd(C_5H_5)_3$	1273-98-9	339.522	red-blue cry	380				s thf
1848	Neon	Ne	7440-01-9	20.180	col gas	-248.59	-246.046	$1.207^{-246.0}$ liq		sl H_2O
1849	Neptunium	Np	7439-99-8	237	silv metal	644	≈3902	20.2		s HCl
1850	Neptunium(IV) oxide	NpO_2	12035-79-9	269	grn cub cry	2547		11.1		
1851	Nickel	Ni	7440-02-0	58.693	wh metal; cub	1455	2913	8.90		i H_2O; sl dil acid
1852	Nickelocene	$Ni(C_5H_5)_2$	1271-28-9	188.879	grn cry	173				
1853	Nickel aluminide (NiAl)	NiAl	12003-78-0	85.675	metallic solid	1638				
1854	Nickel antimonide	NiSb	12035-52-8	180.453	hex cry	1147		8.74		

No.	Name	Formula	CAS Reg. No.	Mol. wt.	Phys. form	mp/°C	bp/°C	ρ/g cm⁻³	Sol. in g/100 g H_2O	Qualitative solubility
1855	Nickel arsenide	NiAs	27016-75-7	133.615	hex cry	967		7.77		
1856	Nickel boride (NiB)	NiB	12007-00-0	69.504	grn refrac solid	1035		7.13		
1857	Nickel boride (Ni_2B)	Ni_2B	12007-01-1	128.198	refrac solid	1125		7.90		
1858	Nickel boride (Ni_3B)	Ni_3B	12007-02-2	186.891	refrac solid	1166		8.17		
1859	Nickel carbonyl [$Ni(CO)_4$]	$Ni(CO)_4$	13463-39-3	170.734	col liq	-19.3	43 (exp 60)	1.31^{25}		i H_2O; s EtOH, bz, ace, ctc
1860	Nickel phosphide	Ni_2P	12035-64-2	148.361	hex cry	1100		7.33		
1861	Nickel silicide [$NiSi_2$]	$NiSi_2$	12201-89-7	114.864	cub cry	993		4.83		
1862	Nickel silicide (Ni_2Si)	Ni_2Si	12059-14-2	145.473	orth cry	1255		7.40		
1863	Nickel subsulfide	Ni_3S_2	12035-72-2	240.210	yel hex cry	789		5.87		
1864	Nickel(II) acetate tetrahydrate	$Ni(C_2H_3O_2)_2 \cdot 4H_2O$	6018-89-9	248.842	grn monocl cry	250 dec		1.74	16^{20}	s H_2O, EtOH
1865	Nickel(II) ammonium chloride hexahydrate	$NH_4NiCl_3 \cdot 6H_2O$	16122-03-5	291.182	grn hyg cry			1.65		s H_2O
1866	Nickel(II) ammonium sulfate	$Ni(NH_4)_2(SO_4)_2$	15699-18-0	286.895	blue-grn cry	250 dec				sl H_2O
1867	Nickel(II) ammonium sulfate hexahydrate	$Ni(NH_4)_2(SO_4)_2 \cdot 6H_2O$	7785-20-8	394.987	blue-grn cry	130 dec		1.92	6.5^{20}	s H_2O; i EtOH
1868	Nickel(II) arsenate octahydrate	$Ni_3(AsO_4)_2 \cdot 8H_2O$	7784-48-7	598.040	yel-grn powder	dec		4.98		i H_2O; s acid
1869	Nickel(II) bromide	$NiBr_2$	13462-88-9	218.501	yel hex cry; hyg	963	subl	5.10	131^{20}	
1870	Nickel(II) bromide trihydrate	$NiBr_2 \cdot 3H_2O$	13462-88-9	272.547	yel-grn hyg cry	200 dec				vs H_2O; s EtOH, eth
1871	Nickel(II) carbonate	$NiCO_3$	3333-67-3	118.702	grn rhomb cry			4.389	0.0043^{20}	s dil acid
1872	Nickel(II) chlorate hexahydrate	$Ni(ClO_3)_2 \cdot 6H_2O$	13477-94-6	333.687	grn cub cry	80 dec		2.07		vs H_2O
1873	Nickel(II) chloride	$NiCl_2$	7718-54-9	129.599	yel hex cry; hyg	1031	985 sp	3.51	67.5^{25}	s EtOH
1874	Nickel(II) chloride hexahydrate	$NiCl_2 \cdot 6H_2O$	7791-20-0	237.690	grn monocl cry				67.5^{25}	s EtOH
1875	Nickel(II) chromate	$NiCrO_4$	14721-18-7	174.687	red solid					sl H_2O
1876	Nickel(II) cyanide tetrahydrate	$Ni(CN)_2 \cdot 4H_2O$	13477-95-7	182.789	grn plates	200 dec				i H_2O; sl dil acid; s NH_4OH
1877	Nickel(II) fluoride	NiF_2	10028-18-9	96.690	yel tetr cry	1380		4.7	2.56^{25}	i EtOH, eth
1878	Nickel(II) fluoride tetrahydrate	$NiF_2 \cdot 4H_2O$	13940-83-5	168.752	grn pow					sl H_2O
1879	Nickel(II) hydroxide	$Ni(OH)_2$	12054-48-7	92.708	grn hex cry	230 dec		4.1	0.00015^{20}	
1880	Nickel(II) hydroxide monohydrate	$Ni(OH)_2 \cdot H_2O$	36897-37-7	110.723	grn powder				0.00015^{20}	s dil acid
1881	Nickel(II) iodate	$Ni(IO_3)_2$	13477-98-0	408.498	yel needles			5.07	1.1^{30}	sl H_2O
1882	Nickel(II) iodate tetrahydrate	$Ni(IO_3)_2 \cdot 4H_2O$	13477-99-1	480.560	yel hex cry	100 dec		5.07		sl H_2O
1883	Nickel(II) iodide	NiI_2	13462-90-3	312.502	blk hex cry; hyg	800	subl	5.22	154^{25}	
1884	Nickel(II) iodide hexahydrate	$NiI_2 \cdot 6H_2O$	7790-34-3	420.593	grn monocl cry; hyg				154^{25}	vs EtOH
1885	Nickel(II) nitrate	$Ni(NO_3)_2$	13138-45-9	182.702	grn cry				99.2^{25}	s EtOH
1886	Nickel(II) nitrate hexahydrate	$Ni(NO_3)_2 \cdot 6H_2O$	13478-00-7	290.794	grn monocl cry; hyg	56 dec		2.05	99.2^{25}	s EtOH
1887	Nickel(II) oxalate dihydrate	$NiC_2O_4 \cdot 2H_2O$	6018-94-6	182.742	grn-wh solid	150 dec			0.0012^{25}	i H_2O; s acid, NH_4OH
1888	Nickel(II) oxide	NiO	1313-99-1	74.692	grn cub cry	1957		6.72		i H_2O; s acid
1889	Nickel(II) perchlorate hexahydrate	$Ni(ClO_4)_2 \cdot 6H_2O$	13637-71-3	365.686	grn hex needles	140			158.8^{25}	s EtOH, ace
1890	Nickel(II) phosphate octahydrate	$Ni_3(PO_4)_2 \cdot 8H_2O$	10381-36-9	510.145	grn plates					s acid
1891	Nickel(II) selenate hexahydrate	$NiSeO_4 \cdot 6H_2O$	15060-62-5	309.74	grn tetr cry			2.314	35.5^{20}	
1892	Nickel(II) selenide	NiSe	1314-05-2	137.65	yel-grn hex cry	980		7.2		
1893	Nickel(II) stannate dihydrate	$NiSnO_3 \cdot 2H_2O$	12035-38-0	261.432	grn pow	120 dec				
1894	Nickel(II) sulfate	$NiSO_4$	7786-81-4	154.756	grn-yel orth cry	840 dec		4.01	40.4^{25}	
1895	Nickel(II) sulfate hexahydrate	$NiSO_4 \cdot 6H_2O$	10101-97-0	262.847	blue-grn tetr cry	≈100 dec		2.07	40.4^{25}	sl EtOH
1896	Nickel(II) sulfate heptahydrate	$NiSO_4 \cdot 7H_2O$	10101-98-1	280.862	grn orth cry			1.98	40.4^{25}	s EtOH
1897	Nickel(II) sulfide	NiS	16812-54-7	90.758	yel hex cry	976		5.5		i H_2O
1898	Nickel(II) thiocyanate	$Ni(SCN)_2$	13689-92-4	174.857	grn pwd				55.0^{25}	
1899	Nickel(II) titanate	$NiTiO_3$	12035-39-1	154.558	brn hex cry			5.0		
1900	Nickel(II,III) sulfide	Ni_3S_4	12137-12-1	304.340	cub cry	995		4.77		
1901	Nickel(III) oxide	Ni_2O_3	1314-06-3	165.385	gray-blk cub cry	≈600 dec				i H_2O; s hot acid
1902	Niobium	Nb	7440-03-1	92.906	gray metal; cub	2477	4741	8.57		i acid
1903	Niobocene dichloride	$Nb(C_5H_5)_2Cl_2$	12793-14-5	293.998	hyg blk cry					sl tol
1904	Niobium boride (NbB)	NbB	12045-19-1	103.717	gray orth cry	2270		7.5		
1905	Niobium boride [NbB_2]	NbB_2	12007-29-3	114.528	gray hex cry	3050		6.97		
1906	Niobium carbide (NbC)	NbC	12069-94-2	104.917	gray cub cry	3608	4300	7.82		i H_2O, acid
1907	Niobium carbide (Nb_2C)	Nb_2C	12011-99-3	197.824	refrac hex cry	3080		7.8		i H_2O
1908	Niobium nitride	NbN	24621-21-4	106.913	gray cry; cub	2050		8.47		i HCl, acid
1909	Niobium phosphide	NbP	12034-66-1	123.880	tetr cry			6.5		
1910	Niobium silicide	$NbSi_2$	12034-80-9	149.077	gray hex cry	1950		5.7		
1911	Niobium(II) oxide	NbO	12034-57-0	108.905	gray cub cry	1937		7.30		
1912	Niobium(III) bromide	$NbBr_3$	15752-41-7	332.618	dark brn solid		400 subl			
1913	Niobium(III) chloride	$NbCl_3$	13569-59-0	199.265	blk solid					
1914	Niobium(III) fluoride	NbF_3	15195-53-6	149.901	blue cub cry			4.2		
1915	Niobium(III) iodide	NbI_3	13870-20-7	473.619	blk solid	510 dec				
1916	Niobium(IV) bromide	$NbBr_4$	13842-75-6	412.522	dark brn cry		300 subl	4.72		reac H_2O
1917	Niobium(IV) chloride	$NbCl_4$	13569-70-5	234.718	viol-blk monocl cry	800 dec	275 subl	3.2		reac H_2O
1918	Niobium(IV) fluoride	NbF_4	13842-88-1	168.900	blk tetr cry; hyg	>350 dec		4.01		
1919	Niobium(IV) iodide	NbI_4	13870-21-8	600.524	gray orth cry	503		5.6		reac H_2O
1920	Niobium(IV) oxide	NbO_2	12034-59-2	124.905	wh tetr cry or powder	1901		5.9		
1921	Niobium(IV) selenide	$NbSe_2$	12034-77-4	250.83	gray hex cry	>1300		6.3		
1922	Niobium(IV) sulfide	NbS_2	12136-97-9	157.036	blk rhomb cry			4.4		
1923	Niobium(IV) telluride	$NbTe_2$	12034-83-2	348.11	hex cry			7.6		
1924	Niobium(V) bromide	$NbBr_5$	13478-45-0	492.426	oran orth cry	254	361.6	4.36		s H_2O, EtOH
1925	Niobium(V) chloride	$NbCl_5$	10026-12-7	270.171	yel monocl cry; hyg	205.8	247.4	2.78		reac H_2O; s HCl, ctc
1926	Niobium(V) dioxyfluoride	NbO_2F	15195-33-2	143.903	wh cub cry			4.0		

No.	Name	Formula	CAS Reg. No.	Mol. wt.	Phys. form	mp/°C	bp/°C	ρ/g cm^{-3}	Sol. in g/100 g H_2O	Qualitative solubility
1927	Niobium(V) ethoxide	$Nb(OC_2H_5)_5$	3236-82-6	318.209	col hyg liq	5	203	1.258		reac H_2O; s peth
1928	Niobium(V) fluoride	NbF_5	7783-68-8	187.898	col monocl cry; hyg	80	234	2.70		reac H_2O; sl CS_2, chl
1929	Niobium(V) iodide	NbI_5	13779-92-5	727.428	yel-blk monocl cry	327		5.32		reac H_2O
1930	Niobium(V) oxide	Nb_2O_5	1313-96-8	265.810	wh orth cry	1512		4.47		i H_2O; s HF
1931	Niobium(V) oxybromide	$NbOBr_3$	14459-75-7	348.617	yel-brn cry	≈320 dec	subl			
1932	Niobium(V) oxychloride	$NbOCl_3$	13597-20-1	215.264	wh tetr cry		subl	3.72		
1933	Nitrogen	N_2	7727-37-9	28.014	col gas	-210.0	-195.795	$0.8061^{-195.8}$ liq		sl H_2O; i EtOH
1934	Nitramide	NO_2NH_2	7782-94-7	62.028	unstab wh cry	72 dec				s H_2O, EtOH, ace, eth; i chl
1935	Nitric acid	HNO_3	7697-37-2	63.013	col liq; hyg	-41.6	83	1.5129^{20}		vs H_2O
1936	Nitrous acid	HNO_2	7782-77-6	47.014	stab only in soln					
1937	Nitrous oxide	N_2O	10024-97-2	44.012	col gas	-90.8	-88.48	0.001799^{25} gas		sl H_2O; s EtOH, eth
1938	Nitric oxide	NO	10102-43-9	30.006	col gas	-163.6	-151.74	0.001226^{25} gas		sl H_2O
1939	Nitrogen dioxide	NO_2	10102-44-0	46.006	brn gas; equil with N_2O_4		see N_2O_4	0.001880^{25} gas		reac H_2O
1940	Nitrogen trioxide	N_2O_3	10544-73-7	76.011	blue solid or liq (low temp)	-101.1	≈3 dec	1.4^2 liq		reac H_2O
1941	Nitrogen tetroxide	N_2O_4	10544-72-6	92.011	col liq; equil with NO_2	-9.3	21.15	1.45^{20}		reac H_2O
1942	Nitrogen pentoxide	N_2O_5	10102-03-1	108.010	col hex cry		33 sp	2.0		s chl; sl ctc
1943	Nitrogen tribromide	NBr_3	15162-90-0	253.719	unstab solid	-100 exp				
1944	Nitrogen trichloride	NCl_3	10025-85-1	120.366	yel oily liq; exp	-40	71	1.653		i H_2O; s CS_2, bz, ctc
1945	Nitrogen trifluoride	NF_3	7783-54-2	71.002	col gas	-206.79	-128.75	0.002902^{25} gas		i H_2O
1946	Nitrogen triiodide	NI_3	13444-85-4	394.720	unstab blk cry; exp					
1947	Nitrogen chloride difluoride	$NClF_2$	13637-87-1	87.457	col gas	-195	-67	0.003575^{25} gas		
1948	Chloramine	NH_2Cl	10599-90-3	51.476	yel liq	-66				s H_2O, EtOH, eth; sl bz, ctc
1949	Fluoramine	NH_2F	15861-05-9	35.021	unstab gas	≈-100		0.001431^{25} gas		
1950	Difluoramine	NHF_2	10405-27-3	53.012	col gas	-116	-23	0.002167^{25} gas		
1951	*cis*-Difluorodiazine	N_2F_2	13812-43-6	66.010	col gas	<-195	-105.75	0.002698^{25} gas		
1952	*trans*-Difluorodiazine	N_2F_2	13776-62-0	66.010	col gas	-172	-111.45	0.002698^{25} gas		
1953	Tetrafluorohydrazine	N_2F_4	10036-47-2	104.007	col gas	-164.5	-74	0.004251^{25} gas		
1954	Nitrosyl bromide	NOBr	13444-87-6	109.910	red gas	-56	≈0	0.004492^{25} gas		reac H_2O
1955	Nitrosyl chloride	NOCl	2696-92-6	65.459	yel gas	-59.6	-5.5	0.002676^{25} gas		reac H_2O
1956	Nitrosyl fluoride	NOF	7789-25-5	49.004	col gas	-132.5	-59.9	0.002003^{25} gas		
1957	Trifluoramine oxide	NOF_3	13847-65-9	87.001	col gas	-161	-87.5	0.003556^{25} gas		
1958	Nitryl chloride	NO_2Cl	13444-90-1	81.459	col gas	-145	-15	0.003330^{25} gas		
1959	Nitryl fluoride	NO_2F	10022-50-1	65.004	col gas	-166	-72.4	0.002657^{25} gas		reac H_2O
1960	Nitrogen selenide	N_4Se_4	12033-88-4	371.87	red monocl cry; hyg	exp		4.2		i H_2O, eth, EtOH; sl bz, CS_2
1961	Nobelium	No	10028-14-5	259	metal	827				
1962	Osmium	Os	7440-04-2	190.23	blue-wh metal; hex	3033	5008	22.587^{20}		s aqua regia
1963	Osmocene	$Os(C_5H_5)_2$	1273-81-0	320.42	col cry	229				
1964	Osmium carbonyl	$Os_3(CO)_{12}$	15696-40-9	906.81	yel cry	224		3.48		
1965	Osmium pentacarbonyl	$Os(CO)_5$	16406-49-8	330.28	col liq	-15	100 dec			s os
1966	Osmium nonacarbonyl	$Os_2(CO)_9$	28411-13-4	632.55	oran-yel cry	65 dec				s hc
1967	Osmium(II) chloride	$OsCl_2$	13444-92-3	261.14	hyg brn solid	>450 dec				s EtOH, eth
1968	Osmium(III) bromide	$OsBr_3$	59201-51-3	429.94	dark gray cry	340 dec				i H_2O, os, acid
1969	Osmium(III) chloride	$OsCl_3$	13444-93-4	296.59	gray cub cry	450 dec				i H_2O, os; s conc acid
1970	Osmium(IV) chloride	$OsCl_4$	10026-01-4	332.04	red-blk orth cry	323 dec		4.38		reac H_2O; i os
1971	Osmium(IV) fluoride	OsF_4	54120-05-7	266.22	yel cry	230				reac H_2O
1972	Osmium(IV) oxide	OsO_2	12036-02-1	222.23	yel-brn tetr cry	500 dec		11.4		i H_2O, acid
1973	Osmium(V) fluoride	OsF_5	31576-40-6	285.22	hyg blue-grn cry	70	233			reac H_2O
1974	Osmium(VI) fluoride	OsF_6	13768-38-2	304.22	yel cub cry	33.4	47.5	4.1		reac H_2O
1975	Osmium(VI) tetrachloride oxide	$OsOCl_4$	36509-15-6	348.04	dark brn hyg cry	32	200			reac H_2O; s hc
1976	Osmium(VIII) oxide	OsO_4	20816-12-0	254.23	yel monocl cry	40.6	131.2	5.1	6.44^{20}	sl H_2O; s ctc, bz, EtOH, eth
1977	Oxygen	O_2	7782-44-7	31.998	col gas	-218.79	-182.962	$1.141^{-183.0}$ liq		sl H_2O, EtOH, os

No.	Name	Formula	CAS Reg. No.	Mol. wt.	Phys. form	mp/°C	bp/°C	ρ/g cm^{-3}	Sol. in g/100 g H_2O	Qualitative solubility
1978	Ozone	O_3	10028-15-6	47.998	blue gas	-193	-111.35	0.00196^{25} gas		sl H_2O
1979	Palladium	Pd	7440-05-3	106.42	silv-wh metal; cub	1554.8	2963	12.0		s aqua regia
1980	Palladium(II) acetate	$Pd(C_2H_3O_2)_2$	3375-31-3	224.51	oran-brn cry	205 dec				i H_2O; s MeCN, eth, ace
1981	Palladium(II) bromide	$PdBr_2$	13444-94-5	266.23	red-blk monocl cry; hyg	250 dec		≈5.2		i H_2O
1982	Palladium(II) chloride	$PdCl_2$	7647-10-1	177.33	red rhomb cry; hyg	679		4.0		s H_2O, EtOH, ace
1983	Palladium(II) chloride dihydrate	$PdCl_2 \cdot 2H_2O$	7647-10-1	213.36	brn cry					s H_2O, EtOH, ace
1984	Palladium(II) cyanide	$Pd(CN)_2$	2035-66-7	158.45	yel solid	dec				
1985	Palladium(II) fluoride	PdF_2	13444-96-7	144.42	viol tetr cry; hyg	952		5.76		reac H_2O
1986	Palladium(II) iodide	PdI_2	7790-38-7	360.23	blk cry	360 dec		6.0		i H_2O, EtOH, eth
1987	Palladium(II) nitrate	$Pd(NO_3)_2$	10102-05-3	230.43	brn hyg cry	dec				sl H_2O; s dil HNO_3
1988	Palladium(II) oxide	PdO	1314-08-5	122.42	grn-blk tetr cry	750 dec		8.3		i H_2O, acid; sl aqua regia
1989	Palladium(II) 2,4-pentanedioate	$Pd(CH_3COCHCOCH_3)_2$	14024-61-4	304.64	oran-yel cry	205 dec				s bz, chl
1990	Palladium(II) sulfate dihydrate	$PdSO_4 \cdot 2H_2O$	13566-03-5	238.51	grn-brn cry	dec				
1991	Palladium(II) sulfide	PdS	12125-22-3	138.49	gray tetr cry			6.7		
1992	*cis*-Dichlorodiamminepalladium(II)	$Pd(NH_3)_2Cl_2$	15684-18-1	211.39	yel pow				0.025^{25}	
1993	*trans*-Dichlorodiamminepalladium(II)	$Pd(NH_3)_2Cl_2$	13782-33-7	211.39	yel solid			2.50		
1994	Phosphorus (white)	P	7723-14-0	30.974	col waxlike cub cry	44.15	280.5	1.823		i H_2O; sl bz, EtOH, chl; s CS_2
1995	Phosphorus (red)	P	7723-14-0	30.974	red-viol amorp powder	579.2	431 sp	2.16		i H_2O, os
1996	Phosphorus (black)	P	7723-14-0	30.974	blk orth cry or amorp solid	610		2.69		i os
1997	Phosphine	PH_3	7803-51-2	33.998	col gas; flam	-133.8	-87.75	0.001390^{25} gas		i H_2O; sl EtOH, eth
1998	Diphosphine	P_2H_4	13445-50-6	65.980	col liq	-99	63.5 dec			reac H_2O
1999	Diphosphorus tetrachloride	P_2Cl_4	13497-91-1	203.760	col oily liq	-28	≈180 dec			
2000	Diphosphorus tetrafluoride	P_2F_4	13824-74-3	137.942	col gas	-86.5	-6.2	0.005638^{25} gas		
2001	Diphosphorus tetraiodide	P_2I_4	13455-00-0	569.566	red tricl needles	125.5	dec	3.89		
2002	Phosphonium chloride	PH_4Cl	24567-53-1	70.459	gas		-27 sp	0.002880^{25} gas		reac H_2O
2003	Phosphonium iodide	PH_4I	12125-09-6	161.910	col tetr cry	18.5	62.5	2.86		reac H_2O, EtOH
2004	Phosphoric acid	H_3PO_4	7664-38-2	97.995	col visc liq	42.4	407		548^{20}	s EtOH
2005	Phosphotungstic acid	$H_3PW_{12}O_{40}$	12067-99-1	2880.05	wh-yel cry	89				vs H_2O; s EtOH, eth
2006	Phosphonic acid	H_3PO_3	13598-36-2	81.996	wh hyg cry	74.4	200	1.65	309^{0}	vs EtOH
2007	Phosphinic acid	HPH_2O_2	6303-21-5	65.997	hyg cry or col oily liq	26.5	130	1.49		vs H_2O, EtOH, eth
2008	Metaphosphoric acid	HPO_3	37267-86-0	79.980	gl solid; hyg					sl H_2O; s EtOH
2009	Hypophosphoric acid	$H_4P_2O_6$	7803-60-3	161.976	col orth cry	73 dec				vs H_2O
2010	Diphosphoric acid	$H_4P_2O_7$	2466-09-3	177.975	wh cry	71.5			709^{23}	
2011	Difluorophosphoric acid	HPO_2F_2	13779-41-4	101.978	col liq	≈-94	110 dec	1.583		reac H_2O
2012	Hexafluorophosphoric acid	HPF_6	16940-81-1	145.972	col oily liq	25 dec				reac H_2O
2013	Fluorophosphonic acid	H_2PFO_3	13537-32-1	99.986	col visc liq	<-70		1.82		vs H_2O
2014	Phosphorus nitride [P_3N_5]	P_3N_5	12136-91-3	162.955	yel-brn solid	800 dec				i H_2O; s os
2015	Phosphorus sesquisulfide	P_4S_3	1314-85-8	220.090	yel-grn orth cry	173	407	2.03		i H_2O; s bz; vs CS_2
2016	Phosphorus heptasulfide	P_4S_7	12037-82-0	348.350	pale yel monocl cry	308	523	2.19		sl CS_2
2017	Phosphonitrilic chloride trimer	$(PNCl_2)_3$	940-71-6	347.659	wh hyg cry	128.8		1.98		reac H_2O
2018	Phosphorus(III) bromide	PBr_3	7789-60-8	270.686	col liq	-41.5	173.2	2.8		reac H_2O, EtOH; s ace, CS_2
2019	Phosphorus(III) dibromide fluoride	PBr_2F	15597-39-4	209.780	col liq	-115	78.5			
2020	Phosphorus(III) bromide difluoride	$PBrF_2$	15597-40-7	148.875	col gas	-133.8	-16.1	0.006085^{25} gas		
2021	Phosphorus(III) chloride	PCl_3	7719-12-2	137.333	col liq	-93	76	1.574		reac H_2O, EtOH; s bz, chl, eth
2022	Phosphorus(III) dichloride fluoride	PCl_2F	15597-63-4	120.878	col gas	-144	13.85	0.004941^{25} gas		
2023	Phosphorus(III) chloride difluoride	$PClF_2$	14335-40-1	104.424	col gas	-164.8	-47.3	0.004268^{25} gas		
2024	Phosphorus(III) fluoride	PF_3	7783-55-3	87.969	col gas	-151.5	-101.8	0.003596^{25} gas		reac H_2O
2025	Phosphorus(III) iodide	PI_3	13455-01-1	411.687	red-oran hex cry; hyg	61.2	227 dec	4.18		reac H_2O; s EtOH
2026	Phosphorus(III) oxide	P_2O_3	1314-24-5	109.946	col monocl cry or liq	23.8	173	2.13		reac H_2O
2027	Tetraphosphorus(III) hexoxide	P_4O_6	12440-00-5	219.891	soft wh cry	23.8	175.4			
2028	Phosphorus(III) selenide	P_2Se_3	1314-86-9	298.83	oran-red cry	245	≈380	1.31		reac H_2O; s bz, ctc, CS_2, ace

Inorganic

No.	Name	Formula	CAS Reg. No.	Mol. wt.	Phys. form	mp/°C	bp/°C	ρ/g cm-3	Sol. in g/100 g H2O	Qualitative solubility
2029	Phosphorus(III) sulfide	P_2S_3	12165-69-4	158.143	yel solid	290	490			reac H_2O; s EtOH, eth, CS_2
2030	Phosphorus(V) bromide	PBr_5	7789-69-7	430.494	yel orth cry, hyg	≈100 dec		3.61		reac H_2O, EtOH; s CS_2, ctc
2031	Phosphorus(V) tetrabromide fluoride	PBr_4F		369.588	pale yel cry	87 dec				
2032	Phosphorus(V) dibromide trifluoride	PBr_2F_3	13445-58-4	247.777	yel-red liq	-20	15 dec			
2033	Phosphorus(V) chloride	PCl_5	10026-13-8	208.239	wh-yel tetr cry; hyg	167 tp	160 sp	2.1		reac H_2O; s CS_2, ctc
2034	Phosphorus(V) tetrachloride fluoride	PCl_4F	13498-11-8	191.784	col liq	-59	30 dec			
2035	Phosphorus(V) trichloride difluoride	PCl_3F_2	13537-23-0	175.330	col liq	-63				
2036	Phosphorus(V) dichloride trifluoride	PCl_2F_3	13454-99-4	158.875	col gas	-125	7.1	0.006494^{25} gas		
2037	Phosphorus(V) chloride tetrafluoride	$PClF_4$	13498-11-8	142.421	col gas	-132	-43.4	0.005821^{25} gas		
2038	Phosphorus(V) fluoride	PF_5	7647-19-0	125.966	col gas	-93.8	-84.6	0.005149^{25} gas		reac H_2O
2039	Phosphorus(V) oxide	P_2O_5	1314-56-3	141.945	wh orth cry; hyg	562	605	2.30		reac H_2O, EtOH
2040	Phosphorus(V) selenide	P_2Se_5	1314-82-5	456.75	blk-purp amorp solid					reac hot H_2O, ctc; i CS_2
2041	Phosphorus(V) sulfide	P_2S_5	1314-80-3	222.273	grn-yel hyg cry	285	515	2.03		reac H_2O; s CS_2
2042	Phosphonic difluoride	POF_2H	14939-34-5	85.978	volatile liq	>-120	≈60 (gas unstab)			
2043	Phosphoryl bromide	$POBr_3$	7789-59-5	286.685	faint oran plates	55	191.7	2.822		reac H_2O; s bz, eth, chl
2044	Phosphoryl dibromide chloride	$POBr_2Cl$	13550-31-7	242.234	yel solid	31	165			
2045	Phosphoryl dibromide fluoride	$POBr_2F$	14014-19-8	225.779	col liq	-117.2	110.1			
2046	Phosphoryl bromide dichloride	$POBrCl_2$	13455-03-3	197.783	col liq	11	136.5	2.104^{14}		
2047	Phosphoryl bromide difluoride	$POBrF_2$	14014-18-7	164.874	col liq	-84.8	31.6			
2048	Phosphoryl bromide chloride fluoride	$POBrClF$	14518-81-1	181.328	col liq		79			
2049	Phosphoryl chloride	$POCl_3$	10025-87-3	153.332	col liq	1.18	105.5	1.645		reac H_2O, EtOH
2050	Phosphoryl dichloride fluoride	$POCl_2F$	13769-76-1	136.877	col liq	-80.1	52.9			
2051	Phosphoryl chloride difluoride	$POClF_2$	13769-75-0	120.423	col gas	-96.4	3.1	0.004922^{25} gas		
2052	Phosphoryl fluoride	POF_3	13478-20-1	103.968	col gas	-39.1 tp	-39.7 sp	0.004250^{25} gas		reac H_2O
2053	Phosphoryl iodide	POI_3	13455-04-4	427.686	viol cry	53				
2054	Phosphorothioc tribromide	$PSBr_3$	3931-89-3	302.751	yel cry	37.8	212 dec	2.85		
2055	Phosphorothioc dibromide fluoride	$PSBr_2F$	13706-10-0	241.845	yel liq	-75.2	125.3			
2056	Phosphorothioc bromide difluoride	$PSBrF_2$	13706-09-7	180.940	yel liq	-136.9	35.5			
2057	Phosphorothioc trichloride	$PSCl_3$	3982-91-0	169.398	fuming liq	-36.2	125	1.635		reac H_2O; s bz, ctc, chl, CS_2
2058	Phosphorothioc dichloride fluoride	$PSCl_2F$	155698-29-6	152.943	col liq	-96.0	64.7			
2059	Phosphorothioc chloride difluoride	$PSClF_2$	2524-02-9	136.489	col gas	-155.2	6.3	0.005579^{25} gas		
2060	Phosphorothioc trifluoride	PSF_3	2404-52-6	120.034	col gas	-148.8	-52.25	0.004906^{25} gas		
2061	Phosphorothioc triiodide	PSI_3	63972-04-3	443.752	yel cry	48	dec			
2062	Platinum	Pt	7440-06-4	195.084	silv-gray metal; cub	1768.2	3825	21.5		i acid; s aqua regia
2063	Hexachloroplatinic acid	H_2PtCl_6	16941-12-1	409.818	hyg yel-brn cry	60				s H_2O, EtOH
2064	Hydrogen hexahydroxyplatinate(IV)	$H_2Pt(OH)_6$	51850-20-5	299.144	yel needles	100 dec				s H_2O, acid, dil alk
2065	Platinum(II) bromide	$PtBr_2$	13455-12-4	354.892	red-brn powder	250 dec		6.65		i H_2O
2066	Platinum(II) chloride	$PtCl_2$	10025-65-7	265.990	grn hex cry	581 dec		6.0		i H_2O, EtOH, eth; s HCl
2067	Platinum(II) cyanide	$Pt(CN)_2$	592-06-3	247.118	pale yel cry					i H_2O, acid, alk
2068	Platinum(II) iodide	PtI_2	7790-39-8	448.893	blk powder	325 dec		6.4		i H_2O
2069	Platinum(II) oxide	PtO	12035-82-4	211.083	blk tetr cry	325 dec		14.1		i H_2O, EtOH; s aqua regia
2070	Platinum(II) sulfide	PtS	12038-20-9	227.149	tetr cry			10.25		
2071	Platinum(III) bromide	$PtBr_3$	25985-07-3	434.796	grn-blk cry	200 dec				
2072	Platinum(III) chloride	$PtCl_3$	25909-39-1	301.443	grn-blk cry	435 dec		5.26		
2073	Platinum(IV) bromide	$PtBr_4$	68938-92-1	514.700	brn-blk cry	180 dec			0.41^{20}	sl EtOH, eth
2074	Platinum(IV) chloride	$PtCl_4$	37773-49-2	336.896	red-brn cub cry	327 dec		4.30	142^{25}	
2075	Platinum(IV) chloride pentahydrate	$PtCl_4 \cdot 5H_2O$	13454-96-1	426.972	red cry			2.43		s H_2O, EtOH
2076	Platinum(IV) fluoride	PtF_4	13455-15-7	271.078	red cry	600				
2077	Platinum(IV) iodide	PtI_4	7790-46-7	702.702	brn-blk powder	130 dec				s H_2O
2078	Platinum(IV) oxide	PtO_2	1314-15-4	227.083	blk hex cry	450		11.8		i H_2O; s conc acid, dil alk
2079	Platinum(IV) sulfide	PtS_2	12038-21-0	259.214	hex cry			7.85		
2080	Platinum(VI) fluoride	PtF_6	13693-05-5	309.074	red cub cry	61.3	69.1	≈4.0		
2081	*cis*-Diamminedichloroplatinum	$Pt(NH_3)_2Cl_2$	15663-27-1	300.051	yel solid	270 dec			0.253^{25}	
2082	*trans*-Diamminedichloroplatinum	$Pt(NH_3)_2Cl_2$	14913-33-8	300.051	pale yel solid	270 dec			0.036^{25}	s DMF, DMSO
2083	Hexachloroplatinic acid hexahydrate	$H_2PtCl_6 \cdot 6H_2O$	16941-12-1	517.909	brn-yel hyg cry	60		2.43	140^{18}	vs EtOH
2084	Platinum silicide	PtSi	12137-83-6	223.170	orth cry	1229		12.4		

Inorganic

No.	Name	Formula	CAS Reg. No.	Mol. wt.	Phys. form	mp/°C	bp/°C	ρ/g cm^{-3}	Sol. in g/100 g H_2O	Qualitative solubility
2085	Plutonium	Pu	7440-07-5	244	silv-wh metal; monocl	640	3228	19.7		
2086	Plutonium nitride	PuN	12033-54-4	258	gray cub cry	2550		14.4		
2087	Plutonium(II) oxide	PuO	12035-83-5	260	cub cry			14.0		
2088	Plutonium(III) bromide	$PuBr_3$	15752-46-2	484	grn orth cry	681		6.75		s H_2O
2089	Plutonium(III) chloride	$PuCl_3$	13569-62-5	350	grn hex cry	760		5.71		s H_2O
2090	Plutonium(III) fluoride	PuF_3	13842-83-6	301	purp hex cry	1396		9.33		i H_2O; sl acid
2091	Plutonium(III) iodide	PuI_3	13813-46-2	625	grn orth cry; hyg	777		6.92		s H_2O
2092	Plutonium(III) oxide	Pu_2O_3	12036-34-9	536	blk cub cry	2085		10.5		
2093	Plutonium(IV) fluoride	PuF_4	13709-56-3	320	red-brn monocl cry	1037		7.1		
2094	Plutonium(IV) oxide	PuO_2	12059-95-9	276	yel-brn cub cry	2390		11.5		
2095	Plutonium(VI) fluoride	PuF_6	13693-06-6	358	red-brn orth cry	51.6		5.08		
2096	Polonium	Po	7440-08-6	209	silv metal; cub	254	962	9.20		
2097	Polonium(IV) chloride	$PoCl_4$	10026-02-5	351	yel hyg cry	≈300	390			s H_2O, EtOH, ace
2098	Polonium(IV) oxide	PoO_2	7446-06-2	241	yel cub cry	500 dec		8.9		
2099	Potassium	K	7440-09-7	39.098	soft silv-wh metal; cub	63.5	759	0.89		reac H_2O
2100	Potassium acetate	$KC_2H_3O_2$	127-08-2	98.142	wh hyg cry	309		1.57	269^{25}	s EtOH; i eth
2101	Potassium aluminate trihydrate	$K_2Al_2O_4 \cdot 3H_2O$	12003-63-3	250.204	wh orth cry			2.13		vs H_2O; i EtOH
2102	Potassium aluminum silicate	$KAlSi_3O_8$	1327-44-2	278.332	col monocl cry			2.56		i H_2O
2103	Potassium aluminum sulfate	$KAl(SO_4)_2$	10043-67-1	258.205	wh hyg powder				5.9^{20}	
2104	Potassium aluminum sulfate dodecahydrate	$KAl(SO_4)_2 \cdot 12H_2O$	7784-24-9	474.389	col cry	≈100 dec		1.72	5.9^{20}	
2105	Potassium amide	KNH_2	17242-52-3	55.121	wh/yel-grn hyg cry	335				reac H_2O, EtOH
2106	Potassium arsenate	K_3AsO_4	13464-36-3	256.215	col cry			2.8	125^{25}	
2107	Potassium arsenite	$KAsO_2$	13464-35-2	146.019	wh hyg powder					s H_2O; sl EtOH
2108	Potassium azide	KN_3	20762-60-1	81.118	tetr cry; exp			2.04	49.7^{17}	
2109	Potassium borohydride	KBH_4	13762-51-1	53.941	wh cub cry	≈500 dec		1.11		s H_2O
2110	Potassium bromate	$KBrO_3$	7758-01-2	167.000	wh hex cry	434 dec		3.27	8.17^{25}	i EtOH
2111	Potassium bromide	KBr	7758-02-3	119.002	col cub cry; hyg	734	1435	2.74	67.8^{25}	sl EtOH
2112	Potassium carbonate	K_2CO_3	584-08-7	138.206	wh monocl cry; hyg	899	dec	2.29	111^{25}	i EtOH
2113	Potassium carbonate sesquihydrate	$K_2CO_3 \cdot 1.5H_2O$	6381-79-9	165.229	granular cry				111^{20}	
2114	Potassium chlorate	$KClO_3$	3811-04-9	122.549	wh monocl cry	357	dec	2.34	8.61^{25}	
2115	Potassium chloride	KCl	7447-40-7	74.551	wh cub cry	771		1.988	35.5^{25}	i eth, ace
2116	Potassium chlorochromate	$KCrO_3Cl$	16037-50-6	174.545	oran cry			2.5		reac H_2O; s ace, acid
2117	Potassium chromate	K_2CrO_4	7789-00-6	194.191	yel orth cry	974		2.73	65.0^{25}	
2118	Potassium citrate monohydrate	$K_3C_6H_5O_7 \cdot H_2O$	6100-05-6	324.410	col hyg cry	180 dec		1.98	172^{20}	vs H_2O; sl EtOH
2119	Potassium cobalt(II) selenate hexahydrate	$K_2Co(SeO_4)_2 \cdot 6H_2O$	28041-86-3	531.14	red monocl cry			2.51		
2120	Potassium cyanate	KCNO	590-28-3	81.115	wh tetr cry	≈700 dec		2.05	75^{25}	sl EtOH
2121	Potassium cyanide	KCN	151-50-8	65.116	wh cub cry; hyg	622		1.55	69.9^{20}	sl EtOH
2122	Potassium cyanoaurite	$KAu(CN)_2$	13967-50-5	288.099	col cry			3.45	14^{20}	s H_2O; sl EtOH; i eth, ace
2123	Potassium dichromate	$K_2Cr_2O_7$	7778-50-9	294.185	oran-red tricl cry	398	≈500 dec	2.68	15.1^{25}	
2124	Potassium dihydrogen arsenate	KH_2AsO_4	7784-41-0	180.034	col cry	288		2.87	19^{6}	i EtOH
2125	Potassium dihydrogen phosphate	KH_2PO_4	7778-77-0	136.085	wh tetr cry	253 dec		2.34	25.0^{25}	sl EtOH
2126	Potassium dihydrogen phosphonate	KH_2PO_3	13977-65-6	120.086	col monocl hyg cry					
2127	Potassium dithionate	$K_2S_2O_6$	13455-20-4	238.323	col hex cry	dec		2.27		sl H_2O; i EtOH
2128	Potassium ferricyanide	$K_3Fe(CN)_6$	13746-66-2	329.244	red cry	dec		1.89	48.8^{25}	
2129	Potassium ferrocyanide trihydrate	$K_4Fe(CN)_6 \cdot 3H_2O$	14459-95-1	422.388	yel monocl cry	60 dec		1.85	36.0^{25}	i EtOH, eth
2130	Potassium fluoride	KF	7789-23-3	58.096	wh cub cry	858	1502	2.48	102^{25}	
2131	Potassium fluoride dihydrate	$KF \cdot 2H_2O$	13455-21-5	94.127	monocl cry	41 dec		2.5	102^{25}	
2132	Potassium fluoroborate	KBF_4	14075-53-7	125.903	col orth cry	570		2.505	0.55^{25}	sl EtOH
2133	Potassium fluorotantalate	K_2TaF_7	16924-00-8	392.134	col cry	730		5.24	0.5^{0}	
2134	Potassium formate	$KCHO_2$	590-29-4	84.116	col hyg cry	167		1.91	331^{18}	
2135	Potassium hexachloroosmate(IV)	K_2OsCl_6	16871-60-6	481.15	red cub cry					vs H_2O; sl EtOH
2136	Potassium hexachloroplatinate	K_2PtCl_6	16921-30-5	485.999	yel-oran cub cry	250 dec		3.50	0.77^{20}	i EtOH
2137	Potassium hexacyanocobaltate	$K_3Co(CN)_6$	13963-58-1	332.332	yel monocl cry	dec		1.91		vs H_2O; i EtOH
2138	Potassium hexafluoromanganate(IV)	K_2MnF_6	16962-31-5	247.125	yel hex cry					reac H_2O
2139	Potassium hexafluorosilicate	K_2SiF_6	16871-90-2	220.273	wh cry	dec		2.27	0.084^{20}	i EtOH
2140	Potassium hexafluorozirconate(IV)	K_2ZrF_6	16923-95-8	283.411	col monocl cry			3.48	0.78^{2}	
2141	Potassium hydride	KH	7693-26-7	40.106	cub cry	619		1.43		reac H_2O
2142	Potassium hydrogen arsenate	K_2HAsO_4	21093-83-4	218.125	col monocl prisms	300 dec			18.7^{6}	i EtOH
2143	Potassium hydrogen carbonate	$KHCO_3$	298-14-6	100.115	col monocl cry	≈100 dec		2.17	36.2^{25}	i EtOH
2144	Potassium hydrogen fluoride	KHF_2	7789-29-9	78.103	col tetr cry	238.8		2.37	39.2^{20}	i EtOH
2145	Potassium hydrogen iodate	$KH(IO_3)_2$	13455-24-8	389.911	col cry	dec			1.3^{15}	sl H_2O; i EtOH
2146	Potassium hydrogen oxalate hemihydrate	$KHC_2O_4 \cdot 0.5H_2O$	127-95-7	137.133	wh cry	dec		2.09	2.5^{20}	sl EtOH
2147	Potassium hydrogen phosphate	K_2HPO_4	7758-11-4	174.176	wh hyg cry	dec			168^{25}	s EtOH
2148	Potassium hydrogen phosphite	K_2HPO_3	13492-26-7	158.177	wh hyg powder	dec			170^{20}	i EtOH
2149	Potassium hydrogen selenite	$KHSeO_3$	7782-70-9	167.06	hyg orth cry	>100 dec				s H_2O; sl EtOH
2150	Potassium hydrogen sulfate	$KHSO_4$	7646-93-7	136.169	wh monocl cry; hyg	≈200		2.32	50.6^{25}	
2151	Potassium hydrogen sulfide	KHS	1310-61-8	72.171	wh hex cry; hyg	≈450		1.69		s H_2O, EtOH

Inorganic

No.	Name	Formula	CAS Reg. No.	Mol. wt.	Phys. form	mp/°C	bp/°C	ρ/g cm⁻³	Sol. in g/100 g H_2O	Qualitative solubility
2152	Potassium hydrogen sulfide hemihydrate	$KHS\cdot0.5H_2O$	1310-61-8	81.179	wh-yel hyg cry	≈175		1.7		vs H_2O, EtOH
2153	Potassium hydrogen sulfite	$KHSO_3$	7773-03-7	120.169	wh cry powder	190 dec			49^{20}	i EtOH
2154	Potassium hydrogen tartrate	$KHC_4H_4O_6$	868-14-4	188.177	wh cry			1.98	0.57^{20}	s acid, alk; i EtOH
2155	Potassium hydroxide	KOH	1310-58-3	56.105	wh rhomb cry; hyg	406	1327	2.044	121^{25}	s EtOH; s MeOH
2156	Potassium hypochlorite	$KOCl$	7778-66-7	90.550	exists only in aq soln					
2157	Potassium phosphinate	KH_2PO_2	7782-87-8	104.087	wh hyg cry	dec				vs H_2O; s EtOH
2158	Potassium iodate	KIO_3	7758-05-6	214.001	wh monocl cry	560 dec		3.89	9.22^{25}	
2159	Potassium iodide	KI	7681-11-0	166.003	col cub cry	681	1323	3.12	148^{25}	sl EtOH
2160	Potassium iron(III) oxalate trihydrate	$K_3Fe(C_2O_4)_3\cdot3H_2O$	5936-11-8	491.243	grn monocl cry	100	230 dec	2.133	4.7^{0}	i EtOH
2161	Potassium manganate	K_2MnO_4	10294-64-1	197.133	grn cry	190 dec				s H_2O; reac HCl
2162	Potassium metaarsenate	$KAsO_3$	19197-73-0	162.018	wh solid	660				
2163	Potassium metabisulfite	$K_2S_2O_5$	16731-55-8	222.324	wh powder	≈150 dec		2.3	49.5^{25}	reac acid; i EtOH
2164	Potassium metaborate	KBO_2	13709-94-9	81.908	wh hex cry	947		≈2.3		
2165	Potassium molybdate	K_2MoO_4	13446-49-6	238.14	wh hyg cry	919		2.3	183^{25}	i EtOH
2166	Potassium niobate	$KNbO_3$	12030-85-2	180.002	wh rhomb cry	≈1100		4.64		i H_2O
2167	Potassium nitrate	KNO_3	7757-79-1	101.103	col orth cry or powder	334	400 dec	2.105	38.3^{25}	i EtOH
2168	Potassium nitrite	KNO_2	7758-09-0	85.104	wh hyg cry	438	537 exp	1.915	312^{25}	sl EtOH
2169	Potassium oxalate	$K_2C_2O_4$	583-52-8	166.216	wh pwd					sl H_2O
2170	Potassium oxalate monohydrate	$K_2C_2O_4\cdot H_2O$	6487-48-5	184.231	col cry	160 dec		2.13	36.4^{20}	
2171	Potassium oxide	K_2O	12136-45-7	94.196	gray cub cry	740		2.35		s H_2O, EtOH, eth
2172	Potassium perbromate	$KBrO_4$	22207-96-1	183.000	wh cry	275 dec			4.21^{25}	
2173	Potassium percarbonate monohydrate	$K_2C_2O_6\cdot H_2O$	589-97-9	216.230	oran or blue pow				6.5^{20}	
2174	Potassium perchlorate	$KClO_4$	7778-74-7	138.549	col orth cry; hyg	525		2.52	2.08^{25}	
2175	Potassium periodate	KIO_4	7790-21-8	230.001	col tetr cry	582	exp	3.618	0.51^{25}	
2176	Potassium permanganate	$KMnO_4$	7722-64-7	158.034	purp orth cry	dec		2.7	7.60^{25}	reac EtOH
2177	Potassium peroxide	K_2O_2	17014-71-0	110.196	yel amorp solid	545				reac H_2O
2178	Potassium persulfate	$K_2S_2O_8$	7727-21-1	270.322	col cry	≈100 dec		2.48	4.7^{20}	
2179	Potassium phosphate	K_3PO_4	7778-53-2	212.266	wh orth cry; hyg	1340		2.564	106^{25}	i EtOH
2180	Potassium pyrophosphate	$K_4P_2O_7$	7320-34-5	330.337	wh cry	1300 dec				s H_2
2181	Potassium pyrophosphate trihydrate	$K_4P_2O_7\cdot3H_2O$	7790-67-2	384.383	col hyg cry	300 dec		2.33		vs H_2O; i EtOH
2182	Potassium pyrosulfate	$K_2S_2O_7$	7790-62-7	254.323	col needles	≈325		2.28		s H_2O
2183	Potassium selenate	K_2SeO_4	7790-59-2	221.16	wh powder			3.07	114^{25}	
2184	Potassium selenide	K_2Se	1312-74-9	157.16	red cub cry; hyg	800		2.29		s H_2O
2185	Potassium selenite	K_2SeO_3	10431-47-7	205.16	wh hyg cry	875 dec			217^{25}	sl EtOH
2186	Potassium silver cyanide	$KAg(CN)_2$	506-61-6	199.000	wh cry					s H_2O
2187	Potassium sodium tartrate tetrahydrate	$KNaC_4H_4O_6\cdot4H_2O$	6381-59-5	282.220	wh cry	≈70	130 anh	1.79		vs H_2O; i EtOH
2188	Potassium stannate trihydrate	$K_2SnO_3\cdot3H_2O$	12142-33-5	298.951	col cry			3.20		vs H_2O; i EtOH
2189	Potassium stearate	$KC_{18}H_{35}O_2$	593-29-3	322.568	wh pow					sl cold H_2O; s hot H_2O, EtOH
2190	Potassium sulfate	K_2SO_4	7778-80-5	174.260	wh orth cry	1069		2.66	12.0^{25}	i EtOH
2191	Potassium sulfide	K_2S	1312-73-8	110.262	red-yel cub cry; hyg	948		1.74		s H_2O, EtOH; i eth
2192	Potassium sulfide pentahydrate	$K_2S\cdot5H_2O$	37248-34-3	200.338	col rhomb cry	60				vs H_2O, EtOH; i eth
2193	Potassium sulfite	K_2SO_3	10117-38-1	158.260	col hex cry				106^{25}	sl EtOH
2194	Potassium sulfite dihydrate	$K_2SO_3\cdot2H_2O$	7790-56-9	194.291	wh monocl cry	dec			107^{20}	sl EtOH; dec dil acid
2195	Potassium superoxide	KO_2	12030-88-5	71.097	yel tetr cry; hyg	535		2.16		reac H_2O
2196	Potassium tellurate(VI) trihydrate	$K_2TeO_4\cdot3H_2O$	15571-91-2	323.84	wh cry powder					s H_2O
2197	Potassium tellurite	K_2TeO_3	7790-58-1	253.80	wh hyg cry	≈460 dec				vs H_2O
2198	Potassium tetraborate pentahydrate	$K_2B_4O_7\cdot5H_2O$	1332-77-0	323.513	wh cry powder				16.5^{30}	sl EtOH
2199	Potassium tetrachloroaurate dihydrate	$KAuCl_4\cdot2H_2O$	13682-61-6	413.908	yel monocl cry					s H_2O, EtOH, eth
2200	Potassium tetrachloroplatinate	K_2PtCl_4	10025-99-7	415.093	pink-red tetr cry	500 dec		3.38		s H_2O; i EtOH
2201	Potassium tetracyanocadmate	$K_2Cd(CN)_4$	14402-75-6	294.678	cub cry	≈450		1.85	25^{20}	sl EtOH
2202	Potassium tetracyanonickelate monohydrate	$K_2[Ni(CN)_4]\cdot H_2O$	14220-17-8	258.975	red-oran cry	100 dec				
2203	Potassium tetracyanoplatinate(II) trihydrate	$K_2Pt(CN)_4\cdot3H_2O$	562-76-5	431.397	col rhomb prisms					s H_2O
2204	Potassium tetracyanozincate	$K_2Zn(CN)_4$	14244-62-3	247.676	cry pow					vs H_2O
2205	Potassium tetraiodomercurate(II)	K_2HgI_4	7783-33-7	786.40	yel hyg cry			4.29		vs H_2O; s EtOH, eth, ace
2206	Potassium thiocyanate	$KSCN$	333-20-0	97.181	col tetr cry; hyg	173	500 dec	1.88	238^{25}	s EtOH
2207	Potassium thiosulfate	$K_2S_2O_3$	10294-66-3	190.325	col hyg cry				165^{25}	i EtOH
2208	Potassium titanate	K_2TiO_3	12030-97-6	174.062	wh orth cry	1515		3.1		reac H_2O
2209	Potassium triiodide monohydrate	$KI_3\cdot H_2O$	7790-42-3	437.827	brn monocl cry; hyg	225 dec		3.5		s H_2O; reac EtOH, eth
2210	Potassium triiodozincate	$KZnI_3$	7790-43-4	485.221	hyg cry					vs H_2O
2211	Potassium thiocarbonate	K_2CS_3	26750-66-3	186.403	yel-red hyg cry					vs H_2O
2212	Potassium tungstate	K_2WO_4	7790-60-5	326.04	hyg cry	921		3.12		vs H_2O; i EtOH
2213	Potassium uranate	$K_2U_2O_7$	7790-63-8	666.251	oran cub cry			6.12		i H_2O; s acid

Inorganic

No.	Name	Formula	CAS Reg. No.	Mol. wt.	Phys. form	mp/°C	bp/°C	ρ/g cm^{-3}	Sol. in g/100 g H_2O	Qualitative solubility
2214	Potassium uranyl nitrate	$K(UO_2)(NO_3)_3$	18078-40-5	495.140	grn-yel cry pow					vs H_2O
2215	Potassium uranyl sulfate dihydrate	$K_2(UO_2)(SO_4)_2 \cdot 2H_2O$	27709-53-1	576.381	grn-yel cry pow	120 dec		3.36		vs H_2O
2216	Potassium zinc sulfate hexahydrate	$K_2Zn(SO_4)_2 \cdot 6H_2O$	13932-17-7	443.823	cry					s H_2O
2217	Potassium zirconium sulfate trihydrate	$K_4Zr(SO_4)_4 \cdot 3H_2O$	53608-79-0	685.914	wh cry pow					sl H_2O
2218	Praseodymium	Pr	7440-10-0	140.908	silv metal; hex	931	3520	6.77		
2219	Praseodymium boride	PrB_6	12008-27-4	205.774	blk cub cry	2610		4.84		
2220	Praseodymium nitride	PrN	25764-09-4	154.915	cub cry			7.46		
2221	Praseodymium silicide	$PrSi_2$	12066-83-0	197.079	tetr cry	1712		5.46		
2222	Praseodymium(II) iodide	PrI_2	65530-47-4	394.717	bronze solid	758				
2223	Praseodymium(III) bromate	$Pr(BrO_3)_3$	15162-93-3	524.615	grn cry					vs H_2O
2224	Praseodymium(III) bromide	$PrBr_3$	13536-53-3	380.620	grn hex cry; hyg	693		5.28		s H_2O
2225	Praseodymium(III) carbonate octahydrate	$Pr_2(CO_3)_3 \cdot 8H_2O$	14948-62-0	605.964	grn silky plates	420 (anh) dec				i H_2O; s acid
2226	Praseodymium(III) chloride	$PrCl_3$	10361-79-2	247.267	grn hex needles; hyg	786		4.0	96.1^{25}	s EtOH
2227	Praseodymium(III) chloride heptahydrate	$PrCl_3 \cdot 7H_2O$	10025-90-8	373.374	grn cry	110 dec			96.1^{25}	s EtOH
2228	Praseodymium(III) fluoride	PrF_3	13709-46-1	197.903	grn hex cry	1399		6.3		
2229	Praseodymium(III) hydroxide	$Pr(OH)_3$	16469-16-2	191.930	grn solid	220 dec		3.7		i H_2O
2230	Praseodymium(III) iodide	PrI_3	13813-23-5	521.621	orth hyg cry	738		≈5.8		s H_2O
2231	Praseodymium(III) nitrate	$Pr(NO_3)_3$	10361-80-5	326.923	pale grn hyg cry				165^{25}	s EtOH
2232	Praseodymium(III) nitrate hexahydrate	$Pr(NO_3)_3 \cdot 6H_2O$	15878-77-0	435.014	grn needles				165^{25}	s EtOH, ace
2233	Praseodymium(III) oxide	Pr_2O_3	12036-32-7	329.813	wh hex cry	2183	3760	6.9		
2234	Praseodymium(III) perchlorate hexahydrate	$Pr(ClO_4)_3 \cdot 6H_2O$	13498-07-2	547.351	hyg grn cry	200 dec				s H_2O, EtOH
2235	Praseodymium(III) sulfate octahydrate	$Pr_2(SO_4)_3 \cdot 8H_2O$	13510-41-3	714.125	grn monocl cry			2.83	17^{20}	s H_2O
2236	Praseodymium(III) sulfide	Pr_2S_3	12038-13-0	378.010	cub cry	1765		5.1		
2237	Praseodymium(III) telluride	Pr_2Te_3	12038-12-9	664.62	cub cry	1500		≈7.0		
2238	Praseodymium(IV) fluoride	PrF_4	15192-24-2	216.902	yel-wh solid	90 dec				
2239	Promethium	Pm	7440-12-2	145	silv metal; hex	1042	≈3000	7.26		
2240	Promethium(III) bromide	$PmBr_3$	14325-78-1	385	red cry	625				s H_2O
2241	Promethium(III) chloride	$PmCl_3$	13779-10-7	251	pale blue hyg cry	655				s H_2O
2242	Promethium(III) fluoride	PmF_3	13709-45-0	202	pink solid	1338				s H_2O
2243	Promethium(III) iodide	PmI_3	13818-73-0	526	red solid	695				
2244	Protactinium	Pa	7440-13-3	231.036	shiny metal; tetr or cub	1572		15.4		
2245	Protactinium(V) chloride	$PaCl_5$	13760-41-3	408.301	yel monocl cry	306		3.74		
2246	Radium	Ra	7440-14-4	226	wh metal; cub	696		5		
2247	Radium bromide	$RaBr_2$	10031-23-9	386	wh orth cry	728		5.79	70.6^{20}	s EtOH
2248	Radium carbonate	$RaCO_3$	7116-98-5	286	wh orth cry					i H_2O
2249	Radium chloride	$RaCl_2$	10025-66-8	297	wh orth cry	1000		4.9	24.5^{20}	s EtOH
2250	Radium fluoride	RaF_2	20610-49-5	264	wh cub cry			6.7		
2251	Radium nitrate	$Ra(NO_3)_2$	10213-12-4	350	cry				13.9	
2252	Radium sulfate	$RaSO_4$	7446-16-4	322	wh cry					i H_2O, acid
2253	Radon	Rn	10043-92-2	222	col gas	-71	-61.7	0.009074^{25} gas		sl H_2O
2254	Rhenium	Re	7440-15-5	186.207	silv-gray metal	3185	5590	20.8		i HCl
2255	Perrhenic acid	$HReO_4$	13768-11-1	251.213	exists only in soln					vs H_2O, os
2256	Rhenium carbonyl	$Re_2(CO)_{10}$	14285-68-8	652.515	yel-wh cry	170 dec		2.87		s os
2257	Rhenium pentacarbonyl bromide	$Re(CO)_5Br$	14220-21-4	406.162	wh cry	90				
2258	Rhenium pentacarbonyl chloride	$Re(CO)_5Cl$	14099-01-5	361.711	wh cry		140 subl			
2259	Rhenium(III) bromide	$ReBr_3$	13569-49-8	425.919	red-brn monocl cry		500 subl	6.10		s ace, MeOH, EtOH
2260	Rhenium(III) chloride	$ReCl_3$	13569-63-6	292.566	red-blk hyg cry	500 dec		4.81		s H_2O
2261	Rhenium(III) iodide	ReI_3	15622-42-1	566.920	blk solid	dec				
2262	Rhenium(IV) chloride	$ReCl_4$	13569-71-6	328.019	purp-blk cry; hyg	300 dec		4.9		
2263	Rhenium(IV) fluoride	ReF_4	15192-42-4	262.201	blue tetr cry		>300 subl	7.49		
2264	Rhenium(IV) oxide	ReO_2	12036-09-8	218.206	gray orth cry	900 dec		11.4		
2265	Rhenium(IV) selenide	$ReSe_2$	12038-64-1	344.13	tricl cry					
2266	Rhenium(IV) silicide	$ReSi_2$	12038-66-3	242.378	refrac solid	2000				
2267	Rhenium(IV) sulfide	ReS_2	12038-63-0	250.337	tricl cry			7.6		
2268	Rhenium(IV) telluride	$ReTe_2$	12067-00-4	441.41	orth cry			8.50		
2269	Rhenium(V) bromide	$ReBr_5$	30937-53-2	585.727	brn solid	110 dec				
2270	Rhenium(V) chloride	$ReCl_5$	39368-69-9	363.472	brn-blk solid	220		4.9		reac H_2O
2271	Rhenium(V) fluoride	ReF_5	30937-52-1	281.199	yel-grn solid	48	221.3			
2272	Rhenium(V) oxide	Re_2O_5	12165-05-8	452.411	blue-blk tetr cry			≈7		
2273	Rhenium(VI) chloride	$ReCl_6$	31234-26-1	398.925	red-grn solid	29				
2274	Rhenium(VI) dioxydifluoride	ReO_2F_2	81155-18-2	256.203	col cry	156				
2275	Rhenium(VI) fluoride	ReF_6	10049-17-9	300.197	yel liq or cub cry	18.5	33.8	4.06 (cry)		s HNO_3
2276	Rhenium(VI) oxide	ReO_3	1314-28-9	234.205	red cub cry	400 dec		6.9		i H_2O, acid, alk
2277	Rhenium(VI) oxytetrachloride	$ReOCl_4$	13814-76-1	344.018	brn cry	29.3	223			reac H_2O
2278	Rhenium(VI) oxytetrafluoride	$ReOF_4$	17026-29-8	278.200	blue solid	108	171.7			
2279	Rhenium(VII) fluoride	ReF_7	17029-21-9	319.196	yel cub cry	48.3	73.7	4.32		
2280	Rhenium(VII) oxide	Re_2O_7	1314-68-7	484.410	yel hyg cry	327	360	6.10		s H_2O, EtOH, eth, diox, py
2281	Rhenium(VII) trioxychloride	ReO_3Cl	7791-09-5	269.658	col liq	4.5	128	3.87		reac H_2O
2282	Rhenium(VII) trioxyfluoride	ReO_3F	42246-24-2	253.203	yel solid	147	164			
2283	Rhenium(VII) dioxytrifluoride	ReO_2F_3	57246-89-6	275.201	yel solid	90	185.4			reac H_2O
2284	Rhenium(VII) oxypentafluoride	$ReOF_5$	23377-53-9	297.198	cream solid	43.8	73.0			
2285	Rhenium(VII) sulfide	Re_2S_7	12038-67-4	596.869	brn-blk tetr cry			4.87		i H_2O

No.	Name	Formula	CAS Reg. No.	Mol. wt.	Phys. form	mp/°C	bp/°C	ρ/g cm^{-3}	Sol. in g/100 g H_2O	Qualitative solubility
2286	Rhodium	Rh	7440-16-6	102.906	silv-wh metal; cub	1963	3695	12.4		i acid, sl aqua regia
2287	Rhodium carbonyl	$Rh_6(CO)_{16}$	28407-51-4	1065.590	red-brn cry	220 dec				
2288	Rhodium carbonyl chloride	$[Rh(CO)_2Cl]_2$	14523-22-9	388.758	red-oran cry	124				s os
2289	Rhodium dodecacarbonyl	$Rh_4(CO)_{12}$	19584-30-6	747.743	red hyg cry	150 dec		2.52		reac H_2O
2290	Rhodium(III) bromide	$RhBr_3$	15608-29-4	342.618	dark brn plates	800 dec		5.56		s H_2O; i acid, os
2291	Rhodium(III) chloride	$RhCl_3$	10049-07-7	209.265	red monocl cry		717	5.38		i H_2O; s alk
2292	Rhodium(III) fluoride	RhF_3	60804-25-3	159.901	red hex cry			5.4		
2293	Rhodium(III) iodide	RhI_3	15492-38-3	483.619	blk monocl cry; hyg			6.4		
2294	Rhodium(III) nitrate	$Rh(NO_3)_3$	10139-58-9	288.921	hyg brn solid	600 dec				i H_2O
2295	Rhodium(III) nitrate dihydrate	$Rh(NO_3)_3 \cdot 2H_2O$	13465-43-5	324.951	blk solid	dec				i H_2O; s aqua regia
2296	Rhodium(III) oxide	Rh_2O_3	12036-35-0	253.809	gray hex cry	1100 dec		8.2		
2297	Rhodium(III) oxide pentahydrate	$Rh_2O_3 \cdot 5H_2O$	39373-27-8	309.010	yel pow	dec				sl H_2O; s acid
2298	Rhodium(III) sulfate	$Rh_2(SO_4)_3$	10489-46-0	493.999	red-yel solid	>500 dec				
2299	Rhodium(IV) oxide	RhO_2	12137-27-8	134.905	blk tetr cry			7.2		
2300	Rhodium(IV) oxide dihydrate	$RhO_2 \cdot 2H_2O$	12137-27-8	170.936	grn solid	dec		8.20		i H_2O, sol HCl, alk
2301	Rhodium(VI) fluoride	RhF_6	13693-07-7	216.896	blk cub cry	≈70		3.1		
2302	Rubidium	Rb	7440-17-7	85.468	soft silv metal; cub	39.30	688	1.53		reac H_2O
2303	Rubidium acetate	$RbC_2H_3O_2$	563-67-7	144.512	wh hyg cry	246				vs H_2O
2304	Rubidium aluminum sulfate	$RbAl(SO_4)_2$	13530-57-9	304.575	hex cry			≈3.1	1.60^{20}	i EtOH
2305	Rubidium aluminum sulfate dodecahydrate	$RbAl(SO_4)_2 \cdot 12H_2O$	7784-29-4	520.759	col cub cry	≈100 dec		≈1.9		s H_2O; i EtOH
2306	Rubidium azide	RbN_3	22756-36-1	127.488	tetr cry; exp	317		2.79	107^{16}	
2307	Rubidium bromate	$RbBrO_3$	13446-70-3	213.370	cub cry	430		3.68	2.95^{25}	
2308	Rubidium bromide	RbBr	7789-39-1	165.372	wh cub cry; hyg	692	1340	3.35	116^{25}	
2309	Rubidium carbonate	Rb_2CO_3	584-09-8	230.945	col monocl cry; hyg	873			223^{20}	
2310	Rubidium chlorate	$RbClO_3$	13446-71-4	168.919	col cry	324	480 dec	3.19	6.63^{25}	sl H_2O
2311	Rubidium chloride	RbCl	7791-11-9	120.921	wh cub cry; hyg	724	1390	2.76	93.9^{25}	sl EtOH
2312	Rubidium chromate	Rb_2CrO_4	13446-72-5	286.930	yel rhom cry			3.518	76.2^{25}	
2313	Rubidium dichromate	$Rb_2Cr_2O_7$	13446-73-6	386.924	red tricl or yel monocl cry			3.1		s H_2O
2314	Rubidium cyanide	RbCN	19073-56-4	111.486	wh cub cry			2.3		s H_2O; i EtOH, eth
2315	Rubidium fluoride	RbF	13446-74-7	104.466	wh cub cry; hyg	795	1410	3.2	300^{20}	i EtOH
2316	Rubidium fluoroborate	$RbBF_4$	18909-68-7	172.273	orth cry	612 dec		2.82		sl H_2O
2317	Rubidium formate	$RbCHO_2$	3495-35-0	130.486	wh hyg cry	dec				
2318	Rubidium hexafluorogermanate	Rb_2GeF_6	16962-48-4	357.57	wh cry	696				s H_2O
2319	Rubidium hydride	RbH	13446-75-8	86.476	wh cub cry; flam	585		2.60		reac H_2O
2320	Rubidium hydrogen carbonate	$RbHCO_3$	19088-74-5	146.485	wh rhomb cry	175 dec			116^{20}	
2321	Rubidium hydrogen fluoride	$RbHF_2$	12280-64-7	124.473	tetr cry	188		3.3		
2322	Rubidium hydrogen sulfate	$RbHSO_4$	15587-72-1	182.539	col monocl cry	208		2.9		s H_2O
2323	Rubidium hydroxide	RbOH	1310-82-3	102.475	gray-wh orth cry; hyg	385		3.2	173^{30}	s EtOH
2324	Rubidium iodate	$RbIO_3$	13446-76-9	260.370	monocl or cub cry	dec		4.33	2.44^{25}	vs HCl
2325	Rubidium iodide	RbI	7790-29-6	212.372	wh cub cry	656	1300	3.55	165^{25}	s EtOH
2326	Rubidium molybdate	Rb_2MoO_4	13718-22-4	330.87	wh cry	958				s H_2O
2327	Rubidium nitrate	$RbNO_3$	13126-12-0	147.473	wh hex cry; hyg	310		3.11	65.0^{25}	vs H_2O
2328	Rubidium nitrite	$RbNO_2$	13825-25-7	131.474	wh cry	422				vs H_2O
2329	Rubidium oxide	Rb_2O	18088-11-4	186.935	yel-brn cub cry; hyg	505		4.0		reac H_2O
2330	Rubidium perchlorate	$RbClO_4$	13510-42-4	184.919	wh hyg cry	550	>550 dec	2.9	1.5^{25}	
2331	Rubidium permanganate	$RbMnO_4$	13465-49-1	204.404	dark purp cry	300 dec		3.24		sl H_2O
2332	Rubidium peroxide	Rb_2O_2	23611-30-5	202.935	wh orth cry	570		3.8		reac H_2O
2333	Rubidium selenide	Rb_2Se	31052-43-4	249.90	wh cub cry	733		3.22		reac H_2O
2334	Rubidium sulfate	Rb_2SO_4	7488-54-2	266.999	wh orth cry	1066		3.61	50.8^{25}	
2335	Rubidium sulfide	Rb_2S	31083-74-6	203.001	wh cub cry	425		2.91		s H_2O
2336	Rubidium superoxide	RbO_2	12137-25-6	117.467	tetr cry	540		≈3.0		
2337	Ruthenium	Ru	7440-18-8	101.07	silv-wh metal; hex	2333	4147	12.1		i acid, aqua regia
2338	Ruthenium dodecacarbonyl	$Ru_3(CO)_{12}$	15243-33-1	639.33	oran cry	150 dec				
2339	Ruthenium pentacarbonyl	$Ru(CO)_5$	16406-48-7	241.12	col liq	-22	50 dec			i H_2O; s EtOH, bz, chl, hc
2340	Ruthenium nonacarbonyl	$Ru_2(CO)_9$	63128-11-0	454.23	stab below -40					s hex
2341	Ruthenium nitrosyl chloride monohydrate	$Ru(NO)Cl_3 \cdot H_2O$	18902-42-6	255.45	hyg red cry					
2342	Hexaammineruthenium(III) chloride	$Ru(NH_3)_6Cl_3$	14282-91-8	309.61	col monocl cry					s H_2O
2343	Ruthenium(III) bromide	$RuBr_3$	14014-88-1	340.78	brn hex cry	500 dec		5.3		i H_2O, acid, EtOH
2344	Ruthenium(III) chloride	$RuCl_3$	10049-08-8	207.43	blk-brn hex cry	≈500 dec		3.1		i H_2O; sl EtOH
2345	Ruthenium(III) fluoride	RuF_3	51621-05-7	158.07	brn rhomb cry	≈600 dec		5.36		i H_2O, dil acid
2346	Ruthenium(III) iodide	RuI_3	13896-65-6	481.78	blk hex cry	300 dec		6.0		sl H_2O
2347	Ruthenium(III) 2,4-pentanedioate	$Ru(CH_3COCHCOCH_3)_3$	14284-93-6	398.39	red-brn cry	230				
2348	Ruthenium(IV) fluoride	RuF_4	71500-16-8	177.06	yel-red cry					reac H_2O
2349	Ruthenium(IV) oxide	RuO_2	12036-10-1	133.07	gray-blk tetr cry	1300 dec		7.05		i H_2O, acid
2350	Ruthenium(V) fluoride	RuF_5	14521-18-7	196.06	grn monocl cry	101	227	3.90		
2351	Ruthenium(VI) fluoride	RuF_6	13693-08-8	215.06	dark brn orth cry	54	200 dec	3.54		reac H_2O

No.	Name	Formula	CAS Reg. No.	Mol. wt.	Phys. form	mp/°C	bp/°C	ρ/g cm⁻³	Sol. in g/100 g H_2O	Qualitative solubility
2352	Ruthenium(VIII) oxide	RuO_4	20427-56-9	165.07	yel monocl prisms	25.4	40	3.29	2.03^{20}	sl H_2O; vs ctc; reac EtOH
2353	Samarium	Sm	7440-19-9	150.36	silv metal; rhomb	1072	1794	7.52		
2354	Samarium boride	SmB_6	12008-30-9	215.23	refrac solid	2580		5.07		
2355	Samarium silicide	$SmSi_2$	12300-22-0	206.53	orth cry			5.14		
2356	Samarium(II) bromide	$SmBr_2$	50801-97-3	310.17	brn cry	669				reac H_2O
2357	Samarium(II) chloride	$SmCl_2$	13874-75-4	221.27	brn cry	855		3.69		reac H_2O
2358	Samarium(II) fluoride	SmF_2	15192-17-3	188.36	purp cry					reac H_2O
2359	Samarium(II) iodide	SmI_2	32248-43-4	404.17	grn cry	520				reac H_2O
2360	Samarium(III) acetate trihydrate	$Sm(C_2H_3O_2)_3{\cdot}3H_2O$	17829-86-6	381.54	hyg yel-wh solid			1.94		s H_2O
2361	Samarium(III) bromate nonahydrate	$Sm(BrO_3)_3{\cdot}9H_2O$	63427-22-5	696.20	pink hex cry	75 dec				vs H_2O; sl EtOH
2362	Samarium(III) bromide	$SmBr_3$	13759-87-0	390.07	yel cry	640				reac H_2O
2363	Samarium(III) carbonate	$Sm_2(CO_3)_3$	5895-47-6	480.75	wh-yel pow	>500 dec				
2364	Samarium(III) chloride	$SmCl_3$	10361-82-7	256.72	yel cry	682		4.46	93.8^{25}	
2365	Samarium(III) chloride hexahydrate	$SmCl_3{\cdot}6H_2O$	13465-55-9	364.81	yel cry	dec		2.38	93.8^{25}	
2366	Samarium(III) fluoride	SmF_3	13765-24-7	207.36	wh cry	1306				reac H_2O
2367	Samarium(III) iodide	SmI_3	13813-25-7	531.07	oran cry	850				reac H_2O
2368	Samarium(III) nitrate	$Sm(NO_3)_3$	10361-83-8	336.38	yel-wh hyg solid				144^{25}	s EtOH
2369	Samarium(III) nitrate hexahydrate	$Sm(NO_3)_3{\cdot}6H_2O$	13759-83-6	444.47	pale yel cry	78				s H_2O, MeOH, ace
2370	Samarium(III) oxide	Sm_2O_3	12060-58-1	348.72	yel-wh cub cry	2335	3780	7.6		
2371	Samarium(III) sulfate octahydrate	$Sm_2(SO_4)_3{\cdot}8H_2O$	13465-58-2	733.03	yel cry			2.93	2.67^{20}	
2372	Samarium(III) sulfide	Sm_2S_3	12067-22-0	396.92	gray-brn cub cry	1720		5.87		
2373	Samarium(III) telluride	Sm_2Te_3	12040-00-5	683.52	orth cry			7.31		
2374	Scandium	Sc	7440-20-2	44.956	silv metal; hex	1541	2836	2.99		
2375	Scandium boride	ScB_2	12007-34-0	66.578	refrac solid	2250		3.17		
2376	Scandium bromide	$ScBr_3$	13465-59-3	284.668	wh hyg cry	969		9.33		s H_2O
2377	Scandium chloride	$ScCl_3$	10361-84-9	151.315	wh hyg cry	967		2.4		s H_2O; i EtOH
2378	Scandium fluoride	ScF_3	13709-47-2	101.951	wh powder	1552				sl H_2O
2379	Scandium hydroxide	$Sc(OH)_3$	17674-34-9	95.978	col amorp solid					i H_2O; s dil acid
2380	Scandium iodide	ScI_3	14474-33-0	425.669	hyg yel cry	953	subl			s H_2O, EtOH, CCD
2381	Scandium nitrate	$Sc(NO_3)_3$	13465-60-6	230.971	wh cry				169^{25}	s EtOH
2382	Scandium oxide	Sc_2O_3	12060-08-1	137.910	wh cub cry	2489		3.864		s conc acid
2383	Scandium sulfate pentahydrate	$Sc_2(SO_4)_3{\cdot}5H_2O$	15292-44-1	468.176	col cry	110 dec				vs H_2O
2384	Scandium sulfide	Sc_2S_3	12166-29-9	186.107	yel orth cry	1775		2.91		
2385	Scandium telluride	Sc_2Te_3	12166-44-8	472.71	blk hex cry			5.29		
2386	Selenium (gray)	Se	7782-49-2	78.96	gray metallic cry; hex	220.8	685	4.809		i H_2O, CS_2
2387	Selenium (α form)	Se	7782-49-2	78.96	red monocl cry	>120 trans gray Se	685	4.39		i H_2O, EtOH; sl eth
2388	Selenium (vitreous)	Se	7782-49-2	78.96	blk amorp solid	180 trans gray Se	685	4.28		i H_2O; sl CS_2
2389	Selenic acid	H_2SeO_4	7783-08-6	144.97	wh hyg solid	58	260 dec	2.95		vs H_2O; reac EtOH
2390	Pentafluoroorthoselenic acid	$HOSeF_5$	38989-47-8	190.96	col solid	38	47			
2391	Selenous acid	H_2SeO_3	7783-00-8	128.97	wh hyg cry	70 dec		3.0		vs H_2O; s EtOH
2392	Selenium dioxide	SeO_2	7446-08-4	110.96	wh tetr needles or powder	360	315 sp	3.95	264^{22}	s EtOH, MeOH; sl ace
2393	Selenium trioxide	SeO_3	13768-86-0	126.96	wh tetr cry; hyg	118	subl	3.44		s H_2O, os
2394	Selenium bromide	Se_2Br_2	7789-52-8	317.73	red liq	5	225 dec	3.60		reac H_2O; s CS_2, chl
2395	Selenium chloride	Se_2Cl_2	10025-68-0	228.83	yel-brn oily liq	-85	127 dec	2.774		reac H_2O; s CS_2, bz, ctc, chl
2396	Selenium tetrabromide	$SeBr_4$	7789-65-3	398.58	oran-red cry	123				reac H_2O; s CS_2, chl
2397	Selenium tetrachloride	$SeCl_4$	10026-03-6	220.77	wh-yel cry	305 tp	191.4 sp	2.6		reac H_2O
2398	Selenium tetrafluoride	SeF_4	13465-66-2	154.95	col liq	-9.5	101.6	2.75		reac H_2O; vs EtOH, eth
2399	Selenium hexafluoride	SeF_6	7783-79-1	192.95	col gas	-34.6 tp	-46.6 sp	0.007887^{25} gas		i H_2O
2400	Selenium chloride pentafluoride	SeF_5Cl	34979-62-9	209.41	col gas	-19	4.5			
2401	Selenium oxybromide	$SeOBr_2$	7789-51-7	254.77	red-yel solid	41.6	220 dec	3.38		reac H_2O; s CS_2, bz, ctc
2402	Selenium oxychloride	$SeOCl_2$	7791-23-3	165.87	col or yel liq	8.5	177	2.44		reac H_2O; s ctc, chl, bz, tol
2403	Selenium oxyfluoride	$SeOF_2$	7783-43-9	132.96	col liq	15	125	2.8		reac H_2O
2404	Selenium oxytetrafluoride	$SeOF_4$	53319-44-1	170.95	unstab col liq	12	65			
2405	Selenium dioxydifluoride	SeO_2F_2	14984-81-7	148.96	col gas	-99.5	-8.4	0.006089^{25} gas		reac H_2O
2406	Selenium monosulfide	SeS	7446-34-6	111.03						
2407	Selenium disulfide	SeS_2	7488-56-4	143.09	red-yel cry	100				i H_2O; s acid
2408	Selenium sulfide [Se_2S_6]	Se_2S_6	75926-26-0	350.31	oran needles	121.5		2.44		s CS_2; sl bz
2409	Selenium sulfide [Se_4S_4]	Se_4S_4	75926-28-2	444.10	red cry	113 dec		3.29		s bz; sl CS_2
2410	Selenium sulfide [Se_6S_2]	Se_6S_2	75926-30-6	537.89	oran cry	121.5				s CS_2
2411	Silicon	Si	7440-21-3	28.085	gray cry or brn amorp solid	1414	3265	2.3296		i H_2O, acid; s alk
2412	Silane	SiH_4	7803-62-5	32.118	col gas; flam	-185	-111.9	0.001313^{25} gas		reac H_2O; i EtOH, bz

Inorganic

No.	Name	Formula	CAS Reg. No.	Mol. wt.	Phys. form	mp/°C	bp/°C	ρ/g cm⁻³	Sol. in g/100 g H_2O	Qualitative solubility
2413	Disilane	Si_2H_6	1590-87-0	62.219	col gas; flam	-129.4	-14.8	0.002543^{25} gas		reac H_2O, ctc, chl; s EtOH, bz
2414	Trisilane	Si_3H_8	7783-26-8	92.321	flam col liq	-117.4	52.9	0.739		reac H_2O
2415	Tetrasilane	Si_4H_{10}	7783-29-1	122.421	col liq; flam	-89.9	108.1	0.792		reac H_2O
2416	2-Silyltrisilane	Si_4H_{10}	13597-87-0	122.421	col liq	-99.4	101.7	0.792		reac H_2O
2417	Cyclopentasilane	Si_5H_{10}	289-22-5	150.507	col liq	-10.5	194.3	0.963		reac H_2O
2418	Pentasilane	Si_5H_{12}	14868-53-2	152.523	col liq	-72.8	153.2	0.827		reac H_2O
2419	2-Silyltetrasilane	Si_5H_{12}	14868-54-3	152.523	col liq	-109.9	146.2	0.820		reac H_2O
2420	2,2-Disilyltrisilane	Si_5H_{12}	15947-57-6	152.523	col liq	-57.8	134.3	0.815		reac H_2O
2421	Cyclohexasilane	Si_6H_{12}	291-59-8	180.608	col liq	16.5	226			reac H_2O
2422	Hexasilane	Si_6H_{14}	14693-61-9	182.624	col liq	-44.7	193.6	0.847		reac H_2O
2423	2-Silylpentasilane	Si_6H_{14}	14868-55-4	182.624	col liq	-78.4	185.2	0.840		
2424	3-Silylpentasilane	Si_6H_{14}	52988-75-7	182.624	col liq	-69	179.5	0.843		reac H_2O
2425	Heptasilane	Si_7H_{16}	14693-65-3	212.726	col liq	-30.1	226.8	0.859		reac H_2O
2426	Bromosilane	SiH_3Br	13465-73-1	111.014	col gas	-94	1.9	0.004538^{25} gas		
2427	Dibromosilane	SiH_2Br_2	13768-94-0	189.910	liq	-70.1	66			
2428	Tribromosilane	$SiHBr_3$	7789-57-3	268.806	flam liq	-73	109	2.7		reac H_2O
2429	Tetrabromosilane	$SiBr_4$	7789-66-4	347.702	col fuming liq	5.39	154	2.8		reac H_2O
2430	Bromotrichlorosilane	$SiBrCl_3$	13465-74-2	214.349	col liq	-62	80.3	1.826		reac H_2O
2431	Dibromodichlorosilane	$SiBr_2Cl_2$	13465-75-3	258.800	col liq	-45.5	104	2.172		reac H_2O
2432	Tribromochlorosilane	$SiBr_3Cl$	13465-76-4	303.251	col liq	-20.8	127	2.497		reac H_2O
2433	Hexabromosilane	Si_2Br_6	13517-13-0	535.595	col cry	95	265			
2434	Octabromotrisilane	Si_3Br_8	54804-32-9	723.489	col liq	46				
2435	Chlorosilane	SiH_3Cl	13465-78-6	66.563	col gas	-118	-30.4	0.002721^{25} gas		
2436	Dichlorosilane	SiH_2Cl_2	4109-96-0	101.008	col gas; flam	-122	8.3	0.004129^{25} gas		reac H_2O
2437	Trichlorosilane	$SiHCl_3$	10025-78-2	135.453	fuming liq	-128.2	33	1.331		reac H_2O
2438	Tetrachlorosilane	$SiCl_4$	10026-04-7	169.898	col fuming liq	-68.74	57.65	1.5		reac H_2O
2439	Chlorotrifluorosilane	$SiClF_3$	14049-36-6	120.534	col gas	-138	-70.0	0.004927^{25} gas		reac H_2O
2440	Dichlorodifluorosilane	$SiCl_2F_2$	18356-71-3	136.989	col gas	-44	-32	0.005599^{25} gas		reac H_2O
2441	Trichlorofluorosilane	$SiCl_3F$	14965-52-7	153.443	col gas		12.25	0.006272^{25} gas		
2442	Trichloroiodosilane	$SiCl_3I$	13465-85-5	261.349	col liq	-60	113.5			reac H_2O
2443	Hexachlorodisilane	Si_2Cl_6	13465-77-5	268.889	col liq	2.5	146			reac H_2O
2444	Octachlorotrisilane	Si_3Cl_8	13596-23-1	367.881	col liq	-67	216			
2445	Fluorosilane	SiH_3F	13537-33-2	50.108	col gas		-98.6	0.002048^{25} gas		
2446	Difluorosilane	SiH_2F_2	13824-36-7	68.099	col gas	-122	-77.8	0.002783^{25} gas		
2447	Trifluorosilane	$SiHF_3$	13465-71-9	86.089	col gas	-131	-95	0.003519^{25} gas		
2448	Tetrafluorosilane	SiF_4	7783-61-1	104.080	col gas	-90.2	-86	0.004254^{25} gas		reac H_2O
2449	Hexafluorodisilane	Si_2F_6	13830-68-7	170.161	col gas	-18.7 tp (780 mm Hg)	-19.1 sp			reac H_2O
2450	Octafluorotrisilane	Si_3F_8	14521-14-3	236.244	col liq	-1.2	42			
2451	Decafluorotetrasilane	Si_4F_{10}	14521-15-4	302.326	col cry	68	85.1			
2452	Iodosilane	SiH_3I	13598-42-0	158.014	col liq	-57	45.6			
2453	Diiodosilane	SiH_2I_2	13760-02-6	283.911	col liq	-1	150			
2454	Triiodosilane	$SiHI_3$	13465-72-0	409.807	liq	8	220 dec			
2455	Tetraiodosilane	SiI_4	13465-84-4	535.704	wh powder	120.5	287.35	4.1		
2456	Hexaiododisilane	Si_2I_6	13510-43-5	817.598	pale yel cry	250				
2457	Disiloxane	$(SiH_3)_2O$	13597-73-4	78.218	gas	-144	-15.2	0.003197^{25} gas		
2458	Hexachlorodisiloxane	$(SiCl_3)_2O$	14986-21-1	284.888	liq	-28	137			
2459	Methylsilane	SiH_3CH_3	992-94-9	46.145	col gas	-156.5	-57.5			
2460	Metasilicic acid	H_2SiO_3	7699-41-4	78.100	wh amorp powder					i H_2O; s HF
2461	Orthosilicic acid	H_4SiO_4	10193-36-9	96.116	exists only in soln					
2462	Fluorosilicic acid	H_2SiF_6	16961-83-4	144.092	stab only in aq soln					s H_2O
2463	Silicon carbide (hexagonal)	SiC	409-21-2	40.097	hard grn-black hex cry	2830		3.16		i H_2O, EtOH
2464	Silicon nitride [Si_3N_4]	Si_3N_4	12033-89-5	140.284	gray refrac solid; hex	1900		3.17		
2465	Silicon monoxide	SiO	10097-28-6	44.085	blk cub cry, stable >1200			2.18		
2466	Silicon dioxide (α-quartz)	SiO_2	14808-60-7	60.085	col hex cry	573 trans beta quartz	2950	2.648		i H_2O, acid; s HF
2467	Silicon dioxide (β-quartz)	SiO_2	14808-60-7	60.085	col hex cry	867 trans tridymite	2950	2.533^{600}		i H_2O, acid; s HF
2468	Silicon dioxide (tridymite)	SiO_2	15468-32-3	60.085	col hex cry	1470 trans cristobalite	2950	2.265		i H_2O, acid; s HF
2469	Silicon dioxide (cristobalite)	SiO_2	14464-46-1	60.085	col hex cry	1722	2950	2.334		i H_2O, acid; s HF
2470	Silicon dioxide (vitreous)	SiO_2	60676-86-0	60.085	col amorp solid	1713	2950	2.196		i H_2O, acid; s HF

No.	Name	Formula	CAS Reg. No.	Mol. wt.	Phys. form	mp/°C	bp/°C	ρ/g cm⁻³	Sol. in g/100 g H_2O	Qualitative solubility
2471	Silicon monosulfide	SiS	12504-41-5	60.151	yel-red hyg powder	1090	940	1.85		reac H_2O
2472	Silicon disulfide	SiS_2	13759-10-9	92.216	wh rhomb cry	1090	subl	2.04		reac H_2O, EtOH; i bz
2473	Silicon tetraacetate	$Si(C_2H_3O_2)_4$	562-90-3	264.262	wh hyg cry	110				reac H_2O; s ace, bz
2474	Silicon tetraboride	SiB_4	12007-81-7	71.330	gray refrac solid	1870 dec		2.4		
2475	Silicotungstic acid	$H_4SiO_4 \cdot (W_3O_9)_4$	12520-88-6	2878.17	hyg yel cry					vs H_2O, EtOH
2476	Silver	Ag	7440-22-4	107.868	silv metal; cub	961.78	2162	10.5		
2477	Silver azide	AgN_3	13863-88-2	149.888	orth cry; exp	≈250 exp		4.9	0.00081^{20}	
2478	Silver subfluoride	Ag_2F	1302-01-8	234.734	yel hex cry	100 dec		8.6		reac H_2O
2479	Silver(I) acetate	$AgC_2H_3O_2$	563-63-3	166.912	wh needles or powder	dec		3.26	1.04^{20}	
2480	Silver(I) acetylide	Ag_2C_2	7659-31-6	239.757	wh powder; exp					
2481	Silver(I) acetylide (AgC_2H)	AgC_2H	13092-75-6	132.897	wh powder; exp					
2482	Silver(I) arsenate	Ag_3AsO_4	13510-44-6	462.524	red cub cry	dec		6.657	0.00085	s NH_4OH
2483	Silver(I) benzoate	$Ag(C_6H_5CO_2)$	532-31-0	228.982	powder				30^{20}	
2484	Silver(I) bromate	$AgBrO_3$	7783-89-3	235.770	wh tetr cry	360 dec		5.21	0.193^{25}	
2485	Silver(I) bromide	$AgBr$	7785-23-1	187.772	yel cub cry	430	1502	6.47	0.000014^{25}	i H_2O, acid, EtOH
2486	Silver(I) carbonate	Ag_2CO_3	534-16-7	275.745	yel monocl cry	218		6.077	0.0036^{20}	s acid
2487	Silver(I) chlorate	$AgClO_3$	7783-92-8	191.319	wh tetr cry	230	270 dec	4.430	17.6^{25}	sl EtOH
2488	Silver(I) chloride	$AgCl$	7783-90-6	143.321	wh cub cry	455	1547	5.56	0.00019^{25}	
2489	Silver(I) chlorite	$AgClO_2$	7783-91-7	175.320	yel cry	105 exp			0.55^{25}	
2490	Silver(I) chromate	Ag_2CrO_4	7784-01-2	331.730	brn-red monocl cry			5.625	0.000014^{0}	
2491	Silver(I) citrate	$Ag_3C_6H_5O_7$	126-45-4	512.705	wh cry powder					i H_2O; s HNO_3
2492	Silver(I) cyanide	$AgCN$	506-64-9	133.886	wh-gray hex cry	320 dec		3.95	0.0000011	i EtOH, dil acid
2493	Silver(I) dichromate	$Ag_2Cr_2O_7$	7784-02-3	431.724	red cry			4.770		sl H_2O
2494	Silver(I) diethyldithiocarbamate	$Ag(C_2H_5)_2NCS_2$	1470-61-7	256.138	powder	173				s py
2495	Silver(I) fluoride	AgF	7775-41-9	126.866	yel-brn cub cry; hyg	435	1159	5.852	172^{20}	
2496	Silver(I) hexafluoroantimonate	$AgSbF_6$	26042-64-8	343.618	powder					
2497	Silver(I) hexafluoroarsenate	$AgAsF_6$	12005-82-2	296.780	powder					
2498	Silver(I) hexafluorophosphate	$AgPF_6$	26042-63-7	252.832	powder	102 dec				
2499	Silver(I) hydrogen fluoride	$AgHF_2$	12249-52-4	146.873	hyg cry	dec				
2500	Silver(I) iodate	$AgIO_3$	7783-97-3	282.770	wh orth cry	>200		5.53	0.053^{25}	
2501	Silver(I) iodide	AgI	7783-96-2	234.772	yel powder; hex	558	1506	5.68	0.000003	i acid
2502	Silver(I) lactate monohydrate	$AgC_3H_5O_3 \cdot H_2O$	128-00-7	214.954	gray cry powder					sl H_2O, EtOH
2503	Silver(I) metaphosphate	$AgPO_3$	13465-96-8	186.840	grn glass	490		6.37		i H_2O; s HNO_3, NH_4OH
2504	Silver(I) molybdate	Ag_2MoO_4	13765-74-7	375.67	yel cub cry	483		6.18		sl H_2O
2505	Silver(I) nitrate	$AgNO_3$	7761-88-8	169.873	col rhomb cry	210	440 dec	4.35	234^{25}	sl EtOH, ace
2506	Silver(I) nitrite	$AgNO_2$	7783-99-5	153.874	yel needles	140 dec		4.453	0.415^{25}	i EtOH; reac acid
2507	Silver(I) oxalate	$Ag_2C_2O_4$	533-51-7	303.755	wh cry powder	140 exp		5.03	0.0043^{20}	
2508	Silver(I) oxide	Ag_2O	20667-12-3	231.735	brn-blk cub cry	827		7.2	0.0025	i EtOH; s acid, alk
2509	Silver(I) perchlorate	$AgClO_4$	7783-93-9	207.319	col cub cry; hyg	486 dec		2.806	558^{25}	s bz, py, os
2510	Silver(I) perchlorate monohydrate	$AgClO_4 \cdot H_2O$	14242-05-8	225.334	hyg wh cry	43 dec			558^{25}	
2511	Silver(I) permanganate	$AgMnO_4$	7783-98-4	226.804	viol monocl cry	dec		4.49	0.91^{18}	reac EtOH
2512	Silver(I) phosphate	Ag_3PO_4	7784-09-0	418.576	yel powder	849		6.37	0.0064	sl dil acid
2513	Silver(I) picrate monohydrate	$AgC_6H_2N_3O_7 \cdot H_2O$	146-84-9	353.979	yel cry					sl H_2O, EtOH; i chl, eth
2514	Silver(I) selenate	Ag_2SeO_4	7784-07-8	358.69	orth cry			5.72	0.118^{20}	
2515	Silver(I) selenide	Ag_2Se	1302-09-6	294.70	gray hex needles	880		8.216		i H_2O
2516	Silver(I) selenite	Ag_2SeO_3	7784-05-6	342.69	needles	530	>550 dec	5.930		sl H_2O; s acid
2517	Silver(I) sulfate	Ag_2SO_4	10294-26-5	311.799	col cry or powder	660		5.45	0.84^{25}	
2518	Silver(I) sulfide	Ag_2S	21548-73-2	247.801	gray-blk orth powder	836		7.23		i H_2O; s acid
2519	Silver(I) sulfite	Ag_2SO_3	13465-98-0	295.799	wh cry	100 dec			0.00046^{20}	s acid, NH_4OH
2520	Silver(I) telluride	Ag_2Te	12002-99-2	343.34	blk orth cry	955		8.4		
2521	Silver(I) tetraiodomercurate(II)	Ag_2HgI_4	7784-03-4	923.94	yel tetr cry	≈40 trans red cub		6.1		i H_2O, dil acid
2522	Silver(I) thiocyanate	$AgSCN$	1701-93-5	165.951	wh powder	dec				i H_2O
2523	Silver(I) thiosulfate	$Ag_2S_2O_3$	23149-52-2	327.864	wh cry	dec				sl H_2O; s NH_4OH
2524	Silver(II) oxide	AgO	1301-96-8	123.867	gray powder; monocl or cub	>100 dec		7.5	0.0027^{25}	s alk; reac acid
2525	Silver(I) tungstate	Ag_2WO_4	13465-93-5	463.57	yel cry	620			0.015	s HNO_3, NH_4OH
2526	Silver(II) fluoride	AgF_2	7783-95-1	145.865	wh or gray hyg cry	690		4.58		reac H_2O
2527	Silver[II] oxide [Ag_2O_2]	Ag_2O_2	25455-73-6	247.735	gray-blk cub cry	>100		7.44		i H_2O; s acid, NH_4OH
2528	Sodium	Na	7440-23-5	22.990	soft silv met; cub	97.794	882.940	0.97		reac H_2O
2529	Sodium acetate	$NaC_2H_3O_2$	127-09-3	82.034	col cry	328.2		1.528	50.4^{25}	
2530	Sodium acetate trihydrate	$NaC_2H_3O_2 \cdot 3H_2O$	6131-90-4	136.079	col cry	58 dec		1.45	50.4^{25}	sl EtOH
2531	Sodium aluminate	$NaAlO_2$	1302-42-7	81.971	wh orth cry; hyg	1650		4.63		vs H_2O; i EtOH
2532	Sodium aluminum hydride	$NaAlH_4$	13770-96-2	54.004	wh hyg solid	174 dec		1.24		i eth; s thf
2533	Sodium aluminum sulfate dodecahydrate	$NaAl(SO_4)_2 \cdot 12H_2O$	10102-71-3	458.281	col cry	≈60		1.61	39.7^{20}	i EtOH
2534	Sodium amide	$NaNH_2$	7782-92-5	39.013	wh-grn orth cry	210	500 dec	1.39		reac H_2O

Inorganic

No.	Name	Formula	CAS Reg. No.	Mol. wt.	Phys. form	mp/°C	bp/°C	ρ/g cm^{-3}	Sol. in g/100 g H_2O	Qualitative solubility
2535	Sodium ammonium phosphate tetrahydrate	$NaNH_4HPO_4 \cdot 4H_2O$	13011-54-6	209.069	monocl cry	≈80 dec		1.54		s H_2O; i EtOH
2536	Sodium arsenate dodecahydrate	$Na_3AsO_4 \cdot 12H_2O$	7778-43-0	424.072	col monocl prism	86 dec				s H_2O; sl EtOH; i eth
2537	Sodium arsenite	$NaAsO_2$	7784-46-5	129.911	wh-gray hyg powder			1.87		vs H_2O; i EtOH
2538	Sodium azide	NaN_3	26628-22-8	65.010	col hex cry	300 dec		1.846	40.8^{20}	sl EtOH; i eth
2539	Sodium borohydride	$NaBH_4$	16940-66-2	37.833	wh cub cry; hyg	≈400 dec		1.07	55^{20}	reac EtOH
2540	Sodium bromate	$NaBrO_3$	7789-38-0	150.892	col cub cry	381		3.34	39.4^{25}	i EtOH
2541	Sodium bromide	NaBr	7647-15-6	102.894	wh cub cry	747	1390	3.200	94.6^{25}	s EtOH
2542	Sodium bromide dihydrate	$NaBr \cdot 2H_2O$	13466-08-5	138.925	wh cry	36 dec		2.18	94.6^{25}	sl EtOH
2543	Sodium carbonate	Na_2CO_3	497-19-8	105.989	wh hyg powder	856		2.54	30.7^{25}	i EtOH
2544	Sodium carbonate monohydrate	$Na_2CO_3 \cdot H_2O$	5968-11-6	124.005	col orth cry	100 dec		2.25	30.7^{25}	i EtOH
2545	Sodium carbonate decahydrate	$Na_2CO_3 \cdot 10H_2O$	6132-02-1	286.142	col cry	34 dec		1.46	30.7^{25}	i EtOH
2546	Sodium chlorate	$NaClO_3$	7775-09-9	106.441	col cub cry	248	630 dec	2.5	100^{25}	sl EtOH
2547	Sodium chloride	NaCl	7647-14-5	58.443	col cub cry	802.018	1465	2.17	36.0^{25}	sl EtOH
2548	Sodium chlorite	$NaClO_2$	7758-19-2	90.442	wh hyg cry	≈180 dec			64^{17}	
2549	Sodium chromate	Na_2CrO_4	7775-11-3	161.974	yel orth cry	794		2.72	87.6^{25}	sl EtOH
2550	Sodium chromate tetrahydrate	$Na_2CrO_4 \cdot 4H_2O$	10034-82-9	234.035	yel hyg cry	dec			87.6^{25}	sl EtOH
2551	Sodium citrate dihydrate	$Na_3C_6H_5O_7 \cdot 2H_2O$	6132-04-3	294.099	wh cry	150 dec				vs H_2O; i EtOH, eth
2552	Sodium citrate pentahydrate	$Na_3C_6H_5O_7 \cdot 5H_2O$	6858-44-2	348.145	hyg col cry	150 dec		1.86	92^{25}	vs H_2O; sl EtOH; i eth
2553	Sodium cyanate	NaCNO	917-61-3	65.007	col needles	550		1.89		s H_2O; sl EtOH; i eth
2554	Sodium cyanide	NaCN	143-33-9	49.008	wh cub cry; hyg	562		1.6	58.2^{20}	sl EtOH
2555	Sodium cyanoborohydride	$NaBH_3(CN)$	25895-60-7	62.843	wh hyg powder	240 dec		1.12		vs H_2O; s thf; sl EtOH; i bz, eth
2556	Sodium dichromate	$Na_2Cr_2O_7$	10588-01-9	261.968	red hyg cry	357	400 dec		187^{25}	
2557	Sodium dichromate dihydrate	$Na_2Cr_2O_7 \cdot 2H_2O$	7789-12-0	297.999	oran-red monocl cry	85 dec		2.35		vs H_2O; S HOAc
2558	Sodium dihydrogen phosphate	NaH_2PO_4	7558-80-7	119.977	col monocl cry	200 dec			94.9^{25}	
2559	Sodium dihydrogen phosphate monohydrate	$NaH_2PO_4 \cdot H_2O$	10049-21-5	137.993	wh hyg cry	100 dec			94.9^{25}	i EtOH
2560	Sodium dihydrogen phosphate dihydrate	$NaH_2PO_4 \cdot 2H_2O$	13472-35-0	156.008	col orth cry	60 dec		1.91	94.9^{25}	i EtOH
2561	Sodium dihydrogen hypophosphate hexahydrate	$Na_2H_2P_2O_6 \cdot 6H_2O$	7782-95-8	314.031	monocl plates	110 dec		1.849	2.0^{25}	i EtOH
2562	Sodium dihydrogen pyrophosphate	$Na_2H_2P_2O_7$	7758-16-9	221.939	wh powder	220 dec		≈1.9		s H_2O
2563	Sodium dithionate	$Na_2S_2O_4$	7775-14-6	174.108	gray-wh powder	52 dec			24.1^{20}	sl EtOH
2564	Sodium dithionate dihydrate	$Na_2S_2O_6 \cdot 2H_2O$	7631-94-9	242.137	col orth cry	110 dec		2.19	15.1^{20}	i EtOH
2565	Sodium ethanolate	NaC_2H_5O	141-52-6	68.050	wh-yel hyg powder	260 dec				reac H_2O; s EtOH
2566	Sodium ferricyanide monohydrate	$Na_3Fe(CN)_6 \cdot H_2O$	14217-21-1	298.933	red hyg cry					s H_2O; i EtOH
2567	Sodium ferrocyanide decahydrate	$Na_4Fe(CN)_6 \cdot 10H_2O$	13601-19-9	484.061	yel monocl cry	≈50 dec		1.46	20^{20}	i os
2568	Sodium fluoride	NaF	7681-49-4	41.988	col cub or tetr cry	996	1704	2.78	4.13^{25}	i EtOH
2569	Sodium fluorophosphate	Na_2PO_3F	10163-15-2	143.950	powder					
2570	Sodium formate	$NaCHO_2$	141-53-7	68.008	wh hyg cry	257.3	dec	1.92	94.9^{25}	sl EtOH
2571	Sodium germanate	Na_2GeO_3	12025-19-3	166.62	wh monocl hyg cry	1083		3.31		
2572	Sodium gold cyanide	$NaAu(CN)_2$	15280-09-8	271.991	wh-yel cry pow					s H_2O, NH_4OH
2573	Sodium gold thiosulfate dihydrate	$Na_3Au(S_2O_3)_2 \cdot 2H_2O$	10233-88-2	526.200	wh needles or prisms	150 dec		3.09		vs H_2O; i EtOH
2574	Sodium hexabromoplatinate(IV) hexahydrate	$Na_2PtBr_6 \cdot 6H_2O$	39277-13-9	828.579	cry					
2575	Sodium hexachloroiridate(IV) hexahydrate	$Na_2IrCl_6 \cdot 6H_2O$	19567-78-3	559.006	cry	600 dec				
2576	Sodium hexachloroplatinate(IV)	Na_2PtCl_6	16923-58-3	453.782	yel hyg cry				53^{16}	s EtOH
2577	Sodium hexachloroplatinate(IV) hexahydrate	$Na_2PtCl_6 \cdot 6H_2O$	16923-58-3	561.873	yel cry	110 dec		2.50	53^{16}	s EtOH; i eth
2578	Sodium hexafluoroaluminate	Na_3AlF_6	13775-53-6	209.941	col monocl cry; trans cub 560	1013		2.97		i H_2O
2579	Sodium hexafluoroantimonate	$NaSbF_6$	16925-25-0	258.740	wh cub cry			3.375	129^{20}	s EtOH, ace
2580	Sodium hexafluorophosphate monohydrate	$NaPF_6 \cdot H_2O$	20644-15-9	185.969	col orth cry			2.369	103^{0}	s EtOH, MeOH, ace
2581	Sodium hexafluorosilicate	Na_2SiF_6	16893-85-9	188.056	wh hex cry	847		2.7	0.67^{20}	i EtOH
2582	Sodium hexanitrocobaltate(III)	$Na_3Co(NO_2)_6$	14649-73-1	403.935	yel-brn cry powder					vs H_2O; sl EtOH
2583	Sodium hydride	NaH	7646-69-7	23.998	silv cub cry; flam	638		1.39		reac H_2O, EtOH
2584	Sodium hydrogen arsenate	Na_2HAsO_4	7778-43-0	185.908	wh powder	≈195 dec			51^{20}	sl EtOH
2585	Sodium hydrogen arsenate heptahydrate	$Na_2HAsO_4 \cdot 7H_2O$	10048-95-0	312.014	wh monocl cry	≈50 dec		1.87	51^{20}	sl EtOH
2586	Sodium hydrogen carbonate	$NaHCO_3$	144-55-8	84.007	wh monocl cry	527		2.20	10.3^{25}	i EtOH
2587	Sodium hydrogen fluoride	$NaHF_2$	1333-83-1	61.995	wh hex cry	>160 dec		2.08	3.25^{20}	
2588	Sodium hydrogen phosphate	Na_2HPO_4	7558-79-4	141.959	wh hyg powder			1.7	11.8^{25}	
2589	Sodium hydrogen phosphate heptahydrate	$Na_2HPO_4 \cdot 7H_2O$	7782-85-6	268.066	col cry			≈1.7	11.8^{25}	i EtOH
2590	Sodium hydrogen phosphate dodecahydrate	$Na_2HPO_4 \cdot 12H_2O$	10039-32-4	358.143	col cry	≈35 dec		≈1.5	11.8^{25}	i EtOH
2591	Sodium hydrogen sulfate	$NaHSO_4$	7681-38-1	120.061	wh hyg cry	≈315		2.43	28.5^{25}	

No.	Name	Formula	CAS Reg. No.	Mol. wt.	Phys. form	mp/°C	bp/°C	ρ/g cm^{-3}	Sol. in g/100 g H_2O	Qualitative solubility
2592	Sodium hydrogen sulfate monohydrate	$NaHSO_4{\cdot}H_2O$	10034-88-5	138.076	wh monocl cry			2.10	28.5^{25}	reac EtOH
2593	Sodium hydrogen sulfide	NaHS	16721-80-5	56.063	col rhomb cry	350		1.79		s H_2O, EtOh, eth
2594	Sodium hydrogen sulfide dihydrate	$NaHS{\cdot}2H_2O$	16721-80-5	92.094	yel hyg needles	55 dec				vs H_2O, EtOH, eth
2595	Sodium hydrogen sulfite	$NaHSO_3$	7631-90-5	104.061	wh cry			1.48		s H_2O; sl EtOH
2596	Sodium hydroxide	NaOH	1310-73-2	39.997	wh orth cry; hyg	323	1388	2.13	100^{25}	s EtOH, MeOH
2597	Sodium hypochlorite	NaClO	7681-52-9	74.442	stab in aq soln	anh form exp			79.9^{25}	
2598	Sodium hypochlorite pentahydrate	$NaOCl{\cdot}5H_2O$	10022-70-5	164.518	pale grn orth cry	18		1.6		s H_2O
2599	Sodium iodate	$NaIO_3$	7681-55-2	197.892	wh orth cry	422		4.28	9.47^{25}	i EtOH
2600	Sodium iodide	NaI	7681-82-5	149.894	wh cub cry; hyg	661	1304	3.67	184^{25}	s EtOH, ace
2601	Sodium iodide dihydrate	$NaI{\cdot}2H_2O$	13517-06-1	185.925	hyg col monocl cry	69 dec		2.45	318^{0}	vs H_2O
2602	Sodium bismuthate	$NaBiO_3$	12232-99-4	279.968	yel-brn hyg cry					i cold H_2O, reac acid
2603	Sodium metabisulfite	$Na_2S_2O_5$	7681-57-4	190.107	wh cry				66.7^{25}	sl EtOH
2604	Sodium metaborate	$NaBO_2$	7775-19-1	65.800	wh hex cry	966	1434	2.46		s H_2O
2605	Sodium metasilicate	Na_2SiO_3	6834-92-0	122.064	wh amorp solid; hyg	1089		2.61		s cold H_2O; reac hot H_2O; i EtOH
2606	Sodium metasilicate pentahydrate	$Na_2SiO_3{\cdot}5H_2O$	13517-24-3	212.140	wh pow	72 dec				s H_2O
2607	Sodium molybdate	Na_2MoO_4	7631-95-0	205.92	col cub cry	687		≈3.5	65.0^{25}	
2608	Sodium molybdate dihydrate	$Na_2MoO_4{\cdot}2H_2O$	10102-40-6	241.95	cry powder	100 dec		≈3.5	65.0^{25}	
2609	Sodium molybdophosphate	$Na_3PO_4{\cdot}12MoO_3$	1313-30-0	1891.20	hyg solid			2.83		vs H_2O, EtOH
2610	Sodium niobate	$NaNbO_3$	12034-09-2	163.894	rhom cry	1422		4.55		i H_2O
2611	Sodium nitrate	$NaNO_3$	7631-99-4	84.995	col hex cry; hyg	306.5		2.261	91.2^{25}	sl EtOH, MeOH
2612	Sodium nitrite	$NaNO_2$	7632-00-0	68.996	wh orth cry; hyg	284	>320 dec	2.17	84.8^{25}	sl EtOH; reac acid
2613	Sodium nitroferricyanide dihydrate	$Na_2[Fe(CN)_5NO]{\cdot}2H_2O$	13755-38-9	297.949	red cry			1.72	40^{16}	sl EtOH
2614	Sodium orthovanadate	Na_3VO_4	13721-39-6	183.909	col hex prisms	860				s H_2O; i EtOH
2615	Sodium oxalate	$Na_2C_2O_4$	62-76-0	133.999	wh powder	≈250 dec		2.34	3.61^{25}	i EtOH
2616	Sodium oxide	Na_2O	1313-59-3	61.979	wh amorp powder	1134		2.27		reac H_2O
2617	Sodium perborate tetrahydrate	$NaBO_3{\cdot}4H_2O$	7632-04-4	153.861	wh cry	60 dec				reac H_2O
2618	Sodium perchlorate	$NaClO_4$	7601-89-0	122.441	wh orth cry; hyg	482 dec		2.52	205^{25}	
2619	Sodium perchlorate monohydrate	$NaClO_4{\cdot}H_2O$	7791-07-3	140.456	wh hyg cry	≈130 dec		2.02	205^{25}	
2620	Sodium periodate	$NaIO_4$	7790-28-5	213.892	wh tetr cry	≈300 dec		3.86	14.4^{25}	s acid
2621	Sodium periodate trihydrate	$NaIO_4{\cdot}3H_2O$	13472-31-6	267.938	wh hex cry	175 dec		3.22	14.4^{25}	
2622	Sodium permanganate trihydrate	$NaMnO_4{\cdot}3H_2O$	10101-50-5	195.972	red-blk hyg cry	170 dec		2.47	144^{20}	reac EtOH
2623	Sodium peroxide	Na_2O_2	1313-60-6	77.979	yel hyg powder	675		2.805		reac H_2O
2624	Sodium perrhenate	$NaReO_4$	13472-33-8	273.195	cry	300		5.39		
2625	Sodium persulfate	$Na_2S_2O_8$	7775-27-1	238.105	wh hyg cry					vs H_2O; reac EtOH
2626	Sodium phosphate	Na_3PO_4	7601-54-9	163.940	col cry	1583		2.54	14.5^{25}	s H_2O
2627	Sodium phosphate dodecahydrate	$Na_3PO_4{\cdot}12H_2O$	10101-89-0	380.124	col hex cry	≈75		1.62	14.4^{25}	i EtOH
2628	Sodium phosphate, chlorinated	$Na_3PO_4{\cdot}NaOCl$	56802-99-4	238.383	wh cry				25^{25}	
2629	Sodium phosphide	Na_3P	12058-85-4	99.943	red solid	>650				reac H_2O
2630	Sodium phosphinate	NaH_2PO_2	7681-53-0	87.979	wh cry				100^{25}	
2631	Sodium phosphinate monohydrate	$NaH_2PO_2{\cdot}H_2O$	10039-56-2	105.994	col hyg cry	310 dec			100^{25}	s EtOH
2632	Sodium phosphonate pentahydrate	$Na_2HPO_3{\cdot}5H_2O$	13517-23-2	216.036	wh hex plates	200 dec			429^{20}	vs H_2O; i EtOH
2633	Sodium pyrophosphate	$Na_4P_2O_7$	7722-88-5	265.902	col cry	988		2.53	7.09^{25}	
2634	Sodium selenate	Na_2SeO_4	13410-01-0	188.94	col orth cry				58.5^{25}	
2635	Sodium selenate decahydrate	$Na_2SeO_4{\cdot}10H_2O$	10102-23-5	369.09	wh cry			1.61	58.5^{25}	
2636	Sodium selenide	Na_2Se	1313-85-5	124.94	amorp solid	>875		2.62		reac H_2O
2637	Sodium selenite	Na_2SeO_3	10102-18-8	172.94	wh tetr cry				89.8^{25}	i EtOH
2638	Sodium selenite pentahydrate	$Na_2SeO_3{\cdot}5H_2O$	26970-82-1	263.02	wh tetr cry	dec				s H_2O; i EtOH
2639	Sodium stannate trihydrate	$Na_2SnO_3{\cdot}3H_2O$	12209-98-2	266.734	col hex cry	140 dec			61^{15}	vs H_2O; i EtOH, ace
2640	Sodium stearate	$NaC_{18}H_{35}O_2$	822-16-2	306.460	wh powder					sl H_2O, EtOH; vs hot H_2O
2641	Sodium succinate hexahydrate	$Na_2C_4H_4O_4{\cdot}6H_2O$	150-90-3	270.144	cry powder	120 dec			20	i EtOH
2642	Sodium sulfate	Na_2SO_4	7757-82-6	142.043	wh orth cry or powder	884		2.7	28.1^{25}	i EtOH
2643	Sodium sulfate heptahydrate	$Na_2SO_4{\cdot}7H_2O$	13472-39-4	268.14	wh cry	dec				vs H_2O
2644	Sodium sulfate decahydrate	$Na_2SO_4{\cdot}10H_2O$	7727-73-3	322.196	col monocl cry	32 dec		1.46	28.1^{25}	i EtOH
2645	Sodium sulfide	Na_2S	1313-82-2	78.045	wh cub cry; hyg	1172		1.856	20.6^{25}	sl EtOH; i eth
2646	Sodium sulfide pentahydrate	$Na_2S{\cdot}5H_2O$	1313-83-3	168.121	col orth cry	120 dec		1.58	20.6^{25}	s EtOH; i eth
2647	Sodium sulfide nonahydrate	$Na_2S{\cdot}9H_2O$	1313-84-4	240.183	wh-yel hyg cry	≈50 dec		1.43	20.6^{25}	sl EtOH; i eth
2648	Sodium sulfite	Na_2SO_3	7757-83-7	126.043	wh hex cry	911		2.63	30.7^{25}	i EtOH
2649	Sodium sulfite heptahydrate	$Na_2SO_3{\cdot}7H_2O$	10102-15-5	252.150	wh monocl cry; unstab			1.56	30.7^{25}	sl EtOH
2650	Sodium superoxide	NaO_2	12034-12-7	54.989	yel cub cry	552		2.2		reac H_2O
2651	Sodium tartrate dihydrate	$Na_2C_4H_4O_6{\cdot}2H_2O$	6106-24-7	230.082				1.545^{25}		s H_2O; i EtOH
2652	Sodium tellurate	Na_2TeO_4	10101-83-4	237.58	wh powder				0.8	
2653	Sodium tellurite	Na_2TeO_3	10102-20-2	221.58	wh rhomb prisms					sl H_2O
2654	Sodium tetraborate	$Na_2B_4O_7$	1330-43-4	201.220	col gl solid; hyg	743	1575	2.4	3.17^{25}	sl MeOH
2655	Sodium tetraborate tetrahydrate	$Na_2B_4O_7{\cdot}4H_2O$	12045-87-3	273.281	wh monocl cry			1.95	3.17^{25}	
2656	Sodium tetraborate pentahydrate	$Na_2B_4O_7{\cdot}5H_2O$	12045-88-4	291.296	hex cry	dec		1.88	3.17^{25}	
2657	Sodium tetraborate decahydrate	$Na_2B_4O_7{\cdot}10H_2O$	1303-96-4	381.373	wh monocl cry	75 dec		1.73	3.17^{25}	i EtOH
2658	Sodium tetrachloroaluminate	$NaAlCl_4$	7784-16-9	191.784	orth cry			2.01		s H_2O

No.	Name	Formula	CAS Reg. No.	Mol. wt.	Phys. form	mp/°C	bp/°C	ρ/g cm⁻³	Sol. in g/100 g H_2O	Qualitative solubility
2659	Sodium tetrachloroaurate(III) dihydrate	$NaAuCl_4 \cdot 2H_2O$	13874-02-7	397.800	oran-yel rhom cry	100 dec			150^{10}	s EtOH, eth
2660	Sodium tetrachloropalladate(II) trihydrate	$Na_2PdCl_4 \cdot 3H_2O$	13820-53-6	348.26	brn-red hyg cry					vs H_2O; s EtOH
2661	Sodium tetrachloroplatinate(II) tetrahydrate	$Na_2PtCl_4 \cdot 4H_2O$	10026-00-3	454.938	red prisms	100				s H_2O, EtOH
2662	Sodium tetrafluoroberyllate	Na_2BeF_4	13871-27-7	130.986	orth cry	575		2.47		sl H_2O
2663	Sodium tetrafluoroborate	$NaBF_4$	13755-29-8	109.795	wh orth prisms	384		2.47	108^{20}	sl EtOH
2664	Sodium thioantimonate nonahydrate	$Na_3SbS_4 \cdot 9H_2O$	10101-91-4	481.127	yel cry	108 dec		1.8	28^{20}	i EtOH
2665	Sodium thiocyanate	NaSCN	540-72-7	81.073	col hyg cry	287			151^{25}	
2666	Sodium thiophosphate dodecahydrate	$Na_3PO_3S \cdot 12H_2O$	10101-88-9	396.190	hex hyg leaflets	60				vs hot H_2O
2667	Sodium thiosulfate	$Na_2S_2O_3$	7772-98-7	158.108	col monocl cry	100 dec		1.69	76.4^{25}	i EtOH
2668	Sodium thiosulfate pentahydrate	$Na_2S_2O_3 \cdot 5H_2O$	10102-17-7	248.184	col cry	≈50 dec		1.69	76.4^{25}	i EtOH
2669	Sodium trimetaphosphate	$Na_3(PO_3)_3$	7785-84-4	305.885	wh cry			2.49	22	
2670	Sodium trimetaphosphate hexahydrate	$Na_3(PO_3)_3 \cdot 6H_2O$	7785-84-4	413.976	tricl-rhom hyg prisms	53		1.786	22	i EtOH
2671	Sodium tripolyphosphate	$Na_5P_3O_{10}$	7758-29-4	367.864	wh hyg powder	622			20^{25}	
2672	Sodium tungstate	Na_2WO_4	13472-45-2	293.82	wh rhom cry	695		4.18	74.2^{25}	
2673	Sodium tungstate dihydrate	$Na_2WO_4 \cdot 2H_2O$	10213-10-2	329.85	wh orth cry	100 dec		3.25	74.2^{25}	i EtOH
2674	Sodium uranate(VI) monohydrate	$Na_2U_2O_7 \cdot H_2O$	13721-34-1	652.049	yel powder					i H_2O; s acid
2675	Sodium vanadate(V)	$NaVO_3$	13718-26-8	121.930	col monocl prisms	630			21^{25}	
2676	Sodium vanadate(V) tetrahydrate	$NaVO_3 \cdot 4H_2O$	13718-26-8	193.992	yel-wh cry powder				21^{25}	
2677	Strontium	Sr	7440-24-6	87.62	silv-wh metal; cub	777	1377	2.64		reac H_2O; s EtOH
2678	Strontium acetate	$Sr(C_2H_3O_2)_2$	543-94-2	205.71	col hyg cry	dec		2.1	40^{25}	vs H_2O
2679	Strontium arsenite tetrahydrate	$Sr(AsO_2)_2 \cdot 4H_2O$	10378-48-0	373.52	wh powder					sl H_2O, EtOH; sol dil acid
2680	Strontium bromate monohydrate	$Sr(BrO_3)_2 \cdot H_2O$	14519-18-7	361.44	yel hyg monocl cry	120 dec		3.773	39.0^{25}	
2681	Strontium bromide	$SrBr_2$	10476-81-0	247.43	wh tetr cry	657		4.216	107^{25}	
2682	Strontium bromide hexahydrate	$SrBr_2 \cdot 6H_2O$	7789-53-9	355.52	col hyg cry	88 dec			107^{25}	s EtOH; i eth
2683	Strontium carbide	SrC_2	12071-29-3	111.64	blk tetr cry	>1700		3.19		i H_2O
2684	Strontium carbonate	$SrCO_3$	1633-05-2	147.63	wh orth cry; hyg	1494		3.785	0.00034^{20}	s dil acid
2685	Strontium chlorate	$Sr(ClO_3)_2$	7791-10-8	254.52	col cry	120 dec		3.15	176^{25}	sl EtOH
2686	Strontium chloride	$SrCl_2$	10476-85-4	158.53	wh cub cry; hyg	874	1250	3.052	54.7^{25}	
2687	Strontium chloride hexahydrate	$SrCl_2 \cdot 6H_2O$	10025-70-4	266.62	col hyg cry	100 dec		1.96	54.7^{25}	s EtOH
2688	Strontium chromate	$SrCrO_4$	7789-06-2	203.61	yel monocl cry	dec		3.9	0.106^{20}	s dil acid
2689	Strontium cyanide tetrahydrate	$Sr(CN)_2 \cdot 4H_2O$	52870-08-3	211.72	wh hyg cry	dec				vs H_2O
2690	Strontium ferrocyanide pentadecahydrate	$SrFe(CN)_6 \cdot 15H_2O$	14654-44-5	569.80	yel monocl cry				50	
2691	Strontium fluoride	SrF_2	7783-48-4	125.62	wh cub cry or powder	1477	2460	4.24	0.021^{25}	s dil acid
2692	Strontium formate	$Sr(CHO_2)_2$	592-89-2	177.66	wh cry	71.9		2.693	9.1^{0}	
2693	Strontium formate dihydrate	$Sr(CHO_2)_2 \cdot 2H_2O$	6160-34-5	213.69	col rhom cry	100 dec		2.25	9.1^{37}	i EtOH, eth
2694	Strontium hexaboride	SrB_6	12046-54-7	152.49	blk cub cry	2235		3.39		i H_2O; s HNO_3
2695	Strontium hydride	SrH_2	13598-33-9	89.64	orth cry	1050		3.26		reac H_2O
2696	Strontium hydroxide	$Sr(OH)_2$	18480-07-4	121.64	col orth cry; hyg	535	710 dec	3.625	2.25^{25}	
2697	Strontium iodate	$Sr(IO_3)_2$	13470-01-4	437.43	tricl cry			5.045	0.165^{25}	
2698	Strontium iodide	SrI_2	10476-86-5	341.43	wh hyg cry	538	1773 dec	4.55	177^{25}	
2699	Strontium iodide hexahydrate	$SrI_2 \cdot 6H_2O$	73796-25-5	449.52	wh-yel hex cry; hyg	120 dec		4.4	177^{25}	s EtOH
2700	Strontium molybdate	$SrMoO_4$	13470-04-7	247.56	wh cry pow	1040		4.54		i H_2O
2701	Strontium niobate	$SrNb_2O_6$	12034-89-8	369.43	monocl cry	1225		5.11		i H_2O
2702	Strontium nitrate	$Sr(NO_3)_2$	10042-76-9	211.63	wh cub cry	570		2.99	80.2^{25}	sl EtOH, ace
2703	Strontium nitride	Sr_3N_2	12033-82-8	290.87	refrac solid	1200				reac H_2O; s HCl
2704	Strontium nitrite	$Sr(NO_2)_2$	13470-06-9	179.63	wh-yel hyg needles	240 dec		2.8	72.1^{30}	s H_2O
2705	Strontium orthosilicate	Sr_2SiO_4	13597-55-2	267.32	orth cry			4.5		
2706	Strontium oxalate monohydrate	$SrC_2O_4 \cdot H_2O$	814-95-9	193.65	cry pow	150 dec			0.00005^{20}	sl dil acid
2707	Strontium oxide	SrO	1314-11-0	103.62	col cub cry	2531		5.1		reac H_2O
2708	Strontium perchlorate	$Sr(ClO_4)_2$	13450-97-0	286.52	col hyg cry				306^{25}	s EtOH, MeOH
2709	Strontium permanganate trihydrate	$Sr(MnO_4)_2 \cdot 3H_2O$	14446-13-0	379.54	purp cub cry	175 dec		2.75	250^{18}	
2710	Strontium peroxide	SrO_2	1314-18-7	119.62	wh tetr cry; unstab	215 dec		4.78		reac H_2O
2711	Strontium phosphate	$Sr_3(PO_4)_2$	7446-28-8	452.80	wh powder				0.000011^{20}	s acid
2712	Strontium selenate	$SrSeO_4$	7446-21-1	230.58	orth cry			4.25	0.115^{20}	s hot HCl
2713	Strontium selenide	SrSe	1315-07-7	166.58	wh cub cry	1600		4.54		
2714	Strontium silicide	$SrSi_2$	12138-28-2	143.79	silv-gray cub cry	1100		3.35		
2715	Strontium sulfate	$SrSO_4$	7759-02-6	183.68	wh orth cry	1606		3.96	0.0135^{25}	i EtOH; sl acid
2716	Strontium sulfide	SrS	1314-96-1	119.69	gray cub cry	2226		3.70		sl H_2O; s acid
2717	Strontium sulfite	$SrSO_3$	13451-02-0	167.68	col cry	dec			0.0015^{25}	s H_2SO_4, HCl
2718	Strontium telluride	SrTe	12040-08-3	215.22	wh cub cry			4.83		
2719	Strontium thiosulfate pentahydrate	$SrS_2O_3 \cdot 5H_2O$	15123-90-7	289.82	monocl needles	100 dec		2.17	36.3^{25}	i EtOH
2720	Strontium titanate	$SrTiO_3$	12060-59-2	183.49	wh cub cry	2080		5.1		i H_2O
2721	Strontium tungstate	$SrWO_4$	13451-05-3	335.46	col tetr cry	dec		6.187	0.14^{15}	i EtOH
2722	Strontium zirconate	$SrZrO_3$	12036-39-4	226.84	col cry	2600				
2723	Sulfur (rhombic)	S	7704-34-9	32.06	yel orth cry	95.2 trans monocl	444.61	2.07		i H_2O; sl EtOH, bz, eth; s CS_2

No.	Name	Formula	CAS Reg. No.	Mol. wt.	Phys. form	mp/°C	bp/°C	ρ/g cm^{-3}	Sol. in g/100 g H_2O	Qualitative solubility
2724	Sulfur (monoclinic)	S	7704-34-9	32.06	yel monocl needles, stable 95.3-120	115.21	444.61	2.00		i H_2O; sl EtOH, bz, eth; s CS_2
2725	Sulfuric acid	H_2SO_4	7664-93-9	98.079	col oily liq	10.31	337	1.8305^{20}		vs H_2O
2726	Peroxysulfuric acid	H_2SO_5	7722-86-3	114.078	wh cry; unstab	45 dec				vs H_2O
2727	Nitrosylsulfuric acid	$HNOSO_4$	7782-78-7	127.077	prisms	73 dec				reac H_2O; s H_2SO_4
2728	Chlorosulfonic acid	$SO_2(OH)Cl$	7790-94-5	116.524	col-yel liq	-80	152	1.75		reac H_2O; s py
2729	Fluorosulfonic acid	$SO_2(OH)F$	7789-21-1	100.069	col liq	-89	163	1.726		reac H_2O
2730	Sulfurous acid	H_2SO_3	7782-99-2	82.079	exists only in aq soln					
2731	Sulfamic acid	H_2NSO_3H	5329-14-6	97.094	orth cry	≈205 dec		2.15	14.7^0	sl ace; i eth
2732	Sulfur dioxide	SO_2	7446-09-5	64.064	col gas	-75.45	-10.02	$1.620^{-75.5}$ liq		s H_2O, EtOh, eth, chl
2733	Sulfur trioxide (α-form)	SO_3	7446-11-9	80.063	wh needles	62.2	subl			reac H_2O
2734	Sulfur trioxide (γ-form)	SO_3	7446-11-9	80.063	col solid or liq	16.8	44.5	1.90		reac H_2O
2735	Sulfur trioxide (β-form)	SO_3	7446-11-9	80.063	wh needles	30.5	44.5			reac H_2O
2736	Sulfur bromide [BrSSBr]	S_2Br_2	13172-31-1	223.938	red oily liq	-46	>25 dec	2.63		reac H_2O
2737	Sulfur chloride [ClSSCl]	S_2Cl_2	10025-67-9	135.036	yel-red oily liq	-77	137	1.69		reac H_2O; s EtOH, bz, eth, ctc
2738	Sulfur fluoride [SSF_2]	SSF_2	16860-99-4	102.127	col gas	-164.6	-10.6	0.004174^{25} gas		reac H_2O
2739	Sulfur fluoride [FSSF]	FSSF	13709-35-8	102.127	col gas	-133	15	0.004174^{25} gas		reac H_2O
2740	Sulfur dichloride	SCl_2	10545-99-0	102.971	red visc liq	-122	59.6	1.62		reac H_2O
2741	Sulfur tetrafluoride	SF_4	7783-60-0	108.059	col gas	-125	-40.45	0.004417^{25} gas		reac H_2O
2742	Sulfur hexafluoride	SF_6	2551-62-4	146.055	col gas	-49.596 tp	-63.8 sp	$1.845^{-49.6}$ liq		sl H_2O; s EtOH
2743	Sulfur bromide pentafluoride	SF_5Br	15607-89-3	206.961	col gas	-79	3.1	0.008459^{25} gas		
2744	Sulfur chloride pentafluoride	SF_5Cl	13780-57-9	162.510	col gas	-64	-19.05	0.006642^{25} gas		
2745	Sulfur decafluoride	S_2F_{10}	5714-22-7	254.114	liq	-52.7	30 dec 150	2.08		i H_2O
2746	Sulfuryl amide	$(NH_2)_2SO_2$	7803-58-9	96.109	orth plates	93	250 dec			vs H_2O; sl EtOH
2747	Sulfuryl chloride	SO_2Cl_2	7791-25-5	134.970	col liq	-51	69.4	1.680		reac H_2O; s bz, tol, eth
2748	Sulfuryl fluoride	SO_2F_2	2699-79-8	102.061	col gas	-135.8	-55.4	0.004172^{25} gas		sl H_2O, EtOH; s tol, ctc
2749	Sulfuryl bromide fluoride	SO_2BrF	13536-61-3	162.966	col liq	-86	41			reac H_2O
2750	Sulfuryl chloride fluoride	SO_2ClF	13637-84-8	118.515	col gas	-124.7	7.1	1.62^0		reac H_2O
2751	Pyrosulfuryl chloride	$S_2O_5Cl_2$	7791-27-7	215.033	col fuming liq	-37	151	1.837		reac H_2O
2752	Thionyl bromide	$SOBr_2$	507-16-4	207.872	yel liq	-50	140			reac H_2O
2753	Thionyl chloride	$SOCl_2$	7719-09-7	118.970	yel fuming liq	-101	75.6	1.631		reac H_2O; s bz, ctc, chl
2754	Thionyl fluoride	SOF_2	7783-42-8	86.061	col gas	-129.5	-43.8	0.003518^{25} gas		reac H_2O; s bz, eth
2755	Sulfur fluoride oxide [SOF_4]	SOF_4	13709-54-1	124.058	col gas	-99.6	-48.5	1.95^{-82}		reac H_2O
2756	Sulfur fluoride hypofluorite	F_5SOF	15179-32-5	162.054	col gas	-86	-35.1	0.006624^{25} gas		
2757	Tetrasulfur tetranitride	S_4N_4	28950-34-7	184.287	yel-oran cry	178.2	subl			i H_2O; reac alk, acid
2758	Tantalum	Ta	7440-25-7	180.948	gray metal; cub	3017	5455	16.4		reac HF
2759	Tantalum aluminide	$TaAl_3$	12004-76-1	261.893	gray refrac powder	≈1400		7.02		i H_2O, acid, alk
2760	Tantalum boride	TaB	12007-07-7	191.759	refrac orth cry	2040		14.2		
2761	Tantalum boride [TaB_2]	TaB_2	12007-35-1	202.570	blk hex cry	3100		11.2		i H_2O, acid, alk
2762	Tantalum carbide	TaC	12070-06-3	192.959	gold-brown powder; cub	3880	4780	14.3		s $HF\text{-}HNO_3$ mixture
2763	Tantalum carbide (Ta_2C)	Ta_2C	12070-07-4	373.907	refrac hex cry	3327		15.1		
2764	Tantalum hydride	TaH	13981-95-8	181.956	gray metallic solid			15.1		i acid
2765	Tantalum nitride	TaN	12033-62-4	194.955	blk hex cry	3090		13.7		i H_2O; sl aqua regia; reac alk
2766	Tantalum silicide	$TaSi_2$	12039-79-1	237.119	gray powder	2200		9.14		
2767	Tantalum(III) bromide	$TaBr_3$	13842-73-4	420.660	gray-grn solid	220 dec				
2768	Tantalum(III) chloride	$TaCl_3$	13569-67-0	287.307	blk-grn solid	440 dec				s H_2O
2769	Tantalum(IV) bromide	$TaBr_4$	13842-76-7	500.564	dark blue solid	392		5.77		reac H_2O
2770	Tantalum(IV) chloride	$TaCl_4$	13569-72-7	322.760	dark grn solid	300 dec	subl	4.35		reac H_2O
2771	Tantalum(IV) iodide	TaI_4	14693-80-2	688.566	gray-blk solid	400 dec				reac H_2O
2772	Tantalum(IV) oxide	TaO_2	12036-14-5	212.947	tetr cry			10.0		
2773	Tantalum(IV) selenide	$TaSe_2$	12039-55-3	338.87	hex cry			6.7		
2774	Tantalum(IV) sulfide	TaS_2	12143-72-5	245.078	blk hex cry	>3000		6.86		i H_2O
2775	Tantalum(IV) telluride	$TaTe_2$	12067-66-2	436.15	monocl cry			9.4		
2776	Tantalum(V) bromide	$TaBr_5$	13451-11-1	580.468	yel cry powder	240	348.8	4.67		
2777	Tantalum(V) chloride	$TaCl_5$	7721-01-9	358.213	yel-wh monocl cry; hyg	216.6	239	3.68		reac H_2O; s EtOH
2778	Tantalum(V) fluoride	TaF_5	7783-71-3	275.940	wh monocl cry; hyg	96.9	229.5	4.74		s H_2O, eth; sl CS_2, ctc
2779	Tantalum(V) iodide	TaI_5	14693-81-3	815.470	blk hex cry; hyg	496	543	5.80		reac H_2O
2780	Tantalum(V) oxide	Ta_2O_5	1314-61-0	441.893	wh rhomb cry or powder	1875		8.24		i H_2O, EtOH, acid; s HF
2781	Technetium	Tc	7440-26-8	98	hex cry	2157	4262	11		

No.	Name	Formula	CAS Reg. No.	Mol. wt.	Phys. form	mp/°C	bp/°C	ρ/g cm⁻³	Sol. in g/100 g H_2O	Qualitative solubility
2782	Technetium(V) fluoride	TcF_5	31052-14-9	193	yel solid	50	dec			
2783	Technetium(VI) fluoride	TcF_6	13842-93-8	212	yel cub cry	37.4	55.3	3.0		
2784	Tellurium	Te	13494-80-9	127.60	gray-wh rhomb cry	449.51	988	6.232		i H_2O, bz, CS_2
2785	Telluric(VI) acid	H_6TeO_6	7803-68-1	229.64	wh monocl cry	136		3.07	50.1^{30}	
2786	Tellurous acid	H_2TeO_3	10049-23-7	177.61	wh cry	40 dec		3.0		sl H_2O; s dil acid, alk
2787	Tellurium dioxide	TeO_2	7446-07-3	159.60	wh orth cry	733	1245	5.9		i H_2O; s alk, acid
2788	Tellurium trioxide	TeO_3	13451-18-8	175.60	yel-oran cry	430		5.07		i H_2O
2789	Tellurium dibromide	$TeBr_2$	7789-54-0	287.41	grn-brn hyg cry	210	339			reac H_2O; s eth; sl chl
2790	Tellurium dichloride	$TeCl_2$	10025-71-5	198.51	blk amorp solid; hyg	208	328	6.9		reac H_2O; i ctc
2791	Tellurium tetrabromide	$TeBr_4$	10031-27-3	447.22	yel-oran monocl cry	380	≈420 dec	4.3		reac H_2O; s eth
2792	Tellurium tetrachloride	$TeCl_4$	10026-07-0	269.41	wh monocl cry; hyg	224	387	3.0		reac H_2O; s EtOH, tol
2793	Tellurium tetrafluoride	TeF_4	15192-26-4	203.59	col cry	129	195 dec			reac H_2O
2794	Tellurium decafluoride	Te_2F_{10}	53214-07-6	445.18	col liq	-33.7	59			
2795	Tellurium tetraiodide	TeI_4	7790-48-9	635.22	blk orth cry	280		5.05		reac H_2O; sl ace
2796	Tellurium hexafluoride	TeF_6	7783-80-4	241.59	col gas	-37.6 tp	-38.9 sp	0.009875^{25} gas		reac H_2O
2797	Terbium	Tb	7440-27-9	158.925	silv metal; hex	1359	3230	8.23		
2798	Terbium nitride	TbN	12033-64-6	172.932	cub cry			9.55		
2799	Terbium silicide	$TbSi_2$	12039-80-4	215.096	orth cry			6.66		
2800	Terbium(III) bromide	$TbBr_3$	14456-47-4	398.637	wh hex cry	830	1490			s H_2O
2801	Terbium(III) chloride	$TbCl_3$	10042-88-3	265.284	wh orth cry; hyg	582		4.35		s H_2O
2802	Terbium(III) chloride hexahydrate	$TbCl_3 \cdot 6H_2O$	13798-24-8	373.375	hyg cry			4.35		vs H_2O
2803	Terbium(III) iodide	TbI_3	13813-40-6	539.638	hex cry; hyg	955		≈5.2		s H_2O
2804	Terbium(III) nitrate	$Tb(NO_3)_3$	10043-27-3	344.940	pink hyg solid				157^{25}	s EtOH
2805	Terbium(III) nitrate hexahydrate	$Tb(NO_3)_3 \cdot 6H_2O$	13451-19-9	453.031	col needles	89				s H_2O, EtOH, ace
2806	Terbium(III) oxide	Tb_2O_3	12036-41-8	365.849	wh cub cry	2303		7.91		
2807	Terbium(III) sulfate octahydrate	$Tb_2(SO_4)_3 \cdot 8H_2O$	13842-67-6	750.161	wh cry	360 dec				sl H_2O
2808	Terbium(III) sulfide	Tb_2S_3	12138-11-3	414.046	cub cry			6.35		
2809	Terbium(III) fluoride	TbF_3	13708-63-9	215.920	wh solid	1175	2280			i H_2O
2810	Terbium(IV) fluoride	TbF_4	36781-15-4	234.919	wh monocl cry	300 dec				i H_2O
2811	Thallium	Tl	7440-28-0	204.38	soft blue-wh metal	304	1473	11.8		i H_2O; reac acid
2812	Thallium(I) acetate	$TlC_2H_3O_2$	563-68-8	263.427	hyg wh cry	131		3.68		s H_2O, EtOH
2813	Thallium(I) azide	TlN_3	13847-66-0	246.403	yel cry	334	exp			s H_2O
2814	Thallium(I) bromate	$TlBrO_3$	14550-84-6	332.285	col needles	120 dec			0.49^{30}	s EtOH
2815	Thallium(I) bromide	TlBr	7789-40-4	284.287	yel cub cry	460	819	7.5	0.059^{20}	
2816	Thallium(I) carbonate	Tl_2CO_3	6533-73-9	468.776	wh monocl cry	273		7.11	4.69^{20}	i EtOH
2817	Thallium(I) chlorate	$TlClO_3$	13453-30-0	287.834	col hex cry	500 dec		5.5	3.92^{20}	
2818	Thallium(I) chloride	TlCl	7791-12-0	239.836	wh cub cry	431	720	7.0	0.33^{20}	i EtOH
2819	Thallium(I) chromate	Tl_2CrO_4	13473-75-1	524.761	yel cry				0.003^{20}	sl acid, alk
2820	Thallium(I) cyanide	TlCN	13453-34-4	230.401	wh hex plates			6.523		s H_2O, acid, EtOH
2821	Thallium(I) ethanolate	TlC_2H_5O	20398-06-5	249.443	cloudy liq	-3	130 dec	3.49		reac H_2O
2822	Thallium(I) fluoride	TlF	7789-27-7	223.381	wh orth cry	326	826	8.36	245^{25}	
2823	Thallium(I) formate	$TlCHO_2$	992-98-3	249.401	hyg col needles	101		4.97		vs H_2O; s MeOH
2824	Thallium(I) hexafluorophosphate	$TlPF_6$	60969-19-9	349.347	wh cub cry			4.6		
2825	Thallium(I) hydroxide	TlOH	12026-06-1	221.390	yel needles	139 dec		7.44	34.3^{18}	
2826	Thallium(I) iodate	$TlIO_3$	14767-09-0	379.285	wh needles				0.058	sl HNO_3
2827	Thallium(I) iodide	TlI	7790-30-9	331.287	yel cry powder	441.7	824	7.1	0.0085^{20}	i EtOH
2828	Thallium(I) molybdate	Tl_2MoO_4	34128-09-1	568.71	yel-wh cub cry					i H_2O
2829	Thallium(I) nitrate	$TlNO_3$	10102-45-1	266.388	wh cry	206	450 dec	5.55	9.55^{20}	s H_2O; i EtOH
2830	Thallium(I) nitrite	$TlNO_2$	13826-63-6	250.389	yel cub cry	186		5.7	32.1^{25}	s H_2O
2831	Thallium(I) oxalate	$Tl_2C_2O_4$	30737-24-7	496.786	wh powder			6.31	1.83^{20}	
2832	Thallium(I) oxide	Tl_2O	1314-12-1	424.766	blk rhomb cry; hyg	579	≈1080	9.52		s H_2O, EtOH
2833	Thallium(I) perchlorate	$TlClO_4$	13453-40-2	303.834	col orth cry	501		4.89	19.7^{30}	
2834	Thallium(I) selenate	Tl_2SeO_4	7446-22-2	551.73	orth cry	>400		6.875	2.8^{20}	i EtOH, eth
2835	Thallium(I) selenide	Tl_2Se	15572-25-5	487.73	gray plates	340				i H_2O, acid
2836	Thallium(I) sulfate	Tl_2SO_4	7446-18-6	504.830	wh rhomb prisms	632		6.77	5.47^{25}	
2837	Thallium(I) sulfide	Tl_2S	1314-97-2	440.832	blue-blk cry	457	1367	8.39	0.02^{20}	sl alk; s acid
2838	Thallium(III) acetate	$Tl(C_2H_3O_2)_3$	2570-63-0	381.514	hyg wh platelets	182 dec				
2839	Thallium(III) bromide tetrahydrate	$TlBr_3 \cdot 4H_2O$	13701-90-1	516.157	yel orth cry			3.65		s H_2O, EtOH
2840	Thallium(III) chloride	$TlCl_3$	13453-32-2	310.742	monocl cry	155		4.7		vs H_2O, EtOH, eth
2841	Thallium(III) chloride tetrahydrate	$TlCl_3 \cdot 4H_2O$	13453-32-2	382.804	orth cry			3.00		s H_2O
2842	Thallium(III) fluoride	TlF_3	7783-57-5	261.378	wh orth cry; hyg	550 dec		8.65		reac H_2O
2843	Thallium(III) nitrate	$Tl(NO_3)_3$	13746-98-0	390.398	col cry					reac H_2O
2844	Thallium(III) oxide	Tl_2O_3	1314-32-5	456.765	brn cub cry	834		10.2		i H_2O; reac acid
2845	Thallium(III) sulfate	$Tl_2(SO_4)_3$	16222-66-5	696.955	col leaflets					reac H_2O
2846	Thallium selenide (TlSe)	TlSe	12039-52-0	283.34	blk solid	330				i H_2O, acid
2847	Thorium	Th	7440-29-1	232.038	soft gray-wh metal; cub	1750	4785	11.7		s acid
2848	Thorium hydride	ThH_2	16689-88-6	234.054	tetr cry			9.5		

No.	Name	Formula	CAS Reg. No.	Mol. wt.	Phys. form	mp/°C	bp/°C	ρ/g cm^{-3}	Sol. in g/100 g H_2O	Qualitative solubility
2849	Thorium boride	ThB_6	12229-63-9	296.904	refrac solid	2450		6.99		
2850	Thorium(IV) bromide	$ThBr_4$	13453-49-1	551.654	wh hyg cry	679			65^{20}	
2851	Thorium carbide	ThC	12012-16-7	244.049	cub cry	2500		10.6		reac H_2O
2852	Thorium dicarbide	ThC_2	12071-31-7	256.059	yel monocl cry	≈2650		9.0		reac H_2O
2853	Thorium(IV) chloride	$ThCl_4$	10026-08-1	373.850	gray-wh tetr needles; hyg	770	921	4.59		s H_2O, EtOH
2854	Thorium(IV) fluoride	ThF_4	13709-59-6	308.032	wh monocl cry; hyg	1110	1680	6.1		
2855	Thorium(IV) iodide	ThI_4	7790-49-0	739.656	wh-yel monocl cry	566	837			
2856	Thorium(IV) nitrate	$Th(NO_3)_4$	13823-29-5	480.058	hyg wh plates	55 dec				vs H_2O, EtOH
2857	Thorium(IV) nitrate tetrahydrate	$Th(NO_3)_4 \cdot 4H_2O$	13470-07-0	552.119	wh hyg cry	500 dec			191^{20}	s EtOH
2858	Thorium nitride	ThN	12033-65-7	246.045	refrac cub cry	2820		11.6		reac H_2O
2859	Thorium(IV) oxide	ThO_2	1314-20-1	264.037	wh cub cry	3350	4400	10.0		i H_2O, alk; sl acid
2860	Thorium(IV) selenide	$ThSe_2$	60763-24-8	389.96	orth cry			8.5		
2861	Thorium orthosilicate	$ThSiO_4$	14553-44-7	324.122	brn tetr cry			6.7		
2862	Thorium silicide	$ThSi_2$	12067-54-8	288.209	tetr cry	1850		7.9		
2863	Thorium(IV) sulfate nonahydrate	$Th(SO_4)_2 \cdot 9H_2O$	10381-37-0	586.301	wh monocl cry	dec		2.8	4.2^{20}	
2864	Thorium(IV) sulfide	ThS_2	12138-07-7	296.168	dark brn cry	1905		7.30		i H_2O; s acid
2865	Thulium	Tm	7440-30-4	168.934	silv metal; hex	1545	1950	9.32		s dil acid
2866	Thulium(II) bromide	$TmBr_2$	64171-97-7	328.742	dark grn solid	619				
2867	Thulium(II) chloride	$TmCl_2$	22852-11-5	239.840	red or grn cry	718				reac H_2O
2868	Thulium(II) iodide	TmI_2	60864-26-8	422.743	blk hyg solid	756				reac H_2O
2869	Thulium(III) bromide	$TmBr_3$	14456-51-0	408.646	wh hyg cry	954				s H_2O
2870	Thulium(III) chloride	$TmCl_3$	13537-18-3	275.293	yel hyg cry	845				s H_2O
2871	Thulium(III) chloride heptahydrate	$TmCl_3 \cdot 7H_2O$	13778-39-7	401.400	hyg cry					s H_2O, EtOH
2872	Thulium(III) fluoride	TmF_3	13760-79-7	225.929	wh cry	1158				s H_2O
2873	Thulium(III) hydroxide	$Tm(OH)_3$	1311-33-7	219.956	wh or grn prec					i H_2O
2874	Thulium(III) iodide	TmI_3	13813-43-9	549.647	yel hyg cry	1021				
2875	Thulium(III) nitrate	$Tm(NO_3)_3$	14985-19-4	354.949	grn hyg solid				212^{25}	s EtOH
2876	Thulium(III) nitrate pentahydrate	$Tm(NO_3)_3 \cdot 5H_2O$	36548-87-5	445.025	grn hyg cry					s H_2O, EtOH, ace
2877	Thulium(III) oxalate hexahydrate	$Tm_2(C_2O_4)_3 \cdot 6H_2O$	26677-68-9	710.016	grn solid	50 dec				s alk oxalates
2878	Thulium(III) oxide	Tm_2O_3	12036-44-1	385.866	grn-wh cub cry	2341	3945	8.6		sl acid
2879	Tin (gray)	Sn	7440-31-5	118.710	cub cry	13.2 trans white Sn	2586	5.769		
2880	Tin (white)	Sn	7440-31-5	118.710	silv tetr cry	231.928	2586	7.287		
2881	Stannane	SnH_4	2406-52-2	122.742	unstab col gas	-146	-51.8	0.005017^{25} gas		
2882	Methylstannane	SnH_3CH_3	1631-78-3	136.769	col gas		1.4	0.005590^{25} gas		reac H_2O
2883	(Dimethylamino)trimethystannane	$Sn(CH_3)_3N(CH_3)_2$	993-50-0	207.890	liq	1	126	1.22		reac H_2O
2884	Tin monophosphide	SnP	25324-56-5	149.684	dull metallic solid	540				
2885	Tin triphosphide	Sn_4P_3	12286-33-8	567.761	wh cry	≈550		5.2		
2886	Tin(II) acetate	$Sn(C_2H_3O_2)_2$	638-39-1	236.799	wh orth cry	183	subl	2.31		s dil HCl
2887	Tin(II) bromide	$SnBr_2$	10031-24-0	278.518	yel powder	232	639	5.12	85^{0}	s EtOH, eth, ace
2888	Tin(II) chloride	$SnCl_2$	7772-99-8	189.616	wh orth cry	247.0	623	3.90	178^{10}	s EtOH, ace, eth; i xyl
2889	Tin(II) chloride dihydrate	$SnCl_2 \cdot 2H_2O$	10025-69-1	225.647	wh monocl cry	37 dec		2.71	178^{10}	s EtOH, NaOH; vs HCl
2890	Tin(II) fluoride	SnF_2	7783-47-3	156.707	wh monocl cry; hyg	215	850	4.57		s H_2O; i EtOH, eth, chl
2891	Tin(II) hexafluorozirconate	$SnZrF_6$	12419-43-1	323.924	cry			4.21		s H_2O
2892	Tin(II) hydroxide	$Sn(OH)_2$	12026-24-3	152.725	wh amorp solid					
2893	Tin(II) iodide	SnI_2	10294-70-9	372.519	red-oran powder	320	714	5.28	0.98^{20}	s bz, chl, CS_2
2894	Tin(II) oxalate	SnC_2O_4	814-94-8	206.729	wh powder	280 dec		3.56		i H_2O; s dil HCl
2895	Tin(II) oxide	SnO	21651-19-4	134.709	blue-blk tetr cry	977		6.45		i H_2O, EtOH; s acid
2896	Tin(II) pyrophosphate	$Sn_2P_2O_7$	15578-26-4	411.363	wh amorp powder	400 dec		4.009		i H_2O; s conc acid
2897	Tin(II) selenide	SnSe	1315-06-6	197.67	gray orth cry	861		6.18		i H_2O; s aqua regia
2898	Tin(II) sulfate	$SnSO_4$	7488-55-3	214.773	wh orth cry	378 dec		4.15	18.8^{19}	
2899	Tin(II) sulfide	SnS	1314-95-0	150.775	gray orth cry	881	1210	5.08		i H_2O; s conc acid
2900	Tin(II) tartrate	$SnC_4H_4O_6$	815-85-0	266.781	wh cry powder					s H_2O, dil HCl
2901	Tin(II) telluride	SnTe	12040-02-7	246.31	gray cub cry	806		6.5		
2902	Tin(IV) bromide	$SnBr_4$	7789-67-5	438.326	wh cry	29.1	205	3.34		vs H_2O; s EtOH
2903	Tin(IV) chloride	$SnCl_4$	7646-78-8	260.522	col fuming liq	-34.07	114.15	2.234		reac H_2O; s EtOH, ctc, bz, ace
2904	Tin(IV) chloride pentahydrate	$SnCl_4 \cdot 5H_2O$	10026-06-9	350.598	wh-yel cry	56 dec		2.04		vs H_2O; s EtOH
2905	Tin(IV) chromate	$Sn(CrO_4)_2$	38455-77-5	350.697	brn-yel cry powder	dec				s H_2O
2906	Tin(IV) fluoride	SnF_4	7783-62-2	194.704	wh tetr cry	442	705 subl	4.78		reac H_2O
2907	Tin(IV) iodide	SnI_4	7790-47-8	626.328	yel-brn cub cry	402	364.35	4.46		reac H_2O; s EtOH, bz, chl, eth
2908	Tin(IV) oxide	SnO_2	18282-10-5	150.709	gray tetr cry	1630		6.85		i H_2O, EtOH; s hot conc alk

No.	Name	Formula	CAS Reg. No.	Mol. wt.	Phys. form	mp/°C	bp/°C	ρ/g cm^{-3}	Sol. in g/100 g H_2O	Qualitative solubility
2909	Tin(IV) selenide	$SnSe_2$	20770-09-6	276.63	red-brn cry	650		≈5.0		i H_2O; s alk, conc acid
2910	Tin(IV) selenite	$Sn(SeO_3)_2$	7446-25-5	372.63	cry powder					i H_2O; s hot HCl
2911	Tin(IV) sulfide	SnS_2	1315-01-1	182.840	gold-yel hex cry	600 dec		4.5		i H_2O; s alk, aqua regia
2912	Titanium	Ti	7440-32-6	47.867	gray metal; hex	1670	3287	4.506		
2913	Titanocene dichloride	$Ti(C_5H_5)_2Cl_2$	1271-19-8	248.959	red cry	289		1.60		sl H_2O, bz; s chl, EtOH, tol
2914	Titanium hydride	TiH_2	7704-98-5	49.883	gray-blk powder	≈450 dec		3.75		i H_2O
2915	Titanium boride	TiB_2	12045-63-5	69.489	gray refrac solid; hex	2920		4.38		
2916	Titanium carbide	TiC	12070-08-5	59.878	cub cry	3067		4.93		i H_2O; s HNO_3
2917	Titanium nitride	TiN	25583-20-4	61.874	yel-brn cub cry	2947		5.21		i H_2O; s aqua regia
2918	Titanium phosphide	TiP	12037-65-9	78.841	gray hex cry	1990		4.08		
2919	Titanium silicide	$TiSi_2$	12039-83-7	104.038	blk orth cry	1500		4.0		i H_2O, acid, alk; s HF
2920	Titanium(II) bromide	$TiBr_2$	13783-04-5	207.675	blk powder	400 dec		4.0		reac H_2O
2921	Titanium(II) chloride	$TiCl_2$	10049-06-6	118.773	blk hex cryc	1035	1500	3.13		reac H_2O; s EtOH; i chl, eth
2922	Titanium(II) iodide	TiI_2	13783-07-8	301.676	blk hex cry	400 dec		5.02		reac H_2O
2923	Titanium(II) oxide	TiO	12137-20-1	63.866	yel cub cry	1770	3227	4.95		
2924	Titanium(II) sulfide	TiS	12039-07-5	79.932	brn hex cry	1927		3.85		i H_2O; s conc acid
2925	Titanium(III) bromide	$TiBr_3$	13135-31-4	287.579	viol hex cry	400 dec				s H_2O
2926	Titanium(III) chloride	$TiCl_3$	7705-07-9	154.226	red-viol hex cry; hyg	425 dec	960	2.64		reac H_2O
2927	Titanium(III) fluoride	TiF_3	13470-08-1	104.862	viol hex cry	950 dec		2.98		i H_2O, dil acid, alk
2928	Titanium(III) iodide	TiI_3	13783-08-9	428.580	viol cry	350 dec				
2929	Titanium(III) oxide	Ti_2O_3	1344-54-3	143.732	blk hex cry	1842		4.486		s hot HF
2930	Titanium(III) sulfate	$Ti_2(SO_4)_3$	10343-61-0	383.922	grn cry					i H_2O, EtOH; s dil HCl
2931	Titanium(III) sulfide	Ti_2S_3	12039-16-6	191.929	blk hex cry			3.56		
2932	Titanium(III,IV) oxide	Ti_3O_5	12065-65-5	223.598	blk monocl cry	1777		4.24		
2933	Titanium(IV) bromide	$TiBr_4$	7789-68-6	367.483	yel-oran cub cry; hyg	38.3	233.5	3.37		reac H_2O
2934	Titanium(IV) chloride	$TiCl_4$	7550-45-0	189.679	col or yel liq	-24.12	136.45	1.73		reac H_2O; s EtOH
2935	Titanium(IV) fluoride	TiF_4	7783-63-3	123.861	wh hyg powder	377	284 subl	2.798		reac H_2O; s EtOH, py
2936	Titanium(IV) iodide	TiI_4	7720-83-4	555.485	red hyg powder	155	377	4.3		reac H_2O
2937	Titanium(IV) oxide (anatase)	TiO_2	1317-70-0	79.866	brn tetr cry	1560		3.9		
2938	Titanium(IV) oxide (brookite)	TiO_2	12188-41-9	79.866	wh orth cry			4.17		
2939	Titanium(IV) oxide (rutile)	TiO_2	1317-80-2	79.866	wh tetr cry	1912	≈3000	4.17		i H_2O, dil acid; s conc acid
2940	Titanium(IV) oxysulfate monohydrate	$TiOSO_4 \cdot H_2O$	13825-74-6	177.944	col orth cry			2.71		reac H_2O
2941	Titanium(IV) sulfate	$Ti(SO_4)_2$	13693-11-3	239.992	wh-yel hyg cry	150 dec				s H_2O
2942	Titanium(IV) sulfide	TiS_2	12039-13-3	111.997	yel-brn hex cry; hyg			3.37		s H_2SO_4
2943	Tungsten	W	7440-33-7	183.84	gray-wh metal; cub	3414	5555	19.3		
2944	Tungstic acid	H_2WO_4	7783-03-1	249.85	yel amorp powder	100 dec		5.5		i H_2O, acid; s alk
2945	Tungsten boride (W_2B)	W_2B	12007-10-2	378.49	refrac blk powder	2740		16.0		i H_2O
2946	Tungsten boride	WB	12007-09-9	194.65	blk refrac powder	2800		15.2		i H_2O
2947	Tungsten boride [W_2B_5]	W_2B_5	12007-98-6	421.74	refrac solid	2370		11.0		i H_2O
2948	Tungsten carbide (W_2C)	W_2C	12070-13-2	379.69	refrac hex cry	≈2800		14.8		i H_2O
2949	Tungsten carbide	WC	12070-12-1	195.85	gray hex cry	2785		15.6		i H_2O; s HNO_3/HF
2950	Tungsten carbonyl	$W(CO)_6$	14040-11-0	351.90	wh cry	170 dec	subl	2.65		i H_2O; s os
2951	Tungsten nitride [WN_2]	WN_2	60922-26-1	211.85	hex cry	600 dec		7.7		
2952	Tungsten nitride (W_2N)	W_2N	12033-72-6	381.69	gray cub cry	dec		17.8		
2953	Tungsten silicide [WSi_2]	WSi_2	12039-88-2	240.01	blue-gray tetr cry	2160		9.3		i H_2O
2954	Tungsten silicide [W_5Si_3]	W_5Si_3	12039-95-1	1003.46	blue-gray refrac solid	2320		14.4		
2955	Tungsten(II) bromide	WBr_2	13470-10-5	343.65	yel powder	400 dec				i H_2O
2956	Tungsten(II) chloride	WCl_2	13470-12-7	254.75	gray solid	500 dec		5.44		sl H_2O
2957	Tungsten(II) iodide	WI_2	13470-17-2	437.65	oran-brn cry	800 dec		6.79		i H_2O
2958	Tungsten(III) bromide	WBr_3	15163-24-3	423.55	blk hex cry	180 dec				i H_2O
2959	Tungsten(III) chloride	WCl_3	20193-56-0	290.20	red solid	550 dec	subl			reac H_2O
2960	Tungsten(III) iodide	WI_3	15513-69-6	564.55	blk solid	dec				i H_2O; s ace; sl EtOH, chl
2961	Tungsten(IV) bromide	WBr_4	14055-81-3	503.46	blk orth cry		240 subl			reac H_2O
2962	Tungsten(IV) chloride	WCl_4	13470-13-8	325.65	blk hyg powder	450 dec		4.62		reac H_2O
2963	Tungsten(IV) fluoride	WF_4	13766-47-7	259.83	red-brn cry	800 dec				reac H_2O; s MeCN; i bz, tol, ctc

Inorganic

No.	Name	Formula	CAS Reg. No.	Mol. wt.	Phys. form	mp/°C	bp/°C	ρ/g cm^{-3}	Sol. in g/100 g H_2O	Qualitative solubility
2964	Tungsten(IV) iodide	WI_4	14055-84-6	691.46	blk cry	dec				reac H_2O; s EtOH; i eth chl
2965	Tungsten(IV) oxide	WO_2	12036-22-5	215.84	brn monocl cry	≈1500 dec	1730	10.8		i H_2O, os
2966	Tungsten(IV) selenide	WSe_2	12067-46-8	341.76	gray hex cry			9.2		
2967	Tungsten(IV) sulfide	WS_2	12138-09-9	247.97	gray hex cry	1250 dec		7.6		i H_2O, HCl, alk
2968	Tungsten(IV) telluride	WTe_2	12067-76-4	439.04	gray orth cry	1020		9.43		
2969	Tungsten(V) bromide	WBr_5	13470-11-6	583.36	brn-blk hyg solid	286	333			reac H_2O
2970	Tungsten(V) chloride	WCl_5	13470-14-9	361.11	blk-grn hyg cry	253	286	3.88		reac H_2O
2971	Tungsten(V) ethanolate	$W(C_2H_5O)_5$	62571-53-3	409.14	powder					s EtAc
2972	Tungsten(V) fluoride	WF_5	19357-83-6	278.83	yel solid	20 dec				reac H_2O
2973	Tungsten(V) oxytribromide	$WOBr_3$	20213-56-3	439.55	dark brn tetr cry			≈5.9		
2974	Tungsten(V) oxytrichloride	$WOCl_3$	14249-98-0	306.20	grn tetr cry			≈4.6		
2975	Tungsten(VI) bromide	WBr_6	13701-86-5	663.26	blue-blk cry	309		6.9		reac H_2O
2976	Tungsten(VI) chloride	WCl_6	13283-01-7	396.56	purp hex cry; hyg	282	337	3.52		reac H_2O; s EtOH, os
2977	Tungsten(VI) dioxydibromide	WO_2Br_2	13520-75-7	375.65	red cry		440 subl			
2978	Tungsten(VI) dioxydichloride	WO_2Cl_2	13520-76-8	286.75	yel orth cry	265		4.67		i H_2O
2979	Tungsten(VI) dioxydiiodide	WO_2I_2	14447-89-3	469.65	grn monocl cry	400 dec		6.39		
2980	Tungsten(VI) fluoride	WF_6	7783-82-6	297.83	yel liq or col gas	1.9	17.1	3.44		reac H_2O; vs ctc, cyhex
2981	Tungsten(VI) oxide	WO_3	1314-35-8	231.84	yel powder	1473	≈1700	7.2		i H_2O, os; sl acid; s alk
2982	Tungsten(VI) oxytetrabromide	$WOBr_4$	13520-77-9	519.46	red tetr cry	277	327	≈5.5		reac H_2O
2983	Tungsten(VI) oxytetrachloride	$WOCl_4$	13520-78-0	341.65	red hyg cry	210	230	11.92		reac H_2O; s bz, CS_2
2984	Tungsten(VI) oxytetrafluoride	WOF_4	13520-79-1	275.83	wh monocl cry	105	185.9	5.07		reac H_2O
2985	Tungsten(VI) sulfide	WS_3	12125-19-8	280.04	brn powder					sl H_2O; s alk
2986	Uranium	U	7440-61-1	238.029	silv-wh orth cry	1135	4131	19.1		
2987	Uranium boride [UB_2]	UB_2	12007-36-2	259.651	refrac solid	2430		12.7		
2988	Uranium boride [UB_4]	UB_4	12007-84-0	281.273	refrac solid	2530		9.32		i H_2O
2989	Uranium carbide	UC	12070-09-6	250.040	gray cub cry	2790				
2990	Uranium carbide [UC_2]	UC_2	12071-33-9	262.050	gray tetr cry	2350	4370	11.3		reac H_2O; sl EtOH
2991	Uranium carbide [U_2C_3]	U_2C_3	12076-62-9	512.090	gray cub cry	≈1700 dec		12.7		
2992	Uranium nitride	UN	25658-43-9	252.036	gray cub cry	2805		14.3		i H_2O
2993	Uranium nitride [U_2N_3]	U_2N_3	12033-83-9	518.078	cub cry	dec		11.3		
2994	Uranium(III) bromide	UBr_3	13470-19-4	477.741	red hyg cry	727				s H_2O
2995	Uranium(III) chloride	UCl_3	10025-93-1	344.388	grn hyg cry	837		5.51		vs H_2O; i bz, ctc
2996	Uranium(III) fluoride	UF_3	13775-06-9	295.024	blk hex cry	1495		8.9		i H_2O; s acid
2997	Uranium(III) hydride	UH_3	13598-56-6	241.053	gray-blk cub cry			11.1		
2998	Uranium(III) iodide	UI_3	13775-18-3	618.742	blk hyg cry	766				s H_2O
2999	Uranium(IV) bromide	UBr_4	13470-20-7	557.645	brn hyg cry	519				s H_2O, EtOH
3000	Uranium(IV) chloride	UCl_4	10026-10-5	379.841	grn octahed cry	590	791	4.72		reac H_2O; s EtOH
3001	Uranium(IV) fluoride	UF_4	10049-14-6	314.023	grn monocl cry	1036	1417	6.7	0.01^{25}	s conc acid, alk
3002	Uranium(IV) iodide	UI_4	13470-22-9	745.647	blk hyg cry	506				s H_2O, EtOH
3003	Uranium(IV) oxide	UO_2	1344-57-6	270.028	brn cub cry	2847		10.97		i H_2O, dil acid; s conc acid
3004	Uranium(IV,V) oxide	U_4O_9	12037-15-9	1096.110	cub cry			11.2		
3005	Uranium(V) bromide	UBr_5	13775-16-1	637.549	brn hyg cry					reac H_2O
3006	Uranium(V) chloride	UCl_5	13470-21-8	415.294	brn hyg cry	287				reac H_2O
3007	Uranium(V) fluoride	UF_5	13775-07-0	333.021	pale blue tetr cry; hyg	348		5.81		s H_2O
3008	Uranium(V,VI) oxide	U_3O_8	1344-59-8	842.082	grn-blk orth cry	1300 dec		8.38		
3009	Uranium(VI) chloride	UCl_6	13763-23-0	450.747	grn hex cry	177		3.6		
3010	Uranium(VI) fluoride	UF_6	7783-81-5	352.019	wh monocl solid	64.06 tp	56.5 sp	5.09		reac H_2O; s ctc, chl
3011	Uranium(VI) oxide	UO_3	1344-58-7	286.027	oran-yel cry			≈7.3		i H_2O; s acid
3012	Uranium(VI) oxide monohydrate	$UO_3 \cdot H_2O$	12326-21-5	304.043	yel orth cry	570 dec		7.05		
3013	Uranium peroxide dihydrate	$UO_4 \cdot 2H_2O$	19525-15-6	338.057	yel hyg cry	115 dec				i H_2O
3014	Uranyl acetate dihydrate	$UO_2(C_2H_3O_2)_2 \cdot 2H_2O$	6159-44-0	424.146	ye cry (HOAc)	80 dec		2.89		sl EtOH
3015	Uranyl chloride	UO_2Cl_2	7791-26-6	340.934	yel orth cry; hyg	577				vs H_2O; s EtOH, ace; i bz
3016	Uranyl fluoride	UO_2F_2	13536-84-0	308.025	yel hyg solid				64.4^{20}	i bz
3017	Uranyl hydrogen phosphate tetrahydrate	$UO_2HPO_4 \cdot 4H_2O$	18433-48-2	438.068	yel cry pow					i H_2O; s acid
3018	Uranyl nitrate	$UO_2(NO_3)_2$	10102-06-4	394.037	yel cry				127^{25}	s eth
3019	Uranyl nitrate hexahydrate	$UO_2(NO_3)_2 \cdot 6H_2O$	13520-83-7	502.129	yel orth cry; hyg	60	118 dec	2.81	127^{25}	s EtOH, eth
3020	Uranyl sulfate	UO_2SO_4	1314-64-3	366.090	yel cry					
3021	Uranyl sulfate trihydrate	$UO_2SO_4 \cdot 3H_2O$	20910-28-5	420.137	yel cry			3.28	152^{16}	sl EtOH
3022	Vanadium	V	7440-62-2	50.942	gray-wh metal; cub	1910	3407	6.0		i H_2O; s acid
3023	Vanadocene	$V(C_5H_5)_2$	1277-47-0	181.128	viol cry; hyg	167				s bz, thf
3024	Vanadocene dichloride	$V(C_5H_5)_2Cl_2$	12083-48-6	252.034	dark grn cry	205 dec				s H_2O, chl, EtOH
3025	Vanadium boride (VB)	VB	12045-27-1	61.753	refrac solid	2250				i H_2O
3026	Vanadium boride [VB_2]	VB_2	12007-37-3	72.564	refrac solid	2450				
3027	Vanadium carbide (VC)	VC	12070-10-9	62.953	refrac blk cry; cub	2810		5.77		i H_2O
3028	Vanadium carbide (V_2C)	V_2C	12012-17-8	113.894	hex cry	2167				
3029	Vanadium carbonyl	$V(CO)_6$	14024-00-1	219.002	blue-grn cry; flam	60 dec	subl			

No.	Name	Formula	CAS Reg. No.	Mol. wt.	Phys. form	mp/°C	bp/°C	ρ/g cm^{-3}	Sol. in g/100 g H_2O	Qualitative solubility
3030	Vanadium nitride	VN	24646-85-3	64.949	blk powder; cub	2050		6.13		i H_2O; s aqua regia
3031	Vanadium silicide [VSi_2]	VSi_2	12039-87-1	107.113	metallic prisms			4.42		s HF
3032	Vanadium silicide (V_3Si)	V_3Si	12039-76-8	180.911	cub cry	1935		5.70		
3033	Vanadium(II) bromide	VBr_2	14890-41-6	210.750	oran-brn hex cry		800 subl	4.58		reac H_2O
3034	Vanadium(II) chloride	VCl_2	10580-52-6	121.848	grn hex plates	1350	910 subl	3.23		reac H_2O; s EtOH, eth
3035	Vanadium(II) fluoride	VF_2	13842-80-3	88.939	blue hyg cry	1490				reac H_2O
3036	Vanadium(II) iodide	VI_2	15513-84-5	304.751	red-viol hex cry		800 subl	5.44		reac H_2O
3037	Vanadium(II) oxide	VO	12035-98-2	66.941	gray-blk cry	1790		5.758		s acid
3038	Vanadium(II) sulfate heptahydrate	$VSO_4 \cdot 7H_2O$	36907-42-3	273.111	viol cry					
3039	Vanadium(III) bromide	VBr_3	13470-26-3	290.654	blk-grn hyg cry	500 dec	subl	4.00		reac H_2O
3040	Vanadium(III) chloride	VCl_3	7718-98-1	157.301	red-viol hex cry; hyg	500 dec		3.00		reac H_2O; s EtOH, eth
3041	Vanadium(III) fluoride	VF_3	10049-12-4	107.937	yel-grn hex cry	1395	subl	3.363		i H_2O, EtOH
3042	Vanadium(III) fluoride trihydrate	$VF_3 \cdot 3H_2O$	10049-12-4	161.983	grn rhomb cry	≈100 dec				sl H_2O
3043	Vanadium(III) iodide	VI_3	15513-94-7	431.655	brn-blk rhomb cry; hyg	300 dec		5.21		reac H_2O
3044	Vanadium(III) oxide	V_2O_3	1314-34-7	149.881	blk powder	1957	≈3000	4.87		i H_2O
3045	Vanadium(III) 2,4-pentanedioate	$V(CH_3COCHCOCH_3)_3$	13476-99-8	348.266	brn cry	≈185	subl	≈1.0		s MeOH, ace, bz chl
3046	Vanadium(III) sulfate	$V_2(SO_4)_3$	13701-70-7	390.071	yel powder	≈400 dec				sl H_2O
3047	Vanadium(III) sulfide	V_2S_3	1315-03-3	198.078	grn-blk powder	dec		4.7		i H_2O; s hot HCl
3048	Vanadium(IV) bromide	VBr_4	13595-30-7	370.558	unstab purp cry	-23 dec				
3049	Vanadium(IV) chloride	VCl_4	7632-51-1	192.754	red-brn liq	-28	151	1.816		reac H_2O; s EtOH, eth
3050	Vanadium(IV) fluoride	VF_4	10049-16-8	126.936	grn hyg powder	325 dec	subl	3.15		vs H_2O
3051	Vanadium(IV) oxide	VO_2	12036-21-4	82.941	blue-blk powder	1545		4.339		i H_2O; s acid, alk
3052	Vanadium(V) fluoride	VF_5	7783-72-4	145.934	col liq	19.5	48.3	2.50		reac H_2O
3053	Vanadium(V) dioxide fluoride	VO_2F	14259-82-6	101.939	brn hyg cry	350 dec				reac H_2O
3054	Vanadium(V) dioxide chloride	VO_2Cl	13759-30-3	118.394	oran hyg cry	180 dec				s thf
3055	Vanadium(V) oxide	V_2O_5	1314-62-1	181.880	yel-brn orth cry	681	1750	3.35	0.07^{25}	s conc acid, alk; i EtOH
3056	Vanadium(V) sulfide	V_2S_5	12138-17-9	262.208	grn-blk pow	dec		3.0		i H_2O; s acid, alk
3057	Vanadyl bromide	VOBr	13520-88-2	146.845	viol cry	480 dec				
3058	Vanadyl chloride	VOCl	13520-87-1	102.394	brn orth cry		127	1.72		
3059	Vanadyl dibromide	$VOBr_2$	13520-89-3	226.749	yel-brn cry	180 dec				
3060	Vanadyl dichloride	$VOCl_2$	10213-09-9	137.847	grn hyg cry	380 dec		2.88		reac H_2O; s EtOH
3061	Vanadyl difluoride	VOF_2	13814-83-0	104.938	yel cry					
3062	Vanadyl selenite hydrate	$VOSeO_3 \cdot H_2O$	133578-89-9	211.92	grn tricl plates			3.506		
3063	Vanadyl sulfate dihydrate	$VOSO_4 \cdot 2H_2O$	27774-13-6	199.035	blue cry powder					s H_2O
3064	Vanadyl tribromide	$VOBr_3$	13520-90-6	306.653	deep red liq	-59	170			reac H_2O
3065	Vanadyl trichloride	$VOCl_3$	7727-18-6	173.300	fuming red-yel liq	-79	127	1.829		reac H_2O; s MeOH, eth, ace
3066	Vanadyl trifluoride	VOF_3	13709-31-4	123.936	yel hyg powder	300	480	2.459		reac H_2O
3067	Water	H_2O	7732-18-5	18.015	col liq	0.00	99.974	0.9970^{25}		vs EtOH, MeOH, ace
3068	Water-d_2	D_2O	7789-20-0	20.027	col liq	3.82	101.42	1.1044^{25}		
3069	Water-t_2	T_2O	14940-65-9	22.032	col liq	4.48	101.51	1.2138^{25}		
3070	Xenon	Xe	7440-63-3	131.293	col gas	-111.75	-108.099	$2.942^{-108.1}$ liq		sl H_2O
3071	Xenon trioxide	XeO_3	13776-58-4	179.291	col orth cry	≈25 exp		4.55		s H_2O
3072	Xenon tetroxide	XeO_4	12340-14-6	195.291	yel solid or col gas; exp	-35.9	≈0 dec			
3073	Xenon difluoride	XeF_2	13709-36-9	169.290	col tetr cry	129.03 tp	114.35 sp	4.32		sl H_2O
3074	Xenon tetrafluoride	XeF_4	13709-61-0	207.287	col monocl cry	117.10 tp	115.75 sp	4.04		reac H_2O
3075	Xenon hexafluoride	XeF_6	13693-09-9	245.283	col monocl cry	49.48	75.6	3.56		reac H_2O
3076	Xenon fluoride oxide	$XeOF_2$	13780-64-8	185.289	yel solid, stab <-25	≈0 exp				reac H_2O
3077	Xenon oxytetrafluoride	$XeOF_4$	13774-85-1	223.286	col liq	-46.2		3.17^{0}		reac H_2O
3078	Xenon dioxydifluoride	XeO_2F_2	13875-06-4	201.289	col orth cry	30.8 exp		4.10		
3079	Xenon difluoride trioxide	XeO_3F_2	15192-14-0	217.288	unstab at r.t.	-54.1	exp			
3080	Xenon pentafluoride hexafluoroarsenate	XeF_5AsF_6	20328-94-3	415.197	wh monocl cry	130.5		3.51		
3081	Xenon pentafluoride hexafluororuthenate	XeF_5RuF_6	39796-98-0	441.35	grn orth cry	152		3.79		
3082	Xenon fluoride hexafluoroantimonate	XeF_5SbF_6	39797-63-2	424.039	yel-grn monocl cry	≈110		3.92		
3083	Xenon fluoride hexafluoroarsenate	$Xe_2F_3AsF_6$	50432-32-1	508.494	yel-grn monocl cry	99		3.62		reac H_2O
3084	Xenon fluoride hexafluororuthenate	$XeFRuF_6$	22527-13-5	365.35	yel-grn monocl cry	110		3.78		
3085	Xenon fluoride undecafluoroantimonate	$XeFSb_2F_{11}$	15364-10-0	602.794	yel monocl cry	63		3.69		
3086	Xenon trifluoride undecafluoroantimonate	$XeF_3Sb_2F_{11}$	35718-37-7	640.791	yel-grn tricl cry	82		3.98		
3087	Ytterbium	Yb	7440-64-4	173.054	silv metal; cub	824	1196	6.90		s dil acid
3088	Ytterbium silicide	$YbSi_2$	12039-89-3	229.21	hex cry			7.54		
3089	Ytterbium(II) bromide	$YbBr_2$	25502-05-0	332.85	yel cry	673				reac H_2O

No.	Name	Formula	CAS Reg. No.	Mol. wt.	Phys. form	mp/°C	bp/°C	ρ/g cm^{-3}	Sol. in g/100 g H_2O	Qualitative solubility
3090	Ytterbium(II) chloride	$YbCl_2$	13874-77-6	243.95	grn cry	721		5.27		reac H_2O
3091	Ytterbium(II) fluoride	YbF_2	15192-18-4	211.04	gray solid	1407				i H_2O
3092	Ytterbium(II) iodide	YbI_2	19357-86-9	426.85	blk cry	772				reac H_2O
3093	Ytterbium(III) acetate tetrahydrate	$Yb(C_2H_3O_2)_3 \cdot 4H_2O$	15280-58-7	422.23	hyg col cry	70 dec		2.09		vs H_2O
3094	Ytterbium(III) bromide	$YbBr_3$	13759-89-2	412.75	col cry	956 dec				s H_2O
3095	Ytterbium(III) chloride	$YbCl_3$	10361-91-8	279.40	wh hyg powder	854				s H_2O
3096	Ytterbium(III) chloride hexahydrate	$YbCl_3 \cdot 6H_2O$	19423-87-1	387.49	grn hyg cry	150 dec		2.57		vs H_2O
3097	Ytterbium(III) fluoride	YbF_3	13760-80-0	230.04	wh cry	1157		8.2		i H_2O
3098	Ytterbium(III) iodide	YbI_3	13813-44-0	553.75	yel cry	700 dec				s H_2O
3099	Ytterbium(III) nitrate	$Yb(NO_3)_3$	13768-67-7	359.06	col hyg solid				239^{25}	s EtOH
3100	Ytterbium(III) oxide	Yb_2O_3	1314-37-0	394.08	col cub cry	2355	4070	9.2		s dil acid
3101	Ytterbium(III) sulfate octahydrate	$Yb_2(SO_4)_3 \cdot 8H_2O$	10034-98-7	778.39	col cry			3.3	38.4^{20}	
3102	Yttrium	Y	7440-65-5	88.906	silv metal; hex	1522	3345	4.47		reac H_2O; s dil acid
3103	Yttrium aluminum oxide	$Y_3Al_5O_{12}$	12005-21-9	593.619	grn cub cry			≈4.5		
3104	Yttrium antimonide	YSb	12186-97-9	210.666	cub cry	2310		5.97		
3105	Yttrium arsenide	YAs	12255-48-0	163.828	cub cry			5.59		
3106	Yttrium boride	YB_6	12008-32-1	153.772	refrac solid	2600		3.72		
3107	Yttrium bromide	YBr_3	13469-98-2	328.618	col hyg cry	904			83.3^{30}	
3108	Yttrium carbide	YC_2	12071-35-1	112.927	refrac solid	≈2400		4.13		
3109	Yttrium carbonate trihydrate	$Y_2(CO_3)_3 \cdot 3H_2O$	5970-44-5	411.885	red-brn powder					i H_2O; s dil acid
3110	Yttrium chloride	YCl_3	10361-92-9	195.265	wh monocl cry; hyg	721	1482	2.61	75.1^{20}	vs H_2O
3111	Yttrium chloride hexahydrate	$YCl_3 \cdot 6H_2O$	10025-94-2	303.356	hyg col cry	100 dec				vs H_2O; s EtOH
3112	Yttrium fluoride	YF_3	13709-49-4	145.901	wh hyg powder	1155		4.0		i H_2O
3113	Yttrium hydroxide	$Y(OH)_3$	16469-22-0	139.928	wh prec or pow	190 dec				i H_2O
3114	Yttrium iodide	YI_3	13470-38-7	469.619	hyg wh-yel cry	997				s H_2O, ace, EtOH
3115	Yttrium iron oxide	$Y_3Fe_5O_{12}$	12063-56-8	737.936	cub cry	1555				
3116	Yttrium nitrate	$Y(NO_3)_3$	10361-93-0	274.921	wh hyg solid				149^{25}	s EtOH
3117	Yttrium nitrate tetrahydrate	$Y(NO_3)_3 \cdot 4H_2O$	13773-69-8	346.982	red-wh prisms			2.68	149^{25}	
3118	Yttrium nitrate hexahydrate	$Y(NO_3)_3 \cdot 6H_2O$	13494-98-9	383.012	hyg cry				149^{25}	
3119	Yttrium oxide	Y_2O_3	1314-36-9	225.810	wh cry; cub	2439		5.03		s dil acid
3120	Yttrium phosphide	YP	12294-01-8	119.880	cub cry			≈4.4		
3121	Yttrium sulfate octahydrate	$Y_2(SO_4)_3 \cdot 8H_2O$	7446-33-5	610.122	red monocl cry			2.6	7.47^{16}	
3122	Yttrium sulfide	Y_2S_3	12039-19-9	274.007	yel cub cry	1925		3.87		
3123	Zinc	Zn	7440-66-6	65.38	blue-wh metal; hex	419.527	907	7.134		s acid, alk
3124	Zinc acetate dihydrate	$Zn(C_2H_3O_2)_2 \cdot 2H_2O$	5970-45-6	219.527	wh powder	237 dec		1.735	30.0^{20}	s EtOH
3125	Zinc ammonium sulfate	$Zn(NH_4)_2(SO_4)_2$	7783-24-6	293.611	wh cry				9.2^{20}	
3126	Zinc antimonide	ZnSb	12039-35-9	187.169	silv-wh orth cry	565		6.33		reac H_2O
3127	Zinc arsenate	$Zn_3(AsO_4)_2$	13464-44-3	474.065	wh powder				0.000078^{20}	s acid, alk
3128	Zinc arsenate octahydrate	$Zn_3(AsO_4)_2 \cdot 8H_2O$	13464-45-4	618.187	wh monocl cry			3.33	0.000078^{20}	s acid, alk
3129	Zinc arsenide	Zn_3As_2	12006-40-5	346.070	powder	1015		5.528		
3130	Zinc arsenite	$Zn(AsO_2)_2$	10326-24-6	279.250	col powder					i H_2O; s acid
3131	Zinc borate	$3ZnO \cdot 2B_2O_3$	27043-84-1	383.466	wh amorp powder			3.64		sl H_2O; s dil acid
3132	Zinc borate hemiheptahydrate	$2ZnO \cdot 3B_2O_3 \cdot 3.5H_2O$	12513-27-8	434.69	wh cry	980		4.22		i H_2O
3133	Zinc borate pentahydrate	$2ZnO \cdot 3B_2O_3 \cdot 5H_2O$	12536-65-1	461.753	wh powder			3.64	0.007^{25}	sl HCl
3134	Zinc bromate hexahydrate	$Zn(BrO_3)_2 \cdot 6H_2O$	13517-27-6	429.305	wh hyg solid	100		2.57		vs H_2O
3135	Zinc bromide	$ZnBr_2$	7699-45-8	225.217	wh hex cry; hyg	402	≈670	4.5	488^{25}	vs EtOH; s eth
3136	Zinc caprylate	$Zn(C_8H_{15}O_2)_2$	557-09-5	351.816	wh hyg cry	136				sl H_2O
3137	Zinc carbonate	$ZnCO_3$	3486-35-9	125.418	wh hex cry	140 dec		4.434	0.000091^{20}	s dil acid, alk
3138	Zinc carbonate hydroxide	$3Zn(OH)_2 \cdot 2ZnCO_3$	12070-69-8	549.107	wh powder					
3139	Zinc chlorate	$Zn(ClO_3)_2$	10361-95-2	232.311	yel hyg cry	60 dec		2.15	200^{20}	
3140	Zinc chloride	$ZnCl_2$	7646-85-7	136.315	wh hyg cry	325	732	2.907	408^{25}	vs H_2O; s EtOH, ace
3141	Zinc chromate	$ZnCrO_4$	13530-65-9	181.403	yel prisms	316		3.40	3.08	s acid; i ace
3142	Zinc chromite	$ZnCr_2O_4$	12018-19-8	233.399	grn cub cry			5.29		
3143	Zinc citrate dihydrate	$Zn_3(C_6H_5O_7)_2 \cdot 2H_2O$	546-46-3	610.456	col powder					sl H_2O; s dil acid, alk
3144	Zinc cyanide	$Zn(CN)_2$	557-21-1	117.443	wh powder			1.852	0.00047^{20}	reac acid
3145	Zinc diethyl	$Zn(C_2H_5)_2$	557-20-0	123.531	col liq	-33.34	133	1.2065^{20}		reac H_2O; msc eth, peth, bz
3146	Zinc dithionate	ZnS_2O_6	7779-86-4	193.537	wh amorp solid	200 dec			40^{20}	
3147	Zinc fluoride	ZnF_2	7783-49-5	103.406	wh tetr needles; hyg	872	1500	4.9	1.55^{25}	sl H_2O
3148	Zinc fluoride tetrahydrate	$ZnF_2 \cdot 4H_2O$	13986-18-0	175.468	wh orth cry			2.30	1.55^{25}	
3149	Zinc fluoroborate hexahydrate	$Zn(BF_4)_2 \cdot 6H_2O$	27860-83-9	347.109	hex cry			2.12		vs H_2O; s EtOH
3150	Zinc formate dihydrate	$Zn(CHO_2)_2 \cdot 2H_2O$	5970-62-7	191.474	wh cry			2.207	5.2^{20}	i EtOH
3151	Zinc hexafluorosilicate hexahydrate	$ZnSiF_6 \cdot 6H_2O$	16871-71-9	315.576	wh cry					s H_2O
3152	Zinc hydroxide	$Zn(OH)_2$	20427-58-1	99.424	col orth cry	125 dec		3.05	0.000042^{20}	
3153	Zinc iodate	$Zn(IO_3)_2$	7790-37-6	415.214	wh cry powder				0.64^{25}	
3154	Zinc iodide	ZnI_2	10139-47-6	319.218	wh-yel hyg cry	450	625	4.74	438^{25}	vs H_2O; s EtOH, eth
3155	Zinc laurate	$Zn(C_{12}H_{23}O_2)_2$	2452-01-9	464.029	wh powder	128				sl H_2O
3156	Zinc molybdate	$ZnMoO_4$	13767-32-3	225.35	wh tetr cry	>700		4.3		i H_2O
3157	Zinc nitrate	$Zn(NO_3)_2$	7779-88-6	189.418	wh powder				120^{25}	
3158	Zinc nitrate hexahydrate	$Zn(NO_3)_2 \cdot 6H_2O$	10196-18-6	297.510	col orth cry	36 dec		2.067	120^{25}	vs EtOH
3159	Zinc nitride	Zn_3N_2	1313-49-1	224.240	blue-gray cub cry	700 dec		6.22		i H_2O
3160	Zinc nitrite	$Zn(NO_2)_2$	10102-02-0	157.420	hyg solid					reac H_2O
3161	Zinc oleate	$Zn(C_{18}H_{33}O_2)_2$	557-07-3	628.316	wh powder	70 dec				i H_2O; s EtOH, eth, bz

No.	Name	Formula	CAS Reg. No.	Mol. wt.	Phys. form	mp/°C	bp/°C	ρ/g cm^{-3}	Sol. in g/100 g H_2O	Qualitative solubility
3162	Zinc oxalate	ZnC_2O_4	547-68-2	153.428	wh pwd				0.0026^{25}	
3163	Zinc oxalate dihydrate	$ZnC_2O_4·2H_2O$	4255-07-6	189.458	wh powder	100 dec		2.56	0.0026^{25}	s dil acid
3164	Zinc tartrate dihydrate	$ZnC_4H_4O_6·2H_2O$	22570-08-7	249.511	wh cry pow	150 dec			0.022^{20}	
3165	Zinc oxide	ZnO	1314-13-2	81.408	wh powder; hex	1974		5.6		i H_2O; s dil acid
3166	Zinc 2,4-pentanedioate	$Zn(CH_3COCHCOCH_3)_2$	14024-63-6	263.625	cry	127.3				sl H_2O; s EtOH, DMSO
3167	Zinc pentanoate dihydrate	$Zn(C_5H_9O_2)_2·2H_2O$	556-38-7	303.687	scales or powder					sl H_2O; reac acid; s EtOH
3168	Zinc perchlorate hexahydrate	$Zn(ClO_4)_2·6H_2O$	10025-64-6	372.402	wh cub cry; hyg	106 dec		2.2	121.3^{25}	s EtOH
3169	Zinc permanganate hexahydrate	$Zn(MnO_4)_2·6H_2O$	23414-72-4	411.372	blk orth cry; hyg			2.45		s H_2O; reac EtOH
3170	Zinc peroxide	ZnO_2	1314-22-3	97.408	yel-wh powder	>150 dec	212 exp	1.57		i H_2O; reac acid, EtOH, ace
3171	Zinc phosphate	$Zn_3(PO_4)_2$	7779-90-0	386.170	wh monocl cry	900		4.0		i H_2O
3172	Zinc phosphate tetrahydrate	$Zn_3(PO_4)_2·4H_2O$	7543-51-3	458.231	col orth cry			3.04		i H_2O, EtOH; s dil acid, alk
3173	Zinc phosphide [Zn_3P_2]	Zn_3P_2	1314-84-7	258.175	gray tetr cry	1160		4.55		i H_2O, EtOH; reac acid; s bz
3174	Zinc pyrophosphate	$Zn_2P_2O_7$	7446-26-6	304.761	wh cry powder			3.75		i H_2O; s dil acid
3175	Zinc selenate pentahydrate	$ZnSeO_4·5H_2O$	13597-54-1	298.44	tricl cry	50 dec		2.59	63.4^{25}	
3176	Zinc selenide	ZnSe	1315-09-9	144.37	yel-red cub cry	>1100	subl	5.65		i H_2O; s dil acid
3177	Zinc orthosilicate	Zn_2SiO_4	13597-65-4	222.902	wh hex cry	1509		4.1		i H_2O, dil acid
3178	Zinc selenite	$ZnSeO_3$	13597-46-1	192.37	wh powder	621				
3179	Zinc stearate	$Zn(C_{18}H_{35}O_2)_2$	557-05-1	632.348	wh powder	130		1.095		i H_2O, EtOH, eth; s bz
3180	Zinc sulfate	$ZnSO_4$	7733-02-0	161.472	col orth cry	680 dec		3.8	57.7^{25}	
3181	Zinc sulfate monohydrate	$ZnSO_4·H_2O$	7446-19-7	179.487	wh monocl cry	238 dec		3.20	57.7^{25}	i EtOH
3182	Zinc sulfate heptahydrate	$ZnSO_4·7H_2O$	7446-20-0	287.578	col orth cry	100 dec		1.97	57.7^{25}	i EtOH
3183	Zinc sulfide (sphalerite)	ZnS	1314-98-3	97.474	gray-wh cub cry	1020 trans wurtzite		4.04		i H_2O, EtOH; s dil acid
3184	Zinc sulfide (wurtzite)	ZnS	1314-98-3	97.474	wh hex cry	1827	subl	4.09		i H_2O; s dil acid
3185	Zinc sulfite dihydrate	$ZnSO_3·2H_2O$	7488-52-0	181.503	wh powder	200 dec			0.224^{25}	i EtOH
3186	Zinc telluride	ZnTe	1315-11-3	193.01	red cub cry	1295		5.9		i H_2O
3187	Zinc thiocyanate	$Zn(SCN)_2$	557-42-6	181.573	wh hyg cry					sl H_2O; s EtOH
3188	Zirconium	Zr	7440-67-7	91.224	gray-wh metal; hex	1854	4406	6.52		s hot conc acid
3189	Zirconocene dichloride	$Zr(C_5H_5)_2Cl_2$	1291-32-3	292.316	col cry	248	150 subl			
3190	Zirconium boride	ZrB_2	12045-64-6	112.846	gray refrac solid; hex	3050		6.17		
3191	Zirconium carbide	ZrC	12020-14-3	103.235	gray refrac solid; cub	3532		6.73		s HF
3192	Zirconium nitride	ZrN	25658-42-8	105.231	yel cub cry	2952		7.09		s conc HF; sl dil acid
3193	Zirconium phosphide	ZrP_2	12037-80-8	153.172	orth cry			≈5.1		
3194	Zirconium silicide	$ZrSi_2$	12039-90-6	147.395	gray powder	1620		4.88		i H_2O, aqua regia; s HF
3195	Zirconium(II) bromide	$ZrBr_2$	24621-17-8	251.032	blue-blk cry	827				
3196	Zirconium(II) chloride	$ZrCl_2$	13762-26-0	162.130	blk cry	722		3.16		reac H_2O
3197	Zirconium(II) fluoride	ZrF_2	13842-94-9	129.221	blk cry	902				
3198	Zirconium(II) hydride	ZrH_2	7704-99-6	93.240	gray tetr cry	800 dec		5.6		i H_2O
3199	Zirconium(II) iodide	ZrI_2	15513-85-6	345.033	blk cry	827				
3200	Zirconium(III) bromide	$ZrBr_3$	24621-18-9	330.936	dark blue cry	727				
3201	Zirconium(III) chloride	$ZrCl_3$	10241-03-9	197.583	dark blue cry	627		3.05		reac H_2O
3202	Zirconium(III) fluoride	ZrF_3	13814-22-7	148.219	blue-grn cry	927		4.26		i H_2O; s acid
3203	Zirconium(III) iodide	ZrI_3	13779-87-8	471.937	dark blue cry	727				
3204	Zirconium(IV) acetate hydroxide	$Zr(C_2H_3O_2)_2(OH)_2$	14311-93-4	243.327	wh amorp solid					s H_2O
3205	Zirconium(IV) ammonium hydroxide carbonate dihydrate	$Zr(NH_4)_3OH(CO_3)_3·2H_2O$	12616-24-9	378.402	prisms; unstab					s H_2O
3206	Zirconium(IV) bromide	$ZrBr_4$	13777-25-8	410.840	wh cub cry	450 tp	360 sp	3.98		reac H_2O
3207	Zirconium(IV) chloride	$ZrCl_4$	10026-11-6	233.036	wh monocl cry; hyg	437 tp	331 sp	2.80		reac H_2O; s EtOH, eth
3208	Zirconium(IV) fluoride	ZrF_4	7783-64-4	167.218	wh monocl cry	910	912 sp	4.43	1.5^{25}	
3209	Zirconium(IV) hydroxide	$Zr(OH)_4$	14475-63-9	159.254	wh amorp powder	dec		3.25		i H_2O; s acid
3210	Zirconium(IV) iodide	ZrI_4	13986-26-0	598.842	yel-oran cub cry	500	431 sp	4.85		vs H_2O
3211	Zirconium(IV) nitrate pentahydrate	$Zr(NO_3)_4·5H_2O$	13746-89-9	429.320	wh hyg cry	100 dec				vs H_2O; s EtOH
3212	Zirconium(IV) orthosilicate	$ZrSiO_4$	10101-52-7	183.308	wh tetr cry	1540 dec		4.6		i H_2O, acid
3213	Zirconium(IV) oxide	ZrO_2	1314-23-4	123.223	wh amorp powder	2710	4300	5.68		i H_2O; sl acid
3214	Zirconium(IV) pyrophosphate	ZrP_2O_7	13565-97-4	265.167	wh refrac solid	1550 dec				i H_2O, dil acid; s HF
3215	Zirconium(IV) sulfate	$Zr(SO_4)_2$	14644-61-2	283.349	wh hyg cry	410 dec		3.22		s H_2O; sl EtOH
3216	Zirconium(IV) sulfate tetrahydrate	$Zr(SO_4)_2·4H_2O$	7446-31-3	355.411	wh tetr cry	100 dec		2.80		vs H_2O
3217	Zirconium(IV) sulfide	ZrS_2	12039-15-5	155.354	red-brn hex cry	1550		3.87		i H_2O
3218	Zirconium(IV) tungstate	$Zr(WO_4)_2$	16853-74-0	586.90	grn pow					
3219	Zirconyl chloride	$ZrOCl_2$	7699-43-6	178.129	wh solid	250 dec				s H_2O, EtOH
3220	Zirconyl chloride octahydrate	$ZrOCl_2·8H_2O$	13520-92-8	322.252	tetr cry	400 dec		1.91		vs H_2O, EtOH

[a] This is the estimated diamond-graphite-liquid carbon triple point (Ref. 12).

FORMULA INDEX OF INORGANIC COMPOUNDS

Inorganic

Inorganic

Inorganic

Inorganic

T_2	1243
T_2O	3069
Ta	2758
$TaAl_3$	2759
TaB	2760
TaB_2	2761
$TaBr_3$	2767
$TaBr_4$	2769
$TaBr_5$	2776
TaC	2762
$TaCl_3$	2768
$TaCl_4$	2770
$TaCl_5$	2777
TaF_5	2778
TaH	2764
TaI_4	2771
TaI_5	2779
TaN	2765
TaO_2	2772
TaS_2	2774
$TaSe_2$	2773
$TaSi_2$	2766
$TaTe_2$	2775
Ta_2C	2763
Ta_2O_5	2780
Tb	2797
$TbBr_3$	2800
$TbCl_3$	2801
$TbCl_3 \cdot 6H_2O$	2802
TbF_3	2809
TbF_4	2810
TbI_3	2803
TbN	2798
$Tb(NO_3)_3$	2804
$Tb(NO_3)_3 \cdot 6H_2O$	2805
$TbSi_2$	2799
Tb_2O_3	2806
Tb_2S_3	2808
$Tb_2(SO_4)_3 \cdot 8H_2O$	2807
Tc	2781
TcF_5	2782
TcF_6	2783
Te	2784
$TeBr_2$	2789
$TeBr_4$	2791
$TeCl_2$	2790
$TeCl_4$	2792
TeF_4	2793
TeF_6	2796
TeI_4	2795
TeO_2	2787
TeO_3	2788
Te_2F_{10}	2794
Th	2847
ThB_6	2849
$ThBr_4$	2850
ThC	2851
ThC_2	2852
$ThCl_4$	2853
ThF_4	2854
ThH_2	2848
ThI_4	2855
ThN	2858
$Th(NO_3)_4$	2856
$Th(NO_3)_4 \cdot 4H_2O$	2857
ThO_2	2859
ThS_2	2864
$Th(SO_4)_2 \cdot 9H_2O$	2863
$ThSe_2$	2860
$ThSiO_4$	2861
$ThSi_2$	2862
Ti	2912
TiI_3	2928
TiB_2	2915
$TiBr_2$	2920
$TiBr_3$	2925
$TiBr_4$	2933
TiC	2916
$Ti(C_5H_5)_2Cl_2$	2913
$TiCl_2$	2921
$TiCl_3$	2926
$TiCl_4$	2934
TiF_3	2927
TiF_4	2935
TiH_2	2914
TiI_2	2922
TiI_4	2936
TiN	2917
TiO	2923
$TiOSO_4 \cdot H_2O$	2940
TiO_2	2937
TiO_2	2938
TiO_2	2939
TiP	2918
TiS	2924
TiS_2	2942
$Ti(SO_4)_2$	2941
$TiSi_2$	2919
Ti_2O_3	2929
Ti_2S_3	2931
$Ti_2(SO_4)_3$	2930
Ti_3O_5	2932
Tl	2811
$TlBr$	2815
$TlBrO_3$	2814
$TlBr_3 \cdot 4H_2O$	2839
$TlCHO_2$	2823
$TlCN$	2820
$TlC_2H_3O_2$	2812
TlC_2H_5O	2821
$Tl(C_2H_3O_2)_3$	2838
$TlCl$	2818
$TlClO_3$	2817
$TlClO_4$	2833
$TlCl_3$	2840
$TlCl_3 \cdot 4H_2O$	2841
TlF	2822
TlF_3	2842
TlI	2827
$TlIO_3$	2826
$TlNO_2$	2830
$TlNO_3$	2829
TlN_3	2813
$Tl(NO_3)_3$	2843
$TlOH$	2825
$TlPF_6$	2824
$TlSe$	2846
Tl_2CO_3	2816
$Tl_2C_2O_4$	2831
Tl_2CrO_4	2819
Tl_2MoO_4	2828
Tl_2O	2832
Tl_2O_3	2844
Tl_2S	2837
Tl_2SO_4	2836
$Tl_2(SO_4)_3$	2845
Tl_2Se	2835
Tl_2SeO_4	2834
Tm	2865
$TmBr_2$	2866
$TmBr_3$	2869
$TmCl_2$	2867
$TmCl_3$	2870
$TmCl_3 \cdot 7H_2O$	2871
TmF_3	2872
TmI_2	2868
TmI_3	2874
$Tm(NO_3)_3$	2875
$Tm(NO_3)_3 \cdot 5H_2O$	2876
$Tm(OH)_3$	2873
$Tm_2(C_2O_4)_3 \cdot 6H_2O$	2877
Tm_2O_3	2878
U	2986
UB_2	2987
UB_4	2988
UBr_3	2994
UBr_4	2999
UBr_5	3005
UC	2989
UC_2	2990
UCl_3	2995
UCl_4	3000
UCl_5	3006
UCl_6	3009
UF_3	2996
UF_4	3001
UF_5	3007
UF_6	3010
UH_3	2997
UI_3	2998
UI_4	3002
UN	2992
UO_2	3003
$UO_2(C_2H_3O_2)_2 \cdot 2H_2O$	3014
UO_2Cl_2	3015
UO_2F_2	3016
$UO_2HPO_4 \cdot 4H_2O$	3017
$UO_2(NO_3)_2$	3018
$UO_2(NO_3)_2 \cdot 6H_2O$	3019
$UO_2(NH_4)_3F_5$	204
UO_2SO_4	3020
$UO_2SO_4 \cdot 3H_2O$	3021
UO_3	3011
$UO_3 \cdot H_2O$	3012
$UO_4 \cdot 2H_2O$	3013
U_2C_3	2991
U_2N_3	2993
U_3O_8	3008
U_4O_9	3004
V	3022
VB	3025
VB_2	3026
VBr_2	3033
VBr_3	3039
VBr_4	3048
VC	3027
$V(CO)_6$	3029
$V(C_5H_5)_2$	3023
$V(C_5H_5)_2Cl_2$	3024
$V(CH_3COCHCOCH_3)_3$	3045
VCl_2	3034
VCl_3	3040
VCl_4	3049
VF_2	3035
VF_3	3041
$VF_3 \cdot 3H_2O$	3042
VF_4	3050
VF_5	3052
VI_2	3036
VI_3	3043
VN	3030
VO	3037
$VOBr$	3057
$VOBr_2$	3059
$VOBr_3$	3064
$VOCl$	3058
$VOCl_2$	3060
$VOCl_3$	3065
VOF_2	3061
VOF_3	3066
$VOSO_4 \cdot 2H_2O$	3063
$VOSeO_3 \cdot H_2O$	3062
VO_2	3051
VO_2Cl	3054
VO_2F	3053
$VSO_4 \cdot 7H_2O$	3038
VSi_2	3031
V_2C	3028
V_2O_3	3044
V_2O_5	3055
V_2S_3	3047
$V_2(SO_4)_3$	3046
V_2S_5	3056
V_3Si	3032
W	2943
WB	2946
WBr_2	2955
WBr_3	2958
WBr_4	2961
WBr_5	2969
WBr_6	2975
WC	2949
$W(CO)_6$	2950
$W(C_2H_5O)_5$	2971
WCl_2	2956
WCl_3	2959
WCl_4	2962
WCl_5	2970
WCl_6	2976
WF_4	2963
WF_5	2972
WF_6	2980
WI_2	2957
WI_3	2960
WI_4	2964
WN_2	2951

PHYSICAL PROPERTIES OF THE RARE EARTH METALS

K. A. Gschneidner

Table 1. Data for the Trivalent Ions of the Rare Earth Elements

Rare earth	Symbol	Atomic no.	Atomic wt.[a]	No. 4f electrons in R^{3+}ion	S^b	L^b	J^b	Spectroscopic ground state
Scandium	Sc	21	44.955908	0				
Yttrium	Y	39	88.90584	0				
Lanthanum	La	57	138.90547	0				
Cerium	Ce	58	140.116	1	1/2	3	5/2	$^2F_{5/2}$
Praseodymium	Pr	59	140.90766	2	1	5	4	$^3H_4$
Neodymium	Nd	60	144.242	3	3/2	6	9/2	$^4I_{9/2}$
Promethium	Pm	61	145[c]	4	2	6	4	$^5I_4$
Samarium	Sm	62	150.36	5	5/2	5	5/2	$^6H_{5/2}$
Europium	Eu	63	151.964	6	3	3	0	$^7F_0$
Gadolinium	Gd	64	157.25	7	7/2	0	7/2	$^8S_{7/2}$
Terbium	Tb	65	158.92535	8	3	3	6	$^7F_6$
Dysprosium	Dy	66	162.500	9	5/2	5	15/2	$^6H_{15/2}$
Holmium	Ho	67	164.93033	10	2	6	8	$^5I_8$
Erbium	Er	68	167.259	11	3/2	6	15/2	$^4I_{15/2}$
Thulium	Tm	69	168.93422	12	1	5	6	$^3H_6$
Ytterbium	Yb	70	173.045	13	1/2	3	7/2	$^2F_{7/2}$
Lutetium	Lu	71	174.9668	14				

[a] 2013 standard atomic weights
[b] In R^{3+} ion
[c] Mass number of longest-lived isotope

Note: For additional information, see Goldschmidt, Z.B., in *Handbook on the Physics and Chemistry of Rare Earths,* Vol. 1, Gschneidner, K.A., Jr. and Eyring, L., Eds., North-Holland Physics, Amsterdam, 1978; DeLaeter, J.R., and Heumann, K.G., *J. Phys. Chem. Ref. Data*, 20, 1313 , 1991; *Pure Appl. Chem.*, 66, 2423, 1994.

Table 2. Crystallographic Data for the Rare Earth Metals at 24 °C (297 K) or Below

Rare earth metal	Crystal structure[a]	a_o/Å	b_o/Å	c_o/Å	Metallic radius CN = 12 (Å)	Atomic volume (cm^3/mol)	Density (g/cm^3)
αSc	hcp	3.3088		5.2680	1.6406	15.039	2.989
αY	hcp	3.6482		5.7318	1.8012	19.893	4.469
αLa	dhcp	3.7740		12.171	1.8791	22.602	6.146
αCe[b]	fcc	4.85[b]			1.72[b]	17.2[b]	8.16[b]
βCe	dhcp	3.6810		11.857	1.8321	20.947	6.689
γCe[c]	fcc	5.1610			1.8247	20.696	6.770
αPr	dhcp	3.6721		11.8326	1.8279	20.803	6.773
αNd	dhcp	3.6582		11.7966	1.8214	20.583	7.008
αPm	dhcp	3.65		11.65	1.811	20.24	7.264
αSm	rhomb[d]	3.6290[d]		26.207	1.8041	20.000	7.520
Eu	bcc	4.5827			2.0418	28.979	5.244
αGd	hcp	3.6336		5.7810	1.8013	19.903	7.901
α′Tb[e]	ortho	3.605[e]	6.244[e]	5.706[e]	1.784[e]	19.34[e]	8.219[e]
αTb	hcp	3.6055		5.6966	1.7833	19.310	8.230
α′Dy[f]	ortho	3.595[f]	6.184[f]	5.678[f]	1.774[f]	19.00[f]	8.551[f]
αDy	hcp	3.5915		5.6501	1.7740	19.004	8.551
Ho	hcp	3.5778		5.6178	1.7661	18.752	8.795
Er	hcp	3.5592		5.5850	1.7566	18.449	9.066
Tm	hcp	3.5375		5.5540	1.7462	18.124	9.321
αYb[g]	hcp	3.8799[g]		6.3859[g]	1.9451[g]	25.067[g]	6.903[g]
βYb	fcc	5.4848			1.9392	24.841	6.966

Rare earth metal	Crystal structure[a]	a_o/Å	b_o/Å	c_o/Å	Metallic radius CN = 12 (Å)	Atomic volume (cm³/mol)	Density (g/cm³)
Lu	hcp	3.5052		5.5494	1.7349	17.779	9.841

[a] hcp = hexagonal close-packed; $P6_3/mmc$, hP2, A3, Mg-type; dhcp = double-c hexagonal close-packed; $P6_3/mmc$, hP4, A3′, αLa-type; fcc = face-centered cubic; $Fm\bar{3}m$, cF4, A1, Cu-type; rhomb = rhombohedral; $R\bar{3}m$, hR3, αSm-type; bcc = body-centered cubic; $Im\bar{3}m$, cI2, A2, W-type; ortho = orthorhombic; Cmcm, oC4, α′ Dy-type

[b] At 77 K (−196 °C)

[c] Equilibrium room temperature (standard state) phase

[d] Rhombohedral is the primitive cell. Lattice parameters given are for the nonprimitive hexagonal cell.

[e] At 220 K (−53 °C)

[f] At 86 K (−187 °C)

[g] At 23 °C

Note: For additional information, see Gschneidner, K.A., Jr. and Calderwood, F.W., in *Handbook on the Physics and Chemistry of Rare Earths,* Vol. 8, Gschneidner, K.A., Jr. and Eyring, L., Eds., North-Holland Physics, Amsterdam, 1986; Gschneidner, K.A., Jr., Pecharsky, V.K., Cho, Jaephil and Martin, S.W., *Scripta Mater*., 34, 1717, 1996.

Table 3. Crystallographic Data for Rare Earth Metals at High Temperature

Rare earth metal	Crystal structure	Lattice parameter (Å)	Temp. (°C)	Metallic radius CN = 8 (Å)	Metallic radius CN = 12 (Å)	Atomic volume (cm³/mol)	Density (g/cm³)
βSc	bcc	3.73[d]	1337	1.62	1.66	15.6	2.88
βY	bcc	4.10[a]	1478	1.78	1.83	20.8	4.28
βLa	fcc	5.303	325		1.875	22.45	6.187
γLa	bcc	4.26	887	1.84	1.90	23.3	5.97
δCe	bcc	4.12	757	1.78	1.84	21.1	6.65
βPr	bcc	4.13	821	1.79	1.84	21.2	6.64
βNd	bcc	4.13	883	1.79	1.84	21.2	6.80
βPm	bcc	4.10[d]	890	1.78	1.83	20.8	6.99
βSm	hcp	[e]	450[b]		1.8176	20.450	7.353
γSm	bcc	4.10[d]	922	1.77	1.82	20.8	7.25
βGd	bcc	4.06	1265	1.76	1.81	20.2	7.80
βTb	bcc	4.07[a]	1289	1.76	1.81	20.3	7.82
βDy	bcc	4.03[a]	1381	1.75	1.80	19.7	8.23
γYb	bcc	4.44	763[c]	1.92	1.98	26.4	6.57

[a] Determined by extrapolation to 0% solute of a vs. composition data for R-Mg alloys at 24 °C and corrected for thermal expansion to temperature given

[b] The hcp phase was stabilized by impurities and the temperature of measurement was below the equilibrium transition temperature (see Table 4)

[c] The bcc phase was stabilized by impurities and the temperature of measurement was below the equilibrium transition temperature (see Table 4)

[d] Estimated

[e] *a* = 3.6630, *c* = 5.8448

Note: The rare earths Eu, Ho, Er, Tm, and Lu are monomorphic. For additional information, see Gschneidner, K.A., Jr. and Calderwood, F.W., in *Handbook on the Physics and Chemistry of Rare Earths*, Vol. 8, Gschneidner, K.A., Jr. and Eyring, L., Eds., North-Holland Physics, Amsterdam, 1986, 1.

Table 4. High Temperature Transition Temperatures and Melting Point of Rare Earth Metals

Rare earth metal	CASRN	Temp. (°C)	Phases (α – β)[a]	Temp. (C°)	Phases (β – γ)[a]	Melting point (°C)
Sc	7440-20-2	1337	hcp ⇌ bcc			1541
Y	7440-65-5	1478	hcp ⇌ bcc			1522
La[b]	7439-91-0	310	dhcp ⇌ fcc	865	fcc ⇌ bcc	920
Ce[c, d]	7440-45-1	139	dhcp ⇌ fcc (β - γ)	726	fcc ⇌ bcc (γ - δ)	799
Pr	7440-10-0	795	dhcp ⇌ bcc			931
Nd	7440-00-8	863	dhcp ⇌ bcc			1016
Pm	7440-12-2	890	dhcp ⇌ bcc			1042
Sm[e]	7440-19-9	734	rhom ⇌ hcp	922	hcp ⇌ bcc	1072
Eu	7440-53-1					822
Gd	7440-54-2	1235	hcp ⇌ bcc			1313
Tb	7440-27-9	1289	hcp ⇌ bcc			1359
Dy	7429-91-6	1381	hcp ⇌ bcc			1412
Ho	7440-60-0					1472
Er	7440-52-0					1529

Rare earth metal	CASRN	Temp. (°C)	Phases (α – β)[a]	Temp. (C°)	Phases (β – γ)[a]	Melting point (°C)
Tm	7440-30-4					1545
Yb	7440-64-4	795	fcc ⇌ bcc (β - γ)			824
Lu	7439-94-3					1663

[a] For all the transformations listed, unless otherwise noted.
[b] On cooling, fcc → dhcp (β → α), 260 °C.
[c] The β ⇌ γ equilibrium transition temperature is 10 ± 5 °C.
[d] On cooling, fcc → dhcp (γ → β), –16 °C.
[e] On cooling, hcp → rhomb (β → α), 727 °C.

Note: For additional information, see Gschneidner, K.A., Jr. and Calderwood, F.W., in *Handbook on the Physics and Chemistry of Rare Earths*, Vol. 8, Gschneidner, K.A., Jr. and Eyring, L., Eds., North-Holland Physics, Amsterdam, 1986; Gschneidner, K.A., Jr., Pecharsky, V.K., Cho, Jaephil and Martin, S.W., *Scripta Mater.*, 34, 1717, 1996.

Table 5. Low Temperature Transition Temperatures of the Rare Earth Metals

Rare earth metal	Cooling transformation	t/°C	T/K	Heating transformation	t/°C	T/K
Ce	γ → β[a]	-16	257	α → β	-148	125
Ce	γ → α	-172	101	α → β + γ	-104	169
Ce	β → α	-228	45	β → γ[a]	139	412
Tb	α → α′	-53	220			
Dy	α → α′	-187	86			
Yb	β → α	-13	260	α → β	7	280

[a] The β ⇌ γ equilibrium transition temperature is 10 ± 5 °C (283 ± 5K).

Note: For additional information, see Beaudry, B.J. and Gschneidner, K.A., Jr., in *Handbook on the Physics and Chemistry of Rare Earths*, Vol. 1, Gschneidner, K.A., Jr. and Eyring, L., Eds., North-Holland Physics, Amsterdam, 1978, 173; Koskenmaki, D.C. and Gschneidner, K.A., Jr., 1978, in *Handbook on the Physics and Chemistry of Rare Earths*, Vol. 1, Gschneidner, K.A., Jr. and Eyring, L., Eds., North-Holland Physics, Amsterdam, 1978, 337; Gschneidner, K.A., Jr., Pecharsky, V.K., Cho, Jaephil and Martin, S.W., *Scripta Mater.*, 34, 1717, 1996.

Table 6. Heat Capacity, Standard Entropy, and Heats of Transformation and Fusion of the Rare Earth Metals

Rare earth metal	Heat capacity at 298 K (J/mol K)	Standard entropy $S°_{298}$ (J/mol K)	trans. 1	$\Delta_{tr}H^1$/kJ mol^{-1}	trans. 2	$\Delta_{tr}H^2$/kJ mol^{-1}	Heat of fusion (kJ/mol)
Sc	25.5	34.6	α ⇌ β	4.00			14.1
Y	26.5	44.4	α ⇌ β	4.99			11.4
La	27.1	56.9	α ⇌ β	0.36	β ⇌ γ	3.12	6.20
Ce	26.9	72.0	β ⇌ γ	0.05	γ ⇌ δ	2.99	5.46
Pr	27.2	73.2	α ⇌ β	3.17			6.89
Nd	27.5	71.5	α ⇌ β	3.03			7.14
Pm	27.3[a]	71.6[a]	α ⇌ β	3.0[a]			7.7[a]
Sm	29.5	69.6	α ⇌ β	0.2[a]	β ⇌ γ	3.11	8.62
Eu	27.7	77.8					9.21
Gd	37.0	68.1	α ⇌ β	3.91			10.0
Tb	28.9	73.2	α ⇌ β	5.02			10.79
Dy	27.7	75.6	α ⇌ β	4.16			11.06
Ho	27.2	75.3					17.0[a]
Er	28.1	73.2					19.9
Tm	27.0	74.0					16.8
Yb	26.7	59.9	β ⇌ γ	1.75			7.66
Lu	26.9	51.0					22[a]

[a] Estimated.

Note: For additional information, see Hultgren, R., Desai, P.D., Hawkins, D.T., Gleiser, M., Kelley, K.K., and Wagman, D.D., *Selected Values of the Thermodynamic Properties of the Elements*, ASM International, Metals Park, Ohio, 1973; Wagman, D.D., Evans, W.H., Parker, V.B., Schumm, R.H., Halow, I., Bailey, S.M., Churney, K.L., and Nuttall, R.L., *The NBS Tables of Chemical Thermodynamic Properties*, *J. Phys. Chem. Ref. Data*, Vol. 11, Suppl. 2, 1982; Amitin, E.B., Bessergenev, W.G., Kovalevskaya, Yu. A., and Paukov, I.E., J. Chem. Thermodyn., 15, 181, 1983; Amitin, E.B., Bessergenev, W.G., Kovalevskaya, Yu. A., and Paukov, I.E., *J. Chem. Thermodyn.*, 15, 181, 1983.

Inorganic

Table 7. Vapor Pressures (Temperature at Which Pressure Reaches Indicated Value), Boiling Points, and Heats of Sublimation of Rare Earth Metals

Rare earth metal	10^{-8} atm (0.001 Pa)	10^{-6} atm (0.101 Pa)	10^{-4} atm (10.1Pa)	10^{-2} atm (1013 Pa)	Boiling point[a] (°C)	Heat of sublimation at 25 °C (kJ/mol)
Sc	1036	1243	1533	1999	2836	377.8
Y	1222	1460	1812	2360	3345	424.7
La	1301	1566	1938	2506	3464	431.0
Ce	1290	1554	1926	2487	3443	422.6
Pr	1083	1333	1701	2305	3520	355.6
Nd	955	1175	1500	2029	3074	327.6
Pm					≈3000	348[b]
Sm	508	642	835	1150	1794	206.7
Eu	399	515	685	964	1529	175.3
Gd	1167	1408	1760	2306	3273	397.5
Tb	1124	1354	1698	2237	3230	388.7
Dy	804	988	1252	1685	2567	290.4
Ho	845	1036	1313	1771	2700	300.8
Er	908	1113	1405	1896	2868	317.1
Tm	599	748	964	1300	1950	232.2
Yb	301	400	541	776	1196	152.1
Lu	1241	1483	1832	2387	3402	427.6

[a] International Temperature Scale of 1990 (ITS-90) values
[b] Estimated

Note: For additional information, see Hultgren, R., Desai, P.D., Hawkins, D.T., Gleiser, M., Kelley, K.K., and Wagman, D.D., *Selected Values of the Thermodynamic Properties of the Elements*, ASM International, Metals Park, Ohio, 1973; Beaudry, B.J. and Gschneidner, K.A., Jr., in *Handbook on the Physics and Chemistry of Rare Earths*, Vol. 1, Gschneidner, K.A., Jr. and Eyring, L., Eds., North-Holland Physics, Amsterdam, 1978, 173.

Inorganic

Table 8. Magnetic Properties of the Rare Earth Metals: Effective Magnetic Moment (μ), Néel Temperature (T_N), Curie Temperature (T_C), and Debye Temperature (θ_D)

Rare earth metal	$\chi_A \times 10^6$ at 298 K (emu/mol)	μ_{theory} (paramag.)	μ_{obs} (paramag.)	μ_{theory} (ferromag.)	μ_{obs} (ferromag.)	Easy axis	T_N/K hex sites	T_N/K cubic sites	T_C/K	θ_D/K ∥c	θ_D/K ⊥c	θ_D/K polycryst. or avg.
αSc	295.2											
αY	187.7											
αLa	95.9											
βLa	105											
γCe	2,270	2.54	2.52		2.14			14.4				-50
βCe	2,500	2.54	2.61		2.14		13.7	12.5				-41
αPr	5,530	3.58	3.56	2.7[c]	3.20	**a**	0.03					0
αNd	5,930	3.62	3.45	2.2[c]	3.27	**b**	19.9	7.5		0	5	3.3
αPm		2.68			2.40							
αSm	1,278[d]	0.85	1.74	0.5[c]	0.71	**a**	109	14.0				
Eu	30,900	7.94	8.48	5.9	7.0	<110>		90.4				100
αGd	185,000[e]	7.94	7.98	7.63	7.0	30° to **c**			293.4	317	317	317
αTb	170,000	9.72	9.77				230.0			195	239	224
α′Tb				9.34	9.0	**b**			219.5			
αDy	98,000	10.64	10.83				180.2			121	169	153
α′Dy				10.33	10.0	**a**			90.5[f]			
Ho	72,900	10.60	11.2	10.34	10.0	**b**	132		19.5	73.0	88.0	83.0
Er	48,000	9.58	9.9	9.1	9.0	30° to **c**	85		18.7	61.7	32.5	42.2
Tm	24,700	7.56	7.61	7.14	7.0	**c**	58		32.0	41.0	-17.0	2.3
βYb	67[d]											
Lu	182.9											

[a] $g[J(J+1)]^{1/2}$.
[b] gJ.
[c] At 38 T and 4.2 K.
[d] At 290 K.
[e] At 350 K.
[f] On cooling T_C = 89.6 K and on warming T_C = 91.5 K.

Note: For additional information, see McEwen, K.A., in *Handbook on the Physics and Chemistry of Rare Earths*, Vol. 1, Gschneidner, K.A., Jr. and Eyring, L., Eds., North-Holland Physics, Amsterdam, 1978, 411; Legvold, S., in *Ferromagnetic Materials*, Vol. 1, Wohlfarth, E.P., Ed., North-Holland Physics, Amsterdam, 1980, 183; Pecharsky, V.K., Gschneidner, K.A., Jr. and Fort, D., *Phys. Rev. B*, 47, 5063, 1993; Pecharsky, V.K., Gschneidner, K.A., Jr. and Fort, D., 1996, to be published; Steward, A.M. and Collocott, S.J., *J. Phys.: Condens. Matter*, 1, 677, 1988.

Table 9. Room Temperature Coefficient of Thermal Expansion (α), Thermal Conductivity (k), Electrical Resistivity (ρ), and Hall Coefficient (R)

Rare earth metal	$\alpha_a \times 10^6$/°C^{-1}	$\alpha_c \times 10^6$/°C^{-1}	$\alpha_{poly} \times 10^6$/°C^{-1}	k/W cm^{-1} K^{-1}	ρ_a/μΩ cm	ρ_c/μΩ cm	ρ_{poly}/μΩ cm	$R_a \times 10^{12}$/V cm A^{-1} Oe^{-1}	$R_c \times 10^{12}$/V cm A^{-1} Oe^{-1}	$R_{poly} \times 10^{12}$/V cm A^{-1} Oe^{-1}
αSc	7.6	15.3	10.2	0.158	70.9	26.9	56.2[a]			-0.13
αY	6.0	19.7	10.6	0.172	72.5	35.5	59.6	-0.27	-1.6	
αLa	4.5	27.2	12.1	0.134			61.5			-0.35
βCe							82.8			
γCe	6.3		6.3	0.113			74.4			1.81
αPr	4.5	11.2	6.7	0.125			70.0			0.709
αNd	7.6	13.5	9.6	0.165			64.3			0.971
αPm	9[b]	16[b]	11[b]	0.15[b]			75[b]			
αSm	9.6	19.0	12.7	0.133			94.0			-0.21
Eu	35.0		35.0	0.139[b]			90.0			24.4
αGd	9.1[c]	10.0[c]	9.4[c]	0.105	135.1	121.7	131.0	-10	-54	4.48[d]
αTb	9.3	12.4	10.3	0.111	123.5	101.5	115.0	-1.0	-3.7	
αDy	7.1	15.6	9.9	0.107	111.0	76.6	92.6	-0.3	-3.7	
Ho	7.0	19.5	11.2	0.162	101.5	60.5	81.4	0.2	-3.2	
Er	7.9	20.9	12.2	0.145	94.5	60.3	86.0	0.3	-3.6	
Tm	8.8	22.2	13.3	0.169	88.0	47.2	67.6			-1.8
βYb	26.3		26.3	0.385			25.0			3.77
Lu	4.8	20.0	9.9	0.164	76.6	34.7	58.2	0.45	-2.6	-0.535

[a] Calculated from single crystal values
[b] Estimated
[c] At 100 °C
[d] At 77 °C

Note: For additional information, see Beaudry, B. J. and Gschneidner, K.A., Jr., in *Handbook on the Physics and Chemistry of Rare Earths*, Vol. 1, Gschneidner, K.A., Jr. and Eyring, L., Eds., North-Holland Physics, Amsterdam, 1978, 173; McEwen, K.A., in *Handbook on the Physics and Chemistry of Rare Earths*, Vol. 1, Gschneidner, K.A., Jr. and Eyring, L., Eds., North-Holland Physics, Amsterdam, 1978, 411.

Table 10. Electronic Specific Heat Constant (γ), Electron–Electron (Coulomb) Coupling Constant (μ*), Electron–Phonon Coupling Constant (λ), Debye Temperature at 0 K (θ_D), and Superconducting Transition Temperature

Rare earth metal	γ/mJ mol^{-1} K^{-2}	μ*	λ	θ_D/K from heat capacity	θ_D/K from elastic constants	Superconducting temperature (K)
αSc	10.334	0.16	0.30	345.3		0.050[a]
αY	7.878	0.15	0.30	244.4	258	1.3[b]
αLa	9.45	0.08	0.76	150	154	5.10
βLa	11.5			140		6.00
αCe	12.8			179		0.022[c]
αPr	20		1.07[d]	155[e]	153	
αNd	[f]		0.86[d]	157[e]	163	
αPm				159[e]		
αSm	8.1[g]		0.81[d]	162[e, f]	169	
Eu	[f]			[f]	118	
αGd	4.48		0.30	169	182	
α′Tb	3.71		0.34[d]	169.6	177	
α′Dy	4.9		0.32[d]	192	183	
Ho	2.1		0.30[d]	175[e]	190	
Er	8.7		0.33[d]	176.9	188	
Tm	[f]		0.36[d]	179[e]	200	
αYb	3.30			117.6	118	
βYb	8.36			109		
Lu	8.194	0.14	0.31	183.2	185	0.022[h]

[a] At 18.6 GPa.
[b] At 11 GPa.
[c] At 2.2 GPa.
[d] Calculated value.
[e] Estimated.
[f] Heat capacity results have been reported, but the resultant γ and θ_D values are unreliable because of the presence of impurities and/or there was no reliable procedure or model to correct for the magn etic contribution to the heat capacity.
[g] ±1.5; based on the values reported for the purer Sm sample (IV).
[h] At 4.5 GPa.

Note: For additional information, see Sundström, L.J., in *Handbook on the Physics and Chemistry of Rare Earths*, Vol. 1, Gschneidner, K.A., Jr., and Eyring, L., Eds., North-Holland Physics, Amsterdam, 1978, 379; Scott, T., in *Handbook on the Physics and Chemistry of Rare Earths*, Vol. 1, Gschneidner, K.A., Jr. and Eyring, L., Eds., North-Holland Physics, Amsterdam, 1978, 591; Probst, C. and Wittig, J., in *Handbook on the Physics and Chemistry of Rare Earths*, Vol. 1, Gschneidner, K.A., Jr. and Eyring, L., Eds., North-Holland Physics, Amsterdam, 1978, 749; Tsang, T.-W.E., Gschneidner, K.A., Jr., Schmidt, F.A., and Thome, D.K., *Phys. Rev., B*, 31, 235, 1985; Collocott, S.J., Hill, R.W. and Stewart, A.M., *J. Phys. F*, 18, L223, 1988; Hill, R.W. and Gschneidner, K.A., Jr., *J. Phys. F*, 18, 2545, 1988; Skriver, H.L. and Mertig, I., *Phys. Rev. B*, 41, 6553, 1990; Collocott, S.J. and Stewart, A.M., *J. Phys.: Condens. Matter*, 4, 6743, 1992; Pecharsky, V.K., Gschneidner, K.A., Jr. and Fort, D., *Phys. Rev. B*, 47, 5063, 1993.

Table 11. Room Temperature Elastic Moduli and Mechanical Properties

Rare earth metal	Young's (elastic) modulus	Shear modulus	Bulk modulus	Poisson's ratio	Yield strength 0.2% offset	Ultimate tensile strength	Uniform elongation (%)	Reduction in area (%)	Recryst. temp. (°C)
Sc	74.4	29.1	56.6	0.279	173[a]	255[a]	5.0[a]	8.0[a]	550
Y	63.5	25.6	41.2	0.243	42	129	34.0		550
αLa	36.6	14.3	27.9	0.280	126[a]	130	7.9[a]		300
βCe					86	138		24.0	
γCe	33.6	13.5	21.5	0.24	28	117	22.0	30.0	325
αPr	37.3	14.8	28.8	0.281	73	147	15.4	67.0	400
αNd	41.4	16.3	31.8	0.281	71	164	25.0	72.0	400
αPm	46[b]	18[b]	33[b]	0.28[b]					400[b]
αSm	49.7	19.5	37.8	0.274	68	156	17.0	29.5	440
Eu	18.2	7.9	8.3	0.152					300
αGd	54.8	21.8	37.9	0.259	15	118	37.0	56.0	500
αTb	55.7	22.1	38.7	0.261					500
αDy	61.4	24.7	40.5	0.247	43	139	30.0	30.0	550
Ho	64.8	26.3	40.2	0.231					520
Er	69.9	28.3	44.4	0.237	60	136	11.5	11.9	520
Tm	74.0	30.5	44.5	0.213					600
βYb	23.9	9.9	30.5	0.207	7	58	43.0	92.0	300
Lu	68.6	27.2	47.6	0.261					600

[a] Value is questionable.
[b] Estimated.

Note: For additional information, see Scott, T., in *Handbook on the Physics and Chemistry of Rare Earths*, Vol. 1, Gschneidner, K.A., Jr. and Eyring, L., Eds., North Holland Physics, Amsterdam, 1978, 591.

Table 12. Liquid Metal Properties near the Melting Point

Rare earth metal	Density (g/cm³)	Surface tension (N/m)	Viscosity (centipoise)	Heat capacity (J/mol K)	Thermal conductivity (W/cm K)	Magnetic susceptibility $\chi \times 10^4$ (emu/mol)	Electrical resistivity (μΩ·cm)	ΔV(l→s)[a] (%)	Spectral emittance[c] (%)	Temp. range (°C)
Sc	2.80	0.954		44.2[b]						
Y	4.24	0.871		43.1					36.8	1522–1647
La	5.96	0.718	2.65	34.3	0.238	1.20	133	-0.6	25.4	920–1287
Ce	6.68	0.706	3.20	37.7	0.210	9.37	130	1.1	32.2	877–1547
Pr	6.59	0.707	2.85	43.0	0.251	17.3	139	-0.02	28.4	931–1537
Nd	6.72	0.687		48.8	0.195	18.7	151	-0.9	39.4	1021–1567
Pm	6.9[b]	0.680[b]		50[b]			160[b]			
Sm	7.16	0.431		50.2[b]		18.3	182	-3.6	43.7	1075
Eu	4.87	0.264		38.1		97	242	-4.8		
Gd	7.4	0.664		37.2	0.149	67	195	-2.0	34.2	1313–1600
Tb	7.65	0.669		46.5		82	193	-3.1		
Dy	8.2	0.648		49.9	0.187	95	210	-4.5	29.7	1412–1437
Ho	8.34	0.650		43.9		88	221	-7.4		
Er	8.6	0.637		38.7		69	226	-9.0	37.2	1529–1587
Tm	9.0[b]			41.4		41	235[b]	-6.9		
Yb	6.21	0.320	2.67	36.8			113	-5.1		
Lu	9.3	0.940		47.9[b]			224	-3.6		

[a] Volume change on freezing
[b] Estimated
[c] At 645 nm wavelength

Note: For additional information, see Van Zytveld, J., in *Handbook on the Physics and Chemistry of Rare Earths*, Vol. 12, Gschneidner, K.A., Jr. and Eyring, L., Eds., North- Holland Physics, Amsterdam, 1989, 357; Stretz, L.A. and Bautista, R.G., in *Temperature, Its Measurement and Control in Science and Industry*, Vol. 4, part I, H.H. Plumb, Ed., Instrument Society of America, Pittsburgh, 1972, 489; King, T.S., Baria, D.N., and Bautista, R.G., *Met. Trans. B*, 7, 411, 1976; Baria, D.N., King, T.S., and Bautista, R.G., *Met. Trans. B*, 7, 577, 1976.

Table 13. Ionization Potentials (eV)

Rare earth	I Neutral atom	II Singly ionized	III Doubly ionized	IV Triply ionized	V Quadruply ionized
Sc	6.56144	12.79967	24.75666	73.4894	91.65
Y	6.217	12.24	20.52	60.597	77.0
La	5.5770	11.060	19.1773	49.95	61.6
Ce	5.5387	10.85	20.198	36.758	65.55
Pr	5.464	10.55	21.624	38.98	57.53
Nd	5.5250	10.73	22.1	40.41	
Pm	5.554	10.90	22.3	41.1	
Sm	5.6437	11.07	23.4	41.4	
Eu	5.6704	11.241	24.92	42.7	
Gd	6.1500	12.09	20.63	44.0	
Tb	5.8639	11.52	21.91	39.79	
Dy	5.9389	11.67	22.8	41.47	
Ho	6.0216	11.80	22.84	42.5	
Er	6.1078	11.93	22.74	42.7	
Tm	6.18431	12.05	23.68	42.7	
Yb	6.25416	12.1761	25.05	43.56	
Lu	5.42585	13.9	20.9594	45.25	66.8

Note: For references, see the table "Ionization Energies of Atoms and Atomic Ions" in Section 10.

Table 14. Effective Ionic Radii (Å)

Rare earth ion	R^{2+} (CN = 6)	R^{2+} (CN = 8)	R^{3+} (CN = 6)	R^{3+} (CN = 8)	R^{3+} (CN = 12)	R^{4+} (CN = 6)	R^{4+} (CN = 8)
Sc			0.745	0.87	1.116		
Y			0.900	1.015	1.220		
La			1.045	1.18	1.320		
Ce			1.010	1.14	1.290	0.80	0.97
Pr			0.997	1.14	1.286	0.78	0.96
Nd			0.983	1.12	1.276		
Pm			0.97	1.10	1.267		
Sm	1.19	1.27	0.958	1.09	1.260		
Eu	1.17	1.25	0.947	1.07	1.252		
Gd			0.938	1.06	1.246		
Tb			0.923	1.04	1.236	0.76	0.88
Dy			0.912	1.03	1.228		
Ho			0.901	1.02	1.221		
Er			0.890	1.00	1.214		
Tm			0.880	0.99	1.207		
Yb	1.00	1.07	0.868	0.98	1.199		
Lu			0.861	0.97	1.194		

Note: Radius of O^{2-} is 1.40 Å; for a coordination number (CN) of 6.

Note: For additional information, see Shannon, R.D. and Prewitt, C.T., *Acta Cryst.*, 25, 925, 1969 and Shannon, R.D. and Prewitt, C.T., *Acta Cryst.*, 26, 1046, 1970.

MELTING, BOILING, TRIPLE, AND CRITICAL POINTS OF THE ELEMENTS

This table summarizes the significant points on the phase diagrams for the elements. When reliable date are available, values are given for the solid-liquid-gas triple-point temperature t_{tp} and pressure p_{tp}; normal melting point at 101.325 kPa pressure t_m; normal boiling point t_b; and critical temperature t_c and pressure p_c. All temperatures are on the ITS-90 scale.

An "sp" notation in the boiling point column indicates a sublimation point, where the vapor pressure of the solid phase reaches 101.325 kPa (1 atm). Transition temperatures between allotropic forms are included for several elements. The notation "trans" in the melting point column indicates the temperature of the transition to the crystalline form immediately below that entry. An asterisk* indicates an extrapolated or estimated value.

The major data sources are listed below.

References

1. The International Temperature Scale of 1990, *Metrologia* 27, 3, 1990; errata in *Metrologia* 27, 107, 1990.
2. Bedford, R. E., Bonnier, G., Maas, H., and Pavese, F., *Metrologia* 33, 133, 1996.
3. Lemmon, E. W., Huber, M. L., and McLinden, M. O., NIST Standard Reference Database 23: Reference Fluid Thermodynamic and Transport Properties-REFPROP, Version 9.0, National Institute of Standards and Technology, Standard Reference Data Program, Gaithersburg, MD, 2007 (http://www.nist.gov/nist23.cfm). [Ar, F, H, He, Kr, N, Ne, O, Xe]
4. Dinsdale, A.T., SGTE Data for Pure Elements, *CALPHAD* 15, 317-425, 1991.
5. *Landolt-Börnstein, Numerical Data and Functional Relationships in Science and Technology, New Series,* IV/19A, "Thermodynamic Properties of Inorganic Materials compiled by SGTE," Springer-Verlag, Heidelberg; Part 1, 1999.
6. Chase, M. W., *NIST-JANAF Thermochemical Tables, Fourth Edition, J. Phys. Chem. Ref. Data,* Monograph No. 9, 1998.
7. Gurvich, L. V., Veyts, I. V., and Alcock, C. B., *Thermodynamic Properties of Individual Substances, Fourth Edition,* Hemisphere Publishing Corp., New York, 1989.
8. Greenwood, N. N., and Earnshaw, A., *Chemistry of the Elements, Second Edition,* Butterworth-Heinemann, Oxford, 1997.
9. Hultgren, R. R., *Selected Values of Thermodynamic Properties of the Elements,* American Society of Metals, 1973.
10. Geiger, F., Busse, C. A., and Loehrke, R. I., *Int. J. Thermophys.* 8, 425, 1987. [Boiling points of In, Ag, Ga, Cu, Sn, and Au]
11. Velasco, S., Roman, F. L., White, J. A., and Mulero, A., *Fluid Phase Equilib.* 244, 11, 2006. [Critical temperatures of Ag, Be, Bi, Ge, Fe, Mn, Te]
12. Gustafson, P., *Carbon* 24, 169, 1986. [Graphite and diamond]
13. Michels, A., and Prins, C., *Physica* 28, 101, 1962. [Triple point of Ar, Kr, Xe]

Element	Formula	t_{tp}/°C	p_{tp}/kPa	t_m/°C	t_b/°C	t_c/°C	p_c/MPa
Actinium	Ac			1050	≈3200		
Aluminum	Al			660.323	2519	6427 *	
Americium	Am			1176	≈2011		
Antimony (gray)	Sb			630.628	1587		
Argon	Ar	-189.3442	68.89	-189.34	-185.848	-122.463	4.863
Arsenic (gray)	As	817	3700	817	616 sp	1400	22.3
Astatine	At			302			
Barium	Ba			727	≈1845		
Berkelium (β form)	Bk			986			
Beryllium	Be			1287	2468	4932 *	
Bismuth	Bi			271.402	1564	4347 *	
Boron	B			2077	4000		
Bromine	Br_2	-7.25	5.879	-7.2	58.8	315	10.34
Cadmium	Cd			321.069	767		
Calcium	Ca			842	1484		
Californium	Cf			900			
Carbon (graphite)	C	4489	10300	4489 tp (10.3 MPa)	3825 sp		
Carbon (diamond)	C			4440 (12.4 GPa)**			
Cerium	Ce			799	3443		
Cesium	Cs			28.5	671	1665	9.4
Chlorine	Cl_2			-101.5	-34.04	143.9	7.991
Chromium	Cr			1907	2671		
Cobalt	Co			1495	2927		
Copper	Cu			1084.62	2560		
Curium	Cm			1345			
Dysprosium	Dy			1412	2567		
Einsteinium	Es			860			
Erbium	Er			1529	2868		
Europium	Eu			822	1529		
Fermium	Fm			1527			
Fluorine	F_2	-219.67	90	-219.67	-188.11	-128.74	5.1724
Francium	Fr			≈21			
Gadolinium	Gd			1313	3273		

Element	Formula	t_{tp}/°C	p_{tp}/kPa	t_m/°C	t_b/°C	t_c/°C	p_c/MPa
Gallium	Ga	29.7666		29.7646	2229		
Germanium	Ge			938.25	2833	9529 *	
Gold	Au			1064.18	2836		
Hafnium	Hf			2233	4600		
Helium	He	-270.973	5.043		-268.928	-267.9550	0.22746
Holmium	Ho			1472	2700		
Hydrogen	H_2	-259.3467	7.041	-259.16	-252.879	-240.212	1.2858
Indium	In	156.5936		156.5985	2027		
Iodine	I_2	113.6	12.11	113.7	184.4	546	
Iridium	Ir			2446	4428		
Iron	Fe			1538	2861	9067 *	
Krypton	Kr	-157.375	73.53	-157.37	-153.415	-63.67	5.525
Lanthanum	La			920	3464		
Lawrencium	Lr			1627			
Lead	Pb			327.462	1749		
Lithium	Li			180.50	1342	2950 *	67 *
Lutetium	Lu			1663	3402		
Magnesium	Mg			650	1090		
Manganese	Mn			1246	2061	4052 *	
Mendelevium	Md			827			
Mercury	Hg	-38.8344		-38.8290	356.619	1491	167
Molybdenum	Mo			2622	4639		
Neodymium	Nd			1016	3074		
Neon	Ne	-248.5939	43.37	-248.59	-246.046	-228.6580	2.6786
Neptunium	Np			644	≈3902		
Nickel	Ni			1455	2913		
Niobium	Nb			2477	4741		
Nitrogen	N_2	-209.999	12.52	-210.0	-195.795	-146.958	3.3958
Nobelium	No			827			
Osmium	Os			3033	5008		
Oxygen	O_2	-218.7916	0.1463	-218.79	-182.962	-118.569	5.0430
Palladium	Pd			1554.8	2963		
Phosphorus (white)	P			44.15	280.5	721	
Phosphorus (red)	P			579.2	431 sp	721	
Platinum	Pt			1768.2	3825		
Plutonium	Pu			640	3228		
Polonium	Po			254	962		
Potassium	K			63.5	759	1950 *	16 *
Praseodymium	Pr			931	3520		
Promethium	Pm			1042	≈3000		
Protactinium	Pa			1572			
Radium	Ra			696			
Radon	Rn			-71	-61.7	104	6.28
Rhenium	Re			3185	5590		
Rhodium	Rh			1963	3695		
Rubidium	Rb	39.26		39.30	688	1820 *	16 *
Ruthenium	Ru			2333	4147		
Samarium	Sm			1072	1794		
Scandium	Sc			1541	2836		
Selenium (vitreous)	Se			180 trans gray Se	685	1493	
Selenium (gray)	Se			220.8	685	1493	27.2
Silicon	Si			1414	3265		
Silver	Ag			961.78	2162	6137 *	
Sodium	Na			97.794	882.940	2300 *	35 *
Strontium	Sr			777	1377		
Sulfur (rhombic)	S			95.2 trans monocl	444.61	1041	20.7
Sulfur (monoclinic)	S			115.21	444.61	1041	
Tantalum	Ta			3017	5455		
Technetium	Tc			2157	4262		
Tellurium	Te			449.51	988	2056 *	
Terbium	Tb			1359	3230		

Element	Formula	t_{tp}/°C	p_{tp}/kPa	t_m/°C	t_b/°C	t_c/°C	p_c/MPa
Thallium	Tl			304	1473		
Thorium	Th			1750	4785		
Thulium	Tm			1545	1950		
Tin (gray)	Sn			13.2 trans white Sn	2586		
Tin (white)	Sn			231.928	2586		
Titanium	Ti			1670	3287		
Tungsten	W			3414	5555		
Uranium	U			1135	4131		
Vanadium	V			1910	3407		
Xenon	Xe	-111.745	81.77	-111.75	-108.099	16.583	5.8420
Ytterbium	Yb			824	1196		
Yttrium	Y			1522	3345		
Zinc	Zn			419.527	907		
Zirconium	Zr			1854	4406		

* Extrapolated or estimated value.

** This is the estimated diamond-graphite-liquid carbon triple point (Ref. 12).

HEAT CAPACITY OF THE ELEMENTS AT 25 °C

This table gives the specific heat capacity (c_p) in J/g K and the molar heat capacity (C_p) in J/mol K at a temperature of 25 °C and a pressure of 100 kPa (1 bar or 0.987 standard atmospheres) for all the elements for which reliable data are available.

Name	State	c_p/J g^{-1} K^{-1}	C_p/J mol^{-1} K^{-1}
Actinium	solid	0.120	27.2
Aluminum	solid	0.897	24.20
Antimony (gray)	solid	0.207	25.23
Argon	gas	0.520	20.786
Arsenic (gray)	solid	0.329	24.64
Barium	solid	0.204	28.07
Beryllium	solid	1.825	16.443
Bismuth	solid	0.122	25.52
Boron	solid	1.026	11.087
Bromine	liquid	0.474	75.69
Cadmium	solid	0.231	26.020
Calcium	solid	0.647	25.929
Carbon (graphite)	solid	0.709	8.517
Cerium	solid	0.192	26.94
Cesium	solid	0.242	32.210
Chlorine	gas	0.479	33.949
Chromium	solid	0.449	23.35
Cobalt	solid	0.421	24.81
Copper	solid	0.385	24.440
Dysprosium	solid	0.173	28.16
Erbium	solid	0.168	28.12
Europium	solid	0.182	27.66
Fluorine	gas	0.824	31.304
Gadolinium	solid	0.235	37.03
Gallium	solid	0.373	26.03
Germanium	solid	0.320	23.222
Gold	solid	0.129	25.418
Hafnium	solid	0.144	25.73
Helium	gas	5.193	20.786
Holmium	solid	0.165	27.15
Hydrogen	gas	14.304	28.836
Indium	solid	0.233	26.74
Iodine	solid	0.214	54.43
Iridium	solid	0.131	25.10
Iron	solid	0.449	25.10
Krypton	gas	0.248	20.786
Lanthanum	solid	0.195	27.11
Lead	solid	0.130	26.84
Lithium	solid	3.582	24.860
Lutetium	solid	0.154	26.86
Magnesium	solid	1.023	24.869
Manganese	solid	0.479	26.32
Mercury	liquid	0.140	27.983
Molybdenum	solid	0.251	24.06
Neodymium	solid	0.190	27.45
Neon	gas	1.030	20.786
Nickel	solid	0.444	26.07
Niobium	solid	0.265	24.60
Nitrogen	gas	1.040	29.124
Osmium	solid	0.130	24.7
Oxygen	gas	0.918	29.378
Palladium	solid	0.244	25.98
Phosphorus (white)	solid	0.769	23.824
Platinum	solid	0.133	25.86
Potassium	solid	0.757	29.600
Praseodymium	solid	0.193	27.20
Radon	gas	0.094	20.786
Rhenium	solid	0.137	25.48
Rhodium	solid	0.243	24.98
Rubidium	solid	0.363	31.060
Ruthenium	solid	0.238	24.06
Samarium	solid	0.196	29.54
Scandium	solid	0.568	25.52
Selenium (gray)	solid	0.321	25.363
Silicon	solid	0.712	19.99
Silver	solid	0.235	25.350
Sodium	solid	1.228	28.230
Strontium	solid	0.306	26.79
Sulfur (rhombic)	solid	0.708	22.70
Tantalum	solid	0.140	25.36
Tellurium	solid	0.202	25.73
Terbium	solid	0.182	28.91
Thallium	solid	0.129	26.32
Thorium	solid	0.118	27.32
Thulium	solid	0.160	27.03
Tin (white)	solid	0.227	26.99
Titanium	solid	0.524	25.060
Tungsten	solid	0.132	24.27
Uranium	solid	0.116	27.665
Vanadium	solid	0.489	24.89
Xenon	gas	0.158	20.786
Ytterbium	solid	0.155	26.74
Yttrium	solid	0.298	26.53
Zinc	solid	0.388	25.390
Zirconium	solid	0.278	25.36

Inorganic

VAPOR PRESSURE OF THE METALLIC ELEMENTS — EQUATIONS

C. B. Alcock

This table gives coefficients in an equation for the vapor pressure of 65 metallic elements in both the solid and liquid states. Vapor pressures in the range 10^{-10} to 10^{2} Pa (10^{-15} to 10^{-3} atm) are covered. The equation is:

for p in atmospheres: $\log(p/\text{atm}) = A + BT^{-1} + C\log T + DT^{-3}$
for p in pascals: $\log(p/\text{Pa}) = 5.006 + A + BT^{-1} + C\log T + DT^{-3}$
for p in torr (mmHg): $\log(p/\text{torr}) = 2.881 + A + BT^{-1} + C\log T + DT^{-3}$

where T is the temperature in K.

This equation reproduces the observed vapor pressures to an accuracy of 5% or better. The metals are listed alphabetically by name, and the melting point is included.

The table following this one gives values of the vapor pressure at several temperatures in the 400 K to 2400 K range, as calculated from these equations.

Reprinted with permission of the publisher, Pergamon Press.

Reference

Alcock, C. B., Itkin, V. P., and Horrigan, M. K., *Canadian Metallurgical Quarterly*, 23, 309, 1984.

Coefficients in Vapor Pressure Equations

Element	Phase	A	B	C	D	Range/K	mp/K
Aluminum	solid	9.459	-17342	-0.7927		298-mp	933
Aluminum	liquid	5.911	-16211			mp-1800	933
Americium	solid	11.311	-15059	-1.3449		298-mp	1449
Barium	solid	12.405	-9690	-2.2890		298-mp	1000
Barium	liquid	4.007	-8163			mp-1200	1000
Beryllium	solid	8.042	-17020	-0.4440		298-mp	1560
Beryllium	liquid	5.786	-15731			mp-1800	1560
Cadmium	solid	5.939	-5799			298-mp	594
Cadmium	liquid	5.242	-5392			mp-650	594
Calcium	solid	10.127	-9517	-1.4030		298-mp	1115
Cerium	solid	6.139	-21752			298-mp	1072
Cerium	liquid	5.611	-21200			mp-2450	1072
Cesium	solid	4.711	-3999			298-mp	302
Cesium	liquid	4.165	-3830			mp-550	302
Chromium	solid	6.800	-20733	0.4391	-0.4094	298-2000	2180
Cobalt	solid	10.976	-22576	-1.0280		298-mp	1768
Cobalt	liquid	6.488	-20578			mp-2150	1768
Copper	solid	9.123	-17748	-0.7317		298-mp	1358
Copper	liquid	5.849	-16415			mp-1850	1358
Curium	solid	8.369	-20364	-0.5770		298-mp	1618
Curium	liquid	5.223	-18292			mp-2200	1618
Dysprosium	solid	9.579	-15336	-1.1114		298-mp	1685
Erbium	solid	9.916	-16642	-1.2154		298-mp	1802
Erbium	liquid	4.668	-14380			mp-1900	1802
Europium	solid	9.240	-9459	-1.1661		298-mp	1095
Gadolinium	solid	8.344	-20861	-0.5775		298-mp	1586
Gadolinium	liquid	5.557	-19389			mp-2250	1586
Gallium	solid	6.657	-14208			298-mp	303
Gallium	liquid	6.754	-13984	-0.3413		mp-1600	303
Gold	solid	9.152	-19343	-0.7479		298-mp	1337
Gold	liquid	5.832	-18024			mp-2050	1337
Hafnium	solid	9.445	-32482	-0.6735		298-mp	2506
Holmium	solid	9.785	-15899	-1.1753		298-mp	1745
Indium	solid	5.991	-12548			298-mp	430
Indium	liquid	5.374	-12276			mp-1500	430
Iridium	solid	10.506	-35099	-0.7500		298-2500	2719
Iron	solid	7.100	-21723	0.4536	-0.5846	298-mp	1811
Iron	liquid	6.347	-19574			mp-2100	1811
Lanthanum	solid	7.463	-22551	-0.3142		298-mp	1193
Lanthanum	liquid	5.911	-21855			mp-2450	1193
Lead	solid	5.643	-10143			298-mp	600
Lead	liquid	4.911	-9701			mp-1200	600
Lithium	solid	5.667	-8310			298-mp	454
Lithium	liquid	5.055	-8023			mp-1000	454

Element	Phase	*A*	*B*	*C*	*D*	Range/K	mp/K
Lutetium	solid	8.793	-22423	-0.6200		298-mp	1936
Lutetium	liquid	5.648	-20302			mp-2350	1936
Magnesium	solid	8.489	-7813	-0.8253		298-mp	923
Manganese	solid	12.805	-15097	-1.7896		298-mp	1519
Mercury	liquid	5.116	-3190			298-400	234
Molybdenum	solid	11.529	-34626	-1.1331		298-2500	2895
Neodymium	solid	8.996	-17264	-0.9519		298-mp	1289
Neodymium	liquid	4.912	-15824			mp-2000	1289
Neptunium	solid	19.643	-24886	-3.9991		298-mp	917
Neptunium	liquid	10.076	-23378	-1.3250		mp-2500	917
Nickel	solid	10.557	-22606	-0.8717		298-mp	1728
Nickel	liquid	6.666	-20765			mp-2150	1728
Niobium	solid	8.882	-37818	-0.2575		298-2500	2750
Osmium	solid	9.419	-41198	-0.3896		298-2500	3306
Palladium	solid	9.502	-19813	-0.9258		298-mp	1828
Palladium	liquid	5.426	-17899			mp-2100	1828
Platinum	solid	4.882	-29387	1.1039	-0.4527	298-mp	2041
Platinum	liquid	6.386	-26856			mp-2500	2041
Plutonium	solid	26.160	-19162	-6.6675		298-600	913
Plutonium	solid	18.858	-18460	-4.4720		500-913	913
Plutonium	liquid	3.666	-16658			mp-2450	913
Potassium	solid	4.961	-4646			298-mp	337
Potassium	liquid	4.402	-4453			mp-600	337
Praseodymium	solid	8.859	-18720	-0.9512		298-mp	1204
Praseodymium	liquid	4.772	-17315			mp-2200	1204
Protactinium	solid	10.552	-34869	-1.0075		298-mp	1845
Protactinium	liquid	6.177	-32874			mp-2500	1845
Rhenium	solid	11.543	-40726	-1.1629		298-2500	3458
Rhodium	solid	10.168	-29010	-0.7068		298-mp	2236
Rhodium	liquid	6.802	-26792			mp-2500	2236
Rubidium	solid	4.857	-4215			298-mp	312
Rubidium	liquid	4.312	-4040			mp-550	312
Ruthenium	solid	9.755	-34154	-0.4723		298-mp	2606
Samarium	solid	9.988	-11034	-1.3287		298-mp	1345
Scandium	solid	6.650	-19721	0.2885	-0.3663	298-mp	1814
Scandium	liquid	5.795	-17681			mp-2000	1814
Silver	solid	9.127	-14999	-0.7845		298-mp	1235
Silver	liquid	5.752	-13827			mp-1600	1235
Sodium	solid	5.298	-5603			298-mp	371
Sodium	liquid	4.704	-5377			mp-700	371
Strontium	solid	9.226	-8572	-1.1926		298-mp	1050
Tantalum	solid	16.807	-41346	-3.2152	0.7437	298-2500	3290
Terbium	solid	9.510	-20457	-0.9247		298-mp	1632
Terbium	liquid	5.411	-18639			mp-2200	1632
Thallium	solid	5.971	-9447			298-mp	577
Thallium	liquid	5.259	-9037			mp-1100	577
Thorium	solid	8.668	-31483	-0.5288		298-mp	2023
Thorium	liquid	-18.453	-24569	6.6473		mp-2500	2023
Thulium	solid	8.882	-12270	-0.9564		298-1400	1818
Tin (white)	solid	6.036	-15710			298-mp	505
Tin (white)	liquid	5.262	-15332			mp-1850	505
Titanium	solid	11.925	-24991	-1.3376		298-mp	1943
Titanium	liquid	6.358	-22747			mp-2400	1943
Tungsten	solid	2.945	-44094	1.3677		298-2350	3687
Tungsten	solid	54.527	-57687	-12.2231		2200-2500	3687
Uranium	solid	0.770	-27729	2.6982	-1.5471	298-mp	1408
Uranium	liquid	20.735	-28776	-4.0962		mp-2500	1408
Vanadium	solid	9.744	-27132	-0.5501		298-mp	2183
Vanadium	liquid	6.929	-25011			mp-2500	2183
Ytterbium	solid	9.111	-8111	-1.0849		298-900	1097
Yttrium	solid	9.735	-22306	-0.8705		298-mp	1795
Yttrium	liquid	5.795	-20341			mp-2300	1795
Zinc	solid	6.102	-6776			298-mp	693
Zinc	liquid	5.378	-6286			mp-750	693
Zirconium	solid	10.008	-31512	-0.7890		298-mp	2127
Zirconium	liquid	6.806	-30295			mp-2500	2127

VAPOR PRESSURE OF THE METALLIC ELEMENTS — DATA

The following values of the vapor pressure of metallic elements are calculated from the equations in the preceding table. All values are given in pascals. For conversion, note that 1 Pa = 7.50 μm Hg = $9.87 \cdot 10^{-6}$ atm.

Vapor Pressure at Selected Temperatures

Element	mp/K	*p*/Pa (400 K)	*p*/Pa (600 K)	*p*/Pa (800 K)	*p*/Pa (1000 K)	*p*/Pa (1200 K)	*p*/Pa (1400 K)	*p*/Pa (1600 K)	*p*/Pa (1800 K)	*p*/Pa (2000 K)	*p*/Pa (2200 K)	*p*/Pa (2400 K)
Aluminum	933			$3.06 \cdot 10^{-10}$	$5.08 \cdot 10^{-6}$	0.00256	0.218	6.10	81.4			
Americium	1449			$3.88 \cdot 10^{-7}$	0.00167	0.423	21.35					
Barium	1000		$7.97 \cdot 10^{-6}$	0.0450	7.11	162						
Beryllium	1560			$3.04 \cdot 10^{-10}$	$4.96 \cdot 10^{-6}$	0.00314	0.312	9.12	113			
Cadmium	594	0.000280	18.2									
Calcium	1115		$2.36 \cdot 10^{-5}$	0.146	25.5							
Cerium	1072				$2.47 \cdot 10^{-11}$	$8.91 \cdot 10^{-8}$	$2.97 \cdot 10^{-5}$	0.00233	0.0691	1.04	9.56	60.8
Cesium	302	0.394										
Chromium	2180				$2.45 \cdot 10^{-8}$	$7.59 \cdot 10^{-5}$	0.0239	1.80	52.1	774		
Cobalt	1768				$2.09 \cdot 10^{-10}$	$1.00 \cdot 10^{-6}$	0.000419	0.0379	1.15	16.0		
Copper	1358			$6.60 \cdot 10^{-11}$	$1.53 \cdot 10^{-6}$	0.00122	0.135	3.94	54.4			
Curium	1618				$1.90 \cdot 10^{-9}$	$4.24 \cdot 10^{-6}$	0.00103	0.0629	1.17	12.1	82.1	
Dysprosium	1685			$1.54 \cdot 10^{-8}$	$8.21 \cdot 10^{-5}$	0.0241	1.362	27.5				
Erbium	1802			$3.90 \cdot 10^{-10}$	$4.30 \cdot 10^{-6}$	0.00205	0.163	4.23	52.5			
Europium	1095		$1.74 \cdot 10^{-5}$	0.109	19.4							
Gadolinium	1586				$5.70 \cdot 10^{-10}$	$1.54 \cdot 10^{-6}$	0.000429	0.0279	0.618	7.39	56.2	
Gallium	303			$1.94 \cdot 10^{-7}$	0.000565	0.114	4.98	84.4				
Gold	1337				$3.72 \cdot 10^{-8}$	$5.44 \cdot 10^{-5}$	0.00920	0.374	6.68	67.0		
Hafnium	2506						$1.35 \cdot 10^{-11}$	$9.81 \cdot 10^{-9}$	$1.63 \cdot 10^{-6}$	$9.69 \cdot 10^{-5}$	0.00272	0.0437
Holmium	1745			$3.20 \cdot 10^{-9}$	$2.32 \cdot 10^{-5}$	0.00837	0.546	12.3				
Indium	430		$8.31 \cdot 10^{-11}$	$1.08 \cdot 10^{-5}$	0.0127	1.413	40.9					
Iridium	2719							$1.48 \cdot 10^{-9}$	$3.72 \cdot 10^{-7}$	$3.06 \cdot 10^{-5}$	0.00112	0.0225
Iron	1811				$5.54 \cdot 10^{-9}$	$2.51 \cdot 10^{-5}$	0.0104	0.961	32.7	36.8		
Lanthanum	1193					$5.09 \cdot 10^{-8}$	$2.02 \cdot 10^{-5}$	0.00181	0.0596	0.976	9.61	64.7
Lead	600		$5.54 \cdot 10^{-7}$	0.00618	1.64	68.1						
Lithium	454	$7.90 \cdot 10^{-11}$	0.000489	1.08	109							
Lutetium	1936				$3.28 \cdot 10^{-11}$	$1.59 \cdot 10^{-7}$	$6.79 \cdot 10^{-5}$	0.00628	0.211	3.18	26.7	
Magnesium	923	$6.53 \cdot 10^{-9}$	0.0152	21.5								
Manganese	1519			$5.55 \cdot 10^{-7}$	0.00221	0.524	24.9					
Mercury	234	140										
Molybdenum	2895							$1.83 \cdot 10^{-9}$	$4.07 \cdot 10^{-7}$	$3.03 \cdot 10^{-5}$	0.00102	0.0189
Neodymium	1289			$4.55 \cdot 10^{-11}$	$7.62 \cdot 10^{-7}$	0.000483	0.0412	1.07	13.4	101		
Neptunium	917					$3.31 \cdot 10^{-9}$	$1.63 \cdot 10^{-6}$	0.000168	0.00604	0.105	1.06	7.28
Nickel	1728				$2.19 \cdot 10^{-10}$	$1.09 \cdot 10^{-6}$	0.000471	0.0438	1.37	19.5		
Niobium	2750							$2.32 \cdot 10^{-11}$	$9.54 \cdot 10^{-9}$	$1.17 \cdot 10^{-6}$	$5.98 \cdot 10^{-5}$	0.00158
Osmium	3306								$1.85 \cdot 10^{-10}$	$3.46 \cdot 10^{-8}$	$2.49 \cdot 10^{-6}$	$8.75 \cdot 10^{-5}$
Palladium	1828				$8.27 \cdot 10^{-9}$	$1.40 \cdot 10^{-5}$	0.00277	0.144	3.07	30.4		
Platinum	2041						$2.34 \cdot 10^{-8}$	$1.14 \cdot 10^{-5}$	0.00143	0.0689	0.153	1.59
Plutonium	913				$1.03 \cdot 10^{-8}$	$6.17 \cdot 10^{-6}$	0.000594	0.0182	0.262	2.20	12.6	53.8
Potassium	337	0.0188	96.9									
Praseodymium	1204				$1.95 \cdot 10^{-8}$	$2.16 \cdot 10^{-5}$	0.00257	0.0904	1.44	13.2	80.8	
Protactinium	1845							$3.44 \cdot 10^{-10}$	$8.06 \cdot 10^{-8}$	$5.57 \cdot 10^{-6}$	0.000174	0.00306
Rhenium	3458								$1.37 \cdot 10^{-10}$	$2.22 \cdot 10^{-8}$	$1.41 \cdot 10^{-6}$	$4.45 \cdot 10^{-5}$
Rhodium	2236						$1.69 \cdot 10^{-8}$	$5.99 \cdot 10^{-6}$	0.000571	0.0217	0.422	4.41
Rubidium	312	0.165										
Ruthenium	2606							$7.96 \cdot 10^{-9}$	$1.77 \cdot 10^{-6}$	0.000133	0.00455	0.0858
Samarium	1345		$8.17 \cdot 10^{-8}$	0.00221	0.942	51.0						
Scandium	1814				$6.31 \cdot 10^{-8}$	0.000129	0.0300	1.80	43.6	91.3		
Silver	1235			$1.27 \cdot 10^{-7}$	0.000603	0.165	7.61	131				
Sodium	371	0.000185	5.60									
Strontium	1050	$4.99 \cdot 10^{-11}$	0.000429	1.134	121							
Tantalum	3290									$3.36 \cdot 10^{-10}$	$1.87 \cdot 10^{-8}$	$5.21 \cdot 10^{-7}$
Terbium	1632				$1.92 \cdot 10^{-9}$	$4.18 \cdot 10^{-6}$	0.000988	0.0585	1.15	12.5	88.0	

Element	mp/K	p/Pa (400 K)	p/Pa (600 K)	p/Pa (800 K)	p/Pa (1000 K)	p/Pa (1200 K)	p/Pa (1400 K)	p/Pa (1600 K)	p/Pa (1800 K)	p/Pa (2000 K)	p/Pa (2200 K)	p/Pa (2400 K)
Thallium	577		$1.59 \cdot 10^{-5}$	0.0931	16.9							
Thorium	2023						$3.33 \cdot 10^{-11}$	$2.00 \cdot 10^{-8}$	$2.89 \cdot 10^{-6}$	0.000154	0.00401	0.0610
Thulium	1818		$6.03 \cdot 10^{-10}$	$5.94 \cdot 10^{-5}$	0.0561	5.22	130					
Tin	505			$1.26 \cdot 10^{-9}$	$8.62 \cdot 10^{-6}$	0.00310	0.207	4.85	56.3			
Titanium	1943					$9.69 \cdot 10^{-9}$	$7.44 \cdot 10^{-6}$	0.00106	0.0493	0.978	10.6	76.9
Tungsten	3687									$2.62 \cdot 10^{-10}$	$3.01 \cdot 10^{-8}$	$1.59 \cdot 10^{-6}$
Uranium	1408					$9.47 \cdot 10^{-10}$	$2.87 \cdot 10^{-6}$	$4.27 \cdot 10^{-6}$	0.000263	0.00678	0.0933	0.803
Vanadium	2183					$2.79 \cdot 10^{-10}$	$4.35 \cdot 10^{-7}$	0.000107	0.00769	0.233	3.68	32.6
Ytterbium	1097	$1.03 \cdot 10^{-9}$	0.00384	6.74								
Yttrium	1795				$6.66 \cdot 10^{-11}$	$2.96 \cdot 10^{-7}$	0.000117	0.0102	0.316	4.27	35.9	
Zinc	693	$1.47 \cdot 10^{-6}$	0.653									
Zirconium	2127						$1.05 \cdot 10^{-10}$	$6.17 \cdot 10^{-8}$	$8.68 \cdot 10^{-6}$	0.000450	0.0110	0.155

DENSITY OF MOLTEN ELEMENTS AND REPRESENTATIVE SALTS

This table lists the liquid density at the melting point, ρ_m, for elements that are solid at room temperature, as well as for some representative salts of these elements. Densities at higher temperatures (up to the t_{max} given in the last column) may be estimated from the equation

$$\rho(t) = \rho_m - k(t - t_m)$$

where t_m is the melting point and k is the derivative of density with respect to temperature. If a value of t_{max} is not given, the equation should not be used to extrapolate more than about 20 °C beyond the melting point.

Data for the elements were selected from the primary literature. The molten salt data were derived from Ref. 1.

References

1. Janz, G. J., Thermodynamic and Transport Properties of Molten Salts: Correlation Equations for Critically Evaluated Density, Surface Tension, Electrical Conductance, and Viscosity Data, *J. Phys. Chem. Ref. Data*, 17, Suppl. 2, 1988.
2. Nasch, P. M., and Steinemann, S. G., *Phys. Chem. Liq.*, 29, 43, 1995.
3. Assael, M. J., Kakosimos, K., Banish, R. M., Brillo, J., Egry, I., Brooks, R., Quested, P. N., Mills, K. C., Nagashima, A., Sato, Y., and Wakeham, W. A., *J. Phys. Chem. Ref. Data* 35, 285, 2006. [Al, Fe]
4. Assael, M. J., Kalyva, A. E., Antoniadis, K. D., Banish, R. M., Egry, I., Wu, J., Kaschnitz, E., and Wakeham, W. A., *J. Phys. Chem. Ref. Data* 39, 033105-1, 2010. [Cu, Sn]

Density of Molten Elements and Salts

Name	Formula	t_m/°C	ρ_m/ g cm^{-3}	k/ g cm^{-3} °C^{-1}	t_{max}/ °C
Aluminum	Al	660.323	2.377	0.000311	917
Aluminum bromide	$AlBr_3$	97.5	2.647	0.002435	267
Aluminum chloride	$AlCl_3$	192.6	1.302	0.002711	296
Aluminum iodide	AlI_3	188.28	3.223	0.0025	240
Antimony (gray)	Sb	630.628	6.53	0.00067	745
Antimony(III) chloride	$SbCl_3$	73.4	2.681	0.002293	77
Antimony(III) iodide	SbI_3	171	4.171	0.002483	322
Antimony(V) chloride	$SbCl_5$	4	2.37	0.001869	77
Arsenic (gray)	As	817	5.22	0.000544	
Barium	Ba	727	3.338	0.000299	1550
Barium bromide	$BaBr_2$	857	3.991	0.000924	900
Barium chloride	$BaCl_2$	961	3.174	0.000681	1081
Barium fluoride	BaF_2	1368	4.14	0.000999	1727
Barium iodide	BaI_2	711	4.26	0.000977	975
Beryllium	Be	1287	1.690	0.00011	
Beryllium chloride	$BeCl_2$	415	1.54	0.0011	473
Beryllium fluoride	BeF_2	552	1.96	0.000015	850
Bismuth	Bi	271.402	10.05	0.00135	800
Bismuth tribromide	$BiBr_3$	219	4.76	0.002637	927
Bismuth trichloride	$BiCl_3$	234	3.916	0.0023	350
Boron	B	2077	2.08		
Cadmium	Cd	321.069	7.996	0.001218	500
Cadmium bromide	$CdBr_2$	568	4.075	0.00108	720
Cadmium chloride	$CdCl_2$	568	3.392	0.00082	807
Cadmium iodide	CdI_2	388	4.396	0.001117	700
Calcium	Ca	842	1.378	0.000230	1484
Calcium bromide	$CaBr_2$	742	3.111	0.0005	791
Calcium chloride	$CaCl_2$	775	2.085	0.000422	950
Calcium fluoride	CaF_2	1418	2.52	0.000391	2027
Calcium iodide	CaI_2	783	3.443	0.000751	1028
Cerium	Ce	799	6.55	0.000710	1460
Cerium(III) chloride	$CeCl_3$	807	3.25	0.00092	950
Cerium(III) fluoride	CeF_3	1430	4.659	0.000936	1927
Cesium	Cs	28.5	1.843	0.000556	510
Cesium bromide	CsBr	636	3.133	0.001223	860
Cesium chloride	CsCl	646	2.79	0.001065	906
Cesium fluoride	CsF	703	3.649	0.001282	912
Cesium iodide	CsI	632	3.197	0.001183	907
Cesium nitrate	$CsNO_3$	409	2.820	0.001166	491
Cesium sulfate	Cs_2SO_4	1005	3.1	0.00095	1530
Chromium	Cr	1907	6.3	0.0011	2100
Cobalt	Co	1495	7.75	0.00165	1580
Copper	Cu	1084.62	7.997	0.000819	2227
Copper(I) chloride	CuCl	423	3.692	0.00076	585
Dysprosium	Dy	1412	8.37	0.00143	1540
Dysprosium(III) chloride	$DyCl_3$	718	3.62	0.00068	987
Erbium	Er	1529	8.86	0.00157	1700
Europium	Eu	822	5.13	0.0028	980
Gadolinium	Gd	1313	7.4		
Gadolinium(III) chloride	$GdCl_3$	602	3.56	0.000671	1007
Gadolinium(III) iodide	GdI_3	930	4.12	0.000908	1032
Gallium	Ga	29.7646	6.08	0.00062	400
Gallium(III) bromide	$GaBr_3$	123	3.116	0.00246	135
Gallium(III) chloride	$GaCl_3$	77.9	2.053	0.002083	141
Gallium(III) iodide	GaI_3	212	3.630	0.002377	252
Germanium	Ge	938.25	5.60	0.00055	1600
Gold	Au	1064.18	17.31	0.001343	1200
Hafnium	Hf	2233	12		
Holmium	Ho	1472	8.34		
Indium	In	156.5985	7.02	0.000836	500
Indium(III) bromide	$InBr_3$	420	3.121	0.0015	528
Indium(III) chloride	$InCl_3$	583	2.140	0.0021	666
Indium(III) iodide	InI_3	207	3.820	0.0015	360
Iridium	Ir	2446	19		
Iron	Fe	1538	7.035	0.000926	2207
Iron(II) chloride	$FeCl_2$	677	2.348	0.000555	877
Lanthanum	La	920	5.94	0.00061	1600
Lanthanum bromide	$LaBr_3$	788	4.933	0.000096	912
Lanthanum chloride	$LaCl_3$	858	3.209	0.000777	973
Lanthanum fluoride	LaF_3	1493	4.589	0.000682	2177
Lanthanum iodide	LaI_3	778	4.29	0.001110	907
Lead	Pb	327.462	10.66	0.00122	700
Lead(II) bromide	$PbBr_2$	371	5.73	0.00165	600
Lead(II) chloride	$PbCl_2$	501	4.951	0.0015	710
Lead(II) iodide	PbI_2	410	5.691	0.001594	697
Lithium	Li	180.50	0.512	0.00052	285
Lithium bromide	LiBr	550	2.528	0.000652	739
Lithium chloride	LiCl	610	1.502	0.000432	781

Name	Formula	t_m/°C	ρ_m/ g cm^{-3}	k/ g cm^{-3} °C^{-1}	t_{max}/ °C
Lithium fluoride	LiF	848.2	1.81	0.000490	1047
Lithium iodide	LiI	469	3.109	0.000917	667
Lithium nitrate	$LiNO_3$	253	1.781	0.000546	441
Lithium sulfate	Li_2SO_4	860	2.003	0.000407	1214
Lutetium	Lu	1663	9.3		
Magnesium	Mg	650	1.584	0.000234	900
Magnesium bromide	$MgBr_2$	711	2.62	0.000478	935
Magnesium chloride	$MgCl_2$	714	1.68	0.000271	826
Magnesium iodide	MgI_2	634	3.05	0.000651	888
Manganese	Mn	1246	5.95	0.00105	1590
Manganese(II) chloride	$MnCl_2$	650	2.353	0.000437	850
Mercury(II) bromide	$HgBr_2$	241	5.126	0.003233	319
Mercury(II) chloride	$HgCl_2$	277	4.368	0.002862	304
Mercury(II) iodide (yellow)	HgI_2	256	5.222	0.003235	354
Molybdenum	Mo	2622	9.33		
Neodymium	Nd	1016	6.89	0.00076	1350
Nickel	Ni	1455	7.81	0.000726	1700
Nickel(II) chloride	$NiCl_2$	1031	2.653	0.00066	1057
Osmium	Os	3033	20		
Palladium	Pd	1554.8	10.38	0.001169	1700
Platinum	Pt	1768.2	19.77	0.0024	2200
Plutonium	Pu	640	16.63	0.001419	950
Potassium	K	63.5	0.828	0.000232	500
Potassium bromide	KBr	734	2.127	0.000825	930
Potassium chloride	KCl	771	1.527	0.000583	939
Potassium fluoride	KF	858	1.910	0.000651	1037
Potassium iodide	KI	681	2.448	0.000956	904
Potassium nitrate	KNO_3	334	1.865	0.000723	457
Praseodymium	Pr	931	6.50	0.00093	1460
Praseodymium(III) chloride	$PrCl_3$	786	3.23	0.00074	977
Rhenium	Re	3185	18.9		
Rhodium	Rh	1963	10.7	0.000895	2200
Rubidium	Rb	39.30	1.46	0.000451	800
Rubidium bromide	RbBr	692	2.715	0.001072	907
Rubidium carbonate	Rb_2CO_3	873	2.84	0.000640	1007
Rubidium chloride	RbCl	724	2.248	0.000883	923
Rubidium fluoride	RbF	795	2.87	0.00102	1067
Rubidium iodide	RbI	656	2.904	0.001143	902
Rubidium nitrate	$RbNO_3$	310	2.519	0.001068	417
Rubidium sulfate	Rb_2SO_4	1066	2.56	0.000665	1545
Ruthenium	Ru	2333	10.65		
Samarium	Sm	1072	7.16		
Scandium	Sc	1541	2.68	0.00054	1680
Selenium (gray)	Se	220.8	3.99		
Silicon	Si	1414	2.57	0.00036	1500
Silver	Ag	961.78	9.320	0.0009	1500
Silver(I) bromide	AgBr	430	5.577	0.001035	667
Silver(I) chloride	AgCl	455	4.83	0.00094	627
Silver(I) iodide	AgI	558	5.58	0.00101	802
Silver(I) nitrate	$AgNO_3$	210	3.970	0.001098	360
Silver(I) sulfate	Ag_2SO_4	660	4.84	0.001089	770
Sodium	Na	97.794	0.927	0.00023	600
Sodium bromide	NaBr	747	2.342	0.000816	945
Sodium carbonate	Na_2CO_3	856	1.972	0.000448	1004
Sodium chloride	NaCl	802.018	1.556	0.000543	1027
Sodium fluoride	NaF	996	1.948	0.000636	1097
Sodium iodide	NaI	661	2.742	0.000949	912
Sodium nitrate	$NaNO_3$	306.5	1.90	0.000715	370
Sodium sulfate	Na_2SO_4	884	2.069	0.000483	1077
Strontium	Sr	777	6.980		
Strontium bromide	$SrBr_2$	657	3.70	0.000745	1004
Strontium chloride	$SrCl_2$	874	2.727	0.000578	1037
Strontium fluoride	SrF_2	1477	3.470	0.000751	1927
Strontium iodide	SrI_2	538	4.085	0.000885	1026
Sulfur (monoclinic)	S	115.21	1.819	0.00080	160
Tantalum	Ta	3017	15		
Tantalum(V) chloride	$TaCl_5$	216.6	2.700	0.004316	457
Tellurium	Te	449.51	5.70	0.00035	600
Terbium	Tb	1359	7.65		
Thallium	Tl	304	11.22	0.00144	600
Thallium(I) bromide	TlBr	460	5.98	0.001755	647
Thallium(I) chloride	TlCl	431	5.628	0.0018	642
Thallium(I) iodide	TlI	441.7	6.15	0.001761	737
Thallium(I) nitrate	$TlNO_3$	206	4.91	0.001873	279
Thallium(I) sulfate	Tl_2SO_4	632	5.62	0.00130	927
Thorium(IV) chloride	$ThCl_4$	770	3.363	0.0014	847
Thorium(IV) fluoride	ThF_4	1110	6.058	0.000759	1378
Thulium	Tm	1545	8.56	0.00050	1675
Tin (white)	Sn	231.928	6.979	0.000652	1650
Tin(II) chloride	$SnCl_2$	247.0	3.36	0.001253	480
Tin(IV) chloride	$SnCl_4$	-34.07	2.37	0.002687	138
Titanium	Ti	1670	4.11		
Titanium(IV) chloride	$TiCl_4$	-24.12	1.807	0.001735	137
Tungsten	W	3414	17.6		
Uranium	U	1135	17.3		
Uranium(III) chloride	UCl_3	837	4.84	0.007943	1057
Uranium(IV) chloride	UCl_4	590	3.572	0.001945	667
Uranium(IV) fluoride	UF_4	1036	6.485	0.000992	1341
Vanadium	V	1910	5.5		
Ytterbium	Yb	824	6.21		
Yttrium	Y	1522	4.18	0.00029	1685
Yttrium chloride	YCl_3	721	2.510	0.0005	845
Zinc	Zn	419.527	6.57	0.0011	700
Zinc bromide	$ZnBr_2$	402	3.47	0.000959	602
Zinc chloride	$ZnCl_2$	325	2.54	0.00053	557
Zinc iodide	ZnI_2	450	3.878	0.00136	588
Zinc sulfate	$ZnSO_4$	680 dec	3.14	0.00047	987
Zirconium	Zr	1854	5.8		
Zirconium(IV) chloride	$ZrCl_4$	437 tp	1.643	0.007464	492

MAGNETIC SUSCEPTIBILITY OF THE ELEMENTS AND INORGANIC COMPOUNDS

When a material is placed in a magnetic field $\boldsymbol{H}$, a magnetization (magnetic moment per unit volume) $\boldsymbol{M}$ is induced in the material which is related to $\boldsymbol{H}$ by $\boldsymbol{M}= \kappa \boldsymbol{H}$, where κ is called the volume susceptibility. Since $\boldsymbol{H}$ and $\boldsymbol{M}$ have the same dimensions, κ is dimensionless. A more useful parameter is the molar susceptibility χ_m, defined by

$$\chi_m = \kappa V_m = \kappa M/\rho$$

where V_m is the molar volume of the substance, M the molar mass, and ρ the mass density. When the cgs system is used, the customary units for χ_m are cm³ mol⁻¹; the corresponding SI units are m³ mol⁻¹.

Substances that have no unpaired electron orbital or spin angular momentum generally have negative values of χ_m and are called diamagnetic. Their molar susceptibility varies only slightly with temperature. Substances with unpaired electrons, which are termed paramagnetic, have positive χ_m and show a much stronger temperature dependence, varying roughly as $1/T$. The net susceptibility of a paramagnetic substance is the sum of the paramagnetic and diamagnetic contributions, but the former almost always dominates. Ferromagnetic substances, indicated here by a footnote, have more complex magnetic behavior.

This table gives values of χ_m for the elements and selected inorganic compounds. All values refer to nominal room temperature (285 to 300 K) unless otherwise indicated. When the phase is indicated as "solid," the most common crystalline form is understood.

Substances are arranged in alphabetical order by the most common name. In keeping with customary practice, the molar susceptibility is given here in units appropriate to the cgs system. These values should be multiplied by 4π to obtain values for use in SI equations (where the magnetic field strength $\boldsymbol{H}$ has units of A m⁻¹).

References

1. *Landolt-Börnstein, Numerical Data and Functional Relationships in Science and Technology, New Series*, II/16, *Diamagnetic Susceptibility*, Springer-Verlag, Heidelberg, 1986.
2. *Landolt-Börnstein, Numerical Data and Functional Relationships in Science and Technology, New Series*, III/19, Subvolumes a to i2, *Magnetic Properties of Metals*, Springer-Verlag, Heidelberg, 1986-1992.
3. *Landolt-Börnstein, Numerical Data and Functional Relationships in Science and Technology, New Series*, II/2, II/8, II/10, II/11, and II/12a, *Coordination and Organometallic Transition Metal Compounds*, Springer-Verlag, Heidelberg, 1966-1984.
4. *Tables de Constantes et Données Numérique, Volume 7, Relaxation Paramagnetique*, Masson, Paris, 1957.

Magnetic Susceptibility of Inorganic Materials

Name	Formula	Phase	χ_m/ 10⁻⁶ cm³ mol⁻¹
Aluminum	Al	solid	16.5
Aluminum fluoride	AlF_3	solid	-13.9
Aluminum oxide (α)	Al_2O_3	solid	-37
Aluminum sulfate	$Al_2(SO_4)_3$	solid	-93
Ammonia	NH_3	gas	-16.3
Ammonia	NH_3	aq. soln.	-18.3
Ammonium acetate	$NH_4C_2H_3O_2$	solid	-41.1
Ammonium bromide	NH_4Br	solid	-47
Ammonium carbonate	$(NH_4)_2CO_3$	solid	-42.5
Ammonium chlorate	NH_4ClO_3	solid	-42.1
Ammonium chloride	NH_4Cl	solid	-36.7
Ammonium fluoride	NH_4F	solid	-23
Ammonium iodate	NH_4IO_3	solid	-62.3
Ammonium iodide	NH_4I	solid	-66
Ammonium nitrate	NH_4NO_3	solid	-33
Ammonium sulfate	$(NH_4)_2SO_4$	solid	-67
Ammonium thiocyanate	NH_4SCN	solid	-48.1
Antimony	Sb	solid	-99
Antimony(III) bromide	$SbBr_3$	solid	-111.4
Antimony(III) chloride	$SbCl_3$	solid	-86.7
Antimony(V) chloride	$SbCl_5$	solid	-120.5
Antimony(III) fluoride	SbF_3	solid	-46
Antimony(III) iodide	SbI_3	solid	-147.2
Antimony(III) oxide (senarmontite)	Sb_2O_3	solid	-69.4
Antimony(III) sulfide	Sb_2S_3	solid	-86
Argon	Ar	gas	-19.32
Arsenic (gray)	As	solid	-5.6
Arsenic(III) bromide	$AsBr_3$	solid	-106
Arsenic(III) chloride	$AsCl_3$	solid	-72.5
Arsenic(III) iodide	AsI_3	solid	-142.2
Arsenic(III) oxide (arsenolite)	As_2O_3	solid	-30.34
Arsenic(III) sulfide	As_2S_3	solid	-70
Arsenic (yellow)	As	solid	-23.2
Arsine	AsH_3	gas	-35.2
Barium	Ba	solid	20.6
Barium bromide	$BaBr_2$	solid	-92
Barium bromide dihydrate	$BaBr_2 \cdot 2H_2O$	solid	-119.3
Barium carbonate	$BaCO_3$	solid	-58.9
Barium chloride	$BaCl_2$	solid	-72.6
Barium chloride dihydrate	$BaCl_2 \cdot 2H_2O$	solid	-100
Barium fluoride	BaF_2	solid	-51
Barium hydroxide	$Ba(OH)_2$	solid	-53.2
Barium iodate	$Ba(IO_3)_2$	solid	-122.5
Barium iodide	BaI_2	solid	-124.4
Barium iodide dihydrate	$BaI_2 \cdot 2H_2O$	solid	-163
Barium nitrate	$Ba(NO_3)_2$	solid	-66.5
Barium oxide	BaO	solid	-29.1
Barium peroxide	BaO_2	solid	-40.6
Barium sulfate	$BaSO_4$	solid	-65.8
Beryllium	Be	solid	-9.0
Beryllium chloride	$BeCl_2$	solid	-26.5
Beryllium hydroxide (α)	$Be(OH)_2$	solid	-23.1
Beryllium oxide	BeO	solid	-11.9
Beryllium sulfate	$BeSO_4$	solid	-37
Bismuth	Bi	solid	-280.1
Bismuth hydroxide	$Bi(OH)_3$	solid	-65.8
Bismuth nitrate pentahydrate	$Bi(NO_3)_3 \cdot 5H_2O$	solid	-159
Bismuth oxide	Bi_2O_3	solid	-83
Bismuth phosphate	$BiPO_4$	solid	-77
Bismuth sulfate	$Bi_2(SO_4)_3$	solid	-199
Bismuth sulfide	Bi_2S_3	solid	-123

Inorganic

Name	Formula	Phase	χ_m/ 10^{-6} cm^3 mol^{-1}
Bismuth tribromide	$BiBr_3$	solid	-147
Bismuth trichloride	$BiCl_3$	solid	-26.5
Bismuth trifluoride	BiF_3	solid	-61.2
Bismuth triiodide	BiI_3	solid	-200.5
Boric acid	H_3BO_3	solid	-34.1
Boron	B	solid	-6.7
Boron oxide	B_2O_3	solid	-38.7
Boron trichloride	BCl_3	gas	-59.9
Bromine	Br_2	liquid	-56.4
Bromine	Br_2	gas	-73.5
Bromine pentafluoride	BrF_5	liquid	-45.1
Bromine trifluoride	BrF_3	gas	-33.9
Cadmium	Cd	solid	-19.7
Cadmium bromide	$CdBr_2$	solid	-87.3
Cadmium bromide tetrahydrate	$CdBr_2·4H_2O$	solid	-131.5
Cadmium carbonate	$CdCO_3$	solid	-46.7
Cadmium chloride	$CdCl_2$	solid	-68.7
Cadmium chromate	$CdCrO_4$	solid	-16.8
Cadmium cyanide	$Cd(CN)_2$	solid	-54
Cadmium fluoride	CdF_2	solid	-40.6
Cadmium hydroxide	$Cd(OH)_2$	solid	-41
Cadmium iodate	$Cd(IO_3)_2$	solid	-108.4
Cadmium iodide	CdI_2	solid	-117.2
Cadmium nitrate	$Cd(NO_3)_2$	solid	-55.1
Cadmium nitrate tetrahydrate	$Cd(NO_3)_2·4H_2O$	solid	-140
Cadmium oxide	CdO	solid	-30
Cadmium sulfate	$CdSO_4$	solid	-59.2
Cadmium sulfide	CdS	solid	-50
Calcium	Ca	solid	40
Calcium bromide	$CaBr_2$	solid	-73.8
Calcium carbonate (calcite)	$CaCO_3$	solid	-38.2
Calcium chloride	$CaCl_2$	solid	-54.7
Calcium fluoride	CaF_2	solid	-28
Calcium hydroxide	$Ca(OH)_2$	solid	-22
Calcium iodate	$Ca(IO_3)_2$	solid	-101.4
Calcium iodide	CaI_2	solid	-109
Calcium oxide	CaO	solid	-15.0
Calcium sulfate	$CaSO_4$	solid	-49.7
Calcium sulfate dihydrate	$CaSO_4·2H_2O$	solid	-74
Carbon (diamond)	C	solid	-5.9
Carbon dioxide	CO_2	gas	-21.0
Carbon (graphite)	C	solid	-6.0
Carbon monoxide	CO	gas	-9.8
Cerium	Ce	β form	2500
Cerium(III) chloride	$CeCl_3$	solid	2490
Cerium(III) fluoride	CeF_3	solid	2190
Cerium(IV) oxide	CeO_2	solid	26
Cerium(IV) sulfate tetrahydrate	$Ce(SO_4)_2·4H_2O$	solid	-97
Cerium(II) sulfide	CeS	solid	2110
Cerium(III) sulfide	Ce_2S_3	solid	5080
Cesium	Cs	solid	29
Cesium bromate	$CsBrO_3$	solid	-75.1
Cesium bromide	CsBr	solid	-67.2
Cesium carbonate	Cs_2CO_3	solid	-103.6
Cesium chlorate	$CsClO_3$	solid	-65
Cesium chloride	CsCl	solid	-56.7
Cesium fluoride	CsF	solid	-44.5
Cesium iodide	CsI	solid	-82.6

Name	Formula	Phase	χ_m/ 10^{-6} cm^3 mol^{-1}
Cesium sulfate	Cs_2SO_4	solid	-116
Cesium superoxide	CsO_2	solid	1534
Chlorine	Cl_2	liquid	-40.4
Chlorine trifluoride	ClF_3	gas	-26.5
Chromium	Cr	solid	167
Chromium(II) chloride	$CrCl_2$	solid	7230
Chromium(III) chloride	$CrCl_3$	solid	6350
Chromium(III) fluoride	CrF_3	solid	4370
Chromium(III) oxide	Cr_2O_3	solid	1960
Chromium(VI) oxide	CrO_3	solid	40
Chromium(III) sulfate	$Cr_2(SO_4)_3$	solid	11800
Cobalt	Co	solid	[a]
Cobalt(II) bromide	$CoBr_2$	solid	13000
Cobalt(II) chloride	$CoCl_2$	solid	12660
Cobalt(II) chloride hexahydrate	$CoCl_2·6H_2O$	solid	9710
Cobalt(II) cyanide	$Co(CN)_2$	solid	3825
Cobalt(II) fluoride	CoF_2	solid	9490
Cobalt(III) fluoride	CoF_3	solid	1900
Cobalt(II) iodide	CoI_2	solid	10760
Cobalt(II,III) oxide	Co_3O_4	solid	7380
Cobalt(III) oxide	Co_2O_3	solid	4560
Cobalt(II) sulfate	$CoSO_4$	solid	10000
Cobalt(II) sulfide	CoS	solid	225
Copper	Cu	solid	-5.46
Copper(I) bromide	CuBr	solid	-49
Copper(II) bromide	$CuBr_2$	solid	685
Copper(I) chloride	CuCl	solid	-40
Copper(II) chloride	$CuCl_2$	solid	1080
Copper(II) chloride dihydrate	$CuCl_2·2H_2O$	solid	1420
Copper(I) cyanide	CuCN	solid	-24
Copper(II) fluoride	CuF_2	solid	1050
Copper(II) fluoride dihydrate	$CuF_2·2H_2O$	solid	1600
Copper(II) hydroxide	$Cu(OH)_2$	solid	1170
Copper(I) iodide	CuI	solid	-63
Copper(II) nitrate hexahydrate	$Cu(NO_3)_2·6H_2O$	solid	1625
Copper(II) nitrate trihydrate	$Cu(NO_3)_2·3H_2O$	solid	1570
Copper(I) oxide	Cu_2O	solid	-20
Copper(II) oxide	CuO	solid	238
Copper(II) sulfate	$CuSO_4$	solid	1330
Copper(II) sulfate pentahydrate	$CuSO_4·5H_2O$	solid	1460
Copper(II) sulfide	CuS	solid	-2.0
Diborane	B_2H_6	gas	-21.0
Disilane	Si_2H_6	gas	-37.3
Dysprosium	Dy	α form	98000
Dysprosium(III) oxide	Dy_2O_3	solid	89600
Dysprosium(III) sulfide	Dy_2S_3	solid	95200
Erbium	Er	solid	48000
Erbium oxide	Er_2O_3	solid	73920
Erbium sulfate octahydrate	$Er_2(SO_4)_3·8H_2O$	solid	74600
Erbium sulfide	Er_2S_3	solid	77200
Europium	Eu	solid	30900
Europium(II) bromide	$EuBr_2$	solid	26800
Europium(II) chloride	$EuCl_2$	solid	26500
Europium(II) fluoride	EuF_2	solid	23750
Europium(II) iodide	EuI_2	solid	26000

Name	Formula	Phase	χ_m/ 10^{-6} cm^3 mol^{-1}
Europium(III) oxide	Eu_2O_3	solid	10100
Europium(III) sulfate	$Eu_2(SO_4)_3$	solid	10400
Europium(II) sulfide	EuS	solid	23800
Gadolinium	Gd	solid, 350 K	185000
Gadolinium(III) chloride	$GdCl_3$	solid	27930
Gadolinium(III) oxide	Gd_2O_3	solid	53200
Gadolinium(III) sulfate octahydrate	$Gd_2(SO_4)_3{\cdot}8H_2O$	solid	53280
Gadolinium(III) sulfide	Gd_2S_3	solid	55500
Gallium	Ga	solid	-21.6
Gallium(III) chloride	$GaCl_3$	solid	-63
Gallium suboxide	Ga_2O	solid	-34
Gallium(II) sulfide	GaS	solid	-23
Gallium(III) sulfide	Ga_2S_3	solid	-80
Germane	GeH_4	gas	-29.7
Germanium	Ge	solid	-11.6
Germanium(IV) chloride	$GeCl_4$	solid	-72
Germanium(IV) fluoride	GeF_4	solid	-50
Germanium(IV) iodide	GeI_4	solid	-171
Germanium(II) oxide	GeO	solid	-28.8
Germanium(IV) oxide	GeO_2	solid	-34.3
Germanium(II) sulfide	GeS	solid	-40.9
Germanium(IV) sulfide	GeS_2	solid	-53.9
Gold	Au	solid	-28
Gold(I) bromide	AuBr	solid	-61
Gold(I) chloride	AuCl	solid	-67
Gold(III) chloride	$AuCl_3$	solid	-112
Gold(I) iodide	AuI	solid	-91
Hafnium	Hf	solid	71
Hafnium(IV) oxide	HfO_2	solid	-23
Helium	He	gas	-2.02
Holmium	Ho	solid	72900
Holmium oxide	Ho_2O_3	solid	88100
Hydrazine	N_2H_4	liquid	-201
Hydrogen	H_2	liquid, 20.3 K	-5.44
Hydrogen	H_2	gas	-3.99
Hydrogen chloride	HCl	liquid	-22.6
Hydrogen chloride	HCl	aq. soln.	-22
Hydrogen fluoride	HF	liquid	-8.6
Hydrogen fluoride	HF	aq. soln.	-9.3
Hydrogen iodide	HI	solid, 195 K	-47.3
Hydrogen iodide	HI	liquid, 233 K	-48.3
Hydrogen iodide	HI	aq. soln.	-50.2
Hydrogen peroxide	H_2O_2	liquid	-17.3
Hydrogen sulfide	H_2S	gas	-25.5
Indium	In	solid	-10.2
Indium(III) bromide	$InBr_3$	solid	-107
Indium(I) chloride	InCl	solid	-30
Indium(II) chloride	$InCl_2$	solid	-56
Indium(III) chloride	$InCl_3$	solid	-86
Indium(III) oxide	In_2O_3	solid	-56
Indium(II) sulfide	InS	solid	-28
Indium(III) sulfide	In_2S_3	solid	-98
Iodic acid	HIO_3	solid	-48
Iodine	I_2	solid	-90
Iodine chloride	ICl	liquid	-54.6
Iodine pentafluoride	IF_5	liquid	-58.1
Iodine pentoxide	I_2O_5	solid	-79.4
Iodine trichloride	ICl_3	solid	-90.2
Iridium	Ir	solid	25
Iridium(III) chloride	$IrCl_3$	solid	-14.4
Iridium(IV) oxide	IrO_2	solid	224
Iron	Fe	solid	[a]
Iron(II) bromide	$FeBr_2$	solid	13600
Iron(II) carbonate	$FeCO_3$	solid	11300
Iron(II) chloride	$FeCl_2$	solid	14750
Iron(III) chloride	$FeCl_3$	solid	13450
Iron(III) chloride hexahydrate	$FeCl_3{\cdot}6H_2O$	solid	15250
Iron(II) chloride tetrahydrate	$FeCl_2{\cdot}4H_2O$	solid	12900
Iron(II) fluoride	FeF_2	solid	9500
Iron(III) fluoride	FeF_3	solid	13760
Iron(III) fluoride trihydrate	$FeF_3{\cdot}3H_2O$	solid	7870
Iron(II) iodide	FeI_2	solid	13600
Iron(III) nitrate nonahydrate	$Fe(NO_3)_3{\cdot}9H_2O$	solid	15200
Iron(II) oxide	FeO	solid	7200
Iron(II) sulfate	$FeSO_4$	solid	12400
Iron(II) sulfate heptahydrate	$FeSO_4{\cdot}7H_2O$	solid	11200
Iron(II) sulfate monohydrate	$FeSO_4{\cdot}H_2O$	solid	10500
Iron(II) sulfide	FeS	solid	1074
Krypton	Kr	gas	-29.0
Lanthanum	La	α form	95.9
Lanthanum oxide	La_2O_3	solid	-78
Lanthanum sulfate nonahydrate	$La_2(SO_4)_3{\cdot}9H_2O$	solid	-262
Lanthanum sulfide	La_2S_3	solid	-37
Lead	Pb	solid	-23
Lead(II) acetate	$Pb(C_2H_3O_2)_2$	solid	-89.1
Lead(II) bromide	$PbBr_2$	solid	-90.6
Lead(II) carbonate	$PbCO_3$	solid	-61.2
Lead(II) chloride	$PbCl_2$	solid	-73.8
Lead(II) chromate	$PbCrO_4$	solid	-18
Lead(II) fluoride	PbF_2	solid	-58.1
Lead(II) iodate	$Pb(IO_3)_2$	solid	-131
Lead(II) iodide	PbI_2	solid	-126.5
Lead(II) nitrate	$Pb(NO_3)_2$	solid	-74
Lead(II) oxide (massicot)	PbO	solid	-42
Lead(II) phosphate	$Pb_3(PO_4)_2$	solid	-182
Lead(II) sulfate	$PbSO_4$	solid	-69.7
Lead(II) sulfide	PbS	solid	-83.6
Lithium	Li	solid	14.2
Lithium bromide	LiBr	solid	-34.3
Lithium carbonate	Li_2CO_3	solid	-27
Lithium chloride	LiCl	solid	-24.3
Lithium fluoride	LiF	solid	-10.1
Lithium hydride	LiH	solid	-4.6
Lithium hydroxide	LiOH	aq. soln.	-12.3
Lithium iodide	LiI	solid	-50
Lithium sulfate	Li_2SO_4	solid	-41.6
Lutetium	Lu	solid	182.9
Magnesium	Mg	solid	13.1
Magnesium bromide	$MgBr_2$	solid	-72
Magnesium carbonate	$MgCO_3$	solid	-32.4
Magnesium chloride	$MgCl_2$	solid	-47.4

Name	Formula	Phase	χ_m/ 10^{-6} cm^3 mol^{-1}
Magnesium fluoride	MgF_2	solid	-22.7
Magnesium hydroxide	$Mg(OH)_2$	solid	-22.1
Magnesium iodide	MgI_2	solid	-111
Magnesium oxide	MgO	solid	-10.2
Magnesium sulfate	$MgSO_4$	solid	-42
Magnesium sulfate heptahydrate	$MgSO_4 \cdot 7H_2O$	solid	-135.7
Magnesium sulfate monohydrate	$MgSO_4 \cdot H_2O$	solid	-61
Manganese	Mn	solid	511
Manganese(II) bromide	$MnBr_2$	solid	13900
Manganese(II) carbonate	$MnCO_3$	solid	11400
Manganese(II) chloride	$MnCl_2$	solid	14350
Manganese(II) chloride tetrahydrate	$MnCl_2 \cdot 4H_2O$	solid	14600
Manganese(II) fluoride	MnF_2	solid	10700
Manganese(III) fluoride	MnF_3	solid	10500
Manganese(II) hydroxide	$Mn(OH)_2$	solid	13500
Manganese(II) iodide	MnI_2	solid	14400
Manganese(II) oxide	MnO	solid	4850
Manganese(II,III) oxide	Mn_3O_4	solid	12400
Manganese(III) oxide	Mn_2O_3	solid	14100
Manganese(IV) oxide	MnO_2	solid	2280
Manganese(II) sulfate	$MnSO_4$	solid	13660
Manganese(II) sulfate monohydrate	$MnSO_4 \cdot H_2O$	solid	14200
Manganese(II) sulfate tetrahydrate	$MnSO_4 \cdot 4H_2O$	solid	14600
Manganese(II) sulfide (α form)	MnS	α form	5630
Manganese(II) sulfide (β form)	MnS	β form	3850
Mercury	Hg	solid, 234 K	-24.1
Mercury	Hg	liquid	-33.5
Mercury(I) bromide	Hg_2Br_2	solid	-105
Mercury(II) bromide	$HgBr_2$	solid	-94.2
Mercury(I) chloride	Hg_2Cl_2	solid	-120
Mercury(II) chloride	$HgCl_2$	solid	-82
Mercury(II) cyanide	$Hg(CN)_2$	solid	-67
Mercury(I) fluoride	Hg_2F_2	solid	-106
Mercury(II) fluoride	HgF_2	solid	-57.3
Mercury(I) iodide	Hg_2I_2	solid	-166
Mercury(II) iodide (red)	HgI_2	solid	-165
Mercury(I) nitrate	$Hg_2(NO_3)_2$	solid	-121
Mercury(II) nitrate	$Hg(NO_3)_2$	solid	-74
Mercury(I) oxide	Hg_2O	solid	-76.3
Mercury(II) oxide	HgO	solid	-46
Mercury(I) sulfate	Hg_2SO_4	solid	-123
Mercury(II) sulfate	$HgSO_4$	solid	-78.1
Mercury(II) sulfide (red)	HgS	solid	-55.4
Mercury(II) thiocyanate	$Hg(SCN)_2$	solid	-96.5
Molybdenum	Mo	solid	72
Molybdenum(III) bromide	$MoBr_3$	solid	525
Molybdenum(IV) bromide	$MoBr_4$	solid	520
Molybdenum(III) chloride	$MoCl_3$	solid	43
Molybdenum(IV) chloride	$MoCl_4$	solid	1750
Molybdenum(V) chloride	$MoCl_5$	solid	990
Molybdenum(VI) fluoride	MoF_6	solid	-26.0
Molybdenum(III) oxide	Mo_2O_3	solid	-42.0
Molybdenum(IV) oxide	MoO_2	solid	41

Name	Formula	Phase	χ_m/ 10^{-6} cm^3 mol^{-1}
Molybdenum(VI) oxide	MoO_3	solid	3
Neodymium	Nd	α form	5930
Neodymium(III) fluoride	NdF_3	solid	4980
Neodymium(III) oxide	Nd_2O_3	solid	10200
Neodymium(III) sulfate	$Nd_2(SO_4)_3$	solid	9990
Neodymium(III) sulfide	Nd_2S_3	solid	5550
Neon	Ne	gas	-6.96
Neptunium	Np	solid	575
Nickel	Ni	solid	[a]
Nickel(II) bromide	$NiBr_2$	solid	5600
Nickel(II) chloride	$NiCl_2$	solid	6145
Nickel(II) chloride hexahydrate	$NiCl_2 \cdot 6H_2O$	solid	4240
Nickel(II) fluoride	NiF_2	solid	2410
Nickel(II) hydroxide	$Ni(OH)_2$	solid	4500
Nickel(II) iodide	NiI_2	solid	3875
Nickel(II) nitrate hexahydrate	$Ni(NO_3)_2 \cdot 6H_2O$	solid	4300
Nickel(II) oxide	NiO	solid	660
Nickel subsulfide	Ni_3S_2	solid	1030
Nickel(II) sulfate	$NiSO_4$	solid	4005
Nickel(II) sulfide	NiS	solid	190
Niobium	Nb	solid	208
Niobium(V) oxide	Nb_2O_5	solid	-10
Nitric acid	HNO_3	liquid	-19.9
Nitric oxide	NO	solid, 90 K	19.8
Nitric oxide	NO	liquid, 118 K	114.2
Nitric oxide	NO	gas	1461
Nitrogen	N_2	gas	-12.0
Nitrogen dioxide	NO_2	gas, 408 K	150
Nitrogen tetroxide	N_2O_4	gas	-23.0
Nitrogen trioxide	N_2O_3	gas	-16
Nitrous oxide	N_2O	gas	-18.9
Osmium	Os	solid	11
Oxygen	O_2	solid, 54 K	10200
Oxygen	O_2	liquid, 90 K	7699
Oxygen	O_2	gas	3415
Ozone	O_3	liquid	6.7
Palladium	Pd	solid	540
Palladium(II) chloride	$PdCl_2$	solid	-38
Phosphine	PH_3	gas	-26.2
Phosphonic acid	H_3PO_3	aq. soln.	-42.5
Phosphoric acid	H_3PO_4	aq. soln.	-43.8
Phosphorus (white)	P	solid	-26.66
Phosphorus (red)	P	solid	-20.77
Phosphorus(III) chloride	PCl_3	liquid	-63.4
Platinum	Pt	solid	193
Platinum(II) chloride	$PtCl_2$	solid	-54
Platinum(III) chloride	$PtCl_3$	solid	-66.7
Platinum(IV) chloride	$PtCl_4$	solid	-93
Platinum(IV) fluoride	PtF_4	solid	445
Plutonium	Pu	solid	525
Plutonium(IV) fluoride	PuF_4	solid	1760
Plutonium(VI) fluoride	PuF_6	solid	173
Plutonium(IV) oxide	PuO_2	solid	730
Potassium	K	solid	20.8
Potassium bromate	$KBrO_3$	solid	-52.6
Potassium bromide	KBr	solid	-49.1
Potassium carbonate	K_2CO_3	solid	-59

Name	Formula	Phase	χ_m/ 10^{-6} cm^3 mol^{-1}
Potassium chlorate	$KClO_3$	solid	-42.8
Potassium chloride	KCl	solid	-38.8
Potassium chromate	K_2CrO_4	solid	-3.9
Potassium cyanide	KCN	solid	-37
Potassium ferricyanide	$K_3Fe(CN)_6$	solid	2290
Potassium ferrocyanide trihydrate	$K_4Fe(CN)_6 \cdot 3H_2O$	solid	-172.3
Potassium fluoride	KF	solid	-23.6
Potassium hydrogen sulfate	$KHSO_4$	solid	-49.8
Potassium hydroxide	KOH	aq. soln.	-22
Potassium iodate	KIO_3	solid	-63.1
Potassium iodide	KI	solid	-63.8
Potassium nitrate	KNO_3	solid	-33.7
Potassium nitrite	KNO_2	solid	-23.3
Potassium permanganate	$KMnO_4$	solid	20
Potassium sulfate	K_2SO_4	solid	-67
Potassium sulfide	K_2S	solid	-60
Potassium superoxide	KO_2	solid	3230
Potassium thiocyanate	$KSCN$	solid	-48
Praseodymium	Pr	α form	5530
Praseodymium(III) chloride	$PrCl_3$	solid	44.5
Praseodymium(III) oxide	Pr_2O_3	solid	8994
Praseodymium(III) sulfide	Pr_2S_3	solid	10770
Protactinium	Pa	solid	277
Rhenium	Re	solid	67
Rhenium(V) chloride	$ReCl_5$	solid	1225
Rhenium(IV) oxide	ReO_2	solid	44
Rhenium(VI) oxide	ReO_3	solid	16
Rhenium(VII) oxide	Re_2O_7	solid	-16
Rhenium(IV) sulfide	ReS_2	solid	38
Rhodium	Rh	solid	102
Rhodium(III) chloride	$RhCl_3$	solid	-7.5
Rhodium(III) oxide	Rh_2O_3	solid	104
Rubidium	Rb	solid	17
Rubidium bromide	$RbBr$	solid	-56.4
Rubidium carbonate	Rb_2CO_3	solid	-75.4
Rubidium chloride	$RbCl$	solid	-46
Rubidium fluoride	RbF	solid	-31.9
Rubidium iodide	RbI	solid	-72.2
Rubidium nitrate	$RbNO_3$	solid	-41
Rubidium sulfate	Rb_2SO_4	solid	-88.4
Rubidium superoxide	RbO_2	solid	1527
Ruthenium	Ru	solid	39
Ruthenium(III) chloride	$RuCl_3$	solid	1998
Ruthenium(IV) oxide	RuO_2	solid	162
Samarium	Sm	α form	1278
Samarium(II) bromide	$SmBr_2$	solid	5337
Samarium(III) bromide	$SmBr_3$	solid	972
Samarium(III) oxide	Sm_2O_3	solid	1988
Samarium(III) sulfate octahydrate	$Sm_2(SO_4)_3 \cdot 8H_2O$	solid	1710
Samarium(III) sulfide	Sm_2S_3	solid	3300
Scandium	Sc	α form	295.2
Selenium (gray)	Se	solid	-25
Selenium bromide	Se_2Br_2	liquid	-113
Selenium chloride	Se_2Cl_2	liquid	-94.8
Selenium dioxide	SeO_2	solid	-27.2
Selenium hexafluoride	SeF_6	gas	-51
Silane	SiH_4	gas	-20.4
Silicon	Si	solid	-3.12
Silicon carbide (hexagonal)	SiC	solid	-12.8
Silicon dioxide (α-quartz)	SiO_2	solid	-29.6
Silver	Ag	solid	-19.5
Silver(I) bromide	$AgBr$	solid	-61
Silver(I) carbonate	Ag_2CO_3	solid	-80.90
Silver(I) chloride	$AgCl$	solid	-49
Silver(I) chromate	Ag_2CrO_4	solid	-40
Silver(I) cyanide	$AgCN$	solid	-43.2
Silver(I) fluoride	AgF	solid	-36.5
Silver(I) iodide	AgI	solid	-80
Silver(I) nitrate	$AgNO_3$	solid	-45.7
Silver(I) nitrite	$AgNO_2$	solid	-42
Silver(I) oxide	Ag_2O	solid	-134
Silver(II) oxide	AgO	solid	-19.6
Silver(I) phosphate	Ag_3PO_4	solid	-120
Silver(I) sulfate	Ag_2SO_4	solid	-92.90
Silver(I) thiocyanate	$AgSCN$	solid	-61.8
Sodium	Na	solid	16
Sodium acetate	$NaC_2H_3O_2$	solid	-37.6
Sodium bromate	$NaBrO_3$	solid	-44.2
Sodium bromide	$NaBr$	solid	-41
Sodium carbonate	Na_2CO_3	solid	-41
Sodium chlorate	$NaClO_3$	solid	-34.7
Sodium chloride	$NaCl$	solid	-30.2
Sodium dichromate	$Na_2Cr_2O_7$	solid	55
Sodium fluoride	NaF	solid	-15.6
Sodium hydrogen phosphate	Na_2HPO_4	solid	-56.6
Sodium hydroxide	$NaOH$	aq. soln.	-15.8
Sodium iodate	$NaIO_3$	solid	-53
Sodium iodide	NaI	solid	-57
Sodium nitrate	$NaNO_3$	solid	-25.6
Sodium nitrite	$NaNO_2$	solid	-14.5
Sodium oxide	Na_2O	solid	-19.8
Sodium peroxide	Na_2O_2	solid	-28.10
Sodium sulfate	Na_2SO_4	solid	-52
Sodium sulfate decahydrate	$Na_2SO_4 \cdot 10H_2O$	solid	-184
Sodium sulfide	Na_2S	solid	-39
Sodium tetraborate	$Na_2B_4O_7$	solid	-85
Stibine	SbH_3	gas	-34.6
Strontium	Sr	solid	92
Strontium bromide	$SrBr_2$	solid	-86.6
Strontium bromide hexahydrate	$SrBr_2 \cdot 6H_2O$	solid	-160
Strontium carbonate	$SrCO_3$	solid	-47
Strontium chlorate	$Sr(ClO_3)_2$	solid	-73
Strontium chloride	$SrCl_2$	solid	-61.5
Strontium chloride hexahydrate	$SrCl_2 \cdot 6H_2O$	solid	-145
Strontium chromate	$SrCrO_4$	solid	-5.1
Strontium fluoride	SrF_2	solid	-37.2
Strontium hydroxide	$Sr(OH)_2$	solid	-40
Strontium iodate	$Sr(IO_3)_2$	solid	-108
Strontium iodide	SrI_2	solid	-112
Strontium nitrate	$Sr(NO_3)_2$	solid	-57.2
Strontium oxide	SrO	solid	-35
Strontium peroxide	SrO_2	solid	-32.3
Strontium sulfate	$SrSO_4$	solid	-57.9
Sulfur (rhombic)	S	solid	-15.5
Sulfur (monoclinic)	S	solid	-14.9
Sulfur chloride [ClSSCl]	S_2Cl_2	liquid	-62.2
Sulfur dichloride	SCl_2	liquid	-49.4

Name	Formula	Phase	χ_m/ 10^{-6} cm³ mol⁻¹
Sulfur dioxide	SO_2	gas	-18.2
Sulfur hexafluoride	SF_6	gas	-44
Sulfuric acid	H_2SO_4	liquid	-39
Sulfur trioxide (α-form)	SO_3	liquid	-28.54
Tantalum	Ta	solid	154
Tantalum(V) chloride	$TaCl_5$	solid	140
Tantalum(V) oxide	Ta_2O_5	solid	-32
Technetium	Tc	solid	115
Tellurium	Te	solid	-38
Tellurium dibromide	$TeBr_2$	solid	-106
Tellurium dichloride	$TeCl_2$	solid	-94
Tellurium hexafluoride	TeF_6	gas	-66
Terbium	Tb	α form	170000
Terbium(III) oxide	Tb_2O_3	solid	78340
Tetrabromosilane	$SiBr_4$	liquid	-126
Tetrachlorosilane	$SiCl_4$	liquid	-87.5
Tetraethylsilane	$(C_2H_5)_4Si$	liquid	-120.2
Tetramethylsilane	$(CH_3)_4Si$	liquid	-74.80
Thallium	Tl	solid	-50
Thallium(I) bromate	$TlBrO_3$	solid	-75.9
Thallium(I) bromide	TlBr	solid	-63.9
Thallium(I) carbonate	Tl_2CO_3	solid	-101.6
Thallium(I) chlorate	$TlClO_3$	solid	-65.5
Thallium(I) chloride	TlCl	solid	-57.8
Thallium(I) chromate	Tl_2CrO_4	solid	-39.3
Thallium(I) cyanide	TlCN	solid	-49
Thallium(I) fluoride	TlF	solid	-44.4
Thallium(I) iodate	$TlIO_3$	solid	-86.8
Thallium(I) iodide	TlI	solid	-82.2
Thallium(I) nitrate	$TlNO_3$	solid	-56.5
Thallium(I) nitrite	$TlNO_2$	solid	-50.8
Thallium(I) sulfate	Tl_2SO_4	solid	-112.6
Thallium(I) sulfide	Tl_2S	solid	-88.8
Thionyl chloride	$SOCl_2$	liquid	-44.3
Thorium	Th	solid	97
Thorium(IV) oxide	ThO_2	solid	-16
Thulium	Tm	solid	24700
Thulium(III) oxide	Tm_2O_3	solid	51444
Tin(IV) bromide	$SnBr_4$	solid	-149
Tin(II) chloride	$SnCl_2$	solid	-69
Tin(IV) chloride	$SnCl_4$	liquid	-115
Tin(II) chloride dihydrate	$SnCl_2{\cdot}2H_2O$	solid	-91.4
Tin (gray)	Sn	solid	-37.4
Tin(II) oxide	SnO	solid	-19
Tin(IV) oxide	SnO_2	solid	-41
Titanium	Ti	solid	151
Titanium(II) bromide	$TiBr_2$	solid	720
Titanium(III) bromide	$TiBr_3$	solid	660
Titanium(II) chloride	$TiCl_2$	solid	484
Titanium(III) chloride	$TiCl_3$	solid	1110
Titanium(IV) chloride	$TiCl_4$	solid	-54
Titanium(III) fluoride	TiF_3	solid	1300
Titanium(II) iodide	TiI_2	solid	1790
Titanium(III) oxide	Ti_2O_3	solid	132
Titanium(IV) oxide (rutile)	TiO_2	solid	5.9
Titanium(II) sulfide	TiS	solid	432
Tungsten	W	solid	53
Tungsten(V) bromide	WBr_5	solid	270
Tungsten carbide	WC	solid	10
Tungsten(II) chloride	WCl_2	solid	-25
Tungsten(V) chloride	WCl_5	solid	387
Tungsten(VI) chloride	WCl_6	solid	-71
Tungsten(VI) fluoride	WF_6	gas	-53
Tungsten(IV) oxide	WO_2	solid	57
Tungsten(VI) oxide	WO_3	solid	-15.8
Tungsten(IV) sulfide	WS_2	solid	5850
Uranium	U	solid	409
Uranium(III) bromide	UBr_3	solid	4740
Uranium(IV) bromide	UBr_4	solid	3530
Uranium(III) chloride	UCl_3	solid	3460
Uranium(IV) chloride	UCl_4	solid	3680
Uranium(IV) fluoride	UF_4	solid	3530
Uranium(VI) fluoride	UF_6	solid	43
Uranium(III) hydride	UH_3	solid	6244
Uranium(III) iodide	UI_3	solid	4460
Uranium(IV) oxide	UO_2	solid	2360
Uranium(VI) oxide	UO_3	solid	128
Vanadium	V	solid	285
Vanadium(II) bromide	VBr_2	solid	3230
Vanadium(III) bromide	VBr_3	solid	2910
Vanadium(II) chloride	VCl_2	solid	2410
Vanadium(III) chloride	VCl_3	solid	3030
Vanadium(IV) chloride	VCl_4	liquid	1215
Vanadium(III) fluoride	VF_3	solid	2757
Vanadium(III) oxide	V_2O_3	solid	1976
Vanadium(IV) oxide	VO_2	solid	99
Vanadium(V) oxide	V_2O_5	solid	128
Vanadium(III) sulfide	V_2S_3	solid	1560
Water	H_2O	solid, 273 K	-12.63
Water	H_2O	liquid, 293 K	-12.96
Water	H_2O	liquid, 373 K	-13.09
Water	H_2O	gas, 373 K	-13.1
Xenon	Xe	gas	-45.5
Ytterbium	Yb	β form	67
Yttrium	Y	α form	187.7
Yttrium oxide	Y_2O_3	solid	44.4
Yttrium sulfide	Y_2S_3	solid	100
Zinc	Zn	solid	-9.15
Zinc carbonate	$ZnCO_3$	solid	-34
Zinc chloride	$ZnCl_2$	solid	-55.33
Zinc cyanide	$Zn(CN)_2$	solid	-46
Zinc fluoride	ZnF_2	solid	-34.3
Zinc hydroxide	$Zn(OH)_2$	solid	-67
Zinc iodide	ZnI_2	solid	-108
Zinc oxide	ZnO	solid	-27.2
Zinc phosphate	$Zn_3(PO_4)_2$	solid	-141
Zinc sulfate	$ZnSO_4$	solid	-47.8
Zinc sulfate heptahydrate	$ZnSO_4{\cdot}7H_2O$	solid	-138
Zinc sulfate monohydrate	$ZnSO_4{\cdot}H_2O$	solid	-63
Zinc sulfide (sphalerite)	ZnS	solid	-25
Zirconium	Zr	solid	120
Zirconium carbide	ZrC	solid	-26
Zirconium(IV) nitrate pentahydrate	$Zr(NO_3)_4{\cdot}5H_2O$	solid	-77
Zirconium(IV) oxide	ZrO_2	solid	-13.8

[a] Ferromagnetic

INDEX OF REFRACTION OF INORGANIC LIQUIDS

This table gives the index of refraction n of several inorganic substances in the liquid state. The temperature in °C is indicated by a superscript on n. The measurements refer to ambient atmospheric pressure except for substances whose normal boiling points are greater than the indicated temperature; in this case the pressure is the saturated vapor pressure of the substance. All values refer to a wavelength of 589 nm unless otherwise indicated in a footnote. Entries are arranged in alphabetical order by chemical name.

Data on the index of refraction at other temperatures and wavelengths may be found in Ref. 1.

References

1. Wohlfarth, C., and Wohlfarth, B., *Landolt-Börnstein, Numerical Data and Functional Relationships in Science and Technology, New Series*, III/38A, Martienssen, W., Editor, Springer-Verlag, Heidelberg, 1996.
2. Francis, A.W., *J. Chem. Eng. Data*, 5, 534, 1960.

Index of Refraction of Inorganic Liquids

Name	Formula	n
Ammonia[b, d]	NH_3	1.3944^{-77}
Antimony(V) chloride	$SbCl_5$	1.5925^{22}
Argon	Ar	1.2312^{-188}
Arsenic(III) chloride	$AsCl_3$	1.604^{16}
Boron tribromide	BBr_3	1.312^{16}
Bromine	Br_2	1.659^{15}
Bromine pentafluoride	BrF_5	1.3529^{25}
Bromine trifluoride	BrF_3	1.4536^{25}
Carbon dioxide	CO_2	1.6630^{24} at 6.3 MPa
Carbon disulfide	CS_2	1.6277^{20}
Carbon oxysulfide	OCS	1.24^{-87}
Carbon suboxide	C_3O_2	1.4538^{0}
Chlorine	Cl_2	1.3834^{20}
Chromium(VI) dichloride dioxide	CrO_2Cl_2	1.524^{23}
Germanium(IV) bromide	$GeBr_4$	1.6269^{26}
Germanium(IV) chloride	$GeCl_4$	1.4614^{25}
Helium[c]	He	1.02451^{-269}
Hydrazine	N_2H_4	1.470^{22}
Hydrogen	H_2	1.1096^{-253}
Hydrogen bromide	HBr	1.325^{10}
Hydrogen chloride[a]	HCl	1.3287^{18}
Hydrogen cyanide	HCN	1.2614^{20}
Hydrogen disulfide	H_2S_2	1.630^{20}
Hydrogen fluoride	HF	1.1574^{25}
Hydrogen iodide	HI	1.466^{16}
Hydrogen peroxide	H_2O_2	1.4061^{28}
Hydrogen sulfide[e]	H_2S	1.460^{-80}
Iron pentacarbonyl	$Fe(CO)_5$	1.5196^{20}
Krypton[c]	Kr	1.3032^{-157}
Nitric acid	HNO_3	1.393^{25}
Nitric oxide	NO	1.330^{-90}
Nitrogen	N_2	1.19876^{-196}
Nitrous oxide	N_2O	1.238^{25} at high pressure
Oxygen	O_2	1.2243^{-183}
Perchloric acid	$HClO_4$	1.3819^{50}
Phosphine	PH_3	1.317^{17}
Phosphorus(III) bromide	PBr_3	1.687^{25}
Phosphorus(III) chloride	PCl_3	1.5122^{21}
Phosphorus(III) oxide	P_2O_3	1.540^{27}
Sulfur (rhombic)	S	1.9170^{125}
Sulfur chloride [ClSSCl]	S_2Cl_2	1.671^{20}
Sulfur dichloride	SCl_2	1.557^{14}
Sulfur dioxide	SO_2	1.3396^{25}
Sulfur hexafluoride	SF_6	1.167^{25}
Sulfuric acid	H_2SO_4	1.4183^{20}
Sulfur trioxide (α-form)	SO_3	1.40965^{20}
Sulfuryl chloride	SO_2Cl_2	1.444^{12}
Tetrabromosilane	$SiBr_4$	1.5685^{31}
Tetrachlorosilane	$SiCl_4$	1.41156^{25}
Thionyl chloride	$SOCl_2$	1.527^{10}
Tin(IV) bromide	$SnBr_4$	1.6628^{31}
Tin(IV) chloride	$SnCl_4$	1.5086^{25}
Titanium(IV) chloride	$TiCl_4$	1.6076^{18}
Water	H_2O	1.33336^{20}
Xenon	Xe	1.3918^{-112}

[a] At 581 nm
[b] At 578 nm
[c] At 546 nm
[d] The value for compressed liquid NH_3 at 20 °C and 589 nm wavelength is 1.3327.
[e] The value for compressed liquid H_2S at 20 °C is 1.3682.

PHYSICAL AND OPTICAL PROPERTIES OF MINERALS

The chemical formula, crystal system, density, hardness, and index of refraction of some common minerals are given in this table. Entries are arranged alphabetically by mineral name. The columns are:

Formula: Chemical formula for a typical sample of the mineral. Composition often varies considerably with the origin of the sample.

Crystal system

Density: Typical density in g/cm^3. Individual samples may vary by a few percent.

Hardness: On the Mohs' scale (range of 1 to 10, with talc = 1 and diamond = 10).

Index of refraction: Values are given for the three coordinate axes in the order of least, intermediate, and greatest index. For cubic crystals there is only a single value. Index values refer to light of wavelength in the neighborhood of 589 nm. See Ref. 1 for details on the axis systems. Variations of several percent, depending on the origin and exact composition of the sample, are common.

References

1. Deer, W. A., Howie, R. A., and Zussman, J., *An Introduction to the Rock-Forming Minerals, 2nd Edition*, Longman Scientific & Technical, Harlow, Essex, 1992.
2. Carmichael, R. S., *Practical Handbook of Physical Properties of Rocks and Minerals*, CRC Press, Boca Raton, FL, 1989.
3. Donnay, J. D. H., and Ondik, H. M., *Crystal Data Determinative Tables, Third Edition, Volume 2, Inorganic Compounds*, Joint Committee on Powder Diffraction Standards, Swarthmore, PA, 1973.

Density, Hardness, and Refractive Index along the Coordinate Axes of Least, Intermediate, and Greatest Index

Name	Formula	Crystal system	$\rho/g\ cm^{-3}$	Hardness	n_α	n_β	n_γ
Acanthite	Ag_2S	orthorhombic	7.2	2.3			
Actinolite	$Ca_2(Mg,Fe)_5Si_8O_{22}(OH,F)_2$	monoclinic	3.23	5.5	1.624	1.655	1.664
Aegirine	$NaFe(SiO_3)_2$	monoclinic	3.58	6	1.763	1.800	1.815
Akermanite	$Ca_2MgSi_2O_7$	tetragonal	2.94	5.5	1.632	1.640	
Alabandite	MnS	cubic	4.0	3.8			
Albite	$NaAlSi_3O_8$	triclinic	2.63	6.3	1.527	1.531	1.538
Allanite	$(Ca,Mn,Ce,La,Y,Th)_2(Fe,Ti)(Al,Fe)O{\cdot}OH(Si_2O_7)(SiO_4)$	monoclinic	3.8	5.8	1.75	1.78	1.80
Allemontite	SbAs	hexagonal	6.0	3.5			
Almandine	$Fe_3Al_2Si_3O_{12}$	cubic	4.32	6.8	1.830		
Altaite	PbTe	cubic	8.16	3			
Aluminite	$Al_2(SO_4)(OH)_4{\cdot}7H_2O$	monoclinic	1.74	1.5	1.459	1.464	1.470
Alunite	$KAl_3(SO_4)_2(OH)_6$	rhombohedral	2.8	3.8	1.572	1.592	
Alunogen	$Al_2(SO_4)_3{\cdot}18H_2O$	monoclinic	1.69	1.8	1.467	1.47	1.478
Amblygonite	$(Li,Na)Al(PO_4)(F,OH)$	triclinic	3.1	5.8	1.591	1.604	1.613
Analcite	$NaAl(SiO_3)_2{\cdot}H_2O$	cubic	2.27	5.5	1.486		
Anatase	TiO_2	tetragonal	4.23	5.8	2.488	2.561	
Andalusite	Al_2OSiO_4	orthorhombic	3.15	7.5	1.635	1.639	1.644
Andesine	$NaAlSi_3O_8{\cdot}CaAl_2Si_2O_8$	triclinic	2.67	6.3	1.550	1.553	1.557
Andorite	$PbAgSb_3S_6$	rhombohedral	2.67	3.3			
Andradite	$Ca_3Fe_2Si_3O_{12}$	cubic	3.86	6.8	1.887		
Anglesite	$PbSO_4$	orthorhombic	6.29	2.8	1.877	1.883	1.894
Anhydrite	$CaSO_4$	orthorhombic	2.96	3.5	1.570	1.575	1.614
Ankerite	$Ca(Fe,Mg,Mn)(CO_3)_2$	rhombohedral	3.0	3.8	1.529	1.720	
Anorthite	$CaAl_2Si_2O_8$	triclinic	2.76	6.3	1.577	1.585	1.590
Anorthoclase	$(Na,K)AlSi_3O_8$	triclinic	2.58	6	1.523	1.528	1.529
Anthophyllite	$Mg_7Si_8O_{22}(OH)_2$	rhombohedral	3.21	5.8	1.645	1.658	1.668
Apatite	$Ca_5(PO_4)_3(OH,F,Cl)$	hexagonal	3.2	5	1.645	1.648	
Apophyllite	$KFCa_4Si_8O_{20}{\cdot}8H_2O$	tetragonal	2.35	4.8	1.535	1.536	
Aragonite	$CaCO_3$	orthorhombic	2.83	3.5	1.531	1.680	1.686
Arcanite	K_2SO_4	orthorhombic	2.66		1.494	1.494	1.497
Argentite	Ag_2S	orthorhombic	7.2	2.3			
Arsenolite	As_2O_3	cubic	3.86	1.5	1.755		
Arsenopyrite	FeAsS	monoclinic	6.1	5.8			
Atacamite	$Cu_2(OH)_3Cl$	rhombohedral	3.76	3.3	1.831	1.861	1.880
Augelite	$Al_2(PO_4)(OH)_3$	monoclinic	2.70	4.8	1.574	1.576	1.588
Augite	$(Ca,Mg,Fe,Ti,Al)_2(Si,Al)_2O_6$	monoclinic	3.38	6	1.703	1.707	1.738
Autunite	$Ca(UO_2)_2(PO_4)_2{\cdot}10H_2O$	tetragonal	3.2	2.3	1.553	1.577	
Axinite	$(Ca,Mn,Fe)_3Al_2BO_3Si_4O_{12}(OH)$	triclinic	3.31	6.8	1.684	1.691	1.694
Azurite	$Cu_3(OH)_2(CO_3)_2$	monoclinic	3.77	3.8	1.730	1.758	1.838

Name	Formula	Crystal system	ρ/g cm^{-3}	Hardness	n_α	n_β	n_γ
Baddeleyite	ZrO_2	monoclinic	5.7	6.5	2.13	2.19	2.20
Barite	$BaSO_4$	orthorhombic	4.49	3.3	1.636	1.637	1.648
Benitoite	$BaTi(SiO_3)_3$	rhombohedral	3.65	6.3	1.757	1.804	
Bertrandite	$Be_4Si_2O_7(OH)_2$	rhombohedral	2.6	6	1.589	1.602	1.613
Beryl	$Be_3Al_2(SiO_3)_6$	hexagonal	2.64	7.8	1.582	1.589	
Beryllonite	$NaBe(PO)_4$	monoclinic	2.81	5.8	1.552	1.558	1.561
Biotite	$K(Mg,Fe)_3AlSi_3O_{10}(OH,F)_2$	monoclinic	3.0	2.8	1.595	1.651	1.651
Bismuthinite	Bi_2S_3	orthorhombic	6.78	2			
Bixbyite	$(Mn,Fe)_2O_3$	cubic	4.95	6.3			
Bloedite	$Na_2Mg(SO_4)_2 \cdot 4H_2O$	monoclinic	2.25	2.8	1.483	1.486	1.487
Boehmite	$AlO(OH)$	orthorhombic	3.44	3.8	1.64	1.65	1.66
Boracite	$Mg_3B_7O_{13}Cl$	rhombohedral	2.94	7.3	1.66	1.66	1.67
Borax	$Na_2B_4O_7 \cdot 10H_2O$	monoclinic	1.73	2.3	1.447	1.469	1.472
Bornite	Cu_5FeS_4	cubic	5.07	3			
Boulangerite	$Pb_5Sb_4S_{11}$	monoclinic	6.1	2.8			
Bournonite	$PbCuSbS_3$	rhombohedral	5.83	2.8			
Braggite	PtS	tetragonal	10.2				
Braunite	$(Mn,Si)_2O_3$	tetragonal	4.78	6.3			
Bravoite	$(Ni,Fe)S_2$	cubic	4.62	5.8			
Breithauptite	$NiSb$	hexagonal	≈8.7	5.5			
Brochantite	$Cu_4SO_4(OH)_6$	monoclinic	3.79	3.8	1.728	1.771	1.800
Bromyrite	$AgBr$	cubic	6.47	2.5	2.253		
Brookite	TiO_2	orthorhombic	4.23	5.8	2.583	2.584	2.700
Brucite	$Mg(OH)_2$	hexagonal	2.37	2.5	1.575	1.59	
Bunsenite	NiO	cubic	6.72	5.5			
Cacoxenite	$Fe_4(PO_4)_3(OH)_3 \cdot 12H_2O$	hexagonal	2.3	3.5	1.580	1.646	
Calcite	$CaCO_3$	hexagonal	2.71	3	1.486	1.658	
Caledonite	$Cu_2Pb_5(SO_4)_3(CO_3)(OH)_6$	rhombohedral	5.76	2.8	1.818	1.866	1.909
Calomel	Hg_2Cl_2	tetragonal	7.16	1.5	1.973	2.656	
Cancrinite	$(Na,Ca,K)_7[Al_6Si_6O_{24}](CO_3,SO_4,Cl,OH)_2 \cdot H_2O$	hexagonal	2.42	5.5	1.495	1.509	
Carbon (diamond)	C	cubic	3.513	10	2.418		
Carnalite	$KMgCl_3 \cdot 6H_2O$	rhombohedral	1.60	2.5	1.466	1.475	1.494
Carnotite	$K_2(UO_2)_2(VO_4)_2 \cdot 3H_2O$	rhombohedral		1.5	1.75	1.92	1.95
Cassiterite	SnO_2	tetragonal	6.85	6.5	2.006	2.097	
Celestite	$SrSO_4$	orthorhombic	3.96	3.3	1.622	1.624	1.631
Celsian	$BaAl_2Si_2O_8$	monoclinic	3.25	6.3	1.583	1.588	1.594
Cerargyrite	$AgCl$	cubic	5.56	2.5	2.071		
Cerussite	$PbCO_3$	orthorhombic	6.6	3.3	1.804	2.076	2.079
Cervantite	Sb_2O_4	orthorhombic	6.64	4.5			
Chabazite	$CaAl_2Si_4O_{12} \cdot 6H_2O$	trig	2.08	4.5	1.482		
Chalcanthite	$CuSO_4 \cdot 5H_2O$	triclinic	2.29	2.5	1.514	1.537	1.543
Chalcocite	Cu_2S	orthorhombic	5.6	2.8			
Chalcopyrite	$CuFeS_2$	tetragonal	4.2	3.8			
Chiolite	$Na_5Al_3F_{14}$	tetragonal	3.00	3.8	1.342	1.349	
Chlorite	$(Mg,Al,Fe)_{12}(Si,Al)_8O_{20}(OH)_{16}$	monoclinic	3.0	2.5	1.61	1.62	1.62
Chloritoid	$FeAl_4O_2(SiO_4)_2(OH)_4$	monoclinic	3.66	6.5	1.717	1.721	1.726
Chondrodite	$2Mg_2SiO_4 \cdot MgF_2$	monoclinic	3.21	6.5	1.604	1.615	1.634
Chromite	$FeCr_2O_4$	cubic	5.0	5.5	2.16		
Chrysoberyl	$BeAl_2O_4$	orthorhombic	3.65	8.5	1.746	1.748	1.756
Chrysocolla	$CuSiO_3 \cdot 2H_2O$	rhombohedral	2.4	2	1.575	1.597	1.598
Cinnabar	HgS	hexagonal	8.17	2.3	2.814	3.143	
Claudetite	As_2O_3	monoclinic	3.74	2.5	1.87	1.92	2.01
Clinohumite	$4Mg_2SiO_4 \cdot MgF_2$	monoclinic	3.21	6	1.633	1.647	1.668
Clinozoisite	$Ca_2Al_3(SiO_4)_3OH$	monoclinic	3.30	6.5	1.693	1.700	1.712
Cobaltite	$CoAsS$	cubic	≈6.1	5.5			
Colemanite	$Ca_2B_6O_{11} \cdot 5H_2O$	monoclinic	2.42	4.5	1.586	1.592	1.614
Columbite	$(Fe,Mn)(Nb,Ta)_2O_6$	rhombohedral	5.20	6			
Connellite	$Cu_{19}(SO_4)Cl_4(OH)_{32} \cdot 3H_2O$	hexagonal	3.36	3	1.731	1.752	
Copiapite	$(Fe,Mg)Fe_4(SO_4)_6(OH)_2 \cdot 20H_2O$	triclinic	2.13	2.8	1.52	1.54	1.59
Coquimbite	$Fe_2(SO_4)_3 \cdot 9H_2O$	hexagonal	2.1	2.5	1.54	1.56	
Cordierite	$Al_3(Mg,Fe)_2Si_5AlO_{18}$	rhombohedral	2.66	7	1.540	1.549	1.553

Name	Formula	Crystal system	ρ/g cm^{-3}	Hardness	n_α	n_β	n_γ
Corundum	Al_2O_3	hexagonal	3.97	9	1.761	1.769	
Cotunnite	$PbCl_2$	orthorhombic	5.98	2.5	2.199	2.217	2.260
Covellite	CuS	hexagonal	4.8	1.8			
Cristobalite	SiO_2	hexagonal	2.33	6.5	1.484	1.487	
Crocoite	$PbCrO_4$	monoclinic	6.12	2.8	2.29	2.36	2.66
Cryolite	Na_3AlF_6	monoclinic	2.97	2.5	1.338	1.338	1.339
Cryolithionite	$Na_3Li_3Al_2F_{12}$	cubic	2.77	2.8	1.340		
Cubanite	$CuFe_2S_3$	rhombohedral	4.11	3.5			
Cummingtonite	$(Mg,Fe,Mn)_7(Si_4O_{11})_2(OH)_2$	monoclinic	3.4	5.5	1.650	1.660	1.676
Cuprite	Cu_2O	cubic	6.0	3.8			
Danburite	$CaB_2Si_2O_8$	rhombohedral	3.0	7	1.63	1.63	1.63
Datolite	$CaBSiO_4(OH)$	monoclinic	2.98	5.3	1.624	1.652	1.668
Daubreelite	Cr_2FeS_4	cubic	3.81				
Derbylite	$Fe_6Ti_6Sb_2O_{23}$	rhombohedral	4.53	5	2.45	2.45	2.51
Diaspore	$AlO(OH)$	orthorhombic	3.4	6.8	1.694	1.715	1.741
Digenite	$Cu_{1.79}S$	cubic	5.55	2.8			
Diopside	$CaMg(SiO_3)_2$	monoclinic	3.30	6	1.680	1.687	1.708
Dioptase	$CuSiO_2(OH)_2$	rhombohedral	3.5	5	1.65	1.70	
Dolomite	$CaMg(CO_3)_2$	rhombohedral	2.86	3.5	1.500	1.679	
Douglasite	$K_2FeCl_4{\cdot}2H_2O$	orthorhombic	2.16		1.488	1.500	
Dyscrasite	Ag_3Sb	rhombohedral	9.74	3.8			
Eddingtonite	$BaAl_2Si_3O_{10}{\cdot}4H_2O$	rhombohedral	2.8		1.541	1.553	1.557
Eglestonite	Hg_4OCl_2	cubic	8.4	2.5	2.49		
Emplectite	$CuBiS_2$	rhombohedral	6.38	2			
Enargite	Cu_3AsS_4	rhombohedral	4.5	3			
Enstatite	$MgSiO_3$	monoclinic	3.19	5.5	1.656	1.662	1.669
Epidote	$Ca_2Al_2FeOH(SiO_4)_3$	monoclinic	3.44	6	1.733	1.755	1.765
Epsomite	$MgSO_4{\cdot}7H_2O$	orthorhombic	1.67	2.3	1.433	1.455	1.461
Erythrite	$(Co,Ni)_3(AsO_4)_2{\cdot}8H_2O$	monoclinic	3.06	2	1.626	1.661	1.699
Eucairite	$AgCuSe$	orthorhombic	7.7	2.5			
Euclasite	$BeAlSiO_4(OH)$	monoclinic	3.1	7.5	1.651	1.655	1.671
Eudialite	$(Na,Ca,Ce)_5(Fe,Mn)(Zr,Ti)(Si_3O_9)_2(OH,Cl)$	hexagonal	3.0	5.5	1.623	1.600	1.615
Eulytite	$Bi_4(SiO_4)_3$	cubic	6.6	4.5	2.05		
Euxenite	$(Y,Ca,Ce,U,Th)(Nb,Ta,Ti)_2O_6$	rhombohedral	5.5	6	2.2		
Fayalite	Fe_2SiO_4	orthorhombic	4.30	6.5	1.827	1.869	1.879
Ferberite	$FeWO_4$	monoclinic	7.51	4.3			
Fergussonite	$(Y,Er,Ce,Fe)(Nb,Ta,Ti)O_4$	tetragonal	5.7	6	2.1		
Fluorite	CaF_2	cubic	3.18	4	1.434		
Forsterite	Mg_2SiO_4	orthorhombic	3.21	7	1.635	1.651	1.670
Franklinite	$ZnFe_2O_4$	cubic	5.21	6	2.36		
Gahnite	$ZnAl_2O_4$	cubic	4.62	7.8	1.805		
Galaxite	$MnAl_2O_4$	cubic	4.04	7.8	1.92		
Galena	PbS	cubic	7.60	2.5	3.91		
Galenabismuthite	$PbBi_2S_4$	rhombohedral	7.04	3			
Ganomalite	$(Ca,Pb)_{10}(OH,Cl)_2(Si_2O_7)_3$	hexagonal	5.6	3.5	1.910	1.945	
Gaylussite	$Na_2Ca(CO_3)_2{\cdot}5H_2O$	monoclinic	1.99	2.8	1.444	1.516	1.523
Gehlenite	$Ca_2Al_2SiO_7$	tetragonal	3.04	5.5	1.658	1.669	
Geikielite	$MgTiO_3$	hexagonal	3.85	5.5	1.95	2.31	
Gibbsite	$Al(OH)_3$	monoclinic	2.42	3	1.57	1.57	1.59
Glauberite	$Na_2Ca(SO_4)_2$	monoclinic	2.80	2.8	1.515	1.535	1.536
Glauconite	$(K,Na,Ca)_{1.6}(Fe,Al,Mg)_{4.0}Si_{7.3}Al_{0.7}O_{20}(OH)_4$	monoclinic	2.7	2	1.60	1.63	1.63
Glaucophane	$Na_2Mg_3Al_2[Si_8O_{22}](OH)_2$	monoclinic	3.19	6	1.634	1.645	1.648
Gmelinite	$(Ca,Na_2)[Al_2Si_4O_{12}]{\cdot}6H_2O$	hexagonal	2.10	4.5	1.477	1.485	
Goethite	$FeO(OH)$	orthorhombic	4.3	5.3	2.268	2.401	2.457
Goslarite	$ZnSO_4{\cdot}7H_2O$	orthorhombic	1.97	2.3	1.457	1.480	1.484
Greenockite	CdS	hexagonal	4.8	3.3	2.506	2.529	
Grossularite	$Ca_3Al_2(SiO_4)_3$	cubic	3.59	6.8	1.734		
Gummite	$UO_3{\cdot}H_2O$	orthorhombic	7.05	3.8			
Gypsum	$CaSO_4{\cdot}2H_2O$	monoclinic	2.32	2	1.520	1.525	1.530
Halite	$NaCl$	cubic	2.17	2	1.544		
Hambergite	$Be_2(OH,F)BO_3$	rhombohedral	2.36	7.5	1.56	1.59	1.63

Name	Formula	Crystal system	ρ/g cm^{-3}	Hardness	n_α	n_β	n_γ
Hanksite	$Na_{22}K(SO_4)_9(CO_3)_2Cl$	hexagonal	2.56	3.3	1.461	1.481	
Harmotome	$Ba[Al_2Si_6O_{16}]\cdot 6H_2O$	monoclinic	2.44	4.5	1.506	1.507	1.511
Hausmannite	Mn_3O_4	tetragonal	4.84	5.5	2.15	2.46	
Haüyne	$(Na,Ca)_{4-8}Al_6Si_6O_{24}(SO_4,S)_{1-2}$	cubic	2.47	5.8	1.502		
Hedenbergite	$CaFe(SiO_3)_2$	monoclinic	3.53	6	1.721	1.727	1.746
Helvite	$Mn_4Be_3Si_3O_{12}S$	cubic	3.32	6	1.739		
Hematite	Fe_2O_3	hexagonal	5.25	6	2.91	3.19	
Hemimorphite	$Zn_4(OH)_2Si_2O_7\cdot H_2O$	rhombohedral	3.45	5	1.614	1.617	1.636
Hercynite	$Fe(AlO_2)_2$	cubic	4.3	7.8	1.835		
Herderite	$CaBe(PO_4)(Fe,OH)$	monoclinic	2.98	5.3	1.592	1.612	1.621
Hessite	Ag_2Te	orthorhombic	8.4	2.5			
Heulandite	$(Ca,Na_2,K_2)[Al_2Si_7O_{18}]\cdot 6H_2O$	monoclinic	2.2	3.8	1.498	1.498	1.506
Hopeite	$Zn_3(PO_4)_2\cdot 4H_2O$	orthorhombic	3.0	3.2	1.58	1.59	1.59
Hornblende	$Ca_2(Mg,Fe)_4Al(Si_7AlO_{22})(OH)_2$	monoclinic	3.24	5.5	1.67	1.67	1.69
Huebnerite	$MnWO_4$	monoclinic	7.2	4.3	2.17	2.22	2.32
Humite	$Mg(OH,F)_2\cdot 3Mg_2SiO_4$	orthorhombic	3.3	6	1.625	1.636	1.657
Huntite	$Mg_3Ca(CO_3)_4$	trigonal-trapezohedral	2.70				
Hydrogrossularite	$Ca_3Al_2Si_2O_8(SiO_4)_{1-m}(OH)_{4m}$	cubic	3.4	6.8	1.70		
Hydromagnesite	$3MgCO_3\cdot Mg(OH)_2\cdot 3H_2O$	monoclinic	2.24	3.5	1.523	1.527	1.545
Illite	$KAl_4[Si_7AlO_{20}](OH)_4$	monoclinic	2.8	1.5	1.56	1.59	1.59
Ilmenite	$FeTiO_3$	rhombohedral	4.72	5.5			
Iodyrite	AgI	hexagonal	5.68	1.5	2.21	2.22	
Jacobsite	$MnFe_2O_4$	cubic	4.87	7.8	2.3		
Jadeite	$NaAl(SiO_3)_2$	monoclinic	3.34	6	1.649	1.654	1.663
Jamesonite	$Pb_4FeSb_6S_{14}$	monoclinic	5.63	2.5			
Jarosite	$KFe_3(SO_4)_2(OH)_6$	rhombohedral	3.09	3	1.715	1.820	
Kainite	$KMg(SO_4)Cl\cdot 3H_2O$	monoclinic	2.15	2.8	1.494	1.505	1.516
Kaliophilite	$KAlSiO_4$	hexagonal	2.61	6	1.532	1.537	
Kaolinite	$Al_2Si_2O_5(OH)_4$	triclinic	2.65	2.3	1.549	1.564	1.565
Kernite	$Na_2B_4O_7\cdot 4H_2O$	monoclinic	1.95	2.5	1.454	1.472	1.488
Kieserite	$MgSO_4\cdot H_2O$	monoclinic	2.57	3.5	1.520	1.533	1.584
Kyanite	Al_2OSiO_4	triclinic	3.59	6.3	1.715	1.722	1.731
Lanarkite	$Pb_2(SO_4)O$	monoclinic	6.92	2.3	1.928	2.007	2.036
Lanthanite	$(La,Ce)_2(CO_3)_3\cdot 8H_2O$	rhombohedral	2.72	2.8	1.52	1.587	1.613
Laumontite	$Ca_4[Al_8Si_{16}O_{48}]\cdot 16H_2O$	monoclinic	2.3	3.3	1.508	1.517	1.519
Laurionite	$Pb(OH)Cl$	rhombohedral	6.24	3.3	2.08	2.12	2.16
Lawsonite	$CaAl_2Si_2O_7(OH)_2\cdot H_2O$	rhombohedral	3.08	6	1.655	1.675	1.685
Lazulite	$(Mg,Fe)Al_2(PO_4)_2(OH)_2$	monoclinic	3.23	5.8	1.615	1.64	1.650
Lazurite	$Na_4SSi_3Al_3O_{12}$	cubic	2.42	5.3	1.500		
Leadhillite	$Pb_4(SO_4)(CO_3)_2(OH)_2$	monoclinic	6.55	2.8	1.87	2.00	2.01
Lepidocrocite	$FeO(OH)$	orthorhombic	4.26	5	1.94	2.20	2.51
Lepidolite	$K_2Al_3Li_2AlSi_7O_{20}(OH)_4$	monoclinic	2.85	3.3	1.536	1.565	1.566
Leucite	$KAlSi_2O_6$	tetragonal	2.49	5.8	1.510		
Levyne	$(Ca,Na_2)Al_2Si_4O_{12}\cdot 6H_2O$	rhombohedral	2.10	4.5	1.496	1.501	
Litharge	PbO	tetragonal	9.35	2	2.535	2.665	
Loellingite	$FeAs_2$	rhombohedral	7.40	5.3			
Maghemite	Fe_2O_3	cubic	4.88	7.8	2.63		
Magnesite	$MgCO_3$	hexagonal	3.05	4	1.536	1.741	
Magnetite	Fe_3O_4	cubic	5.17	6	2.42		
Malachite	$Cu_2(OH)_2CO_3$	monoclinic	4.05	3.8	1.655	1.875	1.909
Manganite	$MnO(OH)$	monoclinic	≈4.3	4	2.25	2.25	2.53
Manganosite	MnO	cubic	5.37	5.5			
Marcasite	FeS_2	cubic	5.02	6.3			
Marialite	$Na_4Al_3Si_9O_{24}Cl$	tetragonal	2.56	5.5	1.541	1.548	
Marshite	CuI	cubic	5.67	2.5	2.346		
Mascagnite	$(NH_4)_2SO_4$	orthorhombic	1.77	2.3	1.520	1.523	1.533
Matlockite	$PbClF$	tetragonal	7.05	2.8	2.006	2.145	
Meionite	$Ca_4Al_6Si_6O_{24}CO_3$	tetragonal	2.78	5.5	1.559	1.595	
Melanterite	$FeSO_4\cdot 7H_2O$	monoclinic	1.89	2	1.47	1.48	1.49
Melilite	$(Ca,Na)_2(Mg,Fe,Al,Si)_3O_7$	tetragonal	3.00	5.5	1.639	1.645	

Name	Formula	Crystal system	ρ/g cm^{-3}	Hardness	n_α	n_β	n_γ
Mellite	$Al_2C_{12}O_{12}\cdot 18H_2O$	tetragonal	1.64	2.3	1.511	1.539	
Mendipite	$Pb_3O_2Cl_2$	rhombohedral	7.24	2.5	2.24	2.27	2.31
Mesolite	$Na_2Ca_2(Al_2Si_3O_{10})_3\cdot 8H_2O$	orthorhombic	2.26	5	1.506		
Metacinnabar	HgS	cubic	7.70	3			
Microcline	$KAlSi_3O_8$	monoclinic	2.56	6.3	1.522	1.526	1.530
Miersite	AgI	hexagonal	5.68	2.5	2.20		
Millerite	NiS	hexagonal	5.5	3.3			
Mimetite	$Pb_5(AsO_4,PO_4)_3Cl$	hexagonal	7.24	3.8	2.128	2.147	
Minium	Pb_3O_4	tetragonal	8.9	2.5			
Mirabilite	$Na_2SO_4\cdot 10H_2O$	monoclinic	1.46	1.8	1.394	1.396	1.398
Moissanite	SiC	hexagonal	3.16	9.5	2.648	2.691	
Molybdenite	MoS_2	hexagonal	5.06	1.3			
Monazite	$(Ce,La,Th)PO_4$	monoclinic	5.2	5	1.787	1.789	1.840
Monetite	$CaHPO_4$	triclinic	2.92	3.5	1.587	1.61	1.640
Monticellite	$CaMgSiO_4$	orthorhombic	3.18	5.5	1.647	1.655	1.664
Montmorillonite	$(0.5Ca,Na)_{0.7}(Al,Mg,Fe)_4[(Si,Al)_8O_{20}](OH)_4\cdot nH_2O$	monoclinic	2.5	1.5	1.55	1.57	1.57
Montroydite	HgO	orthorhombic	11.14	2.5	2.37	2.50	2.65
Mordenite	$(Na,K,Ca)[Al_2Si_{10}O_{24}]\cdot 7H_2O$	orthorhombic	2.13	3.5	1.478	1.480	1.482
Muscovite	$KAl_2AlSi_3O_{10}(OH)_2$	monoclinic	2.83	2.8	1.563	1.596	1.602
Nantokite	CuCl	cubic	4.14	2.5	1.930		
Natrolite	$Na_2Al_2Si_3O_{10}\cdot 2H_2O$	orthorhombic	2.23	5	1.478	1.481	1.491
Nepheline	$Na_3KAl_4Si_4O_{16}$	hexagonal	2.61	5.8	1.534	1.538	
Newberyite	$MgHPO_4\cdot 3H_2O$	orthorhombic	2.13	3.3	1.514	1.517	1.533
Niccolite	NiAs	hexagonal	7.77	5.3			
Norbergite	$Mg(OH,F)_2\cdot Mg_2SiO_4$	orthorhombic	3.21	6.5	1.565	1.573	1.592
Nosean	$Na_8Al_6Si_6O_{24}SO_4$	cubic	2.35	5.5	1.495		
Oldhamite	CaS	cubic	2.59	4	2.137		
Oligoclase	$([NaSi]_{0.9\text{-}0.7}[CaAl]_{0.1\text{-}0.3})AlSi_2O_8$	triclinic	2.64	6.3	1.539	1.543	1.547
Olivenite	$Cu_2(AsO_4)(OH)$	rhombohedral	4.2	3	1.77	1.80	1.85
Olivine	$(Mg,Fe)SiO_4$	rhombohedral	3.81	6.8	1.73	1.76	1.78
Opal	$SiO_2\cdot nH_2O$	amorp	1.9	5	1.44		
Orpiment	As_2S_3	monoclinic	3.46	1.8	2.40	2.81	3.02
Orthoclase	$KAlSi_3O_8$	monoclinic	2.56	6	1.523	1.527	1.531
Orthopyroxene	$(Mg,Fe)SiO_3$	rhombohedral	3.6	5.5	1.709	1.712	1.723
Paragonite	$NaAl_2AlSi_3O_{10}(OH)_2$	monoclinic	2.85	2.5	1.572	1.602	1.605
Parisite	$(Ce,La,Na)FCO_3\cdot CaCO_3$	hexagonal	4.42	4.5	1.672	1.771	
Pectolite	$Ca_2NaH(SiO_3)_3$	triclinic	2.88	4.8	1.603	1.610	1.639
Penfieldite	$Pb_4Cl_6(OH)_2$	hexagonal	6.6		2.13	2.21	
Pentlandite	$Fe_{4.75}Ni_{5.25}S_8$	cubic	4.8	3.8			
Percylite	$PbCuCl_2(OH)_2$	cubic		2.5	2.05		
Periclase	MgO	cubic	3.6	5.5	1.735		
Perovskite	$CaTiO_3$	cubic	3.98	5.5	2.34		
Petalite	$LiAlSi_4O_{10}$	monoclinic	2.42	6.5	1.506	1.511	1.519
Pharmacosiderite	$Fe_3(AsO_4)_2(OH)_3\cdot 5H_2O$	cubic	2.80	2.5	1.690		
Phenakite	Be_2SiO_4	rhombohedral	2.98	7.5	1.654	1.670	
Phillipsite	$K(Ca_{0.5},Na)_2[Al_3Si_5O_{16}]\cdot 6H_2O$	monoclinic	2.2	4.3	1.494	1.497	1.505
Phlogopite	$KMg_3AlSi_3O_{10}(OH)_2$	monoclinic	2.83	2.3	1.560	1.597	1.598
Phosgenite	$Pb_2(CO_3)Cl_2$	tetragonal	6.13	2.5	2.118	2.145	
Piemontite	$Ca_2Al_{1.5}Mn_{1.5}(SiO_4)_3OH$	monoclinic	3.49	6	1.762	1.773	1.796
Pigeonite	$(Mg,Fe,Ca)(Mg,Fe)Si_2O_6$	monoclinic	3.38	6	1.702	1.703	1.728
Pollucite	$CsAlSi_2O_6$	tetragonal	2.9	6.5	1.517		
Polybasite	$(Ag,Cu)_{16}Sb_2S_{11}$	monoclinic	6.1	2.5			
Powellite	$Ca(Mo,W)O_4$	tetragonal	4.35	3.8	1.971	1.980	
Prehnite	$Ca_2Al_2Si_3O_{10}(OH)_2$	rhombohedral	2.93	6.3	1.622	1.628	1.648
Proustite	Ag_3AsS_3	rhombohedral	5.57	2.3	2.792	3.088	
Pseudobrookite	Fe_2TiO_5	rhombohedral	4.36	6	2.38	2.39	2.42
Psilomelane	$BaMn_9O_{16}(OH)_4$	rhombohedral	4.71	5.5			
Pumpellyite	$Ca_2Al_2(Al,Fe,Mg)[Si_2(O,OH)_7](SiO_4)(OH,O)_3$	monoclinic	3.21	5.5	1.688	1.695	1.705
Pyrargyrite	Ag_3SbS_3	rhombohedral	5.85	2.5	2.88	3.08	
Pyrite	FeS_2	cubic	5.02	6.3			

Name	Formula	Crystal system	ρ/g cm^{-3}	Hardness	n_α	n_β	n_γ
Pyrochlore	$NaCaNb_2O_6F$	cubic	5.3	5.3			
Pyrochroite	$Mn(OH)_2$	hexagonal	3.26	2.5	1.68	1.72	
Pyrolusite	MnO_2	tetragonal	5.08	6.3			
Pyromorphite	$Pb_5(PO_4,AsO_4)_3Cl$	hexagonal	7.04	3.8	2.048	2.058	
Pyrope	$Mg_3Al_2Si_3O_{12}$	cubic	3.58	6.8	1.714		
Pyrophyllite	$Al_2Si_4O_{10}(OH)_2$	monoclinic	2.78	1.5	1.545	1.579	1.599
Pyrrhotite	$Fe_{0.885}S$	hexagonal	4.62	4			
Quartz (α)	SiO_2	hexagonal	2.65	7	1.544	1.553	
Rammelsbergite	$NiAs_2$	orthorhombic	7.1	5.8			
Raspite	$PbWO_4$	monoclinic	8.46	2.8	1.27	1.27	1.30
Realgar	As_4S_4	monoclinic	3.5	1.8	2.538	2.684	2.704
Rhodochrosite	$MnCO_3$	hexagonal	3.70	3.8	1.597	1.816	
Rhodonite	$MnSiO_3$	orthorhombic	3.48	6	1.725	1.729	1.737
Riebeckite	$Na_2Fe_5FSi_8O_{22}(OH)_2$	monoclinic	3.3	5	1.675	1.683	1.694
Rutile	TiO_2	tetragonal	4.23	6.2	2.609	2.900	
Safflorite	$(Co,Fe)As_2$	rhombohedral	7.3	4.8			
Samarskite	$(Y,Er,Ce,U,Ca,Fe,Pb,Th)(Nb,Ta,Ti,Sn)_2O_6$	rhombohedral	5.69	5.5	2.200		
Sapphirine	$Mg_2Al_4O_6SiO_4$	monoclinic	3.49	7.5	1.709	1.712	1.715
Scapolite	$(Na,Ca)_4Al_3(Al,Si)_3Si_6O_{24}(Cl,F,OH,CO_3,SO_4)$	tetragonal	2.64	5.5	1.551	1.573	
Scheelite	$CaWO_4$	tetragonal	6.06	4.8	1.920	1.936	
Scolecite	$CaAl_2Si_3O_{10}\cdot 3H_2O$	monoclinic	2.27	5	1.510	1.518	1.519
Scorodite	$Fe(AsO_4)\cdot 2H_2O$	rhombohedral	3.28	3.8	1.784	1.795	1.814
Sellaite	MgF_2	tetragonal	3.15	5	1.378	1.390	
Senarmontite	Sb_2O_3	cubic	5.58	2.3	2.087		
Serpentine	$Mg_3Si_2O_5(OH)_4$	monoclinic	2.55	3	1.55	1.56	1.56
Siderite	$FeCO_3$	hexagonal	3.9	4.3	1.635	1.875	
Sillimanite	Al_2OSiO_4	rhombohedral	3.25	7	1.658	1.660	1.660
Co-Skutterudite	$(Co,Ni)As_3$	cubic	6.8	5.8			
Smithsonite	$ZnCO_3$	rhombohedral	4.4	4.3	1.621	1.848	
Sodalite	$Na_8Al_6Si_6O_{24}Cl_2$	cubic	2.30	5.8	1.485		
Sperrylite	$PtAs_2$	cubic	10.58	6.5			
Spessartite	$Mn_3Al_2Si_3O_{12}$	cubic	4.19	6.8	1.800		
Sphalerite	ZnS	cubic	4.0	3.8	2.369		
Sphene	$CaTiSiO_5$	monoclinic	3.50	5	1.90	1.95	2.03
Spinel	$MgAl_2O_4$	cubic	3.55	7.8	1.719		
Spodumene	$LiAl(SiO_3)_2$	monoclinic	3.13	6.8	1.656	1.662	1.671
Stannite	Cu_2FeSn_4	tetragonal	4.4	4			
Staurolite	$Fe_2Al_9Si_4O_{22}(OH)_2$	monoclinic	3.79	7.5	1.743	1.747	1.755
Stercorite	$Na(NH_4)H(PO_4)\cdot 4H_2O$	triclinic	1.62	2	1.439	1.442	1.469
Stibiotantalite	$Sb(Ta,Nb)O_4$	rhombohedral	6.6	5.5	2.38	2.41	2.46
Stibnite	Sb_2S_3	orthorhombic	4.56	2			
Stilbite	$NaCa_2[Al_5Si_{13}O_{36}]\cdot 14H_2O$	monoclinic	2.2	3.8	1.492	1.499	1.503
Stilpnomelane	$(K,Na,Ca)_{0.6}(Fe,Mg)_6Si_8Al(O,OH)_{27}\cdot 2H_2O$	monoclinic	2.8	3.5	1.585	1.665	1.665
Stolzite	$PbWO_4$	tetragonal	8.2	2.8	2.19	2.27	
Strengite	$FePO_4\cdot 2H_2O$	orthorhombic	2.87	4	1.707	1.719	1.741
Strontianite	$SrCO_3$	orthorhombic	3.5	3.5	1.518	1.666	1.668
Struvite	$Mg(NH_4)(PO_4)\cdot 6H_2O$	rhombohedral	1.71	2	1.495	1.496	1.504
Sulfur (orthorhombic)	S_8	orthorhombic	2.07	2	1.958	2.038	2.245
Sylvanite	$(Ag,Au)Te_2$	monoclinic	8.16	1.8			
Sylvite	KCl	cubic	1.99	2	1.490		
Talc	$3MgO\cdot 4SiO_2\cdot H_2O$	monoclinic	2.71	1	1.545	1.592	1.595
Tantalite	$(Fe,Mn)(Ta,Nb)_2O_6$	rhombohedral	7.95	6.5	2.26	2.32	2.43
Tapiolite	$FeTa_2O_6$	tetragonal	7.9	6.3	2.27	2.42	
Tellurobismuthite	Bi_2Te_3	hexagonal	7.74	1.8			
Terlinguaite	Hg_2OCl	monoclinic	8.73	2.5	2.35	2.64	2.66
Tetrahedrite	$(Cu,Fe)_{12}Sb_4S_{13}$	cubic	4.9	3.8			
Thenardite	Na_2SO_4	orthorhombic	2.7	2.8	1.468	1.475	1.483
Thermonatrite	$Na_2CO_3\cdot H_2O$	orthorhombic	2.25	1.3	1.420	1.506	1.524
Thomsenolite	$NaCaAlF_6\cdot H_2O$	monoclinic	2.98	2	1.407	1.414	1.415
Thorianite	ThO_2	cubic	10.0	6.5	2.200		
Thorite	$ThSiO_4$	tetragonal	6.7	4.8	1.8		

Inorganic

Name	Formula	Crystal system	ρ/g cm^{-3}	Hardness	n_α	n_β	n_γ
Topaz	$Al_2(SiO_4)(F,OH)_2$	rhombohedral	3.53	8	1.618	1.620	1.627
Torbernite	$Cu(UO_2)_2(PO_4)_2{\cdot}8H_2O$	tetragonal	3.22	2.3	1.582	1.592	
Tourmaline	$Na(Mg,Fe,Mn,Li,Al)_3Al_6Si_6O_{18}(BO_3)_3$	rhombohedral	3.14	7	1.62	1.65	
Tremolite	$Ca_2Mg_5Si_8O_{22}(OH)_2$	monoclinic	3.0	5.5	1.599	1.612	1.622
Trevorite	$NiFe_2O_4$	cubic	5.33	7.8	2.3		
Tridymite	SiO_2	hexagonal	2.27	7	1.475	1.476	1.479
Triphyllite-Lithiophyllite	$Li(Fe,Mn)PO_4$	rhombohedral	3.46	4.5	1.68	1.68	1.69
Troegerite	$(UO_2)_3(AsO_4)_2{\cdot}12H_2O$	tetragonal		2.5	1.59	1.630	
Troilite	FeS	hexagonal	4.7	4			
Trona	$Na_3H(CO_3)_2{\cdot}2H_2O$	monoclinic	2.14	2.8	1.412	1.492	1.540
Turquois	$CuAl_6(PO_4)_4(OH)_8{\cdot}4H_2O$	triclinic	2.9	5.3	1.70	1.73	1.75
Ullmannite	NiSbS	cubic	6.65	5.3			
Uraninite	UO_2	cubic	11.0	5.5			
Uvarovite	$Ca_3Cr_2Si_3O_{12}$	cubic	3.83	6.8	1.865		
Valentinite	Sb_2O_3	orthorhombic	5.7	2.8	2.18	2.35	2.35
Vanadinite	$Pb_5(VO_4)_3Cl$	hexagonal	6.8	2.9	2.350	2.416	
Variseite-Strengite	$(Al,Fe)(PO_4){\cdot}2H_2O$	rhombohedral	2.72	4	1.635	1.654	1.668
Vaterite	$CaCO_3$	hexagonal	2.71		1.550	1.645	
Vermiculite	$(Mg,Ca)_{0.7}(Mg,Fe,Al)_6[(Al,Si)_8O_{20}](OH)_4{\cdot}8H_2O$	monoclinic	2.3	1.5	1.542	1.556	1.556
Vesuvianite	$Ca_{10}(Mg,Fe)_2Al_4(Si_2O_7)_2(SiO_4)_5(OH,F)_4$	tetragonal	3.33	6.5	1.72	1.73	
Villiaumite	NaF	cubic	2.78	2.3	1.327		
Vivianite	$Fe_3(PO_4)_2{\cdot}8H_2O$	monoclinic	2.58	1.8	1.598	1.629	1.652
Wagnerite	$Mg_2(PO_4)F$	monoclinic	3.15	5.3	1.568	1.572	1.582
Wavellite	$Al_3(OH)_3(PO_4)_2{\cdot}5H_2O$	rhombohedral	2.36	3.6	1.527	1.535	1.553
Whewellite	$CaC_2O_4{\cdot}H_2O$	cubic	2.2	2.8	1.491	1.554	1.650
Willemite	Zn_2SiO_4	hexagonal	4.1	5.5	1.691	1.719	
Witherite	$BaCO_3$	orthorhombic	4.29	3.5	1.529	1.676	1.677
Wolframite	$Fe_{0.5}Mn_{0.5}WO_4$	monoclinic	7.3	4.3	2.26	2.32	2.42
Wollastonite	$CaSiO_3$	monoclinic	2.92	4.8	1.628	1.639	1.642
Wulfenite	$PbMoO_4$	tetragonal	6.7	2.9	2.283	2.403	
Wurtzite	ZnS	hexagonal	4.09	3.8	2.356	2.378	
Xenotime	YPO_4	tetragonal	4.8	4.5	1.721	1.816	
Zeunerite	$Cu(UO_2)_2(AsO_4)_2{\cdot}10H_2O$	tetragonal			1.606		
Zincite	ZnO	hexagonal	5.6	4	2.013	2.029	
Zircon	$ZrSiO_4$	tetragonal	4.6	7.5	1.94	1.99	
Zoisite	$Ca_2Al_3(SiO_4)_3OH$	rhombohedral	3.26	6	1.695	1.699	1.711

CRYSTALLOGRAPHIC DATA ON MINERALS

This table contains x-ray crystallographic data on about 400 common minerals, as well as selected crystalline elements. Entries are arranged alphabetically by mineral name. The columns are:

Name: Common name of the mineral.
Formula: Chemical formula for a typical sample of the mineral. Composition often varies considerably with the origin of the sample.
Crystal system
Structure type: Prototype for the structural arrangement of the crystallographic cell.
***Z*:** Number of formula units per the unit cell.
***a, b, c*:** Lengths of the cell edges in Å (1 Å = 10^{-8} cm).
***α, β, γ*:** Angles between cell axes in degrees.

References

1. Robie, R.A., Bethke, P.M., and Beardsley, K.M., *U. S. Geological Survey Bulletin 1248*, U. S. Government Printing Office, Washington, D.C.
2. Donnay, J.D.H., and Ondik, H.M., *Crystal Data Determinative Tables, Third Edition, Volume 2, Inorganic Compounds*, Joint Committee on Powder Diffraction Standards, Swarthmore, PA, 1973.
3. Deer, W.A., Howie, R.A., and Zussman, J., *An Introduction to the Rock-Forming Minerals, 2nd Edition*, Longman Scientific & Technical, Harlow, Essex, 1992.

Crystallographic Data on Minerals

Name	Formula	Crystal system	Structure type	*Z*	*a*/Å	*b*/Å	*c*/Å	*α*/°	*β*/°	*γ*/°
Acanthite	Ag_2S	monoclinic		4	4.228	6.928	7.862		99.58	
Acmite	$NaFe(SiO_3)_2$	monoclinic	diopside	4	9.658	8.795	5.294		107.42	
Akermanite	$Ca_2MgSi_2O_7$	tetragonal	melilite	2	7.8435		5.010			
Alabandite	MnS	cubic	rock salt	4	5.223					
Almandine	$Fe_3Al_2Si_3O_{12}$	cubic	garnet	8	11.526					
Altaite	PbTe	cubic	rock salt	4	6.4606					
Aluminum	Al	cubic	copper	4	4.049					
Alunite	$KAl_3(SO_4)_2(OH)_6$	rhombohedral		3	6.982		17.32			
Analcite	$NaAl(SiO_3)_2{\cdot}H_2O$	cubic		16	13.733					
Anatase	TiO_2	tetragonal		4	3.785		9.514			
Andalusite	Al_2OSiO_4	orthorhombic		4	7.7959	7.8983	5.5583			
Andradite	$Ca_3Fe_2Si_3O_{12}$	cubic	garnet	8	12.048					
Anglesite	$PbSO_4$	orthorhombic	barite	4	8.480	5.398	6.958			
Anhydrite	$CaSO_4$	orthorhombic	anhydrite	4	6.991	6.996	6.238			
Annite	$KFe_3[AlSi_3O_{10}](OH)_2$	monoclinic	1M mica	2	10.29	9.33	5.39		105.1	
Anorthite	$CaAl_2Si_2O_8$	triclinic	primitive cell	8	8.177	12.877	14.169	93.17	115.85	91.22
Anthophyllite	$Mg_7Si_8O_{22}(OH)_2$	orthorhombic		4	18.61	18.01	5.24			
Antimony	Sb	rhombohedral	arsenic	6	4.2996		11.2516			
Aragonite	$CaCO_3$	orthorhombic	aragonite	4	5.741	7.968	4.959			
Arcanite	K_2SO_4	orthorhombic	arcanite	4	5.772	10.072	7.483			
Argentite	Ag_2S	cubic		2	4.870					
Argentopyrite	$AgFe_2S_3$	orthorhombic		4	6.64	11.47	6.45			
Arsenic (gray)	As	rhombohedral	arsenic	6	3.760		10.555			
Arsenolite	As_2O_3	cubic	diamond	16	11.074					
Arsenopyrite	FeAsS	triclinic		4	5.760	5.690	5.785	90.00	112.23	90.00
Azurite	$Cu_3(OH)_2(CO_3)_2$	monoclinic		2	5.008	5.844	10.336		92.45	
Baddeleyite	ZrO_2	monoclinic	baddeleyite	4	5.1454	5.2075	5.3107		99.23	
Banalsite	$BaNa_2Al_4Si_4O_{16}$	orthorhombic		4	8.50	9.97	16.72			
Barite	$BaSO_4$	orthorhombic	barite	4	8.878	5.450	7.152			
Berlinite	$AlPO_4$	hexagonal	α-quartz	3	4.942		10.97			
Beryl	$Be_3Al_2(SiO_3)_6$	hexagonal	beryl	2	9.215		9.192			
Berzelianite	Cu_2Se	cubic		4	5.85					
Bismite	Bi_2O_3	monoclinic	pseudo-orth	4	7.48	8.14	5.83		112.9	
Bismuth	Bi	rhombohedral	arsenic	6	4.5367		11.8383			
Bismuthinite	Bi_2S_3	orthorhombic	stibnite	4	11.150	11.300	3.981			
Bixbyite	$(Mn,Fe)_2O_3$	cubic	thallium trioxide	16	9.411					
Boehmite	AlO(OH)	orthorhombic	lepidocrocite	4	2.868	12.227	3.700			
Borax	$Na_2B_4O_7{\cdot}10H_2O$	monoclinic		4	11.858	10.674	12.197		106.68	
Bornite (metastable)	Cu_5FeS_4	cubic		8	10.94					
Braggite	PtS	tetragonal		2	3.4699		6.1098			
Breithauptite	NiSb	hexagonal	niccolite	2	3.942		5.155			
Brochantite	$Cu_4SO_4(OH)_6$	monoclinic		4	13.066	9.85	6.022		103.27	

Name	Formula	Crystal system	Structure type	Z	a/Å	b/Å	c/Å	α/°	β/°	γ/°
Bromellite	BeO	hexagonal	zincite	2	2.6979		4.3772			
Bromyrite	$AgBr$	cubic	rock salt	4	5.7745					
Brookite	TiO_2	orthorhombic		8	5.456	9.182	5.143			
Brucite	$Mg(OH)_2$	hexagonal	cadmium iodide	1	3.147		4.769			
Bunsenite	NiO	cubic	rock salt	4	4.177					
Bustamite	$CaMn(SiO_3)_2$	triclinic		6	7.736	7.157	13.824	90.52	94.58	103.87
Cadmium telluride	$CdTe$	cubic	sphalerite	4	6.4805					
Cadmoselite	$CdSe$	hexagonal	zincite	2	4.2977		7.0021			
Calcite	$CaCO_3$	rhombohedral	calcite	6	4.9899		17.064			
Calomel	Hg_2Cl_2	tetragonal		4	4.478		10.910			
Carbonate-apatite	$Ca_{10}(PO_4)_6CO_3{\cdot}H_2O$	hexagonal	apatite	1	9.436		6.883			
Carbon (diamond)	C	cubic	diamond	8	3.5670					
Carbon (graphite)	C	hexagonal	graphite	4	2.4612		6.7079			
Cassiterite	SnO_2	tetragonal	rutile	2	4.738		3.188			
Cattierite	CoS_2	cubic	pyrite	4	5.5345					
Celestite	$SrSO_4$	orthorhombic	barite	4	8.359	5.352	6.866			
Celsian	$BaAl_2Si_2O_8$	monoclinic		8	8.627	13.045	14.408		115.20	
Cerargyrite	$AgCl$	cubic	rock salt	4	5.5491					
Cerianite	CeO_2	cubic	fluorite	4	5.4110					
Cerussite	$PbCO_3$	orthorhombic	aragonite	4	6.152	8.436	5.195			
Cervantite	Sb_2O_4	cubic		8	10.305					
Chalcanthite	$CuSO_4{\cdot}5H_2O$	triclinic		2	6.1045	10.72	5.949	97.57	107.28	77.43
Chalcocite	Cu_2S	orthorhombic		96	11.881	27.323	13.491			
Chalcopyrite	$CuFeS_2$	tetragonal		4	5.2988		10.434			
Chlorapatite	$Ca_5(PO_4)_3Cl$	hexagonal	apatite	2	9.629		6.777			
Chloritoid	$FeAl_4O_2(SiO_4)_2(OH)_4$	monoclinic		8	9.48	5.48	18.18		101.77	
Chloromagnesite	$MgCl_2$	rhombohedral		3	3.632		17.795			
Chondrodite	$2Mg_2SiO_4{\cdot}MgF_2$	monoclinic		2	7.89	4.743	10.29		109.03	
Chrysoberyl	$BeAl_2O_4$	orthorhombic	olivine	4	5.4756	9.4041	4.4267			
Cinnabar	HgS	hexagonal	cinnabar	3	4.149		9.495			
Claudetite	As_2O_3	monoclinic		4	5.339	12.984	4.5405		94.27	
Clausthalite	$PbSe$	cubic	rock salt	4	6.1255					
Clinoenstatite	$MgSiO_3$	monoclinic		8	9.620	8.825	5.188		108.33	
Clinoferrosilite	$FeSiO_3$	monoclinic		8	9.7085	9.0872	5.2284		108.43	
Clinohumite	$4Mg_2SiO_4{\cdot}MgF_2$	monoclinic		2	13.68	4.75	10.27		100.83	
Clinozoisite	$Ca_2Al_3(SiO_4)_3OH$	monoclinic		2	8.887	5.581	10.14		115.93	
Cobalticalcite	$CoCO_3$	rhombohedral	calcite	6	4.6581		14.958			
Cobaltite	$CoAsS$	cubic	NiSbS	4	5.60					
Cobalt olivine	Co_2SiO_4	orthorhombic	olivine	4	4.782	10.301	6.003			
Cobalt(II) oxide	CoO	cubic	rock salt	4	4.260					
Cobalt(II) sulfide	CoS	cubic	sphalerite	4	5.339					
Cobalt(II) titanate	$CoTiO_3$	rhombohedral	ilmenite	6	5.066		13.918			
Coesite	SiO_2	monoclinic		16	7.152	12.379	7.152		120.00	
Coffinite	$USiO_4$	tetragonal	zircon	4	6.995		6.263			
Colemanite	$Ca_2B_6O_{11}{\cdot}5H_2O$	monoclinic		4	8.743	11.264	6.102		110.12	
Coloradoite	$HgTe$	cubic	sphalerite	4	6.4600					
Copper	Cu	cubic	face-centered cubic	4	3.6150					
Fe-Cordierite	$Fe_2Al_3(AlSi_5O_{18})$	orthorhombic	cordierite	4	9.726	17.065	9.287			
Corundum	Al_2O_3	rhombohedral	corundum	6	4.7591		12.9894			
Cotunnite	$PbCl_2$	orthorhombic		4	4.535	7.62	9.05			
Covellite	CuS	hexagonal		6	3.792		16.34			
Cristobalite (α)	SiO_2	tetragonal		4	4.971		6.918			
Cristobalite (β)	SiO_2	cubic		8	7.1382					
Cryolite	Na_3AlF_6	monoclinic		2	5.40	5.60	7.776		90.18	
Cubanite	$CuFe_2S_3$	orthorhombic		4	6.46	11.12	6.23			
Cummingtonite	$(Mg,Fe,Mn)_7(Si_4O_{11})_2(OH)_2$	monoclinic	tremolite	2	9.522	18.223	5.332		101.92	
Cuprite	Cu_2O	cubic		2	4.2696					
Danburite	$CaB_2Si_2O_8$	orthorhombic		4	8.04	8.77	7.74			
Datolite	$CaBSiO_4(OH)$	monoclinic		4	9.62	7.60	4.84		90.15	

Inorganic

Name	Formula	Crystal system	Structure type	Z	a/Å	b/Å	c/Å	α/°	β/°	γ/°
Daubreeite	$FeCr_2S_4$	cubic	spinel	8	9.966					
Diaspore	$AlO(OH)$	orthorhombic		4	4.401	9.421	2.845			
Dickite	$Al_2Si_2O_5(OH)_4$	monoclinic		4	5.150	8.940	14.736		103.58	
Digenite	$Cu_{1.79}S$	cubic	deformed fluorite	4	5.5695					
Diopside	$CaMg(SiO_3)_2$	monoclinic	diopside	4	9.743	8.923	5.251		105.93	
Dioptase	$CuSiO_2(OH)_2$	rhombohedral	phenacite	18	14.61		7.80			
Dolerophanite	$Cu_2O(SO_4)$	monoclinic		4	8.334	6.312	7.628		108.4	
Dolomite	$CaMg(CO_3)_2$	rhombohedral	calcite	3	4.8079		16.010			
Dravite	$NaMg_3Al_6B_3Si_6O_{27}(OH)_4$	rhombohedral	tourmaline	3	15.942		7.224			
Elbaite	$NaLiAl_{7.67}B_3Si_6O_{27}(OH)_4$	rhombohedral	tourmaline	3	15.842		7.009			
Enargite	Cu_3AsS_4	orthorhombic		2	6.426	7.422	6.144			
Enstatite	$MgSiO_3$	orthorhombic		16	8.829	18.22	5.192			
Epidote	$Ca_2Al_2FeOH(SiO_4)_3$	monoclinic		2	8.89	5.63	10.19		115.40	
Epsomite	$MgSO_4{\cdot}7H_2O$	orthorhombic		4	11.86	11.99	6.858			
Eskolaite	Cr_2O_3	rhombohedral	corundum	6	4.9607		13.599			
Eucairite	$AgCuSe$	orthorhombic		10	4.105	20.35	6.31			
Euclase	$AlBeSiO_4(OH)$	monoclinic		4	4.763	14.29	4.618		100.25	
Famatimite	Cu_3SbS_4	tetragonal		2	5.384		10.770			
Fayalite	Fe_2SiO_4	orthorhombic	olivine	4	4.817	10.477	6.105			
Ferberite	$FeWO_4$	monoclinic	wolframite	2	4.732	5.708	4.965		90.00	
Ferriannite	$KFe_3[FeSi_3O_{10}](OH)_2$	monoclinic		2	5.430	9.404	10.341		100.07	
Ferroselite	$FeSe_2$	orthorhombic	marcasite	2	4.801	5.778	3.587			
Ferrotremolite	$Ca_2Fe_5[Si_8O_{22}](OH)_2$	monoclinic	tremolite	2	9.97	18.34	5.30		104.50	
Fluorapatite	$Ca_5(PO_4)_3F$	hexagonal	apatite	2	9.3684		6.8841			
Fluor-edenite	$NaCa_2Mg_5[AlSi_7O_{22}]F_2$	monoclinic	tremolite	2	9.847	18.00	5.282		104.83	
Fluor-humite	$3Mg_2SiO_4{\cdot}MgF_2$	orthorhombic		4	10.243	20.72	4.735			
Fluorite	CaF_2	cubic	fluorite	4	5.4638					
Fluor-norbergite	$Mg_2SiO_4{\cdot}MgF_2$	orthorhombic		4	8.727	10.271	4.709			
Fluor-phlogopite	$KMg_3[AlSi_3O_{10}]F_2$	monoclinic	1M mica	2	5.299	9.188	10.135		99.92	
Fluor-richterite	$Na_2CaMg_5[Si_8O_{22}]F_2$	monoclinic	tremolite	2	9.823	17.96	5.268		104.33	
Fluor-tremolite	$Ca_2Mg_5[Si_8O_{22}]F_2$	monoclinic	tremolite	2	9.781	18.01	5.267		104.52	
Forsterite	Mg_2SiO_4	orthorhombic	olivine	4	4.758	10.214	5.984			
Frohbergite	$FeTe_2$	orthorhombic	marcasite	2	5.265	6.265	3.869			
Gahnite	$ZnAl_2O_4$	cubic	spinel	8	8.0848					
Galaxite	$MnAl_2O_4$	cubic	spinel	8	8.258					
Galena	PbS	cubic	rock salt	4	5.9360					
Gallium(III) oxide	Ga_2O_3	rhombohedral	corundum	6	4.9793		13.429			
Gehlenite	$Ca_2Al_2SiO_7$	tetragonal	melilite	2	7.690		5.0675			
Fe-Gehlenite	$Ca_2Fe_2SiO_7$	tetragonal	melilite	2	7.54		4.855			
Geikielite	$MgTiO_3$	rhombohedral	ilmenite	6	5.054		13.898			
Gerhardite	$Cu_2(NO_3)(OH)_3$	orthorhombic		4	6.075	13.812	5.592			
Gersdorfite	$NiAsS$	cubic		4	5.693					
Gibbsite	$Al(OH)_3$	monoclinic		8	9.719	5.0705	8.6412		94.57	
Glauchroite	$CaMnSiO_4$	orthorhombic	olivine	4	4.944	11.19	6.529			
Glaucodot	$(Co,Fe)AsS$	orthorhombic		24	6.64	28.39	5.64			
Glaucophane I	$Na_2Mg_3Al_2[Si_8O_{22}](OH)_2$	monoclinic	tremolite	2	9.748	17.915	5.273		102.78	
Glaucophane II	$Na_2Mg_3Al_2[Si_8O_{22}](OH)_2$	monoclinic	tremolite	2	9.663	17.696	5.277		103.67	
Goethite	$FeO(OH)$	orthorhombic		4	4.596	9.957	3.021			
Gold	Au	cubic	face-centered cubic	4	4.0786					
Goldmanite	$Ca_3V_2Si_3O_{12}$	cubic	garnet	8	12.070					
Goslarite	$ZnSO_4{\cdot}7H_2O$	orthorhombic	epsomite	4	11.779	12.050	6.822			
Greenockite	CdS	hexagonal	zincite	2	4.1354		6.7120			
Greigite	Fe_3S_4	cubic	spinel	8	9.876					
Grossularite	$Ca_3Al_2(SiO_4)_3$	cubic	garnet	8	11.851					
Grunerite	$Fe_7[Si_8O_{22}](OH)_2$	monoclinic	tremolite	2	9.572	18.44	5.342		101.77	
Gudmundite	$FeSbS$	monoclinic		8	10.00	5.93	6.73		90.00	
Gypsum	$CaSO_4{\cdot}2H_2O$	monoclinic		4	5.68	15.18	6.29		113.83	
Hafnia	HfO_2	monoclinic	baddeleyite	4	5.1156	5.1722	5.2948		99.18	
Halite	$NaCl$	cubic	rock salt	4	5.6402					
Hambergite	$Be_2(OH,F)BO_3$	orthorhombic		8	9.755	12.201	4.426			

Name	Formula	Crystal system	Structure type	Z	a/Å	b/Å	c/Å	α/°	β/°	γ/°
Hardystonite	$Ca_2ZnSi_2O_7$	tetragonal	melilite	2	7.87		5.01			
Hauerite	MnS_2	cubic	pyrite	4	6.1014					
Hausmannite	Mn_3O_4	tetragonal		8	8.136		9.422			
Hawleyite	CdS	cubic	sphalerite	4	5.833					
Heazelwoodite	Ni_3S_2	rhombohedral		3	5.746		7.134			
Hedenbergite	$CaFe(SiO_3)_2$	monoclinic	diopside	4	9.854	9.024	5.263		104.23	
Hematite	Fe_2O_3	rhombohedral	corundum	6	5.025		13.735			
Hemimorphite	$Zn_4(OH)_2Si_2O_7{\cdot}H_2O$	orthorhombic		2	8.370	10.719	5.120			
Hercynite	$Fe(AlO_2)_2$	cubic	spinel	8	8.150					
Herzenbergite	SnS	orthorhombic	germanium sulfide	4	4.328	11.190	3.978			
Hessite	Ag_2Te	monoclinic		4	8.13	4.48	8.09		111.9	
Hexahydrite	$MgSO_4{\cdot}6H_2O$	monoclinic		8	10.110	7.212	24.41		98.30	
High albite	$NaAlSi_3O_8$	triclinic		4	8.160	12.870	7.106	93.54	116.36	90.19
High argentite	Ag_2S	cubic		4	6.269					
High bornite	Cu_5FeS_4	cubic		1	5.50					
High carnegeite	$NaAlSiO_4$	cubic		4	7.325					
High chalcocite	Cu_2S	hexagonal		2	3.961		6.722			
High clinoenstatite	$MgSiO_3$	triclinic		8	10.000	8.934	5.170	88.27	70.03	91.01
High digenite	Cu_2S	cubic		4	5.725					
High germania	GeO_2	hexagonal	α-quartz	3	4.987		5.652			
High leucite	$KAlSi_2O_6$	cubic		16	13.43					
High naumannite	Ag_2Se	cubic		2	4.993					
High sanidine	$KAlSi_3O_8$	monoclinic		4	8.615	13.031	7.177		115.98	
Huebnerite	$MnWO_4$	monoclinic	wolframite	2	4.834	5.758	4.999		91.18	
Huntite	$Mg_3Ca(CO_3)_4$	rhombohedral	calcite	3	9.498		7.816			
Hydroxylapatite	$Ca_5(PO_4)_3OH$	hexagonal	apatite	2	9.418		6.883			
Ice	H_2O	hexagonal		4	4.5212		7.3666			
Ilmenite	$FeTiO_3$	rhombohedral	ilmenite	6	5.093		14.055			
Indialite	$Mg_2Al_3(AlSi_5O_{18})$	hexagonal	beryl	2	9.7698		9.3517			
Fe-Indialite	$Fe_2Al_3(AlSi_5O_{18})$	hexagonal	beryl	2	9.860		9.285			
Mn-Indialite	$Mn_2Al_3(AlSi_5O_{18})$	hexagonal	beryl	2	9.925		9.297			
Iodargyrite	AgI	hexagonal	zincite	2	4.5955		7.5005			
Iron	Fe	cubic	body-centered cubic	2	2.8664					
Jacobsite	$MnFe_2O_4$	cubic	spinel	8	8.499					
Jadeite	$NaAl(SiO_3)_2$	monoclinic	diopside	4	9.409	8.564	5.220		107.50	
Jalpaite	$Ag_{1.55}Cu_{0.45}S$	tetragonal		16	8.673		11.756			
Johannsenite	$CaMn(SiO_3)_2$	monoclinic	diopside	4	9.83	9.04	5.27		105.00	
Kaliophilite	$KAlSiO_4$	hexagonal		54	26.930		8.522			
Kalsilite	$KAlSiO_4$	hexagonal		2	5.1597		8.7032			
Kaolinite	$Al_2Si_2O_5(OH)_4$	triclinic		2	5.155	8.959	7.407	91.68	104.87	89.93
Karelianite	V_2O_3	rhombohedral	corundum	6	4.952		14.002			
Keatite	SiO_2	tetragonal		12	7.456		8.604			
Kernite	$Na_2B_4O_7{\cdot}4H_2O$	monoclinic		4	7.022	9.151	15.676		108.83	
Kerschsteinite	$CaFeSiO_4$	orthorhombic	olivine	4	4.886	11.146	6.434			
Klockmannite	CuSe	hexagonal	deformed covellite	78	14.206		17.25			
Knebelite	$MnFeSiO_4$	orthorhombic	olivine	4	4.854	10.602	6.162			
Kyanite	Al_2OSiO_4	triclinic		4	7.123	7.848	5.564	89.92	101.25	105.97
Larnite	Ca_2SiO_4	monoclinic		4	5.48	6.76	9.28		94.55	
Laurite	RuS_2	cubic	pyrite	4	5.60					
Lawrencite	$FeCl_2$	rhombohedral		3	3.593		17.58			
Lawsonite	$CaAl_2Si_2O_7(OH)_2{\cdot}H_2O$	orthorhombic		4	8.787	5.836	13.123			
Lead	Pb	cubic	face-centered cubic	4	4.9505					
Leonhardtite	$MgSO_4{\cdot}4H_2O$	monoclinic		4	5.922	13.604	7.905		90.85	
Lepidocrocite	FeO(OH)	orthorhombic		4	3.868	12.525	3.066			
Lepidolite	$K_2Al_3Li_2AlSi_7O_{20}(OH)_4$	monoclinic	2M2 mica	2	9.2	5.3	20.0		98.00	
Leucite	$KAlSi_2O_6$	tetragonal		16	13.074		13.738			
Fe-Leucite	$KFeSi_2O_6$	tetragonal		16	13.205		13.970			
Lime	CaO	cubic	rock salt	4	4.8108					

Name	Formula	Crystal system	Structure type	Z	a/Å	b/Å	c/Å	α/°	β/°	γ/°
Lime olivine	Ca_2SiO_4	orthorhombic	olivine	4	5.091	11.371	6.782			
Linnaeite	Co_3S_4	cubic	spinel	8	9.401					
Litharge	PbO	tetragonal		2	3.9759		5.023			
Loellingite	$FeAs_2$	orthorhombic	marcasite	2	5.300	5.981	2.882			
Low albite	$NaAlSi_3O_8$	triclinic		4	8.139	12.788	7.160	94.27	116.57	87.68
Low bornite	Cu_5FeS_4	tetragonal		16	10.94		21.88			
Low cordierite	$Mg_2Al_3(AlSi_5O_{18})$	orthorhombic		4	9.721	17.062	9.339			
Low germania	GeO_2	tetragonal	rutile	2	4.3963		2.8626			
Low nepheline	$NaAlSiO_4$	hexagonal		8	9.986		8.330			
Luzonite	Cu_3AsS_4	tetragonal		2	5.289		10.440			
Mackinawite	FeS	tetragonal		2	3.675		5.030			
Magnesioriebeckite	$Na_2Mg_3Fe_2[Si_8O_{22}](OH)_2$	monoclinic	tremolite	2	9.733	17.946	5.299		103.30	
Magnesite	$MgCO_3$	rhombohedral	calcite	6	4.6330		15.016			
Magnetite	Fe_3O_4	cubic	spinel	8	8.3940					
Malachite	$Cu_2(OH)_2CO_3$	monoclinic		4	9.502	11.974	3.240		98.75	
Maldonite	Au_2Bi	cubic		8	7.958					
Manganese(II) sulfide (β form)	MnS	cubic	sphalerite	4	5.611					
Manganese(II) sulfide (γ form)	MnS	hexagonal	zincite	2	3.976		6.432			
Manganosite	MnO	cubic	rock salt	4	4.4448					
Marcasite	FeS_2	orthorhombic	marcasite	2	4.443	5.423	3.3876			
Margarite	$CaAl_2[AlSi_2O_{10}](OH)_2$	monoclinic	2M mica	4	5.13	8.92	19.50		95.00	
Marialite	$Na_4Al_3Si_9O_{24}Cl$	tetragonal		2	12.064		7.514			
Marshite	CuI	cubic	sphalerite	4	6.0507					
Mascagnite	$(NH_4)_2SO_4$	orthorhombic	arcanite	4	7.782	5.993	10.636			
Massicot	PbO	orthorhombic		4	5.489	4.755	5.891			
Matlockite	PbClF	tetragonal		2	4.106		7.23			
Maucherite	$Ni_{11}As_8$	tetragonal		4	6.870		21.81			
Meionite	$Ca_4Al_6Si_6O_{24}CO_3$	tetragonal		2	12.174		7.652			
Melanophlogite	SiO_2	cubic	clathrate type	46	13.402					
Melanterite	$FeSO_4{\cdot}7H_2O$	monoclinic		4	14.072	6.503	11.041		105.57	
Melonite	$NiTe_2$	hexagonal	cadmium iodide	1	3.869		5.308			
Metacinnabar	HgS	cubic	sphalerite	4	5.8517					
Miargyrite	$AgSbS_2$	monoclinic		8	12.862	4.111	13.220		98.63	
Microcline	$KAlSi_3O_8$	triclinic		4	8.582	12.964	7.222	90.62	115.92	87.68
Fe-Microcline	$KFeSi_3O_8$	triclinic		4	8.68	13.10	7.340	90.75	116.05	86.23
Miersite	AgI	cubic	sphalerite	4	6.4963					
Millerite	NiS	rhombohedral		9	9.616		3.152			
Minium	Pb_3O_4	tetragonal		4	8.815		6.565			
Minnesotaite	$Fe_3Si_4O_{10}(OH)_2$	monoclinic		4	5.4	9.42	19.4		100.00	
Mirabilite	$Na_2SO_4{\cdot}10H_2O$	monoclinic		4	11.51	10.38	12.83		107.75	
Molybdenite	MoS_2	hexagonal	molybdenite	2	3.1604		12.295			
Molybdenum	Mo	cubic		2	3.1653					
Molybdite	MoO_3	orthorhombic		4	3.962	13.858	3.697			
Monteponite	CdO	cubic	rock salt	4	4.6953					
Monticellite	$CaMgSiO_4$	orthorhombic	olivine	4	4.827	11.084	6.376			
Montroydite	HgO	orthorhombic		4	6.608	5.518	3.519			
Mullite (2:1)	$2Al_2O_3{\cdot}SiO_2$	orthorhombic		6	7.5788	7.6909	2.8883			
Mullite (3:2)	$3Al_2O_3{\cdot}2SiO_2$	orthorhombic		3	7.557	7.6876	2.8842			
Muscovite	$KAl_2AlSi_3O_{10}(OH)_2$	monoclinic	2M2 mica	4	5.203	8.995	20.030		94.47	
Nacrite	$Al_2Si_2O_5(OH)_4$	monoclinic		4	8.909	5.146	15.697		113.70	
Nantokite	CuCl	cubic	sphalerite	4	5.416					
Natroalunite	$NaAl_3(SO_4)_2(OH)_6$	rhombohedral		3	6.974		16.69			
Natrolite	$Na_2Al_2Si_3O_{10}{\cdot}2H_2O$	orthorhombic		8	18.30	18.63	6.60			
Neighborite	$NaMgF_3$	orthorhombic	perovskite	4	5.363	7.676	5.503			
Niccolite	NiAs	hexagonal	niccolite	2	3.618		5.034			
Nickel	Ni	cubic	face-centered cubic	4	3.5238					
Nickel(II) carbonate	$NiCO_3$	rhombohedral	calcite	6	4.5975		14.723			
Nickel olivine	Ni_2SiO_4	orthorhombic	olivine	4	4.727	10.121	5.915			

Name	Formula	Crystal system	Structure type	Z	a/Å	b/Å	c/Å	α/°	β/°	γ/°
Nickel selenide [$NiSe_2$]	$NiSe_2$	cubic	pyrite	4	5.9604					
Niter	KNO_3	orthorhombic	aragonite	4	6.431	9.164	5.414			
Norsethite	$BaMg(CO_3)_2$	rhombohedral	calcite	3	5.020		16.75			
Oldhamite	CaS	cubic	rock salt	4	5.689					
Orpiment	As_2S_3	monoclinic		4	11.49	9.59	4.25		90.45	
Orthoclase	$KAlSi_3O_8$	monoclinic		4	8.562	12.996	7.193		116.02	
Orthoferrosilite	$FeSiO_3$	orthorhombic	enstatite	16	9.080	18.431	5.238			
Otavite	$CdCO_3$	rhombohedral	calcite	6	4.9204		16.298			
Paracelsian	$BaAl_2Si_2O_8$	monoclinic		4	8.58	9.583	9.08		90.00	
Paragonite	$NaAl_2AlSi_3O_{10}(OH)_2$	monoclinic	2M1 mica	4	5.13	8.89	19.32		95.17	
Pararammelsbergite	$NiAs_2$	orthorhombic		8	5.75	5.82	11.428			
Paratellurite	TeO_2	tetragonal		4	4.810		7.613			
Parawollastonite	$CaSiO_3$	monoclinic		12	15.417	7.321	7.066		95.40	
Pectolite	$Ca_2NaH(SiO_3)_3$	triclinic		2	7.99	7.04	7.02	90.05	95.27	102.47
Pentlandite	$Fe_{4.75}Ni_{5.25}S_8$	cubic		4	10.095					
Periclase	MgO	cubic	rock salt	4	4.2117					
Perovskite	$CaTiO_3$	orthorhombic	perovskite	4	5.3670	7.6438	5.4439			
Petalite	$LiAlSi_4O_{10}$	monoclinic		2	11.32	5.14	7.62		105.90	
Petzite	Ag_3AuTe_2	cubic		8	10.38					
Phenacite	Be_2SiO_4	rhombohedral	phenacite	18	12.472		8.252			
Phlogopite	$KMg_3AlSi_3O_{10}(OH)_2$	monoclinic	1M mica	2	5.326	9.210	10.311		100.17	
Picrochromite	$MgCr_2O_4$	cubic	spinel	8	8.333					
Piemontite	$Ca_2Al_{1.5}Mn_{1.5}(SiO_4)_3OH$	monoclinic		2	8.95	5.70	9.41		115.70	
Platinum	Pt	cubic	face-centered cubic	4	3.9231					
Polymidite	Ni_3S_4	cubic	spinel	8	9.480					
Portlandite	$Ca(OH)_2$	hexagonal	cadmium iodide	1	3.5933		4.9086			
Powellite	$Ca(Mo,W)O_4$	tetragonal	scheelite	4	5.226		11.43			
Protoenstatite	$MgSiO_3$	orthorhombic		8	9.25	8.74	5.32			
Proustite	Ag_3AsS_3	rhombohedral		6	10.816		8.6948			
Pseudowollastonite	$CaSiO_3$	triclinic		24	6.90	11.78	19.65	90.00	90.80	90.00
Pyrargyrite	Ag_3SbS_3	rhombohedral		6	11.052		8.7177			
Pyrite	FeS_2	cubic	pyrite	4	5.4175					
Pyrolusite	MnO_2	tetragonal	rutile	2	4.388		2.865			
Pyrope	$Mg_3Al_2Si_3O_{12}$	cubic	garnet	8	11.459					
Pyrophanite	$MnTiO_3$	rhombohedral	ilmenite	6	5.155		14.18			
Pyrophyllite	$Al_2Si_4O_{10}(OH)_2$	monoclinic	2M1 mica	4	5.14	8.90	18.55		99.92	
Pyroxmangite	$MnFe(SiO_3)_2$	triclinic		7	7.56	17.45	6.67	84.00	94.30	113.70
Pyrrhotite	$Fe_{0.885}S$	hexagonal	defect niccolite	2	3.440		5.709			
Quartz (α)	SiO_2	hexagonal		3	4.9136		5.4051			
Quartz (β)	SiO_2	hexagonal		3	4.999		5.4592			
Rammelsbergite	$NiAs_2$	orthorhombic	marcasite	2	4.757	5.797	3.542			
Realgar	As_4S_4	monoclinic		16	9.29	13.53	6.57		106.55	
Retgersite	$NiSO_4{\cdot}4H_2O$	tetragonal		4	6.782		18.28			
Rhodochrosite	$MnCO_3$	rhombohedral	calcite	6	4.7771		15.664			
Rhodonite	$MnSiO_3$	triclinic		10	7.682	11.818	6.707	92.36	93.95	105.66
Riebeckite	$Na_2Fe_5FSi_8O_{22}(OH)_2$	monoclinic	tremolite	2	9.729	18.065	5.334		103.31	
Rutile	TiO_2	tetragonal		2	4.5937		2.9618			
Safflorite	$(Co,Fe)As_2$	orthorhombic	marcasite	2	5.231	5.953	2.962			
Fe-Sanidine	$KFeSi_3O_8$	monoclinic		4	8.689	13.12	7.319		116.10	
Sanmartinite	$ZnWO_4$	monoclinic	wolframite	2	4.691	5.720	4.925		89.36	
Sapphirine	$Mg_2Al_4O_6SiO_4$	monoclinic		8	9.96	28.60	9.85		110.5	
Scacchite	$MnCl_2$	rhombohedral		3	3.711		17.59			
Scheelite	$CaWO_4$	tetragonal	scheelite	4	5.242		11.372			
Schorl	$NaFe_3Al_6B_3Si_6O_{27}(OH)_4$	rhombohedral	tourmaline	3	16.032		7.149			
Selenium (gray)	Se	hexagonal		3	4.3642		4.9588			
Selenolite	SeO_2	tetragonal		8	8.35		5.05			
Sellaite	MgF_2	tetragonal	rutile	2	4.621		3.050			
Senarmontite	Sb_2O_3	cubic	arsenic trioxide	16	11.152					
Shandite	$Ni_3Pb_2S_2$	rhombohedral		3	5.576		13.658			
Shortite	$Na_2Ca_2(CO_3)_3$	orthorhombic		2	4.961	11.03	7.12			

Name	Formula	Crystal system	Structure type	Z	a/Å	b/Å	c/Å	α/°	β/°	γ/°
Siderite	$FeCO_3$	rhombohedral	calcite	6	4.6887		15.373			
Silicon	Si	cubic	diamond	8	5.4305					
Sillimanite	Al_2OSiO_4	orthorhombic		4	7.4843	7.6730	5.7711			
Silver	Ag	cubic	face-centered cubic	4	4.0862					
Silver telluride I	Ag_2Te	cubic		2	5.29					
Silver telluride II	Ag_2Te	cubic		4	6.585					
Fe-Skutterudite	$FeAs_{2.95}$	cubic		8	8.1814					
Ni-Skutterudite	$NiAs_{2.95}$	cubic		8	8.3300					
Smithsonite	$ZnCO_3$	rhombohedral	calcite	6	4.6528		15.025			
Soda niter	$NaNO_3$	rhombohedral	calcite	6	5.0696		16.829			
Sodium melilite	$NaCaAlSi_2O_7$	tetragonal	melilite	2	8.511		4.809			
Sperrylite	$PtAs_2$	cubic	pyrite	4	5.968					
Spessartite	$Mn_3Al_2Si_3O_{12}$	cubic	garnet	8	11.621					
Sphalerite	ZnS	cubic	sphalerite	4	5.4093					
Sphene	$CaTiSiO_5$	monoclinic		4	7.07	8.72	6.56		113.95	
Spinel	$MgAl_2O_4$	cubic	spinel	8	8.080					
Spodumene	$LiAl(SiO_3)_2$	monoclinic	diopside	4	9.451	8.387	5.208		110.07	
Spodumene (β)	$LiAl(SiO_3)_2$	tetragonal		4	7.5332		9.1540			
Staurolite	$Fe_2Al_9Si_4O_{22}(OH)_2$	monoclinic		2	7.90	16.65	5.63		90.00	
Sternbergite	$AgFe_2S_3$	orthorhombic		8	11.60	12.675	6.63			
Stibnite	Sb_2S_3	orthorhombic	stibnite	4	11.229	11.310	3.8389			
Stilleite	ZnSe	cubic	sphalerite	4	5.6685					
Stishovite	SiO_2	tetragonal	rutile	2	4.1790		2.6649			
Stolzite	$PbWO_4$	tetragonal	scheelite	4	5.4616		12.046			
Stromeyerite	$Ag_{0.93}Cu_{1.07}S$	orthorhombic		4	4.066	6.628	7.972			
Strontianite	$SrCO_3$	orthorhombic	aragonite	4	6.029	8.414	5.107			
Sulfur (monoclinic)	S	monoclinic	S8 ring molecules	48	11.04	10.98	10.92		96.73	
Sulfur (rhombohedral)	S_6	rhombohedral	S6 ring molecules	18	10.818		4.280			
Sulfur (orthorhombic)	S_8	orthorhombic	S8 ring molecules	128	10.4646	12.8660	24.4860			
Sylvite	KCl	cubic	rock salt	4	6.2931					
Syngenite	$K_2Ca(SO_4)_2{\cdot}H_2O$	monoclinic		2	9.775	7.156	6.251		104.00	
Synthetic anorthite (hexagonal)	$CaAl_2Si_2O_8$	hexagonal		2	5.10		14.72			
Synthetic anorthite (orthorhombic)	$CaAl_2Si_2O_8$	orthorhombic		2	8.22	8.60	4.83			
Talc	$3MgO{\cdot}4SiO_2{\cdot}H_2O$	monoclinic	2M1 mica	4	5.287	9.158	18.95		99.50	
Tantalum	Ta	cubic	tungsten	2	3.3058					
Teallite	$PbSnS_2$	orthorhombic	germanium sulfide	2	4.266	11.419	4.090			
Tellurite	TeO_2	orthorhombic	tellurite	8	5.607	12.034	5.463			
Tellurium	Te	hexagonal	selenium	3	4.4570		5.9290			
Tellurobismuthite	Bi_2Te_3	rhombohedral		3	4.3835		30.487			
Tennantite	$Cu_{12}As_4S_{13}$	cubic	tetrahedrite	2	10.190					
Tenorite	CuO	monoclinic		4	4.684	3.425	5.129		99.47	
Tephroite	Mn_2SiO_4	orthorhombic	olivine	4	4.871	10.636	6.232			
Tetrahedrite	$(Cu,Fe)_{12}Sb_4S_{13}$	cubic	tetrahedrite	2	10.327					
Thenardite	Na_2SO_4	orthorhombic	thenardite	8	5.863	12.304	9.821			
Thorianite	ThO_2	cubic	fluorite	4	5.5952					
Thorite	$ThSiO_4$	tetragonal	zircon	4	7.143		6.327			
Tiemannite	HgSe	cubic	sphalerite	4	6.0853					
Tin (white)	Sn	tetragonal		4	5.8315		3.1813			
Titanium	Ti	hexagonal		2	2.953		4.729			
Titanium(III) oxide	Ti_2O_3	rhombohedral	corundum	6	5.149		13.642			
Topaz	$Al_2(SiO_4)(F,OH)_2$	orthorhombic		4	8.394	8.792	4.649			
Tremolite	$Ca_2Mg_5Si_8O_{22}(OH)_2$	monoclinic	tremolite	2	9.840	18.052	5.275		104.70	
Trevorite	$NiFe_2O_4$	cubic	spinel	8	8.339					
Tridymite	SiO_2	hexagonal		4	5.0463		8.2563			

Inorganic

Name	Formula	Crystal system	Structure type	*Z*	*a*/Å	*b*/Å	*c*/Å	*α*/°	*β*/°	*γ*/°
Trogtalite	$CoSe_2$	cubic	pyrite	4	5.8588					
Troilite	FeS	hexagonal	niccolite	2	3.446		5.877			
Tschermakite	$CaAl_2SiO_6$	monoclinic	diopside	4	9.615	8.661	5.272		106.12	
Tungsten	W	cubic		2	3.1653					
Tungstenite	WS_2	hexagonal	molybdenite	2	3.154		12.362			
Turquois	$CuAl_6(PO_4)_4(OH)_8{\cdot}4H_2O$	triclinic		1	7.424	7.629	9.910	68.61	69.71	65.08
Umangite	Cu_3Se_2	tetragonal		2	6.402		4.276			
Uraninite	UO_2	cubic	fluorite	4	5.4682					
Ureyite	$NaCr(SiO_3)_2$	monoclinic	diopside	4	9.550	8.712	5.273		107.44	
Uvarovite	$Ca_3Cr_2Si_3O_{12}$	cubic	garnet	8	11.999					
Uvite	$CaMg_4Al_5B_3Si_6O_{27}(OH)_4$	rhombohedral	tourmaline	3	15.86		7.19			
Vaesite	NiS_2	cubic	pyrite	4	5.6873					
Valentinite	Sb_2O_3	orthorhombic	antimony trioxide	4	4.914	12.468	5.421			
Vanthoffite	$MgSO_4{\cdot}3Na_2SO_4$	monoclinic		2	9.797	9.217	8.199		113.50	
Vaterite	$CaCO_3$	hexagonal		6	7.135		8.524			
Villiaumite	NaF	cubic	rock salt	4	4.6342					
Violarite	$FeNi_2S_4$	cubic	spinel	8	9.464					
Willemite	Zn_2SiO_4	rhombohedral	phenacite	18	13.94		9.309			
Witherite	$BaCO_3$	orthorhombic	aragonite	4	6.430	8.904	5.314			
Wolframite	$Fe_{0.5}Mn_{0.5}WO_4$	monoclinic	wolframite	2	4.782	5.731	4.982		90.57	
Wollastonite	$CaSiO_3$	triclinic		6	7.94	7.32	7.07	90.03	95.37	103.43
Wulfenite	$PbMoO_4$	tetragonal	scheelite	4	5.435		12.110			
Wurtzite	ZnS	hexagonal	zincite	2	3.8230		6.2565			
Wustite	$Fe_{0.953}O$	cubic	defect rock salt	4	4.3088					
Xenotime	YPO_4	tetragonal	zircon	4	6.885		5.982			
Zinc	Zn	hexagonal	hexagonal close pack	2	2.665		4.947			
Zincite	ZnO	hexagonal	zincite	2	3.2495		5.2069			
Zinc telluride	$ZnTe$	cubic	sphalerite	4	6.1020					
Zinkosite	$ZnSO_4$	orthorhombic	barite	4	8.588	6.740	4.770			
Zircon	$ZrSiO_4$	tetragonal	zircon	4	6.604		5.979			
Zoisite	$Ca_2Al_3(SiO_4)_3OH$	orthorhombic		4	16.15	5.581	10.06			

Section 5
Thermochemistry, Electrochemistry, and Solution Chemistry

CODATA KEY VALUES FOR THERMODYNAMICS

The Committee on Data for Science and Technology (CODATA) has conducted a project to establish internationally agreed values for the thermodynamic properties of key chemical substances. This table presents the final results of the project. Use of these recommended, internally consistent values is encouraged in the analysis of thermodynamic measurements, data reduction, and preparation of other thermodynamic tables.

The table includes the standard enthalpy of formation at 298.15 K, the entropy at 298.15 K, and the quantity $H°(298.15\ \text{K})–H°(0)$. A value of 0 in the $\Delta_f H°$ column for an element indicates the reference state for that element. The standard state pressure is 100,000 Pa (1 bar). See the reference for information on the dependence of gas-phase entropy on the choice of standard state pressure.

Substances are listed in alphabetical order of their chemical formulas when written in the most common form.

The table is reprinted with permission of CODATA.

Reference

Cox, J. D., Wagman, D. D., and Medvedev, V. A., *CODATA Key Values for Thermodynamics*, Hemisphere Publishing Corp., New York, 1989.

CODATA Key Values for Thermodynamics

Substance	State	$\Delta_f H°(298)$/ kJ mol^{-1}	$S°(298)$/ J mol^{-1} K^{-1}	$H°(298)-H°(0)$/ kJ mol^{-1}
Ag	cr	0	42.55 ± 0.20	5.745 ± 0.020
Ag	g	284.9 ± 0.8	172.997 ± 0.004	6.197 ± 0.001
Ag^+	aq	105.79 ± 0.08	73.45 ± 0.40	
AgCl	cr	-127.01 ± 0.05	96.25 ± 0.20	12.033 ± 0.020
Al	cr	0	28.30 ± 0.10	4.540 ± 0.020
Al	g	330.0 ± 4.0	164.554 ± 0.004	6.919 ± 0.001
Al^{+3}	aq	-538.4 ± 1.5	-325 ± 10	
AlF_3	cr	-1510.4 ± 1.3	66.5 ± 0.5	11.62 ± 0.04
Al_2O_3	cr, corundum	-1675.7 ± 1.3	50.92 ± 0.10	10.016 ± 0.020
Ar	g	0	154.846 ± 0.003	6.197 ± 0.001
B	cr, rhombic	0	5.90 ± 0.08	1.222 ± 0.008
B	g	565 ± 5	153.436 ± 0.015	6.316 ± 0.002
BF_3	g	-1136.0 ± 0.8	254.42 ± 0.20	11.650 ± 0.020
B_2O_3	cr	-1273.5 ± 1.4	53.97 ± 0.30	9.301 ± 0.040
Be	cr	0	9.50 ± 0.08	1.950 ± 0.020
Be	g	324 ± 5	136.275 ± 0.003	6.197 ± 0.001
BeO	cr	-609.4 ± 2.5	13.77 ± 0.04	2.837 ± 0.008
Br^-	aq	-121.41 ± 0.15	82.55 ± 0.20	
Br	g	111.87 ± 0.12	175.018 ± 0.004	6.197 ± 0.001
Br_2	l	0	152.21 ± 0.30	24.52 ± 0.01
Br_2	g	30.91 ± 0.11	245.468 ± 0.005	9.725 ± 0.001
C	cr, graphite	0	5.74 ± 0.10	1.050 ± 0.020
C	g	716.68 ± 0.45	158.100 ± 0.003	6.536 ± 0.001
CO	g	-110.53 ± 0.17	197.660 ± 0.004	8.671 ± 0.001
CO_2	g	-393.51 ± 0.13	213.785 ± 0.010	9.365 ± 0.003
CO_2	aq, undissoc.	-413.26 ± 0.20	119.36 ± 0.60	
CO_3^{-2}	aq	-675.23 ± 0.25	-50.0 ± 1.0	
Ca	cr	0	41.59 ± 0.40	5.736 ± 0.040
Ca	g	177.8 ± 0.8	154.887 ± 0.004	6.197 ± 0.001
Ca^{+2}	aq	-543.0 ± 1.0	-56.2 ± 1.0	
CaO	cr	-634.92 ± 0.90	38.1 ± 0.4	6.75 ± 0.06
Cd	cr	0	51.80 ± 0.15	6.247 ± 0.015
Cd	g	111.80 ± 0.20	167.749 ± 0.004	6.197 ± 0.001
Cd^{+2}	aq	-75.92 ± 0.60	-72.8 ± 1.5	
CdO	cr	-258.35 ± 0.40	54.8 ± 1.5	8.41 ± 0.08
$CdSO_4$•(8/3)H_2O	cr	-1729.30 ± 0.80	229.65 ± 0.40	35.56 ± 0.04
Cl^-	aq	-167.080 ± 0.10	56.60 ± 0.20	
Cl	g	121.301 ± 0.008	165.190 ± 0.004	6.272 ± 0.001
ClO_4^-	aq	-128.10 ± 0.40	184.0 ± 1.5	
Cl_2	g	0	223.081 ± 0.010	9.181 ± 0.001
Cs	cr	0	85.23 ± 0.40	7.711 ± 0.020
Cs	g	76.5 ± 1.0	175.601 ± 0.003	6.197 ± 0.001
Cs^+	aq	-258.00 ± 0.50	132.1 ± 0.5	
Cu	cr	0	33.15 ± 0.08	5.004 ± 0.008
Cu	g	337.4 ± 1.2	166.398 ± 0.004	6.197 ± 0.001
Cu^{+2}	aq	64.9 ± 1.0	-98 ± 4	
$CuSO_4$	cr	-771.4 ± 1.2	109.2 ± 0.4	16.86 ± 0.08
F^-	aq	-335.35 ± 0.65	-13.8 ± 0.8	
F	g	79.38 ± 0.30	158.751 ± 0.004	6.518 ± 0.001
F_2	g	0	202.791 ± 0.005	8.825 ± 0.001
Ge	cr	0	31.09 ± 0.15	4.636 ± 0.020
Ge	g	372 ± 3	167.904 ± 0.005	7.398 ± 0.001
GeF_4	g	-1190.20 ± 0.50	301.9 ± 1.0	17.29 ± 0.10
GeO_2	cr, tetragonal	-580.0 ± 1.0	39.71 ± 0.15	7.230 ± 0.020
H	g	217.998 ± 0.006	114.717 ± 0.002	6.197 ± 0.001
H^+	aq	0	0	
HBr	g	-36.29 ± 0.16	198.700 ± 0.004	8.648 ± 0.001
HCO_3^-	aq	-689.93 ± 0.20	98.4 ± 0.5	
HCl	g	-92.31 ± 0.10	186.902 ± 0.005	8.640 ± 0.001
HF	g	-273.30 ± 0.70	173.779 ± 0.003	8.599 ± 0.001
HI	g	26.50 ± 0.10	206.590 ± 0.004	8.657 ± 0.001
HPO_4^{-2}	aq	-1299.0 ± 1.5	-33.5 ± 1.5	
HSO_4^-	aq	-886.9 ± 1.0	131.7 ± 3.0	
H_2	g	0	130.680 ± 0.003	8.468 ± 0.001
H_2O	l	-285.830 ± 0.040	69.95 ± 0.03	13.273 ± 0.020
H_2O	g	-241.826 ± 0.040	188.835 ± 0.010	9.905 ± 0.005
$H_2PO_4^-$	aq	-1302.6 ± 1.5	92.5 ± 1.5	
H_2S	g	-20.6 ± 0.5	205.81 ± 0.05	9.957 ± 0.010
H_2S	aq, undissoc.	-38.6 ± 1.5	126 ± 5	
H_3BO_3	cr	-1094.8 ± 0.8	89.95 ± 0.60	13.52 ± 0.04
H_3BO_3	aq, undissoc.	-1072.8 ± 0.8	162.4 ± 0.6	
He	g	0	126.153 ± 0.002	6.197 ± 0.001
Hg	l	0	75.90 ± 0.12	9.342 ± 0.008
Hg	g	61.38 ± 0.04	174.971 ± 0.005	6.197 ± 0.001
Hg^{+2}	aq	170.21 ± 0.20	-36.19 ± 0.80	
HgO	cr, red	-90.79 ± 0.12	70.25 ± 0.30	9.117 ± 0.025
Hg_2^{+2}	aq	166.87 ± 0.50	65.74 ± 0.80	
Hg_2Cl_2	cr	-265.37 ± 0.40	191.6 ± 0.8	23.35 ± 0.20
Hg_2SO_4	cr	-743.09 ± 0.40	200.70 ± 0.20	26.070 ± 0.030
I^-	aq	-56.78 ± 0.05	106.45 ± 0.30	
I	g	106.76 ± 0.04	180.787 ± 0.004	6.197 ± 0.001
I_2	cr	0	116.14 ± 0.30	13.196 ± 0.040
I_2	g	62.42 ± 0.08	260.687 ± 0.005	10.116 ± 0.001
K	cr	0	64.68 ± 0.20	7.088 ± 0.020
K	g	89.0 ± 0.8	160.341 ± 0.003	6.197 ± 0.001
K^+	aq	-252.14 ± 0.08	101.20 ± 0.20	
Kr	g	0	164.085 ± 0.003	6.197 ± 0.001
Li	cr	0	29.12 ± 0.20	4.632 ± 0.040
Li	g	159.3 ± 1.0	138.782 ± 0.010	6.197 ± 0.001
Li^+	aq	-278.47 ± 0.08	12.24 ± 0.15	
Mg	cr	0	32.67 ± 0.10	4.998 ± 0.030
Mg	g	147.1 ± 0.8	148.648 ± 0.003	6.197 ± 0.001
Mg^{+2}	aq	-467.0 ± 0.6	-137 ± 4	
MgF_2	cr	-1124.2 ± 1.2	57.2 ± 0.5	9.91 ± 0.06
MgO	cr	-601.60 ± 0.30	26.95 ± 0.15	5.160 ± 0.020

Substance	State	$\Delta_f H°(298)$/ kJ mol^{-1}	$S°(298)$/ J mol^{-1} K^{-1}	$H°(298)-H°(0)$/ kJ mol^{-1}
N	g	472.68 ± 0.40	153.301 ± 0.003	6.197 ± 0.001
NH_3	g	-45.94 ± 0.35	192.77 ± 0.05	10.043 ± 0.010
NH_4^+	aq	-133.26 ± 0.25	111.17 ± 0.40	
NO_3^-	aq	-206.85 ± 0.40	146.70 ± 0.40	
N_2	g	0	191.609 ± 0.004	8.670 ± 0.001
Na	cr	0	51.30 ± 0.20	6.460 ± 0.020
Na	g	107.5 ± 0.7	153.718 ± 0.003	6.197 ± 0.001
Na^+	aq	-240.34 ± 0.06	58.45 ± 0.15	
Ne	g	0	146.328 ± 0.003	6.197 ± 0.001
O	g	249.18 ± 0.10	161.059 ± 0.003	6.725 ± 0.001
OH^-	aq	-230.015 ± 0.040	-10.90 ± 0.20	
O_2	g	0	205.152 ± 0.005	8.680 ± 0.002
P	cr, white	0	41.09 ± 0.25	5.360 ± 0.015
P	g	316.5 ± 1.0	163.199 ± 0.003	6.197 ± 0.001
P_2	g	144.0 ± 2.0	218.123 ± 0.004	8.904 ± 0.001
P_4	g	58.9 ± 0.3	280.01 ± 0.50	14.10 ± 0.20
Pb	cr	0	64.80 ± 0.30	6.870 ± 0.030
Pb	g	195.2 ± 0.8	175.375 ± 0.005	6.197 ± 0.001
Pb^{+2}	aq	0.92 ± 0.25	18.5 ± 1.0	
$PbSO_4$	cr	-919.97 ± 0.40	148.50 ± 0.60	20.050 ± 0.040
Rb	cr	0	76.78 ± 0.30	7.489 ± 0.020
Rb	g	80.9 ± 0.8	170.094 ± 0.003	6.197 ± 0.001
Rb^+	aq	-251.12 ± 0.10	121.75 ± 0.25	
S	cr, rhombic	0	32.054 ± 0.050	4.412 ± 0.006
S	g	277.17 ± 0.15	167.829 ± 0.006	6.657 ± 0.001
SH^-	aq	-16.3 ± 1.5	67 ± 5	
SO_2	g	-296.81 ± 0.20	248.223 ± 0.050	10.549 ± 0.010
SO_4^{-2}	aq	-909.34 ± 0.40	18.50 ± 0.40	
S_2	g	128.60 ± 0.30	228.167 ± 0.010	9.132 ± 0.002
Si	cr	0	18.81 ± 0.08	3.217 ± 0.008
Si	g	450 ± 8	167.981 ± 0.004	7.550 ± 0.001
SiF_4	g	-1615.0 ± 0.8	282.76 ± 0.50	15.36 ± 0.05
SiO_2	cr, alpha quartz	-910.7 ± 1.0	41.46 ± 0.20	6.916 ± 0.020
Sn	cr, white	0	51.18 ± 0.08	6.323 ± 0.008
Sn	g	301.2 ± 1.5	168.492 ± 0.004	6.215 ± 0.001
Sn^{+2}	aq	-8.9 ± 1.0	-16.7 ± 4.0	
SnO	cr, tetragonal	-280.71 ± 0.20	57.17 ± 0.30	8.736 ± 0.020
SnO_2	cr, tetragonal	-577.63 ± 0.20	49.04 ± 0.10	8.384 ± 0.020
Th	cr	0	51.8 ± 0.5	6.35 ± 0.05
Th	g	602 ± 6	190.17 ± 0.05	6.197 ± 0.003
ThO_2	cr	-1226.4 ± 3.5	65.23 ± 0.20	10.560 ± 0.020
Ti	cr	0	30.72 ± 0.10	4.824 ± 0.015
Ti	g	473 ± 3	180.298 ± 0.010	7.539 ± 0.002
$TiCl_4$	g	-763.2 ± 3.0	353.2 ± 4.0	21.5 ± 0.5
TiO_2	cr, rutile	-944.0 ± 0.8	50.62 ± 0.30	8.68 ± 0.05
U	cr	0	50.20 ± 0.20	6.364 ± 0.020
U	g	533 ± 8	199.79 ± 0.10	6.499 ± 0.020
UO_2	cr	-1085.0 ± 1.0	77.03 ± 0.20	11.280 ± 0.020
UO_2^{+2}	aq	-1019.0 ± 1.5	-98.2 ± 3.0	
UO_3	cr, gamma	-1223.8 ± 1.2	96.11 ± 0.40	14.585 ± 0.050
U_3O_8	cr	-3574.8 ± 2.5	282.55 ± 0.50	42.74 ± 0.10
Xe	g	0	169.685 ± 0.003	6.197 ± 0.001
Zn	cr	0	41.63 ± 0.15	5.657 ± 0.020
Zn	g	130.40 ± 0.40	160.990 ± 0.004	6.197 ± 0.001
Zn^{+2}	aq	-153.39 ± 0.20	-109.8 ± 0.5	
ZnO	cr	-350.46 ± 0.27	43.65 ± 0.40	6.933 ± 0.040

STANDARD THERMODYNAMIC PROPERTIES OF CHEMICAL SUBSTANCES

This table gives the standard state chemical thermodynamic properties of about 2500 individual substances in the crystalline, liquid, and gaseous states. The properties tabulated are:

$\Delta_f H°$: Standard molar enthalpy (heat) of formation at 298.15 K in kJ/mol

$\Delta_f G°$: Standard molar Gibbs energy of formation at 298.15 K in kJ/mol

$S°$: Standard molar entropy at 298.15 K in J/mol K

C_p: Molar heat capacity at constant pressure at 298.15 K in J/mol K

The standard state pressure is 100 kPa (1 bar). The standard states are defined for different phases by:

The standard state of a pure gaseous substance is that of the substance as a (hypothetical) ideal gas at the standard state pressure.

The standard state of a pure liquid substance is that of the liquid under the standard state pressure.

The standard state of a pure crystalline substance is that of the crystalline substance under the standard state pressure.

An entry of 0.0 for $\Delta_f H°$ for an element indicates the reference state of that element. See Refs. 1 and 2 for further information on reference states. A blank means no value is available.

The data are derived from the sources listed in the references, from other papers appearing in the *Journal of Physical and Chemical Reference Data*, and from the primary research literature. We are indebted to M. V. Korobov for providing data on fullerene compounds.

References

1. Cox, J. D., Wagman, D. D., and Medvedev, V. A., *CODATA Key Values for Thermodynamics*, Hemisphere Publishing Corp., New York, 1989.
2. Wagman, D. D., Evans, W. H., Parker, V. B., Schumm, R. H., Halow, I., Bailey, S. M., Churney, K. L., and Nuttall, R. L., *The NBS Tables of Chemical Thermodynamic Properties, J. Phys. Chem. Ref. Data*, Vol. 11, Suppl. 2, 1982.
3. Chase, M. W., Davies, C. A., Downey, J. R., Frurip, D. J., McDonald, R. A., and Syverud, A. N., *JANAF Thermochemical Tables, Third Edition, J. Phys. Chem. Ref. Data*, Vol. 14, Suppl. 1, 1985.
4. Chase, M. W., *NIST-JANAF Thermochemical Tables, Fourth Edition, J. Phys. Chem. Ref. Data*, Monograph 9, 1998.
5. Daubert, T. E., Danner, R. P., Sibul, H. M., and Stebbins, C. C., *Physical and Thermodynamic Properties of Pure Compounds: Data Compilation*, extant 1994 (core with 4 supplements), Taylor & Francis, Bristol, PA.
6. Pedley, J. B., Naylor, R. D., and Kirby, S. P., *Thermochemical Data of Organic Compounds, Second Edition*, Chapman & Hall, London, 1986.
7. Pedley, J. B., *Thermochemical Data and Structures of Organic Compounds*, Thermodynamic Research Center, Texas A & M University, College Station, TX, 1994.
8. Domalski, E. S., and Hearing, E. D., Heat Capacities and Entropies of Organic Compounds in the Condensed Phase, Volume III, *J. Phys. Chem. Ref. Data*, 25, 1–525, 1996.
9. Zabransky, M., Ruzicka , V., Majer, V., and Domalski, E. S., *Heat Capacity of Liquids, J. Phys. Chem. Ref. Data*, Monograph No. 6, 1996.
10. Gurvich, L. V., Veyts, I.V., and Alcock, C. B., *Thermodynamic Properties of Individual Substances, Fourth Edition*, Vol. 1, Hemisphere Publishing Corp., New York, 1989.
11. Gurvich, L. V., Veyts, I.V., and Alcock, C. B., *Thermodynamic Properties of Individual Substances, Fourth Edition*, Vol. 3, CRC Press, Boca Raton, FL, 1994.
12. *NIST Chemistry Webbook*, <webbook.nist.gov>.

Standard Thermodynamic Properties of Chemical Substances

Name	Mol. form.	$\Delta_f H°$(c)/ kJ mol^{-1}	$\Delta_f G°$(c)/ kJ mol^{-1}	$S°$(c)/ J mol^{-1} K^{-1}	C_p(c)/ J mol^{-1} K^{-1}	$\Delta_f H°$(l)/ kJ mol^{-1}	$\Delta_f G°$(l)/ kJ mol^{-1}	$S°$(l)/ J mol^{-1} K^{-1}	C_p(l)/ J mol^{-1} K^{-1}	$\Delta_f H°$(g)/ kJ mol^{-1}	$\Delta_f G°$(g)/ kJ mol^{-1}	$S°$(g)/ J mol^{-1} K^{-1}	C_p(g)/ J mol^{-1} K^{-1}
Acenaphthene	$C_{12}H_{10}$	70.3		188.9	190.4					156.0			
Acenaphthylene	$C_{12}H_8$	186.7			166.4					259.7			
Acetaldehyde	C_2H_4O					-192.2	-127.6	160.2	89.0	-166.2	-133.0	263.8	55.3
Acetamide	C_2H_5NO	-317.0		115.0	91.3					-238.3			
Acetanilide	C_8H_9NO	-209.4			179.3								
Acetic acid	$C_2H_4O_2$					-484.3	-389.9	159.8	123.3	-432.2	-374.2	283.5	63.4
Acetic anhydride	$C_4H_6O_3$					-624.4				-572.5			
Acetone	C_3H_6O					-248.4		199.8	126.3	-217.1	-152.7	295.3	74.5
Acetone cyanohydrin	C_4H_7NO					-120.9							
Acetonitrile	C_2H_3N					40.6	86.5	149.6	91.5	74.0	91.9	243.4	52.2
Acetophenone	C_8H_8O					-142.5				-86.7			
Acetyl bromide	C_2H_3BrO					-223.5				-190.4			
Acetyl chloride	C_2H_3ClO					-272.9	-208.0	200.8	117.0	-242.8	-205.8	295.1	67.8
Acetylene	C_2H_2									227.4	209.9	200.9	44.0
Acetyl fluoride	C_2H_3FO					-467.2				-442.1			
Acetyl iodide	C_2H_3IO					-163.5				-126.4			
2-(Acetyloxy)benzoic acid	$C_9H_8O_4$	-815.6											
9-Acridinamine	$C_{13}H_{10}N_2$	159.2											
Acridine	$C_{13}H_9N$	179.4								273.9			
Acrolein	C_3H_4O												71.3
Acrylamide	C_3H_5NO	-212.1			110.6	-224.0				-130.2			
Acrylic acid	$C_3H_4O_2$					-383.8			145.7				
Acrylonitrile	C_3H_3N					147.1				180.6			
Actinium	Ac	0.0		56.5	27.2					406.0	366.0	188.1	20.8
Adenine	$C_5H_5N_5$	96.9			147.0					205.7			
DL-Alanine	$C_3H_7NO_2$	-563.6											

Name	Mol. form.	$\Delta_f H°$(c)/ kJ mol⁻¹	$\Delta_f G°$(c)/ kJ mol⁻¹	$S°$(c)/ J mol⁻¹ K⁻¹	C_p(c)/ J mol⁻¹ K⁻¹	$\Delta_f H°$(l)/ kJ mol⁻¹	$\Delta_f G°$(l)/ kJ mol⁻¹	$S°$(l)/ J mol⁻¹ K⁻¹	C_p(l)/ J mol⁻¹ K⁻¹	$\Delta_f H°$(g)/ kJ mol⁻¹	$\Delta_f G°$(g)/ kJ mol⁻¹	$S°$(g)/ J mol⁻¹ K⁻¹	C_p(g)/ J mol⁻¹ K⁻¹
D-Alanine	$C_3H_7NO_2$	-561.2											
L-Alanine	$C_3H_7NO_2$	-604.0								-465.9			
β-Alanine	$C_3H_7NO_2$	-558.0								-424.0			
Allene	C_3H_4									190.5			
Allyl acetate	$C_5H_8O_2$								184.1				
Allyl alcohol	C_3H_6O					-171.8			138.9	-124.5			
Allylamine	C_3H_7N					-10.0							
Allylcyclopentane	C_8H_{14}					-64.5				-24.1			
Aluminum	Al	0.0		28.3	24.2					330.0	289.4	164.6	21.4
Aluminum borohydride	AlB_3H_{12}					-16.3	145.0	289.1	194.6	13.0	147.0	379.2	
Aluminum bromide	$AlBr_3$	-527.2		180.2	100.6					-425.1			
Aluminum chloride	$AlCl_3$	-704.2	-628.8	109.3	91.1					-583.2			
Aluminum dichloride	$AlCl_2$									-331.0			
Aluminum fluoride	AlF_3	-1510.4	-1431.1	66.5	75.1					-1204.6	-1188.2	277.1	62.6
Aluminum hexabromide	Al_2Br_6									-970.7			
Aluminum hexachloride	Al_2Cl_6									-1290.8	-1220.4	490.0	
Aluminum hexafluoride	Al_2F_6									-2628.0			
Aluminum hexaiodide	Al_2I_6									-516.7			
Aluminum hydride	AlH_3	-46.0		30.0	40.2								
Aluminum iodide	AlI_3	-302.9		195.9	98.7					-289.4		223.6	
Aluminum monobromide	AlBr									-4.0	-42.0	239.5	35.6
Aluminum monochloride	AlCl									-47.7	-74.1	228.1	35.0
Aluminum monofluoride	AlF									-258.2	-283.7	215.0	31.9
Aluminum monohydride	AlH									259.2	231.2	187.9	29.4
Aluminum monoiodide	AlI									65.5			36.0
Aluminum monosulfide	AlS									200.9	150.1	230.6	33.4
Aluminum monoxide	AlO									91.2	65.3	218.4	30.9
Aluminum nitride	AlN	-318.0	-287.0	20.2	30.1								
Aluminum oxide (α)	Al_2O_3	-1675.7	-1582.3	50.9	79.0								
Aluminum oxide (Al_2O)	Al_2O									-130.0	-159.0	259.4	45.7
Aluminum phosphate	AlO_4P	-1733.8	-1617.9	90.8	93.2								
Aluminum phosphide	AlP	-166.5											
Aluminum sulfide	Al_2S_3	-724.0		116.9	105.1								
Americium	Am	0.0											
Amidogen	H_2N									184.9	194.6	195.0	33.9
Aminetrifluoroboron	BF_3H_3N	-1353.9											
2-Aminobiphenyl	$C_{12}H_{11}N$	93.8								184.4			
4-Aminobiphenyl	$C_{12}H_{11}N$	81.0											
4-Aminobutanoic acid	$C_4H_9NO_2$	-581.0								-441.0			
N-(Aminocarbonyl)acetamide	$C_3H_6N_2O_2$	-544.2								-441.2			
6-Aminohexanoic acid	$C_6H_{13}NO_2$	-637.3											
3-Amino-1-nitroguanidine	$CH_5N_5O_2$	22.1											
5-Aminopentanoic acid	$C_5H_{11}NO_2$	-604.1								-460.0			
Amminetrimethylboron	$C_3H_{12}BN$	-284.1	-79.3	218.0									
Ammonia	H_3N									-45.9	-16.4	192.8	35.1
Ammonium azide	H_4N_4	115.5	274.2	112.5									
Ammonium bromide	BrH_4N	-270.8	-175.2	113.0	96.0								
Ammonium chloride	ClH_4N	-314.4	-202.9	94.6	84.1								
Ammonium cyanide	CH_4N_2	0.4			134.0								
Ammonium fluoride	FH_4N	-464.0	-348.7	72.0	65.3								
Ammonium hexafluorosilicate	$F_6H_8N_2Si$	-2681.7	-2365.3	280.2	228.1								
Ammonium hydrogen carbonate	CH_5NO_3	-849.4	-665.9	120.9									
Ammonium hydrogen phosphate	$H_9N_2O_4P$	-1566.9			188.0								
Ammonium hydrogen sulfate	H_5NO_4S	-1027.0											
Ammonium hydrogen sulfite	H_5NO_3S	-768.6											
Ammonium hydroxide	H_5NO					-361.2	-254.0	165.6	154.9				
Ammonium iodide	H_4IN	-201.4	-112.5	117.0									
Ammonium nitrate	$H_4N_2O_3$	-365.6	-183.9	151.1	139.3								
Ammonium nitrite	$H_4N_2O_2$	-256.5											
Ammonium oxalate	$C_2H_8N_2O_4$	-1123.0			226.0								
Ammonium perchlorate	ClH_4NO_4	-295.3	-88.8	186.2									
Ammonium phosphate	$H_{12}N_3O_4P$	-1671.9											
Ammonium sulfate	$H_8N_2O_4S$	-1180.9	-901.7	220.1	187.5								
Aniline	C_6H_7N					31.6			191.9	87.5	-7.0	317.9	107.9
Aniline-2-carboxylic acid	$C_7H_7NO_2$	-401.1								-296.0			
Aniline-3-carboxylic acid	$C_7H_7NO_2$	-417.3								-283.6			

Name	Mol. form.	$\Delta_f H°(c)$/ kJ mol⁻¹	$\Delta_f G°(c)$/ kJ mol⁻¹	$S°(c)$/ J mol⁻¹ K⁻¹	$C_p(c)$/ J mol⁻¹ K⁻¹	$\Delta_f H°(l)$/ kJ mol⁻¹	$\Delta_f G°(l)$/ kJ mol⁻¹	$S°(l)$/ J mol⁻¹ K⁻¹	$C_p(l)$/ J mol⁻¹ K⁻¹	$\Delta_f H°(g)$/ kJ mol⁻¹	$\Delta_f G°(g)$/ kJ mol⁻¹	$S°(g)$/ J mol⁻¹ K⁻¹	$C_p(g)$/ J mol⁻¹ K⁻¹
Aniline-4-carboxylic acid	$C_7H_7NO_2$	-410.0			177.8					-296.7			
Anisole	C_7H_8O					-114.8				-67.9			
Anthracene	$C_{14}H_{10}$	129.2		207.5	210.5					230.9			
9,10-Anthracenedione	$C_{14}H_8O_2$	-188.5								-75.7			
Antimony	Sb	0.0		45.7	25.2					262.3	222.1	180.3	20.8
Antimony(III) bromide	Br_3Sb	-259.4	-239.3	207.1						-194.6	-223.9	372.9	80.2
Antimony(III) chloride	Cl_3Sb	-382.2	-323.7	184.1	107.9								
Antimony(III) fluoride	F_3Sb	-915.5											
Antimony(III) iodide	I_3Sb	-100.4											
Antimony(V) oxide	O_5Sb_2	-971.9	-829.2	125.1									
α-D-Arabinopyranose	$C_5H_{10}O_5$	-1057.9											
D-Arginine	$C_6H_{14}N_4O_2$	-623.5		250.6	232.0								
Argon	Ar									0.0		154.8	20.8
Arsenic (gray)	As	0.0		35.1	24.6					302.5	261.0	174.2	20.8
Arsenic acid	AsH_3O_4	-906.3											
Arsenic(III) bromide	$AsBr_3$	-197.5								-130.0	-159.0	363.9	79.2
Arsenic(III) chloride	$AsCl_3$					-305.0	-259.4	216.3		-261.5	-248.9	327.2	75.7
Arsenic(III) fluoride	AsF_3					-821.3	-774.2	181.2	126.6	-785.8	-770.8	289.1	65.6
Arsenic(III) iodide	AsI_3	-58.2	-59.4	213.1	105.8							388.3	80.6
Arsenic monoxide	AsO									70.0			
Arsenic(V) oxide	As_2O_5	-924.9	-782.3	105.4	116.5								
Arsenic(III) sulfide	As_2S_3	-169.0	-168.6	163.6	116.3								
Arsenic (yellow)	As	14.6											
Arsine	AsH_3									66.4	68.9	222.8	38.1
L-Ascorbic acid	$C_6H_8O_6$	-1164.6											
L-Asparagine	$C_4H_8N_2O_3$	-789.4											
L-Asparagine, monohydrate	$C_4H_{10}N_2O_4$	-1086.6											
L-Aspartic acid	$C_4H_7NO_4$	-973.3											
Astatine	At	0.0											
Atrazine	$C_8H_{14}ClN_5$	-125.4											
2,2'-Azobis[isobutyronitrile]	$C_8H_{12}N_4$	246.0			237.6								0.0
Azobutane	$C_8H_{18}N_2$					-40.1				9.2			
Azopropane	$C_6H_{14}N_2$					11.5				51.3			
trans-Azoxybenzene	$C_{12}H_{10}N_2O$	243.4								342.0			
Azulene	$C_{10}H_8$	212.3								289.1			
Barbituric acid	$C_4H_4N_2O_3$	-634.7											
Barium	Ba	0.0		62.5	28.1					180.0	146.0	170.2	20.8
Barium bromide	$BaBr_2$	-757.3	-736.8	146.0									
Barium carbonate	$CBaO_3$	-1213.0	-1134.4	112.1	86.0								
Barium chloride	$BaCl_2$	-855.0	-806.7	123.7	75.1								
Barium chloride dihydrate	$BaCl_2H_4O_2$	-1456.9	-1293.2	203.0									
Barium fluoride	BaF_2	-1207.1	-1156.8	96.4	71.2								
Barium hydride	BaH_2	-177.0	-138.2	63.0	46.0								
Barium hydroxide	BaH_2O_2	-944.7											
Barium iodide	BaI_2	-602.1											
Barium nitrate	BaN_2O_6	-988.0	-792.6	214.0	151.4								
Barium nitrite	BaN_2O_4	-768.2											
Barium oxide	BaO	-548.0	-520.3	72.1	47.3					-112.0			
Barium sulfate	BaO_4S	-1473.2	-1362.2	132.2	101.8								
Barium sulfide	BaS	-460.0	-456.0	78.2	49.4								
Benzaldehyde	C_7H_6O					-87.0		221.2	172.0	-36.7			
Benzamide	C_7H_7NO	-202.6								-100.9			
Benz[*a*]anthracene	$C_{18}H_{12}$	170.8								293.0			
Benzene	C_6H_6					49.1	124.5	173.4	136.0	82.9	129.7	269.2	82.4
1,2-Benzenediamine	$C_6H_8N_2$	-0.3											
1,3-Benzenediamine	$C_6H_8N_2$	-7.8		154.5	159.6								
1,4-Benzenediamine	$C_6H_8N_2$	3.0											
Benzeneethanol	$C_8H_{10}O$								252.6				
Benzenethiol	C_6H_6S					63.7		222.8	173.2	111.3			
1,2,3-Benzenetriol	$C_6H_6O_3$	-551.1								-434.2			
1,2,4-Benzenetriol	$C_6H_6O_3$	-563.8								-444.0			
1,3,5-Benzenetriol	$C_6H_6O_3$	-584.6								-452.9			
p-Benzidine	$C_{12}H_{12}N_2$	70.7											
Benzil	$C_{14}H_{10}O_2$	-153.9								-55.5			
1*H*-Benzimidazole	$C_7H_6N_2$	79.5								181.7			
Benzoic acid	$C_7H_6O_2$	-385.2		167.6	146.8					-294.0			

Name	Mol. form.	$\Delta_fH°$(c)/ kJ mol^{-1}	$\Delta_fG°$(c)/ kJ mol^{-1}	$S°$(c)/ J mol^{-1} K^{-1}	C_p(c)/ J mol^{-1} K^{-1}	$\Delta_fH°$(l)/ kJ mol^{-1}	$\Delta_fG°$(l)/ kJ mol^{-1}	$S°$(l)/ J mol^{-1} K^{-1}	C_p(l)/ J mol^{-1} K^{-1}	$\Delta_fH°$(g)/ kJ mol^{-1}	$\Delta_fG°$(g)/ kJ mol^{-1}	$S°$(g)/ J mol^{-1} K^{-1}	C_p(g)/ J mol^{-1} K^{-1}
Benzonitrile	C_7H_5N					163.2		209.1	165.2	215.7			
Benzophenone	$C_{13}H_{10}O$	-34.5			224.8					54.9			
Benzo[*a*]pyrene	$C_{20}H_{12}$												254.8
Benzo[*f*]quinoline	$C_{13}H_9N$	150.6								233.7			
p-Benzoquinone	$C_6H_4O_2$	-185.7			129.0					-122.9			
Benzo[*b*]thiophene	C_8H_6S	100.6								166.3			
1*H*-Benzotriazole	$C_6H_5N_3$	236.5								335.5			
Benzoxazole	C_7H_5NO	-24.2								44.8			
Benzoyl chloride	C_7H_5ClO					-158.0				-103.2			
Benzoyl peroxide	$C_{14}H_{10}O_4$	-369.4								-281.7			
Benzyl acetate	$C_9H_{10}O_2$								148.5				
Benzyl alcohol	C_7H_8O					-160.7		216.7	217.9	-100.4			
Benzylamine	C_7H_9N					34.2			207.2	94.4			0.0
N-Benzylaniline	$C_{13}H_{13}N$	101.4											
Berkelium (β form)	Bk	0.0											
Beryllium	Be	0.0		9.5	16.4					324.0	286.6	136.3	20.8
Beryllium bromide	$BeBr_2$	-353.5		108.0	69.4								
Beryllium carbonate	$CBeO_3$	-1025.0		52.0	65.0								
Beryllium chloride	$BeCl_2$	-490.4	-445.6	75.8	62.4								
Beryllium fluoride	BeF_2	-1026.8	-979.4	53.4	51.8								
Beryllium hydroxide (α)	BeH_2O_2	-902.5	-815.0	45.5	62.1								
Beryllium iodide	BeI_2	-192.5		121.0	71.1								
Beryllium oxide	BeO	-609.4	-580.1	13.8	25.6								
Beryllium sulfate	BeO_4S	-1205.2	-1093.8	77.9	85.7								
Beryllium sulfide	BeS	-234.3		34.0	34.0								
9,9'-Bianthracene	$C_{28}H_{18}$	326.2								454.3			
Bicyclo[2.2.1]heptane	C_7H_{12}	-95.1			151.0					-54.8			0.0
1,1'-Bicyclopentyl	$C_{10}H_{18}$					-178.9							
Biphenyl	$C_{12}H_{10}$	99.4		209.4	198.4					181.4			
Bis(2-aminoethyl)amine	$C_4H_{13}N_3$								254.0				
Bis(2-chloroethyl) ether	$C_4H_8Cl_2O$								220.9				
Bis(2-cyanoethyl) sulfide	$C_6H_8N_2S$					96.3							
Bis(2-ethylhexyl) phthalate	$C_{24}H_{38}O_4$								704.7				
2,2-Bis(4-hydroxyphenyl) propane	$C_{15}H_{16}O_2$	-368.6											
Bismuth	Bi	0.0		56.7	25.5					207.1	168.2	187.0	20.8
Bismuth hydroxide	BiH_3O_3	-711.3											
Bismuth oxide	Bi_2O_3	-573.9	-493.7	151.5	113.5								
Bismuth oxychloride	BiClO	-366.9	-322.1	120.5									
Bismuth sulfate	$Bi_2O_{12}S_3$	-2544.3											
Bismuth sulfide	Bi_2S_3	-143.1	-140.6	200.4	122.2								
Bismuth trichloride	$BiCl_3$	-379.1	-315.0	177.0	105.0					-265.7	-256.0	358.9	79.7
Bismuth triiodide	BiI_3		-175.3										
Borane(1)	BH									442.7	412.7	171.8	29.2
Borane(3)	BH_3									89.2	93.3	188.2	36.0
Borane carbonyl	CH_3BO									-111.2	-92.9	249.4	59.5
Borazine	$B_3H_6N_3$					-541.0	-392.7	199.6					
Boric acid	BH_3O_3	-1094.3	-968.9	90.0	86.1					-994.1			
Boron	B	0.0		5.9	11.1					565.0	521.0	153.4	20.8
Boron dioxide	BO_2									-300.4	-305.9	229.6	43.0
Boron monosulfide	BS									342.0	288.8	216.2	30.0
Boron monoxide	BO									25.0	-4.0	203.5	29.2
Boron nitride	BN	-254.4	-228.4	14.8	19.7					647.5	614.5	212.3	29.5
Boron oxide	B_2O_3	-1273.5	-1194.3	54.0	62.8					-843.8	-832.0	279.8	66.9
Boron sulfide	B_2S_3	-240.6		100.0	111.7					67.0			
Boron tribromide	BBr_3					-239.7	-238.5	229.7		-205.6	-232.5	324.2	67.8
Boron trichloride	BCl_3					-427.2	-387.4	206.3	106.7	-403.8	-388.7	290.1	62.7
Boron trifluoride	BF_3									-1136.0	-1119.4	254.4	
Boron triiodide	BI_3									71.1	20.7	349.2	70.8
Bromine	Br_2					0.0		152.2	75.7	30.9	3.1	245.5	36.0
Bromine (atomic)	Br									111.9	82.4	175.0	20.8
Bromine chloride	BrCl									14.6	-1.0	240.1	35.0
Bromine dioxide	BrO_2									152.0	155.0	271.1	45.4
Bromine fluoride	BrF									-93.8	-109.2	229.0	33.0
Bromine monoxide	BrO									125.8	109.6	233.0	34.2
Bromine pentafluoride	BrF_5					-458.6	-351.8	225.1		-428.9	-350.6	320.2	99.6

Name	Mol. form.	$\Delta_f H°$(c)/ kJ mol^{-1}	$\Delta_f G°$(c)/ kJ mol^{-1}	$S°$(c)/ J mol^{-1} K^{-1}	C_p(c)/ J mol^{-1} K^{-1}	$\Delta_f H°$(l)/ kJ mol^{-1}	$\Delta_f G°$(l)/ kJ mol^{-1}	$S°$(l)/ J mol^{-1} K^{-1}	C_p(l)/ J mol^{-1} K^{-1}	$\Delta_f H°$(g)/ kJ mol^{-1}	$\Delta_f G°$(g)/ kJ mol^{-1}	$S°$(g)/ J mol^{-1} K^{-1}	C_p(g)/ J mol^{-1} K^{-1}
Bromine trifluoride	BrF_3					-300.8	-240.5	178.2	124.6	-255.6	-229.4	292.5	66.6
Bromoacetic acid	$C_2H_3BrO_2$									-383.5	-338.3	337.0	80.5
Bromoacetone	C_3H_5BrO									-181.0			
Bromoacetylene	C_2HBr											253.7	55.7
Bromobenzene	C_6H_5Br					60.9		219.2	154.3				
Bromoborane(1)	BBr									238.1	195.4	225.0	32.9
1-Bromobutane	C_4H_9Br					-143.8				-107.1			
2-Bromobutane, (±)-	C_4H_9Br					-154.9				-120.3			
Bromochlorodifluoromethane	$CBrClF_2$											318.5	74.6
1-Bromo-2-chloroethane	C_2H_4BrCl								130.1				
Bromochlorofluoromethane	CHBrClF											304.3	63.2
Bromochloromethane	CH_2BrCl											287.6	52.7
1-Bromo-2-chloro-1,1,2-trifluoroethane	$C_2HBrClF_3$					-675.3				-644.8			
2-Bromo-2-chloro-1,1,1-trifluoroethane	$C_2HBrClF_3$					-720.0				-690.4			
Bromodichlorofluoromethane	$CBrCl_2F$											330.6	80.0
Bromodichloromethane	$CHBrCl_2$											316.4	67.4
Bromodifluoromethane	$CHBrF_2$									-424.9		295.1	58.7
1-Bromododecane	$C_{12}H_{25}Br$					-344.7				-269.9			
Bromoethane	C_2H_5Br					-90.5	-25.8	198.7	100.8	-61.9	-23.9	286.7	64.5
Bromoethene	C_2H_3Br									79.2	81.8	275.8	55.5
Bromofluoromethane	CH_2BrF											276.3	49.2
Bromogermane	$BrGeH_3$											274.8	56.4
1-Bromoheptane	$C_7H_{15}Br$					-218.4				-167.8			
1-Bromohexadecane	$C_{16}H_{33}Br$					-444.5				-350.2			
1-Bromohexane	$C_6H_{13}Br$					-194.2		453.0	204.0	-148.3			
Bromomethane	CH_3Br					-59.8				-35.4	-26.3	246.4	42.4
2-Bromo-2-methylpropane	C_4H_9Br					-164.4				-132.4			
1-Bromooctane	$C_8H_{17}Br$					-245.1				-189.3			
Bromopentafluoroethane	C_2BrF_5									-1064.4			
1-Bromopentane	$C_5H_{11}Br$					-170.2				-128.9			
1-Bromopropane	C_3H_7Br					-121.9				-87.0			
2-Bromopropane	C_3H_7Br					-130.5				-99.4			
cis-1-Bromopropene	C_3H_5Br					7.9				40.8			
3-Bromopropene	C_3H_5Br					12.2				45.2			
Bromosilane	BrH_3Si											262.4	52.8
Bromosilyldyne	BrSi									209.0			38.6
Bromosilylene	BrHSi									-464.4			
N-Bromosuccinimide	$C_4H_4BrNO_2$	-335.9											
4-Bromotoluene	C_7H_7Br					12.0							
Bromotrichloromethane	$CBrCl_3$									-41.1			85.3
Bromotrichlorosilane	$BrCl_3Si$											350.1	90.9
2-Bromo-1,1,1-trifluoroethane	$C_2H_2BrF_3$									-694.5			
Bromotrifluoromethane	$CBrF_3$									-648.3			69.3
Bromotrinitromethane	$CBrN_3O_6$					32.5				80.3			
1,2-Butadiene	C_4H_6					138.6				162.3			
1,3-Butadiene	C_4H_6					88.5		199.0	123.6	110.0			
Butanal	C_4H_8O					-239.2		246.6	163.7	-204.8		343.7	103.4
Butanamide	C_4H_9NO	-364.8								-282.0			
Butane	C_4H_{10}					-147.3			140.9	-125.7			
1,2-Butanediol, (±)-	$C_4H_{10}O_2$					-523.6							
1,3-Butanediol	$C_4H_{10}O_2$					-501.0				-433.2			
1,4-Butanediol	$C_4H_{10}O_2$					-505.3		223.4	200.1	-428.7			
2,3-Butanediol	$C_4H_{10}O_2$					-541.5			213.0	-482.3			
1,4-Butanedithiol	$C_4H_{10}S_2$					-105.7				-50.6			
Butanenitrile	C_4H_7N					-5.8				33.6			
1-Butanethiol	$C_4H_{10}S$					-124.7			171.2	-88.0			
2-Butanethiol	$C_4H_{10}S$					-131.0				-96.9			
Butanoic acid	$C_4H_8O_2$					-533.8		222.2	178.6	-475.9			
Butanoic anhydride	$C_8H_{14}O_3$								283.7				
1-Butanol	$C_4H_{10}O$					-327.3		225.8	177.2	-274.9			
2-Butanol	$C_4H_{10}O$					-342.6		214.9	196.9	-292.8		359.5	112.7
2-Butanone	C_4H_8O					-273.3		239.1	158.7	-238.5		339.9	101.7
trans-2-Butenal	C_4H_6O					-138.7				-100.6			
1-Butene	C_4H_8					-20.8		227.0	118.0	0.1			

Name	Mol. form.	$\Delta_f H°$(c)/ kJ mol⁻¹	$\Delta_f G°$(c)/ kJ mol⁻¹	$S°$(c)/ J mol⁻¹ K⁻¹	C_p(c)/ J mol⁻¹ K⁻¹	$\Delta_f H°$(l)/ kJ mol⁻¹	$\Delta_f G°$(l)/ kJ mol⁻¹	$S°$(l)/ J mol⁻¹ K⁻¹	C_p(l)/ J mol⁻¹ K⁻¹	$\Delta_f H°$(g)/ kJ mol⁻¹	$\Delta_f G°$(g)/ kJ mol⁻¹	$S°$(g)/ J mol⁻¹ K⁻¹	C_p(g)/ J mol⁻¹ K⁻¹
cis-2-Butene	C_4H_8					-29.8		219.9	127.0	-7.1			
trans-2-Butene	C_4H_8					-33.3				-11.4			
trans-2-Butenedinitrile	$C_4H_2N_2$	268.2								340.2			
trans-2-Butenenitrile	C_4H_5N					95.1				134.3			
3-Butenenitrile	C_4H_5N					117.8				159.7			
trans-2-Butenoic acid	$C_4H_6O_2$												
2-Butoxyethanol	$C_6H_{14}O_2$								281.0				
N-Butylacetamide	$C_6H_{13}NO$					-380.9				-305.9			
Butyl acetate	$C_6H_{12}O_2$					-529.2			227.8	-485.3			
tert-Butyl acetate	$C_6H_{12}O_2$					-554.5			231.0	-516.5			
Butyl acrylate	$C_7H_{12}O_2$					-422.6			251.0	-375.3			
Butylamine	$C_4H_{11}N$					-127.6			179.2	-91.9			
sec-Butylamine	$C_4H_{11}N$					-137.5				-104.6			
tert-Butylamine	$C_4H_{11}N$					-150.6			192.1	-121.0			
Butylbenzene	$C_{10}H_{14}$					-63.2		321.2	243.4	-11.8			
sec-Butylbenzene, (±)-	$C_{10}H_{14}$					-66.4				-18.4			
tert-Butylbenzene	$C_{10}H_{14}$					-71.9				-23.0			
Butyl chloroacetate	$C_6H_{11}ClO_2$					-538.4				-487.4			
Butyl 2-chloropropanoate	$C_7H_{13}ClO_2$					-571.7				-517.3			
Butyl 3-chloropropanoate	$C_7H_{13}ClO_2$					-557.9				-502.3			
Butylcyclohexane	$C_{10}H_{20}$					-263.1		345.0	271.0	-213.7			
Butyl dichloroacetate	$C_6H_{10}Cl_2O_2$					-550.1				-497.8			
1-*tert*-Butyl-3,5-dimethylbenzene	$C_{12}H_{18}$					-146.5							
5-Butyldocosane	$C_{26}H_{54}$					-713.5				-587.6			
11-Butyldocosane	$C_{26}H_{54}$					-716.0				-593.4			
Butyl ethyl ether	$C_6H_{14}O$								159.0				
tert-Butyl ethyl ether	$C_6H_{14}O$									-313.9			
Butyl ethyl sulfide	$C_6H_{14}S$					-172.3				-127.8			
Butyl formate	$C_5H_{10}O_2$								200.2				
tert-Butyl hydroperoxide	$C_4H_{10}O_2$					-293.6				-245.9			
tert-Butyl isobutyl ether	$C_8H_{18}O$					-409.1				-369.0			
tert-Butyl isopropyl ether	$C_7H_{16}O$					-392.8				-358.1			
1-*tert*-Butyl-3-methylbenzene	$C_{11}H_{16}$					-109.7							
1-*tert*-Butyl-4-methylbenzene	$C_{11}H_{16}$					-109.7				-57.0			
Butyl methyl ether	$C_5H_{12}O$					-290.6		295.3	192.7	-258.1			
Butyl methyl sulfide	$C_5H_{12}S$					-142.9		307.5	200.9	-102.4			
tert-Butyl methyl sulfide	$C_5H_{12}S$					-157.1		276.1	199.9	-121.3			
Butyl oleate	$C_{22}H_{42}O_2$					-816.9							
Butyl pentanoate	$C_9H_{18}O_2$					-613.3				-560.2			
sec-Butyl pentanoate	$C_9H_{18}O_2$					-624.2				-573.2			
N-Butylpiperidine	$C_9H_{19}N$					-171.8							
Butyl stearate	$C_{22}H_{44}O_2$												
Butyl trichloroacetate	$C_6H_9Cl_3O_2$					-545.8				-492.3			
Butylurea	$C_5H_{12}N_2O$	-419.5											
tert-Butylurea	$C_5H_{12}N_2O$	-417.4											
Butyl vinyl ether	$C_6H_{12}O$					-218.8			232.0	-182.6			
1-Butyne	C_4H_6					141.4				165.2			
2-Butyne	C_4H_6					119.1				145.7			
2-Butynedinitrile	C_4N_2					500.4				529.2			
2-Butynedioic acid	$C_4H_2O_4$	-577.3											
γ-Butyrolactone	$C_4H_6O_2$					-420.9			141.4	-366.5			
Cadmium	Cd	0.0		51.8	26.0					111.8		167.7	20.8
Cadmium bromide	Br_2Cd	-316.2	-296.3	137.2	76.7								
Cadmium carbonate	$CCdO_3$	-750.6	-669.4	92.5									
Cadmium chloride	$CdCl_2$	-391.5	-343.9	115.3	74.7								
Cadmium fluoride	CdF_2	-700.4	-647.7	77.4									
Cadmium hydroxide	CdH_2O_2	-560.7	-473.6	96.0									
Cadmium iodide	CdI_2	-203.3	-201.4	161.1	80.0								
Cadmium oxide	CdO	-258.4	-228.7	54.8	43.4								
Cadmium sulfate	CdO_4S	-933.3	-822.7	123.0	99.6								
Cadmium sulfide	CdS	-161.9	-156.5	64.9									
Cadmium telluride	CdTe	-92.5	-92.0	100.0									
Calcium	Ca	0.0		41.6	25.9					177.8	144.0	154.9	20.8
Calcium bromide	Br_2Ca	-682.8	-663.6	130.0									
Calcium carbide	C_2Ca	-59.8	-64.9	70.0	62.7								
Calcium carbonate (calcite)	$CCaO_3$	-1207.6	-1129.1	91.7	83.5								

Name	Mol. form.	$\Delta_f H°$(c)/ kJ mol^{-1}	$\Delta_f G°$(c)/ kJ mol^{-1}	$S°$(c)/ J mol^{-1} K^{-1}	C_p(c)/ J mol^{-1} K^{-1}	$\Delta_f H°$(l)/ kJ mol^{-1}	$\Delta_f G°$(l)/ kJ mol^{-1}	$S°$(l)/ J mol^{-1} K^{-1}	C_p(l)/ J mol^{-1} K^{-1}	$\Delta_f H°$(g)/ kJ mol^{-1}	$\Delta_f G°$(g)/ kJ mol^{-1}	$S°$(g)/ J mol^{-1} K^{-1}	C_p(g)/ J mol^{-1} K^{-1}
Calcium carbonate (aragonite)	$CCaO_3$	-1207.8	-1128.2	88.0	82.3								
Calcium chloride	$CaCl_2$	-795.4	-748.8	108.4	72.9								
Calcium cyanide	C_2CaN_2	-184.5											
Calcium fluoride	CaF_2	-1228.0	-1175.6	68.5	67.0								
Calcium hydride	CaH_2	-181.5	-142.5	41.4	41.0								
Calcium hydroxide	CaH_2O_2	-985.2	-897.5	83.4	87.5								
Calcium iodide	CaI_2	-533.5	-528.9	142.0									
Calcium nitrate	CaN_2O_6	-938.2	-742.8	193.2	149.4								
Calcium oxalate	C_2CaO_4	-1360.6											
Calcium oxide	CaO	-634.9	-603.3	38.1	42.0								
Calcium phosphate	$Ca_3O_8P_2$	-4120.8	-3884.7	236.0	227.8								
Calcium sulfate	CaO_4S	-1434.5	-1322.0	106.5	99.7								
Calcium sulfide	CaS	-482.4	-477.4	56.5	47.4								
Californium	Cf	0.0											
Camphor, (±)-	$C_{10}H_{16}O$	-319.4			271.2					-267.5			
Caprolactam	$C_6H_{11}NO$	-329.4			156.8					-239.6			
Carbazole	$C_{12}H_9N$	101.7								200.7			
Carbon (diamond)	C	1.9	2.9	2.4	6.1								
Carbon dioxide	CO_2									-393.5	-394.4	213.8	37.1
Carbon diselenide	CSe_2					164.8							
Carbon disulfide	CS_2					89.0	64.6	151.3	76.4	116.7	67.1	237.8	45.4
Carbon [fullerene-C_{60}]	C_{60}	2327.0	2302.0	426.0	520.0					2502.0	2442.0	544.0	512.0
Carbon [fullerene-C_{70}]	C_{70}	2555.0	2537.0	464.0	650.0					2755.0	2692.0	614.0	585.0
Carbon (graphite)	C	0.0		5.7	8.5					716.7	671.3	158.1	20.8
Carbon monosulfide	CS									280.3	228.8	210.6	29.8
Carbon monoxide	CO									-110.5	-137.2	197.7	29.1
Carbon oxysulfide	COS									-142.0	-169.2	231.6	41.5
Carbonyl bromide	CBr_2O					-127.2				-96.2	-110.9	309.1	61.8
Carbonyl chloride	CCl_2O									-219.1	-204.9	283.5	57.7
Carbonyl chloride fluoride	CClFO											276.7	52.4
Carbonyl fluoride	CF_2O									-639.8			46.8
Cerium	Ce	0.0		72.0	26.9					423.0	385.0	191.8	23.1
Cerium(III) bromide	Br_3Ce	-891.4											
Cerium(III) chloride	$CeCl_3$	-1060.5	-984.8	151.0	87.4								
Cerium(III) iodide	CeI_3	-669.3											
Cerium(III) oxide	Ce_2O_3	-1796.2	-1706.2	150.6	114.6								
Cerium(IV) oxide	CeO_2	-1088.7	-1024.6	62.3	61.6								
Cerium(II) sulfide	CeS	-459.4	-451.5	78.2	50.0								
Cesium	Cs	0.0		85.2	32.2					76.5	49.6	175.6	20.8
Cesium amide	CsH_2N	-118.4											
Cesium bromide	BrCs	-405.8	-391.4	113.1	52.9								
Cesium carbonate	CCs_2O_3	-1139.7	-1054.3	204.5	123.9								
Cesium chloride	ClCs	-443.0	-414.5	101.2	52.5								
Cesium fluoride	CsF	-553.5	-525.5	92.8	51.1								
Cesium hydride	CsH	-54.2											
Cesium hydrogen carbonate	$CHCsO_3$	-966.1											
Cesium hydrogen fluoride	CsF_2H	-923.8	-858.9	135.2	87.3								
Cesium hydrogen sulfate	$CsHO_4S$	-1158.1											
Cesium hydroxide	CsHO	-416.2	-371.8	104.2	69.9					-256.0	-256.5	254.8	49.7
Cesium iodide	CsI	-346.6	-340.6	123.1	52.8								
Cesium metaborate	$BCsO_2$	-972.0	-915.0	104.4	80.6								
Cesium nitrate	$CsNO_3$	-506.0	-406.5	155.2									
Cesium oxide	Cs_2O	-345.8	-308.1	146.9	76.0								
Cesium perchlorate	$ClCsO_4$	-443.1	-314.3	175.1	108.3								
Cesium sulfate	Cs_2O_4S	-1443.0	-1323.6	211.9	134.9								
Cesium sulfide	Cs_2S	-359.8											
Cesium sulfite	Cs_2O_3S	-1134.7											
Cesium superoxide	CsO_2	-286.2											
Chlorine	Cl_2									0.0		223.1	33.9
Chlorine (atomic)	Cl									121.3	105.3	165.2	21.8
Chlorine dioxide	ClO_2									102.5	120.5	256.8	42.0
Chlorine fluoride	ClF									-50.3	-51.8	217.9	32.1
Chlorine monoxide	Cl_2O									80.3	97.9	266.2	45.4
Chlorine oxide (ClO)	ClO									101.8	98.1	226.6	31.5
Chlorine superoxide [ClOO]	ClO_2									89.1	105.0	263.7	46.0
Chlorine trifluoride	ClF_3					-189.5				-163.2	-123.0	281.6	63.9

Name	Mol. form.	$\Delta_f H°$(c)/ kJ mol^{-1}	$\Delta_f G°$(c)/ kJ mol^{-1}	$S°$(c)/ J mol^{-1} K^{-1}	C_p(c)/ J mol^{-1} K^{-1}	$\Delta_f H°$(l)/ kJ mol^{-1}	$\Delta_f G°$(l)/ kJ mol^{-1}	$S°$(l)/ J mol^{-1} K^{-1}	C_p(l)/ J mol^{-1} K^{-1}	$\Delta_f H°$(g)/ kJ mol^{-1}	$\Delta_f G°$(g)/ kJ mol^{-1}	$S°$(g)/ J mol^{-1} K^{-1}	C_p(g)/ J mol^{-1} K^{-1}
Chloroacetic acid	$C_2H_3ClO_2$	-509.7								-427.6	-368.5	325.9	78.8
Chloroacetyl chloride	$C_2H_2Cl_2O$					-283.7				-244.8			
Chloroacetylene	C_2HCl											242.0	54.3
2-Chloroaniline	C_6H_6ClN					-4.6							
3-Chloroaniline	C_6H_6ClN					-20.3			198.7				
4-Chloroaniline	C_6H_6ClN	-33.3			147.3								
Chlorobenzene	C_6H_5Cl					11.1			150.1	52.0			
2-Chlorobenzoic acid	$C_7H_5ClO_2$	-404.5								-325.0			
3-Chlorobenzoic acid	$C_7H_5ClO_2$	-424.3								-342.3			
4-Chlorobenzoic acid	$C_7H_5ClO_2$	-428.9			163.2					-341.0			
3-Chlorobenzoyl chloride	$C_7H_4Cl_2O$					-189.7							
Chloroborane(1)	BCl									149.5	120.9	213.2	31.7
1-Chlorobutane	C_4H_9Cl					-188.1				-154.4			
2-Chlorobutane	C_4H_9Cl					-192.8				-161.1			
2-Chlorobutanoic acid	$C_4H_7ClO_2$					-575.5							
3-Chlorobutanoic acid	$C_4H_7ClO_2$					-556.3							
4-Chlorobutanoic acid	$C_4H_7ClO_2$					-566.3							
Chlorocyclohexane	$C_6H_{11}Cl$					-207.2				-163.7			
Chlorodibromomethane	$CHBr_2Cl$											327.7	69.2
1-Chloro-1,1-difluoroethane	$C_2H_3ClF_2$											307.2	82.5
1-Chloro-2,2-difluoroethene	C_2HClF_2									-315.5	-289.1	303.0	72.1
Chlorodifluoromethane	$CHClF_2$									-482.6		280.9	55.9
1-Chlorododecane	$C_{12}H_{25}Cl$					-392.3				-321.1			
Chloroethane	C_2H_5Cl					-136.8	-59.3	190.8	104.3	-112.1	-60.4	276.0	62.8
2-Chloroethanol	C_2H_5ClO					-295.4							
Chloroethene	C_2H_3Cl	-94.1			59.4	14.6				37.2	53.6	264.0	53.7
2-Chloroethyl ethyl ether	C_4H_9ClO					-335.6				-301.3			
2-Chloroethyl vinyl ether	C_4H_7ClO					-208.1				-170.1			
1-Chloro-1-fluoroethane	C_2H_4ClF									-313.4			
Chlorofluoromethane	CH_2ClF											264.4	47.0
Chlorogermane	$ClGeH_3$											263.7	54.7
2-Chlorohexane	$C_6H_{13}Cl$					-246.1				-204.3			
Chloromethane	CH_3Cl									-81.9		234.6	40.8
(Chloromethyl)benzene	C_7H_7Cl					-32.5				18.9			
1-Chloro-3-methylbutane	$C_5H_{11}Cl$					-216.0				-179.7			
2-Chloro-2-methylbutane	$C_5H_{11}Cl$					-235.7				-202.2			
2-Chloro-3-methylbutane	$C_5H_{11}Cl$					-226.6				-185.1			
1-Chloro-2-methylpropane	C_4H_9Cl					-191.1				-159.3			
2-Chloro-2-methylpropane	C_4H_9Cl					-211.3				-182.2			
1-Chloronaphthalene	$C_{10}H_7Cl$					54.6			212.6	119.8			
2-Chloronaphthalene	$C_{10}H_7Cl$	55.4								137.4			
1-Chloro-4-nitrobenzene	$C_6H_4ClNO_2$	-48.7			250.2								
1-Chlorooctadecane	$C_{18}H_{37}Cl$					-544.1				-446.0			
1-Chlorooctane	$C_8H_{17}Cl$					-291.3				-238.9			
Chloropentafluorobenzene	C_6ClF_5	-858.4								-809.3			
Chloropentafluoroethane	C_2ClF_5									-1118.8			184.2
1-Chloropentane	$C_5H_{11}Cl$					-213.2				-174.9			
2-Chlorophenol	C_6H_5ClO								188.7				
3-Chlorophenol	C_6H_5ClO	-206.4				-189.3							
4-Chlorophenol	C_6H_5ClO	-197.7				-181.3							
1-Chloropropane	C_3H_7Cl					-160.5				-131.9			
2-Chloropropane	C_3H_7Cl					-172.3				-144.9			
3-Chloro-1,2-propanediol	$C_3H_7ClO_2$					-525.3							
2-Chloro-1,3-propanediol	$C_3H_7ClO_2$					-517.5							
2-Chloropropanoic acid	$C_3H_5ClO_2$					-522.5				-475.8			
3-Chloropropanoic acid	$C_3H_5ClO_2$	-549.3											
2-Chloropropene	C_3H_5Cl									-21.0			
3-Chloropropene	C_3H_5Cl								125.1				
Chlorosilane	ClH_3Si											250.7	51.0
Chlorosilylidyne	ClSi									189.9			36.9
N-Chlorosuccinimide	$C_4H_4ClNO_2$	-357.9											
2-Chlorotoluene	C_7H_7Cl								166.8				
2-Chloro-1,1,1-trifluoroethane	$C_2H_2ClF_3$											326.5	89.1
Chlorotrifluoroethene	C_2ClF_3					-522.7				-505.5	-523.8	322.1	83.9
Chlorotrifluoromethane	$CClF_3$									-706.3			66.9
Chlorotrinitromethane	$CClN_3O_6$					-27.1				18.4			

Name	Mol. form.	$\Delta_f H°$(c)/ kJ mol^{-1}	$\Delta_f G°$(c)/ kJ mol^{-1}	$S°$(c)/ J mol^{-1} K^{-1}	C_p(c)/ J mol^{-1} K^{-1}	$\Delta_f H°$(l)/ kJ mol^{-1}	$\Delta_f G°$(l)/ kJ mol^{-1}	$S°$(l)/ J mol^{-1} K^{-1}	C_p(l)/ J mol^{-1} K^{-1}	$\Delta_f H°$(g)/ kJ mol^{-1}	$\Delta_f G°$(g)/ kJ mol^{-1}	$S°$(g)/ J mol^{-1} K^{-1}	C_p(g)/ J mol^{-1} K^{-1}
Chloroxyborane	$BClO$									-314.0			
Chromium	Cr	0.0		23.8	23.4					396.6	351.8	174.5	20.8
Chromium(II) bromide	Br_2Cr	-302.1											
Chromium(II) chloride	Cl_2Cr	-395.4	-356.0	115.3	71.2								
Chromium(III) chloride	Cl_3Cr	-556.5	-486.1	123.0	91.8								
Chromium(VI) dichloride dioxide	Cl_2CrO_2					-579.5	-510.8	221.8		-538.1	-501.6	329.8	84.5
Chromium(II) fluoride	CrF_2	-778.0											
Chromium(III) fluoride	CrF_3	-1159.0	-1088.0	93.9	78.7								
Chromium(II) iodide	CrI_2	-156.9											
Chromium(III) iodide	CrI_3	-205.0											
Chromium iron oxide	Cr_2FeO_4	-1444.7	-1343.8	146.0	133.6								
Chromium(II,III) oxide	Cr_3O_4	-1531.0											
Chromium(III) oxide	Cr_2O_3	-1139.7	-1058.1	81.2	118.7								
Chromium(IV) oxide	CrO_2	-598.0											
Chromium(VI) oxide	CrO_3									-292.9		266.2	56.0
Chrysene	$C_{18}H_{12}$	145.3								269.8			
Citric acid	$C_6H_8O_7$	-1543.8											
Cobalt	Co	0.0		30.0	24.8					424.7	380.3	179.5	23.0
Cobalt(II) bromide	Br_2Co	-220.9			79.5								
Cobalt(II) carbonate	$CCoO_3$	-713.0											
Cobalt(II) chloride	Cl_2Co	-312.5	-269.8	109.2	78.5								
Cobalt(II) fluoride	CoF_2	-692.0	-647.2	82.0	68.8								
Cobalt(II) hydroxide	CoH_2O_2	-539.7	-454.3	79.0									
Cobalt(II) iodide	CoI_2	-88.7											
Cobalt(II) nitrate	CoN_2O_6	-420.5											
Cobalt(II) oxide	CoO	-237.9	-214.2	53.0	55.2								
Cobalt(II,III) oxide	Co_3O_4	-891.0	-774.0	102.5	123.4								
Cobalt(II) sulfate	CoO_4S	-888.3	-782.3	118.0									
Cobalt(II) sulfide	CoS	-82.8											
Cobalt(III) sulfide	Co_2S_3	-147.3											
Copper	Cu	0.0		33.2	24.4					337.4	297.7	166.4	20.8
Copper(I) bromide	$BrCu$	-104.6	-100.8	96.1	54.7								
Copper(II) bromide	Br_2Cu	-141.8											
Copper(I) chloride	$ClCu$	-137.2	-119.9	86.2	48.5								
Copper(II) chloride	Cl_2Cu	-220.1	-175.7	108.1	71.9								
Copper(I) cyanide	$CCuN$	96.2	111.3	84.5									
Copper(II) fluoride	CuF_2	-542.7											
Copper(II) hydroxide	CuH_2O_2	-449.8											
Copper(I) iodide	CuI	-67.8	-69.5	96.7	54.1								
Copper(II) nitrate	CuN_2O_6	-302.9											
Copper(I) oxide	Cu_2O	-168.6	-146.0	93.1	63.6								
Copper(II) oxide	CuO	-157.3	-129.7	42.6	42.3								
Copper(II) selenide	$CuSe$	-39.5											
Copper(II) sulfate	CuO_4S	-771.4	-662.2	109.2									
Copper(I) sulfide	Cu_2S	-79.5	-86.2	120.9	76.3								
Copper(II) sulfide	CuS	-53.1	-53.6	66.5	47.8								
Copper(II) tungstate	CuO_4W	-1105.0											
Creatine	$C_4H_9N_3O_2$	-537.2											
Creatinine	$C_4H_7N_3O$	-238.5											
o-Cresol	C_7H_8O	-204.6		165.4	154.6					-128.6			
m-Cresol	C_7H_8O					-194.0		212.6	224.9	-132.3			
p-Cresol	C_7H_8O	-199.3		167.3	150.2					-125.4			
Curium	Cm	0.0											
Cyanamide	CH_2N_2	58.8											
Cyanide	CN									437.6	407.5	202.6	29.2
Cyanogen	C_2N_2					285.9				306.7		241.9	56.8
Cyanogen bromide	$CBrN$	140.5								186.2	165.3	248.3	46.9
Cyanogen chloride	$CClN$					112.1				138.0	131.0	236.2	45.0
Cyanogen fluoride	CFN											224.7	41.8
Cyanogen iodide	CIN	166.2	185.0	96.2						225.5	196.6	256.8	48.3
Cyclobutanamine	C_4H_9N					5.6				41.2			
Cyclobutane	C_4H_8					3.7				27.7			
Cyclobutanecarbonitrile	C_5H_7N					103.0				147.4			
Cyclobutene	C_4H_6									156.7			
Cycloheptane	C_7H_{14}					-156.6				-118.1			
Cyclohexane	C_6H_{12}					-156.4			154.9	-123.4			

Name	Mol. form.	$\Delta_f H°$(c)/ kJ mol^{-1}	$\Delta_f G°$(c)/ kJ mol^{-1}	$S°$(c)/ J mol^{-1} K^{-1}	C_p(c)/ J mol^{-1} K^{-1}	$\Delta_f H°$(l)/ kJ mol^{-1}	$\Delta_f G°$(l)/ kJ mol^{-1}	$S°$(l)/ J mol^{-1} K^{-1}	C_p(l)/ J mol^{-1} K^{-1}	$\Delta_f H°$(g)/ kJ mol^{-1}	$\Delta_f G°$(g)/ kJ mol^{-1}	$S°$(g)/ J mol^{-1} K^{-1}	C_p(g)/ J mol^{-1} K^{-1}
Cyclohexanecarbonitrile	$C_7H_{11}N$					-47.2				4.8			
Cyclohexanethiol	$C_6H_{12}S$					-140.7		255.6	192.6	-96.2			
Cyclohexanol	$C_6H_{12}O$					-348.2			208.2	-286.2			
Cyclohexanone	$C_6H_{10}O$					-271.2			182.2	-226.1			
Cyclohexene	C_6H_{10}					-38.5		214.6	148.3	-5.0			
1-Cyclohexenecarbonitrile	C_7H_9N					48.1				101.6			
Cyclohexylamine	$C_6H_{13}N$					-147.6				-104.0			
Cyclohexylbenzene	$C_{12}H_{16}$					-76.6				-16.7			
Cyclohexylcyclohexane	$C_{12}H_{22}$					-273.7				-215.7			
Cyclooctane	C_8H_{16}					-167.7				-124.4			
1,3-Cyclopentadiene	C_5H_6					105.9				134.3			
Cyclopentane	C_5H_{10}					-105.1		204.5	128.8	-76.4			
Cyclopentanecarbonitrile	C_6H_9N					0.7				44.1			
cis-1,2-Cyclopentanediol	$C_5H_{10}O_2$	-485.0											
trans-1,2-Cyclopentanediol	$C_5H_{10}O_2$	-490.1											
Cyclopentanethiol	$C_5H_{10}S$					-89.5		256.9	165.2	-48.1			
Cyclopentanol	$C_5H_{10}O$					-300.1		204.1	182.5	-242.5		362.9	
Cyclopentanone	C_5H_8O					-235.9				-192.1			
Cyclopentene	C_5H_8					4.3		201.2	122.4	34.0			
1-Cyclopentenecarbonitrile	C_6H_7N					111.5				156.5			
Cyclopentylamine	$C_5H_{11}N$					-95.1		241.0	181.2	-54.9			
Cyclopentyl methyl sulfide	$C_6H_{12}S$					-109.8				-64.7			
Cyclopropane	C_3H_6					35.2				53.3	104.5	237.5	55.6
Cyclopropanecarbonitrile	C_4H_5N					140.8				182.8			
Cyclopropene	C_3H_4									277.1			
Cyclopropylamine	C_3H_7N					45.8		187.7	147.1	77.0			
Cyclopropylbenzene	C_9H_{10}					100.3				150.5			
Cyclotetramethylenetetranitramine	$C_4H_8N_8O_8$									187.9		568.8	275.5
L-Cysteine	$C_3H_7NO_2S$	-534.1											
L-Cystine	$C_6H_{12}N_2O_4S_2$	-1032.7											
Cytosine	$C_4H_5N_3O$	-221.3			132.6								
Decaborane(14)	$B_{10}H_{14}$									47.3	232.8	350.7	186.1
cis-Decahydronaphthalene	$C_{10}H_{18}$					-219.4		265.0	232.0	-169.2			
trans-Decahydronaphthalene	$C_{10}H_{18}$					-230.6		264.9	228.5	-182.1			
Decane	$C_{10}H_{22}$					-300.9			314.4	-249.5			
Decanedioic acid	$C_{10}H_{18}O_4$	-1082.6								-921.9			
1,10-Decanediol	$C_{10}H_{22}O_2$	-678.9											
Decanenitrile	$C_{10}H_{19}N$					-158.4				-91.5			
1-Decanethiol	$C_{10}H_{22}S$	-309.9				-276.5		476.1	350.4	-211.5			
Decanoic acid	$C_{10}H_{20}O_2$	-713.7				-684.3				-594.9			
1-Decanol	$C_{10}H_{22}O$					-478.1			370.6	-396.6			
1-Decene	$C_{10}H_{20}$					-173.8		425.0	300.8	-123.3			
Decylbenzene	$C_{16}H_{26}$					-218.3				-138.6			
Decylcyclopentane	$C_{15}H_{30}$					-367.3							
11-Decylheneicosane	$C_{31}H_{64}$					-848.0				-705.8			
Diacetone alcohol	$C_6H_{12}O_2$								221.3				
Dialuminum	Al_2									485.9	433.3	233.2	36.4
4,4'-Diaminodiphenylmethane	$C_{13}H_{14}N_2$				270.9								
Diantimony	Sb_2									235.6	187.0	254.9	36.4
Diarsenic	As_2									222.2	171.9	239.4	35.0
Diazomethane	CH_2N_2											242.9	52.5
Dibenz[*a,h*]anthracene	$C_{22}H_{14}$												283.9
Dibenzofuran	$C_{12}H_8O$	-5.3								83.4			
Dibenzothiophene	$C_{12}H_8S$	120.0								205.1			
Dibismuth	Bi_2									219.7			36.9
Diborane	B_2H_6									36.4	87.6	232.1	56.7
Diboron	B_2									830.5	774.0	201.9	30.5
Diboron dioxide	B_2O_2									-454.8	-462.3	242.5	57.3
1,2-Dibromobutane	$C_4H_8Br_2$					-142.1				-91.6			
1,3-Dibromobutane	$C_4H_8Br_2$					-148.0							
1,4-Dibromobutane	$C_4H_8Br_2$					-140.3				-87.8			
2,3-Dibromobutane	$C_4H_8Br_2$					-139.6				-102.0			
Dibromochlorofluoromethane	CBr_2ClF											342.8	82.4
1,2-Dibromo-1-chloro-1,2,2-trifluoroethane	$C_2Br_2ClF_3$					-691.7				-656.6			
1,2-Dibromo-1,2-dichloroethane	$C_2H_2Br_2Cl_2$									-36.9			

Thermochem

Name	Mol. form.	$\Delta_fH°$(c)/ kJ mol^{-1}	$\Delta_fG°$(c)/ kJ mol^{-1}	$S°$(c)/ J mol^{-1} K^{-1}	C_p(c)/ J mol^{-1} K^{-1}	$\Delta_fH°$(l)/ kJ mol^{-1}	$\Delta_fG°$(l)/ kJ mol^{-1}	$S°$(l)/ J mol^{-1} K^{-1}	C_p(l)/ J mol^{-1} K^{-1}	$\Delta_fH°$(g)/ kJ mol^{-1}	$\Delta_fG°$(g)/ kJ mol^{-1}	$S°$(g)/ J mol^{-1} K^{-1}	C_p(g)/ J mol^{-1} K^{-1}
Dibromodichloromethane	CBr_2Cl_2											347.8	87.1
Dibromodifluoromethane	CBr_2F_2											325.3	77.0
1,1-Dibromoethane	$C_2H_4Br_2$					-66.2						327.7	80.8
1,2-Dibromoethane	$C_2H_4Br_2$					-79.2		223.3	136.0	-37.5			
cis-1,2-Dibromoethene	$C_2H_2Br_2$											311.3	68.8
trans-1,2-Dibromoethene	$C_2H_2Br_2$											313.5	70.3
Dibromofluoromethane	$CHBr_2F$											316.8	65.1
1,2-Dibromoheptane	$C_7H_{14}Br_2$					-212.3				-157.9			
Dibromomethane	CH_2Br_2											293.2	54.7
2,3-Dibromo-2-methylbutane	$C_5H_{10}Br_2$									-137.6			
1,2-Dibromo-2-methylpropane	$C_4H_8Br_2$					-156.6				-113.3			
1,2-Dibromopropane	$C_3H_6Br_2$					-113.6				-71.6			
Dibromosilane	Br_2H_2Si											309.7	65.5
1,2-Dibromotetrafluoroethane	$C_2Br_2F_4$					-817.7				-789.1			
1,2-Dibutoxyethane	$C_{10}H_{22}O_2$								350.0				
Dibutylamine	$C_8H_{19}N$					-206.0			292.9	-156.6			
1,3-Di-*tert*-butylbenzene	$C_{14}H_{22}$					-188.8							
1,4-Di-*tert*-butylbenzene	$C_{14}H_{22}$	-212.0											
Dibutyl disulfide	$C_8H_{18}S_2$					-222.9				-160.6			
Di-*tert*-butyl disulfide	$C_8H_{18}S_2$					-255.2				-201.0			
cis-1,2-Di-*tert*-butylethene	$C_{10}H_{20}$					-163.6							
Dibutyl ether	$C_8H_{18}O$					-377.9			278.2	-332.8			
Di-*sec*-butyl ether	$C_8H_{18}O$					-401.5				-360.6			
Di-*tert*-butyl ether	$C_8H_{18}O$					-399.6			276.1	-362.0			
1,3-Di-*tert*-butyl-5-methylbenzene	$C_{15}H_{24}$	-245.8											
2,6-Di-*tert*-butyl-4-methylphenol	$C_{15}H_{24}O$	-410.0								-296.9			
Dibutyl phthalate	$C_{16}H_{22}O_4$					-842.6				-750.9			
Dibutyl sebacate	$C_{18}H_{34}O_4$								619.0				
Dibutyl sulfide	$C_8H_{18}S$					-220.7		405.1	284.3	-167.7			
Di-*sec*-butyl sulfide	$C_8H_{18}S$					-220.7				-167.7			
Di-*tert*-butyl sulfide	$C_8H_{18}S$					-232.6				-188.8			
Dibutyl sulfite	$C_8H_{18}O_3S$					-693.1				-625.3			
Dicarbon	C_2									831.9	775.9	199.4	43.2
Dichloroacetic acid	$C_2H_2Cl_2O_2$					-496.3							
Dichloroacetyl chloride	C_2HCl_3O					-280.4				-241.0			
3,4-Dichloroaniline	$C_6H_5Cl_2N$	-89.1											
o-Dichlorobenzene	$C_6H_4Cl_2$					-17.5			162.4	30.2			
m-Dichlorobenzene	$C_6H_4Cl_2$					-20.7				25.7			
p-Dichlorobenzene	$C_6H_4Cl_2$	-42.3		175.4	147.8					22.5			
1,3-Dichlorobutane	$C_4H_8Cl_2$					-237.3				-195.0			
1,4-Dichlorobutane	$C_4H_8Cl_2$					-229.8				-183.4			
Dichlorodifluoromethane	CCl_2F_2									-477.4	-439.4	300.8	72.3
1,1-Dichloroethane	$C_2H_4Cl_2$					-158.4	-73.8	211.8	126.3	-127.7	-70.8	305.1	76.2
1,2-Dichloroethane	$C_2H_4Cl_2$					-166.8			128.4	-126.4		308.4	78.7
1,1-Dichloroethene	$C_2H_2Cl_2$					-23.9	24.1	201.5	111.3	2.8	25.4	289.0	67.1
cis-1,2-Dichloroethene	$C_2H_2Cl_2$					-26.4		198.4	116.4	4.6		289.6	65.1
trans-1,2-Dichloroethene	$C_2H_2Cl_2$					-24.3	27.3	195.9	116.8	5.0	28.6	290.0	66.7
1,1-Dichloro-1-fluoroethane	$C_2H_3Cl_2F$											320.2	88.7
1,1-Dichloro-2-fluoroethene	C_2HCl_2F											313.9	76.5
Dichlorofluoromethane	$CHCl_2F$											293.1	60.9
Dichloromethane	CH_2Cl_2					-124.2		177.8	101.2	-95.4		270.2	51.0
2,4-Dichlorophenol	$C_6H_4Cl_2O$	-226.4								-156.3			
1,2-Dichloropropane, (±)-	$C_3H_6Cl_2$					-198.8			149.1	-162.8			
1,3-Dichloropropane	$C_3H_6Cl_2$					-199.9				-159.2			
2,2-Dichloropropane	$C_3H_6Cl_2$					-205.8				-173.2			
2,3-Dichloro-1-propanol	$C_3H_6Cl_2O$					-381.5				-316.3			
1,3-Dichloro-2-propanol	$C_3H_6Cl_2O$					-385.3				-318.4			
2,3-Dichloropropene	$C_3H_4Cl_2$					-73.3							
Dichlorosilane	Cl_2H_2Si											285.7	60.5
1,2-Dichloro-1,1,2,2-tetrafluoroethane	$C_2Cl_2F_4$					-960.2			111.7	-937.0			
2,2-Dichloro-1,1,1-trifluoroethane	$C_2HCl_2F_3$											352.8	102.5
Dicopper	Cu_2									484.2	431.9	241.6	36.6
Diethanolamine	$C_4H_{11}NO_2$	-493.8			233.5					-397.1			
1,1-Diethoxyethane	$C_6H_{14}O_2$					-491.4				-453.5			

Name	Mol. form.	$\Delta_f H°$(c)/ kJ mol⁻¹	$\Delta_f G°$(c)/ kJ mol⁻¹	$S°$(c)/ J mol⁻¹ K⁻¹	C_p(c)/ J mol⁻¹ K⁻¹	$\Delta_f H°$(l)/ kJ mol⁻¹	$\Delta_f G°$(l)/ kJ mol⁻¹	$S°$(l)/ J mol⁻¹ K⁻¹	C_p(l)/ J mol⁻¹ K⁻¹	$\Delta_f H°$(g)/ kJ mol⁻¹	$\Delta_f G°$(g)/ kJ mol⁻¹	$S°$(g)/ J mol⁻¹ K⁻¹	C_p(g)/ J mol⁻¹ K⁻¹
1,2-Diethoxyethane	$C_6H_{14}O_2$					-451.4			259.4	-408.1			
Diethoxymethane	$C_5H_{12}O_2$					-450.5				-414.7			
2,2-Diethoxypropane	$C_7H_{16}O_2$					-538.9				-506.9			
Diethylamine	$C_4H_{11}N$					-103.7			169.2	-72.2			
Diethylamine hydrochloride	$C_4H_{12}ClN$	-358.6											
2-Diethylaminoethanol	$C_6H_{15}NO$					-305.9							
o-Diethylbenzene	$C_{10}H_{14}$					-68.5							
m-Diethylbenzene	$C_{10}H_{14}$					-73.5							
p-Diethylbenzene	$C_{10}H_{14}$					-72.8							
Diethyl carbonate	$C_5H_{10}O_3$					-681.5				-637.9			
Diethyl 3,5-dimethylpyrrole-2,4-dicarboxylate	$C_{12}H_{17}NO_4$	-916.7											
Diethyl disulfide	$C_4H_{10}S_2$					-120.1		305.0	204.0	-79.4			
Diethylene glycol	$C_4H_{10}O_3$					-628.5			244.8	-571.2			
Diethylene glycol dibutyl ether	$C_{12}H_{26}O_3$								452.0				
Diethylene glycol diethyl ether	$C_8H_{18}O_3$								341.4				
Diethylene glycol dimethyl ether	$C_6H_{14}O_3$								274.1				
Diethylene glycol monobutyl ether	$C_8H_{18}O_3$								354.9				
Diethylene glycol monoethyl ether	$C_6H_{14}O_3$								301.0				
Diethylene glycol monomethyl ether	$C_5H_{12}O_3$								271.1				
Diethyl ether	$C_4H_{10}O$					-279.5		253.5	172.5	-252.1		342.7	119.5
Diethyl malonate	$C_7H_{12}O_4$								285.0				
Diethyl mercury	$C_4H_{10}Hg$					30.1			182.8	75.3			
Diethyl oxalate	$C_6H_{10}O_4$					-805.5				-742.0			
3,3-Diethylpentane	C_9H_{20}					-275.4			278.2	-233.3			
Diethyl phthalate	$C_{12}H_{14}O_4$					-776.6		425.1	366.1	-688.4			
Diethyl sulfate	$C_4H_{10}O_4S$					-813.2				-756.3			
Diethyl sulfide	$C_4H_{10}S$					-119.4		269.3	171.4	-83.5		368.1	117.0
Diethyl sulfite	$C_4H_{10}O_3S$					-600.7				-552.2			
Diethyl sulfoxide	$C_4H_{10}OS$					-268.0				-205.6			
N,N-Diethylurea	$C_5H_{12}N_2O$	-372.2											
Difluoramine	F_2HN											252.8	43.4
Difluorine dioxide	F_2O_2									19.2	58.2	277.2	62.1
Difluoroamidogen	F_2N									43.1	57.8	249.9	41.0
o-Difluorobenzene	$C_6H_4F_2$					-330.0		222.6	159.0	-293.8			
m-Difluorobenzene	$C_6H_4F_2$					-343.9		223.8	159.1	-309.2			
p-Difluorobenzene	$C_6H_4F_2$					-342.3			157.5	-306.7			
cis-Difluorodiazine	F_2N_2									69.5			
trans-Difluorodiazine	F_2N_2									82.0			
1,1-Difluoroethane	$C_2H_4F_2$									-497.0		282.5	67.8
1,1-Difluoroethene	$C_2H_2F_2$									-335.0		266.2	60.1
cis-1,2-Difluoroethene	$C_2H_2F_2$											268.3	58.2
Difluoromethane	CH_2F_2									-452.3		246.7	42.9
Difluorosilylene	F_2Si									-619.0	-628.0	252.7	43.9
Digallium	Ga_2									438.5			
Digermane	Ge_2H_6					137.3				162.3			
Digermanium	Ge_2									473.1	416.3	252.8	35.6
Digold	Au_2									515.1			36.9
Dihydro-5-methyl-2(3*H*)-furanone	$C_5H_8O_2$					-461.3				-406.5			
1,2-Dihydronaphthalene	$C_{10}H_{10}$					71.6							
1,4-Dihydronaphthalene	$C_{10}H_{10}$					84.2							
2,3-Dihydrothiophene	C_4H_6S					52.9				90.7	133.5	303.5	79.8
2,5-Dihydrothiophene	C_4H_6S					47.0				86.9	131.6	297.1	83.3
1,4-Dihydroxy-9,10-anthracenedione	$C_{14}H_8O_4$	-595.8								-471.7			
Diindium	In_2									380.9			
Diiodoacetylene	C_2I_2											313.1	70.3
1,4-Diiodobutane	$C_4H_8I_2$					-30.0							
1,2-Diiodoethane	$C_2H_4I_2$	9.3								75.0			
cis-1,2-Diiodoethene	$C_2H_2I_2$									-207.4			
Diiodomethane	CH_2I_2					68.5	90.4	174.1	134.0	119.5	95.8	309.7	57.7
1,2-Diiodopropane	$C_3H_6I_2$									35.6			

Name	Mol. form.	$\Delta_f H°$(c)/ kJ mol^{-1}	$\Delta_f G°$(c)/ kJ mol^{-1}	$S°$(c)/ J mol^{-1} K^{-1}	C_p(c)/ J mol^{-1} K^{-1}	$\Delta_f H°$(l)/ kJ mol^{-1}	$\Delta_f G°$(l)/ kJ mol^{-1}	$S°$(l)/ J mol^{-1} K^{-1}	C_p(l)/ J mol^{-1} K^{-1}	$\Delta_f H°$(g)/ kJ mol^{-1}	$\Delta_f G°$(g)/ kJ mol^{-1}	$S°$(g)/ J mol^{-1} K^{-1}	C_p(g)/ J mol^{-1} K^{-1}
1,3-Diiodopropane	$C_3H_6I_2$					-9.0							
Diisobutylamine	$C_8H_{19}N$					-218.5				-179.2			
Diisobutyl sulfide	$C_8H_{18}S$					-229.2				-180.5			
Diisopentyl ether	$C_{10}H_{22}O$								379.0				
Diisopentyl sulfide	$C_{10}H_{22}S$					-281.8				-221.5			
Diisopropylamine	$C_6H_{15}N$					-178.5				-143.8			
Diisopropyl ether	$C_6H_{14}O$					-351.5			216.8	-319.2			
Diisopropyl sulfide	$C_6H_{14}S$					-181.6		313.0	232.0	-142.0			
Diketene	$C_4H_4O_2$					-233.1				-190.3			
Dilithium	Li_2									215.9	174.4	197.0	36.1
Dimagnesium	Mg_2									287.7			
Dimercury	Hg_2									108.8	68.2	288.1	37.4
1,2-Dimethoxybenzene	$C_8H_{10}O_2$					-290.3				-223.3			
1,2-Dimethoxyethane	$C_4H_{10}O_2$					-376.6			193.3				
Dimethoxymethane	$C_3H_8O_2$					-377.8		244.0	162.0	-348.5			
1,1-Dimethoxypropane	$C_5H_{12}O_2$					-443.6							
2,2-Dimethoxypropane	$C_5H_{12}O_2$					-459.4				-429.9			
Dimethylacetal	$C_4H_{10}O_2$					-420.6				-389.7			
N,N-Dimethylacetamide	C_4H_9NO					-278.3			175.6	-228.0			
Dimethylamine	C_2H_7N					-43.9	70.0	182.3	137.7	-18.8	68.5	273.1	70.7
Dimethylamine hydrochloride	C_2H_8ClN	-289.3											
2,4-Dimethylaniline	$C_8H_{11}N$					-39.2							
2,5-Dimethylaniline	$C_8H_{11}N$					-38.9							
2,6-Dimethylaniline	$C_8H_{11}N$								238.9				
N,N-Dimethylaniline	$C_8H_{11}N$					46.0				100.5			
2,2-Dimethylbutane	C_6H_{14}					-213.8		272.5	191.9	-185.9			
2,3-Dimethylbutane	C_6H_{14}					-207.4		287.8	189.7	-178.1			
2,3-Dimethyl-2-butanethiol	$C_6H_{14}S$					-187.1				-147.9			
3,3-Dimethyl-2-butanone	$C_6H_{12}O$					-328.6				-290.6			
2,3-Dimethyl-1-butene	C_6H_{12}					-93.2				-62.4			
3,3-Dimethyl-1-butene	C_6H_{12}					-87.5				-60.3			
2,3-Dimethyl-2-butene	C_6H_{12}					-101.4		270.2	174.7	-68.1			
3,3-Dimethyl-1-butyne	C_6H_{10}					78.4							
Dimethyl cadmium	C_2H_6Cd					63.6	139.0	201.9	132.0	101.6	146.9	303.0	
1,1-Dimethylcyclohexane	C_8H_{16}					-218.7		267.2	209.2	-180.9			
cis-1,2-Dimethylcyclohexane	C_8H_{16}					-211.8		274.1	210.2	-172.1			
trans-1,2-Dimethylcyclohexane	C_8H_{16}					-218.2		273.2	209.4	-179.9			
cis-1,3-Dimethylcyclohexane	C_8H_{16}					-222.9		272.6	209.4	-184.6			
trans-1,3-Dimethylcyclohexane	C_8H_{16}					-215.7		276.3	212.8	-176.5			
cis-1,4-Dimethylcyclohexane	C_8H_{16}					-215.6		271.1	212.1	-176.6			
trans-1,4-Dimethylcyclohexane	C_8H_{16}					-222.4		268.0	210.2	-184.5			
1,1-Dimethylcyclopentane	C_7H_{14}					-172.1				-138.2			
cis-1,2-Dimethylcyclopentane	C_7H_{14}					-165.3		269.2		-129.5			
trans-1,2-Dimethylcyclopentane	C_7H_{14}					-171.2				-136.6			
cis-1,3-Dimethylcyclopentane	C_7H_{14}					-170.1				-135.8			
trans-1,3-Dimethylcyclopentane	C_7H_{14}					-168.1				-133.6			
1,1-Dimethylcyclopropane	C_5H_{10}					-33.3				-8.2			
cis-1,2-Dimethylcyclopropane	C_5H_{10}					-26.3							
trans-1,2-Dimethylcyclopropane	C_5H_{10}					-30.7							
Dimethyl disulfide	$C_2H_6S_2$					-62.6		235.4	146.1	-24.7			
N,N-Dimethylethanolamine	$C_4H_{11}NO$					-253.7				-203.6			
Dimethyl ether	C_2H_6O					-203.3				-184.1	-112.6	266.4	64.4
N,N-Dimethylformamide	C_3H_7NO					-239.3			150.6	-192.4			
3,4-Dimethyl-2,5-furandione	$C_6H_6O_3$	-581.4											
Dimethylglyoxime	$C_4H_8N_2O_2$	-199.7											
2,2-Dimethylheptane	C_9H_{20}					-288.1							
2,6-Dimethyl-4-heptanone	$C_9H_{18}O$					-408.5			297.3	-357.6			
2,2-Dimethylhexane	C_8H_{18}					-261.9				-224.5			
2,3-Dimethylhexane	C_8H_{18}					-252.6				-213.8			
2,4-Dimethylhexane	C_8H_{18}					-257.0				-219.2			
2,5-Dimethylhexane	C_8H_{18}					-260.4			249.2	-222.5			
3,3-Dimethylhexane	C_8H_{18}					-257.5			246.6	-219.9			
3,4-Dimethylhexane	C_8H_{18}					-251.8				-212.8			
2,5-Dimethyl-2,5-hexanediol	$C_8H_{18}O_2$	-681.7											
cis-2,2-Dimethyl-3-hexene	C_8H_{16}					-126.4				-89.3			
trans-2,2-Dimethyl-3-hexene	C_8H_{16}					-144.9				-107.7			

Name	Mol. form.	$\Delta_f H°$(c)/ kJ mol⁻¹	$\Delta_f G°$(c)/ kJ mol⁻¹	$S°$(c)/ J mol⁻¹ K⁻¹	C_p(c)/ J mol⁻¹ K⁻¹	$\Delta_f H°$(l)/ kJ mol⁻¹	$\Delta_f G°$(l)/ kJ mol⁻¹	$S°$(l)/ J mol⁻¹ K⁻¹	C_p(l)/ J mol⁻¹ K⁻¹	$\Delta_f H°$(g)/ kJ mol⁻¹	$\Delta_f G°$(g)/ kJ mol⁻¹	$S°$(g)/ J mol⁻¹ K⁻¹	C_p(g)/ J mol⁻¹ K⁻¹
1,1-Dimethylhydrazine	$C_2H_8N_2$					48.9	206.4	198.0	164.1	84.1			
1,2-Dimethylhydrazine	$C_2H_8N_2$					52.7				92.2			
1,1-Dimethylindan	$C_{11}H_{14}$					-53.6				-1.6			
Dimethyl isophthalate	$C_{10}H_{10}O_4$	-730.9											
Dimethyl maleate	$C_6H_8O_4$								263.2				
Dimethyl mercury	C_2H_6Hg					59.8	140.3	209.0		94.4	146.1	306.0	83.3
cis, cis-2,6-Dimethyl-2,4,6-octatriene	$C_{10}H_{16}$					-24.0							
Dimethyl oxalate	$C_4H_6O_4$	-756.3								-708.9			
3,3-Dimethyloxetane	$C_5H_{10}O$					-182.2				-148.2			
2,2-Dimethylpentane	C_7H_{16}					-238.3		300.3	221.1	-205.7			
2,3-Dimethylpentane	C_7H_{16}					-233.1				-198.7			
2,4-Dimethylpentane	C_7H_{16}					-234.6		303.2	224.2	-201.6			
3,3-Dimethylpentane	C_7H_{16}					-234.2				-201.0			
2,2-Dimethyl-3-pentanone	$C_7H_{14}O$					-356.1				-313.6			
2,4-Dimethyl-3-pentanone	$C_7H_{14}O$					-352.9		318.0	233.7	-311.3			
2,4-Dimethyl-1-pentene	C_7H_{14}					-117.0				-83.8			
4,4-Dimethyl-1-pentene	C_7H_{14}					-110.6				-81.6			
2,4-Dimethyl-2-pentene	C_7H_{14}					-123.1				-88.7			
cis-4,4-Dimethyl-2-pentene	C_7H_{14}					-105.3				-72.6			
trans-4,4-Dimethyl-2-pentene	C_7H_{14}					-121.7				-88.8			
Dimethyl phthalate	$C_{10}H_{10}O_4$								303.1				
2,2-Dimethylpropanamide	$C_5H_{11}NO$	-399.7								-313.1			
2,2-Dimethyl-1,3-propanediol	$C_5H_{12}O_2$	-551.2											
2,2-Dimethylpropanenitrile	C_5H_9N					-39.8		232.0	179.4	-2.3			
2,2-Dimethyl-1-propanethiol	$C_5H_{12}S$					-165.4				-129.0			
2,2-Dimethylpropanoic acid	$C_5H_{10}O_2$	-564.5								-491.3			
2,2-Dimethyl-1-propanol	$C_5H_{12}O$					-399.4							
2,3-Dimethylpyridine	C_7H_9N					19.4		243.7	189.5	67.1			
2,4-Dimethylpyridine	C_7H_9N					16.1		248.5	184.8	63.6			
2,5-Dimethylpyridine	C_7H_9N					18.7		248.8	184.7	66.5			
2,6-Dimethylpyridine	C_7H_9N					12.7		244.2	185.2	58.1			
3,4-Dimethylpyridine	C_7H_9N					18.3		240.7	191.8	68.8			
3,5-Dimethylpyridine	C_7H_9N					22.5		241.7	184.5	72.0			
2,4-Dimethylpyrrole	C_6H_9N	-422.3											
2,5-Dimethylpyrrole	C_6H_9N					-16.7				39.8			
Dimethyl sulfate	$C_2H_6O_4S$					-735.5				-687.0			
Dimethyl sulfide	C_2H_6S					-65.3		196.4	118.1	-37.4		286.0	74.1
Dimethyl sulfite	$C_2H_6O_3S$					-523.6				-483.4			
Dimethyl sulfone	$C_2H_6O_2S$	-450.1	-302.4	142.0						-373.1	-272.7	310.6	100.0
Dimethyl sulfoxide	C_2H_6OS					-204.2	-99.9	188.3	153.0	-151.3			
Dimethyl terephthalate	$C_{10}H_{10}O_4$	-732.6			261.1								
N,N-Dimethylurea	$C_3H_8N_2O$	-319.1											
N,N'-Dimethylurea	$C_3H_8N_2O$	-312.1											
Dimethyl zinc	C_2H_6Zn					23.4		201.6	129.2	53.0			
2,3-Dinitroaniline	$C_6H_5N_3O_4$	-11.7											
2,4-Dinitroaniline	$C_6H_5N_3O_4$	-67.8											
2,5-Dinitroaniline	$C_6H_5N_3O_4$	-44.3											
2,6-Dinitroaniline	$C_6H_5N_3O_4$	-50.6											
3,5-Dinitroaniline	$C_6H_5N_3O_4$	-38.9											
1,2-Dinitrobenzene	$C_6H_4N_2O_4$	-2.0			200.4								
1,3-Dinitrobenzene	$C_6H_4N_2O_4$	-27.0			197.5	-36.0							
1,4-Dinitrobenzene	$C_6H_4N_2O_4$	-38.0			200.0								
3,5-Dinitrobenzoic acid	$C_7H_4N_2O_6$	-409.8											
1,4-Dinitrobutane	$C_4H_8N_2O_4$					-237.5							
1,1-Dinitroethane	$C_2H_4N_2O_4$					-148.2							
1,2-Dinitroethane	$C_2H_4N_2O_4$					-165.2							
Dinitromethane	$CH_2N_2O_4$					-104.9				-61.5		358.1	86.4
1,5-Dinitronaphthalene	$C_{10}H_6N_2O_4$	29.8											
1,8-Dinitronaphthalene	$C_{10}H_6N_2O_4$	39.7											
2,4-Dinitrophenol	$C_6H_4N_2O_5$	-232.7								-128.1			
1,1-Dinitropropane	$C_3H_6N_2O_4$					-163.2				-100.7			
1,3-Dinitropropane	$C_3H_6N_2O_4$					-207.1							
2,2-Dinitropropane	$C_3H_6N_2O_4$					-181.2							
1,3-Dioxane	$C_4H_8O_2$					-379.7			143.9	-340.6			

Name	Mol. form.	$\Delta_f H°$(c)/ kJ mol^{-1}	$\Delta_f G°$(c)/ kJ mol^{-1}	$S°$(c)/ J mol^{-1} K^{-1}	C_p(c)/ J mol^{-1} K^{-1}	$\Delta_f H°$(l)/ kJ mol^{-1}	$\Delta_f G°$(l)/ kJ mol^{-1}	$S°$(l)/ J mol^{-1} K^{-1}	C_p(l)/ J mol^{-1} K^{-1}	$\Delta_f H°$(g)/ kJ mol^{-1}	$\Delta_f G°$(g)/ kJ mol^{-1}	$S°$(g)/ J mol^{-1} K^{-1}	C_p(g)/ J mol^{-1} K^{-1}
1,4-Dioxane	$C_4H_8O_2$					-353.9		270.2	152.1	-315.3			
1,3-Dioxolane	$C_3H_6O_2$					-333.5			118.0	-298.0			
1,3-Dioxol-2-one	$C_3H_2O_3$					-459.9				-418.6			
Dipentene	$C_{10}H_{16}$					-50.8			249.4	-2.6			
Dipentyl ether	$C_{10}H_{22}O$								250.0				
Dipentyl sulfide	$C_{10}H_{22}S$					-266.4				-204.9			
Diphenylacetylene	$C_{14}H_{10}$	312.4			225.9								
Diphenylamine	$C_{12}H_{11}N$	130.2								219.3			
9,10-Diphenylanthracene	$C_{26}H_{18}$	308.7								465.6			
1,1-Diphenylethane	$C_{14}H_{14}$					48.7							
1,2-Diphenylethane	$C_{14}H_{14}$	51.5								142.9			
Diphenyl ether	$C_{12}H_{10}O$	-32.1		233.9	216.6	-14.9				52.0			
Diphenylmethane	$C_{13}H_{12}$	71.5		239.3		89.7				139.0			
Diphenyl phthalate	$C_{20}H_{14}O_4$	-489.2											
Diphosphine	H_4P_2					-5.0				20.9			
Diphosphoric acid	$H_4O_7P_2$	-2241.0				-2231.7							
Diphosphorus	P_2									144.0	103.5	218.1	32.1
Dipotassium	K_2									123.7	87.5	249.7	37.9
Dipropylamine	$C_6H_{15}N$					-156.1				-116.0			
Dipropyl disulfide	$C_6H_{14}S_2$					-171.5				-118.3			
Dipropyl ether	$C_6H_{14}O$					-328.8		323.9	221.6	-293.0			
Dipropyl sulfate	$C_6H_{14}O_4S$					-859.0				-792.0			
Dipropyl sulfoxide	$C_6H_{14}OS$					-329.4				-254.9			
2,2'-Dipyrrolylmethane	$C_9H_{10}N_2$	126.2											
Diselenium	Se_2									146.0	96.2	252.0	35.4
Disilane	H_6Si_2									80.3	127.3	272.7	80.8
Disilicon	Si_2									594.0	536.0	229.9	34.4
Disilver	Ag_2									410.0	358.8	257.1	37.0
Disodium	Na_2									142.1	103.9	230.2	37.6
Disulfur	S_2									128.6	79.7	228.2	32.5
Ditellurium	Te_2									168.2	118.0	268.1	36.7
1,3-Dithiane	$C_4H_8S_2$									-10.0	72.4	333.5	110.4
1,4-Dithiane	$C_4H_8S_2$									0.0	84.5	326.2	109.7
1,2-Dithiolane	$C_3H_6S_2$									0.0	47.7	313.5	86.5
1,3-Dithiolane	$C_3H_6S_2$									10.0	54.7	323.3	84.7
Diuron	$C_9H_{10}Cl_2N_2O$	-329.0											
cis-1,2-Divinylcyclobutane	C_8H_{12}					124.3				166.5			
trans-1,2-Divinylcyclobutane	C_8H_{12}					101.3				143.5			
Divinyl ether	C_4H_6O					-39.8				-13.6			
trans-13-Docosenoic acid	$C_{22}H_{42}O_2$	-960.7											
3,9-Dodecadiyne	$C_{12}H_{18}$					197.8							
5,7-Dodecadiyne	$C_{12}H_{18}$					181.5							
Dodecane	$C_{12}H_{26}$					-350.9			375.8	-289.4			
Dodecanedioic acid	$C_{12}H_{22}O_4$	-1130.0								-976.9			
Dodecanoic acid	$C_{12}H_{24}O_2$	-774.6			404.3	-737.9				-642.0			
1-Dodecanol	$C_{12}H_{26}O$					-528.5			438.1	-436.6			
1-Dodecene	$C_{12}H_{24}$					-226.2		484.8	360.7	-165.4			
Dotriacontane	$C_{32}H_{66}$	-968.3								-697.2			
Dysprosium	Dy	0.0		75.6	28.2					290.4	254.4	196.6	20.8
Dysprosium(III) bromide	Br_3Dy	-836.2											
Dysprosium(III) chloride	Cl_3Dy	-1000.0											
Dysprosium(III) iodide	DyI_3	-620.5											
Dysprosium(III) oxide	Dy_2O_3	-1863.1	-1771.5	149.8	116.3								
Eicosanoic acid	$C_{20}H_{40}O_2$	-1011.9			545.1	-940.0				-812.4			
Einsteinium	Es	0.0											
Epichlorohydrin	C_3H_5ClO					-148.4			131.6	-107.8			
1,2-Epoxybutane	C_4H_8O					-168.9		230.9	147.0				
Erbium	Er	0.0		73.2	28.1					317.1	280.7	195.6	20.8
Erbium chloride	Cl_3Er	-998.7			100.0								
Erbium fluoride	ErF_3	-1711.0											
Erbium oxide	Er_2O_3	-1897.9	-1808.7	155.6	108.5								
Ethane	C_2H_6									-84.0	-32.0	229.2	52.5
Ethanedial dioxime	$C_2H_4N_2O_2$	-90.5											
1,2-Ethanediamine	$C_2H_8N_2$					-63.0			172.6	-18.0			
1,2-Ethanediol	$C_2H_6O_2$					-460.0		163.2	148.6	-392.2		303.8	82.7
1,2-Ethanediol, diacetate	$C_6H_{10}O_4$								310.0				

Name	Mol. form.	$\Delta_f H°$(c)/ kJ mol^{-1}	$\Delta_f G°$(c)/ kJ mol^{-1}	$S°$(c)/ J mol^{-1} K^{-1}	C_p(c)/ J mol^{-1} K^{-1}	$\Delta_f H°$(l)/ kJ mol^{-1}	$\Delta_f G°$(l)/ kJ mol^{-1}	$S°$(l)/ J mol^{-1} K^{-1}	C_p(l)/ J mol^{-1} K^{-1}	$\Delta_f H°$(g)/ kJ mol^{-1}	$\Delta_f G°$(g)/ kJ mol^{-1}	$S°$(g)/ J mol^{-1} K^{-1}	C_p(g)/ J mol^{-1} K^{-1}
Ethanedithioamide	$C_2H_4N_2S_2$	-20.8								83.0			
1,2-Ethanedithiol	$C_2H_6S_2$					-54.3				-9.7			
Ethanethiol	C_2H_6S					-73.6	-5.5	207.0	117.9	-46.1	-4.8	296.2	72.7
Ethanol	C_2H_6O					-277.6	-174.8	160.7	112.3	-234.8	-167.9	281.6	65.6
Ethanolamine	C_2H_7NO								195.5				
Ethoxybenzene	$C_8H_{10}O$					-152.6			228.5	-101.6			
2-Ethoxyethanol	$C_4H_{10}O_2$								210.8				
2-Ethoxyethyl acetate	$C_6H_{12}O_3$								376.0				
(Ethoxymethyl)oxirane	$C_5H_{10}O_2$					-296.5							
Ethyl acetate	$C_4H_8O_2$					-479.3		257.7	170.7	-443.6			
Ethyl acetoacetate	$C_6H_{10}O_3$								248.0				
Ethyl acrylate	$C_5H_8O_2$					-370.6				-354.2			
Ethylamine	C_2H_7N					-74.1			130.0	-47.5	36.3	283.8	71.5
N-Ethylaniline	$C_8H_{11}N$					8.2				56.3			
Ethylbenzene	C_8H_{10}					-12.3			183.2	29.9			
Ethyl benzoate	$C_9H_{10}O_2$								246.0				
Ethyl butanoate	$C_6H_{12}O_2$								228.0				
2-Ethyl-1-butene	C_6H_{12}					-87.1				-56.0			
Ethyl *trans*-2-butenoate	$C_6H_{10}O_2$					-420.0				-375.6			
Ethyl carbamate	$C_3H_7NO_2$	-517.1			156.4	-497.3				-446.3			
Ethyl 4-chlorobutanoate	$C_6H_{11}ClO_2$					-566.5				-513.8			
Ethyl chloroformate	$C_3H_5ClO_2$					-505.3				-462.9			
Ethyl cyanoacetate	$C_5H_7NO_2$								220.2				
Ethylcyclobutane	C_6H_{12}					-59.0				-27.5			
Ethylcyclohexane	C_8H_{16}					-212.1		280.9	211.8	-171.5			
Ethylcyclopentane	C_7H_{14}					-163.4		279.9		-126.9			
1-Ethylcyclopentene	C_7H_{12}					-53.3				-19.8			
Ethylcyclopropane	C_5H_{10}					-24.8							
1-Ethyl-2,4-dimethylbenzene	$C_{10}H_{14}$					-84.1							
1-Ethyl-3,5-dimethylbenzene	$C_{10}H_{14}$					-87.8							
2-Ethyl-1,3-dimethylbenzene	$C_{10}H_{14}$					-80.1							
2-Ethyl-1,4-dimethylbenzene	$C_{10}H_{14}$					-84.8							
3-Ethyl-1,2-dimethylbenzene	$C_{10}H_{14}$					-80.5							
4-Ethyl-1,2-dimethylbenzene	$C_{10}H_{14}$					-86.0							
3-Ethyl-2,2-dimethylpentane	C_9H_{20}					-272.7							
3-Ethyl-2,4-dimethylpentane	C_9H_{20}					-269.7							
Ethyl 2,2-dimethylpropanoate	$C_7H_{14}O_2$					-577.2				-536.0			
Ethyl 3,5-dimethylpyrrole-2-carboxylate	$C_9H_{13}NO_2$	-474.5											
Ethyl 2,4-dimethylpyrrole-3-carboxylate	$C_9H_{13}NO_2$	-463.2											
Ethyl 2,5-dimethylpyrrole-3-carboxylate	$C_9H_{13}NO_2$	-478.7											
Ethyl 4,5-dimethylpyrrole-3-carboxylate	$C_9H_{13}NO_2$	-470.3											
Ethylene	C_2H_4									52.4	68.4	219.3	42.9
Ethylene carbonate	$C_3H_4O_3$					-682.8			133.9	-508.4			
Ethylenediaminetetraacetic acid	$C_{10}H_{16}N_2O_8$	-1759.5											
Ethyleneimine	C_2H_5N					91.9				126.5			
Ethyl formate	$C_3H_6O_2$								149.3				
2-Ethylhexanal	$C_8H_{16}O$					-348.5				-299.6			
3-Ethylhexane	C_8H_{18}					-250.4				-210.7			
2-Ethylhexanoic acid	$C_8H_{16}O_2$					-635.1				-559.5			
2-Ethyl-1-hexanol	$C_8H_{18}O$					-432.8		347.0	317.5	-365.3			
Ethylidenecyclohexane	C_8H_{14}					-103.5				-59.5			
Ethyl isocyanide	C_3H_5N					108.6				141.7			
Ethyl isopropyl sulfide	$C_5H_{12}S$					-156.1				-118.3			
Ethyl lactate	$C_5H_{10}O_3$								254.0				
Ethyl 3-methylbutanoate	$C_7H_{14}O_2$					-571.0				-527.0			
2-Ethyl-3-methyl-1-butene	C_7H_{14}					-114.1				-79.5			
1-Ethyl-1-methylcyclopentane	C_8H_{16}					-193.8							
cis-1-Ethyl-2-methylcyclopentane	C_8H_{16}					-190.8							

Name	Mol. form.	$\Delta_fH°$(c)/ kJ mol^{-1}	$\Delta_fG°$(c)/ kJ mol^{-1}	$S°$(c)/ J mol^{-1} K^{-1}	C_p(c)/ J mol^{-1} K^{-1}	$\Delta_fH°$(l)/ kJ mol^{-1}	$\Delta_fG°$(l)/ kJ mol^{-1}	$S°$(l)/ J mol^{-1} K^{-1}	C_p(l)/ J mol^{-1} K^{-1}	$\Delta_fH°$(g)/ kJ mol^{-1}	$\Delta_fG°$(g)/ kJ mol^{-1}	$S°$(g)/ J mol^{-1} K^{-1}	C_p(g)/ J mol^{-1} K^{-1}
trans-1-Ethyl-2-methylcyclopentane	C_8H_{16}					-195.1				-156.2			
cis-1-Ethyl-3-methylcyclopentane	C_8H_{16}					-194.4							
trans-1-Ethyl-3-methylcyclopentane	C_8H_{16}					-196.0							
Ethyl methyl ether	C_3H_8O									-216.4		309.2	93.3
3-Ethyl-2-methylpentane	C_8H_{18}					-249.6				-211.0			
3-Ethyl-3-methylpentane	C_8H_{18}					-252.8				-214.8			
3-Ethyl-2-methyl-1-pentene	C_8H_{16}					-137.9				-100.3			
Ethyl methyl sulfide	C_3H_8S					-91.6		239.1	144.6	-59.6			
Ethyl nitrate	$C_2H_5NO_3$					-190.4				-154.1			
Ethyl nitroacetate	$C_4H_7NO_4$					-487.1							
2-Ethyl-2-nitro-1,3-propanediol	$C_5H_{11}NO_4$	-606.4											
Ethyl *cis*-9-octadecenoate	$C_{20}H_{38}O_2$					-775.8							
Ethyl *trans*-9-octadecenoate	$C_{20}H_{38}O_2$					-773.3							
3-Ethylpentane	C_7H_{16}					-224.9		314.5	219.6	-189.5			
Ethyl pentanoate	$C_7H_{14}O_2$					-553.0				-505.9			
Ethyl 2-pentynoate	$C_7H_{10}O_2$					-301.8				-250.3			
2-Ethylphenol	$C_8H_{10}O$					-208.8				-145.2			
3-Ethylphenol	$C_8H_{10}O$					-214.3				-146.1			
4-Ethylphenol	$C_8H_{10}O$	-224.4			206.9					-144.1			
Ethyl propanoate	$C_5H_{10}O_2$					-502.7				-463.4			
Ethyl propyl ether	$C_5H_{12}O$					-303.6		295.0	197.2	-272.0			
Ethyl propyl sulfide	$C_5H_{12}S$					-144.8		309.5	198.4	-104.8			
Ethyl silicate	$C_8H_{20}O_4Si$							533.1	364.4				
S-Ethyl thioacetate	C_4H_8OS					-268.2				-228.1			
2-Ethyltoluene	C_9H_{12}					-46.4				1.3			
3-Ethyltoluene	C_9H_{12}					-48.7				-1.8			
4-Ethyltoluene	C_9H_{12}					-49.8				-3.2			
3-Ethyl-2,4,5-trimethylpyrrole	$C_9H_{15}N$	-89.2											
N-Ethylurea	$C_3H_8N_2O$	-357.8											
Ethyl vinyl ether	C_4H_8O					-167.4				-140.8			
Ethynylsilane	C_2H_4Si											269.4	72.6
Europium	Eu	0.0		77.8	27.7					175.3	142.2	188.8	20.8
Europium(III) chloride	Cl_3Eu	-936.0											
Europium(II,III) oxide	Eu_3O_4	-2272.0	-2142.0	205.0									
Europium(III) oxide	Eu_2O_3	-1651.4	-1556.8	146.0	122.2								
Fermium	Fm	0.0											
Fluoranthene	$C_{16}H_{10}$	189.9		230.6	230.2					289.0			
9*H*-Fluorene	$C_{13}H_{10}$	89.9		207.3	203.1					175.0			173.1
Fluorine	F_2									0.0		202.8	31.3
Fluorine (atomic)	F									79.4	62.3	158.8	22.7
Fluorine monoxide	F_2O									24.5	41.8	247.5	43.3
Fluorine oxide	FO									109.0	105.3	216.4	32.0
Fluorine superoxide [FOO]	FO_2									25.4	39.4	259.5	44.5
Fluoroacetylene	C_2HF											231.7	52.4
Fluorobenzene	C_6H_5F					-150.6		205.9	146.4	-115.9			
Fluoroborane(1)	BF									-122.2	-149.8	200.5	29.6
Fluoroethane	C_2H_5F											264.5	58.6
Fluoroethene	C_2H_3F									-138.8			
Fluorogermane	$FGeH_3$											252.8	51.6
Fluoromethane	CH_3F											222.9	37.5
Fluorooxyborane	BFO									-607.0			
1-Fluoropropane	C_3H_7F									-285.9			
2-Fluoropropane	C_3H_7F									-293.5			
Fluorosilane	FH_3Si											238.4	47.4
Fluorosilylidyne	FSi									7.1	-24.3	225.8	32.6
4-Fluorotoluene	C_7H_7F					-186.9			171.2	-147.4			
Formaldehyde	CH_2O									-108.6	-102.5	218.8	35.4
Formamide	CH_3NO					-254.0				-193.9			
Formic acid	CH_2O_2					-425.0	-361.4	129.0	99.0	-378.7			
Formyl fluoride	CHFO											246.6	39.9
Francium	Fr	0.0		95.4									
β-D-Fructose	$C_6H_{12}O_6$	-1265.6											
Fumaric acid	$C_4H_4O_4$	-811.7		168.0	142.0					-675.8			

Name	Mol. form.	$\Delta_f H°$(c)/ kJ mol^{-1}	$\Delta_f G°$(c)/ kJ mol^{-1}	$S°$(c)/ J mol^{-1} K^{-1}	C_p(c)/ J mol^{-1} K^{-1}	$\Delta_f H°$(l)/ kJ mol^{-1}	$\Delta_f G°$(l)/ kJ mol^{-1}	$S°$(l)/ J mol^{-1} K^{-1}	C_p(l)/ J mol^{-1} K^{-1}	$\Delta_f H°$(g)/ kJ mol^{-1}	$\Delta_f G°$(g)/ kJ mol^{-1}	$S°$(g)/ J mol^{-1} K^{-1}	C_p(g)/ J mol^{-1} K^{-1}
Furan	C_4H_4O					-62.3		177.0	114.8	-34.8		267.2	65.4
2-Furancarboxylic acid	$C_5H_4O_3$	-498.4								-390.0			
3-(2-Furanyl)-2-propenal	$C_7H_6O_2$	-182.0								-105.9			
Furfural	$C_5H_4O_2$					-201.6			163.2	-151.0			
Furfuryl alcohol	$C_5H_6O_2$					-276.2			204.0	-211.8			
Gadolinium	Gd	0.0		68.1	37.0					397.5	359.8	194.3	27.5
Gadolinium(III) chloride	Cl_3Gd	-1008.0			88.0								
Gadolinium(III) fluoride	F_3Gd									-1297.0			
Gadolinium(III) oxide	Gd_2O_3	-1819.6			106.7								
Galactitol	$C_6H_{14}O_6$					-1317.0							
D-Galactose	$C_6H_{12}O_6$	-1286.3											
Gallium	Ga	0.0	0.0	40.8	26.0	5.6				272.0	233.7	169.0	25.3
Gallium antimonide	GaSb	-41.8	-38.9	76.1	48.5								
Gallium arsenide	AsGa	-71.0	-67.8	64.2	46.2								
Gallium(III) bromide	Br_3Ga	-386.6	-359.8	180.0									
Gallium(III) chloride	Cl_3Ga	-524.7	-454.8	142.0									
Gallium(III) fluoride	F_3Ga	-1163.0	-1085.3	84.0									
Gallium(III) hydroxide	GaH_3O_3	-964.4	-831.3	100.0									
Gallium(III) iodide	GaI_3	-238.9		205.0	100.0								
Gallium monofluoride	FGa									-251.9			33.3
Gallium monoxide	GaO									279.5	253.5	231.1	32.1
Gallium nitride	GaN	-110.5											
Gallium(III) oxide	Ga_2O_3	-1089.1	-998.3	85.0	92.1								
Gallium phosphide	GaP	-88.0											
Gallium suboxide	Ga_2O	-356.0											
Germane	GeH_4									90.8	113.4	217.1	45.0
Germanium	Ge	0.0		31.1	23.3					372.0	331.2	167.9	30.7
Germanium(IV) bromide	Br_4Ge					-347.7	-331.4	280.7		-300.0	-318.0	396.2	101.8
Germanium(IV) chloride	Cl_4Ge					-531.8	-462.7	245.6		-495.8	-457.3	347.7	96.1
Germanium(IV) fluoride	F_4Ge									-1190.2	-1150.0	301.9	
Germanium(IV) iodide	GeI_4	-141.8	-144.3	271.1						-56.9	-106.3	428.9	104.1
Germanium monobromide	BrGe									235.6			37.1
Germanium monochloride	ClGe									155.2	124.2	247.0	36.9
Germanium monofluoride	FGe									-33.4			34.7
Germanium(II) oxide	GeO	-261.9	-237.2	50.0						-46.2	-73.2	224.3	30.9
Germanium(IV) oxide	GeO_2	-580.0	-521.4	39.7	52.1								
Germanium phosphide	GeP	-21.0	-17.0	63.0									
Germanium(II) sulfide	GeS	-69.0	-71.5	71.0						92.0	42.0	234.0	33.7
Germanium(II) telluride	GeTe	20.0											
α-D-Glucose	$C_6H_{12}O_6$	-1273.3											
α-D-Glucose pentaacetate	$C_{16}H_{22}O_{11}$	-2249.4											
β-D-Glucose pentaacetate	$C_{16}H_{22}O_{11}$	-2232.6											
D-Glutamic acid	$C_5H_9NO_4$	-1005.3											
L-Glutamic acid	$C_5H_9NO_4$	-1009.7											
L-Glutamine	$C_5H_{10}N_2O_3$	-826.4											
Glycerol	$C_3H_8O_3$					-669.6		206.3	218.9	-577.9			
Glycerol 1-acetate	$C_5H_{10}O_4$					-909.2							
Glycerol triacetate	$C_9H_{14}O_6$					-1330.8		458.3	384.7	-1245.0			
Glycine	$C_2H_5NO_2$	-528.5								-392.1			
Glycolic acid	$C_2H_4O_3$									-583.0	-504.9	318.6	87.1
N-Glycylglycine	$C_4H_8N_2O_3$	-747.7											
Glyoxal	$C_2H_2O_2$									-212.0	-189.7	272.5	60.6
Gold	Au	0.0		47.4	25.4					366.1	326.3	180.5	20.8
Gold(I) bromide	AuBr	-14.0											
Gold(III) bromide	$AuBr_3$	-53.3											
Gold(I) chloride	AuCl	-34.7											
Gold(III) chloride	$AuCl_3$	-117.6											
Gold(III) fluoride	AuF_3	-363.6											
Gold hydride	AuH									295.0	265.7	211.2	29.2
Gold(I) iodide	AuI	0.0											
Guanidine	CH_5N_3	-56.0											
Guanine	$C_5H_5N_5O$	-183.9											
Hafnium	Hf	0.0		43.6	25.7					619.2	576.5	186.9	20.8
Hafnium(IV) chloride	Cl_4Hf	-990.4	-901.3	190.8	120.5					-884.5			
Hafnium(IV) fluoride	F_4Hf	-1930.5	-1830.4	113.0						-1669.8			
Hafnium(IV) oxide	HfO_2	-1144.7	-1088.2	59.3	60.3								

Name	Mol. form.	$\Delta_f H°$(c)/ kJ mol⁻¹	$\Delta_f G°$(c)/ kJ mol⁻¹	$S°$(c)/ J mol⁻¹ K⁻¹	C_p(c)/ J mol⁻¹ K⁻¹	$\Delta_f H°$(l)/ kJ mol⁻¹	$\Delta_f G°$(l)/ kJ mol⁻¹	$S°$(l)/ J mol⁻¹ K⁻¹	C_p(l)/ J mol⁻¹ K⁻¹	$\Delta_f H°$(g)/ kJ mol⁻¹	$\Delta_f G°$(g)/ kJ mol⁻¹	$S°$(g)/ J mol⁻¹ K⁻¹	C_p(g)/ J mol⁻¹ K⁻¹
Helium	He									0.0		126.2	20.8
Heptadecanoic acid	$C_{17}H_{34}O_2$	-924.4			475.7	-865.6							
Heptanal	$C_7H_{14}O$					-311.5		335.4	230.1	-263.8			
Heptane	C_7H_{16}					-224.2			224.7	-187.6			
1,7-Heptanediol	$C_7H_{16}O_2$					-574.2							
Heptanenitrile	$C_7H_{13}N$					-82.8				-31.0			
1-Heptanethiol	$C_7H_{16}S$					-200.5				-149.9			
Heptanoic acid	$C_7H_{14}O_2$					-610.2			265.4	-536.2			
1-Heptanol	$C_7H_{16}O$					-403.3			272.1	-336.5			
2-Heptanone	$C_7H_{14}O$								232.6				
3-Heptanone	$C_7H_{14}O$									-297.1			
4-Heptanone	$C_7H_{14}O$									-298.3			
1-Heptene	C_7H_{14}					-97.9		327.6	211.8	-62.3			
cis-2-Heptene	C_7H_{14}					-105.1							
trans-2-Heptene	C_7H_{14}					-109.5							
cis-3-Heptene	C_7H_{14}					-104.3							
trans-3-Heptene	C_7H_{14}					-109.3							
Hexaborane(10)	B_6H_{10}					56.3				94.6	211.3	296.8	125.7
Hexabromoethane	C_2Br_6											441.9	139.3
Hexachlorobenzene	C_6Cl_6	-127.6		260.2	201.2					-35.5			
Hexachloro-1,3-butadiene	C_4Cl_6					-24.5							
Hexachloroethane	C_2Cl_6	-202.8		237.3	198.2					-143.6			
Hexadecane	$C_{16}H_{34}$					-456.1			501.6	-374.8			
Hexadecanoic acid	$C_{16}H_{32}O_2$	-891.5		452.4	460.7	-838.1				-737.1			
1-Hexadecanol	$C_{16}H_{34}O$	-686.5			422.0					-517.0			
1-Hexadecene	$C_{16}H_{32}$					-328.7		587.9	488.9	-248.4			
1,5-Hexadiene	C_6H_{10}					54.1				84.2			
1,5-Hexadiyne	C_6H_6					384.2							
Hexafluoroacetylacetone	$C_5H_2F_6O_2$	-2286.7											
Hexafluorobenzene	C_6F_6					-991.3		280.8	221.6	-955.4			
Hexafluorodisilane	F_6Si_2	-2427.0	-2299.7	219.1	129.5					-2383.3	-2307.3	391.0	129.9
Hexafluoroethane	C_2F_6									-1344.2		332.3	106.7
Hexahydro-1,3,5-trinitro-1,3,5-triazine	$C_3H_6N_6O_6$									192.0		482.4	230.2
Hexamethylbenzene	$C_{12}H_{18}$	-162.4		306.3	245.6					-77.4			
Hexamethyldisiloxane	$C_6H_{18}OSi_2$					-815.0	-541.5	433.8	311.4	-777.7	-534.5	535.0	238.5
Hexamethylphosphoric triamide	$C_6H_{18}N_3OP$								321.0				
Hexanal	$C_6H_{12}O$							280.3	210.4				
Hexanamide	$C_6H_{13}NO$	-423.0								-324.2			
Hexane	C_6H_{14}					-198.7			195.6	-166.9			
1,6-Hexanediamine	$C_6H_{16}N_2$	-205.0											
Hexanedinitrile	$C_6H_8N_2$					85.1			128.7	149.5			
1,6-Hexanedioic acid	$C_6H_{10}O_4$	-994.3											
1,2-Hexanediol	$C_6H_{14}O_2$					-577.1				-490.1			
1,6-Hexanediol	$C_6H_{14}O_2$	-569.9				-548.6				-461.2			
1-Hexanethiol	$C_6H_{14}S$					-175.7				-129.9			
Hexanoic acid	$C_6H_{12}O_2$					-583.8				-511.9			
1-Hexanol	$C_6H_{14}O$					-377.5		287.4	240.4	-315.9			
2-Hexanol	$C_6H_{14}O$					-392.0				-333.5			
3-Hexanol	$C_6H_{14}O$					-392.4			286.2				
2-Hexanone	$C_6H_{12}O$					-322.0			213.3	-278.9			
3-Hexanone	$C_6H_{12}O$					-320.2		305.3	216.9	-277.6			
1-Hexene	C_6H_{12}					-74.2		295.2	183.3	-43.5			
cis-2-Hexene	C_6H_{12}					-83.9				-52.3			
trans-2-Hexene	C_6H_{12}					-85.5				-53.9			
cis-3-Hexene	C_6H_{12}					-78.9				-47.6			
trans-3-Hexene	C_6H_{12}					-86.1				-54.4			
Hexyl acetate	$C_8H_{16}O_2$								282.8				
L-Histidine	$C_6H_9N_3O_2$	-466.7											
Holmium	Ho	0.0		75.3	27.2					300.8	264.8	195.6	20.8
Holmium chloride	Cl_3Ho	-1005.4			88.0								
Holmium fluoride	F_3Ho	-1707.0											
Holmium oxide	Ho_2O_3	-1880.7	-1791.1	158.2	115.0								
Hydrazine	H_4N_2					50.6	149.3	121.2	98.9	95.4	159.4	238.5	48.4
Hydrazinecarbothioamide	CH_5N_3S	24.7											
1,2-Hydrazinedicarboxamide	$C_2H_6N_4O_2$	-498.7											

Name	Mol. form.	$\Delta_f H°$(c)/ kJ mol^{-1}	$\Delta_f G°$(c)/ kJ mol^{-1}	$S°$(c)/ J mol^{-1} K^{-1}	C_p(c)/ J mol^{-1} K^{-1}	$\Delta_f H°$(l)/ kJ mol^{-1}	$\Delta_f G°$(l)/ kJ mol^{-1}	$S°$(l)/ J mol^{-1} K^{-1}	C_p(l)/ J mol^{-1} K^{-1}	$\Delta_f H°$(g)/ kJ mol^{-1}	$\Delta_f G°$(g)/ kJ mol^{-1}	$S°$(g)/ J mol^{-1} K^{-1}	C_p(g)/ J mol^{-1} K^{-1}
Hydrazoic acid	HN_3					264.0	327.3	140.6		294.1	328.1	239.0	43.7
Hydrogen (atomic)	H									218.0	203.3	114.7	20.8
Hydrogen	H_2									0.0		130.7	28.8
Hydrogen bromide	BrH									-36.3	-53.4	198.7	29.1
Hydrogen chloride	ClH									-92.3	-95.3	186.9	29.1
Hydrogen cyanide	CHN					108.9	125.0	112.8	70.6	135.1	124.7	201.8	35.9
Hydrogen disulfide	H_2S_2					-18.1			84.1	15.5			51.5
Hydrogen fluoride	FH					-299.8				-273.3	-275.4	173.8	
Hydrogen iodide	HI									26.5	1.7	206.6	29.2
Hydrogen peroxide	H_2O_2					-187.8	-120.4	109.6	89.1	-136.3	-105.6	232.7	43.1
Hydrogen selenide	H_2Se									29.7	15.9	219.0	34.7
Hydrogen sulfide	H_2S									-20.6	-33.4	205.8	34.2
Hydrogen telluride	H_2Te									99.6			
Hydroperoxy	HO_2									10.5	22.6	229.0	34.9
p-Hydroquinone	$C_6H_6O_2$	-364.5			136.0					-265.3			
2-Hydroxybenzoic acid	$C_7H_6O_3$	-589.9								-494.8			
Hydroxyl	HO									39.0	34.2	183.7	29.9
Hydroxylamine	H_3NO	-114.2											
2-(Hydroxymethyl)-2-methyl-1,3-propanediol	$C_5H_{12}O_3$	-744.6											
Hypochlorous acid	ClHO									-78.7	-66.1	236.7	37.2
Hypoxanthine	$C_5H_4N_4O$	-110.8		145.6	134.5								
Imidazole	$C_3H_4N_2$	49.8								132.9			
Imidogen	HN									351.5	345.6	181.2	29.2
Iminodiacetic acid	$C_4H_7NO_4$	-932.6											
Indan	C_9H_{10}					11.5		234.4	190.2	60.3			
1*H*-Indazole	$C_7H_6N_2$	151.9								243.0			
Indene	C_9H_8					110.6		215.3	186.9	163.4			
Indium	In	0.0		57.8	26.7					243.3	208.7	173.8	20.8
Indium antimonide	InSb	-30.5	-25.5	86.2	49.5					344.3			
Indium arsenide	AsIn	-58.6	-53.6	75.7	47.8								
Indium(I) bromide	BrIn	-175.3	-169.0	113.0						-56.9	-94.3	259.5	36.7
Indium(III) bromide	Br_3In	-428.9								-282.0			
Indium(I) chloride	ClIn	-186.2								-75.0			
Indium(III) chloride	Cl_3In	-537.2								-374.0			
Indium(I) fluoride	FIn									-203.4			
Indium(I) iodide	IIn	-116.3	-120.5	130.0						7.5	-37.7	267.3	36.8
Indium(III) iodide	I_3In	-238.0								-120.5			
Indium monoxide	InO									387.0	364.4	236.5	32.6
Indium(III) oxide	In_2O_3	-925.8	-830.7	104.2	92.0								
Indium phosphide	InP	-88.7	-77.0	59.8	45.4								
Indium(II) sulfide	InS	-138.1	-131.8	67.0						238.0			
Indium(III) sulfide	In_2S_3	-427.0	-412.5	163.6	118.0								
Indium(IV) telluride	In_2Te_5	-175.3											
1*H*-Indole	C_8H_7N	86.6								156.5			
1*H*-Indole-2,3-dione	$C_8H_5NO_2$	-268.2											
Iodic acid	HIO_3	-230.1											
Iodine	I_2	0.0		116.1	54.4					62.4	19.3	260.7	36.9
Iodine (atomic)	I									106.8	70.2	180.8	20.8
Iodine bromide	BrI									40.8	3.7	258.8	36.4
Iodine chloride	ClI					-23.9	-13.6	135.1		17.8	-5.5	247.6	35.6
Iodine fluoride	FI									-95.7	-118.5	236.2	33.4
Iodine monoxide	IO									126.0	102.5	239.6	32.9
Iodine pentafluoride	F_5I					-864.8				-822.5	-751.7	327.7	99.2
Iodoacetone	C_3H_5IO									-130.5			
Iodobenzene	C_6H_5I					117.2		205.4	158.7	164.9			
Iodoethane	C_2H_5I					-40.0	14.7	211.7	115.1	-8.1	19.2	306.0	66.9
Iodoethene	C_2H_3I											285.0	57.9
Iodogermane	GeH_3I											283.2	57.5
Iodomethane	CH_3I					-13.6		163.2	126.0	14.4		254.1	44.1
1-Iodo-2-methylpropane	C_4H_9I								162.3				
2-Iodo-2-methylpropane	C_4H_9I					-107.5				-72.1			
1-Iodonaphthalene	$C_{10}H_7I$					161.5				233.8			
2-Iodonaphthalene	$C_{10}H_7I$	144.3								235.1			
1-Iodopropane	C_3H_7I					-66.0				-30.0			
2-Iodopropane	C_3H_7I					-74.8				-40.3			

Name	Mol. form.	$\Delta_fH°$(c)/ kJ mol⁻¹	$\Delta_fG°$(c)/ kJ mol⁻¹	$S°$(c)/ J mol⁻¹ K⁻¹	C_p(c)/ J mol⁻¹ K⁻¹	$\Delta_fH°$(l)/ kJ mol⁻¹	$\Delta_fG°$(l)/ kJ mol⁻¹	$S°$(l)/ J mol⁻¹ K⁻¹	C_p(l)/ J mol⁻¹ K⁻¹	$\Delta_fH°$(g)/ kJ mol⁻¹	$\Delta_fG°$(g)/ kJ mol⁻¹	$S°$(g)/ J mol⁻¹ K⁻¹	C_p(g)/ J mol⁻¹ K⁻¹
3-Iodopropanoic acid	$C_3H_5IO_2$	-460.0											
3-Iodopropene	C_3H_5I					53.7				91.5			
Iodosilane	H_3ISi											270.9	54.4
Iridium	Ir	0.0		35.5	25.1					665.3	617.9	193.6	20.8
Iridium(III) chloride	Cl_3Ir	-245.6											
Iridium(VI) fluoride	F_6Ir	-579.7	-461.6	247.7						-544.0	-460.0	357.8	121.1
Iridium(IV) oxide	IrO_2	-274.1			57.3								
Iridium(III) sulfide	Ir_2S_3	-234.0											
Iridium(IV) sulfide	IrS_2	-138.0											
Iron	Fe	0.0		27.3	25.1					416.3	370.7	180.5	25.7
Iron(II) bromide	Br_2Fe	-249.8	-238.1	140.6									
Iron(III) bromide	Br_3Fe	-268.2											
Iron carbide	CFe_3	25.1	20.1	104.6	105.9								
Iron(II) carbonate	$CFeO_3$	-740.6	-666.7	92.9	82.1								
Iron(II) chloride	Cl_2Fe	-341.8	-302.3	118.0	76.7								
Iron(III) chloride	Cl_3Fe	-399.5	-334.0	142.3	96.7								
Iron disulfide	FeS_2	-178.2	-166.9	52.9	62.2								
Iron(II) fluoride	F_2Fe	-711.3	-668.6	87.0	68.1								
Iron(II) iodide	FeI_2	-113.0											
Iron(III) iodide	FeI_3									71.0			
Iron(II) molybdate	$FeMoO_4$	-1075.0	-975.0	129.3	118.5								
Iron(II) orthosilicate	Fe_2O_4Si	-1479.9	-1379.0	145.2	132.9								
Iron(II) oxide	FeO	-272.0											
Iron(II,III) oxide	Fe_3O_4	-1118.4	-1015.4	146.4	143.4								
Iron(III) oxide	Fe_2O_3	-824.2	-742.2	87.4	103.9								
Iron pentacarbonyl	C_5FeO_5					-774.0	-705.3	338.1	240.6				
Iron(II) sulfate	FeO_4S	-928.4	-820.8	107.5	100.6								
Iron(II) sulfide	FeS	-100.0	-100.4	60.3	50.5								
Iron(II) tungstate	FeO_4W	-1155.0	-1054.0	131.8	114.6								
Isobutanal	C_4H_8O					-247.3				-215.7			
Isobutane	C_4H_{10}					-154.2				-134.2			
Isobutene	C_4H_8					-37.5				-16.9			
Isobutyl acetate	$C_6H_{12}O_2$								233.8				
Isobutylamine	$C_4H_{11}N$					-132.6			183.2	-98.7			
Isobutylbenzene	$C_{10}H_{14}$					-69.8				-21.9			
Isobutyl 2-chloropropanoate	$C_7H_{13}ClO_2$					-603.1				-549.6			
Isobutyl 3-chloropropanoate	$C_7H_{13}ClO_2$					-572.6				-517.3			
Isobutyl isobutanoate	$C_8H_{16}O_2$					-587.4				-542.9			
Isobutyl pentanoate	$C_9H_{18}O_2$					-620.0				-568.6			
Isobutyl trichloroacetate	$C_6H_9Cl_3O_2$					-553.4				-500.2			
Isocyanic acid (HNCO)	CHNO											238.0	44.9
Isocyanomethane	C_2H_3N					130.8	159.5	159.0		163.5	165.7	246.9	52.9
DL-Isoleucine	$C_6H_{13}NO_2$	-635.3											
L-Isoleucine	$C_6H_{13}NO_2$	-637.8											
Isopentane	C_5H_{12}					-178.4		260.4	164.8	-153.6			
Isopentyl acetate	$C_7H_{14}O_2$								248.5				
Isopentyl trichloroacetate	$C_7H_{11}Cl_3O_2$					-580.9				-523.1			
Isophorone	$C_9H_{14}O$								253.5				
Isophthalic acid	$C_8H_6O_4$	-803.0								-696.3			
Isopropyl acetate	$C_5H_{10}O_2$					-518.9			199.4	-481.6			
Isopropylamine	C_3H_9N					-112.3		218.3	163.8	-83.7	32.2	312.2	97.5
Isopropylbenzene	C_9H_{12}					-41.1			210.7	4.0			
Isopropylbenzene hydroperoxide	$C_9H_{12}O_2$					-148.3				-78.4			
1-Isopropyl-2-methylbenzene	$C_{10}H_{14}$					-73.3							
1-Isopropyl-3-methylbenzene	$C_{10}H_{14}$					-78.6							
1-Isopropyl-4-methylbenzene	$C_{10}H_{14}$					-78.0			236.4				
Isopropyl methyl ether	$C_4H_{10}O$					-278.8		253.8	161.9	-252.0			
Isopropyl methyl sulfide	$C_4H_{10}S$					-124.7		263.1	172.4	-90.5			
Isopropyl nitrate	$C_3H_7NO_3$					-229.7				-191.0			
Isopropyl pentanoate	$C_8H_{16}O_2$					-592.2				-544.9			
2-Isopropylphenol	$C_9H_{12}O$					-233.7				-182.2			
3-Isopropylphenol	$C_9H_{12}O$					-252.5				-196.0			
4-Isopropylphenol	$C_9H_{12}O$	-270.0								-175.3			
Isoquinoline	C_9H_7N					144.3		216.0	196.2	204.6			
Isothiocyanic acid	CHNS									127.6	113.0	247.8	46.9
Isoxazole	C_3H_3NO					42.1				78.6			

Thermochem

Name	Mol. form.	$\Delta_f H°$(c)/ kJ mol^{-1}	$\Delta_f G°$(c)/ kJ mol^{-1}	$S°$(c)/ J mol^{-1} K^{-1}	C_p(c)/ J mol^{-1} K^{-1}	$\Delta_f H°$(l)/ kJ mol^{-1}	$\Delta_f G°$(l)/ kJ mol^{-1}	$S°$(l)/ J mol^{-1} K^{-1}	C_p(l)/ J mol^{-1} K^{-1}	$\Delta_f H°$(g)/ kJ mol^{-1}	$\Delta_f G°$(g)/ kJ mol^{-1}	$S°$(g)/ J mol^{-1} K^{-1}	C_p(g)/ J mol^{-1} K^{-1}
Ketene	C_2H_2O					-67.9				-47.5	-48.3	247.6	51.8
Krypton	Kr									0.0		164.1	20.8
β-D-Lactose	$C_{12}H_{22}O_{11}$	-2236.7											
α-Lactose monohydrate	$C_{12}H_{24}O_{12}$	-2484.1											
Lanthanum	La	0.0		56.9	27.1					431.0	393.6	182.4	22.8
Lanthanum chloride	Cl_3La	-1072.2			108.8								
Lanthanum iodide	I_3La	-668.9											
Lanthanum monosulfide	LaS	-456.0	-451.5	73.2	59.0								
Lanthanum oxide	La_2O_3	-1793.7	-1705.8	127.3	108.8								
Lawrencium	Lr	0.0											
Lead	Pb	0.0		64.8	26.8					195.2	162.2	175.4	20.8
Lead(II) bromide	Br_2Pb	-278.7	-261.9	161.5	80.1								
Lead(II) carbonate	CO_3Pb	-699.1	-625.5	131.0	87.4								
Lead(II) chloride	Cl_2Pb	-359.4	-314.1	136.0									
Lead(IV) chloride	Cl_4Pb					-329.3							
Lead(II) chromate	CrO_4Pb	-930.9											
Lead(II) fluoride	F_2Pb	-664.0	-617.1	110.5									
Lead(IV) fluoride	F_4Pb	-941.8											
Lead(II) iodide	I_2Pb	-175.5	-173.6	174.9	77.4								
Lead(II) metasilicate	O_3PbSi	-1145.7	-1062.1	109.6	90.0								
Lead(II) molybdate	MoO_4Pb	-1051.9	-951.4	166.1	119.7								
Lead(II) nitrate	N_2O_6Pb	-451.9											
Lead(II) orthosilicate	O_4Pb_2Si	-1363.1	-1252.6	186.6	137.2								
Lead(II) oxalate	C_2O_4Pb	-851.4	-750.1	146.0	105.4								
Lead(II,II,IV) oxide	O_4Pb_3	-718.4	-601.2	211.3	146.9								
Lead(IV) oxide	O_2Pb	-277.4	-217.3	68.6	64.6								
Lead(II) oxide (litharge)	OPb	-219.0	-188.9	66.5	45.8								
Lead(II) oxide (massicot)	OPb	-217.3	-187.9	68.7	45.8								
Lead(II) selenate	O_4PbSe	-609.2	-504.9	167.8									
Lead(II) selenide	PbSe	-102.9	-101.7	102.5	50.2								
Lead(II) sulfate	O_4PbS	-920.0	-813.0	148.5	103.2								
Lead(II) sulfide	PbS	-100.4	-98.7	91.2	49.5								
Lead(II) sulfite	O_3PbS	-669.9											
Lead(II) telluride	PbTe	-70.7	-69.5	110.0	50.5								
DL-Leucine	$C_6H_{13}NO_2$	-640.6											
D-Leucine	$C_6H_{13}NO_2$	-637.3											
L-Leucine	$C_6H_{13}NO_2$	-637.4			200.1					-486.8			
d-Limonene	$C_{10}H_{16}$					-54.5			249.0				
Lithium	Li	0.0		29.1	24.9					159.3	126.6	138.8	20.8
Lithium aluminum hydride	AlH_4Li	-116.3	-44.7	78.7	83.2								
Lithium amide	H_2LiN	-179.5											
Lithium borohydride	BH_4Li	-190.8	-125.0	75.9	82.6								
Lithium bromide	BrLi	-351.2	-342.0	74.3									
Lithium carbonate	CLi_2O_3	-1215.9	-1132.1	90.4	99.1								
Lithium chloride	ClLi	-408.6	-384.4	59.3	48.0								
Lithium fluoride	FLi	-616.0	-587.7	35.7	41.6								
Lithium hydride	HLi	-90.5	-68.3	20.0	27.9								
Lithium hydroxide	HLiO	-487.5	-441.5	42.8	49.6					-229.0	-234.2	214.4	46.0
Lithium iodide	ILi	-270.4	-270.3	86.8	51.0								
Lithium metaborate	$BLiO_2$	-1032.2	-976.1	51.5	59.8								
Lithium metasilicate	Li_2O_3Si	-1648.1	-1557.2	79.8	99.1								
Lithium nitrate	$LiNO_3$	-483.1	-381.1	90.0									
Lithium nitrite	$LiNO_2$	-372.4	-302.0	96.0									
Lithium oxide	Li_2O	-597.9	-561.2	37.6	54.1								
Lithium perchlorate	$ClLiO_4$	-381.0											
Lithium peroxide	Li_2O_2	-634.3											
Lithium phosphate	Li_3O_4P	-2095.8											
Lithium sulfate	Li_2O_4S	-1436.5	-1321.7	115.1	117.6								
Lithium sulfide	Li_2S	-441.4											
Lutetium	Lu	0.0		51.0	26.9					427.6	387.8	184.8	20.9
Lutetium chloride	Cl_3Lu	-945.6								-649.0			
Lutetium iodide	I_3Lu	-548.0											
Lutetium oxide	Lu_2O_3	-1878.2	-1789.0	110.0	101.8								
DL-Lysine	$C_6H_{14}N_2O_2$	-678.7											
Magnesium	Mg	0.0		32.7	24.9					147.1	112.5	148.6	20.8
Magnesium bromide	Br_2Mg	-524.3	-503.8	117.2									

Name	Mol. form.	$\Delta_f H°$(c)/ kJ mol^{-1}	$\Delta_f G°$(c)/ kJ mol^{-1}	$S°$(c)/ J mol^{-1} K^{-1}	C_p(c)/ J mol^{-1} K^{-1}	$\Delta_f H°$(l)/ kJ mol^{-1}	$\Delta_f G°$(l)/ kJ mol^{-1}	$S°$(l)/ J mol^{-1} K^{-1}	C_p(l)/ J mol^{-1} K^{-1}	$\Delta_f H°$(g)/ kJ mol^{-1}	$\Delta_f G°$(g)/ kJ mol^{-1}	$S°$(g)/ J mol^{-1} K^{-1}	C_p(g)/ J mol^{-1} K^{-1}
Magnesium carbonate	$CMgO_3$	-1095.8	-1012.1	65.7	75.5								
Magnesium chloride	Cl_2Mg	-641.3	-591.8	89.6	71.4								
Magnesium fluoride	F_2Mg	-1124.2	-1071.1	57.2	61.6								
Magnesium hydride	H_2Mg	-75.3	-35.9	31.1	35.4								
Magnesium hydroxide	H_2MgO_2	-924.5	-833.5	63.2	77.0								
Magnesium iodide	I_2Mg	-364.0	-358.2	129.7									
Magnesium nitrate	MgN_2O_6	-790.7	-589.4	164.0	141.9								
Magnesium orthosilicate	Mg_2O_4Si	-2174.0	-2055.1	95.1	118.5								
Magnesium oxalate	C_2MgO_4	-1269.0											
Magnesium oxide	MgO	-601.6	-569.3	27.0	37.2								
Magnesium selenate	MgO_4Se	-968.5											
Magnesium sulfate	MgO_4S	-1284.9	-1170.6	91.6	96.5								
Magnesium sulfide	MgS	-346.0	-341.8	50.3	45.6								
Maleic acid	$C_4H_4O_4$	-789.4		160.8	137.0					-679.4			
Maleic anhydride	$C_4H_2O_3$	-469.8								-398.3			
Malononitrile	$C_3H_2N_2$	186.4								265.5			
Manganese	Mn	0.0		32.0	26.3					280.7	238.5	173.7	20.8
Manganese(II) bromide	Br_2Mn	-384.9											
Manganese(II) carbonate	$CMnO_3$	-894.1	-816.7	85.8	81.5								
Manganese(II) chloride	Cl_2Mn	-481.3	-440.5	118.2	72.9								
Manganese(II) metasilicate	MnO_3Si	-1320.9	-1240.5	89.1	86.4								
Manganese(II) nitrate	MnN_2O_6	-576.3											
Manganese(II) orthosilicate	Mn_2O_4Si	-1730.5	-1632.1	163.2	129.9								
Manganese(II) oxide	MnO	-385.2	-362.9	59.7	45.4								
Manganese(II,III) oxide	Mn_3O_4	-1387.8	-1283.2	155.6	139.7								
Manganese(III) oxide	Mn_2O_3	-959.0	-881.1	110.5	107.7								
Manganese(IV) oxide	MnO_2	-520.0	-465.1	53.1	54.1								
Manganese(II) selenide	MnSe	-106.7	-111.7	90.8	51.0								
Manganese(II) sulfide (α form)	MnS	-214.2	-218.4	78.2	50.0								
D-Mannitol	$C_6H_{14}O_6$					-1314.5							
D-Mannose	$C_6H_{12}O_6$	-1263.0											
Mendelevium	Md	0.0											
Mercapto	HS									142.7	113.3	195.7	32.3
Mercury	Hg					0.0		75.9	28.0	61.4	31.8	175.0	20.8
Mercury(I) bromide	Br_2Hg_2	-206.9	-181.1	218.0									
Mercury(II) bromide	Br_2Hg	-170.7	-153.1	172.0									
Mercury(I) carbonate	CHg_2O_3	-553.5	-468.1	180.0									
Mercury(I) chloride	Cl_2Hg_2	-265.4	-210.7	191.6									
Mercury(II) chloride	Cl_2Hg	-224.3	-178.6	146.0									
Mercury(I) iodide	Hg_2I_2	-121.3	-111.0	233.5									
Mercury(II) iodide (red)	HgI_2	-105.4	-101.7	180.0									
Mercury(II) oxalate	C_2HgO_4	-678.2											
Mercury(II) oxide	HgO	-90.8	-58.5	70.3	44.1								
Mercury(I) sulfate	Hg_2O_4S	-743.1	-625.8	200.7	132.0								
Mercury(II) sulfate	HgO_4S	-707.5											
Mercury(II) sulfide (red)	HgS	-58.2	-50.6	82.4	48.4								
Mercury(II) telluride	HgTe	-42.0											
Mesityl oxide	$C_6H_{10}O$								212.5				
Metaboric acid (β form)	BHO_2	-794.3	-723.4	38.0						-561.9	-551.0	240.1	42.2
Metaphosphoric acid	HO_3P	-948.5											
Metasilicic acid	H_2O_3Si	-1188.7	-1092.4	134.0									
Methacrylic acid	$C_4H_6O_2$								161.1				
Methane	CH_4									-74.6	-50.5	186.3	35.7
Methanethiol	CH_4S					-46.7	-7.7	169.2	90.5	-22.9	-9.3	255.2	50.3
Methanol	CH_4O					-239.2	-166.6	126.8	81.1	-201.0	-162.3	239.9	44.1
L-Methionine	$C_5H_{11}NO_2S$	-577.5								-413.5			
2-Methoxyethanol	$C_3H_8O_2$								171.1				
2-Methoxyethyl acetate	$C_5H_{10}O_3$								310.0				
Methyl	CH_3									145.7	147.9	194.2	38.7
Methyl acetate	$C_3H_6O_2$					-445.9			141.9	-413.3		324.4	86.0
Methyl acetoacetate	$C_5H_8O_3$					-623.2							
Methyl acrylate	$C_4H_6O_2$					-362.2		239.5	158.8	-333.0			
2-Methylacrylonitrile	C_4H_5N								126.3				
Methylamine	CH_5N					-47.3	35.7	150.2	102.1	-22.5	32.7	242.9	50.1
Methylamine hydrochloride	CH_6ClN	-298.1											
2-Methylaniline	C_7H_9N					-6.3				56.4	167.6	351.0	130.2

Name	Mol. form.	$\Delta_f H°$(c)/ kJ mol^{-1}	$\Delta_f G°$(c)/ kJ mol^{-1}	$S°$(c)/ J mol^{-1} K^{-1}	C_p(c)/ J mol^{-1} K^{-1}	$\Delta_f H°$(l)/ kJ mol^{-1}	$\Delta_f G°$(l)/ kJ mol^{-1}	$S°$(l)/ J mol^{-1} K^{-1}	C_p(l)/ J mol^{-1} K^{-1}	$\Delta_f H°$(g)/ kJ mol^{-1}	$\Delta_f G°$(g)/ kJ mol^{-1}	$S°$(g)/ J mol^{-1} K^{-1}	C_p(g)/ J mol^{-1} K^{-1}
3-Methylaniline	C_7H_9N					-8.1				54.6	165.4	352.5	125.5
4-Methylaniline	C_7H_9N	-23.5								55.3	167.7	347.0	126.2
N-Methylaniline	C_7H_9N								207.1				
Methyl benzoate	$C_8H_8O_2$					-343.5			221.3	-287.9			
1-Methylbicyclo[3.1.0]hexane	C_7H_{12}					-33.2				1.7			
3-Methyl-1,2-butadiene	C_5H_8					101.2							
2-Methyl-1,3-butadiene	C_5H_8					48.2		229.3	152.6	75.5			
2-Methyl-1-butanethiol, (+)	$C_5H_{12}S$					-154.4				-114.9			
3-Methyl-1-butanethiol	$C_5H_{12}S$					-154.4				-114.9			
2-Methyl-2-butanethiol	$C_5H_{12}S$					-162.8		290.1	198.1	-127.1			
3-Methyl-2-butanethiol	$C_5H_{12}S$					-158.8				-121.3			
Methyl butanoate	$C_5H_{10}O_2$								198.2				
2-Methylbutanoic acid	$C_5H_{10}O_2$					-554.5							
3-Methylbutanoic acid	$C_5H_{10}O_2$					-561.6				-510.0			
2-Methyl-1-butanol, (±)-	$C_5H_{12}O$					-356.6				-301.4			
3-Methyl-1-butanol	$C_5H_{12}O$					-356.4				-300.7			
2-Methyl-2-butanol	$C_5H_{12}O$					-379.5			247.1	-329.3			
3-Methyl-2-butanol, (±)-	$C_5H_{12}O$					-366.6				-313.5			
3-Methyl-2-butanone	$C_5H_{10}O$					-299.5		268.5	179.9	-262.6			
2-Methyl-1-butene	C_5H_{10}					-61.1		254.0	157.2	-35.2			
3-Methyl-1-butene	C_5H_{10}					-51.5		253.3	156.1	-27.5			
2-Methyl-2-butene	C_5H_{10}					-68.6		251.0	152.8	-41.7			
Methyl *trans*-2-butenoate	$C_5H_8O_2$					-382.9				-341.9			
3-Methylbutyl 2-chloropropanoate	$C_8H_{15}ClO_2$					-627.3				-575.0			
3-Methylbutyl 3-chloropropanoate	$C_8H_{15}ClO_2$					-593.4				-539.4			
Methyl *tert*-butyl ether	$C_5H_{12}O$					-313.6		265.3	187.5	-283.7			
9-Methyl-9*H*-carbazole	$C_{13}H_{11}N$	105.5								201.0			
Methyl chloroacetate	$C_3H_5ClO_2$					-487.0				-444.0			
Methylcyclobutane	C_5H_{10}					-44.5							
Methyl cyclobutanecarboxylate	$C_6H_{10}O_2$					-395.0				-350.2			
Methylcyclohexane	C_7H_{14}					-190.1			184.8	-154.7			
cis-2-Methylcyclohexanol	$C_7H_{14}O$					-390.2				-327.0			
trans-2-Methylcyclohexanol, (±)-	$C_7H_{14}O$					-415.7				-352.5			
cis-3-Methylcyclohexanol, (±)-	$C_7H_{14}O$					-416.1				-350.9			
trans-3-Methylcyclohexanol, (±)-	$C_7H_{14}O$					-394.4				-329.1			
cis-4-Methylcyclohexanol	$C_7H_{14}O$					-413.2				-347.5			
trans-4-Methylcyclohexanol	$C_7H_{14}O$					-433.3				-367.2			
Methylcyclopentane	C_6H_{12}					-137.9				-106.2			
cis-2-Methylcyclopentanol	$C_6H_{12}O$					-345.5							
2-Methylcyclopentanone	$C_6H_{10}O$					-265.2							
1-Methylcyclopentene	C_6H_{10}					-36.4				-3.8			
3-Methylcyclopentene	C_6H_{10}					-23.7				7.4			
4-Methylcyclopentene	C_6H_{10}					-17.6				14.6			
Methylcyclopropane	C_4H_8					1.7							
Methyl decanoate	$C_{11}H_{22}O_2$					-640.5				-573.8			
Methyl 2,2-dimethylpropanoate	$C_6H_{12}O_2$					-530.0			257.9	-491.2			
1-Methyl-2,4-dinitrobenzene	$C_7H_6N_2O_4$	-66.4								33.2			
4-Methyl-1,3-dioxane	$C_5H_{10}O_2$					-416.1				-376.9			
2-Methyl-1,3-dioxolane	$C_4H_8O_2$					-386.9				-352.0			
Methyl dodecanoate	$C_{13}H_{26}O_2$					-693.0				-614.9			
Methylene	CH_2									390.4	372.9	194.9	33.8
Methylenecyclobutane	C_5H_8					93.8				121.6			
Methylenecyclohexane	C_7H_{12}					-61.3				-25.2			
2-Methylenecyclohexanol	$C_7H_{12}O$					-277.6							
N-Methylformamide	C_2H_5NO								123.8				
Methyl formate	$C_2H_4O_2$					-386.1			119.1	-357.4		285.3	64.4
3-Methyl-2,5-furandione	$C_5H_4O_3$					-504.5				-447.2			
Methyl α-D-glucopyranoside	$C_7H_{14}O_6$	-1233.3											
2-Methylheptane	C_8H_{18}					-255.0		356.4	252.0	-215.3			
3-Methylheptane	C_8H_{18}					-252.3		362.6	250.2	-212.5			
4-Methylheptane	C_8H_{18}					-251.6			251.1	-211.9			
Methyl heptanoate	$C_8H_{16}O_2$					-567.1			285.1	-515.5			
2-Methylhexane	C_7H_{16}					-229.5		323.3	222.9	-194.5			
3-Methylhexane	C_7H_{16}					-226.4				-191.3			

Name	Mol. form.	$\Delta_f H°$(c)/ kJ mol⁻¹	$\Delta_f G°$(c)/ kJ mol⁻¹	$S°$(c)/ J mol⁻¹ K⁻¹	C_p(c)/ J mol⁻¹ K⁻¹	$\Delta_f H°$(l)/ kJ mol⁻¹	$\Delta_f G°$(l)/ kJ mol⁻¹	$S°$(l)/ J mol⁻¹ K⁻¹	C_p(l)/ J mol⁻¹ K⁻¹	$\Delta_f H°$(g)/ kJ mol⁻¹	$\Delta_f G°$(g)/ kJ mol⁻¹	$S°$(g)/ J mol⁻¹ K⁻¹	C_p(g)/ J mol⁻¹ K⁻¹
Methyl hexanoate	$C_7H_{14}O_2$					-540.2				-492.2			
5-Methyl-1-hexene	C_7H_{14}					-100.0				-65.7			
cis-3-Methyl-3-hexene	C_7H_{14}					-115.9				-79.4			
trans-3-Methyl-3-hexene	C_7H_{14}					-112.7				-76.8			
Methyl 2-hexynoate	$C_7H_{10}O_2$					-242.7							
Methylhydrazine	CH_6N_2					54.2	180.0	165.9	134.9	94.7	187.0	278.8	71.1
Methylidyne	CH									595.8			
Methyl isocyanate	C_2H_3NO					-92.0							
Methyl isothiocyanate	C_2H_3NS	79.4											
Methyl methacrylate	$C_5H_8O_2$								191.2				
1-Methylnaphthalene	$C_{11}H_{10}$					56.3		254.8	224.4				
2-Methylnaphthalene	$C_{11}H_{10}$	44.9		220.0	196.0					106.7			
Methyl nitrate	CH_3NO_3					-156.3	-43.4	217.1	157.3	-122.0		305.8	76.6
Methyl nitrite	CH_3NO_2									-66.1			
Methyl nitroacetate	$C_3H_5NO_4$					-464.0							
2-Methyl-2-nitropropane	$C_4H_9NO_2$					-217.2				-177.1			
2-Methyl-2-nitro-1-propanol	$C_4H_9NO_3$	-410.1											
2-Methylnonane	$C_{10}H_{22}$					-309.8		420.1	313.3	-260.2			
5-Methylnonane	$C_{10}H_{22}$					-307.9		423.8	314.4	-258.6			
Methyl nonanoate	$C_{10}H_{20}O_2$					-616.2				-554.2			
Methyl *trans*-9-octadecenoate	$C_{19}H_{36}O_2$					-737.0							
Methyl octanoate	$C_9H_{18}O_2$					-590.3				-533.9			
Methyl *cis*-9-octadecenoate	$C_{19}H_{36}O_2$					-734.5				-649.9			
2-Methyl-2-oxazoline	C_4H_7NO					-169.5				-130.5			
Methyloxirane	C_3H_6O					-123.0		196.5	120.4	-94.7		286.9	72.6
Methyl pentadecanoate	$C_{16}H_{32}O_2$					-771.0				-680.0			
2-Methylpentane	C_6H_{14}					-204.6		290.6	193.7	-174.6			
3-Methylpentane	C_6H_{14}					-202.4		292.5	190.7	-171.9			
2-Methyl-2,4-pentanediol	$C_6H_{14}O_2$								336.0				
2-Methyl-2-pentanethiol	$C_6H_{14}S$					-188.3				-148.3			
Methyl pentanoate	$C_6H_{12}O_2$					-514.2			229.3	-471.1			
2-Methyl-1-pentanol	$C_6H_{14}O$								248.0				
3-Methyl-2-pentanol	$C_6H_{14}O$								275.9				
4-Methyl-2-pentanol	$C_6H_{14}O$					-394.7			273.0				
2-Methyl-3-pentanol	$C_6H_{14}O$					-396.4							
3-Methyl-3-pentanol	$C_6H_{14}O$								293.4				
4-Methyl-2-pentanone	$C_6H_{12}O$								213.3				
2-Methyl-3-pentanone	$C_6H_{12}O$					-325.9				-286.0			
2-Methyl-1-pentene	C_6H_{12}					-90.0				-59.4			
3-Methyl-1-pentene	C_6H_{12}					-78.2				-49.5			
4-Methyl-1-pentene	C_6H_{12}					-80.0				-51.3			
2-Methyl-2-pentene	C_6H_{12}					-98.5				-66.9			
3-Methyl-*cis*-2-pentene	C_6H_{12}					-94.5				-62.3			
3-Methyl-*trans*-2-pentene	C_6H_{12}					-94.6				-63.1			
4-Methyl-*cis*-2-pentene	C_6H_{12}					-87.0				-57.5			
4-Methyl-*trans*-2-pentene	C_6H_{12}					-91.6				-61.5			
Methyl pentyl sulfide	$C_6H_{14}S$					-167.1				-121.8			
2-Methylpiperidine, (±)-	$C_6H_{13}N$					-124.9				-84.4			
1-Methyl-2-piperidinone	$C_6H_{11}NO$					-293.0							
2-Methylpropanamide	C_4H_9NO	-368.6								-282.6			
N-Methylpropanamide	C_4H_9NO								179.0				
2-Methyl-1,2-propanediamine	$C_4H_{12}N_2$					-133.9				-90.3			
2-Methyl-1,2-propanediol	$C_4H_{10}O_2$					-539.7							
2-Methylpropanenitrile	C_4H_7N					-13.8				23.4			
2-Methyl-1-propanethiol	$C_4H_{10}S$					-132.0				-97.3			
2-Methyl-2-propanethiol	$C_4H_{10}S$					-140.5				-109.6			
Methyl propanoate	$C_4H_8O_2$								171.2				
2-Methylpropanoic acid	$C_4H_8O_2$								173.0				
2-Methyl-1-propanol	$C_4H_{10}O$					-334.7		214.7	181.5	-283.8			
2-Methyl-2-propanol	$C_4H_{10}O$					-359.2		193.3	218.6	-312.5		326.7	113.6
Methyl propyl ether	$C_4H_{10}O$					-266.0		262.9	165.4	-238.1			
Methyl propyl sulfide	$C_4H_{10}S$					-118.5		272.5	171.6	-82.2			
2-Methylpyridine	C_6H_7N					56.7			158.6	99.2			
3-Methylpyridine	C_6H_7N					61.9		216.3	158.7	106.4			
4-Methylpyridine	C_6H_7N					59.2		209.1	159.0	103.8			
1-Methylpyrrole	C_5H_7N					62.4				103.1			

Thermochem

Name	Mol. form.	$\Delta_f H°$(c)/ kJ mol^{-1}	$\Delta_f G°$(c)/ kJ mol^{-1}	$S°$(c)/ J mol^{-1} K^{-1}	C_p(c)/ J mol^{-1} K^{-1}	$\Delta_f H°$(l)/ kJ mol^{-1}	$\Delta_f G°$(l)/ kJ mol^{-1}	$S°$(l)/ J mol^{-1} K^{-1}	C_p(l)/ J mol^{-1} K^{-1}	$\Delta_f H°$(g)/ kJ mol^{-1}	$\Delta_f G°$(g)/ kJ mol^{-1}	$S°$(g)/ J mol^{-1} K^{-1}	C_p(g)/ J mol^{-1} K^{-1}
2-Methylpyrrole	C_5H_7N					23.3				74.0			
3-Methylpyrrole	C_5H_7N					20.5				70.2			
N-Methyl-2-pyrrolidinone	C_5H_9NO					-262.2			307.8				
Methyl salicylate	$C_8H_8O_3$								249.0				
Methylsilane	CH_6Si											256.5	65.9
Methyl tetradecanoate	$C_{15}H_{30}O_2$					-743.9				-656.9			
4-Methylthiazole	C_4H_5NS					67.9				111.8			
Methylthiirane	C_3H_6S					11.3				45.8			
2-Methylthiophene	C_5H_6S					44.6		218.5	149.8	83.5			
3-Methylthiophene	C_5H_6S					43.1				82.5			
Methyl tridecanoate	$C_{14}H_{28}O_2$					-717.9				-635.3			
Methyl undecanoate	$C_{12}H_{24}O_2$					-665.2				-593.8			
N-Methylurea	$C_2H_6N_2O$	-332.8											
Molybdenum	Mo	0.0		28.7	24.1					658.1	612.5	182.0	20.8
Molybdenum carbonyl	C_6MoO_6	-982.8	-877.7	325.9	242.3					-912.1	-856.0	490.0	205.0
Molybdenum(VI) fluoride	F_6Mo					-1585.5	-1473.0	259.7	169.8	-1557.7	-1472.2	350.5	120.6
Molybdenum(IV) oxide	MoO_2	-588.9	-533.0	46.3	56.0								
Molybdenum(VI) oxide	MoO_3	-745.1	-668.0	77.7	75.0								
Molybdenum silicide (Mo_3Si)	Mo_3Si	-125.2	-125.7	106.3	93.1								
Molybdenum(IV) sulfide	MoS_2	-235.1	-225.9	62.6	63.6								
Morpholine	C_4H_9NO								164.8				
β-Myrcene	$C_{10}H_{16}$					14.5							
Naphthalene	$C_{10}H_8$	78.5	201.6	167.4	165.7					150.6	224.1	333.1	131.9
1-Naphthaleneacetic acid	$C_{12}H_{10}O_2$	-359.2											
2-Naphthaleneacetic acid	$C_{12}H_{10}O_2$	-371.9											
1-Naphthalenecarboxylic acid	$C_{11}H_8O_2$	-333.5								-223.1			
2-Naphthalenecarboxylic acid	$C_{11}H_8O_2$	-346.1								-232.5			
1-Naphthol	$C_{10}H_8O$	-121.5			166.9					-30.4			149.4
2-Naphthol	$C_{10}H_8O$	-124.1		179.0	172.8					-29.9		366.6	147.8
1-Naphthylamine	$C_{10}H_9N$	67.8								132.8			
2-Naphthylamine	$C_{10}H_9N$	60.2								134.3			
Neodymium	Nd	0.0		71.5	27.5					327.6	292.4	189.4	22.1
Neodymium(III) chloride	Cl_3Nd	-1041.0			113.0								
Neodymium(III) fluoride	F_3Nd	-1657.0											
Neodymium(III) oxide	Nd_2O_3	-1807.9	-1720.8	158.6	111.3								
Neon	Ne									0.0		146.3	20.8
Neopentane	C_5H_{12}					-190.2				-168.0			
Nickel	Ni	0.0		29.9	26.1					429.7	384.5	182.2	23.4
Nickel(II) bromide	Br_2Ni	-212.1											
Nickel carbonyl [$Ni(CO)_4$]	C_4NiO_4					-633.0	-588.2	313.4	204.6	-602.9	-587.2	410.6	145.2
Nickel(II) chloride	Cl_2Ni	-305.3	-259.0	97.7	71.7								
Nickel(II) fluoride	F_2Ni	-651.4	-604.1	73.6	64.1								
Nickel(II) hydroxide	H_2NiO_2	-529.7	-447.2	88.0									
Nickel(II) iodide	I_2Ni	-78.2											
Nickel(III) oxide	Ni_2O_3	-489.5											
Nickel(II) sulfate	NiO_4S	-872.9	-759.7	92.0	138.0								
Nickel(II) sulfide	NiS	-82.0	-79.5	53.0	47.1								
Niobium	Nb	0.0		36.4	24.6					725.9	681.1	186.3	30.2
Niobium(V) chloride	Cl_5Nb	-797.5	-683.2	210.5	148.1					-703.7	-646.0	400.6	120.8
Niobium(V) fluoride	F_5Nb	-1813.8	-1699.0	160.2	134.7					-1739.7	-1673.6	321.9	97.1
Niobium(II) oxide	NbO	-405.8	-378.6	48.1	41.3								
Niobium(IV) oxide	NbO_2	-796.2	-740.5	54.5	57.5								
Niobium(V) oxide	Nb_2O_5	-1899.5	-1766.0	137.2	132.1								
Nitramide	$H_2N_2O_2$	-89.5											
Nitric acid	HNO_3					-174.1	-80.7	155.6	109.9	-133.9	-73.5	266.9	54.1
Nitric oxide	NO									91.3	87.6	210.8	29.9
Nitrilotriacetic acid	$C_6H_9NO_6$	-1311.9											
Nitroacetone	$C_3H_5NO_3$					-278.6							
2-Nitroaniline	$C_6H_6N_2O_2$	-26.1			166.0	-9.4				63.8			
3-Nitroaniline	$C_6H_6N_2O_2$	-38.3			158.8	-14.4				58.4			
4-Nitroaniline	$C_6H_6N_2O_2$	-42.0			167.0	-20.7				58.8			
Nitrobenzene	$C_6H_5NO_2$					12.5			185.8	68.5		348.8	120.4
2-Nitrobenzoic acid	$C_7H_5NO_4$	-378.8											
3-Nitrobenzoic acid	$C_7H_5NO_4$	-394.7											
4-Nitrobenzoic acid	$C_7H_5NO_4$	-392.2											
1-Nitrobutane	$C_4H_9NO_2$					-192.5				-143.9		369.9	115.1

Name	Mol. form.	$\Delta_f H°$(c)/ kJ mol⁻¹	$\Delta_f G°$(c)/ kJ mol⁻¹	$S°$(c)/ J mol⁻¹ K⁻¹	C_p(c)/ J mol⁻¹ K⁻¹	$\Delta_f H°$(l)/ kJ mol⁻¹	$\Delta_f G°$(l)/ kJ mol⁻¹	$S°$(l)/ J mol⁻¹ K⁻¹	C_p(l)/ J mol⁻¹ K⁻¹	$\Delta_f H°$(g)/ kJ mol⁻¹	$\Delta_f G°$(g)/ kJ mol⁻¹	$S°$(g)/ J mol⁻¹ K⁻¹	C_p(g)/ J mol⁻¹ K⁻¹
3-Nitro-2-butanol	$C_4H_9NO_3$					-390.0							
1-Nitrodecane	$C_{10}H_{21}NO_2$					-351.5							
N-Nitrodiethylamine	$C_4H_{10}N_2O_2$					-106.2				-53.0			
Nitroethane	$C_2H_5NO_2$					-143.9			134.4	-103.8		320.5	79.0
2-Nitroethanol	$C_2H_5NO_3$					-350.7							
Nitroethene	$C_2H_3NO_2$									33.3		300.5	73.7
2-Nitrofuran	$C_4H_3NO_3$	-104.1								-28.8			
5-Nitro-2-furancarboxylic acid	$C_5H_3NO_5$	-516.8											
Nitrogen	N_2									0.0		191.6	29.1
Nitrogen (atomic)	N									472.7	455.5	153.3	20.8
Nitrogen dioxide	NO_2									33.2	51.3	240.1	37.2
Nitrogen pentoxide	N_2O_5	-43.1	113.9	178.2	143.1					13.3	117.1	355.7	95.3
Nitrogen tetroxide	N_2O_4					-19.5	97.5	209.2	142.7	11.1	99.8	304.4	79.2
Nitrogen trichloride	Cl_3N					230.0							
Nitrogen trifluoride	F_3N									-132.1	-90.6	260.8	53.4
Nitrogen trioxide	N_2O_3					50.3				86.6	142.4	314.7	72.7
Nitroguanidine	$CH_4N_4O_2$	-92.4											
Nitromethane	CH_3NO_2					-112.6	-14.4	171.8	106.6	-80.8		282.9	55.5
(Nitromethyl)benzene	$C_7H_7NO_2$					-22.8				30.7			
1-Nitronaphthalene	$C_{10}H_7NO_2$	42.6								111.2			
1-Nitropentane	$C_5H_{11}NO_2$					-215.4				-164.4		390.9	137.1
2-Nitrophenol	$C_6H_5NO_3$	-202.4											
N-Nitropiperidine	$C_5H_{10}N_2O_2$					-93.0				-44.5			
1-Nitropropane	$C_3H_7NO_2$					-167.2				-124.3		350.0	104.1
2-Nitropropane	$C_3H_7NO_2$					-180.3			170.3	-138.9			
N-Nitrosodiphenylamine	$C_{12}H_{10}N_2O$	227.2											
N-Nitrosopiperidine	$C_5H_{10}N_2O$					-31.1				16.6			
Nitrosyl bromide	BrNO									82.2	82.4	273.7	45.5
Nitrosyl chloride	ClNO									51.7	66.1	261.7	44.7
Nitrosyl fluoride	FNO									-66.5	-51.0	248.1	41.3
2-Nitrotoluene	$C_7H_7NO_2$					-9.7							
3-Nitrotoluene	$C_7H_7NO_2$					-31.5							
4-Nitrotoluene	$C_7H_7NO_2$	-48.1			172.3					31.0			
Nitrous acid	HNO_2									-79.5	-46.0	254.1	45.6
Nitrous oxide	N_2O									81.6	103.7	220.0	38.6
Nitryl chloride	$ClNO_2$									12.6	54.4	272.2	53.2
Nitryl fluoride	FNO_2											260.4	49.8
Nobelium	No	0.0											
Nonaborane(15)	B_9H_{15}									158.4	357.5	364.9	187.0
Nonane	C_9H_{20}					-274.7			284.4	-228.2			
Nonanedioic acid	$C_9H_{16}O_4$	-1054.3											
1,9-Nonanediol	$C_9H_{20}O_2$	-657.6											
Nonanoic acid	$C_9H_{18}O_2$					-659.7			362.4	-577.3			
1-Nonanol	$C_9H_{20}O$					-453.4				-376.5			
2-Nonanone	$C_9H_{18}O$					-397.2				-340.7			
5-Nonanone	$C_9H_{18}O$					-398.2		401.4	303.6	-344.9			
1-Nonyne	C_9H_{16}					16.3				62.3			
L-Norleucine	$C_6H_{13}NO_2$	-639.1											
Octadecane	$C_{18}H_{38}$	-567.4		480.2	485.6					-414.6			
cis-9-Octadecenoic acid	$C_{18}H_{34}O_2$								577.0				
1,7-Octadiyne	C_8H_{10}					334.4							
Octanal	$C_8H_{16}O$									-291.9		365.4	
Octanamide	$C_8H_{17}NO$	-473.2								-362.7			
Octane	C_8H_{18}					-250.1			254.6	-208.5			
1,8-Octanediol	$C_8H_{18}O_2$	-626.6											
Octanenitrile	$C_8H_{15}N$					-107.3				-50.5			
Octanoic acid	$C_8H_{16}O_2$					-636.0			297.9	-554.3			
1-Octanol	$C_8H_{18}O$					-426.5			305.2	-355.6			
2-Octanol	$C_8H_{18}O$								330.1				
2-Octanone	$C_8H_{16}O$								273.3				
1-Octene	C_8H_{16}					-124.5			241.0	-81.3			
cis-2-Octene	C_8H_{16}					-135.7			239.0				
trans-2-Octene	C_8H_{16}					-135.7			239.0				
1-Octen-3-yne	C_8H_{12}					140.7							
Octyldimethylamine	$C_{10}H_{23}N$					-232.8							
Orthosilicic acid	H_4O_4Si	-1481.1	-1332.9	192.0									

Thermochem

Name	Mol. form.	$\Delta_f H°$(c)/ kJ mol⁻¹	$\Delta_f G°$(c)/ kJ mol⁻¹	$S°$(c)/ J mol⁻¹ K⁻¹	C_p(c)/ J mol⁻¹ K⁻¹	$\Delta_f H°$(l)/ kJ mol⁻¹	$\Delta_f G°$(l)/ kJ mol⁻¹	$S°$(l)/ J mol⁻¹ K⁻¹	C_p(l)/ J mol⁻¹ K⁻¹	$\Delta_f H°$(g)/ kJ mol⁻¹	$\Delta_f G°$(g)/ kJ mol⁻¹	$S°$(g)/ J mol⁻¹ K⁻¹	C_p(g)/ J mol⁻¹ K⁻¹
Osmium	Os	0.0		32.6	24.7					791.0	745.0	192.6	20.8
Osmium(III) chloride	Cl_3Os	-190.4											
Osmium(VI) fluoride	F_6Os			246.0								358.1	120.8
Osmium(VIII) oxide	O_4Os	-394.1	-304.9	143.9						-337.2	-292.8	293.8	74.1
Oxalic acid	$C_2H_2O_4$	-829.9		109.8	91.0					-731.8	-662.7	320.6	86.2
Oxalyl chloride	$C_2Cl_2O_2$					-367.6				-335.8			
Oxalyl dihydrazide	$C_2H_6N_4O_2$	-295.2											
Oxamic acid	$C_2H_3NO_3$	-661.2								-552.3			
Oxamide	$C_2H_4N_2O_2$	-504.4								-387.1			
Oxazole	C_3H_3NO					-48.0				-15.5			
Oxetane	C_3H_6O					-110.8				-80.5			
2-Oxetanone	$C_3H_4O_2$					-329.9		175.3	122.1	-282.9			
Oxirane	C_2H_4O					-78.0	-11.8	153.9	88.0	-52.6	-13.0	242.5	47.9
Oxomethyl [HCO]	CHO									43.1	28.0	224.7	34.6
Oxygen	O_2									0.0		205.2	29.4
Oxygen (atomic)	O									249.2	231.7	161.1	21.9
Oxymethurea	$C_3H_8N_2O_3$	-717.0											
Ozone	O_3									142.7	163.2	238.9	39.2
Palladium	Pd	0.0		37.6	26.0					378.2	339.7	167.1	20.8
Palladium(II) oxide	OPd	-85.4			31.4					348.9	325.9	218.0	
Palladium(II) sulfide	PdS	-75.0	-67.0	46.0									
Paraformaldehyde	$(CH_2O)x$	-177.6											
Paraldehyde	$C_6H_{12}O_3$					-673.1				-631.7			
Pentaborane(9)	B_5H_9					42.7	171.8	184.2	151.1	73.2	173.6	280.6	99.6
Pentaborane(11)	B_5H_{11}					73.2				103.3	230.6	321.0	130.3
Pentachloroethane	C_2HCl_5					-187.6			173.8	-142.0			
Pentachlorophenol	C_6HCl_5O	-292.5		253.2	202.0								
Pentadecanoic acid	$C_{15}H_{30}O_2$	-861.7			443.3	-811.7				-699.0			
1-Pentadecanol	$C_{15}H_{32}O$	-658.2											
1,2-Pentadiene	C_5H_8									140.7			
cis-1,3-Pentadiene	C_5H_8									81.4			
trans-1,3-Pentadiene	C_5H_8									76.1			
1,4-Pentadiene	C_5H_8									105.7			
2,3-Pentadiene	C_5H_8									133.1			
Pentaerythritol	$C_5H_{12}O_4$	-920.6								-776.7			
Pentaerythritol tetranitrate	$C_5H_8N_4O_{12}$	-538.6								-387.0		614.7	294.8
Pentafluorobenzene	C_6HF_5	-852.7				-841.8				-806.5			
Pentafluoroethane	C_2HF_5									-1100.4			
Pentafluorophenol	C_6HF_5O	-1024.1				-1007.7							
2,3,4,5,6-Pentafluorotoluene	$C_7H_3F_5$					-883.8		306.4	225.8	-842.7			
Pentamethylbenzene	$C_{11}H_{16}$	-144.6								-67.2			
Pentanal	$C_5H_{10}O$					-267.2				-228.4			
Pentanamide	$C_5H_{11}NO$	-379.5								-290.2			
Pentane	C_5H_{12}					-173.5			167.2	-146.9			
Pentanedioic acid	$C_5H_8O_4$	-960.0											
1,5-Pentanediol	$C_5H_{12}O_2$					-528.8				-450.8			
2,4-Pentanedione	$C_5H_8O_2$					-423.8				-382.0			
Pentanenitrile	C_5H_9N					-33.1				10.5			
1-Pentanethiol	$C_5H_{12}S$					-151.3				-110.0			
Pentanoic acid	$C_5H_{10}O_2$					-559.4		259.8	210.3	-491.9			
1-Pentanol	$C_5H_{12}O$					-351.6			208.1	-294.6			
2-Pentanol	$C_5H_{12}O$					-365.2				-311.0			
3-Pentanol	$C_5H_{12}O$					-368.9			239.7	-314.9			
2-Pentanone	$C_5H_{10}O$					-297.3			184.1	-258.8			
3-Pentanone	$C_5H_{10}O$					-296.5		266.0	190.9	-257.9			
1-Pentene	C_5H_{10}					-46.9		262.6	154.0	-21.1			
cis-2-Pentene	C_5H_{10}					-53.7		258.6	151.7	-27.6			
trans-2-Pentene	C_5H_{10}					-58.2		256.5	157.0	-31.9			
trans-3-Pentenenitrile	C_5H_7N					80.9				125.7			
4-Pentenoic acid	$C_5H_8O_2$					-430.6							
cis-3-Penten-1-yne	C_5H_6					226.5							
trans-3-Penten-1-yne	C_5H_6					228.2							
Pentetic acid	$C_{14}H_{23}N_3O_{10}$	-2225.2											
Pentyl acetate	$C_7H_{14}O_2$								261.0				
Pentylamine	$C_5H_{13}N$								218.0				
Perchloric acid	$ClHO_4$					-40.6							

Thermochem

Name	Mol. form.	$\Delta_f H°$(c)/ kJ mol^{-1}	$\Delta_f G°$(c)/ kJ mol^{-1}	$S°$(c)/ J mol^{-1} K^{-1}	C_p(c)/ J mol^{-1} K^{-1}	$\Delta_f H°$(l)/ kJ mol^{-1}	$\Delta_f G°$(l)/ kJ mol^{-1}	$S°$(l)/ J mol^{-1} K^{-1}	C_p(l)/ J mol^{-1} K^{-1}	$\Delta_f H°$(g)/ kJ mol^{-1}	$\Delta_f G°$(g)/ kJ mol^{-1}	$S°$(g)/ J mol^{-1} K^{-1}	C_p(g)/ J mol^{-1} K^{-1}
Perchloryl fluoride	$ClFO_3$									-23.8	48.2	279.0	64.9
Perfluorobutane	C_4F_{10}								127.2				
Perfluorocyclobutane	C_4F_8									-1542.6			
Perfluorocyclohexane	C_6F_{12}					-2406.3				-2370.4			
Perfluorocyclohexene	C_6F_{10}					-1963.5				-1932.7			
Perfluoroheptane	C_7F_{16}					-3420.0		561.8	419.0	-3383.6			
Perfluoromethylcyclohexane	C_7F_{14}					-2931.1			353.1	-2897.2			
Perfluoropropane	C_3F_8									-1783.2			
Perfluorotoluene	C_7F_8					-1311.1		355.5	262.3				
Peroxyacetic acid	$C_2H_4O_3$												82.4
Perrhenic acid	HO_4Re	-762.3	-656.4	158.2									
Perylene	$C_{20}H_{12}$	182.8		264.6	274.9								
Phenanthrene	$C_{14}H_{10}$	116.2		215.1	220.6					207.5			
9,10-Phenanthrenedione	$C_{14}H_8O_2$	-154.7								-46.6			
Phenanthridine	$C_{13}H_9N$	141.9								240.5			
Phenazine	$C_{12}H_8N_2$	237.0								328.8			
Phenol	C_6H_6O	-165.1		144.0	127.4					-96.4			
L-Phenylalanine	$C_9H_{11}NO_2$	-466.9		213.6	203.0					-312.9			
Phenylhydrazine	$C_6H_8N_2$					141.0			217.0	202.9			
Phenylurea	$C_7H_8N_2O$	-218.6											
Phenyl vinyl ether	C_8H_8O					-26.2				22.7			
Phosphine	H_3P									5.4	13.5	210.2	37.1
Phosphinic acid	H_3O_2P	-604.6				-595.4							
Phosphonic acid	H_3O_3P	-964.4											
Phosphonium chloride	ClH_4P	-145.2											
Phosphoric acid	H_3O_4P	-1284.4	-1124.3	110.5	106.1	-1271.7	-1123.6	150.8	145.0				
Phosphorus (white)	P	0.0		41.1	23.8					316.5	280.1	163.2	20.8
Phosphorus (red)	P	-17.6		22.8	21.2								
Phosphorus (black)	P	-39.3											
Phosphorus(III) bromide	Br_3P					-184.5	-175.7	240.2		-139.3	-162.8	348.1	76.0
Phosphorus(V) bromide	Br_5P	-269.9											
Phosphorus(III) chloride	Cl_3P					-319.7	-272.3	217.1		-287.0	-267.8	311.8	71.8
Phosphorus(V) chloride	Cl_5P	-443.5								-374.9	-305.0	364.6	112.8
Phosphorus dioxide	O_2P									-279.9	-281.6	252.1	39.5
Phosphorus(III) fluoride	F_3P									-958.4	-936.9	273.1	58.7
Phosphorus(V) fluoride	F_5P									-1594.4	-1520.7	300.8	84.8
Phosphorus(III) iodide	I_3P	-45.6										374.4	78.4
Phosphorus monoxide	OP									-28.5	-51.9	222.8	31.8
Phosphorus nitride	NP	-63.0								171.5	149.4	211.1	29.7
Phosphoryl bromide	Br_3OP	-458.6										359.8	89.9
Phosphoryl chloride	Cl_3OP					-597.1	-520.8	222.5	138.8	-558.5	-512.9	325.5	84.9
Phosphoryl fluoride	F_3OP									-1254.3	-1205.8	285.4	68.8
Phthalic acid	$C_8H_6O_4$	-782.0		207.9	188.1								
Phthalic anhydride	$C_8H_4O_3$	-460.1		180.0	160.0					-371.4			
α-Pinene	$C_{10}H_{16}$					-16.4				28.3			
β-Pinene	$C_{10}H_{16}$					-7.7				38.7			
Piperazine	$C_4H_{10}N_2$	-45.6											
2,5-Piperazinedione	$C_4H_6N_2O_2$	-446.5											
Piperidine	$C_5H_{11}N$					-86.4		210.0	179.9	-47.1			
2-Piperidinone	C_5H_9NO	-306.6											
Platinum	Pt	0.0		41.6	25.9					565.3	520.5	192.4	25.5
Platinum(II) bromide	Br_2Pt	-82.0											
Platinum(III) bromide	Br_3Pt	-120.9											
Platinum(IV) bromide	Br_4Pt	-156.5											
Platinum(II) chloride	Cl_2Pt	-123.4											
Platinum(III) chloride	Cl_3Pt	-182.0											
Platinum(IV) chloride	Cl_4Pt	-231.8											
Platinum(VI) fluoride	F_6Pt			235.6								348.3	122.8
Platinum(IV) iodide	I_4Pt	-72.8											
Platinum(II) sulfide	PtS	-81.6	-76.1	55.1	43.4								
Platinum(IV) sulfide	PtS_2	-108.8	-99.6	74.7	65.9								
Plutonium	Pu	0.0											
Polonium	Po	0.0											
Potassium	K	0.0		64.7	29.6					89.0	60.5	160.3	20.8
Potassium acetate	$C_2H_3KO_2$	-723.0											
Potassium aluminum hydride	AlH_4K	-183.7											

Name	Mol. form.	$\Delta_f H°$(c)/ kJ mol^{-1}	$\Delta_f G°$(c)/ kJ mol^{-1}	$S°$(c)/ J mol^{-1} K^{-1}	C_p(c)/ J mol^{-1} K^{-1}	$\Delta_f H°$(l)/ kJ mol^{-1}	$\Delta_f G°$(l)/ kJ mol^{-1}	$S°$(l)/ J mol^{-1} K^{-1}	C_p(l)/ J mol^{-1} K^{-1}	$\Delta_f H°$(g)/ kJ mol^{-1}	$\Delta_f G°$(g)/ kJ mol^{-1}	$S°$(g)/ J mol^{-1} K^{-1}	C_p(g)/ J mol^{-1} K^{-1}
Potassium amide	H_2KN	-128.9											
Potassium borohydride	BH_4K	-227.4	-160.3	106.3	96.1								
Potassium bromate	$BrKO_3$	-360.2	-271.2	149.2	105.2								
Potassium bromide	BrK	-393.8	-380.7	95.9	52.3								
Potassium carbonate	CK_2O_3	-1151.0	-1063.5	155.5	114.4								
Potassium chlorate	$ClKO_3$	-397.7	-296.3	143.1	100.3								
Potassium chloride	ClK	-436.5	-408.5	82.6	51.3					-214.6	-233.3	239.1	36.5
Potassium cyanide	CKN	-113.0	-101.9	128.5	66.3								
Potassium dihydrogen phosphate	H_2KO_4P	-1568.3	-1415.9	134.9	116.6								
Potassium fluoride	FK	-567.3	-537.8	66.6	49.0								
Potassium formate	$CHKO_2$	-679.7											
Potassium hexafluorosilicate	F_6K_2Si	-2956.0	-2798.6	226.0									
Potassium hydride	HK	-57.7											
Potassium hydrogen carbonate	$CHKO_3$	-963.2	-863.5	115.5									
Potassium hydrogen fluoride	F_2HK	-927.7	-859.7	104.3	76.9								
Potassium hydrogen sulfate	HKO_4S	-1160.6	-1031.3	138.1									
Potassium hydroxide	HKO	-424.6	-379.4	81.2	68.9					-232.0	-229.7	238.3	49.2
Potassium iodate	IKO_3	-501.4	-418.4	151.5	106.5								
Potassium iodide	IK	-327.9	-324.9	106.3	52.9								
Potassium metaborate	BKO_2	-981.6	-923.4	80.0	66.7								
Potassium nitrate	KNO_3	-494.6	-394.9	133.1	96.4								
Potassium nitrite	KNO_2	-369.8	-306.6	152.1	107.4								
Potassium oxalate	$C_2K_2O_4$	-1346.0											
Potassium oxide	K_2O	-361.5											
Potassium perbromate	$BrKO_4$	-287.9	-174.4	170.1	120.2								
Potassium perchlorate	$ClKO_4$	-432.8	-303.1	151.0	112.4								
Potassium periodate	IKO_4	-467.2	-361.4	175.7									
Potassium permanganate	$KMnO_4$	-837.2	-737.6	171.7	117.6								
Potassium peroxide	K_2O_2	-494.1	-425.1	102.1									
Potassium phosphate	K_3O_4P	-1950.2											
Potassium sodium	KNa					6.3							
Potassium sulfate	K_2O_4S	-1437.8	-1321.4	175.6	131.5								
Potassium sulfide	K_2S	-380.7	-364.0	105.0									
Potassium superoxide	KO_2	-284.9	-239.4	116.7	77.5								
Potassium thiocyanate	CKNS	-200.2	-178.3	124.3	88.5								
Praseodymium	Pr	0.0		73.2	27.2					355.6	320.9	189.8	21.4
Praseodymium(III) chloride	Cl_3Pr	-1056.9			100.0								
Praseodymium(III) oxide	O_3Pr_2	-1809.6			117.4								
L-Proline	$C_5H_9NO_2$	-515.2								-366.2			
Promethium	Pm	0.0										187.1	24.3
Propanal	C_3H_6O					-215.6				-185.6		304.5	80.7
Propanamide	C_3H_7NO	-338.2								-259.0			
Propane	C_3H_8					-120.9				-103.8	-23.4	270.3	73.6
Propanediamide	$C_3H_6N_2O_2$	-546.1											
1,2-Propanediamine, (±)-	$C_3H_{10}N_2$					-97.8				-53.6			
1,2-Propanediol	$C_3H_8O_2$					-501.0			190.8	-429.8			
1,3-Propanediol	$C_3H_8O_2$					-480.8				-408.0			
1,2-Propanedione	$C_3H_4O_2$					-309.1				-271.0			
1,3-Propanedithiol	$C_3H_8S_2$					-79.4				-29.8			
Propanenitrile	C_3H_5N					15.5			119.3	51.7			
1-Propanethiol	C_3H_8S					-99.9		242.5	144.6	-67.8			
2-Propanethiol	C_3H_8S					-105.9		233.5	145.3	-76.2			
Propanoic acid	$C_3H_6O_2$					-510.7		191.0	152.8	-455.7			
Propanoic anhydride	$C_6H_{10}O_3$					-679.1				-626.5			
1-Propanol	C_3H_8O					-302.6		193.6	143.9	-255.1		322.6	85.6
2-Propanol	C_3H_8O					-318.1		181.1	156.5	-272.6		309.2	89.3
Propene	C_3H_6					4.0				20.0			
trans-1-Propene-1,2-dicarboxylic acid	$C_5H_6O_4$	-824.4											
cis-1-Propene-1,2,3-tricarboxylic acid	$C_6H_6O_6$	-1224.4											
trans-1-Propene-1,2,3-tricarboxylic acid	$C_6H_6O_6$	-1232.7											
Propyl acetate	$C_5H_{10}O_2$								196.2				
Propylamine	C_3H_9N					-101.5			164.1	-70.1	39.9	325.4	91.2
Propylamine hydrochloride	$C_3H_{10}ClN$	-354.7											

Name	Mol. form.	$\Delta_f H°$(c)/ kJ mol^{-1}	$\Delta_f G°$(c)/ kJ mol^{-1}	S°(c)/ J mol^{-1} K^{-1}	C_p(c)/ J mol^{-1} K^{-1}	$\Delta_f H°$(l)/ kJ mol^{-1}	$\Delta_f G°$(l)/ kJ mol^{-1}	S°(l)/ J mol^{-1} K^{-1}	C_p(l)/ J mol^{-1} K^{-1}	$\Delta_f H°$(g)/ kJ mol^{-1}	$\Delta_f G°$(g)/ kJ mol^{-1}	S°(g)/ J mol^{-1} K^{-1}	C_p(g)/ J mol^{-1} K^{-1}
Propylbenzene	C_9H_{12}					-38.3		287.8	214.7	7.9			
Propyl carbamate	$C_4H_9NO_2$	-552.6								-471.4			
Propyl chloroacetate	$C_5H_9ClO_2$					-515.5				-467.0			
Propyl 2-chlorobutanoate	$C_7H_{13}ClO_2$					-630.7				-578.4			
Propyl chlorocarbonate	$C_4H_7ClO_2$					-533.4				-492.7			
Propyl 3-chloropropanoate	$C_6H_{11}ClO_2$					-537.6				-485.7			
Propylcyclohexane	C_9H_{18}					-237.4		311.9	242.0	-192.3			
Propylcyclopentane	C_8H_{16}					-188.8		310.8	216.3	-147.7			
Propylene carbonate	$C_4H_6O_3$					-613.2			218.6	-582.5			
Propyl formate	$C_4H_8O_2$					-500.3				-462.7			
Propyl nitrate	$C_3H_7NO_3$					-214.5				-174.1		362.6	123.2
Propyl pentanoate	$C_8H_{16}O_2$					-583.0				-533.6			
2-Propyn-1-amine	C_3H_5N					205.7							
Propyne	C_3H_4									184.9			
2-Propynoic acid	$C_3H_2O_2$					-193.2							
Protactinium	Pa	0.0		51.9						607.0	563.0	198.1	22.9
Protactinium(IV) bromide	Br_4Pa	-824.0	-787.8	234.0									
Protactinium(IV) chloride	Cl_4Pa	-1043.0	-953.0	192.0									
Protactinium(V) chloride	Cl_5Pa	-1145.0	-1034.0	238.0									
1*H*-Purine	$C_5H_4N_4$	169.4											
Pyrazine	$C_4H_4N_2$	139.8								196.1			
1*H*-Pyrazole	$C_3H_4N_2$	105.4			81.0					179.4			
Pyrene	$C_{16}H_{10}$	125.5		224.9	229.7					225.7			
Pyridazine	$C_4H_4N_2$					224.9				278.3			
Pyridine	C_5H_5N					100.2			132.7	140.4			
3-Pyridinecarboxylic acid	$C_6H_5NO_2$	-344.9								-221.5			
Pyrimidine	$C_4H_4N_2$					145.9				195.7			
Pyrocatechol	$C_6H_6O_2$	-354.1								-267.5			
Pyrrole	C_4H_5N					63.1		156.4	127.7	108.2			
1*H*-Pyrrole-2-carboxaldehyde	C_5H_5NO	-106.4											
Pyrrolidine	C_4H_9N					-41.1		204.1	156.6	-3.6			
2-Pyrrolidone	C_4H_7NO					-286.2							
Quinoline	C_9H_7N					141.2				200.5			
2-Quinolinecarbonitrile	$C_{10}H_6N_2$	246.5											
3-Quinolinecarbonitrile	$C_{10}H_6N_2$	242.3											
2-Quinolinol	C_9H_7NO	-144.9								-25.5			
8-Quinolinol	C_9H_7NO	82.1											
Radium	Ra	0.0		71.0						159.0	130.0	176.5	20.8
Radium nitrate	N_2O_6Ra	-992.0	-796.1	222.0									
Radium oxide	ORa	-523.0											
Radium sulfate	O_4RaS	-1471.1	-1365.6	138.0									
Radon	Rn									0.0		176.2	20.8
Resorcinol	$C_6H_6O_2$	-368.0								-274.7			
Rhenium	Re	0.0		36.9	25.5					769.9	724.6	188.9	20.8
Rhenium(III) bromide	Br_3Re	-167.0											
Rhenium(III) chloride	Cl_3Re	-264.0	-188.0	123.8	92.4								
Rhenium(VII) oxide	O_7Re_2	-1240.1	-1066.0	207.1	166.1					-1100.0	-994.0	452.0	
Rhodium	Rh	0.0		31.5	25.0					556.9	510.8	185.8	21.0
Rhodium(III) chloride	Cl_3Rh	-299.2											
Rhodium monoxide	ORh									385.0			
Rhodium(III) oxide	O_3Rh_2	-343.0			103.8								
D-Ribose	$C_5H_{10}O_5$	-1047.2											
Rubidium	Rb	0.0		76.8	31.1					80.9	53.1	170.1	20.8
Rubidium amide	H_2NRb	-113.0											
Rubidium bromide	BrRb	-394.6	-381.8	110.0	52.8								
Rubidium carbonate	CO_3Rb_2	-1136.0	-1051.0	181.3	117.6								
Rubidium chloride	ClRb	-435.4	-407.8	95.9	52.4								
Rubidium fluoride	FRb	-557.7											
Rubidium hydride	HRb	-52.3											
Rubidium hydrogen fluoride	F_2HRb	-922.6	-855.6	120.1	79.4								
Rubidium hydrogen sulfate	HO_4RbS	-1159.0											
Rubidium hydroxide	HORb	-418.8	-373.9	94.0	69.0					-238.0	-239.1	248.5	49.5
Rubidium iodide	IRb	-333.8	-328.9	118.4	53.2								
Rubidium metaborate	BO_2Rb	-971.0	-913.0	94.3	74.1								
Rubidium nitrate	NO_3Rb	-495.1	-395.8	147.3	102.1								
Rubidium nitrite	NO_2Rb	-367.4	-306.2	172.0									

Name	Mol. form.	$\Delta_f H°$(c)/ kJ mol^{-1}	$\Delta_f G°$(c)/ kJ mol^{-1}	$S°$(c)/ J mol^{-1} K^{-1}	C_p(c)/ J mol^{-1} K^{-1}	$\Delta_f H°$(l)/ kJ mol^{-1}	$\Delta_f G°$(l)/ kJ mol^{-1}	$S°$(l)/ J mol^{-1} K^{-1}	C_p(l)/ J mol^{-1} K^{-1}	$\Delta_f H°$(g)/ kJ mol^{-1}	$\Delta_f G°$(g)/ kJ mol^{-1}	$S°$(g)/ J mol^{-1} K^{-1}	C_p(g)/ J mol^{-1} K^{-1}
Rubidium oxide	ORb_2	-339.0											
Rubidium perchlorate	ClO_4Rb	-437.2	-306.9	161.1									
Rubidium peroxide	O_2Rb_2	-472.0											
Rubidium sulfate	O_4Rb_2S	-1435.6	-1316.9	197.4	134.1								
Rubidium superoxide	O_2Rb	-278.7											
Ruthenium	Ru	0.0		28.5	24.1					642.7	595.8	186.5	21.5
Ruthenium(III) bromide	Br_3Ru	-138.0											
Ruthenium(III) chloride	Cl_3Ru	-205.0											
Ruthenium(III) iodide	I_3Ru	-65.7											
Ruthenium(IV) oxide	O_2Ru	-305.0											
Ruthenium(VIII) oxide	O_4Ru	-239.3	-152.2	146.4									
Salicylaldehyde	$C_7H_6O_2$								222.0				
Salicylaldoxime	$C_7H_7NO_2$	-183.7											
Samarium	Sm	0.0		69.6	29.5					206.7	172.8	183.0	30.4
Samarium(III) chloride	Cl_3Sm	-1025.9											
Samarium(III) fluoride	F_3Sm	-1778.0											
Samarium(III) oxide	O_3Sm_2	-1823.0	-1734.6	151.0	114.5								
Sarcosine	$C_3H_7NO_2$	-513.3								-367.3			
Scandium	Sc	0.0		34.6	25.5					377.8	336.0	174.8	22.1
Scandium bromide	Br_3Sc	-743.1											
Scandium chloride	Cl_3Sc	-925.1											
Scandium fluoride	F_3Sc	-1629.2	-1555.6	92.0						-1247.0	-1234.0	300.5	67.8
Scandium oxide	O_3Sc_2	-1908.8	-1819.4	77.0	94.2								
Selenic acid	H_2O_4Se	-530.1											
Selenium (gray)	Se	0.0		42.4	25.4					227.1	187.0	176.7	20.8
Selenium (α form)	Se	6.7								227.1			
Selenium (vitreous)	Se	5.0								227.1			
Selenium dibromide	Br_2Se									-21.0			
Selenium dioxide	O_2Se	-225.4											
Selenium hexafluoride	F_6Se									-1117.0	-1017.0	313.9	110.5
Selenium monoxide	OSe									53.4	26.8	234.0	31.3
DL-Serine	$C_3H_7NO_3$	-739.0											
L-Serine	$C_3H_7NO_3$	-732.7											
Silane	H_4Si									34.3	56.9	204.6	42.8
Silicon	Si	0.0		18.8	20.0					450.0	405.5	168.0	22.3
Silicon carbide (hexagonal)	CSi	-62.8	-60.2	16.5	26.7								
Silicon carbide (cubic)	CSi	-65.3	-62.8	16.6	26.9								
Silicon dioxide (α-quartz)	O_2Si	-910.7	-856.3	41.5	44.4					-322.0			
Silicon monosulfide	SSi									112.5	60.9	223.7	32.3
Silicon monoxide	OSi									-99.6	-126.4	211.6	29.9
Silicon nitride [Si_3N_4]	N_4Si_3	-743.5	-642.6	101.3									
Silver	Ag	0.0		42.6	25.4					284.9	246.0	173.0	20.8
Silver(I) bromate	$AgBrO_3$	-10.5	71.3	151.9									
Silver(I) bromide	AgBr	-100.4	-96.9	107.1	52.4								
Silver(I) carbonate	CAg_2O_3	-505.8	-436.8	167.4	112.3								
Silver(I) chlorate	$AgClO_3$	-30.3	64.5	142.0									
Silver(I) chloride	AgCl	-127.0	-109.8	96.3	50.8								
Silver(I) chromate	Ag_2CrO_4	-731.7	-641.8	217.6	142.3								
Silver(I) cyanide	CAgN	146.0	156.9	107.2	66.7								
Silver(I) fluoride	AgF	-204.6											
Silver(II) fluoride	AgF_2	-360.0											
Silver(I) iodate	$AgIO_3$	-171.1	-93.7	149.4	102.9								
Silver(I) iodide	AgI	-61.8	-66.2	115.5	56.8								
Silver(I) nitrate	$AgNO_3$	-124.4	-33.4	140.9	93.1								
Silver(I) oxide	Ag_2O	-31.1	-11.2	121.3	65.9								
Silver[II] oxide [Ag_2O_2]	Ag_2O_2	-24.3	27.6	117.0	88.0								
Silver(III) oxide	Ag_2O_3	33.9	121.4	100.0									
Silver(I) perchlorate	$AgClO_4$	-31.1											
Silver(I) sulfate	Ag_2O_4S	-715.9	-618.4	200.4	131.4								
Silver(I) sulfide	Ag_2S	-32.6	-40.7	144.0	76.5								
Silylidyne	HSi									361.0			
Sodium	Na	0.0		51.3	28.2					107.5	77.0	153.7	20.8
Sodium acetate	$C_2H_3NaO_2$	-708.8	-607.2	123.0	79.9								
Sodium aluminum hydride	AlH_4Na	-115.5											
Sodium amide	H_2NNa	-123.8	-64.0	76.9	66.2								
Sodium azide	N_3Na	21.7	93.8	96.9	76.6								

Name	Mol. form.	$\Delta_f H°$(c)/ kJ mol⁻¹	$\Delta_f G°$(c)/ kJ mol⁻¹	$S°$(c)/ J mol⁻¹ K⁻¹	C_p(c)/ J mol⁻¹ K⁻¹	$\Delta_f H°$(l)/ kJ mol⁻¹	$\Delta_f G°$(l)/ kJ mol⁻¹	$S°$(l)/ J mol⁻¹ K⁻¹	C_p(l)/ J mol⁻¹ K⁻¹	$\Delta_f H°$(g)/ kJ mol⁻¹	$\Delta_f G°$(g)/ kJ mol⁻¹	$S°$(g)/ J mol⁻¹ K⁻¹	C_p(g)/ J mol⁻¹ K⁻¹
Sodium borohydride	BH_4Na	-188.6	-123.9	101.3	86.8								
Sodium bromate	$BrNaO_3$	-334.1	-242.6	128.9									
Sodium bromide	$BrNa$	-361.1	-349.0	86.8	51.4					-143.1	-177.1	241.2	36.3
Sodium carbonate	CNa_2O_3	-1130.7	-1044.4	135.0	112.3								
Sodium chlorate	$ClNaO_3$	-365.8	-262.3	123.4									
Sodium chloride	$ClNa$	-411.2	-384.1	72.1	50.5								
Sodium chlorite	$ClNaO_2$	-307.0											
Sodium cyanate	$CNNaO$	-405.4	-358.1	96.7	86.6								
Sodium cyanide	$CNNa$	-87.5	-76.4	115.6	70.4								
Sodium fluoride	FNa	-576.6	-546.3	51.1	46.9								
Sodium formate	$CHNaO_2$	-666.5	-599.9	103.8	82.7								
Sodium hexafluorosilicate	F_6Na_2Si	-2909.6	-2754.2	207.1	187.1								
Sodium hydride	HNa	-56.3	-33.5	40.0	36.4								
Sodium hydrogen carbonate	$CHNaO_3$	-950.8	-851.0	101.7	87.6								
Sodium hydrogen fluoride	F_2HNa	-920.3	-852.2	90.9	75.0								
Sodium hydrogen phosphate	HNa_2O_4P	-1748.1	-1608.2	150.5	135.3								
Sodium hydrogen sulfate	$HNaO_4S$	-1125.5	-992.8	113.0									
Sodium hydroxide	$HNaO$	-425.8	-379.7	64.4	59.5					-191.0	-193.9	229.0	48.0
Sodium iodate	$INaO_3$	-481.8			92.0								
Sodium iodide	INa	-287.8	-286.1	98.5	52.1								
Sodium metaborate	$BNaO_2$	-977.0	-920.7	73.5	65.9								
Sodium metasilicate	Na_2O_3Si	-1554.9	-1462.8	113.9									
Sodium molybdate	$MoNa_2O_4$	-1468.1	-1354.3	159.7	141.7								
Sodium nitrate	$NNaO_3$	-467.9	-367.0	116.5	92.9								
Sodium nitrite	$NNaO_2$	-358.7	-284.6	103.8									
Sodium oxalate	$C_2Na_2O_4$									-1318.0			
Sodium oxide	Na_2O	-414.2	-375.5	75.1	69.1								
Sodium perchlorate	$ClNaO_4$	-383.3	-254.9	142.3									
Sodium periodate	$INaO_4$	-429.3	-323.0	163.0									
Sodium permanganate	$MnNaO_4$	-1156.0											
Sodium peroxide	Na_2O_2	-510.9	-447.7	95.0	89.2								
Sodium sulfate	Na_2O_4S	-1387.1	-1270.2	149.6	128.2								
Sodium sulfide	Na_2S	-364.8	-349.8	83.7									
Sodium sulfite	Na_2O_3S	-1100.8	-1012.5	145.9	120.3								
Sodium superoxide	NaO_2	-260.2	-218.4	115.9	72.1								
Sodium tetraborate	$B_4Na_2O_7$	-3291.1	-3096.0	189.5	186.8								
Sodium tetrafluoroaluminate	AlF_4Na									-1869.0	-1827.5	345.7	105.9
Sodium tetrafluoroborate	BF_4Na	-1844.7	-1750.1	145.3	120.3								
L-Sorbose	$C_6H_{12}O_6$	-1271.5											
Spiro[2.2]pentane	C_5H_8					157.5		193.7	134.5	185.2			
Spiro[5.5]undecane	$C_{11}H_{20}$					-244.5				-188.3			
S-Propyl thioacetate	$C_5H_{10}OS$					-294.5				-250.4			
Stannane	H_4Sn									162.8	188.3	227.7	49.0
Stearic acid	$C_{18}H_{36}O_2$	-947.7			501.5	-884.7				-781.2			
Stibine	H_3Sb									145.1	147.8	232.8	41.1
cis-Stilbene	$C_{14}H_{12}$					183.3				252.3			
trans-Stilbene	$C_{14}H_{12}$	136.9								236.1			
Strontium	Sr	0.0		55.0	26.8					164.4	130.9	164.6	20.8
Strontium bromide	Br_2Sr	-717.6	-697.1	135.1	75.3								
Strontium carbonate	CO_3Sr	-1220.1	-1140.1	97.1	81.4								
Strontium chloride	Cl_2Sr	-828.9	-781.1	114.9	75.6								
Strontium fluoride	F_2Sr	-1216.3	-1164.8	82.1	70.0								
Strontium formate	$C_2H_2O_4Sr$	-1393.3											
Strontium hydride	H_2Sr	-180.3											
Strontium hydroxide	H_2O_2Sr	-959.0											
Strontium iodide	I_2Sr	-558.1			81.6								
Strontium metasilicate	O_3SiSr	-1633.9	-1549.7	96.7	88.5								
Strontium nitrate	N_2O_6Sr	-978.2	-780.0	194.6	149.9								
Strontium nitrite	N_2O_4Sr	-762.3											
Strontium orthosilicate	O_4SiSr_2	-2304.5	-2191.1	153.1	134.3								
Strontium oxide	OSr	-592.0	-561.9	54.4	45.0					1.5			
Strontium selenide	$SeSr$	-385.8											
Strontium sulfate	O_4SSr	-1453.1	-1340.9	117.0									
Strontium sulfide	SSr	-472.4	-467.8	68.2	48.7								
Styrene	C_8H_8					103.8			182.0	147.9			
Succinamide	$C_4H_8N_2O_2$	-581.2											

Name	Mol. form.	$\Delta_fH°$(c)/ kJ mol^{-1}	$\Delta_fG°$(c)/ kJ mol^{-1}	$S°$(c)/ J mol^{-1} K^{-1}	C_p(c)/ J mol^{-1} K^{-1}	$\Delta_fH°$(l)/ kJ mol^{-1}	$\Delta_fG°$(l)/ kJ mol^{-1}	$S°$(l)/ J mol^{-1} K^{-1}	C_p(l)/ J mol^{-1} K^{-1}	$\Delta_fH°$(g)/ kJ mol^{-1}	$\Delta_fG°$(g)/ kJ mol^{-1}	$S°$(g)/ J mol^{-1} K^{-1}	C_p(g)/ J mol^{-1} K^{-1}
Succinic acid	$C_4H_6O_4$	-940.5		167.3	153.1					-823.0			
Succinic anhydride	$C_4H_4O_3$	-608.6								-527.9			
Succinimide	$C_4H_5NO_2$	-459.0								-375.4			
Succinonitrile	$C_4H_4N_2$	139.7		191.6	145.6					209.7			
Sucrose	$C_{12}H_{22}O_{11}$	-2226.1											
Sulfolane	$C_4H_8O_2S$								180.0				
Sulfur (rhombic)	S	0.0		32.1	22.6					277.2	236.7	167.8	23.7
Sulfur (monoclinic)	S	0.3											
Sulfur bromide [BrSSBr]	Br_2S_2					-13.0							
Sulfur chloride [ClSSCl]	Cl_2S_2					-59.4							
Sulfur chloride pentafluoride	ClF_5S					-1065.7							
Sulfur dichloride	Cl_2S					-50.0							
Sulfur dioxide	O_2S					-320.5				-296.8	-300.1	248.2	39.9
Sulfur hexafluoride	F_6S									-1220.5	-1116.5	291.5	97.0
Sulfuric acid	H_2O_4S					-814.0	-690.0	156.9	138.9				
Sulfur monoxide	OS									6.3	-19.9	222.0	30.2
Sulfur tetrafluoride	F_4S									-763.2	-722.0	299.6	77.6
Sulfur trioxide (α-form)	O_3S	-454.5	-374.2	70.7		-441.0	-373.8	113.8		-395.7	-371.1	256.8	50.7
Sulfuryl chloride	Cl_2O_2S					-394.1			134.0	-364.0	-320.0	311.9	77.0
Sulfuryl fluoride	F_2O_2S											284.0	66.0
Tantalum	Ta	0.0		41.5	25.4					782.0	739.3	185.2	20.9
Tantalum(V) bromide	Br_5Ta	-598.3											
Tantalum(V) chloride	Cl_5Ta	-859.0											
Tantalum(V) fluoride	F_5Ta	-1903.6											
Tantalum hydride (Ta_2H)	HTa_2	-32.6	-69.0	79.1	90.8								
Tantalum(V) oxide	O_5Ta_2	-2046.0	-1911.2	143.1	135.1								
Technetium	Tc	0.0								678.0		181.1	20.8
Tellurium	Te	0.0		49.7	25.7					196.7	157.1	182.7	20.8
Tellurium dioxide	O_2Te	-322.6	-270.3	79.5									
Tellurium hexafluoride	F_6Te									-1318.0			
Tellurium tetrabromide	Br_4Te	-190.4											
Tellurium tetrachloride	Cl_4Te	-326.4			138.5								
Terbium	Tb	0.0		73.2	28.9					388.7	349.7	203.6	24.6
Terbium(III) chloride	Cl_3Tb	-997.0											
Terbium(III) oxide	O_3Tb_2	-1865.2			115.9								
Terephthalic acid	$C_8H_6O_4$	-816.1								-717.9			
o-Terphenyl	$C_{18}H_{14}$			298.8	274.8			337.1	369.1				
p-Terphenyl	$C_{18}H_{14}$	163.0		285.6	278.7					279.0			
α-Terpinene	$C_{10}H_{16}$									-20.6			
Tetraborane(10)	B_4H_{10}									66.1	184.3	280.3	93.2
1,1,2,2-Tetrabromoethane	$C_2H_2Br_4$								165.7				
Tetrabromoethene	C_2Br_4											387.1	102.7
Tetrabromomethane	CBr_4	29.4	47.7	212.5	144.3					83.9	67.0	358.1	91.2
Tetrabromosilane	Br_4Si					-457.3	-443.9	277.8		-415.5	-431.8	377.9	97.1
Tetrabutylammonium iodide	$C_{16}H_{36}IN$	-498.6											
Tetrachlorodiborane	B_2Cl_4					-523.0	-464.8	262.3	137.7	-490.4	-460.6	357.4	95.4
1,1,1,2-Tetrachloro-2,2-difluoroethane	$C_2Cl_4F_2$									-489.9	-407.0	382.9	123.4
1,1,2,2-Tetrachloro-1,2-difluoroethane	$C_2Cl_4F_2$								173.6				
1,1,1,2-Tetrachloroethane	$C_2H_2Cl_4$											356.0	102.7
1,1,2,2-Tetrachloroethane	$C_2H_2Cl_4$					-195.0		246.9	162.3	-149.2		362.8	100.8
Tetrachloroethene	C_2Cl_4					-50.6	3.0	266.9	143.4	-10.9			
Tetrachloromethane	CCl_4					-128.2			130.7	-95.7			83.3
1,1,1,3-Tetrachloropropane	$C_3H_4Cl_4$					-208.7							
1,2,2,3-Tetrachloropropane	$C_3H_4Cl_4$					-251.8							
Tetrachlorosilane	Cl_4Si					-687.0	-619.8	239.7	145.3	-657.0	-617.0	330.7	90.3
Tetracyanoethene	C_6N_4	623.8								705.0			
Tetradecanenitrile	$C_{14}H_{27}N$					-260.2				-174.9			
Tetradecanoic acid	$C_{14}H_{28}O_2$	-833.5			432.0	-788.8				-693.7			
1-Tetradecanol	$C_{14}H_{30}O$	-629.6			388.0	-580.6							
Tetraethylammonium bromide	$C_8H_{20}BrN$	-342.7											
Tetraethylene glycol	$C_8H_{18}O_5$					-981.7			428.8	-883.0			
Tetraethyl lead	$C_8H_{20}Pb$					52.7		464.6	307.4	109.6			
Tetraethylsilane	$C_8H_{20}Si$								298.1				
Tetraethylurea	$C_9H_{20}N_2O$					-380.0				-316.4			

Name	Mol. form.	$\Delta_f H°$(c)/ kJ mol^{-1}	$\Delta_f G°$(c)/ kJ mol^{-1}	$S°$(c)/ J mol^{-1} K^{-1}	C_p(c)/ J mol^{-1} K^{-1}	$\Delta_f H°$(l)/ kJ mol^{-1}	$\Delta_f G°$(l)/ kJ mol^{-1}	$S°$(l)/ J mol^{-1} K^{-1}	C_p(l)/ J mol^{-1} K^{-1}	$\Delta_f H°$(g)/ kJ mol^{-1}	$\Delta_f G°$(g)/ kJ mol^{-1}	$S°$(g)/ J mol^{-1} K^{-1}	C_p(g)/ J mol^{-1} K^{-1}
1,2,4,5-Tetrafluorobenzene	$C_6H_2F_4$					-683.8							
Tetrafluorodiborane	B_2F_4									-1440.1	-1410.4	317.3	79.1
Tetrafluoroethene	C_2F_4	-820.5								-658.9		300.1	80.5
Tetrafluorohydrazine	F_4N_2									-8.4	79.9	301.2	79.2
Tetrafluoromethane	CF_4									-933.6		261.6	61.1
2,2,3,3-Tetrafluoro-1-propanol	$C_3H_4F_4O$					-1114.9				-1061.3			
Tetrafluorosilane	F_4Si									-1615.0	-1572.8	282.8	73.6
Tetrahydrofuran	C_4H_8O					-216.2		204.3	124.0	-184.1		302.4	76.3
Tetrahydrofurfuryl alcohol	$C_5H_{10}O_2$					-435.7				-369.1			
1,2,3,4-Tetrahydronaphthalene	$C_{10}H_{12}$					-29.2			217.5	26.0			
Tetrahydropyran	$C_5H_{10}O$					-258.3				-223.4			
Tetrahydro-2*H*-pyran-2-one	$C_5H_8O_2$					-436.7				-379.6			
1,2,5,6-Tetrahydropyridine	C_5H_9N					33.5							
Tetrahydrothiophene	C_4H_8S					-72.9				-34.1	45.8	309.6	92.5
Tetraiodoethene	C_2I_4	305.0											
Tetraiodomethane	CI_4	-392.9								474.0		391.9	95.9
Tetraiodosilane	I_4Si	-189.5											
1-Tetralone	$C_{10}H_{10}O$	-209.6											
Tetramethylammonium bromide	$C_4H_{12}BrN$	-251.0											
Tetramethylammonium chloride	$C_4H_{12}ClN$	-276.4											
Tetramethylammonium iodide	$C_4H_{12}IN$	-203.9											
1,2,4,5-Tetramethylbenzene	$C_{10}H_{14}$	-119.9		245.6	215.1								
2,2,3,3-Tetramethylbutane	C_8H_{18}	-269.0		273.7	239.2					-226.0			
2,2,4,4-Tetramethyl-1,3-cyclobutanedione	$C_8H_{12}O_2$	-379.9								-307.6			
1,1,2,2-Tetramethylcyclopropane	C_7H_{14}					-119.8							
Tetramethyl lead	$C_4H_{12}Pb$					97.9				135.9			
N,N,N',N'-Tetramethylmethanediamine	$C_5H_{14}N_2$					-51.1				-18.2			
2,2,3,3-Tetramethylpentane	C_9H_{20}					-278.3			271.5	-237.1			
2,2,3,4-Tetramethylpentane	C_9H_{20}					-277.7				-236.9			
2,2,4,4-Tetramethylpentane	C_9H_{20}					-280.0			266.3	-241.6			
2,3,3,4-Tetramethylpentane	C_9H_{20}					-277.9				-236.1			
2,2,6,6-Tetramethylpiperidine	$C_9H_{19}N$					-206.9				-159.9			
2,2,6,6-Tetramethyl-4-piperidinone	$C_9H_{17}NO$	-334.2								-273.4			
Tetramethylsilane	$C_4H_{12}Si$					-264.0	-100.0	277.3	204.1	-239.1	-99.9	359.0	143.9
Tetramethylstannane	$C_4H_{12}Sn$					-52.3				-18.8			
Tetramethylthiourea	$C_5H_{12}N_2S$	-38.1								44.9			
Tetramethylurea	$C_5H_{12}N_2O$					-262.2							
Tetranitromethane	CN_4O_8					38.4				82.4		503.7	176.1
Tetraphosphorus	P_4									58.9	24.4	280.0	67.2
Thallium	Tl	0.0		64.2	26.3					182.2	147.4	181.0	20.8
Thallium(I) bromide	BrTl	-173.2	-167.4	120.5						-37.7			
Thallium(I) carbonate	CO_3Tl_2	-700.0	-614.6	155.2									
Thallium(I) chloride	ClTl	-204.1	-184.9	111.3	50.9					-67.8			
Thallium(III) chloride	Cl_3Tl	-315.1											
Thallium(I) fluoride	FTl	-324.7								-182.4			
Thallium(I) hydroxide	HOTl	-238.9	-195.8	88.0									
Thallium(I) iodide	ITl	-123.8	-125.4	127.6						7.1			
Thallium(I) nitrate	NO_3Tl	-243.9	-152.4	160.7	99.5								
Thallium(I) oxide	OTl_2	-178.7	-147.3	126.0									
Thallium(I) selenide	$SeTl_2$	-59.0	-59.0	172.0									
Thallium(I) sulfate	O_4STl_2	-931.8	-830.4	230.5									
Thallium(I) sulfide	STl_2	-97.1	-93.7	151.0									
Thiacyclohexane	$C_5H_{10}S$					-106.3		218.2	163.3	-63.5	53.1	323.0	109.7
Thianthrene	$C_{12}H_8S_2$	182.0								286.0			
Thiepane	$C_6H_{12}S$									-65.8	79.4	363.5	131.3
Thietane	C_3H_6S					24.7		184.9		60.6	107.1	285.0	68.3
Thiirane	C_2H_4S					51.6				82.0	96.8	255.2	53.3
Thiirene	C_2H_2S									300.0	275.8	255.3	54.7
Thioacetamide	C_2H_5NS	-71.7								11.4			
Thioacetic acid	C_2H_4OS					-216.9				-175.1			
Thiolactic acid	$C_3H_6O_2S$					-468.4							
Thionitrosyl fluoride (NSF)	FNS											259.8	44.1
Thionyl chloride	Cl_2OS					-245.6			121.0	-212.5	-198.3	309.8	66.5
Thionyl fluoride	F_2OS											278.7	56.8

Thermochem

Name	Mol. form.	$\Delta_f H°$(c)/ kJ mol^{-1}	$\Delta_f G°$(c)/ kJ mol^{-1}	$S°$(c)/ J mol^{-1} K^{-1}	C_p(c)/ J mol^{-1} K^{-1}	$\Delta_f H°$(l)/ kJ mol^{-1}	$\Delta_f G°$(l)/ kJ mol^{-1}	$S°$(l)/ J mol^{-1} K^{-1}	C_p(l)/ J mol^{-1} K^{-1}	$\Delta_f H°$(g)/ kJ mol^{-1}	$\Delta_f G°$(g)/ kJ mol^{-1}	$S°$(g)/ J mol^{-1} K^{-1}	C_p(g)/ J mol^{-1} K^{-1}
Thiophene	C_4H_4S					80.2		181.2	123.8	114.9	126.1	278.8	72.8
Thiourea	CH_4N_2S	-89.1								22.9			
Thiram	$C_6H_{12}N_2S_4$	40.2			301.7								
Thorium	Th	0.0		51.8	27.3					602.0	560.7	190.2	20.8
Thorium(IV) chloride	Cl_4Th	-1186.2	-1094.1	190.4	120.3					-964.4	-932.0	390.7	107.5
Thorium(III) fluoride	F_3Th									-1166.1	-1160.6	339.2	73.3
Thorium(IV) fluoride	F_4Th	-2097.8	-2003.4	142.0	110.7					-1759.0	-1724.0	341.7	93.0
Thorium hydride	H_2Th	-139.7	-100.0	50.7	36.7								
Thorium(IV) oxide	O_2Th	-1226.4	-1169.2	65.2	61.8								
DL-Threonine	$C_4H_9NO_3$	-758.8											
L-Threonine	$C_4H_9NO_3$	-807.2											
Thulium	Tm	0.0		74.0	27.0					232.2	197.5	190.1	20.8
Thulium(III) chloride	Cl_3Tm	-986.6											
Thulium(III) oxide	O_3Tm_2	-1888.7	-1794.5	139.7	116.7								
Thymine	$C_5H_6N_2O_2$	-462.8			150.8					-328.7			
Thymol	$C_{10}H_{14}O$	-309.7								-218.5			
Tin(II) bromide	Br_2Sn	-243.5											
Tin(IV) bromide	Br_4Sn	-377.4	-350.2	264.4						-314.6	-331.4	411.9	103.4
Tin(II) chloride	Cl_2Sn	-325.1											
Tin(IV) chloride	Cl_4Sn					-511.3	-440.1	258.6	165.3	-471.5	-432.2	365.8	98.3
Tin (gray)	Sn	-2.1	0.1	44.1	25.8								
Tin(II) hydroxide	H_2O_2Sn	-561.1	-491.6	155.0									
Tin(II) iodide	I_2Sn	-143.5											
Tin(IV) iodide	I_4Sn				84.9							446.1	105.4
Tin(II) oxide	OSn	-280.7	-251.9	57.2	44.3					15.1	-8.4	232.1	31.6
Tin(IV) oxide	O_2Sn	-577.6	-515.8	49.0	52.6								
Tin(II) sulfide	SSn	-100.0	-98.3	77.0	49.3								
Tin (white)	Sn	0.0		51.2	27.0					301.2	266.2	168.5	21.3
Titanium	Ti	0.0		30.7	25.1					473.0	428.4	180.3	24.4
Titanium(II) bromide	Br_2Ti	-402.0											
Titanium(III) bromide	Br_3Ti	-548.5	-523.8	176.6	101.7								
Titanium(IV) bromide	Br_4Ti	-616.7	-589.5	243.5	131.5					-549.4	-568.2	398.4	100.8
Titanium(II) chloride	Cl_2Ti	-513.8	-464.4	87.4	69.8								
Titanium(III) chloride	Cl_3Ti	-720.9	-653.5	139.7	97.2								
Titanium(IV) chloride	Cl_4Ti					-804.2	-737.2	252.3	145.2	-763.2	-726.3	353.2	95.4
Titanium(IV) iodide	I_4Ti	-375.7	-371.5	249.4	125.7					-277.8			
Titanium(II) oxide	OTi	-519.7	-495.0	50.0	40.0								
Titanium(III) oxide	O_3Ti_2	-1520.9	-1434.2	78.8	97.4								
Titanium(III,IV) oxide	O_5Ti_3	-2459.4	-2317.4	129.3	154.8								
Titanium(IV) oxide (rutile)	O_2Ti	-944.0	-888.8	50.6	55.0								
Toluene	C_7H_8					12.4			157.3	50.5			
Toluene-2,4-diisocyanate	$C_9H_6N_2O_2$								287.8				
o-Toluic acid	$C_8H_8O_2$	-416.5			174.9								
m-Toluic acid	$C_8H_8O_2$	-426.1			163.6								
p-Toluic acid	$C_8H_8O_2$	-429.2			169.0								
Triacetamide	$C_6H_9NO_3$					-610.5				-550.1			
1*H*-1,2,4-Triazol-3-amine	$C_2H_4N_4$	76.8											
Tribromochloromethane	CBr_3Cl											357.8	89.4
Tribromochlorosilane	Br_3ClSi											377.1	95.3
Tribromofluoromethane	CBr_3F											345.9	84.4
Tribromomethane	$CHBr_3$					-22.3	-5.0	220.9	130.7	23.8	8.0	330.9	71.2
Tribromosilane	Br_3HSi					-355.6	-336.4	248.1		-317.6	-328.5	348.6	80.8
Tributylamine	$C_{12}H_{27}N$					-281.6							
1,3,5-Tri-*tert*-butylbenzene	$C_{18}H_{30}$	-320.0											
Tributyl phosphate	$C_{12}H_{27}O_4P$								379.4				
Trichloroacetaldehyde	C_2HCl_3O					-234.5			151.0	-196.6			
2,2,2-Trichloroacetamide	$C_2H_2Cl_3NO$	-358.0											
Trichloroacetic acid	$C_2HCl_3O_2$	-503.3											
Trichloroacetonitrile	C_2Cl_3N											336.6	96.1
Trichloroacetyl chloride	C_2Cl_4O					-280.8				-239.8			
1,2,3-Trichlorobenzene	$C_6H_3Cl_3$	-70.8								3.8			
1,2,4-Trichlorobenzene	$C_6H_3Cl_3$					-63.1				-8.1			
1,3,5-Trichlorobenzene	$C_6H_3Cl_3$	-78.4								-13.4			
1,1,1-Trichloroethane	$C_2H_3Cl_3$					-177.4		227.4	144.3	-144.4		323.1	93.3
1,1,2-Trichloroethane	$C_2H_3Cl_3$					-190.8		232.6	150.9	-151.3		337.2	89.0
Trichloroethene	C_2HCl_3					-43.6		228.4	124.4	-9.0		324.8	80.3

Name	Mol. form.	$\Delta_f H°$(c)/ kJ mol^{-1}	$\Delta_f G°$(c)/ kJ mol^{-1}	$S°$(c)/ J mol^{-1} K^{-1}	C_p(c)/ J mol^{-1} K^{-1}	$\Delta_f H°$(l)/ kJ mol^{-1}	$\Delta_f G°$(l)/ kJ mol^{-1}	$S°$(l)/ J mol^{-1} K^{-1}	C_p(l)/ J mol^{-1} K^{-1}	$\Delta_f H°$(g)/ kJ mol^{-1}	$\Delta_f G°$(g)/ kJ mol^{-1}	$S°$(g)/ J mol^{-1} K^{-1}	C_p(g)/ J mol^{-1} K^{-1}
Trichlorofluoromethane	CCl_3F					-301.3	-236.8	225.4	121.6	-268.3			78.1
Trichloromethane	$CHCl_3$					-134.1	-73.7	201.7	114.2	-102.7	6.0	295.7	65.7
Trichloromethyl	CCl_3									59.0			
Trichloromethylsilane	CH_3Cl_3Si							262.8	163.1	-528.9		351.1	102.4
1,2,3-Trichloropropane	$C_3H_5Cl_3$					-230.6			183.6	-182.9			
1,2,3-Trichloro-1-propene	$C_3H_3Cl_3$					-101.8							
Trichlorosilane	Cl_3HSi					-539.3	-482.5	227.6		-513.0	-482.0	313.9	75.8
1,1,2-Trichloro-1,2,2-trifluoroethane	$C_2Cl_3F_3$					-745.0			170.1	-716.8			
Tri-*o*-cresyl phosphate	$C_{21}H_{21}O_4P$			570.0	578.0								
Tridecane	$C_{13}H_{28}$								406.7				
Tridecanedioic acid	$C_{13}H_{24}O_4$	-1148.3											
1-Tridecanol	$C_{13}H_{28}O$	-599.4											
1-Tridecene	$C_{13}H_{26}$								391.8				
Triethanolamine	$C_6H_{15}NO_3$	-664.2			389.0					-558.3			
Triethylamine	$C_6H_{15}N$					-127.7			219.9	-92.7			
Triethylborane	$C_6H_{15}B$					-194.6	9.4	336.7	241.2	-157.7	16.1	437.8	
Triethylene glycol	$C_6H_{14}O_4$					-804.3				-725.0			
Trifluoroacetic acid	$C_2HF_3O_2$					-1069.9				-1031.4			
Trifluoroacetonitrile	C_2F_3N									-497.9		298.1	77.9
1,1,1-Trifluoroethane	$C_2H_3F_3$									-744.6		279.9	78.2
1,1,2-Trifluoroethane	$C_2H_3F_3$									-730.7			
2,2,2-Trifluoroethanol	$C_2H_3F_3O$					-932.4				-888.4			
Trifluoroethene	C_2HF_3									-490.5			
1,1,1-Trifluoro-2-iodoethane	$C_2H_2F_3I$									-644.5			
Trifluoroiodomethane	CF_3I									-587.8		307.4	70.9
Trifluoromethane	CHF_3									-695.4		259.7	51.0
Trifluoromethyl	CF_3									-477.0	-464.0	264.5	49.6
(Trifluoromethyl)benzene	$C_7H_5F_3$								188.4				
1,1,1-Trifluoro-2,4-pentanedione	$C_5H_5F_3O_2$					-1040.2				-993.3			
3,3,3-Trifluoro-1-propene	$C_3H_3F_3$									-614.2			
Trifluorosilane	F_3HSi											271.9	60.5
Trigermane	Ge_3H_8					193.7				226.8			
Trihexylamine	$C_{18}H_{39}N$					-433.0							
Trihydro(phosphorus trifluoride) boron	BF_3H_3P									-854.0			
Triiodomethane	CHI_3	-181.1								251.0		356.2	75.0
Trimethyl aluminum	C_3H_9Al					-136.4	-9.9	209.4	155.6	-74.1			
Trimethylamine	C_3H_9N					-45.7		208.5	137.9	-23.6		287.1	91.8
Trimethylamine borane	$C_3H_{12}BN$	-142.5	70.7	187.0									
Trimethylamine hydrochloride	$C_3H_{10}ClN$	-282.9											
1,2,3-Trimethylbenzene	C_9H_{12}					-58.5		267.9	216.4	-9.5			
1,2,4-Trimethylbenzene	C_9H_{12}					-61.8			215.0	-13.8			
1,3,5-Trimethylbenzene	C_9H_{12}					-63.4			209.3	-15.9			
Trimethylborane	C_3H_9B					-143.1	-32.1	238.9		-124.3	-35.9	314.7	88.5
Trimethyl borate	$C_3H_9BO_3$								189.9				
2,2,3-Trimethylbutane	C_7H_{16}					-236.5		292.2	213.5	-204.4			
2,3,3-Trimethyl-1-butene	C_7H_{14}					-117.7				-85.5			
Trimethylchlorosilane	C_3H_9ClSi					-382.8	-246.4	278.2		-352.8	-243.5	369.1	
1α,3α,5β-1,3,5-Trimethylcyclohexane	C_9H_{18}									-212.1			
1,1,2-Trimethylcyclopropane	C_6H_{12}					-96.2							
2,2,3-Trimethylhexane	C_9H_{20}					-282.7							
2,2,4-Trimethylhexane	C_9H_{20}					-282.8							
2,2,5-Trimethylhexane	C_9H_{20}					-293.3							
2,3,3-Trimethylhexane	C_9H_{20}					-281.1							
2,3,5-Trimethylhexane	C_9H_{20}					-284.0				-242.6			
2,4,4-Trimethylhexane	C_9H_{20}					-280.2							
3,3,4-Trimethylhexane	C_9H_{20}					-277.5							
Trimethylolpropane	$C_6H_{14}O_3$	-750.9											
2,2,3-Trimethylpentane	C_8H_{18}					-256.9				-220.0			
2,2,4-Trimethylpentane	C_8H_{18}					-259.2			239.1	-224.0			
2,3,3-Trimethylpentane	C_8H_{18}					-253.5			245.6	-216.3			
2,3,4-Trimethylpentane	C_8H_{18}					-255.0		329.3	247.3	-217.3			
2,2,4-Trimethyl-3-pentanone	$C_8H_{16}O$					-381.6				-338.3			
2,4,4-Trimethyl-1-pentene	C_8H_{16}					-145.9				-110.5			

Name	Mol. form.	$\Delta_f H°$(c)/ kJ mol^{-1}	$\Delta_f G°$(c)/ kJ mol^{-1}	$S°$(c)/ J mol^{-1} K^{-1}	C_p(c)/ J mol^{-1} K^{-1}	$\Delta_f H°$(l)/ kJ mol^{-1}	$\Delta_f G°$(l)/ kJ mol^{-1}	$S°$(l)/ J mol^{-1} K^{-1}	C_p(l)/ J mol^{-1} K^{-1}	$\Delta_f H°$(g)/ kJ mol^{-1}	$\Delta_f G°$(g)/ kJ mol^{-1}	$S°$(g)/ J mol^{-1} K^{-1}	C_p(g)/ J mol^{-1} K^{-1}
2,4,4-Trimethyl-2-pentene	C_8H_{16}					-142.4				-104.9			
Trimethylsilane	$C_3H_{10}Si$											331.0	117.9
Trimethylurea	$C_4H_{10}N_2O$	-330.5											
Trinitroacetonitrile	$C_2N_4O_6$					183.7							
1,3,5-Trinitrobenzene	$C_6H_3N_3O_6$	-37.0			214.6								
2,4,6-Trinitro-1,3-benzenediol	$C_6H_3N_3O_8$	-467.5											
Trinitroglycerol	$C_3H_5N_3O_9$					-370.9				-279.1		545.9	234.2
Trinitromethane	CHN_3O_6					-32.8				-13.4		435.6	134.1
2,4,6-Trinitrophenol	$C_6H_3N_3O_7$	-217.9			239.7								
2,4,6-Trinitrotoluene	$C_7H_5N_3O_6$	-63.2			243.3								
Trioctylamine	$C_{24}H_{51}N$					-585.0							
1,3,5-Trioxane	$C_3H_6O_3$	-522.5		133.0	111.4					-465.9			
Triphenylamine	$C_{18}H_{15}N$	234.7								326.8			
Triphenylmethanol	$C_{19}H_{16}O$	-2.5											
Triphenyl phosphate	$C_{18}H_{15}O_4P$			397.5	356.2								
Triphenylphosphine	$C_{18}H_{15}P$				312.5								
Tripropylamine	$C_9H_{21}N$					-207.1				-161.0			
Tris(hydroxymethyl)methylamine	$C_4H_{11}NO_3$	-717.8											
Trisilane	H_8Si_3					92.5				120.9			
Tris(perfluorobutyl)amine	$C_{12}F_{27}N$								418.4				
1,3,5-Trithiane	$C_3H_6S_3$									80.0	130.4	336.4	111.3
Trithiocarbonic acid	CH_2S_3					24.0							
L-Tryptophan	$C_{11}H_{12}N_2O_2$	-415.3		251.0	238.1								
Tungsten	W	0.0		32.6	24.3					849.4	807.1	174.0	21.3
Tungsten(VI) bromide	Br_6W	-348.5											
Tungsten(VI) chloride	Cl_6W	-602.5								-513.8			
Tungsten(VI) fluoride	F_6W					-1747.7	-1631.4	251.5		-1721.7	-1632.1	341.1	119.0
Tungsten(IV) oxide	O_2W	-589.7	-533.9	50.5	56.1								
Tungsten(VI) oxide	O_3W	-842.9	-764.0	75.9	73.8								
L-Tyrosine	$C_9H_{11}NO_3$	-685.1		214.0	216.4								
Undecane	$C_{11}H_{24}$					-327.2			344.9	-270.8			
1-Undecanol	$C_{11}H_{24}O$					-504.8							
1-Undecene	$C_{11}H_{22}$								344.9				
Uracil	$C_4H_4N_2O_2$	-429.4			120.5					-302.9			
Uranium	U	0.0		50.2	27.7					533.0	488.4	199.8	23.7
Uranium(III) chloride	Cl_3U	-866.5	-799.1	159.0	102.5								
Uranium(IV) chloride	Cl_4U	-1019.2	-930.0	197.1	122.0					-809.6	-786.6	419.0	
Uranium(VI) chloride	Cl_6U	-1092.0	-962.0	285.8	175.7					-1013.0	-928.0	431.0	
Uranium(III) fluoride	F_3U	-1502.1	-1433.4	123.4	95.1					-1058.5	-1051.9	331.9	74.3
Uranium(IV) fluoride	F_4U	-1914.2	-1823.3	151.7	116.0					-1598.7	-1572.7	368.0	91.2
Uranium(VI) fluoride	F_6U	-2197.0	-2068.5	227.6	166.8					-2147.4	-2063.7	377.9	129.6
Uranium(III) hydride	H_3U	-127.2	-72.8	63.7	49.3								
Uranium(II) oxide	OU									21.0			
Uranium(IV) oxide	O_2U	-1085.0	-1031.8	77.0	63.6					-465.7	-471.5	274.6	51.4
Uranium(IV,V) oxide	O_9U_4	-4510.4	-4275.1	334.1	293.3								
Uranium(IV,VI) oxide	O_7U_3	-3427.1	-3242.9	250.5	215.5								
Uranium(V,VI) oxide	O_8U_3	-3574.8	-3369.5	282.6	238.4								
Uranium(VI) oxide	O_3U	-1223.8	-1145.7	96.1	81.7								
Uranyl chloride	Cl_2O_2U	-1243.9	-1146.4	150.5	107.9								
Uranyl fluoride	F_2O_2U	-1653.5	-1557.4	135.6	103.2								
Urea	CH_4N_2O	-333.1								-245.8			
Uric acid	$C_5H_4N_4O_3$	-618.8		173.2	166.1								
DL-Valine	$C_5H_{11}NO_2$	-628.9											
L-Valine	$C_5H_{11}NO_2$	-617.9								-455.1			
Vanadium	V	0.0		28.9	24.9					514.2	754.4	182.3	26.0
Vanadium(IV) bromide	Br_4V									-336.8			
Vanadium(III) chloride	Cl_3V	-580.7	-511.2	131.0	93.2								
Vanadium(IV) chloride	Cl_4V					-569.4	-503.7	255.0		-525.5	-492.0	362.4	96.2
Vanadium(IV) fluoride	F_4V	-1403.3											
Vanadium(V) fluoride	F_5V					-1480.3	-1373.1	175.7		-1433.9	-1369.8	320.9	98.6
Vanadium(IV) iodide	I_4V									-122.6			
Vanadium(II) oxide	OV	-431.8	-404.2	38.9	45.4								
Vanadium(III) oxide	O_3V_2	-1218.8	-1139.3	98.3	103.2								
Vanadium(III,IV) oxide	O_5V_3	-1933.0	-1803.0	163.0									
Vanadium(V) oxide	O_5V_2	-1550.6	-1419.5	131.0	127.7								
Vanadyl chloride	ClOV	-607.0	-556.0	75.0									

Name	Mol. form.	$\Delta_f H°$(c)/ kJ mol^{-1}	$\Delta_f G°$(c)/ kJ mol^{-1}	$S°$(c)/ J mol^{-1} K^{-1}	C_p(c)/ J mol^{-1} K^{-1}	$\Delta_f H°$(l)/ kJ mol^{-1}	$\Delta_f G°$(l)/ kJ mol^{-1}	$S°$(l)/ J mol^{-1} K^{-1}	C_p(l)/ J mol^{-1} K^{-1}	$\Delta_f H°$(g)/ kJ mol^{-1}	$\Delta_f G°$(g)/ kJ mol^{-1}	$S°$(g)/ J mol^{-1} K^{-1}	C_p(g)/ J mol^{-1} K^{-1}
Vanadyl trichloride	Cl_3OV					-734.7	-668.5	244.3		-695.6	-659.3	344.3	89.9
Vinyl acetate	$C_4H_6O_2$					-349.2				-314.4			
Vinylcyclopentane	C_7H_{12}					-34.8							
2-Vinylfuran	C_6H_6O					-10.3				27.8			
Water	H_2O					-285.8	-237.1	70.0	75.3	-241.8	-228.6	188.8	33.6
Xanthine	$C_5H_4N_4O_2$	-379.6		161.1	151.3								
Xanthone	$C_{13}H_8O_2$	-191.5											
Xenon	Xe									0.0		169.7	20.8
Xenon tetrafluoride	F_4Xe	-261.5											
o-Xylene	C_8H_{10}					-24.4			186.1	19.1			
m-Xylene	C_8H_{10}					-25.4			183.0	17.3			
p-Xylene	C_8H_{10}					-24.4			181.5	18.0			
2,3-Xylenol	$C_8H_{10}O$	-241.1								-157.2			
2,4-Xylenol	$C_8H_{10}O$					-228.7				-163.8			
2,5-Xylenol	$C_8H_{10}O$	-246.6								-161.6			
2,6-Xylenol	$C_8H_{10}O$	-237.4								-162.1			
3,4-Xylenol	$C_8H_{10}O$	-242.3								-157.3			
3,5-Xylenol	$C_8H_{10}O$	-244.4								-162.4			
Xylitol	$C_5H_{12}O_5$	-1118.5											
D-Xylose	$C_5H_{10}O_5$	-1057.8											
Ytterbium	Yb	0.0		59.9	26.7					152.3	118.4	173.1	20.8
Ytterbium(III) chloride	Cl_3Yb	-959.8											
Ytterbium(III) oxide	O_3Yb_2	-1814.6	-1726.7	133.1	115.4								
Yttrium	Y	0.0		44.4	26.5					421.3	381.1	179.5	25.9
Yttrium chloride	Cl_3Y	-1000.0								-750.2			75.0
Yttrium fluoride	F_3Y	-1718.8	-1644.7	100.0						-1288.7	-1277.8	311.8	70.3
Yttrium oxide	O_3Y_2	-1905.3	-1816.6	99.1	102.5								
Zinc	Zn	0.0		41.6	25.4					130.4	94.8	161.0	20.8
Zinc bromide	Br_2Zn	-328.7	-312.1	138.5									
Zinc carbonate	CO_3Zn	-812.8	-731.5	82.4	79.7								
Zinc chloride	Cl_2Zn	-415.1	-369.4	111.5	71.3					-266.1			
Zinc fluoride	F_2Zn	-764.4	-713.3	73.7	65.7								
Zinc hydroxide	H_2O_2Zn	-641.9	-553.5	81.2									
Zinc iodide	I_2Zn	-208.0	-209.0	161.1									
Zinc nitrate	N_2O_6Zn	-483.7											
Zinc orthosilicate	O_4SiZn_2	-1636.7	-1523.2	131.4	123.3								
Zinc oxide	OZn	-350.5	-320.5	43.7	40.3								
Zinc selenide	SeZn	-163.0	-163.0	84.0									
Zinc sulfate	O_4SZn	-982.8	-871.5	110.5	99.2								
Zinc sulfide (sphalerite)	SZn	-206.0	-201.3	57.7	46.0								
Zinc sulfide (wurtzite)	SZn	-192.6											
Zirconium	Zr	0.0		39.0	25.4					608.8	566.5	181.4	26.7
Zirconium(IV) bromide	Br_4Zr	-760.7											
Zirconium(II) chloride	Cl_2Zr	-502.0											
Zirconium(IV) chloride	Cl_4Zr	-980.5	-889.9	181.6	119.8								
Zirconium(IV) fluoride	F_4Zr	-1911.3	-1809.9	104.6	103.7								
Zirconium(II) hydride	H_2Zr	-169.0	-128.8	35.0	31.0								
Zirconium(IV) iodide	I_4Zr	-481.6											
Zirconium(IV) orthosilicate	O_4SiZr	-2033.4	-1919.1	84.1	98.7								
Zirconium(IV) oxide	O_2Zr	-1100.6	-1042.8	50.4	56.2								
Zirconium(IV) sulfate	O_8S_2Zr	-2217.1			172.0								
Zirconium titanate	O_4TiZr	-2024.1	-1915.8	116.7	114.0								

THERMODYNAMIC PROPERTIES AS A FUNCTION OF TEMPERATURE

L. V. Gurvich, V. S. Iorish, V. S. Yungman, and O. V. Dorofeeva

The thermodynamic properties $C^\circ_p(T)$, $S^\circ(T)$, $H^\circ(T)$-$H^\circ(T_r)$, -$[G^\circ(T)$-$H^\circ(T_r)]/T$ and formation properties $\Delta_f H^\circ(T)$, $\Delta_f G^\circ(T)$, log $K_f^\circ(T)$ are tabulated as functions of temperature in the range 298.15 to 1500 K for 80 substances in the standard state. The reference temperature, T_r, is equal to 298.15 K. The standard state pressure is taken as 1 bar (100,000 Pa). The tables are presented in the JANAF Thermochemical Tables format (Ref. 2). The numerical data are extracted from IVTANTHERMO databases except for C_2H_4O, C_3H_6O, C_6H_6, C_6H_6O, $C_{10}H_8$, and CH_5N, which are based upon TRC Tables. See the references for information on standard states and other details.

References

1. Gurvich, L. V., Veyts, I. V., and Alcock, C. B., Eds., *Thermodynamic Properties of Individual Substances, 4th ed.*, Hemisphere Publishing Corp., New York, 1989.
2. Chase, M. W., *NIST-JANAF Thermochemical Tables, Fourth Edition*, *J. Phys. Chem. Ref. Data*, Monograph No. 9, 1998.

Order of Listing of Tables

No.	Mol. form.	Name	State
1	Ar	Argon	g
2	Br	Bromine (atomic)	g
3	Br_2	Bromine	g
4	BrH	Hydrogen bromide	g
5	C	Carbon (graphite)	cr
6	C	Carbon (diamond)	cr
7	C_2	Dicarbon	g
8	C_3	Carbon trimer	g
9	CO	Carbon monoxide	g
10	CO_2	Carbon dioxide	g
11	CH_4	Methane	g
12	C_2H_2	Acetylene	g
13	C_2H_4	Ethylene	g
14	C_2H_6	Ethane	g
15	C_3H_6	Cyclopropane	g
16	C_3H_8	Propane	g
17	C_6H_6	Benzene	l
18	C_6H_6	Benzene	g
19	$C_{10}H_8$	Naphthalene	cr, l
20	$C_{10}H_8$	Naphthalene	g
21	CH_2O	Formaldehyde	g
22	CH_4O	Methanol	g
23	C_2H_4O	Acetaldehyde	g
24	C_2H_6O	Ethanol	g
25	$C_2H_4O_2$	Acetic acid	g
26	C_3H_6O	Acetone	g
27	C_6H_6O	Phenol	g
28	CF_4	Tetrafluoromethane	g
29	CHF_3	Trifluoromethane	g
30	$CClF_3$	Chlorotrifluoromethane	g
31	CCl_2F_2	Dichlorodifluoromethane	g
32	$CHClF_2$	Chlorodifluoromethane	g
33	CH_5N	Methylamine	g
34	Cl	Chlorine (atomic)	g
35	Cl_2	Chlorine	g
36	ClH	Hydrogen chloride	g
37	Cu	Copper	cr, l
38	Cu	Copper	g
39	CuO	Copper(II) oxide	cr
40	Cu_2O	Copper(I) oxide	cr
41	Cl_2Cu	Copper(II) chloride	cr, l
42	Cl_2Cu	Copper(II) chloride	g
43	F	Fluorine (atomic)	g
44	F_2	Fluorine	g
45	FH	Hydrogen fluoride	g
46	Ge	Germanium	cr, l
47	Ge	Germanium	g
48	GeO_2	Germanium(IV) oxide	cr, l
49	Cl_4Ge	Germanium(IV) chloride	g
50	H	Hydrogen (atomic)	g
51	H_2	Hydrogen	g
52	HO	Hydroxyl	g
53	H_2O	Water	l
54	H_2O	Water	g
55	I	Iodine (atomic)	g
56	I_2	Iodine	cr, l
57	I_2	Iodine	g
58	HI	Hydrogen iodide	g
59	K	Potassium	cr, l
60	K	Potassium	g
61	K_2O	Potassium oxide	cr, l
62	HKO	Potassium hydroxide	cr, l
63	HKO	Potassium hydroxide	g
64	ClK	Potassium chloride	cr, l
65	ClK	Potassium chloride	g
66	N_2	Nitrogen	g
67	NO	Nitric oxide	g
68	NO_2	Nitrogen dioxide	g
69	H_3N	Ammonia	g
70	O	Oxygen	g
71	O_2	Oxygen	g
72	S	Sulfur (rhombic)	cr, l
73	S	Sulfur (rhombic)	g
74	S_2	Disulfur	g
75	S_8	Sulfur (orthorhombic)	g
76	O_2S	Sulfur dioxide	g
77	Si	Silicon	cr
78	Si	Silicon	g
79	O_2Si	Silicon dioxide (α-quartz)	cr
80	Cl_4Si	Tetrachlorosilane	g

Standard State Thermodynamic Properties

T/K	$C_p°$/ J K⁻¹mol⁻¹	*S*°/ J K⁻¹ mol⁻¹	-(*G*°-*H*°(*T*))/ J K⁻¹ mol⁻¹	(*H*-*H*°(*T*))/ kJ mol⁻¹	$\Delta_f H°$/ kJ mol⁻¹	$\Delta_f G°$/ kJ mol⁻¹	Log K_f
1. ARGON Ar (g)							
298.15	20.786	154.845	154.845	0.000	0.000	0.000	0.000
300	20.786	154.973	154.845	0.038	0.000	0.000	0.000
400	20.786	160.953	155.660	2.117	0.000	0.000	0.000
500	20.786	165.591	157.200	4.196	0.000	0.000	0.000
600	20.786	169.381	158.924	6.274	0.000	0.000	0.000
700	20.786	172.585	160.653	8.353	0.000	0.000	0.000
800	20.786	175.361	162.322	10.431	0.000	0.000	0.000
900	20.786	177.809	163.909	12.510	0.000	0.000	0.000
1000	20.786	179.999	165.410	14.589	0.000	0.000	0.000
1100	20.786	181.980	166.828	16.667	0.000	0.000	0.000
1200	20.786	183.789	168.167	18.746	0.000	0.000	0.000
1300	20.786	185.453	169.434	20.824	0.000	0.000	0.000
1400	20.786	186.993	170.634	22.903	0.000	0.000	0.000
1500	20.786	188.427	171.773	24.982	0.000	0.000	0.000
2. BROMINE Br (g)							
298.15	20.786	175.017	175.017	0.000	111.870	82.379	-14.432
300	20.786	175.146	175.018	0.038	111.838	82.196	-14.311
400	20.787	181.126	175.833	2.117	96.677	75.460	-9.854
500	20.798	185.765	177.373	4.196	96.910	70.129	-7.326
600	20.833	189.559	179.097	6.277	97.131	64.752	-5.637
700	20.908	192.776	180.827	8.364	97.348	59.338	-4.428
800	21.027	195.575	182.499	10.461	97.568	53.893	-3.519
900	21.184	198.061	184.093	12.571	97.796	48.420	-2.810
1000	21.365	200.302	185.604	14.698	98.036	42.921	-2.242
1100	21.559	202.347	187.034	16.844	98.291	37.397	-1.776
1200	21.752	204.231	188.390	19.010	98.560	31.850	-1.386
1300	21.937	205.980	189.676	21.195	98.844	26.279	-1.056
1400	22.107	207.612	190.900	23.397	99.141	20.686	-0.772
1500	22.258	209.142	192.065	25.615	99.449	15.072	-0.525
3. DIBROMINE Br_2 (g)							
298.15	36.057	245.467	245.467	0.000	30.910	3.105	-0.544
300	36.074	245.690	245.468	0.067	30.836	2.933	-0.511
332.25[a]	36.340	249.387	245.671	1.235			
400	36.729	256.169	246.892	3.711	0.000	0.000	0.000
500	37.082	264.406	249.600	7.403	0.000	0.000	0.000
600	37.305	271.188	252.650	11.123	0.000	0.000	0.000
700	37.464	276.951	255.720	14.862	0.000	0.000	0.000
800	37.590	281.962	258.694	18.615	0.000	0.000	0.000
900	37.697	286.396	261.530	22.379	0.000	0.000	0.000
1000	37.793	290.373	264.219	26.154	0.000	0.000	0.000
1100	37.883	293.979	266.763	29.938	0.000	0.000	0.000
1200	37.970	297.279	269.170	33.730	0.000	0.000	0.000
1300	38.060	300.322	271.451	37.532	0.000	0.000	0.000
1400	38.158	303.146	273.615	41.343	0.000	0.000	0.000
1500	38.264	305.782	275.673	45.164	0.000	0.000	0.000
4. HYDROGEN BROMIDE HBr (g)							
298.15	29.141	198.697	198.697	0.000	-36.290	-53.360	9.348
300	29.141	198.878	198.698	0.054	-36.333	-53.466	9.309
400	29.220	207.269	199.842	2.971	-52.109	-55.940	7.305
500	29.454	213.811	202.005	5.903	-52.484	-56.854	5.939
600	29.872	219.216	204.436	8.868	-52.844	-57.694	5.023
700	30.431	223.861	206.886	11.882	-53.168	-58.476	4.363
800	31.063	227.965	209.269	14.957	-53.446	-59.214	3.866
900	31.709	231.661	211.555	18.095	-53.677	-59.921	3.478
1000	32.335	235.035	213.737	21.298	-53.864	-60.604	3.166

T/K	$C_p°$/ J K^{-1}mol^{-1}	$S°$/ J K^{-1} mol^{-1}	-($G°$-$H°(T)$)/ J K^{-1} mol^{-1}	(H-$H°(T)$)/ kJ mol^{-1}	$\Delta_f H°$/ kJ mol^{-1}	$\Delta_f G°$/ kJ mol^{-1}	Log K_f
1100	32.919	238.145	215.816	24.561	-54.012	-61.271	2.909
1200	33.454	241.032	217.799	27.880	-54.129	-61.925	2.696
1300	33.938	243.729	219.691	31.250	-54.220	-62.571	2.514
1400	34.374	246.261	221.499	34.666	-54.291	-63.211	2.358
1500	34.766	248.646	223.230	38.123	-54.348	-63.846	2.223
5. CARBON (GRAPHITE) C (cr; graphite)							
298.15	8.536	5.740	5.740	0.000	0.000	0.000	0.000
300	8.610	5.793	5.740	0.016	0.000	0.000	0.000
400	11.974	8.757	6.122	1.054	0.000	0.000	0.000
500	14.537	11.715	6.946	2.385	0.000	0.000	0.000
600	16.607	14.555	7.979	3.945	0.000	0.000	0.000
700	18.306	17.247	9.113	5.694	0.000	0.000	0.000
800	19.699	19.785	10.290	7.596	0.000	0.000	0.000
900	20.832	22.173	11.479	9.625	0.000	0.000	0.000
1000	21.739	24.417	12.662	11.755	0.000	0.000	0.000
1100	22.452	26.524	13.827	13.966	0.000	0.000	0.000
1200	23.000	28.502	14.968	16.240	0.000	0.000	0.000
1300	23.409	30.360	16.082	18.562	0.000	0.000	0.000
1400	23.707	32.106	17.164	20.918	0.000	0.000	0.000
1500	23.919	33.749	18.216	23.300	0.000	0.000	0.000
6. CARBON (DIAMOND) C (cr; diamond)							
298.15	6.109	2.362	2.362	0.000	1.850	2.857	-0.501
300	6.201	2.400	2.362	0.011	1.846	2.863	-0.499
400	10.321	4.783	2.659	0.850	1.645	3.235	-0.422
500	13.404	7.431	3.347	2.042	1.507	3.649	-0.381
600	15.885	10.102	4.251	3.511	1.415	4.087	-0.356
700	17.930	12.709	5.274	5.205	1.361	4.537	-0.339
800	19.619	15.217	6.361	7.085	1.338	4.993	-0.326
900	21.006	17.611	7.479	9.118	1.343	5.450	-0.316
1000	22.129	19.884	8.607	11.277	1.372	5.905	-0.308
1100	23.020	22.037	9.731	13.536	1.420	6.356	-0.302
1200	23.709	24.071	10.842	15.874	1.484	6.802	-0.296
1300	24.222	25.990	11.934	18.272	1.561	7.242	-0.291
1400	24.585	27.799	13.003	20.714	1.646	7.675	-0.286
1500	24.824	29.504	14.047	23.185	1.735	8.103	-0.282
7. DICARBON C_2 (g)							
298.15	43.548	197.095	197.095	0.000	830.457	775.116	-135.795
300	43.575	197.365	197.096	0.081	830.506	774.772	-134.898
400	42.169	209.809	198.802	4.403	832.751	755.833	-98.700
500	39.529	218.924	201.959	8.483	834.170	736.423	-76.933
600	37.837	225.966	205.395	12.342	834.909	716.795	-62.402
700	36.984	231.726	208.758	16.078	835.148	697.085	-52.016
800	36.621	236.637	211.943	19.755	835.020	677.366	-44.227
900	36.524	240.943	214.931	23.411	834.618	657.681	-38.170
1000	36.569	244.793	217.728	27.065	834.012	638.052	-33.328
1100	36.696	248.284	220.349	30.728	833.252	618.492	-29.369
1200	36.874	251.484	222.812	34.406	832.383	599.006	-26.074
1300	37.089	254.444	225.133	38.104	831.437	579.596	-23.288
1400	37.329	257.201	227.326	41.824	830.445	560.261	-20.903
1500	37.589	259.785	229.405	45.570	829.427	540.997	-18.839
8. TRICARBON C_3 (g)							
298.15	42.202	237.611	237.611	0.000	839.958	774.249	-135.643
300	42.218	237.872	237.611	0.078	839.989	773.841	-134.736
400	43.383	250.164	239.280	4.354	841.149	751.592	-98.147
500	44.883	260.003	242.471	8.766	841.570	729.141	-76.172
600	46.406	268.322	246.104	13.331	841.453	706.659	-61.519
700	47.796	275.582	249.807	18.042	840.919	684.230	-51.057
800	48.997	282.045	253.440	22.884	840.053	661.901	-43.217

T/K	$C_p°$/ J K⁻¹mol⁻¹	$S°$/ J K⁻¹ mol⁻¹	$-(G°-H°(T))$/ J K⁻¹ mol⁻¹	$(H-H°(T))$/ kJ mol⁻¹	$\Delta_f H°$/ kJ mol⁻¹	$\Delta_f G°$/ kJ mol⁻¹	Log K_f
900	50.006	287.876	256.948	27.835	838.919	639.698	-37.127
1000	50.844	293.189	260.310	32.879	837.572	617.633	-32.261
1100	51.535	298.069	263.524	37.999	836.059	595.711	-28.288
1200	52.106	302.578	266.593	43.182	834.420	573.933	-24.982
1300	52.579	306.768	269.524	48.417	832.690	552.295	-22.191
1400	52.974	310.679	272.326	53.695	830.899	530.793	-19.804
1500	53.307	314.346	275.006	59.010	829.068	509.421	-17.739

9. CARBON OXIDE CO (g)

298.15	29.141	197.658	197.658	0.000	-110.530	-137.168	24.031
300	29.142	197.838	197.659	0.054	-110.519	-137.333	23.912
400	29.340	206.243	198.803	2.976	-110.121	-146.341	19.110
500	29.792	212.834	200.973	5.930	-110.027	-155.412	16.236
600	30.440	218.321	203.419	8.941	-110.157	-164.480	14.319
700	31.170	223.067	205.895	12.021	-110.453	-173.513	12.948
800	31.898	227.277	208.309	15.175	-110.870	-182.494	11.915
900	32.573	231.074	210.631	18.399	-111.378	-191.417	11.109
1000	33.178	234.538	212.851	21.687	-111.952	-200.281	10.461
1100	33.709	237.726	214.969	25.032	-112.573	-209.084	9.928
1200	34.169	240.679	216.990	28.426	-113.228	-217.829	9.482
1300	34.568	243.430	218.920	31.864	-113.904	-226.518	9.101
1400	34.914	246.005	220.763	35.338	-114.594	-235.155	8.774
1500	35.213	248.424	222.527	38.845	-115.291	-243.742	8.488

10. CARBON DIOXIDE CO_2 (g)

298.15	37.135	213.783	213.783	0.000	-393.510	-394.373	69.092
300	37.220	214.013	213.784	0.069	-393.511	-394.379	68.667
400	41.328	225.305	215.296	4.004	-393.586	-394.656	51.536
500	44.627	234.895	218.280	8.307	-393.672	-394.914	41.256
600	47.327	243.278	221.762	12.909	-393.791	-395.152	34.401
700	49.569	250.747	225.379	17.758	-393.946	-395.367	29.502
800	51.442	257.492	228.978	22.811	-394.133	-395.558	25.827
900	53.008	263.644	232.493	28.036	-394.343	-395.724	22.967
1000	54.320	269.299	235.895	33.404	-394.568	-395.865	20.678
1100	55.423	274.529	239.172	38.893	-394.801	-395.984	18.803
1200	56.354	279.393	242.324	44.483	-395.035	-396.081	17.241
1300	57.144	283.936	245.352	50.159	-395.265	-396.159	15.918
1400	57.818	288.196	248.261	55.908	-395.488	-396.219	14.783
1500	58.397	292.205	251.059	61.719	-395.702	-396.264	13.799

11. METHANE CH_4 (g)

298.15	35.695	186.369	186.369	0.000	-74.600	-50.530	8.853
300	35.765	186.590	186.370	0.066	-74.656	-50.381	8.772
400	40.631	197.501	187.825	3.871	-77.703	-41.827	5.462
500	46.627	207.202	190.744	8.229	-80.520	-32.525	3.398
600	52.742	216.246	194.248	13.199	-82.969	-22.690	1.975
700	58.603	224.821	198.008	18.769	-85.023	-12.476	0.931
800	64.084	233.008	201.875	24.907	-86.693	-1.993	0.130
900	69.137	240.852	205.773	31.571	-88.006	8.677	-0.504
1000	73.746	248.379	209.660	38.719	-88.996	19.475	-1.017
1100	77.919	255.607	213.511	46.306	-89.698	30.358	-1.442
1200	81.682	262.551	217.310	54.289	-90.145	41.294	-1.797
1300	85.067	269.225	221.048	62.630	-90.367	52.258	-2.100
1400	88.112	275.643	224.720	71.291	-90.390	63.231	-2.359
1500	90.856	281.817	228.322	80.242	-90.237	74.200	-2.584

12. ACETYLENE C_2H_2 (g)

298.15	44.036	200.927	200.927	0.000	227.400	209.879	-36.769
300	44.174	201.199	200.927	0.082	227.397	209.770	-36.524
400	50.388	214.814	202.741	4.829	227.161	203.928	-26.630
500	54.751	226.552	206.357	10.097	226.846	198.154	-20.701
600	58.121	236.842	210.598	15.747	226.445	192.452	-16.754

T/K	$C_p°$/ J K⁻¹mol⁻¹	S°/ J K⁻¹ mol⁻¹	-(G°-H°(T))/ J K⁻¹ mol⁻¹	(H-H°(T))/ kJ mol⁻¹	$\Delta_f H°$/ kJ mol⁻¹	$\Delta_f G°$/ kJ mol⁻¹	Log K_f
700	60.970	246.021	215.014	21.704	225.968	186.823	-13.941
800	63.511	254.331	219.418	27.931	225.436	181.267	-11.835
900	65.831	261.947	223.726	34.399	224.873	175.779	-10.202
1000	67.960	268.995	227.905	41.090	224.300	170.355	-8.898
1100	69.909	275.565	231.942	47.985	223.734	164.988	-7.835
1200	71.686	281.725	235.837	55.067	223.189	159.672	-6.950
1300	73.299	287.528	239.592	62.317	222.676	154.400	-6.204
1400	74.758	293.014	243.214	69.721	222.203	149.166	-5.565
1500	76.077	298.218	246.709	77.264	221.774	143.964	-5.013
13. ETHYLENE C_2H_4 (g)							
298.15	42.883	219.316	219.316	0.000	52.400	68.358	-11.976
300	43.059	219.582	219.317	0.079	52.341	68.457	-11.919
400	53.045	233.327	221.124	4.881	49.254	74.302	-9.703
500	62.479	246.198	224.864	10.667	46.533	80.887	-8.450
600	70.673	258.332	229.441	17.335	44.221	87.982	-7.659
700	77.733	269.770	234.393	24.764	42.278	95.434	-7.121
800	83.868	280.559	239.496	32.851	40.655	103.142	-6.734
900	89.234	290.754	244.630	41.512	39.310	111.036	-6.444
1000	93.939	300.405	249.730	50.675	38.205	119.067	-6.219
1100	98.061	309.556	254.756	60.280	37.310	127.198	-6.040
1200	101.670	318.247	259.688	70.271	36.596	135.402	-5.894
1300	104.829	326.512	264.513	80.599	36.041	143.660	-5.772
1400	107.594	334.384	269.225	91.223	35.623	151.955	-5.669
1500	110.018	341.892	273.821	102.107	35.327	160.275	-5.581
14. ETHANE C_2H_6 (g)							
298.15	52.487	229.161	229.161	0.000	-84.000	-32.015	5.609
300	52.711	229.487	229.162	0.097	-84.094	-31.692	5.518
400	65.459	246.378	231.379	5.999	-88.988	-13.473	1.759
500	77.941	262.344	235.989	13.177	-93.238	5.912	-0.618
600	89.188	277.568	241.660	21.545	-96.779	26.086	-2.271
700	99.136	292.080	247.835	30.972	-99.663	46.800	-3.492
800	107.936	305.904	254.236	41.334	-101.963	67.887	-4.433
900	115.709	319.075	260.715	52.525	-103.754	89.231	-5.179
1000	122.552	331.628	267.183	64.445	-105.105	110.750	-5.785
1100	128.553	343.597	273.590	77.007	-106.082	132.385	-6.286
1200	133.804	355.012	279.904	90.131	-106.741	154.096	-6.708
1300	138.391	365.908	286.103	103.746	-107.131	175.850	-7.066
1400	142.399	376.314	292.178	117.790	-107.292	197.625	-7.373
1500	145.905	386.260	298.121	132.209	-107.260	219.404	-7.640
15. CYCLOPROPANE C_3H_6 (g)							
298.15	55.571	237.488	237.488	0.000	53.300	104.514	-18.310
300	55.941	237.832	237.489	0.103	53.195	104.832	-18.253
400	76.052	256.695	239.924	6.708	47.967	122.857	-16.043
500	93.859	275.637	245.177	15.230	43.730	142.091	-14.844
600	108.542	294.092	251.801	25.374	40.405	162.089	-14.111
700	120.682	311.763	259.115	36.854	37.825	182.583	-13.624
800	130.910	328.564	266.755	49.447	35.854	203.404	-13.281
900	139.658	344.501	274.516	62.987	34.384	224.441	-13.026
1000	147.207	359.616	282.277	77.339	33.334	245.618	-12.830
1100	153.749	373.961	289.965	92.395	32.640	266.883	-12.673
1200	159.432	387.588	297.538	108.060	32.249	288.197	-12.545
1300	164.378	400.549	304.967	124.257	32.119	309.533	-12.437
1400	168.689	412.892	312.239	140.915	32.215	330.870	-12.345
1500	172.453	424.662	319.344	157.976	32.507	352.193	-12.264
16. PROPANE C_3H_8 (g)							
298.15	73.597	270.313	270.313	0.000	-103.847	-23.458	4.110
300	73.931	270.769	270.314	0.136	-103.972	-22.959	3.997
400	94.014	294.739	273.447	8.517	-110.33	15.029	-0.657

T/K	$C_p°$/ J K^{-1}mol^{-1}	$S°$/ J K^{-1} mol^{-1}	$-(G°-H°(T))$/ J K^{-1} mol^{-1}	$(H-H°(T))$/ kJ mol^{-1}	$\Delta_fH°$/ kJ mol^{-1}	$\Delta_fG°$/ kJ mol^{-1}	Log K_f
500	112.591	317.768	280.025	18.872	-115.658	34.507	-3.605
600	128.700	339.753	288.162	30.955	-119.973	64.961	-5.655
700	142.674	360.668	297.039	44.540	-123.384	96.065	-7.168
800	154.766	380.528	306.245	59.427	-126.016	127.603	-8.331
900	165.352	399.381	315.555	75.444	-127.982	159.430	-9.253
1000	174.598	417.293	324.841	92.452	-129.380	191.444	-10.000
1100	182.673	434.321	334.026	110.325	-130.296	223.574	-10.617
1200	189.745	450.526	343.064	128.954	-130.802	255.770	-11.133
1300	195.853	465.961	351.929	148.241	-130.961	287.993	-11.572
1400	201.209	480.675	360.604	168.100	-130.829	320.217	-11.947
1500	205.895	494.721	369.080	188.460	-130.445	352.422	-12.272
17. BENZENE C_6H_6 (l)							
298.15	135.950	173.450	173.450	0.000	49.080	124.521	-21.815
300	136.312	174.292	173.453	0.252	49.077	124.989	-21.762
400	161.793	216.837	179.082	15.102	48.978	150.320	-19.630
500	207.599	257.048	190.639	33.204	50.330	175.559	-18.340
18. BENZENE C_6H_6 (g)							
298.15	82.430	269.190	269.190	0.000	82.880	129.750	-22.731
300	83.020	269.700	269.190	0.153	82.780	130.040	-22.641
400	113.510	297.840	272.823	10.007	77.780	146.570	-19.140
500	139.340	326.050	280.658	22.696	73.740	164.260	-17.160
600	160.090	353.360	290.517	37.706	70.490	182.680	-15.903
700	176.790	379.330	301.360	54.579	67.910	201.590	-15.042
800	190.460	403.860	312.658	72.962	65.910	220.820	-14.418
900	201.840	426.970	324.084	92.597	64.410	240.280	-13.945
1000	211.430	448.740	335.473	113.267	63.340	259.890	-13.575
1100	219.580	469.280	346.710	134.827	62.620	277.640	-13.184
1200	226.540	488.690	357.743	157.137	62.200	299.320	-13.029
1300	232.520	507.070	368.534	180.097	62.000	319.090	-12.821
1400	237.680	524.490	379.056	203.607	61.990	338.870	-12.643
1500	242.140	541.040	389.302	227.607	62.110	358.640	-12.489
19. NAPHTHALENE $C_{10}H_8$ (cr, l)							
298.15	165.720	167.390	167.390	0.000	78.530	201.585	-35.316
300	167.001	168.419	167.393	0.308	78.466	202.349	-35.232
353.43	208.722	198.948	169.833	10.290	96.099	224.543	-33.186
Phase transition: $\Delta_{trs}H$ = 18.980 kJ/mol, $\Delta_{trs}S$ = 53.702 J/K·mol, cr–l							
353.43	217.200	252.650	169.833	29.270	96.099	224.543	-33.186
400	241.577	280.916	181.124	39.917	96.067	241.475	-31.533
470	276.409	322.712	199.114	58.091	97.012	266.859	-29.658
20. NAPHTHALENE $C_{10}H_8$ (g)							
298.15	131.920	333.150	333.150	0.000	150.580	224.100	-39.260
300	132.840	333.970	333.157	0.244	150.450	224.560	-39.098
400	180.070	378.800	338.950	15.940	144.190	250.270	-32.681
500	219.740	423.400	351.400	36.000	139.220	277.340	-28.973
600	251.530	466.380	367.007	59.624	135.350	305.330	-26.581
700	277.010	507.140	384.146	86.096	132.330	333.950	-24.919
800	297.730	545.520	401.935	114.868	130.050	362.920	-23.696
900	314.850	581.610	419.918	145.523	128.430	392.150	-22.759
1000	329.170	615.550	437.806	177.744	127.510	421.700	-22.027
1100	341.240	647.500	455.426	211.281	127.100	450.630	-21.398
1200	351.500	677.650	472.707	245.932	126.960	480.450	-20.913
1300	360.260	706.130	489.568	281.531	127.060	509.770	-20.482
1400	367.780	733.110	506.009	317.941	127.390	539.740	-20.137
1500	374.270	758.720	522.019	355.051	127.920	568.940	-19.812
21. FORMALDEHYDE H_2CO (g)							
298.15	35.387	218.760	218.760	0.000	-108.700	-102.667	17.987
300	35.443	218.979	218.761	0.066	-108.731	-102.630	17.869

T/K	$C_p°$/ J K⁻¹mol⁻¹	$S°$/ J K⁻¹ mol⁻¹	$-(G°-H°(T))$/ J K⁻¹ mol⁻¹	$(H-H°(T))$/ kJ mol⁻¹	$\Delta_f H°$/ kJ mol⁻¹	$\Delta_f G°$/ kJ mol⁻¹	Log K_f
400	39.240	229.665	220.192	3.789	-110.438	-100.340	13.103
500	43.736	238.900	223.028	7.936	-112.073	-97.623	10.198
600	48.181	247.270	226.381	12.534	-113.545	-94.592	8.235
700	52.280	255.011	229.924	17.560	-114.833	-91.328	6.815
800	55.941	262.236	233.517	22.975	-115.942	-87.893	5.739
900	59.156	269.014	237.088	28.734	-116.889	-84.328	4.894
1000	61.951	275.395	240.603	34.792	-117.696	-80.666	4.213
1100	64.368	281.416	244.042	41.111	-118.382	-76.929	3.653
1200	66.453	287.108	247.396	47.655	-118.966	-73.134	3.183
1300	68.251	292.500	250.660	54.392	-119.463	-69.294	2.784
1400	69.803	297.616	253.833	61.297	-119.887	-65.418	2.441
1500	71.146	302.479	256.915	68.346	-120.249	-61.514	2.142
22. METHANOL CH_3OH (g)							
298.15	44.101	239.865	239.865	0.000	-201.000	-162.298	28.434
300	44.219	240.139	239.866	0.082	-201.068	-162.057	28.216
400	51.713	253.845	241.685	4.864	-204.622	-148.509	19.393
500	59.800	266.257	245.374	10.442	-207.750	-134.109	14.010
600	67.294	277.835	249.830	16.803	-210.387	-119.125	10.371
700	73.958	288.719	254.616	23.873	-212.570	-103.737	7.741
800	79.838	298.987	259.526	31.569	-214.350	-88.063	5.750
900	85.025	308.696	264.455	39.817	-215.782	-72.188	4.190
1000	89.597	317.896	269.343	48.553	-216.916	-56.170	2.934
1100	93.624	326.629	274.158	57.718	-217.794	-40.050	1.902
1200	97.165	334.930	278.879	67.262	-218.457	-23.861	1.039
1300	100.277	342.833	283.497	77.137	-218.936	-7.624	0.306
1400	103.014	350.367	288.007	87.304	-219.261	8.644	-0.322
1500	105.422	357.558	292.405	97.729	-219.456	24.930	-0.868
23. ACETALDEHYDE C_2H_4O (g)							
298.15	55.318	263.840	263.840	0.000	-166.190	-133.010	23.302
300	55.510	264.180	263.837	0.103	-166.250	-132.800	23.122
400	66.282	281.620	266.147	6.189	-169.530	-121.130	15.818
500	76.675	297.540	270.850	13.345	-172.420	-108.700	11.356
600	85.942	312.360	276.550	21.486	-174.870	-95.720	8.334
700	94.035	326.230	282.667	30.494	-176.910	-82.350	6.145
800	101.070	339.260	288.938	40.258	-178.570	-68.730	4.487
900	107.190	351.520	295.189	50.698	-179.880	-54.920	3.187
1000	112.490	363.100	301.431	61.669	-180.850	-40.930	2.138
1100	117.080	374.040	307.537	73.153	-181.560	-27.010	1.283
1200	121.060	384.400	313.512	85.065	-182.070	-12.860	0.560
1300	124.500	394.230	319.350	97.344	-182.420	1.240	-0.050
1400	127.490	403.570	325.031	109.954	-182.640	15.470	-0.577
1500	130.090	412.460	330.571	122.834	-182.750	29.580	-1.030
24. ETHANOL C_2H_5OH (g)							
298.15	65.652	281.622	281.622	0.000	-234.800	-167.874	29.410
300	65.926	282.029	281.623	0.122	-234.897	-167.458	29.157
400	81.169	303.076	284.390	7.474	-239.826	-144.216	18.832
500	95.400	322.750	290.115	16.318	-243.940	-119.820	12.517
600	107.656	341.257	297.112	26.487	-247.260	-94.672	8.242
700	118.129	358.659	304.674	37.790	-249.895	-69.023	5.151
800	127.171	375.038	312.456	50.065	-251.951	-43.038	2.810
900	135.049	390.482	320.276	63.185	-253.515	-16.825	0.976
1000	141.934	405.075	328.033	77.042	-254.662	9.539	-0.498
1100	147.958	418.892	335.670	91.543	-255.454	36.000	-1.709
1200	153.232	431.997	343.156	106.609	-255.947	62.520	-2.721
1300	157.849	444.448	350.473	122.168	-256.184	89.070	-3.579
1400	161.896	456.298	357.612	138.160	-256.206	115.630	-4.314
1500	165.447	467.591	364.571	154.531	-256.044	142.185	-4.951

T/K	C_p°/ J K⁻¹mol⁻¹	S°/ J K⁻¹ mol⁻¹	-(G°-H°(T))/ J K⁻¹ mol⁻¹	(H-H°(T))/ kJ mol⁻¹	$\Delta_f H^\circ$/ kJ mol⁻¹	$\Delta_f G^\circ$/ kJ mol⁻¹	Log K_f
25. ACETIC ACID $C_2H_4O_2$ (g)							
298.15	63.438	283.470	283.470	0.000	-432.249	-374.254	65.567
300	63.739	283.863	283.471	0.118	-432.324	-373.893	65.100
400	79.665	304.404	286.164	7.296	-436.006	-353.840	46.206
500	93.926	323.751	291.765	15.993	-438.875	-332.950	34.783
600	106.181	341.988	298.631	26.014	-440.993	-311.554	27.123
700	116.627	359.162	306.064	37.169	-442.466	-289.856	21.629
800	125.501	375.331	313.722	49.287	-443.395	-267.985	17.497
900	132.989	390.558	321.422	62.223	-443.873	-246.026	14.279
1000	139.257	404.904	329.060	75.844	-443.982	-224.034	11.702
1100	144.462	418.429	336.576	90.039	-443.798	-202.046	9.594
1200	148.760	431.189	343.933	104.707	-443.385	-180.086	7.839
1300	152.302	443.240	351.113	119.765	-442.795	-158.167	6.355
1400	155.220	454.637	358.105	135.146	-442.071	-136.299	5.085
1500	157.631	465.432	364.903	150.793	-441.247	-114.486	3.987
26. ACETONE C_3H_6O (g)							
298.15	74.517	295.349	295.349	0.000	-217.150	-152.716	26.757
300	74.810	295.809	295.349	0.138	-217.233	-152.339	26.521
400	91.755	319.658	298.498	8.464	-222.212	-129.913	16.962
500	107.864	341.916	304.988	18.464	-226.522	-106.315	11.107
600	122.047	362.836	312.873	29.978	-230.120	-81.923	7.133
700	134.306	382.627	321.470	42.810	-233.049	-56.986	4.252
800	144.934	401.246	330.265	56.785	-235.350	-31.673	2.068
900	154.097	418.860	339.141	71.747	-237.149	-6.109	0.353
1000	162.046	435.513	347.950	87.563	-238.404	19.707	-1.030
1100	168.908	451.286	356.617	104.136	-239.283	45.396	-2.157
1200	174.891	466.265	365.155	121.332	-239.827	71.463	-3.110
1300	180.079	480.491	373.513	139.072	-240.120	97.362	-3.912
1400	184.556	493.963	381.596	157.314	-240.203	123.470	-4.607
1500	188.447	506.850	389.533	175.975	-240.120	149.369	-5.202
27. PHENOL C_6H_6O (g)							
298.15	103.220	314.810	314.810	0.000	-96.400	-32.630	5.720
300	103.860	315.450	314.810	0.192	-96.490	-32.230	5.610
400	135.790	349.820	319.278	12.217	-100.870	-10.180	1.330
500	161.910	383.040	328.736	27.152	-104.240	12.970	-1.360
600	182.480	414.450	340.430	44.412	-106.810	36.650	-3.190
700	198.840	443.860	353.134	63.508	-108.800	60.750	-4.530
800	212.140	471.310	366.211	84.079	-110.300	85.020	-5.550
900	223.190	496.950	379.327	105.861	-111.370	109.590	-6.360
1000	232.490	520.960	392.302	128.658	-111.990	134.280	-7.010
1100	240.410	543.500	405.033	152.314	-112.280	158.620	-7.530
1200	247.200	564.720	417.468	176.703	-112.390	183.350	-7.980
1300	253.060	584.740	429.568	201.723	-112.330	208.070	-8.360
1400	258.120	603.680	441.331	227.288	-112.120	233.050	-8.700
1500	262.520	621.650	452.767	253.325	-111.780	257.540	-8.970
28. CARBON TETRAFLUORIDE CF_4 (g)							
298.15	61.050	261.455	261.455	0.000	-933.200	-888.518	155.663
300	61.284	261.833	261.456	0.113	-933.219	-888.240	154.654
400	72.399	281.057	264.001	6.822	-933.986	-873.120	114.016
500	80.713	298.153	269.155	14.499	-934.372	-857.852	89.618
600	86.783	313.434	275.284	22.890	-934.490	-842.533	73.348
700	91.212	327.162	281.732	31.801	-934.431	-827.210	61.726
800	94.479	339.566	288.199	41.094	-934.261	-811.903	53.011
900	96.929	350.842	294.542	50.670	-934.024	-796.622	46.234
1000	98.798	361.156	300.695	60.460	-933.745	-781.369	40.814
1100	100.250	370.643	306.629	70.416	-933.442	-766.146	36.381
1200	101.396	379.417	312.334	80.500	-933.125	-750.952	32.688
1300	102.314	387.571	317.811	90.687	-932.800	-735.784	29.564

T/K	$C_p°$/ J K⁻¹mol⁻¹	S°/ J K⁻¹ mol⁻¹	-(G°-H°(T))/ J K⁻¹ mol⁻¹	(H-H°(T))/ kJ mol⁻¹	$\Delta_f H°$/ kJ mol⁻¹	$\Delta_f G°$/ kJ mol⁻¹	Log K_f
1400	103.059	395.181	323.069	100.957	-932.470	-720.641	26.887
1500	103.671	402.313	328.116	111.295	-932.137	-705.522	24.568

29. TRIFLUOROMETHANE CHF_3 (g)

T/K	$C_p°$/ J K⁻¹mol⁻¹	S°/ J K⁻¹ mol⁻¹	-(G°-H°(T))/ J K⁻¹ mol⁻¹	(H-H°(T))/ kJ mol⁻¹	$\Delta_f H°$/ kJ mol⁻¹	$\Delta_f G°$/ kJ mol⁻¹	Log K_f
298.15	51.069	259.675	259.675	0.000	-696.700	-662.237	116.020
300	51.258	259.991	259.676	0.095	-696.735	-662.023	115.267
400	61.148	276.113	261.807	5.722	-698.427	-650.186	84.905
500	69.631	290.700	266.149	12.275	-699.715	-637.969	66.647
600	76.453	304.022	271.368	19.593	-700.634	-625.528	54.456
700	81.868	316.230	276.917	27.519	-701.253	-612.957	45.739
800	86.201	327.455	282.542	35.930	-701.636	-600.315	39.196
900	89.719	337.818	288.116	44.732	-701.832	-587.636	34.105
1000	92.617	347.426	293.572	53.854	-701.879	-574.944	30.032
1100	95.038	356.370	298.879	63.240	-701.805	-562.253	26.699
1200	97.084	364.730	304.022	72.849	-701.629	-549.574	23.922
1300	98.833	372.571	308.997	82.647	-701.368	-536.913	21.573
1400	100.344	379.952	313.804	92.607	-701.033	-524.274	19.561
1500	101.660	386.921	318.449	102.709	-700.635	-511.662	17.817

30. CHLOROTRIFLUOROMETHANE $CClF_3$ (g)

T/K	$C_p°$/ J K⁻¹mol⁻¹	S°/ J K⁻¹ mol⁻¹	-(G°-H°(T))/ J K⁻¹ mol⁻¹	(H-H°(T))/ kJ mol⁻¹	$\Delta_f H°$/ kJ mol⁻¹	$\Delta_f G°$/ kJ mol⁻¹	Log K_f
298.15	66.886	285.419	285.419	0.000	-707.800	-667.238	116.896
300	67.111	285.834	285.421	0.124	-707.810	-666.986	116.131
400	77.528	306.646	288.187	7.383	-708.153	-653.316	85.313
500	85.013	324.797	293.734	15.532	-708.170	-639.599	66.818
600	90.329	340.794	300.271	24.314	-707.975	-625.901	54.489
700	94.132	355.020	307.096	33.547	-707.654	-612.246	45.686
800	96.899	367.780	313.897	43.106	-707.264	-598.642	39.087
900	98.951	379.317	320.536	52.903	-706.837	-585.090	33.957
1000	100.507	389.827	326.947	62.880	-706.396	-571.586	29.856
1100	101.708	399.465	333.108	72.993	-705.950	-558.126	26.503
1200	102.651	408.357	339.013	83.213	-705.505	-544.707	23.710
1300	103.404	416.604	344.668	93.517	-705.064	-531.326	21.349
1400	104.012	424.290	350.084	103.889	-704.628	-517.977	19.326
1500	104.512	431.484	355.273	114.316	-704.196	-504.660	17.574

31. DICHLORODIFLUOROMETHANE CCl_2F_2 (g)

T/K	$C_p°$/ J K⁻¹mol⁻¹	S°/ J K⁻¹ mol⁻¹	-(G°-H°(T))/ J K⁻¹ mol⁻¹	(H-H°(T))/ kJ mol⁻¹	$\Delta_f H°$/ kJ mol⁻¹	$\Delta_f G°$/ kJ mol⁻¹	Log K_f
298.15	72.476	300.903	300.903	0.000	-486.000	-447.030	78.317
300	72.691	301.352	300.905	0.134	-486.002	-446.788	77.792
400	82.408	323.682	303.883	7.919	-485.945	-433.716	56.637
500	89.063	342.833	309.804	16.514	-485.618	-420.692	43.949
600	93.635	359.500	316.729	25.663	-485.136	-407.751	35.497
700	96.832	374.189	323.909	35.196	-484.576	-394.897	29.467
800	99.121	387.276	331.027	44.999	-483.984	-382.126	24.950
900	100.801	399.053	337.942	55.000	-483.388	-369.429	21.441
1000	102.062	409.742	344.596	65.146	-482.800	-356.799	18.637
1100	103.030	419.517	350.969	75.402	-482.226	-344.227	16.346
1200	103.786	428.515	357.061	85.745	-481.667	-331.706	14.439
1300	104.388	436.847	362.882	96.154	-481.121	-319.232	12.827
1400	104.874	444.602	368.445	106.618	-480.588	-306.799	11.447
1500	105.270	451.851	373.767	117.126	-480.065	-294.404	10.252

32. CHLORODIFLUOROMETHANE $CHClF_2$ (g)

T/K	$C_p°$/ J K⁻¹mol⁻¹	S°/ J K⁻¹ mol⁻¹	-(G°-H°(T))/ J K⁻¹ mol⁻¹	(H-H°(T))/ kJ mol⁻¹	$\Delta_f H°$/ kJ mol⁻¹	$\Delta_f G°$/ kJ mol⁻¹	Log K_f
298.15	55.853	280.915	280.915	0.000	-475.000	-443.845	77.759
300	56.039	281.261	280.916	0.104	-475.028	-443.652	77.246
400	65.395	298.701	283.231	6.188	-476.390	-432.978	56.540
500	73.008	314.145	287.898	13.123	-477.398	-422.001	44.086
600	78.940	328.003	293.448	20.733	-478.103	-410.851	35.767
700	83.551	340.533	299.294	28.867	-478.574	-399.603	29.818
800	87.185	351.936	305.172	37.411	-478.870	-388.299	25.353
900	90.100	362.379	310.956	46.280	-479.031	-376.967	21.878
1000	92.475	371.999	316.586	55.413	-479.090	-365.622	19.098
1100	94.433	380.908	322.033	64.761	-479.068	-354.276	16.823

T/K	$C_p°$/ J K^{-1}mol^{-1}	$S°$/ J K^{-1} mol^{-1}	-($G°$-$H°$(T))/ J K^{-1} mol^{-1}	(H-$H°$(T))/ kJ mol^{-1}	$\Delta_f H°$/ kJ mol^{-1}	$\Delta_f G°$/ kJ mol^{-1}	Log K_f
1200	96.066	389.196	327.289	74.289	-478.982	-342.935	14.927
1300	97.438	396.941	332.352	83.966	-478.843	-331.603	13.324
1400	98.601	404.206	337.228	93.769	-478.661	-320.283	11.950
1500	99.593	411.044	341.923	103.681	-478.443	-308.978	10.759
33. METHYLAMINE CH_5N (g)							
298.15	50.053	242.881	242.881	0.000	-22.529	32.734	-5.735
300	50.227	243.196	242.893	0.091	-22.614	33.077	-5.759
400	60.171	258.986	244.975	5.604	-26.846	52.294	-6.829
500	70.057	273.486	249.244	12.121	-30.431	72.510	-7.575
600	78.929	287.063	254.431	19.579	-33.364	93.382	-8.129
700	86.711	299.826	260.008	27.873	-35.712	114.702	-8.559
800	93.545	311.865	265.749	36.893	-37.548	136.316	-8.900
900	99.573	323.239	271.511	46.555	-38.949	158.138	-9.178
1000	104.886	334.006	277.220	56.786	-39.967	180.098	-9.407
1100	109.576	344.233	282.861	67.509	-40.681	201.822	-9.584
1200	113.708	353.944	288.374	78.685	-41.136	224.240	-9.761
1300	117.341	363.190	293.775	90.239	-41.376	246.364	-9.899
1400	120.542	372.012	299.061	102.131	-41.451	268.504	-10.018
1500	123.353	380.426	304.209	114.326	-41.381	290.639	-10.121
34. CHLORINE Cl (g)							
298.15	21.838	165.190	165.190	0.000	121.302	105.306	-18.449
300	21.852	165.325	165.190	0.040	121.311	105.207	-18.318
400	22.467	171.703	166.055	2.259	121.795	99.766	-13.028
500	22.744	176.752	167.708	4.522	122.272	94.203	-9.841
600	22.781	180.905	169.571	6.800	122.734	88.546	-7.709
700	22.692	184.411	171.448	9.074	123.172	82.813	-6.179
800	22.549	187.432	173.261	11.337	123.585	77.019	-5.029
900	22.389	190.079	174.986	13.584	123.971	71.175	-4.131
1000	22.233	192.430	176.615	15.815	124.334	65.289	-3.410
1100	22.089	194.542	178.150	18.031	124.675	59.368	-2.819
1200	21.959	196.458	179.597	20.233	124.996	53.416	-2.325
1300	21.843	198.211	180.963	22.423	125.299	47.439	-1.906
1400	21.742	199.826	182.253	24.602	125.587	41.439	-1.546
1500	21.652	201.323	183.475	26.772	125.861	35.418	-1.233
35. DICHLORINE Cl_2 (g)							
298.15	33.949	223.079	223.079	0.000	0.000	0.000	0.000
300	33.981	223.290	223.080	0.063	0.000	0.000	0.000
400	35.296	233.263	224.431	3.533	0.000	0.000	0.000
500	36.064	241.229	227.021	7.104	0.000	0.000	0.000
600	36.547	247.850	229.956	10.736	0.000	0.000	0.000
700	36.874	253.510	232.926	14.408	0.000	0.000	0.000
800	37.111	258.450	235.815	18.108	0.000	0.000	0.000
900	37.294	262.832	238.578	21.829	0.000	0.000	0.000
1000	37.442	266.769	241.203	25.566	0.000	0.000	0.000
1100	37.567	270.343	243.692	29.316	0.000	0.000	0.000
1200	37.678	273.617	246.052	33.079	0.000	0.000	0.000
1300	37.778	276.637	248.290	36.851	0.000	0.000	0.000
1400	37.872	279.440	250.416	40.634	0.000	0.000	0.000
1500	37.961	282.056	252.439	44.426	0.000	0.000	0.000
36. HYDROGEN CHLORIDE HCl (g)							
298.15	29.136	186.902	186.902	0.000	-92.310	-95.298	16.696
300	29.137	187.082	186.902	0.054	-92.314	-95.317	16.596
400	29.175	195.468	188.045	2.969	-92.587	-96.278	12.573
500	29.304	201.990	190.206	5.892	-92.911	-97.164	10.151
600	29.576	207.354	192.630	8.835	-93.249	-97.983	8.530
700	29.988	211.943	195.069	11.812	-93.577	-98.746	7.368
800	30.500	215.980	197.435	14.836	-93.879	-99.464	6.494
900	31.063	219.604	199.700	17.913	-94.149	-100.145	5.812

T/K	$C_p°$/ J K^{-1}mol^{-1}	$S°$/ J K^{-1} mol^{-1}	-($G°$-$H°(T)$)/ J K^{-1} mol^{-1}	(H-$H°(T)$)/ kJ mol^{-1}	$\Delta_f H°$/ kJ mol^{-1}	$\Delta_f G°$/ kJ mol^{-1}	Log K_f
1000	31.639	222.907	201.858	21.049	-94.384	-100.798	5.265
1100	32.201	225.949	203.912	24.241	-94.587	-101.430	4.816
1200	32.734	228.774	205.867	27.488	-94.760	-102.044	4.442
1300	33.229	231.414	207.732	30.786	-94.908	-102.645	4.124
1400	33.684	233.893	209.513	34.132	-95.035	-103.235	3.852
1500	34.100	236.232	211.217	37.522	-95.146	-103.817	3.615

37. COPPER Cu (cr, l)

298.15	24.440	33.150	33.150	0.000	0.000	0.000	0.000
300	24.460	33.301	33.150	0.045	0.000	0.000	0.000
400	25.339	40.467	34.122	2.538	0.000	0.000	0.000
500	25.966	46.192	35.982	5.105	0.000	0.000	0.000
600	26.479	50.973	38.093	7.728	0.000	0.000	0.000
700	26.953	55.090	40.234	10.399	0.000	0.000	0.000
800	27.448	58.721	42.322	13.119	0.000	0.000	0.000
900	28.014	61.986	44.328	15.891	0.000	0.000	0.000
1000	28.700	64.971	46.245	18.726	0.000	0.000	0.000
1100	29.553	67.745	48.075	21.637	0.000	0.000	0.000
1200	30.617	70.361	49.824	24.644	0.000	0.000	0.000
1300	31.940	72.862	51.501	27.769	0.000	0.000	0.000
1358	32.844	74.275	52.443	29.647	0.000	0.000	0.000

Phase transition: $\Delta_{trs}H$ = 13.141 kJ/mol, $\Delta_{trs}S$ = 9.676 J/K·mol, cr–l

1358	32.800	83.951	52.443	42.788	0.000	0.000	0.000
1400	32.800	84.950	53.403	44.166	0.000	0.000	0.000
1500	32.800	87.213	55.583	47.446	0.000	0.000	0.000

38. COPPER Cu (g)

298.15	20.786	166.397	166.397	0.000	337.600	297.873	-52.185
300	20.786	166.525	166.397	0.038	337.594	297.626	-51.821
400	20.786	172.505	167.213	2.117	337.179	284.364	-37.134
500	20.786	177.143	168.752	4.196	336.691	271.215	-28.333
600	20.786	180.933	170.476	6.274	336.147	258.170	-22.475
700	20.786	184.137	172.205	8.353	335.554	245.221	-18.298
800	20.786	186.913	173.874	10.431	334.913	232.359	-15.171
900	20.786	189.361	175.461	12.510	334.219	219.581	-12.744
1000	20.786	191.551	176.963	14.589	333.463	206.883	-10.806
1100	20.788	193.532	178.380	16.667	332.631	194.265	-9.225
1200	20.793	195.341	179.719	18.746	331.703	181.726	-7.910
1300	20.803	197.006	180.986	20.826	330.657	169.270	-6.801
1400	20.823	198.548	182.186	22.907	316.342	157.305	-5.869
1500	20.856	199.986	183.325	24.991	315.146	145.987	-5.084

39. COPPER OXIDE CuO (cr)

298.15	42.300	42.740	42.740	0.000	-162.000	-134.277	23.524
300	42.417	43.002	42.741	0.078	-161.994	-134.105	23.349
400	46.783	55.878	44.467	4.564	-161.487	-124.876	16.307
500	49.190	66.596	47.852	9.372	-160.775	-115.803	12.098
600	50.827	75.717	51.755	14.377	-159.973	-106.883	9.305
700	52.099	83.651	55.757	19.526	-159.124	-98.102	7.320
800	53.178	90.680	59.691	24.791	-158.247	-89.444	5.840
900	54.144	97.000	63.491	30.158	-157.356	-80.897	4.695
1000	55.040	102.751	67.134	35.617	-156.462	-72.450	3.784
1100	55.890	108.037	70.615	41.164	-155.582	-64.091	3.043
1200	56.709	112.936	73.941	46.794	-154.733	-55.812	2.429
1300	57.507	117.507	77.118	52.505	-153.940	-47.601	1.913
1400	58.288	121.797	80.158	58.295	-166.354	-39.043	1.457
1500	59.057	125.845	83.070	64.163	-165.589	-29.975	1.044

40. DICOPPER OXIDE Cu_2O (cr)

298.15	62.600	92.550	92.550	0.000	-173.100	-150.344	26.339
300	62.721	92.938	92.551	0.116	-173.102	-150.203	26.152

T/K	C_p°/ J K^{-1}mol^{-1}	S°/ J K^{-1} mol^{-1}	$-(G^\circ$-$H^\circ(T))$/ J K^{-1} mol^{-1}	$(H$-$H^\circ(T))$/ kJ mol^{-1}	Δ_fH°/ kJ mol^{-1}	Δ_fG°/ kJ mol^{-1}	Log K_f
400	67.587	111.712	95.078	6.654	-173.036	-142.572	18.618
500	70.784	127.155	99.995	13.580	-172.772	-134.984	14.101
600	73.323	140.291	105.643	20.789	-172.389	-127.460	11.096
700	75.552	151.764	111.429	28.235	-171.914	-120.009	8.955
800	77.616	161.989	117.121	35.894	-171.363	-112.631	7.354
900	79.584	171.245	122.629	43.755	-170.750	-105.325	6.113
1000	81.492	179.729	127.920	51.809	-170.097	-98.091	5.124
1100	83.360	187.584	132.992	60.052	-169.431	-90.922	4.317
1200	85.202	194.917	137.850	68.480	-168.791	-83.814	3.648
1300	87.026	201.808	142.507	77.092	-168.223	-76.756	3.084
1400	88.836	208.324	146.978	85.885	-194.030	-68.926	2.572
1500	90.636	214.515	151.276	94.858	-193.438	-60.010	2.090

41. COPPER DICHLORIDE $CuCl_2$ (cr, l)

298.15	71.880	108.070	108.070	0.000	-218.000	-173.826	30.453
300	71.998	108.515	108.071	0.133	-217.975	-173.552	30.218
400	76.338	129.899	110.957	7.577	-216.494	-158.962	20.758
500	78.654	147.204	116.532	15.336	-214.873	-144.765	15.123
600	80.175	161.687	122.884	23.282	-213.182	-130.901	11.396
675	81.056	171.183	127.732	29.329	-211.185	-120.693	9.340

Phase transition: $\Delta_{trs}H$ = 0.700 kJ/mol, $\Delta_{trs}S$ = 1.037 J/K·mol, crII–crI

675	82.400	172.220	127.732	30.029	-211.185	-120.693	9.340
700	82.400	175.216	129.375	32.089	-210.719	-117.350	8.757
800	82.400	186.219	135.808	40.329	-208.898	-104.137	6.799
871	82.400	193.226	140.207	46.179	-192.649	-94.893	5.691

Phase transition: $\Delta_{trs}H$ = 15.001 kJ/mol, $\Delta_{trs}S$ = 17.221 J/K·mol, crI–l

871	100.000	210.447	140.207	61.180	-192.649	-94.893	5.691
900	100.000	213.723	142.523	64.080	-191.640	-91.655	5.319
1000	100.000	224.259	150.179	74.080	-188.212	-80.730	4.217
1100	100.000	233.790	157.353	84.080	-184.873	-70.144	3.331
1130.75	100.000	236.547	159.470	87.155	-183.867	-66.951	3.093

42. COPPER DICHLORIDE $CuCl_2$ (g)

298.15	56.814	278.418	278.418	0.000	-43.268	-49.883	8.739
300	56.869	278.769	278.419	0.105	-43.271	-49.924	8.692
400	58.992	295.456	280.679	5.911	-43.428	-52.119	6.806
500	60.111	308.752	285.010	11.871	-43.606	-54.271	5.670
600	60.761	319.774	289.911	17.918	-43.814	-56.385	4.909
700	61.168	329.173	294.865	24.015	-44.060	-58.462	4.362
800	61.439	337.360	299.677	30.147	-44.349	-60.500	3.950
900	61.630	344.608	304.274	36.301	-44.688	-62.499	3.627
1000	61.776	351.109	308.638	42.471	-45.088	-64.457	3.367
1100	61.900	357.003	312.771	48.655	-45.566	-66.372	3.152
1200	62.022	362.394	316.685	54.851	-46.139	-68.239	2.970
1300	62.159	367.364	320.395	61.060	-46.829	-70.053	2.815
1400	62.325	371.976	323.916	67.284	-60.784	-71.404	2.664
1500	62.531	376.283	327.265	73.526	-61.613	-72.133	2.512

43. FLUORINE F (g)

298.15	22.746	158.750	158.750	0.000	79.380	62.280	-10.911
300	22.742	158.891	158.750	0.042	79.393	62.173	-10.825
400	22.432	165.394	159.639	2.302	80.043	56.332	-7.356
500	22.100	170.363	161.307	4.528	80.587	50.340	-5.259
600	21.832	174.368	163.161	6.724	81.046	44.246	-3.852
700	21.629	177.717	165.008	8.897	81.442	38.081	-2.842
800	21.475	180.595	166.780	11.052	81.792	31.862	-2.080
900	21.357	183.117	168.458	13.193	82.106	25.601	-1.486
1000	21.266	185.362	170.039	15.324	82.391	19.308	-1.009
1100	21.194	187.386	171.525	17.447	82.654	12.986	-0.617
1200	21.137	189.227	172.925	19.563	82.897	6.642	-0.289

T/K	$C_p°$/ J K⁻¹mol⁻¹	S°/ J K⁻¹ mol⁻¹	-(G°-H°(T))/ J K⁻¹ mol⁻¹	(H-H°(T))/ kJ mol⁻¹	$\Delta_fH°$/ kJ mol⁻¹	$\Delta_fG°$/ kJ mol⁻¹	Log K_f
1300	21.091	190.917	174.245	21.675	83.123	0.278	-0.011
1400	21.054	192.479	175.492	23.782	83.335	-6.103	0.228
1500	21.022	193.930	176.673	25.886	83.533	-12.498	0.435
44. DIFLUORINE F_2 (g)							
298.15	31.304	202.790	202.790	0.000	0.000	0.000	0.000
300	31.337	202.984	202.790	0.058	0.000	0.000	0.000
400	32.995	212.233	204.040	3.277	0.000	0.000	0.000
500	34.258	219.739	206.453	6.643	0.000	0.000	0.000
600	35.171	226.070	209.208	10.117	0.000	0.000	0.000
700	35.839	231.545	212.017	13.669	0.000	0.000	0.000
800	36.343	236.365	214.765	17.279	0.000	0.000	0.000
900	36.740	240.669	217.409	20.934	0.000	0.000	0.000
1000	37.065	244.557	219.932	24.625	0.000	0.000	0.000
1100	37.342	248.103	222.334	28.346	0.000	0.000	0.000
1200	37.588	251.363	224.619	32.093	0.000	0.000	0.000
1300	37.811	254.381	226.794	35.863	0.000	0.000	0.000
1400	38.019	257.191	228.866	39.654	0.000	0.000	0.000
1500	38.214	259.820	230.843	43.466	0.000	0.000	0.000
45. HYDROGEN FLUORIDE HF (g)							
298.15	29.137	173.776	173.776	0.000	-273.300	-275.399	48.248
300	29.137	173.956	173.776	0.054	-273.302	-275.412	47.953
400	29.149	182.340	174.919	2.968	-273.450	-276.096	36.054
500	29.172	188.846	177.078	5.884	-273.679	-276.733	28.910
600	29.230	194.169	179.496	8.804	-273.961	-277.318	24.142
700	29.350	198.683	181.923	11.732	-274.277	-277.852	20.733
800	29.549	202.614	184.269	14.676	-274.614	-278.340	18.174
900	29.827	206.110	186.505	17.645	-274.961	-278.785	16.180
1000	30.169	209.270	188.626	20.644	-275.309	-279.191	14.583
1100	30.558	212.163	190.636	23.680	-275.652	-279.563	13.275
1200	30.974	214.840	192.543	26.756	-275.988	-279.904	12.184
1300	31.403	217.336	194.355	29.875	-276.315	-280.217	11.259
1400	31.831	219.679	196.081	33.037	-276.631	-280.505	10.466
1500	32.250	221.889	197.729	36.241	-276.937	-280.771	9.777
46. GERMANIUM Ge (cr, l)							
298.15	23.222	31.090	31.090	0.000	0.000	0.000	0.000
300	23.249	31.234	31.090	0.043	0.000	0.000	0.000
400	24.310	38.083	32.017	2.426	0.000	0.000	0.000
500	24.962	43.582	33.798	4.892	0.000	0.000	0.000
600	25.452	48.178	35.822	7.414	0.000	0.000	0.000
700	25.867	52.133	37.876	9.980	0.000	0.000	0.000
800	26.240	55.612	39.880	12.586	0.000	0.000	0.000
900	26.591	58.723	41.804	15.227	0.000	0.000	0.000
1000	26.926	61.542	43.639	17.903	0.000	0.000	0.000
1100	27.252	64.124	45.386	20.612	0.000	0.000	0.000
1200	27.571	66.509	47.048	23.353	0.000	0.000	0.000
1211.4	27.608	66.770	47.232	23.668	0.000	0.000	0.000
Phase transition: $\Delta_{trs}H$ = 37.030 kJ/mol, $\Delta_{trs}S$ = 30.568 J/K·mol, cr–l							
1211.4	27.600	97.338	47.232	60.698	0.000	0.000	0.000
1300	27.600	99.286	50.714	63.143	0.000	0.000	0.000
1400	27.600	101.331	54.258	65.903	0.000	0.000	0.000
1500	27.600	103.236	57.460	68.663	0.000	0.000	0.000
47. GERMANIUM Ge (g)							
298.15	30.733	167.903	167.903	0.000	367.800	327.009	-57.290
300	30.757	168.094	167.904	0.057	367.814	326.756	-56.893
400	31.071	177.025	169.119	3.162	368.536	312.959	-40.868
500	30.360	183.893	171.415	6.239	369.147	298.991	-31.235
600	29.265	189.334	173.965	9.222	369.608	284.914	-24.804

T/K	$C_p°$/ J K^{-1}mol^{-1}	$S°$/ J K^{-1} mol^{-1}	-($G°$-$H°(T)$)/ J K^{-1} mol^{-1}	(H-$H°(T)$)/ kJ mol^{-1}	$\Delta_f H°$/ kJ mol^{-1}	$\Delta_f G°$/ kJ mol^{-1}	Log K_f
700	28.102	193.758	176.487	12.090	369.910	270.773	-20.205
800	27.029	197.439	178.882	14.845	370.060	256.598	-16.754
900	26.108	200.567	181.122	17.501	370.073	242.414	-14.069
1000	25.349	203.277	183.205	20.072	369.969	228.234	-11.922
1100	24.741	205.664	185.141	22.575	369.763	214.069	-10.165
1200	24.264	207.795	186.941	25.025	369.471	199.928	-8.703
1300	23.898	209.722	188.621	27.432	332.088	188.521	-7.575
1400	23.624	211.483	190.192	29.807	331.704	177.492	-6.622
1500	23.426	213.105	191.666	32.159	331.296	166.491	-5.798

48. GERMANIUM DIOXIDE GeO_2 (cr, l)

T/K	$C_p°$/ J K^{-1}mol^{-1}	$S°$/ J K^{-1} mol^{-1}	-($G°$-$H°(T)$)/ J K^{-1} mol^{-1}	(H-$H°(T)$)/ kJ mol^{-1}	$\Delta_f H°$/ kJ mol^{-1}	$\Delta_f G°$/ kJ mol^{-1}	Log K_f
298.15	50.166	39.710	39.710	0.000	-580.200	-521.605	91.382
300	50.475	40.021	39.711	0.093	-580.204	-521.242	90.755
400	61.281	56.248	41.850	5.759	-579.893	-501.610	65.503
500	66.273	70.519	46.191	12.164	-579.013	-482.134	50.368
600	69.089	82.872	51.299	18.943	-577.915	-462.859	40.295
700	70.974	93.671	56.597	25.952	-576.729	-443.776	33.115
800	72.449	103.247	61.841	33.125	-575.498	-424.866	27.741
900	73.764	111.857	66.928	40.436	-574.235	-406.113	23.570
1000	75.049	119.696	71.819	47.877	-572.934	-387.502	20.241
1100	76.378	126.910	76.504	55.447	-571.582	-369.024	17.523
1200	77.796	133.616	80.987	63.155	-570.166	-350.671	15.264
1300	79.332	139.903	85.279	71.010	-605.685	-329.732	13.249
1308	79.460	140.390	85.615	71.646	-584.059	-328.034	13.100

Phase transition: $\Delta_{trs}H$ = 21.500 kJ/mol, $\Delta_{trs}S$ = 16.437 J/K·mol, crII–crI

T/K	$C_p°$/ J K^{-1}mol^{-1}	$S°$/ J K^{-1} mol^{-1}	-($G°$-$H°(T)$)/ J K^{-1} mol^{-1}	(H-$H°(T)$)/ kJ mol^{-1}	$\Delta_f H°$/ kJ mol^{-1}	$\Delta_f G°$/ kJ mol^{-1}	Log K_f
1308	80.075	156.827	85.615	93.146	-584.059	-328.034	13.100
1388	81.297	161.617	89.858	99.601	-565.504	-312.415	11.757

Phase transition: $\Delta_{trs}H$ = 17.200 kJ/mol, $\Delta_{trs}S$ = 12.392 J/K·mol, crI–l

T/K	$C_p°$/ J K^{-1}mol^{-1}	$S°$/ J K^{-1} mol^{-1}	-($G°$-$H°(T)$)/ J K^{-1} mol^{-1}	(H-$H°(T)$)/ kJ mol^{-1}	$\Delta_f H°$/ kJ mol^{-1}	$\Delta_f G°$/ kJ mol^{-1}	Log K_f
1388	78.500	174.009	89.858	116.801	-565.504	-312.415	11.757
1400	78.500	174.685	90.582	117.743	-565.328	-310.228	11.575
1500	78.500	180.100	96.372	125.593	-563.882	-292.057	10.170

49. GERMANIUM TETRACHLORIDE $GeCl_4$ (g)

T/K	$C_p°$/ J K^{-1}mol^{-1}	$S°$/ J K^{-1} mol^{-1}	-($G°$-$H°(T)$)/ J K^{-1} mol^{-1}	(H-$H°(T)$)/ kJ mol^{-1}	$\Delta_f H°$/ kJ mol^{-1}	$\Delta_f G°$/ kJ mol^{-1}	Log K_f
298.15	95.918	348.393	348.393	0.000	-500.000	-461.582	80.866
300	96.041	348.987	348.395	0.178	-499.991	-461.343	80.326
400	100.750	377.342	352.229	10.045	-499.447	-448.540	58.573
500	103.206	400.114	359.604	20.255	-498.845	-435.882	45.536
600	104.624	419.067	367.980	30.652	-498.234	-423.347	36.855
700	105.509	435.266	376.463	41.162	-497.634	-410.914	30.662
800	106.096	449.396	384.715	51.744	-497.057	-398.565	26.023
900	106.504	461.917	392.611	62.375	-496.509	-386.287	22.419
1000	106.799	473.155	400.113	73.041	-495.993	-374.068	19.539
1100	107.020	483.344	407.224	83.733	-495.512	-361.899	17.185
1200	107.189	492.664	413.961	94.444	-495.067	-349.772	15.225
1300	107.320	501.249	420.349	105.169	-531.677	-334.973	13.459
1400	107.425	509.206	426.416	115.907	-531.265	-319.857	11.934
1500	107.509	516.621	432.185	126.654	-530.861	-304.771	10.613

50. HYDROGEN H (g)

T/K	$C_p°$/ J K^{-1}mol^{-1}	$S°$/ J K^{-1} mol^{-1}	-($G°$-$H°(T)$)/ J K^{-1} mol^{-1}	(H-$H°(T)$)/ kJ mol^{-1}	$\Delta_f H°$/ kJ mol^{-1}	$\Delta_f G°$/ kJ mol^{-1}	Log K_f
298.15	20.786	114.716	114.716	0.000	217.998	203.276	-35.613
300	20.786	114.845	114.716	0.038	218.010	203.185	-35.377
400	20.786	120.824	115.532	2.117	218.635	198.149	-25.875
500	20.786	125.463	117.071	4.196	219.253	192.956	-20.158
600	20.786	129.252	118.795	6.274	219.867	187.639	-16.335
700	20.786	132.457	120.524	8.353	220.476	182.219	-13.597
800	20.786	135.232	122.193	10.431	221.079	176.712	-11.538
900	20.786	137.680	123.780	12.510	221.670	171.131	-9.932
1000	20.786	139.870	125.282	14.589	222.247	165.485	-8.644
1100	20.786	141.852	126.700	16.667	222.806	159.781	-7.587
1200	20.786	143.660	128.039	18.746	223.345	154.028	-6.705

T/K	$C_p°$/ J K^{-1}mol^{-1}	$S°$/ J K^{-1} mol^{-1}	-($G°$-$H°(T)$)/ J K^{-1} mol^{-1}	(H-$H°(T)$)/ kJ mol^{-1}	$\Delta_f H°$/ kJ mol^{-1}	$\Delta_f G°$/ kJ mol^{-1}	Log K_f
1300	20.786	145.324	129.305	20.824	223.864	148.230	-5.956
1400	20.786	146.864	130.505	22.903	224.360	142.393	-5.313
1500	20.786	148.298	131.644	24.982	224.835	136.522	-4.754
51. DIHYDROGEN H_2 (g)							
298.15	28.836	130.680	130.680	0.000	0.000	0.000	0.000
300	28.849	130.858	130.680	0.053	0.000	0.000	0.000
400	29.181	139.217	131.818	2.960	0.000	0.000	0.000
500	29.260	145.738	133.974	5.882	0.000	0.000	0.000
600	29.327	151.078	136.393	8.811	0.000	0.000	0.000
700	29.440	155.607	138.822	11.749	0.000	0.000	0.000
800	29.623	159.549	141.172	14.702	0.000	0.000	0.000
900	29.880	163.052	143.412	17.676	0.000	0.000	0.000
1000	30.204	166.217	145.537	20.680	0.000	0.000	0.000
1100	30.580	169.113	147.550	23.719	0.000	0.000	0.000
1200	30.991	171.791	149.460	26.797	0.000	0.000	0.000
1300	31.422	174.288	151.275	29.918	0.000	0.000	0.000
1400	31.860	176.633	153.003	33.082	0.000	0.000	0.000
1500	32.296	178.846	154.653	36.290	0.000	0.000	0.000
52. HYDROXYL OH (g)							
298.15	29.886	183.737	183.737	0.000	39.349	34.631	-6.067
300	29.879	183.922	183.738	0.055	39.350	34.602	-6.025
400	29.604	192.476	184.906	3.028	39.384	33.012	-4.311
500	29.495	199.067	187.104	5.982	39.347	31.422	-3.283
600	29.513	204.445	189.560	8.931	39.252	29.845	-2.598
700	29.655	209.003	192.020	11.888	39.113	28.287	-2.111
800	29.914	212.979	194.396	14.866	38.945	26.752	-1.747
900	30.265	216.522	196.661	17.874	38.763	25.239	-1.465
1000	30.682	219.731	198.810	20.921	38.577	23.746	-1.240
1100	31.135	222.677	200.848	24.012	38.393	22.272	-1.058
1200	31.603	225.406	202.782	27.149	38.215	20.814	-0.906
1300	32.069	227.954	204.621	30.332	38.046	19.371	-0.778
1400	32.522	230.347	206.374	33.562	37.886	17.941	-0.669
1500	32.956	232.606	208.048	36.836	37.735	16.521	-0.575
53. WATER H_2O (l)							
298.15	75.300	69.950	69.950	0.000	-285.830	-237.141	41.546
300	75.281	70.416	69.951	0.139	-285.771	-236.839	41.237
373.21	76.079	86.896	71.715	5.666	-283.454	-225.160	31.513
54. WATER H_2O (g)							
298.15	33.598	188.832	188.832	0.000	-241.826	-228.582	40.046
300	33.606	189.040	188.833	0.062	-241.844	-228.500	39.785
400	34.283	198.791	190.158	3.453	-242.845	-223.900	29.238
500	35.259	206.542	192.685	6.929	-243.822	-219.050	22.884
600	36.371	213.067	195.552	10.509	-244.751	-214.008	18.631
700	37.557	218.762	198.469	14.205	-245.620	-208.814	15.582
800	38.800	223.858	201.329	18.023	-246.424	-203.501	13.287
900	40.084	228.501	204.094	21.966	-247.158	-198.091	11.497
1000	41.385	232.792	206.752	26.040	-247.820	-192.603	10.060
1100	42.675	236.797	209.303	30.243	-248.410	-187.052	8.882
1200	43.932	240.565	211.753	34.574	-248.933	-181.450	7.898
1300	45.138	244.129	214.108	39.028	-249.392	-175.807	7.064
1400	46.281	247.516	216.374	43.599	-249.792	-170.132	6.348
1500	47.356	250.746	218.559	48.282	-250.139	-164.429	5.726
55. IODINE I (g)							
298.15	20.786	180.787	180.787	0.000	106.760	70.172	-12.294
300	20.786	180.915	180.787	0.038	106.748	69.945	-12.178
400	20.786	186.895	181.602	2.117	97.974	58.060	-7.582
500	20.786	191.533	183.142	4.196	75.988	50.202	-5.244

T/K	$C_p°$/ J K⁻¹mol⁻¹	$S°$/ J K⁻¹ mol⁻¹	$-(G°-H°(T))$/ J K⁻¹ mol⁻¹	$(H-H°(T))$/ kJ mol⁻¹	$\Delta_f H°$/ kJ mol⁻¹	$\Delta_f G°$/ kJ mol⁻¹	Log K_f
600	20.786	195.323	184.866	6.274	76.190	45.025	-3.920
700	20.786	198.527	186.594	8.353	76.385	39.816	-2.971
800	20.787	201.303	188.263	10.432	76.574	34.579	-2.258
900	20.789	203.751	189.851	12.510	76.757	29.319	-1.702
1000	20.795	205.942	191.352	14.589	76.936	24.038	-1.256
1100	20.806	207.924	192.770	16.669	77.109	18.740	-0.890
1200	20.824	209.735	194.110	18.751	77.277	13.426	-0.584
1300	20.851	211.403	195.377	20.835	77.440	8.098	-0.325
1400	20.889	212.950	196.577	22.921	77.596	2.758	-0.103
1500	20.936	214.392	197.717	25.013	77.745	-2.592	0.090

56. DIIODINE I_2 (cr, l)

T/K	$C_p°$/ J K⁻¹mol⁻¹	$S°$/ J K⁻¹ mol⁻¹	$-(G°-H°(T))$/ J K⁻¹ mol⁻¹	$(H-H°(T))$/ kJ mol⁻¹	$\Delta_f H°$/ kJ mol⁻¹	$\Delta_f G°$/ kJ mol⁻¹	Log K_f
298.15	54.440	116.139	116.139	0.000	0.000	0.000	0.000
300	54.518	116.476	116.140	0.101	0.000	0.000	0.000
386.75	61.531	131.039	117.884	5.088	0.000	0.000	0.000
Phase transition: $\Delta_{trs}H$ = 15.665 kJ/mol, $\Delta_{trs}S$ = 40.504 J/K·mol, cr–l							
386.75	79.555	171.543	117.884	20.753	0.000	0.000	0.000
400	79.555	174.223	119.706	21.807	0.000	0.000	0.000
457.67	79.555	184.938	127.266	26.395	0.000	0.000	0.000

57. DIIODINE I_2 (g)

T/K	$C_p°$/ J K⁻¹mol⁻¹	$S°$/ J K⁻¹ mol⁻¹	$-(G°-H°(T))$/ J K⁻¹ mol⁻¹	$(H-H°(T))$/ kJ mol⁻¹	$\Delta_f H°$/ kJ mol⁻¹	$\Delta_f G°$/ kJ mol⁻¹	Log K_f
298.15	36.887	260.685	260.685	0.000	62.420	19.324	-3.385
300	36.897	260.913	260.685	0.068	62.387	19.056	-3.318
400	37.256	271.584	262.138	3.778	44.391	5.447	-0.711
457.67[a]	37.385	276.610	263.652	5.931			
500	37.464	279.921	264.891	7.515	0.000	0.000	0.000
600	37.613	286.765	267.983	11.269	0.000	0.000	0.000
700	37.735	292.573	271.092	15.037	0.000	0.000	0.000
800	37.847	297.619	274.099	18.816	0.000	0.000	0.000
900	37.956	302.083	276.965	22.606	0.000	0.000	0.000
1000	38.070	306.088	279.681	26.407	0.000	0.000	0.000
1100	38.196	309.722	282.249	30.220	0.000	0.000	0.000
1200	38.341	313.052	284.679	34.047	0.000	0.000	0.000
1300	38.514	316.127	286.981	37.890	0.000	0.000	0.000
1400	38.719	318.989	289.166	41.751	0.000	0.000	0.000
1500	38.959	321.668	291.245	45.635	0.000	0.000	0.000

58. HYDROGEN IODIDE HI (g)

T/K	$C_p°$/ J K⁻¹mol⁻¹	$S°$/ J K⁻¹ mol⁻¹	$-(G°-H°(T))$/ J K⁻¹ mol⁻¹	$(H-H°(T))$/ kJ mol⁻¹	$\Delta_f H°$/ kJ mol⁻¹	$\Delta_f G°$/ kJ mol⁻¹	Log K_f
298.15	29.157	206.589	206.589	0.000	26.500	1.700	-0.298
300	29.158	206.769	206.589	0.054	26.477	1.546	-0.269
400	29.329	215.176	207.734	2.977	17.093	-6.289	0.821
500	29.738	221.760	209.904	5.928	-5.481	-9.946	1.039
600	30.351	227.233	212.348	8.931	-5.819	-10.806	0.941
700	31.070	231.965	214.820	12.002	-6.101	-11.614	0.867
800	31.807	236.162	217.230	15.145	-6.323	-12.386	0.809
900	32.511	239.950	219.548	18.362	-6.489	-13.133	0.762
1000	33.156	243.409	221.763	21.646	-6.608	-13.865	0.724
1100	33.735	246.597	223.878	24.991	-6.689	-14.586	0.693
1200	34.249	249.555	225.896	28.391	-6.741	-15.302	0.666
1300	34.703	252.314	227.823	31.839	-6.775	-16.014	0.643
1400	35.106	254.901	229.666	35.330	-6.797	-16.723	0.624
1500	35.463	257.336	231.430	38.858	-6.814	-17.432	0.607

59. POTASSIUM K (cr, l)

T/K	$C_p°$/ J K⁻¹mol⁻¹	$S°$/ J K⁻¹ mol⁻¹	$-(G°-H°(T))$/ J K⁻¹ mol⁻¹	$(H-H°(T))$/ kJ mol⁻¹	$\Delta_f H°$/ kJ mol⁻¹	$\Delta_f G°$/ kJ mol⁻¹	Log K_f
298.15	29.600	64.680	64.680	0.000	0.000	0.000	0.000
300	29.671	64.863	64.681	0.055	0.000	0.000	0.000
336.86	32.130	68.422	64.896	1.188	0.000	0.000	0.000
Phase transition: $\Delta_{trs}H$ = 2.321 kJ/mol, $\Delta_{trs}S$ = 6.891 J/K·mol, cr–l							
336.86	32.129	75.313	64.896	3.509	0.000	0.000	0.000
400	31.552	80.784	66.986	5.519	0.000	0.000	0.000
500	30.741	87.734	70.469	8.632	0.000	0.000	0.000

T/K	$C_p°$/ J K^{-1} mol^{-1}	$S°$/ J K^{-1} mol^{-1}	$-(G°-H°(T))$/ J K^{-1} mol^{-1}	$(H-H°(T))$/ kJ mol^{-1}	$\Delta_f H°$/ kJ mol^{-1}	$\Delta_f G°$/ kJ mol^{-1}	Log K_f
600	30.158	93.283	73.824	11.675	0.000	0.000	0.000
700	29.851	97.905	76.943	14.673	0.000	0.000	0.000
800	29.838	101.887	79.818	17.655	0.000	0.000	0.000
900	30.130	105.415	82.470	20.651	0.000	0.000	0.000
1000	30.730	108.618	84.927	23.691	0.000	0.000	0.000
1039.4	31.053	109.812	85.847	24.908	0.000	0.000	0.000
60. POTASSIUM K (g)							
298.15	20.786	160.340	160.340	0.000	89.000	60.479	-10.596
300	20.786	160.468	160.340	0.038	88.984	60.302	-10.499
400	20.786	166.448	161.155	2.117	85.598	51.332	-6.703
500	20.786	171.086	162.695	4.196	84.563	42.887	-4.480
600	20.786	174.876	164.419	6.274	83.599	34.643	-3.016
700	20.786	178.080	166.148	8.353	82.680	26.557	-1.982
800	20.786	180.856	167.817	10.431	81.776	18.601	-1.215
900	20.786	183.304	169.404	12.510	80.859	10.759	-0.624
1000	20.786	185.494	170.905	14.589	79.897	3.021	-0.158
1039.4[a]	20.786	186.297	171.474	15.408			
1100	20.786	187.475	172.323	16.667	0.000	0.000	0.000
1200	20.786	189.284	173.662	18.746	0.000	0.000	0.000
1300	20.789	190.948	174.929	20.825	0.000	0.000	0.000
1400	20.793	192.489	176.129	22.904	0.000	0.000	0.000
1500	20.801	193.923	177.268	24.983	0.000	0.000	0.000
61. DIPOTASSIUM OXIDE K_2O (cr, l)							
298.15	72.000	96.000	96.000	0.000	-361.700	-321.171	56.267
300	72.130	96.446	96.001	0.133	-361.704	-320.920	55.876
400	79.154	118.158	98.914	7.698	-366.554	-306.416	40.013
500	86.178	136.575	104.647	15.964	-366.043	-291.423	30.444
590	92.500	151.348	110.662	24.005	-364.204	-278.079	24.619
Phase transition: $\Delta_{trs}H$ = 0.700 kJ/mol, $\Delta_{trs}S$ = 1.186 J/K·mol, crIII–crII							
590	100.000	152.534	110.662	24.705	-364.204	-278.079	24.619
600	100.000	154.215	111.374	25.705	-363.968	-276.621	24.082
645	100.000	161.447	114.618	30.205	-358.901	-270.109	21.874
Phase transition: $\Delta_{trs}H$ = 4.000 kJ/mol, $\Delta_{trs}S$ = 6.202 J/K·mol, crII–crI							
645	100.000	167.649	114.618	34.205	-358.901	-270.109	21.874
700	100.000	175.832	119.111	39.705	-357.592	-262.592	19.595
800	100.000	189.185	127.054	49.705	-355.224	-249.183	16.270
900	100.000	200.963	134.625	59.705	-352.919	-236.067	13.701
1000	100.000	211.499	141.794	69.705	-350.732	-223.202	11.659
1013	100.000	212.791	142.697	71.005	-323.459	-221.546	11.424
Phase transition: $\Delta_{trs}H$ = 27.000 kJ/mol, $\Delta_{trs}S$ = 26.654 J/K·mol, crI–l							
1013	100.000	239.444	142.697	98.005	-323.459	-221.546	11.424
1100	100.000	247.684	150.679	106.705	-479.439	-203.633	9.670
1200	100.000	256.385	159.131	116.705	-475.371	-178.740	7.780
1300	100.000	264.389	166.924	126.705	-471.321	-154.185	6.195
1400	100.000	271.800	174.154	136.705	-467.287	-129.941	4.848
1500	100.000	278.699	180.896	146.705	-463.268	-105.986	3.691
62. POTASSIUM HYDROXIDE KOH (cr, l)							
298.15	64.900	78.870	78.870	0.000	-424.580	-378.747	66.354
300	65.038	79.272	78.871	0.120	-424.569	-378.463	65.895
400	72.519	99.007	81.512	6.998	-426.094	-362.765	47.372
500	80.000	115.993	86.745	14.624	-424.572	-347.093	36.260
520	81.496	119.159	87.931	16.239	-417.725	-344.002	34.555
Phase transition: $\Delta_{trs}H$ = 6.450 kJ/mol, $\Delta_{trs}S$ = 12.404 J/K·mol, crII–crI							
520	79.000	131.563	87.931	22.689	-417.725	-344.002	34.555
600	79.000	142.868	94.520	29.009	-416.274	-332.766	28.969
678	79.000	152.523	100.649	35.171	-405.464	-321.998	24.807

T/K	$C_p°$/ J K^{-1}mol^{-1}	$S°$/ J K^{-1} mol^{-1}	-($G°$-$H°$(T))/ J K^{-1} mol^{-1}	(H-$H°$(T))/ kJ mol^{-1}	$\Delta_f H°$/ kJ mol^{-1}	$\Delta_f G°$/ kJ mol^{-1}	Log K_f
Phase transition: $\Delta_{trs}H$ = 9.400 kJ/mol, $\Delta_{trs}S$ = 13.865 J/K·mol, crl–l							
678	83.000	166.388	100.649	44.571	-405.464	-321.998	24.807
700	83.000	169.038	102.757	46.397	-404.981	-319.297	23.826
800	83.000	180.121	111.750	54.697	-402.808	-307.206	20.058
900	83.000	189.897	119.901	62.997	-400.694	-295.383	17.143
1000	83.000	198.642	127.345	71.297	-398.668	-283.791	14.824
1100	83.000	206.553	134.192	79.597	-475.618	-267.780	12.716
1200	83.000	213.775	140.527	87.897	-472.711	-249.014	10.839
1300	83.000	220.418	146.421	96.197	-469.843	-230.490	9.261
1400	83.000	226.569	151.929	104.497	-467.011	-212.184	7.917
1500	83.000	232.296	157.098	112.797	-464.217	-194.080	6.758
63. POTASSIUM HYDROXIDE KOH (g)							
298.15	49.184	238.283	238.283	0.000	-227.989	-229.685	40.239
300	49.236	238.588	238.284	0.091	-228.007	-229.696	39.993
400	51.178	253.053	240.243	5.124	-231.377	-229.667	29.991
500	52.178	264.591	243.998	10.296	-232.309	-229.129	23.937
600	52.804	274.163	248.251	15.547	-233.145	-228.413	19.885
700	53.296	282.340	252.551	20.853	-233.934	-227.562	16.981
800	53.758	289.487	256.730	26.206	-234.708	-226.599	14.795
900	54.229	295.846	260.730	31.605	-235.495	-225.538	13.090
1000	54.713	301.585	264.533	37.052	-236.322	-224.388	11.721
1100	55.203	306.823	268.143	42.548	-316.077	-218.535	10.377
1200	55.686	311.647	271.570	48.092	-315.925	-209.674	9.127
1300	56.153	316.122	274.827	53.684	-315.764	-200.826	8.069
1400	56.598	320.300	277.927	59.322	-315.595	-191.991	7.163
1500	57.016	324.220	280.884	65.003	-315.420	-183.169	6.378
64. POTASSIUM CHLORIDE KCl (cr, l)							
298.15	51.300	82.570	82.570	0.000	-436.490	-408.568	71.579
300	51.333	82.887	82.571	0.095	-436.481	-408.395	71.107
400	52.977	97.886	84.605	5.312	-438.463	-398.651	52.058
500	54.448	109.867	88.498	10.685	-437.990	-388.749	40.612
600	55.885	119.921	92.919	16.201	-437.332	-378.960	32.991
700	57.425	128.649	97.413	21.865	-436.502	-369.295	27.557
800	59.205	136.430	101.812	27.694	-435.505	-359.760	23.490
900	61.361	143.523	106.058	33.719	-434.337	-350.360	20.334
1000	64.032	150.121	110.138	39.983	-432.981	-341.100	17.817
1044	65.405	152.908	111.882	42.830	-485.450	-336.720	16.847
Phase transition: $\Delta_{trs}H$ = 26.320 kJ/mol, $\Delta_{trs}S$ = 25.210 J/K·mol, cr–l							
1044	72.000	178.118	111.882	69.150	-485.450	-336.720	16.847
1100	72.000	181.880	115.351	73.182	-483.633	-328.790	15.613
1200	72.000	188.145	121.160	80.382	-480.393	-314.856	13.705
1300	72.000	193.908	126.537	87.582	-477.158	-301.192	12.102
1400	72.000	199.244	131.542	94.782	-473.928	-287.778	10.737
1500	72.000	204.211	136.223	101.982	-470.704	-274.594	9.562
65. POTASSIUM CHLORIDE KCl (g)							
298.15	36.505	239.091	239.091	0.000	-214.575	-233.320	40.876
300	36.518	239.317	239.092	0.068	-214.594	-233.436	40.644
400	37.066	249.904	240.532	3.749	-218.112	-239.107	31.224
500	37.384	258.212	243.267	7.473	-219.287	-244.219	25.513
600	37.597	265.048	246.344	11.222	-220.396	-249.100	21.686
700	37.769	270.857	249.441	14.991	-221.461	-253.799	18.938
800	37.907	275.910	252.441	18.775	-222.509	-258.347	16.868
900	38.041	280.382	255.302	22.572	-223.568	-262.764	15.250
1000	38.162	284.397	258.014	26.383	-224.667	-267.061	13.950
1100	38.279	288.039	260.581	30.205	-304.696	-266.627	12.661
1200	38.401	291.375	263.010	34.039	-304.821	-263.161	11.455
1300	38.518	294.454	265.312	37.885	-304.941	-259.684	10.434

T/K	$C_p°$/ J K⁻¹mol⁻¹	S°/ J K⁻¹ mol⁻¹	-(G°-H°(T))/ J K⁻¹ mol⁻¹	(H-H°(T))/ kJ mol⁻¹	$\Delta_f H°$/ kJ mol⁻¹	$\Delta_f G°$/ kJ mol⁻¹	Log K_f
1400	38.639	297.313	267.496	41.743	-305.053	-256.199	9.559
1500	38.761	299.983	269.574	45.613	-305.159	-252.706	8.800
66. DINITROGEN N_2 (g)							
298.15	29.124	191.608	191.608	0.000	0.000	0.000	0.000
300	29.125	191.788	191.608	0.054	0.000	0.000	0.000
400	29.249	200.180	192.752	2.971	0.000	0.000	0.000
500	29.580	206.738	194.916	5.911	0.000	0.000	0.000
600	30.109	212.175	197.352	8.894	0.000	0.000	0.000
700	30.754	216.864	199.812	11.936	0.000	0.000	0.000
800	31.433	221.015	202.208	15.046	0.000	0.000	0.000
900	32.090	224.756	204.509	18.222	0.000	0.000	0.000
1000	32.696	228.169	206.706	21.462	0.000	0.000	0.000
1100	33.241	231.311	208.802	24.759	0.000	0.000	0.000
1200	33.723	234.224	210.801	28.108	0.000	0.000	0.000
1300	34.147	236.941	212.708	31.502	0.000	0.000	0.000
1400	34.517	239.485	214.531	34.936	0.000	0.000	0.000
1500	34.842	241.878	216.275	38.404	0.000	0.000	0.000
67. NITRIC OXIDE NO (g)							
298.15	29.862	210.745	210.745	0.000	91.277	87.590	-15.345
300	29.858	210.930	210.746	0.055	91.278	87.567	-15.247
400	29.954	219.519	211.916	3.041	91.320	86.323	-11.272
500	30.493	226.255	214.133	6.061	91.340	85.071	-8.887
600	31.243	231.879	216.635	9.147	91.354	83.816	-7.297
700	32.031	236.754	219.168	12.310	91.369	82.558	-6.160
800	32.770	241.081	221.642	15.551	91.386	81.298	-5.308
900	33.425	244.979	224.022	18.862	91.405	80.036	-4.645
1000	33.990	248.531	226.298	22.233	91.426	78.772	-4.115
1100	34.473	251.794	228.469	25.657	91.445	77.505	-3.680
1200	34.883	254.811	230.540	29.125	91.464	76.237	-3.318
1300	35.234	257.618	232.516	32.632	91.481	74.967	-3.012
1400	35.533	260.240	234.404	36.170	91.495	73.697	-2.750
1500	35.792	262.700	236.209	39.737	91.506	72.425	-2.522
68. NITROGEN DIOXIDE NO_2 (g)							
298.15	37.178	240.166	240.166	0.000	34.193	52.316	-9.165
300	37.236	240.397	240.167	0.069	34.181	52.429	-9.129
400	40.513	251.554	241.666	3.955	33.637	58.600	-7.652
500	43.664	260.939	244.605	8.167	33.319	64.882	-6.778
600	46.383	269.147	248.026	12.673	33.174	71.211	-6.199
700	48.612	276.471	251.575	17.427	33.151	77.553	-5.787
800	50.405	283.083	255.107	22.381	33.213	83.893	-5.478
900	51.844	289.106	258.555	27.496	33.3340	90.221	-5.236
1000	53.007	294.631	261.891	32.741	33.495	96.534	-5.042
1100	53.956	299.729	265.102	38.090	33.686	102.828	-4.883
1200	54.741	304.459	268.187	43.526	33.898	109.105	-4.749
1300	55.399	308.867	271.148	49.034	34.124	115.363	-4.635
1400	55.960	312.994	273.992	54.603	34.360	121.603	-4.537
1500	56.446	316.871	276.722	60.224	34.604	127.827	-4.451
69. AMMONIA NH_3 (g)							
298.15	35.630	192.768	192.768	0.000	-45.940	-16.407	2.874
300	35.678	192.989	192.769	0.066	-45.981	-16.223	2.825
400	38.674	203.647	194.202	3.778	-48.087	-5.980	0.781
500	41.994	212.633	197.011	7.811	-49.908	4.764	-0.498
600	45.229	220.578	200.289	12.174	-51.430	15.846	-1.379
700	48.269	227.781	203.709	16.850	-52.682	27.161	-2.027
800	51.112	234.414	207.138	21.821	-53.695	38.639	-2.523
900	53.769	240.589	210.516	27.066	-54.499	50.231	-2.915
1000	56.244	246.384	213.816	32.569	-55.122	61.903	-3.233
1100	58.535	251.854	217.027	38.309	-55.589	73.629	-3.496
1200	60.644	257.039	220.147	44.270	-55.920	85.392	-3.717

T/K	$C_p°$/ J K⁻¹mol⁻¹	$S°$/ J K⁻¹ mol⁻¹	-($G°$-$H°(T)$)/ J K⁻¹ mol⁻¹	(H-$H°(T)$)/ kJ mol⁻¹	$\Delta_f H°$/ kJ mol⁻¹	$\Delta_f G°$/ kJ mol⁻¹	Log K_f
1300	62.576	261.970	223.176	50.432	-56.136	97.177	-3.905
1400	64.339	266.673	226.117	56.779	-56.251	108.975	-4.066
1500	65.945	271.168	228.971	63.295	-56.282	120.779	-4.206
70. OXYGEN O (g)							
298.15	21.911	161.058	161.058	0.000	249.180	231.743	-40.600
300	21.901	161.194	161.059	0.041	249.193	231.635	-40.331
400	21.482	167.430	161.912	2.207	249.874	225.677	-29.470
500	21.257	172.197	163.511	4.343	250.481	219.556	-22.937
600	21.124	176.060	165.290	6.462	251.019	213.319	-18.571
700	21.040	179.310	167.067	8.570	251.500	206.997	-15.446
800	20.984	182.115	168.777	10.671	251.932	200.610	-13.098
900	20.944	184.584	170.399	12.767	252.325	194.171	-11.269
1000	20.915	186.789	171.930	14.860	252.686	187.689	-9.804
1100	20.893	188.782	173.372	16.950	253.022	181.173	-8.603
1200	20.877	190.599	174.733	19.039	253.335	174.628	-7.601
1300	20.864	192.270	176.019	21.126	253.630	168.057	-6.753
1400	20.853	193.815	177.236	23.212	253.908	161.463	-6.024
1500	20.845	195.254	178.389	25.296	254.171	154.851	-5.392
71. DIOXYGEN O_2 (g)							
298.15	29.378	205.148	205.148	0.000	0.000	0.000	0.000
300	29.387	205.330	205.148	0.054	0.000	0.000	0.000
400	30.109	213.873	206.308	3.026	0.000	0.000	0.000
500	31.094	220.695	208.525	6.085	0.000	0.000	0.000
600	32.095	226.454	211.045	9.245	0.000	0.000	0.000
700	32.987	231.470	213.612	12.500	0.000	0.000	0.000
800	33.741	235.925	216.128	15.838	0.000	0.000	0.000
900	34.365	239.937	218.554	19.244	0.000	0.000	0.000
1000	34.881	243.585	220.878	22.707	0.000	0.000	0.000
1100	35.314	246.930	223.096	26.217	0.000	0.000	0.000
1200	35.683	250.019	225.213	29.768	0.000	0.000	0.000
1300	36.006	252.888	227.233	33.352	0.000	0.000	0.000
1400	36.297	255.568	229.162	36.968	0.000	0.000	0.000
1500	36.567	258.081	231.007	40.611	0.000	0.000	0.000
72. SULFUR S (cr, l)							
298.15	22.690	32.070	32.070	0.000	0.000	0.000	0.000
300	22.737	32.210	32.070	0.042	0.000	0.000	0.000
368.3	24.237	37.030	32.554	1.649	0.000	0.000	0.000
Phase transition: $\Delta_{trs}H$ = 0.401 kJ/mol, $\Delta_{trs}S$ = 1.089 J/K·mol, crII–crI							
368.3	24.773	38.119	32.553	2.050	0.000	0.000	0.000
388.36	25.180	39.444	32.875	2.551	0.000	0.000	0.000
Phase transition: $\Delta_{trs}H$ = 1.722 kJ/mol, $\Delta_{trs}S$ = 4.431 J/K·mol, crI–l							
388.36	31.710	43.875	32.872	4.273	0.000	0.000	0.000
400	32.369	44.824	33.206	4.647	0.000	0.000	0.000
500	38.026	53.578	36.411	8.584	0.000	0.000	0.000
600	34.371	60.116	39.842	12.164	0.000	0.000	0.000
700	32.451	65.278	43.120	15.511	0.000	0.000	0.000
800	32.000	69.557	46.163	18.715	0.000	0.000	0.000
882.38	32.000	72.693	48.496	21.351	0.000	0.000	0.000
73. SULFUR S (g)							
298.15	23.673	167.828	167.828	0.000	277.180	236.704	-41.469
300	23.669	167.974	167.828	0.044	277.182	236.453	-41.170
400	23.233	174.730	168.752	2.391	274.924	222.962	-29.115
500	22.741	179.860	170.482	4.689	273.286	210.145	-21.953
600	22.338	183.969	172.398	6.942	271.958	197.646	-17.206
700	22.031	187.388	174.302	9.160	270.829	185.352	-13.831
800	21.800	190.314	176.125	11.351	269.816	173.210	-11.309
900	21.624	192.871	177.847	13.522	215.723	162.258	-9.417
1000	21.489	195.142	179.465	15.677	216.018	156.301	-8.164

T/K	$C_p°$/ J K⁻¹mol⁻¹	$S°$/ J K⁻¹ mol⁻¹	-($G°$-$H°(T)$)/ J K⁻¹ mol⁻¹	(H-$H°(T)$)/ kJ mol⁻¹	$\Delta_f H°$/ kJ mol⁻¹	$\Delta_f G°$/ kJ mol⁻¹	Log K_f
1100	21.386	197.185	180.985	17.821	216.284	150.317	-7.138
1200	21.307	199.043	182.413	19.955	216.525	144.309	-6.282
1300	21.249	200.746	183.759	22.083	216.743	138.282	-5.556
1400	21.209	202.319	185.029	24.206	216.940	132.239	-4.934
1500	21.186	203.781	186.231	26.325	217.119	126.182	-4.394
74. DISULFUR S2 (g)							
298.15	32.505	228.165	228.165	0.000	128.600	79.696	-13.962
300	32.540	228.366	228.165	0.060	128.576	79.393	-13.823
400	34.108	237.956	229.462	3.398	122.703	63.380	-8.276
500	35.133	245.686	231.959	6.863	118.296	49.031	-5.122
600	35.815	252.156	234.800	10.413	114.685	35.530	-3.093
700	36.305	257.715	237.686	14.020	111.599	22.588	-1.685
800	36.697	262.589	240.501	17.671	108.841	10.060	-0.657
882.38[a]	36.985	266.200	242.734	20.706			
900	37.045	266.932	243.201	21.358	0.000	0.000	0.000
1000	37.377	270.852	245.773	25.079	0.000	0.000	0.000
1100	37.704	274.430	248.218	28.833	0.000	0.000	0.000
1200	38.030	277.725	250.541	32.620	0.000	0.000	0.000
1300	38.353	280.781	252.751	36.439	0.000	0.000	0.000
1400	38.669	283.635	254.856	40.290	0.000	0.000	0.000
1500	38.976	286.314	256.865	44.173	0.000	0.000	0.000
75. OCTASULFUR S8 (g)							
298.15	156.500	432.536	432.536	0.000	101.277	48.810	-8.551
300	156.768	433.505	432.539	0.290	101.231	48.484	-8.442
400	167.125	480.190	438.834	16.542	80.642	32.003	-4.179
500	173.181	518.176	451.022	33.577	66.185	21.409	-2.237
600	177.936	550.180	464.951	51.137	55.101	13.549	-1.180
700	182.441	577.948	479.152	69.157	46.349	7.343	-0.548
800	186.764	602.596	493.071	87.620	39.177	2.263	-0.148
900	190.595	624.821	506.495	106.494	-392.062	6.554	-0.380
1000	193.618	645.067	519.355	125.712	-387.728	50.614	-2.644
1100	195.684	663.625	531.639	145.185	-383.272	94.233	-4.475
1200	196.825	680.707	543.359	164.817	-378.786	137.444	-5.983
1300	197.195	696.480	554.539	184.524	-374.356	180.283	-7.244
1400	196.988	711.089	565.206	204.237	-370.048	222.785	-8.312
1500	196.396	724.662	575.389	223.909	-365.905	264.984	-9.227
76. SULFUR DIOXIDE SO2 (g)							
298.15	39.842	248.219	248.219	0.000	-296.810	-300.090	52.574
300	39.909	248.466	248.220	0.074	-296.833	-300.110	52.253
400	43.427	260.435	249.828	4.243	-300.240	-300.935	39.298
500	46.490	270.465	252.978	8.744	-302.735	-300.831	31.427
600	48.938	279.167	256.634	13.520	-304.699	-300.258	26.139
700	50.829	286.859	260.413	18.513	-306.308	-299.386	22.340
800	52.282	293.746	264.157	23.671	-307.691	-298.302	19.477
900	53.407	299.971	267.796	28.958	-362.075	-295.987	17.178
1000	54.290	305.646	271.301	34.345	-362.012	-288.647	15.077
1100	54.993	310.855	274.664	39.810	-361.934	-281.314	13.358
1200	55.564	315.665	277.882	45.339	-361.849	-273.989	11.926
1300	56.033	320.131	280.963	50.920	-361.763	-266.671	10.715
1400	56.426	324.299	283.911	56.543	-361.680	-259.359	9.677
1500	56.759	328.203	286.735	62.203	-361.605	-252.053	8.777
77. SILICON Si (cr)							
298.15	19.789	18.810	18.810	0.000	0.000	0.000	0.000
300	19.855	18.933	18.810	0.037	0.000	0.000	0.000
400	22.301	25.023	19.624	2.160	0.000	0.000	0.000
500	23.610	30.152	21.231	4.461	0.000	0.000	0.000
600	24.472	34.537	23.092	6.867	0.000	0.000	0.000
700	25.124	38.361	25.006	9.348	0.000	0.000	0.000
800	25.662	41.752	26.891	11.888	0.000	0.000	0.000

T/K	C_p°/ J K^{-1}mol^{-1}	S°/ J K^{-1} mol^{-1}	$-(G^\circ-H^\circ(T))$/ J K^{-1} mol^{-1}	$(H-H^\circ(T))$/ kJ mol^{-1}	$\Delta_f H^\circ$/ kJ mol^{-1}	$\Delta_f G^\circ$/ kJ mol^{-1}	Log K_f
900	26.135	44.802	28.715	14.478	0.000	0.000	0.000
1000	26.568	47.578	30.464	17.114	0.000	0.000	0.000
1100	26.974	50.130	32.138	19.791	0.000	0.000	0.000
1200	27.362	52.493	33.737	22.508	0.000	0.000	0.000
1300	27.737	54.698	35.265	25.263	0.000	0.000	0.000
1400	28.103	56.767	36.728	28.055	0.000	0.000	0.000
1500	28.462	58.719	38.130	30.883	0.000	0.000	0.000
78. SILICON Si (g)							
298.15	22.251	167.980	167.980	0.000	450.000	405.525	-71.045
300	22.234	168.117	167.980	0.041	450.004	405.249	-70.559
400	21.613	174.416	168.843	2.229	450.070	390.312	-50.969
500	21.316	179.204	170.456	4.374	449.913	375.388	-39.216
600	21.153	183.074	172.246	6.497	449.630	360.508	-31.385
700	21.057	186.327	174.032	8.607	449.259	345.682	-25.795
800	21.000	189.135	175.748	10.709	448.821	330.915	-21.606
900	20.971	191.606	177.375	12.808	448.329	316.205	-18.352
1000	20.968	193.815	178.911	14.904	447.791	301.553	-15.751
1100	20.989	195.815	180.358	17.002	447.211	286.957	-13.626
1200	21.033	197.643	181.723	19.103	446.595	272.416	-11.858
1300	21.099	199.329	183.014	21.209	445.946	257.927	-10.364
1400	21.183	200.895	184.236	23.323	445.268	243.489	-9.085
1500	21.282	202.360	185.396	25.446	444.563	229.101	-7.978
79. SILICON DIOXIDE SiO2 (cr)							
298.15	44.602	41.460	41.460	0.000	-910.700	-856.288	150.016
300	44.712	41.736	41.461	0.083	-910.708	-855.951	149.032
400	53.477	55.744	43.311	4.973	-910.912	-837.651	109.385
500	60.533	68.505	47.094	10.705	-910.540	-819.369	85.598
600	64.452	79.919	51.633	16.971	-909.841	-801.197	69.749
700	68.234	90.114	56.414	23.590	-908.958	-783.157	58.439
800	76.224	99.674	61.226	30.758	-907.668	-765.265	49.966
848	82.967	104.298	63.533	34.569	-906.310	-756.747	46.613
Phase transition: $\Delta_{trs}H$ = 0.411 kJ/mol, $\Delta_{trs}S$ = 0.484 J/K·mol, crII–crII′							
848	67.446	104.782	63.532	34.980	-906.310	-756.747	46.613
900	67.953	108.811	66.033	38.500	-905.922	-747.587	43.388
1000	68.941	116.021	70.676	45.345	-905.176	-730.034	38.133
1100	69.940	122.639	75.104	52.289	-904.420	-712.557	33.836
1200	70.947	128.768	79.323	59.333	-901.382	-695.148	30.259
Phase transition: $\Delta_{trs}H$ = 2.261 kJ/mol, $\Delta_{trs}S$ = 1.883 J/K·mol, crII′–crI							
1200	71.199	130.651	79.323	61.594	-901.382	-695.148	30.259
1300	71.743	136.372	83.494	68.742	-900.574	-677.994	27.242
1400	72.249	141.707	87.463	75.941	-899.782	-660.903	24.658
1500	72.739	146.709	91.248	83.191	-899.004	-643.867	22.421
80. SILICON TETRACHLORIDE $SiCl_4$ (g)							
298.15	90.404	331.446	331.446	0.000	-662.200	-622.390	109.039
300	90.562	332.006	331.448	0.167	-662.195	-622.143	108.323
400	96.893	359.019	335.088	9.572	-661.853	-608.841	79.505
500	100.449	381.058	342.147	19.456	-661.413	-595.637	62.225
600	102.587	399.576	350.216	29.616	-660.924	-582.527	50.713
700	103.954	415.500	358.432	39.948	-660.417	-569.501	42.496
800	104.875	429.445	366.455	50.392	-659.912	-556.548	36.338
900	105.523	441.837	374.155	60.914	-659.422	-543.657	31.553
1000	105.995	452.981	381.490	71.491	-658.954	-530.819	27.727
1100	106.349	463.101	388.456	82.109	-658.515	-518.027	24.599
1200	106.620	472.366	395.068	92.758	-658.107	-505.274	21.994
1300	106.834	480.909	401.347	103.431	-657.735	-492.553	19.791
1400	107.003	488.833	407.316	114.123	-657.400	-479.860	17.904
1500	107.141	496.220	413.000	124.830	-657.104	-467.189	16.269

[a] Boiling point at 1 bar pressure

THERMODYNAMIC PROPERTIES OF AQUEOUS IONS

This table contains standard state thermodynamic properties of positive and negative ions in aqueous solution. It includes enthalpy and Gibbs energy of formation, entropy, and heat capacity, and thus serves as a companion to the preceding table, "Standard Thermodynamic Properties of Chemical Substances." The standard state is the hypothetical ideal solution with molality m = 1 mol/kg (mean ionic molality $m_{\pm}$ in the case of a species that is assumed to dissociate at infinite dilution). Further details on conventions may be found in Ref. 1.

All values refer to standard conditions of 25 °C and 100 kPa pressure. Cations are listed first in alphabetical order by formula, followed by anions.

References

1. Wagman, D. D., Evans, W. H., Parker, V. B., Schumm, R. H., Halow, I., Bailey, S. M., Churney, K. L., and Nuttall, R. L., *The NBS Tables of Chemical Thermodynamic Properties, J. Phys. Chem. Ref. Data*, Vol. 11, Suppl. 2, 1982.
2. Zemaitis, J. F., Clark, D. M., Rafal, M., and Scrivner, N. C., *Handbook of Aqueous Electrolyte Thermodynamics*, American Institute of Chemical Engineers, New York, 1986.

Thermodynamic Properties of Aqueous Ions

Formula	$\Delta_f H°$(aq)/ kJ mol^{-1}	$\Delta_f G°$(aq)/ kJ mol^{-1}	$S°$(aq)/ J mol^{-1} K^{-1}	C_p(aq)/ J mol^{-1} K^{-1}
Cations				
Ag^{+}	105.6	77.1	72.7	21.8
Al^{+3}	-531.0	-485.0	-321.7	
$AlOH^{+2}$		-694.1		
$BaOH^{+}$		-730.5		
Ba^{+2}	-537.6	-560.8	9.6	
Be^{+2}	-382.8	-379.7	-129.7	
Bi^{+3}		82.8		
$BiOH^{+2}$		-146.4		
Ca^{+2}	-542.8	-553.6	-53.1	
$CaOH^{+}$		-718.4		
Cd^{+2}	-75.9	-77.6	-73.2	
$CdOH^{+}$		-261.1		
Ce^{+3}	-696.2	-672.0	-205.0	
Ce^{+4}	-537.2	-503.8	-301.0	
Co^{+2}	-58.2	-54.4	-113.0	
Co^{+3}	92.0	134.0	-305.0	
Cr^{+2}	-143.5			
Cs^{+}	-258.3	-292.0	133.1	-10.5
Cu^{+}	71.7	50.0	40.6	
Cu^{+2}	64.8	65.5	-99.6	
Dy^{+3}	-699.0	-665.0	-231.0	21.0
Er^{+3}	-705.4	-669.1	-244.3	21.0
Eu^{+2}	-527.0	-540.2	-8.0	
Eu^{+3}	-605.0	-574.1	-222.0	8.0
Fe^{+2}	-89.1	-78.9	-137.7	
Fe^{+3}	-48.5	-4.7	-315.9	
$FeOH^{+}$	-324.7	-277.4	-29.0	
$FeOH^{+2}$	-290.8	-229.4	-142.0	
$Fe(OH)_2^{+}$		-438.0		
Ga^{+2}		-88.0		
Ga^{+3}	-211.7	-159.0	-331.0	
$GaOH^{+2}$		-380.3		
$Ga(OH)_2^{+}$		-597.4		
Gd^{+3}	-686.0	-661.0	-205.9	
H^{+}	0	0	0	0
Hg^{+2}	171.1	164.4	-32.2	
$HgOH^{+}$	-84.5	-52.3	71.0	
Hg_2^{+2}	172.4	153.5	84.5	
Ho^{+3}	-705.0	-673.7	-226.8	17.0
In^{+}		-12.1		
$InOH^{+2}$	-370.3	-313.0	-88.0	
$In(OH)_2^{+}$	-619.0	-525.0	25.0	
In^{+2}		-50.7		
In^{+3}	-105.0	-98.0	-151.0	
K^{+}	-252.4	-283.3	102.5	21.8
La^{+3}	-707.1	-683.7	-217.6	-13.0
Li^{+}	-278.5	-293.3	13.4	68.6
Lu^{+3}	-665.0	-628.0	-264.0	25.0
LuF^{+2}		-931.4		
Mg^{+2}	-466.9	-454.8	-138.1	
$MgOH^{+}$		-626.7		
Mn^{+2}	-220.8	-228.1	-73.6	50.0
$MnOH^{+}$	-450.6	-405.0	-17.0	
NH_4^{+}	-132.5	-79.3	113.4	79.9
$N_2H_5^{+}$	-7.5	82.5	151.0	70.3
Na^{+}	-240.1	-261.9	59.0	46.4
Nd^{+3}	-696.2	-671.6	-206.7	-21.0
Ni^{+2}	-54.0	-45.6	-128.9	
$NiOH^{+}$	-287.9	-227.6	-71.0	
PH_4^{+}		92.1		
Pa^{+4}	-619.0			
Pb^{+2}	-1.7	-24.4	10.5	
$PbOH^{+}$		-226.3		
Pd^{+2}	149.0	176.5	-184.0	
Po^{+2}		71.0		
Po^{+4}		293.0		
Pr^{+3}	-704.6	-679.1	-209.0	-29.0
Pt^{+2}		254.8		
Ra^{+2}	-527.6	-561.5	54.0	
Rb^{+}	-251.2	-284.0	121.5	
Re^{+}		-33.0		
Sc^{+3}	-614.2	-586.6	-255.0	
$ScOH^{+2}$	-861.5	-801.2	-134.0	
Sm^{+2}		-497.5		
Sm^{+3}	-691.6	-666.6	-211.7	-21.0
Sn^{+2}	-8.8	-27.2	-17.0	
$SnOH^{+}$	-286.2	-254.8	50.0	
Sr^{+2}	-545.8	-559.5	-32.6	
$SrOH^{+}$		-721.3		
Tb^{+3}	-682.8	-651.9	-226.0	17.0
$Te(OH)_3^{+}$	-608.4	-496.1	111.7	
Th^{+4}	-769.0	-705.1	-422.6	
$Th(OH)^{+3}$	-1030.1	-920.5	-343.0	

Formula	$\Delta_f H°$(aq)/ kJ mol^{-1}	$\Delta_f G°$(aq)/ kJ mol^{-1}	$S°$(aq)/ J mol^{-1} K^{-1}	C_p(aq)/ J mol^{-1} K^{-1}
$Th(OH)_2^{+2}$	-1282.4	-1140.9	-218.0	
Tl^{+}	5.4	-32.4	125.5	
Tl^{+3}	196.6	214.6	-192.0	
$TlOH^{+2}$		-15.9		
$Tl(OH)_2^{+}$		-244.7		
Tm^{+3}	-697.9	-662.0	-243.0	25.0
U^{+3}	-489.1	-476.2	-188.0	
U^{+4}	-591.2	-531.9	-410.0	
Y^{+3}	-723.4	-693.8	-251.0	
$Y(OH)^{+2}$		-879.1		
$Y_2(OH)_2^{+4}$		-1780.3		
Yb^{+2}		-527.0		
Yb^{+3}	-674.5	-644.0	-238.0	25.0
Zn^{+2}	-153.9	-147.1	-112.1	46.0
$ZnOH^{+}$		-330.1		
Anions				
AlO_2^{-}	-930.9	-830.9	-36.8	
$Al(OH)_4^{-}$	-1502.5	-1305.3	102.9	
AsO_2^{-}	-429.0	-350.0	40.6	
AsO_4^{-3}	-888.1	-648.4	-162.8	
BF_4^{-}	-1574.9	-1486.9	180.0	
BH_4^{-}	48.2	114.4	110.5	
BO_2^{-}	-772.4	-678.9	-37.2	
$B_4O_7^{-2}$		-2604.8		
BeO_2^{-2}	-790.8	-640.1	-159.0	
Br^{-}	-121.6	-104.0	82.4	-141.8
BrO^{-}	-94.1	-33.4	42.0	
BrO_3^{-}	-67.1	18.6	161.7	
BrO_4^{-}	13.0	118.1	199.6	
$CHOO^{-}$	-425.6	-351.0	92.0	-87.9
CH_3COO^{-}	-486.0	-369.3	86.6	-6.3
CN^{-}	150.6	172.4	94.1	
CO_3^{-2}	-677.1	-527.8	-56.9	
$C_2O_4^{-2}$	-825.1	-673.9	45.6	
$C_2O_4H^{-}$	-818.4	-698.3	149.4	
Cl^{-}	-167.2	-131.2	56.5	-136.4
ClO^{-}	-107.1	-36.8	42.0	
ClO_2^{-}	-66.5	17.2	101.3	
ClO_3^{-}	-104.0	-8.0	162.3	
ClO_4^{-}	-129.3	-8.5	182.0	
CrO_4^{-2}	-881.2	-727.8	50.2	
$Cr_2O_7^{-2}$	-1490.3	-1301.1	261.9	
F^{-}	-332.6	-278.8	-13.8	-106.7
$Fe(CN)_6^{-3}$	561.9	729.4	270.3	
$Fe(CN)_6^{-4}$	455.6	695.1	95.0	
$HB_4O_7^{-}$		-2685.1		
HCO_3^{-}	-692.0	-586.8	91.2	
HF_2^{-}	-649.9	-578.1	92.5	
HPO_3F^{-}		-1198.2		
HPO_4^{-2}	-1292.1	-1089.2	-33.5	
$HP_2O_7^{-3}$	-2274.8	-1972.2	46.0	
HSO_3^{-}	-626.2	-527.7	139.7	
HSO_4^{-}	-887.3	-755.9	131.8	-84.0
$HS_2O_4^{-}$		-614.5		
HSe^{-}	15.9	44.0	79.0	
$HSeO_3^{-}$	-514.6	-411.5	135.1	
$HSeO_4^{-}$	-581.6	-452.2	149.4	
$H_2AsO_3^{-}$	-714.8	-587.1	110.5	
$H_2AsO_4^{-}$	-909.6	-753.2	117.0	
$H_2PO_4^{-}$	-1296.3	-1130.2	90.4	
$H_2P_2O_7^{-2}$	-2278.6	-2010.2	163.0	
I^{-}	-55.2	-51.6	111.3	-142.3
IO^{-}	-107.5	-38.5	-5.4	
IO_3^{-}	-221.3	-128.0	118.4	
IO_4^{-}	-151.5	-58.5	222.0	
MnO_4^{-}	-541.4	-447.2	191.2	-82.0
MnO_4^{-2}	-653.0	-500.7	59.0	
MoO_4^{-2}	-997.9	-836.3	27.2	
NO_2^{-}	-104.6	-32.2	123.0	-97.5
NO_3^{-}	-207.4	-111.3	146.4	-86.6
N_3^{-}	275.1	348.2	107.9	
OCN^{-}	-146.0	-97.4	106.7	
OH^{-}	-230.0	-157.2	-10.8	-148.5
PO_4^{-3}	-1277.4	-1018.7	-220.5	
$P_2O_7^{-4}$	-2271.1	-1919.0	-117.0	
Re^{-}	46.0	10.1	230.0	
S^{-2}	33.1	85.8	-14.6	
SCN^{-}	76.4	92.7	144.3	-40.2
SH^{-}	-17.6	12.1	62.8	
SO_3^{-2}	-635.5	-486.5	-29.0	
SO_4^{-2}	-909.3	-744.5	20.1	-293.0
S_2^{-2}	30.1	79.5	28.5	
$S_2O_3^{-2}$	-652.3	-522.5	67.0	
$S_2O_4^{-2}$	-753.5	-600.3	92.0	
$S_2O_8^{-2}$	-1344.7	-1114.9	244.3	
Se^{-2}		129.3		
SeO_3^{-2}	-509.2	-369.8	13.0	
SeO_4^{-2}	-599.1	-441.3	54.0	
VO_3^{-}	-888.3	-783.6	50.0	
VO_4^{-3}		-899.0		
WO_4^{-2}	-1075.7			

HEAT OF COMBUSTION

The heat of combustion of a substance at 25 °C can be calculated from the enthalpy of formation ($\Delta_f H°$) data in the table "Standard Thermodynamic Properties of Chemical Substances" in this section. We can write the general combustion reaction as

$$X + O_2 \rightarrow CO_2(g) + H_2O\ (l) + \text{other products}$$

For a compound containing only carbon, hydrogen, and oxygen, the reaction is simply

$$C_aH_bO_c + \left(a + \frac{1}{4}b - \frac{1}{2}c\right) O_2 \rightarrow a\,CO_2\ (g) + \frac{1}{2}b\ H_2O\ (l)$$

and the standard heat of combustion $\Delta_c H°$, which is defined as the negative of the enthalpy change for the reaction (i.e., the heat released in the combustion process), is given by

$$\Delta_c H^o = -a\Delta_f H^o(CO_2, g) - \frac{1}{2}b\Delta_f H^o(H_2O, l) + \Delta_f H^o(C_aH_bO_c)$$
$$= 393.51a + 142.915b + \Delta_f H^o(C_aH_bO_c)$$

This equation applies if the reactants start in their standard states (25 °C and one atmosphere pressure) and the products return to the same conditions. The same equation applies to a compound containing another element if that element ends in its standard reference state (e.g., nitrogen, if the product is N_2); in general, however, the exact products containing the other elements must be known in order to calculate the heat of combustion.

The following table gives the standard heat of combustion calculated in this manner for a few representative substances. Data for a much longer list of substances is given in the Internet edition of the *CRC Handbook of Chemistry and Physics.*

Heat of Combustion of Selected Substances

Name	Mol. form.	State	$\Delta_c H°$/ kJ mol^{-1}
Acetaldehyde	C_2H_4O	liq	1167
Acetamide	C_2H_5NO	cry	1185
Acetic acid	$C_2H_4O_2$	liq	874
Acetone	C_3H_6O	liq	1790
Acetonitrile	C_2H_3N	liq	1256
Acetylene	C_2H_2	gas	1300
L-Alanine	$C_3H_7NO_2$	cry	1577
Ammonia	H_3N	gas	383
Aniline	C_6H_7N	liq	3393
Anthracene	$C_{14}H_{10}$	cry	7068
Benzene	C_6H_6	liq	3268
Benzoic acid	$C_7H_6O_2$	cry	3228.2
1,3-Butadiene	C_4H_6	gas	2542
Butane	C_4H_{10}	gas	2878
2-Butanone	C_4H_8O	liq	2444
1-Butene	C_4H_8	gas	2718
cis-2-Butene	C_4H_8	gas	2710
trans-2-Butene	C_4H_8	gas	2706
Carbon (graphite)	C	cry	394
Carbon monoxide	CO	gas	283
Cyclobutene	C_4H_6	gas	2588
Cyclohexane	C_6H_{12}	liq	3920
Cyclopropane	C_3H_6	gas	2091
Decane	$C_{10}H_{22}$	liq	6778
Diethyl ether	$C_4H_{10}O$	liq	2724
Dimethyl ether	C_2H_6O	gas	1460
Ethane	C_2H_6	gas	1561
1,2-Ethanediol	$C_2H_6O_2$	liq	1185
Ethanol	C_2H_6O	liq	1367
Ethyl acetate	$C_4H_8O_2$	liq	2238
Ethylene	C_2H_4	gas	1411
Formaldehyde	CH_2O	gas	571
Formic acid	CH_2O_2	liq	254
Glycerol	$C_3H_8O_3$	liq	1654
Heptane	C_7H_{16}	liq	4817
Hexane	C_6H_{14}	liq	4163
Hydrazine	H_4N_2	liq	622
Hydrogen	H_2	gas	286
Hydrogen cyanide	CHN	gas	672
Ketene	C_2H_2O	gas	1025
Methane	CH_4	gas	891
Methanol	CH_4O	liq	726
Methyl acetate	$C_3H_6O_2$	liq	1592
Methylamine	CH_5N	gas	1086
Methyl formate	$C_2H_4O_2$	liq	973
Naphthalene	$C_{10}H_8$	cry	5157
Nitric oxide	NO	gas	91
Nitrobenzene	$C_6H_5NO_2$	liq	3088
Nitromethane	CH_3NO_2	liq	710
Nitrous oxide	N_2O	gas	82
Nonane	C_9H_{20}	liq	6125
Octane	C_8H_{18}	liq	5470
Pentane	C_5H_{12}	liq	3509
1-Pentanol	$C_5H_{12}O$	liq	3331
Phenanthrene	$C_{14}H_{10}$	cry	7055
Phenol	C_6H_6O	cry	3054
Propanal	C_3H_6O	liq	1822
Propane	C_3H_8	gas	2220
Propanenitrile	C_3H_5N	liq	1911
1-Propanol	C_3H_8O	liq	2021
2-Propanol	C_3H_8O	liq	2006
Propene	C_3H_6	gas	2058
Pyridine	C_5H_5N	liq	2782
Toluene	C_7H_8	liq	3910
Trimethylamine	C_3H_9N	gas	2443
2,4,6-Trinitrotoluene	$C_7H_5N_3O_6$	cry	3406
Urea	CH_4N_2O	cry	632.7

ENERGY CONTENT OF FUELS

Several fuels are compared in this table with respect to their energy content per unit mass and the amount of CO_2 released per unit of available energy. The energy content is taken to be the negative of the standard enthalpy of combustion (see the table "Heat of Combustion" in this section for more details). The energy is assumed to be released by combustion with oxygen at normal atmospheric pressure, with products of gaseous CO_2 and liquid H_2O at room temperature. This quantity is often called the "gross heat of combustion" to distinguish it from the "net heat of combustion," for which the water remains in the gas state. The latter quantity is typically 5% to 10% less than the values given here.

The energy content is given both in SI units of MJ/kg and conventional units of BTU/lb. Values for the fossil fuels and other materials are typical; individual samples show wide variations.

The last column gives the grams of carbon released as carbon dioxide per megajoule of energy. Examination of the table shows that the minimum CO_2 release occurs for fuels that have a high ratio of hydrogen to carbon. Furthermore, fuels containing oxygen have a lower energy content and higher CO_2 release than hydrocarbons with the same number of carbon atoms.

References

1. Domalski, E. S., Jobe, T. L., and Milne, T. A., *Thermodynamic Data for Biomass Conversion and Waste Incineration,* SERI/SP-271-2839, Solar Technical Information Program, U. S. Department of Energy, September 1986; see also NBSIR 78-1479, National Bureau of Standards, August 1978.
2. Green, D. W., and Ackers, D. E, *Perry's Chemical Engineers' Handbook, Eighth Edition,* McGraw-Hill, New York, 2007.
3. *Transportation Energy Data Book,* U. S. Department of Energy, May 2007, http://cta.ornl.gov/data/index.shtml.
4. *Chemical Composition of Natural Gas,* Union Gas Limited, http://www.uniongas.com/aboutus/aboutng/composition.asp.

Fuel	Energy content		g of C
	MJ/kg	10^3 BTU/lb	per MJ
Pure compounds			
Hydrogen	141.8	61.0	0.0
Methane	55.5	23.9	13.5
Ethane	51.9	22.3	15.4
Propane	50.3	21.7	16.2
Hexane	48.3	20.8	17.3
Heptane	48.1	20.7	17.5
Octane	47.9	20.6	17.6
Methanol	22.7	9.7	16.5
Ethanol	29.7	12.8	17.6
1-Propanol	33.6	14.5	17.8
1-Butanol	36.1	15.5	18.0
1-Octanol	40.7	17.5	18.1
Methyl *tert*-butyl ether	38.2	16.4	17.8
Fossil fuels			
Natural gas[a]	54.0	23.2	13.9
Gasoline	46.5	20.0	17.6
Kerosene	46.4	20.0	18.5
Fuel oil	40.9	17.6	21.3
Coal, high bituminous	36.3	15.6	23.5
Coal, low bituminous	28.9	12.4	26.3
Coal, anthracite	34.6	14.9	27.3
Other materials			
Wood, oak	18.9	8.1	25.3
Wood, locust	19.7	8.5	25.7
Wood, Ponderosa pine	20.0	8.6	24.6
Wood, redwood	20.7	8.9	24.4
Charcoal, wood	34.7	14.9	26.8
Newsprint	18.6	8.0	26.5
Cellulose	17.3	7.5	25.6
Grass (lawn clippings)	19.3	8.3	24.9

[a] Assumed to be 95% methane, 2.5% ethane, and 2.5% inert compounds; however, the actual composition varies widely (see Ref. 4).

IONIZATION CONSTANT OF WATER

Serguei N. Lvov and Allan H. Harvey

This table gives values of $pK_w = -\log_{10}(K_w)$, where K_w is the equilibrium constant of the reaction $2H_2O = H_3O^+(aq) + OH^-(aq)$. K_w is defined as

$$K_w = a_{H_3O^+} a_{OH^-} / a_{H_2O}^2$$

where a_i is the dimensionless activity of species i. The activities are on the molality basis for ions and mole fraction basis for water molecules. It is assumed that the activity of $H_3O^+(aq)$ is the same as the activity of $H^+(aq)$, so that K_w is numerically equal to

$$a_{H^+} a_{OH^-} / a_{H_2O}^2$$

the equilibrium constant for the ionization reaction of water, $H_2O = H^+(aq) + OH^-(aq)$, that is most commonly used in the literature. Values in the table are calculated using an analytical equation given in Refs. 1 and 2 where K_w is presented as a function of temperature and density from 0 °C to 800 °C and 0 g cm^{-3} to 1.25 g cm^{-3}.

References

1. International Association for the Properties of Water and Steam, *Release on the Ionization Constant of H_2O* (2007), available from <www.iapws.org>.
2. Bandura, A.V., and Lvov, S. N., *J. Phys. Chem. Ref. Data* 35, 15, 2006.

Ionization Constant of Water — Values of pK_w as a Function of Temperature and Pressure

p/MPa	0 °C	25 °C	50 °C	75 °C	100 °C	150 °C	200 °C	250 °C	300 °C	400 °C	600 °C	700 °C	800 °C
0.1[a]	14.946	13.995	13.264	12.696	12.252[a]	11.641[a]	11.310[a]	11.205[a]	11.339[a]	47.961	46.384	43.925	40.785
25	14.848	13.908	13.181	12.613	12.165	11.543	11.189	11.050	11.125	16.566	19.425	19.829	20.113
50	14.754	13.824	13.102	12.533	12.084	11.450	11.076	10.898	10.893	11.557	15.621	16.279	16.693
75	14.665	13.745	13.026	12.458	12.006	11.364	10.974	10.769	10.715	11.045	13.507	14.301	14.791
100	14.580	13.668	12.953	12.385	11.933	11.283	10.880	10.655	10.568	10.744	12.296	13.040	13.544
150	14.422	13.524	12.815	12.249	11.795	11.135	10.713	10.458	10.327	10.345	11.117	11.613	12.032
200	14.278	13.390	12.687	12.123	11.668	11.000	10.564	10.289	10.131	10.063	10.513	10.853	11.171
250	14.145	13.265	12.567	12.004	11.549	10.876	10.430	10.140	9.963	9.839	10.112	10.360	10.609
300	14.021	13.148	12.453	11.892	11.437	10.760	10.306	10.005	9.814	9.651	9.810	9.998	10.199
350	13.906	13.037	12.346	11.786	11.331	10.651	10.191	9.881	9.679	9.487	9.567	9.712	9.877
400	13.797	12.932	12.243	11.685	11.230	10.548	10.083	9.766	9.555	9.341	9.361	9.475	9.613
500	13.595	12.736	12.052	11.496	11.042	10.356	9.884	9.557	9.332	9.086	9.024	9.094	9.191
600	13.411	12.556	11.875	11.322	10.868	10.181	9.703	9.369	9.135	8.866	8.749	8.790	8.861
700	13.240	12.389	11.710	11.159	10.705	10.018	9.537	9.197	8.956	8.670	8.514	8.536	8.587
800	13.080	12.233	11.556	11.006	10.553	9.865	9.381	9.037	8.791	8.493	8.308	8.314	8.352
900	12.930	12.085	11.410	10.861	10.410	9.721	9.236	8.888	8.638	8.330	8.122	8.117	8.144
1000	12.788	11.946	11.272	10.725	10.273	9.585	9.098	8.748	8.495	8.178	7.952	7.939	7.957

[a] *Note*: Pressure for first row is 0.1 MPa at $t < 100$ °C and $t \geq 400$ °C, or p_s (saturated liquid) for 100 °C $\leq t \leq$ 350 °C.

Thermochem

IONIZATION CONSTANT OF NORMAL AND HEAVY WATER

Serguei N. Lvov and Allan H. Harvey

This table gives the ionization constant for liquid H_2O and D_2O at temperatures from 0 °C to 100 °C at the saturated vapor pressure. The quantity tabulated is $pK_w = -\log_{10}(K_w)$, where

$$K_w = a_{H_3O^+}a_{OH^-} / a_{H_2O}^2$$

for H_2O and

$$K_w = a_{D_3O^+}a_{OD^-} / a_{D_2O}^2$$

for D_2O. Values in the table are calculated using analytical equations given in Refs. 1 and 2 for H_2O and Ref. 3 for D_2O.

References

1. International Association for the Properties of Water and Steam, *Release on the Ionization Constant of H2O* (2007), available from <www.iapws.org>.
2. Bandura, A. V., and Lvov, S. N., *J. Phys. Chem. Ref. Data* 35, 15, 2006.
3. Mesmer, R. E., and Herting, D. L., *J. Solution Chem.* 7, 901, 1978.

Ionization Constant of Normal and Heavy Water — Values of $pK_w = -\log_{10} K_w$ of Normal and Heavy Water

t/°C	$pK_w(H_2O)$	$pK_w(D_2O)$
0	14.947	15.972
5	14.734	15.743
10	14.534	15.527
15	14.344	15.324
20	14.165	15.132
25	13.995	14.951
30	13.833	14.779
35	13.680	14.616
40	13.535	14.462
45	13.396	14.316
50	13.265	14.176
55	13.140	14.044
60	13.020	13.918
65	12.907	13.798
70	12.799	13.683
75	12.696	13.574
80	12.598	13.470
85	12.505	13.371
90	12.417	13.276
95	12.332	13.186
100	12.252	13.099

ELECTRICAL CONDUCTIVITY OF WATER

This table gives the electrical conductivity of highly purified water over a range of temperature and pressure. The first column of conductivity data refers to water at its own vapor pressure. Equations for calculating the conductivity at any temperature and pressure may be found in the reference.

Reference

Marshall, W. L., *J. Chem. Eng. Data* 32, 221, 1987.

Electrical Conductivity of Water in μS/cm at the Indicated Pressure and Temperature

t/°C	κ(sat. vap.)/ μS cm^{-1}	κ(50 MPa)/ μS cm^{-1}	κ(100 MPa)/ μS cm^{-1}	κ(200 MPa)/ μS cm^{-1}	κ(400 MPa)/ μS cm^{-1}	κ(600 MPa)/ μS cm^{-1}
0	0.0115	0.0150	0.0189	0.0275	0.0458	0.0667
25	0.0550	0.0686	0.0836	0.117	0.194	0.291
100	0.765	0.942	1.13	1.53	2.45	3.51
200	2.99	4.08	5.22	7.65	13.1	19.5
300	2.41	4.87	7.80	14.1	28.9	46.5
400		1.17	4.91	14.3	39.2	71.3
600			0.134	4.65	33.8	85.7

ELECTRICAL CONDUCTIVITY OF AQUEOUS SOLUTIONS

The following table gives the electrical conductivity of aqueous solutions of some acids, bases, and salts as a function of concentration. All values refer to 20 °C. The conductivity κ (often called specific conductance in older literature) is the reciprocal of the resistivity. The molar conductivity Λ is related to this by $\Lambda = \kappa/c$, where c is the amount-of-substance concentration of the electrolyte. Thus if κ has units of millisiemens per centimeter (mS/cm), as in this table, and c is expressed in mol/L, then Λ has units of S cm^2 mol^{-1}. For these electrolytes the concentration c corresponding to the mass percent values given here can be found in the table "Concentrative Properties of Aqueous Solutions" in this section.

References

1. *CRC Handbook of Chemistry, and Physics, 70th Edition*, Weast, R. C., Ed., CRC Press, Boca Raton, FL, 1989, p. D-221.
2. Wolf, A. V., *Aqueous Solutions and Body Fluids*, Harper and Row, New York, 1966.

Electrical Conductivity κ of Aqueous Solutions in mS/cm at 20 °C for the Indicated Concentration in Mass Percent

Solute	Formula	κ(0.5%)	κ(1%)	κ(2%)	κ(5%)	κ(10%)	κ(15%)	κ(20%)	κ(25%)	κ(30%)	κ(40%)	κ(50%)
Acetic acid	CH_3COOH	0.3	0.6	0.8	1.2	1.5	1.7	1.7	1.6	1.4	1.1	0.8
Ammonia	NH_3	0.5	0.7	1.0	1.1	1.0	0.7	0.5	0.4			
Ammonium chloride	NH_4Cl	10.5	20.4	40.3	95.3	180						
Ammonium sulfate	$(NH_4)_2SO_4$	7.4	14.2	25.7	57.4	105	147	185	215			
Barium chloride	$BaCl_2$	4.7	9.1	17.4	40.4	76.7	109.0	137.0				
Calcium chloride	$CaCl_2$	8.1	15.7	29.4	67.0	117	157	177	183	172	106	
Cesium chloride	CsCl	3.8	7.4	13.8	32.9	65.8	102	142				
Citric acid	$C(OH)(COOH)_3$	1.2	2.1	3.0	4.7	6.2	7.0	7.2	7.1			
Copper(II) sulfate	$CuSO_4$	2.9	5.4	9.3	19.0	32.2	42.3					
Formic acid	HCOOH	1.4	2.4	3.5	5.6	7.8	9.0	9.9	10.4	10.5	9.9	8.6
Hydrogen chloride	HCl	45.1	92.9	183								
Lithium chloride	LiCl	10.1	19.0	34.9	76.4	127	155	170	165	146		
Magnesium chloride	$MgCl_2$	8.6	16.6	31.2	66.9	108	129	134	122	98		
Magnesium sulfate	$MgSO_4$	4.1	7.6	13.3	27.4	42.7	54.2	51.1	44.1			
Manganese(II) sulfate	$MnSO_4$		6.2	10.6	21.6	34.5	43.7	47.6				
Nitric acid	HNO_3	28.4	56.1	108								
Oxalic acid	$H_2C_2O_4$	14.0	21.8	35.3	65.6							
Phosphoric acid	H_3PO_4	5.5	10.1	16.2	31.5	59.4	88.4	118	146	173	209	
Potassium bromide	KBr	5.2	10.2	19.5	47.7	95.6	144	194				
Potassium carbonate	K_2CO_3	7.0	13.6	25.4	58.0	109	152	188	223			
Potassium chloride	KCl	8.2	15.7	29.5	71.9	143	208					
Potassium dihydrogen phosphate	KH_2PO_4	3.0	5.9	11.0	25.0	44.6						
Potassium hydrogen carbonate	$KHCO_3$	4.6	8.9	17.0	38.8	72.4	101	128				
Potassium hydrogen phosphate	K_2HPO_4	5.2	9.9	18.3	40.3							
Potassium hydroxide	KOH	20.0	38.5	75.0	178							
Potassium iodide	KI	3.8	7.5	14.2	35.2	71.8	110	188	224			
Potassium nitrate	KNO_3	5.5	10.7	20.1	47.0	87.3	124	157	182			
Potassium permanganate	$KMnO_4$	3.5	6.9	13.0	30.5							
Potassium sulfate	K_2SO_4	5.8	11.2	21.0	48.0	88.6						
Silver(I) nitrate	$AgNO_3$	3.1	6.1	12.0	26.7	49.8	72.0	92.8	112	129	162	
Sodium acetate	$NaC_2H_3O_2$	3.9	7.6	14.4	30.9	53.4	64.1	69.3	69.2	64.3		
Sodium bromide	NaBr	5.0	9.7	18.4	44.0	84.6	122	157	191	216		
Sodium carbonate	Na_2CO_3	7.0	13.1	23.3	47.0	74.4	88.6					
Sodium chloride	NaCl	8.2	16.0	30.2	70.1	126	171	204	222			
Sodium dihydrogen phosphate	NaH_2PO_4	2.2	4.4	9.1	21.0	33.2	43.3	49.6	53.1	54.0	46.1	
Sodium hydrogen carbonate	$NaHCO_3$	4.2	8.2	15.0	31.4							
Sodium hydrogen phosphate	Na_2HPO_4	4.6	8.7	15.6	31.4							
Sodium hydroxide	NaOH	24.8	48.6	93.1	206							
Sodium nitrate	$NaNO_3$	5.4	10.6	20.4	46.2	82.6	111	134	152	165	178	
Sodium phosphate	Na_3PO_4	7.3	14.1	22.7	43.5							
Sodium sulfate	Na_2SO_4	5.9	11.2	19.8	42.7	71.3	91.1	109				
Sodium thiosulfate	$Na_2S_2O_3$	5.7	10.7	19.5	43.3	76.7	104	123	134	136	118	
Strontium chloride	$SrCl_2$	5.9	11.4	22.0	49.1	91.5	127	153	168	178		
Sulfuric acid	H_2SO_4	24.3	47.8	92	211							
Trichloroacetic acid	CCl_3COOH	10.3	19.6	37.2	84.7	148	193	221				
Trisodium citrate			7.4	12.8	26.2	42.1	52.0	57.1	57.3	53.5		
Zinc sulfate	$ZnSO_4$	2.8	5.4	10.0	20.5	33.7	43.3					

STANDARD KCl SOLUTIONS FOR CALIBRATING CONDUCTIVITY CELLS

This table presents recommended electrolytic conductivity (κ) values for aqueous potassium chloride solutions with molalities of 0.01 mol/kg, 0.1 mol/kg and 1.0 mol/kg at temperatures from 0 °C to 50 °C. The values, which are based on measurements at the National Institute of Standards and Technology, provide primary standards for the calibration of conductivity cells. The measurements at 0.01 and 0.1 molal are described in Ref. 1, while those at 1.0 molal are in Ref. 2. Temperatures are given on the ITS-90 scale. The uncertainty in the conductivity is about 0.03% for the 0.01 molal values and about 0.04% for the 0.1 and 1.0 molal values. The conductivity of water saturated with atmospheric CO_2 is given in the last column. These values were subtracted from the original measurements to give the values in the second, third, and fourth columns. All κ values are given in units of 10^{-4} S/m (numerically equal to μS/cm).

The assistance of Kenneth W. Pratt is appreciated.

References

1. Wu, Y. C., Koch, W. F., and Pratt, K. W., *J. Res. Natl. Inst. Stand. Technol.* 96, 191, 1991.
2. Wu, Y. C., Koch, W. F., Feng, D., Holland, L. A., Juhasz, E., Arvay, E., and Tomek, A., *J. Res. Natl. Inst. Stand. Technol.* 99, 241, 1994.
3. Pratt, K. W., Koch, W. F., Wu, Y. C., and Berezansky, P. A., *Pure Appl. Chem.* 73, 1783, 2001.

Conductivity of Aqueous Potassium Chloride at Various Concentrations and Temperatures

t/°C	$10^4\kappa$(0.01 m)/ S m^{-1}	$10^4\kappa$(0.1 m)/ S m^{-1}	$10^4\kappa$(1.0 m)/ S m^{-1}	$10^4\kappa(H_2O)$/ S m^{-1}
0	772.92	7116.85	63488	0.58
5	890.96	8183.70	72030	0.68
10	1013.95	9291.72	80844	0.79
15	1141.45	10437.1	89900	0.89
18	1219.93	11140.6		0.95
20	1273.03	11615.9	99170	0.99
25	1408.23	12824.6	108620	1.10
30	1546.63	14059.2	118240	1.20
35	1687.79	15316.0	127970	1.30
40	1831.27	16591.0	137810	1.40
45	1976.62	17880.6	147720	1.51
50	2123.43	19180.9	157670	1.61

MOLAR CONDUCTIVITY OF AQUEOUS HF, HCl, HBr, AND HI

The molar conductivity Λ of an electrolyte solution is defined as the conductivity divided by amount-of-substance concentration. The customary unit is S cm^2 mol^{-1} (i.e., Ω^{-1} cm^2 mol^{-1}). The first table gives the molar conductivity of the hydrohalogen acids at 25 °C as a function of the concentration. The second and third tables give both the temperature and concentration dependence of Λ for HCl and HBr. More extensive tables and mathematical representations may be found in the reference.

Reference

Hamer, W. J., and DeWane, H. J., *Electrolytic Conductance and the Conductances of the Hydrohalogen Acids in Water,* Natl. Stand. Ref. Data Sys.- Natl. Bur. Standards (U.S.), No. 33, 1970.

Molar Conductivity Λ of Aqueous Hydrohalogen Acids at 25 °C in Units of S cm^2 mol^{-1} for the Specified Concentrations Expressed as Molarity (mol/L)

	inf. dil.	0.0001 M	0.001 M	0.005 M	0.01 M	0.05 M	0.1 M	0.5 M	1.0 M	2.0 M	3.0 M	4.0 M	5.0 M	6.0 M	7.0 M	8.0 M	9.0 M	10.0 M
HF	405.1			128.1	96.1	50.1	39.1	26.3	24.3									
HCl	426.1	424.5	421.2	415.7	411.9	398.9	391.1	360.7	332.2	281.4	237.6	200.0	167.4	139.7	116.9	98.2	83.1	70.7
HBr	427.7	425.9	422.9	417.6	413.7	400.4	391.9	361.9	334.5	281.7	236.8	199.4	166.5	138.2	114.2	94.4		
HI	426.4	424.6	421.7	416.4	412.8	400.8	394.0	369.8	343.9	288.9	237.9	195.1	160.4	131.7	105.7			

Temperature and Concentration Dependence of Molar Conductivity Λ/S cm^2 mol^{-1} for HCl

c/mol L^{-1}	−20 °C	−10 °C	0 °C	10 °C	20 °C	30 °C	40 °C	50 °C
0.5			228.7	283.0	336.4	386.8	436.9	482.4
1.0			211.7	261.6	312.2	359.0	402.9	445.3
1.5			196.2	241.5	287.5	331.1	371.6	410.8
2.0			182.0	222.7	262.9	303.3	342.4	378.2
2.5		131.7	168.5	205.1	239.8	277.0	315.2	347.6
3.0		120.8	154.6	188.5	219.3	253.3	289.3	319.0
3.5	85.5	111.3	139.6	172.2	201.6	232.9	263.9	292.1
4.0	79.3	102.7	129.2	158.1	185.6	214.2	242.2	268.2
4.5	73.7	94.9	119.5	145.4	170.6	196.6	222.5	246.7
5.0	68.5	87.8	110.3	133.5	156.6	180.2	204.1	226.5
5.5	63.6	81.1	101.7	122.5	143.6	165.0	187.1	207.7
6.0	58.9	74.9	93.7	112.3	131.5	151.0	171.3	190.3
6.5	54.4	69.1	86.2	103.0	120.4	138.2	156.9	174.3
7.0	50.2	63.7	79.3	94.4	110.2	126.4	143.3	159.7
7.5	46.3	58.6	73.0	86.5	100.9	115.7	131.6	146.2
8.0	42.7	54.0	67.1	79.4	92.4	106.1	120.6	134.0
8.5	39.4	49.8	61.7	72.9	84.7	97.3	110.7	123.0
9.0	36.4	45.9	56.8	67.1	77.8	89.4	101.7	112.9
9.5	33.6	42.3	52.3	61.8	71.5	82.3	93.6	103.9
10.0	31.2	39.1	48.2	57.0	65.8	75.9	86.3	95.7
10.5	28.9	36.1	44.5	52.7	60.7	70.1	79.6	88.4
11.0	26.8	33.4	41.1	48.8	56.1	64.9	73.6	81.7
11.5	24.9	31.0	38.0	45.3	51.9	60.1	68.0	75.6
12.0	23.1	28.7	35.3	42.0	48.0	55.6	62.8	70.0
12.5	21.4	26.7	32.7	39.0	44.4	51.4	57.9	64.8

Temperature and Concentration Dependence of Molar Conductivity Λ/S cm^2 mol^{-1} for HBr

c/mol L^{-1}	−20 °C	−10 °C	0 °C	10 °C	20 °C	30 °C	40 °C	50 °C
0.5			240.9	295.9	347.0	398.9	453.6	496.8
1.0			229.6	276.0	329.0	380.4	418.6	465.2
1.5			209.5	254.9	298.9	340.6	381.8	421.4
2.0		150.8	188.6	231.3	271.8	314.1	350.5	387.4
2.5		136.8	171.7	208.3	244.8	281.7	316.0	349.1
3.0		125.7	157.2	189.5	222.2	255.0	287.8	318.6
3.5		116.1	144.1	174.6	203.2	234.4	263.7	291.9
4.0	84.0	107.5	132.3	160.2	186.8	214.2	239.7	266.9
4.5	78.0	99.0	123.0	146.4	171.2	195.1	218.8	242.6
5.0	72.3	91.4	112.6	134.0	155.7	178.2	199.6	221.3
5.5	67.0	84.2	103.1	122.7	142.1	162.8	181.4	201.8
6.0	61.8	77.2	94.3	112.0	129.6	148.0	165.4	183.4
6.5	56.8	70.7	86.0	102.0	118.0	134.1	150.5	166.3
7.0	51.9	64.6	78.4	92.6	107.1	121.4	136.3	150.8

EQUIVALENT CONDUCTIVITY OF ELECTROLYTES IN AQUEOUS SOLUTION

Petr Vanýsek

This table gives the equivalent (molar) conductivity Λ at 25 °C for some common electrolytes in aqueous solution at concentrations up to 0.1 M (0.1 mol/L). The units of Λ are cm² S mol⁻¹ (10^{-4} m² S mol⁻¹).

For very dilute solutions, the equivalent conductivity for any electrolyte of concentration c can be approximately calculated using the Debye–Hückel–Onsager equation, which can be written for a symmetrical (equal charge on cation and anion) electrolyte as

$$\Lambda = \Lambda^\circ - (A + B\Lambda^\circ)c^{1/2}$$

For a solution at 25 °C and both cation and anion with charge |1|, the constants are A = 60.20 and B = 0.229. Λ° can be found from the "inf. dil." column in the table below or calculated from the ion values in the more extensive table "Ionic Conductivity and Diffusion at Infinite Dilution." The equation is reliable for c <0.001 mol/L; with higher concentration the error increases.

Equivalent Conductivity of Electrolytes in Aqueous Solutions at 25 °C as a Function of Concentration

Species	Λ°(inf. dil.)/ S cm² mol⁻¹	Λ(0.0005 M)/ S cm² mol⁻¹	Λ(0.001 M)/ S cm² mol⁻¹	Λ(0.005 M)/ S cm² mol⁻¹	Λ(0.01 M)/ S cm² mol⁻¹	Λ(0.02 M)/ S cm² mol⁻¹	Λ(0.05 M)/ S cm² mol⁻¹	Λ(0.1 M)/ S cm² mol⁻¹
$AgNO_3$	133.29	131.29	130.45	127.14	124.70	121.35	115.18	109.09
$1/2BaCl_2$	139.91	135.89	134.27	127.96	123.88	119.03	111.42	105.14
$1/2CaCl_2$	135.77	131.86	130.30	124.19	120.30	115.59	108.42	102.41
$1/2Ca(OH)_2$	258			233	226	214		
$CuSO_4$	133.6	121.6	115.20	94.02	83.08	72.16	59.02	50.55
HCl	425.95	422.53	421.15	415.59	411.80	407.04	398.89	391.13
KBr	151.9	149.8	148.9	146.02	143.36	140.41	135.61	131.32
KCl	149.79	147.74	146.88	143.48	141.20	138.27	133.30	128.90
$KClO_4$	139.97	138.69	137.80	134.09	131.39	127.86	121.56	115.14
$1/3K_3Fe(CN)_6$	174.5	166.4	163.1	150.7				
$1/4K_4Fe(CN)_6$	184		167.16	146.02	134.76	122.76	107.65	97.82
$KHCO_3$	117.94	116.04	115.28	112.18	110.03	107.17		
KI	150.31	148.2	143.32	144.30	142.11	139.38	134.90	131.05
KIO_4	127.86	125.74	124.88	121.18	118.45	114.08	106.67	98.2
$KMnO_4$	134.8	132.7	131.9		126.5			113
KNO_3	144.89	142.70	141.77	138.41	132.75	132.34	126.25	120.34
KOH	271.5		234	230	228		219	213
$KReO_4$	128.20	126.03	125.12	121.31	118.49	114.49	106.40	97.40
$1/3LaCl_3$	145.9	139.6	137.0	127.5	121.8	115.3	106.2	99.1
LiCl	114.97	113.09	112.34	109.35	107.27	104.60	100.06	95.81
$LiClO_4$	105.93	104.13	103.39	100.52	98.56	96.13	92.15	88.52
$1/2MgCl_2$	129.34	125.55	124.15	118.25	114.49	109.99	103.03	97.05
NH_4Cl	149.6	147.5	146.7	143.9	141.21	138.25	133.22	128.69
$NaC_2H_3O_2$	91.0	89.2	88.5	85.68	83.72	81.20	76.88	72.76
Na picrate	80.45	78.7	78.6	75.7	73.7		66.3	61.8
NaCl	126.39	124.44	123.68	120.59	118.45	115.70	111.01	106.69
$NaClO_4$	117.42	115.58	114.82	111.70	109.54	106.91	102.35	98.38
NaI	126.88	125.30	124.19	121.19	119.18	116.64	112.73	108.73
NaOH	247.7	245.5	244.6	240.7	237.9			
$1/2Na_2SO_4$	129.8	125.68	124.09	117.09	112.38	106.73	97.70	89.94
$1/2SrCl_2$	135.73	131.84	130.27	124.18	120.23	115.48	108.20	102.14
$ZnSO_4$	132.7	121.3	114.47	95.44	84.87	74.20	61.17	52.61

IONIC CONDUCTIVITY AND DIFFUSION AT INFINITE DILUTION

Petr Vanýsek

This table gives the molar (equivalent) conductivity λ for common ions at infinite dilution. All values refer to aqueous solutions at 25 °C. It also lists the diffusion coefficient D of the ion in dilute aqueous solution, which is related to λ through the equation

$$D = (RT/F^2)(\lambda/|z|)$$

where R is the molar gas constant, T the temperature, F the Faraday constant, and z the charge on the ion. The variation with temperature is fairly sharp; for typical ions, λ and D increase by 2 to 3% per degree as the temperature increases from 25 °C.

The diffusion coefficient for a salt, D_{salt}, may be calculated from the D_+ and D_- values of the constituent ions by the relation

$$D_{salt} = \frac{(z_+ + |z_-|)D_+D_-}{z_+D_+ + |z_-|D_-}$$

For solutions of simple, pure electrolytes (one positive and one negative ionic species), such as NaCl, equivalent ionic conductivity $\Lambda°$, which is the molar conductivity per unit concentration of charge, is defined as

$$\Lambda° = \Lambda_+ + \Lambda_-$$

where Λ_+ and Λ_- are equivalent ionic conductivities of the cation and anion. The more general formula is

$$\Lambda° = \nu_+\Lambda_+ + \nu_-\Lambda_-$$

where ν_+ and ν_- refer to the number of moles of cations and anions, respectively, to which one mole of the electrolyte gives rise in solution.

References

1. Gray, D. E., Ed., *American Institute of Physics Handbook*, McGraw-Hill, New York, 1972, 2–226.
2. Robinson, R. A., and Stokes, R. H., *Electrolyte Solutions*, Butterworths, London, 1959.
3. Lobo, V. M. M., and Quaresma, J. L., *Handbook of Electrolyte Solutions*, Physical Science Data Series 41, Elsevier, Amsterdam, 1989.
4. Conway, B. E., *Electrochemical Data*, Elsevier, Amsterdam, 1952.
5. Milazzo, G., *Electrochemistry: Theoretical Principles and Practical Applications*, Elsevier, Amsterdam, 1963.

Ionic Conductivity and Diffusion at Infinite Dilution

Ion	$\Lambda_\pm$/ 10^{-4} m² S mol⁻¹	D/ 10^{-5} cm² s⁻¹
Inorganic Cations		
Ag^+	61.9	1.648
$1/3Al^{3+}$	61	0.541
$1/2Ba^{2+}$	63.6	0.847
$1/2Be^{2+}$	45	0.599
$1/2Ca^{2+}$	59.47	0.792
$1/2Cd^{2+}$	54	0.719
$1/3Ce^{3+}$	69.8	0.620
$1/2Co^{2+}$	55	0.732
$1/3[Co(NH_3)_6]^{3+}$	101.9	0.904
$1/3[Co(en)_3]^{3+}$	74.7	0.663
$1/6[Co_2(trien)_3]^{6+}$	69	0.306
$1/3Cr^{3+}$	67	0.595
Cs^+	77.2	2.056
$1/2Cu^{2+}$	53.6	0.714
D^+	249.9	6.655
$1/3Dy^{3+}$	65.6	0.582
$1/3Er^{3+}$	65.9	0.585
$1/3Eu^{3+}$	67.8	0.602
$1/2Fe^{2+}$	54	0.719
$1/3Fe^{3+}$	68	0.604
$1/3Gd^{3+}$	67.3	0.597
H^+	349.65	9.311
$1/2Hg^{2+}$	68.6	0.913
$1/2Hg^{2+}$	63.6	0.847
$1/3Ho^{3+}$	66.3	0.589
K^+	73.48	1.957
$1/3La^{3+}$	69.7	0.619
Li^+	38.66	1.029
$1/2Mg^{2+}$	53.0	0.706
$1/2Mn^{2+}$	53.5	0.712
NH_4^+	73.5	1.957
$N_2H_5^+$	59	1.571
Na^+	50.08	1.334
$1/3Nd^{3+}$	69.4	0.616
$1/2Ni^{2+}$	49.6	0.661
$1/4[Ni_2(trien)_3]^{4+}$	52	0.346
$1/2Pb^{2+}$	71	0.945
$1/3Pr^{3+}$	69.5	0.617
$1/2Ra^{2+}$	66.8	0.889
Rb^+	77.8	2.072
$1/3Sc^{3+}$	64.7	0.574
$1/3Sm^{3+}$	68.5	0.608
$1/2Sr^{2+}$	59.4	0.791
Tl^+	74.7	1.989
$1/3Tm^{3+}$	65.4	0.581
$1/2UO_2^{2+}$	32	0.426
$1/3Y^{3+}$	62	0.550
$1/3Yb^{3+}$	65.6	0.582
$1/2Zn^{2+}$	52.8	0.703
Inorganic Anions		
$Au(CN)_2^-$	50	1.331
$Au(CN)_4^-$	36	0.959
$B(C_6H_5)_4^-$	21	0.559
Br^-	78.1	2.080
Br_3^-	43	1.145
BrO_3^-	55.7	1.483
CN^-	78	2.077
CNO^-	64.6	1.720
$1/2CO_3^{2-}$	69.3	0.923
Cl^-	76.31	2.032
ClO_2^-	52	1.385

Ion	$\Lambda_\pm$/ 10^{-4} m^2 S mol^{-1}	D/ 10^{-5} cm^2 s^{-1}
ClO_3^-	64.6	1.720
ClO_4^-	67.3	1.792
$1/3[Co(CN)_6]^{3-}$	98.9	0.878
$1/2CrO_4^{2-}$	85	1.132
F^-	55.4	1.475
$1/4[Fe(CN)_6]^{4-}$	110.4	0.735
$1/3[Fe(CN)_6]^{3-}$	100.9	0.896
$H_2AsO_4^-$	34	0.905
HCO_3^-	44.5	1.185
HF_2^-	75	1.997
$1/2HPO_4^{2-}$	57	0.759
$H_2PO_4^-$	36	0.959
$H_2PO_2^-$	46	1.225
HS^-	65	1.731
HSO_3^-	58	1.545
HSO_4^-	52	1.385
$H_2SbO_4^-$	31	0.825
I^-	76.8	2.045
IO_3^-	40.5	1.078
IO_4^-	54.5	1.451
MnO_4^-	61.3	1.632
$1/2MoO_4^{2-}$	74.5	1.984
$N(CN)_2^-$	54.5	1.451
NO_2^-	71.8	1.912
NO_3^-	71.42	1.902
$NH_2SO_3^-$	48.3	1.286
N_3^-	69	1.837
OCN^-	64.6	1.720
OD^-	119	3.169
OH^-	198	5.273
PF_6^-	56.9	1.515
$1/2PO_3F^{2-}$	63.3	0.843
$1/3PO_4^{3-}$	92.8	0.824
$1/4P_2O_7^{4-}$	96	0.639
$1/3P_3O_9^{3-}$	83.6	0.742
$1/5P_3O_{10}^{5-}$	109	0.581
ReO_4^-	54.9	1.462
SCN^-	66	1.758
$1/2SO_3^{2-}$	72	0.959
$1/2SO_4^{2-}$	80.0	1.065
$1/2S_2O_3^{2-}$	85.0	1.132
$1/2S_2O_4^{2-}$	66.5	0.885
$1/2S_2O_6^{2-}$	93	1.238
$1/2S_2O_8^{2-}$	86	1.145
$Sb(OH)_6^-$	31.9	0.849
$SeCN^-$	64.7	1.723
$1/2SeO_4^{2-}$	75.7	1.008
$1/2WO_4^{2-}$	69	0.919
Organic Cations		
Benzyltrimethylammonium$^+$	34.6	0.921
Butyltrimethylammonium$^+$	33.6	0.895
Decylpyridinium$^+$	29.5	0.786
Decyltrimethylammonium$^+$	24.4	0.650
Diethylammonium$^+$	42.0	1.118
Dimethylammonium$^+$	51.8	1.379
Dipropylammonium$^+$	30.1	0.802
Dodecylammonium$^+$	23.8	0.634
Dodecyltrimethylammonium$^+$	22.6	0.602
Ethanolammonium$^+$	42.2	1.124
Ethylammonium$^+$	47.2	1.257

Ion	$\Lambda_\pm$/ 10^{-4} m^2 S mol^{-1}	D/ 10^{-5} cm^2 s^{-1}
Ethyltrimethylammonium$^+$	40.5	1.078
Hexadecyltrimethylammonium$^+$	20.9	0.557
Hexyltrimethylammonium$^+$	29.6	0.788
Histidyl$^+$	23.0	0.612
Hydroxyethyltrimethylarsonium$^+$	39.4	1.049
Isobutylammonium	38	1.012
Methylammonium$^+$	58.7	1.563
Octadecylpyridinium$^+$	20	0.533
Octadecyltributylammonium$^+$	16.6	0.442
Octadecyltriethylammonium$^+$	17.9	0.477
Octadecyltrimethylammonium$^+$	19.9	0.530
Octadecyltripropylammonium$^+$	17.2	0.458
Octyltrimethylammonium$^+$	26.5	0.706
Pentylammonium$^+$	37	0.985
Piperidinium$^+$	37.2	0.991
Propylammonium$^+$	40.8	1.086
Pyrilammonium$^+$	24.3	0.647
Tetrabutylammonium$^+$	19.5	0.519
Tetradecyltrimethylammonium$^+$	21.5	0.573
Tetraethylammonium$^+$	32.6	0.868
Tetramethylammonium$^+$	44.9	1.196
Tetraisopentylammonium$^+$	17.9	0.477
Tetrapentylammmonium$^+$	17.5	0.466
Tetrapropylammonium$^+$	23.4	0.623
Triethylammonium$^+$	34.3	0.913
Triethylsulfonium$^+$	36.1	0.961
Trimethylammonium$^+$	47.23	1.258
Trimethylhexylammonium$^+$	34.6	0.921
Trimethylsulfonium$^+$	51.4	1.369
Tripropylammonium$^+$	26.1	0.695
Organic Anions		
Acetate$^-$	40.9	1.089
p-Anisate$^-$	29.0	0.772
1/2Azelate^{2-}	40.6	0.541
Benzoate$^-$	32.4	0.863
Bromoacetate$^-$	39.2	1.044
Bromobenzoate$^-$	30	0.799
Butyrate$^-$	32.6	0.868
Chloroacetate$^-$	39.8	1.060
m-Chlorobenzoate$^-$	31	0.825
o-Chlorobenzoate$^-$	30.2	0.804
1/3Citrate^{3-}	70.2	0.623
Crotonate$^-$	33.2	0.884
Cyanoacetate$^-$	43.4	1.156
Cyclohexane carboxylate$^-$	28.7	0.764
1/2 1,1-Cyclopropanedicarboxylate^{2-}	53.4	0.711
Decylsulfate$^-$	26	0.692
Dichloroacetate$^-$	38.3	1.020
1/2Diethylbarbiturate^{2-}	26.3	0.350
Dihydrogencitrate$^-$	30	0.799
1/2Dimethylmalonate^{2-}	49.4	0.658
3,5-Dinitrobenzoate$^-$	28.3	0.754
Dodecylsulfate$^-$	24	0.639
Ethylmalonate$^-$	49.3	1.313
Ethylsulfate$^-$	39.6	1.055
Fluoroacetate$^-$	44.4	1.182
Fluorobenzoate$^-$	33	0.879
Formate$^-$	54.6	1.454
1/2Fumarate^{2-}	61.8	0.823
1/2Glutarate^{2-}	52.6	0.700

Ion	$\Lambda_\pm$/ 10^{-4} m² S mol⁻¹	D/ 10^{-5} cm² s⁻¹
Hydrogenoxalate⁻	40.2	1.070
Isovalerate⁻	32.7	0.871
Iodoacetate⁻	40.6	1.081
Lactate⁻	38.8	1.033
1/2Malate²⁻	58.8	0.783
1/2Maleate²⁻	61.9	0.824
1/2Malonate²⁻	63.5	0.845
Methylsulfate⁻	48.8	1.299
Naphthylacetate⁻	28.4	0.756
1/2Oxalate²⁻	74.11	0.987
Octylsulfate⁻	29	0.772
Phenylacetate⁻	30.6	0.815
1/2*o*-Phthalate²⁻	52.3	0.696
1/2*m*-Phthalate²⁻	54.7	0.728
Picrate⁻	30.37	0.809
Pivalate⁻	31.9	0.849
Propionate⁻	35.8	0.953
Propylsulfate⁻	37.1	0.988
Salicylate⁻	36	0.959
1/2Suberate²⁻	36	0.479
1/2Succinate²⁻	58.8	0.783
p-Sulfonate	29.3	0.780
1/2Tartarate²⁻	59.6	0.794
Trichloroacetate⁻	35	0.932

Thermochem

ELECTROCHEMICAL SERIES

Petr Vanýsek

There are three tables for the electrochemical series. Each table lists standard reduction potentials, E° values, at 298.15 K (25 °C), and at a pressure of 101.325 kPa (1 atm) (not the standard pressure of 1 bar). The activity of all soluble species is assumed to be 1.000 mol/L. This is particularly important when pH (H^+ or OH^-) takes part in the equilibrium. Table 1 is an alphabetical listing of the elements, according to the symbol of the elements. Thus, data for silver (Ag) precede those for aluminum (Al). Table 2 lists only those reduction reactions that have E° values positive with respect to the standard hydrogen electrode. In Table 2, the reactions are listed in the order of increasing positive potential, and they range from 0.0000 V to +3.4 V. Table 3 lists only those reduction potentials that have E° negative with respect to the standard hydrogen electrode. In Table 3, the reactions are listed in the order of decreasing potential and range from 0.0000 V to −4.10 V. The reliability of the potentials is not the same for all the data. Typically, the values with fewer significant figures have lower reliability. The values of reduction potentials, in particular those of less common reactions, are not definite; they are subject to occasional revisions.

Abbreviations: ac = acetate; bipy = 2,2'-dipyridine, or bipyridine; en = ethylenediamine; phen = 1,10-phenanthroline.

References

1. Milazzo, G., Caroli, S., and Sharma, V. K., *Tables of Standard Electrode Potentials, 4th Edition*, Wiley, Chichester, 1978.
2. Bard, A. J., Parsons, R., and Jordan, J., *Standard Potentials in Aqueous Solutions*, Marcel Dekker, New York, 1985.
3. Bratsch, S. G., *J. Phys. Chem. Ref. Data*, 18, 1–21, 1989.

TABLE 1. Electrochemical Series - Alphabetical Listing

Reaction	E°/V
$Ac^{3+} + 3\,e \rightleftharpoons Ac$	-2.20
$Ag^+ + e \rightleftharpoons Ag$	0.7996
$Ag^{2+} + e \rightleftharpoons Ag^+$	1.980
$Ag(ac) + e \rightleftharpoons Ag + (ac)^-$	0.643
$AgBr + e \rightleftharpoons Ag + Br^-$	0.07133
$AgBrO_3 + e \rightleftharpoons Ag + BrO_3^-$	0.546
$Ag_2C_2O_4 + 2\,e \rightleftharpoons 2\,Ag + C_2O_4^{2-}$	0.4647
$AgCl + e \rightleftharpoons Ag + Cl^-$	0.22233
$AgCN + e \rightleftharpoons Ag + CN^-$	-0.017
$Ag_2CO_3 + 2\,e \rightleftharpoons 2\,Ag + CO_3^{2-}$	0.47
$Ag_2CrO_4 + 2\,e \rightleftharpoons 2\,Ag + CrO_4^{2-}$	0.4470
$AgF + e \rightleftharpoons Ag + F^-$	0.779
$Ag_4[Fe(CN)_6] + 4e \rightleftharpoons 4Ag + [Fe(CN)_6]^{4-}$	0.1478
$AgI + e \rightleftharpoons Ag + I^-$	-0.15224
$AgIO_3 + e \rightleftharpoons Ag + IO_3^-$	0.354
$Ag_2MoO_4 + 2\,e \rightleftharpoons 2\,Ag + MoO_4^{2-}$	0.4573
$AgNO_2 + e \rightleftharpoons Ag + 2\,NO_2^-$	0.564
$Ag_2O + H_2O + 2\,e \rightleftharpoons 2\,Ag + 2\,OH^-$	0.342
$Ag_2O_3 + H_2O + 2\,e \rightleftharpoons 2\,AgO + 2\,OH^-$	0.739
$Ag^{3+} + 2\,e \rightleftharpoons Ag^+$	1.9
$Ag^{3+} + e \rightleftharpoons Ag^{2+}$	1.8
$Ag_2O_2 + 4\,H^+ + e \rightleftharpoons 2\,Ag + 2\,H_2O$	1.802
$2\,AgO + H_2O + 2\,e \rightleftharpoons Ag_2O + 2\,OH^-$	0.607
$AgOCN + e \rightleftharpoons Ag + OCN^-$	0.41
$Ag_2S + 2\,e \rightleftharpoons 2\,Ag + S^{2-}$	-0.691
$Ag_2S + 2\,H^+ + 2\,e \rightleftharpoons 2\,Ag + H_2S$	-0.0366
$AgSCN + e \rightleftharpoons Ag + SCN^-$	0.08951
$Ag_2SeO_3 + 2\,e \rightleftharpoons 2\,Ag + SeO_4^{2-}$	0.3629
$Ag_2SO_4 + 2\,e \rightleftharpoons 2\,Ag + SO_4^{2-}$	0.654
$Ag_2WO_4 + 2\,e \rightleftharpoons 2\,Ag + WO_4^{2-}$	0.4660
$Al^{3+} + 3\,e \rightleftharpoons Al$	-1.676
$Al(OH)_3 + 3\,e \rightleftharpoons Al + 3\,OH^-$	-2.30
$Al(OH)_4^- + 3\,e \rightleftharpoons Al + 4\,OH^-$	-2.310
$H_2AlO_3^- + H_2O + 3\,e \rightleftharpoons Al + 4\,OH^-$	-2.33
$AlF_6^{3-} + 3\,e \rightleftharpoons Al + 6\,F^-$	-2.069
$Am^{4+} + e \rightleftharpoons Am^{3+}$	2.60
$Am^{2+} + 2\,e \rightleftharpoons Am$	-1.9
$Am^{3+} + 3\,e \rightleftharpoons Am$	-2.048
$Am^{3+} + e \rightleftharpoons Am^{2+}$	-2.3
$As + 3\,H^+ + 3\,e \rightleftharpoons AsH_3$	-0.608
$As_2O_3 + 6\,H^+ + 6\,e \rightleftharpoons 2\,As + 3\,H_2O$	0.234
$HAsO_2 + 3\,H^+ + 3\,e \rightleftharpoons As + 2\,H_2O$	0.248
$AsO_2^- + 2\,H_2O + 3\,e \rightleftharpoons As + 4\,OH^-$	-0.68
$H_3AsO_4 + 2\,H^+ + 2\,e \rightleftharpoons HAsO_2 + 2\,H_2O$	0.560
$AsO_4^{3-} + 2\,H_2O + 2\,e \rightleftharpoons AsO_2^- + 4\,OH^-$	-0.71
$At_2 + 2\,e \rightleftharpoons 2\,At^-$	0.2
$Au^+ + e \rightleftharpoons Au$	1.692
$Au^{3+} + 2\,e \rightleftharpoons Au^+$	1.401
$Au^{3+} + 3\,e \rightleftharpoons Au$	1.498
$Au^{2+} + e^- \rightleftharpoons Au^+$	1.8
$AuOH^{2+} + H^+ + 2\,e \rightleftharpoons Au^+ + H_2O$	1.32
$AuBr_2^- + e \rightleftharpoons Au + 2\,Br^-$	0.959
$AuBr_4^- + 3\,e \rightleftharpoons Au + 4\,Br^-$	0.854
$AuCl_4^- + 3\,e \rightleftharpoons Au + 4\,Cl^-$	1.002
$Au(OH)_3 + 3\,H^+ + 3\,e \rightleftharpoons Au + 3\,H_2O$	1.45
$H_2BO_3^- + 5\,H_2O + 8\,e \rightleftharpoons BH_4^- + 8\,OH^-$	-1.24
$H_2BO_3^- + H_2O + 3\,e \rightleftharpoons B + 4\,OH^-$	-1.79
$H_3BO_3 + 3\,H^+ + 3\,e \rightleftharpoons B + 3\,H_2O$	-0.8698
$B(OH)_3 + 7\,H^+ + 8\,e \rightleftharpoons BH_4^- + 3\,H_2O$	-0.481
$Ba^{2+} + 2\,e \rightleftharpoons Ba$	-2.912
$Ba^{2+} + 2\,e \rightleftharpoons Ba(Hg)$	-1.570
$Ba(OH)_2 + 2\,e \rightleftharpoons Ba + 2\,OH^-$	-2.99
$Be^{2+} + 2\,e \rightleftharpoons Be$	-1.847
$Be_2O_3^{2-} + 3\,H_2O + 4\,e \rightleftharpoons 2\,Be + 6\,OH^-$	-2.63
p–benzoquinone + $2\,H^+ + 2\,e \rightleftharpoons$ hydroquinone	0.6992
$Bi^+ + e \rightleftharpoons Bi$	0.5
$Bi^{3+} + 3\,e \rightleftharpoons Bi$	0.308
$Bi^{3+} + 2\,e \rightleftharpoons Bi^+$	0.2
$Bi + 3\,H^+ + 3\,e \rightleftharpoons BiH_3$	-0.8
$BiCl_4^- + 3\,e \rightleftharpoons Bi + 4\,Cl^-$	0.16
$Bi_2O_3 + 3\,H_2O + 6\,e \rightleftharpoons 2\,Bi + 6\,OH^-$	-0.46
$Bi_2O_4 + 4\,H^+ + 2\,e \rightleftharpoons 2\,BiO^+ + 2\,H_2O$	1.593
$BiO^+ + 2\,H^+ + 3\,e \rightleftharpoons Bi + H_2O$	0.320
$BiOCl + 2\,H^+ + 3\,e \rightleftharpoons Bi + Cl^- + H_2O$	0.1583
$Bk^{4+} + e \rightleftharpoons Bk^{3+}$	1.67
$Bk^{2+} + 2\,e \rightleftharpoons Bk$	-1.6
$Bk^{3+} + e \rightleftharpoons Bk^{2+}$	-2.8
$Br_2(aq) + 2\,e \rightleftharpoons 2\,Br^-$	1.0873
$Br_2(l) + 2\,e \rightleftharpoons 2\,Br^-$	1.066
$HBrO + H^+ + 2\,e \rightleftharpoons Br^- + H_2O$	1.331
$HBrO + H^+ + e \rightleftharpoons 1/2\,Br_2(aq) + H_2O$	1.574
$HBrO + H^+ + e \rightleftharpoons 1/2\,Br_2(l) + H_2O$	1.596
$BrO^- + H_2O + 2\,e \rightleftharpoons Br^- + 2\,OH^-$	0.761
$BrO_3^- + 6\,H^+ + 5\,e \rightleftharpoons 1/2\,Br_2 + 3\,H_2O$	1.482
$BrO_3^- + 6\,H^+ + 6\,e \rightleftharpoons Br^- + 3\,H_2O$	1.423
$BrO_3^- + 3\,H_2O + 6\,e \rightleftharpoons Br^- + 6\,OH^-$	0.61
$(CN)_2 + 2\,H^+ + 2\,e \rightleftharpoons 2\,HCN$	0.373
$2\,HCNO + 2\,H^+ + 2\,e \rightleftharpoons (CN)_2 + 2\,H_2O$	0.330
$(CNS)_2 + 2\,e \rightleftharpoons 2\,CNS^-$	0.77
$CO_2 + 2\,H^+ + 2\,e \rightleftharpoons HCOOH$	-0.199
$Ca^+ + e \rightleftharpoons Ca$	-3.80
$Ca^{2+} + 2\,e \rightleftharpoons Ca$	-2.868
$Ca(OH)_2 + 2\,e \rightleftharpoons Ca + 2\,OH^-$	-3.02
Calomel electrode, 1 molal KCl	0.2800
Calomel electrode, 1 molar KCl (NCE)	0.2801
Calomel electrode, 0.1 molar KCl	0.3337
Calomel electrode, saturated KCl (SCE)	0.2412
Calomel electrode, saturated NaCl (SSCE)	0.2360
$Cd^{2+} + 2\,e \rightleftharpoons Cd$	-0.4030
$Cd^{2+} + 2\,e \rightleftharpoons Cd(Hg)$	-0.3521
$Cd(OH)_2 + 2\,e \rightleftharpoons Cd(Hg) + 2\,OH^-$	-0.809
$CdSO_4 + 2\,e \rightleftharpoons Cd + SO_4^{2-}$	-0.246

Reaction	E°/V
$Cd(OH)_4^{2-} + 2 e \rightleftharpoons Cd + 4 OH^-$	-0.658
$CdO + H_2O + 2 e \rightleftharpoons Cd + 2 OH^-$	-0.783
$Ce^{3+} + 3 e \rightleftharpoons Ce$	-2.336
$Ce^{3+} + 3 e \rightleftharpoons Ce(Hg)$	-1.4373
$Ce^{4+} + e \rightleftharpoons Ce^{3+}$	1.72
$CeOH^{3+} + H^+ + e \rightleftharpoons Ce^{3+} + H_2O$	1.715
$Cf^{4+} + e \rightleftharpoons Cf^{3+}$	3.3
$Cf^{3+} + e \rightleftharpoons Cf^{2+}$	-1.6
$Cf^{3+} + 3 e \rightleftharpoons Cf$	-1.94
$Cf^{2+} + 2 e \rightleftharpoons Cf$	-2.12
$Cl_2(g) + 2 e \rightleftharpoons 2 Cl^-$	1.35827
$HClO + H^+ + e \rightleftharpoons 1/2\, Cl_2 + H_2O$	1.611
$HClO + H^+ + 2 e \rightleftharpoons Cl^- + H_2O$	1.482
$ClO^- + H_2O + 2 e \rightleftharpoons Cl^- + 2 OH^-$	0.81
$ClO_2 + H^+ + e \rightleftharpoons HClO_2$	1.277
$HClO_2 + 2 H^+ + 2 e \rightleftharpoons HClO + H_2O$	1.645
$HClO_2 + 3 H^+ + 3 e \rightleftharpoons 1/2\, Cl_2 + 2 H_2O$	1.628
$HClO_2 + 3 H^+ + 4 e \rightleftharpoons Cl^- + 2 H_2O$	1.570
$ClO_2^- + H_2O + 2 e \rightleftharpoons ClO^- + 2 OH^-$	0.66
$ClO_2^- + 2 H_2O + 4 e \rightleftharpoons Cl^- + 4 OH^-$	0.76
$ClO_2(aq) + e \rightleftharpoons ClO_2^-$	0.954
$ClO_3^- + 2 H^+ + e \rightleftharpoons ClO_2 + H_2O$	1.152
$ClO_3^- + 3 H^+ + 2 e \rightleftharpoons HClO_2 + H_2O$	1.214
$ClO_3^- + 6 H^+ + 5 e \rightleftharpoons 1/2\, Cl_2 + 3 H_2O$	1.47
$ClO_3^- + 6 H^+ + 6 e \rightleftharpoons Cl^- + 3 H_2O$	1.451
$ClO_3^- + H_2O + 2 e \rightleftharpoons ClO_2^- + 2 OH^-$	0.33
$ClO_3^- + 3 H_2O + 6 e \rightleftharpoons Cl^- + 6 OH^-$	0.62
$ClO_4^- + 2 H^+ + 2 e \rightleftharpoons ClO_3^-H_2O$	1.189
$ClO_4^- + 8 H^+ + 7 e \rightleftharpoons 1/2\, Cl_2 + 4 H_2O$	1.39
$ClO_4^- + 8 H^+ + 8 e \rightleftharpoons Cl^- + 4 H_2O$	1.389
$ClO_4^- + H_2O + 2 e \rightleftharpoons ClO_3^- + 2 OH^-$	0.36
$Cm^{4+} + e \rightleftharpoons Cm^{3+}$	3.0
$Cm^{3+} + 3 e \rightleftharpoons Cm$	-2.04
$Co^{2+} + 2 e \rightleftharpoons Co$	-0.28
$Co^{3+} + e \rightleftharpoons Co^{2+}$	1.92
$[Co(NH_3)_6]^{3+} + e \rightleftharpoons [Co(NH_3)_6]^{2+}$	0.108
$Co(OH)_2 + 2 e \rightleftharpoons Co + 2 OH^-$	-0.73
$Co(OH)_3 + e \rightleftharpoons Co(OH)_2 + OH^-$	0.17
$Cr^{2+} + 2 e \rightleftharpoons Cr$	-0.913
$Cr^{3+} + e \rightleftharpoons Cr^{2+}$	-0.407
$Cr^{3+} + 3 e \rightleftharpoons Cr$	-0.744
$Cr_2O_7^{2-} + 14 H^+ + 6 e \rightleftharpoons 2 Cr^{3+} + 7 H_2O$	1.36
$CrO_2^- + 2 H_2O + 3 e \rightleftharpoons Cr + 4 OH^-$	-1.2
$HCrO_4^- + 7 H^+ + 3 e \rightleftharpoons Cr^{3+} + 4 H_2O$	1.350
$CrO_2 + 4 H^+ + e \rightleftharpoons Cr^{3+} + 2H_2O$	1.48
$Cr(V) + e \rightleftharpoons Cr(IV)$	1.34
$CrO_4^{2-} + 4H_2O + 3e \rightleftharpoons Cr(OH)_3 + 5OH^-$	-0.13
$Cr(OH)_3 + 3 e \rightleftharpoons Cr + 3 OH^-$	-1.48
$Cs^+ + e \rightleftharpoons Cs$	-3.026
$Cu^+ + e \rightleftharpoons Cu$	0.521
$Cu^{2+} + e \rightleftharpoons Cu^+$	0.153
$Cu^{2+} + 2 e \rightleftharpoons Cu$	0.3419
$Cu^{2+} + 2 e \rightleftharpoons Cu(Hg)$	0.345
$Cu^{3+} + e \rightleftharpoons Cu^{2+}$	2.4
$Cu_2O_3 + 6 H^+ + 2e \rightleftharpoons 2Cu^{2+} + 3 H_2O$	2.0
$Cu^{2+} + 2 CN^- + e \rightleftharpoons [Cu(CN)_2]^-$	1.103
$CuI_2^- + e \rightleftharpoons Cu + 2 I^-$	0.00
$Cu_2O + H_2O + 2 e \rightleftharpoons 2 Cu + 2 OH^-$	-0.360
$Cu(OH)_2 + 2 e \rightleftharpoons Cu + 2 OH^-$	-0.222
$2 Cu(OH)_2 + 2 e \rightleftharpoons Cu_2O + 2 OH^- + H_2O$	-0.080
$2 D^+ + 2 e \rightleftharpoons D_2$	-0.013
$Dy^{2+} + 2 e \rightleftharpoons Dy$	-2.2
$Dy^{3+} + 3 e \rightleftharpoons Dy$	-2.295
$Dy^{3+} + e \rightleftharpoons Dy^{2+}$	-2.6
$Er^{2+} + 2 e \rightleftharpoons Er$	-2.0
$Er^{3+} + 3 e \rightleftharpoons Er$	-2.331
$Er^{3+} + e \rightleftharpoons Er^{2+}$	-3.0
$Es^{3+} + e \rightleftharpoons Es^{2+}$	-1.3
$Es^{3+} + 3 e \rightleftharpoons Es$	-1.91
$Es^{2+} + 2 e \rightleftharpoons Es$	-2.23
$Eu^{2+} + 2 e \rightleftharpoons Eu$	-2.812
$Eu^{3+} + 3 e \rightleftharpoons Eu$	-1.991
$Eu^{3+} + e \rightleftharpoons Eu^{2+}$	-0.36
$F_2 + 2 H^+ + 2 e \rightleftharpoons 2 HF$	3.053
$F_2 + 2 e \rightleftharpoons 2 F^-$	2.866
$F_2O + 2 H^+ + 4 e \rightleftharpoons H_2O + 2 F^-$	2.153
$Fe^{2+} + 2 e \rightleftharpoons Fe$	-0.447
$Fe^{3+} + 3 e \rightleftharpoons Fe$	-0.037
$Fe^{3+} + e \rightleftharpoons Fe^{2+}$	0.771
$2 HFeO_4^- + 8 H^+ + 6 e \rightleftharpoons Fe_2O_3 + 5 H_2O$	2.09
$HFeO_4^- + 4 H^+ + 3 e \rightleftharpoons FeOOH + 2 H_2O$	2.08
$HFeO_4^- + 7 H^+ + 3 e \rightleftharpoons Fe^{3+} + 4 H_2O$	2.07
$Fe_2O_3 + 4 H^+ + 2 e \rightleftharpoons 2 FeOH^+ + H_2O$	0.16
$[Fe(CN)_6]^{3-} + e \rightleftharpoons [Fe(CN)_6]^{4-}$	0.358
$FeO_4^{2-} + 8 H^+ + 3 e\ Fe^{3+} + 4 H_2O$	2.20
$[Fe(bipy)_2]^{3+} + e \rightleftharpoons Fe(bipy)_2]^{2+}$	0.78
$[Fe(bipy)_3]^{3+} + e \rightleftharpoons Fe(bipy)_3]^{2+}$	1.03
$Fe(OH)_3 + e \rightleftharpoons Fe(OH)_2 + OH^-$	-0.56
$[Fe(phen)_3]^{3+} + e \rightleftharpoons [Fe(phen)_3]^{2+}$	1.147
$[Fe(phen)_3]^{3+} + e \rightleftharpoons [Fe(phen)_3]^{2+}$(1 molar H_2SO_4)	1.06
$[Ferricinium]^+ + e \rightleftharpoons$ ferrocene	0.400
$Fm^{3+} + e \rightleftharpoons Fm^{2+}$	-1.1
$Fm^{3+} + 3 e \rightleftharpoons Fm$	-1.89
$Fm^{2+} + 2 e \rightleftharpoons Fm$	-2.30
$Fr^+ + e \rightleftharpoons Fr$	-2.9
$Ga^{3+} + 3 e \rightleftharpoons Ga$	-0.549
$Ga^+ + e \rightleftharpoons Ga$	-0.2
$GaOH^{2+} + H^+ + 3 e \rightleftharpoons Ga + H_2O$	-0.498
$H_2GaO^-_3 + H_2O + 3 e \rightleftharpoons Ga + 4 OH^-$	-1.219
$Gd^{3+} + 3 e \rightleftharpoons Gd$	-2.279
$Ge^{2+} + 2 e \rightleftharpoons Ge$	0.24
$Ge^{4+} + 4 e \rightleftharpoons Ge$	0.124
$Ge^{4+} + 2 e \rightleftharpoons Ge^{2+}$	0.00
$GeO_2 + 2 H^+ + 2 e \rightleftharpoons GeO + H_2O$	-0.118
$H_2GeO_3 + 4 H^+ + 4 e \rightleftharpoons Ge + 3 H_2O$	-0.182
$2 H^+ + 2 e \rightleftharpoons H_2$	0.00000
$H_2 + 2 e \rightleftharpoons 2 H^-$	-2.23
$HO_2 + H^+ + e \rightleftharpoons H_2O_2$	1.495
$2 H_2O + 2 e \rightleftharpoons H_2 + 2 OH^-$	-0.8277
$H_2O_2 + 2 H^+ + 2 e \rightleftharpoons 2 H_2O$	1.776
$Hf^{4+} + 4 e \rightleftharpoons Hf$	-1.55
$HfO^{2+} + 2 H^+ + 4 e \rightleftharpoons Hf + H_2O$	-1.724
$HfO_2 + 4 H^+ + 4 e \rightleftharpoons Hf + 2 H_2O$	-1.505
$HfO(OH)_2 + H_2O + 4 e \rightleftharpoons Hf + 4 OH^-$	-2.50
$Hg^{2+} + 2 e \rightleftharpoons Hg$	0.851
$2 Hg^{2+} + 2 e \rightleftharpoons Hg_2^{2+}$	0.920
$Hg_2^{2+} + 2 e \rightleftharpoons 2 Hg$	0.7973
$Hg_2(ac)_2 + 2 e \rightleftharpoons 2 Hg + 2(ac)^-$	0.51163
$Hg_2Br_2 + 2 e \rightleftharpoons 2 Hg + 2 Br^-$	0.13923
$Hg_2Cl_2 + 2 e \rightleftharpoons 2 Hg + 2 Cl^-$	0.26808
$Hg_2HPO_4 + 2 e \rightleftharpoons 2 Hg + HPO_4^{2-}$	0.6359
$Hg_2I_2 + 2 e \rightleftharpoons 2 Hg + 2 I^-$	-0.0405
$Hg_2O + H_2O + 2 e \rightleftharpoons 2 Hg + 2 OH^-$	0.123
$HgO + H_2O + 2 e \rightleftharpoons Hg + 2 OH^-$	0.0977
$Hg(OH)_2 + 2 H^+ + 2 e \rightleftharpoons Hg + 2 H_2O$	1.034
$Hg_2SO_4 + 2 e \rightleftharpoons 2 Hg + SO_4^{2-}$	0.6125
$Ho^{2+} + 2 e \rightleftharpoons Ho$	-2.1
$Ho^{3+} + 3 e \rightleftharpoons Ho$	-2.33
$Ho^{3+} + e \rightleftharpoons Ho^{2+}$	-2.8
$I_2 + 2 e \rightleftharpoons 2 I^-$	0.5355
$I_3^- + 2 e \rightleftharpoons 3 I^-$	0.536
$H_3IO_6^{2-} + 2 e \rightleftharpoons IO^-_3 + 3 OH^-$	0.7
$H_5IO_6 + H^+ + 2 e \rightleftharpoons IO_3^- + 3 H_2O$	1.601
$2 HIO + 2 H^+ + 2 e \rightleftharpoons I_2 + 2 H_2O$	1.439
$HIO + H^+ + 2 e \rightleftharpoons I^- + H_2O$	0.987
$IO^- + H_2O + 2 e \rightleftharpoons I^- + 2 OH^-$	0.485
$2 IO_3^- + 12 H^+ + 10 e \rightleftharpoons I_2 + 6 H_2O$	1.195
$IO_3^- + 6 H^+ + 6 e \rightleftharpoons I^- + 3 H_2O$	1.085
$IO_3^- + 2 H_2O + 4 e \rightleftharpoons IO^- + 4 OH^-$	0.15
$IO_3^- + 3 H_2O + 6 e \rightleftharpoons IO^- + 6 OH^-$	0.26
$In^+ + e \rightleftharpoons In$	-0.14
$In^{2+} + e \rightleftharpoons In^+$	-0.40
$In^{3+} + e \rightleftharpoons In^{2+}$	-0.49
$In^{3+} + 2 e \rightleftharpoons In^+$	-0.443
$In^{3+} + 3 e \rightleftharpoons In$	-0.3382
$In(OH)_3 + 3 e \rightleftharpoons In + 3 OH^-$	-0.99
$In(OH)_4^- + 3 e \rightleftharpoons In + 4 OH^-$	-1.007
$In_2O_3 + 3 H_2O + 6 e \rightleftharpoons 2 In + 6 OH^-$	-1.034
$Ir^{3+} + 3 e \rightleftharpoons Ir$	1.156
$[IrCl_6]^{2-} + e \rightleftharpoons [IrCl_6]^{3-}$	0.8665
$[IrCl_6]^{3-} + 3 e \rightleftharpoons Ir + 6 Cl^-$	0.77
$Ir_2O_3 + 3 H_2O + 6 e \rightleftharpoons 2 Ir + 6 OH^-$	0.098
$K^+ + e \rightleftharpoons K$	-2.931
$La^{3+} + 3 e \rightleftharpoons La$	-2.379
$La(OH)_3 + 3 e \rightleftharpoons La + 3 OH^-$	-2.90
$Li^+ + e \rightleftharpoons Li$	-3.0401
$Lr^{3+} + 3 e \rightleftharpoons Lr$	-1.96
$Lu^{3+} + 3 e \rightleftharpoons Lu$	-2.28
$Md^{3+} + e \rightleftharpoons Md^{2+}$	-0.1
$Md^{3+} + 3 e \rightleftharpoons Md$	-1.65
$Md^{2+} + 2 e \rightleftharpoons Md$	-2.40
$Mg^+ + e \rightleftharpoons Mg$	-2.70
$Mg^{2+} + 2 e \rightleftharpoons Mg$	-2.372
$Mg(OH)_2 + 2 e \rightleftharpoons Mg + 2 OH^-$	-2.690
$Mn^{2+} + 2 e \rightleftharpoons Mn$	-1.185
$Mn^{3+} + e \rightleftharpoons Mn^{2+}$	1.5415
$MnO_2 + 4 H^+ + 2 e \rightleftharpoons Mn^{2+} + 2 H_2O$	1.224
$MnO_4^- + e \rightleftharpoons MnO_4^{2-}$	0.558
$MnO_4^- + 4 H^+ + 3 e \rightleftharpoons MnO_2 + 2 H_2O$	1.679
$MnO_4^- + 8 H^+ + 5 e \rightleftharpoons Mn^{2+} + 4 H_2O$	1.507
$MnO_4^- + 2 H_2O + 3 e \rightleftharpoons MnO_2 + 4 OH^-$	0.595
$MnO_4^- + 4 H_2O + 5 e \rightleftharpoons Mn(OH)_2 + 6 OH^-$	0.34
$MnO_4^{2-} + 2 H_2O + 2 e \rightleftharpoons MnO_2 + 4 OH^-$	0.60

Reaction	$E°/V$
$Mn(OH)_2 + 2 e \rightleftharpoons Mn + 2 OH^-$	-1.56
$Mn(OH)_3 + e \rightleftharpoons Mn(OH)_2 + OH^-$	0.15
$Mn_2O_3 + 6 H^+ + e \rightleftharpoons 2 Mn^{2+} + 3 H_2O$	1.485
$Mo^{3+} + 3 e \rightleftharpoons Mo$	-0.200
$MoO_2 + 4 H^+ + 4 e \rightleftharpoons Mo + 4 H_2O$	-0.152
$H_3Mo_7O_{24}^{3-} + 45 H^+ + 42 e \rightleftharpoons 7 Mo + 24 H_2O$	0.082
$MoO_3 + 6 H^+ + 6 e \rightleftharpoons Mo + 3 H_2O$	0.075
$N_2 + 2 H_2O + 6 H^+ + 6 e \rightleftharpoons 2 NH_4OH$	0.092
$3 N_2 + 2 H^+ + 2 e \rightleftharpoons 2 HN_3$	-3.09
$N_5^+ + 3 H^+ + 2 e \rightleftharpoons 2 NH_4^+$	1.275
$N_2O + 2 H^+ + 2 e \rightleftharpoons N_2 + H_2O$	1.766
$H_2N_2O_2 + 2 H^+ + 2 e \rightleftharpoons N_2 + 2 H_2O$	2.65
$N_2O_4 + 2 e \rightleftharpoons 2 NO_2^-$	0.867
$N_2O_4 + 2 H^+ + 2 e \rightleftharpoons 2 NHO_2$	1.065
$N_2O_4 + 4 H^+ + 4 e \rightleftharpoons 2 NO + 2 H_2O$	1.035
$2 NH_3OH^+ + H^+ + 2 e \rightleftharpoons N_2H_5^+ + 2 H_2O$	1.42
$2 NO + 2 H^+ + 2 e \rightleftharpoons N_2O + H_2O$	1.591
$2 NO + H_2O + 2 e \rightleftharpoons N_2O + 2 OH^-$	0.76
$HNO_2 + H^+ + e \rightleftharpoons NO + H_2O$	0.983
$2 HNO_2 + 4 H^+ + 4 e \rightleftharpoons H_2N_2O_2 + 2 H_2O$	0.86
$2 HNO_2 + 4 H^+ + 4 e \rightleftharpoons N_2O + 3 H_2O$	1.297
$NO_2^- + H_2O + e \rightleftharpoons NO + 2 OH^-$	-0.46
$2 NO_2^- + 2 H_2O + 4 e \rightleftharpoons N_2O_2^{2-} + 4 OH^-$	-0.18
$2 NO_2^- + 3 H_2O + 4 e \rightleftharpoons N_2O + 6 OH^-$	0.15
$NO_3^- + 3 H^+ + 2 e \rightleftharpoons HNO_2 + H_2O$	0.934
$NO_3^- + 4 H^+ + 3 e \rightleftharpoons NO + 2 H_2O$	0.957
$2 NO_3^- + 4 H^+ + 2 e \rightleftharpoons N_2O_4 + 2 H_2O$	0.803
$NO_3^- + H_2O + 2 e \rightleftharpoons NO_2^- + 2 OH^-$	0.01
$2 NO_3^- + 2 H_2O + 2 e \rightleftharpoons N_2O_4 + 4 OH^-$	-0.85
$Na^+ + e \rightleftharpoons Na$	-2.71
$Nb^{3+} + 3 e \rightleftharpoons Nb$	-1.099
$NbO_2 + 2 H^+ + 2 e \rightleftharpoons NbO + H_2O$	-0.646
$NbO_2 + 4 H^+ + 4 e \rightleftharpoons Nb + 2 H_2O$	-0.690
$NbO + 2 H^+ + 2 e \rightleftharpoons Nb + H_2O$	-0.733
$Nb_2O_5 + 10 H^+ + 10 e \rightleftharpoons 2 Nb + 5 H_2O$	-0.644
$Nd^{3+} + 3 e \rightleftharpoons Nd$	-2.323
$Nd^{2+} + 2 e \rightleftharpoons Nd$	-2.1
$Nd^{3+} + e \rightleftharpoons Nd^{2+}$	-2.7
$Ni^{2+} + 2 e \rightleftharpoons Ni$	-0.257
$Ni(OH)_2 + 2 e \rightleftharpoons Ni + 2 OH^-$	-0.72
$NiO_2 + 4 H^+ + 2 e \rightleftharpoons Ni^{2+} + 2 H_2O$	1.678
$NiO_2 + 2 H_2O + 2 e \rightleftharpoons Ni(OH)_2 + 2 OH^-$	-0.490
$No^{3+} + e \rightleftharpoons No^{2+}$	1.4
$No^{3+} + 3 e \rightleftharpoons No$	-1.20
$No^{2+} + 2 e \rightleftharpoons No$	-2.50
$Np^{3+} + 3 e \rightleftharpoons Np$	-1.856
$Np^{4+} + e \rightleftharpoons Np^{3+}$	0.147
$NpO_2 + H_2O + H^+ + e \rightleftharpoons Np(OH)_3$	-0.962
$O_2 + 2 H^+ + 2 e \rightleftharpoons H_2O_2$	0.695
$O_2 + 4 H^+ + 4 e \rightleftharpoons 2 H_2O$	1.229
$O_2 + H_2O + 2 e \rightleftharpoons HO_2^- + OH^-$	-0.076
$O_2 + 2 H_2O + 2 e \rightleftharpoons H_2O_2 + 2 OH^-$	-0.146
$O_2 + 2 H_2O + 4 e \rightleftharpoons 4 OH^-$	0.401
$O_3 + 2 H^+ + 2 e \rightleftharpoons O_2 + H_2O$	2.076
$O_3 + H_2O + 2 e \rightleftharpoons O_2 + 2 OH^-$	1.24
$O(g) + 2 H^+ + 2 e \rightleftharpoons H_2O$	2.421
$OH + e \rightleftharpoons OH^-$	2.02
$HO_2^- + H_2O + 2 e \rightleftharpoons 3 OH^-$	0.878
$OsO_4 + 8 H^+ + 8 e \rightleftharpoons Os + 4 H_2O$	0.838
$OsO_4 + 4 H^+ + 4 e \rightleftharpoons OsO_2 + 2 H_2O$	1.02
$[Os(bipy)_2]^{3+} + e \rightleftharpoons [Os(bipy)_2]^{2+}$	0.81
$[Os(bipy)_3]^{3+} + e \rightleftharpoons [Os(bipy)_3]^{2+}$	0.80
$P(red) + 3 H^+ + 3 e \rightleftharpoons PH_3(g)$	-0.111
$P(white) + 3 H^+ + 3 e \rightleftharpoons PH_3(g)$	-0.063
$P + 3 H_2O + 3 e \rightleftharpoons PH_3(g) + 3 OH^-$	-0.87
$H_2P_2^- + e \rightleftharpoons P + 2 OH^-$	-1.82
$H_3PO_2 + H^+ + e \rightleftharpoons P + 2 H_2O$	-0.508
$H_3PO_3 + 2 H^+ + 2 e \rightleftharpoons H_3PO_2 + H_2O$	-0.499
$H_3PO_3 + 3 H^+ + 3 e \rightleftharpoons P + 3 H_2O$	-0.454
$HPO_3^{2-} + 2 H_2O + 2 e \rightleftharpoons H_2PO_2^- + 3 OH^-$	-1.65
$HPO_3^{2-} + 2 H_2O + 3 e \rightleftharpoons P + 5 OH^-$	-1.71
$H_3PO_4 + 2 H^+ + 2 e \rightleftharpoons H_3PO_3 + H_2O$	-0.276
$PO_4^{3-} + 2 H_2O + 2 e \rightleftharpoons HPO_3^{2-} + 3 OH^-$	-1.05
$Pa^{3+} + 3 e \rightleftharpoons Pa$	-1.34
$Pa^{4+} + 4 e \rightleftharpoons Pa$	-1.49
$Pa^{4+} + e \rightleftharpoons Pa^{3+}$	-1.9
$Pb^{2+} + 2 e \rightleftharpoons Pb$	-0.1262
$Pb^{2+} + 2 e \rightleftharpoons Pb(Hg)$	-0.1205
$PbBr_2 + 2 e \rightleftharpoons Pb + 2 Br^-$	-0.284
$PbCl_2 + 2 e \rightleftharpoons Pb + 2 Cl^-$	-0.2675
$PbF_2 + 2 e \rightleftharpoons Pb + 2 F^-$	-0.3444
$PbHPO_4 + 2 e \rightleftharpoons Pb + HPO_4^{2-}$	-0.465
$PbI_2 + 2 e \rightleftharpoons Pb + 2 I^-$	-0.365
$PbO + H_2O + 2 e \rightleftharpoons Pb + 2 OH^-$	-0.580
$PbO_2 + 4 H^+ + 2 e \rightleftharpoons Pb^{2+} + 2 H_2O$	1.455
$HPbO_2^- + H_2O + 2 e \rightleftharpoons Pb + 3 OH^-$	-0.537
$PbO_2 + H_2O + 2 e \rightleftharpoons PbO + 2 OH^-$	0.247
$PbO_2 + SO_4^{2-} + 4 H^+ + 2 e \rightleftharpoons PbSO_4 + 2 H_2O$	1.6913
$PbSO_4 + 2 e \rightleftharpoons Pb + SO_4^{2-}$	-0.3588
$PbSO_4 + 2 e \rightleftharpoons Pb(Hg) + SO_4^{2-}$	-0.3505
$Pd^{2+} + 2 e \rightleftharpoons Pd$	0.951
$[PdCl_4]^{2-} + 2 e \rightleftharpoons Pd + 4 Cl^-$	0.591
$[PdCl_6]^{2-} + 2 e \rightleftharpoons [PdCl_4]^{2-} + 2 Cl^-$	1.288
$Pd(OH)_2 + 2 e \rightleftharpoons Pd + 2 OH^-$	0.07
$Pm^{2+} + 2 e \rightleftharpoons Pm$	-2.2
$Pm^{3+} + 3 e \rightleftharpoons Pm$	-2.30
$Pm^{3+} + e \rightleftharpoons Pm^{2+}$	-2.6
$Po^{2+} + 2 e \rightleftharpoons Po$	0.368
$PoO_2 + 4 H^+ + 2 e \rightleftharpoons Po^{2+} + 2 H_2O$	1.095
$Pr^{4+} + e \rightleftharpoons Pr^{3+}$	3.2
$Pr^{2+} + 2 e \rightleftharpoons Pr$	-2.0
$Pr^{3+} + 3 e \rightleftharpoons Pr$	-2.353
$Pr^{3+} + e \rightleftharpoons Pr^{2+}$	-3.1
$Pt^{2+} + 2 e \rightleftharpoons Pt$	1.18
$[PtCl_4]^{2-} + 2 e \rightleftharpoons Pt + 4 Cl^-$	0.755
$[PtCl_6]^{2-} + 2 e \rightleftharpoons [PtCl_4]^{2-} + 2 Cl^-$	0.68
$Pt(OH)_2 + 2 e \rightleftharpoons Pt + 2 OH^-$	0.14
$PtO_3 + 2 H^+ + 2 e \rightleftharpoons PtO_2 + H_2O$	1.7
$PtO_3 + 4 H^+ + 2 e \rightleftharpoons Pt(OH)_2^{2+} + H_2O$	1.5
$PtOH^+ + H^+ + 2 e \rightleftharpoons Pt + H_2O$	1.2
$PtO_2 + 2 H^+ + 2 e \rightleftharpoons PtO + H_2O$	1.01
$PtO_2 + 4 H^+ + 4 e \rightleftharpoons Pt + 2 H_2O$	1.00
$Pu^{3+} + 3 e \rightleftharpoons Pu$	-2.031
$Pu^{4+} + e \rightleftharpoons Pu^{3+}$	1.006
$Pu^{5+} + e \rightleftharpoons Pu^{4+}$	1.099
$PuO_2(OH)_2 + 2 H^+ + 2 e \rightleftharpoons Pu(OH)_4$	1.325
$PuO_2(OH)_2 + H^+ + e \rightleftharpoons PuO_2OH + H_2O$	1.062
$Ra^{2+} + 2 e \rightleftharpoons Ra$	-2.8
$Rb^+ + e \rightleftharpoons Rb$	-2.98
$Re^{3+} + 3 e \rightleftharpoons Re$	0.300
$ReO_4^- + 4 H^+ + 3 e \rightleftharpoons ReO_2 + 2 H_2O$	0.510
$ReO_2 + 4 H^+ + 4 e \rightleftharpoons Re + 2 H_2O$	0.2513
$ReO_4^- + 2 H^+ + e \rightleftharpoons ReO_3 + H_2O$	0.768
$ReO_4^- + 4 H_2O + 7 e \rightleftharpoons Re + 8 OH^-$	-0.604
$ReO_4^- + 8 H^+ + 7 e \rightleftharpoons Re + 4 H_2O$	0.34
$Rh^+ + e \rightleftharpoons Rh$	0.600
$Rh^{3+} + 3 e \rightleftharpoons Rh$	0.758
$[RhCl_6]^{3-} + 3 e \rightleftharpoons Rh + 6 Cl^-$	0.431
$RhOH^{2+} + H^+ + 3 e \rightleftharpoons Rh + H_2O$	0.83
$Ru^{2+} + 2 e \rightleftharpoons Ru$	0.455
$Ru^{3+} + e \rightleftharpoons Ru^{2+}$	0.2487
$RuO_2 + 4 H^+ + 2 e \rightleftharpoons Ru^{2+} + 2 H_2O$	1.120
$RuO_4^- + e \rightleftharpoons RuO_4^{2-}$	0.59
$RuO_4 + e \rightleftharpoons RuO_4^-$	1.00
$RuO_4 + 6 H^+ + 4 e \rightleftharpoons Ru(OH)_2^{2+} + 2 H_2O$	1.40
$RuO_4 + 8 H^+ + 8 e \rightleftharpoons Ru + 4 H_2O$	1.038
$[Ru(bipy)_3]^{3+} + e \rightleftharpoons [Ru(bipy)_3]^{2+}$	1.24
$[Ru(H_2O)_6]^{3+} + e \rightleftharpoons [Ru(H_2O)_6]^{2+}$	0.23
$[Ru(NH_3)_6]^{3+} + e \rightleftharpoons [Ru(NH_3)_6]^{2+}$	0.10
$[Ru(en)_3]^{3+} + e \rightleftharpoons [Ru(en)_3]^{2+}$	0.210
$[Ru(CN)_6]^{3-} + e \rightleftharpoons [Ru(CN)_6]^{4-}$	0.86
$S + 2 e \rightleftharpoons S^{2-}$	-0.47627
$S + 2H^+ + 2 e \rightleftharpoons H_2S(aq)$	0.142
$S + H_2O + 2 e \rightleftharpoons SH^- + OH^-$	-0.478
$2 S + 2 e \rightleftharpoons S_2^{2-}$	-0.42836
$S_2O_6^{2-} + 4 H^+ + 2 e \rightleftharpoons 2 H_2SO_3$	0.564
$S_2O_8^{2-} + 2 e \rightleftharpoons 2 SO_4^{2-}$	2.010
$S_2O_8^{2-} + 2 H^+ + 2 e \rightleftharpoons 2 HSO_4^-$	2.123
$S_4O_6^{2-} + 2 e \rightleftharpoons 2 S_2O_3^{2-}$	0.08
$2 H_2SO_3 + H^+ + 2 e \rightleftharpoons HS_2O_4^- + 2 H_2O$	-0.056
$H_2SO_3 + 4 H^+ + 4 e \rightleftharpoons S + 3 H_2O$	0.449
$2 SO_3^{2-} + 2 H_2O + 2 e \rightleftharpoons S_2O_4^{2-} + 4 OH^-$	-1.12
$2 SO_3^{2-} + 3 H_2O + 4 e \rightleftharpoons S_2O_3^{2-} + 6 OH^-$	-0.571
$SO_4^{2-} + 4 H^+ + 2 e \rightleftharpoons H_2SO_3 + H_2O$	0.172
$2 SO_4^{2-} + 4 H^+ + 2 e \rightleftharpoons S_2O_6^{2-} + H_2O$	-0.22
$SO_4^{2-} + H_2O + 2 e \rightleftharpoons SO_3^{2-} + 2 OH^-$	-0.93
$Sb + 3 H^+ + 3 e \rightleftharpoons SbH_3$	-0.510
$Sb_2O_3 + 6 H^+ + 6 e \rightleftharpoons 2 Sb + 3 H_2O$	0.152
$Sb_2O_5(senarmontite) + 4 H^+ + 4 e \rightleftharpoons Sb_2O_3 + 2 H_2O$	0.671
$Sb_2O_5(valentinite) + 4 H^+ + 4 e \rightleftharpoons Sb_2O_3 + 2 H_2O$	0.649
$Sb_2O_5 + 6 H^+ + 4 e \rightleftharpoons 2 SbO^+ + 3 H_2O$	0.581
$SbO^+ + 2 H^+ + 3 e \rightleftharpoons Sb + 2 H_2O$	0.212
$SbO_2^- + 2 H_2O + 3 e \rightleftharpoons Sb + 4 OH^-$	-0.66
$SbO_3^- + H_2O + 2 e \rightleftharpoons SbO_2^- + 2 OH^-$	-0.59
$Sc^{3+} + 3 e \rightleftharpoons Sc$	-2.077
$Se + 2 e \rightleftharpoons Se^{2-}$	-0.670
$Se + 2 H^+ + 2 e \rightleftharpoons H_2Se(aq)$	-0.399
$H_2SeO_3 + 4 H^+ + 4 e \rightleftharpoons Se + 3 H_2O$	0.74
$Se + 2 H^+ + 2 e \rightleftharpoons H_2Se$	-0.082
$SeO_3^{2-} + 3 H_2O + 4 e \rightleftharpoons Se + 6 OH^-$	-0.366
$SeO_4^{2-} + 4 H^+ + 2 e \rightleftharpoons H_2SeO_3 + H_2O$	1.151

Reaction	$E°$/V
$SeO_4^{2-} + H_2O + 2e \rightleftharpoons SeO_3^{2-} + 2OH^-$	0.05
$SiF_6^{2-} + 4e \rightleftharpoons Si + 6F^-$	-1.24
$SiO + 2H^+ + 2e \rightleftharpoons Si + H_2O$	-0.8
SiO_2(quartz) $+ 4H^+ + 4e \rightleftharpoons Si + 2H_2O$	0.857
$SiO_3^{2-} + 3H_2O + 4e \rightleftharpoons Si + 6OH^-$	-1.697
$Sm^{3+} + e \rightleftharpoons Sm^{2+}$	-1.55
$Sm^{3+} + 3e \rightleftharpoons Sm$	-2.304
$Sm^{2+} + 2e \rightleftharpoons Sm$	-2.68
$Sn^{2+} + 2e \rightleftharpoons Sn$	-0.1375
$Sn^{4+} + 2e \rightarrow Sn^{2+}$	0.151
$Sn(OH)_3^+ + 3H^+ + 2e \rightleftharpoons Sn^{2+} + 3H_2O$	0.142
$SnO_2 + 4H^+ + 2e \rightleftharpoons Sn^{2+} + 2H_2O$	-0.094
$SnO_2 + 4H^+ + 4e \rightleftharpoons Sn + 2H_2O$	-0.117
$SnO_2 + 3H^+ + 2e \rightleftharpoons SnOH^+ + H_2O$	-0.194
$SnO_2 + 2H_2O + 4e \rightleftharpoons Sn + 4OH^-$	-0.945
$HSnO_2^- + H_2O + 2e \rightleftharpoons Sn + 3OH^-$	-0.909
$Sn(OH)_6^{2-} + 2e \rightleftharpoons HSnO_2^- + 3OH^- + H_2O$	-0.93
$Sr^+ + e \rightleftharpoons Sr$	-4.10
$Sr^{2+} + 2e \rightleftharpoons Sr$	-2.899
$Sr^{2+} + 2e \rightleftharpoons Sr(Hg)$	-1.793
$Sr(OH)_2 + 2e \rightleftharpoons Sr + 2OH^-$	-2.88
$Ta_2O_5 + 10H^+ + 10e \rightleftharpoons 2Ta + 5H_2O$	-0.750
$Ta^{3+} + 3e \rightleftharpoons Ta$	-0.6
$Tc^{2+} + 2e \rightleftharpoons Tc$	0.400
$TcO_4^- + 4H^+ + 3e \rightleftharpoons TcO_2 + 2H_2O$	0.738
$Tc^{3+} + e \rightleftharpoons Tc^{2+}$	0.3
$TcO_4^- + 8H^+ + 7e \rightleftharpoons Tc + 4H_2O$	0.472
$Tb^{4+} + e \rightleftharpoons Tb^{3+}$	3.1
$Tb^{3+} + 3e \rightleftharpoons Tb$	-2.28
$Te + 2e \rightleftharpoons Te^{2-}$	-1.143
$Te + 2H^+ + 2e \rightleftharpoons H_2Te$	-0.793
$Te^{4+} + 4e \rightleftharpoons Te$	0.568

Reaction	$E°$/V
$TeO_2 + 4H^+ + 4e \rightleftharpoons Te + 2H_2O$	0.593
$TeO_3^{2-} + 3H_2O + 4e \rightleftharpoons Te + 6OH^-$	-0.57
$TeO_4^- + 8H^+ + 7e \rightleftharpoons Te + 4H_2O$	0.472
$H_6TeO_6 + 2H^+ + 2e \rightleftharpoons TeO_2 + 4H_2O$	1.02
$Th^{4+} + 4e \rightleftharpoons Th$	-1.899
$ThO_2 + 4H^+ + 4e \rightleftharpoons Th + 2H_2O$	-1.789
$Th(OH)_4 + 4e \rightleftharpoons Th + 4OH^-$	-2.48
$Ti^{2+} + 2e \rightleftharpoons Ti$	-1.628
$Ti^{3+} + e \rightleftharpoons Ti^{2+}$	-0.369
$TiO_2 + 4H^+ + 2e \rightleftharpoons Ti^{2+} + 2H_2O$	-0.502
$Ti^{3+} + 3e \rightleftharpoons Ti$	-1.209
$TiOH^{3+} + H^+ + e \rightleftharpoons Ti^{3+} + H_2O$	-0.055
$Tl^+ + e \rightleftharpoons Tl$	-0.336
$Tl^+ + e \rightleftharpoons Tl(Hg)$	-0.3338
$Tl^{3+} + 2e \rightleftharpoons Tl^+$	1.252
$Tl^{3+} + 3e \rightleftharpoons Tl$	0.741
$TlBr + e \rightleftharpoons Tl + Br^-$	-0.658
$TlCl + e \rightleftharpoons Tl + Cl^-$	-0.5568
$TlI + e \rightleftharpoons Tl + I^-$	-0.752
$Tl_2O_3 + 3H_2O + 4e \rightleftharpoons 2Tl^+ + 6OH^-$	0.02
$TlOH + e \rightleftharpoons Tl + OH^-$	-0.34
$Tl(OH)_3 + 2e \rightleftharpoons TlOH + 2OH^-$	-0.05
$Tl_2SO_4 + 2e \rightleftharpoons Tl + SO_4^{2-}$	-0.4360
$Tm^{3+} + e \rightleftharpoons Tm^{2+}$	-2.2
$Tm^{3+} + 3e \rightleftharpoons Tm$	-2.319
$Tm^{2+} + 2e \rightleftharpoons Tm$	-2.4
$U^{3+} + 3e \rightleftharpoons U$	-1.66
$U^{4+} + e \rightleftharpoons U^{3+}$	-0.52
$UO_2^+ + 4H^+ + e \rightleftharpoons U^{4+} + 2H_2O$	0.612
$UO_2^{2+} + e \rightleftharpoons UO_2^+$	0.16
$UO_2^{2+} + 4H^+ + 2e \rightleftharpoons U^{4+} + 2H_2O$	0.327
$UO_2^{2+} + 4H^+ + 6e \rightleftharpoons U + 2H_2O$	-1.444
$V^{2+} + 2e \rightleftharpoons V$	-1.175

Reaction	$E°$/V
$V^{3+} + e \rightleftharpoons V^{2+}$	-0.255
$VO^{2+} + 2H^+ + e \rightleftharpoons V^{3+} + H_2O$	0.337
$VO_2^+ + 2H^+ + e \rightleftharpoons VO^{2+} + H_2O$	0.991
$V_2O_5 + 6H^+ + 2e \rightleftharpoons 2VO^{2+} + 3H_2O$	0.957
$V_2O_5 + 10H^+ + 10e \rightleftharpoons 2V + 5H_2O$	-0.242
$V(OH)_4^+ + 2H^+ + e \rightleftharpoons VO^{2+} + 3H_2O$	1.00
$V(OH)_4^+ + 4H^+ + 5e \rightleftharpoons V + 4H_2O$	-0.254
$[V(phen)_3]^{3+} + e \rightleftharpoons [V(phen)_3]^{2+}$	0.14
$W^{3+} + 3e \rightleftharpoons W$	0.1
$W_2O_5 + 2H^+ + 2e \rightleftharpoons 2WO_2 + H_2O$	-0.031
$WO_2 + 4H^+ + 4e \rightleftharpoons W + 2H_2O$	-0.119
$WO_3 + 6H^+ + 6e \rightleftharpoons W + 3H_2O$	-0.090
$WO_3 + 2H^+ + 2e \rightleftharpoons WO_2 + H_2O$	0.036
$2WO_3 + 2H^+ + 2e \rightleftharpoons W_2O_5 + H_2O$	-0.029
$H_4XeO_6 + 2H^+ + 2e \rightleftharpoons XeO_3 + 3H_2O$	2.42
$XeO_3 + 6H^+ + 6e \rightleftharpoons Xe + 3H_2O$	2.10
$XeF + e \rightleftharpoons Xe + F^-$	3.4
$Y^{3+} + 3e \rightleftharpoons Y$	-2.372
$Yb^{3+} + e \rightleftharpoons Yb^{2+}$	-1.05
$Yb^{3+} + 3e \rightleftharpoons Yb$	-2.19
$Yb^{2+} + 2e \rightleftharpoons Yb$	-2.76
$Zn^{2+} + 2e \rightleftharpoons Zn$	-0.7618
$Zn^{2+} + 2e \rightleftharpoons Zn(Hg)$	-0.7628
$ZnO_2^{2-} + 2H_2O + 2e \rightleftharpoons Zn + 4OH^-$	-1.215
$ZnSO_4 \cdot 7H_2O + 2e = Zn(Hg) + SO_4^{2-} + 7H_2O$ (Saturated $ZnSO_4$)	-0.7993
$ZnOH^+ + H^+ + 2e \rightleftharpoons Zn + H_2O$	-0.497
$Zn(OH)_4^{2-} + 2e \rightleftharpoons Zn + 4OH^-$	-1.199
$Zn(OH)_2 + 2e \rightleftharpoons Zn + 2OH^-$	-1.249
$ZnO + H_2O + 2e \rightleftharpoons Zn + 2OH^-$	-1.260
$ZrO_2 + 4H^+ + 4e \rightleftharpoons Zr + 2H_2O$	-1.553
$ZrO(OH)_2 + H_2O + 4e \rightleftharpoons Zr + 4OH^-$	-2.36
$Zr^{4+} + 4e \rightleftharpoons Zr$	-1.45

TABLE 2. Reduction Reactions Having $E°$ Values More Positive Than That of the Standard Hydrogen Electrode

Reaction	$E°$/V
$2H^+ + 2e \rightleftharpoons H_2$	0.00000
$CuI_2^- + e \rightleftharpoons Cu + 2I^-$	0.00
$Ge^{4+} + 2e \rightleftharpoons Ge^{2+}$	0.00
$NO_3^- + H_2O + 2e \rightleftharpoons NO_2^- + 2OH^-$	0.01
$Tl_2O_3 + 3H_2O + 4e \rightleftharpoons 2Tl^+ + 6OH^-$	0.02
$SeO_4^{2-} + H_2O + 2e \rightleftharpoons SeO_3^{2-} + 2OH^-$	0.05
$WO_3 + 2H^+ + 2e \rightleftharpoons WO_2 + H_2O$	0.036
$Pd(OH)_2 + 2e \rightleftharpoons Pd + 2OH^-$	0.07
$AgBr + e \rightleftharpoons Ag + Br^-$	0.07133
$MoO_3 + 6H^+ + 6e \rightleftharpoons Mo + 3H_2O$	0.075
$S_4O_6^{2-} + 2e \rightleftharpoons 2S_2O_3^{2-}$	0.08
$H_3Mo_7O_{24}^{3-} + 45H^+ + 42e \rightleftharpoons 7Mo + 24H_2O$	0.082
$AgSCN + e \rightleftharpoons Ag + SCN^-$	0.8951
$N_2 + 2H_2O + 6H^+ + 6e \rightleftharpoons 2NH_4OH$	0.092
$HgO + H_2O + 2e \rightleftharpoons Hg + 2OH^-$	0.0977
$Ir_2O_3 + 3H_2O + 6e \rightleftharpoons 2Ir + 6OH^-$	0.098
$2NO + 2e \rightleftharpoons N_2O_2^{2-}$	0.10
$[Ru(NH_3)_6]^{3+} + e \rightleftharpoons [Ru(NH_3)_6]^{2+}$	0.10
$W^{3+} + 3e \rightleftharpoons W$	0.1
$[Co(NH_3)_6]^{3+} + e \rightleftharpoons [Co(NH_3)_6]^{2+}$	0.108

Reaction	$E°$/V
$Hg_2O + H_2O + 2e \rightleftharpoons 2Hg + 2OH^-$	0.123
$Ge^{4+} + 4e \rightleftharpoons Ge$	0.124
$Hg_2Br_2 + 2e \rightleftharpoons 2Hg + 2Br^-$	0.13923
$Pt(OH)_2 + 2e \rightleftharpoons Pt + 2OH^-$	0.14
$[V(phen)_3]^{3+} + e \rightleftharpoons [V(phen)_3]^{2+}$	0.14
$S + 2H^+ + 2e \rightleftharpoons H_2S(aq)$	0.142
$Sn(OH)_3^+ + 3H^+ + 2e \rightleftharpoons Sn^{2+} + 3H_2O$	0.142
$Np^{4+} + e \rightleftharpoons Np^{3+}$	0.147
$Ag_4[Fe(CN)_6] + 4e \rightleftharpoons 4Ag + [Fe(CN)_6]^{4-}$	0.1478
$IO_3^- + 2H_2O + 4e \rightleftharpoons IO^- + 4OH^-$	0.15
$Mn(OH)_3 + e \rightleftharpoons Mn(OH)_2 + OH^-$	0.15
$2NO_2^- + 3H_2O + 4e \rightleftharpoons N_2O + 6OH^-$	0.15
$Sn^{4+} + 2e \rightleftharpoons Sn^{2+}$	0.151
$Sb_2O_3 + 6H^+ + 6e \rightleftharpoons 2Sb + 3H_2O$	0.152
$Cu^{2+} + e \rightleftharpoons Cu^+$	0.153
$BiOCl + 2H^+ + 3e \rightleftharpoons Bi + Cl^- + H_2O$	0.1583
$BiCl_4^- + 3e \rightleftharpoons Bi + 4Cl^-$	0.16
$Fe_2O_3 + 4H^+ + 2e \rightleftharpoons 2FeOH^+ + H_2O$	0.16
$UO_2^{2+} + e = UO_2^+$	0.16
$Co(OH)_3 + e \rightleftharpoons Co(OH)_2 + OH^-$	0.17

Reaction	$E°$/V
$SO_4^{2-} + 4H^+ + 2e \rightleftharpoons H_2SO_3 + H_2O$	0.172
$Bi^{3+} + 2e \rightleftharpoons Bi^+$	0.2
$At_2 + 2e \rightleftharpoons 2At^-$	0.2
$[Ru(en)_3]^{3+} + e \rightleftharpoons [Ru(en)_3]^{2+}$	0.210
$SbO^+ + 2H^+ + 3e \rightleftharpoons Sb + 2H_2O$	0.212
$AgCl + e \rightleftharpoons Ag + Cl^-$	0.22233
$[Ru(H_2O)_6]^{3+} + e \rightleftharpoons [Ru(H_2O)_6]^{2+}$	0.23
$As_2O_3 + 6H^+ + 6e \rightleftharpoons 2As + 3H_2O$	0.234
Calomel electrode, saturated NaCl (SSCE)	0.2360
$Ge^{2+} + 2e \rightleftharpoons Ge$	0.24
$Ru^{3+} + e \rightleftharpoons Ru^{2+}$	0.24
Calomel electrode, saturated KCl	0.2412
$PbO_2 + H_2O + 2e \rightleftharpoons PbO + 2OH^-$	0.247
$HAsO_2 + 3H^+ + 3e \rightleftharpoons As + 2H_2O$	0.248
$Ru^{3+} + e \rightarrow Ru^{2+}$	0.2487
$ReO_2 + 4H^+ + 4e \rightleftharpoons Re + 2H_2O$	0.2513
$IO_3^- + 3H_2O + 6e \rightleftharpoons I^- + OH^-$	0.26
$Hg_2Cl_2 + 2e \rightleftharpoons 2Hg + 2Cl^-$	0.26808
Calomel electrode, 1 molal KCl	0.2800

Reaction	$E°/V$
Calomel electrode, 1 molar KCl (NCE)	0.2801
$Re^{3+} + 3 e \rightleftharpoons Re$	0.300
$Tc^{3+} + e \rightleftharpoons Tc^{2+}$	0.3
$Bi^{3+} + 3 e \rightleftharpoons Bi$	0.308
$BiO^{+} + 2 H^{+} + 3 e \rightleftharpoons Bi + H_2O$	0.320
$UO_2^{2+} + 4 H^{+} + 2 e \rightleftharpoons U^{4+} + 2 H_2O$	0.327
$ClO_3^{-} + H_2O + 2 e \rightleftharpoons ClO_2^{-} + 2 OH^{-}$	0.33
$2 HCNO + 2 H^{+} + 2 e \rightleftharpoons (CN)_2 + 2 H_2O$	0.330
Calomel electrode, 0.1 molar KCl	0.3337
$VO^{2+} + 2 H^{+} + e \rightleftharpoons V^{3+} + H_2O$	0.337
$MnO_4^{-} + 4 H_2O + 5 e \rightleftharpoons Mn(OH)_2 + 6 OH^{-}$	0.34
$ReO_4^{-} + 8 H^{+} + 7 e \rightleftharpoons Re + 4 H_2O$	0.34
$Cu^{2+} + 2 e \rightleftharpoons Cu$	0.3419
$Ag_2O + H_2O + 2 e \rightleftharpoons 2 Ag + 2 OH^{-}$	0.342
$Cu^{2+} + 2 e \rightleftharpoons Cu(Hg)$	0.345
$AgIO_3 + e \rightleftharpoons Ag + IO_3^{-}$	0.354
$[Fe(CN)_6]^{3-} + e \rightleftharpoons [Fe(CN)_6]^{4-}$	0.358
$ClO_4^{-} + H_2O + 2 e \rightleftharpoons ClO^{-}_3 + 2 OH^{-}$	0.36
$Ag_2SeO_3 + 2 e \rightleftharpoons 2 Ag + SeO_3^{2-}$	0.3629
$Po^{2+} + 2 e \rightleftharpoons Po$	0.368
$(CN)_2 + 2 H^{+} + 2 e \rightleftharpoons 2 HCN$	0.373
$[Ferricinium]^{+} + e \rightleftharpoons$ ferrocene	0.400
$Tc^{2+} + 2 e \rightleftharpoons Tc$	0.400
$O_2 + 2 H_2O + 4 e \rightleftharpoons 4 OH^{-}$	0.401
$AgOCN + e \rightleftharpoons Ag + OCN^{-}$	0.41
$[RhCl_6]^{3-} + 3 e \rightleftharpoons Rh + 6 Cl^{-}$	0.431
$Ag_2CrO_4 + 2 e \rightleftharpoons 2 Ag + CrO_4^{2-}$	0.4470
$H_2SO_3 + 4 H^{+} + 4 e \rightleftharpoons S + 3 H_2O$	0.449
$Ru^{2+} + 2 e \rightleftharpoons Ru$	0.455
$Ag_2MoO_4 + 2 e \rightleftharpoons 2 Ag + MoO_4^{2-}$	0.4573
$Ag_2C_2O_4 + 2 e \rightleftharpoons 2 Ag + C_2O_4^{2-}$	0.4647
$Ag_2WO_4 + 2 e \rightleftharpoons 2 Ag + WO_4^{2-}$	0.4660
$Ag_2CO_3 + 2 e \rightleftharpoons 2 Ag + CO_3^{2-}$	0.47
$TcO_4^{-} + 8 H^{+} + 7 e \rightleftharpoons Tc + 4 H_2O$	0.472
$TeO_4^{-} + 8 H^{+} + 7 e \rightleftharpoons Te + 4 H_2O$	0.472
$IO^{-} + H_2O + 2 e \rightleftharpoons I^{-} + 2 OH^{-}$	0.485
$NiO_2 + 2 H_2O + 2 e \rightleftharpoons Ni(OH)_2 + 2 OH^{-}$	0.490
$Bi^{+} + e \rightleftharpoons Bi$	0.5
$ReO_4^{-} + 4 H^{+} + 3 e \rightleftharpoons ReO_2 + 2 H_2O$	0.510
$Hg_2(ac)_2 + 2 e \rightleftharpoons 2 Hg + 2(ac)^{-}$	0.51163
$Cu^{+} + e \rightleftharpoons Cu$	0.521
$I_2 + 2 e \rightleftharpoons 2 I^{-}$	0.5355
$I_3^{-} + 2 e \rightleftharpoons 3 I^{-}$	0.536
$AgBrO_3 + e \rightleftharpoons Ag + BrO_3^{-}$	0.546
$MnO_4^{-} + e \rightleftharpoons MnO_4^{2-}$	0.558
$H_3AsO_4 + 2 H^{+} + 2 e \rightleftharpoons HAsO_2 + 2 H_2O$	0.560
$S_2O_6^{2-} + 4 H^{+} + 2 e \rightleftharpoons 2 H_2SO_3$	0.564
$AgNO_2 + e \rightleftharpoons Ag + NO_2^{-}$	0.564
$Te^{4+} + 4 e \rightleftharpoons Te$	0.568
$Sb_2O_5 + 6 H^{+} + 4 e \rightleftharpoons 2 SbO^{+} + 3 H_2O$	0.581
$RuO_4^{-} + e \rightleftharpoons RuO_4^{2-}$	0.59
$[PdCl_4]^{2-} + 2 e \rightleftharpoons Pd + 4 Cl^{-}$	0.591
$TeO_2 + 4 H^{+} + 4 e \rightleftharpoons Te + 2 H_2O$	0.593
$MnO_4^{-} + 2 H_2O + 3 e \rightleftharpoons MnO_2 + 4 OH^{-}$	0.595
$Rh^{2+} + 2 e \rightleftharpoons Rh$	0.600
$Rh^{+} + e \rightleftharpoons Rh$	0.600

Reaction	$E°/V$
$MnO_4^{2-} + 2 H_2O + 2 e \rightleftharpoons MnO_2 + 4 OH^{-}$	0.60
$2 AgO + H_2O + 2 e \rightleftharpoons Ag_2O + 2 OH^{-}$	0.607
$BrO_3^{-} + 3 H_2O + 6 e \rightleftharpoons Br^{-} + 6 OH^{-}$	0.61
$UO_2^{+} + 4 H^{+} + e \rightleftharpoons U^{4+} + 2 H_2O$	0.612
$Hg_2SO_4 + 2 e \rightleftharpoons 2 Hg + SO_4^{2-}$	0.6125
$ClO_3^{-} + 3 H_2O + 6 e \rightleftharpoons Cl^{-} + 6 OH^{-}$	0.62
$Hg_2HPO_4 + 2 e \rightleftharpoons 2 Hg + HPO_4^{2-}$	0.6359
$Ag(ac) + e \rightleftharpoons Ag + (ac)^{-}$	0.643
Sb_2O_5(valentinite) $+ 4 H^{+} + 4 e \rightleftharpoons Sb_2O_3 + 2 H_2O$	0.649
$Ag_2SO_4 + 2 e \rightleftharpoons 2 Ag + SO_4^{2-}$	0.654
$ClO_2^{-} + H_2O + 2 e \rightleftharpoons ClO^{-} + 2 OH^{-}$	0.66
Sb_2O_5(senarmontite) $+ 4 H^{+} + 4 e \rightleftharpoons Sb_2O_5 + 2 H_2O$	0.671
$[PtCl_6]^{2-} + 2 e \rightleftharpoons [PtCl_4]^{2-} + 2 Cl^{-}$	0.68
$O_2 + 2 H^{+} + 2 e \rightleftharpoons H_2O_2$	0.695
p–benzoquinone $+ 2 H^{+} + 2 e \rightleftharpoons$ hydroquinone	0.6992
$H_3IO_6^{2-} + 2 e \rightleftharpoons IO_3^{-} + 3 OH^{-}$	0.7
$TcO_4^{-} + 4 H^{+} + 3 e \rightleftharpoons TcO_2 + 2 H_2O$	0.738
$Ag_2O_3 + H_2O + 2 e \rightleftharpoons 2 AgO + 2 OH^{-}$	0.739
$Tl^{3+} + 3 e \rightleftharpoons Tl$	0.741
$[PtCl_4]^{2-} + 2 e \rightleftharpoons Pt + 4 Cl^{-}$	0.755
$Rh^{3+} + 3 e \rightleftharpoons Rh$	0.758
$ClO_2^{-} + 2 H_2O + 4 e \rightleftharpoons Cl^{-} + 4 OH^{-}$	0.76
$2 NO + H_2O + 2 e \rightleftharpoons N_2O + 2 OH^{-}$	0.76
$BrO^{-} + H_2O + 2 e \rightleftharpoons Br^{-} + 2 OH^{-}$	0.761
$ReO_4^{-} + 2 H^{+} + e \rightleftharpoons ReO_3 + H_2O$	0.768
$(CNS)_2 + 2 e \rightleftharpoons 2 CNS^{-}$	0.77
$[IrCl_6]^{3-} + 3 e \rightleftharpoons Ir + 6 Cl^{-}$	0.77
$Fe^{3+} + e \rightleftharpoons Fe^{2+}$	0.771
$AgF + e \rightleftharpoons Ag + F^{-}$	0.779
$[Fe(bipy)_2]^{3+} + e \rightleftharpoons [Fe(bipy)_2]^{2+}$	0.78
$Hg_2^{2+} + 2 e \rightleftharpoons 2 Hg$	0.7973
$Ag^{+} + e \rightleftharpoons Ag$	0.7996
$[Os(bipy)_3]^{3+} + e \rightleftharpoons [Os(bipy)_3]^{2+}$	0.80
$2 NO_3^{-} + 4 H^{+} + 2 e \rightleftharpoons N_2O_4 + 2 H_2O$	0.803
$[Os(bipy)_2]^{3+} + e \rightleftharpoons [Os(bipy)_2]^{2+}$	0.81
$RhOH^{2+} + H + 3 e \rightleftharpoons Rh + H_2O$	0.83
$OsO_4 + 8 H^{+} + 8 e \rightleftharpoons Os + 4 H_2O$	0.838
$ClO^{-} + H_2O + 2 e \rightleftharpoons Cl^{-} + 2 OH^{-}$	0.841
$Hg^{2+} + 2 e \rightleftharpoons Hg$	0.851
$AuBr_4^{-} + 3 e \rightleftharpoons Au + 4 Br^{-}$	0.854
SiO_2(quartz) $+ 4 H^{+} + 4 e \rightleftharpoons Si + 2 H_2O$	0.857
$2 HNO_2 + 4 H^{+} + 4 e \rightleftharpoons H_2N_2O_2 + H_2O$	0.86
$[Ru(CN)_6]^{3-} + e \rightleftharpoons [Ru(CN)_6]^{4-}$	0.86
$[IrCl_6]^{2-} + e \rightleftharpoons [IrCl_6]^{3-}$	0.8665
$N_2O_4 + 2 e \rightleftharpoons 2 NO_2^{-}$	0.867
$HO_2^{-} + H_2O + 2 e \rightleftharpoons 3 OH^{-}$	0.878
$2 Hg^{2+} + 2 e \rightleftharpoons Hg_2^{2+}$	0.920
$NO_3^{-} + 3 H^{+} + 2 e \rightleftharpoons HNO_2 + H_2O$	0.934
$Pd^{2+} + 2 e \rightleftharpoons Pd$	0.951
$ClO_2(aq) + e \rightleftharpoons ClO_2^{-}$	0.954
$NO_3^{-} + 4 H^{+} + 3 e \rightleftharpoons NO + 2 H_2O$	0.957
$V_2O_5 + 6 H^{+} + 2 e \rightleftharpoons 2 VO^{2+} + 3 H_2O$	0.957
$AuBr_2^{-} + e \rightleftharpoons Au + 2 Br^{-}$	0.959
$HNO_2 + H^{+} + e \rightleftharpoons NO + H_2O$	0.983
$HIO + H^{+} + 2 e \rightleftharpoons I^{-} + H_2O$	0.987
$VO_2^{+} + 2 H^{+} + e \rightleftharpoons VO^{2+} + H_2O$	0.991

Reaction	$E°/V$
$PtO_2 + 4 H^{+} + 4 e \rightleftharpoons Pt + 2 H_2O$	1.00
$RuO_4 + e \rightleftharpoons RuO_4^{-}$	1.00
$V(OH)_4^{+} + 2 H^{+} + e \rightleftharpoons VO^{2+} + 3 H_2O$	1.00
$AuCl_4^{-} + 3 e \rightleftharpoons Au + 4 Cl^{-}$	1.002
$Pu^{4+} + e \rightleftharpoons Pu^{3+}$	1.006
$PtO_2 + 2 H^{+} + 2 e \rightleftharpoons PtO + H_2O$	1.01
$OsO_4 + 4 H + 4 e \rightleftharpoons OsO_2 + 2 H_2O$	1.02
$H_6TeO_6 + 2 H^{+} + 2 e \rightleftharpoons TeO_2 + 4 H_2O$	1.02
$[Fe(bipy)_3]^{3+} + e \rightleftharpoons [Fe(bipy)_3]^{2+}$	1.03
$Hg(OH)_2 + 2 H^{+} + 2 e \rightleftharpoons Hg + 2 H_2O$	1.034
$N_2O_4 + 4 H^{+} + 4 e \rightleftharpoons 2 NO + 2 H_2O$	1.035
$RuO_4 + 8 H^{+} + 8 e \rightleftharpoons Ru + 4H_2O$	1.038
$[Fe(phen)_3]^{3+} + e \rightleftharpoons [Fe(phen)_3]^{2+}$(1 molar H_2SO_4)	1.06
$PuO_2(OH)_2 + H^{+} + e \rightleftharpoons PuO_2OH + H_2O$	1.062
$N_2O_4 + 2 H^{+} + 2 e \rightleftharpoons 2 HNO_2$	1.065
$Br_2(l) + 2 e \rightleftharpoons 2Br^{-}$	1.066
$IO_3^{-} + 6 H^{+} + 6 e \rightleftharpoons I^{-} + 3 H_2O$	1.085
$Br_2(aq) + 2 e \rightleftharpoons 2Br^{-}$	1.0873
$PoO_3 + 4 H^{+} + 3 e \rightleftharpoons Po^{2+} + 2 H_2O$	1.095
$Pu^{5+} + e \rightleftharpoons Pu^{4+}$	1.099
$Cu^{2+} + 2 CN^{-} + e \rightleftharpoons [Cu(CN)_2]^{-}$	1.103
$RuO_2 + 4 H^{+} + 2 e \rightleftharpoons Ru^{2+} + 2 H_2O$	1.120
$[Fe(phen)_3]^{3+} + e \rightleftharpoons [Fe(phen)_3]^{2+}$	1.147
$SeO_4^{2-} + 4 H^{+} + 2 e \rightleftharpoons H_2SeO_3 + H_2O$	1.151
$ClO_3^{-} + 2 H^{+} + e \rightleftharpoons ClO_2 + H_2O$	1.152
$Ir^{3+} + 3 e \rightharpoonup Ir$	1.156
$Pt^{2+} + 2 e \rightleftharpoons Pt$	1.18
$ClO_4^{-} + 2 H^{+} + 2 e \rightleftharpoons ClO_3^{-} + H_2O$	1.189
$2 IO_3^{-} + 12 H^{+} + 10 e \rightleftharpoons I_2 + 6 H_2O$	1.195
$PtOH^{+} + H^{+} + 2 e \rightleftharpoons Pt + H_2O$	1.2
$ClO_3^{-} + 3 H^{+} + 2 e \rightleftharpoons HClO_2 + H_2O$	1.214
$MnO_2 + 4 H^{+} + 2 e \rightleftharpoons Mn^{2+} + 2 H_2O$	1.224
$O_2 + 4 H^{+} + 4 e \rightleftharpoons 2 H_2O$	1.229
$O_3 + H_2O + 2 e \rightleftharpoons O_2 + 2 OH^{-}$	1.24
$[Ru(bipy)_3]^{3+} + e \rightleftharpoons [Ru(bipy)_3]^{2+}$	1.24
$Tl^{3+} + 2 e \rightleftharpoons Tl^{+}$	1.252
$N_2H_5^{+} + 3 H^{+} + 2 e \rightleftharpoons 2 NH_4^{+}$	1.275
$ClO_2 + H^{+} + e \rightleftharpoons HClO_2$	1.277
$[PdCl_6]^{2-} + 2 e \rightleftharpoons [PdCl_4]^{2-} + 2 Cl^{-}$	1.288
$2 HNO_2 + 4 H^{+} + 4 e \rightleftharpoons N_2O + 3 H_2O$	1.297
$AuOH^{2+} + H^{+} + 2 e \rightleftharpoons Au^{+} + H_2O$	1.32
$PuO_2(OH)_2 + 2 H^{-} + 2 e \rightleftharpoons Pu(OH)_4$	1.325
$HBrO + H^{+} + 2 e \rightleftharpoons Br^{-} + H_2O$	1.331
$Cr(V) + e \rightleftharpoons Cr(IV)$	1.34
$HCrO_4^{-} + 7 H^{+} + 3 e \rightleftharpoons Cr^{3+} + 4 H_2O$	1.350
$Cl_2(g) + 2 e \rightleftharpoons 2Cl^{-}$	1.35827
$Cr_2O_7^{2-} + 14 H^{+} + 6 e \rightleftharpoons 2 Cr^{3+} + 7 H_2O$	1.36
$ClO_4^{-} + 8 H^{+} + 8 e \rightleftharpoons Cl^{-} + 4 H_2O$	1.389
$ClO_4^{-} + 8 H^{+} + 7 e \rightleftharpoons 1/2 Cl_2 + 4 H_2O$	1.39
$No^{3+} + e \rightleftharpoons No^{2+}$	1.4
$RuO_4 + 6 H^{+} + 4 e \rightleftharpoons Ru(OH)_2^{2+} + 2 H_2O$	1.40
$Au^{3+} + 2 e \rightleftharpoons Au^{+}$	1.401
$2 NH_3OH^{+} + H^{+} + 2 e \rightleftharpoons N_2H_5^{+} + 2 H_2O$	1.42
$BrO_3^{-} + 6 H^{+} + 6 e \rightleftharpoons Br^{-} + 3 H_2O$	1.423
$2 HIO + 2 H^{+} + 2 e \rightleftharpoons I_2 + 2 H_2O$	1.439

Reaction	$E°/V$
$Au(OH)_3 + 3 H^+ + 3 e \rightleftharpoons Au^- + 3 H_2O$	1.45
$3IO_3^- + 6 H^+ + 6 e \rightleftharpoons Cl^- + 3 H_2O$	1.451
$PbO_2 + 4 H^+ + 2 e \rightleftharpoons Pb^{2+} + 2 H_2O$	1.455
$ClO_3^- + 6 H^+ + 5 e \rightleftharpoons 1/2 Cl_2 + 3 H_2O$	1.47
$CrO_2 + 4 H^+ + e \rightleftharpoons Cr^{3+} + 2 H_2O$	1.48
$BrO_3^- + 6 H^+ + 5 e \rightleftharpoons 1/2 Br_2 + 3 H_2O$	1.482
$HClO + H^+ + 2 e \rightleftharpoons Cl^- + H_2O$	1.482
$Mn_2O_3 + 6 H^+ + e \rightleftharpoons 2 Mn^{2+} + 3 H_2O$	1.485
$HO_2 + H^+ + e \rightleftharpoons H_2O_2$	1.495
$Au^{3+} + 3 e \rightleftharpoons Au$	1.498
$PtO_3 + 4 H^+ + 2 e \rightleftharpoons Pt(OH)_2^{2+} + H_2O$	1.5
$MnO_4^- + 8 H^+ + 5 e \rightleftharpoons Mn^{2+} + 4 H_2O$	1.507
$Mn^{3+} + e \rightleftharpoons Mn^{2-}$	1.5415
$HClO_2 + 3 H^+ + 4 e \rightleftharpoons Cl^- + 2 H_2O$	1.570
$HBrO + H^+ + e \rightleftharpoons 1/2 Br_2(aq) + H_2O$	1.574
$2 NO + 2 H^+ + 2 e \rightleftharpoons N_2O + H_2O$	1.591
$Bi_2O_4 + 4 H^+ + 2 e \rightleftharpoons 2 BiO^+ + 2 H_2O$	1.593
$HBrO + H^+ + e \rightleftharpoons 1/2 Br_2(l) + H_2O$	1.596
$H_5IO_6 + H^+ + 2 e \rightleftharpoons IO_3^- + 3 H_2O$	1.601
$HClO + H^+ + e \rightleftharpoons 1/2 Cl_2 + H_2O$	1.611
$HClO_2 + 3 H^+ + 3 e \rightleftharpoons 1/2 Cl_2 + 2 H_2O$	1.628
$HClO_2 + 2 H^+ + 2 e \rightleftharpoons HClO + H_2O$	1.645
$Bk^{4+} + e \rightleftharpoons Bk^{3+}$	1.67
$NiO_2 + 4 H^+ + 2 e \rightleftharpoons Ni^{2+} + 2 H_2O$	1.678
$MnO_4^- + 4 H^+ + 3 e \rightarrow MnO_2 + 2 H_2O$	1.679
$PbO_2 + SO_4^{2-} + 4 H^+ + 2 e \rightleftharpoons PbSO_4 + 2 H_2O$	1.6913
$Au^+ + e \rightleftharpoons Au$	1.692
$PtO_3 + 2 H^+ + 2 e \rightleftharpoons PtO_2 + H_2O$	1.7
$CeOH^{3+} + H^+ + e \rightleftharpoons Ce^{3+} + H_2O$	1.715
$Ce^{4+} + e \rightleftharpoons Ce^{3+}$	1.72
$N_2O + 2 H^+ + 2 e \rightleftharpoons N_2 + H_2O$	1.766
$H_2O_2 + 2 H^+ + 2 e \rightleftharpoons 2 H_2O$	1.776
$Ag^{3+} + e \rightleftharpoons Ag^{2+}$	1.8
$Au^{2+} + e \rightleftharpoons Au^+$	1.8
$Ag_2O_2 + 4 H^+ + e \rightleftharpoons 2 Ag + 2 H_2O$	1.802
$Co^{3+} + e \rightleftharpoons Co^{2-}$(2 molar H_2SO_4)	1.83
$Ag^{3+} + 2 e \rightleftharpoons Ag^+$	1.9
$Co^{3+} + e \rightleftharpoons Co^{2+}$	1.92
$Ag^{2+} + e \rightleftharpoons Ag^+$	1.980
$Cu_2O_3 + 6 H^+ + 2 e \rightleftharpoons 2 Cu^{2+} + 3 H_2O$	2.0
$S_2O_8^{2-} + 2 e \rightleftharpoons 2 SO_4^{2-}$	2.010
$OH + e \rightleftharpoons OH^-$	2.02
$HFeO_4^- + 7 H^+ + 3 e \rightleftharpoons Fe^{3+} + 4 H_2O$	2.07
$O_3 + 2 H^+ + 2 e \rightleftharpoons O_2 + H_2O$	2.076
$HFeO_4^- + 4 H^+ + 3 e \rightleftharpoons FeOOH + 2 H_2O$	2.08
$2 HFeO_4^- + 8 H^+ + 6 e \rightleftharpoons Fe_2O_3 + 5 H_2O$	2.09
$XeO_3 + 6 H^+ + 6 e \rightleftharpoons Xe + 3 H_2O$	2.10
$S_2O_8^{2-} + 2 H^+ + 2 e \rightleftharpoons 2 HSO_4^-$	2.123
$F_2O + 2 H^+ + 4 e \rightleftharpoons H_2O + 2 F^-$	2.153
$FeO_4^{2-} + 8 H^+ + 3 e \rightleftharpoons Fe^{3+} + 4 H_2O$	2.20
$Cu^{3+} + e \rightleftharpoons Cu^{2+}$	2.4
$H_4XeO_6 + 2 H^+ + 2 e \rightleftharpoons XeO_3 + 3 H_2O$	2.42
$O(g) + 2 H^+ + 2 e \rightleftharpoons H_2O$	2.421
$Am^{4+} + e \rightleftharpoons Am^{3+}$	2.60
$H_2N_2O_2 + 2 H^+ + 2 e \rightleftharpoons N_2 + 2 H_2O$	2.65
$F_2 + 2 e \rightleftharpoons 2 F^-$	2.866
$Cm^{4+} + e \rightleftharpoons Cm^{3+}$	3.0
$F_2 + 2 H^+ + 2 e \rightleftharpoons 2 HF$	3.053
$Tb^{4+} + e \rightleftharpoons Tb^{3+}$	3.1
$Pr^{4+} + e \rightleftharpoons Pr^{3+}$	3.2
$Cf^{4+} + e \rightleftharpoons Cf^{3+}$	3.3
$XeF + e \rightleftharpoons Xe + F^-$	3.4

TABLE 3. Reduction Reactions Having $E°$ Values More Negative Than That of the Standard Hydrogen Electrode

Reaction	$E°/V$
$2 H^+ + 2 e \rightleftharpoons H_2$	0.00000
$2 D^+ + 2 e \rightleftharpoons D_2$	-0.013
$AgCN + e \rightleftharpoons Ag + CN^-$	-0.017
$2 WO_3 + 2 H^+ + 2 e \rightleftharpoons W_2O_5 + H_2O$	-0.029
$W_2O_5 + 2 H^+ + 2 e \rightleftharpoons 2 WO_2 + H_2O$	-0.031
$Ag_2S + 2 H^+ + 2 e \rightleftharpoons 2 Ag + H_2S$	-0.0366
$Fe^{3+} + 3 e \rightleftharpoons Fe$	-0.037
$Hg_2I_2 + 2 e \rightleftharpoons 2 Hg + 2 I^-$	-0.0405
$Tl(OH)_3 + 2 e \rightleftharpoons TlOH + 2 OH^-$	-0.05
$TiOH^{3+} + H^+ + e \rightleftharpoons Ti^{3+} + H_2O$	-0.055
$2 H_2SO_3 + H^+ + 2 e \rightleftharpoons HS_2O_4^- + 2 H_2O$	-0.056
$P(white) + 3 H^+ + 3 e \rightleftharpoons PH_3(g)$	-0.063
$O_2 + H_2O + 2 e \rightleftharpoons HO_2^- + OH^-$	-0.076
$2 Cu(OH)_2 + 2 e \rightleftharpoons Cu_2O + 2 OH^- + H_2O$	-0.080
$Se + 2 H^+ + 2 e \rightleftharpoons H_2Se$	-0.082
$WO_3 + 6 H^+ + 6 e \rightleftharpoons W + 3 H_2O$	-0.090
$SnO_2 + 4 H^+ + 2 e \rightleftharpoons Sn^{2+} + 2 H_2O$	-0.094
$Md^{3+} + e \rightleftharpoons Md^{2+}$	-0.1
$P(red) + 3 H^+ + 3 e \rightleftharpoons PH_3(g)$	-0.111
$SnO_2 + 4 H^+ + 4 e \rightleftharpoons Sn + 2 H_2O$	-0.117
$GeO_2 + 2 H^+ + 2 e \rightleftharpoons GeO + H_2O$	-0.118
$WO_2 + 4 H^+ + 4 e \rightleftharpoons W + 2 H_2O$	-0.119
$Pb^{2+} + 2 e \rightleftharpoons Pb(Hg)$	-0.1205
$Pb^{2+} + 2 e \rightleftharpoons Pb$	-0.1262
$CrO_4^{2-} + 4H_2O + 3e \rightleftharpoons Cr(OH)_3 + 5OH^-$	-0.13
$Sn^{2-} + 2 e \rightleftharpoons Sn$	-0.1375
$In^+ + e \rightleftharpoons In$	-0.14
$O_2 + 2 H_2O + 2 e \rightleftharpoons H_2O_2 + 2 OH^-$	-0.146
$MoO_2 + 4 H^+ + 4 e \rightleftharpoons Mo + 4 H_2O$	-0.152
$AgI + e \rightleftharpoons Ag + I^-$	-0.15224
$2 NO_2^- + 2 H_2O + 4 e \rightleftharpoons N_2O_2^{2-} + 4 OH^-$	-0.18
$H_2GeO_3 + 4 H^+ + 4 e \rightleftharpoons Ge + 3 H_2O$	-0.182
$SnO_2 + 3 H^+ + 2 e \rightleftharpoons SnOH^+ + H_2O$	-0.194
$CO_2 + 2 H^+ + 2 e \rightleftharpoons HCOOH$	-0.199
$Mo^{3+} + 3 e \rightleftharpoons Mo$	-0.200
$Ga^+ + e \rightleftharpoons Ga$	-0.2
$2 SO_2^{2-} + 4 H^+ + 2 e \rightleftharpoons S_2O_6^{2-} + H_2O$	-0.22
$Cu(OH)_2 + 2 e \rightleftharpoons Cu + 2 OH^-$	-0.222
$V_2O_5 + 10 H^+ + 10 e \rightleftharpoons 2 V + 5 H_2O$	-0.242
$CdSO_4 + 2 e \rightleftharpoons Cd + SO_4^{2-}$	-0.246
$V(OH)_4^+ + 4 H^+ + 5 e \rightleftharpoons V + 4 H_2O$	-0.254
$V^{3+} + e \rightleftharpoons V^{2+}$	-0.255
$Ni^{2+} + 2 e \rightleftharpoons Ni$	-0.257
$PbCl_2 + 2 e \rightleftharpoons Pb + 2 Cl^-$	-0.2675
$H_3PO_4 + 2 H^+ + 2 e \rightleftharpoons H_3PO_3 + H_2O$	-0.276
$Co^{2+} + 2 e \rightleftharpoons Co$	-0.28
$PbBr_2 + 2 e \rightleftharpoons Pb + 2 Br^-$	-0.284
$Tl^+ + e \rightleftharpoons Tl(Hg)$	-0.3338
$Tl^+ + e \rightleftharpoons Tl$	-0.336
$In^{3+} + 3 e \rightleftharpoons In$	-0.3382
$TlOH + e \rightleftharpoons Tl + OH^-$	-0.34
$PbF_2 + 2 e \rightleftharpoons Pb + 2 F^-$	-0.3444
$PbSO_4 + 2 e \rightleftharpoons Pb(Hg) + SO_4^{2-}$	-0.3505
$Cd^{2+} + 2 e \rightleftharpoons Cd(Hg)$	-0.3521
$PbSO_4 + 2 e \rightleftharpoons Pb + SO_4^{2-}$	-0.3588
$Cu_2O + H_2O + 2 e \rightleftharpoons 2 Cu + 2 OH^-$	-0.360
$Eu^{3+} + e \rightleftharpoons Eu^{2+}$	-0.36
$PbI_2 + 2 e \rightleftharpoons Pb + 2 I^-$	-0.365
$SeO_3^{2-} + 3 H_2O + 4 e \rightleftharpoons Se + 6 OH^-$	-0.366
$Ti^{3+} + e \rightleftharpoons Ti^{2+}$	-0.369
$Se + 2 H^+ + 2 e \rightleftharpoons H_2Se(aq)$	-0.399
$In^{2+} + e \rightleftharpoons In^+$	-0.40
$Cd^{2+} + 2 e \rightleftharpoons Cd$	-0.4030
$Cr^{3+} + e \rightleftharpoons Cr^{2+}$	-0.407
$2 S + 2 e \rightleftharpoons S_2^{2-}$	-0.42836
$Tl_2SO_4 + 2 e \rightleftharpoons Tl + SO_4^{2-}$	-0.4360
$In^{3+} + 2 e \rightleftharpoons In^+$	-0.443
$Fe^{2+} + 2 e \rightleftharpoons Fe$	-0.447
$H_3PO_3 + 3 H^+ + 3 e \rightleftharpoons P + 3 H_2O$	-0.454
$Bi_2O_3 + 3 H_2O + 6 e \rightleftharpoons 2 Bi + 6 OH^-$	-0.46
$NO_2^- + H_2O + e \rightleftharpoons NO + 2 OH$	-0.46
$PbHPO_4 + 2 e \rightleftharpoons Pb + HPO_4^{2-}$	-0.465
$S + 2 e \rightleftharpoons S^{2-}$	-0.47627
$S + H_2O + 2 e \rightleftharpoons HS^- + OH^-$	-0.478
$B(OH)_3 + 7 H^+ + 8 e \rightleftharpoons BH_4^- + 3 H_2O$	-0.481
$In^{3+} + e \rightleftharpoons In^{2+}$	-0.49
$ZnOH^+ + H^+ + 2 e \rightleftharpoons Zn + H_2O$	-0.497
$GaOH^{2+} + H^+ + 3 e \rightleftharpoons Ga + H_2O$	-0.498
$H_3PO_3 + 2 H^+ + 2 e \rightleftharpoons H_3PO_2 + H_2O$	-0.499
$TiO_2 + 4 H^+ + 2 e \rightleftharpoons Ti^{2+} + 2 H_2O$	-0.502
$H_3PO_2 + H^+ + e \rightleftharpoons P + 2 H_2O$	-0.508
$Sb + 3 H^+ + 3 e \rightleftharpoons SbH_3$	-0.510
$U^{4+} + e \rightleftharpoons U^{3+}$	-0.52
$HPbO_2^- + H_2O + 2 e \rightleftharpoons Pb + 3 OH^-$	-0.537
$Ga^{3+} + 3 e \rightleftharpoons Ga$	-0.549
$TlCl + e \rightleftharpoons Tl + Cl^-$	-0.5568
$Fe(OH)_3 + e \rightleftharpoons Fe(OH)_2 + OH^-$	-0.56
$TeO_3^{2-} + 3 H_2O + 4 e \rightleftharpoons Te + 6 OH^-$	-0.57
$2 SO_3^{2-} + 3 H_2O + 4 e \rightleftharpoons S_2O_3^{2-} + 6 OH^-$	-0.571
$PbO + H_2O + 2 e \rightleftharpoons Pb + 2 OH^-$	-0.580
$SbO_3^- + H_2O + 2 e \rightleftharpoons SbO_2^- + 2 OH^-$	-0.59
$Ta^{3+} + 3 e \rightleftharpoons Ta$	-0.6
$ReO_2^- + 4 H_2O + 7 e \rightleftharpoons Re + 8 OH^-$	-0.604
$As + 3 H^+ + 3 e \rightleftharpoons AsH_3$	-0.608

Reaction	E°/V
$Nb_2O_5 + 10\ H^+ + 10\ e \rightleftharpoons 2\ Nb + 5\ H_2O$	-0.644
$NbO_2 + 2\ H^+ + 2\ e \rightleftharpoons NbO + H_2O$	-0.646
$Cd(OH)_4^{2-} + 2\ e \rightleftharpoons Cd + 4\ OH^-$	-0.658
$TlBr + e \rightleftharpoons Tl + Br^-$	-0.658
$SbO_2^- + 2\ H_2O + 3\ e \rightleftharpoons Sb + 4\ OH^-$	-0.66
$Se + 2\ e \rightleftharpoons Se^{2-}$	-0.670
$AsO_2^- + 2\ H_2O + 3\ e \rightleftharpoons As + 4\ OH^-$	-0.68
$NbO_2 + 4\ H^+ + 4\ e \rightleftharpoons Nb + 2\ H_2O$	-0.690
$Ag_2S + 2\ e \rightleftharpoons 2\ Ag + S^{2-}$	-0.691
$AsO_4^{3-} + 2\ H_2O + 2\ e \rightleftharpoons AsO_2^- + 4\ OH^-$	-0.71
$Ni(OH)_2 + 2\ e \rightleftharpoons Ni + 2\ OH^-$	-0.72
$Co(OH)_2 + 2\ e \rightleftharpoons Co + 2\ OH^-$	-0.73
$NbO + 2\ H^+ + 2\ e \rightleftharpoons Nb + H_2O$	-0.733
$H_2SeO_3 + 4\ H^+ + 4\ e \rightleftharpoons Se + 3\ H_2O$	-0.74
$Cr^{3+} + 3\ e \rightleftharpoons Cr$	-0.744
$Ta_2O_5 + 10\ H^+ + 10\ e \rightleftharpoons 2\ Ta + 5\ H_2O$	-0.750
$TlI + e \rightleftharpoons Tl + I^-$	-0.752
$Zn^{2+} + 2\ e \rightleftharpoons Zn$	-0.7618
$Zn^{2+} + 2\ e \rightleftharpoons Zn(Hg)$	-0.7628
$CdO + H_2O + 2\ e \rightleftharpoons Cd + 2\ OH^-$	-0.783
$Te + 2\ H^+ + 2\ e \rightleftharpoons H_2Te$	-0.793
$ZnSO_4 7H_2O + 2\ e \rightleftharpoons Zn(Hg) + SO_4^{2-} + 7\ H_2O$ (Saturated $ZnSO_4$)	-0.7993
$Bi + 3\ H^+ + 3\ e \rightleftharpoons BiH_3$	-0.8
$SiO + 2\ H^+ + 2\ e \rightleftharpoons Si + H_2O$	-0.8
$Cd(OH)_2 + 2\ e \rightleftharpoons Cd(Hg) + 2\ OH^-$	-0.809
$2\ H_2O + 2\ e \rightleftharpoons H_2 + 2\ OH^-$	-0.8277
$2\ NO^-_3 + 2\ H_2O + 2\ e \rightleftharpoons N_2O_4 + 4\ OH^-$	-0.85
$H_3BO_3 + 3\ H^+ + 3\ e \rightleftharpoons B + 3\ H_2O$	-0.8698
$P + 3\ H_2O + 3\ e \rightleftharpoons PH_3(g) + 3\ OH^-$	-0.87
$HSnO_2^- + H_2O + 2\ e \rightleftharpoons Sn + 3\ OH^-$	-0.909
$Cr^{2+} + 2\ e \rightleftharpoons Cr$	-0.913
$SO_4^{2-} + H_2O + 2\ e \rightleftharpoons SO_3^{2-} + 2\ OH^-$	-0.93
$Sn(OH)_6^{2-} + 2\ e \rightleftharpoons HSnO_2^- + 3\ OH^- + H_2O$	-0.93
$SnO_2 + 2\ H_2O + 4\ e \rightleftharpoons Sn + 4\ OH^-$	-0.945
$In(OH)_3 + 3\ e \rightleftharpoons In + 3\ OH^-$	-0.99
$NpO_2 + H_2O + H^+ + e \rightleftharpoons Np(OH)_3$	-0.962
$In(OH)_4^- + 3\ e \rightleftharpoons In + 4\ OH^-$	-1.007
$In_2O_3 + 3\ H_2O + 6\ e \rightleftharpoons 2\ In + 6\ OH^-$	-1.034
$PO_4^{3-} + 2\ H_2O + 2\ e \rightleftharpoons HPO_3^{2-} + 3\ OH^-$	-1.05
$Yb^{3+} + e \rightleftharpoons Yb^{2+}$	-1.05
$Nb^{3+} + 3\ e \rightleftharpoons Nb$	-1.099
$Fm^{3+} + e \rightleftharpoons Fm^{2+}$	-1.1
$2\ SO_3^{2-} + 2\ H_2O + 2\ e \rightleftharpoons S_2O_4^{2-} + 4\ OH^-$	-1.12
$Te + 2\ e \rightleftharpoons Te^{2-}$	-1.143
$V^{2+} + 2\ e \rightleftharpoons V$	-1.175
$Mn^{2+} + 2\ e \rightleftharpoons Mn$	-1.185
$Zn(OH)_4^{2-} + 2\ e \rightleftharpoons Zn + 4\ OH^-$	-1.199
$CrO_2 + 2\ H_2O + 3\ e \rightleftharpoons Cr + 4\ OH^-$	-1.2
$No^{3+} + 3\ e \rightleftharpoons No$	-1.20
$Ti^{3+} + 3\ e \rightleftharpoons Ti$	-1.209
$ZnO_2^- + 2\ H_2O + 2\ e \rightleftharpoons Zn + 4\ OH^-$	-1.215
$H_2GaO_3^- + H_2O + 3\ e \rightleftharpoons Ga + 4\ OH^-$	-1.219
$H_2BO_3^- + 5\ H_2O + 8\ e \rightleftharpoons BH_4^- + 8\ OH^-$	-1.24

Reaction	E°/V
$SiF_6^{2-} + 4\ e \rightleftharpoons Si + 6\ F^-$	-1.24
$Zn(OH)_2 + 2\ e \rightleftharpoons Zn + 2\ OH^-$	-1.249
$ZnO + H_2O + 2\ e \rightleftharpoons Zn + 2\ OH^-$	-1.260
$Es^{3+} + e \rightleftharpoons Es^{2+}$	-1.3
$Pa^{3+} + 3\ e \rightleftharpoons Pa$	-1.34
$Ce^{3+} + 3\ e \rightleftharpoons Ce(Hg)$	-1.4373
$UO_2^{2+} + 4\ H^+ + 6\ e \rightleftharpoons U + 2\ H_2O$	-1.444
$Zr^{4+} + 4\ e \rightleftharpoons Zr$	-1.45
$Cr(OH)_3 + 3\ e \rightleftharpoons Cr + 3\ OH^-$	-1.48
$Pa^{4+} + 4\ e \rightleftharpoons Pa$	-1.49
$HfO_2 + 4\ H^+ + 4\ e \rightleftharpoons Hf + 2\ H_2O$	-1.505
$Hf^{4+} + 4\ e \rightleftharpoons Hf$	-1.55
$Sm^{3+} + e \rightleftharpoons Sm^{2+}$	-1.55
$ZrO_2 + 4\ H^+ + 4\ e \rightleftharpoons Zr + 2\ H_2O$	-1.553
$Mn(OH)_2 + 2\ e \rightleftharpoons Mn + 2\ OH^-$	-1.56
$Ba^{2+} + 2\ e \rightleftharpoons Ba(Hg)$	-1.570
$Bk^{2+} + 2\ e \rightleftharpoons Bk$	-1.6
$Cf^{3+} + e \rightleftharpoons Cf^{2+}$	-1.6
$Ti^{2+} + 2\ e \rightleftharpoons Ti$	-1.628
$Md^{3+} + 3\ e \rightleftharpoons Md$	-1.65
$HPO_3^{2-} + 2H_2O + 2e \rightleftharpoons H_2PO_2^- + 3OH^-$	-1.65
$U^{3+} + 3\ e \rightleftharpoons U$	-1.66
$Al^{3+} + 3\ e \rightleftharpoons Al$	-1.676
$SiO_3^{2-} + H_2O + 4\ e \rightleftharpoons Si + 6\ OH^-$	-1.697
$HPO_3^{2-} + 2\ H_2O + 3\ e \rightleftharpoons P + 5\ OH^-$	-1.71
$HfO^{2+} + 2\ H^+ + 4\ e \rightleftharpoons Hf + H_2O$	-1.724
$ThO_2 + 4\ H^+ + 4\ e \rightleftharpoons Th + 2\ H_2O$	-1.789
$H_2BO_3^- + H_2O + 3\ e \rightleftharpoons B + 4\ OH^-$	-1.79
$Sr^{2+} + 2\ e \rightleftharpoons Sr(Hg)$	-1.793
$H_2PO^-_2 + e \rightleftharpoons P + 2\ OH^-$	-1.82
$Be^{2+} + 2\ e \rightleftharpoons Be$	-1.847
$Np^{3+} + 3\ e \rightleftharpoons Np$	-1.856
$Fm^{3+} + 3\ e \rightleftharpoons Fm$	-1.89
$Th^{4+} + 4\ e \rightleftharpoons Th$	-1.899
$Am^{2+} + 2\ e \rightleftharpoons Am$	-1.9
$Pa^{4+} + e \rightleftharpoons Pa^{3+}$	-1.9
$Es^{3+} + 3\ e \rightleftharpoons Es$	-1.91
$Cf^{3+} + 3\ e \rightleftharpoons Cf$	-1.94
$Lr^{3+} + 3\ e \rightleftharpoons Lr$	-1.96
$Eu^{3+} + 3\ e \rightleftharpoons Eu$	-1.991
$Er^{2+} + 2\ e \rightleftharpoons Er$	-2.0
$Pr^{2+} + 2\ e \rightleftharpoons Pr$	-2.0
$Pu^{3+} + 3\ e \rightleftharpoons Pu$	-2.031
$Cm^{3+} + 3\ e \rightleftharpoons Cm$	-2.04
$Am^{3+} + 3\ e \rightleftharpoons Am$	-2.048
$AlF_6^{3-} + 3\ e \rightleftharpoons Al + 6\ F^-$	-2.069
$Sc^{3+} + 3\ e \rightleftharpoons Sc$	-2.077
$Ho^{2+} + 2\ e \rightleftharpoons Ho$	-2.1
$Nd^{2+} + 2\ e \rightleftharpoons Nd$	-2.1
$Cf^{2+} + 2\ e \rightleftharpoons Cf$	-2.12
$Yb^{3+} + 3\ e \rightleftharpoons Yb$	-2.19
$Ac^{3+} + 3\ e \rightleftharpoons Ac$	-2.20
$Dy^{2+} + 2\ e \rightleftharpoons Dy$	-2.2
$Tm^{3+} + e \rightleftharpoons Tm^{2+}$	-2.2
$Pm^{2+} + 2\ e \rightleftharpoons Pm$	-2.2
$Es^{2+} + 2\ e \rightleftharpoons Es$	-2.23
$H_2 + 2\ e \rightleftharpoons 2\ H^-$	-2.23

Reaction	E°/V
$Gd^{3+} + 3\ e \rightleftharpoons Gd$	-2.279
$Tb^{3+} + 3\ e \rightleftharpoons Tb$	-2.28
$Lu^{3+} + 3\ e \rightleftharpoons Lu$	-2.28
$Dy^{3+} + 3\ e \rightleftharpoons Dy$	-2.295
$Am^{3+} + e \rightleftharpoons Am^{2+}$	-2.3
$Fm^{2+} + 2\ e \rightleftharpoons Fm$	-2.30
$Pm^{3+} + 3\ e \rightleftharpoons Pm$	-2.30
$Al(OH)_3 + 3\ e \rightleftharpoons Al + 3\ OH^-$	-2.30
$Sm^{3+} + 3\ e \rightleftharpoons Sm$	-2.304
$Al(OH)^- + 3\ e \rightleftharpoons Al + 4\ OH^-$	-2.310
$Tm^{3+} + 3\ e \rightleftharpoons Tm$	-2.319
$Nd^{3+} + 3\ e \rightleftharpoons Nd$	-2.323
$H_2AlO_3^- + H_2O + 3\ e \rightleftharpoons Al + 4\ OH^-$	-2.33
$Ho^{3+} + 3\ e \rightleftharpoons Ho$	-2.33
$Er^{3+} + 3\ e \rightleftharpoons Er$	-2.331
$Ce^{3+} + 3\ e \rightleftharpoons Ce$	-2.336
$Pr^{3+} + 3\ e \rightleftharpoons Pr$	-2.353
$ZrO(OH)_2 + H_2O + 4\ e \rightleftharpoons Zr + 4\ OH^-$	-2.36
$Mg^{2+} + 2\ e \rightleftharpoons Mg$	-2.372
$Y^{3+} + 3\ e \rightleftharpoons Y$	-2.372
$La^{3+} + 3\ e \rightleftharpoons La$	-2.379
$Tm^{2+} + 2\ e \rightleftharpoons Tm$	-2.4
$Md^{2+} + 2\ e \rightleftharpoons Md$	-2.40
$Th(OH)_4 + 4\ e \rightleftharpoons Th + 4\ OH^-$	-2.48
$HfO(OH)_2 + H_2O + 4\ e \rightleftharpoons Hf + 4\ OH^-$	-2.50
$No^{2+} + 2\ e \rightleftharpoons No$	-2.50
$Dy^{3+} + e \rightleftharpoons Dy^{2+}$	-2.6
$Pm^{3+} + e \rightleftharpoons Pm^{2+}$	-2.6
$Be_2O_3^{2-} + 3\ H_2O + 4\ e \rightleftharpoons 2\ Be + 6\ OH^-$	-2.63
$Sm^{2+} + 2\ e \rightleftharpoons Sm$	-2.68
$Mg(OH)_2 + 2\ e \rightleftharpoons Mg + 2\ OH^-$	-2.690
$Nd^{3+} + e \rightleftharpoons Nd^{2+}$	-2.7
$Mg^+ + e \rightleftharpoons Mg$	-2.70
$Na^+ + e \rightleftharpoons Na$	-2.71
$Yb^{2+} + 2\ e \rightleftharpoons Yb$	-2.76
$Bk^{3+} + e \rightleftharpoons Bk^{2+}$	-2.8
$Ho^{3+} + e \rightleftharpoons Ho^{2+}$	-2.8
$Ra^{2+} + 2\ e \rightleftharpoons Ra$	-2.8
$Eu^{2+} + 2\ e \rightleftharpoons Eu$	-2.812
$Ca^{2+} + 2\ e \rightleftharpoons Ca$	-2.868
$Sr(OH)_2 + 2\ e \rightleftharpoons Sr + 2\ OH^-$	-2.88
$Sr^{2+} + 2\ e \rightleftharpoons Sr$	-2.899
$Fr^+ + e \rightleftharpoons Fr$	-2.9
$La(OH)_3 + 3\ e \rightleftharpoons La + 3\ OH^-$	-2.90
$Ba^{2+} + 2\ e \rightleftharpoons Ba$	-2.912
$K^+ + e \rightleftharpoons K$	-2.931
$Rb^+ + e \rightleftharpoons Rb$	-2.98
$Ba(OH)_2 + 2\ e \rightleftharpoons Ba + 2\ OH^-$	-2.99
$Er^{3+} + e \rightleftharpoons Er^{2+}$	-3.0
$Ca(OH)_2 + 2\ e \rightleftharpoons Ca + 2\ OH^-$	-3.02
$Cs^+ + e \rightleftharpoons Cs$	-3.026
$Li^+ + e \rightleftharpoons Li$	-3.0401
$3\ N_2 + 2\ H^+ + 2\ e \rightleftharpoons 2\ HN_3$	-3.09
$Pr^{3+} + e \rightleftharpoons Pr^{2+}$	-3.1
$Ca^+ + e \rightleftharpoons Ca$	-3.80
$Sr^+ + e \rightleftharpoons Sr$	-4.10

REDUCTION AND OXIDATION POTENTIALS FOR CERTAIN ION RADICALS

Petr Vanýsek

There are two tables for ion radicals. The first table lists reduction potentials for organic compounds that produce anion radicals during reduction, a process described as $A + e^- \rightleftharpoons A^{-\cdot}$. The second table lists oxidation potentials for organic compounds that produce cation radicals during oxidation, a process described as $A \rightleftharpoons A^{+\cdot} + e^-$. To obtain reduction potential for a reverse reaction, the sign for the potential is changed.

Unlike the table of the Electrochemical Series, which lists *standard* potentials, values for radicals are experimental values with experimental conditions given in the second column. Because the measurements leading to potentials for ion radicals are very dependent on conditions, an attempt to report standard potentials for radicals would serve no useful purpose. For the same reason, the potentials are also reported as experimental values, usually a half-wave potential ($E_{1/2}$ in polarography) or a peak potential (E_p in cyclic voltammetry). Unless otherwise stated, the values are reported vs. SCE (saturated calomel electrode). To obtain a value vs. normal hydrogen electrode, 0.241 V has to be added to the SCE values. All the ion radicals chosen for inclusion in the tables result from electrochemically reversible reactions. More detailed data on ion radicals can be found in the *Encyclopedia of Electrochemistry of Elements*, (A. J. Bard, Ed.), Vols. XI and XII in particular, Marcel Dekker, New York, 1978.

Abbreviations are:

CV: cyclic voltammetry
DMF: *N,N*-Dimethylformamide
***E* swp:** potential sweep
$E°$: standard potential
E_p: peak potential
$E_{p/2}$: half-peak potential
$E_{1/2}$: half-wave potential
***M*:** mol/L
MeCN: acetonitrile
pol: polarography
rot Pt dsk: rotated Pt disk
SCE: saturated calomel electrode
$TBABF_4$: tetrabutylammonium tetrafluoroborate
TBAI: tetrabutylammonium iodide
TBAP: tetrabutylammonium perchlorate
TEABr: tetraethylammonium bromide
TEAP: tetraethylammonium perchlorate
THF: tetrahydrofuran
$TPACF_3SO_3$: tetrapropylammonium trifluoromethanesulfite
TPAP: tetrapropylammonium perchlorate
wr: wire

Reduction Potentials (Products Are Anion Radicals)

Substance	Conditions/electrode/ technique	Potential V (vs. SCE)
Acetone	DMF, 0.1 *M* TEABr/Hg/pol	$E_{1/2} = -2.84$
1-Naphthyphenylacetylene	DMF, 0.03 *M* TBAI/Hg/pol	$E_{1/2} = -1.91$
1-Naphthalenecarboxyaldehyde	-/Hg/pol	$E_{1/2} = -0.91$
2-Naphthalenecarboxyaldehyde	-/Hg/pol	$E_{1/2} = -0.96$
2-Phenanthrenecarboxaldehyde	-/Hg/pol	$E_{1/2} = -1.00$
3-Phenanthrenecarboxaldehyde	-/Hg/pol	$E_{1/2} = -0.94$
9-Phenanthrenecarboxaldehyde	-/Hg/pol	$E_{1/2} = -0.83$
1-Anthracenecarboxaldehyde	-/Hg/pol	$E_{1/2} = -0.75$
1-Pyrenecarboxaldehyde	-/Hg/pol	$E_{1/2} = -0.76$
2-Pyrenecarboxaldehyde	-/Hg/pol	$E_{1/2} = -1.00$
Anthracene	DMF, 0.1 *M* TBAP/Pt dsk/CV	$E_p = -2.00$
	DMF, 0.5 *M* $TBABF_4$/Hg/CV	$E_{1/2} = -1.93$
	MeCN, 0.1 *M* TEAP/Hg/CV	$E_{1/2} = -2.07$
	DMF, 0.1 *M* TBAI/Hg/pol	$E_{1/2} = -1.92$
9,10-Dimethylanthracene	DMF, 0.1 *M* TBAP/Pt/CV	$E_p = -2.08$
	MeCN, 0.1 *M* TBAP/Pt/CV	$E_p = -2.10$
1-Phenylanthracene	DMF, 0.5 *M* $TBABF_2$/Hg/CV	$E_{1/2} = -1.91$
	DMF, 0.1 *M* TBAI/Hg/pol	$E_{1/2} = -1.878$
2-Phenylanthracene	DMF, 0.1 *M* TBAI/Hg/pol	$E_{1/2} = -1.875$
8-Phenylanthracene	DMF, 0.5 *M* $TBABF_4$/Hg/CV	$E_{1/2} = -1.91$
9-Phenylanthracene	DMF, 0.5 *M* $TBABF_4$/Hg/CV	$E_{1/2} = -1.93$
	DMF, 0.1 *M* TBAI/Hg/pol	$E_{1/2} = -1.863$
1,8-Diphenylanthracene	DMF, 0.5 *M* $TBABF_4$/Hg/CV	$E_{1/2} = -1.88$
1,9-Diphenylanthracene	DMF, 0.1 *M* TBAI/Hg/pol	$E_{1/2} = -1.846$
1,10-Diphenylanthracene	DMF, 0.1 *M* TBAI/Hg/pol	$E_{1/2} = -1.786$
8,9-Diphenylanthracene	DMF, 0.5 *M* $TBABF_4$/Hg/CV	$E_{1/2} = -1.90$
9,10-Diphenylanthracene	MeCN, 0.1 *M* TBAP/rot Pt/*E* swp	$E_{1/2} = -1.83$
	DMF, 0.1 *M* TBAI/Hg/pol	$E_{1/2} = -1.835$
1,8,9-Triphenylanthracene	DMF, 0.5 *M* $TBABF_4$/Hg/CV	$E_{1/2} = -1.85$
1,8,10-Triphenylanthracene	DMF, 0.5 *M* $TBABF_4$/Hg/CV	$E_{1/2} = -1.81$
9,10-Dibiphenylanthracene	MeCN, 0.1 *M* TBAP/rot Pt/*E* swp	$E_{1/2} = -1.94$
Benz(a)anthracene	MeCN, 0.1 *M* TEAP/Hg/CV	$E_{1/2} = -2.11$
	MeCN, 0.1 *M* TEAP/Hg/pol	$E_{1/2} = -2.40$[a]
Azulene	DMF, 0.1 *M* TBAI/Hg/pol	$E_{1/2} = -1.10$[c]
Annulene	DMF, 0.5 *M* TBAP 0 °C/Hg/pol	$E_{1/2} = -1.23$
Benzaldehyde	DMF, 0.1 *M* TBAP/Hg/pol	$E_{1/2} = -1.67$
Benzil	DMSO, 0.1 *M* TBAP/Hg/pol	$E_{1/2} = -1.04$

Substance	Conditions/electrode/technique	Potential V (vs. SCE)
Benzophenone	-/Hg/pol	$E_{1/2} = -1.80$
	DMF/Pt dsk/CV	$E° = -1.72$
Chrysene	MeCN, 0.1 *M* TEAP/Hg/pol	$E_{1/2} = -2.73$[a]
Fluoranthrene	DMF, 0.1 *M* TBAP/Pt dsk/CV	$E_p = -1.76$
Cyclohexanone	DMF, 0.1 *M* TEABr/Hg/pol	$E_{1/2} = -2.79$
5,5-Dimethyl-3-phenyl-2-cyclohexen-1-one	DMF, 0.5 *M*/Hg/pol	$E_{1/2} = -1.71$
1,2,3-Indanetrione hydrate (ninhydrin)	DMF, 0.2 *M* $NaNO_3$/Hg/pol	$E_{1/2} = -0.039$
Naphthacene	DMF, 0.1 *M* TBAI/Hg/pol	$E_{1/2} = -1.53$
Naphthalene	DMF, 0.1 *M* TBAP/Pt dsk/CV	$E_p = -2.55$
	DMF, 0.5 *M* $TBABF_4$/Hg/CV	$E_{1/2} = -2.56$
	DMF, MeCN, 0.1 *M* TEAP/Hg/CV	$E_{1/2} = -2.63$
	DMF, 0.1 *M* TBAI/Hg/pol	$E_{1/2} = -2.50$
1-Phenylnaphthalene	DMF, 0.5 *M* $TBABF_4$/Hg/CV	$E_{1/2} = -2.36$
1,2-Diphenylnaphthalene	DMF, 0.5 *M* $TBABF_4$/Hg/CV	$E_{1/2} = -2.25$
Cyclopentanone	DMF, 0.1 *M* TEABr/Hg/pol	$E_{1/2} = -2.82$
Phenanthrene	MeCN, 0.1 *M* TBAP/Pt wr/CV	$E_{1/2} = -2.47$
	MeCN, 0.1 *M* TEAP/Hg/pol	$E_{1/2} = -2.88$[a]
Pentacene	THF, 0.1 *M* TBAP/rot Pt dsk/*E* swp	$E_{1/2} = -1.40$
Perylene	MeCN, 0.1 *M* TEAP/Hg/CV	$E_{1/2} = -1.73$
1,3-Diphenyl-1,3-propanedione	DMSO, 0.2 *M* TBAP/Hg/CV	$E_{1/2} = -1.42$
2,2-Dimethyl-1,3-diphenyl-1,3 propanedione	DMSO, TBAP/Hg/CV	$E_{1/2} = -1.80$
Pyrene	DMF, 0.1 *M* TBAP/Pt/CV	$E_p = -2.14$
	MeCN, 0.1 *M* TEAP/Hg/pol	$E_{1/2} = -2.49$[a]
Diphenylsulfone	DMF, TEABr	$E_{1/2} = -2.16$
Triphenylene	MeCN, 0.1 *M* TEAP/Hg/pol	$E_{1/2} = -2.87$[a]
9,10-Anthraquinone	DMF, 0.5 *M* TBAP, 20°/Pt dsk/CV	$E_{1/2} = -1.01$
1,4-Benzoquinone	MeCN, 0.1 *M* TEAP/Pt/CV	$E_p = -0.54$
1,4-Naphthohydroquinone, dipotassium salt	DMF, 0.5 *M* TBAP, 20°/Pt dsk/CV	$E_{1/2} = -1.55$
Rubrene	DMF, 0.1 *M* TBAP/Pt dsk/CV	$E_p = -1.48$
	DMF, 0.1 *M* TBAI/Hg/pol	$E_{1/2} = -1.410$
Benzocyclooctatetraene	THF, 0.1 *M* TBAP/Hg/pol	$E_{1/2} = -2.13$
sym-Dibenzocyclo-octatetraene	THF, 0.1 *M* TBAP/Hg/pol	$E_{1/2} = -2.29$
Ubiquinone-6	MeCN, 0.1 *M* TEAP/Pt/CV	$E_p = -1.05$[d]
(9-Phenyl-fluorenyl)$^+$	10.2 *M* H_2SO_4/Hg/CV	$E_p = -0.01$[b]
(Triphenylcyclopropenyl)$^+$	MeCN, 0.1 *M* TEAP/Hg/CV	$E_p = -1.87$
(Triphenylmethyl)$^+$	MeCN, 0.1 *M* TBAP/Hg/pol	$E_{1/2} = 0.27$
	H_2SO_4, 10.2 *M*/Hg/CV	$E_p = -0.58$[b]
(Tribiphenylmethyl)$^+$	MeCN, 0.1 *M* TBAP/Hg/pol	$E_{1/2} = 0.19$
(Tri-4-*t*-butyl-5-phenylmethyl)$^+$	MeCN, 0.1 *M* TBAP/Hg/pol	$E_{1/2} = 0.13$
(Tri-4-isopropylphenylmethyl)$^+$	MeCN, 0.1 *M* TBAP/Hg/pol	$E_{1/2} = 0.07$
(Tri-4-methylphenylmethyl)$^+$	MeCN, 0.1 *M* TBAP/Hg/pol	$E_{1/2} = 0.05$
(Tri-4-cyclopropylphenylmethyl)$^+$	MeCN, 0.1 *M* TBAP/Hg/pol	$E_{1/2} = 0.01$
(Tropylium)+	MeCN, 0.1 *M* TBAP/Hg/pol	$E_{1/2} = -0.17$
	DMF, 0.15 *M* TBAI/Hg/pol	$E_{1/2} = -1.55$
	DMF, 0.15 *M* TBAI/Hg/pol	$E_{1/2} = -1.55$
	DMF, 0.15 *M* TBAI/Hg/pol	$E_{1/2} = -1.57$
	DMF, 0.15 *M* TBAI/Hg/pol	$E_{1/2} = -1.60$
	DMF, 0.15 *M* TBAI/Hg/pol	$E_{1/2} = -1.87$
	DMF, 0.15 *M* TBAI/Hg/pol	$E_{1/2} = -1.96$
	DMF, 0.15 *M* TBAI/Hg/pol	$E_{1/2} = -2.05$

[a] vs. 0.01 M Ag/$AgClO_4$
[b] vs. Hg/Hg_2SO_4, 17 M H_2SO_4
[c] vs. Hg pool
[d] vs. Ag/0.01 M Ag+

Oxidation Potentials (Products Are Cation Radicals)

Substance	Conditions/electrode/technique	Potential V (vs. SCE)
Anthracene	CH_2Cl_2, 0.2 *M* $TBABF_4$, −70 °C/Pt dsk/CV	$E_p = +0.73$[e]
9,10-Dimethylanthracene	MeCN, 0.1 *M* $LiClO_4$/Pt wr/CV	$E_p = +1.0$
9,10-Dipropylanthracene	MeCN, 0.1 *M* TEAP/Pt/CV	$E_p = +1.08$
1,8-Diphenylanthracene	CH_2Cl_2, 0.2 *M* $TPrACF_3SO_3$/rot Pt wr/*E* swp	$E_{1/2} = +1.34$
8,9-Diphenylanthracene	CH_2Cl_2, 0.2 *M* $TPrACF_3SO_3$/rot Pt wr/*E* swp	$E_{1/2} = +1.30$
9,10-Diphenylanthracene	MeCN/Pt/CV	$E_p = +1.22$
Perylene	MeCN, 0.1 *M* TBAP/Pt/CV	$E_p = +1.34$
Pyrene	DMF, 0.1 *M* TBAP/Pt dsk/CV	$E_p = +1.25$
Rubrene	DMF, 0.1 *M* TBAP/Pt dsk/CV	$E_p = +1.10$
Tetracene	CH_2Cl_2, 0.2 *M* $TBABF_4$, −70 °C/Pt wr/CV	$E_p = +0.35$[e]
1,4-Dithiabenzene	MeCN, 0.1 *M* TEAP/Pt dsk/rot	$E_{1/2} = +0.69$
1,4-Dithianaphthalene	MeCN, 0.1 *M* TEAP/Pt dsk/rot	$E_{1/2} = +0.80$
Thianthrene	0.1 *M* TPAP/Pt/CV	$E_{1/2} = +1.28$

[e] vs. Ag/saturated $AgNO_3$

DISSOCIATION CONSTANTS OF INORGANIC ACIDS AND BASES

The data in this table are presented as values of pK_a, defined as the negative logarithm of the acid dissociation constant K_a for the reaction

$$BH \rightleftharpoons B^- + H^+$$

Thus $pK_a = -\log K_a$, and the hydrogen ion concentration $[H^+]$ can be calculated from

$$K_a = \frac{[H^+][B^-]}{[BH]}$$

In the case of bases, the entry in the table is for the conjugate acid; e.g., ammonium ion for ammonia. The OH^- concentration in the system

$$NH_3 + H_2O \rightleftharpoons NH_4^+ + OH^-$$

can be calculated from the equation

$$K_b = K_{water}/K_a = \frac{[OH^-][NH_4^+]}{[NH_3]}$$

where $K_{water} = 1.01 \times 10^{-14}$ at 25 °C. Note that $pK_a + pK_b = pK_{water}$.

All values refer to dilute aqueous solutions at zero ionic strength at the temperature indicated. The table is arranged alphabetically by compound name.

Reference

1. Perrin, D. D., *Ionization Constants of Inorganic Acids and Bases in Aqueous Solution, Second Edition*, Pergamon, Oxford, 1982.

Acid Dissociation Constant (as Negative Logarithm) for Inorganic Compounds at Specified Temperatures

Name	Formula	Step	t/°C	pK_a
Aluminum ion $[Al^{+3}]$	Al^{+3}		25	5.0
Ammonia	NH_3		25	9.25
Arsenic acid	H_3AsO_4	1	25	2.26
		2	25	6.76
		3	25	11.29
Arsenious acid	H_3AsO_3		25	9.29
Barium ion $[Ba^{+2}]$	Ba^{+2}		25	13.4
Boric acid	H_3BO_3	1	20	9.27
		2	20	14[a]
Calcium ion $[Ca^{+2}]$	Ca^{+2}		25	12.6
Carbonic acid	H_2CO_3	1	25	6.35
		2	25	10.33
Chlorous acid	$HClO_2$		25	1.94
Chromic acid	H_2CrO_4	1	25	0.74
		2	25	6.49
Cyanic acid	HOCN		25	3.46
Diphosphoric acid	$H_4P_2O_7$	1	25	0.91
		2	25	2.10
		3	25	6.70
		4	25	9.32
Germanic acid	H_2GeO_3	1	25	9.01
		2	25	12.3
Hydrazine	N_2H_4		25	8.1
Hydrazoic acid	HN_3		25	4.6
Hydrogen cyanide	HCN		25	9.21
Hydrogen fluoride	HF		25	3.20
Hydrogen peroxide	H_2O_2		25	11.62
Hydrogen selenide	H_2Se	1	25	3.89
		2	25	11.0
Hydrogen sulfide	H_2S	1	25	7.05
		2	25	19
Hydrogen telluride	H_2Te	1	18	2.6
		2	25	11
Hydroxylamine	H_2NOH		25	5.94
Hypobromous acid	HOBr		25	8.55
Hypochlorous acid	HOCl		25	7.40
Hypoiodous acid	HIO		25	10.5
Iodic acid	HIO_3		25	0.78
Lithium ion $[Li^+]$	Li^+		25	13.8
Magnesium ion $[Mg^{+2}]$	Mg^{+2}		25	11.4
Nitrous acid	HNO_2		25	3.25
Orthosilicic acid	H_4SiO_4	1	30	9.9
		2	30	11.8
		3	30	12
		4	30	12
Perchloric acid	$HClO_4$		20	-1.6
Periodic acid	HIO_4		25	1.64
Phosphonic acid	H_3PO_3	1	20	1.3
		2	20	6.70
Phosphoric acid	H_3PO_4	1	25	2.16
		2	25	7.21
		3	25	12.32
Selenic acid	H_2SeO_4	2	25	1.7
Selenous acid	H_2SeO_3	1	25	2.62
		2	25	8.32
Sodium ion $[Na^+]$	Na^+		25	14.8
Strontium ion $[Sr^{+2}]$	Sr^{+2}		25	13.2
Sulfamic acid	H_2NSO_3H		25	1.05
Sulfuric acid	H_2SO_4	2	25	1.99
Sulfurous acid	H_2SO_3	1	25	1.85
		2	25	7.2
Telluric(VI) acid	H_6TeO_6	1	18	7.68
		2	18	11.0
Tellurous acid	H_2TeO_3	1	25	6.27
		2	25	8.43
Tetrafluoroboric acid	HBF_4		25	0.5
Thiocyanic acid	HCNS		25	-1.8
Water	H_2O		25	13.995

[a] Lower limit

DISSOCIATION CONSTANTS OF ORGANIC ACIDS AND BASES

This table lists the dissociation (ionization) constants of over 1070 organic acids, bases, and amphoteric compounds. All data apply to dilute aqueous solutions and are presented as values of pK_a, which is defined as the negative of the logarithm of the equilibrium constant K_a for the reaction

$$HA \rightleftharpoons H^+ + A^-$$

i.e.,

$$K_a = [H^+][A^-]/[HA]$$

where $[H^+]$, etc., represent the concentrations of the respective species in mol/L. It follows that $pK_a = pH + \log[HA] - \log[A^-]$, so that a solution with 50% dissociation has pH equal to the pK_a of the acid.

Data for bases are presented as pK_a values for the conjugate acid, i.e., for the reaction

$$BH^+ \rightleftharpoons H^+ + B$$

In older literature, an ionization constant K_b was used for the reaction $B + H_2O \rightleftharpoons BH^+ + OH^-$. This is related to K_a by

$$pK_a + pK_b = pK_{water} = 14.00 \text{ (at 25 °C)}$$

Compounds are listed alphabetically by name.

References

1. Perrin, D. D., *Dissociation Constants of Organic Bases in Aqueous Solution*, Butterworths, London, 1965; Supplement, 1972.
2. Serjeant, E. P., and Dempsey, B., *Ionization Constants of Organic Acids in Aqueous Solution*, Pergamon, Oxford, 1979.
3. Albert, A., "Ionization Constants of Heterocyclic Substances", in Katritzky, A. R., Ed., *Physical Methods in Heterocyclic Chemistry*, Academic Press, New York, 1963.
4. Sober, H. A., Ed., *CRC Handbook of Biochemistry*, CRC Press, Boca Raton, FL, 1968.
5. Perrin, D. D., Dempsey, B., and Serjeant, E. P., *pK_a Prediction for Organic Acids and Bases*, Chapman and Hall, London, 1981.
6. Albert, A., and Serjeant, E. P., *The Determination of Ionization Constants, Third Edition*, Chapman and Hall, London, 1984.
7. O'Neil, M. J., Ed., *The Merck Index, 14th Edition*, Merck & Co., Whitehouse Station, NJ, 2006.

Acid Dissociation Constants (as Negative Logarithms) of Organic Compounds at Specified Temperatures

Name	Mol. form.	Step	t/°C	pK_a
Acetaldehyde	C_2H_4O		25	13.57
Acetamide	C_2H_5NO		25	15.1
Acetanilide	C_8H_9NO		25	0.5
Acetazolamide	$C_4H_6N_4O_3S_2$			7.2
Acetic acid	$C_2H_4O_2$		25	4.756
Acetoacetic acid	$C_4H_6O_3$		25	3.6
Acetohydroxamic acid	$C_2H_5NO_2$			8.70
N-Acetylglycine	$C_4H_7NO_3$		25	3.67
2-(Acetyloxy)benzoic acid	$C_9H_8O_4$		25	3.48
Aconine	$C_{25}H_{41}NO_9$			9.52
Aconitine	$C_{34}H_{47}NO_{11}$			5.88
9-Acridinamine	$C_{13}H_{10}N_2$		20	9.99
Acridine	$C_{13}H_9N$		20	5.58
3,6-Acridinediamine	$C_{13}H_{11}N_3$		20	9.65
Acrylic acid	$C_3H_4O_2$		25	4.25
Adenine	$C_5H_5N_5$	1		4.3
		2		9.83
Adenosine	$C_{10}H_{13}N_5O_4$	1	25	3.6
		2	25	12.4
Adenosine 5'-monophosphate	$C_{10}H_{14}N_5O_7P$	1		3.8
		2		6.2
Adipamic acid	$C_6H_{11}NO_3$		25	4.63
Agaritine	$C_{12}H_{17}N_3O_4$	1		3.4
		2		8.86
L-Alanine	$C_3H_7NO_2$	1	25	2.34
		2	25	9.87
β-Alanine	$C_3H_7NO_2$	1	25	3.55
		2	25	10.24
Allantoin	$C_4H_6N_4O_3$		25	8.96
Allopurinol	$C_5H_4N_4O$			10.2
Alloxanic acid	$C_4H_4N_2O_5$		25	6.64
Allyl alcohol	C_3H_6O		25	15.5
Allylamine	C_3H_7N		25	9.49
N-Allylaniline	$C_9H_{11}N$		25	4.17
Aminoacetonitrile	$C_2H_4N_2$		25	5.34
2-Aminoadipic acid	$C_6H_{11}NO_4$	1	25	2.14
		2	25	4.21
		3	25	9.77
4-Aminoazobenzene	$C_{12}H_{11}N_3$		25	2.82
2-Aminobenzenesulfonic acid	$C_6H_7NO_3S$		25	2.46
3-Aminobenzenesulfonic acid	$C_6H_7NO_3S$		25	3.74
4-Aminobenzenesulfonic acid	$C_6H_7NO_3S$		25	3.23
2-Aminobenzonitrile	$C_7H_6N_2$		25	0.77
3-Aminobenzonitrile	$C_7H_6N_2$		25	2.75
4-Aminobenzonitrile	$C_7H_6N_2$		25	1.74
2-Aminobiphenyl	$C_{12}H_{11}N$		25	3.83
3-Aminobiphenyl	$C_{12}H_{11}N$		18	4.25
4-Aminobiphenyl	$C_{12}H_{11}N$		18	4.35
DL-2-Aminobutanoic acid	$C_4H_9NO_2$	1	25	2.29
		2	25	9.83
4-Aminobutanoic acid	$C_4H_9NO_2$	1	25	4.031
		2	25	10.556
4-(2-Aminoethyl)phenol	$C_8H_{11}NO$	1	25	9.74
		2	25	10.52
6-Aminohexanoic acid	$C_6H_{13}NO_2$	1	25	4.37
		2	25	10.80
4-Amino-2-hydroxybenzoic acid	$C_7H_7NO_3$			3.25
2-Amino-4-hydroxypteridine	$C_6H_5N_5O$	1	20	2.27
		2	20	7.96
2-Amino-2-methylpropanoic acid	$C_4H_9NO_2$	1	25	2.36
		2	25	10.21
5-Amino-1-naphthol	$C_{10}H_9NO$		25	3.97
5-Amino-4-oxopentanoic acid	$C_5H_9NO_3$	1	25	4.05

Name	Mol. form.	Step	t/°C	pK_a
		2	25	8.90
5-Aminopentanoic acid	$C_5H_{11}NO_2$	1	25	4.27
		2	25	10.77
2-Aminophenol	C_6H_7NO	1	20	4.78
		2	20	9.97
3-Aminophenol	C_6H_7NO	1	20	4.37
		2	20	9.82
4-Aminophenol	C_6H_7NO	1	25	5.48
		2	25	10.30
3-Aminopropanenitrile	$C_3H_6N_2$		20	7.80
4-Amino-3,5,6-trichloro-2-pyridinecarbox	$C_6H_3Cl_3N_2O_2$			3.6
Amiodarone	$C_{25}H_{29}I_2NO_3$		25	6.56
Amitriptyline	$C_{20}H_{23}N$			9.4
Amobarbital	$C_{11}H_{18}N_2O_3$		25	8.0
Aniline	C_6H_7N		25	4.87
Aniline-2-carboxylic acid	$C_7H_7NO_2$	1	25	2.17
		2	25	4.85
Aniline-3-carboxylic acid	$C_7H_7NO_2$	1	25	3.07
		2	25	4.79
Aniline-4-carboxylic acid	$C_7H_7NO_2$	1	25	2.50
		2	25	4.87
Apomorphine	$C_{17}H_{17}NO_2$	1		7.0
		2		8.92
Arecoline	$C_8H_{13}NO_2$			6.84
L-Arginine	$C_6H_{14}N_4O_2$	1	25	1.82
		2	25	8.99
		3	25	12.5
L-Argininosuccinic acid	$C_{10}H_{18}N_4O_5$	1	25	1.62
		2	25	2.70
		3	25	4.26
		4	25	9.58
L-Ascorbic acid	$C_6H_8O_6$	1	25	4.04
		2	16	11.7
L-Asparagine	$C_4H_8N_2O_3$	1	20	2.1
		2	20	8.80
L-Aspartic acid	$C_4H_7NO_4$	1	25	1.99
		2	25	3.90
		3	25	9.90
Aspergillic acid	$C_{12}H_{20}N_2O_2$			5.5
Atenolol	$C_{14}H_{22}N_2O_3$			9.6
Atisine	$C_{22}H_{33}NO_2$			12.2
Azaserine	$C_5H_7N_3O_4$			8.55
Azathioprine	$C_9H_7N_7O_2S$			8.2
6-Azauridine	$C_8H_{11}N_3O_6$			6.70
Azetidine	C_3H_7N		25	11.29
Barbital	$C_8H_{12}N_2O_3$		25	7.43
Barbituric acid	$C_4H_4N_2O_3$		25	4.01
Benzaldehyde	C_7H_6O		25	14.90
Benzamide	C_7H_7NO		25	13[a]
Benzeneacetic acid	$C_8H_8O_2$		25	4.31
Benzeneboronic acid	$C_6H_7BO_2$			8.83
Benzenebutanoic acid	$C_{10}H_{12}O_2$		25	4.76
1,2-Benzenediamine	$C_6H_8N_2$	1	20	4.57
		2	20	0.80
1,3-Benzenediamine	$C_6H_8N_2$	1	20	5.11
		2	20	2.50
1,4-Benzenediamine	$C_6H_8N_2$	1	20	6.31
		2	20	2.97
Benzeneethanamine	$C_8H_{11}N$		25	9.83
Benzenemethanethiol	C_7H_8S		25	9.43
Benzenepropanoic acid	$C_9H_{10}O_2$		25	4.66
Benzenesulfinic acid	$C_6H_6O_2S$		20	1.3
Benzenesulfonic acid	$C_6H_6O_3S$		25	0.70
Benzenethiol	C_6H_6S		25	6.62
p-Benzidine	$C_{12}H_{12}N_2$	1	20	4.65
		2	20	3.43
1*H*-Benzimidazole	$C_7H_6N_2$		25	5.53
Benzoic acid	$C_7H_6O_2$		25	4.204
1*H*-Benzotriazole	$C_6H_5N_3$		20	1.6
N-Benzoylglycine	$C_9H_9NO_3$		25	3.62
Benzpiperylon	$C_{22}H_{25}N_3O$	1		6.73
		2		9.13
Benzylamine	C_7H_9N		25	9.34
4-Benzylaniline	$C_{13}H_{13}N$		25	2.17
2-Benzylpyridine	$C_{12}H_{11}N$		25	5.13
Betaine	$C_5H_{11}NO_2$		0	1.83
Biguanide	$C_2H_7N_5$	1		11.52
		2		2.93
N,N-Bis(2-hydroxyethyl)glycine	$C_6H_{13}NO_4$	2	20	8.35
Bithionol	$C_{12}H_6Cl_4O_2S$	1		4.82
		2		10.50
Bornylamine	$C_{10}H_{19}N$		25	10.17
Bromadiolone	$C_{30}H_{23}BrO_4$		21	4.04
Bromoacetic acid	$C_2H_3BrO_2$		25	2.90
2-Bromoaniline	C_6H_6BrN		25	2.53
3-Bromoaniline	C_6H_6BrN		25	3.53
4-Bromoaniline	C_6H_6BrN		25	3.89
2-Bromobenzoic acid	$C_7H_5BrO_2$		25	2.85
3-Bromobenzoic acid	$C_7H_5BrO_2$		25	3.81
4-Bromobenzoic acid	$C_7H_5BrO_2$		25	3.96
Bromocresol Green	$C_{21}H_{14}Br_4O_5S$			4.7
Bromocresol Purple	$C_{21}H_{16}Br_2O_5S$			6.3
4-Bromo-*N,N*-dimethylaniline	$C_8H_{10}BrN$		25	4.23
2-Bromophenol	C_6H_5BrO		25	8.45
3-Bromophenol	C_6H_5BrO		25	9.03
4-Bromophenol	C_6H_5BrO		25	9.37
Bromophenol Blue	$C_{19}H_{10}Br_4O_5S$			4.0
3-Bromopropanoic acid	$C_3H_5BrO_2$		25	4.00
3-Bromopyridine	C_5H_4BrN		25	2.84
3-Bromoquinoline	C_9H_6BrN		25	2.69
Bromothymol Blue	$C_{27}H_{28}Br_2O_5S$			7.0
Brucine	$C_{23}H_{26}N_2O_4$	1		6.04
		2		11.07
1,4-Butanediamine	$C_4H_{12}N_2$	1	25	10.80
		2	25	9.63
1,2,3,4-Butanetetrol	$C_4H_{10}O_4$			13.9
Butanoic acid	$C_4H_8O_2$		25	4.83
trans-2-Butenoic acid	$C_4H_6O_2$		25	4.69
3-Butenoic acid	$C_4H_6O_2$		25	4.34
Butylamine	$C_4H_{11}N$		25	10.60
sec-Butylamine	$C_4H_{11}N$		25	10.56
tert-Butylamine	$C_4H_{11}N$		25	10.68
N-*tert*-Butylaniline	$C_{10}H_{15}N$		25	7.00
2-*tert*-Butylbenzoic acid	$C_{11}H_{14}O_2$		25	3.54
3-*tert*-Butylbenzoic acid	$C_{11}H_{14}O_2$		25	4.20
4-*tert*-Butylbenzoic acid	$C_{11}H_{14}O_2$		25	4.38
Butylcyclohexylamine	$C_{10}H_{21}N$		25	11.23
2-*tert*-Butylphenol	$C_{10}H_{14}O$		25	10.62
3-*tert*-Butylphenol	$C_{10}H_{14}O$		25	10.12
4-*tert*-Butylphenol	$C_{10}H_{14}O$		25	10.23
N-Butylpiperidine	$C_9H_{19}N$		23	10.47

Name	Mol. form.	Step	t/°C	pK_a
Butylpropanedioic acid	$C_7H_{12}O_4$	1	5	2.96
2-Butynoic acid	$C_4H_4O_2$		25	2.62
Captopril	$C_9H_{15}NO_3S$	1		3.7
		2		9.8
Carbamodithioic acid	CH_3NS_2		25	2.95
Carbendazim	$C_9H_9N_3O_2$			4.48
L-γ-Carboxyglutamic acid	$C_6H_9NO_6$	1	25	1.7
		2	25	3.2
		3	25	4.75
		4	25	9.9
Carnitine	$C_7H_{15}NO_3$		25	3.80
Carnosine	$C_9H_{14}N_4O_3$	1	20	2.73
		2	20	6.87
		3	20	9.73
Cephalexin	$C_{16}H_{17}N_3O_4S$	1		5.2
		2		7.3
Cephaloridine	$C_{19}H_{17}N_3O_4S_2$			3.2
Cephradine	$C_{16}H_{19}N_3O_4S$	1		2.63
		2		7.27
Chloroacetic acid	$C_2H_3ClO_2$		25	2.87
2-Chloroaniline	C_6H_6ClN		25	2.66
3-Chloroaniline	C_6H_6ClN		25	3.52
4-Chloroaniline	C_6H_6ClN		25	3.98
2-Chlorobenzeneacetic acid	$C_8H_7ClO_2$		25	4.07
3-Chlorobenzeneacetic acid	$C_8H_7ClO_2$		25	4.14
4-Chlorobenzeneacetic acid	$C_8H_7ClO_2$		25	4.19
2-Chlorobenzoic acid	$C_7H_5ClO_2$		25	2.90
3-Chlorobenzoic acid	$C_7H_5ClO_2$		25	3.84
4-Chlorobenzoic acid	$C_7H_5ClO_2$		25	4.00
2-Chlorobutanoic acid	$C_4H_7ClO_2$			2.86
3-Chlorobutanoic acid	$C_4H_7ClO_2$			4.05
4-Chlorobutanoic acid	$C_4H_7ClO_2$			4.52
trans-o-Chlorocinnamic acid	$C_9H_7ClO_2$		25	4.23
trans-m-Chlorocinnamic acid	$C_9H_7ClO_2$		25	4.29
trans-p-Chlorocinnamic acid	$C_9H_7ClO_2$		25	4.41
Chlorodiazepoxide	$C_{16}H_{14}ClN_3O$			4.8
3-Chloro-*N,N*-dimethylaniline	$C_8H_{10}ClN$		20	3.83
4-Chloro-*N,N*-dimethylaniline	$C_8H_{10}ClN$		20	4.39
2-Chlorophenol	C_6H_5ClO		25	8.56
3-Chlorophenol	C_6H_5ClO		25	9.12
4-Chlorophenol	C_6H_5ClO		25	9.41
2-Chlorophenoxyacetic acid	$C_8H_7ClO_3$		25	3.05
3-Chlorophenoxyacetic acid	$C_8H_7ClO_3$		25	3.10
3-(2-Chlorophenyl)propanoic acid	$C_9H_9ClO_2$		25	4.58
3-(3-Chlorophenyl)propanoic acid	$C_9H_9ClO_2$		25	4.59
3-(4-Chlorophenyl)propanoic acid	$C_9H_9ClO_2$		25	4.61
2-Chloropropanoic acid	$C_3H_5ClO_2$		25	2.83
3-Chloropropanoic acid	$C_3H_5ClO_2$		25	3.98
2-Chloropyridine	C_5H_4ClN		25	0.49
3-Chloropyridine	C_5H_4ClN		25	2.81
4-Chloropyridine	C_5H_4ClN		25	3.83
Chlorothiazide	$C_7H_6ClN_3O_4S_2$	1		6.85
		2		9.45
Cholic acid	$C_{24}H_{40}O_5$		20	4.98
Choline	$C_5H_{14}NO$		25	13.9
Cinchonidine	$C_{19}H_{22}N_2O$	1		5.80
		2		10.03
Cinchonine	$C_{19}H_{22}N_2O$	1		5.85
		2		9.92

Name	Mol. form.	Step	t/°C	pK_a
cis-Cinnamic acid	$C_9H_8O_2$		25	3.88
trans-Cinnamic acid	$C_9H_8O_2$		25	4.44
Cinnoline	$C_8H_6N_2$		20	2.37
Citric acid	$C_6H_8O_7$	1	25	3.13
		2	25	4.76
		3	25	6.40
Citrulline	$C_6H_{13}N_3O_3$	1	25	2.43
		2	25	9.69
Clindamycin	$C_{18}H_{33}ClN_2O_5S$			7.6
Clonazepam	$C_{15}H_{10}ClN_3O_3$	1		1.5
		2		10.5
Clozapine	$C_{18}H_{19}ClN_4$	1		3.70
		2		7.60
Codeine	$C_{18}H_{21}NO_3$			8.21
Colchicine	$C_{22}H_{25}NO_6$		20	12.36
Creatine	$C_4H_9N_3O_2$	1	25	2.63
		2	25	14.3
Creatinine	$C_4H_7N_3O$	1	25	4.8
		2		9.2
o-Cresol	C_7H_8O		25	10.29
m-Cresol	C_7H_8O		25	10.09
p-Cresol	C_7H_8O		25	10.26
o-Cresolphthalein	$C_{22}H_{18}O_4$			9.4
Cresol Red	$C_{21}H_{18}O_5S$			8.3
Cupreine	$C_{19}H_{22}N_2O_2$			6.57
Cyanamide	CH_2N_2		29	1.1
Cyanic acid	CHNO		25	3.7
Cyanoacetic acid	$C_3H_3NO_2$		25	2.47
3-Cyanobenzoic acid	$C_8H_5NO_2$		25	3.60
4-Cyanobenzoic acid	$C_8H_5NO_2$		25	3.55
4-Cyanobutanoic acid	$C_5H_7NO_2$		25	2.42
(2-Cyanophenoxy)acetic acid	$C_9H_7NO_3$		25	2.98
(3-Cyanophenoxy)acetic acid	$C_9H_7NO_3$		25	3.03
(4-Cyanophenoxy)acetic acid	$C_9H_7NO_3$		25	2.93
Cyanuric acid	$C_3H_3N_3O_3$	1		6.88
		2		11.40
		3		13.5
Cyclohexanecarboxylic acid	$C_7H_{12}O_2$		25	4.91
cis-1,2-Cyclohexanediamine	$C_6H_{14}N_2$	1	20	9.93
		2	20	6.13
trans-1,2-Cyclohexanediamine	$C_6H_{14}N_2$	1	20	9.94
		2	20	6.47
1,3-Cyclohexanedione	$C_6H_8O_2$		25	5.26
Cyclohexylamine	$C_6H_{13}N$		25	10.64
Cyclopentanecarboxylic acid	$C_6H_{10}O_2$		25	4.99
Cyclopropanecarboxylic acid	$C_4H_6O_2$		25	4.83
1,1-Cyclopropanedicarboxylic acid	$C_5H_6O_4$	1	25	1.82
		2	25	7.43
Cysteamine	C_2H_7NS	1	25	8.27
		2	25	10.53
DL-Cysteic acid	$C_3H_7NO_5S$	1	25	1.3
		2	25	1.9
		3	25	8.70
L-Cysteine	$C_3H_7NO_2S$	1	25	1.5
		2	25	8.7
		3	25	10.2
L-Cystine	$C_6H_{12}N_2O_4S_2$	1		1
		2		2.1
		3		8.02
		4		8.71

Thermochem

Name	Mol. form.	Step	t/°C	pK_a
Cytidine	$C_9H_{13}N_3O_5$	1		4.22
		2		12.5
3'-Cytidylic acid	$C_9H_{14}N_3O_8P$	1		0.8
		2		4.28
		3		6.0
Cytisine	$C_{11}H_{14}N_2O$	1		6.11
		2		13.08
Cytosine	$C_4H_5N_3O$	1		4.60
		2		12.16
Decanedioic acid	$C_{10}H_{18}O_4$	1		4.59
		2		5.59
Decylamine	$C_{10}H_{23}N$		25	10.64
D-2-Deoxyribose	$C_5H_{10}O_4$		25	12.61
2,4-Diaminobutanoic acid	$C_4H_{10}N_2O_2$	1	25	1.85
		2	25	8.24
		3	25	10.44
1,3-Diamino-2-propanol	$C_3H_{10}N_2O$	1	20	9.69
		2	20	7.93
Dibenzepin	$C_{18}H_{21}N_3O$			8.25
3,5-Dibromoaniline	$C_6H_5Br_2N$		25	2.34
3,5-Dibromo-4-hydroxybenzonitrile	$C_7H_3Br_2NO$			4.06
3,5-Dibromo-L-tyrosine	$C_9H_9Br_2NO_3$	1		2.17
		2		6.45
		3		7.60
Dibutylamine	$C_8H_{19}N$		21	11.25
2,6-Di-*tert*-butylpyridine	$C_{13}H_{21}N$			3.58
Dichloroacetic acid	$C_2H_2Cl_2O_2$		25	1.35
2,4-Dichloroaniline	$C_6H_5Cl_2N$		22	2.05
2,3-Dichlorophenol	$C_6H_4Cl_2O$		25	7.44
Dicyclohexylamine	$C_{12}H_{23}N$			10.4
2',3'-Dideoxyinosine	$C_{10}H_{12}N_4O_3$			9.12
Diethylamine	$C_4H_{11}N$		25	10.84
N,N-Diethylaniline	$C_{10}H_{15}N$		25	6.57
Diethylmethylamine	$C_5H_{13}N$		25	10.35
N,N-Diethyl-2-methylaniline	$C_{11}H_{17}N$		25	7.24
1,7-Dihydro-6*H*-purine-6-thione	$C_5H_4N_4S$	1		7.77
		2		11.17
2,5-Dihydroxybenzeneacetic acid	$C_8H_8O_4$		25	4.40
2,4-Dihydroxybenzoic acid	$C_7H_6O_4$	1	25	3.11
		2	25	8.55
		3	25	14.0
2,5-Dihydroxybenzoic acid	$C_7H_6O_4$	1	25	2.97
3,4-Dihydroxybenzoic acid	$C_7H_6O_4$	1	25	4.48
		2	25	8.83
		3	25	12.6
3,5-Dihydroxybenzoic acid	$C_7H_6O_4$	1	25	4.04
3,12-Dihydroxycholan-24-oic acid, (3α,5β,12α)	$C_{24}H_{40}O_4$		20	5.15
Dihydroxytartaric acid	$C_4H_6O_8$		25	1.92
3,5-Diiodo-L-tyrosine	$C_9H_9I_2NO_3$	1	25	2.12
		2	25	5.32
		3	25	9.48
Diisopropylamine	$C_6H_{15}N$		25	11.05
Dimethylamine	C_2H_7N		25	10.73
4-(Dimethylamino)benzoic acid	$C_9H_{11}NO_2$	1		6.03
		2		11.49
2-Dimethylaminopurine	$C_7H_9N_5$	1	20	4.00
		2	20	10.24
2,6-Dimethylaniline	$C_8H_{11}N$		25	3.89
N,N-Dimethylaniline	$C_8H_{11}N$		25	5.07
Dimethylarsinic acid	$C_2H_7AsO_2$	1	25	1.57
		2	25	6.27
3,5-Dimethylbenzoic acid	$C_9H_{10}O_2$		25	4.32
5,5-Dimethyl-1,3-cyclohexanedione	$C_8H_{12}O_2$		25	5.15
2,2-Dimethyl-1,3-dioxane-4,6-dione	$C_6H_8O_4$			5.1
N,N-Dimethylglycine	$C_4H_9NO_2$		25	9.89
2,4-Dimethyl-1*H*-imidazole	$C_5H_8N_2$		25	8.36
Dimethylmalonic acid	$C_5H_8O_4$		25	3.15
N,N-Dimethyl-1-naphthylamine	$C_{12}H_{13}N$		25	4.83
N,N-Dimethyl-2-naphthylamine	$C_{12}H_{13}N$		25	4.566
N,N-Dimethyl-3-nitroaniline	$C_8H_{10}N_2O_2$		25	2.62
5,5-Dimethyl-2,4-oxazolidinedione	$C_5H_7NO_3$		37	6.13
cis-2,5-Dimethylpiperazine	$C_6H_{14}N_2$	1	25	9.66
		2	25	5.20
1,2-Dimethylpiperidine, (±)-	$C_7H_{15}N$		25	10.22
2,2-Dimethylpropanoic acid	$C_5H_{10}O_2$		20	5.03
2,2-Dimethylpropylamine	$C_5H_{13}N$		25	10.15
2,3-Dimethylpyridine	C_7H_9N		25	6.57
2,4-Dimethylpyridine	C_7H_9N		25	6.99
2,5-Dimethylpyridine	C_7H_9N		25	6.40
2,6-Dimethylpyridine	C_7H_9N		25	6.65
3,4-Dimethylpyridine	C_7H_9N		25	6.46
3,5-Dimethylpyridine	C_7H_9N		25	6.15
4,6-Dimethyl-2-pyrimidinamine	$C_6H_9N_3$		20	4.82
1,2-Dimethylpyrrolidine	$C_6H_{13}N$		26	10.20
2,4-Dinitrobenzoic acid	$C_7H_4N_2O_6$		25	1.43
2,4-Dinitrophenol	$C_6H_4N_2O_5$		25	4.07
2,5-Dinitrophenol	$C_6H_4N_2O_5$		15	5.15
Dinoseb	$C_{10}H_{12}N_2O_5$			4.62
Diphenylamine	$C_{12}H_{11}N$		25	0.79
4-[(Dipropylamino)sulfonyl] benzoic acid	$C_{13}H_{19}NO_4S$			5.8
Dodecylamine	$C_{12}H_{27}N$		25	10.63
Dopamine	$C_8H_{11}NO_2$	1	25	8.9
		2	25	10.6
Droperidol	$C_{22}H_{22}FN_3O_2$			7.64
Emetine	$C_{29}H_{40}N_2O_4$	1		5.77
		2		6.64
Ephedrine, (1*S*,2*R*)	$C_{10}H_{15}NO$		10	10.139
Ephedrine, (1*R*,2*S*)	$C_{10}H_{15}NO$		10	9.958
Epinephrine	$C_9H_{13}NO_3$	1	25	8.66
		2	25	9.95
Ergometrinine	$C_{19}H_{23}N_3O_2$			7.3
Ergonovine	$C_{19}H_{23}N_3O_2$			6.8
Erythromycin	$C_{37}H_{67}NO_{13}$			8.8
Ethacrynic acid	$C_{13}H_{12}Cl_2O_4$			3.50
1,2-Ethanediamine	$C_2H_8N_2$	1	25	9.92
		2	25	6.86
1,2-Ethanediol	$C_2H_6O_2$		25	15.1
Ethanimidamide	$C_2H_6N_2$		25	12.1
Ethanol	C_2H_6O		25	15.5
Ethanolamine	C_2H_7NO		25	9.50
Ethoxyacetic acid	$C_4H_8O_3$		18	3.65
2-Ethoxyaniline	$C_8H_{11}NO$		28	4.43
3-Ethoxyaniline	$C_8H_{11}NO$		25	4.18
4-Ethoxyaniline	$C_8H_{11}NO$		28	5.20
Ethyl acetoacetate	$C_6H_{10}O_3$		25	10.68
Ethylamine	C_2H_7N		25	10.65

Name	Mol. form.	Step	t/°C	pK_a
Ethyl 4-aminobenzoate	$C_9H_{11}NO_2$			2.5
N-Ethylaniline	$C_8H_{11}N$		25	5.12
2-Ethyl-1*H*-benzimidazole	$C_9H_{10}N_2$		25	6.18
O-Ethyl *S*-[2-(diisopropylamino) ethyl] methylphosphonothioate	$C_{11}H_{26}NO_2PS$			7.9
Ethyleneimine	C_2H_5N		25	8.04
α-Ethylglutamic acid	$C_7H_{13}NO_4$	1	25	3.846
		2	25	7.838
N-Ethylmorpholine	$C_6H_{13}NO$		25	7.67
1-Ethylpiperidine	$C_7H_{15}N$		23	10.45
2-Ethylpyridine	C_7H_9N		25	5.89
Etoposide	$C_{29}H_{32}O_{13}$			9.8
Fluoroacetic acid	$C_2H_3FO_2$		25	2.59
2-Fluoroaniline	C_6H_6FN		25	3.20
3-Fluoroaniline	C_6H_6FN		25	3.59
4-Fluoroaniline	C_6H_6FN		25	4.65
2-Fluorobenzoic acid	$C_7H_5FO_2$		25	3.27
3-Fluorobenzoic acid	$C_7H_5FO_2$		25	3.86
4-Fluorobenzoic acid	$C_7H_5FO_2$		25	4.15
5-Fluorocytosine	$C_4H_4FN_3O$			3.26
2-Fluorophenol	C_6H_5FO		25	8.73
3-Fluorophenol	C_6H_5FO		25	9.29
4-Fluorophenol	C_6H_5FO		25	9.89
2-Fluoropyridine	C_5H_4FN		25	-0.44
Folinic acid	$C_{20}H_{23}N_7O_7$	1		3.1
		2		4.8
		3		10.4
Formaldehyde	CH_2O		25	13.27
Formic acid	CH_2O_2		25	3.75
3-Formylbenzoic acid	$C_8H_6O_3$		25	3.84
4-Formylbenzoic acid	$C_8H_6O_3$		25	3.77
β-D-Fructose	$C_6H_{12}O_6$		25	12.27
Fumaric acid	$C_4H_4O_4$	1	25	3.02
		2	25	4.38
2-Furancarboxylic acid	$C_5H_4O_3$		25	3.16
3-Furancarboxylic acid	$C_5H_4O_3$		25	3.9
Furethidine	$C_{21}H_{31}NO_4$			7.48
Gibberellic acid	$C_{19}H_{22}O_6$			4.0
α-D-Glucose	$C_6H_{12}O_6$		25	12.46
L-Glutamic acid	$C_5H_9NO_4$	1	25	2.13
		2	25	4.31
		3		9.67
L-Glutamine	$C_5H_{10}N_2O_3$	1	25	2.17
		2	25	9.13
Glutathione	$C_{10}H_{17}N_3O_6S$	1	25	2.12
		2	25	3.59
		3	25	8.75
		4	25	9.65
Glyceric acid	$C_3H_6O_4$		25	3.52
Glycerol	$C_3H_8O_3$		25	14.15
Glycine	$C_2H_5NO_2$	1	25	2.35
		2	25	9.78
Glycocholic acid	$C_{26}H_{43}NO_6$			4.4
Glycolic acid	$C_2H_4O_3$		25	3.83
Glycylalanine	$C_5H_{10}N_2O_3$		25	3.15
L-Glycylasparagine	$C_6H_{11}N_3O_4$	1	25	2.942
		2	18	8.44
N-Glycylglycine	$C_4H_8N_2O_3$	1	25	3.14
		2		8.17
N-(*N*-Glycylglycyl)glycine	$C_6H_{11}N_3O_4$	1	25	3.225
		2	25	8.09
N-Glycyl-L-leucine	$C_8H_{16}N_2O_3$		25	3.18
N-Glycylserine, (*DL*)-	$C_5H_{10}N_2O_4$	1	25	2.98
		2	25	8.38
Glyoxylic acid	$C_2H_2O_3$		25	3.18
Guanidine	CH_5N_3		25	13.6
Guanidinoacetic acid	$C_3H_7N_3O_2$		25	2.82
Guanine	$C_5H_5N_5O$		40	9.92
Haloperidol	$C_{21}H_{23}ClFNO_2$			8.3
Harmaline	$C_{13}H_{14}N_2O$			4.2
Harmine	$C_{13}H_{12}N_2O$			7.70
2-Heptanamine	$C_7H_{17}N$		19	10.7
Heptanedioic acid	$C_7H_{12}O_4$	1	25	4.71
		2	25	5.58
Heptanoic acid	$C_7H_{14}O_2$		25	4.89
Heptylamine	$C_7H_{17}N$		25	10.67
Hexadecylamine	$C_{16}H_{35}N$		25	10.61
2,4-Hexadienoic acid	$C_6H_8O_2$		25	4.76
Hexamethyldisilazane	$C_6H_{19}NSi_2$			7.55
1,6-Hexanediamine	$C_6H_{16}N_2$	1	0	11.86
		2	0	10.76
1,6-Hexanedioic acid	$C_6H_{10}O_4$	1	18	4.41
		2	18	5.41
Hexanoic acid	$C_6H_{12}O_2$		25	4.85
Hexylamine	$C_6H_{15}N$		25	10.56
Histamine	$C_5H_9N_3$	1	25	6.04
		2	25	9.75
L-Histidine	$C_6H_9N_3O_2$	1	25	1.80
		2	25	6.04
		3	25	9.33
DL-Homocysteine	$C_4H_9NO_2S$	1	25	2.22
		2	25	8.87
		3	25	10.86
Homocystine	$C_8H_{16}N_2O_4S_2$	1	25	1.59
		2	25	2.54
		3	25	8.52
		4	25	9.44
L-Homoserine	$C_4H_9NO_3$	1	25	2.71
		2	25	9.62
Hydrastine	$C_{21}H_{21}NO_6$			7.8
Hydrastinine	$C_{11}H_{13}NO_3$			11.38
Hydrochlorothiazide	$C_7H_8ClN_3O_4S_2$	1		7.9
		2		9.2
Hydroflumethiazide	$C_8H_8F_3N_3O_4S_2$	1		8.9
		2		9.7
Hydroquinine	$C_{20}H_{26}N_2O_2$			5.33
p-Hydroquinone	$C_6H_6O_2$	1	25	9.85
		2	25	11.4
3-Hydroxybenzaldehyde	$C_7H_6O_2$		25	8.98
4-Hydroxybenzaldehyde	$C_7H_6O_2$		25	7.61
α-Hydroxybenzeneaceticacid,(±)-	$C_8H_8O_3$		25	3.37
3-Hydroxybenzenesulfonic acid	$C_6H_6O_4S$		25	9.07
4-Hydroxybenzenesulfonic acid	$C_6H_6O_4S$		25	9.11
2-Hydroxybenzoic acid	$C_7H_6O_3$	1	20	2.98
		2	20	13.6
3-Hydroxybenzoic acid	$C_7H_6O_3$	1	25	4.08
		2	19	9.92
4-Hydroxybenzoic acid	$C_7H_6O_3$	1	25	4.57
		2	25	9.46
2-Hydroxybenzonitrile	C_7H_5NO		25	6.86
3-Hydroxybenzonitrile	C_7H_5NO		25	8.61
4-Hydroxybenzonitrile	C_7H_5NO		25	7.97

Name	Mol. form.	Step	t/°C	pK_a
2-Hydroxybiphenyl	$C_{12}H_{10}O$		25	10.01
3-Hydroxybiphenyl	$C_{12}H_{10}O$		25	9.64
4-Hydroxybiphenyl	$C_{12}H_{10}O$		25	9.55
3-Hydroxybutanoic acid, (±)-	$C_4H_8O_3$		25	4.70
4-Hydroxybutanoic acid	$C_4H_8O_3$		25	4.72
1-Hydroxy-1,1-diphosphonoethane	$C_2H_8O_7P_2$	1		1.35
		2		2.87
		3		7.03
		4		11.3
5-Hydroxy-2-(hydroxymethyl)-4*H*-pyran-4-one	$C_6H_6O_4$			7.9
α-Hydroxy-α-methylbenzeneaceti	$C_9H_{10}O_3$		25	3.47
α-Hydroxy-α-phenylbenzeneacetic acid	$C_{14}H_{12}O_3$		25	3.04
1-(2-Hydroxyphenyl)ethanone	$C_8H_8O_2$		25	10.06
1-(3-Hydroxyphenyl)ethanone	$C_8H_8O_2$		25	9.19
1-(4-Hydroxyphenyl)ethanone	$C_8H_8O_2$		25	8.05
trans-4-Hydroxy-L-proline	$C_5H_9NO_3$	1	25	1.82
		2	25	9.66
Hydroxypropanedioic acid	$C_3H_4O_5$	1		2.42
		2		4.54
3-Hydroxypropanoic acid	$C_3H_6O_3$		25	4.51
5-Hydroxytryptamine	$C_{10}H_{12}N_2O$	1	25	9.8
		2	25	11.1
Hyoscyamine	$C_{17}H_{23}NO_3$		21	9.7
Hypoxanthine	$C_5H_4N_4O$		25	8.7
Imazapyr	$C_{13}H_{15}N_3O_3$	1		1.9
		2		3.6
Imazethapyr	$C_{15}H_{19}N_3O_3$	1		2.1
		2		3.9
Imidazole	$C_3H_4N_2$		25	6.99
Iminodiacetic acid	$C_4H_7NO_4$	1		2.98
		2		9.89
1-Indanamine	$C_9H_{11}N$		22	9.21
1*H*-Indole-3-acetic acid	$C_{10}H_9NO_2$			4.75
Indomethacin	$C_{19}H_{16}ClNO_4$			4.5
Iocetamic acid	$C_{12}H_{13}I_3N_2O_3$			4
Iodoacetic acid	$C_2H_3IO_2$		25	3.18
2-Iodoaniline	C_6H_6IN		25	2.54
3-Iodoaniline	C_6H_6IN		25	3.58
4-Iodoaniline	C_6H_6IN		25	3.81
2-Iodobenzoic acid	$C_7H_5IO_2$		25	2.86
3-Iodobenzoic acid	$C_7H_5IO_2$		25	3.87
4-Iodobenzoic acid	$C_7H_5IO_2$		25	4.00
2-Iodophenol	C_6H_5IO		25	8.51
3-Iodophenol	C_6H_5IO		25	9.03
4-Iodophenol	C_6H_5IO		25	9.33
L-3-Iodotyrosine	$C_9H_{10}INO_3$	1	25	2.2
		2	25	8.7
		3	25	9.1
Iopanoic acid	$C_{11}H_{12}I_3NO_2$			4.8
Isocitric acid	$C_6H_8O_7$	1	25	3.29
		2	25	4.71
		3	25	6.40
L-Isoleucine	$C_6H_{13}NO_2$	1	25	2.32
		2	25	9.76
Isophthalic acid	$C_8H_6O_4$	1	25	3.70
		2	25	4.60
Isopropylamine	C_3H_9N		25	10.63
N-Isopropylaniline	$C_9H_{13}N$		25	5.77
Isoproterenol	$C_{11}H_{17}NO_3$			8.64
1-Isoquinolinamine	$C_9H_8N_2$		20	7.62
3-Isoquinolinamine	$C_9H_8N_2$		20	5.05
Isoquinoline	C_9H_7N		20	5.40
7-Isoquinolinol	C_9H_7NO	1	20	5.68
		2	20	8.90
Isoxazole	C_3H_3NO		25	-2.0
Ketamine	$C_{13}H_{16}ClNO$			7.5
D-Lactic acid	$C_3H_6O_3$		25	3.86
L-Leucine	$C_6H_{13}NO_2$	1	25	2.33
		2	25	9.74
N-Leucylglycine	$C_8H_{16}N_2O_3$	1	25	3.25
		2	25	8.2
Levodopa	$C_9H_{11}NO_4$	1	25	2.32
		2	25	8.72
		3	25	9.96
		4	25	11.79
Lisinopril	$C_{21}H_{35}N_3O_7$	1		2.5
		2		4.0
		3		6.7
		4		10.1
Lycodine	$C_{16}H_{22}N_2$	1		3.97
		2		8.08
Lysergic acid	$C_{16}H_{16}N_2O_2$	1		3.44
		2		7.68
L-Lysine	$C_6H_{14}N_2O_2$	1	25	2.16
		2	25	9.06
		3	25	10.54
Maleic acid	$C_4H_4O_4$	1	25	1.92
		2	25	6.23
Malic acid	$C_4H_6O_5$	1	25	3.40
		2	25	5.11
Malonic acid	$C_3H_4O_4$	1	25	2.85
		2	25	5.70
α-Maltose	$C_{12}H_{22}O_{11}$		21	12.05
D-Mannitol	$C_6H_{14}O_6$		18	13.5
D-Mannose	$C_6H_{12}O_6$		25	12.08
Mefenamic acid	$C_{15}H_{15}NO_2$			4.2
Mefluidide	$C_{11}H_{13}F_3N_2O_3S$			4.6
2-Mercaptoethanol	C_2H_6OS		25	9.72
Methanethiol	CH_4S		25	10.33
Methanol	CH_4O		25	15.5
Metharbital	$C_9H_{14}N_2O_3$			8.45
Methazolamide	$C_5H_8N_4O_3S_2$			7.30
L-Methionine	$C_5H_{11}NO_2S$	1	25	2.13
		2	25	9.27
Methoxamine hydrochloride	$C_{11}H_{18}ClNO_3$		25	9.2
2-Methoxyaniline	C_7H_9NO		25	4.53
3-Methoxyaniline	C_7H_9NO		25	4.20
4-Methoxyaniline	C_7H_9NO		25	5.36
2-Methoxybenzoic acid	$C_8H_8O_3$		25	4.08
3-Methoxybenzoic acid	$C_8H_8O_3$		25	4.10
4-Methoxybenzoic acid	$C_8H_8O_3$		25	4.50
2-Methoxyethanol	$C_3H_8O_2$		25	14.8
2-Methoxyethylamine	C_3H_9NO		25	9.40
2-(2-Methoxyethyl)pyridine	$C_8H_{11}NO$			5.5
2-Methoxyphenol	$C_7H_8O_2$		25	9.98
3-Methoxyphenol	$C_7H_8O_2$		25	9.65
4-Methoxyphenol	$C_7H_8O_2$		25	10.21
2-Methoxypyridine	C_6H_7NO		20	3.28
3-Methoxypyridine	C_6H_7NO		25	4.78

Thermochem

Name	Mol. form.	Step	t/°C	pK_a
4-Methoxypyridine	C_6H_7NO		25	6.58
6-Methoxyquinoline	$C_{10}H_9NO$		20	5.03
Methyclothiazide	$C_9H_{11}Cl_2N_3O_4S_2$			9.4
Methylamine	CH_5N		25	10.66
2-(Methylamino)benzoic acid	$C_8H_9NO_2$		25	5.34
3-(Methylamino)benzoic acid	$C_8H_9NO_2$		25	5.10
4-(Methylamino)benzoic acid	$C_8H_9NO_2$		25	5.04
2-Methylaniline	C_7H_9N		25	4.45
3-Methylaniline	C_7H_9N		25	4.71
4-Methylaniline	C_7H_9N		25	5.08
N-Methylaniline	C_7H_9N		25	4.85
α-Methylbenzeneacetic acid, (±)-	$C_9H_{10}O_2$		25	4.64
2-Methyl-1*H*-benzimidazole	$C_8H_8N_2$		25	6.19
3-Methyl-1-butanamine	$C_5H_{13}N$		25	10.60
2-Methyl-2-butanamine	$C_5H_{13}N$		19	10.85
2-Methylbutanoic acid	$C_5H_{10}O_2$		25	4.80
3-Methylbutanoic acid	$C_5H_{10}O_2$		25	4.77
trans-o-Methylcinnamic acid	$C_{10}H_{10}O_2$		25	4.50
trans-m-Methylcinnamic acid	$C_{10}H_{10}O_2$		25	4.44
trans-p-Methylcinnamic acid	$C_{10}H_{10}O_2$		25	4.56
α-Methylenebenzeneacetic acid	$C_9H_8O_2$			4.35
Methyl α-*D*-glucopyranoside	$C_7H_{14}O_6$		25	13.71
N-Methyl-2-heptanamine	$C_8H_{19}N$		17	10.99
L-1-Methylhistidine	$C_7H_{11}N_3O_2$	1	25	1.69
		2	25	6.48
		3	25	8.85
L-3-Methylhistidine	$C_7H_{11}N_3O_2$	1	25	1.92
		2	25	6.56
		3	25	8.73
O-Methylhydroxylamine	CH_5NO			12.5
1-Methylimidazol	$C_4H_6N_2$		25	6.95
Methylmalonic acid	$C_4H_6O_4$	1	25	3.07
		2	25	5.76
4-Methylmorpholine	$C_5H_{11}NO$		25	7.38
Methyl-1-naphthylamine	$C_{11}H_{11}N$		27	3.67
3-Methylpentanedioic acid	$C_6H_{10}O_4$		25	4.24
4-Methylpentanoic acid	$C_6H_{12}O_2$		18	4.84
Methylphenidate	$C_{14}H_{19}NO_2$			8.9
1-Methylpiperidine	$C_6H_{13}N$		25	10.38
2-Methylpropanoic acid	$C_4H_8O_2$		20	4.84
3-Methylpyrazinamine	$C_5H_7N_3$		25	3.39
2-Methylpyrazine	$C_5H_6N_2$		27	1.45
N-Methyl-4-pyridinamine	$C_6H_8N_2$		20	9.65
2-Methylpyridine	C_6H_7N		25	6.00
3-Methylpyridine	C_6H_7N		25	5.70
4-Methylpyridine	C_6H_7N		25	5.99
N-Methylpyrrolidine	$C_5H_{11}N$		25	10.46
2-Methylquinoline	$C_{10}H_9N$		20	5.83
4-Methylquinoline	$C_{10}H_9N$		20	5.67
5-Methylquinoline	$C_{10}H_9N$		20	5.20
Methyl Red	$C_{15}H_{15}N_3O_2$	1		2.5
		2		9.5
Methylsuccinic acid	$C_5H_8O_4$	1	25	4.13
		2	25	5.64
(Methylthio)acetic acid	$C_3H_6O_2S$		25	3.66
2-(Methylthio)aniline	C_7H_9NS		25	3.45
4-(Methylthio)aniline	C_7H_9NS		25	4.35
4-(Methylthio)phenol	C_7H_8OS		25	9.53
6-Methyl-1,2,4-triazine-3,5(2*H*,4*H*)-dione	$C_4H_5N_3O_2$			7.6
Minoxidil	$C_9H_{15}N_5O$			4.61
Morphine	$C_{17}H_{19}NO_3$	1	25	8.21
		2	20	9.85
Morpholine	C_4H_9NO		25	8.50
Mycophenolic acid	$C_{17}H_{20}O_6$			4.5
Nadolol	$C_{17}H_{27}NO_4$			9.67
1-Naphthalenecarboxylic acid	$C_{11}H_8O_2$		25	3.69
2-Naphthalenecarboxylic acid	$C_{11}H_8O_2$		25	4.16
1-Naphthol	$C_{10}H_8O$		25	9.39
2-Naphthol	$C_{10}H_8O$		25	9.63
1-Naphthylamine	$C_{10}H_9N$		25	3.92
2-Naphthylamine	$C_{10}H_9N$		25	4.16
Neobornylamine	$C_{10}H_{19}N$		25	10.01
Neutral Red	$C_{15}H_{17}ClN_4$			6.7
L-Nicotine	$C_{10}H_{14}N_2$	1		8.02
		2		3.12
Nitrilotriacetic acid	$C_6H_9NO_6$	1	20	3.03
		2	20	3.07
		3	20	10.70
Nitroacetic acid	$C_2H_3NO_4$		24	1.48
2-Nitroaniline	$C_6H_6N_2O_2$		25	-0.25
3-Nitroaniline	$C_6H_6N_2O_2$		25	2.46
4-Nitroaniline	$C_6H_6N_2O_2$		25	1.02
Nitrobenzene	$C_6H_5NO_2$		0	3.98
2-Nitrobenzeneacetic acid	$C_8H_7NO_4$		25	4.00
3-Nitrobenzeneacetic acid	$C_8H_7NO_4$		25	3.97
4-Nitrobenzeneacetic acid	$C_8H_7NO_4$		25	3.85
2-Nitrobenzoic acid	$C_7H_5NO_4$		25	2.17
3-Nitrobenzoic acid	$C_7H_5NO_4$		25	3.46
4-Nitrobenzoic acid	$C_7H_5NO_4$		25	3.43
Nitroethane	$C_2H_5NO_2$		25	8.46
Nitrofurantoin	$C_8H_6N_4O_5$			7.2
Nitromethane	CH_3NO_2		25	10.21
2-Nitrophenol	$C_6H_5NO_3$		25	7.23
3-Nitrophenol	$C_6H_5NO_3$		25	8.36
4-Nitrophenol	$C_6H_5NO_3$		25	7.15
3-(2-Nitrophenyl)propanoic acid	$C_9H_9NO_4$		25	4.50
3-(4-Nitrophenyl)propanoic acid	$C_9H_9NO_4$		25	4.47
4-Nitropyridine	$C_5H_4N_2O_2$		25	1.61
5-Nitropyrimidinamine	$C_4H_4N_4O_2$		20	0.35
Nizatidine	$C_{12}H_{21}N_5O_2S_2$	1		2.1
		2		6.8
Nonanedioic acid	$C_9H_{16}O_4$	1	25	4.53
		2	25	5.33
Nonanoic acid	$C_9H_{18}O_2$		25	4.96
Nonylamine	$C_9H_{21}N$		25	10.64
Norepinephrine	$C_8H_{11}NO_3$	1	25	8.64
		2	25	9.70
L-Norleucine	$C_6H_{13}NO_2$	1	25	2.34
		2	25	9.83
DL-Norvaline	$C_5H_{11}NO_2$	1		2.36
		2		9.72
L-Norvaline	$C_5H_{11}NO_2$	1	25	2.32
		2	25	9.81
Noscapine	$C_{22}H_{23}NO_7$			7.8
Novobiocin	$C_{31}H_{36}N_2O_{11}$	1		4.3
		2		9.1
cis,cis-9,12-Octadecadienoic acid	$C_{18}H_{32}O_2$		25	4.77
Octadecylamine	$C_{18}H_{39}N$		25	10.60
1,8-Octanediamine	$C_8H_{20}N_2$	1	20	11.00

Name	Mol. form.	Step	t/°C	pK_a
		2	20	10.1
Octanedioic acid	$C_8H_{14}O_4$	1	25	4.52
Octanoic acid	$C_8H_{16}O_2$		25	4.89
Octylamine	$C_8H_{19}N$		25	10.65
Oleanolic acid	$C_{30}H_{48}O_3$			2.52
L-Ornithine	$C_5H_{12}N_2O_2$	1	25	1.71
		2	25	8.69
		3	25	10.76
Oxalic acid	$C_2H_2O_4$	1	25	1.25
		2	25	3.81
Oxaloacetic acid	$C_4H_4O_5$	1	25	2.55
		2	25	4.37
		3	25	13.03
Oxazole	C_3H_3NO		33	0.8
2-Oxobutanoic acid	$C_4H_6O_3$		25	2.50
2-Oxoglutaric acid	$C_5H_6O_5$	1	25	2.47
		2	25	4.68
Papaverine	$C_{20}H_{21}NO_4$			6.4
Pentadecylamine	$C_{15}H_{33}N$		25	10.61
Pentafluorobenzoic acid	$C_7HF_5O_2$		25	1.75
1,2,2,6,6-Pentamethylpiperidine	$C_{10}H_{21}N$		30	11.25
3-Pentanamine	$C_5H_{13}N$		17	10.59
1,5-Pentanediamine	$C_5H_{14}N_2$	1	25	10.05
		2	25	10.93
Pentanedioic acid	$C_5H_8O_4$	1	18	4.32
		2	25	5.42
Pentanoic acid	$C_5H_{10}O_2$		20	4.83
trans-3-Pentenoic acid	$C_5H_8O_2$		25	4.51
Pentostatin	$C_{11}H_{16}N_4O_4$			5.2
Pentylamine	$C_5H_{13}N$		25	10.63
Perfluidone	$C_{14}H_{12}F_3NO_4S_2$			2.5
1*H*-Perimidine	$C_{11}H_8N_2$		20	6.35
Phenanthridine	$C_{13}H_9N$		20	5.58
1,10-Phenanthroline	$C_{12}H_8N_2$		25	4.84
Phenazine	$C_{12}H_8N_2$		20	1.20
Phenobarbital	$C_{12}H_{12}N_2O_3$	1		7.3
		2		11.8
Phenol	C_6H_6O		25	9.99
Phenolphthalein	$C_{20}H_{14}O_4$		25	9.7
Phenol Red	$C_{19}H_{14}O_5S$			7.9
Phenoxyacetic acid	$C_8H_8O_3$		25	3.17
2-Phenoxybenzoic acid	$C_{13}H_{10}O_3$		25	3.53
3-Phenoxybenzoic acid	$C_{13}H_{10}O_3$		25	3.95
4-Phenoxybenzoic acid	$C_{13}H_{10}O_3$		25	4.57
2-(3-Phenoxyphenyl)propanoic acid, (±)-	$C_{15}H_{14}O_3$			4.5
L-Phenylalanine	$C_9H_{11}NO_2$	1	25	2.20
		2	25	9.31
α-Phenylbenzeneacetic acid	$C_{14}H_{12}O_2$		25	3.94
2-Phenylbenzimidazole	$C_{13}H_{10}N_2$	1	25	5.23
		2	25	11.91
2-Phenylbenzoic acid	$C_{13}H_{10}O_2$		25	3.46
Phenyl biguanide	$C_8H_{11}N_5$	1		10.76
		2		2.13
Phenylbutazone	$C_{19}H_{20}N_2O_2$			4.5
N-Phenylglycine	$C_8H_9NO_2$	1	25	1.83
		2		4.39
Phenylhydrazine	$C_6H_8N_2$		15	8.79
Phenylpropanolamine hydrochloride	$C_9H_{14}ClNO$			9.44
Phthalazine	$C_8H_6N_2$		20	3.47
Phthalic acid	$C_8H_6O_4$	1	25	2.943
		2	25	5.432
Physostigmine	$C_{15}H_{21}N_3O_2$	1		6.12
		2		12.24
Pilocarpine	$C_{11}H_{16}N_2O_2$	1	25	1.6
		2	25	6.9
Piperazine	$C_4H_{10}N_2$	1	25	9.73
		2	25	5.33
Piperidine	$C_5H_{11}N$		25	11.123
2-Piperidinecarboxylic acid	$C_6H_{11}NO_2$	1	25	2.28
		2	25	10.72
Piperine	$C_{17}H_{19}NO_3$		18	12.22
L-Proline	$C_5H_9NO_2$	1	25	1.95
		2	25	10.64
1,2-Propanediamine, (±)-	$C_3H_{10}N_2$	1	25	9.82
		2	25	6.61
1,3-Propanediamine	$C_3H_{10}N_2$	1	25	10.55
		2	25	8.88
1,2,3-Propanetriamine	$C_3H_{11}N_3$	1	20	9.59
		2	20	7.95
Propanoic acid	$C_3H_6O_2$		25	4.87
2-Propanone oxime	C_3H_7NO		25	12.42
Propargyl alcohol	C_3H_4O		25	13.6
trans-1-Propene-1,2-dicarboxylic	$C_5H_6O_4$	1	25	3.09
		2	25	4.75
1-Propene-2,3-dicarboxylic acid	$C_5H_6O_4$	1	25	3.85
		2	25	5.45
cis-1-Propene-1,2,3-tricarboxylic	$C_6H_6O_6$		25	1.95
trans-1-Propene-1,2,3-tricarboxylic acid	$C_6H_6O_6$	1	25	2.80
		2	25	4.46
Propylamine	C_3H_9N		25	10.54
N-Propylglycine	$C_5H_{11}NO_2$	1	25	2.35
		2	25	10.19
2-Propylpentanoic acid	$C_8H_{16}O_2$			4.6
2-Propylpiperidine, (*S*)-	$C_8H_{17}N$			10.9
trans-6-Propyl-3-piperidinol,(3S)-	$C_8H_{17}NO$			10.3
Propyl 3,4,5-trihydroxybenzoate	$C_{10}H_{12}O_5$			8.11
2-Propynoic acid	$C_3H_2O_2$		25	1.84
Protriptyline	$C_{19}H_{21}N$			8.2
Pseudotropine	$C_8H_{15}NO$		15	3.80
Pteridine	$C_6H_4N_4$		20	4.05
1*H*-Purine	$C_5H_4N_4$	1	20	2.30
		2	20	8.96
Pyrazine	$C_4H_4N_2$		20	0.65
Pyrazinecarboxamide	$C_5H_5N_3O$			0.5
1*H*-Pyrazole	$C_3H_4N_2$		25	2.49
Pyridazine	$C_4H_4N_2$		20	2.24
2-Pyridinamine	$C_5H_6N_2$		20	6.82
3-Pyridinamine	$C_5H_6N_2$		25	6.04
4-Pyridinamine	$C_5H_6N_2$		25	9.11
Pyridine	C_5H_5N		25	5.23
2-Pyridinecarboxaldehyde	C_6H_5NO		25	12.68
4-Pyridinecarboxaldehyde	C_6H_5NO		30	12.05
2-Pyridinecarboxaldehyde oxime	$C_6H_6N_2O$	1	20	3.59
		2	20	10.18
3-Pyridinecarboxamide	$C_6H_6N_2O$		20	3.3
2-Pyridinecarboxylic acid	$C_6H_5NO_2$	1	20	0.99
		2	20	5.39
3-Pyridinecarboxylic acid	$C_6H_5NO_2$	1	25	2.00

Name	Mol. form.	Step	t/°C	pK_a
		2	25	4.82
4-Pyridinecarboxylic acid	$C_6H_5NO_2$	1	25	1.77
		2	25	4.84
2,5-Pyridinediamine	$C_5H_7N_3$		20	6.48
2,3-Pyridinedicarboxylic acid	$C_7H_5NO_4$	1	25	2.43
		2	25	4.78
2,4-Pyridinedicarboxylic acid	$C_7H_5NO_4$	1	25	2.15
2,6-Pyridinedicarboxylic acid	$C_7H_5NO_4$	1	25	2.16
		2	25	4.76
3,5-Pyridinedicarboxylic acid	$C_7H_5NO_4$	1	25	2.80
Pyridine-1-oxide	C_5H_5NO		24	0.79
2-Pyridinol	C_5H_5NO	1	20	0.75
		2	20	11.65
3-Pyridinol	C_5H_5NO	1	20	4.79
		2	20	8.75
4-Pyridinol	C_5H_5NO	1	20	3.20
		2	20	11.12
2(1*H*)-Pyridinone	C_5H_5NO	1	20	0.75
		2	20	11.65
2-Pyrimidinamine	$C_4H_5N_3$		20	3.45
4-Pyrimidinamine	$C_4H_5N_3$		20	5.71
Pyrimidine	$C_4H_4N_2$		20	1.23
2,4,6-Pyrimidinetriamine	$C_4H_7N_5$		20	6.84
Pyrocatechol	$C_6H_6O_2$	1	25	9.34
		2	25	12.6
L-Pyroglutamic acid	$C_5H_7NO_3$		25	3.32
Pyrrole	C_4H_5N		25	-3.8
1*H*-Pyrrole-2-carboxylic acid	$C_5H_5NO_2$		20	4.45
1*H*-Pyrrole-3-carboxylic acid	$C_5H_5NO_2$		20	5.00
Pyrrolidine	C_4H_9N		25	11.31
Pyruvic acid	$C_3H_4O_3$		25	2.39
Quinazoline	$C_8H_6N_2$		29	3.43
Quinidine	$C_{20}H_{24}N_2O_2$	1	20	5.4
		2	20	10.0
Quinine	$C_{20}H_{24}N_2O_2$	1	25	8.52
		2	25	4.13
2-Quinolinamine	$C_9H_8N_2$		20	7.34
3-Quinolinamine	$C_9H_8N_2$		20	4.91
4-Quinolinamine	$C_9H_8N_2$		20	9.17
Quinoline	C_9H_7N		20	4.90
8-Quinolinecarboxylic acid	$C_{10}H_7NO_2$		25	1.82
2-Quinolinol	C_9H_7NO	1	20	-0.31
		2	20	11.76
3-Quinolinol	C_9H_7NO	1	20	4.28
		2	20	8.08
4-Quinolinol	C_9H_7NO	1	20	2.23
		2	20	11.28
6-Quinolinol	C_9H_7NO	1	20	5.15
		2	20	8.90
8-Quinolinol	C_9H_7NO	1	25	4.91
		2	25	9.81
Quinoxaline	$C_8H_6N_2$		20	0.56
Reserpine	$C_{33}H_{40}N_2O_9$			6.6
Resorcinol	$C_6H_6O_2$	1	25	9.32
		2	25	11.1
Riboflavin	$C_{17}H_{20}N_4O_6$	1		1.7
		2	25	9.69
L-Ribose	$C_5H_{10}O_5$		25	12.22
Rifampin	$C_{43}H_{58}N_4O_{12}$	1		1.7
		2		7.9
Saccharin	$C_7H_5NO_3S$		18	11.68
Salicylaldehyde	$C_7H_6O_2$		25	8.37
Sarcosine	$C_3H_7NO_2$	1	25	2.21
		2	25	10.1
L-Serine	$C_3H_7NO_3$	1	25	2.19
		2	25	9.21
Solanine	$C_{45}H_{73}NO_{15}$		15	6.66
Sparteine	$C_{15}H_{26}N_2$	1	20	2.24
		2	20	9.46
Strychnine	$C_{21}H_{22}N_2O_2$		25	8.26
Succinic acid	$C_4H_6O_4$	1	25	4.21
		2	25	5.64
Succinimide	$C_4H_5NO_2$		25	9.62
Sucrose	$C_{12}H_{22}O_{11}$		25	12.7
Sulfabenzamide	$C_{13}H_{12}N_2O_3S$		25	4.57
Sulfacytine	$C_{12}H_{14}N_4O_3S$			6.9
Sulfamethazine	$C_{12}H_{14}N_4O_2S$	1		7.4
		2		2.65
Sulfamethoxypyridazine	$C_{11}H_{12}N_4O_3S$			6.7
Sulfathiazole	$C_9H_9N_3O_2S_2$			7.2
Sulfisoxazole	$C_{11}H_{13}N_3O_3S$			5
DL-Tartaric acid	$C_4H_6O_6$	1	25	3.03
		2	25	4.37
meso-Tartaric acid	$C_4H_6O_6$	1	25	3.17
		2	25	4.91
L-Tartaric acid	$C_4H_6O_6$	1	25	2.98
		2	25	4.34
Taurine	$C_2H_7NO_3S$	1	25	1.5
		2	25	9.06
Taurocholic acid	$C_{26}H_{45}NO_7S$			1.4
Teniposide	$C_{32}H_{32}O_{13}S$			10.13
Terephthalic acid	$C_8H_6O_4$	1	25	3.54
		2	25	4.34
Tetradecylamine	$C_{14}H_{31}N$		25	10.62
5,6,7,8-Tetrahydro-2-naphthol	$C_{10}H_{12}O$		25	10.48
N,N,N',N'-Tetramethyl-1,2-ethanediamine	$C_6H_{16}N_2$	1	25	10.40
		2	25	8.26
2,2,6,6-Tetramethylpiperidine	$C_9H_{19}N$		25	11.07
Tetramethylurea	$C_5H_{12}N_2O$			2
Tetrodotoxin	$C_{11}H_{17}N_3O_8$			8.76
Thebaine	$C_{19}H_{21}NO_3$		15	6.05
Theobromine	$C_7H_8N_4O_2$		18	7.89
Theophylline	$C_7H_8N_4O_2$	1	25	8.77
2-Thiazolamine	$C_3H_4N_2S$		20	5.36
Thiazole	C_3H_3NS		25	2.52
Thioacetic acid	C_2H_4OS		25	3.33
Thioctic acid	$C_8H_{14}O_2S_2$			5.4
Thioglycolic acid	$C_2H_4O_2S$		25	3.68
2-Thiophenecarboxylic acid	$C_5H_4O_2S$		25	3.49
3-Thiophenecarboxylic acid	$C_5H_4O_2S$		25	4.1
Thiourea	CH_4N_2S		25	-1
L-Threonine	$C_4H_9NO_3$	1	25	2.09
		2	25	9.10
Thymine	$C_5H_6N_2O_2$		25	9.94
L-Thyroxine	$C_{15}H_{11}I_4NO_4$	1	25	2.2
		2	25	6.45
		3	25	10.1
Tolazamide	$C_{14}H_{21}N_3O_3S$		25	3.6
o-Toluic acid	$C_8H_8O_2$		25	3.91
m-Toluic acid	$C_8H_8O_2$		25	4.25
p-Toluic acid	$C_8H_8O_2$		25	4.37

Name	Mol. form.	Step	t/°C	pK_a
1,3,5-Triazine-2,4,6-triamine	$C_3H_6N_6$		25	5.00
1*H*-1,2,3-Triazole	$C_2H_3N_3$		20	1.17
1*H*-1,2,4-Triazole	$C_2H_3N_3$		20	2.27
Trichloroacetaldehyde	C_2HCl_3O		25	10.04
Trichloroacetic acid	$C_2HCl_3O_2$		20	0.66
2,2,2-Trichloroethanol	$C_2H_3Cl_3O$		25	12.24
Triclopyr	$C_7H_4Cl_3NO_3$			2.68
(Tridecyl)amine	$C_{13}H_{29}N$		25	10.63
Triethanolamine	$C_6H_{15}NO_3$		25	7.76
Triethylamine	$C_6H_{15}N$		25	10.75
Triethylenediamine	$C_6H_{12}N_2$	1		3.0
		2		8.7
Trifluoroacetic acid	$C_2HF_3O_2$		25	0.52
2,2,2-Trifluoroethanol	$C_2H_3F_3O$		25	12.37
3-(Trifluoromethyl)aniline	$C_7H_6F_3N$		25	3.49
4-(Trifluoromethyl)aniline	$C_7H_6F_3N$		25	2.45
2-(Trifluoromethyl)phenol	$C_7H_5F_3O$		25	8.95
3-(Trifluoromethyl)phenol	$C_7H_5F_3O$		25	8.68
2,4,6-Trihydroxybenzoic acid	$C_7H_6O_5$		25	1.68
3,4,5-Trihydroxybenzoic acid	$C_7H_6O_5$		25	4.41
Trimethoprim	$C_{14}H_{18}N_4O_3$			6.6
Trimethylamine	C_3H_9N		25	9.80
Trimethylamine oxide	C_3H_9NO		20	4.65
2,2,4-Trimethylpiperidine	$C_8H_{17}N$		30	11.04
2,4,6-Trimethylpyridine	$C_8H_{11}N$		25	7.43
2,4,6-Trinitrobenzoic acid	$C_7H_3N_3O_8$		25	0.65
2,4,6-Trinitrophenol	$C_6H_3N_3O_7$		24	0.42
Tris(hydroxymethyl) methylamine	$C_4H_{11}NO_3$		20	8.3
Tropacocaine	$C_{15}H_{19}NO_2$		15	4.32
Tropine	$C_8H_{15}NO$		15	3.80
Tryptamine	$C_{10}H_{12}N_2$		25	10.2
L-Tryptophan	$C_{11}H_{12}N_2O_2$	1	25	2.46
		2	25	9.41
L-Tyrosine	$C_9H_{11}NO_3$	1	25	2.20
		2	25	9.11
		3	25	10.1
Tyrosineamide	$C_9H_{12}N_2O_2$		25	7.33
Undecylamine	$C_{11}H_{25}N$		25	10.63
Uracil	$C_4H_4N_2O_2$		25	9.45
Urea	CH_4N_2O		25	0.10
Uric acid	$C_5H_4N_4O_3$		12	3.89
5'-Uridylic acid	$C_9H_{13}N_2O_9P$	1		6.4
		2		9.5
L-Valine	$C_5H_{11}NO_2$	1	25	2.29
		2	25	9.74
Valium	$C_{16}H_{13}ClN_2O$			3.4
Verapamil	$C_{27}H_{38}N_2O_4$			8.6
Veratridine	$C_{36}H_{51}NO_{11}$			9.54
Vinblastine	$C_{46}H_{58}N_4O_9$	1		5.4
		2		7.4
Vincristine	$C_{46}H_{56}N_4O_{10}$			5.4
Xanthopterin	$C_6H_5N_5O_2$	2	20	6.59
		3	20	9.31
D-Xylose	$C_5H_{10}O_5$		18	12.14

[a] Approximate value

Thermochem

ACTIVITY COEFFICIENTS OF ACIDS, BASES, AND SALTS

Petr Vanýsek

This table gives mean activity coefficients γ at 25 °C for molalities in the range 0.1 to 1.0 mol/kg. See the following table "Mean Activity Coefficients of Electrolytes as a Function of Concentration" for definitions, references, and data over a wider concentration range. Compounds are listed alphabetically by formula.

Activity Coefficients of Acids, Bases, and Salts at 25 °C and the Indicated Molalities

Formula	γ(0.1 m)	γ(0.2 m)	γ(0.3 m)	γ(0.4 m)	γ(0.5 m)	γ(0.6 m)	γ(0.7 m)	γ(0.8 m)	γ(0.9 m)	γ(1.0 m)
$AgNO_3$	0.734	0.657	0.606	0.567	0.536	0.509	0.485	0.464	0.446	0.429
$AlCl_3$	0.337	0.305	0.302	0.313	0.331	0.356	0.388	0.429	0.479	0.539
$Al_2(SO_4)_3$	0.035	0.0225	0.0176	0.0153	0.0143	0.014	0.0142	0.0149	0.0159	0.0175
$BaCl_2$	0.500	0.444	0.419	0.405	0.397	0.391	0.391	0.391	0.392	0.395
$BeSO_4$	0.150	0.109	0.0885	0.0769	0.0692	0.0639	0.0600	0.0570	0.0546	0.0530
$CaCl_2$	0.518	0.472	0.455	0.448	0.448	0.453	0.460	0.470	0.484	0.500
$CdCl_2$	0.2280	0.1638	0.1329	0.1139	0.1006	0.0905	0.0827	0.0765	0.0713	0.0669
$Cd(NO_3)_2$	0.513	0.464	0.442	0.430	0.425	0.423	0.423	0.425	0.428	0.433
$CdSO_4$	0.150	0.103	0.0822	0.0699	0.0615	0.0553	0.0505	0.0468	0.0438	0.0415
$CoCl_2$	0.522	0.479	0.463	0.459	0.462	0.470	0.479	0.492	0.511	0.531
$CrCl_3$	0.331	0.298	0.294	0.300	0.314	0.335	0.362	0.397	0.436	0.481
$Cr(NO_3)_3$	0.319	0.285	0.279	0.281	0.291	0.304	0.322	0.344	0.371	0.401
$Cr_2(SO_4)_3$	0.0458	0.0300	0.0238	0.0207	0.0190	0.0182	0.0181	0.0185	0.0194	0.0208
CsBr	0.754	0.694	0.654	0.626	0.603	0.586	0.571	0.558	0.547	0.538
$CsC_2H_3O_2$	0.799	0.771	0.761	0.759	0.762	0.768	0.776	0.783	0.792	0.802
CsCl	0.756	0.694	0.656	0.628	0.606	0.589	0.575	0.563	0.553	0.544
CsI	0.754	0.692	0.651	0.621	0.599	0.581	0.567	0.554	0.543	0.533
$CsNO_3$	0.733	0.655	0.602	0.561	0.528	0.501	0.478	0.458	0.439	0.422
CsOH	0.795	0.761	0.744	0.739	0.739	0.742	0.748	0.754	0.762	0.771
Cs_2SO_4	0.456	0.382	0.338	0.311	0.291	0.274	0.262	0.251	0.242	0.235
$CuCl_2$	0.508	0.455	0.429	0.417	0.411	0.409	0.409	0.410	0.413	0.417
$Cu(NO_3)_2$	0.511	0.460	0.439	0.429	0.426	0.427	0.431	0.437	0.445	0.455
$CuSO_4$	0.150	0.104	0.0829	0.0704	0.0620	0.0559	0.0512	0.0475	0.0446	0.0423
$FeCl_2$	0.5185	0.473	0.454	0.448	0.450	0.454	0.463	0.473	0.488	0.506
HBr	0.805	0.782	0.777	0.781	0.789	0.801	0.815	0.832	0.850	0.871
HCl	0.796	0.767	0.756	0.755	0.757	0.763	0.772	0.783	0.795	0.809
$HClO_4$	0.803	0.778	0.768	0.766	0.769	0.776	0.785	0.795	0.808	0.823
HI	0.818	0.807	0.811	0.823	0.839	0.860	0.883	0.908	0.935	0.963
HNO_3	0.791	0.754	0.735	0.725	0.720	0.717	0.717	0.718	0.721	0.724
H_2SO_4	0.2655	0.2090	0.1826		0.1557		0.1417			0.1316
KBr	0.772	0.722	0.693	0.673	0.657	0.646	0.636	0.629	0.622	0.617
$KC_2H_3O_2$	0.796	0.766	0.754	0.750	0.751	0.754	0.759	0.766	0.774	0.783
KCl	0.770	0.718	0.688	0.666	0.649	0.637	0.626	0.618	0.610	0.604
$KClO_3$	0.749	0.681	0.635	0.599	0.568	0.541	0.518			
KF	0.775	0.727	0.700	0.682	0.670	0.661	0.654	0.650	0.646	0.645
KH_2PO_4	0.731	0.653	0.602	0.561	0.529	0.501	0.477	0.456	0.438	0.421
KI	0.778	0.733	0.707	0.689	0.676	0.667	0.660	0.654	0.649	0.645
KNO_3	0.739	0.663	0.614	0.576	0.545	0.519	0.496	0.476	0.459	0.443
KOH	0.798	0.760	0.742	0.734	0.732	0.733	0.736	0.742	0.749	0.756
KSCN	0.769	0.716	0.685	0.663	0.646	0.633	0.623	0.614	0.606	0.599
K_2CrO_4	0.456	0.382	0.340	0.313	0.292	0.276	0.263	0.253	0.243	0.235
K_2SO_4	0.441	0.360	0.316	0.286	0.264	0.246	0.232			
$K_3Fe(CN)_6$	0.268	0.212	0.184	0.167	0.155	0.146	0.140	0.135	0.131	0.128
$K_4Fe(CN)_6$	0.139	0.0993	0.0808	0.0693	0.0614	0.0556	0.0512	0.0479	0.0454	
LiBr	0.796	0.766	0.756	0.752	0.753	0.758	0.767	0.777	0.789	0.803
$LiC_2H_3O_2$	0.784	0.742	0.721	0.709	0.700	0.691	0.689	0.688	0.688	0.689
LiCl	0.790	0.757	0.744	0.740	0.739	0.743	0.748	0.755	0.764	0.774
$LiClO_4$	0.812	0.794	0.792	0.798	0.808	0.820	0.834	0.852	0.869	0.887

Formula	γ(0.1 m)	γ(0.2 m)	γ(0.3 m)	γ(0.4 m)	γ(0.5 m)	γ(0.6 m)	γ(0.7 m)	γ(0.8 m)	γ(0.9 m)	γ(1.0 m)
LiI	0.815	0.802	0.804	0.813	0.824	0.838	0.852	0.870	0.888	0.910
$LiNO_3$	0.788	0.752	0.736	0.728	0.726	0.727	0.729	0.733	0.737	0.743
LiOH	0.760	0.702	0.665	0.638	0.617	0.599	0.585	0.573	0.563	0.554
Li_2SO_4	0.468	0.398	0.361	0.337	0.319	0.307	0.297	0.289	0.282	0.277
$MgCl_2$	0.529	0.489	0.477	0.475	0.481	0.491	0.506	0.522	0.544	0.570
$MgSO_4$	0.150	0.107	0.0874	0.0756	0.0675	0.0616	0.0571	0.0536	0.0508	0.0485
$MnCl_2$	0.516	0.469	0.450	0.442	0.440	0.443	0.448	0.455	0.466	0.479
$MnSO_4$	0.150	0.105	0.0848	0.0725	0.0640	0.0578	0.0530	0.0493	0.0463	0.0439
NH_4Cl	0.770	0.718	0.687	0.665	0.649	0.636	0.625	0.617	0.609	0.603
NH_4NO_3	0.740	0.677	0.636	0.606	0.582	0.562	0.545	0.530	0.516	0.504
$(NH_4)_2SO_4$	0.439	0.356	0.311	0.280	0.257	0.240	0.226	0.214	0.205	0.196
NaBr	0.782	0.741	0.719	0.704	0.697	0.692	0.689	0.687	0.687	0.687
$NaC_2H_3O_2$	0.791	0.757	0.744	0.737	0.735	0.736	0.740	0.745	0.752	0.757
NaCl	0.778	0.735	0.710	0.693	0.681	0.673	0.667	0.662	0.659	0.657
$NaClO_3$	0.772	0.720	0.688	0.664	0.645	0.630	0.617	0.606	0.597	0.589
$NaClO_4$	0.775	0.729	0.701	0.683	0.668	0.656	0.648	0.641	0.635	0.629
NaF	0.765	0.710	0.676	0.651	0.632	0.616	0.603	0.592	0.582	0.573
NaH_2PO_4	0.744	0.675	0.629	0.593	0.563	0.539	0.517	0.499	0.483	0.468
NaI	0.787	0.751	0.735	0.727	0.723	0.723	0.724	0.727	0.731	0.736
$NaNO_3$	0.762	0.703	0.666	0.638	0.617	0.599	0.583	0.570	0.558	0.548
NaOH	0.766	0.727	0.708	0.697	0.690	0.685	0.681	0.679	0.678	0.678
NaSCN	0.787	0.750	0.731	0.720	0.715	0.712	0.710	0.710	0.711	0.712
Na_2CrO_4	0.464	0.394	0.353	0.327	0.307	0.292	0.280	0.269	0.261	0.253
Na_2SO_4	0.445	0.365	0.320	0.289	0.266	0.248	0.233	0.221	0.210	0.201
$NiCl_2$	0.522	0.479	0.463	0.460	0.464	0.471	0.482	0.496	0.515	0.563
$NiSO_4$	0.150	0.105	0.0841	0.0713	0.0627	0.0562	0.0515	0.0478	0.0448	0.0425
$Pb(NO_3)_2$	0.395	0.308	0.260	0.228	0.205	0.187	0.172	0.160	0.150	0.141
RbBr	0.763	0.706	0.673	0.650	0.632	0.617	0.605	0.595	0.586	0.578
$RbC_2H_3O_2$	0.796	0.767	0.756	0.753	0.755	0.759	0.766	0.773	0.782	0.792
RbCl	0.764	0.709	0.675	0.652	0.634	0.620	0.608	0.599	0.590	0.583
RbI	0.762	0.705	0.671	0.647	0.629	0.614	0.602	0.591	0.583	0.575
$RbNO_3$	0.734	0.658	0.606	0.565	0.534	0.508	0.485	0.465	0.446	0.430
Rb_2SO_4	0.451	0.374	0.331	0.301	0.279	0.263	0.249	0.238	0.228	0.219
$SrCl_2$	0.511	0.462	0.442	0.433	0.430	0.431	0.434	0.441	0.449	0.461
$TlClO_4$	0.730	0.652	0.599	0.527						
$TlNO_3$	0.702	0.606	0.545	0.500						
UO_2Cl_2	0.544	0.510	0.520	0.505	0.517	0.532	0.549	0.571	0.595	0.620
UO_2SO_4	0.150	0.102	0.0807	0.0689	0.0611	0.0566	0.0515	0.0483	0.0458	0.0439
$ZnCl_2$	0.515	0.462	0.432	0.411	0.394	0.380	0.369	0.357	0.348	0.339
$Zn(NO_3)_2$	0.531	0.489	0.474	0.469	0.473	0.480	0.489	0.501	0.518	0.535
$ZnSO_4$	0.150	0.10	0.0835	0.0714	0.0630	0.0569	0.0523	0.0487	0.0458	0.0435

MEAN ACTIVITY COEFFICIENTS OF ELECTROLYTES AS A FUNCTION OF CONCENTRATION

The mean activity coefficient γ_{mean} of an electrolyte X_aY_b is defined as

$$\gamma = (\gamma_+^a \gamma_-^b)^{1/(a+b)}$$

where γ_+ and γ_- are activity coefficients of the individual ions (which cannot be directly measured). This table gives the mean activity coefficients of about 100 electrolytes in aqueous solution as a function of concentration, expressed in molality terms. All values refer to a temperature of 25 °C. Substances are arranged alphabetically in order of formula.

References

1. Hamer, W. J., and Wu, Y. C., *J. Phys.Chem. Ref. Data,* 1, 1047, 1972.
2. Staples, B. R., *J. Phys. Chem. Ref. Data,* 6, 385, 1977; 10, 767, 1981; 10, 779, 1981.
3. Goldberg, R. N. et al., *J. Phys. Chem. Ref. Data,* 7, 263, 1978; 8, 923, 1979; 8, 1005, 1979; 10, 1, 1981; 10, 671, 1981.

Mean Activity Coefficients of Electrolytes at 25 °C

Electrolyte	m/ mol kg^{-1}	γ_{mean}
$AgNO_3$	0.001	0.964
	0.002	0.950
	0.005	0.924
	0.010	0.896
	0.020	0.859
	0.050	0.794
	0.100	0.732
	0.200	0.656
	0.500	0.536
	1.000	0.430
	2.000	0.316
	5.000	0.181
	10.000	0.108
	15.000	0.085
$BaBr_2$	0.001	0.881
	0.002	0.850
	0.005	0.785
	0.010	0.727
	0.020	0.661
	0.050	0.573
	0.100	0.517
	0.200	0.463
	0.500	0.435
	1.000	0.470
	2.000	0.654
$BaCl_2$	0.001	0.887
	0.002	0.849
	0.005	0.782
	0.010	0.721
	0.020	0.653
	0.050	0.559
	0.100	0.492
	0.200	0.436
	0.500	0.391
	1.000	0.393
BaI_2	0.001	0.890
	0.002	0.853
	0.005	0.792
	0.010	0.737
	0.020	0.678
	0.050	0.600
	0.100	0.551
	0.200	0.520
	0.500	0.536
	1.000	0.664
	2.000	1.242
$CaBr_2$	0.001	0.890
	0.002	0.853
	0.005	0.791
	0.010	0.735
	0.020	0.674
	0.050	0.594
	0.100	0.540
	0.200	0.502
	0.500	0.500
	1.000	0.604
	2.000	1.125
	5.000	18.7
$CaCl_2$	0.001	0.888
	0.002	0.851
	0.005	0.787
	0.010	0.727
	0.020	0.664
	0.050	0.577
	0.100	0.517
	0.200	0.469
	0.500	0.444
	1.000	0.495
	2.000	0.784
	5.000	5.907
	10.000	43.1
CaI_2	0.001	0.890
	0.002	0.853
	0.005	0.791
	0.010	0.736
	0.020	0.677
	0.050	0.600
	0.100	0.552
	0.200	0.524
	0.500	0.554
	1.000	0.729
$Cd(NO_2)_2$	0.001	0.881
	0.002	0.837
	0.005	0.759
	0.010	0.681
	0.020	0.589
	0.050	0.451
	0.100	0.344
	0.200	0.247
	0.500	0.148
	1.000	0.098
	2.000	0.069
	5.000	0.054
$Cd(NO_3)_2$	0.001	0.888
	0.002	0.851
	0.005	0.787
	0.010	0.728
	0.020	0.664
	0.050	0.576
	0.100	0.515
	0.200	0.465
	0.500	0.428
	1.000	0.437
	2.000	0.517
$CoBr_2$	0.001	0.890
	0.002	0.854
	0.005	0.794
	0.010	0.740
	0.020	0.681
	0.050	0.605
	0.100	0.556
	0.200	0.523
	0.500	0.538
	1.000	0.685
	2.000	1.421
	5.000	13.9
$CoCl_2$	0.001	0.889
	0.002	0.852
	0.005	0.789
	0.010	0.732
	0.020	0.670
	0.050	0.586
	0.100	0.528
	0.200	0.483
	0.500	0.465
	1.000	0.532

Electrolyte	m/mol kg^{-1}	γ_{mean}
	2.000	0.864
CoI_2	0.001	0.887
	0.002	0.849
	0.005	0.783
	0.010	0.724
	0.020	0.661
	0.050	0.582
	0.100	0.540
	0.200	0.527
	0.500	0.596
	1.000	0.845
	2.000	2.287
	5.000	55.3
	10.000	196
$Co(NO_3)_2$	0.001	0.888
	0.002	0.850
	0.005	0.786
	0.010	0.728
	0.020	0.663
	0.050	0.576
	0.100	0.516
	0.200	0.469
	0.500	0.446
	1.000	0.492
	2.000	0.722
	5.000	3.338
CsBr	0.001	0.965
	0.002	0.951
	0.005	0.925
	0.010	0.898
	0.020	0.864
	0.050	0.806
	0.100	0.752
	0.200	0.691
	0.500	0.605
	1.000	0.540
	2.000	0.485
	5.000	0.454
CsCl	0.001	0.965
	0.002	0.951
	0.005	0.925
	0.010	0.898
	0.020	0.864
	0.050	0.805
	0.100	0.751
	0.200	0.691
	0.500	0.607
	1.000	0.546
	2.000	0.496
	5.000	0.474
	10.000	0.508
CsF	0.001	0.965
	0.002	0.952
	0.005	0.929
	0.010	0.905
	0.020	0.876
	0.050	0.830
	0.100	0.792
	0.200	0.755
	0.500	0.721
	1.000	0.726
	2.000	0.803
CsI	0.001	0.965
	0.002	0.951
	0.005	0.925
	0.010	0.898
	0.020	0.863
	0.050	0.804
	0.100	0.749
	0.200	0.688
	0.500	0.601
	1.000	0.534
	2.000	0.470
$CsNO_3$	0.001	0.964
	0.002	0.951
	0.005	0.924
	0.010	0.897
	0.020	0.860
	0.050	0.796
	0.100	0.733
	0.200	0.655
	0.500	0.529
	1.000	0.421
CsOH	0.001	0.966
	0.002	0.953
	0.005	0.930
	0.010	0.906
	0.020	0.878
	0.050	0.836
	0.100	0.802
	0.200	0.772
	0.500	0.755
	1.000	0.782
Cs_2SO_4	0.001	0.885
	0.002	0.845
	0.005	0.775
	0.010	0.709
	0.020	0.634
	0.050	0.526
	0.100	0.444
	0.200	0.369
	0.500	0.285
	1.000	0.233
$CuBr_2$	0.001	0.889
	0.002	0.853
	0.005	0.791
	0.010	0.735
	0.020	0.674
	0.050	0.594
	0.100	0.541
	0.200	0.504
	0.500	0.503
	1.000	0.591
	2.000	0.859
$CuCl_2$	0.001	0.887
	0.002	0.849
	0.005	0.783
	0.010	0.722
	0.020	0.654
	0.050	0.561
	0.100	0.495
	0.200	0.441
	0.500	0.401
	1.000	0.405
	2.000	0.453
	5.000	0.601
$Cu(ClO_4)_2$	0.001	0.890
	0.002	0.854
	0.005	0.795
	0.010	0.741
	0.020	0.685
	0.050	0.613
	0.100	0.572
	0.200	0.553
	0.500	0.617
	1.000	0.892
	2.000	2.445
$Cu(NO_3)_2$	0.001	0.888
	0.002	0.851
	0.005	0.787
	0.010	0.729
	0.020	0.664
	0.050	0.577
	0.100	0.516
	0.200	0.466
	0.500	0.431
	1.000	0.456
	2.000	0.615
	5.000	2.083
$FeCl_2$	0.001	0.888
	0.002	0.850
	0.005	0.785
	0.010	0.725
	0.020	0.659
	0.050	0.570
	0.100	0.509
	0.200	0.462
	0.500	0.443
	1.000	0.500
	2.000	0.782
HBr	0.001	0.966
	0.002	0.953
	0.005	0.930
	0.010	0.907
	0.020	0.879
	0.050	0.837
	0.100	0.806
	0.200	0.783
	0.500	0.790
	1.000	0.872
	2.000	1.167
	5.000	3.800
	10.000	33.4
HCl	0.001	0.965
	0.002	0.952
	0.005	0.929
	0.010	0.905
	0.020	0.876
	0.050	0.832
	0.100	0.797

Electrolyte	*m*/mol kg⁻¹	γ_{mean}
	0.200	0.768
	0.500	0.759
	1.000	0.811
	2.000	1.009
	5.000	2.380
	10.000	10.4
$HClO_4$	0.001	0.966
	0.002	0.953
	0.005	0.929
	0.010	0.906
	0.020	0.878
	0.050	0.836
	0.100	0.803
	0.200	0.776
	0.500	0.769
	1.000	0.826
	2.000	1.055
	5.000	3.100
	10.000	30.8
	15.000	323
HF	0.001	0.551
	0.002	0.429
	0.005	0.302
	0.010	0.225
	0.020	0.163
	0.050	0.106
	0.100	0.0766
	0.200	0.0550
	0.500	0.0352
	1.000	0.0249
	2.000	0.0175
	5.000	0.0110
	10.000	0.0085
	15.000	0.0077
	20.000	0.0075
HI	0.001	0.966
	0.002	0.953
	0.005	0.931
	0.010	0.909
	0.020	0.884
	0.050	0.847
	0.100	0.823
	0.200	0.811
	0.500	0.845
	1.000	0.969
	2.000	1.363
	5.000	4.760
	10.000	49.100
HNO_3	0.001	0.965
	0.002	0.952
	0.005	0.929
	0.010	0.905
	0.020	0.875
	0.050	0.829
	0.100	0.792
	0.200	0.756
	0.500	0.725
	1.000	0.730
	2.000	0.788
	5.000	1.063

Electrolyte	*m*/mol kg⁻¹	γ_{mean}
	10.000	1.644
	15.000	2.212
	20.000	2.607
H_2SO_4	0.001	0.804
	0.002	0.740
	0.005	0.634
	0.010	0.542
	0.020	0.445
	0.050	0.325
	0.100	0.251
	0.200	0.195
	0.500	0.146
	1.000	0.125
	2.000	0.119
	5.000	0.197
	10.000	0.527
	15.000	1.077
	20.000	1.701
KBr	0.001	0.965
	0.002	0.952
	0.005	0.927
	0.010	0.902
	0.020	0.870
	0.050	0.817
	0.100	0.771
	0.200	0.722
	0.500	0.658
	1.000	0.617
	2.000	0.593
	5.000	0.626
KCl	0.001	0.965
	0.002	0.951
	0.005	0.927
	0.010	0.901
	0.020	0.869
	0.050	0.816
	0.100	0.768
	0.200	0.717
	0.500	0.649
	1.000	0.604
	2.000	0.573
	5.000	0.593
$KClO_3$	0.001	0.965
	0.002	0.951
	0.005	0.926
	0.010	0.899
	0.020	0.865
	0.050	0.805
	0.100	0.749
	0.200	0.681
	0.500	0.569
KF	0.001	0.965
	0.002	0.952
	0.005	0.927
	0.010	0.902
	0.020	0.870
	0.050	0.818
	0.100	0.773
	0.200	0.726
	0.500	0.670

Electrolyte	*m*/mol kg⁻¹	γ_{mean}
	1.000	0.645
	2.000	0.658
	5.000	0.871
	10.000	1.715
	15.000	3.120
KH_2PO_4	0.001	0.964
	0.002	0.950
	0.005	0.924
	0.010	0.896
	0.020	0.859
	0.050	0.793
	0.100	0.730
	0.200	0.652
	0.500	0.529
	1.000	0.422
KI	0.001	0.965
	0.002	0.952
	0.005	0.927
	0.010	0.902
	0.020	0.871
	0.050	0.820
	0.100	0.776
	0.200	0.731
	0.500	0.676
	1.000	0.646
	2.000	0.638
KNO_3	0.001	0.964
	0.002	0.950
	0.005	0.924
	0.010	0.896
	0.020	0.860
	0.050	0.797
	0.100	0.735
	0.200	0.662
	0.500	0.546
	1.000	0.444
	2.000	0.332
KOH	0.001	0.965
	0.002	0.952
	0.005	0.927
	0.010	0.902
	0.020	0.871
	0.050	0.821
	0.100	0.779
	0.200	0.740
	0.500	0.710
	1.000	0.733
	2.000	0.860
	5.000	1.697
	10.000	6.110
	15.000	19.9
	20.000	46.4
KSCN	0.001	0.965
	0.002	0.951
	0.005	0.927
	0.010	0.901
	0.020	0.869
	0.050	0.815
	0.100	0.768
	0.200	0.716

Electrolyte	m/ mol kg⁻¹	γ_{mean}
	0.500	0.647
	1.000	0.598
	2.000	0.556
	5.000	0.525
K_2CrO_4	0.001	0.886
	0.002	0.847
	0.005	0.779
	0.010	0.715
	0.020	0.643
	0.050	0.539
	0.100	0.460
	0.200	0.385
	0.500	0.296
	1.000	0.239
	2.000	0.199
K_2HPO_4	0.001	0.886
	0.002	0.847
	0.005	0.779
	0.010	0.715
	0.020	0.643
	0.050	0.538
	0.100	0.457
	0.200	0.379
	0.500	0.283
K_2SO_4	0.001	0.885
	0.002	0.844
	0.005	0.772
	0.010	0.704
	0.020	0.625
	0.050	0.511
	0.100	0.424
	0.200	0.343
	0.500	0.251
LiBr	0.001	0.965
	0.002	0.952
	0.005	0.929
	0.010	0.905
	0.020	0.877
	0.050	0.832
	0.100	0.797
	0.200	0.767
	0.500	0.754
	1.000	0.803
	2.000	1.012
	5.000	2.696
	10.000	20.0
	15.000	147
	20.000	486
LiCl	0.001	0.965
	0.002	0.952
	0.005	0.928
	0.010	0.904
	0.020	0.874
	0.050	0.827
	0.100	0.789
	0.200	0.756
	0.500	0.739
	1.000	0.775
	2.000	0.924
	5.000	2.000

Electrolyte	m/ mol kg⁻¹	γ_{mean}
	10.000	9.600
	15.000	30.9
$LiClO_4$	0.001	0.966
	0.002	0.953
	0.005	0.931
	0.010	0.908
	0.020	0.882
	0.050	0.843
	0.100	0.815
	0.200	0.795
	0.500	0.806
	1.000	0.887
	2.000	1.161
LiI	0.001	0.966
	0.002	0.953
	0.005	0.930
	0.010	0.908
	0.020	0.882
	0.050	0.843
	0.100	0.817
	0.200	0.802
	0.500	0.824
	1.000	0.912
	2.000	1.197
$LiNO_3$	0.001	0.965
	0.002	0.952
	0.005	0.928
	0.010	0.904
	0.020	0.874
	0.050	0.827
	0.100	0.788
	0.200	0.753
	0.500	0.726
	1.000	0.743
	2.000	0.837
	5.000	1.298
	10.000	2.500
	15.000	3.960
	20.000	4.970
LiOH	0.001	0.964
	0.002	0.950
	0.005	0.923
	0.010	0.895
	0.020	0.858
	0.050	0.794
	0.100	0.735
	0.200	0.668
	0.500	0.579
	1.000	0.522
	2.000	0.484
	5.000	0.493
Li_2SO_4	0.001	0.887
	0.002	0.847
	0.005	0.780
	0.010	0.716
	0.020	0.645
	0.050	0.544
	0.100	0.469
	0.200	0.400
	0.500	0.325

Electrolyte	m/ mol kg⁻¹	γ_{mean}
	1.000	0.284
	2.000	0.270
$MgBr_2$	0.001	0.889
	0.002	0.852
	0.005	0.790
	0.010	0.733
	0.020	0.672
	0.050	0.593
	0.100	0.543
	0.200	0.512
	0.500	0.540
	1.000	0.715
	2.000	1.590
	5.000	36.1
$MgCl_2$	0.001	0.889
	0.002	0.852
	0.005	0.790
	0.010	0.734
	0.020	0.672
	0.050	0.590
	0.100	0.535
	0.200	0.493
	0.500	0.485
	1.000	0.577
	2.000	1.065
	5.000	14.40
MgI_2	0.001	0.889
	0.002	0.853
	0.005	0.791
	0.010	0.736
	0.020	0.677
	0.050	0.602
	0.100	0.556
	0.200	0.535
	0.500	0.594
	1.000	0.858
	2.000	2.326
	5.000	109.8
$MnBr_2$	0.001	0.889
	0.002	0.853
	0.005	0.791
	0.010	0.735
	0.020	0.674
	0.050	0.595
	0.100	0.543
	0.200	0.508
	0.500	0.519
	1.000	0.650
	2.000	1.224
	5.000	6.697
$MnCl_2$	0.001	0.888
	0.002	0.850
	0.005	0.786
	0.010	0.727
	0.020	0.662
	0.050	0.574
	0.100	0.513
	0.200	0.464
	0.500	0.437
	1.000	0.477

Electrolyte	m/ mol kg⁻¹	γ_{mean}
	2.000	0.661
	5.000	1.539
$Mn(ClO_4)_2$	0.001	0.892
	0.002	0.858
	0.005	0.801
	0.010	0.752
	0.020	0.700
	0.050	0.637
	0.100	0.604
	0.200	0.596
	0.500	0.686
	1.000	1.030
	2.000	3.072
NH_4Cl	0.001	0.965
	0.002	0.952
	0.005	0.927
	0.010	0.901
	0.020	0.869
	0.050	0.816
	0.100	0.769
	0.200	0.718
	0.500	0.649
	1.000	0.603
	2.000	0.569
	5.000	0.563
NH_4ClO_4	0.001	0.964
	0.002	0.950
	0.005	0.924
	0.010	0.895
	0.020	0.859
	0.050	0.794
	0.100	0.734
	0.200	0.663
	0.500	0.560
	1.000	0.479
	2.000	0.399
NH_4NO_3	0.001	0.964
	0.002	0.951
	0.005	0.925
	0.010	0.897
	0.020	0.862
	0.050	0.801
	0.100	0.744
	0.200	0.678
	0.500	0.582
	1.000	0.502
	2.000	0.419
	5.000	0.303
	10.000	0.220
	15.000	0.179
	20.000	0.154
$(NH_4)_2HPO_4$	0.001	0.882
	0.002	0.839
	0.005	0.763
	0.010	0.688
	0.020	0.600
	0.050	0.469
	0.100	0.367
	0.200	0.273
	0.500	0.171

Electrolyte	m/ mol kg⁻¹	γ_{mean}
	1.000	0.114
	2.000	0.074
NaBr	0.001	0.965
	0.002	0.952
	0.005	0.928
	0.010	0.903
	0.020	0.873
	0.050	0.824
	0.100	0.783
	0.200	0.742
	0.500	0.697
	1.000	0.687
	2.000	0.730
	5.000	1.083
$NaBrO_3$	0.001	0.965
	0.002	0.951
	0.005	0.926
	0.010	0.900
	0.020	0.867
	0.050	0.811
	0.100	0.759
	0.200	0.698
	0.500	0.605
	1.000	0.528
	2.000	0.449
NaCl	0.001	0.965
	0.002	0.952
	0.005	0.928
	0.010	0.903
	0.020	0.872
	0.050	0.822
	0.100	0.779
	0.200	0.734
	0.500	0.681
	1.000	0.657
	2.000	0.668
	5.000	0.874
$NaClO_3$	0.001	0.965
	0.002	0.952
	0.005	0.927
	0.010	0.902
	0.020	0.870
	0.050	0.818
	0.100	0.771
	0.200	0.719
	0.500	0.646
	1.000	0.590
	2.000	0.537
$NaClO_4$	0.001	0.965
	0.002	0.952
	0.005	0.928
	0.010	0.903
	0.020	0.872
	0.050	0.821
	0.100	0.777
	0.200	0.729
	0.500	0.668
	1.000	0.630
	2.000	0.608
	5.000	0.648

Electrolyte	m/ mol kg⁻¹	γ_{mean}
NaF	0.001	0.965
	0.002	0.951
	0.005	0.926
	0.010	0.901
	0.020	0.868
	0.050	0.813
	0.100	0.764
	0.200	0.710
	0.500	0.633
	1.000	0.573
NaI	0.001	0.965
	0.002	0.952
	0.005	0.928
	0.010	0.904
	0.020	0.874
	0.050	0.827
	0.100	0.789
	0.200	0.753
	0.500	0.722
	1.000	0.734
	2.000	0.823
	5.000	1.402
	10.000	4.011
$NaNO_3$	0.001	0.965
	0.002	0.951
	0.005	0.926
	0.010	0.900
	0.020	0.866
	0.050	0.810
	0.100	0.759
	0.200	0.701
	0.500	0.617
	1.000	0.550
	2.000	0.480
	5.000	0.388
	10.000	0.329
NaOH	0.001	0.965
	0.002	0.952
	0.005	0.927
	0.010	0.902
	0.020	0.870
	0.050	0.819
	0.100	0.775
	0.200	0.731
	0.500	0.685
	1.000	0.674
	2.000	0.714
	5.000	1.076
	10.000	3.258
	15.000	9.796
	20.000	19.410
Na_2CO_3	0.001	0.887
	0.002	0.847
	0.005	0.780
	0.010	0.716
	0.020	0.644
	0.050	0.541
	0.100	0.462
	0.200	0.385
	0.500	0.292

Electrolyte	m/ mol kg^{-1}	γ_{mean}
	1.000	0.229
	2.000	0.182
Na_2CrO_4	0.001	0.887
	0.002	0.849
	0.005	0.783
	0.010	0.722
	0.020	0.653
	0.050	0.554
	0.100	0.479
	0.200	0.406
	0.500	0.318
	1.000	0.261
	2.000	0.231
Na_2HPO_4	0.001	0.887
	0.002	0.848
	0.005	0.780
	0.010	0.717
	0.020	0.644
	0.050	0.539
	0.100	0.456
	0.200	0.373
	0.500	0.266
	1.000	0.191
	2.000	0.133
Na_2SO_3	0.001	0.887
	0.002	0.847
	0.005	0.779
	0.010	0.716
	0.020	0.644
	0.050	0.540
	0.100	0.462
	0.200	0.386
	0.500	0.296
	1.000	0.237
	2.000	0.196
Na_2SO_4	0.001	0.886
	0.002	0.846
	0.005	0.777
	0.010	0.712
	0.020	0.637
	0.050	0.529
	0.100	0.446
	0.200	0.366
	0.500	0.268
	1.000	0.204
	2.000	0.155
Na_2WO_4	0.001	0.886
	0.002	0.846
	0.005	0.777
	0.010	0.712
	0.020	0.638
	0.050	0.534
	0.100	0.457
	0.200	0.388
	0.500	0.320
	1.000	0.291
	2.000	0.291
$NiBr_2$	0.001	0.889
	0.002	0.853
	0.005	0.791

Electrolyte	m/ mol kg^{-1}	γ_{mean}
	0.010	0.735
	0.020	0.675
	0.050	0.596
	0.100	0.546
	0.200	0.514
	0.500	0.535
	1.000	0.692
	2.000	1.476
$NiCl_2$	0.001	0.889
	0.002	0.852
	0.005	0.789
	0.010	0.732
	0.020	0.669
	0.050	0.584
	0.100	0.527
	0.200	0.482
	0.500	0.465
	1.000	0.538
	2.000	0.915
	5.000	4.785
$Ni(ClO_4)_2$	0.001	0.891
	0.002	0.855
	0.005	0.797
	0.010	0.745
	0.020	0.690
	0.050	0.621
	0.100	0.582
	0.200	0.567
	0.500	0.639
	1.000	0.946
	2.000	2.812
$Ni(NO_3)_2$	0.001	0.889
	0.002	0.851
	0.005	0.787
	0.010	0.730
	0.020	0.666
	0.050	0.581
	0.100	0.524
	0.200	0.481
	0.500	0.467
	1.000	0.528
	2.000	0.797
$Pb(ClO_4)_2$	0.001	0.889
	0.002	0.851
	0.005	0.787
	0.010	0.729
	0.020	0.666
	0.050	0.580
	0.100	0.522
	0.200	0.476
	0.500	0.458
	1.000	0.516
	2.000	0.799
	5.000	4.043
	10.000	33.8
$Pb(NO_3)_2$	0.001	0.882
	0.002	0.840
	0.005	0.764
	0.010	0.690
	0.020	0.604

Electrolyte	m/ mol kg^{-1}	γ_{mean}
	0.050	0.476
	0.100	0.379
	0.200	0.291
	0.500	0.195
	1.000	0.136
RbBr	0.001	0.965
	0.002	0.951
	0.005	0.926
	0.010	0.900
	0.020	0.866
	0.050	0.811
	0.100	0.760
	0.200	0.705
	0.500	0.630
	1.000	0.578
	2.000	0.535
	5.000	0.514
RbCl	0.001	0.965
	0.002	0.951
	0.005	0.926
	0.010	0.900
	0.020	0.867
	0.050	0.811
	0.100	0.761
	0.200	0.707
	0.500	0.633
	1.000	0.583
	2.000	0.546
	5.000	0.544
RbF	0.001	0.965
	0.002	0.952
	0.005	0.927
	0.010	0.902
	0.020	0.871
	0.050	0.821
	0.100	0.780
	0.200	0.739
	0.500	0.701
	1.000	0.697
	2.000	0.724
RbI	0.001	0.965
	0.002	0.951
	0.005	0.926
	0.010	0.900
	0.020	0.866
	0.050	0.810
	0.100	0.759
	0.200	0.703
	0.500	0.627
	1.000	0.574
	2.000	0.532
	5.000	0.517
$RbNO_3$	0.001	0.964
	0.002	0.950
	0.005	0.924
	0.010	0.896
	0.020	0.859
	0.050	0.795
	0.100	0.733
	0.200	0.657

Electrolyte	m/ mol kg⁻¹	γ_{mean}
	0.500	0.536
	1.000	0.430
	2.000	0.320
Rb_2SO_4	0.001	0.886
	0.002	0.845
	0.005	0.776
	0.010	0.710
	0.020	0.635
	0.050	0.526
	0.100	0.443
	0.200	0.365
	0.500	0.274
	1.000	0.217
$SrBr_2$	0.001	0.889
	0.002	0.852
	0.005	0.790
	0.010	0.734
	0.020	0.673
	0.050	0.591
	0.100	0.535
	0.200	0.492
	0.500	0.476
	1.000	0.545
	2.000	0.921
$SrCl_2$	0.001	0.888
	0.002	0.850
	0.005	0.785
	0.010	0.725
	0.020	0.659
	0.050	0.569
	0.100	0.506
	0.200	0.455
	0.500	0.421
	1.000	0.451
	2.000	0.650
SrI_2	0.001	0.890
	0.002	0.854
	0.005	0.793

Electrolyte	m/ mol kg⁻¹	γ_{mean}
	0.010	0.740
	0.020	0.681
	0.050	0.606
	0.100	0.557
	0.200	0.526
	0.500	0.542
	1.000	0.686
UO_2Cl_2	0.001	0.888
	0.002	0.851
	0.005	0.787
	0.010	0.729
	0.020	0.666
	0.050	0.583
	0.100	0.529
	0.200	0.493
	0.500	0.501
	1.000	0.601
	2.000	0.948
$UO_2(NO_3)_2$	0.001	0.888
	0.002	0.849
	0.005	0.784
	0.010	0.726
	0.020	0.663
	0.050	0.583
	0.100	0.535
	0.200	0.509
	0.500	0.532
	1.000	0.673
	2.000	1.223
	5.000	3.020
$ZnBr_2$	0.001	0.890
	0.002	0.854
	0.005	0.794
	0.010	0.741
	0.020	0.683
	0.050	0.606
	0.100	0.553
	0.200	0.515

Electrolyte	m/ mol kg⁻¹	γ_{mean}
	0.500	0.516
	1.000	0.558
	2.000	0.578
	5.000	0.788
	10.000	2.317
	15.000	5.381
	20.000	7.965
$ZnCl_2$	0.001	0.887
	0.002	0.847
	0.005	0.781
	0.010	0.719
	0.020	0.652
	0.050	0.561
	0.100	0.499
	0.200	0.447
	0.500	0.384
	1.000	0.330
	2.000	0.283
	5.000	0.342
	10.000	0.876
	15.000	1.914
	20.000	2.968
ZnI_2	0.001	0.893
	0.002	0.859
	0.005	0.804
	0.010	0.757
	0.020	0.708
	0.050	0.644
	0.100	0.601
	0.200	0.574
	0.500	0.635
	1.000	0.836
	2.000	1.062
	5.000	1.546
	10.000	4.698

[a] The anion is $H_2PO_4^-$.
[b] The anion is HPO_4^{-2}.

ENTHALPY OF DILUTION OF ACIDS

The quantity given in this table is $-\Delta_{dil}H$, the negative of the enthalpy (heat) of dilution to infinite dilution for aqueous solutions of several common acids; i.e., the negative of the enthalpy change when a solution of molality m at a temperature of 25 °C is diluted with an infinite amount of water. The tabulated numbers thus represent the heat produced (or, if the value is negative, the heat absorbed) when the acid is diluted from the initial molality m to infinite dilution.

It is sometimes useful to have the dilution ratio, which is the number of moles of water that must be added to one mole of acid to produce a solution of desired molality. This may be calucalated from the relation: dilution ratio = 55.506/m, where m is in mol/kg.

Reference

Parker, V. B., *Thermal Properties of Aqueous Uni-Univalent Electrolytes,* Natl. Stand. Ref. Data Ser. - Natl. Bur. Stand. (U.S.) 2, U.S. Government Printing Office, 1965.

Enthalpy of Dilution ($-\Delta_{dil}H$ in kJ/mol) at 25 °C for the Specified Molality m in mol/kg

Formula	0.000111 m	0.00111 m	0.01110 m	0.05 m	0.1 m	0.2 m	0.5 m	1 m	2 m	3 m	4 m	5 m	9 m	20 m	55.506 m
HF	1.255	5.439	9.874	12.24	12.80	13.09	13.20	13.30	13.40	13.45	13.53	13.62	13.81	14.88	
HCl	0.021	0.067	0.197	0.406	0.556	0.761	1.172	1.695	2.623	3.506	4.402	5.318	9.213	19.87	45.61
HBr	0.021	0.054	0.184	0.372	0.498	0.649	0.941	1.314	1.996	2.611	3.330	4.113	7.719	19.92	48.83
HI	0.021	0.050	0.172	0.339	0.439	0.536	0.711	0.933	1.318	1.787	2.460	3.197	6.569	21.71	
HNO_3	0.021	0.063	0.176	0.305	0.372	0.439	0.498	0.506	0.527	0.665	0.958	1.310	3.368	9.498	19.73
$HClO_4$	0.021	0.059	0.167	0.259	0.272	0.247	0.075	-0.201	-0.623	-0.782	-0.787	-0.628	1.280	13.81	
HCOOH	0.038	0.084	0.109	0.121	0.134	0.146	0.176	0.226	0.276	0.289	0.289	0.289	0.230	0.038	0.046
CH_3COOH	0.167	0.222	0.255	0.272	0.289	0.331	0.406	0.544	0.803	1.025	1.218	1.393	1.782	2.075	2.167

ENTHALPY OF SOLUTION OF ELECTROLYTES

This table gives the molar enthalpy (heat) of solution at infinite dilution for some common uni-univalent electrolytes. This is the enthalpy change when 1 mol of solute in its standard state is dissolved in an infinite amount of water. Values are given in kilojoules per mole at 25 °C.

Reference

Parker, V. B., *Thermal Properties of Uni-Univalent Electrolytes*, Natl. Stand. Ref. Data Series - Natl. Bur. Stand. (U.S.), No. 2, 1965.

Solute	State	$\Delta_{sol}H$/ kJ mol^{-1}
Acids		
HF	gas	-61.50
HCl	gas	-74.84
HBr	gas	-85.14
HI	gas	-81.67
$HClO_4$	liquid	-88.76
$HClO_4 \cdot H_2O$	crystal	-32.95
HIO_3	crystal	8.79
HNO_3	liquid	-33.28
HCOOH	liquid	-0.86
CH_3COOH	liquid	-1.51
Silver Compounds		
$AgClO_4$	crystal	7.36
$AgNO_2$	crystal	36.94
$AgNO_3$	crystal	22.59
Cesium Compounds		
CsBr	crystal	25.98
$CsBrO_3$	crystal	50.46
CsCl	crystal	17.78
$CsClO_4$	crystal	55.44
CsF	crystal	-36.86
$CsF \cdot H_2O$	crystal	-10.46
$CsF \cdot 1.5H_2O$	crystal	-5.44
CsI	crystal	33.35
$CsNO_3$	crystal	40.00
CsOH	crystal	-71.55
$CsOH \cdot H_2O$	crystal	-20.50
Potassium Compounds		
KBr	crystal	19.87
$KBrO_3$	crystal	41.13
KCN	crystal	11.72
KCNO	crystal	20.25
$KC_2H_3O_2$	crystal	-15.33
KCl	crystal	17.22
$KClO_3$	crystal	41.38
$KClO_4$	crystal	51.04
KF	crystal	-17.73
$KF \cdot 2H_2O$	crystal	6.97
KI	crystal	20.33
KIO_3	crystal	27.74
$KMnO_4$	crystal	43.56
KNO_2	crystal	13.35
KNO_3	crystal	34.89
KOH	crystal	-57.61
$KOH \cdot H_2O$	crystal	-14.64
$KOH \cdot 1.5H_2O$	crystal	-10.46
KSCN	crystal	24.23
Lithium Compounds		
LiBr	crystal	-48.83
$LiBr \cdot H_2O$	crystal	-23.26
$LiBr \cdot 2H_2O$	crystal	-9.41
$LiBrO_3$	crystal	1.42
LiCl	crystal	-37.03
$LiCl \cdot H_2O$	crystal	-19.08
$LiClO_4$	crystal	-26.55
$LiClO_4 \cdot 3H_2O$	crystal	32.61
LiF	crystal	4.73
LiI	crystal	-63.30
$LiI \cdot H_2O$	crystal	-29.66
$LiI \cdot 2H_2O$	crystal	-14.77
$LiI \cdot 3H_2O$	crystal	0.59
$LiNO_2$	crystal	-11.00
$LiNO_2 \cdot H_2O$	crystal	7.03
$LiNO_3$	crystal	-2.51
LiOH	crystal	-23.56
$LiOH \cdot H_2O$	crystal	-6.69
Ammonium Compounds		
NH_3	gas	-30.50
NH_4Br	crystal	16.78
NH_4Cl	crystal	14.78
NH_4I	crystal	13.72
NH_4CN	crystal	17.57
$NH_4C_2H_3O_2$	crystal	-2.38
NH_4ClO_4	crystal	33.47
NH_4IO_3	crystal	31.80
NH_4NO_2	crystal	19.25
NH_4NO_3	crystal	25.69
NH_4SCN	crystal	22.59
$N(CH_3)_4Br$	crystal	24.27
$N(CH_3)_4Cl$	crystal	4.08
$N(CH_3)_4I$	crystal	42.07
$(CH_3)_3N \cdot HCl$	crystal	1.46
$CH_3NH_2 \cdot HCl$	crystal	5.77
Sodium Compounds		
NaBr	crystal	-0.60
$NaBr \cdot 2H_2O$	crystal	18.64
$NaBrO_3$	crystal	26.90
NaCN	crystal	1.21
$NaCN \cdot 0.5H_2O$	crystal	3.31
$NaCN \cdot 2H_2O$	crystal	18.58
NaCNO	crystal	19.20
$NaC_2H_3O_2$	crystal	-17.32
$NaC_2H_3O_2 \cdot 3H_2O$	crystal	19.66
NaCl	crystal	3.88
$NaClO_2$	crystal	0.33
$NaClO_2 \cdot 3H_2O$	crystal	28.58
$NaClO_3$	crystal	21.72
$NaClO_4$	crystal	13.88
$NaClO_4 \cdot H_2O$	crystal	22.51
NaF	crystal	0.91
NaI	crystal	-7.53
$NaI \cdot 2H_2O$	crystal	16.13
$NaIO_3$	crystal	20.29
$NaNO_2$	crystal	13.89
$NaNO_3$	crystal	20.50
NaOH	crystal	-44.51
$NaOH \cdot H_2O$	crystal	-21.41
NaSCN	crystal	6.83
Rubidium Compounds		
RbBr	crystal	21.88
$RbBrO_3$	crystal	48.95
RbCl	crystal	17.28
$RbClO_3$	crystal	47.74
$RbClO_4$	crystal	56.74
RbF	crystal	-26.11
$RbF \cdot H_2O$	crystal	-0.42
$RbF \cdot 1.5H_2O$	crystal	1.34
RbI	crystal	25.10
$RbNO_3$	crystal	36.48
RbOH	crystal	-62.34
$RbOH \cdot H_2O$	crystal	-17.99
$RbOH \cdot 2H_2O$	crystal	0.88

ENTHALPY OF HYDRATION OF GASES

The molar enthalpy of hydration $\Delta_{hyd}H^{\infty}$ is defined as the enthalpy change when one mole of an ideal gas is dissolved in an infinite amount of water. Another term for this quantity is enthalpy (heat) of solvation in water at infinite dilution. The enthalpy of hydration influences the distribution of a volatile compound between the aqueous solution phase and air and is thus important in fields such as environmental science, geochemistry, and chemical engineering. It is related to the enthalpy of solution of the liquid or solid phase, $\Delta_{sol}H^{\infty}$, by

$$\Delta_{hyd}H^{\infty} = \Delta_{sol}H^{\infty} - \Delta_{vap}H^{\circ}$$

where $\Delta_{vap}H^{\circ}$ is the molar enthalpy of vaporization. This table gives the molar enthalpy of hydration for a number of common substances at 298.15 K.

References

1. Plyasunov, A. V., and Shock, E. L., *J. Chem. Eng. Data* [Hydrocarbons, alcohols, and ketones] 46, 1016, 2001.
2. Plyasunov, A. V., Plyasunova, N. V., and Shock, E. L., *J. Chem. Eng. Data* [Esters] 49, 1152, 2004.
3. Plyasunov, A. V., Plyasunova, N. V., and Shock, E. L., *J. Chem. Eng. Data* [Ethers, diethers, and polyethers] 51, 276, 2006.
4. Plyasunova, N. V., Plyasunov, A. V., and Shock, E. L., *J. Chem. Eng. Data* [Thiols, sulfides, and polysulfides] 50, 246, 2005.
5. Plyasunov, A. V., Plyasunova, N. V., and Shock, E. L., *J. Chem. Eng. Data* [Nitriles and dinitriles] 51, 1481, 2006.
6. Plyasunov, A. V., and Shock, E. L., *Geochim. Cosmochim. Acta* [Inorganics and halogen compounds; temperature dependence] 67, 4981, 2003.
7. Kühne, R., Ebert, R-U, and Schüürmann, G., *Environ. Sci. Technol.* [Compilation; temperature dependence] 39, 6705, 2005.
8. Mintz, C., Clark, M., Acree, W. E., and Abraham, M. H., *J. Chem. Inf. Model.* [Compilation and modelling] 47, 115, 2007.

Enthalpy of Hydration for Gases at 298.15 K

Name	Mol. form.	$\Delta_{hyd}H^{\infty}$/ kJ mol^{-1}
Acenaphthene	$C_{12}H_{10}$	-52.1
Acetic acid	$C_2H_4O_2$	-52.8
Acetone	C_3H_6O	-39.7
Acetonitrile	C_2H_3N	-34.9
Acetophenone	C_8H_8O	-53.3
Ammonia	H_3N	-35.4
Aniline	C_6H_7N	-56.5
Anisole	C_7H_8O	-41.4
Argon	Ar	-12.2
Benzaldehyde	C_7H_6O	-42.1
Benzene	C_6H_6	-28.1
Benzonitrile	C_7H_5N	-48.5
Benzyl alcohol	C_7H_8O	-66.9
Biphenyl	$C_{12}H_{10}$	-47.2
Bromobenzene	C_6H_5Br	-33.5
Bromodichloromethane	$CHBrCl_2$	-28.9
Bromoethane	C_2H_5Br	-29.5
Bromomethane	CH_3Br	-23.8
2-Bromo-2-methylpropane	C_4H_9Br	-25.4
1,3-Butadiene	C_4H_6	-31.4
Butane	C_4H_{10}	-24.8
1,4-Butanediamine	$C_4H_{12}N_2$	-91.6
1,4-Butanediol	$C_4H_{10}O_2$	-89.6
Butanenitrile	C_4H_7N	-42.1
1,2,3,4-Butanetetrol	$C_4H_{10}O_4$	-114
1-Butanethiol	$C_4H_{10}S$	-36.3
Butanoic acid	$C_4H_8O_2$	-59.5
1-Butanol	$C_4H_{10}O$	-61.9
2-Butanol	$C_4H_{10}O$	-62.7
2-Butanone	C_4H_8O	-41.9
1-Butene	C_4H_8	-24.1
2-Butoxyethanol	$C_6H_{14}O_2$	-73.6
Butyl acetate	$C_6H_{12}O_2$	-52.7
sec-Butyl acetate	$C_6H_{12}O_2$	-51.9
tert-Butyl acetate	$C_6H_{12}O_2$	-46.2
Butylamine	$C_4H_{11}N$	-59.2
sec-Butylamine	$C_4H_{11}N$	-57.1
tert-Butylamine	$C_4H_{11}N$	-59.0
Butylbenzene	$C_{10}H_{14}$	-38.5
Butyl butanoate	$C_8H_{16}O_2$	-63.5
Butyl ethyl ether	$C_6H_{14}O$	-48.4
tert-Butyl ethyl ether	$C_6H_{14}O$	-53.4
4-*tert*-Butylphenol	$C_{10}H_{14}O$	-63.8
Butyl propanoate	$C_7H_{14}O_2$	-57.8
1-Butyne	C_4H_6	-13.5
Carbon dioxide	CO_2	-17.9
Carbon monoxide	CO	-11.1
Chlorine	Cl_2	-23.4
Chlorine dioxide	ClO_2	-27.8
Chlorobenzene	C_6H_5Cl	-30.6
2-Chlorobiphenyl	$C_{12}H_9Cl$	-42.8
1-Chlorobutane	C_4H_9Cl	-28.2
2-Chlorobutane	C_4H_9Cl	-34.6
Chlorodibromomethane	$CHBr_2Cl$	-33.3
Chlorodifluoromethane	$CHClF_2$	-22.8
2-Chloro-2-(difluoromethoxy)-1,1,1-trifluoroethane	$C_3H_2ClF_5O$	-35.3
Chloroethane	C_2H_5Cl	-22.0
Chlorofluoromethane	CH_2ClF	-21.7
1-Chlorohexane	$C_6H_{13}Cl$	-34.5
Chloromethane	CH_3Cl	-20.2
1-Chloropentane	$C_5H_{11}Cl$	-34.1
3-Chlorophenol	C_6H_5ClO	-50.3
1-Chloropropane	C_3H_7Cl	-27.0
2-Chloropyridine	C_5H_4ClN	-42.6
3-Chloropyridine	C_5H_4ClN	-46.2
2-Chlorotoluene	C_7H_7Cl	-38.3
3-Chlorotoluene	C_7H_7Cl	-37.0
4-Chlorotoluene	C_7H_7Cl	-33.3
o-Cresol	C_7H_8O	-64.8
m-Cresol	C_7H_8O	-58.7
p-Cresol	C_7H_8O	-61.3

Name	Mol. form.	$\Delta_{hyd}H^{\infty}$/ kJ mol^{-1}
Cycloheptanol	$C_7H_{14}O$	-74.6
Cyclohexane	C_6H_{12}	-30.0
cis-1,2-Cyclohexanediol	$C_6H_{12}O_2$	-82.4
Cyclohexanol	$C_6H_{12}O$	-70.7
Cyclohexanone	$C_6H_{10}O$	-49.8
Cyclohexene	C_6H_{10}	-27.3
Cyclooctane	C_8H_{16}	-39.0
cis-Cyclooctene	C_8H_{14}	-45.5
Cyclopentane	C_5H_{10}	-30.3
Cyclopentanol	$C_5H_{10}O$	-58.5
Cyclopentanone	C_5H_8O	-44.3
Cyclopropane	C_3H_6	-15.4
Dibromomethane	CH_2Br_2	-33.0
Dibutylamine	$C_8H_{19}N$	-59.3
Dibutyl ether	$C_8H_{18}O$	-55.8
o-Dichlorobenzene	$C_6H_4Cl_2$	-37.3
m-Dichlorobenzene	$C_6H_4Cl_2$	-35.3
p-Dichlorobenzene	$C_6H_4Cl_2$	-28.4
2,3-Dichloro-1,1'-biphenyl	$C_{12}H_8Cl_2$	-45.6
2,4-Dichloro-1,1'-biphenyl	$C_{12}H_8Cl_2$	-43.0
2,5-Dichlorobiphenyl	$C_{12}H_8Cl_2$	-45.6
2,4'-Dichloro-1,1'-biphenyl	$C_{12}H_8Cl_2$	-44.2
Dichlorodifluoromethane	CCl_2F_2	-26.0
2,2-Dichloro-1,1-difluoro-1-methoxyethane	$C_3H_4Cl_2F_2O$	-30.4
1,1-Dichloroethane	$C_2H_4Cl_2$	-30.3
1,2-Dichloroethane	$C_2H_4Cl_2$	-27.9
1,1-Dichloroethene	$C_2H_2Cl_2$	-28.5
cis-1,2-Dichloroethene	$C_2H_2Cl_2$	-26.9
trans-1,2-Dichloroethene	$C_2H_2Cl_2$	-29.3
Dichloromethane	CH_2Cl_2	-30.3
1,2-Dichloropropane, (±)-	$C_3H_6Cl_2$	-31.1
1,3-Dichloropropane	$C_3H_6Cl_2$	-29.7
1,2-Dichloro-1,1,2,2-tetrafluoroethane	$C_2Cl_2F_4$	-20.2
1,2-Diethoxyethane	$C_6H_{14}O_2$	-71.9
Diethylamine	$C_4H_{11}N$	-64.3
N,N-Diethylaniline	$C_{10}H_{15}N$	-45.7
p-Diethylbenzene	$C_{10}H_{14}$	-46.4
Diethylene glycol dimethyl ether	$C_6H_{14}O_3$	-96.2
Diethyl ether	$C_4H_{10}O$	-46.4
Diethyl sulfide	$C_4H_{10}S$	-40.2
1,1-Difluoroethane	$C_2H_4F_2$	-20.7
Difluoromethane	CH_2F_2	-17.2
Diiodomethane	CH_2I_2	-41.6
Diisopropyl ether	$C_6H_{14}O$	-51.7
1,2-Dimethoxyethane	$C_4H_{10}O_2$	-59.3
Dimethylamine	C_2H_7N	-53.1
2,4-Dimethylaniline	$C_8H_{11}N$	-58.7
2,5-Dimethylaniline	$C_8H_{11}N$	-61.5
2,6-Dimethylaniline	$C_8H_{11}N$	-60.5
N,N-Dimethylaniline	$C_8H_{11}N$	-49.6
2,3-Dimethylbutane	C_6H_{14}	-32.4
3,3-Dimethyl-2-butanone	$C_6H_{12}O$	-47.5
cis-1,2-Dimethylcyclohexane	C_8H_{16}	-38.3
trans-1,2-Dimethylcyclohexane	C_8H_{16}	-36.1
Dimethyl ether	C_2H_6O	-34.0
N,N-Dimethylformamide	C_3H_7NO	-62.9
2,4-Dimethyl-3-pentanone	$C_7H_{14}O$	-54.0
2,3-Dimethylpyridine	C_7H_9N	-57.7
2,4-Dimethylpyridine	C_7H_9N	-60.7
2,5-Dimethylpyridine	C_7H_9N	-54.9
2,6-Dimethylpyridine	C_7H_9N	-52.3
3,4-Dimethylpyridine	C_7H_9N	-50.5
3,5-Dimethylpyridine	C_7H_9N	-51.3
Dimethyl sulfide	C_2H_6S	-31.5
Dimethyl sulfoxide	C_2H_6OS	-71.9
2,5-Dimethyltetrahydrofuran	$C_6H_{12}O$	-56.3
1,4-Dioxane	$C_4H_8O_2$	-48.4
1,2-Dipropoxyethane	$C_8H_{18}O_2$	-76.8
Dipropylamine	$C_6H_{15}N$	-65.2
Dipropyl ether	$C_6H_{14}O$	-49.9
Dipropyl sulfide	$C_6H_{14}S$	-47.7
1-Dodecanol	$C_{12}H_{26}O$	-81.9
Ethane	C_2H_6	-17.9
1,2-Ethanediamine	$C_2H_8N_2$	-76.1
1,2-Ethanediol	$C_2H_6O_2$	-77.3
Ethanethiol	C_2H_6S	-28.9
Ethanol	C_2H_6O	-50.6
2-Ethoxyethanol	$C_4H_{10}O_2$	-66.4
1-Ethoxy-2-methoxyethane	$C_5H_{12}O_2$	-66.1
Ethyl acetate	$C_4H_8O_2$	-45.3
Ethylamine	C_2H_7N	-53.7
2-Ethylaniline	$C_8H_{11}N$	-59.7
4-Ethylaniline	$C_8H_{11}N$	-65.0
Ethylbenzene	C_8H_{10}	-39.4
Ethyl butanoate	$C_6H_{12}O_2$	-52.7
Ethylcyclohexane	C_8H_{16}	-36.8
Ethyl 2,2-dimethylpropanoate	$C_7H_{14}O_2$	-50.3
Ethylene	C_2H_4	-13.7
Ethyl formate	$C_3H_6O_2$	-38.1
Ethyl hexanoate	$C_8H_{16}O_2$	-60.2
Ethyl 2-methylbutanoate	$C_7H_{14}O_2$	-55.4
Ethyl 3-methylbutanoate	$C_7H_{14}O_2$	-56.0
Ethyl 2-methylpropanoate	$C_6H_{12}O_2$	-51.3
Ethyl pentanoate	$C_7H_{14}O_2$	-56.5
Ethyl propanoate	$C_5H_{10}O_2$	-49.5
2-Ethylpyridine	C_7H_9N	-55.7
3-Ethylpyridine	C_7H_9N	-53.5
4-Ethylpyridine	C_7H_9N	-52.2
9*H*-Fluorene	$C_{13}H_{10}$	-42.7
Fluorobenzene	C_6H_5F	-29.3
Fluoromethane	CH_3F	-16.1
Glycerol	$C_3H_8O_3$	-103.5
Helium	He	-0.67
1,1,1,2,3,3,3-Heptafluoropropane	C_3HF_7	-24.8
Heptanal	$C_7H_{14}O$	-56.6
Heptane	C_7H_{16}	-34.0
1-Heptanol	$C_7H_{16}O$	-72.1
2-Heptanol, (±)-	$C_7H_{16}O$	-72.6
4-Heptanol	$C_7H_{16}O$	-75.3
2-Heptanone	$C_7H_{14}O$	-54.9
4-Heptanone	$C_7H_{14}O$	-58.1
1,1,1,3,3,3-Hexafluoro-2-propanol	$C_3H_2F_6O$	-57.1
Hexanal	$C_6H_{12}O$	-55.2
Hexane	C_6H_{14}	-31.9
Hexanedinitrile	$C_6H_8N_2$	-66.6
1-Hexanol	$C_6H_{14}O$	-67.4
3-Hexanol	$C_6H_{14}O$	-69.6
2-Hexanone	$C_6H_{12}O$	-48.9
3-Hexanone	$C_6H_{12}O$	-46.0
1-Hexene	C_6H_{12}	-30.4

Name	Mol. form.	$\Delta_{hyd}H^{\infty}$/kJ mol^{-1}
Hexyl acetate	$C_8H_{16}O_2$	-60.8
Hexylamine	$C_6H_{15}N$	-65.9
Hexylbenzene	$C_{12}H_{18}$	-52.7
Hydrogen	H_2	-0.402
Hydrogen selenide	H_2Se	-15.7
Hydrogen sulfide	H_2S	-18.0
3-Hydroxybenzaldehyde	$C_7H_6O_2$	-70.7
3-Hydroxybenzonitrile	C_7H_5NO	-70.7
4-Hydroxybenzonitrile	C_7H_5NO	-70.3
Iodoethane	C_2H_5I	-31.7
Iodomethane	CH_3I	-28.2
1-Iodopropane	C_3H_7I	-35.3
2-Iodopropane	C_3H_7I	-36.6
Isobutanal	C_4H_8O	-40.0
Isobutane	C_4H_{10}	-21.7
Isobutene	C_4H_8	-22.7
Isobutyl acetate	$C_6H_{12}O_2$	-51.8
Isobutyl formate	$C_5H_{10}O_2$	-43.0
Isobutyl isobutanoate	$C_8H_{16}O_2$	-55.3
Isobutyl propanoate	$C_7H_{14}O_2$	-54.7
Isopentyl acetate	$C_7H_{14}O_2$	-53.8
Isopentyl formate	$C_6H_{12}O_2$	-47.7
Isophorone	$C_9H_{14}O$	-59.1
Isopropyl acetate	$C_5H_{10}O_2$	-46.8
Isopropylamine	C_3H_9N	-55.0
Isopropylbenzene	C_9H_{12}	-33.7
Isopropyl formate	$C_4H_8O_2$	-43.0
1-Isopropyl-4-methylbenzene	$C_{10}H_{14}$	-34.6
Krypton	Kr	-15.6
Methane	CH_4	-12.0
Methanethiol	CH_4S	-24.4
Methanol	CH_4O	-52.0
2-Methoxyethanol	$C_3H_8O_2$	-60.4
2-Methoxy-2-methylbutane	$C_6H_{14}O$	-52.5
2-Methoxyphenol	$C_7H_8O_2$	-62.6
Methyl acetate	$C_3H_6O_2$	-40.1
Methylamine	CH_5N	-45.3
Methyl benzoate	$C_8H_8O_2$	-50.3
Methyl butanoate	$C_5H_{10}O_2$	-47.5
3-Methyl-1-butanol	$C_5H_{12}O$	-66.0
2-Methyl-2-butanol	$C_5H_{12}O$	-68.4
3-Methyl-2-butanone	$C_5H_{10}O$	-57.6
2-Methyl-2-butene	C_5H_{10}	-26.6
Methyl *tert*-butyl ether	$C_5H_{12}O$	-48.7
Methyl 2,2-dimethylpropanoate	$C_6H_{12}O_2$	-46.2
Methyl formate	$C_2H_4O_2$	-32.0
Methyl hexanoate	$C_7H_{14}O_2$	-54.7
Methyl isobutanoate	$C_5H_{10}O_2$	-46.0
4-Methylmorpholine	$C_5H_{11}NO$	-68.7
1-Methylnaphthalene	$C_{11}H_{10}$	-45.0
2-Methylnaphthalene	$C_{11}H_{10}$	-44.9
2-Methylpentane	C_6H_{14}	-30.5
3-Methylpentane	C_6H_{14}	-36.8
Methyl pentanoate	$C_6H_{12}O_2$	-50.4
4-Methyl-2-pentanol	$C_6H_{14}O$	-69.9
4-Methyl-2-pentanone	$C_6H_{12}O$	-44.6
1-Methylpiperidine	$C_6H_{13}N$	-65.8
2-Methylpropanenitrile	C_4H_7N	-40.0
Methyl propanoate	$C_4H_8O_2$	-44.5
2-Methyl-1-propanol	$C_4H_{10}O$	-60.2
2-Methyl-2-propanol	$C_4H_{10}O$	-62.9
Methyl propyl ether	$C_4H_{10}O$	-38.0
2-Methylpyridine	C_6H_7N	-50.3
3-Methylpyridine	C_6H_7N	-50.3
4-Methylpyridine	C_6H_7N	-51.8
N-Methylpyrrolidine	$C_5H_{11}N$	-63.4
2-Methyltetrahydrofuran	$C_5H_{10}O$	-51.4
Morpholine	C_4H_9NO	-69.5
Naphthalene	$C_{10}H_8$	-42.8
Neon	Ne	-3.90
Neopentane	C_5H_{12}	-23.4
Nitric oxide	NO	-11.9
Nitrobenzene	$C_6H_5NO_2$	-43.8
Nitroethane	$C_2H_5NO_2$	-32.5
Nitrogen	N_2	-1.04
Nitromethane	CH_3NO_2	-35.7
2-Nitrophenol	$C_6H_5NO_3$	-49.8
3-Nitrophenol	$C_6H_5NO_3$	-67.7
4-Nitrophenol	$C_6H_5NO_3$	-68.6
1-Nitropropane	$C_3H_7NO_2$	-34.4
2-Nitropropane	$C_3H_7NO_2$	-34.1
2-Nitrotoluene	$C_7H_7NO_2$	-46.4
3-Nitrotoluene	$C_7H_7NO_2$	-38.5
Nitrous oxide	N_2O	-19.8
2-Nonanone	$C_9H_{18}O$	-65.3
5-Nonanone	$C_9H_{18}O$	-62.8
Octanal	$C_8H_{16}O$	-48.8
Octane	C_8H_{18}	-36.0
1-Octanol	$C_8H_{18}O$	-74.1
2-Octanone	$C_8H_{16}O$	-58.3
1-Octene	C_8H_{16}	-39.2
Octylamine	$C_8H_{19}N$	-52.3
Oxygen	O_2	-1.20
Pentachlorobenzene	C_6HCl_5	-39.9
Pentafluoroethane	C_2HF_5	-21.5
2,2,3,3,3-Pentafluoro-1-propanol	$C_3H_3F_5O$	-51.9
Pentanal	$C_5H_{10}O$	-42.9
Pentane	C_5H_{12}	-28.3
1,5-Pentanediamine	$C_5H_{14}N_2$	-95.1
Pentanedinitrile	$C_5H_6N_2$	-63.5
Pentanenitrile	C_5H_9N	-45.6
1-Pentanol	$C_5H_{12}O$	-61.9
2-Pentanol	$C_5H_{12}O$	-63.3
3-Pentanol	$C_5H_{12}O$	-59.6
2-Pentanone	$C_5H_{10}O$	-45.3
3-Pentanone	$C_5H_{10}O$	-49.6
Pentyl acetate	$C_7H_{14}O_2$	-55.3
Pentylamine	$C_5H_{13}N$	-62.1
Pentylbenzene	$C_{11}H_{16}$	-49.5
Pentyl formate	$C_6H_{12}O_2$	-48.1
Perfluoropropene	C_3F_6	-17.4
Phenol	C_6H_6O	-57.7
1-Phenyl-1-propanone	$C_9H_{10}O$	-61.9
Piperidine	$C_5H_{11}N$	-65.4
Propanal	C_3H_6O	-39.4
Propanamide	C_3H_7NO	-73.4
Propane	C_3H_8	-20.4
1,3-Propanediamine	$C_3H_{10}N_2$	-85.6
1,3-Propanediol	$C_3H_8O_2$	-81.1
Propanenitrile	C_3H_5N	-39.5

Name	Mol. form.	$\Delta_{hyd}H^{\infty}$/ kJ mol⁻¹
1-Propanethiol	C_3H_8S	-30.2
Propanoic acid	$C_3H_6O_2$	-56.5
1-Propanol	C_3H_8O	-59.9
2-Propanol	C_3H_8O	-58.2
Propene	C_3H_6	-21.6
2-Propoxyethanol	$C_5H_{12}O_2$	-69.6
Propyl acetate	$C_5H_{10}O_2$	-48.7
Propylamine	C_3H_9N	-56.0
Propylbenzene	C_9H_{12}	-36.4
Propyl butanoate	$C_7H_{14}O_2$	-54.9
Propyl formate	$C_4H_8O_2$	-40.5
Propyl propanoate	$C_6H_{12}O_2$	-51.2
Propyne	C_3H_4	-15.6
Pyridine	C_5H_5N	-42.1
Quinoline	C_9H_7N	-58.2
Radon	Rn	-24.0
Styrene	C_8H_8	-28.4
Succinonitrile	$C_4H_4N_2$	-58.2
Sulfur hexafluoride	F_6S	-20.7
1,2,3,4-Tetrachlorobenzene	$C_6H_2Cl_4$	-35.0
1,1,1,2-Tetrachloroethane	$C_2H_2Cl_4$	-36.2
1,1,2,2-Tetrachloroethane	$C_2H_2Cl_4$	-34.8
Tetrachloroethene	C_2Cl_4	-41.5
Tetrachloromethane	CCl_4	-30.5
Tetraethylene glycol dimethyl ether	$C_{10}H_{22}O_5$	-126
1,1,1,2-Tetrafluoroethane	$C_2H_2F_4$	-22.2
Tetrafluoroethene	C_2F_4	-15.1
Tetrafluoromethane	CF_4	-13.5
2,2,3,3-Tetrafluoro-1-propanol	$C_3H_4F_4O$	-57.9
Tetrahydrofuran	C_4H_8O	-47.3
Tetrahydropyran	$C_5H_{10}O$	-48.9
Thiophene	C_4H_4S	-29.9
Toluene	C_7H_8	-32.4
Tribromomethane	$CHBr_3$	-35.8
1,2,3-Trichlorobenzene	$C_6H_3Cl_3$	-32.6
1,3,5-Trichlorobenzene	$C_6H_3Cl_3$	-34.2
1,1,1-Trichloroethane	$C_2H_3Cl_3$	-28.7
1,1,2-Trichloroethane	$C_2H_3Cl_3$	-32.5
Trichloroethene	C_2HCl_3	-32.2
Trichlorofluoromethane	CCl_3F	-19.8
Trichloromethane	$CHCl_3$	-33.5
1,1,2-Trichloro-1,2,2-trifluoroethane	$C_2Cl_3F_3$	-28.8
Triethylamine	$C_6H_{15}N$	-69.7
Triethylene glycol dimethyl ether	$C_8H_{18}O_4$	-102.4
2,2,2-Trifluoroethanol	$C_2H_3F_3O$	-50.2
Trifluoromethane	CHF_3	-22.6
1,1,1-Trifluoro-2-propanol, (±)-	$C_3H_5F_3O$	-53.5
Trimethylamine	C_3H_9N	-52.7
1,2,3-Trimethylbenzene	C_9H_{12}	-37.4
1,2,4-Trimethylbenzene	C_9H_{12}	-36.6
1,3,5-Trimethylbenzene	C_9H_{12}	-39.1
2,2,4-Trimethylpentane	C_8H_{18}	-31.0
2,3,4-Trimethylpentane	C_8H_{18}	-38.5
Xenon	Xe	-19.4
o-Xylene	C_8H_{10}	-37.7
m-Xylene	C_8H_{10}	-38.6
p-Xylene	C_8H_{10}	-34.8

pH SCALE FOR AQUEOUS SOLUTIONS

A. K. Covington

A Working Party of IUPAC, after extensive considerations over five years, has produced a report (1) which sets pH firmly within the International System of Units (SI). A summary of these important developments is given below.

The concept of pH is unique amongst the commonly encountered physicochemical quantities in that, in terms of its definition,

$$pH = -\lg a_H \tag{1}$$

it involves a single ion quantity, the activity of the hydrogen ion, which is immeasurable by any thermodynamically valid method and requires a convention for its evaluation.

pH was originally defined by Sørensen (2) in terms of the concentration of hydrogen ions (in modern nomenclature) as pH $= -\lg (c_H/c^o)$ where c_H is the hydrogen ion concentration in mol dm^{-3}, and $c^o = 1$ mol dm^{-3} is the standard amount concentration. Subsequently (3), it was accepted as more satisfactory to define pH in terms of the relative activity of hydrogen ions in solution

$$pH = -\lg a_H = -\lg (m_H \gamma_H / m^o) \tag{2}$$

where a_H is the relative (molality basis) activity and γ_H is the molal activity coefficient of the hydrogen ion H^+ at the molality m_H, and m^o the standard molality. The quantity pH is intended to be a measure of the activity of hydrogen ions in solution. However, since it is defined in terms of a quantity that cannot be measured by a thermodynamically valid method, eqn. (2) can only be considered a *notional definition* of pH.

pH being a single ion quantity, it is not determinable in terms of a fundamental (or base) unit of any measurement system, and there is difficulty providing a proper basis for the traceability of pH measurements. A satisfactory approach is now available in that pH determinations can be incorporated into the International System (SI) if they can be traced to measurements made using a method that fulfils the definition of a 'primary method of measurement' (4).

The essential feature of a primary method is that it must operate according to a well-defined measurement equation in which all of the variables can be determined experimentally in terms of SI units. Any limitation in the determination of the experimental variables, or in the theory, must be included within the estimated uncertainty of the method if traceability to the SI is to be established. If a convention were used without an estimate of its uncertainty, true traceability to SI would not be established. The electrochemical cell without liquid junction, known as the Harned cell (5), fulfills the definition of a primary method for the measurement of the acidity function, $p(a_H\gamma_{Cl})$, and subsequently of the pH of buffer solutions.

The Harned cell is written as

$$\text{Pt} \mid H_2 \mid \text{buffer S, } Cl^- \mid \text{AgCl} \mid \text{Ag} \qquad \text{(Cell I)}$$

and contains a standard buffer, S, with chloride ions, as potassium or sodium chloride, added in order to use the silver–silver chloride electrode as reference electrode. The application of the Nernst equation to the spontaneous cell reaction of Cell I:

$$\tfrac{1}{2} H_2 + AgCl \rightarrow Ag(s) + H^+ + Cl^-$$

yields the potential difference E_I of the cell (corrected to 1 atm (101.325 kPa), the partial pressure of hydrogen gas used in electrochemistry in preference to 100 kPa) as

$$E_I = E^o - (RT/F)\ln 10 \lg [(m_H\gamma_H/m^o)(m_{Cl}\gamma_{Cl}/m^o)] \tag{3}$$

which can be rearranged, since $a_H = m_H\gamma_H/m^o$, to give the acidity function

$$p(a_H\gamma_{Cl}) = -\lg(a_H\gamma_{Cl}) = (E_I - E^o)/[(RT/F)\ln 10] + \lg(m_{Cl}/m^o) \tag{4}$$

where E^o is the standard potential difference of the cell, and hence of the silver–silver chloride electrode, and γ_{Cl} is the activity coefficient of the chloride ion.

The standard potential difference of the silver–silver chloride electrode, E^o, is determined from a Harned cell in which only HCl is present at a fixed molality (e.g., $m = 0.01$ mol kg^{-1})

$$\text{Pt} \mid H_2 \mid \text{HCl}(m) \mid \text{AgCl} \mid \text{Ag} \qquad \text{(Cell Ia)}$$

The application of the Nernst equation to the HCl cell (Ia) gives

$$E_{Ia} = E^o - (2RT/F)\ln 10 \lg[(m_{HCl}/m^o)(\gamma_{\pm HCl})] \tag{5}$$

where E_{Ia} has been corrected to 1 atmosphere partial pressure of hydrogen gas (101.325 kPa) and $\gamma_{\pm HCl}$ is the mean ionic activity coefficient of HCl.

Values of the activity coefficient ($\gamma_{\pm HCl}$) at molality 0.01 mol kg^{-1} and various temperatures were given by Bates and Robinson (6). The standard potential difference depends on the method of preparation of the electrodes, but individual determinations of the activity coefficient of HCl at 0.01 mol kg^{-1} are more uniform than values of E^o. Hence the practical determination of the potential difference of the cell with HCl at 0.01 mol kg^{-1} is recommended at 298.15 K at which the mean ionic activity coefficient is 0.904. (It is unnecessary to repeat the measurement of E^o at other temperatures but simply to correct published smoothed values by the observed difference in E^o at 298.15 K.)

In national metrology institutes (NMIs), measurements of Cells I and Ia are often done simultaneously in a thermostat bath. Subtracting eqn. (5) from (3) gives

$$\Delta E = E_I - E_{Ia} = -(RT/F)\ln 10\{\lg[(m_H\gamma_H/m^o)(m_{Cl}\gamma_{Cl}/m^o)] - \lg[(m_{HCl}/m^o)^2\gamma^2_{\pm HCl}]\} \tag{6}$$

which is independent of the standard potential difference. Therefore, the subsequently calculated pH does not depend on the standard potential difference and hence does not depend on the assumption that the standard potential of the hydrogen electrode is zero at all temperatures. Therefore, the Harned cell gives an exact comparison between hydrogen ion activities at different temperatures.

The quantity $p(a_H\gamma_{Cl}) = -\lg (a_H\gamma_{Cl})$, on the left-hand side of (4), is called the acidity function (5). To obtain the quantity pH according to eqn. (2) from the acidity function, it is necessary to evaluate $\lg \gamma_{Cl}$ independently. This is done in two steps: (i) the value of $\lg (a_H\gamma_{Cl})$ at zero chloride molality, $\lg (a_H\gamma_{Cl})^o$, is evaluated and (ii) a value for the activity of the chloride ion γ^o_{Cl}, at zero chloride molality (sometimes referred to as the limiting or 'trace' activity coefficient) is calculated using the Bates-Guggenheim convention (7). The value of $\lg (a_H\gamma_{Cl})^o$ corresponding to zero chloride molality is determined by linear extrapolation of measurements using Harned cells with at least three added molalities of sodium or potassium chloride ($I < 0.1$ mol kg^{-1}).

Thermochem

The value of lg $(a_H\gamma_{Cl})^o$ corresponding to zero chloride molality is determined by linear extrapolation of measurements using Harned cells with at least three added molalities of sodium or potassium chloride (I < 0.1 mol kg^{-1}) in accord with eqn. (7):

$$-\lg(a_H\gamma_{Cl}) = -\lg(a_H\gamma_{Cl})^o + Sm_{Cl} \quad (7)$$

where S is an empirical, temperature dependent, constant.

The Bates-Guggenheim convention (7) assumes that the trace activity coefficient of the chloride ion γ^o_{Cl} is given by

$$\lg \gamma^o_{Cl} = -A\,I^{1/2}/(1 + Ba\,I^{1/2}) \quad (8)$$

where A is the Debye-Hückel temperature dependent constant (limiting slope), a is the *mean* distance of closest approach of the ions (ion size parameter), Ba is set equal to 1.5 (mol kg^{-1})$^{-1/2}$ at all temperatures in the range 5–50 °C, and I is the ionic strength of the buffer (which for its evaluation requires knowledge of appropriate acid dissociation constants).

The various stages in the assignment of primary standard pH values are combined in eqn. (9), which is derived from eqns. (4), (5) and (8)

$$\mathrm{pH(PS)} = \lim m_{Cl\to 0}\,\{(E_I - E^o)/[(RT/F)\ln 10] + \lg(m_{Cl}/m^o)\} - AI^{1/2}/[1 + 1.5\,(I/m^o)^{1/2}] \quad (9)$$

In order for a particular buffer solution to be considered a primary buffer solution, it must be of the "highest metrological" quality (4) in accordance with the definition of a primary standard. It is recommended that it have the following attributes (9):

1. High buffer value in the range 0.016–0.07 (mol OH$^-$)/pH.
2. Small dilution value at half concentration (change in pH with change in buffer concentration) in the range 0.01–0.20.
3. Small dependence of pH on temperature less than ± 0.01 K^{-1}.
4. Low residual liquid junction potential < 0.01 in pH.
5. Ionic strength ≤0.1 mol kg^{-1} to permit applicability of Bates-Guggenheim convention.
6. NMI certificate for specific batch.
7. Reproducible purity of preparation (lot to lot differences of $|\Delta\mathrm{pH(PS)}| < 0.003$).
8. Long-term stability of stored solid material.

Values for the above and other important parameters for the primary and secondary buffer materials are given in Table 1.

Primary Standard Buffers

As there can be significant variations in the purity of samples of a buffer of the same nominal chemical composition, it is essential that the primary buffer material used has been certified with values that have been measured with Cell I. The Harned cell is used by many national metrological institutes for accurate measurements of pH of buffer solutions.

Typical values of the pH(PS) of the seven solutions from the six accepted primary standard reference buffers, which meet the conditions stated above, are listed in Table 2. Batch-to-batch variations in purity can result in changes in the pH value of samples of at most 0.003. The typical values in Table 2 should not be used in place of the certified value (from a Harned cell measurement) for a specific batch of buffer material.

The required attributes listed above effectively limit the range of primary buffers available to between pH 3 and 10 (at 25 °C). Calcium hydroxide and potassium tetraoxalate are excluded because the contribution of hydroxide or hydrogen ions to the ionic strength is significant. Also excluded are the nitrogen bases of the type BH$^+$ (such as tris(hydroxymethyl)aminomethane and piperazine phosphate) and the zwitterionic buffers (e.g., HEPES and MOPS (10)). These do not comply because either the Bates-Guggenheim convention is not applicable, or the liquid junction potentials are high. This means the choice of primary standards is restricted to buffers derived from oxy-carbon, -phosphorus, -boron and mono, di- and tri-protic carboxylic acids. The uncertainties (11) associated with Harned cell measurements are calculated (1) to be 0.004 in pH at NMIs, with typical variation between batches of primary standard buffers of 0.003.

Secondary Standards

Substances that do not fulfill all the criteria for primary standards, but to which pH values can be assigned using Cell I are considered to be secondary standards (Table 3). Reasons for their exclusion as primary standards include difficulties in achieving consistent and suitable chemical quality (e.g. acetic acid is a liquid), suspected high liquid junction potential, or inappropriateness of the Bates-Guggenheim convention (e.g., other charge-type buffers). The uncertainty is higher (e.g., 0.01) for biological buffers. Certain other substances, which cannot be used in cells containing hydrogen gas electrodes, are also classed as secondary standards.

Calibration Procedures

(a) One-point calibration
A single point calibration is insufficient to determine both slope and one-point parameters. The theoretical value for the slope can be assumed but the practical slope may be up to 5% lower. Alternatively, a value for the practical slope can be assumed from the manufacturer's prior calibration. The one-point calibration therefore yields only an estimate of pH(X). Since both parameters may change with age of the electrodes, this is not a reliable procedure.

(b) Two-point calibration [target uncertainty: 0.02–0.03 at 25 °C]
In the majority of practical applications, glass electrodes cells are calibrated by a two-point calibration, or bracketing, procedure using two standard buffer solutions, with pH values, pH(S_1) and pH(S_2), bracketing the unknown pH(X). Bracketing is often taken to mean that the pH(S_1) and pH(S_2) buffers selected from Table 2 should be those that are immediately above and below pH(X). This may not be appropriate in all situations and choice of a wider range may be better.

(c) Multi-point calibration [target uncertainty: 0.01–0.03 at 25 °C].
Multi-point calibration is carried out using up to five standard buffers. The use of more than five points yields no significant improvement in the statistical information obtainable.
Details of uncertainty computations (11) have been given (1).

Measurement of pH and Choice of pH Standard Solutions

1a) If pH is not required to better than ±0.05 any pH standard solution may be selected.

1b) If pH is required to ±0.002 and interpretation in terms of hydrogen ion concentration or activity is desired, choose a standard solution, pH(PS), to match X as closely as possible in terms of pH, composition and ionic strength.

2) Alternatively, a bracketing procedure may be adopted whereby two standard solutions are chosen whose pH values, pH(S1), pH(S2) are on either side of pH(X). Then if the corresponding potential difference measurements are E(S1), E(S2), E(X), then pH(X) is obtained from

$$\mathrm{pH(X)} = \mathrm{pH(S1)} + [E(\mathrm{X}) - E(\mathrm{S1})]/\%k$$

where $\%k = 100[E(\mathrm{S2}) - E(\mathrm{S1})]/[\mathrm{pH(S2)} - \mathrm{pH(S1)}]$ is the apparent percentage slope. This procedure is very easily done on some pH meters simply by adjusting downwards the slope factor control with the electrodes in S2. The purpose of the bracketing procedure is to compensate for deficiencies in the electrodes and measuring system.

Information to Be Given about the Measurement of pH(X)

The standard solutions selected for calibration of the pH meter system should be reported with the measurement as follows:

System calibrated with pH(S) = ... at ... K.

System calibrated with two primary standards, pH(PS1) = ... and pH(PS2) = ... at .... K.

System calibrated with n standards, pH(S1) = ..., pH(S2) = ... etc. at .... K.

Interpretation of pH(X) in Terms of Hydrogen Ion Concentration

The defined pH has no simple interpretation in terms of hydrogen ion concentration but the mean ionic activity coefficient of a typical 1:1 electrolyte can be used to obtain hydrogen ion concentration subject to an uncertainty of 3.9% in concentration, corresponding to 0.02 in pH.

References

1. Buck, R.P., Rondinini, S., Covington, A.K., Baucke, F.G.K., Brett, C.M.A., Camoes, M.F.C., Milton, M.J.T., Mussini, T., Naumann, R., Pratt, K.W., Spitzer, P., and Wilson, G.S. *Pure Appl. Chem.*, 74, 2105, 2002.
2. Sørensen, S.P.L. *Comp. Rend. Trav. Lab. Carlsberg*, 8, 1, 1909.
3. Sørensen, S.P.L., and Linderstrøm-Lang, K.L., *Comp. Rend. Trav. Lab. Carlsberg*, 15, 1924.
4. BIPM, *Com. Cons. Quantité de Matière* 4, 1998. See also M.J.T. Milton and T.J. Quinn, *Metrologia* 38, 289, 2001.
5. Harned H.S., and Owen, B.B., *The Physical Chemistry of Electrolytic Solutions*, Ch 14, Reinhold, New York, 1958.
6. Bates R.G., and Robinson, R.A., *J. Soln. Chem.* 9, 455, 1980.
7. Bates R.G., and Guggenheim, E.A., *Pure Appl. Chem.* 1, 163, 1960.
8. *International Vocabulary of Basic and General Terms in Metrology* (VIM), Beuth, Berlin, 2nd Edn. 1994.
9. Bates, R.G. *Determination of pH*, Wiley, New York, 1973.
10. Good, N.E. et al., *Biochem. J.* 5, 467, 1966.
11. *Guide to the Expression of Uncertainty* (GUM), BIPM, IEC, IFCC, ISO, IUPAC, IUPAP, OIML, 1993.

TABLE 1. Summary of Useful Properties of Some Primary and Secondary Standard Buffer Substances and Solutions

Salt or solid substance	Formula	Molality/ mol kg^{-1}	Molar mass/ g mol^{-1}	Density/ g/mL	Amount conc. at 20 °C/ mol dm^{-3}	Mass/g to make 1 dm^3	Dilution value $\Delta pH_{1/2}$	Buffer value (β)/ mol OH$^-$ dm^{-3}	pH Temperature coefficient/ K^{-1}
Potassium tetroxalate dihydrate	$KH_3C_4O_8{\cdot}2H_2O$	0.1	254.191	1.0091	0.09875	25.101			
Potassium tetraoxalate dihydrate	$KH_3C_4O_8{\cdot}2H_2O$	0.05	254.191	1.0032	0.04965	12.620	0.186	0.070	0.001
Potassium hydrogen tartrate (sat at 25 °C)	$KHC_4H_4O_6$	0.0341	188.18	1.0036	0.034	6.4	0.049	0.027	–0.0014
Potassium dihydrogen citrate	$KH_2C_6H_5O_7$	0.05	230.22	1.0029	0.04958	11.41	0.024	0.034	–0.022
Potassium hydrogen phthalate	$KHC_8H_4O_4$	0.05	204.44	1.0017	0.04958	10.12	0.052	0.016	0.00012
Disodium hydrogen orthophosphate +	Na_2HPO_4	0.025	141.958	1.0038	0.02492	3.5379	0.080	0.029	–0.0028
potassium dihydrogen orthophosphate	KH_2PO_4	0.025	136.085			3.3912			
Disodium hydrogen orthophosphate +	Na_2HPO_4	0.03043	141.959	1.0020	0.08665	4.302	0.07	0.016	–0.0028
potassium dihydrogen orthophosphate	KH_2PO_4	0.00869	136.085		0.03032	1.179			
Disodium tetraborate decahydrate	$Na_2B_4O_7{\cdot}10H_2O$	0.05	381.367	1.0075	0.04985	19.012			
Disodium tetraborate decahydrate	$Na_2B_4O_7{\cdot}10H_2O$	0.01	381.367	1.0001	0.00998	3.806	0.01	0.020	–0.0082
Sodium hydrogen carbonate +	$NaHCO_3$	0.025	84.01	1.0013	0.02492	2.092	0.079	0.029	–0.0096
sodium carbonate	Na_2CO_3	0.025	105.99			2.640			
Calcium hydroxide (sat. at 25 °C)	$Ca(OH)_2$	0.0203	74.09	0.9991	0.02025	1.5	–0.28	0.09	–0.033

TABLE 2. Typical Values of pH(PS) for Primary Standards at 0–50 °C

	Temperature in °C										
Primary standards (PS)	0	5	10	15	20	25	30	35	37	40	50
Sat. potassium hydrogen tartrate (at 25 °C)						3.557	3.552	3.549	3.548	3.547	3.549
0.05 mol kg^{-1} potassium dihydrogen citrate	3.863	3.840	3.820	3.802	3.788	3.776	3.766	3.759	3.756	3.754	3.749
0.05 mol kg^{-1} potassium hydrogen phthalate	4.000	3.998	3.997	3.998	4.000	4.005	4.011	4.018	4.022	4.027	4.050
0.025 mol kg^{-1} disodium hydrogen phosphate + 0.025 mol kg^{-1} potassium dihydrogen phosphate	6.984	6.951	6.923	6.900	6.881	6.865	6.853	6.844	6.841	6.838	6.833
0.03043 mol kg^{-1} disodium hydrogen phosphate + 0.008695 mol kg^{-1} potassium dihydrogen phosphate	7.534	7.500	7.472	7.448	7.429	7.413	7.400	7.389	7.386	7.380	7.367
0.01 mol kg^{-1} disodium tetraborate	9.464	9.395	9.332	9.276	9.225	9.180	9.139	9.102	9.088	9.068	9.011
0.025 mol kg^{-1} sodium hydrogen carbonate + 0.025 mol kg^{-1} sodium carbonate	10.317	10.245	10.179	10.118	10.062	10.012	9.966	9.926	9.910	9.889	9.828

TABLE 3. Values of pH(SS) of Some Secondary Standards from Harned Cell I Measurements

	Temperature in °C									
Secondary standards	0	5	10	15	20	25	30	37	40	50
0.05 mol kg^{-1} potassium tetroxalate[a]	1.67	1.67	1.67	1.67	1.68	1.68	1.68	1.69	1.69	1.71
0.05 mol kg^{-1} sodium hydrogen diglycolate[b]		3.47	3.47	3.48	3.48	3.49	3.50	3.52	3.53	3.56
0.1 mol dm^{-3} acetic acid + 0.1 mol dm^{-3} sodium acetate	4.68	4.67	4.67	4.66	4.66	4.65	4.65	4.66	4.66	4.68
0.01 mol dm^{-3} acetic acid + 0.1 mol dm^{-3} sodium acetate	4.74	4.73	4.73	4.72	4.72	4.72	4.72	4.73	4.73	4.75
0.02 mol kg^{-1} piperazine phosphate[c]	6.58	6.51	6.45	6.39	6.34	6.29	6.24	6.16	6.14	6.06
0.05 mol kg^{-1} tris hydrochloride + 0.01667 mol kg^{-1} tris[c]	8.47	8.30	8.14	7.99	7.84	7.70	7.56	7.38	7.31	7.07
0.05 mol kg^{-1} disodium tetraborate	9.51	9.43	9.36	9.30	9.25	9.19	9.15	9.09	9.07	9.01
Saturated (at 25 °C) calcium hydroxide	13.42	13.21	13.00	12.81	12.63	12.45	12.29	12.07	11.98	11.71

[a] Potassium trihydrogen dioxalate ($KH_3C_4O_8$)
[b] Sodium hydrogen 2,2 -oxydiacetate
[c] 2-Amino-2-(hydroxymethyl)-1,3 propanediol or tris(hydroxymethyl)aminomethane

BUFFER SOLUTIONS GIVING ROUND VALUES OF pH AT 25 °C

This tables gives a series of buffer solutions A, B, ··· J that produce round values of pH from 1.0 to 13.0. The value of x to be used for each pH value is given. The buffers are:

A: 25 mL of 0.2 molar KCl + x mL of 0.2 molar HCl.

B: 50 mL of 0.1 molar potassium hydrogen phthalate + x mL of 0.1 molar HCl.

C: 50 mL of 0.1 molar potassium hydrogen phthalate + x mL of 0.1 molar NaOH.

D: 50 mL of 0.1 molar potassium dihydrogen phosphate + x mL of 0.1 molar NaOH.

E: 50 mL of 0.1 molar tris(hydroxymethyl)aminomethane + x mL of 0.1 molar HCl.

F: 50 mL of 0.025 molar borax + x mL of 0.1 molar HCl.

G: 50 mL of 0.025 molar borax + x mL of 0.1 molar NaOH.

H: 50 mL of 0.05 molar sodium bicarbonate + x mL of 0.1 molar NaOH.

I: 50 mL of 0.05 molar disodium hydrogen phosphate + x mL of 0.1 molar NaOH.

J: 25 mL of 0.2 molar KCl + x mL of 0.2 molar NaOH.

Final volume of mixtures = 100 mL.

References

1. Bower, V. E., and Bates, R. G., *J. Res. Natl. Bur. Stand.*, 55, 197, 1955 (A–D).
2. Bates, R. G., and Bower, V. E., *Anal. Chem.*, 28, 1322, 1956 (E–J).

Value of x Needed to Produce the Indicated pH for Each Buffer Solution

A	x	B	x	C	x	D	x	E	x	F	x	G	x	H	x	I	x	J	x
1.0	67.0	**2.2**	49.5	**4.1**	1.3	**5.8**	3.6	**7.0**	46.6	**8.0**	20.5	**9.2**	0.9	**9.6**	5.0	**10.9**	3.3	**12.0**	6.0
1.1	52.8	**2.3**	45.8	**4.2**	3.0	**5.9**	4.6	**7.1**	45.7	**8.1**	19.7	**9.3**	3.6	**9.7**	6.2	**11.0**	4.1	**12.1**	8.0
1.2	42.5	**2.4**	42.2	**4.3**	4.7	**6.0**	5.6	**7.2**	44.7	**8.2**	18.8	**9.4**	6.2	**9.8**	7.6	**11.1**	5.1	**12.2**	10.2
1.3	33.6	**2.5**	38.8	**4.4**	6.6	**6.1**	6.8	**7.3**	43.4	**8.3**	17.7	**9.5**	8.8	**9.9**	9.1	**11.2**	6.3	**12.3**	12.8
1.4	26.6	**2.6**	35.4	**4.5**	8.7	**6.2**	8.1	**7.4**	42.0	**8.4**	16.6	**9.6**	11.1	**10.0**	10.7	**11.3**	7.6	**12.4**	16.2
1.5	20.7	**2.7**	32.1	**4.6**	11.1	**6.3**	9.7	**7.5**	40.3	**8.5**	15.2	**9.7**	13.1	**10.1**	12.2	**11.4**	9.1	**12.5**	20.4
1.6	16.2	**2.8**	28.9	**4.7**	13.6	**6.4**	11.6	**7.6**	38.5	**8.6**	13.5	**9.8**	15.0	**10.2**	13.8	**11.5**	11.1	**12.6**	25.6
1.7	13.0	**2.9**	25.7	**4.8**	16.5	**6.5**	13.9	**7.7**	36.6	**8.7**	11.6	**9.9**	16.7	**10.3**	15.2	**11.6**	13.5	**12.7**	32.2
1.8	10.2	**3.0**	22.3	**4.9**	19.4	**6.6**	16.4	**7.8**	34.5	**8.8**	9.6	**10.0**	18.3	**10.4**	16.5	**11.7**	16.2	**12.8**	41.2
1.9	8.1	**3.1**	18.8	**5.0**	22.6	**6.7**	19.3	**7.9**	32.0	**8.9**	7.1	**10.1**	19.5	**10.5**	17.8	**11.8**	19.4	**12.9**	53.0
2.0	6.5	**3.2**	15.7	**5.1**	25.5	**6.8**	22.4	**8.0**	29.2	**9.0**	4.6	**10.2**	20.5	**10.6**	19.1	**11.9**	23.0	**13.0**	66.0
2.1	5.10	**3.3**	12.9	**5.2**	28.8	**6.9**	25.9	**8.1**	26.2	**9.1**	2.0	**10.3**	21.3	**10.7**	20.2	**12.0**	26.9		
2.2	3.9	**3.4**	10.4	**5.3**	31.6	**7.0**	29.1	**8.2**	22.9			**10.4**	22.1	**10.8**	21.2				
		3.5	8.2	**5.4**	34.1	**7.1**	32.1	**8.3**	19.9			**10.5**	22.7	**10.9**	22.0				
		3.6	6.3	**5.5**	36.6	**7.2**	34.7	**8.4**	17.2			**10.6**	23.3	**11.0**	22.7				
		3.7	4.5	**5.6**	38.8	**7.3**	37.0	**8.5**	14.7			**10.7**	23.8						
		3.8	2.9	**5.7**	40.6	**7.4**	39.1	**8.6**	12.2			**10.8**	24.25						
		3.9	1.4	**5.8**	42.3	**7.5**	40.9	**8.7**	10.3										
		4.0	0.1	**5.9**	43.7	**7.6**	42.4	**8.8**	8.5										
						7.7	43.5	**8.9**	7.0										
						7.8	44.5	**9.0**	5.7										
						7.9	45.3												
						8.0	46.1												

CONCENTRATIVE PROPERTIES OF AQUEOUS SOLUTIONS

Density, Refractive Index, Freezing Point Depression, and Viscosity

This table gives properties of aqueous solutions of 66 substances as a function of concentration. All data refer to a temperature of 20 °C. The properties are:

Mass %: Mass of solute divided by total mass of solution, expressed as percent.
m: Molality (moles of solute per kg of water).
c: Molarity (moles of solute per liter of solution).
ρ: Density of solution in g/cm³.
n: Index of refraction, relative to air, at a wavelength of 589 nm (sodium D line); the index of pure water at 20 °C is 1.3330.
Δ: Freezing point depression in °C relative to pure water.
η: Absolute (dynamic) viscosity in mPa s (equal to centipoise, cP); the viscosity of pure water at 20 °C is 1.002 mPa s.

Density data for aqueous solutions over a wider range of temperatures and pressures (and for additional compounds not in this table) may be found in Ref. 2.

References

1. Wolf, A. V., *Aqueous Solutions and Body Fluids*, Hoeber, 1966.
2. Söhnel, O., and Novotny, P., *Densities of Aqueous Solutions of Inorganic Substances*, Elsevier, Amsterdam, 1985.

Concentrative Properties of Aqueous Solutions: Density, Refractive Index, Freezing Point Depression, and Viscosity

Mass %	m/mol kg⁻¹	c/mol L⁻¹	ρ/g cm⁻³	n	Δ/°C	η/mPa s
Acetic acid						
0.5	0.084	0.083	0.9989	1.3334	0.16	1.012
1.0	0.168	0.166	0.9996	1.3337	0.32	1.022
2.0	0.340	0.333	1.0011	1.3345	0.63	1.042
3.0	0.515	0.501	1.0025	1.3352	0.94	1.063
4.0	0.694	0.669	1.0038	1.3359	1.26	1.084
5.0	0.876	0.837	1.0052	1.3366	1.58	1.105
6.0	1.063	1.006	1.0066	1.3373	1.90	1.125
7.0	1.253	1.175	1.0080	1.3381	2.23	1.143
8.0	1.448	1.345	1.0093	1.3388	2.56	1.162
9.0	1.647	1.515	1.0107	1.3395	2.89	1.186
10.0	1.850	1.685	1.0121	1.3402	3.23	1.210
11.0	2.058	1.856	1.0134	1.3409	3.57	1.231
12.0	2.271	2.028	1.0147	1.3416	3.91	1.253
13.0	2.488	2.200	1.0161	1.3423	4.26	1.275
14.0	2.711	2.372	1.0174	1.3430	4.61	1.298
15.0	2.939	2.545	1.0187	1.3437	4.97	1.320
16.0	3.172	2.718	1.0200	1.3444	5.33	1.341
17.0	3.411	2.891	1.0213	1.3451	5.69	1.360
18.0	3.655	3.065	1.0225	1.3458	6.06	1.380
19.0	3.906	3.239	1.0238	1.3465	6.43	1.405
20.0	4.163	3.414	1.0250	1.3472	6.81	1.431
22.0	4.697	3.764	1.0275	1.3485	7.57	1.478
24.0	5.259	4.116	1.0299	1.3498	8.36	1.525
26.0	5.851	4.470	1.0323	1.3512	9.17	1.572
28.0	6.476	4.824	1.0346	1.3525	10.00	1.613
30.0	7.137	5.180	1.0369	1.3537	10.84	1.669
32.0	7.837	5.537	1.0391	1.3550	11.70	1.715
34.0	8.579	5.896	1.0413	1.3562	12.55	1.762
36.0	9.367	6.255	1.0434	1.3574	13.38	1.812
38.0	10.207	6.615	1.0454	1.3586		1.852
40.0	11.102	6.977	1.0474	1.3598		1.912
42.0	12.059	7.339	1.0493	1.3610		1.960
44.0	13.084	7.701	1.0510	1.3621		2.007
46.0	14.186	8.065	1.0528	1.3632		2.052
48.0	15.372	8.429	1.0545	1.3642		2.110
50.0	16.653	8.794	1.0562	1.3653		2.158
52.0	18.041	9.159	1.0577	1.3663		2.212
54.0	19.549	9.525	1.0592	1.3673		2.265
56.0	21.194	9.890	1.0605	1.3682		2.308
58.0	22.997	10.256	1.0618	1.3691		2.360
60.0	24.979	10.620	1.0629	1.3700		2.409
62.0	27.170	10.986	1.0640	1.3708		2.456
64.0	29.605	11.351	1.0650	1.3716		2.502
66.0	32.326	11.715	1.0659	1.3724		2.553
68.0	35.387	12.080	1.0668	1.3732		2.594
70.0	38.857	12.441	1.0673	1.3738		2.629
72.0	42.821	12.801	1.0676	1.3745		2.662
74.0	47.396	13.159	1.0678	1.3751		2.687
76.0	52.734	13.517	1.0680	1.3757		2.714
78.0	59.042	13.874	1.0681	1.3762		2.720
80.0	66.611	14.228	1.0680	1.3767		2.720
82.0	75.863	14.580	1.0677	1.3770		2.696
84.0	87.427	14.930	1.0673	1.3773		2.658
86.0	102.296	15.275	1.0666	1.3774		2.596
88.0	122.120	15.619	1.0658	1.3774		2.511
90.0	149.875	15.953	1.0644	1.3771		2.386
92.0	191.507	16.284	1.0629	1.3766		2.240
94.0	260.894	16.602	1.0606	1.3759		2.036
96.0	399.667	16.911	1.0578	1.3748		1.813
98.0	815.987	17.198	1.0538	1.3734		1.535
100.0		17.447	1.0477	1.3716		1.223
Acetone						
0.5	0.087	0.086	0.9975	1.3334	0.16	1.013
1.0	0.174	0.172	0.9968	1.3337	0.32	1.024
1.5	0.262	0.257	0.9961	1.3341	0.48	1.035
2.0	0.351	0.343	0.9954	1.3344	0.65	1.047
2.5	0.441	0.428	0.9947	1.3348	0.81	1.059
3.0	0.533	0.513	0.9940	1.3352	0.97	1.072
3.5	0.624	0.599	0.9933	1.3355	1.13	1.085
4.0	0.717	0.684	0.9926	1.3359	1.30	1.099
4.5	0.811	0.769	0.9919	1.3363	1.46	1.112
5.0	0.906	0.853	0.9912	1.3366	1.63	1.125
6.0	1.099	1.023	0.9899	1.3373	1.96	1.150
7.0	1.296	1.191	0.9886	1.3381	2.29	1.174
8.0	1.497	1.360	0.9874	1.3388	2.62	1.198
9.0	1.703	1.528	0.9861	1.3395	2.95	1.221
10.0	1.913	1.696	0.9849	1.3402	3.29	1.244

Ammonia

Mass %	m/mol kg^{-1}	c/mol L^{-1}	ρ/g cm^{-3}	n	Δ/°C	η/mPa s
0.5	0.295	0.292	0.9960	1.3332	0.55	1.009
1.0	0.593	0.584	0.9938	1.3335	1.14	1.015
2.0	1.198	1.162	0.9895	1.3339	2.32	1.029
3.0	1.816	1.736	0.9853	1.3344	3.53	1.043
4.0	2.447	2.304	0.9811	1.3349	4.78	1.057
5.0	3.090	2.868	0.9770	1.3354	6.08	1.071
6.0	3.748	3.428	0.9730	1.3359	7.43	1.085
7.0	4.420	3.983	0.9690	1.3365	8.95	1.099
8.0	5.106	4.533	0.9651	1.3370	10.34	1.113
9.0	5.807	5.080	0.9613	1.3376	11.90	1.127
10.0	6.524	5.622	0.9575	1.3381	13.55	1.141
11.0	7.257	6.160	0.9538	1.3387	15.29	1.155
12.0	8.007	6.695	0.9502	1.3393	17.13	1.169
13.0	8.774	7.226	0.9466	1.3398	19.07	1.182
14.0	9.558	7.753	0.9431	1.3404	21.13	1.195
15.0	10.362	8.275	0.9396	1.3410	23.32	1.207
16.0	11.184	8.794	0.9361	1.3416	25.63	1.218
17.0	12.026	9.310	0.9327	1.3422	28.09	1.228
18.0	12.889	9.823	0.9294	1.3428	30.70	1.237
19.0	13.773	10.332	0.9261	1.3434	33.47	1.245
20.0	14.679	10.837	0.9228	1.3440	36.42	1.254
22.0	16.561	11.838	0.9164	1.3453	43.36	1.268
24.0	18.542	12.826	0.9102	1.3465	51.38	1.280
26.0	20.630	13.801	0.9040	1.3477	60.77	1.288
28.0	22.834	14.764	0.8980	1.3490	71.66	
30.0	25.164	15.713	0.8920	1.3502	84.06	

Ammonium chloride

Mass %	m/mol kg^{-1}	c/mol L^{-1}	ρ/g cm^{-3}	n	Δ/°C	η/mPa s
0.5	0.094	0.093	0.9998	1.3340	0.32	0.999
1.0	0.189	0.187	1.0014	1.3349	0.64	0.996
2.0	0.382	0.376	1.0045	1.3369	1.27	0.992
3.0	0.578	0.565	1.0076	1.3388	1.91	0.988
4.0	0.779	0.756	1.0107	1.3407	2.57	0.985
5.0	0.984	0.948	1.0138	1.3426	3.25	0.982
6.0	1.193	1.141	1.0168	1.3445	3.94	0.979
7.0	1.407	1.335	1.0198	1.3464	4.66	0.976
8.0	1.626	1.529	1.0227	1.3483	5.40	0.974
9.0	1.849	1.726	1.0257	1.3502	6.16	0.972
10.0	2.077	1.923	1.0286	1.3521	6.95	0.970
11.0	2.311	2.121	1.0315	1.3540	7.76	0.969
12.0	2.549	2.320	1.0344	1.3559	8.60	0.969
13.0	2.793	2.521	1.0373	1.3578	9.47	0.969
14.0	3.043	2.722	1.0401	1.3596		0.969
15.0	3.299	2.924	1.0429	1.3615		0.970
16.0	3.561	3.128	1.0457	1.3634		0.971
17.0	3.829	3.332	1.0485	1.3652		0.972
18.0	4.104	3.537	1.0512	1.3671		0.973
19.0	4.385	3.744	1.0540	1.3689		0.975
20.0	4.674	3.951	1.0567	1.3708		0.978
22.0	5.273	4.368	1.0621	1.3745		0.986
24.0	5.903	4.789	1.0674	1.3782		0.996

Ammonium sulfate

Mass %	m/mol kg^{-1}	c/mol L^{-1}	ρ/g cm^{-3}	n	Δ/°C	η/mPa s
0.5	0.038	0.038	1.0012	1.3338	0.17	1.008
1.0	0.076	0.076	1.0042	1.3346	0.33	1.014
2.0	0.154	0.153	1.0101	1.3363	0.63	1.027
3.0	0.234	0.231	1.0160	1.3379	0.92	1.041
4.0	0.315	0.309	1.0220	1.3395	1.21	1.057
5.0	0.398	0.389	1.0279	1.3411	1.49	1.073
6.0	0.483	0.469	1.0338	1.3428	1.77	1.090
7.0	0.570	0.551	1.0397	1.3444	2.05	1.108
8.0	0.658	0.633	1.0456	1.3460	2.33	1.127
9.0	0.748	0.716	1.0515	1.3476	2.61	1.147
10.0	0.841	0.800	1.0574	1.3492	2.89	1.168
12.0	1.032	0.971	1.0691	1.3523	3.47	1.210
14.0	1.232	1.145	1.0808	1.3555	4.07	1.256
16.0	1.441	1.323	1.0924	1.3586	4.69	1.305
18.0	1.661	1.504	1.1039	1.3616		1.359
20.0	1.892	1.688	1.1154	1.3647		1.421
22.0	2.134	1.876	1.1269	1.3677		1.490
24.0	2.390	2.067	1.1383	1.3707		1.566
26.0	2.659	2.262	1.1496	1.3737		1.650
28.0	2.943	2.460	1.1609	1.3766		1.743
30.0	3.243	2.661	1.1721	1.3795		1.847
32.0	3.561	2.866	1.1833	1.3824		1.961
34.0	3.898	3.073	1.1945	1.3853		2.086
36.0	4.257	3.284	1.2056	1.3881		2.222
38.0	4.638	3.499	1.2166	1.3909		2.371
40.0	5.045	3.716	1.2277	1.3938		2.530

Barium chloride

Mass %	m/mol kg^{-1}	c/mol L^{-1}	ρ/g cm^{-3}	n	Δ/°C	η/mPa s
0.5	0.024	0.024	1.0026	1.3337	0.12	1.009
1.0	0.049	0.048	1.0070	1.3345	0.23	1.016
2.0	0.098	0.098	1.0159	1.3360	0.46	1.026
3.0	0.149	0.148	1.0249	1.3375	0.69	1.037
4.0	0.200	0.199	1.0341	1.3391	0.93	1.049
5.0	0.253	0.251	1.0434	1.3406	1.18	1.062
6.0	0.307	0.303	1.0528	1.3422	1.44	1.075
7.0	0.361	0.357	1.0624	1.3438	1.70	1.087
8.0	0.418	0.412	1.0721	1.3454	1.98	1.101
9.0	0.475	0.468	1.0820	1.3470	2.27	1.114
10.0	0.534	0.524	1.0921	1.3487	2.58	1.129
12.0	0.655	0.641	1.1128	1.3520	3.22	1.161
14.0	0.782	0.763	1.1342	1.3555	3.92	1.195
16.0	0.915	0.889	1.1564	1.3591	4.69	1.234
18.0	1.054	1.019	1.1793	1.3627		1.277
20.0	1.201	1.156	1.2031	1.3664		1.325
22.0	1.355	1.297	1.2277	1.3703		1.378
24.0	1.517	1.444	1.2531	1.3741		1.437
26.0	1.687	1.597	1.2793	1.3781		1.503

Calcium chloride

Mass %	m/mol kg^{-1}	c/mol L^{-1}	ρ/g cm^{-3}	n	Δ/°C	η/mPa s
0.5	0.045	0.045	1.0024	1.3342	0.22	1.015
1.0	0.091	0.091	1.0065	1.3354	0.44	1.028
2.0	0.184	0.183	1.0148	1.3378	0.88	1.050
3.0	0.279	0.277	1.0232	1.3402	1.33	1.078
4.0	0.375	0.372	1.0316	1.3426	1.82	1.110
5.0	0.474	0.469	1.0401	1.3451	2.35	1.143
6.0	0.575	0.567	1.0486	1.3475	2.93	1.175
7.0	0.678	0.667	1.0572	1.3500	3.57	1.208
8.0	0.784	0.768	1.0659	1.3525	4.28	1.242
9.0	0.891	0.872	1.0747	1.3549	5.04	1.279
10.0	1.001	0.976	1.0835	1.3575	5.86	1.319
11.0	1.114	1.083	1.0923	1.3600	6.74	1.362
12.0	1.229	1.191	1.1014	1.3625	7.70	1.408
13.0	1.346	1.301	1.1105	1.3651	8.72	1.457
14.0	1.467	1.413	1.1198	1.3677	9.83	1.508
15.0	1.590	1.526	1.1292	1.3704	11.01	1.564
16.0	1.716	1.641	1.1386	1.3730	12.28	1.625
17.0	1.846	1.759	1.1482	1.3757	13.65	1.691
18.0	1.978	1.878	1.1579	1.3784	15.11	1.764
19.0	2.114	1.999	1.1677	1.3812	16.70	1.843

Mass %	m/ mol kg⁻¹	c/mol L⁻¹	ρ/g cm⁻³	n	Δ/°C	η/mPa s
20.0	2.253	2.122	1.1775	1.3839	18.30	1.930
22.0	2.541	2.374	1.1976	1.3895	21.70	2.127
24.0	2.845	2.634	1.2180	1.3951	25.30	2.356
26.0	3.166	2.902	1.2388	1.4008	29.70	2.645
28.0	3.504	3.179	1.2600	1.4066	34.70	3.000
30.0	3.862	3.464	1.2816	1.4124	41.00	3.467
32.0	4.240	3.759	1.3036	1.4183	49.70	4.035
34.0	4.642	4.062	1.3260	1.4242		4.820
36.0	5.068	4.375	1.3488	1.4301		5.807
38.0	5.522	4.698	1.3720	1.4361		7.321
40.0	6.007	5.030	1.3957	1.4420		8.997
Cesium chloride						
0.5	0.030	0.030	1.0020	1.3334	0.10	1.000
1.0	0.060	0.060	1.0058	1.3337	0.20	0.997
1.5	0.090	0.090	1.0097	1.3341	0.30	0.994
2.0	0.121	0.120	1.0135	1.3345	0.40	0.992
2.5	0.152	0.151	1.0174	1.3349	0.51	0.990
3.0	0.184	0.182	1.0214	1.3353	0.61	0.988
3.5	0.215	0.213	1.0253	1.3357	0.71	0.986
4.0	0.247	0.245	1.0293	1.3361	0.81	0.984
4.5	0.280	0.276	1.0334	1.3365	0.92	0.982
5.0	0.313	0.308	1.0374	1.3369	1.02	0.980
5.5	0.346	0.340	1.0415	1.3373	1.12	0.978
6.0	0.379	0.373	1.0456	1.3377	1.22	0.977
6.5	0.413	0.405	1.0498	1.3382	1.33	0.975
7.0	0.447	0.438	1.0540	1.3386	1.43	0.974
7.5	0.482	0.471	1.0582	1.3390	1.53	0.973
8.0	0.516	0.505	1.0625	1.3394	1.64	0.971
8.5	0.552	0.539	1.0668	1.3399	1.74	0.970
9.0	0.587	0.573	1.0711	1.3403	1.85	0.969
9.5	0.624	0.607	1.0754	1.3407	1.95	0.968
10.0	0.660	0.641	1.0798	1.3412	2.06	0.966
11.0	0.734	0.711	1.0887	1.3421	2.28	0.963
12.0	0.810	0.782	1.0978	1.3430	2.51	0.961
13.0	0.888	0.855	1.1070	1.3439	2.74	0.958
14.0	0.967	0.928	1.1163	1.3448	2.97	0.955
15.0	1.048	1.003	1.1258	1.3458	3.21	0.953
16.0	1.131	1.079	1.1355	1.3468	3.46	0.950
17.0	1.217	1.156	1.1453	1.3477	3.71	0.947
18.0	1.304	1.235	1.1552	1.3487	3.96	0.945
19.0	1.393	1.315	1.1653	1.3497	4.22	0.942
20.0	1.485	1.397	1.1756	1.3507	4.49	0.939
22.0	1.675	1.564	1.1967	1.3528		0.934
24.0	1.876	1.737	1.2185	1.3550		0.930
26.0	2.087	1.917	1.2411	1.3572		0.926
28.0	2.310	2.103	1.2644	1.3594		0.924
30.0	2.546	2.296	1.2885	1.3617		0.922
32.0	2.795	2.497	1.3135	1.3641		0.922
34.0	3.060	2.705	1.3393	1.3666		0.924
36.0	3.341	2.921	1.3661	1.3691		0.926
38.0	3.640	3.146	1.3938	1.3717		0.930
40.0	3.960	3.380	1.4226	1.3744		0.934
42.0	4.301	3.624	1.4525	1.3771		0.940
44.0	4.667	3.877	1.4835	1.3800		0.947
46.0	5.060	4.142	1.5158	1.3829		0.956
48.0	5.483	4.418	1.5495	1.3860		0.967
50.0	5.940	4.706	1.5846	1.3892		0.981
52.0	6.435	5.007	1.6212	1.3925		1.000
54.0	6.973	5.323	1.6596	1.3960		1.023
56.0	7.560	5.654	1.6999	1.3996		1.050
58.0	8.202	6.002	1.7422	1.4035		1.080
60.0	8.910	6.368	1.7868	1.4076		1.120
62.0	9.691	6.754	1.8340	1.4120		1.172
64.0	10.560	7.163	1.8842	1.4167		1.238
Citric acid						
0.5	0.026	0.026	1.0002	1.3336	0.05	1.013
1.0	0.053	0.052	1.0022	1.3343	0.11	1.024
2.0	0.106	0.105	1.0063	1.3356	0.21	1.048
3.0	0.161	0.158	1.0105	1.3368	0.32	1.073
4.0	0.217	0.211	1.0147	1.3381	0.43	1.098
5.0	0.274	0.265	1.0189	1.3394	0.54	1.125
6.0	0.332	0.320	1.0232	1.3407	0.65	1.153
7.0	0.392	0.374	1.0274	1.3420	0.76	1.183
8.0	0.453	0.430	1.0316	1.3433	0.88	1.214
9.0	0.515	0.485	1.0359	1.3446	1.00	1.247
10.0	0.578	0.541	1.0402	1.3459	1.12	1.283
12.0	0.710	0.655	1.0490	1.3486	1.38	1.357
14.0	0.847	0.771	1.0580	1.3514	1.66	1.436
16.0	0.991	0.889	1.0672	1.3541	1.95	1.525
18.0	1.143	1.008	1.0764	1.3569	2.26	1.625
20.0	1.301	1.130	1.0858	1.3598	2.57	1.740
22.0	1.468	1.254	1.0953	1.3626	2.88	1.872
24.0	1.644	1.380	1.1049	1.3655	3.21	2.017
26.0	1.829	1.508	1.1147	1.3684	3.55	2.178
28.0	2.024	1.639	1.1246	1.3714	3.89	2.356
30.0	2.231	1.772	1.1346	1.3744	4.25	2.549
Copper(II) sulfate						
0.5	0.031	0.031	1.0033	1.3339	0.08	1.017
1.0	0.063	0.063	1.0085	1.3348	0.14	1.036
2.0	0.128	0.128	1.0190	1.3367	0.26	1.084
3.0	0.194	0.194	1.0296	1.3386	0.37	1.129
4.0	0.261	0.261	1.0403	1.3405	0.48	1.173
5.0	0.330	0.329	1.0511	1.3424	0.59	1.221
6.0	0.400	0.399	1.0620	1.3443	0.70	1.276
7.0	0.472	0.471	1.0730	1.3462	0.82	1.336
8.0	0.545	0.543	1.0842	1.3481	0.93	1.400
9.0	0.620	0.618	1.0955	1.3501	1.05	1.469
10.0	0.696	0.694	1.1070	1.3520	1.18	1.543
11.0	0.774	0.771	1.1186	1.3540	1.31	1.620
12.0	0.854	0.850	1.1304	1.3560	1.45	1.701
13.0	0.936	0.930	1.1424	1.3581	1.60	1.790
14.0	1.020	1.013	1.1545	1.3601	1.75	1.889
15.0	1.106	1.097	1.1669	1.3622		2.004
16.0	1.193	1.182	1.1796	1.3644		2.136
17.0	1.283	1.270	1.1926	1.3666		2.285
18.0	1.375	1.360	1.2059	1.3689		2.449
1,2-Ethanediol						
0.5	0.081	0.080	0.9988	1.3335	0.15	1.010
1.0	0.163	0.161	0.9995	1.3339	0.30	1.020
2.0	0.329	0.322	1.0007	1.3348	0.61	1.048
3.0	0.498	0.484	1.0019	1.3358	0.92	1.074
4.0	0.671	0.646	1.0032	1.3367	1.24	1.099
5.0	0.848	0.809	1.0044	1.3377	1.58	1.125
6.0	1.028	0.972	1.0057	1.3386	1.91	1.153
7.0	1.213	1.136	1.0070	1.3396	2.26	1.182
8.0	1.401	1.299	1.0082	1.3405	2.62	1.212
9.0	1.593	1.464	1.0095	1.3415	2.99	1.243
10.0	1.790	1.628	1.0108	1.3425	3.37	1.277
12.0	2.197	1.959	1.0134	1.3444	4.16	1.348

Mass %	m/mol kg⁻¹	c/mol L⁻¹	ρ/g cm⁻³	n	Δ/°C	η/mPa s
14.0	2.623	2.292	1.0161	1.3464	5.01	1.424
16.0	3.069	2.626	1.0188	1.3484	5.91	1.500
18.0	3.537	2.962	1.0214	1.3503	6.89	1.578
20.0	4.028	3.300	1.0241	1.3523	7.93	1.661
24.0	5.088	3.981	1.0296	1.3564	10.28	1.843
28.0	6.265	4.669	1.0350	1.3605	13.03	2.047
32.0	7.582	5.364	1.0405	1.3646	16.23	2.280
36.0	9.062	6.067	1.0460	1.3687	19.82	2.537
40.0	10.741	6.776	1.0514	1.3728	23.84	2.832
44.0	12.659	7.491	1.0567	1.3769	28.32	3.166
48.0	14.872	8.212	1.0619	1.3811	33.30	3.544
52.0	17.453	8.939	1.0670	1.3851	38.81	3.981
56.0	20.505	9.671	1.0719	1.3892	44.83	4.475
60.0	24.166	10.406	1.0765	1.3931	51.23	5.026

Ethanol

Mass %	m/mol kg⁻¹	c/mol L⁻¹	ρ/g cm⁻³	n	Δ/°C	η/mPa s
0.5	0.109	0.108	0.9973	1.3333	0.20	1.023
1.0	0.219	0.216	0.9963	1.3336	0.40	1.046
2.0	0.443	0.432	0.9945	1.3342	0.81	1.095
3.0	0.671	0.646	0.9927	1.3348	1.23	1.140
4.0	0.904	0.860	0.9910	1.3354	1.65	1.183
5.0	1.142	1.074	0.9893	1.3360	2.09	1.228
6.0	1.385	1.286	0.9878	1.3367	2.54	1.279
7.0	1.634	1.498	0.9862	1.3374	2.99	1.331
8.0	1.887	1.710	0.9847	1.3381	3.47	1.385
9.0	2.147	1.921	0.9833	1.3388	3.96	1.442
10.0	2.412	2.131	0.9819	1.3395	4.47	1.501
11.0	2.683	2.341	0.9805	1.3403	5.00	1.563
12.0	2.960	2.551	0.9792	1.3410	5.56	1.627
13.0	3.243	2.759	0.9778	1.3417	6.13	1.694
14.0	3.534	2.967	0.9765	1.3425	6.73	1.761
15.0	3.830	3.175	0.9752	1.3432	7.36	1.826
16.0	4.134	3.382	0.9739	1.3440	8.01	1.890
17.0	4.446	3.589	0.9726	1.3447	8.69	1.955
18.0	4.765	3.795	0.9713	1.3455	9.40	2.019
19.0	5.092	4.000	0.9700	1.3462	10.14	2.081
20.0	5.427	4.205	0.9687	1.3469	10.92	2.142
22.0	6.122	4.613	0.9660	1.3484	12.60	2.259
24.0	6.855	5.018	0.9632	1.3498	14.47	2.370
26.0	7.626	5.419	0.9602	1.3511	16.41	2.476
28.0	8.441	5.817	0.9571	1.3524	18.43	2.581
30.0	9.303	6.212	0.9539	1.3535	20.47	2.667
32.0	10.215	6.601	0.9504	1.3546	22.44	2.726
34.0	11.182	6.987	0.9468	1.3557	24.27	2.768
36.0	12.210	7.370	0.9431	1.3566	25.98	2.803
38.0	13.304	7.747	0.9392	1.3575	27.62	2.829
40.0	14.471	8.120	0.9352	1.3583	29.26	2.846
42.0	15.718	8.488	0.9311	1.3590	30.98	2.852
44.0	17.055	8.853	0.9269	1.3598	32.68	2.850
46.0	18.490	9.213	0.9227	1.3604	34.36	2.843
48.0	20.036	9.568	0.9183	1.3610	36.04	2.832
50.0	21.706	9.919	0.9139	1.3616	37.67	2.813
52.0	23.515	10.266	0.9095	1.3621	39.20	2.789
54.0	25.481	10.607	0.9049	1.3626	40.65	2.754
56.0	27.626	10.945	0.9004	1.3630	42.06	2.701
58.0	29.975	11.278	0.8958	1.3634	43.49	2.632
60.0	32.559	11.605	0.8911	1.3638	44.93	2.547
62.0	35.415	11.930	0.8865	1.3641	46.28	2.479
64.0	38.589	12.250	0.8818	1.3644	47.52	2.415
66.0	42.135	12.565	0.8771	1.3647	48.64	2.347
68.0	46.125	12.877	0.8724	1.3650	49.52	2.281
70.0	50.648	13.183	0.8676	1.3652		2.214
72.0	55.816	13.486	0.8629	1.3654		2.148
74.0	61.779	13.783	0.8581	1.3655		2.082
76.0	68.736	14.077	0.8533	1.3657		2.015
78.0	76.958	14.366	0.8485	1.3657		1.948
80.0	86.824	14.649	0.8436	1.3658		1.881
82.0	98.883	14.928	0.8387	1.3657		1.808
84.0	113.957	15.197	0.8335	1.3656		1.741
86.0	133.337	15.464	0.8284	1.3655		1.674
88.0	159.178	15.724	0.8232	1.3653		1.606
90.0	195.355	15.980	0.8180	1.3650		1.542
92.0	249.620	16.225	0.8125	1.3646		1.475
94.0	340.062	16.466	0.8070	1.3642		1.407
96.0	520.946	16.697	0.8013	1.3636		1.342
98.0		16.920	0.7954	1.3630		1.273
100.0		17.133	0.7893	1.3614		1.203

Ethylenediaminetetraacetic acid, disodium salt, dihydrate

Mass %	m/mol kg⁻¹	c/mol L⁻¹	ρ/g cm⁻³	n	Δ/°C	η/mPa s
0.5	0.015	0.015	1.0009	1.3339	0.07	1.017
1.0	0.030	0.030	1.0036	1.3348	0.14	1.032
1.5	0.045	0.045	1.0062	1.3356	0.21	1.046
2.0	0.061	0.060	1.0089	1.3365	0.27	1.062
2.5	0.076	0.075	1.0115	1.3374	0.33	1.077
3.0	0.092	0.090	1.0142	1.3383	0.40	1.093
3.5	0.108	0.106	1.0169	1.3392	0.46	1.109
4.0	0.124	0.121	1.0196	1.3400	0.52	1.125
4.5	0.140	0.137	1.0223	1.3409	0.58	1.142
5.0	0.157	0.152	1.0250	1.3418	0.65	1.160
5.5	0.173	0.168	1.0277	1.3427	0.71	1.178
6.0	0.190	0.184	1.0305	1.3436	0.77	1.197

Formic acid

Mass %	m/mol kg⁻¹	c/mol L⁻¹	ρ/g cm⁻³	n	Δ/°C	η/mPa s
0.5	0.109	0.109	0.9994	1.3333	0.21	1.006
1.0	0.219	0.217	1.0006	1.3336	0.42	1.011
2.0	0.443	0.436	1.0029	1.3342	0.82	1.017
3.0	0.672	0.655	1.0053	1.3348	1.24	1.195
4.0	0.905	0.876	1.0077	1.3354	1.67	1.032
5.0	1.143	1.097	1.0102	1.3359	2.10	1.039
6.0	1.387	1.320	1.0126	1.3365	2.53	1.046
7.0	1.635	1.544	1.0150	1.3371	2.97	1.052
8.0	1.889	1.768	1.0175	1.3376	3.40	1.058
9.0	2.149	1.994	1.0199	1.3382	3.84	1.064
10.0	2.414	2.221	1.0224	1.3387	4.27	1.070
12.0	2.962	2.678	1.0273	1.3397	5.19	1.082
14.0	3.537	3.139	1.0322	1.3408	6.11	1.094
16.0	4.138	3.605	1.0371	1.3418	7.06	1.106
18.0	4.769	4.074	1.0419	1.3428	8.08	1.119
20.0	5.431	4.548	1.0467	1.3437	9.11	1.132
24.0	6.861	5.507	1.0562	1.3456	11.10	1.156
28.0	8.449	6.481	1.0654	1.3475	13.10	1.179
32.0	10.224	7.471	1.0746	1.3493	15.28	1.203
36.0	12.220	8.477	1.0839	1.3511	17.65	1.227
40.0	14.483	9.502	1.0935	1.3529	20.18	1.254
44.0	17.070	10.529	1.1015	1.3547	22.93	1.281
48.0	20.054	11.572	1.1097	1.3565	26.06	1.309
52.0	23.535	12.633	1.1183	1.3581	29.69	1.340
56.0	27.650	13.715	1.1273	1.3597	33.81	1.374
60.0	32.587	14.813	1.1364	1.3612	38.26	1.410
64.0	38.622	15.928	1.1456	1.3626	43.02	1.449
68.0	46.166	17.054	1.1544	1.3641		1.490
70.0	50.692	17.619	1.1586	1.3648		1.511

β-D-Fructose

Mass %	m/mol kg^{-1}	c/mol L^{-1}	ρ/g cm^{-3}	n	Δ/°C	η/mPa s
0.5	0.028	0.028	1.0002	1.3337	0.05	1.015
1.0	0.056	0.056	1.0021	1.3344	0.10	1.028
2.0	0.113	0.112	1.0061	1.3358	0.21	1.054
3.0	0.172	0.168	1.0101	1.3373	0.32	1.080
4.0	0.231	0.225	1.0140	1.3387	0.43	1.106
5.0	0.292	0.283	1.0181	1.3402	0.54	1.134
6.0	0.354	0.340	1.0221	1.3417	0.66	1.165
7.0	0.418	0.399	1.0262	1.3431	0.78	1.198
8.0	0.483	0.458	1.0303	1.3446	0.90	1.232
9.0	0.549	0.517	1.0344	1.3461	1.03	1.270
10.0	0.617	0.576	1.0385	1.3476	1.16	1.309
11.0	0.686	0.637	1.0427	1.3492	1.29	1.349
12.0	0.757	0.697	1.0469	1.3507	1.43	1.391
13.0	0.829	0.759	1.0512	1.3522	1.57	1.435
14.0	0.904	0.820	1.0554	1.3538	1.71	1.483
15.0	0.980	0.882	1.0597	1.3554	1.86	1.533
16.0	1.057	0.945	1.0640	1.3569	2.01	1.587
17.0	1.137	1.008	1.0684	1.3585	2.16	1.643
18.0	1.218	1.072	1.0728	1.3601	2.32	1.703
19.0	1.302	1.136	1.0772	1.3618	2.48	1.768
20.0	1.388	1.201	1.0816	1.3634	2.64	1.837
22.0	1.566	1.332	1.0906	1.3667	3.05	1.986
24.0	1.753	1.465	1.0996	1.3700	3.43	2.154
26.0	1.950	1.600	1.1089	1.3734	3.82	2.348
28.0	2.159	1.738	1.1182	1.3768	4.20	2.562
30.0	2.379	1.878	1.1276	1.3803		2.817
32.0	2.612	2.020	1.1372	1.3839		3.112
34.0	2.859	2.164	1.1469	1.3874		3.462
36.0	3.122	2.312	1.1568	1.3911		3.899
38.0	3.402	2.461	1.1668	1.3948		4.418
40.0	3.700	2.613	1.1769	1.3985		5.046
42.0	4.019	2.767	1.1871	1.4023		5.773
44.0	4.361	2.925	1.1975	1.4062		6.644
46.0	4.728	3.084	1.2080	1.4101		7.753
48.0	5.124	3.247	1.2187	1.4141		9.060

α-D-Glucose

Mass %	m/mol kg^{-1}	c/mol L^{-1}	ρ/g cm^{-3}	n	Δ/°C	η/mPa s
0.5	0.028	0.028	1.0001	1.3337	0.05	1.010
1.0	0.056	0.056	1.0020	1.3344	0.11	1.021
2.0	0.113	0.112	1.0058	1.3358	0.21	1.052
3.0	0.172	0.168	1.0097	1.3373	0.32	1.083
4.0	0.231	0.225	1.0136	1.3387	0.43	1.113
5.0	0.292	0.282	1.0175	1.3402	0.55	1.145
6.0	0.354	0.340	1.0214	1.3417	0.67	1.179
7.0	0.418	0.398	1.0254	1.3432	0.79	1.214
8.0	0.483	0.457	1.0294	1.3447	0.91	1.250
9.0	0.549	0.516	1.0334	1.3462	1.04	1.289
10.0	0.617	0.576	1.0375	1.3477	1.17	1.330
11.0	0.686	0.636	1.0416	1.3492	1.30	1.372
12.0	0.757	0.697	1.0457	1.3508	1.44	1.416
13.0	0.829	0.758	1.0498	1.3523	1.59	1.462
14.0	0.904	0.819	1.0540	1.3539	1.73	1.512
15.0	0.980	0.881	1.0582	1.3555	1.88	1.566
16.0	1.057	0.944	1.0624	1.3571	2.03	1.625
17.0	1.137	1.007	1.0667	1.3587	2.19	1.688
18.0	1.218	1.070	1.0710	1.3603	2.35	1.757
19.0	1.302	1.134	1.0753	1.3619	2.52	1.829
20.0	1.388	1.199	1.0797	1.3635	2.70	1.904
22.0	1.566	1.329	1.0884	1.3668	3.07	2.063
24.0	1.753	1.462	1.0973	1.3702	3.48	2.242
26.0	1.950	1.597	1.1063	1.3736	3.90	2.458
28.0	2.159	1.734	1.1154	1.3770	4.34	2.707
30.0	2.379	1.873	1.1246	1.3805	4.79	2.998
32.0	2.612	2.014	1.1340	1.3840		3.324
34.0	2.859	2.158	1.1434	1.3876		3.704
36.0	3.122	2.304	1.1529	1.3912		4.193
38.0	3.402	2.452	1.1626	1.3949		4.786
40.0	3.700	2.603	1.1724	1.3986		5.493
42.0	4.019	2.756	1.1823	1.4024		6.288
44.0	4.361	2.912	1.1924	1.4062		7.235
46.0	4.728	3.071	1.2026	1.4101		8.454
48.0	5.124	3.232	1.2130	1.4141		9.883
50.0	5.551	3.396	1.2235	1.4181		11.884
52.0	6.013	3.562	1.2342	1.4222		14.489
54.0	6.516	3.732	1.2451	1.4263		17.916
56.0	7.064	3.905	1.2562	1.4306		22.886
58.0	7.665	4.081	1.2676	1.4349		29.389
60.0	8.326	4.261	1.2793	1.4394		37.445

Glycerol

Mass %	m/mol kg^{-1}	c/mol L^{-1}	ρ/g cm^{-3}	n	Δ/°C	η/mPa s
0.5	0.055	0.054	0.9994	1.3336	0.07	1.022
1.0	0.110	0.109	1.0005	1.3342	0.18	1.034
2.0	0.222	0.218	1.0028	1.3353	0.41	1.060
3.0	0.336	0.327	1.0051	1.3365	0.63	1.088
4.0	0.452	0.438	1.0074	1.3376	0.85	1.116
5.0	0.572	0.548	1.0097	1.3388	1.08	1.145
6.0	0.693	0.659	1.0120	1.3400	1.32	1.176
7.0	0.817	0.771	1.0144	1.3412	1.56	1.207
8.0	0.944	0.883	1.0167	1.3424	1.81	1.240
9.0	1.074	0.996	1.0191	1.3436	2.06	1.275
10.0	1.207	1.109	1.0215	1.3448	2.32	1.310
12.0	1.481	1.337	1.0262	1.3472	2.88	1.386
14.0	1.768	1.568	1.0311	1.3496	3.47	1.469
16.0	2.068	1.800	1.0360	1.3521	4.09	1.560
18.0	2.384	2.035	1.0409	1.3547	4.76	1.658
20.0	2.715	2.271	1.0459	1.3572	5.46	1.766
24.0	3.429	2.752	1.0561	1.3624	7.01	2.01
28.0	4.223	3.242	1.0664	1.3676	8.77	2.32
32.0	5.110	3.742	1.0770	1.3730	10.74	2.69
36.0	6.108	4.252	1.0876	1.3785	12.96	3.15
40.0	7.239	4.771	1.0984	1.3841	15.50	3.73
44.0	8.532	5.300	1.1092	1.3897		4.48
48.0	10.024	5.838	1.1200	1.3954		5.45
52.0	11.764	6.385	1.1308	1.4011		6.73
56.0	13.820	6.944	1.1419	1.4069		8.47
60.0	16.288	7.512	1.1530	1.4129		10.9
64.0	19.305	8.092	1.1643	1.4189		14.3
68.0	23.075	8.680	1.1755	1.4249		19.4
72.0	27.923	9.277	1.1866	1.4310		27.2
76.0	34.387	9.884	1.1976	1.4370		39.6
80.0	43.436	10.498	1.2085	1.4431		60.6
84.0	57.009	11.121	1.2192	1.4492		98
88.0	79.632	11.753	1.2299	1.4553		170
92.0	124.878	12.392	1.2404	1.4613		319
96.0	260.615	13.039	1.2508	1.4674		648
100.0		13.694	1.2611	1.4735		1460

Hydrogen chloride

Mass %	m/mol kg^{-1}	c/mol L^{-1}	ρ/g cm^{-3}	n	Δ/°C	η/mPa s
0.5	0.138	0.137	1.0007	1.3341	0.49	1.008
1.0	0.277	0.275	1.0031	1.3353	0.99	1.015
2.0	0.560	0.553	1.0081	1.3376	2.08	1.029

Mass %	m/ mol kg⁻¹	c/mol L⁻¹	ρ/g cm⁻³	n	Δ/°C	η/mPa s
3.0	0.848	0.833	1.0130	1.3399	3.28	1.044
4.0	1.143	1.117	1.0179	1.3422	4.58	1.059
5.0	1.444	1.403	1.0228	1.3445	5.98	1.075
6.0	1.751	1.691	1.0278	1.3468	7.52	1.091
7.0	2.064	1.983	1.0327	1.3491	9.22	1.108
8.0	2.385	2.277	1.0377	1.3515	11.10	1.125
9.0	2.713	2.574	1.0426	1.3538	13.15	1.143
10.0	3.047	2.873	1.0476	1.3561	15.40	1.161
11.0	3.390	3.176	1.0526	1.3584	17.85	1.180
12.0	3.740	3.481	1.0576	1.3607	20.51	1.199
13.0	4.098	3.789	1.0626	1.3630		1.219
14.0	4.465	4.099	1.0676	1.3653		1.239
15.0	4.840	4.413	1.0726	1.3676		1.261
16.0	5.224	4.729	1.0777	1.3700		1.282
17.0	5.617	5.049	1.0828	1.3723		1.304
18.0	6.020	5.370	1.0878	1.3746		1.326
19.0	6.433	5.695	1.0929	1.3769		1.350
20.0	6.857	6.023	1.0980	1.3792		1.374
22.0	7.736	6.687	1.1083	1.3838		1.426
24.0	8.661	7.362	1.1185	1.3884		1.483
26.0	9.636	8.049	1.1288	1.3930		1.547
28.0	10.666	8.748	1.1391	1.3976		1.620
30.0	11.754	9.456	1.1492	1.4020		1.705
32.0	12.907	10.175	1.1594	1.4066		1.799
34.0	14.129	10.904	1.1693	1.4112		1.900
36.0	15.427	11.642	1.1791	1.4158		2.002
38.0	16.810	12.388	1.1886	1.4204		2.105
40.0	18.284	13.140	1.1977	1.4250		

Iron(III) chloride

Mass %	m/ mol kg⁻¹	c/mol L⁻¹	ρ/g cm⁻³	n	Δ/°C	η/mPa s
0.5	0.031	0.031	1.0025	1.3344	0.21	1.024
1.0	0.062	0.062	1.0068	1.3358	0.39	1.047
2.0	0.126	0.125	1.0153	1.3386	0.75	1.093
3.0	0.191	0.189	1.0238	1.3413	1.15	1.139
4.0	0.257	0.255	1.0323	1.3441	1.56	1.187
5.0	0.324	0.321	1.0408	1.3468	2.00	1.238
6.0	0.394	0.388	1.0493	1.3496	2.48	1.292
7.0	0.464	0.457	1.0580	1.3524	2.99	1.350
8.0	0.536	0.526	1.0668	1.3552	3.57	1.412
9.0	0.610	0.597	1.0760	1.3581	4.19	1.480
10.0	0.685	0.669	1.0853	1.3611	4.85	1.553
12.0	0.841	0.817	1.1040	1.3670	6.38	1.707
14.0	1.004	0.969	1.1228	1.3730	8.22	1.879
16.0	1.174	1.126	1.1420		10.45	2.080
18.0	1.353	1.289	1.1615		13.08	2.311
20.0	1.541	1.457	1.1816		16.14	2.570
24.0	1.947	1.810	1.2234		23.79	3.178
28.0	2.398	2.189	1.2679		33.61	4.038
32.0	2.901	2.595	1.3153		49.16	5.274
36.0	3.468	3.030	1.3654			7.130
40.0	4.110	3.496	1.4176			9.674

D-Lactic acid

Mass %	m/ mol kg⁻¹	c/mol L⁻¹	ρ/g cm⁻³	n	Δ/°C	η/mPa s
0.5	0.056	0.055	0.9992	1.3335	0.10	1.014
1.0	0.112	0.111	1.0002	1.3340	0.19	1.027
2.0	0.227	0.223	1.0023	1.3350	0.38	1.056
3.0	0.343	0.334	1.0043	1.3360	0.57	1.084
4.0	0.463	0.447	1.0065	1.3370	0.76	1.110
5.0	0.584	0.560	1.0086	1.3380	0.95	1.138
6.0	0.709	0.673	1.0108	1.3390	1.16	1.167
7.0	0.836	0.787	1.0131	1.3400	1.36	1.198
8.0	0.965	0.902	1.0153	1.3410	1.57	1.229
9.0	1.098	1.017	1.0176	1.3420	1.79	1.262
10.0	1.233	1.132	1.0199	1.3430	2.02	1.296
12.0	1.514	1.365	1.0246	1.3450	2.49	1.366
14.0	1.807	1.600	1.0294	1.3470	2.99	1.441
16.0	2.115	1.837	1.0342	1.3491	3.48	1.522
18.0	2.437	2.076	1.0390	1.3511	3.96	1.607
20.0	2.775	2.318	1.0439	1.3532	4.44	1.699
24.0	3.506	2.807	1.0536	1.3573		1.902
28.0	4.317	3.305	1.0632	1.3615		2.136
32.0	5.224	3.811	1.0728	1.3657		2.414
36.0	6.244	4.325	1.0822	1.3700		2.730
40.0	7.401	4.847	1.0915	1.3743		3.114
44.0	8.722	5.377	1.1008	1.3786		3.566
48.0	10.247	5.917	1.1105	1.3828		4.106
52.0	12.026	6.466	1.1201	1.3871		4.789
56.0	14.129	7.023	1.1297	1.3914		5.579
60.0	16.652	7.588	1.1392	1.3958		6.679
64.0	19.736	8.161	1.1486	1.4001		8.024
68.0	23.590	8.741	1.1579	1.4045		9.863
72.0	28.546	9.328	1.1670	1.4088		12.866
76.0	35.154	9.922	1.1760	1.4131		16.974
80.0	44.405	10.522	1.1848	1.4173		22.164

β-D-Lactose

Mass %	m/ mol kg⁻¹	c/mol L⁻¹	ρ/g cm⁻³	n	Δ/°C	η/mPa s
0.5	0.015	0.015	1.0002	1.3337	0.03	1.013
1.0	0.030	0.029	1.0021	1.3345	0.06	1.026
1.5	0.044	0.044	1.0041	1.3352	0.08	1.041
2.0	0.060	0.059	1.0061	1.3359	0.11	1.058
2.5	0.075	0.074	1.0081	1.3367	0.14	1.074
3.0	0.090	0.089	1.0102	1.3375	0.17	1.089
3.5	0.106	0.103	1.0122	1.3382	0.20	1.105
4.0	0.122	0.119	1.0143	1.3390	0.23	1.120
4.5	0.138	0.134	1.0163	1.3398	0.26	1.137
5.0	0.154	0.149	1.0184	1.3406	0.29	1.154
5.5	0.170	0.164	1.0204	1.3413	0.32	1.172
6.0	0.186	0.179	1.0225	1.3421	0.35	1.191
6.5	0.203	0.195	1.0246	1.3429	0.39	1.211
7.0	0.220	0.210	1.0267	1.3437	0.42	1.232
7.5	0.237	0.225	1.0287	1.3445	0.46	1.254
8.0	0.254	0.241	1.0308	1.3453	0.50	1.276
8.5	0.271	0.256	1.0329	1.3460		1.298
9.0	0.289	0.272	1.0349	1.3468		1.321
9.5	0.307	0.288	1.0370	1.3476		1.345
10.0	0.325	0.304	1.0390	1.3484		1.370
11.0	0.361	0.335	1.0432	1.3500		1.421
12.0	0.398	0.367	1.0473	1.3515		1.476
13.0	0.437	0.399	1.0515	1.3531		1.533
14.0	0.476	0.432	1.0558	1.3548		1.593
15.0	0.516	0.465	1.0602	1.3564		1.657
16.0	0.556	0.498	1.0648	1.3582		1.724
17.0	0.598	0.531	1.0696	1.3600		1.795
18.0	0.641	0.565	1.0746	1.3619		1.869

Lithiun chloride

Mass %	m/ mol kg⁻¹	c/mol L⁻¹	ρ/g cm⁻³	n	Δ/°C	η/mPa s
0.5	0.119	0.118	1.0012	1.3341	0.42	1.019
1.0	0.238	0.237	1.0041	1.3351	0.84	1.037
2.0	0.481	0.476	1.0099	1.3373	1.72	1.072
3.0	0.730	0.719	1.0157	1.3394	2.68	1.108
4.0	0.983	0.964	1.0215	1.3415	3.73	1.146
5.0	1.241	1.211	1.0272	1.3436	4.86	1.185

Mass %	m/mol kg^{-1}	c/mol L^{-1}	ρ/g cm^{-3}	n	Δ/°C	η/mPa s
6.0	1.506	1.462	1.0330	1.3457	6.14	1.226
7.0	1.775	1.715	1.0387	1.3478	7.56	1.269
8.0	2.051	1.971	1.0444	1.3499	9.11	1.313
9.0	2.333	2.230	1.0502	1.3520	10.79	1.360
10.0	2.621	2.491	1.0560	1.3541	12.61	1.411
12.0	3.217	3.022	1.0675	1.3583	16.59	1.522
14.0	3.840	3.564	1.0792	1.3625	21.04	1.647
16.0	4.493	4.118	1.0910	1.3668		1.787
18.0	5.178	4.683	1.1029	1.3711		1.942
20.0	5.897	5.260	1.1150	1.3755		2.128
22.0	6.653	5.851	1.1274	1.3799		2.341
24.0	7.449	6.453	1.1399	1.3844		2.600
26.0	8.288	7.069	1.1527	1.3890		2.925
28.0	9.173	7.700	1.1658	1.3936		3.318
30.0	10.109	8.344	1.1791	1.3983		3.785

Magnesium chloride

Mass %	m/mol kg^{-1}	c/mol L^{-1}	ρ/g cm^{-3}	n	Δ/°C	η/mPa s
0.5	0.053	0.053	1.0022	1.3343	0.26	1.024
1.0	0.106	0.106	1.0062	1.3356	0.52	1.046
2.0	0.214	0.213	1.0144	1.3381	1.06	1.091
3.0	0.325	0.322	1.0226	1.3406	1.65	1.139
4.0	0.438	0.433	1.0309	1.3432	2.30	1.188
5.0	0.553	0.546	1.0394	1.3457	3.01	1.241
6.0	0.670	0.660	1.0479	1.3483		1.298
7.0	0.791	0.777	1.0564	1.3508		1.358
8.0	0.913	0.895	1.0651	1.3534		1.423
9.0	1.039	1.015	1.0738	1.3560		1.493
10.0	1.167	1.137	1.0826	1.3587		1.570
12.0	1.432	1.387	1.1005	1.3641		1.745
14.0	1.710	1.645	1.1189	1.3695		1.956
16.0	2.001	1.911	1.1372	1.3749		2.207
18.0	2.306	2.184	1.1553	1.3804		2.507
20.0	2.626	2.467	1.1742	1.3859		2.867
22.0	2.962	2.758	1.1938	1.3915		3.323
24.0	3.317	3.060	1.2140	1.3972		3.917
26.0	3.690	3.371	1.2346	1.4030		4.694
28.0	4.085	3.692	1.2555	1.4089		5.709
30.0	4.501	4.022	1.2763	1.4148		7.017

Magnesium sulfate

Mass %	m/mol kg^{-1}	c/mol L^{-1}	ρ/g cm^{-3}	n	Δ/°C	η/mPa s
0.5	0.042	0.042	1.0033	1.3340	0.10	1.027
1.0	0.084	0.084	1.0084	1.3350	0.19	1.054
2.0	0.170	0.169	1.0186	1.3371	0.36	1.112
3.0	0.257	0.256	1.0289	1.3391	0.52	1.177
4.0	0.346	0.345	1.0392	1.3411	0.69	1.249
5.0	0.437	0.436	1.0497	1.3431	0.87	1.328
6.0	0.530	0.528	1.0602	1.3451	1.05	1.411
7.0	0.625	0.623	1.0708	1.3471	1.24	1.498
8.0	0.722	0.719	1.0816	1.3492	1.43	1.593
9.0	0.822	0.817	1.0924	1.3512	1.64	1.702
10.0	0.923	0.917	1.1034	1.3532	1.85	1.829
12.0	1.133	1.122	1.1257	1.3572	2.31	2.104
14.0	1.352	1.336	1.1484	1.3613	2.86	2.412
16.0	1.582	1.557	1.1717	1.3654	3.67	2.809
18.0	1.824	1.788	1.1955	1.3694		3.360
20.0	2.077	2.027	1.2198	1.3735		4.147
22.0	2.343	2.275	1.2447	1.3776		5.199
24.0	2.624	2.532	1.2701	1.3817		6.498
26.0	2.919	2.800	1.2961	1.3858		8.066

α-Maltose

Mass %	m/mol kg^{-1}	c/mol L^{-1}	ρ/g cm^{-3}	n	Δ/°C	η/mPa s
0.5	0.015	0.015	1.0003	1.3337	0.03	1.016
1.0	0.030	0.029	1.0023	1.3345	0.06	1.030
2.0	0.060	0.059	1.0063	1.3359	0.11	1.060
3.0	0.090	0.089	1.0104	1.3374	0.17	1.092
4.0	0.122	0.119	1.0144	1.3389	0.23	1.126
5.0	0.154	0.149	1.0184	1.3404	0.29	1.162
6.0	0.186	0.179	1.0224	1.3420	0.35	1.200
7.0	0.220	0.210	1.0265	1.3435	0.42	1.239
8.0	0.254	0.241	1.0305	1.3450	0.48	1.281
9.0	0.289	0.272	1.0345	1.3466	0.55	1.325
10.0	0.325	0.303	1.0385	1.3482	0.62	1.372
11.0	0.361	0.335	1.0425	1.3497	0.69	1.422
12.0	0.398	0.367	1.0465	1.3513	0.77	1.474
13.0	0.437	0.399	1.0505	1.3529	0.84	1.530
14.0	0.476	0.431	1.0545	1.3546	0.92	1.588
15.0	0.516	0.464	1.0585	1.3562	1.00	1.649
16.0	0.556	0.497	1.0629	1.3578	1.08	1.715
17.0	0.598	0.530	1.0672	1.3595	1.17	1.784
18.0	0.641	0.564	1.0716	1.3612	1.25	1.859
19.0	0.685	0.597	1.0759	1.3628	1.34	1.940
20.0	0.730	0.631	1.0801	1.3644	1.43	2.021
22.0	0.824	0.700	1.0894	1.3678	1.64	2.216
24.0	0.923	0.770	1.0984	1.3714	1.85	2.463
26.0	1.026	0.842	1.1080	1.3749	2.08	2.753
28.0	1.136	0.914	1.1171	1.3785	2.34	3.066
30.0	1.252	0.988	1.1269	1.3821	2.62	3.427
32.0	1.375	1.063	1.1367	1.3858	2.93	3.918
34.0	1.505	1.139	1.1463	1.3896	3.25	4.447
36.0	1.643	1.216	1.1561	1.3935	3.60	5.050
38.0	1.791	1.295	1.1663	1.3974	3.99	5.832
40.0	1.948	1.375	1.1769	1.4013	4.41	6.926
42.0	2.116	1.457	1.1878	1.4051	4.88	8.191
44.0	2.295	1.540	1.1979	1.4094	5.35	9.649
46.0	2.489	1.624	1.2084	1.4136		11.473
48.0	2.697	1.710	1.2194	1.4177		14.148
50.0	2.921	1.797	1.2304	1.4217		17.786
52.0	3.165	1.886	1.2416	1.4260		22.034
54.0	3.429	1.976	1.2528	1.4308		28.757
56.0	3.718	2.068	1.2638	1.4350		38.226
58.0	4.034	2.159	1.2740	1.4394		49.298
60.0	4.382	2.253	1.2855	1.4440		

Manganese(II) sulfate

Mass %	m/mol kg^{-1}	c/mol L^{-1}	ρ/g cm^{-3}	n	Δ/°C	η/mPa s
1.0	0.067	0.067	1.0080	1.3348	0.16	1.046
2.0	0.135	0.135	1.0178	1.3366	0.31	1.090
3.0	0.205	0.204	1.0277	1.3384	0.44	1.137
4.0	0.276	0.275	1.0378	1.3402	0.57	1.187
5.0	0.349	0.347	1.0480	1.3420	0.70	1.242
6.0	0.423	0.421	1.0583	1.3438	0.84	1.301
7.0	0.498	0.495	1.0688	1.3457	0.98	1.363
8.0	0.576	0.572	1.0794	1.3475	1.12	1.431
9.0	0.655	0.650	1.0902	1.3494	1.28	1.505
10.0	0.736	0.729	1.1012	1.3513	1.44	1.587
11.0	0.819	0.810	1.1123	1.3532	1.61	1.678
12.0	0.903	0.893	1.1236	1.3551	1.80	1.779
13.0	0.990	0.977	1.1351	1.3570	2.00	1.887
14.0	1.078	1.063	1.1467	1.3589	2.21	2.005
15.0	1.169	1.151	1.1585	1.3609	2.43	2.133
16.0	1.261	1.240	1.1705	1.3629	2.67	2.272
17.0	1.356	1.331	1.1827	1.3648	2.92	2.420
18.0	1.454	1.424	1.1950	1.3668	3.19	2.580
19.0	1.553	1.519	1.2075	1.3688	3.49	2.752

Mass %	m/ mol kg⁻¹	c/mol L⁻¹	ρ/g cm⁻³	n	Δ/°C	η/mPa s
20.0	1.656	1.616	1.2203	1.3708	3.80	2.938
D-Mannitol						
0.5	0.028	0.027	1.0000	1.3337	0.05	1.019
1.0	0.055	0.055	1.0017	1.3345	0.10	1.032
1.5	0.084	0.083	1.0035	1.3352	0.16	1.044
2.0	0.112	0.110	1.0053	1.3359	0.21	1.057
2.5	0.141	0.138	1.0070	1.3367	0.26	1.069
3.0	0.170	0.166	1.0088	1.3374	0.32	1.081
3.5	0.199	0.194	1.0106	1.3381	0.37	1.094
4.0	0.229	0.222	1.0124	1.3389	0.43	1.107
4.5	0.259	0.251	1.0141	1.3396	0.48	1.121
5.0	0.289	0.279	1.0159	1.3403	0.54	1.135
5.5	0.319	0.307	1.0177	1.3411	0.60	1.150
6.0	0.350	0.336	1.0195	1.3418	0.66	1.166
6.5	0.382	0.364	1.0212	1.3425	0.72	1.183
7.0	0.413	0.393	1.0230	1.3433	0.77	1.200
7.5	0.445	0.422	1.0248	1.3440	0.84	1.218
8.0	0.477	0.451	1.0266	1.3447	0.90	1.236
8.5	0.510	0.480	1.0284	1.3455	0.96	1.256
9.0	0.543	0.509	1.0302	1.3462	1.02	1.275
9.5	0.576	0.538	1.0320	1.3469	1.08	1.294
10.0	0.610	0.567	1.0338	1.3477	1.15	1.314
11.0	0.678	0.626	1.0375	1.3491	1.28	1.355
12.0	0.749	0.686	1.0412	1.3506	1.41	1.398
13.0	0.820	0.746	1.0450	1.3521	1.55	1.443
14.0	0.894	0.806	1.0489	1.3536	1.69	1.489
15.0	0.969	0.867	1.0529	1.3552	1.84	1.537
Methanol						
0.5	0.157	0.156	0.9973	1.3331	0.28	1.022
1.0	0.315	0.311	0.9964	1.3332	0.56	1.040
2.0	0.637	0.621	0.9947	1.3334	1.14	1.070
3.0	0.965	0.930	0.9930	1.3336	1.75	1.100
4.0	1.300	1.238	0.9913	1.3339	2.37	1.131
5.0	1.643	1.544	0.9896	1.3341	3.02	1.163
6.0	1.992	1.850	0.9880	1.3343	3.71	1.196
7.0	2.349	2.155	0.9864	1.3346	4.41	1.229
8.0	2.714	2.459	0.9848	1.3348	5.13	1.264
9.0	3.087	2.762	0.9832	1.3351	5.85	1.297
10.0	3.468	3.064	0.9816	1.3354	6.60	1.329
11.0	3.858	3.365	0.9801	1.3356	7.36	1.360
12.0	4.256	3.665	0.9785	1.3359	8.14	1.389
13.0	4.664	3.964	0.9770	1.3362	8.93	1.418
14.0	5.081	4.262	0.9755	1.3365	9.72	1.446
15.0	5.508	4.560	0.9740	1.3367	10.53	1.474
16.0	5.945	4.856	0.9725	1.3370	11.36	1.501
17.0	6.393	5.152	0.9710	1.3373	12.23	1.528
18.0	6.851	5.447	0.9695	1.3376	13.13	1.554
19.0	7.321	5.740	0.9680	1.3379	14.06	1.579
20.0	7.803	6.034	0.9666	1.3381	15.02	1.604
22.0	8.803	6.616	0.9636	1.3387	16.98	1.652
24.0	9.856	7.196	0.9606	1.3392	19.04	1.697
26.0	10.966	7.771	0.9576	1.3397	21.23	1.735
28.0	12.138	8.341	0.9545	1.3402	23.59	1.769
30.0	13.376	8.908	0.9514	1.3407	25.91	1.795
32.0	14.688	9.470	0.9482	1.3411	28.15	1.814
34.0	16.078	10.028	0.9450	1.3415	30.48	1.827
36.0	17.556	10.580	0.9416	1.3419	32.97	1.835
38.0	19.129	11.127	0.9382	1.3422	35.60	1.839
40.0	20.807	11.669	0.9347	1.3425	38.60	1.837
42.0	22.601	12.205	0.9311	1.3427	41.50	1.831
44.0	24.523	12.734	0.9273	1.3429	44.50	1.821
46.0	26.587	13.259	0.9235	1.3430	47.80	1.805
48.0	28.810	13.777	0.9196	1.3431	51.20	1.785
50.0	31.211	14.288	0.9156	1.3431	54.50	1.761
52.0	33.812	14.792	0.9114	1.3431	58.10	1.735
54.0	36.639	15.290	0.9072	1.3430	62.00	1.708
56.0	39.723	15.783	0.9030	1.3429	66.00	1.676
58.0	43.101	16.269	0.8987	1.3428	70.00	1.641
60.0	46.816	16.749	0.8944	1.3426	74.50	1.600
62.0	50.923	17.224	0.8901	1.3425	79.30	1.553
64.0	55.486	17.690	0.8856	1.3422	84.40	1.503
66.0	60.586	18.148	0.8810	1.3419	89.60	1.456
68.0	66.323	18.598	0.8763	1.3415	96.30	1.413
70.0	72.826	19.040	0.8715	1.3411		1.368
72.0	80.257	19.476	0.8667	1.3407		1.318
74.0	88.831	19.904	0.8618	1.3402		1.267
76.0	98.835	20.324	0.8568	1.3397		1.218
78.0	110.657	20.737	0.8518	1.3391		1.172
80.0	124.844	21.144	0.8468	1.3385		1.128
82.0	142.183	21.539	0.8416	1.3379		1.074
84.0	163.858	21.931	0.8365	1.3372		1.023
86.0	191.725	22.311	0.8312	1.3365		0.970
88.0	228.881	22.684	0.8259	1.3357		0.916
90.0	280.899	23.045	0.8204	1.3348		0.861
92.0	358.926	23.396	0.8148	1.3339		0.806
94.0	488.972	23.732	0.8089	1.3328		0.749
96.0	749.064	24.072	0.8034	1.3316		0.695
98.0		24.396	0.7976	1.3304		0.639
100.0		24.710	0.7917	1.3290		0.586
Nitric acid						
0.5	0.080	0.079	1.0009	1.3336	0.28	1.004
1.0	0.160	0.159	1.0037	1.3343	0.56	1.005
2.0	0.324	0.320	1.0091	1.3356	1.12	1.007
3.0	0.491	0.483	1.0146	1.3368	1.70	1.010
4.0	0.661	0.648	1.0202	1.3381	2.32	1.014
5.0	0.835	0.814	1.0257	1.3394	2.96	1.018
6.0	1.013	0.982	1.0314	1.3407	3.63	1.022
7.0	1.194	1.152	1.0370	1.3421	4.33	1.027
8.0	1.380	1.324	1.0427	1.3434	5.05	1.032
9.0	1.570	1.498	1.0485	1.3447	5.81	1.038
10.0	1.763	1.673	1.0543	1.3460	6.60	1.044
11.0	1.961	1.851	1.0602	1.3474	7.42	1.051
12.0	2.164	2.030	1.0660	1.3487	8.27	1.058
13.0	2.371	2.212	1.0720	1.3500	9.15	1.066
14.0	2.583	2.395	1.0780	1.3514	10.08	1.075
15.0	2.801	2.580	1.0840	1.3527	11.04	1.084
16.0	3.023	2.768	1.0901	1.3541	12.04	1.094
17.0	3.250	2.958	1.0963	1.3555	13.08	1.105
18.0	3.484	3.149	1.1025	1.3569	14.16	1.116
19.0	3.723	3.343	1.1087	1.3582	15.30	1.128
20.0	3.967	3.539	1.1150	1.3596		1.141
22.0	4.476	3.937	1.1277	1.3624		1.169
24.0	5.011	4.344	1.1406	1.3652		1.199
26.0	5.576	4.760	1.1536	1.3680		1.233
28.0	6.172	5.185	1.1668	1.3708		1.271
30.0	6.801	5.618	1.1801	1.3736		1.311
32.0	7.468	6.060	1.1934	1.3763		1.354
34.0	8.175	6.512	1.2068	1.3790		1.400
36.0	8.927	6.971	1.2202	1.3817		1.450

Mass %	m/mol kg⁻¹	c/mol L⁻¹	ρ/g cm⁻³	n	Δ/°C	η/mPa s
38.0	9.727	7.439	1.2335	1.3842		1.504
40.0	10.580	7.913	1.2466	1.3867		1.561
Oxalic acid						
0.5	0.056	0.056	1.0006	1.3336	0.16	1.013
1.0	0.112	0.111	1.0030	1.3342	0.30	1.023
1.5	0.169	0.167	1.0054	1.3347	0.44	1.033
2.0	0.227	0.224	1.0079	1.3353	0.57	1.044
2.5	0.285	0.281	1.0103	1.3359	0.71	1.055
3.0	0.343	0.337	1.0126	1.3364	0.84	1.065
3.5	0.403	0.395	1.0150	1.3370	0.97	1.076
4.0	0.463	0.452	1.0174	1.3375	1.09	1.086
4.5	0.523	0.510	1.0197	1.3381		1.097
5.0	0.585	0.568	1.0220	1.3386		1.108
5.5	0.646	0.626	1.0244	1.3392		1.118
6.0	0.709	0.684	1.0265	1.3397		1.129
6.5	0.772	0.743	1.0288	1.3402		1.140
7.0	0.836	0.802	1.0310	1.3407		1.150
7.5	0.901	0.861	1.0332	1.3413		1.161
8.0	0.966	0.920	1.0355	1.3418		1.172
Phosphoric acid						
0.5	0.051	0.051	1.0010	1.3335	0.12	1.010
1.0	0.103	0.102	1.0038	1.3340	0.24	1.020
2.0	0.208	0.206	1.0092	1.3349	0.46	1.050
3.0	0.316	0.311	1.0146	1.3358	0.69	1.079
4.0	0.425	0.416	1.0200	1.3367	0.93	1.108
5.0	0.537	0.523	1.0254	1.3376	1.16	1.138
6.0	0.651	0.631	1.0309	1.3385	1.38	1.169
7.0	0.768	0.740	1.0363	1.3394	1.62	1.200
8.0	0.887	0.850	1.0418	1.3403	1.88	1.232
9.0	1.009	0.962	1.0474	1.3413	2.16	1.267
10.0	1.134	1.075	1.0531	1.3422	2.45	1.303
11.0	1.261	1.189	1.0589	1.3431	2.72	1.341
12.0	1.392	1.304	1.0647	1.3441	3.01	1.382
13.0	1.525	1.420	1.0705	1.3450	3.38	1.424
14.0	1.661	1.538	1.0765	1.3460	3.76	1.469
15.0	1.801	1.657	1.0825	1.3470	4.08	1.516
16.0	1.944	1.777	1.0885	1.3480	4.45	1.565
17.0	2.090	1.899	1.0947	1.3489	4.82	1.616
18.0	2.240	2.022	1.1009	1.3500	5.25	1.671
19.0	2.394	2.147	1.1071	1.3510	5.72	1.727
20.0	2.551	2.273	1.1135	1.3520	6.23	1.788
22.0	2.878	2.529	1.1263	1.3540	7.38	1.914
24.0	3.223	2.791	1.1395	1.3561	8.69	2.049
26.0	3.585	3.059	1.1528	1.3582	10.12	2.198
28.0	3.968	3.333	1.1665	1.3604	11.64	2.365
30.0	4.373	3.614	1.1804	1.3625	13.23	2.553
32.0	4.802	3.901	1.1945	1.3647	14.94	2.766
34.0	5.257	4.194	1.2089	1.3669	16.81	3.001
36.0	5.740	4.495	1.2236	1.3691	18.85	3.260
38.0	6.254	4.803	1.2385	1.3713	21.09	3.544
40.0	6.803	5.117	1.2536	1.3735	23.58	3.856
Potassium bromide						
0.5	0.042	0.042	1.0018	1.3336	0.15	1.000
1.0	0.085	0.084	1.0054	1.3342	0.29	0.998
2.0	0.171	0.170	1.0127	1.3354	0.59	0.994
3.0	0.260	0.257	1.0200	1.3366	0.88	0.990
4.0	0.350	0.345	1.0275	1.3379	1.18	0.985
5.0	0.442	0.435	1.0350	1.3391	1.48	0.981
6.0	0.536	0.526	1.0426	1.3403	1.78	0.977

Mass %	m/mol kg⁻¹	c/mol L⁻¹	ρ/g cm⁻³	n	Δ/°C	η/mPa s
7.0	0.633	0.618	1.0503	1.3416	2.10	0.974
8.0	0.731	0.711	1.0581	1.3429	2.42	0.970
9.0	0.831	0.806	1.0660	1.3441	2.74	0.967
10.0	0.934	0.903	1.0740	1.3454	3.07	0.964
11.0	1.039	1.000	1.0821	1.3467	3.42	0.961
12.0	1.146	1.099	1.0903	1.3481	3.76	0.958
13.0	1.256	1.200	1.0986	1.3494	4.12	0.956
14.0	1.368	1.302	1.1070	1.3507	4.49	0.953
15.0	1.483	1.406	1.1155	1.3521	4.86	0.951
16.0	1.601	1.512	1.1242	1.3535	5.25	0.949
17.0	1.721	1.619	1.1330	1.3548	5.64	0.948
18.0	1.845	1.727	1.1419	1.3562	6.04	0.946
19.0	1.971	1.838	1.1509	1.3577	6.46	0.945
20.0	2.101	1.950	1.1601	1.3591	6.88	0.944
22.0	2.370	2.179	1.1788	1.3620	7.76	0.943
24.0	2.654	2.416	1.1980	1.3650	8.70	0.943
26.0	2.952	2.661	1.2179	1.3680	9.68	0.944
28.0	3.268	2.914	1.2383	1.3711	10.72	0.947
30.0	3.601	3.175	1.2593	1.3743	11.82	0.952
32.0	3.954	3.445	1.2810	1.3776	12.98	0.959
34.0	4.329	3.724	1.3033	1.3809		0.968
36.0	4.727	4.012	1.3263	1.3843		0.979
38.0	5.150	4.311	1.3501	1.3878		0.993
40.0	5.602	4.620	1.3746	1.3914		1.010
Potassium carbonate						
0.5	0.036	0.036	1.0027	1.3339	0.18	1.013
1.0	0.073	0.073	1.0072	1.3347	0.34	1.025
2.0	0.148	0.147	1.0163	1.3365	0.66	1.048
3.0	0.224	0.223	1.0254	1.3382	0.99	1.071
4.0	0.301	0.299	1.0345	1.3399	1.32	1.094
5.0	0.381	0.378	1.0437	1.3416	1.67	1.119
6.0	0.462	0.457	1.0529	1.3433	2.03	1.146
7.0	0.545	0.538	1.0622	1.3450	2.40	1.174
8.0	0.629	0.620	1.0715	1.3467	2.77	1.204
9.0	0.716	0.704	1.0809	1.3484	3.17	1.235
10.0	0.804	0.789	1.0904	1.3501	3.57	1.269
12.0	0.987	0.963	1.1095	1.3535	4.45	1.339
14.0	1.178	1.144	1.1291	1.3569	5.39	1.414
16.0	1.378	1.330	1.1490	1.3603	6.42	1.497
18.0	1.588	1.523	1.1692	1.3637	7.55	1.594
20.0	1.809	1.722	1.1898	1.3671	8.82	1.707
24.0	2.285	2.139	1.2320	1.3739	11.96	1.978
28.0	2.814	2.584	1.2755	1.3807	16.01	2.331
32.0	3.405	3.057	1.3204	1.3874	21.46	2.834
36.0	4.070	3.559	1.3665	1.3940	28.58	3.503
40.0	4.824	4.093	1.4142	1.4006	37.55	4.360
44.0	5.685	4.659	1.4633	1.4071		5.720
48.0	6.679	5.259	1.5142	1.4136		7.764
50.0	7.236	5.573	1.5404	1.4168		9.369
Potassium chloride						
0.5	0.067	0.067	1.0014	1.3337	0.23	1.000
1.0	0.135	0.135	1.0046	1.3343	0.46	0.999
2.0	0.274	0.271	1.0110	1.3357	0.92	0.999
3.0	0.415	0.409	1.0174	1.3371	1.38	0.998
4.0	0.559	0.549	1.0239	1.3384	1.85	0.997
5.0	0.706	0.691	1.0304	1.3398	2.32	0.996
6.0	0.856	0.835	1.0369	1.3411	2.80	0.994
7.0	1.010	0.980	1.0434	1.3425	3.29	0.992
8.0	1.166	1.127	1.0500	1.3438	3.80	0.990

Mass %	m/ mol kg⁻¹	c/mol L⁻¹	ρ/g cm⁻³	n	Δ/°C	η/mPa s
9.0	1.327	1.276	1.0566	1.3452	4.30	0.989
10.0	1.490	1.426	1.0633	1.3466	4.81	0.988
11.0	1.658	1.579	1.0700	1.3479	5.33	0.989
12.0	1.829	1.733	1.0768	1.3493	5.88	0.990
13.0	2.004	1.890	1.0836	1.3507	6.45	0.992
14.0	2.184	2.048	1.0905	1.3521		0.994
15.0	2.367	2.208	1.0974	1.3535		0.997
16.0	2.555	2.370	1.1043	1.3549		0.999
17.0	2.747	2.534	1.1114	1.3563		1.001
18.0	2.944	2.701	1.1185	1.3577		1.004
19.0	3.146	2.869	1.1256	1.3592		1.007
20.0	3.353	3.039	1.1328	1.3606		1.012
22.0	3.783	3.386	1.1474	1.3635		1.024
24.0	4.236	3.742	1.1623	1.3665		1.040
Potassium dihydrogen phosphate						
0.5	0.037	0.037	1.0018	1.3336	0.13	1.010
1.0	0.074	0.074	1.0053	1.3342	0.25	1.019
1.5	0.112	0.111	1.0089	1.3348	0.37	1.028
2.0	0.150	0.149	1.0125	1.3354	0.49	1.038
2.5	0.188	0.187	1.0161	1.3359	0.60	1.048
3.0	0.227	0.225	1.0197	1.3365	0.72	1.060
3.5	0.267	0.263	1.0233	1.3371	0.84	1.071
4.0	0.306	0.302	1.0269	1.3377	0.96	1.083
4.5	0.346	0.341	1.0306	1.3382	1.08	1.096
5.0	0.387	0.380	1.0342	1.3388	1.19	1.108
5.5	0.428	0.419	1.0378	1.3394	1.30	1.121
6.0	0.469	0.459	1.0414	1.3400	1.41	1.133
6.5	0.511	0.499	1.0450	1.3405	1.52	1.146
7.0	0.553	0.539	1.0486	1.3411	1.63	1.160
7.5	0.596	0.580	1.0522	1.3417	1.74	1.173
8.0	0.639	0.621	1.0558	1.3422	1.84	1.187
8.5	0.683	0.662	1.0594	1.3428	1.94	1.201
9.0	0.727	0.703	1.0630	1.3434	2.04	1.215
9.5	0.771	0.745	1.0667	1.3439	2.14	1.230
10.0	0.816	0.786	1.0703	1.3445	2.23	1.245
Potassium hydrogen carbonate						
0.5	0.050	0.050	1.0014	1.3335	0.18	1.009
1.0	0.101	0.100	1.0046	1.3341	0.34	1.015
2.0	0.204	0.202	1.0114	1.3353	0.67	1.027
3.0	0.309	0.305	1.0181	1.3365	0.98	1.040
4.0	0.416	0.409	1.0247	1.3376	1.29	1.053
5.0	0.526	0.515	1.0310	1.3386	1.60	1.067
6.0	0.638	0.622	1.0379	1.3397	1.91	1.081
7.0	0.752	0.730	1.0446	1.3409	2.22	1.096
8.0	0.869	0.840	1.0514	1.3419	2.53	1.112
9.0	0.988	0.951	1.0581	1.3430	2.84	1.128
10.0	1.110	1.064	1.0650	1.3441	3.16	1.145
12.0	1.362	1.293	1.0788	1.3462	3.79	1.183
14.0	1.626	1.528	1.0929	1.3484	4.41	1.224
16.0	1.903	1.770	1.1073	1.3506		1.270
18.0	2.193	2.017	1.1221	1.3528		1.319
20.0	2.497	2.272	1.1372	1.3550		1.373
22.0	2.817	2.533	1.1527	1.3572		1.432
24.0	3.154	2.801	1.1685	1.3595		1.497
Potassium hydrogen phosphate						
0.5	0.029	0.029	1.0025	1.3338	0.13	1.013
1.0	0.058	0.058	1.0068	1.3345	0.25	1.023
1.5	0.087	0.087	1.0110	1.3353	0.37	1.034
2.0	0.117	0.117	1.0153	1.3361	0.49	1.046

Mass %	m/ mol kg⁻¹	c/mol L⁻¹	ρ/g cm⁻³	n	Δ/°C	η/mPa s
2.5	0.147	0.146	1.0195	1.3368	0.61	1.057
3.0	0.178	0.176	1.0238	1.3376	0.73	1.069
3.5	0.208	0.207	1.0281	1.3384	0.86	1.081
4.0	0.239	0.237	1.0324	1.3392	0.97	1.094
4.5	0.271	0.268	1.0368	1.3399	1.10	1.107
5.0	0.302	0.299	1.0412	1.3407	1.22	1.120
5.5	0.334	0.330	1.0456	1.3415	1.34	1.133
6.0	0.366	0.362	1.0500	1.3422	1.46	1.147
6.5	0.399	0.394	1.0545	1.3430	1.58	1.162
7.0	0.432	0.426	1.0590	1.3438	1.70	1.177
7.5	0.466	0.458	1.0635	1.3445	1.82	1.193
8.0	0.499	0.491	1.0680	1.3453	1.95	1.209
Potassium hydroxide						
0.5	0.090	0.089	1.0025	1.3340	0.30	1.010
1.0	0.180	0.179	1.0068	1.3350	0.61	1.019
2.0	0.364	0.362	1.0155	1.3369	1.24	1.038
3.0	0.551	0.548	1.0242	1.3388	1.89	1.058
4.0	0.743	0.736	1.0330	1.3408	2.57	1.079
5.0	0.938	0.929	1.0419	1.3427	3.36	1.102
6.0	1.138	1.124	1.0509	1.3445	4.14	1.126
7.0	1.342	1.322	1.0599	1.3464	4.92	1.151
8.0	1.550	1.524	1.0690	1.3483		1.177
9.0	1.763	1.729	1.0781	1.3502		1.205
10.0	1.980	1.938	1.0873	1.3520		1.233
11.0	2.203	2.150	1.0966	1.3539		1.264
12.0	2.431	2.365	1.1059	1.3558		1.294
13.0	2.663	2.584	1.1153	1.3576		1.327
14.0	2.902	2.806	1.1246	1.3595		1.361
15.0	3.145	3.032	1.1341	1.3614		1.397
16.0	3.395	3.261	1.1435	1.3632		1.436
17.0	3.651	3.494	1.1531	1.3651		1.477
18.0	3.913	3.730	1.1626	1.3670		1.521
19.0	4.181	3.970	1.1722	1.3688		1.568
20.0	4.456	4.213	1.1818	1.3707		1.619
22.0	5.027	4.711	1.2014	1.3744		1.732
24.0	5.629	5.223	1.2210	1.3781		1.861
26.0	6.262	5.750	1.2408	1.3818		2.006
28.0	6.931	6.293	1.2609	1.3854		2.170
30.0	7.639	6.851	1.2813	1.3889		2.357
32.0	8.388	7.426	1.3020	1.3923		2.570
34.0	9.182	8.017	1.3230	1.3957		2.820
36.0	10.026	8.626	1.3444	1.3993		3.111
38.0	10.924	9.253	1.3661	1.4030		3.460
40.0	11.882	9.896	1.3881	1.4068		3.879
42.0	12.907	10.558	1.4104	1.4106		4.389
44.0	14.004	11.239	1.4331	1.4143		5.013
46.0	15.183	11.938	1.4560	1.4179		5.781
48.0	16.453	12.654	1.4791	1.4214		6.726
50.0	17.824	13.389	1.5024	1.4247		7.892
Potassium iodide						
0.5	0.030	0.030	1.0019	1.3337	0.11	0.999
1.0	0.061	0.061	1.0056	1.3343	0.22	0.997
2.0	0.123	0.122	1.0131	1.3357	0.43	0.991
3.0	0.186	0.184	1.0206	1.3370	0.64	0.986
4.0	0.251	0.248	1.0282	1.3384	0.86	0.981
5.0	0.317	0.312	1.0360	1.3397	1.08	0.976
6.0	0.385	0.377	1.0438	1.3411	1.30	0.969
7.0	0.453	0.443	1.0517	1.3425	1.53	0.963
8.0	0.524	0.511	1.0598	1.3440	1.77	0.957

Mass %	m/ mol kg⁻¹	c/mol L⁻¹	ρ/g cm⁻³	n	Δ/°C	η/mPa s
9.0	0.596	0.579	1.0679	1.3454	2.01	0.951
10.0	0.669	0.648	1.0762	1.3469	2.26	0.946
12.0	0.821	0.790	1.0931	1.3498	2.77	0.937
14.0	0.981	0.937	1.1105	1.3529	3.30	0.929
16.0	1.147	1.088	1.1284	1.3560	3.87	0.921
18.0	1.322	1.244	1.1469	1.3593	4.46	0.915
20.0	1.506	1.405	1.1659	1.3626	5.09	0.910
22.0	1.699	1.571	1.1856	1.3661	5.76	0.905
24.0	1.902	1.744	1.2060	1.3696	6.46	0.901
26.0	2.117	1.922	1.2270	1.3733	7.21	0.898
28.0	2.343	2.106	1.2487	1.3771	8.01	0.895
30.0	2.582	2.297	1.2712	1.3810	8.86	0.892
32.0	2.835	2.495	1.2944	1.3851	9.76	0.891
34.0	3.103	2.700	1.3185	1.3893	10.72	0.890
36.0	3.388	2.913	1.3434	1.3936	11.73	0.890
38.0	3.692	3.134	1.3692	1.3981	12.81	0.893
40.0	4.016	3.364	1.3959	1.4027	13.97	0.897
Potassium nitrate						
0.5	0.050	0.050	1.0014	1.3335	0.17	0.999
1.0	0.100	0.099	1.0045	1.3339	0.33	0.996
2.0	0.202	0.200	1.0108	1.3349	0.64	0.990
3.0	0.306	0.302	1.0171	1.3358	0.94	0.986
4.0	0.412	0.405	1.0234	1.3368	1.22	0.983
5.0	0.521	0.509	1.0298	1.3377	1.50	0.980
6.0	0.631	0.615	1.0363	1.3386	1.76	0.977
7.0	0.744	0.722	1.0428	1.3396	2.02	0.975
8.0	0.860	0.830	1.0494	1.3405	2.27	0.973
9.0	0.978	0.940	1.0560	1.3415	2.52	0.971
10.0	1.099	1.051	1.0627	1.3425	2.75	0.970
12.0	1.349	1.277	1.0762	1.3444		0.970
14.0	1.610	1.509	1.0899	1.3463		0.972
16.0	1.884	1.747	1.1039	1.3482		0.976
18.0	2.171	1.991	1.1181	1.3502		0.982
20.0	2.473	2.240	1.1326	1.3521		0.990
22.0	2.790	2.497	1.1473	1.3541		0.999
24.0	3.123	2.759	1.1623	1.3561		1.010
Potassium permanganate						
0.5	0.032	0.032	1.0017		0.11	1.001
1.0	0.064	0.064	1.0051		0.22	1.000
1.5	0.096	0.096	1.0085		0.32	0.999
2.0	0.129	0.128	1.0118		0.43	0.998
2.5	0.162	0.161	1.0152			0.996
3.0	0.196	0.193	1.0186			0.995
3.5	0.230	0.226	1.0220			0.994
4.0	0.264	0.260	1.0254			0.992
4.5	0.298	0.293	1.0288			0.991
5.0	0.333	0.327	1.0322			0.989
5.5	0.368	0.360	1.0356			0.987
6.0	0.404	0.394	1.0390			0.985
Potassium sulfate						
0.5	0.029	0.029	1.0022	1.3336	0.14	1.006
1.0	0.058	0.058	1.0062	1.3343	0.26	1.011
2.0	0.117	0.116	1.0143	1.3355	0.50	1.021
2.5	0.147	0.146	1.0183	1.3362	0.61	1.027
3.0	0.177	0.176	1.0224	1.3368	0.73	1.033
3.5	0.208	0.206	1.0265	1.3374	0.84	1.039
4.0	0.239	0.237	1.0306	1.3380	0.95	1.045
4.5	0.270	0.267	1.0347	1.3386	1.06	1.051
5.0	0.302	0.298	1.0388	1.3393	1.17	1.058

Mass %	m/ mol kg⁻¹	c/mol L⁻¹	ρ/g cm⁻³	n	Δ/°C	η/mPa s
6.0	0.366	0.360	1.0470	1.3405		1.072
7.0	0.432	0.424	1.0553	1.3417		1.087
8.0	0.499	0.488	1.0637	1.3428		1.102
9.0	0.568	0.554	1.0721	1.3440		1.117
10.0	0.638	0.620	1.0806	1.3452		1.132
1-Propanol						
1.0	0.168	0.166	0.9963	1.3339	0.31	1.051
2.0	0.340	0.331	0.9946	1.3348	0.61	1.100
3.0	0.515	0.496	0.9928	1.3357	0.93	1.152
4.0	0.693	0.660	0.9911	1.3366	1.24	1.208
5.0	0.876	0.823	0.9896	1.3376	1.57	1.267
6.0	1.062	0.987	0.9882	1.3385	1.91	1.325
7.0	1.252	1.149	0.9868	1.3394	2.26	1.387
8.0	1.447	1.312	0.9855	1.3404	2.61	1.449
9.0	1.646	1.474	0.9842	1.3414	2.99	1.514
10.0	1.849	1.635	0.9829	1.3423	3.36	1.577
12.0	2.269	1.958	0.9804	1.3442	4.09	1.710
14.0	2.709	2.278	0.9779	1.3460	4.91	1.849
16.0	3.169	2.595	0.9749	1.3477	5.78	1.986
18.0	3.652	2.911	0.9719	1.3494	6.67	2.106
20.0	4.160	3.223	0.9686	1.3510	7.76	2.218
24.0	5.254	3.838	0.9612	1.3539	9.12	2.432
28.0	6.471	4.441	0.9533	1.3566	10.17	2.612
32.0	7.830	5.033	0.9452	1.3592	10.66	2.765
36.0	9.359	5.613	0.9370	1.3614		2.900
40.0	11.093	6.182	0.9288	1.3635		3.010
60.0	24.958	8.860	0.8875	1.3734		3.186
80.0	66.556	11.275	0.8470	1.3812		2.822
100.0		13.368	0.8034	1.3852		2.227
2-Propanol						
1.0	0.168	0.166	0.9960	1.3338	0.30	1.056
2.0	0.340	0.331	0.9939	1.3346	0.60	1.112
3.0	0.515	0.495	0.9920	1.3355	0.93	1.166
4.0	0.693	0.659	0.9902	1.3364	1.26	1.225
5.0	0.876	0.822	0.9884	1.3373	1.61	1.287
6.0	1.062	0.985	0.9871	1.3382	1.96	1.352
7.0	1.252	1.148	0.9855	1.3392	2.32	1.417
8.0	1.447	1.310	0.9843	1.3400	2.68	1.485
9.0	1.646	1.472	0.9831	1.3410	3.06	1.553
10.0	1.849	1.633	0.9816	1.3420	3.48	1.629
12.0	2.269	1.955	0.9793	1.3439	4.43	1.794
14.0	2.709	2.276	0.9772	1.3459	5.29	1.970
16.0	3.169	2.596	0.9751	1.3478	6.36	2.160
18.0	3.652	2.913	0.9725	1.3496	7.40	2.352
20.0	4.160	3.227	0.9696	1.3514	8.52	2.550
40.0	11.093	6.191	0.9302	1.3642		
60.0	24.958	8.809	0.8824	1.3717		
80.0	66.556	11.103	0.8341	1.3742		
100.0		13.058	0.7848	1.3776		
Silver nitrate						
0.5	0.030	0.030	1.0027	1.3336	0.10	1.003
1.0	0.059	0.059	1.0070	1.3342	0.20	1.005
2.0	0.120	0.120	1.0154	1.3352	0.40	1.009
3.0	0.182	0.181	1.0239	1.3363	0.59	1.013
4.0	0.245	0.243	1.0327	1.3374	0.78	1.016
5.0	0.310	0.307	1.0417	1.3385	0.96	1.020
6.0	0.376	0.371	1.0506	1.3396	1.15	1.024
7.0	0.443	0.437	1.0597	1.3407	1.33	1.027
8.0	0.512	0.503	1.0690	1.3419	1.51	1.031

Mass %	m/mol kg⁻¹	c/mol L⁻¹	ρ/g cm⁻³	n	Δ/°C	η/mPa s
9.0	0.582	0.571	1.0785	1.3431	1.69	1.035
10.0	0.654	0.641	1.0882	1.3443	1.87	1.039
12.0	0.803	0.783	1.1079	1.3467	2.21	1.049
14.0	0.958	0.930	1.1284	1.3493	2.55	1.060
16.0	1.121	1.083	1.1496	1.3519	2.86	1.072
18.0	1.292	1.241	1.1715	1.3546		1.086
20.0	1.472	1.406	1.1942	1.3574		1.101
22.0	1.660	1.577	1.2177	1.3602		1.117
24.0	1.859	1.755	1.2420	1.3632		1.135
26.0	2.068	1.940	1.2672	1.3662		1.154
28.0	2.289	2.132	1.2933	1.3694		1.176
30.0	2.523	2.332	1.3204	1.3726		1.200
32.0	2.770	2.541	1.3487	1.3760		1.227
34.0	3.033	2.758	1.3780	1.3795		1.257
36.0	3.311	2.985	1.4087	1.3832		1.290
38.0	3.608	3.223	1.4407	1.3871		1.326
40.0	3.925	3.472	1.4743	1.3911		1.366

Sodium acetate

Mass %	m/mol kg⁻¹	c/mol L⁻¹	ρ/g cm⁻³	n	Δ/°C	η/mPa s
0.5	0.061	0.061	1.0008	1.3337	0.22	1.021
1.0	0.123	0.122	1.0034	1.3344	0.43	1.040
2.0	0.249	0.246	1.0085	1.3358	0.88	1.080
3.0	0.377	0.371	1.0135	1.3372	1.34	1.124
4.0	0.508	0.497	1.0184	1.3386	1.82	1.171
5.0	0.642	0.624	1.0234	1.3400	2.32	1.222
6.0	0.778	0.752	1.0283	1.3414	2.85	1.278
7.0	0.918	0.882	1.0334	1.3428	3.40	1.337
8.0	1.060	1.013	1.0386	1.3442	3.98	1.401
9.0	1.206	1.145	1.0440	1.3456	4.57	1.468
10.0	1.354	1.279	1.0495	1.3470		1.539
12.0	1.662	1.552	1.0607	1.3498		1.688
14.0	1.984	1.829	1.0718	1.3526		1.855
16.0	2.322	2.112	1.0830	1.3554		2.054
18.0	2.676	2.400	1.0940	1.3583		2.284
20.0	3.047	2.694	1.1050	1.3611		2.567
22.0	3.438	2.993	1.1159	1.3639		2.948
24.0	3.849	3.297	1.1268	1.3666		3.400
26.0	4.283	3.606	1.1377	1.3693		3.877
28.0	4.741	3.921	1.1488	1.3720		4.388
30.0	5.224	4.243	1.1602	1.3748		4.940

Sodium bromide

Mass %	m/mol kg⁻¹	c/mol L⁻¹	ρ/g cm⁻³	n	Δ/°C	η/mPa s
0.5	0.049	0.049	1.0021	1.3337	0.17	1.004
1.0	0.098	0.098	1.0060	1.3344	0.34	1.007
2.0	0.198	0.197	1.0139	1.3358	0.69	1.012
3.0	0.301	0.298	1.0218	1.3372	1.04	1.017
4.0	0.405	0.400	1.0298	1.3386	1.39	1.022
5.0	0.512	0.504	1.0380	1.3401	1.76	1.028
6.0	0.620	0.610	1.0462	1.3415	2.14	1.034
7.0	0.732	0.717	1.0546	1.3430	2.53	1.040
8.0	0.845	0.826	1.0630	1.3445	2.93	1.046
9.0	0.961	0.937	1.0716	1.3460	3.34	1.053
10.0	1.080	1.050	1.0803	1.3475	3.77	1.060
11.0	1.201	1.164	1.0892	1.3491	4.21	1.068
12.0	1.325	1.281	1.0981	1.3506	4.67	1.077
13.0	1.452	1.399	1.1072	1.3522	5.15	1.086
14.0	1.582	1.519	1.1164	1.3538	5.65	1.096
15.0	1.715	1.641	1.1257	1.3554	6.18	1.107
16.0	1.851	1.765	1.1352	1.3570	6.74	1.119
17.0	1.991	1.891	1.1448	1.3587	7.32	1.131
18.0	2.133	2.020	1.1546	1.3604		1.144

Mass %	m/mol kg⁻¹	c/mol L⁻¹	ρ/g cm⁻³	n	Δ/°C	η/mPa s
19.0	2.280	2.150	1.1645	1.3621		1.159
20.0	2.430	2.283	1.1745	1.3638		1.174
22.0	2.741	2.555	1.1951	1.3673		1.207
24.0	3.069	2.837	1.2163	1.3708		1.244
26.0	3.415	3.129	1.2382	1.3745		1.287
28.0	3.780	3.431	1.2608	1.3783		1.336
30.0	4.165	3.744	1.2842	1.3822		1.395
32.0	4.574	4.069	1.3083	1.3862		1.465
34.0	5.007	4.406	1.3333	1.3903		1.546
36.0	5.467	4.755	1.3592	1.3946		1.639
38.0	5.957	5.119	1.3860	1.3990		1.745
40.0	6.479	5.496	1.4138	1.4035		1.866

Sodium carbonate

Mass %	m/mol kg⁻¹	c/mol L⁻¹	ρ/g cm⁻³	n	Δ/°C	η/mPa s
0.5	0.047	0.047	1.0034	1.3341	0.22	1.025
1.0	0.095	0.095	1.0086	1.3352	0.43	1.049
2.0	0.193	0.192	1.0190	1.3375	0.75	1.102
3.0	0.292	0.291	1.0294	1.3397	1.08	1.159
4.0	0.393	0.392	1.0398	1.3419	1.42	1.222
5.0	0.497	0.495	1.0502	1.3440	1.77	1.292
6.0	0.602	0.600	1.0606	1.3462	2.13	1.367
7.0	0.710	0.707	1.0711	1.3483		1.448
8.0	0.820	0.816	1.0816	1.3504		1.538
9.0	0.933	0.927	1.0922	1.3525		1.638
10.0	1.048	1.041	1.1029	1.3547		1.754
11.0	1.166	1.156	1.1136	1.3568		1.884
12.0	1.287	1.273	1.1244	1.3589		2.028
13.0	1.410	1.392	1.1353	1.3610		2.186
14.0	1.536	1.514	1.1463	1.3631		2.361
15.0	1.665	1.638	1.1574	1.3652		2.551

Sodium chloride

Mass %	m/mol kg⁻¹	c/mol L⁻¹	ρ/g cm⁻³	n	Δ/°C	η/mPa s
0.1	0.017	0.017	0.9989	1.3332	0.06	1.004
0.2	0.034	0.034	0.9997	1.3333	0.12	1.006
0.3	0.051	0.051	1.0004	1.3335	0.18	1.008
0.4	0.069	0.069	1.0011	1.3337	0.24	1.009
0.5	0.086	0.086	1.0018	1.3339	0.30	1.011
1.0	0.173	0.172	1.0053	1.3347	0.59	1.020
1.5	0.261	0.259	1.0089	1.3356	0.89	1.028
2.0	0.349	0.346	1.0125	1.3365	1.19	1.036
2.5	0.439	0.435	1.0160	1.3374	1.49	1.044
3.0	0.529	0.523	1.0196	1.3383	1.79	1.052
3.5	0.621	0.613	1.0232	1.3391	2.10	1.060
4.0	0.713	0.703	1.0268	1.3400	2.41	1.068
4.5	0.806	0.793	1.0304	1.3409	2.73	1.076
5.0	0.901	0.885	1.0340	1.3418	3.05	1.085
6.0	1.092	1.069	1.0413	1.3435	3.70	1.104
7.0	1.288	1.256	1.0486	1.3453	4.38	1.124
8.0	1.488	1.445	1.0559	1.3470	5.08	1.145
9.0	1.692	1.637	1.0633	1.3488	5.81	1.168
10.0	1.901	1.832	1.0707	1.3505	6.56	1.193
11.0	2.115	2.029	1.0781	1.3523	7.35	1.220
12.0	2.333	2.229	1.0857	1.3541	8.18	1.250
13.0	2.557	2.432	1.0932	1.3558	9.04	1.283
14.0	2.785	2.637	1.1008	1.3576	9.94	1.317
15.0	3.020	2.845	1.1085	1.3594	10.89	1.352
16.0	3.259	3.056	1.1162	1.3612	11.89	1.388
17.0	3.505	3.270	1.1240	1.3630	12.94	1.424
18.0	3.756	3.486	1.1319	1.3648	14.04	1.463
19.0	4.014	3.706	1.1398	1.3666	15.22	1.507
20.0	4.278	3.928	1.1478	1.3684	16.46	1.557

Mass %	m/ mol kg⁻¹	c/mol L⁻¹	ρ/g cm⁻³	n	Δ/°C	η/mPa s
21.0	4.548	4.153	1.1558	1.3702	17.78	1.614
22.0	4.826	4.382	1.1640	1.3721	19.18	1.676
23.0	5.111	4.613	1.1721	1.3739	20.67	1.745
24.0	5.403	4.847	1.1804	1.3757		1.821
25.0	5.704	5.085	1.1887	1.3776		1.902
26.0	6.012	5.326	1.1972	1.3795		1.990

Sodium citrate dihydrate

Mass %	m/ mol kg⁻¹	c/mol L⁻¹	ρ/g cm⁻³	n	Δ/°C	η/mPa s
1.0	0.039	0.039	1.0049	1.3348	0.20	1.043
2.0	0.079	0.078	1.0120	1.3366	0.39	1.081
3.0	0.120	0.118	1.0186	1.3383	0.59	1.122
4.0	0.161	0.159	1.0260	1.3401	0.79	1.166
5.0	0.204	0.200	1.0331	1.3419	0.97	1.210
6.0	0.247	0.242	1.0405	1.3437	1.17	1.263
7.0	0.292	0.284	1.0482	1.3455	1.36	1.314
8.0	0.337	0.327	1.0557	1.3473	1.57	1.371
9.0	0.383	0.371	1.0632	1.3491	1.77	1.427
10.0	0.431	0.415	1.0708	1.3509	1.96	1.499
12.0	0.528	0.505	1.0861	1.3546	2.38	1.649
14.0	0.631	0.598	1.1019	1.3583	2.82	1.832
16.0	0.738	0.693	1.1173	1.3618	3.27	2.045
18.0	0.851	0.790	1.1327	1.3656	3.82	2.290
20.0	0.969	0.891	1.1492	1.3693	4.39	2.596
24.0	1.224	1.099	1.1813	1.3767		3.409
28.0	1.507	1.318	1.2151	1.3845		4.586
32.0	1.823	1.548	1.2487	1.3923		6.541
36.0	2.180	1.792	1.2843	1.4001		9.788

Sodium dihydrogen phosphate

Mass %	m/ mol kg⁻¹	c/mol L⁻¹	ρ/g cm⁻³	n	Δ/°C	η/mPa s
0.5	0.042	0.042	1.0019	1.3336	0.14	1.018
1.0	0.084	0.084	1.0056	1.3343	0.28	1.035
1.5	0.127	0.126	1.0094	1.3349	0.42	1.051
2.0	0.170	0.169	1.0131	1.3356	0.56	1.068
2.5	0.214	0.212	1.0168	1.3362	0.70	1.085
3.0	0.258	0.255	1.0206	1.3369	0.84	1.103
3.5	0.302	0.299	1.0244	1.3375	0.98	1.121
4.0	0.347	0.343	1.0281	1.3382	1.12	1.140
4.5	0.393	0.387	1.0319	1.3388	1.25	1.160
5.0	0.439	0.432	1.0358	1.3395	1.39	1.180
5.5	0.485	0.477	1.0396	1.3401	1.52	1.201
6.0	0.532	0.522	1.0434	1.3408	1.65	1.223
6.5	0.579	0.567	1.0473	1.3414	1.77	1.245
7.0	0.627	0.613	1.0511	1.3421	1.89	1.270
7.5	0.676	0.660	1.0550	1.3427	2.00	1.294
8.0	0.725	0.706	1.0589	1.3434	2.12	1.319
8.5	0.774	0.753	1.0628	1.3440	2.23	1.345
9.0	0.824	0.800	1.0668	1.3447	2.35	1.371
9.5	0.875	0.848	1.0707	1.3453	2.47	1.399
10.0	0.926	0.896	1.0747	1.3460	2.58	1.428
11.0	1.030	0.993	1.0826	1.3473	2.82	1.488
12.0	1.137	1.091	1.0907	1.3486	3.06	1.552
13.0	1.246	1.191	1.0988	1.3499	3.29	1.620
14.0	1.357	1.292	1.1070	1.3512	3.53	1.694
15.0	1.471	1.394	1.1152	1.3525	3.78	1.775
16.0	1.588	1.499	1.1236	1.3538	4.03	1.861
17.0	1.707	1.604	1.1320	1.3552	4.29	1.952
18.0	1.830	1.711	1.1404	1.3565	4.55	2.050
19.0	1.955	1.820	1.1490	1.3578	4.82	2.159
20.0	2.084	1.930	1.1576	1.3592	5.10	2.283
22.0	2.351	2.155	1.1752	1.3618		2.550
24.0	2.632	2.387	1.1931	1.3646		2.850

Mass %	m/ mol kg⁻¹	c/mol L⁻¹	ρ/g cm⁻³	n	Δ/°C	η/mPa s
26.0	2.929	2.625	1.2113	1.3673		3.214
28.0	3.242	2.870	1.2299	1.3700		3.682
30.0	3.572	3.123	1.2488	1.3728		4.300
32.0	3.923	3.383	1.2682	1.3756		5.079
34.0	4.294	3.650	1.2879	1.3784		6.008
36.0	4.689	3.925	1.3080	1.3812		7.098
38.0	5.109	4.208	1.3285	1.3840		8.363
40.0	5.557	4.499	1.3493	1.3869		9.814

Sodium hydrogen carbonate

Mass %	m/ mol kg⁻¹	c/mol L⁻¹	ρ/g cm⁻³	n	Δ/°C	η/mPa s
0.5	0.060	0.060	1.0018	1.3337	0.20	1.015
1.0	0.120	0.120	1.0054	1.3344	0.40	1.028
1.5	0.181	0.180	1.0089	1.3351	0.59	1.042
2.0	0.243	0.241	1.0125	1.3357	0.78	1.057
2.5	0.305	0.302	1.0160	1.3364	0.98	1.071
3.0	0.368	0.364	1.0196	1.3370	1.16	1.086
3.5	0.432	0.426	1.0231	1.3377	1.35	1.102
4.0	0.496	0.489	1.0266	1.3383	1.54	1.118
4.5	0.561	0.552	1.0301	1.3390	1.72	1.134
5.0	0.627	0.615	1.0337	1.3396	1.90	1.151
5.5	0.693	0.679	1.0372	1.3403	2.08	1.168
6.0	0.760	0.743	1.0408	1.3409	2.26	1.185

Sodium hydrogen phosphate

Mass %	m/ mol kg⁻¹	c/mol L⁻¹	ρ/g cm⁻³	n	Δ/°C	η/mPa s
0.5	0.035	0.035	1.0032	1.3340	0.17	1.021
1.0	0.071	0.071	1.0082	1.3349	0.32	1.042
1.5	0.107	0.107	1.0131	1.3358	0.46	1.064
2.0	0.144	0.143	1.0180	1.3368		1.088
2.5	0.181	0.180	1.0229	1.3377		1.113
3.0	0.218	0.217	1.0279	1.3386		1.138
3.5	0.255	0.255	1.0328	1.3396		1.165
4.0	0.293	0.292	1.0378	1.3405		1.193
4.5	0.332	0.331	1.0428	1.3414		1.223
5.0	0.371	0.369	1.0478	1.3424		1.254
5.5	0.410	0.408	1.0528	1.3433		1.286

Sodium hydroxide

Mass %	m/ mol kg⁻¹	c/mol L⁻¹	ρ/g cm⁻³	n	Δ/°C	η/mPa s
0.5	0.126	0.125	1.0039	1.3344	0.43	1.027
1.0	0.253	0.252	1.0095	1.3358	0.86	1.054
2.0	0.510	0.510	1.0207	1.3386	1.74	1.112
3.0	0.773	0.774	1.0318	1.3414	2.64	1.176
4.0	1.042	1.043	1.0428	1.3441	3.59	1.248
5.0	1.316	1.317	1.0538	1.3467	4.57	1.329
6.0	1.596	1.597	1.0648	1.3494	5.60	1.416
7.0	1.882	1.883	1.0758	1.3520	6.69	1.510
8.0	2.174	2.174	1.0869	1.3546	7.87	1.616
9.0	2.473	2.470	1.0979	1.3572	9.12	1.737
10.0	2.778	2.772	1.1089	1.3597	10.47	1.882
11.0	3.090	3.080	1.1199	1.3623	11.89	2.039
12.0	3.409	3.393	1.1309	1.3648	13.42	2.201
13.0	3.736	3.711	1.1419	1.3673	15.04	2.376
14.0	4.070	4.036	1.1530	1.3697	16.76	2.568
15.0	4.412	4.365	1.1640	1.3722		2.789
16.0	4.762	4.701	1.1751	1.3746		3.043
17.0	5.121	5.041	1.1861	1.3770		3.344
18.0	5.488	5.387	1.1971	1.3793		3.698
19.0	5.865	5.739	1.2082	1.3817		4.119
20.0	6.250	6.096	1.2192	1.3840		4.619
22.0	7.052	6.827	1.2412	1.3885		5.765
24.0	7.895	7.579	1.2631	1.3929		7.100
26.0	8.784	8.352	1.2848	1.3971		8.744
28.0	9.723	9.145	1.3064	1.4012		10.832

Mass %	m/mol kg⁻¹	c/mol L⁻¹	ρ/g cm⁻³	n	Δ/°C	η/mPa s
30.0	10.715	9.958	1.3277	1.4051		13.517
32.0	11.766	10.791	1.3488	1.4088		16.844
34.0	12.880	11.643	1.3697	1.4123		20.751
36.0	14.064	12.512	1.3901	1.4156		25.290
38.0	15.324	13.398	1.4102	1.4186		30.461
40.0	16.668	14.300	1.4299	1.4215		36.312

Sodium nitrate

Mass %	m/mol kg⁻¹	c/mol L⁻¹	ρ/g cm⁻³	n	Δ/°C	η/mPa s
0.5	0.059	0.059	1.0016	1.3336	0.20	1.004
1.0	0.119	0.118	1.0050	1.3341	0.40	1.007
2.0	0.240	0.238	1.0117	1.3353	0.79	1.012
3.0	0.364	0.359	1.0185	1.3364	1.18	1.018
4.0	0.490	0.483	1.0254	1.3375	1.56	1.025
5.0	0.619	0.607	1.0322	1.3387	1.94	1.032
6.0	0.751	0.734	1.0392	1.3398	2.32	1.040
7.0	0.886	0.862	1.0462	1.3409	2.70	1.049
8.0	1.023	0.991	1.0532	1.3421	3.08	1.059
9.0	1.164	1.123	1.0603	1.3432	3.46	1.069
10.0	1.307	1.256	1.0674	1.3443	3.84	1.081
12.0	1.604	1.527	1.0819	1.3466	4.60	1.107
14.0	1.915	1.806	1.0967	1.3489	5.37	1.138
18.0	2.583	2.387	1.1272	1.3536	6.98	1.215
20.0	2.941	2.689	1.1429	1.3559	7.81	1.263
24.0	3.715	3.318	1.1752	1.3607	9.52	1.377
28.0	4.575	3.981	1.2085	1.3654	11.28	1.522
30.0	5.042	4.326	1.2256	1.3678		1.609
34.0	6.061	5.044	1.2610	1.3726		1.818
40.0	7.844	6.200	1.3175	1.3802		2.226

Sodium phosphate

Mass %	m/mol kg⁻¹	c/mol L⁻¹	ρ/g cm⁻³	n	Δ/°C	η/mPa s
0.5	0.031	0.031	1.0042	1.3343	0.19	1.033
1.0	0.062	0.062	1.0100	1.3356	0.37	1.064
1.5	0.093	0.093	1.0158	1.3369	0.53	1.094
2.0	0.124	0.125	1.0216	1.3381	0.67	1.126
2.5	0.156	0.157	1.0275	1.3394	0.79	1.161
3.0	0.189	0.189	1.0335	1.3406		1.198
3.5	0.221	0.222	1.0395	1.3419		1.238
4.0	0.254	0.255	1.0456	1.3432		1.281
4.5	0.287	0.289	1.0517	1.3444		1.327
5.0	0.321	0.323	1.0579	1.3457		1.375
5.5	0.355	0.357	1.0642	1.3470		1.426
6.0	0.389	0.392	1.0705	1.3482		1.480
6.5	0.424	0.427	1.0768	1.3495		1.538
7.0	0.459	0.462	1.0832	1.3507		1.598
7.5	0.495	0.498	1.0896	1.3519		1.662
8.0	0.530	0.535	1.0961	1.3532		1.729

Sodium sulfate

Mass %	m/mol kg⁻¹	c/mol L⁻¹	ρ/g cm⁻³	n	Δ/°C	η/mPa s
0.5	0.035	0.035	1.0027	1.3338	0.17	1.013
1.0	0.071	0.071	1.0071	1.3345	0.32	1.026
2.0	0.144	0.143	1.0161	1.3360	0.61	1.058
3.0	0.218	0.217	1.0252	1.3376	0.87	1.091
4.0	0.293	0.291	1.0343	1.3391	1.13	1.126
5.0	0.371	0.367	1.0436	1.3406	1.36	1.163
6.0	0.449	0.445	1.0526	1.3420	1.56	1.202
7.0	0.530	0.523	1.0619	1.3435		1.244
8.0	0.612	0.603	1.0713	1.3449		1.289
9.0	0.696	0.685	1.0808	1.3464		1.337
10.0	0.782	0.768	1.0905	1.3479		1.390
11.0	0.870	0.852	1.1002	1.3494		1.447
12.0	0.960	0.938	1.1101	1.3509		1.508
13.0	1.052	1.025	1.1201	1.3524		1.574
14.0	1.146	1.114	1.1301	1.3539		1.646
15.0	1.242	1.204	1.1402	1.3553		1.725
16.0	1.341	1.296	1.1503	1.3567		1.812
17.0	1.442	1.389	1.1604	1.3581		1.905
18.0	1.545	1.483	1.1705	1.3595		2.005
19.0	1.651	1.579	1.1806	1.3608		2.112
20.0	1.760	1.677	1.1907	1.3620		2.227
22.0	1.986	1.875	1.2106	1.3643		2.481

Sodium thiosulfate

Mass %	m/mol kg⁻¹	c/mol L⁻¹	ρ/g cm⁻³	n	Δ/°C	η/mPa s
0.5	0.032	0.032	1.0024	1.3340	0.14	1.012
1.0	0.064	0.064	1.0065	1.3351	0.28	1.023
2.0	0.129	0.128	1.0148	1.3371	0.57	1.044
3.0	0.196	0.194	1.0231	1.3392	0.84	1.066
4.0	0.264	0.261	1.0315	1.3413	1.09	1.090
5.0	0.333	0.329	1.0399	1.3434	1.34	1.115
6.0	0.404	0.398	1.0483	1.3454	1.59	1.141
7.0	0.476	0.468	1.0568	1.3475	1.83	1.169
8.0	0.550	0.539	1.0654	1.3496	2.06	1.199
9.0	0.626	0.611	1.0740	1.3517	2.30	1.231
10.0	0.703	0.685	1.0827	1.3538	2.55	1.267
12.0	0.862	0.835	1.1003	1.3581	3.06	1.345
14.0	1.030	0.990	1.1182	1.3624	3.60	1.435
16.0	1.205	1.150	1.1365	1.3667	4.17	1.537
18.0	1.388	1.315	1.1551	1.3711	4.76	1.657
20.0	1.581	1.485	1.1740	1.3756	5.37	1.798
22.0	1.784	1.660	1.1932	1.3801		1.958
24.0	1.997	1.841	1.2128	1.3847		2.141
26.0	2.222	2.027	1.2328	1.3893		2.356
28.0	2.460	2.219	1.2532	1.3940		2.596
30.0	2.711	2.417	1.2739	1.3987		2.903
32.0	2.976	2.621	1.2950	1.4035		3.305
34.0	3.258	2.831	1.3164	1.4084		3.792
36.0	3.558	3.047	1.3382	1.4132		4.359
38.0	3.876	3.269	1.3603	1.4181		5.011
40.0	4.216	3.498	1.3827	1.4229		5.758

Strontium chloride

Mass %	m/mol kg⁻¹	c/mol L⁻¹	ρ/g cm⁻³	n	Δ/°C	η/mPa s
0.5	0.032	0.032	1.0027	1.3339	0.16	1.012
1.0	0.064	0.064	1.0071	1.3348	0.31	1.021
2.0	0.129	0.128	1.0161	1.3366	0.62	1.039
3.0	0.195	0.194	1.0252	1.3384	0.93	1.057
4.0	0.263	0.261	1.0344	1.3402	1.26	1.076
5.0	0.332	0.329	1.0437	1.3421	1.61	1.096
6.0	0.403	0.399	1.0532	1.3440	1.98	1.116
7.0	0.475	0.469	1.0628	1.3459	2.38	1.136
8.0	0.549	0.541	1.0726	1.3478	2.80	1.157
9.0	0.624	0.615	1.0825	1.3498	3.25	1.180
10.0	0.701	0.689	1.0925	1.3518	3.74	1.204
12.0	0.860	0.843	1.1131	1.3558	4.81	1.258
14.0	1.027	1.002	1.1342	1.3599	6.03	1.317
16.0	1.202	1.167	1.1558	1.3641	7.41	1.383
18.0	1.385	1.338	1.1780	1.3684	8.98	1.460
20.0	1.577	1.515	1.2008	1.3728	10.74	1.549
22.0	1.779	1.699	1.2241	1.3772	12.74	1.650
24.0	1.992	1.890	1.2481	1.3817	14.99	1.765
26.0	2.216	2.087	1.2728	1.3864		1.897
28.0	2.453	2.293	1.2983	1.3911		2.056
30.0	2.703	2.507	1.3248	1.3961		2.245
32.0	2.968	2.730	1.3523	1.4013		2.527
34.0	3.250	2.962	1.3811	1.4067		2.846

Mass %	m/mol kg^{-1}	c/mol L^{-1}	ρ/g cm^{-3}	n	Δ/°C	η/mPa s
36.0	3.548	3.205	1.4114	1.4124		3.206

Sucrose

Mass %	m/mol kg^{-1}	c/mol L^{-1}	ρ/g cm^{-3}	n	Δ/°C	η/mPa s
0.5	0.015	0.015	1.0002	1.3337	0.03	1.015
1.0	0.030	0.029	1.0021	1.3344	0.06	1.028
2.0	0.060	0.059	1.0060	1.3359	0.11	1.055
3.0	0.090	0.089	1.0099	1.3373	0.17	1.084
4.0	0.122	0.118	1.0139	1.3388	0.23	1.114
5.0	0.154	0.149	1.0178	1.3403	0.29	1.146
6.0	0.186	0.179	1.0218	1.3418	0.35	1.179
7.0	0.220	0.210	1.0259	1.3433	0.42	1.215
8.0	0.254	0.241	1.0299	1.3448	0.49	1.254
9.0	0.289	0.272	1.0340	1.3463	0.55	1.294
10.0	0.325	0.303	1.0381	1.3478	0.63	1.336
11.0	0.361	0.335	1.0423	1.3494	0.70	1.381
12.0	0.398	0.367	1.0465	1.3509	0.77	1.429
13.0	0.437	0.399	1.0507	1.3525	0.85	1.480
14.0	0.476	0.431	1.0549	1.3541	0.93	1.534
15.0	0.516	0.464	1.0592	1.3557	1.01	1.592
16.0	0.556	0.497	1.0635	1.3573	1.10	1.653
17.0	0.598	0.530	1.0678	1.3589	1.19	1.719
18.0	0.641	0.564	1.0722	1.3606	1.27	1.790
19.0	0.685	0.598	1.0766	1.3622	1.37	1.865
20.0	0.730	0.632	1.0810	1.3639	1.47	1.945
22.0	0.824	0.700	1.0899	1.3672	1.67	2.124
24.0	0.923	0.771	1.0990	1.3706	1.89	2.331
26.0	1.026	0.842	1.1082	1.3741	2.12	2.573
28.0	1.136	0.914	1.1175	1.3776	2.37	2.855
30.0	1.252	0.988	1.1270	1.3812	2.64	3.187
32.0	1.375	1.063	1.1366	1.3848	2.94	3.762
34.0	1.505	1.139	1.1464	1.3885	3.27	4.052
36.0	1.643	1.216	1.1562	1.3922	3.63	4.621
38.0	1.791	1.295	1.1663	1.3960	4.02	5.315
40.0	1.948	1.375	1.1765	1.3999	4.45	6.162
42.0	2.116	1.456	1.1868	1.4038	4.93	7.234
44.0	2.295	1.539	1.1972	1.4078		8.596
46.0	2.489	1.623	1.2079	1.4118		10.301
48.0	2.697	1.709	1.2186	1.4159		12.515
50.0	2.921	1.796	1.2295	1.4201		15.431
60.0	4.382	2.255	1.2864	1.4419		58.487
70.0	6.817	2.755	1.3472	1.4654		481.561
80.0	11.686	3.299	1.4117	1.4906		
84.0	15.337	3.530	1.4383	1.5010		

Sulfuric acid

Mass %	m/mol kg^{-1}	c/mol L^{-1}	ρ/g cm^{-3}	n	Δ/°C	η/mPa s
0.5	0.051	0.051	1.0016	1.3336	0.21	1.010
1.0	0.103	0.102	1.0049	1.3342	0.42	1.019
2.0	0.208	0.206	1.0116	1.3355	0.80	1.036
3.0	0.315	0.311	1.0183	1.3367	1.17	1.059
4.0	0.425	0.418	1.0250	1.3379	1.60	1.085
5.0	0.537	0.526	1.0318	1.3391	2.05	1.112
6.0	0.651	0.635	1.0385	1.3403	2.50	1.136
7.0	0.767	0.746	1.0453	1.3415	2.95	1.159
8.0	0.887	0.858	1.0522	1.3427	3.49	1.182
9.0	1.008	0.972	1.0591	1.3439	4.08	1.206
10.0	1.133	1.087	1.0661	1.3451	4.64	1.230
11.0	1.260	1.204	1.0731	1.3463	5.25	1.256
12.0	1.390	1.322	1.0802	1.3475	5.93	1.282
13.0	1.524	1.441	1.0874	1.3488	6.67	1.309
14.0	1.660	1.563	1.0947	1.3500	7.49	1.337
15.0	1.799	1.685	1.1020	1.3513	8.35	1.367
16.0	1.942	1.810	1.1094	1.3525	9.26	1.399
17.0	2.088	1.936	1.1169	1.3538	10.23	1.434
18.0	2.238	2.064	1.1245	1.3551	11.29	1.470
19.0	2.392	2.193	1.1321	1.3563	12.43	1.508
20.0	2.549	2.324	1.1398	1.3576	13.64	1.546
22.0	2.876	2.592	1.1554	1.3602	16.48	1.624
24.0	3.220	2.866	1.1714	1.3628	19.85	1.706
26.0	3.582	3.147	1.1872	1.3653	24.29	1.797
28.0	3.965	3.435	1.2031	1.3677	29.65	1.894
30.0	4.370	3.729	1.2191	1.3701	36.21	2.001
32.0	4.798	4.030	1.2353	1.3725	44.76	2.122
34.0	5.252	4.339	1.2518	1.3749	55.28	2.255
36.0	5.735	4.656	1.2685	1.3773		2.392
38.0	6.249	4.981	1.2855	1.3797		2.533
40.0	6.797	5.313	1.3028	1.3821		2.690
42.0	7.383	5.655	1.3205	1.3846		2.872
44.0	8.011	6.005	1.3386	1.3870		3.073
46.0	8.685	6.364	1.3570	1.3895		3.299
48.0	9.411	6.734	1.3759	1.3920		3.546
50.0	10.196	7.113	1.3952	1.3945		3.826
52.0	11.045	7.502	1.4149	1.3971		4.142
54.0	11.969	7.901	1.4351	1.3997		4.499
56.0	12.976	8.312	1.4558	1.4024		4.906
58.0	14.080	8.734	1.4770	1.4050		5.354
60.0	15.294	9.168	1.4987	1.4077		5.917
62.0	16.635	9.608	1.5200			
64.0	18.126	10.063	1.5421			
66.0	19.792	10.529	1.5646			
68.0	21.666	11.006	1.5874			
70.0	23.790	11.494	1.6105			
72.0	26.218	11.994	1.6338			
74.0	29.019	12.505	1.6574			
76.0	32.287	13.026	1.6810			
78.0	36.149	13.554	1.7043			
80.0	40.783	14.088	1.7272			
82.0	46.447	14.623	1.7491			
84.0	53.528	15.153	1.7693			
86.0	62.631	15.671	1.7872			
88.0	74.769	16.170	1.8022			
90.0	91.762	16.649	1.8144			
92.0	117.251	17.109	1.8240			
94.0	159.734	17.550	1.8312			
96.0	244.698	17.966	1.8355			
98.0	499.592	18.346	1.8361			
100.0		18.663	1.8305			

Trichloroacetic acid

Mass %	m/mol kg^{-1}	c/mol L^{-1}	ρ/g cm^{-3}	n	Δ/°C	η/mPa s
0.5	0.031	0.031	1.0008	1.3337	0.11	1.011
1.0	0.062	0.061	1.0034	1.3343	0.21	1.021
2.0	0.125	0.123	1.0083	1.3356	0.42	1.044
3.0	0.189	0.186	1.0133	1.3369	0.64	1.069
4.0	0.255	0.249	1.0182	1.3381	0.86	1.096
5.0	0.322	0.313	1.0230	1.3394	1.08	1.123
6.0	0.391	0.377	1.0279	1.3406	1.30	1.150
7.0	0.461	0.442	1.0328	1.3418	1.53	1.177
8.0	0.532	0.508	1.0378	1.3431	1.76	1.204
9.0	0.605	0.574	1.0428	1.3444	1.99	1.233
10.0	0.680	0.641	1.0479	1.3456	2.23	1.263
12.0	0.835	0.777	1.0583	1.3483	2.73	1.326
14.0	0.996	0.916	1.0692	1.3510	3.26	1.393
16.0	1.166	1.058	1.0806	1.3539	3.82	1.462

Mass %	m/mol kg^{-1}	c/mol L^{-1}	ρ/g cm^{-3}	n	Δ/°C	η/mPa s
18.0	1.343	1.203	1.0921	1.3568		1.533
20.0	1.530	1.351	1.1035	1.3597		1.608
24.0	1.933	1.654	1.1260	1.3652		1.768
28.0	2.380	1.968	1.1485	1.3705		1.935
32.0	2.880	2.294	1.1713	1.3759		2.118
36.0	3.443	2.632	1.1947	1.3813		2.320
40.0	4.080	2.984	1.2188	1.3868		1.543
44.0	4.809	3.349	1.2435	1.3923		2.797
48.0	5.650	3.726	1.2682	1.3977		3.076
50.0	6.120	3.918	1.2803	1.4003		3.225

Tris(hydroxymethyl)methylamine

Mass %	m/mol kg^{-1}	c/mol L^{-1}	ρ/g cm^{-3}	n	Δ/°C	η/mPa s
0.5	0.041	0.041	0.9994	1.3337	0.08	1.014
1.0	0.083	0.083	1.0006	1.3344	0.16	1.027
2.0	0.168	0.166	1.0030	1.3359	0.31	1.054
3.0	0.255	0.249	1.0054	1.3374	0.47	1.083
4.0	0.344	0.333	1.0078	1.3388	0.64	1.115
5.0	0.434	0.417	1.0103	1.3403	0.80	1.148
6.0	0.527	0.502	1.0128	1.3418	0.97	1.182
7.0	0.621	0.587	1.0153	1.3433	1.15	1.218
8.0	0.718	0.672	1.0179	1.3448	1.33	1.256
9.0	0.816	0.758	1.0204	1.3463	1.51	1.295
10.0	0.917	0.844	1.0230	1.3478	1.70	1.337
12.0	1.126	1.019	1.0282	1.3508	2.08	1.427
14.0	1.344	1.194	1.0335	1.3539	2.47	1.527
16.0	1.572	1.372	1.0389	1.3570	2.90	1.642
18.0	1.812	1.552	1.0443	1.3601	3.36	1.772
20.0	2.064	1.733	1.0498	1.3633	3.85	1.920
22.0	2.328	1.917	1.0554	1.3665		2.083
24.0	2.607	2.102	1.0610	1.3697		2.261
26.0	2.900	2.289	1.0666	1.3730		2.464
28.0	3.210	2.478	1.0723	1.3763		2.705
30.0	3.538	2.670	1.0781	1.3797		2.998
32.0	3.885	2.863	1.0839	1.3831		3.344
34.0	4.253	3.058	1.0897	1.3865		3.736
36.0	4.643	3.256	1.0956	1.3900		4.177
38.0	5.059	3.456	1.1016	1.3935		4.667
40.0	5.503	3.657	1.1076	1.3970		5.208

Urea

Mass %	m/mol kg^{-1}	c/mol L^{-1}	ρ/g cm^{-3}	n	Δ/°C	η/mPa s
0.5	0.084	0.083	0.9995	1.3337	0.16	1.007
1.0	0.168	0.167	1.0007	1.3344	0.31	1.010
2.0	0.340	0.334	1.0033	1.3358	0.62	1.012
3.0	0.515	0.502	1.0058	1.3372	0.93	1.017
4.0	0.694	0.672	1.0085	1.3387	1.24	1.025
5.0	0.876	0.842	1.0111	1.3401	1.55	1.033
6.0	1.063	1.013	1.0138	1.3416	1.88	1.041

Mass %	m/mol kg^{-1}	c/mol L^{-1}	ρ/g cm^{-3}	n	Δ/°C	η/mPa s
7.0	1.253	1.185	1.0165	1.3431	2.22	1.049
8.0	1.448	1.358	1.0192	1.3446	2.56	1.057
9.0	1.647	1.531	1.0220	1.3461	2.91	1.065
10.0	1.850	1.706	1.0248	1.3476	3.26	1.074
11.0	2.058	1.882	1.0276	1.3491	3.61	1.083
12.0	2.270	2.059	1.0304	1.3506	3.95	1.091
13.0	2.488	2.236	1.0332	1.3521	4.30	1.100
14.0	2.710	2.415	1.0360	1.3537	4.66	1.109
15.0	2.938	2.594	1.0388	1.3552	5.02	1.119
16.0	3.171	2.775	1.0417	1.3568	5.40	1.130
17.0	3.410	2.956	1.0445	1.3583	5.79	1.141
18.0	3.655	3.139	1.0473	1.3599	6.19	1.153
19.0	3.906	3.322	1.0502	1.3614	6.59	1.165
20.0	4.163	3.506	1.0530	1.3629	7.00	1.178
22.0	4.696	3.878	1.0586	1.3661	7.81	1.205
24.0	5.258	4.253	1.0643	1.3692	8.64	1.235
26.0	5.850	4.632	1.0699	1.3723	9.52	1.266
28.0	6.475	5.014	1.0756	1.3754	10.45	1.298
30.0	7.136	5.401	1.0812	1.3785	11.40	1.332
32.0	7.835	5.791	1.0869	1.3817	12.34	1.371
34.0	8.577	6.185	1.0926	1.3848	13.27	1.413
36.0	9.366	6.584	1.0984	1.3881	14.20	1.459
38.0	10.205	6.988	1.1044	1.3913	15.11	1.509
40.0	11.100	7.397	1.1106	1.3947	15.99	1.565
42.0	12.057	7.812	1.1171	1.3982	16.83	1.629
44.0	13.082	8.234	1.1239	1.4018	17.62	1.700
46.0	14.183	8.665	1.1313	1.4056		1.780

Zinc sulfate

Mass %	m/mol kg^{-1}	c/mol L^{-1}	ρ/g cm^{-3}	n	Δ/°C	η/mPa s
0.5	0.031	0.031	1.0034	1.3339	0.08	1.021
1.0	0.063	0.062	1.0085	1.3348	0.15	1.040
2.0	0.126	0.126	1.0190	1.3366	0.28	1.081
3.0	0.192	0.191	1.0296	1.3384	0.41	1.126
4.0	0.258	0.258	1.0403	1.3403	0.53	1.175
5.0	0.326	0.326	1.0511	1.3421	0.65	1.227
6.0	0.395	0.395	1.0620	1.3439	0.77	1.283
7.0	0.466	0.465	1.0730	1.3457	0.89	1.341
8.0	0.539	0.537	1.0842	1.3475	1.01	1.403
9.0	0.613	0.611	1.0956	1.3494	1.14	1.470
10.0	0.688	0.686	1.1071	1.3513	1.27	1.545
11.0	0.766	0.762	1.1188	1.3532	1.41	1.627
12.0	0.845	0.840	1.1308	1.3551	1.55	1.716
13.0	0.926	0.920	1.1429	1.3570	1.71	1.814
14.0	1.008	1.002	1.1553	1.3590	1.89	1.918
15.0	1.093	1.085	1.1679	1.3610	2.09	2.031
16.0	1.180	1.170	1.1806	1.3630	2.31	2.152

SOLUBILITY OF SELECTED GASES IN WATER

L. H. Gevantman

The values in this table are taken almost exclusively from the International Union of Pure and Applied Chemistry "Solubility Data Series." Unless noted, they comprise evaluated data fitted to a smoothing equation. The data at each temperature are then derived from the smoothing equation which expresses the mole fraction solubility X_1 of the gas in solution as:

$$\ln X_1 = A + B/T^* + C \ln T^*$$

where

$$T^* = T/100 \text{ K}$$

All values refer to a partial pressure of the gas of 101.325 kPa (one atmosphere).

The equation constants, the standard deviation for $\ln X_1$ (except where noted), and the temperature range over which the equation applies are given in the column headed Equation constants. There are two exceptions. The equation for methane has an added term, DT^*. The equation for H_2Se and H_2S takes the form,

$$\ln X_1 = A + B/T + C \ln T + DT$$

where T is the temperature in kelvin.

Solubilities given for those gases that react with water, namely ozone, nitrogen oxides, chlorine and its oxides, carbon dioxide, hydrogen sulfide, hydrogen selenide, and sulfur dioxide, are recorded as bulk solubilities; i.e., all chemical species of the gas and its reaction products with water are included.

Solubility of Gases in Water — Constants in the Solubility Equation and Solubility Values at Selected Temperatures

T/K	Solubility (X_1)	Equation constants	Ref.
Hydrogen (H_2) M_r = 2.01588			
288.15	1.510×10^{-5}	$A = -48.1611$	1
293.15	1.455×10^{-5}	$B = 55.2845$	1
298.15	1.411×10^{-5}	$C = 16.8893$	1
303.15	1.377×10^{-5}	Std. dev. = ±0.54%	1
308.15	1.350×10^{-5}	Temp. range = 273.15—353.15	1
Deuterium (D_2) M_r = 4.0282			
283.15	$1.675 \times 10^{-5} \pm 0.57\%$	Averaged experimental values	1
288.15	$1.595 \times 10^{-5} \pm 0.57\%$		1
293.15	$1.512 \times 10^{-5} \pm 0.78\%$	Temp. range = 278.15—303.15	1
298.15	$1.460 \times 10^{-5} \pm 0.52\%$		1
303.15	$1.395 \times 10^{-5} \pm 0.37\%$		1
Helium (He) A_r = 4.0026			
288.15	7.123×10^{-6}	$A = -41.4611$	2
293.15	7.044×10^{-6}	$B = 42.5962$	2
298.15	6.997×10^{-6}	$C = 14.0094$	2
303.15	6.978×10^{-6}	Std. dev. = ±0.54%	2
308.15	6.987×10^{-6}	Temp. range = 273.15—348.15	2
Neon (Ne) A_r = 20.1797			
288.15	8.702×10^{-6}	$A = -52.8573$	2
293.15	8.395×10^{-6}	$B = 61.0494$	2
298.15	8.152×10^{-6}	$C = 18.9157$	2
303.15	7.966×10^{-6}	Std. dev. = ±0.47%	2
308.15	7.829×10^{-6}	Temp. range = 273.15—348.15	2
Argon (Ar) A_r = 39.948			
288.15	3.025×10^{-5}	$A = -57.6661$	3
293.15	2.748×10^{-5}	$B = 74.7627$	3
298.15	2.519×10^{-5}	$C = 20.1398$	3
303.15	2.328×10^{-5}	Std. dev. = ±0.26%	3
308.15	2.169×10^{-5}	Temp. range = 273.15—348.15	3
Krypton (Kr) A_r = 83.80			
288.15	5.696×10^{-5}	$A = -66.9928$	4
293.15	5.041×10^{-5}	$B = 91.0166$	4
298.15	4.512×10^{-5}	$C = 24.2207$	4
303.15	4.079×10^{-5}	Std. dev. = ±0.32%	4
308.15	3.725×10^{-5}	Temp. range = 273.15—353.15	4
Xenon (Xe) A_r = 131.29			
288.15	10.519×10^{-5}	$A = -74.7398$	4
293.15	9.051×10^{-5}	$B = 105.210$	4
298.15	7.890×10^{-5}	$C = 27.4664$	4
303.15	6.961×10^{-5}	Std. dev. = ±0.35%	4
308.15	6.212×10^{-5}	Temp. range = 273.15—348.15	4
Radon (^{222}Rn) A_r = 222			
288.15	2.299×10^{-4}	$A = -90.5481$	4
293.15	1.945×10^{-4}	$B = 130.026$	4
298.15	1.671×10^{-4}	$C = 35.0047$	4
303.15	1.457×10^{-4}	Std. dev. = ±1.02%	4
308.15	1.288×10^{-4}	Temp. range = 273.15—373.15	4
Oxygen (O_2) M_r = 31.9988			
288.15	2.756×10^{-5}	$A = -66.7354$	5
293.15	2.501×10^{-5}	$B = 87.4755$	5
298.15	2.293×10^{-5}	$C = 24.4526$	5
303.15	2.122×10^{-5}	Std. dev. = ±0.36%	5
308.15	1.982×10^{-5}	Temp. range = 273.15—348.15	5
Ozone (O_3) M_r = 47.9982			
293.15	$1.885 \times 10^{-6} \pm 10\%$	Derived from Henry's Law Constant Equation	5
		pH = 7.0	5
Nitrogen (N_2) M_r = 28.0134			
288.15	1.386×10^{-5}	$A = -67.3877$	6
293.15	1.274×10^{-5}	$B = 86.3213$	6
298.15	1.183×10^{-5}	$C = 24.7981$	6
303.15	1.108×10^{-5}	Std. dev. = ±0.72%	6
308.15	1.047×10^{-5}	Temp. range = 273.15—348.15	6
Nitrous oxide (N_2O) M_r = 44.0129			
288.15	5.948×10^{-4}	$A = -60.7467$	7
293.15	5.068×10^{-4}	$B = 88.8280$	7
298.15	4.367×10^{-4}	$C = 21.2531$	7
303.15	3.805×10^{-4}	Std. dev. = ±1.2%	7
308.15	3.348×10^{-4}	Temp. range = 273.15—313.15	7

T/K	Solubility (X_1)	Equation constants	Ref.
Nitric oxide (NO) M_r = 30.0061			
288.15	4.163×10^{-5}	A = −62.8086	7
293.15	3.786×10^{-5}	B = 82.3420	7
298.15	3.477×10^{-5}	C = 22.8155	7
303.15	3.222×10^{-5}	Std. dev. = ±0.76%	7
308.15	3.012×10^{-5}	Temp. range = 273.15—358.15	7
Carbon monoxide (CO) M_r = 28.0104			
288.15	2.095×10^{-5}	Derived from Henry's Law Constant Equation	8
293.15	1.918×10^{-5}		8
298.15	1.774×10^{-5}	Std. dev. = ±0.043%	8
303.15	1.657×10^{-5}	Temp. range = 273.15—328.15	8
308.15	1.562×10^{-5}		8
Carbon dioxide (CO_2) M_r = 44.0098			
288.15	8.21×10^{-4}	Derived from Henry's Law Constant Equation	9
293.15	7.07×10^{-4}		9
298.15	6.15×10^{-4}	Std. dev. = ±1.1%	9
303.15	5.41×10^{-4}	Temp. range = 273.15—353.15	9
308.15	4.80×10^{-4}		9
Hydrogen selenide (H_2Se) M_r = 80.976			
288.15	1.80×10^{-3}	A = 9.15	10
298.15	1.49×10^{-3}	B = 974	10
308.15	1.24×10^{-3}	C = −3.542	10
		D = 0.0042	10
		Std. dev. = $\pm 2.3 \times 10^{-5}$	10
		Temp. range = 288.15—343.15	10
Hydrogen sulfide (H_2S) M_r = 34.082			
288.15	2.335×10^{-3}	A = −24.912	10
293.15	2.075×10^{-3}	B = 3477	10
298.15	1.85×10^{-3}	C = 0.3993	10
303.15	1.66×10^{-3}	D = 0.0157	10
308.15	1.51×10^{-3}	Std. dev. = $\pm 6.5 \times 10^{-5}$	10
		Temp. range = 283.15—603.15	11
Sulfur dioxide (SO_2) M_r = 64.0648			
288.15	3.45×10^{-2}	A = −25.2629	11
293.15	2.90×10^{-2}	B = 45.7552	11
298.15	2.46×10^{-2}	C = 5.6855	11
303.15	2.10×10^{-2}	Std. dev. = ±1.8%	11
308.15	1.80×10^{-2}	Temp. range = 278.15—328.15	11
Chlorine (Cl_2) M_r = 70.9054			
283.15	$2.48 \times 10^{-3} \pm 2\%$	Experimental data	11
293.15	$1.88 \times 10^{-3} \pm 2\%$	Temp. range = 283.15—333.15	11
303.15	$1.50 \times 10^{-3} \pm 2\%$		11
313.15	$1.23 \times 10^{-3} \pm 2\%$		11
Chlorine monoxide (Cl_2O) M_r = 86.9048			
273.15	$5.25 \times 10^{-1} \pm 1\%$	Experimental data	11
276.61	$4.54 \times 10^{-1} \pm 1\%$	Temp. range = 273.15—293.15	11
283.15	$4.273 \times 10^{-1} \pm 1\%$		11
293.15	$3.353 \times 10^{-1} \pm 1\%$		11
Chlorine dioxide (ClO_2) M_r = 67.4515			
288.15	2.67×10^{-2}	A = 7.9163	11
293.15	2.20×10^{-2}	B = 0.4791	11
298.15	1.823×10^{-2}	C = 11.0593	11
303.15	1.513×10^{-2}	Std. dev. = ±4.6%	11
308.15	1.259×10^{-2}	Temp. range = 283.15—333.15	11
Methane (CH_4) M_r = 16.0428			
288.15	3.122×10^{-5}	A = −115.6477	12
293.15	2.806×10^{-5}	B = 155.5756	12
298.15	2.552×10^{-5}	C = 65.2553	12
303.15	2.346×10^{-5}	D = −6.1698	12
308.15	2.180×10^{-5}	Std. dev. = ±0.056%	12
		Temp. range = 273.15—328.15	12
Ethane (C_2H_6) M_r = 30.0696			
288.15	4.556×10^{-5}	A = −90.8225	13
293.15	3.907×10^{-5}	B = 126.9559	13
298.15	3.401×10^{-5}	C = 34.7413	13
303.15	3.002×10^{-5}	Std. dev. = ±0.13%	13
308.15	2.686×10^{-5}	Temp. range = 273.15—323.15	13
Propane (C_3H_8) M_r = 44.097			
288.15	3.813×10^{-5}	A = −102.044	14
293.15	3.200×10^{-5}	B = 144.345	14
298.15	2.732×10^{-5}	C = 39.4740	14
303.15	2.370×10^{-5}	Std. dev. = ±0.012%	14
308.15	2.088×10^{-5}	Temp. range = 273.15—347.15	14
Butane (C_4H_{10}) M_r = 58.123			
288.15	3.274×10^{-5}	A = −102.029	14
293.15	2.687×10^{-5}	B = 146.040	14
298.15	2.244×10^{-5}	C = 38.7599	14
303.15	1.906×10^{-5}	Std. dev. = ±0.026%	14
308.15	1.645×10^{-5}	Temp. range = 273.15—349.15	14
Isobutane (C_4H_{10}) M_r = 58.123			
288.15	2.333×10^{-5}	A = −129.714	14
293.15	1.947×10^{-5}	B = 183.044	14
298.15	1.659×10^{-5}	C = 53.4651	14
303.15	1.443×10^{-5}	Std. dev. = ±0.034%	14
308.15	1.278×10^{-5}	Temp. range = 278.15—318.15	14

References

1. C. L. Young, Ed., *IUPAC Solubility Data Series*, Vol. 5/6, Hydrogen and Deuterium, Pergamon Press, Oxford, England, 1981.
2. H. L. Clever, Ed., *IUPAC Solubility Data Series*, Vol. 1, Helium and Neon, Pergamon Press, Oxford, England, 1979.
3. H. L. Clever, Ed., *IUPAC Solubility Data Series*, Vol. 4, Argon, Pergamon Press, Oxford, England, 1980.
4. H. L. Clever, Ed., *IUPAC Solubility Data Series*, Vol. 2, Krypton, Xenon and Radon, Pergamon Press, Oxford, England, 1979.
5. R. Battino, Ed., *IUPAC Solubility Data Series*, Vol. 7, Oxygen and Ozone, Pergamon Press, Oxford, England, 1981.
6. R. Battino, Ed., *IUPAC Solubility Data Series*, Vol. 10, Nitrogen and Air, Pergamon Press, Oxford, England, 1982.
7. C. L. Young, Ed., *IUPAC Solubility Data Series*, Vol. 8, Oxides of Nitrogen, Pergamon Press, Oxford, England, 1981.
8. R. W. Cargill, Ed., *IUPAC Solubility Data Series*, Vol. 43, Carbon Monoxide, Pergamon Press, Oxford, England, 1990.
9. R. Crovetto, Evaluation of Solubility Data for the System CO_2-H_2O, *J. Phys. Chem. Ref. Data*, 20, 575, 1991.
10. P. G. T. Fogg and C. L. Young, Eds., *IUPAC Solubility Data Series*, Vol. 32, Hydrogen Sulfide, Deuterium Sulfide, and Hydrogen Selenide, Pergamon Press, Oxford, England, 1988.
11. C. L. Young, Ed., *IUPAC Solubility Data Series*, Vol. 12, Sulfur Dioxide, Chlorine, Fluorine and Chlorine Oxides, Pergamon Press, Oxford, England, 1983.
12. H. L. Clever and C. L. Young, Eds., *IUPAC Solubility Data Series*, Vol. 27/28, Methane, Pergamon Press, Oxford, England, 1987.
13. W. Hayduk, Ed., *IUPAC Solubility Data Series*, Vol. 9, Ethane, Pergamon Press, Oxford, England, 1982.
14. W. Hayduk, Ed., *IUPAC Solubility Data Series*, Vol. 24, Propane, Butane and 2-Methylpropane, Pergamon Press, Oxford, England, 1986.

SOLUBILITY OF CARBON DIOXIDE IN WATER AT VARIOUS TEMPERATURES AND PRESSURES

The solubility of CO_2 in water, expressed as mole fraction of CO_2 in the liquid phase, is given for pressures up to atmospheric and temperatures of 0 to 100 °C. Note that 1 standard atmosphere equals 101.325 kPa. The references give data over a wider range of temperature and pressure. The estimated uncertainty is about 2%.

References

1. Carroll, J. J., Slupsky, J. D., and Mather, A. E., *J. Phys. Chem. Ref. Data*, 20, 1201, 1991.
2. Fernandez-Prini, R. and Crovetto, R., *J.Phys. Chem. Ref. Data*, 18, 1231, 1989.
3. Crovetto, R., *J. Phys. Chem. Ref. Data*, 20, 575,1991.

Solubility of CO_2 in Water - 1000 × Mole Fraction of CO_2 in Liquid Phase as a Function of CO_2 Partial Pressure

t/°C	5 kPa	10 kPa	20 kPa	30 kPa	40 kPa	50 kPa	100 kPa
0	0.067	0.135	0.269	0.404	0.538	0.671	1.337
5	0.056	0.113	0.226	0.338	0.451	0.564	1.123
10	0.048	0.096	0.191	0.287	0.382	0.477	0.950
15	0.041	0.082	0.164	0.245	0.327	0.409	0.814
20	0.035	0.071	0.141	0.212	0.283	0.353	0.704
25	0.031	0.062	0.123	0.185	0.247	0.308	0.614
30	0.027	0.054	0.109	0.163	0.218	0.271	0.541
35	0.024	0.048	0.097	0.145	0.193	0.242	0.481
40	0.022	0.043	0.087	0.130	0.173	0.216	0.431
45	0.020	0.039	0.078	0.117	0.156	0.196	0.389
50	0.018	0.036	0.071	0.107	0.142	0.178	0.354
55	0.016	0.033	0.065	0.098	0.131	0.163	0.325
60	0.015	0.030	0.060	0.090	0.121	0.150	0.300
65	0.014	0.028	0.056	0.084	0.112	0.140	0.279
70	0.013	0.026	0.052	0.079	0.105	0.131	0.261
75	0.012	0.025	0.049	0.074	0.099	0.123	0.245
80	0.012	0.023	0.047	0.070	0.093	0.116	0.232
85	0.011	0.022	0.044	0.067	0.089	0.111	0.221
90	0.011	0.021	0.042	0.064	0.085	0.106	0.211
95	0.010	0.020	0.041	0.061	0.082	0.102	0.203
100	0.010	0.020	0.039	0.059	0.079	0.098	0.196

AQUEOUS SOLUBILITY AND HENRY'S LAW CONSTANTS OF ORGANIC COMPOUNDS

The solubility in water of about 1300 organic compounds, including many compounds of environmental interest, is tabulated here. When data are available, values are given at several temperatures between 0 °C and 100 °C. Solids, liquids, and gases are included; additional data on gases can be found in the table "Solubility of Selected Gases in Water" in this section.

Solubility of solids is defined as the concentration of the compound in a solution that is in equilibrium with the solid phase at the specified temperature and one atmosphere pressure. For liquids whose water mixtures separate into two phases, the solubility given here is the concentration of the specified compound in the water-rich phase at equilibrium. In the case of gases (i.e., compounds whose vapor pressure at the specified temperature exceeds one atmosphere) the solubility is defined here as the concentration in the water phase when the partial pressure of the compound above the solution is 101.325 kPa (1 atm). Values for gases are marked with an asterisk.

The solubility values in this table are expressed as mass percent of solute, $s = 100w_2$, where the mass fraction w_2 is defined as

$$w_2 = m_2/(m_1 + m_2)$$

where m_2 is the mass of solute and m_1 the mass of water. For convenience, the solubility expressed in grams of solute that will dissolve in 1 kilogram of water is tabulated in the adjacent column to mass percent. For compounds with low solubility (e.g., $s < 1\%$), that column is, to a high approximation, numerically identical to the solubility expressed in grams of solute per liter of solution.

The mass fraction w_2 is related to other common measures of solubility as follows:

Molality: $m_2 = 1000\, w_2/M_2(1 - w_2)$
Molarity: $c_2 = 1000\rho w_2/M_2$
Mole fraction: $x_2 = (w_2/M_2)/\{(w_2/M_2) + (1 - w_2)/M_1\}$
Mass of solute per 100 g of H_2O: $100w_2/(1 - w_2)$
Mass of solute per liter of solution: $1000\rho w_2$

Here, M_2 is the molar mass of the solute, M_1 = 18.015 g/mol is the molar mass of water, and ρ is the density of the solution in g/mL.

Data have been selected from evaluated sources wherever possible, in particular the *IUPAC Solubility Data Series*. Many values come from experimental measurements reported in the *Journal of Chemical and Engineering Data* and the *Journal of Chemical Thermodynamics*, as well as critical review papers in the *Journal of Physical and Chemical Reference Data*. The primary source for each value is listed in the column following the solubility values; additional references of interest are sometimes given. Many of the references contain solubility data at other temperatures and pH values and in the presence of other compounds. The user is cautioned that wide variations of data are found in the literature for the lower solubility compounds. The references should be consulted for more information on these compounds.

The table also contains values of the Henry's Law constant k_H, which provides a measure of the partition of a substance between the atmosphere and the aqueous phase. Here, k_H is defined as the limit of p_2/c_2 as the concentration approaches zero, where p_2 is the partial pressure of the solute above the solution and c_2 is the concentration in the solution at equilibrium (other formulations of Henry's Law are often used; see Ref. 5). The values of k_H listed here are based on direct experimental measurement whenever available, but many of them are simply calculated as the ratio of the pure compound vapor pressure to the solubility. This approximation is reliable only for compounds of very low solubility. In fact, values of k_H found in the literature frequently differ by a factor of two or three, and variations over an order of magnitude are not unusual (Ref. 5). Therefore, the data given here should be taken only as a rough indication of the true Henry's Law constant, which is difficult to measure precisely.

All values of k_H refer to 25 °C. If the vapor pressure of the compound at 25 °C is greater than one atmosphere, it can be assumed that the k_H value has been calculated as $101.325/c_2$. The source of the Henry's Law data is given in the last column. The air-water partition coefficient (i.e., ratio of air concentration to water concentration when both are expressed in the same units) is equal to k_H/RT or $k_H/2.48$ in the units used here.

Compounds are listed alphabetically by systematic name.

References

1. *Solubility Data Series, International Union of Pure and Applied Chemistry, Vol. 15*, Pergamon Press, Oxford, 1982.
2. *Solubility Data Series, International Union of Pure and Applied Chemistry, Vol. 20*, Pergamon Press, Oxford, 1985.
3. *Solubility Data Series, International Union of Pure and Applied Chemistry, Vol. 37*, Pergamon Press, Oxford, 1988.
4. *Solubility Data Series, International Union of Pure and Applied Chemistry, Vol. 38*, Pergamon Press, Oxford, 1988.
5. Mackay, D., and Shiu, W. Y., *J. Phys. Chem. Ref. Data*, 10, 1175, 1981.
6. Pearlman, R. S., and Yalkowsky, S. H., *J. Phys. Chem. Ref. Data*, 13, 975, 1984.
7. Shiu, W. Y., and Mackay, D., *J. Phys. Chem. Ref. Data*, 15, 911, 1986.
8. Varhanickova, D., Lee, S. C., Shiu, W. Y., and Mackay, D., *J. Chem. Eng. Data*, 40, 620, 1995.
9. Miller, M. M., Ghodbane, S., Wasik, S. P., Tewari, Y. B., and Martire, D. E., *J. Chem. Eng. Data*, 29, 184, 1984.
10. Riddick, J. A., Bunger, W. B., and Sakano, T. K., *Organic Solvents, Fourth Edition*, John Wiley & Sons, New York, 1986.
11. Mackay, D., Shiu, W. Y., and Ma, K. C., *Illustrated Handbook of Physical-Chemical Properties and Environmental Fate for Organic Chemicals, Vol. I*, Lewis Publishers/CRC Press, Boca Raton, FL, 1992.
12. Mackay, D., Shiu, W. Y., and Ma, K. C., *Illustrated Handbook of Physical-Chemical Properties and Environmental Fate for Organic Chemicals, Vol. II*, Lewis Publishers/CRC Press, Boca Raton, FL, 1992.
13. Mackay, D., Shiu, W. Y., and Ma, K. C., *Illustrated Handbook of Physical-Chemical Properties and Environmental Fate for Organic Chemicals, Vol. III*, Lewis Publishers/CRC Press, Boca Raton, FL, 1993.
14. Horvath, A. L., *Halogenated Hydrocarbons*, Marcel Dekker, New York, 1982.
15. Howard, P. H., *Handbook of Environmental Fate and Exposure Data for Organic Chemicals, Vol. I*, Lewis Publishers/CRC Press, Boca Raton, FL, 1989.
16. Howard, P. H., *Handbook of Environmental Fate and Exposure Data for Organic Chemicals, Vol. II*, Lewis Publishers/CRC Press, Boca Raton, FL, 1990.
17. Banergee, S., Yalkowsky, S. H., and Valvani, S. C., *Environ. Sci. Technol.*, 14, 1227, 1980.
18. Gevantman, L. H., in *CRC Handbook of Chemistry and Physics, 90th Edition*, pp. 8–80, CRC Press, Boca Raton, FL, 2009.
19. Wilhelm, E., Battino, R., and Wilcock, R. J., *Chem. Rev.* 77, 219, 1977.
20. Stephenson, R. M., *J. Chem. Eng. Data*, 37, 80, 1992.

21. Stephenson, R. M., Stuart, J., and Tabak, M., *J. Chem. Eng. Data*, 29, 287, 1984.
22. Shiu, W.-Y., and Ma, K.-C, *J. Phys. Chem. Ref. Data*, 29, 41, 2000.
23. Lun, R., Varhanickova, D., Shiu, W.-Y., and Mackay, D., *J. Chem. Eng. Data*, 42, 951 (1997).
24. Huang, G.-L., Xiao, H., Chi, J., Shiu, W.-Y., and Mackay, D., *J. Chem. Eng. Data*, 45, 411, 2000.
25. Horvath, A. L., Getzen, F. W., and Maczynska, Z., *J. Phys. Chem. Ref. Data*, 28, 395, 2000 [IUPAC No. 67].
26. Dawson, R. M. C., Elliott, D. C., Elliott, W. H., and Jones, K. M., *Data for Biochemical Research*, Third Edition, Clarendon Press, Oxford, 1986.
27. Stephen, H., and Stephen, T., *Solubilities of Organic and Inorganic Compounds*, MacMillan, New York, 1963.
28. Shiu, W.-Y., and Mackay, D., *J. Chem. Eng. Data* 42, 27, 1997.
29. Hinz, H.-J., ed., *Thermodynamic Data for Biochemistry and Biotechnology*, Springer-Verlag, Berlin, 1986.
30. Budavari, S., ed., *The Merck Index, Twelfth Edition*, Merck & Co., Rahway, NJ, 1996.
31. Bamford, H. A., Poster, D. L., and Baker, J. E., *J. Chem. Eng. Data*, 45, 1069, 2000.
32. Lide, D. R., and Milne, G. W. A., *Handbook of Data on Organic Compounds, Third Edition*, CRC Press, Boca Raton, FL, 1994.
33. Apelblat, A., and Manzurola, E., *J. Chem. Thermodynamics* 21, 1005, 1989.
34. Apelblat, A., and Manzurola, E., *J. Chem. Thermodynamics* 22, 289, 1990.
35. Horvath, A. L., and Getzen, F. W., *J. Phys. Chem. Ref. Data* 28, 649, 1999 [IUPAC No. 68].
36. Sazonov, V. P., Marsh, K. N., and Hefter, G. T., *J. Phys. Chem. Ref. Data* 29, 1165, 2000 [IUPAC No. 71].
37. Verbruggen, E. M. J., Hermens, J. L. M., and Tolls, J., *J. Phys. Chem. Ref. Data* 29, 1435, 2000.
38. Sazonov, V. P., Shaw, D. G., and Marsh, K. N., *J. Phys. Chem. Ref. Data* 31, 1, 2002 [IUPAC No. 77].
39. Sazonov, V. P., and Shaw, D. G., *J. Phys. Chem. Ref. Data* 31, 989, 2002 [IUPAC No. 78].
40. Yalkowsky, S. H., and He, Y., *Handbook of Aqueous Solubility Data*, CRC Press, Boca Raton, FL, 2003.
41. Shiu, W.-Y., and Ma, K.-C., *J. Phys. Chem. Ref. Data* 29, 387, 2000.
42. Shaw, D. G., and Maczynski, A., *J. Phys. Chem. Ref. Data* 35, 687, 2006 [IUPAC No. 81, Part 11].
43. Nordstrom, F. L., and Rasmuson, A. C., *J. Chem. Eng. Data* 51, 1668, 2006.
44. Nordstrom, F. L., and Rasmuson, A. C., *J. Chem. Eng. Data* 51, 1775, 2006.
45. Sapoundjiev, D., Lorenz, H,. and Seidel-Morgenstern, A., *J. Chem. Eng. Data* 51, 1562, 2006.
46. Marche, C., Ferronato, C., and Jose, J., *J. Chem. Eng. Data* 48, 967, 2003.
47. Lu, J., Wang, X., Yang, X., and Ching, C., *J. Chem. Eng. Data* 51, 1593, 2006.
48. Achard, C., Jaoui, M., Schwing, M., and Rogalski, M., *J. Chem. Eng. Data* 41, 504, 1996.
49. Shareef, A., et al., *J. Chem. Eng. Data* 51, 879, 2006.
50. Clever, H. L., et al., *J. Phys. Chem. Ref. Data* 34, 201, 2005 [IUPAC No. 80].
51. Jaoui, M., Achard, C., and Rogalski, M., *J. Chem. Eng. Data* 47, 297, 2002.
52. Fichan, I., Larroche, C., and Gros, J. B., *J. Chem. Eng. Data* 44, 56, 1999.
53. Freire, M. G., et al., *J. Chem. Eng. Data* 50, 237, 2005.
54. Domanska, U., and Kozlowska, M. K., *J. Chem. Eng. Data* 47, 456, 2002.
55. Phelan, J. M., and Barnett, J. L., *J. Chem. Eng. Data* 46, 375, 2001.
56. Long, B-W., Wang, L-S., and Wu, J-S., *J. Chem. Eng. Data* 50, 136, 2005.
57. Marche, C., Ferronato, C., and Jose, J., *J. Chem. Eng. Data* 49, 937, 2004.
58. Oleszek-Kudlak, S., Shibata, E., and Nakamura, T., *J. Chem. Eng. Data* 49, 570, 2004.
59. Lynch, J. C., et al., *J. Chem. Eng. Data* 46, 1549, 2001.
60. Xiao, H., Li, N., and Wania, F., *J. Chem. Eng. Data* 49, 173, 2004.
61. Ma, J. H. Y., Hung, H., Shiu, W-Y., and Mackay, D., *J. Chem. Eng. Data* 46, 619, 2001.
62. Carta, R., and Tola, G., *J. Chem. Eng. Data* 41, 414, 1996; 44, 563, 1999.
63. Kao, H. D., et al., *Pharm. Res.* 17, 978, 2000.
64. Heric, E. L., and Langford, R. E., *J. Chem. Eng. Data* 17, 471, 1972.
65. Marche, C., Delépine, H., Ferronato, C., and Jose, J., *J. Chem. Eng. Data* 48, 398, 2003.
66. Wang, L-C, and Wang, F-A, *J. Chem. Eng. Data* 49, 155, 2004.
67. Shen, L, and Wania, F., *J. Chem. Eng. Data* 50, 742, 2005.
68. Oleszek-Kudlak, S., Shibata, E., and Nakamura, T., *J. Chem. Eng. Data* 52, 1824, 2007.
69. Zhao, H-K., Li, R-R, Ji, H-Z, Zhang, D-S., Tang, C., and Yang, L-Q., *J. Chem. Eng. Data* 52, 2072, 2007.
70. Yang, X., Wang, X., and Ching, C. B., *J. Chem. Eng. Data* 53, 1133, 2008.
71. Liu, L., and Chen, J., *J. Chem. Eng. Data* 53, 1649, 2008.
72. Szterner, P., *J. Chem. Eng. Data* 53, 1738, 2008.
73. Kong, M-Z., Shi, X-H, Cao, Y-C., and Zhou, C-R., *J. Chem. Eng. Data* 53, 615, 2008.
74. Daneshfar, A., Ghaziaskar, H. S., and Homayoun, N., *J. Chem. Eng. Data* 53, 776, 2008.
75. Manzurola, E., and Apelblat, A., *J. Chem. Thermodynamics* 34, 1127, 2002.
76. Apelblat, A., Manzurola, E., and Balal, N. A., *J. Chem. Thermodynamics* 38, 565, 2006.
77. Apelblat, A., and Mishelevich, A., *J. Chem. Thermodynamics* 40, 897, 2008.
78. Góral, M., Wiśniewska-Goclowska, B., and Mączyński, A., *J. Phys. Chem. Ref. Data* 35, 1391, 2006.
79. Mączyński, A., Shaw, D. G., Góral, M., and Wiśniewska-Goclowska, B., *J. Phys. Chem. Ref. Data* 37, 1119, 2008.
80. Mączyński, A., Shaw, D. G., Góral, M., and Wiśniewska-Goclowska, B., *J. Phys. Chem. Ref. Data* 37, 1147, 2008.
81. Mączyński, A., Shaw, D. G., Góral, M., and Wiśniewska-Goclowska, B., *J. Phys. Chem. Ref. Data* 37, 1169, 2008.
82. Mączyński, A., Shaw, D. G., Góral, M., and Wiśniewska-Goclowska, B., *J. Phys. Chem. Ref. Data* 37, 1517, 2008.
83. Mączyński, A., Shaw, D. G., Góral, M., and Wiśniewska-Goclowska, B., *J. Phys. Chem. Ref. Data* 37, 1575, 2008.
84. Mączyński, A., Shaw, D. G., Góral, M., and Wiśniewska-Goclowska, B., *J. Phys. Chem. Ref. Data* 37, 1611, 2008.
85. Luning Prak, D. J., and O'Sullivam, D. W., *J. Chem. Eng. Data* 51, 448, 2006.
86. Shiu, W.-Y., Wania, F., Hung, H., and Mackay, D., *J. Chem. Eng. Data* 42, 293, 1997.
87. Mączyński, A., and Shaw, D. G., *J. Phys. Chem. Ref. Data* 36, 59, 2007.
88. Mączyński, A., and Shaw, D. G., *J. Phys. Chem. Ref. Data* 36, 133, 2007.
89. Góral, M., Wiśniewska-Goclowska, B., and Mączyński, A., *J. Phys. Chem. Ref. Data* 33, 1159, 2004.
90. Heric, E. L., and Langford, R. E., *J. Chem. Eng. Data* 17, 209, 1972.

Aqueous Solubility and Henry's Law Constants for Organic Compounds

Name	Mol. form.	CAS Reg. No.	Mol. wt.	t/°C	s/mass%	s/g kg^{-1} H_2O	Ref.	k_H/ kPa m^3 mol^{-1}	Ref.
Acenaphthene	$C_{12}H_{10}$	83-32-9	154.207	0	0.00015	0.0015	4		
				25	0.000380	0.00380	22	0.01217	22
				50	0.00092	0.0092	4		
Acenaphthylene	$C_{12}H_8$	208-96-8	152.192	20	0.0016	0.016	28	0.012	28
Acephate	$C_4H_{10}NO_3PS$	30560-19-1	183.166	20	≈28	≈390	40		
Acetamide	C_2H_5NO	60-35-5	59.067	20	40.8	689	10		
Acetanilide	C_8H_9NO	103-84-4	135.163	20	0.52	5.2	27		
				70	2.7	28	27		
Acetazolamide	$C_4H_6N_4O_3S_2$	59-66-5	222.246	30	0.10	1.0	40		
Acetohexamide	$C_{15}H_{20}N_2O_4S$	968-81-0	324.396	37	0.0013	0.013	40		
Acetonitrile	C_2H_3N	75-05-8	41.052	-10	31.7	464	39		
				-3	40.5	681	39		
Acetophenone	C_8H_8O	98-86-2	120.149	20	0.67	6.7	84	0.00108	28
				50	0.81	8.2	84	0.00108	28
				80	1.16	11.7	84	0.00108	28
Acetylene	C_2H_2	74-86-2	26.037	25	0.108	1.08	19		
2-(Acetyloxy)benzoic acid	$C_9H_8O_4$	50-78-2	180.158		0.25	2.5	27		
2-(Acetyloxy)-5-bromobenzoic acid	$C_9H_7BrO_4$	1503-53-3	259.054		0.07	0.7	30		
Acridine	$C_{13}H_9N$	260-94-6	179.217	25	0.00466	0.0466	6		
Acrolein	C_3H_4O	107-02-8	56.063	20	20.8	263	10		
Acrylamide	C_3H_5NO	79-06-1	71.078	20	≈27	≈370	40		
Acrylonitrile	C_3H_3N	107-13-1	53.063	20	7.35	79.3	10		
Adenine	$C_5H_5N_5$	73-24-5	135.128	25	0.104	1.04	29		
Adenosine	$C_{10}H_{13}N_5O_4$	58-61-7	267.242	25	0.51	5.1	29		
Alachlor	$C_{14}H_{20}ClNO_2$	15972-60-8	269.768	23	0.024	0.24	40		
L-Alanine	$C_3H_7NO_2$	56-41-7	89.094	25	14.30	167	26		
β-Alanine	$C_3H_7NO_2$	107-95-9	89.094	25	47.1	890	26		
Aldicarb	$C_7H_{14}N_2O_2S$	116-06-3	190.263	20	0.60	6.0	40		
Aldrin	$C_{12}H_8Cl_6$	309-00-2	364.910	25	0.00002	0.0002	67		
Allopurinol	$C_5H_4N_4O$	315-30-0	136.112	25	0.057	0.57	40		
Ametryn	$C_9H_{17}N_5S$	834-12-8	227.330	20	0.0190	0.190	40		
2-Amino-9,10-anthracenedione	$C_{14}H_9NO_2$	117-79-3	223.227	25	0.000016	0.00016	40		
4-Aminoazobenzene	$C_{12}H_{11}N_3$	60-09-3	197.235	25	0.0030	0.030	40		
				97	0.068	0.68	40		
4-Aminobenzenesulfonamide	$C_6H_8N_2O_2S$	63-74-1	172.205	20	0.71	7.2	40		
4-Aminobenzenesulfonic acid	$C_6H_7NO_3S$	121-57-3	173.190	7	0.59	5.9	27		
DL-2-Aminobutanoic acid	$C_4H_9NO_2$	2835-81-6	103.120	25	17.4	211	26		
DL-3-Aminobutanoic acid	$C_4H_9NO_2$	2835-82-7	103.120	25	55.6	1250	26		
4-Amino-*N*-[(butylamino)carbonyl] benzenesulfonamide	$C_{11}H_{17}N_3O_3S$	339-43-5	271.336	37	0.053	0.53	40		
3-Amino-2,5-dichlorobenzoic acid	$C_7H_5Cl_2NO_2$	133-90-4	206.027	25	0.070	0.70	40		
4-(2-Aminoethyl)phenol	$C_8H_{11}NO$	51-67-2	137.179	15	1.03	10.4	40		
6-Aminohexanoic acid	$C_6H_{13}NO_2$	60-32-2	131.173	25	46	852	29		
4-Amino-2-hydroxybenzoic acid	$C_7H_7NO_3$	65-49-6	153.136	20	0.20	2.0	40		
2-Amino-2-methylpropanoic acid	$C_4H_9NO_2$	62-57-7	103.120	25	12.1	138	26		
2-Aminophenol	C_6H_7NO	95-55-6	109.126	20	1.92	19.6	40		
3-Aminophenol	C_6H_7NO	591-27-5	109.126	20	2.56	26.3	40		
				70	≈24	≈320	40		
4-Aminophenol	C_6H_7NO	123-30-8	109.126	20	1.55	15.7	40		
Aminopyrine	$C_{13}H_{17}N_3O$	58-15-1	231.293	25	4.8	50	40		
Amitriptyline	$C_{20}H_{23}N$	50-48-6	277.404	24	0.00097	0.0097	40		
Amobarbital	$C_{11}H_{18}N_2O_3$	57-43-2	226.272	25	0.06	0.6	40		
Anilazine	$C_9H_5Cl_3N_4$	101-05-3	275.522	20	0.001	0.01	40		
Aniline	C_6H_7N	62-53-3	93.127	25	3.38	35.0	10	14	15
Aniline-2-carboxylic acid	$C_7H_7NO_2$	118-92-3	137.137	20	0.349	3.49	40		
Aniline-4-carboxylic acid	$C_7H_7NO_2$	150-13-0	137.137	25	0.54	5.4	40		
Aniline hydrochloride	C_6H_8ClN	142-04-1	129.588	15	15.1	178	27		
Anisole	C_7H_8O	100-66-3	108.138	20	0.203	2.03	20	0.025	13
				40	0.184	1.84	20	0.025	13

Name	Mol. form.	CAS Reg. No.	Mol. wt.	t/°C	s/mass%	s/g kg^{-1} H_2O	Ref.	k_H/ kPa m^3 mol^{-1}	Ref.
				81	0.294	2.95	20	0.025	13
Anthracene	$C_{14}H_{10}$	120-12-7	178.229	0	0.0000022	0.000022	42,4		
				25	0.0000044	0.000044	42,22	0.00396	22
				50	0.000029	0.00029	42		
9,10-Anthracenedione	$C_{14}H_8O_2$	84-65-1	208.213	25	0.00014	0.0014	40		
Apomorphine	$C_{17}H_{17}NO_2$	58-00-4	267.323	25	2.0	20	40		
L-Arginine	$C_6H_{14}N_4O_2$	74-79-3	174.201	25	15.44	183	26		
L-Ascorbic acid	$C_6H_8O_6$	50-81-7	176.124	25	25.2	337	33		
				50	41.0	695	33		
L-Asparagine	$C_4H_8N_2O_3$	70-47-3	132.118	25	2.45	25.1	26		
L-Aspartic acid	$C_4H_7NO_4$	56-84-8	133.104	10	0.29	2.9	77		
				25	0.49	4.9	77		
				50	1.31	13.3	77		
Atrazine	$C_8H_{14}ClN_5$	1912-24-9	215.684	25	0.007	0.07	26		
Atropine	$C_{17}H_{23}NO_3$	51-55-8	289.370	20	0.3	3	40		
Azinphos-methyl	$C_{10}H_{12}N_3O_3PS_2$	86-50-0	317.324	20	0.00209	0.0209	40		
trans-Azobenzene	$C_{12}H_{10}N_2$	17082-12-1	182.220	20	0.03	0.3	27		
Bayleton	$C_{14}H_{16}ClN_3O_2$	43121-43-3	293.749	20	0.026	0.26	40		
Bendiocarb	$C_{11}H_{13}NO_4$	22781-23-3	223.226	25	0.004	0.04	40		
Bentazon	$C_{10}H_{12}N_2O_3S$	25057-89-0	240.278	20	0.050	0.50	40		
Benzaldehyde	C_7H_6O	100-52-7	106.122	20	0.3	3	10		
Benzamide	C_7H_7NO	55-21-0	121.137	12	0.577	5.77	27		
Benz[*a*]anthracene	$C_{18}H_{12}$	56-55-3	228.288	10	0.00000038	0.0000038	42		
				25	0.00000093	0.0000093	42,22	0.00058	22
Benzene	C_6H_6	71-43-2	78.112	10	0.174	1.74	22		
				20	0.177	1.77	22		
				30	0.183	1.83	22		
				40	0.192	1.92	22		
				50	0.206	2.06	22		
				70	0.249	2.50	65		
				101	0.398	4.00	65		
Benzeneacetic acid	$C_8H_8O_2$	103-82-2	136.149	25	1.71	17.4	27		
1,2-Benzenediamine	$C_6H_8N_2$	95-54-5	108.141	20	3.02	31.1	40		
1,3-Benzenediamine	$C_6H_8N_2$	108-45-2	108.141	20	3.48	36.1	40		
1,4-Benzenediamine	$C_6H_8N_2$	106-50-3	108.141	24	3.45	35.7	40		
1,2-Benzenedicarboxamide	$C_8H_8N_2O_2$	88-96-0	164.162	30	0.59	5.9	40		
Benzeneethanol	$C_8H_{10}O$	60-12-8	122.164	25	1.72	17.5	40		
Benzenehexacarboxylic acid	$C_{12}H_6O_{12}$	517-60-2	342.169	25	49.3	972	76		
Benzenepentacarboxylic acid	$C_{11}H_6O_{10}$	1585-40-6	298.160	10	11.9	135	76		
				25	21.1	267	76		
				50	36.2	567	76		
1,2,3,4-Benzenetetracarboxylic acid	$C_{10}H_6O_8$	476-73-3	254.150	10	11.0	124	76		
				25	20.9	264	76		
				50	39.5	653	76		
1,2,3,5-Benzenetetracarboxylic acid	$C_{10}H_6O_8$	479-47-0	254.150	10	7.50	81.1	76		
				25	10.1	112	76		
				50	15.8	188	76		
1,2,4,5-Benzenetetracarboxylic acid	$C_{10}H_6O_8$	89-05-4	254.150	10	0.51	5.1	76		
				25	1.06	10.7	76		
				50	3.82	39.7	76		
1,2,3-Benzenetricarboxylic acid	$C_9H_6O_6$	569-51-7	210.140	10	2.39	24.5	76		
				25	4.78	50.2	76		
				50	17.4	211	76		
1,2,4-Benzenetricarboxylic acid	$C_9H_6O_6$	528-44-9	210.140	10	1.02	10.3	76		
				25	1.92	19.6	76		
				50	5.45	57.6	76		
1,3,5-Benzenetricarboxylic acid	$C_9H_6O_6$	554-95-0	210.140	10	0.110	1.10	76		
				25	0.207	2.07	76		
				50	0.598	6.02	76		
1,2,3-Benzenetriol	$C_6H_6O_3$	87-66-1	126.110	25	38.5	626	27		

Name	Mol. form.	CAS Reg. No.	Mol. wt.	t/°C	s/mass%	s/g kg^{-1} H_2O	Ref.	k_H/kPa m^3 mol^{-1}	Ref.
1,3,5-Benzenetriol	$C_6H_6O_3$	108-73-6	126.110	20	1.12	11.3	27		
p-Benzidine	$C_{12}H_{12}N_2$	92-87-5	184.236	24	0.0360	0.360	40		
1*H*-Benzimidazole	$C_7H_6N_2$	51-17-2	118.136	15	0.33	3.3	54		
				20	0.201	2.01	6		
1,3-Benzodioxole-5-carboxaldehyde	$C_8H_6O_3$	120-57-0	150.132	20	0.35	3.5	40		
Benzo[*b*]fluoranthene	$C_{20}H_{12}$	205-99-2	252.309	20	0.0000002	0.000002	40		
Benzo[*k*]fluoranthene	$C_{20}H_{12}$	207-08-9	252.309		0.00000008	0.0000008	40		
11*H*-Benzo[*a*]fluorene	$C_{17}H_{12}$	238-84-6	216.277	25	0.0000045	0.000045	42,4		
11*H*-Benzo[b]fluorene	$C_{17}H_{12}$	243-17-4	216.277	25	0.0000002	0.000002	42,4		
Benzoic acid	$C_7H_6O_2$	65-85-0	122.122	10	0.209	2.09	76		
				25	0.343	3.44	76		
				50	0.842	8.49	76		
Benzoin	$C_{14}H_{12}O_2$	579-44-2	212.244	25	0.03	0.3	40		
Benzonitrile	C_7H_5N	100-47-0	103.122	25	0.2	2	10		
Benzo[ghi]perylene	$C_{22}H_{12}$	191-24-2	276.330	25	0.000000026	0.00000026	42,4	0.000075	12
Benzophenone	$C_{13}H_{10}O$	119-61-9	182.217	20	0.0075	0.075	40		
2*H*-1-Benzopyran-2-one	$C_9H_6O_2$	91-64-5	146.143	20	0.190	1.90	40		
				60	0.69	6.9	40		
Benzo[*a*]pyrene	$C_{20}H_{12}$	50-32-8	252.309	25	0.00000043	0.0000043	42,22	0.0000465	22
Benzo[e]pyrene	$C_{20}H_{12}$	192-97-2	252.309	8	0.00000032	0.0000032	42		
				17	0.00000044	0.0000044	42,22	0.0000467	22
				25	0.00000048	0.0000048	42		
Benzo[f]quinoline	$C_{13}H_9N$	85-02-9	179.217	25	0.0079	0.079	6		
p-Benzoquinone	$C_6H_4O_2$	106-51-4	108.095	25	1.36	13.8	27		
Benzo[*b*]thiophene	C_8H_6S	95-15-8	134.199	20	0.0130	0.130	6		
Benzo[b]triphenylene	$C_{22}H_{14}$	215-58-7	278.346	25	0.0000027	0.000027	4		
Benzoxazole	C_7H_5NO	273-53-0	119.121	20	0.834	8.34	6		
N-Benzoylglycine	$C_9H_9NO_3$	495-69-2	179.172	25	0.37	3.7	29		
Benzoyl peroxide	$C_{14}H_{10}O_4$	94-36-0	242.227	20	0.000016	0.00016	40		
N-Benzoyl-L-phenylalanine	$C_{16}H_{15}NO_3$	2566-22-5	269.295	25	0.085	0.85	29		
Benzyl acetate	$C_9H_{10}O_2$	140-11-4	150.174	25	0.150	1.50	40		
Benzyl alcohol	C_7H_8O	100-51-6	108.138	20	0.08	0.8	10		
Benzyl formate	$C_8H_8O_2$	104-57-4	136.149	20	1.07	10.8	20		
				80	1.43	14.5	20		
Bifenthrin	$C_{23}H_{22}ClF_3O_2$	82657-04-3	422.868	25	0.00001	0.0001	32		
Biotin	$C_{10}H_{16}N_2O_3S$	58-85-5	244.310	25	0.035	0.35	40		
Biphenyl	$C_{12}H_{10}$	92-52-4	154.207	0	0.000272	0.00272	4		
				25	0.00054	0.0054	58,22	0.028	22
				50	0.0022	0.022	4		
2,2'-Bipyridine	$C_{10}H_8N_2$	366-18-7	156.184	25	0.61	6.1	40		
2,2'-Biquinoline	$C_{18}H_{12}N_2$	119-91-5	256.301	24	0.000102	0.00102	6		
Bis(4-aminophenyl) sulfone	$C_{12}H_{12}N_2O_2S$	80-08-0	248.300	25	0.016	0.16	40		
Bis(2-chloroethyl) ether	$C_4H_8Cl_2O$	111-44-4	143.012	20	1.04	10.5	20	0.003	13
				81	1.26	12.8	20		
1,1-Bis(4-chlorophenyl)-2,2,2-trichloroethanol	$C_{14}H_9Cl_5O$	115-32-2	370.485	25	0.00013	0.0013	40		
Bis(2-ethylhexyl) phthalate	$C_{24}H_{38}O_4$	117-81-7	390.557	25	0.000027	0.00027	40		
2,2-Bis(4-hydroxyphenyl)propane	$C_{15}H_{16}O_2$	80-05-7	228.287	25	0.0300	0.30	49		
1,3-Bis(trifluoromethyl)benzene	$C_8H_4F_6$	402-31-3	214.108	25	0.0041	0.041	2		
Borneol, (-)-	$C_{10}H_{18}O$	464-45-9	154.249	25	0.046	0.46	52		
Bromacil	$C_9H_{13}BrN_2O_2$	314-40-9	261.115	25	0.082	0.82	40		
Bromobenzene	C_6H_5Br	108-86-1	157.008	10	0.0387	0.387	2		
				25	0.0445	0.445	2	0.25	28
				40	0.0516	0.516	2		
2-Bromobenzoic acid	$C_7H_5BrO_2$	88-65-3	201.018	25	0.185	1.85	27		
3-Bromobenzoic acid	$C_7H_5BrO_2$	585-76-2	201.018	25	0.040	0.40	27		
4-Bromobenzoic acid	$C_7H_5BrO_2$	586-76-5	201.018	25	0.0056	0.056	27		
1-Bromobutane	C_4H_9Br	109-65-9	137.018	25	0.087	0.87	35	1.2	13
4-Bromo-1-butene	C_4H_7Br	5162-44-7	135.003	25	0.076	0.76	35		
1-Bromo-2-chlorobenzene	C_6H_4BrCl	694-80-4	191.453	25	0.0124	0.124	2		
1-Bromo-3-chlorobenzene	C_6H_4BrCl	108-37-2	191.453	25	0.0118	0.118	2		

Name	Mol. form.	CAS Reg. No.	Mol. wt.	t/°C	s/mass%	s/g kg⁻¹ H₂O	Ref.	k_H/kPa m³ mol⁻¹	Ref.
1-Bromo-4-chlorobenzene	C_6H_4BrCl	106-39-8	191.453	25	0.00442	0.0442	2		
1-Bromo-2-chloroethane	C_2H_4BrCl	107-04-0	143.410	30	0.683	6.83	25		
Bromochloromethane	CH_2BrCl	74-97-5	129.384	25	1.7	17	10	0.18	13
1-Bromo-3-chloropropane	C_3H_6BrCl	109-70-6	157.437	25	0.223	2.23	35		
2-Bromo-2-chloro-1,1,1-trifluoroethane	$C_2HBrClF_3$	151-67-7	197.381	10	0.52	5.2	25		
				25	0.41	4.1	25		
				40	0.40	4.0	25		
Bromodichloromethane	$CHBrCl_2$	75-27-4	163.829	30	0.300	3.00	40		
Bromoethane	C_2H_5Br	74-96-4	108.965	0	1.05	10.6	25		
				25	0.90	9.0	25	1.23	13
1-Bromoheptane	$C_7H_{15}Br$	629-04-9	179.098	25	0.00067	0.0067	35		
1-Bromohexane	$C_6H_{13}Br$	111-25-1	165.071	25	0.00258	0.0258	35		
1-Bromo-4-iodobenzene	C_6H_4BrI	589-87-7	282.904	25	0.000794	0.00794	2		
Bromomethane	CH_3Br	74-83-9	94.939	20	1.80	18.3	5	0.63	13
1-Bromo-3-methylbutane	$C_5H_{11}Br$	107-82-4	151.045	16	0.020	0.20	35		
1-Bromo-2-methylpropane	C_4H_9Br	78-77-3	137.018	18	0.051	0.51	35		
1-Bromooctane	$C_8H_{17}Br$	111-83-1	193.125	25	0.000167	0.00167	35		
1-Bromopentane	$C_5H_{11}Br$	110-53-2	151.045	25	0.0127	0.127	35		
4-Bromophenol	C_6H_5BrO	106-41-2	173.007	25	1.86	19.0	2		
1-Bromopropane	C_3H_7Br	106-94-5	122.992	0	0.298	2.98	35		
				25	0.234	2.34	35	3.8	13
2-Bromopropane	C_3H_7Br	75-26-3	122.992	20	0.32	3.2	35	1.27	13
3-Bromopropene	C_3H_5Br	106-95-6	120.976	25	0.38	3.8	35		
4-Bromotoluene	C_7H_7Br	106-38-7	171.035	25	0.011	0.11	2		
Bromotrifluoromethane	$CBrF_3$	75-63-8	148.910	25	0.032*	0.32*	14		
5-Bromouracil	$C_4H_3BrN_2O_2$	51-20-7	190.983	25	0.288	2.89	72		
Brucine	$C_{23}H_{26}N_2O_4$	357-57-3	394.463	20	0.012	0.12	27		
1,3-Butadiene	C_4H_6	106-99-0	54.091	25	0.0735*	0.735*	5	20.7	13
Butanal	C_4H_8O	123-72-8	72.106	25	7.1	76	10		
Butanamide	C_4H_9NO	541-35-5	87.120	25	≈19	≈230	40		
Butane	C_4H_{10}	106-97-8	58.122	25	0.00724*	0.0724*	18	95.9	5
2,3-Butanedione	$C_4H_6O_2$	431-03-8	86.090	20	31.7	464	20		
				80	21.8	279	20		
Butanenitrile	C_4H_7N	109-74-0	69.106	20	3.3	34	10		
1,2,3,4-Butanetetrol	$C_4H_{10}O_4$	149-32-6	122.120	20	38.0	613	27		
1-Butanethiol	$C_4H_{10}S$	109-79-5	90.187	20	0.0597	0.597	10		
1-Butanol	$C_4H_{10}O$	71-36-3	74.121	0	10.5	117	78,1		
				25	7.3	79	78,1		
				50	6.4	68	78,1		
				100	8.8	96	78		
2-Butanol	$C_4H_{10}O$	78-92-2	74.121	10	23.9	314	1,87		
				25	18.1	221	1,87		
				50	14.0	163	1,87		
2-Butanone	C_4H_8O	78-93-3	72.106	0	35.9	560	82		
				25	25.6	344	82		
				40	21.5	274	82		
				70	18.1	221	20		
				100	19.3	239	82		
trans-2-Butenal	C_4H_6O	123-73-9	70.090	20	15.6	185	10		
1-Butene	C_4H_8	106-98-9	56.107	25	0.0222*	0.222*	5	25.6	13
trans-2-Butenoic acid	$C_4H_6O_2$	107-93-7	86.090	20	7.1	76	26		
cis-2-Buten-1-ol	C_4H_8O	4088-60-2	72.106	20	16.6	199	10		
3-Buten-2-one	C_4H_6O	78-94-4	70.090	28	54.3	1190	82		
				50	35.6	553	82		
				80	37.6	603	82		
Butyl acetate	$C_6H_{12}O_2$	123-86-4	116.158	20	0.68	6.8	10		
sec-Butyl acetate	$C_6H_{12}O_2$	105-46-4	116.158	20	0.62	6.2	10		
Butyl 4-aminobenzoate	$C_{11}H_{15}NO_2$	94-25-7	193.243	25	0.018	0.18	40		
Butylbenzene	$C_{10}H_{14}$	104-51-8	134.218	25	0.00138	0.0138	22,89	1.33	22
sec-Butylbenzene, (±)-	$C_{10}H_{14}$	36383-15-0	134.218	25	0.0014	0.014	4,89	1.89	11

Name	Mol. form.	CAS Reg. No.	Mol. wt.	t/°C	s/mass%	s/g kg⁻¹ H₂O	Ref.	k_H/ kPa m³ mol⁻¹	Ref.
tert-Butylbenzene	$C_{10}H_{14}$	98-06-6	134.218	25	0.0032	0.032	4	1.28	11
Butyl ethyl ether	$C_6H_{14}O$	628-81-9	102.174	20	0.65	6.5	20		
				70	0.39	3.9	20		
Butyl 4-hydroxybenzoate	$C_{11}H_{14}O_3$	94-26-8	194.227	25	0.020	0.20	40		
Butyl methyl ether	$C_5H_{12}O$	628-28-4	88.148	0	2.51	25.7	79		
				25	0.89	9.0	79		
4-*tert*-Butylphenol	$C_{10}H_{14}O$	98-54-4	150.217	25	0.058	0.58	40		
Butyl propanoate	$C_7H_{14}O_2$	590-01-2	130.185	22	0.572	5.72	27		
Butyl stearate	$C_{22}H_{44}O_2$	123-95-5	340.583	25	0.2	2	10		
Butyl vinyl ether	$C_6H_{12}O$	111-34-2	100.158	20	0.3	3	10		
1-Butyne	C_4H_6	107-00-6	54.091	25	0.287*	2.87*	5	1.91	5
Caffeine	$C_8H_{10}N_4O_2$	58-08-2	194.191	25	2.12	21.7	29		
Camphor, (+)	$C_{10}H_{16}O$	464-49-3	152.233	20	0.01	0.1	10		
trans-Camphoric acid, (±)-	$C_{10}H_{16}O_4$	560-08-7	200.232	25	0.8	8	27		
Cantharidin	$C_{10}H_{12}O_4$	56-25-7	196.200	20	0.003	0.03	40		
Caprolactam	$C_6H_{11}NO$	105-60-2	113.157	25	84.0	5250	10		
Captafol	$C_{10}H_9Cl_4NO_2S$	191906	349.061	20	0.000142	0.00142	40		
Captan	$C_9H_8Cl_3NO_2S$	133-06-2	300.590	20	0.00005	0.0005	40		
Carbaryl	$C_{12}H_{11}NO_2$	63-25-2	201.221	20	0.0102	0.102	40		
Carbazole	$C_{12}H_9N$	86-74-8	167.206	22	0.000120	0.00120	6		
Carbofuran	$C_{12}H_{15}NO_3$	1563-66-2	221.252	20	0.032	0.32	40		
Carbon dioxide	CO_2	124-38-9	44.010	25	0.150*	1.50*	18		
Carbon disulfide	CS_2	75-15-0	76.141	20	0.210	2.10	10		
Carbon monoxide	CO	630-08-0	28.010	25	0.00276*	0.0276*	18		
Carboxin	$C_{12}H_{13}NO_2S$	5234-68-4	235.302	25	0.017	0.17	40		
Carminic acid	$C_{22}H_{20}O_{13}$	1260-17-9	492.386	20	0.13	1.3	40		
Carnosine	$C_9H_{14}N_4O_3$	305-84-0	226.232	25	24.4	323	26		
Carvenol	$C_{10}H_{16}O$	99-48-9	152.233	25	0.29	2.9	52		
Carvenone, (*S*)-	$C_{10}H_{16}O$	10395-45-6	152.233	15	0.22	2.2	27		
Carvone, (±)-	$C_{10}H_{14}O$	22327-39-5	150.217	15	0.13	1.3	27		
(*S*)-Carvone	$C_{10}H_{14}O$	2244-16-8	150.217	25	0.13	1.3	52		
Cephalexin	$C_{16}H_{17}N_3O_4S$	15686-71-2	347.389	25	1.2	12	40		
Chloramphenicol	$C_{11}H_{12}Cl_2N_2O_5$	56-75-7	323.129	25	0.38	3.8	40		
Chlordane	$C_{10}H_6Cl_8$	57-74-9	409.779	25	0.00006	0.0006	67		
2-Chloroaniline	C_6H_6ClN	95-51-2	127.572	25	0.876	8.76	10		
3-Chloroaniline	C_6H_6ClN	108-42-9	127.572	20	0.54	5.4	40		
4-Chloroaniline	C_6H_6ClN	106-47-8	127.572	20	0.275	2.75	40		
Chlorobenzene	C_6H_5Cl	108-90-7	112.557	5	0.050	0.50	61		
				25	0.050	0.50	61		
				45	0.055	0.55	61,2		
Chlorobenzilate	$C_{16}H_{14}Cl_2O_3$	510-15-6	325.186	20	0.001	0.01	32		
2-Chlorobenzoic acid	$C_7H_5ClO_2$	118-91-2	156.567	25	0.209	2.09	27		
3-Chlorobenzoic acid	$C_7H_5ClO_2$	535-80-8	156.567	25	0.040	0.40	27		
4-Chlorobenzoic acid	$C_7H_5ClO_2$	74-11-3	156.567	25	0.072	0.72	27		
2-Chlorobiphenyl	$C_{12}H_9Cl$	2051-60-7	188.652	25	0.00055	0.0055	7	0.0701	7
1-Chlorobutane	C_4H_9Cl	109-69-3	92.567	1	0.062	0.62	35		
				25	0.087	0.87	35	1.54	13
2-Chlorobutane	C_4H_9Cl	53178-20-4	92.567	0	0.107	1.07	35		
				25	0.092	0.92	35		
3-Chloro-2-butanone	C_4H_7ClO	4091-39-8	106.551	19	2.80	28.8	20		
				92	3.38	35.0	20		
Chlorodiazepoxide	$C_{16}H_{14}ClN_3O$	58-25-3	299.754	20	0.2	2	40		
Chlorodibromomethane	$CHBr_2Cl$	124-48-1	208.280	30	0.251	2.51	40		
Chlorodifluoromethane	$CHClF_2$	75-45-6	86.469	25	0.30*	3.0*	10	3	13
4-Chloro-2,5-dimethylphenol	C_8H_9ClO	1124-06-7	156.609	25	0.89	8.9	2		
4-Chloro-2,6-dimethylphenol	C_8H_9ClO	1123-63-3	156.609	25	0.52	5.2	2		
4-Chloro-3,5-dimethylphenol	C_8H_9ClO	88-04-0	156.609	25	0.34	3.4	2		
1-Chloro-2,4-dinitrobenzene	$C_6H_3ClN_2O_4$	97-00-7	202.552	25	0.00092	0.0092	40		
Chloroethane	C_2H_5Cl	75-00-3	64.514	0	0.45	4.5	25		
				25	0.67*	6.7*	25	1.02	13

Thermochem

Name	Mol. form.	CAS Reg. No.	Mol. wt.	t/°C	s/mass%	s/g kg⁻¹ H₂O	Ref.	k_H/ kPa m³ mol⁻¹	Ref.
Chloroethene	C_2H_3Cl	75-01-4	62.498	25	0.27*	2.7*	5	2.68	13
1-Chloro-2-fluorobenzene	C_6H_4ClF	348-51-6	130.547	25	0.0502	0.502	40		
Chlorofluoromethane	CH_2ClF	593-70-4	68.478	25	1.05*	10.6*	14		
1-Chloroheptane	$C_7H_{15}Cl$	629-06-1	134.647	25	0.00136	0.0136	35		
1-Chlorohexane	$C_6H_{13}Cl$	544-10-5	120.620	5	0.0047	0.047	35		
				25	0.0064	0.064	35		
2-Chloro-4-hydroxy-5-methoxybenzaldehyde	$C_8H_7ClO_3$	18268-76-3	186.593	25	0.013	0.13	8		
3-Chloro-4-hydroxy-5-methoxybenzaldehyde	$C_8H_7ClO_3$	19463-48-0	186.593	25	0.093	0.93	8		
1-Chloro-2-iodobenzene	C_6H_4ClI	615-41-8	238.453	25	0.00689	0.0689	2		
1-Chloro-3-iodobenzene	C_6H_4ClI	625-99-0	238.453	25	0.00674	0.0674	2		
1-Chloro-4-iodobenzene	C_6H_4ClI	637-87-6	238.453	25	0.00311	0.0311	2		
Chloromethane	CH_3Cl	74-87-3	50.488	25	0.535*	5.35*	5	0.98	13
1-Chloro-2-methoxyethane	C_3H_7ClO	627-42-9	94.540	20	7.79	84.5	20		
				70	6.31	67.3	20		
(Chloromethyl)benzene	C_7H_7Cl	100-44-7	126.584	20	0.0493	0.493	10		
3-(Chloromethyl)heptane	$C_8H_{17}Cl$	123-04-6	148.674	20	0.01	0.1	10		
2-Chloro-6-methylphenol	C_7H_7ClO	87-64-9	142.583	25	0.36	3.6	2		
4-Chloro-2-methylphenol	C_7H_7ClO	1570-64-5	142.583	25	0.68	6.8	2		
4-Chloro-3-methylphenol	C_7H_7ClO	59-50-7	142.583	25	0.40	4.0	2		
(4-Chloro-2-methylphenoxy)acetic acid	$C_9H_9ClO_3$	94-74-6	200.618	25	0.117	1.17	40		
1-Chloro-2-methylpropane	C_4H_9Cl	513-36-0	92.567	25	0.92	9.2	35		
2-Chloro-2-methylpropane	C_4H_9Cl	507-20-0	92.567	15	0.29	2.9	35		
1-Chloro-2-methylpropene	C_4H_7Cl	513-37-1	90.552	25	0.916	9.16	5	0.12	5
1-Chloronaphthalene	$C_{10}H_7Cl$	90-13-1	162.616	25	0.00224	0.0224	5	0.0363	28
2-Chloronaphthalene	$C_{10}H_7Cl$	91-58-7	162.616	25	0.00117	0.0117	5	0.0335	28
1-Chloro-2-nitrobenzene	$C_6H_4ClNO_2$	88-73-3	157.555	20	0.0441	0.441	40		
1-Chloro-3-nitrobenzene	$C_6H_4ClNO_2$	121-73-3	157.555	20	0.0273	0.273	40		
1-Chloro-4-nitrobenzene	$C_6H_4ClNO_2$	100-00-5	157.555	20	0.0453	0.453	40		
3-Chloro-2-nitrobenzoic acid	$C_7H_4ClNO_4$	4771-47-5	201.565	25	0.047	0.47	27		
5-Chloro-2-nitrobenzoic acid	$C_7H_4ClNO_4$	2516-95-2	201.565	25	0.96	9.6	27		
1-Chlorooctane	$C_8H_{17}Cl$	111-85-3	148.674	25	0.0345	0.345	35		
Chloropentafluoroethane	C_2ClF_5	76-15-3	154.466	25	0.006*	0.06*	10	260	13
1-Chloropentane	$C_5H_{11}Cl$	543-59-9	106.594	5	0.020	0.20	35		
				25	0.0201	0.201	35	2.37	13
3-Chloropentane	$C_5H_{11}Cl$	616-20-6	106.594	25	0.025	0.25	35		
5-Chloro-2-pentanone	C_5H_9ClO	5891-21-4	120.577	22	4.7	49	20		
				71	13.5	156	20		
2-Chlorophenol	C_6H_5ClO	95-57-8	128.556	25	2.27	23.2	48,51,2		
3-Chlorophenol	C_6H_5ClO	108-43-0	128.556	25	2.2	22	2		
4-Chlorophenol	C_6H_5ClO	106-48-9	128.556	25	2.55	26.2	48,51,2		
N'-(4-Chlorophenyl)-N,N-dimethylurea	$C_9H_{11}ClN_2O$	150-68-5	198.648	25	0.023	0.23	26		
1-Chloropropane	C_3H_7Cl	540-54-5	78.541	25	0.250	2.50	35	1.41	13
2-Chloropropane	C_3H_7Cl	75-29-6	78.541	0	0.44	4.4	35		
				20	0.30	3.0	35		
3-Chloropropene	C_3H_5Cl	107-05-1	76.525	25	0.40	4.0	35	1.1	5
3-Chloropropene	C_3H_5Cl	107-05-1	76.525	50	0.13	1.3	35		
Chloropropham	$C_{10}H_{12}ClNO_2$	101-21-3	213.661	25	0.0080	0.080	40		
1-Chlorotetradecane	$C_{14}H_{29}Cl$	2425-54-9	232.833	25	0.0232	0.232	35		
Chlorothalonil	$C_8Cl_4N_2$	1897-45-6	265.911	25	0.00006	0.0006	40		
Chlorothiazide	$C_7H_6ClN_3O_4S_2$	58-94-6	295.724	25	0.0283	0.283	40		
2-Chlorotoluene	C_7H_7Cl	95-49-8	126.584	25	0.0117	0.117	61		
3-Chlorotoluene	C_7H_7Cl	108-41-8	126.584	25	0.0117	0.117	61		
4-Chlorotoluene	C_7H_7Cl	106-43-4	126.584	25	0.0123	0.123	61		
Chlorotrifluoromethane	$CClF_3$	75-72-9	104.459	25	0.009*	0.09*	10	6.9	13
3-Chloro-1,1,1-trifluoropropane	$C_3H_4ClF_3$	460-35-5	132.512	20	0.133	1.33	35		
2-Chloro-1,3,5-trinitrobenzene	$C_6H_2ClN_3O_6$	88-88-0	247.549	15	0.018	0.18	40		
5-Chlorouracil	$C_4H_3ClN_2O_2$	1820-81-1	146.532	25	0.250	2.51	72		

Name	Mol. form.	CAS Reg. No.	Mol. wt.	t/°C	s/mass%	s/g kg^{-1} H_2O	Ref.	k_H/ kPa m^3 mol^{-1}	Ref.
Chlorpyrifos	$C_9H_{11}Cl_3NO_3PS$	2921-88-2	350.586	20	0.000073	0.00073	40		
Chlorsulfuron	$C_{12}H_{12}ClN_5O_4S$	64902-72-3	357.773	25	2.71	27.9	32		
Cholic acid	$C_{24}H_{40}O_5$	81-25-4	408.572	20	0.028	0.28	26		
Chrysene	$C_{18}H_{12}$	218-01-9	228.288	7	0.00000007	0.0000007	42		
				25	0.00000019	0.0000019	42,22	0.000065	22
trans-Cinnamaldehyde	C_9H_8O	14371-10-9	132.159	25	0.135	1.35	40		
trans-Cinnamic acid	$C_9H_8O_2$	140-10-3	148.159	20	0.1	1	26		
				98	0.59	5.9	26		
Citric acid	$C_6H_8O_7$	77-92-9	192.124	20	59	1440	26		
Clopyralid	$C_6H_3Cl_2NO_2$	1702-17-6	192.000	20	0.1	1	40		
Clorophene	$C_{13}H_{11}ClO$	120-32-1	218.678	20	0.42	4.2	40		
Cocaine	$C_{17}H_{21}NO_4$	50-36-2	303.354	25	0.17	1.7	27		
Codeine	$C_{18}H_{21}NO_3$	76-57-3	299.365	25	0.79	7.9	27		
Colchicine	$C_{22}H_{25}NO_6$	64-86-8	399.437	20	4	42	26		
Coronene	$C_{24}H_{12}$	191-07-1	300.352	25	0.000000014	0.00000014	42,4		
Creatine	$C_4H_9N_3O_2$	57-00-1	131.133	25	1.6	16	26		
o-Cresol	C_7H_8O	95-48-7	108.138	40	3.08	31.8	10		
m-Cresol	C_7H_8O	108-39-4	108.138	40	2.51	25.7	10		
p-Cresol	C_7H_8O	106-44-5	108.138	40	2.26	23.1	10		
Crufomate	$C_{12}H_{19}ClNO_3P$	299-86-5	291.711	20	0.50	5.0	40		
Cyanazine	$C_9H_{13}ClN_6$	21725-46-2	240.692	25	0.0171	0.171	40		
2-Cyanoacetamide	$C_3H_4N_2O$	107-91-5	84.076	20	11.5	130	40		
Cyanogen	C_2N_2	460-19-5	52.034	25	0.8*	8*	30		
Cyanogen chloride	CClN	506-77-4	61.471	0	5.7	60	40		
Cyanoguanidine	$C_2H_4N_4$	461-58-5	84.080	25	3.8	40	40		
Cyanuric acid	$C_3H_3N_3O_3$	108-80-5	129.074	25	0.259	2.59	40		
Cycloheptane	C_7H_{14}	291-64-5	98.186	25	0.0030	0.030	3	9.59	13
Cycloheptanone	$C_7H_{12}O$	502-42-1	112.169	20	3.61	37.5	20		
				92	2.82	29.0	20		
1,3,5-Cycloheptatriene	C_7H_8	544-25-2	92.139	25	0.064	0.64	3	0.47	13
Cycloheptene	C_7H_{12}	628-92-2	96.170	25	0.0066	0.066	3	4.9	13
1,4-Cyclohexadiene	C_6H_8	628-41-1	80.128	25	0.08	0.8	3	1.03	13
Cyclohexane	C_6H_{12}	110-82-7	84.159	25	0.0058	0.058	3	19.4	13
				70	0.0092	0.092	65		
				100	0.0163	0.163	65		
Cyclohexanecarboxylic acid	$C_7H_{12}O_2$	98-89-5	128.169	15	0.201	2.01	27		
Cyclohexanol	$C_6H_{12}O$	108-93-0	100.158	10	4.62	48.4	1		
				25	3.8	40	1		
				40	3.30	34.1	1		
Cyclohexanone	$C_6H_{10}O$	108-94-1	98.142	10	12.2	139	83		
				25	9.5	105	83		
				50	7.6	82	83		
				80	6.8	73	20		
Cyclohexanone oxime	$C_6H_{11}NO$	100-64-1	113.157	25	1.57	16.0	40		
Cyclohexene	C_6H_{10}	110-83-8	82.143	25	0.016	0.16	3	4.57	13
Cyclohexyl butanoate	$C_{10}H_{18}O_2$	1551-44-6	170.249	20	0.11	1.1	20		
				90	0.09	0.90	20		
Cyclooctane	C_8H_{16}	292-64-8	112.213	25	0.00079	0.0079	4	10.7	13
1,3-Cyclopentadiene	C_5H_6	542-92-7	66.102	25	0.068	0.68	3		
Cyclopentane	C_5H_{10}	287-92-3	70.133	25	0.0157	0.157	3	19.1	13
Cyclopentanol	$C_5H_{10}O$	96-41-3	86.132	19	10.6	119	88		
				50	8.3	91	88		
				90	9.2	101	88		
Cyclopentanone	C_5H_8O	120-92-3	84.117	0	37.7	605	20		
				20	31.0	449	20		
				80	24.8	330	20		
Cyclopentene	C_5H_8	142-29-0	68.118	25	0.054	0.54	3	6.56	13
Cyclopropane	C_3H_6	75-19-4	42.080	25	0.0484*	0.484*	19		
Cyfluthrin	$C_{22}H_{18}Cl_2FNO_3$	68359-37-5	434.287	20	0.0000002	0.000002	32		
Cygon	$C_5H_{12}NO_3PS_2$	60-51-5	229.258	20	2.6	27	40		

Name	Mol. form.	CAS Reg. No.	Mol. wt.	t/°C	s/mass%	s/g kg⁻¹ H₂O	Ref.	k_H/ kPa m³ mol⁻¹	Ref.
Cyhalothrin	$C_{23}H_{19}ClF_3NO_3$	91465-08-6	449.850	20	0.0000005	0.000005	32		
Cypermethrin	$C_{22}H_{19}Cl_2NO_3$	52315-07-8	416.297	20	0.000001	0.00001	32		
L-Cystine	$C_6H_{12}N_2O_4S_2$	56-89-3	240.300	25	0.0166	0.166	62		
Cytisine	$C_{11}H_{14}N_2O$	485-35-8	190.241	16	≈30	≈430	40		
Cytosine	$C_4H_5N_3O$	71-30-7	111.102	25	0.73	7.3	29		
Daminozide	$C_6H_{12}N_2O_3$	1596-84-5	160.170	25	9.1	100	40		
Dazomet	$C_5H_{10}N_2S_2$	533-74-4	162.276	25	0.12	1.2	40		
Decabromobiphenyl ether	$C_{12}Br_{10}O$	1163-19-5	959.167	25	0.0000025	0.000025	40		
Decachlorobiphenyl	$C_{12}Cl_{10}$	2051-24-3	498.658	25	0.00000000012	0.0000000012	7	0.0208	7
cis-Decahydronaphthalene	$C_{10}H_{18}$	493-01-6	138.250	25	0.000089	0.00089	37		
trans-Decahydronaphthalene	$C_{10}H_{18}$	493-02-7	138.250	25	0.000089	0.00089	4	3	13
Decane	$C_{10}H_{22}$	124-18-5	142.282	0	0.0000015	0.000015	4	479	13
Decanedioic acid	$C_{10}H_{18}O_4$	111-20-6	202.248	20	0.10	1.0	40		
Decanoic acid	$C_{10}H_{20}O_2$	334-48-5	172.265	20	0.015	0.15	26		
1-Decanol	$C_{10}H_{22}O$	112-30-1	158.281	25	0.0037	0.037	1		
2-Decanone	$C_{10}H_{20}O$	693-54-9	156.265	25	0.0079	0.079	84		
4-Decanone	$C_{10}H_{20}O$	624-16-8	156.265	20	0.0238	0.238	20		
				80	0.0064	0.064	20		
1-Decene	$C_{10}H_{20}$	872-05-9	140.266	25	0.00057	0.0057	4		
2'-Deoxyadenosine	$C_{10}H_{13}N_5O_3$	958-09-8	251.242	25	0.67	6.7	29		
Deoxycholic acid	$C_{24}H_{40}O_4$	83-44-3	392.573	20	0.001	0.01	40		
Dexamethasone	$C_{22}H_{29}FO_5$	50-02-2	392.460	25	0.009	0.09	40		
Dibenz[a,j]acridine	$C_{21}H_{13}N$	224-42-0	279.335	25	0.000016	0.00016	6		
Dibenz[a,h]anthracene	$C_{22}H_{14}$	53-70-3	278.346	25	0.00000005	0.0000005	42,4		
Dibenz[a,j]anthracene	$C_{22}H_{14}$	224-41-9	278.346	27	0.0000012	0.000012	42,4		
13*H*-Dibenzo[a,i]carbazole	$C_{20}H_{13}N$	239-64-5	267.324	24	0.00000104	0.0000104	6		
Dibenzo[b,e][1,4]dioxin	$C_{12}H_8O_2$	262-12-4	184.191	25	0.000126	0.00126	68		
Dibenzofuran	$C_{12}H_8O$	132-64-9	168.191	25	0.000475	0.00475	41	0.011	12
Dibenzothiophene	$C_{12}H_8S$	132-65-0	184.257	25	0.000103	0.00103	6		
Dibenzyl ether	$C_{14}H_{14}O$	103-50-4	198.260	35	0.0040	0.040	10		
o-Dibromobenzene	$C_6H_4Br_2$	583-53-9	235.904	25	0.00748	0.0748	2		
m-Dibromobenzene	$C_6H_4Br_2$	108-36-1	235.904	25	0.0064	0.064	2		
p-Dibromobenzene	$C_6H_4Br_2$	106-37-6	235.904	25	0.0020	0.020	2		
1,4-Dibromobutane	$C_4H_8Br_2$	110-52-1	215.915	25	0.035	0.35	35		
1,2-Dibromo-1-chloroethane	$C_2H_3Br_2Cl$	598-20-9	222.306	20	0.060	0.60	25		
1,2-Dibromo-3-chloropropane	$C_3H_5Br_2Cl$	96-12-8	236.333	20	0.123	1.23	35		
1,2-Dibromo-1,2-dichloroethane	$C_2H_2Br_2Cl_2$	683-68-1	256.751	20	0.070	0.70	25		
1,2-Dibromoethane	$C_2H_4Br_2$	106-93-4	187.861	20	0.412	4.14	20		
				50	0.493	4.95	20	0.066	13
				80	0.572	5.75	20		
1,2-Dibromo-1,1,2,3,3,3-hexafluoropropane	$C_3Br_2F_6$	661-95-0	309.830	21	0.0068	0.068	35		
3,5-Dibromo-4-hydroxybenzonitrile	$C_7H_3Br_2NO$	1689-84-5	276.913	25	0.013	0.13	40		
Dibromomethane	CH_2Br_2	74-95-3	173.835	20	1.28	13.0	20	0.086	13
				90	1.51	15.3	20		
2,4-Dibromophenol	$C_6H_4Br_2O$	615-58-7	251.903	25	0.2	2	2		
1,2-Dibromopropane	$C_3H_6Br_2$	78-75-1	201.888	25	0.143	1.43	10		
1,3-Dibromopropane	$C_3H_6Br_2$	109-64-8	201.888	25	0.169	1.69	35		
1,2-Dibromotetrafluoroethane	$C_2Br_2F_4$	124-73-2	259.823	25	0.00030	0.0030	25		
Dibutylamine	$C_8H_{19}N$	111-92-2	129.244	20	0.47	4.7	10		
Dibutyl ether	$C_8H_{18}O$	142-96-1	130.228	0	0.040	0.40	20	0.48	13
				20	0.023	0.23	20	0.48	13
				90	0.010	0.10	20		
Dibutyl phthalate	$C_{16}H_{22}O_4$	84-74-2	278.344	25	0.00112	0.0112	15		
Dibutyl sebacate	$C_{18}H_{34}O_4$	109-43-3	314.461	20	0.004	0.04	10		
o-Dichlorobenzene	$C_6H_4Cl_2$	95-50-1	147.002	5	0.012	0.12	61,58,2		
				25	0.015	0.15	61,58,2		
				45	0.020	0.20	61,58,2		
m-Dichlorobenzene	$C_6H_4Cl_2$	541-73-1	147.002	10	0.0103	0.103	41,2		
				25	0.0120	0.120	41,2	0.376	11
				45	0.0141	0.141	61,2		

Name	Mol. form.	CAS Reg. No.	Mol. wt.	t/°C	s/mass%	s/g kg^{-1} H_2O	Ref.	k_H/ kPa m^3 mol^{-1}	Ref.
p-Dichlorobenzene	$C_6H_4Cl_2$	106-46-7	147.002	10	0.00512	0.0512	2		
				25	0.0080	0.080	41	0.244	28
				50	0.0167	0.167	2		
3,5-Dichloro-1,2-benzenediol	$C_6H_4Cl_2O_2$	13673-92-2	179.001	25	0.78	7.8	8		
4,5-Dichloro-1,2-benzenediol	$C_6H_4Cl_2O_2$	3428-24-8	179.001	25	1.19	12.0	8		
3,3'-Dichloro-*p*-benzidine	$C_{12}H_{10}Cl_2N_2$	91-94-1	253.126	25	0.00031	0.0031	40		
2,5-Dichlorobiphenyl	$C_{12}H_8Cl_2$	34883-39-1	223.098	25	0.0002	0.002	7	0.0201	7
2,6-Dichlorobiphenyl	$C_{12}H_8Cl_2$	33146-45-1	223.098	25	0.00014	0.0014	7		
1,1-Dichloro-2,2-bis(*p*-chlorophenyl) ethane	$C_{14}H_{10}Cl_4$	72-54-8	320.041	25	0.000009	0.00009	40		
				45	0.000024	0.00024	40		
1,1-Dichlorobutane	$C_4H_8Cl_2$	541-33-3	127.013	25	0.050	0.50	35		
1,4-Dichlorobutane	$C_4H_8Cl_2$	110-56-5	127.013	25	0.16	1.6	35		
2,3-Dichlorobutane, (±)-	$C_4H_8Cl_2$	2211-67-8	127.013	20	0.056	0.56	35		
2,7-Dichlorodibenzo-*p*-dioxin	$C_{12}H_6Cl_2O_2$	33857-26-0	253.081	25	4.1E-07	0.0000041	68		
1,2-Dichloro-1,1-difluoroethane	$C_2H_2Cl_2F_2$	1649-08-7	134.940	24	0.49	4.9	25		
Dichlorodifluoromethane	CCl_2F_2	75-71-8	120.914	20	0.028*	0.28*	5	41	13
1,3-Dichloro-5,5-dimethyl hydantoin	$C_5H_6Cl_2N_2O_2$	118-52-5	197.019	20	0.050	0.50	40		
1,1-Dichloroethane	$C_2H_4Cl_2$	75-34-3	98.959	0	0.62	6.2	25		
				25	0.50	5.0	25	0.63	13
				50	0.50	5.0	25		
1,2-Dichloroethane	$C_2H_4Cl_2$	107-06-2	98.959	0	0.92	9.2	25		
				25	0.86	8.6	25	0.14	13
				50	1.05	10.6	25		
				100	2.17	22.2	25		
1,1-Dichloroethene	$C_2H_2Cl_2$	75-35-4	96.943	5	0.310	3.10	25		
				25	0.242	2.42	25	2.62	13
				50	0.225	2.25	25		
				90	0.355	3.55	25		
cis-1,2-Dichloroethene	$C_2H_2Cl_2$	156-59-2	96.943	10	0.76	7.6	25		
				25	0.64	6.4	25	0.46	13
				40	0.66	6.6	25		
trans-1,2-Dichloroethene	$C_2H_2Cl_2$	156-60-5	96.943	10	0.53	5.3	25		
				25	0.45	4.5	25	0.96	13
				40	0.41	4.1	25		
1,1-Dichloro-1-fluoroethane	$C_2H_3Cl_2F$	1717-00-6	116.949	25	0.042	0.42	25		
Dichlorofluoromethane	$CHCl_2F$	75-43-4	102.923	25	0.95*	9.5*	10		
1,2-Dichloro-1,1,2,3,3,3-hexafluoropropane	$C_3Cl_2F_6$	661-97-2	220.928	21	0.0096	0.096	35		
1,4-Dichloro-5-isopropyl-2-methylbenzene	$C_{10}H_{12}Cl_2$	81686-41-1	203.108	25	0.00049	0.0049	23		
Dichloromethane	CH_2Cl_2	75-09-2	84.933	25	1.73	17.6	20	0.3	13
3,6-Dichloro-2-methoxybenzoic acid	$C_8H_6Cl_2O_3$	1918-00-9	221.038	25	0.45	4.5	40		
(Dichloromethyl)benzene	$C_7H_6Cl_2$	98-87-3	161.029	30	0.025	0.25	10		
2,3-Dichloro-2-methylbutane	$C_5H_{10}Cl_2$	507-45-9	141.038	25	0.029	0.29	35		
2,4-Dichloro-6-methylphenol	$C_7H_6Cl_2O$	1570-65-6	177.028	25	0.0283	0.283	2		
2,6-Dichloro-4-methylphenol	$C_7H_6Cl_2O$	194649	177.028	25	0.0673	0.673	2		
2,3-Dichloro-1,4-naphthalenedione	$C_{10}H_4Cl_2O_2$	117-80-6	227.044	25	0.00001	0.0001	40		
1,2-Dichloro-4-nitrobenzene	$C_6H_3Cl_2NO_2$	99-54-7	192.000	20	0.0121	0.121	40		
1,2-Dichloropentane	$C_5H_{10}Cl_2$	1674-33-5	141.038	25	0.029	0.29	35		
1,5-Dichloropentane	$C_5H_{10}Cl_2$	628-76-2	141.038	19	0.02	0.2	35		
2,3-Dichloropentane	$C_5H_{10}Cl_2$	600-11-3	141.038	25	0.029	0.29	35		
Dichlorophene	$C_{13}H_{10}Cl_2O_2$	97-23-4	269.123	25	0.003	0.03	40		
2,3-Dichlorophenol	$C_6H_4Cl_2O$	576-24-9	163.001	25	0.82	8.3	40		
2,4-Dichlorophenol	$C_6H_4Cl_2O$	120-83-2	163.001	25	0.55	5.5	48,51,24		
2,6-Dichlorophenol	$C_6H_4Cl_2O$	87-65-0	163.001	25	0.262	2.62	40		
(2,4-Dichlorophenoxy)acetic acid	$C_8H_6Cl_2O_3$	94-75-7	221.038	25	0.07	0.7	40		
4-(2,4-Dichlorophenoxy)butanoic acid	$C_{10}H_{10}Cl_2O_3$	94-82-6	249.090	25	0.0046	0.046	40		
2-(2,4-Dichlorophenoxy)propanoic acid	$C_9H_8Cl_2O_3$	120-36-5	235.064	25	0.083	0.83	40		
1,2-Dichloropropane, (±)-	$C_3H_6Cl_2$	78-87-5	112.986	5	0.270	2.70	35		

Name	Mol. form.	CAS Reg. No.	Mol. wt.	t/°C	s/mass%	s/g kg^{-1} H_2O	Ref.	k_H/ kPa m^3 mol^{-1}	Ref.
				25	0.274	2.74	35	0.29	13
				40	0.297	2.97	35		
1,3-Dichloropropane	$C_3H_6Cl_2$	142-28-9	112.986	5	0.218	2.18	35		
				25	0.280	2.80	35		
cis-1,3-Dichloropropene	$C_3H_4Cl_2$	10061-01-5	110.970	20	0.27	2.7	5	0.24	5
trans-1,3-Dichloropropene	$C_3H_4Cl_2$	10061-02-6	110.970	20	0.28	2.8	5	0.18	5
2,3-Dichloropropene	$C_3H_4Cl_2$	78-88-6	110.970	25	0.215	2.15	5	0.36	5
1,2-Dichloro-1,1,2,2-tetrafluoroethane	$C_2Cl_2F_4$	76-14-2	170.921	25	0.013*	0.13*	10	127	13
2,4-Dichlorotoluene	$C_7H_6Cl_2$	95-73-8	161.029	25	0.00260	0.0260	61		
2,6-Dichlorotoluene	$C_7H_6Cl_2$	118-69-4	161.029	25	0.00233	0.0233	61		
2,2-Dichloro-1,1,1-trifluoroethane	$C_2HCl_2F_3$	306-83-2	152.930	25	0.46	4.6	25		
Diclofop-methyl	$C_{16}H_{14}Cl_2O_4$	51338-27-3	341.186	20	0.0003	0.003	32		
Dieldrin	$C_{12}H_8Cl_6O$	60-57-1	380.909	25	0.000020	0.00020	67		
Diethanolamine	$C_4H_{11}NO_2$	111-42-2	105.136	20	95.4	20700	10		
1,1-Diethoxyethane	$C_6H_{14}O_2$	105-57-7	118.174	25	5	53	10		
1,2-Diethoxyethane	$C_6H_{14}O_2$	629-14-1	118.174	20	21.0	266	10		
2-(Diethylamino)-*N*-(2,6-dimethylphenyl)acetamide	$C_{14}H_{22}N_2O$	137-58-6	234.337	25	0.38	3.8	40		
o-Diethylbenzene	$C_{10}H_{14}$	135-01-3	134.218	20	0.0071	0.071	40		
p-Diethylbenzene	$C_{10}H_{14}$	105-05-5	134.218	20	0.0025	0.025	40		
Diethyl carbonate	$C_5H_{10}O_3$	105-58-8	118.131	20	1.8	18	40		
Diethyl ether	$C_4H_{10}O$	60-29-7	74.121	0	12.5	143	79	0.088	13
				25	5.9	63	79	0.088	13
				38	4.6	48	79	0.088	13
				82	3.1	32	79	0.088	13
Diethyl glutarate	$C_9H_{16}O_4$	818-38-2	188.221	30	1.20	12.1	20		
				91	0.91	9.2	20		
Diethyl maleate	$C_8H_{12}O_4$	141-05-9	172.179	20	1.56	15.8	20		
				91	1.75	17.8	20		
Diethyl malonate	$C_7H_{12}O_4$	105-53-3	160.168	20	2.26	23.1	20		
				91	2.47	25.3	20		
Diethyl phthalate	$C_{12}H_{14}O_4$	84-66-2	222.237	25	0.12	1.2	40		
trans-Diethylstilbestrol	$C_{18}H_{20}O_2$	56-53-1	268.351	20	0.01	0.1	40		
Diethyl succinate	$C_8H_{14}O_4$	123-25-1	174.195	20	0.19	1.9	40		
Diethyl sulfide	$C_4H_{10}S$	352-93-2	90.187	25	0.307	3.07	40		
Diflubenzuron	$C_{14}H_9ClF_2N_2O_2$	35367-38-5	310.683	20	0.00002	0.0002	40		
o-Difluorobenzene	$C_6H_4F_2$	367-11-3	114.093	25	0.114	1.14	2		
m-Difluorobenzene	$C_6H_4F_2$	372-18-9	114.093	25	0.114	1.14	2		
p-Difluorobenzene	$C_6H_4F_2$	540-36-3	114.093	25	0.122	1.22	2		
1,1-Difluoroethane	$C_2H_4F_2$	75-37-6	66.050	20	0.29*	2.9*	50		
Digitoxin	$C_{41}H_{64}O_{13}$	71-63-6	764.939	25	0.0004	0.004	40		
Diglycolic acid	$C_4H_6O_5$	110-99-6	134.088	24	40.0	667	34		
				50	59.9	1490	34		
Digoxin	$C_{41}H_{64}O_{14}$	20830-75-5	780.939	25	0.0059	0.059	40		
Dihexyl ether	$C_{12}H_{26}O$	112-58-3	186.333	20	0.019	0.19	20		
				90	0.019	0.19	20		
1,2-Dihydrobenz[j]aceanthrylene	$C_{20}H_{14}$	479-23-2	254.325	27	0.00000035	0.0000035	42,6		
1,3-Dihydro-2*H*-benzimidazoL-2-one	$C_7H_6N_2O$	615-16-7	134.135	24	0.37	3.7	54		
1,2-Dihydro-3-methylbenz[j]aceanthrylene	$C_{21}H_{16}$	56-49-5	268.352	25	0.00000022	0.0000022	42,6		
				27	0.00000028	0.0000028	42		
2,3-Dihydro-6-propyL-2-thioxo-4(1H)-pyrimidinone	$C_7H_{10}N_2OS$	51-52-5	170.231	25	0.120	1.20	40		
1,7-Dihydro-6*H*-purine-6-thione	$C_5H_4N_4S$	50-44-2	152.178	25	0.0124	0.124	40		
3,4-Dihydro-2*H*-pyran	C_5H_8O	110-87-2	84.117	20	1.04	10.5	20		
				82	2.26	23.1	20		
1,4-Dihydroxy-9,10-anthracenedione	$C_{14}H_8O_4$	81-64-1	240.212	25	0.0000096	0.000096	40		
3,4-Dihydroxybenzoic acid	$C_7H_6O_4$	99-50-3	154.121	14	1.8	18	26		
				80	21.3	271	26		
17,21-Dihydroxypregna-1,4-diene-3,11,20-trione	$C_{21}H_{26}O_5$	53-03-2	358.428	25	0.012	0.12	40		

Name	Mol. form.	CAS Reg. No.	Mol. wt.	t/°C	s/mass%	s/g kg^{-1} H_2O	Ref.	k_H/ kPa m^3 mol^{-1}	Ref.
17,21-Dihydroxypregn-4-ene-3,11,20-trione	$C_{21}H_{28}O_5$	53-06-5	360.444	25	0.028	0.28	30		
o-Diiodobenzene	$C_6H_4I_2$	615-42-9	329.905	25	0.00192	0.0192	2		
m-Diiodobenzene	$C_6H_4I_2$	626-00-6	329.905	25	0.000185	0.00185	2		
p-Diiodobenzene	$C_6H_4I_2$	624-38-4	329.905	25	0.000893	0.00893	2		
cis-1,2-Diiodoethene	$C_2H_2I_2$	590-26-1	279.846	25	0.046	0.46	25		
trans-1,2-Diiodoethene	$C_2H_2I_2$	590-27-2	279.846	25	0.015	0.15	25		
Diiodomethane	CH_2I_2	75-11-6	267.836	30	0.124	1.24	10	0.032	13
3,5-Diiodo-L-tyrosine	$C_9H_9I_2NO_3$	300-39-0	432.981	25	0.062	0.62	26		
Diisopentyl ether	$C_{10}H_{22}O$	544-01-4	158.281	20	0.02	0.2	10		
Diisopropyl ether	$C_6H_{14}O$	108-20-3	102.174	20	0.79	8.0	20	0.26	13
				61	0.22	2.2	20		
1,2-Dimethoxybenzene	$C_8H_{10}O_2$	91-16-7	138.164	20	0.716	7.21	20		
				92	1.073	10.85*	20		
3,3'-Dimethoxybenzidine	$C_{14}H_{16}N_2O_2$	119-90-4	244.289	25	0.006	0.06	40		
Dimethoxymethane	$C_3H_8O_2$	109-87-5	76.095	16	24.4	323	10		
4-(Dimethylamino)azobenzene	$C_{14}H_{15}N_3$	60-11-7	225.289	20	0.00014	0.0014	40		
2',3-Dimethyl-4-aminoazobenzene	$C_{14}H_{15}N_3$	97-56-3	225.289	37	0.0007	0.007	40		
2,5-Dimethylaniline	$C_8H_{11}N$	95-78-3	121.180	20	0.66	6.6	27		
N,N-Dimethylaniline	$C_8H_{11}N$	121-69-7	121.180	25	0.111	1.11	40		
9,10-Dimethylanthracene	$C_{16}H_{14}$	781-43-1	206.282	25	0.0000056	0.000056	42,4		
Dimethylarsinic acid	$C_2H_7AsO_2$	75-60-5	137.998	25	≈41	≈700	40		
7,12-Dimethylbenz[a]anthracene	$C_{20}H_{16}$	57-97-6	256.341	25	0.0000061	0.000061	42		
2,2-Dimethylbutane	C_6H_{14}	75-83-2	86.175	25	0.0021	0.021	3	199	13
2,3-Dimethylbutane	C_6H_{14}	79-29-8	86.175	25	0.0021	0.021	3	144	13
2,2-Dimethyl-1-butanol	$C_6H_{14}O$	1185-33-7	102.174	25	0.78	7.9	78,1		
2,3-Dimethyl-2-butanol	$C_6H_{14}O$	594-60-5	102.174	25	4.2	44	1		
3,3-Dimethyl-2-butanol, (±)-	$C_6H_{14}O$	20281-91-8	102.174	25	2.4	25	1		
3,3-Dimethyl-2-butanone	$C_6H_{12}O$	75-97-8	100.158	0	2.92	30.1	83		
				19	1.97	20.1	20		
				25	1.85	18.8	83		
				50	1.46	14.8	83		
				90	1.14	11.5	20		
2,3-Dimethyl-1-butene	C_6H_{12}	563-78-0	84.159	30	0.046	0.46	3		
cis-1,2-Dimethylcyclohexane	C_8H_{16}	112134	112.213	25	0.00060	0.0060	4	36	5
trans-1,2-Dimethylcyclohexane	C_8H_{16}	6876-23-9	112.213	30	0.00050	0.0050	57,4	88.2	5
				100	0.00293	0.0293	57,4		
Dimethyl ether	C_2H_6O	115-10-6	46.068	25	35.3*	546	79		
				50	29.2*	412	79		
Dimethylglyoxime	$C_4H_8N_2O_2$	95-45-4	116.119	20	0.06	0.6	40		
3,5-Dimethyl-4-heptanol	$C_9H_{20}O$	19549-79-2	144.254	15	0.072	0.72	1		
2,6-Dimethyl-4-heptanone	$C_9H_{18}O$	108-83-8	142.238	21	0.045	0.45	20		
				91	0.037	0.37	20		
1,2-Dimethyl-1*H*-imidazole	$C_5H_8N_2$	1739-84-0	96.131	19	94.3	16500	54		
Dimethyl maleate	$C_6H_8O_4$	624-48-6	144.126	25	8.0	87	10		
Dimethyl malonate	$C_5H_8O_4$	108-59-8	132.116	19	14.9	175	20		
				90	29.8	425	20		
1,3-Dimethylnaphthalene	$C_{12}H_{12}$	575-41-7	156.223	25	0.0008	0.008	4		
1,4-Dimethylnaphthalene	$C_{12}H_{12}$	571-58-4	156.223	25	0.00114	0.0114	4		
1,5-Dimethylnaphthalene	$C_{12}H_{12}$	571-61-9	156.223	25	0.00031	0.0031	4	0.036	28
2,3-Dimethylnaphthalene	$C_{12}H_{12}$	581-40-8	156.223	25	0.00025	0.0025	4		
2,6-Dimethylnaphthalene	$C_{12}H_{12}$	581-42-0	156.223	25	0.00017	0.0017	4		
Dimethyl oxalate	$C_4H_6O_4$	553-90-2	118.089	20	5.82	61.8	27		
2,2-Dimethylpentane	C_7H_{16}	590-35-2	100.202	25	0.00044	0.0044	3	318	5
2,3-Dimethylpentane	C_7H_{16}	565-59-3	100.202	25	0.00052	0.0052	3	175	5
2,4-Dimethylpentane	C_7H_{16}	108-08-7	100.202	25	0.00042	0.0042	3	323	13
3,3-Dimethylpentane	C_7H_{16}	562-49-2	100.202	25	0.00059	0.0059	3	186	5
2,3-Dimethyl-2-pentanol	$C_7H_{16}O$	4911-70-0	116.201	25	1.5	15	1		
2,4-Dimethyl-2-pentanol	$C_7H_{16}O$	625-06-9	116.201	25	1.3	13	1		
2,2-Dimethyl-3-pentanol	$C_7H_{16}O$	3970-62-5	116.201	25	0.82	8.2	1		
2,3-Dimethyl-3-pentanol	$C_7H_{16}O$	595-41-5	116.201	25	1.6	16	1		

Name	Mol. form.	CAS Reg. No.	Mol. wt.	t/°C	s/mass%	s/g kg^{-1} H_2O	Ref.	k_H/ kPa m^3 mol^{-1}	Ref.
2,4-Dimethyl-3-pentanol	$C_7H_{16}O$	600-36-2	116.201	25	0.70	7.0	1		
2,4-Dimethyl-3-pentanone	$C_7H_{14}O$	565-80-0	114.185	20	0.52	5.2	20		
				90	0.30	3.0	20		
N,N-Dimethyl-*N'*-phenylurea	$C_9H_{12}N_2O$	101-42-8	164.203	25	0.32	3.2	40		
Dimethyl phthalate	$C_{10}H_{10}O_4$	131-11-3	194.184	25	0.40	4.0	15		
2,2-Dimethyl-1-propanol	$C_5H_{12}O$	75-84-3	88.148	12	3.87	40.3	78,1		
				25	3.26	33.7	78,1		
				80	2.84	29.2	78,1		
4-(1,1-Dimethylpropyl)phenol	$C_{11}H_{16}O$	80-46-6	164.244	25	0.017	0.17	40		
Dimethyl succinate	$C_6H_{10}O_4$	106-65-0	146.141	21	12.4	142	20		
				92	17.1	206	20		
Dimethyl sulfate	$C_2H_6O_4S$	77-78-1	126.132	18	2.7	28	27		
Dimethyl sulfide	C_2H_6S	75-18-3	62.134	25	2	20	10		
Dimethyl sulfoxide	C_2H_6OS	67-68-5	78.133	25	25.3	339	10		
Dimethyl terephthalate	$C_{10}H_{10}O_4$	120-61-6	194.184	25	0.00328	0.0328	40		
Dimethyl tetrachloroterephthalate	$C_{10}H_6Cl_4O_4$	1861-32-1	331.965	25	0.00005	0.0005	40		
N,N-Dimethyl-N'-[3-(trifluoromethyl) phenyl]urea	$C_{10}H_{11}F_3N_2O$	2164-17-2	232.201	20	0.0105	0.105	40		
2,4-Dinitroaniline	$C_6H_5N_3O_4$	97-02-9	183.122	25	0.0078	0.078	40		
1,2-Dinitrobenzene	$C_6H_4N_2O_4$	528-29-0	168.107	20	0.21	2.1	27		
1,3-Dinitrobenzene	$C_6H_4N_2O_4$	99-65-0	168.107	20	2.09	21.3	27		
1,4-Dinitrobenzene	$C_6H_4N_2O_4$	100-25-4	168.107	20	1.30	13.2	27		
3,5-Dinitrobenzoic acid	$C_7H_4N_2O_6$	99-34-3	212.116	25	0.134	1.34	27		
2,4-Dinitrophenol	$C_6H_4N_2O_5$	51-28-5	184.106	25	0.069	0.69	48,51		
				35	0.098	0.98	48,51		
Dipentyl ether	$C_{10}H_{22}O$	693-65-2	158.281	25	0.11	1.1	81		
Diphenamid	$C_{16}H_{17}NO$	957-51-7	239.312	27	0.026	0.26	32		
Diphenylamine	$C_{12}H_{11}N$	122-39-4	169.222	20	0.0055	0.055	40		
				50	0.0058	0.058	40		
1,2-Diphenylethane	$C_{14}H_{14}$	103-29-7	182.261	25	0.00044	0.0044	6	0.017	12
Diphenyl ether	$C_{12}H_{10}O$	101-84-8	170.206	25	0.0018	0.0180	6	0.027	13
Diphenylmethane	$C_{13}H_{12}$	101-81-5	168.234	25	0.00014	0.0014	42,4	0.001	12
Diphenyl phthalate	$C_{20}H_{14}O_4$	84-62-8	318.323	24	0.000008	0.00008	40		
1,3-DiphenyL-1-triazene	$C_{12}H_{11}N_3$	136-35-6	197.235	20	0.050	0.50	40		
N,N'-Diphenylurea	$C_{13}H_{12}N_2O$	102-07-8	212.246	20	0.015	0.15	40		
Dipropylamine	$C_6H_{15}N$	142-84-7	101.190	20	2.5	26	10		
Dipropyl ether	$C_6H_{14}O$	111-43-3	102.174	0	2.67	27.4	80	0.26	13
				25	0.91	9.2	80	0.26	13
Diuron	$C_9H_{10}Cl_2N_2O$	330-54-1	233.093	25	0.0042	0.042	40		
Docosane	$C_{22}H_{46}$	629-97-0	310.600	22	0.0000006	0.000006	37		
Dodecane	$C_{12}H_{26}$	112-40-3	170.334	25	0.00000037	0.0000037	4	750	5
Dodecanedioic acid	$C_{12}H_{22}O_4$	693-23-2	230.301	20	0.004	0.04	40		
Dodecanoic acid	$C_{12}H_{24}O_2$	143-07-7	200.318	20	0.0055	0.055	26		
1-Dodecanol	$C_{12}H_{26}O$	112-53-8	186.333	25	0.0004	0.004	1		
Droperidol	$C_{22}H_{22}FN_3O_2$	548-73-2	379.427	30	0.00041	0.0041	40		
Eicosane	$C_{20}H_{42}$	112-95-8	282.547	25	0.00000019	0.0000019	42,4		
Emetine	$C_{29}H_{40}N_2O_4$	483-18-1	480.639	15	0.096	0.96	40		
Endrin	$C_{12}H_8Cl_6O$	72-20-8	380.909	25	0.000025	0.00025	67		
Ephedrine, (1R,2S)	$C_{10}H_{15}NO$	299-42-3	165.232	25	0.57	5.7	40		
Epichlorohydrin	C_3H_5ClO	13403-37-7	92.524	20	6.58	70.4	10	0.003	13
				65	7.2	78	40		
Epinephrine	$C_9H_{13}NO_3$	51-43-4	183.204	20	0.018	0.18	40		
1,2-Epoxy-4-(epoxyethyl)cyclohexane	$C_8H_{12}O_2$	106-87-6	140.180	20	13.4	155	40		
2,3-Epoxy-α-pinane	$C_{10}H_{16}O$	1686-14-2	152.233	25	0.039	0.39	52		
Erythromycin	$C_{37}H_{67}NO_{13}$	114-07-8	733.927	30	0.12	1.2	40		
				80	0.04	0.4	40		
Estra-1,3,5(10)-triene-3,17-diol (17β)	$C_{18}H_{24}O_2$	50-28-2	272.383	25	0.000151	0.00151	49		
Estrone	$C_{18}H_{22}O_2$	53-16-7	270.367	25	0.000130	0.00130	49		
Ethane	C_2H_6	74-84-0	30.069	25	0.00568*	0.0568*	18	50.6	5
1,2-Ethanediol, diacetate	$C_6H_{10}O_4$	111-55-7	146.141	25	13.3	153	40		
Ethinylestradiol	$C_{20}H_{24}O_2$	57-63-6	296.404	25	0.000921	0.00921	49		

Name	Mol. form.	CAS Reg. No.	Mol. wt.	t/°C	s/mass%	s/g kg⁻¹ H₂O	Ref.	k_H/ kPa m³ mol⁻¹	Ref.
Ethoxybenzene	$C_8H_{10}O$	103-73-1	122.164	25	0.12	1.2	10		
2-Ethoxyethyl acetate	$C_6H_{12}O_3$	111-15-9	132.157	20	14	163	30		
N-(4-Ethoxyphenyl)acetamide	$C_{10}H_{13}NO_2$	62-44-2	179.216	25	0.0502	0.502	40		
Ethyl acetate	$C_4H_8O_2$	141-78-6	88.106	25	8.08	87.9	10		
Ethyl acetoacetate	$C_6H_{10}O_3$	141-97-9	130.141	25	12	136	10		
Ethyl acrylate	$C_5H_8O_2$	140-88-5	100.117	25	1.50	15.2	10		
Ethylbenzene	C_8H_{10}	100-41-4	106.165	0	0.020	0.20	4,89		
				25	0.0161	0.161	22,89	0.843	22
				40	0.0200	0.200	4,89		
Ethyl benzoate	$C_9H_{10}O_2$	93-89-0	150.174	25	0.083	0.83	20		
Ethyl butanoate	$C_6H_{12}O_2$	105-54-4	116.158	20	0.49	4.9	10		
2-Ethyl-1-butanol	$C_6H_{14}O$	97-95-0	102.174	20	0.92	9.3	78		
				50	0.80	8.1	78		
Ethyl carbamate	$C_3H_7NO_2$	51-79-6	89.094	15	48	920	27		
Ethyl cyanoacetate	$C_5H_7NO_2$	105-56-6	113.116	20	25.9	350	10		
Ethylcyclohexane	C_8H_{16}	1678-91-7	112.213	30	0.00061	0.0061	57,4		
				100	0.00212	0.0212	57,4		
Ethylcyclopentane	C_7H_{14}	1640-89-7	98.186	20	0.012	0.12	3		
Ethyl decanoate	$C_{12}H_{24}O_2$	110-38-3	200.318	20	0.0015	0.015	27		
Ethylene	C_2H_4	74-85-1	28.053	25	0.01336*	0.1336*	19	21.7	5
Ethyleneimine	C_2H_5N	151-56-4	43.068	20	0.90	9.1	40		
Ethyl formate	$C_3H_6O_2$	109-94-4	74.079	25	11.8	134	10		
Ethyl heptanoate	$C_9H_{18}O_2$	106-30-9	158.238	20	0.029	0.29	27		
Ethyl hexanoate	$C_8H_{16}O_2$	123-66-0	144.212	20	0.063	0.63	27		
2-Ethyl-1-hexanol	$C_8H_{18}O$	104-76-7	130.228	25	0.071	0.71	78		
				50	0.074	0.74	78		
2-Ethylhexylamine	$C_8H_{19}N$	104-75-6	129.244	20	0.25	2.5	10		
Ethyl 4-hydroxybenzoate	$C_9H_{10}O_3$	120-47-8	166.173	25	0.0080	0.080	40		
Ethyl isopropyl ether	$C_5H_{12}O$	625-54-7	88.148	25	0.52	5.2	79		
Ethyl 2-methylbutanoate, (+)	$C_7H_{14}O_2$	10307-61-6	130.185	19	0.257	2.58	20		
				91	0.151	1.51	20		
Ethyl 3-methylbutanoate	$C_7H_{14}O_2$	108-64-5	130.185	20	0.2	2	10		
Ethyl *N*-methylcarbamate	$C_4H_9NO_2$	105-40-8	103.120	15	69	2230	27		
1-Ethylnaphthalene	$C_{12}H_{12}$	1127-76-0	156.223	25	0.00101	0.0101	4	0.039	12
2-Ethylnaphthalene	$C_{12}H_{12}$	939-27-5	156.223	25	0.00080	0.0080	4	0.078	12
O-Ethyl *O*-*p*-nitrophenyl benzenethiophosphonate	$C_{14}H_{14}NO_4PS$	2104-64-5	323.304	22	0.00031	0.0031	40		
N-Ethyl-*N*-nitrosourea	$C_3H_7N_3O_2$	759-73-9	117.107	20	1.3	13	40		
Ethyl nonanoate	$C_{11}H_{22}O_2$	123-29-5	186.292	20	0.003	0.03	27		
Ethyl octanoate	$C_{10}H_{20}O_2$	106-32-1	172.265	20	0.007	0.07	27		
Ethyl pentanoate	$C_7H_{14}O_2$	539-82-2	130.185	25	0.3	3	27		
3-Ethyl-3-pentanol	$C_7H_{16}O$	597-49-9	116.201	25	1.7	17	1		
4-Ethylphenol	$C_8H_{10}O$	123-07-9	122.164	20	0.59	5.9	40		
Ethyl propanoate	$C_5H_{10}O_2$	105-37-3	102.132	20	1.92	19.6	10		
Ethyl *N*-propylcarbamate	$C_6H_{13}NO_2$	623-85-8	131.173	15	7.70	83.4	27		
Ethyl propyl ether	$C_5H_{12}O$	628-32-0	88.148	25	1.87	19.2	79		
2-Ethyltoluene	C_9H_{12}	611-14-3	120.191	25	0.0075	0.075	89,5	0.529	13
4-Ethyltoluene	C_9H_{12}	622-96-8	120.191	25	0.0094	0.094	5	0.5	13
Ethyl vinyl ether	C_4H_8O	109-92-2	72.106	20	0.9	9	10		
Etoposide	$C_{29}H_{32}O_{13}$	33419-42-0	588.556	20	0.02	0.2	40		
Eucalyptol	$C_{10}H_{18}O$	470-82-6	154.249	21	0.379	3.79	40		
				50	0.170	1.70	40		
Fenamiphos	$C_{13}H_{22}NO_3PS$	22224-92-6	303.358	20	0.0329	0.329	40		
Fenbutatin oxide	$C_{60}H_{78}OSn_2$	13356-08-6	1052.68	23	0.0000005	0.000005	32		
α-Fenchol, (+)-	$C_{10}H_{18}O$	115823	154.249	25	0.083	0.83	52		
(±)-Fenchone	$C_{10}H_{16}O$	18492-37-0	152.233	20	0.2	2	84		
Fenoxycarb	$C_{17}H_{19}NO_4$	79127-80-3	301.338	20	0.0006	0.006	32		
Ferbam	$C_9H_{18}FeN_3S_6$	14484-64-1	416.494	20	0.013	0.13	40		
Fluoranthene	$C_{16}H_{10}$	206-44-0	202.250	20	0.000017	0.00017	42		
Fluoranthene	$C_{16}H_{10}$	206-44-0	202.250	25	0.000021	0.00021	42,22	0.00096	22
9*H*-Fluorene	$C_{13}H_{10}$	86-73-7	166.218	0	0.00007	0.0007	42,4		

Name	Mol. form.	CAS Reg. No.	Mol. wt.	t/°C	s/mass%	s/g kg^{-1} H_2O	Ref.	k_H/kPa m^3 mol^{-1}	Ref.
				25	0.00019	0.0019	42,22	0.00787	22
				50	0.00063	0.0063	42,4		
Fluorescein	$C_{20}H_{12}O_5$	153954	332.306	20	0.005	0.05	27		
Fluorobenzene	C_6H_5F	462-06-6	96.102	19	0.170	1.70	20	0.7	11
				80	0.188	1.88	20	0.7	11
2-Fluorobenzoic acid	$C_7H_5FO_2$	445-29-4	140.112	25	0.72	7.2	27		
3-Fluorobenzoic acid	$C_7H_5FO_2$	455-38-9	140.112	25	0.15	1.5	27		
4-Fluorobenzoic acid	$C_7H_5FO_2$	456-22-4	140.112	25	0.12	1.2	27		
Fluoroethane	C_2H_5F	353-36-6	48.059	25	0.216*	2.16*	14		
Fluoromethane	CH_3F	593-53-3	34.033	0	0.420*	4.20*	50		
				25	0.201*	2.01*	50		
				80	0.082*	0.82*	50		
1-Fluoropropane	C_3H_7F	460-13-9	62.086	14	0.386*	3.86*	14		
2-Fluoropropane	C_3H_7F	420-26-8	62.086	15	0.366*	3.66*	14		
5-Fluorouracil	$C_4H_3FN_2O_2$	51-21-8	130.077	25	1.77	18.0	72		
Folic acid	$C_{19}H_{19}N_7O_6$	59-30-3	441.397	0	0.001	0.01	26		
				100	0.05	0.5	26		
Folpet	$C_9H_4Cl_3NO_2S$	133-07-3	296.558	20	0.00010	0.0010	40		
β-D-Fructose	$C_6H_{12}O_6$	53188-23-1	180.155	20	≈31	≈450	40		
Furan	C_4H_4O	110-00-9	68.074	25	1	10	10	0.54	13
2-Furancarboxylic acid	$C_5H_4O_3$	88-14-2	112.084	25	4.76	50.0	33		
				50	25.2	337	33		
Furfural	$C_5H_4O_2$	98-01-1	96.085	20	8.2	89	10		
Galactaric acid	$C_6H_{10}O_8$	526-99-8	210.138	14	0.33	3.3	40		
D-Galactose	$C_6H_{12}O_6$	59-23-4	180.155	20	40.6	684	27		
D-Glucitol	$C_6H_{14}O_6$	50-70-4	182.171	20	≈41	≈700	40		
α-D-Glucose	$C_6H_{12}O_6$	26655-34-5	180.155	15	45.0	818	27		
				30	54.6	1200	27		
				80	81.5	4400	27		
DL-Glutamic acid	$C_5H_9NO_4$	617-65-2	147.130	25	2.30	23.5	29		
L-Glutamic acid	$C_5H_9NO_4$	56-86-0	147.130	10	0.444	4.46	75		
				25	0.824	8.31	75		
				50	2.13	21.8	75		
L-Glutamine	$C_5H_{10}N_2O_3$	56-85-9	146.144	25	4.0	42	26		
Glycerol triacetate	$C_9H_{14}O_6$	102-76-1	218.203	25	5.8	62	10		
Glycine	$C_2H_5NO_2$	56-40-6	75.067	25	18.5	227	70		
				36	22.1	284	70		
				50	26.1	353	70		
Glycolic acid	$C_2H_4O_3$	79-14-1	76.051	25	71.2	2470	34		
				55	77.9	3520	34		
N-Glycylglycine	$C_4H_8N_2O_3$	556-50-3	132.118	25	18.8	232	47,29		
Glyphosate	$C_3H_8NO_5P$	1071-83-6	169.074	25	1.2	12	32		
Guanidinoacetic acid	$C_3H_7N_3O_2$	352-97-6	117.107	25	0.5	5	26		
Guanine	$C_5H_5N_5O$	73-40-5	151.127	25	0.0068	0.068	29		
Guanosine	$C_{10}H_{13}N_5O_5$	118-00-3	283.241	25	0.0500	0.500	29		
Haloperidol	$C_{21}H_{23}ClFNO_2$	52-86-8	375.865	30	0.0003	0.003	40		
Heptachlor	$C_{10}H_5Cl_7$	76-44-8	373.318	25	0.000018	0.00018	67		
2,2',3,3',4,4',6-Heptachlorobiphenyl	$C_{12}H_3Cl_7$	52663-71-5	395.323	25	0.0000002	0.000002	7	0.0054	7
Heptadecanoic acid	$C_{17}H_{34}O_2$	506-12-7	270.451	20	0.00042	0.0042	26		
1,6-Heptadiyne	C_7H_8	2396-63-6	92.139	25	0.125	1.25	3		
Heptanal	$C_7H_{14}O$	111-71-7	114.185	11	0.124	1.24	27		
Heptane	C_7H_{16}	142-82-5	100.202	25	0.000242	0.00242	46		
				50	0.000341	0.00341	46	209	13
				75	0.000570	0.00570	46		
				100	0.00108	0.0108	46		
Heptanedioic acid	$C_7H_{12}O_4$	111-16-0	160.168	25	6.347	67.77	33		
				50	42.80	748	33		
Heptanoic acid	$C_7H_{14}O_2$	111-14-8	130.185	15	0.24	2.4	27		
1-Heptanol	$C_7H_{16}O$	111-70-6	116.201	0	0.236	2.37	78		
				25	0.164	1.64	78,1	0.00562	28

Name	Mol. form.	CAS Reg. No.	Mol. wt.	t/°C	s/mass%	s/g kg^{-1} H_2O	Ref.	k_H/ kPa m^3 mol^{-1}	Ref.
				50	0.164	1.64	78,1		
				90	0.245	2.46	78		
2-Heptanol, (±)-	$C_7H_{16}O$	52390-72-4	116.201	30	0.33	3.3	1		
3-Heptanol, (*S*)-	$C_7H_{16}O$	26549-25-7	116.201	25	0.43	4.3	1		
4-Heptanol	$C_7H_{16}O$	589-55-9	116.201	25	0.47	4.7	1		
2-Heptanone	$C_7H_{14}O$	110-43-0	114.185	25	0.435	4.37	20	0.0171	28
				90	0.353	3.53	20	0.0171	28
3-Heptanone	$C_7H_{14}O$	106-35-4	114.185	20	0.479	4.81	20		
				90	0.309	3.10	20		
4-Heptanone	$C_7H_{14}O$	123-19-3	114.185	20	0.457	4.57	20		
				90	0.316	3.16	20		
1-Heptene	C_7H_{14}	592-76-7	98.186	25	0.032	0.32	3	40.3	13
trans-2-Heptene	C_7H_{14}	14686-13-6	98.186	25	0.015	0.15	3	42.2	13
Heptyl butanoate	$C_{11}H_{22}O_2$	5870-93-9	186.292	20	0.028	0.28	20		
				80	0.020	0.20	20		
1-Heptyne	C_7H_{12}	628-71-7	96.170	25	0.0094	0.094	3	4.47	13
Hesperetin	$C_{16}H_{14}O_6$	520-33-2	302.278	15	0.00004	0.0004	71		
				25	0.00014	0.0014	71		
				35	0.00052	0.0052	71		
Hexachlorobenzene	C_6Cl_6	118-74-1	284.782	25	0.00000096	0.0000096	58	0.131	11
				35	0.0000018	0.000018	58		
				55	0.0000038	0.000038	58		
2,2',3,3',4,4'-Hexachlorobiphenyl	$C_{12}H_4Cl_6$	38380-07-3	360.878	25	0.00000006	0.0000006	7	0.0354	31
2,2',4,4',6,6'-Hexachlorobiphenyl	$C_{12}H_4Cl_6$	33979-03-2	360.878	25	0.0000003	0.000003	41	0.818	7
2,2',3,3',6,6'-Hexachlorobiphenyl	$C_{12}H_4Cl_6$	38411-22-2	360.878	25	0.0000004	0.000004	41		
Hexachloro-1,3-butadiene	C_4Cl_6	87-68-3	260.761	25	0.41	4.1	35		
1,2,3,4,5,6-Hexachlorocyclohexane, (1α,2α,3β,4α,5α,6β)	$C_6H_6Cl_6$	58-89-9	290.830	25	0.00078	0.0078	60		
				45	0.0015	0.015	60		
1,2,3,4,5,6-Hexachlorocyclohexane, (1α,2α,3β,4α,5β,6β)	$C_6H_6Cl_6$	319-84-6	290.830	25	0.00018	0.0018	60		
1,2,3,4,5,6-Hexachlorocyclohexane, (1α,2β,3α,4β,5α,6β)	$C_6H_6Cl_6$	319-85-7	290.830	25	0.00002	0.0002	60		
Hexachloroethane	C_2Cl_6	67-72-1	236.739	25	0.005	0.05	25	0.85	13
Hexachloropropene	C_3Cl_6	1888-71-7	248.750	20	0.00118	0.0118	35		
Hexacosafluorododecane	$C_{12}F_{26}$	307-59-5	638.086	20	0.00000096	0.0000096	35		
Hexacosane	$C_{26}H_{54}$	630-01-3	366.707	25	0.00000017	0.0000017	42,37		
Hexadecane	$C_{16}H_{34}$	544-76-3	226.441	25	0.0000004	0.000004	42,37		
Hexadecanoic acid	$C_{16}H_{32}O_2$	57-10-3	256.424	20	0.00072	0.0072	26		
1-Hexadecanol	$C_{16}H_{34}O$	36653-82-4	242.440	25	0.000003	0.00003	1		
1,5-Hexadiene	C_6H_{10}	592-42-7	82.143	25	0.017	0.17	3		
Hexafluorobenzene	C_6F_6	392-56-3	186.054	8	0.0778	0.778	53		
				28	0.0616	0.616	53		
				67	0.0636	0.636	53		
Hexahydro-1,3,5-trinitro-1,3,5-triazine	$C_3H_6N_6O_6$	121-82-4	222.116	3	0.0014	0.014	59		
				20	0.0037	0.037	59		
				25	0.0060	0.060	17		
				34	0.0086	0.086	59		
Hexamethylenetetramine	$C_6H_{12}N_4$	100-97-0	140.186	12	44.8	812	27		
Hexane	C_6H_{14}	110-54-3	86.175	25	0.00098	0.0098	46		
				50	0.00114	0.0114	46		
				75	0.00167	0.0167	46	183	13
				100	0.00291	0.0291	46		
1,6-Hexanediamine	$C_6H_{16}N_2$	124-09-4	116.204	5	≈42	≈720	40		
Hexanedinitrile	$C_6H_8N_2$	111-69-3	108.141	20	0.80	8.0	16		
1,6-Hexanedioic acid	$C_6H_{10}O_4$	124-04-9	146.141	15	1.48	15.0	26		
				100	61.5	1600	26		
Hexanoic acid	$C_6H_{12}O_2$	142-62-1	116.158	25	1.01	10.2	64		
				35	1.09	11.0	64		
				60	1.16	11.7	26		

Name	Mol. form.	CAS Reg. No.	Mol. wt.	t/°C	s/mass%	s/g kg^{-1} H_2O	Ref.	k_H/kPa m^3 mol^{-1}	Ref.
1-Hexanol	$C_6H_{14}O$	111-27-3	102.174	0	0.79	7.9	1		
				10	0.70	7.0	78		
				25	0.59	5.9	78,1		
				50	0.55	5.5	78,1		
2-Hexanol	$C_6H_{14}O$	20281-86-1	102.174	25	1.4	14	1		
3-Hexanol	$C_6H_{14}O$	17015-11-1	102.174	25	1.6	16	1		
2-Hexanone	$C_6H_{12}O$	591-78-6	100.158	10	1.91	19.5	83		
				25	1.49	15.1	83		
				50	1.17	11.8	83		
3-Hexanone	$C_6H_{12}O$	589-38-8	100.158	25	1.47	14.9	83		
Hexatriacontane	$C_{36}H_{74}$	630-06-8	506.973	25	0.00000017	0.0000017	42,37		
Hexazinone	$C_{12}H_{20}N_4O_2$	51235-04-2	252.313	25	3.2	33	40		
1-Hexene	C_6H_{12}	592-41-6	84.159	25	0.0053	0.053	3	41.8	5
trans-2-Hexene	C_6H_{12}	4050-45-7	84.159	25	0.0067	0.067	3		
1-Hexen-3-ol	$C_6H_{12}O$	4798-44-1	100.158	25	2.52	25.9	1		
4-Hexen-2-ol	$C_6H_{12}O$	52387-50-5	100.158	25	3.81	39.6	1		
Hexyl acetate	$C_8H_{16}O_2$	142-92-7	144.212	20	0.02	0.2	10		
sec-Hexyl acetate	$C_8H_{16}O_2$	108-84-9	144.212	20	0.13	1.3	10		
Hexylbenzene	$C_{12}H_{18}$	1077-16-3	162.271	25	0.00021	0.0021	4		
4-Hexyl-1,3-benzenediol	$C_{12}H_{18}O_2$	136-77-6	194.270	18	0.05	0.5	40		
Hexyl butanoate	$C_{10}H_{20}O_2$	2639-63-6	172.265	29	0.021	0.21	20		
1-Hexyne	C_6H_{10}	693-02-7	82.143	25	0.036	0.36	3	4.14	13
L-Histidine	$C_6H_9N_3O_2$	71-00-1	155.154	25	4.17	43.5	26		
Homocystine	$C_8H_{16}N_2O_4S_2$	870-93-9	268.354	25	0.02	0.2	26		
L-Homoserine	$C_4H_9NO_3$	672-15-1	119.119	25	52.4	1100	26		
Hydramethylnon	$C_{25}H_{24}F_6N_4$	67485-29-4	494.476	20	0.0000006	0.000006	32		
Hydrochlorothiazide	$C_7H_8ClN_3O_4S_2$	58-93-5	297.740	25	0.007	0.07	40		
Hydrocortisone	$C_{21}H_{30}O_5$	50-23-7	362.460	25	0.029	0.29	40		
Hydroflumethiazide	$C_8H_8F_3N_3O_4S_2$	135-09-1	331.293	37	0.068	0.68	40		
p-Hydroquinone	$C_6H_6O_2$	123-31-9	110.111	25	7.42	80.1	27		
17-Hydroxyandrost-4-en-3-one, (17β)	$C_{19}H_{28}O_2$	58-22-0	288.424	25	0.0024	0.024	40		
4-Hydroxybenzaldehyde	$C_7H_6O_2$	123-08-0	122.122	30	1.27	12.9	40		
2-Hydroxybenzamide	$C_7H_7NO_2$	65-45-2	137.137	10	0.122	1.22	44		
				25	0.241	2.42	44		
				50	0.737	7.42	44		
α-Hydroxybenzeneacetic acid, (±)-	$C_8H_8O_3$	611-72-3	152.148	25	11.3	127	27		
2-Hydroxybenzoic acid	$C_7H_6O_3$	69-72-7	138.121	10	0.119	1.19	43,33		
				25	0.189	1.89	43,33		
				50	0.521	5.24	43,33		
4-Hydroxybenzoic acid	$C_7H_6O_3$	99-96-7	138.121	15	0.8	8	26		
				75	2.5	26	27		
2-Hydroxybiphenyl	$C_{12}H_{10}O$	90-43-7	170.206	25	0.07	0.7	40		
4-Hydroxybiphenyl	$C_{12}H_{10}O$	92-69-3	170.206	25	0.0056	0.056	40		
4-Hydroxy-3-methoxybenzaldehyde	$C_8H_8O_3$	121-33-5	152.148	25	0.247	2.47	8		
3-Hydroxy-4-oxo-4*H*-pyran-2,6-dicarboxylic acid	$C_7H_4O_7$	497-59-6	200.103	25	0.84	8.4	27		
N-(4-Hydroxyphenyl)acetamide	$C_8H_9NO_2$	103-90-2	151.163	25	1.3	13	40		
trans-4-Hydroxy-L-proline	$C_5H_9NO_3$	51-35-4	131.130	25	26.5	361	26		
Hyoscyamine	$C_{17}H_{23}NO_3$	101-31-5	289.370	20	0.36	3.6	40		
Hypoxanthine	$C_5H_4N_4O$	68-94-0	136.112	25	0.070	0.70	29		
Ibuprofen	$C_{13}H_{18}O_2$	15687-27-1	206.281	25	0.0011	0.011	40		
				60	0.0048	0.048	40		
Imazaquin	$C_{17}H_{17}N_3O_3$	81335-37-7	311.335	20	0.009	0.09	32		
Imidacloprid	$C_9H_{10}ClN_5O_2$	105827-78-9	255.66	30	0.038	0.38	73		
				51	0.117	1.17	73		
Imidazole	$C_3H_4N_2$	288-32-4	68.077	19	67.3	2060	54		
2,4-Imidazolidinedione	$C_3H_4N_2O_2$	461-72-3	100.076	25	3.93	40.9	29		
Imidodicarbonic diamide	$C_2H_5N_3O_2$	108-19-0	103.080	15	1.5	15	40		
Iminodiacetic acid	$C_4H_7NO_4$	142-73-4	133.104	5	2.32	23.8	40		
Indan	C_9H_{10}	496-11-7	118.175	25	0.010	0.10	4		

Name	Mol. form.	CAS Reg. No.	Mol. wt.	t/°C	s/mass%	s/g kg^{-1} H_2O	Ref.	k_H/ kPa m^3 mol^{-1}	Ref.
1*H*-Indazole	$C_7H_6N_2$	271-44-3	118.136	20	0.0827	0.827	6		
Indeno[1,2,3-cd]pyrene	$C_{22}H_{12}$	193-39-5	276.330	20	0.00000002	0.0000002	40		
1*H*-Indole	C_8H_7N	120-72-9	117.149	20	0.187	1.87	6		
Indomethacin	$C_{19}H_{16}ClNO_4$	53-86-1	357.788	25	0.001	0.01	40		
Inosine	$C_{10}H_{12}N_4O_5$	58-63-9	268.226	20	1.6	16	29		
Iodobenzene	C_6H_5I	591-50-4	204.008	10	0.0193	0.193	2		
				25	0.0226	0.226	2	0.078	11
				45	0.0279	0.279	2		
2-Iodobenzoic acid	$C_7H_5IO_2$	88-67-5	248.018	25	0.095	0.95	27		
3-Iodobenzoic acid	$C_7H_5IO_2$	618-51-9	248.018	25	0.016	0.16	27		
4-Iodobenzoic acid	$C_7H_5IO_2$	619-58-9	248.018	25	0.0027	0.027	27		
1-Iodobutane	C_4H_9I	542-69-8	184.018	17	0.021	0.21	10	1.87	13
Iodoethane	C_2H_5I	75-03-6	155.965	0	0.44	4.4	25		
				25	0.40	4.0	25	0.52	13
1-Iodoheptane	$C_7H_{15}I$	4282-40-0	226.098	25	0.00035	0.0035	35		
Iodomethane	CH_3I	74-88-4	141.939	20	1.4	14	10	0.54	13
1-Iodopropane	C_3H_7I	107-08-4	169.992	0	0.114	1.14	35		
				20	0.100	1.00	35	0.93	13
2-Iodopropane	C_3H_7I	75-30-9	169.992	0	0.167	1.67	35		
				20	0.140	1.40	35		
5-Iodouracil	$C_4H_3IN_2O_2$	696-07-1	237.983	25	0.49	4.9	72		
trans-β-Ionone	$C_{13}H_{20}O$	79-77-6	192.297	25	0.017	0.17	52		
Iopanoic acid	$C_{11}H_{12}I_3NO_2$	96-83-3	570.932	37	0.034	0.34	40		
Iprodione	$C_{13}H_{13}Cl_2N_3O_3$	36734-19-7	330.166	20	0.0013	0.013	40		
Isobutanal	C_4H_8O	78-84-2	72.106	20	9.1	100	10		
Isobutane	C_4H_{10}	75-28-5	58.122	25	0.00535*	0.0535*	18	120	5
Isobutene	C_4H_8	115-11-7	56.107	25	0.0263*	0.263*	5	21.6	13
Isobutyl acetate	$C_6H_{12}O_2$	110-19-0	116.158	20	0.63	6.3	10		
Isobutylbenzene	$C_{10}H_{14}$	538-93-2	134.218	25	0.0010	0.010	4	3.32	11
Isobutyl formate	$C_5H_{10}O_2$	542-55-2	102.132	22	1.0	10	10		
Isobutyl isobutanoate	$C_8H_{16}O_2$	97-85-8	144.212	20	0.5	5	10		
Isobutyl propanoate	$C_7H_{14}O_2$	540-42-1	130.185	19	0.225	2.26	20		
				91	0.142	1.42	20		
Isoguanine	$C_5H_5N_5O$	3373-53-3	151.127	25	0.006	0.06	26		
1*H*-Isoindole-1,3(2H)-dione	$C_8H_5NO_2$	85-41-6	147.132	25	0.036	0.36	40		
L-Isoleucine	$C_6H_{13}NO_2$	73-32-5	131.173	25	3.31	34.2	26		
Isoniazid	$C_6H_7N_3O$	54-85-3	137.139	25	11.0	124	40		
Isopentane	C_5H_{12}	78-78-4	72.149	25	0.00485	0.0485	3	479	13
Isopentyl acetate	$C_7H_{14}O_2$	123-92-2	130.185	20	0.2	2	10		
Isopentyl formate	$C_6H_{12}O_2$	110-45-2	116.158	22	0.3	3	27		
Isophorone	$C_9H_{14}O$	78-59-1	138.206	20	1.57	16.0	20		
				80	1.27	12.9	20		
Isophthalic acid	$C_8H_6O_4$	121-91-5	166.132	10	0.0062	0.062	76		
				25	0.0154	0.154	56		
				50	0.0395	0.395	56		
				80	0.123	1.23	56		
Isopropenylbenzene	C_9H_{10}	98-83-9	118.175	20	0.0116	0.116	40		
Isopropyl acetate	$C_5H_{10}O_2$	108-21-4	102.132	20	2.9	30	10		
Isopropylbenzene	C_9H_{12}	98-82-8	120.191	25	0.0050	0.050	22	1.466	22
1-Isopropyl-2-methylbenzene	$C_{10}H_{14}$	527-84-4	134.218	25	0.00482	0.0482	23		
1-Isopropyl-3-methylbenzene	$C_{10}H_{14}$	535-77-3	134.218	25	0.00425	0.0425	23		
1-Isopropyl-4-methylbenzene	$C_{10}H_{14}$	99-87-6	134.218	25	0.0051	0.051	23	0.8	5
Isopropyl phenylcarbamate	$C_{10}H_{13}NO_2$	122-42-9	179.216	20	0.01	0.1	40		
Isoquinoline	C_9H_7N	119-65-3	129.159	20	0.452	4.52	6		
Isosorbide dinitrate	$C_6H_8N_2O_8$	87-33-2	236.136	25	0.055	0.55	40		
Kepone	$C_{10}Cl_{10}O$	143-50-0	490.636	100	0.4	4	40		
L-Lanthionine	$C_6H_{12}N_2O_4S$	922-55-4	208.235	25	0.15	1.5	26		
Lasiocarpine	$C_{21}H_{33}NO_7$	303-34-4	411.490	20	0.67	6.7	40		
L-Leucine	$C_6H_{13}NO_2$	61-90-5	131.173	25	2.32	23.8	62		
Levodopa	$C_9H_{11}NO_4$	59-92-7	197.188	20	0.165	1.65	63		

Name	Mol. form.	CAS Reg. No.	Mol. wt.	t/°C	s/mass%	s/g kg^{-1} H_2O	Ref.	k_H/ kPa m^3 mol^{-1}	Ref.
d-Limonene	$C_{10}H_{16}$	5989-27-5	136.234	0	0.001	0.01	4		
				25	0.0020	0.020	52		
Linalol	$C_{10}H_{18}O$	22564-99-4	154.249	25	0.156	1.56	52		
Linuron	$C_9H_{10}Cl_2N_2O_2$	330-55-2	249.093	25	0.0075	0.075	40		
L-Lysine	$C_6H_{14}N_2O_2$	56-87-1	146.187	25	0.58	5.8	26		
Maleic acid	$C_4H_4O_4$	110-16-7	116.073	25	44.1	789	26		
Malic acid	$C_4H_6O_5$	617-48-1	134.088	26	59	1440	26		
Malonic acid	$C_3H_4O_4$	141-82-2	104.062	0	37.9	610	26		
				20	42.4	736	26		
				50	48.1	927	26		
Malononitrile	$C_3H_2N_2$	109-77-3	66.061	20	10.6	119	40		
α-Maltose	$C_{12}H_{22}O_{11}$	4482-75-1	342.296	20	51.9	1080	27		
D-Mannitol	$C_6H_{14}O_6$	69-65-8	182.171	25	17.7	215	27		
Mefenamic acid	$C_{15}H_{15}NO_2$	61-68-7	241.286	20	0.0026	0.026	40		
Melphalan	$C_{13}H_{18}Cl_2N_2O_2$	148-82-3	305.200	30	0.44	4.4	40		
Mercury(II) phenyl acetate	$C_8H_8HgO_2$	62-38-4	336.74	20	0.2	2	30		
Mesityl oxide	$C_6H_{10}O$	141-79-7	98.142	20	2.8	29	83		
Methacrylic acid	$C_4H_6O_2$	79-41-4	86.090	20	8.9	98	10		
Methane	CH_4	74-82-8	16.043	25	0.00227*	0.0227*	18	67.4	5
Methazolamide	$C_5H_8N_4O_3S_2$	554-57-4	236.273	15	0.0472	0.472	40		
Methazole	$C_9H_6Cl_2N_2O_3$	20354-26-1	261.061	24	0.00015	0.0015	40		
Methidathion	$C_6H_{11}N_2O_4PS_3$	950-37-8	302.330	20	0.0187	0.187	40		
L-Methionine	$C_5H_{11}NO_2S$	63-68-3	149.212	25	5.3	56	26		
Methomyl	$C_5H_{10}N_2O_2S$	16752-77-5	162.210	25	5.5	58	40		
Methoxsalen	$C_{12}H_8O_4$	298-81-7	216.190	30	0.0048	0.048	40		
2-Methoxyaniline	C_7H_9NO	90-04-0	123.152	25	1.24	12.6	40		
4-Methoxyaniline	C_7H_9NO	104-94-9	123.152	20	1.14	11.5	40		
4-Methoxybenzaldehyde	$C_8H_8O_2$	123-11-5	136.149	25	0.429	4.29	40		
4-Methoxybenzoic acid	$C_8H_8O_3$	100-09-4	152.148	25	0.023	0.23	27		
Methoxychlor	$C_{16}H_{15}Cl_3O_2$	72-43-5	345.648	25	0.000005	0.00005	40		
2-Methoxy-2-methylbutane	$C_6H_{14}O$	994-05-8	102.174	20	1.10	11.1	20		
				79	0.36	3.6	20		
4-Methoxyphenol	$C_7H_8O_2$	150-76-5	124.138	20	2.51	25.7	40		
Methyclothiazide	$C_9H_{11}Cl_2N_3O_4S_2$	135-07-9	360.237	20	0.005	0.05	40		
Methyl acetate	$C_3H_6O_2$	79-20-9	74.079	20	24.5	325	10		
Methyl acrylate	$C_4H_6O_2$	96-33-3	86.090	25	4.94	52.0	10		
2-Methylacrylonitrile	C_4H_5N	126-98-7	67.090	20	2.57	26.4	10		
2-Methylaniline	C_7H_9N	95-53-4	107.153	20	1.66	16.9	10		
4-Methylaniline	C_7H_9N	106-49-0	107.153	21	7.35	79.3	10		
N-Methylaniline	C_7H_9N	100-61-8	107.153	25	0.56	5.6	40		
2-Methylanthracene	$C_{15}H_{12}$	613-12-7	192.256	6	0.0000007	0.000007	42		
				25	0.0000021	0.000021	42,22		
9-Methylanthracene	$C_{15}H_{12}$	779-02-2	192.256	25	0.000026	0.00026	42,4		
9-Methylbenz[a]anthracene	$C_{19}H_{14}$	2381-16-0	242.314	27	0.0000066	0.000066	42,4		
10-Methylbenz[a]anthracene	$C_{19}H_{14}$	2381-15-9	242.314	25	0.0000055	0.000055	42,4		
2-Methylbenzenesulfonamide	$C_7H_9NO_2S$	88-19-7	171.217	25	0.162	1.62	27		
3-Methylbenzenesulfonamide	$C_7H_9NO_2S$	1899-94-1	171.217	25	0.78	7.8	27		
4-Methylbenzenesulfonamide	$C_7H_9NO_2S$	70-55-3	171.217	25	0.316	3.16	27		
2-Methyl-1*H*-benzimidazole	$C_8H_8N_2$	615-15-6	132.163	20	0.145	1.45	6		
Methyl benzoate	$C_8H_8O_2$	93-58-3	136.149	20	0.21	2.1	10		
2-Methyl-1,3-butadiene	C_5H_8	78-79-5	68.118	25	0.061	0.61	3	7.78	5
				50	0.076*	0.76*	3		
Methyl butanoate	$C_5H_{10}O_2$	623-42-7	102.132		1.6	16	30		
3-Methylbutanoic acid	$C_5H_{10}O_2$	503-74-2	102.132	20	4.0	42	26		
2-Methyl-1-butanol, (±)-	$C_5H_{12}O$	34713-94-5	88.148	10	3.38	35.0	78		
				25	2.75	28.3	78		
				50	2.35	24.1	78		
3-Methyl-1-butanol	$C_5H_{12}O$	123-51-3	88.148	10	3.17	32.7	78,1		
				25	2.59	26.6	78,1		
				70	2.24	22.9	78,1		

Name	Mol. form.	CAS Reg. No.	Mol. wt.	t/°C	s/mass%	s/g kg^{-1} H_2O	Ref.	k_H/ kPa m^3 mol^{-1}	Ref.
2-Methyl-2-butanol	$C_5H_{12}O$	75-85-4	88.148	25	11.0	124	88,1		
				60	6.6	71	88,1		
3-Methyl-2-butanol, (±)-	$C_5H_{12}O$	70116-68-6	88.148	25	5.6	59	1		
3-Methyl-2-butanone	$C_5H_{10}O$	563-80-4	86.132	0	9.4	104	82		
				25	6.1	65	82		
				40	5.2	55	82		
3-Methyl-1-butene	C_5H_{10}	563-45-1	70.133	25	0.013*	0.13*	3	54.7	5
2-Methyl-2-butene	C_5H_{10}	513-35-9	70.133	25	0.041	0.41	3		
2-Methyl-3-buten-2-ol	$C_5H_{10}O$	115-18-4	86.132	18	27.4	377	88		
				29	18.4	225	88		
Methyl *tert*-butyl ether	$C_5H_{12}O$	1634-04-4	88.148	0	7.72	83.7	79		
				25	3.25	33.6	79		
				35	2.56	26.3	79		
				70	1.64	16.7	79		
Methyl carbamate	$C_2H_5NO_2$	598-55-0	75.067	15	69	2230	27		
5-Methylchrysene	$C_{19}H_{14}$	3697-24-3	242.314	27	0.0000062	0.000062	42,4		
Methylcyclohexane	C_7H_{14}	108-87-2	98.186	26	0.00161	0.0161	3	43.3	13
				100	0.00548	0.0548	3		
2-Methylcyclohexanone, (±)-	$C_7H_{12}O$	24965-84-2	112.169	0	2.93	30.2	84		
				20	1.98	20.2	20		
				31	1.72	17.5	84		
				60	1.44	14.6	84		
				90	1.54	15.6	20		
4-Methylcyclohexanone	$C_7H_{12}O$	589-92-4	112.169	20	2.43	24.9	20		
				80	1.95	19.9	20		
1-Methylcyclohexene	C_7H_{12}	591-49-1	96.170	25	0.0052	0.052	3		
Methylcyclopentane	C_6H_{12}	96-37-7	84.159	25	0.0043	0.043	3	36.7	5
5-Methylcytosine	$C_5H_7N_3O$	554-01-8	125.129	25	0.45	4.5	26		
1-Methyl-2,4-dinitrobenzene	$C_7H_6N_2O_4$	121-14-2	182.134	12	0.0130	0.130	55		
				32	0.0270	0.270	85		
				62	0.098	0.98	85		
2-Methyl-4,6-dinitrophenol	$C_7H_6N_2O_5$	534-52-1	198.133		0.0130	0.130	40		
Methyl formate	$C_2H_4O_2$	107-31-3	60.052	25	23	300	10		
3-Methylheptane	C_8H_{18}	589-81-1	114.229	25	0.000079	0.00079	4	376	5
2-Methyl-2-heptanol	$C_8H_{18}O$	625-25-2	130.228	30	0.25	2.5	1		
5-Methyl-3-heptanone	$C_8H_{16}O$	541-85-5	128.212	20	0.192	1.92	20		
				90	0.131	1.31	20		
2-Methylhexane	C_7H_{16}	591-76-4	100.202	25	0.00025	0.0025	3	346	5
3-Methylhexane	C_7H_{16}	78918-91-9	100.202	25	0.00026	0.0026	3	249	13
2-Methyl-2-hexanol	$C_7H_{16}O$	625-23-0	116.201	25	1.0	10	1		
5-Methyl-2-hexanol	$C_7H_{16}O$	627-59-8	116.201	25	0.49	4.9	1		
3-Methyl-3-hexanol	$C_7H_{16}O$	597-96-6	116.201	25	1.2	12	1		
5-Methyl-2-hexanone	$C_7H_{14}O$	110-12-3	114.185	19	0.537	5.40	20		
				90	0.417	4.19	20		
5-Methyl-3-hexanone	$C_7H_{14}O$	623-56-3	114.185	20	0.47	4.7	20		
				81	0.32	3.2	20		
Methyl 4-hydroxybenzoate	$C_8H_8O_3$	99-76-3	152.148	25	0.24	2.4	40		
2-Methyl-1*H*-imidazole	$C_4H_6N_2$	693-98-1	82.104	18	23.2	302	54		
3-Methyl-1*H*-indole	C_9H_9N	83-34-1	131.174	20	0.050	0.50	6		
3-Methylisoquinoline	$C_{10}H_9N$	1125-80-0	143.185	20	0.092	0.92	6		
Methyl isothiocyanate	C_2H_3NS	556-61-6	73.117	20	0.75	7.6	40		
Methylmalonic acid	$C_4H_6O_4$	516-05-2	118.089	0	30.1	431	26		
				20	40	670	26		
Methyl methacrylate	$C_5H_8O_2$	80-62-6	100.117	20	1.56	15.8	10		
2-Methyl-3-(2-methylphenyl)-4(3H)-quinazolinone	$C_{16}H_{14}N_2O$	72-44-6	250.294	23	0.03	0.3	40		
1-Methylnaphthalene	$C_{11}H_{10}$	90-12-0	142.197	25	0.00281	0.0281	22	0.045	22
2-Methylnaphthalene	$C_{11}H_{10}$	91-57-6	142.197	25	0.0025	0.025	4	0.051	12
2-Methyl-1,4-naphthalenedione	$C_{11}H_8O_2$	58-27-5	172.181	25	0.016	0.16	40		
N-Methyl-*N*-nitrosourea	$C_2H_5N_3O_2$	684-93-5	103.080	14	2.3	24	40		
4-Methyloctane	C_9H_{20}	2216-34-4	128.255	25	0.0000115	0.000115	4	1000	5

Name	Mol. form.	CAS Reg. No.	Mol. wt.	t/°C	s/mass%	s/g kg⁻¹ H_2O	Ref.	k_H/ kPa m³ mol⁻¹	Ref.
Methyloxirane	C_3H_6O	75-56-9	58.079	20	40.5	681	10	0.0087	13
Methyl parathion	$C_8H_{10}NO_5PS$	298-00-0	263.208	10	0.00218	0.0218	40		
				20	0.00380	0.0380	40		
				30	0.0059	0.059	40		
2-Methylpentane	C_6H_{14}	107-83-5	86.175	25	0.00137	0.0137	3	176	13
3-Methylpentane	C_6H_{14}	96-14-0	86.175	25	0.00129	0.0129	3	170	13
2-Methyl-1-pentanol	$C_6H_{14}O$	105-30-6	102.174	25	0.76	7.7	78,1		
				50	0.70	7.0	78		
4-Methyl-1-pentanol	$C_6H_{14}O$	626-89-1	102.174	25	0.76	7.6	1		
2-Methyl-2-pentanol	$C_6H_{14}O$	590-36-3	102.174	25	3.2	33	1		
3-Methyl-2-pentanol	$C_6H_{14}O$	565-60-6	102.174	25	1.9	19	1		
4-Methyl-2-pentanol	$C_6H_{14}O$	108-11-2	102.174	27	1.5	15	1		
2-Methyl-3-pentanol	$C_6H_{14}O$	565-67-3	102.174	25	2.0	20	1		
3-Methyl-3-pentanol	$C_6H_{14}O$	77-74-7	102.174	25	4.3	45	1		
4-Methyl-2-pentanone	$C_6H_{12}O$	108-10-1	100.158	0	2.92	30.1	83		
				25	1.85	18.8	83		
				50	1.46	14.8	83		
2-Methyl-3-pentanone	$C_6H_{12}O$	565-69-5	100.158	25	1.5	15	83		
2-Methyl-1-pentene	C_6H_{12}	763-29-1	84.159	25	0.0078	0.078	3	28.1	5
4-Methyl-1-pentene	C_6H_{12}	691-37-2	84.159	25	0.0048	0.048	3	63.2	5
1-Methylphenanthrene	$C_{15}H_{12}$	832-69-9	192.256	7	0.0000095	0.000095	42		
				25	0.0000269	0.000269	42,4		
Methylprednisolone	$C_{22}H_{30}O_5$	83-43-2	374.470	25	0.012	0.12	40		
Methyl propanoate	$C_4H_8O_2$	554-12-1	88.106		6	60	30		
2-Methylpropanoic acid	$C_4H_8O_2$	79-31-2	88.106	20	22.8	295	10		
2-Methyl-1-propanol	$C_4H_{10}O$	78-83-1	74.121	0	12.2	139	78,1		
				25	8.1	88	78,1	0.00273	28
				50	7.0	70	78,1		
Methyl propyl ether	$C_4H_{10}O$	557-17-5	74.121	0	5.4	57	79		
				25	3.0	31	79		
2-Methyl-2-propyL-1,3-propanediol dicarbamate	$C_9H_{18}N_2O_4$	57-53-4	218.250	25	0.33	3.3	40		
Methyl salicylate	$C_8H_8O_3$	119-36-8	152.148	30	0.74	7.4	10		
17-Methyltestosterone	$C_{20}H_{30}O_2$	58-18-4	302.451	25	0.0033	0.033	40		
2-Methyltetrahydrofuran	$C_5H_{10}O$	96-47-9	86.132	19	14.4	168	20	0.67	13
				71	6.0	64	20		
N-Methyl-*N*,2,4,6-tetranitroaniline	$C_7H_5N_5O_8$	479-45-8	287.144	20	0.0074	0.074	40		
Methylthiouracil	$C_5H_6N_2OS$	56-04-2	142.179	25	0.0533	0.533	40		
1-Methyl-2,3,4-trinitrobenzene	$C_7H_5N_3O_6$	602-29-9	227.131	14	0.0091	0.091	85,59		
				23	0.0116	0.116	85,59		
				61	0.0643	0.643	85,59		
Metronidazole	$C_6H_9N_3O_3$	443-48-1	171.153	20	0.93	9.4	40		
Mirex	$C_{10}Cl_{12}$	2385-85-5	545.543	25	0.0000085	0.000085	40		
Morphine	$C_{17}H_{19}NO_3$	57-27-2	285.338	20	0.015	0.15	27		
β-Myrcene	$C_{10}H_{16}$	123-35-3	136.234	25	0.030	0.30	52		
Naphthacene	$C_{18}H_{12}$	92-24-0	228.288	25	0.00000007	0.0000007	42,4	0.000004	12
Naphthalene	$C_{10}H_8$	91-20-3	128.171	10	0.0019	0.019	4		
				25	0.00316	0.0316	22	0.043	22
				50	0.0082	0.082	4		
1-Naphthaleneacetic acid	$C_{12}H_{10}O_2$	86-87-3	186.206	25	0.0415	0.415	40		
1-Naphthalenecarboxylic acid	$C_{11}H_8O_2$	86-55-5	172.181	25	0.0058	0.058	27		
1-Naphthalenylthiourea	$C_{11}H_{10}N_2S$	86-88-4	202.275	20	0.06	0.6	40		
1-Naphthol	$C_{10}H_8O$	90-15-3	144.170	20	0.111	1.11	40		
2-Naphthol	$C_{10}H_8O$	135-19-3	144.170	20	0.064	0.64	40		
				80	0.67	6.7	40		
1-Naphthylamine	$C_{10}H_9N$	134-32-7	143.185	20	0.17	1.7	40		
2-Naphthylamine	$C_{10}H_9N$	91-59-8	143.185	20	0.0189	0.189	40		
Narceine	$C_{23}H_{27}NO_8$	131-28-2	445.462	13	0.078	0.78	27		
Neopentane	C_5H_{12}	463-82-1	72.149	25	0.00332*	0.0332*	3	220	13
Nitrapyrin	$C_6H_3Cl_4N$	1929-82-4	230.907	20	0.0040	0.040	40		
2-Nitroaniline	$C_6H_6N_2O_2$	88-74-4	138.124	30	1.47	14.9	27		

Name	Mol. form.	CAS Reg. No.	Mol. wt.	t/°C	s/mass%	s/g kg^{-1} H_2O	Ref.	k_H/ kPa m^3 mol^{-1}	Ref.
3-Nitroaniline	$C_6H_6N_2O_2$	99-09-2	138.124	30	0.121	1.21	27		
4-Nitroaniline	$C_6H_6N_2O_2$	100-01-6	138.124	30	0.073	0.73	27		
2-Nitroanisole	$C_7H_7NO_3$	91-23-6	153.136	30	0.169	1.69	10		
4-Nitroanisole	$C_7H_7NO_3$	100-17-4	153.136	30	0.059	0.59	27		
3-Nitrobenzaldehyde	$C_7H_5NO_3$	99-61-6	151.120	25	0.16	1.6	27		
4-Nitrobenzaldehyde	$C_7H_5NO_3$	555-16-8	151.120	25	0.23	2.3	27		
Nitrobenzene	$C_6H_5NO_2$	98-95-3	123.110	25	0.21	2.1	17		
3-Nitro-1,2-benzenedicarboxylic acid	$C_8H_5NO_6$	603-11-2	211.129	25	1.63	16.6	69		
				40	2.97	30.6	69		
4-Nitro-1,2-benzenedicarboxylic acid	$C_8H_5NO_6$	610-27-5	211.129	25	60.9	1560	69		
				40	68.0	2125	69		
2-Nitrobenzoic acid	$C_7H_5NO_4$	552-16-9	167.120	25	0.55	5.5	40		
3-Nitrobenzoic acid	$C_7H_5NO_4$	121-92-6	167.120	10	0.197	1.97	75		
				25	0.313	3.14	75		
				50	0.90	9.1	75		
4-Nitrobenzoic acid	$C_7H_5NO_4$	62-23-7	167.120	25	0.0422	0.422	40		
Nitroethane	$C_2H_5NO_2$	79-24-3	75.067	25	4.4	46	38		
				50	5.3	56	38		
Nitrofen	$C_{12}H_7Cl_2NO_3$	1836-75-5	284.095	22	0.00095	0.0095	40		
Nitrofurantoin	$C_8H_6N_4O_5$	67-20-9	238.158	30	0.011	0.11	40		
Nitrofurazone	$C_6H_6N_4O_4$	59-87-0	198.137	20	0.0238	0.238	40		
Nitroguanidine	$CH_4N_4O_2$	556-88-7	104.069	25	1.2	12	40		
Nitromethane	CH_3NO_2	75-52-5	61.041	0	9.2	101	36		
				25	11.0	124	36		
				50	14.8	174	36		
1-Nitronaphthalene	$C_{10}H_7NO_2$	86-57-7	173.169	18	0.005	0.05	40		
2-Nitrophenol	$C_6H_5NO_3$	88-75-5	139.109	25	0.170	1.70	48,51		
3-Nitrophenol	$C_6H_5NO_3$	554-84-7	139.109	20	2.14	21.9	27		
4-Nitrophenol	$C_6H_5NO_3$	100-02-7	139.109	20	1.56	15.8	48,51		
1-Nitropropane	$C_3H_7NO_2$	108-03-2	89.094	25	1.54	15.6	38		
				90	2.29	23.4	20		
2-Nitropropane	$C_3H_7NO_2$	79-46-9	89.094	25	1.75	17.8	38		
				90	2.36	24.2	20		
N-Nitrosodiethylamine	$C_4H_{10}N_2O$	55-18-5	102.134	24	9.6	106	40		
N-Nitrosodiphenylamine	$C_{12}H_{10}N_2O$	86-30-6	198.219	25	0.0035	0.035	17		
2-Nitrotoluene	$C_7H_7NO_2$	88-72-2	137.137	30	0.065	0.65	27		
3-Nitrotoluene	$C_7H_7NO_2$	99-08-1	137.137	30	0.050	0.50	27		
4-Nitrotoluene	$C_7H_7NO_2$	99-99-0	137.137	30	0.044	0.44	27		
2,2',3,3',4,5,5',6,6'-Nonachlorobiphenyl	$C_{12}HCl_9$	52663-77-1	464.213	25	0.0000000018	0.000000018	7		
1,8-Nonadiyne	C_9H_{12}	2396-65-8	120.191	25	0.0125	0.125	4		
Nonane	C_9H_{20}	111-84-2	128.255	25	0.000017	0.00017	4	333	13
				50	0.000022	0.00022	4		
Nonanedioic acid	$C_9H_{16}O_4$	123-99-9	188.221	25	0.1780	1.780	34		
				65	1.322	13.40	34		
Nonanoic acid	$C_9H_{18}O_2$	112-05-0	158.238	20	0.0284	0.284	26		
1-Nonanol	$C_9H_{20}O$	143-08-8	144.254	25	0.0129	0.129	78,1		
				90	0.0291	0.291	78		
2-Nonanol, (±)-	$C_9H_{20}O$	74683-66-2	144.254	15	0.026	0.26	1		
3-Nonanol, (±)-	$C_9H_{20}O$	74742-08-8	144.254	15	0.032	0.32	1		
4-Nonanol	$C_9H_{20}O$	52708-03-9	144.254	15	0.0026	0.026	1		
5-Nonanol	$C_9H_{20}O$	623-93-8	144.254	15	0.0032	0.032	1		
2-Nonanone	$C_9H_{18}O$	821-55-6	142.238	20	0.038	0.38	20		
				70	0.034	0.34	20		
3-Nonanone	$C_9H_{18}O$	925-78-0	142.238	30	0.056	0.56	20		
				80	0.046	0.46	20		
5-Nonanone	$C_9H_{18}O$	502-56-7	142.238	20	0.054	0.54	20		
				80	0.029	0.29	20		
1-Nonene	C_9H_{18}	124-11-8	126.239	25	0.000112	0.00112	40		
Nonyl formate	$C_{10}H_{20}O_2$	5451-92-3	172.265	10	0.012	0.12	20		
				90	0.039	0.39	20		

Name	Mol. form.	CAS Reg. No.	Mol. wt.	t/°C	s/mass%	s/g kg^{-1} H_2O	Ref.	k_H/ kPa m^3 mol^{-1}	Ref.
4-Nonylphenol	$C_{15}H_{24}O$	104-40-5	220.351	25	0.000636	0.00636	40		
1-Nonyne	C_9H_{16}	9/3/3452	124.223	25	0.00072	0.0072	4		
Norethisterone	$C_{20}H_{26}O_2$	68-22-4	298.419	25	0.00063	0.0063	40		
Norflurazon	$C_{12}H_9ClF_3N_3O$	27314-13-2	303.666	25	0.0028	0.028	40		
L-Norleucine	$C_6H_{13}NO_2$	327-57-1	131.173	25	1.5	15	26		
L-Norvaline	$C_5H_{11}NO_2$	6600-40-4	117.147	25	9.7	107	26		
Noscapine	$C_{22}H_{23}NO_7$	128-62-1	413.421	25	0.03	0.3	40		
2,2',3,3',5,5',6,6'-Octachlorobiphenyl	$C_{12}H_2Cl_8$	2136-99-4	429.768	25	0.00000015	0.0000015	41	0.0381	7
Octachlorodibenzo-*p*-dioxin	$C_{12}Cl_8O_2$	3268-87-9	459.751	25	2.3E-11	2.3E-10	68		
Octachloro-1,3-pentadiene	C_5Cl_8	1888-73-9	343.678	20	0.000020	0.00020	35		
Octacosane	$C_{28}H_{58}$	630-02-4	394.761	22	0.0000006	0.000006	37		
Octadecane	$C_{18}H_{38}$	593-45-3	254.495	25	0.00000021	0.0000021	42,37		
Octadecanoic acid	$C_{18}H_{36}O_2$	57-11-4	284.478	20	0.00029	0.0029	26		
1-Octadecanol	$C_{18}H_{38}O$	112-92-5	270.494	34	0.000011	0.00011	1		
Octane	C_8H_{18}	111-65-9	114.229	25	0.000073	0.00073	46	311	13
				50	0.000102	0.00102	47		
				75	0.000179	0.00179	46		
				100	0.000377	0.00377	46		
Octanedioic acid	$C_8H_{14}O_4$	505-48-6	174.195	25	0.242	2.43	34		
				50	0.557	5.570	34		
Octanoic acid	$C_8H_{16}O_2$	124-07-2	144.212	25	0.080	0.80	26		
1-Octanol	$C_8H_{18}O$	111-87-5	130.228	25	0.0460	0.460	78		
				60	0.0536	0.536	78		
2-Octanol	$C_8H_{18}O$	4128-31-8	130.228	25	0.4	4	1		
2-Octanone	$C_8H_{16}O$	111-13-7	128.212	20	0.134	1.34	84		
				50	0.098	0.98	84		
				80	0.091	0.91	84		
3-Octanone	$C_8H_{16}O$	106-68-3	128.212	20	0.137	1.37	20		
				91	0.106	1.06	20		
1-Octene	C_8H_{16}	111-66-0	112.213	25	0.00027	0.0027	4	96.3	13
Octyl acetate	$C_{10}H_{20}O_2$	112-14-1	172.265	19	0.020	0.20	20		
				92	0.012	0.12	20		
1-Octyne	C_8H_{14}	629-05-0	110.197	25	0.0024	0.024	4	7.87	13
Orotic acid	$C_5H_4N_2O_4$	65-86-1	156.097	18	0.18	1.8	26		
Oryzalin	$C_{12}H_{18}N_4O_6S$	19044-88-3	346.359	25	0.00024	0.0024	40		
Ouabain	$C_{29}H_{44}O_{12}$	630-60-4	584.652	25	1.3	13	40		
Oxalic acid	$C_2H_2O_4$	144-62-7	90.035	20	8.69	95.2	27		
				80	45.8	845	27		
Oxamyl	$C_7H_{13}N_3O_3S$	23135-22-0	219.261	25	≈21	≈270	40		
4-Oxopentanoic acid	$C_5H_8O_3$	123-76-2	116.116	10	63.6	1750	34		
				25	84.0	5250	34		
4-Oxo-4*H*-pyran-2,6-dicarboxylic acid	$C_7H_4O_6$	99-32-1	184.103	25	1.45	14.7	27		
Papaverine	$C_{20}H_{21}NO_4$	58-74-2	339.386	37	0.0037	0.037	40		
Paraldehyde	$C_6H_{12}O_3$	123-63-7	132.157	25	11	124	30		
Parathion	$C_{10}H_{14}NO_5PS$	56-38-2	291.261	20	0.00129	0.0129	40		
Pendimethalin	$C_{13}H_{19}N_3O_4$	40487-42-1	281.308	20	0.00003	0.0003	40		
Pentachlorobenzene	C_6HCl_5	608-93-5	250.337	25	0.000050	0.00050	41	0.085	11
2,3,4,5,6-Pentachlorobiphenyl	$C_{12}H_5Cl_5$	18259-05-7	326.433	25	0.0000008	0.000008	7		
2,2',4,5,5'-Pentachlorobiphenyl	$C_{12}H_5Cl_5$	37680-73-2	326.433	25	0.000001	0.00001	7	0.0421	31
Pentachloroethane	C_2HCl_5	76-01-7	202.294	25	0.049	0.49	25	0.25	13
Pentachloronitrobenzene	$C_6Cl_5NO_2$	82-68-8	295.335	20	0.000044	0.00044	40		
Pentachlorophenol	C_6HCl_5O	87-86-5	266.336	25	0.0021	0.021	48,51,24		
2,3,4,5,6-Pentachlorotoluene	$C_7H_3Cl_5$	877-11-2	264.364	25	0.0000028	0.000028	61		
Pentadecanoic acid	$C_{15}H_{30}O_2$	1002-84-2	242.398	20	0.0012	0.012	26		
1-Pentadecanol	$C_{15}H_{32}O$	629-76-5	228.414	25	0.000010	0.00010	1		
1,4-Pentadiene	C_5H_8	591-93-5	68.118	25	0.056	0.56	3	12	5
Pentaerythritol	$C_5H_{12}O_4$	115-77-5	136.147	15	5.3	56	30		
Pentaerythritol tetranitrate	$C_5H_8N_4O_{12}$	78-11-5	316.138	20	0.0002	0.002	40		
Pentanal	$C_5H_{10}O$	110-62-3	86.132	25	1.2	12	40		
Pentane	C_5H_{12}	109-66-0	72.149	25	0.0041	0.041	3	128	13

Name	Mol. form.	CAS Reg. No.	Mol. wt.	t/°C	s/mass%	s/g kg^{-1} H$_2$O	Ref.	k_H/ kPa m^3 mol^{-1}	Ref.
Pentanedioic acid	$C_5H_8O_4$	110-94-1	132.116	25	58.3	1400	33		
				50	78.1	3570	33		
2,4-Pentanedione	$C_5H_8O_2$	123-54-6	100.117	20	16.1	192	20		
				80	32.2	475	20		
Pentanoic acid	$C_5H_{10}O_2$	109-52-4	102.132	16	3.6	37	26		
				25	4.32	45.2	90		
				35	5.26	55.5	90		
1-Pentanol	$C_5H_{12}O$	71-41-0	88.148	0	3.23	33.4	78,1		
				25	2.14	21.9	78,1		
				50	1.83	18.6	78,1		
				90	2.12	21.7	78		
2-Pentanol	$C_5H_{12}O$	6032-29-7	88.148	25	4.3	45	21		
3-Pentanol	$C_5H_{12}O$	584-02-1	88.148	25	5.6	59	21		
2-Pentanone	$C_5H_{10}O$	107-87-9	86.132	0	8.7	95	20	0.00847	28
				25	5.5	58	20	0.00847	28
				80	3.8	40	20	0.00847	28
3-Pentanone	$C_5H_{10}O$	96-22-0	86.132	0	7.6	82	82		
				25	4.9	52	82		
3-Pentanone	$C_5H_{10}O$	96-22-0	86.132	80	3.6	37	82		
1-Pentene	C_5H_{10}	109-67-1	70.133	25	0.0148	0.148	3	40.3	5
cis-2-Pentene	C_5H_{10}	627-20-3	70.133	25	0.0203	0.203	3	22.8	5
Pentyl acetate	$C_7H_{14}O_2$	628-63-7	130.185	20	0.17	1.7	10		
sec-Pentyl acetate (*S*)-	$C_7H_{14}O_2$	55621-90-4	130.185	25	0.2	2	27		
Pentylbenzene	$C_{11}H_{16}$	538-68-1	148.245	25	0.00043	0.0043	89,5	1.69	11
Pentylcyclopentane	$C_{10}H_{20}$	3741-00-2	140.266	25	0.0000115	0.000115	4	185	5
Pentyl propanoate	$C_8H_{16}O_2$	624-54-4	144.212	20	0.1	1	27		
1-Pentyne	C_5H_8	627-19-0	68.118	25	0.157	1.57	3	2.5	5
Perfluorocyclobutane	C_4F_8	115-25-3	200.030	5	0.00638*	0.0638*	50		
				25	0.00247*	0.0247*	50		
				45	0.00158*	0.0158*	50		
Perfluorodecane	$C_{10}F_{22}$	307-45-9	538.072	20	0.000031	0.00031	35		
Perfluoroheptane	C_7F_{16}	335-57-9	388.049	25	0.0000013	0.000013	35		
Perfluorohexane	C_6F_{14}	355-42-0	338.042	25	0.0000098	0.000098	35		
Perfluoro-2-methylpentane	C_6F_{14}	355-04-4	338.042	25	0.000017	0.00017	35		
Perfluorooctane	C_8F_{18}	307-34-6	438.057	25	0.00000017	0.0000017	35		
Perfluoropentane	C_5F_{12}	678-26-2	288.035	25	0.00012	0.0012	35		
Perfluoropropane	C_3F_8	76-19-7	188.019	15	0.0015*	0.015*	14		
Perfluoropropene	C_3F_6	116-15-4	150.022	25	0.0194*	0.194*	14		
Permethrin	$C_{21}H_{20}Cl_2O_3$	52645-53-1	391.288	20	0.00002	0.0002	32		
Perylene	$C_{20}H_{12}$	198-55-0	252.309	25	0.00000004	0.0000004	42,4	0.000003	12
Phenanthrene	$C_{14}H_{10}$	85-01-8	178.229	0	0.000039	0.00039	42		
				10	0.000047	0.00047	42,4		
				25	0.00012	0.0012	42,22	0.00324	22
				50	0.00042	0.0042	42,4		
Phenmedipham	$C_{16}H_{16}N_2O_4$	13684-63-4	300.309	25	0.00047	0.0047	32		
Phenobarbital	$C_{12}H_{12}N_2O_3$	50-06-6	232.234	25	0.12	1.2	40		
				45	0.26	2.6	40		
Phenol	C_6H_6O	108-95-2	94.111	15	7.60	82.3	48,51		
				25	8.40	91.7	48,51		
				35	9.31	102.7	48,51		
Phenolphthalein	$C_{20}H_{14}O_4$	77-09-8	318.323	20	0.018	0.18	27		
10*H*-Phenothiazine	$C_{12}H_9NS$	92-84-2	199.271	25	0.00016	0.0016	40		
2-Phenoxyethanol	$C_8H_{10}O_2$	122-99-6	138.164	20	2.53	26.0	40		
Phenyl acetate	$C_8H_8O_2$	122-79-2	136.149	20	0.59	5.9	20		
				91	0.91	9.2	20		
DL-Phenylalanine	$C_9H_{11}NO_2$	150-30-1	165.189	25	1.40	14.2	29		
L-Phenylalanine	$C_9H_{11}NO_2$	63-91-2	165.189	25	2.71	27.9	26		
Phenylbutazone	$C_{19}H_{20}N_2O_2$	50-33-9	308.374	25	0.0034	0.034	40		
1-Phenyl-1-propanone	$C_9H_{10}O$	93-55-0	134.174	19	0.32	3.2	20		
				80	0.24	2.4	20		

Name	Mol. form.	CAS Reg. No.	Mol. wt.	t/°C	s/mass%	s/g kg^{-1} H_2O	Ref.	k_H/ kPa m^3 mol^{-1}	Ref.
Phenylthiourea	$C_7H_8N_2S$	103-85-5	152.217	25	2.55	26.2	27		
Phenytoin	$C_{15}H_{12}N_2O_2$	57-41-0	252.268	37	0.0038	0.038	40		
Phosalone	$C_{12}H_{15}ClNO_4PS_2$	2310-17-0	367.808	20	0.00026	0.0026	40		
Phosmet	$C_{11}H_{12}NO_4PS_2$	732-11-6	317.321	25	0.0025	0.025	40		
Phthalic acid	$C_8H_6O_4$	88-99-3	166.132	10	0.464	4.66	76		
				25	0.719	7.24	76		
				50	1.76	17.9	76		
				65	3.57	37.0	33		
Phthalic anhydride	$C_8H_4O_3$	85-44-9	148.116	27	0.62	6.20	40		
Picene	$C_{22}H_{14}$	213-46-7	278.346	27	0.00000025	0.0000025	42,4		
α-Pinene, (-)	$C_{10}H_{16}$	7785-26-4	136.234	25	0.00050	0.0050	52		
β-Pinene, (1S)-	$C_{10}H_{16}$	18172-67-3	136.234	25	0.00110	0.0110	52		
2,5-Piperazinedione	$C_4H_6N_2O_2$	106-57-0	114.103	25	1.64	16.7	29		
2-PivaloyL-1,3-indandione	$C_{14}H_{14}O_3$	83-26-1	230.259	25	0.0018	0.018	40		
Prednisolone	$C_{21}H_{28}O_5$	50-24-8	360.444	25	0.03	0.3	40		
Progesterone	$C_{21}H_{30}O_2$	57-83-0	314.462	25	0.00088	0.0088	40		
				41	0.00206	0.0206	40		
L-Proline	$C_5H_9NO_2$	147-85-3	115.131	25	61.9	1625	26		
Prometone	$C_{10}H_{19}N_5O$	1610-18-0	225.291	20	0.075	0.75	40		
Prometryn	$C_{10}H_{19}N_5S$	7287-19-6	241.357	20	0.0048	0.048	32		
Propachlor	$C_{11}H_{14}ClNO$	1918-16-7	211.688	20	0.07	0.7	40		
Propanal	C_3H_6O	123-38-6	58.079	25	30.6	441	10		
Propane	C_3H_8	74-98-6	44.096	25	0.00669*	0.0669*	18	71.6	5
Propanenitrile	C_3H_5N	107-12-0	55.079	25	10.3	115	10		
Propanil	$C_9H_9Cl_2NO$	709-98-8	218.079	20	0.013	0.13	40		
Propazine	$C_9H_{16}ClN_5$	139-40-2	229.710	20	0.00086	0.0086	40		
Propene	C_3H_6	115-07-1	42.080	25	0.0200*	0.200*	5	21.3	5
1-Propene-2,3-dicarboxylic acid	$C_5H_6O_4$	97-65-4	130.100	20	7.7	83	26		
trans-1-Propene-1,2,3-tricarboxylic acid	$C_6H_6O_6$	4023-65-8	174.108	25	20.9	264	26		
				90	52.5	1105	26		
Propoxur	$C_{11}H_{15}NO_3$	114-26-1	209.242	20	0.193	1.93	40		
Propyl acetate	$C_5H_{10}O_2$	109-60-4	102.132	20	2.3	24	10		
Propylbenzene	C_9H_{12}	103-65-1	120.191	25	0.0052	0.052	22	1.041	22
Propyl butanoate	$C_7H_{14}O_2$	105-66-8	130.185	17	0.162	1.62	27		
Propylcyclopentane	C_8H_{16}	2040-96-2	112.213	25	0.00020	0.0020	4	90.2	5
Propyl formate	$C_4H_8O_2$	110-74-7	88.106	22	2.05	20.9	10		
Propyl 4-hydroxybenzoate	$C_{10}H_{12}O_3$	94-13-3	180.200	25	0.04	0.4	40		
Propyl propanoate	$C_6H_{12}O_2$	106-36-5	116.158	25	0.6	6	27		
Propyne	C_3H_4	74-99-7	40.064	25	0.364*	3.64*	5	1.11	5
Propyzamide	$C_{12}H_{11}Cl_2NO$	23950-58-5	256.127	25	0.0015	0.015	32		
Pyrene	$C_{16}H_{10}$	129-00-0	202.250	0	0.0000049	0.000049	42		
				15	0.0000069	0.000069	42		
				25	0.0000139	0.000139	42,22	0.00092	22
				50	0.000053	0.00053	42,4		
				75	0.000231	0.00231	42		
3-Pyridinecarboxamide	$C_6H_6N_2O$	98-92-0	122.124	20	≈33	≈490	40		
3-Pyridinecarboxylic acid	$C_6H_5NO_2$	59-67-6	123.110	24	1.63	16.6	66		
				52	3.40	35.2	66		
				72	5.20	54.9	66		
Pyrocatechol	$C_6H_6O_2$	120-80-9	110.111	20	31.1	451	27		
Pyrrole	C_4H_5N	109-97-7	67.090	25	4.5	47	10		
Quinic acid	$C_7H_{12}O_6$	77-95-2	192.166	9	29	410	26		
Quinidine	$C_{20}H_{24}N_2O_2$	56-54-2	324.417	20	0.020	0.20	27		
Quinine	$C_{20}H_{24}N_2O_2$	130-95-0	324.417	25	0.057	0.57	27		
Quinoline	C_9H_7N	91-22-5	129.159	20	0.633	6.33	6		
8-Quinolinol	C_9H_7NO	148-24-3	145.158	25	0.065	0.65	40		
Quinoxaline	$C_8H_6N_2$	91-19-0	130.147	50	54	1170	6		
Raffinose	$C_{18}H_{32}O_{16}$	512-69-6	504.437	20	12.5	143	27		
Reserpine	$C_{33}H_{40}N_2O_9$	50-55-5	608.679	30	0.0073	0.073	40		
Resorcinol	$C_6H_6O_2$	108-46-3	110.111	20	63.7	1750	27		

Name	Mol. form.	CAS Reg. No.	Mol. wt.	t/°C	s/mass%	s/g kg^{-1} H_2O	Ref.	k_H/ kPa m^3 mol^{-1}	Ref.
Riboflavin	$C_{17}H_{20}N_4O_6$	83-88-5	376.364	25	0.0075	0.075	40		
Ronnel	$C_8H_8Cl_3O_3PS$	299-84-3	321.546	20	0.00011	0.0011	40		
Rotenone	$C_{23}H_{22}O_6$	83-79-4	394.417	25	0.000017	0.00017	40		
Saccharin	$C_7H_5NO_3S$	81-07-2	183.185	25	0.40	4.0	27		
				100	4.0	42	27		
Salicylaldehyde	$C_7H_6O_2$	90-02-8	122.122	86	1.68	17.1	10		
Sarcosine	$C_3H_7NO_2$	107-97-1	89.094	25	30.0	429	26		
L-Serine	$C_3H_7NO_3$	56-45-1	105.093	25	20	250	26		
Shikimic acid	$C_7H_{10}O_5$	138-59-0	174.151		15	176	26		
Silvex	$C_9H_7Cl_3O_3$	93-72-1	269.509	25	0.014	0.14	40		
Solanine	$C_{45}H_{73}NO_{15}$	20562-02-1	868.060	15	0.0026	0.026	40		
L-Sorbose	$C_6H_{12}O_6$	87-79-6	180.155	17	≈26	≈350	40		
trans-Stilbene	$C_{14}H_{12}$	103-30-0	180.245	25	0.000029	0.00029	42,4	0.04	12
Streptozotocin	$C_8H_{15}N_3O_7$	18883-66-4	265.221	25	0.50	5.0	40		
Strychnine	$C_{21}H_{22}N_2O_2$	57-24-9	334.412	20	0.013	0.13	27		
Styrene	C_8H_8	100-42-5	104.150	25	0.032	0.32	22	0.286	22
				50	0.046	0.46	4,89	0.3	13
Succinamide	$C_4H_8N_2O_2$	110-14-5	116.119	50	18.4	225	27		
Succinic acid	$C_4H_6O_4$	110-15-6	118.089	25	7.71	83.5	27		
				100	55	1220	27		
Succinonitrile	$C_4H_4N_2$	110-61-2	80.088	25	11.5	130	10		
Sucrose	$C_{12}H_{22}O_{11}$	57-50-1	342.296	20	67.1	2040	27		
			342.296	50	72.3	2610	27		
			342.296	100	83.0	4880	27		
Sulfamethazine	$C_{12}H_{14}N_4O_2S$	57-68-1	278.330	20	0.053	0.53	40		
Sulfamethoxazole	$C_{10}H_{11}N_3O_3S$	723-46-6	253.277	25	0.0281	0.281	40		
Sulfathiazole	$C_9H_9N_3O_2S_2$	72-14-0	255.316	20	0.048	0.48	40		
Sulfisoxazole	$C_{11}H_{13}N_3O_3S$	127-69-5	267.304	37	0.03	0.3	40		
DL-Tartaric acid	$C_4H_6O_6$	133-37-9	150.087	0	8.95	98.3	26		
				20	17.1	206	26		
				100	65	1860	26		
L-Tartaric acid	$C_4H_6O_6$	87-69-4	150.087	20	58	1380	26		
				100	77	3350	26		
Tebuthiuron	$C_9H_{16}N_4OS$	34014-18-1	228.314	20	0.23	2.3	40		
Terbacil	$C_9H_{13}ClN_2O_2$	5902-51-2	216.664	25	0.071	0.71	40		
Terephthalic acid	$C_8H_6O_4$	100-21-0	166.132	10	0.0082	0.082	76		
				25	0.0065	0.065	76		
				50	0.0074	0.074	76		
o-Terphenyl	$C_{18}H_{14}$	84-15-1	230.304	25	0.000124	0.00124	42,40		
m-Terphenyl	$C_{18}H_{14}$	92-06-8	230.304	25	0.000152	0.00152	42,40		
p-Terphenyl	$C_{18}H_{14}$	92-94-4	230.304	25	0.00000180	0.000018	42,40		
α-Terpineol	$C_{10}H_{18}O$	12/2/2438	154.249	25	0.189	1.89	52		
1,2,4,5-Tetrabromobenzene	$C_6H_2Br_4$	636-28-2	393.696	25	0.00000434	0.0000434	2		
1,1,2,2-Tetrabromoethane	$C_2H_2Br_4$	79-27-6	345.653	0	0.052	0.52	25		
				25	0.068	0.68	25		
				50	0.106	1.06	25		
				100	0.307	3.07	25		
Tetrabromomethane	CBr_4	558-13-4	331.627	30	0.024	0.24	14		
1,2,3,4-Tetrachlorobenzene	$C_6H_2Cl_4$	634-66-2	215.892	25	0.0007	0.007	41	0.144	11
1,2,3,5-Tetrachlorobenzene	$C_6H_2Cl_4$	634-90-2	215.892	25	0.00035	0.0035	41	0.59	11
1,2,4,5-Tetrachlorobenzene	$C_6H_2Cl_4$	95-94-3	215.892	25	0.000060	0.00060	41	0.122	11
3,4,5,6-Tetrachloro-1,2-benzenediol	$C_6H_2Cl_4O_2$	1198-55-6	247.891	25	0.071	0.71	8		
2,2',4',5-Tetrachlorobiphenyl	$C_{12}H_6Cl_4$	41464-40-8	291.988	25	0.0000016	0.000016	9		
2,3,4,5-Tetrachlorobiphenyl	$C_{12}H_6Cl_4$	33284-53-6	291.988	25	0.000002	0.00002	7		
2,3,5,6-Tetrachloro-2,5-cyclohexadiene-1,4-dione	$C_6Cl_4O_2$	118-75-2	245.875	20	0.025	0.25	40		
2,3,7,8-Tetrachlorodibenzo-*p*-dioxin	$C_{12}H_4Cl_4O_2$	1746-01-6	321.971	22	0.0000000019	0.000000019	40		
1,1,2,2-Tetrachloro-1,2-difluoroethane	$C_2Cl_4F_2$	76-12-0	203.830	27	0.016	0.16	25		
1,1,1,2-Tetrachloroethane	$C_2H_2Cl_4$	630-20-6	167.849	0	0.120	1.20	25		
				25	0.107	1.07	25	0.24	13
				50	0.123	1.23	25		

Name	Mol. form.	CAS Reg. No.	Mol. wt.	t/°C	s/mass%	s/g kg⁻¹ H_2O	Ref.	k_H/ kPa m³ mol⁻¹	Ref.
1,1,2,2-Tetrachloroethane	$C_2H_2Cl_4$	79-34-5	167.849	5	0.302	3.02	25		
				25	0.283	2.83	25	0.026	13
				50	0.318	3.18	25		
Tetrachloroethene	C_2Cl_4	127-18-4	165.833	0	0.0273	0.273	20		
				20	0.0286	0.286	20	1.73	13
				80	0.0380	0.380	20		
Tetrachloromethane	CCl_4	56-23-5	153.823	25	0.065	0.65	20	2.99	13
				75	0.115	1.15	20	2.99	13
2,3,4,6-Tetrachloro-5-methylphenol	$C_7H_4Cl_4O$	10460-33-0	245.918	25	0.00061	0.0061	2		
2,3,4,6-Tetrachlorophenol	$C_6H_2Cl_4O$	58-90-2	231.891	25	0.017	0.17	24		
1,1,1,3-Tetrachloro-2,2,3,3-tetrafluoropropane	$C_3Cl_4F_4$	2268-46-4	253.838	21	0.0052	0.052	35		
Tetracosane	$C_{24}H_{50}$	646-31-1	338.654	22	0.0000004	0.000004	37		
Tetradecane	$C_{14}H_{30}$	629-59-4	198.388	25	0.00000023	0.0000023	42,5		
Tetradecanoic acid	$C_{14}H_{28}O_2$	544-63-8	228.371	20	0.0020	0.020	26		
1-Tetradecanol	$C_{14}H_{30}O$	112-72-1	214.387	25	0.000031	0.00031	1		
Tetraethylsilane	$C_8H_{20}Si$	631-36-7	144.331	25	0.0000325	0.000325	10		
Tetrafluoroethene	C_2F_4	116-14-3	100.015	25	0.0158*	0.158*	19,50		
Tetrafluoroethene	C_2F_4	116-14-3	100.015	70	0.0090*	0.090*	50		
Tetrafluoromethane	CF_4	75-73-0	88.005	0	0.00390*	0.0390*	50		
				25	0.00185*	0.0185*	50,19		
				50	0.00134*	0.0134*	50		
Tetrahydro-2,5-dimethoxyfuran	$C_6H_{12}O_3$	696-59-3	132.157	21	32	470	20		
				90	19	235	20		
1,2,3,4-Tetrahydronaphthalene	$C_{10}H_{12}$	119-64-2	132.202	20	0.0045	0.045	40		
Tetrahydropyran	$C_5H_{10}O$	142-68-7	86.132	20	8.57	93.7	20		
				81	4.29	44.8	20		
1,2,4,5-Tetramethylbenzene	$C_{10}H_{14}$	95-93-2	134.218	25	0.000348	0.00348	4	2.55	11
N,N,N′,N′-Tetramethyl-4,4′-diaminobenzophenone	$C_{17}H_{20}N_2O$	90-94-8	268.353	20	0.04	0.4	40		
Tetramethylsilane	$C_4H_{12}Si$	75-76-3	88.224	25	0.00196	0.0196	10		
Theophylline	$C_7H_8N_4O_2$	58-55-9	180.165	20	0.52	5.2	29		
Thioacetamide	C_2H_5NS	62-55-5	75.133	25	12.3	140	40		
Thiourea	CH_4N_2S	62-56-6	76.121	20	10.6	119	40		
				80	≈37	≈590	40		
2-Thioxo-4-thiazolidinone	$C_3H_3NOS_2$	141-84-4	133.192	25	0.225	2.25	40		
Thiram	$C_6H_{12}N_2S_4$	137-26-8	240.432	20	0.003	0.03	40		
DL-Threonine	$C_4H_9NO_3$	80-68-2	119.119	10	14.34	167	45		
				20	15.69	186	45		
				40	19.84	248	45		
L-Threonine	$C_4H_9NO_3$	72-19-5	119.119	10	7.34	79.2	45		
				20	8.31	90.6	45		
				40	10.78	121	45		
Thymidine	$C_{10}H_{14}N_2O_5$	50-89-5	242.228	25	5.1	54	29		
Thymine	$C_5H_6N_2O_2$	65-71-4	126.114	25	0.35	3.5	29		
Thymol	$C_{10}H_{14}O$	89-83-8	150.217		0.1	1	30		
Tolazamide	$C_{14}H_{21}N_3O_3S$	1156-19-0	311.400	30	0.0065	0.065	40		
Tolbutamide	$C_{12}H_{18}N_2O_3S$	64-77-7	270.347	25	0.011	0.11	40		
o-Tolidine	$C_{14}H_{16}N_2$	119-93-7	212.290	25	0.13	1.3	40		
Toluene	C_7H_8	108-88-3	92.139	5	0.054	0.54	61		
				25	0.0519	0.519	61,22	0.66	22
				45	0.063	0.63	61		
				90	0.12	1.2	22		
p-Toluenesulfonic acid	$C_7H_8O_3S$	104-15-4	172.202	40	≈33	≈490	40		
o-Toluic acid	$C_8H_8O_2$	118-90-1	136.149	25	0.118	1.18	27		
m-Toluic acid	$C_8H_8O_2$	99-04-7	136.149	25	0.098	0.98	27		
p-Toluic acid	$C_8H_8O_2$	99-94-5	136.149	10	0.030	0.30	75		
				25	0.036	0.36	75		
				50	0.089	0.89	75		
1,3,5-Triazine-2,4,6-triamine	$C_3H_6N_6$	108-78-1	126.120	20	0.323	3.23	40		
				95	4.2	44	40		

Name	Mol. form.	CAS Reg. No.	Mol. wt.	t/°C	s/mass%	s/g kg⁻¹ H_2O	Ref.	k_H/ kPa m^3 mol^{-1}	Ref.
1*H*-1,2,4-TriazoL-3-amine	$C_2H_4N_4$	61-82-5	84.080	23	22	280	26		
1,2,4-Tribromobenzene	$C_6H_3Br_3$	615-54-3	314.800	25	0.0010	0.010	2		
1,3,5-Tribromobenzene	$C_6H_3Br_3$	626-39-1	314.800	25	0.0000789	0.000789	2		
1,1,2-Tribromoethane	$C_2H_3Br_3$	78-74-0	266.757	20	0.050	0.50	25		
Tribromofluoromethane	CBr_3F	353-54-8	270.721	25	0.040	0.40	14		
Tribromomethane	$CHBr_3$	75-25-2	252.731	25	0.30	3.0	5	0.047	13
2,4,6-Tribromophenol	$C_6H_3Br_3O$	118-79-6	330.799	15	0.0007	0.007	2		
Tributylamine	$C_{12}H_{27}N$	102-82-9	185.349	25	0.0142	0.142	40		
Tributyl phosphate	$C_{12}H_{27}O_4P$	126-73-8	266.313	25	0.039	0.39	10		
Tributyrin	$C_{15}H_{26}O_6$	60-01-5	302.363	20	0.010	0.10	40		
Trichloroacetaldehyde	C_2HCl_3O	75-87-6	147.387	25	≈39	≈640	40		
Trichloroacetic acid	$C_2HCl_3O_2$	76-03-9	163.387	25	92.3	11990	27		
1,2,3-Trichlorobenzene	$C_6H_3Cl_3$	87-61-6	181.447	25	0.0021	0.021	41	0.242	11
1,2,4-Trichlorobenzene	$C_6H_3Cl_3$	120-82-1	181.447	15	0.0029	0.029	61		
				25	0.0037	0.037	61,41	0.277	11
				45	0.0047	0.047	61		
1,3,5-Trichlorobenzene	$C_6H_3Cl_3$	108-70-3	181.447	25	0.0008	0.008	41	1.1	11
3,4,5-Trichloro-1,2-benzenediol	$C_6H_3Cl_3O_2$	56961-20-7	213.446	25	0.051	0.51	8		
2,4,5-Trichlorobiphenyl	$C_{12}H_7Cl_3$	15862-07-4	257.543	25	0.000014	0.00014	7	0.0379	31
2,4,6-Trichlorobiphenyl	$C_{12}H_7Cl_3$	35693-92-6	257.543	25	0.00002	0.0002	7	0.0495	7
1,1,1-Trichloro-2,2-bis(4-chlorophenyl)ethane	$C_{14}H_9Cl_5$	50-29-3	354.486	25	0.0000004	0.000004	67		
2,4,6-Trichloro-3,5-dimethylphenol	$C_8H_7Cl_3O$	6972-47-0	225.500	25	0.00050	0.0050	2		
1,1,1-Trichloroethane	$C_2H_3Cl_3$	71-55-6	133.404	0	0.134	1.34	25		
				25	0.129	1.29	25	1.76	13
				50	0.138	1.38	25		
1,1,2-Trichloroethane	$C_2H_3Cl_3$	79-00-5	133.404	0	0.425	4.25	25		
				25	0.459	4.59	25	0.092	13
				50	0.536	5.36	25		
Trichloroethene	C_2HCl_3	79-01-6	131.388	0	0.145	1.45	25		
				25	0.128	1.28	25	1.03	13
				60	0.133	1.33	25		
Trichlorofluoromethane	CCl_3F	75-69-4	137.368	20	0.11	1.1	5	10.2	13
Trichloromethane	$CHCl_3$	67-66-3	119.378	25	0.80	8.0	20	0.43	13
				59	0.79	7.9	20	0.43	13
1,2,4-Trichloro-5-methylbenzene	$C_7H_5Cl_3$	6639-30-1	195.474	25	0.00023	0.0023	61		
(Trichloromethyl)benzene	$C_7H_5Cl_3$	98-07-7	195.474	5	0.0053	0.053	10		
2,4,6-Trichloro-3-methylphenol	$C_7H_5Cl_3O$	551-76-8	211.473	25	0.0112	0.112	2		
Trichloronitromethane	CCl_3NO_2	76-06-2	164.376	0	0.227	2.27	40		
				25	0.162	1.62	40		
1,1,1-Trichloro-2,2,3,3,3-pentafluoropropane	$C_3Cl_3F_5$	4259-43-2	237.383	21	0.0058	0.058	35		
2,4,5-Trichlorophenol	$C_6H_3Cl_3O$	95-95-4	197.446	25	0.1	1	2		
2,4,6-Trichlorophenol	$C_6H_3Cl_3O$	88-06-2	197.446	25	0.069	0.69	48,51,24		
2,4,5-Trichlorophenoxyacetic acid	$C_8H_5Cl_3O_3$	93-76-5	255.483	25	0.028	0.28	40		
1,2,3-Trichloropropane	$C_3H_5Cl_3$	96-18-4	147.431	10	0.14	1.4	35		
				25	0.20	2.0	35	0.038	13
1,1,2-Trichloro-1,2,2-trifluoroethane	$C_2Cl_3F_3$	76-13-1	187.375	25	0.017	0.17	25	32	13
Tri-*p*-cresyl phosphate	$C_{21}H_{21}O_4P$	78-32-0	368.363	25	0.00004	0.0004	40		
Tridecane	$C_{13}H_{28}$	629-50-5	184.361	25	0.000000033	0.00000033	37		
Tridecanoic acid	$C_{13}H_{26}O_2$	638-53-9	214.344	20	0.0033	0.033	26		
Triethylamine	$C_6H_{15}N$	121-44-8	101.190	20	5.5	58	10		
Triethylamine hydrochloride	$C_6H_{16}ClN$	554-68-7	137.651	25	57.8	1370	27		
Trifluoromethane	CHF_3	75-46-7	70.014	25	0.15*	1.5*	50,14		
3,4,5-Trihydroxybenzoic acid	$C_7H_6O_5$	149-91-7	170.120	25	1.52	15.4	74		
				50	3.82	39.7	74		
				100	25.0	333	27		
Triiodomethane	CHI_3	75-47-8	393.732	25	0.012	0.12	14		
Trimethoprim	$C_{14}H_{18}N_4O_3$	738-70-5	290.318	25	0.04	0.4	40		
1,2,3-Trimethylbenzene	C_9H_{12}	526-73-8	120.191	25	0.0070	0.070	22	0.343	22
1,2,4-Trimethylbenzene	C_9H_{12}	95-63-6	120.191	25	0.0057	0.057	22	0.569	22

Name	Mol. form.	CAS Reg. No.	Mol. wt.	t/°C	s/mass%	s/g kg^{-1} H_2O	Ref.	k_H/ kPa m^3 mol^{-1}	Ref.
1,3,5-Trimethylbenzene	C_9H_{12}	108-67-8	120.191	25	0.0050	0.050	22	0.781	22
2,3,3-Trimethyl-2-butanol	$C_7H_{16}O$	594-83-2	116.201	40	2.2	22	1		
1,1,3-Trimethylcyclohexane	C_9H_{18}	3073-66-3	126.239	25	0.000177	0.00177	4	105	13
1,1,3-Trimethylcyclopentane	C_8H_{16}	4516-69-2	112.213	25	0.00037	0.0037	4	159	5
2,2,5-Trimethylhexane	C_9H_{20}	3522-94-9	128.255	25	0.00008	0.0008	4	246	13
1,4,5-Trimethylnaphthalene	$C_{13}H_{14}$	2131-41-1	170.250	25	0.00021	0.0021	42,4		
2,6,8-Trimethyl-4-nonanone	$C_{12}H_{24}O$	123-18-2	184.318	10	0.012	0.12	20		
				80	0.014	0.14	20		
2,2,4-Trimethylpentane	C_8H_{18}	540-84-1	114.229	25	0.00022	0.0022	4	307	13
2,3,4-Trimethylpentane	C_8H_{18}	565-75-3	114.229	25	0.00018	0.0018	4	206	13
Trimethyl phosphate	$C_3H_9O_4P$	512-56-1	140.074	25	≈33	≈490	40		
1,3,5-Trinitrobenzene	$C_6H_3N_3O_6$	99-35-4	213.104	15	0.028	0.28	40		
2,4,6-Trinitrobenzoic acid	$C_7H_3N_3O_8$	129-66-8	257.114	23	1.97	20.1	40		
Trinitroglycerol	$C_3H_5N_3O_9$	55-63-0	227.087	25	0.13	1.3	40		
Trinitroglycerol	$C_3H_5N_3O_9$	55-63-0	227.087	80	0.34	3.4	40		
2,4,6-Trinitrophenol	$C_6H_3N_3O_7$	88-89-1	229.104	25	1.25	12.7	40		
				90	4.9	52	40		
2,4,6-Trinitrotoluene	$C_7H_5N_3O_6$	118-96-7	227.131	20	0.012	0.12	40		
				100	0.15	1.5	40		
2,4,6-Trinitro-*N*-(2,4,6-trinitrophenyl) aniline	$C_{12}H_5N_7O_{12}$	131-73-7	439.208	17	0.0060	0.060	40		
1,3,5-Trioxane	$C_3H_6O_3$	110-88-3	90.078	25	17.4	211	30		
Triphenylene	$C_{18}H_{12}$	217-59-4	228.288	25	0.0000043	0.000043	42,4	0.00001	12
Triphenyl phosphate	$C_{18}H_{15}O_4P$	115-86-6	326.283	24	0.000073	0.00073	40		
Triphenyltin hydroxide	$C_{18}H_{16}OSn$	76-87-9	367.029	20	0.0001	0.001	32		
Tris(hydroxymethyl)methylamine	$C_4H_{11}NO_3$	77-86-1	121.135	25	≈41	≈700	40		
L-Tryptophan	$C_{11}H_{12}N_2O_2$	73-22-3	204.225	25	1.30	13.2	26		
DL-Tyrosine	$C_9H_{11}NO_3$	556-03-6	181.188	25	0.35	3.5	30		
L-Tyrosine	$C_9H_{11}NO_3$	60-18-4	181.188	25	0.0507	0.507	62		
Undecane	$C_{11}H_{24}$	1120-21-4	156.309	25	0.0000004	0.000004	37		
Uracil	$C_4H_4N_2O_2$	66-22-8	112.087	25	0.460	4.62	72		
Urea	CH_4N_2O	57-13-6	60.055	5	44	790	26		
				25	54.4	1200	26		
Uric acid	$C_5H_4N_4O_3$	69-93-2	168.111	20	0.002	0.02	26		
L-Valine	$C_5H_{11}NO_2$	72-18-4	117.147	25	8.13	88.5	26		
Valium	$C_{16}H_{13}ClN_2O$	439-14-5	284.739	25	0.005	0.05	40		
Vidarabine	$C_{10}H_{15}N_5O_5$	5536-17-4	285.257	20	0.051	0.51	40		
Vinclozolin	$C_{12}H_9Cl_2NO_3$	50471-44-8	286.110	20	0.1	1	32		
Vinyl acetate	$C_4H_6O_2$	108-05-4	86.090	20	2.0	20	10		
4-Vinylcyclohexene	C_8H_{12}	100-40-3	108.181	25	0.005	0.05	4		
Warfarin	$C_{19}H_{16}O_4$	81-81-2	308.328	20	0.004	0.04	40		
Xanthine	$C_5H_4N_4O_2$	69-89-6	152.112	20	0.05	0.5	26		
o-Xylene	C_8H_{10}	95-47-6	106.165	25	0.0171	0.171	22	0.551	22
				45	0.021	0.21	4		
m-Xylene	C_8H_{10}	108-38-3	106.165	0	0.0203	0.203	4		
				25	0.0161	0.161	22	0.73	22
				40	0.022	0.22	4		
p-Xylene	C_8H_{10}	106-42-3	106.165	0	0.0160	0.160	4		
				25	0.0181	0.181	22	0.69	22
				40	0.022	0.22	4		
2,3-Xylenol	$C_8H_{10}O$	526-75-0	122.164	25	0.457	4.57	40		
2,4-Xylenol	$C_8H_{10}O$	105-67-9	122.164	25	0.787	7.87	10		
2,5-Xylenol	$C_8H_{10}O$	95-87-4	122.164	25	0.354	3.54	40		
2,6-Xylenol	$C_8H_{10}O$	576-26-1	122.164	25	0.60	6.0	40		
3,4-Xylenol	$C_8H_{10}O$	95-65-8	122.164	25	0.477	4.77	40		
3,5-Xylenol	$C_8H_{10}O$	108-68-9	122.164	29	0.62	6.2	10		
D-Xylose	$C_5H_{10}O_5$	58-86-6	150.130	25	≈30	≈430	40		
Ziram	$C_6H_{12}N_2S_4Zn$	137-30-4	305.841	20	0.0065	0.065	40		

* Indicates a value of *s* for a gas at a partial pressure of 101.325 kPa (1 atm) in equilibrium with the solution.

AQUEOUS SOLUBILITY OF INORGANIC COMPOUNDS AT VARIOUS TEMPERATURES

The solubility of over 300 common inorganic compounds in water is tabulated here as a function of temperature. Solubility is defined as the concentration of the compound in a solution that is in equilibrium with a solid phase at the specified temperature. In this table the solid phase is generally the most stable crystalline phase at the temperature in question. An asterisk * on solubility values in adjacent columns indicates that the solid phase changes between those two temperatures (usually from one hydrated phase to another or from a hydrate to the anhydrous solid). In such cases the slope of the solubility vs. temperature curve may show a discontinuity.

All solubility values are expressed as mass percent of solute, $100 \times w_2$, where

$$w_2 = m_2/(m_1 + m_2)$$

and m_2 is the mass of solute and m_1 the mass of water. This quantity is related to other common measures of solubility as follows:

Molarity: $c_2 = 1000\rho w_2/M_2$
Molality: $m_2 = 1000w_2/M_2(1\text{-}w_2)$
Mole fraction: $x_2 = (w_2/M_2)/\{(w_2/M_2) + (1\text{-}w_2)/M_1\}$
Mass of solute per 100 g of H_2O: $r_2 = 100w_2/(1\text{-}w_2)$

Here M_2 is the molar mass of the solute and M_1 = 18.015 g/mol is the molar mass of water; ρ is the density of the solution in g cm^{-3}.

The data in the table have been derived from the references indicated; in many cases the data have been refitted or interpolated in order to present solubility at rounded values of temperature. Where available, values were taken from the IUPAC *Solubility Data Series* (Ref. 1) or related papers in the *Journal of Physical and Chemical Reference Data* (Refs. 2 to 5), which present carefully evaluated data.

The solubility of sparingly soluble compounds that do not appear in this table may be calculated from the data in the table "Solubility Product Constants." Solubility of inorganic gases may be found in the table "Solubility of Selected Gases in Water."

Compounds are listed alphabetically by formula.

References

1. *Solubility Data Series*, International Union of Pure and Applied Chemistry. Volumes 1 to 53 were published by Pergamon Press, Oxford, from 1979 to 1994; subsequent volumes were published by Oxford University Press, Oxford. The number following the colon is the volume number in the series. Current reports in the series appear in the *Journal of Physical and Chemical Reference Data.*
2. Clever, H. L., and Johnston, F. J., *J. Phys. Chem. Ref. Data*, 9, 751, 1980.
3. Marcus, Y., *J. Phys. Chem. Ref. Data*, 9, 1307, 1980.
4. Clever, H. L., Johnson, S. A., and Derrick, M. E., *J. Phys. Chem. Ref. Data*, 14, 631, 1985.
5. Clever, H. L., Johnson, S. A., and Derrick, M. E., *J. Phys. Chem. Ref. Data*, 21, 941, 1992.
6. Söhnel, O., and Novotny, P., *Densities of Aqueous Solutions of Inorganic Substances*, Elsevier, Amsterdam, 1985.
7. Krumgalz, B.S., *Mineral Solubility in Water at Various Temperatures*, Israel Oceanographic and Limnological Research Ltd., Haifa, 1994.
8. Potter, R. W., and Clynne, M. A., *J. Research U.S. Geological Survey*, 6, 701, 1978; Clynne, M. A., and Potter, R. W., *J. Chem. Eng. Data*, 24, 338, 1979.
9. Marshal, W. L., and Slusher, R., *J. Phys. Chem.*, 70, 4015, 1966; Knacke, O., and Gans, W., *Zeit. Phys. Chem.*, NF, 104, 41, 1977.
10. Stephen, H., and Stephen, T., *Solubilities of Inorganic and Organic Compounds, Vol. 1*, Macmillan, New York, 1963.

Aqueous Solubility of Inorganic Compounds in Mass% as a Function of Temperature

Formula	0 °C	10 °C	20 °C	25 °C	30 °C	40 °C	50 °C	60 °C	70 °C	80 °C	90 °C	100 °C	Ref.
$AgBrO_3$				0.193							1.32		7
$AgClO_2$	0.17	0.31	0.47	0.55	0.64	0.82	1.02	1.22	1.44	1.66	1.88	2.11	7
$AgClO_3$				15									7
$AgClO_4$	81.6	83.0	84.2	84.8	85.3	86.3	86.9	87.5	87.9	88.3	88.6	88.8	6
$AgNO_2$	0.155			0.413									7
$AgNO_3$	55.9	62.3	67.8	70.1	72.3	76.1	79.2	81.7	83.8	85.4	86.7	87.8	6
Ag_2SO_4	0.56	0.67	0.78	0.83	0.88	0.97	1.05	1.13	1.20	1.26	1.32	1.39	7
$AlCl_3$	30.84	30.91	31.03	31.10	31.18	31.37	31.60	31.87	32.17	32.51	32.90	33.32	7
$Al(ClO_4)_3$	54.9										64.4		7
AlF_3	0.25	0.34	0.44	0.50	0.56	0.68	0.81	0.96	1.11	1.28	1.45	1.64	7
$Al(NO_3)_3$	37.0	38.2	39.9	40.8	42.0	44.5	47.3	50.4	53.8*			61.5*	6
$Al_2(SO_4)_3$	27.5			27.8	28.2	29.2	30.7	32.6	34.9	37.6	40.7	44.2	7
As_2O_3	1.19	1.48	1.80	2.01	2.27	2.86	3.43	4.11	4.89	5.77	6.72	7.71	10
$BaBr_2$	47.6	48.5	49.5	50.0	50.4	51.4	52.5	53.5	54.5	55.5	56.6	57.6	6
$Ba(BrO_3)_2$	0.285	0.442	0.656	0.788	0.935	1.30	1.74	2.27	2.90	3.61	4.40	5.25	1:14
$Ba(C_2H_3O_2)_2$	37.0			44.2									7
$BaCl_2$	23.30	24.88	26.33	27.03	27.70	29.00	30.27	31.53	32.81	34.14	35.54	37.05	8
$Ba(ClO_2)_2$	30.5			31.3								44.7	7
$Ba(ClO_3)_2$	16.90	21.23	23.66	27.50	29.43	33.16	36.69	40.05	43.04	45.90	48.70	51.17	1:14
$Ba(ClO_4)_2$	67.30	70.96	74.30	75.75	77.05	79.23	80.92	82.21	83.16	83.88	84.43	84.90	7
BaF_2		0.158		0.161									7
BaI_2	62.5	64.7	67.3	68.8	69.1	69.5	70.1	70.7	71.3	72.0	72.7	73.4	6
$Ba(IO_3)_2$	0.0182	0.0262	0.0342	0.0396	0.045*	0.058*	0.073	0.090	0.109	0.131	0.156	0.182	1:14

Formula	0 °C	10 °C	20 °C	25 °C	30 °C	40 °C	50 °C	60 °C	70 °C	80 °C	90 °C	100 °C	Ref.
$Ba(NO_2)_2$	31.1	36.6	41.8	44.3	46.8	51.6	56.2	60.5	64.6	68.5	72.1	75.6	10
$Ba(NO_3)_2$	4.7	6.3	8.2	9.3	10.2	12.4	14.7	17.0	19.3	21.5	23.5	25.5	6
$Ba(OH)_2$	1.67			4.68	8.4	19	33	52	74	100			7
BaS	2.79	4.78	6.97	8.21	9.58	12.67	16.18	20.05	24.19	28.55	33.04	37.61	7
$BaSO_3$				0.0011									1:26
$Ba(SCN)_2$				62.6									7
$BeCl_2$	40.5			41.7									7
$Be(ClO_4)_2$				59.5									7
$BeSO_4$	26.69	27.58	28.61	29.22	29.90	31.51	33.39	35.50	37.78	40.21	42.72	45.28	7
$CaBr_2$	55	56	59	61	63	68	71	73					10
$CaCl_2$	36.70	39.19	42.13	44.83*	49.12*	52.85*	56.05*	56.73	57.44	58.21	59.04	59.94	8
$Ca(ClO_3)_2$	63.2	64.2	65.5	66.3	67.2	69.0	71.0	73.2	75.5*	77.4*	77.7	78.0	1:14
$Ca(ClO_4)_2$				65.3									7
CaF_2	0.0013			0.0016									10
CaI_2	64.6	66.0	67.6	68.3	69.0	70.8	72.4	74.0	76.0	78.0	79.6	81.0	7
$Ca(IO_3)_2$	0.082	0.155	0.243	0.305	0.384*	0.517*	0.590	0.652	0.811*	0.665*	0.668		1:14
$Ca(NO_2)_2$	38.6	39.5	44.5	48.6									7
$Ca(NO_3)_2$	50.1	53.1	56.7	59.0	60.9	65.4	77.8	78.1	78.2	78.3	78.4	78.5	6
$CaSO_3$			0.0059	0.0054	0.0049	0.0041	0.0035	0.0030	0.0026	0.0023	0.0020	0.0019	1:26
$CaSO_4$	0.174	0.191	0.202	0.205	0.208	0.210	0.207	0.201	0.193	0.184	0.173	0.163	9
$CdBr_2$	36.0	43.0	49.9	53.4	56.4	60.3*	60.3*	60.5	60.7	60.9	61.3	61.6	6
CdC_2O_4				0.0060									5
$CdCl_2$	47.2	50.1	53.2	54.6	56.3*	57.3*	57.5	57.8	58.1	58.51	58.98	59.5	6
$Cd(ClO_4)_2$				58.7								66.9	7
CdF_2		5.82	4.65	4.18	3.76								5
CdI_2	44.1	44.9	45.8	46.3	46.8	47.9	49.0	50.2	51.5	52.7	54.1	55.4	6
$Cd(IO_3)_2$				0.091									5
$Cd(NO_3)_2$	55.4	57.1	59.6	61.0	62.8	66.5	70.6	86.1	86.5	86.8	87.1	87.4	6
$CdSO_4$	43.1	43.1	43.2	43.4	43.6	44.1	43.5	42.5	41.4	40.2	38.5	36.7	6
$CdSeO_4$	42.04	40.59	39.02	38.18	37.29	35.35	33.15	30.65	27.84	24.69	21.24	17.49	5
$Ce(NO_3)_3$	57.99	59.80	61.89	63.05	64.31*	67.0*	68.6	71.1*	74.9*	79.2	80.9	83.1	1:13
$CoCl_2$	30.30	32.60	34.87	35.99	37.10	39.27	41.38	43.46	45.50	47.51	49.51	51.50	7
$Co(ClO_4)_2$	50.0			53.0									7
CoF_2				1.4									7
CoI_2	58.00	61.78	65.35	66.99	68.51	71.17	73.41	75.29	76.89	78.28	79.52	80.70	7
$Co(NO_2)_2$	0.076			0.49									7
$Co(NO_3)_2$	45.5	47.0	49.4	50.8	52.4	56.0	60.1	62.6	64.9	67.7			6
$CoSO_4$	19.9	23.0	26.1	27.7	29.2	32.3	34.4	35.9	35.5	33.2	30.6	27.8	6
$Co(SCN)_2$				50.7									7
CrO_3	62.2	62.3	62.6	62.8	63.0	63.5	64.1	64.7	65.5	66.2	67.1	67.9	6
$CsBr$				55.2									7
$CsBrO_3$	1.16	1.93	3.01	3.69	4.46	6.32	8.60	11.32	14.45	17.96	21.83	25.98	1:30
$CsCl$	61.83	63.48	64.96	65.64	66.29	67.50	68.60	69.61	70.54	71.40	72.21	72.96	1:47
$CsClO_3$	2.40	3.87	5.94	7.22	8.69	12.15	16.33	21.14	26.45	32.10	37.89	43.42	1:30
$CsClO_4$	0.79	1.01	1.51	1.96	2.57	4.28	6.55	9.29	12.41	15.80	19.39	23.07	7
CsI	30.9	37.2	43.2	45.9	48.6	53.3	57.3	60.7	63.6	65.9	67.7	69.2	6
$CsIO_3$	1.08	1.58	2.21	2.59	3.02	3.96	5.06	6.29	7.70	9.20	10.79	12.45	1:30
$CsNO_3$	8.46	13.0	18.6	21.8	25.1	32.0	39.0	45.7	51.9	57.3	62.1	66.2	6
$CsOH$					75								7
Cs_2SO_4	62.6	63.4	64.1	64.5	64.8	65.5	66.1	66.7	67.3	67.8	68.3	68.8	6
$CuBr_2$				55.8									7
$CuCl_2$	40.8	41.7	42.6	43.1	43.7	44.8	46.0	47.2	48.5	49.9	51.3	52.7	6
$Cu(ClO_4)_2$	54.3				59.3								7
CuF_2				0.075									7
$Cu(NO_3)_2$	45.2	49.8	56.3	59.2	61.1	62.0	63.1	64.5	65.9	67.5	69.2	71.0	6
$CuSO_4$	12.4	14.4	16.7	18.0	19.3	22.2	25.4	28.8	32.4	36.3	40.3	43.5	6
$CuSeO_4$	10.6			16.0									7
$Dy(NO_3)_3$	58.79	59.99	61.49	62.35	63.29	65.43	68.04	71.58					1:13
$Er(NO_3)_3$	61.58	63.15	64.84	65.75	66.69	68.70	70.96	73.64	77.75				1:13
$Eu(NO_3)_3$	55.2	56.7	58.5	59.4	60.4	62.5	64.6						1:13
$FeBr_2$				54.6*								64.8*	7

Formula	0 °C	10 °C	20 °C	25 °C	30 °C	40 °C	50 °C	60 °C	70 °C	80 °C	90 °C	100 °C	Ref.
$FeCl_2$	33.2*			39.4*								48.7*	7
$Fe(ClO_4)_2$	63.39			67.76									7
$FeCl_3$	42.7	44.9	47.9	47.7	51.6	74.8	76.7	84.6	84.3	84.3	84.4	84.7	6
FeF_3				5.59									7
$Fe(NO_3)_2$	41.44			46.67									7
$Fe(NO_3)_3$	40.15			46.57									7
$FeSO_4$	13.5	17.0	20.8	22.8	24.8	28.8	32.8	35.5	33.6	30.4	27.1	24.0	6
$Gd(NO_3)_3$	56.3	57.7	59.2	60.1	61.0	62.9	65.2	67.9	71.5				1:13
HIO_3	73.45	74.10	74.98	75.48	76.03	77.20	78.46	79.78	81.13	82.48	83.82	85.14	1:30
H_3BO_3	2.61	3.57	4.77	5.48	6.27	8.10	10.3	12.9	15.9	19.3	23.1	27.3	6
$HgBr_2$	0.26	0.37	0.52	0.61	0.72	0.96	1.26	1.63	2.08	2.61	3.23	3.95	4
$Hg(CN)_2$	6.57	7.83	9.33	10.2	11.1	13.1	15.5	18.2	21.2	24.6	28.3	32.3	6
$HgCl_2$	4.24	5.05	6.17	6.81	7.62	9.53	12.02	15.18	19.16	24.06	29.90	36.62	4
HgI_2			0.0041	0.0055	0.0072	0.0122	0.0199						4
$Hg(SCN)_2$				0.070									4
Hg_2Cl_2				0.0004									3
$Hg_2(ClO_4)_2$	73.8			79.8*								85.3*	7
Hg_2SO_4	0.038	0.043	0.048	0.051	0.054	0.059	0.065	0.070	0.076	0.082	0.088	0.093	4
$Ho(NO_3)_3$				63.8									1:13
KBF_4	0.28	0.34	0.45	0.55	0.75	1.38	2.09	2.82	3.58	4.34	5.12	5.90	10
KBr	35.0	37.3	39.4	40.4	41.4	43.2	44.8	46.2	47.6	48.8	49.8	50.8	6
$KBrO_3$	2.97	4.48	6.42	7.55	8.79	11.57	14.71	18.14	21.79	25.57	29.42	33.28	1:30
$KC_2H_3O_2$	68.40	70.29	72.09	72.92	73.70	75.08	76.27	77.31	78.22	79.04	79.80	80.55	7
KCl	21.74	23.61	25.39	26.22	27.04	28.59	30.04	31.40	32.66	33.86	34.99	36.05	1:47
$KClO_3$	3.03	4.67	6.74	7.93	9.21	12.06	15.26	18.78	22.65	26.88	31.53	36.65	1:30
$KClO_4$	0.70	1.10	1.67	2.04	2.47	3.54	4.94	6.74	8.99	11.71	14.94	18.67	6
KF	30.90	39.8	47.3	50.41	53.2					60.0			7
$KHCO_3$	18.62	21.73	24.92	26.6	28.13	31.32	34.46	37.51	40.45				6
$KHSO_4$	27.1	29.7	32.3	33.6	35.0	37.8	40.5	43.4	46.2	49.02	51.82	54.6	6
KH_2PO_4	11.74	14.91	18.25	19.97	21.77	25.28	28.95	32.76	36.75	40.96	45.41	50.12	1:31
KI	56.0	57.6	59.0	59.7	60.4	61.6	62.8	63.8	64.8	65.7	66.6	67.4	6
KIO_3	4.53	5.96	7.57	8.44	9.34	11.09	13.22	15.29	17.41	19.58	21.78	24.03	1:30
KIO_4	0.16	0.22	0.37	0.51	0.70	1.24	1.96	2.83	3.82	4.89	6.02	7.17	7
$KMnO_4$	2.74	4.12	5.96	7.06	8.28	11.11	14.42	18.16					6
KNO_2	73.7	74.6	75.3	75.7	76.0	76.7	77.4	78.0	78.5	79.1	79.6	80.1	6
KNO_3	12.0	17.6	24.2	27.7	31.3	38.6	45.7	52.2	58.0	63.0	67.3	70.8	6
KOH	48.7	50.8	53.2	54.7	56.1	57.9	58.6	59.5	60.6	61.8	63.1	64.6	6
KSCN	63.8	66.4	69.1	70.4	71.6	74.1	76.5	78.9	81.1	83.3	85.3	87.3	6
K_2CO_3	51.3	51.7	52.3	52.7	53.1	54.0	54.9	56.0	57.2	58.4	59.6	61.0	6
K_2CrO_4	37.1	38.1	38.9	39.4	39.8	40.5	41.3	41.9	42.6	43.2	43.8	44.3	6
$K_2Cr_2O_7$	4.30	7.12	10.9	13.1	15.5	20.8	26.3	31.7	36.9	41.5	45.5	48.9	6
K_2HAsO_4	48.5*			63.6*								79.8*	7
K_2HPO_4	57.0	59.1	61.5	62.7	64.1	67.7*		72.7*					1:31
K_2MoO_4				64.7							66.5		7
K_2SO_3	51.30	51.39	51.49	51.55	51.62	51.76	51.93	52.11	52.32	52.54	52.79	53.06	1:26
K_2SO_4	7.11	8.46	9.95	10.7	11.4	12.9	14.2	15.5	16.7	17.7	18.6	19.3	6
$K_2S_2O_3$	49.0*			62.3*							75.7*		7
$K_2S_2O_5$	22.1	26.7	31.1	33.1	35.2	39.0	42.6	46.0	49.1	52.0	54.6		1:26
K_2SeO_3	68.4*			68.5*								68.5*	7
K_2SeO_4	52.70	52.93	53.17	53.30	53.43	53.70	53.99	54.30	54.61	54.94	55.26	55.60	7
K_3AsO_4	51.5*			55.6*								73*	7
$K_3Fe(CN)_6$	23.9	27.6	31.1	32.8	34.3	37.2	39.6	41.7	43.5	45.0	46.1	47.0	6
K_3PO_4	44.3			51.4									7
$K_4Fe(CN)_6$	12.5	17.3	22.0	23.9	25.6	29.2	32.5	35.5	38.2	40.6	41.4	43.1	6
$LaCl_3$	49.0	48.5	48.6	48.9	49.3	50.5	52.1	54.0	56.3	58.9	61.7		6
$La(NO_3)_3$	55.0	56.9	58.9	60.0	61.1	63.6	66.3	69.9*	74.1*				1:13
LiBr	58.4	60.1	62.7	64.4	65.9	67.8	68.3	69.0	69.8	70.7	71.7	72.8	6
$LiBrO_3$	61.03	62.62	64.44	65.44	66.51	68.90	71.68*	73.24*	74.43	75.66	76.93	78.32	1:30
$LiC_2H_3O_2$	23.76	26.49	29.42	31.02	32.72	36.48	40.65	45.15	49.93	54.91	60.04	65.26	7
LiCl	40.45	42.46*	45.29*	45.81	46.25	47.30	48.47	49.78	51.27	52.98	54.98*	56.34*	1:47
$LiClO_3$	73.2	75.6*	80.8*	82.1	83.4	85.9*	87.1*	88.2	89.6	91.3	93.4	95.7	1:30

Formula	0 °C	10 °C	20 °C	25 °C	30 °C	40 °C	50 °C	60 °C	70 °C	80 °C	90 °C	100 °C	Ref.
$LiClO_4$	30.1	32.6	35.5	37.0	38.6	41.9	45.5	49.2	53.2	57.2	61.3	71.4	6
LiF	0.120	0.126	0.131	0.134									7
LiH_2PO_4	55.8												7
LiI	59.4	60.5	61.7	62.3	63.0	64.3	65.8	67.3	68.8	81.3	81.7	82.6	6
$LiIO_3$				43.8									1:30
$LiNO_2$	41	45	49	51	53	56	60	63	66	68			10
$LiNO_3$	34.8	37.6	42.7	50.5	57.9	60.1	62.2	64.0	65.7	67.2	68.5	69.7	6
LiOH	10.8	10.8	11.0	11.1	11.3	11.7	12.2	12.7	13.4	14.2	15.1	16.1	6
LiSCN				54.5									7
Li_2CO_3	1.54	1.43	1.33	1.28	1.24	1.15	1.07	0.99	0.92	0.85	0.78	0.72	7
$Li_2C_2O_4$				5.87									7
Li_2HPO_3	9.07	8.40	7.77	7.47	7.18	6.64	6.16	5.71	5.30	4.91	4.53	4.16	7
Li_2SO_4	26.3	25.9	25.6	25.5	25.3	25.0	24.8	24.5	24.3	24.0	23.8	23.6	6
Li_3PO_4				0.027									1:31
$Lu(NO_3)_3$				71.1									1:13
$MgBr_2$	49.3	49.8	50.3	50.6	50.9	51.5	52.1	52.8	53.5	54.2	55.0	55.7	6
$Mg(BrO_3)_2$	43.0	45.2	48.0	49.4	51.0	54.3	57.9	61.6	65.3	69.0*	70.9*	71.7	1:14
MgC_2O_4				0.038									7
$Mg(C_2H_3O_2)_2$	36.18	37.55	38.92	39.61									7
$MgCl_2$	33.96	34.85	35.58	35.90	36.20	36.77	37.34	37.97	38.71	39.62	40.75	42.15	8
$Mg(ClO_3)_2$	53.35	54.40	56.81	58.66	60.91*	65.46*	67.33	69.27	71.01	72.44	73.48		1:14
$Mg(ClO_4)_2$	47.8	48.7	49.6	50.1	50.5	51.3	52.1						6
$MgCrO_4$	32.06*			35.39*									7
$MgCr_2O_7$				58.9						67.0			7
MgF_2				0.013									7
MgI_2	54.7	56.1	58.2	59.4	60.8	63.9	65.0	65.0	65.0	65.0	65.1	65.2	6
$Mg(IO_3)_2$	3.19*	6.70*	7.92	8.52	9.11	10.45	11.99	13.7	15.6	17.6	19.6		1:14
$Mg(NO_2)_2$				47									7
$Mg(NO_3)_2$	38.4	39.5	40.8	41.6	42.4	44.1	45.9	47.9	50.0	52.2	70.6	72.0	6
$MgSO_3$	0.32	0.37	0.46	0.52	0.61	0.87*	0.85*	0.76	0.69	0.64	0.62	0.60	1:26
$MgSO_4$	18.2	21.7	25.1	26.3	28.2	30.9	33.4	35.6	36.9	35.9	34.7	33.3	6
MgS_2O_3	30.7			34.1									7
$MgSeO_4$	31.4*			35.7*								47*	7
$MnBr_2$	56.00	57.72	59.39	60.19	60.96	62.41	63.75	65.01	66.19	67.32	68.42	69.50	7
$MnCl_2$	38.7	40.6	42.5	43.6	44.7	47.0	49.4	54.1	54.7	55.2	55.7	56.1	6
MnF_2	0.80*			1.01*								0.48	7
$Mn(IO_3)_2$				0.27							0.34		7
$Mn(NO_3)_2$	50.5			61.7									7
$MnSO_4$	34.6	37.3	38.6	38.9	38.9	37.7	36.3	34.6	32.8	30.8	28.8	26.7	6
NH_4Br	37.5	40.2	42.7	43.9	45.1	47.3	49.4	51.3	53.0	54.6	56.1	57.4	7
NH_4Cl	22.92	25.12	27.27	28.34	29.39	31.46	33.50	35.49	37.46	39.40	41.33	43.24	1:47
NH_4ClO_4	10.8	14.1	17.8	19.7	21.7	25.8	29.8	33.6	37.3	40.7	43.8	46.6	6
NH_4F	41.7	43.2	44.7	45.5	46.3	47.8	49.3	50.9	52.5	54.1			7
NH_4HCO_3	10.6	13.7	17.6	19.9	22.4	27.9	34.2	41.4	49.3	58.1	67.6	78.0	7
$NH_4H_2AsO_4$	25.2	29.0	32.7	34.5	36.3	39.7	43.1	46.2	49.3	52.2	55.0		7
$NH_4H_2PO_4$	17.8	22.0	26.4	28.8	31.2	36.2	41.6	47.2	53.0	59.2	65.7	72.4	7
NH_4I	60.7	62.1	63.4	64.0	64.6	65.8	66.8	67.8	68.7	69.6	70.4	71.1	6
NH_4IO_3				3.70	4.20	5.64	7.63						1:30
NH_4NO_2	55.7	59.0	64.9	68.8									7
NH_4NO_3	54.0	60.1	65.5	68.0	70.3	74.3	77.7	80.8	83.4	85.8	88.2	90.3	6
NH_4SCN				64.4					81.1				7
$(NH_4)_2C_2O_4$	2.31	3.11	4.25	4.94	5.73	7.56	9.73	12.2	15.1	18.3	21.8	25.7	7
$(NH_4)_2HPO_4$	36.4	38.2	40.0	41.0	42.0	44.1	46.2	48.5	50.9	53.3	55.9	58.6	7
$(NH_4)_2SO_3$	32.2	34.9	37.7	39.1	40.6	43.7	47.0	50.6	54.5	58.9			1:26
$(NH_4)_2SO_4$	41.3	42.1	42.9	43.3	43.8	44.7	45.6	46.6	47.5	48.5	49.5	50.5	6
$(NH_4)_2S_2O_5$	65.5	67.9	69.8	70.5	71.3	72.3	72.9	73.1					1:26
$(NH_4)_2S_2O_8$	37.00	40.45	43.84	45.49	47.11	50.25	53.28	56.23	59.13	62.00			7
$(NH_4)_2SeO_3$	49.0	51.1	53.4	54.7	56.0	58.9	62.0	65.4	69.1				7
$(NH_4)_2SeO_4$				54.02									7
$(NH_4)_3PO_4$				15.5									7
NaBr	44.4	45.9	47.7	48.6	49.6	51.6	53.7	54.1	54.3	54.5	54.7	54.9	6

Thermochem

Formula	0 °C	10 °C	20 °C	25 °C	30 °C	40 °C	50 °C	60 °C	70 °C	80 °C	90 °C	100 °C	Ref.
$NaBrO_3$	20.0	23.22	26.65	28.28	29.86	32.83	35.55	38.05	40.37	42.52			1:30
$NaCHO_2$	30.8	37.9	45.7	48.7	50.6	52.0	53.5	55.0					6
$NaC_2H_3O_2$	26.5	28.8	31.8	33.5	35.5	39.9	45.1	58.3	59.3	60.5	61.7	62.9	6
$NaCl$	26.28	26.32	26.41	26.45	26.52	26.67	26.84	27.03	27.25	27.50	27.78	28.05	1:47
$NaClO$	22.7			44.4									7
$NaClO_2$				97.0*				95.3*					7
$NaClO_3$	44.27	46.67	49.3	50.1	51.2	53.6	55.5	57.0	58.5	60.5	63.3	67.1	1:30
$NaClO_4$	61.9	64.1	66.2	67.2	68.3	70.4	72.5	74.1	74.7	75.4	76.1	76.7	6
NaF	3.52	3.72	3.89	3.97	4.05	4.20	4.34	4.46	4.57	4.66	4.75	4.82	6
$NaHCO_3$	6.48	7.59	8.73	9.32	9.91	11.13	12.40	13.70	15.02	16.37	17.73	19.10	7
$NaHSO_4$				22.2								33.3	10
NaH_2PO_4	36.54	41.07	46.00	48.68	51.54	57.89*	61.7*	62.3*	65.9	68.7			1:31
NaI	61.2	62.4	63.9	64.8	65.7	67.7	69.8	72.0	74.7	74.8	74.9	75.1	6
$NaIO_3$	2.43	4.40	7.78*	8.65*	9.60	11.67	13.99	16.52	19.25*	21.1*	22.9	24.7	1:30
$NaIO_4$				12.62									7
$NaNO_2$	41.9	43.4	45.1	45.9	46.8	48.7	50.7	52.8	55.0	57.2	59.5	61.8	6
$NaNO_3$	42.2	44.4	46.6	47.7	48.8	51.0	53.2	55.3	57.5	59.6	61.7	63.8	6
$NaOH$	30	39	46	50	53	58	63	67	71	74	76	79	10
$NaSCN$		52.9	57.1	60.2	62.7	63.5	64.2	65.0	65.9	66.9	67.9	69.0	6
$Na_2B_4O_7$	1.23	1.71	2.50	3.07	3.82	6.02	9.7	14.9	17.1	19.9	23.5	28.0	6
Na_2CO_3	6.44	10.8	17.9	23.5	28.7	32.8	32.2	31.7	31.3	31.1	30.9	30.9	6
$Na_2C_2O_4$	2.62	2.95	3.30	3.48	3.65	4.00	4.36	4.71	5.06	5.41	5.75	6.08	6
Na_2CrO_4	22.6	32.3	44.6	46.7	46.9	48.9	51.0	53.4	55.3	55.5	55.8	56.1	6
$Na_2Cr_2O_7$	62.1	63.1	64.4	65.2	66.1	68.0	70.1	72.3	74.6	77.0	79.6	80.7	6
Na_2HAsO_4	5.6*			29.3*								67*	7
Na_2HPO_4	1.66	4.19	7.51	10.55	16.34*	35.17*	44.64*	45.20	46.81	48.78	50.52	51.53	1:31
Na_2MoO_4	30.6	38.8	39.4	39.4	39.8	40.3	41.0	41.7	42.6	43.5	44.5	45.5	6
Na_2S	11.1	13.2	15.7	17.1	18.6	22.1	26.7	28.1	30.2	33.0	36.4	41.0	6
Na_2SO_3	12.0	16.1	20.9	23.5	26.3*	27.3*	25.9	24.8	23.7	22.8	22.1	21.5	1:26
Na_2SO_4			16.13	21.94	29.22*	32.35*	31.55	30.90	30.39	30.02	29.79	29.67	8
$Na_2S_2O_3$	33.1	36.3	40.6	43.3	45.9	52.0	62.3	65.7	68.8	69.4	70.1	71.0	6
$Na_2S_2O_5$		38.4	39.5	40.0	40.6	41.8	43.0	44.2	45.5	46.8	48.1	49.5	1:26
Na_2SeO_3				47.3*								45*	7
Na_2SeO_4	11.7			36.9*								42.1*	7
Na_2WO_4	41.6	41.9	42.3	42.6	42.9	43.6	44.4	45.3	46.2	47.3	48.4	49.5	6
Na_3PO_4	4.28	7.30	10.8	12.6	14.1	16.6	22.9	28.4	32.4	37.6	40.4	43.5	6
$Na_4P_2O_7$	2.23	3.28	4.81	6.62	7.00	10.10	14.38	20.07	27.31	36.03	32.37	30.67	6
$NdCl_3$	49.0	49.3	49.7	50.0	50.4	51.2	52.2	53.3	54.5	55.8	57.1	58.5	6
$Nd(NO_3)_3$	55.76	57.49	59.37	60.38	61.43	63.69	66.27	69.47					1:13
$NiCl_2$	34.7	36.1	38.5	40.3	41.7	42.1	43.2	45.0	46.1	46.2	46.4	46.6	6
$Ni(ClO_4)_2$	51.1			52.8									7
NiF_2				2.50							2.52		7
NiI_2	55.40	57.68	59.78	60.69	61.50	62.80	63.73	64.38	64.80	65.09	65.30		7
$Ni(NO_3)_2$	44.1	46.0	48.4	49.8	51.3	54.6	58.3	61.0	63.1	65.6	67.9	69.0	6
$NiSO_4$	21.4	24.4	27.4	28.8	30.3*	32.0*	34.1	35.8	37.7	39.9	42.3	44.8	6
$Ni(SCN)_2$				35.48									7
$NiSeO_4$	21.6		26.2*									45.6*	7
$PbBr_2$	0.449	0.620	0.841	0.966	1.118	1.46	1.89						2
$PbCl_2$	0.66	0.81	0.98	1.07	1.17	1.39	1.64	1.93	2.24	2.60	2.99	3.42	2
$Pb(ClO_4)_2$				81.5									7
PbF_2		0.0603	0.0649	0.0670	0.0693								2
PbI_2	0.041	0.052	0.067	0.076	0.086	0.112	0.144	0.187	0.243	0.315			2
$Pb(IO_3)_2$				0.0025									7
$Pb(NO_3)_2$	28.46	32.13	35.67	37.38	39.05	42.22	45.17	47.90	50.42	52.72	54.82	56.75	2
$PbSO_4$	0.0033	0.0038	0.0042	0.0044	0.0047	0.0052	0.0058						2
$PrCl_3$	48.0	48.1	48.6	49.0	49.5	50.8	52.3	54.1	56.1	58.3			6
$Pr(NO_3)_3$	57.50	59.20	61.16	62.24	63.40*	65.7*	67.8	70.2	73.4				1:13
$RbBr$	47.4	50.1	52.6	53.8	54.9	57.0	58.8	60.6	62.1	63.5	64.8	65.9	6
$RbBrO_3$	0.97	1.55	2.36	2.87	3.45	4.87	6.64	8.78	11.29	14.15	17.32	20.76	1:30
$RbCl$	43.58	45.65	47.53	48.42	49.27	50.86	52.34	53.67	54.92	56.08	57.16	58.15	1:47
$RbClO_3$	2.10	3.38	5.14	6.22	7.45	10.35	13.85	17.93	22.53	27.57	32.96	38.60	1:30

Formula	0 °C	10 °C	20 °C	25 °C	30 °C	40 °C	50 °C	60 °C	70 °C	80 °C	90 °C	100 °C	Ref.
$RbClO_4$	1			1.5								17	7
RbF			75										7
$RbHCO_3$			53.7										7
RbI	55.8	58.6	61.1	62.3	63.4	65.4	67.2	68.8	70.3	71.6	72.7	73.8	6
$RbIO_3$	1.09	1.53	2.07	2.38	2.74	3.52	4.41	5.42	6.52	7.74	9.00	10.36	1:30
$RbNO_3$	16.4	25.0	34.6	39.4	44.2	53.1	60.8	67.2	72.2	76.1	79.0	81.2	6
$RbOH$					63.4								7
Rb_2CrO_4	38.27			43.26									7
Rb_2SO_4	27.3	30.0	32.5	33.7	34.8	36.9	38.7	40.3	41.8	43.0	44.1	44.9	6
$SbCl_3$	85.7			90.8									7
SbF_3	79.4			83.1									7
$Sc(NO_3)_3$	57.0	59.3	61.6	62.8	63.9	66.2	68.5						1:13
$SmCl_3$		48.0	48.2	48.4	48.6	49.2	50.0						6
$Sm(NO_3)_3$	54.83	56.33	58.08	59.05	60.08	62.38	65.05*	68.1*	70.8	74.2			1:13
$SnCl_2$	46	64											7
SnI_2			0.97									3.87	7
$SrBr_2$	46.0	48.3	50.6	51.7	52.9	55.2	57.6	59.9	62.3	64.6	66.8	69.0	6
$Sr(BrO_3)_2$	18.53	22.00	25.39	27.02	28.59	31.55	34.21	36.57	38.64*	40.2*	40.8	41.0	1:14
$SrCl_2$	31.94	32.93	34.43	35.37	36.43	38.93	41.94	45.44*	46.81*	47.69	48.70	49.87	8
$Sr(ClO_2)_2$	13.0	13.6	14.1	14.3	14.5	14.9	15.3	15.6	15.9				7
$Sr(ClO_3)_2$	63.29	63.42	63.64	63.77	63.93	64.29	64.70	65.16	65.65	66.18	66.74	67.31	1:14
$Sr(ClO_4)_2$	70.04*			75.35*		78.44*							7
SrF_2	0.011			0.021									7
SrI_2	62.5	62.8	63.5	63.9	64.5	65.8	67.3	69.0	70.8	72.7	74.7	79.2	6
$Sr(IO_3)_2$	0.102	0.126	0.152	0.165	0.179	0.206	0.233	0.259	0.284	0.307	0.328	0.346	1:14
$Sr(MnO_4)_2$	2.5												7
$Sr(NO_2)_2$					41.9	44.3						58.6	7
$Sr(NO_3)_2$	28.2	34.6	41.0	44.5	47.0	47.4	47.9	48.4	48.9	49.5	50.1	50.7	6
$Sr(OH)_2$	0.9			2.2									7
$SrSO_3$				0.0015									1:26
$SrSO_4$				0.0135									7
SrS_2O_3	8.8	13.2	17.7	20.0	22.2	26.8							7
$Tb(NO_3)_3$			60.6	61.02									1:13
Tl_2SO_4	2.65	3.56	4.61	5.19	5.80	7.09	8.46	9.89	11.33	12.77	14.18	15.53	6
$Tm(NO_3)_3$				67.9									1:13
$UO_2(NO_3)_2$	49.52	51.82	54.42	55.85	57.55	61.59	67.07						1:??
$Y(NO_3)_3$	55.57	56.93	58.75	59.86	61.11*	63.3*	64.9	67.9	72.5				1:13
$Yb(NO_3)_3$				70.5									1:13
$ZnBr_2$	79.3	80.1	81.8	83.0	84.1	85.6	85.8	86.1	86.3	86.6	86.8	87.1	6
ZnC_2O_4		0.0010	0.0019	0.0026									5
$ZnCl_2$		76.6	79.0	80.3	81.4	81.8	82.4	83.0	83.7	84.4	85.2	86.0	6
$Zn(ClO_4)_2$	44.29*			46.27*			48.70						7
ZnF_2				1.53									5
ZnI_2	81.1	81.2	81.3	81.4	81.5	81.7	82.0	82.3	82.6	83.0	83.3	83.7	6
$Zn(IO_3)_2$			0.58	0.64	0.69	0.77	0.82						5
$Zn(NO_3)_2$	47.8	50.8	54.4	54.6	58.5	79.1	80.1	87.5	89.9				6
$ZnSO_3$			0.1786	0.1790	0.1794	0.1803	0.1812						5
$ZnSO_4$	29.1	32.0	35.0	36.6	38.2	41.3	43.0	42.1	41.0	39.9	38.8	37.6	6
$ZnSeO_4$	33.06	34.98	37.38	38.79	40.34								5

* Solid phase changes between these temperatures.

OCTANOL-WATER PARTITION COEFFICIENTS

The octanol–water partition coefficient, *P*, is a widely used parameter for correlating biological effects of organic substances. It is a property of the two-phase system in which water and 1-octanol are in equilibrium at a fixed temperature and the substance is distributed between the water-rich and octanol-rich phases. *P* is defined as the ratio of the equilibrium concentration of the substance in the octanol-rich phase to that in the water-rich phase, in the limit of zero concentration. In general, *P* tends to be large for compounds with extended non-polar structures (such as long chain or multi-ring hydrocarbons) and small for compounds with highly polar groups. Thus *P* (or, in its more common form of expression, log *P*) provides a measure of the lipophilic vs. hydrophilic nature of a compound, which is an important consideration in assessing the potential toxicity. A discussion of methods of measurement and accuracy considerations for log *P* may be found in Ref. 1.

This table gives selected values of log *P* for about 450 organic compounds, including many of environmental importance. All values refer to a nominal temperature of 25 °C. The source of each value is indicated in the last column. These references contain data on many more compounds than are included here.

Compounds are listed alphabetically by systematic name.

References

1. Sangster, J., *J. Phys. Chem. Ref. Data*, 18, 1111, 1989.
2. Mackay, D., Shiu, W. Y., and Ma, K. C., *Illustrated Handbook of Physical-Chemical Properties and Environmental Fate for Organic Chemicals*, Lewis Publishers/CRC Press, Boca Raton, FL, 1992.
3. Shiu, W. Y., and Mackay, D., *J. Phys. Chem. Ref. Data*, 15, 911, 1986.
4. Pinsuwan, S., Li, L., and Yalkowsky, S. H., *J. Chem. Eng. Data*, 40, 623, 1995.
5. *Solubility Data Series, International Union of Pure and Applied Chemistry, Vol. 20*, Pergamon Press, Oxford, 1985.
6. *Solubility Data Series, International Union of Pure and Applied Chemistry, Vol. 38*, Pergamon Press, Oxford, 1985.
7. Miller, M. M., Ghodbane, S., Wasik, S. P., Tewari, Y. B., and Martire, D. E., *J. Chem. Eng. Data*, 29, 184, 1984.

Octanol-Water Partition Coefficients

Name	Mol. form.	log *P*	Ref.
Acenaphthene	$C_{12}H_{10}$	3.96	4
Acetaldehyde	C_2H_4O	0.45	1
Acetamide	C_2H_5NO	-1.26	1
Acetic acid	$C_2H_4O_2$	-0.17	1
Acetone	C_3H_6O	-0.24	1
Acetonitrile	C_2H_3N	-0.34	1
Acetophenone	C_8H_8O	1.63	1
Acridine	$C_{13}H_9N$	3.40	1
Acrolein	C_3H_4O	-0.01	1
Acrylamide	C_3H_5NO	-0.78	1
Acrylonitrile	C_3H_3N	0.25	1
Allyl alcohol	C_3H_6O	0.17	1
Allylamine	C_3H_7N	0.03	1
Aniline	C_6H_7N	0.90	1
Anisole	C_7H_8O	2.11	1
Anthracene	$C_{14}H_{10}$	4.56	4
trans-Azobenzene	$C_{12}H_{10}N_2$	3.82	1
Azulene	$C_{10}H_8$	3.22	1
Benzaldehyde	C_7H_6O	1.48	1
Benz[*a*]anthracene	$C_{18}H_{12}$	5.91	1
Benzene	C_6H_6	2.13	1
Benzeneacetaldehyde	C_8H_8O	1.78	1
Benzeneacetic acid	$C_8H_8O_2$	1.41	1
Benzeneacetonitrile	C_8H_7N	1.56	1
Benzeneethanamine	$C_8H_{11}N$	1.41	1
Benzeneethanol	$C_8H_{10}O$	1.36	1
Benzenepropanenitrile	C_9H_9N	1.72	1
Benzenepropanol	$C_9H_{12}O$	1.88	1
Benzenethiol	C_6H_6S	2.52	1
11*H*-Benzo[*a*]fluorene	$C_{17}H_{12}$	5.40	1
11*H*-Benzo[*b*]fluorene	$C_{17}H_{12}$	5.75	1
Benzofuran	C_8H_6O	2.67	1
Benzoic acid	$C_7H_6O_2$	1.88	4
Benzonitrile	C_7H_5N	1.56	1
Benzo[*ghi*]perylene	$C_{22}H_{12}$	6.90	1
Benzophenone	$C_{13}H_{10}O$	3.18	1
Benzo[*a*]pyrene	$C_{20}H_{12}$	6.20	4
Benzo[*b*]thiophene	C_8H_6S	3.12	1
Benzyl acetate	$C_9H_{10}O_2$	1.96	1
Benzyl alcohol	C_7H_8O	1.05	1
Benzylamine	C_7H_9N	1.09	1
Benzyl benzoate	$C_{14}H_{12}O_2$	3.97	1
Benzyl methyl ether	$C_8H_{10}O$	1.35	1
Benzyl phenyl ether	$C_{13}H_{12}O$	3.79	1
Biphenyl	$C_{12}H_{10}$	3.76	6
Bis(2-chloroethyl) ether	$C_4H_8Cl_2O$	1.12	2
Bromobenzene	C_6H_5Br	2.99	2
2-Bromobenzoic acid	$C_7H_5BrO_2$	2.20	4
3-Bromobenzoic acid	$C_7H_5BrO_2$	2.87	4
4-Bromobenzoic acid	$C_7H_5BrO_2$	2.86	4
1-Bromobutane	C_4H_9Br	2.75	1
Bromochloromethane	CH_2BrCl	1.41	2
Bromocyclohexane	$C_6H_{11}Br$	3.20	1
Bromoethane	C_2H_5Br	1.6	2
1-Bromoheptane	$C_7H_{15}Br$	4.36	1
1-Bromohexane	$C_6H_{13}Br$	3.80	1
Bromomethane	CH_3Br	1.19	2
(Bromomethyl)benzene	C_7H_7Br	2.92	1
1-Bromooctane	$C_8H_{17}Br$	4.89	1
1-Bromopentane	$C_5H_{11}Br$	3.37	1
1-Bromopropane	C_3H_7Br	2.1	2
2-Bromopropane	C_3H_7Br	1.9	2
3-Bromopropene	C_3H_5Br	1.79	1
1,3-Butadiene	C_4H_6	1.99	1
Butanal	C_4H_8O	0.88	1
Butanamide	C_4H_9NO	-0.21	1
Butanenitrile	C_4H_7N	0.60	1
1-Butanethiol	$C_4H_{10}S$	2.28	1
Butanoic acid	$C_4H_8O_2$	0.79	1
1-Butanol	$C_4H_{10}O$	0.84	1
2-Butanol	$C_4H_{10}O$	0.65	1
2-Butanone	C_4H_8O	0.29	1
cis-2-Butene	C_4H_8	2.33	1
trans-2-Butene	C_4H_8	2.31	1

Name	Mol. form.	log *P*	Ref.
Butyl acetate	$C_6H_{12}O_2$	1.82	1
Butylamine	$C_4H_{11}N$	0.86	1
tert-Butylamine	$C_4H_{11}N$	0.40	1
Butylbenzene	$C_{10}H_{14}$	4.26	1
tert-Butylbenzene	$C_{10}H_{14}$	4.11	1
Butyl methacrylate	$C_8H_{14}O_2$	2.88	1
4-Butylphenol	$C_{10}H_{14}O$	3.65	1
2-Butyne	C_4H_6	1.46	1
Carbazole	$C_{12}H_9N$	3.72	1
Chlorobenzene	C_6H_5Cl	2.84	1
2-Chlorobiphenyl	$C_{12}H_9Cl$	4.52	1
3-Chlorobiphenyl	$C_{12}H_9Cl$	4.58	1
4-Chlorobiphenyl	$C_{12}H_9Cl$	4.61	1
1-Chlorobutane	C_4H_9Cl	2.64	2
Chloroethane	C_2H_5Cl	1.43	2
Chloroethene	C_2H_3Cl	1.38	2
1-Chloroheptane	$C_7H_{15}Cl$	4.15	1
Chloromethane	CH_3Cl	0.91	2
(Chloromethyl)benzene	C_7H_7Cl	2.30	1
1-Chloronaphthalene	$C_{10}H_7Cl$	3.90	1
2-Chloronaphthalene	$C_{10}H_7Cl$	3.98	1
1-Chloropropane	C_3H_7Cl	2.04	1
2-Chloropropane	C_3H_7Cl	1.90	1
2-Chlorotoluene	C_7H_7Cl	3.42	1
3-Chlorotoluene	C_7H_7Cl	3.28	1
4-Chlorotoluene	C_7H_7Cl	3.33	1
Chrysene	$C_{18}H_{12}$	5.73	4
trans-Cinnamic acid	$C_9H_8O_2$	2.13	1
Coronene	$C_{24}H_{12}$	6.05	4
o-Cresol	C_7H_8O	1.98	1
m-Cresol	C_7H_8O	1.98	1
p-Cresol	C_7H_8O	1.97	1
Cyclododecanone	$C_{12}H_{22}O$	4.10	1
1,3,5-Cycloheptatriene	C_7H_8	2.63	2
1,4-Cyclohexadiene	C_6H_8	2.3	2
Cyclohexane	C_6H_{12}	3.44	1
Cyclohexanol	$C_6H_{12}O$	1.23	1
Cyclohexanone	$C_6H_{10}O$	0.81	1
Cyclohexene	C_6H_{10}	2.86	1
2-Cyclohexen-1-one	C_6H_8O	0.61	1
Cyclohexylamine	$C_6H_{13}N$	1.49	1
Cyclooctane	C_8H_{16}	4.45	2
Cyclopentane	C_5H_{10}	3.00	1
Decachlorobiphenyl	$C_{12}Cl_{10}$	8.26	3
Decane	$C_{10}H_{22}$	6.25	1
Decanoic acid	$C_{10}H_{20}O_2$	4.09	1
1-Decanol	$C_{10}H_{22}O$	4.57	1
2-Decanone	$C_{10}H_{20}O$	3.77	1
Dibenzofuran	$C_{12}H_8O$	4.12	1
Dibromomethane	CH_2Br_2	2.3	2
Dibutyl ether	$C_8H_{18}O$	3.21	1
o-Dichlorobenzene	$C_6H_4Cl_2$	3.38	5
m-Dichlorobenzene	$C_6H_4Cl_2$	3.48	5
p-Dichlorobenzene	$C_6H_4Cl_2$	3.38	5
2,5-Dichlorobiphenyl	$C_{12}H_8Cl_2$	5.10	3
2,6-Dichlorobiphenyl	$C_{12}H_8Cl_2$	5.00	3
Dichlorodifluoromethane	CCl_2F_2	2.16	2
1,1-Dichloroethane	$C_2H_4Cl_2$	1.79	2
1,2-Dichloroethane	$C_2H_4Cl_2$	1.48	2
1,1-Dichloroethene	$C_2H_2Cl_2$	2.13	2
cis-1,2-Dichloroethene	$C_2H_2Cl_2$	1.86	2
trans-1,2-Dichloroethene	$C_2H_2Cl_2$	1.93	2
Dichloromethane	CH_2Cl_2	1.25	2
2,4-Dichlorophenol	$C_6H_4Cl_2O$	3.23	4
1,2-Dichloropropane, (±)-	$C_3H_6Cl_2$	2.0	2
cis-1,3-Dichloropropene	$C_3H_4Cl_2$	2.03	2
Diethylamine	$C_4H_{11}N$	0.58	1
Diethyl carbonate	$C_5H_{10}O_3$	1.21	1
Diethyl ether	$C_4H_{10}O$	0.89	1
Diethyl sulfide	$C_4H_{10}S$	1.95	1
Difluoromethane	CH_2F_2	0.20	1
2,3-Dihydrobenzofuran	C_8H_8O	2.14	1
2,5-Dihydrofuran	C_4H_6O	0.46	1
1,2-Dihydro-3-methylbenz[*j*]aceanthrylene	$C_{21}H_{16}$	6.75	1
Diiodomethane	CH_2I_2	2.5	2
Diisopropyl ether	$C_6H_{14}O$	1.52	1
N,N-Dimethylacetamide	C_4H_9NO	-0.77	1
Dimethylamine	C_2H_7N	-0.38	1
N,N-Dimethylaniline	$C_8H_{11}N$	2.31	1
9,10-Dimethylanthracene	$C_{16}H_{14}$	5.69	1
N,N-Dimethylbenzylamine	$C_9H_{13}N$	1.98	1
4,4'-Dimethylbiphenyl	$C_{14}H_{14}$	5.09	1
2,2-Dimethylbutane	C_6H_{14}	3.82	1
2,3-Dimethylbutane	C_6H_{14}	3.85	2
3,3-Dimethyl-2-butanol, (±)-	$C_6H_{14}O$	1.48	1
Dimethyl ether	C_2H_6O	0.10	1
N,N-Dimethylformamide	C_3H_7NO	-1.01	1
1,2-Dimethylnaphthalene	$C_{12}H_{12}$	4.31	1
1,4-Dimethylnaphthalene	$C_{12}H_{12}$	4.37	1
2,2-Dimethyl-1-propanol	$C_5H_{12}O$	1.31	1
Dimethyl sulfone	$C_2H_6O_2S$	-1.41	1
Dimethyl sulfoxide	C_2H_6OS	-1.35	1
Diphenylamine	$C_{12}H_{11}N$	3.44	4
1,2-Diphenylethane	$C_{14}H_{14}$	4.70	1
Diphenyl ether	$C_{12}H_{10}O$	4.21	1
Diphenylmethane	$C_{13}H_{12}$	4.14	1
Diphenylmethanol	$C_{13}H_{12}O$	2.67	1
Diphenyl sulfide	$C_{12}H_{10}S$	4.45	1
Dipropylamine	$C_6H_{15}N$	1.67	1
Dipropyl ether	$C_6H_{14}O$	2.03	1
Dodecanoic acid	$C_{12}H_{24}O_2$	4.6	1
1-Dodecanol	$C_{12}H_{26}O$	5.13	1
Eicosanoic acid	$C_{20}H_{40}O_2$	9.29	1
5,8,11,14-Eicosatetraenoic acid, (all-*cis*)	$C_{20}H_{32}O_2$	6.98	1
Epichlorohydrin	C_3H_5ClO	0.30	2
Ethanol	C_2H_6O	-0.30	1
Ethoxybenzene	$C_8H_{10}O$	2.51	1
Ethyl acetate	$C_4H_8O_2$	0.73	1
Ethyl acrylate	$C_5H_8O_2$	1.32	1
Ethylamine	C_2H_7N	-0.13	1
4-Ethylaniline	$C_8H_{11}N$	1.96	1
Ethylbenzene	C_8H_{10}	3.15	1
Ethyl benzoate	$C_9H_{10}O_2$	2.64	1
2-Ethylfuran	C_6H_8O	2.40	1
Ethyl methacrylate	$C_6H_{10}O_2$	1.94	1
Ethylmethylamine	C_3H_9N	0.15	1
1-Ethylnaphthalene	$C_{12}H_{12}$	4.40	1
2-Ethylphenol	$C_8H_{10}O$	2.47	1
3-Ethylphenol	$C_8H_{10}O$	2.50	1
4-Ethylphenol	$C_8H_{10}O$	2.50	1
Ethyl propanoate	$C_5H_{10}O_2$	1.21	1
2-Ethyltoluene	C_9H_{12}	3.53	1
4-Ethyltoluene	C_9H_{12}	3.63	2
Ethyl vinyl ether	C_4H_8O	1.04	1

Name	Mol. form.	log *P*	Ref.
Fluoranthene	$C_{16}H_{10}$	5.07	4
9*H*-Fluorene	$C_{13}H_{10}$	4.20	4
9*H*-Fluoren-9-one	$C_{13}H_8O$	3.58	1
Fluorobenzene	C_6H_5F	2.27	2
1-Fluorobutane	C_4H_9F	2.58	1
Fluoromethane	CH_3F	0.51	1
1-Fluoropentane	$C_5H_{11}F$	2.33	1
Formaldehyde	CH_2O	0.35	1
Formamide	CH_3NO	-1.51	1
Formic acid	CH_2O_2	-0.54	1
Furan	C_4H_4O	1.34	1
2,2',3,3',4,4',6-Heptachlorobiphenyl	$C_{12}H_3Cl_7$	6.70	3
Heptane	C_7H_{16}	4.50	1
1-Heptanol	$C_7H_{16}O$	2.62	1
2-Heptanol, (±)-	$C_7H_{16}O$	2.31	1
3-Heptanol, (*S*)-	$C_7H_{16}O$	2.24	1
4-Heptanol	$C_7H_{16}O$	2.22	1
2-Heptanone	$C_7H_{14}O$	1.98	1
1-Heptene	C_7H_{14}	3.99	1
Heptylamine	$C_7H_{17}N$	2.57	1
Hexachlorobenzene	C_6Cl_6	5.47	5
2,2',3,3',4,4'-Hexachlorobiphenyl	$C_{12}H_4Cl_6$	7.00	3
2,2',4,4',6,6'-Hexachlorobiphenyl	$C_{12}H_4Cl_6$	7.00	3
2,2',3,3',6,6'-Hexachlorobiphenyl	$C_{12}H_4Cl_6$	6.70	3
Hexachloroethane	C_2Cl_6	4.00	4
Hexadecanoic acid	$C_{16}H_{32}O_2$	7.17	1
1,5-Hexadiene	C_6H_{10}	2.8	2
Hexamethylbenzene	$C_{12}H_{18}$	4.69	4
Hexanal	$C_6H_{12}O$	1.78	1
Hexane	C_6H_{14}	4.00	1
Hexanenitrile	$C_6H_{11}N$	1.66	1
Hexanoic acid	$C_6H_{12}O_2$	1.92	1
1-Hexanol	$C_6H_{14}O$	2.03	1
2-Hexanol	$C_6H_{14}O$	1.76	1
3-Hexanol	$C_6H_{14}O$	1.65	1
2-Hexanone	$C_6H_{12}O$	1.38	1
1-Hexene	C_6H_{12}	3.40	1
5-Hexen-2-one	$C_6H_{10}O$	1.02	1
Hexylamine	$C_6H_{15}N$	2.06	1
Hexylbenzene	$C_{12}H_{18}$	5.52	1
1-Hexyne	C_6H_{10}	2.73	2
5-Hexyn-2-one	C_6H_8O	0.58	1
2-Hydroxybenzoic acid	$C_7H_6O_3$	2.20	4
Indan	C_9H_{10}	3.33	1
Indene	C_9H_8	2.92	1
1*H*-Indole	C_8H_7N	2.14	1
Iodobenzene	C_6H_5I	3.28	2
1-Iodobutane	C_4H_9I	3	2
Iodoethane	C_2H_5I	2	2
1-Iodoheptane	$C_7H_{15}I$	4.70	1
Iodomethane	CH_3I	1.5	2
1-Iodopropane	C_3H_7I	2.5	2
Isobutane	C_4H_{10}	2.8	2
Isobutene	C_4H_8	2.35	1
Isobutylbenzene	$C_{10}H_{14}$	4.01	2
Isopropylamine	C_3H_9N	0.26	1
Isopropylbenzene	C_9H_{12}	3.66	1
Isopropyl benzoate	$C_{10}H_{12}O_2$	3.18	1
1-Isopropyl-4-methylbenzene	$C_{10}H_{14}$	4.10	1
Isoquinoline	C_9H_7N	2.08	1
Methacrylic acid	$C_4H_6O_2$	0.93	1
Methanol	CH_4O	-0.74	1
N-Methylacetamide	C_3H_7NO	-1.05	1
Methyl acetate	$C_3H_6O_2$	0.18	1
4-Methylacetophenone	$C_9H_{10}O$	2.19	1
Methyl acrylate	$C_4H_6O_2$	0.80	1
Methylamine	CH_5N	-0.57	1
2-Methylaniline	C_7H_9N	1.32	1
3-Methylaniline	C_7H_9N	1.40	1
4-Methylaniline	C_7H_9N	1.39	1
N-Methylaniline	C_7H_9N	1.66	1
2-Methylanisole	$C_8H_{10}O$	2.74	1
3-Methylanisole	$C_8H_{10}O$	2.66	1
4-Methylanisole	$C_8H_{10}O$	2.81	1
2-Methylanthracene	$C_{15}H_{12}$	5.15	2
9-Methylanthracene	$C_{15}H_{12}$	5.07	1
2-Methylbenzaldehyde	C_8H_8O	2.26	1
α-Methylbenzeneacetic acid, (±)-	$C_9H_{10}O_2$	1.80	1
α-Methylbenzenemethanol	$C_8H_{10}O$	1.42	1
3-Methylbenzenemethanol	$C_8H_{10}O$	1.60	1
4-Methylbenzenemethanol	$C_8H_{10}O$	1.58	1
Methyl benzoate	$C_8H_8O_2$	2.20	1
4-Methylbiphenyl	$C_{13}H_{12}$	4.63	1
3-Methyl-1-butanol	$C_5H_{12}O$	1.28	1
2-Methyl-2-butanol	$C_5H_{12}O$	0.89	1
3-Methyl-2-butanol, (±)-	$C_5H_{12}O$	1.28	1
3-Methyl-2-butanone	$C_5H_{10}O$	0.56	1
Methyl *tert*-butyl ether	$C_5H_{12}O$	0.94	1
Methylcyclohexane	C_7H_{14}	3.88	1
Methylcyclopentane	C_6H_{12}	3.37	2
Methyl decanoate	$C_{11}H_{22}O_2$	4.41	1
1-Methyl-9*H*-fluorene	$C_{14}H_{12}$	4.97	1
2-Methylfuran	C_5H_6O	1.85	1
5-Methyl-2-hexanone	$C_7H_{14}O$	1.88	1
Methyl methacrylate	$C_5H_8O_2$	1.38	1
1-Methylnaphthalene	$C_{11}H_{10}$	3.87	1
2-Methylnaphthalene	$C_{11}H_{10}$	4.00	1
5-Methyl-2-octanone	$C_9H_{18}O$	2.92	1
Methyloxirane	C_3H_6O	0.03	1
3-Methylpentane	C_6H_{14}	3.60	2
4-Methyl-2-pentanone	$C_6H_{12}O$	1.31	1
4-Methyl-1-pentene	C_6H_{12}	2.5	2
1-Methylphenanthrene	$C_{15}H_{12}$	5.14	2
4-Methylphenyl acetate	$C_9H_{10}O_2$	2.11	1
2-Methyl-1-propanol	$C_4H_{10}O$	0.76	1
2-Methyl-2-propanol	$C_4H_{10}O$	0.35	1
2-Methylpyridine	C_6H_7N	1.11	1
3-Methylpyridine	C_6H_7N	1.20	1
4-Methylpyridine	C_6H_7N	1.22	1
1-Methylpyrrole	C_5H_7N	1.21	1
2-Methyltetrahydrofuran	$C_5H_{10}O$	1.85	2
Naphthacene	$C_{18}H_{12}$	5.76	1
Naphthalene	$C_{10}H_8$	3.34	4
1-Naphthol	$C_{10}H_8O$	2.84	1
2-Naphthol	$C_{10}H_8O$	2.70	1
Neopentane	C_5H_{12}	3.11	1
Nitrobenzene	$C_6H_5NO_2$	1.85	1
1-Nitrobutane	$C_4H_9NO_2$	1.47	1
Nitroethane	$C_2H_5NO_2$	0.18	1
Nitromethane	CH_3NO_2	-0.33	1
1-Nitropentane	$C_5H_{11}NO_2$	2.01	1
1-Nitropropane	$C_3H_7NO_2$	0.87	1
4-Nitrotoluene	$C_7H_7NO_2$	2.42	1
2,2',3,3',4,5,5',6,6'-Nonachlorobiphenyl	$C_{12}HCl_9$	8.16	3
Nonane	C_9H_{20}	5.65	1
1-Nonanol	$C_9H_{20}O$	4.02	1

Name	Mol. form.	log P	Ref.
2-Nonanone	$C_9H_{18}O$	3.16	1
1-Nonene	C_9H_{18}	5.15	1
2,2',3,3',5,5',6,6'-Octachlorobiphenyl	$C_{12}H_2Cl_8$	7.10	3
cis,cis-9,12-Octadecadienoic acid	$C_{18}H_{32}O_2$	7.05	1
cis,cis,cis-9,12,15-Octadecatrienoic acid	$C_{18}H_{30}O_2$	6.46	1
cis-9-Octadecenoic acid	$C_{18}H_{34}O_2$	7.64	1
Octane	C_8H_{18}	5.15	1
Octanenitrile	$C_8H_{15}N$	2.75	1
Octanoic acid	$C_8H_{16}O_2$	3.05	1
1-Octanol	$C_8H_{18}O$	3.07	1
2-Octanol	$C_8H_{18}O$	2.90	1
4-Octanol	$C_8H_{18}O$	2.68	1
2-Octanone	$C_8H_{16}O$	2.37	1
1-Octene	C_8H_{16}	4.57	1
Octylbenzene	$C_{14}H_{22}$	6.30	1
Oxirane	C_2H_4O	-0.30	1
Pentachlorobenzene	C_6HCl_5	5.03	5
2,3,4,5,6-Pentachlorobiphenyl	$C_{12}H_5Cl_5$	6.30	3
2,2',4,5,5'-Pentachlorobiphenyl	$C_{12}H_5Cl_5$	6.40	3
Pentachloroethane	C_2HCl_5	2.89	2
Pentachlorophenol	C_6HCl_5O	5.07	4
1,4-Pentadiene	C_5H_8	2.48	1
Pentamethylbenzene	$C_{11}H_{16}$	4.56	1
Pentane	C_5H_{12}	3.45	1
Pentanenitrile	C_5H_9N	0.94	1
Pentanoic acid	$C_5H_{10}O_2$	1.39	1
1-Pentanol	$C_5H_{12}O$	1.51	1
2-Pentanol	$C_5H_{12}O$	1.25	1
3-Pentanol	$C_5H_{12}O$	1.21	1
2-Pentanone	$C_5H_{10}O$	0.84	1
3-Pentanone	$C_5H_{10}O$	0.82	1
1-Pentene	C_5H_{10}	2.2	2
Pentylamine	$C_5H_{13}N$	1.49	1
Pentylbenzene	$C_{11}H_{16}$	4.90	1
1-Pentyne	C_5H_8	1.98	1
Perylene	$C_{20}H_{12}$	6.25	1
Phenanthrene	$C_{14}H_{10}$	4.52	4
Phenol	C_6H_6O	1.48	4
Phenyl acetate	$C_8H_8O_2$	1.49	1
2-Phenylacetophenone	$C_{14}H_{12}O$	3.18	1
Phenylacetylene	C_8H_6	2.40	1
N-Phenylbenzamide	$C_{13}H_{11}NO$	2.62	1
Phenyl benzoate	$C_{13}H_{10}O_2$	3.59	1
4-Phenylcyclohexanone	$C_{12}H_{14}O$	2.45	1
Phenyl formate	$C_7H_6O_2$	1.26	1
Phenyloxirane	C_8H_8O	1.61	1
1-Phenyl-1-propanone	$C_9H_{10}O$	2.19	1
1-Phenyl-2-propanone	$C_9H_{10}O$	1.44	1
1-Phenyl-2-propylamine, (±)-	$C_9H_{13}N$	1.76	1
4-Phenylpyridine	$C_{11}H_9N$	2.59	1
Piperidine	$C_5H_{11}N$	0.84	1
Propanal	C_3H_6O	0.59	1
Propanenitrile	C_3H_5N	0.16	1
1-Propanethiol	C_3H_8S	1.81	1
Propanoic acid	$C_3H_6O_2$	0.33	1
1-Propanol	C_3H_8O	0.25	1
2-Propanol	C_3H_8O	0.05	1
Propargyl alcohol	C_3H_4O	-0.38	1
Propyl acetate	$C_5H_{10}O_2$	1.24	1
Propylamine	C_3H_9N	0.48	1
Propylbenzene	C_9H_{12}	3.69	1
Propyl formate	$C_4H_8O_2$	0.83	1
2-Propylphenol	$C_9H_{12}O$	2.93	1
4-Propylphenol	$C_9H_{12}O$	3.20	1
Pyrene	$C_{16}H_{10}$	5.08	4
Pyridine	C_5H_5N	0.65	1
Pyrrole	C_4H_5N	0.75	1
Pyrrolidine	C_4H_9N	0.46	1
Quinoline	C_9H_7N	2.03	1
Stearic acid	$C_{18}H_{36}O_2$	8.23	1
trans-Stilbene	$C_{14}H_{12}$	4.81	1
Styrene	C_8H_8	3.05	1
1,2,3,4-Tetrachlorobenzene	$C_6H_2Cl_4$	4.55	5
1,2,3,5-Tetrachlorobenzene	$C_6H_2Cl_4$	4.65	5
1,2,4,5-Tetrachlorobenzene	$C_6H_2Cl_4$	4.51	5
2,2',4',5-Tetrachlorobiphenyl	$C_{12}H_6Cl_4$	5.73	7
2,3,4,5-Tetrachlorobiphenyl	$C_{12}H_6Cl_4$	5.72	3
1,1,2,2-Tetrachloroethane	$C_2H_2Cl_4$	2.39	2
Tetrachloroethene	C_2Cl_4	2.88	2
Tetrachloromethane	CCl_4	2.64	2
Tetradecanoic acid	$C_{14}H_{28}O_2$	6.1	1
Tetrahydrofuran	C_4H_8O	0.46	1
Tetrahydropyran	$C_5H_{10}O$	0.82	1
1,2,3,4-Tetramethylbenzene	$C_{10}H_{14}$	4.00	1
1,2,3,5-Tetramethylbenzene	$C_{10}H_{14}$	4.10	1
1,2,4,5-Tetramethylbenzene	$C_{10}H_{14}$	4.10	2
Thiophene	C_4H_4S	1.81	1
Toluene	C_7H_8	2.73	1
o-Toluic acid	$C_8H_8O_2$	2.32	4
m-Toluic acid	$C_8H_8O_2$	2.37	1
p-Toluic acid	$C_8H_8O_2$	2.34	1
Tribromomethane	$CHBr_3$	2.38	2
1,2,3-Trichlorobenzene	$C_6H_3Cl_3$	4.04	5
1,2,4-Trichlorobenzene	$C_6H_3Cl_3$	3.98	5
1,3,5-Trichlorobenzene	$C_6H_3Cl_3$	4.02	5
2,4,5-Trichlorobiphenyl	$C_{12}H_7Cl_3$	5.60	3
2,4,6-Trichlorobiphenyl	$C_{12}H_7Cl_3$	5.47	3
1,1,1-Trichloroethane	$C_2H_3Cl_3$	2.49	2
1,1,2-Trichloroethane	$C_2H_3Cl_3$	2.38	2
Trichloroethene	C_2HCl_3	2.53	2
Trichlorofluoromethane	CCl_3F	2.53	2
Trichloromethane	$CHCl_3$	1.97	2
1,2,3-Trichloropropane	$C_3H_5Cl_3$	2.63	2
1,1,2-Trichloro-1,2,2-trifluoroethane	$C_2Cl_3F_3$	3.16	2
Triethylamine	$C_6H_{15}N$	1.45	1
Trimethylamine	C_3H_9N	0.16	1
1,2,3-Trimethylbenzene	C_9H_{12}	3.60	1
1,2,4-Trimethylbenzene	C_9H_{12}	3.63	1
1,3,5-Trimethylbenzene	C_9H_{12}	3.42	1
2,3,6-Trimethylphenol	$C_9H_{12}O$	2.67	1
2,4,6-Trimethylphenol	$C_9H_{12}O$	2.46	1
Triphenylamine	$C_{18}H_{15}N$	5.74	1
Triphenylene	$C_{18}H_{12}$	5.49	4
Triphenylmethanol	$C_{19}H_{16}O$	3.68	1
Tripropylamine	$C_9H_{21}N$	2.79	1
2-Undecanone	$C_{11}H_{22}O$	4.09	1
Vinyl acetate	$C_4H_6O_2$	0.73	1
o-Xylene	C_8H_{10}	3.12	1
m-Xylene	C_8H_{10}	3.20	1
p-Xylene	C_8H_{10}	3.15	1
2,4-Xylenol	$C_8H_{10}O$	2.35	1
2,5-Xylenol	$C_8H_{10}O$	2.34	1
2,6-Xylenol	$C_8H_{10}O$	2.36	1
3,4-Xylenol	$C_8H_{10}O$	3.23	1
3,5-Xylenol	$C_8H_{10}O$	2.35	1

SOLUBILITY PRODUCT CONSTANTS OF INORGANIC SALTS

The solubility product constant K_{sp} is a useful parameter for calculating the aqueous solubility of sparingly soluble compounds under various conditions. It may be determined by direct measurement or calculated from the standard Gibbs energies of formation $\Delta_f G°$ of the species involved at their standard states. Thus if $K_{sp} = [M^+]^m \cdot [A^-]^n$ is the equilibrium constant for the reaction

$$M_mA_n(s) \rightleftharpoons mM^+(aq) + nA^-(aq),$$

where M_mA_n is the slightly soluble substance and M^+ and A^- are the ions produced in solution by the dissociation of M_mA_n, then the Gibbs energy change is

$$\Delta G° = m\,\Delta_f G°(M^+,aq) + n\,\Delta_f G°(A^-,aq) - \Delta_f G°(M_mA_n, s)$$

The solubility product constant is calculated from the equation

$$\ln K_{sp} = -\Delta G°/RT$$

The table gives selected values of K_{sp} at 25 °C. Many of these have been calculated from standard state thermodynamic data in Refs. 1 and 2; other values are taken from publications of the IUPAC Solubility Data Project (Refs. 3 to 7).

The above formulation is not convenient for treating sulfides because the S^{-2} ion is usually not present in significant concentrations (see Ref. 8). This is due to the hydrolysis reaction

$$S^{-2} + H_2O \rightleftharpoons HS^- + OH^-$$

which is strongly shifted to the right except in very basic solutions. Furthermore, the equilibrium constant for this reaction, which depends on the second ionization constant of H_2S, is poorly known. Therefore it is more useful in the case of sulfides to define a different solubility product K_{spa} based on the reaction

$$M_mS_n(s) + 2H^+ \rightleftharpoons mM^+ + nH_2S(aq)$$

Values of K_{spa} taken from Ref. 8 are given for several sulfides in the last column. Additional discussion of sulfide equilibria may be found in Refs. 7 and 9. Compounds are listed alphabetically by name.

References

1. Wagman, D. D., Evans, W. H., Parker, V. B., Schumm, R. H., Halow, I., Bailey, S. M., Churney, K. L., and Nuttall, R. L., *The NBS Tables of Chemical Thermodynamic Properties, J. Phys. Chem. Ref. Data*, Vol. 11, Suppl. 2, 1982.
2. Garvin, D., Parker, V. B., and White, H. J., *CODATA Thermodynamic Tables*, Hemisphere, New York, 1987.
3. *Solubility Data Series* (53 Volumes), International Union of Pure and Applied Chemistry, Pergamon Press, Oxford, 1979–1992.
4. Clever, H. L., and Johnston, F. J., *J. Phys. Chem. Ref. Data*, 9, 751, 1980.
5. Marcus, Y., *J. Phys. Chem. Ref. Data*, 9, 1307, 1980.
6. Clever, H. L., Johnson, S. A., and Derrick, M. E., *J. Phys. Chem. Ref. Data*, 14, 631, 1985.
7. Clever, H. L., Johnson, S. A., and Derrick, M. E., *J. Phys. Chem. Ref. Data*, 21, 941, 1992.
8. Myers, R. J., *J. Chem. Educ.*, 63, 687, 1986.
9. Licht, S., *J. Electrochem. Soc.*, 135, 2971, 1988.

Solubility Product Constants of Inorganic Salts

Name	Formula	K_{sp}	K_{spa}
Aluminum phosphate	$AlPO_4$	$9.84 \cdot 10^{-21}$	
Barium bromate	$Ba(BrO_3)_2$	$2.43 \cdot 10^{-4}$	
Barium carbonate	$BaCO_3$	$2.58 \cdot 10^{-9}$	
Barium chromate(VI)	$BaCrO_4$	$1.17 \cdot 10^{-10}$	
Barium fluoride	BaF_2	$1.84 \cdot 10^{-7}$	
Barium hydroxide octahydrate	$Ba(OH)_2 \cdot 8H_2O$	$2.55 \cdot 10^{-4}$	
Barium iodate	$Ba(IO_3)_2$	$4.01 \cdot 10^{-9}$	
Barium iodate monohydrate	$Ba(IO_3)_2 \cdot H_2O$	$1.67 \cdot 10^{-9}$	
Barium molybdate	$BaMoO_4$	$3.54 \cdot 10^{-8}$	
Barium selenate	$BaSeO_4$	$3.40 \cdot 10^{-8}$	
Barium sulfate	$BaSO_4$	$1.08 \cdot 10^{-10}$	
Barium sulfite	$BaSO_3$	$5.0 \cdot 10^{-10}$	
Beryllium hydroxide (α)	$Be(OH)_2$	$6.92 \cdot 10^{-22}$	
Bismuth arsenate	$BiAsO_4$	$4.43 \cdot 10^{-10}$	
Bismuth triiodide	BiI_3	$7.71 \cdot 10^{-19}$	
Cadmium arsenate	$Cd_3(AsO_4)_2$	$2.2 \cdot 10^{-33}$	
Cadmium carbonate	$CdCO_3$	$1.0 \cdot 10^{-12}$	
Cadmium fluoride	CdF_2	$6.44 \cdot 10^{-3}$	
Cadmium hydroxide	$Cd(OH)_2$	$7.2 \cdot 10^{-15}$	
Cadmium iodate	$Cd(IO_3)_2$	$2.5 \cdot 10^{-8}$	
Cadmium oxalate trihydrate	$CdC_2O_4 \cdot 3H_2O$	$1.42 \cdot 10^{-8}$	
Cadmium phosphate	$Cd_3(PO_4)_2$	$2.53 \cdot 10^{-33}$	
Cadmium sulfide	CdS		$8 \cdot 10^{-7}$
Calcium carbonate (calcite)	$CaCO_3$	$3.36 \cdot 10^{-9}$	
Calcium fluoride	CaF_2	$3.45 \cdot 10^{-11}$	
Calcium hydroxide	$Ca(OH)_2$	$5.02 \cdot 10^{-6}$	
Calcium iodate	$Ca(IO_3)_2$	$6.47 \cdot 10^{-6}$	
Calcium iodate hexahydrate	$Ca(IO_3)_2 \cdot 6H_2O$	$7.10 \cdot 10^{-7}$	
Calcium molybdate	$CaMoO_4$	$1.46 \cdot 10^{-8}$	
Calcium oxalate monohydrate	$CaC_2O_4 \cdot H_2O$	$2.32 \cdot 10^{-9}$	
Calcium phosphate	$Ca_3(PO_4)_2$	$2.07 \cdot 10^{-33}$	
Calcium sulfate	$CaSO_4$	$4.93 \cdot 10^{-5}$	
Calcium sulfate dihydrate	$CaSO_4 \cdot 2H_2O$	$3.14 \cdot 10^{-5}$	
Calcium sulfite hemihydrate	$CaSO_3 \cdot 0.5H_2O$	$3.1 \cdot 10^{-7}$	
Cesium perchlorate	$CsClO_4$	$3.95 \cdot 10^{-3}$	
Cesium periodate	$CsIO_4$	$5.16 \cdot 10^{-6}$	
Cobalt(II) arsenate	$Co_3(AsO_4)_2$	$6.80 \cdot 10^{-29}$	
Cobalt(II) hydroxide	$Co(OH)_2$	$5.92 \cdot 10^{-15}$	
Cobalt(II) iodate dihydrate	$Co(IO_3)_2 \cdot 2H_2O$	$1.21 \cdot 10^{-2}$	
Cobalt(II) phosphate	$Co_3(PO_4)_2$	$2.05 \cdot 10^{-35}$	
Copper(II) arsenate	$Cu_3(AsO_4)_2$	$7.95 \cdot 10^{-36}$	
Copper(I) bromide	$CuBr$	$6.27 \cdot 10^{-9}$	
Copper(I) chloride	$CuCl$	$1.72 \cdot 10^{-7}$	
Copper(I) cyanide	$CuCN$	$3.47 \cdot 10^{-20}$	
Copper(II) iodate monohydrate	$Cu(IO_3)_2 \cdot H_2O$	$6.94 \cdot 10^{-8}$	
Copper(I) iodide	CuI	$1.27 \cdot 10^{-12}$	
Copper(II) oxalate	CuC_2O_4	$4.43 \cdot 10^{-10}$	
Copper(II) phosphate	$Cu_3(PO_4)_2$	$1.40 \cdot 10^{-37}$	
Copper(II) sulfide	CuS		$6 \cdot 10^{-16}$
Copper(I) thiocyanate	$CuSCN$	$1.77 \cdot 10^{-13}$	

Name	Formula	K_{sp}	K_{spa}
Europium(III) hydroxide	$Eu(OH)_3$	$9.38 \cdot 10^{-27}$	
Gallium(III) hydroxide	$Ga(OH)_3$	$7.28 \cdot 10^{-36}$	
Iron(II) carbonate	$FeCO_3$	$3.13 \cdot 10^{-11}$	
Iron(II) fluoride	FeF_2	$2.36 \cdot 10^{-6}$	
Iron(II) hydroxide	$Fe(OH)_2$	$4.87 \cdot 10^{-17}$	
Iron(III) hydroxide	$Fe(OH)_3$	$2.79 \cdot 10^{-39}$	
Iron(III) phosphate dihydrate	$FePO_4 \cdot 2H_2O$	$9.91 \cdot 10^{-16}$	
Iron(II) sulfide	FeS		$6 \cdot 10^{2}$
Lanthanum iodate	$La(IO_3)_3$	$7.50 \cdot 10^{-12}$	
Lead(II) bromide	$PbBr_2$	$6.60 \cdot 10^{-6}$	
Lead(II) carbonate	$PbCO_3$	$7.40 \cdot 10^{-14}$	
Lead(II) chloride	$PbCl_2$	$1.70 \cdot 10^{-5}$	
Lead(II) fluoride	PbF_2	$3.3 \cdot 10^{-8}$	
Lead(II) hydroxide	$Pb(OH)_2$	$1.43 \cdot 10^{-20}$	
Lead(II) iodate	$Pb(IO_3)_2$	$3.69 \cdot 10^{-13}$	
Lead(II) iodide	PbI_2	$9.8 \cdot 10^{-9}$	
Lead(II) selenate	$PbSeO_4$	$1.37 \cdot 10^{-7}$	
Lead(II) sulfate	$PbSO_4$	$2.53 \cdot 10^{-8}$	
Lead(II) sulfide	PbS		$3 \cdot 10^{-7}$
Lithium carbonate	Li_2CO_3	$8.15 \cdot 10^{-4}$	
Lithium fluoride	LiF	$1.84 \cdot 10^{-3}$	
Lithium phosphate	Li_3PO_4	$2.37 \cdot 10^{-11}$	
Magnesium carbonate	$MgCO_3$	$6.82 \cdot 10^{-6}$	
Magnesium carbonate pentahydrate	$MgCO_3 \cdot 5H_2O$	$3.79 \cdot 10^{-6}$	
Magnesium carbonate trihydrate	$MgCO_3 \cdot 3H_2O$	$2.38 \cdot 10^{-6}$	
Magnesium fluoride	MgF_2	$5.16 \cdot 10^{-11}$	
Magnesium hydroxide	$Mg(OH)_2$	$5.61 \cdot 10^{-12}$	
Magnesium oxalate dihydrate	$MgC_2O_4 \cdot 2H_2O$	$4.83 \cdot 10^{-6}$	
Magnesium phosphate	$Mg_3(PO_4)_2$	$1.04 \cdot 10^{-24}$	
Manganese(II) carbonate	$MnCO_3$	$2.24 \cdot 10^{-11}$	
Manganese(II) iodate	$Mn(IO_3)_2$	$4.37 \cdot 10^{-7}$	
Manganese(II) oxalate dihydrate	$MnC_2O_4 \cdot 2H_2O$	$1.70 \cdot 10^{-7}$	
Manganese(II) sulfide (α form)	MnS		$3 \cdot 10^{7}$
Mercury(I) bromide	Hg_2Br_2	$6.40 \cdot 10^{-23}$	
Mercury(II) bromide	$HgBr_2$	$6.2 \cdot 10^{-20}$	
Mercury(I) carbonate	Hg_2CO_3	$3.6 \cdot 10^{-17}$	
Mercury(I) chloride	Hg_2Cl_2	$1.43 \cdot 10^{-18}$	
Mercury(I) fluoride	Hg_2F_2	$3.10 \cdot 10^{-6}$	
Mercury(I) iodide	Hg_2I_2	$5.2 \cdot 10^{-29}$	
Mercury(II) iodide (red)	HgI_2	$2.9 \cdot 10^{-29}$	
Mercury(I) oxalate	$Hg_2C_2O_4$	$1.75 \cdot 10^{-13}$	
Mercury(I) sulfate	Hg_2SO_4	$6.5 \cdot 10^{-7}$	
Mercury(II) sulfide (red)	HgS		$4 \cdot 10^{-33}$
Mercury(II) sulfide (black)	HgS		$2 \cdot 10^{-32}$
Mercury(I) thiocyanate	$Hg_2(SCN)_2$	$3.2 \cdot 10^{-20}$	
Neodymium carbonate	$Nd_2(CO_3)_3$	$1.08 \cdot 10^{-33}$	
Nickel(II) carbonate	$NiCO_3$	$1.42 \cdot 10^{-7}$	
Nickel(II) hydroxide	$Ni(OH)_2$	$5.48 \cdot 10^{-16}$	
Nickel(II) iodate	$Ni(IO_3)_2$	$4.71 \cdot 10^{-5}$	
Nickel(II) phosphate	$Ni_3(PO_4)_2$	$4.74 \cdot 10^{-32}$	
Palladium(II) thiocyanate	$Pd(SCN)_2$	$4.39 \cdot 10^{-23}$	
Potassium hexachloroplatinate	K_2PtCl_6	$7.48 \cdot 10^{-6}$	
Potassium perchlorate	$KClO_4$	$1.05 \cdot 10^{-2}$	
Potassium periodate	KIO_4	$3.71 \cdot 10^{-4}$	
Praseodymium(III) hydroxide	$Pr(OH)_3$	$3.39 \cdot 10^{-24}$	
Radium iodate	$Ra(IO_3)_2$	$1.16 \cdot 10^{-9}$	
Radium sulfate	$RaSO_4$	$3.66 \cdot 10^{-11}$	
Rubidium perchlorate	$RbClO_4$	$3.00 \cdot 10^{-3}$	
Scandium fluoride	ScF_3	$5.81 \cdot 10^{-24}$	
Scandium hydroxide	$Sc(OH)_3$	$2.22 \cdot 10^{-31}$	
Silver(I) acetate	$AgC_2H_3O_2$	$1.94 \cdot 10^{-3}$	
Silver(I) arsenate	Ag_3AsO_4	$1.03 \cdot 10^{-22}$	
Silver(I) bromate	$AgBrO_3$	$5.38 \cdot 10^{-5}$	
Silver(I) bromide	$AgBr$	$5.35 \cdot 10^{-13}$	
Silver(I) carbonate	Ag_2CO_3	$8.46 \cdot 10^{-12}$	
Silver(I) chloride	$AgCl$	$1.77 \cdot 10^{-10}$	
Silver(I) chromate	Ag_2CrO_4	$1.12 \cdot 10^{-12}$	
Silver(I) cyanide	$AgCN$	$5.97 \cdot 10^{-17}$	
Silver(I) iodate	$AgIO_3$	$3.17 \cdot 10^{-8}$	
Silver(I) iodide	AgI	$8.52 \cdot 10^{-17}$	
Silver(I) oxalate	$Ag_2C_2O_4$	$5.40 \cdot 10^{-12}$	
Silver(I) phosphate	Ag_3PO_4	$8.89 \cdot 10^{-17}$	
Silver(I) sulfate	Ag_2SO_4	$1.20 \cdot 10^{-5}$	
Silver(I) sulfide	Ag_2S		$6 \cdot 10^{-30}$
Silver(I) sulfite	Ag_2SO_3	$1.50 \cdot 10^{-14}$	
Silver(I) thiocyanate	$AgSCN$	$1.03 \cdot 10^{-12}$	
Strontium arsenate	$Sr_3(AsO_4)_2$	$4.29 \cdot 10^{-19}$	
Strontium carbonate	$SrCO_3$	$5.60 \cdot 10^{-10}$	
Strontium fluoride	SrF_2	$4.33 \cdot 10^{-9}$	
Strontium iodate	$Sr(IO_3)_2$	$1.14 \cdot 10^{-7}$	
Strontium iodate hexahydrate	$Sr(IO_3)_2 \cdot 6H_2O$	$4.55 \cdot 10^{-7}$	
Strontium iodate monohydrate	$Sr(IO_3)_2 \cdot H_2O$	$3.77 \cdot 10^{-7}$	
Strontium sulfate	$SrSO_4$	$3.44 \cdot 10^{-7}$	
Thallium(I) bromate	$TlBrO_3$	$1.10 \cdot 10^{-4}$	
Thallium(I) bromide	$TlBr$	$3.71 \cdot 10^{-6}$	
Thallium(I) chloride	$TlCl$	$1.86 \cdot 10^{-4}$	
Thallium(I) chromate	Tl_2CrO_4	$8.67 \cdot 10^{-13}$	
Thallium(III) hydroxide	$Tl(OH)_3$	$1.68 \cdot 10^{-44}$	
Thallium(I) iodate	$TlIO_3$	$3.12 \cdot 10^{-6}$	
Thallium(I) iodide	TlI	$5.54 \cdot 10^{-8}$	
Thallium(I) thiocyanate	$TlSCN$	$1.57 \cdot 10^{-4}$	
Tin(II) hydroxide	$Sn(OH)_2$	$5.45 \cdot 10^{-27}$	
Tin(II) sulfide	SnS		$1 \cdot 10^{-5}$
Yttrium carbonate	$Y_2(CO_3)_3$	$1.03 \cdot 10^{-31}$	
Yttrium fluoride	YF_3	$8.62 \cdot 10^{-21}$	
Yttrium hydroxide	$Y(OH)_3$	$1.00 \cdot 10^{-22}$	
Yttrium iodate	$Y(IO_3)_3$	$1.12 \cdot 10^{-10}$	
Zinc arsenate	$Zn_3(AsO_4)_2$	$2.8 \cdot 10^{-28}$	
Zinc carbonate	$ZnCO_3$	$1.46 \cdot 10^{-10}$	
Zinc carbonate monohydrate	$ZnCO_3 \cdot H_2O$	$5.42 \cdot 10^{-11}$	
Zinc fluoride	ZnF_2	$3.04 \cdot 10^{-2}$	
Zinc hydroxide	$Zn(OH)_2$	$3 \cdot 10^{-17}$	
Zinc iodate dihydrate	$Zn(IO_3)_2 \cdot 2H_2O$	$4.1 \cdot 10^{-6}$	
Zinc oxalate dihydrate	$ZnC_2O_4 \cdot 2H_2O$	$1.38 \cdot 10^{-9}$	
Zinc selenide	$ZnSe$	$3.6 \cdot 10^{-26}$	
Zinc selenite monohydrate	$ZnSeO_3 \cdot H_2O$	$1.59 \cdot 10^{-7}$	
Zinc sulfide (sphalerite)	ZnS		$2 \cdot 10^{-4}$
Zinc sulfide (wurtzite)	ZnS		$3 \cdot 10^{-2}$

SOLUBILITY OF COMMON SALTS AT AMBIENT TEMPERATURES

This table gives the aqueous solubility of selected salts at temperatures from 10 °C to 40 °C. Values are given in molality terms.

References

1. Apelblat, A., *J. Chem. Thermodynamics*, 24, 619, 1992.
2. Apelblat, A., *J. Chem. Thermodynamics*, 25, 63, 1993.
3. Apelblat, A., *J. Chem. Thermodynamics*, 25, 1513, 1993.
4. Apelblat, A. and Korin, E., *J. Chem. Thermodynamics*, 30, 59, 1998.

Solubility of Common Salts (in mol/kg) at the Indicated Temperature

Salt	Formula	Mol. wt.	10 °C	15 °C	20 °C	25 °C	30 °C	35 °C	40 °C	Ref.
Ammonium chloride	NH_4Cl	53.492	6.199	6.566	6.943	7.331				2
Ammonium nitrate	NH_4NO_3	80.043	18.809	21.163	23.721	26.496				2
Ammonium sulfate	$(NH_4)_2SO_4$	132.140	5.494	5.589	5.688	5.790	5.896	6.005		3
Barium chloride	$BaCl_2$	208.233	1.603	1.659	1.716	1.774	1.834	1.895	1.958	1
Calcium nitrate	$Ca(NO_3)_2$	164.087	6.896	7.398	7.986	8.675	9.480	10.421		1
Copper(II) sulfate	$CuSO_4$	159.609	1.055	1.153	1.260	1.376	1.502	1.639		3
Iron(II) sulfate	$FeSO_4$	151.908	1.352	1.533	1.729	1.940	2.165	2.405		3
Lithium chloride	LiCl	42.394	19.296	19.456	19.670	19.935				2
Magnesium nitrate	$Mg(NO_3)_2$	148.314	4.403	4.523	4.656	4.800	4.958	5.130	5.314	1
Manganese(II) chloride	$MnCl_2$	125.844	5.421	5.644	5.884	6.143	6.422	6.721		3
Potassium bromide	KBr	119.002	5.002	5.237	5.471	5.703	5.932	6.157		3
Potassium carbonate	K_2CO_3	138.206	7.756	7.846	7.948	8.063	8.191	8.331	8.483	1
Potassium iodate	KIO_3	214.001	0.291	0.333	0.378	0.426	0.478	0.534	0.593	4
Rubidium chloride	RbCl	120.921	6.911	7.180	7.449	7.717	7.986	8.253	8.520	4
Sodium bromide	NaBr	102.894	8.258	8.546	8.856	9.191	9.550	9.937	10.351	4
Sodium chloride	NaCl	58.443	6.110	6.121	6.136	6.153	6.174	6.197	6.222	4
Sodium nitrate	$NaNO_3$	84.995	9.395	9.819	10.261	10.723	11.204	11.706	12.230	4
Sodium nitrite	$NaNO_2$	68.996	11.111	11.484	11.883	12.310	12.766	13.253	13.772	4
Zinc sulfate	$ZnSO_4$	161.472	2.911	3.116	3.336	3.573	3.827	4.099	4.194	1

SOLUBILITY OF HYDROCARBONS IN SEAWATER

Concern about pollution of the oceans has stimulated measurements of the solubility of organic compounds in seawater. This table gives the solubility of several hydrocarbons in seawater. The data are derived from a review in the IUPAC Solubility Data Series (Ref. 1).

Solubility is expressed in this table as parts per million by mass, i.e.,

$$S/\text{ppm(mass)} = 10^6 \times w_2 = 10^6 \times m_2/(m_1 + m_2)$$

where m_1 and m_2 are the masses of solvent (seawater) and solute, respectively, under saturation conditions, and w_2 is the mass fraction. Because the solubilities in this table are very low, the value of S is effectively the mass of hydrocarbon in grams per 1000 kg of seawater.

The temperature and salinity of each measurement are given in the table. Salinity is a standardized measure of the concentration of dissolved salts, as explained in the table "Properties of Seawater" in Section 14. Salinity values in the open oceans at mid-latitude typically fall between 34 and 36.

Ref. 1 gives details of the method of measurement and an indication of the reliability of the measurements.

Reference

1. Shaw, David G., and Maczynski, A., IUPAC-NIST Solubility Data Series 81. Hydrocarbons with Water and Seawater — Revised and Updated. Part 12. C_5-C_{26} Hydrocarbons with Seawater, *J. Phys. Chem. Ref. Data* 35, 785, 2006.

Name	Mol. form.	Salinity	t/°C	S/ppm (mass)
Acenaphthene	$C_{12}H_{10}$	35	15	0.21
Acenaphthene	$C_{12}H_{10}$	35	25	1.8
Anthracene	$C_{14}H_{10}$	35	25	0.031
Benz[a]anthracene	$C_{18}H_{12}$	35	25	0.0056
Benzene	C_6H_6	34.4	0	1320
Benzene	C_6H_6	35	25	1360
Benzo[ghi]perylene	$C_{22}H_{12}$	6	25	0.00021
Benzo[a]pyrene	$C_{20}H_{12}$	6	25	0.00013
Benzo[e]pyrene	$C_{20}H_{12}$	30	25	0.0033
Benzo[b]triphenylene	$C_{22}H_{14}$	6	25	0.027
Biphenyl	$C_{12}H_{10}$	35	25	4.76
Butylbenzene	$C_{10}H_{14}$	34.5	25	7.1
sec-Butylbenzene	$C_{10}H_{14}$	34.5	25	12
tert-Butylbenzene	$C_{10}H_{14}$	34.5	25	21
Chrysene	$C_{18}H_{12}$	35	25	0.0011
Dibenz[a,h]anthracene	$C_{22}H_{14}$	6	25	0.021
Dibenz[a,j]anthracene	$C_{22}H_{14}$	6	25	0.010
Dodecane	$C_{12}H_{26}$	35	25	0.0029
Eicosane	$C_{20}H_{42}$	35	25	0.0008
Ethylbenzene	C_8H_{10}	34.4	0	140
Ethylbenzene	C_8H_{10}	34.4	10	129
Ethylbenzene	C_8H_{10}	34.4	25	111
Fluoranthene	$C_{16}H_{10}$	35	25	0.124
9H-Fluorene	$C_{13}H_{10}$	35	25	1.2
Heptane	C_7H_{16}	6	25	10.3
Hexacosane	$C_{26}H_{54}$	35	25	0.0001
Hexadecane	$C_{16}H_{34}$	35	25	0.0004
Hexane	C_6H_{14}	35.3	25	7.9
Isopropylbenzene	C_9H_{12}	34.5	25	43
2-Methylanthracene	$C_{15}H_{12}$	35	25	0.013
Methylcyclopentane	C_6H_{12}	34.5	25	29
1-Methylnaphthalene	$C_{11}H_{10}$	30	25	23
1-Methylphenanthrene	$C_{15}H_{12}$	35	25	0.20
Naphthalene	$C_{10}H_8$	35	25	22.8
Nonane	C_9H_{20}	6	25	0.43
Octadecane	$C_{18}H_{38}$	35	25	0.0008
Pentane	C_5H_{12}	34.5	25	28
Phenanthrene	$C_{14}H_{10}$	34	25	0.69
Pyrene	$C_{16}H_{10}$	35	25	0.086
Tetradecane	$C_{14}H_{30}$	35	25	0.0017
Toluene	C_7H_8	34.4	0	450
Toluene	C_7H_8	35	25	387
1,2,3-Trimethylbenzene	C_9H_{12}	34.5	25	49
1,2,4-Trimethylbenzene	C_9H_{12}	34.5	25	40
1,3,5-Trimethylbenzene	C_9H_{12}	34.5	25	31
Undecane	$C_{11}H_{24}$	6	25	0.01
o-Xylene	C_8H_{10}	34.5	25	130
m-Xylene	C_8H_{10}	34.5	25	106
p-Xylene	C_8H_{10}	34.5	25	111

SOLUBILITY OF ORGANIC COMPOUNDS IN PRESSURIZED HOT WATER

Liquid water at elevated temperatures and pressures, but still in the subcritical region, is of interest as a solvent in various laboratory and industrial processes. In effect, this means water at a temperature between about 100 °C and 373 °C, the critical temperature, and at pressures up to 400 bar or greater. Because the dielectric constant of water decreases with increasing temperature, the solubility of many compounds, especially non-polar compounds, increases dramatically at higher temperature. The fact that solubility can be fine-tuned by controlling temperature and pressure makes pressurized hot water a useful tool in various extraction and reaction processes.

This table gives a sample of the variations of solubility with temperature and pressure for several compounds, mostly hydrocarbons. The solubility S is expressed in both mole fraction of solute, x_2, and mass percent, $100w_2$, where w_2 is the mass fraction. More information is available in the references. Compounds are listed alphabetically by name.

References

1. *Solubility Data Series, International Union of Pure and Applied Chemistry, Vol. 38*, Pergamon Press, Oxford, 1988.
2. Shaw, D. G., and Maczynski, A., *J. Phys. Chem. Ref. Data* 35, 687, 2006.
3. Stephenson, R. M., *J. Chem. Eng. Data* 37, 80, 1992.
4. Lun, R., Varhanickova, D., Shiu, W.-Y., and Mackay, D., *J. Chem. Eng. Data* 42, 951, 1997.
5. Miller, D. J., et al., *J. Chem. Eng. Data* 43, 1043, 1998.
6. Miller, D. J., and Hawthorne, S. B., *J. Chem. Eng. Data* 45, 78, 2000.
7. Ma, J. H. Y., Hung, H., Shiu, W-Y., and Mackay, D., *J. Chem. Eng. Data* 46, 619, 2001.
8. Marche, C., Ferronato, C., and Jose, J., *J. Chem. Eng. Data* 48, 967, 2003.
9. Oleszek-Kudlak, S., Shibata, E., and Nakamura, T., *J. Chem. Eng. Data* 49, 570, 2004.
10. Marche, C., Ferronato, C., and Jose, J., *J. Chem. Eng. Data* 49, 937, 2004.
11. Andersson, T. A., Hartonen, K. M., and Riekkola, M-L., *J. Chem. Eng. Data* 50, 1177, 2005.
12. Karasek, P., Planeta, J., and Roth, M., *J. Chem. Eng. Data* 51, 616, 2006.
13. Shiu, W.-Y., and Ma, K.-C, *J. Phys. Chem. Ref. Data* 29, 41, 2000.

Aqueous Solubility of Organic Compounds at Various Pressures and Temperatures

Mol. form.	t/°C	p/bar	$S/10^3x_2$	S/mass %	Ref.
Acenaphthene					
$C_{12}H_{10}$	25	1	0.000444	0.000380	13
	250	50	1.25	1.06	11
Anthracene					
$C_{14}H_{10}$	25	1	0.0000074	0.0000044	2
	50	50	0.000017	0.000017	5
	100	45	0.00032	0.00032	5
	100	39	0.000457	0.00045	12
	150	50	0.0102	0.0101	11
	200	77	0.13	0.13	12
	250	50	0.497	0.49	11
	300	100	3.78	3.62	11
Benz[a]anthracene					
$C_{18}H_{12}$	25	1	0.00000073	0.00000093	2
	60	50	0.00000846	0.0000107	12
	100	50	0.000113	0.000143	12
	120	52	0.000418	0.00053	12
	150	49	0.00296	0.00375	12
Benzene					
C_6H_6	25	1	0.40	0.178	13
	25	65	0.40	0.173	6
	25	400	0.33	0.143	6
	50	65	0.47	0.203	6
	100	65	0.89	0.38	6
	150	65	2.2	0.95	6
	200	65	5.0	2.13	6
	200	400	4.1	1.75	6
Carbazole					
$C_{12}H_9N$	25	1	0.00013	0.00012	5
	25	54	0.00011	0.000102	5
	50	56	0.00045	0.00042	5
	100	54	0.0099	0.0092	5
	150	54	0.162	0.150	5
	200	52	1.9	1.74	5
Chrysene					
$C_{18}H_{12}$	25	1	0.00000016	0.00000019	2
	25	32	0.00000063	0.0000008	5
	50	36	0.000001	0.0000013	5
	100	38	0.000013	0.000016	5
	150	43	0.00060	0.00076	5
	200	45	0.0158	0.020	5
	225	62	0.0758	0.096	5
***o*-Dichlorobenzene**					
$C_6H_4Cl_2$	25	1	0.018	0.0094	9
	50	65	0.023	0.019	6
	100	65	0.055	0.045	6
	150	65	0.18	0.15	6
	200	65	0.57	0.46	6
***trans*-1,2-Dimethylcyclohexane**					
C_8H_{16}	25	1	0.008	0.00050	10
	101	7	0.0047	0.0029	10
	131	7	0.0108	0.0067	10
	151	7	0.0223	0.0139	10
	170	7	0.0356	0.0222	10
Ethylcyclohexane					
C_8H_{16}	25	1	0.00098	0.00061	10
	100	7	0.00340	0.00212	10
	131	7	0.0085	0.0053	10
	151	7	0.01665	0.0104	10
	171	7	0.0334	0.0208	10
Heptane					
C_7H_{16}	25	1	0.0004352	0.000242	8
	50	7	0.000613	0.00034096	8
	100	7	0.001938	0.00108	8

Mol. form.	t/°C	p/bar	$S/10^3x_2$	S/mass %	Ref.
	125	7	0.00400	0.00222	8
	150	7	0.00878	0.00488	8
	170	7	0.01701	0.00946	8
Hexane					
C_6H_{14}	25	1	0.002045	0.00098	8
	100	7	0.006074	0.0029	8
	125	7	0.01192	0.0057	8
	150	7	0.02555	0.0122	8
	170	7	0.04935	0.0236	8
1-Isopropyl-4-methylbenzene					
$C_{10}H_{14}$	25	1	0.0030	0.0051	4
	50	60	0.0040	0.0030	6
	100	60	0.011	0.0082	6
	150	60	0.043	0.032	6
	200	60	0.20	0.15	6
Methylcyclohexane					
C_7H_{14}	25	1	0.00293	0.00151	10
	100	7	0.01006	0.0055	10
	131	7	0.0244	0.0133	10
	151	7	0.0423	0.0231	10
	171	7	0.0708	0.0386	10
Naphthalene					
$C_{10}H_8$	25	1	0.00444	0.00316	13
	40	50	0.00692	0.0049	12
	50	50	0.0114	0.0081	12
	65	50	0.0264	0.0188	12
	75	50	0.0435	0.0309	12
Octane					
C_8H_{18}	25	1	0.0001158	0.000073	8
	100	7	0.0005943	0.000377	8
	125	7	0.0014163	0.000898	8
	150	7	0.0036957	0.00234	8
	170	7	0.0083483	0.00529	8
	200	65	0.029	0.018	6
Perylene					
$C_{20}H_{12}$	25	1	0.00000003	0.00000004	2
	50	50	0.00000029	0.0000004	5
	100	45	0.00000210	0.00000294	5
	150	47	0.000120	0.000168	5
	200	48	0.0050	0.0070	5
Pyrene					
$C_{16}H_{10}$	25	1	0.000012	0.0000139	2
	100	50	0.000637	0.00072	11
	100	200	0.00078	0.00087	5
	140	50	0.0054	0.0061	11
	200	50	0.0492	0.055	11
	250	50	0.205	0.23	11
	300	50	1.41	1.56	11
***p*-Terphenyl**					
$C_{18}H_{14}$	25	1	0.00000141	0.00000180	2
	100	49	0.0000219	0.000028	12
	140	51	0.000372	0.000476	12
	180	55	0.00626	0.0080	12
	200	53	0.0241	0.0308	12
	210	54	0.0393	0.0502	12
Tetrachloroethene					
C_2Cl_4	25	1	0.0285	0.0286	3
	50	65	0.027	0.025	6
	100	65	0.059	0.054	6
	150	65	0.18	0.17	6
	200	65	0.59	0.54	6
Toluene					
C_7H_8	25	1	0.107	0.0519	7,13
	50	50	0.125	0.064	6
	100	50	0.27	0.138	6
	150	50	0.66	0.337	6
	200	50	1.9	0.96	6
2,2,4-Trimethylpentane					
C_8H_{18}	25	1	0.00035	0.00022	1
	50	65	0.00052	0.00033	6
	100	65	0.0020	0.00127	6
	150	65	0.0102	0.0065	6
	200	65	0.061	0.0387	6
Triphenylene					
$C_{18}H_{12}$	25	1	0.0000034	0.0000043	2
	100	51	0.0000899	0.000114	12
	140	50	0.00126	0.00160	12
	180	64	0.0123	0.0156	12
	195	60	0.0283	0.0359	12
***m*-Xylene**					
C_8H_{10}	25	1	0.028	0.0161	13
	50	60	0.036	0.021	6
	100	60	0.085	0.050	6
	150	60	0.27	0.159	6
	200	60	0.88	0.516	6

SOLUBILITY CHART FOR INORGANIC SALTS

Abbreviations:

W: soluble in water
A: insoluble in water but soluble in acids
w: sparingly soluble in water but soluble in acids
a: insoluble in water and only sparingly soluble in acids
I: insoluble in water and acids
d: decomposes in water

Solubility Chart for Inorganic Salts

No.	Anion	Ag	Al	Au (I)	Au (II)	Ba	Bi	Ca	Cd	Co	Cr	Cu	Fe (II)	Fe (III)	H
1	Acetate —($C_2H_3O_2$)	w Ag(—)	W Al(—)$_3$	W	W	W Ba(—)$_2$	W Bi(—)$_3$	W Ca(—)$_2$	W Cd(—)$_2$	W Co(—)$_2$	W Cr(—)$_3$	W Cu(—)$_2$	Fe(—)$_2$	Fe$_2$(—)$_6$	W $C_2H_4O_2$
2	Arsenate —(AsO_4)	A Ag$_3$(—)	a Al(—)			w Ba$_3$(—)$_2$	A Bi(—)	w Ca$_3$(—)$_2$	A Cd$_3$(—)$_2$	A Co$_3$(—)$_2$		A Cu$_3$(—)$_2$	A Fe$_3$(—)$_2$	A Fe(—)	W H_3AsO_4
3	Arsenite —(AsO_3)	A Ag$_3$H(—)						w Ca$_3$(—)$_2$		A Co$_3$H$_6$(—)$_4$		A CuH(—)			
4	Benzoate —($C_7H_5O_2$)	w Ag(—)				W Ba(—)$_2$	A Bi(—)$_3$	W Ca(—)$_2$	W Cd(—)$_2$	W Co(—)$_2$		w Cu(—)$_2$	W Fe(—)$_2$	A Fe$_2$(—)$_6$	W $C_7H_6O_2$
5	Bromide	a AgBr	W $AlBr_2$	w AuBr	W $AuBr_3$	W $BaBr_2$	d $BiBr_3$	W $CaBr_2$	W $CdBr_2$	W $CoBr_2$	W(I) $CrBr_3$*	W $CuBr_2$	W $FeBr_2$	W $FeBr_3$	W HBr
6	Carbonate	A Ag_2CO_3				w $BaCO_3$		w $CaCO_3$	A $CdCO_3$	A $CoCO_3$	W $CrCO_3$		w $FeCO_3$		
7	Chlorate —(ClO_3)	W Ag(—)	W Al(—)$_3$			W Ba(—)$_2$	W Bi(—)$_3$	W Ca(—)$_2$	W Cd(—)$_2$	W Co(—)$_2$		W Cu(—)$_2$	W Fe(—)$_2$	W Fe(—)$_3$	W $HClO_3$
8	Chloride	a AgCl	W $AlCl_3$	w AuCl	W $AuCl_3$	W $BaCl_2$	d $BiCl_3$	W $CaCl_2$	W $CdCl_2$	W $CoCl_2$	I $CrCl_3$	W $CuCl_2$	W $FeCl_2$	W $FeCl_3$	W HCl
9	Chromate —(CrO_4)	w Ag$_2$(—)				A Ba(—)		W Ca(—)	A Cd(—)	A Co(—)				A Fe$_2$(—)$_3$	
10	Citrate —($C_6H_5O_7$)	w Ag$_3$(—)	W Al(—)			w Ba$_3$(—)$_2$	A Bi(—)	w Ca$_3$(—)$_2$	A Cd$_3$(—)$_2$	w Co$_3$(—)$_2$				W Fe(—)	W $C_6H_8O_7$
11	Cyanide	a AgCN		w AuCN	W Au(CN)$_3$	W Ba(CN)$_2$	w Bi(CN)$_3$	W Ca(CN)$_2$	W Cd(CN)$_2$	A Co(CN)$_2$	A Cr(CN)$_3$	A Cu(CN)$_2$	a Fe(CN)$_2$		W HCN
12	Ferricyanide —(Fe(CN)$_6$)	I Ag$_3$(—)				w Ba$_3$(—)$_2$		W Ca$_3$(—)$_2$	A Cd$_3$(—)$_2$	I Co$_3$(—)$_2$		I Cu$_3$(—)$_2$	I Fe$_3$(—)$_2$		W H$_3$(—)
13	Ferrocyanide —(Fe(CN)$_6$)	I Ag$_4$(—)	w Al$_4$(—)$_3$			W Ba$_2$(—)		W Ca$_2$(—)	A Cd$_2$(—)	I Co$_2$(—)		I Cu$_2$(—)	I Fe$_2$(—)	a Fe$_4$(—)$_3$	W H$_4$(—)
14	Fluoride	W AgF	W AlF_3			w BaF_2	W BiF_3	w CaF_2	W CdF_2	W CoF_2	W(a) CrF_3*	w CuF_2	w FeF_2	w FeF_3	W HF
15	Formate —(CHO_2)	W Ag(—)	W Al(—)$_3$			W Ba(—)$_2$	W Bi(—)$_3$	W Ca(—)$_2$	W Cd(—)$_2$	W Co(—)$_2$		W Cu(—)$_2$	W Fe(—)$_2$	W Fe(—)$_3$	W CH_2O_2
16	Hydroxide		A Al(OH)$_3$	W AuOH	A Au(OH)$_3$	W Ba(OH)$_2$	A Bi(OH)$_3$	W Ca(OH)$_2$	A Cd(OH)$_2$	A Co(OH)$_2$	A Cr(OH)$_3$	A Cu(OH)$_2$	A Fe(OH)$_2$	A Fe(OH)$_3$	
17	Iodide	I AgI	W AlI_3	a AuI	a AuI_3	W BaI_2	A BiI_3	W CaI_2	W CdI_2	W CoI_2	W CrI_3	a CuI	W FeI_2	W FeI_3	W HI
18	Nitrate	W $AgNO_3$	W Al(NO$_3$)$_3$			W Ba(NO$_3$)$_2$	d Bi(NO$_3$)$_3$	W Ca(NO$_3$)$_2$	W Cd(NO$_3$)$_2$	W Co(NO$_3$)$_2$	W Cr(NO$_3$)$_3$	W Cu(NO$_3$)$_2$	W Fe(NO$_3$)$_2$	W Fe(NO)$_3$	W HNO_3
19	Oxalate —(C_2O_4)	a Ag$_2$(—)	A Al$_2$(—)$_3$			w Ba(—)	A Bi$_2$(—)$_3$	A Ca(—)	w Cd(—)	A Co(—)	W Cr(—)	A Cu(—)	A Fe(—)	W Fe$_2$(—)$_3$	W $C_2H_2O_4$
20	Oxide	w Ag_2O	a Al_2O_3	Au_2O	A Au_2O_3	W BaO	A Bi_2O_3	w CaO	A CdO	A CoO	a Cr_2O_3	A CuO	A FeO	A Fe_2O_3	W H_2O_2
21	Phosphate	A Ag_3PO_4	A $AlPO_4$		H_3PO_4	A Ba$_3$(PO$_4$)$_2$	A $BiPO_4$	w Ca$_3$(PO$_4$)$_2$	A Cd$_3$(PO$_4$)$_2$	A Co$_3$(PO$_4$)$_2$	w Cr$_2$(PO$_4$)$_2$	A Cu$_3$(PO$_4$)$_2$	A $FePO_4$	w	W Fe$_3$(PO$_4$)$_2$
22	Silicate, —(SiO_3)		I Al$_2$(—)$_3$			W Ba(—)		w Ca(—)	A Cd(—)	A Co_2SiO_4		A Cu(—)			I H_2SiO_3
23	Sulfate	w Ag_2SO_4	W Al$_2$(SO$_4$)$_3$			a $BaSO_4$	d Bi$_2$(SO$_4$)$_3$	w $CaSO_4$	W $CdSO_4$	W $CoSO_4$	W(I) Cr$_2$(SO$_4$)$_3$*	W $CuSO_4$	W $FeSO_4$	w Fe(SO$_4$)$_3$	W H_2SO_4
24	Sulfide	A Ag_2S	d Al_2S_3	I Au_2S	I Au_2S_3	d BaS	A Bi_2S_3	w CaS	A CdS	A CoS	d Cr_2S_3	A CuS	A FeS	d Fe_2S_3	W H_2S
25	Tartrate —($C_4H_4O_6$)	A Ag$_2$(—)	w Al$_2$(—)$_3$	I	I	w Ba(—)	A Bi$_2$(—)$_3$	w Ca(—)	A Cd(—)	A Co(—)	d	A Cu(—)	A Fe(—)	d Fe$_2$(—)$_3$	W $C_4H_6O_6$
26	Thiocyanide	I AgCNS				W Ba(CNS)$_2$		W Ca(CNS)		W Co(CNS)$_2$		d CuCNS	W Fe(CNS)$_2$	W Fe(CNS)$_3$	W CNSH

* Indicates two modifications of the salt.

Solubility Chart for Inorganic Salts

No.	Hg (I)	Hg (II)	K	Mg	Mn	NH_4	Na	Ni	Pb	Pt	Sb	Sn (II)	Sn (IV)	Sr	Zn
1	w Hg(—)	W $Hg(—)_2$	W K(—)	W $Mg(—)_2$	W $Mn(—)_2$	W $NH_4(—)$	w Na(—)	W $Ni(—)_2$	W $Pb(—)_2$			d $Sn(—)_2$	W $Sn(—)_4$	W $Sr(—)_2$	W $Zn(—)_2$
2	A $Hg_3(—)$	w $Hg_3(—)_2$	W $K_3(—)$	A $Mg_3(—)$	w MnH(—)	W $(NH_4)_3(—)$	W $Na_3(—)$	A $Ni_3(—)_2$	A PbH(—)		A Sb(—)			w SrH(—)	A $Zn_3(—)_2$
3	A $Hg_3(—)$	A $Hg_3(—)$	W K_3AsO_3	W $Mg_3(—)_2$	A $Mn_3H_6(—)_4$	W NH_4AsO_2	W $Na_2H(—)$	A $Ni_3H_6(—)_4$			A Sb(—)	A $Sn_3(—)_2$		w $Sr_3(—)_2$	
4	A $Hg_2(—)_2$	w $Hg(—)_2$	W K(—)	W $Mg(—)_2$	W $Mn(—)_2$	W $NH_4(—)$	W Na(—)	w $Ni(—)_2$	w $Pb(—)_2$						W $Zn(—)_2$
5	A HgBr	W $HgBr_2$	W KBr	W $MgBr_2$	W $MnBr_2$	W NH_4Br	W NaBr	W $NiBr_2$	W $PbBr_2$	w $PtBr_4$	d $SbBr_3$	W $SnBr_2$	W $SnBr_4$	W $SrBr_2$	W $ZnBr_2$
6	A Hg_2CO_3		W K_2CO_3	w $MgCO_3$	w $MnCO_3$	W $(NH_4)_2CO_3$	W Na_2CO_3	w $NiCO_3$	A $PbCO_3$					w $SrCO_3$	w $ZnCO_3$
7	W Hg(—)	W $Hg(—)_2$	W K(—)	W $Mg(—)_2$	W $Mn(—)_2$	W $NH_4(—)$	W Na(—)	W $Ni(—)_2$	W $Pb(—)_2$			W $Sn(—)_2$		W $Sr(—)_2$	W $Zn(—)_2$
8	a HgCl	W $HgCl_2$	W KCl	W $MgCl_2$	W $MnCl_2$	W NH_4Cl	W NaCl	W iCl_2	W $PbCl_2$	W $PtCl_4$	W $SbCl_3$	W $SnCl_2$	W $SnCl_4$	W $SrCl_2$	W $ZnCl_2$
9	w $Hg_2(—)$	w Hg(—)	W $K_2(—)$	W Mg(—)		W $(NH_4)_2(—)$	W $Na_2(—)$	A Ni(—)	A Pb(—)			A Sn(—)	W $Sn(—)_2$	w Sr(—)	w Zn(—)
10	w $Hg_3(—)$		W $K_3(—)$	W $Mg_3(—)_2$	w MnH(—)	W $NH_4)_3(—)$	W $Na_3(—)$	W $Ni_3(—)_2$	W $Pb_3(—)_2$					A SrH(—)	w $Zn_3(—)_2$
11	A HgCN	W $Hg(CN)_2$	W KCN	W $Mg(CN)_2$		W NH_4CN	W NaCN	a $Ni(CN)_2$	w $Pb(CN)_2$	I $Pt(CN)_2$				W $Sr(CN)_2$	A $Zn(CN)_2$
12		A $Hg_3(—)_2$	W $K_3(—)$	W $Mg_3(—)_2$		W $NH_4)_3(—)$	W $Na_3(—)$	I $Ni_3(—)_2$	w $Pb_3(—)_2$			A $Sn_3(—)_2$		W $Sr_3(—)_2$	A $Zn_3(—)_2$
13		I $Hg_2(—)$	W $K_4(—)$	W $Mg_2(—)$	A $Mn_2(—)$	W $NH_4)_4(—)$	W $Na_4(—)$	I $Ni_2(—)$	a $Pb_2(—)$			a $Sn_2(—)$		W $Sr_2(—)$	I $Zn_2(—)$
14	d HgF	d HgF_2	W KF	w MgF_2	A MnF_2	W NH_4F	W NaF	w NiF_2	w PbF_2	W PtF_4	W SbF_3	W SnF_2	W SnF_4	w SrF_2	w ZnF_2
15	w Hg(—)	W $Hg(—)_2$	W K(—)	W $Mg(—)_2$	W $Mn(—)_2$	W $NH_4(—)$	W Na(—)	W $Ni(—)_2$	W $Pb(—)_2$					W $Sr(—)_2$	W $Zn(—)_2$
16		A $Hg(OH)_2$	W KOH	A $Mg(OH)_2$	A $Mn(OH)_2$	W NH_4OH	W NaOH	w $Ni(OH)_2$	w $Pb(OH)_2$	A $Pt(OH)_4$		A $Sn(OH)_2$	w $Sn(OH)_4$	W $Sr(OH)_2$	A $Zn(OH)_2$
17	A HgI	w HgI_2	W KI	W MgI_2	W MnI_2	W NH_4I	W NaI	W NiI_2	w PbI_2	I PtI_2	d SbI_3	W SnI_2	d SnI_4	W SrI_2	W ZnI_2
18	W $HgNO_3$	W $Hg(NO_3)_2$	W KNO_3	W $Mg(NO_3)_2$	W $Mn(NO_3)_2$	W NH_4NO_3	W $NaNO_3$	W $Ni(NO_3)_2$	W $Pb(NO_3)_2$	W $Pt(NO_3)_4$		d $Sn(NO_3)_2$		W $Sr(NO_3)_2$	W $Zn(NO_3)_2$
19	a $Hg_2(—)$	A Hg(—)	W $K_2(—)$	w Mg(—)	w Mn(—)	W $(NH_4)_2(—)$	W $Na_2(—)$	A Ni(—)	A Pb(—)			A Sn(—)		w Sr(—)	A Zn(—)
20	A Hg_2O	w HgO	W K_2O	A MgO	A MnO		d Na_2O	A NiO	w PbO	A PtO	w Sb_2O_3	A SnO	A SnO_2	W SrO	w ZnO
21	A Hg_3PO_4	A $Hg_3(PO_4)_2$	W K_3PO_4	w $Mg_2(PO_4)_2$	w $Mn_3(PO_4)_2$	W $NH_4H_2PO_4$	W Na_3PO_4	A $Ni_3(PO_4)_2$	A $Pb_3(PO_4)_2$			A $Sn_3(PO_4)$		A $Sr_3(PO_4)_2$	A $Zn_3(PO_4)_2$
22			W $K_2(—)$	A Mg(—)	I Mn(—)		W $Na_2(—)$		A Pb(—)					A Sr(—)	A Zn(—)
23	w Hg_2SO_4	d $HgSO_4$	W K_2SO_4	W $MgSO_4$	W $MnSO_4$	W $(NH_4)_2SO_4$	W Na_2SO_4	W $NiSO_4$	w $PbSO_4$	W $Pt(SO_4)_2$	A $Sb_2(SO_4)_3$	W $SnSO_4$	W $Sn(SO_4)_2$	w $SrSO_4$	W $ZnSO_4$
24	I Hg_2S	I HgS	W K_2S	d MgS	A MnS	W $(NH_4)_2S$	W Na_2S	A NiS	A PbS	I PtS	A Sb_2S_3	A SnS	A SnS_2	W SrS	A ZnS
25	I $Hg_2(—)$	I	W $K_2(—)$	w Mg(—)	w Mn(—)	W $(NH_4)_2(—)$	W $Na_2(—)$	A Ni(—)	A Pb(—)	I	W $Sb_2(—)_3$	A Sn(—)	A	W Sr(—)	A Zn(—)
26	A HgCNS	w $Hg(CNS)_2$	W KCNS	W $Mg(CNS)_2$	W $Mn(CNS)_2$	W NH_4CNS	W NaCNS		w $Pb(CNS)_2$					w $Sr(CNS)_2$	W $Zn(CNS)_2$

* Indicates two modifications of the salt.

Section 6
Fluid Properties

THERMOPHYSICAL PROPERTIES OF WATER AND STEAM

Eric W. Lemmon and Allan H. Harvey

These tables summarize the thermophysical properties of water and steam at equilibrium as accepted by the International Association for the Properties of Water and Steam (http://www.iapws.org) for general and scientific use. The thermodynamic properties are calculated from the equation of state of Wagner and Pruß (Ref. 6). The reference state for these tables is the liquid at the triple point, at which the internal energy and entropy are taken as zero. These tables refer to states at 1 bar (100 KPa) pressure with temperatures in °C in the first section; liquid and gaseous states at equilibrium as a function of temperature in the second section; and properties along isobars in the third section. The tabulated properties are pressure (*P*), density (*ρ*), enthalpy (*H*), entropy (*S*), isochoric heat capacity (C_v), isobaric heat capacity (C_p), speed of sound (*u*), viscosity (*η*), thermal conductivity (*λ*), and static dielectric constant (*D*). In the saturation tables, the first line of identical temperatures is for the liquid state and the second line is for the vapor state. A duplicate entry in the isobar section indicates a phase transition (liquid-vapor) at that temperature; property values are then given for both phases. These are identified by the high densities in the liquid and the low densities in the vapor. The temperature scale is ITS-90. Additional calculations at state points not listed below can be obtained by using the NIST Standard Reference Data program REFPROP (Ref. 5) or the water-specific program steam (Ref. 2).

The uncertainty in density of the equation of state is 0.0001% at 1 atm in the liquid phase, and 0.001% at other liquid states at pressures up to 10 MPa and temperatures to 423 K. In the vapor phase, the uncertainty is 0.05% or less. The uncertainties rise at higher temperatures and/or pressures, but are generally less than 0.1% in density except at extreme conditions. The uncertainty in pressure in the critical region is 0.1%. The uncertainty of the speed of sound is 0.15% in the vapor and 0.1% or less in the liquid, and increases near the critical region and at high temperatures and pressures. The uncertainty in the isobaric heat capacity is 0.2% in the vapor and 0.1% in the liquid, with increasing values in the critical region and at high pressures. The uncertainties of saturation conditions are 0.025% in vapor pressure, 0.0025% in saturated liquid density, and 0.1% in saturated vapor density. The uncertainties in the saturated densities increase substantially as the critical region is approached.

The uncertainty in density of the equation of state is 0.0001% at 1 atm in the liquid phase, and 0.001% at other liquid states at pressures up to 10 MPa and temperatures to 423 K. In the vapor phase, the uncertainty is 0.05% or less. The uncertainties rise at higher temperatures and/or pressures, but are generally less than 0.1% in density except at extreme conditions. The uncertainty in pressure in the critical region is 0.1%. The uncertainty of the speed of sound is 0.15% in the vapor and 0.1% or less in the liquid, and increases near the critical region and at high temperatures and pressures. The uncertainty in the isobaric heat capacity is 0.2% in the vapor and 0.1% in the liquid, with increasing values in the critical region and at high pressures. The uncertainties of saturation conditions are 0.025% in vapor pressure, 0.0025% in saturated liquid density, and 0.1% in saturated vapor density. The uncertainties in the saturated densities increase substantially as the critical region is approached.

References

1. Fernández, D.P., Goodwin, A.R.H., Lemmon, E.W., Levelt Sengers, J.M.H., and Williams, R.C., A Formulation for the Static Permittivity of Water and Steam at Temperatures from 238 K to 873 K at Pressures up to 1200 MPa, Including Derivatives and Debye-Hückel Coefficients, *J. Phys. Chem. Ref. Data* 26, 1125, 1997.
2. Harvey, A.H., and Lemmon, E.W., NIST Standard Reference Database 10: NIST/ASME Steam Properties, Version 3.0, National Institute of Standards and Technology, Standard Reference Data Program, Gaithersburg, Maryland, 20013 (http://www.nist.gov/srd/nist10.cfm).
3. Huber, M.L., Perkins, R.A., Laesecke, A., Friend, D.G., Sengers, J.V., Assael, M.J., Metaxa, I.M., Vogel, E., Mareš, R., and Miyagawa, K., New International Formulation for the Viscosity of Water, *J. Phys. Chem. Ref. Data* 38, 101, 2009.
4. Huber, M.L., Perkins, R.A., Friend, D.G., Sengers, J.V., Asseal, M.J., Metaxa, I.N., Miyagawa, K., Hellmann, R., and Vogel, E., New International Formulation for the Thermal Conductivity of H_2O, *J. Phys. Chem. Ref. Data* 41, 033102, 2012.
5. Lemmon, E.W., Huber, M.L., and McLinden, M.O., NIST Standard Reference Database 23: Reference Fluid Thermodynamic and Transport Properties-REFPROP, Version 9.1, National Institute of Standards and Technology, Standard Reference Data Program, Gaithersburg, Maryland, 2013 (http://www.nist.gov/srd/nist23.cfm).
6. Wagner, W. and Pruß, A., The IAPWS Formulation 1995 For the Thermodynamic Properties of Ordinary Water Substance for General and Scientific Use, *J. Phys. Chem. Ref. Data* 31, 387, 2002.

Thermophysical Properties of H_2O from the Triple Point to the Critical Point

t/°C	*P*/MPa	*ρ*/kg m^{-3}	*H*/kJ kg^{-1}	*S*/ kJ kg^{-1} K^{-1}	C_v/ kJ kg^{-1} K^{-1}	C_p/ kJ kg^{-1} K^{-1}	*u*/m s^{-1}	*D*	*η*/μPa s	*λ*/ mW m^{-1} K^{-1}
***P* = 0.1 MPa (1 bar)**										
0.01	0.1	999.84	0.10186	0.000007	4.2170	4.2194	1402.4	87.899	1791.1	555.67
10	0.1	999.70	42.118	0.15108	4.1906	4.1952	1447.3	83.974	1305.9	578.78
20	0.1	998.21	84.006	0.29646	4.1567	4.1841	1482.3	80.223	1001.6	598.01
25	0.1	997.05	104.92	0.36720	4.1376	4.1813	1496.7	78.408	890.02	606.52
30	0.1	995.65	125.82	0.43673	4.1172	4.1798	1509.2	76.634	797.22	614.39
40	0.1	992.22	167.62	0.57237	4.0734	4.1794	1528.9	73.201	652.73	628.48
50	0.1	988.03	209.42	0.70377	4.0262	4.1813	1542.6	69.916	546.52	640.62
60	0.1	983.20	251.25	0.83125	3.9765	4.1850	1551.0	66.774	466.03	651.00

t/°C	P/MPa	ρ/kg m^{-3}	H/kJ kg^{-1}	S/ kJ kg^{-1} K^{-1}	C_v/ kJ kg^{-1} K^{-1}	C_p/ kJ kg^{-1} K^{-1}	u/m s^{-1}	D	η/μPa s	λ/ mW m^{-1} K^{-1}
70	0.1	977.76	293.12	0.95509	3.9251	4.1901	1554.7	63.770	403.55	659.76
80	0.1	971.79	335.05	1.0755	3.8728	4.1968	1554.4	60.898	354.05	666.99
90	0.1	965.31	377.06	1.1928	3.8204	4.2052	1550.4	58.152	314.17	672.79
99.606	0.1	958.63	417.50	1.3028	3.7702	4.2152	1543.5	55.628	282.75	677.06
99.606	0.1	0.59034	2674.9	7.3588	1.5548	2.0784	471.99	1.0058	12.218	24.532
100	0.1	0.58967	2675.8	7.3610	1.5535	2.0766	472.28	1.0058	12.234	24.564
Saturation										
0.01	0.000612	999.79	0.000612	0.0	4.2174	4.2199	1402.3	87.895	1791.4	555.60
0.01	0.000612	0.0048546	2500.9	9.1555	1.4184	1.8844	409.00	1.00006	8.9458	16.761
10	0.0012282	999.65	42.021	0.15109	4.1910	4.1955	1447.1	83.971	1306.0	578.71
10	0.0012282	0.0094071	2519.2	8.8998	1.4269	1.8947	416.17	1.00012	9.2384	17.412
20	0.0023393	998.16	83.914	0.29648	4.1570	4.1844	1482.2	80.219	1001.6	597.95
20	0.0023393	0.017314	2537.4	8.6660	1.4359	1.9059	423.18	1.00021	9.5441	18.087
25	0.0031699	997.00	104.83	0.36722	4.1379	4.1816	1496.5	78.405	890.04	606.46
25	0.0031699	0.023075	2546.5	8.5566	1.4405	1.9118	426.63	1.00028	9.7009	18.433
30	0.0042470	995.61	125.73	0.43675	4.1175	4.1801	1509.0	76.630	797.22	614.34
30	0.0042470	0.030415	2555.5	8.4520	1.4452	1.9180	430.03	1.00036	9.8602	18.786
40	0.0073849	992.18	167.53	0.57240	4.0737	4.1796	1528.7	73.197	652.72	628.44
40	0.0073849	0.051242	2573.5	8.2555	1.4552	1.9314	436.71	1.00059	10.185	19.509
50	0.012352	988.00	209.34	0.70381	4.0264	4.1815	1542.4	69.913	546.50	640.57
50	0.012352	0.083147	2591.3	8.0748	1.4663	1.9468	443.21	1.00094	10.516	20.261
60	0.019946	983.16	251.18	0.83129	3.9767	4.1851	1550.8	66.772	466.02	650.96
60	0.019946	0.13043	2608.8	7.9081	1.4789	1.9648	449.50	1.0014	10.854	21.043
70	0.031201	977.73	293.07	0.95513	3.9252	4.1902	1554.6	63.768	403.53	659.72
70	0.031201	0.19843	2626.1	7.7540	1.4937	1.9862	455.57	1.0021	11.195	21.860
80	0.047414	971.77	335.01	1.0756	3.8729	4.1969	1554.3	60.896	354.04	666.97
80	0.047414	0.29367	2643.0	7.6111	1.5111	2.0120	461.39	1.0030	11.539	22.717
90	0.070182	965.30	377.04	1.1929	3.8204	4.2053	1550.4	58.151	314.17	672.77
90	0.070182	0.42390	2659.5	7.4781	1.5316	2.0429	466.94	1.0043	11.885	23.618
100	0.10142	958.35	419.17	1.3072	3.7682	4.2157	1543.2	55.527	281.58	677.21
100	0.10142	0.59817	2675.6	7.3541	1.5558	2.0800	472.20	1.0059	12.232	24.570
110	0.14338	950.95	461.42	1.4188	3.7167	4.2283	1532.9	53.018	254.61	680.35
110	0.14338	0.82693	2691.1	7.2381	1.5843	2.1244	477.13	1.0080	12.580	25.579
120	0.19867	943.11	503.81	1.5279	3.6662	4.2435	1519.9	50.620	232.03	682.24
120	0.19867	1.1221	2705.9	7.1291	1.6177	2.1770	481.73	1.0105	12.927	26.652
140	0.36154	926.13	589.16	1.7392	3.5694	4.2826	1486.2	46.131	196.64	682.53
140	0.36154	1.9667	2733.4	6.9293	1.7002	2.3109	489.82	1.0177	13.618	29.016
160	0.61823	907.45	675.47	1.9426	3.4788	4.3354	1443.2	42.018	170.43	678.73
160	0.61823	3.2596	2757.4	6.7491	1.8044	2.4883	496.29	1.0282	14.304	31.721
180	1.0028	887.00	763.05	2.1392	3.3949	4.4050	1391.7	38.235	150.38	671.28
180	1.0028	5.1588	2777.2	6.5840	1.9279	2.7129	501.04	1.0431	14.985	34.832
200	1.5549	864.66	852.27	2.3305	3.3179	4.4958	1332.1	34.742	134.58	660.01
200	1.5549	7.8610	2792.0	6.4302	2.0666	2.9895	503.92	1.0636	15.666	38.426
220	2.3196	840.22	943.58	2.5177	3.2479	4.6146	1264.5	31.495	121.77	645.26
220	2.3196	11.615	2800.9	6.2840	2.2172	3.3289	504.77	1.0915	16.354	42.606
240	3.3469	813.37	1037.6	2.7020	3.1850	4.7719	1189.0	28.455	111.06	627.17
240	3.3469	16.749	2803.0	6.1423	2.3794	3.7537	503.32	1.1292	17.062	47.525
260	4.6923	783.63	1135.0	2.8849	3.1301	4.9856	1105.3	25.580	101.81	605.78
260	4.6923	23.712	2796.6	6.0016	2.5555	4.3075	499.21	1.1802	17.810	53.446
280	6.4166	750.28	1236.9	3.0685	3.0849	5.2889	1012.6	22.824	93.550	581.03
280	6.4166	33.165	2779.9	5.8579	2.7503	5.0731	491.93	1.2505	18.630	60.878
300	8.5879	712.14	1345.0	3.2552	3.0530	5.7504	909.40	20.135	85.855	552.65
300	8.5879	46.168	2749.6	5.7059	2.9708	6.2197	480.73	1.3504	19.580	70.900
320	11.284	667.09	1462.2	3.4494	3.0428	6.5373	793.16	17.440	78.310	520.02
320	11.284	64.638	2700.6	5.5372	3.2276	8.1589	464.43	1.5012	20.773	86.156
340	14.601	610.67	1594.5	3.6601	3.0781	8.2080	658.27	14.606	70.331	481.93
340	14.601	92.759	2621.8	5.3356	3.5430	12.236	440.72	1.7555	22.477	114.52
360	18.666	527.59	1761.7	3.9167	3.2972	15.004	479.74	11.225	60.306	439.16
360	18.666	143.90	2481.5	5.0536	4.0068	27.356	402.37	2.3096	25.638	191.44
373.946	22.064	322.00	2084.3	4.4070				5.3606		

Thermophysical Properties of H_2O along Isobars from 0.1 MPa to 100 MPa

T/K	P/MPa	ρ/kg m^{-3}	H/kJ kg^{-1}	S/kJ kg^{-1} K^{-1}	C_v/ kJ kg^{-1} K^{-1}	C_p/ kJ kg^{-1} K^{-1}	u/m s^{-1}	D	η/μPa s	λ/ mW m^{-1} K^{-1}
P = 0.1 MPa (1 bar)										
273.16	0.1	999.84	0.10186	0.000007	4.2170	4.2194	1402.4	87.899	1791.1	555.67
280	0.1	999.91	28.894	0.10411	4.1998	4.2009	1434.3	85.192	1433.6	571.98
300	0.1	996.56	112.65	0.39306	4.1302	4.1806	1501.5	77.747	853.74	609.50
320	0.1	989.43	196.25	0.66281	4.0414	4.1805	1538.9	70.935	576.73	637.00
340	0.1	979.54	279.93	0.91646	3.9414	4.1883	1554.0	64.702	421.63	657.17
360	0.1	967.40	363.82	1.1562	3.8369	4.2023	1552.1	59.004	325.86	671.11
372.756	0.1	958.63	417.50	1.3028	3.7702	4.2152	1543.5	55.628	282.75	677.06
372.756	0.1	0.59034	2674.9	7.3588	1.5548	2.0784	471.99	1.0058	12.218	24.532
380	0.1	0.57824	2689.9	7.3986	1.5356	2.0507	477.08	1.0056	12.498	25.134
400	0.1	0.54761	2730.4	7.5025	1.5082	2.0078	490.31	1.0051	13.278	26.825
450	0.1	0.48458	2829.7	7.7365	1.4943	1.9752	520.60	1.0040	15.267	31.269
500	0.1	0.43514	2928.6	7.9447	1.5082	1.9813	548.31	1.0033	17.299	36.032
550	0.1	0.39507	3028.1	8.1344	1.5319	2.0010	574.19	1.0027	19.356	41.093
600	0.1	0.36185	3128.8	8.3096	1.5600	2.0268	598.61	1.0023	21.425	46.424
650	0.1	0.33384	3230.8	8.4730	1.5903	2.0557	621.79	1.0020	23.496	51.996
700	0.1	0.30988	3334.4	8.6264	1.6222	2.0867	643.92	1.0017	25.562	57.782
750	0.1	0.28915	3439.5	8.7715	1.6553	2.1191	665.11	1.0015	27.617	63.759
800	0.1	0.27102	3546.3	8.9093	1.6892	2.1525	685.47	1.0013	29.657	69.908
900	0.1	0.24085	3765.0	9.1668	1.7589	2.2216	724.03	1.0011	33.680	82.647
1000	0.1	0.21673	3990.7	9.4045	1.8297	2.2921	760.17	1.00088	37.615	95.877
1100	0.1	0.19701	4223.4	9.6263	1.9000	2.3621	794.33	1.00074	41.453	109.50
1200	0.1	0.18058	4463.0	9.8348	1.9682	2.4302	826.85	1.00063	45.192	123.44
P = 1 MPa (10 bar)										
273.16	1.0	1000.3	1.0180	0.000066	4.2127	4.2150	1403.9	87.937	1789.1	556.36
280	1.0	1000.3	29.783	0.10407	4.1960	4.1973	1435.7	85.228	1432.5	572.59
300	1.0	996.96	113.48	0.39281	4.1272	4.1781	1503.0	77.781	853.66	610.00
320	1.0	989.82	197.03	0.66242	4.0390	4.1784	1540.5	70.967	576.89	637.47
340	1.0	979.93	280.67	0.91594	3.9395	4.1863	1555.7	64.734	421.86	657.64
360	1.0	967.81	364.52	1.1556	3.8353	4.2004	1553.9	59.035	326.10	671.61
380	1.0	953.74	448.73	1.3832	3.7315	4.2220	1538.4	53.827	262.82	680.00
400	1.0	937.87	533.47	1.6005	3.6315	4.2535	1511.3	49.065	218.82	683.32
450	1.0	890.39	749.20	2.1086	3.4076	4.3924	1400.6	38.814	153.23	672.77
453.028	1.0	887.13	762.52	2.1381	3.3954	4.4045	1392.0	38.258	150.49	671.33
453.028	1.0	5.1450	2777.1	6.5850	1.9271	2.7114	501.02	1.0430	14.981	34.812
500	1.0	4.5323	2891.2	6.8250	1.6699	2.2795	535.74	1.0344	17.054	38.473
550	1.0	4.0581	3001.8	7.0359	1.6159	2.1647	565.75	1.0282	19.215	42.908
600	1.0	3.6871	3109.0	7.2224	1.6098	2.1292	592.58	1.0236	21.349	47.802
650	1.0	3.3843	3215.2	7.3925	1.6227	2.1254	617.34	1.0202	23.462	53.075
700	1.0	3.1305	3321.7	7.5504	1.6447	2.1368	640.55	1.0174	25.555	58.664
750	1.0	2.9140	3429.0	7.6984	1.6715	2.1566	662.53	1.0153	27.628	64.516
800	1.0	2.7265	3537.5	7.8384	1.7014	2.1816	683.48	1.0135	29.680	70.591
900	1.0	2.4174	3758.5	8.0986	1.7663	2.2402	722.85	1.0108	33.716	83.281
1000	1.0	2.1723	3985.7	8.3380	1.8346	2.3048	759.50	1.0088	37.655	96.534
1100	1.0	1.9729	4219.5	8.5608	1.9034	2.3713	794.01	1.0074	41.494	110.21
1200	1.0	1.8074	4460.0	8.7699	1.9708	2.4371	826.77	1.0063	45.231	124.21
P = 10 MPa (100 bar)										
273.16	10.0	1004.8	10.111	0.00049	4.1721	4.1726	1418.4	88.311	1770.0	563.05
280	10.0	1004.7	38.613	0.10355	4.1593	4.1622	1450.3	85.588	1422.2	578.60
300	10.0	1001.0	121.73	0.39029	4.0984	4.1536	1518.2	78.113	852.99	614.97
320	10.0	993.70	204.84	0.65846	4.0157	4.1580	1556.4	71.284	578.57	642.13
340	10.0	983.84	288.08	0.91079	3.9205	4.1672	1572.6	65.044	424.17	662.33
360	10.0	971.85	371.56	1.1493	3.8198	4.1810	1572.0	59.345	328.53	676.49
380	10.0	957.99	455.37	1.3759	3.7188	4.2013	1558.0	54.140	265.22	685.19
400	10.0	942.42	539.67	1.5921	3.6210	4.2302	1532.7	49.385	221.17	688.88
450	10.0	896.16	753.94	2.0967	3.4010	4.3553	1428.4	39.172	155.48	679.41
500	10.0	838.02	977.18	2.5669	3.2211	4.6022	1271.3	30.794	119.83	646.41

T/K	P/MPa	ρ/kg m⁻³	H/kJ kg⁻¹	S/kJ kg⁻¹ K⁻¹	C_v/ kJ kg⁻¹ K⁻¹	C_p/ kJ kg⁻¹ K⁻¹	u/m s⁻¹	D	η/μPa s	λ/ mW m⁻¹ K⁻¹
550	10.0	761.82	1218.8	3.0270	3.0865	5.1407	1054.6	23.531	96.080	590.37
584.147	10.0	688.42	1408.1	3.3606	3.0438	6.1237	847.33	18.660	81.718	535.29
584.147	10.0	55.463	2725.5	5.6160	3.1065	7.1408	472.51	1.4248	20.194	78.344
600	10.0	49.773	2820.0	5.7756	2.6239	5.1365	503.34	1.3649	21.017	72.247
650	10.0	40.479	3022.6	6.1009	2.1103	3.3968	562.10	1.2672	23.472	68.284
700	10.0	35.355	3177.4	6.3305	1.9338	2.8741	602.20	1.2145	25.773	69.997
750	10.0	31.810	3314.6	6.5200	1.8625	2.6452	634.58	1.1793	27.973	73.764
800	10.0	29.107	3443.7	6.6867	1.8367	2.5313	662.61	1.1536	30.101	78.664
900	10.0	25.123	3691.6	6.9787	1.8439	2.4458	710.98	1.1182	34.201	90.420
1000	10.0	22.241	3935.5	7.2357	1.8843	2.4397	753.03	1.0948	38.144	103.67
1100	10.0	20.017	4180.6	7.4693	1.9377	2.4661	791.02	1.0782	41.960	117.74
1200	10.0	18.230	4429.2	7.6855	1.9957	2.5070	826.16	1.0659	45.663	132.27
P = 100 MPa (1000 bar)										
273.16	100.0	1045.3	95.444	-0.0083717	3.8761	3.9053	1575.5	91.834	1660.1	616.62
280	100.0	1043.6	122.26	0.088571	3.8869	3.9328	1603.8	88.976	1367.4	627.72
300	100.0	1037.2	201.44	0.36171	3.8751	3.9798	1667.9	81.216	859.19	657.72
320	100.0	1028.9	281.30	0.61941	3.8289	4.0043	1707.7	74.222	599.18	683.70
340	100.0	1019.0	361.55	0.86265	3.7637	4.0194	1728.7	67.891	448.03	705.08
360	100.0	1007.8	442.05	1.0927	3.6883	4.0309	1735.4	62.150	352.49	721.60
380	100.0	995.37	522.79	1.3110	3.6089	4.0427	1730.9	56.937	288.37	733.29
400	100.0	981.82	603.78	1.5187	3.5293	4.0569	1717.3	52.201	243.33	740.39
450	100.0	943.51	807.84	1.9993	3.3430	4.1105	1652.8	42.149	175.71	740.46
500	100.0	899.21	1015.4	2.4366	3.1820	4.1968	1555.7	34.149	139.57	720.25
550	100.0	848.78	1228.2	2.8421	3.0452	4.3234	1435.5	27.667	117.36	684.16
600	100.0	791.49	1448.6	3.2256	2.9295	4.5019	1300.4	22.290	101.85	634.92
650	100.0	726.21	1679.5	3.5952	2.8329	4.7503	1158.6	17.717	89.647	575.50
700	100.0	651.77	1925.0	3.9589	2.7538	5.0832	1020.0	13.754	79.123	508.87
750	100.0	568.52	2188.5	4.3223	2.6866	5.4492	898.84	10.336	69.732	438.73
800	100.0	482.23	2466.5	4.6811	2.6169	5.6108	813.97	7.5622	61.842	371.17
900	100.0	343.61	3000.1	5.3104	2.4386	4.8879	765.30	4.2835	52.771	275.39
1000	100.0	265.45	3440.1	5.7749	2.2950	3.9788	792.50	2.9559	50.506	237.77
1100	100.0	220.62	3809.8	6.1276	2.2276	3.4715	832.67	2.3472	51.089	232.34
1200	100.0	191.53	4142.5	6.4172	2.2098	3.2098	872.28	2.0111	52.802	238.54

VAPOR PRESSURE AND OTHER SATURATION PROPERTIES OF WATER

Eric W. Lemmon

This table summarizes the vapor pressure, enthalpy (heat) of vaporization, and surface tension γ of water as accepted by the International Association for the Properties of Water and Steam (www.iapws.org) for general and scientific use. The vapor pressure and heat of vaporization are calculated from the equation of state of Wagner and Pruss (Ref. 1). The temperature scale is ITS-90. Additional calculations at state points not listed below can be obtained by using the NIST Standard Reference Data program REFPROP (www.nist.gov/srd/nist23.cfm) or the water-specific program Steam Properties (www.nist.gov/srd/nist10.cfm).

References

1. Wagner, W. and Pruss, A., The IAPWS Formulation 1995 for the Thermodynamic Properties of Ordinary Water Substance for General and Scientific Use, *J. Phys. Chem. Ref. Data*, 31, 387, 2002.
2. International Association for the Properties of Water and Steam, Release on the surface tension of ordinary water substance, Physical Chemistry of Aqueous Systems: Proceedings of the 12th International Conference on the Properties of Water and Steam, Orlando, Florida, September 11–16, 1994, pp. A139–A142.

Vapor Pressure, Enthalpy of Vaporization, and Surface Tension of Water

t/°C	P/kPa	$\Delta_{vap}H$/kJ kg^{-1}	γ/mN m^{-1}
0.01	0.61165	2500.9	75.65
2	0.70599	2496.2	75.37
4	0.81355	2491.4	75.08
6	0.93536	2486.7	74.80
8	1.0730	2481.9	74.51
10	1.2282	2477.2	74.22
12	1.4028	2472.5	73.93
14	1.5990	2467.7	73.63
16	1.8188	2463.0	73.34
18	2.0647	2458.3	73.04
20	2.3393	2453.5	72.74
22	2.6453	2448.8	72.43
24	2.9858	2444.0	72.13
25	3.1699	2441.7	71.97
26	3.3639	2439.3	71.82
28	3.7831	2434.6	71.51
30	4.2470	2429.8	71.19
32	4.7596	2425.1	70.88
34	5.3251	2420.3	70.56
36	5.9479	2415.5	70.24
38	6.6328	2410.8	69.92
40	7.3849	2406.0	69.60
42	8.2096	2401.2	69.27
44	9.1124	2396.4	68.94
46	10.099	2391.6	68.61
48	11.177	2386.8	68.28
50	12.352	2381.9	67.94
52	13.631	2377.1	67.61
54	15.022	2372.3	67.27
56	16.533	2367.4	66.93
58	18.171	2362.5	66.58
60	19.946	2357.7	66.24
62	21.867	2352.8	65.89
64	23.943	2347.8	65.54
66	26.183	2342.9	65.19
68	28.599	2338.0	64.84
70	31.201	2333.0	64.48
72	34.000	2328.1	64.12
74	37.009	2323.1	63.76
76	40.239	2318.1	63.40
78	43.703	2313.0	63.04
80	47.414	2308.0	62.67
82	51.387	2302.9	62.31
84	55.635	2297.9	61.94
86	60.173	2292.8	61.56
88	65.017	2287.6	61.19
90	70.182	2282.5	60.82
92	75.684	2277.3	60.44
94	81.541	2272.1	60.06
96	87.771	2266.9	59.68
98	94.390	2261.7	59.30
100	101.42	2256.4	58.91
102	108.87	2251.1	58.53
104	116.78	2245.8	58.14
106	125.15	2240.4	57.75
108	134.01	2235.1	57.36
110	143.38	2229.6	56.96
112	153.28	2224.2	56.57
114	163.74	2218.7	56.17
116	174.77	2213.2	55.77
118	186.41	2207.7	55.37
120	198.67	2202.1	54.97
122	211.59	2196.5	54.56
124	225.18	2190.9	54.16
126	239.47	2185.2	53.75
128	254.50	2179.5	53.34
130	270.28	2173.7	52.93
132	286.85	2167.9	52.52
134	304.23	2162.1	52.11
136	322.45	2156.2	51.69
138	341.54	2150.3	51.27
140	361.54	2144.3	50.86
142	382.47	2138.3	50.44
144	404.37	2132.2	50.01
146	427.26	2126.1	49.59
148	451.18	2119.9	49.17
150	476.16	2113.7	48.74
152	502.25	2107.5	48.31
154	529.46	2101.2	47.89
156	557.84	2094.8	47.46
158	587.42	2088.4	47.02
160	618.23	2082.0	46.59
162	650.33	2075.5	46.16
164	683.73	2068.9	45.72
166	718.48	2062.3	45.28
168	754.62	2055.6	44.85
170	792.19	2048.8	44.41

t/°C	P/kPa	$\Delta_{vap}H$/kJ kg^{-1}	γ/mN m^{-1}
172	831.22	2042.0	43.97
174	871.76	2035.1	43.52
176	913.84	2028.2	43.08
178	957.51	2021.2	42.64
180	1002.8	2014.2	42.19
182	1049.8	2007.0	41.74
184	1098.5	1999.8	41.30
186	1148.9	1992.6	40.85
188	1201.1	1985.3	40.40
190	1255.2	1977.9	39.95
192	1311.2	1970.4	39.49
194	1369.1	1962.8	39.04
196	1429.0	1955.2	38.59
198	1490.9	1947.5	38.13
200	1554.9	1939.7	37.67
202	1621.0	1931.9	37.22
204	1689.3	1923.9	36.76
206	1759.8	1915.9	36.30
208	1832.6	1907.8	35.84
210	1907.7	1899.6	35.38
212	1985.1	1891.4	34.92
214	2065.0	1883.0	34.46
216	2147.3	1874.6	33.99
218	2232.2	1866.0	33.53
220	2319.6	1857.4	33.07
222	2409.6	1848.6	32.60
224	2502.3	1839.8	32.14
226	2597.8	1830.9	31.67
228	2696.0	1821.8	31.20
230	2797.1	1812.7	30.74
232	2901.0	1803.5	30.27
234	3008.0	1794.1	29.80
236	3117.9	1784.7	29.33
238	3230.8	1775.1	28.86
240	3346.9	1765.4	28.39
242	3466.2	1755.6	27.92
244	3588.7	1745.7	27.45
246	3714.5	1735.6	26.98
248	3843.6	1725.5	26.51
250	3976.2	1715.2	26.04
252	4112.2	1704.7	25.57
254	4251.8	1694.2	25.10
256	4394.9	1683.5	24.63
258	4541.7	1672.6	24.16
260	4692.3	1661.6	23.69
262	4846.6	1650.5	23.22
264	5004.7	1639.2	22.75
266	5166.8	1627.8	22.28
268	5332.9	1616.2	21.81
270	5503.0	1604.4	21.34
272	5677.2	1592.5	20.87
274	5855.6	1580.4	20.40
276	6038.3	1568.1	19.93
278	6225.2	1555.6	19.46
280	6416.6	1543.0	18.99
282	6612.4	1530.1	18.53
284	6812.8	1517.1	18.06
286	7017.7	1503.8	17.59
288	7227.4	1490.4	17.13
290	7441.8	1476.7	16.66
292	7661.0	1462.7	16.20
294	7885.2	1448.6	15.74
296	8114.3	1434.2	15.28
298	8348.5	1419.5	14.82
300	8587.9	1404.6	14.36
302	8832.5	1389.4	13.90
304	9082.4	1374.0	13.45
306	9337.8	1358.2	12.99
308	9598.6	1342.1	12.54
310	9865.1	1325.7	12.09
312	10137	1309.0	11.64
314	10415	1291.9	11.19
316	10699	1274.5	10.75
318	10989	1256.6	10.30
320	11284	1238.4	9.86
322	11586	1219.7	9.43
324	11895	1200.6	8.99
326	12209	1180.9	8.56
328	12530	1160.8	8.13
330	12858	1140.2	7.70
332	13193	1118.9	7.28
334	13534	1097.1	6.86
336	13882	1074.6	6.44
338	14238	1051.3	6.03
340	14601	1027.3	5.63
342	14971	1002.5	5.22
344	15349	976.7	4.83
346	15734	949.9	4.43
348	16128	922.0	4.05
350	16529	892.7	3.67
352	16939	862.1	3.29
354	17358	829.8	2.93
356	17785	795.5	2.57
358	18221	759.0	2.22
360	18666	719.8	1.88
362	19121	677.3	1.55
364	19585	630.5	1.23
366	20060	578.2	0.93
368	20546	517.8	0.65
370	21044	443.8	0.39
372	21554	340.3	0.16
373.95	22064	0.0	0.0

STANDARD DENSITY OF WATER

This table gives the density ρ of water in the temperature range from 0 °C to 100 °C at a pressure of 101325 Pa (one standard atmosphere). Temperatures are given on the ITS-90 scale. From 0 °C to 40 °C the values are taken from the publication in Reference 1 and refer to standard mean ocean water (SMOW), free from dissolved salts and gases. SMOW is a standard water sample of high purity and known isotopic composition. Methods of correcting for different isotopic compositions are discussed in Ref. 2. The remaining values are calculated from the NIST REFPROP program, Ref. 3, which obtains thermodynamic properties from the equation of state of Wagner and Pruss given in Ref. 4.

References

1. Tanaka, M., Girard, G., Davis, R., Peuto, A., and Bignell, N., *Metrologia* 38, 301, 2001.
2. Marsh, K. N., Ed., *Recommended Reference Materials for the Realization of Physicochemical Properties,* Blackwell Scientific Publications, Oxford, 1987.
3. Lemmon, E.W., Huber, M.L., and McLinden, M.O., NIST Standard Reference Database 23: Reference Fluid Thermodynamic and Transport Properties-REFPROP, Version 9.0, National Institute of Standards and Technology, Standard Reference Data Program, Gaithersburg, Maryland, 2010 (www.nist.gov/srd/nist23.cfm).
4. Wagner, W., and Pruss, A., *J. Phys. Chem. Ref. Data* 31, 387, 2002.

t/°C	ρ/g cm^{-3}	t/°C	ρ/g cm^{-3}	t/°C	ρ/g cm^{-3}	t/°C	ρ/g cm^{-3}	t/°C	ρ/g cm^{-3}
0.1	0.9998495	4.1	0.9999748	8.1	0.9998452	12.1	0.9994890	16.1	0.9989296
0.2	0.9998560	4.2	0.9999746	8.2	0.9998389	12.2	0.9994774	16.2	0.9989132
0.3	0.9998624	4.3	0.9999742	8.3	0.9998325	12.3	0.9994657	16.3	0.9988967
0.4	0.9998685	4.4	0.9999736	8.4	0.9998260	12.4	0.9994539	16.4	0.9988800
0.5	0.9998745	4.5	0.9999728	8.5	0.9998193	12.5	0.9994419	16.5	0.9988633
0.6	0.9998803	4.6	0.9999719	8.6	0.9998125	12.6	0.9994298	16.6	0.9988464
0.7	0.9998859	4.7	0.9999709	8.7	0.9998056	12.7	0.9994176	16.7	0.9988294
0.8	0.9998913	4.8	0.9999697	8.8	0.9997985	12.8	0.9994052	16.8	0.9988123
0.9	0.9998966	4.9	0.9999683	8.9	0.9997912	12.9	0.9993927	16.9	0.9987951
1.0	0.9999017	5.0	0.9999668	9.0	0.9997839	13.0	0.9993801	17.0	0.9987778
1.1	0.9999066	5.1	0.9999651	9.1	0.9997764	13.1	0.9993674	17.1	0.9987603
1.2	0.9999113	5.2	0.9999633	9.2	0.9997687	13.2	0.9993546	17.2	0.9987428
1.3	0.9999158	5.3	0.9999613	9.3	0.9997610	13.3	0.9993416	17.3	0.9987251
1.4	0.9999202	5.4	0.9999592	9.4	0.9997530	13.4	0.9993285	17.4	0.9987073
1.5	0.9999244	5.5	0.9999569	9.5	0.9997450	13.5	0.9993153	17.5	0.9986895
1.6	0.9999285	5.6	0.9999544	9.6	0.9997368	13.6	0.9993020	17.6	0.9986715
1.7	0.9999323	5.7	0.9999518	9.7	0.9997285	13.7	0.9992885	17.7	0.9986534
1.8	0.9999360	5.8	0.9999491	9.8	0.9997200	13.8	0.9992749	17.8	0.9986351
1.9	0.9999396	5.9	0.9999462	9.9	0.9997114	13.9	0.9992612	17.9	0.9986168
2.0	0.9999429	6.0	0.9999431	10.0	0.9997027	14.0	0.9992474	18.0	0.9985984
2.1	0.9999461	6.1	0.9999400	10.1	0.9996938	14.1	0.9992335	18.1	0.9985798
2.2	0.9999491	6.2	0.9999366	10.2	0.9996848	14.2	0.9992194	18.2	0.9985611
2.3	0.9999519	6.3	0.9999331	10.3	0.9996757	14.3	0.9992052	18.3	0.9985424
2.4	0.9999546	6.4	0.9999295	10.4	0.9996665	14.4	0.9991909	18.4	0.9985235
2.5	0.9999571	6.5	0.9999257	10.5	0.9996571	14.5	0.9991765	18.5	0.9985045
2.6	0.9999595	6.6	0.9999217	10.6	0.9996475	14.6	0.9991619	18.6	0.9984854
2.7	0.9999616	6.7	0.9999176	10.7	0.9996379	14.7	0.9991473	18.7	0.9984662
2.8	0.9999636	6.8	0.9999134	10.8	0.9996281	14.8	0.9991325	18.8	0.9984469
2.9	0.9999655	6.9	0.9999090	10.9	0.9996182	14.9	0.9991176	18.9	0.9984275
3.0	0.9999672	7.0	0.9999045	11.0	0.9996081	15.0	0.9991026	19.0	0.9984079
3.1	0.9999687	7.1	0.9998998	11.1	0.9995979	15.1	0.9990874	19.1	0.9983883
3.2	0.9999700	7.2	0.9998950	11.2	0.9995876	15.2	0.9990722	19.2	0.9983686
3.3	0.9999712	7.3	0.9998900	11.3	0.9995772	15.3	0.9990568	19.3	0.9983487
3.4	0.9999722	7.4	0.9998849	11.4	0.9995666	15.4	0.9990413	19.4	0.9983287
3.5	0.9999731	7.5	0.9998797	11.5	0.9995559	15.5	0.9990257	19.5	0.9983087
3.6	0.9999738	7.6	0.9998743	11.6	0.9995451	15.6	0.9990100	19.6	0.9982885
3.7	0.9999743	7.7	0.9998687	11.7	0.9995341	15.7	0.9989942	19.7	0.9982682
3.8	0.9999747	7.8	0.9998631	11.8	0.9995230	15.8	0.9989782	19.8	0.9982478
3.9	0.9999749	7.9	0.9998572	11.9	0.9995118	15.9	0.9989621	19.9	0.9982273
4.0	0.9999749	8.0	0.9998513	12.0	0.9995005	16.0	0.9989459	20.0	0.9982067

t/°C	ρ/g cm^{-3}	t/°C	ρ/g cm^{-3}	t/°C	ρ/g cm^{-3}	t/°C	ρ/g cm^{-3}	t/°C	ρ/g cm^{-3}
20.1	0.9981860	25.3	0.9969696	30.5	0.9954967	35.7	0.9937899	49.0	0.98848
20.2	0.9981652	25.4	0.9969436	30.6	0.9954660	35.8	0.9937549	50.0	0.98804
20.3	0.9981443	25.5	0.9969176	30.7	0.9954352	35.9	0.9937199	51.0	0.98758
20.4	0.9981233	25.6	0.9968914	30.8	0.9954044	36.0	0.9936847	52.0	0.98712
20.5	0.9981022	25.7	0.9968651	30.9	0.9953734	36.1	0.9936495	53.0	0.98665
20.6	0.9980810	25.8	0.9968387	31.0	0.9953424	36.2	0.9936142	54.0	0.98617
20.7	0.9980596	25.9	0.9968123	31.1	0.9953113	36.3	0.9935788	55.0	0.98569
20.8	0.9980382	26.0	0.9967857	31.2	0.9952801	36.4	0.9935434	56.0	0.98521
20.9	0.9980167	26.1	0.9967591	31.3	0.9952488	36.5	0.9935078	57.0	0.98471
21.0	0.9979950	26.2	0.9967324	31.4	0.9952175	36.6	0.9934722	58.0	0.98421
21.1	0.9979733	26.3	0.9967055	31.5	0.9951860	36.7	0.9934365	59.0	0.98371
21.2	0.9979514	26.4	0.9966786	31.6	0.9951545	36.8	0.9934007	60.0	0.98320
21.3	0.9979295	26.5	0.9966516	31.7	0.9951228	36.9	0.9933649	61.0	0.98268
21.4	0.9979074	26.6	0.9966245	31.8	0.9950911	37.0	0.9933290	62.0	0.98216
21.5	0.9978853	26.7	0.9965973	31.9	0.9950593	37.1	0.9932929	63.0	0.98163
21.6	0.9978630	26.8	0.9965700	32.0	0.9950275	37.2	0.9932569	64.0	0.98109
21.7	0.9978407	26.9	0.9965426	32.1	0.9949955	37.3	0.9932207	65.0	0.98055
21.8	0.9978182	27.0	0.9965151	32.2	0.9949635	37.4	0.9931844	66.0	0.98000
21.9	0.9977956	27.1	0.9964875	32.3	0.9949313	37.5	0.9931481	67.0	0.97945
22.0	0.9977730	27.2	0.9964599	32.4	0.9948991	37.6	0.9931117	68.0	0.97890
22.1	0.9977502	27.3	0.9964321	32.5	0.9948668	37.7	0.9930753	69.0	0.97833
22.2	0.9977273	27.4	0.9964043	32.6	0.9948344	37.8	0.9930387	70.0	0.97776
22.3	0.9977044	27.5	0.9963763	32.7	0.9948020	37.9	0.9930021	71.0	0.97719
22.4	0.9976813	27.6	0.9963483	32.8	0.9947694	38.0	0.9929654	72.0	0.97661
22.5	0.9976582	27.7	0.9963202	32.9	0.9947368	38.1	0.9929286	73.0	0.97603
22.6	0.9976349	27.8	0.9962920	33.0	0.9947041	38.2	0.9928917	74.0	0.97544
22.7	0.9976115	27.9	0.9962637	33.1	0.9946713	38.3	0.9928548	75.0	0.97484
22.8	0.9975881	28.0	0.9962353	33.2	0.9946384	38.4	0.9928178	76.0	0.97424
22.9	0.9975645	28.1	0.9962068	33.3	0.9946055	38.5	0.9927807	77.0	0.97364
23.0	0.9975408	28.2	0.9961783	33.4	0.9945724	38.6	0.9927435	78.0	0.97303
23.1	0.9975171	28.3	0.9961496	33.5	0.9945393	38.7	0.9927063	79.0	0.97241
23.2	0.9974932	28.4	0.9961208	33.6	0.9945061	38.8	0.9926689	80.0	0.97179
23.3	0.9974692	28.5	0.9960920	33.7	0.9944728	38.9	0.9926316	81.0	0.97116
23.4	0.9974452	28.6	0.9960631	33.8	0.9944394	39.0	0.9925941	82.0	0.97053
23.5	0.9974210	28.7	0.9960341	33.9	0.9944060	39.1	0.9925565	83.0	0.96990
23.6	0.9973968	28.8	0.9960050	34.0	0.9943724	39.2	0.9925189	84.0	0.96926
23.7	0.9973724	28.9	0.9959758	34.1	0.9943388	39.3	0.9924812	85.0	0.96861
23.8	0.9973480	29.0	0.9959465	34.2	0.9943051	39.4	0.9924434	86.0	0.96796
23.9	0.9973234	29.1	0.9959171	34.3	0.9942713	39.5	0.9924056	87.0	0.96731
24.0	0.9972988	29.2	0.9958876	34.4	0.9942375	39.6	0.9923677	88.0	0.96664
24.1	0.9972740	29.3	0.9958581	34.5	0.9942035	39.7	0.9923297	89.0	0.96598
24.2	0.9972492	29.4	0.9958285	34.6	0.9941695	39.8	0.9922916	90.0	0.96531
24.3	0.9972243	29.5	0.9957987	34.7	0.9941354	39.9	0.9922534	91.0	0.96463
24.4	0.9971992	29.6	0.9957689	34.8	0.9941012	40.0	0.9922152	92.0	0.96396
24.5	0.9971741	29.7	0.9957390	34.9	0.9940669	41.0	0.99183	93.0	0.96327
24.6	0.9971489	29.8	0.9957090	35.0	0.9940326	42.0	0.99144	94.0	0.96258
24.7	0.9971236	29.9	0.9956790	35.1	0.9939982	43.0	0.99104	95.0	0.96189
24.8	0.9970981	30.0	0.9956488	35.2	0.9939637	44.0	0.99063	96.0	0.96119
24.9	0.9970726	30.1	0.9956185	35.3	0.9939291	45.0	0.99021	97.0	0.96049
25.0	0.9970470	30.2	0.9955882	35.4	0.9938944	46.0	0.98979	98.0	0.95978
25.1	0.9970213	30.3	0.9955578	35.5	0.9938597	47.0	0.98936	99.0	0.95907
25.2	0.9969955	30.4	0.9955273	35.6	0.9938248	48.0	0.98893	99.974	0.95837

FIXED-POINT PROPERTIES OF H_2O AND D_2O

Allan H. Harvey

Temperatures are given on the ITS-90 scale.

References

1. International Association for the Properties of Water and Steam (IAPWS), *Guideline on the Use of Fundamental Physical Constants and Basic Constants of Water* (2001; 2008 update), available from http://www.iapws.org
2. IAPWS, *Revised Release on the Pressure along the Melting and Sublimation Curves of Ordinary Water Substance* (2011), available from http://www.iapwsâœ.org
3. IAPWS, *IAPWS Release on the Values of Temperature, Pressure, and Density of Ordinary and Heavy Water Substances at their Respective Critical Points* (1992), available from http://www.iapws.org
4. Wagner, W., and Pruß, A., *J. Phys. Chem. Ref. Data* 31, 387, 2002.
5. Hill, P. G., MacMillan, R. D. C., and Lee, V., *J. Phys. Chem. Ref. Data* 11, 1, 1982.
6. Guildner, L. A., Johnson, D. P., and Jones, F. E., *J. Res. Nat. Bur. Stand.* 80A, 505, 1976.
7. Harvey, A. H., and Lemmon, E. W., *J. Phys. Chem. Ref. Data* 31, 173, 2002.
8. Harvey, A. H., Peskin, A. P., and Klein, S. A., NIST Standard Reference Database 10: *NIST/ASME Steam Properties*, Version 2.22, National Institute of Standards and Technology, Standard Reference Data Program, Gaithersburg, Maryland, 2008 (http://www.nist.gov/srd/nist10.cfm).
9. Lemmon, E. W., Huber, M. L., and McLinden, M. O., NIST Standard Reference Database 23: *Reference Fluid Thermodynamic and Transport Properties-REFPROP*, Version 9.0, National Institute of Standards and Technology, Standard Reference Data Program, Gaithersburg, Maryland, 2010 (http://www.nist.gov/srd/nist23.cfm).

Fixed-Point Properties of H_2O and D_2O

	Unit	H_2O	D_2O
Molar mass	g mol^{-1}	18.015268	20.02751
Melting point (101.325 kPa)	°C	0.0025	3.81
Boiling point (101.325 kPa)	°C	99.974	101.40
Triple-point temperature	°C	0.01[a]	3.82
Triple-point pressure	Pa	611.657	661
Triple-point density (liquid)	g cm^{-3}	0.99979	1.1055
Critical temperature	°C	373.946	370.697
Critical pressure	MPa	22.064	21.671
Critical density	g cm^{-3}	0.322	0.356
Maximum density (101.325 kPa)	g cm^{-3}	0.999975	1.1060
Temperature of maximum density	°C	3.98	11.2

[a] Exact

PROPERTIES OF SATURATED LIQUID D_2O

Allan H. Harvey

Properties of saturated liquid heavy water, D_2O, are given in this table as a function of temperature from the melting point to the critical point. The vapor pressure was calculated from the formulation of Harvey and Lemmon (Ref. 2). The other properties were generated from the NIST REFPROP program (Ref. 1) and are consistent with formulations adopted for general and scientific use by The International Association for the Properties of Water and Steam (IAPWS). The background for the equation of state used for density and heat capacity is given by Hill et al. (Ref. 3), and the background for the transport property correlations is given by Matsunaga and Nagashima (Ref. 4). The unpublished surface tension correlation and the other IAPWS formulations may be found on the IAPWS Web site (http://www.iapws.org). The temperature scale is ITS-90. Additional calculations at state points not listed below can be obtained by using the REFPROP program (http://www.nist.gov/srd/nist23.htm). The properties are

P: vapor pressure
ρ: density
C_p: isobaric heat capacity
η: viscosity
λ: thermal conductivity
σ: surface tension

References

1. Lemmon, E.W., Huber, M.L., and McLinden, M.O., NIST Standard Reference Database 23: Reference Fluid Thermodynamic and Transport Properties-REFPROP, Version 9.0, National Institute of Standards and Technology, Standard Reference Data Program, Gaithersburg, MD, 2010 (www.nist.gov/srd/nist23.cfm).
2. Harvey, A. H. and Lemmon, E. W., *J. Phys. Chem. Ref. Data* 31, 173, 2002.
3. Hill, P. G., MacMillan, R. D. C., and Lee, V., *J. Phys. Chem. Ref. Data* 11, 1, 1982.
4. Matsunaga, N. and Nagashima, A., *J. Phys. Chem. Ref. Data* 12, 933, 1983.

t/°C	P/kPa	ρ/kg m^{-3}	C_p/kJ kg^{-1} K^{-1}	η/mPa s	λ/W m^{-1} K^{-1}	σ/mN m^{-1}
3.82	0.661	1105.5	4.211	2.086	0.565	74.93
10	1.026	1106.0	4.231	1.679	0.575	74.06
20	1.999	1105.3	4.243	1.247	0.589	72.61
30	3.702	1103.2	4.240	0.972	0.600	71.09
40	6.550	1099.9	4.232	0.785	0.610	69.52
50	11.12	1095.6	4.220	0.651	0.618	67.89
60	18.20	1090.5	4.207	0.552	0.625	66.21
70	28.81	1084.6	4.194	0.476	0.629	64.47
80	44.24	1078.1	4.181	0.416	0.633	62.67
90	66.09	1071.0	4.170	0.368	0.635	60.82
100	96.30	1063.3	4.161	0.329	0.636	58.93
110	137.1	1055.1	4.155	0.296	0.636	56.98
120	191.3	1046.4	4.153	0.269	0.635	54.99
130	261.8	1037.2	4.155	0.247	0.632	52.95
140	352.0	1027.5	4.162	0.227	0.629	50.87
150	465.8	1017.2	4.174	0.210	0.625	48.75
160	607.3	1006.5	4.190	0.195	0.620	46.59
170	781.1	995.2	4.212	0.182	0.614	44.39
180	992.0	983.4	4.240	0.171	0.607	42.16
190	1246	971.1	4.273	0.161	0.600	39.90
200	1547	958.2	4.313	0.152	0.592	37.61
210	1903	944.7	4.360	0.144	0.583	35.29
220	2319	930.5	4.415	0.136	0.574	32.95
230	2802	915.7	4.479	0.130	0.563	30.59
240	3359	900.1	4.554	0.123	0.553	28.22
250	3998	883.7	4.643	0.118	0.541	25.84
260	4725	866.4	4.750	0.112	0.529	23.45
270	5550	848.0	4.880	0.107	0.516	21.07
280	6480	828.4	5.038	0.103	0.502	18.69
290	7525	807.5	5.237	0.098	0.488	16.33
300	8694	784.8	5.490	0.094	0.473	13.99
310	9998	760.1	5.823	0.089	0.458	11.68

t/°C	P/kPa	ρ/kg m^{-3}	C_p/kJ kg^{-1} K^{-1}	η/mPa s	λ/W m^{-1} K^{-1}	σ/mN m^{-1}
320	11449	732.7	6.281	0.085	0.442	9.428
330	13059	702.0	6.952	0.080	0.425	7.238
340	14845	666.5	8.041	0.075	0.408	5.141
350	16824	623.3	10.17	0.069	0.391	3.173
360	19024	565.1	16.41	0.062	0.382	1.405
370	21487	430.7	268.8	0.046	0.548	0.0467
370.697	21671	356.0				0

Fluids

PROPERTIES OF ICE AND SUPERCOOLED WATER

Allan H. Harvey

The common form of ice at ambient pressure is hexagonal ice, designated as ice Ih (see phase diagram in Section 12). The data given here refer to that form, at standard atmospheric pressure (101.325 kPa). Data have been taken from the references indicated, which in most cases are formulations based on critical evaluation of available experimental data. Most properties are sensitive to the method of preparation of the sample, because air and other gases are sometimes occluded. For this reason, there is often disagreement among values in the literature. For all properties except the dielectric constant of ice, the cited reference contains information on the uncertainty of the property.

Phase Transition Properties (Ref. 2)

$\Delta_{fus}H(0\ °C) = 333.4\ J\ g^{-1}$
$\Delta_{subl}H(0\ °C) = 2834\ J\ g^{-1}$

Properties of Ice Ih and Supercooled Water

ρ: mass density (Refs. 2, 3)
α_V: cubic expansion coefficient, $\alpha_V = -(1/V)(\partial V/\partial T)_p$ (Refs. 2, 3)
κ_s: isentropic compressibility, $\kappa_s = -(1/V)(\partial V/\partial p)_s$ (Refs. 2, 3)
c_p: specific heat capacity at constant pressure (Refs. 2, 3)
λ: thermal conductivity (Refs. 4, 5)
ε: static dielectric constant (relative permittivity) (Ref. 6)

	t/°C	ρ/g cm^{-3}	$10^3 \cdot \alpha_V$/K^{-1}	κ_S/GPa^{-1}	c_p/J g^{-1} K^{-1}	λ/W m^{-1} K^{-1}	ε
Ice	0	0.9167	0.160	0.114	2.10	2.16	91.2
Ice	-10	0.9182	0.155	0.113	2.02	2.26	95.1
Ice	-20	0.9196	0.150	0.111	1.95	2.38	99.4
Ice	-30	0.9209	0.144	0.110	1.88	2.50	104.1
Ice	-40	0.9222	0.138	0.108	1.80	2.63	109.2
Ice	-50	0.9235	0.132	0.107	1.73	2.77	115.0
Ice	-60	0.9247	0.126	0.106	1.66	2.93	121.4
Ice	-80	0.9269	0.111	0.103	1.52	3.27	136.6
Ice	-100	0.9288	0.095	0.101	1.38	3.69	
Ice	-120	0.9304	0.078	0.099	1.25	4.2	
Ice	-140	0.9317	0.060	0.097	1.11	4.9	
Ice	-160	0.9326	0.041	0.096	0.97	5.7	
Ice	-180	0.9332	0.025	0.095	0.82	7.0	
Ice	-200	0.9336	0.013	0.095	0.65	8.9	
Ice	-220	0.9337	0.0050	0.095	0.47	12.2	
Ice	-240	0.9338	0.0012	0.095	0.27	20	
Ice	-260	0.9338	0.00008	0.095	0.036		
Supercooled water							
Water	0	0.9998					
Water	-5	0.9993					
Water	-10	0.9981					
Water	-15	0.9963					
Water	-20	0.9936					
Water	-25	0.9896					
Water	-30	0.9838					

References

1. Wagner, W., and Pruß, A., *J. Phys. Chem. Ref. Data* 31, 387, 2002.
2. Feistel, R., and Wagner, W., *J. Phys. Chem. Ref. Data* 35, 1021, 2006.
3. International Association for the Properties of Water and Steam (IAPWS), *Revised Release on the Equation of State 2006 for H_2O Ice Ih* (2009), available from <www.iapws.org>.
4. Andersson, O., and Inaba, A., *Phys. Chem. Chem. Phys.* 7, 1441, 2005.
5. Slack, G.A., *Phys. Rev. B* 22, 3065, 1980.
6. Wörz, O., and Cole, R.H., *J. Chem. Phys.* 51, 1546, 1969.

VAPOR PRESSURE OF ICE

The values of the vapor (sublimation) pressure of ice Ih were calculated from the equation recommended by the International Association for the Properties of Water and Steam (IAPWS) in 2008. See Refs. 1 and 2 for details on the uncertainty in different temperature ranges. The first entry in the table is the triple point of water, whose pressure has an uncertainty of 0.010 Pa.

References

1. International Association for the Properties of Water and Steam (IAPWS), *Revised Release on the Pressure along the Melting and Sublimation Curves of Ordinary Water Substance* (2011), available from http://www.iapws.org.
2. Wagner, W., Riethmann, T., Feistel, R., and Harvey, A. H., New Equations for the Sublimation Pressure and Melting Pressure of H_2O Ice Ih, *J. Phys. Chem. Ref. Data* 40 143103, 2011.

Vapor Pressure for Temperature in °C

t/°C	*p*/Pa	*t*/°C	*p*/Pa	*t*/°C	*p*/Pa
0.01	611.657	-17	137.22	-65	0.541
0	611.15	-18	124.90	-70	0.262
-1	562.66	-19	113.60	-75	0.122
-2	517.70	-20	103.24	-80	5.48×10^{-2}
-3	476.04	-22	85.08	-90	9.68×10^{-3}
-4	437.45	-24	69.89	-100	1.40×10^{-3}
-5	401.74	-26	57.24	-110	1.61×10^{-4}
-6	368.71	-28	46.72	-120	1.41×10^{-5}
-7	338.17	-30	38.01	-130	8.75×10^{-7}
-8	309.95	-32	30.81	-140	3.62×10^{-8}
-9	283.91	-34	24.89	-150	9.00×10^{-10}
-10	259.87	-36	20.04	-160	1.18×10^{-11}
-11	237.71	-38	16.07	-170	6.80×10^{-14}
-12	217.29	-40	12.84	-180	1.32×10^{-16}
-13	198.49	-45	7.203	-190	5.89×10^{-20}
-14	181.19	-50	3.938	-200	3.31×10^{-24}
-15	165.27	-55	2.094	-210	8.86×10^{-30}
-16	150.65	-60	1.081	-220	2.06×10^{-37}

Vapor Pressure for Temperature in K

T/K	*p*/Pa	*T*/K	*p*/Pa	*T*/K	*p*/Pa
273.16	611.657	256	135.30	190	3.24×10^{-2}
273.15	611.15	254	111.98	180	5.39×10^{-3}
273	603.65	252	92.40	170	7.30×10^{-4}
272	555.70	250	76.01	160	7.73×10^{-5}
271	511.25	248	62.33	150	6.10×10^{-6}
270	470.06	246	50.95	140	3.37×10^{-7}
269	431.92	244	41.51	130	1.20×10^{-8}
268	396.62	242	33.70	120	2.48×10^{-10}
267	363.97	240	27.27	110	2.57×10^{-12}
266	333.79	235	15.81	100	1.09×10^{-14}
265	305.91	230	8.947	90	1.39×10^{-17}
264	280.18	225	4.939	80	3.50×10^{-21}
263	256.43	220	2.654	70	8.58×10^{-26}
262	234.54	215	1.386	60	6.51×10^{-32}
261	214.37	210	0.702	50	1.93×10^{-40}
260	195.80	205	0.344		
258	163.00	200	0.163		

MELTING POINT OF ICE AS A FUNCTION OF PRESSURE

This table gives the melting temperature of ice at various pressures, calculated from the equation for the ice Ih – liquid water phase boundary recommended by the International Association for the Properties of Water and Steam (IAPWS) in 2008. See Refs. 1 and 2 for information on the solid/liquid transitions for high-pressure forms of ice. IAPWS gives the following locations for the triple points where equilibrium exists among two ice forms and liquid water:

Phase transition	*p*/MPa	*t*/°C
ice I – ice III	208.566 MPa	−21.985 °C
ice III – ice V	350.1 MPa	−16.986 °C
ice V – ice VI	632.4 MPa	0.16 °C
ice VI – ice VII	2216 MPa	81.85 °C

References

1. International Association for the Properties of Water and Steam (IAPWS), *Revised Release on the Pressure along the Melting and Sublimation Curves of Ordinary Water Substance* (2011), available from http://www.iapws.org.
2. Wagner, W., Riethmann, T., Feistel, R., and Harvey, A. H., New Equations for the Sublimation Pressure and Melting Pressure of H_2O Ice Ih, *J. Phys. Chem. Ref. Data*, 40, 143103, 2011.

Melting Temperature of Ice Ih as a Function of Pressure

p/MPa	*t*/°C
0.000 612	0.01
0.1	0.0026
1	-0.064
2	-0.14
5	-0.37
10	-0.75
15	-1.14
20	-1.54
30	-2.36
40	-3.21
50	-4.09
60	-5.00
70	-5.94
80	-6.91
90	-7.91
100	-8.94
120	-11.09
140	-13.35
160	-15.73
180	-18.22
200	-20.83

Fluids

PERMITTIVITY (DIELECTRIC CONSTANT) OF WATER AT VARIOUS FREQUENCIES

The permittivity of liquid water in the radiofrequency and microwave regions can be represented by the Debye equation (Refs. 1 and 2):

$$\varepsilon' = \varepsilon_\infty + \frac{\varepsilon_s - \varepsilon_\infty}{1+\omega^2\tau^2}$$

$$\varepsilon'' = \frac{(\varepsilon_s - \varepsilon_\infty)\omega\tau}{1+\omega^2\tau^2}$$

where $\varepsilon = \varepsilon' + i\,\varepsilon''$ is the (complex) relative permittivity (i.e., the absolute permittivity divided by the permittivity of free space $\varepsilon_0 = 8.854{\cdot}10^{-12}$ F m^{-1}). Here ε_s is the static permittivity (see Ref. 3 and the table "Thermophysical Properties of Water and Steam" in this section); ε_∞ is a parameter describing the permittivity in the high frequency limit; τ is the relaxation time for molecular orientation; and $\omega = 2\pi f$ is the angular frequency. The values in this table have been calculated from parameters given in Ref. 2:

	0 °C	25 °C	50 °C
ε_∞	5.7	5.2	4.0
τ/ps	17.67	8.27	4.75

Other useful quantities that can be calculated from the values in the table are the loss tangent:

$$\tan\delta = \varepsilon''/\varepsilon'$$

and the absorption coefficient α which describes the power attenuation per unit length ($P = P_0 e^{-\alpha l}$):

$$\alpha = \frac{\pi f \varepsilon''}{c\sqrt{\varepsilon'}}$$

and c is the speed of light. The last equation is valid when $\varepsilon''/\varepsilon' << 1$.

References

1. Fernández, D. P., Mulev, Y., Goodwin, A. R. H., and Levelt Sengers, J. M. H., *J. Phys. Chem. Ref. Data*, 24, 33, 1995.
2. Kaatze, U., *J. Chem. Eng. Data*, 34, 371, 1989.
3. Archer, D. G., and Wang, P., *J. Phys. Chem. Ref. Data*, 19, 371, 1990.

Frequency	ε' (0 °C)	ε'' (0 °C)	ε' (25 °C)	ε'' (25 °C)	ε' (50 °C)	ε'' (50 °C)
0	87.90	0.00	78.36	0.00	69.88	0.00
1 kHz	87.90	0.00	78.36	0.00	69.88	0.00
1 MHz	87.90	0.01	78.36	0.00	69.88	0.00
10 MHz	87.90	0.09	78.36	0.04	69.88	0.02
100 MHz	87.89	0.91	78.36	0.38	69.88	0.20
200 MHz	87.86	1.82	78.35	0.76	69.88	0.39
500 MHz	87.65	4.55	78.31	1.90	69.87	0.98
1 GHz	86.90	9.01	78.16	3.79	69.82	1.96
2 GHz	84.04	17.39	77.58	7.52	69.65	3.92
3 GHz	79.69	24.64	76.62	11.13	69.36	5.85
4 GHz	74.36	30.49	75.33	14.58	68.95	7.75
5 GHz	68.54	34.88	73.73	17.81	68.45	9.62
10 GHz	42.52	40.88	62.81	29.93	64.49	18.05
20 GHz	19.56	30.78	40.37	36.55	52.57	28.99
30 GHz	12.50	22.64	26.53	33.25	40.57	32.74
40 GHz	9.67	17.62	18.95	28.58	31.17	32.43
50 GHz	8.28	14.34	14.64	24.53	24.42	30.47

THERMOPHYSICAL PROPERTIES OF AIR

Eric W. Lemmon

These tables summarize the thermophysical properties of air in the liquid and gaseous states as calculated from the pseudo-pure fluid equation of state of Lemmon et al. (Ref. 1). The first table refers to liquid and gaseous air at equilibrium as a function of temperature. The tabulated properties are the bubble-point pressure (i.e., pressure at which boiling begins as the pressure of the liquid is lowered); the dew-point pressure (i.e., pressure at which condensation begins as the pressure of the gas is raised); density (ρ); enthalpy (H); entropy (S); isochoric heat capacity (C_v); isobaric heat capacity (C_p); speed of sound (u); viscosity (η); and thermal conductivity (λ). The first line of identical temperatures is the bubble point (liquid) and the second line is the dew point (vapor). The normal boiling point of air, i.e., the temperature at which the bubble-point pressure reaches 1 standard atmosphere (1.01325 bar), is 78.90 K (-194.25 °C).

The second table gives the properties of air along various isobars. An entry with non-integer temperatures in the isobar section indicates a phase transition (liquid–vapor) at these temperatures; property values are then given for both phases. These are identified by the high densities in the liquid and the low densities in the vapor. Additional calculations at state points not listed below can be obtained by using the NIST program REFPROP (http://www.nist.gov/srd/nist23.cfm).

In the range from the solidification point to 873 K at pressures to 70 MPa, the estimated uncertainty of density values calculated with the equation of state is 0.1%. The estimated uncertainty of calculated speed-of-sound values is 0.2%, and that for calculated heat capacities is 1%. At temperatures above 873 K and 70 MPa, the estimated uncertainty of calculated density values is 0.5%, increasing to 1.0% at 2000 K and 2000 MPa.

References

1. Lemmon, E.W., Jacobsen, R.T, Penoncello, S.G., and Friend, D.G., Thermodynamic Properties of Air and Mixtures of Nitrogen, Argon, and Oxygen from 60 to 2000 K at Pressures to 2000 MPa, *J. Phys. Chem. Ref. Data*, 29, 331, 2000.
2. Lemmon, E. W. and Jacobsen, R. T, Viscosity and Thermal Conductivity Equations for Nitrogen, Oxygen, Argon, and Air, *Int. J. Thermophys.*, 25, 21, 2004.
3. Lemmon, E.W., Huber, M.L., and McLinden, M.O., NIST Standard Reference Database 23: Reference Fluid Thermodynamic and Transport Properties-REFPROP, Version 9.0, National Institute of Standards and Technology, Standard Reference Data Program, Gaithersburg, MD, 2010 <http://www.nist.gov/srd/nist23.cfm>.

Thermophysical Properties of Air along the Boiling and Condensation Curves

T/ K	P/ MPa	ρ/ kg m^{-3}	H/ kJ kg^{-1}	S/ kJ kg^{-1} K^{-1}	C_v/ kJ kg^{-1} K^{-1}	C_p/ kJ kg^{-1} K^{-1}	u/ m s^{-1}	η/ µPa s	λ/ mW m^{-1} K^{-1}
59.75	0.005265	957.6	-36.66	-0.5306	1.174	1.901	1030	376.6	171.4
59.75	0.002432	0.1421	185.5	3.340	0.7184	1.009	154.8	4.220	5.294
60	0.005546	956.5	-36.19	-0.5226	1.173	1.901	1028	371.9	171.0
60	0.002584	0.1504	185.8	3.326	0.7186	1.009	155.1	4.238	5.320
62	0.008270	948.2	-32.38	-0.4603	1.157	1.901	1012	336.9	167.8
62	0.004111	0.2318	187.7	3.225	0.7198	1.012	157.6	4.386	5.529
64	0.01200	939.9	-28.58	-0.3999	1.143	1.902	995.8	306.3	164.5
64	0.006325	0.3460	189.6	3.132	0.7212	1.015	160.0	4.532	5.739
66	0.01699	931.5	-24.77	-0.3414	1.129	1.903	979.1	279.4	161.3
66	0.009442	0.5018	191.5	3.047	0.7230	1.019	162.3	4.679	5.950
68	0.02352	923.0	-20.95	-0.2846	1.115	1.906	962.2	255.7	158.0
68	0.01371	0.7089	193.4	2.968	0.7252	1.024	164.5	4.825	6.162
70	0.03191	914.4	-17.13	-0.2293	1.102	1.908	945.1	234.8	154.7
70	0.01943	0.9785	195.2	2.896	0.7277	1.030	166.7	4.970	6.376
72	0.04250	905.7	-13.31	-0.1756	1.090	1.912	927.7	216.3	151.4
72	0.02692	1.322	197.0	2.828	0.7305	1.037	168.7	5.115	6.592
74	0.05566	897.0	-9.468	-0.1232	1.078	1.917	910.0	199.9	148.1
74	0.03655	1.753	198.7	2.766	0.7338	1.046	170.6	5.260	6.810
76	0.07179	888.1	-5.617	-0.07209	1.067	1.923	892.1	185.2	144.8
76	0.04870	2.285	200.4	2.708	0.7375	1.055	172.5	5.405	7.031
78	0.09129	879.1	-1.751	-0.02217	1.056	1.930	873.9	172.1	141.5
78	0.06381	2.933	202.0	2.653	0.7416	1.066	174.2	5.549	7.256
80	0.1146	870.0	2.132	0.02665	1.045	1.938	855.4	160.4	138.2
80	0.08232	3.711	203.6	2.602	0.7460	1.078	175.8	5.694	7.485
82	0.1422	860.7	6.036	0.07444	1.035	1.948	836.7	149.8	134.8
82	0.1047	4.635	205.1	2.554	0.7510	1.092	177.4	5.839	7.719
84	0.1745	851.3	9.962	0.1213	1.025	1.959	817.6	140.2	131.4
84	0.1315	5.724	206.5	2.509	0.7563	1.108	178.8	5.984	7.959
86	0.2121	841.7	13.91	0.1673	1.016	1.972	798.2	131.5	128.1
86	0.1631	6.993	207.8	2.466	0.7620	1.125	180.0	6.131	8.206

T/ K	P/ MPa	ρ/ kg m^{-3}	H/ kJ kg^{-1}	S/ kJ kg^{-1} K^{-1}	C_v/ kJ kg^{-1} K^{-1}	C_p/ kJ kg^{-1} K^{-1}	u/ m s^{-1}	η/ μPa s	λ/ mW m^{-1} K^{-1}
88	0.2553	832.0	17.90	0.2125	1.007	1.986	778.6	123.6	124.8
88	0.2002	8.464	209.1	2.425	0.7682	1.144	181.2	6.278	8.461
90	0.3048	822.0	21.91	0.2569	0.9984	2.003	758.5	116.4	121.4
90	0.2432	10.16	210.3	2.386	0.7748	1.166	182.2	6.427	8.725
92	0.3609	811.8	25.97	0.3007	0.9902	2.022	738.2	109.7	118.0
92	0.2927	12.09	211.4	2.349	0.7817	1.190	183.1	6.578	9.001
94	0.4243	801.4	30.06	0.3439	0.9825	2.044	717.5	103.6	114.6
94	0.3493	14.29	212.3	2.313	0.7891	1.217	183.8	6.732	9.289
96	0.4954	790.7	34.21	0.3866	0.9752	2.069	696.5	97.88	111.2
96	0.4136	16.78	213.2	2.279	0.7969	1.248	184.5	6.889	9.593
98	0.5749	779.7	38.41	0.4288	0.9684	2.098	675.0	92.57	107.8
98	0.4861	19.60	214.0	2.246	0.8052	1.282	184.9	7.050	9.915
100	0.6631	768.4	42.66	0.4707	0.9619	2.131	653.3	87.61	104.4
100	0.5674	22.76	214.6	2.213	0.8138	1.320	185.3	7.215	10.26
102	0.7608	756.7	46.98	0.5122	0.9560	2.168	631.1	82.94	101.0
102	0.6582	26.32	215.1	2.182	0.8230	1.363	185.5	7.387	10.63
104	0.8684	744.6	51.38	0.5535	0.9505	2.212	608.5	78.53	97.62
104	0.7590	30.31	215.5	2.151	0.8326	1.413	185.6	7.566	11.02
106	0.9864	732.1	55.86	0.5947	0.9456	2.262	585.5	74.35	94.25
106	0.8706	34.78	215.7	2.120	0.8429	1.470	185.5	7.754	11.46
108	1.116	719.1	60.44	0.6358	0.9412	2.321	562.1	70.36	90.89
108	0.9934	39.79	215.8	2.089	0.8537	1.536	185.2	7.952	11.94
110	1.256	705.5	65.12	0.6769	0.9375	2.390	538.2	66.54	87.55
110	1.128	45.41	215.6	2.059	0.8653	1.614	184.9	8.163	12.47
112	1.409	691.2	69.93	0.7182	0.9345	2.472	513.9	62.87	84.24
112	1.276	51.73	215.2	2.028	0.8777	1.708	184.3	8.391	13.07
114	1.575	676.2	74.87	0.7598	0.9324	2.571	489.0	59.31	80.96
114	1.437	58.84	214.6	1.997	0.8912	1.821	183.6	8.637	13.76
116	1.755	660.3	79.98	0.8019	0.9312	2.693	463.5	55.85	77.72
116	1.612	66.88	213.8	1.965	0.9059	1.961	182.7	8.909	14.56
118	1.948	643.4	85.29	0.8447	0.9312	2.847	437.3	52.47	74.52
118	1.801	76.04	212.6	1.932	0.9220	2.139	181.7	9.210	15.50
120	2.156	625.1	90.83	0.8885	0.9327	3.048	410.2	49.13	71.36
120	2.007	86.55	211.0	1.898	0.9402	2.374	180.4	9.552	16.63
122	2.379	605.3	96.66	0.9338	0.9363	3.323	382.0	45.81	68.24
122	2.229	98.76	208.9	1.861	0.9608	2.694	179.1	9.946	18.04
124	2.617	583.3	102.9	0.9811	0.9427	3.723	352.3	42.46	65.17
124	2.468	113.2	206.3	1.821	0.9847	3.157	177.5	10.41	19.85
126	2.872	558.3	109.7	1.032	0.9537	4.367	320.4	39.01	62.18
126	2.727	130.6	202.9	1.777	1.013	3.882	175.8	10.98	22.29
128	3.143	528.3	117.3	1.088	0.9728	5.589	285.0	35.33	59.44
128	3.006	152.6	198.3	1.725	1.049	5.166	174.0	11.72	25.84
130	3.429	488.3	126.7	1.157	1.010	8.849	243.7	31.07	58.05
130	3.308	182.7	191.7	1.660	1.096	8.033	171.9	12.77	31.81
132	3.723	411.2	142.6	1.273	1.117	35.04	189.1	24.47	67.80
132	3.646	235.4	179.7	1.556	1.168	20.65	169.4	14.80	47.00
132.63	3.785	302.6	164.5	1.437				17.83	

Thermophysical Properties of Air along Various Isobars

T/ K	ρ/ kg m^{-3}	H/ kJ kg^{-1}	S/ kJ kg^{-1} K^{-1}	C_v/ kJ kg^{-1} K^{-1}	C_p/ kJ kg^{-1} K^{-1}	u/ m s^{-1}	η/ μPa s	λ/ mW m^{-1} K^{-1}
P = 0.1 MPa (1 bar)								
60	956.7	-36.11	-0.5230	1.173	1.901	1029	372.4	171.1
78.79	875.5	-0.2237	-0.002818	1.051	1.933	866.7	167.4	140.2
81.61	4.442	204.8	2.563	0.7500	1.089	177.1	5.811	7.673
100	3.557	224.3	2.779	0.7282	1.040	198.2	7.107	9.469
120	2.938	244.9	2.966	0.7211	1.022	218.3	8.457	11.38
140	2.507	265.2	3.123	0.7184	1.014	236.4	9.750	13.24
160	2.188	285.5	3.258	0.7172	1.011	253.2	10.99	15.05

T/ K	ρ/ kg m⁻³	H/ kJ kg⁻¹	S/ kJ kg⁻¹ K⁻¹	C_v/ kJ kg⁻¹ K⁻¹	C_p/ kJ kg⁻¹ K⁻¹	u/ m s⁻¹	η/ μPa s	λ/ mW m⁻¹ K⁻¹
180	1.942	305.6	3.377	0.7166	1.008	268.8	12.18	16.80
200	1.746	325.8	3.483	0.7163	1.007	283.5	13.33	18.50
220	1.586	345.9	3.579	0.7163	1.006	297.4	14.44	20.16
240	1.453	366.0	3.667	0.7164	1.006	310.7	15.51	21.77
260	1.341	386.2	3.747	0.7168	1.006	323.4	16.55	23.35
280	1.245	406.3	3.822	0.7173	1.006	335.6	17.56	24.88
300	1.161	426.4	3.891	0.7181	1.007	347.4	18.54	26.38
320	1.089	446.5	3.956	0.7192	1.007	358.7	19.49	27.85
340	1.024	466.7	4.018	0.7206	1.009	369.6	20.41	29.29
360	0.9674	486.9	4.075	0.7223	1.010	380.3	21.32	30.71
380	0.9164	507.1	4.130	0.7243	1.012	390.5	22.20	32.09
400	0.8706	527.4	4.182	0.7266	1.014	400.5	23.06	33.45
500	0.6964	629.5	4.410	0.7426	1.030	446.4	27.09	39.94
600	0.5803	733.6	4.599	0.7641	1.051	487.1	30.77	46.01
700	0.4974	839.9	4.763	0.7879	1.075	523.9	34.18	51.76
800	0.4352	948.6	4.908	0.8117	1.099	557.8	37.37	57.25
900	0.3869	1060	5.039	0.8340	1.121	589.6	40.39	62.54
1000	0.3482	1173	5.158	0.8540	1.141	619.6	43.28	67.68
***P* = 0.5 MPa (5 bar)**								
60	957.3	-35.80	-0.5248	1.173	1.900	1031	374.6	171.4
80	870.9	2.387	0.02430	1.046	1.934	858.5	161.4	138.6
96.12	790.0	34.46	0.3892	0.9748	2.071	695.2	97.55	111.0
98.36	20.14	214.1	2.240	0.8067	1.288	185.0	7.079	9.974
100	19.65	216.2	2.261	0.7967	1.261	187.4	7.192	10.10
120	15.48	239.6	2.475	0.7461	1.115	212.4	8.542	11.81
140	12.94	261.4	2.643	0.7311	1.068	232.8	9.834	13.58
160	11.17	282.5	2.784	0.7245	1.045	250.9	11.07	15.32
180	9.842	303.3	2.906	0.7213	1.032	267.4	12.26	17.04
200	8.811	323.8	3.014	0.7195	1.025	282.6	13.41	18.71
220	7.981	344.3	3.112	0.7186	1.020	297.0	14.51	20.34
240	7.297	364.6	3.200	0.7182	1.017	310.6	15.58	21.94
260	6.724	385.0	3.282	0.7182	1.015	323.5	16.62	23.50
280	6.235	405.2	3.357	0.7185	1.014	335.9	17.62	25.02
300	5.813	425.5	3.427	0.7191	1.013	347.8	18.60	26.51
320	5.446	445.8	3.492	0.7200	1.013	359.3	19.54	27.97
340	5.123	466.0	3.553	0.7213	1.013	370.3	20.47	29.41
360	4.836	486.3	3.611	0.7229	1.014	381.0	21.37	30.81
380	4.580	506.6	3.666	0.7248	1.016	391.3	22.24	32.19
400	4.350	526.9	3.718	0.7271	1.018	401.3	23.10	33.55
500	3.477	629.3	3.947	0.7429	1.032	447.3	27.13	40.02
600	2.897	733.5	4.137	0.7643	1.053	488.0	30.80	46.07
700	2.483	840.0	4.301	0.7881	1.076	524.8	34.20	51.80
800	2.173	948.8	4.446	0.8119	1.100	558.7	37.40	57.29
900	1.932	1060	4.577	0.8341	1.122	590.5	40.42	62.58
1000	1.739	1173	4.696	0.8542	1.142	620.4	43.30	67.71
***P* = 1 MPa (10 bar)**								
60	958.0	-35.42	-0.5271	1.174	1.898	1033	377.2	171.7
80	872.2	2.720	0.02129	1.047	1.930	862.5	162.8	139.1
100	770.1	42.76	0.4673	0.9623	2.119	658.2	88.33	105.0
106.22	730.7	56.36	0.5992	0.9451	2.268	583.0	73.90	93.88
108.10	40.07	215.8	2.088	0.8543	1.540	185.2	7.963	11.96
120	33.48	232.3	2.233	0.7844	1.285	204.0	8.718	12.59
140	27.02	256.3	2.419	0.7481	1.148	228.2	9.978	14.11
160	22.94	278.7	2.568	0.7341	1.093	248.1	11.20	15.74
180	20.03	300.2	2.695	0.7273	1.065	265.7	12.38	17.38
200	17.83	321.3	2.806	0.7236	1.048	281.7	13.51	19.00
220	16.09	342.2	2.906	0.7216	1.038	296.6	14.61	20.60
240	14.68	362.9	2.996	0.7204	1.031	310.6	15.67	22.17
260	13.50	383.5	3.078	0.7199	1.026	323.8	16.70	23.70
280	12.50	403.9	3.154	0.7199	1.023	336.4	17.70	25.21

T/ K	ρ/ kg m^{-3}	H/ kJ kg^{-1}	S/ kJ kg^{-1} K^{-1}	C_v/ kJ kg^{-1} K^{-1}	C_p/ kJ kg^{-1} K^{-1}	u/ m s^{-1}	η/ μPa s	λ/ mW m^{-1} K^{-1}
300	11.64	424.4	3.224	0.7203	1.021	348.4	18.67	26.68
320	10.90	444.8	3.290	0.7211	1.020	360.0	19.62	28.13
340	10.25	465.2	3.352	0.7222	1.019	371.1	20.54	29.55
360	9.668	485.6	3.410	0.7237	1.019	381.9	21.43	30.95
380	9.153	506.0	3.465	0.7255	1.020	392.3	22.31	32.32
400	8.690	526.4	3.518	0.7277	1.022	402.3	23.16	33.67
500	6.943	629.1	3.747	0.7434	1.034	448.5	27.18	40.11
600	5.784	733.5	3.937	0.7646	1.054	489.2	30.84	46.15
700	4.957	840.0	4.101	0.7884	1.077	526.0	34.24	51.87
800	4.338	948.9	4.247	0.8121	1.100	559.9	37.43	57.35
900	3.857	1060	4.378	0.8343	1.122	591.5	40.45	62.63
1000	3.472	1173	4.497	0.8543	1.142	621.5	43.33	67.75

P = 2 MPa (20 bar)

T/ K	ρ/ kg m^{-3}	H/ kJ kg^{-1}	S/ kJ kg^{-1} K^{-1}	C_v/ kJ kg^{-1} K^{-1}	C_p/ kJ kg^{-1} K^{-1}	u/ m s^{-1}	η/ μPa s	λ/ mW m^{-1} K^{-1}
60.11[a]	959.1	-34.44	-0.5282	1.175	1.895	1037	380.5	172.2
80	874.6	3.390	0.01535	1.048	1.921	870.2	165.5	140.1
100	775.0	43.09	0.4576	0.9636	2.086	672.5	90.43	106.6
118.52	638.8	86.69	0.8559	0.9314	2.894	430.4	51.60	73.71
119.94	86.20	211.0	1.899	0.9396	2.365	180.5	9.540	16.59
120	86.03	211.2	1.900	0.9382	2.354	180.7	9.542	16.57
140	59.88	244.8	2.161	0.7883	1.387	218.3	10.44	15.68
160	48.61	270.5	2.333	0.7545	1.213	242.7	11.55	16.80
180	41.54	294.0	2.471	0.7396	1.139	262.8	12.67	18.21
200	36.52	316.3	2.589	0.7319	1.100	280.3	13.77	19.69
220	32.70	338.0	2.692	0.7275	1.076	296.2	14.84	21.19
240	29.66	359.4	2.785	0.7249	1.060	310.8	15.88	22.68
260	27.18	380.5	2.870	0.7235	1.050	324.6	16.89	24.16
280	25.11	401.4	2.947	0.7228	1.042	337.6	17.88	25.62
300	23.34	422.2	3.019	0.7227	1.037	349.9	18.84	27.06
320	21.82	442.9	3.086	0.7231	1.033	361.7	19.77	28.47
340	20.49	463.5	3.148	0.7240	1.031	373.0	20.68	29.87
360	19.32	484.1	3.207	0.7253	1.030	383.9	21.57	31.24
380	18.27	504.7	3.263	0.7269	1.029	394.4	22.44	32.59
400	17.34	525.3	3.315	0.7290	1.029	404.5	23.29	33.93
500	13.84	628.6	3.546	0.7442	1.039	450.8	27.28	40.31
600	11.52	733.4	3.737	0.7653	1.057	491.5	30.93	46.30
700	9.878	840.2	3.902	0.7889	1.079	528.3	34.32	52.00
800	8.646	949.3	4.047	0.8125	1.102	562.1	37.49	57.46
900	7.689	1061	4.178	0.8346	1.123	593.7	40.50	62.73
1000	6.923	1174	4.298	0.8546	1.143	623.6	43.38	67.84

P = 5 MPa (50 bar)

T/ K	ρ/ kg m^{-3}	H/ kJ kg^{-1}	S/ kJ kg^{-1} K^{-1}	C_v/ kJ kg^{-1} K^{-1}	C_p/ kJ kg^{-1} K^{-1}	u/ m s^{-1}	η/ μPa s	λ/ mW m^{-1} K^{-1}
60.64[a]	961.4	-31.10	-0.5246	1.176	1.886	1048	386.2	173.3
80	881.7	5.437	-0.001766	1.054	1.898	892.1	173.5	143.0
100	788.3	44.27	0.4311	0.9681	2.009	710.6	96.44	111.1
120	665.1	87.67	0.8256	0.9194	2.429	496.8	57.06	79.02
140	321.3	172.5	1.467	1.049	8.515	199.5	19.25	43.18
160	151.4	240.7	1.930	0.8269	1.916	231.8	13.80	22.96
180	116.7	273.4	2.123	0.7785	1.453	258.2	14.14	22.02
200	97.93	300.6	2.267	0.7569	1.288	279.4	14.90	22.54
220	85.43	325.5	2.385	0.7451	1.205	297.7	15.77	23.49
240	76.25	349.0	2.488	0.7381	1.155	314.0	16.68	24.62
260	69.12	371.8	2.579	0.7338	1.123	328.8	17.60	25.83
280	63.36	394.0	2.661	0.7312	1.101	342.7	18.52	27.09
300	58.59	415.8	2.736	0.7297	1.085	355.6	19.42	28.39
320	54.55	437.4	2.806	0.7291	1.074	367.9	20.31	29.69
340	51.07	458.8	2.871	0.7292	1.065	379.6	21.18	30.98
360	48.05	480.1	2.932	0.7298	1.059	390.7	22.04	32.27
380	45.38	501.2	2.989	0.7310	1.055	401.4	22.88	33.55
400	43.01	522.3	3.043	0.7327	1.052	411.7	23.70	34.81
500	34.21	627.4	3.277	0.7466	1.052	458.3	27.61	40.97
600	28.48	733.2	3.470	0.7671	1.066	498.9	31.20	46.83

T/ K	ρ/ kg m⁻³	H/ kJ kg⁻¹	S/ kJ kg⁻¹ K⁻¹	C_v/ kJ kg⁻¹ K⁻¹	C_p/ kJ kg⁻¹ K⁻¹	u/ m s⁻¹	η/ μPa s	λ/ mW m⁻¹ K⁻¹
700	24.42	840.8	3.636	0.7903	1.085	535.4	34.55	52.43
800	21.39	950.4	3.782	0.8136	1.106	569.0	37.69	57.83
900	19.03	1062	3.914	0.8356	1.127	600.3	40.68	63.04
1000	17.14	1176	4.034	0.8554	1.146	629.9	43.54	68.12
***P* = 10 MPa (100 bar)**								
61.52[a]	965.2	-25.55	-0.5187	1.177	1.871	1064	395.1	175.0
80	892.4	8.950	-0.02831	1.063	1.868	925.2	186.8	147.5
100	806.9	46.73	0.3930	0.9767	1.924	763.5	105.8	117.8
120	706.1	86.65	0.7565	0.9192	2.094	591.1	67.00	89.01
140	573.7	132.1	1.106	0.8916	2.517	418.1	41.82	63.61
160	397.0	188.0	1.479	0.8787	2.882	297.6	24.99	43.38
180	273.3	238.0	1.774	0.8267	2.098	283.1	19.31	32.89
200	214.1	274.8	1.968	0.7908	1.646	296.3	18.16	29.63
220	179.9	305.4	2.114	0.7701	1.434	312.4	18.17	28.80
240	157.0	332.8	2.233	0.7574	1.317	328.0	18.60	28.89
260	140.2	358.3	2.336	0.7492	1.245	342.7	19.20	29.38
280	127.3	382.7	2.426	0.7439	1.196	356.5	19.90	30.15
300	116.9	406.3	2.507	0.7404	1.162	369.5	20.64	31.12
320	108.3	429.3	2.582	0.7383	1.138	381.8	21.40	32.15
340	101.0	451.8	2.650	0.7372	1.120	393.5	22.17	33.22
360	94.80	474.1	2.714	0.7369	1.106	404.6	22.95	34.33
380	89.36	496.1	2.773	0.7374	1.096	415.3	23.72	35.45
400	84.57	517.9	2.829	0.7384	1.088	425.6	24.49	36.58
500	67.06	625.7	3.070	0.7505	1.073	471.8	28.19	42.26
600	55.82	733.3	3.266	0.7699	1.080	511.9	31.67	47.84
700	47.90	842.0	3.433	0.7926	1.095	547.8	34.94	53.26
800	42.00	952.4	3.581	0.8155	1.113	580.8	38.04	58.52
900	37.42	1065	3.713	0.8372	1.132	611.6	40.99	63.64
1000	33.75	1179	3.833	0.8568	1.150	640.7	43.81	68.64
***P* = 20 MPa (200 bar)**								
63.24[a]	972.3	-14.49	-0.5069	1.180	1.847	1094	411.4	178.4
80	911.3	16.25	-0.07568	1.081	1.825	982.1	213.1	155.4
100	836.2	52.71	0.3311	0.9948	1.827	846.0	123.1	128.7
120	755.9	89.57	0.6670	0.9334	1.864	711.9	82.28	103.9
140	668.7	127.4	0.9589	0.8906	1.926	589.6	58.52	82.08
160	575.8	166.5	1.220	0.8606	1.977	491.8	43.00	65.18
180	484.9	206.0	1.452	0.8365	1.946	428.0	33.22	53.43
200	407.6	243.6	1.651	0.8143	1.808	397.2	27.80	46.02
220	348.5	278.1	1.815	0.7956	1.642	387.6	25.12	41.70
240	304.6	309.5	1.952	0.7812	1.506	388.7	23.92	39.30
260	271.5	338.6	2.068	0.7705	1.405	394.7	23.48	37.98
280	245.7	365.9	2.169	0.7627	1.331	402.8	23.47	37.46
300	225.0	392.0	2.259	0.7572	1.277	412.0	23.70	37.57
320	208.0	417.1	2.340	0.7532	1.236	421.5	24.08	37.94
340	193.7	441.5	2.414	0.7506	1.204	431.1	24.56	38.47
360	181.5	465.3	2.482	0.7491	1.180	440.6	25.10	39.13
380	170.9	488.7	2.546	0.7485	1.161	450.0	25.68	39.88
400	161.7	511.7	2.605	0.7486	1.146	459.2	26.29	40.69
500	128.3	624.0	2.856	0.7575	1.108	501.8	29.49	45.27
600	107.0	734.4	3.057	0.7753	1.103	539.5	32.69	50.19
700	92.10	845.1	3.227	0.7969	1.112	573.7	35.78	55.18
800	80.99	957.0	3.377	0.8191	1.126	605.2	38.75	60.13
900	72.34	1070	3.510	0.8402	1.142	634.8	41.61	65.02
1000	65.41	1185	3.631	0.8594	1.157	662.8	44.36	69.84
***P* = 50 MPa (500 bar)**								
68.21[a]	991.5	18.30	-0.4728	1.192	1.793	1173	450.9	187.5
80	955.5	39.24	-0.1896	1.128	1.761	1112	292.7	174.1
100	895.9	73.98	0.1982	1.043	1.715	1012	171.6	152.8
120	837.8	107.9	0.5077	0.9811	1.680	920.9	118.9	133.2

Fluids

T/ K	ρ/ kg m^{-3}	H/ kJ kg^{-1}	S/ kJ kg^{-1} K^{-1}	C_v/ kJ kg^{-1} K^{-1}	C_p/ kJ kg^{-1} K^{-1}	u/ m s^{-1}	η/ μPa s	λ/ mW m^{-1} K^{-1}
140	781.1	141.2	0.7643	0.9342	1.649	839.4	90.60	115.9
160	726.2	173.9	0.9826	0.8983	1.619	770.2	72.86	101.0
180	673.9	205.9	1.171	0.8706	1.586	714.1	60.73	88.73
200	624.9	237.3	1.337	0.8488	1.549	670.6	52.14	79.05
220	579.8	267.9	1.482	0.8313	1.509	638.1	45.98	71.65
240	539.0	297.7	1.612	0.8171	1.467	614.7	41.60	66.11
260	502.4	326.6	1.728	0.8057	1.425	598.5	38.49	61.97
280	470.0	354.7	1.832	0.7964	1.385	588.0	36.32	58.99
300	441.2	382.0	1.926	0.7890	1.349	581.7	34.82	57.07
320	415.6	408.7	2.012	0.7832	1.316	578.4	33.83	55.77
340	392.9	434.7	2.091	0.7787	1.287	577.4	33.19	54.93
360	372.7	460.2	2.164	0.7754	1.262	578.1	32.82	54.41
380	354.6	485.2	2.232	0.7731	1.241	580.0	32.66	54.15
400	338.2	509.8	2.295	0.7717	1.223	582.9	32.64	54.07
500	276.1	629.0	2.561	0.7749	1.168	604.1	33.85	55.45
600	234.6	744.6	2.772	0.7890	1.148	629.5	36.00	58.34
700	204.6	859.2	2.948	0.8080	1.147	655.5	38.46	61.93
800	181.8	974.2	3.102	0.8285	1.154	681.0	41.00	65.86
900	163.8	1090	3.238	0.8483	1.164	705.9	43.55	69.98
1000	149.2	1207	3.362	0.8665	1.175	730.1	46.07	74.20

[a] Freezing point for the liquid state.

THERMOPHYSICAL PROPERTIES OF FLUIDS

Eric W. Lemmon

These tables give thermodynamic and transport properties of a variety of fluids, as generated from the equations of state presented in the references below. The properties tabulated are pressure (P), density (ρ), enthalpy (H), entropy (S), isochoric heat capacity (C_v), isobaric heat capacity (C_p), speed of sound (u), viscosity (η), thermal conductivity (λ), and static dielectric constant (D). All extensive properties are given on a mass basis. Not all properties are included for every substance. The references should be consulted for information on the uncertainties.

Values are given first along the saturation line. The first two points are the properties at the triple point. The final line gives the properties at the critical point. Two lines are given for each temperature (except at the critical point); the first line gives the values of the liquid phase (note the high density) and the second line gives the values of the vapor phase (lower densities). Following the saturation tables, values are given as a function of temperature for several isobars. A duplicate entry in the isobar section indicates a phase transition (liquid–vapor) at that temperature; property values are then given for both phases. The phase can be determined by noting the sharp decrease in density between two successive temperature entries; all lines above this point refer to the liquid phase, and all lines below refer to the gas phase. If there is no sharp discontinuity in density, all data in the table refer to the supercritical region (i.e., the isobar is above the critical pressure). If the first temperature in the isobars is not an integer, this state point refers to the properties of the liquid at the melting line.

All temperatures are given on ITS-90, except those for oxygen (Ref. 16) and helium (Ref. 13), where the source equations of state still use the IPTS-68 temperature scale. Nitrogen, oxygen, and argon use a reference state based on zero enthalpy in the gas phase at 0 K; parahydrogen, helium, methane, and ethane use a reference state of zero enthalpy and entropy at the saturated liquid state at the normal boiling point; and propane and carbon dioxide use a reference state of 200 kJ kg^{-1} and 1 kJ kg^{-1} K^{-1}, respectively, at −40 °C. Additional calculations at state points not listed below and for fluids not contained here can be obtained at the NIST Chemistry WebBook website <http://webbook.nist.gov/chemistry/fluid/>, or by using the NIST Standard Reference Data program REFPROP (Ref. 9).

References

1. Arp, V.D., McCarty, R.D., and Friend, D.G., Thermophysical Properties of Helium-4 from 0.8 to 1500 K with Pressures to 2000 MPa, NIST Technical Note 1334 (revised), 1998.
2. Buecker, D. and Wagner, W., A Reference Equation of State for the Thermodynamic Properties of Ethane for Temperatures from the Melting Line to 675 K and Pressures up to 900 MPa, *J. Phys. Chem. Ref. Data,* 35, 205, 2006.
3. Friend, D.G., Ely, J.F., and Ingham, H., Tables for the Thermophysical Properties of Methane, NIST Technical Note 1325, 1989.
4. Friend, D.G., Ingham, H., and Ely, J.F., Thermophysical Properties of Ethane, *J. Phys. Chem. Ref. Data,* 20, 275, 1991.
5. Hands, B.A. and Arp, V.D., A Correlation of Thermal Conductivity Data for Helium, *Cryogenics*, 21, 697, 1981.
6. Harvey, A.H. and Lemmon, E.W., Method for Estimating the Dielectric Constant of Natural Gas Mixtures, *Int. J. Thermophys.*, 26, 31, 2005.
7. Leachman, J.W., Jacobsen, R.T, Lemmon, E.W., and Penoncello, S.G., Fundamental Equations of State for Parahydrogen, Normal Hydrogen, and Orthohydrogen, *J. Phys. Chem. Ref. Data,* 38, 721, 2009.
8. Lemmon, E.W. and Jacobsen, R.T, Viscosity and Thermal Conductivity Equations for Nitrogen, Oxygen, Argon, and Air, *Int. J. Thermophys.*, 25, 21, 2004.
9. Lemmon, E.W., Huber, M.L., and McLinden, M.O., NIST Standard Reference Database 23: Reference Fluid Thermodynamic and Transport Properties-REFPROP, Version 9.0, National Institute Of Standards and Technology, Standard Reference Data Program, Gaithersburg, MD, 2010 (http://www.nist.gov/srd/nist23.cfm).
10. Lemmon, E.W., McLinden, M.O., and Friend, D.G., Thermophysical Properties of Fluid Systems, NIST Chemistry WebBook, NIST Standard Reference Database Number 69, Linstrom, P.J. and Mallard, W.G., Eds., National Institute of Standards and Technology, Gaithersburg, MD.
11. Lemmon, E.W., McLinden, M.O., and Wagner, W., Thermodynamic Properties of Propane. III. A Reference Equation of State for Temperatures from the Melting Line to 650 K and Pressures up to 1000 MPa, submitted to *J. Chem. Eng. Data,* 2009.
12. Marsh, K., Perkins, R., and Ramires, M.L.V., Measurement and Correlation of the Thermal Conductivity of Propane from 86 to 600 K at Pressures to 70 MPa, *J. Chem. Eng. Data,* 47, 932, 2002.
13. McCarty, R.D. and Arp, V.D., A New Wide Range Equation of State for Helium, *Adv. Cryo. Eng.*, 35, 1465, 1990.
14. McCarty, R.D. and Weber, L.A., Thermophysical Properties of Parahydrogen from the Freezing Liquid Line to 5000 R for Pressures to 10,000 psia, Natl. Bur. Stand., Tech. Note 617, 1972.
15. Quinones-Cisneros, S.E., Huber, M.L., and Deiters, U.K., Reference Correlation for the Viscosity of Methane, private communication.
16. Schmidt, R. and Wagner, W., A New Form of the Equation of State for Pure Substances and Its Application to Oxygen, *Fluid Phase Equilib.*, 19, 175, 1985.
17. Setzmann, U. and Wagner, W., A New Equation of State and Tables of Thermodynamic Properties for Methane Covering the Range from the Melting Line to 625 K at Pressures up to 1000 MPa, *J. Phys. Chem. Ref. Data,* 20, 1061, 1991.
18. Span, R. and Wagner, W., A New Equation of State for Carbon Dioxide Covering the Fluid Region from the Triple-Point Temperature to 1100 K at Pressures up to 800 MPa, *J. Phys. Chem. Ref. Data,* 25, 1509, 1996.
19. Span, R., Lemmon, E.W., Jacobsen, R.T, Wagner, W., and Yokozeki, A., A Reference Equation of State for the Thermodynamic Properties of Nitrogen for Temperatures from 63.151 to 1000 K and Pressures to 2200 MPa, *J. Phys. Chem. Ref. Data,* 29, 1361, 2000.
20. Tegeler, Ch., Span, R., and Wagner, W., A New Equation of State for Argon Covering the Fluid Region for Temperatures from the Melting Line to 700 K at Pressures up to 1000 MPa, *J. Phys. Chem. Ref. Data,* 28, 779, 1999.
21. Vogel, E., Kuechenmeister, C., Bich, E., and Laesecke, A., Reference Correlation of the Viscosity of Propane, *J. Phys. Chem. Ref. Data,* 27, 947, 1998.

Thermophysical Properties of Nitrogen (N_2)

T / K	P / MPa	ρ / kg m^{-3}	H / kJ kg^{-1}	S / kJ kg^{-1} K^{-1}	C_v / kJ kg^{-1} K^{-1}	C_p / kJ kg^{-1} K^{-1}	u / m s^{-1}	D	η / μPa s	λ / mW m^{-1} K^{-1}
Saturation										
63.15	0.01252	867.2	-150.7	2.426	1.176	2.000	995.3	1.47003	311.6	173.2
63.15	0.01252	0.6743	64.78	5.838	0.7499	1.058	161.1	1.00032	4.376	5.621
70	0.03854	838.5	-137.0	2.632	1.130	2.014	925.7	1.45241	220.2	159.5
70	0.03854	1.896	71.10	5.605	0.7580	1.082	168.4	1.00089	4.883	6.355
80	0.1369	793.9	-116.6	2.903	1.069	2.056	824.4	1.42541	145.1	139.5
80	0.1369	6.089	79.10	5.349	0.7773	1.145	176.7	1.00286	5.652	7.506
90	0.3605	745.0	-95.52	3.147	1.020	2.141	719.0	1.39622	102.8	119.8
90	0.3605	15.08	84.97	5.153	0.8078	1.266	181.8	1.00710	6.482	8.868
100	0.7783	689.4	-73.21	3.376	0.9832	2.318	605.2	1.36351	75.76	100.1
100	0.7783	31.96	87.77	4.986	0.8548	1.503	183.3	1.01510	7.429	10.73
110	1.466	621.5	-48.49	3.601	0.9667	2.743	476.4	1.32430	55.99	80.44
110	1.466	62.58	85.84	4.823	0.9284	2.062	180.8	1.02974	8.626	13.83
120	2.511	523.4	-17.87	3.851	1.011	4.508	317.3	1.26895	38.43	61.01
120	2.511	125.1	74.17	4.618	1.099	4.631	172.6	1.06012	10.62	21.72
126.19	3.396	313.3	29.23	4.215				1.15559	18.30	
***P* = 0.1 MPa (1 bar)**										
63.17	0.1	867.3	-150.6	2.426	1.176	2.000	995.6	1.47007	311.6	173.3
77.24	0.1	806.6	-122.2	2.831	1.085	2.041	852.5	1.43304	161.4	145.0
77.24	0.1	4.556	77.07	5.412	0.7710	1.123	174.7	1.00214	5.435	7.174
80	0.1	4.379	80.15	5.451	0.7666	1.112	178.3	1.00206	5.623	7.443
100	0.1	3.437	101.9	5.694	0.7514	1.071	201.6	1.00162	6.958	9.381
120	0.1	2.840	123.2	5.888	0.7466	1.057	222.0	1.00133	8.244	11.27
140	0.1	2.424	144.2	6.050	0.7447	1.050	240.4	1.00114	9.480	13.11
160	0.1	2.116	165.2	6.190	0.7438	1.047	257.3	1.00099	10.67	14.89
180	0.1	1.878	186.1	6.313	0.7433	1.045	273.2	1.00088	11.81	16.61
200	0.1	1.688	207.0	6.423	0.7430	1.043	288.1	1.00079	12.91	18.28
220	0.1	1.534	227.9	6.523	0.7429	1.043	302.3	1.00072	13.97	19.90
240	0.1	1.405	248.7	6.613	0.7428	1.042	315.8	1.00066	15.00	21.48
260	0.1	1.297	269.5	6.697	0.7428	1.042	328.7	1.00061	15.99	23.01
280	0.1	1.204	290.4	6.774	0.7429	1.041	341.2	1.00057	16.96	24.51
300	0.1	1.123	311.2	6.846	0.7432	1.041	353.2	1.00053	17.89	25.97
320	0.1	1.053	332.0	6.913	0.7436	1.042	364.7	1.00049	18.80	27.39
340	0.1	0.9909	352.9	6.976	0.7441	1.042	375.9	1.00047	19.68	28.79
360	0.1	0.9357	373.7	7.036	0.7450	1.043	386.8	1.00044	20.55	30.15
400	0.1	0.8421	415.5	7.146	0.7475	1.045	407.5	1.00040	22.21	32.81
500	0.1	0.6736	520.5	7.380	0.7592	1.056	454.6	1.00032	26.06	39.04
600	0.1	0.5613	627.0	7.574	0.7781	1.075	496.3	1.00026	29.58	44.84
700	0.1	0.4811	735.6	7.741	0.8011	1.098	533.9	1.00023	32.83	50.31
800	0.1	0.4210	846.6	7.890	0.8254	1.122	568.4	1.00020	35.89	55.51
900	0.1	0.3742	960.0	8.023	0.8488	1.146	600.7	1.00018	38.78	60.52
1000	0.1	0.3368	1076	8.145	0.8705	1.167	631.1	1.00016	41.54	65.36
***P* = 1 MPa (10 bar)**										
63.37	1.0	868.0	-149.5	2.427	1.177	1.995	998.9	1.47049	311.8	173.6
80	1.0	796.3	-116.0	2.896	1.071	2.044	832.3	1.42683	147.3	140.5
100	1.0	690.8	-73.18	3.373	0.9833	2.305	609.4	1.36432	76.26	100.6
103.75	1.0	665.8	-64.33	3.460	0.9739	2.431	559.2	1.34984	67.78	92.74
103.75	1.0	41.33	87.73	4.926	0.8786	1.652	182.8	1.01956	7.835	11.67
120	1.0	32.10	110.9	5.134	0.7980	1.297	208.9	1.01517	8.714	12.49
140	1.0	26.01	135.5	5.324	0.7701	1.177	232.8	1.01227	9.844	14.00
160	1.0	22.11	158.5	5.477	0.7588	1.127	252.9	1.01043	10.96	15.60
180	1.0	19.33	180.7	5.608	0.7531	1.101	270.6	1.00911	12.06	17.21
200	1.0	17.21	202.6	5.723	0.7500	1.085	286.9	1.00811	13.13	18.81
220	1.0	15.54	224.2	5.826	0.7480	1.075	301.9	1.00732	14.16	20.37
240	1.0	14.17	245.6	5.919	0.7468	1.067	316.1	1.00667	15.17	21.90
260	1.0	13.04	266.9	6.005	0.7460	1.062	329.5	1.00614	16.14	23.39
280	1.0	12.07	288.1	6.083	0.7456	1.059	342.4	1.00569	17.09	24.86
300	1.0	11.25	309.2	6.156	0.7454	1.056	354.6	1.00530	18.01	26.29

T / K	*P* / MPa	*ρ* / kg m^{-3}	*H* / kJ kg^{-1}	*S* / kJ kg^{-1} K^{-1}	C_v / kJ kg^{-1} K^{-1}	C_p / kJ kg^{-1} K^{-1}	*u* / m s^{-1}	*D*	*η* / μPa s	*λ* / mW m^{-1} K^{-1}
320	1.0	10.53	330.3	6.224	0.7455	1.054	366.4	1.00496	18.91	27.69
340	1.0	9.901	351.4	6.288	0.7458	1.053	377.8	1.00466	19.79	29.07
360	1.0	9.343	372.4	6.348	0.7464	1.052	388.8	1.00440	20.64	30.42
400	1.0	8.399	414.5	6.459	0.7486	1.052	409.7	1.00395	22.29	33.04
500	1.0	6.711	520.1	6.695	0.7600	1.061	457.0	1.00316	26.13	39.23
600	1.0	5.592	627.0	6.889	0.7786	1.078	498.7	1.00263	29.63	44.99
700	1.0	4.793	735.8	7.057	0.8015	1.100	536.2	1.00226	32.87	50.43
800	1.0	4.195	847.0	7.206	0.8257	1.124	570.7	1.00197	35.92	55.63
900	1.0	3.730	960.6	7.339	0.8492	1.147	602.9	1.00176	38.81	60.62
1000	1.0	3.358	1076	7.461	0.8708	1.168	633.2	1.00158	41.57	65.45
***P* = 10 MPa (100 bar)**										
65.32	10.0	875.0	-138.3	2.441	1.186	1.955	1031	1.47447	314.0	177.1
80	10.0	818.4	-109.6	2.837	1.093	1.960	904.1	1.43984	169.0	150.4
100	10.0	733.6	-69.89	3.280	0.9997	2.022	734.2	1.38918	93.65	115.9
120	10.0	632.9	-27.82	3.663	0.9401	2.218	559.6	1.33060	58.57	84.11
140	10.0	499.8	20.66	4.036	0.9127	2.676	392.0	1.25566	36.11	57.60
160	10.0	344.2	76.44	4.408	0.8849	2.668	300.8	1.17162	22.40	40.20
180	10.0	248.7	122.6	4.681	0.8407	1.990	294.7	1.12203	18.50	32.06
200	10.0	199.4	158.4	4.870	0.8107	1.628	307.7	1.09701	17.70	29.38
220	10.0	169.4	188.9	5.015	0.7927	1.445	323.2	1.08198	17.74	28.72
240	10.0	148.8	216.7	5.136	0.7813	1.340	338.4	1.07172	18.13	28.87
260	10.0	133.4	242.8	5.240	0.7737	1.273	352.9	1.06415	18.67	29.39
280	10.0	121.4	267.7	5.333	0.7685	1.227	366.6	1.05827	19.29	30.21
300	10.0	111.7	291.9	5.417	0.7649	1.195	379.5	1.05351	19.96	31.14
320	10.0	103.6	315.6	5.493	0.7623	1.171	391.8	1.04958	20.66	32.13
340	10.0	96.80	338.8	5.563	0.7606	1.153	403.6	1.04625	21.37	33.16
360	10.0	90.89	361.7	5.629	0.7596	1.138	414.8	1.04338	22.08	34.21
400	10.0	81.19	406.8	5.748	0.7593	1.119	436.0	1.03869	23.51	36.35
500	10.0	64.50	517.5	5.995	0.7672	1.100	483.1	1.03065	26.99	41.73
600	10.0	53.74	627.6	6.195	0.7841	1.104	523.9	1.02549	30.28	47.00
700	10.0	46.15	738.7	6.367	0.8059	1.118	560.4	1.02187	33.40	52.10
800	10.0	40.49	851.4	6.517	0.8294	1.137	593.8	1.01917	36.35	57.04
900	10.0	36.08	966.1	6.652	0.8523	1.157	625.0	1.01708	39.17	61.85
1000	10.0	32.56	1083	6.775	0.8735	1.176	654.4	1.01540	41.88	66.54

Thermophysical Properties of Oxygen (O_2)

T / K	*P* / MPa	*ρ* / kg m^{-3}	*H* / kJ kg^{-1}	*S* / kJ kg^{-1} K^{-1}	C_v / kJ kg^{-1} K^{-1}	C_p / kJ kg^{-1} K^{-1}	*u* / m s^{-1}	*D*	*η* / μPa s	*λ* / mW m^{-1} K^{-1}
Saturation										
54.36	0.0001463	1306	-193.6	2.092	1.195	1.673	1123	1.56799	773.6	201.9
54.36	0.0001463	0.01036	49.11	6.557	0.6638	0.9260	140.3	1.00000	4.096	4.420
60	0.0007258	1282	-184.2	2.257	1.089	1.673	1127	1.55615	578.1	193.9
60	0.0007258	0.04659	54.19	6.230	0.6817	0.9475	147.0	1.00002	4.553	4.984
70	0.006262	1237	-167.4	2.516	1.017	1.678	1066	1.53399	371.8	179.7
70	0.006262	0.3457	63.09	5.809	0.7052	0.9780	158.1	1.00013	5.356	5.992
80	0.03012	1190	-150.6	2.740	0.9697	1.682	987.4	1.51120	261.2	165.4
80	0.03012	1.468	71.69	5.519	0.6950	0.9743	168.4	1.00054	6.149	7.028
90	0.09935	1142	-133.7	2.938	0.9296	1.699	905.9	1.48766	195.6	151.0
90	0.09935	4.387	79.55	5.308	0.6758	0.9705	177.3	1.00163	6.936	8.124
100	0.2540	1091	-116.4	3.118	0.8949	1.738	822.2	1.46295	152.6	136.6
100	0.2540	10.42	86.16	5.144	0.6752	1.006	184.1	1.00387	7.728	9.336
110	0.5434	1035	-98.64	3.286	0.8658	1.807	734.8	1.43650	121.5	121.9
110	0.5434	21.28	91.05	5.010	0.6988	1.101	188.1	1.00791	8.547	10.75
120	1.022	973.9	-79.90	3.444	0.8430	1.927	641.5	1.40746	97.43	107.2
120	1.022	39.31	93.75	4.892	0.7415	1.276	189.4	1.01465	9.427	12.51
130	1.749	902.5	-59.66	3.600	0.8293	2.153	539.5	1.37432	77.57	92.63
130	1.749	68.37	93.47	4.778	0.8002	1.600	187.8	1.02558	10.45	14.94
140	2.788	813.2	-36.70	3.761	0.8323	2.691	423.1	1.33363	60.22	78.22
140	2.788	116.8	88.47	4.655	0.8834	2.370	182.8	1.04395	11.82	18.98

T / K	P / MPa	ρ / kg m^{-3}	H / kJ kg^{-1}	S / kJ kg^{-1} K^{-1}	C_v / kJ kg^{-1} K^{-1}	C_p / kJ kg^{-1} K^{-1}	u / m s^{-1}	D	η / μPa s	λ / mW m^{-1} K^{-1}
150	4.219	675.5	-6.671	3.955	0.9057	5.464	273.8	1.27248	42.90	64.19
150	4.219	214.9	72.56	4.483	1.049	6.625	172.8	1.08192	14.72	29.67
154.58	5.043	436.1	32.42	4.201				1.17084	24.84	
P = 0.1 MPa (1 bar)										
54.37	0.1	1306	-193.5	2.092	1.195	1.673	1124	1.56802	773.8	201.9
60	0.1	1282	-184.1	2.257	1.089	1.673	1128	1.55621	578.5	194.0
80	0.1	1191	-150.6	2.739	0.9699	1.681	987.7	1.51126	261.4	165.5
90.06	0.1	1142	-133.6	2.939	0.9293	1.699	905.4	1.48751	195.3	151.0
90.06	0.1	4.413	79.60	5.307	0.6757	0.9705	177.4	1.00164	6.940	8.131
100	0.1	3.941	88.99	5.405	0.6527	0.9352	188.4	1.00146	7.712	9.085
120	0.1	3.252	107.6	5.575	0.6543	0.9280	207.3	1.00121	9.219	10.99
140	0.1	2.774	126.1	5.718	0.6530	0.9218	224.6	1.00103	10.67	12.86
160	0.1	2.420	144.5	5.841	0.6519	0.9180	240.5	1.00090	12.07	14.69
180	0.1	2.147	162.8	5.949	0.6513	0.9158	255.4	1.00080	13.41	16.48
200	0.1	1.930	181.1	6.045	0.6512	0.9146	269.3	1.00072	14.72	18.24
220	0.1	1.753	199.4	6.132	0.6516	0.9142	282.6	1.00065	15.98	19.95
240	0.1	1.606	217.7	6.212	0.6525	0.9146	295.2	1.00060	17.20	21.64
260	0.1	1.482	236.0	6.285	0.6539	0.9156	307.2	1.00055	18.38	23.28
280	0.1	1.376	254.3	6.353	0.6560	0.9174	318.7	1.00051	19.53	24.90
300	0.1	1.284	272.7	6.416	0.6587	0.9199	329.7	1.00048	20.65	26.49
320	0.1	1.203	291.1	6.476	0.6621	0.9231	340.3	1.00045	21.74	28.04
340	0.1	1.132	309.6	6.532	0.6661	0.9269	350.5	1.00042	22.80	29.58
360	0.1	1.069	328.2	6.585	0.6707	0.9313	360.4	1.00040	23.84	31.08
400	0.1	0.9622	365.7	6.684	0.6812	0.9416	379.0	1.00036	25.84	34.03
500	0.1	0.7696	461.3	6.897	0.7119	0.9722	421.3	1.00029	30.49	41.05
600	0.1	0.6413	560.1	7.077	0.7432	1.003	458.9	1.00024	34.73	47.66
700	0.1	0.5497	661.9	7.234	0.7710	1.031	493.3	1.00020	38.65	53.97
800	0.1	0.4809	766.2	7.373	0.7945	1.055	525.4	1.00018	42.33	60.02
900	0.1	0.4275	872.6	7.498	0.8140	1.074	555.6	1.00016	45.81	65.87
1000	0.1	0.3848	980.9	7.612	0.8301	1.090	584.3	1.00014	49.12	71.55
P = 1 MPa (10 bar)										
54.47	1.0	1307	-192.8	2.093	1.190	1.669	1127	1.56831	775.0	202.1
60	1.0	1283	-183.6	2.254	1.089	1.671	1130	1.55674	582.7	194.3
80	1.0	1192	-150.1	2.736	0.9716	1.679	991.1	1.51204	263.6	166.0
100	1.0	1093	-116.1	3.115	0.8964	1.731	826.8	1.46397	153.9	137.2
119.62	1.0	976.3	-80.64	3.438	0.8438	1.921	645.2	1.40862	98.25	107.8
119.62	1.0	38.46	93.70	4.896	0.7396	1.268	189.4	1.01433	9.392	12.43
120	1.0	38.25	94.18	4.900	0.7355	1.257	190.0	1.01425	9.421	12.45
140	1.0	30.39	116.8	5.075	0.6851	1.065	214.2	1.01131	10.89	13.90
160	1.0	25.65	137.5	5.213	0.6708	1.005	233.7	1.00954	12.29	15.54
180	1.0	22.33	157.2	5.329	0.6631	0.9743	250.8	1.00831	13.64	17.21
200	1.0	19.84	176.5	5.431	0.6590	0.9568	266.3	1.00738	14.94	18.88
220	1.0	17.88	195.5	5.521	0.6571	0.9462	280.6	1.00665	16.20	20.53
240	1.0	16.30	214.4	5.603	0.6566	0.9397	294.0	1.00606	17.41	22.15
260	1.0	14.98	233.1	5.678	0.6572	0.9359	306.6	1.00557	18.59	23.76
280	1.0	13.86	251.8	5.748	0.6586	0.9342	318.5	1.00515	19.73	25.34
300	1.0	12.91	270.5	5.812	0.6609	0.9340	329.9	1.00480	20.85	26.89
320	1.0	12.08	289.2	5.872	0.6640	0.9351	340.8	1.00449	21.93	28.43
340	1.0	11.35	307.9	5.929	0.6677	0.9373	351.2	1.00422	22.99	29.93
360	1.0	10.71	326.7	5.983	0.6720	0.9404	361.2	1.00398	24.02	31.42
400	1.0	9.623	364.5	6.082	0.6823	0.9487	380.1	1.00358	26.01	34.33
500	1.0	7.683	460.7	6.297	0.7126	0.9763	422.7	1.00286	30.63	41.29
600	1.0	6.398	559.8	6.478	0.7436	1.006	460.4	1.00238	34.85	47.86
700	1.0	5.483	661.8	6.635	0.7713	1.033	494.9	1.00204	38.77	54.14
800	1.0	4.798	766.2	6.774	0.7948	1.056	526.9	1.00179	42.43	60.17
900	1.0	4.265	872.8	6.900	0.8142	1.075	557.1	1.00159	45.90	66.00
1000	1.0	3.839	981.1	7.014	0.8303	1.091	585.8	1.00143	49.20	71.67
P = 10 MPa (100 bar)										
55.50	10.0	1312	-185.4	2.103	1.149	1.640	1158	1.57109	786.0	203.9
60	10.0	1294	-178.0	2.231	1.092	1.653	1155	1.56193	625.3	197.9

Fluids

T / K	P / MPa	ρ / kg m^{-3}	H / kJ kg^{-1}	S / kJ kg^{-1} K^{-1}	C_v / kJ kg^{-1} K^{-1}	C_p / kJ kg^{-1} K^{-1}	u / m s^{-1}	D	η / μPa s	λ / mW m^{-1} K^{-1}
80	10.0	1207	-144.8	2.708	0.9885	1.654	1023	1.51940	285.2	171.5
100	10.0	1116	-111.6	3.078	0.9136	1.672	877.1	1.47515	169.5	144.8
120	10.0	1015	-77.51	3.389	0.8571	1.755	727.0	1.42679	112.4	118.1
140	10.0	892.5	-40.64	3.673	0.8191	1.966	564.5	1.36967	75.70	91.70
160	10.0	716.0	4.057	3.970	0.8137	2.659	379.7	1.29022	47.96	66.52
180	10.0	423.3	70.35	4.360	0.8161	3.312	250.9	1.16550	26.07	42.24
200	10.0	277.6	119.7	4.621	0.7506	1.894	259.0	1.10662	20.74	31.77
220	10.0	219.9	152.6	4.778	0.7152	1.462	278.2	1.08384	20.24	29.46
240	10.0	186.7	179.8	4.896	0.6973	1.280	295.8	1.07090	20.67	29.19
260	10.0	164.2	204.3	4.995	0.6878	1.183	311.6	1.06220	21.40	29.67
280	10.0	147.6	227.3	5.080	0.6830	1.124	326.1	1.05578	22.25	30.48
300	10.0	134.6	249.4	5.156	0.6809	1.086	339.4	1.05078	23.15	31.47
320	10.0	124.0	270.8	5.225	0.6809	1.060	351.7	1.04675	24.07	32.56
340	10.0	115.2	291.9	5.289	0.6823	1.042	363.3	1.04339	25.00	33.74
360	10.0	107.8	312.6	5.348	0.6848	1.030	374.2	1.04054	25.92	34.96
400	10.0	95.67	353.5	5.456	0.6923	1.017	394.5	1.03595	27.74	37.45
500	10.0	75.32	454.8	5.682	0.7187	1.015	438.7	1.02825	32.07	43.71
600	10.0	62.44	557.1	5.869	0.7477	1.031	476.8	1.02339	36.10	49.86
700	10.0	53.46	661.2	6.029	0.7743	1.050	511.2	1.02002	39.87	55.84
800	10.0	46.79	767.1	6.170	0.7970	1.069	543.1	1.01752	43.43	61.66
900	10.0	41.63	874.8	6.297	0.8159	1.085	572.9	1.01558	46.80	67.32
1000	10.0	37.51	984.0	6.412	0.8316	1.098	601.2	1.01404	50.03	72.85

Thermophysical Properties of Parahydrogen (H_2)

T / K	P / MPa	ρ / kg m^{-3}	H / kJ kg^{-1}	S / kJ kg^{-1} K^{-1}	C_v / kJ kg^{-1} K^{-1}	C_p / kJ kg^{-1} K^{-1}	u / m s^{-1}	D	η / μPa s	λ / mW m^{-1} K^{-1}
Saturation										
13.80	0.007041	76.98	-53.74	-3.084	5.131	6.924	1263	1.25267	25.90	75.27
13.80	0.007041	0.1255	396.3	29.52	6.226	10.53	305.6	1.00038	0.6507	10.46
14	0.007884	76.82	-52.36	-2.986	5.158	6.981	1257	1.25208	25.21	76.86
14	0.007884	0.1388	398.1	29.19	6.236	10.56	307.6	1.00042	0.6669	10.65
16	0.02155	75.13	-37.60	-2.013	5.334	7.659	1210	1.24585	19.81	90.02
16	0.02155	0.3377	415.8	26.33	6.323	10.87	325.4	1.00102	0.8106	12.60
18	0.04815	73.25	-21.18	-1.068	5.472	8.513	1168	1.23903	16.22	98.39
18	0.04815	0.6880	431.5	24.08	6.387	11.29	340.7	1.00207	0.9355	14.60
20	0.09341	71.14	-2.692	-0.1281	5.637	9.569	1119	1.23147	13.63	103.0
20	0.09341	1.244	444.5	22.23	6.450	11.92	353.5	1.00375	1.057	16.71
22	0.1635	68.74	18.32	0.8244	5.811	10.86	1059	1.22301	11.65	104.9
22	0.1635	2.071	454.2	20.64	6.541	12.88	363.7	1.00625	1.182	18.98
24	0.2648	66.01	42.37	1.805	5.974	12.52	989.9	1.21345	10.06	104.7
24	0.2648	3.255	459.7	19.19	6.681	14.40	371.3	1.00984	1.315	21.54
26	0.4038	62.83	70.23	2.832	6.120	14.83	909.3	1.20242	8.725	102.6
26	0.4038	4.924	460.0	17.82	6.890	16.90	376.1	1.01491	1.463	24.53
28	0.5875	58.98	103.2	3.941	6.266	18.57	813.3	1.18922	7.533	98.79
28	0.5875	7.300	453.4	16.45	7.188	21.53	378.2	1.02216	1.634	28.26
30	0.8232	53.98	144.2	5.211	6.472	26.65	693.0	1.17223	6.390	92.78
30	0.8232	10.87	435.7	14.93	7.625	32.58	377.2	1.03313	1.865	33.47
32	1.120	45.90	204.1	6.945	7.010	68.19	522.6	1.14520	5.076	82.53
32	1.120	17.49	392.3	12.83	8.336	92.39	372.0	1.05368	2.337	43.17
32.94	1.286	31.32	295.6	9.625				1.09756	3.512	
***P* = 0.1 MPa (1 bar)**										
20	0.1	71.14	-2.629	-0.1296	5.637	9.566	1119	1.23150	13.63	103.1
20.23	0.1	70.88	-0.4432	-0.02096	5.657	9.702	1112	1.23055	13.38	103.4
20.23	0.1	1.323	445.8	22.04	6.458	12.01	354.8	1.00399	1.071	16.95
40	0.1	0.6157	661.5	29.52	6.218	10.57	521.4	1.00186	2.039	31.85
60	0.1	0.4059	873.1	33.81	6.503	10.72	635.7	1.00122	2.865	45.53
80	0.1	0.3035	1096	37.02	7.598	11.77	713.9	1.00091	3.566	61.36
100	0.1	0.2425	1348	39.82	9.277	13.43	772.5	1.00073	4.190	80.43
120	0.1	0.2020	1633	42.41	10.83	14.98	827.4	1.00061	4.763	101.4

T / K	P / MPa	ρ / kg m⁻³	H / kJ kg⁻¹	S / kJ kg⁻¹ K⁻¹	C_v / kJ kg⁻¹ K⁻¹	C_p / kJ kg⁻¹ K⁻¹	u / m s⁻¹	D	η / μPa s	λ / mW m⁻¹ K⁻¹
140	0.1	0.1731	1943	44.80	11.82	15.96	883.4	1.00052	5.301	119.7
160	0.1	0.1514	2267	46.96	12.21	16.35	940.4	1.00046	5.814	134.8
180	0.1	0.1346	2594	48.89	12.19	16.32	997.7	1.00041	6.306	146.5
200	0.1	0.1211	2918	50.60	11.95	16.08	1054	1.00037	6.780	155.6
220	0.1	0.1101	3237	52.11	11.63	15.76	1110	1.00033	7.239	163.4
240	0.1	0.1010	3549	53.47	11.33	15.46	1163	1.00030	7.684	170.6
260	0.1	0.09319	3856	54.70	11.07	15.20	1214	1.00028	8.118	177.7
280	0.1	0.08654	4157	55.82	10.87	15.00	1263	1.00026	8.540	184.9
300	0.1	0.08077	4456	56.85	10.72	14.85	1310	1.00024	8.953	192.3
320	0.1	0.07572	4752	57.80	10.61	14.74	1355	1.00023	9.357	200.0
340	0.1	0.07127	5045	58.69	10.53	14.66	1398	1.00022	9.752	208.0
360	0.1	0.06731	5338	59.53	10.48	14.61	1439	1.00020	10.14	216.4
400	0.1	0.06058	5921	61.06	10.43	14.55	1518	1.00018	10.89	233.1
500	0.1	0.04847	7374	64.31	10.40	14.53	1698	1.00015	12.67	275.0
600	0.1	0.04040	8828	66.96	10.42	14.55	1859	1.00012	14.34	317.6
700	0.1	0.03463	10290	69.20	10.48	14.60	2006	1.00011	15.92	360.8
800	0.1	0.03030	11750	71.16	10.57	14.70	2142	1.00009	17.43	404.7
900	0.1	0.02693	13230	72.90	10.71	14.83	2268	1.00008	18.89	449.2
1000	0.1	0.02424	14720	74.47	10.87	14.99	2386	1.00007	20.30	494.3
***P* = 1 MPa (10 bar)**										
20	1.0	72.29	6.046	-0.3233	5.634	9.187	1164	1.23552	14.53	105.1
31.24	1.0	49.65	177.4	6.181	6.718	40.77	596.5	1.15769	5.627	87.27
31.24	1.0	14.31	414.3	13.76	8.017	52.90	374.7	1.04376	2.096	38.44
40	1.0	7.270	590.6	18.85	6.455	14.12	497.2	1.02207	2.228	37.28
60	1.0	4.228	838.4	23.90	6.588	11.70	634.2	1.01279	2.976	48.40
80	1.0	3.072	1075	27.31	7.645	12.25	717.9	1.00929	3.648	63.68
100	1.0	2.429	1335	30.19	9.307	13.72	778.4	1.00734	4.255	82.51
120	1.0	2.015	1624	32.83	10.85	15.17	834.2	1.00609	4.817	103.1
140	1.0	1.723	1938	35.24	11.84	16.09	890.7	1.00520	5.348	121.2
160	1.0	1.506	2264	37.42	12.23	16.45	948.1	1.00455	5.854	136.1
180	1.0	1.338	2593	39.36	12.20	16.40	1006	1.00404	6.341	147.6
200	1.0	1.204	2918	41.07	11.96	16.14	1062	1.00364	6.812	156.7
220	1.0	1.095	3238	42.60	11.64	15.81	1117	1.00331	7.268	164.3
240	1.0	1.004	3551	43.96	11.34	15.50	1171	1.00303	7.711	171.5
260	1.0	0.9267	3858	45.19	11.08	15.24	1222	1.00280	8.143	178.6
280	1.0	0.8607	4161	46.31	10.88	15.03	1271	1.00260	8.563	185.7
300	1.0	0.8035	4460	47.34	10.73	14.87	1317	1.00243	8.975	193.0
320	1.0	0.7534	4756	48.30	10.62	14.76	1362	1.00228	9.377	200.7
340	1.0	0.7093	5050	49.19	10.54	14.68	1405	1.00215	9.771	208.7
360	1.0	0.6700	5343	50.03	10.49	14.62	1446	1.00203	10.16	217.0
400	1.0	0.6033	5927	51.56	10.43	14.56	1525	1.00183	10.91	233.7
500	1.0	0.4830	7381	54.81	10.41	14.53	1704	1.00146	12.68	275.5
600	1.0	0.4027	8835	57.46	10.43	14.55	1865	1.00122	14.35	318.0
700	1.0	0.3454	10290	59.71	10.48	14.61	2011	1.00105	15.93	361.2
800	1.0	0.3023	11760	61.66	10.58	14.70	2147	1.00092	17.44	405.0
900	1.0	0.2688	13230	63.40	10.71	14.83	2272	1.00082	18.90	449.5
1000	1.0	0.2420	14720	64.97	10.87	15.00	2390	1.00074	20.30	494.6
***P* = 10 MPa (100 bar)**										
20	10.0	79.91	98.92	-1.567	5.529	7.566	1467	1.26244	23.45	120.3
40	10.0	63.19	310.5	5.532	6.665	13.45	1154	1.20345	8.791	121.8
60	10.0	42.86	617.0	11.71	6.960	16.19	923.2	1.13508	5.594	100.6
80	10.0	29.87	933.6	16.27	7.952	15.47	884.9	1.09289	5.034	96.74
100	10.0	23.01	1244	19.73	9.546	15.76	904.3	1.07104	5.169	106.3
120	10.0	18.87	1567	22.67	11.05	16.56	942.4	1.05803	5.497	121.6
140	10.0	16.08	1905	25.27	12.00	17.11	990.1	1.04932	5.888	136.6
160	10.0	14.06	2249	27.57	12.37	17.22	1042	1.04303	6.303	149.4
180	10.0	12.51	2592	29.59	12.32	17.01	1096	1.03824	6.727	159.4
200	10.0	11.28	2928	31.36	12.07	16.63	1150	1.03445	7.151	167.3
220	10.0	10.28	3257	32.93	11.74	16.22	1203	1.03138	7.571	173.9
240	10.0	9.453	3577	34.32	11.43	15.84	1254	1.02883	7.985	180.4

T / K	P / MPa	ρ / kg m⁻³	H / kJ kg⁻¹	S / kJ kg⁻¹ K⁻¹	C_v / kJ kg⁻¹ K⁻¹	C_p / kJ kg⁻¹ K⁻¹	u / m s⁻¹	D	η / µPa s	λ / mW m⁻¹ K⁻¹
260	10.0	8.751	3891	35.58	11.17	15.52	1303	1.02667	8.393	186.8
280	10.0	8.149	4199	36.72	10.96	15.27	1350	1.02483	8.795	193.3
300	10.0	7.626	4502	37.77	10.80	15.08	1394	1.02323	9.189	200.2
320	10.0	7.167	4802	38.73	10.68	14.94	1437	1.02183	9.577	207.4
340	10.0	6.762	5100	39.64	10.60	14.84	1478	1.02059	9.959	215.0
360	10.0	6.400	5396	40.48	10.55	14.77	1518	1.01948	10.33	223.0
400	10.0	5.784	5985	42.03	10.49	14.68	1593	1.01761	11.07	239.2
500	10.0	4.665	7447	45.30	10.45	14.60	1764	1.01420	12.81	280.0
600	10.0	3.911	8907	47.96	10.46	14.59	1919	1.01191	14.46	321.9
700	10.0	3.368	10370	50.21	10.51	14.63	2062	1.01026	16.02	364.7
800	10.0	2.957	11830	52.17	10.60	14.72	2193	1.00902	17.52	408.2
900	10.0	2.636	13310	53.91	10.73	14.84	2315	1.00805	18.97	452.4
1000	10.0	2.378	14800	55.48	10.89	15.00	2430	1.00727	20.37	497.3

Thermophysical Properties of Helium (^{4}He)

T / K	P / MPa	ρ / kg m⁻³	H / kJ kg⁻¹	S / kJ kg⁻¹ K⁻¹	C_v / kJ kg⁻¹ K⁻¹	C_p / kJ kg⁻¹ K⁻¹	u / m s⁻¹	D	η / µPa s	λ / mW m⁻¹ K⁻¹
Saturation										
2.177	0.004856	146.2	-7.501	-2.177	6.287	6.318	216.8	1.07842	3.596	13.52
2.177	0.004856	1.146	15.73	8.494	3.470	6.061	83.22	1.00045	0.5376	3.977
2.5	0.01000	145.0	-6.180	-1.620	2.521	2.700	216.0	1.07491	3.746	14.89
2.5	0.01000	2.116	16.97	7.641	3.486	6.258	87.62	1.00082	0.6382	4.765
3	0.02373	141.3	-4.869	-1.177	2.061	2.598	214.3	1.06975	3.694	16.55
3	0.02373	4.428	18.66	6.666	3.452	6.592	93.04	1.00173	0.7961	5.893
3.5	0.04663	136.1	-3.258	-0.7324	2.332	3.414	202.4	1.06471	3.511	17.75
3.5	0.04663	8.022	19.95	5.898	3.378	7.122	96.94	1.00315	0.9640	7.060
4	0.08100	128.9	-1.164	-0.2439	2.504	4.523	185.7	1.05941	3.278	18.47
4	0.08100	13.41	20.68	5.217	3.284	8.179	99.47	1.00529	1.151	8.356
4.5	0.1292	118.8	1.586	0.3106	2.601	6.776	164.0	1.05319	3.010	18.77
4.5	0.1292	21.76	20.52	4.519	3.179	10.96	100.8	1.00865	1.372	10.04
5	0.1945	100.8	5.854	1.081	2.709	20.24	133.0	1.04367	2.634	19.28
5	0.1945	38.21	18.18	3.547	3.053	29.09	101.6	1.01543	1.687	13.60
5.195	0.2275	69.64	11.87	2.183				1.02904	2.129	
***P* = 0.1 MPa (1 bar)**										
20	0.1	2.408	108.5	13.94	3.121	5.250	264.3	1.00093	3.582	26.20
40	0.1	1.200	212.9	17.56	3.119	5.206	373.4	1.00047	5.542	40.44
60	0.1	0.8006	317.0	19.67	3.118	5.198	456.9	1.00031	7.116	52.55
80	0.1	0.6007	420.9	21.16	3.117	5.195	527.2	1.00023	8.503	63.52
100	0.1	0.4807	524.8	22.32	3.117	5.194	589.2	1.00019	9.778	73.71
120	0.1	0.4007	628.7	23.27	3.117	5.194	645.3	1.00016	10.79	83.33
140	0.1	0.3435	732.6	24.07	3.116	5.193	696.9	1.00013	11.94	92.50
160	0.1	0.3006	836.4	24.76	3.116	5.193	744.9	1.00012	13.05	101.3
180	0.1	0.2672	940.3	25.37	3.116	5.193	790.0	1.00010	14.11	109.8
200	0.1	0.2405	1044	25.92	3.116	5.193	832.7	1.00009	15.14	118.0
220	0.1	0.2187	1148	26.42	3.116	5.193	873.3	1.00008	16.14	126.0
240	0.1	0.2005	1252	26.87	3.116	5.193	912.0	1.00008	17.12	133.7
260	0.1	0.1851	1356	27.28	3.116	5.193	949.2	1.00007	18.08	141.3
280	0.1	0.1718	1460	27.67	3.116	5.193	985.0	1.00007	19.01	148.7
300	0.1	0.1604	1563	28.03	3.116	5.193	1020	1.00006	19.93	156.0
320	0.1	0.1504	1667	28.36	3.116	5.193	1053	1.00006	20.83	163.1
340	0.1	0.1415	1771	28.68	3.116	5.193	1085	1.00005	21.72	170.1
360	0.1	0.1337	1875	28.97	3.116	5.193	1117	1.00005	22.59	177.0
400	0.1	0.1203	2083	29.52	3.116	5.193	1177	1.00005	24.29	190.4
500	0.1	0.09626	2602	30.68	3.116	5.193	1316	1.00004	28.36	222.3
600	0.1	0.08022	3121	31.63	3.116	5.193	1442	1.00003	32.22	252.4
700	0.1	0.06876	3641	32.43	3.116	5.193	1557	1.00003	35.89	281.1
800	0.1	0.06017	4160	33.12	3.116	5.193	1664	1.00002	39.43	308.5
900	0.1	0.05348	4679	33.73	3.116	5.193	1765	1.00002	42.85	335.0
1000	0.1	0.04814	5199	34.28	3.116	5.193	1861	1.00002	46.16	360.6

T / K	P / MPa	ρ / kg m⁻³	H / kJ kg⁻¹	S / kJ kg⁻¹ K⁻¹	C_v / kJ kg⁻¹ K⁻¹	C_p / kJ kg⁻¹ K⁻¹	u / m s⁻¹	D	η / μPa s	λ / mW m⁻¹ K⁻¹
P = 1 MPa (10 bar)										
20	1.0	24.02	104.1	8.936	3.156	5.728	276.4	1.00938	3.933	29.04
40	1.0	11.72	213.2	12.73	3.143	5.317	384.9	1.00455	5.740	42.01
60	1.0	7.850	318.6	14.87	3.133	5.242	466.8	1.00305	7.269	53.72
80	1.0	5.912	423.1	16.37	3.128	5.217	535.8	1.00229	8.633	64.50
100	1.0	4.745	527.4	17.54	3.125	5.206	596.9	1.00184	9.892	74.60
120	1.0	3.963	631.4	18.48	3.123	5.200	652.2	1.00154	10.90	84.16
140	1.0	3.403	735.4	19.29	3.121	5.197	703.2	1.00132	12.04	93.29
160	1.0	2.982	839.3	19.98	3.120	5.195	750.7	1.00116	13.13	102.1
180	1.0	2.653	943.2	20.59	3.120	5.194	795.4	1.00103	14.19	110.5
200	1.0	2.390	1047	21.14	3.119	5.194	837.7	1.00093	15.21	118.7
220	1.0	2.174	1151	21.63	3.119	5.193	878.0	1.00084	16.20	126.7
240	1.0	1.994	1255	22.09	3.118	5.193	916.5	1.00077	17.17	134.4
260	1.0	1.842	1359	22.50	3.118	5.193	953.5	1.00071	18.12	142.0
280	1.0	1.711	1463	22.89	3.118	5.192	989.1	1.00066	19.05	149.4
300	1.0	1.597	1566	23.24	3.118	5.192	1023	1.00062	19.96	156.6
320	1.0	1.498	1670	23.58	3.117	5.192	1057	1.00058	20.86	163.8
340	1.0	1.410	1774	23.89	3.117	5.192	1089	1.00055	21.74	170.7
360	1.0	1.332	1878	24.19	3.117	5.192	1120	1.00052	22.61	177.6
400	1.0	1.200	2086	24.74	3.117	5.192	1180	1.00047	24.32	191.0
500	1.0	0.9604	2605	25.90	3.117	5.192	1319	1.00037	28.38	222.9
600	1.0	0.8007	3124	26.84	3.116	5.192	1444	1.00031	32.23	253.0
700	1.0	0.6866	3643	27.64	3.116	5.192	1559	1.00027	35.91	281.6
800	1.0	0.6009	4163	28.34	3.116	5.192	1666	1.00023	39.44	309.1
900	1.0	0.5342	4682	28.95	3.116	5.193	1767	1.00021	42.86	335.5
1000	1.0	0.4809	5201	29.50	3.116	5.193	1863	1.00019	46.17	361.1
P = 10 MPa (100 bar)										
20	10.0	147.2	111.0	3.756	3.286	5.413	497.6	1.06056	6.928	58.72
40	10.0	91.33	225.2	7.703	3.305	5.721	512.2	1.03614	7.483	59.90
60	10.0	65.36	337.4	9.980	3.251	5.506	566.3	1.02560	8.638	68.12
80	10.0	51.09	446.0	11.54	3.215	5.376	620.9	1.01993	9.802	76.83
100	10.0	42.05	552.8	12.74	3.192	5.303	672.3	1.01637	10.93	85.42
120	10.0	35.77	658.4	13.70	3.176	5.261	720.4	1.01391	11.82	93.82
140	10.0	31.14	763.3	14.51	3.165	5.235	765.6	1.01210	12.88	102.0
160	10.0	27.58	867.8	15.21	3.157	5.219	808.5	1.01071	13.89	110.1
180	10.0	24.75	972.1	15.82	3.151	5.208	849.2	1.00961	14.87	118.0
200	10.0	22.46	1076	16.37	3.146	5.201	888.2	1.00872	15.82	125.7
220	10.0	20.55	1180	16.86	3.142	5.196	925.6	1.00798	16.75	133.3
240	10.0	18.94	1284	17.31	3.139	5.193	961.6	1.00735	17.65	140.7
260	10.0	17.57	1388	17.73	3.137	5.190	996.4	1.00682	18.54	148.0
280	10.0	16.38	1492	18.11	3.135	5.188	1030	1.00635	19.41	155.2
300	10.0	15.34	1595	18.47	3.133	5.187	1063	1.00595	20.26	162.3
320	10.0	14.43	1699	18.81	3.132	5.186	1094	1.00560	21.14	169.2
340	10.0	13.62	1803	19.12	3.130	5.186	1125	1.00528	22.01	176.0
360	10.0	12.89	1907	19.42	3.129	5.185	1155	1.00500	22.86	182.8
400	10.0	11.65	2114	19.96	3.127	5.185	1213	1.00452	24.54	196.0
500	10.0	9.387	2632	21.12	3.124	5.185	1347	1.00364	28.56	227.6
600	10.0	7.860	3151	22.07	3.122	5.186	1469	1.00305	32.38	257.4
700	10.0	6.760	3670	22.87	3.121	5.186	1582	1.00262	36.04	285.9
800	10.0	5.930	4188	23.56	3.120	5.187	1687	1.00230	39.56	313.2
900	10.0	5.281	4707	24.17	3.119	5.188	1786	1.00205	42.96	339.6
1000	10.0	4.760	5226	24.72	3.119	5.188	1880	1.00185	46.26	365.1

Thermophysical Properties of Argon (Ar)

T / K	P / MPa	ρ / kg m⁻³	H / kJ kg⁻¹	S / kJ kg⁻¹ K⁻¹	C_v / kJ kg⁻¹ K⁻¹	C_p / kJ kg⁻¹ K⁻¹	u / m s⁻¹	D	η / μPa s	λ / mW m⁻¹ K⁻¹
Saturation										
83.81	0.06889	1417	-121.4	1.329	0.5496	1.116	862.4	1.51232	290.2	133.6
83.81	0.06889	4.055	42.28	3.283	0.3247	0.5550	168.1	1.00126	6.856	5.359

T / K	P / MPa	ρ / kg m^{-3}	H / kJ kg^{-1}	S / kJ kg^{-1} K^{-1}	C_v / kJ kg^{-1} K^{-1}	C_p / kJ kg^{-1} K^{-1}	u / m s^{-1}	D	η / μPa s	λ / mW m^{-1} K^{-1}
90	0.1335	1379	-114.5	1.409	0.5268	1.121	819.5	1.49650	240.0	124.5
90	0.1335	7.436	44.57	3.176	0.3309	0.5757	172.8	1.00231	7.413	5.835
100	0.3238	1314	-103.1	1.528	0.4976	1.154	746.9	1.46993	181.3	110.2
100	0.3238	16.86	47.40	3.032	0.3445	0.6269	178.9	1.00525	8.349	6.689
110	0.6653	1243	-91.13	1.639	0.4747	1.218	669.2	1.44136	140.4	96.41
110	0.6653	33.29	48.84	2.911	0.3633	0.7122	183.0	1.01039	9.366	7.732
120	1.213	1163	-78.35	1.746	0.4576	1.332	584.2	1.40965	110.2	83.13
120	1.213	60.14	48.41	2.802	0.3893	0.8627	185.1	1.01884	10.54	9.154
130	2.025	1068	-64.16	1.854	0.4492	1.564	487.9	1.37273	85.86	70.43
130	2.025	103.6	45.30	2.696	0.4275	1.172	184.8	1.03259	12.03	11.45
140	3.168	943.7	-47.16	1.971	0.4598	2.225	371.6	1.32519	63.62	58.06
140	3.168	178.9	37.47	2.576	0.4940	2.104	181.5	1.05677	14.32	16.39
150	4.735	680.4	-17.88	2.159	0.7060	23.58	174.7	1.22816	36.78	57.63
150	4.735	394.5	11.52	2.355	0.8218	35.47	157.0	1.12827	21.18	55.88
150.69	4.863	535.6	-4.332	2.248				1.17684	27.63	
***P* = 0.1 MPa (1 bar)**										
83.81	0.1	1417	-121.4	1.329	0.5496	1.116	862.5	1.51233	290.2	133.6
87.18	0.1	1396	-117.7	1.373	0.5366	1.117	839.2	1.50375	261.3	128.6
87.18	0.1	5.704	43.57	3.223	0.3279	0.5654	170.8	1.00178	7.157	5.614
100	0.1	4.915	50.69	3.299	0.3206	0.5470	184.2	1.00153	8.234	6.450
120	0.1	4.058	61.49	3.398	0.3161	0.5347	202.8	1.00126	9.878	7.735
140	0.1	3.461	72.12	3.480	0.3144	0.5293	219.6	1.00108	11.48	8.987
160	0.1	3.020	82.68	3.550	0.3135	0.5264	235.1	1.00094	13.03	10.21
180	0.1	2.680	93.19	3.612	0.3131	0.5247	249.6	1.00083	14.53	11.39
200	0.1	2.409	103.7	3.667	0.3128	0.5236	263.2	1.00075	16.00	12.54
220	0.1	2.189	114.1	3.717	0.3127	0.5229	276.2	1.00068	17.42	13.66
240	0.1	2.005	124.6	3.762	0.3125	0.5224	288.5	1.00062	18.80	14.74
260	0.1	1.850	135.0	3.804	0.3125	0.5220	300.3	1.00058	20.15	15.80
280	0.1	1.717	145.5	3.843	0.3124	0.5217	311.7	1.00053	21.46	16.83
300	0.1	1.603	155.9	3.879	0.3124	0.5215	322.7	1.00050	22.74	17.84
320	0.1	1.502	166.3	3.913	0.3124	0.5214	333.3	1.00047	23.99	18.82
340	0.1	1.414	176.8	3.944	0.3123	0.5212	343.5	1.00044	25.21	19.77
360	0.1	1.335	187.2	3.974	0.3123	0.5211	353.5	1.00042	26.40	20.71
400	0.1	1.201	208.0	4.029	0.3123	0.5209	372.7	1.00037	28.70	22.52
500	0.1	0.9608	260.1	4.145	0.3123	0.5207	416.6	1.00030	34.08	26.73
600	0.1	0.8006	312.2	4.240	0.3122	0.5206	456.4	1.00025	39.00	30.57
700	0.1	0.6862	364.2	4.320	0.3122	0.5205	493.0	1.00021	43.56	34.13
800	0.1	0.6004	416.3	4.390	0.3122	0.5205	527.0	1.00019	47.82	37.46
900	0.1	0.5337	468.3	4.451	0.3122	0.5204	558.9	1.00017	51.85	40.60
1000	0.1	0.4803	520.3	4.506	0.3122	0.5204	589.1	1.00015	55.69	43.58
***P* = 1 MPa (10 bar)**										
84.04	1.0	1418	-120.8	1.330	0.5498	1.113	865.3	1.51276	291.1	134.0
100	1.0	1316	-102.8	1.525	0.4984	1.148	751.6	1.47104	183.1	110.9
116.60	1.0	1191	-82.82	1.710	0.4627	1.285	614.1	1.42088	119.7	87.58
116.60	1.0	49.55	48.81	2.839	0.3795	0.8007	184.6	1.01550	10.12	8.608
120	1.0	47.20	51.45	2.861	0.3682	0.7559	189.3	1.01476	10.37	8.732
140	1.0	37.75	65.11	2.967	0.3375	0.6350	212.1	1.01179	11.87	9.704
160	1.0	31.93	77.31	3.048	0.3266	0.5907	230.5	1.00997	13.36	10.80
180	1.0	27.83	88.89	3.116	0.3214	0.5686	246.8	1.00868	14.83	11.90
200	1.0	24.74	100.1	3.176	0.3185	0.5556	261.6	1.00772	16.26	13.00
220	1.0	22.31	111.1	3.228	0.3168	0.5473	275.3	1.00696	17.65	14.07
240	1.0	20.33	122.0	3.275	0.3157	0.5417	288.3	1.00634	19.01	15.12
260	1.0	18.69	132.8	3.319	0.3149	0.5376	300.5	1.00583	20.34	16.15
280	1.0	17.30	143.5	3.358	0.3144	0.5346	312.2	1.00539	21.64	17.16
300	1.0	16.11	154.2	3.395	0.3140	0.5324	323.4	1.00502	22.90	18.14
320	1.0	15.08	164.8	3.429	0.3137	0.5306	334.2	1.00470	24.14	19.10
340	1.0	14.17	175.4	3.462	0.3135	0.5292	344.7	1.00441	25.35	20.04
360	1.0	13.37	186.0	3.492	0.3133	0.5280	354.7	1.00416	26.53	20.96
400	1.0	12.01	207.1	3.547	0.3131	0.5263	374.1	1.00374	28.82	22.75
500	1.0	9.593	259.6	3.664	0.3127	0.5239	418.3	1.00299	34.16	26.92

T / K	P / MPa	ρ / kg m⁻³	H / kJ kg⁻¹	S / kJ kg⁻¹ K⁻¹	C_v / kJ kg⁻¹ K⁻¹	C_p / kJ kg⁻¹ K⁻¹	u / m s⁻¹	D	η / μPa s	λ / mW m⁻¹ K⁻¹
600	1.0	7.988	311.9	3.760	0.3126	0.5227	458.1	1.00249	39.06	30.73
700	1.0	6.846	364.1	3.840	0.3125	0.5220	494.6	1.00213	43.61	34.27
800	1.0	5.990	416.3	3.910	0.3124	0.5215	528.6	1.00186	47.87	37.58
900	1.0	5.325	468.4	3.971	0.3124	0.5212	560.5	1.00166	51.90	40.71
1000	1.0	4.793	520.6	4.026	0.3124	0.5210	590.7	1.00149	55.72	43.68
P = 10 MPa (100 bar)										
86.27	10.0	1428	-114.2	1.333	0.5519	1.085	891.3	1.51685	300.0	137.1
100	10.0	1349	-99.27	1.493	0.5089	1.093	805.9	1.48430	205.8	118.6
120	10.0	1222	-76.82	1.698	0.4650	1.163	674.3	1.43296	130.0	93.55
140	10.0	1066	-52.01	1.888	0.4374	1.349	527.4	1.37161	85.36	70.81
160	10.0	833.6	-20.15	2.100	0.4356	1.970	357.8	1.28395	50.84	49.64
180	10.0	491.9	26.15	2.373	0.4254	2.091	259.3	1.16159	27.88	30.35
200	10.0	337.7	57.55	2.539	0.3815	1.215	267.7	1.10910	23.43	23.57
220	10.0	270.9	78.52	2.639	0.3585	0.9257	283.7	1.08686	22.90	21.77
240	10.0	231.2	95.59	2.713	0.3457	0.7960	299.0	1.07382	23.22	21.31
260	10.0	204.0	110.7	2.774	0.3379	0.7240	313.1	1.06491	23.87	21.39
280	10.0	183.6	124.7	2.826	0.3327	0.6789	326.1	1.05829	24.68	21.73
300	10.0	167.6	138.0	2.872	0.3290	0.6481	338.4	1.05312	25.58	22.19
320	10.0	154.6	150.7	2.913	0.3264	0.6261	350.0	1.04892	26.53	22.78
340	10.0	143.7	163.1	2.950	0.3243	0.6095	361.0	1.04542	27.51	23.43
360	10.0	134.4	175.1	2.985	0.3228	0.5967	371.6	1.04245	28.49	24.11
400	10.0	119.4	198.6	3.046	0.3205	0.5784	391.5	1.03765	30.48	25.50
500	10.0	94.09	255.0	3.173	0.3174	0.5540	436.1	1.02957	35.35	29.04
600	10.0	78.03	309.8	3.272	0.3159	0.5423	475.7	1.02448	39.97	32.48
700	10.0	66.80	363.7	3.355	0.3151	0.5357	511.7	1.02093	44.34	35.76
800	10.0	58.47	417.0	3.427	0.3145	0.5317	545.1	1.01830	48.47	38.88
900	10.0	52.03	470.0	3.489	0.3142	0.5290	576.4	1.01628	52.40	41.86
1000	10.0	46.88	522.8	3.545	0.3139	0.5271	606.0	1.01466	56.16	44.71

Thermophysical Properties of Methane (CH_4)

T / K	P / MPa	ρ / kg m⁻³	H / kJ kg⁻¹	S / kJ kg⁻¹ K⁻¹	C_v / kJ kg⁻¹ K⁻¹	C_p / kJ kg⁻¹ K⁻¹	u / m s⁻¹	D	η / μPa s	λ / mW m⁻¹ K⁻¹
Saturation										
90.69	0.01170	451.5	-71.82	-7099	2.168	3.368	1539	1.67721	193.4	211.2
90.69	0.01170	0.2507	472.4	5.291	1.574	2.110	249.1	1.00031	3.585	8.776
100	0.03438	438.9	-40.27	-3793	2.114	3.408	1452	1.65478	151.1	199.6
100	0.03438	0.6746	490.2	4.925	1.589	2.146	260.1	1.00082	3.914	9.910
110	0.08813	424.8	-5.813	-5217	2.064	3.469	1355	1.62995	121.0	186.1
110	0.08813	1.598	508.0	4.619	1.611	2.205	270.0	1.00196	4.273	11.22
120	0.1914	409.9	29.41	0.2521	2.020	3.549	1253	1.60408	98.68	172.0
120	0.1914	3.262	524.0	4.374	1.639	2.293	277.8	1.00399	4.636	12.66
130	0.3673	394.0	65.63	0.5385	1.980	3.658	1148	1.57683	81.29	157.7
130	0.3673	5.980	537.7	4.170	1.674	2.421	283.1	1.00733	5.008	14.28
140	0.6412	376.9	103.2	0.8116	1.945	3.813	1038	1.54771	67.35	143.5
140	0.6412	10.15	548.3	3.991	1.717	2.611	285.9	1.01248	5.398	16.15
150	1.040	357.9	142.6	1.076	1.919	4.047	920.8	1.51599	55.99	129.2
150	1.040	16.33	555.2	3.827	1.773	2.908	286.0	1.02013	5.827	18.38
160	1.592	336.3	184.8	1.338	1.904	4.435	795.4	1.48045	46.55	115.0
160	1.592	25.38	557.1	3.664	1.847	3.419	283.0	1.03144	6.335	21.23
170	2.328	310.5	231.2	1.605	1.910	5.187	657.5	1.43870	38.45	100.5
170	2.328	38.97	551.5	3.490	1.956	4.459	276.7	1.04861	7.012	25.34
180	3.285	276.2	285.9	1.899	1.967	7.292	497.0	1.38453	30.92	85.50
180	3.285	61.38	532.8	3.271	2.140	7.574	266.0	1.07739	8.127	33.50
190	4.519	200.8	378.3	2.369	2.602	94.01	250.3	1.27031	20.34	94.07
190	4.519	125.2	459.0	2.794	2.855	140.8	238.5	1.16274	12.24	120.5
190.56	4.599	162.7	415.6	2.562				1.21521	15.91	
P = 0.1 MPa (1 bar)										
90.72	0.1	451.5	-71.60	-7097	2.168	3.367	1539	1.67726	193.6	211.2
100	0.1	438.9	-40.17	-3798	2.114	3.408	1453	1.65487	151.2	199.6

T / K	P / MPa	ρ / kg m⁻³	H / kJ kg⁻¹	S / kJ kg⁻¹ K⁻¹	C_v / kJ kg⁻¹ K⁻¹	C_p / kJ kg⁻¹ K⁻¹	u / m s⁻¹	D	η / μPa s	λ / mW m⁻¹ K⁻¹
111.51	0.1	422.6	-5573	-4967	2.057	3.480	1340	1.62612	117.2	184.0
111.51	0.1	1.795	510.6	4.579	1.615	2.216	271.3	1.00220	4.327	11.43
120	0.1	1.655	529.2	4.740	1.594	2.174	282.8	1.00202	4.668	12.35
140	0.1	1.403	572.1	5.071	1.574	2.129	307.6	1.00172	5.449	14.65
160	0.1	1.221	614.5	5.354	1.568	2.110	330.0	1.00149	6.213	16.98
180	0.1	1.081	656.6	5.602	1.568	2.104	350.7	1.00132	6.963	19.32
200	0.1	0.9709	698.7	5.824	1.574	2.106	370.0	1.00119	7.697	21.65
220	0.1	0.8812	740.9	6.025	1.587	2.115	388.0	1.00108	8.416	23.99
240	0.1	0.8069	783.4	6.209	1.607	2.133	404.9	1.00099	9.117	26.39
260	0.1	0.7442	826.3	6.381	1.634	2.159	420.8	1.00091	9.803	28.88
280	0.1	0.6906	869.8	6.542	1.670	2.194	435.7	1.00085	10.47	31.47
300	0.1	0.6443	914.1	6.695	1.713	2.236	449.7	1.00079	11.13	34.19
320	0.1	0.6038	959.3	6.841	1.763	2.285	463.1	1.00074	11.77	37.04
340	0.1	0.5681	1006	6.981	1.819	2.340	475.7	1.00070	12.39	40.03
360	0.1	0.5364	1053	7.117	1.880	2.401	487.8	1.00066	13.00	43.15
400	0.1	0.4826	1152	7.376	2.013	2.534	510.6	1.00059	14.18	49.80
500	0.1	0.3859	1423	7.981	2.381	2.901	561.9	1.00047	16.92	68.34
600	0.1	0.3215	1732	8.543	2.754	3.273	608.0	1.00039	19.41	88.80
700	0.1	0.2756	2077	9.074	3.110	3.629	650.8	1.00034	21.68	110.4
800	0.1	0.2411	2457	9.581	3.441	3.959	690.9	1.00030	23.76	132.5
900	0.1	0.2143	2868	10.06	3.744	4.263	729.0	1.00026	25.70	154.7
1000	0.1	0.1929	3308	10.53	4.020	4.538	765.2	1.00024	27.49	176.7
P = 1 MPa (10 bar)										
90.95	1.0	451.8	-69.37	-7070	2.169	3.363	1543	1.67772	195.7	211.7
100	1.0	439.6	-38.76	-3862	2.116	3.401	1460	1.65604	153.4	200.5
120	1.0	410.8	30.49	0.2447	2.022	3.536	1262	1.60560	99.86	173.0
140	1.0	377.5	103.5	0.8069	1.946	3.798	1044	1.54879	67.81	144.0
149.14	1.0	359.6	139.2	1.054	1.920	4.023	931.2	1.51885	56.88	130.4
149.14	1.0	15.70	554.8	3.841	1.767	2.876	286.1	1.01935	5.788	18.17
160	1.0	13.97	584.2	4.031	1.682	2.588	305.2	1.01720	6.287	18.79
180	1.0	11.81	633.6	4.322	1.631	2.380	333.7	1.01453	7.089	20.65
200	1.0	10.33	680.2	4.568	1.613	2.289	357.8	1.01269	7.838	22.74
220	1.0	9.217	725.5	4.784	1.613	2.248	379.1	1.01133	8.558	24.90
240	1.0	8.344	770.3	4.978	1.626	2.234	398.2	1.01025	9.258	27.19
260	1.0	7.634	815.0	5.157	1.648	2.238	415.8	1.00938	9.939	29.60
280	1.0	7.043	859.9	5.324	1.681	2.258	432.0	1.00865	10.60	32.12
300	1.0	6.542	905.4	5.480	1.722	2.289	447.0	1.00803	11.25	34.79
320	1.0	6.110	951.5	5.629	1.770	2.330	461.1	1.00750	11.89	37.59
340	1.0	5.733	998.6	5.772	1.825	2.379	474.4	1.00704	12.51	40.54
360	1.0	5.402	1047	5.910	1.885	2.434	487.0	1.00663	13.11	43.64
400	1.0	4.846	1147	6.172	2.017	2.559	510.6	1.00595	14.28	50.23
500	1.0	3.860	1420	6.781	2.383	2.915	563.0	1.00474	17.01	68.68
600	1.0	3.210	1730	7.345	2.755	3.282	609.7	1.00394	19.48	89.08
700	1.0	2.750	2076	7.878	3.110	3.635	652.8	1.00338	21.74	110.6
800	1.0	2.405	2456	8.385	3.441	3.964	693.1	1.00296	23.82	132.7
900	1.0	2.137	2868	8.870	3.744	4.266	731.2	1.00263	25.75	154.9
1000	1.0	1.924	3308	9.334	4.020	4.541	767.5	1.00237	27.54	176.8
P = 10 MPa (100 bar)										
93.22	10.0	454.5	-47.07	-6806	2.176	3.325	1582	1.68223	212.8	216.9
100	10.0	446.0	-24.48	-4466	2.139	3.344	1526	1.66701	174.8	209.0
120	10.0	419.8	43.14	0.1696	2.047	3.424	1352	1.62085	112.1	183.3
140	10.0	391.2	112.8	0.7062	1.969	3.552	1171	1.57167	78.30	157.0
160	10.0	358.8	185.8	1.194	1.909	3.777	981.8	1.51718	56.70	131.5
180	10.0	319.6	265.3	1.661	1.873	4.226	780.8	1.45298	41.52	107.0
200	10.0	266.2	358.9	2.153	1.878	5.304	567.9	1.36870	29.55	83.93
220	10.0	187.6	482.7	2.742	1.903	6.713	404.4	1.25087	19.37	63.29
240	10.0	128.4	599.6	3.252	1.847	4.853	383.6	1.16707	14.89	49.74
260	10.0	101.0	683.8	3.589	1.803	3.730	402.8	1.12974	13.80	44.85
280	10.0	85.51	752.9	3.846	1.793	3.242	424.5	1.10902	13.61	43.81
300	10.0	75.18	815.1	4.060	1.807	3.002	444.5	1.09539	13.75	44.37

T / K	P / MPa	ρ / kg m⁻³	H / kJ kg⁻¹	S / kJ kg⁻¹ K⁻¹	C_v / kJ kg⁻¹ K⁻¹	C_p / kJ kg⁻¹ K⁻¹	u / m s⁻¹	D	η / μPa s	λ / mW m⁻¹ K⁻¹
320	10.0	67.61	873.8	4.250	1.838	2.878	462.8	1.08549	14.04	45.79
340	10.0	61.75	930.6	4.422	1.880	2.817	479.5	1.07786	14.41	47.75
360	10.0	57.01	986.7	4.582	1.931	2.795	494.8	1.07173	14.82	50.09
400	10.0	49.74	1099	4.877	2.051	2.819	522.6	1.06238	15.72	55.61
500	10.0	38.32	1391	5.529	2.401	3.055	581.0	1.04783	18.06	72.56
600	10.0	31.47	1712	6.113	2.766	3.370	630.6	1.03918	20.32	92.15
700	10.0	26.82	2066	6.657	3.117	3.696	675.2	1.03334	22.44	113.2
800	10.0	23.41	2451	7.171	3.445	4.008	716.3	1.02907	24.43	134.9
900	10.0	20.79	2866	7.660	3.747	4.300	754.7	1.02581	26.28	156.8
1000	10.0	18.72	3310	8.128	4.022	4.568	791.1	1.02323	28.02	178.5

Thermophysical Properties of Ethane (C_2H_6)

T / K	P / MPa	ρ / kg m⁻³	H / kJ kg⁻¹	S / kJ kg⁻¹ K⁻¹	C_v / kJ kg⁻¹ K⁻¹	C_p / kJ kg⁻¹ K⁻¹	u / m s⁻¹	D	η / μPa s	λ / mW m⁻¹ K⁻¹
Saturation										
90.37	0.0000011	651.5	-219.2	-1.655	1.605	2.326	2009	1.94483	1281	255.6
90.37	0.0000011	0.0000457	375.6	4.927	0.8916	1.168	180.9	1.00000	3.043	2.908
100	0.0000111	640.9	-197.0	-1.422	1.541	2.283	1938	1.92579	873.2	247.8
100	0.0000111	0.0004007	386.9	4.418	0.9107	1.187	189.9	1.00000	3.316	3.456
120	0.0003523	618.9	-151.5	-1.007	1.478	2.280	1794	1.88657	485.8	229.9
120	0.0003523	0.01062	411.0	3.681	0.9528	1.230	206.9	1.00001	3.896	4.657
140	0.003814	596.6	-105.6	-0.6532	1.450	2.311	1649	1.84725	319.8	210.6
140	0.003814	0.09880	435.6	3.213	1.003	1.284	222.0	1.00011	4.496	5.966
160	0.02141	573.6	-58.95	-0.3418	1.436	2.357	1501	1.80737	231.2	190.8
160	0.02141	0.4890	460.3	2.904	1.048	1.338	235.1	1.00054	5.110	7.422
180	0.07864	549.5	-11.13	-0.06082	1.434	2.421	1350	1.76641	176.2	171.3
180	0.07864	1.625	484.2	2.691	1.098	1.409	245.5	1.00181	5.739	9.079
200	0.2172	524.0	38.30	0.1981	1.444	2.512	1196	1.72364	138.3	152.6
200	0.2172	4.170	506.2	2.538	1.179	1.537	252.3	1.00465	6.391	11.01
220	0.4920	496.3	90.01	0.4419	1.468	2.645	1037	1.67809	110.1	134.6
220	0.4920	9.017	525.5	2.421	1.280	1.720	254.6	1.01008	7.089	13.32
240	0.9668	465.3	145.0	0.6767	1.507	2.847	873.3	1.62822	87.80	117.5
240	0.9668	17.43	540.9	2.326	1.388	1.976	252.1	1.01957	7.881	16.18
260	1.712	429.1	204.9	0.9095	1.566	3.195	700.5	1.57121	69.30	101.2
260	1.712	31.58	550.2	2.238	1.516	2.418	243.8	1.03567	8.876	20.02
280	2.807	382.7	273.1	1.152	1.654	3.987	512.4	1.50037	52.84	85.43
280	2.807	56.37	548.6	2.136	1.696	3.522	228.1	1.06440	10.37	26.33
300	4.357	303.5	364.4	1.451	1.912	10.02	274.9	1.38458	34.97	71.49
300	4.357	114.5	514.1	1.950	2.089	13.30	200.5	1.13418	14.02	47.46
305.32	4.872	206.2	439.0	1.690				1.25115	21.80	
P = 0.1 MPa (1 bar)										
90.38	0.1	651.5	-219.0	-1.655	1.605	2.326	2009	1.94485	1281	255.6
100	0.1	641.0	-196.9	-1.422	1.541	2.283	1939	1.92586	873.9	247.9
120	0.1	619.0	-151.4	-1.007	1.478	2.279	1795	1.88665	486.1	230.0
140	0.1	596.6	-105.5	-0.6535	1.450	2.311	1650	1.84735	320.0	210.6
160	0.1	573.6	-58.85	-0.3420	1.436	2.357	1502	1.80747	231.3	190.9
180	0.1	549.5	-11.10	-0.06091	1.434	2.421	1351	1.76644	176.2	171.4
184.33	0.1	544.1	-0.5954	-0.003215	1.435	2.438	1317	1.75734	166.9	167.2
184.33	0.1	2.030	489.1	2.654	1.113	1.432	247.3	1.00226	5.877	9.471
200	0.1	1.856	511.8	2.771	1.149	1.457	257.9	1.00207	6.368	10.72
220	0.1	1.676	541.3	2.912	1.200	1.499	270.3	1.00187	6.991	12.47
240	0.1	1.529	571.8	3.045	1.260	1.554	281.8	1.00170	7.608	14.39
260	0.1	1.407	603.5	3.172	1.328	1.617	292.5	1.00157	8.217	16.48
280	0.1	1.303	636.5	3.294	1.401	1.688	302.6	1.00145	8.817	18.76
300	0.1	1.214	671.1	3.413	1.479	1.764	312.2	1.00135	9.408	21.22
320	0.1	1.137	707.1	3.530	1.560	1.844	321.4	1.00127	9.989	23.85
340	0.1	1.069	744.8	3.644	1.644	1.927	330.3	1.00119	10.56	26.65
360	0.1	1.009	784.2	3.756	1.729	2.011	338.8	1.00112	11.12	29.61
400	0.1	0.9067	868.1	3.977	1.901	2.181	355.2	1.00101	12.21	35.96

T / K	P / MPa	ρ / kg m⁻³	H / kJ kg⁻¹	S / kJ kg⁻¹ K⁻¹	C_v / kJ kg⁻¹ K⁻¹	C_p / kJ kg⁻¹ K⁻¹	u / m s⁻¹	D	η / μPa s	λ / mW m⁻¹ K⁻¹
500	0.1	0.7242	1107	4.509	2.315	2.594	393.1	1.00081	14.78	53.77
600	0.1	0.6031	1386	5.015	2.689	2.967	427.6	1.00067	17.13	73.34
700	0.1	0.5167	1699	5.498	3.021	3.298	459.6	1.00058	19.32	93.86
800	0.1	0.4520	2044	5.958	3.314	3.591	489.6	1.00050	21.36	114.8
900	0.1	0.4018	2416	6.396	3.572	3.849	517.9	1.00045	23.28	135.9
1000	0.1	0.3616	2813	6.814	3.799	4.076	544.8	1.00040	25.10	156.8
P = 1 MPa (10 bar)										
90.53	1.0	651.7	-217.5	-1.653	1.605	2.324	2012	1.94510	1283	255.9
100	1.0	641.3	-195.7	-1.424	1.542	2.282	1942	1.92644	880.2	248.3
120	1.0	619.4	-150.2	-1.010	1.480	2.278	1799	1.88737	489.1	230.5
140	1.0	597.2	-104.4	-0.6563	1.452	2.309	1654	1.84822	322.0	211.2
160	1.0	574.3	-57.80	-0.3452	1.437	2.354	1508	1.80855	232.8	191.6
180	1.0	550.4	-10.13	-0.06458	1.435	2.416	1358	1.76780	177.5	172.1
200	1.0	524.9	39.03	0.1943	1.445	2.505	1204	1.72518	139.3	153.3
220	1.0	497.1	90.36	0.4388	1.468	2.637	1044	1.67944	110.7	135.2
240	1.0	465.4	145.0	0.6764	1.507	2.846	873.9	1.62834	87.85	117.6
241.10	1.0	463.5	148.1	0.6894	1.510	2.861	864.0	1.62531	86.70	116.6
241.10	1.0	18.04	541.6	2.321	1.395	1.994	251.8	1.02025	7.929	16.36
260	1.0	15.93	577.9	2.466	1.400	1.880	268.9	1.01786	8.485	17.92
280	1.0	14.32	615.3	2.605	1.449	1.869	283.8	1.01604	9.072	19.91
300	1.0	13.07	652.9	2.735	1.513	1.898	296.9	1.01463	9.652	22.18
320	1.0	12.05	691.3	2.859	1.585	1.948	308.7	1.01349	10.22	24.67
340	1.0	11.21	730.9	2.979	1.663	2.010	319.6	1.01254	10.79	27.36
360	1.0	10.49	771.8	3.095	1.745	2.079	329.9	1.01173	11.34	30.24
400	1.0	9.312	857.9	3.322	1.911	2.230	348.8	1.01041	12.42	36.46
500	1.0	7.323	1101	3.862	2.321	2.620	390.2	1.00818	14.95	54.08
600	1.0	6.058	1381	4.372	2.693	2.983	426.6	1.00677	17.29	73.56
700	1.0	5.173	1696	4.857	3.023	3.309	459.7	1.00578	19.46	94.03
800	1.0	4.518	2042	5.318	3.315	3.598	490.3	1.00505	21.48	115.0
900	1.0	4.011	2415	5.757	3.573	3.855	519.0	1.00448	23.39	136.0
1000	1.0	3.608	2812	6.175	3.800	4.081	546.2	1.00403	25.20	156.9
P = 10 MPa (100 bar)										
91.96	10.0	653.3	-202.5	-1.640	1.605	2.310	2034	1.94751	1310	258.6
100	10.0	644.8	-184.1	-1.448	1.553	2.275	1976	1.93215	947.3	252.5
120	10.0	623.6	-138.8	-1.035	1.490	2.266	1839	1.89429	520.6	235.4
140	10.0	602.2	-93.25	-0.6839	1.463	2.290	1701	1.85661	342.0	216.9
160	10.0	580.4	-47.10	-0.3758	1.449	2.326	1564	1.81879	248.1	197.9
180	10.0	558.1	-0.1206	-0.09918	1.447	2.374	1425	1.78049	190.5	179.3
200	10.0	534.8	47.98	0.1542	1.456	2.439	1287	1.74125	151.3	161.4
220	10.0	510.2	97.60	0.3905	1.478	2.527	1147	1.70050	122.4	144.4
240	10.0	483.8	149.3	0.6152	1.511	2.646	1008	1.65744	99.96	128.5
260	10.0	454.6	203.7	0.8331	1.557	2.809	866.9	1.61088	81.74	113.6
280	10.0	421.4	262.1	1.049	1.616	3.044	724.4	1.55888	66.38	99.57
300	10.0	381.3	326.4	1.271	1.689	3.419	579.3	1.49786	52.87	86.39
320	10.0	328.8	400.9	1.511	1.783	4.109	434.8	1.42048	40.36	73.81
340	10.0	255.9	493.3	1.791	1.888	5.070	319.9	1.31781	28.77	62.21
360	10.0	187.0	591.3	2.071	1.941	4.452	286.8	1.22573	21.62	53.36
400	10.0	126.2	738.6	2.460	2.025	3.205	310.4	1.14850	17.94	48.70
500	10.0	80.57	1035	3.123	2.369	2.933	377.8	1.09293	17.95	59.52
600	10.0	62.41	1338	3.675	2.722	3.150	427.4	1.07144	19.50	76.88
700	10.0	51.79	1666	4.180	3.043	3.415	467.8	1.05903	21.26	96.40
800	10.0	44.59	2021	4.653	3.330	3.672	502.9	1.05068	23.01	116.8
900	10.0	39.30	2400	5.100	3.584	3.909	534.3	1.04457	24.73	137.5
1000	10.0	35.20	2802	5.523	3.809	4.122	563.2	1.03988	26.40	158.2

Thermophysical Properties of Propane (C_3H_8)

T / K	P / MPa	ρ / kg m⁻³	H / kJ kg⁻¹	S / kJ kg⁻¹ K⁻¹	C_v / kJ kg⁻¹ K⁻¹	C_p / kJ kg⁻¹ K⁻¹	u / m s⁻¹	D	η / μPa s	λ / mW m⁻¹ K⁻¹
Saturation										
85.53	0.17×10⁻⁹	733.1	-196.6	-1.396	1.355	1.916	2136	2.08838	10780	207.9

T / K	P / MPa	ρ / kg m^{-3}	H / kJ kg^{-1}	S / kJ kg^{-1} K^{-1}	C_v / kJ kg^{-1} K^{-1}	C_p / kJ kg^{-1} K^{-1}	u / m s^{-1}	D	η / μPa s	λ / mW m^{-1} K^{-1}
85.53	0.17×10^{-9}	0.11×10^{-7}	366.3	5.186	0.6907	0.8792	143.3	1.00000	2.641	1.706
100	0.25×10^{-7}	718.1	-168.8	-1.095	1.341	1.930	2038	2.05908	3774	203.2
100	0.25×10^{-7}	0.0000013	379.4	4.387	0.7475	0.9361	153.7	1.00000	2.979	2.417
120	0.0000030	697.8	-129.9	-0.7411	1.335	1.957	1901	2.01930	1500	194.4
120	0.0000030	0.0001310	398.9	3.666	0.8210	1.010	166.8	1.00000	3.467	3.500
140	0.0000790	677.6	-90.51	-0.4372	1.335	1.988	1767	1.98025	822.2	183.9
140	0.0000790	0.002993	419.7	3.207	0.8878	1.076	178.9	1.00000	3.970	4.699
160	0.0008502	657.3	-50.38	-0.1693	1.342	2.025	1633	1.94163	535.0	172.3
160	0.0008502	0.02821	441.8	2.907	0.9514	1.141	190.0	1.00003	4.483	6.013
180	0.005068	636.6	-9.433	0.07174	1.356	2.070	1499	1.90310	382.2	160.2
180	0.005068	0.1499	464.9	2.707	1.017	1.209	200.1	1.00016	5.000	7.439
200	0.02019	615.4	32.53	0.2926	1.383	2.127	1366	1.86434	288.0	147.8
200	0.02019	0.5417	488.6	2.573	1.088	1.287	208.7	1.00059	5.515	8.971
220	0.06057	593.4	75.80	0.4984	1.421	2.199	1233	1.82499	224.2	135.7
220	0.06057	1.499	512.7	2.484	1.169	1.381	215.6	1.00164	6.026	10.61
240	0.1480	570.4	120.7	0.6932	1.469	2.289	1102	1.78460	178.4	124.0
240	0.1480	3.438	536.6	2.426	1.259	1.494	220.1	1.00376	6.540	12.38
260	0.3107	545.8	167.7	0.8801	1.528	2.403	971.7	1.74256	143.9	112.9
260	0.3107	6.903	560.1	2.389	1.357	1.630	221.9	1.00755	7.073	14.32
280	0.5817	519.2	217.3	1.062	1.596	2.547	840.3	1.69800	117.1	102.5
280	0.5817	12.62	582.3	2.365	1.466	1.803	220.4	1.01383	7.655	16.54
300	0.9977	489.4	270.2	1.241	1.675	2.740	706.8	1.64961	95.28	92.86
300	0.9977	21.63	602.6	2.349	1.588	2.041	214.8	1.02381	8.340	19.24
320	1.599	454.9	327.3	1.421	1.764	3.028	569.1	1.59504	76.78	83.94
320	1.599	35.74	619.5	2.334	1.729	2.416	204.4	1.03957	9.230	22.78
340	2.431	411.8	390.9	1.608	1.872	3.585	422.5	1.52907	59.98	75.57
340	2.431	58.88	629.8	2.311	1.898	3.197	187.4	1.06579	10.57	28.20
360	3.555	345.6	468.2	1.820	2.049	5.984	251.0	1.43242	42.27	67.72
360	3.555	105.4	622.4	2.249	2.183	7.111	161.1	1.11995	13.38	41.36
369.89	4.251	220.5	555.2	2.052				1.26277	22.89	
***P* = 0.1 MPa (1 bar)**										
85.53	0.1	733.1	-196.5	-1.396	1.355	1.916	2137	2.08840	10790	207.9
100	0.1	718.2	-168.7	-1.096	1.341	1.930	2038	2.05912	3778	203.2
120	0.1	697.8	-129.8	-0.7413	1.335	1.957	1902	2.01936	1502	194.5
140	0.1	677.6	-90.39	-0.4374	1.335	1.988	1767	1.98032	822.9	184.0
160	0.1	657.3	-50.27	-0.1695	1.342	2.025	1633	1.94172	535.5	172.4
180	0.1	636.7	-9.328	0.07149	1.357	2.070	1500	1.90320	382.5	160.2
200	0.1	615.5	32.62	0.2924	1.383	2.127	1366	1.86444	288.2	147.9
220	0.1	593.5	75.84	0.4983	1.421	2.198	1234	1.82505	224.3	135.7
230.74	0.1	581.2	99.69	0.6042	1.446	2.245	1163	1.80347	197.9	129.4
230.74	0.1	2.387	525.6	2.450	1.216	1.439	218.3	1.00261	6.301	11.54
240	0.1	2.284	539.0	2.507	1.249	1.467	222.9	1.00249	6.559	12.40
260	0.1	2.092	569.0	2.627	1.324	1.535	232.3	1.00228	7.110	14.32
280	0.1	1.932	600.5	2.744	1.405	1.610	241.1	1.00211	7.656	16.36
300	0.1	1.796	633.5	2.857	1.490	1.692	249.4	1.00196	8.196	18.51
320	0.1	1.679	668.2	2.969	1.578	1.777	257.3	1.00183	8.732	20.78
340	0.1	1.576	704.6	3.080	1.668	1.865	264.9	1.00172	9.262	23.16
360	0.1	1.486	742.8	3.189	1.758	1.954	272.3	1.00162	9.787	25.66
400	0.1	1.334	824.5	3.404	1.937	2.130	286.2	1.00145	10.82	30.99
500	0.1	1.064	1059	3.925	2.356	2.547	318.3	1.00116	13.29	46.36
600	0.1	0.8852	1332	4.422	2.722	2.912	347.4	1.00096	15.59	64.63
700	0.1	0.7581	1640	4.896	3.038	3.228	374.2	1.00082	17.70	85.81
800	0.1	0.6631	1976	5.345	3.313	3.502	399.3	1.00072	19.62	109.9
900	0.1	0.5892	2339	5.772	3.553	3.743	422.8	1.00064	21.36	136.9
1000	0.1	0.5302	2724	6.177	3.764	3.953	445.1	1.00058	22.93	166.8
***P* = 1 MPa (10 bar)**										
85.62	1.0	733.3	-195.3	-1.396	1.356	1.916	2139	2.08856	10840	208.1
100	1.0	718.5	-167.6	-1.097	1.342	1.930	2041	2.05953	3816	203.5
120	1.0	698.2	-128.8	-0.7432	1.336	1.956	1905	2.01987	1515	194.8
140	1.0	678.1	-89.34	-0.4394	1.336	1.987	1771	1.98096	829.3	184.3

T / K	P / MPa	ρ / kg m^{-3}	H / kJ kg^{-1}	S / kJ kg^{-1} K^{-1}	C_v / kJ kg^{-1} K^{-1}	C_p / kJ kg^{-1} K^{-1}	u / m s^{-1}	D	η / μPa s	λ / mW m^{-1} K^{-1}
160	1.0	657.8	-49.24	-0.1717	1.343	2.024	1638	1.94249	539.4	172.8
180	1.0	637.3	-8.331	0.06918	1.358	2.068	1505	1.90415	385.3	160.7
200	1.0	616.2	33.57	0.2899	1.384	2.124	1372	1.86560	290.4	148.4
220	1.0	594.4	76.72	0.4955	1.422	2.194	1241	1.82648	226.2	136.3
240	1.0	571.5	121.5	0.6901	1.470	2.284	1111	1.78630	180.0	124.6
260	1.0	547.0	168.2	0.8771	1.528	2.395	980.1	1.74433	145.2	113.5
280	1.0	520.1	217.5	1.060	1.596	2.539	846.7	1.69945	117.9	102.9
300	1.0	489.5	270.2	1.241	1.675	2.740	706.9	1.64962	95.28	92.86
300.09	1.0	489.3	270.4	1.242	1.675	2.741	706.2	1.64937	95.19	92.82
300.09	1.0	21.68	602.7	2.349	1.588	2.042	214.8	1.02386	8.344	19.25
320	1.0	19.34	642.9	2.479	1.646	2.009	229.9	1.02125	8.877	21.49
340	1.0	17.61	683.2	2.601	1.717	2.032	242.4	1.01931	9.414	23.91
360	1.0	16.25	724.3	2.719	1.795	2.081	253.5	1.01779	9.948	26.46
400	1.0	14.17	810.1	2.944	1.958	2.211	272.6	1.01549	11.00	31.92
500	1.0	10.93	1050	3.478	2.363	2.583	311.6	1.01191	13.48	47.56
600	1.0	8.964	1326	3.980	2.725	2.933	344.1	1.00976	15.76	66.04
700	1.0	7.623	1635	4.456	3.040	3.242	372.9	1.00830	17.86	87.38
800	1.0	6.641	1973	4.907	3.315	3.512	399.1	1.00723	19.76	111.6
900	1.0	5.889	2336	5.334	3.555	3.750	423.4	1.00641	21.49	138.7
1000	1.0	5.292	2722	5.741	3.765	3.959	446.3	1.00576	23.04	168.7
P = 10 MPa (100 bar)										
86.45	10.0	735.2	-182.9	-1.394	1.362	1.914	2159	2.09013	11310	209.8
100	10.0	721.5	-156.9	-1.115	1.350	1.927	2070	2.06349	4212	205.7
120	10.0	701.8	-118.1	-0.7615	1.344	1.951	1939	2.02490	1649	197.6
140	10.0	682.2	-78.80	-0.4586	1.345	1.979	1809	1.98713	894.7	187.7
160	10.0	662.7	-38.90	-0.1922	1.352	2.012	1682	1.94997	579.3	176.7
180	10.0	643.1	1.726	0.04696	1.367	2.052	1556	1.91318	413.7	165.2
200	10.0	623.2	43.23	0.2656	1.393	2.101	1432	1.87653	312.8	153.5
220	10.0	602.8	85.84	0.4686	1.430	2.162	1310	1.83978	245.3	142.0
240	10.0	581.9	129.8	0.6598	1.478	2.236	1192	1.80267	197.1	130.9
260	10.0	560.1	175.4	0.8422	1.535	2.325	1077	1.76484	161.3	120.5
280	10.0	537.1	222.9	1.018	1.600	2.429	963.4	1.72590	133.7	110.7
300	10.0	512.7	272.7	1.190	1.673	2.550	852.7	1.68529	111.7	101.8
320	10.0	486.1	325.0	1.359	1.751	2.692	744.0	1.64228	93.65	93.68
340	10.0	456.7	380.5	1.527	1.833	2.865	637.1	1.59580	78.30	86.37
360	10.0	423.2	439.9	1.697	1.921	3.085	532.1	1.54422	64.77	79.79
400	10.0	334.5	576.0	2.054	2.111	3.790	339.0	1.41464	41.09	68.25
500	10.0	143.3	947.1	2.887	2.438	3.221	275.9	1.16393	19.83	58.85
600	10.0	99.02	1263	3.462	2.755	3.183	330.8	1.11103	19.67	73.95
700	10.0	78.82	1591	3.967	3.058	3.386	372.8	1.08759	20.88	95.32
800	10.0	66.46	1940	4.434	3.328	3.610	407.1	1.07350	22.27	120.1
900	10.0	57.89	2312	4.871	3.565	3.821	436.7	1.06382	23.62	148.0
1000	10.0	51.49	2704	5.284	3.774	4.013	463.1	1.05665	24.89	178.8

Thermophysical Properties of Carbon Dioxide (CO_2)

T / K	P / MPa	ρ / kg m^{-3}	H / kJ kg^{-1}	S / kJ kg^{-1} K^{-1}	C_v / kJ kg^{-1} K^{-1}	C_p / kJ kg^{-1} K^{-1}	u / m s^{-1}	D	η / μPa s	λ / mW m^{-1} K^{-1}
Saturation										
216.59	0.5180	1178	80.04	0.5213	0.9747	1.953	975.8	1.75696	256.7	180.6
216.59	0.5180	13.76	430.4	2.139	0.6292	0.9087	222.8	1.00694	10.95	11.01
220	0.5991	1166	86.73	0.5517	0.9698	1.962	951.2	1.74853	242.0	176.2
220	0.5991	15.82	431.6	2.119	0.6389	0.9303	223.1	1.00798	11.14	11.30
230	0.8929	1129	106.6	0.6387	0.9567	1.997	879.1	1.72239	204.2	163.3
230	0.8929	23.27	434.6	2.065	0.6700	1.005	223.6	1.01178	11.69	12.22
240	1.282	1089	126.8	0.7235	0.9454	2.051	806.4	1.69398	173.0	150.7
240	1.282	33.30	436.5	2.014	0.7053	1.103	223.0	1.01692	12.27	13.30
250	1.785	1046	147.7	0.8068	0.9364	2.132	731.8	1.66298	146.7	138.5
250	1.785	46.64	437.0	1.964	0.7459	1.237	221.2	1.02382	12.90	14.61
260	2.419	998.9	169.4	0.8895	0.9323	2.255	652.6	1.62876	124.4	126.3
260	2.419	64.42	435.9	1.914	0.7943	1.429	218.2	1.03309	13.61	16.31

T/ K	P/ MPa	ρ/ kg m⁻³	H/ kJ kg⁻¹	S/ kJ kg⁻¹ K⁻¹	C_v/ kJ kg⁻¹ K⁻¹	C_p/ kJ kg⁻¹ K⁻¹	u/ m s⁻¹	D	η/ μPa s	λ/ mW m⁻¹ K⁻¹
270	3.203	945.8	192.4	0.9732	0.9396	2.453	565.5	1.59026	105.0	114.3
270	3.203	88.37	432.6	1.863	0.8517	1.731	213.8	1.04574	14.47	18.69
280	4.161	883.6	217.3	1.060	0.9605	2.814	471.5	1.54547	87.73	102.0
280	4.161	121.7	425.9	1.805	0.9232	2.277	207.7	1.06363	15.60	22.47
290	5.318	804.7	245.6	1.154	0.9937	3.676	371.9	1.48961	71.41	89.55
290	5.318	172.0	413.8	1.734	1.026	3.614	199.4	1.09112	17.36	29.82
300	6.713	679.2	283.4	1.276	1.120	8.698	245.7	1.40346	53.11	80.59
300	6.713	268.6	387.1	1.622	1.248	11.92	185.3	1.14588	21.31	53.69
304.13	7.377	467.6	332.2	1.434				1.26600	33.04	
P = 0.1 MPa (1 bar)										
220	0.1	2.439	442.2	2.492	0.5791	0.7807	233.4	1.00122	11.06	10.90
240	0.1	2.228	458.0	2.561	0.5981	0.7962	243.2	1.00112	12.07	12.24
260	0.1	2.052	474.1	2.625	0.6184	0.8142	252.3	1.00103	13.06	13.68
280	0.1	1.902	490.6	2.686	0.6390	0.8333	261.0	1.00095	14.05	15.20
300	0.1	1.773	507.4	2.745	0.6593	0.8525	269.4	1.00089	15.02	16.79
320	0.1	1.661	524.7	2.800	0.6791	0.8715	277.4	1.00083	15.98	18.42
340	0.1	1.562	542.3	2.854	0.6982	0.8900	285.2	1.00078	16.93	20.09
360	0.1	1.474	560.3	2.905	0.7165	0.9079	292.8	1.00074	17.87	21.77
400	0.1	1.326	597.3	3.002	0.7510	0.9417	307.3	1.00066	19.70	25.14
500	0.1	1.059	695.2	3.221	0.8255	1.015	340.6	1.00053	24.02	33.49
600	0.1	0.8824	799.9	3.411	0.8867	1.076	370.8	1.00044	28.00	41.55
700	0.1	0.7562	910.2	3.581	0.9375	1.127	398.7	1.00038	31.68	49.30
800	0.1	0.6616	1025	3.735	0.9800	1.169	424.7	1.00033	35.09	56.71
900	0.1	0.5880	1144	3.874	1.015	1.205	449.2	1.00029	38.27	63.80
1000	0.1	0.5292	1266	4.003	1.045	1.234	472.4	1.00027	41.26	70.57
P = 1 MPa (10 bar)										
216.70	1.0	1179	80.37	0.5210	0.9751	1.950	977.8	1.75735	257.2	180.9
220	1.0	1167	86.83	0.5506	0.9703	1.959	953.6	1.74911	242.8	176.5
233.03	1.0	1117	112.7	0.6646	0.9530	2.011	857.2	1.71404	194.1	159.5
233.03	1.0	26.01	435.3	2.049	0.6803	1.032	223.5	1.01318	11.86	12.53
240	1.0	24.86	442.4	2.079	0.6733	0.9991	228.5	1.01259	12.20	12.94
260	1.0	22.21	461.7	2.157	0.6656	0.9450	241.2	1.01123	13.18	14.26
280	1.0	20.20	480.4	2.226	0.6709	0.9252	252.3	1.01019	14.15	15.71
300	1.0	18.58	498.8	2.289	0.6822	0.9209	262.4	1.00937	15.11	17.25
320	1.0	17.23	517.3	2.349	0.6960	0.9243	271.8	1.00868	16.07	18.84
340	1.0	16.09	535.8	2.405	0.7111	0.9320	280.6	1.00810	17.01	20.47
360	1.0	15.11	554.6	2.459	0.7266	0.9421	289.0	1.00760	17.94	22.12
400	1.0	13.48	592.7	2.559	0.7575	0.9655	304.7	1.00677	19.76	25.46
500	1.0	10.66	692.4	2.781	0.8282	1.027	339.8	1.00535	24.06	33.74
600	1.0	8.845	798.0	2.974	0.8881	1.083	370.9	1.00444	28.04	41.76
700	1.0	7.564	908.8	3.144	0.9384	1.132	399.2	1.00379	31.71	49.47
800	1.0	6.610	1024	3.298	0.9805	1.173	425.5	1.00332	35.12	56.86
900	1.0	5.872	1143	3.438	1.016	1.207	450.2	1.00295	38.30	63.93
1000	1.0	5.283	1265	3.567	1.046	1.236	473.6	1.00265	41.28	70.69
P = 10 MPa (100 bar)										
218.60	10.0	1190	86.78	0.5155	0.9827	1.902	1012	1.76433	266.3	185.4
220	10.0	1186	89.44	0.5277	0.9806	1.904	1003	1.76113	260.2	183.6
240	10.0	1115	127.9	0.6949	0.9536	1.949	870.9	1.71160	189.4	159.2
260	10.0	1035	167.8	0.8545	0.9340	2.052	735.5	1.65372	139.5	136.0
280	10.0	938.2	210.8	1.014	0.9263	2.280	588.7	1.58337	102.2	113.1
300	10.0	801.6	261.8	1.189	0.9496	2.991	414.3	1.48612	71.03	88.91
320	10.0	448.3	362.9	1.514	1.058	7.617	219.1	1.25290	32.39	60.44
340	10.0	258.6	443.4	1.759	0.9025	2.402	238.1	1.13908	22.80	35.70
360	10.0	208.2	483.0	1.873	0.8526	1.707	257.8	1.11030	21.83	31.88
400	10.0	161.5	542.1	2.029	0.8287	1.333	286.9	1.08425	22.23	31.58
500	10.0	113.1	663.8	2.301	0.8552	1.162	337.4	1.05801	25.39	37.26
600	10.0	89.94	779.3	2.511	0.9017	1.156	375.4	1.04580	28.94	44.39
700	10.0	75.49	895.9	2.691	0.9465	1.178	407.4	1.03828	32.39	51.59
800	10.0	65.35	1015	2.850	0.9859	1.205	435.8	1.03305	35.66	58.65
900	10.0	57.76	1137	2.993	1.020	1.231	461.8	1.02916	38.75	65.48
1000	10.0	51.82	1261	3.124	1.048	1.255	485.9	1.02613	41.66	72.05

THERMOPHYSICAL PROPERTIES OF SELECTED FLUIDS AT SATURATION

Eric W. Lemmon

These tables give thermodynamic and transport properties of a variety of fluids, as generated from the equations of state presented in the references below. The properties tabulated are pressure (*P*), density (ρ), enthalpy (*H*), entropy (*S*), isochoric heat capacity (C_v), isobaric heat capacity (C_p), speed of sound (*u*), viscosity (η), thermal conductivity (λ), and static dielectric constant (*D*). All extensive properties are given on a mass basis. Not all properties are included for every substance. The references should be consulted for information on the uncertainties. Values are given along the saturation line. The first two lines give the properties at the triple point (except for R-1234yf). The final line gives the properties at the critical point. Two lines are given for each temperature (except at the critical point); the first line gives the values of the liquid phase (note the high density), and the second line gives the values of the vapor phase (at low densities). Ammonia uses a reference state based on zero energy and entropy at the triple point; benzene, carbon monoxide, ethylene, hydrogen sulfide, sulfur dioxide, and toluene use a reference state of zero enthalpy and entropy at the saturated liquid state at the normal boiling point; butane, isobutane, ethanol, propylene, R-134a, R-1234yf, and sulfur hexafluoride use a reference state of 200 kJ kg^{-1} and 1 kJ kg^{-1} K^{-1} at −40 °C. Additional calculations at state points not listed below and for fluids not contained here can be obtained at the NIST Chemistry WebBook Web site (http://webbook.nist.gov/chemistry/fluid/), or by using the NIST Standard Reference Data program REFPROP (http://www.nist.gov/srd/nist23.cfm).

References

1. Bücker, D., and Wagner, W., Reference Equations of State for the Thermodynamic Properties of Fluid Phase n-Butane and Isobutane, *J. Phys. Chem. Ref. Data* 35, 929, 2006.
2. Schroeder, J.A., A New Fundamental Equation for Ethanol, Master's Thesis, University of Idaho, Moscow, Idaho, 2011.
3. Fenghour, A., Wakeham, W. A., Vesovic, V., Watson, J. T. R., Millat, J., and Vogel, E., The Viscosity of Ammonia, *J. Phys. Chem. Ref. Data* 24, 1649, 1995.
4. Guder, C., and Wagner, W., A Reference Equation of State for the Thermodynamic Properties of Sulfur Hexafluoride (SF_6) for Temperatures from the Melting Line to 625 K and Pressures up to 150 MPa, *J. Phys. Chem. Ref. Data* 38, 33, 2009.
5. Harvey, A. H., and Lemmon, E. W., Method for Estimating the Dielectric Constant of Natural Gas Mixtures, *Int. J. Thermophys.* 26, 31, 2005.
6. Holland, P. M., Eaton, B. E., and Hanley, H. J. M., A Correlation of the Viscosity and Thermal Conductivity Data of Gaseous and Liquid Ethylene, *J. Phys. Chem. Ref. Data* 12, 917, 1983.
7. Huber, M. L., Lasescke, A., and Perkins, R. A., Model for the Viscosity and Thermal Conductivity of Refrigerants, Including a New Correlation for the Viscosity of R134a, *Ind. Eng. Chem. Res.* 42, 3163, 2003.
8. Kiselev, S. B., Ely, J. F., Abdulagatov, I. M., and Huber, M. L., Generalized SAFT-DFT/DMT Model for the Thermodynamic, Interfacial, and Transport Properties of Associating Fluids: Application for n-Alkanols, *Ind. Eng. Chem. Res.* 44, 6916, 2005.
9. Lemmon, E. W., and Span, R., Short Fundamental Equations of State for 20 Industrial Fluids, *J. Chem. Eng. Data* 51, 785, 2006.
10. Lemmon, E.W., Overhoff, U., McLinden, M.O., and Wagner, W., A Reference Equation of State for Propylene for Temperatures from the Melting Line to 575 K and Pressures up to 1000 MPa, submitted to *J. Phys. Chem. Ref. Data*, 2013.
11. Perkins, R. A, Ramires, M. L. V., Nieto de Castro, C. A., and Cusco, L., Measurement and Correlation of the Thermal Conductivity of Butane from 135 K to 600 K at Pressures to 70 MPa, *J. Chem. Eng. Data* 47, 1263, 2002.
12. Perkins, R. A., Laesecke, A., Howley, J., Ramires, M. L. V., Gurova, A. N., and Cusco, L., Experimental Thermal Conductivity Values for the IUPAC Round-Robin Sample of 1,1,1,2-Tetrafluoroethane (R134a), NISTIR, 2000.
13. Perkins, R. A., Measurement and Correlation of the Thermal Conductivity of Isobutane from 114 K to 600 K at Pressures to 70 MPa, *J. Chem. Eng. Data* 47, 1272, 2002.
14. Thol, M., Lemmon, E.W., and Span, R., Equation of State for Benzene for Temperatures from the Melting Line up to 725 K with Pressures up to 500 MPa, *High Temp.-High Press.* 41, 81, 2012.
15. Schmidt, K. A. G., Carroll, J. J., Quiñones-Cisneros, S. E., and Kvamme, B., Hydrogen Sulphide Viscosity Model, Proceedings of the 86th Annual GPA Convention, San Antonio, Texas, 2007.
16. Smukala, J., Span, R., and Wagner, W., New Equation of State for Ethylene Covering the Fluid Region for Temperatures from the Melting Line to 450 K at Pressures up to 300 MPa, *J. Phys. Chem. Ref. Data* 29, 1053, 2000.
17. Tillner-Roth, R. and Baehr, H. D., An International Standard Formulation for the Thermodynamic Properties of 1,1,1,2-Tetrafluoroethane (HFC-134a) for Temperatures from 170 K to 455 K and Pressures up to 70 MPa, *J. Phys. Chem. Ref. Data* 23, 657, 1994.
18. Tillner-Roth, R., Harms-Watzenberg, F., and Baehr, H. D., A New Fundamental Equation for Ammonia, *DKV-Tagungsbericht* 20, 167, 1993.
19. Tufeu, R., Ivanov, D. Y., Garrabos, Y., and Le Neindre, B., Thermal Conductivity of Ammonia in a Large Temperature and Pressure Range Including the Critical Region, *Ber. Bunsenges. Phys. Chem.* 88, 422, 1984.
20. Vogel, E., Kuechenmeister, C., and Bich, E., Viscosity Correlation for Isobutane over Wide Ranges of the Fluid Region, *Int. J. Thermophys.* 21, 343, 2000.
21. Vogel, E., Kuechenmeister, C., and Bich, E., Viscosity Correlation for n-Butane in the Fluid Region, *High Temp.-High Press.* 31, 173, 1999.
22. Richter, M., McLinden, M. O., and Lemmon, E. W., Thermodynamic Properties of 2,3,3,3-Tetrafluoroprop-1-ene (R1234yf): p-ρ-T Measurements and an Equation of State, submitted to *J. Chem. Eng. Data*, 2011.
23. Assael, M.J., Mihailidou, E., Huber, M.L., and Perkins, R.A., Reference Correlation of the Thermal Conductivity of Benzene from the Triple Point to 725 K and up to 500 MPa, submitted to *J. Phys. Chem. Ref. Data*, 2013.

Ammonia (NH_3)

T/ K	*P*/ MPa	ρ/ kg m^{-3}	*H*/ kJ kg^{-1}	*S*/ kJ kg^{-1} K^{-1}	C_v/ kJ kg^{-1} K^{-1}	C_p/ kJ kg^{-1} K^{-1}	*u*/ m s^{-1}	η/ µPa s	λ/ mW m^{-1} K^{-1}
195.50	0.006091	732.9	0.008311	0.	2.934	4.202	2124.	559.6	819.0
195.50	0.006091	0.06409	1484.	7.593	1.557	2.063	354.1	6.840	19.64
200.00	0.008651	728.1	19.00	0.09601	2.926	4.227	2080.	507.3	803.1

Fluids

T/ K	P/ MPa	ρ/ kg m⁻³	H/ kJ kg⁻¹	S/ kJ kg⁻¹ K⁻¹	C_v/ kJ kg⁻¹ K⁻¹	C_p/ kJ kg⁻¹ K⁻¹	u/ m s⁻¹	η/ μPa s	λ/ mW m⁻¹ K⁻¹
200.00	0.008651	0.08908	1493.	7.465	1.565	2.075	357.9	6.952	19.68
220.00	0.03379	705.8	104.7	0.5042	2.889	4.342	1914.	346.7	733.2
220.00	0.03379	0.3188	1529.	6.978	1.620	2.160	373.4	7.485	20.13
240.00	0.1022	681.8	192.7	0.8866	2.855	4.449	1767.	254.9	665.1
240.00	0.1022	0.8969	1562.	6.591	1.705	2.298	386.3	8.059	20.98
260.00	0.2553	656.2	282.8	1.246	2.821	4.548	1625.	197.3	600.1
260.00	0.2553	2.115	1590.	6.274	1.823	2.503	396.2	8.656	22.26
280.00	0.5509	629.1	375.0	1.586	2.790	4.656	1481.	158.1	538.5
280.00	0.5509	4.382	1612.	6.005	1.972	2.788	402.6	9.266	24.03
300.00	1.062	600.0	469.7	1.910	2.762	4.800	1333.	129.3	480.3
300.00	1.062	8.251	1628.	5.770	2.148	3.177	405.0	9.894	26.41
320.00	1.873	568.2	567.9	2.222	2.743	5.018	1178.	106.9	424.8
320.00	1.873	14.51	1634.	5.554	2.349	3.718	402.7	10.56	29.57
340.00	3.080	532.4	671.3	2.529	2.739	5.385	1012.	88.55	371.5
340.00	3.080	24.40	1629.	5.346	2.574	4.530	395.1	11.33	33.94
360.00	4.793	490.3	783.0	2.838	2.764	6.082	830.6	72.80	319.2
360.00	4.793	40.19	1608.	5.129	2.832	5.954	380.8	12.35	40.75
380.00	7.140	436.1	910.3	3.169	2.853	7.818	628.8	58.31	266.6
380.00	7.140	67.37	1558.	4.874	3.147	9.395	358.0	14.02	54.56
400.00	10.30	344.6	1085.	3.595	3.177	22.73	384.6	41.80	216.0
400.00	10.30	131.1	1432.	4.462	3.598	34.92	318.2	18.53	113.5
405.40	11.33	225.0	1262.	4.026				27.04	

Benzene (C_6H_6)

T/ K	P/ MPa	ρ/ kg m⁻³	H/ kJ kg⁻¹	S/ kJ kg⁻¹ K⁻¹	C_v/ kJ kg⁻¹ K⁻¹	C_p/ kJ kg⁻¹ K⁻¹	u/ m s⁻¹	η/ mPa s	λ/ mW m⁻¹ K⁻¹
278.67	0.004784	894.1	−133.4	−0.4229	1.162	1.688	1399.	828.1	148.0
278.67	0.004784	0.1618	314.3	1.184	0.8586	0.9667	182.1	6.960	9.130
280.00	0.005139	892.7	−131.2	−0.4149	1.165	1.691	1392.	810.4	147.5
280.00	0.005139	0.1731	315.6	1.181	0.8640	0.9722	182.5	6.994	9.222
300.00	0.01382	871.5	−96.89	−0.2966	1.216	1.740	1292.	596.2	140.5
300.00	0.01382	0.4361	335.3	1.144	0.9469	1.057	187.3	7.509	10.70
320.00	0.03207	850.0	−61.51	−0.1825	1.271	1.798	1197.	452.5	133.9
320.00	0.03207	0.9556	356.4	1.123	1.032	1.145	191.6	8.031	12.34
340.00	0.06619	828.1	−24.91	−0.07174	1.330	1.861	1106.	352.7	127.6
340.00	0.06619	1.876	378.5	1.115	1.117	1.237	195.1	8.564	14.12
360.00	0.1243	805.8	13.02	0.03642	1.389	1.928	1019.	281.1	121.5
360.00	0.1243	3.376	401.6	1.116	1.203	1.331	197.6	9.113	16.04
380.00	0.2161	782.7	52.34	0.1424	1.449	1.999	934.3	228.4	115.6
380.00	0.2161	5.672	425.5	1.124	1.289	1.429	199.1	9.685	18.08
400.00	0.3526	758.6	93.14	0.2466	1.508	2.075	852.2	188.5	109.7
400.00	0.3526	9.019	449.9	1.139	1.374	1.532	199.2	10.30	20.26
450.00	0.9715	692.7	202.3	0.5014	1.655	2.293	651.1	122.5	95.27
450.00	0.9715	24.30	512.0	1.190	1.583	1.822	192.6	12.22	26.44
500.00	2.165	610.5	324.0	0.7538	1.799	2.645	441.3	82.34	80.99
500.00	2.165	57.61	570.2	1.246	1.788	2.285	171.6	15.62	34.53
550.00	4.223	466.0	472.6	1.029	1.983	4.852	185.7	52.37	69.19
550.00	4.223	153.7	599.6	1.260	2.047	5.771	124.2	25.18	52.86
562.02	4.906	304.8	550.8	1.166					

Butane (C_4H_{10})

T/ K	P/ MPa	ρ/ kg m⁻³	H/ kJ kg⁻¹	S/ kJ kg⁻¹ K⁻¹	C_v/ kJ kg⁻¹ K⁻¹	C_p/ kJ kg⁻¹ K⁻¹	u/ m s⁻¹	D	η/ μPa s	λ/ mW m⁻¹ K⁻¹
134.90	$0.6657 \cdot 10^{-6}$	735.0	−89.82	−0.4651	1.441	1.973	1827.	2.03981	2304.	176.6
134.90	$0.6657 \cdot 10^{-6}$	0.3450E-10⁻⁴	406.1	3.211	0.9634	1.106	148.9	1.00000	3.321	4.855

Fluids

T/ K	P/ MPa	ρ/ kg m^{-3}	H/ kJ kg^{-1}	S/ kJ kg^{-1} K^{-1}	C_v/ kJ kg^{-1} K^{-1}	C_p/ kJ kg^{-1} K^{-1}	u/ m s^{-1}	D	η/ μPa s	λ/ mW m^{-1} K^{-1}
150.00	$0.8573\cdot10^{-5}$	720.9	−59.93	−0.2550	1.442	1.986	1730.	2.01487	1370.	171.2
150.00	$0.8573\cdot10^{-5}$	$0.3995E\cdot10^{-3}$	423.2	2.966	1.015	1.158	156.5	1.00000	3.709	5.579
200.00	0.001939	674.0	41.02	0.3252	1.473	2.062	1438.	1.93301	496.5	149.4
200.00	0.001939	0.06792	484.8	2.544	1.174	1.319	178.9	1.00007	4.979	8.497
220.00	0.007805	654.8	82.75	0.5240	1.505	2.113	1326.	1.90007	373.4	139.9
220.00	0.007805	0.2496	511.3	2.472	1.244	1.392	186.4	1.00027	5.477	9.884
240.00	0.02409	635.1	125.6	0.7105	1.548	2.176	1216.	1.86666	290.9	130.4
240.00	0.02409	0.7120	538.6	2.431	1.322	1.476	192.9	1.00076	5.966	11.39
260.00	0.06098	614.6	170.0	0.8875	1.601	2.254	1106.	1.83250	232.3	121.2
260.00	0.06098	1.688	566.6	2.413	1.408	1.573	197.9	1.00180	6.450	13.03
280.00	0.1328	593.3	216.0	1.058	1.662	2.345	998.3	1.79722	189.0	112.3
280.00	0.1328	3.490	595.0	2.411	1.501	1.684	201.1	1.00373	6.935	14.82
300.00	0.2576	570.7	264.0	1.222	1.729	2.451	890.9	1.76038	155.6	103.9
300.00	0.2576	6.516	623.6	2.421	1.602	1.811	202.2	1.00697	7.436	16.78
320.00	0.4562	546.4	314.3	1.384	1.803	2.577	783.6	1.72136	129.1	96.10
320.00	0.4562	11.29	651.9	2.439	1.708	1.960	200.7	1.01209	7.978	19.00
340.00	0.7520	519.7	367.4	1.543	1.881	2.729	675.4	1.67929	107.2	88.88
340.00	0.7520	18.52	679.5	2.461	1.820	2.142	196.2	1.01990	8.601	21.58
360.00	1.170	489.6	423.8	1.702	1.965	2.926	564.3	1.63273	88.54	82.27
360.00	1.170	29.35	705.6	2.484	1.935	2.379	187.9	1.03167	9.369	24.72
380.00	1.740	454.2	484.3	1.862	2.059	3.221	447.1	1.57904	71.86	76.22
380.00	1.740	45.85	728.7	2.505	2.058	2.754	174.7	1.04981	10.41	28.81
400.00	2.495	408.5	551.2	2.029	2.173	3.838	318.4	1.51174	55.96	70.60
400.00	2.495	73.08	745.2	2.514	2.210	3.623	154.8	1.08025	12.03	35.03
420.00	3.490	327.8	635.2	2.227	2.357	8.852	165.6	1.39801	37.39	67.19
420.00	3.490	135.0	739.9	2.476	2.444	10.72	124.5	1.15196	15.96	53.10
425.13	3.796	228.0	693.9	2.363				1.26627	23.90	

Carbon monoxide (CO)

T/ K	P/ MPa	ρ/ kg m^{-3}	H/ kJ kg^{-1}	S/ kJ kg^{-1} K^{-1}	C_v/ kJ kg^{-1} K^{-1}	C_p/ kJ kg^{-1} K^{-1}	u/ m s^{-1}	η/ μPa s	λ/ mW m^{-1} K^{-1}
68.16	0.01554	849.5	−28.96	−0.3863	1.262	2.157	998.2	274.2	180.3
68.16	0.01554	0.7761	203.0	3.017	0.7529	1.063	167.2	4.637	6.687
70.00	0.02105	842.1	−24.99	−0.3290	1.243	2.150	980.5	252.2	175.5
70.00	0.02105	1.027	204.7	2.953	0.7553	1.069	169.2	4.777	6.884
80.00	0.08374	800.3	−3.524	−0.04334	1.154	2.142	883.4	178.0	151.4
80.00	0.08374	3.658	213.4	2.668	0.7747	1.120	178.4	5.608	7.982
90.00	0.2385	755.4	18.18	0.2099	1.088	2.188	783.6	139.5	129.2
90.00	0.2385	9.660	220.4	2.457	0.8073	1.216	184.7	6.618	9.129
100.00	0.5444	705.4	40.71	0.4428	1.037	2.306	678.7	114.0	108.3
100.00	0.5444	21.20	224.9	2.285	0.8544	1.389	187.5	7.924	10.37
110.00	1.067	647.4	64.89	0.6657	1.002	2.558	564.7	93.40	88.95
110.00	1.067	41.74	226.0	2.130	0.9190	1.725	186.7	9.722	11.83
120.00	1.877	574.6	92.30	0.8924	0.9951	3.202	433.1	73.95	71.30
120.00	1.877	78.77	221.2	1.967	1.016	2.601	182.0	12.51	14.10
130.00	3.065	456.2	129.3	1.169	1.110	8.070	254.0	51.35	53.11
130.00	3.065	164.8	201.1	1.721	1.235	9.853	171.9	18.94	21.85
132.86	3.494	303.9	164.7	1.429				32.12	

Ethanol (C_2H_6O)

T/ K	P/ MPa	ρ/ kg m^{-3}	H/ kJ kg^{-1}	S/ kJ kg^{-1} K^{-1}	C_v/ kJ kg^{-1} K^{-1}	C_p/ kJ kg^{-1} K^{-1}	u/ m s^{-1}	η/ μPa s	λ/ mW m^{-1} K^{-1}
160.00	$0.9015\cdot10^{-9}$	907.9	−429.4	−1.713	1.323	1.804	1701.		
160.00	$0.9015\cdot10^{-9}$	$0.3122\cdot10^{-7}$	612.0	4.795	0.8595	1.040	186.9		11.66
180.00	$0.4812\cdot10^{-7}$	887.9	−392.0	−1.493	1.503	1.918	1608.		210.0

T/ K	P/ MPa	ρ/ kg m⁻³	H/ kJ kg⁻¹	S/ kJ kg⁻¹ K⁻¹	C_v/ kJ kg⁻¹ K⁻¹	C_p/ kJ kg⁻¹ K⁻¹	u/ m s⁻¹	η/ μPa s	λ/ mW m⁻¹ K⁻¹
180.00	0.4812·10⁻⁷	0.1481·10⁻⁵	633.2	4.203	0.9026	1.083	197.4		12.17
200.00	0.1102·10⁻⁵	869.5	−353.2	−1.289	1.555	1.954	1528.		199.2
200.00	0.1102·10⁻⁵	0.3052·10⁻⁴	655.3	3.754	0.9487	1.129	207.3		12.80
220.00	0.1369·10⁻⁴	851.8	−313.7	−1.101	1.613	2.001	1443.		190.0
220.00	0.1369·10⁻⁴	0.3449·10⁻³	678.4	3.409	0.9990	1.179	216.5		13.55
240.00	0.1080·10⁻³	834.5	−273.0	−0.9234	1.697	2.077	1358.		182.2
240.00	0.1080·10⁻³	0.002495	702.5	3.141	1.054	1.235	225.2		14.44
260.00	0.6023·10⁻³	817.5	−230.5	−0.7533	1.802	2.179	1277.	2465.	175.4
260.00	0.6023·10⁻³	0.01285	727.6	2.932	1.115	1.296	233.4	7.586	15.46
280.00	0.002560	800.5	−185.7	−0.5875	1.922	2.302	1204.	1565.	169.6
280.00	0.002560	0.05078	753.6	2.767	1.181	1.365	241.1	8.211	16.61
300.00	0.008768	783.5	−138.2	−0.4238	2.056	2.449	1135.	1044.	164.5
300.00	0.008768	0.1629	780.4	2.638	1.254	1.444	248.1	8.829	17.90
320.00	0.02524	765.9	−87.56	−0.2604	2.201	2.620	1069.	726.6	159.8
320.00	0.02524	0.4426	807.5	2.537	1.336	1.535	254.3	9.439	19.34
340.00	0.06303	747.6	−33.25	−0.09599	2.350	2.812	1005.	524.3	155.5
340.00	0.06303	1.052	834.4	2.456	1.428	1.645	259.5	10.04	20.97
360.00	0.1400	727.9	25.12	0.07045	2.497	3.023	938.0	389.6	151.4
360.00	0.1400	2.245	860.5	2.391	1.532	1.778	263.3	10.63	22.84
380.00	0.2821	706.4	87.96	0.2397	2.639	3.257	867.4	296.0	147.2
380.00	0.2821	4.393	885.0	2.337	1.651	1.943	265.5	11.22	25.04
400.00	0.5237	682.1	155.8	0.4127	2.771	3.516	790.5	228.1	143.0
400.00	0.5237	8.010	907.4	2.292	1.786	2.150	265.6	11.82	27.75
420.00	0.9064	654.2	228.9	0.5896	2.880	3.790	706.3	177.1	138.5
420.00	0.9064	13.80	926.8	2.251	1.938	2.418	263.3	12.44	31.24
440.00	1.478	621.8	307.2	0.7697	2.961	4.060	615.1	137.9	134.0
440.00	1.478	22.77	942.3	2.213	2.111	2.785	258.1	13.13	36.00
460.00	2.289	583.9	390.6	0.9519	3.027	4.365	517.8	107.6	130.0
460.00	2.289	36.47	952.3	2.173	2.308	3.339	249.3	13.96	42.94
480.00	3.398	537.8	480.3	1.139	3.113	4.959	408.1	83.55	127.4
480.00	3.398	57.90	953.1	2.124	2.544	4.381	235.0	15.11	54.07
500.00	4.872	467.4	586.1	1.348	3.273	7.648	266.0	60.98	129.7
500.00	4.872	96.62	932.3	2.041	2.866	8.052	210.1	17.23	76.63
514.71	6.268	273.2	758.7	1.680					

Ethylene (C_2H_4)

T/ K	P/ MPa	ρ/ kg m⁻³	H/ kJ kg⁻¹	S/ kJ kg⁻¹ K⁻¹	C_v/ kJ kg⁻¹ K⁻¹	C_p/ kJ kg⁻¹ K⁻¹	u/ m s⁻¹	D	η/ μPa s	λ/ mW m⁻¹ K⁻¹
103.99	0.1220·10⁻³	654.6	−158.1	−1.179	1.622	2.429	1767.	2.00849	685.7	270.6
103.99	0.1220·10⁻³	0.003958	409.4	4.279	0.8901	1.187	202.7	1.00000	0.7727	6.801
120.00	0.001368	634.2	−119.2	−0.8308	1.553	2.427	1660.	1.96600	427.5	248.8
120.00	0.001368	0.03852	428.3	3.731	0.8937	1.192	217.5	1.00004	3.339	6.780
140.00	0.01185	608.0	−70.81	−0.4581	1.465	2.408	1521.	1.91418	277.6	222.3
140.00	0.01185	0.2876	451.3	3.271	0.9064	1.212	233.9	1.00033	4.841	7.892
160.00	0.05623	580.9	−22.66	−0.1372	1.395	2.407	1377.	1.86255	200.1	197.5
160.00	0.05623	1.212	473.0	2.961	0.9336	1.260	247.4	1.00139	5.700	9.115
180.00	0.1818	552.2	25.87	0.1473	1.346	2.441	1227.	1.80986	153.0	174.6
180.00	0.1818	3.589	492.3	2.739	0.9783	1.347	257.0	1.00413	6.401	10.36
200.00	0.4555	521.2	75.68	0.4070	1.321	2.529	1070.	1.75467	120.5	153.7
200.00	0.4555	8.494	508.1	2.569	1.043	1.492	261.9	1.00979	7.133	11.86
220.00	0.9566	486.7	128.0	0.6516	1.320	2.700	903.5	1.69488	95.68	134.5
220.00	0.9566	17.45	519.1	2.429	1.132	1.738	261.5	1.02023	8.007	13.94
240.00	1.773	446.1	184.9	0.8911	1.344	3.043	724.4	1.62679	74.96	116.1
240.00	1.773	33.07	522.7	2.299	1.254	2.211	254.9	1.03865	9.162	17.14
260.00	3.003	393.5	250.4	1.141	1.407	3.946	524.1	1.54141	55.95	97.24

T/ K	P/ MPa	ρ/ kg m⁻³	H/ kJ kg⁻¹	S/ kJ kg⁻¹ K⁻¹	C_v/ kJ kg⁻¹ K⁻¹	C_p/ kJ kg⁻¹ K⁻¹	u/ m s⁻¹	D	η/ μPa s	λ/ mW m⁻¹ K⁻¹
260.00	3.003	61.54	513.5	2.153	1.439	3.512	240.5	1.07303	10.96	23.22
265.00	3.390	376.5	269.4	1.210	1.437	4.473	467.2	1.51452	51.11	92.17
265.00	3.390	72.62	507.3	2.108	1.506	4.308	235.3	1.08667	11.63	26.05
270.00	3.812	356.4	290.3	1.284	1.484	5.409	405.0	1.48318	46.03	86.86
270.00	3.812	86.80	498.3	2.054	1.592	5.766	229.2	1.10433	12.51	30.45
275.00	4.275	330.8	314.6	1.368	1.566	7.588	334.7	1.44393	40.40	81.58
275.00	4.275	106.5	484.3	1.985	1.716	9.293	221.6	1.12922	13.76	39.17
280.00	4.784	290.7	347.8	1.481	1.778	19.56	246.7	1.38366	33.08	83.42
280.00	4.784	140.7	457.5	1.873	1.981	29.26	208.9	1.17365	16.18	71.72
282.35	5.042	214.2	399.4	1.661				1.27358	22.88	

Hydrogen sulfide (H_2S)

T/ K	P/ MPa	ρ/ kg m⁻³	H/ kJ kg⁻¹	S/ kJ kg⁻¹ K⁻¹	C_v/ kJ kg⁻¹ K⁻¹	C_p/ kJ kg⁻¹ K⁻¹	u/ m s⁻¹	η/ μPa s	λ/ mW m⁻¹ K⁻¹
187.70	0.02326	992.3	−50.47	−0.2520	1.302	2.020	1438.	543.7	249.9
187.70	0.02326	0.5120	524.5	2.811	0.7437	0.9976	245.8	7.673	6.240
200.00	0.05034	971.5	−25.72	−0.1244	1.263	2.003	1373.	378.5	236.7
200.00	0.05034	1.046	535.5	2.682	0.7507	1.012	252.8	8.188	7.174
220.00	0.1437	936.5	14.28	0.06577	1.209	1.995	1268.	265.5	215.8
220.00	0.1437	2.758	552.2	2.511	0.7671	1.047	262.6	9.055	8.740
240.00	0.3377	899.8	54.38	0.2393	1.164	2.009	1162.	211.3	195.6
240.00	0.3377	6.093	566.9	2.375	0.7898	1.102	270.1	9.954	10.42
260.00	0.6875	860.6	95.09	0.4006	1.128	2.050	1054.	175.0	176.1
260.00	0.6875	11.87	579.1	2.262	0.8190	1.183	275.0	10.88	12.32
280.00	1.256	818.0	137.0	0.5533	1.099	2.127	942.1	146.2	157.6
280.00	1.256	21.16	588.0	2.164	0.8549	1.302	277.1	11.83	14.58
300.00	2.110	770.5	180.8	0.7008	1.078	2.262	825.0	121.9	139.7
300.00	2.110	35.48	592.7	2.074	0.8981	1.488	276.1	12.80	17.48
300.00	2.110	770.5	180.8	0.7008	1.078	2.262	825.0	121.9	139.7
300.00	2.110	35.48	592.7	2.074	0.8981	1.488	276.1	12.80	17.48
310.00	2.667	744.3	203.8	0.7738	1.070	2.365	763.8	110.9	131.0
310.00	2.667	45.26	592.9	2.029	0.9230	1.626	274.3	13.31	19.32
320.00	3.323	715.7	227.8	0.8472	1.064	2.509	700.2	100.6	122.5
320.00	3.323	57.40	591.4	1.983	0.9509	1.816	271.6	13.85	21.59
330.00	4.089	684.1	253.2	0.9218	1.062	2.719	633.5	90.67	114.0
330.00	4.089	72.67	587.6	1.935	0.9825	2.095	268.0	14.44	24.50
340.00	4.976	648.3	280.4	0.9989	1.064	3.055	562.6	81.07	105.6
340.00	4.976	92.34	580.8	1.883	1.020	2.548	263.4	15.13	28.43
350.00	5.997	605.8	310.4	1.081	1.074	3.678	485.6	71.43	97.33
350.00	5.997	118.9	569.3	1.821	1.065	3.409	257.7	16.04	34.18
360.00	7.171	551.1	345.6	1.175	1.099	5.271	398.9	61.01	89.85
360.00	7.171	158.3	549.8	1.742	1.126	5.653	250.8	17.54	43.94
370.00	8.529	457.9	397.2	1.309	1.176	18.59	292.8	46.39	88.51
370.00	8.529	238.3	506.6	1.604	1.225	23.66	242.8	21.88	72.03
373.10	9.000	347.3	450.5	1.449				31.95	

Isobutane (C_4H_{10})

T/ K	P/ MPa	ρ/ kg m⁻³	H/ kJ kg⁻¹	S/ kJ kg⁻¹ K⁻¹	C_v/ kJ kg⁻¹ K⁻¹	C_p/ kJ kg⁻¹ K⁻¹	u/ m s⁻¹	D	η/ μPa s	λ/ mW m⁻¹ K⁻¹
113.73	$0.2289 \cdot 10^{-7}$	740.3	−112.4	−0.6798	1.174	1.689	2000.	2.10871	8767.	157.9
113.73	$0.2289 \cdot 10^{-7}$	$0.1407 \cdot 10^{-5}$	368.3	3.547	0.7366	0.8796	139.4	1.00000	2.848	2.272
150.00	$0.2388 \cdot 10^{-4}$	706.0	−49.02	−0.1971	1.251	1.805	1714.	2.03196	1748.	148.4
150.00	$0.2388 \cdot 10^{-4}$	0.001113	403.1	2.817	0.8944	1.037	157.8	1.00000	3.799	4.461
200.00	0.003814	657.7	45.22	0.3440	1.363	1.968	1389.	1.93554	547.1	129.1
200.00	0.003814	0.1338	459.6	2.416	1.095	1.241	179.4	1.00015	5.080	8.027

T/ K	P/ MPa	ρ/ kg m⁻³	H/ kJ kg⁻¹	S/ kJ kg⁻¹ K⁻¹	C_v/ kJ kg⁻¹ K⁻¹	C_p/ kJ kg⁻¹ K⁻¹	u/ m s⁻¹	D	η/ μPa s	λ/ mW m⁻¹ K⁻¹
220.00	0.01402	637.7	85.31	0.5349	1.415	2.041	1269.	1.89795	395.7	120.6
220.00	0.01402	0.4500	484.5	2.349	1.180	1.330	186.5	1.00049	5.576	9.620
240.00	0.04022	616.9	127.0	0.7159	1.475	2.123	1151.	1.86024	298.3	112.1
240.00	0.04022	1.197	510.3	2.313	1.270	1.428	192.3	1.00130	6.063	11.30
260.00	0.09588	595.4	170.4	0.8893	1.541	2.215	1036.	1.82195	231.6	103.9
260.00	0.09588	2.685	536.7	2.298	1.366	1.538	196.3	1.00291	6.545	13.07
280.00	0.1988	572.6	215.8	1.057	1.613	2.320	922.8	1.78256	183.8	95.99
280.00	0.1988	5.316	563.5	2.299	1.469	1.664	198.3	1.00575	7.033	14.96
300.00	0.3700	548.3	263.5	1.220	1.690	2.442	810.3	1.74142	148.2	88.60
300.00	0.3700	9.610	590.4	2.310	1.578	1.810	197.7	1.01041	7.546	17.02
320.00	0.6333	521.8	313.9	1.381	1.773	2.589	697.4	1.69761	120.7	81.78
320.00	0.6333	16.27	616.7	2.328	1.694	1.985	194.2	1.01765	8.118	19.37
340.00	1.015	492.1	367.4	1.541	1.862	2.777	582.7	1.64968	98.39	75.56
340.00	1.015	26.36	642.0	2.349	1.814	2.209	186.9	1.02871	8.813	22.21
360.00	1.543	457.2	425.0	1.703	1.958	3.052	463.6	1.59499	79.37	69.94
360.00	1.543	41.79	664.7	2.368	1.939	2.548	175.0	1.04579	9.758	25.97
380.00	2.252	412.8	488.6	1.870	2.069	3.588	335.1	1.52752	61.79	64.93
380.00	2.252	66.87	681.9	2.379	2.101	3.298	156.7	1.07403	11.28	31.85
400.00	3.186	341.0	565.4	2.060	2.250	6.349	184.4	1.42315	42.50	61.90
400.00	3.186	118.4	683.3	2.355	2.354	7.555	128.9	1.13392	14.76	46.45
407.81	3.629	225.5	633.9	2.226				1.26683	24.42	

Propylene (C_3H_6)

T/ K	P/ MPa	ρ/ kg m⁻³	H/ kJ kg⁻¹	S/ kJ kg⁻¹ K⁻¹	C_v/ kJ kg⁻¹ K⁻¹	C_p/ kJ kg⁻¹ K⁻¹	u/ m s⁻¹	η/ μPa s	λ/ mW m⁻¹ K⁻¹
87.95	$0.7472 \cdot 10^{-9}$	768.1	−199.4	−1.420	1.569	2.182	2013.	9889.	192.9
87.95	$0.7472 \cdot 10^{-9}$	$0.4299 \cdot 10^{-7}$	386.1	5.237	0.7037	0.9013	149.2	2.589	3.110
100.00	$0.4115 \cdot 10^{-7}$	754.0	−173.4	−1.143	1.508	2.141	1952.	3842.	191.0
100.00	$0.4115 \cdot 10^{-7}$	$0.2082 \cdot 10^{-5}$	397.2	4.563	0.7331	0.9307	158.4	2.900	3.585
150.00	$0.4110 \cdot 10^{-3}$	697.6	−68.75	−0.2932	1.360	2.064	1667.	595.9	178.6
150.00	$0.4110 \cdot 10^{-3}$	0.01387	446.6	3.143	0.8545	1.052	191.0	4.250	5.884
200.00	0.02675	640.8	35.18	0.3042	1.338	2.114	1352.	265.2	160.2
200.00	0.02675	0.6864	501.0	2.633	1.003	1.213	215.6	5.647	8.831
220.00	0.07823	616.9	78.03	0.5080	1.359	2.170	1222.	208.4	151.6
220.00	0.07823	1.855	523.0	2.530	1.077	1.303	222.4	6.204	10.21
240.00	0.1872	591.8	122.3	0.6996	1.393	2.249	1091.	167.8	142.3
240.00	0.1872	4.179	544.5	2.459	1.162	1.416	226.7	6.773	11.76
260.00	0.3862	565.0	168.4	0.8828	1.437	2.354	959.8	137.4	132.6
260.00	0.3862	8.277	565.0	2.408	1.256	1.557	228.1	7.386	13.59
280.00	0.7128	535.8	216.9	1.060	1.490	2.493	827.5	113.7	122.4
280.00	0.7128	14.99	583.8	2.371	1.360	1.740	225.9	8.098	15.91
300.00	1.209	503.1	268.7	1.236	1.553	2.691	692.3	94.26	111.8
300.00	1.209	25.57	600.0	2.340	1.478	2.002	219.5	9.001	19.10
320.00	1.919	464.5	324.9	1.412	1.627	3.014	550.7	77.41	100.9
320.00	1.919	42.31	611.7	2.308	1.611	2.439	208.1	10.28	23.92
340.00	2.897	414.7	388.4	1.598	1.725	3.751	396.4	61.38	89.42
340.00	2.897	70.66	614.7	2.263	1.772	3.485	190.2	12.37	32.30
360.00	4.219	325.7	473.1	1.829	1.955	10.53	206.3	41.61	80.20
360.00	4.219	138.1	588.4	2.149	2.090	14.37	161.9	17.78	57.59
364.21	4.555	229.6	531.1	1.986					

R–134a (1,1,1,2–tetrafluoroethane, CF_3CH_2F)

T/ K	P/ MPa	ρ/ kg m⁻³	H/ kJ kg⁻¹	S/ kJ kg⁻¹ K⁻¹	C_v/ kJ kg⁻¹ K⁻¹	C_p/ kJ kg⁻¹ K⁻¹	u/ m s⁻¹	η/ μPa s	λ/ mW m⁻¹ K⁻¹
169.85	$0.3896 \cdot 10^{-3}$	1591.	71.46	0.4126	0.7922	1.184	1120.	2154.	145.2

T/ K	P/ MPa	ρ/ kg m^{-3}	H/ kJ kg^{-1}	S/ kJ kg^{-1} K^{-1}	C_v/ kJ kg^{-1} K^{-1}	C_p/ kJ kg^{-1} K^{-1}	u/ m s^{-1}	η/ μPa s	λ/ mW m^{-1} K^{-1}
169.85	0.3896·10^{-3}	0.02817	334.9	1.964	0.5030	0.5853	126.8	6.829	3.080
180.00	0.001128	1564.	83.48	0.4814	0.7912	1.187	1068.	1479.	139.1
180.00	0.001128	0.07701	340.9	1.911	0.5267	0.6097	130.1	7.232	3.893
200.00	0.006313	1510.	107.4	0.6073	0.8016	1.206	967.6	867.3	127.7
200.00	0.006313	0.3898	353.1	1.836	0.5732	0.6586	136.0	8.015	5.498
220.00	0.02443	1455.	131.8	0.7235	0.8193	1.233	869.9	582.2	117.2
220.00	0.02443	1.385	365.7	1.787	0.6204	0.7109	141.0	8.779	7.108
240.00	0.07248	1398.	156.8	0.8321	0.8403	1.267	775.0	420.2	107.3
240.00	0.07248	3.837	378.3	1.755	0.6700	0.7705	144.7	9.521	8.732
260.00	0.1768	1337.	182.6	0.9349	0.8631	1.308	682.1	316.6	97.92
260.00	0.1768	8.905	390.8	1.736	0.7234	0.8418	146.8	10.25	10.39
280.00	0.3727	1272.	209.3	1.033	0.8877	1.361	590.2	244.3	88.99
280.00	0.3727	18.23	402.5	1.724	0.7810	0.9296	146.6	10.98	12.12
300.00	0.7028	1200.	237.2	1.129	0.9144	1.432	497.9	190.5	80.34
300.00	0.7028	34.19	413.3	1.716	0.8426	1.044	143.9	11.77	14.01
320.00	1.217	1117.	266.8	1.223	0.9443	1.543	404.0	147.8	71.78
320.00	1.217	60.71	422.3	1.709	0.9093	1.211	137.9	12.73	16.30
340.00	1.972	1015.	298.9	1.318	0.9802	1.751	306.4	111.8	63.08
340.00	1.972	105.7	428.2	1.698	0.9852	1.524	127.6	14.16	19.71
360.00	3.040	870.1	336.1	1.421	1.039	2.437	196.0	78.15	54.06
360.00	3.040	193.6	427.1	1.674	1.085	2.606	111.2	17.14	27.37
374.21	4.059	511.9	389.6	1.562				34.69	

R-1234yf (2,3,3,3-Tetrafluoroprop-1-ene, $CF_3CF{=}CH_2$)

T/ K	P/ MPa	ρ/ kg m^{-3}	H/ kJ kg^{-1}	S/ kJ kg^{-1} K^{-1}	C_v/ kJ kg^{-1} K^{-1}	C_p/ kJ kg^{-1} K^{-1}	u/ m s^{-1}
240.00	0.08603	1273.	159.1	0.8412	0.7990	1.178	704.6
240.00	0.08603	5.129	341.2	1.600	0.7096	0.8004	134.5
250.00	0.1327	1245.	171.0	0.8898	0.8205	1.210	659.9
250.00	0.1327	7.713	347.9	1.597	0.7374	0.8352	135.4
260.00	0.1972	1216.	183.3	0.9378	0.8411	1.243	616.0
260.00	0.1972	11.23	354.6	1.597	0.7656	0.8725	135.8
270.00	0.2834	1186.	195.9	0.9852	0.8607	1.278	572.8
270.00	0.2834	15.89	361.2	1.597	0.7940	0.9127	135.6
280.00	0.3959	1154.	208.9	1.032	0.8794	1.315	529.7
280.00	0.3959	22.00	367.7	1.599	0.8227	0.9569	134.7
290.00	0.5393	1121.	222.3	1.079	0.8973	1.356	486.6
290.00	0.5393	29.88	374.0	1.602	0.8514	1.006	133.2
300.00	0.7187	1085.	236.1	1.125	0.9147	1.401	443.1
300.00	0.7187	39.99	379.9	1.604	0.8808	1.065	130.9
310.00	0.9394	1047.	250.3	1.171	0.9321	1.454	399.2
310.00	0.9394	52.95	385.5	1.607	0.9122	1.139	127.7
320.00	1.207	1005.	265.1	1.217	0.9496	1.521	354.6
320.00	1.207	69.65	390.5	1.609	0.9467	1.238	123.6
330.00	1.528	957.6	280.5	1.263	0.9667	1.617	308.2
330.00	1.528	91.48	394.8	1.610	0.9847	1.382	118.3
340.00	1.910	902.8	296.8	1.311	0.9855	1.768	257.2
340.00	1.910	120.8	397.9	1.608	1.027	1.614	111.7
350.00	2.361	834.9	314.5	1.360	1.015	2.061	200.6
350.00	2.361	162.8	399.2	1.602	1.077	2.087	103.4
360.00	2.893	738.9	335.0	1.416	1.074	3.029	138.4
360.00	2.893	232.4	396.3	1.587	1.146	3.724	92.64
367.85	3.382	475.6	369.6	1.509			

Sulfur hexafluoride (SF_6)

T/ K	P/ MPa	ρ/ kg m^{-3}	H/ kJ kg^{-1}	S/ kJ kg^{-1} K^{-1}	C_v/ kJ kg^{-1} K^{-1}	C_p/ kJ kg^{-1} K^{-1}	u/ m s^{-1}
223.56	0.2314	1845.	154.1	0.8175	0.5275	0.8371	552.3

T/ K	P/ MPa	ρ/ kg m⁻³	H/ kJ kg⁻¹	S/ kJ kg⁻¹ K⁻¹	C_v/ kJ kg⁻¹ K⁻¹	C_p/ kJ kg⁻¹ K⁻¹	u/ m s⁻¹
223.56	0.2314	19.56	264.7	1.312	0.4833	0.5631	112.8
230.00	0.3013	1813.	159.5	0.8414	0.5407	0.8573	521.5
230.00	0.3013	25.14	267.4	1.310	0.4996	0.5851	112.8
240.00	0.4401	1761.	168.3	0.8784	0.5604	0.8902	474.7
240.00	0.4401	36.21	271.6	1.309	0.5249	0.6219	112.3
250.00	0.6220	1705.	177.4	0.9151	0.5796	0.9264	428.6
250.00	0.6220	50.87	275.6	1.308	0.5506	0.6636	111.0
260.00	0.8546	1646.	186.9	0.9517	0.5986	0.9681	382.8
260.00	0.8546	70.10	279.5	1.308	0.5762	0.7121	109.0
270.00	1.146	1580.	196.8	0.9884	0.6176	1.019	336.9
270.00	1.146	95.32	283.0	1.308	0.6026	0.7739	106.1
280.00	1.506	1507.	207.2	1.025	0.6372	1.087	290.2
280.00	1.506	128.7	286.1	1.307	0.6327	0.8623	102.1
290.00	1.942	1423.	218.3	1.063	0.6589	1.190	241.6
290.00	1.942	173.9	288.4	1.305	0.6650	0.9997	96.92
300.00	2.468	1319.	230.3	1.103	0.6863	1.380	189.8
300.00	2.468	238.4	289.6	1.300	0.7029	1.272	90.15
310.00	3.098	1175.	244.3	1.147	0.7308	1.961	131.5
310.00	3.098	344.1	288.2	1.289	0.7618	2.195	81.33
318.72	3.755	742.3	269.5	1.225			

Sulfur dioxide (SO_2)

T/ K	P/ MPa	ρ/ kg m⁻³	H/ kJ kg⁻¹	S/ kJ kg⁻¹ K⁻¹	C_v/ kJ kg⁻¹ K⁻¹	C_p/ kJ kg⁻¹ K⁻¹	u/ m s⁻¹
197.70	0.001660	1620.	−89.30	−0.3900	0.8776	1.376	1362.
197.70	0.001660	0.06483	355.0	1.857	0.4424	0.5746	182.2
200.00	0.002026	1615.	−86.14	−0.3741	0.8756	1.374	1350.
200.00	0.002026	0.07823	356.3	1.838	0.4443	0.5768	183.2
220.00	0.009334	1567.	−58.76	−0.2437	0.8578	1.364	1253.
220.00	0.009334	0.3289	367.1	1.692	0.4638	0.6006	191.1
240.00	0.03199	1519.	−31.52	−0.1252	0.8406	1.360	1160.
240.00	0.03199	1.041	377.7	1.580	0.4885	0.6330	198.1
260.00	0.08791	1469.	−4.277	−0.01632	0.8248	1.363	1070.
260.00	0.08791	2.677	387.6	1.491	0.5180	0.6750	203.9
280.00	0.2043	1417.	23.14	0.08495	0.8112	1.375	980.3
280.00	0.2043	5.897	396.6	1.419	0.5514	0.7274	208.4
300.00	0.4172	1362.	50.92	0.1802	0.7999	1.398	891.0
300.00	0.4172	11.57	404.6	1.359	0.5875	0.7915	211.3
320.00	0.7702	1304.	79.31	0.2710	0.7910	1.436	801.0
320.00	0.7702	20.83	411.1	1.308	0.6253	0.8712	212.4
340.00	1.313	1240.	108.6	0.3586	0.7849	1.495	709.5
340.00	1.313	35.23	415.7	1.262	0.6648	0.9756	211.5
360.00	2.099	1168.	139.3	0.4445	0.7819	1.590	615.4
360.00	2.099	57.07	417.8	1.218	0.7065	1.126	208.2
380.00	3.190	1086.	172.2	0.5306	0.7832	1.759	516.7
380.00	3.190	90.35	416.1	1.173	0.7529	1.380	202.2
400.00	4.656	983.7	208.7	0.6205	0.7922	2.134	409.4
400.00	4.656	143.8	408.2	1.119	0.8091	1.942	192.9
420.00	6.590	832.0	254.3	0.7264	0.8262	3.867	279.4
420.00	6.590	248.6	385.7	1.039	0.8916	4.570	179.3
430.64	7.884	525.0	320.7	0.8778			

Toluene (CH_3–C_6H_5)

T/ K	P/ MPa	ρ/ kg m^{-3}	H/ kJ kg^{-1}	S/ kJ kg^{-1} K^{-1}	C_v/ kJ kg^{-1} K^{-1}	C_p/ kJ kg^{-1} K^{-1}	u/ m s^{-1}
178.00	$0.3939 \cdot 10^{-7}$	974.8	−344.9	−1.261	1.024	1.472	1888.
178.00	$0.3939 \cdot 10^{-7}$	$0.2453 \cdot 10^{-5}$	147.1	1.503	0.5940	0.6843	136.0
180.00	$0.5534 \cdot 10^{-7}$	972.9	−342.0	−1.244	1.024	1.472	1877.
180.00	$0.5534 \cdot 10^{-7}$	$0.3407 \cdot 10^{-5}$	148.5	1.480	0.6004	0.6907	136.7
200.00	$0.1083 \cdot 10^{-5}$	953.5	−312.5	−1.089	1.039	1.479	1768.
200.00	$0.1083 \cdot 10^{-5}$	$0.6002 \cdot 10^{-4}$	163.0	1.288	0.6668	0.7570	143.1
220.00	$0.1148 \cdot 10^{-4}$	934.6	−282.7	−0.9471	1.068	1.504	1665.
220.00	$0.1148 \cdot 10^{-4}$	$0.5783 \cdot 10^{-3}$	178.8	1.151	0.7373	0.8275	149.3
240.00	$0.7754 \cdot 10^{-4}$	916.0	−252.2	−0.8147	1.108	1.542	1566.
240.00	$0.7754 \cdot 10^{-4}$	0.003581	196.1	1.053	0.8111	0.9014	155.1
260.00	$0.3731 \cdot 10^{-3}$	897.5	−220.9	−0.6895	1.156	1.590	1472.
260.00	$0.3731 \cdot 10^{-3}$	0.01591	214.8	0.9865	0.8876	0.9781	160.7
280.00	0.001383	879.0	−188.6	−0.5697	1.210	1.646	1381.
280.00	0.001383	0.05483	235.0	0.9433	0.9660	1.057	166.0
300.00	0.004177	860.4	−155.1	−0.4541	1.268	1.707	1294.
300.00	0.004177	0.1549	256.7	0.9184	1.046	1.138	171.0
320.00	0.01073	841.7	−120.3	−0.3418	1.329	1.773	1211.
320.00	0.01073	0.3745	279.7	0.9080	1.126	1.221	175.5
340.00	0.02417	822.6	−84.12	−0.2323	1.391	1.842	1130.
340.00	0.02417	0.7995	303.9	0.9090	1.208	1.305	179.4
360.00	0.04898	803.1	−46.54	−0.1250	1.455	1.915	1051.
360.00	0.04898	1.545	329.3	0.9190	1.289	1.391	182.7
380.00	0.09099	783.0	−7.483	−0.01956	1.518	1.989	974.1
380.00	0.09099	2.757	355.7	0.9361	1.370	1.479	185.1
400.00	0.1573	762.2	33.10	0.08428	1.581	2.066	898.1
400.00	0.1573	4.612	382.9	0.9587	1.450	1.570	186.4
450.00	0.4862	705.7	141.6	0.3385	1.737	2.272	709.6
450.00	0.4862	13.62	453.2	1.031	1.649	1.814	184.2
500.00	1.177	638.1	261.2	0.5882	1.887	2.529	516.5
500.00	1.177	33.64	523.7	1.113	1.846	2.132	170.9
550.00	2.428	544.8	395.8	0.8406	2.043	3.029	306.4
550.00	2.428	80.11	585.4	1.185	2.057	2.873	140.1
591.75	4.126	292.0	565.8	1.130			

VIRIAL COEFFICIENTS OF SELECTED GASES

Henry V. Kehiaian

This table gives second virial coefficients of about 110 inorganic and organic gases as a function of temperature. Selected data from the literature have been fitted by least squares to the equation

$$B / \text{cm}^3\text{mol}^{-1} = \sum_{i=1}^{n} a(i)[(T_0 / T) - 1]^{i-1}$$

where T_0 = 298.15 K. The table gives values of B at fixed temperature increments, as calculated from this smoothing equation. The first row (lowest temperature) for each compound includes the coefficients $a(i)$ for that compound. Compounds are listed by name.

The equation may be used with the tabulated coefficients for interpolation within the indicated temperature range. It should not be used for extrapolation beyond this range. A useful compilation of virial coefficient data from the literature may be found in the reference.

Reference

Dymond, J. H., and Smith, E. B,, *The Virial Coefficients of Pure Gases and Mixtures, A Critical Compilation*, Oxford University Press, Oxford, 1980.

Name	Mol. form.	T/K	B/cm³ mol⁻¹	$a(1)$	$a(2)$	$a(3)$	$a(4)$	$a(5)$
Acetaldehyde	C_2H_4O	290	-1352	-1217	-4647	-5725		
Acetaldehyde	C_2H_4O	320	-927					
Acetaldehyde	C_2H_4O	350	-654					
Acetaldehyde	C_2H_4O	380	-482					
Acetaldehyde	C_2H_4O	410	-375					
Acetaldehyde	C_2H_4O	440	-314					
Acetaldehyde	C_2H_4O	470	-283					
Acetone	C_3H_6O	300	-1996	-2051	-8903	-18056	-16448	
Acetone	C_3H_6O	320	-1522					
Acetone	C_3H_6O	340	-1198					
Acetone	C_3H_6O	360	-971					
Acetone	C_3H_6O	380	-806					
Acetone	C_3H_6O	400	-683					
Acetone	C_3H_6O	420	-586					
Acetone	C_3H_6O	440	-506					
Acetone	C_3H_6O	460	-437					
Acetone	C_3H_6O	480	-375					
Acetonitrile	C_2H_3N	330	-3468	-5840	-29175	-47611		
Acetonitrile	C_2H_3N	340	-2971					
Acetonitrile	C_2H_3N	350	-2563					
Acetonitrile	C_2H_3N	360	-2233					
Acetonitrile	C_2H_3N	370	-1970					
Acetonitrile	C_2H_3N	380	-1765					
Acetonitrile	C_2H_3N	390	-1610					
Acetonitrile	C_2H_3N	400	-1499					
Acetonitrile	C_2H_3N	410	-1425					
Acetylene	C_2H_2	200	-573	-216	-375	-716		
Acetylene	C_2H_2	210	-500					
Acetylene	C_2H_2	220	-440					
Acetylene	C_2H_2	230	-390					
Acetylene	C_2H_2	240	-349					
Acetylene	C_2H_2	250	-315					
Acetylene	C_2H_2	260	-287					
Acetylene	C_2H_2	270	-263					
Ammonia	H_3N	290	-302	-271	-1022	-2715	-4189	
Ammonia	H_3N	300	-265					
Ammonia	H_3N	310	-236					
Ammonia	H_3N	320	-213					
Ammonia	H_3N	330	-194					
Ammonia	H_3N	340	-179					
Ammonia	H_3N	350	-166					
Ammonia	H_3N	360	-154					

Name	Mol. form.	T/K	B/cm³ mol⁻¹	$a(1)$	$a(2)$	$a(3)$	$a(4)$	$a(5)$
Ammonia	H_3N	370	-144					
Ammonia	H_3N	380	-135					
Ammonia	H_3N	400	-118					
Ammonia	H_3N	420	-101					
Argon	Ar	100	-184	-16	-60	-9.7	-1.5	
Argon	Ar	120	-131					
Argon	Ar	140	-98					
Argon	Ar	160	-76					
Argon	Ar	180	-60					
Argon	Ar	200	-48					
Argon	Ar	300	-16					
Argon	Ar	400	-1					
Argon	Ar	500	7					
Argon	Ar	600	12					
Argon	Ar	700	15					
Argon	Ar	800	18					
Argon	Ar	900	20					
Argon	Ar	1000	22					
Benzene	C_6H_6	290	-1588	-1477	-3851	-3683	-1423	
Benzene	C_6H_6	300	-1454					
Benzene	C_6H_6	310	-1335					
Benzene	C_6H_6	320	-1231					
Benzene	C_6H_6	330	-1139					
Benzene	C_6H_6	340	-1056					
Benzene	C_6H_6	350	-983					
Benzene	C_6H_6	400	-712					
Benzene	C_6H_6	450	-542					
Benzene	C_6H_6	500	-429					
Benzene	C_6H_6	550	-349					
Benzene	C_6H_6	600	-291					
Boron trifluoride	BF_3	200	-338	-106	-330	-251	-80	
Boron trifluoride	BF_3	240	-202					
Boron trifluoride	BF_3	280	-129					
Boron trifluoride	BF_3	320	-85					
Boron trifluoride	BF_3	360	-56					
Boron trifluoride	BF_3	400	-37					
Boron trifluoride	BF_3	440	-23					
Bromomethane	CH_3Br	280	-645	-559	-1324			
Bromomethane	CH_3Br	290	-596					
Bromomethane	CH_3Br	300	-551					
Bromomethane	CH_3Br	310	-509					
Bromomethane	CH_3Br	320	-469					
Bromomethane	CH_3Br	340	-396					
Bromomethane	CH_3Br	360	-332					
Bromomethane	CH_3Br	380	-274					
Butane	C_4H_{10}	250	-1170	-735	-1835	-1922	-1330	
Butane	C_4H_{10}	280	-863					
Butane	C_4H_{10}	310	-668					
Butane	C_4H_{10}	340	-536					
Butane	C_4H_{10}	370	-442					
Butane	C_4H_{10}	400	-371					
Butane	C_4H_{10}	430	-315					
Butane	C_4H_{10}	460	-270					
Butane	C_4H_{10}	490	-232					
Butane	C_4H_{10}	520	-199					
Butane	C_4H_{10}	550	-171					
1-Butanol	$C_4H_{10}O$	350	-1693	-2629	-6315			
1-Butanol	$C_4H_{10}O$	360	-1544					
1-Butanol	$C_4H_{10}O$	370	-1402					
1-Butanol	$C_4H_{10}O$	380	-1268					
1-Butanol	$C_4H_{10}O$	390	-1141					

Name	Mol. form.	T/K	B/cm³ mol⁻¹	a(1)	a(2)	a(3)	a(4)	a(5)
1-Butanol	$C_4H_{10}O$	400	-1021					
1-Butanol	$C_4H_{10}O$	420	-796					
1-Butanol	$C_4H_{10}O$	440	-593					
2-Butanol	$C_4H_{10}O$	380	-1110	-2232	-5209			
2-Butanol	$C_4H_{10}O$	390	-1005					
2-Butanol	$C_4H_{10}O$	400	-906					
2-Butanol	$C_4H_{10}O$	410	-811					
2-Butanol	$C_4H_{10}O$	420	-721					
2-Butanone	C_4H_8O	310	-2056	-2282	-5907			
2-Butanone	C_4H_8O	320	-1878					
2-Butanone	C_4H_8O	330	-1712					
2-Butanone	C_4H_8O	340	-1555					
2-Butanone	C_4H_8O	350	-1407					
2-Butanone	C_4H_8O	360	-1267					
2-Butanone	C_4H_8O	370	-1135					
1-Butene	C_4H_8	300	-624	-633	-1442	-932		
1-Butene	C_4H_8	320	-539					
1-Butene	C_4H_8	340	-470					
1-Butene	C_4H_8	360	-413					
1-Butene	C_4H_8	380	-366					
1-Butene	C_4H_8	400	-327					
1-Butene	C_4H_8	420	-294					
Carbon dioxide	CO_2	220	-244	-127	-288	-118		
Carbon dioxide	CO_2	240	-204					
Carbon dioxide	CO_2	260	-172					
Carbon dioxide	CO_2	280	-146					
Carbon dioxide	CO_2	300	-126					
Carbon dioxide	CO_2	320	-108					
Carbon dioxide	CO_2	340	-94					
Carbon dioxide	CO_2	360	-81					
Carbon dioxide	CO_2	380	-71					
Carbon dioxide	CO_2	400	-62					
Carbon dioxide	CO_2	500	-30					
Carbon dioxide	CO_2	600	-13					
Carbon dioxide	CO_2	700	-1					
Carbon dioxide	CO_2	800	7					
Carbon dioxide	CO_2	900	12					
Carbon dioxide	CO_2	1000	16					
Carbon dioxide	CO_2	1100	19					
Carbon disulfide	CS_2	280	-932	-807	-1829	-1371		
Carbon disulfide	CS_2	310	-740					
Carbon disulfide	CS_2	340	-603					
Carbon disulfide	CS_2	370	-504					
Carbon disulfide	CS_2	400	-431					
Carbon disulfide	CS_2	430	-375					
Carbon monoxide	CO	210	-36	-9	-58	-18		
Carbon monoxide	CO	240	-24					
Carbon monoxide	CO	270	-15					
Carbon monoxide	CO	300	-8					
Carbon monoxide	CO	330	-3					
Carbon monoxide	CO	360	1					
Carbon monoxide	CO	420	7					
Carbon monoxide	CO	480	11					
Chlorine	Cl_2	210	-508	-303	-555	9	329	68
Chlorine	Cl_2	220	-483					
Chlorine	Cl_2	230	-457					
Chlorine	Cl_2	240	-432					
Chlorine	Cl_2	250	-407					
Chlorine	Cl_2	260	-383					
Chlorine	Cl_2	270	-360					
Chlorine	Cl_2	280	-339					

Name	Mol. form.	*T*/K	*B*/cm³ mol⁻¹	*a*(1)	*a*(2)	*a*(3)	*a*(4)	*a*(5)
Chlorine	Cl_2	290	-318					
Chlorine	Cl_2	300	-299					
Chlorine	Cl_2	350	-221					
Chlorine	Cl_2	400	-166					
Chlorine	Cl_2	450	-126					
Chlorine	Cl_2	500	-97					
Chlorine	Cl_2	600	-59					
Chlorine	Cl_2	700	-36					
Chlorine	Cl_2	800	-22					
Chlorine	Cl_2	900	-12					
1-Chlorobutane	C_4H_9Cl	330	-1224	-1643	-4897	-6178	-3718	
1-Chlorobutane	C_4H_9Cl	370	-898					
1-Chlorobutane	C_4H_9Cl	410	-691					
1-Chlorobutane	C_4H_9Cl	450	-551					
1-Chlorobutane	C_4H_9Cl	490	-449					
1-Chlorobutane	C_4H_9Cl	530	-371					
1-Chlorobutane	C_4H_9Cl	570	-309					
Chlorodifluoromethane	$CHClF_2$	300	-343	-347	-575	187		
Chlorodifluoromethane	$CHClF_2$	325	-298					
Chlorodifluoromethane	$CHClF_2$	350	-257					
Chlorodifluoromethane	$CHClF_2$	375	-221					
Chlorodifluoromethane	$CHClF_2$	400	-188					
Chlorodifluoromethane	$CHClF_2$	425	-158					
Chloroethane	C_2H_5Cl	320	-634	-777	-2205	-1764		
Chloroethane	C_2H_5Cl	360	-450					
Chloroethane	C_2H_5Cl	400	-330					
Chloroethane	C_2H_5Cl	440	-249					
Chloroethane	C_2H_5Cl	480	-195					
Chloroethane	C_2H_5Cl	520	-157					
Chloroethane	C_2H_5Cl	560	-131					
Chloroethane	C_2H_5Cl	600	-114					
Chloromethane	CH_3Cl	280	-466	-407	-887	-385		
Chloromethane	CH_3Cl	300	-402					
Chloromethane	CH_3Cl	320	-348					
Chloromethane	CH_3Cl	340	-304					
Chloromethane	CH_3Cl	360	-266					
Chloromethane	CH_3Cl	380	-234					
Chloromethane	CH_3Cl	400	-206					
Chloromethane	CH_3Cl	420	-182					
Chloromethane	CH_3Cl	440	-161					
Chloromethane	CH_3Cl	460	-142					
Chloromethane	CH_3Cl	480	-126					
Chloromethane	CH_3Cl	500	-112					
Chloromethane	CH_3Cl	600	-58					
1-Chloropropane	C_3H_7Cl	310	-1001	-1121	-3271	-3786	-1974	
1-Chloropropane	C_3H_7Cl	340	-772					
1-Chloropropane	C_3H_7Cl	370	-614					
1-Chloropropane	C_3H_7Cl	400	-501					
1-Chloropropane	C_3H_7Cl	430	-417					
1-Chloropropane	C_3H_7Cl	460	-352					
1-Chloropropane	C_3H_7Cl	490	-302					
1-Chloropropane	C_3H_7Cl	520	-261					
1-Chloropropane	C_3H_7Cl	550	-227					
1-Chloropropane	C_3H_7Cl	580	-198					
Chlorotrifluoromethane	$CClF_3$	240	-369	-223	-504	-340	-291	
Chlorotrifluoromethane	$CClF_3$	290	-237					
Chlorotrifluoromethane	$CClF_3$	340	-165					
Chlorotrifluoromethane	$CClF_3$	390	-119					
Chlorotrifluoromethane	$CClF_3$	440	-86					
Chlorotrifluoromethane	$CClF_3$	490	-60					
Chlorotrifluoromethane	$CClF_3$	540	-39					

Name	Mol. form.	T/K	B/cm³ mol⁻¹	a(1)	a(2)	a(3)	a(4)	a(5)
Cyclohexane	C_6H_{12}	300	-1698	-1733	-5618	-9486	-7936	
Cyclohexane	C_6H_{12}	320	-1391					
Cyclohexane	C_6H_{12}	340	-1170					
Cyclohexane	C_6H_{12}	360	-1007					
Cyclohexane	C_6H_{12}	380	-883					
Cyclohexane	C_6H_{12}	400	-786					
Cyclohexane	C_6H_{12}	420	-707					
Cyclohexane	C_6H_{12}	440	-641					
Cyclohexane	C_6H_{12}	460	-584					
Cyclohexane	C_6H_{12}	480	-534					
Cyclohexane	C_6H_{12}	500	-488					
Cyclohexane	C_6H_{12}	520	-446					
Cyclohexane	C_6H_{12}	540	-406					
Cyclohexane	C_6H_{12}	560	-368					
Cyclopentane	C_5H_{10}	300	-1049	-1062	-2116			
Cyclopentane	C_5H_{10}	305	-1015					
Cyclopentane	C_5H_{10}	310	-981					
Cyclopentane	C_5H_{10}	315	-949					
Cyclopentane	C_5H_{10}	320	-918					
Cyclopropane	C_3H_6	300	-383	-388	-861	-538		
Cyclopropane	C_3H_6	310	-356					
Cyclopropane	C_3H_6	320	-332					
Cyclopropane	C_3H_6	330	-310					
Cyclopropane	C_3H_6	340	-290					
Cyclopropane	C_3H_6	350	-272					
Cyclopropane	C_3H_6	360	-256					
Cyclopropane	C_3H_6	370	-241					
Cyclopropane	C_3H_6	380	-227					
Cyclopropane	C_3H_6	390	-215					
Cyclopropane	C_3H_6	400	-204					
Dichlorodifluoromethane	CCl_2F_2	250	-769	-486	-1217	-1188	-698	
Dichlorodifluoromethane	CCl_2F_2	280	-570					
Dichlorodifluoromethane	CCl_2F_2	310	-441					
Dichlorodifluoromethane	CCl_2F_2	340	-353					
Dichlorodifluoromethane	CCl_2F_2	370	-289					
Dichlorodifluoromethane	CCl_2F_2	400	-241					
Dichlorodifluoromethane	CCl_2F_2	430	-204					
Dichlorodifluoromethane	CCl_2F_2	460	-174					
1,2-Dichloroethane	$C_2H_4Cl_2$	370	-812	-1362	-3240	-2100		
1,2-Dichloroethane	$C_2H_4Cl_2$	390	-716					
1,2-Dichloroethane	$C_2H_4Cl_2$	410	-635					
1,2-Dichloroethane	$C_2H_4Cl_2$	430	-566					
1,2-Dichloroethane	$C_2H_4Cl_2$	450	-508					
1,2-Dichloroethane	$C_2H_4Cl_2$	470	-458					
1,2-Dichloroethane	$C_2H_4Cl_2$	490	-416					
1,2-Dichloroethane	$C_2H_4Cl_2$	510	-379					
1,2-Dichloroethane	$C_2H_4Cl_2$	530	-347					
1,2-Dichloroethane	$C_2H_4Cl_2$	550	-319					
1,2-Dichloroethane	$C_2H_4Cl_2$	570	-295					
Dichlorofluoromethane	$CHCl_2F$	250	-728	-562	-862			
Dichlorofluoromethane	$CHCl_2F$	275	-634					
Dichlorofluoromethane	$CHCl_2F$	300	-557					
Dichlorofluoromethane	$CHCl_2F$	325	-491					
Dichlorofluoromethane	$CHCl_2F$	350	-434					
Dichlorofluoromethane	$CHCl_2F$	375	-385					
Dichlorofluoromethane	$CHCl_2F$	400	-343					
Dichlorofluoromethane	$CHCl_2F$	425	-305					
Dichlorofluoromethane	$CHCl_2F$	450	-271					
Dichloromethane	CH_2Cl_2	320	-706	-913	-3371	-5013		
Dichloromethane	CH_2Cl_2	330	-634					
Dichloromethane	CH_2Cl_2	340	-574					

Name	Mol. form.	*T*/K	*B*/cm³ mol⁻¹	*a*(1)	*a*(2)	*a*(3)	*a*(4)	*a*(5)
Dichloromethane	CH_2Cl_2	350	-524					
Dichloromethane	CH_2Cl_2	360	-482					
Dichloromethane	CH_2Cl_2	370	-447					
Dichloromethane	CH_2Cl_2	380	-420					
Dichloromethane	CH_2Cl_2	400	-380					
Dichloromethane	CH_2Cl_2	420	-357					
1,2-Dichloro-1,1,2,2-tetrafluoroethane	$C_2Cl_2F_4$	300	-801	-812	-1773	-963		
1,2-Dichloro-1,1,2,2-tetrafluoroethane	$C_2Cl_2F_4$	320	-695					
1,2-Dichloro-1,1,2,2-tetrafluoroethane	$C_2Cl_2F_4$	340	-608					
1,2-Dichloro-1,1,2,2-tetrafluoroethane	$C_2Cl_2F_4$	360	-536					
1,2-Dichloro-1,1,2,2-tetrafluoroethane	$C_2Cl_2F_4$	380	-475					
1,2-Dichloro-1,1,2,2-tetrafluoroethane	$C_2Cl_2F_4$	400	-423					
1,2-Dichloro-1,1,2,2-tetrafluoroethane	$C_2Cl_2F_4$	420	-379					
1,2-Dichloro-1,1,2,2-tetrafluoroethane	$C_2Cl_2F_4$	440	-341					
1,2-Dichloro-1,1,2,2-tetrafluoroethane	$C_2Cl_2F_4$	460	-307					
1,2-Dichloro-1,1,2,2-tetrafluoroethane	$C_2Cl_2F_4$	480	-279					
1,2-Dichloro-1,1,2,2-tetrafluoroethane	$C_2Cl_2F_4$	500	-253					
Diethylamine	$C_4H_{11}N$	320	-1228	-1522	-5204	-15047	-28835	
Diethylamine	$C_4H_{11}N$	330	-1134					
Diethylamine	$C_4H_{11}N$	340	-1056					
Diethylamine	$C_4H_{11}N$	350	-988					
Diethylamine	$C_4H_{11}N$	360	-926					
Diethylamine	$C_4H_{11}N$	370	-868					
Diethylamine	$C_4H_{11}N$	380	-812					
Diethylamine	$C_4H_{11}N$	390	-755					
Diethylamine	$C_4H_{11}N$	400	-697					
Diethyl ether	$C_4H_{10}O$	280	-1550	-1226	-4458	-7746	-10005	
Diethyl ether	$C_4H_{10}O$	300	-1199					
Diethyl ether	$C_4H_{10}O$	320	-954					
Diethyl ether	$C_4H_{10}O$	340	-776					
Diethyl ether	$C_4H_{10}O$	360	-638					
Diethyl ether	$C_4H_{10}O$	380	-525					
Diethyl ether	$C_4H_{10}O$	400	-428					
Diethyl ether	$C_4H_{10}O$	420	-340					
Difluoromethane	CH_2F_2	280	-375	-321	-754	-1300		
Difluoromethane	CH_2F_2	290	-343					
Difluoromethane	CH_2F_2	300	-316					
Difluoromethane	CH_2F_2	310	-294					
Difluoromethane	CH_2F_2	320	-275					
Difluoromethane	CH_2F_2	330	-260					
Difluoromethane	CH_2F_2	340	-248					
Difluoromethane	CH_2F_2	350	-238					
Dimethylamine	C_2H_7N	310	-606	-662	-1504	-667		
Dimethylamine	C_2H_7N	320	-563					
Dimethylamine	C_2H_7N	330	-523					
Dimethylamine	C_2H_7N	340	-487					
Dimethylamine	C_2H_7N	350	-454					
Dimethylamine	C_2H_7N	360	-423					
Dimethylamine	C_2H_7N	370	-395					
Dimethylamine	C_2H_7N	380	-369					
Dimethylamine	C_2H_7N	390	-345					
Dimethylamine	C_2H_7N	400	-322					
Dimethyl ether	C_2H_6O	275	-536	-455	-965			
Dimethyl ether	C_2H_6O	280	-517					
Dimethyl ether	C_2H_6O	285	-499					
Dimethyl ether	C_2H_6O	290	-482					
Dimethyl ether	C_2H_6O	295	-465					
Dimethyl ether	C_2H_6O	300	-449					
Dimethyl ether	C_2H_6O	305	-433					
Dimethyl ether	C_2H_6O	310	-418					
Ethane	C_2H_6	200	-409	-184	-376	-143	-54	

Name	Mol. form.	T/K	B/cm³ mol⁻¹	$a(1)$	$a(2)$	$a(3)$	$a(4)$	$a(5)$
Ethane	C_2H_6	220	-337					
Ethane	C_2H_6	240	-284					
Ethane	C_2H_6	260	-242					
Ethane	C_2H_6	280	-209					
Ethane	C_2H_6	300	-181					
Ethane	C_2H_6	320	-159					
Ethane	C_2H_6	340	-140					
Ethane	C_2H_6	360	-123					
Ethane	C_2H_6	380	-109					
Ethane	C_2H_6	400	-96					
Ethane	C_2H_6	500	-52					
Ethane	C_2H_6	600	-24					
Ethanol	C_2H_6O	320	-2710	-4475	-29719	-56716		
Ethanol	C_2H_6O	330	-2135					
Ethanol	C_2H_6O	340	-1676					
Ethanol	C_2H_6O	350	-1317					
Ethanol	C_2H_6O	360	-1043					
Ethanol	C_2H_6O	370	-843					
Ethanol	C_2H_6O	380	-705					
Ethanol	C_2H_6O	390	-622					
Ethyl acetate	$C_4H_8O_2$	330	-1543	-2272	-8818	-13130		
Ethyl acetate	$C_4H_8O_2$	340	-1385					
Ethyl acetate	$C_4H_8O_2$	350	-1254					
Ethyl acetate	$C_4H_8O_2$	360	-1144					
Ethyl acetate	$C_4H_8O_2$	370	-1055					
Ethyl acetate	$C_4H_8O_2$	380	-982					
Ethyl acetate	$C_4H_8O_2$	390	-923					
Ethyl acetate	$C_4H_8O_2$	400	-878					
Ethylamine	C_2H_7N	300	-773	-785	-2012	-1397		
Ethylamine	C_2H_7N	310	-710					
Ethylamine	C_2H_7N	320	-654					
Ethylamine	C_2H_7N	330	-604					
Ethylamine	C_2H_7N	340	-558					
Ethylamine	C_2H_7N	350	-517					
Ethylamine	C_2H_7N	360	-480					
Ethylamine	C_2H_7N	370	-447					
Ethylamine	C_2H_7N	380	-416					
Ethylamine	C_2H_7N	390	-389					
Ethylamine	C_2H_7N	400	-363					
Ethylene	C_2H_4	240	-218	-140	-296	-101		
Ethylene	C_2H_4	270	-172					
Ethylene	C_2H_4	300	-139					
Ethylene	C_2H_4	330	-113					
Ethylene	C_2H_4	360	-92					
Ethylene	C_2H_4	390	-76					
Ethylene	C_2H_4	420	-63					
Ethylene	C_2H_4	450	-52					
Ethyl formate	$C_3H_6O_2$	330	-1003	-1371	-4231	-4312		
Ethyl formate	$C_3H_6O_2$	340	-916					
Ethyl formate	$C_3H_6O_2$	350	-839					
Ethyl formate	$C_3H_6O_2$	360	-771					
Ethyl formate	$C_3H_6O_2$	370	-712					
Ethyl formate	$C_3H_6O_2$	380	-660					
Ethyl formate	$C_3H_6O_2$	390	-614					
Fluorine	F_2	80	-378	8.5	-163.2	84.0	-27.9	
Fluorine	F_2	110	-165					
Fluorine	F_2	140	-109					
Fluorine	F_2	170	-79					
Fluorine	F_2	200	-55					
Fluorine	F_2	230	-33					
Fluorine	F_2	260	-14					

Name	Mol. form.	*T*/K	*B*/cm³ mol⁻¹	*a*(1)	*a*(2)	*a*(3)	*a*(4)	*a*(5)
Fluoromethane	CH_3F	280	-244	-209	-525	-365		
Fluoromethane	CH_3F	300	-205					
Fluoromethane	CH_3F	320	-174					
Fluoromethane	CH_3F	340	-150					
Fluoromethane	CH_3F	360	-129					
Fluoromethane	CH_3F	380	-112					
Fluoromethane	CH_3F	400	-99					
Fluoromethane	CH_3F	420	-87					
Helium	He	2	-172	12.44	-1.25			
Helium	He	6	-48					
Helium	He	10	-24					
Helium	He	14	-13					
Helium	He	18	-7					
Helium	He	22	-3					
Helium	He	26	-1					
Helium	He	30	1					
Helium	He	50	6					
Helium	He	70	8					
Helium	He	90	10					
Helium	He	110	10					
Helium	He	150	11					
Helium	He	250	12					
Helium	He	650	13					
Helium	He	700	13					
Heptane	C_7H_{16}	300	-2782	-2834	-8523	-10068	-5051	
Heptane	C_7H_{16}	320	-2297					
Heptane	C_7H_{16}	340	-1928					
Heptane	C_7H_{16}	360	-1641					
Heptane	C_7H_{16}	380	-1415					
Heptane	C_7H_{16}	400	-1233					
Heptane	C_7H_{16}	420	-1085					
Heptane	C_7H_{16}	440	-963					
Heptane	C_7H_{16}	460	-862					
Heptane	C_7H_{16}	480	-775					
Heptane	C_7H_{16}	500	-702					
Heptane	C_7H_{16}	540	-583					
Heptane	C_7H_{16}	580	-490					
Heptane	C_7H_{16}	620	-416					
Heptane	C_7H_{16}	660	-355					
Heptane	C_7H_{16}	700	-304					
1-Heptene	C_7H_{14}	340	-1781	-2491	-6230	-3780		
1-Heptene	C_7H_{14}	350	-1651					
1-Heptene	C_7H_{14}	360	-1532					
1-Heptene	C_7H_{14}	370	-1424					
1-Heptene	C_7H_{14}	380	-1324					
1-Heptene	C_7H_{14}	390	-1233					
1-Heptene	C_7H_{14}	400	-1150					
1-Heptene	C_7H_{14}	410	-1073					
Hexane	C_6H_{14}	300	-1920	-1961	-6691	-13167	-15273	
Hexane	C_6H_{14}	310	-1724					
Hexane	C_6H_{14}	320	-1561					
Hexane	C_6H_{14}	330	-1424					
Hexane	C_6H_{14}	340	-1309					
Hexane	C_6H_{14}	350	-1209					
Hexane	C_6H_{14}	360	-1123					
Hexane	C_6H_{14}	370	-1046					
Hexane	C_6H_{14}	380	-978					
Hexane	C_6H_{14}	390	-916					
Hexane	C_6H_{14}	400	-859					
Hexane	C_6H_{14}	410	-806					
Hexane	C_6H_{14}	430	-707					

Name	Mol. form.	*T*/K	*B*/cm³ mol⁻¹	*a*(1)	*a*(2)	*a*(3)	*a*(4)	*a*(5)
Hexane	C_6H_{14}	450	-616					
Hydrogen	H_2	15	-230	15.4	-9.0	-0.21		
Hydrogen	H_2	20	-151					
Hydrogen	H_2	25	-108					
Hydrogen	H_2	30	-82					
Hydrogen	H_2	35	-64					
Hydrogen	H_2	40	-52					
Hydrogen	H_2	45	-42					
Hydrogen	H_2	50	-35					
Hydrogen	H_2	60	-24					
Hydrogen	H_2	70	-16					
Hydrogen	H_2	80	-11					
Hydrogen	H_2	90	-7					
Hydrogen	H_2	100	-3					
Hydrogen	H_2	200	11					
Hydrogen	H_2	300	15					
Hydrogen	H_2	400	18					
Hydrogen chloride	ClH	190	-451	-144	-325	-277	-170	
Hydrogen chloride	ClH	230	-269					
Hydrogen chloride	ClH	270	-181					
Hydrogen chloride	ClH	310	-132					
Hydrogen chloride	ClH	350	-102					
Hydrogen chloride	ClH	390	-81					
Hydrogen chloride	ClH	430	-66					
Hydrogen chloride	ClH	470	-54					
Iodine pentafluoride	F_5I	320	-2540	-3077	-8474	-9116		
Iodine pentafluoride	F_5I	330	-2344					
Iodine pentafluoride	F_5I	340	-2172					
Iodine pentafluoride	F_5I	350	-2021					
Iodine pentafluoride	F_5I	360	-1890					
Iodine pentafluoride	F_5I	370	-1775					
Iodine pentafluoride	F_5I	380	-1674					
Iodine pentafluoride	F_5I	390	-1587					
Iodine pentafluoride	F_5I	400	-1510					
Iodine pentafluoride	F_5I	410	-1443					
Iodomethane	CH_3I	310	-725	-844	-3353	-6590		
Iodomethane	CH_3I	320	-646					
Iodomethane	CH_3I	330	-582					
Iodomethane	CH_3I	340	-531					
Iodomethane	CH_3I	350	-492					
Iodomethane	CH_3I	360	-462					
Iodomethane	CH_3I	370	-441					
Iodomethane	CH_3I	380	-427					
Isobutane	C_4H_{10}	270	-900	-707	-1719	-1282		
Isobutane	C_4H_{10}	300	-697					
Isobutane	C_4H_{10}	330	-553					
Isobutane	C_4H_{10}	360	-450					
Isobutane	C_4H_{10}	390	-374					
Isobutane	C_4H_{10}	420	-317					
Isobutane	C_4H_{10}	450	-273					
Isobutane	C_4H_{10}	480	-240					
Isobutane	C_4H_{10}	510	-215					
Isopentane	C_5H_{12}	280	-1263	-1095	-2503	-1534		
Isopentane	C_5H_{12}	290	-1166					
Isopentane	C_5H_{12}	300	-1079					
Isopentane	C_5H_{12}	310	-1001					
Isopentane	C_5H_{12}	320	-931					
Isopentane	C_5H_{12}	330	-867					
Isopentane	C_5H_{12}	340	-810					
Isopentane	C_5H_{12}	350	-757					
Isopentane	C_5H_{12}	400	-557					

Name	Mol. form.	T/K	B/cm^3 mol^{-1}	$a(1)$	$a(2)$	$a(3)$	$a(4)$	$a(5)$
Isopentane	C_5H_{12}	450	-424					
Krypton	Kr	110	-363	-51	-118	-29	-5	
Krypton	Kr	120	-307					
Krypton	Kr	130	-263					
Krypton	Kr	140	-229					
Krypton	Kr	150	-201					
Krypton	Kr	160	-178					
Krypton	Kr	170	-159					
Krypton	Kr	180	-143					
Krypton	Kr	190	-129					
Krypton	Kr	200	-117					
Krypton	Kr	250	-75					
Krypton	Kr	300	-51					
Krypton	Kr	400	-23					
Krypton	Kr	500	-8					
Krypton	Kr	600	2					
Krypton	Kr	700	8					
Methane	CH_4	110	-328	-43	-114	-19	-7	
Methane	CH_4	120	-276					
Methane	CH_4	130	-237					
Methane	CH_4	140	-206					
Methane	CH_4	150	-181					
Methane	CH_4	160	-160					
Methane	CH_4	170	-143					
Methane	CH_4	180	-128					
Methane	CH_4	190	-116					
Methane	CH_4	200	-105					
Methane	CH_4	250	-66					
Methane	CH_4	300	-43					
Methane	CH_4	350	-27					
Methane	CH_4	400	-16					
Methane	CH_4	500	0					
Methane	CH_4	600	10					
Methanol	CH_4O	320	-1431	-1752	-4694			
Methanol	CH_4O	330	-1299					
Methanol	CH_4O	340	-1174					
Methanol	CH_4O	350	-1056					
Methanol	CH_4O	360	-945					
Methanol	CH_4O	370	-840					
Methanol	CH_4O	380	-741					
Methanol	CH_4O	390	-646					
Methanol	CH_4O	400	-557					
Methyl acetate	$C_3H_6O_2$	320	-1320	-1709	-6348	-9650		
Methyl acetate	$C_3H_6O_2$	330	-1186					
Methyl acetate	$C_3H_6O_2$	340	-1074					
Methyl acetate	$C_3H_6O_2$	350	-980					
Methyl acetate	$C_3H_6O_2$	360	-903					
Methyl acetate	$C_3H_6O_2$	370	-840					
Methyl acetate	$C_3H_6O_2$	380	-789					
Methyl acetate	$C_3H_6O_2$	390	-749					
Methylamine	CH_5N	300	-451	-459	-1191	-995		
Methylamine	CH_5N	325	-367					
Methylamine	CH_5N	350	-304					
Methylamine	CH_5N	375	-257					
Methylamine	CH_5N	400	-220					
Methylamine	CH_5N	425	-192					
Methylamine	CH_5N	450	-170					
Methylamine	CH_5N	500	-140					
Methylamine	CH_5N	550	-122					
Methylcyclopentane	C_6H_{12}	305	-1447	-1512	-2910			
Methylcyclopentane	C_6H_{12}	315	-1357					

Name	Mol. form.	*T*/K	*B*/cm³ mol⁻¹	*a*(1)	*a*(2)	*a*(3)	*a*(4)	*a*(5)
Methylcyclopentane	C_6H_{12}	325	-1272					
Methylcyclopentane	C_6H_{12}	335	-1192					
Methylcyclopentane	C_6H_{12}	345	-1117					
Methyl formate	$C_2H_4O_2$	320	-821	-1035	-3425	-4203		
Methyl formate	$C_2H_4O_2$	330	-744					
Methyl formate	$C_2H_4O_2$	340	-677					
Methyl formate	$C_2H_4O_2$	350	-620					
Methyl formate	$C_2H_4O_2$	360	-571					
Methyl formate	$C_2H_4O_2$	370	-528					
Methyl formate	$C_2H_4O_2$	380	-492					
Methyl formate	$C_2H_4O_2$	390	-461					
Methyl formate	$C_2H_4O_2$	400	-435					
Methyl propanoate	$C_4H_8O_2$	330	-1588	-2216	-7339	-8658		
Methyl propanoate	$C_4H_8O_2$	340	-1444					
Methyl propanoate	$C_4H_8O_2$	350	-1319					
Methyl propanoate	$C_4H_8O_2$	360	-1211					
Methyl propanoate	$C_4H_8O_2$	370	-1117					
Methyl propanoate	$C_4H_8O_2$	380	-1037					
Methyl propanoate	$C_4H_8O_2$	390	-968					
Methyl propanoate	$C_4H_8O_2$	400	-908					
2-Methyl-1-propanol	$C_4H_{10}O$	390	-1076	-2269	-5065			
2-Methyl-1-propanol	$C_4H_{10}O$	400	-979					
2-Methyl-1-propanol	$C_4H_{10}O$	410	-887					
2-Methyl-1-propanol	$C_4H_{10}O$	420	-800					
2-Methyl-1-propanol	$C_4H_{10}O$	430	-716					
2-Methyl-1-propanol	$C_4H_{10}O$	440	-636					
2-Methyl-2-propanol	$C_4H_{10}O$	380	-924	-1952	-4775			
2-Methyl-2-propanol	$C_4H_{10}O$	390	-827					
2-Methyl-2-propanol	$C_4H_{10}O$	400	-736					
2-Methyl-2-propanol	$C_4H_{10}O$	410	-649					
2-Methyl-2-propanol	$C_4H_{10}O$	420	-567					
2-Methylpyridine	C_6H_7N	360	-1656	-2940	-8813	-7809		
2-Methylpyridine	C_6H_7N	370	-1523					
2-Methylpyridine	C_6H_7N	380	-1404					
2-Methylpyridine	C_6H_7N	390	-1297					
2-Methylpyridine	C_6H_7N	400	-1202					
2-Methylpyridine	C_6H_7N	410	-1117					
2-Methylpyridine	C_6H_7N	420	-1040					
2-Methylpyridine	C_6H_7N	430	-972					
3-Methylpyridine	C_6H_7N	380	-1819	-6304	-30415	-44549		
3-Methylpyridine	C_6H_7N	390	-1612					
3-Methylpyridine	C_6H_7N	400	-1448					
3-Methylpyridine	C_6H_7N	410	-1322					
3-Methylpyridine	C_6H_7N	420	-1230					
3-Methylpyridine	C_6H_7N	430	-1166					
4-Methylpyridine	C_6H_7N	380	-1787	-6553	-32873	-49874		
4-Methylpyridine	C_6H_7N	390	-1578					
4-Methylpyridine	C_6H_7N	400	-1417					
4-Methylpyridine	C_6H_7N	410	-1297					
4-Methylpyridine	C_6H_7N	420	-1214					
4-Methylpyridine	C_6H_7N	430	-1163					
Molybdenum(VI) fluoride	F_6Mo	300	-896	-914	-2922	-4778		
Molybdenum(VI) fluoride	F_6Mo	310	-810					
Molybdenum(VI) fluoride	F_6Mo	320	-737					
Molybdenum(VI) fluoride	F_6Mo	330	-677					
Molybdenum(VI) fluoride	F_6Mo	340	-627					
Molybdenum(VI) fluoride	F_6Mo	350	-586					
Molybdenum(VI) fluoride	F_6Mo	360	-553					
Molybdenum(VI) fluoride	F_6Mo	370	-527					
Molybdenum(VI) fluoride	F_6Mo	380	-506					
Molybdenum(VI) fluoride	F_6Mo	390	-491					

Name	Mol. form.	*T*/K	*B*/cm³ mol⁻¹	*a*(1)	*a*(2)	*a*(3)	*a*(4)	*a*(5)
Neon	Ne	60	-25	10.8	-7.5	-0.4		
Neon	Ne	80	-13					
Neon	Ne	100	-6					
Neon	Ne	120	-1					
Neon	Ne	140	2					
Neon	Ne	160	4					
Neon	Ne	180	6					
Neon	Ne	200	7					
Neon	Ne	300	11					
Neon	Ne	400	13					
Neon	Ne	500	14					
Neon	Ne	600	15					
Neopentane	C_5H_{12}	300	-916	-931	-2387	-2641	-1810	
Neopentane	C_5H_{12}	310	-843					
Neopentane	C_5H_{12}	320	-780					
Neopentane	C_5H_{12}	330	-724					
Neopentane	C_5H_{12}	340	-674					
Neopentane	C_5H_{12}	350	-629					
Neopentane	C_5H_{12}	360	-590					
Neopentane	C_5H_{12}	370	-554					
Neopentane	C_5H_{12}	380	-521					
Neopentane	C_5H_{12}	390	-492					
Neopentane	C_5H_{12}	400	-464					
Neopentane	C_5H_{12}	450	-357					
Neopentane	C_5H_{12}	500	-279					
Neopentane	C_5H_{12}	550	-218					
Nitric oxide	NO	120	-232	-12	-119	89	-73	
Nitric oxide	NO	130	-176					
Nitric oxide	NO	140	-138					
Nitric oxide	NO	150	-113					
Nitric oxide	NO	160	-96					
Nitric oxide	NO	170	-83					
Nitric oxide	NO	180	-73					
Nitric oxide	NO	190	-65					
Nitric oxide	NO	200	-58					
Nitric oxide	NO	210	-52					
Nitric oxide	NO	230	-42					
Nitric oxide	NO	250	-32					
Nitric oxide	NO	270	-24					
Nitrogen	N_2	75	-274	-4.3	-55.7	-11.8		
Nitrogen	N_2	100	-161					
Nitrogen	N_2	125	-104					
Nitrogen	N_2	150	-71					
Nitrogen	N_2	175	-49					
Nitrogen	N_2	200	-34					
Nitrogen	N_2	225	-24					
Nitrogen	N_2	250	-15					
Nitrogen	N_2	300	-4					
Nitrogen	N_2	400	9					
Nitrogen	N_2	500	16					
Nitrogen	N_2	600	21					
Nitrogen	N_2	700	24					
Nitrous oxide	N_2O	240	-219	-130	-307	-248		
Nitrous oxide	N_2O	260	-181					
Nitrous oxide	N_2O	280	-151					
Nitrous oxide	N_2O	300	-128					
Nitrous oxide	N_2O	320	-110					
Nitrous oxide	N_2O	340	-96					
Nitrous oxide	N_2O	360	-85					
Nitrous oxide	N_2O	380	-76					
Nitrous oxide	N_2O	400	-68					

Name	Mol. form.	T/K	B/cm³ mol⁻¹	$a(1)$	$a(2)$	$a(3)$	$a(4)$	$a(5)$
Octane	C_8H_{18}	300	-4042	-4123	-13120	-16408	-8580	
Octane	C_8H_{18}	350	-2511					
Octane	C_8H_{18}	400	-1704					
Octane	C_8H_{18}	450	-1234					
Octane	C_8H_{18}	500	-936					
Octane	C_8H_{18}	550	-732					
Octane	C_8H_{18}	600	-583					
Octane	C_8H_{18}	650	-468					
Octane	C_8H_{18}	700	-375					
1-Octene	C_8H_{16}	360	-2147	-3273	-6557			
1-Octene	C_8H_{16}	370	-2000					
1-Octene	C_8H_{16}	380	-1861					
1-Octene	C_8H_{16}	390	-1729					
1-Octene	C_8H_{16}	400	-1604					
1-Octene	C_8H_{16}	410	-1485					
Oxygen	O_2	90	-241	-16	-62	-8	-3	
Oxygen	O_2	110	-161					
Oxygen	O_2	130	-117					
Oxygen	O_2	150	-88					
Oxygen	O_2	170	-69					
Oxygen	O_2	190	-55					
Oxygen	O_2	210	-44					
Oxygen	O_2	230	-36					
Oxygen	O_2	250	-29					
Oxygen	O_2	270	-23					
Oxygen	O_2	290	-18					
Oxygen	O_2	310	-14					
Oxygen	O_2	330	-10					
Oxygen	O_2	350	-7					
Oxygen	O_2	400	-1					
Pentane	C_5H_{12}	300	-1234	-1254	-3345	-2726		
Pentane	C_5H_{12}	310	-1130					
Pentane	C_5H_{12}	320	-1038					
Pentane	C_5H_{12}	330	-957					
Pentane	C_5H_{12}	340	-884					
Pentane	C_5H_{12}	350	-818					
Pentane	C_5H_{12}	400	-579					
Pentane	C_5H_{12}	450	-436					
Pentane	C_5H_{12}	500	-348					
Pentane	C_5H_{12}	550	-294					
2-Pentanone	$C_5H_{10}O$	330	-2850	-4962	-26372	-46537		
2-Pentanone	$C_5H_{10}O$	340	-2420					
2-Pentanone	$C_5H_{10}O$	350	-2076					
2-Pentanone	$C_5H_{10}O$	360	-1804					
2-Pentanone	$C_5H_{10}O$	370	-1595					
2-Pentanone	$C_5H_{10}O$	380	-1440					
2-Pentanone	$C_5H_{10}O$	390	-1332					
1-Pentene	C_5H_{10}	310	-966	-1055	-2377	-1189		
1-Pentene	C_5H_{10}	320	-898					
1-Pentene	C_5H_{10}	330	-836					
1-Pentene	C_5H_{10}	340	-780					
1-Pentene	C_5H_{10}	350	-729					
1-Pentene	C_5H_{10}	360	-681					
1-Pentene	C_5H_{10}	370	-638					
1-Pentene	C_5H_{10}	380	-598					
1-Pentene	C_5H_{10}	390	-561					
1-Pentene	C_5H_{10}	400	-527					
1-Pentene	C_5H_{10}	410	-495					
Phosphine	H_3P	190	-457	-146	-733	1022	-1220	
Phosphine	H_3P	200	-404					
Phosphine	H_3P	210	-364					

Name	Mol. form.	T/K	B/cm³ mol⁻¹	a(1)	a(2)	a(3)	a(4)	a(5)
Phosphine	H_3P	220	-332					
Phosphine	H_3P	230	-305					
Phosphine	H_3P	240	-281					
Phosphine	H_3P	250	-258					
Phosphine	H_3P	260	-235					
Phosphine	H_3P	270	-213					
Phosphine	H_3P	280	-190					
Phosphine	H_3P	290	-166					
Phosphorus(V) fluoride	F_5P	320	-162	-186	-345			
Phosphorus(V) fluoride	F_5P	340	-143					
Phosphorus(V) fluoride	F_5P	360	-127					
Phosphorus(V) fluoride	F_5P	380	-112					
Phosphorus(V) fluoride	F_5P	400	-98					
Phosphorus(V) fluoride	F_5P	420	-86					
Phosphorus(V) fluoride	F_5P	440	-75					
Phosphorus(V) fluoride	F_5P	460	-64					
Propane	C_3H_8	240	-641	-386	-844	-720	-574	
Propane	C_3H_8	260	-527					
Propane	C_3H_8	280	-444					
Propane	C_3H_8	300	-381					
Propane	C_3H_8	320	-331					
Propane	C_3H_8	340	-292					
Propane	C_3H_8	360	-259					
Propane	C_3H_8	380	-232					
Propane	C_3H_8	400	-208					
Propane	C_3H_8	440	-169					
Propane	C_3H_8	480	-138					
Propane	C_3H_8	520	-112					
Propane	C_3H_8	560	-90					
1-Propanol	C_3H_8O	380	-873	-2690	-12040	-16738		
1-Propanol	C_3H_8O	385	-826					
1-Propanol	C_3H_8O	390	-783					
1-Propanol	C_3H_8O	395	-744					
1-Propanol	C_3H_8O	400	-709					
1-Propanol	C_3H_8O	405	-679					
1-Propanol	C_3H_8O	410	-651					
1-Propanol	C_3H_8O	415	-627					
1-Propanol	C_3H_8O	420	-606					
2-Propanol	C_3H_8O	380	-821	-3165	-16092	-24197		
2-Propanol	C_3H_8O	385	-766					
2-Propanol	C_3H_8O	390	-717					
2-Propanol	C_3H_8O	395	-674					
2-Propanol	C_3H_8O	400	-636					
2-Propanol	C_3H_8O	405	-604					
2-Propanol	C_3H_8O	410	-576					
2-Propanol	C_3H_8O	415	-552					
2-Propanol	C_3H_8O	420	-533					
Propene	C_3H_6	280	-395	-347	-727	-325		
Propene	C_3H_6	300	-342					
Propene	C_3H_6	320	-299					
Propene	C_3H_6	340	-262					
Propene	C_3H_6	360	-232					
Propene	C_3H_6	380	-205					
Propene	C_3H_6	400	-183					
Propene	C_3H_6	420	-163					
Propene	C_3H_6	440	-146					
Propene	C_3H_6	460	-131					
Propene	C_3H_6	480	-118					
Propene	C_3H_6	500	-106					
Propyl formate	$C_4H_8O_2$	330	-1496	-2118	-7299	-8851		
Propyl formate	$C_4H_8O_2$	340	-1354					

Name	Mol. form.	T/K	B/cm³ mol⁻¹	$a(1)$	$a(2)$	$a(3)$	$a(4)$	$a(5)$
Propyl formate	$C_4H_8O_2$	350	-1231					
Propyl formate	$C_4H_8O_2$	360	-1126					
Propyl formate	$C_4H_8O_2$	370	-1035					
Propyl formate	$C_4H_8O_2$	380	-957					
Propyl formate	$C_4H_8O_2$	390	-890					
Propyl formate	$C_4H_8O_2$	400	-834					
Pyridine	C_5H_5N	350	-1257	-1765	-3431			
Pyridine	C_5H_5N	360	-1176					
Pyridine	C_5H_5N	370	-1099					
Pyridine	C_5H_5N	380	-1026					
Pyridine	C_5H_5N	390	-957					
Pyridine	C_5H_5N	400	-892					
Pyridine	C_5H_5N	420	-770					
Pyridine	C_5H_5N	440	-659					
Sulfur dioxide	O_2S	290	-465	-430	-1193	-1029		
Sulfur dioxide	O_2S	320	-354					
Sulfur dioxide	O_2S	350	-276					
Sulfur dioxide	O_2S	380	-221					
Sulfur dioxide	O_2S	410	-181					
Sulfur dioxide	O_2S	440	-153					
Sulfur dioxide	O_2S	470	-132					
Sulfur hexafluoride	F_6S	200	-685	-279	-647	-335	-72	
Sulfur hexafluoride	F_6S	250	-416					
Sulfur hexafluoride	F_6S	300	-275					
Sulfur hexafluoride	F_6S	350	-190					
Sulfur hexafluoride	F_6S	400	-135					
Sulfur hexafluoride	F_6S	450	-96					
Sulfur hexafluoride	F_6S	500	-68					
Tetrachloromethane	CCl_4	320	-1345	-1600	-4059	-4653		
Tetrachloromethane	CCl_4	340	-1171					
Tetrachloromethane	CCl_4	360	-1040					
Tetrachloromethane	CCl_4	380	-942					
Tetrachloromethane	CCl_4	400	-868					
Tetrachloromethane	CCl_4	420	-814					
Tetrafluoromethane	CF_4	250	-137	-88	-238	-70		
Tetrafluoromethane	CF_4	300	-87					
Tetrafluoromethane	CF_4	350	-55					
Tetrafluoromethane	CF_4	400	-32					
Tetrafluoromethane	CF_4	450	-16					
Tetrafluoromethane	CF_4	500	-4					
Tetrafluoromethane	CF_4	600	14					
Tetrafluoromethane	CF_4	700	25					
Tetrafluoromethane	CF_4	800	33					
Tetrafluorosilane	F_4Si	210	-268	-138	-312			
Tetrafluorosilane	F_4Si	240	-213					
Tetrafluorosilane	F_4Si	270	-170					
Tetrafluorosilane	F_4Si	300	-136					
Tetrafluorosilane	F_4Si	330	-108					
Tetrafluorosilane	F_4Si	360	-84					
Tetrafluorosilane	F_4Si	390	-64					
Tetrafluorosilane	F_4Si	420	-47					
Tetrafluorosilane	F_4Si	450	-32					
Toluene	C_7H_8	350	-1641	-2620	-7548	-6349		
Toluene	C_7H_8	360	-1511					
Toluene	C_7H_8	370	-1394					
Toluene	C_7H_8	380	-1289					
Toluene	C_7H_8	390	-1195					
Toluene	C_7H_8	400	-1110					
Toluene	C_7H_8	410	-1034					
Toluene	C_7H_8	420	-965					
Toluene	C_7H_8	430	-903					

Name	Mol. form.	T/K	B/cm³ mol⁻¹	a(1)	a(2)	a(3)	a(4)	a(5)
Trichlorofluoromethane	CCl_3F	240	-1140	-786	-1428	-142		
Trichlorofluoromethane	CCl_3F	280	-879					
Trichlorofluoromethane	CCl_3F	320	-689					
Trichlorofluoromethane	CCl_3F	360	-545					
Trichlorofluoromethane	CCl_3F	400	-431					
Trichlorofluoromethane	CCl_3F	440	-340					
Trichlorofluoromethane	CCl_3F	480	-265					
Trichloromethane	$CHCl_3$	320	-1001	-1193	-2936	-1751		
Trichloromethane	$CHCl_3$	330	-926					
Trichloromethane	$CHCl_3$	340	-858					
Trichloromethane	$CHCl_3$	350	-797					
Trichloromethane	$CHCl_3$	360	-740					
Trichloromethane	$CHCl_3$	370	-689					
Trichloromethane	$CHCl_3$	380	-642					
Trichloromethane	$CHCl_3$	390	-599					
Trichloromethane	$CHCl_3$	400	-559					
1,1,2-Trichloro-1,2,2-trifluoroethane	$C_2Cl_3F_3$	290	-1041	-999	-1479			
1,1,2-Trichloro-1,2,2-trifluoroethane	$C_2Cl_3F_3$	310	-943					
1,1,2-Trichloro-1,2,2-trifluoroethane	$C_2Cl_3F_3$	330	-856					
1,1,2-Trichloro-1,2,2-trifluoroethane	$C_2Cl_3F_3$	350	-780					
1,1,2-Trichloro-1,2,2-trifluoroethane	$C_2Cl_3F_3$	370	-712					
1,1,2-Trichloro-1,2,2-trifluoroethane	$C_2Cl_3F_3$	390	-651					
1,1,2-Trichloro-1,2,2-trifluoroethane	$C_2Cl_3F_3$	410	-596					
1,1,2-Trichloro-1,2,2-trifluoroethane	$C_2Cl_3F_3$	430	-546					
1,1,2-Trichloro-1,2,2-trifluoroethane	$C_2Cl_3F_3$	450	-500					
Triethylamine	$C_6H_{15}N$	330	-1562	-2061	-5735	-5899		
Triethylamine	$C_6H_{15}N$	340	-1444					
Triethylamine	$C_6H_{15}N$	350	-1340					
Triethylamine	$C_6H_{15}N$	360	-1249					
Triethylamine	$C_6H_{15}N$	370	-1169					
Triethylamine	$C_6H_{15}N$	380	-1099					
Triethylamine	$C_6H_{15}N$	390	-1037					
Triethylamine	$C_6H_{15}N$	400	-983					
Trifluoromethane	CHF_3	200	-433	-177	-399	-250		
Trifluoromethane	CHF_3	220	-350					
Trifluoromethane	CHF_3	240	-288					
Trifluoromethane	CHF_3	260	-241					
Trifluoromethane	CHF_3	280	-204					
Trifluoromethane	CHF_3	300	-174					
Trifluoromethane	CHF_3	320	-151					
Trifluoromethane	CHF_3	340	-132					
Trifluoromethane	CHF_3	360	-116					
Trifluoromethane	CHF_3	380	-103					
Trifluoromethane	CHF_3	400	-91					
Trimethylamine	C_3H_9N	310	-675	-737	-1669	-986		
Trimethylamine	C_3H_9N	320	-628					
Trimethylamine	C_3H_9N	330	-585					
Trimethylamine	C_3H_9N	340	-547					
Trimethylamine	C_3H_9N	350	-512					
Trimethylamine	C_3H_9N	360	-480					
Trimethylamine	C_3H_9N	370	-450					
Tungsten(VI) fluoride	F_6W	320	-641	-719	-1143			
Tungsten(VI) fluoride	F_6W	340	-578					
Tungsten(VI) fluoride	F_6W	360	-523					
Tungsten(VI) fluoride	F_6W	380	-473					
Tungsten(VI) fluoride	F_6W	400	-428					
Tungsten(VI) fluoride	F_6W	420	-387					
Tungsten(VI) fluoride	F_6W	440	-350					
Tungsten(VI) fluoride	F_6W	460	-317					
Uranium(VI) fluoride	F_6U	320	-1030	-1204	-2690	-2144		
Uranium(VI) fluoride	F_6U	340	-905					

Name	Mol. form.	*T*/K	*B*/cm³ mol⁻¹	*a*(1)	*a*(2)	*a*(3)	*a*(4)	*a*(5)
Uranium(VI) fluoride	F_6U	360	-805					
Uranium(VI) fluoride	F_6U	380	-724					
Uranium(VI) fluoride	F_6U	400	-658					
Uranium(VI) fluoride	F_6U	420	-604					
Uranium(VI) fluoride	F_6U	440	-560					
Water	H_2O	300	-1126	-1158	-5157	-10301	-10597	-4415
Water	H_2O	320	-850					
Water	H_2O	340	-660					
Water	H_2O	360	-526					
Water	H_2O	380	-428					
Water	H_2O	400	-356					
Water	H_2O	420	-301					
Water	H_2O	440	-258					
Water	H_2O	460	-224					
Water	H_2O	480	-197					
Water	H_2O	500	-175					
Water	H_2O	600	-104					
Water	H_2O	700	-67					
Water	H_2O	800	-44					
Water	H_2O	900	-30					
Water	H_2O	1000	-20					
Water	H_2O	1100	-14					
Water	H_2O	1200	-11					
Xenon	Xe	160	-421	-130	-262	-87		
Xenon	Xe	170	-377					
Xenon	Xe	180	-340					
Xenon	Xe	190	-307					
Xenon	Xe	200	-280					
Xenon	Xe	210	-255					
Xenon	Xe	220	-234					
Xenon	Xe	230	-215					
Xenon	Xe	240	-199					
Xenon	Xe	250	-184					
Xenon	Xe	300	-129					
Xenon	Xe	350	-93					
Xenon	Xe	400	-69					
Xenon	Xe	450	-39					
Xenon	Xe	600	-21					
Xenon	Xe	650	-14					
o-Xylene	C_8H_{10}	380	-2046	-5632	-22873	-28900		
o-Xylene	C_8H_{10}	390	-1848					
o-Xylene	C_8H_{10}	400	-1681					
o-Xylene	C_8H_{10}	410	-1543					
o-Xylene	C_8H_{10}	420	-1428					
o-Xylene	C_8H_{10}	430	-1335					
o-Xylene	C_8H_{10}	440	-1261					
m-Xylene	C_8H_{10}	380	-2082	-5808	-23244	-27607		
m-Xylene	C_8H_{10}	390	-1865					
m-Xylene	C_8H_{10}	400	-1679					
m-Xylene	C_8H_{10}	410	-1521					
m-Xylene	C_8H_{10}	420	-1388					
m-Xylene	C_8H_{10}	430	-1276					
m-Xylene	C_8H_{10}	440	-1184					
p-Xylene	C_8H_{10}	380	-2043	-4921	-16843	-16159		
p-Xylene	C_8H_{10}	390	-1851					
p-Xylene	C_8H_{10}	400	-1680					
p-Xylene	C_8H_{10}	410	-1529					
p-Xylene	C_8H_{10}	420	-1395					
p-Xylene	C_8H_{10}	430	-1276					
p-Xylene	C_8H_{10}	440	-1171					

MEAN FREE PATH AND RELATED PROPERTIES OF GASES

In the simplest version of the kinetic theory of gases, molecules are treated as hard spheres of diameter d which make binary collisions only. In this approximation the mean distance traveled by a molecule between successive collisions, the mean free path l, is related to the collision diameter by:

$$l = \frac{kT}{\pi\sqrt{2}Pd^2}$$

where P is the pressure, T the absolute temperature, and k the Boltzmann constant. At standard conditions (P = 100,000 Pa and T = 298.15 K) this relation becomes:

$$l = \frac{9.27 \cdot 10^{27}}{d^2}$$

where l and d are in meters.

Using the same model and the same standard pressure, the collision diameter can be calculated from the viscosity η by the kinetic theory relation:

$$\eta = \frac{2.67 \cdot 10^{-20}(MT)^{1/2}}{d^2}$$

where η is in units of μPa s and M is the molar mass in g/mol. Kinetic theory also gives a relation for the mean velocity $\bar{v}$ of molecules of mass m:

$$\bar{v} = \left(\frac{8kT}{\pi m}\right)^{1/2} = 145.5(T/M)^{1/2}\,\mathrm{m/s}$$

Finally, the mean time τ between collisions can be calculated from the relation $\tau\bar{v} = l$.

The table below gives values of l, $\bar{v}$, and τ for some common gases at 25 °C and atmospheric pressure, as well as the value of d, all calculated from measured gas viscosities (see Refs. 2 and 3 and the table "Viscosity of Gases" in this section). It is seen from the above equations that the mean free path varies directly with T and inversely with P, while the mean velocity varies as the square root of T and, in this approximation, is independent of P.

A more accurate model, in which molecular interactions are described by a Lennard-Jones potential, gives mean free path values about 5% lower than this table (see Ref. 4).

References

1. Reid, R. C., Prausnitz, J. M., and Poling, B. E., *The Properties of Gases and Liquids, Fourth Edition,* McGraw-Hill, New York, 1987.
2. Lide, D. R., and Kehiaian, H. V., *CRC Handbook of Thermophysical and Thermochemical Data,* CRC Press, Boca Raton, FL, 1994.
3. Vargaftik, N. B., *Tables of Thermophysical Properties of Liquids and Gases, Second Edition,* John Wiley, New York, 1975.
4. Kaye, G. W. C., and Laby, T. H., *Tables of Physical and Chemical Constants, 15th Edition,* Longman, London, 1986.

Gas	Formula	$10^{10} \cdot d$/m	$10^8 \cdot l$/m	$\bar{v}$/m s^{-1}	τ/ps
Air		3.66	6.91	467	148
Ammonia	NH_3	4.32	4.97	609	82
Argon	Ar	3.58	7.22	397	182
Carbon dioxide	CO_2	4.53	4.51	379	119
Helium	He	2.15	20.0	1256	159
Hydrogen	H_2	2.71	12.6	1769	71
Krypton	Kr	4.08	5.58	274	203
Neon	Ne	2.54	14.3	559	256
Nitrogen	N_2	3.70	6.76	475	142
Oxygen	O_2	3.55	7.36	444	166
Xenon	Xe	4.78	4.05	219	185

Fluids

INFLUENCE OF PRESSURE ON FREEZING POINTS

This table illustrates the variation of the freezing point of representative types of liquids with pressure. Substances are listed in alphabetical order. Note that 1 MPa = 0.01 kbar = 9.87 atm.

References

1. Isaacs, N. S., *Liquid Phase High Pressure Chemistry*, John Wiley, New York, 1981.
2. Merrill, L., *J. Phys. Chem. Ref. Data*, 6, 1205, 1977; 11, 1005, 1982.

Freezing Point in °C at the Indicated Pressure

Substance	Mol. form.	0.1013 MPa	100 MPa	1000 MPa
Acetic acid	$C_2H_4O_2$	17	37	
Acetophenone	C_8H_8O	19.4	41.2	
Aniline	C_6H_7N	-6.0	13.5	140
Benzene	C_6H_6	5.538	33.4	
Benzonitrile	C_7H_5N	-12.82	7.6	
Benzyl alcohol	C_7H_8O	-15.5	0.2	
Bromobenzene	C_6H_5Br	-30.74	-12	108
Bromoethane	C_2H_5Br	-118.4	-108	
1-Bromonaphthalene	$C_{10}H_7Br$	6.1	6.1	
1-Bromopropane	C_3H_7Br	-110.1	-98	
4-Bromotoluene	C_7H_7Br	26.2	56.7	
Butanoic acid	$C_4H_8O_2$	-5.12	13.8	
1-Butanol	$C_4H_{10}O$	-88.60	-77.2	
Carbon disulfide	CS_2	-111.7	-98	
Chlorobenzene	C_6H_5Cl	-45.2	-28	84
4-Chlorotoluene	C_7H_7Cl	7.4	33.1	
o-Cresol	C_7H_8O	31.0	47.7	
m-Cresol	C_7H_8O	12.2	25.6	
p-Cresol	C_7H_8O	34.77	56.2	
Cyclohexane	C_6H_{12}	6.7	32.5	
Cyclohexanol	$C_6H_{12}O$	26	62.3	
1,2-Dibromoethane	$C_2H_4Br_2$	9.8	34.0	
p-Dichlorobenzene	$C_6H_4Cl_2$	53.1	79.1	
Dichloromethane	CH_2Cl_2	-94.9	-83	
N,N-Dimethylaniline	$C_8H_{11}N$	2.1	26.3	
1,4-Dioxane	$C_4H_8O_2$	11.75	23	
Ethanol	C_2H_6O	-114.14	-108	
Formamide	CH_3NO	2.57	10.8	
Formic acid	CH_2O_2	8.3	20.6	
Furan	C_4H_4O	-85.58	-73	
Hexamethyldisiloxane	$C_6H_{18}OSi_2$	-68.2	-37	
(-)-Menthol	$C_{10}H_{20}O$	42.1	60	
Methyl benzoate	$C_8H_8O_2$	-12.35	31.8	
2-Methyl-2-butanol	$C_5H_{12}O$	-8.7	13.4	
2-Methyl-2-propanol	$C_4H_{10}O$	25.81	58.1	
Naphthalene	$C_{10}H_8$	80.22	115.7	
Nitrobenzene	$C_6H_5NO_2$	5.65	13.5	
3-Nitrotoluene	$C_7H_7NO_2$	15.9	40.6	
Pentachloroethane	C_2HCl_5	-29.0	-6.3	
Potassium	K	63.5	78	170
Potassium chloride	ClK	771		945
Propanoic acid	$C_3H_6O_2$	-20.5	-1.2	
Silver(I) chloride	AgCl	455		545
Sodium	Na	97.794	106	167
Sodium chloride	ClNa	802.018		997
Sodium fluoride	FNa	996		1115
Tetrachloromethane	CCl_4	-22.8	14.2	
Tribromomethane	$CHBr_3$	8.69	31.5	
Trichloromethane	$CHCl_3$	-63.3	-45.2	
Water	H_2O	0.00	-9.0	
o-Xylene	C_8H_{10}	-25.16	-3.5	
m-Xylene	C_8H_{10}	-47.85	-25.2	
p-Xylene	C_8H_{10}	13.3	46.0	

CRITICAL CONSTANTS OF ORGANIC COMPOUNDS

Chris D. Muzny, Vladimir Diky, Andrei Kazakov, Robert D. Chirico, and Michael Frenkel

The parameters of the liquid-gas critical point are important constants in determining the behavior of fluids. This table lists the critical temperature, pressure, and molar volume, as well as the normal boiling point, for over 850 organic substances. The properties and their units are:

T_b: Normal boiling point in K at a pressure of 101.325 kPa (1 atmosphere); "sp" following the value indicates a sublimation point (temperature at which the solid is in equilibrium with the gas at a pressure of 101.325 kPa)

T_c: Critical temperature in K

P_c: Critical pressure in MPa

V_c: Critical molar volume in cm^3 mol^{-1}

The listed values of the critical constants are critically evaluated using the NIST ThermoData Engine, TDE (Ref. 1), designed to implement the dynamic data evaluation concept (Refs. 2–5). This concept requires large electronic databases capable of storing essentially all relevant experimental data known to date with detailed descriptions of metadata and uncertainties. The combination of these electronic databases with expert-system software, designed to automatically generate recommended property values based on available experimental and predicted data, leads to the ability to produce critically evaluated data dynamically or "to order." The evaluated data have been generated only for compounds for which experimental data for critical properties are available. Group contribution methods such as Joback-Reed (Ref. 6), Constantinou-Gani (Ref. 7), Marrero-Pardillo (Ref. 8), and Wilson-Jasperson (Ref. 9) as well as quantitative structure-property relationship (QSPR) methods (Ref. 5) were used within the TDE environment to validate available experimental data. Each recommended value in the table is characterized with a combined expanded uncertainty (Ref. 10) (level of confidence, approximately 95%) listed in parentheses. Only references to original experimental data actually used by TDE to generate critically evaluated data are indicated for each compound. Compounds are listed alphabetically by name.

The values of the normal boiling temperatures provided in the table along with the combined expanded uncertainities listed in parentheses have also been critically evaluated using TDE. Additional details on the determination of the normal boiling temperatures using TDE can be found in the "Physical Constants of Organic Compounds" table in Section 3. The remaining values of the normal boiling temperatures (without uncertainities) are taken from the same table.

Critical Constants

Name	Mol. form.	T_b/K	T_c/K	P_c/MPa	V_c/cm³ mol⁻¹	Ref. (see text)
Acetaldehyde	C_2H_4O	294.0(6)	462(8)	7.5(10)	154(5)	11-13
Acetic acid	$C_2H_4O_2$	391.1(2)	593(2)	5.79(3)	171(2)	14-21
Acetic anhydride	$C_4H_6O_3$	412.7(3)	606(1)	4.00(8)	294(12)	22
Acetone	C_3H_6O	329.23(7)	508.1(2)	4.7(1)	221(20)	18, 23-31
Acetonitrile	C_2H_3N	354.8(2)	545.47(7)	4.88(5)	173(59)	32-40
Acetophenone	C_8H_8O	475.3(2)	709.5(7)	4.01(5)	373(40)	11, 41
Acetylene	C_2H_2	188.5 sp	308.4(4)	6.24(4)	119(11)	42-48
Acrylonitrile	C_3H_3N	350.4(2)	540(2)	4.6(1)	211(10)	49
Allene	C_3H_4	238.4(3)	394(4)	6.5(7)	167(8)	50
Allyl alcohol	C_3H_6O	370.1(5)	539.8(6)	5.76(4)	222(9)	51
Allylamine	C_3H_7N	327(2)	540.0(7)	4.83(3)	217(11)	52
Allyl ethyl ether	$C_5H_{10}O$	338(4)	518(10)		320(10)	21
2-Aminobiphenyl	$C_{12}H_{11}N$	571.5(2)	838(2)	3.52(3)	548(99)	53
2-(2-Aminoethoxy)ethanol	$C_4H_{11}NO_2$	496.3(1)	721(4)	4.88(10)	333(19)	54
N-(2-Aminoethyl)ethanolamine	$C_4H_{12}N_2O$	515(5)	739(2)	4.53(9)	340(17)	55
Aniline	C_6H_7N	457.3(4)	704(7)	5.3(1)	291(3)	39, 40, 58-60
Anisole	C_7H_8O	426.8(2)	646.1(2)	4.2(1)	355(12)	25, 39, 40, 49
Benzene	C_6H_6	353.23(7)	562.0(1)	4.90(2)	257(11)	19, 27, 28, 34, 61-97
Benzeneacetic acid	$C_8H_8O_2$	541(2)	766(8)	3.9(3)	372(16)	98
Benzenebutanoic acid	$C_{10}H_{12}O_2$	569(2)	783(8)	3.2(2)	493(18)	98
Benzenebutanol	$C_{10}H_{14}O$	537(3)	746(14)	3.1(2)	493(19)	584
Benzeneethanol	$C_8H_{10}O$	493(3)	724(4)	4.0(2)	390(15)	99
Benzeneheptanoic acid	$C_{13}H_{18}O_2$	585	798(8)	2.5(3)	662(21)	98
Benzenehexanoic acid	$C_{12}H_{16}O_2$	574	794(8)	2.6(3)	611(20)	98
Benzenepentanoic acid	$C_{11}H_{14}O_2$	583(1)	790(8)	3.1(2)	526(19)	98
Benzenepropanoic acid	$C_9H_{10}O_2$	557(2)	776(8)	3.5(2)	440(17)	98
Benzenepropanol	$C_9H_{12}O$	514(4)	732(14)	3.4(5)	450(14)	584
Benzonitrile	C_7H_5N	464(1)	691(9)	4.22(4)	348(7)	39, 40, 101, 102
Benzophenone	$C_{13}H_{10}O$	579.1(2)	830(2)	3.0(1)	568(44)	101
Benzo[*b*]thiophene	C_8H_6S	494.1(4)	764(2)	4.68(4)	359(59)	100

Fluids

Name	Mol. form.	T_b/K	T_c/K	P_c/MPa	V_c/cm³ mol⁻¹	Ref. (see text)
Benzyl alcohol	C_7H_8O	478.5(2)	715(3)	4.3(2)	333(13)	103
[1,1'-Bicyclohexyl]-2-one	$C_{12}H_{20}O$	537	787(70)		584(20)	104
1,1'-Bicyclopentyl	$C_{10}H_{18}$	463.61(3)	690(2)	3.27(3)	497(41)	105
Biphenyl	$C_{12}H_{10}$	528.4(3)	773(5)	3.43(6)	481(69)	39, 106-109
Bis(2-aminoethyl)amine	$C_4H_{13}N_3$	479.7(3)	710(2)	4.43(7)	350(21)	55
1,1-Bis(difluoromethoxy)-1,2,2,2-tetrafluoroethane	$C_4H_2F_8O_2$	319.78	450(1)	2.40(8)	410(4)	110
Bis(difluoromethyl) ether	$C_2H_2F_4O$	278.7(4)	420.2(1)	4.16(6)	223(13)	111
Bis(2-ethylhexyl) phthalate	$C_{24}H_{38}O_4$	657	835(9)	1.1(2)	1495(27)	112
Bis(2-hydroxyethyl)methylamine	$C_5H_{13}NO_2$	518(1)	742(4)	4.2(4)	404(17)	54
Bis(2,2,2-trifluoroethyl) ether	$C_4H_4F_6O$	336.91	476.31(9)	2.78(5)	365(5)	113
Bis(trimethylsilyl)methane	$C_7H_{20}Si_2$	406	573.9(3)			114
Bromochlorodifluoromethane	$CBrClF_2$	269.3(7)	428(12)	4.31(5)	229(16)	115
Bromodifluoromethane	$CHBrF_2$	257.6(5)	412.0(3)	5.2(1)	173(18)	116
Bromoethane	C_2H_5Br	311.4(6)	503.9(4)	6.2(1)	214(10)	117, 118
1-Bromo-2-fluorobenzene	C_6H_4BrF	427	669.6(6)	4.3(6)	342(18)	119
1-Bromo-3-fluorobenzene	C_6H_4BrF	423	652.0(4)	4.2(6)	337(18)	119
1-Bromo-4-fluorobenzene	C_6H_4BrF	423(2)	654.8(4)	4.2(2)	338(18)	119
1-Bromopropane	C_3H_7Br	344.0(2)	536.9(1)	4.33(6)	271(6)	120
Bromotrifluoromethane	$CBrF_3$	215.4(4)	340.06(5)	3.96(1)	199(6)	121
1-Bromo-2-(trifluoromethyl)benzene	$C_7H_4BrF_3$	440.7	656.5(4)	3.3(8)	415(24)	119
1-Bromo-3-(trifluoromethyl)benzene	$C_7H_4BrF_3$	424.7	627.1(4)	3.2(7)	413(24)	119
1-Bromo-4-(trifluoromethyl)benzene	$C_7H_4BrF_3$	433	629.8(4)	3.2(8)	413(24)	119
1,3-Butadiene	C_4H_6	268.6(2)	425(1)	4.35(7)	221(23)	122, 123
Butanal	C_4H_8O	348.0(2)	537(2)		258(9)	104, 124
Butane	C_4H_{10}	272.7(5)	425.2(1)	3.79(1)	257(4)	29, 125-139
1,4-Butanediamine	$C_4H_{12}N_2$	429(10)	651(7)	4.5(5)	317(14)	140
1,2-Butanediol, (±)-	$C_4H_{10}O_2$	469.57(6)	680(2)	5.4(1)	298(12)	141
1,3-Butanediol	$C_4H_{10}O_2$	481.4(1)	679(17)	4.7(1)	302(67)	54, 141
1,4-Butanediol	$C_4H_{10}O_2$	502.7(4)	724(4)	5.5(2)	307(14)	49, 55
Butanenitrile	C_4H_7N	390.8(4)	585.40(7)	3.82(5)	265(6)	35, 39, 40, 142
1-Butanethiol	$C_4H_{10}S$	371.6(5)	570.1(6)	4.01(2)	324(12)	143, 144
Butanoic acid	$C_4H_8O_2$	436.9(1)	623(6)	4.0(3)	292(10)	145-147
1-Butanol	$C_4H_{10}O$	390.8(2)	563.0(4)	4.43(7)	280(14)	34, 80, 148-155
2-Butanol	$C_4H_{10}O$	372.6(2)	535(4)	4.2(1)	269(4)	148, 149, 153, 156
2-Butanone	C_4H_8O	352.8(2)	537(1)	4.18(2)	274(30)	25, 30, 33, 34, 38, 157
1-Butene	C_4H_8	266.9(2)	419.3(1)	4.00(5)	236(14)	123, 158-161
cis-2-Butene	C_4H_8	276.87(8)	435.7(2)	4.23(2)	235(4)	86, 123, 158
trans-2-Butene	C_4H_8	274.03(9)	428.6(1)	4.03(2)	238(4)	86, 123, 158
2-Butoxyethanol	$C_6H_{14}O_2$	444(2)	633.9(10)	3.3(1)	424(15)	11, 41
1-*tert*-Butoxy-2-ethoxyethane	$C_8H_{18}O_2$	421.3	585(3)	2.5(4)	546(14)	162
2-Butoxyethyl acetate	$C_8H_{16}O_3$	464.3(9)	640(2)	2.7(2)	551(21)	144, 163
1-*tert*-Butoxy-2-methoxyethane	$C_7H_{16}O_2$	404	574(1)	2.8(7)	480(13)	162
1-Butoxy-2-propanol	$C_7H_{16}O_2$	445(3)	625(1)	2.7(1)	479(20)	32
Butyl acetate	$C_6H_{12}O_2$	399.2(1)	578(10)	3.16(6)	403(6)	162, 164-167
sec-Butyl acetate	$C_6H_{12}O_2$	381(4)	571.1(5)	3.01(10)	398(14)	164, 165
tert-Butyl acetate	$C_6H_{12}O_2$	371.1(10)	541(4)	3.0(1)	399(7)	54
Butyl acrylate	$C_7H_{12}O_2$	419.8(6)	597.4(6)	2.76(3)	445(7)	51
Butylamine	$C_4H_{11}N$	350.2(2)	531.9(2)	4.20(4)	291(13)	168
sec-Butylamine	$C_4H_{11}N$	335.86(8)	514.3(2)	4.0(2)	284(11)	168
tert-Butylamine	$C_4H_{11}N$	317.17(7)	483.7(6)	3.85(6)	293(23)	75
Butylbenzene	$C_{10}H_{14}$	456.5(3)	660.5(1)	2.89(3)	498(18)	79, 86, 88
sec-Butylbenzene, (±)-	$C_{10}H_{14}$	446.5(4)	652(1)	2.94(3)	488(39)	58
tert-Butylbenzene	$C_{10}H_{14}$	442.3(3)	648(1)	3.00(3)	474(30)	58, 64
Butyl benzoate	$C_{11}H_{14}O_2$	522(3)	725(14)	2.4(3)	594(10)	169
Butyl butanoate	$C_8H_{16}O_2$	438.10(1)	612(3)	2.4(2)	550(9)	162
Butylcyclohexane	$C_{10}H_{20}$	454.1(6)	653.1(4)	2.57(7)	547(14)	170, 171
tert-Butylcyclohexane	$C_{10}H_{20}$	444.8(4)	652.0(4)	2.82(9)	537(15)	170
tert-Butyl ethyl ether	$C_6H_{14}O$	345.9(1)	509(2)	3.0(2)	394(4)	172
Butyl methyl ether	$C_5H_{12}O$	343.3(3)	512.7(1)	3.37(2)	340(2)	25, 173, 174
Butyl propanoate	$C_7H_{14}O_2$	418.3(1)	594(1)	2.8(2)	464(10)	162

Fluids

Name	Mol. form.	T_b/K	T_c/K	P_c/MPa	V_c/cm³ mol⁻¹	Ref. (see text)
Butyl silicate	$C_{16}H_{36}O_4Si$	529	682(14)			57
Butyl vinyl ether	$C_6H_{12}O$	367(1)	540(1)	3.12(5)	379(10)	175
γ-Butyrolactone	$C_4H_6O_2$	477.8(4)	731(1)	5(1)	246(18)	49
Chlorobenzene	C_6H_5Cl	404.8(2)	632.4(1)	4.5(1)		19, 33, 34
1-Chlorobutane	C_4H_9Cl	351.6(2)	539.2(6)	4.1(2)	303(11)	119
2-Chlorobutane	C_4H_9Cl	344(8)	518.6(6)	3.4(2)	307(14)	119
1-Chloro-2,4-difluorobenzene	$C_6H_3ClF_2$	400	609.6(4)	4.0(7)	333(16)	119
1-Chloro-2,5-difluorobenzene	$C_6H_3ClF_2$	401	612.5(4)	4.0(7)	333(16)	119
1-Chloro-3,4-difluorobenzene	$C_6H_3ClF_2$	400	609.2(4)	4.0(6)	333(16)	119
1-Chloro-3,5-difluorobenzene	$C_6H_3ClF_2$	391.8	592.0(4)	3.9(7)	327(16)	119
1-Chloro-1,1-difluoroethane	$C_2H_3ClF_2$	264.03(7)	410.31(5)	4.06(3)	230(6)	113, 176-178
1-Chloro-2,2-difluoroethene	C_2HClF_2	254.4(5)	400.5(7)	4.54(7)	197(6)	179
Chlorodifluoromethane	$CHClF_2$	232.4(5)	369.30(5)	4.98(1)	165(2)	180-191
2-Chloro-2-(difluoromethoxy)-1,1,1-trifluoroethane	$C_3H_2ClF_5O$	322.5	467.8(6)	3.05(3)	316(24)	192
Chloroethane	C_2H_5Cl	285.5(2)	460.3(4)	5.24(4)	198(11)	193
Chloroethene	C_2H_3Cl	259.4(3)	425(5)	5.60(3)	171(9)	194
1-Chloro-2-fluorobenzene	C_6H_4ClF	410.8	633.8(4)	4.3(6)	319(21)	119
1-Chloro-3-fluorobenzene	C_6H_4ClF	401(25)	615.9(4)	4.2(6)	324(21)	119
1-Chloro-4-fluorobenzene	C_6H_4ClF	403	620.1(4)	4.2(4)	322(18)	119
1-Chloroheptane	$C_7H_{15}Cl$	432(2)	614(8)	3.1(6)	492(14)	119
1-Chlorohexane	$C_6H_{13}Cl$	408.2(5)	599(3)	3.3(3)	422(12)	119
Chloromethane	CH_3Cl	249.1(3)	416.24(4)	6.72(3)	136(2)	195, 196
2-Chloro-2-methylbutane	$C_5H_{11}Cl$	358(1)	509.1(6)	3.2(5)	397(15)	119
3-Chloro-3-methylpentane	$C_6H_{13}Cl$	389	528(3)		414(14)	119
2-Chloro-2-methylpropane	C_4H_9Cl	324.1(5)	497.8(1)	3.7(4)	308(13)	34
1-Chlorooctane	$C_8H_{17}Cl$	456(3)	643(2)	2.5(4)	543(13)	119
Chloropentafluoroacetone	C_3ClF_5O	281.0(9)	410.6(1)	2.89(1)	277(21)	197
Chloropentafluorobenzene	C_6ClF_5	391.11	570(1)	3.2(2)	367(25)	198
Chloropentafluoroethane	C_2ClF_5	234.0(2)	353.0(2)	3.141(10)	255(4)	199, 200
1-Chloropentane	$C_5H_{11}Cl$	381.1(3)	571.2(4)	3.3(2)	361(12)	119
1-Chloropropane	C_3H_7Cl	319.4(5)	503.3(4)	4.56(4)	268(24)	118, 193, 201, 202
2-Chloropropane	C_3H_7Cl	308.2(6)	482.4(4)	4.25(4)	245(16)	201, 202
1-Chloro-1,2,2,2-tetrafluoroethane	C_2HClF_4	261.19(9)	395.43(6)	3.62(1)	244(4)	203, 204
4-Chlorotoluene	C_7H_7Cl	435.0(2)	615.9(5)	2.33(9)	377(16)	88
2-Chloro-1,1,1-trifluoroethane	$C_2H_2ClF_3$	279.2(6)	425.0(2)	4.02(2)	232(6)	205
Chlorotrifluoroethene	C_2ClF_3	244.9(3)	380.1(1)	3.95(3)	214(12)	206, 207
2-Chloro-1,1,2-trifluoroethyl difluoromethyl ether	$C_3H_2ClF_5O$	330.0	475.0(6)	2.98(3)	343(25)	192
Chlorotrifluoromethane	$CClF_3$	191.67	301.9(2)	3.89(1)	180.3(1)	191, 208-216
o-Cresol	C_7H_8O	464.2(1)	697.6(2)	4.2(2)	336(12)	217, 218
m-Cresol	C_7H_8O	475.4(1)	705.8(4)	4.4(2)	337(12)	39, 40, 217, 218
p-Cresol	C_7H_8O	475.1(1)	704.6(3)	4.1(2)	349(13)	217, 218
Cyanogen	C_2N_2	252.1	397(3)	6.2(4)	149(8)	219
Cycloheptane	C_7H_{14}	392.0(2)	604.2(1)	3.85(4)	361(12)	34, 220, 221
Cyclohexane	C_6H_{12}	353.9(7)	553.4(3)	4.07(1)	307(12)	19, 34, 78, 82, 84, 88, 90, 144, 163, 170, 171, 220, 222-231
Cyclohexanol	$C_6H_{12}O$	434.1(2)	647.1(3)	4.3(1)	334(33)	49, 232, 233
Cyclohexanone	$C_6H_{10}O$	428.6(1)	665(1)	4.61(9)	354(12)	162, 166
Cyclohexene	C_6H_{10}	356.1(2)	560.45(5)	4.43(8)	290(20)	86, 201, 202
Cyclohexylamine	$C_6H_{13}N$	406.8(5)	626.8(9)	3.9(6)	349(15)	170
Cyclooctane	C_8H_{16}	424.3(1)	647.2(4)	3.55(6)	417(28)	34, 170, 220, 221
Cyclopentane	C_5H_{10}	322.4(1)	511.7(2)	4.51(7)	264(10)	34, 78, 90, 234-236
Cyclopentanol	$C_5H_{10}O$	413.6(2)	619(1)	4.9(1)	288(19)	233
Cyclopentanone	C_5H_8O	403.7(2)	624(2)	4.59(5)	276(17)	233
Cyclopentene	C_5H_8	317.4(2)	506.1(2)	4.78(5)	252(16)	90, 144, 163, 201, 202
Cyclopropane	C_3H_6	242(2)	398.2(4)	5.58(2)	164(9)	237, 238
2,2',3,3',4,4',5,5',6,6'-Decafluoro-1,1'-biphenyl	$C_{12}F_{10}$	480(2)	640(4)	2.3(3)	641(40)	239
cis-Decahydronaphthalene	$C_{10}H_{18}$	469.0(3)	702(1)	3.2(3)	492(19)	241
trans-Decahydronaphthalene	$C_{10}H_{18}$	460.5(2)	687(1)	3.1(1)	499(19)	241
Decamethylcyclopentasiloxane	$C_{10}H_{30}O_5Si_5$	486(3)	617.4(3)	1.04(2)	1201(8)	34, 114

Name	Mol. form.	T_b/K	T_c/K	P_c/MPa	V_c/cm^3 mol^{-1}	Ref. (see text)
Decanal	$C_{10}H_{20}O$	485(3)	674(1)	2.6(3)	601(14)	163, 242
Decane	$C_{10}H_{22}$	447.3(1)	618.1(9)	2.10(3)	621(35)	16, 33, 65, 78, 86, 131, 243-252
1,10-Decanediamine	$C_{10}H_{24}N_2$	535(12)	736(8)	2.4(3)	654(28)	140
Decanedioic acid	$C_{10}H_{18}O_4$	647(5)	845(13)	2.5(1)	724(21)	253
Decanoic acid	$C_{10}H_{20}O_2$	543(1)	724(5)		638(24)	145, 146
1-Decanol	$C_{10}H_{22}O$	502(3)	690(10)	2.3(1)	624(87)	149, 152, 254, 255
2-Decanol	$C_{10}H_{22}O$	484	668.5(3)	2.3(5)	646(13)	255
3-Decanol	$C_{10}H_{22}O$	490(7)	666.1(3)	2.3(3)	643(13)	255
4-Decanol	$C_{10}H_{22}O$	487(3)	663.7(3)	2.3(1)	643(13)	255
5-Decanol	$C_{10}H_{22}O$	489(5)	663.2(4)	2.3(4)	646(13)	255
2-Decanone	$C_{10}H_{20}O$	484(3)	671.8(5)	2.2(3)	625(25)	256
3-Decanone	$C_{10}H_{20}O$	485(4)	668(1)	2.2(2)	628(15)	256
4-Decanone	$C_{10}H_{20}O$	479.7	662.9(5)	2.2(2)	636(18)	256
5-Decanone	$C_{10}H_{20}O$	477	661.0(4)	2.2(2)	628(25)	256
1-Decene	$C_{10}H_{20}$	444(1)	616.0(3)	2.16(5)	594(3)	257
Decylbenzene	$C_{16}H_{26}$	571(1)	752(8)	1.72(10)	879(29)	258
Decyl silicate	$C_{40}H_{84}O_4Si$		849(16)			57
Dibenzofuran	$C_{12}H_8O$	558.4(3)	824(2)	3.37(3)	494(32)	259
Dibenzothiophene	$C_{12}H_8S$	604.8(4)	897(2)	3.9(2)		260
1,2-Dibromo-1-chloro-1,2,2-trifluoroethane	$C_2Br_2ClF_3$	366.0(2)	560.6(2)	3.61(2)	368(4)	261
1,4-Dibromooctafluorobutane	$C_4Br_2F_8$	371(25)	532(2)	2.4(3)	452(29)	198
Dibutylamine	$C_8H_{19}N$	435(2)	607.5(2)	3.11(3)	532(21)	168
1,4-Di-*tert*-butylbenzene	$C_{14}H_{22}$	510.5(5)	708(2)	2.23(1)	732(70)	232
Dibutyl ether	$C_8H_{18}O$	414.8(3)	584.1(2)	2.4(2)	521(12)	262
Dibutyl phthalate	$C_{16}H_{22}O_4$	611(9)	797(9)	1.6(3)	954(18)	112
m-Dichlorobenzene	$C_6H_4Cl_2$	445(2)	685.7(4)	4.2(2)	366(22)	119
p-Dichlorobenzene	$C_6H_4Cl_2$	447.1(2)	669(5)	3.54(7)	364(22)	263
Dichlorodiethylsilane	$C_4H_{10}Cl_2Si$	403(2)	595.7(6)	3.06(3)	455(4)	264
Dichlorodifluoromethane	CCl_2F_2	243.4(1)	384.9(2)	4.12(1)	218(36)	191, 216, 265
Dichlorodimethylsilane	$C_2H_6Cl_2Si$	343.7(5)	520.3(6)	3.49(3)	350(5)	266
1,1-Dichloroethane	$C_2H_4Cl_2$	329.5(7)	523.4(1)	5.1(5)	248(12)	267
1,2-Dichloroethane	$C_2H_4Cl_2$	356.6(1)	561.5(4)	5.4(1)	225(8)	33, 34, 268, 269
cis-1,2-Dichloroethene	$C_2H_2Cl_2$	333(2)	535.8(4)	5.4(3)	220(15)	34
trans-1,2-Dichloroethene	$C_2H_2Cl_2$	320.79(8)	515.5(2)	5.3(2)	216(14)	33, 34
1,1-Dichloro-1-fluoroethane	$C_2H_3Cl_2F$	305.20(9)	477.3(1)	4.20(2)	253.7(6)	178, 270
Dichlorofluoromethane	$CHCl_2F$	282.1	451.6(4)	5.20(1)	196(1)	271
1,2-Dichloro-1,1,2,3,3,3-hexafluoropropane	$C_3Cl_2F_6$	307(2)	451.8(1)	2.63(7)	365(48)	272
Dichloromethane	CH_2Cl_2	313.0(3)	508.0(2)	6.35(5)	177(13)	273
1,2-Dichloropropane, (±)-	$C_3H_6Cl_2$	369.6	578(2)	4.63(6)	292(6)	119, 232
1,3-Dichloropropane	$C_3H_6Cl_2$	394.0(3)	615(3)	4.7(5)	299(16)	119
1,1-Dichloro-1,2,2,2-tetrafluoroethane	$C_2Cl_2F_4$	276(1)	418.6(8)	3.31(3)	294(8)	179
1,2-Dichloro-1,1,2,2-tetrafluoroethane	$C_2Cl_2F_4$	276.8(5)	418.74(6)	3.25(2)	295(3)	191, 274
1,2-Dichloro-1,1,2-trifluoroethane	$C_2HCl_2F_3$	303.2(1)	461.6(1)	3.77(8)	283.2(5)	178
2,2-Dichloro-1,1,1-trifluoroethane	$C_2HCl_2F_3$	301.0(6)	456.8(2)	3.67(1)	278(2)	182, 275-280
Didecyl phthalate	$C_{28}H_{46}O_4$	736(4)	870(10)	0.94(5)	1807(27)	112
1,1-Diethoxyethane	$C_6H_{14}O_2$	375(2)	539.7(4)	3.22(8)	426(12)	166
1,2-Diethoxyethane	$C_6H_{14}O_2$	393.8(7)	542(3)	2.14(2)	432(11)	162
Diethoxymethane	$C_5H_{12}O_2$	359(2)	532(1)	3.4(5)	370(10)	162
Diethylamine	$C_4H_{11}N$	328.6(1)	499.5(4)	3.75(2)	304(32)	38, 117, 133, 281, 282
p-Diethylbenzene	$C_{10}H_{14}$	457(1)	657.90(5)	2.80(8)	494(12)	79, 86
Diethylene glycol	$C_4H_{10}O_3$	518.7(2)	753(4)	4.8(2)	325(19)	283
Diethylene glycol diethyl ether	$C_8H_{18}O_3$	458(4)	612(10)	2.4(7)	587(18)	162
Diethylene glycol dimethyl ether	$C_6H_{14}O_3$	435(2)	617(4)	3.0(6)	450(16)	162
Diethylene glycol monobutyl ether	$C_8H_{18}O_3$	505(4)	692(3)	2.8(6)	546(26)	41
Diethylene glycol monobutyl ether acetate	$C_{10}H_{20}O_4$	521(2)	694(2)	2.15(5)	627(20)	55
Diethylene glycol monoethyl ether	$C_6H_{14}O_3$	475(3)	670(4)	3.2(1)	427(23)	49
Diethylene glycol monoethyl ether acetate	$C_8H_{16}O_4$	491(1)	670(12)	2.50(6)	524(18)	49, 55
Diethylene glycol monomethyl ether	$C_5H_{12}O_3$	467(2)	672(2)	3.7(2)	378(25)	49
Diethylene glycol monopropyl ether	$C_7H_{16}O_3$	488.0(4)	680(2)	3.05(7)	495(18)	41, 104
Diethyl ether	$C_4H_{10}O$	307.6(5)	466.8(3)	3.64(1)	280(5)	19, 77, 95, 155, 196, 284-301

Name	Mol. form.	T_b/K	T_c/K	P_c/MPa	V_c/cm^3 mol^{-1}	Ref. (see text)
Diethyl oxalate	$C_6H_{10}O_4$	459(1)	618(2)	2.14(2)	464(37)	101
Diethyl phthalate	$C_{12}H_{14}O_4$	571(2)	776(9)	2.2(2)	687(15)	112
Diethyl succinate	$C_8H_{14}O_4$	490(1)	663(30)	2.26(2)	567(44)	101, 302
Diethyl sulfide	$C_4H_{10}S$	365.3(2)	557.5(10)	4.0(1)	322(8)	32, 303, 304
o-Difluorobenzene	$C_6H_4F_2$	367.1(5)	566.0(4)	4.28(7)	290(21)	119
m-Difluorobenzene	$C_6H_4F_2$	356.2(5)	548.4(4)	4.20(7)	289(21)	119
p-Difluorobenzene	$C_6H_4F_2$	362.1(3)	556.9(4)	4.28(7)	297(22)	119
1,1-Difluoroethane	$C_2H_4F_2$	249.13	386.4(1)	4.52(1)	178(2)	178, 180, 185, 305-307
1,1-Difluoroethene	$C_2H_2F_2$	187.7(8)	302.9(6)	4.48(5)	155(4)	179, 215
2,2-Difluoroethylbis(trifluoromethyl)amine	$C_4H_3F_8N$	324.6	460.20(9)	2.64(1)	375(1)	308
Difluoromethane	CH_2F_2	221.50(7)	351.28(3)	5.79(1)	121(4)	306, 309-316
3-Difluoromethoxy-1,1,1,2,2-pentafluoropropane	$C_4H_3F_7O$	319.09	455.1(1)	2.77(2)	363(1)	113
2-(Difluoromethoxy)-1,1,1-trifluoroethane	$C_3H_3F_5O$	302.4(2)	444.9(3)	3.43(1)	291(19)	113
2,4-Difluorotoluene	$C_7H_6F_2$	390	581.4(4)	3.7(4)	340(21)	119
2,5-Difluorotoluene	$C_7H_6F_2$	391	587.8(4)	3.8(5)	341(21)	119
2,6-Difluorotoluene	$C_7H_6F_2$	385	581.8(4)	3.7(4)	341(21)	119
3,4-Difluorotoluene	$C_7H_6F_2$	385	598.5(5)	3.8(6)	342(22)	119
Diheptyl phthalate	$C_{22}H_{34}O_4$	633	830(9)	1.24(8)	1153(22)	112
Dihexyl phthalate	$C_{20}H_{30}O_4$	652(5)	817(9)	1.3(1)	1061(22)	112
3,4-Dihydro-2*H*-pyran	C_5H_8O	358.7(2)	561(2)	4.63(8)	268(34)	75
Diisobutylamine	$C_8H_{19}N$	412.8	584.4(2)	3.20(6)	518(22)	168
Diisopropylamine	$C_6H_{15}N$	357(3)	523.1(2)	3.02(2)	407(18)	168
1,4-Diisopropylbenzene	$C_{12}H_{18}$	483.5(2)	675(1)	2.30(4)	610(65)	317
Diisopropyl ether	$C_6H_{14}O$	341.6(2)	500.2(7)	2.85(4)	386(5)	25, 290, 318, 319
1,2-Dimethoxyethane	$C_4H_{10}O_2$	358.2(1)	539(4)	3.91(5)	305(9)	162, 166, 175, 290
Dimethoxymethane	$C_3H_8O_2$	315.5(2)	488(11)	4.0(2)	259(10)	75, 162, 320
1,2-Dimethoxypropane	$C_5H_{12}O_2$	369	543(1)	3.4(6)	356(12)	162
2,2-Dimethoxypropane	$C_5H_{12}O_2$	350.6(7)	510(3)		360(13)	162
Dimethylamine	C_2H_7N	280.5(4)	437.5(4)	5.34(5)	188(13)	193, 322, 323
N,N-Dimethylaniline	$C_8H_{11}N$	466(1)	687.7(6)	3.63(9)	407(15)	39, 40
2,2-Dimethylbutane	C_6H_{14}	322.9(2)	489.1(5)	3.10(1)	364(1)	86, 324-328
2,3-Dimethylbutane	C_6H_{14}	331.2(3)	500.2(3)	3.13(1)	358(1)	19, 86, 252, 324-330
3,3-Dimethyl-2-butanone	$C_6H_{12}O$	379.3(2)	570.9(3)	3.67(3)	383(6)	164, 165
2,3-Dimethyl-1-butene	C_6H_{12}	328.74(4)	497.7(9)	3.31(1)	346(9)	170
3,3-Dimethyl-1-butene	C_6H_{12}	314.39(4)	477.4(9)	3.18(2)	348(10)	170
2,3-Dimethyl-2-butene	C_6H_{12}	346.34(6)	521.0(9)	3.4(1)	344(7)	170
Dimethyl carbonate	$C_3H_6O_3$	363.26(9)	557(1)	4.8(2)		331, 332
cis-1,3-Dimethylcyclohexane	C_8H_{16}	397.6(6)	587.7(5)	2.88(1)	429(10)	64
cis-1,4-Dimethylcyclohexane	C_8H_{16}	397.5(7)	603.2(3)	3.44(2)	434(7)	164, 165
trans-1,4-Dimethylcyclohexane	C_8H_{16}	392.5(5)	588(2)	3.04(1)	439(18)	74
Dimethyl disulfide	$C_2H_6S_2$	382.87(8)	608(4)	5.1(1)	266(8)	54
Dimethyl ether	C_2H_6O	248.4(2)	400.1(8)	5.31(3)	171(3)	125, 174, 186, 333-343
N,N-Dimethylformamide	C_3H_7NO	426.0(5)	649.6(8)	4.4(1)	262(9)	11, 344
Dimethyl glutarate	$C_7H_{12}O_4$	489(4)	682(14)	2.8(4)	488(14)	321
2,2-Dimethylheptane	C_9H_{20}	406(1)	576.7(5)	2.35(7)	546(12)	345
Dimethyl heptanedioate	$C_9H_{16}O_4$	518(2)	711(14)	2.4(1)	608(14)	321
2,2-Dimethylhexane	C_8H_{18}	380.0(4)	549.9(4)	2.53(3)	481(10)	346
2,3-Dimethylhexane	C_8H_{18}	388.8(5)	563.5(4)	2.63(2)	466(16)	346
2,4-Dimethylhexane	C_8H_{18}	382.6(4)	553(3)	2.55(2)	480(39)	330, 346
2,5-Dimethylhexane	C_8H_{18}	382.3(7)	550.0(3)	2.49(2)	485(20)	19, 346
3,3-Dimethylhexane	C_8H_{18}	385.1(6)	562.0(4)	2.65(2)	450(17)	346
3,4-Dimethylhexane	C_8H_{18}	390.9(4)	568.8(4)	2.69(2)	467(21)	346
Dimethyl 1,6-hexanedioate	$C_8H_{14}O_4$	504(3)	692(14)	2.5(5)	561(13)	321
Dimethyl malonate	$C_5H_8O_4$	454.3(6)	647(1)	3.5(1)	368(13)	317, 321
2,7-Dimethylnaphthalene	$C_{12}H_{12}$	535.6(3)	775(2)	3.02(3)	515(45)	347
Dimethyl octanedioate	$C_{10}H_{18}O_4$	532(8)	723(14)	2.3(3)	672(14)	321
Dimethyl oxalate	$C_4H_6O_4$	436.6(5)	632(14)	4.0(3)	315(12)	321, 348
2,2-Dimethyloxirane	C_4H_8O	324(2)	500(4)	4.4(1)	327(31)	54
2,2-Dimethylpentane	C_7H_{16}	352.4(3)	520.6(6)	2.77(4)	409(12)	346, 349
2,3-Dimethylpentane	C_7H_{16}	363.0(6)	537.5(6)	2.92(8)	394(23)	346, 349, 350

Name	Mol. form.	T_b/K	T_c/K	P_c/MPa	V_c/cm³ mol⁻¹	Ref. (see text)
2,4-Dimethylpentane	C_7H_{16}	353.6(5)	520.0(7)	2.74(6)	415(19)	346, 349
3,3-Dimethylpentane	C_7H_{16}	359.2(6)	536.4(4)	2.94(2)	413(24)	346
2,3-Dimethyl-1-pentene	C_7H_{14}	357(1)	534(4)	2.9(8)	400(11)	170
4,4-Dimethyl-1-pentene	C_7H_{14}	345.7(2)	516(4)	2.91(1)	406(11)	170
Dimethyl phthalate	$C_{10}H_{10}O_4$	555.9(2)	772(9)	2.76(8)	557(17)	112
2,3-Dimethylpyridine	C_7H_9N	434.3(4)	655.5(3)	4.03(1)	337(28)	86, 351
2,4-Dimethylpyridine	C_7H_9N	431.6(3)	647.1(9)	3.83(2)	363(34)	90, 352
2,5-Dimethylpyridine	C_7H_9N	430.15(5)	644.2(3)	4.11(4)	371(45)	86
2,6-Dimethylpyridine	C_7H_9N	417.2(1)	623.8(2)	3.80(4)	353(43)	90
3,4-Dimethylpyridine	C_7H_9N	452.3(3)	683.8(4)	4.06(1)	353(13)	86, 351, 352
3,5-Dimethylpyridine	C_7H_9N	445.1(1)	667.3(3)	3.84(2)	369(12)	86, 351
2,6-Dimethylquinoline	$C_{11}H_{11}N$	541.3(4)	786(2)	3.27(2)	505(24)	529
Dimethyl sebacate	$C_{12}H_{22}O_4$	562(3)	742(14)	2.1(2)	695(14)	321
Dimethyl succinate	$C_6H_{10}O_4$	470(1)	662(14)	3.5(2)	426(16)	321
Dimethyl sulfide	C_2H_6S	310.47(5)	503.0(3)	5.40(6)	201(9)	52, 117, 304
Dimethyl sulfoxide	C_2H_6OS	465.1(9)	707(1)	4.6(7)	228(7)	23
1,3-Dimethyl-1,1,3,3-tetraphenyldisiloxane	$C_{26}H_{26}OSi_2$	701	893(9)	1.38(10)	1300(89)	353
1,3-Dimethyltricyclo[3.3.1.1^{3,7}]decane	$C_{12}H_{20}$	476.53	708(2)	2.86(1)	595(115)	141
Dinonyl phthalate	$C_{26}H_{42}O_4$	686	858(9)	1.0(3)	1652(26)	112
Dioctyl phthalate	$C_{24}H_{38}O_4$	688(4)	840(9)	1.1(1)	1510(22)	112
1,4-Dioxane	$C_4H_8O_2$	374.4(3)	587.3(1)	5.2(2)	251(5)	34, 290, 354
Dipentyl phthalate	$C_{18}H_{26}O_4$	614(40)	811(9)		957(21)	112
Diphenyl ether	$C_{12}H_{10}O$	531.2(1)	766.9(8)	3.10(4)	526(23)	25, 355
Diphenylmethane	$C_{13}H_{12}$	537.4(3)	776(9)	3.02(7)		39, 302, 356, 357
Dipropylamine	$C_6H_{15}N$	380.7(9)	555.8(1)	3.6(1)	414(15)	168
Dipropylene glycol	$C_6H_{14}O_3$	504(2)	705(4)	3.4(1)	444(18)	54
Dipropyl ether	$C_6H_{14}O$	363.3(3)	531(2)	2.92(5)	402(8)	25, 52
Dipropyl phthalate	$C_{14}H_{18}O_4$	592(2)	784(9)	1.9(1)	816(20)	112
Diundecyl phthalate	$C_{30}H_{50}O_4$	711(25)	886(10)	0.89(10)	1590(25)	112
Docosane	$C_{22}H_{46}$	642(5)	786(6)	1.0(1)	1434(50)	62, 358, 359
Docosanoic acid	$C_{22}H_{44}O_2$	693(3)	837(8)	1.11(8)	1485(29)	360
1-Docosanol	$C_{22}H_{46}O$	680(12)	827(8)		1243(20)	361
1,2,2,3,3,4,4,5,5,6,6,7-Dodecafluoro-1-heptanol	$C_7H_4F_{12}O$	444.8	589(5)	2.0(2)	620(34)	362
Dodecane	$C_{12}H_{26}$	489.5(2)	658.8(9)	1.80(9)	747(18)	34, 86, 118, 243, 244, 363, 364
1,12-Dodecanediamine	$C_{12}H_{28}N_2$	572(13)	767(8)	2.0(3)	765(37)	140
Dodecanedioic acid	$C_{12}H_{22}O_4$	621(10)	859(13)	2.1(2)	730(19)	253
1-Dodecanethiol	$C_{12}H_{26}S$	550(3)	734(4)	1.81(10)	726(25)	99
Dodecanoic acid	$C_{12}H_{24}O_2$	572(1)	743(7)	1.9(2)	787(19)	360
1-Dodecanol	$C_{12}H_{26}O$	537.3(3)	719.4(6)	2.02(5)	805(15)	149
2-Dodecanone	$C_{12}H_{24}O$	520(6)	702(4)		742(20)	256
3-Dodecanone	$C_{12}H_{24}O$	523	701(2)	1.9(2)	680(19)	256
4-Dodecanone	$C_{12}H_{24}O$	528	697(2)	1.9(4)	672(19)	256
5-Dodecanone	$C_{12}H_{24}O$	521	695(5)	1.9(4)	678(19)	256
6-Dodecanone	$C_{12}H_{24}O$	522	694(2)	1.9(3)	677(19)	256
1-Dodecene	$C_{12}H_{24}$	486.6(9)	657.6(6)	1.88(1)	710(15)	257
Eicosane	$C_{20}H_{42}$	617.3(9)	768(6)	1.08(5)	1325(48)	358, 359
Eicosanoic acid	$C_{20}H_{40}O_2$	673(6)	820(8)	1.2(1)	1346(27)	360
1-Eicosanol	$C_{20}H_{42}O$	629	808(8)	1.1(2)	1130(18)	361
1-Eicosene	$C_{20}H_{40}$	614	772(15)	1.1(3)	1213(36)	365
Ethane	C_2H_6	184.6(4)	305.36(4)	4.88(1)	146(3)	44, 48, 92, 131, 132, 134, 214, 215, 229, 246, 366-393
1,2-Ethanediamine	$C_2H_8N_2$	390.1(5)	613.1(3)	6.71(4)	204(13)	49
1,2-Ethanediol	$C_2H_6O_2$	470.7(1)	719(5)	8.1(4)	180(11)	11, 32, 41, 394
1,1-Ethanediol, diacetate	$C_6H_{10}O_4$	441(3)	618(4)	2.9(1)	457(13)	54
Ethanethiol	C_2H_6S	308.2(1)	498.7(3)	5.53(8)	208(10)	304
Ethanol	C_2H_6O	351.39(9)	515(1)	6.25(4)	169(4)	19, 25, 34, 80, 85, 128, 148-150, 152, 222, 395-409
Ethanolamine	C_2H_7NO	443.5(4)	671(3)	8.0(5)	207(13)	41
Ethoxybenzene	$C_8H_{10}O$	443.0(2)	647(2)	3.45(5)	407(16)	39, 40
2-Ethoxyethyl acetate	$C_6H_{12}O_3$	429.8(4)	609(2)	3.07(3)	443(38)	141, 144, 163

Name	Mol. form.	T_b/K	T_c/K	P_c/MPa	V_c/cm³ mol⁻¹	Ref. (see text)
2-Ethoxy-2-methylbutane	$C_7H_{16}O$	374.7(4)	546(2)	2.83(9)	448(29)	410
1-Ethoxy-1,1,2,2,3,3,4,4,4-nonafluorobutane	$C_6H_5F_9O$	350.04	482.0(1)	1.98(1)	518(2)	113
Ethyl acetate	$C_4H_8O_2$	350.3(2)	523.27(7)	3.88(2)	288(19)	19, 23, 400, 411-413
Ethylamine	C_2H_7N	289.8(2)	456.5(9)	5.6(1)	183(22)	193, 414
Ethylbenzene	C_8H_{10}	409.4(4)	617.1(1)	3.61(1)	365(61)	68, 78, 79, 86, 88, 172, 345
Ethyl benzoate	$C_9H_{10}O_2$	485.7(2)	700(14)	3.01(5)	470(12)	169
Ethyl butanoate	$C_6H_{12}O_2$	394.3(4)	566.1(1)	3.2(3)	421(30)	415, 416
Ethyl *trans*-2-butenoate	$C_6H_{10}O_2$	413(5)	599(10)		382(7)	21
Ethylcyclohexane	C_8H_{16}	405.0(4)	606.9(4)	3.27(4)	431(12)	170, 171
Ethylcyclopentane	C_7H_{14}	376.7(6)	569.48(5)	3.40(7)	377(3)	236
Ethyl 2,2-dimethylpropanoate	$C_7H_{14}O_2$	391.5(4)	566(2)	2.88(2)	461(8)	417
Ethylene	C_2H_4	169.4(3)	282.35(3)	5.06(1)	130.9(2)	61, 246, 371, 418-436
Ethyl 3-ethoxypropanoate	$C_7H_{14}O_3$	441(2)	621(3)	2.7(1)	478(19)	11, 41
Ethyl formate	$C_3H_6O_2$	327.24(10)	508.5(5)	4.78(3)		413, 415
Ethyl heptanoate	$C_9H_{18}O_2$	461(2)	634(1)	2.3(2)	587(8)	162
3-Ethylhexane	C_8H_{18}	391.7(5)	565.5(4)	2.61(2)	450(15)	346
Ethyl hexanoate	$C_8H_{16}O_2$	438(1)	615(1)	2.6(2)	528(8)	162
2-Ethylhexanoic acid	$C_8H_{16}O_2$	500.7(1)	674(1)	2.75(6)	543(8)	144, 163
2-Ethyl-1-hexanol	$C_8H_{18}O$	459.4(2)	640.2(3)	3.0(2)	508(29)	437
2-Ethylhexyl acetate	$C_{10}H_{20}O_2$	473(1)	642(2)	2.02(1)	644(10)	410
Ethyl 3-methylbutanoate	$C_7H_{14}O_2$	408(3)	584(6)		463(9)	147, 162
Ethyl methyl ether	C_3H_8O	279(2)	437.8(2)	4.39(6)	219(5)	174, 304, 438
3-Ethyl-2-methylpentane	C_8H_{18}	388.8(6)	567.1(4)	2.70(2)	442(22)	346
3-Ethyl-3-methylpentane	C_8H_{18}	391.4(9)	576.5(4)	2.77(1)	463(13)	346
Ethyl 2-methylpropanoate	$C_6H_{12}O_2$	384(2)	554(4)	3.1(3)	421(76)	415
Ethyl methyl sulfide	C_3H_8S	339.8(3)	533(10)	4.62(3)	260(6)	303
Ethyl nonanoate	$C_{11}H_{22}O_2$	497(5)	664(1)	2.0(4)	715(8)	162
Ethyl octanoate	$C_{10}H_{20}O_2$	479(1)	652(12)		657(8)	147, 162
3-Ethylpentane	C_7H_{16}	366.6(4)	540.7(4)	2.90(3)	412(14)	346, 349
Ethyl pentanoate	$C_7H_{14}O_2$	415(3)	593(1)	2.8(4)	466(10)	162
2-Ethylphenol	$C_8H_{10}O$	477.7(1)	703(1)	3.7(3)	388(15)	218
3-Ethylphenol	$C_8H_{10}O$	491.6(1)	716(1)	3.8(3)	393(15)	218
4-Ethylphenol	$C_8H_{10}O$	491.12(6)	716(1)	3.1(6)	395(15)	218
Ethyl propanoate	$C_5H_{10}O_2$	372.1(2)	547(1)	3.37(5)	343(23)	19, 64, 162, 413, 415
Ethyl propyl ether	$C_5H_{12}O$	336(3)	500.2(4)	3.37(1)	343(44)	25, 304
Ethyl silicate	$C_8H_{20}O_4Si$	441(1)	587(12)	2.0(4)	701.3(2)	57
S-Ethyl thioacetate	C_4H_8OS	387(3)	590.5(2)	4.1(1)	320(10)	49
4-Ethyltoluene	C_9H_{12}	435.2(6)	640.2(5)	3.23(4)	446(12)	439, 440
Ethyl vinyl ether	C_4H_8O	309(2)	475(2)	4.06(4)	262(8)	290
Fluorobenzene	C_6H_5F	357.9(3)	560.10(7)	4.55(1)	272(9)	86, 441
Fluoroethane	C_2H_5F	235.5(3)	375.2(2)	5.02(1)	164(3)	206, 442, 443
Fluoromethane	CH_3F	194.84	317.42(1)	5.88(1)	112.4(1)	444, 445
2-Fluorotoluene	C_7H_7F	387(2)	591.2(4)	3.9(3)	323(47)	119
3-Fluorotoluene	C_7H_7F	389(2)	591.8(4)	3.9(3)	332(17)	119
4-Fluorotoluene	C_7H_7F	389.8(4)	592.1(8)	3.85(1)	332(17)	119
Formic acid	CH_2O_2	374	588(10)		115.9(1)	146
Furan	C_4H_4O	304.5(2)	490.2(2)	5.43(8)	218(3)	241, 290
Glycerol	$C_3H_8O_3$	562(3)	850(9)	7.6(8)	251(15)	394
Heneicosane	$C_{21}H_{44}$	632(6)	778(8)	1.0(1)	1366(48)	359
Heptadecane	$C_{17}H_{36}$	576(2)	736(1)	1.33(7)	1081(28)	359
Heptadecanoic acid	$C_{17}H_{34}O_2$	635(4)	792(8)	1.4(1)	1130(24)	360
1-Heptadecanol	$C_{17}H_{36}O$	597	780(8)	1.4(1)	1097(18)	361
1-Heptadecene	$C_{17}H_{34}$	574(3)	734(7)	1.34(8)	1053(35)	365
2,2,3,3,5,5,6-Heptafluoro-1,4-dioxane	$C_4HF_7O_2$	312.6	453(1)	2.86(6)	359(4)	110
1,1,1,2,2,3,3-Heptafluoropentan-4-one	$C_5H_3F_7O$	337.5	476.55(8)	2.57(1)	394.2(7)	308
1,1,1,2,3,3,3-Heptafluoropropane	C_3HF_7	256.81	375.0(1)	2.93(1)	299(7)	446-449
1,1,1,2,4,4,4-Heptafluoro-2-trifluoromethoxybutane	$C_5H_2F_{10}O$	322.73	447(1)	2.15(6)	465(5)	110
1,1,1,2,2,3,3-Heptafluoro-3-(trifluoromethoxy)propane	$C_4F_{10}O$	280.1	391.7(7)	1.89(5)	431(31)	450
2,2,4,4,6,8,8-Heptamethylnonane	$C_{16}H_{34}$	519(1)	692(4)	1.53(1)	957(27)	451

Name	Mol. form.	T_b/K	T_c/K	P_c/MPa	V_c/cm^3 mol^{-1}	Ref. (see text)
1,1,1,3,5,5,5-Heptamethyltrisiloxane	$C_7H_{22}O_2Si_3$	416(1)	553.4(6)	1.48(2)	828(3)	452
Heptanal	$C_7H_{14}O$	426(3)	616.8(4)	3.2(2)	434(7)	242
2-Heptanamine	$C_7H_{17}N$	414(4)	598(2)	2.9(3)	455(18)	170
Heptane	C_7H_{16}	371.53(7)	540.1(2)	2.74(1)	428(15)	34, 36, 62, 75, 78, 86, 131, 133-135, 157, 243, 244, 254, 286, 324, 346, 359, 394, 432, 453- 463
Heptanedioic acid	$C_7H_{12}O_4$	615.2	842(13)	3.3(2)	463(15)	253
Heptanoic acid	$C_7H_{14}O_2$	495(2)	678(2)	3.0(3)	476(5)	145, 146, 464
1-Heptanol	$C_7H_{16}O$	451(1)	632.4(6)	3.1(2)	430(9)	149, 254, 465
2-Heptanol, (±)-	$C_7H_{16}O$	432	608.4(6)	3.0(1)	442(2)	149, 465
3-Heptanol, (*S*)-	$C_7H_{16}O$	436(2)	605.4(3)	3.1(4)	451(3)	465
4-Heptanol	$C_7H_{16}O$	434(2)	602.6(3)	3.1(6)	455(4)	465
2-Heptanone	$C_7H_{14}O$	424.2(3)	611.4(2)	2.98(4)	436(4)	25, 172, 466
3-Heptanone	$C_7H_{14}O$	419(2)	606.6(2)	3.0(1)	433(5)	466
4-Heptanone	$C_7H_{14}O$	417(1)	602.0(2)	3.0(3)	434(5)	466
1-Heptene	C_7H_{14}	367(1)	537.3(3)	2.85(2)	409(2)	51, 52, 86, 124, 257, 467
cis-2-Heptene	C_7H_{14}	370(2)	548.5(6)	3.0(3)	410(7)	170
trans-2-Heptene	C_7H_{14}	371(2)	542.8(4)	3.0(2)	410(7)	170
trans-3-Heptene	C_7H_{14}	369(2)	538.6(7)	3.0(2)	411(7)	170
Heptylbenzene	$C_{13}H_{20}$	515(4)	708(7)	2.1(2)	680(16)	258
Heptyl silicate	$C_{28}H_{60}O_4Si$		778(16)			57
Hexacosane	$C_{26}H_{54}$	688(11)	816(8)	0.8(2)	1740(59)	358
Hexadecane	$C_{16}H_{34}$	560.1(7)	722.2(8)	1.4(2)	1009(53)	34, 244, 245
Hexadecanoic acid	$C_{16}H_{32}O_2$	624(6)	785(8)	1.5(2)	1059(23)	360
1-Hexadecanol	$C_{16}H_{34}O$	598(2)	770(8)	1.47(10)	1019(17)	361
1-Hexadecene	$C_{16}H_{32}$	558(1)	718(7)	1.4(2)	986(33)	365
Hexaethyldisiloxane	$C_{12}H_{30}OSi_2$	525(7)	692.9(1)		955(2)	34
Hexafluoroacetylacetone	$C_5H_2F_6O_2$	342(2)	485.1(5)	2.9(2)	313(49)	468
Hexafluorobenzene	C_6F_6	353.4(2)	516.4(5)	3.28(1)	337(4)	74, 198, 286, 469-473
2,2,4,4,5,5-Hexafluoro-1,3-dioxolane	$C_3F_6O_2$	251.0	368.1(7)	2.72(4)	293(30)	116
Hexafluoroethane	C_2F_6	195.1(1)	292.9(2)	3.03(1)	223(3)	172, 474-477
1,1,1,2,3,3-Hexafluoro-3-(2,2,3,3,3-pentafluoropropoxy)propane	$C_6H_3F_{11}O$	360.64	486.48(7)	1.95(1)	529(2)	113
1,1,1,2,3,3-Hexafluoropropane	$C_3H_2F_6$	279.35	412.40(6)	3.42(1)	270(5)	203, 447, 448, 478, 479
1,1,1,3,3,3-Hexafluoropropane	$C_3H_2F_6$	271.8(2)	398.07(6)	3.18(1)	262(18)	203
1,1,1,2,3,3-Hexafluoro-3-(2,2,3,3-tetrafluoropropoxy)propane	$C_6H_4F_{10}O$	379.07	516.2(3)	2.2(2)	543(34)	113
1,1,1,2,3,3-Hexafluoro-3-(2,2,2-trifluoroethoxy)propane	$C_5H_3F_9O$	345.87	475.74(9)	2.23(2)	455(2)	113
1,1,1,3,3,3-Hexafluoro-2-trifluoromethyl-2-methoxypropane	$C_5H_3F_9O$	327	463(1)	2.37(7)	448(5)	110
Hexamethylbenzene	$C_{12}H_{18}$	541(3)	758(2)	2.6(4)	581(15)	25
1,1,1,5,5,5-Hexamethyl-3,3-bis[(trimethylsilyl)oxy]trisiloxane	$C_{12}H_{36}O_4Si_5$	494.7	622.6(2)	1.03(2)	1323(90)	34
Hexamethyldisiloxane	$C_6H_{18}OSi_2$	373.7(3)	518.7(6)	1.95(2)	629(15)	480
2,6,10,15,19,23-Hexamethyltetracosane	$C_{30}H_{62}$	693(6)	796(2)	0.60(4)	2060(70)	481
Hexanal	$C_6H_{12}O$	402.8(4)	592(3)	3.4(2)	378(7)	163, 242
Hexane	C_6H_{14}	341.87(6)	507.5(1)	3.03(1)	366.0(8)	19, 29, 33, 34, 38, 52, 77, 78, 84, 86, 96, 106, 124, 131, 133, 134, 243, 244, 254, 281, 286, 318, 324-328, 401, 456, 458, 470, 482-488
1,6-Hexanediamine	$C_6H_{16}N_2$	470(2)	685(7)	3.6(5)	446(17)	140
1,6-Hexanedioic acid	$C_6H_{10}O_4$	610.7	841(13)	3.8(3)	449(17)	253
1,6-Hexanediol	$C_6H_{14}O_2$	481	741(10)	4.1(1)	404(13)	489
Hexanenitrile	$C_6H_{11}N$	436.7(3)	633.8(2)	2.99(6)	378(8)	35
Hexanoic acid	$C_6H_{12}O_2$	478.1(6)	661(7)		413(15)	145, 146, 360, 464
1-Hexanol	$C_6H_{14}O$	430.1(7)	611.0(4)	3.40(9)	381(30)	34, 149, 151, 152, 254, 255, 456
2-Hexanol	$C_6H_{14}O$	411(6)	585.9(5)	3.3(3)	406(8)	149, 255, 437
3-Hexanol	$C_6H_{14}O$	416(2)	582.4(4)	3.3(1)	378(14)	64, 254, 255
2-Hexanone	$C_6H_{12}O$	400.8(1)	586.7(5)	3.31(4)	377(4)	25, 172, 466
3-Hexanone	$C_6H_{12}O$	396.7(3)	583.1(5)	3.32(1)	378(4)	25, 466
Hexatriacontane	$C_{36}H_{74}$	777(7)	872(9)	0.47(7)	2711(2)	358
1-Hexene	C_6H_{12}	336.6(1)	504.1(9)	3.20(3)	381(9)	86, 201, 202, 252, 257, 490

Name	Mol. form.	T_b/K	T_c/K	P_c/MPa	V_c/cm³ mol⁻¹	Ref. (see text)
cis-2-Hexene	C_6H_{12}	342.1(5)	513.4(9)	3.34(6)	347(7)	170
trans-2-Hexene	C_6H_{12}	341.00(9)	509.0(7)	3.16(1)	353(7)	170
cis-3-Hexene	C_6H_{12}	339.6(5)	510(1)	3.29(1)	351(7)	170
trans-3-Hexene	C_6H_{12}	340.21(9)	507(2)	3.18(1)	352(7)	170
5-Hexen-2-one	$C_6H_{10}O$	402.3(5)	593.5(6)	3.51(4)	359(10)	51
Hexyl acetate	$C_8H_{16}O_2$	444.3(7)	618(1)	2.5(1)	526(8)	162
Hexylamine	$C_6H_{15}N$	405(1)	592.3(7)	3.4(3)	402(15)	170
Hexylbenzene	$C_{12}H_{18}$	499(2)	695(7)	2.4(2)	620(14)	258
Hexyl benzoate	$C_{13}H_{18}O_2$	545	748(14)	2.0(3)	658(14)	169
Hexyl silicate	$C_{24}H_{52}O_4Si$		757(16)			57
Indan	C_9H_{10}	451.0(4)	684.8(4)	3.95(3)	385(22)	25
Isobutanal	C_4H_8O	337.3(2)	543.6(6)	5.12(8)	283(10)	164, 165
Isobutane	C_4H_{10}	261.5(5)	407.84(7)	3.64(2)	256(7)	125, 132, 491, 492
Isobutene	C_4H_8	266.2(2)	418.0(3)	4.00(4)	240(3)	123, 158, 493, 494
Isobutyl acetate	$C_6H_{12}O_2$	390.1(6)	562(2)	2.97(6)	369(37)	166, 201, 202, 415
Isobutylbenzene	$C_{10}H_{14}$	445.9(4)	650(3)	3.0(2)	493(15)	96
Isobutyl butanoate	$C_8H_{16}O_2$	430(1)	611(6)	2.5(3)	524(9)	147
Isobutylcyclohexane	$C_{10}H_{20}$	444.5	642.1(6)	2.61(7)	550(14)	170
Isobutyl formate	$C_5H_{10}O_2$	371.6(3)	551(4)	3.9(4)	359(25)	415
Isobutyl isobutanoate	$C_8H_{16}O_2$	421(3)	602(6)		530(9)	147
Isobutyl 3-methylbutanoate	$C_9H_{18}O_2$	442(3)	621(6)		581(9)	147
Isobutyl propanoate	$C_7H_{14}O_2$	409(2)	586(8)		462(9)	162, 167
Isopentane	C_5H_{12}	300.98(6)	460.37(9)	3.35(6)	313(17)	19, 86, 96, 127, 495
Isopentyl acetate	$C_7H_{14}O_2$	414.8(7)	586.1(4)	2.76(7)	464(9)	166
Isopentyl butanoate	$C_9H_{18}O_2$	458.0(3)	619(6)		595(10)	147
Isopentyl nitrite	$C_5H_{11}NO_2$	372(3)	626(16)	5.07(4)	386.2(1)	52
Isopentyl propanoate	$C_8H_{16}O_2$	446(4)	611(6)		523(9)	147
Isopropyl acetate	$C_5H_{10}O_2$	361.8(2)	531.1(6)	3.31(4)	343(4)	64, 166, 172, 411, 490
Isopropylamine	C_3H_9N	305.0(2)	472.2(9)	4.55(7)	231(5)	75, 170
Isopropylbenzene	C_9H_{12}	425.6(2)	631(1)	3.2(1)	423(5)	79, 90, 96, 172
Isopropylcyclohexane	C_9H_{18}	427.6(4)	632.2(4)	3.1(2)	484(14)	170
Isopropyl formate	$C_4H_8O_2$	341(2)	534.6(5)	3.95(3)	294(11)	164, 165
1-Isopropyl-4-methylbenzene	$C_{10}H_{14}$	450(2)	654(8)	2.8(1)	495(16)	96, 147
Isopropyl methyl ether	$C_4H_{10}O$	304.0(5)	464.4(2)	3.76(1)	287(11)	25
Isoquinoline	C_9H_7N	516.4(6)	803(8)	5.07(3)	380(17)	218
d-Limonene	$C_{10}H_{16}$	450.8(5)	653(2)	2.81(2)	498(11)	496
(-)-Menthol	$C_{10}H_{20}O$	487(12)	694(5)	2.7(5)	539(14)	
Mesityl oxide	$C_6H_{10}O$	402.9(4)	605(2)	3.85(2)	353(27)	497
Methane	CH_4	111.7(2)	190.56(2)	4.60(1)	99(3)	132, 134, 498-508
Methane-d_4	CD_4		189.2(6)		98(3)	507
Methanethiol	CH_4S	279.2(1)	469.9(3)	7.24(9)	148(4)	304
Methanol	CH_4O	337.7(7)	512.7(6)	8.01(3)	117(4)	19, 38, 67, 80, 85, 87, 131, 148, 152, 155, 196, 405, 458, 482, 509-516
1-Methoxy-2,4-dimethylbenzene	$C_9H_{12}O$	465	682(4)	3.2(7)	451(18)	162
2-Methoxy-1,4-dimethylbenzene	$C_9H_{12}O$	467	677(1)	3.2(7)	451(15)	162
2-Methoxyethanol	$C_3H_8O_2$	397.5(1)	598(1)	5.28(8)	263(6)	49
2-Methoxyethyl acetate	$C_5H_{10}O_3$	415(3)	603(3)	3.6(5)	368(15)	162
4-Methoxy-1,1,1,2,2,3,3-heptafluorobutane	$C_5H_5F_7O$	344.13	481.5(2)	2.38(1)	431(2)	113
1-Methoxy-1,1,2,2,3,3-hexafluoropropane	$C_4H_4F_6O$	341.02	487.0(3)	2.9(1)	370(25)	113
2-Methoxy-2-methylbutane	$C_6H_{14}O$	359.6(1)	536(2)	3.23(9)	372(19)	172, 410, 517
5-Methoxy-1,1,2,2,3,3,4,4-octafluoropentane	$C_6H_6F_8O$	395.83	546.1(3)	2.40(7)	493(30)	113
1-Methoxy-2-propanol	$C_4H_{10}O_2$	393.2(6)	579.8(3)	4.11(4)	304(12)	49
2-Methoxy-1-propene	C_4H_8O	308.9(3)	478.5(6)	4.2(3)	257(13)	518
Methyl acetate	$C_3H_6O_2$	329.9(2)	506.7(4)	4.73(7)	227(22)	19, 411-413
Methylamine	CH_5N	266.8(3)	430.6(6)	7.61(9)	139(1)	130, 193, 322, 323
2-Methylaniline	C_7H_9N	473.2(4)	710(1)	3.6(1)	377(52)	519
3-Methylaniline	C_7H_9N	476.5(5)	709(10)	4.6(5)	346(17)	106
4-Methylaniline	C_7H_9N	474(1)	667(10)	3.3(7)	334(16)	106
N-Methylaniline	C_7H_9N	470(1)	702(5)	5.2(6)	347(16)	117

Name	Mol. form.	T_b/K	T_c/K	P_c/MPa	V_c/cm³ mol⁻¹	Ref. (see text)
2-Methylanisole	$C_8H_{10}O$	446(2)	662(1)	3.6(3)	397(18)	162
3-Methylanisole	$C_8H_{10}O$	450(2)	665(1)	3.6(3)	449(22)	162
4-Methylanisole	$C_8H_{10}O$	448(2)	667(1)	3.6(4)	396(15)	162
α-Methylbenzenemethanol	$C_8H_{10}O$	478(4)	699(5)	3.8(7)	399(16)	32
Methyl benzoate	$C_8H_8O_2$	472(2)	702(1)	3.8(1)	408(11)	517
2-Methylbutanal, (±)-	$C_5H_{10}O$	363	531.6(1)	4.04(3)	318(10)	124
Methyl butanoate	$C_5H_{10}O_2$	375.1(1)	554.4(1)	3.49(8)	341(20)	19, 413, 416
3-Methylbutanoic acid	$C_5H_{10}O_2$	449.7(2)	629(1)	3.4(2)	355(10)	146
2-Methyl-1-butanol, (±)-	$C_5H_{12}O$	402.2(4)	575.4(5)	3.9(1)	342(8)	254
3-Methyl-1-butanol	$C_5H_{12}O$	404.0(3)	579(2)	3.9(3)	335(7)	21, 91, 222, 254, 511, 520
2-Methyl-2-butanol	$C_5H_{12}O$	375.6	544(1)	3.71(5)	326(9)	147, 254
3-Methyl-2-butanol, (±)-	$C_5H_{12}O$	386.9(4)	556.1(5)	3.9(4)	336(9)	254
3-Methyl-2-butanone	$C_5H_{10}O$	367.4(2)	553.1(3)	3.83(10)	321(33)	157, 166
3-Methyl-1-butene	C_5H_{10}	293.3(2)	452.7(5)	3.51(4)	305(8)	490
2-Methyl-2-butene	C_5H_{10}	311.7(4)	470(1)	3.4(1)	299(7)	521
Methyl *tert*-butyl ether	$C_5H_{12}O$	328.3(1)	497.0(6)	3.41(5)	335(10)	25, 440
Methylcyclohexane	C_7H_{14}	374.1(1)	572.3(2)	3.48(9)	368(3)	34, 74, 78, 86, 88, 171, 235, 236
Methylcyclopentane	C_6H_{12}	345.0(2)	532.78(5)	3.79(5)	322(2)	78, 235, 236
2-Methylcyclopentanone	$C_6H_{10}O$	413(3)	631(2)	4.0(6)	328(17)	162
2-Methyl-*N,N*-dimethylaniline	$C_9H_{13}N$	458(2)	668.0(7)	3.12(8)	466(15)	39, 40
Methyl dodecanoate	$C_{13}H_{26}O_2$	541(2)	712(5)	1.4(4)	842(9)	218
1,1'-Methylenebis[(1-methylethyl)benzene]	$C_{19}H_{24}$	592	795(8)	1.6(1)	871(30)	394
Methyl formate	$C_2H_4O_2$	304.8(3)	487.16(10)	6.01(1)	172(6)	19, 412, 413
2-Methylfuran	C_5H_6O	337.1(2)	528(3)	4.77(8)	252(3)	290
2-Methylheptane	C_8H_{18}	390.8(9)	559.6(1)	2.50(2)	487(12)	86, 346, 522
3-Methylheptane	C_8H_{18}	392.1(6)	563.7(4)	2.54(2)	463(12)	346
4-Methylheptane	C_8H_{18}	390.9(5)	561.7(4)	2.54(2)	480(14)	346
Methyl heptanoate	$C_8H_{16}O_2$	442.9(4)	628(2)	2.6(4)	521(8)	162
4-Methyl-3-heptanol	$C_8H_{18}O$	430(2)	623.5(7)	2.8(4)	505(13)	437
5-Methyl-3-heptanol	$C_8H_{18}O$	427(2)	621.2(3)	2.8(3)	493(13)	437
2-Methyl-3-heptanone	$C_8H_{16}O$	431	615(1)	2.7(3)	487(11)	162
5-Methyl-3-heptanone	$C_8H_{16}O$	432(4)	619(4)	2.7(7)	484(11)	162
2-Methyl-1-heptene	C_8H_{16}	392(2)	567.5(9)	2.6(2)	466(10)	170
2-Methyl-2-heptene	C_8H_{16}	395(2)	569(1)	2.6(4)	465(11)	170
2-Methylhexane	C_7H_{16}	363.2(8)	530.4(1)	2.73(3)	420(15)	86, 346, 349, 522
3-Methylhexane	C_7H_{16}	365	535.4(5)	2.82(6)	405(19)	346, 349
5-Methyl-2-hexanone	$C_7H_{14}O$	412(2)	604(1)	2.9(3)	434(13)	162
2-Methyl-3-hexanone	$C_7H_{14}O$	407(3)	593(1)	2.9(4)	428(13)	162
2-Methyl-1-hexene	C_7H_{14}	365(2)	542(1)	2.9(3)	407(8)	170
5-Methyl-1-hexene	C_7H_{14}	358(1)	528.7(4)	2.9(2)	410(9)	170
N-Methylhexylamine	$C_7H_{17}N$	418	592(1)	2.8(8)	458(20)	170
Methyl isobutanoate	$C_5H_{10}O_2$	365(1)	540.7(5)	3.43(1)	341(42)	19, 413
Methyl methacrylate	$C_5H_8O_2$	373.8(2)	540.3(6)	2.97(6)	320(6)	51
1-Methylnaphthalene	$C_{11}H_{10}$	517.6(9)	771(5)	3.56(7)	479(22)	86, 218, 523
2-Methylnaphthalene	$C_{11}H_{10}$	514.3(3)	761(3)	3.37(6)	464(20)	218
2-Methyloctane	C_9H_{20}	416(1)	582.8(2)	2.30(2)	547(17)	345, 522
Methyloxirane	C_3H_6O	308	488.11(8)	5.44(2)		290, 524
Methyl pentafluoroethyl ether	$C_3H_3F_5O$	278.8(9)	406.81(5)	2.89(1)	301(5)	176, 448, 525
2-Methylpentane	C_6H_{14}	333.36(9)	497.9(2)	3.03(1)	371(2)	86, 127, 252, 325-328, 487, 522
3-Methylpentane	C_6H_{14}	336.5(5)	504.6(2)	3.12(1)	368.7(3)	252, 324-328, 526
Methyl pentanoate	$C_6H_{12}O_2$	400.51(6)	588.9(3)	3.20(5)	398(6)	164, 165
2-Methyl-1-pentanol	$C_6H_{14}O$	430(6)	604.4(5)	3.4(2)	410(8)	254
4-Methyl-1-pentanol	$C_6H_{14}O$	424(2)	603.5(7)	3.4(4)	406(7)	437
2-Methyl-2-pentanol	$C_6H_{14}O$	394(1)	559.5(7)	3.6(4)	410(11)	437
4-Methyl-2-pentanol	$C_6H_{14}O$	405.2(5)	574.4(5)		389(9)	437
2-Methyl-3-pentanol	$C_6H_{14}O$	401.1(2)	576(1)	3.5(1)	380(9)	254
3-Methyl-3-pentanol	$C_6H_{14}O$	402(4)	575.6(6)	3.5(2)	376(10)	254
4-Methyl-2-pentanone	$C_6H_{12}O$	388.9(2)	575.4(10)	3.4(1)	378(12)	166, 527
4-Methyl-1-pentene	C_6H_{12}	327(2)	493.1(5)	3.18(7)		170, 497
2-Methyl-2-pentene	C_6H_{12}	340.5(5)	509.3(5)	3.26(1)	348(7)	170
4-Methyl-*cis*-2-pentene	C_6H_{12}	329.6(1)	496.3(7)	3.24(1)	350(7)	170

Name	Mol. form.	T_b/K	T_c/K	P_c/MPa	V_c/cm³ mol⁻¹	Ref. (see text)
Methyl pentyl ether	$C_6H_{14}O$	372(3)	546.5(2)	3.04(10)	395(14)	173, 174
2-Methyl-1,3-propanediol	$C_4H_{10}O_2$	494(4)	708(2)	5.4(4)	300(12)	55
Methyl propanoate	$C_4H_8O_2$	351.8(2)	530.57(10)	4.0(2)		19, 412, 413, 415
2-Methylpropanoic acid	$C_4H_8O_2$	427.6(2)	605(2)	3.7(3)	296(10)	146
2-Methyl-1-propanol	$C_4H_{10}O$	380.99(7)	548(2)	4.30(4)	274(17)	25, 91, 148, 153, 155
2-Methyl-2-propanol	$C_4H_{10}O$	355.5(1)	506.2(1)	3.98(7)	283(4)	153
Methyl propyl ether	$C_4H_{10}O$	311.7(10)	476.2(2)	3.80(1)	281(7)	25
2-Methylpyridine	C_6H_7N	402.6(2)	622(1)	4.62(4)		75, 528
3-Methylpyridine	C_6H_7N	417.3(1)	644.8(6)	4.63(3)		90, 528
4-Methylpyridine	C_6H_7N	418.5(1)	645.8(5)	4.68(4)		75, 90
N-Methyl-2-pyrrolidinone	C_5H_9NO	477.4(3)	721.7(4)	4.5(4)	330(12)	144, 242
2-Methylquinoline	$C_{10}H_9N$	520.6(4)	778(2)	3.91(2)	447(49)	530
8-Methylquinoline	$C_{10}H_9N$	520.6(7)	787(2)	4.22(2)		530
Methyl salicylate	$C_8H_8O_3$	495.8(5)	709(30)	4.4(7)	436(17)	302
Methyl silicate	$C_4H_{12}O_4Si$	393.3(7)	558(12)	2.8(6)	464(22)	57
2-Methyltetrahydrofuran	$C_5H_{10}O$	353(1)	537(2)	3.74(6)	292(4)	290
(Methylthio)benzene	C_7H_8S	467.5(2)	706(4)	4.1(1)	374(13)	99
Methyl trifluoromethyl ether	$C_2H_3F_3O$	248.0(7)	377.92(6)	3.64(3)	219(2)	447, 525
Methyltris(trimethylsiloxy)silane	$C_{10}H_{30}O_3Si_4$	464.4	597.4(2)	1.23(2)	1089(74)	34
4-Morpholinecarboxaldehyde	$C_5H_9NO_2$	511(1)	779(4)	5.0(4)	326(14)	54
Naphthalene	$C_{10}H_8$	491.2(1)	748.3(4)	4.06(4)	408(21)	79, 82, 86, 241, 251, 355, 451, 531
Neopentane	C_5H_{12}	282.65(6)	433.71(1)	3.20(1)	311.6(7)	532
Nitromethane	CH_3NO_2	374.34(10)	588(3)	6.0(2)	175(2)	533, 534
Nonadecane	$C_{19}H_{40}$	603(3)	756(5)	1.16(7)	1216(43)	358, 359
1-Nonadecene	$C_{19}H_{38}$	604(17)	755(8)	1.2(2)	1196(36)	365
1,1,1,2,2,3,3,4,4-Nonafluorohexan-5-one	$C_6H_3F_9O$	360.47	498.97(8)	2.20(2)	504(2)	308
Nonanal	$C_9H_{18}O$	468(3)	658(2)	2.7(1)	546(10)	242
Nonane	C_9H_{20}	424.0(2)	594.2(5)	2.29(5)	547(23)	34, 36, 78, 86, 131, 243-245, 247, 249, 251, 454, 456, 535, 536
1,9-Nonanediamine	$C_9H_{22}N_2$	531.9	726(7)	2.6(3)	600(23)	140
Nonanedioic acid	$C_9H_{16}O_4$	630.3	844(13)	2.7(2)	586(17)	253
Nonanoic acid	$C_9H_{18}O_2$	529(1)	712(3)		592(16)	146
1-Nonanol	$C_9H_{20}O$	486.9(4)	670.6(5)	2.54(7)	555(75)	149, 152, 254, 255
2-Nonanol, (±)-	$C_9H_{20}O$	466.7	649(1)	2.53(10)	575(11)	149, 255
3-Nonanol, (±)-	$C_9H_{20}O$	468	648.0(3)	2.5(3)	577(12)	255
4-Nonanol	$C_9H_{20}O$	465.7	645.1(3)	2.5(3)	577(12)	255
2-Nonanone	$C_9H_{18}O$	467(1)	652.1(7)	2.5(1)	560(8)	49, 466
3-Nonanone	$C_9H_{18}O$	460(4)	648(4)	2.4(6)	560(7)	466
4-Nonanone	$C_9H_{18}O$	461(4)	643.7(3)	2.4(3)	560(7)	466
5-Nonanone	$C_9H_{18}O$	461.6(3)	641.4(3)	2.35(2)	560(7)	466
1-Nonene	C_9H_{18}	420.1(6)	594(1)	2.38(1)	529(2)	257
Nonyl silicate	$C_{36}H_{76}O_4Si$		830(16)			57
Octacosane	$C_{28}H_{58}$	705(6)	824(8)	0.8(1)	1916(65)	358
Octadecane	$C_{18}H_{38}$	589(2)	748(1)	1.3(1)	1167(41)	244
1-Octadecanol	$C_{18}H_{38}O$	624(2)	790(8)	1.28(10)	1157(18)	361
1-Octadecene	$C_{18}H_{36}$	588.7(10)	748(8)	1.3(1)	1119(34)	365
1,1,1,2,2,3,3,4-Octafluorobutane	$C_4H_2F_8$	300.62	432.0(1)	2.80(2)	360(22)	537
1,2,2,3,3,4,4,5-Octafluoro-1-pentanol	$C_5H_4F_8O$	413.2	571(1)	2.9(1)	440(24)	362
Octafluorotetrahydrofuran	C_4F_8O	272.3	399.6(7)	2.68(9)	350(30)	450
Octamethylcyclotetrasiloxane	$C_8H_{24}O_4Si_4$	448.6(9)	585.8(9)	1.33(1)	1006(67)	114, 538
Octamethyltrisiloxane	$C_8H_{24}O_2Si_3$	425.7(8)	564.1(2)	1.42(1)	882(16)	539
Octanal	$C_8H_{16}O$	447(3)	639.3(3)	3.0(3)	489(7)	163, 242
Octane	C_8H_{18}	398.77(10)	568.7(1)	2.48(1)	490(22)	11, 19, 33, 34, 36, 78, 83, 86, 91, 96, 131, 133, 134, 243-245, 247, 249, 318, 319, 324, 346, 465, 469, 535, 540, 541
1,8-Octanediamine	$C_8H_{20}N_2$	498.8	712(7)	2.8(3)	547(20)	140
Octanedioic acid	$C_8H_{14}O_4$	618.7	843(13)	3.0(2)	520(16)	253
Octanenitrile	$C_8H_{15}N$	475(3)	674.4(4)	2.85(3)	494(10)	35
Octanoic acid	$C_8H_{16}O_2$	513(1)	694(1)	2.9(3)	522(19)	145, 146, 360, 464
1-Octanol	$C_8H_{18}O$	467.9(8)	651(2)	2.80(7)	490(47)	25, 149, 152, 254, 255
2-Octanol	$C_8H_{18}O$	452(7)	629.5(9)	2.75(4)	519(10)	149, 254, 255

Fluids

Name	Mol. form.	T_b/K	T_c/K	P_c/MPa	V_c/cm³ mol⁻¹	Ref. (see text)
3-Octanol	$C_8H_{18}O$	457(6)	628.4(3)	2.8(4)	515(10)	255
4-Octanol	$C_8H_{18}O$	449.5	625.1(3)	2.8(3)	516(10)	255
2-Octanone	$C_8H_{16}O$	446(3)	632.7(2)	2.7(5)	497(6)	466
3-Octanone	$C_8H_{16}O$	439(4)	627.7(2)	2.7(3)	497(6)	466
4-Octanone	$C_8H_{16}O$	439(3)	623.8(2)	2.7(3)	497(6)	466
1-Octene	C_8H_{16}	394.5(2)	566.58(5)	2.68(2)	464(2)	86, 257
trans-2-Octene	C_8H_{16}	398.1(5)	569.8(4)	2.58(9)	471(9)	170
trans-4-Octene	C_8H_{16}	395.6(5)	566(1)	2.55(6)	472(9)	170
Octylamine	$C_8H_{19}N$	451.8(2)	641(1)	2.82(3)	494(41)	542
Octylbenzene	$C_{14}H_{22}$	536(2)	725(7)	2.0(2)	746(17)	258
Octyl silicate	$C_{32}H_{68}O_4Si$		812(16)			57
Oxazole	C_3H_3NO	342.7(2)	551(4)	6.8(2)	185(23)	54
Oxirane	C_2H_4O	283.6(1)	469(1)	7.2(2)	138(4)	543, 544
Paraldehyde	$C_6H_{12}O_3$	397(2)	563(10)		410(15)	12
Pentacene	$C_{22}H_{14}$		1115(47)		806(23)	545
1*H*-Pentadecafluoroheptane	C_7HF_{15}	368(2)	495.8(7)	1.7(5)	644(38)	546
Pentadecane	$C_{15}H_{32}$	543.8(4)	707(2)	1.54(9)	938(36)	243, 244, 245, 247, 249, 363, 547
Pentadecanoic acid	$C_{15}H_{30}O_2$	612(4)	777(8)	1.6(2)	1002(22)	360
1-Pentadecanol	$C_{15}H_{32}O$	591(2)	757(8)	1.6(2)	961(16)	361
1-Pentadecene	$C_{15}H_{30}$	541.6(4)	705(7)	1.56(5)	933(30)	365
Pentafluorobenzene	C_6HF_5	358(3)	530.93(5)	3.53(1)	322(22)	472, 548
3,3,4,4,4-Pentafluorobutan-2-one	$C_4H_3F_5O$	314.36	453(1)	2.90(6)	333(4)	110
Pentafluoroethane	C_2HF_5	225.06	339.2(2)	3.63(1)	210(3)	180, 309, 312, 315, 447, 448, 549-554
1,1,1,2,2-Pentafluoropentan-3-one	$C_5H_5F_5O$	335.24	475.5(1)	2.64(1)	356(1)	308
1,1,1,2,2-Pentafluoropropane	$C_3H_3F_5$	255.2(3)	380.1(4)	3.14(2)	273(3)	555
1,1,1,3,3-Pentafluoropropane	$C_3H_3F_5$	288.29	427.20(7)	3.66(2)	262(14)	203
1,1,2,2,3-Pentafluoropropane	$C_3H_3F_5$	298.28	447.57(6)	3.96(2)	258(13)	203
1,1,1,2,2-Pentafluoro-3-(1,1,2,2-tetrafluoroethoxy)propane	$C_5H_3F_9O$	343.5	473.0(1)	2.24(1)	457(2)	113
Pentafluoro(trifluoromethoxy)ethane	C_3F_8O	249.5	356.8(1)	2.4(5)	319(5)	442
Pentanal	$C_5H_{10}O$	376(2)	567(3)	3.1(3)	313(11)	104, 124
Pentane	C_5H_{12}	309.21(7)	469.7(1)	3.37(1)	310(1)	19, 34, 36, 52, 64, 77, 84-86, 127, 131, 133, 134, 201, 202, 243, 244, 254, 286, 324, 359, 394, 456, 490, 556-562
Pentanedioic acid	$C_5H_8O_4$	546(10)	840(13)	4.3(5)	343(13)	253
Pentanenitrile	C_5H_9N	413(1)	610.3(2)	3.58(5)	320(8)	35
Pentanoic acid	$C_5H_{10}O_2$	459.3(3)	639(2)	3.6(1)	347(15)	11, 41, 145, 146, 464
1-Pentanol	$C_5H_{12}O$	410.8(4)	587.9(4)	3.9(3)	331(9)	25, 34, 149, 329, 456, 465, 563
2-Pentanol	$C_5H_{12}O$	392.3(5)	560.4(2)	4.2(4)	340(3)	149, 254, 465
3-Pentanol	$C_5H_{12}O$	396(2)	559.6(3)	4.9(9)	325(2)	465
2-Pentanone	$C_5H_{10}O$	375.4(1)	561.0(2)	3.70(6)	324(4)	25, 172
3-Pentanone	$C_5H_{10}O$	375.1(1)	561.4(2)	3.73(7)	319(11)	25
1-Pentene	C_5H_{10}	303.2(3)	464.74(5)	3.55(2)	301.0(1)	86, 470, 559
cis-2-Pentene	C_5H_{10}	310.1(2)	474.9(4)	3.69(2)	301(7)	564
Pentyl acetate	$C_7H_{14}O_2$	422.6(3)	600(2)	2.79(3)		141, 162, 166
Pentylbenzene	$C_{11}H_{16}$	476(3)	675(7)	2.6(3)	559(12)	258
Pentyl benzoate	$C_{12}H_{16}O_2$	533(3)	736(14)	2.2(2)	661(14)	169
Pentyl formate	$C_6H_{12}O_2$	399(3)	576(4)	3.5(8)	453(35)	415
Pentyl silicate	$C_{20}H_{44}O_4Si$		714(14)			57
Perfluoroacetone	C_3F_6O	245.8(4)	357.2(1)	2.85(1)	329(7)	197, 470
Perfluorobutane	C_4F_{10}	271.1(8)	386.3(2)	2.33(2)	380(18)	565, 566
Perfluorocyclobutane	C_4F_8	267.24	388.4(1)	2.78(1)	316(8)	265, 470, 567, 568
Perfluorocyclohexane	C_6F_{12}	326.0 sp	457.1(5)	2.24(3)	424(32)	569
Perfluorocyclohexene	C_6F_{10}	324.9(1)	461.7(7)	2.6(1)	434(28)	546
Perfluorodecane	$C_{10}F_{22}$	408(3)	542.4(4)	1.45(3)	892(7)	240, 570
Perfluorodimethoxymethane	$C_3F_8O_2$	263.2(7)	372.4(2)	2.34(1)	370(30)	116, 571
Perfluoro-2,3-dimethylbutane	C_6F_{14}	333.0(3)	463.0(1)	1.95(3)	523(18)	240, 572
Perfluoroethyl ethyl ether	$C_4H_5F_5O$	301(3)	431.23(8)	2.53(1)	366(3)	176
Perfluoroethyl 2,2,2-trifluoroethyl ether	$C_4H_2F_8O$	301.04	421.68(8)	2.33(1)	409(3)	176
Perfluoroheptane	C_7F_{16}	355.7(2)	477(3)	1.63(1)	603(14)	232, 566, 570, 573-576

Fluids

Name	Mol. form.	T_b/K	T_c/K	P_c/MPa	V_c/cm³ mol⁻¹	Ref. (see text)
Perfluoro-1-heptene	C_7F_{14}	354(4)	478.2(7)	1.7(3)	555(34)	546
Perfluorohexane	C_6F_{14}	330.4(2)	451(3)	1.88(2)	552(77)	198, 470, 546, 570, 572, 573, 577
1*H*-Perfluorohexane	C_6HF_{13}	345	471.8(7)	2.0(3)	504(32)	546
Perfluoro-1-hexene	C_6F_{12}	330.2	454.3(7)		462(31)	546
Perfluoroisobutane	C_4F_{10}	273	395.4(7)		396(22)	578
Perfluoroisopropyl methyl ether	$C_4H_3F_7O$	302(1)	433.30(8)	2.55(1)	369(3)	176
Perfluoromethylcyclohexane	C_7F_{14}	349.5(2)	486.5(10)	2.02(1)	561(4)	325, 326, 327, 566, 569
Perfluoromethylcyclopentane	C_6F_{12}	321.58	451.43(5)	2.17(1)	419(37)	579
Perfluoro-2-methylpentane	C_6F_{14}	330.8(3)	454.6(2)	1.87(2)	585(10)	572, 580
Perfluoro-3-methylpentane	C_6F_{14}	331(9)	450(1)	1.69(1)	511(36)	572
Perfluoronaphthalene	$C_{10}F_8$	473(8)	673(1)	2.9(6)	464(29)	546
Perfluorononane	C_9F_{20}	390(3)	524.0(1)	1.56(4)	846(48)	570
Perfluorooctane	C_8F_{18}	378(2)	502.3(1)	1.66(2)	738(17)	34, 240, 570, 573
Perfluorooxetane	C_3F_6O	244.6(6)	361.8(5)	3.10(2)	274(24)	116, 571
Perfluoropentane	C_5F_{12}	302.4(2)	421.8(1)	2.04(2)	463(7)	570, 573
1*H*-Perfluoropentane	C_5HF_{11}	319	443.9(7)	2.2(3)	413(28)	546
Perfluoropropane	C_3F_8	236.4(3)	345.03(8)	2.67(1)	301(12)	187, 305, 470, 581
Perfluoropropyl methyl ether	$C_4H_3F_7O$	307(1)	437.7(1)	2.48(1)	382(3)	176
Perfluorotoluene	C_7F_8	377.8(3)	534.4(2)	2.70(2)	423(26)	25
Perfluorovaleric acid	$C_5HF_9O_2$	415.4	545.5(2)	2.10(5)	496(29)	583
Phenol	C_6H_6O	455.0(1)	694.3(1)	5.5(2)	283(10)	218, 584
Phenyl acetate	$C_8H_8O_2$	468(1)	686(2)	3.60(6)	407(12)	55
Phenyl isocyanate	C_7H_5NO	439.5(4)	657(10)	3.6(1)	342(15)	54
α-Pinene, (-)	$C_{10}H_{16}$	429.2	644(2)	3.4(2)	472(10)	496
Piperazine	$C_4H_{10}N_2$	421.78(5)	660(5)	5.4(4)	283(16)	331, 489
Piperidine	$C_5H_{11}N$	379.34(9)	594.14(5)	4.7(1)	294(10)	39, 218
Propanal	C_3H_6O	321.2(2)	503.7(8)	5.04(3)	218(9)	104, 124, 163, 242
Propane	C_3H_8	231.04(9)	369.9(1)	4.25(1)	199(6)	29, 125, 127, 131-134, 252, 322, 323, 336, 371, 387, 470, 486, 509, 540, 585-596
1,3-Propanediamine	$C_3H_{10}N_2$	412.4(7)	632(7)	5.7(6)	257(13)	140
1,2-Propanediol	$C_3H_8O_2$	460.5(2)	676(1)	5.9(2)	237(10)	32
1,3-Propanediol	$C_3H_8O_2$	487.9(3)	718(2)	6.7(2)	255(12)	55
Propanenitrile	C_3H_5N	370.5(4)	561.3(2)	4.26(7)	211(7)	35
1-Propanethiol	C_3H_8S	340.9(1)	536.6(6)	4.7(1)	286(11)	143, 144
Propanoic acid	$C_3H_6O_2$	414.7(2)	603(3)	4.5(7)	232(12)	39, 145, 146, 493, 597, 598
1-Propanol	C_3H_8O	370.19(9)	536.8(2)	5.1(1)	220(22)	19, 34, 80, 148-150, 152, 153, 599, 600
2-Propanol	C_3H_8O	355.36(9)	508.3(2)	4.7(1)	226(1)	25, 91, 148, 149, 153, 324, 440, 520, 601, 602
Propene	C_3H_6	225.6(1)	364.9(5)	4.59(2)	184(11)	67, 123, 172, 181, 427, 603-610
2-Propoxyethanol	$C_5H_{12}O_2$	425(3)	614.7(7)	3.65(9)	364(13)	11, 41
1-Propoxy-2-propanol	$C_6H_{14}O_2$	423.4(7)	605(1)	3.1(1)	417(16)	32
Propyl acetate	$C_5H_{10}O_2$	374.2(2)	549.69(8)	3.37(7)	346(24)	19, 411, 413, 415
Propylamine	C_3H_9N	320.36(8)	499.2(4)	4.77(6)	230(12)	170, 193
Propylbenzene	C_9H_{12}	432.4(5)	638.3(1)	3.20(2)	441(6)	79, 86, 88
Propyl benzoate	$C_{10}H_{12}O_2$	504(2)	710(14)	2.6(3)	530(10)	169
Propyl butanoate	$C_7H_{14}O_2$	417(2)	593(1)	2.72(6)	463(10)	162, 166
Propylcyclohexane	C_9H_{18}	429.9(3)	630.8(9)	2.87(4)	489(13)	170, 171
Propylene carbonate	$C_4H_6O_3$	514.8(7)	763(2)	4.1(2)	256.5(5)	55
1,2-Propylene glycol 1-*tert*-butyl ether	$C_7H_{16}O_2$	425.3	601(4)	2.7(1)	468(15)	99
1,2-Propylene glycol monomethyl ether acetate	$C_6H_{12}O_3$	419.2(4)	598(1)	3.1(2)	432(16)	41, 144
Propyl formate	$C_4H_8O_2$	353.8(2)	538.1(1)	4.07(2)	281(31)	19, 412, 413, 415
Propyl isobutanoate	$C_7H_{14}O_2$	407(4)	582(11)		463(9)	162, 167
Propyl 3-methylbutanoate	$C_8H_{16}O_2$	428(3)	609(6)		523(9)	147
Propyl propanoate	$C_6H_{12}O_2$	395.4(1)	569(3)	3.1(1)	403(6)	162, 166
Propyl silicate	$C_{12}H_{28}O_4Si$	499	649(12)			57
Propyne	C_3H_4	250.0	402(2)	5.63(6)	160(10)	605, 611
Pyrazine	$C_4H_4N_2$	389.5(1)	627(1)	6.49(3)	225(16)	612
Pyridine	C_5H_5N	388.4(1)	619(2)	5.63(7)	248(12)	16, 17, 60, 90, 246, 290, 520, 613
Pyrrole	C_4H_5N	402.89(4)	639.7(2)	8.0(2)	222(15)	241

Name	Mol. form.	T_b/K	T_c/K	P_c/MPa	V_c/cm³ mol⁻¹	Ref. (see text)
Pyrrolidine	C_4H_9N	359.8(1)	568.6(2)	5.69(8)	259(3)	241, 290
Quinoline	C_9H_7N	510.3(5)	782(3)	4.75(10)	382(17)	218
Resorcinol	$C_6H_6O_2$	553(2)	836(10)	6.3(3)	292(10)	489
Stearic acid	$C_{18}H_{36}O_2$	644(3)	803(8)	1.3(2)	1251(27)	360
Styrene	C_8H_8	418.5(6)	635(2)	3.9(2)	357(15)	481
Succinic acid	$C_4H_6O_4$	507(3)	851(20)		308(21)	253
o-Terphenyl	$C_{18}H_{14}$	610(5)	857(6)	2.9(1)	737(37)	108
m-Terphenyl	$C_{18}H_{14}$	648(1)	883(7)	2.2(2)	747(37)	108, 358
p-Terphenyl	$C_{18}H_{14}$	649	913(22)	2.5(5)	713(37)	108, 614
Tetrachloromethane	CCl_4	349.9(2)	556.5(3)	4.57(7)	276(9)	19, 27, 469, 615-619
Tetracosane	$C_{24}H_{50}$	664(5)	800(5)	0.9(1)	1585(55)	358, 359
Tetradecamethylcycloheptasiloxane	$C_{14}H_{42}O_7Si_7$	548.5	683.2(2)	0.99(2)	1634(110)	34
Tetradecane	$C_{14}H_{30}$	526.7(4)	693(1)	1.56(8)	870(49)	34, 244
Tetradecanedioic acid	$C_{14}H_{26}O_4$	639(10)		1.9(2)		253
Tetradecanoic acid	$C_{14}H_{28}O_2$	599(1)	763(8)	1.6(2)	921(20)	360
1-Tetradecanol	$C_{14}H_{30}O$	569.0(4)	743(7)	1.70(4)	887(15)	361
2-Tetradecanone	$C_{14}H_{28}O$	562(6)	728(9)		896(26)	256
3-Tetradecanone	$C_{14}H_{28}O$	552	727(6)		896(26)	256
4-Tetradecanone	$C_{14}H_{28}O$	552	725(6)		900(27)	256
7-Tetradecanone	$C_{14}H_{28}O$	552	723(8)		904(27)	256
1-Tetradecene	$C_{14}H_{28}$	524.3(4)	691(7)	1.58(7)	851(24)	365
Tetraethylene glycol	$C_8H_{18}O_5$	588(7)	800(30)	2.8(7)	608(31)	283
Tetraethylsilane	$C_8H_{20}Si$	426.6(7)	606(2)	2.297(10)	596.4(2)	620
1,2,3,4-Tetrafluorobenzene	$C_6H_2F_4$	367.5(8)	550.8(2)	3.791(10)	312(22)	25
1,2,3,5-Tetrafluorobenzene	$C_6H_2F_4$	357.5(8)	535.2(2)	3.75(1)	311(22)	25
1,2,4,5-Tetrafluorobenzene	$C_6H_2F_4$	363.4(3)	543.3(2)	3.80(1)	309(22)	25
1,1,2,2-Tetrafluoro-2-(2,2-difluoroethoxy) ethane	$C_4H_4F_6O$	352.13	501.08(8)	3.09(2)	356(1)	113
1,1,1,2-Tetrafluoroethane	$C_2H_2F_4$	247.1(1)	374.2(2)	4.06(1)	200(2)	113, 182, 183, 188, 278-280, 305, 313, 315, 447, 448, 478, 621-631
1,1,2,2-Tetrafluoroethane	$C_2H_2F_4$	253(1)	391.75(8)	4.61(1)	192(1)	178, 632
Tetrafluoroethene	C_2F_4	197(1)	307(1)	3.94(5)	183(3)	633, 634
1,2,2,2-Tetrafluoroethyl difluoromethyl ether	$C_3H_2F_6O$	296(2)	428.95(8)	3.05(1)	315(2)	176
1,1,2,2-Tetrafluoroethyl 1,1,1-trifluoroethyl ether	$C_4H_3F_7O$	329.37	463.89(7)	2.71(1)	373(1)	113
Tetrafluoromethane	CF_4	145.3(1)	227.54(3)	3.73(3)	140(1)	506, 635
1,1,2,2-Tetrafluoro-3-methoxypropane	$C_4H_6F_4O$	347.5	505.4(1)	3.28(1)	331(1)	113
1,2,2,3-Tetrafluoro-1-propanol	$C_3H_4F_4O$	386.5	554(2)	3.3(2)	280(15)	362
1,1,2,2-Tetrafluoro-3-(1,1,2,2-tetrafluoroethoxy)propane	$C_5H_4F_8O$	366.32	510.07(8)	2.58(1)	440(2)	113
3,4,4,4-Tetrafluoro-3-trifluoromethylbutan-2-one	$C_5H_3F_7O$	328.76	468(1)	2.50(6)	409(5)	110
4,4,5,5-Tetrafluoro-2-trifluoromethyl-1,3-dioxolane	$C_4HF_7O_2$	304.7	435(1)	2.62(7)	376(4)	110
Tetrahydrofuran	C_4H_8O	339.2(1)	540(1)	5.29(6)	223(2)	23, 241, 290
1,2,3,4-Tetrahydronaphthalene	$C_{10}H_{12}$	480.4(3)	720(1)	3.6(1)	431(40)	144, 242, 636
Tetrahydropyran	$C_5H_{10}O$	361.2(4)	572.0(3)	4.8(2)	278(18)	75
Tetrahydrothiophene	C_4H_8S	394.3(2)	632.0(2)	5.4(6)	276(19)	143, 241
1,2,4,5-Tetraisopropylbenzene	$C_{18}H_{30}$	532	703(1)	1.65(2)	983(83)	317
1,2,4,5-Tetramethylbenzene	$C_{10}H_{14}$	470(1)	676(2)	2.9(3)	489(11)	39
1,1,3,3-Tetramethyl-1,3-diphenyldisiloxane	$C_{16}H_{22}OSi_2$	565	750(8)			353
2,2,3,3-Tetramethylhexane	$C_{10}H_{22}$	433(2)	623.0(5)	2.51(6)	574(14)	536
2,2,5,5-Tetramethylhexane	$C_{10}H_{22}$	410(2)	581.4(5)	2.19(1)	600(14)	536
2,2,3,3-Tetramethylpentane	C_9H_{20}	413.4(4)	607.5(5)	2.74(3)	514(15)	536
2,2,3,4-Tetramethylpentane	C_9H_{20}	406.2(8)	592.6(5)	2.60(3)	517(17)	536
2,2,4,4-Tetramethylpentane	C_9H_{20}	395.4(10)	574.6(5)	2.49(1)	532(16)	536
2,3,3,4-Tetramethylpentane	C_9H_{20}	414.7(7)	607.5(5)	2.72(4)	517(17)	536
Tetramethylsilane	$C_4H_{12}Si$	299.9(5)	449(2)	2.82(1)	362(7)	637-639
Tetramethylstannane	$C_4H_{12}Sn$	350.0(10)	521.77(5)	2.98(1)		640, 641
Thiacyclohexane	$C_5H_{10}S$	414.88(4)	684(44)	6.50(7)	284(11)	104

Fluids

Name	Mol. form.	T_b/K	T_c/K	P_c/MPa	V_c/cm³ mol⁻¹	Ref. (see text)
Thiobis(trifluoromethane)	C_2F_6S	251.3	376.8(1)	3.2(5)	216(19)	442
Thiophene	C_4H_4S	357.3(1)	579.4(2)	5.7(2)	230(3)	241, 290
Thymol	$C_{10}H_{14}O$	506(3)	698(10)		528(21)	302
Toluene	C_7H_8	383.75(7)	591.9(2)	4.13(2)	314(7)	34, 63, 68, 74, 78, 79, 84, 86, 88, 90, 96, 163, 172, 258, 481, 636, 642-646
Triacontane	$C_{30}H_{62}$	724(7)	843(8)	0.6(1)	2055(69)	358
Tribromomethane	$CHBr_3$	422.4(5)	682(1)	5.8(2)	261(12)	647
Trichloroacetyl chloride	C_2Cl_4O	391.4(3)	604(2)	4.21(3)	331(54)	410
Trichloroethylsilane	$C_2H_5Cl_3Si$	371.9(7)	559.9(6)	3.34(4)	403(5)	266
Trichlorofluoromethane	CCl_3F	296.9(6)	471.1(2)	4.40(3)	248.0(9)	33, 34, 191, 271, 571
Trichloromethane	$CHCl_3$	334.4(1)	536.0(4)	5.5(2)	237(7)	18, 27, 28, 117
Trichloromethylsilane	CH_3Cl_3Si	339(2)	517.7(3)	3.52(3)	329(9)	264, 648
1,3,5-Trichloro-2,4,6-trifluorobenzene	$C_6Cl_3F_3$	472(27)	684.7(4)	3.3(1)	443(27)	25
1,1,2-Trichloro-1,2,2-trifluoroethane	$C_2Cl_3F_3$	320.8(2)	487.4(2)	3.40(2)	325(1)	191, 265, 271, 649
Tricosane	$C_{23}H_{48}$	654(9)	790(8)	0.9(1)	1527(53)	359
Tridecane	$C_{13}H_{28}$	508.6(4)	676(1)	1.68(4)	824(30)	34, 244, 249
1-Tridecanol	$C_{13}H_{28}O$	560(8)	732(7)	1.8(2)	828(14)	361
2-Tridecanone	$C_{13}H_{26}O$	541(1)	717(6)	1.8(2)	820(24)	256
3-Tridecanone	$C_{13}H_{26}O$	539	716(5)	1.7(5)	823(24)	256
4-Tridecanone	$C_{13}H_{26}O$	539	712(6)	1.7(5)	823(24)	256
5-Tridecanone	$C_{13}H_{26}O$	539	710(8)	1.7(5)	826(17)	256
6-Tridecanone	$C_{13}H_{26}O$	539	709(5)	1.7(5)	826(24)	256
7-Tridecanone	$C_{13}H_{26}O$	539(9)	708(5)	1.7(5)	830(24)	256
1-Tridecene	$C_{13}H_{26}$	506.0(7)	673(7)	1.74(5)	770(17)	365
Tridecylbenzene	$C_{19}H_{32}$	613(4)	790(8)	1.5(1)	1079(43)	258
Triethylamine	$C_6H_{15}N$	362.0(2)	535.6(3)	3.1(3)	392(25)	117, 286
1,3,5-Triethylbenzene	$C_{12}H_{18}$	489.0(9)	679(2)	2.32(1)	624(60)	332
Triethylene glycol	$C_6H_{14}O_4$	561.8(2)	775(30)	3.3(2)	454(25)	283
Trifluoroacetonitrile	C_2F_3N	204.4(8)	311.1(4)	3.61(4)	202(4)	470
1,2,3-Trifluorobenzene	$C_6H_3F_3$	368	560.3(4)	4.1(4)	296(20)	119
1,2,4-Trifluorobenzene	$C_6H_3F_3$	363	551.1(4)	4.1(6)	297(20)	119
1,3,5-Trifluorobenzene	$C_6H_3F_3$	350.2(5)	530.9(4)	3.8(2)	300(20)	119
1,1,1-Trifluoroethane	$C_2H_3F_3$	226.0(1)	345.89(7)	3.77(1)	195(15)	179, 203, 309, 478, 479, 549, 624, 650-653
2,2,2-Trifluoroethanol	$C_2H_3F_3O$	347.0(3)	498.57(5)	4.81(1)	211(12)	654, 655
2,2,2-Trifluoroethyl methyl ether	$C_3H_5F_3O$	304.77	448.98(8)	3.51(6)	277(3)	176
Trifluoroiodomethane	CF_3I	251.4(6)	396.44(6)	3.95(1)	231(3)	656-659
Trifluoromethane	CHF_3	191.2(1)	299.00(2)	4.82(1)	133(1)	310, 606, 660-664
Trifluoromethyl difluoromethyl ether	C_2HF_5O	238.2(2)	354.49(6)	3.36(2)	226(20)	203
(Trifluoromethyl)sulfur pentafluoride	CF_8S	252.4	381.2(1)	3.4(1)	284(4)	442
Trifluoromethyl 1,1,2,2-tetrafluoroethyl ether	C_3HF_7O	269.9(10)	387.8(5)	2.65(1)	341(22)	116, 571
Trifluoromethyl 1,2,2,2-tetrafluoroethyl ether	C_3HF_7O	263.6	377.26(6)	2.62(1)	321(2)	525
3,3,3-Trifluoro-1-propene	$C_3H_3F_3$	246(4)	378.6(5)	3.61(8)	229(14)	440
Trimethylamine	C_3H_9N	276.0(2)	433.0(6)	4.08(4)	254(6)	322, 323, 665
1,2,3-Trimethylbenzene	C_9H_{12}	449.2(4)	664.4(1)	3.45(3)	423(11)	79
1,2,4-Trimethylbenzene	C_9H_{12}	442.6(3)	649.1(1)	3.3(1)	436(12)	79, 86, 96, 251
1,3,5-Trimethylbenzene	C_9H_{12}	437.9(3)	637.31(10)	3.13(5)	435(12)	79
3,7,7-Trimethyl-bicyclo[4.1.0]hept-3-ene	$C_{10}H_{16}$	445	658(2)	2.9(5)	487(10)	496
2,2,3-Trimethylbutane	C_7H_{16}	354.0(1)	531.3(5)	2.96(3)	401(13)	346, 349
Trimethylchlorosilane	C_3H_9ClSi	330.8(4)	497.7(6)	3.20(3)	366(6)	266
1α,3α,5β-1,3,5-Trimethylcyclohexane	C_9H_{18}	414(2)	602(2)	2.6(3)	494(14)	74
3,3,5-Trimethylheptane	$C_{10}H_{22}$	430(3)	609.5(5)	2.32(5)	583(18)	536
2,2,5-Trimethylhexane	C_9H_{20}	397(2)	570(2)	2.46(3)	547(18)	350
2,2,3-Trimethylpentane	C_8H_{18}	383.0(4)	563.5(4)	2.73(2)	442(16)	346
2,2,4-Trimethylpentane	C_8H_{18}	372.4(2)	543.9(4)	2.57(2)	475(20)	33, 34, 86, 159, 346, 666, 667
2,3,3-Trimethylpentane	C_8H_{18}	387.9(3)	573.5(4)	2.82(3)	454(14)	346
2,3,4-Trimethylpentane	C_8H_{18}	386.6(3)	566.4(4)	2.72(2)	462(12)	346
cis-Tri(methylphenyl)trisiloxane	$C_{21}H_{24}O_3Si_3$		824(8)			353

Name	Mol. form.	T_b/K	T_c/K	P_c/MPa	V_c/cm³ mol⁻¹	Ref. (see text)
2,4,6-Trimethyl-2,4,6-triphenylcyclotrisiloxane	$C_{21}H_{24}O_3Si_3$		839(8)			353
Tris(perfluorobutyl)amine	$C_{12}F_{27}N$	451(2)	566(4)	1.24(9)	1196(66)	582
Undecafluorocyclohexane	C_6HF_{11}	335.2	477.7(7)			546
Undecane	$C_{11}H_{24}$	469.1(3)	638.8(2)	2.01(3)	683(20)	34, 86, 131, 243, 249, 363, 535
Undecanoic acid	$C_{11}H_{22}O_2$	553	728(7)	2.1(2)	741(20)	360
1-Undecanol	$C_{11}H_{24}O$	519(2)	703.0(6)	2.15(7)	707(12)	149
2-Undecanone	$C_{11}H_{22}O$	506.3(3)	688(2)	2.08(1)	692(20)	256
3-Undecanone	$C_{11}H_{22}O$	500	685(2)	2.0(4)	692(20)	256
4-Undecanone	$C_{11}H_{22}O$	501	681(2)	2.0(2)	692(20)	256
5-Undecanone	$C_{11}H_{22}O$	500	679(2)	2.0(2)	692(20)	256
6-Undecanone	$C_{11}H_{22}O$	500.6(5)	678(2)	2.02(1)	692(20)	256
Undecylbenzene	$C_{17}H_{28}$	585(3)	763(8)	1.6(1)	946(35)	258
Vinyl acetate	$C_4H_6O_2$	345.8(3)	519.2(2)	4.17(3)	269(7)	156, 440
o-Xylene	C_8H_{10}	417.6(4)	630.26(10)	3.74(1)	372(40)	34, 68, 78, 79, 88, 669
m-Xylene	C_8H_{10}	412.3(4)	616.9(3)	3.54(1)	377(7)	34, 68, 79, 88, 90, 106, 668
p-Xylene	C_8H_{10}	411.5(5)	616.17(9)	3.55(2)	372(35)	34, 68, 74, 79, 88, 90, 669
2,3-Xylenol	$C_8H_{10}O$	490.03(5)	723(1)	4.1(3)	397(15)	218
2,4-Xylenol	$C_8H_{10}O$	484.09(3)	708(1)	3.5(3)	389(15)	218
2,5-Xylenol	$C_8H_{10}O$	484.29(8)	707(1)	3.9(1)	397(15)	25
2,6-Xylenol	$C_8H_{10}O$	474.18(5)	701(1)	3.8(1)	396(15)	218
3,4-Xylenol	$C_8H_{10}O$	500.46(5)	730(1)	4.9(5)	388(15)	218
3,5-Xylenol	$C_8H_{10}O$	494.86(5)	716(1)	3.8(2)	396(15)	218

References

1. Frenkel, M., Chirico, R. D., Diky, V. V., Kazakov, A., and Muzny, C. D., *ThermoData Engine*. NIST Standard Reference Database 103b, Version 4.0 (Pure Compounds, Binary Mixtures, and Chemical Reactions, TDE-SOURCE Version 4.3), National Institute of Standards and Technology, Gaithersburg, MD – Boulder, CO, 2009, http://www.nist.gov/srd/nist103b.cfmhttp://www.nist.gov/srd/nist103b.cfm.
2. Frenkel, M., Chirico, R. D., Diky, V., Yan, X., Dong, Q., and Muzny, C., *J. Chem. Inf. Model.* 45, 816, 2005.
3. Diky, V., Muzny, C. D., Lemmon, E. W., Chirico, R. D., and Frenkel, M., *J. Chem. Inf. Model.* 47, 1713, 2007.
4. Diky, V., Chirico, R. D., Kazakov, A. F., Muzny, C., and Frenkel, M., *J. Chem. Inf. Model.* 49, 503, 2009.
5. Diky, V., Chirico, R. D., Kazakov, A. F., Muzny, C., and Frenkel, M., *J. Chem. Inf. Model.* 49, 2883, 2009.
6. Joback, K. G., and Reid, R. C., *Chem. Eng. Commun.* 57, 233, 1987.
7. Constantinou, L., and Gani, R., *AIChE J.* 40, 1697, 1994.
8. Marrero-Morejon, J., and Pardillo-Fontdevila, E., *AIChE J.* 45, 615, 1999.
9. Wilson, G. M., and Jasperson, L. V., AIChE Meeting, New Orleans, LA, 1996.
10. Chirico, R. D., Frenkel, M., Diky, V. V., Marsh, K. N., and Wilhoit, R. C., *J. Chem. Eng. Data* 48, 1344, 2003.
11. Teja, A. S., and Anselme, M. J., *AIChE Symp. Ser.* 86 (279), 115, 1990.
12. Hollmann, R., *Z. Phys. Chem., Stoechiom. Verwandtschaftsl.* 43, 129, 1903.
13. Van der Waals, J. D., *Continuity of Gas and Liquid Data*, 1st edition, Leipzig, p. 168, 1881.
14. Vandana, V., and Teja, A. S., *Fluid Phase Equilib.* 103, 113, 1995.
15. Ambrose, D., Ellender, J. H., Sprake, C. H. S., and Townsend, R., *J. Chem. Thermodyn.* 9, 735, 1977.
16. Kreglewski, A., *Rocz. Chem.* 31, 1001, 1957.
17. Swietoslawski, W., and Kreglewski, A., *Bull. Acad. Pol. Sci., Cl. 3* 2, 77, 1954.
18. Swietoslawski, W., and Kreglewski, A., *Bull. Acad. Pol. Sci., Cl. 3* 2, 187, 1954.
19. Young, S., *Sci. Proc. R. Dublin Soc.* 12, 374, 1910.
20. Young, S., *J. Chem. Soc.* 59, 903, 1891.
21. Pawlewski, B., *Ber. Dtsch. Chem. Ges.* 16, 2633, 1883.
22. Ambrose, D., and Ghiassee, N. B., *J. Chem. Thermodyn.* 19, 911, 1987.
23. Sassa, Y., Konishi, R., and Katayama, T., *J. Chem. Eng. Data* 19, 44, 1974.
24. Ambrose, D., Sprake, C. H. S., and Townsend, R., *J. Chem. Thermodyn.* 6, 693, 1974.
25. Ambrose, D., Broderick, B. E., and Townsend, R., *J. Appl. Chem. Biotechnol.* 24, 359, 1974.
26. Campbell, A. N., and Musbally, G. M., *Can. J. Chem.* 48, 3173, 1970.
27. Campbell, A. N., and Chatterjee, R. M., *Can. J. Chem.* 47, 3893, 1969.
28. Campbell, A. N., and Chatterjee, R. M., *Can. J. Chem.* 46, 575, 1968.
29. Kay, W. B., *J. Phys. Chem.* 68, 827, 1964.
30. Rosenbaum, M., M.S. Thesis, Univ. Texas, Austin, TX, 1951.
31. Kuenen, J. P., and Robson, W. G., *Philos. Mag.* 3, 622, 1902.
32. VonNiederhausern, D. M., Wilson, L. C., Giles, N. F., and Wilson, G. M., *J. Chem. Eng. Data* 45, 154, 2000.
33. Christou, G., Young, C. L., and Svejda, P., *Ber. Bunsen-Ges. Phys. Chem.* 95, 510, 1991.
34. Christou, G., Ph.D. Dissertation, Univ. Melbourne, 1988.
35. Castillo-Lopez, N., and Trejo Rodriguez, A., *J. Chem. Thermodyn.* 19, 671, 1987.
36. Trejo Rodriguez, A., and McLure, I. A., *Fluid Phase Equilib.* 12, 297, 1983.
37. Trejo Rodriguez, A., and McLure, I. A., *J. Chem. Thermodyn.* 11, 1113, 1979.
38. Khera, R., Ph.D. Thesis, Ohio State Univ., Columbus, OH, 1968.
39. Guye, P. A., and Mallet, E., *Arch. Sci. Phys. Nat.* 13, 274, 1902.
40. Guye, P. A., and Mallet, E., *C. R. Hebd. Seances Acad. Sci.* 133, 168, 1902.
41. Teja, A. S., and Rosenthal, D. J., *Experimental Results for Phase Equilibria and Pure Component Properties*, DIPPR DATA Series No. 1, p. 96, 1991.
42. Goloborod'ko, N. P., and Khodeeva, S. M., *Russ. J. Phys. Chem.* (*Engl. Transl.*) 46, 235, 1972.
43. Mislavskaya, V. S., and Khodeeva, S. M., *Zh. Fiz. Khim.* 43, 2367, 1969.
44. Khodeeva, S. M., *Russ. J. Phys. Chem.* (*Engl. Transl.*) 40, 1061, 1966.
45. Ambrose, D., and Townsend, R., *Trans. Faraday Soc.* 60, 1025, 1964.
46. Ambrose, D., *Trans. Faraday Soc.* 52, 772, 1956.

47. McIntosh, D., *J. Phys. Chem.* 11, 306, 1907.
48. Kuenen, J. P., *Philos. Mag.* 44, 174, 1897.
49. Wilson, L. C., Wilson, H. L., Wilding, W. V., and Wilson, G. M., *J. Chem. Eng. Data* 41, 1252, 1996.
50. Lespieau, R., and Chavanne, G., *C. R. Hebd. Seances Acad. Sci.* 140, 1035, 1905.
51. Wang, X., Jia, Q., Gao, J., Xia, S., and Ma, P. S., *J. Chem. Ind. Eng. (China)* 56, 1385, 2005.
52. Liang, Y.-H., Ma, P. S., and Zhang, H., *J. Chem. Ind. Eng.* (*China*) 51, 243, 2000.
53. Steele, W. V., Chirico, R. D., Knipmeyer, S. E., and Nguyen, A., *J. Chem. Thermodyn.* 23, 957, 1991.
54. VonNiederhausern, D. M., Wilson, G. M., and Giles, N. F., *J. Chem. Eng. Data* 51, 1990, 2006.
55. Wilson, G. M., VonNiederhausern, D. M., and Giles, N. F., *J. Chem. Eng. Data* 47, 761, 2002.
56. Chashkin, Yu. R., Gorbunova, V. G., and Voronel, A. V., *Zh. Exp. Teor. Fiz.* 49, 432, 1965.
57. Nikitin, E. D., and Popov, A. P., *J. Chem. Eng. Data* 53, 1371, 2008.
58. Steele, W. V., Chirico, R. D., Knipmeyer, S. E., and Nguyen, A., *J. Chem. Eng. Data* 47, 648–666, 2002.
59. Lagutkin, O. D., and Kuropatkin, E. I., *Zh. Fiz. Khim.* 55, 1329, 1981.
60. Livingston, J., Morgan, R., and Higgins, E., *Z. Phys. Chem., Stoechiom. Verwandtschaftsl.* 64, 170, 1908.
61. Liu, T., Fu, J., Wang, K., Gao, Y., and Yuan, W., *J. Chem. Eng. Data* 46, 809, 2001.
62. Nikitin, E. D., Pavlov, P. A., and Skutin, M., *Fluid Phase Equilib.* 161, 119, 1999.
63. Chirico, R. D., and Steele, W. V., *Ind. Eng. Chem. Res.* 33, 157, 1994.
64. Zhang, J., Zhao, X., and Ma, P., *Huagong Xuebao* 43, 105, 1992.
65. Knipmeyer, S. E., Archer, D. G., Chirico, R. D., Gammon, B. E., Hossenlopp, I. A., Nguyen, A., Smith, N. K., Steele, W. V., and Strube, M. M., *Fluid Phase Equilib.* 52, 185, 1989.
66. Goodwin, R. D., *J. Phys. Chem. Ref. Data* 17, 1541, 1988.
67. Brunner, E., *J. Chem. Thermodyn.* 20, 1397, 1988.
68. Ambrose, D., *J. Chem. Thermodyn.* 19, 1007, 1987.
69. Kay, W. B., and Kreglewski, A., *Fluid Phase Equilib.* 11, 251, 1983.
70. Hales, J. L., and Gundry, H. A., *J. Phys. E* 16, 91, 1983.
71. Hugill, J. A., and McGlashan, M. L., *J. Chem. Thermodyn.* 13, 429, 1981.
72. Ewing, M. B., McGlashan, M. L., and Tzias, P., *J. Chem. Thermodyn.* 13, 527, 1981.
73. Akhundov, T. S., and Abdullaev, F. G., *Izv. Vyssh. Uchebn. Zaved., Neft Gaz* 20, 73, 1977.
74. Powell, R. J., Swinton, F. L., and Young, C. L., *J. Chem. Thermodyn.* 2, 105, 1970.
75. Kobe, K. A., and Mathews, J. F., *J. Chem. Eng. Data* 15, 182, 1970.
76. Artyukhovskaya, L. M., Shimanskaya, E. T., and Shimanskii, Yu. I., *Ukr. Fiz. Zh.* (*Ukr. Ed.*) 15, 1974, 1970.
77. Skripov, V. P., and Sinitsyn, E. N., *Zh. Fiz. Khim.* 42, 309, 1968.
78. Kay, W. B., and Hissong, D. W., *Proc. - Am. Pet. Inst., Div. Refin.* 47, 653, 1967.
79. Ambrose, D., Broderick, B. E., and Townsend, R., *J. Chem. Soc. A* 633, 1967.
80. Skaates, J. M., and Kay, W. B., *Chem. Eng. Sci.* 19, 431, 1964.
81. Makhan'ko, I. G., and Nozdrev, V. F., *Akust. Zh.* 10, 249, 1964.
82. Cheng, D. C. H., *Chem. Eng. Sci.* 18, 715, 1963.
83. Connolly, J. F., and Kandalic, G. A., *J. Chem. Eng. Data* 7, 137, 1962.
84. Partington, E. J., Rowlinson, J. S., and Weston, J. F., *Trans. Faraday Soc.* 56, 479, 1960.
85. McCracken, P. G., Storvick, T. S., and Smith, J. M., *J. Chem. Eng. Data* 5, 130, 1960.
86. Ambrose, D., Cox, J. D., and Townsend, R., *Trans. Faraday Soc.* 56, 1452, 1960.
87. Krichevskii, I. R., Khazanova, N. E., and Linshits, L. R., *Tr. GIAP* No. 9, 40, 1959.
88. Simon, M., *Bull. Soc. Chim. Belg.* 66, 375, 1957.
89. Krichevskii, I. R., Khazanova, N. E., and Linshits, L. R., *Zh.Fiz. Khim.* 31, 2711, 1957.
90. Ambrose, D., and Grant, D. G., *Trans. Faraday Soc.* 53, 771, 1957.
91. Kreglewski, A., *Rocz. Chem.* 29, 754, 1955.
92. Kay, W. B., and Nevens, T. D., *Chem. Eng. Prog., Symp. Ser.* 48, 108, 1952.
93. Bender, P., Furukawa, G. T., and Hyndman, J. R., *Ind. Eng. Chem.* 44, 387, 1952.
94. Gornowski, E. J., Amick, E. H., and Hixson, A. N., *Ind. Eng. Chem.* 39, 1348, 1947.
95. Schamhardt, H. O., Thesis, Amsterdam, the Netherlands, 1908.
96. Altschul, M., *Z. Phys. Chem., Stoechiom. Verwandtschaftsl.* 11, 577, 1893.
97. Young, S., *J. Chem. Soc., Trans.* 55, 486, 1889.
98. Nikitin, E. D., Popov, A. P., and Yatluk, Y. G., *J. Chem. Eng. Data* 51, 1335, 2006.
99. VonNiederhausern, D. M., Wilson, G. M., and Giles, N. F., *J. Chem. Eng. Data* 51, 1982, 2006.
100. Chirico, R. D., Knipmeyer, S. E., Nguyen, A., and Steele, W. V., *J. Chem. Thermodyn.* 23, 759, 1991.
101. Steele, W. V., Chirico, R. D., Hossenlopp, I. A., Knipmeyer, S. E., Nguyen, A., and Smith, N. K., *Experimental Results for DIPPR 1990–91 Projects on Phase Equilibria and Pure Component Properties*, DIPPR Data Ser. No. 2, p. 188, 1994.
102. Guseinov, S. O., Naziev, Y. M., Farzaliev, B. I., and Movsunov, T. G., *Izv. Vyssh. Uchebn. Zaved., Neft Gaz* 21, 48, 1978.
103. Ambrose, D., and Ghiassee, N. B., *J. Chem. Thermodyn.* 22, 307, 1990.
104. Anselme, M. J., and Teja, A. S., *AIChE Symp. Ser.* 86 (279), 128, 1990.
105. Chirico, R. D., and Steele, W. V., *J. Chem. Thermodyn.* 36, 633, 2004.
106. Glaser, F., and Ruland, H., *Chem.-Ing.-Tech.* 29, 772, 1957.
107. Chirico, R. D., Knipmeyer, S. E., Nguyen, A., and Steele, W. V., *J. Chem. Thermodyn.* 21, 1307, 1989.
108. Reiter, R. W., *NASA Document* N63, 1963.
109. Ellard, J. A., and Yanko, W. H., *U. S. A. E. C. Rep.* IDO-11008, 1963.
110. Sako, T., Yasumoto, M., Nakazawa, N., and Kamizawa, C., *J. Chem. Eng. Data* 46, 1078, 2001.
111. Defibaugh, D. R., Gillis, K. A., Moldover, M. R., Morrison, G., and Schmidt, J. W., *Fluid Phase Equilib.* 81, 285, 1992.
112. Nikitin, E. D., Popov, A. P., and Yatluk, Y. G., *J. Chem. Eng. Data* 51, 1326, 2006.
113. Yasumoto, M., Yamada, Y., Murata, J., Urata, S., and Otake, K., *J. Chem. Eng. Data* 48, 1368, 2003.
114. McLure, I. A., and Neville, J. F., *J. Chem. Thermodyn.* 14, 385, 1982.
115. Badylkes, S., *Kholod. Tekh.* 43, 18, 1966.
116. Salvi-Narkhede, M., Wang, B. -H., Adcock, J. I., and Van Hook, W. A., *J. Chem. Thermodyn.* 24, 1065, 1992.
117. Herz, W., and Neukirch, E., *Z. Phys. Chem., Stoechiom. Verwandtschaftsl.* 104, 433, 1923.
118. Adamenko, I. I., and Chernyavskaya, I. A., *Ukr. Fiz. Zh.* 11, 336, 1966.
119. Morton, D. W., Lui, M. P. W., Tran, C. A., and Young, C. L., *J. Chem. Eng. Data* 45, 437, 2000.
120. Li, Y., Ma, P., and Ruan, Y., *Shiyou Huagong* 22, 322, 1993.
121. Higashi, Y., Uematsu, M., and Watanabe, K., *Bull. JSME* 28, 2660, 1985.
122. Scott, R. B., Meyers, C. H., Rands, R. D., Brickwedde, F. G., and Bekkedahl, N., *J. Res. Natl. Bur. Stand.* (*U.S.*) 35, 39, 1945.
123. Cragoe, C. S., *Natl. Bur. Stand.* (*U.S.*) LC-736, 1943.
124. Ma, P. S., Gao, J., and Xia, S., *Chin. J. Chem. Eng.* 10, 473, 2002.
125. Yasumoto, M., Uchida, Y., Ochi, K., Furuya, T., and Otake, K., *J. Chem. Eng. Data* 50, 596, 2005.
126. Warowny, W., *J. Chem. Eng. Data* 41, 689, 1996.
127. Holcomb, C. D., Magee, J. W., and Haynes, W. M., *Research Report RR-147*, Gas Processors Association Project 916, Tulsa, OK, 1995.
128. Deak, A., Victorov, A. I., and De Loos, T. W., *Fluid Phase Equilib.* 107, 277, 1995.
129. Vasserman, A. A., Khasilev, I. P., and Cymarnyi, V. A., *Deposited Doc. VNIIKI*, Doc. No. 604-kk, 1989.
130. Li, L., and Kiran, E., *J. Chem. Eng. Data* 33, 342, 1988.
131. Brunner, E., *J. Chem. Thermodyn.* 20, 273, 1988.

132. Younglove, B. A., and Ely, J. F., *J. Phys. Chem. Ref. Data* 16, 577, 1987.
133. Kreglewski, A., and Kay, W. B., *J. Phys. Chem.* 73, 3359, 1969.
134. Golubev, I. F., and Agaev, N. A., *Dokl. Akad. Nauk SSSR* 151(4), 875, 1963.
135. Kay, W. B., *Ind. Eng. Chem.* 33, 590, 1941.
136. Sage, B. H., Hicks, B. L., and Lascey, W. N., *Ind. Eng. Chem.* 32, 1085, 1940.
137. Kay, W. B., *Ind. Eng. Chem.* 32, 353, 1940.
138. Beattie, J. A., Simard, G. L., and Su, G.-J., *J. Am. Chem. Soc.* 61, 24, 1939.
139. Kuenen, J. P., *Commun. Kamerlingh Onnes Lab., Univ. Leiden*, No. 125, 1, 1911.
140. Nikitin, E. D., Popov, A. P., and Yatluk, Y. G., *J. Chem. Eng. Data* 51, 609, 2006.
141. Steele, W. V., Chirico, R. D., Knipmeyer, S. E., and Nguyen, A., *J. Chem. Eng. Data* 41, 1255, 1996.
142. Eliosa, G., Murrieta-Guevara, F., Reza, J., and Trejo Rodriguez, A., *Fluid Phase Equilib.* 61, 99, 1990.
143. Tsonopoulos, C., and Ambrose, D., *J. Chem. Eng. Data* 46, 480, 2001.
144. Teja, A. S., and Anselme, M. J., *AIChE Symp. Ser.* 86 (279), 122, 1990.
145. Gude, M. T., Mendez-Santiago, J., and Teja, A. S., *J. Chem. Eng. Data* 42, 278, 1997.
146. Ambrose, D., and Ghiassee, N. B., *J. Chem. Thermodyn.* 19, 505, 1987.
147. Brown, J. C., *J. Chem. Soc., Trans.* 89, 311, 1906.
148. Lydersen, A. L., and Tsochev, V., *Chem. Eng. Technol.* 13, 125, 1990.
149. Rosenthal, D. J., and Teja, A. S., *Ind. Eng. Chem. Res.* 28, 1693, 1989.
150. Christou, G., and Young, C. L., *Int. DATA Ser.*, Sel. Data Mixtures, Ser. A, 14(4), 245, 1986.
151. Naumova, A. A., Tyvina, T. N., and Fokina, V. V., *Zh. Prikl. Khim.* 1980, 1667, 1980.
152. Efremov, Yu. V., *Zh. Fiz. Khim.* 40, 1240, 1966.
153. Ambrose, D., and Townsend, R., *J. Chem. Soc.* 54, 3614, 1963.
154. Singh, R., and Shemilt, L. W., *J. Chem. Phys.* 23, 1370, 1955.
155. Kay, W. B., and Donham, W. E., *Chem. Eng. Sci.* 4, 1, 1955.
156. Stevens, R. M. M., Van Roermund, J. C., Jager, M. D., De Loos, T. W., and De Swaan Arons, J., *Fluid Phase Equilib.*, 138, 159, 1997.
157. Kobe, K. A., Crawford, H. R., and Stephenson, R. W., *Ind. Eng. Chem.* 47, 1767, 1955.
158. Ihmels, E. C., Fischer, K., and Gmehling, J., *Fluid Phase Equilib.* 228–229, 155, 2005.
159. Li, J., Qin, Z., Wang, G., Dong, M., and Wang, J., *J. Chem. Eng. Data* 52, 1736, 2007.
160. Beattie, J. A., and Marple, S., *J. Am. Chem. Soc.* 72, 1449, 1950.
161. Olds, R. H., Sage, B. H., and Lacey, W. N., *Ind. Eng. Chem.* 38, 301, 1946.
162. Morton, D. W., Lui, M., and Young, C. L., *J. Chem. Thermodyn.* 31, 675, 1999.
163. Teja, A. S., and Rosenthal, D. J., *AIChE Symp. Ser.* 86 (279), 133, 1990.
164. Ma, F., Wang, J., and Ruan, Y., *J. Chem. Eng. Chin. Univ.* 9, 62, 1995.
165. Ma, P., and Ruan, Y., *Gaoxiao Huaxue Gongcheng Xuebao* 9, 62, 1995.
166. Quadri, S. K., and Kudchadker, A. P., *J. Chem. Thermodyn.* 23, 129, 1991.
167. Pawlewski, B., *Ber. Dtsch. Chem. Ges.* 15, 2460, 1882.
168. Toczylkin. L. S., and Young, C. L., *J. Chem. Thermodyn.* 12, 365, 1980.
169. Nikitin, E. D., and Popov, A. P., *J. Chem. Eng. Data* 52, 1336, 2007.
170. Morton, D. W., Lui, M. P. W., Tran, C. A., and Young, C. L., *J. Chem. Eng. Data* 49, 283, 2004.
171. Nikitin, E. D., Popov, A. P., and Bogatishcheva, N. S., *J. Chem. Eng. Data* 48, 1137, 2003.
172. Wilson, L. C., Wilding, W. V., Wilson, H. L., and Wilson, G. M., *J. Chem. Eng. Data* 40, 765, 1995.
173. Zawisza, A. C., and Glowka, S., *Bull. Acad. Pol. Sci., Ser. Sci. Chim.* 19, 191, 1971.
174. Osipiuk, B., and Stryjek, R., *Bull. Acad. Pol. Sci., Ser. Sci. Chim.* 18, 289, 1970.
175. Steele, W. V., Chirico, R. D., Knipmeyer, S. E., Nguyen, A., and Smith, N. K., *J. Chem. Eng. Data* 41, 1285, 1996.
176. Sako, T., Sato, M., Nakazawa, N., Oowa, M., Yasumoto, M., Ito, H., And Yamashita, S., *J. Chem. Eng. Data* 41, 802, 1996.
177. Tanikawa, S., Tatoh, J., Maezawa, Y., Sato, H., and Watanabe, K., *J. Chem. Eng. Data* 37, 74, 1992.
178. Chae, H. B., Schmidt, J. W., and Moldover, M. R., *J. Phys. Chem.* 94, 8840, 1990.
179. Mears, W. H., Stahl, R. F., Orfeo, S. R., Shair, R. C., Kells, L. F., Thompson, W., and McCann, H., *Ind. Eng. Chem.* 47, 1449, 1955.
180. Yata, J., Hori, M., Kawakatsu, H., and Minamiyama, T., *Int. J. Thermophys.* 17, 65, 1996.
181. Zhao, X., and Ma, P. S., *Chin. J. Chem. Eng.* 3, 233, 1995.
182. Nishiumi, H., Kohmatsu, S., Yokoyama, T., and Konda, A., *Fluid Phase Equilib.* 104, 131, 1995.
183. Economou, I. G., Peters, C. J., Florusse, L. J., and De Swaan Arons, J., *Fluid Phase Equilib.* 111, 239, 1995.
184. Nishiumi, H., Komatsu, M., Yokoyama, T., and Kohmatsu, S., *Fluid Phase Equilib.* 83, 109, 1993.
185. Wang, J., Liu, Z. G., Tan, L. C., and Yin, J. M., *Fluid Phase Equilib.* 80, 203, 1992.
186. Noles, J. R., and Zollweg, J. A., *J. Chem. Eng. Data* 37, 306, 1992.
187. Leu, A. D., and Robinson, D. B., *J. Chem. Eng. Data* 37, 7, 1992.
188. Goodwin, A. R. H., Defibaugh, D. R., and Weber, L. A., *Int. J. Thermophys.* 13, 837, 1992.
189. Zhimai, H., and Jianfen, H., *Gongcheng Rewuli Xuebao* 10, 233, 1989.
190. He, Z., Zhang, Y., and Hong, J., *The Second Asian Thermophysical Properties Conference*, Hunan University of Science and Technology, Guangzhou, China, p. 519, 1989.
191. *Chemicals and Plastics Physical Properties*, Union Carbide Corp. (Product Bulletin), 1968.
192. Ambrose, D., and Ghiassee, N. B., *J. Chem. Thermodyn.* 20, 765, 1988.
193. Berthoud, A., *J. Chim. Phys. Phys.-Chim. Biol.* 15, 3, 1917.
194. Cullick, A. S., and Ely, J. F., *J. Chem. Eng. Data* 27, 276, 1982.
195. Mansoorian, H., Hall, K. R., Holste, J. C., and Eubank, P. T., *J. Chem. Thermodyn.* 13, 1001, 1981.
196. Centnerszwer, M., *Z. Phys. Chem., Stoechiom. Verwandtschaftsl.* 49, 199, 1904.
197. Murphy, K. P., *J. Chem. Eng. Data* 9, 259, 1964.
198. Skripov, V. P., and Muratov, G. N., *Russ. J. Phys. Chem. (Engl. Transl.)* 51, 806, 1977.
199. Yada, N., Uematsu, M., and Watanabe, K., *Nippon Kikai Gakkai Ronbunshu, B-hen* 55, 2426, 1989.
200. Mears, W. H., Rosenthal. E., and Sinka, J. V., *J. Chem. Eng. Data* 11, 338, 1966.
201. Ma, P., Fang, Z., Zhang, J., and Ruan, Y., *J. Chem. Eng. Chin. Univ.* 6, 112, 1992.
202. Ma, P., Fang, Z., Zhang, J., and Ruan, Y., *Gaoxiao Huaxue Gongcheng Xuebao* 6, 112, 1992.
203. Schmidt, J. W., Carrillo-Nava, E., and Moldover, M. R., *Fluid Phase Equilib.* 122, 187, 1996.
204. Fukushima, M., and Watanabe, N., *Nippon Reito Kyokai Ronbunshu* 10, 75, 1993.
205. Liu, Z., Liang, D., He, M., Ju, B., and Yin, J., *Gongchengrewuli Xuebao* 18, 261, 1997.
206. Booth, H. S., and Swinehart, C. F., *J. Am. Chem. Soc.* 57, 1337, 1935.
207. Oliver, G. D., Grisard, J. W., and Cunningham, C. W., *J. Am. Chem. Soc.* 73, 5719, 1951.
208. Weber, L. A., *J. Chem. Eng. Data* 34, 171, 1989.
209. Shavandrin, A. M., and Li, S. A., *Inzh.-Fiz. Zh.* 37, 830, 1979.
210. Vitkalov, V. S., Kolpakov, Y. D., and Skripov, V. P., *Zh. Fiz. Khim.* 50, 2336, 1976.
211. Oguchi, K., Tanishita, I., Watanabe, K., Yamaguchi, T., and Sasayama, A., *Bull. JSME* 18, 1456, 1975.
212. Muratov, G. N., and Skripov, V. P., *Zh. Fiz. Khim.* 49, 2148, 1975.

213. Levelt Sengers, J. M. H., Straub, J., and Vincentini-Missoni, M., *J. Chem. Phys.* 54, 5034, 1971.
214. Tsiklis, D. S., and Prokhorov, V. M., *Dokl. Akad. Nauk SSSR* 174, 470, 1967.
215. Tsiklis, D. S., and Prokhorov, V. M., *Zh. Fiz. Khim.* 41, 2195, 1967.
216. Michels, A., Wassenaar, T., Wolkers, G. J., Prins, Chr., and van de Klundert, L., *J. Chem. Eng. Data* 11, 449, 1966.
217. Delaunois, C., *Ann. Mines Belg.* No. 1, 9, 1968.
218. Ambrose, D., *Trans. Faraday Soc.* 59, 1988, 1963.
219. Dewar, J., *Philos. Mag.* 18, 210, 1884.
220. Young, C. L., *Aust. J. Chem.* 25, 1625, 1972.
221. Hicks, C. P., and Young, C. L., *Trans. Faraday Soc.* 67, 1605, 1971.
222. Fischer, R., and Reichel, T., *Mikrochem. Ver. Mikrochim. Acta* 31, 102, 1943.
223. Zhang, R., Qin, Z., Wang, G., Dong, M., Hou, X., and Wang, J., *J. Chem. Eng. Data* 50, 1414, 2005.
224. Hugill, J. A., and McGlashan, M. L., *J. Chem. Thermodyn.* 10, 95, 1978.
225. Naziev, Y. M., Abasov, A. A., Nurberdiev, A. A., and Shakhverdiev, A. N., *Zh. Fiz. Khim.* 68, 434, 1974.
226. Krichevskii, I. R., and Sorina, G. A., *Zh. Fiz. Khim.* 34, 1420, 1960.
227. Richardson, M. J., and Rowlinson, J. S., *Trans. Faraday Soc.* 53, 1586, 1959.
228. Reamer, H. H., Sage, B. H., and Lacey, W. N., *Chem. Eng. Data Ser.* 3, 240, 1958.
229. Kay, W. B., and Albert, R. E., *Ind. Eng. Chem.* 48, 422, 1956.
230. Rotinyantz, L., and Nagornov, N. N., *Z. Phys. Chem., Abt. A* 169, 20, 1934.
231. Young, S., and Fortey, E. C., *J. Chem. Soc., Trans.* 75, 873, 1899.
232. Steele, W. V., Chirico, R. D., Knipmeyer, S. E., and Nguyen, A., *J. Chem. Eng. Data* 42, 1021, 1997.
233. Ambrose, D., and Ghiassee, N. B., *J. Chem. Thermodyn.* 19, 903, 1987.
234. Alekhin, O. D., Krupskii, N. P., and Minchenko, Y. B., *Ukr. Fiz. Zh. (Ukr. Edc.)* 15, 509, 1970.
235. Kudchadker, A. P., Alani, G. H., and Zwolinski, B. J., *Chem. Rev.* 68, 659, 1968.
236. Kay, W. B., *J. Am. Chem. Soc.* 69, 1273, 1947.
237. Lin, D. C. K., Silberberg, I. H., and McKetta, J. J., *J. Chem. Eng. Data* 15, 483, 1970.
238. Booth, H. S., and Morris, W. C., *J. Phys. Chem.* 62, 875, 1958.
239. Grzyll, L. R., Ramos, C., and Back, D. D., *J. Chem. Eng. Data* 41, 446, 1996.
240. Ermakov, G. V., and Skripov, V. P., *Zh. Fiz. Khim.* 43, 1308, 1969.
241. Cheng, D. C. H., McCoubrey, J. C., and Phillips, D. G., *Trans. Faraday Soc.* 58, 224, 1962.
242. Gude, M. T., and Teja, A. S., *Experimental Results for DIPPR 1990-91 Projects on Phase Equilibria and Pure Component Properties*, DIPPR Data Series No. 2, p. 174, 1994.
243. Anselme, M. J., Gude, M., and Teja, A. S., *Fluid Phase Equilib.* 57, 317, 1990.
244. Rosenthal, D. J., and Teja, A. S., *AIChE J.* 35, 1829, 1989.
245. Smith, R. L., Teja, A. S., and Kay, W. B., *AIChE J.* 33, 232, 1987.
246. Brunner, E., *J. Chem. Thermodyn.* 19, 823, 1987.
247. Smith, R. L., Anselme, M., and Teja, A. S., *Proc. World Congress III Chem. Eng.*, Tokyo, Vol. II, p. 135, 1986.
248. Gehrig, M., and Lentz, H., *J. Chem. Thermodyn.* 15, 1159, 1983.
249. Mogollon, E., Kay, W. B., and Teja, A. S., *Ind. Eng. Chem. Fundam.* 21, 173, 1982.
250. Cholpan, P. F., Sperkach, V. S., and Garkusha, L. N., *Fiz. Zhidk. Sostoyaniya* 9, 79, 1981.
251. Kay, W. B., and Pak, S. C., *J. Chem. Thermodyn.* 12, 673, 1980.
252. Chun, S. W., Ph.D. Thesis, Ohio State Univ., Columbus, OH, 1964.
253. Nikitin, E. D., Popov, A. P., Bogatishcheva, N. S., and Yatluk, Y. G., *J. Chem. Eng. Data* 49, 1515, 2004.
254. Quadri, S. K., Khilar, K. C., Kudchadker, A. P., and Patni, M. J., *J. Chem. Thermodyn.* 23, 67, 1991.
255. Anselme, M. J., and Teja, A. S., *Fluid Phase Equilib.* 40, 127, 1988.
256. Pulliam, M. K., Gude, M. T., and Teja, A. S., *J. Chem. Eng. Data* 40, 455, 1995.
257. Gude, M. T., Rosenthal, D. J., and Teja, A. S., *Fluid Phase Equilib.* 70, 55, 1991.
258. Nikitin, E. D., Popov, A. P., Bogatishcheva, N. S., and Yatluk, Y. G., *J. Chem. Eng. Data* 47, 1012, 2002.
259. Chirico, R. D., Gammon, B. E., Knipmeyer, S. E., Nguyen, A., Strube, M. M., Tsonopoulos, C., and Steele, W. V., *J. Chem. Thermodyn.* 22, 1075, 1990.
260. Chirico, R. D., Knipmeyer, S. E., Nguyen, A., and Steele, W. V., *J. Chem. Thermodyn.* 23, 431, 1991.
261. Nisel'son, L. A., Tret'yakova, K. V., Yatko, M. E., Tsirut, E. K., and Antonova, N. P., *Thermophysical Properties of Matter and Substances*, Vol. 4, Rabinovich, V. A., Ed., Amerind Pub., New Delhi, p. 132, 1975.
262. Toczylkin. L. S., and Young, C. L., *J. Chem. Thermodyn.* 12, 355, 1980.
263. Golik, A. Z., and Ravikovich, S. D., *Zh. Fiz. Khim.* 23, 86, 1949.
264. Stepanov, N. G., *Russ. J. Phys. Chem.* (*Engl. Transl.*) 46, 464, 1972.
265. Krauss, R., and Stephan, K., *J. Phys. Chem. Ref. Data* 18, 43, 1989.
266. Stepanov, N. G., and Nozdrev, V. F., *Russ. J. Phys. Chem.* (*Engl. Transl.*) 42, 1300, 1968.
267. Garcia-Sanchez, F., and Trejo Rodriguez, A., *J. Chem. Thermodyn.* 19, 359, 1987.
268. Garcia-Sanchez, F., and Trejo Rodriguez, A., *J. Chem. Thermodyn.* 17, 981, 1985.
269. Hojendahl, K., *Mat.-Fys. Medd. - K. Dan. Vidensk. Selsk.* 24, 1, 1946.
270. Duarte-Garza, H. A., Hwang, C.-A., Kellerman, S. A., Miller, R. C., Hall, K. R., Holste, J. C., Marsh, K. N., and Gammon, B. E., *J. Chem. Eng. Data* 42, 497, 1997.
271. Benning, A. F., and McHarness, R. C., *Ind. Eng. Chem.* 32, 814, 1940.
272. Gorchakovskii, V. K., Zadov, V. E., and Podvezennyi, V. N., *Inzh.-Fiz. Zh.* 59, 122, 1990.
273. Garcia-Sanchez, F., Romero-Martinez, A., and Trejo Rodriguez, A., *J. Chem. Thermodyn.* 21, 823, 1989.
274. Higashi, Y., Uematsu, M., and Watanabe, K., *Bull. JSME* 28, 2968, 1985.
275. Piao, C. C., Sato, H., and Watanabe, K., *J. Chem. Eng. Data* 36, 398, 1991.
276. Weber, L. A., and Levelt Sengers, J. M. H., *Fluid Phase Equilib.* 55, 241, 1990.
277. Tanikawa, S., Kabata, Y., Sato, H., and Watanabe, K., *J. Chem. Eng. Data* 35, 381, 1990.
278. Fukushima, M., Watanabe, N., and Kamimura, T., *Nippon Reito Kyokai Ronbunshu* 7, 243, 1990.
279. Fukushima, M., Watanabe, N., and Kamimura, T., *Nippon Reito Kyokai Ronbunshu* 7, 189, 1990.
280. Yamashita, T., Kubota, H., Tanaka, Y., Makita, T., and Kashiwagi, H., *Proc. 10th Symp. Thermophys. Prop.*, Japan, pp. 75–78, 1989.
281. Mandlekar, A. V., Kay, W. B., Smith, R. L., and Teja, A. S., *Fluid Phase Equilib.* 23, 79, 1985.
282. Herz, W., *Z. Anorg. Allg. Chem.* 149, 230, 1925.
283. Nikitin, E. D., Pavlov, P. A., and Popov, A. P., *J. Chem. Thermodyn.* 27, 43, 1995.
284. Ratzsh, M. T., *Z. Phys. Chem. Leipzig* 243, 212, 1970.
285. Schmidt, G. C., *Justus Liebigs Ann. Chem.* 266, 266, 1891.
286. Young, C. L., *Int. DATA Ser., Sel. Data Mixtures, Ser. A*, No. 1, 66, 1975.
287. Ambrose, D., Sprake, C. H. S., and Townsend, R., *J. Chem. Thermodyn.* 4, 247, 1972.
288. Zawisza, A. C., *Bull. Acad. Pol. Sci., Ser. Sci. Chim.* 15, 291, 1967.
289. Stryjek, R., and Kreglewski, A., *Bull. Acad. Pol. Sci., Ser. Sci. Chim.* 13, 201, 1965.
290. Kobe, K. A., Ravicz, A. E., and Vohra, S. P., *J. Chem. Eng. Data* 1, 50, 1956.
291. Schroeer, E., *Z. Phys. Chem., Abt. A* 140, 379, 1929.
292. Schroeer, E., *Z. Phys. Chem., Abt. A* 140, 241, 1929.
293. Wilip, J., *Eesti Vabariigi Tartu Ulik. Toim. A* 6 (2), 1924.
294. Audant, *C. R. Hebd. Seances Acad. Sci.* 170, 1573, 1920.
295. Prins, A., and Scheffer, F. E. C., *J. Phys. Chem.* 84, 827, 1913.

296. Travers, M. W., and Usher, F. L., *Z. Phys. Chem., Stoechiom. Verwandtschaftsl.* 57, 365, 1906.
297. Centerszwer, M., and Pakalneet, A., *Z. Phys. Chem., Stoechiom. Verwandtschaftsl.* 55, 303, 1906.
298. Smits, A., *Z. Phys. Chem., Stoechiom. Verwandtschaftsl.* 52, 587, 1905.
299. Galitzine, B., and Wilip, J., *Bull. Acad. Pet.* 11, No. 3, 117, 1901.
300. De Vries, E. C., *Arch. Neerl. Sci. Exactes Nat.* 28, 215, 1895.
301. Ramsay, W., and Young, S., *Philos. Trans. R. Soc. London, A* 178, 57, 1887.
302. Radice, G., Ph.D. Thesis, Univ. of Geneve, 1899.
303. Vespigniani, G. R., *Gazz. Chim. Ital.* 33, 73, 1903.
304. Berthoud, A., and Brum, R., *J. Chim. Phys. Phys.-Chim. Biol.* 21, 143, 1924.
305. Grebenkov, A. J., Zhelezny, V. P., Klepatsky, P. M., Beljajeva, O. V., Chernjak, Y. A., Kotelevsky, Y. G., and Timofejev, B. D., *Int. J. Thermophys.* 17, 535, 1996.
306. Holcomb, C. D., Niesen, V. G., Van Poolen, L. J., and Outcalt, S. L., *Fluid Phase Equilib.* 91, 145, 1993.
307. Higashi, Y., Ashizawa, M., Kabata, Y., Majima, T., Uematsu, M., and Watanabe, K., *JSME Int. J.* 30, 1106, 1987.
308. Otake, K., Yasumoto, M., Yamada, Y., Murata, J., and Urata, S., *J. Chem. Eng. Data* 48, 1380, 2003.
309. Pitschmann, M., and Straub, J., *Int. J. Thermophys.* 23, 877, 2002.
310. Diefenbacher, A., and Tuerk, M., *J. Chem. Thermodyn.* 31, 905, 1999.
311. Shi, L., Zhu, M., Han, L., Duan, Y., Sun, L;, and Fu, Y.-D., *Science in China, Ser. E* 41, 435, 1998.
312. Kuwabara, S., Aoyama, H., Sato, H., and Watanabe, K., *J. Chem. Eng. Data* 40, 112, 1995.
313. Higashi, Y., *Int. J. Thermophys.* 16, 1175, 1995.
314. Fu, Y. D., Han, L.-Z., and Zhu, M.-S., *Fluid Phase Equilib.* 111, 273, 1995.
315. Higashi, Y., *Int. J. Refrig.* 17, 524, 1994.
316. Malbrunot, P. F., Meunier, P. A., Scatena, G. M., Mears, W. H., Murphy, K. P., and Sinka, J. V., *J. Chem. Eng. Data* 13, 16, 1968.
317. Steele, W. V., Chirico, R. D., Cowell, A. B., Knipmeyer, S. E., and Nguyen, A., *J. Chem. Eng. Data* 47, 725, 2002.
318. Young, C. L., *Int. DATA Ser., Sel. Data Mixtures, Ser. A*, No. 1, 159, 1975.
319. Durig, J. R., and Li, Y. S., *J. Chem. Phys.* 63, 4110, 1975.
320. Bourgou, A., *Bull. Soc. Chim. Belg.* 33, 101, 1924.
321. Verevkin, S. P., Kozlova, S. A., Emel'yanenko, V. N., Nikitin, E. D., Popov, A. P., and Krasnykh, E. L., *J. Chem. Eng. Data* 51, 1896, 2006.
322. Kay, W. B., and Young, C. L., *Int. DATA Ser., Sel. Data Mixtures, Ser. A*, No. 2, 154, 1974.
323. Weaver, D. L., M.S. Thesis, Ohio State Univ., Columbus, OH, 1973.
324. Young, C. L., *Int. DATA Ser., Sel. Data Mixtures, Ser. A*, No. 1, 47, 1974.
325. Genco, J. M., Teja, A. S., and Kay, W. B., *J. Chem. Eng. Data* 25, 350, 1980.
326. Kay, W. B., and Young, C. L., *Int. DATA Ser., Sel. Data Mixtures, Ser. A*, No. 1, 52, 1975.
327. Genco, J. M., Ph.D. Thesis, Ohio State Univ., Columbus, OH, 1965.
328. Kay, W. B., *J. Am. Chem. Soc.* 68, 1336, 1946.
329. Quadri, S. K., and Kudchadker, A. P., *J. Chem. Thermodyn.* 24, 473, 1992.
330. Young, S., and Fortey, E. C., *J. Chem. Soc.* 35, 1126, 1879.
331. Steele, W. V., Chirico, R. D., Knipmeyer, S. E., Nguyen, A., and Smith, N. K., *J. Chem. Eng. Data* 42, 1037, 1997.
332. Steele, W. V., Chirico, R. D., Knipmeyer, S. E., and Nguyen, A., *J. Chem. Eng. Data* 42, 1008, 1997.
333. Ihmels, E. C. C., and Lemmon, E. W., *Fluid Phase Equilib.* 260, 36, 2007.
334. Wu, J., Liu, Z., Wang, B., and Pan, J., *J. Chem. Eng. Data* 49, 704, 2004.
335. Noles, J. R., and Zollweg, J. A., *Fluid Phase Equilib.* 66, 275, 1991.
336. Glowka, S., *Bull. Acad. Pol. Sci., Ser. Sci. Chim.* 20, 163, 1972.
337. Zawisza, A. C., and Glowka, S., *Bull. Acad. Pol. Sci., Ser. Sci. Chim.* 18, 549, 1970.
338. Edwards, J., and Maass, O., *Can. J. Res., Sect. A* 12, 357, 1935.
339. Tapp, J. S., Steacie, E. W. R., and Maass, O., *Can. J. Res.* 9, 217, 1933.
340. Cardoso, E., and Coppola, A. A., J. *Chim. Phys. Phys.-Chim. Biol.* 20, 337, 1923.
341. Cardoso, E., and Bruno, A., *J. Chim. Phys. Phys.-Chim. Biol.* 20, 347, 1923.
342. Briner, E., and Cardoso, E., *J. Chim. Phys. Phys.-Chim. Biol.* 6, 641, 1908.
343. Briner, E., and Cardoso, E., *C. R. Hebd. Seances Acad. Sci.* 144, 911, 1907.
344. Bogoslovskii, V. E., Mikhalyuk, G. I., and Shamolin, A. I., *Zh. Prikl. Khim. (Leningrad)* 45, 1154, 1972.
345. Kay, W. B., and Hissong, D. W., *Proc. - Am. Pet. Inst., Div. Refin.* 49, 13, 1969.
346. McMicking, J. H., and Kay, W. B., *Proc., Am. Pet. Inst., Sect. 3* 45, 75, 1965.
347. Chirico, R. D., Knipmeyer, S. E., Nguyen, A., and Steele, W. V., *J. Chem. Thermodyn.* 25, 1461, 1993.
348. Stern, S. A., and Kay, W. B., *J. Phys. Chem.* 61, 374, 1957.
349. Edgar, G., and Calingaert, G., *J. Am. Chem. Soc.* 51, 1540, 1929.
350. Francis, A. W., *Ind. Eng. Chem.* 49, 1779, 1957.
351. Cox, J. D., *Trans. Faraday Soc.* 56, 959, 1960.
352. Steele, W. V., Chirico, R. D., Nguyen, A., and Knipmeyer, S. E., *J. Chem. Thermodyn.* 27, 311, 1995.
353. Nikitin, E. D., Pavlov, P. A., and Popov, A. P., *J. Chem. Thermodyn.* 26, 1047, 1994.
354. Cristou, G., Young, C. L., and Svejda, P., *Fluid Phase Equilib.* 67, 45, 1991.
355. Zhuravlev, D. I., *Zh. Fiz. Khim.* 9, 875, 1937.
356. Chirico, R. D., and Steele, W. V., *J. Chem. Eng. Data* 50, 1052, 2005.
357. Smith, R. L., Ph.D. Dissertation, Georgia Institute of Technology 1985.
358. Nikitin, E. D., Pavlov, P. A., and Popov, A. P., *Fluid Phase Equilib.* 141, 155, 1997.
359. Nikitin, E. D., Pavlov, P. A., and Bessonova, N. V., *J. Chem. Thermodyn.* 26, 177, 1994.
360. Nikitin, E. D., Pavlov, P. A., and Popov, A. P., *Fluid Phase Equilib.* 189, 151, 2001.
361. Nikitin, E. D., Pavlov, P. A., and Popov, A. P., *Fluid Phase Equilib.* 149, 223, 1998.
362. Sinicyn, E. N., Mikhalevich, L. A., and Yankovskaya, O. P., *Deposited Doc. VINITI*, Doc. No. 2510-V90, 1990.
363. Teja, A. S., Gude, M., and Rosenthal, D. J., *Fluid Phase Equilib.* 52, 193, 1989.
364. Beale, E. S., and Docksey, P., *J. Inst. Pet.* 21, 860, 1935.
365. Nikitin, E. D., and Popov, A. P., *Fluid Phase Equilib.* 166, 237, 1999.
366. Horstman, S., Fischer, K., Gmehling, J., and Kolar, P., *J. Chem. Thermodyn.* 32, 451, 2000.
367. Colgate, S. O., Sivaraman, A., and Dejsupa, C., *Fluid Phase Equilib.* 76, 175, 1992.
368. Friend, D. G., Ingham, H., and Ely, J. F., *J. Phys. Chem. Ref. Data* 20, 275, 1991.
369. Jangkamolkulchai, A., and Luks, K. D., *J. Chem. Eng. Data* 34, 92, 1989.
370. Calado, J. C. G., Chang, E., Clancy, P., and Streett, W. B., *J. Phys. Chem.* 91, 3914, 1987.
371. Brunner, E., *J. Chem. Thermodyn.* 17, 871, 1985.
372. Morrison, G., and Kincaid, J. M., *AIChE J.* 30, 257, 1984.
373. Sychev, V. V., Vasserman, A. A., Kozlov, A. D., Zagoruchenko, V. A., Spiridonov, G. A., and Tsymarny, V. A., *Thermodynamic Properties of Ethane*, Standards Publishing House, Moscow, 1982.
374. Morrison, G., *J. Phys. Chem.* 85, 759, 1981.
375. Bulavin, L. A., and Shimanskii, Yu. I., *Zh. Eksp. Teor. Fiz.* 29, 482, 1979.
376. Strumpf, H. J., Collings, A. F., and Pings, C. J., *J. Chem. Phys.* 60, 3109, 1974.
377. Burton, M., and Balzarini, D., *Can. J. Phys.* 52, 2011, 1974.
378. Douslin, D. R., and Harrison, R. H., *J. Chem. Thermodyn.* 5, 491, 1973.

379. Berestov, A. T., Giterman, M. S., and Shmakov, N. G., *Sov. Phys. - JETP (Engl. Transl.)* 37, 1128, 1973.
380. Efremova, G. D., and Shvarts, A. V., *Russ. J. Phys. Chem. (Engl. Transl.)* 46, 237, 1972.
381. Miniovich, V. M., and Sorina, G. A., *Russ. J. Phys. Chem. (Engl. Transl.)* 45, 306, 1971.
382. Khazanova, N. E., and Sominskaya, E. E., *Russ. J. Phys. Chem. (Engl. Transl.)* 45, 88, 1971.
383. Bulavin, L. A., Ostanevich, Yu M., Simkina, A. P., and Stelkov, A. V., *Ukr. Fiz. Zh. (Ukr. Ed.)* 16, 90, 1971.
384. Chashkin, Yu. R., Smirnov, V. A., and Voronel, A. V., *Teplofiz. Svoistva Veshchestv Mater.* 2, 139, 1970.
385. Sliwinski, P., *Z. Phys. Chem. (Munich)* 63, 263, 1969.
386. Khazanova, N. E., Lesnevskaya, L. S., and Zakharova, A. V., *Khim. Prom-st. (Moscow)* 42, 364, 1966.
387. Matschke, D. E., and Thodos, G., *J. Chem. Eng. Data* 7, 232, 1962.
388. Schmidt, E., and Thomas, W., *Forsch. Geb. Ingenieurw.* 20B, 161, 1954.
389. Kay, W. B., and Brice, D. B., *Ind. Eng. Chem.* 45, 615, 1953.
390. Murray, F. E., and Mason, S. G., *Can. J. Chem.* 30, 550, 1952.
391. Mason, S. G., Naldrett, S. N., and Maass, O., *Can. J. Res., Sect.* B 18, 103, 1940.
392. Beattie, J. A., Su, G.-J., and Simard, G. L., *J. Am. Chem. Soc.* 61, 924, 1939.
393. Price, T. W., *J. Chem. Soc.* 107, 188, 1915.
394. Nikitin, E. D., Pavlov, P. A., and Skripov, P. V., *J. Chem. Thermodyn.* 25, 869, 1993.
395. Marshall, W. L., and Jones, E. V., *J. Inorg. Nucl. Chem.* 36, 2319, 1974.
396. Sajotschewsky, W., *Beibl. Ann. Phys.* 3, 741, 1879.
397. Mocharnyuk, R. F., *Zh. Obshch. Khim.* 30, 1098, 1960.
398. Golik, A. Z., Ravikovich, S. D., and Orishchenko, A. V., *Ukr. Khim. Zh. (Russ. Ed.)* 21, 167, 1955.
399. Polikhronidi, N. G., Abdulagatov, I. M., Stepanov, G. V., and Batyrova, R. G., *J. Supercrit. Fluids* 43, 1, 2007.
400. Hu, T., Qin, Z., Wang, G., Hou, X., and Wang, J., *J. Chem. Eng. Data* 49, 1809, 2004.
401. Sauermann, P., Holzapfel, K., Oprzynski, J., Kohler, F., Poot, W., and De Loos, T. W., *Fluid Phase Equilib.* 112, 249, 1995.
402. Mousa, A. H. N., *J. Chem. Eng. Jpn.* 20, 635, 1987.
403. Wilson, K. S., Lindley, D. D., Kay, W. B., and Hershey, H. C., *J. Chem. Eng. Data* 29, 243, 1984.
404. Hentze, G., *Thermochim. Acta* 20, 27, 1977.
405. Nozdrev, V. F., *Akust. Zh.* 2, 209, 1956.
406. Griswold, J., Haney, J. D., and Klein, V. A., *Ind. Eng. Chem.* 35, 701, 1943.
407. Battelli, A., *Mem. Torino, Ser.* 2 44, 57, 1893.
408. Ramsay, W., and Young, S., *Philos. Trans. R. Soc. London* 177, 123, 1886.
409. Strauss, O, *Beibl. Ann. Phys.* 6, 282, 1882.
410. Steele, W. V., Chirico, R. D., Knipmeyer, S. E., and Nguyen, A., *Experimental Results for DIPPR 1990–91 Projects on Phase Equilibria and Pure Component Properties*, DIPPR Data Series No. 2, 154, 1994.
411. Ambrose, D., Ellender, J. H., Gundry, H. A., Lee, D. A., and Townsend, R., *J. Chem. Thermodyn.* 13, 795, 1981.
412. Lambert, J. D., Clarke, J. S., Duke, J. F., Hicks, C. L., Lawrence, S. D., Morris, D. M., and Shone, M. G. T., *Proc. R. Soc. London, A* 249, 414, 1959.
413. Young, S., and Thomas, G. L., *J. Chem. Soc.* 63, 1191, 1893.
414. Pohland, E., and Mehl, W., *Z. Phys. Chem., Abt.* A 164, 48, 1933.
415. Nadezhdin, A., *Rep. Phys.* 23, 708, 1887.
416. Guseinov, K. D., and Zhabbarov, O., *Izv. Vyssh. Uchebn. Zaved. Neft Gaz* 2, 76, 1975.
417. Steele, W. V., Chirico, R. D., Cowell, A. B., Knipmeyer, S. E., and Nguyen, A., *J. Chem. Eng. Data* 47, 700, 2002.
418. Nowak, P., Kleinrahm, R., and Wagner, W., *J. Chem. Thermodyn.* 28, 1441, 1996.
419. Hasch, B. M., and McHugh, M. A., *Fluid Phase Equilib.* 64, 251, 1991.
420. Jahangiri, M., Jacobsen, R. T., Stewart, R. B., and McCarty, R. D., *J. Phys. Chem. Ref. Data* 15, 593, 1986.
421. Younglove, B. A., *J. Phys. Chem. Ref. Data*, Vol. 11, suppl. No. 1, Am. Chem. Soc., Washington, DC, 1982.
422. McCarty, R. D., and Jacobsen, R. T., *NBS Tech. Note (U.S.)* 1045, 1981.
423. Thomas, W., and Zander, M., *Int. J. Thermophys* 1, 383, 1980.
424. Hastings, J. R., Levelt Sengers, J. M. H., and Balfour, F. W., *J. Chem. Thermodyn.* 12, 1009, 1980.
425. Hastings, J. R., and Levelt Sengers, J. M. H., *Proc. 7th Symp. Thermophys. Prop.*, Cezairliyan, A., Ed., ASME, New York, p. 794, 1977.
426. Douslin, D. R., and Harrison, R. H., *J. Chem. Thermodyn.* 8, 301, 1976.
427. Bender, E., *Cryogenics* 15, 667, 1975.
428. Moldover, M. R., *J. Chem. Phys.* 61, 1766, 1974.
429. Angus, S., Armstrong, B., and de Reuck, K. M., *International Thermodynamic Tables of the Fluid State - 2 Ethylene*, Butterworths, London, 1974.
430. Zernov, V. S., Kogan, V. B., and Lyubetskii, S. G., *Zh. Prikl. Khim. (Leningrad)* 44, 1819, 1971.
431. Shim, J., and Kohn, J. P., *J. Chem. Eng. Data* 9, 1, 1964.
432. Kay, W. B., *Ind. Eng. Chem.* 40, 1459, 1948.
433. Diepen, G. A. M., and Scheffer, F. E. C., *J. Am. Chem. Soc.* 70, 4081, 1948.
434. Naldrett, S. N., and Maass, O., *Can. J. Res., Sect. B* 18, 118, 1940.
435. Dacey, J. R., McIntosh, R. L., and Maass, O., *Can. J. Res., Sect. B* 17, 206, 1939.
436. Maass, O., and Geddes, A. L., *Philos. Trans. R. Soc. London, A* 236, 303, 1937.
437. Lawrenson, I. J., and Lee, D. A., *J. Chem. Thermodyn.* 10, 1111, 1978.
438. Zawisza, A. C., and Glowka, S., *Bull. Acad. Pol. Sci., Ser. Sci. Chim.* 18, 555, 1970.
439. Lyons, R. L., M.S. Thesis, Pennsylvania State Univ., University Park, PA, 1985.
440. Daubert, T. E., Jalowka, J. W., and Goren, V., *AIChE Symp. Ser.* 83 (256), 128, 1987.
441. Douslin, D. R., Moore, R. T., Dawson, J. P., and Waddington, G., *J. Am. Chem. Soc.* 80, 2031, 1958.
442. Beyerlein, A. L., Desmarteau, D. D., Kul, I., and Zhao, G., *Fluid Phase Equilib.* 150, 287, 1998.
443. Parthasarathy, S., *Proc. - Indian Acad. Sci., Sect.* A 2, 497, 1935.
444. Bominaar, S. A. R. C., Trappeniers, N. J., and Biswas, S. N., *J. Phys. Chem.* 94, 1097, 1990.
445. Bominaar, S. A. R. C., Biswas, S. N., Trappeniers, N. J., and Ten Seldam, C. A., *J. Chem. Thermodyn.* 19, 959, 1987.
446. Froba, A. P., Botero, C., and Leipertz, A., *Int. J. Thermophys.* 27, 1609, 2006.
447. Uchida, Y., Yasumoto, M., Yamada, Y., Ochi, K., Furuya, T., and Otake, K., *J. Chem. Eng. Data* 49, 1615, 2004.
448. Otake, K., Uchida, Y., Yasumoto, M., Yamada, Y., Furuya, T., and Ochi, K., *J. Chem. Eng. Data* 49, 1643, 2004.
449. Hu, P., and Chen, Z. S., *Fluid Phase Equilib.* 221, 7, 2004.
450. Salvi-Narkhede, M., Adcock, J. L., Gakh, A., and Van Hook, W. A., *J. Chem. Thermodyn.* 25, 643, 1993.
451. Ambrose, D., and Ghiassee, N. B., *J. Chem. Thermodyn.* 20, 1231, 1988.
452. Myers, J. E., Hershey, H. C., and Kay, W. B., *J. Chem. Thermodyn.* 11, 1019, 1979.
453. Golik, A. Z., and Adamenko, I. I., *Ukr. Fiz. Zh. (Ukr. Ed.)* 10, 443, 1965.
454. Golik, A. Z., and Ivanova, I. I., *Zh. Fiz. Khim.* 36, 1768, 1962.
455. Kay, W. B., *Ind. Eng. Chem.* 30, 459, 1938.
456. Christou, G., Sadus, R. J., and Young, C. L., *Fluid Phase Equilib.* 67, 259, 1991.
457. Kurumov, D. S., Grigor'ev, B. A., and Vasil'ev, Yu. L., *Teplofiz. Svoistva Veshchestv Mater.* No. 27, 101, 1989.
458. De Loos, T. W., Poot, W., and De Swaan Arons, J., *Fluid Phase Equilib.* 42, 209, 1988.
459. Artyukhovskaya, L. M., Shimanskaya, E. T., and Shimanskii, Yu I., *Opt. Spektrosk.* 37, 935, 1974.
460. Artyukhovskaya, L. M., Shimanskaya, E. T., and Shimanskii, Yu I., *Sov. Phys. - JETP (Engl. Transl.)* 37, 848, 1973.

461. Artyukhovskaya, L. M., Shimanskaya, E. T., and Shimanskii, Yu I., *Zh. Eksp. Teor. Fiz.* 63, 2159, 1972.
462. Smith, L. B., Beattie, J. A., and Kay, W. C., *J. Am. Chem. Soc.* 59, 1587, 1937.
463. Beattie, J. A., and Kay, W. C., *J. Am. Chem. Soc.* 59, 1586, 1937.
464. Rosenthal, D. J., Gude, M. T., Teja, A. S., and Mendez-Santiago, J., *Fluid Phase Equilib.* 135, 89, 1997.
465. Smith, R. L., Anselme, M. J., and Teja, A. S., *Fluid Phase Equilib.* 31, 161, 1986.
466. Pulliam, M. K., Gude, M. T., and Teja, A. S., *Experimental Results For DIPPR 1990–91 Projects On Phase Equilibria and Pure Component Properties*, DIPPR Data Ser. No. 2, p. 184, 1994.
467. Naziev, Y. M., and Abasov, A. A., *Izv. Vyssh. Uchebn. Zaved., Neft Gaz* 12, 81, 1969.
468. Mousa, A. H. N., *J. Chem. Eng. Data* 26, 248, 1981.
469. Hicks, C. P., and Young, C. L., *Chem. Rev.* 75, 119, 1975.
470. Mousa, A. H. N., Kay, W. B., and Kreglewski, A., *J. Chem. Thermodyn.* 4, 301, 1972.
471. Douslin, D. R., Harrison, R. H., and Moore, R. T., *J. Chem. Thermodyn.* 1, 305, 1969.
472. Evans, F. D., and Tiley, P. F., *J. Chem. Soc. B* 134, 1966.
473. Counsell, J. F., Green, J. H. S., Hales, J. L., and Martin, J. F., *Trans. Faraday Soc.* 61, 212, 1965.
474. Saikawa, K., Kijima, J., Uematsu, M., and Watanabe, K., *J. Chem. Eng. Data* 24, 165, 1979.
475. Kijima, J., Saikawa, K., Watanabe, K., Oguchi, K., and Tanishita, I., *Proc. 7th Symp. Thermophys. Prop.*, Cezairliyan, A., Ed., ASME, New York, p. 480, 1977.
476. Kim, K. Y., Ph.D. Dissertation, Univ. Michigan, Ann Arbor, MI, 1974.
477. Swarts, F., *Bull. Soc. Chim. Belg.* 42, 114, 1933.
478. Aoyama, H., Kishizawa, G., Sato, H., and Watanabe, K., *J. Chem. Eng. Data* 41, 1046, 1996.
479. Aoyama, H., Sato, H., and Watanabe, K., *Sixteenth Japan Symposium on Thermophysical Properties*, Hiroshima, p. 173, 1995.
480. McLure, I. A., and Dickinson, E., *J. Chem. Thermodyn.* 8, 93, 1976.
481. VonNiederhausern, D. M., Wilson, G. M., and Giles, N. F., *J. Chem. Eng. Data* 45, 157, 2000.
482. Liu, J., Qin, Z., Wang, G., Hou, X., and Wang, J., *J. Chem. Eng. Data* 48, 1610, 2003.
483. Gude, M. T., and Teja, A. S., *Fluid Phase Equilib.* 83, 139, 1993.
484. Grigor'ev, B. A., Rastorguev, Yu. L., Gerasimov, A. A., Kurumov, D. S., and Plotnikov, S. A., *Int. J. Thermophys.* 9, 439, 1988.
485. Zawisza, A., *J. Chem. Thermodyn.* 17, 941, 1985.
486. Mousa, A. H. N., *J. Chem. Thermodyn.* 9, 1063, 1977.
487. Young, C. L., *Int. DATA Ser., Sel. Data Mixtures, Ser. A*, No. 1, 157, 1975.
488. Nichols, W. B., Reamer, H. H., and Sage, B. H., *AIChE J.* 3, 262, 1957.
489. VonNiederhausern, D. M., Wilson, G. M., and Giles, N. F., *J. Chem. Eng. Data* 51, 1986, 2006.
490. Ma, P., Ma, Y., and Zhang, J., *Gaoxiao Huaxue Gongcheng Xuebao* 5, 175, 1991.
491. Masui, G., Honda, Y., and Uematsu, M., *J. Chem. Thermodyn.* 38, 1711, 2006.
492. Goodwin, R. D., and Haynes, W. M., *NBS Tech. Note (U.S.)* No. 1051, 1982.
493. Pryanikova, R. O., Plenkina, R. M., Kuzyakina, N. V., and Markina, I. A., *Khim. Prom-st. (Moscow)*, 13. 1987.
494. Beattie, J. A., Ingersoll, H. G., and Stockmayer, W. H., *J. Am. Chem. Soc.* 64, 546, 1942.
495. Vohra, S. P., and Kobe, K. A., *J. Chem. Eng. Data* 4, 329, 1959.
496. Smith, R. L., Negishi, E., Arai, K., and Saito, S., *J. Chem. Eng. Jpn.* 23, 99, 1990.
497. Steele, W. V., Chirico, R. D., Cowell, A. B., Knipmeyer, S. E., and Nguyen, A., *J. Chem. Eng. Data* 42, 1053, 1997.
498. Setzmann, U., and Wagner, W., *J. Phys. Chem. Ref. Data* 20, 1061, 1991.
499. Friend, D. G., Ely, J. F., and Ingham, H., *J. Phys. Chem. Ref. Data* 18, 583, 1989.
500. Kleinrahm, R., and Wagner, W., *J. Chem. Thermodyn.* 18, 739, 1986.
501. Calado, J. C. G., Dieters, U., and Strett, W. B., *J. Chem. Soc., Faraday Trans. 1* 77, 2503, 1981.
502. Angus, S., Armstrong, B., and De Reuck, K. M., *International Thermodynamic Tables Of The Fluid State - 5 Methane*, Pergamon, Oxford, 1978.
503. Goodwin, R. D., *NBS Tech. Note (U.S.)* No. 653, 1974.
504. Gielen, H., Jansoone, F., and Verbeke, O. B., *J. Chem. Phys.* 59, 5763, 1973.
505. Jansoone, V., Gielen, H., De Boelpaep, J., and Verbeke, O. B., *Physica (Amsterdam)* 46, 213, 1970.
506. Terry, M. J., Lynch, J. T., Bunclark, M., Mansell, K. R., and Staveley, L. A. K., *J. Chem. Thermodyn.* 1, 413, 1969.
507. Grigor, A. F., and Steele, W. A., *J. Chem. Phys.* 48, 1032, 1968.
508. Keyes, F. G., Taylor, R. S., and Smith, L. B., *J. Math. Phys. (Cambridge, Mass.)* 1, 211, 1922.
509. Kuenen, J. P., *Philos. Mag.* 6, 637, 1903.
510. Crismer, L., *Bull. Soc. Chim. Belg.* 18, 18, 1904.
511. Schmidt, G. C., *Z. Phys. Chem., Stoechiom. Verwandtschaftsl.* 8, 628, 1891.
512. Polikhronidi, N. G., Radzhabova, L. M., Rasulov, A. R., and Stepanov, G. V., *High Temp. (Engl. Transl.)* 44, 512, 2006.
513. Francesconi, Artur Zaghini, Lentz, H., and Franck, E. U., *J. Phys. Chem.* 85, 3303, 1981.
514. Swami, D. R., Kumarkrishna Rao, V. N., and Narasinga Rao, N., *Trans., Indian Inst. Chem. Eng.* 9, 32, 1956.
515. Salzwedel, E., *Ann. Phys. (Leipzig)* 5, 853, 1930.
516. Ramsay, W., and Young, S., *Philos. Trans. R. Soc. London, A* 178, 313, 1887.
517. Steele, W. V., Chirico, R. D., Cowell, A. B., Knipmeyer, S. E., and Nguyen, A., *J. Chem. Eng. Data* 47, 667, 2002.
518. Gurarii, L. L., Kuleshov, G. G., Baglai, A. K., and Petrashkevich, R. I., *Khim.-Farm. Zh.* 21, 247, 1987.
519. Guseinov, S. O., Farzaliev, B. I., and Naziev, Y. M., *Izv. Vyssh. Uchebn. Zaved., Neft Gaz* 22, 52, 1979.
520. Kreglewski, A., *Bull. Acad. Pol. Sci., Cl.* 3 2, 191, 1954.
521. Kiyama, R., Suzuki, K., and Ikegami, T., *Rev. Phys. Chem. Jpn.* 21, 50, 1951.
522. Abara, J. A., Jennings, D. W., Kay, W. B., and Teja, A. S., *J. Chem. Eng. Data* 33, 242, 1988.
523. Wilson, G. M., Johnston, R. H., Hwang, S.-C., and Tsonopoulos, C., *Ind. Eng. Chem. Process Des. Dev.* 20, 94, 1981.
524. Rutenberg, O. L., and Shakhova, S. F., *Russ. J. Phys. Chem. (Engl. Transl.)* 47, 124, 1973.
525. Yasumoto, M., Uchida, Y., Ochi, K., Furuya, T., Shono, A., and Otake, K., *J. Chem. Eng. Data* 52, 1726, 2007.
526. Day, H. O., and Felsing, W. A., *J. Am. Chem. Soc.* 74, 1951, 1952.
527. Ambrose, D., and Ghiassee, N. B., *J. Chem. Thermodyn.* 20, 767, 1988.
528. Chirico, R. D., Knipmeyer, S. E., Nguyen, A., and Steele, W. V., *J. Chem. Thermodyn.* 31, 339, 1999.
529. Chirico, R. D., Johnson, R. D. I., and Steele, W. V., *J. Chem. Thermodyn.* 39, 698, 2007.
530. Chirico, R. D., and Steele, W. V., *J. Chem. Eng. Data* 50, 697, 2005.
531. Schroeer, E., *Z. Phys. Chem., Abt. B* 49, 271, 1941.
532. Dawson, P. P., Silberberg, I. H., and McKetta, J. J., *J. Chem. Eng. Data* 18, 7, 1973.
533. Ambrose, D., Counsell, J. F., and Hicks, C. P., *J. Chem. Thermodyn.* 10, 771, 1978.
534. Griffin, D. N., *J. Am. Chem. Soc.* 71, 1423, 1949.
535. Matzik, I., and Schneider, G. M., *Ber. Bunsen-Ges. Phys. Chem.* 89, 551, 1985.
536. Ambrose, D., and Townsend, R., *Trans. Faraday Soc.* 64, 2622, 1968.
537. Defibaugh, D. R., Carrillo-Nava, E., Hurly, J. J., Moldover, M. R., Schmidt, J. W., and Weber, L. A., *J. Chem. Eng. Data* 42, 488, 1997.
538. Young, C. L., *J. Chem. Thermodyn.* 4, 65, 1972.
539. Lindley, D. D., and Hershey, H. C., *Fluid Phase Equilib.* 55, 109, 1990.
540. Kreglewski, A., *Bull. Acad. Pol. Sci., Cl.* 3 5, 323, 1957.

541. Young, S., *J. Chem. Soc.* 77, 1145, 1900.
542. Steele, W. V., Chirico, R. D., Knipmeyer, S. E., Nguyen, A., Smith, N. K., and Tasker, I. R., *J. Chem. Eng. Data* 41, 1269, 1996.
543. Post, R. G., *Unpublished Rep., Chem. Eng. No.* 362, Univ. Texas, Austin, TX, 1950.
544. Hess, L. G., and Tilton, V. V., *Ind. Eng. Chem.* 42, 1251, 1950.
545. Thodos, G., *AIChE J.* 3, 428, 1957.
546. Cheng, D. C. H., and McCoubrey, J. C., *J. Chem. Soc.* 4993, 1963.
547. Teja, A. S., and Smith, R. L., *AIChE J.* 33, 1560, 1987.
548. Ambrose, D., and Sprake, C. H. S., *J. Chem. Soc. A* 1263, 1971.
549. Yata, J., Hori, M., Kohno, K., and Minamiyama, T., *High Temp. - High Press.* 29, 19, 1997.
550. Duarte-Garza, H. A., Stouffer, C. E., Hall, K. R., Hall, K. R., Holste, J. C., Marsh, K. N., and Gammon, B. E., *J. Chem. Eng. Data* 42, 745, 1997.
551. Ye, F., Sato, H., and Watanabe, K., *J. Chem. Eng. Data* 40, 148, 1995.
552. Tsvetkov, O. B., Kletskii, A. V., Laptev, Yu. A., Asambaev, A. J., and Zausaev, I. A., *Int. J. Thermophys.* 16, 1185, 1995.
553. Sagawa, T., Sato, H., and Watanabe, K., *High Temp. - High Press.* 26, 193, 1994.
554. Wilson, L. C., Wilding, W. V., Wilson, G. M., Rowley, R. L., Felix, V. M., and Chisolm-Carter, T., *Fluid Phase Equilib.* 80, 167, 1992.
555. Shank, R. L., *J. Chem. Eng. Data* 12, 474, 1967.
556. Gude, M. T., and Teja, A. S., *AIChE Symp. Ser.* 90 (298), 14, 1994.
557. Grigor'ev, B. A., Rastorguev, Yu. L., Kurumov, D. S., Gerasimov, A. A., Kharin, V. E., and Plotnikov, S. A., *Int. J. Thermophys.* 11, 487, 1990.
558. Kratzke, H., *AIChE J.* 31, 693, 1985.
559. Wolfe, D., Kay, W. B., and Teja, A. S., *J. Chem. Eng. Data* 28, 319, 1983.
560. Artyukhovskaya, L. M., Shimanskaya, E. T., and Shimanskii, Yu I., *Sov. Phys. - JETP* (*Engl. Transl.*) 59, 375, 1970.
561. Beattie, J. A., Levine, S. W., and Douslin, D. R., *J. Am. Chem. Soc.* 73, 4431, 1951.
562. Sage, B. H., and Lacey, W. N., *Ind. Eng. Chem.* 32, 992, 1940.
563. Quadri, S. K., and Kudchadker, A. P., *AIChE Symp. Ser.* 298, 1, 1994.
564. Lenoir, J. M., Rebert, C. J., and Hipkin, H. G., *J. Chem. Eng. Data* 16, 401, 1971.
565. Brown, J. A., and Mears, W. H., *J. Phys. Chem.* 62, 960, 1958.
566. Fowler, R. D., Hamilton, J. M., Kasper, J. S., Weber, C. E., Burford, W. B., and Anderson, H. C., *Ind. Eng. Chem.* 39, 375, 1947.
567. Martin, J. J., *J. Chem. Eng. Data* 7, 68, 1962.
568. Douslin, D. R., Moore, R. T., and Waddington, G., *J. Phys. Chem.* 63, 1959, 1959.
569. Rowlinson, J. S., and Thacker, R., *Trans. Faraday Soc.* 53, 1, 1957.
570. Ermakov, G. V., and Skripov, V. P., *Russ. J. Phys. Chem.* (*Engl. Transl.*) 41, 39, 1967.
571. Wang, B. -H., Adcock, J. L., Mathur, S. B., and Van Hook, W. A., *J. Chem. Thermodyn.* 23, 699, 1991.
572. Taylor, Z. L., and Reed, T. M., *AIChE J.* 16, 738, 1970.
573. Vandana, V., Rosenthal, D. J., and Teja, A. S., *Fluid Phase Equilib.* 99, 209, 1994.
574. Milton, H. T., and Oliver, G. D., *J. Am. Chem. Soc.* 74, 3951, 1952.
575. Oliver, G. D., and Grisard, J. W., *J. Am. Chem. Soc.* 73, 1688, 1951.
576. Oliver, G. D., Blumkin, S., and Cunningham, C. W., *J. Am. Chem. Soc.* 73, 5722, 1951.
577. Mousa, A. H. N., *J. Chem. Eng. Data* 23, 133, 1978.
578. McLure, I. A., Trejo Rodriguez, A., and Soares, V. A. M., *J. Chem. Thermodyn.* 14, 402, 1982.
579. Ewing, M. B., and Sanchez Ochoa, J. C., *J. Chem. Thermodyn.* 30, 189, 1998.
580. Ernst, G., Gurtner, J., and Wirbser, H., *J. Chem. Thermodyn.* 29, 1125, 1997.
581. Brown, J. A., *J. Chem. Eng. Data* 8, 106, 1963.
582. Young, C. L., *Int. DATA Ser., Sel. Data Mixtures, Ser. A*, No. 4, 291, 1985.
583. Sinitsyn, E. N., Mikhalevich, L. A., Biryukova, L. V., Danilov, N. N., Muratov, G. N., and Fedorov, A. P., *Deposited Doc. VINITI*, Doc. No. 1516-80, 1980.
584. Nikitin, E. D., Popov, A. P., and Yatluk, Y. G., *J. Chem. Eng. Data* 52, 315, 2007.
585. Honda, Y., Sato, T., and Uematsu, M., *J. Chem. Thermodyn.* 40, 208, 2008.
586. Jou, F.-Y., Carroll, J. J., and Mather, A. E., *Fluid Phase Equilib.* 109, 235, 1995.
587. Sychev, V. V., Vasserman, A. A., Kozlov, A. D., and Tsymarny, V. A., *Thermodynamic Properties Of Propane*, Standards Publishing House, Moscow, 1989.
588. Goodwin, R. D., and Haynes, W. M., *NBS Monogr.* (*U.S.*) No. 170, 1982.
589. Barber, J. R., Kay, W. B., and Teja, A. S., *AIChE J.* 28, 134, 1982.
590. Yesavage, V. F., Katz, D. L., and Powers, J. E., *J. Chem. Eng. Data* 14, 197, 1969.
591. Clegg, H. P., and Rowlinson, J. S., *Trans. Faraday Soc.* 51, 1333, 1955.
592. Kay, W. B., and Rambosek, G. M., *Ind. Eng. Chem.* 45, 221, 1953.
593. Meyers, C. H., *J. Res. Natl. Bur. Stand.* (U.S.) 29, 157, 1942.
594. Meyer, R. E., Ph.D. Thesis, Pennsylvania State Univ., University Park, PA, 1941.
595. Deschner, W. W., and Brown, G. G., *Ind. Eng. Chem.* 32, 836, 1940.
596. Beattie, J. A., Poffenberger, N., and Hadlock, C., *J. Chem. Phys.* 3, 96, 1935.
597. Efremova, G. D., and Sokolova, E. S., *Russ. J. Phys. Chem.* (*Engl. Transl.*) 46, 1084, 1972.
598. Anonymous, B., *International Critical Tables of Numerical Data, Phys., Chem. Technol.*, Vol. III, Washburn, E. W., Ed., McGraw-Hill, New York, 1928.
599. Kuenen, J. P., and Robson, W. G., *Philos. Mag.* 4, 116, 1902.
600. Ramsay, W., and Young, S., *Philos. Trans. R. Soc. London*, A 180, 137, 1889.
601. Oh, B. C., Lee, S., Seo, J., and Kim, H., *J. Chem. Eng. Data* 49, 221, 2004.
602. Ambrose, D., Counsell, J. F., Lawrenson, I. J., and Lewis, G. B., *J. Chem. Thermodyn.* 10, 1033, 1978.
603. Seibert, F. M., and Burrell, G. A., *J. Am. Chem. Soc.* 37, 2683, 1915.
604. Lu, H., Newitt, D. M., and Ruhemann, M., *Proc. R. Soc. London, A* 178, 506, 1941.
605. Maass, O., and Wright, C. H., *J. Am. Chem. Soc.* 43, 1098, 1921.
606. Ohgaki, K., Umezono, S., and Katayama, T., *J. Supercrit. Fluids* 3, 78, 1990.
607. Marchman, H., Prengle, H. W., and Motard, R. L., *Ind. Eng. Chem.* 41, 2658, 1949.
608. Farrington, P. S., and Sage, B. H., *Ind. Eng. Chem.* 41, 1734, 1949.
609. Vaughan, W. E., and Graves, N. R., *Ind. Eng. Chem.* 32, 1252, 1940.
610. Winkler, C. A., and Maass, O., *Can. J. Res.* 9, 613, 1933.
611. Vohra, S. P., Kang, T.-L., Kobe, K. A., and McKetta, J. J., *J. Chem. Eng. Data* 7, 150, 1962.
612. Steele, W. V., Chirico, R. D., Knipmeyer, S. E., and Nguyen, A., *J. Chem. Eng. Data* 47, 689, 2002.
613. Chirico, R. D., Steele, W. V., Nguyen, A., Klots, T. D., and Knipmeyer, S. E., *J. Chem. Thermodyn.* 28, 797, 1996.
614. Mandel, H., and Ewbank, N., *Atomics International* NAA-S-R-5129, 1960.
615. Gallant, R. W., *Hydrocarbon Process.* 45, 161, 1966.
616. Altunin, V. V., Geller, V. Z., Kremenvskaya, E. A., Perel'shtein, I. I., and Petrov, E. K., *Thermophysical Properties of Freons, Methane Ser.*, Part 2, Vol. 9, NSRDS-USSR, Selover, T. B., Ed., Hemisphere, New York, 1987.
617. Toczylkin. L. S., and Young, C. L., *Aust. J. Chem.* 30, 1591, 1977.
618. Kordes, E., . *Elektrochem.* 58, 76, 1954.
619. Lewis, D. T., *J. Appl. Chem.* 3, 154, 1953.
620. Steele, W. V., Chirico, R. D., Nguyen, A., Hossenlopp, I. A., and Smith, N. K., *DIPPR Data Ser.* 1, 101, 1991.
621. Poot, W., and De Loos, T. W., *Fluid Phase Equilib.* 222–223, 255, 2004.
622. Poot, W., and De Loos, T. W., *Fluid Phase Equilib.* 210, 69, 2003.
623. Yata, J., Hori, M., Niki, M., Isono, Y., and Yanagitani, Y., *Fluid Phase Equilib.* 174, 221, 2000.
624. Fujiwara, K., Nakamura, S., and Noguchi, M., *J. Chem. Eng. Data* 43, 55, 1998.
625. Morrison, G., and Ward, D., *Fluid Phase Equilib.* 62, 65, 1991.

626. Piao, C. C., Sato, H., and Watanabe, K., *ASHRAE Trans.* 96, 132, 1990.
627. Piao, C. C., Sato, H., and Watanabe, K., *ASHRAE Trans.* 41, 132, 1989.
628. Kubota, H., Yamashita, T., Tanaka, Y., and Makita, T., *Int. J. Thermophys.* 10, 629, 1989.
629. Kabata, Y., Tanikawa, S., Uematsu, M., and Watanabe, K., *Int. J. Thermophys.* 10, 605, 1989.
630. Basu, R. S., and Wilson, D. P., *Int. J. Thermophys.* 10, 591, 1989.
631. Wilson, D. P., and Basu, R. S., *ASHRAE Trans.* 94, 2095, 1988.
632. Tatoh, J., Kuwabara, S., Sato, H., and Watanabe, K., *J. Chem. Eng. Data* 38, 116, 1993.
633. Lebedeva, E. S., and Khodeeva, S. M., *Zh. Fiz. Khim.* 41, 2081, 1967.
634. Renfrew, M. M., and Lewis, E. E., *Ind. Eng. Chem.* 38, 870, 1946.
635. Chari, N. C., Ph.D. Dissertation, Univ. Michigan, Ann Arbor, MI, 1960.
636. Steele, W. V., Chirico, R. D., Knipmeyer, S. E., and Smith, N. K., *Report*, NIPPR-360, NTIS Order No. DE89000709, Dec. 1988.
637. McGlashan, M. L., and McKinnon, I. R., *J. Chem. Thermodyn.* 9, 1205, 1977.
638. Cipollint, N. E., and Allen, A. O., *J. Chem. Phys.* 67, 131, 1977.
639. Hicks, C. P., and Young, C. L., *J. Chem. Soc., Faraday Trans.* 1 72, 122, 1976.
640. Hugill, J. A., and McGlashan, M. L., *J. Chem. Thermodyn.* 10, 85, 1978.
641. Bendtsen, J., *J. Raman Spectrosc.* 6, 306, 1977.
642. Abdulagatov, I. M., Polikhronidi, N. G., Bruno, T. J., Batyrova, R. G., and Stepanov, G. V., *Fluid Phase Equilib.* 263, 71, 2008.
643. Polikhronidi, N. G., Abdulagatov, I. M., Magee, J. W., and Batyrova, R. G., *J. Chem. Eng. Data* 46, 1064, 2001.
644. Goodwin, R. D., *J. Phys. Chem. Ref. Data* 18, 1565, 1989.
645. Akhundov, T. S., and Abdullaev, F. G., *Izv. Vyssh. Uchebn. Zaved., Neft Gaz* 12, 44, 1969.
646. Krase, N. W., and Goodman, J. B., *Ind. Eng. Chem.* 22, 13, 1930.
647. Buchowski, H., Janaszewski, B., and Teperek, J., *Bull. Acad. Pol. Sci., Ser. Sci. Chim.* 14, 403, 1966.
648. Sokolova, T. D., Prokof'eva, N. K., and Nisel'son, L. A., *Russ. J. Phys. Chem.* (*Engl. Transl.*) 47, 154, 1973.
649. Mastroianni, M. J., Stahl, R. F., and Sheldon, P. N., *J. Chem. Eng. Data* 23, 113, 1978.
650. Weber, L. A., and Defibaugh, D. R., *J. Chem. Eng. Data* 41, 1477, 1996.
651. Higashi, Y., and Ikeda, T., *Fluid Phase Equilib.* 125, 139, 1996.
652. Wang, H., Ma, Y., Lu, C., and Tian, Y., *J. Eng. Thermophys.* 14, 122, 1993.
653. Fukushima, M., *Nippon Reito Kyokai Ronbunshu* 10, 87, 1993.
654. Bier, K., Turk, M., and Zhai, J., *Vapour Pressure of Trifluoroethanol*, Insitut für Technische Thermodynamik und Kaltetechnik, Universität Karlsruhe (TH), D 7500 Karlsruhe I, FRG, 1991.
655. Bier, K., Tuerk, M., and Zhai, J., *Proc. Int. Inst. Ref., Comm. B1 Meet.*, Herzlia, Israel, pp. 129–139, 1990.
656. Zhang, C., Duan, Y. -Y., Shi, L., Zhu, M. -S., and Han, L. -Z., *J. Tsinghua Univ.* (*Sci. & Technol.*) 40, 77, 2000.
657. Duan, Y. -Y., Shi, L., Zhu, M. -S., and Han, L. -Z., *J. Tsinghua Univ.* (*Sci. & Technol.*) 40, 60, 2000.
658. Duan, Y. -Y., Shi, L., Sun, L. -Q., Zhu, M. -S., and Han, L. -Z., *Int. J. Thermophys.* 21, 393, 2000.
659. Duan, Y. -Y., Shi, L., Zhu, M. -S., and Han, L. -Z., *J. Chem. Eng. Data* 44, 501, 1999.
660. Khodeeva, S. M., and Gubochkin, I. V., *Russ. J. Phys. Chem.* (*Engl. Transl.*) 51, 998, 1977.
661. Diefenbacher, A., Crone, M., and Turk, M., *J. Chem. Thermodyn.* 30, 481, 1998.
662. Hori, K., Okazaki, S., Uematsu, M., and Watanabe, K., *Proc. 8th Symp. Thermophys. Prop.*, Vol. II, Sengers, J. V., Ed., ASME, New York, pp. 370–376, 1982.
663. Wagner, W., *Kaeltetech.-Klim.* 20, 238, 1968.
664. Hou, Y.-C., and Martin, J. J., *AIChE J.* 5, 125, 1959.
665. Day, H. O., and Felsing, W. A., *J. Am. Chem. Soc.* 72, 1698, 1950.
666. Kay, W. B., and Warzel, F. M., *Ind. Eng. Chem.* 43, 1150, 1951.
667. Beattie, J. A., and Edwards, D. G., *J. Am. Chem. Soc.* 70, 3382, 1948.
668. Akhundov, T. S., and Asadullaeva, N. N., *Izv. Vyssh. Uchebn. Zaved., Neft Gaz* 11, 83, 1968.
669. Akhundov, T. S., and Imanov, Sh. Yu., *Teplofiz. Svoistva Zhidk.* 48, 1970.

CRITICAL CONSTANTS OF INORGANIC COMPOUNDS

The parameters of the liquid-gas critical point are important constants in determining the behavior of fluids. This table lists the critical temperature, pressure, and molar volume, as well as the normal boiling point, for over 140 inorganic substances. The properties and their units are:

T_b: Normal boiling point in K at a pressure of 101.325 kPa (1 atmosphere); an "sp" following the value indicates a sublimation point (temperature at which the solid is in equilibrium with the gas at a pressure of 101.325 kPa)

T_c: Critical temperature in K

P_c: Critical pressure in MPa

V_c: Critical molar volume in cm³ mol⁻¹

The number of digits given for T_b, T_c, and P_c indicates the estimated accuracy of these quantities; however, values of T_c greater than 750 K may be in error by 10 K or more. Although most V_c values are given to three figures, they cannot be assumed accurate to better than a few percent.

All values are experimentally determined except for a few values, indicated by an asterisk*, that are based on extrapolations. Methods of measurement are described and critiqued in Ref. 1. Compounds are listed alphabetically by name.

Name	Formula	T_b/K	T_c/K	P_c/MPa	V_c/cm³ mol⁻¹	Ref.
Aluminum	Al	2792	6700 *			2
Aluminum bromide	$AlBr_3$	528	763	2.89	310	3
Aluminum chloride	$AlCl_3$	453 sp	620	2.63	257	3
Aluminum iodide	AlI_3	655	983		408	3
Ammonia	NH_3	239.82	405.56	11.357	69.8	3,4
Ammonium chloride	NH_4Cl	611 sp	1155	163.5		3
Antimony(III) bromide	$SbBr_3$	561	904		300	3
Antimony(III) chloride	$SbCl_3$	493.5	794		272	3
Antimony(III) iodide	SbI_3	673	1102			3
Argon	Ar	87.302	150.687	4.863	75	3
Arsenic (gray)	As	889 sp	1673	22.3	35	3
Arsenic(III) chloride	$AsCl_3$	403	654		252	3
Arsine	AsH_3	210.7	373.1			3
Beryllium	Be	2741	5205 *			5
Bismuth	Bi	1837	4620 *			5
Bismuth tribromide	$BiBr_3$	735	1220		301	3
Bismuth trichloride	$BiCl_3$	714	1179	12.0	261	3
Boron tribromide	BBr_3	364.5	581		272	3
Boron trichloride	BCl_3	285.7	455	3.87	239	3
Boron trifluoride	BF_3	173.3	260.8	4.98	115	3
Boron triiodide	BI_3	482.7	773		356	3
Bromine	Br_2	332.0	588	10.34	127	3
Carbon dioxide	CO_2	194.686 sp	304.13	7.375	94	6
Carbon disulfide	CS_2	319.4	552	7.90	173	3
Carbon monoxide	CO	81.64	132.86	3.494	93	3,7,8
Carbon oxysulfide	OCS	223.0	375	5.88	137	3,7
Cesium	Cs	944	1938	9.4	341	9
Chlorine	Cl_2	239.11	417.0	7.991	123	3
Chlorine pentafluoride	ClF_5	260.1	416	5.27	233	3
Chlorotrifluorosilane	$SiClF_3$	203.2	307.7	3.46		3
Diborane	B_2H_6	180.66	289.8	4.05		3
Dichlorodifluorosilane	$SiCl_2F_2$	241	369.0	3.5		3
Difluoramine	NHF_2	250	403			3
cis-Difluorodiazine	N_2F_2	167.40	272	7.09		3
trans-Difluorodiazine	N_2F_2	161.70	260	5.57		3
Fluorine	F_2	85.04	144.41	5.1724	66	3
Fluorine monoxide	F_2O	128.9	215			3
Gallium(III) bromide	$GaBr_3$	552	806.7		303	3
Gallium(III) chloride	$GaCl_3$	474	694		263	3
Gallium(III) iodide	GaI_3	613	951		395	3
Germane	GeH_4	185.1	312.2	4.95	147	3
Germanium	Ge	3106	9802 *			5
Germanium(IV) bromide	$GeBr_4$	459.50	718		392	3
Germanium(IV) chloride	$GeCl_4$	359.70	553.2	3.861	330	3

Name	Formula	T_b/K	T_c/K	P_c/MPa	V_c/cm³ mol⁻¹	Ref.
Germanium(IV) iodide	GeI_4	621	973		500	3
Hafnium(IV) bromide	$HfBr_4$	596 sp	746		415	3
Hafnium(IV) chloride	$HfCl_4$	590 sp	725.7	5.42	314	3
Hafnium(IV) iodide	HfI_4	667 sp	916		528	3
Helium	He	4.222	5.1953	0.22746	57	3
Hydrazine	N_2H_4	386.70	653	14.7		3
Hydrogen	H_2	20.271	32.938	1.2858	65	3
Hydrogen bromide	HBr	206.77	363.2	8.55		3
Hydrogen chloride	HCl	188	324.7	8.31	81	3
Hydrogen fluoride	HF	293	461	6.48	69	3
Hydrogen iodide	HI	237.60	424.0	8.31		3
Hydrogen peroxide	H_2O_2	423.4	728 *	22 *		10
Hydrogen selenide	H_2Se	231.90	411	8.92		3
Hydrogen sulfide	H_2S	213.60	373.1	9.00	99	3,7
Iodine	I_2	457.6	819		155	3
Iodine bromide	IBr	389 dec	719		139	3
Iron	Fe	3134	9340 *			5
Krypton	Kr	119.735	209.48	5.525	91	3,7
Lithium	Li	1615	3223 *	67 *	66	11
Manganese	Mn	2334	4325 *			5
Mercury	Hg	629.769	1764	167	43	3,12
Mercury(II) bromide	$HgBr_2$	591	1012			3
Mercury(II) chloride	$HgCl_2$	577	973		174	3
Mercury(II) iodide (yellow)	HgI_2	624	1072			3
Molybdenum(V) chloride	$MoCl_5$	541	850		369	3
Molybdenum(VI) fluoride	MoF_6	307.2	473	4.75	226	3
Neon	Ne	27.104	44.4918	2.6786	42	3
Niobium(V) chloride	$NbCl_5$	520.6	803.5	4.88	397	3
Niobium(V) fluoride	NbF_5	507	737	6.28	155	3
Nitric oxide	NO	121.41	180	6.48	58	3
Nitrogen	N_2	77.355	126.192	3.3958	90	3
Nitrogen chloride difluoride	$NClF_2$	206	337.5	5.15		3
Nitrogen tetroxide	N_2O_4	294.30	431	10.1	167	3
Nitrogen trifluoride	NF_3	144.40	234.0	4.46	126	3
Nitrosyl chloride	NOCl	267.7	440			3
Nitrous oxide	N_2O	184.67	309.52	7.245	97	3,7
Nitryl fluoride	NO_2F	200.8	349.5			3
Osmium(VIII) oxide	OsO_4	404.4	678			3
Oxygen	O_2	90.188	154.581	5.0430	73	3
Ozone	O_3	161.80	261.1	5.57	89	3
Perchloryl fluoride	ClO_3F	226.40	368.4	5.37	161	3
Phosphine	PH_3	185.40	324.5	6.54		3
Phosphonium chloride	PH_4Cl	246 sp	322.3	7.37		3
Phosphorothioc chloride difluoride	$PSClF_2$	279.5	439.2	4.14		3
Phosphorothioc trifluoride	PSF_3	220.90	346.0	3.82		3
Phosphorus (white)	P	553.7	994			3
Phosphorus(III) bromide	PBr_3	446.4	711		300	3
Phosphorus(III) chloride	PCl_3	349	563		264	3
Phosphorus(V) chloride	PCl_5	433 sp	646			3
Phosphorus(III) chloride difluoride	$PClF_2$	225.9	362.4	4.52		3
Phosphorus(III) dichloride fluoride	PCl_2F	287.00	463.0	4.96		3
Phosphorus(III) fluoride	PF_3	171.4	271.2	4.33		3
Potassium	K	1032	2223 *	16 *	209	11
Radon	Rn	211.5	377	6.28		3
Rhenium(VII) oxide	Re_2O_7	633	942		334	3
Rhenium(VI) oxytetrachloride	$ReOCl_4$	496	781		362	3
Rubidium	Rb	961	2093 *	16 *	247	11
Selenium (gray)	Se	958	1766	27.2		3
Selenium hexafluoride	SeF_6	226.6 sp	345.5			3
Selenium oxychloride	$SeOCl_2$	450	730	7.09	235	3
Silver	Ag	2435	6410 *			5

Name	Formula	T_b/K	T_c/K	P_c/MPa	V_c/cm³ mol⁻¹	Ref.
Sodium	Na	1156.090	2573 *	35 *	116	11
Sulfur (rhombic)	S	717.76	1314	20.7	57.0	3
Sulfur chloride pentafluoride	SF_5Cl	254.10	390.9			3
Sulfur dioxide	SO_2	263.13	430.64	7.884	122	3,7
Sulfur hexafluoride	SF_6	209.4 sp	318.723	3.77	197	3,14
Sulfur tetrafluoride	SF_4	232.70	364			3
Sulfur trioxide (β-form)	SO_3	317.7	491.0	8.2	127	3
Tantalum(V) bromide	$TaBr_5$	622.0	974		461	3
Tantalum(V) chloride	$TaCl_5$	512	767		402	3
Tellurium	Te	1261	2329 *			5
Tellurium hexafluoride	TeF_6	234.3 sp	356			3
Tellurium tetrachloride	$TeCl_4$	660	1002	8.56	310	3
Tetrabromosilane	$SiBr_4$	427	663		382	3
Tetrachlorosilane	$SiCl_4$	330.80	508.1	3.593	326	3
Tetrafluorohydrazine	N_2F_4	199	309	3.75		3
Tetrafluorosilane	SiF_4	187	259.0	3.72		3
Tetraiodosilane	SiI_4	560.50	944		558	3
Tin(IV) bromide	$SnBr_4$	478	744		417	3
Tin(IV) chloride	$SnCl_4$	387.30	591.9	3.75	351	3
Tin(IV) iodide	SnI_4	637.50	968		531	3
Titanium(IV) bromide	$TiBr_4$	506.7	795.7		391	3
Titanium(IV) chloride	$TiCl_4$	409.60	638	4.66	339	3
Titanium(IV) iodide	TiI_4	650	1040		505	3
Tribromosilane	$SiHBr_3$	382	610.0		305	3
Trichlorofluorosilane	$SiCl_3F$	285.40	438.6	3.58		3
Trichlorosilane	$SiHCl_3$	306	479		268	3
Trifluoramine oxide	NOF_3	185.7	303	6.43	147	3
Tungsten(VI) chloride	WCl_6	610	923		422	3
Tungsten(VI) fluoride	WF_6	290.3	444	4.34	233	3
Tungsten(VI) oxytetrachloride	$WOCl_4$	503	782		338	3
Uranium(VI) fluoride	UF_6	329.7 sp	505.8	4.66	250	3
Vanadyl chloride	VOCl	400	636		171	3
Water	H_2O	373.124	647.10	22.06	56	3
Xenon	Xe	165.051	289.733	5.8420	118	7,15
Xenon difluoride	XeF_2	387.50 sp	631	9.32	148	3
Xenon tetrafluoride	XeF_4	388.90 sp	612	7.04	188	3
Zirconium(IV) bromide	$ZrBr_4$	633 sp	805		424	3
Zirconium(IV) chloride	$ZrCl_4$	604 sp	778	5.77	319	3
Zirconium(IV) iodide	ZrI_4	704 sp	960		530	3

References

1. Ambrose, D., and Young, C. L., *J. Chem. Eng. Data* 40, 345, 1995.
2. Morel, V., Bultel, A., and Chéron, B. G., Int. *J. Thermophys.* 30, 1853, 2009.
3. Ambrose, D., "Vapor-Liquid Constants of Fluids," in *Handbook of the Thermodynamics of Organic Compounds*, Stevenson, R. M., and Malanowski, S., Eds., Elsevier, New York, 1987.
4. Sato, M., Masui, G., and Uematsu, M., *J. Chem. Thermodyn.* 37, 931, 2005.
5. Velasco, S., Roman, F. L., White, J. A., and Mulero, A., *Fluid Phase Equilib.* 244, 11, 2006.
6. Nowak, P., Tielkes, T., Kleinrahm, R., and Wagner, W., *J. Chem. Thermodyn.* 29, 885, 1997.
7. Lemmon, E. W., and Span, R., *J. Chem. Eng. Data* 51, 785, 2006.
8. Goodwin, R. D., *J. Phys. Chem. Ref. Data* 14, 849, 1985.
9. Vargaftik, N. B., *Int.J. Thermophys.* 11, 467, 1990.
10. Nikitin, E. D., Pavlov, P. A., Popov, A. P., and Nikitina, H. E., *J. Chem. Thermodyn.* 27, 945, 1995.
11. Dillon, I. G., Nelson, P. A., and Swanson, B. S., *J. Chem. Phys.* 44, 4229, 1966.
12. Huber, M. L., Laesecke, A., and Friend, D. G., *The Vapor Pressure of Mercury,* NISTIR 6643, National Institute of Standards and Technology, Boulder, CO, March 2006; *Ind. Eng. Chem. Res.* 45, 7351, 2006.
13. Rau, H., Kutty, T. R. N., and Guedes de Carvalho, J. R. F., *J. Chem. Thermodyn.* 5, 291, 1973.
14. Funke, M., Kleinrahm, R., and Wagner, W., *J. Chem. Thermodyn.* 34, 717, 2002.
15. Sifner, O., and Klomfar, J., *J. Phys. Chem. Ref. Data* 23, 63, 1994.

SUBLIMATION PRESSURE OF SOLIDS

This table gives the sublimation (vapor) pressure of some representative solids as a function of temperature. Entries include simple inorganic and organic substances in their solid phase below room temperature, as well as polycyclic organic compounds which show measurable sublimation pressure only at elevated temperatures. Substances are listed by molecular formula in the Hill order, with inorganic compounds preceeding compounds containing carbon. Values marked by * represent the solid–liquid–gas triple point. Note that some pressure values are in pascals (Pa) and others are in kilopascals (kPa). For conversion, 1 kPa = 7.506 mmHg = 0.0098692 atm.

References

1. Lide, D. R. and Kehiaian, H. V., *CRC Handbook of Thermophysical and Thermochemical Data*, CRC Press, Boca Raton, FL, 1994.
2. *TRC Thermodynamic Tables*, Thermodynamic Research Center, Texas A&M University, College Station, TX.
3. Oja, V. and Suuberg, E. M., *J. Chem. Eng. Data*, 43, 486, 1998.
4. Guder, C., and Wagner, W., *J. Phys. Chem. Ref. Data* 38, 33, 2009. [sulfur hexafluoride]
5. Lemmon, E. W. and Span, R., *J. Chem. Eng. Data*, 51 785, 2006. [krypton, xenon]
6. Setzmann, U., and Wagner, W., *J. Phys. Chem. Ref. Data* 20, 1061, 1991. [methane]
7. Span, R., and Wagner, W., *J. Phys. Chem. Ref. Data* 25, 1509, 1996. [carbon dioxide]
8. Katti, R. S., Jacobsen, R. T, Stewart, R. B., and Jahangiri, M., *Adv. Cryo. Eng.* 31, 1189, 1986. [neon]
9. Span, R., Lemmon, E. W., Jacobsen, R. T, Wagner, W., and Yokozeki, A., *J. Phys. Chem. Ref. Data* 29, 1361, 2000. [nitrogen]

Sublimation Pressure of Solids

Ar	*T*/K	55	60	65	70	75	80	83.81*	
Argon	*p*/kPa	0.2	0.8	2.8	7.7	18.7	40.7	68.89*	
BrH	*T*/K	135	140	150	160	170	180	185.1*	
Hydrogen bromide	*p*/kPa	0.1	0.3	1.1	3.3	8.7	20.1	27.4*	
Br_2	*T*/K	170	180	190	200	210	220	230	240
Bromine	*p*/Pa	0.069	0.416	2.04	8.45	30.3	96.0	273	710
ClH	*T*/K	120	130	140	150	155	159.0*		
Hydrogen chloride	*p*/kPa	0.1	0.5	1.9	5.8	9.5	13.5*		
Cl_2	*T*/K	120	130	140	150	160	172.17*		
Chlorine	*p*/Pa	0.144	1.52	11.2	63.1	283	1392*		
F_4Si	*T*/K	130	140	150	160	170	175	180	186.3*
Tetrafluorosilane	*p*/kPa	0.2	0.9	3.9	14.0	43.8	74.2	122.4	220.8*
F_6S	*T*/K	150	165	180	190	200	210	220	223.554*
Sulfur hexafluoride	*p*/kPa	0.4566	2.596	10.94	25.07	52.71	103.0	189.0	232.7*
HI	*T*/K	160	170	180	190	200	210	220	222.35*
Hydrogen iodide	*p*/kPa	0.2	0.8	2.2	5.3	11.7	23.6	44.1	49.3*
H_2O	*T*/K	190	210	225	240	250	260	270	273.16*
Water	*p*/Pa	0.032	0.702	4.942	27.28	76.04	195.8	470.1	611.655*
H_2S	*T*/K	140	150	160	165	170	175	180	187.55*
Hydrogen sulfide	*p*/kPa	0.2	0.6	1.9	3.2	5.2	8.3	12.7	23.18*
H_3N	*T*/K	160	170	180	190	195	195.4*		
Ammonia	*p*/kPa	0.1	0.4	1.2	3.5	5.8	6.077*		
I_2	*T*/K	240	250	260	270	280	290	300	310
Iodine	*p*/Pa	0.081	0.297	0.971	2.89	7.92	20.1	47.9	107
Kr	*T*/K	80	90	95	100	105	110	115.775*	
Krypton	*p*/kPa	0.4160	2.670	5.840	11.81	22.35	39.39	73.53*	
N_2	*T*/K	50	55	60	63.151*				
Nitrogen	*p*/kPa	0.4005	1.800	6.298	12.52*				
Ne	*T*/K	13	15	17	19	21	22	23	24.5561*
Neon	*p*/kPa	0.003362	0.04915	0.3823	1.930	7.160	12.61	21.15	43.37*
NO	*T*/K	85	90	95	100	105	109.5*		
Nitric oxide	*p*/kPa	0.1	0.4	1.3	3.8	10.0	21.916*		
Xe	*T*/K	110	120	130	140	150	155	160	161.405*
Xenon	*p*/kPa	0.3433	1.298	4.343	12.83	33.32	50.96	74.58	81.77*
CHN	*T*/K	200	210	220	230	240	250	255	259.83*
Hydrogen cyanide	*p*/kPa	0.2	0.4	1.0	2.2	4.8	9.7	13.6	18.62*
CH_4	*T*/K	65	70	75	80	85	90	90.694*	
Methane	*p*/kPa	0.07307	0.2627	0.7965	2.102	4.949	10.59	11.696*	

CO	T/K	50	55	60	65	68.13*			
Carbon monoxide	p/kPa	0.1	0.6	2.6	8.2	15.42*			
CO_2	T/K	130	140	155	170	185	194.7	205	216.592*
Carbon dioxide	p/kPa	0.03087	0.1836	1.666	9.960	43.93	104.0	227.0	518.5*
C_2Cl_6	T/K	275	300	325	350	375	400	425	459.9*
Hexachloroethane	p/Pa	0.004	0.056	0.383	1.62	5.30	14.8	36.4	107.4*
C_2H_2	T/K	130	140	150	160	170	180	190	192.4*
Acetylene	p/kPa	0.2	0.7	2.6	7.8	20.6	49.0	106.3	126.0*
$C_2H_4O_2$	T/K	250	260	270	280	289.7*			
Acetic acid	p/kPa	0.092	0.199	0.406	0.79	1.277*			
C_5H_{12}	T/K	200	210	220	230	240	250	255	256.58*
Neopentane	p/kPa	0.7	1.6	3.6	7.3	13.9	24.8	32.4	35.8*
$C_6H_6Cl_6$	T/K	300	320	330	340	350	360	370	380
1,2,3,4,5,6-Hexachlorocyclohexane (Lindane)	p/Pa	0.01	0.13	0.39	1.04	2.66	6.42	14.8	32.7
$C_6H_6O_2$	T/K	330	340	350	360	370	380		
Resorcinol	p/Pa	1.03	2.78	7.09	17.2	39.6	87.6		
$C_6H_6O_2$	T/K	350	360	370	380	390	400		
p-Hydroquinone	p/Pa	1.20	3.18	7.96	19.0	43.4	95.1		
$C_{10}H_8$	T/K	250	270	280	290	300	310	330	353.43*
Naphthalene	p/Pa	0.036	0.514	1.662	4.918	13.43	34.15	182.9	999.6*
$C_{12}H_8N_2$	T/K	290	300	310	320				
Phenazine	p/Pa	0.0013	0.0046	0.0150	0.0448				
$C_{12}H_8O$	T/K	300	310	320	330	340	350		
Dibenzofuran	p/Pa	0.408	1.21	3.35	8.71	21.4	50.0		
$C_{12}H_9N$	T/K	350	355	360					
Carbazole	p/Pa	0.086	0.140	0.245					
$C_{13}H_7NO_2$	T/K	330	340	350	360	370	380		
Benz[*g*]isoquinoline-5,10-dione	p/Pa	0.006	0.018	0.053	0.148	0.394	0.994		
$C_{13}H_8O$	T/K	330	340	350					
1*H*-Phenalen-1-one	p/Pa	0.040	0.113	0.302					
$C_{13}H_8O_2$	T/K	400	410	420	430				
3-Hydroxy-1*H*-phenalen-1-one	p/Pa	0.006	0.018	0.053	0.144				
$C_{13}H_9N$	T/K	290	300	310	320				
Acridine	p/Pa	0.0024	0.0085	0.0278	0.0845				
$C_{13}H_9N$	T/K	310	320	330	340				
Phenanthridine	p/Pa	0.020	0.066	0.206	0.603				
$C_{14}H_{10}$	T/K	320	330	340	350	360	370	380	390
Anthracene	p/Pa	0.014	0.043	0.125	0.342	1.01	2.38	5.35	11.5
$C_{14}H_{10}$	T/K	300	310	320	330	340	350	360	
Phenanthrene	p/Pa	0.025	0.085	0.270	0.796	2.02	4.89	11.2	
$C_{16}H_{10}$	T/K	320	330	340	350	360	370	380	390
Pyrene	p/Pa	0.008	0.024	0.073	0.208	0.556	1.32	2.86	6.30
$C_{16}H_{10}O$	T/K	360	370	380	390	400			
1-Pyrenol	p/Pa	0.005	0.016	0.047	0.135	0.364			
$C_{16}H_{12}S$	T/K	330	340	350	360	370	380	390	
Benzo[*b*]naphtho(2,1-d)thiophene	p/Pa	0.001	0.004	0.012	0.036	0.098	0.255	0.631	
$C_{17}H_{12}$	T/K	340	350	360	370	380	390	400	
11*H*-Benzo[*b*]fluorene	p/Pa	0.003	0.009	0.029	0.085	0.235	0.619	1.55	
$C_{18}H_{10}O_4$	T/K	420	430	440	450				
6,11-Dihydroxy-5,12-naphthacenedione	p/Pa	0.008	0.022	0.055	0.131				
$C_{18}H_{12}$	T/K	390	400	410	420				
Chrysene	p/Pa	0.087	0.221	0.539	1.26				
$C_{18}H_{12}$	T/K	390	400	410	420	430	440	450	460
Naphthacene	p/Pa	0.005	0.014	0.035	0.084	0.194	0.432	0.928	1.929
$C_{20}H_{12}$	T/K	390	400	410	420	430			
Perylene	p/Pa	0.006	0.015	0.040	0.102	0.246			
$C_{22}H_{14}$	T/K	450	460	470	480	490			
Pentacene	p/Pa	0.002	0.006	0.013	0.031	0.069			
$C_{24}H_{12}$	T/K	430	440	450	460	470	480	490	500
Coronene	p/Pa	0.004	0.010	0.021	0.046	0.097	0.197	0.389	0.747

* Solid-liquid-gas triple point.

VAPOR PRESSURE

David R. Lide

This table gives vapor pressure data for about 1800 substances. In order to accommodate elements and compounds ranging from refractory to highly volatile in a single table, the temperature at which the vapor pressure reaches specified pressure values is listed. The pressure values run in decade steps from 1 Pa (about 7.5 μm Hg) to 100 kPa (about 750 mm Hg). All temperatures are given in °C. The symbol "s" following a value indicates that the substance is a solid at that temperature.

The data used in preparing the table came from a large number of sources; the main references used for each substance are indicated. Since the data were refit in most cases, values appearing in this table may not be identical with values in the source cited. The temperature entry in the 100 kPa column is close to, but not identical with, the normal boiling point (which is defined as the temperature at which the vapor pressure reaches 101.325 kPa). Although some temperatures are quoted to 0.1 °C, uncertainties of several degrees should generally be assumed.

More extensive and detailed vapor pressure data on selected important substances appear in other tables in this section of the Handbook. These substance names are flagged by the following footnotes:

a: More detailed data on this compound can be found in "Vapor Pressure of Fluids at Temperatures below 300 K" in this section.

b: See also "Recommended Data for Vapor Pressure Calibration" in this Section.

c: See also "Vapor Pressure of Ice" and "Vapor Pressure and Other Saturation Properties of Water" in this Section.

The following footnotes appear on individual temperature values:

d: Calculated from ideal gas thermodynamic functions (Ref. 8).

e: Value obtained by extrapolation (usually with an Antoine equation) beyond the region for which experimental measurements were available; thus subject to greater uncertainty.

References

1. Lide, D.R., and Kehiaian, H.V., CRC Handbook of *Thermophysical and Thermochemical Data*, CRC Press, Boca Raton, FL, 1994.
2. Stull, D., in *American Institute of Physics Handbook, Third Edition*, Gray, D.E., Ed., McGraw Hill, New York, 1972.
3. Hultgren, R., Desai, P.D., Hawkins, D.T., Gleiser, M., Kelley, K.K., and Wagman, D.D., *Selected Values of Thermodynamic Properties of the Elements*, American Society for Metals, Metals Park, OH, 1973.
4. Stull, D., *Ind. Eng. Chem.*, 39, 517, 1947.
5. *TRCVP, Vapor Pressure Database, Version 2.2P*, Thermodynamic Research Center, Texas A&M University, College Station, TX.
6. *TRC Thermodynamic Tables*, Thermodynamic Research Center, Texas A&M University, College Station, TX.
7. Ohe, S., *Computer Aided Data Book of Vapor Pressure*, Data Book Publishing Co., Tokyo, 1976.
8. Chase, M.W., Davies, C.A., Downey, J.R., Frurip, D.J., McDonald, R.A., and Syverud, A.N., *JANAF Thermochemical Tables, Third Edition, J. Phys. Chem. Ref. Data*, Vol. 14, Suppl. 1, 1985.
9. Barin, I., *Thermochemical Data of Pure Substances*, VCH Publishers, New York, 1993.
10. Jacobsen, R.T., et al, *International Thermodynamic Tables of the Fluid State, No. 10. Ethylene*, Blackwell Scientific Publications, Oxford, 1988.
11. Wakeham, W.A., *International Thermodynamic Tables of the Fluid State, No. 12. Methanol*, Blackwell Scientific Publications, Oxford, 1993.
12. Janz, G.J., *Molten Salts Handbook*, Academic Press, New York, 1967.
13. Ohse, R.W. *Handbook of Thermodynamic and Transport Properties of Alkali Metals*, Blackwell Scientific Publications, Oxford, 1994.
14. Gschneidner, K.A., in *CRC Handbook of Chemistry and Physics, 77th Edition*, p. 4–112, CRC Press, Boca Raton, FL, 1996.
15. Leider, H.R., Krikorian, O.H., and Young, D.A., *Carbon*, 11, 555, 1973.
16. Ruzicka, K., and Majer, V., *J. Phys. Chem. Ref. Data*, 23, 1, 1994.
17. Tillner–Roth, R., and Baehr, H.D., *J. Phys. Chem. Ref. Data*, 23, 657, 1994.
18. Younglove, B.A., and McLinden, M.O., *J. Phys. Chem. Ref. Data*, 23, 731, 1994.
19. Outcalt, S.L., and McLinden, M.O., *J. Phys. Chem. Ref. Data*, 25, 605, 1996.
20. Weber, L.A., and Defibaugh, D.R., *J. Chem. Eng. Data*, 41, 382, 1996.
21. Rodrigues, M.F., and Bernardo-Gil, M.G., *J. Chem. Eng. Data*, 41, 581, 1996.
22. Piacente, V., Gigli, G., Scardala, P., and Giustini, A., *J. Phys. Chem.*, 100, 9815, 1996.
23. Barton, J.L., and Bloom, H., *J. Phys. Chem.*, 60, 1413, 1956.
24. Sense, K.A., Alexander, C.A., Bowman, R.E., and Filbert, R.B., *J. Phys. Chem.*, 61, 337, 1957.
25. Ewing, C.T., and Stern, K.H., *J. Phys. Chem.* 78, 1998, 1974.
26. Cady, G.H., and Hargreaves, G.B., *J. Chem. Soc.*, 1563, 1961; 1568, 1961.
27. Skudlarski, K., Dudek, J., and Kapala, J., *J. Chem. Thermodynamics*, 19, 857, 1987.
28. Wagner, W., and de Reuck, K.M., *International Thermodynamic Tables of the Fluid State, No. 9. Oxygen*, Blackwell Scientific Publications, Oxford, 1987.
29. Marsh, K.N., Ed., *Recommended Reference Materials for the Realization of Physicochemical Properties*, Blackwell Scientific Publications, Oxford, 1987.
30. Alcock, C.B., Itkin, V.P., and Horrigan, M.K., *Canadian Metallurgical Quarterly*, 23, 309, 1984.
31. Stewart, R.B., and Jacobsen, R.T., *J. Phys. Chem. Ref. Data*, 18, 639, 1989.
32. Sifner, O., and Klomfar, J., *J. Phys. Chem. Ref. Data*, 23, 63, 1994.
33. Bah, A., and Dupont–Pavlovsky, N., *J. Chem. Eng. Data*, 40, 869, 1995.
34. Behrens, R.G., and Rosenblatt, G., *J. Chem. Thermodynamics*, 4, 175, 1972.
35. Behrens, R.G., and Rosenblatt, G., *J. Chem. Thermodynamics*, 5, 173, 1973.
36. Haar, L., Gallagher, J.S., and Kell, G.S., *NBS/NRC Steam Tables*, Hemisphere Publishing Corp., New York, 1984.
37. Wagner, W., Saul, A., and Pruss, A., *J. Phys. Chem. Ref. Data*, 23, 515. 1994.
38. Behrens, R.G., Lemons, R.S., and Rosenblatt, G., *J. Chem. Thermodynamics*, 6, 457, 1974.
39. Boublik, T., Fried, V., and Hala, E., *The Vapor Pressure of Pure Substances, Second Edition*, Elsevier, Amsterdam, 1984.
40. Goodwin, R.D., *J. Phys. Chem. Ref. Data*, 14, 849, 1985.
41. Younglove, B.A., and Ely, J.F., *J. Phys. Chem. Ref. Data*, 16, 577, 1987.

Temperature at Which Vapor Pressure Equals the Indicated Value

Name	Mol. form.	*t*/°C for 1 Pa	*t*/°C for 10 Pa	*t*/°C for 100 Pa	*t*/°C for 1 kPa	*t*/°C for 10 kPa	*t*/°C for 100 kPa	Ref.
Acenaphthene	$C_{12}H_{10}$				126.2	187[e]	276[e]	1
Acenaphthylene	$C_{12}H_8$	24 s	49.8 s	80.6 s				5
Acetaldehyde	C_2H_4O		-105[e]	-87[e]	-62.8	-29.4	20.0	5
Acetamide	C_2H_5NO	16.7 s	39.1 s	65.2 s	102.8	150.8	218.2	5
Acetic acid	$C_2H_4O_2$	-42.8 s	-26.7 s	-8 s	14.2 s	55.9	117.5	1,5
Acetic anhydride	$C_4H_6O_3$	-44[e]	-25[e]	-1[e]	31[e]	75.1	139.7	1
1-Acetonaphthone	$C_{12}H_{10}O$	37[e]	69[e]	107.0	154.6	215.2	294.9	5
2-Acetonaphthone	$C_{12}H_{10}O$	48.3 s		118.7	163.0	221.1	300.3	5
Acetone	C_3H_6O	-95	-81.8	-62.8	-35.6	1.3	55.7	1,5
Acetonitrile	C_2H_3N				-20[e]	21.4	81.2	1
Acetophenone	C_8H_8O			36[e]	73[e]	125.3	201.5	5
Acetyl bromide	C_2H_3BrO	-78[e]	-65[e]	-49[e]	-25[e]	13.9	84[e]	5
Acetyl chloride	C_2H_3ClO	-100[e]	-85[e]	-66[e]	-40[e]	-3.6	50.4	1
Acetylene[a]	C_2H_2			-146.6 s	-130.7 s	-110.6 s	-84.8 s	5
Acetyl fluoride	C_2H_3FO					-64.1	17.0	5
Acetyl iodide	C_2H_3IO				-0.6	47[e]	107.0	5
Acridine	$C_{13}H_9N$			124.4	176.2	246.0	345.4	5
Acrolein	C_3H_4O		-87[e]	-67[e]	-40[e]	-3.0	52.8	1
Acrylamide	C_3H_5NO			109.6	161[e]			5
Acrylic acid	$C_3H_4O_2$				35[e]	78.0	140.7	1
Acrylonitrile	C_3H_3N		-72[e]	-50[e]	-22[e]	17.7	77.0	1
Allene[a]	C_3H_4		-129[e]	-118[e]	-101.4	-76.7	-34.7	5
Allyl alcohol	C_3H_6O	-63[e]	-48[e]	-21.9	6.8	44.5	96.2	5
Allylamine	C_3H_7N		-88[e]	-65[e]	-37[e]	0.4	52[e]	5
Allyl ethyl ether	$C_5H_{10}O$			-56[e]	-28.7	9.8	67.2	5
Allyl glycidyl ether	$C_6H_{10}O_2$				40.1	85.7	152.8	5
Allyl isothiocyanate	C_4H_5NS	-45[e]	-27[e]	-3[e]	32.1	89[e]	198[e]	5
4-Allyl-2-methoxyphenol	$C_{10}H_{12}O_2$	9[e]	37[e]	72[e]	115.9	173.8	252.9	5
Allyltrichlorosilane	$C_3H_5Cl_3Si$					53.0	116.5	5
Aluminum	Al	1209	1359	1544	1781	2091	2517	2
Aluminum borohydride	AlB_3H_{12}				-46.8	-9.4	45.5	4
Aluminum chloride	$AlCl_3$	58.4 s	76.5 s	97.1 s	120.7 s	148.2 s	180.5 s	4
Aluminum fluoride	AlF_3	744 s	819 s	906 s	1008 s	1130 s	1276 s	8
Aluminum iodide	AlI_3				218	285	385	4
Aluminum oxide (α)	Al_2O_3			2122	2351	2629	2975	4
1-Amino-2-propanol	C_3H_9NO			18[e]	53.2	98.2	157.9	5
Ammonia[a]	H_3N	-139 s	-127 s	-112 s	-94.5 s	-71.3	-33.6	1,5,6
Ammonium bromide	BrH_4N	121 s	154 s	195 s	246 s	310.4 s	395.1 s	5
Ammonium chloride	ClH_4N	91 s	121 s	159 s	204.7 s	263.1 s	339.5 s	5
Ammonium iodide	H_4IN	125 s	159 s	201 s	253 s	318.4 s	405.2 s	5
Aniline	C_6H_7N		-2.5	26.7	63.5	112.5	183.5	1,5
Anisole	C_7H_8O		-21[e]	4[e]	38[e]	84[e]	153.2	1,5
Anthracene	$C_{14}H_{10}$	89.2 s	125.9 s	151.5 s	165 s	238.8	340.2	1,5
Antimony	Sb	534 s	603 s	738	946	1218	1585	2,3
Antimony(III) bromide	Br_3Sb				136.5	196.9	286.5	1
Antimony(III) iodide	I_3Sb				214.9	292.0	401.2	4
Antimony(III) oxide (valentinite)	O_3Sb_2	426.1 s	478 s	539 s	610 s	907	1420	4,35
Argon[a]	Ar		-226.4 s	-220.3 s	-212.4 s	-201.7 s	-186.0	1,5,31
Arsenic (gray)	As	280 s	323 s	373 s	433 s	508 s	601 s	3
Arsenic(III) chloride	$AsCl_3$			-8[e]	21.3	63.1	129.4	1
Arsenic(III) fluoride	AsF_3					8.1	56.0	4
Arsenic(III) iodide	AsI_3				187	261	367[e]	7
Arsenic(III) oxide (arsenolite)	As_2O_3	133.7 s	163.0 s	196.8 s	236.2 s	283.0		34
Astatine	At	88 s	119 s	156 s	202 s	258 s	334	2
trans-Azobenzene	$C_{12}H_{10}N_2$			98.1	144.8	206.7	292.7	4
Azulene	$C_{10}H_8$	24.1 s	46 s	71.5 s	103.3	162.6	244.0	5
Barium	Ba	638 s	765	912	1115	1413	1897	9
Benzaldehyde	C_7H_6O		-9[e]	19[e]	54.6	104.6	178.3	1
Benzanthrone	$C_{17}H_{10}O$		184[e]	229.3	290.3	377.2	511[e]	5

Name	Mol. form.	t/°C for 1 Pa	t/°C for 10 Pa	t/°C for 100 Pa	t/°C for 1 kPa	t/°C for 10 kPa	t/°C for 100 kPa	Ref.
Benzene[b]	C_6H_6			-40 s	-15.1 s	20.0	79.7	1,5
Benzeneacetonitrile	C_8H_7N	-3[e]	23[e]	55.3	97.4	153.7	233.1	5
1,3-Benzenediamine	$C_6H_8N_2$			94.5	140.2	200.8	285.0	5
Benzeneethanol	$C_8H_{10}O$	2[e]	25[e]	54[e]	92[e]	143.6	217.7	5
Benzenethiol	C_6H_6S		-15[e]	12[e]	47[e]	96.0	168.6	5
1,2,3-Benzenetriol	$C_6H_6O_3$				162.0	222.8	308.3	5
Benzil	$C_{14}H_{10}O_2$			123	175	246	346	4
Benzofuran	C_8H_6O		-16[e]	12[e]	47.9	97.7	170.7	5
Benzoin	$C_{14}H_{12}O_2$				181	248	342	4
Benzonitrile	C_7H_5N		-6[e]	23.9	63.1	115.7	190.0	5
p-Benzoquinone	$C_6H_4O_2$	-4.1 s	17.8 s	43.5 s	74.3 s	111.6 s		5
Benzoyl bromide	C_7H_5BrO	-15[e]	11[e]	42.6	83.9	139.5	218.0	5
Benzoyl chloride	C_7H_5ClO			27.5	67.0	120.4	196.7	5
Benzyl acetate	$C_9H_{10}O_2$	-11[e]	15[e]	46.6	86.9	139.5	211[e]	5
Benzyl alcohol	C_7H_8O	8[e]	28[e]	54[e]	88[e]	134.7	204.9	1
Benzylamine	C_7H_9N			25.6	62.6	112.7	183.9	5
Benzyl ethyl ether	$C_9H_{12}O$		-10[e]	20.4	59.3	111.3	184.5	5
Beryllium	Be	1189 s	1335	1518	1750	2054	2469	2
Beryllium bromide	$BeBr_2$	203 s	240 s	283 s	335 s	397 s	473 s	4
Beryllium chloride	$BeCl_2$	196 s	237 s	284 s	339 s	402 s	487	4
Beryllium fluoride	BeF_2		686[e]	767[e]	869	999	1172[e]	7
Beryllium iodide	BeI_2	188 s	229 s	276 s	333 s	402 s	487	4
Bicyclo[4.1.0]heptane	C_7H_{12}					49.9	116.3	5
Biphenyl	$C_{12}H_{10}$			69.0	111.1	169.5	254.7	1
Bis(2-aminoethyl)amine	$C_4H_{13}N_3$	-10[e]	13[e]	43[e]	80[e]	129.6	198[e]	5
Bis(2-chloroethyl) ether	$C_4H_8Cl_2O$	-32[e]	-9[e]	19.8	56.9	106.9	177.9	5
Bis(2-ethylhexyl) phthalate	$C_{24}H_{38}O_4$	122.0	153.2	189.2	231.3	281.1	341.1	5
Bis(2-hydroxyethyl) sulfide	$C_4H_{10}O_2S$			31[e]	114.2		282.0	5
Bismuth	Bi	668	768	892	1052	1265	1562	2
Bismuth tribromide	$BiBr_3$			217 s	273[d]	348[d]	455[d]	4,9
Bismuth trichloride	$BiCl_3$				248.9	328.6	438.7	1,4
Borane carbonyl	CH_3BO				-124	-99	-64	4
Boron	B	2075	2289	2549	2868	3272	3799	2
Boron tribromide	BBr_3			-45[e]	-15[e]	27.5	90.4	1
Boron trichloride[a]	BCl_3			-94.0	-70.5	-37.4	12.3	4
Boron trifluoride[a]	BF_3	-173.9 s	-166.0 s	-156.0 s	-143.0 s	-125.9	-101.1	4
Bromine[a]	Br_2	-87.7 s	-71.8 s	-52.7 s	-29.3 s	2.5	58.4	1
Bromobenzene	C_6H_5Br		-25[e]	1[e]	34.9	83.1	155.4	1
1-Bromobutane	C_4H_9Br	-68.4	-53.9	-34.1	-5.4	37.6	101.1	1,5
2-Bromobutane, (±)-	C_4H_9Br	-86[e]	-68[e]	-46[e]	-16[e]	26.6	90.7	5
trans-1-Bromo-1-butene	C_4H_7Br	-87[e]	-68[e]	-43.3	-11.4	31.9	94.4	5
2-Bromo-1-butene	C_4H_7Br	-87[e]	-70[e]	-48[e]	-20[e]	20.7	80.6	5
cis-2-Bromo-2-butene	C_4H_7Br	-90[e]	-72[e]	-49.0	-18.5	23.5	85.2	5
trans-2-Bromo-2-butene	C_4H_7Br	-86[e]	-67[e]	-43.4	-12.0	31.0	93.5	5
Bromochlorodifluoromethane	$CBrClF_2$	-136[e]	-123[e]	-106[e]	-83.4	-51.8	-4.3	1
1-Bromo-2-chloroethane	C_2H_4BrCl				-0.4	41.7	105.7	6
Bromochloromethane	CH_2BrCl	-83[e]	-69[e]	-50[e]	-25[e]	11.4	67.7	1
1-Bromo-3-chloropropane	C_3H_6BrCl	-51[e]	-31[e]	-6[e]	28[e]	74.1	142.9	5
2-Bromo-2-chloro-1,1,1-trifluoroethane	$C_2HBrClF_3$				-41.4	-4.8	49.8	1
1-Bromodecane	$C_{10}H_{21}Br$	9[e]	33[e]	63[e]	104[e]	159.2	240.0	5
Bromodifluoromethane	$CHBrF_2$		-128 s	-111.4 s	-89.7 s	-59.7 s	-16 s	5
1-Bromododecane	$C_{12}H_{25}Br$	31[e]	57[e]	90[e]	132[e]	190.8	275.3	5
Bromoethane	C_2H_5Br	-111[e]	-96[e]	-77[e]	-51.3	-15.5	38.0	5
Bromoethene	C_2H_3Br	-124[e]	-110[e]	-92[e]	-68[e]	-34.5	15.4	5
(2-Bromoethyl)cyclohexane	$C_8H_{15}Br$	-14[e]	8[e]	36.9	75.3	129.7	212.5	5
1-Bromoheptane	$C_7H_{15}Br$	-30[e]	-9[e]	18[e]	54[e]	104.4	178.4	5
1-Bromohexane	$C_6H_{13}Br$	-45[e]	-25[e]	2[e]	36[e]	83.7	154.8	5
1-Bromo-4-isopropylbenzene	$C_9H_{11}Br$	-8[e]	15[e]	45[e]	84[e]	138.1	218.5	5
Bromomethane	CH_3Br				-77[e]	-44.3	3.3	1
(Bromomethyl)benzene	C_7H_7Br			25.4	66.8	121.7	198.3	5

Name	Mol. form.	t/°C for 1 Pa	t/°C for 10 Pa	t/°C for 100 Pa	t/°C for 1 kPa	t/°C for 10 kPa	t/°C for 100 kPa	Ref.
1-Bromo-3-methylbutane	$C_5H_{11}Br$	-67[e]	-49[e]	-25[e]	8[e]	52.4	119.9	5
1-Bromo-2-methylpropane	C_4H_9Br	-85[e]	-68[e]	-46[e]	-16[e]	26.8	91.1	5
2-Bromo-2-methylpropane	C_4H_9Br					11.7	72.4	1,5
1-Bromonaphthalene	$C_{10}H_7Br$	17[e]	45[e]	80.3	126.7	189.8	280.5	5
1-Bromooctane	$C_8H_{17}Br$	-17[e]	6[e]	34[e]	72[e]	123.8	200.3	5
Bromopentafluorobenzene	C_6BrF_5			-10[e]	23[e]	68[e]	136.0	5
1-Bromopentane	$C_5H_{11}Br$	-60[e]	-41[e]	-16[e]	16[e]	61.5	129.1	5
2-Bromopentane	$C_5H_{11}Br$	-69[e]	-51[e]	-27[e]	5[e]	49.7	116.9	5
3-Bromopentane	$C_5H_{11}Br$	-68[e]	-50[e]	-26[e]	6[e]	50.8	118.1	5
1-Bromopropane	C_3H_7Br	-95[e]	-78[e]	-57[e]	-28[e]	11.6	70.6	1
2-Bromopropane	C_3H_7Br		-84[e]	-65[e]	-39.6	-1.7	59.1	1,5
cis-1-Bromopropene	C_3H_5Br	-100[e]	-84[e]	-64[e]	-37[e]	1.0	57.4	5
2-Bromopropene	C_3H_5Br	-112[e]	-95[e]	-75[e]	-47[e]	-9[e]	48.0	5
3-Bromopropene	C_3H_5Br	-98[e]	-80[e]	-58[e]	-28[e]	12[e]	69.6	5
Bromosilane	BrH_3Si				-81.0	-47.3	2.2	4
2-Bromotoluene	C_7H_7Br		-10[e]	17[e]	54[e]	104.8	181.1	5
3-Bromotoluene	C_7H_7Br	-34[e]	-11[e]	19.4	58.1	109.9	183.1	5
4-Bromotoluene	C_7H_7Br				57[e]	107.8	183.8	5
Bromotrichloromethane	$CBrCl_3$				-6[e]	38.9	104.4	5
Bromotrifluoromethane[a]	$CBrF_3$	-168[e]	-156[e]	-142[e]	-122.8	-96.6	-58.1	5
1,2-Butadiene	C_4H_6	-132[e]	-117[e]	-98[e]	-72.8	-38.9	10.5	5
1,3-Butadiene[a]	C_4H_6			-106[e]	-83[e]	-51.9	-4.7	1
Butanal	C_4H_8O	-88[e]	-72[e]	-50[e]	-22[e]	16.6	74.5	1,5
Butane[a]	C_4H_{10}	-134.3	-121.0	-103.9	-81.1	-49.1	-0.8	1,41
1,3-Butanediol	$C_4H_{10}O_2$	-4[e]	23[e]	55[e]	94[e]	142.9	206.1	5
1,4-Butanediol	$C_4H_{10}O_2$		45[e]	77[e]	116[e]	164.7	227.6	5
2,3-Butanediol	$C_4H_{10}O_2$		15[e]	43[e]	77[e]	121.2	180.3	5
2,3-Butanedione	$C_4H_6O_2$					30.7	84.8	5
1,4-Butanedithiol	$C_4H_{10}S_2$	-17[e]	5[e]	32[e]	69.1	119.9	195.1	5
Butanenitrile	C_4H_7N	-67[e]	-48[e]	-24[e]	8[e]	52.3	117.2	1
1-Butanethiol	$C_4H_{10}S$	-77[e]	-59[e]	-37[e]	-6[e]	35.4	98.0	5
2-Butanethiol	$C_4H_{10}S$	-86[e]	-69[e]	-47[e]	-17[e]	23.4	84.5	5
Butanoic acid	$C_4H_8O_2$			12.9	52.2	101.4	163.3	1,5
Butanoic anhydride	$C_8H_{14}O_3$	-28[e]	-2[e]	30[e]	71[e]	123.8	196.5	5
1-Butanol	$C_4H_{10}O$	-37[e]	-20[e]	0[e]	28[e]	64[e]	117.4	1
2-Butanol	$C_4H_{10}O$	-50[e]	-34[e]	-14[e]	12.6	48.2	99.2	1,5
2-Butanone	C_4H_8O	-85[e]	-68[e]	-46[e]	-18.1	21.2	79.2	1
2-Butanone oxime	C_4H_9NO		-18[e]	7[e]	38.9	81.9	142.9	5
trans-2-Butenal	C_4H_6O	-74[e]	-56[e]	-33[e]	-3[e]	39.7	102.4	5
1-Butene	C_4H_8	-139.0	-125.2	-107.8	-85.3	-53.7	-6.6	1,5
cis-2-Butene	C_4H_8	-131.2	-117.4	-99.8	-76.7	-44.8	3.4	1,5
trans-2-Butene	C_4H_8			-102[e]	-80[e]	-47.6	0.6	1
cis-2-Butene-1,4-diol	$C_4H_8O_2$	17[e]	44[e]	77[e]	117.4	168.5	234.9	5
trans-2-Butenedioyl dichloride	$C_4H_2Cl_2O_2$			8.0	45.6	94.3	159.8	5
3-Butenenitrile	C_4H_5N	-67[e]	-48[e]	-23.1	9.3	53.7	118.4	5
cis-2-Butenoic acid	$C_4H_6O_2$			30[e]	63[e]	106.7	168.9	5
trans-2-Butenoic acid	$C_4H_6O_2$				74[e]	120.8	184.9	5
3-Butenoic acid	$C_4H_6O_2$	-19[e]	2[e]	27[e]	61[e]	105.6	168.6	5
3-Buten-2-one	C_4H_6O					21[e]	81.0	5
1-Buten-3-yne	C_4H_4			-96.1	-73.4	-41.8	4.9	5
2-Butoxyethanol	$C_6H_{14}O_2$	-31[e]	-8[e]	20[e]	55[e]	103.2	170.2	5
Butyl acetate	$C_6H_{12}O_2$	-63[e]	-43[e]	-19[e]	14[e]	61.0	125.6	1,5
Butyl acrylate	$C_7H_{12}O_2$	-52[e]	-31[e]	-4.5	30.4	78.0	146.9	5
Butylamine	$C_4H_{11}N$			-46[e]	-18.1	20.0	75.9	5
sec-Butylamine	$C_4H_{11}N$			-55[e]	-29.1	7.5	62.3	5
tert-Butylamine	$C_4H_{11}N$			-67[e]	-42.4	-8.1	43.7	5
N-Butylaniline	$C_{10}H_{15}N$	11[e]	35[e]	66[e]	106[e]	160.9	241.0	5
Butylbenzene	$C_{10}H_{14}$	-28[e]	-7[e]	21[e]	56.9	107.6	182.8	1,5
sec-Butylbenzene, (±)-	$C_{10}H_{14}$	-35[e]	-14[e]	13[e]	48[e]	98.3	172.8	5
tert-Butylbenzene	$C_{10}H_{14}$	-37[e]	-16[e]	10[e]	46[e]	94.9	168.6	5

Name	Mol. form.	t/°C for 1 Pa	t/°C for 10 Pa	t/°C for 100 Pa	t/°C for 1 kPa	t/°C for 10 kPa	t/°C for 100 kPa	Ref.
Butyl benzoate	$C_{11}H_{14}O_2$	6[e]	34[e]	67.9	110.3	165[e]	237[e]	5
Butylcyclohexane	$C_{10}H_{20}$	-31[e]	-9[e]	18[e]	54[e]	104.7	180.4	5
tert-Butylcyclohexane	$C_{10}H_{20}$	-39[e]	-18[e]	9[e]	45[e]	95.3	171.1	5
Butylcyclopentane	C_9H_{18}	-45[e]	-24[e]	1[e]	36[e]	84[e]	156.1	5
Butylethylamine	$C_6H_{15}N$				6.1	47.7	107.0	5
Butyl ethyl ether	$C_6H_{14}O$	-78[e]	-61[e]	-39[e]	-10[e]	31.0	91.9	1
tert-Butyl ethyl ether	$C_6H_{14}O$	-90[e]	-74[e]	-53[e]	-24.6	14.4	72.6	5
Butyl ethyl sulfide	$C_6H_{14}S$	-49[e]	-30[e]	-5[e]	29[e]	74.8	143.8	5
Butyl formate	$C_5H_{10}O_2$			-29[e]	2[e]	44.4	105.7	5
Butyl methacrylate	$C_8H_{14}O_2$				47[e]	93.3	159.0	5
1-*tert*-Butyl-4-methylbenzene	$C_{11}H_{16}$	-24[e]	-2[e]	27[e]	64.1	115.5	190.8	5
Butyl methyl ether	$C_5H_{12}O$			-54[e]	-27[e]	12[e]	69.8	1
Butyl methyl sulfide	$C_5H_{12}S$		-43[e]	-19[e]	13[e]	57[e]	123.0	1
tert-Butyl methyl sulfide	$C_5H_{12}S$				-7.8	34.7	98.4	5
1-Butylnaphthalene	$C_{14}H_{16}$	67[e]	82[e]	103[e]	135[e]	186.7	288.6	5
2-Butylnaphthalene	$C_{14}H_{16}$	44[e]	67[e]	98[e]	139[e]	197.5	287.4	5
Butyl oleate	$C_{22}H_{42}O_2$	95.5	124.2	158[e]	198[e]	245[e]	304[e]	5
2-Butylphenol	$C_{10}H_{14}O$	7[e]	31[e]	61[e]	101[e]	155.2	234.4	5
Butyl phenyl ether	$C_{10}H_{14}O$	-16[e]	8[e]	38[e]	77[e]	131.3	209.7	5
Butyl stearate	$C_{22}H_{44}O_2$	99.6	128[e]	162[e]	201[e]	249[e]	307[e]	5
Butyltrichlorosilane	$C_4H_9Cl_3Si$					77.2	148.4	5
Butyl vinyl ether	$C_6H_{12}O$	-87[e]	-67[e]	-42[e]	-9.3	33.6	93.2	5
1-Butyne	C_4H_6	-125[e]	-111[e]	-94[e]	-71.2	-39.4	7.8	1
2-Butyne	C_4H_6		-89.2 s	-73.8 s	-53.5 s	-23.9	26.6	5
γ-Butyrolactone	$C_4H_6O_2$		-17[e]	24[e]	72[e]	130.2	203[e]	5
Cadmium	Cd	257 s	310 s	381	472	594	767	2
Cadmium bromide	Br_2Cd	373 s	435 s	509 s				27
Cadmium chloride	$CdCl_2$	412 s	471 s	541 s	634	768	959	12,23,27
Cadmium fluoride	CdF_2				1257	1461	1742	4
Cadmium iodide	CdI_2	296 s	344 s	406	498	622	795	4,27
Cadmium oxide	CdO	770 s	866 s	983 s	1128 s	1314 s	1558 s	4
Calcium	Ca	591 s	683 s	798 s	954	1170	1482	2
Camphene	$C_{10}H_{16}$					90.7	160.1	4
Camphor, (+)	$C_{10}H_{16}O$	-15.8 s	10 s	41.5 s	80.8 s	131.4 s	207.6	5
Caprolactam	$C_6H_{11}NO$	36.8 s	58.9 s	86.6 s			270	5
Carbazole	$C_{12}H_9N$					254.7	354.0	5
Carbon dioxide[a, b]	CO_2	-159.1 s	-148.9 s	-136.7 s	-121.6 s	-103.1 s	-78.6 s	5
Carbon diselenide	CSe_2			-24[e]	9.4	56.2	127[e]	1
Carbon disulfide	CS_2		-96[e]	-76[e]	-49[e]	-10.9	45.9	1
Carbon [fullerene-C_{70}]	C_{70}	598 s	662 s					22
Carbon (graphite)	C		2566 s	2775 s	3016 s	3299 s	3635 s	15
Carbon monoxide[a]	CO			-223 s	-216.5 s	-207.2 s	-191.7	40
Carbon oxyselenide	COSe			-120	-98	-67	-22	4
Carbon oxysulfide[a]	COS			-136[e]	-117[e]	-90.0	-50.4	1
Carbonyl chloride	CCl_2O	-127[e]	-113[e]	-96[e]	-73[e]	-40.6	7.2	5
Carbonyl dicyanide	C_3N_2O				-21.7	15.3	65.2	5
Cerium	Ce	1719	1921	2169	2481	2886	3432	14
Cesium	Cs	144.5	195.6	260.9	350.0	477.1	667.0	13,30
Cesium bromide	BrCs	531 s	601 s	701[d]	834[d]	1019[d]	1293[e]	9
Cesium chloride	ClCs			730	864	1043	1297	4
Cesium fluoride	CsF				825	999	1249	4
Cesium iodide	CsI	523 s	595 s	692	854	1029	1278	4,25
Chlorine[a]	Cl_2	-145 s	-133.7 s	-120.2 s	-103.6 s	-76.1	-34.2	1
Chlorine dioxide[a]	ClO_2					-34.3	10.5	5
Chlorine fluoride[a]	ClF				-144.4	-122.6	-90.2	5
Chlorine pentafluoride	ClF_5				-88[e]	-59	-14	7
Chlorine trifluoride	ClF_3				-63.7	-33.0	11.4	5
Chloroacetic acid	$C_2H_3ClO_2$				78.4	123.9	188.9	1
Chloroacetyl chloride	$C_2H_2Cl_2O$			-23.7	5.6	46.1	105.6	5
2-Chloroaniline	C_6H_6ClN		10[e]	39.0	75.2	131.4	208.3	5

Name	Mol. form.	t/°C for 1 Pa	t/°C for 10 Pa	t/°C for 100 Pa	t/°C for 1 kPa	t/°C for 10 kPa	t/°C for 100 kPa	Ref.
3-Chloroaniline	C_6H_6ClN	-5^e	19.7	49.4	94.2	162^e	1069^e	5
2-Chloroanisole	C_7H_7ClO	-22^e	2^e	33^e	72^e	125.2	201^e	5
Chlorobenzene	C_6H_5Cl		-43^e	-17^e	16.8	62.9	131.3	1,5
2-Chlorobenzoyl chloride	$C_7H_4Cl_2O$				93^e	149^e	237.0	5
3-Chlorobenzoyl chloride	$C_7H_4Cl_2O$				87.8	147^e	225.0	5
2-Chloro-1,3-butadiene	C_4H_5Cl	-113^e	-95^e	-71^e	-41^e	0.3	59.0	5
1-Chlorobutane	C_4H_9Cl	-87^e	-71^e	-49^e	-21^e	18.4	78.1	1
2-Chlorobutane	C_4H_9Cl	-96^e	-80^e	-59^e	-31.0	8.5	67.9	1
3-Chloro-1-butene	C_4H_7Cl			-64^e	-36^e	4^e	63.6	5
cis-2-Chloro-2-butene	C_4H_7Cl	-100^e	-83^e	-62^e	-34^e	6^e	66.4	5
trans-2-Chloro-2-butene	C_4H_7Cl	-102^e	-86^e	-65^e	-37^e	3^e	62.2	5
Chlorocyclohexane	$C_6H_{11}Cl$		-35^e	-9^e	25^e	71.6	142.1	5
1-Chlorodecane	$C_{10}H_{21}Cl$	2^e	25^e	54^e	92^e	145.7	225.3	5
1-Chloro-1,1-difluoroethane	$C_2H_3ClF_2$		-123^e	-107^e	-85.3	-55.4	-10.5	5
Chlorodifluoromethane[a]	$CHClF_2$	-152^e	-141^e	-126^e	-107.1	-80.5	-41.1	5
1-Chloro-2,2-dimethylpropane	$C_5H_{11}Cl$				-17^e	23.5	83.9	5
1-Chlorododecane	$C_{12}H_{25}Cl$	27^e	51^e	81^e	122^e	178.7	262.6	5
Chloroethane	C_2H_5Cl	-126^e	-112^e	-94^e	-70^e	-37.0	12.0	1
2-Chloroethanol	C_2H_5ClO	-61^e	-39^e	-12^e	23^e	67.1	127.3	5
Chloroethene	C_2H_3Cl	-139^e	-127^e	-110^e	-89^e	-59.0	-14.1	1
1-Chloro-2-ethylbenzene	C_8H_9Cl	-30^e	-9^e	18^e	54^e	103.7	177.9	5
1-Chloro-4-ethylbenzene	C_8H_9Cl	-27^e	-6^e	22^e	58^e	108.7	183.9	5
1-Chloro-1-fluoroethane	C_2H_4ClF				-69.9	-36.1	15.8	5
Chlorofluoromethane	CH_2ClF		-124^e	-108^e	-86.2	-55.7	-9.4	5
1-Chloroheptane	$C_7H_{15}Cl$	-39^e	-19^e	7^e	41^e	88.6	159.9	5
1-Chlorohexane	$C_6H_{13}Cl$	-55^e	-36^e	-11^e	21^e	66.7	134.6	5
1-Chloro-2-isopropylbenzene	$C_9H_{11}Cl$	-23^e	-1^e	27^e	64^e	114.6	190.5	5
1-Chloro-4-isopropylbenzene	$C_9H_{11}Cl$		3^e	31^e	69^e	120.5	197.8	5
Chloromethane[a]	CH_3Cl	-140.2 s	-128.6 s	-114.7 s	-96^e	-67.1	-24.4	1,33
(Chloromethyl)benzene	C_7H_7Cl	-34^e	-11^e	17.7	55.4	106.3	178.9	5
2-Chloro-2-methylbutane	$C_5H_{11}Cl$			-52^e	-21^e	21.8	85.2	5
3-(Chloromethyl)heptane	$C_8H_{17}Cl$					100.3	172.4	5
Chloromethyl methyl ether	C_2H_5ClO	-96^e	-80^e	-59^e	-32^e	6^e	61^e	5
1-Chloro-2-methylpropane	C_4H_9Cl	-94^e	-78^e	-56.6	-28.7	10.2	68.5	5
2-Chloro-2-methylpropane	C_4H_9Cl					-4.2	50.3	5
3-Chloro-2-methylpropene	C_4H_7Cl		-75^e	-54^e	-25^e	13.8	71.5	5
Chloromethylsilane	CH_5ClSi	-129^e	-115^e	-97.9	-74.4	-41.5	8.3	5
1-Chloronaphthalene	$C_{10}H_7Cl$	14^e	39^e	70.5	112.8	171.6	258.6	5
1-Chloro-4-nitrobenzene	$C_6H_4ClNO_2$	15.4 s	35.8 s		97^e	156.0	238^e	5
1-Chloro-2-nitro-4-(trifluoromethyl)benzene	$C_7H_3ClF_3NO_2$	3^e	26^e	55^e	92.8	145.2	222.0	5
1-Chlorononane	$C_9H_{19}Cl$	-11^e	11^e	39^e	76^e	127.8	204.7	5
1-Chlorooctane	$C_8H_{17}Cl$	-25^e	-4^e	23^e	59^e	108.8	182.9	5
Chloropentafluoroacetone	C_3ClF_5O	-122^e	-109^e	-93^e	-71^e	-39.4	7.4	5
Chloropentafluorobenzene	C_6ClF_5		-44^e	-21^e	11^e	53.8	117.6	1
Chloropentafluoroethane	C_2ClF_5					-80.3	-39.4	1
1-Chloropentane	$C_5H_{11}Cl$	-73^e	-55^e	-32^e	-1^e	42.5	107.9	5
2-Chloropentane, (+)	$C_5H_{11}Cl$	-80^e	-62^e	-39^e	-9^e	33.2	96.1	5
3-Chloropentane	$C_5H_{11}Cl$	-77^e	-60^e	-37^e	-7^e	34.9	97.3	5
2-Chlorophenol	C_6H_5ClO				45.8	97.9	173.9	5
3-Chlorophenol	C_6H_5ClO			39.7	80.2	135.1	213.4	5
4-Chlorophenol	C_6H_5ClO			45.0	86.5	142.0	219.9	5
1-Chloropropane	C_3H_7Cl	-106^e	-90^e	-71^e	-44.5	-8.1	46.2	1
2-Chloropropane	C_3H_7Cl		-91^e	-74^e	-51.1	-17.8	35.4	1,5
2-Chloro-1-propanol	C_3H_7ClO				23^e	63.8	125.7	5
cis-1-Chloropropene	C_3H_5Cl	-114^e	-100^e	-81^e	-55^e	-20.1	32.4	5
trans-1-Chloropropene	C_3H_5Cl		-97^e	-77^e	-52^e	-16.2	37.0	5
2-Chloropropene	C_3H_5Cl	-120^e	-106^e	-87^e	-63^e	-28.7	22.3	5
3-Chloropropene	C_3H_5Cl	-107^e	-92^e	-72.4	-46.3	-9.8	44.6	5
2-Chloropyridine	C_5H_4ClN			7.4	45.8	97.3	169.9	5
2-Chlorostyrene	C_8H_7Cl	-33^e	-10^e	20^e	58^e	110.8	188^e	5

Name	Mol. form.	t/°C for 1 Pa	t/°C for 10 Pa	t/°C for 100 Pa	t/°C for 1 kPa	t/°C for 10 kPa	t/°C for 100 kPa	Ref.
Chlorosulfonic acid	$ClHO_3S$	-40[e]	-20[e]	5[e]	38.7	85.0	153.6	5
1-Chloro-1,1,2,2-tetrafluoroethane	C_2HClF_4			-110[e]	-87.6	-57.0	-12.1	5
2-Chlorothiophene	C_4H_3ClS		-62[e]	-35[e]	2[e]	51.8	123[e]	5
2-Chlorotoluene	C_7H_7Cl		-24[e]	3[e]	38[e]	86.3	158.7	1,5
3-Chlorotoluene	C_7H_7Cl	-41[e]	-21[e]	6[e]	41[e]	89[e]	161.8	5
4-Chlorotoluene	C_7H_7Cl				40[e]	88.9	161.5	1,5
Chlorotrifluoroethene	C_2ClF_3	-146[e]	-134[e]	-119[e]	-99[e]	-71[e]	-28.4	1
Chlorotrifluoromethane[a]	$CClF_3$	-176[e]	-167[e]	-155[e]	-139[e]	-116[e]	-81.7	5
1-Chloro-2-(trifluoromethyl)benzene	$C_7H_4ClF_3$			1[e]	34.5	81.8	151.8	5
1-Chloro-3-(trifluoromethyl)benzene	$C_7H_4ClF_3$	-53[e]	-34[e]	-9[e]	24.2	69.8	137.2	5
1-Chloro-4-(trifluoromethyl)benzene	$C_7H_4ClF_3$			-9[e]	24.2	70.4	138.1	5
3-Chloro-1,1,1-trifluoropropane	$C_3H_4ClF_3$	-102[e]	-87[e]	-68[e]	-43[e]	-8[e]	45.3	5
Chromium	Cr	1383 s	1534 s	1718 s	1950	2257	2669	2
Cobalt	Co	1517	1687	1892	2150	2482	2925	2
Cobalt(II) chloride	Cl_2Co					818	1048	4
Copper	Cu	1236	1388	1577	1816	2131	2563	2
Copper(I) chloride	ClCu		459	543	675	914	1477	4
Copper(I) iodide	CuI				636	864	1331	4
o-Cresol	C_7H_8O	-6.4 s	12.8 s	40.2	72.3	120.3	190.5	1,5
m-Cresol	C_7H_8O	20.8	33.6	52.4	82.6	130.6	201.8	1,5
p-Cresol	C_7H_8O	-0.2 s	20.7 s	52.7	83.1	130.7	201.5	1,5
Cyanic acid	CHNO			-81.1	-56.8	-23.9	23[e]	5
Cyanoacetylene	C_3HN			-58.7 s	-35.6 s	-7 s	42.0	5
Cyanogen	C_2N_2	-127 s	-114.1 s	-98.5 s	-79.2 s	-54.9 s	-21.4	5
Cyanogen bromide	CBrN				-13 s	17.7 s	61.0	1
Cyanogen chloride	CClN		-94.6 s	-78.1 s	-57 s	-29 s	13.0	5
Cyanogen fluoride	CFN		-135 s	-121.2 s	-104.1 s	-82.8 s	-46.2	1,5
Cyanogen iodide	CIN						153.8	5
Cyclobutane	C_4H_8				-71.8	-38.1	12.1	5
Cyclobutanone	C_4H_6O			-34[e]	-4[e]	37.1	97[e]	5
Cyclodecane	$C_{10}H_{20}$			29[e]	68[e]	121.3	201.8	1
1,5,9-Cyclododecatriene	$C_{12}H_{18}$	-14[e]	11[e]	44[e]	87[e]	145.0	229.8	5
Cycloheptane	C_7H_{14}				6[e]	51.1	118.4	1
Cycloheptanone	$C_7H_{12}O$			18[e]	53.7	104.0	178.7	5
Cycloheptene	C_7H_{12}			-30.0	3.4	47.5	108[e]	5
1,3-Cyclohexadiene	C_6H_8	-88[e]	-71[e]	-50[e]	-21[e]	19[e]	79.9	5
1,4-Cyclohexadiene	C_6H_8				-15[e]	27.3	85.0	5
Cyclohexane	C_6H_{12}	-85.6 s	-68.9 s	-47.6 s	-19.8 s	19.3	80.4	1,5
Cyclohexanethiol	$C_6H_{12}S$					84.8	158.3	5
Cyclohexanol	$C_6H_{12}O$			34[e]	61[e]	99.2	160.7	1
Cyclohexanone	$C_6H_{10}O$		-25[e]	1[e]	36[e]	84[e]	155.2	1
Cyclohexene	C_6H_{10}	-87[e]	-70[e]	-49[e]	-19[e]	21[e]	82.6	1
Cyclohexyl acetate	$C_8H_{14}O_2$					103.1	172.9	5
Cyclohexylamine	$C_6H_{13}N$			-9[e]	22[e]	66.6	133.5	1
Cyclohexylbenzene	$C_{12}H_{16}$		28[e]	58[e]	98[e]	154.7	239.5	5
Cyclohexylcyclohexane	$C_{12}H_{22}$		20[e]	53.1	96.0	154.1	237.2	5
cis,cis-1,5-Cyclooctadiene	C_8H_{12}		-37[e]	-8[e]	30[e]	80.2	150[e]	5
Cyclooctane	C_8H_{16}				30[e]	78[e]	150.7	1
1,3,5,7-Cyclooctatetraene	C_8H_8				24.3	71.0	140.1	5
1,3-Cyclopentadiene	C_5H_6			-77[e]	-51[e]	-14[e]	39.8	5
Cyclopentane	C_5H_{10}			-77.0	-45.4	-7.1	48.8	5
Cyclopentanethiol	$C_5H_{10}S$				18[e]	64[e]	131.7	5
Cyclopentanol	$C_5H_{10}O$		-13[e]	11.5	42.2	82.5	140.0	5
Cyclopentanone	C_5H_8O		-39[e]	-14[e]	19[e]	64[e]	130.3	1
Cyclopentene	C_5H_8	-109[e]	-94[e]	-74[e]	-48[e]	-11.1	43.8	5
Cyclopentylamine	$C_5H_{11}N$	-66[e]	-48[e]	-26[e]	4[e]	45.8	108[e]	5
Cyclopropane	C_3H_6			-124[e]	-104[e]	-75.7	-33.1	1
Cyclopropyl methyl ketone	C_5H_8O		-57[e]	-31[e]	3[e]	49[e]	112[e]	5
cis-Decahydronaphthalene	$C_{10}H_{18}$	-26[e]	-4[e]	24[e]	62.4	115.5	195.3	1
trans-Decahydronaphthalene	$C_{10}H_{18}$		-10[e]	18[e]	55.3	107.9	186.8	1

Name	Mol. form.	t/°C for 1 Pa	t/°C for 10 Pa	t/°C for 100 Pa	t/°C for 1 kPa	t/°C for 10 kPa	t/°C for 100 kPa	Ref.
Decamethylcyclopentasiloxane	$C_{10}H_{30}O_5Si_5$	-2[e]	19[e]	46[e]	82[e]	132.9	210.4	5
Decamethyltetrasiloxane	$C_{10}H_{30}O_3Si_4$	-31[e]	-6[e]	26[e]	66.8	118.8	193.9	5
Decanal	$C_{10}H_{20}O$		16[e]	47.2	86.3	137.7	208.0	5
Decane	$C_{10}H_{22}$		-10.6	16.7	52.3	101.1	173.7	16
Decanedioic acid	$C_{10}H_{18}O_4$	125.9 s						5
Decanenitrile	$C_{10}H_{19}N$	13[e]	36[e]	66[e]	105.8	160.6	241.6	5
1-Decanethiol	$C_{10}H_{22}S$	11[e]	34[e]	64[e]	103[e]	157.5	238.6	5
Decanoic acid	$C_{10}H_{20}O_2$	58[e]	80[e]	108[e]	145[e]	195.2	269.5	5
1-Decanol	$C_{10}H_{22}O$	30[e]	50[e]	75[e]	109[e]	157.3	230.6	1,39
4-Decanol	$C_{10}H_{22}O$	18[e]	37[e]	61[e]	93[e]	139[e]	210[e]	5
1-Decene	$C_{10}H_{20}$	-35.5	-13.7	13.7	49.0	97.9	170.1	1,5
Decyl acetate	$C_{12}H_{24}O_2$	12[e]	40[e]	74[e]	115.1	168.1	238[e]	5
Decylcyclopentane	$C_{15}H_{30}$	37[e]	61[e]	93[e]	134[e]	192.5	278.8	5
1-Decyne	$C_{10}H_{18}$	-34[e]	-13[e]	14[e]	51[e]	100.3	173.5	5
Diacetone alcohol	$C_6H_{12}O_2$	-41[e]	-17[e]	13[e]	50.1	98.5	164[e]	5
Diallyl sulfide	$C_6H_{10}S$	-58[e]	-38[e]	-12.4	21.7	68.8	138.1	5
Dibenzylamine	$C_{14}H_{15}N$	48[e]	77[e]	113.1	158.9	218.5	299.4	5
Diborane	B_2H_6			-162[e]	-147.0	-125.8	-92.6	1
m-Dibromobenzene	$C_6H_4Br_2$	-7[e]	16[e]	44[e]	83[e]	137.0	218.2	5
1,2-Dibromobutane	$C_4H_8Br_2$	-54[e]	-30[e]	0.4	39.6	92.1	166.1	5
1,4-Dibromobutane	$C_4H_8Br_2$	-13[e]	9[e]	37[e]	74[e]	124.0	196.5	5
1,2-Dibromo-1-chloro-1,2,2-trifluoroethane	$C_2Br_2ClF_3$						92.3	5
1,2-Dibromo-1,1-dichloroethane	$C_2H_2Br_2Cl_2$					103.6	177.8	5
1,2-Dibromo-1,2-dichloroethane	$C_2H_2Br_2Cl_2$		-11[e]	22[e]	64.1	119[e]	193[e]	5
Dibromodifluoromethane	CBr_2F_2		-110[e]	-91[e]	-66[e]	-30[e]	22.5	1
1,1-Dibromoethane	$C_2H_4Br_2$		-49[e]	-26[e]	5[e]	46.4	107.6	5
1,2-Dibromoethane	$C_2H_4Br_2$				18[e]	62.2	130.9	1
cis-1,2-Dibromoethene	$C_2H_2Br_2$		-45[e]	-21[e]	10[e]	52.2	114.8	1
trans-1,2-Dibromoethene	$C_2H_2Br_2$				-4[e]	42.2	107.4	5
Dibromomethane	CH_2Br_2			-37[e]	-7[e]	35.2	96.5	5
1,5-Dibromopentane	$C_5H_{10}Br_2$	1[e]	25[e]	54[e]	93[e]	145.6	221.8	5
1,2-Dibromopropane	$C_3H_6Br_2$	-46[e]	-26[e]	-2[e]	31[e]	75.3	139.5	5
1,3-Dibromopropane	$C_3H_6Br_2$	-30[e]	-9[e]	17[e]	52[e]	98.7	166.8	5
1,2-Dibromotetrafluoroethane	$C_2Br_2F_4$		-97[e]	-75[e]	-46[e]	-7.2	47.1	5
1,2-Dibutoxyethane	$C_{10}H_{22}O_2$	0[e]	20[e]	44[e]	78.4	127.1	202.9	5
Dibutylamine	$C_8H_{19}N$	-37[e]	-16[e]	10[e]	44[e]	90.8	159.1	5
Dibutyl ether	$C_8H_{18}O$	-55[e]	-35[e]	-8[e]	26[e]	73.0	141.2	5
Di-*sec*-butyl ether	$C_8H_{18}O$			-19[e]	12.1	55.4	120.6	5
Di-*tert*-butyl ether	$C_8H_{18}O$			-33[e]	-2[e]	41.7	106.8	1
Dibutyl maleate	$C_{12}H_{20}O_4$	12.3	50.4	94.0	144.2	203[e]	272[e]	5
Di-*tert*-butyl peroxide	$C_8H_{18}O_2$			-26[e]	4.3	46.6	110.5	5
Dibutyl phthalate	$C_{16}H_{22}O_4$		104.0	142.7	191.5	254.5	339.4	4
Dibutyl sulfide	$C_8H_{18}S$	-22[e]	0[e]	27[e]	63[e]	113.5	188.4	5
1,1-Dichloroacetone	$C_3H_4Cl_2O$				1[e]	47.8	118.0	5
o-Dichlorobenzene	$C_6H_4Cl_2$		-13[e]	16.3	53.9	104.6	180.0	1,5
m-Dichlorobenzene	$C_6H_4Cl_2$		-22[e]	8.0	46.7	97.8	172.5	1,5
p-Dichlorobenzene	$C_6H_4Cl_2$	-45.5 s	-21.8 s	8 s	46.7 s	99.0	173.6	1,5
1,1-Dichlorobutane	$C_4H_8Cl_2$			-25[e]	6[e]	49.3	113.4	5
1,2-Dichlorobutane	$C_4H_8Cl_2$			-28.4	5.8	53.1	123.1	5
1,4-Dichlorobutane	$C_4H_8Cl_2$		-26[e]	0[e]	35[e]	82.4	153.4	5
2,2-Dichlorobutane	$C_4H_8Cl_2$		-58[e]	-35[e]	-5[e]	37.8	102.1	5
1,1-Dichlorocyclohexane	$C_6H_{10}Cl_2$	-39[e]	-19[e]	8[e]	43[e]	93.5	170.5	5
cis-1,2-Dichlorocyclohexane	$C_6H_{10}Cl_2$			27[e]	69[e]	125.7	206.2	5
1,2-Dichloro-1,1-difluoroethane	$C_2H_2Cl_2F_2$	-101[e]	-87[e]	-68[e]	-42.2	-6.8	46.3	5
Dichlorodifluoromethane[a]	CCl_2F_2	-150[e]	-138[e]	-122[e]	-101.8	-73.1	-30.0	5
2,2'-Dichlorodiisopropyl ether	$C_6H_{12}Cl_2O$		-1[e]	27.3	63.4	112.3	182.1	5
Dichlorodimethylsilane	$C_2H_6Cl_2Si$					11.1	70.1	5
1,1-Dichloroethane	$C_2H_4Cl_2$		-84[e]	-64[e]	-36.7	1.0	56.9	1
1,2-Dichloroethane	$C_2H_4Cl_2$				-16.4	23.7	83.1	1
1,1-Dichloroethene	$C_2H_2Cl_2$	-116[e]	-101[e]	-82[e]	-57[e]	-21.4	31.2	1

Name	Mol. form.	t/°C for 1 Pa	t/°C for 10 Pa	t/°C for 100 Pa	t/°C for 1 kPa	t/°C for 10 kPa	t/°C for 100 kPa	Ref.
cis-1,2-Dichloroethene	$C_2H_2Cl_2$			-62[e]	-34[e]	3.8	60.3	1
trans-1,2-Dichloroethene	$C_2H_2Cl_2$				-44[e]	-7.5	47.3	1
1,1-Dichloro-1-fluoroethane	$C_2H_3Cl_2F$		-101[e]	-83[e]	-57.9	-22.7	31.4	5
1,2-Dichloro-1-fluoroethane	$C_2H_3Cl_2F$			-50[e]	-23.8	14.1	73.4	5
Dichlorofluoromethane	$CHCl_2F$	-76[e]	-70[e]	-61[e]	-49[e]	-28.7	8.6	1
1,5-Dichloro-1,1,3,3,5,5-hexamethyltrisiloxane	$C_6H_{18}Cl_2O_2Si_3$	-29[e]	-7[e]	22.2	59.7	110.5	183.4	5
1,2-Dichlorohexane	$C_6H_{12}Cl_2$				49[e]	98.1	171.7	5
Dichloromethane[a]	CH_2Cl_2		-92[e]	-73[e]	-48[e]	-12.5	39.3	1
(Dichloromethyl)benzene	$C_7H_6Cl_2$			31	72	130	213	4
Dichloromethylphenylsilane	$C_7H_8Cl_2Si$			32.4	71.8	126.0	205.0	5
Dichloromethylsilane	CH_4Cl_2Si			-77[e]	-51[e]	-14[e]	40.5	1
1,2-Dichloropentane	$C_5H_{10}Cl_2$				30[e]	77.4	147.8	5
1,5-Dichloropentane	$C_5H_{10}Cl_2$	-31[e]	-10[e]	17[e]	54[e]	104.1	178.9	5
Dichlorophenylarsine	$C_6H_5AsCl_2$	6.9	35.2	70[e]	113[e]	170[e]	245[e]	5
1,1-Dichloropropane	$C_3H_6Cl_2$				-14[e]	27.0	87.7	5
1,2-Dichloropropane, (±)-	$C_3H_6Cl_2$	-78[e]	-61[e]	-38.1	-8.1	33.7	95.9	5
1,3-Dichloropropane	$C_3H_6Cl_2$	-65[e]	-46[e]	-22[e]	10[e]	54.0	119.9	5
2,2-Dichloropropane	$C_3H_6Cl_2$				-28[e]	10.8	68.9	5
1,3-Dichloro-2-propanol	$C_3H_6Cl_2O$			21.8	59.0	107.6	173.9	5
1,1-Dichloro-1,2,2,2-tetrafluoroethane	$C_2Cl_2F_4$					-45.4	2.7	5
1,2-Dichloro-1,1,2,2-tetrafluoroethane	$C_2Cl_2F_4$				-76.8	-44.9	3.2	5
1,3-Dichloro-1,1,3,3-tetramethyldisiloxane	$C_4H_{12}Cl_2OSi_2$		-33[e]	-9[e]	23.8	69.1	136.5	5
2,5-Dichlorothiophene	$C_4H_2Cl_2S$			-20[e]	22[e]	81.4	171[e]	5
2,4-Dichlorotoluene	$C_7H_6Cl_2$		6[e]	33[e]	68.3	119.5	199.1	5
3,4-Dichlorotoluene	$C_7H_6Cl_2$	-13[e]	9[e]	38[e]	76[e]	129.3	208.4	5
2,2-Dichloro-1,1,1-trifluoroethane	$C_2HCl_2F_3$		-101.0	-82.2	-57.4	-23.3	26.7	18
Diethanolamine	$C_4H_{11}NO_2$	53[e]	77[e]	107[e]	146[e]	197.3	268[e]	5
Diethoxydimethylsilane	$C_6H_{16}O_2Si$	-62[e]	-44[e]	-21.2	9.1	51.0	113.0	5
1,1-Diethoxyethane	$C_6H_{14}O_2$	-68[e]	-49[e]	-26[e]	3.7	44.2	101.9	5
1,2-Diethoxyethane	$C_6H_{14}O_2$		-59[e]	-35.3	-2.8	44.4	118.8	5
Diethoxymethane	$C_5H_{12}O_2$		-65[e]	-43[e]	-14[e]	27.3	87.7	5
Diethylamine	$C_4H_{11}N$			-46[e]	-26[e]	5[e]	55.2	1
2-Diethylaminoethanol	$C_6H_{15}NO$					97[e]	160.6	5
N,N-Diethylaniline	$C_{10}H_{15}N$	-11[e]	14[e]	44.3	84.2	138.4	216.3	5
o-Diethylbenzene	$C_{10}H_{14}$	-28[e]	-6[e]	21[e]	58[e]	107.9	182.9	5
m-Diethylbenzene	$C_{10}H_{14}$	-28[e]	-7[e]	20[e]	56[e]	106.2	180.6	5
p-Diethylbenzene	$C_{10}H_{14}$	-28[e]	-6[e]	21[e]	57[e]	108.1	183.3	5
Diethyl carbonate	$C_5H_{10}O_3$		-42[e]	-17[e]	17[e]	61.6	125.9	5
Diethyl disulfide	$C_4H_{10}S_2$	-46[e]	-26[e]	0[e]	35[e]	82.4	153.5	5
Diethylene glycol	$C_4H_{10}O_3$	35[e]	58[e]	86[e]	123[e]	173.6	245.2	1
Diethylene glycol dibutyl ether	$C_{12}H_{26}O_3$	5[e]	34.4	70.2	115.3	174.1	253.8	5
Diethylene glycol diethyl ether	$C_8H_{18}O_3$	-32[e]	-7[e]	25[e]	64.9	117.1	189[e]	5
Diethylene glycol dimethyl ether	$C_6H_{14}O_3$	-42[e]	-20[e]	8.3	44.3	92.3	159.4	5
Diethylene glycol monobutyl ether	$C_8H_{18}O_3$	14[e]	37[e]	66.8	104.9	153[e]	230.4	5
Diethylene glycol monobutyl ether acetate	$C_{10}H_{20}O_4$	6[e]	34[e]	69[e]	112.6	169.2	245.4	5
Diethylene glycol monoethyl ether	$C_6H_{14}O_3$			40[e]	80.3	132.4	201.4	5
Diethylene glycol monoethyl ether acetate	$C_8H_{16}O_4$	-16[e]	10.6	43.9	86.2	141.3	216.6	5
Diethylene glycol monomethyl ether	$C_5H_{12}O_3$		12[e]	40[e]	76[e]	124.2	193.7	1
Diethyl ether	$C_4H_{10}O$	-111[e]	-96[e]	-77[e]	-52.6	-17.8	34.1	1
Diethyl glutarate	$C_9H_{16}O_4$	-1[e]	26[e]	60.2	103.3	159.6	236.5	5
Diethyl hexanedioate	$C_{10}H_{18}O_4$	4[e]	35[e]	72[e]	116.6	171.2	239.5	5
Diethyl maleate	$C_8H_{12}O_4$	-6[e]	20[e]	52.2	93.5	148.4	224.8	5
Diethyl malonate	$C_7H_{12}O_4$	-23[e]	4[e]	36.0	76.4	128.5	198.3	5
1,3-Diethyl-5-methylbenzene	$C_{11}H_{16}$	-26[e]	-1[e]	29.5	69.5	123.5	200.2	5
Diethyl oxalate	$C_6H_{10}O_4$	-5[e]	18[e]	44.9	79.4	124.3	185.2	5
3,3-Diethylpentane	C_9H_{20}			-9[e]	26[e]	73.7	145.7	1
Diethylperoxide	$C_4H_{10}O_2$				-39[e]	3.6	65.0	5
Diethyl phthalate	$C_{12}H_{14}O_4$	12[e]	51[e]	96[e]	150.5	215.9	296.2	5
N,N-Diethyl-1,3-propanediamine	$C_7H_{18}N_2$				50.1	99.9	167.7	5
Diethyl sebacate	$C_{14}H_{26}O_4$		83[e]	120	166	225	305	4

Name	Mol. form.	*t*/°C for 1 Pa	*t*/°C for 10 Pa	*t*/°C for 100 Pa	*t*/°C for 1 kPa	*t*/°C for 10 kPa	*t*/°C for 100 kPa	Ref.
Diethyl succinate	$C_8H_{14}O_4$	-6[e]	20[e]	51.0	91.1	143.7	216.1	5
Diethyl sulfate	$C_4H_{10}O_4S$		3[e]	36[e]	79[e]	134[e]	208.3	5
Diethyl sulfide	$C_4H_{10}S$	-80[e]	-62[e]	-40[e]	-10.8	30.3	91.7	1
1,1-Difluoroethane	$C_2H_4F_2$			-115.2	-94.6	-66.1	-24.3	19
Difluoromethane[a]	CH_2F_2	-156.7	-145.8	-131.9	-113.6	-88.6	-51.9	1
3,4-Dihydro-2*H*-pyran	C_5H_8O				-22[e]	22.0	84.9	5
Diiodomethane	CH_2I_2			17[e]	55[e]	106.1	181.6	5
Diiodosilane	H_2I_2Si				11.8	70.5	149.4	4
Diisobutylamine	$C_8H_{19}N$	-57[e]	-36[e]	-9.0	25.5	72.2	139.0	5
Diisopentyl ether	$C_{10}H_{22}O$			14.0	51.5	101.8	172.8	5
Diisopentyl sulfide	$C_{10}H_{22}S$			7[e]	82[e]	118[e]	139[e]	5
Diisopropylamine	$C_6H_{15}N$			-47[e]	-17.5	23.5	84.0	5
1,2-Diisopropylbenzene	$C_{12}H_{18}$	-14[e]	9[e]	37[e]	74[e]	125.9	203.2	5
1,3-Diisopropylbenzene	$C_{12}H_{18}$	-14[e]	8[e]	36[e]	74[e]	125.5	202.6	5
1,4-Diisopropylbenzene	$C_{12}H_{18}$	-6[e]	18[e]	49[e]	90[e]	148.8	238[e]	5
Diisopropyl ether	$C_6H_{14}O$		-76[e]	-55[e]	-28[e]	11[e]	68.1	1
Diisopropyl sulfide	$C_6H_{14}S$	-65[e]	-47[e]	-23[e]	9[e]	53.1	119.6	5
Diketene	$C_4H_4O_2$				19.3	63.3	126[e]	5
1,3-Dimethoxybenzene	$C_8H_{10}O_2$	18[e]	34[e]	56[e]	86.7	135.5	223[e]	5
Dimethoxyborane	$C_2H_7BO_2$	-116[e]	-101.9	-83.5	-59.2	-25.4	25[e]	5
1,2-Dimethoxyethane	$C_4H_{10}O_2$			-44[e]	-15[e]	25.2	85.2	1
Dimethoxymethane	$C_3H_8O_2$	-93[e]	-81[e]	-64[e]	-42[e]	-9.3	41.7	5
Dimethylacetal	$C_4H_{10}O_2$	-89[e]	-74[e]	-55[e]	-29[e]	7.7	64.1	5
N,*N*-Dimethylacetamide	C_4H_9NO	-8[e]	8[e]	28.0	56.4	98.2	165.7	1
Dimethylamine	C_2H_7N			-88[e]	-66.9	-37.2	6.6	1
(Dimethylamino)dimethylborane	$C_4H_{12}BN$		-81[e]	-60.1	-31.9	7.0	64.2	5
3-(Dimethylamino)propanenitrile	$C_5H_{10}N_2$				51.1	101.8	171.4	5
2,4-Dimethylaniline	$C_8H_{11}N$	-2[e]	21[e]	51[e]	88[e]	139.1	210.9	5
2,6-Dimethylaniline	$C_8H_{11}N$			37[e]	80[e]	137.7	217.7	5
N,*N*-Dimethylaniline	$C_8H_{11}N$			28[e]	66[e]	118.1	193.6	1
2,4-Dimethylbenzaldehyde	$C_9H_{10}O$	-3[e]	23[e]	54[e]	93.2	144.6	214.5	5
2,3-Dimethyl-1,3-butadiene	C_6H_{10}			-59[e]	-30[e]	9.7	68.1	5
2,2-Dimethylbutane	C_6H_{14}		-90[e]	-71.5	-45.5	-7.7	49.4	1
2,3-Dimethylbutane	C_6H_{14}	-103[e]	-87[e]	-66[e]	-39.0	-0.4	57.6	1
3,3-Dimethyl-1-butanol	$C_6H_{14}O$	-37[e]	-16[e]	9[e]	42[e]	84.3	142.5	5
2,3-Dimethyl-2-butanol	$C_6H_{14}O$			-5[e]	23[e]	61.3	118.2	5
3,3-Dimethyl-2-butanone	$C_6H_{12}O$			-30[e]	0[e]	42.5	105.7	1
2,3-Dimethyl-1-butene	C_6H_{12}	-103[e]	-87[e]	-67[e]	-39.9	-1.9	55.2	5
3,3-Dimethyl-1-butene	C_6H_{12}	-110[e]	-95[e]	-76[e]	-50.8	-14.5	40.8	5
2,3-Dimethyl-2-butene	C_6H_{12}		-75[e]	-54[e]	-25[e]	14[e]	72.9	1
1,1-Dimethylcyclohexane	C_8H_{16}			-27[e]	5[e]	50.6	119.1	5
cis-1,2-Dimethylcyclohexane	C_8H_{16}		-44[e]	-20[e]	14[e]	59.7	129.2	5
trans-1,2-Dimethylcyclohexane	C_8H_{16}	-68[e]	-49[e]	-25[e]	8[e]	53.9	122.9	5
cis-1,3-Dimethylcyclohexane	C_8H_{16}	-68[e]	-48[e]	-23[e]	10[e]	55.6	123.1	5
trans-1,3-Dimethylcyclohexane	C_8H_{16}	-62[e]	-45[e]	-23[e]	8[e]	51.5	120.9	5
cis-1,4-Dimethylcyclohexane	C_8H_{16}	-66[e]	-47[e]	-23[e]	10[e]	55.3	123.8	5
trans-1,4-Dimethylcyclohexane	C_8H_{16}			-27[e]	5[e]	50.6	118.9	5
1,1-Dimethylcyclopentane	C_7H_{14}		-69[e]	-47[e]	-17[e]	24.8	87.4	5
cis-1,2-Dimethylcyclopentane	C_7H_{14}			-38[e]	-8[e]	34.9	99.0	5
trans-1,2-Dimethylcyclopentane	C_7H_{14}	-83[e]	-66[e]	-43[e]	-13[e]	28.4	91.4	5
cis-1,3-Dimethylcyclopentane	C_7H_{14}	-84[e]	-66[e]	-44[e]	-14[e]	28.2	91.1	5
trans-1,3-Dimethylcyclopentane	C_7H_{14}	-84[e]	-67[e]	-44[e]	-14[e]	27.4	90.3	5
1,2-Dimethylcyclopentene	C_7H_{12}	-75[e]	-57[e]	-34[e]	-3[e]	40.2	105.3	5
1,5-Dimethylcyclopentene	C_7H_{12}	-77[e]	-59[e]	-36[e]	-5.5	37.3	101.5	5
cis-1,2-Dimethylcyclopropane	C_5H_{10}	-118[e]	-103[e]	-83[e]	-57[e]	-20[e]	36.6	5
trans-1,2-Dimethylcyclopropane	C_5H_{10}	-122[e]	-108[e]	-89[e]	-63[e]	-27[e]	27.8	5
Dimethyl disulfide	$C_2H_6S_2$	-71[e]	-53[e]	-29[e]	1.7	45.0	109.3	5
N,*N*-Dimethylethanolamine	$C_4H_{11}NO$	-52[e]	-31[e]	-6[e]	27[e]	70.9	133[e]	5
Dimethyl ether[a]	C_2H_6O		-135[e]	-118[e]	-96.8	-67.6	-25.1	1,5
N,*N*-Dimethylformamide	C_3H_7NO	-39[e]	-20[e]	5[e]	38.0	83.9	152.6	1

Name	Mol. form.	t/°C for 1 Pa	t/°C for 10 Pa	t/°C for 100 Pa	t/°C for 1 kPa	t/°C for 10 kPa	t/°C for 100 kPa	Ref.
Dimethyl glutarate	$C_7H_{12}O_4$	-11^e	15^e	47^e	87.7	139.8	209.5	5
2,2-Dimethylheptane	C_9H_{20}	-58^e	-39^e	-15^e	18^e	63.6	132.3	5
2,3-Dimethylheptane	C_9H_{20}	-53^e	-33^e	-9^e	25^e	70.8	140.0	5
2,6-Dimethylheptane	C_9H_{20}	-55^e	-36^e	-12^e	21^e	66.4	134.7	5
2,6-Dimethyl-4-heptanone	$C_9H_{18}O$	-32^e	-12^e	14^e	48^e	96.2	167.7	5
2,5-Dimethyl-1,5-hexadiene	C_8H_{14}	-38^e	-26^e	-10^e	14^e	50.8	115.1	5
2,2-Dimethylhexane	C_8H_{18}	-73^e	-55^e	-32^e	-1.5	41.6	106.4	5
2,3-Dimethylhexane	C_8H_{18}				5^e	49.2	115.1	5
2,4-Dimethylhexane	C_8H_{18}				0.6	43.9	109.0	5
2,5-Dimethylhexane	C_8H_{18}	-71^e	-53^e	-30^e	0.7	43.8	108.6	5
3,3-Dimethylhexane	C_8H_{18}	-72^e	-54^e	-30^e	1.4	45.4	111.5	5
3,4-Dimethylhexane	C_8H_{18}				7^e	50.9	117.3	5
Dimethyl 1,6-hexanedioate	$C_8H_{14}O_4$		28^e	61^e	103^e	156.1	227.3	5
2,3-Dimethyl-2-hexene	C_8H_{16}	-65^e	-47^e	-23^e	10^e	54.3	121.3	5
cis-2,2-Dimethyl-3-hexene	C_8H_{16}	-74^e	-56^e	-33^e	-3^e	40.1	105.0	5
1,1-Dimethylhydrazine	$C_2H_8N_2$			-52^e	-25.6	10.5	63^e	5
1,2-Dimethylhydrazine	$C_2H_8N_2$		-49^e	-33^e	-9^e	26.4	88^e	1
Dimethyl isophthalate	$C_{10}H_{10}O_4$			85^e	129.5	189.2	273^e	5
2,4-Dimethyl-3-isopropylpentane	$C_{10}H_{22}$	-46^e	-26^e	0^e	35^e	83.2	156.5	5
Dimethyl maleate	$C_6H_8O_4$		5^e	36^e	76^e	127.3	197^e	5
Dimethyl malonate	$C_5H_8O_4$	-22^e	1^e	30.0	66.7	114.7	180.2	5
Dimethyl mercury	C_2H_6Hg				-13.5	29.0	92.1	5
1,2-Dimethylnaphthalene	$C_{12}H_{12}$	26^e	51^e	82^e	123^e	180.5	265.7	5
2,7-Dimethylnaphthalene	$C_{12}H_{12}$	31.5 s	53.1 s	78.8 s	115.9	175^e	260^e	5
2,4-Dimethyloctane	$C_{10}H_{22}$				38^e	84.9	155.4	5
2,7-Dimethyloctane	$C_{10}H_{22}$	-39^e	-19^e	7^e	41^e	88.4	159.4	5
Dimethyl oxalate	$C_4H_6O_4$				50.5	98.1	163.0	5
2,2-Dimethylpentane	C_7H_{16}	-90^e	-73^e	-52^e	-22.9	17.6	78.8	1
2,3-Dimethylpentane	C_7H_{16}	-87^e	-68.4	-45.3	-14.9	26.8	89.3	5
2,4-Dimethylpentane	C_7H_{16}	-89^e	-72^e	-50^e	-21.3	19.2	80.1	1
3,3-Dimethylpentane	C_7H_{16}	-88^e	-71^e	-49^e	-18.8	22.9	85.6	1
2,2-Dimethyl-3-pentanol	$C_7H_{16}O$			9^e	35^e	73.1	135.5	5
2,4-Dimethyl-3-pentanone	$C_7H_{14}O$	-61^e	-42^e	-18^e	14^e	58.5	124.8	1
2,3-Dimethyl-1-pentene	C_7H_{14}	-85^e	-68^e	-46^e	-17^e	23.4	83.8	5
2,4-Dimethyl-1-pentene	C_7H_{14}	-88^e	-71^e	-50^e	-21^e	20.0	81.2	5
3,3-Dimethyl-1-pentene	C_7H_{14}	-87^e	-71^e	-50^e	-21^e	18.1	77.1	5
4,4-Dimethyl-1-pentene	C_7H_{14}	-94^e	-78^e	-57^e	-28^e	11.5	72.1	5
2,3-Dimethyl-2-pentene	C_7H_{14}	-79^e	-62^e	-39^e	-9^e	33.5	96.9	5
2,4-Dimethyl-2-pentene	C_7H_{14}	-84^e	-68^e	-46^e	-18^e	22.6	82.9	5
cis-3,4-Dimethyl-2-pentene	C_7H_{14}	-83^e	-65^e	-43^e	-14^e	27.2	88.8	5
trans-3,4-Dimethyl-2-pentene	C_7H_{14}	-82^e	-64^e	-42^e	-13^e	29.0	91.1	5
cis-4,4-Dimethyl-2-pentene	C_7H_{14}	-90^e	-73^e	-51^e	-22^e	18.6	80.0	5
trans-4,4-Dimethyl-2-pentene	C_7H_{14}	-90^e	-73^e	-52^e	-23^e	16.6	76.3	5
4,4-Dimethyl-1-pentyne	C_7H_{12}		-73^e	-52^e	-24^e	15.9	75.6	5
4,4-Dimethyl-2-pentyne	C_7H_{12}		-70^e	-48^e	-19^e	21.4	82.6	5
Dimethyl phthalate	$C_{10}H_{10}O_4$	27^e	56^e	92.7	137.8	195.8	272.7	5
2,2-Dimethylpropanenitrile	C_5H_9N					41.1	104.8	5
2,2-Dimethyl-1-propanol	$C_5H_{12}O$					59.2	112.7	5
2,3-Dimethylpyridine	C_7H_9N				42^e	89.9	160.6	5
2,4-Dimethylpyridine	C_7H_9N		-25^e	3.7	40.0	87.5	157.9	1,5
2,5-Dimethylpyridine	C_7H_9N			4^e	39^e	86.2	156.6	1
2,6-Dimethylpyridine	C_7H_9N			-3^e	29.9	75.8	143.6	1
3,4-Dimethylpyridine	C_7H_9N		-9^e	19^e	55^e	104.8	178.6	5
3,5-Dimethylpyridine	C_7H_9N			11^e	48^e	98^e	171.5	1
Dimethyl sebacate	$C_{12}H_{22}O_4$		53^e	97	150	214	293	4
Dimethyl succinate	$C_6H_{10}O_4$			30^e	70.4	123.3	195.4	5
Dimethyl sulfide	C_2H_6S		-96^e	-77^e	-51.2	-16.0	37.0	1,5
Dimethyl sulfone	$C_2H_6O_2S$				109^e	166.8	248.9	5
Dimethyl sulfoxide	C_2H_6OS			27.4	65.0	115.9	188.6	1
Dimethyl terephthalate	$C_{10}H_{10}O_4$	56.6 s	79.4 s	106.1 s	137.9 s	197.9	282^e	5

Name	Mol. form.	t/°C for 1 Pa	t/°C for 10 Pa	t/°C for 100 Pa	t/°C for 1 kPa	t/°C for 10 kPa	t/°C for 100 kPa	Ref.
2,5-Dimethylthiophene	C_6H_8S		-43[e]	-16[e]	20[e]	67.5	134.8	5
1,1-Dinitropropane	$C_3H_6N_2O_4$	-9[e]	12[e]	39[e]	73.2	120[e]	187[e]	5
Dioctyl phthalate	$C_{24}H_{38}O_4$	130[e]	163.7	203.8	252[e]	311[e]	385[e]	5
1,3-Dioxane	$C_4H_8O_2$			-37[e]	-3[e]	43.4	106.0	5
1,4-Dioxane	$C_4H_8O_2$					39.6	101.0	1
1,3-Dioxolane	$C_3H_6O_2$		-72[e]	-50[e]	-22[e]	17.0	75.3	1
Dipentene	$C_{10}H_{16}$	-42[e]	-19[e]	10.6	48.7	100.2	173.9	5
Dipentylamine	$C_{10}H_{23}N$				77[e]	127.7	202.0	5
Dipentyl ether	$C_{10}H_{22}O$	-31[e]	-8[e]	22[e]	60[e]	111.6	186.2	5
Diphenylamine	$C_{12}H_{11}N$	48 s		102.8	150.5	213.7	301.4	5
1,1-Diphenylethane	$C_{14}H_{14}$	19[e]	47[e]	82.0	125.3	181[e]	254[e]	5
Diphenyl ether	$C_{12}H_{10}O$		44[e]	75[e]	116[e]	173[e]	257.4	5
Diphenylmethane	$C_{13}H_{12}$		45[e]	77[e]	119.3	177.7	263.6	1,5
Diphenyl sulfide	$C_{12}H_{10}S$	20[e]	51[e]	88.7	137.5	202.2	291.8	5
1,2-Dipropoxyethane	$C_8H_{18}O_2$			-44.2	-2.0	63.6	179.2	5
Dipropylamine	$C_6H_{15}N$		-48[e]	-25[e]	6[e]	47.5	108.8	5
Dipropylene glycol	$C_6H_{14}O_3$				110[e]	162.6	231.4	5
Dipropyl ether	$C_6H_{14}O$	-80[e]	-63[e]	-41[e]	-12[e]	28.8	89.7	1
Dipropyl oxalate	$C_8H_{14}O_4$	-4[e]	20[e]	49.9	88.6	140.4	213.0	5
Dipropyl succinate	$C_{10}H_{18}O_4$	11[e]	38[e]	72.1	115.4	172.3	250.4	5
Dipropyl sulfide	$C_6H_{14}S$	-50[e]	-30[e]	-6[e]	28[e]	73.6	142.4	5
m-Divinylbenzene	$C_{10}H_{10}$	-29[e]	-4[e]	27.1	67.6	122.1	199[e]	5
Divinyl ether	C_4H_6O		-99[e]	-80[e]	-56[e]	-22.1	28.0	5
Docosane	$C_{22}H_{46}$	83.5	115.0	154.0	203.6	274.8	368.0	5
Docosanoic acid	$C_{22}H_{44}O_2$	145.4	176.5	213.7	259.3	316.2	390[e]	5
cis-13-Docosenoic acid	$C_{22}H_{42}O_2$	126[e]	160[e]	199.4	247.4	306.5	381.1	5
trans-13-Docosenoic acid	$C_{22}H_{42}O_2$	134[e]	166[e]	203.6	249.8	307.6	382.0	5
Dodecamethylcyclohexasiloxane	$C_{12}H_{36}O_6Si_6$	18[e]	41[e]	69[e]	108[e]	162.2	244.7	5
Dodecanal	$C_{12}H_{24}O$			70[e]	116.2	175.9	256.6	5
Dodecane	$C_{12}H_{26}$	-5.4	18.2	47.6	85.8	138.2	215.8	16
Dodecanenitrile	$C_{12}H_{23}N$	36[e]	60[e]	92[e]	133[e]	190.5	275.5	5
Dodecanoic acid	$C_{12}H_{24}O_2$	78[e]	100[e]	128[e]	166[e]	219.1	298.1	5
1-Dodecanol	$C_{12}H_{26}O$				133[e]	185.0	264.1	1
1-Dodecene	$C_{12}H_{24}$	-8.3	15.2	44.8	82.9	135.4	212.8	5
1-Dodecyne	$C_{12}H_{22}$	-11[e]	13[e]	43[e]	82[e]	135.8	214.4	5
Dysprosium	Dy	1105 s	1250 s	1431[d]	1681[d]	2031[d]	2558[d]	3
Eicosamethylnonasiloxane	$C_{20}H_{60}O_8Si_9$			141[e]	183.1	236.7	307.1	5
Eicosane	$C_{20}H_{42}$	80.4	108.9	144.2	189.8	252.1	344[e]	16
1-Eicosanol	$C_{20}H_{42}O$	119[e]	143[e]	173[e]	213[e]	270.0	355.1	5
Epichlorohydrin	C_3H_5ClO			-21[e]	11[e]	53.8	115.5	5
1,2-Epoxybutane	C_4H_8O	-135[e]	-114[e]	-87[e]	-53[e]	-5.5	62.1	5
Erbium	Er	1231 s	1390 s	1612[d]	1890[d]	2279[d]	2859[d]	3
Ethane[a]	C_2H_6	-183.3 s	-173.2	-161.3	-145.3	-122.8	-88.8	41
1,2-Ethanediamine	$C_2H_8N_2$				17.0	57.5	116.6	1,5
1,2-Ethanediol	$C_2H_6O_2$	2[e]	24[e]	51.1	86.1	132.5	196.9	1
1,2-Ethanediol, diacetate	$C_6H_{10}O_4$	-17[e]	6[e]	35.0	71.9	121.1	190.0	5
1,2-Ethanediol, dinitrate	$C_2H_4N_2O_6$	4[e]	25.6	51.0	81[e]	117[e]	162[e]	5
Ethanethiol	C_2H_6S	-112[e]	-97[e]	-78[e]	-53[e]	-18[e]	34.7	1
Ethanol	C_2H_6O	-73[e]	-56[e]	-34[e]	-7[e]	29.2	78.0	1,5
Ethanolamine	C_2H_7NO		11[e]	35[e]	66.2	109.0	170.6	1
2-Ethoxyaniline	$C_8H_{11}NO$	0[e]	27[e]	60[e]	102.2	156.0	228.1	5
Ethoxybenzene	$C_8H_{10}O$		-9[e]	17[e]	51[e]	99[e]	169.3	5
2-Ethoxyethanol	$C_4H_{10}O_2$	-49[e]	-29[e]	-3[e]	30[e]	73.6	135.3	1
2-Ethoxyethyl acetate	$C_6H_{12}O_3$	-25[e]	-8[e]	14[e]	44.6	88.0	155.6	5
Ethyl acetate	$C_4H_8O_2$	-83[e]	-66[e]	-45[e]	-18[e]	20.4	76.8	1
Ethyl acetoacetate	$C_6H_{10}O_3$	-25[e]	-3[e]	25.7	62.3	111.3	180.2	5
Ethyl acrylate	$C_5H_8O_2$		-55[e]	-32.7	-2.8	38.5	99.2	5
Ethylamine	C_2H_7N			-71[e]	-53[e]	-27[e]	16.4	1
4-Ethylaniline	$C_8H_{11}N$	-2[e]	21[e]	49[e]	87[e]	139.4	216.7	5
N-Ethylaniline	$C_8H_{11}N$	-15[e]	8[e]	38[e]	76.4	128.8	204.2	5

Name	Mol. form.	*t*/°C for 1 Pa	*t*/°C for 10 Pa	*t*/°C for 100 Pa	*t*/°C for 1 kPa	*t*/°C for 10 kPa	*t*/°C for 100 kPa	Ref.
Ethylbenzene	C_8H_{10}	-56.2	-36.8	-12.0	21.1	67.1	135.7	1
Ethyl benzoate	$C_9H_{10}O_2$	-18[e]	8[e]	39[e]	80.1	135.1	212.8	5
Ethyl butanoate	$C_6H_{12}O_2$	-49[e]	-34[e]	-14[e]	14.3	55.2	121.1	5
2-Ethylbutanoic acid	$C_6H_{12}O_2$	-9[e]	16[e]	46[e]	83[e]	130.7	192.5	5
2-Ethyl-1-butanol	$C_6H_{14}O$		-5[e]	17[e]	46[e]	85.7	146.1	5
2-Ethyl-1-butene	C_6H_{12}	-98[e]	-81[e]	-60[e]	-32[e]	6.6	64.3	5
Ethyl chloroacetate	$C_4H_7ClO_2$			-2.6	32.6	79.1	143.8	5
Ethyl 2-chloropropanoate	$C_5H_9ClO_2$			1.4	36.4	82.5	146.0	5
Ethyl *trans*-cinnamate	$C_{11}H_{12}O_2$			79	125	187	271	4
Ethyl cyanoacetate	$C_5H_7NO_2$	16[e]	39[e]	67.0	102.1	146.7	205.6	5
Ethylcyclobutane	C_6H_{12}	-99[e]	-82[e]	-61[e]	-32[e]	9[e]	70.2	5
Ethylcyclohexane	C_8H_{16}	-61[e]	-42[e]	-17[e]	15.8	61.9	131.3	5
1-Ethylcyclohexene	C_8H_{14}	-55[e]	-35[e]	-11[e]	22[e]	68[e]	136.5	5
Ethylcyclopentane	C_7H_{14}	-76[e]	-59[e]	-35[e]	-5[e]	38.4	103.0	5
1-Ethylcyclopentene	C_7H_{12}	-75[e]	-57[e]	-34[e]	-3[e]	40.7	105.8	5
Ethylcyclopropane	C_5H_{10}	-118[e]	-102[e]	-83[e]	-57[e]	-20[e]	35.5	5
Ethyl decanoate	$C_{12}H_{24}O_2$	8[e]	35[e]	69[e]	111.8	166.1	238[e]	5
Ethyl dichloroacetate	$C_4H_6Cl_2O_2$			2.6	40.1	89.1	156.3	5
Ethyl diethylmalonate	$C_{11}H_{20}O_4$			74[e]	105[e]	149.4	219[e]	5
Ethyldifluoroarsine	$C_2H_5AsF_2$			-36[e]	-6.0	35.0	93.1	5
1-Ethyl-2,4-dimethylbenzene	$C_{10}H_{14}$	-25[e]	-4[e]	24[e]	61[e]	112.2	187.9	5
1-Ethyl-3,5-dimethylbenzene	$C_{10}H_{14}$	-28[e]	-6[e]	21[e]	58[e]	108.3	183.2	5
2-Ethyl-1,3-dimethylbenzene	$C_{10}H_{14}$		-2[e]	26[e]	63[e]	113.7	189.5	5
2-Ethyl-1,4-dimethylbenzene	$C_{10}H_{14}$	-27[e]	-5[e]	23[e]	60[e]	110.6	186.4	5
3-Ethyl-1,2-dimethylbenzene	$C_{10}H_{14}$	-22[e]	0[e]	28[e]	66[e]	117.2	193.4	5
4-Ethyl-1,2-dimethylbenzene	$C_{10}H_{14}$	-24[e]	-2[e]	26[e]	63[e]	113.6	189.2	5
3-Ethyl-2,4-dimethylpentane	C_9H_{20}	-58[e]	-38[e]	-13[e]	20[e]	66.7	136.2	5
Ethylene[a]	C_2H_4				-155.6	-135.1	-104.0	1,10
Ethylene carbonate	$C_3H_4O_3$	12.7 s	37[e]				247	5
Ethyleneimine	C_2H_5N		-74[e]	-55[e]	-30[e]	4.1	55[e]	5
Ethyl formate	$C_3H_6O_2$		-80[e]	-61[e]	-35[e]	1[e]	54.0	1
3-Ethylhexane	C_8H_{18}				8[e]	52.1	118.1	5
Ethyl hexanoate	$C_8H_{16}O_2$	-31[e]	-9[e]	18.7	53.9	100.7	166.2	5
2-Ethylhexanoic acid	$C_8H_{16}O_2$				108[e]	159.6	226.6	5
2-Ethyl-1-hexanol	$C_8H_{18}O$			45[e]	75[e]	118.3	184.2	1
2-Ethylhexyl acetate	$C_{10}H_{20}O_2$	-11[e]	5[e]	26[e]	57.6	107.1	197.2	5
Ethyl hydroperoxide	$C_2H_6O_2$	-70[e]	-49[e]	-25[e]	6.8	47.0	101[e]	5
Ethyl isopropyl sulfide	$C_5H_{12}S$	-72[e]	-54[e]	-31[e]	0[e]	42.7	106.9	5
Ethyl isothiocyanate	C_3H_5NS				17.4	66[e]	136[e]	5
Ethyl levulinate	$C_7H_{12}O_3$		17[e]	45.3	82.6	133.2	205.7	5
Ethyl methacrylate	$C_6H_{10}O_2$				8[e]	53.2	116.8	5
Ethyl 3-methylbutanoate	$C_7H_{14}O_2$	-57[e]	-36[e]	-10[e]	23.9	69.5	134.4	5
trans-1-Ethyl-4-methylcyclohexane	C_9H_{18}	-53[e]	-33[e]	-8[e]	25[e]	71.8	141.5	5
1-Ethyl-1-methylcyclopentane	C_8H_{16}	-67[e]	-49[e]	-24[e]	8[e]	53.2	121.0	5
cis-1-Ethyl-2-methylcyclopentane	C_8H_{16}	-63[e]	-44[e]	-19[e]	13.3	59.1	127.6	5
1-Ethyl-1-methylcyclopropane	C_6H_{12}	-105[e]	-89[e]	-69[e]	-41[e]	-3[e]	56.3	5
Ethyl methyl ether	C_3H_8O	-98[e]	-89[e]	-77[e]	-60[e]	-34.8	7.0	5
3-Ethyl-4-methylhexane	C_9H_{20}			-9[e]	24[e]	70.6	139.9	5
3-Ethyl-2-methylpentane	C_8H_{18}	-69[e]	-50[e]	-27[e]	5[e]	48.9	115.2	5
3-Ethyl-3-methylpentane	C_8H_{18}	-70[e]	-51[e]	-27[e]	5[e]	50.2	117.8	5
Ethyl 2-methylpropanoate	$C_6H_{12}O_2$	-65[e]	-47[e]	-24.6	5.4	47.3	109.8	5
Ethyl methyl sulfide	C_3H_8S	-94[e]	-78[e]	-57[e]	-29.7	8.8	66.3	1
1-Ethylnaphthalene	$C_{12}H_{12}$	16[e]	41[e]	72[e]	114[e]	171.8	257.7	5
2-Ethylnaphthalene	$C_{12}H_{12}$	14[e]	39[e]	71[e]	113[e]	171.2	257.3	5
Ethyl nitrate	$C_2H_5NO_3$	-81[e]	-63[e]	-41[e]	-12[e]	28.2	87[e]	1
1-Ethyl-4-nitrobenzene	$C_8H_9NO_2$	10[e]	36[e]	69[e]	111.6	168[e]	245[e]	5
Ethyl octanoate	$C_{10}H_{20}O_2$	-17[e]	9[e]	41[e]	81.4	133.2	203[e]	5
3-Ethylpentane	C_7H_{16}	-81[e]	-63[e]	-41[e]	-11[e]	30.5	93.1	1
3-Ethyl-1-pentene	C_7H_{14}	-85[e]	-68[e]	-46[e]	-17[e]	23.2	83.7	5
2-Ethylphenol	$C_8H_{10}O$		16.9	44.5	81.1	130.9	204.0	5

Name	Mol. form.	t/°C for 1 Pa	t/°C for 10 Pa	t/°C for 100 Pa	t/°C for 1 kPa	t/°C for 10 kPa	t/°C for 100 kPa	Ref.
3-Ethylphenol	$C_8H_{10}O$	5.6	29.2	57.5	91.9	144.8	217.9	5
4-Ethylphenol	$C_8H_{10}O$			60[e]	95.5	144.6	217.5	5
Ethyl phenylacetate	$C_{10}H_{12}O_2$	-9[e]	19[e]	52[e]	95[e]	150.2	225[e]	5
5-Ethyl-2-picoline	$C_8H_{11}N$	-33[e]	-9.3	20[e]			178.0	5
Ethyl propanoate	$C_5H_{10}O_2$	-69[e]	-52[e]	-30[e]	-1[e]	38.9	98.7	1
Ethyl propyl ether	$C_5H_{12}O$	-92[e]	-77[e]	-57[e]	-30.5	6.7	63.4	1,5
Ethyl propyl sulfide	$C_5H_{12}S$	-64[e]	-46[e]	-23[e]	9[e]	52.7	118.0	5
2-Ethylpyridine	C_7H_9N	-46[e]	-26[e]	-1[e]	33[e]	79.3	149.0	5
3-Ethylpyridine	C_7H_9N	-38[e]	-17[e]	9[e]	44[e]	92.7	166.5	5
4-Ethylpyridine	C_7H_9N	-35[e]	-15[e]	11[e]	46[e]	94.4	168.6	5
Ethyl silicate	$C_8H_{20}O_4Si$	-77[e]	-52[e]	-21[e]	21.6	80.5	164.1	5
2-Ethylstyrene	$C_{10}H_{12}$	-31[e]	-8[e]	21[e]	60[e]	111.7	187[e]	5
3-Ethylstyrene	$C_{10}H_{12}$	-28[e]	-5.3	24.1	62.6	116[e]	193[e]	5
4-Ethylstyrene	$C_{10}H_{12}$	-31[e]	-8.2	21.3	60.5	115[e]	196[e]	5
Ethyl thiocyanate	C_3H_5NS	-39[e]	-20[e]	4[e]	35[e]	79.1	143.4	5
2-Ethyltoluene	C_9H_{12}	-40[e]	-19[e]	8[e]	43[e]	92.1	164.7	5
3-Ethyltoluene	C_9H_{12}	-42[e]	-21[e]	5[e]	40.4	88.9	160.8	5
4-Ethyltoluene	C_9H_{12}	-41[e]	-21[e]	6[e]	41[e]	89.2	161.5	5
Ethyl trichloroacetate	$C_4H_5Cl_3O_2$			15.3	51.9	100.1	166.6	5
1-Ethyl-2,4,5-trimethylbenzene	$C_{11}H_{16}$	-13[e]	11[e]	40[e]	79.4	132.1	207.7	5
2-Ethyl-1,3,5-trimethylbenzene	$C_{11}H_{16}$		6[e]	36[e]	75.7	129.6	207.6	5
Ethyl 10-undecenoate	$C_{13}H_{24}O_2$	32[e]	55[e]	86[e]	125.2	179.5	258.4	5
Ethyl vinyl ether	C_4H_8O		-102[e]	-81[e]	-53.1	-16.5	34.7	5
Eucalyptol	$C_{10}H_{18}O$			10.6	48.5	100.3	175.4	5
Europium	Eu	590 s	684 s	799 s	961	1179	1523	14
9H-Fluorene	$C_{13}H_{10}$	48.4 s			137.4	205.4	295[e]	5
Fluorine[a]	F_2	-235 s	-229.5 s	-222.9 s	-214.8	-204.3	-188.3	1,5
Fluorine monoxide[a]	F_2O	-211.7	-204.7	-195.9	-184.2	-168.2	-144.9	5
Fluorine nitrate	FNO_3	-160[e]	-149[e]	-135[e]	-115.1	-87.4	-45.0	5
Fluorobenzene	C_6H_5F				-16.9	24.2	84.4	1
1-Fluorobutane	C_4H_9F	-114[e]	-99[e]	-80[e]	-55[e]	-20.0	32.1	5
2-Fluorobutane	C_4H_9F	-117[e]	-103[e]	-85[e]	-60.7	-26.7	24.7	5
1-Fluorodecane	$C_{10}H_{21}F$	-22[e]	0[e]	27[e]	64[e]	113.3	185.7	5
Fluoroethane	C_2H_5F		-142[e]	-127[e]	-106.3	-78.7	-37.9	1
2-Fluoroethanol	C_2H_5FO			-22[e]	8.3	47.5	99[e]	5
Fluoroethene	C_2H_3F			-153.3	-135.2	-109.9	-72.2	5
1-Fluoroheptane	$C_7H_{15}F$	-64[e]	-45[e]	-22[e]	10[e]	53.3	117.4	5
1-Fluorohexane	$C_6H_{13}F$	-80[e]	-62[e]	-40[e]	-11[e]	30.4	91.1	5
Fluoromethane[a]	CH_3F				-130[e]	-111[e]	-78.6	1
1-Fluorooctane	$C_8H_{17}F$				29[e]	74.6	141.8	5
1-Fluoropentane	$C_5H_{11}F$	-97[e]	-80[e]	-60[e]	-32[e]	5.7	62.4	5
1-Fluoropropane	C_3H_7F	-133[e]	-120[e]	-103[e]	-80.7	-49.4	-2.8	5
Fluorosulfonic acid	FHO_3S	-14[e]	4[e]	28[e]	59.1	101.3	162.2	5
2-Fluorotoluene	C_7H_7F		-50[e]	-26[e]	5[e]	49.0	113.9	5
3-Fluorotoluene	C_7H_7F	-67[e]	-48[e]	-25[e]	7[e]	51.0	116.1	5
4-Fluorotoluene	C_7H_7F		-48[e]	-24[e]	7[e]	51[e]	116.2	5
1-Fluoro-4-(trifluoromethyl)benzene	$C_7H_4F_4$			-38[e]	-6[e]	38.6	102.3	5
Formaldehyde[a]	CH_2O				-91[e]	-61.7	-19.3	1
Formamide	CH_3NO		22[e]	53[e]	93[e]	145.0	218[e]	5
Formic acid	CH_2O_2	-56 s	-40.4 s	-22.3 s	-0.8 s	37.0	100.2	1,5
Francium	Fr	131[e]	181[e]	246[e]	335[e]	465[e]	673[e]	2
Fumaric acid	$C_4H_4O_4$	123.9 s	150 s	180 s				5
Furan	C_4H_4O			-78[e]	-54[e]	-20[e]	31.0	1
Furfural	$C_5H_4O_2$	-26[e]	-8[e]	16[e]	47[e]	92.4	161.4	1
Furfuryl alcohol	$C_5H_6O_2$	-30[e]	-5[e]	25[e]	62.6	109.3	169.7	5
Gadolinium	Gd	1563[d]	1755[d]	1994[d]	2300[d]	2703[d]	3262[d]	3
Gallium	Ga	1037	1175	1347	1565	1852	2245	2
Geraniol	$C_{10}H_{18}O$	4[e]	31[e]	63.2	104.3	157.7	229.6	5
Geranyl acetate	$C_{12}H_{20}O_2$			67.7	110.8	166.9	242.9	5
Germanium	Ge	1371	1541	1750	2014	2360	2831	2

Name	Mol. form.	t/°C for 1 Pa	t/°C for 10 Pa	t/°C for 100 Pa	t/°C for 1 kPa	t/°C for 10 kPa	t/°C for 100 kPa	Ref.
Germanium(IV) bromide	Br_4Ge				51	105	188	4
Glycerol	$C_3H_8O_3$	96[e]	113[e]	136[e]	168[e]	213.4	287[e]	1
Glycerol triacetate	$C_9H_{14}O_6$	37.6	62[e]	90[e]	124[e]	165[e]	214[e]	5
Glycolic acid	$C_2H_4O_3$						99.9	5
Gold	Au	1373	1541	1748	2008	2347	2805	2
Hafnium	Hf	2416	2681	3004	3406	3921	4603	9
Helium[a]	He					-270.6	-268.9	2
Heneicosane	$C_{21}H_{44}$	82.3	113.5	152.2	201.6	263.8	355.9	5
Heptacosane	$C_{27}H_{56}$	136.7	168.8	206.5	255.8	323.3	421.2	5
Heptadecane	$C_{17}H_{36}$	51.5	78.5	112.0	155.3	214.5	302[e]	16
1-Heptadecanol	$C_{17}H_{36}O$	94[e]	117[e]	146[e]	185[e]	240.1	323.3	5
Heptanal	$C_7H_{14}O$	-41[e]	-21[e]	4[e]	37[e]	83.7	152.3	5
Heptane	C_7H_{16}	-78.6	-60.2	-37.0	-6.6	35.4	98.0	16
1-Heptanethiol	$C_7H_{16}S$	-30[e]	-9[e]	18[e]	53[e]	102.7	176.4	5
Heptanoic acid	$C_7H_{14}O_2$	24[e]	46[e]	72[e]	107[e]	154.6	222.6	5
1-Heptanol	$C_7H_{16}O$		17[e]	40[e]	70.1	112.5	176[e]	1
2-Heptanol, (±)-	$C_7H_{16}O$	-9[e]	7[e]	27[e]	55.0	95.2	158.7	5
3-Heptanol, (*S*)-	$C_7H_{16}O$	-8[e]	7[e]	27[e]	54.5	93.9	156.3	5
4-Heptanol	$C_7H_{16}O$	-16[e]	1[e]	22[e]	51[e]	91.9	154.6	5
2-Heptanone	$C_7H_{14}O$		-22[e]	3[e]	36[e]	82.2	150.6	1
3-Heptanone	$C_7H_{14}O$		-28[e]	0[e]	36[e]	83.2	147.0	5
4-Heptanone	$C_7H_{14}O$	-27[e]	-6[e]	18.8	50.2	90.3	143.4	5
Heptanoyl chloride	$C_7H_{13}ClO$	-17[e]	4[e]	29.4	59.7	96.9	144.0	5
1-Heptene	C_7H_{14}	-82.1	-63.8	-40.6	-10.7	31.1	93.2	1,5
cis-2-Heptene	C_7H_{14}	-79[e]	-61[e]	-38[e]	-8[e]	34.3	98.0	5
trans-2-Heptene	C_7H_{14}	-79[e]	-61[e]	-39[e]	-8[e]	34.0	97.5	5
cis-3-Heptene	C_7H_{14}	-80[e]	-62[e]	-40[e]	-10[e]	32.3	95.3	5
trans-3-Heptene	C_7H_{14}	-80[e]	-62[e]	-40[e]	-10[e]	32.2	95.2	5
Heptyl acetate	$C_9H_{18}O_2$	-16[e]	6[e]	34[e]	70[e]	119.9	191.9	5
Heptylamine	$C_7H_{17}N$			5[e]	39[e]	86.7	156.4	5
Heptylbenzene	$C_{13}H_{20}$	12[e]	36[e]	66[e]	107[e]	162.7	246.2	5
Heptyl butanoate	$C_{11}H_{22}O_2$	2[e]	29[e]	62[e]	102.6	155.1	224.7	5
Heptylcyclohexane	$C_{13}H_{26}$	11[e]	34[e]	65[e]	105[e]	160.9	244.3	5
Heptylcyclopentane	$C_{12}H_{24}$	-1[e]	22[e]	51[e]	90[e]	143.5	223.5	5
1-Heptyne	C_7H_{12}	-75[e]	-57[e]	-35[e]	-5[e]	37.1	99.5	5
2-Heptyne	C_7H_{12}		-51[e]	-27[e]	4[e]	46.9	111.5	5
3-Heptyne	C_7H_{12}	-71[e]	-53[e]	-31[e]	0[e]	42.7	106.4	5
Hexachloro-1,3-butadiene	C_4Cl_6	-1[e]	22[e]	50[e]	86.7	137.0	209.7	5
Hexachloroethane	C_2Cl_6	-7.6 s	9.9 s	33.6 s	67.7 s	116.9 s	184.2 s	5
Hexachloropropene	C_3Cl_6	-12[e]	11[e]	40[e]	79[e]	132.8	213.6	5
Hexacosane	$C_{26}H_{54}$	125.1	158.8	200.1	252.1	314.3	411.3	5
Hexadecane	$C_{16}H_{34}$	41.1	67.4	100.3	142.7	200.7	286.3	16
Hexadecanoic acid	$C_{16}H_{32}O_2$		136[e]	165[e]	205[e]	261.9	350.2	5
1-Hexadecanol	$C_{16}H_{34}O$	99.5	130.6	171.9	175[e]	229.0	311.7	5
1-Hexadecene	$C_{16}H_{32}$	38.4	65.0	98.1	140.5	198.8	284.3	5
Hexadecylamine	$C_{16}H_{35}N$	63[e]	91[e]	126[e]	171[e]	232.6	320.5	5
trans-1,3-Hexadiene	C_6H_{10}	-86[e]	-70[e]	-51[e]	-24[e]	14[e]	72[e]	5
trans-1,4-Hexadiene	C_6H_{10}	-98[e]	-81[e]	-60[e]	-33[e]	7[e]	65[e]	5
1,5-Hexadiene	C_6H_{10}	-99[e]	-84[e]	-64[e]	-37[e]	0.9	59.2	5
cis,cis-2,4-Hexadiene	C_6H_{10}					18[e]	79.6	5
trans,cis-2,4-Hexadiene	C_6H_{10}	-89[e]	-73[e]	-52[e]	-23[e]	18[e]	79.6	5
trans,trans-2,4-Hexadiene	C_6H_{10}				-23[e]	18[e]	79.6	5
1,5-Hexadien-3-yne	C_6H_6	-82[e]	-66[e]	-44.3	-16.0	23.7	83.6	5
Hexaethylbenzene	$C_{18}H_{30}$				144.1	206.8	297.5	5
Hexafluorobenzene	C_6F_6		-56.9 s	-36 s	-11.5 s	22.6	79.9	1,5
Hexafluoroethane[b]	C_2F_6			-155.2 s	-137.5 s	-113.4 s	-78.4 s	1,5
1,1,1,3,3,3-Hexafluoro-2-propanol	$C_3H_2F_6O$					12.7	57.1	5
Hexamethylbenzene	$C_{12}H_{18}$	46.3 s	72.5 s	81.7 s	121.8 s	178.3	263.7	5
Hexamethyldisiloxane	$C_6H_{18}OSi_2$		-56[e]	-34[e]	-5[e]	37.1	100.1	5
2,6,10,15,19,23-Hexamethyltetracosane	$C_{30}H_{62}$	66[e]	84[e]	105.8	131.9	163.7	203.2	5

Fluids

Name	Mol. form.	t/°C for 1 Pa	t/°C for 10 Pa	t/°C for 100 Pa	t/°C for 1 kPa	t/°C for 10 kPa	t/°C for 100 kPa	Ref.
Hexanal	$C_6H_{12}O$	-56[e]	-37[e]	-13[e]	19[e]	62.6	127.8	5
Hexane	C_6H_{14}	-96.4 s	-79.2	-57.6	-29.3	9.8	68.3	16
1,6-Hexanediamine	$C_6H_{16}N_2$				76.0	128.2	199.0	5
Hexanedinitrile	$C_6H_8N_2$	30[e]	61[e]	100[e]	148.6	211.8	297[e]	5
Hexanenitrile	$C_6H_{11}N$	-40[e]	-19[e]	8[e]	43[e]	91.5	163.2	1,5
1-Hexanethiol	$C_6H_{14}S$	-45[e]	-25[e]	1[e]	35[e]	81.7	152.2	5
2-Hexanethiol	$C_6H_{14}S$	-50[e]	-32[e]	-8[e]	25[e]	69.9	138.4	5
1,2,6-Hexanetriol	$C_6H_{14}O_3$	92[e]	114.8	146.0	191[e]			5
Hexanoic acid	$C_6H_{12}O_2$		33[e]	59[e]	93[e]	139.3	204.5	1
1-Hexanol	$C_6H_{14}O$		5[e]	28[e]	56.8	97.3	157.1	1
2-Hexanol	$C_6H_{14}O$	-28[e]	-10[e]	12[e]	41.4	81.5	139.6	1
3-Hexanol	$C_6H_{14}O$	-43[e]	-23[e]	1[e]	33[e]	75.4	135.1	1
2-Hexanone	$C_6H_{12}O$	-43[e]	-21[e]	4.2	34.5	61.9	127.2	1,5
3-Hexanone	$C_6H_{12}O$		-40[e]	-16[e]	15[e]	58.5	123.1	1
cis-1,3,5-Hexatriene	C_6H_8					21[e]	78[e]	5
1-Hexene	C_6H_{12}	-99.8	-82.8	-61.4	-33.7	5.2	63.1	1,5
cis-2-Hexene	C_6H_{12}	-97[e]	-80[e]	-58[e]	-30[e]	9.9	68.5	5
trans-2-Hexene	C_6H_{12}	-94[e]	-78[e]	-57[e]	-30[e]	9.3	67.5	5
cis-3-Hexene	C_6H_{12}	-96[e]	-79[e]	-59[e]	-30.8	7.9	66.0	5
trans-3-Hexene	C_6H_{12}	-95[e]	-79[e]	-58[e]	-30.0	8.8	66.7	5
Hexyl acetate	$C_8H_{16}O_2$	-37[e]	-13[e]	16[e]	52.8	100.4	164[e]	5
Hexylamine	$C_6H_{15}N$			-10[e]	22[e]	66.0	130.6	5
Hexylbenzene	$C_{12}H_{18}$	-2[e]	22[e]	51[e]	90[e]	144.5	225.5	5
Hexylcyclohexane	$C_{12}H_{24}$	-3[e]	20[e]	50[e]	89[e]	143.1	224.2	5
Hexylcyclopentane	$C_{11}H_{22}$	-15[e]	7[e]	36[e]	73[e]	125.0	202.5	5
2-(Hexyloxy)ethanol	$C_8H_{18}O_2$	-13[e]	14[e]	46[e]	86[e]	137.7	206.9	5
1-Hexyne	C_6H_{10}	-91[e]	-75[e]	-54[e]	-26[e]	12.8	71.0	5
2-Hexyne	C_6H_{10}	-84[e]	-67[e]	-46[e]	-17[e]	23.6	84.1	5
3-Hexyne	C_6H_{10}	-86[e]	-69[e]	-48[e]	-19.1	21.0	81.0	1,5
Holmium	Ho	1159 s	1311 s	1502[d]	1767[d]	2137[d]	2691[d]	3
Hydrazine	H_4N_2				14.7	55.6	113[e]	5
Hydrazoic acid	HN_3			-79[e]	-54[e]	-18.0	35.7	5
Hydrogen[a]	H_2					-258.6	-252.8	1
Hydrogen bromide[a]	BrH		-153.3 s	-140.4 s	-123.8 s	-101.5 s	-67.0	5
Hydrogen chloride[a]	ClH				-138.2 s	-118.0	-85.2	1,5
Hydrogen cyanide[a]	CHN			-77 s	-52.6 s	-22.7 s	25.4	1,5
Hydrogen disulfide	H_2S_2				-27[e]	12.2	70.7	5
Hydrogen fluoride[a]	FH				-71.1	-33.7	19.2	1,5
Hydrogen iodide[a]	HI	-146 s	-135.2 s	-120.8 s	-101.9 s	-75.9 s	-35.9	5
Hydrogen peroxide	H_2O_2			13[e]	45[e]	89.0	149.8	5
Hydrogen selenide	H_2Se	-145 s	-134 s	-120 s	-102.8 s	-78.9 s	-41.5	5
Hydrogen sulfide[a]	H_2S		-149 s	-136 s	-118.9 s	-95.9 s	-60.5	1,5
Hydrogen telluride	H_2Te					-46.6	-2.3	5
Hydroxylamine	H_3NO				43.7	73.3	109.8	4
3-Hydroxypropanenitrile	C_3H_5NO	-11[e]	18[e]	53[e]	96.1	150.3	220.8	5
Indan	C_9H_{10}	-33[e]	-12[e]	16[e]	52[e]	102.3	177.5	1
Indene	C_9H_8			12[e]	53.0	106.8	181.0	5
Indium	In	923	1052	1212	1417	1689	2067	2
Indium(III) bromide	Br_3In			304.6 s	328.7 s	364.8 s		1
1*H*-Indole	C_8H_7N	20.6 s	44.5 s				254.0	5
Iodine	I_2	-12.8 s	9.3 s	35.9 s	68.7 s	108 s	184.0	1,2
Iodobenzene	C_6H_5I	-30[e]	-7[e]	20.9	58.5	110.6	187.8	1
1-Iodobutane	C_4H_9I	-62[e]	-43[e]	-19[e]	14[e]	60.5	130.0	5
2-Iodobutane, (±)-	C_4H_9I	-70[e]	-51[e]	-27[e]	5[e]	50[e]	119.5	5
Iodoethane	C_2H_5I	-94[e]	-78[e]	-56[e]	-27.9	11.9	71.9	5
Iodoethene	C_2H_3I				-41[e]	-3[e]	55.6	5
1-Iodoheptane	$C_7H_{15}I$	-19[e]	3[e]	32[e]	71[e]	123.8	203.4	5
1-Iodohexane	$C_6H_{13}I$	-33[e]	-11[e]	16[e]	53[e]	104.0	180.8	5
Iodomethane	CH_3I				-49[e]	-12.4	42.1	1
1-Iodo-3-methylbutane	$C_5H_{11}I$		-34[e]	-6.6	28.8	77.3	147.8	5

Name	Mol. form.	t/°C for 1 Pa	t/°C for 10 Pa	t/°C for 100 Pa	t/°C for 1 kPa	t/°C for 10 kPa	t/°C for 100 kPa	Ref.
1-Iodo-2-methylpropane	C_4H_9I		-47^{e}	-21.4	12.0	56.8	120.0	5
2-Iodo-2-methylpropane	C_4H_9I	-75.1 s	-58.8 s	-39.5 s	-5.2	41^{e}	100.0	5
1-Iodooctane	$C_8H_{17}I$	-6^{e}	18^{e}	48^{e}	87^{e}	142.5	224.5	5
1-Iodopentane	$C_5H_{11}I$	-47^{e}	-27^{e}	-1^{e}	34^{e}	83.0	156.5	5
1-Iodopropane	C_3H_7I	-78^{e}	-60^{e}	-37^{e}	-6^{e}	36.9	102.0	5
2-Iodopropane	C_3H_7I	-89^{e}	-71^{e}	-47^{e}	-16.3	26.5	89.2	5
3-Iodopropene	C_3H_5I	-80^{e}	-62^{e}	-39^{e}	-8^{e}	36^{e}	101.5	5
Iodosilane	H_3ISi				-47.7	-10.1	45.2	4
2-Iodothiophene	C_4H_3IS			-25^{e}	23^{e}	94.9	181.0	5
Iridium	Ir	2440 s	2684	2979	3341	3796	4386	2
Iridium(VI) fluoride	F_6Ir	-88 s	-71 s	-51 s	-27 s	3.8 s	53.1	26
Iron	Fe	1455 s	1617	1818	2073	2406	2859	2
Iron(II) chloride	Cl_2Fe				685	821	1025	4
Iron(III) chloride	Cl_3Fe	118 s	153 s	190 s	229 s	268 s	319	4
Iron pentacarbonyl	C_5FeO_5				0	44	105	4
Isobutanal	C_4H_8O			-56^{e}	-29^{e}	8^{e}	63.8	1
Isobutane[a]	C_4H_{10}		-129.0	-113.0	-90.9	-59.4	-12.0	1,41
Isobutene	C_4H_8	-139.1	-125.5	-108.2	-85.5	-54.5	-7.3	1,5
Isobutyl acetate	$C_6H_{12}O_2$	-63^{e}	-45^{e}	-21^{e}	10^{e}	53.4	116^{e}	5
Isobutylamine	$C_4H_{11}N$	-85^{e}	-70^{e}	-50^{e}	-24.5	12.0	67.3	5
Isobutylbenzene	$C_{10}H_{14}$	-36^{e}	-15^{e}	12^{e}	47.9	97.8	172.3	5
Isobutylcyclohexane	$C_{10}H_{20}$	-37^{e}	-16^{e}	10^{e}	46^{e}	95.9	170.8	5
Isobutylcyclopentane	C_9H_{18}	-105^{e}	-88^{e}	-64^{e}	-28^{e}	31^{e}	147.0	5
Isobutyl formate	$C_5H_{10}O_2$	-69^{e}	-53^{e}	-31^{e}	-3^{e}	37.4	97.6	5
Isobutyl isobutanoate	$C_8H_{16}O_2$	-47^{e}	-26^{e}	0.4	34.8	81.1	147.0	5
Isobutyl 3-methylbutanoate	$C_9H_{18}O_2$			11.3	48.3	97.9	168.3	5
Isobutyl nitrate	$C_4H_9NO_3$			-18^{e}	15.1	59.2	123.0	5
Isobutyl propanoate	$C_7H_{14}O_2$	-35^{e}	-19^{e}	2^{e}	31^{e}	72.0	136.1	5
Isobutyl vinyl ether	$C_6H_{12}O$	-87^{e}	-68^{e}	-44^{e}	-13^{e}	26.5	80.7	5
Isoeugenol	$C_{10}H_{12}O_2$				125^{e}	185.3	267.1	5
Isopentane	C_5H_{12}	-119^{e}	-105^{e}	-86^{e}	-61^{e}	-26^{e}	27.5	1
Isopentyl acetate	$C_7H_{14}O_2$	-51^{e}	-30^{e}	-4^{e}	30.3	76.2	141.4	5
Isopentyl butanoate	$C_9H_{18}O_2$				55^{e}	105.6	178.4	5
Isopentyl formate	$C_6H_{12}O_2$	-60^{e}	-41^{e}	-17^{e}	15^{e}	59.1	124^{e}	5
Isopentyl isopentanoate	$C_{10}H_{20}O_2$			22^{e}	62.8	116.9	193.6	5
Isopentyl propanoate	$C_8H_{16}O_2$			3.1	40.7	90.6	159.8	5
Isophorone	$C_9H_{14}O$		1^{e}	33.1	75.1	132.4	215.1	5
Isopropenylbenzene	C_9H_{10}			3.2	41.5	92.8	164.9	5
p-Isopropenylisopropylbenzene	$C_{12}H_{16}$	-11^{e}	15^{e}	46^{e}	87^{e}	142.4	221^{e}	5
Isopropyl acetate	$C_5H_{10}O_2$		-61^{e}	-40^{e}	-11^{e}	29.8	88.2	5
Isopropylamine	C_3H_9N		-91^{e}	-74^{e}	-50.4	-17.6	31.5	1,5
4-Isopropylbenzaldehyde	$C_{10}H_{12}O$			54.1	96.0	152.2	231.5	5
Isopropylbenzene	C_9H_{12}	-46^{e}	-26^{e}	-1^{e}	33^{e}	80.9	152.0	1
Isopropyl chloroacetate	$C_5H_9ClO_2$			-2^{e}	35.0	83.3	148.1	5
Isopropylcyclohexane	C_9H_{18}	-48^{e}	-28^{e}	-2^{e}	33^{e}	81.3	154.0	5
Isopropylcyclopentane	C_8H_{16}	-65^{e}	-46^{e}	-21^{e}	12^{e}	57.3	125.9	5
Isopropylcyclopropane	C_6H_{12}	-104^{e}	-88^{e}	-68^{e}	-40^{e}	-1^{e}	57.9	5
Isopropyl formate	$C_4H_8O_2$	-80^{e}	-65^{e}	-47^{e}	-22.2	13.2	67.7	5
Isopropyl isobutanoate	$C_7H_{14}O_2$		-44^{e}	-19.7	12.2	56.0	120.1	5
5-Isopropyl-2-methylaniline	$C_{10}H_{15}N$	19^{e}	43^{e}	72^{e}	107.4	150^{e}	204^{e}	5
1-Isopropyl-2-methylbenzene	$C_{10}H_{14}$	-39^{e}	-16^{e}	13^{e}	51^{e}	103.1	177.8	5
1-Isopropyl-3-methylbenzene	$C_{10}H_{14}$	-34^{e}	-13^{e}	14^{e}	50^{e}	99.9	174.6	5
1-Isopropyl-4-methylbenzene	$C_{10}H_{14}$	-33^{e}	-12^{e}	16^{e}	52^{e}	102.2	176.6	5
Isopropyl methyl ether	$C_4H_{10}O$				-56^{e}	-21.2	30.4	5
Isopropyl methyl sulfide	$C_4H_{10}S$	-85^{e}	-68^{e}	-46^{e}	-17^{e}	23.4	84.3	5
1-Isopropylnaphthalene	$C_{13}H_{14}$	27^{e}	51^{e}	82^{e}	123.2	180.8	267.3	5
Isopropyl propyl sulfide	$C_6H_{14}S$				18.5	63.8	131.6	5
4-Isopropylstyrene	$C_{11}H_{14}$	-25^{e}	-1^{e}	30.2	70.3	124.5	202.1	5
Isoquinoline	C_9H_7N		30.2	60.7	101.3	157.9	242.7	1,5
Ketene	C_2H_2O		-151^{e}	-135^{e}	-115^{e}	-88.2	-50.0	1

Name	Mol. form.	*t*/°C for 1 Pa	*t*/°C for 10 Pa	*t*/°C for 100 Pa	*t*/°C for 1 kPa	*t*/°C for 10 kPa	*t*/°C for 100 kPa	Ref.
Krypton[a]	Kr	-214.0 s	-208.0 s	-199.4 s	-188.9 s	-174.6 s	-153.6	5
Lanthanum	La	1732[d]	1935[d]	2185[d]	2499[d]	2905[d]	3453[d]	3
Lead	Pb	705	815	956	1139	1387	1754	2
Lead(II) bromide	Br_2Pb	374	431	502	597	726	914	4
Lead(II) chloride	Cl_2Pb			541[e]	637	765	949	23
Lead(II) fluoride	F_2Pb				865	1054	1292	4
Lead(II) iodide	I_2Pb			470	558	682	869	4
Lead(II) oxide (massicot)	OPb	724	816	928	1065	1241	1471	4
Lead(II) sulfide	PbS	656 s	741 s	838 s	953 s	1088 s	1280	4
d-Limonene	$C_{10}H_{16}$	-45[e]	-21[e]	9.1	48.0	100.4	174.5	5
l-Limonene	$C_{10}H_{16}$	-33[e]	-12[e]	16[e]	52.0	102.3	177.0	21
Lithium	Li	524.3	612.3	722.1	871.2	1064.3	1337.1	13,30
Lithium bromide	BrLi		630	733	868	1049	1308	4
Lithium chloride	ClLi		649[d]	761[d]	905[d]	1101[d]	1381[d]	8
Lithium fluoride	FLi	801 s	896	1024	1188	1395	1672	4,12,25
Lithium iodide	ILi	545	619	710	824	972	1170	4
Lutetium	Lu	1633 s	1829.8	2072.8	2380[d]	2799[d]	3390[d]	3
Magnesium	Mg	428 s	500 s	588 s	698	859	1088	2
Magnesium chloride	Cl_2Mg			762	908	1111	1414	4
Maleic anhydride	$C_4H_2O_3$				73.7	127.9	201.7	5
Manganese	Mn	955 s	1074 s	1220 s	1418	1682	2060	2
Manganese(II) chloride	Cl_2Mn				760	933	1189	4
Mercury[b]	Hg	42.0	76.6	120.0	175.6	250.3	355.9	29,30
Mercury(II) bromide	Br_2Hg	71 s	98 s	132 s	174 s	227 s	318	4
Mercury(II) chloride	Cl_2Hg	64.4 s	94.7 s	130.8 s	174.5 s	228.5 s	304.0	4
Mercury(II) iodide (red)	HgI_2	85.1 s	115.6 s	152.4 s	197.8 s	255.1 s	353.6	4
Mesityl oxide	$C_6H_{10}O$	-56[e]	-37[e]	-13[e]	19[e]	63.5	129.3	5
Methacrylic acid	$C_4H_6O_2$			22[e]	56[e]	99.9	161.5	5
Methane[a]	CH_4	-220 s	-214.2 s	-206.8 s	-197 s	-183.6 s	-161.7	5,41
Methanethiol	CH_4S		-115[e]	-97[e]	-74[e]	-41.7	5.7	1
Methanol[a]	CH_4O	-87[e]	-69[e]	-47.5	-20.4	15.2	64.2	11
4-Methoxybenzaldehyde	$C_8H_8O_2$	9[e]	35[e]	68.1	110.8	167.9	248.5	5
2-Methoxyethanol	$C_3H_8O_2$	-57[e]	-37[e]	-12[e]	21[e]	63.8	124.3	1
2-Methoxyethyl acetate	$C_5H_{10}O_3$	-47[e]	-26[e]	0[e]	34[e]	79.4	144.1	5
4-Methoxy-4-methyl-2-pentanone	$C_7H_{14}O_2$				43[e]	89.8	160[e]	5
1-Methoxy-4-(2-propenyl)benzene	$C_{10}H_{12}O$			48.5	88.0	140.7	214.6	5
N-Methylacetamide	C_3H_7NO	-13.3 s	13 s	43[e]	83.8	136.1	206.3	5
Methyl acetate	$C_3H_6O_2$	-95[e]	-79[e]	-59[e]	-33[e]	3.3	56.6	1
Methyl acetoacetate	$C_5H_8O_3$				50.1	101.1	171.3	5
Methyl acrylate	$C_4H_6O_2$		-71[e]	-48[e]	-18[e]	22[e]	79.9	5
2-Methylacrylonitrile	C_4H_5N				-12[e]	29.0	89.8	5
Methylamine	CH_5N				-76.7	-48.1	-6.6	1
2-Methylaniline	C_7H_9N	1.0	18.8	42.6	76.1	125.6	199.9	1,5
3-Methylaniline	C_7H_9N	3.8	22.0	46.2	80.1	128.8	202.9	1,5
4-Methylaniline	C_7H_9N				77.1	126.2	199.9	5
N-Methylaniline	C_7H_9N	-16[e]	6[e]	34[e]	70.3	121.1	195.8	1
4-Methyl-1,3-benzenediamine	$C_7H_{10}N_2$			100.4	145.3	202.9	279.5	5
3-Methylbenzenethiol	C_7H_8S		0[e]	29[e]	66[e]	117.9	194.6	5
Methyl benzoate	$C_8H_8O_2$		-1[e]	29[e]	68[e]	121.2	198.9	5
2-Methylbenzonitrile	C_8H_7N		1[e]	32.1	72.2	126.6	204.7	5
4-Methylbenzonitrile	C_8H_7N			40.1	78.7	134.3	221.3	5
1-Methylbicyclo[3.1.0]hexane	C_7H_{12}					29.8	92.6	5
3-Methyl-1,2-butadiene	C_5H_8	-111[e]	-95[e]	-75[e]	-49.2	-13.1	40.4	5
2-Methyl-1,3-butadiene	C_5H_8	-115[e]	-100[e]	-81[e]	-55.4	-19.7	33.7	1,5
2-Methyl-1-butanethiol, (+)	$C_5H_{12}S$				8.0	52.3	118.5	5
3-Methyl-1-butanethiol	$C_5H_{12}S$				7.8	51.9	117.9	5
2-Methyl-2-butanethiol	$C_5H_{12}S$				-8.0	34.6	98.7	5
Methyl butanoate	$C_5H_{10}O_2$	-68[e]	-50[e]	-28[e]	0.9	41.7	102.3	5
2-Methylbutanoic acid	$C_5H_{10}O_2$	-10[e]	10[e]	36[e]	69[e]	112.8	175.2	5
3-Methylbutanoic acid	$C_5H_{10}O_2$	-15.8	4[e]	30.0	64.7	110.6	176.1	5

Name	Mol. form.	t/°C for 1 Pa	t/°C for 10 Pa	t/°C for 100 Pa	t/°C for 1 kPa	t/°C for 10 kPa	t/°C for 100 kPa	Ref.
2-Methyl-1-butanol, (±)-	$C_5H_{12}O$	-27^e	-11^e	9^e	36.2	73.4	128.3	1
3-Methyl-1-butanol	$C_5H_{12}O$	-22^e	-7^e	13^e	39.1	75.7	130.1	5
2-Methyl-2-butanol	$C_5H_{12}O$			-5^e	17.7	50.6	101.7	1,5
3-Methyl-2-butanol, (±)-	$C_5H_{12}O$			-3^e	22.7	58.2	111.1	5
3-Methyl-2-butanone	$C_5H_{10}O$	-69^e	-54^e	-34^e	-6.9	32.2	94.0	1,5
2-Methyl-1-butene	C_5H_{10}	-117.7	-102.2	-82.7	-57.2	-21.9	30.8	1,5
3-Methyl-1-butene	C_5H_{10}	-125.0	-110.1	-91.2	-66.7	-32.1	19.7	1,5
2-Methyl-2-butene	C_5H_{10}	-113.4	-97.6	-77.7	-51.6	-15.8	38.2	1,5
3-Methyl-3-buten-2-one	C_5H_8O			-35^e	-5^e	36.0	97.3	5
3-Methylbutyl benzoate	$C_{12}H_{16}O_2$			66^e	115.0	177.7	261.4	5
Methyl *tert*-butyl ether	$C_5H_{12}O$			-66^e	-39^e	-2^e	54.8	1
3-Methylbutyl nitrate	$C_5H_{11}NO_3$		-26^e	1.0	35.5	81.7	147.0	5
3-Methyl-1-butyne	C_5H_8			-82^e	-57.5	-23.1	28.6	5
Methyl chloroacetate	$C_3H_5ClO_2$		-28^e	-5^e	25^e	66.9	129.1	5
Methyl cyanoacetate	$C_4H_5NO_2$	-3^e	19^e	48^e	84^e	134.0	204.6	5
Methylcyclohexane	C_7H_{14}	-79^e	-62^e	-39^e	-7.9	35.5	100.5	1
1-Methylcyclohexene	C_7H_{12}	-72^e	-53^e	-30^e	1^e	45^e	109.8	5
4-Methylcyclohexene	C_7H_{12}	-76^e	-59^e	-36^e	-5^e	37.9	102.3	5
Methylcyclopentane	C_6H_{12}	-97^e	-80^e	-58^e	-28.8	11.6	71.4	1,5
Methylcyclopropane	C_4H_8	-130^e	-116^e	-99.3	-76.3	-44.2	4.2	5
2-Methyldecane	$C_{11}H_{24}$	-20^e	1^e	28^e	64^e	114.0	188.7	5
3-Methyldecane	$C_{11}H_{24}$	-35^e	-10^e	22^e	61.9	115.6	190.4	5
4-Methyldecane	$C_{11}H_{24}$	-38^e	-12^e	20^e	60.8	113.9	186.4	5
Methyl decanoate	$C_{11}H_{22}O_2$	10^e	33^e	62^e	100.9	154.0	232^e	5
Methyl dichloroacetate	$C_3H_4Cl_2O_2$	-44^e	-25^e	0^e	33^e	77.7	142.3	5
Methyldifluoroarsine	CH_3AsF_2				-15^e	22.1	76.1	5
2-Methyl-*N,N*-dimethylaniline	$C_9H_{13}N$	-25^e	-3^e	24.4	60.6	110.7	184.5	5
Methyl dimethylthioborane	C_3H_9BS			-62^e	-30.4	11.4	70.7	5
Methyldiphenylamine	$C_{13}H_{13}N$	35^e	63^e	98.4	143.1	201.6	281.6	5
Methyl dodecanoate	$C_{13}H_{26}O_2$	38^e	61^e	90^e	130^e	184.9	269^e	5
Methylenecyclohexane	C_7H_{12}	-76^e	-58^e	-35^e	-5^e	38^e	103.0	5
N-Methylformamide	C_2H_5NO		13^e	41^e	78^e	127.9	199.1	1
Methyl formate	$C_2H_4O_2$		-95^e	-76^e	-51.8	-18.1	31.4	5
2-Methylfuran	C_5H_6O			-66^e	-35^e	6^e	64.5	1
2-Methylheptane	C_8H_{18}	-69^e	-49.1	-24.5	7.6	51.6	117.2	1,5
3-Methylheptane	C_8H_{18}	-67^e	-48.1	-23.6	8.5	52.7	118.5	1,5
4-Methylheptane	C_8H_{18}	-65^e	-47^e	-24^e	7.8	51.6	117.2	5
Methyl heptanoate	$C_8H_{16}O_2$	-30^e	-9^e	19^e	54.2	102.4	172^e	5
3-Methyl-3-heptanol	$C_8H_{18}O$	-13^e	4^e	26^e	55^e	96.3	160.3	5
4-Methyl-3-heptanol	$C_8H_{18}O$	-52^e	-28^e	1^e	39^e	87.6	155.0	5
5-Methyl-3-heptanol	$C_8H_{18}O$	-35^e	-16^e	8^e	40^e	84.8	153.0	5
4-Methyl-4-heptanol	$C_8H_{18}O$	-17^e	1^e	24^e	55^e	97.2	160.7	5
2-Methyl-1-heptene	C_8H_{16}	-66^e	-48^e	-24^e	8^e	52.3	118.7	5
Methyl hexadecanoate	$C_{17}H_{34}O_2$	65^e	93	129	177			4
2-Methylhexane	C_7H_{16}	-82^e	-65^e	-43^e	-13^e	27.8	89.7	1
3-Methylhexane	C_7H_{16}	-81^e	-64^e	-42^e	-12^e	29.2	91.5	1
Methyl hexanoate	$C_7H_{14}O_2$	-47^e	-26^e	2^e	36.6	83.3	149^e	5
5-Methyl-2-hexanone	$C_7H_{14}O$		-27^e	-2^e	31.0	76.6	144.4	5
2-Methyl-1-hexene	C_7H_{14}	-81^e	-64^e	-42^e	-12^e	29.3	91.6	5
4-Methyl-1-hexene	C_7H_{14}	-84^e	-67^e	-45^e	-16^e	25.3	86.3	5
2-Methyl-2-hexene	C_7H_{14}	-80^e	-63^e	-40^e	-10^e	32.0	95.0	5
cis-3-Methyl-2-hexene	C_7H_{14}	-79^e	-62^e	-39^e	-9^e	33.4	96.8	5
trans-4-Methyl-2-hexene	C_7H_{14}	-83^e	-66^e	-44^e	-15^e	25.9	87.1	5
trans-5-Methyl-2-hexene	C_7H_{14}	-83^e	-66^e	-44^e	-15^e	26.3	87.7	5
trans-2-Methyl-3-hexene	C_7H_{14}	-84^e	-67^e	-45^e	-16^e	24.6	85.5	5
5-Methyl-1-hexyne	C_7H_{12}	-80^e	-62^e	-40^e	-11^e	30.1	91.4	5
5-Methyl-2-hexyne	C_7H_{12}	-75^e	-57^e	-34^e	-4^e	38.6	102.0	5
2-Methyl-3-hexyne	C_7H_{12}	-78^e	-61^e	-39^e	-9^e	32.6	94.8	5
Methylhydrazine	CH_6N_2			-31^e	4.7	32.9	91^e	1
Methyl isobutanoate	$C_5H_{10}O_2$	-83^e	-65^e	-41^e	-11^e	31^e	92.1	5

Name	Mol. form.	t/°C for 1 Pa	t/°C for 10 Pa	t/°C for 100 Pa	t/°C for 1 kPa	t/°C for 10 kPa	t/°C for 100 kPa	Ref.
Methyl isocyanate	C_2H_3NO				-43.5	-10.2	38.8	1
Methyl isopentanoate	$C_6H_{12}O_2$					53.3	116.3	5
Methyl methacrylate	$C_5H_8O_2$			-31[e]	-1[e]	39.7	100.0	1
1-Methylnaphthalene	$C_{11}H_{10}$	5[e]	29[e]	60[e]	102[e]	159.1	244.1	1
2-Methylnaphthalene	$C_{11}H_{10}$			57[e]	99[e]	156.0	240.5	1
Methyl nitrate	CH_3NO_3		-75[e]	-55[e]	-27[e]	9.8	63[e]	5
Methyl 2-nitrobenzoate	$C_8H_7NO_4$	17[e]	49[e]	89[e]	140[e]	208[e]	302[e]	5
2-Methylnonane	$C_{10}H_{22}$	-34[e]	-14[e]	12[e]	47[e]	94.8	166.5	5
3-Methylnonane	$C_{10}H_{22}$	-34[e]	-14[e]	12[e]	47[e]	95.1	167.3	5
4-Methylnonane	$C_{10}H_{22}$	-36[e]	-16[e]	10[e]	45[e]	93.1	165.2	5
5-Methylnonane	$C_{10}H_{22}$	-36[e]	-16[e]	10[e]	45[e]	92.6	164.6	5
Methyl *cis*-9-octadecenoate	$C_{19}H_{36}O_2$	85[e]	114[e]	149.7	195.6	256[e]	340[e]	5
2-Methyloctane	C_9H_{20}	-49[e]	-30[e]	-5[e]	28[e]	73.9	142.8	5
3-Methyloctane	C_9H_{20}	-49[e]	-29[e]	-5[e]	29[e]	74.7	143.7	5
4-Methyloctane	C_9H_{20}	-50[e]	-30[e]	-6[e]	27[e]	73.2	141.9	5
Methyl octanoate	$C_9H_{18}O_2$	-26[e]	-9[e]	13[e]	40[e]	76[e]	127.9	5
2-Methyl-1-octene	C_9H_{18}	-53[e]	-34[e]	-9[e]	25[e]	72[e]	144.1	5
Methyloxirane	C_3H_6O	-109[e]	-95[e]	-76[e]	-51.5	-17.2	33.9	5
cis-2-Methyl-1,3-pentadiene	C_6H_{10}	-92[e]	-75[e]	-54[e]	-26[e]	14[e]	75.6	5
2-Methylpentane	C_6H_{14}	-100[e]	-84[e]	-64[e]	-36[e]	2[e]	59.9	1
3-Methylpentane	C_6H_{14}	-99[e]	-83[e]	-62[e]	-34.3	4.6	62.9	1
2-Methyl-2,4-pentanediol	$C_6H_{14}O_2$	-8[e]	17[e]	48[e]	86[e]	134.4	197.5	5
4-Methylpentanenitrile	$C_6H_{11}N$		-50[e]	-20[e]	20[e]	75.2	155.2	5
Methyl pentanoate	$C_6H_{12}O_2$				19.2	63.7	127.4	5
4-Methylpentanoic acid	$C_6H_{12}O_2$	36[e]	49[e]	67.1	92.9	133.6	206.8	5
2-Methyl-1-pentanol	$C_6H_{14}O$			14[e]	45.9	88.3	147.6	5
4-Methyl-1-pentanol	$C_6H_{14}O$			24[e]	53[e]	92.4	151.4	5
2-Methyl-2-pentanol	$C_6H_{14}O$	-29[e]	-15[e]	3[e]	27.1	63.0	120.9	5
3-Methyl-2-pentanol	$C_6H_{14}O$				36.5	76.1	133.8	5
4-Methyl-2-pentanol	$C_6H_{14}O$	-43[e]	-24[e]	0[e]	30[e]	71.9	131.3	5
2-Methyl-3-pentanol	$C_6H_{14}O$				29.8	68.8	126.0	5
3-Methyl-3-pentanol	$C_6H_{14}O$		-23[e]	-4[e]	22.9	61.1	121.1	5
3-Methyl-2-pentanone, (±)-	$C_6H_{12}O$				8.5	52.7	117.0	5
4-Methyl-2-pentanone	$C_6H_{12}O$	-61[e]	-43[e]	-21[e]	9[e]	51.5	116.1	5
2-Methyl-3-pentanone	$C_6H_{12}O$					50.2	113.0	5
2-Methyl-1-pentene	C_6H_{12}	-98[e]	-82[e]	-62[e]	-34.2	4.1	61.7	5
3-Methyl-1-pentene	C_6H_{12}	-104[e]	-88[e]	-68[e]	-41.5	-3.6	53.8	5
4-Methyl-1-pentene	C_6H_{12}	-105[e]	-89[e]	-69[e]	-41.6	-3.6	53.5	5
2-Methyl-2-pentene	C_6H_{12}	-95[e]	-78[e]	-58[e]	-30[e]	9.0	66.9	5
3-Methyl-*cis*-2-pentene	C_6H_{12}	-95[e]	-79[e]	-58[e]	-30[e]	8.9	67.3	5
3-Methyl-*trans*-2-pentene	C_6H_{12}	-93[e]	-77[e]	-55[e]	-27.4	11.7	70.0	5
4-Methyl-*cis*-2-pentene	C_6H_{12}	-102[e]	-86[e]	-66[e]	-38.7	-0.9	56.0	5
4-Methyl-*trans*-2-pentene	C_6H_{12}	-100[e]	-84[e]	-64[e]	-36.8	1.2	58.2	5
4-Methyl-4-penten-2-one	$C_6H_{10}O$	-59[e]	-41[e]	-17[e]	14[e]	57.0	121.0	5
4-Methyl-1-pentyne	C_6H_{10}	-97[e]	-81[e]	-61[e]	-34[e]	4.1	60.7	5
4-Methyl-2-pentyne	C_6H_{10}	-91[e]	-74[e]	-54[e]	-26[e]	13.8	72.7	5
N-Methylpropanamide	C_4H_9NO				81.1	105[e]		5
2-Methyl-1-propanethiol	$C_4H_{10}S$		-66[e]	-44[e]	-15[e]	26.5	88.1	5
2-Methyl-2-propanethiol	$C_4H_{10}S$					5.8	63.8	5
Methyl propanoate	$C_4H_8O_2$	-80[e]	-64[e]	-43[e]	-15.8	22.2	79.0	1
2-Methylpropanoic acid	$C_4H_8O_2$	-30.1	-8.2	18.1	50.5	92.9	154.0	5
2-Methyl-1-propanol	$C_4H_{10}O$	-39[e]	-24[e]	-5[e]	20.9	56.0	107.6	1,5
2-Methyl-2-propanol	$C_4H_{10}O$					34.4	82.1	1,5
2-Methyl-2-propenoyl chloride	C_4H_5ClO		-57[e]	-35[e]	-5[e]	36.4	98.2	5
1-Methyl-2-propylbenzene	$C_{10}H_{14}$	-27[e]	-6[e]	22[e]	58.2	108.9	184.3	5
1-Methyl-3-propylbenzene	$C_{10}H_{14}$	-29[e]	-8[e]	20[e]	56.1	106.5	181.3	5
1-Methyl-4-propylbenzene	$C_{10}H_{14}$	-29[e]	-7[e]	20[e]	56.6	107.4	182.8	5
cis-1-Methyl-2-propylcyclopentane	C_9H_{18}	-52[e]	-33[e]	-7[e]	28[e]	77[e]	152.0	5
trans-1-Methyl-2-propylcyclopentane	C_9H_{18}	-56[e]	-36[e]	-11[e]	23[e]	72[e]	145.8	5
Methyl propyl ether	$C_4H_{10}O$				-40[e]	-11.3	38.7	5

Name	Mol. form.	*t*/°C for 1 Pa	*t*/°C for 10 Pa	*t*/°C for 100 Pa	*t*/°C for 1 kPa	*t*/°C for 10 kPa	*t*/°C for 100 kPa	Ref.
Methyl propyl sulfide	$C_4H_{10}S$	-78[e]	-61[e]	-38[e]	-8[e]	33.1	95.1	5
2-Methylpyridine	C_6H_7N	-56.5	-37.8	-13.9	18.3	62.9	129.0	1,5
3-Methylpyridine	C_6H_7N			-5[e]	28.8	75.2	143.7	1
4-Methylpyridine	C_6H_7N	-58.2 s	-43.1 s	-3.9 s	29.6	76.1	144.9	1,5
1-Methylpyrrole	C_5H_7N				8[e]	49.9	112.3	5
N-Methylpyrrolidine	$C_5H_{11}N$				-23[e]	18.5	78[e]	5
N-Methyl-2-pyrrolidinone	C_5H_9NO	1[e]	24[e]	53.1	92.3	147.2	229[e]	5
2-Methylquinoline	$C_{10}H_9N$	5.3	31.9	63.8	102.9	165.8	247.2	5
4-Methylquinoline	$C_{10}H_9N$	29[e]	54[e]	85[e]	127[e]	183.0	265.1	5
6-Methylquinoline	$C_{10}H_9N$	27[e]	51[e]	81[e]	122[e]	179.2	264.5	5
8-Methylquinoline	$C_{10}H_9N$	15[e]	40[e]	70[e]	111[e]	166.1	247.3	5
Methyl salicylate	$C_8H_8O_3$	-1[e]	22[e]	51[e]	88.8	141.8	219.9	5
Methylsilane	CH_6Si			-144[e]	-124.6	-97.5	-57.5	5
Methyl silicate	$C_4H_{12}O_4Si$				14.4	59.3	119.7	5
Methyl silyl ether	CH_6OSi				-90.2	-61.8	-18[e]	1
Methyl tetradecanoate	$C_{15}H_{30}O_2$		75[e]	110	155	214	295	4
2-Methyltetrahydrofuran	$C_5H_{10}O$				-20[e]	19.7	79.8	5
4-Methylthiazole	C_4H_5NS						67.0	5
Methyl thiocyanate	C_2H_3NS			-18.4	16.2	63.5	132.5	5
2-Methylthiophene	C_5H_6S		-58[e]	-32[e]	2[e]	47.9	112.2	1
3-Methylthiophene	C_5H_6S		-53[e]	-28[e]	6[e]	50.6	115.1	1
Methyl 10-undecenoate	$C_{12}H_{22}O_2$	10[e]	38[e]	73[e]	116[e]	172.2	247.1	5
Methyl vinyl ether	C_3H_6O			-114[e]	-89[e]	-52.7	4.6	1
Molybdenum	Mo	2469 s	2721	3039	3434	3939	4606	2
Molybdenum carbonyl	C_6MoO_6		17.4 s	42.8 s	73.1 s	109.9 s	155.4 s	5
Molybdenum(V) fluoride	F_5Mo				86.6	140.3	213[e]	26
Molybdenum(VI) fluoride	F_6Mo	-98 s	-82 s	-64 s	-41.2 s	-13.4 s	33.5	26
Molybdenum(VI) oxide	MoO_3				801	935	1151	4
Molybdenum(VI) oxytetrafluoride	F_4MoO	-21 s	3 s	33 s	69.3 s	117.3	184.1	26
Morpholine	C_4H_9NO				21[e]	64.5	128.5	1
β-Myrcene	$C_{10}H_{16}$			9.4	47.3	98.3	171.0	5
Myristicin	$C_{11}H_{12}O_3$	23[e]	53[e]	88.9	135.2	196.0	279.4	5
Naphthalene[b]	$C_{10}H_8$	3.2 s	24.1 s	49.3 s	80.7	135.6	217.5	1,5
1-Naphthalenecarboxylic acid	$C_{11}H_8O_2$				191.9	239.3	299.6	5
2-Naphthalenecarboxylic acid	$C_{11}H_8O_2$				197.9	246.0	308.1	5
1-Naphthol	$C_{10}H_8O$				137.2	196.7	281.8	5
2-Naphthol	$C_{10}H_8O$				140.7	200.5	286.8	5
1-Naphthylamine	$C_{10}H_9N$		62[e]	99.0	146.9	210.7	300.1	5
2-Naphthylamine	$C_{10}H_9N$	36.3 s	65.9 s	103 s	150.9	215.1	305.5	5
Neodymium	Nd	1322.3	1501.2	1725.3	2023[d]	2442[d]	3063[d]	3
Neon[a]	Ne	-261 s	-260 s	-258 s	-255 s	-252 s	-246.1	2
Neopentane[a]	C_5H_{12}		-107.5 s	-90.8 s	-68.8 s	-38.5 s	9.2	1,5
Nickel	Ni	1510	1677	1881	2137	2468	2911	2
Nickel carbonyl [Ni(CO)$_4$]	C_4NiO_4					-12	42	4
Nickel(II) chloride	Cl_2Ni	534 s	592 s	662 s	747 s	852 s	985 s	4
Niobium	Nb	2669	2934	3251	3637	4120	4740	2
Niobium(V) fluoride	F_5Nb				80	140	224	4
Nitric acid	HNO_3			-37[e]	-9[e]	28.4	82.2	5
Nitric oxide[a]	NO	-201 s	-195 s	-188 s	-179.3 s	-168.1 s	-151.9	5
4-Nitroaniline	$C_6H_6N_2O_2$	87.8 s			192.0	252.6	331.2	5
2-Nitroanisole	$C_7H_7NO_3$	15[e]	45[e]	82[e]	129[e]	189.4	271.8	5
Nitrobenzene	$C_6H_5NO_2$		10[e]	40[e]	78[e]	132[e]	210.3	1
Nitroethane	$C_2H_5NO_2$	-61[e]	-44[e]	-21[e]	8.3	50.1	113.5	5
Nitrogen[a]	N_2	-236 s	-232 s	-226.8 s	-220.2 s	-211.1 s	-195.9	1,5
Nitrogen pentoxide	N_2O_5	-71 s	-56 s	-40 s	-19.9 s	3.9 s	33.2	5
Nitrogen tetroxide	N_2O_4	-92 s	-78 s	-61 s	-41.1 s	-16.6 s	28.7	5
Nitrogen trichloride	Cl_3N				-25[e]	13.2	70.6	5
Nitrogen trifluoride[a]	F_3N	-201[e]	-194[e]	-185[e]	-172.8	-155.5	-129.2	5
Nitromethane	CH_3NO_2				-2[e]	40[e]	100.8	1
4-Nitrophenol	$C_6H_5NO_3$	72.6 s	97.4 s					5

Name	Mol. form.	t/°C for 1 Pa	t/°C for 10 Pa	t/°C for 100 Pa	t/°C for 1 kPa	t/°C for 10 kPa	t/°C for 100 kPa	Ref.
1-Nitropropane	$C_3H_7NO_2$	-56[e]	-37[e]	-13[e]	20[e]	64.8	130.8	1
2-Nitropropane	$C_3H_7NO_2$		-48[e]	-22[e]	10.7	55.6	119.8	1
N-Nitrosodimethylamine	$C_2H_6N_2O$				30.7	80.5	149.8	5
Nitrosyl chloride	ClNO		-116 s	-100 s	-78.7 s	-50.2	-5.7	5
Nitrosyl fluoride	FNO			-131[e]	-116.1	-94.3	-60.1	5
2-Nitrotoluene	$C_7H_7NO_2$	23[e]	40[e]	62[e]	94[e]	141.9	221.9	5
3-Nitrotoluene	$C_7H_7NO_2$			45[e]	89.7	148.7	231.3	5
1-Nitro-3-(trifluoromethyl)benzene	$C_7H_4F_3NO_2$		11[e]	39[e]	76.2	127.3	202.2	5
Nitrous oxide[a]	N_2O	-167 s	-157 s	-145.4 s	-131.1 s	-112.9 s	-88.7	5
Nitryl chloride	$ClNO_2$	-121[e]	-113[e]	-102[e]	-86.1	-60.9	-15.7	5
Nitryl fluoride	FNO_2		-156[e]	-144[e]	-128.1	-106.0	-72.6	5
Nonacosane	$C_{29}H_{60}$	148.2	182.8	221.2	271.5	340.2	439.7	5
Nonadecane	$C_{19}H_{40}$	71.1	99.1	133.8	178.8	240.1	330[e]	16
Nonanal	$C_9H_{18}O$		-3[e]	27.4	65.5	115.6	184.6	5
Nonane	C_9H_{20}	-46.8	-26.0	0.0	34.0	80.8	150.3	16
Nonanenitrile	$C_9H_{17}N$	-3[e]	21[e]	50.9	90.7	145.4	225.1	5
1-Nonanethiol	$C_9H_{20}S$	-2[e]	21[e]	49[e]	87[e]	140.4	219.2	5
Nonanoic acid	$C_9H_{18}O_2$	48[e]	69[e]	97[e]	133[e]	182.7	255.1	5
1-Nonanol	$C_9H_{20}O$		40[e]	64[e]	96.9	141.0	213.0	5,39
3-Nonanol, (±)-	$C_9H_{20}O$		24[e]	47[e]	78[e]	123.0	194.2	5
4-Nonanol	$C_9H_{20}O$			45[e]	76.4	121.3	192.0	5
5-Nonanol	$C_9H_{20}O$	13[e]	31[e]	54[e]	84.5	128.1	194.7	5
2-Nonanone	$C_9H_{18}O$		8[e]	35[e]	71[e]	121.0	194.0	5
5-Nonanone	$C_9H_{18}O$			-1[e]	39.1	94[e]	188[e]	5
1-Nonene	C_9H_{18}	-50.1	-29.4	-3.3	30.4	77.1	146.4	1,5
Nonylamine	$C_9H_{21}N$		9[e]	37[e]	75[e]	126.2	202.1	5
Nonylbenzene	$C_{15}H_{24}$	33.0	58.9	92.0	135.4	193.7	281.4	5
Nonylcyclohexane	$C_{15}H_{30}$	35[e]	60[e]	92[e]	134[e]	193.4	280.9	5
Nonylcyclopentane	$C_{14}H_{28}$	25[e]	49[e]	80[e]	120[e]	177.2	261.5	5
2,5-Norbornadiene	C_7H_8				-15[e]	27.4	91[e]	5
Octacosane	$C_{28}H_{58}$	136.5	169.8	210.9	263.1	332.0	430.6	5
Octadecane	$C_{18}H_{38}$	61.5	89.0	123.1	167.3	227.6	316[e]	16
Octadecanoic acid	$C_{18}H_{36}O_2$		153[e]	183[e]	223[e]	281.6	374.5	5
1-Octadecanol	$C_{18}H_{38}O$	106[e]	130[e]	160[e]	200.5	257.3	343.0	5
cis-9-Octadecenoic acid	$C_{18}H_{34}O_2$	94[e]	126[e]	165.5	214.5	277.0	359.7	5
trans-9-Octadecenoic acid	$C_{18}H_{34}O_2$		124[e]	166	216	280	361	4
Octanal	$C_8H_{16}O$			6[e]	45.7	97.8	170.2	5
Octane	C_8H_{18}		-42.6	-17.9	14.4	58.9	125.3	16
Octanenitrile	$C_8H_{15}N$	-15[e]	8[e]	37[e]	75[e]	127.7	204.4	5
1-Octanethiol	$C_8H_{18}S$	-15[e]	6[e]	34[e]	71[e]	122.1	198.5	5
Octanoic acid	$C_8H_{16}O_2$	37[e]	58[e]	85[e]	120[e]	165.5	238.4	1,5
1-Octanol	$C_8H_{18}O$	12[e]	30[e]	53[e]	84[e]	128.2	194.8	1,39
2-Octanol	$C_8H_{18}O$			40[e]	69.9	112.5	179.4	1,39
3-Octanol	$C_8H_{18}O$	12[e]	24[e]	40[e]	64[e]	102.8	174.1	1
4-Octanol	$C_8H_{18}O$			40[e]	66.9	107.3	176.0	1,39
2-Octanone	$C_8H_{16}O$		-3[e]	23[e]	57[e]	103.8	172.1	5
3-Octanone	$C_8H_{16}O$			8[e]	47.7	97[e]	161[e]	5
Octanoyl chloride	$C_8H_{15}ClO$	1[e]	22[e]	46[e]	74.7	109[e]	150[e]	5
1-Octene	C_8H_{16}	-65.7	-46.1	-21.4	10.5	54.9	120.9	1,5
cis-2-Octene	C_8H_{16}	-59[e]	-41[e]	-17[e]	15[e]	59[e]	125.2	5
trans-2-Octene	C_8H_{16}	-59[e]	-41[e]	-17[e]	14[e]	59[e]	124.5	5
cis-3-Octene	C_8H_{16}	-65[e]	-46[e]	-22[e]	10[e]	55.1	122.4	5
trans-3-Octene	C_8H_{16}	-61[e]	-43[e]	-19[e]	13[e]	57[e]	122.8	5
cis-4-Octene	C_8H_{16}	-63[e]	-44[e]	-20[e]	11[e]	56[e]	122.1	5
trans-4-Octene	C_8H_{16}	-65[e]	-46[e]	-22[e]	10[e]	54.6	121.8	5
Octyl acetate	$C_{10}H_{20}O_2$	-26[e]	-3[e]	27[e]	66.3	120.0	198.2	5
Octylbenzene	$C_{14}H_{22}$	20.1	46.2	79.1	121.9	178.1	263.8	5
Octylcyclohexane	$C_{14}H_{28}$	16.9	44.3	77.8	120.0	177.6	263.2	5
Octylcyclopentane	$C_{13}H_{26}$	13[e]	36[e]	66[e]	106[e]	160.9	243.1	5
1-Octyne	C_8H_{14}	-59[e]	-40[e]	-16[e]	16[e]	60.3	125.8	1

Name	Mol. form.	t/°C for 1 Pa	t/°C for 10 Pa	t/°C for 100 Pa	t/°C for 1 kPa	t/°C for 10 kPa	t/°C for 100 kPa	Ref.
2-Octyne	C_8H_{14}	-52[e]	-33[e]	-8[e]	25[e]	70.6	137.8	1
3-Octyne	C_8H_{14}	-55[e]	-35[e]	-11[e]	22[e]	66.8	132.8	1
4-Octyne	C_8H_{14}	-56[e]	-36[e]	-12[e]	21[e]	65.6	131.4	1
Osmium	Os	2887 s	3150	3478	3875	4365	4983	2
Osmium(V) fluoride	F_5Os			74.1	113.2	162.3	226[e]	26
Osmium(VI) fluoride	F_6Os	-89 s	-73 s	-54 s	-30.6 s	-1.7 s	47.4	26
2-Oxetanone	$C_3H_4O_2$		-21[e]	8[e]	45.5	93.8	159.3	5
Oxirane	C_2H_4O		-111[e]	-93[e]	-70[e]	-37.0	10.2	1
Oxygen[a]	O_2				-211.9	-200.5	-183.1	1,28
Ozone[a]	O_3	-189[e]	-182[e]	-172[e]	-158[e]	-139.7	-111.5	5
Palladium	Pd	1448 s	1624	1844	2122	2480	2961	2
Paraldehyde	$C_6H_{12}O_3$				17[e]	62.2	124[e]	5
Pentaborane(9)	B_5H_9				-34.8	3.8	57.6	4
Pentachloroethane	C_2HCl_5		-23[e]	3[e]	37.4	86.0	159.4	1
Pentacosane	$C_{25}H_{52}$	119.7	152.7	193.2	244.4	305.0	401.1	5
1*H*-Pentadecafluoroheptane	C_7HF_{15}				-7[e]	35.9	96.0	5
Pentadecane	$C_{15}H_{32}$	30.5	56.1	88.1	129.6	186.3	270.1	16
1,2-Pentadiene	C_5H_8	-109[e]	-93[e]	-73[e]	-46.1	-9.7	44.5	5
cis-1,3-Pentadiene	C_5H_8	-109[e]	-93[e]	-73[e]	-47.0	-10.5	43.7	1,5
trans-1,3-Pentadiene	C_5H_8			-75[e]	-49.0	-13[e]	42[e]	5
1,4-Pentadiene	C_5H_8	-120[e]	-105[e]	-86[e]	-60.9	-26.2	25.6	5
2,3-Pentadiene	C_5H_8	-106[e]	-90[e]	-70[e]	-42.9	-6.3	47.9	5
Pentafluorobenzene	C_6HF_5			-41[e]	-13[e]	27[e]	85.3	5
Pentafluorophenol	C_6HF_5O				39[e]	82[e]	145.2	5
1,1,1,2,2-Pentafluoropropane	$C_3H_3F_5$					-60[e]	-17.9	5
2,3,4,5,6-Pentafluorotoluene	$C_7H_3F_5$			-20[e]	11[e]	53.6	117.0	5
2,2,3,3,4-Pentamethylpentane	$C_{10}H_{22}$		-24[e]	3[e]	39[e]	89.1	165.5	5
2,2,3,4,4-Pentamethylpentane	$C_{10}H_{22}$		-29[e]	-3[e]	33[e]	82.8	158.7	5
Pentanal	$C_5H_{10}O$	-71[e]	-53[e]	-31[e]	-1[e]	40.8	102.6	5
Pentane[b]	C_5H_{12}	-115.5	-99.8	-80.0	-54.0	-18.1	35.7	16
Pentanedinitrile	$C_5H_6N_2$	24.1	52[e]	85[e]	126[e]	178[e]	245[e]	5
Pentanedioic acid	$C_5H_8O_4$		121[e]	153.2	191.9	240.3	302.5	5
1,5-Pentanediol	$C_5H_{12}O_2$	25[e]	52[e]	85[e]	125[e]	175.1	238.9	5
2,4-Pentanedione	$C_5H_8O_2$			-5[e]	24.7	67.8	137.4	1
Pentanenitrile	C_5H_9N	-54[e]	-34[e]	-8[e]	26[e]	72.2	140.9	1
1-Pentanethiol	$C_5H_{12}S$	-60[e]	-41[e]	-17[e]	15[e]	60[e]	126.2	1
2-Pentanethiol	$C_5H_{12}S$	-70[e]	-52[e]	-28[e]	3[e]	46.6	111.9	5
3-Pentanethiol	$C_5H_{12}S$	-70[e]	-51[e]	-28[e]	4[e]	47.7	113.4	5
Pentanoic acid	$C_5H_{10}O_2$	-7.4	15.3	42.7	76.3	122.1	185.7	5
1-Pentanol	$C_5H_{12}O$	-27[e]	-10[e]	12[e]	41[e]	79.8	137.4	5
2-Pentanol	$C_5H_{12}O$	-35[e]	-19[e]	1[e]	28.0	64.9	118.7	1
3-Pentanol	$C_5H_{12}O$	-41[e]	-25[e]	-4[e]	24[e]	61.1	114.9	5
2-Pentanone	$C_5H_{10}O$				-1[e]	40.3	101.9	1,5
3-Pentanone	$C_5H_{10}O$			-31[e]	-1[e]	40[e]	101.6	1
1-Pentene	C_5H_{10}	-118.9	-103.4	-84.0	-58.8	-23.3	29.6	1,5
cis-2-Pentene	C_5H_{10}	-113.8	-98.1	-78.4	-52.7	-16.8	36.6	1,5
trans-2-Pentene	C_5H_{10}	-114.5	-98.9	-79.1	-53.3	-17.5	36.0	1,5
4-Pentenoic acid	$C_5H_8O_2$	0[e]	19[e]	44[e]	77[e]	122.0	187.5	5
Pentyl acetate	$C_7H_{14}O_2$	-58[e]	-39[e]	-14[e]	20[e]	70.1	149[e]	5
Pentylamine	$C_5H_{13}N$		-52[e]	-29[e]	1[e]	42.8	104.0	5
Pentylbenzene	$C_{11}H_{16}$	-14[e]	8[e]	37[e]	74[e]	126.7	204.9	5
Pentylcyclohexane	$C_{11}H_{22}$	-17[e]	6[e]	34[e]	72[e]	124.2	202.7	5
1-Pentylnaphthalene	$C_{15}H_{18}$	34[e]	62[e]	96[e]	141.3	202.2	289[e]	5
1-Pentyne	C_5H_8			-75[e]	-49.1	-13.5	39.9	5
2-Pentyne	C_5H_8	-100[e]	-85[e]	-65[e]	-37.9	-0.5	55.7	5
Perfluoroacetone	C_3F_6O			-113[e]	-94[e]	-67.8	-27.6	5
Perfluorobutane	C_4F_{10}		-122[e]	-105[e]	-82[e]	-49.8	-2.5	1,5
Perfluorocyclobutane	C_4F_8						-6.2	1
Perfluorocyclohexane	C_6F_{12}				-46.2 s	-7.6 s	48.9 s	5
Perfluorodecane	$C_{10}F_{22}$					52[e]	132.9	5

Fluids

Name	Mol. form.	*t*/°C for 1 Pa	*t*/°C for 10 Pa	*t*/°C for 100 Pa	*t*/°C for 1 kPa	*t*/°C for 10 kPa	*t*/°C for 100 kPa	Ref.
Perfluoro-2,3-dimethylbutane	C_6F_{14}					4.3	59.3	5
Perfluoroheptane	C_7F_{16}		-62[e]	-41[e]	-14[e]	24.7	82.1	1
Perfluorohexane	C_6F_{14}		-75[e]	-57[e]	-32[e]	2.8	56.8	5
Perfluoromethylcyclohexane	C_7F_{14}				-21[e]	18[e]	75.9	1
Perfluoro-2-methylpentane	C_6F_{14}				-33[e]	2.9	57.1	5
Perfluoro-3-methylpentane	C_6F_{14}	-95[e]	-80[e]	-60[e]	-34[e]	2.8	57.9	5
Perfluoronaphthalene	$C_{10}F_8$	5.2 s	25.1 s	48.1 s				5
Perfluorononane	C_9F_{20}					40[e]	114.7	5
Perfluorooctane	C_8F_{18}				5[e]	45.0	105.6	5
Perfluoropentane	C_5F_{12}				-54.7	-20.9	28.6	5
Perfluoropropane	C_3F_8		-139[e]	-124[e]	-105[e]	-77.5	-37.0	1
Perfluoropropene	C_3F_6	-150[e]	-138[e]	-122[e]	-101[e]	-72[e]	-30.6	5
Peroxyacetic acid	$C_2H_4O_3$				14.4	55.3	109.7	5
β-Phellandrene	$C_{10}H_{16}$			16[e]	53.2	104[e]	171.0	5
Phenanthrene	$C_{14}H_{10}$	53 s	83 s	120.8	170.4	238.4	337.7	5
Phenanthridine	$C_{13}H_9N$	79 s						5
Phenol	C_6H_6O	-9.7 s	9.6 s	34.1 s	68.9	113.7	181.4	1,5
2-Phenoxyethanol	$C_8H_{10}O_2$	21[e]	46[e]	75.9	115.4	168.7	244.8	5
Phenyl acetate	$C_8H_8O_2$		3[e]	33.1	72.2	123.9	195.5	5
Phenyl benzoate	$C_{13}H_{10}O_2$			102.3	151.4	217.9	313.3	5
2-Phenylethyl acetate	$C_{10}H_{12}O_2$	-4[e]	22[e]	54[e]	96[e]	152.3	232.0	5
Phenylhydrazine	$C_6H_8N_2$		38[e]	69[e]	109[e]	163.9	242.5	5
Phenyl isopropyl ether	$C_9H_{12}O$	-20[e]	-1[e]	23[e]	56[e]	103.7	176.9	5
Phenyl isothiocyanate	C_7H_5NS				79.4	105[e]	117[e]	5
1-Phenyl-2-propylamine, (±)-	$C_9H_{13}N$			33[e]	70.1	118[e]	202.0	5
Phenyl propyl ether	$C_9H_{12}O$		-10[e]	21[e]	61[e]	113.9	189.3	5
Phenyl salicylate	$C_{13}H_{10}O_3$				166.0	224.8	312.4	5
Phosphine[a]	H_3P	-182 s	-173 s	-161 s	-145 s	-122.7	-88.0	5
Phosphorus (white)	P	6 s	34 s	69	115	180	276	3,9
Phosphorus (red)	P	182 s	216 s	256 s	303 s	362 s	431 s	2,3
Phosphorus(III) bromide	Br_3P		-23[e]	5[e]	42.3	94.6	172.6	5
Phosphorus(V) bromide	Br_5P		-19 s	4 s	31 s	65.5 s	110.1	5
Phosphorus(III) chloride	Cl_3P	-93[e]	-77[e]	-55[e]	-26.0	14.5	75.7	5
Phosphorus(V) chloride	Cl_5P	-2 s	19 s	44 s	74 s	111.4 s	158.9 s	5
Phosphorus(III) chloride difluoride	ClF_2P				-119.5	-91.1	-47.6	5
Phosphorus(III) dichloride fluoride	Cl_2FP				-71.1	-37.4	13.5	5
Phosphorus(V) dichloride trifluoride	Cl_2F_3P		-120[e]	-101[e]	-77.1	-44.3	3[e]	7
Phosphorus(III) fluoride[a]	F_3P				-152[e]	-132.6	-101.4	5
Phosphorus(V) fluoride	F_5P	-157 s	-148 s	-137 s	-124.5 s	-108.6 s	-84.8	5
Phosphorus(III) oxide	O_3P_2				47.3	100.3	172.8	4
Phosphorus(V) oxide	O_5P_2	285 s	328 s	377.5 s	434.4 s	500.5 s	591	4
Phosphoryl bromide	Br_3OP				64[e]	115.5	191.4	5
Phosphoryl chloride	Cl_3OP					39.9	105.0	5
Phosphoryl fluoride	F_3OP	-124 s	-113 s	-100 s	-83.7 s	-64.1 s	-39.7 s	5
Phthalic anhydride	$C_8H_4O_3$	48.2 s	72.4 s			192.7	284.2	5
α-Pinene	$C_{10}H_{16}$	-48[e]	-27[e]	-1[e]	33.6	82.2	155.1	21
β-Pinene	$C_{10}H_{16}$	-43[e]	-22[e]	5.0	40.6	90.5	165.5	21
Piperidine	$C_5H_{11}N$				2[e]	43.3	105.8	5
Platinum	Pt	2057	2277[e]	2542	2870	3283	3821	2
Plutonium	Pu	1483	1680	1925	2238	2653	3226	2
Polonium	Po				573[e]	730.2	963.3	5
Polonium(IV) chloride	Cl_4Po					300.6	389.4	5
Potassium	K	200.2	256.5	328	424	559	756.2	13,30
Potassium bromide	BrK	597 s	674 s	773				25
Potassium chloride	ClK	625 s	704 s	804	945	1137	1411	12,23,25
Potassium fluoride	FK			869	1017	1216	1499	4
Potassium hydroxide	HKO	520[e]	601[e]	704	842	1035	1325	4
Potassium iodide	IK			731	866	1052	1322	4
Praseodymium	Pr	1497.7	1699.4	1954[d]	2298[d]	2781[d]	3506[d]	3
Propanal	C_3H_6O			-69[e]	-42[e]	-6[e]	47.7	1

Name	Mol. form.	t/°C for 1 Pa	t/°C for 10 Pa	t/°C for 100 Pa	t/°C for 1 kPa	t/°C for 10 kPa	t/°C for 100 kPa	Ref.
Propane[a]	C_3H_8	-156.9	-145.6	-130.9	-111.4	-83.8	-42.3	1,41
1,2-Propanediamine, (±)-	$C_3H_{10}N_2$		-35.4	-12.0	18.8	61[e]	119[e]	5
1,2-Propanediol	$C_3H_8O_2$	-11[e]	13[e]	42[e]	78[e]	125.0	187.2	5
1,3-Propanediol	$C_3H_8O_2$	4[e]	30[e]	62[e]	101[e]	149.9	214.0	5
1,3-Propanedithiol	$C_3H_8S_2$	-53[e]	-28[e]	3[e]	43[e]	97[e]	172.4	5
Propanenitrile	C_3H_5N	-69.4	-55.3	-36.0	-7.9	35.2	97.4	1,5
1-Propanethiol	C_3H_8S	-94[e]	-78[e]	-57[e]	-29.1	9.6	67.4	1,5
2-Propanethiol	C_3H_8S	-102[e]	-87[e]	-67[e]	-41[e]	-3[e]	52.2	1
Propanoic acid	$C_3H_6O_2$			0[e]	35.1	79.9	140.8	1,5
Propanoic anhydride	$C_6H_{10}O_3$	-32[e]	-15[e]	6[e]	36[e]	77.6	142.9	5
1-Propanol	C_3H_8O	-54[e]	-38[e]	-16[e]	10[e]	47[e]	96.9	1,5
2-Propanol	C_3H_8O	-65[e]	-49[e]	-28[e]	-1.3	33.6	82.0	1,5
Propene[a]	C_3H_6	-160.6	-149.0	-134.3	-114.9	-88.2	-47.9	1,5
cis-1-Propenylbenzene	C_9H_{10}	-38[e]	-15.4	13.3	51.4	103.7	178.4	5
trans-1-Propenylbenzene	C_9H_{10}		-16[e]	13.3	51.6	103.7	178.4	5
2-Propoxyethanol	$C_5H_{12}O_2$				40[e]	85.6	149.3	5
Propyl acetate	$C_5H_{10}O_2$	-69[e]	-51[e]	-29[e]	0[e]	40.9	101.2	1
Propylamine	C_3H_9N		-81[e]	-63[e]	-38.3	-4.1	46.9	1,5
Propylbenzene	C_9H_{12}	-43[e]	-23[e]	4[e]	38[e]	86.7	158.8	1
Propyl benzoate	$C_{10}H_{12}O_2$	-8[e]	18[e]	50.2	92.3	149.2	230.5	5
Propyl butanoate	$C_7H_{14}O_2$	-35[e]	-19[e]	3[e]	32.0	74.9	142.8	5
Propylcyclohexane	C_9H_{18}	-46[e]	-26[e]	0[e]	35.1	83.6	156.2	5
Propylcyclopentane	C_8H_{16}	-60[e]	-41[e]	-16[e]	16.5	62.1	130.5	5
1-Propylcyclopentanol	$C_8H_{16}O$	9[e]	24[e]	43[e]	69.0	108.4	173.5	5
Propylene carbonate	$C_4H_6O_3$	-40[e]	-5[e]	43[e]	112[e]	220[e]	410[e]	5
Propyl formate	$C_4H_8O_2$	-78[e]	-62[e]	-42[e]	-15.1	23.0	80.4	1,5
Propyl hexanoate	$C_9H_{18}O_2$	-26[e]	-2[e]	28[e]	65.1	113.4	178[e]	5
Propyl isobutanoate	$C_7H_{14}O_2$		-28[e]	-5.7	24.5	67.5	133.3	5
Propyl methacrylate	$C_7H_{12}O_2$				26[e]	73.8	139.7	5
Propyl 3-methylbutanoate	$C_8H_{16}O_2$			1.8	38.9	87.9	155.6	5
Propyl nitrate	$C_3H_7NO_3$			-23.9	6.1	48.1	111[e]	5
Propyl octanoate	$C_{11}H_{22}O_2$	-2[e]	23[e]	55[e]	94.0	145.2	215[e]	5
Propyl propanoate	$C_6H_{12}O_2$	-62[e]	-42[e]	-18[e]	14[e]	58.3	122.0	5
Propyne	C_3H_4				-94[e]	-65.3	-23.2	1
Pulegone	$C_{10}H_{16}O$	37[e]	49.1	66.4	92.2	135.1	220.2	5
Pyridine	C_5H_5N			-23[e]	8[e]	51.0	114.9	1
Pyrrole	C_4H_5N			-8[e]	24[e]	66.7	129.4	1
Pyrrolidine	C_4H_9N		-59[e]	-38[e]	-10[e]	28.5	86.2	1
Quinoline	C_9H_7N	-1.3	23.7	55.4	96.8	153.4	236.5	1,5
Radium	Ra	546 s	633 s	764	936	1173	1526	2
Radon[a]	Rn	-163 s	-152 s	-139 s	-121.4 s	-97.6 s	-62.3	5
Rhenium	Re	3030 s	3341	3736	4227	4854	5681	2
Rhenium(VI) dioxydifluoride	F_2O_2Re				89.2	131.9	185[e]	26
Rhenium(V) fluoride	F_5Re			58.8	99.5	152[e]	221[e]	26
Rhenium(VI) fluoride	F_6Re	-97 s	-82 s	-63 s	-40.2 s	-11.9 s	33.4	26
Rhenium(VII) oxide	O_7Re_2	147 s	176 s	208 s	244 s	284 s	362	4
Rhenium(VII) oxypentafluoride	F_5ORe	-103 s	-84 s	-59 s	-28 s	13.7 s	72.8	26
Rhenium(VI) oxytetrafluoride	F_4ORe	5 s	26 s	50.7 s	80.1 s	117.1	171.2	26
Rhodium	Rh	2015	2223	2476	2790	3132	3724	2
Rubidium	Rb	160.4	212.5	278.9	368	496.1	685.3	13,30
Rubidium bromide	BrRb			766	903	1087	1350	4
Rubidium chloride	ClRb			777	916	1105	1379	4
Rubidium fluoride	FRb			910	1001	1145	1409	4,12
Rubidium iodide	IRb			733	866	1045	1302	4
Ruthenium	Ru	2315 s	2538	2814	3151	3572	4115	2
Salicylaldehyde	$C_7H_6O_2$		-1[e]	29[e]	68[e]	120.7	196.2	5
Samarium	Sm	728 s	833 s	967 s	1148[d]	1402[d]	1788[d]	3
Scandium	Sc	1372 s	1531 s	1733[d]	1993[d]	2340[d]	2828[d]	3
Selenium	Se	227	279	344	431	540	685	3
Selenium dioxide	O_2Se	124.5 s	153.9 s	188 s	228 s	275 s	315 s	38

Name	Mol. form.	t/°C for 1 Pa	t/°C for 10 Pa	t/°C for 100 Pa	t/°C for 1 kPa	t/°C for 10 kPa	t/°C for 100 kPa	Ref.
Selenium hexafluoride	F_6Se	-143 s	-132 s	-118 s	-100.7 s	-77.8 s	-46.5 s	5
Selenium tetrachloride	Cl_4Se	23 s	45 s	71 s	102 s	141.4 s	191.1 s	5
Selenium tetrafluoride	F_4Se				13.6	51.6	104.7	5
Silane[a]	H_4Si			-181	-165.4	-143.7	-111.8	4
Silicon	Si	1635	1829	2066	2363	2748	3264	2
Silicon dioxide (α-quartz)	O_2Si	1966[d]	2149[d]	2368[d]				8
Silver	Ag	1010	1140	1302	1509	1782	2160	2
Silver(I) bromide	AgBr	569[d]	656[d]	765[d]	905[d]	1093[d]	1359[d]	9
Silver(I) chloride	AgCl	670	769	873	1052	1264	1561	4
Silver(I) iodide	AgI	594	686	803	959	1177	1503	4
Sodium	Na	280.6	344.2	424.3	529	673	880.2	13,30
Sodium bromide	BrNa			791	931	1120	1389	4
Sodium chloride	ClNa	653 s	733 s	835	987	1182	1461	12,23,25
Sodium cyanide	CNNa		672[e]	798	961	1182	1497	4
Sodium fluoride	FNa		920 s	1058	1218	1426	1702	4,12,24
Sodium hydroxide	HNaO	513	605	722	874	1080	1377	4
Sodium iodide	INa			753	883	1058	1301	4
Spiro[2.2]pentane	C_5H_8	-110[e]	-95[e]	-76[e]	-51[e]	-15[e]	38.6	5
Stearaldehyde	$C_{18}H_{36}O$			142[e]	186[e]	246.9	336.7	5
cis-Stilbene	$C_{14}H_{12}$	26[e]	54[e]	88[e]	130.4	183[e]	253[e]	5
trans-Stilbene	$C_{14}H_{12}$				155.6	218.1	305.8	5
Strontium	Sr	523 s	609 s	717 s	866	1072	1373	2
Strontium oxide	OSr	1789 s	1903 s	2047 s	2235 s	2488 s		4
Styrene	C_8H_8		-31[e]	-5[e]	28.6	75.4	144.7	1
Succinic anhydride	$C_4H_4O_3$				121[e]	180.8	260.8	5
Succinonitrile	$C_4H_4N_2$	24.8 s					266.0	5
Sulfolane	$C_4H_8O_2S$		49[e]	87[e]	135[e]	198.0	283.5	5
Sulfur (rhombic)	S	102 s	135	176	235	318	444	3
Sulfur bromide [BrSSBr]	Br_2S_2	-7[e]	15[e]	42[e]	78.4	128.1	200.9	5
Sulfur chloride [ClSSCl]	Cl_2S_2	-55[e]	-36[e]	-12[e]	21.0	67.2	137.1	5
Sulfur decafluoride	$F_{10}S_2$					-22.0	28.5	5
Sulfur dichloride	Cl_2S	-76[e]	-61[e]	-41[e]	-16.7	15.3	58.7	5
Sulfur dioxide[a]	O_2S			-98 s	-80 s	-52.2	-10.3	1,5
Sulfur hexafluoride[a]	F_6S	-158 s	-147 s	-133.6 s	-116.6 s	-94.4 s	-64.1 s	5
Sulfuric acid	H_2O_4S	72	103	140	187	248	330	4
Sulfur tetrafluoride	F_4S				-110.0	-82.1	-40.3	5
Sulfur trioxide (α-form)	O_3S				-20 s	6.6 s	44.5	5
Sulfuryl chloride	Cl_2O_2S				-27[e]	11.8	69.0	5
Tantalum	Ta	3024	3324	3684	4122	4666	5361	2
Tantalum(V) fluoride	F_5Ta					119	229	4
Technetium	Tc	2454[e]	2725[e]	3051[e]	3453[e]	3961[e]	4621[e]	2
Tellurium	Te			502[e]	615[e]	768.8	992.4	5
Tellurium hexafluoride	F_6Te	-142 s	-130 s	-115 s	-96 s	-71.8 s	-39.1 s	5
Tellurium tetrachloride	Cl_4Te				237[e]	299.4	387.8	5
Terbium	Tb	1516.1	1706.1	1928[d]	2232[d]	2640[d]	3218[d]	3
o-Terphenyl	$C_{18}H_{14}$	66[e]	94[e]	129[e]	176[e]	241.3	336.3	5
m-Terphenyl	$C_{18}H_{14}$	87[e]	118[e]	156[e]	206.6	275.3	374.6	5
p-Terphenyl	$C_{18}H_{14}$	127.1 s	154.7 s		217.2	284.0	383.0	5
α-Terpineol	$C_{10}H_{18}O$			48	89	142	217	4
Terpinolene	$C_{10}H_{16}$			26.5	64.9	115.4	184.6	5
1,1,2,2-Tetrabromoethane	$C_2H_2Br_4$	14[e]	38[e]	69[e]	109[e]	163.7	242.9	5
Tetrabromoethene	C_2Br_4		-54.5 s	-31.7 s	-3.5 s	32.2 s	226.0	5
Tetrabromomethane	CBr_4			25.6 s	65.8 s	111.6	188.9	5
1,1,1,2-Tetrachloro-2,2-difluoroethane	$C_2Cl_4F_2$				-7[e]	31.0	91.1	5
1,1,2,2-Tetrachloro-1,2-difluoroethane	$C_2Cl_4F_2$					32.3	92.5	1
1,1,1,2-Tetrachloroethane	$C_2H_2Cl_4$	-58[e]	-40[e]	-15[e]	17[e]	62.2	129.7	1
1,1,2,2-Tetrachloroethane	$C_2H_2Cl_4$		-22[e]	1[e]	32.4	76.9	144.7	1
Tetrachloroethene	C_2Cl_4			-22[e]	10[e]	54.4	120.7	1
Tetrachloromethane[a]	CCl_4	-79.4 s	-70.8 s	-53.5 s	-24.4 s	15.8	76.2	1,5
1,1,1,2-Tetrachloropropane	$C_3H_4Cl_4$	-48[e]	-28[e]	-2[e]	32[e]	79.1	149.5	5

Name	Mol. form.	t/°C for 1 Pa	t/°C for 10 Pa	t/°C for 100 Pa	t/°C for 1 kPa	t/°C for 10 kPa	t/°C for 100 kPa	Ref.
Tetrachlorosilane[a]	Cl_4Si				-39[e]	0[e]	57.3	1
Tetracosane	$C_{24}H_{50}$	115.0	148.1	188.5	239.1	295.4	390.6	5
Tetradecamethylhexasiloxane	$C_{14}H_{42}O_5Si_6$	6[e]	36[e]	72[e]	117[e]	176.0	259.1	5
Tetradecane	$C_{14}H_{30}$	19.1	44.1	75.3	115.7	171.1	253.0	16
Tetradecanenitrile	$C_{14}H_{27}N$	52[e]	79[e]	114.0	159.0	219.7	306.3	5
Tetradecanoic acid	$C_{14}H_{28}O_2$	96[e]	118[e]	147[e]	186[e]	241.3	325.6	5
1-Tetradecanol	$C_{14}H_{30}O$	80.0	110.5	149.6	152[e]	205.3	286.7	5
1-Tetradecene	$C_{14}H_{28}$	16.1	41.3	72.7	113.2	168.7	250.6	5
Tetradecylamine	$C_{14}H_{31}N$			104[e]	147[e]	206.1	290.9	5
Tetraethylene glycol	$C_8H_{18}O_5$	89[e]	117[e]	151.1	192.2	242.9	307.3	5
Tetraethylene glycol dimethyl ether	$C_{10}H_{22}O_5$				138[e]	200.9	275.3	5
Tetraethylsilane	$C_8H_{20}Si$			-6.5	30.5	80.6	152.6	5
1,2,3,4-Tetrafluorobenzene	$C_6H_2F_4$			-36[e]	-7[e]	33.8	94.0	1
1,2,3,5-Tetrafluorobenzene	$C_6H_2F_4$			-43[e]	-14[e]	25.5	84.1	1
1,2,4,5-Tetrafluorobenzene	$C_6H_2F_4$					30.7	89.9	1
Tetrafluorodiborane	B_2F_4						-34	1
1,1,2,2-Tetrafluoro-1,2-dinitroethane	$C_2F_4N_2O_4$				-30[e]	6.4	59.5	5
1,1,1,2-Tetrafluoroethane	$C_2H_2F_4$				-94.3	-66.8	-26.4	17
1,1,2,2-Tetrafluoroethane	$C_2H_2F_4$				-96.0	-66.9	-23.3	5
Tetrafluoroethene	C_2F_4				-132.3	-109.7	-75.8	1
Tetrafluoromethane[a]	CF_4	-199.9 s	-193 s	-183.9 s	-171.6	-153.9	-128.3	1,5
2,2,3,3-Tetrafluoro-1-propanol	$C_3H_4F_4O$			-10[e]	17[e]	53.9	107.2	5
Tetrafluorosilane[a]	F_4Si	-166 s	-157 s	-145.6 s	-132.3 s	-115.7 s	-94.9 s	4,7
cis-Tetrahydro-2,5-dimethylthiophene	$C_6H_{12}S$	-53[e]	-34[e]	-8[e]	25[e]	72.0	142.1	5
Tetrahydrofuran	C_4H_8O	-94[e]	-78[e]	-57.3	-29.8	9[e]	65.6	1
Tetrahydrofurfuryl alcohol	$C_5H_{10}O_2$	-40[e]	-16[e]	15[e]	55[e]	106[e]	176.8	5
1,2,3,4-Tetrahydro-5-methylnaphthalene	$C_{11}H_{14}$	9[e]	31[e]	60[e]	99[e]	153.1	233.8	5
1,2,3,4-Tetrahydro-6-methylnaphthalene	$C_{11}H_{14}$	17[e]	36[e]	62[e]	97[e]	147.8	228.5	5
Tetrahydro-3-methyl-2*H*-thiopyran	$C_6H_{12}S$	-48[e]	-27[e]	0[e]	35[e]	84.1	157.5	5
1,2,3,4-Tetrahydronaphthalene	$C_{10}H_{12}$	-21[e]	3[e]	33.2	74.1	127.4	207.8	5
Tetrahydropyran	$C_5H_{10}O$				-15[e]	26.0	88[e]	5
Tetrahydro-2*H*-pyran-2-one	$C_5H_8O_2$		5[e]	35.1	74.4	128.3	207.0	5
Tetrahydrothiophene	C_4H_8S	-66[e]	-47[e]	-23[e]	9.4	54.1	120.5	1
1,2,3,4-Tetramethylbenzene	$C_{10}H_{14}$		7[e]	36[e]	74[e]	126.6	204.5	5
1,2,3,5-Tetramethylbenzene	$C_{10}H_{14}$	-19[e]	3[e]	32[e]	69[e]	120.9	197.5	5
1,2,4,5-Tetramethylbenzene	$C_{10}H_{14}$					119.9	196.3	5
2,2,3,3-Tetramethylbutane	C_8H_{18}	-62.5 s	-44 s	-20.9 s	8.9 s	48.8 s	105.8	5
1,1,3,3-Tetramethylcyclopentane	C_9H_{18}	-72[e]	-54[e]	-30[e]	2[e]	47[e]	117.4	5
2,2,3,3-Tetramethylhexane	$C_{10}H_{22}$	-46[e]	-25[e]	1[e]	36[e]	85.6	159.8	5
2,2,5,5-Tetramethylhexane	$C_{10}H_{22}$			-10[e]	22[e]	68.3	137.0	5
2,2,3,3-Tetramethylpentane	C_9H_{20}				21[e]	68.5	139.8	1
2,2,3,4-Tetramethylpentane	C_9H_{20}	-61[e]	-42[e]	-17[e]	16[e]	62.5	132.6	1
2,2,4,4-Tetramethylpentane	C_9H_{20}		-49[e]	-25[e]	8[e]	53.2	121.8	1
2,3,3,4-Tetramethylpentane	C_9H_{20}	-57[e]	-37[e]	-12[e]	22[e]	69.7	141.1	1
2,2,4,4-Tetramethyl-3-pentanol	$C_9H_{20}O$				58	100	167	5
Tetramethylsilane	$C_4H_{12}Si$			-83[e]	-59[e]	-25[e]	26.7	5
Tetramethylstannane	$C_4H_{12}Sn$			-55.0	-25.6	16.6	77.7	5
Tetramethylurea	$C_5H_{12}N_2O$			20.7	58.0	106.7	179.5	5
Tetranitromethane	CN_4O_8				18.0	61.8	124[e]	5
Thallium	Tl	609	704	824	979	1188	1485	2
Thallium(I) bromide	BrTl				509	635	817	4
Thallium(I) chloride	ClTl				504	626	806	4
Thallium(I) iodide	ITl				520	644	821	4
Thiacyclohexane	$C_5H_{10}S$				24[e]	71.1	141.2	5
Thiazole	C_3H_3NS					54.4	117.8	5
Thietane	C_3H_6S		-62[e]	-40[e]	-9[e]	32.5	94.5	5
Thionyl bromide	Br_2OS	-49[e]	-29[e]	-5[e]	27.8	72.9	139.6	5
Thionyl chloride	Cl_2OS	-99[e]	-81[e]	-58[e]	-27.1	14.6	75.2	5
Thionyl fluoride	F_2OS			-124[e]	-106.5	-81.5	-44.1	5
Thiophene	C_4H_4S				-17[e]	23.7	83.7	5

Name	Mol. form.	t/°C for 1 Pa	t/°C for 10 Pa	t/°C for 100 Pa	t/°C for 1 kPa	t/°C for 10 kPa	t/°C for 100 kPa	Ref.
Thorium	Th	2360	2634	2975	3410	3986	4782	2
Thulium	Tm	844 s	962 s	1108 s	1297 s	1548[d]	1944[d]	3
Thymol	$C_{10}H_{14}O$	18.9 s	37.9 s	59.5	101.2	155.0	230.4	5
Tin	Sn	1224	1384	1582	1834	2165	2620	2
Tin(IV) bromide	Br_4Sn				67	122	204	4
Tin(II) chloride	Cl_2Sn		253	308	381	479	622	4
Tin(IV) iodide	I_4Sn				167.1	242.7	347.7	4
Titanium	Ti	1709	1898	2130[e]	2419	2791	3285	2
Toluene	C_7H_8	-78.1	-57.1	-31.3	1.5	45.2	110.1	5
Toluene-2,4-diisocyanate	$C_9H_6N_2O_2$		39[e]	72[e]	113.9	169.7	247[e]	5
Tribromoacetaldehyde	C_2HBr_3O			15.0	52.7	103.0	173.5	5
1,2,3-Tribromobutane	$C_4H_7Br_3$	0[e]	23[e]	53[e]	91[e]	143.7	219.5	5
1,2,4-Tribromobutane	$C_4H_7Br_3$	-3[e]	20[e]	49[e]	87[e]	139.4	214.5	5
1,1,2-Tribromoethane	$C_2H_3Br_3$	-18[e]	4[e]	32[e]	68[e]	117.1	188.4	5
Tribromomethane	$CHBr_3$				30.5	78.3	148.8	1
Tributylamine	$C_{12}H_{27}N$	-26[e]	1[e]	35[e]	77.7	134.5	213.4	5
Tributyl phosphate	$C_{12}H_{27}O_4P$					205[e]	288.3	5
Trichloroacetaldehyde	C_2HCl_3O			-41.6	-9.8	33.8	97.4	5
Trichloroacetic acid	$C_2HCl_3O_2$				83.8	130.0	197.2	1,5
Trichloroacetonitrile	C_2Cl_3N				-16[e]	25.3	85.1	1
Trichloroacetyl chloride	C_2Cl_4O			-25[e]	7[e]	51.7	117.8	1,5
1,1,1-Trichloroethane	$C_2H_3Cl_3$				-25.3	14.2	73.7	5
1,1,2-Trichloroethane	$C_2H_3Cl_3$			-23[e]	7[e]	49.9	113.4	1
Trichloroethene	C_2HCl_3	-74[e]	-59[e]	-39[e]	-12[e]	26.7	86.8	1
Trichloroethoxysilane	$C_2H_5Cl_3OSi$	-78[e]	-60[e]	-36.0	-4.6	38.7	102.0	5
Trichloroethylsilane	$C_2H_5Cl_3Si$	-79[e]	-61[e]	-38[e]	-8[e]	34.9	98.7	5
Trichlorofluoromethane[a]	CCl_3F		-107[e]	-89[e]	-63[e]	-28.5	23.3	1,5
Trichloromethane[a]	$CHCl_3$			-61[e]	-34[e]	4.3	60.8	1
(Trichloromethyl)benzene	$C_7H_5Cl_3$		9[e]	40.6	81.5	136.2	213.0	5
Trichloromethylsilane	CH_3Cl_3Si		-83[e]	-61[e]	-33[e]	7[e]	65.7	1
Trichloronitromethane	CCl_3NO_2		-59[e]	-30[e]	4.4	47.8	112.0	5
2,4,6-Trichlorophenol	$C_6H_3Cl_3O$			71.8	114.0	169.5	245.7	5
Trichlorophenylsilane	$C_6H_5Cl_3Si$			33[e]	70.2	122.6	201[e]	5
1,1,3-Trichloropropane	$C_3H_5Cl_3$	-51[e]	-31[e]	-5[e]	28[e]	75.3	145.1	5
1,2,3-Trichloropropane	$C_3H_5Cl_3$			2[e]	37[e]	84.9	156.3	5
Trichlorosilane	Cl_3HSi			-81[e]	-56[e]	-21[e]	31.6	7
1,3,5-Trichloro-2,4,6-trifluorobenzene	$C_6Cl_3F_3$	-19[e]	4[e]	32[e]	70[e]	121.7	197.9	1
1,1,1-Trichloro-2,2,2-trifluoroethane	$C_2Cl_3F_3$						45.6	1,5
1,1,2-Trichloro-1,2,2-trifluoroethane	$C_2Cl_3F_3$					-8.2	47.3	1,5
Tricosane	$C_{23}H_{48}$	102.9	135.1	174.8	221[e]	285.3	379.5	5
Tri-*o*-cresyl phosphate	$C_{21}H_{21}O_4P$	119.0	156.1	201.0	256.3	326.3	418[e]	5
Tri-*m*-cresyl phosphate	$C_{21}H_{21}O_4P$	147.8	177.3	211.4	251.3	298[e]	355[e]	5
Tri-*p*-cresyl phosphate	$C_{21}H_{21}O_4P$	140.6	174[e]	214[e]	262[e]	320[e]	392[e]	5
Tridecane	$C_{13}H_{28}$	7.2	31.5	61.8	101.1	155.1	234.9	16
Tridecanoic acid	$C_{13}H_{26}O_2$	87[e]	109[e]	138[e]	176[e]	230.3	311.5	5
1-Tridecanol	$C_{13}H_{28}O$	71.6	101.0	103[e]	140[e]	192.3	273.1	5
1-Tridecene	$C_{13}H_{26}$	4.1	28.5	59.0	98.3	152.5	232.3	5
Triethanolamine	$C_6H_{15}NO_3$	75[e]	108[e]	148[e]	196[e]	256.7	334[e]	5
Triethylamine	$C_6H_{15}N$	-58[e]	-45[e]	-29[e]	-5[e]	29.9	88.5	1
Triethylene glycol	$C_6H_{14}O_4$	44[e]	74[e]	109.0	152.6	207.2	277.9	5
Triethyl phosphate	$C_6H_{15}O_4P$			34	76	132	211	4
Trifluoroacetic acid	$C_2HF_3O_2$					16.8	71.4	1,5
Trifluoroacetic acid anhydride	$C_4F_6O_3$			-63[e]	-39[e]	-7.1	38.8	5
Trifluoroacetonitrile	C_2F_3N				-126.1	-102.5	-67.8	1
1,3,5-Trifluorobenzene	$C_6H_3F_3$					18.2	75.0	5
1,1,1-Trifluoroethane	$C_2H_3F_3$				-113[e]	-86.6	-47.8	1
2,2,2-Trifluoroethanol	$C_2H_3F_3O$			-33[e]	-8[e]	26.0	74[e]	5
Trifluoromethane[a]	CHF_3			-152[e]	-136[e]	-114.4	-82.3	1
(Trifluoromethyl)benzene	$C_7H_5F_3$				-3[e]	39[e]	101.6	5
Trifluoromethyl difluoromethyl ether	C_2HF_5O	-147[e]	-136[e]	-121[e]	-102[e]	-75.0	-35.4	20

Name	Mol. form.	*t*/°C for 1 Pa	*t*/°C for 10 Pa	*t*/°C for 100 Pa	*t*/°C for 1 kPa	*t*/°C for 10 kPa	*t*/°C for 100 kPa	Ref.
Triiodomethane	CHI_3	51.1 s	82.7 s	121[e]			218.0	5
Triisobutylamine	$C_{12}H_{27}N$		1[e]	28.9	64.9	112.5	178.5	5
Triisopropyl borate	$C_9H_{21}BO_3$					73.1	139.0	5
Trimethylamine	C_3H_9N		-114[e]	-97[e]	-75.0	-43.8	2.6	1,5
2,4,6-Trimethylaniline	$C_9H_{13}N$	12[e]	36[e]	66[e]	104.1	154.9	226[e]	5
Trimethylarsine	C_3H_9As			-74[e]	-45[e]	-5.4	52.0	5
1,2,3-Trimethylbenzene	C_9H_{12}		-12[e]	15[e]	52[e]	101.5	175.6	1
1,2,4-Trimethylbenzene	C_9H_{12}	-37[e]	-16[e]	11[e]	47[e]	95.9	168.9	1
1,3,5-Trimethylbenzene	C_9H_{12}	-39[e]	-18[e]	9[e]	43.7	92.4	164.3	1
Trimethyl borate	$C_3H_9BO_3$				-14[e]	15.6	67.9	5
2,2,3-Trimethylbutane	C_7H_{16}				-23.2	18.1	80.4	5
2,3,3-Trimethyl-1-butene	C_7H_{14}	-91[e]	-75[e]	-53[e]	-24.2	16.3	77.5	5
Trimethylchlorosilane	C_3H_9ClSi				-37.8	0.4	57.3	5
1,1,2-Trimethylcyclohexane	C_9H_{18}			-12[e]	23[e]	71.5	145.5	5
1,1,3-Trimethylcyclohexane	C_9H_{18}	-60[e]	-41[e]	-16[e]	18[e]	65.2	136.1	5
1α,2β,4β-1,2,4-Trimethylcyclohexane	C_9H_{18}	-71[e]	-50[e]	-22[e]	15[e]	65.7	140.7	5
1α,3α,5β-1,3,5-Trimethylcyclohexane	C_9H_{18}	-72[e]	-50[e]	-22[e]	14[e]	65.1	140.0	5
1,1,2-Trimethylcyclopentane	C_8H_{16}				2[e]	46.2	113.2	5
1,1,3-Trimethylcyclopentane	C_8H_{16}	-77[e]	-59[e]	-36[e]	-5[e]	38.7	104.4	5
1α,2α,4β-1,2,4-Trimethylcyclopentane	C_8H_{16}	-70[e]	-52[e]	-28[e]	4[e]	48.9	116.2	5
1α,2β,4α-1,2,4-Trimethylcyclopentane	C_8H_{16}	-74[e]	-56[e]	-33[e]	-1[e]	42.8	108.8	5
1,1,2-Trimethylcyclopropane	C_6H_{12}	-109[e]	-94[e]	-73[e]	-46[e]	-7[e]	52.0	5
2,2,6-Trimethylheptane	$C_{10}H_{22}$	-46[e]	-27[e]	-2[e]	32[e]	78.5	148.4	5
3,3,5-Trimethylheptane	$C_{10}H_{22}$			0[e]	35[e]	82.7	155.2	5
2,2,4-Trimethylhexane	C_9H_{20}	-66.1	-46.4	-21.3	11.8	57.7	126.0	5
2,2,5-Trimethylhexane	C_9H_{20}	-65.1	-45.8	-21.2	11.2	56.2	123.7	1,5
2,3,3-Trimethylhexane	C_9H_{20}	-58[e]	-38[e]	-13[e]	20[e]	66.7	137.2	5
2,3,5-Trimethylhexane	C_9H_{20}	-60[e]	-41[e]	-16[e]	17[e]	62.3	130.9	5
2,4,4-Trimethylhexane	C_9H_{20}	-62[e]	-43[e]	-18[e]	15[e]	61.0	130.2	5
3,3,4-Trimethylhexane	C_9H_{20}	-53[e]	-33[e]	-7[e]	28[e]	76.3	148.9	5
2,4,7-Trimethyloctane	$C_{11}H_{24}$				43[e]	94[e]	170.4	5
Trimethylolpropane	$C_6H_{14}O_3$	73[e]	98[e]	128[e]	167.8	220.5	295[e]	5
2,2,3-Trimethylpentane	C_8H_{18}	-74[e]	-56[e]	-32[e]	-0.8	43.1	109.4	5
2,2,4-Trimethylpentane	C_8H_{18}	-81.9	-63.4	-39.8	-8.9	34.0	98.8	5
2,3,3-Trimethylpentane	C_8H_{18}	-72[e]	-54[e]	-30[e]	2.1	46.9	114.3	5
2,3,4-Trimethylpentane	C_8H_{18}	-74[e]	-54.5	-30.0	2.2	46.7	113.1	1,5
2,4,4-Trimethyl-2-pentanol	$C_8H_{18}O$		-7[e]	13[e]	40[e]	79.8	146.1	5
2,2,4-Trimethyl-3-pentanol	$C_8H_{18}O$	-2[e]	9[e]	24[e]	47[e]	82.6	150.4	5
2,2,4-Trimethyl-3-pentanone	$C_8H_{16}O$			11.3	42.1	81.7	134.6	5
2,3,3-Trimethyl-1-pentene	C_8H_{16}		-53[e]	-30[e]	1[e]	43.8	107.9	5
2,4,4-Trimethyl-1-pentene	C_8H_{16}	-79[e]	-61[e]	-38[e]	-7[e]	36.2	101.0	5
2,3,4-Trimethyl-2-pentene	C_8H_{16}	-68[e]	-49[e]	-26[e]	6[e]	50.0	115.8	5
2,4,4-Trimethyl-2-pentene	C_8H_{16}	-73[e]	-56[e]	-33[e]	-2[e]	40.4	104.5	5
Trimethyl phosphate	$C_3H_9O_4P$	-31[e]	-7[e]	23.6	62.8	116.0	192.0	5
Trimethylphosphine	C_3H_9P			-81[e]	-53[e]	-15.0	37.1	5
Trimethylstibine	C_3H_9Sb			-56[e]	-23.8	19[e]	80[e]	5
Trinitroglycerol	$C_3H_5N_3O_9$	48.6	75.7	118[e]	191[e]	353[e]	1007[e]	5
1,3,5-Trioxane	$C_3H_6O_3$					53[e]	113.7	1
Triphenylmethane	$C_{19}H_{16}$	81 s		112[e]	175[e]	254.6	360.0	5
Tripropylamine	$C_9H_{21}N$	-39[e]	-18[e]	8[e]	42[e]	88.2	156.0	5
Tris(perfluorobutyl)amine	$C_{12}F_{27}N$		3[e]	29.0	63.3	109.9	176.8	5
Tungsten	W	3204 s	3500	3864	4306	4854	5550	2
Tungsten(VI) fluoride	F_6W	-107 s	-92 s	-74 s	-52.1 s	-24.8 s	16.9	26
Tungsten(VI) oxytetrafluoride	F_4OW	2 s	25 s	52.1 s	84.3 s	126.7	185.4	26
Undecane	$C_{11}H_{24}$	-18.4	4.3	32.6	69.5	120.2	195.4	16
Undecanenitrile	$C_{11}H_{21}N$			78.6	120.3	177.3	259.9	5
1-Undecanethiol	$C_{11}H_{24}S$	23[e]	47[e]	77[e]	118[e]	173.6	256.8	5
Undecanoic acid	$C_{11}H_{22}O_2$	68[e]	90[e]	118[e]	156[e]	207.2	283.6	5
1-Undecanol	$C_{11}H_{24}O$	52.2	80.0	82[e]	118[e]	167.6	244.1	5
2-Undecanone	$C_{11}H_{22}O$	17[e]	37[e]	64.3	103.0	153.6	232.6	1,5

Name	Mol. form.	t/°C for 1 Pa	t/°C for 10 Pa	t/°C for 100 Pa	t/°C for 1 kPa	t/°C for 10 kPa	t/°C for 100 kPa	Ref.
6-Undecanone	$C_{11}H_{22}O$		28^{e}	57^{e}	95^{e}	148.4	226.9	1
1-Undecene	$C_{11}H_{22}$	-21.6	1.2	29.7	66.4	117.1	192.2	5
cis-2-Undecene	$C_{11}H_{22}$	-14^{e}	7^{e}	34^{e}	70.2	120.6	196^{e}	5
trans-2-Undecene	$C_{11}H_{22}$	-14^{e}	7^{e}	33^{e}	69.3	119.6	195^{e}	5
cis-4-Undecene	$C_{11}H_{22}$	-19^{e}	3^{e}	30^{e}	66.6	117.1	192^{e}	5
trans-4-Undecene	$C_{11}H_{22}$	-17^{e}	4^{e}	31^{e}	67.1	117.4	193^{e}	5
cis-5-Undecene	$C_{11}H_{22}$	-19^{e}	2^{e}	30^{e}	66.2	116.7	191^{e}	5
trans-5-Undecene	$C_{11}H_{22}$	-18^{e}	3^{e}	31^{e}	67.0	117.4	192^{e}	5
10-Undecenoic acid	$C_{11}H_{20}O_2$	35^{e}	67^{e}	105^{e}	150.0	205.4	274.5	5
1-Undecyne	$C_{11}H_{20}$	-22^{e}	0^{e}	29^{e}	67^{e}	118.5	194.5	5
2-Undecyne	$C_{11}H_{20}$	-17^{e}	6^{e}	35^{e}	74^{e}	127.4	205.4	5
Uranium	U	2052	2291	2586	2961	3454	4129	2
Vanadium	V	1828 s	2016	2250	2541	2914	3406	2
Vinyl acetate	$C_4H_6O_2$	-88^{e}	-71^{e}	-50^{e}	-22^{e}	16.2	72.2	1
Vinyl butanoate	$C_6H_{10}O_2$					53^{e}	114.5	5
4-Vinylcyclohexene	C_8H_{12}	-62^{e}	-43^{e}	-19^{e}	14.1	59.9	129^{e}	5
Vinyl formate	$C_3H_4O_2$			-58^{e}	-34^{e}	-1.6	46.2	1
Vinyl propanoate	$C_5H_8O_2$					31.2	94^{e}	5
Water[b,c]	H_2O	-60.7 s	-42.2 s	-20.3 s	7.0	45.8	99.6	36,37
Xenon[a]	Xe	-190 s	-181 s	-170 s	-155.8 s	-136.6 s	-108.4	5,32
Xenon difluoride	F_2Xe			2.9 s	31.8 s	67.9 s	114 s	1,5
o-Xylene	C_8H_{10}			-7^{e}	27^{e}	74.2	143.9	1
m-Xylene	C_8H_{10}		-35^{e}	-10^{e}	23.4	69.8	138.7	1
p-Xylene	C_8H_{10}				22.4	68.9	137.9	1
2,3-Xylenol	$C_8H_{10}O$	14.3 s	34.3 s	57.2 s	91.4	141.7	216.4	1,5
2,4-Xylenol	$C_8H_{10}O$			50.2	85.5	137.2	210.5	1,5
2,5-Xylenol	$C_8H_{10}O$	13.4 s	33.2 s	55.9 s	87.4	137.0	210.6	5
2,6-Xylenol	$C_8H_{10}O$	-3.1 s	16.7 s	39.6 s	75.3	125.9	200.6	1,5
3,4-Xylenol	$C_8H_{10}O$	19.7 s	40.2 s	63.7 s	102.1	152.3	226.4	1,5
3,5-Xylenol	$C_8H_{10}O$	16.5 s	37.2 s	61.1 s	98.0	147.9	221.3	1,5
Ytterbium	Yb	463 s	540 s	637 s	774 s	993^{d}	1192^{d}	3
Yttrium	Y	1610.1	1802.3	2047^{d}	2354^{d}	2763^{d}	3334^{d}	3
Zinc	Zn	337 s	397 s	477	579	717	912^{e}	2
Zinc chloride	Cl_2Zn	305^{d}	356^{d}	419^{d}	497^{d}	596^{d}	726^{d}	4,9,12
Zinc fluoride	F_2Zn	731 s	813 s	911^{d}	1048^{d}	1237^{d}	1503^{d}	9
Zinc iodide	I_2Zn	301 s	351 s	409 s	488^{d}	598^{d}	750^{d}	9
Zirconium	Zr	2366	2618	2924	3302	3780	4405	2
Zirconium(IV) bromide	Br_4Zr	136 s	167 s	203 s	245 s	295 s	356 s	4
Zirconium(IV) chloride	Cl_4Zr	117 s	146 s	181 s	222 s	272 s	336 s	9
Zirconium(IV) iodide	I_4Zr	187 s	220 s	259 s	305 s	361 s	430 s	4

[a] More detailed data on this compound can be found in "Vapor Pressure of Fluids at Temperatures below 300 K" in this section.
[b] See also "Recommended Data for Vapor-Pressure Calibration" in this section.
[c] See also "Vapor Pressure of Ice" and "Vapor Pressure and Other Saturation Properties of Water" in this section.
[d] Value was calculated from ideal gas thermodynamic functions (Ref. 8).
[e] Value obtained by extrapolation beyond the region where experimental values exist.
[s] Solid phase value.

VAPOR PRESSURE OF FLUIDS AT TEMPERATURES BELOW 300 K

This table gives vapor pressures of 65 important fluids in the temperature range 2 K to 300 K. The data have been taken from evaluated sources, as indicated in the references. The symbol s appearing as a superscript on a vapor pressure value indicates that the substance is solid at that temperature.

Pressures are given in kilopascals (kPa). Note that:

1 kPa = 7.50062 Torr
100 kPa = 1 bar
101.325 kPa = 1 atm

References

1. Ortiz-Vega, D. O., Hall, K. R., Holste, J. C., Arp, V. D., and Lemmon, E. W., to be published in *J. Phys. Chem. Ref. Data*, 2016. [helium]
2. Leachman, J. W., Jacobsen, R. T, Penoncello, S. G., and Lemmon, E. W., J. Phys. Chem. Ref. Data 38, 721, 2009. [parahydrogen]
3. Katti, R. S., Jacobsen, R. T, Stewart, R. B., and Jahangiri, M., *Adv. Cryo. Eng.* 31, 1189, 1986. [neon]
4. Tegeler, Ch., Span, R., and Wagner, W., *J. Phys. Chem. Ref. Data* 28, 779, 1999. [argon]
5. DIPPR Data Compilation of Pure Compounds Properties, Design Institute for Physical Properties Data, American Institute of Chemical Engineers, 1987. [boron trichloride, boron trifluoride, hydrogen bromide, bromine, chlorine fluoride, hydrogen chloride, chlorine dioxide, silicon tetrachloride, hydrogen fluoride, difluorine oxide, phosphorus trifluoride, silicon tetrafluoride, hydrogen iodide, phosphine, silane, nitric oxide, ozone, bromotrifluoromethane, tetrachloromethane, trichloromethane, hydrogen cyanide, dichloromethane, formaldehyde, chloromethane, acetylene, propadiene, buta-1,3-diene]
6. Angus, S., Armstrong, B., de Reuck, K.M., and Craven, W., *International Tables for the Fluid State: Chlorine*, Pergamon Press, Oxford, 1985. [chlorine]
7. de Reuck, K. M., *International Thermodynamic Tables of the Fluid State-11: Fluorine*, International Union of Pure and Applied Chemistry, Pergamon Press, Oxford, 1990. [fluorine]
8. Younglove, B. A., *J. Phys. Chem. Ref. Data*, Vol. 11, Suppl. 1, 1, 1982. [nitrogen trifluoride]
9. Guder, C., and Wagner, W., *J. Phys. Chem. Ref. Data* 38, 33, 2009. [sulfur hexafluoride]
10. Lemmon, E. W. and Span, R., *J. Chem. Eng. Data*, 51 785, 2006. [hydrogen sulfide, krypton, nitrous oxide, sulfur dioxide, xenon, carbon monoxide, carbon oxysulfide, R41-fluoromethane, neopentane]
11. Tillner-Roth, R., Harms-Watzenberg, F., and Baehr, H. D., Eine neue Fundamentalgleichung fuer Ammoniak, *DKV-Tagungsbericht* 20, 167, 1993. [ammonia]
12. Span, R., Lemmon, E. W., Jacobsen, R. T, Wagner, W., and Yokozeki, A., *J. Phys. Chem. Ref. Data* 29, 1361, 2000. [nitrogen]
13. Schmidt, R., and Wagner, W., *Fluid Phase Equilib.* 19, 175, 1985. [oxygen]
14. Magee, J. W., Outcalt, S. L., and Ely, J. F., *Int. J. Thermophys.* 21, 1097, 2000. [R13-chlorotrifluoromethane]
15. Marx, V., Pruss, A., and Wagner, W., Duesseldorf, VDI Verlag, Series 19 (Waermetechnik/Kaeltetechnik), No. 57, 1992. [R12-dichlorodifluoromethane]
16. Jacobsen, R. T, Penoncello, S. G., and Lemmon, E. W., *Fluid Phase Equilib.* 80, 45, 1992. [R11-trichlorofluoromethane]
17. Platzer, B., Polt, A., and Maurer, G., *Thermophysical Properties of Refrigerants*, Springer-Verlag, Berlin, 1990. [R14-tetrafluoromethane]
18. Span, R., and Wagner, W., *J. Phys. Chem. Ref. Data* 25, 1509, 1996. [carbon dioxide]
19. Kamei, A., Beyerlein, S. W., and Jacobsen, R. T, *Int. J. Thermophys.* 16, 1155, 1995. [R22- chlorodifluoromethane]
20. Penoncello, S. G., Lemmon, E. W., Jacobsen, R. T, and Shan, Z., *J. Phys. Chem. Ref. Data* 32, 1473, 2003. [R23-trifluoromethane]
21. Tillner-Roth, R., and Yokozeki, A., *J. Phys. Chem. Ref. Data* 26, 1273, 1997. [R32-difluoromethane]
22. Setzmann, U., and Wagner, W., *J. Phys. Chem. Ref. Data* 20, 1061, 1991. [methane]
23. de Reuck, K. M., and Craven, R. J. B., *International Thermodynamic Tables of the Fluid State -12: Methanol* , IUPAC, Blackwell Scientific Publications, London, 1993. [methanol]
24. Smukala, J., Span, R., and Wagner, W., *J. Phys. Chem. Ref. Data* 29, 1053, 2000. [ethylene]
25. Buecker, D., and Wagner, W., *J. Phys. Chem. Ref. Data* 35, 205, 2006. [ethane]
26. Wu, J., Zhou, Y., and Lemmon, E. W., *J. Phys. Chem. Ref. Data* 40, 1, 2011. [dimethyl ether]
27. Lemmon, E. W., Overhoff, U., McLinden, M. O., and Wagner, W., to be published in *J. Phys. Chem. Ref. Data*, 2015. [propylene]
28. Lemmon, E. W., McLinden, M. O., and Wagner, W., *J. Chem. Eng. Data* 54, 3141, 2009. [propane]
29. Buecker, D., and Wagner, W., *J. Phys. Chem. Ref. Data* 35, 929, 2006. [butane, isobutane]
30. Span, R., and Wagner, W., *Int. J. Thermophys.* 24, 41, 2003. [pentane]
31. TRC Thermodynamic Tables: Non-Hydrocarbons, Thermodynamic Research Center, Texas A&M University, College Station, Texas, 1985. [nitric oxide, radon]
32. Stevenson, R.M., and Malanowski, S., *Handbook of the Thermodynamics of Organic Compounds*, Elsevier, New York, 1987. [hydrogen cyanide, acetylene]

***T*/K**	***p*/kPa**
Acetylene	**HC≡CH**
130	0.1^s
135	0.3^s
140	0.7^s
145	1.3^s
150	2.6^s
155	4.6^s
160	7.8^s
165	12.8^s
170	20.6^s
175	32.2^s
180	49.0^s
185	72.9^s
190	106^s
195	146
200	190
205	244
210	309
215	385
220	475
225	579
230	699
235	837
240	993
245	1170
250	1370
255	1593
260	1843
265	2121
270	2429
275	2771
280	3150
285	3567
290	4028
295	4535
300	5093
Allene	**CH_2=C=CH_2**
150	0.1
155	0.2
160	0.3
165	0.6
170	1.0
175	1.7
180	2.7
185	4.1
190	6.1
195	8.9
200	12.5
205	17.4
210	23.7
215	31.6
220	41.4
225	53.5
230	68.2

T/K	p/kPa
235	85.8
240	107
245	131
250	160
255	193
260	230
265	273
270	322
275	376
280	438
285	506
290	582
295	666
300	759

Ammonia NH_3

T/K	p/kPa
160	0.1^{s}
165	0.2^{s}
170	0.3^{s}
175	0.6^{s}
180	1.2^{s}
185	2.1^{s}
190	3.5^{s}
195	5.8^{s}
200	8.651
205	12.51
210	17.74
215	24.69
220	33.79
225	45.51
230	60.41
235	79.09
240	102.2
245	130.6
250	164.9
255	206.2
260	255.3
265	313.2
270	381.1
275	459.9
280	550.9
285	655.3
290	774.4
295	909.4
300	1062

Argon Ar

T/K	p/kPa
50	0.03714^{s}
55	0.2025^{s}
60	0.8322^{s}
65	2.752^{s}
70	7.670^{s}
75	18.65^{s}
80	40.57^{s}
85	78.90
90	133.5
95	213.0
100	323.8
105	472.2
110	665.3
115	909.8
120	1213
125	1582
130	2025
135	2551
140	3168
145	3890
150	4735

Boron trichloride BCl_3

T/K	p/kPa
180	0.1
185	0.2
190	0.3
195	0.5
200	0.8
205	1.2
210	1.8
215	2.6
220	3.8
225	5.2
230	7.2
235	9.7
240	12.9
245	17.0
250	22.0
255	28.1
260	35.6
265	44.5
270	55.1
275	67.6
280	82.2
285	99.1
290	119
295	141
300	166

Boron trifluoride BF_3

T/K	p/kPa
145	7.7
150	13.4
155	22.3
160	35.2
165	53.7
170	79.1
175	113
180	157
185	214
190	285
195	372
200	479
205	608
210	762
215	944
220	1160
225	1413
230	1709
235	2056
240	2460
245	2913
250	3481
255	4123
260	4874

Bromine Br_2

T/K	p/kPa
220	0.1^{s}
225	0.2^{s}
230	0.3^{s}
235	0.4^{s}
240	0.7^{s}
245	1.1^{s}
250	1.7^{s}
255	2.6^{s}
260	3.8^{s}
265	5.5^{s}
270	7.3
275	9.5
280	12.3
285	15.6
290	19.7
295	24.6
300	30.5

Bromotrifluoromethane CF_3Br

T/K	p/kPa
135	0.1
140	0.3
145	0.5
150	0.9
155	1.5
160	2.5
165	3.9
170	5.9
175	8.8
180	12.8
185	18.1
190	25.1
195	34.1
200	45.6
205	60.0
210	77.8
215	99.5
220	126
225	157
230	194
235	237
240	287
245	344
250	410
255	485
260	570
265	665
270	771
275	889
280	1021
285	1166
290	1325
295	1501
300	1692

1,3-Butadiene C_4H_6

T/K	p/kPa
170	0.1
175	0.2
180	0.4
185	0.6
190	1.0
195	1.5
200	2.3
205	3.4
210	4.8
215	6.7
220	9.2
225	12.5
230	16.7
235	21.9
240	28.4
245	36.3
250	46.0
255	57.6
260	71.3
265	87.6
270	107
275	129
280	154
285	184
290	217
295	255
300	297

Butane C_4H_{10}

T/K	p/kPa
170	0.1
175	0.2
180	0.3
185	0.5
190	0.8
195	1.3
200	1.9
205	2.8
210	4.0
215	5.7
220	7.8
225	10.6
230	14.1
235	18.5
240	24.1
245	30.9
250	39.1
255	49.1
260	61.0
265	75.0
270	91.5
275	111
280	133
285	159
290	188
295	221
300	258

Carbon dioxide CO_2

T/K	p/kPa
135	0.1^{s}
140	0.2^{s}
145	0.4^{s}
150	0.8^{s}
155	1.7^{s}
160	3.1^{s}
165	5.7^{s}
170	9.9^{s}
175	16.8^{s}
180	27.6^{s}
185	44.0^{s}

T/K	p/kPa
190	68.4^{s}
195	104^{s}
200	155^{s}
205	227^{s}
210	327^{s}
215	465^{s}
220	599
225	734
230	893
235	1075
240	1283
245	1519
250	1785
255	2085
260	2419
265	2790
270	3203
275	3658
280	4161
285	4714
290	5318
295	5984
300	6713

Carbon monoxide CO

T/K	p/kPa
50	0.1^{s}
55	0.6^{s}
60	2.6^{s}
65	8.2^{s}
70	21.0
75	44.4
80	83.7
85	147
90	239
95	371
100	545
105	771
110	1067
115	1428
120	1877
125	2400
130	3064

Carbon oxysulfide OCS

T/K	p/kPa
140	0.1
145	0.2
150	0.4
155	0.8
160	1.3
165	2.2
170	3.4
175	5.2
180	7.8
185	11.3
190	15.9
195	22.1
200	30.0
205	40.1
210	52.7
215	68.2
220	87.2
225	110
230	137
235	169
240	207
245	250
250	301
255	358
260	423
265	497
270	580
275	673
280	777
285	892
290	1019
295	1159
300	1313

Chlorine Cl_2

T/K	p/kPa
175	1.8
180	2.8
185	4.2
190	6.1
195	8.7
200	12.3
205	16.9
210	22.9
215	30.5
220	40.1
225	51.9
230	66.4
235	84.0
240	105
245	130
250	160
255	194
260	234
265	280
270	332
275	392
280	459
285	535
290	619
295	714
300	818

Chlorine dioxide ClO_2

T/K	p/kPa
195	0.1
200	0.3
205	0.5
210	0.9
215	1.4
220	2.3
225	3.5
230	5.3
235	7.6
240	10.8
245	14.9
250	20.1
255	26.6
260	34.6
265	44.4
270	56.1
275	69.9
280	86.2
285	105
290	127
295	151
300	179

Chlorine flouride ClF

T/K	p/kPa
115	0.1
120	0.3
125	0.6
130	1.2
135	2.1
140	3.6
145	6.0
150	9.5
155	14.6
160	21.8
165	31.7
170	44.8
175	62.0
180	84.2
185	112
190	147
195	190
200	242
205	304
210	378
215	464
220	564
225	680
230	812
235	961
240	1130
245	1319
250	1529
255	1762
260	2019
265	2301
270	2608
275	2941
280	3303
285	3693
290	4111
295	4560
300	5039

Chlorodifluoromethane $CHClF_2$

T/K	p/kPa
150	0.1
155	0.3
160	0.5
165	0.8
170	1.4
175	2.3
180	3.6
185	5.5
190	8.1
195	11.8
200	16.7
205	23.1
210	31.5
215	42.1
220	55.3
225	71.7
230	91.6
235	116
240	144
245	178
250	218
255	264
260	317
265	377
270	446
275	525
280	613
285	711
290	821
295	944
300	1080

Chloromethane CH_3Cl

T/K	p/kPa
185	2.1
190	3.1
195	4.6
200	6.7
205	9.5
210	13.1
215	17.9
220	24.0
225	31.8
230	41.4
235	53.3
240	67.7
245	85.1
250	106
255	131
260	159
265	193
270	232
275	277
280	327
285	385
290	450
295	524
300	606

Chlorotrifluoromethane CF_3Cl

T/K	p/kPa
100	0.002689
105	0.008350
110	0.02310
115	0.05783
120	0.1327
125	0.2823
130	0.5620
135	1.055
140	1.880
145	3.199
150	5.226
155	8.231
160	12.54
165	18.56
170	26.74
175	37.61

T/K	p/kPa
180	51.76
185	69.84
190	92.56
195	120.7
200	155.0
205	196.3
210	245.5
215	303.6
220	371.4
225	450.0
230	540.3
235	643.4
240	760.2
245	892.0
250	1040
255	1205
260	1388
265	1591
270	1815
275	2061
280	2332
285	2628
290	2954
295	3311
300	3706

Dichlorodifluoromethane CF_2Cl_2

T/K	p/kPa
125	0.001647
130	0.004242
135	0.01009
140	0.02238
145	0.04661
150	0.09183
155	0.1721
160	0.3083
165	0.5303
170	0.8793
175	1.410
180	2.194
185	3.322
190	4.902
195	7.069
200	9.981
205	13.82
210	18.79
215	25.14
220	33.12
225	43.02
230	55.15
235	69.85
240	87.48
245	108.4
250	133.1
255	161.9
260	195.2
265	233.6
270	277.5
275	327.3
280	383.6
285	446.9
290	517.7

T/K	p/kPa
295	596.6
300	683.9

Difluoromethane CH_2F_2

T/K	p/kPa
200	0.1
205	0.2
210	0.3
215	0.4
220	0.6
225	0.9
230	1.4
235	2.0
240	2.8
245	3.8
250	5.3
255	7.1
260	9.5
265	12.4
270	16.1
275	20.7
280	26.3
285	33.0
290	41.1
295	50.8
300	62.1
140	0.1
145	0.2
150	0.3
155	0.6
160	1.0
165	1.7
170	2.8
175	4.4
180	6.8
185	10.2
190	14.8
195	21.2
200	29.5
205	40.5
210	54.5
215	72.1
220	94.1
225	121
230	154
235	193
240	240
245	295
250	360
255	434
260	521
265	620
270	732
275	860
280	1004
285	1165
290	1346
295	1547
300	1770

Dimethyl ether CH_3OCH_3

T/K	p/kPa
155	0.1
160	0.2
165	0.3
170	0.5
175	0.9
180	1.4
185	2.1
190	3.2
195	4.7
200	6.8
205	9.6
210	13.3
215	18.1
220	24.3
225	32.1
230	41.9
235	53.9
240	68.6
245	86.3
250	108
255	133
260	162
265	197
270	237
275	283
280	335
285	395
290	463
295	538
300	623

Ethane C_2H_6

T/K	p/kPa
115	0.1
120	0.4
125	0.7
130	1.3
135	2.2
140	3.8
145	6.0
150	9.7
155	15.0
160	21.5
165	31.0
170	42.9
175	59.0
180	78.7
185	104
190	135
195	172
200	217
205	271
210	334
215	407
220	492
225	590
230	700
235	826
240	967
245	1125
250	1301
255	1496
260	1712
265	1949
270	2210
275	2495
280	2806
285	3146
290	3515
295	3917
300	4355

Ethylene $CH_2{=}CH_2$

T/K	p/kPa
110	0.33
115	0.70
120	1.38
125	2.7
130	4.5
135	7.7
140	11.9
145	18.3
150	27.5
155	39.9
160	56.4
165	77.9
170	105
175	140
180	182
185	234
190	296
195	369
200	456
205	557
210	673
215	806
220	958
225	1128
230	1321
235	1535
240	1774
245	2039
250	2331
255	2652
260	3005
265	3391
270	3813
275	4275

Fluorine F_2

T/K	p/kPa
55	0.382
60	1.49
65	4.61
70	12.0
75	27.1
80	54.7
85	101
90	173
95	279
100	428
105	628
110	889
115	1220
120	1630
125	2140
130	2750
135	3470

T/K	p/kPa
140	4340

Fluorine monoxide F_2O

T/K	p/kPa
75	0.1
80	0.2
85	0.5
90	1.2
95	2.6
100	5.3
105	10.1
110	18.0
115	30.5
120	49.3
125	76.7
130	115
135	168
140	237
145	328
150	444
155	588
160	766
165	981
170	1238
175	1541
180	1895
185	2303
190	2771
195	3302
200	3899
205	4567
210	5308

Fluoromethane CH_3F

T/K	p/kPa
135	0.6
140	1.2
145	2.1
150	3.6
155	5.9
160	9.3
165	14.1
170	20.9
175	29.9
180	42.0
185	57.6
190	77.4
195	102
200	133
205	171
210	216
215	270
220	333
225	408
230	495
235	595
240	711
245	843
250	993
255	1163
260	1355
265	1571
270	1813
275	2084
280	2387
285	2724
290	3099
295	3516
300	3978

Formaldehyde HCHO

T/K	p/kPa
185	1.3
190	2.0
195	3.0
200	4.4
205	6.4
210	9.1
215	12.7
220	17.4
225	23.4
230	31.0
235	40.6
240	52.5
245	67.0
250	84.6
255	106
260	131
265	161
270	196
275	236
280	283
285	337
290	399
295	470
300	549

Helium He

T/K	p/kPa
2.2	5.332
2.3	6.725
2.4	8.351
2.5	10.23
2.6	12.38
2.7	14.81
2.8	17.56
2.9	20.64
3.0	24.06
3.1	27.85
3.2	32.02
3.3	36.60
3.4	41.60
3.5	47.04
3.6	52.94
3.7	59.31
3.8	66.19
3.9	73.58
4.0	81.51
4.1	90.00
4.2	99.08
4.3	108.8
4.4	119.1
4.5	130.1
4.6	141.7
4.7	154.1
4.8	167.3
4.9	181.3
5.0	196.2
5.1	212.1

Hydrogen H_2

T/K	p/kPa
14.0	7.884
14.5	10.38
15.0	13.43
15.5	17.13
16.0	21.55
16.5	26.77
17.0	32.89
17.5	39.98
18.0	48.15
18.5	57.48
19.0	68.07
19.5	80.02
20.0	93.41
20.5	108.4
21.0	125.0
21.5	143.3
22.0	163.5
22.5	185.6
23.0	209.8
23.5	236.2
24.0	264.8
24.5	295.7
25.0	329.2
25.5	365.2
26.0	403.8
26.5	445.3
27.0	489.6
27.5	537.0
28.0	587.5
28.5	641.2
29.0	698.3
29.5	758.9
30.0	823.2
30.5	891.2
31.0	963.3
31.5	1040
32.0	1120
32.5	1206

Hydrogen bromide HBr

T/K	p/kPa
135	0.1^s
140	0.3^s
145	0.6^s
150	1.1
155	1.9
160	3.3
165	5.4
170	8.7^s
175	13.4^s
180	20.1^s
185	29.5^s
190	37.9
195	51.8
200	69.5
205	91.8
210	119
215	153
220	194
225	242
230	299
235	366
240	443
245	532
250	633
255	748
260	878
265	1023
270	1185
275	1364
280	1562
285	1780
290	2018
295	2278
300	2561

Hydrogen chloride HCl

T/K	p/kPa
120	0.1
125	0.3
130	0.5
135	1.0
140	1.9
145	3.4
150	5.8
155	9.5
160	14.7
165	22.0
170	31.9
175	45.1
180	62.5
185	84.7
190	113
195	148
200	190
205	242
210	304
215	377
220	463
225	563
230	678
235	811
240	961
245	1132
250	1325
255	1542
260	1784
265	2054
270	2354
275	2686
280	3053
285	3457
290	3901
295	4388
300	4921

Hydrogen cyanide HCN

T/K	p/kPa
200	0.1^s
205	0.2^s
210	0.4^s
215	0.6^s
220	1.0^s
225	1.5^s

T/K	p/kPa
230	2.2^s
235	3.3^s
240	4.7^s
245	6.8^s
250	9.7^s
255	13.6^s
260	18.8
265	24.1
270	30.5
275	38.3
280	47.7
285	58.8
290	72.1
295	87.6
300	105.9

Hydrogen fluoride HF

T/K	p/kPa
190	0.3
195	0.5
200	0.8
205	1.2
210	1.7
215	2.3
220	3.2
225	4.4
230	5.9
235	7.9
240	10.3
245	13.4
250	17.2
255	21.8
260	27.4
265	34.2
270	42.2
275	51.8
280	63.1
285	76.3
290	91.7
295	110
300	130

Hydrogen iodide HI

T/K	p/kPa
155	0.1
160	0.2
165	0.4
170	0.8
175	1.3^s
180	2.2^s
185	3.4^s
190	5.3^s
195	8.0^s
200	11.7^s
205	16.8^s
210	23.6^s
215	32.5^s
220	44.0^s
225	56.2
230	71.4
235	89.7
240	112
245	137
250	168
255	203
260	244
265	290
270	343
275	404
280	472
285	548
290	633
295	727
300	831

Hydrogen sulfide H_2S

T/K	p/kPa
135	0.1^s
140	0.2^s
145	0.3^s
150	0.6^s
155	1.1^s
160	1.9^s
165	3.2
170	5.2
175	8.3
180	12.7
185	18.9^s
190	27.11
195	37.27
200	50.34
205	66.87
210	87.47
215	112.8
220	143.7
225	180.7
230	224.9
235	276.9
240	337.7
245	408.2
250	489.3
255	582.1
260	687.5
265	806.5
270	940.2
275	1090
280	1256
285	1440
290	1643
295	1866
300	2110

Isobutane C_4H_{10}

T/K	p/kPa
160	0.1
165	0.1
170	0.3
175	0.4
180	0.7
185	1.1
190	1.7
195	2.5
200	3.7
205	5.3
210	7.4
215	10.2
220	13.8
225	18.3
230	24.0
235	31.1
240	39.8
245	50.3
250	62.9
255	77.8
260	95.4
265	116
270	140
275	167
280	198
285	234
290	274
295	319
300	370

Krypton Kr

T/K	p/kPa
60	0.001574^s
65	0.008754^s
70	0.03811^s
75	0.1364^s
80	0.4160^s
85	1.113^s
90	2.670^s
95	5.840^s
100	11.81^s
105	22.35^s
110	39.89^s
115	67.71^s
120	103.4
125	150.1
130	211.2
135	289.6
140	387.8
145	508.7
150	655.1
155	830.1
160	1036
165	1277
170	1556
175	1875
180	2239
185	2651
190	3114
195	3634
200	4216
205	4869

Methane CH_4

T/K	p/kPa
65	0.1
70	0.3
75	0.8
80	2.1
85	4.9
90	10.6
95	20.0
100	34.5
105	57.0
110	88.4
115	133
120	192
125	269
130	368
135	491
140	642
145	824
150	1041
155	1297
160	1594
165	1937
170	2331
175	2779
180	3288
185	3865
190	4520

Methanol CH_3OH

T/K	p/kPa
230	0.1
235	0.2
240	0.4
245	0.5
250	0.8
255	1.2
260	1.7
265	2.4
270	3.3
275	4.5
280	6.2
285	8.3
290	11.0
295	14.4
300	18.7

Neon Ne

T/K	p/kPa
25	50.92
26	71.61
27	98.17
28	131.6
29	172.9
30	223.1
31	283.2
32	354.5
33	437.8
34	534.4
35	645.4
36	772.0
37	915.4
38	1077
39	1258
40	1460
41	1685
42	1935
43	2212
44	2517

Neopentane $C(CH_3)_4$

T/K	p/kPa
185	0.1^s
190	0.2^s
195	0.4^s
200	0.7^s
205	1.1^s
210	1.6^s
215	2.4^s
220	3.6^s

Fluids

T/K	p/kPa
225	5.2^{s}
230	7.3^{s}
235	10.2^{s}
240	13.9^{s}
245	18.7^{s}
250	24.8^{s}
255	32.4^{s}
260	41.6
265	51.4
270	63.0
275	76.6
280	92.3
285	111
290	131
295	155
300	182

Nitric oxide NO

T/K	p/kPa
85	0.1^{s}
90	0.4^{s}
95	1.3^{s}
100	3.8^{s}
105	10.0^{s}
110	23.5
115	46.8
120	86.5
125	151
130	248
135	391
140	592
145	867
150	1231
155	1703
160	2302
165	3050
170	3971
175	5089
180	6433

Nitrogen N_2

T/K	p/kPa
50	0.4005^{s}
55	1.800^{s}
60	6.298^{s}
65	17.40
70	38.54
75	76.04
80	136.9
85	228.9
90	360.5
95	540.5
100	778.3
105	1083
110	1466
115	1937
120	2511
125	3207

Nitrogen trifluoride NF_3

T/K	p/kPa
85	0.0530
90	0.156
95	0.406
100	0.946
105	2.02
110	3.98
115	7.33
120	12.8
125	21.1
130	33.5
135	51.0
140	75.2
145	108
150	150
155	204
160	272
165	356
170	458
175	580
180	725
185	895
190	1090
195	1320
200	1580
205	1870
210	2210
215	2580
220	3000
225	3470
230	3990

Nitrous oxide N_2O

T/K	p/kPa
115	0.1
120	0.1
125	0.3
130	0.7
135	1.3
140	2.5
145	4.3
150	7.1
155	11.4
160	17.6
165	26.4
170	38.5
175	54.7
180	75.9
185	103.3
190	137.8
195	180.9
200	233.7
205	297.7
210	374.3
215	465.1
220	571.6
225	695.4
230	838.3
235	1002
240	1188
245	1398
250	1634
255	1898
260	2192
265	2518
270	2877
275	3273
280	3707
285	4182
290	4701
295	5268
300	5887

Oxygen O_2

T/K	p/kPa
50	0.03096^{s}
55	0.1786
60	0.7258
65	2.335
70	6.262
75	14.55
80	30.12
85	56.83
90	99.35
95	163.1
100	254.0
105	378.5
110	543.4
115	755.6
120	1022
125	1351
130	1749
135	2225
140	2788
145	3448
150	4219

Ozone O_3

T/K	p/kPa
100	0.1
105	0.2
110	0.4
115	1.0
120	2.0
125	3.8
130	6.8
135	11.5
140	18.7
145	29.1
150	43.7
155	63.6
160	89.9
165	124
170	168
175	222
180	289
185	367
190	468
195	584
200	721
205	881
210	1068
215	1285
220	1536
225	1824
230	2155
235	2534
240	2968
245	3464
250	4031
255	4678
260	5417

Pentane C_5H_{12}

T/K	p/kPa
220	1.0
225	1.5
230	2.1
235	3.0
240	4.2
245	5.7
250	7.6
255	10.0
260	13.0
265	16.6
270	21.1
275	26.6
280	33.1
285	40.8
290	50.0
295	60.7
300	73.2

Phosphine PH_3

T/K	p/kPa
110	0.1
115	0.2
120	0.4
125	0.7
130	1.3
135	2.3
140	3.9
145	6.2
150	9.6
155	14.5
160	21.1
165	30.0
170	41.6
175	56.6
180	75.6
185	99.2
190	128
195	163
200	205
205	254
210	312
215	379
220	456
225	544
230	644
235	756
240	881
245	1019
250	1172
255	1341
260	1525
265	1725
270	1942
275	2176
280	2428
285	2699
290	2987
295	3295
300	3621

Phosphorus trifluoride PF_3

T/K	p/kPa
105	0.1
110	0.2
115	0.5
120	1.0
125	1.9
130	3.5
135	5.9
140	9.5
145	14.9
150	22.5
155	33.1
160	47.3
165	66.0
170	90.1
175	121
180	159
185	206
190	262
195	330
200	410
205	503
210	611
215	736
220	877
225	1037
230	1217
235	1418
240	1640
245	1885
250	2154
255	2448
260	2767
265	3112

Propane C_3H_8

T/K	p/kPa
160	0.8
165	1.4
170	2.2
175	3.3
180	5.0
185	7.3
190	10.5
195	15.0
200	20.1
205	27.0
210	36.0
215	47.0
220	60.0
225	77.0
230	97.0
235	120
240	148
245	180
250	218
255	261
260	311
265	367
270	431
275	502
280	582
285	671
290	769
295	878
300	998

Propene $CH_3CH_2{=}CH_2$

T/K	p/kPa
140	0.1
145	0.2
150	0.4
155	0.7
160	1.2
165	2.0
170	3.1
175	4.7
180	7.0
185	10.1
190	14.2
195	19.7
200	26.9
205	35.9
210	47.3
215	61.3
220	78.5
225	99.2
230	124
235	153
240	188
245	228
250	274
255	327
260	387
265	456
270	533
275	619
280	715
285	822
290	940
295	1069
300	1212

Radon Rn

T/K	p/kPa
130	0.1
135	0.3
140	0.5
145	0.9
150	1.5
155	2.4
160	3.8
165	5.8
170	8.6
175	12.5
180	17.7
185	24.5
190	33.2
195	44.4
200	58.2
205	75.3
210	96.0
215	121
220	151
225	185

Silane SiH_3

T/K	p/kPa
95	0.1
100	0.2
105	0.4
110	1.0
115	1.9
120	3.5
125	6.1
130	10.0
135	15.8
140	24.1
145	35.3
150	50.3
155	69.8
160	94.6
165	126
170	164
175	210
180	265
185	331
190	408
195	498
200	602
205	722
210	859
215	1017
220	1196
225	1398
230	1628
235	1888
240	2180
245	2509
250	2880
255	3296
260	3763
265	4288

Sulfur dioxide SO_2

T/K	p/kPa
170	0.1
175	0.2
180	0.3
185	0.5
190	0.8
195	1.3
200	2.026
205	3.068
210	4.539
215	6.574
220	9.334
225	13.01
230	17.84
235	24.06
240	31.99
245	41.95
250	54.31
255	69.48
260	87.91
265	110.1
270	136.5
275	167.7
280	204.3
285	246.9
290	296.1
295	352.7
300	417.2

Sulfur hexafluoride SF_6

T/K	p/kPa
115	0.001259^s
120	0.003635^s
125	0.009614^s
130	0.02353^s
135	0.05379^s
140	0.1157^s
145	0.2355^s
150	0.4566^s
155	0.8471^s
160	1.510^s
165	2.596^s
170	4.317^s
175	6.968^s
180	10.94^s
185	16.75^s
190	25.07^s
195	36.71^s
200	52.71^s
205	74.31^s
210	103.0
215	140.6
220	189.0
225	245.9
230	301.3
235	365.8
240	440.1
245	525.2
250	622.0
255	731.5
260	854.6
265	992.6
270	1146
275	1317
280	1506
285	1714
290	1942
295	2193
300	2468

Tetrachloromethane CCl_4

T/K	p/kPa
255	1.5
260	2.1
265	2.8
270	3.7
275	4.9
280	6.4
285	8.2
290	10.5
295	13.2
300	16.5

Tetrachlorosilane $SiCl_4$

T/K	p/kPa
210	0.1
215	0.2
220	0.3
225	0.5

T/K	p/kPa
230	0.7
235	1.0
240	1.5
245	2.0
250	2.8
255	3.8
260	5.0
265	6.6
270	8.6
275	11.1
280	14.2
285	17.9
290	22.3
295	27.7
300	34.0

Tetrafluoromethane CF_4

T/K	p/kPa
90	0.1
95	0.3
100	0.8
105	1.7
110	3.4
115	6.5
120	11.5
125	19.3
130	30.8
135	47.4
140	70.2
145	101
150	141
155	191
160	254
165	332
170	425
175	537
180	669
185	824
190	1005
195	1216
200	1460
205	1743
210	2073
215	2457
220	2907
225	3438

Tetrafluorosilane SiF_4

T/K	p/kPa
125	0.1
130	0.2[s]
135	0.4[s]
140	0.9[s]
145	1.9[s]
150	3.8[s]
155	7.5[s]
160	14.0[s]
165	25.2[s]
170	43.8[s]
175	74.2[s]
180	122[s]
185	197[s]
190	280
195	376
200	488
205	618
210	766
215	932
220	1117
225	1324
230	1555
235	1816
240	2111
245	2449
250	2841
255	3301

Trichlorofluoromethane CCl_3F

T/K	p/kPa
190	0.2
195	0.3
200	0.4
205	0.6
210	1.0
215	1.4
220	2.0
225	2.9
230	4.1
235	5.6
240	7.6
245	10.1
250	13.3
255	17.2
260	22.1
265	28.0
270	35.1
275	43.7
280	53.8
285	65.7
290	79.6
295	95.6
300	114.1

Trichloromethane $CHCl_3$

T/K	p/kPa
215	0.1
220	0.2
225	0.3
230	0.4
235	0.7
240	1.0
245	1.4
250	2.0
255	2.7
260	3.7
265	5.0
270	6.6
275	8.7
280	11.3
285	14.4
290	18.3
295	22.9
300	28.5

Trifluoromethane CHF_3

T/K	p/kPa
120	0.1
125	0.2
130	0.4
135	0.7
140	1.4
145	2.5
150	4.3
155	7.1
160	11.1
165	17.0
170	25.3
175	36.5
180	51.4
185	70.9
190	95.8
195	127
200	166
205	214
210	271
215	340
220	421
225	516
230	626
235	754
240	900
245	1067
250	1257
255	1472
260	1713
265	1984
270	2287
275	2624
280	3000
285	3418
290	3881
295	4393

Xenon Xe

T/K	p/kPa
100	0.08037[s]
105	0.1686[s]
110	0.3433[s]
115	0.6778[s]
120	1.298[s]
125	2.411[s]
130	4.343[s]
135	7.583[s]
140	12.83[s]
145	21.02[s]
150	33.32[s]
155	50.96[s]
160	74.58[s]
165	101.0
170	133.4
175	173.3
180	221.5
185	279.3
190	347.7
195	427.9
200	520.9
205	628.0
210	750.2
215	888.9
220	1045
225	1220
230	1416
235	1632
240	1871
245	2134
250	2423
255	2738
260	3082
265	3456
270	3862
275	4303
280	4782
285	5303

[s] Solid

VAPOR PRESSURE OF SATURATED SALT SOLUTIONS

This table gives the vapor pressure of water above saturated solutions of some common salts at ambient temperatures. Data on pure water are given on the last line for comparison.

The references provide additional information on water activity, osmotic coefficient, and enthalpy of vaporization.

References

1. Apelblat, A., *J. Chem. Thermodynamics*, 24, 619, 1992.
2. Apelblat, A., *J. Chem. Thermodynamics*, 25, 63, 1993.
3. Apelblat, A., *J. Chem. Thermodynamics*, 25, 1513, 1993.
4. Apelblat, A. and Korin, E., *J. Chem. Thermodynamics*, 30, 59, 1998.

Vapor Pressure of Saturated Solution at the Indicated Temperature

Name	Formula	Vap. pres. (10 °C)/kPa	Vap. pres. (15 °C)/kPa	Vap. pres. (20 °C)/kPa	Vap. pres. (25 °C)/kPa	Vap. pres. (30 °C)/kPa	Vap. pres. (35 °C)/kPa	Vap. pres. (40 °C)/kPa	Ref.
Ammonium chloride	NH_4Cl	0.971	1.328	1.836	2.481				2
Ammonium nitrate	NH_4NO_3	0.853	1.152	1.524	1.972				2
Ammonium sulfate	$(NH_4)_2SO_4$	0.901	1.319	1.871	2.573	3.439	4.474		3
Barium chloride	$BaCl_2$	0.971	1.443	2.073	2.887	3.903	5.133	6.576	1
Calcium nitrate	$Ca(NO_3)_2$	0.701	1.015	1.381	1.772	2.154	2.487		1
Copper(II) sulfate	$CuSO_4$	1.113	1.574	2.189	2.996	4.037	5.363		3
Iron(II) sulfate	$FeSO_4$	0.978	1.516	2.208	3.035	3.950	4.884		3
Lithium chloride	$LiCl$	0.128	0.193	0.279	0.384				2
Magnesium nitrate	$Mg(NO_3)_2$	0.726	0.999	1.339	1.749	2.231	2.782	3.397	1
Manganese(II) chloride	$MnCl_2$	0.697	1.064	1.515	2.020	2.535	3.002		3
Potassium bromide	KBr	0.953	1.338	1.853	2.533	3.419	4.563		3
Potassium carbonate	K_2CO_3	0.541	0.802	1.134	1.536	1.997	2.499	3.016	1
Potassium iodate	KIO_3	1.100	1.564	2.177	2.970	3.979	5.236	6.778	4
Rubidium chloride	$RbCl$	0.862	1.215	1.684	2.298	3.088	4.089	5.343	4
Sodium bromide	$NaBr$	0.722	1.004	1.376	1.858	2.475	3.255	4.229	4
Sodium chloride	$NaCl$	0.921	1.285	1.768	2.401	3.218	4.262	5.581	4
Sodium nitrate	$NaNO_3$	0.884	1.244	1.719	2.335	3.121	4.109	5.333	4
Sodium nitrite	$NaNO_2$	0.703	0.994	1.381	1.888	2.540	3.368	4.403	4
Zinc sulfate	$ZnSO_4$	0.945	1.401	1.986	2.698	3.523	4.431	5.382	1
Water	H_2O	1.228	1.706	2.339	3.169	4.246	5.627	7.381	

RECOMMENDED DATA FOR VAPOR-PRESSURE CALIBRATION

Eric W. Lemmon

These precise vapor pressure values are recommended as secondary standards. Values are given in kPa (1 kPa = 0.0098692 atm = 7.5006 Torr). The four tables below give vapor pressure values in the temperature range 180 K to 500 K.

References for water and CO_2 are given in the "Thermophysical Properties of Fluids" table.

References

1. Huber, M. L., Laesecke, A., and Friend, D. G., Correlation for the Vapor Pressure of Mercury, *Ind. Eng. Chem. Res.* 45, 7351, 2006.
2. Ruzicka, K., Fulem, M., and Ruzicka, V., Recommended Vapor Pressure of Solid Naphthalene, *J. Chem. Eng. Data*, 50, 1956, 2005.
3. Span, R., and Wagner, W., Equations of State for Technical Applications. II. Results for Nonpolar Fluids, *Int. J. Thermophys.* 24, 41, 2003. [pentane]
4. Thol, M., Lemmon, E.W., and Span, R., Equation of State for Benzene for Temperatures from the Melting Line up to 725 K and Pressures up to 500 MPa, *High Temp. – High Press.* 41, 81, 2012.

Vapor Pressure in kPa for Temperature Range 180 K to 280 K

Compound	State	180 K	190 K	200 K	210 K	220 K	230 K	240 K	250 K	260 K	270 K	280 K
Naphthalene	sol	3.04×10^{-11}	4.33×10^{-10}	4.69×10^{-9}	4.02×10^{-8}	2.82×10^{-7}	1.66×10^{-6}	8.37×10^{-6}	3.69×10^{-5}	1.446×10^{-4}	5.09×10^{-4}	0.00163
Ice	sol			0.00016	0.0007	0.0027	0.0089	0.0273	0.0760	0.1958	0.4701	
Water	liq											0.992
Benzene	liq											5.139
Carbon dioxide	sol	27.56	68.34	155.03	327.09							
Pentane	liq								7.59	12.96	21.14	33.10

Vapor Pressure in kPa for Temperature Range 290 K to 390 K

Compound	State	290 K	300 K	310 K	320 K	330 K	340 K	350 K	360 K	370 K	380 K	390 K
Naphthalene	sol	0.004798	0.01308	0.03328	0.07956	0.1797	0.3854	0.7884				
Water	liq	1.920	3.537	6.231	10.546	17.213	27.188	41.682	62.194	90.535	128.85	179.64
Benzene	liq	8.602	13.82	21.40	32.07	46.68	66.19	91.67	124.3	165.3	216.1	278.0
Pentane	liq	50.00	73.17	104.07	144.3	195.7	260.2	339.6	436.1	551.8	689.2	850.5

Vapor Pressure in kPa for Temperature Range 400 K to 500 K

Compound	State	400 K	410 K	420 K	430 K	440 K	450 K	460 K	470 K	480 K	490 K	500 K
Mercury	liq	0.139	0.216	0.330	0.494	0.726	1.048	1.489	2.082	2.870	3.905	5.245
Water	liq	245.77	330.45	437.30	570.26	733.67	932.20	1170.9	1455.1	1790.5	2183.1	2639.2
Benzene	liq	352.6	441.3	545.9	668.0	809.2	971.5	1156	1366	1603	1868	2165
Pentane	liq	1038	1256	1505	1790	2115	2484	2903				

Vapor Pressure in kPa for Temperature Range 510 K to 600 K

Compound	State	510 K	520 K	530 K	540 K	550 K	560 K	570 K	580 K	590 K	600 K
Mercury	liq	6.962	9.139	11.871	15.267	19.452	24.564	30.76	38.22	47.12	57.69
Benzene	liq	2495	2863	3270	3721	4223	4784				
Water	liq	3165.5	3769.0	4456.9	5236.9	6117.2	7106.2	8213.2	9448.0	10821	12345

ENTHALPY OF VAPORIZATION

The molar enthalpy (heat) of vaporization $\Delta_{vap}H$, which is defined as the enthalpy change in the conversion of one mole of liquid to gas at constant temperature, is tabulated here for about 950 inorganic and organic compounds. Values are given, when available, both at the normal boiling point t_b, referred to a pressure of 101.325 kPa (760 mmHg), and at 25 °C.

The values in this table were measured either by calorimetric techniques or by application of the Claperyon equation to the variation of vapor pressure with temperature. See Ref. 1 for a discussion of the accuracy of different experimental techniques and methods of estimating enthalpy of vaporization at other temperatures. Several of the references present empirical techniques for correlating enthalpy of vaporization with molecular structure.

Compounds are listed by systematic name.

References

1. Majer, V., and Svoboda, V., *Enthalpies of Vaporization of Organic Compounds*, Blackwell Scientific Publications, Oxford, 1985.
2. Chase, M. W., Davies, C. A., Downey, J. R., Frurip, D. J., McDonald, R. A., and Syverud, A. N., *JANAF Thermochemical Tables, Third Edition, J. Phys. Chem. Ref. Data*, 14, Suppl. 1, 1985.
3. *Landolt-Börnstein, Numerical Data and Functional Relationships in Science and Technology, Sixth Edition*, II/4, *Caloric Quantities of State*, Springer-Verlag, Heidelberg, 1961.
4. Daubert, T. E., Danner, R. P., Sibul, H. M., and Stebbins, C. C., *Physical and Thermodynamic Properties of Pure Compounds: Data Compilation*, extant 1994 (core with 4 supplements), Taylor & Francis, Bristol, PA.
5. Ruzicka, K., and Majer, V., Simultaneous Treatment of Vapor Pressures and Related Thermal Data Between the Triple and Normal Boiling Temperatures for *n*-Alkanes C5–C20, *J. Phys. Chem. Ref. Data*, 23, 1, 1994.
6. Verevkin, S. P., Thermochemistry of Amines: Experimental Standard Molar Enthalpies of Formation of Some Aliphatic and Aromatic Amines, *J. Chem. Thermodynamics*, 29, 891, 1997.
7. Cady, G. H., and Hargreaves, G. B., The Vapor Pressure of Some Heavy Transition Metal Hexafluorides, *J. Chem. Soc.*, 1563, 1578, 1961.
8. Steele, W. V., Chirico, R. D., Knipmeyer, S. E., and Nguyen, A., *J. Chem. Eng. Data* 41, 1255, 1996.
9. Nichols, G., et al., *J. Chem. Eng. Data* 51, 475, 2006.
10. Dias, A. M. A., et al., *J. Chem. Eng. Data* 50, 1328, 2005.
11. Umnahanant, p., et al., *J. Chem. Eng. Data* 51, 2246, 2006.
12. Raganov, G. N., Pisarev, P. N., and Emel'yanenko, V. N., *J. Chem. Eng. Data* 50, 1114, 2005.
13. Verevkin, S. P., et al., *J. Chem. Eng. Data* 51, 1896, 2006.
14. Verevkin, S. P., *J. Chem. Thermodynamics* 38, 1111, 2006.
15. Lide, D. R., *Handbook of Organic Solvents*, CRC Press, Boca Raton,1995; Riddick, J. A., Bunger, W. B., and Sakano, T. K., *Organic Solvents*, Fourth Edition, John Wiley & Sons, New York, 1986.
16. Steele, W. V., Chirico, R. D., Nguyen, A., and Knipmeyer, S. E., *J. Chem. Thermodynamics* 27, 322, 1995.
17. Ribeiro da Silva, M., et al., *J. Chem. Thermodynamics* 27, 565, 1995.
18. Kabo, G. J., et al., *J. Chem. Thermodynamics* 27, 953, 1995.
19. Duarte-Garza, H. A., et al., *J. Chem. Eng. Data* 42, 497, 1997.
20. Steele, W. V., and Chirico, R. D., *J. Phys. Chem. Ref. Data* 22, 377, 1993.
21. Linstrom, P. J., and Mallard, W. G., Editors, *NIST Chemistry WebBook*, NIST Standard Reference Database No. 69, June 2005, National Institute of Standards and Technology, Gaithersburg, MD, <http://webbook.nist.gov>.
22. Weber, L. A. and Defibaugh, D. R., *J. Chem. Eng. Data* 41, 1477, 1996.

Enthalpy of Vaporization at the Boiling Point and at 25 °C

Name	Mol. form.	t_b/°C	$\Delta_{vap}H(t_b)$/ kJ mol^{-1}	$\Delta_{vap}H$(25°C)/ kJ mol^{-1}	Ref.
Acetaldehyde	C_2H_4O	20.8	25.76	25.47	1
Acetic acid	$C_2H_4O_2$	117.9	23.70	23.36	1
Acetic anhydride	$C_4H_6O_3$	139.5	38.2		
Acetone	C_3H_6O	56.08	29.10	30.99	1
Acetonitrile	C_2H_3N	81.6	29.75	32.94	1
Acetophenone	C_8H_8O	202.1	43.98	55.40	8
Acrolein	C_3H_4O	52.3	28.3		
Acrylonitrile	C_3H_3N	77.2	32.6		
Allyl acetate	$C_5H_8O_2$	104	36.3		
Allyl alcohol	C_3H_6O	96.9	40.0		
Aluminum	Al	2519	294		
Aluminum borohydride	AlB_3H_{12}	44.5	30		
Aluminum bromide	$AlBr_3$	255	23.5		
Aluminum iodide	AlI_3	382	32.2		
2-Amino-2-methyl-1-propanol	$C_4H_{11}NO$	163.8	50.6		
Ammonia	H_3N	-33.33	23.33	19.86	
Aniline	C_6H_7N	184.1	42.44	55.83	1
Anisole	C_7H_8O	153.6	38.97	46.90	1
Antimony(III) bromide	Br_3Sb	288	59		
Antimony(III) chloride	Cl_3Sb	220.3	45.19		
Antimony(III) iodide	I_3Sb	400	68.6		
Argon	Ar	-185.848	6.43		
Arsenic(III) bromide	$AsBr_3$	221	41.8		
Arsenic(III) chloride	$AsCl_3$	130	35.01		
Arsenic(III) fluoride	AsF_3	57.13	29.7		
Arsenic(V) fluoride	AsF_5	-52.8	20.8		

Fluids

Name	Mol. form.	t_b/°C	$\Delta_{vap}H(t_b)$/ kJ mol^{-1}	$\Delta_{vap}H$(25°C)/ kJ mol^{-1}	Ref.
Arsenic(III) iodide	AsI_3	424	59.3		
Arsine	AsH_3	-62.5	16.69		
Azobutane	$C_8H_{18}N_2$			49.31	1
Azopropane	$C_6H_{14}N_2$	114		39.88	1
Barium	Ba	≈1845	140		
Benzaldehyde	C_7H_6O	178.7	42.5		
Benzene	C_6H_6	80.08	30.72	33.83	1
Benzenethiol	C_6H_6S	169.1	39.93	47.56	1
Benzonitrile	C_7H_5N	191	45.9		
Benzyl acetate	$C_9H_{10}O_2$	215	49.4		
Benzyl alcohol	C_7H_8O	205.3	50.48		1
Benzylamine	C_7H_9N	185		60.16	1
N-Benzylaniline	$C_{13}H_{13}N$	306.5		79.6	6
Benzyl benzoate	$C_{14}H_{12}O_2$	321.3	53.6		
Beryllium chloride	$BeCl_2$	482	105		
Beryllium iodide	BeI_2	590	70.5		
Bis(2-chloroethyl) ether	$C_4H_8Cl_2O$	178	45.2		
Bis(ethoxymethyl) ether	$C_6H_{14}O_3$	136	36.17	44.69	1
Bismuth	Bi	1564	151		
Bismuth tribromide	$BiBr_3$	462	75.4		
Bismuth trichloride	$BiCl_3$	441	72.61		
Boron	B	4000	480		
Boron tribromide	BBr_3	91.3	30.5		
Boron trichloride	BCl_3	12.5	23.77	23.1	
Boron trifluoride	BF_3	-99.9	19.33		
Boron triiodide	BI_3	209.5	40.5		
Bromine	Br_2	58.8	29.96	30.91	
Bromine fluoride	BrF	≈20 dec	25.1		
Bromine pentafluoride	BrF_5	41.3	30.6		
Bromine trifluoride	BrF_3	125.8	47.57		
Bromobenzene	C_6H_5Br	155.9		44.54	1
1-Bromobutane	C_4H_9Br	101.4	32.51	36.64	1
2-Bromobutane, (±)-	C_4H_9Br	91	30.77	34.41	1
Bromochloromethane	CH_2BrCl	67.9	30.0		
2-Bromo-2-chloro-1,1,1-trifluoroethane	$C_2HBrClF_3$	50	28.08	29.61	1
Bromoethane	C_2H_5Br	38.2	27.04	28.03	1
Bromoethene	C_2H_3Br	16	23.4		
1-Bromoheptane	$C_7H_{15}Br$	179		50.60	1
1-Bromohexane	$C_6H_{13}Br$	156		45.89	1
Bromomethane	CH_3Br	3.4	23.91	22.81	1
1-Bromo-2-methylpropane	C_4H_9Br	91.3	31.33	34.82	1
2-Bromo-2-methylpropane	C_4H_9Br	73.3	29.23	31.81	1
1-Bromonaphthalene	$C_{10}H_7Br$	280	39.3		
1-Bromooctane	$C_8H_{17}Br$	199		55.77	1
1-Bromopentane	$C_5H_{11}Br$	126	35.01	41.28	1
1-Bromopropane	C_3H_7Br	70.8	29.84	32.01	1
2-Bromopropane	C_3H_7Br	59.34	28.33	30.17	1
3-Bromopropene	C_3H_5Br	70.1	30.24	32.73	1
Bromosilane	BrH_3Si	1.9	24.4		
1,2-Butadiene	C_4H_6	11.0	24.02	23.21	1
1,3-Butadiene	C_4H_6	-4.6	22.47	20.86	1
Butanal	C_4H_8O	74.8	31.5		
Butane	C_4H_{10}	-0.5	22.44	21.02	1
1,2-Butanediol, (±)-	$C_4H_{10}O_2$	196.42	52.84	71.55	8
1,3-Butanediol	$C_4H_{10}O_2$	208.2	54.31	74.46	8
1,4-Butanediol	$C_4H_{10}O_2$	229.5		77.1	11
1,4-Butanedithiol	$C_4H_{10}S_2$	195.5		55.10	
Butanenitrile	C_4H_7N	117.6	33.68	39.33	1
1-Butanethiol	$C_4H_{10}S$	98.4	32.23	36.63	1
2-Butanethiol	$C_4H_{10}S$	85.0	30.59	33.99	1
Butanoic acid	$C_4H_8O_2$	163.7		40.45	1
Butanoic anhydride	$C_8H_{14}O_3$	195	50.0		

Name	Mol. form.	t_b/°C	$\Delta_{vap}H(t_b)$/ kJ mol⁻¹	$\Delta_{vap}H$(25°C)/ kJ mol⁻¹	Ref.
1-Butanol	$C_4H_{10}O$	117.6	43.29	52.35	1
2-Butanol	$C_4H_{10}O$	99.4	40.75	49.72	1
2-Butanone	C_4H_8O	79.6	31.30	34.79	1
1-Butene	C_4H_8	-6.3	22.07	20.22	1
cis-2-Butene	C_4H_8	3.72	23.34	22.16	1
trans-2-Butene	C_4H_8	0.88	22.72	21.40	1
2-Butoxyethanol	$C_6H_{14}O_2$	171		56.59	1
Butyl acetate	$C_6H_{12}O_2$	126.0	36.28	43.86	1
tert-Butyl acetate	$C_6H_{12}O_2$	97.9	33.07	38.03	1
Butylamine	$C_4H_{11}N$	77.0	31.81	35.72	1
sec-Butylamine	$C_4H_{11}N$	62.71	29.92	32.85	1
tert-Butylamine	$C_4H_{11}N$	44.02	28.27	29.64	1
Butylbenzene	$C_{10}H_{14}$	183.3	38.87	50.8	14
sec-Butylbenzene, (±)-	$C_{10}H_{14}$	173.3		48.1	14
tert-Butylbenzene	$C_{10}H_{14}$	169.1		47.6	14
Butylcyclohexane	$C_{10}H_{20}$	180.9		49.36	1
Butylcyclopentane	C_9H_{18}	156	36.16	45.89	1
Butylethylamine	$C_6H_{15}N$	104.8	33.97	40.15	1
Butyl ethyl ether	$C_6H_{14}O$	89	31.63	36.32	1
Butyl ethyl sulfide	$C_6H_{14}S$	144.2	37.01	44.51	1
Butyl formate	$C_5H_{10}O_2$	106.1	36.58	41.11	1
tert-Butyl isobutyl ether	$C_8H_{18}O$	112.9	33.11	40.5	
Butyl methyl ether	$C_5H_{12}O$	70.1	29.55	32.37	1
sec-Butyl methyl ether	$C_5H_{12}O$	59.1	28.09	30.23	1
Butyl methyl sulfide	$C_5H_{12}S$	123.4	34.47	40.46	1
tert-Butyl methyl sulfide	$C_5H_{12}S$	98.9	31.47	35.84	1
Butyl propyl ether	$C_7H_{16}O$	117	33.72	40.22	1
Butyl vinyl ether	$C_6H_{12}O$	94	31.58	36.17	1
1-Butyne	C_4H_6	8.1	24.52	23.35	1
2-Butyne-1,4-diol	$C_4H_6O_2$	238		81.5	11
γ-Butyrolactone	$C_4H_6O_2$	204.6	52.2		
Cadmium	Cd	767	99.87		
Cadmium bromide	Br_2Cd	863	115		
Cadmium chloride	$CdCl_2$	964	124.3		
Cadmium fluoride	CdF_2	1750	214		
Cadmium iodide	CdI_2	744	115		
Camphor, (+)	$C_{10}H_{16}O$	209	59.5		
Carbon disulfide	CS_2	46.2	26.74	27.51	1
Carbon monoxide	CO	-191.51	6.04		1
Chlorine	Cl_2	-34.04	20.41	17.65	
Chlorine dioxide	ClO_2	11	30		
Chlorine fluoride	ClF	-101.1	24		
Chlorine monoxide	Cl_2O	2.2	25.9		
Chlorine trifluoride	ClF_3	11.75	27.53		
2-Chloroaniline	C_6H_6ClN	209	44.4		
Chlorobenzene	C_6H_5Cl	131.6	35.19	40.97	1
1-Chlorobutane	C_4H_9Cl	78.4	30.39	33.51	1
2-Chlorobutane	C_4H_9Cl	71	29.17	31.53	1
Chlorodifluoromethane	$CHClF_2$	-40.8	20.2		
Chloroethane	C_2H_5Cl	12.3	24.65		1
2-Chloroethanol	C_2H_5ClO	126	41.4		
Chloroethene	C_2H_3Cl	-13.8	20.8		
1-Chloroheptane	$C_7H_{15}Cl$	159		47.66	1
1-Chlorohexane	$C_6H_{13}Cl$	135.0	35.67	42.83	1
Chloromethane	CH_3Cl	-24.1	21.40	18.92	1
1-Chloro-3-methylbutane	$C_5H_{11}Cl$	99	32.02	36.24	1
1-Chloro-2-methylpropane	C_4H_9Cl	69	29.22	31.67	1
2-Chloro-2-methylpropane	C_4H_9Cl	50.9	27.55	28.98	1
1-Chloronaphthalene	$C_{10}H_7Cl$	259	52.1		
1-Chlorooctane	$C_8H_{17}Cl$	183		52.42	1
Chloropentafluorobenzene	C_6ClF_5	117.96	34.76	41.07	1
Chloropentafluoroethane	C_2ClF_5	-39.2	19.41		1

Fluids

Name	Mol. form.	t_b/°C	$\Delta_{vap}H(t_b)$/ kJ mol⁻¹	$\Delta_{vap}H$(25°C)/ kJ mol⁻¹	Ref.
1-Chloropentane	$C_5H_{11}Cl$	107.9	33.15	38.24	1
2-Chloropentane, (+)	$C_5H_{11}Cl$	96.3	31.79	36.03	1
1-Chloropropane	C_3H_7Cl	46.2	27.18	28.35	1
2-Chloropropane	C_3H_7Cl	35.0	26.30	26.90	1
3-Chloropropene	C_3H_5Cl	44.8	29.0		
Chlorosilane	ClH_3Si	-30.4	21		
2-Chlorotoluene	C_7H_7Cl	158.8	37.5		
4-Chlorotoluene	C_7H_7Cl	161.8	38.7		
Chlorotrifluoromethane	$CClF_3$	-81.48	15.8		
Chlorotrifluorosilane	ClF_3Si	-70.0	18.7		
Cholesterol	$C_{27}H_{46}O$	459		148.0	9
Chromium(II) chloride	Cl_2Cr	1120	197		
Chromium(VI) dichloride dioxide	Cl_2CrO_2	117	35.1		
o-Cresol	C_7H_8O	191.0	45.19		1
m-Cresol	C_7H_8O	202.2	47.40	61.71	1
p-Cresol	C_7H_8O	201.9	47.45		1
Cyanogen	C_2N_2	-21.1	23.33	19.75	1
Cyclobutane	C_4H_8	12.5	24.19	23.51	1
Cyclobutanecarbonitrile	C_5H_7N	148	36.88	44.34	1
Cyclohexane	C_6H_{12}	80.7	29.97	33.01	1
Cyclohexanecarbonitrile	$C_7H_{11}N$	188		51.92	1
Cyclohexanethiol	$C_6H_{12}S$	158.8	37.06	44.57	1
Cyclohexanol	$C_6H_{12}O$	160.9		62.01	1
Cyclohexanone	$C_6H_{10}O$	155.4		45.06	1
Cyclohexene	C_6H_{10}	82.9	30.46	33.47	1
1-Cyclohexenecarbonitrile	C_7H_9N			53.55	1
Cyclohexylamine	$C_6H_{13}N$	133.6	36.14	43.67	1
Cyclohexylbenzene	$C_{12}H_{16}$	239		60.8	14
Cyclohexylcyclohexane	$C_{12}H_{22}$	239		57.98	1
Cyclopentane	C_5H_{10}	49.2	27.30	28.52	1
Cyclopentanecarbonitrile	C_6H_9N	169		43.43	1
Cyclopentanethiol	$C_5H_{10}S$	132.2	35.32	41.42	1
Cyclopentanol	$C_5H_{10}O$	140.4		57.05	
Cyclopentanone	C_5H_8O	130.5	36.35	42.72	1
1-Cyclopentenecarbonitrile	C_6H_7N			44.98	1
Cyclopropane	C_3H_6	-31	20.05	16.93	1
Cyclopropanecarbonitrile	C_4H_5N	135	35.55	41.94	1
Cyclopropylbenzene	C_9H_{10}	172		50.22	1
Cyclopropyl methyl ketone	C_5H_8O	111	34.07	39.41	1
cis-Decahydronaphthalene	$C_{10}H_{18}$	195.8	41.0		
trans-Decahydronaphthalene	$C_{10}H_{18}$	187.3	40.2		
Decane	$C_{10}H_{22}$	174.1	39.58	51.42	
1,10-Decanediol	$C_{10}H_{22}O_2$			120.0	11
Decanenitrile	$C_{10}H_{19}N$	241		66.84	
1-Decanethiol	$C_{10}H_{22}S$	240		65.48	1
1-Decanol	$C_{10}H_{22}O$	229		81.50	1
1-Decene	$C_{10}H_{20}$	171		50.43	1
Decylbenzene	$C_{16}H_{26}$	298		78.2	14
Diborane	B_2H_6	-92.49	14.28		
1,4-Dibromobutane	$C_4H_8Br_2$	197		53.09	1
1,2-Dibromo-1-chloro-1,2,2-trifluoroethane	$C_2Br_2ClF_3$	92.8	31.17	35.04	1
1,2-Dibromoethane	$C_2H_4Br_2$	131.3	34.77	41.73	1
Dibromomethane	CH_2Br_2	97.0	32.92	36.97	1
1,2-Dibromopropane	$C_3H_6Br_2$	140	35.61	41.67	1
1,3-Dibromopropane	$C_3H_6Br_2$	164		47.45	1
Dibromosilane	Br_2H_2Si	66	31		
1,2-Dibromotetrafluoroethane	$C_2Br_2F_4$	47.1	27.03	28.39	1
Dibutylamine	$C_8H_{19}N$	162	38.44	49.45	1
Dibutyl ether	$C_8H_{18}O$	141.6	36.49	44.97	1
Di-*sec*-butyl ether	$C_8H_{18}O$	121.9	34.06	40.84	1
Di-*tert*-butyl ether	$C_8H_{18}O$	107.1	32.15	37.61	1
Dibutyl phthalate	$C_{16}H_{22}O_4$	338	79.2		

Name	Mol. form.	t_b/°C	$\Delta_{vap}H(t_b)$/ kJ mol^{-1}	$\Delta_{vap}H$(25°C)/ kJ mol^{-1}	Ref.
Dibutyl sulfide	$C_8H_{18}S$	168		52.96	1
Di-*tert*-butyl sulfide	$C_8H_{18}S$	152.3	33.26	43.76	1
o-Dichlorobenzene	$C_6H_4Cl_2$	180.2	39.66	50.21	1
m-Dichlorobenzene	$C_6H_4Cl_2$	172	38.62	48.58	1
p-Dichlorobenzene	$C_6H_4Cl_2$	173.9	38.79	49.0	1
1,2-Dichlorobutane	$C_4H_8Cl_2$	123.9	33.90	39.58	1
1,4-Dichlorobutane	$C_4H_8Cl_2$	155		46.36	1
Dichlorodifluoromethane	CCl_2F_2	-29.8	20.1		
Dichlorodifluorosilane	Cl_2F_2Si	-32	21.2		
1,1-Dichloroethane	$C_2H_4Cl_2$	56.3	28.85	30.62	1
1,2-Dichloroethane	$C_2H_4Cl_2$	83.4	31.98	35.16	1
1,1-Dichloroethene	$C_2H_2Cl_2$	31.6	26.14	26.48	1
cis-1,2-Dichloroethene	$C_2H_2Cl_2$	60	30.2		
trans-1,2-Dichloroethene	$C_2H_2Cl_2$	47.64	28.9		
1,1-Dichloro-1-fluoroethane	$C_2H_3Cl_2F$	32.05	26.06	26.48	
Dichlorofluoromethane	$CHCl_2F$	8.9	25.2		
1,2-Dichloro-1,1,2,3,3,3-hexafluoropropane	$C_3Cl_2F_6$	34	26.28	26.93	1
1,2-Dichlorohexane	$C_6H_{12}Cl_2$	172.2		48.16	1
Dichloromethane	CH_2Cl_2	39.8	28.06	28.82	1
1,2-Dichloropentane	$C_5H_{10}Cl_2$	148.2	36.45	43.89	1
1,5-Dichloropentane	$C_5H_{10}Cl_2$	182.9		50.71	1
1,3-Dichloropropane	$C_3H_6Cl_2$	120.8	35.18	40.75	1
Dichlorosilane	Cl_2H_2Si	8.3	25	24.2	
1,2-Dichloro-1,1,2,2-tetrafluoroethane	$C_2Cl_2F_4$	3.6	23.3		
Dicyclopropyl ketone	$C_7H_{10}O$	161		53.70	1
Diethanolamine	$C_4H_{11}NO_2$	271.2	65.2		
1,1-Diethoxyethane	$C_6H_{14}O_2$	102	36.28	43.20	1
1,2-Diethoxyethane	$C_6H_{14}O_2$	120.6	36.28	43.20	1
Diethoxymethane	$C_5H_{12}O_2$	86	31.33	35.65	1
Diethylamine	$C_4H_{11}N$	55.4	29.06	31.31	1
Diethyl carbonate	$C_5H_{10}O_3$	125.9		43.60	1
Diethyl disulfide	$C_4H_{10}S_2$	154.0	37.58	45.18	1
Diethylene glycol	$C_4H_{10}O_3$	245.5	52.3		
Diethylene glycol diethyl ether	$C_8H_{18}O_3$	185		58.40	1
Diethylene glycol dimethyl ether	$C_6H_{14}O_3$	162	36.17	44.69	1
Diethylene glycol monoethyl ether	$C_6H_{14}O_3$	202	47.5		
Diethylene glycol monomethyl ether	$C_5H_{12}O_3$	194	46.6		
Diethyl ether	$C_4H_{10}O$	34.4	26.52	27.10	1
Diethyl malonate	$C_7H_{12}O_4$	200	54.8		
Diethyl oxalate	$C_6H_{10}O_4$	186	42.0		
3,3-Diethylpentane	C_9H_{20}	146.2	34.61	42.0	1
Diethyl sulfide	$C_4H_{10}S$	92.1	31.77	35.80	1
Difluorine dioxide	F_2O_2	-57 extrap	19.1		
o-Difluorobenzene	$C_6H_4F_2$	93.9	32.21	36.18	1
m-Difluorobenzene	$C_6H_4F_2$	83.0	31.10	34.59	1
p-Difluorobenzene	$C_6H_4F_2$	88.9	31.77	35.54	1
1,1-Difluoroethane	$C_2H_4F_2$	-24.02	21.56	19.08	1
Difluorosilane	F_2H_2Si	-77.8	16.3		
Digermane	Ge_2H_6	29	25.1		
2,3-Dihydrothiophene	C_4H_6S	112	33.24	37.74	1
2,5-Dihydrothiophene	C_4H_6S	122.4	34.83	39.95	1
Diiodomethane	CH_2I_2	182	42.5		
Diisobutyl sulfide	$C_8H_{18}S$	173		48.71	1
Diisopentyl ether	$C_{10}H_{22}O$	172	35.1		
Diisopropylamine	$C_6H_{15}N$	84	30.40	34.61	1
Diisopropyl ether	$C_6H_{14}O$	68.4	29.10	32.12	1
Diisopropyl sulfide	$C_6H_{14}S$	120.0	33.80	39.60	1
Diketene	$C_4H_4O_2$	127.0	36.80	42.89	1
1,2-Dimethoxyethane	$C_4H_{10}O_2$	85.0	32.42	36.39	1
N,N-Dimethylacetamide	C_4H_9NO	165.9		50.24	1
Dimethylamine	C_2H_7N	7.3	26.40	25.05	1
2,4-Dimethylaniline	$C_8H_{11}N$	215		61.3	6

Name	Mol. form.	t_b/°C	$\Delta_{vap}H(t_b)$/ kJ mol^{-1}	$\Delta_{vap}H$(25°C)/ kJ mol^{-1}	Ref.
2,5-Dimethylaniline	$C_8H_{11}N$	214		61.7	6
N,N-Dimethylaniline	$C_8H_{11}N$	193		52.83	1
2,2-Dimethylbutane	C_6H_{14}	49.7	26.31	27.68	1
2,3-Dimethylbutane	C_6H_{14}	58.0	27.38	29.12	1
2,3-Dimethyl-2-butanethiol	$C_6H_{14}S$	126.1		39.3	
3,3-Dimethyl-2-butanone	$C_6H_{12}O$	106.1	33.39	37.91	1
2,3-Dimethyl-1-butene	C_6H_{12}	55.59		29.18	1
3,3-Dimethyl-1-butene	C_6H_{12}	41.24		26.61	
2,3-Dimethyl-2-butene	C_6H_{12}	73.19	29.64	32.51	1
1,1-Dimethylcyclohexane	C_8H_{16}	119.5	32.51	37.92	1
cis-1,2-Dimethylcyclohexane	C_8H_{16}	129.7	33.47	39.70	1
trans-1,2-Dimethylcyclohexane	C_8H_{16}	123.4	32.96	38.36	1
cis-1,3-Dimethylcyclohexane	C_8H_{16}	124.4	32.91	38.26	1
trans-1,3-Dimethylcyclohexane	C_8H_{16}	120.1	33.39	39.16	1
cis-1,4-Dimethylcyclohexane	C_8H_{16}	124.3	33.28	39.02	1
trans-1,4-Dimethylcyclohexane	C_8H_{16}	119.3	32.56	37.90	1
cis-1,3-Dimethylcyclopentane	C_7H_{14}	91.7	30.40	34.20	1
Dimethyl disulfide	$C_2H_6S_2$	109.72	33.78	37.86	1
Dimethyl ether	C_2H_6O	-24.8	21.51	18.51	1
N,N-Dimethylformamide	C_3H_7NO	152.8		46.89	1
Dimethyl glutarate	$C_7H_{12}O_4$	216		65.7	13
Dimethyl heptanedioate	$C_9H_{16}O_4$	245		73.5	13
2,6-Dimethyl-4-heptanol	$C_9H_{20}O$	193		65.17	12
2,6-Dimethyl-4-heptanone	$C_9H_{18}O$	157		50.92	1
2,2-Dimethylhexane	C_8H_{18}	106.8	32.07	37.28	1
2,3-Dimethylhexane	C_8H_{18}	115.6	33.17	38.78	1
2,4-Dimethylhexane	C_8H_{18}	109.4	32.51	37.76	1
2,5-Dimethylhexane	C_8H_{18}	109.1	32.54	37.85	1
3,3-Dimethylhexane	C_8H_{18}	111.9	32.31	37.53	1
3,4-Dimethylhexane	C_8H_{18}	117.7	33.24	38.97	1
Dimethyl 1,6-hexanedioate	$C_8H_{14}O_4$	231		69.0	13
2,5-Dimethyl-2,5-hexanediol	$C_8H_{18}O_2$	233		85.2	11
cis-2,2-Dimethyl-3-hexene	C_8H_{16}	105.4		36.86	
trans-2,2-Dimethyl-3-hexene	C_8H_{16}	100.9		37.03	
2,5-Dimethyl-3-hexyne-2,5-diol	$C_8H_{14}O_2$	206		82.8	11
1,1-Dimethylhydrazine	$C_2H_8N_2$	62.4	32.55	35.0	1
Dimethyl malonate	$C_5H_8O_4$	181.1		57.5	13
Dimethyl nonanedioate	$C_{11}H_{20}O_4$			82.3	13
2,4-Dimethyloctane	$C_{10}H_{22}$	154	36.47	47.13	1
Dimethyl octanedioate	$C_{10}H_{18}O_4$	259		78.1	13
Dimethyl oxalate	$C_4H_6O_4$	163.4		54.7	13
3,3-Dimethyloxetane	$C_5H_{10}O$	78	30.85	33.94	1
2,2-Dimethylpentane	C_7H_{16}	79.2	29.23	32.42	1
2,3-Dimethylpentane	C_7H_{16}	89.8	30.46	34.26	1
2,4-Dimethylpentane	C_7H_{16}	80.4	29.55	32.88	1
3,3-Dimethylpentane	C_7H_{16}	86.0	29.62	33.03	1
2,2-Dimethyl-3-pentanone	$C_7H_{14}O$	125	36.09	42.34	1
2,4-Dimethyl-3-pentanone	$C_7H_{14}O$	125.2	34.64	41.51	1
2,4-Dimethyl-1-pentene	C_7H_{14}	81.6		33.03	
4,4-Dimethyl-1-pentene	C_7H_{14}	72.5		31.13	
2,4-Dimethyl-2-pentene	C_7H_{14}	83.3		34.19	
cis-4,4-Dimethyl-2-pentene	C_7H_{14}	80.4		32.56	
trans-4,4-Dimethyl-2-pentene	C_7H_{14}	76.7		32.81	
2,2-Dimethylpropanenitrile	C_5H_9N	105.2	32.40	37.35	1
2,2-Dimethyl-1-propanethiol	$C_5H_{12}S$	103.6		36.4	
2,3-Dimethylpyridine	C_7H_9N	161.1	39.08	47.82	
2,4-Dimethylpyridine	C_7H_9N	158.4	38.53	47.49	
2,5-Dimethylpyridine	C_7H_9N	157.00	38.68	47.04	
2,6-Dimethylpyridine	C_7H_9N	144.0	37.46	45.34	
3,4-Dimethylpyridine	C_7H_9N	179.1	39.99	50.50	
3,5-Dimethylpyridine	C_7H_9N	171.9	39.46	49.33	
Dimethyl sebacate	$C_{12}H_{22}O_4$	289		86.4	13

Name	Mol. form.	t_b/°C	$\Delta_{vap}H(t_b)$/ kJ mol^{-1}	$\Delta_{vap}H$(25°C)/ kJ mol^{-1}	Ref.
Dimethyl succinate	$C_6H_{10}O_4$	197		61.0	13
Dimethyl sulfide	C_2H_6S	37.32	27.0	27.65	1
Dimethyl sulfoxide	C_2H_6OS	191.9	43.1		
1,3-Dioxane	$C_4H_8O_2$	105	34.37	39.09	1
1,4-Dioxane	$C_4H_8O_2$	101.2	34.16	38.60	1
Diphenyl ether	$C_{12}H_{10}O$	258.0	48.2		
Diphosphine	H_4P_2	63.5 dec	28.8		
1,2-Dipropoxyethane	$C_8H_{18}O_2$	161		50.62	1
Dipropylamine	$C_6H_{15}N$	107.5	33.47	40.04	1
Dipropyl ether	$C_6H_{14}O$	90.1	31.31	35.69	1
Dipropyl sulfide	$C_6H_{14}S$	142.8	36.60	44.21	1
Disilane	H_6Si_2	-14.8	21.2		
1-Docosanol	$C_{22}H_{46}O$	407		135.9	9
Dodecane	$C_{12}H_{26}$	216.3	44.09	61.52	1
1,12-Dodecanediol	$C_{12}H_{26}O_2$			135	11
Dodecanenitrile	$C_{12}H_{23}N$	276.1		76.12	1
1-Dodecanol	$C_{12}H_{26}O$	264.1		90.8	9
1-Dodecene	$C_{12}H_{24}$	213.4		60.78	1
Dodecylbenzene	$C_{18}H_{30}$	329		86.6	14
Eicosane	$C_{20}H_{42}$	344.1	58.49	101.81	5
1-Eicosanol	$C_{20}H_{42}O$	356		125.9	9
1,2-Epoxybutane	C_4H_8O	63.4	30.3		
Ethane	C_2H_6	-88.6	14.69	5.16	1
1,2-Ethanediamine	$C_2H_8N_2$	116.9	37.98	44.98	1
1,2-Ethanediol	$C_2H_6O_2$	197.5	50.5	63.9	11
1,2-Ethanediol, diacetate	$C_6H_{10}O_4$	184		61.44	1
1,2-Ethanedithiol	$C_2H_6S_2$	144	37.93	44.68	
Ethanethiol	C_2H_6S	35.0	26.79	27.30	1
Ethanol	C_2H_6O	78.24	38.56	42.32	1
Ethanolamine	C_2H_7NO	170.3	49.83		1
Ethoxybenzene	$C_8H_{10}O$	169.8		51.04	1
2-Ethoxyethanol	$C_4H_{10}O_2$	134.7	39.22	48.21	1
2-Ethoxyethyl acetate	$C_6H_{12}O_3$	156.6	40.76	52.61	8
1-Ethoxy-2-methoxyethane	$C_5H_{12}O_2$	88	34.33	39.83	1
N-Ethylacetamide	C_4H_9NO	205		64.89	1
Ethyl acetate	$C_4H_8O_2$	77.1	31.94	35.60	1
Ethyl acrylate	$C_5H_8O_2$	98.9	34.7		
N-Ethylaniline	$C_8H_{11}N$	204		58.3	6
Ethylbenzene	C_8H_{10}	136.2	35.57	42.24	1
Ethyl butanoate	$C_6H_{12}O_2$	121.1	35.47	42.68	1
2-Ethyl-1-butanol	$C_6H_{14}O$	155	43.2		
2-Ethyl-1-butene	C_6H_{12}	64.7		31.13	1
Ethyl chloroacetate	$C_4H_7ClO_2$	144	40.43	49.47	1
Ethylcyclobutane	C_6H_{12}	70	28.67	31.24	1
Ethylcyclohexane	C_8H_{16}	131.8	34.04	40.56	1
Ethylcyclopentane	C_7H_{14}	103.5	31.96	36.40	1
Ethyl dichloroacetate	$C_4H_6Cl_2O_2$	155		50.60	1
Ethyl 2,2-dimethylpropanoate	$C_7H_{14}O_2$	118.3	34.51	41.25	1
Ethylene	C_2H_4	-103.8	13.53		1
N-Ethylformamide	C_3H_7NO	198		58.44	1
Ethyl formate	$C_3H_6O_2$	54.09	29.91	31.96	1
3-Ethylhexane	C_8H_{18}	118.5	33.59	39.64	1
Ethyl hexanoate	$C_8H_{16}O_2$	165		51.72	1
2-Ethylhexanoic acid	$C_8H_{16}O_2$	227.5		75.60	1
2-Ethyl-1-hexanol	$C_8H_{18}O$	186.2	54.2	68.51	12
2-Ethylhexyl acetate	$C_{10}H_{20}O_2$	200	43.5		
2-Ethylhexylamine	$C_8H_{19}N$	172	40.0		
Ethylisopropylamine	$C_5H_{13}N$	68	29.94	33.13	1
Ethyl isopropyl ether	$C_5H_{12}O$	54	28.21	30.08	1
Ethyl isopropyl sulfide	$C_5H_{12}S$	107.3	32.74	37.78	1
Ethyl 3-methylbutanoate	$C_7H_{14}O_2$	135	37.0		
2-Ethyl-3-methyl-1-butene	C_7H_{14}	86.3		34.35	

Fluids

Name	Mol. form.	t_b/°C	$\Delta_{vap}H(t_b)$/ kJ mol^{-1}	$\Delta_{vap}H$(25°C)/ kJ mol^{-1}	Ref.
1-Ethyl-1-methylcyclopentane	C_8H_{16}	121.5	33.20	38.85	1
3-Ethyl-2-methylpentane	C_8H_{18}	115.6	32.93	38.52	1
3-Ethyl-3-methylpentane	C_8H_{18}	118.2	32.78	37.99	1
3-Ethyl-2-methyl-1-pentene	C_8H_{16}	109.2		37.27	
Ethyl 2-methylpropanoate	$C_6H_{12}O_2$	111	33.67	39.83	1
Ethyl methyl sulfide	C_3H_8S	66.6	29.53	31.85	1
3-Ethylpentane	C_7H_{16}	93.4	31.12	35.22	1
Ethyl pentanoate	$C_7H_{14}O_2$	142	36.96	47.01	1
3-Ethyl-3-pentanol	$C_7H_{16}O$	142		57.34	12
Ethyl pentyl ether	$C_7H_{16}O$	118	34.41	41.01	1
Ethyl propanoate	$C_5H_{10}O_2$	98.9	33.88	39.21	1
Ethyl propyl ether	$C_5H_{12}O$	63	28.94	31.43	1
Ethyl propyl sulfide	$C_5H_{12}S$	118.5	34.24	39.97	1
Ethyl trichloroacetate	$C_4H_5Cl_3O_2$	167		50.97	1
Ethyl vinyl ether	C_4H_8O	36	26.2		
Fluorine	F_2	-188.11	6.62		
Fluorine monoxide	F_2O	-144.3	11.09		
Fluorobenzene	C_6H_5F	84.7	31.19	34.58	1
1-Fluorooctane	$C_8H_{17}F$	146	40.43	49.65	1
Fluorosilane	FH_3Si	-98.6	18.8		
2-Fluorotoluene	C_7H_7F	114	35.4		
4-Fluorotoluene	C_7H_7F	116.6	34.08	39.42	1
Formamide	CH_3NO	217		60.15	1
Formic acid	CH_2O_2	101	22.69	20.10	1
Furan	C_4H_4O	31.3	27.10	27.45	1
Furfural	$C_5H_4O_2$	161.5	43.2		
Furfuryl alcohol	$C_5H_6O_2$	168	53.6		
Gallium	Ga	2229	254		
Gallium(III) bromide	Br_3Ga	279	38.9		
Gallium(III) chloride	Cl_3Ga	201	23.9		
Gallium(III) iodide	GaI_3	340	56.5		
Germane	GeH_4	-88.1	14.06		
Germanium	Ge	2833	334		
Germanium(IV) bromide	Br_4Ge	186.35	41.4		
Germanium(IV) chloride	Cl_4Ge	86.55	27.9		
Glycerol	$C_3H_8O_3$	289	61.0		
Glycerol triacetate	$C_9H_{14}O_6$	259		85.74	1
Gold	Au	2836	324		
Helium	He	-268.928	0.08		
Heptadecane	$C_{17}H_{36}$	303	53.58	86.47	1
1-Heptadecanol	$C_{17}H_{36}O$	324		112.5	9
6-Heptadecanol	$C_{17}H_{36}O$			108.6	9
7-Heptadecanol	$C_{17}H_{36}O$			108.2	9
9-Heptadecanol	$C_{17}H_{36}O$			108.5	9
Heptane	C_7H_{16}	98.38	31.77	36.57	1
1,7-Heptanediol	$C_7H_{16}O_2$	262		97.9	11
1-Heptanol	$C_7H_{16}O$	178		66.81	1
3-Heptanol, (*S*)-	$C_7H_{16}O$	163	42.5		
2-Heptanone	$C_7H_{14}O$	151.0		47.24	1
1-Heptene	C_7H_{14}	94		35.49	1
cis-2-Heptene	C_7H_{14}	97		36.26	
trans-2-Heptene	C_7H_{14}	98		36.27	
cis-3-Heptene	C_7H_{14}	96		35.81	
trans-3-Heptene	C_7H_{14}	96		35.84	
Heptylamine	$C_7H_{17}N$	153		49.96	1
Heptylbenzene	$C_{13}H_{20}$	242		64.2	14
1-Hexacosanol	$C_{26}H_{54}O$			153.7	9
Hexadecane	$C_{16}H_{34}$	286.9	51.84	81.35	1
1,16-Hexadecanediol	$C_{16}H_{34}O_2$			163	11
1-Hexadecanol	$C_{16}H_{34}O$	325		107.7	9
1-Hexadecene	$C_{16}H_{32}$	285		80.25	1
Hexadecylbenzene	$C_{22}H_{38}$	385		104.8	14

Name	Mol. form.	t_b/°C	$\Delta_{vap}H(t_b)$/ kJ mol⁻¹	$\Delta_{vap}H$(25°C)/ kJ mol⁻¹	Ref.
Hexafluoroacetylacetone	$C_5H_2F_6O_2$	69	27.05	30.58	1
Hexafluorobenzene	C_6F_6	80.2	31.66	35.71	1
Hexafluoroethane	C_2F_6	-78.1	16.15		1
Hexane	C_6H_{14}	68.72	28.85	31.56	1
1,6-Hexanediol	$C_6H_{14}O_2$	208		90.2	10,11
Hexanenitrile	$C_6H_{11}N$	163.5		47.91	1
1-Hexanol	$C_6H_{14}O$	156.9	44.50	61.61	1
2-Hexanol	$C_6H_{14}O$	138	41.01	58.46	1
2-Hexanone	$C_6H_{12}O$	127.6	36.35	43.14	1
3-Hexanone	$C_6H_{12}O$	123.5	35.36	42.47	1
1-Hexene	C_6H_{12}	63.4		30.61	1
cis-2-Hexene	C_6H_{12}	68.9		32.19	1
trans-2-Hexene	C_6H_{12}	67.85		31.60	1
cis-3-Hexene	C_6H_{12}	66.4		31.23	
trans-3-Hexene	C_6H_{12}	67.06		31.55	
Hexylamine	$C_6H_{15}N$	132	36.54	45.10	1
Hexylbenzene	$C_{12}H_{18}$	226		60.4	14
Hexyl methyl ether	$C_7H_{16}O$	125	34.93	42.07	1
Hydrazine	H_4N_2	113.55	41.8	44.7	
Hydrazoic acid	HN_3	35.7	30.5		
Hydrogen	H_2	-252.879	0.90		
Hydrogen bromide	BrH	-66.38		12.69	
Hydrogen chloride	ClH	-85	16.15	9.08	
Hydrogen disulfide	H_2S_2	70.7		33.78	
Hydrogen iodide	HI	-35.55	19.76	17.36	
Hydrogen peroxide	H_2O_2	150.2		51.6	
Hydrogen selenide	H_2Se	-41.25	19.7		
Hydrogen sulfide	H_2S	-59.55	18.67	14.08	
Hydrogen telluride	H_2Te	-2	19.2		
Indan	C_9H_{10}	177.8	39.63	48.79	1
Indium(I) bromide	BrIn	656	92		
Indium(I) iodide	IIn	712	90.8		
Iodine	I_2	184.4	41.57		
Iodine pentafluoride	F_5I	100.5	41.3		
Iodobenzene	C_6H_5I	188.5	39.5		
1-Iodobutane	C_4H_9I	130	34.66	40.63	1
2-Iodobutane, (±)-	C_4H_9I	118	33.27	38.46	1
Iodoethane	C_2H_5I	72	29.44	31.93	1
1-Iodohexane	$C_6H_{13}I$	182		49.75	1
Iodomethane	CH_3I	42.4	27.34	27.97	1
1-Iodo-2-methylpropane	C_4H_9I	121	33.54	38.83	1
2-Iodo-2-methylpropane	C_4H_9I	98	31.43	35.41	1
1-Iodopentane	$C_5H_{11}I$	156		45.27	1
1-Iodopropane	C_3H_7I	102	32.08	36.25	1
2-Iodopropane	C_3H_7I	89	30.68	34.06	1
Iridium(VI) fluoride	F_6Ir	53.6	30.9		7
Isobutane	C_4H_{10}	-11.7	21.30	19.23	1
Isobutyl acetate	$C_6H_{12}O_2$	116.9	35.9		
Isobutylamine	$C_4H_{11}N$	68.8	30.61	33.85	1
Isobutylbenzene	$C_{10}H_{14}$	172.7		48.0	14
Isobutyl formate	$C_5H_{10}O_2$	98.4	33.6		
Isobutyl isobutanoate	$C_8H_{16}O_2$	148	38.2		
Isobutyl methyl ether	$C_5H_{12}O$	58.6	28.02	30.13	1
Isopentane	C_5H_{12}	27.83	24.69	24.85	1
Isopentyl acetate	$C_7H_{14}O_2$	141.6	37.5		
Isopentyl isopentanoate	$C_{10}H_{20}O_2$	190	45.9		
Isopropyl acetate	$C_5H_{10}O_2$	88.6	32.93	37.20	1
Isopropylamine	C_3H_9N	31.8	27.83	28.36	1
Isopropylbenzene	C_9H_{12}	152.4		45.13	1
Isopropylcyclohexane	C_9H_{18}	154.4		44.02	1
Isopropylcyclopentane	C_8H_{16}	126.4	33.56	39.44	1
Isopropylmethylamine	$C_4H_{11}N$	53.3	28.71	30.69	1

Name	Mol. form.	t_b/°C	$\Delta_{vap}H(t_b)$/ kJ mol^{-1}	$\Delta_{vap}H$(25°C)/ kJ mol^{-1}	Ref.
1-Isopropyl-4-methylbenzene	$C_{10}H_{14}$	177	38.2		
Isopropyl methyl ether	$C_4H_{10}O$	30.8	26.05	26.41	1
Isopropyl methyl sulfide	$C_4H_{10}S$	84.7	30.71	34.15	1
Isopropylpropylamine	$C_6H_{15}N$	96.2	32.14	37.23	1
Isopropyl propyl sulfide	$C_6H_{14}S$	132.0	35.11	41.78	1
Isoquinoline	C_9H_7N	243.2	49.0	60.26	
Krypton	Kr	-153.415	9.08		
Lead	Pb	1749	179.5		
Lead(II) bromide	Br_2Pb	892	133		
Lead(II) chloride	Cl_2Pb	951	127		
Lead(II) fluoride	F_2Pb	1293	160.4		
Lead(II) iodide	I_2Pb	872 dec	104		
Lithium fluoride	FLi	1673	147		
Lithium hydroxide	HLiO	1626	188		
Mercury	Hg	356.619	59.11		
Mercury(II) bromide	Br_2Hg	318	58.89		
Mercury(II) chloride	Cl_2Hg	304	58.9		
Mercury(II) iodide (yellow)	HgI_2	351	59.2		
Mesityl oxide	$C_6H_{10}O$	129.7	36.1		
Methane	CH_4	-161.5	8.19		
Methanol	CH_4O	64.5	35.21	37.43	1
2-Methoxyethanol	$C_3H_8O_2$	124.3	37.54	45.17	1
2-Methoxyethyl acetate	$C_5H_{10}O_3$	142	43.9		
Methyl acetate	$C_3H_6O_2$	56.7	30.32	32.29	1
Methyl acrylate	$C_4H_6O_2$	80.1	33.1		
2-Methylacrylonitrile	C_4H_5N	90	31.8		
Methylamine	CH_5N	-6.4	25.60	23.37	1
2-Methylaniline	C_7H_9N	200.0	44.6		
3-Methylaniline	C_7H_9N	203.3	44.9		
4-Methylaniline	C_7H_9N	201	44.3		
Methyl benzoate	$C_8H_8O_2$	199		55.57	1
1-Methylbicyclo[3.1.0]hexane	C_7H_{12}	88	31.07	34.77	1
3-Methylbutanenitrile	C_5H_9N	129	35.10	41.64	1
2-Methyl-1-butanethiol, (+)	$C_5H_{12}S$	119.0	33.79	39.45	1
2-Methyl-2-butanethiol	$C_5H_{12}S$	99.1	31.37	35.67	1
3-Methyl-2-butanethiol	$C_5H_{12}S$	109.8		37.5	
Methyl butanoate	$C_5H_{10}O_2$	101.9	33.79	39.28	1
2-Methylbutanoic acid	$C_5H_{10}O_2$	177		46.91	1
2-Methyl-1-butanol, (±)-	$C_5H_{12}O$	129.0		55.16	1
3-Methyl-1-butanol	$C_5H_{12}O$	130.8	44.07	55.61	1
2-Methyl-2-butanol	$C_5H_{12}O$	102.4	39.04	50.10	1
3-Methyl-2-butanol, (±)-	$C_5H_{12}O$	113.7		53.0	
3-Methyl-2-butanone	$C_5H_{10}O$	94.2	32.35	36.78	1
2-Methyl-1-butene	C_5H_{10}	31.1	25.50	25.92	
3-Methyl-1-butene	C_5H_{10}	20.1		23.77	
2-Methyl-2-butene	C_5H_{10}	38.5	26.31	27.06	1
(1-Methylbutyl)benzene	$C_{11}H_{16}$	199		53.0	14
Methyl *tert*-butyl ether	$C_5H_{12}O$	55.1	27.94	29.82	1
Methyl chloroacetate	$C_3H_5ClO_2$	130	39.23	46.73	1
Methyl cyanoacetate	$C_4H_5NO_2$	203	48.2		
Methyl cyclobutanecarboxylate	$C_6H_{10}O_2$	135.5	37.13	44.72	1
Methylcyclohexane	C_7H_{14}	100.9	31.27	35.36	1
1-Methylcyclohexanol	$C_7H_{14}O$	155	79.0		
cis-2-Methylcyclohexanol	$C_7H_{14}O$	165	48.5		
trans-2-Methylcyclohexanol, (±)-	$C_7H_{14}O$	168.4	53.0		
Methylcyclopentane	C_6H_{12}	71.8	29.08	31.64	1
Methyl cyclopropanecarboxylate	$C_5H_8O_2$	112	35.25	41.27	1
2-Methyldecane	$C_{11}H_{24}$	189.2	40.25	54.28	1
4-Methyldecane	$C_{11}H_{24}$	188	40.70	53.76	1
Methyl dichloroacetate	$C_3H_4Cl_2O_2$	127	39.28	47.72	1
Methyl 2,2-dimethylpropanoate	$C_6H_{12}O_2$	102	33.42	38.76	1
Methyl dodecanoate	$C_{13}H_{26}O_2$	268		77.17	1

Name	Mol. form.	t_b/°C	$\Delta_{vap}H(t_b)$/ kJ mol^{-1}	$\Delta_{vap}H$(25°C)/ kJ mol^{-1}	Ref.
N-Methylformamide	C_2H_5NO	186		56.19	1
Methyl formate	$C_2H_4O_2$	31.6	27.92	28.35	1
2-Methylheptane	C_8H_{18}	117.6	33.26	39.67	1
3-Methylheptane	C_8H_{18}	118.9	33.66	39.83	1
4-Methylheptane	C_8H_{18}	117.7	33.35	39.69	1
Methyl heptanoate	$C_8H_{16}O_2$	169.7		51.62	1
2-Methyl-2-heptanol	$C_8H_{18}O$	173		62.87	12
2-Methylhexane	C_7H_{16}	90.0	30.62	34.87	1
3-Methylhexane	C_7H_{16}	92	30.9		
Methyl hexanoate	$C_7H_{14}O_2$	151	38.55	48.04	1
2-Methyl-2-hexanol	$C_7H_{16}O$	142		58.57	12
5-Methyl-3-hexanol, (±)-	$C_7H_{16}O$	147		59.82	12
cis-3-Methyl-3-hexene	C_7H_{14}	95.3		36.31	1
trans-3-Methyl-3-hexene	C_7H_{14}	94		35.70	1
Methylhydrazine	CH_6N_2	83	36.12	40.37	1
Methyl isobutanoate	$C_5H_{10}O_2$	92	32.61	37.32	1
Methyl methacrylate	$C_5H_8O_2$	100.6	36.0		
1-Methylnaphthalene	$C_{11}H_{10}$	244.4	45.5		
2-Methylnonane	$C_{10}H_{22}$	167	38.23	49.63	1
3-Methylnonane	$C_{10}H_{22}$	168	38.26	49.71	1
5-Methylnonane	$C_{10}H_{22}$	164	38.14	49.34	1
Methyl octanoate	$C_9H_{18}O_2$	194.1		56.41	1
Methyloxirane	C_3H_6O	35	27.35	27.89	1
2-Methylpentane	C_6H_{14}	60.21	27.79	29.89	1
3-Methylpentane	C_6H_{14}	63.3	28.06	30.28	1
2-Methyl-2,4-pentanediol	$C_6H_{14}O_2$	197.9	57.3		
Methyl pentanoate	$C_6H_{12}O_2$	127.36	35.36	43.10	1
2-Methyl-1-pentanol	$C_6H_{14}O$	157	50.2		
4-Methyl-1-pentanol	$C_6H_{14}O$	151	44.46	60.47	1
2-Methyl-2-pentanol	$C_6H_{14}O$	121	39.59	54.77	1
4-Methyl-2-pentanol	$C_6H_{14}O$	132.0	44.2		
3-Methyl-2-pentanone, (±)-	$C_6H_{12}O$	117	34.16	40.53	1
4-Methyl-2-pentanone	$C_6H_{12}O$	115.7	34.49	40.61	1
2-Methyl-3-pentanone	$C_6H_{12}O$	118	33.84	39.79	1
2-Methyl-1-pentene	C_6H_{12}	62.1		30.48	1
3-Methyl-1-pentene	C_6H_{12}	54		28.62	1
4-Methyl-1-pentene	C_6H_{12}	54		28.71	1
2-Methyl-2-pentene	C_6H_{12}	67.3		31.60	1
3-Methyl-*cis*-2-pentene	C_6H_{12}	70.4		32.09	1
3-Methyl-*trans*-2-pentene	C_6H_{12}	67.67		31.35	1
4-Methyl-*cis*-2-pentene	C_6H_{12}	56.4		29.48	1
4-Methyl-*trans*-2-pentene	C_6H_{12}	58.58		29.97	1
Methyl pentyl ether	$C_6H_{14}O$	99	32.02	36.85	1
Methyl pentyl sulfide	$C_6H_{14}S$	148	37.41	45.24	1
2-Methylpropanenitrile	C_4H_7N	102	32.39	37.13	1
2-Methyl-1-propanethiol	$C_4H_{10}S$	88.5	31.01	34.63	1
2-Methyl-2-propanethiol	$C_4H_{10}S$	64.2	28.45	30.78	1
Methyl propanoate	$C_4H_8O_2$	78.6	32.24	35.85	1
2-Methylpropanoic acid	$C_4H_8O_2$	154.4		35.30	1
2-Methyl-1-propanol	$C_4H_{10}O$	107.84	41.82	50.82	1
2-Methyl-2-propanol	$C_4H_{10}O$	82.3	39.07	46.69	1
Methyl propyl ether	$C_4H_{10}O$	38.5	26.75	27.60	1
Methyl propyl sulfide	$C_4H_{10}S$	95.5	32.08	36.24	1
2-Methylpyridine	C_6H_7N	129.4	36.17	42.48	1
3-Methylpyridine	C_6H_7N	144.1	37.35	44.44	1
4-Methylpyridine	C_6H_7N	145.3	37.51	44.56	1
2-Methylquinoline	$C_{10}H_9N$	247.4		66.1	
4-Methylquinoline	$C_{10}H_9N$	266		67.6	
6-Methylquinoline	$C_{10}H_9N$	265		67.7	
8-Methylquinoline	$C_{10}H_9N$	247.4		65.7	
Methyl salicylate	$C_8H_8O_3$	222.6	46.7		
4-Methylthiazole	C_4H_5NS	133.2	37.58	43.85	1

Fluids

Name	Mol. form.	t_b/°C	$\Delta_{vap}H(t_b)$/ kJ mol^{-1}	$\Delta_{vap}H$(25°C)/ kJ mol^{-1}	Ref.
2-Methylthiophene	C_5H_6S	112.5	33.90	38.87	1
3-Methylthiophene	C_5H_6S	115.4	34.24	39.43	1
Methyl trichloroacetate	$C_3H_3Cl_3O_2$	152		48.33	1
Molybdenum(V) chloride	Cl_5Mo	268	62.8		
Molybdenum(V) fluoride	F_5Mo	213.6	51.8		7
Molybdenum(VI) fluoride	F_6Mo	34.0	29.0		7
Molybdenum(VI) oxide	MoO_3	1155	138		
Molybdenum(VI) oxytetrafluoride	F_4MoO	186.0	50.6		7
Morpholine	C_4H_9NO	128.2	37.1		
Naphthalene	$C_{10}H_8$	218.0	43.2		
Neon	Ne	-246.046	1.71		
Neopentane	C_5H_{12}	9.50	22.74	21.84	1
Niobium(V) chloride	Cl_5Nb	247.4	52.7		
Niobium(V) fluoride	F_5Nb	234	52.3		
Nitric acid	HNO_3	83		39.1	
Nitric oxide	NO	-151.74	13.83		
Nitrobenzene	$C_6H_5NO_2$	210.7		55.01	1
Nitroethane	$C_2H_5NO_2$	114.1	38.0		
Nitrogen	N_2	-195.795	5.57		
Nitrogen tetroxide	N_2O_4	21.15	38.12		
Nitrogen trifluoride	F_3N	-128.75	11.56		
Nitromethane	CH_3NO_2	101.19	33.99	38.27	1
1-Nitropropane	$C_3H_7NO_2$	131.2	38.5		
2-Nitropropane	$C_3H_7NO_2$	120.2	36.8		
Nitrosyl chloride	ClNO	-5.5	25.78		
Nitrosyl fluoride	FNO	-59.9	19.28		
Nitrous oxide	N_2O	-88.48	16.53		
Nitryl chloride	$ClNO_2$	-15	25.7		
Nitryl fluoride	FNO_2	-72.4	18.05		
Nonadecane	$C_{19}H_{40}$	330	56.93	96.4	5
Nonane	C_9H_{20}	150.8	37.18	46.55	
1,9-Nonanediol	$C_9H_{20}O_2$			112.5	11
1-Nonanol	$C_9H_{20}O$	213.7		76.86	1
2-Nonanone	$C_9H_{18}O$	194		56.44	1
5-Nonanone	$C_9H_{18}O$	188.4		53.30	1
Nonylbenzene	$C_{15}H_{24}$	280		74.1	14
Octadecane	$C_{18}H_{38}$	316	55.23	91.44	5
1-Octadecanol	$C_{18}H_{38}O$	351		116.8	9
cis-9-Octadecenoic acid	$C_{18}H_{34}O_2$	360	67.4		
Octane	C_8H_{18}	125.62	34.41	41.49	1
1,8-Octanediol	$C_8H_{18}O_2$			104.9	11
Octanenitrile	$C_8H_{15}N$	202		56.80	1
Octanoic acid	$C_8H_{16}O_2$	240	58.5		
1-Octanol	$C_8H_{18}O$	194.7		70.98	1
2-Octanol	$C_8H_{18}O$	179	44.4		
1-Octene	C_8H_{16}	121.3	34.07	40.34	
Octylbenzene	$C_{14}H_{22}$	263		69.1	14
1-Octyne	C_8H_{14}	126.2	35.83	42.30	1
2-Octyne	C_8H_{14}	138.0	37.26	44.49	1
3-Octyne	C_8H_{14}	135.6	36.94	43.92	1
4-Octyne	C_8H_{14}	133.5	36.0	42.73	1
Osmium(V) fluoride	F_5Os	233	65.6		7
Osmium(VI) fluoride	F_6Os	47.5	28.1		7
Oxetane	C_3H_6O	47.6	28.67	29.85	1
2-Oxetanone	$C_3H_4O_2$	161		47.03	1
Oxirane	C_2H_4O	10.4	25.54	24.75	1
Oxygen	O_2	-182.962	6.82		
Pentaborane(11)	B_5H_{11}	65	31.8		
Pentachloroethane	C_2HCl_5	161	36.9		
Pentadecane	$C_{15}H_{32}$	270.6	50.08	76.77	
1,15-Pentadecanediol	$C_{15}H_{32}O_2$			139	11
1-Pentadecanol	$C_{15}H_{32}O$	318		103.5	9

Name	Mol. form.	t_b/°C	$\Delta_{vap}H(t_b)$/ kJ mol^{-1}	$\Delta_{vap}H(25°C)$/ kJ mol^{-1}	Ref.
Pentadecylbenzene	$C_{21}H_{36}$	373		100.3	14
Pentafluorobenzene	C_6HF_5	85	32.15	36.27	1
2,3,4,5,6-Pentafluorotoluene	$C_7H_3F_5$	117	34.75	41.12	1
2,2,4,6,6-Pentamethylheptane	$C_{12}H_{26}$	178		48.97	1
Pentane	C_5H_{12}	36.06	25.79	26.43	1
1,5-Pentanediol	$C_5H_{12}O_2$	241	60.7	83.0	11
2,4-Pentanedione	$C_5H_8O_2$	140.7	34.30	41.77	1
Pentanenitrile	C_5H_9N	140	36.09	43.60	1
1-Pentanethiol	$C_5H_{12}S$	126.6	34.88	41.24	1
Pentanoic acid	$C_5H_{10}O_2$	186.1	44.1		
1-Pentanol	$C_5H_{12}O$	137.6	44.36	57.02	1
2-Pentanol	$C_5H_{12}O$	119.1	41.40	54.21	1
3-Pentanol	$C_5H_{12}O$	123		54.0	1
2-Pentanone	$C_5H_{10}O$	102.2	33.44	38.40	1
3-Pentanone	$C_5H_{10}O$	101.9	33.45	38.52	1
1-Pentene	C_5H_{10}	30.0	25.20	25.47	1
cis-2-Pentene	C_5H_{10}	36.9		26.86	1
trans-2-Pentene	C_5H_{10}	36.3		26.76	1
trans-3-Pentenenitrile	C_5H_7N	144	37.09	44.77	1
Pentyl acetate	$C_7H_{14}O_2$	149.4	38.42	48.56	8
Pentylamine	$C_5H_{13}N$	104.7	34.01	40.08	1
Pentylbenzene	$C_{11}H_{16}$	203		55.1	14
Pentylcyclohexane	$C_{11}H_{22}$	204		53.88	1
Perchloryl fluoride	$ClFO_3$	-46.75	19.33		
Perfluorobutane	C_4F_{10}	-2.1	22.9		
Perfluorocyclobutane	C_4F_8	-5.91	23.2		
Perfluorodecalin	$C_{10}F_{18}$	143		41.54	10
Perfluorohexane	C_6F_{14}	57.2		32.47	10
Perfluorononane	C_9F_{20}	117		45.27	10
Perfluorooctane	C_8F_{18}	105	33.38	41.13	1
Perfluorotoluene	C_7F_8	104.6		40.52	10
Phenanthrene	$C_{14}H_{10}$	338.4		75.50	1
Phenol	C_6H_6O	181.8	45.69	57.82	1
Phosphine	H_3P	-87.75	14.6		
Phosphorothioc trifluoride	F_3PS	-52.25	19.6		
Phosphorus (white)	P	280.5	12.4	14.2	
Phosphorus(III) bromide	Br_3P	173.2	38.8		
Phosphorus(III) chloride	Cl_3P	76	30.5	32.1	
Phosphorus(III) chloride difluoride	ClF_2P	-47.3	17.6		
Phosphorus(III) dichloride fluoride	Cl_2FP	13.85	24.9		
Phosphorus(III) fluoride	F_3P	-101.8	16.5		
Phosphorus(V) fluoride	F_5P	-84.6	17.2		
Phosphorus(III) iodide	I_3P	227 dec	43.9		
Phosphoryl bromide	Br_3OP	191.7	38		
Phosphoryl chloride	Cl_3OP	105.5	34.35	38.6	
Piperidine	$C_5H_{11}N$	106.19		39.29	1
Propanal	C_3H_6O	48.0	28.31	29.62	1
Propane	C_3H_8	-42.11	19.04	14.79	1
1,3-Propanediamine	$C_3H_{10}N_2$	139.2	40.85	50.16	1
1,2-Propanediol	$C_3H_8O_2$	187.3	52.4		
1,3-Propanediol	$C_3H_8O_2$	214.7	57.9	69.8	11
1,3-Propanedithiol	$C_3H_8S_2$	172.9		49.66	1
Propanenitrile	C_3H_5N	97.3	31.81	36.03	1
1-Propanethiol	C_3H_8S	67.7	29.54	31.89	1
2-Propanethiol	C_3H_8S	52.6	27.91	29.45	1
Propanoic acid	$C_3H_6O_2$	141.5		32.14	1
Propanoic anhydride	$C_6H_{10}O_3$	168	41.7		
1-Propanol	C_3H_8O	97.04	41.44	47.45	1
2-Propanol	C_3H_8O	82.21	39.85	45.39	1
Propene	C_3H_6	-47.6	18.42	14.24	1
2-Propoxyethanol	$C_5H_{12}O_2$	152	41.40	52.12	1
Propyl acetate	$C_5H_{10}O_2$	101.0	33.92	39.72	1

Name	Mol. form.	t_b/°C	$\Delta_{vap}H(t_b)$/ kJ mol^{-1}	$\Delta_{vap}H$(25°C)/ kJ mol^{-1}	Ref.
Propylamine	C_3H_9N	47.21	29.55	31.27	1
Propylbenzene	C_9H_{12}	159.2		46.22	1
Propylcyclohexane	C_9H_{18}	156.7		45.08	1
Propylcyclopentane	C_8H_{16}	130.9	34.70	41.08	1
Propyl formate	$C_4H_8O_2$	80.6	33.61	37.53	1
Propyl propanoate	$C_6H_{12}O_2$	122.2	35.54	43.45	1
Pyridazine	$C_4H_4N_2$	208		53.47	1
Pyridine	C_5H_5N	115.2	35.09	40.21	1
Pyrimidine	$C_4H_4N_2$	123.8	43.09	49.79	1
Pyrrole	C_4H_5N	129.74	38.75	45.09	1
Pyrrolidine	C_4H_9N	86.6	33.01	37.52	1
Quinoline	C_9H_7N	237.1	49.7	59.30	1
Rhenium(VII) dioxytrifluoride	F_3O_2Re	185.4	65.7		7
Rhenium(V) fluoride	F_5Re	221.3	58.1		7
Rhenium(VI) fluoride	F_6Re	33.8	28.7		7
Rhenium(VI) oxytetrafluoride	F_4ORe	171.7	61.0		7
Salicylaldehyde	$C_7H_6O_2$	208	38.2		
Selenium (α form)	Se	685	95.48		
Selenium tetrafluoride	F_4Se	101.6	47.2		
Silane	H_4Si	-111.9	12.1		
Silver(I) bromide	AgBr	1502	198		
Silver(I) chloride	AgCl	1547	199		
Silver(I) iodide	AgI	1506	143.9		
Sodium hydroxide	HNaO	1388	175		
Spiro[2.2]pentane	C_5H_8	39.0	26.76	27.49	1
Stannane	H_4Sn	-51.8	19.05		
Stibine	H_3Sb	-17	21.3		
Styrene	C_8H_8	145.3	38.7		
Succinonitrile	$C_4H_4N_2$	266	48.5		
Sulfur (rhombic)	S	444.61	45		
Sulfur dioxide	O_2S	-10.02	24.94	22.92	
Sulfur hexafluoride	F_6S	-63.8 sp		8.99	
Sulfur tetrafluoride	F_4S	-40.45	26.44		
Sulfur trioxide (α-form)	O_3S	subl		43.14	
Sulfuryl chloride	Cl_2O_2S	69.4	31.4	30.1	
Tantalum(V) bromide	Br_5Ta	348.8	62.3		
Tantalum(V) chloride	Cl_5Ta	239	54.8		
Tantalum(V) fluoride	F_5Ta	229.5	56.9		
Tellurium	Te	988	114.1		
Tellurium tetrachloride	Cl_4Te	387	77		
Tetraborane(10)	B_4H_{10}	18	27.1		
1,1,2,2-Tetrabromoethane	$C_2H_2Br_4$	248	48.7		
Tetrabromosilane	Br_4Si	154	37.9		
1,1,2,2-Tetrachloroethane	$C_2H_2Cl_4$	146.0	37.64	45.71	1
Tetrachloroethene	C_2Cl_4	121.2	34.68	39.68	1
Tetrachloromethane	CCl_4	76.7	29.82	32.43	1
Tetrachlorosilane	Cl_4Si	57.65	28.7	29.7	
Tetradecane	$C_{14}H_{30}$	253.5	48.16	71.73	
1,14-Tetradecanediol	$C_{14}H_{30}O_2$			149.7	11
Tetradecanenitrile	$C_{14}H_{27}N$			85.29	1
1-Tetradecanol	$C_{14}H_{30}O$	295.8		98.9	9
Tetradecylbenzene	$C_{20}H_{34}$	347		95.8	14
Tetrafluorodiborane	B_2F_4	-34.0	28		
Tetrafluorohydrazine	F_4N_2	-74	13.27		
Tetrahydrofuran	C_4H_8O	66.0	29.81	31.99	1
Tetrahydrofurfuryl alcohol	$C_5H_{10}O_2$	176.3	45.2		
1,2,3,4-Tetrahydronaphthalene	$C_{10}H_{12}$	207.2	43.9		
Tetrahydropyran	$C_5H_{10}O$	88.0	31.17	34.58	1
Tetrahydrothiophene	C_4H_8S	121.1	34.66	39.43	1
Tetraiodosilane	I_4Si	287.35	50.2		
2,2,3,3-Tetramethylbutane	C_8H_{18}	106.32		42.90	1
2,2,4,4-Tetramethylpentane	C_9H_{20}	122.2	32.51	38.49	1

Name	Mol. form.	t_b/°C	$\Delta_{vap}H(t_b)$/ kJ mol^{-1}	$\Delta_{vap}H(25°C)$/ kJ mol^{-1}	Ref.
Tetranitromethane	CN_4O_8	125.6	40.74	49.93	1
Thallium(I) bromide	BrTl	819	99.56		
Thallium(I) chloride	ClTl	720	102.2		
Thallium(I) iodide	ITl	824	104.7		
Thallium(I) sulfide	STl_2	1367	154		
Thiacyclohexane	$C_5H_{10}S$	141.73	35.96	42.58	1
Thietane	C_3H_6S	95.0	32.32	35.97	1
Thionitrosyl fluoride (NSF)	FNS	4.8	22.2		
Thionyl chloride	Cl_2OS	75.6	31.7	31	
Thionyl fluoride	F_2OS	-43.8	21.8		
Thiophene	C_4H_4S	84.1	31.48	34.70	1
Thorium(IV) chloride	Cl_4Th	921	146.4		
Thorium(IV) fluoride	F_4Th	1680	258		
Tin(II) bromide	Br_2Sn	639	102		
Tin(IV) bromide	Br_4Sn	205	43.5		
Tin(II) chloride	Cl_2Sn	623	86.8		
Tin(IV) chloride	Cl_4Sn	114.15	34.9		
Tin(II) iodide	I_2Sn	714	105		
Tin(IV) iodide	I_4Sn	364.35	56.9		
Titanium(IV) bromide	Br_4Ti	233.5	44.37		
Titanium(II) chloride	Cl_2Ti	1500	232		
Titanium(III) chloride	Cl_3Ti	960	124		
Titanium(IV) chloride	Cl_4Ti	136.45	36.2		
Titanium(IV) iodide	I_4Ti	377	58.4		
Toluene	C_7H_8	110.60	33.18	38.01	1
Triacetamide	$C_6H_9NO_3$			60.41	1
Tribromomethane	$CHBr_3$	149.2	39.66	46.05	1
Tribromosilane	Br_3HSi	109	34.8		
Tributylamine	$C_{12}H_{27}N$	207	46.9		
Tributyl borate	$C_{12}H_{27}BO_3$	233.8	56.1		
1,1,1-Trichloroethane	$C_2H_3Cl_3$	74.02	29.86	32.50	1
1,1,2-Trichloroethane	$C_2H_3Cl_3$	113	34.82	40.24	1
Trichloroethene	C_2HCl_3	86.8	31.40	34.54	1
Trichlorofluoromethane	CCl_3F	23.7	25.1		
Trichloromethane	$CHCl_3$	61.2	29.24	31.28	1
1,2,3-Trichloropropane	$C_3H_5Cl_3$	158	37.1		
Trichlorosilane	Cl_3HSi	33		25.7	
1,1,1-Trichloro-2,2,2-trifluoroethane	$C_2Cl_3F_3$	46	26.85	28.08	1
1,1,2-Trichloro-1,2,2-trifluoroethane	$C_2Cl_3F_3$	47.6	27.04	28.40	1
Tridecane	$C_{13}H_{28}$	235.4	46.20	66.68	
1,13-Tridecanediol	$C_{13}H_{28}O_2$			133	11
1-Tridecanol	$C_{13}H_{28}O$	287		94.7	9
Tridecylbenzene	$C_{19}H_{32}$	340		91.8	14
Triethylamine	$C_6H_{15}N$	88.8	31.01	34.84	1
Triethylene glycol	$C_6H_{14}O_4$	288.6	71.4		
Trifluoroacetic acid	$C_2HF_3O_2$	72	33.3		
1,1,1-Trifluoroethane	$C_2H_3F_3$	-47.2	18.99		
(Trifluoromethyl)benzene	$C_7H_5F_3$	102.0	32.63	37.60	1
Trifluorosilane	F_3HSi	-95	16.2		
Trigermane	Ge_3H_8	110.5	32.2		
Trimethylamine	C_3H_9N	2.8	22.94	21.66	1
1,2,3-Trimethylbenzene	C_9H_{12}	176.0		49.05	1
1,2,4-Trimethylbenzene	C_9H_{12}	169.4		47.93	1
1,3,5-Trimethylbenzene	C_9H_{12}	164.7		47.50	1
2,2,3-Trimethylbutane	C_7H_{16}	80.8	28.90	32.05	1
2,3,3-Trimethyl-1-butene	C_7H_{14}	77.8		32.09	
2,2,5-Trimethylhexane	C_9H_{20}	124	33.65	40.16	1
2,3,5-Trimethylhexane	C_9H_{20}	131	34.43	41.41	1
3,5,5-Trimethyl-1-hexanol	$C_9H_{20}O$	193		67.86	12
2,4,7-Trimethyloctane	$C_{11}H_{24}$	170	38.22	49.91	1
2,2,3-Trimethylpentane	C_8H_{18}	109.8	31.94	36.91	1
2,2,4-Trimethylpentane	C_8H_{18}	99.2	30.79	35.14	1

Name	Mol. form.	t_b/°C	$\Delta_{vap}H(t_b)$/ kJ mol^{-1}	$\Delta_{vap}H$(25°C)/ kJ mol^{-1}	Ref.
2,3,3-Trimethylpentane	C_8H_{18}	114.7	32.12	37.27	1
2,3,4-Trimethylpentane	C_8H_{18}	113.4	32.36	37.75	1
2,2,4-Trimethyl-3-pentanone	$C_8H_{16}O$	146	35.64	43.30	1
2,4,4-Trimethyl-1-pentene	C_8H_{16}	101.3		35.59	
2,4,4-Trimethyl-2-pentene	C_8H_{16}	104.9		37.23	
2,3,6-Trimethylpyridine	$C_8H_{11}N$	170	39.95	50.61	1
2,4,6-Trimethylpyridine	$C_8H_{11}N$	170	39.87	50.33	1
Trisilane	H_8Si_3	52.9	28.5		
Tris(perfluorobutyl)amine	$C_{12}F_{27}N$	178	46.4		
Tungsten(VI) chloride	Cl_6W	337	52.7		
Tungsten(VI) fluoride	F_6W	17.1	26.5		7
Tungsten(VI) oxytetrachloride	Cl_4OW	230	67.8		
Tungsten(VI) oxytetrafluoride	F_4OW	185.9	59.5		7
Undecane	$C_{11}H_{24}$	195.9	41.91	56.58	
1,11-Undecanediol	$C_{11}H_{24}O_2$			132	11
Undecanenitrile	$C_{11}H_{21}N$	265		71.14	1
1-Undecanol	$C_{11}H_{24}O$	246		85.8	9
Undecylbenzene	$C_{17}H_{28}$	312		82.4	14
Vanadium(IV) chloride	Cl_4V	151	41.4	42.5	
Vanadium(V) fluoride	F_5V	48.3	44.52		
Vanadyl trichloride	Cl_3OV	127	36.78		
Vinyl acetate	$C_4H_6O_2$	72.6	34.6		
Water	H_2O	99.974	40.65	43.98	
Xenon	Xe	-108.099	12.57		
o-Xylene	C_8H_{10}	144.4	36.24	43.43	1
m-Xylene	C_8H_{10}	139.1	35.66	42.65	1
p-Xylene	C_8H_{10}	138.3	35.67	42.40	1
2,4-Xylenol	$C_8H_{10}O$	210.94		64.96	1
2,5-Xylenol	$C_8H_{10}O$	211.14	46.9		
2,6-Xylenol	$C_8H_{10}O$	201.03		75.31	1
3,4-Xylenol	$C_8H_{10}O$	227.31		85.03	1
3,5-Xylenol	$C_8H_{10}O$	221.71		82.01	1
Zinc bromide	Br_2Zn	≈670	118		
Zinc chloride	Cl_2Zn	732	126		
Zinc fluoride	F_2Zn	1500	190.1		

ENTHALPY OF FUSION

This table lists the molar enthalpy (heat) of fusion, $\Delta_{fus}H$, of over 1100 inorganic and organic compounds. All values refer to the enthalpy change at equilibrium between the liquid phase and the most stable solid phase at the phase transition temperature. Most values of $\Delta_{fus}H$ are given at the normal melting point t_m. However, a "tp" following the entry in the melting point column indicates a triple-point temperature, where the solid, liquid, and gas phases are in equilibrium. Temperatures are given on the ITS–90 scale.

Substances are listed by name, either an IUPAC systematic name or, in the case of drugs and other complex compounds, a common synonym. The molecular formula in the Hill convention is included.

References

1. Chase, M. W., Davies, C. A., Downey, J. R., Frurip, D. J., McDonald, R. A., and Syverud, A. N., *JANAF Thermochemical Tables, Third Edition, J. Phys. Chem. Ref. Data*, Vol. 14, Suppl. 1, 1985.
2. Chase, M. W., *NIST–JANAF Thermochemical Tables, Fourth Edition, J. Phys. Chem. Ref. Data*, Monograph No. 9, 1998.
3. Gurvich, L. V., Veyts, I. V., and Alcock, C. B., *Thermodynamic Properties of Individual Substances, Fourth Edition;* Vol. 2, Hemisphere Publishing Corp., New York, 1991; Vol. 3, CRC Press, Boca Raton, FL, 1994.
4. Dinsdale, A. T., "SGTE Data for Pure Elements", *CALPHAD*, 15, 317–425, 1991.
5. *Landolt–Börnstein, Numerical Data and Functional Relationships in Science and Technology, New Series,* IV/8A, "Enthalpies of Fusion and Transition of Organic Compounds," Springer–Verlag, Heidelberg, 1995.
6. *Landolt–Börnstein, Numerical Data and Functional Relationships in Science and Technology, New Series,* IV/19A, "Thermodynamic Properties of Inorganic Materials compiled by SGTE," Springer–Verlag, Heidelberg; Part 1, 1999; Part 2; 1999; Part 3, 2000; Part 4, 2001.
7. Janz, G. J., et al., *Physical Properties Data Compilations Relevant to Energy Storage. II. Molten Salts,* Nat. Stand. Ref. Data Sys.– Nat. Bur. Standards (U.S.), No. 61, Part 2, 1979.
8. Dirand, M., Bouroukba, M., Chevallier, V., Petitjean, D., Behar, E., and Ruffier–Meray, V., "Normal Alkanes, Multialkane Synthetic Model Mixtures, and Real Petroleum Waxes: Crystallographic Structures, Thermodynamic Properties, and Crystallization," *J. Chem. Eng. Data*, 47, 115–143, 2002.
9. Linstrom, P. J., and Mallard, W. G., Editors, *NIST Chemistry WebBook,* NIST Standard Reference Database No. 69, June 2005, National Institute of Standards and Technology, Gaithersburg, MD, <http://webbook.nist.gov>.
10. Thermodynamic Research Center, National Institute of Standards and Technology, *TRC Thermodynamic Tables*, <http://trc.nist.gov>.
11. Sangster, J., "Phase Diagrams and Thermodynamic Properties of Binary Systems of Drugs," *J. Phys. Chem. Ref. Data* 28, 889, 1999.
12. Chevallier, V., et al., *J. Chem. Eng. Data* 46, 1114, 2001.
13. Dirand, M., et al., *J. Chem. Eng. Data* 47, 115, 2002.
14. Rycerz, L., and Gaune-Escard, M., *J. Chem. Eng. Data* 49, 1078, 2004.
15. Ott, J. B., and Goates, J. R., *J. Chem. Eng. Data* 41, 669, 1996.

Molar Enthalpy of Fusion

Name	Mol. form.	t_m/°C	$\Delta_{fus}H$/ kJ mol^{-1}
Acenaphthene	$C_{12}H_{10}$	93	21.49
Acenaphthylene	$C_{12}H_8$	89.4	6.9
Acetaldehyde	C_2H_4O	-123.4	2.31
Acetamide	C_2H_5NO	80.16	15.59
Acetanilide	C_8H_9NO	114.35	21.3
Acetic acid	$C_2H_4O_2$	17	11.73
Acetic anhydride	$C_4H_6O_3$	-73.4	10.5
Acetone	C_3H_6O	-94.9	5.77
Acetonitrile	C_2H_3N	-44	8.16
Acrylic acid	$C_3H_4O_2$	13.56	9.51
Acrylonitrile	C_3H_3N	-83.51	6.23
Actinium	Ac	1050	12.0
Allene	C_3H_4	-136.4	4.40
Aluminum	Al	660.323	10.71
Aluminum bromide	$AlBr_3$	97.5	11.25
Aluminum chloride	$AlCl_3$	192.6	35.35
Aluminum fluoride	AlF_3	2250 tp (220 MPa)	98
Aluminum iodide	AlI_3	188.28	15.90
Aluminum oxide (α)	Al_2O_3	2053	111.1
Aluminum sulfide	Al_2S_3	1100	66
Americium	Am	1176	14.39
N-[(4-Aminophenyl)sulfonyl] acetamide	$C_8H_{10}N_2O_3S$	182.0	22.4
3-Amino-1-propanol	C_3H_9NO	12.1	19.7
4-Amino-*N*-2-pyridinylbenzenesulfonamide	$C_{11}H_{11}N_3O_2S$	190	34.4
4-Amino-*N*-2-pyrimidinylbenzenesulfonamide	$C_{10}H_{10}N_4O_2S$	261	42.6
Aminopyrine	$C_{13}H_{17}N_3O$	107.5	27.6
Ammonia	H_3N	-77.65	5.66
Ammonium chloride	ClH_4N	520.1 tp (dec)	10.6
Ammonium fluoride	FH_4N	238	12.6
Ammonium iodide	H_4IN	551 dec	21
Ammonium nitrate	$H_4N_2O_3$	169.7	5.86
Ampyrone	$C_{11}H_{13}N_3O$	109	24.9
Aniline	C_6H_7N	-6.0	10.54
Aniline-2-carboxylic acid	$C_7H_7NO_2$	144.6	20.5
Aniline-4-carboxylic acid	$C_7H_7NO_2$	188.2	22.5
Anisole	C_7H_8O	-37.3	12.9
Anthracene	$C_{14}H_{10}$	216	29.4
Antimony (gray)	Sb	630.628	19.79
Antimony(III) bromide	Br_3Sb	97	14.6
Antimony(III) chloride	Cl_3Sb	73.4	12.97
Antimony(III) fluoride	F_3Sb	287	22.8
Antimony(III) iodide	I_3Sb	171	22.8
Antimony(III) oxide (valentinite)	O_3Sb_2	655	54
Antimony(III) sulfide	S_3Sb_2	550	47.9
Argon	Ar	-189.34	1.18
Arsenic (gray)	As	817	24.44
Arsenic(III) bromide	$AsBr_3$	31.1	11.7
Arsenic(III) chloride	$AsCl_3$	-16	10.1
Arsenic(III) fluoride	AsF_3	-5.9	10.4
Arsenic(III) iodide	AsI_3	141	21.8

Name	Mol. form.	t_m/°C	$\Delta_{fus}H$/ kJ mol^{-1}
Arsenic(V) oxide	As_2O_5	730	60
Arsenic(III) oxide (claudetite)	As_2O_3	314	18
Arsenic(III) selenide	As_2Se_3	377	40.8
Arsenic sulfide	As_4S_4	307	25.4
Arsenic(III) sulfide	As_2S_3	312	28.7
Arsenic(III) telluride	As_2Te_3	375	46.0
trans-Azobenzene	$C_{12}H_{10}N_2$	67.88	22.52
trans-Azoxybenzene	$C_{12}H_{10}N_2O$	34.6	17.9
Barbital	$C_8H_{12}N_2O_3$	189	24.7
Barium	Ba	727	7.12
Barium bromide	$BaBr_2$	857	32.2
Barium carbonate	$CBaO_3$	1555	40
Barium chloride	$BaCl_2$	961	15.85
Barium fluoride	BaF_2	1368	23.36
Barium hydride	BaH_2	1200	25
Barium hydroxide	BaH_2O_2	408	16
Barium iodide	BaI_2	711	26.5
Barium oxide	BaO	1973	46
Barium sulfate	BaO_4S	1580	40
Barium sulfide	BaS	2227	63
Benzaldehyde	C_7H_6O	-57.12	9.32
Benzamide	C_7H_7NO	128	19.5
Benz[*a*]anthracene	$C_{18}H_{12}$	160	21.4
Benzene	C_6H_6	5.538	9.87
Benzeneacetic acid	$C_8H_8O_2$	76.7	16.3
1,2-Benzenediamine	$C_6H_8N_2$	103	23.1
1,3-Benzenediamine	$C_6H_8N_2$	65.5	15.57
1,4-Benzenediamine	$C_6H_8N_2$	140.3	23.8
Benzenethiol	C_6H_6S	-14.87	11.48
p-Benzidine	$C_{12}H_{12}N_2$	127.0	19.1
Benzil	$C_{14}H_{10}O_2$	94.84	23.5
Benzoic acid	$C_7H_6O_2$	122.352	18.02
Benzonitrile	C_7H_5N	-12.82	9.1
Benzo[*c*]phenanthrene	$C_{18}H_{12}$	67	16.3
Benzophenone	$C_{13}H_{10}O$	48.0	18.19
Benzo[*a*]pyrene	$C_{20}H_{12}$	179	17.3
Benzo[*e*]pyrene	$C_{20}H_{12}$	180	16.6
p-Benzoquinone	$C_6H_4O_2$	113	18.5
Benzoyl chloride	C_7H_5ClO	-0.5	19.2
Benzyl alcohol	C_7H_8O	-15.5	8.97
Beryllium	Be	1287	7.895
Beryllium bromide	$BeBr_2$	508	18
Beryllium carbide	CBe_2	2127	75.3
Beryllium chloride	$BeCl_2$	415	8.66
Beryllium fluoride	BeF_2	552	4.77
Beryllium iodide	BeI_2	480	20.92
Beryllium nitride	Be_3N_2	2200	111
Beryllium oxide	BeO	2578	86
Beryllium sulfate	BeO_4S	1127	6
2,2'-Binaphthalene	$C_{20}H_{14}$	188.1	38.9
Biphenyl	$C_{12}H_{10}$	68.93	18.57
Bismuth	Bi	271.402	11.106
Bismuth oxide	Bi_2O_3	825	14.7
Bismuth sulfide	Bi_2S_3	777	78.2
Bismuth tribromide	$BiBr_3$	219	21.7
Bismuth trichloride	$BiCl_3$	234	23.6
Bismuth trifluoride	BiF_3	649	21.6
Bismuth triiodide	BiI_3	408.6	39.1
Boric acid	BH_3O_3	170.9	22.3
Boron	B	2077	50.2

Name	Mol. form.	t_m/°C	$\Delta_{fus}H$/ kJ mol^{-1}
Boron nitride	BN	2967	81
Boron oxide	B_2O_3	450	24.56
Boron sulfide	B_2S_3	563	48.12
Boron trichloride	BCl_3	-107.3	2.10
Boron trifluoride	BF_3	-126.8	4.20
Bromine	Br_2	-7.2	10.57
Bromine pentafluoride	BrF_5	-60.5	5.67
Bromobenzene	C_6H_5Br	-30.74	10.70
1-Bromobutane	C_4H_9Br	-112.5	9.23
2-Bromobutane, (±)-	C_4H_9Br	-112.6	6.89
Bromoethane	C_2H_5Br	-118.4	7.47
Bromoethene	C_2H_3Br	-139.5	5.12
1-Bromoheptane	$C_7H_{15}Br$	-56.1	21.8
1-Bromohexane	$C_6H_{13}Br$	-84.9	18.1
Bromomethane	CH_3Br	-93.7	5.98
1-Bromonaphthalene	$C_{10}H_7Br$	6.1	15.2
2-Bromonaphthalene	$C_{10}H_7Br$	58	14.4
1-Bromooctane	$C_8H_{17}Br$	-55.0	24.7
1-Bromopentane	$C_5H_{11}Br$	-88.0	14.37
1-Bromopropane	C_3H_7Br	-110.1	6.44
2-Bromopropane	C_3H_7Br	-88.9	6.53
Bromotrichloromethane	$CBrCl_3$	-5.6	2.53
1,2-Butadiene	C_4H_6	-136.20	6.96
1,3-Butadiene	C_4H_6	-108.9	7.98
Butanal	C_4H_8O	-96.86	10.77
Butane	C_4H_{10}	-138.2	4.66
1,4-Butanediol	$C_4H_{10}O_2$	20.43	18.70
1-Butanethiol	$C_4H_{10}S$	-115.66	10.46
Butanoic acid	$C_4H_8O_2$	-5.12	11.59
1-Butanol	$C_4H_{10}O$	-88.60	9.37
2-Butanol	$C_4H_{10}O$	-88.44	5.97
2-Butanone	C_4H_8O	-86.67	8.39
1-Butene	C_4H_8	-185.33	3.96
cis-2-Butene	C_4H_8	-138.89	7.31
trans-2-Butene	C_4H_8	-105.52	9.76
cis-2-Butenoic acid	$C_4H_6O_2$	15	12.6
trans-2-Butenoic acid	$C_4H_6O_2$	71.3	13.0
tert-Butylamine	$C_4H_{11}N$	-66.92	0.882
Butylbenzene	$C_{10}H_{14}$	-87.81	11.22
Butylcyclohexane	$C_{10}H_{20}$	-74.68	14.16
Butyl methyl ether	$C_5H_{12}O$	-115.7	10.85
1-Butyne	C_4H_6	-125.7	6.03
2-Butyne	C_4H_6	-32.2	9.23
γ-Butyrolactone	$C_4H_6O_2$	-43.36	9.57
Cadmium	Cd	321.069	6.21
Cadmium bromide	Br_2Cd	568	33.35
Cadmium chloride	$CdCl_2$	568	48.58
Cadmium fluoride	CdF_2	1075	22.6
Cadmium iodide	CdI_2	388	15.3
Cadmium nitrate	CdN_2O_6	360	18.3
Caffeine	$C_8H_{10}N_4O_2$	236.1	22.0
Calcium	Ca	842	8.54
Calcium bromide	Br_2Ca	742	29.1
Calcium carbonate (calcite)	$CCaO_3$	800	36
Calcium chloride	$CaCl_2$	775	28.05
Calcium fluoride	CaF_2	1418	30
Calcium hydride	CaH_2	1000	6.7
Calcium iodide	CaI_2	783	41.8
Calcium nitrate	CaN_2O_6	561	23.4
Calcium oxide	CaO	2613	80

Name	Mol. form.	t_m/°C	$\Delta_{fus}H$/ kJ mol^{-1}
Calcium sulfate	CaO_4S	1460	28
Calcium sulfide	CaS	2524	70
Carbazole	$C_{12}H_9N$	245	24.1
Carbon dioxide	CO_2	-56.561 tp	9.02
Carbon diselenide	CSe_2	-43.6	6.36
Carbon disulfide	CS_2	-111.7	4.39
Carbon (graphite)	C	4489 tp (10.3 MPa)	117.4
Carbon monoxide	CO	-205.1	0.833
Carbon oxysulfide	COS	-138.8	4.73
Carbonyl chloride	CCl_2O	-127.77	5.74
Cerium	Ce	799	5.460
Cerium(III) bromide	Br_3Ce	732	51.9
Cerium(III) chloride	$CeCl_3$	807	53.1
Cerium(III) fluoride	CeF_3	1430	55.6
Cerium(III) iodide	CeI_3	760	51.0
Cerium(III) oxide	Ce_2O_3	2250	120
Cerium(IV) oxide	CeO_2	2480	80
Cesium	Cs	28.5	2.09
Cesium carbonate	CCs_2O_3	793	31
Cesium chloride	ClCs	646	20.4
Cesium chromate	$CrCs_2O_4$	963	35.3
Cesium fluoride	CsF	703	21.7
Cesium hydride	CsH	528	15
Cesium hydroxide	CsHO	342.3	7.78
Cesium iodide	CsI	632	25.7
Cesium metaborate	$BCsO_2$	732	27
Cesium molybdate	Cs_2MoO_4	956.3	31.8
Cesium nitrate	$CsNO_3$	409	13.8
Cesium nitrite	$CsNO_2$	406	10.9
Cesium oxide	Cs_2O	495	20
Cesium peroxide	Cs_2O_2	594	22
Cesium sulfate	Cs_2O_4S	1005	35.7
Chlorine	Cl_2	-101.5	6.40
Chloroacetic acid	$C_2H_3ClO_2$	62.0	12.28
2-Chloroaniline	C_6H_6ClN	-2.3	11.9
3-Chloroaniline	C_6H_6ClN	-10.3	10.15
4-Chloroaniline	C_6H_6ClN	70.4	20.0
Chlorobenzene	C_6H_5Cl	-45.2	9.6
2-Chlorobenzoic acid	$C_7H_5ClO_2$	140.4	25.6
Chlorocyclohexane	$C_6H_{11}Cl$	-45	2.043
Chlorodifluoromethane	$CHClF_2$	-157.41	4.12
Chloroethane	C_2H_5Cl	-138.4	4.45
Chloroethene	C_2H_3Cl	-153.84	4.92
Chloromethane	CH_3Cl	-97.6	6.43
2-Chloro-2-methylpropane	C_4H_9Cl	-25.60	2.07
1-Chloronaphthalene	$C_{10}H_7Cl$	-6.0	12.9
2-Chloronaphthalene	$C_{10}H_7Cl$	58.02	14.0
1-Chloro-2-nitrobenzene	$C_6H_4ClNO_2$	32.1	17.9
1-Chloro-3-nitrobenzene	$C_6H_4ClNO_2$	43.6	19.4
1-Chloro-4-nitrobenzene	$C_6H_4ClNO_2$	82.2	14.1
Chloropentafluoroethane	C_2ClF_5	-99.4	1.86
2-Chlorophenol	C_6H_5ClO	8	13.0
3-Chlorophenol	C_6H_5ClO	32.5	14.9
4-Chlorophenol	C_6H_5ClO	43.1	14.1
1-Chloropropane	C_3H_7Cl	-122.9	5.54
2-Chloropropane	C_3H_7Cl	-117.1	7.39
2-Chlorotoluene	C_7H_7Cl	-35.9	9.6
Chlorotrifluoroethene	C_2ClF_3	-158.14	5.55
Chromium	Cr	1907	21.00
Chromium(III) bromide	Br_3Cr	812	60
Chromium(II) bromide	Br_2Cr	842	45
Chromium(II) chloride	Cl_2Cr	824	45.0
Chromium(III) chloride	Cl_3Cr	827	60
Chromium(II) fluoride	CrF_2	894	34
Chromium(III) fluoride	CrF_3	1425	66
Chromium(II) iodide	CrI_2	867	46
Chromium(III) iodide	CrI_3	857	61
Chromium(III) oxide	Cr_2O_3	2432	125
Chromium(VI) oxide	CrO_3	197	14.2
Chromium(II) sulfide	CrS	1567	25.5
Chrysene	$C_{18}H_{12}$	255.0	26.2
Cobalt	Co	1495	16.20
Cobalt(II) bromide	Br_2Co	678	43
Cobalt(II) chloride	Cl_2Co	737	46.0
Cobalt(II) fluoride	CoF_2	1127	58.1
Cobalt(II) iodide	CoI_2	520	35
Cobalt(II) selenite	CoO_3Se	659	16.3
Cobalt(II) sulfide	CoS	1117	30
Copper	Cu	1084.62	13.26
Copper(I) bromide	BrCu	483	5.1
Copper(I) chloride	ClCu	423	7.08
Copper(II) chloride	Cl_2Cu	598	15.0
Copper(II) fluoride	CuF_2	836	55
Copper(I) iodide	CuI	591	7.93
Copper(I) oxide	Cu_2O	1244	65.6
Copper(II) oxide	CuO	1227	49
Copper(I) sulfide	Cu_2S	1129	9.62
Coronene	$C_{24}H_{12}$	437.3	19.2
o-Cresol	C_7H_8O	31.0	15.82
m-Cresol	C_7H_8O	12.2	10.71
p-Cresol	C_7H_8O	34.77	12.71
Curium	Cm	1345	14.64
Cyanamide	CH_2N_2	45.55	7.27
Cyanogen	C_2N_2	-27.83	8.11
Cyclobutane	C_4H_8	-90.7	1.09
Cycloheptane	C_7H_{14}	-8.0	1.88
Cycloheptanol	$C_7H_{14}O$	7.15	1.60
Cyclohexane	C_6H_{12}	6.7	2.68
Cyclohexanol	$C_6H_{12}O$	26	1.78
Cyclohexanone	$C_6H_{10}O$	-27.93	1.328
Cyclohexene	C_6H_{10}	-103.5	3.29
Cyclohexylamine	$C_6H_{13}N$	-17.7	17.5
Cyclohexylbenzene	$C_{12}H_{16}$	7.02	15.6
Cyclooctane	C_8H_{16}	14.82	2.41
Cyclopentane	C_5H_{10}	-93.4	0.61
Cyclopentanol	$C_5H_{10}O$	-17	1.535
Cyclopentene	C_5H_8	-135.02	3.36
Cyclopentylamine	$C_5H_{11}N$	-82.69	8.31
Cyclopropane	C_3H_6	-127.6	5.44
Cyclopropylamine	C_3H_7N	-35.38	13.18
Decaborane(14)	$B_{10}H_{14}$	98.78	21.97
cis-Decahydronaphthalene	$C_{10}H_{18}$	-42.9	9.49
trans-Decahydronaphthalene	$C_{10}H_{18}$	-30.35	14.41
Decanal	$C_{10}H_{20}O$	-3.9	34.5
Decane	$C_{10}H_{22}$	-29.61	28.72
Decanedioic acid	$C_{10}H_{18}O_4$	131	40.8
Decanoic acid	$C_{10}H_{20}O_2$	31.39	27.8
1-Decanol	$C_{10}H_{22}O$	7	43
1-Decene	$C_{10}H_{20}$	-66.21	13.81
5,5-Diallyl-2,4,6(1*H*,3*H*,5*H*)-pyrimidinetrione	$C_{10}H_{12}N_2O_3$	172	32.3

Fluids

Name	Mol. form.	t_m/°C	$\Delta_{fus}H$/ kJ mol^{-1}
1,2-Dibromoethane	$C_2H_4Br_2$	9.8	10.89
1,2-Dibromopropane	$C_3H_6Br_2$	-55.4	8.94
1,3-Dibromopropane	$C_3H_6Br_2$	-35	14.6
1,2-Dibromotetrafluoroethane	$C_2Br_2F_4$	-110.1	7.04
o-Dichlorobenzene	$C_6H_4Cl_2$	-17.0	12.4
m-Dichlorobenzene	$C_6H_4Cl_2$	-24.8	12.6
p-Dichlorobenzene	$C_6H_4Cl_2$	53.1	18.19
1,1-Dichloroethane	$C_2H_4Cl_2$	-96.93	7.87
1,2-Dichloroethane	$C_2H_4Cl_2$	-35.6	8.84
1,1-Dichloroethene	$C_2H_2Cl_2$	-122.5	6.51
cis-1,2-Dichloroethene	$C_2H_2Cl_2$	-80.0	7.2
Dichloromethane	CH_2Cl_2	-94.9	4.60
1,2-Dichloropropane, (±)-	$C_3H_6Cl_2$	-100.53	6.40
2,2-Dichloropropane	$C_3H_6Cl_2$	-33.9	2.30
1,2-Dichloro-1,1,2,2-tetrafluoroethane	$C_2Cl_2F_4$	-92.52	1.51
Diethyl ether	$C_4H_{10}O$	-116.22	7.19
3,3-Diethylpentane	C_9H_{20}	-33.04	10.09
Diethyl sulfide	$C_4H_{10}S$	-103.9	10.90
o-Difluorobenzene	$C_6H_4F_2$	-47.1	11.05
m-Difluorobenzene	$C_6H_4F_2$	-69.11	8.58
1,2-Dihydro-1,5-dimethyl-2-phenyl-3*H*-pyrazol-3-one	$C_{11}H_{12}N_2O$	108.0	27.3
Diisopropyl ether	$C_6H_{14}O$	-85.37	12.04
1,2-Dimethoxyethane	$C_4H_{10}O_2$	-69.0	12.6
Dimethoxymethane	$C_3H_8O_2$	-105.11	8.33
Dimethylamine	C_2H_7N	-93	5.94
2,2-Dimethylbutane	C_6H_{14}	-99.0	0.58
2,3-Dimethylbutane	C_6H_{14}	-128.1	0.79
2,3-Dimethyl-2-butene	C_6H_{12}	-74.3	6.45
1,1-Dimethylcyclohexane	C_8H_{16}	-33.31	2.07
cis-1,2-Dimethylcyclohexane	C_8H_{16}	-49.83	1.64
trans-1,2-Dimethylcyclohexane	C_8H_{16}	-88.12	10.49
cis-1,3-Dimethylcyclohexane	C_8H_{16}	-75.51	10.82
trans-1,3-Dimethylcyclohexane	C_8H_{16}	-90.05	9.87
cis-1,4-Dimethylcyclohexane	C_8H_{16}	-87.4	9.31
trans-1,4-Dimethylcyclohexane	C_8H_{16}	-36.9	12.33
Dimethyl disulfide	$C_2H_6S_2$	-84.67	9.19
Dimethyl ether	C_2H_6O	-141.49	4.94
N,N-Dimethylformamide	C_3H_7NO	-60.3	7.90
1,1-Dimethylhydrazine	$C_2H_8N_2$	-57.15	10.07
1,2-Dimethylhydrazine	$C_2H_8N_2$	-8.86	13.64
Dimethyl oxalate	$C_4H_6O_4$	51	21.1
2,2-Dimethylpentane	C_7H_{16}	-123.71	5.82
2,4-Dimethylpentane	C_7H_{16}	-119.16	6.85
3,3-Dimethylpentane	C_7H_{16}	-134.4	6.85
Dimethyl sulfide	C_2H_6S	-98.26	7.99
Dimethyl sulfone	$C_2H_6O_2S$	108.83	18.30
Dimethyl sulfoxide	C_2H_6OS	18.52	14.37
N,N-Dimethylurea	$C_3H_8N_2O$	181.2	23.0
N,N'-Dimethylurea	$C_3H_8N_2O$	106	13.0
Dimethyl zinc	C_2H_6Zn	-43.01	6.83
1,4-Dioxane	$C_4H_8O_2$	11.75	12.84
1,3-Dioxolane	$C_3H_6O_2$	-97.21	6.57
Diphenylamine	$C_{12}H_{11}N$	53.2	18.5
Diphenyl ether	$C_{12}H_{10}O$	26.865	17.22
Diphenylmethane	$C_{13}H_{12}$	25.22	18.6
Dipropyl ether	$C_6H_{14}O$	-114.8	10.8
Divinyl ether	C_4H_6O	-101	7.9
Docosane	$C_{22}H_{46}$	43.8	48.8
Dodecane	$C_{12}H_{26}$	-9.55	36.8
Dodecanoic acid	$C_{12}H_{24}O_2$	43.82	36.3
1-Dodecanol	$C_{12}H_{26}O$	24.2	40.2
1-Dodecene	$C_{12}H_{24}$	-35.19	19.9
Dotriacontane	$C_{32}H_{66}$	69.7	75.8
Dysprosium	Dy	1412	11.35
Dysprosium(III) fluoride	DyF_3	1157	58.6
Dysprosium(III) oxide	Dy_2O_3	2408	120
Eicosane	$C_{20}H_{42}$	36.48	69.9
1-Eicosanol	$C_{20}H_{42}O$	63.9	42
Einsteinium	Es	860	9.41
Erbium	Er	1529	19.90
Erbium chloride	Cl_3Er	776	32.6
Erbium fluoride	ErF_3	1146	28.2
Erbium oxide	Er_2O_3	2418	130
Estra-1,3,5(10)-triene-3,17-diol 3-benzoate, (17β)	$C_{25}H_{28}O_3$	198	41.8
Ethane	C_2H_6	-182.77	2.72 [a]
1,2-Ethanediamine	$C_2H_8N_2$	11.14	22.58
1,2-Ethanediol	$C_2H_6O_2$	-13	9.96
Ethanethiol	C_2H_6S	-147.89	4.98
Ethanol	C_2H_6O	-114.14	4.931
Ethinylestradiol	$C_{20}H_{24}O_2$	186	27.9
N-(4-Ethoxyphenyl)acetamide	$C_{10}H_{13}NO_2$	135	33.0
Ethyl acetate	$C_4H_8O_2$	-83.8	10.48
Ethyl 4-aminobenzoate	$C_9H_{11}NO_2$	89.6	22.3
Ethylbenzene	C_8H_{10}	-94.95	9.18
Ethylcyclohexane	C_8H_{16}	-111.28	8.33
Ethylene	C_2H_4	-169.15	3.35
Ethyl methyl sulfide	C_3H_8S	-105.89	9.76
3-Ethylpentane	C_7H_{16}	-118.55	9.55
2-Ethyltoluene	C_9H_{12}	-80.7	9.96
3-Ethyltoluene	C_9H_{12}	-95.7	7.6
4-Ethyltoluene	C_9H_{12}	-62.7	12.7
Europium	Eu	822	9.21
Europium(II) bromide	Br_2Eu	683	25.1
Europium(III) chloride	Cl_3Eu	623	33.1
Europium(III) fluoride	EuF_3	647	6.40
Europium(II) oxide	EuO	1967	40
Europium(III) oxide	Eu_2O_3	2350	117
Fluoranthene	$C_{16}H_{10}$	110.2	18.69
9*H*-Fluorene	$C_{13}H_{10}$	114.76	19.58
Fluorine	F_2	-219.67	0.51
Fluorobenzene	C_6H_5F	-42.18	11.31
Formamide	CH_3NO	2.57	8.44
Formic acid	CH_2O_2	8.3	12.68
Furan	C_4H_4O	-85.58	3.80
Furfural	$C_5H_4O_2$	-38.3	14.37
Furfuryl alcohol	$C_5H_6O_2$	-14.5	13.13
Gadolinium	Gd	1313	9.67
Gadolinium(III) bromide	Br_3Gd	785	38.1
Gadolinium(III) chloride	Cl_3Gd	602	40.6
Gadolinium(III) fluoride	F_3Gd	1232	52.4
Gadolinium(III) iodide	GdI_3	930	54.0
Gadolinium(III) oxide	Gd_2O_3	2425	60
Gallium	Ga	29.7646	5.585
Gallium antimonide	GaSb	712	25.1
Gallium arsenide	AsGa	1238	87.64
Gallium(III) bromide	Br_3Ga	123	11.7
Gallium(III) chloride	Cl_3Ga	77.9	11.51
Gallium(III) iodide	GaI_3	212	12.9
Gallium(III) oxide	Ga_2O_3	1807	100

Name	Mol. form.	t_m/°C	$\Delta_{fus}H$/ kJ mol⁻¹
Germanium	Ge	938.25	36.94
Germanium(IV) bromide	Br_4Ge	26.1	12
Germanium(II) iodide	GeI_2	428	33.3
Germanium(IV) iodide	GeI_4	146	19.1
Germanium(IV) oxide	GeO_2	1116	12.6
Germanium(II) selenide	GeSe	675	24.7
Germanium(II) sulfide	GeS	658	21.3
Germanium(IV) sulfide	GeS_2	840	16.3
Germanium(II) telluride	GeTe	724	47.3
Glycerol	$C_3H_8O_3$	18.2	18.3
Gold	Au	1064.18	12.55
Hafnium	Hf	2233	27.20
Hafnium nitride	HfN	3310	62.8
Hafnium(IV) oxide	HfO_2	2800	96
Heneicosane	$C_{21}H_{44}$	40.4	45.21
Heptacosane	$C_{27}H_{56}$	58.8	61.9
Heptadecane	$C_{17}H_{36}$	21.97	40.16
Heptanal	$C_7H_{14}O$	-43.94	23.2
Heptane	C_7H_{16}	-90.549	14.03
Heptanoic acid	$C_7H_{14}O_2$	-7.17	15.13
1-Heptanol	$C_7H_{16}O$	-33.2	18.17
1-Heptene	C_7H_{14}	-118.83	12.41
Hexachlorobenzene	C_6Cl_6	230	25.2
Hexachloroethane	C_2Cl_6	186.8	9.75
Hexacontane	$C_{60}H_{122}$	100	193.2
Hexacosane	$C_{26}H_{54}$	56.09	60.0
Hexadecane	$C_{16}H_{34}$	18.18	53.36
Hexadecanoic acid	$C_{16}H_{32}O_2$	62.49	53.7
1-Hexadecanol	$C_{16}H_{34}O$	49.30	33.6
Hexafluorobenzene	C_6F_6	5.10	11.59
Hexafluoroethane	C_2F_6	-100.02	2.69
Hexamethylbenzene	$C_{12}H_{18}$	165.6	20.6
Hexanal	$C_6H_{12}O$	-58.2	13.3
Hexane	C_6H_{14}	-95.27	13.08
1,6-Hexanedioic acid	$C_6H_{10}O_4$	151.5	36.3
1,6-Hexanediol	$C_6H_{14}O_2$	41.5	22.2
1-Hexanol	$C_6H_{14}O$	-46.4	15.38
2-Hexanone	$C_6H_{12}O$	-55.45	14.9
3-Hexanone	$C_6H_{12}O$	-55.4	13.49
Hexatetracontane	$C_{46}H_{94}$	87.6	176.0
Hexatriacontane	$C_{36}H_{74}$	75.81	87.7
1-Hexene	C_6H_{12}	-139.76	9.35
cis-2-Hexene	C_6H_{12}	-141.12	8.88
Holmium	Ho	1472	11.76
Holmium bromide	Br_3Ho	919	50.1
Holmium chloride	Cl_3Ho	720	30.5
Holmium fluoride	F_3Ho	1143	56.3
Holmium oxide	Ho_2O_3	2415	130
Hydrazine	H_4N_2	1.54	12.66
Hydrogen	H_2	-259.16	0.12
Hydrogen bromide	BrH	-86.80	2.41
Hydrogen chloride	ClH	-114.17	2.00
Hydrogen cyanide	CHN	-13.28	8.41
Hydrogen fluoride	FH	-83.36	4.58
Hydrogen iodide	HI	-50.76	2.87
Hydrogen peroxide	H_2O_2	-0.43	12.50
Hydrogen sulfide	H_2S	-85.5	2.38
p-Hydroquinone	$C_6H_6O_2$	173	26.8
2-Hydroxybenzoic acid	$C_7H_6O_3$	158.6	14.2
N-(4-Hydroxyphenyl)acetamide	$C_8H_9NO_2$	168.0	30.5

Name	Mol. form.	t_m/°C	$\Delta_{fus}H$/ kJ mol⁻¹
Imidazole	$C_3H_4N_2$	89.52	12.82
Indan	C_9H_{10}	-51.34	8.60
Indene	C_9H_8	-1.45	10.20
Indium	In	156.5985	3.291
Indium antimonide	InSb	524	47.7
Indium arsenide	AsIn	942	77.0
Indium(I) bromide	BrIn	285	24.3
Indium(III) bromide	Br_3In	420	26
Indium(I) chloride	ClIn	225	9.20
Indium(III) chloride	Cl_3In	583	27
Indium(III) fluoride	F_3In	1172	64
Indium(II) iodide	I_2In	155	1.29
Indium(I) iodide	IIn	364.4	17.26
Indium(III) iodide	I_3In	207	18.48
Indium(III) oxide	In_2O_3	1912	105
Indium(II) sulfide	InS	692	36.0
Indomethacin	$C_{19}H_{16}ClNO_4$	160	36.9
Iodine	I_2	113.7	15.52
Iodine chloride	ClI	27.38	11.6
Iodobenzene	C_6H_5I	-30.7	9.75
Iridium	Ir	2446	41.12
Iridium(VI) fluoride	F_6Ir	44	8.40
Iron	Fe	1538	13.81
Iron boride (FeB)	BFe	1658	62.66
Iron(II) bromide	Br_2Fe	691	43.0
Iron(II) chloride	Cl_2Fe	677	42.83
Iron(III) chloride	Cl_3Fe	307.6	40
Iron(II) fluoride	F_2Fe	1100	50
Iron(III) fluoride	F_3Fe	367	0.58
Iron(II) iodide	FeI_2	594	39
Iron(II) oxide	FeO	1377	24.1
Iron(II,III) oxide	Fe_3O_4	1597	138
Iron(III) oxide	Fe_2O_3	1539	87
Iron sodium oxide	$FeNaO_2$	1347	49.4
Iron(II) sulfide	FeS	1188	31.5
Isobutane	C_4H_{10}	-159.38	4.54
Isobutene	C_4H_8	-140.7	5.92
Isopentane	C_5H_{12}	-159.8	5.15
Isopropylamine	C_3H_9N	-95.119	7.33
Isopropylbenzene	C_9H_{12}	-96.01	7.33
1-Isopropyl-4-methylbenzene	$C_{10}H_{14}$	-68.1	9.66
Isoquinoline	C_9H_7N	26.46	13.54
Khellin	$C_{14}H_{12}O_5$	154 dec	32.3
Krypton	Kr	-157.37	1.64
Lanthanum	La	920	6.20
Lanthanum bromide	Br_3La	788	54.0
Lanthanum chloride	Cl_3La	858	54.4
Lanthanum fluoride	F_3La	1493	50.2
Lanthanum iodide	I_3La	778	56.1
Lead	Pb	327.462	4.774
Lead(II) bromide	Br_2Pb	371	16.44
Lead(II) chloride	Cl_2Pb	501	21.88
Lead(II) fluoride	F_2Pb	830	14.7
Lead(II) iodide	I_2Pb	410	23.4
Lead(II) oxide (massicot)	OPb	887	25.6
Lead(II) sulfate	O_4PbS	1087	40.2
Lead(II) sulfide	PbS	1113	49.4
Lithium	Li	180.50	3.00
Lithium aluminate	$AlLiO_2$	1610	87.9
Lithium bromide	BrLi	550	17.66

Name	Mol. form.	t_m/°C	$\Delta_{fus}H$/ kJ mol^{-1}
Lithium carbonate	CLi_2O_3	732	44.8
Lithium chloride	$ClLi$	610	19.8
Lithium chromate	$CrLi_2O_4$	482	30.5
Lithium fluoride	FLi	848.2	27.09
Lithium hexafluoroaluminate	AlF_6Li_3	785	86.19
Lithium hydride	HLi	692	21.8
Lithium hydride-*d*	DLi	694	22
Lithium hydroxide	$HLiO$	473	20.9
Lithium iodide	ILi	469	14.6
Lithium metasilicate	Li_2O_3Si	1201	28
Lithium nitrate	$LiNO_3$	253	26.7
Lithium nitrite	$LiNO_2$	222	9.2
Lithium perchlorate	$ClLiO_4$	236	29.3
Lithium sulfate	Li_2O_4S	860	9.00
Lutetium	Lu	1663	18.65
Lutetium oxide	Lu_2O_3	2490	133
Magnesium	Mg	650	8.48
Magnesium bromide	Br_2Mg	711	39.3
Magnesium carbonate	$CMgO_3$	990	59
Magnesium chloride	Cl_2Mg	714	43.1
Magnesium fluoride	F_2Mg	1263	58.7
Magnesium hydride	H_2Mg	327	14
Magnesium iodide	I_2Mg	634	26
Magnesium orthosilicate	Mg_2O_4Si	1897	71
Magnesium oxide	MgO	2825	77
Magnesium phosphate	$Mg_3O_8P_2$	1348	121
Magnesium sulfate	MgO_4S	1137	14.6
Magnesium sulfide	MgS	2226	63
Magnesium tetraboride	B_4Mg	727	0.0
Maleic anhydride	$C_4H_2O_3$	52.56	13.60
Manganese	Mn	1246	12.91
Manganese(II) bromide	Br_2Mn	698	33.5
Manganese(II) chloride	Cl_2Mn	650	30.7
Manganese(II) fluoride	F_2Mn	900	30
Manganese(II) iodide	I_2Mn	638	41.8
Manganese(II) oxide	MnO	1842	43.9
Manganese(II) sulfide (α form)	MnS	1530	26.1
Mercury	Hg	-38.8290	2.295
Mercury(II) bromide	Br_2Hg	241	17.9
Mercury(II) chloride	Cl_2Hg	277	19.41
Mercury(II) fluoride	F_2Hg	645 dec	23.0
Mercury(I) iodide	Hg_2I_2	290	31.4
Mercury(II) iodide (yellow)	HgI_2	256	15.6
Mercury(II) sulfide (black)	HgS	820	40
Metaboric acid (γ form)	BHO_2	236	14.3
Methane	CH_4	-182.475	0.94
Methanethiol	CH_4S	-122.98	5.91
Methanol	CH_4O	-97.5	3.215
Methyl acetate	$C_3H_6O_2$	-98.2	7.49
Methylamine	CH_5N	-93.42	6.13
2-Methylaniline	C_7H_9N	-14.41	11.66
3-Methylaniline	C_7H_9N	-30.8	7.9
4-Methylaniline	C_7H_9N	43.3	18.9
Methyl benzoate	$C_8H_8O_2$	-12.35	9.74
2-Methyl-1,3-butadiene	C_5H_8	-146.1	4.93
2-Methyl-2-butanol	$C_5H_{12}O$	-8.7	4.46
3-Methyl-2-butanone	$C_5H_{10}O$	-93.13	9.34
2-Methyl-1-butene	C_5H_{10}	-137.53	7.91
3-Methyl-1-butene	C_5H_{10}	-168.41	5.36
2-Methyl-2-butene	C_5H_{10}	-133.72	7.60

Name	Mol. form.	t_m/°C	$\Delta_{fus}H$/ kJ mol^{-1}
Methyl *tert*-butyl ether	$C_5H_{12}O$	-108.6	7.60
Methylcyclohexane	C_7H_{14}	-126.6	6.75
Methylcyclopentane	C_6H_{12}	-142.419	6.93
Methylcyclopropane	C_4H_8	-177.2	2.8
2-Methylfuran	C_5H_6O	-91.2	8.55
2-Methylheptane	C_8H_{18}	-109	11.92
3-Methylheptane	C_8H_{18}	-120.48	11.69
4-Methylheptane	C_8H_{18}	-121.0	10.8
2-Methylhexane	C_7H_{16}	-118.23	9.19
Methylhydrazine	CH_6N_2	-52.3	10.42
Methyl methacrylate	$C_5H_8O_2$	-47.55	14.4
1-Methylnaphthalene	$C_{11}H_{10}$	-30.43	6.95
2-Methylnaphthalene	$C_{11}H_{10}$	34.6	12.13
Methyl nitrate	CH_3NO_3	-82.9	8.24
Methyloxirane	C_3H_6O	-111.9	6.53
2-Methylpentane	C_6H_{14}	-153.60	6.27
3-Methylpentane	C_6H_{14}	-162.89	5.30
2-Methyl-1-propanol	$C_4H_{10}O$	-101.96	6.32
2-Methyl-2-propanol	$C_4H_{10}O$	25.81	6.70
2-Methylpyridine	C_6H_7N	-66.65	9.72
3-Methylpyridine	C_6H_7N	-18.1	14.18
4-Methylpyridine	C_6H_7N	3.68	12.58
N-Methylurea	$C_2H_6N_2O$	101.3	14.0
Molybdenum	Mo	2622	37.48
Molybdenum boride [Mo_2B_5]	B_5Mo_2	2210	226
Molybdenum(IV) chloride	Cl_4Mo	317	16.7
Molybdenum(V) chloride	Cl_5Mo	194	19
Molybdenum(VI) dioxydichloride	Cl_2MoO_2	176	17.0
Molybdenum(V) fluoride	F_5Mo	45.67	6.1
Molybdenum(VI) fluoride	F_6Mo	17.5	4.33
Molybdenum monoboride	BMo	2600	55.23
Molybdenum(VI) oxide	MoO_3	802	48.7
Molybdenum(VI) oxytetrachloride	Cl_4MoO	105	14.3
Molybdenum(VI) oxytetrafluoride	F_4MoO	97.2	4
Molybdenum(V) oxytrichloride	Cl_3MoO	310	22
Molybdenum(III) sulfide	Mo_2S_3	1807	0.13
Morpholine	C_4H_9NO	-4.8	14.5
Naphthalene	$C_{10}H_8$	80.22	19.01
1-Naphthol	$C_{10}H_8O$	95.1	23.1
2-Naphthol	$C_{10}H_8O$	122	18.1
Neodymium	Nd	1016	7.14
Neodymium(III) bromide	Br_3Nd	682	45.3
Neodymium(III) chloride	Cl_3Nd	759	48.5
Neodymium(III) fluoride	F_3Nd	1377	54.8
Neodymium(III) iodide	I_3Nd	787	41.5
Neon	Ne	-248.59	0.328
Neopentane	C_5H_{12}	-16.37	3.10
Neptunium	Np	644	3.20
Nickel	Ni	1455	17.48
Nickel boride (Ni_2B)	BNi_2	1125	42.15
Nickel boride (Ni_3B)	BNi_3	1166	72.28
Nickel(II) bromide	Br_2Ni	963	56
Nickel(II) chloride	Cl_2Ni	1031	77.9
Nickel disulfide	NiS_2	1007	65.7
Nickel(II) fluoride	F_2Ni	1380	69
Nickel(II) iodide	I_2Ni	800	48
Nickel(II) oxide	NiO	1957	50.7
Nickel subsulfide	Ni_3S_2	789	19.7
Nickel(II) sulfide	NiS	976	30.1
Niobium	Nb	2477	30

Name	Mol. form.	t_m/°C	$\Delta_{fus}H$/ kJ mol⁻¹
Niobium(V) bromide	Br_5Nb	254	24.0
Niobium(V) chloride	Cl_5Nb	205.8	33.9
Niobium(V) fluoride	F_5Nb	80	12.2
Niobium(V) iodide	I_5Nb	327	37.7
Niobium nitride	NNb	2050	46.0
Niobium(II) oxide	NbO	1937	85.4
Niobium(IV) oxide	NbO_2	1901	92
Niobium(V) oxide	Nb_2O_5	1512	104.3
Nitric acid	HNO_3	-41.6	10.5
Nitric oxide	NO	-163.6	2.30
2-Nitroaniline	$C_6H_6N_2O_2$	71	16.1
3-Nitroaniline	$C_6H_6N_2O_2$	112	23.6
4-Nitroaniline	$C_6H_6N_2O_2$	147.7	21.2
Nitrobenzene	$C_6H_5NO_2$	5.65	12.12
Nitroethane	$C_2H_5NO_2$	-89.42	9.85
Nitrogen	N_2	-210.0	0.71
Nitrogen tetroxide	N_2O_4	-9.3	14.65
Nitromethane	CH_3NO_2	-28.7	9.70
2-Nitrophenol	$C_6H_5NO_3$	44.9	17.7
3-Nitrophenol	$C_6H_5NO_3$	95	20.6
4-Nitrophenol	$C_6H_5NO_3$	113.8	18.8
Nitrosobenzene	C_6H_5NO	67.8	31.0
4-Nitrotoluene	$C_7H_7NO_2$	51.7	16.81
Nitrous oxide	N_2O	-90.8	6.54
Nonacosane	$C_{29}H_{60}$	63.7	66.9
Nonadecane	$C_{19}H_{40}$	31.5	45.8
Nonanal	$C_9H_{18}O$	-19.3	30.5
Nonane	C_9H_{20}	-53.47	15.47
Nonanoic acid	$C_9H_{18}O_2$	12.38	19.82
5-Nonanone	$C_9H_{18}O$	-3.84	24.93
Octacosane	$C_{28}H_{58}$	61.3	65.1
Octadecane	$C_{18}H_{38}$	28.17	61.7
1-Octadecanol	$C_{18}H_{38}O$	58.0	45
Octane	C_8H_{18}	-56.73	20.73
Octanoic acid	$C_8H_{16}O_2$	16.51	21.35
1-Octanol	$C_8H_{18}O$	-14.7	23.7
Octatriacontane	$C_{38}H_{78}$	78.6	133.2
1-Octene	C_8H_{16}	-101.66	15.31
Osmium	Os	3033	57.85
Osmium(VIII) oxide	O_4Os	40.6	14.3
2-Oxepanone	$C_6H_{10}O_2$	-1.02	13.83
Oxetane	C_3H_6O	-97	6.5
Oxirane	C_2H_4O	-112.46	5.17
4-Oxopentanoic acid	$C_5H_8O_3$	33.0	9.22
Oxygen	O_2	-218.79	0.44
Palladium	Pd	1554.8	16.74
Palladium(II) chloride	Cl_2Pd	679	18.41
Paraldehyde	$C_6H_{12}O_3$	12	13.5
Pentachloroethane	C_2HCl_5	-29.0	11.3
Pentacontane	$C_{50}H_{102}$	91.7	162.4
Pentacosane	$C_{25}H_{52}$	53.3	56.9
Pentadecane	$C_{15}H_{32}$	9.95	34.6
cis-1,3-Pentadiene	C_5H_8	-140.81	5.64
trans-1,3-Pentadiene	C_5H_8	-87.5	7.14
1,4-Pentadiene	C_5H_8	-148.3	6.12
Pentaerythritol	$C_5H_{12}O_4$	258	4.8
Pentafluorobenzene	C_6HF_5	-47.3	10.87
Pentafluorophenol	C_6HF_5O	32.80	16.41
2,3,4,5,6-Pentafluorotoluene	$C_7H_3F_5$	-29.79	13.1
Pentane	C_5H_{12}	-129.67	8.40
Pentanedioic acid	$C_5H_8O_4$	97.9	20.3
Pentanenitrile	C_5H_9N	-96.2	9
1-Pentanethiol	$C_5H_{12}S$	-75.69	17.53
Pentanoic acid	$C_5H_{10}O_2$	-33.63	14.16
1-Pentanol	$C_5H_{12}O$	-77.58	10.50
2-Pentanone	$C_5H_{10}O$	-76.83	10.63
3-Pentanone	$C_5H_{10}O$	-38.98	11.59
Pentatriacontane	$C_{35}H_{72}$	74.4	86.3
1-Pentene	C_5H_{10}	-165.13	5.94
cis-2-Pentene	C_5H_{10}	-151.35	7.11
trans-2-Pentene	C_5H_{10}	-140.20	8.35
Perfluoroacetone	C_3F_6O	-125.45	8.38
Perfluorobutane	C_4F_{10}	-129	7.66
Perfluorocyclobutane	C_4F_8	-40.16	2.77
Perfluoroheptane	C_7F_{16}	-51.3	6.95
Perfluorohexane	C_6F_{14}	-86.1	6.84
Perfluoropropane	C_3F_8	-147.7	0.477
Perfluorotoluene	C_7F_8	-65.48	11.54
Perylene	$C_{20}H_{12}$	276	31.9
Phenanthrene	$C_{14}H_{10}$	99	16.46
Phenobarbital	$C_{12}H_{12}N_2O_3$	176	27.8
Phenol	C_6H_6O	40.89	11.51
α-Phenylbenzeneacetic acid	$C_{14}H_{12}O_2$	147.26	31.3
Phenylbutazone	$C_{19}H_{20}N_2O_2$	104	27.7
Phenylhydrazine	$C_6H_8N_2$	20	14.05
Phosphinic acid	H_3O_2P	26.5	9.7
Phosphonic acid	H_3O_3P	74.4	12.8
Phosphoric acid	H_3O_4P	42.4	13.4
Phosphorus (white)	P	44.15	0.659
Phosphorus (red)	P	579.2	18.54
Phosphorus(III) chloride	Cl_3P	-93	7.10
Phosphorus heptasulfide	P_4S_7	308	36.6
Phosphorus(V) oxide	O_5P_2	562	27.2
Phosphorus sesquisulfide	P_4S_3	173	20.1
Phosphoryl chloride	Cl_3OP	1.18	13.1
Piperidine	$C_5H_{11}N$	-11.05	14.85
Platinum	Pt	1768.2	22.175
Plutonium	Pu	640	2.824
Plutonium(III) bromide	Br_3Pu	681	58.6
Plutonium(III) chloride	Cl_3Pu	760	63.6
Plutonium(III) fluoride	F_3Pu	1396	59.8
Plutonium(IV) fluoride	F_4Pu	1037	42.7
Plutonium(VI) fluoride	F_6Pu	51.6	18.6
Plutonium(III) iodide	I_3Pu	777	50.2
Plutonium(III) oxide	O_3Pu_2	2085	113
Plutonium(IV) oxide	O_2Pu	2390	67
Polonium	Po	254	10.0
Potassium	K	63.5	2.335
Potassium acetate	$C_2H_3KO_2$	309	7.65
Potassium aluminate	$AlKO_2$	1713	82
Potassium bromide	BrK	734	25.52
Potassium carbonate	CK_2O_3	899	27.6
Potassium chloride	ClK	771	26.28
Potassium chromate	CrK_2O_4	974	33.0
Potassium cyanide	CKN	622	14.6
Potassium fluoride	FK	858	27.2
Potassium fluoroborate	BF_4K	570	17.66
Potassium hydride	HK	619	21
Potassium hydrogen fluoride	F_2HK	238.8	6.62
Potassium hydroxide	HKO	406	7.90

Name	Mol. form.	t_m/°C	$\Delta_{fus}H$/ kJ mol^{-1}
Potassium iodide	IK	681	24.0
Potassium metaborate	BKO_2	947	31.38
Potassium nitrate	KNO_3	334	9.6
Potassium nitrite	KNO_2	438	16.7
Potassium oxide	K_2O	740	27
Potassium peroxide	K_2O_2	545	20.5
Potassium sulfate	K_2O_4S	1069	36.6
Potassium sulfide	K_2S	948	16.15
Potassium superoxide	KO_2	535	20.6
Praseodymium	Pr	931	6.89
Praseodymium(III) bromide	Br_3Pr	693	47.3
Praseodymium(III) chloride	Cl_3Pr	786	50.6
Praseodymium(III) fluoride	F_3Pr	1399	57.3
Praseodymium(III) iodide	I_3Pr	738	53.1
Propane	C_3H_8	-187.62	3.50
1,3-Propanediol	$C_3H_8O_2$	-27.6	7.1
Propanenitrile	C_3H_5N	-93	5.03
1-Propanethiol	C_3H_8S	-113.12	5.48
2-Propanethiol	C_3H_8S	-130.50	5.74
Propanoic acid	$C_3H_6O_2$	-20.5	10.66
1-Propanol	C_3H_8O	-124.39	5.37
2-Propanol	C_3H_8O	-87.91	5.41
Propene	C_3H_6	-185.19	3.003
Propylamine	C_3H_9N	-84.78	10.97
Propylbenzene	C_9H_{12}	-99.52	9.27
Propylcyclohexane	C_9H_{18}	-94.86	10.37
Protactinium	Pa	1572	12.34
Pyrazine	$C_4H_4N_2$	52.30	12.9
1*H*-Pyrazole	$C_3H_4N_2$	59.9	14.0
Pyrene	$C_{16}H_{10}$	150.62	17.36
Pyridine	C_5H_5N	-41.63	8.28
3-Pyridinecarboxamide	$C_6H_6N_2O$	128.8	23.2
Pyrocatechol	$C_6H_6O_2$	104.6	22.8
Pyrrole	C_4H_5N	-23.39	7.91
Pyrrolidine	C_4H_9N	-57.79	8.58
Quinoline	C_9H_7N	-14.78	10.66
Radium	Ra	696	7.7
Resorcinol	$C_6H_6O_2$	109.8	20.4
Rhenium	Re	3185	34.08
Rhenium(VII) oxide	O_7Re_2	327	65.7
Rhodium	Rh	1963	26.59
Rubidium	Rb	39.30	2.19
Rubidium bromide	BrRb	692	23.3
Rubidium carbonate	CO_3Rb_2	873	30
Rubidium chloride	ClRb	724	24.4
Rubidium fluoride	FRb	795	25.8
Rubidium hydride	HRb	585	22
Rubidium hydroxide	HORb	385	8.0
Rubidium iodide	IRb	656	22.1
Rubidium metaborate	BO_2Rb	860	31
Rubidium nitrate	NO_3Rb	310	4.6
Rubidium nitrite	NO_2Rb	422	12.1
Rubidium oxide	ORb_2	505	20
Rubidium peroxide	O_2Rb_2	570	21
Rubidium sulfate	O_4Rb_2S	1066	37.3
Rubidium superoxide	O_2Rb	540	21
Ruthenium	Ru	2333	38.59
Ruthenium(V) fluoride	F_5Ru	101	74.5
Samarium	Sm	1072	8.62
Samarium(III) oxide	O_3Sm_2	2335	119
Scandium	Sc	1541	14.10
Scandium chloride	Cl_3Sc	967	67.4
Scandium fluoride	F_3Sc	1552	62.6
Scandium oxide	O_3Sc_2	2489	127
Selenium (gray)	Se	220.8	6.69
Selenium dioxide	O_2Se	360	17.6
Silicon	Si	1414	50.21
Silicon dioxide (cristobalite)	O_2Si	1722	9.6
Silicon monosulfide	SSi	1090	31
Silver	Ag	961.78	11.30
Silver(I) bromide	AgBr	430	9.163
Silver(I) chloride	AgCl	455	13.054
Silver(I) iodide	AgI	558	9.414
Silver(I) nitrate	$AgNO_3$	210	11.72
Silver(I) oxide	Ag_2O	827	15
Silver(I) sulfate	Ag_2O_4S	660	17.99
Silver(I) sulfide	Ag_2S	836	7.9
Sodium	Na	97.794	2.60
Sodium acetate	$C_2H_3NaO_2$	328.2	17.9
Sodium bromate	$BrNaO_3$	381	28.11
Sodium bromide	BrNa	747	26.23
Sodium carbonate	CNa_2O_3	856	29.7
Sodium chlorate	$ClNaO_3$	248	22.6
Sodium chloride	ClNa	802.018	28.16
Sodium chromate	$CrNa_2O_4$	794	24.7
Sodium cyanide	CNNa	562	8.79
Sodium fluoride	FNa	996	33.35
Sodium formate	$CHNaO_2$	257.3	17.7
Sodium hexafluoroaluminate	AlF_6Na_3	1013	114.4
Sodium hexafluorosilicate	F_6Na_2Si	847	99.6
Sodium hydride	HNa	638	26
Sodium hydrogen carbonate	$CHNaO_3$	527	25
Sodium hydroxide	HNaO	323	6.60
Sodium iodate	$INaO_3$	422	35.1
Sodium iodide	INa	661	23.7
Sodium metaborate	$BNaO_2$	966	36.2
Sodium metasilicate	Na_2O_3Si	1089	51.8
Sodium nitrate	$NNaO_3$	306.5	15.5
Sodium nitrite	$NNaO_2$	284	14.9
Sodium oxide	Na_2O	1134	47.7
Sodium peroxide	Na_2O_2	675	24.5
Sodium sulfate	Na_2O_4S	884	23.85
Sodium sulfide	Na_2S	1172	19
Sodium sulfite	Na_2O_3S	911	25.9
Spiro[2.2]pentane	C_5H_8	-106.98	6.43
Stearic acid	$C_{18}H_{36}O_2$	69.3	61.2
trans-Stilbene	$C_{14}H_{12}$	124.82	27.7
Strontium	Sr	777	7.43
Strontium bromide	Br_2Sr	657	10.5
Strontium carbonate	CO_3Sr	1494	40
Strontium chloride	Cl_2Sr	874	16.22
Strontium fluoride	F_2Sr	1477	29.7
Strontium hydride	H_2Sr	1050	23
Strontium hydroxide	H_2O_2Sr	535	23
Strontium iodide	I_2Sr	538	19.7
Strontium nitrate	N_2O_6Sr	570	44.6
Strontium oxide	OSr	2531	81
Strontium sulfate	O_4SSr	1606	36
Strontium sulfide	SSr	2226	63
Styrene	C_8H_8	-30.65	10.9

Name	Mol. form.	t_m/°C	$\Delta_{fus}H$/ kJ mol⁻¹
Succinic acid	$C_4H_6O_4$	185	32.4
Succinic anhydride	$C_4H_4O_3$	119.5	20.4
Succinonitrile	$C_4H_4N_2$	57.985	3.70
Sulfamerazine	$C_{11}H_{12}N_4O_2S$	238	38.7
Sulfamethoxazole	$C_{10}H_{11}N_3O_3S$	167	32.2
Sulfamethoxypyridazine	$C_{11}H_{12}N_4O_3S$	182.5	31.3
Sulfathiazole	$C_9H_9N_3O_2S_2$	200	26.4
Sulfisoxazole	$C_{11}H_{13}N_3O_3S$	195.0	30.2
Sulfur (monoclinic)	S	115.21	1.721
Sulfur hexafluoride	F_6S	-49.596 tp	5.02
Sulfuric acid	H_2O_4S	10.31	10.71
Sulfur trioxide (γ-form)	O_3S	16.8	8.60
Tantalum	Ta	3017	36.57
Tantalum boride [TaB_2]	B_2Ta	3100	83.68
Tantalum(V) bromide	Br_5Ta	240	37.7
Tantalum(V) chloride	Cl_5Ta	216.6	35.1
Tantalum(V) fluoride	F_5Ta	96.9	12
Tantalum(V) iodide	I_5Ta	496	7.74
Tantalum nitride	NTa	3090	6.7
Tantalum nitride (Ta2N)	NTa_2	2727	92.0
Tantalum(V) oxide	O_5Ta_2	1875	120
Technetium	Tc	2157	33.29
Tellurium	Te	449.51	17.38
Tellurium dioxide	O_2Te	733	28.9
Tellurium tetrabromide	Br_4Te	380	24.7
Tellurium tetrachloride	Cl_4Te	224	18.9
Terbium	Tb	1359	10.15
Terbium(III) bromide	Br_3Tb	830	31.5
Terbium(III) chloride	Cl_3Tb	582	19.5
o-Terphenyl	$C_{18}H_{14}$	56.19	17.19
p-Terphenyl	$C_{18}H_{14}$	213.8	35.3
Tetrabromomethane	CBr_4	90	3.76
1,1,2,2-Tetrachloro-1,2-difluoroethane	$C_2Cl_4F_2$	26.54	3.67
1,1,2,2-Tetrachloroethane	$C_2H_2Cl_4$	-42.4	9.17
Tetrachloroethene	C_2Cl_4	-22.2	10.88
Tetrachloromethane	CCl_4	-22.8	2.56
Tetrachlorosilane	Cl_4Si	-68.74	7.60
Tetracontane	$C_{40}H_{82}$	81.4	135.5
Tetracosane	$C_{24}H_{50}$	50.3	54.4
Tetradecane	$C_{14}H_{30}$	5.87	45.07
Tetradecanoic acid	$C_{14}H_{28}O_2$	54.16	45.1
1-Tetradecanol	$C_{14}H_{30}O$	37.7	25.1 [a]
1,2,3,5-Tetrafluorobenzene	$C_6H_2F_4$	-46.2	6.36
1,2,4,5-Tetrafluorobenzene	$C_6H_2F_4$	3.88	15.05
Tetrafluoroethene	C_2F_4	-131.14	7.72
Tetrafluoromethane	CF_4	-183.58	0.704
Tetrahydrofuran	C_4H_8O	-108.38	8.54
Tetrahydropyran	$C_5H_{10}O$	-49.1	1.8
Tetrahydrothiophene	C_4H_8S	-96.13	7.35
Tetraiodosilane	I_4Si	120.5	19.7
1,2,4,5-Tetramethylbenzene	$C_{10}H_{14}$	79.2	21
Tetramethyl lead	$C_4H_{12}Pb$	-30.2	10.80
2,2,3,3-Tetramethylpentane	C_9H_{20}	-9.75	2.33
2,2,4,4-Tetramethylpentane	C_9H_{20}	-66.53	9.74
Tetramethylsilane	$C_4H_{12}Si$	-99.063	6.87
Tetramethylstannane	$C_4H_{12}Sn$	-55.09	9.30
Tetratetracontane	$C_{44}H_{90}$	86.0	149.6
Tetratriacontane	$C_{34}H_{70}$	72.8	79.4
1*H*-Tetrazole	CH_2N_4	156.9	18.2
Thallium	Tl	304	4.142
Thallium(I) bromide	BrTl	460	16.4
Thallium(I) carbonate	CO_3Tl_2	273	18
Thallium(I) chloride	ClTl	431	15.56
Thallium(I) fluoride	FTl	326	13.87
Thallium(I) formate	CHO_2Tl	101	10.9
Thallium(I) iodide	ITl	441.7	14.7
Thallium(I) nitrate	NO_3Tl	206	9.6
Thallium(I) oxide	OTl_2	579	30.3
Thallium(III) oxide	O_3Tl_2	834	53
Thallium(I) sulfate	O_4STl_2	632	23.8
Thallium(I) sulfide	STl_2	457	23.0
Thiazole	C_3H_3NS	-33.61	9.57
Thietane	C_3H_6S	-73.19	8.25
Thiophene	C_4H_4S	-38.12	5.07
Thiourea	CH_4N_2S	176	14.0
Thorium	Th	1750	13.81
Thorium(IV) bromide	Br_4Th	679	54.4
Thorium(IV) chloride	Cl_4Th	770	43.9
Thorium(IV) fluoride	F_4Th	1110	41.8
Thorium(IV) iodide	I_4Th	566	48.1
Thorium(IV) oxide	O_2Th	3350	90
Thulium	Tm	1545	16.84
Thulium(III) chloride	Cl_3Tm	845	34.9
Thulium(III) fluoride	F_3Tm	1158	28.9
Thymol	$C_{10}H_{14}O$	49.6	21.3
Tin(II) bromide	Br_2Sn	232	18.0
Tin(IV) bromide	Br_4Sn	29.1	12.2
Tin(II) chloride	Cl_2Sn	247.0	14.52
Tin(IV) chloride	Cl_4Sn	-34.07	9.20
Tin(II) fluoride	F_2Sn	215	10.5
Tin(IV) fluoride	F_4Sn	442	27.6
Tin(II) iodide	I_2Sn	320	18.0
Tin(IV) iodide	I_4Sn	402	0.16
Tin(II) oxide	OSn	977	27.7
Tin(IV) oxide	O_2Sn	1630	23.4
Tin(II) sulfide	SSn	881	31.6
Tin(II) telluride	SnTe	806	45.2
Tin (white)	Sn	231.928	7.15
Titanium	Ti	1670	14.15
Titanium boride	B_2Ti	2920	100.4
Titanium(IV) bromide	Br_4Ti	38.3	12.9
Titanium(II) chloride	Cl_2Ti	1035	34.3
Titanium(IV) chloride	Cl_4Ti	-24.12	9.97
Titanium(IV) fluoride	F_4Ti	377	41
Titanium(IV) iodide	I_4Ti	155	19.8
Titanium nitride	NTi	2947	66.9
Titanium(III) oxide	O_3Ti_2	1842	104.6
Titanium(IV) oxide (rutile)	O_2Ti	1912	68
Titanium(II) sulfide	STi	1927	32
Toluene	C_7H_8	-95.0	6.64
o-Toluic acid	$C_8H_8O_2$	103.4	19.5
m-Toluic acid	$C_8H_8O_2$	109.3	15.7
p-Toluic acid	$C_8H_8O_2$	180	22.7
Triacontane	$C_{30}H_{62}$	65.9	68.3
1,3,5-Triazine	$C_3H_3N_3$	80.3	14.56
Tribromomethane	$CHBr_3$	8.69	11.05
Trichloroacetic acid	$C_2HCl_3O_2$	59.1	5.90
1,2,3-Trichlorobenzene	$C_6H_3Cl_3$	53	17.9
1,2,4-Trichlorobenzene	$C_6H_3Cl_3$	17.0	16.4
1,3,5-Trichlorobenzene	$C_6H_3Cl_3$	62.8	18.1

Name	Mol. form.	t_m/°C	$\Delta_{fus}H$/ kJ mol-1
1,1,1-Trichloroethane	$C_2H_3Cl_3$	-30	2.35
1,1,2-Trichloroethane	$C_2H_3Cl_3$	-36.3	11.46
Trichloroethene	C_2HCl_3	-84.7	8.45
Trichlorofluoromethane	CCl_3F	-110.44	6.89
Trichloromethane	$CHCl_3$	-63.3	9.5
1,1,2-Trichloro-1,2,2-trifluoroethane	$C_2Cl_3F_3$		2.47
Tricosane	$C_{23}H_{48}$	47.4	50.86
Tridecane	$C_{13}H_{28}$	-5.35	28.50
1-Tridecanol	$C_{13}H_{28}O$	31	41.4
1,1,1-Trifluoroethane	$C_2H_3F_3$	-111.6	6.19
Trifluoromethane	CHF_3	-155.18	4.06
Triiodomethane	CHI_3	120	16.44
Trimethoprim	$C_{14}H_{18}N_4O_3$	199	49.4
Trimethylamine	C_3H_9N	-117.1	7
1,2,3-Trimethylbenzene	C_9H_{12}	-25.32	8.18
1,2,4-Trimethylbenzene	C_9H_{12}	-43.8	13.19
1,3,5-Trimethylbenzene	C_9H_{12}	-44.69	9.51
2,2,3-Trimethylbutane	C_7H_{16}	-24.56	2.26
2,2,4-Trimethylpentane	C_8H_{18}	-107.36	9.20
1,3,5-Trinitrobenzene	$C_6H_3N_3O_6$	121.3	15.4
Trinitroglycerol	$C_3H_5N_3O_9$	12.8	21.87
2,4,6-Trinitrotoluene	$C_7H_5N_3O_6$	80.9	22.9
1,3,5-Trioxane	$C_3H_6O_3$	60	15.11
Triphenylamine	$C_{18}H_{15}N$	126.5	24.9
Triphenylene	$C_{18}H_{12}$	197.82	24.74
Tritriacontane	$C_{33}H_{68}$	71.2	79.5
Tungsten	W	3414	52.31
Tungsten boride	BW	2800	80
Tungsten boride (W_2B)	BW_2	2740	117
Tungsten boride [W_2B_5]	B_5W_2	2370	240
Tungsten(V) bromide	Br_5W	286	17.2
Tungsten(V) chloride	Cl_5W	253	20.6
Tungsten(VI) chloride	Cl_6W	282	6.69
Tungsten(VI) fluoride	F_6W	1.9	4.10
Tungsten(VI) oxide	O_3W	1473	73
Tungsten(VI) oxytetrachloride	Cl_4OW	210	18.8
Tungsten(VI) oxytetrafluoride	F_4OW	105	6
Undecane	$C_{11}H_{24}$	-25.54	22.2
Uranium	U	1135	9.14
Uranium(III) bromide	Br_3U	727	43.9
Uranium(IV) bromide	Br_4U	519	55.2
Uranium(IV) chloride	Cl_4U	590	44.8
Uranium(III) fluoride	F_3U	1495	36.8
Uranium(IV) fluoride	F_4U	1036	47
Uranium(V) fluoride	F_5U	348	35
Uranium(VI) fluoride	F_6U	64.06 tp	19.2
Uranium(IV) iodide	I_4U	506	42.1
Uranium(IV) oxide	O_2U	2847	74.2
Uranyl chloride	Cl_2O_2U	577	44.06
Urea	CH_4N_2O	132.4	13.9
Vanadium	V	1910	21.5
Vanadium(II) chloride	Cl_2V	1350	35.0
Vanadium(IV) chloride	Cl_4V	-28	2.30
Vanadium(II) fluoride	F_2V	1490	44
Vanadium(III) fluoride	F_3V	1395	57
Vanadium(V) fluoride	F_5V	19.5	49.96
Vanadium(II) oxide	OV	1790	50
Vanadium(III) oxide	O_3V_2	1957	140
Vanadium(IV) oxide	O_2V	1545	56.0
Vanadium(V) oxide	O_5V_2	681	64
Water	H_2O	0.00	6.01
Xenon	Xe	-111.75	2.27
Xenon difluoride	F_2Xe	129.03 tp	16.8
Xenon hexafluoride	F_6Xe	49.48	5.74
Xenon tetrafluoride	F_4Xe	117.10 tp	16.3
o-Xylene	C_8H_{10}	-25.16	13.6
m-Xylene	C_8H_{10}	-47.85	11.6
p-Xylene	C_8H_{10}	13.3	17.12
2,3-Xylenol	$C_8H_{10}O$	72.7	21.0
2,5-Xylenol	$C_8H_{10}O$	74.9	23.4
2,6-Xylenol	$C_8H_{10}O$	45.4	18.9
3,4-Xylenol	$C_8H_{10}O$	65.1	18.1
3,5-Xylenol	$C_8H_{10}O$	63.4	17.4
Ytterbium	Yb	824	7.66
Ytterbium(III) chloride	Cl_3Yb	854	35.4
Yttrium	Y	1522	11.39
Yttrium chloride	Cl_3Y	721	31.5
Yttrium fluoride	F_3Y	1155	27.9
Yttrium oxide	O_3Y_2	2439	81
Zinc	Zn	419.527	7.068
Zinc bromide	Br_2Zn	402	15.7
Zinc chloride	Cl_2Zn	325	10.30
Zinc fluoride	F_2Zn	872	40
Zinc iodide	I_2Zn	450	17
Zinc oxide	OZn	1974	70
Zinc phosphide [ZnP_2]	P_2Zn	980	92.9
Zinc selenite	O_3SeZn	621	46.4
Zinc sulfide (wurtzite)	SZn	1827	30
Zinc telluride	TeZn	1295	63
Zirconium	Zr	1854	21.00
Zirconium boride	B_2Zr	3050	104.6
Zirconium(IV) bromide	Br_4Zr	450 tp	30
Zirconium(II) bromide	Br_2Zr	827	28
Zirconium(III) bromide	Br_3Zr	727	33
Zirconium(III) chloride	Cl_3Zr	627	30
Zirconium(II) chloride	Cl_2Zr	722	27.0
Zirconium(IV) chloride	Cl_4Zr	437 tp	29
Zirconium(IV) fluoride	F_4Zr	910	61
Zirconium(II) fluoride	F_2Zr	902	37.7
Zirconium(III) fluoride	F_3Zr	927	50
Zirconium(IV) iodide	I_4Zr	500	32
Zirconium(II) iodide	I_2Zr	827	28
Zirconium(III) iodide	I_3Zr	727	33
Zirconium nitride	NZr	2952	67.4
Zirconium(IV) oxide	O_2Zr	2710	90
Zirconium(IV) sulfide	S_2Zr	1550	45

[a] The value of $\Delta_{fus}H$ includes the enthalpy of transition between two crystalline phases whose transformation occurs within 1 °C of the melting point.

COMPRESSIBILITY AND EXPANSION COEFFICIENTS OF LIQUIDS

This table gives data on the variation of the density of some common liquids with pressure and temperature. The pressure dependence is described to first order by the isothermal compressibility coefficient κ defined as

$$\kappa_T = -(1/V)\,(\partial V/\partial P)_T$$

where V is the volume and P the pressure, and the temperature dependence by the cubic expansion coefficient α,

$$\alpha_V = (1/V)\,(\partial V/\partial T)_P$$

Substances are listed alphabetically by name. More precise data on the variation of density with temperature over a wide temperature range can be found in Reference 1.

References

1. Lide, D. R., and Kehiaian, H. V., *CRC Handbook of Thermophysical and Thermochemical Data*, CRC Press, Boca Raton, FL, 1994.
2. Le Neindre, B., *Effets des Hautes et Très Hautes Pressions*, in *Techniques de l'Ingénieur*, Paris, 1991.
3. *Landolt-Börnstein, Numerical Data and Functional Relationships in Science and Technology, New Series*, IV/4, *High-Pressure Properties of Matter*, Springer-Verlag, Heidelberg, 1980.
4. Riddick, J.A., Bunger, W.B., and Sakano, T.K., *Organic Solvents, Fourth Edition*, John Wiley & Sons, New York, 1986.
5. Isaacs, N. S., *Liquid Phase High Pressure Chemistry*, John Wiley, New York, 1981.

Compressibility and Expansion Coefficients of Liquids

Name	Mol. form.	t/°C	$\kappa_T \cdot 10^4$/ MPa^{-1}	$\alpha_V \cdot 10^3$/ °C^{-1}
Acetic acid	$C_2H_4O_2$	20	9.08	1.08
Acetic acid	$C_2H_4O_2$	80	13.7	1.38
Acetone	C_3H_6O	20	12.62	1.46
Acetone	C_3H_6O	40	15.6	1.57
Aniline	C_6H_7N	20	4.53	0.81
Aniline	C_6H_7N	80	6.32	0.91
Anisole	C_7H_8O	20	6.60	0.951
Benzene	C_6H_6	25	9.66	1.14
Benzene	C_6H_6	45	11.28	1.21
Bromobenzene	C_6H_5Br	20	6.46	0.86
1-Bromobutane	C_4H_9Br	20		1.13
1-Bromobutane	C_4H_9Br	25	10.26	
Bromoethane	C_2H_5Br	20	11.53	1.31
1-Bromopentane	$C_5H_{11}Br$	0	8.42	
1-Bromopentane	$C_5H_{11}Br$	25		1.04
1-Bromopropane	C_3H_7Br	0	10.22	
1-Bromopropane	C_3H_7Br	25		1.2
1-Butanol	$C_4H_{10}O$	0	8.10	1.12
Carbon disulfide	CS_2	20	9.38	1.12
Carbon disulfide	CS_2	35		1.16
Carbon disulfide	CS_2	40	10.6	
Chlorobenzene	C_6H_5Cl	20	7.45	0.94
1-Chloropropane	C_3H_7Cl	0	12.09	
1-Chloropropane	C_3H_7Cl	20		1.4
Cycloheptane	C_7H_{14}	20	9.22	
Cyclohexane	C_6H_{12}	20	11.30	1.15
Cyclohexane	C_6H_{12}	60	15.2	1.29
Cyclooctane	C_8H_{16}	20	8.03	
Cyclopentane	C_5H_{10}	20	13.31	1.35
Decane	$C_{10}H_{22}$	25	10.94	1.02
Dibromomethane	CH_2Br_2	27	6.85	
Dibutyl phthalate	$C_{16}H_{22}O_4$	0	5.0	
Dibutyl phthalate	$C_{16}H_{22}O_4$	25		0.86
1,1-Dichloroethane	$C_2H_4Cl_2$	20	7.97	
1,1-Dichloroethane	$C_2H_4Cl_2$	25		0.93
1,2-Dichloroethane	$C_2H_4Cl_2$	20		1.14
1,2-Dichloroethane	$C_2H_4Cl_2$	30	8.46	
trans-1,2-Dichloroethene	$C_2H_2Cl_2$	25	11.2	1.36
Dichloromethane	CH_2Cl_2	25	10.3	1.39
Diethylene glycol	$C_4H_{10}O_3$	0	3.34	
Diethylene glycol	$C_4H_{10}O_3$	20		0.635
Diethyl ether	$C_4H_{10}O$	20	18.65	1.65
Diethyl ether	$C_4H_{10}O$	30	20.85	1.72
2,3-Dimethylbutane	C_6H_{14}	20	17.97	
2,3-Dimethylbutane	C_6H_{14}	25		1.39
Dodecane	$C_{12}H_{26}$	25	9.88	0.93
1,2-Ethanediol	$C_2H_6O_2$	20	3.64	0.626
Ethanol	C_2H_6O	20	11.19	1.40
Ethanol	C_2H_6O	70	15.93	1.67
Ethyl acetate	$C_4H_8O_2$	20	11.32	1.35
Ethyl acetate	$C_4H_8O_2$	60	16.2	1.54
Glycerol	$C_3H_8O_3$	0	2.54	
Glycerol	$C_3H_8O_3$	20		0.520
Glycerol triacetate	$C_9H_{14}O_6$	0	4.49	
Glycerol triacetate	$C_9H_{14}O_6$	25		0.94
Heptane	C_7H_{16}	25	14.38	1.26
Hexadecane	$C_{16}H_{34}$	25	8.57	
Hexadecane	$C_{16}H_{34}$	45	9.78	
Hexane	C_6H_{14}	25	16.69	1.41
Hexane	C_6H_{14}	45	20.27	1.52
1-Hexanol	$C_6H_{14}O$	25	8.24	1.03
1-Iodobutane	C_4H_9I	0	7.73	
1-Iodobutane	C_4H_9I	25		1.02
Iodoethane	C_2H_5I	20	9.82	
Iodoethane	C_2H_5I	25		1.17
Iodomethane	CH_3I	25		1.26
Iodomethane	CH_3I	27	10.3	
1-Iodopentane	$C_5H_{11}I$	0	7.56	
1-Iodopropane	C_3H_7I	0	10.22	
1-Iodopropane	C_3H_7I	25		1.09
Mercury	Hg	20	0.401	1.811
Methanol	CH_4O	20	12.14	1.49
Methanol	CH_4O	40	13.83	1.59
Methyl *cis*-9-octadecenoate	$C_{19}H_{36}O_2$	0	6.18	
Methyl *cis*-9-octadecenoate	$C_{19}H_{36}O_2$	60		0.85

Name	Mol. form.	t/°C	$\kappa_T \cdot 10^4$/ MPa^{-1}	$\alpha_V \cdot 10^3$/ °C^{-1}
2-Methylpentane	C_6H_{14}	0	13.97	
2-Methylpentane	C_6H_{14}	25		1.43
3-Methylpentane	C_6H_{14}	0	14.57	
3-Methylpentane	C_6H_{14}	25		1.40
Nitrobenzene	$C_6H_5NO_2$	20	4.93	
Nitrobenzene	$C_6H_5NO_2$	25		0.833
Nonane	C_9H_{20}	25	11.75	1.08
Octane	C_8H_{18}	25	12.82	1.16
Octane	C_8H_{18}	45	15.06	1.23
1-Octanol	$C_8H_{18}O$	25	7.64	0.827
Pentadecane	$C_{15}H_{32}$	25	8.82	
Pentane	C_5H_{12}	25	21.80	1.64
1-Pentanol	$C_5H_{12}O$	0	7.71	1.02
Phenol	C_6H_6O	60	6.05	0.82
Phosphorus(III) chloride	Cl_3P	20	9.45	1.9
1,2-Propanediol	$C_3H_8O_2$	0	4.45	
1,2-Propanediol	$C_3H_8O_2$	20		0.695
1,3-Propanediol	$C_3H_8O_2$	0	4.09	
1,3-Propanediol	$C_3H_8O_2$	20		0.61
1-Propanol	C_3H_8O	0	8.43	1.22
2-Propanol	C_3H_8O	40	13.32	1.55
Tetrachloroethene	C_2Cl_4	25	7.56	1.02
Tetrachloromethane	CCl_4	20	10.50	1.14
Tetrachloromethane	CCl_4	40	12.20	1.21
Tetrachloromethane	CCl_4	70	15.6	1.33
Tetradecane	$C_{14}H_{30}$	25	9.10	0.87
Toluene	C_7H_8	20	8.96	1.05
Toluene	C_7H_8	50	11.0	1.13
Tribromomethane	$CHBr_3$	25		0.91
Tribromomethane	$CHBr_3$	50	8.76	
Trichloroethene	C_2HCl_3	25	8.57	1.17
Trichloromethane	$CHCl_3$	20	9.96	1.21
Trichloromethane	$CHCl_3$	50	12.9	1.33
Tridecane	$C_{13}H_{28}$	25	9.48	0.90
Triethanolamine	$C_6H_{15}NO_3$	0	3.61	
Triethanolamine	$C_6H_{15}NO_3$	55		0.53
1,3,5-Trimethylbenzene	C_9H_{12}	25	8.14	0.94
Undecane	$C_{11}H_{24}$	25	10.31	0.97
Water	H_2O	20	4.591	0.206
Water	H_2O	25	4.524	0.256
Water	H_2O	30	4.475	0.302
o-Xylene	C_8H_{10}	25	8.10	0.96
m-Xylene	C_8H_{10}	20	8.46	0.99
p-Xylene	C_8H_{10}	25	8.59	1.00

TEMPERATURE AND PRESSURE DEPENDENCE OF LIQUID DENSITY

Ivan Cibulka

This table records parameters of the Tait equation (Refs. 1,2) that gives the ratio between the density at pressure P, $\rho(T,P)$, relative to the density at a reference pressure, $\rho(T,P_{ref})$, at the same temperature T:

$$\frac{\rho(T,P)}{\rho(T,P_{ref})} = \frac{1}{1 - C(T)\ln\dfrac{B(T)+P}{B(T)+P_{ref}}},$$

$$C(T) = a_1 + b_1(T/K) + c_1(T/K)^2, \qquad (1)$$

$$B(T)/\mathrm{MPa} = a_2 + a_3(T/K) + a_4(T/K)^2 + a_5(T/K)^3 + a_6(T/K)^4$$

Parameters a_i (i = 1, ..., 6) and b_1, c_1 were adjusted to selected experimental data using the weighted least-squares method. Parameters b_1 and c_1 are zero for most substances, and therefore their values are given in footnotes for those few substances where the statistical significance was justified. The reference pressure is P_{ref} = 0.101325 MPa at temperatures either at or below the normal boiling point temperature (T_{nbp}) and $P_{ref} = P_{sat}(T)$ (saturated vapor pressure) at temperatures $T > T_{nbp}$. Ranges of validity of the equation (T_{min}, T_{max}, P_{max}) are derived from ranges of experimental data; the minimal pressure of validity is taken as P_{ref}, *i.e.*, interpolation between P_{ref} and lowest experimental pressure is allowed. The upper limit of application is the freezing line (if not limited by the ranges of validity). To avoid any large-scale extrapolation, the validity ranges are rectangular areas ($T_{max} - T_{min}$)P_{max}. If in particular temperature interval(s) the maximum experimental pressure exceeded the given value of P_{max}, then the maximum pressure given in the table is denoted by (r) which means that the validity range given in the table is a rectangular subset of the non-rectangular experimental T, P range. In a few cases P_{max} is given as a ratio where the first value corresponds to T_{min} and the second one to T_{max}, *i.e.*, the validity range has approximately a trapezoidal shape. Values of parameters were taken from the papers (Refs. 3–10) where detailed information on the fits, experimental data, and application ranges is available. The numerical values of the parameters are different from those reported in papers (Refs. 3–10) since the forms of polynomials $C(T)$ and $B(T)$ differ. Besides, the parameters recorded in the table below must not necessarily correspond to those in Refs. 3–10 since some fits were updated using newly published experimental data.

In the second line for each substance, the parameters of the smoothing function for the density at reference pressure ρ (T,P_{ref}) are given. The functions are either the polynomial expansion,

$$\rho(T,P_{ref})/(\mathrm{kg\,m^{-3}}) = \sum_{i=1}^{N_p} a_i (T/K)^{(i-1)} \qquad (2)$$

or the expansion,

$$\rho(T,P_{ref}) = \rho_c\left[1 + \sum_{i=1}^{N_p} a_i (1-T_r)^{(i/3)}\right], \quad T_r = T/T_c \qquad (3)$$

where N_p is the number of adjustable parameters. Values of the critical density ρ_c and the critical temperature T_c used for the fits using Eq. 3 are also recorded in the table. Parameters were mostly taken from Refs. 3–10, those for 1-alkanols C_1 to C_{10} and n-alkanes C_5 to C_{16} are from Ref. 11. Data used for the fits were predominantly recommended values published in the TRC Thermodynamic Tables (Refs. 12,13), sometimes combined with the original experimental data or, in a few cases, the original experimental data were correlated.

RMSD is a relative root-mean-square deviation (in per cent) between experimental values of density and those calculated from the particular function (Tait Eq. 1 or Eqs. 2, 3);

$$RMSD/\% = 100\left[\frac{1}{N}\sum_{i=1}^{N}\left(\frac{\rho_{exp}-\rho_{calc}}{\rho_{exp}}\right)^2\right]^{1/2}$$

where N is the number of experimental values included in the fit.

If the maximum temperature T_{max} of validity of the Tait equation is greater than the normal boiling point temperature T_{nbp}, then parameters of the Wagner equation in the form of either

$$P_{sat}(T) = P_c \exp\left[\frac{a_1(1-T_r) + a_2(1-T_r)^{1.5} + a_3(1-T_r)^{2.5} + a_4(1-T_r)^5}{T_r}\right] \qquad (4)$$

or

$$P_{sat}(T) = P_c \exp\left[\frac{a_1(1-T_r) + a_2(1-T_r)^{1.5} + a_3(1-T_r)^3 + a_4(1-T_r)^6}{T_r}\right] \qquad (5)$$

where $T_r = T/T_c$ are recorded in the third line for each substance. Values of the critical pressure P_c and critical temperature T_c used in Eqs. 4 and 5 are also recorded in the table (values of the critical temperature may differ a little from those recorded for the function, Eq. 3). Parameters of Eqs. 4 and 5 were taken mostly from the papers by McGarry (Ref. 14) and Ambrose and Walton (Ref. 15); in a few cases, the fits were performed using original experimental data or in combination with the recommended values from the TRC Thermodynamic Tables (Refs. 12,13).

The two right-hand-most columns gives values of the isothermal compressibility coefficient, $\kappa_T = -(1/V)(\partial V/\partial P)_T = (1/\rho)(\partial\rho/\partial P)_T$, and the isobaric cubic expansion coefficient, $\alpha_p = (1/V)(\partial V/\partial T)_p = -(1/\rho)(\partial\rho/\partial T)_p$, calculated for T = 298.15 K and P = 0.101325 MPa from the Tait Eq. 1 and from the ρ (T, P_{ref}) equation, respectively. In a very few cases when the lower temperature limit of the Tait equation T_{min} is greater than 298.15 K, the extrapolated values of isothermal compressibility are given.

References

1. Tait, P. G., in *Physics and Chemistry of the Voyage of H.M.S. Challenger*, Vol. II, Part IV, Thomson, C. W., and Murray, J., Eds., H.M.S.O., London, 1889.
2. Tamman, G., *Z. Phys. Chem.* 17, 620, 1895.

3. Cibulka, I., and Ziková, M., *J. Chem. Eng. Data* 39, 876, 1994.
4. Cibulka, I., and Hnědkovský, L., *J. Chem. Eng. Data* 41, 657, 1996.
5. Cibulka, I., Hnědkovský, L., and Takagi, T., *J. Chem. Eng. Data* 42, 2, 1997.
6. Cibulka, I., Hnědkovský, L., and Takagi, T., *J. Chem. Eng. Data* 42, 415, 1997.
7. Cibulka, I., and Takagi, T., *J. Chem. Eng. Data* 44, 411, 1999.
8. Cibulka, I., and Takagi, T., *J. Chem. Eng. Data* 44, 1105, 1999.
9. Cibulka, I., Takagi, T., and Růžička, K., *J. Chem. Eng. Data* 46, 2, 2001.
10. Cibulka, I., and Takagi, T., *J. Chem. Eng. Data* 47, 1037, 2002.
11. Cibulka, I., *Fluid Phase Equilib.* 89, 1, 1993.
12. TRC Thermodynamic Tables, Hydrocarbons, Thermodynamics Research Center (TRC), NIST, Thermophysical Properties Division, Boulder, Colorado.
13. TRC Thermodynamic Tables, Non-Hydrocarbons, Thermodynamics Research Center (TRC), NIST, Thermophysical Properties Division, Boulder, Colorado.
14. McGarry, J., *Ind. Eng. Chem., Process Des. Develop.* 22, 313, 1983.
15. Ambrose, D., and Walton, J., *Pure Appl. Chem.* 61, 1395, 1989.

Eq.	a_1	a_2	a_3	a_4	a_5	a_6	T_{min}/T_{max} K	P_{max} MPa	T_c K	P_c MPa	ρ_c kg m^{-3}	RMSD %	T_{nbp} K	κ_T GPa^{-1}	α_P kK^{-1}
CCl_4 Tetrachloromethane (Ref. 9)															
1	$9.33340\cdot10^{-2}$	$1.11363\cdot10^{3}$	−8.68453	$2.80698\cdot10^{-2}$	$-4.22880\cdot10^{-5}$	$2.37923\cdot10^{-8}$	273/413	51/388				0.040	349.9	1.074	1.209
3	1.58994	2.51946	−5.82313	6.96793	−2.51359		253/554		556.40		557.33	0.042			
5	−7.07139	1.71497	−2.89930	−2.49466			250/556		556.40	4.551					
$CHBr_3$ Tribromomethane (Ref. 9)															
1	$1.03492\cdot10^{-1}$	$2.64208\cdot10^{2}$	$-4.57399\cdot10^{-1}$				323/368	150/343				0.058	422.3	0.809	0.907
2	$3.55953\cdot10^{3}$	−1.96212	$-1.08712\cdot10^{-3}$				283/403					0.001			
$CHCl_3$ Trichloromethane (Ref. 9)															
1	$9.57210\cdot10^{-2}$	$4.79593\cdot10^{2}$	−1.84011	$1.81340\cdot10^{-3}$			273/348	100(r)				0.031	334.4	1.037	1.274
3	3.56339	−3.86051	3.35636				213/333		536.40		499.49	0.043			
5	−6.95546	1.16625	−2.13970	−3.44421			215/536		536.40	5.366					
CH_2Cl_2 Dichloromethane (Ref. 9)															
1	$9.76370\cdot10^{-2}$	$5.24365\cdot10^{2}$	−2.06633	$2.09494\cdot10^{-3}$			293/423	100(r)				0.091	313.4	1.032	1.428
3	3.00368	−2.19763	2.34269				178/383		510.00		440.07	0.014			
5	−7.35739	2.17546	−4.07038	3.50701			233/510		510.00	6.300					
CH_3I Iodomethane (Ref. 9)															
1	$9.54770\cdot10^{-2}$	$5.36810\cdot10^{2}$	−2.25115	$2.53188\cdot10^{-3}$			253/313	160				0.038	315.6	1.052	1.255
2	$3.48981\cdot10^{3}$	−7.47709	$1.83592\cdot10^{-2}$	$-2.36742\cdot10^{-5}$			213/313					0.011			
CH_4O Methanol (Ref. 3)															
1[a]	$1.15068\cdot10^{-1}$	$6.49718\cdot10^{2}$	−4.34583	$1.30722\cdot10^{-2}$	$-2.00292\cdot10^{-5}$	$1.20566\cdot10^{-8}$	183/483	104(r)				0.059	337.7	1.231	1.201
3	2.62781	−4.04742	$1.58343\cdot10$	$-2.25066\cdot10$	$1.09160\cdot10$	$3.04774\cdot10^{-1}$	175/509		512.60		272.00	0.180			
4	−8.63571	1.17982	−2.47900	−1.02400			175/513		512.64	8.092					
C_2Cl_4 Tetrachloroethylene (Ref. 9)															
1	$1.01727\cdot10^{-1}$	$1.36610\cdot10^{2}$					298/298	101				0.007	394.5	0.744	1.012
3	1.10232	1.45142					298/313		620.20		571.84	0.019			
C_2HCl_3 Trichloroethylene (Ref. 9)															
1	$1.00987\cdot10^{-1}$	$1.16005\cdot10^{2}$					298/298	101				0.010	360.4	0.870	1.142
3	1.30631	1.33512					291/315		571.00		513.24	0.008			
$C_2H_4Cl_2$ 1,1-Dichloroethane (Ref. 9)															
1	$9.79930\cdot10^{-2}$	$4.40906\cdot10^{2}$	−1.67041	$1.60492\cdot10^{-3}$			298/398	101(r)				0.028	330.4	1.144	1.320
3	2.81773	−2.41755	2.40906				263/398		523.00		419.32	0.084			
4	−9.51254	7.49332	−7.87028	$6.32540\cdot10^{-1}$			320/450		523.00	5.070					
$C_2H_4Cl_2$ 1,2-Dichloroethane (Ref. 9)															
1	$9.60440\cdot10^{-2}$	$5.39476\cdot10^{2}$	−1.94358	$1.79758\cdot10^{-3}$			278/398	101(r)				0.027	356.7	0.801	1.158
3	2.77660	−2.33774	2.31530				263/398		561.60		439.82	0.063			
5	−7.36864	1.76727	−3.34295	−1.43530			260/566		566.00	5.362					
$C_2H_4O_2$ Acetic acid (Ref. 5)															
1	$9.44550\cdot10^{-2}$	$7.22131\cdot10^{2}$	−3.33202	$4.23588\cdot10^{-3}$			293/328	25/253				0.040	391.1	0.897	1.085
3	4.74861	$-1.76644\cdot10$	$4.44807\cdot10$	$-4.82535\cdot10$	$1.98308\cdot10$		293/493		592.71		351.19	0.003			
C_2H_5Br Bromoethane (Ref. 9)															
1	$9.43490\cdot10^{-2}$	$7.70215\cdot10^{2}$	−4.06470	$5.75736\cdot10^{-3}$			253/313	157(r)				0.055	311.6	1.344	1.475
3	1.50516	1.35063					253/313		503.90		506.82	0.058			
5	−9.14807	5.49831	−6.68657	6.27287			301/504		503.80	6.232					
C_2H_6O Ethanol (Ref. 3)															
1[b]	$1.09012\cdot10^{-1}$	$5.22523\cdot10^{2}$	−2.99367	$8.25136\cdot10^{-3}$	$-1.31014\cdot10^{-5}$	$8.74008\cdot10^{-9}$	193/480	200(r)				0.055	351.4	1.117	1.097
3[c]	$-9.92640\cdot10^{-1}$	$3.80287\cdot10$	$-1.81117\cdot10^{2}$	$4.45905\cdot10^{2}$	$-5.88518\cdot10^{2}$	$3.93920\cdot10^{2}$	159/508		513.88		276.00	0.160			
4	−8.68587	1.17831	−4.87620	1.58800			159/514		513.92	6.132					
$C_2H_6O_2$ Ethylene glycol (Ref. 6)															
1	$9.50140\cdot10^{-2}$	$6.74427\cdot10^{2}$	−1.77594	$1.27583\cdot10^{-3}$			298/378	100(r)				0.031	470.5	0.368	0.639
3	1.77482	1.11208					298/378		790.00		333.70	0.022			
C_3H_6O Acetone (Ref. 5)															
1	$9.92390\cdot10^{-2}$	$6.22264\cdot10^{2}$	−2.94315	$3.73329\cdot10^{-3}$			278/323	392				0.032	329.2	1.293	1.457
3	1.83805	$5.82830\cdot10^{-1}$	$-1.00291\cdot10^{-1}$	$5.70410\cdot10^{-1}$			179/506		508.10		277.90	0.025			

Eq.	a_1	a_2	a_3	a_4	a_5	a_6	T_{min}/T_{max} K	P_{max} MPa	T_c K	P_c MPa	ρ_c kg m^{-3}	RMSD %	T_{nbp} K	κ_T GPa^{-1}	α_p kK^{-1}
C_3H_7Br 1-Bromopropane (Ref. 9)															
1	$8.88710\cdot10^{-2}$	$4.24716\cdot10^{2}$	−1.81352	$2.55049\cdot10^{-3}$	$-1.23227\cdot10^{-6}$		280/500	100				0.094	344.0	1.137	1.251
3	1.65835	1.17536					280/340		536.00		455.53	0.009			
4	−9.68919	7.16793	−7.05282	1.00727			300/536		536.00	4.750					
C_3H_7Cl 1-Chloropropane (Ref. 9)															
1	$1.07245\cdot10^{-1}$	$2.78606\cdot10^{2}$	$-5.17553\cdot10^{-2}$	$-3.83546\cdot10^{-3}$	$5.85824\cdot10^{-6}$		273/368	98(r)				0.078	319.6	1.382	1.462
3	4.15272	−5.73954	4.74431				253/343		503.00		309.22	0.015			
5	−7.55764	2.60153	−5.06041	3.31163			248/503		503.00	4.580					
C_3H_7I 1-Iodopropane (Ref. 9)															
1	$1.01044\cdot10^{-1}$	$3.01114\cdot10^{2}$	$-6.61560\cdot10^{-1}$	$8.74342\cdot10^{-5}$			273/368	1177				0.070	375.7	0.904	1.072
2	$2.22323\cdot10^{3}$	−1.38867	$-7.96985\cdot10^{-4}$				288/361					0.196			
C_3H_8O 1-Propanol (Ref. 3)															
1[d]	$9.60290\cdot10^{-2}$	$3.10861\cdot10^{2}$	$-8.35531\cdot10^{-1}$	$4.27987\cdot10^{-5}$	$7.78234\cdot10^{-7}$		170/524	49(r)				0.109	370.3	1.007	0.999
3	$9.40534\cdot10^{-1}$	$1.29442\cdot10$	$-5.39519\cdot10$	$1.13639\cdot10^{2}$	$-1.13866\cdot10^{2}$	$4.33832\cdot10$	153/530		536.74		274.00	0.160			
4	−8.53706	1.96214	−7.69180	2.94500			147/537		536.78	5.168					
C_3H_8O 2-Propanol (Ref. 6)															
1	$8.90020\cdot10^{-2}$	$1.87411\cdot10^{2}$	$-2.36391\cdot10^{-1}$	$-4.24787\cdot10^{-4}$			273/400	50(r)				0.054	355.4	1.123	1.099
3	$6.24542\cdot10^{-1}$	6.18938	−6.79039	2.49009			243/430		508.30		273.16	0.093			
5	−8.16927	$-9.43213\cdot10^{-2}$	−8.10040	7.85000			250/508		508.30	4.742					
$C_3H_8O_2$ 1,2-Propanediol (Ref. 6)															
1	$8.38940\cdot10^{-2}$	$9.09931\cdot10^{2}$	−4.03075	$5.24295\cdot10^{-3}$			273/368	200				0.019	460.2	0.481	0.714
2	$1.18071\cdot10^{3}$	$-2.56645\cdot10^{-1}$	$-8.05917\cdot10^{-4}$				283/363					0.017			
$C_3H_8O_2$ 1,3-Propanediol (Ref. 6)															
1	$9.10310\cdot10^{-2}$	$5.95160\cdot10^{2}$	−1.64714	$1.36994\cdot10^{-3}$			273/368	200				0.067	487.6	0.403	0.593
3	3.64841	−2.06696	1.22738				283/363		685.70		318.39	0.002			
$C_3H_8O_3$ Glycerol (Ref. 6)															
1	$1.14255\cdot10^{-1}$	$9.25959\cdot10^{2}$	−1.58140	$4.48909\cdot10^{-4}$			223/368	686				0.061	563.1	0.231	0.486
3	6.94960	−9.25236	5.43535				214/364		800.00		351.51	0.069			
$C_4H_8O_2$ Ethyl acetate (Ref. 5)															
1	$8.94090\cdot10^{-2}$	$1.05394\cdot10^{3}$	−7.58700	$1.98689\cdot10^{-2}$	$-1.82587\cdot10^{-5}$		253/343	49(r)				0.106	350.3	1.204	1.425
3	1.69043	2.64150	−4.70476	3.52368			253/473		523.20		308.06	0.015			
C_4H_9Br 1-Bromobutane (Ref. 9)															
1	$9.76600\cdot10^{-2}$	$4.97301\cdot10^{2}$	−2.02541	$2.24791\cdot10^{-3}$			273/368	1177				0.074	374.8	1.046	1.119
3	1.59525	1.23746					293/323		572.00		421.60	0.003			
$C_4H_{10}O$ 1-Butanol (Ref. 3)															
1[e]	$1.01542\cdot10^{-1}$	$5.01888\cdot10^{2}$	−2.41576	$5.35877\cdot10^{-3}$	$-7.31982\cdot10^{-6}$	$4.60628\cdot10^{-9}$	195/524	49(r)				0.053	390.9	0.914	0.949
3	−3.57379	$5.02450\cdot10$	$-1.75934\cdot10^{2}$	$3.08588\cdot10^{2}$	$-2.65628\cdot10^{2}$	$8.94997\cdot10$	186/559		563.01		271.00	0.352			
4	−8.40615	2.23010	−8.24860	$-7.11000\cdot10^{-1}$			185/563		563.05	4.424					
$C_4H_{10}O$ Diethyl ether (Ref. 5)															
1	$9.70250\cdot10^{-2}$	$3.95413\cdot10^{2}$	−1.68339	$1.82398\cdot10^{-3}$			293/353	981				0.239	307.6	1.740	1.657
3	3.31808	−8.55179	$2.09100\cdot10$	$-2.10672\cdot10$	8.26525		140/430		466.74		264.72	0.027			
5	−7.29916	1.24828	−2.91931	−3.36740			250/467		466.74	3.646					
C_5H_{10} Cyclopentane (Ref. 8)															
1	$8.85580\cdot10^{-2}$	$8.94968\cdot10^{2}$	−7.76576	$3.01636\cdot10^{-2}$	$-5.82645\cdot10^{-5}$	$4.43996\cdot10^{-8}$	193/353	48/196				0.050	322.4	1.308	1.325
3	1.68431	$6.19856\cdot10^{-1}$	$4.90670\cdot10^{-2}$	$2.19121\cdot10^{-1}$			179/508		511.70		275.04	0.031			
5	−6.51809	$3.84422\cdot10^{-1}$	−1.11706	−4.50275			289/512		511.60	4.509					
$C_5H_{11}Br$ 1-Bromopentane (Ref. 9)															
1	$8.98660\cdot10^{-2}$	$5.28042\cdot10^{2}$	−2.41620	$3.92466\cdot10^{-3}$	$-2.31231\cdot10^{-6}$		283/523	98				0.061	402.7	0.943	1.076
2	$1.60209\cdot10^{3}$	−1.30486					283/363					0.006			
4	−8.59341	3.46288	−3.06661	−4.07997			294/606		605.90	3.900					
C_5H_{12} Pentane (Ref. 4)															
1[f]	$1.00620\cdot10^{-1}$	$6.52014\cdot10^{2}$	−4.94167	$1.54854\cdot10^{-2}$	$-2.36811\cdot10^{-5}$	$1.44530\cdot10^{-8}$	173/373	87/285				0.119	309.2	2.133	1.610
3	1.17756	3.89157	−5.50896	3.29181			143/444		469.80		232.00	0.154			
4	−7.30698	1.75845	−2.16290	−2.91300			143/470		469.80	3.375					
$C_5H_{12}O$ 1-Pentanol (Ref. 3)															
1	$9.52000\cdot10^{-2}$	$3.00019\cdot10^{2}$	$-5.62481\cdot10^{-1}$	$-5.98193\cdot10^{-4}$	$1.15744\cdot10^{-6}$		233/550	59(r)				0.149	411.2	0.866	0.905
3	−2.61129	$4.23375\cdot10$	$-1.49640\cdot10^{2}$	$2.62210\cdot10^{2}$	$-2.23090\cdot10^{2}$	$7.37879\cdot10$	195/585		588.11		270.00	0.353			
4	−8.98005	3.91624	−9.90810	−2.19100			196/588		588.15	3.909					
C_6H_5Br Bromobenzene (Ref. 9)															
1	$9.73880\cdot10^{-2}$	$4.87590\cdot10^{2}$	−1.49313	$1.16283\cdot10^{-3}$			278/358	100(r)				0.043	429.2	0.668	0.901
3	2.84948	−2.30055	2.31394				298/358		670.00		484.60	0.002			
C_6H_5Cl Chlorobenzene (Ref. 9)															
1[g]	$1.06896\cdot10^{-1}$	$5.95178\cdot10^{2}$	−2.52143	$4.02563\cdot10^{-3}$	$-3.10524\cdot10^{-6}$	$1.06813\cdot10^{-9}$	278/583	50(r)				0.057	404.9	0.752	0.969
3	$1.17137\cdot10^{-1}$	$1.23712\cdot10$	$-2.84464\cdot10$	$2.94048\cdot10$	$-1.07075\cdot10$		253/630		632.40		365.45	0.008			
5	−7.58700	2.26551	−4.09118	$1.70377\cdot10^{-1}$			335/632		632.40	4.519					

Fluids

PROPERTIES OF CRYOGENIC FLUIDS

Eric W. Lemmon

This table gives physical and thermodynamic properties of ten cryogenic fluids. The properties are:

M	Molar mass
T_t	Triple-point temperature
P_t	Triple-point pressure
ρ_t (l)	Liquid density at the triple point
$\Delta_{fus}H$ @ T_t	Enthalpy of fusion at the triple point
T_b	Normal boiling point at a pressure of 101.325 kPa (760 mmHg)
$\Delta_{vap}H$ @ T_b	Enthalpy of vaporization at the normal boiling point
ρ @ T_b	Density at the normal boiling point for the liquid (l) or vapor (v)
C_p @ T_b	Heat capacity at constant pressure at the normal boiling point for the liquid (l) or vapor (v)
u @ T_b	Speed of sound at the normal boiling point for the liquid (l) or vapor (v)
T_c	Critical temperature
P_c	Critical pressure
ρ_c	Critical density

All properties except the heat of fusion are calculated from the REFPROP program, see Ref. 4. Temperatures are listed on the ITS-90 scale, except those values for neon and oxygen that are obtained from older equations of state based on the IPTS-68 temperature scale. The references for all fluids except air, helium, neon, krypton, and xenon are given in the "Thermophysical Properties of Fluids" table. The properties of hydrogen are given for the para form of the molecule.

The triple-point temperature of air is the solidification temperature of the liquid (see Ref. 2 for details). The boiling-point temperature for air is the bubble-point temperature (i.e., the temperature at which boiling begins as the pressure of the liquid is lowered). The dew-point (vapor) properties of air at 101.325 kPa are calculated at a temperature of 81.72 K; the liquid and vapor properties of these two state points are not in equilibrium. The triple-point properties of helium are given at the temperature of the lambda line (change from normal-to-superfluid helium) for the saturated-liquid state.

References

1. Katti, R., Jacobsen, R. T, Stewart, R. B., and Jahangiri, M., Thermodynamic Properties of Neon for Temperatures from the Triple Point to 700 K at Pressures to 700 MPa, *Adv. Cryo. Eng.* 31, 1189, 1986.
2. Lemmon, E. W., Jacobsen, R. T, Penoncello, S. G., and Friend, D. G., Thermodynamic Properties of Air and Mixtures of Nitrogen, Argon, and Oxygen from 60 to 2000 K at Pressures to 2000 MPa, *J. Phys. Chem. Ref. Data* 29, 331, 2000.
3. Lemmon, E. W., and Span, R., Short Fundamental Equations of State for 20 Industrial Fluids, *J. Chem. Eng. Data* 51, 785, 2006.
4. Lemmon, E. W., Huber, M. L., and McLinden, M. O., NIST Standard Reference Database 23: Reference Fluid Thermodynamic and Transport Properties-REFPROP, Version 9.0, National Institute of Standards and Technology, Standard Reference Data Program, Gaithersburg, Maryland, 2010 (www.nist.gov/srd/nist23.cfm).
5. Ortiz-Vega, D. O., Hall, K. R., Arp, V. D., and Lemmon, E. W., Interim Helium Equation, to be published in *Int. J. Thermophys.*, 2011.

Property	Units	Air	N_2	O_2	H_2	He	Ne	Ar	Kr	Xe	CH_4
M	g mol^{-1}	28.9655	28.0134	31.9988	2.01588	4.0026	20.1797	39.948	83.798	131.293	16.0428
T_t	K	59.75	63.151	54.361	13.8033	2.1768	24.556	83.8058	115.775	161.405	90.6941
P_t	kPa	5.265	12.52	0.1463	7.041	5.043	43.37	68.89	73.53	81.77	11.70
ρ_t(l)	g cm^{-3}	0.9578	0.8672	1.306	0.07698	0.1459	1.252	1.417	2.447	2.966	0.4515
$\Delta_{fus}H$ @ T_t	J g^{-1}		25.3	13.7	59.5		16.8	28.0	16.3	13.8	58.41
T_b	K	78.903	77.355	90.188	20.271	4.222	27.104	87.302	119.735	165.051	111.667
$\Delta_{vap}H$ @ T_b	J g^{-1}	204.8	199.2	213.1	446.1	20.91	85.75	161.1	107.1	95.59	510.8
ρ(l) @ T_b	g cm^{-3}	0.8752	0.8061	1.141	0.07083	0.1250	1.207	1.395	2.417	2.942	0.4224
ρ(v) @ T_b	g dm^{-3}	4.497	4.612	4.467	1.339	16.70	9.577	5.774	8.818	10.01	1.816
C_p(l) @ T_b	J g^{-1} K^{-1}	1.933	2.041	1.699	9.729	5.105	1.862	1.117	0.5198	0.3393	3.481
C_p(v) @ T_b	J g^{-1} K^{-1}	1.090	1.124	0.9707	12.03	9.327	1.415	0.5658	0.2748	0.1751	2.218
u(l) @ T_b	m s^{-1}	865.5	851.4	904.3	1111	177.3	594.1	838.3	682.5	643.1	1338
u(v) @ T_b	m s^{-1}	177.1	174.8	177.5	355.0	101.5	126.6	170.9	137.9	129.4	271.5
T_c	K	132.5306	126.192	154.581	32.938	5.1953	44.4918	150.687	209.48	289.733	190.564
P_c	MPa	3.7860	3.3958	5.0430	1.2858	0.22746	2.6786	4.8630	5.5250	5.8420	4.5992
ρ_c	g cm^{-3}	0.3426	0.3133	0.4361	0.03132	0.06958	0.4819	0.5356	0.9092	1.103	0.1627

PROPERTIES OF LIQUID HELIUM

The following data were obtained by a critical evaluation of all existing experimental measurements on liquid helium, using a fitting procedure described in the reference. All values refer to liquid helium at saturated vapor pressure; temperatures are on the ITS-90 scale. Several properties show a singularity at the lambda point (2.1768 K).

- p: vapor pressure
- ρ: density
- C_s: heat capacity
- $\Delta_{vap}H$: molar enthalpy of vaporization
- ε: relative permittivity (dielectric constant)
- σ: surface tension
- α_V: cubic expansion coefficient
- η: viscosity
- λ: thermal conductivity

Reference

Donnelly, R. J., and Barenghi, C. F., *J. Phys. Chem. Ref. Data*, 27, 1217, 1998.

Properties of Liquid Helium

T/K	p/kPa	ρ/g cm^{-3}	C_s/J mol^{-1} K^{-1}	$\Delta_{vap}H$/J mol^{-1}	ε	σ/mN m^{-1}	$10^3\alpha_V$/K^{-1}	η/µPa s	λ/W cm^{-1} K^{-1}
0.0		0.1451397	0	59.83	1.057255		0.000		
0.5		0.1451377	0.010	70.24	1.057254	0.3530	0.107		
1.0	0.01558	0.1451183	0.415	80.33	1.057246	0.3471	0.309	3.873	
1.5	0.4715	0.1451646	4.468	89.35	1.057265	0.3322	-2.36	1.346	
2.0	3.130	0.1456217	21.28	93.07	1.057449	0.3021	-12.2	1.468	
2.5	10.23	0.1448402	9.083	92.50	1.057135	0.2623	39.4	3.259	0.1497
3.0	24.05	0.1412269	9.944	94.11	1.055683	0.2161	61.5	3.517	0.1717
3.5	47.05	0.1360736	12.37	92.84	1.053615	0.1626	88.7	3.509	0.1868
4.0	81.62	0.1289745	15.96	87.00	1.050770	0.1095	129	3.319	0.1965
4.5	130.3	0.1188552	21.8	75.86	1.046725	0.0609	211		
5.0	196.0		44.7	47.67		0.0157			

Fluids

PROPERTIES OF REFRIGERANTS

This table gives physical properties and safe exposure limits for compounds that have been used as working fluids in traditional refrigeration systems or are under consideration as replacements in newer systems. Some are also used as solvents and blowing agents. Many of the compounds listed are believed to be less harmful to the environment than the traditional halocarbon refrigerants.

Compounds are listed by chemical name. Their ASHRAE standard refrigerant designations (Ref. 1) appear in the second column. The ASHRAE codes are often prefixed by symbols such as CFC- (for chlorofluorocarbon), HCFC- (for hydrochlorofluorocarbon), or simply R- (for refrigerant). The "R" number assigned to refrigerants is specified by ANSI/ASHRAE Standard 34. This system is most useful for the hydrocarbons and halocarbons with one to three carbons; for such molecules the chemical composition can be determined from the number and vice versa. The first digit on the far right is the number of fluorine atoms in the compound. The second digit from the right is one more than the number of hydrogen atoms. The third digit from the right is one less than the number of carbon atoms; for single-carbon compounds, this digit is omitted. The fourth digit from the right is equal to the number of unsaturated carbon–carbon bonds; for saturated compounds, this digit is omitted. The number of bromine and iodine atoms is indicated, if needed, by appending "Bn" or "In" to the digits specified by the above rules, where "n" is the number of bromine or iodine atoms. All atoms not specified by the above are assumed to be chlorine. Appended lowercase letter(s) designate different isomers. Additional rules are used to specify cyclic compounds, ethers, inorganic fluids (R700- and R7000-series), miscellaneous organic compounds (R600-series), and blends (R400- and R500-series).

The properties tabulated are

t_m: normal melting point in °C (tp indicates a triple point)

t_b: normal boiling point in °C at 101.325 kPa or 760 mmHg (sp indicates a sublimation point)

t_c: critical temperature in °C

TWA: Threshold limit value, expressed as the time-weighted average over an 8-hr workday and 40-hr workweek, for safe exposure to the vapor; expressed as parts per million by volume.

STEL: Short-term exposure limit, which should not be exceeded for more than 15 min.

Further references and additional data on the critical properties may be found in the tables "Critical Constants of Organic Compounds" and "Critical Constants of Inorganic Compounds" in this section. Details on threshold limits are given in Section 16.

References

1. ASHRAE (2007). ANSI/ASHRAE Standard 34-2007 Designation and Safety Classification of Refrigerants, American Society of Heating, Refrigerating and Air-Conditioning Engineers, Atlanta, GA.
2. *ASHRAE Fundamentals Handbook 2001*, Chapter 19, Refrigerants, American Society of Heating, Refrigerating, and Air-Conditioning Engineers, Atlanta, GA, 2001.
3. Platzer, B., Polt, A., and Mauer, G., *Thermophysical Properties of Refrigerants*, Springer, Berlin, 1990.
4. Sako, T., Sato, M., Nakazawa, N., Oowa, M., Yasumoto, M., Ito, H., and Yamashita, S., *J. Chem. Eng. Data* 41, 802, 1996.
5. Schmidt, J. W., Carrillo-Nava, E., and Moldover, M. R., *Fluid Phase Equilib.* 122, 187, 1996.
6. Salvi-Narkhede, M., Wang, B-H., Adcock, J. L., and Van Hook, W. A., *J. Chem. Thermodyn.* 24, 1065, 1992.
7. Fialho, P. S., and Nieto de Castro, C. A., *Int. J. Thermophys.* 21, 385, 2000.
8. Daubert, T. E., Danner, R. P., Sibul, H. M., and Stebbins, C. C., *Physical and Thermodynamic Properties of Pure Compounds: Data Compilation*, extant 2002 (core with supplements), Taylor & Francis, Bristol, PA.
9. McLinden, M.O., Lemmon, E.W., and Huber, M.L., The REFPROP Database for the Thermophysical Properties of Refrigerants, 21st International Congress of Refrigeration, Washington D.C., International Institute of Refrigeration, Paper ICR0443, 2003.
10. Lemmon, E.W., Huber, M.L., and McLinden, M.O., NIST Standard Reference Database 23: Reference Fluid Thermodynamic and Transport Properties - REFPROP, Version 9.0, National Institute of Standards and Technology, Standard Reference Data Program, Gaithersburg, MD, 2010, <http://www.nist.gov/srd/nist23.cfm>.

Properties of Refrigerants

Name	ASHRAE code	Mol. form.	CAS Reg. No.	t_m/°C	t_b/°C	t_c/°C	p_c/MPa	TWA/ ppm	STEL/ ppm
Ammonia	717	H_3N	7664-41-7	-77.65	-33.33	132.41	11.357	25	35
Bis(difluoromethyl) ether	E134	$C_2H_2F_4O$	1691-17-4		5.5	147.1	4.16		
Bromochlorodifluoromethane	12B1	$CBrClF_2$	353-59-3	-159.5	-3.9	155	4.31		
Bromodifluoromethane	22B1	$CHBrF_2$	1511-62-2	-145	-15.6	138.9	5.2		
Bromotrifluoromethane	13B1	$CBrF_3$	75-63-8	-174.4	-57.8	66.91	3.96	1000	
Butane	600	C_4H_{10}	106-97-8	-138.2	-0.5	152.1	3.79		1000
Carbon dioxide	744	CO_2	124-38-9	-56.561 tp	-78.464 sp	30.98	7.375	5000	30000
1-Chloro-1,1-difluoroethane	142b	$C_2H_3ClF_2$	75-68-3	-130.43	-9.12	137.16	4.06		
1-Chloro-2,2-difluoroethane	142	$C_2H_3ClF_2$	338-65-8		35				
Chlorodifluoromethane	22	$CHClF_2$	75-45-6	-157.41	-40.8	96.15	4.98	1000	
Chloroethane	160	C_2H_5Cl	75-00-3	-138.4	12.3	187.2	5.24	100	
Chloroethene	1140	C_2H_3Cl	75-01-4	-153.84	-13.8	152	5.60	1	
1-Chloro-1-fluoroethane	151a	C_2H_4ClF	1615-75-4		16				
1-Chloro-2-fluoroethane	151	C_2H_4ClF	762-50-5		53.1				
Chlorofluoromethane	31	CH_2ClF	593-70-4	-135.1	-9.1	154			
1-Chloro-1,2,2,3,3,4,4-heptafluorocyclobutane	C317	C_4ClF_7	377-41-3	-39.1	25				

Name	ASHRAE code	Mol. form.	CAS Reg. No.	t_m/°C	t_b/°C	t_c/°C	p_c/MPa	TWA/ ppm	STEL/ ppm
Chloromethane	40	CH_3Cl	74-87-3	-97.6	-24.1	143.09	6.72	50	100
Chloropentafluoroethane	115	C_2ClF_5	76-15-3	-99.4	-39.2	79.9	3.141	1000	
1-Chloro-1,1,2,2-tetrafluoroethane	124a	C_2HClF_4	354-25-6	-117	-13	126.8			
1-Chloro-1,2,2,2-tetrafluoroethane	124	C_2HClF_4	2837-89-0	-199.15	-11.96	122.28	3.62		
1-Chloro-1,1,2-trifluoroethane	133b	$C_2H_2ClF_3$	421-04-5		16				
1-Chloro-1,2,2-trifluoroethane	133	$C_2H_2ClF_3$	431-07-2		17.3				
2-Chloro-1,1,1-trifluoroethane	133a	$C_2H_2ClF_3$	75-88-7	-105.5	6.0	151.9	4.02		
Chlorotrifluoroethene	1113	C_2ClF_3	79-38-9	-158.14	-28.3	107	3.95		
Chlorotrifluoromethane	13	$CClF_3$	75-72-9	-181.2	-81.48	28.8	3.89		
Dibromodifluoromethane	12B2	CBr_2F_2	75-61-6	-110.1	22.79	198.2		100	
1,2-Dibromotetrafluoroethane	114B2	$C_2Br_2F_4$	124-73-2	-110.1	47.1	214.7			
1,2-Dichloro-1,1-difluoroethane	132b	$C_2H_2Cl_2F_2$	1649-08-7	-101.2	47				
1,2-Dichloro-1,2-difluoroethane	132	$C_2H_2Cl_2F_2$	431-06-1	-101.2	59.6				
1,1-Dichloro-2,2-difluoroethene	1112a	$C_2Cl_2F_2$	79-35-6	-116	19				
Dichlorodifluoromethane	12	CCl_2F_2	75-71-8	-157.05	-29.8	111.8	4.12	1000	
1,1-Dichloroethane	150a	$C_2H_4Cl_2$	75-34-3	-96.93	56.3	250.3	5.1	100	
1,2-Dichloroethane	150	$C_2H_4Cl_2$	107-06-2	-35.6	83.4	288.4	5.4	10	
trans-1,2-Dichloroethene	1130	$C_2H_2Cl_2$	156-60-5	-49.8	47.64	242.4	5.3	200	
1,1-Dichloro-1-fluoroethane	141b	$C_2H_3Cl_2F$	1717-00-6	-103.5	32.05	204.2	4.20		
1,2-Dichloro-1-fluoroethane	141	$C_2H_3Cl_2F$	430-57-9	-60	74				
Dichlorofluoromethane	21	$CHCl_2F$	75-43-4	-130.35	8.9	178.5	5.20	10	
1,2-Dichloro-1,2,3,3,4,4-hexafluorocyclobutane	C316	$C_4Cl_2F_6$	356-18-3	-24.2	59.5	224			
1,3-Dichloro-1,1,2,2,3,3-hexafluoropropane	216ca	$C_3Cl_2F_6$	662-01-1	-125.4	35.7	180			
Dichloromethane	30	CH_2Cl_2	75-09-2	-94.9	39.8	234.9	6.35	50	
1,1-Dichloro-1,2,2,2-tetrafluoroethane	114a	$C_2Cl_2F_4$	374-07-2	-56.6	3	145.5	3.31		
1,2-Dichloro-1,1,2,2-tetrafluoroethane	114	$C_2Cl_2F_4$	76-14-2	-92.52	3.6	145.59	3.25	1000	
1,2-Dichloro-1,1,2-trifluoroethane	123a	$C_2HCl_2F_3$	354-23-4	-78	30.0	188.5	3.77		
2,2-Dichloro-1,1,1-trifluoroethane	123	$C_2HCl_2F_3$	306-83-2	-107.15	27.8	183.7	3.67		
Diethyl ether	610	$C_4H_{10}O$	60-29-7	-116.22	34.4	193.7	3.64	400	500
1,1-Difluoroethane	152a	$C_2H_4F_2$	75-37-6	-118.59	-24.02	113.3	4.52		
1,2-Difluoroethane	152	$C_2H_4F_2$	624-72-6		26				
1,1-Difluoroethene	1132a	$C_2H_2F_2$	75-38-7	-144	-85.5	29.8	4.48	500	
Difluoromethane	32	CH_2F_2	75-10-5	-136.8	-51.65	78.13	5.79		
2-(Difluoromethoxy)-1,1,1-trifluoroethane	245mf	$C_3H_3F_5O$	1885-48-9		29.2	171.8	3.43		
Difluoromethyl 1,1,2-trifluoroethyl ether	245qc	$C_3H_3F_5O$	69948-24-9		43.1				
Dimethyl ether	E170	C_2H_6O	115-10-6	-141.49	-24.8	127	5.31		
Ethane	170	C_2H_6	74-84-0	-182.77	-88.6	32.21	4.88		
Ethylene	1150	C_2H_4	74-85-1	-169.15	-103.8	9.20	5.06	200	
Fluoroethane	161	C_2H_5F	353-36-6	-143.2	-37.7	102.1	5.02		
Fluoroethene	1141	C_2H_3F	75-02-5	-160.5	-72	54.8		1	
Fluoromethane	41	CH_3F	593-53-3	-143.33	-78.31	44.27	5.88		
1,1,1,2,3,3,3-Heptafluoropropane	227ea	C_3HF_7	431-89-0	-126.80	-16.34	101.9	2.93		
Hexachloroethane	110	C_2Cl_6	67-72-1	186.8	184.7 sp	422		1	
Hexafluoroethane	116	C_2F_6	76-16-4	-100.02	-78.1	19.8	3.03		
1,1,1,2,3,3-Hexafluoropropane	236ea	$C_3H_2F_6$	431-63-0		6.20	139.25	3.42		
1,1,1,3,3,3-Hexafluoropropane	236fa	$C_3H_2F_6$	690-39-1	-93.63	-1.4	124.92	3.18		
Isobutane	600a	C_4H_{10}	75-28-5	-159.38	-11.7	134.69	3.64		1000
Methane	50	CH_4	74-82-8	-182.475	-161.5	-82.59	4.60		
Methyl formate	611	$C_2H_4O_2$	107-31-3	-99.7	31.6	214.01	6.01	50	100
Methyl pentafluoroethyl ether	245mc	$C_3H_3F_5O$	22410-44-2		5.6	133.66	2.89		
Methyl 1,1,2,2-tetrafluoroethyl ether	254pc	$C_3H_4F_4O$	425-88-7	-107	37.1				
Methyl trifluoromethyl ether	143m	$C_2H_3F_3O$	421-14-7	-149.2	-25.2	104.77	3.64		
Pentachloroethane	120	C_2HCl_5	76-01-7	-29.0	161				
Pentachlorofluoroethane	111	C_2Cl_5F	354-56-3	101	138				
Pentafluoroethane	125	C_2HF_5	354-33-6	-100.63	-48.09	66.1	3.63		
1,1,1,2,2-Pentafluoropropane	245cb	$C_3H_3F_5$	1814-88-6		-18.0	107	3.14		
1,1,1,3,3-Pentafluoropropane	245fa	$C_3H_3F_5$	460-73-1	-102.10	15.14	154.05	3.66		
1,1,2,2,3-Pentafluoropropane	245ca	$C_3H_3F_5$	679-86-7		25.13	174.42	3.96		
Perfluorocyclobutane	C318	C_4F_8	115-25-3	-40.16	-5.91	115.3	2.78		
Perfluoroisopropyl methyl ether	347mmy	$C_4H_3F_7O$	22052-84-2		29	160.15	2.55		

Fluids

Name	ASHRAE code	Mol. form.	CAS Reg. No.	t_m/°C	t_b/°C	t_c/°C	p_c/MPa	TWA/ ppm	STEL/ ppm
Perfluoropropane	218	C_3F_8	76-19-7	-147.7	-36.8	71.88	2.67		
Perfluoropropyl methyl ether	347mcc	$C_4H_3F_7O$	375-03-1		34	164.6	2.48		
Propane	290	C_3H_8	74-98-6	-187.62	-42.11	96.8	4.25		
Propene	1270	C_3H_6	115-07-1	-185.19	-47.6	91.8	4.59	500	
Sulfur dioxide	764	O_2S	7446-09-5	-75.45	-10.02	157.49	7.884		0.25
1,1,1,2-Tetrachloro-2,2-difluoroethane	112a	$C_2Cl_4F_2$	76-11-9	41.0	96			100	
1,1,2,2-Tetrachloro-1,2-difluoroethane	112	$C_2Cl_4F_2$	76-12-0	26.54	92.83	278		50	
1,1,2,2-Tetrachloroethane	130	$C_2H_2Cl_4$	79-34-5	-42.4	146.0	388		1	
1,1,1,2-Tetrachloro-2-fluoroethane	121a	C_2HCl_4F	354-11-0	-95.3	117.1				
1,1,2,2-Tetrachloro-1-fluoroethane	121	C_2HCl_4F	354-14-3	-82.6	116.7				
Tetrachloromethane	10	CCl_4	56-23-5	-22.8	76.7	283.4	4.57	5	10
1,1,1,2-Tetrafluoroethane	134a	$C_2H_2F_4$	811-97-2	-103.30	-26.1	101.1	4.06		
1,1,2,2-Tetrafluoroethane	134	$C_2H_2F_4$	359-35-3	-89	-20	118.60	4.61		
Tetrafluoroethene	1114	C_2F_4	116-14-3	-131.14	-76	34	3.94	2	
1,2,2,2-Tetrafluoroethyl difluoromethyl ether	236me	$C_3H_2F_6O$	57041-67-5		23	155.80	3.05		
Tetrafluoromethane	14	CF_4	75-73-0	-183.58	-127.9	-45.61	3.73		
2,3,3,3-Tetrafluoropropene	1234yf	$C_3H_2F_4$	754-12-1		-29.5	94.8			
1,1,1-Trichloro-2,2-difluoroethane	122b	$C_2HCl_3F_2$	354-12-1		75				
1,2,2-Trichloro-1,1-difluoroethane	122	$C_2HCl_3F_2$	354-21-2	-150.1	71.9				
1,2,2-Trichloro-1,2-difluoroethane	122a	$C_2HCl_3F_2$	354-15-4	-174	73.2				
1,1,1-Trichloroethane	140a	$C_2H_3Cl_3$	71-55-6	-30	74.02	272		350	450
1,1,2-Trichloroethane	140	$C_2H_3Cl_3$	79-00-5	-36.3	113	329		10	
Trichloroethene	1120	C_2HCl_3	79-01-6	-84.7	86.8	271.1		10	25
1,1,2-Trichloro-2-fluoroethane	131	$C_2H_2Cl_3F$	359-28-4		102.4				
Trichlorofluoromethane	11	CCl_3F	75-69-4	-110.44	23.7	198	4.40	1000	
Trichloromethane	20	$CHCl_3$	67-66-3	-63.3	61.2	262.9	5.5	10	
1,1,1-Trichloro-2,2,2-trifluoroethane	113a	$C_2Cl_3F_3$	354-58-5	14.37	46	209.8			
1,1,2-Trichloro-1,2,2-trifluoroethane	113	$C_2Cl_3F_3$	76-13-1		47.6	214.3	3.40	1000	1250
1,1,1-Trifluoroethane	143a	$C_2H_3F_3$	420-46-2	-111.6	-47.2	72.74	3.77		
1,1,2-Trifluoroethane	143	$C_2H_3F_3$	430-66-0	-84	3.5	156.7			
2,2,2-Trifluoroethyl methyl ether	E143a	$C_3H_5F_3O$	460-43-5		31.62	175.83	3.51		
Trifluoromethane	23	CHF_3	75-46-7	-155.18	-82.0	25.85	4.82		
Trifluoromethyl difluoromethyl ether	E125	C_2HF_5O	3822-68-2	-157	-35.0	81.34	3.36		
Trifluoromethyl 1,1,2,2-tetrafluoroethyl ether	227ca2	C_3HF_7O	2356-61-8	-141	-3.3	114.7	2.65		
Trifluoromethyl 1,2,2,2-tetrafluoroethyl ether	227me	C_3HF_7O	2356-62-9		-9.6	104.11	2.62		

PROPERTIES OF GAS CLATHRATE HYDRATES

Carolyn A. Koh, M. Naveed Khan, and E. Dendy Sloan

Gas clathrate hydrates (also known as gas hydrates) are crystalline inclusion compounds composed of hydrogen-bonded water cavities (host) which encage small gas (guest) molecules. Generally, a maximum of one guest molecule occupies each water cavity. Typical guest molecules that form gas hydrates are methane, ethane, carbon dioxide, and propane (see gas hydrate phase equilibria data in Table II). The structural and physical properties of gas hydrates are given in Tables Ia and Ib. Data have been taken from the references indicated.

Table Ia. Gas Hydrate Structural Properties (All values are from Ref. 1 unless noted otherwise)

Structure	sI Small	sI Large	sII Small	sII Large	sH Small	sH Medium	sH Large
Crystal system	Cubic		Cubic		Hexagonal		
Space group	Pm3n (No. 223)[b]		Fd3m (No. 227)[b]		P6/mmm (No. 191)[b]		
Lattice description	Primitive		Face centered		Hexagonal		
Lattice parameters[a]	a = 12 Å $\alpha = \beta = \gamma = 90°$		a = 17.3 Å $\alpha = \beta = \gamma = 90°$		a = 12.2 Å, c = 10.1 Å $\alpha = \beta = 90°, \gamma = 120°$		
Ideal unit cell formula	$6(5^{12}6^2)\cdot 2(5^{12})\cdot 46H_2O$		$8(5^{12}6^4)\cdot 16(5^{12})\cdot 136H_2O$		$1(5^{12}6^8)\cdot 3(5^{12})\cdot 2(4^35^66^3)\cdot 34H_2O$		
Cavity	Small	Large	Small	Large	Small	Medium	Large
Description	5^{12}	$5^{12}6^2$	5^{12}	$5^{12}6^4$	5^{12}	$4^35^66^3$	$5^{12}6^8$
Number of cavities/unit cell	2	6	16	8	3	2	1
Average cavity radius (Å)[c]	3.95	4.33	3.91	4.73	3.94[d]	4.04[d]	5.79[d]
H_2O molecules/cavity[e]	20	24	20	28	20	20	36

[a] Lattice parameters are a function of temperature, pressure, and guest composition. Typical average values given.
[b] Space group reference numbers from the International Tables of Crystallography.
[c] The average cavity radius will vary with temperature, pressure, and guest composition.
[d] From the atomic coordinates measured using single crystal x-ray diffraction on 2,2-dimethylpentane·$5(Xe,H_2S)\cdot 34H_2O$ at 173 K (Ref. 2). The Rietveld refinement package, GSAS was used to determine the atomic distances for each cage oxygen to the cage center.
[e] Number of oxygen atoms at the periphery of each cavity.

Table Ib. Physical Properties of sI, sII Hydrates Compared to Ice Ih (All values are from Refs. 1, 3, 4, and 5 unless noted otherwise)

Property	Ice Ih	sI	sII
Dielectric constant at 273 K	94	~58	~58
H_2O reorientation time at 273 K (μs)	21	~10	~10
H_2O diffusion jump time (μs)	2.7	>200	>200
Isothermal Young's modulus at 268 K (10^6 Pa)	9.5	8.4[est]	8.2[est]
Poisson's ratio	0.3301[f]	0.31403[f]	0.31119[f]
Bulk modulus (GPa)	9.097[f]	8.762[f]	8.482[f]
Shear modulus (GPa)	3.488[f]	3.574[f]	3.6663[f]
Compressional velocity, V_p (m/s)	3870.1[f]	3778[f]	3821.8[f]
Shear velocity, V_s (m/s)	1949[f]	1963.6	2001.14[g]
Linear thermal expansion at 200 K (K^{-1})	56×10^{-6}	77×10^{-6}	52×10^{-6}
Thermal conductivity ($W\ m^{-1}\ K^{-1}$) at 263 K	2.18±0.01[h]	0.51±0.01[h]	0.50±0.01[h]
Adiabatic bulk compression at 273 K (GPa)	12	14[est]	14[est]
Heat capacity ($J\ kg^{-1}\ K^{-1}$)	1700±200[h]	2080	2130±40[h]
Refractive index (632.8 nm, –3 °C)	1.3082 (Ref. 9)	1.346 (Ref. 9)	1.350 (Ref. 9)
Density (g/cm^3)	0.91[j]	0.94	1.291[k]

[f] At 253–268 K, 22.4–32.8 MPa (ice, Ih), 258–288 K, 27.1–62.1 MPa (CH_4, sI), 258–288 K, 30.5–91.6 MPa (CH_4–C_2H_6, sII), Ref. 6.
[g] At 258–288 K, 26.6–62.1 MPa, Ref. 7.
[h] At 248–268 K (ice, Ih), 253–288 K (CH_4, sI), 248–265.5 K (THF, sII), Ref. 8.
[j] Fractional occupancy (calculated from a theoretical model) in small (S) and large (L) cavities: sI = CH_4: 0.87 (S) and CH_4: 0.973 (L); sII = CH_4: 0.672 (S), 0.057 (L); C_2H_6: 0.096 (L) only; C_3H_8: 0.84 (L) only.
[k] Calculated for 2,2-dimethylpentane $5(Xe,H_2S)\,34H_2O$, Ref. 2; est = estimated.

References for Table I

1. Sloan, E.D. and Koh, C.A., *Clathrate Hydrates of Natural Gases*, 3rd Edition, CRC Press, 2008.
2. Udachin, K.A., Ratcliffe, C.I., Enright, G.D., and Ripmeester, J.A., *Supramol. Chem.*, 8, 173, 1997.
3. Davidson, D.W., *Natural Gas Hydrates* (Cox, J.L., Ed.) Butterworths, Boston, 1, 1983.
4. Davidson, D.W., Handa, Y.P., and Ripmeester, J.A., *J. Phys. Chem.*, 90, 6549, 1986.
5. Ripmeester, J.A., Ratcliffe, C.I., Klug, D.D., and Tse, J.S., in *Proc. First International Conference on Natural Gas Hydrates*, (Sloan, E.D., Happel, J., and Hnatow, M.A., eds.) *Annals of the New York Academy of Sciences*, 715, 161, 1994.
6. Helgerud, M.B., Circone, S., Stern, L., Kirby, S., and Lorenson, T.D., in *Proc. Fourth International Conference on Gas Hydrates*, Yokohama, May 19–23, 2002, 716, 2002.
7. Helgerud, M.B., Waite, W.F., Kirby, S.H., and Nur, A., *Can. J. Phys.*, 81, 47, 2003.
8. Waite, W.F., Gilbert, L.Y., Winters, W.J., and Mason, D.H., in *Proc. Fifth International Conference on Gas Hydrates*, Trondheim, Norway, June 13–16, Paper 5042, 2005.
9. Bylov, M. and Rasmussen, P., *Chem. Eng. Sci.*, 52, 3295, 1997.

Table II: Phase Equilibria Data of Gas Clathrate Hydrates

This table gives measured phase equilibria data of pure sI and sII gas clathrate hydrates (see Table I for gas hydrate structure and physical property data). The temperature and pressure conditions at which gas hydrates are stable are listed here for typical guest molecules (Tables IIa–IIg). For example, data for methane hydrate (Table IIa) show that at 277.1 K, methane hydrate will dissociate at pressures below 3.81 MPa. In addition to small hydrocarbons (methane, ethane, and propane) and carbon dioxide, the nitrogen (N_2), xenon, and hydrogen sulfide hydrate systems have been measured extensively, and the hydrate phase equilibria data are available at temperatures and pressures as low as about 211.2 K and 6.5 kPa to as high as about 340.15 K and 1500 kPa.

Table IIa. Methane Hydrate (Ref. 1)

T/K	*P*/MPa
***Ice-Hydrate-Vapor* (Ref. 2)**	
262.4	1.79
264.2	1.9
266.5	2.08
268.6	2.22
270.9	2.39
***Liquid Water-Hydrate-Vapor* (Ref. 2)**	
273.7	2.77
274.3	2.9
275.4	3.24
275.9	3.42
275.9	3.43
277.1	3.81
279.3	4.77
280.4	5.35
280.9	5.71
281.5	6.06
282.6	6.77
284.3	8.12
285.9	9.78
***Liquid Water-Hydrate-Vapor* (Ref. 3)**	
295.7	33.99
295.9	35.3
301.0	64.81
302.0	77.5

T/K	*P*/MPa
***Liquid Water-Hydrate-Vapor* (Ref. 4)**	
285.7	9.62
286.3	10.31
286.1	10.1
285.7	9.62
289.0	13.96
292.1	21.13
295.9	34.75
298.7	48.68
300.9	62.4
301.6	68.09
***Liquid Water-Hydrate-Vapor* (Ref. 5)**	
275.4	2.87
276.2	3.37
277.2	3.9
278.2	4.5
279.2	4.9
281.2	6.1
***Ice-Hydrate-Vapor* (Ref. 6)**	
190.2	0.08251
198.2	0.1314
208.2	0.222
218.2	0.3571
243.2	0.955
262.4	1.798

Table IIb. Ethane Hydrate

T/K	*P*/kPa
***Ice-Hydrate-Vapor* (Ref. 7)**	
260.8	294
260.9	290
269.3	441
***Liquid Water-Hydrate-Vapor* (Ref. 7)**	
273.4	545
275.4	669
277.6	876
279.1	1048
281.1	1317
282.8	1641
284.4	2137
284.6	2055
285.8	2537
287.0	3054
***Liquid Water-Hydrate-Liquid Ethane* (Ref. 7)**	
287.7	4909
287.8	3413
287.8	4289
288.1	3716
288.1	6840
288.2	4944
288.2	5082
288.3	4358
288.4	6840
***Ice-Hydrate-Vapor* (Ref. 1)**	
263.6	313
266.5	357
269.3	405
272.0	457
***Liquid Water-Hydrate-Vapor* (Ref. 2)**	
273.7	510
273.7	503
274.8	579
275.9	662
277.6	814
278.7	931
279.3	1007
279.8	1083
280.4	1165
280.4	1165
280.9	1255
281.5	1345
282.1	1448
282.6	1558
283.2	1689
284.3	1986
285.4	2303
285.4	2310
286.5	2730
***Liquid Water-Hydrate-Vapor* (Ref. 8)**	
277.5	780
278.1	840
279.9	1040
281.5	1380
283.3	1660
284.5	2100
286.5	2620

Table IIc. Propane Hydrate

T/K	*P*/kPa
***Ice-Hydrate-Vapor* (Ref. 1)**	
261.2	100
264.2	115
267.4	132
267.6	135
269.8	149
272.2	167
272.9	172
***Liquid Water-Hydrate-Vapor* (Ref. 2)**	
273.7	183
274.8	232
275.4	270
275.9	301
277.1	386
***Ice-Hydrate-Vapor* (Ref. 9)**	
247.9	48.2
251.4	58.3
251.6	58.3
255.4	69.6
258.2	81.1
260.8	90.5
260.9	94.5
262.1	99.4
***Liquid Water-Hydrate-Vapor* (Ref. 10)**	
276.77	0.368
277.01	0.377
277.22	0.405
277.36	0.425
277.44	0.433
277.87	0.473
278.01	0.527
278.55	0.547
***Liquid Water-Hydrate-Liquid Propane* (Ref. 10)**	
278.71	0.643
278.75	0.893
278.75	1.393
278.75	1.891
278.78	1.893
278.80	2.391
278.80	2.891
278.79	2.893
278.75	3.891
278.77	3.391
278.81	4.391
278.79	5.892
278.86	6.392
278.88	6.892
278.80	8.393
278.84	8.893
278.89	9.893

Table IId. Carbon Dioxide Hydrate (Ref. 1)

T/K	*P*/kPa
***Liquid Water-Hydrate-Vapor* (Ref. 11)**	
279.6	2.74
282.1	4.01
282.8	4.36
***Liquid Water-Hydrate-Liquid Carbon Dioxide* (Ref. 11)**	
282.9	5.03
282.9	5.62
283.1	6.47
283.2	9.01
283.6	11.98
283.9	14.36
***Liquid Water-Hydrate-Vapor* (Ref. 12)**	
276.52	1.82
277.85	1.95
278.52	2.21
279.49	2.62
280.44	2.88
281.49	3.35
281.97	3.68
282.00	3.69
282.45	3.85
282.50	4.01
***Liquid Water-Hydrate-Liquid Carbon Dioxide* (Ref. 12)**	
283.33	5.97
283.36	7.35

Table IIe. Nitrogen (N_2) Hydrate

T/K	*P*/kPa
***Liquid Water-Hydrate-Vapor* (Ref. 13)**	
272.0	14480
272.6	15300
272.8	15910
273.8	15910
273.2	16010
273.2	16310
273.4	16620
274.0	17530
274.2	17730
274.8	19150
274.8	19250
275.2	19660
275.6	20670
275.8	21580
276.2	22390
276.6	23100
277.2	24830
278.2	27360
278.2	27970
278.6	28270
279.2	29890
279.2	30300
280.2	33940
281.2	37490
281.6	38610
282.2	41440
283.2	45900
284.2	50660
284.6	52290
285.2	55430
286.2	61400
287.2	67790
287.8	71230
288.4	74580
289.2	81470
290.2	89370
290.6	92210
291.0	95860
***Liquid Water-Hydrate-Vapor* (Ref. 14)**	
277.6	24930
282.2	36820
286.7	63710
291.6	101980
293.0	115490
294.3	128800
296.6	153480
297.7	169270
298.8	192370
299.7	207780
300.6	219600
302.6	268320
304.7	317650
305.5	328890
***Liquid Water-Hydrate-Vapor* (Ref. 15)**	
273.2	16270
273.7	17130
274.9	19130
276.5	23690
277.4	25200
278.6	28610
279.3	30270
281.1	35160
***Liquid Water-Hydrate-Vapor* (Ref. 16)**	
285.63	55000
287.87	59000
289.47	78000
290.80	88000
291.96	101000
292.90	113000
294.60	127000
295.61	139000
296.62	152000
297.32	162000
297.86	169000
298.47	180000
299.31	195000
299.92	206000
300.49	219000
301.86	240000
302.64	256000
303.08	265000
303.82	280000
304.23	294000
304.43	304000
304.56	306000
305.46	324000
305.85	331000
306.26	342000
306.74	354000
307.21	373000
307.50	383000
308.09	398000
308.57	412000
308.82	420000
309.21	431000
309.43	439000

Table IIf. Xenon Hydrate

T/K	P/kPa
***Liquid Water-Hydrate-Vapor* (Ref. 17)**	
211.2	6.5
223.0	13.5
229.4	19.4
238.2	30.8
253.3	64.8
268.2	121.3
***Liquid Water-Hydrate-Vapor* (Ref. 18)**	
273.15	153.976
275.15	179.301
277.15	220.834
278.15	252.237
279.15	283.64
281.15	343.407
283.15	418.369
285.15	497.383
***Liquid Water-Hydrate-Vapor* (Ref. 19)**	
286.65	675
295.35	1780
300.15	2380
300.65	2680
302.15	3060
303.55	3560
305.45	4060
307.15	5100
307.35	5400
308.45	6100
309.05	6600
309.75	6900
310.15	7800
310.35	8620
310.55	9060
311.25	11200
311.75	13400
311.95	13900
312.15	16200
312.55	17700
312.95	20000
313.15	17000
316.65	37500
319.15	60000
320.45	63000
323.75	90000
327.15	125000
328.75	145000
330.15	159000
331.85	184000
333.15	195000
333.75	215000
334.95	220000
335.65	234000
336.85	250000
337.15	258000
349.45	549000
349.15	548000
348.95	555800
349.15	560000
349.65	579600
349.65	587100
350.95	600000
313.45	21900
313.95	25000
314.65	28000
315.15	30000
316.75	35000
317.75	45000
317.65	50000
319.35	57500
322.25	75000
324.95	100000
326.15	113000
327.05	118000
327.35	125000
328.05	130000
328.35	137000
329.65	140000
329.65	150000
331.35	159300
331.15	165000
331.15	170500
331.45	175000
331.65	180000
332.55	182500
333.35	190600
334.15	200000
333.95	207000
334.55	210000
334.85	216000
335.35	217400
335.35	220000
335.55	228000
335.95	230000
335.65	234000
336.95	240000
336.65	241300
337.65	250000
336.95	251700
337.35	255000
337.95	264400
338.95	287500
338.65	303000
340.15	308000
339.65	310000
351.65	645000
351.85	660000
352.15	684000
352.35	700000
352.35	725000
352.65	730000
352.65	750000
352.25	760000
352.75	780000
352.65	800000
352.65	820000
352.65	850000
351.85	875000
352.65	880000
352.35	900000
352.65	920000
351.65	937000
352.15	950000
351.35	965000
350.95	970000
351.95	980000
350.65	1000000
351.65	1000000
350.35	1016000
351.25	1025000
350.65	1070000
349.85	1080000
349.05	1100000
348.75	1133000
348.05	1162000
348.45	1200000
347.25	1230000
346.15	1250000
345.85	1270000
345.55	1290000
344.95	1300000
344.35	1325000
344.45	1340000
343.85	1355000
343.25	1377000
343.15	1390000
342.95	1400000
342.55	1408000
297.90	426000
314.00	479000
316.40	493000
320.20	502000
324.40	520000
331.60	544000
331.70	548000
334.60	561000
339.65	315000
340.65	333000
341.65	347000
341.95	355400
342.95	376200
343.35	392000
343.65	395600
343.15	1420000
342.15	1425000
341.85	1445000
340.85	1463000
340.95	1475000
341.15	1490000
340.15	1500000
339.85	1500000
343.15	398000
344.65	409000
344.35	415700
344.95	421700
344.95	448500
346.15	453000
346.35	474500
345.95	488000
346.95	488700
347.65	498000

T/K	P/kPa
346.65	499000
346.65	503000
347.95	506600
348.15	523000
348.15	529700
348.65	540000
344.7	591000
353.2	624000
354.4	629000
356.6	641000

Liquid Water-Hydrate-Vapor **(Ref. 20)**

T/K	P/kPa
283.36	5140
285.37	5360
287.37	5580
288.85	5750
289.36	5810
289.66	5850
289.76	5860
290.46	930
292.47	1110
294.50	1340
295.91	1540
296.91	1710
298.92	2090
300.92	2560
302.92	3160
304.90	3960
306.13	4520
307.93	5710
308.76	6510
309.63	7560
310.09	8510
310.99	11560
311.40	13100
312.11	16180
313.23	21530
314.02	25580
314.87	30360
315.84	36230
317.60	47400
318.96	56930
320.56	69020

Table IIg. Hydrogen Sulfide Hydrate

Ice-Hydrate-Vapor **(Ref. 21)**

T/K	P/kPa
250.5	34
255.4	44
258.2	50
260.9	57
263.7	64
265.3	69
266.5	72
269.3	81
272.1	90
272.8	93

Liquid Water-Hydrate-Vapor **(Ref. 21)**

T/K	P/kPa
272.8	93[Q1]
277.6	157
283.2	280
285.2	345
288.7	499
291.8	689
294.3	890
295.7	1034
298.5	1379
299.8	1596
300.5	1724
302.1	2068
302.7	2239[Q2]

Liquid Water-Hydrate-Liquid Hydrogen Sulfide **(Ref. 21)**

T/K	P/kPa
302.7	2239
302.8	3447
303.1	6895
303.2	7826
303.4	10342
303.7	13790
303.7	14190
304.0	17237
304.3	20685
304.3	20954
304.6	24132
304.8	27580
304.8	27842
305.1	31027
305.3	34475
305.4	35068

Liquid Water-Hydrate-Vapor **(Ref. 22)**

T/K	P/kPa
283.2	310
291.2	710
299.7	1496
302.7	2241

Liquid Water-Hydrate-Liquid Hydrogen Sulfide **(Ref. 23)**

T/K	P/kPa
298.0	2030
299.0	2050
299.1	2060
299.4	2080
299.6	2080
299.7	2090
299.8	2090
300.1	2090
300.4	2120
300.8	2110
301.0	2130
301.1	2150
301.2	2170
301.2	2150
301.4	2170
301.4	2180
301.6	2200
301.6	2220
302.6	2240

Liquid Water-Hydrate-Vapor **(Ref. 23)**

T/K	P/kPa
298.6	1610
298.8	1620
299.0	1710
299.1	1680
299.2	1700
299.4	1700
299.8	1750
299.8	1770
300.1	1810
300.2	1850
300.4	1870
300.7	1970
300.8	2070

Liquid Water-Hydrate-Vapor **(Ref. 24)**

T/K	P/kPa
277.7	164
279.0	188
280.5	218
281.7	247
283.1	283
284.6	332
286.1	380
287.4	458
288.6	514
290.0	578
291.3	672
292.5	764
294.0	865
295.3	988
296.6	1131
297.5	1237
298.8	1425
300.4	1692
301.3	1861

Liquid Water-Hydrate-Vapor **(Ref. 25)**

T/K	P/kPa
273.6	107.89
274.1	112.08
274.6	117.72
275.1	123.81
275.6	132.39
276.1	137.31
276.6	146.44
277.1	151.72
277.6	161.49
278.1	167.42
278.6	177.48
279.1	185.06
279.6	195.91
280.1	204.31
280.6	216.08
281.1	226.05
281.6	238.58
282.1	249.94
282.6	264.69

T/K	P/kPa	T/K	P/kPa	T/K	P/kPa
283.1	276.82	290.0	560.39	297.0	1167.65
283.6	291.62	290.5	585.85	297.5	1236.72
284.1	305.08	291.0	620.23	298.0	1306.68
284.6	322.07	291.5	650.69	298.5	1381.8
285.1	338.16	292.0	681.87	299.0	1460.44
285.6	356.2	292.5	723.21	299.5	1548.55
286.1	372.71	293.0	762.36	300.0	1639.3
286.6	393.14	293.5	802.82	300.5	1738.63
287.1	412.55	294.0	846.56	301.0	1845.91
287.6	432.16	294.5	892.44	301.5	1961.13
288.1	453.87	295.0	940.69		
288.6	478.05	295.5	991.56		
289.0	503.19	296.0	1046.1		
289.6	529.25	296.5	1104.06		

Q1 Lower Quadruple Point
Q2 Upper Quadruple Point

References for Table II

1. Sloan, E.D. and Koh, C.A., *Clathrate Hydrates of Natural Gases*, 3rd Edition, CRC Press, 2008.
2. Deaton, W.M. and Frost, E.M., Jr., *Gas Hydrates and Their Relation to the Operation of Natural-Gas Pipe Lines*, U.S. Bureau of Mines Monograph 8, p. 101, 1946.
3. Kobayashi, R. and Katz, D.L., *Trans AIME*, 186, 66, 1949.
4. McLeod, H.O. and Campbell, J.M., *J. Petl Tech.*, 222, 590, 1961.
5. Thakore, J.L. and Holder, G.D., *Ind. Eng Chem. Res.*, 26, 462, 1987.
6. Makogon, T.Y. and Sloan, E.D., *J. Chem. Eng. Data*, 39, 351, 1994.
7. Roberts, O.L., Brownscombe, E.R., and Howe, L.S., *Oil Gas J.*, 39, 37, 1940.
8. Holder, G.D. and Grigoriou, G.C., *J. Chem. Thermodyn.*, 12, 1093, 1980.
9. Holder, G.D. and Godbole, S.P., *AIChE J.*, 28, 930, 1982.
10. Mooijer-van den Heuvel, M.M., Peters, C.J., and de Swaan Arons, J., *Fluid Phase Equilib.*, 193, 245, 2002.
11. Ng, H.-J. and Robinson, D.B., *Fluid Phase Equilib.*, 21, 145, 1985.
12. Mooijer-van den Heuvel, M.M., Witteman, R., and Peters, C.J., *Fluid Phase Equilib.*, 182, 97, 2001.
13. Van Cleeff, A., and Diepen,G. A. M., *Recueil des Travaux Chimiques des Pays-Bas*, 79, 582, 1960.
14. Marshall, D. R., Saito, S., and Kobayashi, R., *AIChE Journal*, 10, 202, 1964.
15. Jhaveri, J. and Robinson, D. B., *The Canadian Journal of Chemical Engineering*, 43, 75, 1965.
16. Sugahara, K. et al., *Journal of Supramolecular Chemistry*, 2, 365, 2002.
17. Barrer, R. M. and Edge, A. V. J., *Proceedings of the Royal Society of London A: Mathematical, Physical and Engineering Sciences*, 300, 1, 1967.
18. Ewing, G. J., and Ionescu, L. G., *Journal of Chemical and Engineering Data*, 19, 367, 1974.
19. Dyadin, Y. A. et al., *Journal of Inclusion Phenomena and Molecular Recognition in Chemistry*, 28, 271, 1997.
20. Ohgaki, K. et al., *Fluid Phase Equilibria*, 175, 1, 2000.
21. Selleck, F. T., Carmichael, L.T., and Sage, B.H., *Industrial and Engineering Chemistry*, 44, 2219, 1952.
22. Bond, D. C. and Russell, N.B., *Transactions of the AIME*, 179, 192, 1949.
23. Carroll, J. J. and Mather, A. E., *The Canadian Journal of Chemical Engineering*, 69, 1206, 1991.
24. Mohammadi, A. H. and Richon, D., *Journal of Chemical and Engineering Data*, 54, 2338, 2009.
25. Ward, Z. T. et al., *Journal of Chemical and Engineering Data*, 60, 403, 2014.

Table III. Phase Equilibria Data for Ternary Gas Clathrate Hydrates

Structure H Hydrate phase equilibria data for multi-component systems are reported in Table III for various ternary systems involving methane as a co-guest and methylcyclohexane as an sH hydrate former. The structure H (CH_4+MCH+H_2O) ternary system has been measured extensively, and experimental data are available at temperature and pressure ranges of 251.50-290.62 K and 519-11933 kPa, respectively.

Ternary Methane+MCH (Methylcyclohexane)+Water *Structure H* Hydrate

T/K	P/kPa
***Liquid Water-Methane-MCH Hydrate-Vapor* (Ref. 1)**	
275.9	2078
277.4	2482
279.2	3088
280.8	2795
***Liquid Water-Methane-MCH Hydrate-Vapor* (Ref. 2)**	
275.65	1599
277.65	2137
279.45	2688
281.15	3357
***Liquid Water-Methane-MCH Hydrate-Vapor* (Ref. 3)**	
282.65	3990
284.15	4860
286.45	6470
287.65	7610
289.15	8820
290.25	10500
***Liquid Water-Methane-MCH Hydrate-Vapor* (Ref. 4)**	
277.1	2041
279.9	2951
282.2	3937
283.4	4606
287.1	7391
***Liquid Water-Methane-MCH Hydrate-Vapor* (Ref. 5)**	
279.48	2650
280.49	2930
281.43	3170
282.42	3870
283.39	4370
284.44	4810
284.95	5200

T/K	P/kPa
285.44	5500
285.95	5960
286.49	6340
286.96	6750
287.46	7340
288.40	8330
288.97	9130
***Liquid Water-Methane-MCH Hydrate-Vapor* (Ref. 6)**	
273.60	1354
274.99	1639
277.48	2156
279.95	3086
282.29	4084
285.19	5862
287.59	8076
289.48	10228
290.62	11933
***Liquid Water-Methane-MCH Hydrate-Vapor* (Ref. 7)**	
274.09	1420
274.78	1540
275.28	1660

T/K	P/kPa
275.79	1750
276.26	1870
276.79	1990
277.26	2110
277.8	2250
278.30	2390
279.25	2700
280.26	3050
281.27	3450
282.28	3900
283.29	4430
284.30	5030
285.29	5720
286.28	6500
287.25	7420
288.29	8480
***Liquid Water-Methane-MCH Hydrate-Vapor* (Ref. 8)**	
279.48	2650
280.49	2930
281.43	3170
282.42	3870
283.39	4370

T/K	P/kPa
284.44	4810
284.95	5200
285.44	5500
285.95	5960
286.49	6340
286.96	6750
287.46	7340
288.40	8330
288.97	9130
***Liquid Water-Methane-MCH Hydrate-Vapor* (Ref. 9)**	
251.50	519
253.15	559
255.70	619
258.10	686
261.00	774
264.00	873
267.00	984
269.05	1063
271.00	1145
272.60	1213

References for Table III

1. Becke, P. Kessel, D., and Rahimian, I., *European Petroleum Conference, Society of Petroleum Engineers*, 1992.
2. Mehta, A. P., and Sloan Jr., E. D., *Journal of Chemical and Engineering Data*, 38, 580, 1993.
3. Thomas, M. and Behar, E., *Proceedings of the Annual Convention-Gas Processors Association*, 1994.
4. Tohidi, B. et al., *International Conference on Natural Gas Hydrates*, 1996.
5. Mooijer-van den Heuvel, M. M., Peters, C. J., and de Swaan A. J., *Fluid Phase Equilibria*, 172, 73, 2000.
6. Sun, Z.-G. et al., *Fluid Phase Equilibria*, 198, 293, 2002.
7. Nakamura, T. et al., *Chemical Engineering Science*, 58, 269, 2003.
8. Mooijer-van den Heuvel, M. M. "Phase behavior and structural aspects of ternary clathrate hydrate systems: the role of additives," *Diss. Ph. D. Thesis, Technische Universiteit Delft*, 2004.
9. Ohmura, R. et al., *Journal of Chemical and Engineering Data*, 50, 993, 2005.

Table IV. Phase Equilibria Data of Multi-Component Gas Clathrate Hydrates

Gas hydrate phase equilibria data for multi-component systems, which includes CH_4+C_3H_8+Water, CH_4+CO_2+Water, CH_4+H_2S+Water and CH_4+C_2H_6+Water, are reported in Tables IVa–IVd for various concentrations, temperatures, and pressure ranges.

Table IVa. Ternary CH_4+C_3H_8+Water Hydrate Phase Equilibria

T/K	P/kPa	x_{CH4}
***Liquid Water-Hydrate-Vapor* (Ref. 1)**		
274.8	1627	0.99
277.6	2248	0.99
277.6	2255	0.99
280.4	3123	0.99
283.0	4357	0.99
274.8	1151	0.974
277.6	1593	0.974
280.4	2193	0.974
283.2	3013	0.974
274.8	814	0.952
277.6	1138	0.952
280.4	1586	0.952
283.2	2227	0.952
274.8	552	0.883
277.6	779	0.883
280.4	1110	0.883

T/K	P/kPa	x_{CH4}
283.2	1558	0.883
274.8	365	0.712
277.6	538	0.712
280.4	800	0.712
280.4	800	0.712
283.2	1151	0.712
274.8	272	0.362
277.6	436	0.362
280.4	687	0.362
***Liquid Water-Hydrate-Vapor* (Ref. 2)**		
290.5	6929	0.965
290.7	6929	0.965
293.3	10446	0.965
294.5	13893	0.965
296.6	20857	0.965
299.1	34508	0.965
301.6	48367	0.965

T/K	P/kPa	x_{CH4}
303.7	62294	0.965
303.7	62467	0.965
304.4	68982	0.965
293.1	7412	0.945
292.8	7412	0.945
296.2	13893	0.945
298.5	23615	0.945
300.6	34577	0.945
302.7	48367	0.945
304.9	62225	0.945
***Liquid Water-Hydrate-Vapor* (Ref. 3)**		
274.88	263	0.2375
276.45	350	0.2375
277.74	443	0.2375
279.11	560	0.2375
280.23	689	0.2375
274.45	270	0.371

T/K	P/kPa	x_{CH4}
275.91	343	0.371
277.08	419	0.371
278.65	536	0.371
280.16	691	0.371
282.31	945	0.371
279.80	3032	0.0072
279.91	6996	0.0072
280.04	10271	0.0072
280.21	15250	0.0072
280.49	1834	0.0092
280.59	3166	0.0092
280.71	6989	0.0092
280.91	11002	0.0092
281.11	15394	0.0092
282.35	1756	0.022
282.57	3170	0.022
282.88	6003	0.022
285.09	2060	0.0446
285.30	3611	0.0446
285.69	6590	0.0446
286.25	10140	0.0446
286.67	14326	0.0446
288.01	2811	0.078
288.27	4804	0.078
288.72	7362	0.078
289.27	11496	0.078
289.84	15339	0.078
290.37	3363	0.137
290.66	5507	0.137
290.91	6362	0.137

T/K	P/kPa	x_{CH4}
291.56	9375	0.137
291.66	10389	0.137
292.10	13133	0.137
292.73	16939	0.137
294.24	5259	0.257
294.85	7810	0.257
295.62	10816	0.257
296.39	13795	0.257
297.15	17063	0.257
297.14	8431	0.4823
297.93	11078	0.4823
298.66	13988	0.4823
299.67	17491	0.4823
298.09	10430	0.5635
298.98	13181	0.5635
299.69	15381	0.5635
300.22	17408	0.5635
298.25	10706	0.594
298.94	12623	0.594
299.40	14526	0.594
300.20	17291	0.594
298.02	10600	0.651
298.93	13891	0.651
300.12	17125	0.651
***Liquid Water-Hydrate-Vapor* (Ref. 4)**		
278.15	1306	0.956
278.15	1144	0.947
278.15	848	0.894
278.15	630	0.768
278.15	496	0.53

T/K	P/kPa	x_{CH4}
278.15	489	0.51
278.15	479	0.502
278.15	475	0.468
278.15	479	0.412
278.15	458	0.394
278.15	458	0.39
278.15	479	0.03
278.15	480	0.026
278.15	509	0
***Liquid Water-Hydrate-Vapor* (Ref. 5)**		
274.55	443	0.8813
277.20	645	0.8813
280.45	956	0.8813
283.65	1452	0.8813
289.25	2638	0.8813
291.05	3607	0.8813
291.32	3799	0.8813
295.15	6630	0.8813
295.71	7306	0.8813
296.91	8276	0.8813
299.35	15216	0.8813
***Liquid Water-Hydrate-Vapor* (Ref. 6)**		
278.09	1416	0.9707
283.52	2640	0.9707
288.80	5080	0.9707
292.87	9196	0.9707
295.33	14966	0.9707
296.59	19814	0.9707
297.53	24363	0.9707

Table IVb. Ternary CH_4+CO_2+Water Hydrate

T/K	P/kPa	x_{CO2}
***Liquid Water-Hydrate-Vapor* (Ref. 7)**		
277.0	2841	0.539
278.9	3461	0.539
278.9	3434	0.545
280.9	4240	0.545
282.9	5171	0.545
284.7	6467	0.545
275.5	1993	0.777
279.2	3082	0.777
276.4	3199	0.274
278.4	3951	0.274
281.0	5102	0.274
283.8	6895	0.274
279.5	2999	0.824
282.2	4275	0.824
283.8	5274	0.824
285.5	6895	0.824
285.7	6998	0.824
***Liquid Water-Hydrate-Vapor* (Ref. 8)**		
273.7	2520	0.1
275.8	3100	0.09
277.8	3830	0.08

T/K	P/kPa	x_{CO2}
280.2	4910	0.08
283.2	6800	0.08
285.1	8400	0.08
287.2	10760	0.09
274.6	2590	0.14
276.9	3240	0.13
279.1	4180	0.13
281.6	5380	0.13
284.0	7170	0.13
286.1	9240	0.12
287.4	10950	0.13
273.8	2120	0.25
279.4	3960	0.22
283.4	6230	0.22
285.2	7750	0.21
287.6	10440	0.25
273.7	1810	0.44
276.9	2630	0.42
280.7	4030	0.40
283.1	5430	0.39
285.1	6940	0.39
287.4	9780	0.39
275.6	1990	0.50

T/K	P/kPa	x_{CO2}
278.5	2980	0.47
280.9	4140	0.40
281.8	4470	0.41
285.1	6840	0.44
287.4	9590	0.45
274.6	1660	0.73
276.4	2080	0.70
278.2	2580	0.68
280.2	3280	0.68
282.0	4120	0.67
273.7	1450	0.79
275.9	1880	0.78
277.8	2370	0.76
279.6	2970	0.75
281.6	3790	0.74
282.7	4370	0.85
***Liquid Water-Hydrate-Vapor* (Ref. 9)**		
271.25	1271	0.77
271.41	1434	0.50
271.37	2022	0.25
271.29	1294	0.77
271.44	1450	0.50
271.37	2033	0.25

Table IVc. Ternary CH_4+H_2S+Water Hydrate

T/K	*P*/kPa	x_{CH4}
***Liquid Water-Hydrate-Vapor* (Ref. 10)**		
288.7	4830	0.93
284.3	2590	0.91
282.3	3030	0.937
287.1	4790	0.935
290.1	6790	0.93
279.3	2210	0.935
290.1	6380	0.930
278.7	2830	0.970
282.9	4270	0.969
287.6	6650	0.9705
276.5	2030	0.961
278.4	3240	0.990
282.3	4620	0.9896
284.8	6690	0.9894
287.6	2100	0.780
295.4	5070	0.790
279.8	1030	0.780
281.5	2070	0.905
287.3	3590	0.890
292.1	6000	0.885
***Liquid Water-Hydrate-Vapor* (Ref. 11)**		
283.31	1998.734	0.8866
289.726	4110.473	0.8863
292.068	5990.616	0.8985
288.62	3586.468	0.8833
290.932	4763.755	0.8914
293.224	6869.082	0.8916

References for Table IV

1. Deaton, W. M., and Frost Jr. E. M., Gas hydrates and their relation to the operation of natural-gas pipe lines, *No. BM-Mon-8. Bureau of Mines,* Amarillo, TX (USA), 1946.
2. McLeod Jr, H. O., and Campbell J. M., *Journal of Petroleum Technology,* 13, 590, 1961.
3. Verma, V. K., Hand, J. H., and Katz D. L.. *GVC/AIChE Joint Meeting,* 10, 1975.
4. Thakore, J. L. and Holder G. D., *Industrial and Engineering Chemistry Research,* 26, 462, 1987.
5. Song, K. Y. and Kobayashi, R., *Fluid Phase Equilibria,* 47, 295, 1989.
6. Nixdorf, J. and Oellrich L. R., *Fluid Phase Equilibria,* 139, 325, 1997.
7. Unruh, C. H. and Katz D. L., *Journal of Petroleum Technology,* 1, 83, 1949.
8. Adisasmito, S., Frank III, R. J., and Sloan Jr., E. D., *Journal of Chemical and Engineering Data,* 36, 68, 1991.
9. Hachikubo, A. et al., *Fourth International Conference on Gas Hydrates,* Japan, 2002.
10. Bond, D. C. and Russell, N. B., *Transactions of the AIME,* 179, 192, 1949.
11. Ward, Z. T., "Phase equilibria of gas hydrates containing hydrogen sulfide and carbon dioxide," *Diss. Colorado School of Mines,* 2015.
12. McLeod Jr., H. O. and Campbell J. M.. *Journal of Petroleum Technology,* 13, 590, 1961.
13. Subramanian, S. et al., *Chemical Engineering Science,,* 55, 1981, 2000.

Table IVd. Ternary CH_4+C_2H_6+Water Hydrate

T/K	*P*/kPa	x_{C2H6}
***Liquid Water-Hydrate-Vapor* (Ref. 1)**		
274.8	2861	0.012
277.6	3806	0.012
280.4	5088	0.012
274.8	2365	0.022
277.6	3227	0.022
280.4	4413	0.022
282.6	5667	0.022
283.2	6088	0.022
274.8	2158	0.029
277.6	2958	0.029
280.4	4033	0.029
274.8	1841	0.05
274.8	1841	0.05
277.6	2530	0.05
280.4	3447	0.05
283.2	4771	0.05
274.8	1524	0.096
277.6	2096	0.096
280.4	2889	0.096
283.2	3964	0.096
274.8	945	0.436
277.6	1289	0.436
280.4	1758	0.436
283.2	2434	0.436
***Liquid Water-Hydrate-Vapor* (Ref. 12)**		
302	68430	0.054
301.2	62225	0.054
299.1	48229	0.054
296.6	34439	0.054
293.6	24235	0.054
289.7	13893	0.054
287.9	10446	0.054
284.9	6929	0.054
304.1	68568	0.191
303.1	61949	0.191
301.3	48643	0.191
299	35611	0.191
296.4	23477	0.191
293.3	13893	0.191
291.7	10446	0.191
288.8	6998	0.191
***Liquid Water-Hydrate-Vapor* (Ref. 13)**		
274.2	883	0.419
274.2	958	0.31
274.2	972	0.292
274.2	986	0.27
274.2	1165	0.156
274.2	1448	0.082

Table V. Hydrate Phase Equilibria for Inhibited Gas Hydrate Systems

Tables Va-Vc provide a brief review of hydrate phase equilibria for inhibited gas hydrate systems. The methane hydrate phase equilibria data in the presence of sodium chloride (NaCl), methanol (MeOH), and 1,2-ethanediol (monoethylene glycol (MEG)) are included.

Table Va. Methane Hydrate Phase Equilibria in the Presence of Sodium Chloride

T/K	P/kPa	x_{NaCl}
Liquid Water-Hydrate-Vapor **(Ref. 1)**		
272.7	1416	0.03986
275.3	2640	0.03986
277.7	5080	0.03986
269.2	9196	0.05978
271.4	14966	0.05978
275.1	19814	0.05978
Liquid Water-Hydrate-Vapor **(Ref. 2)**		
265.9	3778	0.2 (wt)
267.8	4626	0.2 (wt)
272.3	6178	0.2 (wt)
272.3	7191	0.2 (wt)
275.7	11094	0.2 (wt)
276.4	10818	0.2 (wt)
276.3	13659	0.2 (wt)
Liquid Water-Hydrate-Vapor **(Ref. 3)**		
280.66	6600	0.02001
286.00	13710	0.02001
289.42	21380	0.02001
291.71	28180	0.02001
293.42	35940	0.02001
293.51	34840	0.02001
295.62	45520	0.02001
296.30	48750	0.02001
297.42	56200	0.02001
298.42	63250	0.02001
299.06	67810	0.02001
279.16	7510	0.03611
284.53	14070	0.03611
287.50	23380	0.03611
288.30	23880	0.03611
289.23	27190	0.03611
290.55	34690	0.03611
292.15	41850	0.03611
293.37	49000	0.03611
294.58	57630	0.03611
295.47	64600	0.03611
296.03	71560	0.03611
274.40	7920	0.05994
280.25	14320	0.05994
282.33	20740	0.05994
284.67	29120	0.05994
286.23	35930	0.05994
287.47	42930	0.05994
288.42	49530	0.05994
289.21	56470	0.05994
290.37	64140	0.05994
291.00	70560	0.05994
270.66	7850	0.08014
275.22	14940	0.08014
278.04	22980	0.08014
279.20	28180	0.08014
281.29	37860	0.08014
282.39	42260	0.08014
283.49	50570	0.08014
284.36	58850	0.08014
284.92	64030	0.08014
285.76	71300	0.08014

Table Vb. Methane Hydrate Phase Equilibria in the Presence of Methanol (MeOH)

T/K	P/kPa	f_{MeOH}
Liquid Water-Hydrate-Vapor **(Ref. 2)**		
266.23	2140	0.1
271.24	3410	0.1
275.87	5630	0.1
280.31	9070	0.1
283.67	13320	0.1
286.40	18820	0.1
263.34	2830	0.2
267.51	4200	0.2
270.08	5610	0.2
273.55	8410	0.2
277.56	13300	0.2
280.17	18750	0.2
Liquid Water-Hydrate-Vapor **(Ref. 4)**		
233.05	1470	0.50
240.05	2950	0.50
247.35	7240	0.50
250.35	10540	0.50
255.25	16980	0.50
250.85	2380	0.35
256.25	3690	0.35
260.25	6810	0.35
264.55	10160	0.35
267.75	13680	0.35
268.45	17220	0.35
270.05	20510	0.35
Liquid Water-Hydrate-Vapor **(Ref. 5)**		
277.85	4200	0.2
279.65	5000	0
278.65	16800	0.2
291.45	18000	0
282.15	24000	0.2
293.85	25500	0
296.75	35000	0
285.25	35900	0.2
296.25	36100	0
297.25	40500	0
297.85	43200	0
286.95	50500	0.2
299.55	54000	0
288.45	57000	0.2
289.00	57600	0.2
289.05	62000	0.2
301.35	62500	0
288.95	62500	0.2
301.55	65000	0
291.35	75700	0.2
303.35	76500	0
305.15	92000	0
294.05	92500	0.2
294.45	96200	0.2
295.15	100000	0.2

Table Vc. Methane Hydrate Phase Equilibria in the Presence of 1,2-Ethanediol (MEG)

T/K	P/kPa	f_{MEG}
Liquid Water-Hydrate-Vapor **(Ref. 4)**		
270.24	2420	0.1
273.49	3400	0.1
280.19	6530	0.1
287.10	15610	0.1
267.59	3770	0.3
269.73	4930	0.3
274.36	7860	0.3
280.14	16140	0.3
279.89	16380	0.3
263.43	9890	0.5
266.32	14080	0.5
266.48	15240	0.5

References for Table V

1. Kharrat, M. and Dalmazzone, D., *The Journal of Chemical Thermodynamics*, 35, 1489, 2003.
2. Sloan Jr., E. D. and Koh, C., *Clathrate Hydrates of Natural Gases*, CRC Press, 2007.
3. Jager, M. D., and Sloan Jr, E. D., *Fluid Phase Equilibria*, 185, 89, 2001.
4. Robinson, D. B., *Journal of Canadian Petroleum Technology*, 25 (04), 1986.
5. Blanc, C. and Tornier-Lasserve, J., *World Oil (USA)*, 211 (5), 1990.

IONIC LIQUIDS

Eugene Paulechka and Chris Muzny

Ionic liquids are a class of organic salts with melting points below 100 °C. Many of these salts are liquid at room temperature. This table lists four thermophysical properties for 153 ionic liquids. The properties and their units are:

- **Mol. form.**: molecular formula in the Hill convention
- **CAS Reg. No.**: Chemical Abstracts Service Registry Number
- **Mol. wt.**: molecular weight (relative molar mass)
- t_m: normal melting point in °C; the notation "gl" indicates a glass–liquid transition
- ρ: density in g cm^{-3} at 25 °C
- η: viscosity in mPa s at 25 °C
- C_p: heat capacity in J·K^{-1}·mol^{-1} at 25 °C; the superscripts indicate the temperature in °C, if different from 25 °C

The original experimental data used in this table can be found in the ILThermo database (Ref. 1). The melting points, densities, and viscosities are provided if at least three independent and consistent data sources are available for a compound. A few exceptions were made if the melting point (triple-point temperature) was determined by adiabatic calorimetry. If several polymorphs that melt at different temperatures are known for a compound, the highest melting point was specified. The glass-liquid transition temperatures are those determined by differential scanning calorimetry at a typical heating rate of 10 K·min^{-1}.

The listed values of the thermophysical properties were critically evaluated using the NIST ThermoData Engine (Ref. 2). For the heat capacity, the procedures described in Ref. 3 were additionally used to validate experimental data. Each recommended value in the table is characterized with an expanded uncertainty (level of confidence approximately 95%, k = 2) listed in parentheses. Compounds are listed alphabetically by name.

References

1. Kazakov, A., Magee, J.W., Chirico, R.D., Paulechka, E., Diky, V., Muzny, C.D., Kroenlein, K., and Frenkel, M., *NIST Standard Reference Database 147: NIST Ionic Liquids Database - (ILThermo), Version 2.0*, National Institute of Standards and Technology, Gaithersburg MD, 2016, http://ilthermo.boulder.nist.gov/
2. Diky, V., Chirico, R. D., Frenkel, M., Bazyleva, A., Magee, J. W., Paulechka, E., Kazakov, A. F., Lemmon, E. W., Muzny, C. D., Smolyanitsky, A. Y., Townsend, S., and Kroenlein, K., *NIST Standard Reference Database 103b: ThermoData Engine - Pure Compounds, Binary Mixtures, Ternary Mixtures, and Chemical Reactions, Version 10.1*,
3. Paulechka, E., *J. Phys. Chem. Ref. Data*,39, 033108, 2010.

Properties of Ionic Liquids

Name	Mol. form.	CAS Reg. No.	Mol. wt.	t_m/°C	ρ/g cm^{-3}	η/mPa s	C_p/ J K^{-1} mol^{-1}
1-Benzyl-3-methylimidazolium bis(trifluoromethylsulfonyl)imide	$C_{13}H_{13}F_6N_3O_4S_2$	433337-24-7	453.37				599(2)
1-Benzyl-3-methylimidazolium chloride	$C_{11}H_{13}ClN_2$	36443-80-8	208.69				338(1)
1-Benzyl-3-methylimidazolium tetrafluoroborate	$C_{11}H_{13}BF_4N_2$	500996-04-3	260.04				417(4)^{72}
1-Benzyl-3-methylimidazolium 1,1,2,2-tetrafluoroethanesulfonate	$C_{13}H_{14}F_4N_2O_3S$		354.32				510(2)
1-Butyl-2,3-dimethylimidazolium bis(trifluoromethylsulfonyl)imide	$C_{11}H_{17}F_6N_3O_4S_2$	350493-08-2	433.38		1.4188(4)	102(2)	
1-Butyl-2,3-dimethylimidazolium tetrafluoroborate	$C_9H_{17}N_2BF_4$	402846-78-0	240.05		1.1932(1)		
2-Butyl-1-ethylpyridinium bis(trifluoromethylsulfonyl)imide	$C_{13}H_{18}F_6N_2O_4S_2$	1642855-33-1	444.41				624(2)
1-Butyl-3-methylimidazolium acetate	$C_{10}H_{18}N_2O_2$	284049-75-8	198.27		1.0526(5)		383(2)
1-Butyl-3-methylimidazolium bis(pentafluoroethylsulfonyl)imide	$C_{12}H_{15}F_{10}N_3O_4S_2$	254731-29-8	519.37		1.514(1)	116(5)	
1-Butyl-3-methylimidazolium bis(trifluoromethylsulfonyl)imide	$C_{10}H_{15}F_6N_3O_4S_2$	174899-83-3	419.36	-2.93(4)	1.4357(9)	50.9(0.9)	566(1)
1-Butyl-3-methylimidazolium bromide	$C_8H_{15}BrN_2$	85100-77-2	219.13	78.2(2)	1.294(3)		311(1)
1-Butyl-3-methylimidazolium chloride	$C_8H_{15}ClN_2$	79917-90-1	174.67	67.85(1)	1.082(2)		317(5)
1-Butyl-3-methylimidazolium dibutyl phosphate	$C_{16}H_{33}N_2O_4P$	663199-28-8	348.42		1.045(2)		
1-Butyl-3-methylimidazolium dicyanamide	$C_{10}H_{15}N_5$	448245-52-1	205.27	-2.3(1)	1.0605(9)	30.1(0.6)	376(2)
1-Butyl-3-methylimidazolium dimethyl phosphate	$C_{10}H_{21}N_2O_4P$	891772-94-4	264.26		1.160(2)		
1-Butyl-3-methylimidazolium hexafluorophosphate	$C_8H_{15}F_6N_2P$	174501-64-5	284.18	11.40(6)	1.3654(9)	249(15)	408(2)
1-Butyl-3-methylimidazolium iodide	$C_8H_{15}IN_2$	65039-05-6	266.13	18.77(4)	1.4809(9)		310(1)
1-Butyl-3-methylimidazolium methyl sulfate	$C_9H_{18}N_2O_4S$	401788-98-5	250.31		1.2081(8)	216(11)	
1-Butyl-3-methylimidazolium nitrate	$C_8H_{15}N_3O_3$	179075-88-8	201.23	36.01(7)	1.157(1)		354(1)
1-Butyl-3-methylimidazolium octyl sulfate	$C_{16}H_{32}N_2O_4S$	445473-58-5	348.50		1.067(1)		
1-Butyl-3-methylimidazolium tetrafluoroborate	$C_8H_{15}BF_4N_2$	174501-65-6	226.02	-85(2) gl	1.2013(5)	104(4)	367(1)
1-Butyl-3-methylimidazolium thiocyanate	$C_9H_{15}N_3S$	344790-87-0	197.30		1.0699(6)	56(2)	
1-Butyl-3-methylimidazolium tosylate	$C_{15}H_{22}N_2O_3S$	410522-18-8	310.41	71(1)			548(5)^{77}
1-Butyl-3-methylimidazolium tricyanomethide	$C_{12}H_{15}N_5$	878027-73-7	229.29		1.047(5)	27.8(0.4)	
1-Butyl-3-methylimidazolium trifluoroacetate	$C_{10}H_{15}F_3N_2O_2$	174899-94-6	252.24	23.28(9)	1.216(1)		408(2)
1-Butyl-3-methylimidazolium trifluoromethanesulfonate	$C_9H_{15}F_3N_2O_3S$	174899-66-2	288.28	17.83(4)	1.298(1)	89(3)	428(2)
1-Butyl-3-methylimidazolium tris(pentafluoroethyl) trifluorophosphate	$C_{14}H_{15}F_{18}N_2P$	917762-91-5	584.23		1.6245(7)		
1-Butyl-3-methylpyridinium bis(trifluoromethylsulfonyl)imide	$C_{12}H_{16}F_6N_2O_4S_2$	344790-86-9	430.38	-84(3) gl	1.4145(7)	64.5(0.6)	578(2)
1-Butyl-4-methylpyridinium bis(trifluoromethylsulfonyl)imide	$C_{12}H_{16}F_6N_2O_4S_2$	475681-62-0	430.38		1.417(2)	56(1)	
1-Butyl-3-methylpyridinium dicyanamide	$C_{12}H_{16}N_4$	712355-12-9	216.29		1.0494(5)		
1-Butyl-2-methylpyridinium tetrafluoroborate	$C_{10}H_{16}BF_4N$	286453-46-1	237.05		1.2012(5)		
1-Butyl-3-methylpyridinium tetrafluoroborate	$C_{10}H_{16}BF_4N$	597581-48-1	237.05		1.1825(3)	177(5)	388(12)

Name	Mol. form.	CAS Reg. No.	Mol. wt.	t_m/°C	ρ/g cm^{-3}	η/mPa s	C_p/ J K^{-1} mol^{-1}
1-Butyl-4-methylpyridinium tetrafluoroborate	$C_{10}H_{16}BF_4N$	343952-33-0	237.05		1.1825(5)	201(9)	
1-Butyl-3-methylpyridinium trifluoromethanesulfonate	$C_{11}H_{16}F_3NO_3S$	857841-32-8	299.31		1.279(1)		
1-Butyl-1-methylpyrrolidinium bis(trifluoromethylsulfonyl)imide	$C_{11}H_{20}F_6N_2O_4S_2$	223437-11-4	422.40	-7.4(1)	1.3945(4)	77(1)	586(2)
1-Butyl-1-methylpyrrolidinium dicyanamide	$C_{11}H_{20}N_4$	370865-80-8	208.31		1.0135(9)	33.8(0.7)	
1-Butyl-1-methylpyrrolidinium tetracyanoborate	$C_{13}H_{20}BN_5$	1266721-18-9	257.15		0.9761(4)		
1-Butyl-1-methylpyrrolidinium trifluoromethanesulfonate	$C_{10}H_{20}F_3NO_3S$	367522-96-1	291.33		1.2527(3)	166(5)	
1-Butyl-1-methylpyrrolidinium tris(pentafluoroethyl) trifluorophosphate	$C_{15}H_{20}F_{18}NP$	851856-47-8	587.27		1.5833(9)	219(8)	
1-Butylpyridinium bis(trifluoromethylsulfonyl)imide	$C_{11}H_{14}F_6N_2O_4S_2$	187863-42-9	416.35		1.448(1)	60(1)	
1-Butylpyridinium tetrafluoroborate	$C_9H_{14}BF_4N$	203389-28-0	223.02		1.2137(3)	167(8)	
Butyltrimethylammonium bis(trifluoromethylsulfonyl)imide	$C_9H_{18}F_6N_2O_4S_2$	258273-75-5	396.36	17.08(5)	1.3922(7)	105(3)	559(2)
2-Decyl-1-ethylpyridinium bis(trifluoromethylsulfonyl)imide	$C_{19}H_{30}F_6N_2O_4S_2$	1642855-44-4	528.57				811(2)
1-Decyl-3-methylimidazolium bis(trifluoromethylsulfonyl)imide	$C_{16}H_{27}F_6N_3O_4S_2$	433337-23-6	503.52	4.18(5)	1.278(1)	119(4)	755(2)
1-Decyl-3-methylimidazolium hexafluorophosphate	$C_{14}H_{27}F_6N_2P$	362043-46-7	368.35				616(6)[42]
1-Decyl-3-methylimidazolium tetracyanoborate	$C_{18}H_{27}BN_6$	1201894-90-7	338.27		0.9625(2)		
Decyltriethylammonium bis(trifluoromethylsulfonyl)imide	$C_{18}H_{36}F_6N_2O_4S_2$	1190369-93-7	522.60				842(5)
1,3-Dibutylimidazolium bis(trifluoromethylsulfonyl)imide	$C_{13}H_{21}F_6N_3O_4S_2$	749921-07-1	461.44				657(2)
1,3-Didecylimidazolium bis(trifluoromethylsulfonyl)imide	$C_{25}H_{45}F_6N_3O_4S_2$	1453194-51-8	629.76				1027(3)
1,3-Diethylimidazolium bis(trifluoromethylsulfonyl)imide	$C_9H_{13}F_6N_3O_4S_2$	174899-88-8	405.33				533(2)
1,2-Diethylpyridinium bis(trifluoromethylsulfonyl)imide	$C_{11}H_{14}F_6N_2O_4S_2$	1642855-30-8	416.35				566(2)
1,3-Diheptylimidazolium bis(trifluoromethylsulfonyl)imide	$C_{19}H_{33}F_6N_3O_4S_2$	1453194-49-4	545.60				839(3)
1,3-Dihexylimidazolium bis(trifluoromethylsulfonyl)imide	$C_{17}H_{29}F_6N_3O_4S_2$	1394227-32-7	517.54				779(2)
1,3-Dimethylimidazolium bis(trifluoromethylsulfonyl)imide	$C_7H_9F_6N_3O_4S_2$	174899-81-1	377.27		1.569(1)		472(1)
1,3-Dimethylimidazolium dimethyl phosphate	$C_7H_{15}N_2O_4P$	654058-04-5	222.18		1.260(1)	276(19)	
1,3-Dimethylimidazolium methyl sulfate	$C_6H_{12}N_2O_4S$	97345-90-9	208.23		1.329(3)	72(1)	
1,3-Dinonylimidazolium bis(trifluoromethylsulfonyl)imide	$C_{23}H_{41}F_6N_3O_4S_2$	1453194-50-7	601.71				966(3)
1,3-Dioctylimidazolium bis(trifluoromethylsulfonyl)imide	$C_{21}H_{37}F_6N_3O_4S_2$	220749-78-0	573.65				906(3)
1,3-Dipentylimidazolium bis(trifluoromethylsulfonyl)imide	$C_{15}H_{25}F_6N_3O_4S_2$	1138216-85-9	489.49				718(2)
1,3-Dipropylimidazolium bis(trifluoromethylsulfonyl)imide	$C_{11}H_{17}F_6N_3O_4S_2$	1138216-84-8	433.38				595(2)
1-Dodecyl-3-methylimidazolium bis(trifluoromethylsulfonyl)imide	$C_{18}H_{31}F_6N_3O_4S_2$	404001-48-5	531.57				820(2)
1-Dodecyl-3-methylimidazolium hexafluorophosphate	$C_{16}H_{31}F_6N_2P$	219947-93-0	396.40				681(7)[62]
Dodecyltriethylammonium bis(trifluoromethylsulfonyl)imide	$C_{20}H_{40}F_6N_2O_4S_2$	876048-22-5	550.66				905(5)
1-Ethyl-2-heptylpyridium bis(trifluoromethylsulfonyl)imide	$C_{16}H_{24}F_6N_2O_4S_2$	1642855-39-7	486.49				718(2)
1-Ethyl-2-hexylpyridinium bis(trifluoromethylsulfonyl)imide	$C_{15}H_{22}F_6N_2O_4S_2$	1642855-37-5	472.46				686(2)
1-Ethyl-3-methylimidazolium acetate	$C_8H_{14}N_2O_2$	143314-17-4	170.21		1.098(1)	137(5)	322(1)
1-Ethyl-3-methylimidazolium bis(pentafluoroethylsulfonyl)imide	$C_{10}H_{11}F_{10}N_3O_4S_2$	216299-76-2	491.32		1.592(2)		
1-Ethyl-3-methylimidazolium bis(trifluoromethylsulfonyl)imide	$C_8H_{11}F_6N_3O_4S_2$	174899-82-2	391.30	-1.71(7)	1.5183(4)	32.2(0.6)	506(2)
1-Ethyl-3-methylimidazolium bromide	$C_6H_{11}BrN_2$	65039-08-9	191.07	76.8(2)			265(1)[77]
1-Ethyl-3-methylimidazolium dicyanamide	$C_8H_{11}N_5$	370865-89-7	177.21		1.105(4)	16(2)	315(1)
1-Ethyl-3-methylimidazolium diethyl phosphate	$C_{10}H_{21}N_2O_4P$	848641-69-0	264.26		1.146(1)		
1-Ethyl-3-methylimidazolium dimethyl phosphate	$C_8H_{17}N_2O_4P$	945611-27-8	236.21		1.219(1)		412(1)
1-Ethyl-3-methylimidazolium ethyl sulfate	$C_8H_{16}N_2O_4S$	342573-75-5	236.29		1.2368(4)	97(2)	383(2)
1-Ethyl-3-methylimidazolium hexafluorophosphate	$C_6H_{11}F_6N_2P$	155371-19-0	256.13	60(1)			362(4)[67]
1-Ethyl-3-methylimidazolium hexyl sulfate	$C_{12}H_{24}N_2O_4S$	942916-86-1	292.39		1.1303(6)		
1-Ethyl-3-methylimidazolium hydrogen sulfate	$C_6H_{12}N_2O_4S$	412009-61-1	208.23		1.3658(4)		
1-Ethyl-3-methylimidazolium iodide	$C_6H_{11}IN_2$	35935-34-3	238.07	77(3)			
1-Ethyl-3-methylimidazolium methanesulfonate	$C_7H_{14}N_2O_3S$	145022-45-3	206.26		1.2417(7)	160(5)	346(1)
1-Ethyl-3-methylimidazolium 2-(2-methoxyethoxy)ethyl sulfate	$C_{11}H_{22}N_2O_6S$	790663-77-3	310.37		1.236(1)		
1-Ethyl-3-methylimidazolium methyl phosphonate	$C_7H_{15}N_2O_3P$	81994-80-1	206.18		1.192(1)		
1-Ethyl-3-methylimidazolium methyl sulfate	$C_7H_{14}N_2O_4S$	516474-01-4	222.26		1.284(1)	78(7)	
1-Ethyl-3-methylimidazolium octyl sulfate	$C_{14}H_{28}N_2O_4S$	790663-79-5	320.45		1.0948(7)	568(21)	
1-Ethyl-3-methylimidazolium tetracyanoborate	$C_{10}H_{11}BN_6$	742099-80-5	226.05		1.0364(2)	18.4(0.3)	407(1)
1-Ethyl-3-methylimidazolium tetrafluoroborate	$C_6H_{11}BF_4N_2$	143314-16-3	197.97	14(1)	1.281(1)	38(1)	305(2)
1-Ethyl-3-methylimidazolium thiocyanate	$C_7H_{11}N_3S$	331717-63-6	169.25	-95(3) gl	1.1164(7)	24.4(0.3)	281(1)
1-Ethyl-3-methylimidazolium tosylate	$C_{13}H_{18}N_2O_3S$	328090-25-1	282.36		1.221(2)		
1-Ethyl-3-methylimidazolium tricyanomethanide	$C_{10}H_{11}N_5$	666823-18-3	201.23		1.0814(3)	14.5(0.2)	
1-Ethyl-3-methylimidazolium trifluoromethanesulfonate	$C_7H_{11}F_3N_2O_3S$	145022-44-2	260.23				363(1)
1-Ethyl-3-methylimidazolium tris(pentafluoroethyl) trifluorophosphate	$C_{12}H_{11}F_{18}N_2P$	377739-43-0	556.17		1.7087(6)	59(1)	
1-Ethyl-3-methylpyridinium ethyl sulfate	$C_{10}H_{17}NO_4S$	872672-50-9	247.31		1.2211(5)		
1-Ethyl-1-methylpyrrolidinium bis(trifluoromethylsulfonyl)imide	$C_9H_{16}F_6N_2O_4S_2$	223436-99-5	394.35	90(1)			
1-Ethyl-2-nonylpyridinium bis(trifluoromethylsulfonyl)imide	$C_{18}H_{28}F_6N_2O_4S_2$	1642855-43-3	514.54				779(2)
1-Ethyl-2-octylpyridinium bis(trifluoromethylsulfonyl)imide	$C_{17}H_{26}F_6N_2O_4S_2$	1642855-41-1	500.51				749(2)
1-Ethyl-2-pentylpyridinium bis(trifluoromethylsulfonyl)imide	$C_{14}H_{20}F_6N_2O_4S_2$	1642855-35-3	458.43				653(2)
1-Ethyl-3-propylimidazolium bis(trifluoromethylsulfonyl)imide	$C_{10}H_{15}F_6N_3O_4S_2$	347882-21-7	419.36				565(2)
1-Ethyl-2-propylpyridinium bis(trifluoromethylsulfonyl)imide	$C_{12}H_{16}F_6N_2O_4S_2$	1642855-31-9	430.38				594(2)

Name	Mol. form.	CAS Reg. No.	Mol. wt.	t_m/°C	ρ/g cm^{-3}	η/mPa s	C_p/ J K^{-1} mol^{-1}
1-Ethylpyridinium bis(trifluoromethylsulfonyl)imide	$C_9H_{10}F_6N_2O_4S_2$	712354-97-7	388.30		1.5359(4)	40.3(0.7)	
1-Heptyl-3-methylimidazolium bis(trifluoromethylsulfonyl)imide	$C_{13}H_{21}F_6N_3O_4S_2$	425382-14-5	461.44		1.3449(6)		659(2)
1-Heptyl-3-methylimidazolium hexafluorophosphate	$C_{11}H_{21}F_6N_2P$	357915-04-9	326.27		1.2623(4)		500(5)
1-Hexadecyl-3-methylimidazolium bis(trifluoromethylsulfonyl)imide	$C_{22}H_{39}F_6N_3O_4S_2$	404001-50-9	587.68	46.10(5)			972(4)[47]
1-Hexyl-3-methylimidazolium acetate	$C_{12}H_{22}N_2O_2$	888320-05-6	226.32				441(4)
1-Hexyl-3-methylimidazolium bis(trifluoromethylsulfonyl)imide	$C_{12}H_{19}F_6N_3O_4S_2$	382150-50-7	447.41	-1.1(1)	1.3719(3)	69.5(0.9)	629(2)
1-Hexyl-3-methylimidazolium bromide	$C_{10}H_{19}BrN_2$	85100-78-3	247.18		1.2239(7)		
1-Hexyl-3-methylimidazolium chloride	$C_{10}H_{19}ClN_2$	171058-17-6	202.73		1.0408(9)		
1-Hexyl-3-methylimidazolium dicyanamide	$C_{12}H_{19}N_5$	927902-57-6	233.32		1.0288(7)		
1-Hexyl-3-methylimidazolium hexafluorophosphate	$C_{10}H_{19}F_6N_2P$	304680-35-1	312.24		1.293(7)	470(23)	469(5)
1-Hexyl-3-methylimidazolium iodide	$C_{10}H_{19}IN_2$	178631-05-5	294.18		1.384(2)		
1-Hexyl-3-methylimidazolium tetracyanoborate	$C_{14}H_{19}BN_6$	1240857-50-4	282.16		0.9900(3)	49.8(0.5)	
1-Hexyl-3-methylimidazolium tetrafluoroborate	$C_{10}H_{19}BF_4N_2$	244193-50-8	254.08		1.1457(7)		431(1)
1-Hexyl-3-methylimidazolium trifluoromethanesulfonate	$C_{11}H_{19}F_3N_2O_3S$	460345-16-8	316.34		1.238(1)		
1-Hexyl-3-methylimidazolium bis(pentafluoroethyl) trifluorophosphate	$C_{16}H_{19}F_{18}N_2P$	713512-19-7	612.28		1.5506(6)		
1-Hexyl-3-methylpyridinium bis(trifluoromethylsulfonyl)imide	$C_{14}H_{20}F_6N_2O_4S_2$	547718-92-3	458.43			85(2)	
1-Hexylpyridinium bis(trifluoromethylsulfonyl)imide	$C_{13}H_{18}F_6N_2O_4S_2$	460983-97-5	444.41		1.388(2)	91(6)	
1-Hexylpyridinium tetrafluoroborate	$C_{11}H_{18}BF_4N$	474368-70-2	251.07		1.155(1)		
1-(2-Hydroxyethyl)-3-methylimidazolium tris(pentafluoroethyl) trifluorophosphate	$C_{12}H_{11}F_{18}N_2OP$	1187194-46-2	572.17		1.7652(5)		
1-Isobutyl-3-methylimidazolium bis(trifluoromethylsulfonyl)imide	$C_{10}H_{15}F_6N_3O_4S_2$	174899-85-5	419.36				565(2)
1-Isobutyl-1-methylpiperidinium bis(trifluoromethylsulfonyl)imide	$C_{12}H_{22}F_6N_2O_4S_2$	1566595-48-9	436.43				608(2)
1-Isobutyl-3-methylpyridinium bis(trifluoromethylsulfonyl)imide	$C_{12}H_{16}F_6N_2O_4S_2$	1621008-58-9	430.38				579(2)
1-Isobutyl-1-methylpyrrolidinium bis(trifluoromethylsulfonyl)imide	$C_{11}H_{20}F_6N_2O_4S_2$	1111670-82-6	422.40				582(2)
1-(2-Methoxyethyl)-1-methylpyrrolidinium bis(trifluoromethylsulfonyl)imide	$C_{10}H_{18}F_6N_2O_5S_2$	757240-24-7	424.37		1.4548(8)		
1-(2-Methoxyethyl)-1-methylpyrrolidinium tris(pentafluoroethyl) trifluorophosphate	$C_{14}H_{18}F_{18}NOP$	1195983-48-2	589.24		1.6306(6)		
1-Methyl-3-nonylimidazolium bis(trifluoromethylsulfonyl)imide	$C_{15}H_{25}F_6N_3O_4S_2$	433337-21-4	489.49		1.2985(7)		
1-Methyl-3-nonylimidazolium hexafluorophosphate	$C_{13}H_{25}F_6N_2P$	343952-29-4	354.32		1.2127(5)		569(6)
1-Methyl-3-octylimidazolium bis(trifluoromethylsulfonyl)imide	$C_{14}H_{23}F_6N_3O_4S_2$	178631-04-4	475.46	-9.19(4)	1.3206(6)	90(3)	691(2)
1-Methyl-3-octylimidazolium bromide	$C_{12}H_{23}BrN_2$	61545-99-1	275.23		1.175(1)		
1-Methyl-3-octylimidazolium chloride	$C_{12}H_{23}ClN_2$	64697-40-1	230.78		1.010(1)		
1-Methyl-3-octylimidazolium hexafluorophosphate	$C_{12}H_{23}F_6N_2P$	304680-36-2	340.29		1.236(1)	723(19)	536(5)
1-Methyl-3-octylimidazolium tetrafluoroborate	$C_{12}H_{23}BF_4N_2$	244193-52-0	282.13	-27.34(4)	1.102(1)	340(8)	498(2)
1-Methyl-3-octylimidazolium trifluoromethanesulfonate	$C_{13}H_{23}F_3N_2O_3S$	403842-84-2	344.39		1.1932(2)		
1-Methyl-3-pentylimidazolium bis(trifluoromethylsulfonyl)imide	$C_{11}H_{17}F_6N_3O_4S_2$	280779-53-5	433.38		1.4036(6)		596(2)
1-Methyl-3-pentylimidazolium hexafluorophosphate	$C_9H_{17}F_6N_2P$	280779-52-4	298.21		1.3276(5)		437(4)
1-Methyl-3-pentylimidazolium tetrafluoroborate	$C_9H_{17}BF_4N_2$	244193-49-5	240.05		1.1768(9)		
1-Methyl-3-propylimidazolium bis(trifluoromethylsulfonyl)imide	$C_9H_{13}F_6N_3O_4S_2$	216299-72-8	405.33	-87(1) gl	1.4747(6)	44(1)	535(2)
1-Methyl-3-propylimidazolium bromide	$C_7H_{13}BrN_2$	85100-76-1	205.10	36.39(5)			281(1)
1-Methyl-1-propylpiperidinium bis(trifluoromethylsulfonyl)imide	$C_{11}H_{20}F_6N_2O_4S_2$	608140-12-1	422.40		1.4101(8)		
3-Methyl-1-propylpyridinium bis(trifluoromethylsulfonyl)imide	$C_{11}H_{14}F_6N_2O_4S_2$	817575-06-7	416.35		1.4481(6)	54.4(0.9)	
1-Methyl-1-propylpyrrolidinium bis(trifluoromethylsulfonyl)imide	$C_{10}H_{18}F_6N_2O_4S_2$	223437-05-6	408.37	11(1)	1.4276(5)	59(4)	554(6)
1-Methyl-3-tetradecylimidazolium bis(trifluoromethylsulfonyl)imide	$C_{20}H_{35}F_6N_3O_4S_2$	404001-49-6	559.63	35.57(5)			897(4)[37]
Methyltrioctylammonium bis(trifluoromethylsulfonyl)imide	$C_{27}H_{54}F_6N_2O_4S_2$	375395-33-8	648.85		1.103(2)		
Methyltrioctylammonium chloride	$C_{25}H_{54}ClN$	5137-55-3	404.16		0.887(2)		
1-Methyl-3-propylimidazolium tetrafluoroborate	$C_7H_{13}BF_4N_2$	244193-48-4	212.00		1.2356(9)		
Tributylethylphosphonium diethyl phosphate	$C_{18}H_{42}O_4P_2$	20445-94-7	384.48		1.0086(9)		
Triethylhexylammonium bis(trifluoromethylsulfonyl)imide	$C_{14}H_{28}F_6N_2O_4S_2$	210230-46-9	466.50		1.2887(3)	182(8)	713(4)
Triethyloctylammonium bis(trifluoromethylsulfonyl)imide	$C_{16}H_{32}F_6N_2O_4S_2$	210230-48-1	494.55	-77(3) gl	1.2506(6)	226(12)	775(5)
Triethyltetradecylammonium bis(trifluoromethylsulfonyl)imide	$C_{22}H_{44}F_6N_2O_4S_2$	1190369-94-8	578.71				972(6)
Tributylmethylammonium bis(trifluoromethylsulfonyl)imide	$C_{15}H_{30}F_6N_2O_4S_2$	405514-94-5	480.52		1.2626(6)		
Trihexyl(tetradecyl)phosphonium bis(trifluoromethylsulfonyl)imide	$C_{34}H_{68}F_6NO_4PS_2$	460092-03-9	764.00		1.0672(8)		
Trihexyl(tetradecyl)phosphonium bis(2,4,4-trimethylpentyl) phosphinate	$C_{48}H_{102}O_2P_2$	465527-59-7	773.29		0.8856(8)		
Trihexyl(tetradecyl)phosphonium bromide	$C_{32}H_{68}BrP$	654057-97-3	563.77		0.956(2)		
Trihexyl(tetradecyl)phosphonium chloride	$C_{32}H_{68}ClP$	258864-54-9	519.32		0.8906(6)		
Trihexyl(tetradecyl)phosphonium dicyanamide	$C_{34}H_{68}N_3P$	701921-71-3	549.91		0.8987(6)	405(24)	
Trihexyl(tetradecyl)phosphonium tris(pentafluoroethyl) trifluorophosphate	$C_{38}H_{68}F_{18}P_2$	883860-35-3	928.87		1.1826(8)	346(12)	
Triisobutylmethylphosphonium *p*-toluenesulfonate	$C_{20}H_{37}O_3PS$	344774-05-6	388.55		1.0731(6)		

DENSITY AND SPECIFIC VOLUME OF MERCURY

The data in this table have been adjusted to the ITS-90 temperature scale. The uncertainty in density values is 0.0003 g/mL between –20 and –10 °C; 0.0001 or less between –10 and 200 °C; and 0.0002 between 200 and 300 °C.

Reference

Ambrose, D., *Metrologia*, 27, 245, 1990.

t/°C	ρ/g cm^{-3}	v/cm^3 kg^{-1}	t/°C	ρ/g cm^{-3}	v/cm^3 kg^{-1}	t/°C	ρ/g cm^{-3}	v/cm^3 kg^{-1}
–20	13.64461	73.2890	30	13.52134	73.9572	80	13.39971	74.6285
–19	13.64212	73.3024	31	13.51889	73.9705	81	13.39729	74.6420
–18	13.63964	73.3157	32	13.51645	73.9839	82	13.39487	74.6554
–17	13.63716	73.3291	33	13.51400	73.9973	83	13.39245	74.6689
–16	13.63468	73.3424	34	13.51156	74.0107	84	13.39003	74.6824
–15	13.63220	73.3558	35	13.50911	74.0241	85	13.38762	74.6959
–14	13.62972	73.3691	36	13.50667	74.0375	86	13.38520	74.7094
–13	13.62724	73.3824	37	13.50422	74.0509	87	13.38278	74.7229
–12	13.62476	73.3958	38	13.50178	74.0643	88	13.38037	74.7364
–11	13.62228	73.4091	39	13.49934	74.0777	89	13.37795	74.7498
–10	13.61981	73.4225	40	13.49690	74.0911	90	13.37554	74.7633
–9	13.61733	73.4358	41	13.49446	74.1045	91	13.37313	74.7768
–8	13.61485	73.4492	42	13.49202	74.1179	92	13.37071	74.7903
–7	13.61238	73.4625	43	13.48958	74.1313	93	13.36830	74.8038
–6	13.60991	73.4759	44	13.48714	74.1447	94	13.36589	74.8173
–5	13.60743	73.4892	45	13.48470	74.1581	95	13.36347	74.8308
–4	13.60496	73.5026	46	13.48226	74.1715	96	13.36106	74.8443
–3	13.60249	73.5160	47	13.47982	74.1850	97	13.35865	74.8579
–2	13.60002	73.5293	48	13.47739	74.1984	98	13.35624	74.8714
–1	13.59755	73.5427	49	13.47495	74.2118	99	13.35383	74.8849
0	13.59508	73.5560	50	13.47251	74.2252	100	13.35142	74.8984
1	13.59261	73.5694	51	13.47008	74.2386	110	13.3273	75.0337
2	13.59014	73.5827	52	13.46765	74.2520	120	13.3033	75.1693
3	13.58768	73.5961	53	13.46521	74.2655	130	13.2793	75.3052
4	13.58521	73.6095	54	13.46278	74.2789	140	13.2553	75.4413
5	13.58275	73.6228	55	13.46035	74.2923	150	13.2314	75.5778
6	13.58028	73.6362	56	13.45791	74.3057	160	13.2075	75.7147
7	13.57782	73.6495	57	13.45548	74.3192	170	13.1836	75.8519
8	13.57535	73.6629	58	13.45305	74.3326	180	13.1597	75.9895
9	13.57289	73.6763	59	13.45062	74.3460	190	13.1359	76.1274
10	13.57043	73.6896	60	13.44819	74.3594	200	13.1120	76.2659
11	13.56797	73.7030	61	13.44576	74.3729	210	13.0882	76.4047
12	13.56551	73.7164	62	13.44333	74.3863	220	13.0644	76.5440
13	13.56305	73.7297	63	13.44090	74.3998	230	13.0406	76.6838
14	13.56059	73.7431	64	13.43848	74.4132	240	13.0167	76.8241
15	13.55813	73.7565	65	13.43605	74.4266	250	12.9929	76.9650
16	13.55567	73.7698	66	13.43362	74.4401	260	12.9691	77.1064
17	13.55322	73.7832	67	13.43120	74.4535	270	12.9453	77.2484
18	13.55076	73.7966	68	13.42877	74.4670	280	12.9214	77.3909
19	13.54831	73.8100	69	13.42635	74.4804	290	12.8975	77.5341
20	13.54585	73.8233	70	13.42392	74.4939	300	12.8736	77.6779
21	13.54340	73.8367	71	13.42150	74.5073			
22	13.54094	73.8501	72	13.41908	74.5208			
23	13.53849	73.8635	73	13.41665	74.5342			
24	13.53604	73.8769	74	13.41423	74.5477			
25	13.53359	73.8902	75	13.41181	74.5612			
26	13.53114	73.9036	76	13.40939	74.5746			
27	13.52869	73.9170	77	13.40697	74.5881			
28	13.52624	73.9304	78	13.40455	74.6016			
29	13.52379	73.9438	79	13.40213	74.6150			

THERMAL PROPERTIES OF MERCURY

Lev R. Fokin

This table gives thermal properties of solid, liquid, and gaseous mercury over a range of temperature. The properties tabulated are:

- Molar heat capacity: C_p/J mol^{-1} K^{-1} of the solid (Ref. 1, 2), liquid (Ref. 3), and gas (Ref. 3)
- Cubic thermal expansion coefficient: $10^4 \times \alpha$/K^{-1} of the liquid (Ref. 4)
- Isothermal compressibility coefficient: $10^6 \times \kappa$/bar^{-1} of the liquid (Ref. 4)
- Speed of sound: u/m s^{-1} in the liquid (Ref. 4)

The last three properties are defined as follows (v is the specific volume):

$$\alpha = \frac{1}{v}\left(\frac{\partial v}{\partial T}\right)_p \qquad \kappa_T = -\frac{1}{v}\left(\frac{\partial v}{\partial P}\right)_T \qquad u^2 = \left(\frac{\partial P}{\partial \rho}\right)_s \qquad \rho = v^{-1}$$

References

1. Busey, R. H., and Giaque, W. F., *J. Am. Chem. Soc.*, 75, 806, 1953.
2. Amitin, Lebedeva, and Paukov, *Rus. J. Phys. Chem.*, 2666, 1979.
3. Douglas, T. B., Ball, A. T., and Ginnings, D. C., *J. Res. Natl. Bur. Stands.*, 46, 334, 1951.
4. Vukalovich, M. P., et al., *Thermophysical Properties of Mercury*, Moscow Standard Press, 1971.

t/°C	C_p(s)/ J mol^{-1} K^{-1}	C_p(l)/ J mol^{-1} K^{-1}	C_p(g)/ J mol^{-1} K^{-1}	$10^4 \times \alpha$(l)/ K^{-1}	$10^6 \times \kappa_T$(l)/ bar^{-1} (1 bar)	$10^6 \times \kappa_T$(l)/ bar^{-1} (1 kbar)	u(l)/ m s^{-1}
-268.99[a]	0.99						
-268.99[b]	0.97		20.786				
-268.15	1.6		20.786				
-263.15	4.6		20.786				
-258.15	7.6		20.786				
-253.15	10.33		20.786				
-248.15	12.74		20.786				
-243.15	14.78		20.786				
-233.15	17.90		20.786				
-223.15	19.94		20.786				
-213.15	21.40		20.786				
-203.15	22.42		20.786				
-193.15	23.16		20.786				
-183.15	23.76		20.786				
-173.15	24.24		20.786				
-153.15	25.00		20.786				
-133.15	25.61		20.786				
-113.15	26.15		20.786				
-93.15	26.69		20.786				
-73.15	27.28		20.786				
-53.15	27.96		20.786				
-38.83	28.5	28.2746	20.786				
-20		28.1466	20.786	1.818	3.83		1470
0		28.0190	20.786	1.8144	3.918	3.78	1460.8
20		27.9002	20.786	1.8110	4.013	3.87	1451.4
25		27.8717	20.786				
40		27.7897	20.786	1.8083	4.109	3.96	1442.0
60		27.6880	20.786	1.8064	4.207		1432.7
80		27.5952	20.786	1.8053	4.308	4.14	1423.4
100		27.5106	20.786	1.8051	4.41		1414.1
120		27.4349	20.786	1.8058	4.513	4.33	1404.7
140		27.3675	20.786	1.8074	4.622		1395.4
160		27.3090	20.786	1.8100	4.731	4.53	1386.1
180		27.2588	20.790	1.8136	4.844		1376.7
200		27.2169	20.790	1.818	4.96		1367
220		27.1834	20.794				
240		27.1583	20.794				
250				1.834	5.26		1344
260		27.1412	20.799				
280		27.1320	20.807				
300		27.1303	20.815	1.856	5.59		1321
320		27.1366	20.824				
340		27.1500	20.836				
356.73		27.1677	20.849				
360		27.1709	20.853				
380		27.1981	20.870				
400		27.2324	20.891				
420		27.2738	20.916				
440		27.3207	20.941				
460		27.3742	20.974				
480		27.4332	21.008				
500		27.4985	21.046				

[a] Superconducting state
[b] Normal state

MELTING CURVE OF MERCURY

The solid–liquid phase boundary of mercury provides a convenient means of calibrating pressures up to 1 GPa in the neighborhood of room temperature. The best representation of this curve is given by:

$$p/\text{GPa} = 1.932845 \times 10^{-2}d + 1.8333 \times 10^{-6}d^2 + 5.9791 \times 10^{-8}d^3$$

where $d = t/°\text{C} - 38.8344$. Temperature is on the ITS-90 scale, and the relation is valid for pressures up to about 1.2 GPa. The following table is calculated from this equation.

Melting Curve of Mercury

t/°C	p/GPa	p/GPa	t/°C
-38.83	0.000	0.010	-38.32
-35.00	0.074	0.020	-37.80
-30.00	0.171	0.050	-36.25
-25.00	0.268	0.100	-33.66
-20.00	0.365	0.200	-28.50
-15.00	0.463	0.300	-23.35
-10.00	0.560	0.400	-18.21
-5.00	0.658	0.500	-13.08
0.00	0.757	0.600	-7.97
5.00	0.856	0.700	-2.88
10.00	0.955	0.800	2.18
15.00	1.055	0.900	7.23
20.00	1.156	1.000	12.24
25.00	1.257	1.100	17.23
30.00	1.359	1.200	22.19

Reference

Molinar, G. F., Bean, V., Houck, J., and Welch, B., *Metrologia* 16, 21, 1980; 28, 353, 1991.

VAPOR PRESSURE OF MERCURY

The following table gives the vapor pressure of mercury in kilopascals (100 kPa = 1 bar) from the triple point (234.3156 K) to the critical point (1764 K). The data are generated from the formulation of Huber, Laesecke, and Friend in Reference 1, which is based on a critical evaluation of all the published data on mercury vapor pressure and related thermodynamic properties. The estimated uncertainty in the vapor pressure is:

−38 to −10 °C	3%
0 to 130 °C	1%
140 to 350 °C	0.15%
360 to 620 °C	0.5%
630 to 1491 °C	5%

Most of the entries in this table carry one significant figure beyond the estimated accuracy.

Note that the table refers to mercury vapor in equilibrium with liquid mercury, in the absence of air or other gases.

References

1. Huber, M. L., Laesecke, A., and Friend, D. G., *The Vapor Pressure of Mercury*, NISTIR 6643, National Institute of Standards and Technology, Boulder, CO, March 2006.
2. Huber, M. L., Laesecke, A., and Friend, D. G., *Ind. Eng. Chem. Res.* 45, 7351, 2006.
3. Vargaftik, N. B., Vinogradov, Y. K., and Yargin, V. S., *Handbook of Physical Properties of Liquids and Gases, Third Edition*, Begell House, New York, 1996.

t/°C	p/kPa	t/°C	p/kPa	t/°C	p/kPa	t/°C	p/kPa
−38.83	2.985×10^{-7}	38	0.0007350	300	32.965	720	6254
−30	9.451×10^{-7}	39	0.0007929	310	40.856	730	6718
−20	3.160×10^{-6}	40	0.0008551	320	50.260	740	7205
−10	9.625×10^{-6}	41	0.0009216	330	61.396	750	7718
0	2.699×10^{-5}	42	0.0009928	340	74.498	760	8258
1	2.979×10^{-5}	43	0.001069	350	89.823	770	8824
2	3.287×10^{-5}	44	0.001151	360	107.65	780	9417
3	3.623×10^{-5}	45	0.001238	370	128.26	790	10040
4	3.991×10^{-5}	46	0.001331	380	151.99	800	10690
5	4.393×10^{-5}	47	0.001430	390	179.17	810	11370
6	4.833×10^{-5}	48	0.001537	400	210.15	820	12080
7	5.312×10^{-5}	49	0.001650	410	245.32	830	12820
8	5.836×10^{-5}	50	0.001771	420	285.07	840	13600
9	6.406×10^{-5}	55	0.002506	430	329.82	850	14410
10	7.028×10^{-5}	60	0.003508	440	380.00	860	15250
11	7.705×10^{-5}	65	0.004862	450	436.07	870	16120
12	8.441×10^{-5}	70	0.006673	460	498.51	880	17030
13	9.242×10^{-5}	75	0.009075	470	567.81	890	17980
14	0.0001011	80	0.01223	480	644.46	900	18960
15	0.0001106	85	0.01635	490	729.01	910	19980
16	0.0001208	90	0.02167	500	821.99	920	21040
17	0.0001320	95	0.02850	510	923.96	930	22140
18	0.0001440	100	0.03721	520	1035.5	940	23270
19	0.0001571	110	0.06209	530	1157.2	950	24450
20	0.0001713	120	0.1009	540	1289.6	960	25670
21	0.0001866	130	0.1599	550	1433.3	970	26930
22	0.0002032	140	0.2478	560	1589.1	980	28230
23	0.0002211	150	0.3759	570	1757.4	990	29580
24	0.0002404	160	0.5592	580	1939	1000	30970
25	0.0002613	170	0.8168	590	2135	1050	38600
26	0.0002839	180	1.1728	600	2345	1100	47450
27	0.0003082	190	1.6573	610	2570	1150	57590
28	0.0003344	200	2.3071	620	2811	1200	69100
29	0.0003627	210	3.1670	630	3069	1250	82100
30	0.0003931	220	4.2906	640	3344	1300	96600
31	0.0004259	230	5.7414	650	3637	1350	112700
32	0.0004611	240	7.5939	660	3949	1400	130000
33	0.0004990	250	9.9347	670	4281	1450	150000
34	0.0005398	260	12.863	680	4632	1491	167000
35	0.0005835	270	16.494	690	5005		
36	0.0006305	280	20.955	700	5399		
37	0.0006809	290	26.392	710	5815		

SURFACE TENSION OF COMMON LIQUIDS

Ian H. Bell

The surface tension γ of about 230 liquids is tabulated here as a function of temperature. Values of γ are given in units of millinewtons per meter (mN m^{-1}), which is equivalent to dyn cm^{-1} in cgs units. The uncertainty of the values is 0.1 to 0.2 mN m^{-1} or less in most cases. Values at temperatures between the points tabulated can be obtained by linear interpolation to a good approximation. Substances are listed alphabetically by name. A more extensive compilation of surface tension may be found in Ref. 1.

Most of the values from Ref. 1 were obtained by capiliary measurements in which the liquid phase is in contact with its vapor plus air at ambient pressure. Values from Refs. 3-6 refer to measurements in which the liquid and vapor phases are in equlibrium, in the absence of other gases. Details may be found in the references.

References

1. Jasper, J. J., *J. Phys. Chem. Ref. Data*, 1, 841, 1972.
2. Kahl, H., Wadewitz, T., and Winkelmann, J., *J. Chem. Eng. Data*, 48, 580, 2003.
3. Mulero, A., Cachadiña., I, and Parra, M. I., *J. Phys. Chem. Ref. Data*, 41, 043105, 2012.
4. Mulero, A. and Cachadiña, I., *J. Phys. Chem. Ref. Data*, 43, 023104, 2014.
5. Mulero, A., Cachadiña, I., and Sanjuán, E. L., *J. Phys. Chem. Ref. Data*, 44, 033104, 2015.
6. Cachadiña, I., Mulero, A., and Tian, J., *J. Phys. Chem. Ref. Data*, 44, 023104, 2015.

Surface Tension at Various Temperatures

Name	Synonym	Mol. form.	γ(10 °C)/ mN m^{-1}	γ(25 °C)/ mN m^{-1}	γ(50 °C)/ mN m^{-1}	γ(75 °C)/ mN m^{-1}	γ(100 °C)/ mN m^{-1}	Ref.
Acetaldehyde	Ethanal	C_2H_4O	22.54	20.50	17.10			1
Acetic acid	Ethanoic acid	$C_2H_4O_2$		27.10	24.61	22.13		1
Acetic anhydride	Acetyl acetate	$C_4H_6O_3$	34.08	31.93	28.34	24.75	21.16	1
Acetone	2-Propanone	C_3H_6O	24.60	22.71	19.60			2,3
Acetonitrile	Methyl cyanide	C_2H_3N		28.66	25.51			1
Acetophenone	Methyl phenyl ketone	C_8H_8O		39.04	36.15	33.27		1
Allyl alcohol	2-Propen-1-ol	C_3H_6O	26.63	25.30	23.06	20.79		5
Ammonia	R-717	H_3N	24.04	20.58	14.94	9.61	4.76	6
Aniline	Benzenamine	C_6H_7N		42.12	39.41	36.69		1
Anisole	Methoxybenzene	C_7H_8O		35.10	32.09	29.08		1
Benzaldehyde	Benzenecarboxaldehyde	C_7H_6O	39.63	38.00	35.27	32.55	29.82	1
Benzene	[6]Annulene	C_6H_6		28.21	24.93	21.73		3
Benzonitrile	Phenyl cyanide	C_7H_5N		38.79	35.90	33.00		1
Benzyl alcohol	Benzenemethanol	C_7H_8O				31.37	27.67	5
Benzylamine	Benzenemethanamine	C_7H_9N		39.30	36.27	33.23		1
Benzyl benzoate	Benzyl benzenecarboxylate	$C_{14}H_{12}O_2$	44.47	42.82	40.06	37.31	34.55	1
Bromine		Br_2	43.68	40.95	36.40			1
Bromobenzene	Phenyl bromide	C_6H_5Br	36.98	35.24	32.34	29.44	26.54	1
1-Bromobutane	Butyl bromide	C_4H_9Br	27.58	25.90	23.08	20.27	17.45	1
Bromoethane	Ethyl bromide	C_2H_5Br	25.36	23.62				1
1-Bromopropane	Propyl bromide	C_3H_7Br	27.08	25.26	22.21			1
2-Bromopropane	Isopropyl bromide	C_3H_7Br	25.03	23.25	20.30			1
3-Bromopropene	Allyl bromide	C_3H_5Br		26.31	23.17			1
Bromotrifluoromethane	Halon-1301	$CBrF_3$	5.51	3.75	1.19			6
Butanenitrile	Propyl cyanide	C_4H_7N		26.92	24.33	21.73		1
Butanoic acid	Butyric acid	$C_4H_8O_2$		26.05	23.75	21.45		1
1-Butanol	Butyl alcohol	$C_4H_{10}O$	25.28	24.13	22.13	20.03	17.83	5
2-Butanol	*sec*-Butyl alcohol	$C_4H_{10}O$	24.24	23.00	20.91	18.79	16.62	5
2-Butanone	Methyl ethyl ketone	C_4H_8O		23.97	21.16			1
2-Butoxyethanol	Ethylene glycol monobutyl ether	$C_6H_{14}O_2$	27.36	26.14	24.10	22.06	20.02	1
Butyl acetate		$C_6H_{12}O_2$	26.48	24.88	22.21	19.54	16.87	1
Butylamine	1-Butanamine	$C_4H_{11}N$		23.44	20.63			1
tert-Butylamine	2-Methyl-2-propanamine	$C_4H_{11}N$		16.87				1
Butyl formate		$C_5H_{10}O_2$	26.05	24.52	21.95	19.39	16.82	1
Carbon dioxide	Carbonic anhydride	CO_2	2.76	0.56				6
Carbon disulfide	Carbon bisulfide	CS_2	33.81	31.58	27.87			1
Chlorobenzene	Phenyl chloride	C_6H_5Cl	34.78	32.99	30.02	27.04	24.06	1
1-Chlorobutane	Butyl chloride	C_4H_9Cl	24.85	23.18	20.39			1
Chlorodifluoromethane	HCFC-22	$CHClF_2$	10.29	8.11	4.72	1.75		6
1-Chlorohexane	Hexyl chloride	$C_6H_{13}Cl$	27.28	25.73	23.13	20.54	17.94	1
Chloropentafluoroethane	CFC-115	C_2ClF_5	6.37	4.74	2.23	0.21		6
1-Chloropentane	Pentyl chloride	$C_5H_{11}Cl$	26.01	24.40	21.71	19.02	16.33	1

Name	Synonym	Mol. form.	γ(10 °C)/ mN m^{-1}	γ(25 °C)/ mN m^{-1}	γ(50 °C)/ mN m^{-1}	γ(75 °C)/ mN m^{-1}	γ(100 °C)/ mN m^{-1}	Ref.
2-Chlorophenol		C_6H_5ClO		39.70	36.89	34.09	31.28	1
3-Chlorophenol		C_6H_5ClO		41.18	38.66	36.13	33.61	1
1-Chloropropane	Propyl chloride	C_3H_7Cl	23.16	21.30				1
2-Chloropropane	Isopropyl chloride	C_3H_7Cl	20.49	19.16				1
3-Chloropropene	Allyl chloride	C_3H_5Cl		23.14				1
o-Cresol	2-Methylphenol	C_7H_8O		36.83	34.39	31.92	29.40	5
m-Cresol	3-Methylphenol	C_7H_8O		35.77	33.42	31.08	28.76	5
Cyclohexane	Hexahydrobenzene	C_6H_{12}	26.24	24.42	21.44	18.55		2,3
Cyclohexanol	Cyclohexyl alcohol	$C_6H_{12}O$		33.25	30.56	28.06	25.71	5
Cyclohexanone	Pimelic ketone	$C_6H_{10}O$	36.43	34.57	31.46	28.36	25.25	1
Cyclohexene	Tetrahydrobenzene	C_6H_{10}	28.01	26.17	23.12			1
Cyclohexylamine	Cyclohexanamine	$C_6H_{13}N$		31.22	28.25	25.28		1
Cyclopentane	Pentamethylene	C_5H_{10}	24.01	21.85	18.38			4
Cyclopentanone	Adipic ketone	C_5H_8O	34.45	32.80	30.05	27.30	24.55	1
Cyclopentene		C_5H_8	24.45	22.20				1
Decane		$C_{10}H_{22}$	24.81	23.39	21.05	18.78	16.56	3
1-Decanol	Capric alcohol	$C_{10}H_{22}O$	28.96	28.02	26.41	24.73	23.01	5
1,2-Dibromoethane	Ethylene dibromide	$C_2H_4Br_2$		39.55	36.25	32.95		1
Dibromomethane	Methylene bromide	CH_2Br_2		39.38	35.43	31.55		1
Dibutylamine	*N*-Butylbutanamine	$C_8H_{19}N$		24.12	21.74	19.36		6
m-Dichlorobenzene	1,3-Dichlorobenzene	$C_6H_4Cl_2$	37.15	35.43	32.57	29.70	26.83	1
Dichlorodifluoromethane	CFC-12	CCl_2F_2	10.50	8.58	5.58	2.89	0.68	3
1,1-Dichloroethane	Ethylidene dichloride	$C_2H_4Cl_2$		24.06				6
1,2-Dichloroethane	Ethylene dichloride	$C_2H_4Cl_2$		31.86	28.29	24.72		1
Dichloromethane	Methylene chloride	CH_2Cl_2		27.20				1,6
1,2-Dichloropropane, (±)-	Propylene dichloride	$C_3H_6Cl_2$		28.32	25.22	22.12		1
1,2-Dichloro-1,1,2,2-tetrafluoroethane	CFC-114	$C_2Cl_2F_4$	12.65	10.94	8.20	5.61	3.23	6
1,1-Diethoxyethane	Acetal	$C_6H_{14}O_2$		20.89	18.31	15.74		1
Diethylamine	*N*-Ethylethanamine	$C_4H_{11}N$		19.85				1
Diethylene glycol	Diglycol	$C_4H_{10}O_3$		44.82	42.78	40.72	38.63	5
Diethyl ether	Ethyl ether	$C_4H_{10}O$		16.5				1,4
Diethyl sulfide	Ethyl sulfide	$C_4H_{10}S$	26.22	24.57	21.8			1
1,1-Difluoroethane	Ethylidene difluoride	$C_2H_4F_2$	11.91	9.89	6.64	3.59	0.95	6
Difluoromethane	Methylene fluoride	CH_2F_2	9.29	6.85	3.10	0.04		6
Diisobutylamine	2-Methyl-*N*-(2-methylpropyl)-1-propanamine	$C_8H_{19}N$		21.72	19.44	17.16		1
Diisopropylamine	*N*-Isopropyl-2-propanamine	$C_6H_{15}N$		19.14	16.45			1
Diisopropyl ether	Isopropyl ether	$C_6H_{14}O$		17.27	14.65			1
Dimethylamine	*N*-Methylmethanamine	C_2H_7N		26.34				1
N,N-Dimethylaniline	*N,N*-Dimethylbenzenamine	$C_8H_{11}N$		35.52	32.9	30.27		1
Dimethyl disulfide	Methyl disulfide	$C_2H_6S_2$		33.39	30.04			1
N,N-Dimethylformamide	DMF	C_3H_7NO	37.56	35.74	32.70	29.66	26.62	2
2,5-Dimethylhexane	Biisobutyl	C_8H_{18}	20.77	19.40	17.12	14.84	12.56	1
2,3-Dimethylpyridine	2,3-Lutidine	C_7H_9N		32.71	30.04	27.36		1
Dimethyl sulfide	2-Thiapropane	C_2H_6S	25.27	24.06				1
Dimethyl sulfoxide	DMSO	C_2H_6OS		42.92	40.06			1
1,4-Dioxane	1,4-Dioxacyclohexane	$C_4H_8O_2$		32.75	29.28	25.80	22.32	1
Diphenyl ether	Oxybisbenzene	$C_{12}H_{10}O$		26.75	24.80			1
Dipropylamine	*N*-Propyl-1-propanamine	$C_6H_{15}N$		22.31	19.75	17.20		1
Dodecane		$C_{12}H_{26}$	26.32	24.93	22.70	20.55	18.49	3
Epichlorohydrin	(Chloromethyl)oxirane	C_3H_5ClO	38.4	36.36	32.96	29.56	26.16	1
Ethane		C_2H_6	2.03	0.48				3
1,2-Ethanediol	Ethylene glycol	$C_2H_6O_2$		48.02	45.86	43.66	41.42	5
Ethanethiol	Ethyl mercaptan	C_2H_6S		23.08				1
Ethanol	Ethyl alcohol	C_2H_6O	23.12	21.91	19.85			1,5
Ethanolamine	Glycinol	C_2H_7NO		48.32	45.53	42.73		1
Ethoxybenzene	Phenetole	$C_8H_{10}O$		32.41	29.65	26.89		1
2-Ethoxyethanol	Ethylene glycol monoethyl ether	$C_4H_{10}O_2$		28.35	26.11	23.86	21.62	1
Ethyl acetate		$C_4H_8O_2$	25.13	23.39	20.49	17.58	14.68	1
Ethylamine	Ethanamine	C_2H_7N		19.20				1,6
N-Ethylaniline		$C_8H_{11}N$		36.33	33.65	30.98		1
Ethylbenzene	Phenylethane	C_8H_{10}	30.16	28.52	25.82	23.16	20.56	4
Ethyl butanoate	Ethyl butyrate	$C_6H_{12}O_2$	25.51	23.94	21.33	18.71	16.10	1

Name	Synonym	Mol. form.	γ(10 °C)/ mN m⁻¹	γ(25 °C)/ mN m⁻¹	γ(50 °C)/ mN m⁻¹	γ(75 °C)/ mN m⁻¹	γ(100 °C)/ mN m⁻¹	Ref.
Ethylcyclohexane		C_8H_{16}	26.73	25.15	22.51			1
Ethyl formate		$C_3H_6O_2$	25.16	23.18				1
Ethyl propanoate	Ethyl propionate	$C_5H_{10}O_2$	25.55	23.80	20.88	17.96		1
Fluorobenzene	Phenyl fluoride	C_6H_5F	28.47	26.66	23.65	20.64		1
Formamide	Methanamide	CH_3NO		57.03	54.92	52.82	50.71	1
Formic acid	Methanoic acid	CH_2O_2		37.13	34.38	31.64		1
Furfural	2-Furaldehyde	$C_5H_4O_2$	45.08	43.09	39.78	36.46	33.14	1
1,1,1,2,3,3,3-Heptafluoropropane	HFC-227ea	C_3HF_7	8.79	7.04	4.30	1.87	0.05	6
Heptane		C_7H_{16}	21.23	19.73	17.28	14.88		2,3
Heptanoic acid	Enanthic acid	$C_7H_{14}O_2$		27.76	25.64			1
1-Heptanol	Heptyl alcohol	$C_7H_{16}O$	27.57	26.44	24.52	22.58	20.60	5
2-Heptanone	Methyl pentyl ketone	$C_7H_{14}O$		26.12	23.48			1
1-Heptene		C_7H_{14}	21.29	19.80	17.33	14.85		1
Hexadecane	Cetane	$C_{16}H_{34}$		27.05	24.91	22.78	20.64	1
Hexafluoroethane	Perfluoroethane	C_2F_6	0.64					6
1,1,1,3,3,3-Hexafluoropropane	HFC-236fa	$C_3H_2F_6$	11.43	9.61	6.71	4.05	1.70	6
Hexane		C_6H_{14}	19.46	17.88	15.30			3
Hexanedinitrile	Adiponitrile	$C_6H_8N_2$		45.45	43.02	40.58		1
Hexanenitrile	Capronitrile	$C_6H_{11}N$		27.37	25.11	22.84		1
1-Hexanol	Caproyl alcohol	$C_6H_{14}O$		25.77	23.77	21.79	19.80	5
2-Hexanone	Butyl methyl ketone	$C_6H_{12}O$		25.45	22.72			1
1-Hexene		C_6H_{12}	19.44	17.90	15.33			1
Hydrazine		H_4N_2		66.39				1
Iodobenzene	Phenyl iodide	C_6H_5I	40.40	38.71	35.91	33.10	30.29	1
1-Iodobutane	Butyl iodide	C_4H_9I	29.79	28.24	25.67	23.09	20.51	1
Iodoethane	Ethyl iodide	C_2H_5I	30.38	28.46	25.24			1
Iodomethane	Methyl iodide	CH_3I	32.19	30.34				1
Isobutane	2-Methylpropane	C_4H_{10}	11.69	10.00	7.28	4.70	2.37	3
Isobutyl acetate	2-Methylpropyl acetate	$C_6H_{12}O_2$	24.58	23.06	20.53	17.99	15.46	1
Isobutylamine	2-Methyl-1-propanamine	$C_4H_{11}N$		21.75	19.02			1
Isopropyl acetate	1-Methylethyl acetate	$C_5H_{10}O_2$	23.37	21.76	19.08	16.40		1
Isopropylbenzene	Cumene	C_9H_{12}	29.27	27.69	25.05	22.42	19.78	1
Mercury	Quicksilver	Hg	488.55	485.48	480.36	475.23	470.11	1
Methanol	Methyl alcohol	CH_4O	23.47	22.17	19.98			1,5
2-Methoxyethanol	Ethylene glycol monomethyl ether	$C_3H_8O_2$	32.32	30.84	28.38	25.92	23.46	1
Methyl acetate		$C_3H_6O_2$	26.66	24.73	21.51			1
Methylamine	Methanamine	CH_5N		19.09				6
N-Methylaniline	*N*-Methylbenzenamine	C_7H_9N		36.90	34.47	32.05		1
Methyl benzoate	Methyl benzenecarboxylate	$C_8H_8O_2$		37.17	34.25	31.32		1
Methyl butanoate		$C_5H_{10}O_2$	26.34	24.62	21.76	18.89	16.03	1
3-Methyl-1-butanol	Isopentyl alcohol	$C_5H_{12}O$	24.95	23.73	21.69	19.64	17.58	5
2-Methyl-2-butene		C_5H_{10}	18.61	17.15				1
Methylcyclohexane		C_7H_{14}	24.94	23.27	20.55			4
Methylcyclopentane		C_6H_{12}	23.47	21.72	18.82			1
Methyl formate		$C_2H_4O_2$	26.72	24.36	20.43	16.50	12.57	1
Methyl hexadecanoate	Methyl palmitate	$C_{17}H_{34}O_2$			27.76	25.44	23.23	4
3-Methylhexane		C_7H_{16}	20.76	19.31	16.88	14.46		1
Methyl octadecanoate	Methyl stearate	$C_{19}H_{38}O_2$			28.41	26.17	24.03	4
2-Methylpentane	Isohexane	C_6H_{14}	18.4	16.87	14.38			3
3-Methylpentane		C_6H_{14}	19.2	17.61	14.96			1
Methyl propanoate	Methyl propionate	$C_4H_8O_2$	26.32	24.44	21.29			1
2-Methyl-2-propanol	*tert*-Butyl alcohol	$C_4H_{10}O$		20.01	17.75			5
2-Methylpyridine	2-Picoline	C_6H_7N		33.00	29.90	26.79		1
N-Methyl-2-pyrrolidinone	1-Methyl-2-pyrrolidinone	C_5H_9NO	41.94	40.21	37.33	34.45	31.57	2
Methyl salicylate	Methyl 2-hydroxybenzoate	$C_8H_8O_3$	40.98	39.22	36.28	33.35	30.41	1
Nitrobenzene		$C_6H_5NO_2$			40.56	37.66	34.77	1
Nitroethane		$C_2H_5NO_2$	34.02	32.13	29.00			1
Nitromethane	Nitrocarbol	CH_3NO_2	39.04	36.53	32.33			1
2-Nitropropane	Isonitropropane	$C_3H_7NO_2$	31.02	29.29	26.39			1
Nonane		C_9H_{20}	23.82	22.38	20.03	17.73	15.49	3
1-Nonanol	Nonyl alcohol	$C_9H_{20}O$	28.60	27.59	25.86	24.08	22.25	5
5-Nonanone	Dibutyl ketone	$C_9H_{18}O$		26.28	23.85			1
Octadecane		$C_{18}H_{38}$		27.87	25.77	23.66	21.55	1

Name	Synonym	Mol. form.	γ(10 °C)/ mN m^{-1}	γ(25 °C)/ mN m^{-1}	γ(50 °C)/ mN m^{-1}	γ(75 °C)/ mN m^{-1}	γ(100 °C)/ mN m^{-1}	Ref.
Octane		C_8H_{18}	22.66	21.17	18.77	16.45	14.21	3
1-Octanol	Capryl alcohol	$C_8H_{18}O$	28.11	26.96	25.07			5
Paraldehyde	2,4,6-Trimethyl-1,3,5-trioxane	$C_6H_{12}O_3$	27.22	25.63	22.97	20.32	17.66	1
Pentachloroethane	Refrigerant 120	C_2HCl_5		34.11	31.21	28.24		6
Pentanal	Valeraldehyde	$C_5H_{10}O$	26.95	25.44	22.91			1
Pentane		C_5H_{12}	17.12	15.45				3
1-Pentanol	Amyl alcohol	$C_5H_{12}O$	26.19	25.01	23.01	20.98	18.89	5
2-Pentanol	*sec*-Amyl alcohol	$C_5H_{12}O$	25.01	23.46	20.91	18.41	15.94	5
2-Pentanone	Methyl propyl ketone	$C_5H_{10}O$		23.25	21.62			1
3-Pentanone	Diethyl ketone	$C_5H_{10}O$		24.74	22.13			1
1-Pentene	α-Amylene	C_5H_{10}	17.10	15.45				1
Pentyl acetate	Amyl acetate	$C_7H_{14}O_2$	26.67	25.17	22.69	20.20	17.72	1
Pentylamine	Amylamine	$C_5H_{13}N$		24.69	22.14	19.58		1
Perfluoropropane	FC-218	C_3F_8	5.32	3.77	1.45			6
Phenol	Hydroxybenzene	C_6H_6O			37.69	35.05	32.42	5
Phosphorus(III) chloride	Phosphorus trichloride	Cl_3P		27.98	24.81			1
Phosphoryl chloride	Phosphorus oxychloride	Cl_3OP		32.03	28.85	25.66		1
Piperidine	Azacyclohexane	$C_5H_{11}N$	30.64	28.91	26.03	23.14	20.26	1
Propane	LPG	C_3H_8	8.87	7.02	4.14	1.61		3
1,2-Propanediol	1,2-Propylene glycol	$C_3H_8O_2$	36.85	35.99	34.41	32.65	30.69	5
Propanenitrile	Ethyl cyanide	C_3H_5N		26.75	23.87			1
1-Propanethiol	Propyl mercaptan	C_3H_8S		24.20	21.02			1
2-Propanethiol	Isopropyl mercaptan	C_3H_8S		21.33	18.39			1
Propanoic acid	Propionic acid	$C_3H_6O_2$		26.20	23.72	21.23		1
1-Propanol	Propyl alcohol	C_3H_8O	24.59	23.37	21.32	19.20		5
2-Propanol	Isopropyl alcohol	C_3H_8O	22.06	20.92	18.97	16.95		5
Propyl acetate		$C_5H_{10}O_2$	25.48	23.80	21.00	18.20	15.40	1
Propylamine	1-Propanamine	C_3H_9N		21.75				1
Pyridazine	1,2-Diazabenzene	$C_4H_4N_2$	49.51	47.96	45.37	42.78	40.19	1
Pyridine	Azine	C_5H_5N		36.56	33.29	30.03		1
Pyrimidine	1,3-Diazine	$C_4H_4N_2$		30.33	27.80	25.28	22.75	1
Pyrrole	Imidole	C_4H_5N	38.71	37.06	34.31			1
Pyrrolidine	Azacyclopentane	C_4H_9N	30.58	29.23	26.98			1
Quinoline	1-Azanaphthalene	C_9H_7N	44.19	42.59	39.94	37.28	34.62	1
Sulfur hexafluoride		F_6S	3.29	1.63				3
Sulfuryl chloride		Cl_2O_2S		28.78				1
1,1,2,2-Tetrachloroethane	Acetylene tetrachloride	$C_2H_2Cl_4$		35.58	32.41	29.24	26.07	1
Tetrachloromethane	Carbon tetrachloride	CCl_4		26.18	23.23	20.18	17.08	6
Tetrachlorosilane	Silicon tetrachloride	Cl_4Si	19.78	18.29	15.80			1
Tetradecane		$C_{14}H_{30}$	27.43	26.13	23.96	21.78	19.61	1
1,1,1,2-Tetrafluoroethane	HFC-134a	$C_2H_2F_4$	10.08	8.07	4.92	2.14	0.04	6
1,2,3,4-Tetrahydronaphthalene	Tetralin	$C_{10}H_{12}$		33.17	30.78	28.40		1
Thiophene	Thiofuran	C_4H_4S		30.68	27.36			1
Toluene	Methylbenzene	C_7H_8	29.76	27.91	24.88	21.93	19.07	2,3
Tribromomethane	Bromoform	$CHBr_3$		44.84	41.61	38.33		6
Tributylamine	*N,N*-Dibutyl-1-butanamine	$C_{12}H_{27}N$		24.39	22.32	20.24		1
1,1,1-Trichloroethane	Methyl chloroform	$C_2H_3Cl_3$		25.18	22.07			1
1,1,2-Trichloroethane	Vinyl trichloride	$C_2H_3Cl_3$		34.02	30.65	27.27	23.89	1
Trichloromethane	Chloroform	$CHCl_3$		26.65	23.42	20.15		6
Tridecane		$C_{13}H_{28}$	26.86	25.55	23.37	21.19	19.01	1
Triethylamine	*N,N*-Diethylethanamine	$C_6H_{15}N$		20.22	17.74			1
Trifluoroacetic acid		$C_2HF_3O_2$		13.53	11.42			1
Trifluoromethane	Fluoroform	CHF_3	1.44	0.03				6
Trimethylamine	*N,N*-Dimethylmethanamine	C_3H_9N		13.41				1
1,2,4-Trimethylbenzene	Pseudocumene	C_9H_{12}	30.74	29.20	26.64	24.07	21.51	1
1,3,5-Trimethylbenzene	Mesitylene	C_9H_{12}	28.89	27.55	25.31	23.07	20.82	1
Undecane	Hendecane	$C_{11}H_{24}$	25.56	24.21	21.96	19.70	17.45	1
Water		H_2O	74.29	72.06	68.02	63.63	58.92	3
Water-d_2	Heavy water	D_2O	72.90	70.95	67.33	63.25	58.75	3
Xenon		Xe	0.44					3
o-Xylene	1,2-Dimethylbenzene	C_8H_{10}	31.15	29.51	26.81	24.16	21.56	4
m-Xylene	1,3-Dimethylbenzene	C_8H_{10}	29.79	28.12	25.38	22.70	20.08	4
p-Xylene	1,4-Dimethylbenzene	C_8H_{10}		27.80	25.18	22.61	20.08	4

SURFACE TENSION OF AQUEOUS MIXTURES

The composition dependence of the surface tension of binary mixtures of several compounds with water is given in this table. The data are tabulated as a function of the mass percent of the non-aqueous component. Data for methanol, ethanol, 1-propanol, and 2-propanol are taken from Reference 1, which also gives values at other temperatures.

References

1. Vazquez, G., Alvarez, E., and Navaza, J. M., *J. Chem. Eng. Data*, 40, 611, 1995.
2. *Landolt-Börnstein, Numerical Data and Functional Relationships in Science and Technology, New Series*, IV/16, *Surface Tension*, Springer-Verlag, Heidelberg, 1997.

Surface Tension in mN/m for the Specified Mass %

Compound	*t*/°C	0%	10%	20%	30%	40%	50%	60%	70%	80%	90%	100%
Acetic acid	30	71.2	51.4	43.3	41.2	38.2	37.4	36.1	33.5	31.5	30.2	26.3
Acetone	25	72.0	44.9	40.5	36.7	33.0	30.1	29.4	29.4	27.6	24.5	23.1
Acetonitrile	20	72.8	48.5	40.2	34.1	31.6	30.6	30.0	29.6	29.1	28.7	28.4
1,2-Butanediol	25	72.0	66.1	60.4	55.1	50.1	45.6	43.3	41.9	40.8	39.2	35.8
1,3-Butanediol	30	71.2	58.1	51.6	48.7	45.8	43.9	42.4	41.2	40.0	39.0	37.0
1,4-Butanediol	30	71.2	61.2	56.9	54.2	52.0	50.7	49.5	47.9	46.6	45.2	43.8
Butanoic acid	30	71.2	42.4	37.5	35.5	34.8	32.2	30.8	29.2	27.4	26.3	25.5
2-Butanone	20	72.8	41.6	32.2				25.2				24.6
γ-Butyrolactone	30	71.2	64	58	53	50	48	46	45	44	42.8	42.7
Chloroacetic acid	25	72.0	59.8	53.6	51.3	49.7	48.3	47.5	46.1			
Diethanolamine	25	72.0	66.8	63.2	60.7	58.8	57.2	55.7	54.3	52.7	50.6	47.2
N,N-Dimethylacetamide	25	72.0	72.0	72.0	72.4	73.5	74.9	75.4	73.0	65.7	54.7	36.4
N,N-Dimethylformamide	25	72.0	65.4	59.2	53.8	49.6	47.3	46.9	44.9	42.3	38.4	35.2
1,4-Dioxane	25	72.0					41.2	39.6	37.9	36.2	34.5	33.7
Ethanol	25	72.01	47.53	37.97	32.98	30.16	27.96	26.23	25.01	23.82	22.72	21.82
Ethylene glycol	20	72.8	68.5	64.9	61.9		57.0					48.2
Formic acid	20	72.8	66	60	55.7	52.2	50.3	48.8	47.1	44.7	40.9	38.0
Glycerol	25	72.0	70.5	69.5	68.5	67.9	67.4	66.9	66.5	65.7	64.5	62.5
Methanol	25	72.01	56.18	47.21	41.09	36.51	32.86	29.83	27.48	25.54	23.93	22.51
Morpholine	20	72.8	65.1	60.7	58.9	56.7	53.0	49.6	47.0	43.7	41.8	38.7
Nitric acid	20	72.8	71.9	70.7	68.9	66.6	63.8	60.6	56.8	52.6	47.9	42.6
Propanoic acid	30	71.2	46.6	42.2	37.7	35.6	33.1	31.7	30.2	28.2	27.4	25.8
1-Propanol	25	72.01	34.32	27.84	25.98	25.26	24.80	24.49	24.08	23.86	23.59	23.28
2-Propanol	25	72.01	40.42	30.57	26.82	25.27	24.26	23.51	22.68	22.14	21.69	21.22
1,2-Propylene glycol	30	71.2	60.5	54.9	50.7	47.2	44.5	41.5	38.6	37.6	36.3	35.5
1,3-Propylene glycol	30	71.2	62.6	58.8	55.7	53.8	52.8	51.7	50.8	49.6	48.2	47.0
Pyridine	25	72.0	52.8	51.2	48.0	46.8	46.6	45.8	45.0	43.6	40.9	37.0
Sulfolane	20	72.8					62.5	61.6	59.6	57.1	54.9	50.9
Sulfuric acid	50	67.9	73.5	75.1	73.6	71.2	68.0	64.1	60.0	56.4	53.6	51.7
Trichloroacetaldehyde	25	72.0	56.7	51.0	46.7	44.1	43.0	42.5	41.5	38.9	34.7	29.4
Trichloroacetic acid	25	72.0	55.8	46.5	42.8	41.6	40.6	39.4	38.3	37.4	36.5	

SURFACE ACTIVE CHEMICALS (SURFACTANTS)

Thomas J. Bruno and Paris D. N. Svoronos

The following table provides the structure and properties of common surface active chemicals or surfactants (Refs. 1-3). These reagents are used industrially to decrease surface tension in many different applications, and they are used in the laboratory in many analytical and biochemical procedures. In this way, the table complements many of the tables presented in Sections 7 and 8, in which these surfactants are incorporated in reagents. The table is arranged alphabetically according to chemical name, although the most common generic name or abbreviation is provided in bold type (unless the chemical name is also the most common name). The surfactant class is provided (anionic, cationic, or nonionic). Anionic surfactants are molecules in which the hydrophilic moiety is a negatively charged group such as a sulfonate, sulfate, or carboxylate. Cationic surfactants are molecules in which the hydrophilic moiety is a positively charged group such as a quaternary ammonium ion. Nonionic surfactants are molecules in which the hydrophilic moiety is uncharged, such as an ethoxylate. Another classification exists, though not represented here, called amphoteric surfactants, in which the ionic character is pH dependent.

Where available, the melting and normal boiling temperatures are provided; if decomposition occurs before this state point is reached, this is indicated by the notation "dec." The measured density of either the liquid or a solution is provided, along with temperature, where possible. The predicted density, while available for some surfactants in which measurements are unavailable, is not provided here.

Where available, the critical micelle concentration (CMC), the concentration of the surfactant above which micelles spontaneously form, is provided. The preferred unit is mmol/L, but in the case of mixtures, this is provided as a percent or ppm, usually on the basis of mass (mass/mass). The CMC is dependent on temperature and the ionic strength of the solution, and is also dependent to some extent on the measurement technique. For anionic and cationic surfactants, the CMC is reduced by increasing ionic strength, but temperature has a minor effect. For nonionic surfactants, the CMC is relatively insensitive to ionic strength, but increases with increases in temperature. For the CMC values given here, the temperature is provided where possible. If not specified, the temperature is to be regarded as ambient.

Where available, the aggregation number is provided (Refs. 1, 4, and 5). This is the mean number of surfactant molecules present in a micelle after the CMC has been reached. It is measured using luminescent probes by varying surfactant concentration, and it is dependent on temperature and the concentration of any organic or ionic species present. Finally, the hydrophilic lipophilic balance (HLB) is provided where available. This is a fit-for-purpose property rather than a fundamental property which has evolved in definition and determination since being devised, a matter beyond the scope of this entry. The HLB is a numerical scale between 0 and 20 descriptive of the tendency of a surfactant to be either hydrophilic or hydrophobic. A relatively high HLB (greater than 10) indicates hydrophilicity, or good water or polar solvent solubility. A relatively low HLB (lower than 10) indicates lipophilicity, or good solubility in nonpolar solvents such as oils or organics. The HLB changes with concentration and mixture composition.

References

1. *Surfactant Basics*: Dow Answer Center, <http://dowac.custhelp.com/app/answers/detail/a_id/11357/~/dow-surfactants---design-for-the-environment-and-cleangredients%C2%AE-status>
2. Lange, R. K., *Surfactants: A Practical Handbook*, Hanser Publishers, Munich, 1999.
3. Porter, M. R., *Handbook of Surfactants*, Blackie Academic and Professional,.London, 1994.
4. Alargova, R. G., Kochijashky, I. I., Sierra, M. L., and Zana, R., Micelle aggregarion numbers and surfactants in aqueous solutions: a comparison between the results from steady state and time resolved fluorescence quenching, *Langmuir* 14, 5412-5418, 1998.
5. *Technical Information Surface Chemistry*, Akso Nobel, <http://surface.akzonobel.com>, 2015.

Name	Synonym	Mol. form.	Mol. wt.	bp/°C	mp/°C	Den./ g cm^{-3}	CMC/ mmol L^{-1}	Aggr. no.	HLB	Sol. H_2O
Anionic surfactants										
Abietic acid	Sylvic acid	$C_{20}H_{30}O_2$	302.451	439.5	173.5	1.06^{25}	2 at 25 °C			Insoluble
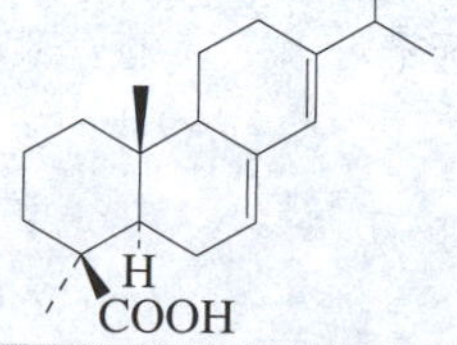							**Soluble in:** ethanol, acetone, ether, chloroform, and benzene			**Notes:** Natural product is primary component of resin acid of pine wood; used as bow rosin; used at high pH.
Bis(2-ethylhexyl) sodium sulfosuccinate	Docusate sodium	$C_{20}H_{37}NaO_7S$	444.559	dec	155	1.1^{25}				15 g/L at 25 °C; 23 g/L at 40 °C; 30 g/L at 50 °C; 55 g/L at 70 °C
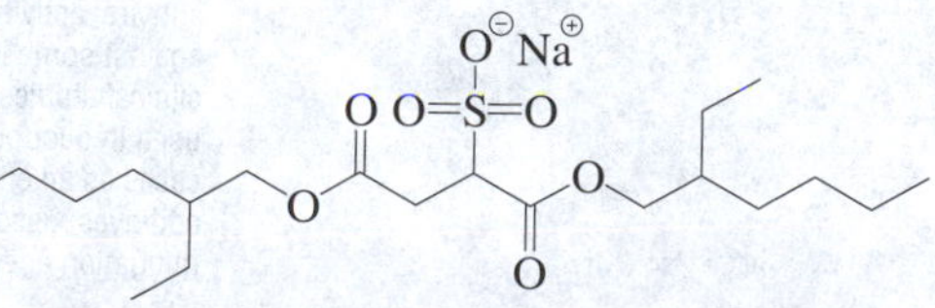							**Soluble in:** ethanol (20 g/L), chloroform (300 g/L), diethyl ether (300 g/L), petroleum ether (unlimited), glycerol, carbon tetrachloride, xylene, petroleum ether, acetone, and vegetable oils			**Notes:** Stable in acid and neutral solutions; hydrolyzes in basic media; used as emulsifier. Dispersant, emulsifier, wetting agent, and laxative to treat constipation and earwax removal.

Fluids

Name	Synonym	Mol. form.	Mol. wt.	bp/°C	mp/°C	Den./ g cm^{-3}	CMC/ mmol L^{-1}	Aggr. no.	HLB	Sol. H_2O
Deoxycholic acid	3,12-Dihydroxycholan-24-oic acid, (3α,5β,12α)	$C_{24}H_{40}O_4$	392.573	dec	177		5 at 25 °C	22	17.6	0.24 mg/mL at 15 °C
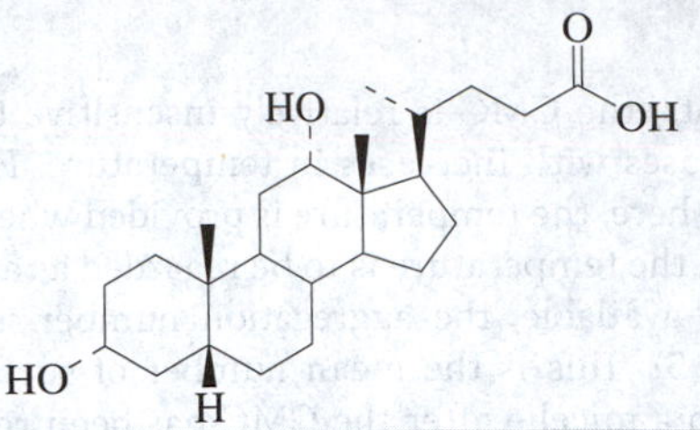							**Soluble in:** ethanol (1 g/100 mL at 20 °C), acetone, ether, and chloroform			**Notes:** Readily dialyzable; used in a modified procedure to recover 40% - 80% of a protein; pK_a = 6.58; λ_{max} = 310 nm; specific rotation +55°; forms molecular coordination compounds; complexes with fatty acids; sodium salt is much more soluble (333 mg/mL in water at 15 °C) but its aqueous solutions precipitate at pH < 5.
Perfluorobutanesulfonic acid	PFBS	$C_4HF_9O_3S$	300.1	211	≈80	1.811[25]				reacts violently
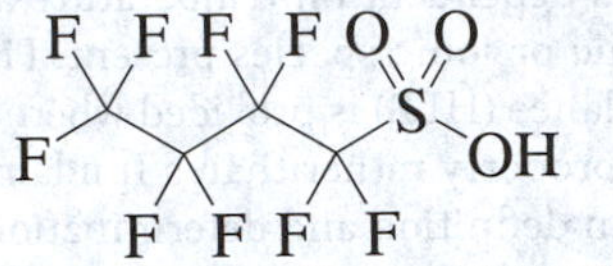										**Notes:** Used as a replacement for perfluorooctanesulfonic acid stain repellent.
Sodium dodecyl sulfate	Sodium lauryl sulfate	$C_{12}H_{25}NaO_4S$	288.38	dec	205	1.06[25]	8.2 at 25 °C	62	40	200 mg/mL at 20 °C
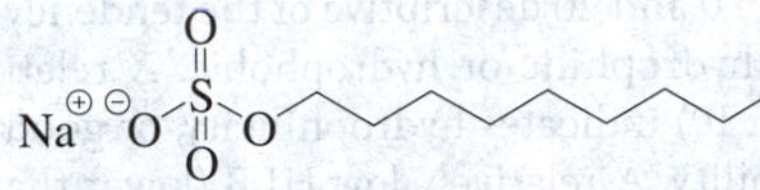							**Soluble in:** ethanol			**Notes:** Aid in lysing cells during DNA extraction and for unraveling proteins in SDS-PAGE; low pK_a; can be used in pH as low as 4; can be prone to foaming.
Cationic surfactant										
Cetyltrimethylammonium bromide	CTAB	$C_{19}H_{42}N$	364.45	dec	240		1 at 25 °C	75-170	≈10	0.3 g/100 mL at 20 °C
							**Soluble in:** alcohols, slightly soluble in acetone; insoluble in diethyl ether and benzene			**Notes:** Aid in high molecular mass DNA isolation and PCR analysis; titrant for perchlorate; phase transfer catalyst in arene and heterocycle reductions.
Ionic (amphiphilic) surfactant										
Sodium lauroyl sarcosinate	Gardol	$C_{15}H_{28}NNaO_3$	293.378	dec	140					7.0% - 8.6% mass/mass
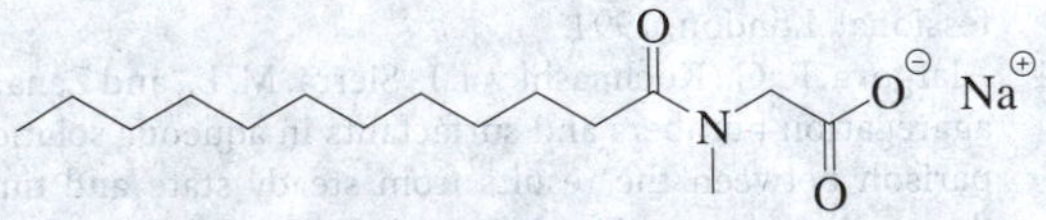										**Notes:** Amphiphilic surfactant used to inhibit D transcription, as a foam stabilizer; lubricant; corrosion inhibitor; bacteriostat; used in emulsion polymerization.
Nonionic surfactants										
Dodecyl-β-D-maltoside	DDM	$C_{24}H_{46}O_{11}$	510.62	dec	225	1.26[25]	0.15 - 0.17 at 25 °C	70-140	≈10	soluble
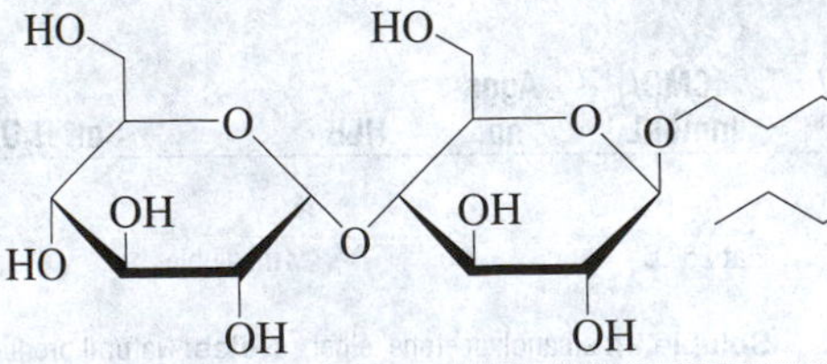										**Notes:** Used to aid in the solubilization and isolation of hydrophobic membrane proteins to preserve their activity.
Ethoxylated tall oil	Ethofat 242/25		945.0	dec		1.08[25a]	4[a] at 25 °C		12.2	soluble
										Notes: Derived from tall oil obtained from paper processing, can be used at wide range of pH.
Glycerol1-dodecanoate,(±)-	Glycerol 1-laurate (±)-	$C_{15}H_{30}O_4$	274.397	397	63.2					12.67 mg/L at 25 °C
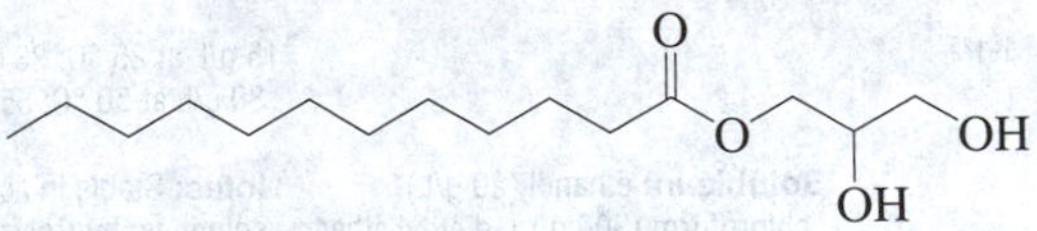							**Soluble in:** methanol and chloroform (50 mg/L)			**Notes:** White fluffy semisolid; has shown both antibacterial and antiviral activity *in vitro*; used against some infections but clinical studies are incomplete; used in deodorants and in some cases as an emulsifier in food additives; used as a methane mitigation agent in ruminants.

Name	Synonym	Mol. form.	Mol. wt.	bp/°C	mp/°C	Den./ g cm^{-3}	CMC/ mmol L^{-1}	Aggr. no.	HLB	Sol. H_2O
Octyl-β-D-1-thioglucopyranoside	OTG	$C_{14}H_{28}O_5S$	308.43	dec	128		9 at 25 °C			slightly soluble
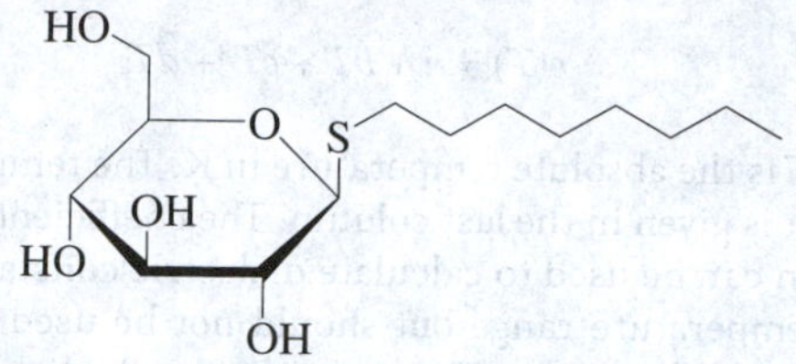							**Soluble in:** ethanol (50 mg/mL)			**Notes:** Colorless powder; used for cell lysis and for solubilizing proteins without denaturing them; dialyzable.
Polyethylene glycol *tert*-octylphenyl ether	Triton X-100		647.0	270	6	1.07^{25}	0.2 - 0.9 at 25 °C	140	13.5	soluble
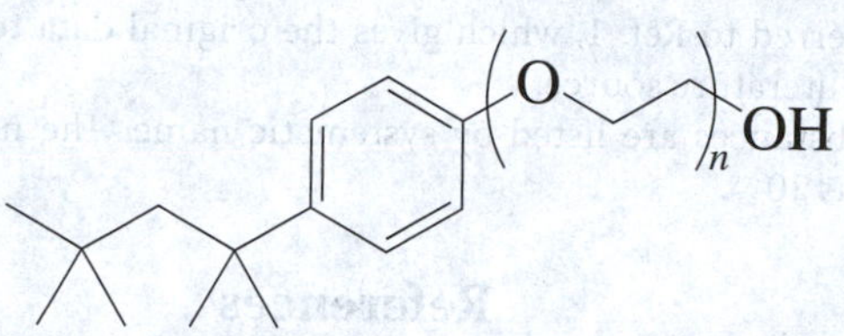							**Soluble in:** benzene, toluene, xylene, trichloroethylene, ethylene glycol, ethyl ether, ethanol, isopropanol, and ethylene dichloride			**Notes:** Nonionic surfactant and emulsifier widely used for solubilizing membrane proteins and lysing cells; no antimicrobial properties; absorbs in the UV, so can interfere with protein quantitation.
Polyoxyethylene (20) sorbitan monolaurate	Tween 20	$C_{58}H_{114}O_{26}$	1227.54	dec		1.06^{25}	0.06 at 25 °C	62	16.7	soluble
										Notes: Washing agent in immunoassay; lysing mammalian cells; used industrially in cleaning applications.
Polyoxyethylene (20) sorbitan monooleate	Tween 80	$C_{64}H_{124}O_{26}$	1310.0	dec		1.07^{25}			13.4-15	soluble
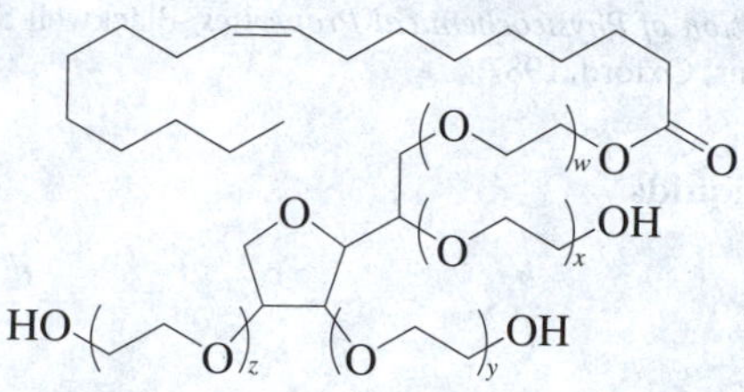 $w + x + y + z = 20$										**Notes:** Used to stabilize aqueous drug formulations; emulsifier for drugs; surfactant used in cosmetics and soaps, and in mouthwash.
Tergitol	Secondary alcohol ethoxylate			dec		1.006^{20}	52 ppm (mass/mass)			soluble
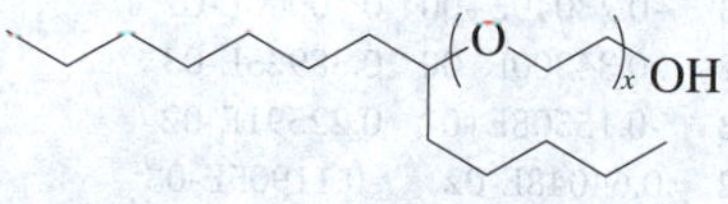							**Soluble in:** chlorinated solvents and most polar organics			**Notes:** Pale yellow liquid; pH (1% aqueous solution) 7.1; chemically stable in the presence of dilute acids, bases, and salts; compatible with anionic, cationic, and other nonionic surfactants; used in chemical analysis, high-performance cleaners and agrochemicals; member of the large Tergitol family of surfactants.

[a] Typical value

PERMITTIVITY (DIELECTRIC CONSTANT) OF LIQUIDS

Christian Wohlfarth

The permittivity of a substance (often called the dielectric constant) is the ratio of the electric displacement D to the electric field strength E when an external field is applied to the substance. The quantity tabulated here is the relative permittivity, which is the ratio of the actual permittivity to the permittivity of a vacuum; it is a dimensionless number.

The table gives the static relative permittivity ε, i.e., the relative permittivity measured in static fields or at low frequencies where no relaxation effects occur. The third column of the table lists the value of ε at the temperature in °C specified by the superscript. When available, a temperature close to 20 °C was chosen, or (as it is the case for many of the substances included here) ε is given at the only temperature for which data are available.

The static permittivity refers to nominal atmospheric pressure as long as the corresponding temperature is below the normal boiling point. Otherwise, at temperatures above the normal boiling point, the pressure is understood to be the saturated vapor pressure of the substance.

For substances where information on the temperature dependence of the permittivity is available, the table gives the coefficients of a simple polynomial fitting of permittivity to temperature with an equation of the form

$$\varepsilon(T) = a + bT + cT^2 + dT^3$$

where T is the absolute temperature in K. The temperature range of the fit is given in the last column. The coefficients of the fitting equation can be used to calculate dielectric constants within the fitted temperature range but should not be used for extrapolation outside this range. The user who needs dielectric constant data with more accuracy than can be provided by this equation is referred to Ref. 1, which gives the original data together with their literature source.

Substances are listed by systematic name. The notation E-xy means 10^{-xy}.

References

1. Wohlfarth, Ch., Static Dielectric Constants of Pure Liquids and Binary Liquid Mixtures, *Landolt-Börnstein, Numerical Data and Functional Relationships in Science and Technology, New Series*, Editor in Chief, O. Madelung, Group IV, Macroscopic and Technical Properties of Matter, Volume 6, Springer-Verlag, Berlin, 1991.
2. Marsh, K. N., Ed., *Recommended Reference Materials for the Realization of Physicochemical Properties*, Blackwell Scientific Publications, Oxford, 1987.

Dielectric Constants of Liquids

Name	Mol. form.	ε	a	b	c	d	Range/K
Acetaldehyde	C_2H_4O	21.0^{18}					
Acetaldoxime	C_2H_5NO	4.70^{25}					
Acetamide	C_2H_5NO	67.6^{91}	-0.20055E+03	0.15515E+01	-0.22392E-02		364 - 448
Acetic acid	$C_2H_4O_2$	6.20^{20}	-0.15731E+02	0.12662E+00	-0.17738E-03		293 - 363
Acetic anhydride	$C_4H_6O_3$	22.45^{20}					
2-Acetonaphthone	$C_{12}H_{10}O$	13.03^{60}	0.14538E+03	-0.73040E+00	0.10000E-02		333 - 363
Acetone	C_3H_6O	21.01^{20}	0.88157E+02	-0.34300E+00	0.38925E-03		273 - 323
Acetonitrile	C_2H_3N	36.64^{20}	0.29724E+03	-0.15508E+01	0.22591E-02		288 - 333
Acetophenone	C_8H_8O	17.44^{25}	0.26099E+02	0.64048E-02	-0.11905E-03		298 - 333
4-Acetylanisole	$C_9H_{10}O_2$	17.3^{40}					
Acetyl chloride	C_2H_3ClO	15.8^{22}					
Acetylene	C_2H_2	2.4841^{-78}					
N-Acetylethanolamine	$C_4H_9NO_2$	96.6^{25}	0.37016E+03	-0.13113E+01	0.13214E-02		298 - 348
2-(Acetyloxy)benzoic acid	$C_9H_8O_4$	6.55^{60}	0.69994E+01	-0.14553E-02			333 - 416
4-Acetylthioanisole	$C_9H_{10}OS$	11.34^{82}					
Acrylonitrile	C_3H_3N	33.0^{20}	0.11109E+03	-0.36806E+00	0.34879E-03		233 - 413
Allene	C_3H_4	2.025^{-4}	0.26049E+01	-0.44147E-03	-0.63420E-05		156 - 269
Allyl alcohol	C_3H_6O	19.7^{20}	0.62714E+02	-0.14771E+00	0.37879E-05		213 - 303
Allylbenzene	C_9H_{10}	2.63^{20}					
Allyl isocyanate	C_4H_5NO	15.15^{15}	0.34299E+02	-0.66444E-01			288 - 333
4-Allyl-2-methoxyphenol	$C_{10}H_{12}O_2$	9.55^{20}	0.52377E+02	-0.24380E+00	0.33333E-03		273 - 323
Aluminum bromide	$AlBr_3$	3.38^{100}					
Ammonia	H_3N	16.61^{20}	0.66756E+02	-0.24696E+00	0.25913E-03		238 - 323
Aniline	C_6H_7N	7.06^{20}	0.89534E+01	0.38990E-02	-0.36310E-04		293 - 413
Anisole	C_7H_8O	4.30^{21}	0.10887E+02	-0.32372E-01	0.33629E-04		294 - 413
Anthracene	$C_{14}H_{10}$	2.649^{229}	0.20571E+02	-0.69169E-01	0.66667E-04		502 - 516
Antimony(V) chloride	Cl_5Sb	3.222^{20}	0.45413E+01	-0.45078E-02			276 - 320
Argon	Ar	1.3247^{-133}	0.12408E+01	0.68755E-02	-0.45344E-04		87 - 149
Arsine	AsH_3	2.40^{-72}	0.37674E+01	-0.97454E-02	0.14537E-04		157 - 201
trans-Azoxybenzene	$C_{12}H_{10}N_2O$	5.2^{38}					
Benzaldehyde	C_7H_6O	17.85^{20}	0.35046E+02	-0.61271E-01	0.16222E-04		301 - 346

Name	Mol. form.	ε	a	b	c	d	Range/K
Benzene	C_6H_6	2.2825^{20}	0.26706E+01	-0.91648E-03	-0.14257E-05		293 - 513
Benzeneacetic acid	$C_8H_8O_2$	3.47^{80}	0.24104E+01	0.30000E-02			353 - 393
Benzeneacetonitrile	C_8H_7N	17.87^{26}	0.82175E+02	-0.37416E+00	0.53220E-03		299 - 343
Benzeneethanol	$C_8H_{10}O$	12.31^{20}	0.12170E+03	-0.63124E+00	0.87776E-03		278 - 333
Benzenemethanethiol	C_7H_8S	4.705^{25}	0.16628E+02	-0.68276E-01	0.94636E-04		298 - 358
Benzenepropanethiol	$C_9H_{12}S$	4.36^{30}	0.82411E+01	-0.15034E-01	0.73617E-05		303 - 358
Benzenepropanol	$C_9H_{12}O$	11.97^{20}	0.94482E+02	-0.45540E+00	0.59307E-03		213 - 303
Benzenesulfonyl chloride	$C_6H_5ClO_2S$	28.90^{50}	0.83886E+02	-0.23405E+00	0.19713E-03		323 - 473
Benzenethiol	C_6H_6S	4.26^{30}	0.57155E+01	-0.70336E-02	0.73617E-05		303 - 358
Benzil	$C_{14}H_{10}O_2$	13.04^{95}	-0.23599E+02	0.22715E+00	-0.34667E-03		368 - 393
Benzonitrile	C_7H_5N	25.9^{20}	0.57605E+02	-0.13354E+00	0.87767E-04		273 - 453
Benzophenone	$C_{13}H_{10}O$	12.62^{27}	0.34130E+02	-0.10249E+00	0.10268E-03		300 - 420
2*H*-1-Benzopyran-2-one	$C_9H_6O_2$	34.04^{70}	0.11311E+03	-0.33804E+00	0.31324E-03		343 - 423
Benzoyl bromide	C_7H_5BrO	21.33^{20}	0.84231E+02	-0.31089E+00	0.32857E-03		283 - 313
Benzoyl chloride	C_7H_5ClO	23.0^{20}					
Benzoyl fluoride	C_7H_5FO	22.7^{20}					
Benzyl acetate	$C_9H_{10}O_2$	5.34^{30}	0.11727E+02	-0.30869E-01	0.32340E-04		303 - 358
Benzyl alcohol	C_7H_8O	11.916^{30}	0.13661E+03	-0.72127E+00	0.10225E-02		303 - 333
Benzylamine	C_7H_9N	5.18^{20}					
Benzyl benzoate	$C_{14}H_{12}O_2$	5.26^{30}	0.76856E+01	-0.80000E-02	-0.80361E-15		303 - 358
Benzyl butanoate	$C_{11}H_{14}O_2$	4.55^{28}					
Benzylethylamine	$C_9H_{13}N$	4.3^{20}					
Benzyl ethyl ether	$C_9H_{12}O$	3.90^{25}					
Benzyl formate	$C_8H_8O_2$	6.34^{30}	0.26162E+02	-0.11026E+00	0.14787E-03		303 - 358
Benzyl nitrite	$C_7H_7NO_2$	7.78^{25}					
Benzyl phenyl ether	$C_{13}H_{12}O$	3.748^{40}					
Benzyl propanoate	$C_{10}H_{12}O_2$	5.11^{30}	0.42301E+01	0.13962E-01	-0.36426E-04		303 - 358
Benzyl salicylate	$C_{14}H_{12}O_3$	4.12^{28}					
Biphenyl	$C_{12}H_{10}$	2.53^{75}	0.26869E+01	0.63072E-03	-0.30995E-05		348 - 428
Bis(2-aminoethyl)amine	$C_4H_{13}N_3$	12.62^{20}	0.57840E+02	-0.23873E+00	0.28841E-03		213 - 333
N,N'-Bis(2-aminoethyl)-1,2-ethanediamine	$C_6H_{18}N_4$	10.76^{20}	0.50699E+02	-0.21730E+00	0.27582E-03		213 - 333
Bis(2-chloroethyl) ether	$C_4H_8Cl_2O$	21.20^{20}					
Bis(chloromethyl) ether	$C_2H_4Cl_2O$	3.51^{20}					
Bis(2-hydroxyethyl) sulfide	$C_4H_{10}O_2S$	28.61^{20}	0.13128E+03	-0.52719E+00	0.60465E-03		253 - 333
1,3-Bis(trifluoromethyl)benzene	$C_8H_4F_6$	5.98^{30}					
l-Bornyl acetate	$C_{12}H_{20}O_2$	4.46^{30}	0.60791E+01	0.98200E-02	-0.50000E-04		303 - 323
Boron tribromide	BBr_3	2.58^{0}					
Bromine	Br_2	3.1484^{25}	0.32701E+01	-0.12535E-03			273 - 327
Bromine pentafluoride	BrF_5	7.91^{25}	0.11428E+02	-0.11822E-01			262 - 298
Bromine trifluoride	BrF_3	106.8^{25}					
3-Bromoaniline	C_6H_6BrN	13.0^{20}					
2-Bromoanisole	C_7H_7BrO	8.96^{30}	0.12023E+02	-0.59116E-02	-0.13787E-04		303 - 358
4-Bromoanisole	C_7H_7BrO	7.40^{30}	0.74367E+01	0.12648E-01	-0.42128E-04		303 - 358
Bromobenzene	C_6H_5Br	5.45^{20}	0.94100E+01	-0.12537E-01	-0.31127E-05		234 - 333
1-Bromobutane	C_4H_9Br	7.315^{10}	0.22542E+02	-0.79306E-01	0.89867E-04		183 - 363
2-Bromobutane, (±)-	C_4H_9Br	8.64^{25}	0.18461E+02	-0.32933E-01			274 - 328
2-Bromobutanoic acid, (±)-	$C_4H_7BrO_2$	7.2^{20}					
cis-2-Bromo-2-butene	C_4H_7Br	5.38^{20}					
trans-2-Bromo-2-butene	C_4H_7Br	6.76^{20}					
Bromochlorodifluoromethane	$CBrClF_2$	3.920^{-150}	0.52442E+01	-0.11000E-01			123 - 223
1-Bromo-2-chloroethane	C_2H_4BrCl	7.41^{10}	0.19493E+02	-0.59054E-01	0.58036E-04		263 - 363
Bromocyclohexane	$C_6H_{11}Br$	8.0026^{30}					
1-Bromodecane	$C_{10}H_{21}Br$	4.44^{25}	0.11202E+02	-0.33491E-01	0.36314E-04		274 - 328
1-Bromododecane	$C_{12}H_{25}Br$	4.07^{25}	0.86103E+01	-0.20891E-01	0.18994E-04		274 - 328
Bromoethane	C_2H_5Br	9.01^{25}	0.28473E+02	-0.85495E-01	0.67971E-04		243 - 308
1-Bromo-2-ethoxybenzene	C_8H_9BrO	7.04^{40}	0.23146E+02	-0.75753E-01	0.77778E-04		313 - 358
1-Bromo-2-ethylbenzene	C_8H_9Br	5.55^{25}					
1-Bromo-3-ethylbenzene	C_8H_9Br	5.56^{25}					
1-Bromo-4-ethylbenzene	C_8H_9Br	5.42^{25}					
1-Bromo-2-fluorobenzene	C_6H_4BrF	4.72^{25}					

Name	Mol. form.	ε	a	b	c	d	Range/K
1-Bromo-3-fluorobenzene	C_6H_4BrF	4.85^{25}					
1-Bromo-4-fluorobenzene	C_6H_4BrF	2.60^{25}					
1-Bromoheptane	$C_7H_{15}Br$	5.255^{30}	0.15289E+02	-0.50621E-01	0.57753E-04		203 - 343
2-Bromoheptane	$C_7H_{15}Br$	6.46^{22}					
4-Bromoheptane	$C_7H_{15}Br$	6.81^{22}					
1-Bromohexadecane	$C_{16}H_{33}Br$	3.68^{25}	0.58668E+01	-0.73333E-02	-0.52666E-14		298 - 328
1-Bromohexane	$C_6H_{13}Br$	5.82^{25}	0.15233E+02	-0.44385E-01	0.43039E-04		274 - 328
Bromomethane	CH_3Br	9.71^{3}	0.40580E+02	-0.18418E+00	0.26219E-03		195 - 276
(Bromomethyl)benzene	C_7H_7Br	6.658^{20}	0.18482E+02	-0.57207E-01	0.57321E-04		273 - 323
1-Bromo-3-methylbutane	$C_5H_{11}Br$	6.33^{18}	0.27743E+02	-0.13927E+00	0.22627E-03		123 - 292
2-Bromo-2-methylbutane	$C_5H_{11}Br$	9.21^{25}					
1-Bromo-2-methylpropane	C_4H_9Br	7.70^{0}	0.37558E+02	-0.20571E+00	0.35496E-03		112 - 273
2-Bromo-2-methylpropane	C_4H_9Br	10.98^{20}	0.35085E+02	-0.14075E+00	0.19960E-03		258 - 293
1-Bromonaphthalene	$C_{10}H_7Br$	4.768^{25}	0.10561E+02	-0.27671E-01	0.27655E-04		293 - 323
1-Bromo-3-nitrobenzene	$C_6H_4BrNO_2$	20.2^{55}	0.81413E+02	-0.27645E+00	0.27367E-03		328 - 413
1-Bromononane	$C_9H_{19}Br$	4.74^{25}	0.79870E+01	-0.10488E-01	-0.13450E-05		274 - 328
1-Bromooctadecane	$C_{18}H_{37}Br$	3.53^{30}	0.46790E+01	-0.30355E-02	-0.24798E-05		303 - 332
1-Bromooctane	$C_8H_{17}Br$	5.0957^{20}	0.12404E+02	-0.35050E-01	0.34542E-04		283 - 353
2-Bromooctane, (±)-	$C_8H_{17}Br$	5.44^{20}					
1-Bromopentadecane	$C_{15}H_{31}Br$	3.88^{20}					
1-Bromopentane	$C_5H_{11}Br$	6.31^{26}	0.20954E+02	-0.78743E-01	0.98908E-04		183 - 328
3-Bromopentane	$C_5H_{11}Br$	8.37^{25}					
1-Bromopropane	C_3H_7Br	8.09^{20}	0.17769E+02	-0.32599E-01			274 - 328
2-Bromopropane	C_3H_7Br	9.46^{20}	0.26195E+02	-0.72995E-01	0.55454E-04		186 - 328
2-Bromopropanoic acid, (±)-	$C_3H_5BrO_2$	11.0^{21}					
3-Bromopropene	C_3H_5Br	7.0^{20}					
(3-Bromopropyl)benzene	$C_9H_{11}Br$	5.41^{29}	0.11360E+02	-0.27471E-01	0.25775E-04		302 - 358
2-Bromopyridine	C_5H_4BrN	23.18^{25}	0.73391E+02	-0.23678E+00	0.22930E-03		298 - 398
1-Bromotetradecane	$C_{14}H_{29}Br$	3.84^{20}	0.10058E+02	-0.33905E-01	0.43528E-04		274 - 328
2-Bromotoluene	C_7H_7Br	4.641^{20}	0.10229E+02	-0.25050E-01	0.20357E-04		273 - 323
3-Bromotoluene	C_7H_7Br	5.566^{20}	0.11522E+02	-0.24946E-01	0.15714E-04		273 - 323
4-Bromotoluene	C_7H_7Br	5.503^{20}	0.10014E+02	-0.13918E-01	-0.50000E-05		273 - 293
Bromotrichloromethane	$CBrCl_3$	2.405^{20}	0.29249E+01	-0.17650E-02			273 - 333
1-Bromotridecane	$C_{13}H_{27}Br$	4.19^{8}					
Bromotrifluoromethane	$CBrF_3$	3.730^{-150}	0.54154E+01	-0.13680E-01			123 - 173
1-Bromoundecane	$C_{11}H_{23}Br$	4.61^{-1}					
1,3-Butadiene	C_4H_6	2.050^{-8}	0.27674E+01	-0.26738E-02			185 - 265
Butanal	C_4H_8O	13.45^{25}					
Butane	C_4H_{10}	1.7697^{22}	0.22379E+01	-0.13884E-02	-0.66711E-06		135 - 303
1,2-Butanediol, (±)-	$C_4H_{10}O_2$	22.4^{25}	0.63702E+02	-0.13807E+00			278 - 323
1,3-Butanediol	$C_4H_{10}O_2$	28.8^{25}	0.72883E+02	-0.14770E+00			278 - 323
1,4-Butanediol	$C_4H_{10}O_2$	31.9^{25}	0.13079E+03	-0.46985E+00	0.46320E-03		288 - 328
2,3-Butanedione	$C_4H_6O_2$	4.04^{25}	0.46907E+01	-0.22302E-02			278 - 348
Butanenitrile	C_4H_7N	24.83^{20}	0.53884E+02	-0.99257E-01			293 - 333
1,2,3,4-Butanetetrol	$C_4H_{10}O_4$	28.2^{120}					
1-Butanethiol	$C_4H_{10}S$	5.204^{15}	0.11201E+02	-0.20767E-01			273 - 318
2-Butanethiol	$C_4H_{10}S$	5.645^{15}	0.10866E+02	-0.17993E-01			273 - 318
Butanoic acid	$C_4H_8O_2$	2.98^{14}	0.15010E+01	0.50046E-02			287 - 403
Butanoic anhydride	$C_8H_{14}O_3$	12.8^{20}					
1-Butanol	$C_4H_{10}O$	17.84^{20}	0.10578E+03	-0.50587E+00	0.84733E-03	-0.48841E-06	193 - 553
2-Butanol	$C_4H_{10}O$	17.26^{20}	0.13850E+03	-0.75146E+00	0.14086E-02	-0.89512E-06	172 - 533
2-Butanone	C_4H_8O	18.56^{20}	0.15457E+02	0.90152E-01	-0.27100E-03		293 - 333
2-Butanone oxime	C_4H_9NO	3.4^{20}					
1-Butene	C_4H_8	2.2195^{-53}	0.29354E+01	-0.32580E-02			220 - 250
cis-2-Butene	C_4H_8	1.960^{23}	0.28802E+01	-0.31064E-02			197 - 296
Butoxyacetylene	$C_6H_{10}O$	6.62^{25}					
N-Butylacetamide	$C_6H_{13}NO$	104.0^{20}	0.70739E+03	-0.37369E+01	0.71585E-02	-0.48716E-05	253 - 493
Butyl acetate	$C_6H_{12}O_2$	5.07^{20}	0.13825E+02	-0.43994E-01	0.48214E-04		253 - 353
sec-Butyl acetate	$C_6H_{12}O_2$	5.135^{20}	0.12427E+02	-0.32035E-01	0.24286E-04		273 - 323
tert-Butyl acetate	$C_6H_{12}O_2$	5.672^{20}	0.55435E+02	-0.30494E+00	0.46107E-03		273 - 323
tert-Butylacetic acid	$C_6H_{12}O_2$	2.85^{23}					

Name	Mol. form.	ε	a	b	c	d	Range/K
Butyl acrylate	$C_7H_{12}O_2$	5.25^{28}	0.38296E+02	-0.19109E+00	0.27006E-03		301 - 343
Butylamine	$C_4H_{11}N$	4.71^{20}	0.13322E+02	-0.44176E-01	0.50250E-04		223 - 333
Butylbenzene	$C_{10}H_{14}$	2.359^{20}					
sec-Butylbenzene, (±)-	$C_{10}H_{14}$	2.357^{20}	0.28348E+01	-0.68586E-03	-0.32143E-05		273 - 323
tert-Butylbenzene	$C_{10}H_{14}$	2.359^{20}	0.27924E+01	-0.38350E-03	-0.37500E-05		273 - 323
Butyl benzoate	$C_{11}H_{14}O_2$	5.52^{30}	0.77854E+01	-0.34972E-02	-0.13149E-04		303 - 358
Butyl butanoate	$C_8H_{16}O_2$	4.39^{25}	0.79684E+01	-0.12000E-01	0.15266E-13		298 - 318
Butyl formate	$C_5H_{10}O_2$	6.10^{30}	0.21532E+02	-0.84106E-01	0.10952E-03		288 - 323
Butyl nitrate	$C_4H_9NO_3$	13.10^{20}					
tert-Butyl nitrite	$C_4H_9NO_2$	11.47^{25}					
5-Butylnonane	$C_{13}H_{28}$	2.0319^{20}					
2-Butyl-1-octanol	$C_{12}H_{26}O$	3.28^{90}					
Butyl oleate	$C_{22}H_{42}O_2$	4.00^{25}					
N-Butyl-*N*-phenylacetamide	$C_{12}H_{17}NO$	11.66^{25}					
Butyl phenyl ether	$C_{10}H_{14}O$	3.734^{20}					
Butyl propanoate	$C_7H_{14}O_2$	4.838^{20}					
Butyl stearate	$C_{22}H_{44}O_2$	3.120^{25}	0.73894E+02	-0.46261E+00	0.75500E-03		298 - 343
Butyl thiophene-2-carboxylate	$C_9H_{12}O_2S$	6.40^{20}					
Butyl trichloroacetate	$C_6H_9Cl_3O_2$	7.480^{20}					
γ-Butyrolactone	$C_4H_6O_2$	39.0^{20}					
Carbon dioxide	CO_2	1.4492^{22}	0.79062E+00	0.10639E-01	-0.28510E-04		220 - 300
Carbon disulfide	CS_2	2.6320^{20}	0.45024E+01	-0.12054E-01	0.19147E-04		154 - 319
Carbon oxyselenide	COSe	3.47^{10}	0.48740E+01	-0.49425E-02			219 - 283
Carbon oxysulfide	COS	4.47^{-88}	0.84702E+01	-0.21488E-01			143 - 185
Carbonyl chloride	CCl_2O	4.30^{22}					
Carvenone, (*S*)-	$C_{10}H_{16}O$	18.8^{20}					
Chlorine	Cl_2	2.147^{-65}	0.29440E+01	-0.44649E-02	0.30388E-05		208 - 240
Chlorine pentafluoride	ClF_5	4.28^{-80}	0.78192E+01	-0.20860E-01	0.13132E-04		193 - 256
Chlorine trifluoride	ClF_3	4.394^{20}	0.96716E+01	-0.18000E-01			273 - 313
Chloroacetic acid	$C_2H_3ClO_2$	12.35^{65}	0.17310E+02	-0.14674E-01			338 - 393
2-Chloroaniline	C_6H_6ClN	13.40^{20}					
3-Chloroaniline	C_6H_6ClN	13.3^{20}					
4-Chloroanisole	C_7H_7ClO	7.84^{20}	0.64019E+01	0.30560E-01	-0.87500E-04		293 - 333
Chlorobenzene	C_6H_5Cl	5.6895^{20}	0.19471E+02	-0.70786E-01	0.82466E-04		293 - 430
4-Chlorobenzenethiol	C_6H_5ClS	3.59^{65}					
2 Chlorobornane	$C_{10}H_{17}Cl$	5.21^{95}					
2-Chloro-1,3-butadiene	C_4H_5Cl	4.914^{20}					
1-Chlorobutane	C_4H_9Cl	7.276^{20}	0.13565E+02	-0.10161E-01	-0.38750E-04		273 - 323
2-Chlorobutane	C_4H_9Cl	8.564^{20}	0.30376E+02	-0.11377E+00	0.13429E-03		273 - 323
Chlorocyclohexane	$C_6H_{11}Cl$	7.9505^{30}					
1-Chlorodecane	$C_{10}H_{21}Cl$	4.581^{20}	0.68741E+01	-0.12210E-02	-0.22500E-04		293 - 323
4-Chloro-1,3-dioxolan-2-one	$C_3H_3ClO_3$	62.0^{40}					
1-Chlorododecane	$C_{12}H_{25}Cl$	4.17^{25}	0.10002E+02	-0.27798E-01	0.27559E-04		274 - 328
Chloroethane	C_2H_5Cl	9.45^{20}	0.60693E+02	-0.31290E+00	0.47154E-03		237 - 293
2-Chloroethanol	C_2H_5ClO	25.80^{20}	0.11155E+03	-0.30149E+00			140 - 175
1-Chloro-2-ethylbenzene	C_8H_9Cl	4.36^{25}					
1-Chloro-3-ethylbenzene	C_8H_9Cl	5.18^{25}					
1-Chloro-4-ethylbenzene	C_8H_9Cl	5.16^{25}					
2-Chloroethyl isocyanate	C_3H_4ClNO	29.1^{15}	0.64311E+02	-0.12217E+00			288 - 403
1-Chloro-2-fluorobenzene	C_6H_4ClF	6.10^{25}					
1-Chloro-3-fluorobenzene	C_6H_4ClF	4.96^{25}					
1-Chloro-4-fluorobenzene	C_6H_4ClF	3.34^{25}					
1-Chloroheptane	$C_7H_{15}Cl$	5.521^{20}	0.14279E+02	-0.39431E-01	0.32321E-04		273 - 323
2-Chloroheptane	$C_7H_{15}Cl$	6.52^{22}					
3-Chloroheptane	$C_7H_{15}Cl$	6.70^{22}					
4-Chloroheptane	$C_7H_{15}Cl$	6.54^{22}					
1-Chlorohexane	$C_6H_{13}Cl$	6.104^{20}	0.15994E+02	-0.43647E-01	0.33393E-04		273 - 323
6-Chloro-1-hexanol	$C_6H_{13}ClO$	21.6^{-31}	-0.73364E+01	0.46377E+00	-0.14202E-02		195 - 242
Chloromethane	CH_3Cl	10.0^{22}	0.42775E+02	-0.16175E+00	0.17108E-03		190 - 392
(Chloromethyl)benzene	C_7H_7Cl	6.854^{20}	0.17108E+02	-0.45285E-01	0.35000E-04		273 - 323
1-Chloro-3-methylbutane	$C_5H_{11}Cl$	6.10^{19}	0.22228E+02	-0.93189E-01	0.12991E-03		171 - 297

Name	Mol. form.	*ε*	*a*	*b*	*c*	*d*	Range/K
2-Chloro-2-methylbutane	$C_5H_{11}Cl$	12.31^{-50}	0.55104E+02	-0.29866E+00	0.47840E-03		201 - 223
1-Chloro-4-methyl-2-nitrobenzene	$C_7H_6ClNO_2$	28.07^{28}					
1-Chloro-2-methylpropane	C_4H_9Cl	7.027^{20}	0.14945E+02	-0.33747E-01	0.23036E-04		273 - 323
2-Chloro-2-methylpropane	C_4H_9Cl	9.663^{20}	0.35077E+02	-0.12867E+00	0.14304E-03		273 - 323
1-Chloronaphthalene	$C_{10}H_7Cl$	5.04^{25}	0.84861E+01	-0.12357E-01	0.26899E-05		274 - 328
1-Chloro-2-nitrobenzene	$C_6H_4ClNO_2$	37.7^{50}	0.16800E+03	-0.59708E+00	0.59957E-03		323 - 436
1-Chloro-3-nitrobenzene	$C_6H_4ClNO_2$	20.9^{50}	0.77193E+02	-0.25118E+00	0.23798E-03		323 - 433
1-Chloro-4-nitrobenzene	$C_6H_4ClNO_2$	8.09^{120}					
2-Chloro-2-nitropropane	$C_3H_6ClNO_2$	31.90^{-23}					
1-Chlorononane	$C_9H_{19}Cl$	4.803^{20}	0.95528E+01	-0.16200E-01	-0.16365E-13		293 - 323
1-Chlorooctane	$C_8H_{17}Cl$	5.05^{25}	0.11346E+02	-0.25120E-01	0.13450E-04		274 - 328
2-Chlorooctane	$C_8H_{17}Cl$	5.42^{20}					
1-Chloropentane	$C_5H_{11}Cl$	6.654^{20}	0.18626E+02	-0.54719E-01	0.47143E-04		273 - 323
2-Chlorophenol	C_6H_5ClO	7.40^{23}	0.29755E+02	-0.11256E+00	0.12390E-03		296 - 448
3-Chlorophenol	C_6H_5ClO	6.255^{20}					
4-Chlorophenol	C_6H_5ClO	11.18^{41}	0.31997E+02	-0.94241E-01	0.88392E-04		314 - 453
4-Chlorophenyl isocyanate	C_7H_4ClNO	3.177^{15}	0.40896E+01	-0.31667E-02			288 - 348
1-Chloropropane	C_3H_7Cl	8.588^{20}	0.21214E+02	-0.43130E-01			273 - 313
3-Chloro-1,2-propanediol	$C_3H_7ClO_2$	31.0^{20}					
3-Chloro-1,2-propanediol dinitrate	$C_3H_5ClN_2O_6$	17.50^{20}					
3-Chloro-1-propanol	C_3H_7ClO	36.0^{-58}	0.12436E+03	-0.60841E+00	0.92060E-03		145 - 215
1-Chloro-2-propanol	C_3H_7ClO	59.0^{-120}	-0.19169E+02	0.13605E+01	-0.55567E-02		153 - 177
2-Chloropropene	C_3H_5Cl	8.92^{26}					
3-Chloropropene	C_3H_5Cl	8.2^{20}					
2-Chloropyridine	C_5H_4ClN	27.32^{25}	0.98702E+02	-0.34237E+00	0.34502E-03		298 - 398
2-Chlorotoluene	C_7H_7Cl	4.721^{20}	0.11507E+02	-0.31148E-01	0.27143E-04		273 - 323
3-Chlorotoluene	C_7H_7Cl	5.763^{20}	0.13921E+02	-0.37186E-01	0.31786E-04		273 - 323
4-Chlorotoluene	C_7H_7Cl	6.25^{20}	0.20265E+01	0.40060E-01	-0.87500E-04		293 - 333
Chlorotrifluoromethane	$CClF_3$	3.010^{-150}	0.43677E+01	-0.11020E-01			123 - 173
3-Chloro-1,1,1-trifluoropropane	$C_3H_4ClF_3$	7.32^{22}	0.22361E+02	-0.68840E-01	0.60594E-04		275 - 313
trans-Cinnamaldehyde	C_9H_8O	17.72^{33}	0.41837E+02	-0.11060E+00	0.10401E-03		306 - 354
o-Cresol	C_7H_8O	6.76^{25}	0.21633E+02	-0.71069E-01	0.70590E-04		298 - 453
m-Cresol	C_7H_8O	12.44^{25}	0.81716E+02	-0.35039E+00	0.39878E-03		274 - 463
p-Cresol	C_7H_8O	13.05^{25}	0.70253E+02	-0.28870E+00	0.31979E-03		298 - 453
Cyanoacetic acid	$C_3H_3NO_2$	33.4^{4}					
Cyanoacetylene	C_3HN	72.3^{19}	0.91803E+03	-0.49149E+01	0.69104E-02		281 - 314
Cyclobutanone	C_4H_6O	14.27^{25}	0.43974E+02	-0.15712E+00	0.19264E-03		220 - 317
Cyclododecanone	$C_{12}H_{22}O$	11.4^{30}	0.39327E+02	-0.13248E+00	0.13298E-03		303 - 423
Cycloheptane	C_7H_{14}	2.0784^{20}	0.25136E+01	-0.15089E-02	0.84915E-07		278 - 333
Cycloheptanone	$C_7H_{12}O$	13.16^{25}	0.17511E+03	-0.11221E+01	0.19417E-02		258 - 298
Cycloheptene	C_7H_{12}	2.265^{22}	0.32309E+01	-0.42373E-02	0.32572E-05		227 - 363
1,3-Cyclohexadiene	C_6H_8	2.68^{-89}					
1,4-Cyclohexadiene	C_6H_8	2.211^{23}	0.27459E+01	-0.16975E-02	-0.36461E-06		232 - 356
Cyclohexane	C_6H_{12}	2.0243^{20}	0.24293E+01	-0.12095E-02	-0.58741E-06		283 - 333
Cyclohexanecarboxylic acid	$C_7H_{12}O_2$	2.67^{31}					
1,4-Cyclohexanedione	$C_6H_8O_2$	4.40^{78}					
Cyclohexanemethanol	$C_7H_{14}O$	9.70^{60}	0.10164E+03	-0.45839E+00	0.54762E-03		333 - 368
Cyclohexanethiol	$C_6H_{12}S$	5.420^{25}					
Cyclohexanol	$C_6H_{12}O$	16.40^{20}	0.10173E+03	-0.43072E+00	0.47926E-03		293 - 423
Cyclohexanone	$C_6H_{10}O$	16.1^{20}	0.41577E+02	-0.11463E+00	0.92454E-04		253 - 423
Cyclohexanone oxime	$C_6H_{11}NO$	3.04^{89}					
Cyclohexene	C_6H_{10}	2.2176^{20}	0.30598E+01	-0.39841E-02	0.37554E-05		141 - 313
Cyclohexyl acetate	$C_8H_{14}O_2$	5.08^{20}					
Cyclohexylamine	$C_6H_{13}N$	4.547^{20}					
Cyclohexyl butanoate	$C_{10}H_{18}O_2$	4.58^{20}					
Cyclohexyl formate	$C_7H_{12}O_2$	6.47^{20}					
2-Cyclohexylphenol	$C_{12}H_{16}O$	3.97^{55}					
4-Cyclohexylphenol	$C_{12}H_{16}O$	4.42^{131}					
Cyclohexyl propanoate	$C_9H_{16}O_2$	4.82^{20}					
Cyclooctane	C_8H_{16}	2.116^{22}	0.25036E+01	-0.12460E-02	-0.23175E-06		295 - 411
cis-Cyclooctene	C_8H_{14}	2.306^{23}	0.31115E+01	-0.32058E-02	0.16713E-05		269 - 406

Name	Mol. form.	ε	a	b	c	d	Range/K
Cyclopentane	C_5H_{10}	1.9687^{20}	0.24287E+01	-0.15304E-02	-0.13095E-06		278 - 313
Cyclopentanecarbonitrile	C_6H_9N	22.68^{20}	0.69830E+02	-0.25303E+00	0.31491E-03		201 - 293
Cyclopentanol	$C_5H_{10}O$	18.5^{15}	0.10565E+03	-0.44244E+00	0.48657E-03		258 - 323
Cyclopentanone	C_5H_8O	13.58^{25}	0.24083E+02	-0.30286E-01	-0.16802E-04		219 - 298
Cyclopentene	C_5H_8	2.083^{22}	0.28177E+01	-0.27597E-02	0.89346E-06		171 - 319
cis-Decahydronaphthalene	$C_{10}H_{18}$	2.219^{20}	0.25410E+01	-0.11420E-02	0.15092E-06		293 - 373
trans-Decahydronaphthalene	$C_{10}H_{18}$	2.184^{20}	0.26615E+01	-0.21241E-02	0.16864E-05		293 - 373
Decamethylcyclopentasiloxane	$C_{10}H_{30}O_5Si_5$	2.50^{20}					
Decamethyltetrasiloxane	$C_{10}H_{30}O_3Si_4$	2.370^{20}					
Decane	$C_{10}H_{22}$	1.9853^{20}	0.24054E+01	-0.15445E-02	0.44643E-06		253 - 393
1-Decanol	$C_{10}H_{22}O$	7.93^{20}	0.47195E+02	-0.20740E+00	0.24942E-03		293 - 343
2-Decanol	$C_{10}H_{22}O$	5.82^{25}	0.13621E+03	-0.81000E+00	0.12500E-02		288 - 308
3-Decanol	$C_{10}H_{22}O$	4.05^{25}	0.52090E+02	-0.31020E+00	0.50000E-03		288 - 308
4-Decanol	$C_{10}H_{22}O$	3.42^{25}	-0.11260E+02	0.93960E-01	-0.15000E-03		288 - 308
5-Decanol	$C_{10}H_{22}O$	3.24^{25}	-0.25832E+01	0.31456E-01	-0.40000E-04		288 - 308
2-Decanone	$C_{10}H_{20}O$	8.3^{14}					
1-Decene	$C_{10}H_{20}$	2.136^{20}	0.19091E+01	0.33442E-02	-0.87500E-05		273 - 323
cis-5-Decene	$C_{10}H_{20}$	2.071^{25}					
trans-5-Decene	$C_{10}H_{20}$	2.030^{25}					
Decyl acetate	$C_{12}H_{24}O_2$	3.75^{20}					
Decylamine	$C_{10}H_{23}N$	3.31^{20}	0.61497E+01	-0.12801E-01	0.10606E-04		293 - 373
Diacetone alcohol	$C_6H_{12}O_2$	18.2^{25}					
Dibenzofuran	$C_{12}H_8O$	3.00^{100}					
Dibenzylamine	$C_{14}H_{15}N$	3.446^{20}					
Dibenzyl ether	$C_{14}H_{14}O$	3.821^{20}	0.80154E+01	-0.20536E-01	0.21250E-04		293 - 333
Diborane	B_2H_6	1.8725^{-93}	0.23848E+01	-0.29501E-02	0.64189E-06		108 - 181
o-Dibromobenzene	$C_6H_4Br_2$	7.86^{20}	-0.81849E-02	0.62671E-01	-0.12222E-03		293 - 353
m-Dibromobenzene	$C_6H_4Br_2$	4.81^{20}	0.93214E+01	-0.20273E-01	0.16667E-04		293 - 353
p-Dibromobenzene	$C_6H_4Br_2$	2.57^{95}					
1,2-Dibromobutane	$C_4H_8Br_2$	4.74^{20}	0.11199E+03	-0.63334E+00	0.91250E-03		293 - 333
1,3-Dibromobutane	$C_4H_8Br_2$	9.14^{20}	0.34031E+02	-0.13254E+00	0.16250E-03		293 - 333
1,4-Dibromobutane	$C_4H_8Br_2$	8.68^{30}	0.20944E+02	-0.55620E-01	0.50000E-04		303 - 333
2,3-Dibromobutane	$C_4H_8Br_2$	6.245^{25}	0.23849E+02	-0.96300E-01	0.12500E-03		293 - 333
1,10-Dibromodecane	$C_{10}H_{20}Br_2$	6.56^{30}	0.17350E+02	-0.50328E-01	0.48633E-04		303 - 368
Dibromodichloromethane	CBr_2Cl_2	2.542^{25}	0.32330E+01	-0.23162E-02			298 - 333
Dibromodifluoromethane	CBr_2F_2	2.939^{0}	0.67296E+01	-0.22133E-01	0.30213E-04		139 - 273
1,2-Dibromoethane	$C_2H_4Br_2$	4.9612^{20}	0.67142E+01	-0.59800E-02			293 - 313
cis-1,2-Dibromoethene	$C_2H_2Br_2$	7.08^{25}					
trans-1,2-Dibromoethene	$C_2H_2Br_2$	2.88^{25}					
1,2-Dibromoheptane	$C_7H_{14}Br_2$	3.77^{25}					
2,3-Dibromoheptane	$C_7H_{14}Br_2$	5.08^{25}					
3,4-Dibromoheptane	$C_7H_{14}Br_2$	4.70^{25}					
1,6-Dibromohexane	$C_6H_{12}Br_2$	8.52^{25}	-0.55185E+01	0.11746E+00	-0.23658E-03		274 - 328
3,4-Dibromohexane	$C_6H_{12}Br_2$	6.732^{25}					
Dibromomethane	CH_2Br_2	7.77^{10}	0.18060E+02	-0.36333E-01			283 - 313
1,2-Dibromo-2-methylpropane	$C_4H_8Br_2$	4.1^{20}					
1,9-Dibromononane	$C_9H_{18}Br_2$	7.153^{20}	0.18931E+02	-0.57764E-01	0.60000E-04		293 - 343
1,8-Dibromooctane	$C_8H_{16}Br_2$	7.43^{25}	0.94117E+00	0.61520E-01	-0.13333E-03		298 - 328
1,2-Dibromopentane	$C_5H_{10}Br_2$	4.39^{25}					
1,4-Dibromopentane	$C_5H_{10}Br_2$	9.05^{20}	0.26443E+02	-0.88640E-01	0.10000E-03		293 - 333
1,5-Dibromopentane	$C_5H_{10}Br_2$	9.14^{30}	0.38192E+02	-0.15648E+00	0.20000E-03		303 - 333
1,2-Dibromopropane	$C_3H_6Br_2$	4.60^{10}	0.54973E+01	-0.31695E-02			283 - 333
1,3-Dibromopropane	$C_3H_6Br_2$	9.482^{20}	0.29193E+02	-0.94450E-01	0.92800E-04		293 - 368
1,2-Dibromotetrafluoroethane	$C_2Br_2F_4$	2.34^{25}					
N,N-Dibutylacetamide	$C_{10}H_{21}NO$	19.1^{20}					
Dibutylamine	$C_8H_{19}N$	2.765^{20}	0.52504E+01	-0.10538E-01	0.71485E-05		243 - 323
Dibutyl ether	$C_8H_{18}O$	3.0830^{20}	0.65383E+01	-0.16172E-01	0.14969E-04		293 - 314
N,N-Dibutylformamide	$C_9H_{19}NO$	18.4^{20}					
Di-*tert*-butyl ketone	$C_9H_{18}O$	10.0^{14}					
Dibutyl phthalate	$C_{16}H_{22}O_4$	6.58^{20}	0.12444E+02	-0.20000E-01			293 - 333
Dibutyl sebacate	$C_{18}H_{34}O_4$	4.54^{20}					

Name	Mol. form.	*ε*	*a*	*b*	*c*	*d*	Range/K
Dibutyl sulfide	$C_8H_{18}S$	4.29^{25}					
Dibutyl sulfone	$C_8H_{18}O_2S$	25.72^{50}	0.66248E+02	-0.16417E+00	0.12001E-03		323 - 398
Dibutyl sulfoxide	$C_8H_{18}OS$	24.73^{40}	0.67156E+02	-0.16448E+00	0.92275E-04		313 - 393
Dibutyl tartrate	$C_{12}H_{22}O_6$	9.4^{41}					
Dichloroacetic acid	$C_2H_2Cl_2O_2$	8.33^{20}	0.11014E+02	-0.10859E-01	0.49242E-05		284 - 363
Dichloroacetic anhydride	$C_4H_2Cl_4O_3$	15.8^{25}					
1,1-Dichloroacetone	$C_3H_4Cl_2O$	14.6^{20}					
o-Dichlorobenzene	$C_6H_4Cl_2$	10.12^{20}	0.13629E+02	0.10622E-02	-0.44444E-04		293 - 353
m-Dichlorobenzene	$C_6H_4Cl_2$	5.02^{20}	0.77565E+01	-0.93333E-02	-0.26880E-14		293 - 353
p-Dichlorobenzene	$C_6H_4Cl_2$	2.3943^{55}	0.26999E+01	-0.35325E-03	-0.17619E-05		328 - 363
1,2-Dichlorobutane	$C_4H_8Cl_2$	7.74^{20}	0.31925E+02	-0.13232E+00	0.17007E-03		293 - 356
1,4-Dichlorobutane	$C_4H_8Cl_2$	9.30^{35}	0.59766E+01	0.49300E-01	-0.12500E-03		308 - 338
1,10-Dichlorodecane	$C_{10}H_{20}Cl_2$	6.68^{35}	-0.57423E+01	0.94220E-01	-0.17500E-03		308 - 338
Dichlorodifluoromethane	CCl_2F_2	3.500^{-150}	0.46984E+01	-0.97600E-02			123 - 223
1,1-Dichloroethane	$C_2H_4Cl_2$	10.10^{25}	0.24429E+02	-0.48000E-01			288 - 318
1,2-Dichloroethane	$C_2H_4Cl_2$	10.42^{20}	0.24404E+02	-0.47892E-01			293 - 343
1,1-Dichloroethene	$C_2H_2Cl_2$	4.60^{20}					
cis-1,2-Dichloroethene	$C_2H_2Cl_2$	9.20^{25}					
trans-1,2-Dichloroethene	$C_2H_2Cl_2$	2.14^{20}					
1,7-Dichloroheptane	$C_7H_{14}Cl_2$	8.34^{25}					
1,6-Dichlorohexane	$C_6H_{12}Cl_2$	8.60^{35}	0.11277E+02	0.67200E-02	-0.50000E-04		308 - 338
Dichloromethane	CH_2Cl_2	8.93^{25}	0.40452E+02	-0.17748E+00	0.23942E-03		184 - 306
(Dichloromethyl)benzene	$C_7H_6Cl_2$	6.9^{20}					
1,2-Dichloro-2-methylpropane	$C_4H_8Cl_2$	7.15^{23}	0.39429E+02	-0.20028E+00	0.30917E-03		165 - 296
1,1-Dichloro-1-nitroethane	$C_2H_3Cl_2NO_2$	16.3^{30}	0.37576E+02	-0.70400E-01			303 - 333
1,8-Dichlorooctane	$C_8H_{16}Cl_2$	7.64^{25}					
1,2-Dichloropentane	$C_5H_{10}Cl_2$	6.89^{20}	0.19016E+02	-0.57954E-01	0.56801E-04		293 - 356
1,5-Dichloropentane	$C_5H_{10}Cl_2$	9.92^{25}					
1,2-Dichloropropane, (±)-	$C_3H_6Cl_2$	8.37^{20}	0.18915E+02	-0.35907E-01			281 - 323
1,3-Dichloropropane	$C_3H_6Cl_2$	10.27^{30}	0.21609E+02	-0.37333E-01			303 - 333
2,2-Dichloropropane	$C_3H_6Cl_2$	11.37^{20}	0.32421E+02	-0.72188E-01			245 - 293
2,5-Dichlorostyrene	$C_8H_6Cl_2$	2.58^{25}					
1,2-Dichloro-1,1,2,2-tetrafluoroethane	$C_2Cl_2F_4$	2.4842^{0}	0.36663E+01	-0.42271E-02	-0.36255E-06		193 - 273
2,4-Dichlorotoluene	$C_7H_6Cl_2$	5.68^{28}					
2,6-Dichlorotoluene	$C_7H_6Cl_2$	3.36^{28}					
3,4-Dichlorotoluene	$C_7H_6Cl_2$	9.39^{28}					
Dicyclohexyl ether	$C_{12}H_{22}O$	3.45^{20}	0.95324E+01	-0.31740E-01	0.37500E-04		293 - 333
Dicyclohexyl hexanedioate	$C_{18}H_{30}O_4$	4.84^{35}					
Dicyclopentadiene	$C_{10}H_{12}$	2.43^{40}	0.30564E+01	-0.20000E-02	0.82443E-15		313 - 373
Didecyl ether	$C_{20}H_{42}O$	2.644^{20}	0.41465E+01	-0.62240E-02	0.37500E-05		293 - 333
Diethanolamine	$C_4H_{11}NO_2$	25.75^{20}	0.73435E+02	-0.21377E+00	0.17500E-03		273 - 323
Diethoxydimethylsilane	$C_6H_{16}O_2Si$	3.216^{25}					
1,2-Diethoxyethane	$C_6H_{14}O_2$	3.90^{20}	0.99099E+01	-0.33403E-01	0.44048E-04		223 - 303
Diethoxymethane	$C_5H_{12}O_2$	2.527^{20}	0.25294E+01	0.73988E-04	-0.28331E-06		227 - 293
N,N-Diethylacetamide	$C_6H_{13}NO$	32.1^{20}					
Diethylamine	$C_4H_{11}N$	3.680^{20}	0.26462E+02	-0.13750E+00	0.20373E-03		243 - 323
N,N-Diethylaniline	$C_{10}H_{15}N$	5.15^{30}	0.50773E+01	0.15399E-01	-0.50000E-04		303 - 328
o-Diethylbenzene	$C_{10}H_{14}$	2.594^{20}					
m-Diethylbenzene	$C_{10}H_{14}$	2.369^{20}					
p-Diethylbenzene	$C_{10}H_{14}$	2.259^{20}					
Diethyl carbonate	$C_5H_{10}O_3$	2.820^{24}					
Diethylene glycol	$C_4H_{10}O_3$	31.82^{20}	0.13973E+03	-0.54725E+00	0.61149E-03		288 - 343
Diethylene glycol dimethyl ether	$C_6H_{14}O_3$	7.23^{25}	0.28291E+02	-0.11236E+00	0.14000E-03		298 - 333
Diethyl ether	$C_4H_{10}O$	4.2666^{20}	0.79725E+01	-0.12519E-01			283 - 301
N,N-Diethylformamide	$C_5H_{11}NO$	29.6^{20}					
Diethyl fumarate	$C_8H_{12}O_4$	6.56^{23}					
Diethyl glutarate	$C_9H_{16}O_4$	6.659^{30}					
Diethyl hexanedioate	$C_{10}H_{18}O_4$	6.109^{20}	0.14824E+02	-0.40749E-01	0.37600E-04		293 - 343
Diethyl maleate	$C_8H_{12}O_4$	7.560^{25}	0.13953E+02	-0.21969E-01	0.17817E-05		298 - 343
Diethyl malonate	$C_7H_{12}O_4$	7.550^{31}	0.14809E+02	-0.31207E-01	0.24066E-04		304 - 393
1,3-Diethyl-5-methylbenzene	$C_{11}H_{16}$	2.264^{20}					

Fluids

Name	Mol. form.	*ε*	*a*	*b*	*c*	*d*	Range/K
Diethyl nonanedioate	$C_{13}H_{24}O_4$	5.133^{30}					
Diethyl oxalate	$C_6H_{10}O_4$	8.266^{20}	0.21938E+02	-0.66226E-01	0.66800E-04		293 - 368
Diethyl phthalate	$C_{12}H_{14}O_4$	7.86^{20}					
Diethyl sebacate	$C_{14}H_{26}O_4$	4.995^{30}	0.39143E+02	-0.20965E+00	0.32000E-03		303 - 313
Diethylsilane	$C_4H_{12}Si$	2.544^{20}					
Diethyl succinate	$C_8H_{14}O_4$	6.098^{20}	0.80213E+01	0.11810E-02	-0.26400E-04		293 - 343
Diethyl sulfate	$C_4H_{10}O_4S$	29.2^{20}					
Diethyl sulfide	$C_4H_{10}S$	5.723^{25}					
Diethyl sulfite	$C_4H_{10}O_3S$	15.6^{20}					
o-Difluorobenzene	$C_6H_4F_2$	13.38^{28}	0.59107E+02	-0.23611E+00	0.27987E-03		273 - 323
m-Difluorobenzene	$C_6H_4F_2$	5.01^{28}	0.14448E+02	-0.46982E-01	0.51948E-04		273 - 323
Difluoromethane	CH_2F_2	53.74^{-121}	0.19428E+03	-0.12939E+01	0.24280E-02		152 - 224
Dihexyl phthalate	$C_{20}H_{30}O_4$	5.62^{20}					
o-Diiodobenzene	$C_6H_4I_2$	5.41^{50}	0.31150E+02	-0.14428E+00	0.20000E-03		323 - 353
m-Diiodobenzene	$C_6H_4I_2$	4.11^{50}					
p-Diiodobenzene	$C_6H_4I_2$	2.88^{120}					
cis-1,2-Diiodoethene	$C_2H_2I_2$	4.46^{72}					
Diiodomethane	CH_2I_2	5.32^{25}					
Diisobutyl hexanedioate	$C_{14}H_{26}O_4$	5.19^{20}					
Diisopentyl ether	$C_{10}H_{22}O$	2.817^{20}	0.44690E+01	-0.63710E-02	0.25000E-05		293 - 323
Diisopropyl ether	$C_6H_{14}O$	3.805^{30}					
Diisopropyl oxalate	$C_8H_{14}O_4$	6.403^{20}	0.10709E+02	-0.16328E-01	0.56000E-05		293 - 368
1,2-Dimethoxybenzene	$C_8H_{10}O_2$	4.45^{20}	0.74604E+01	-0.13445E-01	0.10737E-04		293 - 443
1,3-Dimethoxybenzene	$C_8H_{10}O_2$	5.363^{25}	0.11911E+02	-0.30804E-01	0.29643E-04		298 - 358
1,4-Dimethoxybenzene	$C_8H_{10}O_2$	5.60^{61}	0.11289E+02	-0.20765E-01	0.11987E-04		334 - 463
Dimethoxydimethylsilane	$C_4H_{12}O_2Si$	3.663^{25}					
1,2-Dimethoxyethane	$C_4H_{10}O_2$	7.30^{24}	0.48832E+02	-0.24218E+00	0.34413E-03		256 - 318
Dimethoxymethane	$C_3H_8O_2$	2.644^{20}	0.25877E+01	-0.93019E-03	0.38472E-05		171 - 293
N,N-Dimethylacetamide	C_4H_9NO	38.85^{21}	0.15420E+03	-0.57506E+00	0.61911E-03		294 - 433
N,N-Dimethylaniline	$C_8H_{11}N$	4.90^{25}	0.84052E+01	-0.13549E-01	0.62835E-05		289 - 453
2,6-Dimethylanisole	$C_9H_{12}O$	3.780^{20}	0.76700E+01	-0.18298E-01	0.17143E-04		293 - 333
3,5-Dimethylanisole	$C_9H_{12}O$	3.711^{20}	0.54981E+01	-0.56651E-02	-0.14286E-05		293 - 333
N,N-Dimethylbenzamide	$C_9H_{11}NO$	20.77^{45}	0.76725E+02	-0.26908E+00	0.29409E-03		318 - 443
α,α-Dimethylbenzenemethanol	$C_9H_{12}O$	5.61^{30}	0.57072E+01	0.86568E-02	-0.29580E-04		303 - 373
2,3-Dimethyl-1,3-butadiene	C_6H_{10}	2.102^{20}	0.26258E+01	-0.17990E-02	0.12035E-06		223 - 323
2,2-Dimethylbutane	C_6H_{14}	1.869^{20}	0.22740E+01	-0.96229E-03	-0.14286E-05		273 - 313
2,3-Dimethylbutane	C_6H_{14}	1.889^{20}	0.24305E+01	-0.20081E-02	0.53571E-06		273 - 323
2,2-Dimethyl-1-butanol	$C_6H_{14}O$	10.5^{20}	0.14054E+03	-0.72925E+00	0.97821E-03		243 - 393
3,3-Dimethyl-2-butanone	$C_6H_{12}O$	12.73^{20}	0.66857E+02	-0.28552E+00	0.34422E-03		243 - 293
Dimethyl carbonate	$C_3H_6O_3$	3.087^{25}					
1,2-Dimethylcyclohexene	C_8H_{14}	2.144^{23}	0.26443E+01	-0.17973E-02	0.35815E-06		211 - 374
1,3-Dimethylcyclohexene	C_8H_{14}	2.182^{23}	0.29951E+01	-0.34615E-02	0.24026E-05		213 - 373
Dimethyldiphenoxysilane	$C_{14}H_{16}O_2Si$	3.500^{25}	0.51669E+01	-0.77001E-02	0.70156E-05		283 - 353
Dimethyl disulfide	$C_2H_6S_2$	9.6^{25}	0.19109E+02	-0.32000E-01			298 - 323
Dimethyl ether	C_2H_6O	6.18^{-15}	0.22389E+02	-0.86524E-01	0.91291E-04		155 - 258
N,N-Dimethylformamide	C_3H_7NO	38.25^{20}	0.15364E+03	-0.60367E+00	0.71505E-03		213 - 353
Dimethyl glutarate	$C_7H_{12}O_4$	7.87^{20}	0.20697E+02	-0.57794E-01	0.48405E-04		293 - 433
2,4-Dimethylheptane	C_9H_{20}	1.89^{20}					
2,5-Dimethylheptane	C_9H_{20}	1.89^{20}					
2,6-Dimethylheptane	C_9H_{20}	1.987^{20}					
2,6-Dimethyl-4-heptanone	$C_9H_{18}O$	9.91^{20}	0.33178E+02	-0.11290E+00	0.11454E-03		273 - 393
2,2-Dimethylhexane	C_8H_{18}	1.9498^{20}					
2,5-Dimethylhexane	C_8H_{18}	1.9619^{21}	0.25821E+01	-0.26804E-02	0.19404E-05		294 - 324
3,3-Dimethylhexane	C_8H_{18}	1.9645^{20}					
3,4-Dimethylhexane	C_8H_{18}	1.9814^{19}	0.26849E+01	-0.33712E-02	0.32949E-05		292 - 324
Dimethyl 1,6-hexanedioate	$C_8H_{14}O_4$	6.84^{20}	0.11739E+02	-0.17281E-01	0.11447E-05		293 - 433
2,2-Dimethyl-1-hexanol	$C_8H_{18}O$	4.50^{20}	0.91244E+01	-0.21785E-01	0.21018E-04		283 - 393
2,5-Dimethyl-2-hexene	C_8H_{16}	2.431^{20}					
Dimethyl malonate	$C_5H_8O_4$	9.82^{20}	0.26470E+02	-0.76656E-01	0.67888E-04		293 - 433
1,6-Dimethylnaphthalene	$C_{12}H_{12}$	2.7250^{20}					
2,7-Dimethyloctane	$C_{10}H_{22}$	1.98^{20}					

Name	Mol. form.	ε	*a*	*b*	*c*	*d*	Range/K
2,2-Dimethyloctanoic acid	$C_{10}H_{20}O_2$	2.8^{23}					
2,2-Dimethyl-1-octanol	$C_{10}H_{22}O$	7.86^{20}	0.69536E+02	-0.34596E+00	0.46250E-03		293 - 333
2,2-Dimethylpentane	C_7H_{16}	1.915^{20}	0.23414E+01	-0.14362E-02	-0.51322E-07		153 - 353
2,3-Dimethylpentane	C_7H_{16}	1.929^{20}	0.25637E+01	-0.26328E-02	0.16071E-05		273 - 323
2,4-Dimethylpentane	C_7H_{16}	1.902^{20}	0.23979E+01	-0.17436E-02	0.17857E-06		273 - 323
3,3-Dimethylpentane	C_7H_{16}	1.9419^{18}	0.24007E+01	-0.16802E-02	0.36069E-06		291 - 322
2,2-Dimethyl-1-pentanol	$C_7H_{16}O$	6.020^{20}	0.37318E+02	-0.17095E+00	0.22022E-03		283 - 393
Dimethyl phthalate	$C_{10}H_{10}O_4$	8.66^{20}					
2,2-Dimethylpropanal	$C_5H_{10}O$	9.051^{20}	0.18645E+02	-0.32395E-01	-0.16157E-05		280 - 333
2,2-Dimethylpropanamide	$C_5H_{11}NO$	20.13^{25}	0.10400E+03	-0.46017E+00	0.60000E-03		298 - 328
2,2-Dimethylpropanenitrile	C_5H_9N	21.1^{20}	0.58418E+02	-0.16884E+00	0.14131E-03		293 - 453
2,2-Dimethyl-1-propanol	$C_5H_{12}O$	8.35^{60}	0.92350E+02	-0.41870E+00	0.50000E-03		333 - 373
2,5-Dimethylpyrazine	$C_6H_8N_2$	2.436^{20}					
2,6-Dimethylpyrazine	$C_6H_8N_2$	2.653^{35}					
2,4-Dimethylpyridine	C_7H_9N	9.60^{20}	0.25895E+02	-0.73900E-01	0.62500E-04		293 - 333
2,6-Dimethylpyridine	C_7H_9N	7.33^{20}	0.17714E+02	-0.39080E-01	0.12500E-04		293 - 333
2,6-Dimethylpyridine-1-oxide	C_7H_9NO	46.11^{25}	0.22765E+03	-0.90760E+00	0.10011E-02		298 - 398
Dimethyl succinate	$C_6H_{10}O_4$	7.19^{20}	0.13551E+02	-0.23109E-01	0.55440E-05		293 - 433
Dimethyl sulfate	$C_2H_6O_4S$	55.0^{25}					
Dimethyl sulfide	C_2H_6S	6.70^{21}					
Dimethyl sulfone	$C_2H_6O_2S$	47.39^{110}	0.10830E+03	-0.15900E+00			383 - 398
Dimethyl sulfoxide	C_2H_6OS	47.24^{20}	0.38478E+02	0.16939E+00	-0.47423E-03		288 - 343
1,3-Dinitrobenzene	$C_6H_4N_2O_4$	22.9^{92}	0.10406E+03	-0.34133E+00	0.32609E-03		365 - 413
2,2-Dinitropropane	$C_3H_6N_2O_4$	42.4^{52}					
Dioctyl phthalate	$C_{24}H_{38}O_4$	5.22^{20}					
Dioctyl sebacate	$C_{26}H_{50}O_4$	4.01^{26}					
1,4-Dioxane	$C_4H_8O_2$	2.2189^{20}	0.27299E+01	-0.17440E-02			293 - 313
Dipentyl ether	$C_{10}H_{22}O$	2.798^{25}					
Dipentyl phthalate	$C_{18}H_{26}O_4$	6.00^{20}					
Dipentyl sulfide	$C_{10}H_{22}S$	3.826^{25}					
Dipentyl sulfoxide	$C_{10}H_{22}OS$	18.8^{75}					
Diphenylamine	$C_{12}H_{11}N$	3.73^{50}					
1,2-Diphenylethane	$C_{14}H_{14}$	2.47^{58}	0.31178E+01	-0.21572E-02	0.59800E-06		331 - 451
Diphenyl ether	$C_{12}H_{10}O$	3.726^{10}					
Diphenylmethane	$C_{13}H_{12}$	2.540^{30}	0.30638E+01	-0.17286E-02			303 - 333
Diphenyl sulfide	$C_{12}H_{10}S$	5.43^{25}					
Diphenyl sulfone	$C_{12}H_{10}O_2S$	21.1^{133}					
Diphenyl sulfoxide	$C_{12}H_{10}OS$	16.6^{72}					
Dipropylamine	$C_6H_{15}N$	2.923^{20}	0.11376E+02	-0.49796E-01	0.71792E-04		243 - 323
Dipropyl ether	$C_6H_{14}O$	3.38^{24}	0.14600E+02	-0.72670E-01	0.11742E-03		161 - 297
Dipropyl sulfone	$C_6H_{14}O_2S$	32.62^{30}	0.70195E+02	-0.15008E+00	0.86506E-04		303 - 398
Dipropyl sulfoxide	$C_6H_{14}OS$	30.37^{30}	0.84868E+02	-0.23486E+00	0.18198E-03		303 - 373
Divinyl ether	C_4H_6O	3.94^{15}					
Docosane	$C_{22}H_{46}$	2.0840^{20}					
1-Docosanol	$C_{22}H_{46}O$	2.94^{75}	0.82062E+01	-0.25069E-01	0.28571E-04		348 - 373
Dodecane	$C_{12}H_{26}$	2.0120^{20}	0.23697E+01	-0.12200E-02	-0.36375E-16		283 - 363
1-Dodecanol	$C_{12}H_{26}O$	5.82^{30}	0.18518E+02	-0.44859E-01	0.99900E-05		303 - 358
1-Dodecene	$C_{12}H_{24}$	2.152^{20}	0.22581E+01	0.11106E-02	-0.50000E-05		273 - 323
Dodecyl acetate	$C_{14}H_{28}O_2$	3.6^{20}					
Dodecylamine	$C_{12}H_{27}N$	3.07^{30}	0.27999E+01	0.44810E-02	-0.11905E-04		303 - 373
Eicosamethylnonasiloxane	$C_{20}H_{60}O_8Si_9$	2.645^{20}	0.57840E+01	-0.16568E-01	0.20000E-04		293 - 323
1-Eicosanol	$C_{20}H_{42}O$	3.13^{65}	0.21700E+01	0.12497E-01	-0.28571E-04		338 - 363
Epichlorohydrin	C_3H_5ClO	22.6^{20}					
Ethane	C_2H_6	1.9356^{-178}	0.20185E+01	-0.51493E-03	-0.48148E-05		95 - 295
1,2-Ethanediamine	$C_2H_8N_2$	13.82^{20}	0.48922E+02	-0.17021E+00	0.17262E-03		273 - 333
1,2-Ethanediol	$C_2H_6O_2$	41.4^{20}	0.14355E+03	-0.48573E+00	0.46703E-03		293 - 423
1,2-Ethanediol, diacetate	$C_6H_{10}O_4$	7.7^{17}	0.25093E+02	-0.95171E-01	0.12224E-03		223 - 290
1,2-Ethanediol, dihexadecanoate	$C_{34}H_{66}O_4$	2.89^{75}					
1,2-Ethanediol, dinitrate	$C_2H_4N_2O_6$	28.26^{20}					
1,2-Ethanediol, distearate	$C_{38}H_{74}O_4$	2.79^{80}					
1,2-Ethanediol, ditetradecanoate	$C_{30}H_{58}O_4$	2.98^{70}					

Name	Mol. form.	ε	a	b	c	d	Range/K
1,2-Ethanediol, monoacetate	$C_4H_8O_3$	12.95^{30}					
1,2-Ethanediol, monosulfite	$C_2H_4O_3S$	39.6^{25}	0.85483E+02	-0.15400E+00			298 - 328
1,2-Ethanedithiol	$C_2H_6S_2$	7.26^{20}	0.11228E+02	-0.13500E-01			293 - 333
Ethanethiol	C_2H_6S	6.667^{25}					
Ethanol	C_2H_6O	25.3^{20}	0.15145E+03	-0.87020E+00	0.19570E-02	-0.15512E-05	163 - 523
Ethanolamine	C_2H_7NO	31.94^{20}	0.14890E+03	-0.62491E+00	0.77143E-03		253 - 293
Ethoxyacetylene	C_4H_6O	8.05^{25}					
4-Ethoxyaniline	$C_8H_{11}NO$	7.43^{25}					
Ethoxybenzene	$C_8H_{10}O$	4.216^{20}	-0.15043E+02	0.13752E+00	-0.24500E-03		293 - 313
2-Ethoxyethanol	$C_4H_{10}O_2$	13.38^{25}					
2-Ethoxyethyl acetate	$C_6H_{12}O_3$	7.567^{30}	0.23290E+02	-0.71566E-01	0.65000E-04		303 - 323
1-Ethoxynaphthalene	$C_{12}H_{12}O$	3.3^{19}					
Ethoxytrimethylsilane	$C_5H_{14}OSi$	3.013^{25}					
N-Ethylacetamide	C_4H_9NO	135.0^{20}	0.74494E+03	-0.31400E+01	0.36131E-02		213 - 353
Ethyl acetate	$C_4H_8O_2$	6.0814^{20}	0.15646E+02	-0.44066E-01	0.39137E-04		293 - 433
Ethyl acetoacetate	$C_6H_{10}O_3$	14.0^{20}					
Ethyl acrylate	$C_5H_8O_2$	6.05^{30}	0.47827E+02	-0.24394E+00	0.35000E-03		303 - 343
Ethylamine	C_2H_7N	8.7^{0}	0.30163E+02	-0.79000E-01			233 - 273
Ethyl 2-aminobenzoate	$C_9H_{11}NO_2$	4.14^{25}					
4-Ethylaniline	$C_8H_{11}N$	4.84^{25}					
N-Ethylaniline	$C_8H_{11}N$	5.87^{20}					
N-Ethylbenzamide	$C_9H_{11}NO$	42.6^{80}	-0.20109E+03	0.17866E+01	-0.31065E-02		353 - 389
Ethylbenzene	C_8H_{10}	2.4463^{20}	0.35969E+01	-0.53169E-02	0.47500E-05		293 - 323
α-Ethylbenzenemethanol	$C_9H_{12}O$	6.68^{20}	0.44520E+02	-0.21505E+00	0.29443E-03		233 - 373
Ethyl benzoate	$C_9H_{10}O_2$	6.20^{20}	0.18216E+02	-0.62361E-01	0.72884E-04		288 - 343
Ethyl benzoylacetate	$C_{11}H_{12}O_3$	13.50^{30}	0.93644E+01	0.74280E-01	-0.20000E-03		303 - 323
Ethyl bromoacetate	$C_4H_7BrO_2$	9.75^{30}	0.15627E+02	-0.19600E-01			303 - 333
Ethyl 2-bromobutanoate	$C_6H_{11}BrO_2$	8.57^{30}	0.49005E+02	-0.23193E+00	0.32500E-03		303 - 333
Ethyl 2-bromo-2-methylpropanoate	$C_6H_{11}BrO_2$	8.55^{30}	0.77044E+02	-0.40784E+00	0.60000E-03		303 - 333
Ethyl 2-bromopropanoate	$C_5H_9BrO_2$	9.4^{20}					
Ethyl butanoate	$C_6H_{12}O_2$	5.18^{28}	0.48698E+02	-0.25660E+00	0.37237E-03		301 - 343
2-Ethylbutanoic acid	$C_6H_{12}O_2$	2.72^{23}					
2-Ethyl-1-butanol	$C_6H_{14}O$	6.19^{89}					
Ethyl 2-butenoate	$C_6H_{10}O_2$	5.4^{20}					
Ethyl carbamate	$C_3H_7NO_2$	14.14^{55}	0.32431E+02	-0.65097E-01	0.28571E-04		328 - 368
Ethyl chloroformate	$C_3H_5ClO_2$	9.736^{36}	0.15356E+02	-0.18250E-01			309 - 349
Ethyl 2-chloropropanoate	$C_5H_9ClO_2$	11.95^{30}	0.25965E+02	-0.46250E-01			303 - 343
Ethyl 3-chloropropanoate	$C_5H_9ClO_2$	10.19^{30}	0.21951E+02	-0.38750E-01			303 - 343
Ethyl *trans*-cinnamate	$C_{11}H_{12}O_2$	5.63^{20}					
Ethyl cyanoacetate	$C_5H_7NO_2$	31.62^{-10}					
Ethylcyclobutane	C_6H_{12}	1.965^{20}					
Ethyl cyclohexanecarboxylate	$C_9H_{16}O_2$	4.64^{20}					
Ethylcyclopropane	C_5H_{10}	1.933^{20}					
Ethyl decanoate	$C_{12}H_{24}O_2$	3.75^{20}	0.70969E+01	-0.15080E-01	0.12500E-04		293 - 353
1-Ethyl-3,5-dimethylbenzene	$C_{10}H_{14}$	2.275^{20}					
Ethyl dodecanoate	$C_{14}H_{28}O_2$	3.94^{0}					
Ethylene	C_2H_4	1.4833^{-3}	0.13546E+01	0.62614E-02	-0.21374E-04		200 - 270
Ethylene carbonate	$C_3H_4O_3$	89.78^{40}	0.20746E+03	-0.37610E+00			313 - 343
Ethyleneimine	C_2H_5N	18.3^{25}	0.61405E+02	-0.14474E+00			273 - 298
N-Ethylformamide	C_3H_7NO	102.7^{25}	0.64764E+03	-0.28499E+01	0.34286E-02		298 - 338
Ethyl formate	$C_3H_6O_2$	8.57^{15}	0.15884E+02	-0.25333E-01			288 - 318
2-Ethylheptanoic acid	$C_9H_{18}O_2$	1.98^{20}					
Ethyl hexadecanoate	$C_{18}H_{36}O_2$	3.07^{30}	0.57938E+01	-0.12294E-01	0.10919E-04		303 - 455
3-Ethylhexane	C_8H_{18}	1.9617^{20}					
2-Ethyl-1,3-hexanediol	$C_8H_{18}O_2$	18.73^{20}	0.57919E+02	-0.17128E+00	0.12949E-03		233 - 333
Ethyl hexanoate	$C_8H_{16}O_2$	4.45^{20}	0.11007E+02	-0.32800E-01	0.35714E-04		253 - 353
2-Ethylhexanoic acid	$C_8H_{16}O_2$	2.64^{23}					
2-Ethyl-1-hexanol	$C_8H_{18}O$	7.58^{25}	0.86074E+02	-0.42636E+00	0.55078E-03		208 - 318
Ethyl isocyanate	C_3H_5NO	19.7^{20}					
Ethyl isopentyl ether	$C_7H_{16}O$	3.955^{20}	0.66541E+01	-0.55450E-02	-0.12500E-04		293 - 323
Ethyl isothiocyanate	C_3H_5NS	19.6^{20}					

Name	Mol. form.	*ε*	*a*	*b*	*c*	*d*	Range/K
Ethyl lactate	$C_5H_{10}O_3$	15.4^{30}	0.31225E+02	-0.43531E-01	-0.28571E-04		273 - 373
Ethyl methacrylate	$C_6H_{10}O_2$	5.68^{30}	0.40962E+02	-0.20520E+00	0.29286E-03		303 - 343
Ethyl 3-methylbutanoate	$C_7H_{14}O_2$	4.71^{20}					
Ethyl *N*-methylcarbamate	$C_4H_9NO_2$	21.10^{25}	0.11477E+03	-0.47568E+00	0.54127E-03		298 - 373
Ethyl methyl carbonate	$C_4H_8O_3$	2.985^{20}					
3-Ethyl-3-methylpentane	C_8H_{18}	1.9869^{18}	0.25983E+01	-0.28027E-02	0.24195E-05		292 - 324
Ethyl nitrate	$C_2H_5NO_3$	19.7^{20}					
1-Ethyl-2-nitrobenzene	$C_8H_9NO_2$	21.9^{0}					
Ethyl octadecanoate	$C_{20}H_{40}O_2$	2.958^{40}	0.70930E+01	-0.19081E-01	0.19555E-04		331 - 440
Ethyl *cis*-9-octadecenoate	$C_{20}H_{38}O_2$	3.17^{28}	0.57033E+01	-0.11223E-01	0.93447E-05		301 - 423
3-Ethylpentane	C_7H_{16}	1.942^{20}	0.23771E+01	-0.15140E-02	0.10093E-06		163 - 363
Ethyl pentanoate	$C_7H_{14}O_2$	4.71^{18}					
3-Ethyl-3-pentanol	$C_7H_{16}O$	3.158^{20}					
3-Ethyl-2-pentene	C_7H_{14}	2.051^{20}					
Ethyl pentyl ether	$C_7H_{16}O$	3.6^{23}					
Ethyl phenylacetate	$C_{10}H_{12}O_2$	5.320^{20}					
Ethyl phenyl sulfone	$C_8H_{10}O_2S$	39.0^{75}					
Ethyl propanoate	$C_5H_{10}O_2$	5.76^{20}					
2-Ethylpyridine	C_7H_9N	8.33^{20}	0.36397E+02	-0.15070E+00	0.18750E-03		293 - 333
4-Ethylpyridine	C_7H_9N	10.98^{20}	-0.73831E+01	0.14326E+00	-0.27500E-03		293 - 333
Ethyl 4-pyridinecarboxylate	$C_8H_9NO_2$	8.95^{20}					
Ethyl salicylate	$C_9H_{10}O_3$	8.48^{35}	0.18910E+02	-0.35623E-01	0.46529E-05		225 - 321
Ethyl silicate	$C_8H_{20}O_4Si$	2.50^{20}					
4-Ethylstyrene	$C_{10}H_{12}$	3.350^{25}					
Ethyl tetradecanoate	$C_{16}H_{32}O_2$	3.50^{20}	0.52642E+01	-0.60000E-02	-0.47358E-15		293 - 353
(Ethylthio)benzene	$C_8H_{10}S$	4.95^{25}					
Ethyl thiophene-2-carboxylate	$C_7H_8O_2S$	6.18^{20}					
2-Ethyltoluene	C_9H_{12}	2.595^{20}					
3-Ethyltoluene	C_9H_{12}	2.365^{20}					
4-Ethyltoluene	C_9H_{12}	2.265^{20}					
Ethyl trichloroacetate	$C_4H_5Cl_3O_2$	8.428^{20}					
Ethyl undecanoate	$C_{13}H_{26}O_2$	3.55^{20}					
Eucalyptol	$C_{10}H_{18}O$	4.57^{25}					
(±)-Fenchone	$C_{10}H_{16}O$	12.8^{21}					
Fluorine	F_2	1.4913^{-220}	0.14144E+01	0.26387E-02	-0.28356E-04		54 - 144
Fluorobenzene	C_6H_5F	5.465^{20}					
1-Fluoro-2-iodobenzene	C_6H_4FI	8.22^{25}					
1-Fluoro-4-iodobenzene	C_6H_4FI	3.12^{25}					
Fluoromethane	CH_3F	51.0^{-142}	0.11338E+03	-0.63979E+00	0.96983E-03		150 - 299
1-Fluorooctane	$C_8H_{17}F$	3.89^{20}					
1-Fluoropentane	$C_5H_{11}F$	3.931^{20}					
2-Fluorotoluene	C_7H_7F	4.23^{25}					
3-Fluorotoluene	C_7H_7F	5.41^{25}					
4-Fluorotoluene	C_7H_7F	5.88^{25}					
Formamide	CH_3NO	111.0^{20}	0.26076E+03	-0.61145E+00	0.34296E-03		278 - 333
Formic acid	CH_2O_2	51.1^{25}	0.14040E+03	-0.24673E+00	-0.17151E-03		287 - 358
Furan	C_4H_4O	2.94^{25}	0.13636E+01	0.12864E-01	-0.22701E-04		188 - 277
Furfural	$C_5H_4O_2$	42.1^{20}					
Furfuryl alcohol	$C_5H_6O_2$	16.85^{25}					
Germanium(IV) bromide	Br_4Ge	2.955^{27}	0.34450E+01	-0.16083E-02			300 - 316
Germanium(IV) chloride	Cl_4Ge	2.463^{0}	-0.55078E+01	0.64881E-01	-0.13091E-03		246 - 273
D-Glucitol	$C_6H_{14}O_6$	35.5^{80}					
Glycerol	$C_3H_8O_3$	46.53^{20}	0.77503E+02	-0.37984E-01	-0.23107E-03		288 - 343
Glycerol 1-acetate	$C_5H_{10}O_4$	38.57^{-31}	0.10653E+03	-0.26439E+00	-0.62371E-04		215 - 242
Glycerol 1,3-diacetate	$C_7H_{12}O_5$	9.80^{15}	0.28321E+02	-0.89073E-01	0.86891E-04		258 - 374
Glycerol triacetate	$C_9H_{14}O_6$	7.11^{20}	0.17819E+02	-0.53656E-01	0.57759E-04		219 - 304
Glycerol trielaidate	$C_{57}H_{104}O_6$	2.980^{40}					
Glycerol trihexanoate	$C_{21}H_{38}O_6$	4.476^{20}					
Glycerol trilaurate	$C_{39}H_{74}O_6$	3.287^{40}					
Glycerol trioctanoate	$C_{27}H_{50}O_6$	3.931^{20}					
Glycerol trioleate	$C_{57}H_{104}O_6$	3.109^{20}					

Name	Mol. form.	ε	a	b	c	d	Range/K
Glycerol tripalmitate	$C_{51}H_{98}O_6$	2.901^{55}	-0.29131E+01	0.32206E-01	-0.44154E-04		328 - 393
Glycerol tristearate	$C_{57}H_{110}O_6$	2.740^{80}					
Helium	He	1.0555^{-271}	0.10640E+01	-0.35584E-02			2 - 4
Heptadecane	$C_{17}H_{36}$	2.0578^{20}	0.23627E+01	-0.10400E-02	-0.10397E-12		293 - 308
1-Heptadecanol	$C_{17}H_{36}O$	3.41^{60}					
9-Heptadecanone	$C_{17}H_{34}O$	5.43^{55}	0.44176E+02	-0.21183E+00	0.28571E-03		328 - 363
1,6-Heptadiene	C_7H_{12}	2.161^{20}	0.30815E+01	-0.36095E-02	0.16354E-05		184 - 293
2,2,3,3,4,4,4-Heptafluoro-1-butanol	$C_4H_3F_7O$	14.4^{25}					
Heptanal	$C_7H_{14}O$	9.07^{22}					
Heptane	C_7H_{16}	1.9209^{20}	0.24740E+01	-0.22577E-02	0.12428E-05		273 - 373
1-Heptanethiol	$C_7H_{16}S$	4.194^{20}	0.71333E+01	-0.97320E-02	-0.12500E-05		273 - 333
Heptanoic acid	$C_7H_{14}O_2$	3.04^{15}	0.36423E+01	-0.31996E-02	0.39362E-05		288 - 423
1-Heptanol	$C_7H_{16}O$	11.75^{20}	0.60662E+02	-0.24049E+00	0.25155E-03		239 - 513
2-Heptanol, (±)-	$C_7H_{16}O$	9.72^{21}	0.10050E+03	-0.49793E+00	0.64504E-03		207 - 365
3-Heptanol, (*S*)-	$C_7H_{16}O$	7.07^{23}	0.19586E+03	-0.11465E+01	0.17175E-02		248 - 349
4-Heptanol	$C_7H_{16}O$	6.18^{23}	0.28995E+03	-0.18499E+01	0.30109E-02		270 - 301
2-Heptanone	$C_7H_{14}O$	11.95^{20}	0.38348E+02	-0.12531E+00	0.12005E-03		253 - 413
3-Heptanone	$C_7H_{14}O$	12.7^{20}					
4-Heptanone	$C_7H_{14}O$	12.60^{20}	0.41520E+02	-0.13839E+00	0.13497E-03		253 - 393
1-Heptene	C_7H_{14}	2.092^{20}	0.21755E+01	0.13896E-02	-0.57049E-05		273 - 323
Heptyl acetate	$C_9H_{18}O_2$	4.2^{20}					
Heptylamine	$C_7H_{17}N$	3.81^{20}	0.87794E+01	-0.24363E-01	0.25325E-04		253 - 373
Heptylbenzene	$C_{13}H_{20}$	2.26^{20}					
Hexachloroacetone	C_3Cl_6O	3.925^{19}	0.76423E+01	-0.15838E-01	0.10618E-04		269 - 303
Hexachloro-1,3-butadiene	C_4Cl_6	2.55^{20}					
Hexadecane	$C_{16}H_{34}$	2.0460^{20}	0.23861E+01	-0.11600E-02	0.25555E-15		293 - 363
Hexadecanoic acid	$C_{16}H_{32}O_2$	2.417^{65}					
1-Hexadecanol	$C_{16}H_{34}O$	3.69^{60}	0.85935E+01	-0.14714E-01	-0.45533E-13		333 - 363
Hexadecyl acetate	$C_{18}H_{36}O_2$	3.19^{35}	0.47310E+01	-0.50000E-02	0.41338E-14		308 - 348
Hexadecylamine	$C_{16}H_{35}N$	2.71^{55}					
Hexadecyl stearate	$C_{34}H_{68}O_2$	2.61^{60}					
1,5-Hexadiene	C_6H_{10}	2.125^{21}	0.30014E+01	-0.28668E-02	-0.31026E-06		151 - 294
cis,cis-2,4-Hexadiene	C_6H_{10}	2.163^{24}	0.27284E+01	-0.17178E-02	-0.62926E-06		234 - 351
trans,trans-2,4-Hexadiene	C_6H_{10}	2.123^{24}	0.26774E+01	-0.16977E-02	-0.55637E-06		232 - 353
Hexaethyldisiloxane	$C_{12}H_{30}OSi_2$	2.259^{25}	0.36559E+01	-0.72406E-02	0.85714E-05		298 - 333
Hexafluorobenzene	C_6F_6	2.029^{25}	0.24041E+01	-0.83086E-03	-0.14286E-05		298 - 338
1,1,1,3,3,3-Hexafluoro-2-propanol	$C_3H_2F_6O$	16.70^{20}					
Hexamethylbenzene	$C_{12}H_{18}$	2.172^{176}	0.35710E+01	-0.46912E-02	0.35088E-05		449 - 489
Hexamethylcyclotrisiloxane	$C_6H_{18}O_3Si_3$	2.139^{70}					
Hexamethyldisilazane	$C_6H_{19}NSi_2$	2.273^{21}	0.23358E+01	0.16127E-02	-0.62078E-05		294 - 333
Hexamethyldisiloxane	$C_6H_{18}OSi_2$	2.179^{20}	0.34537E+01	-0.61530E-02	0.61544E-05		213 - 313
Hexamethylene diisocyanate	$C_8H_{12}N_2O_2$	14.41^{15}	0.26715E+02	-0.42696E-01			288 - 403
Hexamethylphosphoric triamide	$C_6H_{18}N_3OP$	31.3^{20}	0.95666E+02	-0.29769E+00	0.26407E-03		283 - 363
2,6,10,15,19,23-Hexamethyltetracosane	$C_{30}H_{62}$	1.9106^{100}					
Hexane	C_6H_{14}	1.8865^{20}	0.19768E+01	0.70933E-03	-0.34470E-05		293 - 473
Hexanenitrile	$C_6H_{11}N$	17.26^{25}					
1-Hexanethiol	$C_6H_{14}S$	4.436^{20}	0.11774E+02	-0.37298E-01	0.41875E-04		273 - 333
1,2,6-Hexanetriol	$C_6H_{14}O_3$	31.5^{12}	0.26127E+03	-0.14552E+01	0.22765E-02		261 - 285
Hexanoic acid	$C_6H_{12}O_2$	2.600^{25}	0.21730E+01	0.14840E-02	-0.16526E-06		298 - 433
1-Hexanol	$C_6H_{14}O$	13.03^{20}	0.62744E+02	-0.24214E+00	0.24704E-03		233 - 513
2-Hexanol	$C_6H_{14}O$	11.06^{25}					
3-Hexanol	$C_6H_{14}O$	9.66^{25}					
2-Hexanone	$C_6H_{12}O$	14.56^{20}	0.70378E+02	-0.29385E+00	0.35289E-03		243 - 293
1-Hexene	C_6H_{12}	2.077^{21}	0.31476E+01	-0.50003E-02	0.46673E-05		149 - 294
trans-2-Hexene	C_6H_{12}	1.978^{22}	0.24338E+01	-0.11323E-02	-0.13720E-05		157 - 295
cis-3-Hexene	C_6H_{12}	2.069^{23}	0.30691E+01	-0.45458E-02	0.39898E-05		155 - 296
trans-3-Hexene	C_6H_{12}	1.954^{20}					
Hexyl acetate	$C_8H_{16}O_2$	4.42^{20}					
Hexylamine	$C_6H_{15}N$	4.08^{20}	0.80244E+01	-0.16627E-01	0.10874E-04		253 - 373
Hexylbenzene	$C_{12}H_{18}$	2.3^{20}					
Hexyl benzoate	$C_{13}H_{18}O_2$	4.80^{20}					

Name	Mol. form.	ε	*a*	*b*	*c*	*d*	Range/K
1-Hexyne	C_6H_{10}	2.621^{23}	0.58591E+01	-0.17099E-01	0.20856E-04		184 - 296
Hydrazine	H_4N_2	51.7^{25}	0.22061E+03	-0.89633E+00	0.11066E-02		278 - 323
Hydrogen	H_2	1.2792^{-260}	0.13327E+01	-0.51946E-02			14 - 19
Hydrogen bromide	BrH	8.23^{-86}					
Hydrogen chloride	ClH	14.3^{-114}	0.47316E+02	-0.28455E+00	0.48650E-03		159 - 258
Hydrogen cyanide	CHN	114.9^{20}	0.37331E+04	-0.23180E+02	0.36963E-01		258 - 299
Hydrogen fluoride	FH	83.6^{0}	0.50352E+03	-0.19297E+01	0.14372E-02		200 - 273
Hydrogen iodide	HI	3.87^{-53}	0.51557E+03	-0.44552E+01	0.96795E-02		220 - 236
Hydrogen peroxide	H_2O_2	74.6^{17}	0.48511E+03	-0.23145E+01	0.31020E-02		233 - 303
Hydrogen sulfide	H_2S	5.93^{10}	0.14736E+02	-0.33675E-01	0.96740E-05		212 - 363
(Hydroxyacetyl)benzene	$C_8H_8O_2$	21.33^{25}	0.42286E+02	-0.69215E-01	-0.35714E-05		298 - 368
2-Hydroxybutanoic acid, (±)-	$C_4H_8O_3$	37.7^{23}					
3-Hydroxybutanoic acid, (±)-	$C_4H_8O_3$	31.5^{23}					
3-Hydroxypropanoic acid	$C_3H_6O_3$	30.0^{23}					
Iodine	I_2	11.08^{118}	0.64730E+02	-0.29266E+00	0.39759E-03		391 - 441
Iodine heptafluoride	F_7I	1.75^{25}					
Iodine pentafluoride	F_5I	37.13^{20}	0.95184E+02	-0.19800E+00			273 - 313
Iodobenzene	C_6H_5I	4.59^{20}	0.89442E+01	-0.20008E-01	0.17641E-04		243 - 323
1-Iodobutane	C_4H_9I	6.27^{20}	0.16493E+02	-0.50262E-01	0.52485E-04		293 - 323
2-Iodobutane, (±)-	C_4H_9I	7.873^{20}	0.10883E+02	-0.14680E-02	-0.30000E-04		293 - 323
1-Iodododecane	$C_{12}H_{25}I$	3.91^{25}	0.34641E+01	0.97404E-02	-0.27602E-04		293 - 323
Iodoethane	C_2H_5I	7.82^{20}	0.25598E+02	-0.94367E-01	0.11424E-03		183 - 343
1-Iodoheptane	$C_7H_{15}I$	4.92^{25}	0.11856E+02	-0.33493E-01	0.34368E-04		294 - 323
3-Iodoheptane	$C_7H_{15}I$	6.39^{22}					
1-Iodohexadecane	$C_{16}H_{33}I$	3.57^{20}	0.79531E+01	-0.22859E-01	0.26955E-04		293 - 323
1-Iodohexane	$C_6H_{13}I$	5.35^{20}	0.16685E+02	-0.61309E-01	0.77262E-04		293 - 323
Iodomethane	CH_3I	6.97^{20}	0.24264E+02	-0.93914E-01	0.11926E-03		223 - 303
1-Iodo-3-methylbutane	$C_5H_{11}I$	5.6^{19}					
2-Iodo-2-methylbutane	$C_5H_{11}I$	8.192^{20}					
2-Iodo-2-methylpropane	C_4H_9I	6.65^{10}	0.76780E+01	0.69900E-02	-0.37500E-04		283 - 323
1-Iodooctane	$C_8H_{17}I$	4.67^{20}	0.12452E+02	-0.41229E-01	0.50108E-04		233 - 313
1-Iodopentane	$C_5H_{11}I$	5.78^{20}	0.15753E+02	-0.50543E-01	0.56401E-04		293 - 323
3-Iodopentane	$C_5H_{11}I$	7.432^{20}					
1-Iodopropane	C_3H_7I	7.07^{20}	0.13744E+02	-0.22745E-01			293 - 323
2-Iodopropane	C_3H_7I	8.19^{25}					
3-Iodopropene	C_3H_5I	6.1^{19}					
4-Iodotoluene	C_7H_7I	4.4^{35}					
trans-α-Ionone, (±)-	$C_{13}H_{20}O$	10.78^{19}					
trans-β-Ionone	$C_{13}H_{20}O$	11.66^{24}					
Iron pentacarbonyl	C_5FeO_5	2.602^{20}					
Isobutane	C_4H_{10}	1.7518^{22}	0.23295E+01	-0.19953E-02	0.14197E-06		115 - 303
Isobutene	C_4H_8	2.1225^{16}	0.33701E+01	-0.43295E-02			220 - 289
Isobutyl acetate	$C_6H_{12}O_2$	5.068^{20}	0.14323E+02	-0.46048E-01	0.49286E-04		273 - 323
Isobutylbenzene	$C_{10}H_{14}$	2.318^{20}	0.28055E+01	-0.92614E-03	-0.25000E-05		273 - 323
Isobutyl benzoate	$C_{11}H_{14}O_2$	5.39^{18}					
Isobutyl chlorocarbonate	$C_5H_9ClO_2$	9.1^{20}					
Isobutyl formate	$C_5H_{10}O_2$	6.41^{20}					
Isobutyl isocyanate	C_5H_9NO	11.638^{20}	0.38026E+02	-0.12714E+00	0.12679E-03		293 - 353
Isobutyl pentanoate	$C_9H_{18}O_2$	3.8^{19}					
Isobutyl trichloroacetate	$C_6H_9Cl_3O_2$	7.667^{20}					
Isobutyl vinyl ether	$C_6H_{12}O$	3.34^{20}	0.48060E+01	-0.50000E-02	-0.41495E-14		293 - 323
Isopentane	C_5H_{12}	1.845^{20}	0.22384E+01	-0.12985E-02	-0.16182E-06		143 - 293
Isopentyl acetate	$C_7H_{14}O_2$	4.72^{20}					
Isopentyl butanoate	$C_9H_{18}O_2$	4.0^{20}					
Isopentyl formate	$C_6H_{12}O_2$	5.44^{15}	0.29257E+02	-0.14028E+00	0.20000E-03		288 - 323
Isopentyl isopentanoate	$C_{10}H_{20}O_2$	4.39^{15}	0.14698E+02	-0.57726E-01	0.76190E-04		288 - 323
Isopentyl lactate	$C_8H_{16}O_3$	11.2^{0}	0.48649E+02	-0.21253E+00	0.27619E-03		273 - 373
Isopentyl pentanoate	$C_{10}H_{20}O_2$	3.6^{19}					
Isopentyl propanoate	$C_8H_{16}O_2$	5.21^{0}	0.17665E+02	-0.71718E-01	0.95635E-04		273 - 373
Isopentyl salicylate	$C_{12}H_{16}O_3$	7.26^{20}	0.13129E+02	-0.19190E-01	-0.36060E-05		225 - 397
Isopentyl trichloroacetate	$C_7H_{11}Cl_3O_2$	7.287^{20}					

Name	Mol. form.	ε	*a*	*b*	*c*	*d*	Range/K
Isopropenylbenzene	C_9H_{10}	2.28^{20}					
Isopropylamine	C_3H_9N	5.6268^{20}	0.40429E+02	-0.21441E+00	0.32634E-03		213 - 298
4-Isopropylbenzaldehyde	$C_{10}H_{12}O$	10.68^{15}					
Isopropylbenzene	C_9H_{12}	2.381^{20}	0.31149E+01	-0.30801E-02	0.19643E-05		273 - 323
1-Isopropyl-4-methylbenzene	$C_{10}H_{14}$	2.2322^{25}	0.25266E+01	-0.25121E-03	-0.24867E-05		277 - 333
Isopropyl nitrite	$C_3H_7NO_2$	13.92^{-13}	0.74578E+02	-0.38283E+00	0.57071E-03		150 - 300
Isoquinoline	C_9H_7N	11.0^{25}	0.14412E+03	-0.79935E+00	0.11839E-02		298 - 323
Krypton	Kr	1.664^{-153}					
Lead(IV) chloride	Cl_4Pb	2.78^{20}					
d-Limonene	$C_{10}H_{16}$	2.3746^{25}					
l-Limonene	$C_{10}H_{16}$	2.3738^{25}					
Maleic anhydride	$C_4H_2O_3$	52.75^{53}					
Manganese(VII) oxide	Mn_2O_7	3.28^{20}	0.37655E+01	-0.16463E-02			283 - 312
D-Mannitol	$C_6H_{14}O_6$	24.6^{170}					
Menthol	$C_{10}H_{20}O$	3.90^{36}	0.68202E+01	-0.15894E-01	0.20837E-04		309 - 358
Mesityl oxide	$C_6H_{10}O$	15.6^{0}					
Methane	CH_4	1.6761^{-182}	0.15996E+01	0.27434E-02	-0.22086E-04		91 - 184
Methanesulfonyl chloride	CH_3ClO_2S	34.0^{20}	0.10384E+03	-0.33838E+00	0.34156E-03		293 - 373
Methanol	CH_4O	33.0^{20}	0.19341E+03	-0.92211E+00	0.12839E-02		177 - 293
Methan-d_1-ol	CH_3DO	31.68^{24}	0.20839E+03	-0.10318E+01	0.14740E-02		176 - 298
2-Methoxyaniline	C_7H_9NO	5.230^{30}	0.79911E+01	-0.92183E-02	0.37879E-06		303 - 393
3-Methoxyaniline	C_7H_9NO	8.76^{25}	0.28179E+02	-0.97840E-01	0.11027E-03		289 - 393
4-Methoxyaniline	C_7H_9NO	7.85^{60}	0.30149E+02	-0.10523E+00	0.11467E-03		333 - 453
4-Methoxybenzaldehyde	$C_8H_8O_2$	22.0^{30}					
4-Methoxybenzenesulfonyl chloride	$C_7H_7ClO_3S$	27.2^{41}					
2-Methoxyethanol	$C_3H_8O_2$	17.2^{25}	0.11803E+03	-0.58000E+00	0.81001E-03		254 - 318
1-Methoxynaphthalene	$C_{11}H_{10}O$	4.020^{20}	0.71885E+01	-0.14838E-01	0.13750E-04		293 - 333
2-Methoxynaphthalene	$C_{11}H_{10}O$	3.563^{80}	0.56702E+01	-0.69754E-02	0.28571E-05		353 - 373
2-Methoxyphenol	$C_7H_8O_2$	11.95^{25}	0.31751E+02	-0.88173E-01	0.72953E-04		291 - 448
3-Methoxyphenol	$C_7H_8O_2$	11.59^{25}	0.37279E+02	-0.12113E+00	0.11698E-03		298 - 433
4-Methoxyphenol	$C_7H_8O_2$	11.05^{61}	0.39483E+02	-0.12142E+00	0.10841E-03		334 - 453
4-Methoxyphenyl isocyanate	$C_8H_7NO_2$	10.26^{60}	0.20780E+02	-0.31571E-01			333 - 403
N-Methylacetamide	C_3H_7NO	179.0^{30}	0.15975E+04	-0.90451E+01	0.18345E-01	-0.12998E-04	303 - 473
Methyl acetate	$C_3H_6O_2$	7.07^{15}	0.13190E+02	-0.21226E-01			276 - 318
Methyl 2-(acetyloxy)benzoate	$C_{10}H_{10}O_4$	5.31^{56}	0.19579E+02	-0.69970E-01	0.80889E-04		329 - 371
Methyl acrylate	$C_4H_6O_2$	7.03^{30}	0.11968E+02	-0.16500E-01			303 - 333
Methylamine	CH_5N	16.7^{-58}	0.34398E+02	-0.73630E-01	-0.41279E-04		198 - 258
Methyl 2-aminobenzoate	$C_8H_9NO_2$	21.9^{25}					
2-Methylaniline	C_7H_9N	6.138^{25}	0.10988E+02	-0.18976E-01	0.91958E-05		298 - 398
3-Methylaniline	C_7H_9N	5.816^{25}	0.13477E+02	-0.35551E-01	0.33135E-04		298 - 398
4-Methylaniline	C_7H_9N	5.058^{60}	0.78897E+01	-0.10196E-01	0.51190E-05		333 - 403
N-Methylaniline	C_7H_9N	5.96^{20}					
2-Methylanisole	$C_8H_{10}O$	3.502^{20}	0.50825E+01	-0.62297E-02	0.28571E-05		293 - 333
3-Methylanisole	$C_8H_{10}O$	3.967^{20}	0.12830E+02	-0.49701E-01	0.66429E-04		293 - 333
4-Methylanisole	$C_8H_{10}O$	3.914^{20}	0.86608E+01	-0.23510E-01	0.25000E-04		293 - 333
α-Methylbenzenemethanol	$C_8H_{10}O$	8.77^{20}	0.32971E+02	-0.12042E+00	0.12809E-03		293 - 423
4-Methylbenzenethiol	C_7H_8S	4.74^{50}	0.87052E+01	-0.15347E-01	0.95238E-05		323 - 358
Methyl benzoate	$C_8H_8O_2$	6.642^{30}	0.17486E+02	-0.51027E-01	0.50222E-04		303 - 393
Methyl 3-bromopropanoate	$C_4H_7BrO_2$	5.81^{30}	0.36001E+01	0.72500E-02			303 - 343
2-Methyl-1,3-butadiene	C_5H_8	2.098^{20}	0.28170E+01	-0.23147E-02	-0.43975E-06		198 - 293
2-Methyl-2-butanethiol	$C_5H_{12}S$	5.087^{20}	0.15116E+02	-0.50700E-01	0.56250E-04		273 - 333
Methyl butanoate	$C_5H_{10}O_2$	5.48^{28}	0.38604E+02	-0.19171E+00	0.27128E-03		301 - 343
2-Methyl-1-butanol, (±)-	$C_5H_{12}O$	15.63^{25}	0.14020E+02	0.13948E+00	-0.45000E-03		288 - 318
3-Methyl-1-butanol	$C_5H_{12}O$	15.63^{20}	0.79733E+02	-0.31272E+00	0.32014E-03		173 - 513
2-Methyl-2-butanol	$C_5H_{12}O$	5.78^{25}	0.11662E+03	-0.69756E+00	0.10920E-02		268 - 318
3-Methyl-2-butanol, (±)-	$C_5H_{12}O$	12.1^{25}					
3-Methyl-2-butanone	$C_5H_{10}O$	10.37^{20}	0.30695E+02	-0.10962E+00	0.13810E-03		293 - 328
2-Methyl-1-butene	C_5H_{10}	2.180^{20}					
2-Methyl-2-butene	C_5H_{10}	1.979^{23}	0.26064E+01	-0.19578E-02	-0.53908E-06		225 - 296
Methyl *trans*-2-butenoate	$C_5H_8O_2$	6.6645^{20}					
Methyl carbamate	$C_2H_5NO_2$	18.48^{55}	0.36773E+02	-0.55700E-01			328 - 368

Name	Mol. form.	ε	a	b	c	d	Range/K
Methyl chloroacetate	$C_3H_5ClO_2$	12.0^{20}					
Methyl 4-chlorobutanoate	$C_5H_9ClO_2$	9.51^{30}	0.17127E+02	-0.25000E-01			303 - 343
Methyl 2-chloropropanoate	$C_4H_7ClO_2$	11.45^{30}	0.22449E+02	-0.36250E-01			303 - 343
Methylcyclohexane	C_7H_{14}	2.024^{20}					
Methyl cyclohexanecarboxylate	$C_8H_{14}O_2$	4.87^{20}					
cis-2-Methylcyclohexanol	$C_7H_{14}O$	9.375^{20}	0.17315E+03	-0.98794E+00	0.14634E-02		273 - 323
cis-3-Methylcyclohexanol, (±)-	$C_7H_{14}O$	13.79^{20}	0.65896E+02	-0.21954E+00	0.14107E-03		273 - 323
cis-4-Methylcyclohexanol	$C_7H_{14}O$	13.45^{20}	0.65021E+02	-0.22896E+00	0.17946E-03		273 - 323
2-Methylcyclohexanone, (±)-	$C_7H_{12}O$	14.0^{20}					
3-Methylcyclohexanone, (±)-	$C_7H_{12}O$	12.4^{20}					
4-Methylcyclohexanone	$C_7H_{12}O$	12.35^{20}					
Methylcyclopentane	C_6H_{12}	1.9853^{20}	0.21587E+01	-0.22450E-03	-0.12500E-05		293 - 323
1-Methylcyclopentanol	$C_6H_{12}O$	7.11^{37}	0.75444E+02	-0.36617E+00	0.47021E-03		310 - 333
2-Methyl-*N*,*N*-dimethylaniline	$C_9H_{13}N$	3.4^{20}					
4-Methyl-*N*,*N*-dimethylaniline	$C_9H_{13}N$	3.9^{20}					
Methyl dodecanoate	$C_{13}H_{26}O_2$	3.539^{20}					
N-Methylformamide	C_2H_5NO	189.0^{20}	0.10383E+04	-0.43165E+01	0.48398E-02		276 - 353
Methyl formate	$C_2H_4O_2$	9.20^{15}	0.19699E+02	-0.36429E-01			288 - 302
2-Methylfuran	C_5H_6O	2.76^{20}					
Methyl heptadecanoate	$C_{18}H_{36}O_2$	3.07^{40}					
2-Methylheptane	C_8H_{18}	1.9519^{20}					
Methyl heptanoate	$C_8H_{16}O_2$	4.355^{20}					
2-Methyl-1-heptanol, (±)-	$C_8H_{18}O$	5.16^{20}	0.61698E+02	-0.33647E+00	0.49066E-03		236 - 328
3-Methyl-1-heptanol	$C_8H_{18}O$	2.884^{17}	0.84687E+01	-0.33712E-01	0.49793E-04		241 - 316
4-Methyl-1-heptanol	$C_8H_{18}O$	4.63^{17}	0.48612E+02	-0.26773E+00	0.39972E-03		237 - 332
5-Methyl-1-heptanol, (±)-	$C_8H_{18}O$	7.68^{17}	0.54581E+02	-0.24772E+00	0.29734E-03		235 - 328
6-Methyl-1-heptanol	$C_8H_{18}O$	10.54^{17}	0.57997E+02	-0.23517E+00	0.24663E-03		265 - 328
2-Methyl-2-heptanol	$C_8H_{18}O$	3.43^{19}					
3-Methyl-2-heptanol	$C_8H_{18}O$	7.47^{16}	0.39178E+02	-0.17976E+00	0.24218E-03		229 - 329
4-Methyl-2-heptanol	$C_8H_{18}O$	3.59^{17}	0.39715E+02	-0.23115E+00	0.36771E-03		240 - 333
5-Methyl-2-heptanol	$C_8H_{18}O$	7.5^{5}	0.68568E+02	-0.40706E+00	0.67433E-03		230 - 279
6-Methyl-2-heptanol	$C_8H_{18}O$	6.41^{17}	0.77520E+02	-0.41724E+00	0.59448E-03		239 - 329
2-Methyl-3-heptanol, (±)-	$C_8H_{18}O$	3.260^{20}	-0.59739E+01	0.56700E-01	-0.83125E-04		343 - 403
3-Methyl-3-heptanol	$C_8H_{18}O$	3.013^{20}	-0.38440E+01	0.42327E-01	-0.61250E-04		343 - 403
4-Methyl-3-heptanol	$C_8H_{18}O$	3.312^{20}	-0.48003E+01	0.50740E-01	-0.75000E-04		343 - 403
5-Methyl-3-heptanol	$C_8H_{18}O$	3.832^{20}	0.61967E+01	-0.63750E-02			343 - 383
6-Methyl-3-heptanol, (±)-	$C_8H_{18}O$	4.992^{20}	0.23037E+02	-0.98029E-01	0.12479E-03		283 - 383
2-Methyl-4-heptanol	$C_8H_{18}O$	3.338^{23}	0.42102E+00	0.10427E-01	-0.20438E-05		230 - 333
3-Methyl-4-heptanol	$C_8H_{18}O$	7.46^{17}	0.33354E+02	-0.14077E+00	0.17750E-03		230 - 330
4-Methyl-4-heptanol	$C_8H_{18}O$	2.902^{23}					
cis-3-Methyl-2-heptene	C_8H_{16}	2.436^{20}					
2-Methylheptyl acetate, (±)-	$C_{10}H_{20}O_2$	4.27^{15}	0.23285E+02	-0.11538E+00	0.17143E-03		288 - 323
Methyl hexadecanoate	$C_{17}H_{34}O_2$	3.124^{40}					
2-Methylhexane	C_7H_{16}	1.9221^{20}	0.24759E+01	-0.22535E-02	0.12500E-05		293 - 323
3-Methylhexane	C_7H_{16}	1.920^{20}	0.27089E+01	-0.37908E-02	0.37500E-05		273 - 323
Methyl hexanedioate	$C_7H_{12}O_4$	6.69^{20}	0.11962E+02	-0.23973E-01	0.20608E-04		293 - 433
Methyl hexanoate	$C_7H_{14}O_2$	4.615^{20}					
2-Methyl-2-hexanol	$C_7H_{16}O$	3.257^{24}					
3-Methyl-2-hexanol	$C_7H_{16}O$	4.990^{24}	0.59724E+02	-0.32417E+00	0.47058E-03		244 - 372
3-Methyl-3-hexanol	$C_7H_{16}O$	3.248^{25}					
5-Methyl-2-hexanone	$C_7H_{14}O$	13.53^{20}	0.52353E+02	-0.17695E+00	0.15195E-03		293 - 333
2-Methyl-2-hexene	C_7H_{14}	2.962^{20}					
Methyl isocyanate	C_2H_3NO	21.75^{16}					
Methyl methacrylate	$C_5H_8O_2$	6.32^{30}	0.32098E+02	-0.14568E+00	0.20000E-03		303 - 343
Methyl 2-methoxybenzoate	$C_9H_{10}O_3$	7.7^{21}					
Methyl 4-methylbenzoate	$C_9H_{10}O_2$	4.3^{33}					
1-Methylnaphthalene	$C_{11}H_{10}$	2.915^{20}	0.45126E+01	-0.76480E-02	0.75000E-05		293 - 333
2-Methylnaphthalene	$C_{11}H_{10}$	2.747^{40}					
Methyl nitrate	CH_3NO_3	23.9^{20}					
Methyl nitrite	CH_3NO_2	20.77^{-73}	0.11071E+03	-0.73428E+00	0.14054E-02		110 - 260
Methyl 2-nitrobenzoate	$C_8H_7NO_4$	27.76^{27}					

Name	Mol. form.	*ε*	*a*	*b*	*c*	*d*	Range/K
Methyl nonadecanoate	$C_{20}H_{40}O_2$	2.982^{40}					
Methyl nonanoate	$C_{10}H_{20}O_2$	3.943^{20}					
5-Methyl-4-nonene	$C_{10}H_{20}$	2.175^{20}					
Methyl *cis,cis*-9,12-octadecadienoate	$C_{19}H_{34}O_2$	3.466^{20}					
Methyl octadecanoate	$C_{19}H_{38}O_2$	3.021^{40}					
Methyl *cis,cis,cis*-9,12,15-octadecatrienoate	$C_{19}H_{32}O_2$	3.355^{20}					
Methyl *cis*-9-octadecenoate	$C_{19}H_{36}O_2$	3.211^{20}					
2-Methyloctane	C_9H_{20}	1.967^{20}					
4-Methyloctane	C_9H_{20}	1.967^{20}					
Methyl octanoate	$C_9H_{18}O_2$	4.101^{20}					
2-Methyloctanoic acid	$C_9H_{18}O_2$	2.39^{20}					
Methyl pentadecanoate	$C_{16}H_{32}O_2$	3.296^{20}					
cis-2-Methyl-1,3-pentadiene	C_6H_{10}	2.422^{25}					
3-Methyl-1,3-pentadiene	C_6H_{10}	2.426^{25}					
4-Methyl-1,3-pentadiene	C_6H_{10}	2.599^{20}	0.51328E+01	-0.12774E-01	0.14215E-04		198 - 323
2-Methylpentane	C_6H_{14}	1.886^{20}	0.20745E+01	0.50871E-03	-0.39286E-05		273 - 323
3-Methylpentane	C_6H_{14}	1.886^{20}	0.24739E+01	-0.23190E-02	0.10714E-05		273 - 323
2-Methyl-2,4-pentanediol	$C_6H_{14}O_2$	25.86^{20}	0.14531E+03	-0.65285E+00	0.83503E-03		203 - 333
4-Methylpentanenitrile	$C_6H_{11}N$	17.5^{22}					
Methyl pentanoate	$C_6H_{12}O_2$	4.992^{20}					
3-Methyl-1-pentanol, (±)-	$C_6H_{14}O$	15.2^{25}					
3-Methyl-3-pentanol	$C_6H_{14}O$	4.322^{20}					
4-Methyl-2-pentanone	$C_6H_{12}O$	13.11^{20}	0.36341E+02	-0.97119E-01	0.61896E-04		204 - 373
1-Methyl-1-phenylhydrazine	$C_7H_{10}N_2$	7.3^{19}					
Methyl phenyl sulfone	$C_7H_8O_2S$	37.9^{100}					
N-Methylpropanamide	C_4H_9NO	170.0^{20}					
2-Methylpropanenitrile	C_4H_7N	24.42^{20}	0.52554E+02	-0.96000E-01			293 - 313
2-Methyl-1-propanethiol	$C_4H_{10}S$	4.961^{25}					
2-Methyl-2-propanethiol	$C_4H_{10}S$	5.475^{20}	0.10597E+02	-0.17500E-01			283 - 313
Methyl propanoate	$C_4H_8O_2$	6.200^{20}	0.12798E+02	-0.22540E-01			293 - 333
2-Methylpropanoic acid	$C_4H_8O_2$	2.58^{20}					
2-Methylpropanoic anhydride	$C_8H_{14}O_3$	13.6^{19}					
2-Methyl-1-propanol	$C_4H_{10}O$	17.93^{20}	0.10762E+03	-0.51398E+00	0.83702E-03	-0.45299E-06	173 - 533
2-Methyl-2-propanol	$C_4H_{10}O$	12.47^{25}	0.22541E+03	-0.14990E+01	0.34050E-02	-0.25968E-05	298 - 503
2-Methylpyridine	C_6H_7N	10.18^{20}	0.34560E+02	-0.11980E+00	0.12500E-03		293 - 333
3-Methylpyridine	C_6H_7N	11.10^{30}	0.19643E+03	-0.11167E+01	0.16667E-02		303 - 333
4-Methylpyridine	C_6H_7N	12.2^{20}	0.33765E+02	-0.10113E+00	0.93860E-04		274 - 333
2-Methylpyridine-1-oxide	C_6H_7NO	36.4^{50}	0.11705E+03	-0.35301E+00	0.32000E-03		323 - 398
3-Methylpyridine-1-oxide	C_6H_7NO	28.26^{45}	0.59851E+02	-0.12682E+00	0.86622E-04		318 - 398
N-Methylpyrrolidine	$C_5H_{11}N$	32.2^{25}					
N-Methyl-2-pyrrolidinone	C_5H_9NO	32.55^{20}					
2-Methylquinoline	$C_{10}H_9N$	7.24^{20}	0.11688E+02	-0.78400E-02	-0.25000E-04		293 - 333
4-Methylquinoline	$C_{10}H_9N$	9.31^{20}	0.17788E+02	-0.32580E-01	0.12500E-04		293 - 333
6-Methylquinoline	$C_{10}H_9N$	8.48^{20}	0.21696E+02	-0.63400E-01	0.62500E-04		293 - 333
8-Methylquinoline	$C_{10}H_9N$	6.58^{20}	0.19356E+02	-0.61900E-01	0.62500E-04		293 - 333
Methyl salicylate	$C_8H_8O_3$	8.80^{41}	0.20501E+02	-0.39045E-01	0.68298E-05		223 - 398
Methyl silicate	$C_4H_{12}O_4Si$	6.0^{20}					
3-Methyl sulfolane	$C_5H_{10}O_2S$	29.4^{25}	0.53158E+02	-0.93730E-01	0.47275E-04		298 - 398
Methyl tetradecanoate	$C_{15}H_{30}O_2$	3.352^{20}					
2-Methyltetrahydrofuran	$C_5H_{10}O$	6.97^{25}					
(Methylthio)benzene	C_7H_8S	4.88^{30}	0.21841E+02	-0.97630E-01	0.13750E-03		303 - 343
Methyl tridecanoate	$C_{14}H_{28}O_2$	3.442^{20}					
Methyltriphenoxysilane	$C_{19}H_{18}O_3Si$	3.628^{25}					
Methyl undecanoate	$C_{12}H_{24}O_2$	3.671^{20}					
Monomethyl glutarate	$C_6H_{10}O_4$	8.37^{20}	0.16779E+02	-0.39839E-01	0.38095E-04		293 - 363
Morpholine	C_4H_9NO	7.42^{25}					
β-Myrcene	$C_{10}H_{16}$	2.3^{25}					
Naphthalene	$C_{10}H_8$	2.54^{90}					
N-1-Naphthalenylacetamide	$C_{12}H_{11}NO$	24.3^{160}	0.84739E+02	-0.12391E+00	-0.35714E-04		433 - 533
1-Naphthol	$C_{10}H_8O$	5.03^{100}	0.16489E+02	-0.46700E-01	0.42857E-04		373 - 453
2-Naphthol	$C_{10}H_8O$	4.95^{140}	0.92865E+01	-0.10500E-01	0.42501E-15		413 - 453

Name	Mol. form.	*ε*	*a*	*b*	*c*	*d*	Range/K
1-Naphthylamine	$C_{10}H_9N$	5.20^{60}	0.10577E+02	-0.22114E-01	0.17857E-04		333 - 453
2-Naphthylamine	$C_{10}H_9N$	5.26^{120}	0.19722E+02	-0.60679E-01	0.60714E-04		393 - 473
1-Naphthyl 2-hydroxybenzoate	$C_{17}H_{12}O_3$	6.30^{20}	0.11229E+02	-0.18857E-01	0.70332E-05		293 - 353
Neon	Ne	1.1907^{-247}	0.12667E+01	-0.29064E-02			26 - 29
Neopentane	C_5H_{12}	1.769^{23}	0.10949E+02	-0.63057E-01	0.10835E-03		251 - 296
L-Nicotine	$C_{10}H_{14}N_2$	8.937^{20}	0.21347E+02	-0.57177E-01	0.50655E-04		293 - 363
Nitric oxide	NO	2.00^{-149}					
2-Nitroaniline	$C_6H_6N_2O_2$	47.3^{80}	0.18900E+03	-0.56977E+00	0.47484E-03		353 - 468
3-Nitroaniline	$C_6H_6N_2O_2$	35.6^{125}	0.20352E+03	-0.66582E+00	0.61310E-03		398 - 468
4-Nitroaniline	$C_6H_6N_2O_2$	78.5^{155}	0.48673E+03	-0.15040E+01	0.12857E-02		428 - 468
2-Nitroanisole	$C_7H_7NO_3$	45.75^{20}	0.16684E+03	-0.58196E+00	0.57382E-03		293 - 423
3-Nitroanisole	$C_7H_7NO_3$	25.7^{45}	0.65402E+02	-0.16460E+00	0.12560E-03		318 - 443
4-Nitroanisole	$C_7H_7NO_3$	26.95^{65}	0.59811E+02	-0.10955E+00	0.36042E-04		338 - 443
Nitrobenzene	$C_6H_5NO_2$	35.6^{20}	0.11212E+03	-0.35211E+00	0.31128E-03		279 - 533
Nitroethane	$C_2H_5NO_2$	29.11^{15}	0.57406E+02	-0.97657E-01			276 - 333
Nitrogen	N_2	1.4680^{-210}	0.12550E+01	0.67949E-02	-0.56704E-04		63 - 126
Nitrogen tetroxide	N_2O_4	2.44^{20}	0.28212E+01	-0.13000E-02			253 - 293
Nitrogen trioxide	N_2O_3	31.13^{-70}	0.92287E+02	-0.43306E+00	0.65000E-03		203 - 243
Nitromethane	CH_3NO_2	37.27^{20}	0.11227E+03	-0.35591E+00	0.34206E-03		288 - 343
1-Nitronaphthalene	$C_{10}H_7NO_2$	19.68^{60}	0.36267E+02	-0.41283E-01	-0.25595E-04		333 - 403
1-Nitrooctane	$C_8H_{17}NO_2$	11.46^{20}					
2-Nitrophenol	$C_6H_5NO_3$	16.50^{50}	0.33827E+02	-0.62123E-01	0.26774E-04		323 - 453
3-Nitrophenol	$C_6H_5NO_3$	35.45^{100}	0.18967E+03	-0.66144E+00	0.66532E-03		373 - 458
4-Nitrophenol	$C_6H_5NO_3$	42.20^{120}	0.22901E+03	-0.74264E+00	0.68006E-03		393 - 463
1-Nitropropane	$C_3H_7NO_2$	24.70^{15}	0.94999E+02	-0.38358E+00	0.48480E-03		276 - 333
2-Nitropropane	$C_3H_7NO_2$	26.74^{15}	0.60138E+02	-0.11566E+00			276 - 303
Nitrosyl bromide	BrNO	13.4^{15}					
Nitrosyl chloride	ClNO	18.2^{12}					
4-Nitrothioanisole	$C_7H_7NO_2S$	21.7^{73}					
2-Nitrotoluene	$C_7H_7NO_2$	26.26^{20}	0.10420E+03	-0.41726E+00	0.51607E-03		273 - 323
3-Nitrotoluene	$C_7H_7NO_2$	24.95^{30}	0.62492E+02	-0.16235E+00	0.12844E-03		303 - 403
4-Nitrotoluene	$C_7H_7NO_2$	22.2^{58}					
Nonadecane	$C_{19}H_{40}$	2.0706^{20}					
10-Nonadecanone	$C_{19}H_{38}O$	5.37^{80}					
Nonane	C_9H_{20}	1.9722^{20}	0.23894E+01	-0.14830E-02	0.14881E-06		253 - 393
Nonanenitrile	$C_9H_{17}N$	12.08^{20}					
Nonanoic acid	$C_9H_{18}O_2$	2.475^{22}	0.25039E+01	0.67274E-03	-0.24180E-05		295 - 365
1-Nonanol	$C_9H_{20}O$	8.83^{20}	0.97467E+02	-0.51103E+00	0.71429E-03		288 - 343
2-Nonanol, (±)-	$C_9H_{20}O$	6.66^{25}	0.10136E+03	-0.55612E+00	0.80000E-03		288 - 308
3-Nonanol, (±)-	$C_9H_{20}O$	4.49^{25}	0.55214E+02	-0.31920E+00	0.50000E-03		288 - 308
4-Nonanol	$C_9H_{20}O$	3.69^{25}	0.27954E+01	0.30000E-02	-0.52375E-13		288 - 308
5-Nonanol	$C_9H_{20}O$	3.54^{25}	-0.25463E+01	0.35320E-01	-0.50000E-04		288 - 308
2-Nonanone	$C_9H_{18}O$	9.14^{22}					
5-Nonanone	$C_9H_{18}O$	10.6^{20}					
1-Nonene	C_9H_{18}	2.180^{20}	0.22710E+01	0.15797E-02	-0.64286E-05		273 - 323
2-Nonenoic acid	$C_9H_{16}O_2$	2.5^{23}					
Nonyl acetate	$C_{11}H_{22}O_2$	3.87^{20}					
Nonylamine	$C_9H_{21}N$	3.42^{20}	0.53575E+01	-0.71982E-02	0.19481E-05		293 - 373
cis,cis-9,12-Octadecadienoic acid	$C_{18}H_{32}O_2$	2.754^{20}	0.32073E+01	-0.15477E-02			275 - 368
1-Octadecanol	$C_{18}H_{38}O$	3.38^{60}	0.73784E+01	-0.12000E-01	-0.22871E-13		333 - 363
cis,cis,cis-9,12,15-Octadecatrienoic acid	$C_{18}H_{30}O_2$	2.825^{20}	0.33867E+01	-0.19181E-02			274 - 368
cis-9-Octadecenoic acid	$C_{18}H_{34}O_2$	2.336^{20}	0.25385E+01	-0.69448E-03			275 - 368
Octadecyl acetate	$C_{20}H_{40}O_2$	3.07^{35}	0.44569E+01	-0.45000E-02	0.33923E-14		308 - 348
Octadecylamine	$C_{18}H_{39}N$	2.67^{53}					
1,7-Octadiene	C_8H_{14}	2.186^{20}	0.28376E+01	-0.17442E-02	-0.16141E-05		214 - 293
2,2,3,3,4,4,5,5-Octafluoro-1-pentanol	$C_5H_4F_8O$	15.30^{25}					
Octamethylcyclotetrasiloxane	$C_8H_{24}O_4Si_4$	2.390^{23}	0.36286E+01	-0.56885E-02	0.50874E-05		296 - 333
Octane	C_8H_{18}	1.948^{20}	0.22590E+01	-0.84212E-03	-0.75758E-06		233 - 393
Octanenitrile	$C_8H_{15}N$	13.90^{20}					
1-Octanethiol	$C_8H_{18}S$	3.949^{20}	0.63667E+01	-0.87920E-02	0.18750E-05		273 - 333
Octanoic acid	$C_8H_{16}O_2$	2.85^{15}	0.29391E+01	-0.38721E-03			288 - 423

Name	Mol. form.	*ε*	*a*	*b*	*c*	*d*	Range/K
1-Octanol	$C_8H_{18}O$	10.30^{20}	0.51647E+02	-0.20371E+00	0.21320E-03		258 - 513
2-Octanol	$C_8H_{18}O$	8.13^{20}	0.63760E+02	-0.27643E+00	0.31075E-03		213 - 513
3-Octanol	$C_8H_{18}O$	5.55^{20}	0.12505E+03	-0.70646E+00	0.10245E-02		223 - 383
4-Octanol	$C_8H_{18}O$	4.48^{20}	0.51049E+02	-0.26664E+00	0.37280E-03		243 - 403
2-Octanone	$C_8H_{16}O$	9.51^{20}	-0.16219E+02	0.18799E+00	-0.34156E-03		293 - 333
3-Octanone	$C_8H_{16}O$	10.50^{30}					
1-Octene	C_8H_{16}	2.113^{20}	0.24348E+01	0.34200E-03	-0.50000E-05		273 - 323
cis-3-Octene	C_8H_{16}	2.062^{25}					
trans-3-Octene	C_8H_{16}	2.002^{25}					
cis-4-Octene	C_8H_{16}	2.053^{25}					
trans-4-Octene	C_8H_{16}	2.004^{25}					
Octyl acetate	$C_{10}H_{20}O_2$	4.18^{15}	-0.34691E+01	0.58106E-01	-0.10952E-03		288 - 323
Octylamine	$C_8H_{19}N$	3.58^{20}	0.77931E+01	-0.20015E-01	0.19347E-04		273 - 373
Octylbenzene	$C_{14}H_{22}$	2.26^{20}					
Oxalyl chloride	$C_2Cl_2O_2$	3.470^{21}					
Oxirane	C_2H_4O	12.42^{20}	0.52661E+02	-0.21337E+00	0.25947E-03		293 - 243
Oxygen	O_2	1.5684^{-219}	0.15434E+01	0.14615E-02	-0.21964E-04		55 - 154
Ozone	O_3	4.75^{-183}	0.86344E+01	-0.54807E-01	0.12596E-03		90 - 185
Pentaborane(9)	B_5H_9	21.1^{25}	0.40952E+03	-0.24414E+01	0.38225E-02		226 - 298
Pentachloroethane	C_2HCl_5	3.716^{25}	0.65972E+01	-0.96800E-02			298 - 338
2,3,4,5,6-Pentachlorotoluene	$C_7H_3Cl_5$	4.8^{20}					
Pentadecane	$C_{15}H_{32}$	2.0391^{20}	0.23792E+01	-0.11600E-02	-0.71069E-16		283 - 363
1-Pentadecanol	$C_{15}H_{32}O$	3.70^{60}					
Pentadecylamine	$C_{15}H_{33}N$	2.85^{40}					
cis-1,3-Pentadiene	C_5H_8	2.319^{25}					
1,4-Pentadiene	C_5H_8	2.054^{21}	0.29994E+01	-0.34578E-02	0.85300E-06		178 - 294
Pentamethylbenzene	$C_{11}H_{16}$	2.358^{61}	0.30196E+01	-0.22619E-02	0.83831E-06		334 - 413
Pentanal	$C_5H_{10}O$	10.00^{20}					
Pentane	C_5H_{12}	1.8371^{20}					
1,2-Pentanediol, (±)-	$C_5H_{12}O_2$	17.31^{24}	0.18436E+03	-0.10682E+01	0.17037E-02		197 - 297
1,4-Pentanediol	$C_5H_{12}O_2$	26.74^{23}	0.13568E+03	-0.59198E+00	0.75398E-03		193 - 318
1,5-Pentanediol	$C_5H_{12}O_2$	26.2^{20}	0.11858E+03	-0.45920E+00	0.49341E-03		243 - 343
2,3-Pentanediol	$C_5H_{12}O_2$	17.37^{24}	0.95876E+02	-0.46463E+00	0.67434E-03		238 - 297
2,4-Pentanediol	$C_5H_{12}O_2$	24.69^{21}	0.11914E+03	-0.52569E+00	0.69607E-03		224 - 294
2,4-Pentanedione	$C_5H_8O_2$	26.524^{30}					
Pentanenitrile	C_5H_9N	20.04^{20}	0.55793E+02	-0.15750E+00	0.12432E-03		183 - 333
1-Pentanethiol	$C_5H_{12}S$	4.847^{20}	0.71131E+01	-0.30228E-02	-0.16414E-04		273 - 333
Pentanoic acid	$C_5H_{10}O_2$	2.661^{21}	0.33491E+01	-0.75156E-02	0.17820E-04		250 - 344
1-Pentanol	$C_5H_{12}O$	15.13^{25}	0.73397E+02	-0.28165E+00	0.28427E-03		213 - 513
2-Pentanol	$C_5H_{12}O$	13.71^{25}	0.16437E+03	-0.86506E+00	0.11955E-02		273 - 323
3-Pentanol	$C_5H_{12}O$	13.35^{25}	0.12838E+03	-0.60980E+00	0.75000E-03		288 - 318
2-Pentanone	$C_5H_{10}O$	15.45^{20}	0.40893E+02	-0.10423E+00	0.60557E-04		204 - 353
3-Pentanone	$C_5H_{10}O$	17.00^{20}	0.12690E+02	0.95177E-01	-0.27321E-03		233 - 353
2-Pentanone oxime	$C_5H_{11}NO$	3.3^{20}					
1-Pentene	C_5H_{10}	2.011^{20}	-0.11438E+01	0.25420E-01	-0.50000E-04		273 - 293
Pentyl acetate	$C_7H_{14}O_2$	4.79^{20}	0.12091E+02	-0.36536E-01	0.39732E-04		253 - 353
Pentylamine	$C_5H_{13}N$	4.27^{20}	0.11274E+02	-0.34965E-01	0.37706E-04		223 - 353
Pentyl benzoate	$C_{12}H_{16}O_2$	5.07^{20}					
Pentyl butanoate	$C_9H_{18}O_2$	4.08^{28}	0.59029E+01	-0.49905E-02	-0.34292E-05		301 - 343
Pentyl cinnamate	$C_{14}H_{18}O_2$	4.89^{20}					
Pentyl formate	$C_6H_{12}O_2$	5.7^{19}					
Pentyl hexanoate	$C_{11}H_{22}O_2$	4.22^{15}	0.83503E+01	-0.18449E-01	0.14286E-04		288 - 323
Pentyl nitrite	$C_5H_{11}NO_2$	7.21^{25}					
Pentyl pentanoate	$C_{10}H_{20}O_2$	4.076^{32}	0.77641E+01	-0.14335E-01	0.73740E-05		306 - 393
Pentyl propanoate	$C_8H_{16}O_2$	4.552^{20}					
Pentyl salicylate	$C_{12}H_{16}O_3$	6.25^{28}					
Perchloryl fluoride	$ClFO_3$	2.194^{-123}	0.23808E+01	-0.38629E-03	-0.57143E-05		125 - 150
Perfluoroacetone	C_3F_6O	2.104^{-71}	0.34809E+01	-0.92883E-02	0.12282E-04		151 - 238
Perfluoroheptane	C_7F_{16}	1.847^{16}					
Perfluorohexane	C_6F_{14}	1.76^{25}					
Perfluoromethylcyclohexane	C_7F_{14}	1.82^{25}					

Name	Mol. form.	*ε*	*a*	*b*	*c*	*d*	Range/K
Phenanthrene	$C_{14}H_{10}$	2.72^{110}					
Phenol	C_6H_6O	12.40^{30}	0.63391E+02	-0.24988E+00	0.26930E-03		303 - 433
Phenoxyacetylene	C_8H_6O	4.76^{25}					
Phenyl acetate	$C_8H_8O_2$	5.403^{25}	0.11327E+02	-0.26707E-01	0.22938E-04		298 - 404
Phenylacetylene	C_8H_6	2.98^{25}					
Phenyl 2-(acetyloxy)benzoate	$C_{15}H_{12}O_4$	4.33^{111}					
Phenyl butanoate	$C_{10}H_{12}O_2$	4.48^{20}					
2-Phenylethyl acetate	$C_{10}H_{12}O_2$	4.93^{24}					
Phenylhydrazine	$C_6H_8N_2$	7.15^{20}					
Phenyl isocyanate	C_7H_5NO	8.940^{20}	0.17541E+02	-0.29790E-01	0.15476E-05		293 - 353
Phenyl laurate	$C_{18}H_{28}O_2$	3.28^{20}					
1-Phenyl-2-methyl-2-propanol	$C_{10}H_{14}O$	5.71^{25}	0.21922E+02	-0.84231E-01	0.99475E-04		298 - 423
Phenyl pentanoate	$C_{11}H_{14}O_2$	4.30^{20}					
Phenyl propanoate	$C_9H_{10}O_2$	4.77^{20}					
1-Phenyl-2-propanol	$C_9H_{12}O$	9.35^{20}	0.10762E+03	-0.56026E+00	0.76915E-03		233 - 373
Phenyl salicylate	$C_{13}H_{10}O_3$	6.92^{17}	0.26545E+02	-0.11180E+00	0.15220E-03		290 - 358
Phenyl silicate	$C_{24}H_{20}O_4Si$	3.4915^{60}					
Phosphorothioc trichloride	Cl_3PS	4.94^{25}					
Phosphorus (white)	P	4.096^{34}	0.79018E+00	0.23911E-01	-0.42826E-04		307 - 358
Phosphorus(III) chloride	Cl_3P	3.498^{17}	0.59098E+01	-0.83322E-02			290 - 333
Phosphorus(V) chloride	Cl_5P	2.85^{160}					
Phosphorus(V) dichloride trifluoride	Cl_2F_3P	2.8129^{-45}	0.46501E+01	-0.80358E-02			172 - 229
Phosphorus(V) tetrachloride fluoride	Cl_4FP	2.6499^{-1}	0.33503E+01	-0.29651E-02			244 - 273
Phosphorus(V) trichloride difluoride	Cl_3F_2P	2.3752^{-5}	0.28905E+01	-0.19228E-02			215 - 268
Phosphoryl chloride	Cl_3OP	14.1^{20}					
Pinane	$C_{10}H_{18}$	2.1456^{25}					
α-Pinene	$C_{10}H_{16}$	2.1787^{25}					
β-Pinene	$C_{10}H_{16}$	2.4970^{25}					
Piperidine	$C_5H_{11}N$	4.33^{20}	0.82317E+01	-0.11229E-01	-0.71429E-05		293 - 333
Propanal	C_3H_6O	18.5^{17}					
Propane	C_3H_8	1.6678^{20}	0.22883E+01	-0.23276E-02	0.84710E-06		90 - 300
1,2-Propanediol	$C_3H_8O_2$	27.5^{30}	0.24546E+03	-0.15738E+01	0.38068E-02	-0.32544E-05	193 - 403
1,3-Propanediol	$C_3H_8O_2$	35.1^{20}	0.11365E+03	-0.36680E+00	0.33766E-03		288 - 328
1,2-Propanedithiol	$C_3H_8S_2$	7.24^{20}	0.14667E+02	-0.32660E-01	0.25000E-04		293 - 333
1,3-Propanedithiol	$C_3H_8S_2$	8.11^{30}	0.66607E+01	0.31310E-01	-0.87500E-04		303 - 343
Propanenitrile	C_3H_5N	29.7^{20}	0.82222E+02	-0.22937E+00	0.17424E-03		213 - 473
1-Propanethiol	C_3H_8S	5.937^{15}	0.11602E+02	-0.19580E-01			273 - 318
2-Propanethiol	C_3H_8S	5.952^{25}					
Propanoic acid	$C_3H_6O_2$	3.44^{25}	0.18793E+01	0.46841E-02	0.19983E-05		289 - 408
Propanoic anhydride	$C_6H_{10}O_3$	18.30^{20}					
1-Propanol	C_3H_8O	20.8^{20}	0.98045E+02	-0.36860E+00	0.36422E-03		193 - 493
2-Propanol	C_3H_8O	20.18^{20}	0.10416E+03	-0.41011E+00	0.42049E-03		193 - 493
Propargyl alcohol	C_3H_4O	20.8^{20}	0.99895E+02	-0.38911E+00	0.40776E-03		213 - 293
Propene	C_3H_6	2.1365^{-53}	0.29623E+01	-0.37564E-02			220 - 250
1-Propenylbenzene (unspecified isomer)	C_9H_{10}	2.73^{20}					
Propyl acetate	$C_5H_{10}O_2$	5.62^{20}	0.17677E+02	-0.61404E-01	0.69196E-04		253 - 353
Propylamine	C_3H_9N	5.08^{23}	0.17719E+02	-0.59022E-01	0.54780E-04		204 - 296
N-Propylaniline	$C_9H_{13}N$	5.48^{20}					
Propylbenzene	C_9H_{12}	2.370^{20}	0.26933E+01	0.21679E-03	-0.44643E-05		273 - 323
Propyl benzoate	$C_{10}H_{12}O_2$	5.78^{30}	0.10927E+02	-0.20535E-01	0.11745E-04		303 - 358
Propyl butanoate	$C_7H_{14}O_2$	4.3^{20}					
Propyl carbamate	$C_4H_9NO_2$	12.06^{65}	0.24356E+02	-0.36400E-01			338 - 378
Propyl chlorocarbonate	$C_4H_7ClO_2$	11.2^{20}					
Propyl *trans*-cinnamate	$C_{12}H_{14}O_2$	5.45^{20}					
Propylene carbonate	$C_4H_6O_3$	66.14^{20}	0.15940E+03	-0.39530E+00	0.26284E-03		273 - 333
Propyl formate	$C_4H_8O_2$	6.92^{30}					
4-Propylheptane	$C_{10}H_{22}$	1.9955^{20}					
Propyl nitrite	$C_3H_7NO_2$	12.35^{-23}	0.70552E+02	-0.40362E+00	0.66687E-03		110 - 310
Propyl pentanoate	$C_8H_{16}O_2$	4.0^{19}					
N-Propylpropanamide	$C_6H_{13}NO$	118.1^{25}	0.58846E+03	-0.22012E+01	0.20870E-02		298 - 328
Propyl propanoate	$C_6H_{12}O_2$	5.249^{20}					

Name	Mol. form.	ε	a	b	c	d	Range/K
Propyl silicate	$C_{12}H_{28}O_4Si$	3.21^{25}					
Propyl trichloroacetate	$C_5H_7Cl_3O_2$	8.32^{25}					
Propyne	C_3H_4	3.218^{-27}	0.60871E+01	-0.11730E-01			185 - 246
Pyrazine	$C_4H_4N_2$	2.80^{50}					
Pyridine	C_5H_5N	13.260^{20}	0.43991E+02	-0.15150E+00	0.15925E-03		293 - 323
2-Pyridinecarbonitrile	$C_6H_4N_2$	93.77^{30}	0.45596E+03	-0.17746E+01	0.19105E-02		303 - 398
3-Pyridinecarbonitrile	$C_6H_4N_2$	20.54^{50}	0.60484E+02	-0.17280E+00	0.15218E-03		323 - 398
4-Pyridinecarbonitrile	$C_6H_4N_2$	5.23^{80}	0.12533E+02	-0.30115E-01	0.26674E-04		353 - 398
Pyridine-1-oxide	C_5H_5NO	35.94^{70}	0.20878E+02	0.16450E+00	-0.35269E-03		343 - 398
Pyrocatechol	$C_6H_6O_2$	17.57^{115}	0.74930E+02	-0.22142E+00	0.18919E-03		388 - 463
Pyrrole	C_4H_5N	8.00^{20}	0.12672E+02	-0.14075E-01	-0.62671E-05		293 - 357
Pyrrolidine	C_4H_9N	8.30^{20}	0.38191E+02	-0.15462E+00	0.17941E-03		274 - 333
2-Pyrrolidone	C_4H_7NO	28.18^{25}	0.11054E+03	-0.47945E+00	0.68182E-03		298 - 338
Quinoline	C_9H_7N	9.16^{20}	0.33432E+02	-0.13497E+00	0.17788E-03		258 - 323
Resorcinol	$C_6H_6O_2$	13.55^{120}	0.30252E+02	-0.56443E-01	0.35578E-04		393 - 463
Salicylaldehyde	$C_7H_6O_2$	18.35^{20}	0.51315E+02	-0.15379E+00	0.14111E-03		289 - 453
Selenium (gray)	Se	5.44^{237}	0.67569E+01	-0.25829E-02			511 - 575
Selenium oxychloride	Cl_2OSe	46.2^{20}					
Stearic acid	$C_{18}H_{36}O_2$	2.314^{20}	0.27159E+01	-0.13300E-02			293 - 373
Styrene	C_8H_8	2.4737^{20}	0.44473E+01	-0.11422E-01	0.16000E-04		293 - 313
Succinonitrile	$C_4H_4N_2$	62.6^{25}	0.17724E+03	-0.54654E+00	0.54046E-03		236 - 351
N-Sulfinylaniline	C_6H_5NOS	6.97^{25}					
Sulfur (rhombic)	S	3.4991^{134}	0.51651E+01	-0.77381E-02	0.89120E-05		407 - 479
Sulfur chloride [ClSSCl]	Cl_2S_2	4.79^{15}					
Sulfur decafluoride	$F_{10}S_2$	2.0202^{20}					
Sulfur dichloride	Cl_2S	2.915^{25}					
Sulfur dioxide	O_2S	14.3^{20}	0.52045E+02	-0.16125E+00	0.11042E-03		213 - 449
Sulfur hexafluoride	F_6S	1.81^{-50}					
Sulfur trioxide (α-form)	O_3S	3.11^{18}					
Sulfuryl chloride	Cl_2O_2S	9.1^{20}					
α-Terpinene	$C_{10}H_{16}$	2.4526^{25}					
γ-Terpinene	$C_{10}H_{16}$	2.2738^{25}					
Terpinolene	$C_{10}H_{16}$	2.2918^{25}					
1,1,2,2-Tetrabromoethane	$C_2H_2Br_4$	6.72^{30}	0.16246E+02	-0.31500E-01			303 - 333
Tetrabutylstannane	$C_{16}H_{36}Sn$	9.74^{20}	0.56115E+02	-0.24812E+00	0.30682E-03		293 - 313
1,1,2,2-Tetrachloro-1,2-difluoroethane	$C_2Cl_4F_2$	2.52^{35}					
1,2,3,4-Tetrachloro-5,6-dimethylbenzene	$C_8H_6Cl_4$	8.0^{20}					
1,2,3,5-Tetrachloro-4,6-dimethylbenzene	$C_8H_6Cl_4$	5.4^{20}					
1,1,1,2-Tetrachloroethane	$C_2H_2Cl_4$	9.22^{-66}	0.19606E+02	-0.49847E-01			207 - 233
1,1,2,2-Tetrachloroethane	$C_2H_2Cl_4$	8.50^{20}					
Tetrachloroethene	C_2Cl_4	2.268^{30}					
Tetrachloromethane	CCl_4	2.2379^{20}	0.28280E+01	-0.20339E-02	0.71795E-07		283 - 333
Tetrachlorosilane	Cl_4Si	2.248^{0}	0.58041E+01	-0.27129E-01	0.51678E-04		207 - 273
Tetradecane	$C_{14}H_{30}$	2.0343^{20}	0.23832E+01	-0.11900E-02	-0.51229E-16		283 - 363
1-Tetradecanol	$C_{14}H_{30}O$	4.42^{45}	0.12272E+02	-0.24667E-01	-0.13168E-13		318 - 358
Tetradecylamine	$C_{14}H_{31}N$	2.90^{39}					
Tetraethylene glycol	$C_8H_{18}O_5$	20.44^{20}	0.83547E+02	-0.31691E+00	0.34689E-03		253 - 333
Tetraethylene glycol dimethyl ether	$C_{10}H_{22}O_5$	7.68^{25}					
Tetraethylenepentamine	$C_8H_{23}N_5$	9.40^{20}	0.40553E+02	-0.16681E+00	0.20659E-03		213 - 333
Tetraethylsilane	$C_8H_{20}Si$	2.090^{20}					
Tetraethylstannane	$C_8H_{20}Sn$	2.241^{20}					
Tetraethylurea	$C_9H_{20}N_2O$	14.29^{24}	0.52820E+02	-0.18790E+00	0.19580E-03		205 - 411
Tetrafluoromethane	CF_4	1.685^{-147}	0.20350E+01	-0.27616E-02			126 - 142
2,2,3,3-Tetrafluoro-1-propanol	$C_3H_4F_4O$	21.03^{25}					
Tetrahydrofuran	C_4H_8O	7.52^{22}	0.30739E+02	-0.12946E+00	0.17195E-03		224 - 295
Tetrahydrofurfuryl alcohol	$C_5H_{10}O_2$	13.48^{30}					
1,2,3,4-Tetrahydronaphthalene	$C_{10}H_{12}$	2.771^{25}	0.29172E+01	0.12832E-02	-0.59453E-05		298 - 343
1,2,3,4-Tetrahydro-2-naphthol	$C_{10}H_{12}O$	11.70^{20}	0.98978E+02	-0.48267E+00	0.63008E-03		293 - 363
Tetrahydropyran	$C_5H_{10}O$	5.66^{20}	0.19793E+02	-0.76071E-01	0.94852E-04		234 - 333
Tetrakis(methylthio)methane	$C_5H_{12}S_4$	2.818^{70}					
Tetramethoxymethane	$C_5H_{12}O_4$	2.40^{20}					

Name	Mol. form.	ε	a	b	c	d	Range/K
1,2,3,4-Tetramethylbenzene	$C_{10}H_{14}$	2.538^{23}	0.33822E+01	-0.33630E-02	0.17475E-05		273 - 412
1,2,4,5-Tetramethylbenzene	$C_{10}H_{14}$	2.223^{83}	0.26834E+01	-0.10327E-02	-0.73533E-06		356 - 430
1,1,3,3-Tetramethylguanidine	$C_5H_{13}N_3$	11.5^{25}					
Tetramethylsilane	$C_4H_{12}Si$	1.921^{20}					
Tetramethylurea	$C_5H_{12}N_2O$	23.10^{20}					
Tetranitromethane	CN_4O_8	2.317^{20}					
Tetrapropylstannane	$C_{12}H_{28}Sn$	2.267^{20}					
Thioacetic acid	C_2H_4OS	14.30^{25}					
Thionyl bromide	Br_2OS	9.06^{20}					
Thionyl chloride	Cl_2OS	8.675^{25}					
Thiophene	C_4H_4S	2.739^{20}	0.32941E+01	-0.19019E-02			253 - 293
Thymol	$C_{10}H_{14}O$	4.259^{60}					
Tin(IV) bromide	Br_4Sn	3.169^{30}	0.50001E+01	-0.60383E-02			304 - 316
Tin(IV) chloride	Cl_4Sn	3.014^{0}	0.43951E+01	-0.48805E-02			234 - 273
Titanium(IV) chloride	Cl_4Ti	2.843^{-16}	0.33668E+01	-0.19675E-02			237 - 257
Toluene	C_7H_8	2.379^{23}	0.32584E+01	-0.34410E-02	0.15937E-05		207 - 316
Toluene-2,4-diisocyanate	$C_9H_6N_2O_2$	8.433^{20}	0.22174E+02	-0.66982E-01	0.68571E-04		293 - 353
p-Toluenesulfonyl chloride	$C_7H_7ClO_2S$	22.6^{70}					
Triacontane	$C_{30}H_{62}$	1.9112^{100}					
Tribromoacetaldehyde	C_2HBr_3O	7.6^{20}					
Tribromochloromethane	CBr_3Cl	2.601^{60}					
Tribromofluoromethane	CBr_3F	3.00^{20}	0.53203E+01	-0.11061E-01	0.10688E-04		206 - 323
Tribromomethane	$CHBr_3$	4.404^{10}	0.71707E+01	-0.98000E-02			283 - 343
Tribromonitromethane	CBr_3NO_2	9.034^{25}	0.16079E+02	-0.23630E-01			298 - 328
1,2,3-Tribromopropane	$C_3H_5Br_3$	6.00^{30}	0.11024E+02	-0.16596E-01			303 - 358
Tributylamine	$C_{12}H_{27}N$	2.340^{20}	0.19846E+01	0.28108E-02	-0.54545E-05		233 - 293
Tributyl borate	$C_{12}H_{27}BO_3$	2.23^{20}					
Tributyl phosphate	$C_{12}H_{27}O_4P$	8.34^{20}	0.26304E+02	-0.88480E-01	0.92857E-04		293 - 373
Tributyrin	$C_{15}H_{26}O_6$	5.72^{10}	0.13152E+02	-0.36684E-01	0.36795E-04		199 - 283
Trichloroacetaldehyde	C_2HCl_3O	6.8^{25}					
Trichloroacetic acid	$C_2HCl_3O_2$	4.34^{60}	0.13412E+01	0.90000E-02	-0.24130E-14		333 - 393
Trichloroacetic anhydride	$C_4Cl_6O_3$	5.0^{25}					
Trichloroacetonitrile	C_2Cl_3N	7.85^{19}					
1,2,2-Trichloro-1,1-difluoroethane	$C_2HCl_3F_2$	4.01^{30}	0.75423E+01	-0.11667E-01			303 - 333
1,1,1-Trichloroethane	$C_2H_3Cl_3$	7.243^{20}	0.27705E+02	-0.10621E+00	0.12424E-03		258 - 318
1,1,2-Trichloroethane	$C_2H_3Cl_3$	7.1937^{25}	0.17147E+02	-0.33371E-01			288 - 318
Trichloroethene	C_2HCl_3	3.390^{28}	0.58319E+01	-0.80828E-02			302 - 338
Trichlorofluoromethane	CCl_3F	3.00^{20}	0.53203E+01	-0.11061E-01	0.10688E-04		206 - 323
Trichloromethane	$CHCl_3$	4.8069^{20}	0.15115E+02	-0.51830E-01	0.56803E-04		218 - 323
Trichloromethane-*d*	$CDCl_3$	4.67^{25}					
Trichloronitromethane	CCl_3NO_2	7.319^{20}	0.14403E+02	-0.24178E-01			276 - 333
1,2,3-Trichloropropane	$C_3H_5Cl_3$	7.5^{20}					
Tri-*o*-cresyl phosphate	$C_{21}H_{21}O_4P$	6.7^{25}					
Tridecane	$C_{13}H_{28}$	2.0213^{20}	0.23731E+01	-0.12000E-02	-0.21841E-15		283 - 363
1-Tridecanol	$C_{13}H_{28}O$	4.02^{60}					
7-Tridecanone	$C_{13}H_{26}O$	7.6^{30}					
1-Tridecene	$C_{13}H_{26}$	2.139^{20}	0.14154E+01	0.66514E-02	-0.14286E-04		273 - 323
Triethoxymethane	$C_7H_{16}O_3$	4.779^{20}					
Triethoxymethylsilane	$C_7H_{18}O_3Si$	3.845^{25}					
Triethylamine	$C_6H_{15}N$	2.418^{20}	0.29205E+01	-0.14007E-02	-0.13469E-05		233 - 323
1,3,5-Triethylbenzene	$C_{12}H_{18}$	2.256^{20}					
Triethylborane	$C_6H_{15}B$	1.974^{20}					
Triethylene glycol	$C_6H_{14}O_4$	23.69^{20}	0.91845E+02	-0.33827E+00	0.36062E-03		253 - 333
Triethylene glycol dimethyl ether	$C_8H_{18}O_4$	7.62^{25}					
Triethyl phosphate	$C_6H_{15}O_4P$	13.20^{25}	0.61230E+02	-0.26047E+00	0.33333E-03		298 - 333
Triethylphosphine oxide	$C_6H_{15}OP$	35.5^{50}					
Triethylphosphine sulfide	$C_6H_{15}PS$	39.0^{98}					
Triethylsilane	$C_6H_{16}Si$	2.323^{20}					
Trifluoroacetic acid	$C_2HF_3O_2$	8.42^{20}	0.21652E+02	-0.68146E-01	0.78571E-04		263 - 323
Trifluoroacetic acid anhydride	$C_4F_6O_3$	2.7^{25}					
2,2,2-Trifluoroethanol	$C_2H_3F_3O$	27.68^{20}	0.90593E+02	-0.21421E+00			293 - 318

Fluids

Name	Mol. form.	*ε*	*a*	*b*	*c*	*d*	Range/K
Trifluoromethane	CHF_3	5.2^{21}	0.11442E+03	-0.75600E+00	0.13562E-02		130 - 263
(Trifluoromethyl)benzene	$C_7H_5F_3$	9.22^{25}					
Trihexyl borate	$C_{18}H_{39}BO_3$	2.22^{20}					
Triisopentylamine	$C_{15}H_{33}N$	2.29^{21}					
Trimethoxymethylsilane	$C_4H_{12}O_3Si$	4.9^{25}					
Trimethylamine	C_3H_9N	2.440^{25}	0.39745E+01	-0.51331E-02			273 - 298
1,2,3-Trimethylbenzene	C_9H_{12}	2.656^{20}	0.76006E+01	-0.29118E-01	0.41786E-04		273 - 323
1,2,4-Trimethylbenzene	C_9H_{12}	2.377^{20}	0.31517E+01	-0.30634E-02	0.14286E-05		273 - 323
1,3,5-Trimethylbenzene	C_9H_{12}	2.279^{20}	0.38998E+01	-0.88072E-02	0.11149E-04		288 - 358
Trimethyl borate	$C_3H_9BO_3$	2.2762^{20}					
2,2,3-Trimethylbutane	C_7H_{16}	1.930^{20}					
Trimethylchlorosilane	C_3H_9ClSi	10.21^{0}	-0.19492E+02	0.29806E+00	-0.69284E-03		223 - 273
2,4,6-Trimethyl-3-heptene (unspecified isomer)	$C_{10}H_{20}$	2.293^{20}					
2,2,3-Trimethylpentane	C_8H_{18}	1.960^{20}					
2,2,4-Trimethylpentane	C_8H_{18}	1.943^{20}	0.23677E+01	-0.14768E-02	0.94261E-07		173 - 373
2,3,3-Trimethylpentane	C_8H_{18}	1.9780^{20}					
2,3,4-Trimethylpentane	C_8H_{18}	1.9738^{20}					
2,4,4-Trimethyl-1-pentene	C_8H_{16}	2.0908^{25}					
Trimethylphenoxysilane	$C_9H_{14}OSi$	3.3953^{25}					
Trimethylphenylsilane	$C_9H_{14}Si$	2.3533^{25}	0.21463E+01	0.32711E-02	-0.86264E-05		288 - 323
Trimethyl phosphate	$C_3H_9O_4P$	20.6^{20}					
2,4,6-Trimethylpyridine	$C_8H_{11}N$	7.807^{25}	0.20990E+02	-0.57419E-01	0.44286E-04		298 - 358
Trinitroglycerol	$C_3H_5N_3O_9$	19.25^{20}					
2,4,6-Trinitrophenol	$C_6H_3N_3O_7$	4.0^{21}					
1,3,5-Trioxane	$C_3H_6O_3$	15.55^{65}					
Triphenylmethane	$C_{19}H_{16}$	2.46^{94}	0.40201E+01	-0.66507E-02	0.65329E-05		367 - 448
Tripropylamine	$C_9H_{21}N$	2.380^{20}	0.33380E+01	-0.86332E-02	0.18322E-04		243 - 293
Tripropylborane	$C_9H_{21}B$	2.026^{20}					
Tripropyl phosphate	$C_9H_{21}O_4P$	10.93^{20}	0.33166E+02	-0.10514E+00	0.10000E-03		293 - 373
Tris(perfluorobutyl)amine	$C_{12}F_{27}N$	2.15^{20}					
Undecane	$C_{11}H_{24}$	1.9972^{20}	0.23637E+01	-0.12500E-02	-0.85869E-16		283 - 363
1-Undecanol	$C_{11}H_{24}O$	5.98^{40}					
2-Undecanone	$C_{11}H_{22}O$	8.3^{12}					
1-Undecene	$C_{11}H_{22}$	2.137^{20}	0.22132E+01	0.13121E-02	-0.53571E-05		273 - 323
Undecylamine	$C_{11}H_{25}N$	3.25^{20}	0.54945E+01	-0.96161E-02	0.66017E-05		293 - 373
Vanadium(IV) chloride	Cl_4V	3.05^{25}					
Vanadyl tribromide	Br_3OV	3.6^{25}	0.61112E+01	-0.84211E-02			203 - 298
Vanadyl trichloride	Cl_3OV	3.4^{25}					
2-Vinylpyridine	C_7H_7N	9.126^{20}					
4-Vinylpyridine	C_7H_7N	10.50^{20}					
Water	H_2O	80.1^{20}	0.24921E+03	-0.79069E+00	0.72997E-03		273 - 372
Xenon	Xe	1.880^{-112}					
Xenon hexafluoride	F_6Xe	4.10^{55}					
o-Xylene	C_8H_{10}	2.562^{20}	0.36163E+01	-0.40177E-02	0.14286E-05		273 - 323
m-Xylene	C_8H_{10}	2.359^{20}	0.28421E+01	-0.10191E-02	-0.21429E-05		273 - 323
p-Xylene	C_8H_{10}	2.2735^{20}	0.23140E+01	0.97221E-03	-0.37500E-05		293 - 363
2,3-Xylenol	$C_8H_{10}O$	4.81^{70}	0.14399E+02	-0.41438E-01	0.39244E-04		343 - 433
2,4-Xylenol	$C_8H_{10}O$	5.060^{30}	0.22125E+02	-0.85543E-01	0.96548E-04		303 - 363
2,5-Xylenol	$C_8H_{10}O$	5.36^{65}	0.18049E+02	-0.54991E-01	0.51656E-04		338 - 455
2,6-Xylenol	$C_8H_{10}O$	4.90^{40}	0.12284E+02	-0.32996E-01	0.29867E-04		313 - 453
3,4-Xylenol	$C_8H_{10}O$	9.02^{60}	0.54423E+02	-0.21153E+00	0.22508E-03		333 - 453
3,5-Xylenol	$C_8H_{10}O$	9.06^{50}	0.54251E+02	-0.21647E+00	0.23542E-03		323 - 453
Xylitol	$C_5H_{12}O_5$	40.0^{20}					

PERMITTIVITY (DIELECTRIC CONSTANT) OF GASES

This table gives the relative permittivity ε (often called the dielectric constant) of some common gases at a temperature of 20 °C and pressure of one atmosphere (101.325 kPa). Values of the permanent dipole moment μ in Debye units (1 D = 3.33564×10^{-30} C m) are also included.

The density dependence of the permittivity is given by the equation

$$\frac{\varepsilon - 1}{\varepsilon + 2} = \rho_m \left(\frac{4\pi N\alpha}{3} + \frac{4\pi N\mu^2}{9kT} \right)$$

where ρ_m is the molar density, N is Avogadro's number, k is the Boltzmann constant, T is the temperature, and α is the molecular polarizability. Therefore, in regions where the gas can be considered ideal, $\varepsilon - 1$ is approximately proportional to the pressure at constant temperature. For nonpolar gases ($\mu = 0$), $\varepsilon - 1$ is inversely proportional to temperature at constant pressure.

The number of significant figures indicates the accuracy of the values given. The values of ε for air, Ar, H_2, He, N_2, O_2, and CO_2 are recommended as reference values; these are accurate to 1 ppm or better.

The second part of the table gives the permittivity of water vapor in equilibrium with liquid water as a function of temperature (derived from Ref. 4).

References

1. Maryott, A. A., and Buckley, F., *Table of Dielectric Constants and Electric Dipole Moments of Substances in the Gaseous State*, National Bureau of Standards Circular 537, 1953.
2. Harvey, A. H., and Lemmon, E. W., *Int. J. Thermophys.* 26, 31, 2005 [for nonpolar gases and light hydrocarbons]
3. *Landolt-Börnstein, Numerical Data and Functional Relationships in Science and Technology*, New Series, Group IV, Vol. 4, Springer-Verlag, Heidelberg, 1980 (for data at high pressures).
4. Fernández, D. P., Goodwin, A. R. H., Lemmon, E. W., Levelt Sengers, J. M. H., and Williams, R. C., *J. Phys. Chem. Ref. Data* 26, 1125, 1997 [for water vapor]

Permittivity of Selected Gases at 20 °C and 1 Atmosphere

Name	Formula	ε	μ/D
Compounds not containing carbon			
Air[a]		1.0005360	0
Ammonia	NH_3	1.00622	1.4718
Argon	Ar	1.0005169	0
Boron trifluoride	BF_3	1.0011	0
Hydrogen	H_2	1.0002532	0
Hydrogen bromide	HBr	1.00279	0.8272
Hydrogen chloride	HCl	1.00390	1.1086
Hydrogen iodide	HI	1.00214	0.448
Hydrogen sulfide	H_2S	1.00344	0.97833
Helium	He	1.0000645	0
Krypton	Kr	1.000784	0
Nitric oxide	NO	1.00060	0.15872
Nitrogen	N_2	1.0005474	0
Nitrogen trifluoride	NF_3	1.0013	0.235
Nitrous oxide	N_2O	1.00104	0.16083
Neon	Ne	1.000124	0
Oxygen	O_2	1.0004941	0
Ozone	O_3	1.0017	0.53373
Sulfur dioxide	SO_2	1.00825	1.63305
Sulfur hexafluoride	SF_6	1.00200	0
Xenon	Xe	1.00127	0
Compounds containing carbon			
Acetylene	HC≡CH	1.00124	0
Butane	C_4H_{10}	1.00266	0.05
Bromomethane	CH_3Br	1.01028	1.8203
Carbon dioxide	CO_2	1.0009217	0
Carbon monoxide	CO	1.00065	0.10980
Chloroethane	C_2H_5Cl	1.01325	2.05
Chloroethene	CH_2=CHCl	1.0075	1.45
Chloromethane	CH_3Cl	1.01080	1.8963
Cyclopropane	C_3H_6	1.00178	0
Dimethyl ether	CH_3OCH_3	1.0062	1.30
Ethane	C_2H_6	1.001403	0
Ethylene	CH_2=CH_2	1.00135	0
Fluoromethane	CH_3F	1.00973	1.858
Iodomethane	CH_3I	1.00914	1.6406
Isobutane	C_4H_{10}	1.00268	0.132
Methane	CH_4	1.0008181	0
Propane	C_3H_8	1.002032	0.084
Propene	CH_3CH=CH_2	1.00228	0.366
Tetrafluoromethane	CF_4	1.00121	0

[a] Dry, CO_2 free

Permittivity of Saturated Water Vapor

t/°C	ε
0	1.000064
10	1.000120
20	1.000214
30	1.000363
40	1.000594
50	1.000935
60	1.00143
70	1.00211
80	1.00304
90	1.00428
100	1.00589

AZEOTROPIC DATA FOR BINARY MIXTURES

J. Gmehling, J. Menke, J. Krafczyk, K. Fischer, J.-C. Fontaine, and H. V. Kehiaian

Binary homogeneous (single-phase) liquid mixtures having an extremum (maximum or minimum) vapor pressure P at constant temperature T, as a function of composition, are called azeotropic mixtures, or simply azeotropes. The composition is usually expressed as mole fractions, where x_1 for component 1 in the liquid phase and y_1 for component 1 in the vapor phase are identical. Mixtures that do not show a maximum or minimum are called zeotropic. A maximum (minimum) of the $P(x_1)$ or $P(y_1)$ curves corresponds to a minimum (maximum) of the boiling temperature T at constant P, plotted as a function of x_1 or y_1 [see $T(x_1)$ and $T(y_1)$ curves, Types I and III, in Fig. 1]. Azeotropes in which the pressure is a maximum (temperature is a minimum) are often called positive azeotropes, while pressure-minimum (temperature-maximum) azeotropes are called negative azeotropes. The coordinates of an azeotropic point are the azeotropic temperature T_{Az}, pressure P_{Az}, and the vapor-phase composition $y_{1,Az}$, which is the same as the liquid-phase composition $x_{1,Az}$.

In the two-phase liquid-liquid region of partially miscible (heterogeneous) mixtures, the vapor pressure at constant T (or the boiling temperature at constant P) is independent of the global composition x_1 of the two coexisting liquid phases between the equilibrium compositions x_1' and x_1'' ($x_1' < x_1''$).

The constant vapor pressure (boiling temperature) above the two-phase region of certain partially miscible mixtures is usually larger (smaller) than the vapor pressure (boiling temperature) at any other liquid-phase composition in the homogeneous region. In this case, the vapor-phase composition is inside the miscibility gap. Mixtures of this type are called heteroazeotropic mixtures, or simply heteroazeotropes. (Fig. 1, Type II), as opposed to the other types of azeotropes, called homoazeotropes.

Only in a few cases partially miscible mixtures present a positive or negative azeotropic point in the single-phase region, outside the miscibility gap, similar to the azeotropic points of homogeneous mixtures (Fig. 1, Types IV and VI).

A few binary mixtures, for example the system perfluorobenzene + benzene, may present two azeotropic points at constant temperature (pressure), a positive and a negative one. They are called double azeotropic mixtures, or simply double azeotropes. (Fig. 1, Type V).

The knowledge of the occurrence of azeotropic points in binary and higher systems is of special importance for the design of distillation processes. The number of theoretical stages of a distillation column required for the separation depends on the separation factor α_{12}, i.e., the ratio of the K_i-factors ($K_i = y_i/x_i$) of the components i (i = 1, 2). The required separation factor can be calculated with the following simplified relation (Ref. 1)

$$\alpha_{12} = K_1/K_2 = (y_1/x_1)/(y_2/x_2) = (\gamma_1 P_1^s)/(\gamma_2 P_2^s) \quad (1)$$

where γ_i is the activity coefficient of component i in the liquid phase and P_i^s is the vapor pressure of the pure component i.

In distillation processes, only the difference between the separation factor and unity ($\alpha_{12} - 1$) can be exploited for the separation. If the separation factor is close to unity, a large number of theoretical stages is required for the separation. If the binary system to be separated shows an azeotropic point ($\alpha_{12} = 1$), the separation is impossible by ordinary distillation, even with an infinitely large number of stages.

Following eq. (1) azeotropic behavior will always occur in homogeneous binary systems when the vapor pressure ratio P_1^s/P_2^s is equal to the ratio of the activity coefficients γ_2/γ_1.

Various thermodynamic methods based on g^E -models (Wilson, NRTL, UNIQUAC) or group contribution methods (UNIFAC, modified UNIFAC, ASOG, PSRK) can be used for either calculating or predicting the required activity coefficients for the components under given conditions of temperature and composition (Ref. 2).

Because of the importance of azeotropic data for the design of distillation processes, compilations have been available in book form for quite some time (Refs. 3-7). The most recent printed data collection was published in 1994 (Ref 8). A revised and extended version appeared in 2004 (Ref. 9).

A collection of approximately 47,400 zeotropic and azeotropic data sets, compiled from 6600 references, are stored in a comprehensive computerized data bank (Ref. 10). The references from the above-mentioned compilations and from the vapor-liquid equilibrium part of the Dortmund Data Bank (Ref. 11) were supplemented by references found from CAS online searches, private communications, data from industry, etc. Over 24,000 zeotropic data and over 20,000 azeotropic data are available for binary systems. Nearly 90% of the binary azeotropic data show a pressure maximum. In most cases (ca. 90%) these are homogeneous azeotropes, and in approximately 7–8% of the cases heterogeneous azeotropes are reported. Less than 10% of the data stored show a pressure minimum. Approximately 21,000 of the data sets stored were published after 1970.

The table below provides information about azeotropes for selected binary systems. Mixtures are listed alphabetically by the name of the first component, followed by the name of the second component. The azeotropic coordinates: temperature T_{Az}, pressure P_{Az}, and vapor-phase composition $y_{1,Az}$ are given for each mixture. The explanation of the type of azeotrope is given by the following codes:

O: homogeneous azeotrope in a completely miscible system
L: homogeneous azeotrope in a partially miscible system
E: heterogeneous azeotrope
X: pressure maximum
N: pressure minimum
D: double azeotrope
C: system contains a supercritical compound

References

1. Gmehling, J. and Brehm, A., *Grundoperationen*, Thieme-Verlag, Stuttgart, 1996.
2. Gmehling, J. and Kolbe, B., *Thermodynamik*, VCH-Verlag, Weinheim, 1992.
3. Lecat, M., *Doctoral Dissertation*, 1908.
4. Lecat, M., *L'Azeotropisme*, Monograph, L'Auteur, Brussel, 1918.
5. Lecat, M., *Tables Azeotropiques*, Monograph, Lamertin, Brussel 1949.
6. Ogorodnikov, S. K., Lesteva, T. M., and Kogan V. B., *Azeotropic Mixtures*, Khimia, Leningrad, 1971.
7. Horsley, L. H., *Azeotropic Data III*, American Chemical Society, Washington, DC, 1973.
8. Gmehling, J., Menke, J., Krafczyk, J., and Fischer, K., *Azeotropic Data*, 2 Volumes, VCH Verlag, Weinheim, 1994.
9. Gmehling, J., Menke, J., Krafczyk, J., and Fischer, K., *Azeotropic Data*, 2nd Ed., 3 Volumes, VCH Verlag, Weinheim, 2004.
10. Gmehling, J., Menke, J., Krafczyk, J., and Fischer, K., *A Data Bank for Azeotropic Data, Status and Applications, Fluid Phase Equilib.* 103, 51, 1995.
11. Dortmund Data Bank, <http://www.ddbst.de>.

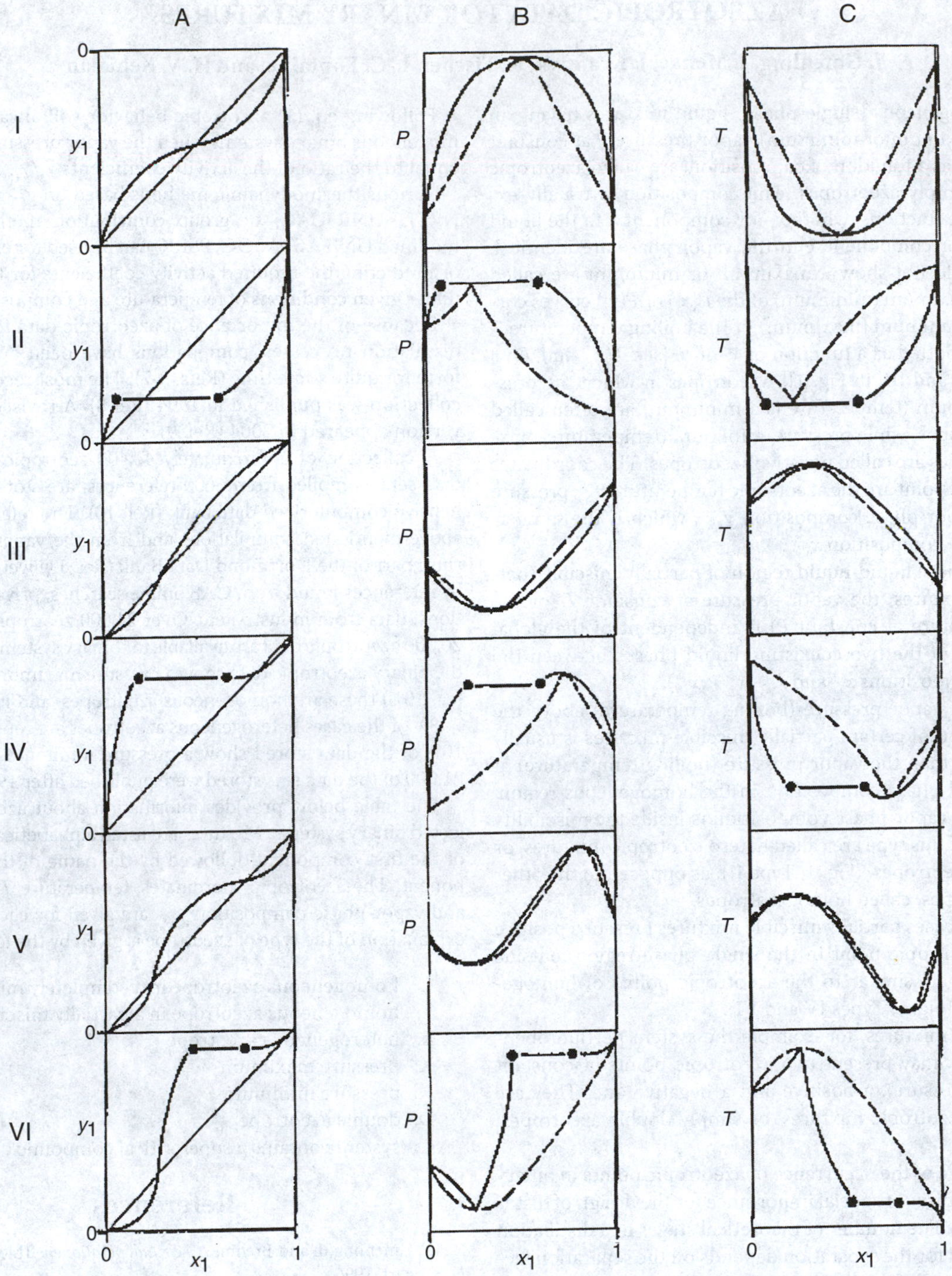

FIGURE 1 Different types of binary azeotropic systems: I — homogeneous pressure-maximum azeotrope in a completely miscible system (OX); II — heterogeneous pressure-maximum azeotrope (EX); III — homogeneous pressure-minimum azeotrope in a completely miscible system (ON); IV — homogeneous pressure-maximum azeotrope in a partially miscible system (LX); V–D: double azeotrope (OND, OXD); VI — homogeneous pressure-minimum azeotrope in a partially miscible system (LN). A — $y_1(x_1)$; B — $P(x_1)$ and $P(y_1)$; C — $T(x_1)$ and $T(y_1)$. Continuous line — (x_1); Dashed line — (y_1).

Azeotropic Temperature, Pressure, and Composition

Component 1	Mol. form. 1	Component 2	Mol. form. 2	T_{AZ}/K	$y_{1,AZ}$	P_{AZ}/kPa	Type
Acetaldehyde	C_2H_4O	1,3-Butadiene	C_4H_6	268.15	0.0520	101.33	OX
Acetaldehyde	C_2H_4O	2-Methyl-1,3-butadiene	C_5H_8	292.23	0.8140	101.33	OX
Acetic acid	$C_2H_4O_2$	Decane	$C_{10}H_{22}$	390.05	0.9250	101.33	OX
Acetic acid	$C_2H_4O_2$	2,4-Dimethylpyridine	C_7H_9N	435.45	0.3022	101.33	ON
Acetic acid	$C_2H_4O_2$	Heptane	C_7H_{16}	364.95	0.4490	101.33	OX
Acetic acid	$C_2H_4O_2$	Hexane	C_6H_{14}	341.40	0.0839	101.33	OX

Component 1	Mol. form. 1	Component 2	Mol. form. 2	T_{AZ}/K	$y_{1,AZ}$	P_{AZ}/kPa	Type
Acetic acid	$C_2H_4O_2$	3-Methyl-2-butanol, (±)-	$C_5H_{12}O$	392.65	0.7210	101.33	ON
Acetic acid	$C_2H_4O_2$	2-Methylpyridine	C_6H_7N	417.27	0.5120	101.33	ON
Acetic acid	$C_2H_4O_2$	Nonane	C_9H_{20}	386.05	0.8250	101.33	OX
Acetic acid	$C_2H_4O_2$	Octane	C_8H_{18}	378.85	0.6870	101.33	OX
Acetic acid	$C_2H_4O_2$	Pyridine	C_5H_5N	411.25	0.5780	101.33	ON
Acetic acid	$C_2H_4O_2$	Undecane	$C_{11}H_{24}$	391.15	0.9720	101.33	OX
Acetic acid	$C_2H_4O_2$	Vinyl butanoate	$C_6H_{10}O_2$	386.45	0.5750	101.33	OX
Acetic acid	$C_2H_4O_2$	*o*-Xylene	C_8H_{10}	389.75	0.8640	101.33	OX
Acetic acid	$C_2H_4O_2$	*p*-Xylene	C_8H_{10}	388.40	0.8200	101.33	OX
Acetic anhydride	$C_4H_6O_3$	Octane	C_8H_{18}	397.65	0.3500	129.80	OX
Acetic anhydride	$C_4H_6O_3$	1-Octene	C_8H_{16}	367.53	0.2840	53.88	OX
Acetone	C_3H_6O	1-Bromopropane	C_3H_7Br	328.75	0.9915	99.75	OX
Acetone	C_3H_6O	2-Chloro-2-methylpropane	C_4H_9Cl	322.05	0.1944	102.11	OX
Acetone	C_3H_6O	Cyclohexane	C_6H_{12}	330.05	0.7590	109.32	OX
Acetone	C_3H_6O	Diisopropyl ether	$C_6H_{14}O$	327.10	0.7424	100.17	OX
Acetone	C_3H_6O	Hexane	C_6H_{14}	322.95	0.6480	101.33	OX
Acetone	C_3H_6O	1-Hexene	C_6H_{12}	323.35	0.5973	101.40	OX
Acetone	C_3H_6O	Isopentane	C_5H_{12}	298.75	0.1730	101.33	OX
Acetone	C_3H_6O	Methanol	CH_4O	328.29	0.7600	101.33	OX
Acetone	C_3H_6O	Methyl acetate	$C_3H_6O_2$	328.85	0.6470	101.33	OX
Acetone	C_3H_6O	2-Methyl-1,3-butadiene	C_5H_8	306.95	0.0610	101.33	OX
Acetone	C_3H_6O	2-Methyl-1-butene	C_5H_{10}	303.25	0.1400	101.33	OX
Acetone	C_3H_6O	2-Methyl-2-butene	C_5H_{10}	308.75	0.2440	101.33	OX
Acetone	C_3H_6O	Methyl *tert*-butyl ether	$C_5H_{12}O$	324.35	0.4824	102.19	OX
Acetone	C_3H_6O	1-Methylcyclobutene	C_5H_8	307.75	0.2220	101.33	OX
Acetone	C_3H_6O	Methylcyclohexane	C_7H_{14}	318.15	0.9500	68.66	OX
Acetone	C_3H_6O	Methylenecyclobutane	C_5H_8	311.25	0.2800	101.33	OX
Acetone	C_3H_6O	2-Methyl-1-pentene	C_6H_{12}	333.40	0.5793	140.60	OX
Acetone	C_3H_6O	Tetrachloromethane	CCl_4	341.25	0.9663	149.93	OX
Acetone	C_3H_6O	Tetrahydrofuran	C_4H_8O	328.85	0.9603	100.35	OX
Acetone	C_3H_6O	Trichloromethane	$CHCl_3$	337.58	0.3602	101.33	ON
Acetone	C_3H_6O	Triethylamine	$C_6H_{15}N$	318.15	0.9800	68.13	OX
Acetonitrile	C_2H_3N	Benzene	C_6H_6	328.15	0.4560	54.65	OX
Acetonitrile	C_2H_3N	2-Butanone	C_4H_8O	352.15	0.3195	101.15	OX
Acetonitrile	C_2H_3N	1-Decene	$C_{10}H_{20}$	354.55	0.9924	100.51	OX
Acetonitrile	C_2H_3N	2-Ethoxy-2-methylbutane	$C_7H_{16}O$	348.85	0.7219	98.99	OX
Acetonitrile	C_2H_3N	Isopentane	C_5H_{12}	298.45	0.1040	101.33	EX
Acetonitrile	C_2H_3N	2-Methoxy-2-methylbutane	$C_6H_{14}O$	346.13	0.5835	100.56	OX
Acetonitrile	C_2H_3N	2-Methyl-1,3-butadiene	C_5H_8	306.75	0.0410	101.33	OX
Acetonitrile	C_2H_3N	2-Methyl-2-butene	C_5H_{10}	308.95	0.1320	101.33	OX
Acetonitrile	C_2H_3N	Methylenecyclobutane	C_5H_8	312.45	0.1450	101.33	OX
Acetonitrile	C_2H_3N	Methyl methacrylate	$C_5H_8O_2$	355.25	0.9866	102.07	OX
Acetonitrile	C_2H_3N	2-Methyl-2-propanol	$C_4H_{10}O$	333.15	0.6200	56.93	OX
Acetonitrile	C_2H_3N	1-Pentene	C_5H_{10}	301.85	0.0830	101.33	OX
Acetonitrile	C_2H_3N	2-Propanol	C_3H_8O	348.15	0.5287	100.81	OX
Acetonitrile	C_2H_3N	Tetrachlorosilane	Cl_4Si	321.05	0.3100	101.33	EX
Acetonitrile	C_2H_3N	Tetrahydrofuran	C_4H_8O	338.95	0.0784	101.13	OX
Acetonitrile	C_2H_3N	Vinyl acetate	$C_4H_6O_2$	344.65	0.1948	98.33	OX
Acetonitrile	C_2H_3N	Water	H_2O	349.95	0.6900	101.33	OX
Acetophenone	C_8H_8O	Water	H_2O	371.15	0.0325	101.19	EX
Acrolein	C_3H_4O	2-Methyl-1,3-butadiene	C_5H_8	306.45	0.1980	101.33	OX
Acrolein	C_3H_4O	Water	H_2O	325.45	0.9270	101.33	LX
Acrylonitrile	C_3H_3N	Benzene	C_6H_6	347.45	0.5575	101.46	OX
Acrylonitrile	C_3H_3N	Cyclohexane	C_6H_{12}	337.75	0.4836	101.94	OX
Acrylonitrile	C_3H_3N	Ethanol	C_2H_6O	343.95	0.5560	101.33	OX
Acrylonitrile	C_3H_3N	Hexane	C_6H_{14}	330.90	0.4048	101.05	OX
Acrylonitrile	C_3H_3N	Methylenecyclobutane	C_5H_8	313.80	0.1275	101.33	OX
Acrylonitrile	C_3H_3N	Water	H_2O	344.05	0.7150	101.33	EX
Allyl alcohol	C_3H_6O	Benzene	C_6H_6	349.90	0.2203	101.33	OX
Allyl alcohol	C_3H_6O	Cyclohexane	C_6H_{12}	333.15	0.2790	63.98	OX
Allyl alcohol	C_3H_6O	Ethyl propanoate	$C_5H_{10}O_2$	367.65	0.5597	99.79	OX
Allyl alcohol	C_3H_6O	Water	H_2O	361.15	0.4438	101.33	OX
Aluminum chloride	$AlCl_3$	Phosphoryl chloride	Cl_3OP	660.15	0.5150	101.33	ONC

Component 1	Mol. form. 1	Component 2	Mol. form. 2	T_{AZ}/K	$y_{1,AZ}$	P_{AZ}/kPa	Type
Aniline	C_6H_7N	Butylbenzene	$C_{10}H_{14}$	448.65	0.4993	101.33	OX
Aniline	C_6H_7N	*sec*-Butylbenzene, (±)-	$C_{10}H_{14}$	443.15	0.3021	101.33	OX
Aniline	C_6H_7N	*tert*-Butylbenzene	$C_{10}H_{14}$	438.25	0.2104	101.33	OX
Aniline	C_6H_7N	1-Butyl-2-methylbenzene	$C_{11}H_{16}$	458.15	0.8091	101.33	OX
Aniline	C_6H_7N	1-Butyl-3-methylbenzene	$C_{11}H_{16}$	456.55	0.7661	101.33	OX
Aniline	C_6H_7N	1-Butyl-4-methylbenzene	$C_{11}H_{16}$	457.05	0.7807	101.33	OX
Aniline	C_6H_7N	*o*-Cresol	C_7H_8O	464.29	0.0953	101.33	ON
Aniline	C_6H_7N	Decane	$C_{10}H_{22}$	440.43	0.4660	101.33	OX
Aniline	C_6H_7N	*o*-Diethylbenzene	$C_{10}H_{14}$	448.75	0.5024	101.33	OX
Aniline	C_6H_7N	*m*-Diethylbenzene	$C_{10}H_{14}$	447.45	0.4584	101.33	OX
Aniline	C_6H_7N	*p*-Diethylbenzene	$C_{10}H_{14}$	448.85	0.5086	101.33	OX
Aniline	C_6H_7N	Dodecane	$C_{12}H_{26}$	453.52	0.8220	101.33	OX
Aniline	C_6H_7N	1-Ethyl-3,5-dimethylbenzene	$C_{10}H_{14}$	449.65	0.5310	101.33	OX
Aniline	C_6H_7N	2-Ethyl-1,4-dimethylbenzene	$C_{10}H_{14}$	449.65	0.5310	101.33	OX
Aniline	C_6H_7N	Isobutylbenzene	$C_{10}H_{14}$	442.75	0.2890	101.33	OX
Aniline	C_6H_7N	1-Isopropyl-2-methylbenzene	$C_{10}H_{14}$	445.95	0.4052	101.33	OX
Aniline	C_6H_7N	1-Isopropyl-3-methylbenzene	$C_{10}H_{14}$	444.15	0.3419	101.33	OX
Aniline	C_6H_7N	1-Isopropyl-4-methylbenzene	$C_{10}H_{14}$	445.35	0.3829	101.33	OX
Aniline	C_6H_7N	1-Methyl-2-propylbenzene	$C_{10}H_{14}$	449.45	0.5270	101.33	OX
Aniline	C_6H_7N	1-Methyl-3-propylbenzene	$C_{10}H_{14}$	447.85	0.4711	101.33	OX
Aniline	C_6H_7N	1-Methyl-4-propylbenzene	$C_{10}H_{14}$	448.75	0.5035	101.33	OX
Aniline	C_6H_7N	Nonane	C_9H_{20}	422.35	0.1770	101.33	OX
Aniline	C_6H_7N	Phenol	C_6H_6O	459.09	0.6116	101.33	ON
Aniline	C_6H_7N	Tetradecane	$C_{14}H_{30}$	457.05	0.9770	101.33	OX
Aniline	C_6H_7N	1,2,3,5-Tetramethylbenzene	$C_{10}H_{14}$	456.55	0.7504	101.33	OX
Aniline	C_6H_7N	1,2,4,5-Tetramethylbenzene	$C_{10}H_{14}$	455.36	0.7349	101.33	OX
Aniline	C_6H_7N	Tridecane	$C_{13}H_{28}$	456.22	0.9300	101.33	OX
Aniline	C_6H_7N	1,2,3-Trimethylbenzene	C_9H_{12}	444.65	0.3331	101.33	OX
Aniline	C_6H_7N	1,2,4-Trimethylbenzene	C_9H_{12}	441.80	0.1850	101.33	OX
Aniline	C_6H_7N	1,3,5-Trimethylbenzene	C_9H_{12}	437.68	0.1071	101.33	OX
Aniline	C_6H_7N	Undecane	$C_{11}H_{24}$	449.05	0.6970	101.33	OX
Aniline	C_6H_7N	Water	H_2O	372.55	0.0420	101.33	EX
Benzaldehyde	C_7H_6O	Phenol	C_6H_6O	447.00	0.3999	73.00	ON
Benzene	C_6H_6	2-Butanone	C_4H_8O	351.53	0.5210	101.33	OX
Benzene	C_6H_6	Butylamine	$C_4H_{11}N$	343.15	0.3000	80.89	OX
Benzene	C_6H_6	Cyclohexane	C_6H_{12}	353.15	0.5460	109.18	OX
Benzene	C_6H_6	Dimethyl carbonate	$C_3H_6O_3$	353.50	0.8634	100.48	OX
Benzene	C_6H_6	Ethanol	C_2H_6O	341.25	0.5400	101.33	OX
Benzene	C_6H_6	Ethyl acetate	$C_4H_8O_2$	350.55	0.0547	102.45	OX
Benzene	C_6H_6	Heptane	C_7H_{16}	353.25	0.9922	101.32	OX
Benzene	C_6H_6	Hexafluorobenzene	C_6F_6	353.60	0.2400	101.33	OND
Benzene	C_6H_6	Hexafluorobenzene	C_6F_6	352.50	0.8168	101.33	OXD
Benzene	C_6H_6	Hexane	C_6H_{14}	341.45	0.0500	101.33	OX
Benzene	C_6H_6	Methanol	CH_4O	331.56	0.3910	101.33	OX
Benzene	C_6H_6	2-Methyl-2-butanol	$C_5H_{12}O$	352.35	0.8500	101.33	OX
Benzene	C_6H_6	Methylcyclopentane	C_6H_{12}	333.15	0.1390	69.93	OX
Benzene	C_6H_6	2-Methyl-1-propanol	$C_4H_{10}O$	352.45	0.9220	101.33	OX
Benzene	C_6H_6	1-Propanol	C_3H_8O	350.20	0.7940	101.33	OX
Benzene	C_6H_6	2-Propanol	C_3H_8O	345.03	0.6040	101.33	OX
Benzene	C_6H_6	Propyl formate	$C_4H_8O_2$	343.15	0.6230	76.08	OX
Benzene	C_6H_6	2,2,4-Trimethylpentane	C_8H_{18}	353.25	0.9751	101.32	OX
Benzene	C_6H_6	Tris(perfluoroethyl)amine	$C_6F_{15}N$	329.95	0.4100	101.33	EX
Benzene	C_6H_6	Water	H_2O	342.35	0.7020	101.33	EX
Benzonitrile	C_7H_5N	*o*-Cresol	C_7H_8O	468.91	0.5100	101.33	ON
Benzonitrile	C_7H_5N	*m*-Cresol	C_7H_8O	476.10	0.1441	101.33	ON
Benzonitrile	C_7H_5N	*p*-Cresol	C_7H_8O	476.95	0.0898	101.33	ON
Benzonitrile	C_7H_5N	Phenol	C_6H_6O	465.11	0.7655	101.33	ON
Benzonitrile	C_7H_5N	2,6-Xylenol	$C_8H_{10}O$	477.15	0.0807	101.33	ON
Benzyl alcohol	C_7H_8O	Decane	$C_{10}H_{22}$	445.75	0.2490	101.33	OX
Benzyl alcohol	C_7H_8O	Water	H_2O	373.05	0.0160	101.33	EX
Bis(2,2,2-trifluoroethyl) ether	$C_4H_4F_6O$	Ethanol	C_2H_6O	331.90	0.7160	101.30	OX
Bis(2,2,2-trifluoroethyl) ether	$C_4H_4F_6O$	Methanol	CH_4O	326.28	0.5550	101.30	OX
Bis(2,2,2-trifluoroethyl) ether	$C_4H_4F_6O$	1-Propanol	C_3H_8O	336.22	0.8900	101.30	OX

Component 1	Mol. form. 1	Component 2	Mol. form. 2	T_{AZ}/K	$y_{1,AZ}$	P_{AZ}/kPa	Type
Bis(2,2,2-trifluoroethyl) ether	$C_4H_4F_6O$	2-Propanol	C_3H_8O	334.16	0.7770	101.30	OX
Bromobenzene	C_6H_5Br	Cyclohexanol	$C_6H_{12}O$	403.15	0.7390	52.45	OX
1-Bromobutane	C_4H_9Br	Water	H_2O	353.95	0.5050	101.33	EX
2-Bromo-2-chloro-1,1,1-trifluoroethane	$C_2HBrClF_3$	Diethyl ether	$C_4H_{10}O$	323.65	0.7200	93.33	ON
2-Bromo-2-chloro-1,1,1-trifluoroethane	$C_2HBrClF_3$	Methanol	CH_4O	317.25	0.8110	93.33	OX
Bromoethane	C_2H_5Br	Isopentane	C_5H_{12}	300.55	0.2180	101.33	OX
Bromoethane	C_2H_5Br	Methanol	CH_4O	308.05	0.8390	101.33	OX
Bromoethane	C_2H_5Br	2-Methyl-2-butene	C_5H_{10}	308.55	0.5110	101.33	OX
Bromoethane	C_2H_5Br	Methyl formate	$C_2H_4O_2$	303.05	0.2640	101.33	OX
1-Bromo-2-methylpropane	C_4H_9Br	Water	H_2O	348.45	0.6270	101.33	EX
1-Bromopropane	C_3H_7Br	Cyclohexane	C_6H_{12}	343.35	0.9219	98.84	OX
1-Bromopropane	C_3H_7Br	Methyl acetate	$C_3H_6O_2$	329.60	0.0273	99.56	OX
1-Bromopropane	C_3H_7Br	Nitromethane	CH_3NO_2	343.25	0.8980	99.82	OX
1-Bromopropane	C_3H_7Br	2-Propanol	C_3H_8O	339.15	0.7349	99.97	OX
1-Bromopropane	C_3H_7Br	Water	H_2O	336.35	0.7790	101.33	EX
2-Bromopropane	C_3H_7Br	Ethyl formate	$C_3H_6O_2$	326.15	0.2910	101.33	OX
1,3-Butadiene	C_4H_6	*cis*-2-Butene	C_4H_8	267.59	0.7650	101.33	OX
Butanal	C_4H_8O	Ethanol	C_2H_6O	345.45	0.6310	101.33	OX
Butanal	C_4H_8O	2-Methyl-1-pentene	C_6H_{12}	334.15	0.2293	101.48	OX
1,4-Butanediol	$C_4H_{10}O_2$	1-Pentadecanol	$C_{15}H_{32}O$	502.75	0.9980	101.33	OX
2,3-Butanedione	$C_4H_6O_2$	1-Propanol	C_3H_8O	359.30	0.6400	100.67	OX
2,3-Butanedione	$C_4H_6O_2$	2-Propanol	C_3H_8O	350.85	0.3546	100.95	OX
2,3-Butanedione	$C_4H_6O_2$	Toluene	C_7H_8	362.70	0.9513	101.34	OX
Butanoic acid	$C_4H_8O_2$	Butyl butanoate	$C_8H_{16}O_2$	434.60	0.6532	93.33	OXD
Butanoic acid	$C_4H_8O_2$	Butyl butanoate	$C_8H_{16}O_2$	434.78	0.8639	93.33	OND
Butanoic acid	$C_4H_8O_2$	Pyridine	C_5H_5N	436.35	0.9117	101.33	ON
Butanoic acid	$C_4H_8O_2$	Undecane	$C_{11}H_{24}$	435.55	0.9060	101.33	OX
Butanoic acid	$C_4H_8O_2$	Water	H_2O	372.95	0.0441	101.33	OX
1-Butanol	$C_4H_{10}O$	Butyl acetate	$C_6H_{12}O_2$	389.97	0.7700	101.33	OX
1-Butanol	$C_4H_{10}O$	1-Butylcyclopentene	C_9H_{16}	356.70	0.8450	79.99	OX
1-Butanol	$C_4H_{10}O$	Chlorobenzene	C_6H_5Cl	388.25	0.6950	101.33	OX
1-Butanol	$C_4H_{10}O$	Cyclohexane	C_6H_{12}	352.68	0.0787	101.33	OX
1-Butanol	$C_4H_{10}O$	Dibutyl ether	$C_8H_{18}O$	390.59	0.8754	101.33	OX
1-Butanol	$C_4H_{10}O$	Diethyl carbonate	$C_5H_{10}O_3$	370.85	0.6346	53.20	OX
1-Butanol	$C_4H_{10}O$	3-Ethylcyclopentene	C_7H_{12}	367.65	0.1900	101.33	OX
1-Butanol	$C_4H_{10}O$	Heptane	C_7H_{16}	366.55	0.2272	101.38	OX
1-Butanol	$C_4H_{10}O$	Hexane	C_6H_{14}	341.35	0.0370	101.33	OX
1-Butanol	$C_4H_{10}O$	Isobutyl acetate	$C_6H_{12}O_2$	387.15	0.5980	101.33	OX
1-Butanol	$C_4H_{10}O$	Nonane	C_9H_{20}	389.05	0.8128	101.33	OX
1-Butanol	$C_4H_{10}O$	1-Nonyne	C_9H_{16}	390.60	0.9400	101.33	OX
1-Butanol	$C_4H_{10}O$	Octane	C_8H_{18}	383.15	0.5500	102.79	OX
1-Butanol	$C_4H_{10}O$	1-Octene	C_8H_{16}	363.45	0.4530	53.33	OX
1-Butanol	$C_4H_{10}O$	*cis*-4-Octene	C_8H_{16}	382.35	0.5300	101.33	OX
1-Butanol	$C_4H_{10}O$	*trans*-4-Octene	C_8H_{16}	382.15	0.5310	101.33	OX
1-Butanol	$C_4H_{10}O$	1-Octyne	C_8H_{14}	386.50	0.6200	101.33	OX
1-Butanol	$C_4H_{10}O$	2-Octyne	C_8H_{14}	398.30	0.7910	101.33	OX
1-Butanol	$C_4H_{10}O$	Pyridine	C_5H_5N	392.00	0.7050	101.33	ON
1-Butanol	$C_4H_{10}O$	Styrene	C_8H_8	388.71	0.8923	98.39	OX
1-Butanol	$C_4H_{10}O$	Tetrachloromethane	CCl_4	349.71	0.0500	101.33	OX
1-Butanol	$C_4H_{10}O$	Toluene	C_7H_8	378.85	0.3320	101.33	OX
1-Butanol	$C_4H_{10}O$	Water	H_2O	365.45	0.2460	101.33	EX
1-Butanol	$C_4H_{10}O$	*o*-Xylene	C_8H_{10}	388.05	0.8671	100.13	OX
1-Butanol	$C_4H_{10}O$	*m*-Xylene	C_8H_{10}	387.75	0.7865	101.46	OX
1-Butanol	$C_4H_{10}O$	*p*-Xylene	C_8H_{10}	387.85	0.7823	99.73	OX
2-Butanol	$C_4H_{10}O$	Cyclohexane	C_6H_{12}	349.90	0.1892	101.02	OX
2-Butanol	$C_4H_{10}O$	Cyclohexene	C_6H_{10}	352.75	0.2046	101.25	OX
2-Butanol	$C_4H_{10}O$	1,4-Dioxane	$C_4H_8O_2$	371.75	0.5268	100.77	OX
2-Butanol	$C_4H_{10}O$	2-Ethoxy-2-methylbutane	$C_7H_{16}O$	367.75	0.4931	102.89	OX
2-Butanol	$C_4H_{10}O$	Heptane	C_7H_{16}	361.95	0.4116	102.70	OX
2-Butanol	$C_4H_{10}O$	Hexane	C_6H_{14}	348.15	0.1010	128.66	OX
2-Butanol	$C_4H_{10}O$	2-Methoxy-2-methylbutane	$C_6H_{14}O$	359.15	0.0991	102.12	OX
2-Butanol	$C_4H_{10}O$	Octane	C_8H_{18}	371.05	0.8001	101.30	OX
2-Butanol	$C_4H_{10}O$	3-Pentanone	$C_5H_{10}O$	370.50	0.6075	99.98	OX

Fluids

Component 1	Mol. form. 1	Component 2	Mol. form. 2	T_{AZ}/K	$y_{1,AZ}$	P_{AZ}/kPa	Type
2-Butanol	$C_4H_{10}O$	Toluene	C_7H_8	353.44	0.5550	56.67	OX
2-Butanol	$C_4H_{10}O$	Water	H_2O	360.50	0.3800	101.33	LX
2-Butanol	$C_4H_{10}O$	*m*-Xylene	C_8H_{10}	369.85	0.9717	101.06	OX
2-Butanol	$C_4H_{10}O$	*p*-Xylene	C_8H_{10}	369.55	0.9646	101.46	OX
2-Butanone	C_4H_8O	Cyclohexene	C_6H_{10}	343.29	0.5110	89.35	OX
2-Butanone	C_4H_8O	Diisopropyl ether	$C_6H_{14}O$	340.55	0.1938	101.56	OX
2-Butanone	C_4H_8O	Dipropyl ether	$C_6H_{14}O$	351.40	0.7785	100.88	OX
2-Butanone	C_4H_8O	Ethanol	C_2H_6O	347.15	0.4920	101.33	OX
2-Butanone	C_4H_8O	Ethyl acetate	$C_4H_8O_2$	349.55	0.1700	101.33	OX
2-Butanone	C_4H_8O	Heptane	C_7H_{16}	350.15	0.7670	101.33	OX
2-Butanone	C_4H_8O	Hexane	C_6H_{14}	337.15	0.3280	101.33	OX
2-Butanone	C_4H_8O	1-Hexene	C_6H_{12}	334.75	0.1760	100.58	OX
2-Butanone	C_4H_8O	Methanol	CH_4O	323.15	0.1980	58.80	OX
2-Butanone	C_4H_8O	Methylcyclohexane	C_7H_{14}	350.50	0.7984	98.93	OX
2-Butanone	C_4H_8O	2-Propanol	C_3H_8O	350.55	0.6170	101.33	OX
2-Butanone	C_4H_8O	Tetrachloromethane	CCl_4	346.99	0.3370	101.33	OX
2-Butanone	C_4H_8O	Water	H_2O	346.54	0.6520	101.33	LX
trans-2-Butenal	C_4H_6O	Octane	C_8H_{18}	353.15	0.4950	97.86	OX
trans-2-Butenal	C_4H_6O	Tetrachloromethane	CCl_4	348.15	0.3500	97.86	OX
trans-2-Butenal	C_4H_6O	Toluene	C_7H_8	374.15	0.5950	97.86	OX
trans-2-Butenal	C_4H_6O	Trichloroethene	C_2HCl_3	360.15	0.1000	97.86	OX
trans-2-Butenal	C_4H_6O	Trichloromethane	$CHCl_3$	329.15	0.0050	97.86	OX
cis-2-Butenenitrile	C_4H_5N	Water	H_2O	358.45	0.6168	101.33	EX
trans-2-Butenenitrile	C_4H_5N	Water	H_2O	363.05	0.3157	101.33	EX
Butyl acetate	$C_6H_{12}O_2$	1-Octene	C_8H_{16}	393.00	0.3030	101.33	OX
Butyl acetate	$C_6H_{12}O_2$	Water	H_2O	363.35	0.2987	101.33	EX
Butylamine	$C_4H_{11}N$	Ethanol	C_2H_6O	354.99	0.4100	101.33	ON
Butylamine	$C_4H_{11}N$	Water	H_2O	349.85	0.9300	101.33	OX
Butylbenzene	$C_{10}H_{14}$	Phenol	C_6H_6O	447.05	0.4465	101.33	OX
sec-Butylbenzene, (±)-	$C_{10}H_{14}$	*o*-Cresol	C_7H_8O	444.65	0.9062	101.33	OX
sec-Butylbenzene, (±)-	$C_{10}H_{14}$	*m*-Cresol	C_7H_8O	445.85	0.9864	101.33	OX
sec-Butylbenzene, (±)-	$C_{10}H_{14}$	*p*-Cresol	C_7H_8O	446.05	0.9814	101.33	OX
sec-Butylbenzene, (±)-	$C_{10}H_{14}$	Phenol	C_6H_6O	441.15	0.6871	101.33	OX
tert-Butylbenzene	$C_{10}H_{14}$	Phenol	C_6H_6O	439.95	0.7227	101.33	OX
Butyl butanoate	$C_8H_{16}O_2$	Water	H_2O	369.85	0.0890	101.33	EX
Butyl ethyl ether	$C_6H_{14}O$	Methanol	CH_4O	335.00	0.1990	98.84	OX
Butyl ethyl ether	$C_6H_{14}O$	Water	H_2O	349.85	0.5930	101.33	EX
tert-Butyl ethyl ether	$C_6H_{14}O$	Ethanol	C_2H_6O	339.95	0.6272	101.72	OX
tert-Butyl ethyl ether	$C_6H_{14}O$	Methanol	CH_4O	330.95	0.3998	101.54	OX
tert-Butyl ethyl ether	$C_6H_{14}O$	2-Methyl-2-propanol	$C_4H_{10}O$	342.85	0.7488	101.44	OX
Butyl formate	$C_5H_{10}O_2$	Formic acid	CH_2O_2	372.15	0.1300	101.33	OX
Butyl formate	$C_5H_{10}O_2$	Water	H_2O	356.95	0.4640	101.33	EX
tert-Butyl isopropyl ether	$C_7H_{16}O$	2-Methyl-2-propanol	$C_4H_{10}O$	350.90	0.4610	102.94	OX
tert-Butyl isopropyl ether	$C_7H_{16}O$	2-Propanol	C_3H_8O	349.95	0.4694	102.70	OX
1-Butyl-2-methylbenzene	$C_{11}H_{16}$	Phenol	C_6H_6O	455.55	0.1496	101.33	OX
1-Butyl-3-methylbenzene	$C_{11}H_{16}$	Phenol	C_6H_6O	454.15	0.1901	101.33	OX
1-Butyl-4-methylbenzene	$C_{11}H_{16}$	Phenol	C_6H_6O	454.55	0.1770	101.33	OX
Butyl methyl ether	$C_5H_{12}O$	Methanol	CH_4O	330.00	0.4485	100.08	OX
2-*sec*-Butyl-4-methylphenol	$C_{11}H_{16}O$	Quinoline	C_9H_7N	516.10	0.4861	101.33	ON
2-*tert*-Butyl-5-methylphenol	$C_{11}H_{16}O$	Quinoline	C_9H_7N	515.45	0.4146	101.33	ON
2-Butylphenol	$C_{10}H_{14}O$	Quinoline	C_9H_7N	515.70	0.4650	101.33	ON
2-*sec*-Butylphenol	$C_{10}H_{14}O$	Quinoline	C_9H_7N	514.70	0.3661	101.33	ON
4-*sec*-Butylphenol	$C_{10}H_{14}O$	Quinoline	C_9H_7N	516.45	0.5449	101.33	ON
2-*tert*-Butylphenol	$C_{10}H_{14}O$	Quinoline	C_9H_7N	513.70	0.2701	101.33	ON
3-*tert*-Butylphenol	$C_{10}H_{14}O$	Quinoline	C_9H_7N	517.05	0.5685	101.33	ON
Butyl propanoate	$C_7H_{14}O_2$	Water	H_2O	367.95	0.1660	101.33	EX
Carbon disulfide	CS_2	Methanol	CH_4O	310.65	0.7000	101.33	LX
1-Chlorobutane	C_4H_9Cl	Cyclohexane	C_6H_{12}	348.31	0.5800	95.85	OX
1-Chloro-2-methylpropane	C_4H_9Cl	Water	H_2O	333.95	0.8030	101.33	LX
1-Chloropropane	C_3H_7Cl	Methanol	CH_4O	313.35	0.7500	101.59	OX
3-Chloropropene	C_3H_5Cl	Hydrogen cyanide	CHN	296.45	0.1984	101.33	OX
3-Chloropropene	C_3H_5Cl	Methanol	CH_4O	312.15	0.7430	100.39	OX
o-Cresol	C_7H_8O	Decane	$C_{10}H_{22}$	433.15	0.3100	78.71	OX

Component 1	Mol. form. 1	Component 2	Mol. form. 2	T_{AZ}/K	$y_{1,AZ}$	P_{AZ}/kPa	Type
o-Cresol	C_7H_8O	*p*-Diethylbenzene	$C_{10}H_{14}$	453.10	0.2694	101.33	OX
o-Cresol	C_7H_8O	Dodecane	$C_{12}H_{26}$	458.15	0.8466	93.55	OX
o-Cresol	C_7H_8O	1,2-Ethanediol	$C_2H_6O_2$	462.67	0.6203	101.33	OX
o-Cresol	C_7H_8O	1,2,4,5-Tetramethylbenzene	$C_{10}H_{14}$	462.37	0.6273	101.33	OX
o-Cresol	C_7H_8O	2,4,6-Trimethylpyridine	$C_8H_{11}N$	470.35	0.6561	101.33	ON
o-Cresol	C_7H_8O	Undecane	$C_{11}H_{24}$	433.15	0.5800	56.40	OX
o-Cresol	C_7H_8O	1-Undecene	$C_{11}H_{22}$	448.15	0.5516	83.07	OX
m-Cresol	C_7H_8O	Decane	$C_{10}H_{22}$	433.15	0.2170	75.85	OX
m-Cresol	C_7H_8O	*p*-Diethylbenzene	$C_{10}H_{14}$	454.10	0.1010	101.33	OX
m-Cresol	C_7H_8O	2,6-Dimethylpyridine	C_7H_9N	475.66	0.9869	101.32	ON
m-Cresol	C_7H_8O	3-Methylpyridine	C_6H_7N	477.01	0.8444	101.32	ON
m-Cresol	C_7H_8O	4-Methylpyridine	C_6H_7N	477.74	0.8178	101.32	ON
m-Cresol	C_7H_8O	Naphthalene	$C_{10}H_8$	474.65	0.9680	101.33	OX
m-Cresol	C_7H_8O	Nonane	C_9H_{20}	413.15	0.0400	76.54	OX
m-Cresol	C_7H_8O	Quinoline	C_9H_7N	511.20	0.0356	101.33	ON
m-Cresol	C_7H_8O	1,2,3,4-Tetrahydronaphthalene	$C_{10}H_{12}$	468.45	0.5900	93.10	OX
m-Cresol	C_7H_8O	1,2,4,5-Tetramethylbenzene	$C_{10}H_{14}$	466.87	0.3591	101.33	OX
p-Cresol	C_7H_8O	*p*-Diethylbenzene	$C_{10}H_{14}$	454.50	0.1105	101.33	OX
p-Cresol	C_7H_8O	Naphthalene	$C_{10}H_8$	474.55	0.9414	101.33	OX
Cyclohexane	C_6H_{12}	Diethoxymethane	$C_5H_{12}O_2$	353.21	0.8226	101.39	OX
Cyclohexane	C_6H_{12}	Dimethyl carbonate	$C_3H_6O_3$	346.95	0.6220	101.49	OX
Cyclohexane	C_6H_{12}	Ethanol	C_2H_6O	337.95	0.5460	102.26	OX
Cyclohexane	C_6H_{12}	Ethyl acetate	$C_4H_8O_2$	345.00	0.4610	102.45	OX
Cyclohexane	C_6H_{12}	Ethyl formate	$C_3H_6O_2$	323.15	0.1790	91.46	OX
Cyclohexane	C_6H_{12}	Methanol	CH_4O	328.75	0.3910	106.66	OX
Cyclohexane	C_6H_{12}	Methyl acetate	$C_3H_6O_2$	328.65	0.2000	101.33	OX
Cyclohexane	C_6H_{12}	2-Methyl-2-butanol	$C_5H_{12}O$	351.95	0.8900	101.33	OX
Cyclohexane	C_6H_{12}	3-Methyl-3-buten-1-ol	$C_5H_{10}O$	352.65	0.9785	101.10	OX
Cyclohexane	C_6H_{12}	2-Methyl-3-buten-2-ol	$C_5H_{10}O$	350.15	0.8096	101.20	OX
Cyclohexane	C_6H_{12}	2-Methyl-1-propanol	$C_4H_{10}O$	351.35	0.8675	101.45	OX
Cyclohexane	C_6H_{12}	1-Propanol	C_3H_8O	347.68	0.7510	101.33	OX
Cyclohexane	C_6H_{12}	2-Propanol	C_3H_8O	342.75	0.5950	101.33	OX
Cyclohexane	C_6H_{12}	Propyl acetate	$C_5H_{10}O_2$	353.15	0.9402	100.43	OX
Cyclohexane	C_6H_{12}	Trichloroethene	C_2HCl_3	353.40	0.9025	101.32	OX
Cyclohexane	C_6H_{12}	Tris(perfluoroethyl)amine	$C_6F_{15}N$	329.35	0.4310	101.33	EX
Cyclohexane	C_6H_{12}	Vinyl acetate	$C_4H_6O_2$	340.45	0.3800	101.33	OX
Cyclohexanol	$C_6H_{12}O$	Nonane	C_9H_{20}	410.20	0.3350	79.99	OX
Cyclohexanol	$C_6H_{12}O$	*o*-Xylene	C_8H_{10}	415.95	0.1426	101.33	OX
Cyclohexanol	$C_6H_{12}O$	*m*-Xylene	C_8H_{10}	411.85	0.0503	101.33	OX
Cyclohexanol	$C_6H_{12}O$	*p*-Xylene	C_8H_{10}	410.95	0.0505	101.33	OX
Cyclohexanone	$C_6H_{10}O$	3-Methyl-1-butanol	$C_5H_{12}O$	404.87	0.0906	101.33	OX
Cyclohexanone	$C_6H_{10}O$	1-Pentanol	$C_5H_{12}O$	392.37	0.0252	53.32	OX
Cyclohexanone	$C_6H_{10}O$	Water	H_2O	369.45	0.1306	101.33	EX
Cyclohexene	C_6H_{10}	1,4-Dioxane	$C_4H_8O_2$	355.75	0.8935	101.44	OX
Cyclohexene	C_6H_{10}	Ethyl acetate	$C_4H_8O_2$	347.45	0.3817	100.87	OX
Cyclohexene	C_6H_{10}	Methyl acetate	$C_3H_6O_2$	330.35	0.0879	102.87	OX
Cyclohexene	C_6H_{10}	2-Methyl-1-propanol	$C_4H_{10}O$	353.75	0.8637	100.31	OX
Cyclohexene	C_6H_{10}	2-Methyl-2-propanol	$C_4H_{10}O$	346.00	0.5828	99.61	OX
Cyclohexene	C_6H_{10}	2-Propanol	C_3H_8O	344.65	0.5729	101.40	OX
Cyclohexene	C_6H_{10}	Water	H_2O	343.95	0.6910	101.33	EX
Cyclohexylamine	$C_6H_{13}N$	Water	H_2O	369.55	0.1308	101.33	OX
1,3-Cyclopentadiene	C_5H_6	Dimethoxymethane	$C_3H_8O_2$	313.65	0.6650	101.33	OX
1,3-Cyclopentadiene	C_5H_6	Methanol	CH_4O	309.05	0.7880	101.33	OX
1,3-Cyclopentadiene	C_5H_6	2-Methyl-2-butene	C_5H_{10}	310.85	0.3000	101.33	OX
Cyclopentane	C_5H_{10}	Ethanol	C_2H_6O	323.44	0.8200	121.00	OX
Cyclopentanone	C_5H_8O	3-Methyl-1-butanol	$C_5H_{12}O$	402.02	0.5944	101.33	OX
Cyclopentanone	C_5H_8O	1-Pentanol	$C_5H_{12}O$	403.84	0.9196	101.33	OX
Cyclopentene	C_5H_8	Ethanol	C_2H_6O	323.40	0.8560	134.00	OX
Cyclopropyl methyl ketone	C_5H_8O	Water	H_2O	361.65	0.2940	101.19	EX
trans-Decahydronaphthalene	$C_{10}H_{18}$	Phenol	C_6H_6O	443.15	0.4581	99.85	OX
Decane	$C_{10}H_{22}$	4,4-Dimethyl-1,3-dioxane	$C_6H_{12}O_2$	405.35	0.0001	100.60	OX
Decane	$C_{10}H_{22}$	1-Heptanol	$C_7H_{16}O$	438.75	0.5692	101.33	OX
Decane	$C_{10}H_{22}$	1-Hexanol	$C_6H_{14}O$	427.05	0.2877	101.33	OX

Component 1	Mol. form. 1	Component 2	Mol. form. 2	T_{AZ}/K	$y_{1,AZ}$	P_{AZ}/kPa	Type
Decane	$C_{10}H_{22}$	2-Methylaniline	C_7H_9N	446.91	0.8230	101.33	OX
Decane	$C_{10}H_{22}$	Nitromethane	CH_3NO_2	371.96	0.0761	99.73	EX
Decane	$C_{10}H_{22}$	1-Octanol	$C_8H_{18}O$	446.45	0.8971	101.33	OX
Decane	$C_{10}H_{22}$	1-Pentanol	$C_5H_{12}O$	410.65	0.0779	101.33	OX
Decane	$C_{10}H_{22}$	Phenol	C_6H_6O	434.15	0.5550	101.32	OX
Decane	$C_{10}H_{22}$	1,2,3-Trimethylbenzene	C_9H_{12}	433.35	0.5990	72.54	OX
Decane	$C_{10}H_{22}$	1,2,4-Trimethylbenzene	C_9H_{12}	433.35	0.1400	80.25	OX
Decane	$C_{10}H_{22}$	Water	H_2O	370.75	0.0820	101.33	EX
1-Decanol	$C_{10}H_{22}O$	Dodecane	$C_{12}H_{26}$	489.25	0.1068	101.33	OX
1-Decanol	$C_{10}H_{22}O$	Water	H_2O	373.13	0.0135	101.33	EX
Diacetone alcohol	$C_6H_{12}O_2$	Water	H_2O	370.00	0.0100	90.79	OX
Dibutylamine	$C_8H_{19}N$	Water	H_2O	370.05	0.1150	101.33	EX
Dibutyl ether	$C_8H_{18}O$	Water	H_2O	368.65	0.2372	101.33	EX
1,1-Dichloroethane	$C_2H_4Cl_2$	Hexane	C_6H_{14}	329.30	0.8025	101.21	OX
1,1-Dichloroethane	$C_2H_4Cl_2$	2-Propanol	C_3H_8O	329.55	0.8928	101.60	OX
1,2-Dichloroethane	$C_2H_4Cl_2$	Formic acid	CH_2O_2	350.17	0.5725	101.33	OX
1,2-Dichloroethane	$C_2H_4Cl_2$	Germanium(IV) chloride	Cl_4Ge	350.75	0.5370	101.33	OX
1,2-Dichloroethane	$C_2H_4Cl_2$	Methylcyclohexane	C_7H_{14}	354.65	0.8036	101.21	OX
1,2-Dichloroethane	$C_2H_4Cl_2$	2-Methyl-1-propanol	$C_4H_{10}O$	356.05	0.9173	101.26	OX
1,2-Dichloroethane	$C_2H_4Cl_2$	2-Methyl-2-propanol	$C_4H_{10}O$	349.45	0.5336	101.43	OX
1,2-Dichloroethane	$C_2H_4Cl_2$	2-Propanol	C_3H_8O	347.25	0.5258	100.32	OX
1,2-Dichloroethane	$C_2H_4Cl_2$	1,1,2,2-Tetrachloro-1,2-difluoroethane	$C_2Cl_4F_2$	353.80	0.7300	101.33	OX
1,2-Dichloroethane	$C_2H_4Cl_2$	Trichloroethene	C_2HCl_3	355.35	0.6676	101.36	OX
1,2-Dichloroethane	$C_2H_4Cl_2$	2,2,4-Trimethylpentane	C_8H_{18}	343.15	0.7600	73.13	OX
1,2-Dichloroethane	$C_2H_4Cl_2$	Water	H_2O	345.43	0.6430	101.33	EX
trans-1,2-Dichloroethene	$C_2H_2Cl_2$	1,1,1,2,3,3-Hexafluoro-3-(2,2,2-trifluoroethoxy)propane	$C_5H_3F_9O$	318.50	0.8390	101.30	OX
Dichloromethane	CH_2Cl_2	Ethanol	C_2H_6O	312.05	0.9600	101.33	OX
Diethoxymethane	$C_5H_{12}O_2$	Dimethyl carbonate	$C_3H_6O_3$	358.71	0.5563	100.42	OX
Diethoxymethane	$C_5H_{12}O_2$	Ethanol	C_2H_6O	348.30	0.3503	102.35	OX
Diethoxymethane	$C_5H_{12}O_2$	Hexane	C_6H_{14}	361.27	0.9101	102.30	OX
Diethoxymethane	$C_5H_{12}O_2$	Methanol	CH_4O	336.03	0.1873	101.52	OX
Diethoxymethane	$C_5H_{12}O_2$	1-Propanol	C_3H_8O	359.01	0.7680	99.43	OX
Diethoxymethane	$C_5H_{12}O_2$	2-Propanol	C_3H_8O	351.45	0.3893	98.61	OX
o-Diethylbenzene	$C_{10}H_{14}$	Phenol	C_6H_6O	447.15	0.4435	101.33	OX
m-Diethylbenzene	$C_{10}H_{14}$	Phenol	C_6H_6O	445.95	0.4848	101.33	OX
p-Diethylbenzene	$C_{10}H_{14}$	Phenol	C_6H_6O	446.45	0.5295	101.33	OX
Diethylene glycol	$C_4H_{10}O_3$	1-Nonanol	$C_9H_{20}O$	486.65	0.0095	101.33	OX
Diethylene glycol dimethyl ether	$C_6H_{14}O_3$	Water	H_2O	372.70	0.0321	101.33	OX
Diethylene glycol monobutyl ether	$C_8H_{18}O_3$	1,2-Ethanediol	$C_2H_6O_2$	469.15	0.0898	101.33	OX
Diethylene glycol monoethyl ether	$C_6H_{14}O_3$	1,2-Ethanediol	$C_2H_6O_2$	467.15	0.3520	101.33	OX
Diethylene glycol monoethyl ether	$C_6H_{14}O_3$	1,3-Propanediol	$C_3H_8O_2$	459.25	0.0650	101.33	OX
Diethylene glycol monoisobutyl ether	$C_8H_{18}O_3$	1,2-Ethanediol	$C_2H_6O_2$	467.55	0.1645	101.33	OX
Diethylene glycol monoisopropyl ether	$C_7H_{16}O_3$	1,2-Ethanediol	$C_2H_6O_2$	466.35	0.3036	101.33	OX
Diethylene glycol monomethyl ether	$C_5H_{12}O_3$	1,2-Ethanediol	$C_2H_6O_2$	463.95	0.5612	101.33	OX
Diethylene glycol monomethyl ether	$C_5H_{12}O_3$	1,3-Propanediol	$C_3H_8O_2$	455.25	0.3700	101.33	OX
Diethylene glycol monopropyl ether	$C_7H_{16}O_3$	1,2-Ethanediol	$C_2H_6O_2$	468.55	0.1552	101.33	OX
Diethyl ether	$C_4H_{10}O$	Methanol	CH_4O	305.15	0.9500	93.33	OX
Diethyl ether	$C_4H_{10}O$	Methyl formate	$C_2H_4O_2$	301.55	0.3970	101.33	OX
Diethyl ether	$C_4H_{10}O$	Pentane	C_5H_{12}	306.85	0.5500	101.33	OX
Dihydromethylpyran (unspecified isomer)	$C_6H_{10}O$	Toluene	C_7H_8	381.85	0.0207	101.30	OX
Dihydromethylpyran (unspecified isomer)	$C_6H_{10}O$	Water	H_2O	360.75	0.4159	100.93	EX
Diisopropylamine	$C_6H_{15}N$	2-Propanol	C_3H_8O	352.94	0.5110	101.33	OX
Diisopropylamine	$C_6H_{15}N$	Water	H_2O	347.25	0.6346	101.33	EX
Diisopropyl ether	$C_6H_{14}O$	Methanol	CH_4O	330.00	0.4610	101.61	OX
Diisopropyl ether	$C_6H_{14}O$	2-Methyl-2-propanol	$C_4H_{10}O$	340.45	0.8942	101.72	OX
Diisopropyl ether	$C_6H_{14}O$	2-Propanol	C_3H_8O	340.00	0.7950	103.36	OX
2,2-Dimethoxybutane	$C_6H_{14}O_2$	Toluene	C_7H_8	380.15	0.9180	101.44	OX
1,2-Dimethoxyethane	$C_4H_{10}O_2$	Methylcyclohexane	C_7H_{14}	350.00	0.8190	79.42	OX
Dimethoxymethane	$C_3H_8O_2$	2-Methyl-1,3-butadiene	C_5H_8	306.80	0.0160	101.33	OX
Dimethoxymethane	$C_3H_8O_2$	Methylenecyclobutane	C_5H_8	310.35	0.4630	101.33	OX

Component 1	Mol. form. 1	Component 2	Mol. form. 2	T_{AZ}/K	$y_{1,AZ}$	P_{AZ}/kPa	Type
Dimethoxymethane	$C_3H_8O_2$	Water	H_2O	315.05	0.9731	101.38	LX
2,2-Dimethoxypropane	$C_5H_{12}O_2$	Methanol	CH_4O	334.15	0.2750	100.00	OX
Dimethylacetal	$C_4H_{10}O_2$	Methanol	CH_4O	330.35	0.5300	101.33	OX
N,N-Dimethylacetamide	C_4H_9NO	Ethylbenzene	C_8H_{10}	408.95	0.0037	101.70	OX
N,N-Dimethylacetamide	C_4H_9NO	*o*-Xylene	C_8H_{10}	416.95	0.0591	103.40	OX
2,4-Dimethylaniline	$C_8H_{11}N$	Dodecane	$C_{12}H_{26}$	482.95	0.4520	101.33	OX
2,4-Dimethylaniline	$C_8H_{11}N$	Tetradecane	$C_{14}H_{30}$	490.53	0.9840	101.33	OX
2,4-Dimethylaniline	$C_8H_{11}N$	Tridecane	$C_{13}H_{28}$	488.43	0.7880	101.33	OX
2,4-Dimethylaniline	$C_8H_{11}N$	Undecane	$C_{11}H_{24}$	468.13	0.1490	101.33	OX
α,α-Dimethylbenzenemethanol	$C_9H_{12}O$	Water	H_2O	371.25	0.0282	101.33	EX
2,3-Dimethylbutane	C_6H_{14}	Methanol	CH_4O	313.15	0.6380	85.50	OX
Dimethyl carbonate	$C_3H_6O_3$	Dipropyl ether	$C_6H_{14}O$	356.45	0.5044	100.73	OX
Dimethyl carbonate	$C_3H_6O_3$	Heptane	C_7H_{16}	355.15	0.5930	99.67	OX
Dimethyl carbonate	$C_3H_6O_3$	Hexane	C_6H_{14}	338.15	0.2540	98.46	OX
Dimethyl carbonate	$C_3H_6O_3$	Methanol	CH_4O	337.25	0.1496	102.52	OX
Dimethyl carbonate	$C_3H_6O_3$	Methylcyclopentane	C_6H_{12}	342.35	0.2680	103.46	OX
4,4-Dimethyl-1,3-dioxane	$C_6H_{12}O_2$	3-Methyl-3-buten-1-ol	$C_5H_{10}O$	403.05	0.2410	102.26	OX
4,4-Dimethyl-1,3-dioxane	$C_6H_{12}O_2$	Nonane	C_9H_{20}	402.15	0.8864	101.30	OX
4,4-Dimethyl-1,3-dioxane	$C_6H_{12}O_2$	Octane	C_8H_{18}	393.95	0.3343	101.20	OX
4,4-Dimethyl-1,3-dioxane	$C_6H_{12}O_2$	1-Propanol	C_3H_8O	368.20	0.0403	101.30	OX
4,4-Dimethyl-1,3-dioxane	$C_6H_{12}O_2$	Water	H_2O	366.00	0.2221	101.33	EX
4,4-Dimethyl-1,3-dioxane	$C_6H_{12}O_2$	*o*-Xylene	C_8H_{10}	404.65	0.9662	101.30	OX
cis-4,5-Dimethyl-1,3-dioxane	$C_6H_{12}O_2$	Water	H_2O	365.05	0.2034	101.50	EX
N,N-Dimethylformamide	C_3H_7NO	1,4-Dimethyl-4-vinylcyclohexene	$C_{10}H_{16}$	415.65	0.5880	101.33	OX
N,N-Dimethylformamide	C_3H_7NO	Heptane	C_7H_{16}	370.15	0.0800	101.33	OX
N,N-Dimethylformamide	C_3H_7NO	1-Methyl-3-(1-methylethylidene) cyclohexene	$C_{10}H_{16}$	419.05	0.7250	101.33	OX
2,6-Dimethylpyridine	C_7H_9N	3-Methylpyridine	C_6H_7N	416.64	0.7060	101.33	OX
2,6-Dimethylpyridine	C_7H_9N	4-Methylpyridine	C_6H_7N	417.08	0.8000	101.32	OX
2,6-Dimethylpyridine	C_7H_9N	Phenol	C_6H_6O	459.32	0.2461	101.32	ON
2,6-Dimethylpyridine	C_7H_9N	Water	H_2O	369.17	0.1353	101.33	EX
2,3-Dimethylquinoline	$C_{11}H_{11}N$	3,4-Xylenol	$C_8H_{10}O$	521.60	0.7887	101.33	ON
2,3-Dimethylquinoline	$C_{11}H_{11}N$	3,5-Xylenol	$C_8H_{10}O$	520.70	0.9470	101.33	ON
1,4-Dioxane	$C_4H_8O_2$	Ethanol	C_2H_6O	351.33	0.0520	101.33	OX
1,4-Dioxane	$C_4H_8O_2$	2-Ethoxy-2-methylbutane	$C_7H_{16}O$	369.15	0.5452	100.27	OX
1,4-Dioxane	$C_4H_8O_2$	Heptane	C_7H_{16}	364.30	0.4868	101.06	OX
1,4-Dioxane	$C_4H_8O_2$	2-Methyl-2-butanol	$C_5H_{12}O$	373.75	0.8119	99.62	OX
1,4-Dioxane	$C_4H_8O_2$	Methylcyclopentane	C_6H_{12}	343.85	0.0538	99.79	OX
1,4-Dioxane	$C_4H_8O_2$	Nitromethane	CH_3NO_2	373.25	0.5899	101.48	OX
1,4-Dioxane	$C_4H_8O_2$	1-Propanol	C_3H_8O	365.30	0.3582	101.30	OX
1,4-Dioxane	$C_4H_8O_2$	Propyl acetate	$C_5H_{10}O_2$	373.35	0.6334	101.13	OX
1,4-Dioxane	$C_4H_8O_2$	Triethylamine	$C_6H_{15}N$	343.15	0.2500	56.80	OX
1,4-Dioxane	$C_4H_8O_2$	Water	H_2O	360.65	0.4720	101.33	OX
1,3-Dioxolane	$C_3H_6O_2$	Methanol	CH_4O	334.66	0.3090	101.30	OX
1,3-Dioxolane	$C_3H_6O_2$	Water	H_2O	344.95	0.7480	101.30	OX
Dipropylamine	$C_6H_{15}N$	Water	H_2O	359.00	0.3954	101.33	EX
Dipropylene glycol monobutyl ether	$C_{10}H_{22}O_3$	1,2-Ethanediol	$C_2H_6O_2$	465.75	0.1870	101.33	OX
Dipropylene glycol monobutyl ether	$C_{10}H_{22}O_3$	1,2-Propanediol	$C_3H_8O_2$	459.65	0.0279	101.33	OX
Dipropylene glycol monoethyl ether	$C_8H_{18}O_3$	1,2-Ethanediol	$C_2H_6O_2$	458.65	0.5200	101.33	OX
Dipropylene glycol monoethyl ether	$C_8H_{18}O_3$	1,2-Propanediol	$C_3H_8O_2$	458.75	0.2222	101.33	OX
Dipropylene glycol monoisobutyl ether	$C_{10}H_{22}O_3$	1,2-Propanediol	$C_3H_8O_2$	459.05	0.0745	101.33	OX
Dipropylene glycol monoisopropyl ether	$C_9H_{20}O_3$	1,2-Propanediol	$C_3H_8O_2$	458.95	0.1870	101.33	OX
Dipropylene glycol monopropyl ether	$C_9H_{20}O_3$	1,2-Ethanediol	$C_2H_6O_2$	463.15	0.3410	101.33	OX
Dipropylene glycol monopropyl ether	$C_9H_{20}O_3$	1,2-Propanediol	$C_3H_8O_2$	458.95	0.0990	101.33	OX
Dodecane	$C_{12}H_{26}$	2-Methylaniline	C_7H_9N	468.90	0.2350	101.33	OX
Dodecane	$C_{12}H_{26}$	Nitromethane	CH_3NO_2	373.75	0.0154	99.73	EX
Dodecane	$C_{12}H_{26}$	1-Nonanol	$C_9H_{20}O$	480.65	0.4765	101.33	OX
Dodecane	$C_{12}H_{26}$	1-Octanol	$C_8H_{18}O$	466.95	0.1164	101.33	OX
Dodecane	$C_{12}H_{26}$	Phenol	C_6H_6O	450.73	0.2100	101.32	OX
1,2-Ethanediamine	$C_2H_8N_2$	Water	H_2O	391.85	0.5550	101.33	ON
1,2-Ethanediol	$C_2H_6O_2$	5-Methyl-1-heptanol, (±)-	$C_8H_{18}O$	457.65	0.3500	101.33	OX
1,2-Ethanediol	$C_2H_6O_2$	2,4,6-Trimethylpyridine	$C_8H_{11}N$	443.65	0.1734	101.33	OX
Ethanol	C_2H_6O	2-Ethoxy-2-methylbutane	$C_7H_{16}O$	349.35	0.7644	101.54	OX

Component 1	Mol. form. 1	Component 2	Mol. form. 2	T_{AZ}/K	$y_{1,AZ}$	P_{AZ}/kPa	Type
Ethanol	C_2H_6O	Ethyl acetate	$C_4H_8O_2$	344.85	0.4590	101.33	OX
Ethanol	C_2H_6O	Fluorobenzene	C_6H_5F	343.85	0.4752	101.54	OX
Ethanol	C_2H_6O	1,1,1,2,3,3-Hexafluoro-3-(2,2,2-trifluoroethoxy)propane	$C_5H_3F_9O$	337.88	0.3980	101.30	OX
Ethanol	C_2H_6O	Hexane	C_6H_{14}	331.65	0.3410	101.33	OX
Ethanol	C_2H_6O	Isopentane	C_5H_{12}	299.95	0.0540	101.33	OX
Ethanol	C_2H_6O	Isopropyl acetate	$C_5H_{10}O_2$	349.85	0.7010	101.33	OX
Ethanol	C_2H_6O	2-Methoxy-2-methylbutane	$C_6H_{14}O$	346.81	0.5820	101.32	OX
Ethanol	C_2H_6O	Methyl acetate	$C_3H_6O_2$	329.79	0.0362	101.33	OX
Ethanol	C_2H_6O	2-Methyl-1,3-butadiene	C_5H_8	305.95	0.1500	101.33	OX
Ethanol	C_2H_6O	Methyl butanoate	$C_5H_{10}O_2$	346.30	0.8800	83.88	OX
Ethanol	C_2H_6O	3-Methyl-2-butanone	$C_5H_{10}O$	350.85	0.8250	101.33	OX
Ethanol	C_2H_6O	2-Methyl-2-butene	C_5H_{10}	309.79	0.0795	101.33	OX
Ethanol	C_2H_6O	Methyl *tert*-butyl ether	$C_5H_{12}O$	327.75	0.0380	101.33	OX
Ethanol	C_2H_6O	Methyl propanoate	$C_4H_8O_2$	346.30	0.5140	103.91	OX
Ethanol	C_2H_6O	Nitromethane	CH_3NO_2	333.15	0.7150	53.61	OX
Ethanol	C_2H_6O	Nonane	C_9H_{20}	351.35	0.9400	101.33	OX
Ethanol	C_2H_6O	Octane	C_8H_{18}	349.85	0.8250	101.33	OX
Ethanol	C_2H_6O	Pentane	C_5H_{12}	307.15	0.0537	101.33	OX
Ethanol	C_2H_6O	2-Pentanone	$C_5H_{10}O$	351.15	0.9779	100.50	OX
Ethanol	C_2H_6O	3-Pentanone	$C_5H_{10}O$	351.33	0.9590	101.33	OX
Ethanol	C_2H_6O	Tetrachloromethane	CCl_4	338.19	0.3860	101.33	OX
Ethanol	C_2H_6O	1,1,2,2-Tetrafluoroethyl 1,1,1-trifluoroethyl ether	$C_4H_3F_7O$	326.67	0.2000	101.30	OX
Ethanol	C_2H_6O	Tetrahydrofuran	C_4H_8O	344.95	0.1290	125.00	OX
Ethanol	C_2H_6O	Toluene	C_7H_8	349.75	0.8152	101.33	OX
Ethanol	C_2H_6O	Trichloroethene	C_2HCl_3	343.85	0.5259	101.33	OX
Ethanol	C_2H_6O	Trichloromethane	$CHCl_3$	332.45	0.1590	101.33	OX
Ethanol	C_2H_6O	1,1,2-Trichloro-1,2,2-trifluoroethane	$C_2Cl_3F_3$	317.75	0.1544	101.42	OX
Ethanol	C_2H_6O	2,2,3-Trimethyloxirane	$C_5H_{10}O$	343.45	0.2930	101.33	OX
Ethanol	C_2H_6O	2,2,4-Trimethylpentane	C_8H_{18}	344.42	0.6450	101.33	OX
Ethanol	C_2H_6O	Water	H_2O	351.25	0.8970	101.33	OX
2-Ethoxyethanol	$C_4H_{10}O_2$	Ethylbenzene	C_8H_{10}	401.05	0.4632	100.94	OX
2-Ethoxyethanol	$C_4H_{10}O_2$	Styrene	C_8H_8	405.75	0.6438	101.33	OX
2-Ethoxyethanol	$C_4H_{10}O_2$	*o*-Xylene	C_8H_{10}	404.95	0.5965	101.36	OX
2-Ethoxyethanol	$C_4H_{10}O_2$	*m*-Xylene	C_8H_{10}	401.75	0.5159	101.33	OX
2-Ethoxyethanol	$C_4H_{10}O_2$	*p*-Xylene	C_8H_{10}	402.55	0.5042	102.19	OX
2-Ethoxy-2-methylbutane	$C_7H_{16}O$	1,4-Dioxane	$C_4H_8O_2$	369.15	0.4548	100.27	OX
2-Ethoxy-2-methylbutane	$C_7H_{16}O$	Methanol	CH_4O	335.15	0.1264	97.28	OX
2-Ethoxy-2-methylbutane	$C_7H_{16}O$	2-Methyl-2-butanol	$C_5H_{12}O$	369.85	0.6096	100.52	OX
2-Ethoxy-2-methylbutane	$C_7H_{16}O$	3-Pentanone	$C_5H_{10}O$	371.15	0.5236	100.21	OX
2-Ethoxy-2-methylbutane	$C_7H_{16}O$	Propyl acetate	$C_5H_{10}O_2$	370.95	0.3471	100.03	OX
Ethyl acetate	$C_4H_8O_2$	Hexane	C_6H_{14}	338.00	0.3430	101.32	OX
Ethyl acetate	$C_4H_8O_2$	1-Hexene	C_6H_{12}	333.15	0.1230	91.47	OX
Ethyl acetate	$C_4H_8O_2$	Methanol	CH_4O	335.66	0.2880	101.33	OX
Ethyl acetate	$C_4H_8O_2$	Methylcyclohexane	C_7H_{14}	349.90	0.9001	101.83	OX
Ethyl acetate	$C_4H_8O_2$	2-Methyl-2-propanol	$C_4H_{10}O$	349.75	0.7778	101.28	OX
Ethyl acetate	$C_4H_8O_2$	Tetrachloromethane	CCl_4	347.95	0.4300	101.33	OX
Ethyl acetate	$C_4H_8O_2$	Water	H_2O	343.55	0.7010	101.33	EX
Ethylbenzene	C_8H_{10}	2-Methyl-1-butanol, (±)-	$C_5H_{12}O$	398.75	0.4343	99.46	OX
Ethylbenzene	C_8H_{10}	Water	H_2O	364.15	0.2779	101.33	EX
1-Ethyl-3,5-dimethylbenzene	$C_{10}H_{14}$	Phenol	C_6H_6O	447.95	0.4160	101.33	OX
2-Ethyl-1,4-dimethylbenzene	$C_{10}H_{14}$	Phenol	C_6H_6O	447.95	0.4160	101.33	OX
4-Ethyl-1,3-dioxane	$C_6H_{12}O_2$	Water	H_2O	365.75	0.2743	101.30	EX
Ethyl formate	$C_3H_6O_2$	Methanol	CH_4O	318.15	0.7000	81.34	OX
Ethyl formate	$C_3H_6O_2$	Water	H_2O	325.75	0.9300	101.33	EX
2-Ethyl-4-methylphenol	$C_9H_{12}O$	Naphthalene	$C_{10}H_8$	488.20	0.2218	101.33	OX
2-Ethyl-5-methylphenol	$C_9H_{12}O$	Naphthalene	$C_{10}H_8$	489.45	0.1710	101.33	OX
2-Ethylphenol	$C_8H_{10}O$	Naphthalene	$C_{10}H_8$	478.35	0.8005	101.33	OX
2-Ethylphenol	$C_8H_{10}O$	Quinoline	C_9H_7N	511.75	0.1041	101.33	ON
2-Ethylphenol	$C_8H_{10}O$	1,2,4,5-Tetramethylbenzene	$C_{10}H_{14}$	471.45	0.3707	101.33	OX
3-Ethylphenol	$C_8H_{10}O$	Naphthalene	$C_{10}H_8$	483.45	0.5551	101.33	OX
3-Ethylphenol	$C_8H_{10}O$	Quinoline	C_9H_7N	512.70	0.2089	101.33	ON

Component 1	Mol. form. 1	Component 2	Mol. form. 2	T_{AZ}/K	$y_{1,AZ}$	P_{AZ}/kPa	Type
3-Ethylphenol	$C_8H_{10}O$	1,2,4,5-Tetramethylbenzene	$C_{10}H_{14}$	475.95	0.1249	101.33	OX
4-Ethylphenol	$C_8H_{10}O$	Naphthalene	$C_{10}H_8$	486.10	0.3762	101.33	OX
4-Ethylphenol	$C_8H_{10}O$	Quinoline	C_9H_7N	513.45	0.2832	101.33	ON
Ethyl propyl ether	$C_5H_{12}O$	Methanol	CH_4O	330.00	0.5950	112.25	OX
Fluorobenzene	C_6H_5F	Methanol	CH_4O	333.35	0.3375	101.62	OX
Fluorobenzene	C_6H_5F	2-Propanol	C_3H_8O	347.75	0.5334	101.25	OX
Formaldehyde	CH_2O	Water	H_2O	355.75	0.0700	53.33	OX
Formic acid	CH_2O_2	Water	H_2O	380.35	0.5728	101.33	ON
Formic acid	CH_2O_2	*m*-Xylene	C_8H_{10}	365.95	0.8545	101.33	EX
Heptane	C_7H_{16}	Methanol	CH_4O	331.95	0.2721	101.33	OX
Heptane	C_7H_{16}	Methyl acetate	$C_3H_6O_2$	323.15	0.0430	79.48	OX
Heptane	C_7H_{16}	3-Methyl-1-butanol	$C_5H_{12}O$	368.15	0.8984	95.06	OX
Heptane	C_7H_{16}	2-Methyl-2-butanol	$C_5H_{12}O$	348.15	0.6860	56.83	OX
Heptane	C_7H_{16}	3-Methyl-3-buten-1-ol	$C_5H_{10}O$	370.00	0.7900	101.30	OX
Heptane	C_7H_{16}	Methyl methacrylate	$C_5H_8O_2$	366.35	0.5403	99.94	OX
Heptane	C_7H_{16}	Nitroethane	$C_2H_5NO_2$	362.95	0.6480	101.33	OX
Heptane	C_7H_{16}	Nitromethane	CH_3NO_2	353.25	0.5210	101.33	EX
Heptane	C_7H_{16}	1-Nitropropane	$C_3H_7NO_2$	369.25	0.8370	101.33	OX
Heptane	C_7H_{16}	2-Nitropropane	$C_3H_7NO_2$	367.55	0.7080	101.33	OX
Heptane	C_7H_{16}	1-Pentanol	$C_5H_{12}O$	371.45	0.9424	101.33	OX
Heptane	C_7H_{16}	3-Pentanol	$C_5H_{12}O$	368.15	0.7999	98.62	OX
Heptane	C_7H_{16}	Perfluoroheptane	C_7F_{16}	328.16	0.3900	53.60	OX
Heptane	C_7H_{16}	1-Propanol	C_3H_8O	357.65	0.5170	101.33	OX
Heptane	C_7H_{16}	2-Propanol	C_3H_8O	349.55	0.3977	101.33	OX
Heptane	C_7H_{16}	Propyl acetate	$C_5H_{10}O_2$	366.75	0.5785	101.38	OX
Heptane	C_7H_{16}	Pyridine	C_5H_5N	368.61	0.6998	101.33	OX
Heptane	C_7H_{16}	Water	H_2O	352.35	0.5490	101.33	EX
1-Heptanol	$C_7H_{16}O$	Nonane	C_9H_{20}	423.45	0.1071	101.33	OX
1-Heptanol	$C_7H_{16}O$	Undecane	$C_{11}H_{24}$	447.85	0.8014	101.33	OX
1-Heptanol	$C_7H_{16}O$	Water	H_2O	371.99	0.0297	101.33	EX
1-Heptene	C_7H_{14}	Water	H_2O	350.20	0.5900	101.33	EX
Hexafluorobenzene	C_6F_6	Methanol	CH_4O	318.15	0.3900	61.73	OX
1,1,1,2,3,3-Hexafluoro-3-(2,2,2-trifluoroethoxy)propane	$C_5H_3F_9O$	Methanol	CH_4O	330.67	0.4400	101.30	OX
1,1,1,2,3,3-Hexafluoro-3-(2,2,2-trifluoroethoxy)propane	$C_5H_3F_9O$	2-Propanol	C_3H_8O	341.23	0.6580	101.30	OX
Hexane	C_6H_{14}	Methanol	CH_4O	333.15	0.4840	149.64	OX
Hexane	C_6H_{14}	Methyl acetate	$C_3H_6O_2$	326.65	0.3410	106.66	OX
Hexane	C_6H_{14}	2-Methyl-2-butanol	$C_5H_{12}O$	339.06	0.9564	93.55	OX
Hexane	C_6H_{14}	Methyl formate	$C_2H_4O_2$	302.65	0.1510	101.33	OX
Hexane	C_6H_{14}	2-Methyl-2-propanol	$C_4H_{10}O$	337.70	0.7498	101.30	OX
Hexane	C_6H_{14}	1-Propanol	C_3H_8O	348.15	0.8100	137.23	OX
Hexane	C_6H_{14}	2-Propanol	C_3H_8O	338.15	0.7100	112.66	OX
Hexane	C_6H_{14}	Tetrahydrofuran	C_4H_8O	323.15	0.4100	65.83	OX
Hexane	C_6H_{14}	Trichloromethane	$CHCl_3$	333.45	0.2160	101.33	OX
Hexane	C_6H_{14}	Tris(perfluoroethyl)amine	$C_6F_{15}N$	327.65	0.5160	101.33	OX
Hexane	C_6H_{14}	Vinyl acetate	$C_4H_6O_2$	335.25	0.5550	101.33	OX
Hexane	C_6H_{14}	Water	H_2O	334.75	0.7890	101.33	EX
1-Hexanol	$C_6H_{14}O$	Nonane	C_9H_{20}	416.95	0.3649	101.33	OX
1-Hexanol	$C_6H_{14}O$	Octane	C_8H_{18}	398.55	0.0886	101.33	OX
1-Hexanol	$C_6H_{14}O$	Water	H_2O	367.89	0.0568	101.33	EX
1-Hexene	C_6H_{12}	Methyl acetate	$C_3H_6O_2$	323.15	0.3660	92.08	OX
1-Hexene	C_6H_{12}	2-Methyl-3-buten-2-ol	$C_5H_{10}O$	336.55	0.9521	101.30	OX
1-Hexene	C_6H_{12}	2-Methyl-2-propanol	$C_4H_{10}O$	333.25	0.7350	101.30	OX
1-Hexene	C_6H_{12}	Water	H_2O	318.15	0.8490	63.35	EX
Hydrogen chloride	ClH	Water	H_2O	389.34	0.1083	133.32	ONC
Hydrogen fluoride	FH	Trichlorofluoromethane	CCl_3F	283.15	0.7840	129.45	EX
Hydrogen fluoride	FH	Water	H_2O	382.15	0.3508	101.33	ON
Isobutanal	C_4H_8O	Water	H_2O	332.80	0.8302	100.99	EX
Isobutyl acetate	$C_6H_{12}O_2$	Water	H_2O	361.05	0.3560	101.33	EX
Isobutylbenzene	$C_{10}H_{14}$	Phenol	C_6H_6O	441.75	0.6478	101.33	OX
Isobutyl formate	$C_5H_{10}O_2$	2-Methyl-1-propanol	$C_4H_{10}O$	370.90	0.8070	101.33	OX
Isobutyl formate	$C_5H_{10}O_2$	Water	H_2O	352.75	0.5540	101.33	EX

Component 1	Mol. form. 1	Component 2	Mol. form. 2	T_{AZ}/K	$y_{1,AZ}$	P_{AZ}/kPa	Type
4-Isobutylphenol	$C_{10}H_{14}O$	Quinoline	C_9H_7N	515.95	0.4939	101.33	ON
Isopentane	C_5H_{12}	Methanol	CH_4O	297.05	0.9070	101.33	OX
Isopentane	C_5H_{12}	2-Methyl-1-buten-3-yne	C_5H_6	299.35	0.6380	101.33	OX
Isopentane	C_5H_{12}	3-Methyl-1-butyne	C_5H_8	297.15	0.4350	101.33	OX
Isopentane	C_5H_{12}	Methyl formate	$C_2H_4O_2$	291.55	0.5080	101.33	OX
Isopentane	C_5H_{12}	2-Propanol	C_3H_8O	298.15	0.8630	101.33	OX
Isopentane	C_5H_{12}	Tris(perfluoroethyl)amine	$C_6F_{15}N$	299.65	0.9020	101.33	OX
Isopentyl acetate	$C_7H_{14}O_2$	3-Methyl-1-butanol	$C_5H_{12}O$	403.95	0.0100	101.33	OX
Isopentyl acetate	$C_7H_{14}O_2$	1-Pentanol	$C_5H_{12}O$	407.45	0.4000	101.33	OX
Isopentyl acetate	$C_7H_{14}O_2$	Water	H_2O	367.05	0.2010	101.46	EX
Isopropenylbenzene	C_9H_{10}	Water	H_2O	369.95	0.1120	101.33	EX
Isopropyl acetate	$C_5H_{10}O_2$	Water	H_2O	349.75	0.6040	101.33	EX
Isopropylbenzene	C_9H_{12}	Water	H_2O	368.15	0.1660	101.33	EX
1-Isopropyl-2-methylbenzene	$C_{10}H_{14}$	Phenol	C_6H_6O	443.75	0.5357	101.33	OX
1-Isopropyl-3-methylbenzene	$C_{10}H_{14}$	Phenol	C_6H_6O	443.05	0.5973	101.33	OX
1-Isopropyl-4-methylbenzene	$C_{10}H_{14}$	Phenol	C_6H_6O	443.65	0.5570	101.33	OX
2-Isopropylphenol	$C_9H_{12}O$	Naphthalene	$C_{10}H_8$	483.15	0.5102	101.33	OX
2-Isopropylphenol	$C_9H_{12}O$	Quinoline	C_9H_7N	512.75	0.1985	101.33	ON
2-Isopropylphenol	$C_9H_{12}O$	1,2,4,5-Tetramethylbenzene	$C_{10}H_{14}$	476.25	0.1036	101.33	OX
3-Isopropylphenol	$C_9H_{12}O$	Quinoline	C_9H_7N	514.70	0.3891	101.33	ON
Isoquinoline	C_9H_7N	2-Methylnaphthalene	$C_{11}H_{10}$	513.90	0.2074	101.33	OX
Isoquinoline	C_9H_7N	3,4-Xylenol	$C_8H_{10}O$	519.75	0.7045	101.33	ON
Isoquinoline	C_9H_7N	3,5-Xylenol	$C_8H_{10}O$	518.05	0.8085	101.33	ON
Methacrylic acid	$C_4H_6O_2$	Water	H_2O	372.25	0.0536	98.93	OX
Methanol	CH_4O	1,3-Cyclopentadiene	C_5H_6	309.05	0.2120	101.33	OX
Methanol	CH_4O	2-Methoxy-2-methylbutane	$C_6H_{14}O$	335.55	0.7735	101.69	OX
Methanol	CH_4O	Methyl acetate	$C_3H_6O_2$	328.15	0.3480	107.19	OX
Methanol	CH_4O	2-Methyl-1,3-butadiene	C_5H_8	303.55	0.1670	101.33	OX
Methanol	CH_4O	2-Methyl-1-butene	C_5H_{10}	300.55	0.1720	101.33	OX
Methanol	CH_4O	3-Methyl-1-butene	C_5H_{10}	291.05	0.0890	101.33	OX
Methanol	CH_4O	2-Methyl-2-butene	C_5H_{10}	306.25	0.2160	101.33	OX
Methanol	CH_4O	Methyl *tert*-butyl ether	$C_5H_{12}O$	325.00	0.3140	103.15	OX
Methanol	CH_4O	1-Methylcyclobutene	C_5H_8	304.85	0.1900	101.33	OX
Methanol	CH_4O	Methylcyclohexane	C_7H_{14}	333.15	0.7520	102.87	EX
Methanol	CH_4O	2-Methyl-1-pentene	C_6H_{12}	330.00	0.4517	141.80	OX
Methanol	CH_4O	Nonane	C_9H_{20}	337.25	0.9526	101.33	OX
Methanol	CH_4O	Octane	C_8H_{18}	335.55	0.8830	101.33	LX
Methanol	CH_4O	*cis*-1,3-Pentadiene	C_5H_8	311.10	0.2300	101.33	OX
Methanol	CH_4O	*trans*-1,3-Pentadiene	C_5H_8	309.65	0.2110	101.33	OX
Methanol	CH_4O	Pentane	C_5H_{12}	303.20	0.1930	101.30	OX
Methanol	CH_4O	1-Pentene	C_5H_{10}	300.05	0.1469	102.47	OX
Methanol	CH_4O	Tetrahydrofuran	C_4H_8O	332.75	0.5040	101.33	OX
Methanol	CH_4O	*N,N,N',N'*-Tetramethylmethanediamine	$C_5H_{14}N_2$	335.15	0.7670	101.33	OX
Methanol	CH_4O	Toluene	C_7H_8	336.65	0.8820	101.33	OX
Methanol	CH_4O	Trichloromethane	$CHCl_3$	328.15	0.3520	107.99	OX
Methanol	CH_4O	2,2,3-Trimethyloxirane	$C_5H_{10}O$	334.95	0.6590	101.33	OX
Methanol	CH_4O	Vinyl acetate	$C_4H_6O_2$	332.05	0.6182	101.33	OX
2-Methoxyethanol	$C_3H_8O_2$	1-Octene	C_8H_{16}	380.75	0.4700	101.33	OX
2-Methoxyethanol	$C_3H_8O_2$	*cis*-4-Octene	C_8H_{16}	381.25	0.4900	101.33	OX
2-Methoxyethanol	$C_3H_8O_2$	*trans*-4-Octene	C_8H_{16}	381.05	0.4900	101.33	OX
2-Methoxyethanol	$C_3H_8O_2$	Styrene	C_8H_8	393.95	0.7787	98.93	OX
2-Methoxyethanol	$C_3H_8O_2$	Water	H_2O	372.65	0.0559	99.99	OX
2-Methoxyethanol	$C_3H_8O_2$	*o*-Xylene	C_8H_{10}	392.65	0.7127	98.79	OX
2-Methoxyethanol	$C_3H_8O_2$	*m*-Xylene	C_8H_{10}	392.15	0.6397	99.73	OX
2-Methoxyethanol	$C_3H_8O_2$	*p*-Xylene	C_8H_{10}	392.65	0.6303	99.99	OX
2-Methoxy-2-methylbutane	$C_6H_{14}O$	2-Methyl-2-propanol	$C_4H_{10}O$	353.20	0.4383	101.80	OX
Methyl acetate	$C_3H_6O_2$	Methylcyclopentane	C_6H_{12}	325.85	0.6917	99.50	OX
Methyl acetate	$C_3H_6O_2$	2-Methyl-1-pentene	C_6H_{12}	325.15	0.5931	100.38	OX
Methyl acetate	$C_3H_6O_2$	Trichloromethane	$CHCl_3$	337.51	0.3240	101.33	ON
Methyl acetate	$C_3H_6O_2$	Water	H_2O	330.05	0.8940	103.62	LX
2-Methylaniline	C_7H_9N	Tridecane	$C_{13}H_{28}$	472.55	0.9070	101.33	OX
2-Methylaniline	C_7H_9N	Undecane	$C_{11}H_{24}$	461.40	0.4930	101.33	OX
2-Methyl-1,3-butadiene	C_5H_8	2-Methyl-1-buten-3-yne	C_5H_6	305.88	0.2790	101.33	OX

Fluids

Component 1	Mol. form. 1	Component 2	Mol. form. 2	T_{AZ}/K	$y_{1,AZ}$	P_{AZ}/kPa	Type
2-Methyl-1,3-butadiene	C_5H_8	Methyl formate	$C_2H_4O_2$	298.90	0.4850	101.33	OX
2-Methyl-1,3-butadiene	C_5H_8	Pentane	C_5H_{12}	310.55	0.7421	114.66	OX
2-Methyl-1,3-butadiene	C_5H_8	Propanal	C_3H_6O	306.35	0.8300	101.33	OX
2-Methyl-1,3-butadiene	C_5H_8	2-Propanol	C_3H_8O	307.05	0.9850	101.33	OX
2-Methyl-1,3-butadiene	C_5H_8	Tris(perfluoroethyl)amine	$C_6F_{15}N$	303.35	0.8200	101.33	EX
2-Methyl-1,3-butadiene	C_5H_8	Water	H_2O	305.85	0.9480	101.33	EX
2-Methyl-1-butanol, (±)-	$C_5H_{12}O$	*o*-Xylene	C_8H_{10}	402.05	0.7417	101.87	OX
2-Methyl-1-butanol, (±)-	$C_5H_{12}O$	*m*-Xylene	C_8H_{10}	400.65	0.6316	101.85	OX
2-Methyl-1-butanol, (±)-	$C_5H_{12}O$	*p*-Xylene	C_8H_{10}	400.15	0.6273	101.07	OX
3-Methyl-1-butanol	$C_5H_{12}O$	Toluene	C_7H_8	383.15	0.1250	101.33	OX
3-Methyl-1-butanol	$C_5H_{12}O$	Water	H_2O	367.97	0.1735	101.33	EX
2-Methyl-2-butanol	$C_5H_{12}O$	Methylcyclohexane	C_7H_{14}	366.60	0.3965	99.87	OX
2-Methyl-2-butanol	$C_5H_{12}O$	Methylcyclopentane	C_6H_{12}	344.75	0.0551	101.80	OX
2-Methyl-2-butanol	$C_5H_{12}O$	Water	H_2O	360.85	0.3645	101.75	EX
3-Methyl-2-butanone	$C_5H_{10}O$	2-Propanol	C_3H_8O	354.75	0.1500	101.33	OX
2-Methyl-1-butene	C_5H_{10}	2-Methyl-1-buten-3-yne	C_5H_6	303.15	0.6550	101.33	OX
2-Methyl-1-butene	C_5H_{10}	Tris(perfluoroethyl)amine	$C_6F_{15}N$	301.95	0.8450	101.33	OX
2-Methyl-2-butene	C_5H_{10}	Methyl formate	$C_2H_4O_2$	297.75	0.4240	101.33	OX
2-Methyl-2-butene	C_5H_{10}	Nitromethane	CH_3NO_2	311.15	0.9430	101.33	LX
2-Methyl-2-butene	C_5H_{10}	1-Pentyne	C_5H_8	310.95	0.6700	101.33	OX
2-Methyl-2-butene	C_5H_{10}	2-Propanol	C_3H_8O	310.95	0.9540	101.33	OX
2-Methyl-2-butene	C_5H_{10}	Tris(perfluoroethyl)amine	$C_6F_{15}N$	307.65	0.8170	101.33	OX
2-Methyl-2-butene	C_5H_{10}	Water	H_2O	309.75	0.9350	101.33	EX
3-Methyl-3-buten-1-ol	$C_5H_{10}O$	Toluene	C_7H_8	381.55	0.2391	101.60	OX
3-Methyl-2-buten-1-ol	$C_5H_{10}O$	Water	H_2O	369.55	0.0859	101.33	EX
3-Methyl-3-buten-1-ol	$C_5H_{10}O$	Water	H_2O	333.15	0.1320	101.33	EX
2-Methyl-3-buten-2-ol	$C_5H_{10}O$	Tetrachloromethane	CCl_4	348.45	0.0991	101.06	OX
2-Methyl-3-buten-2-ol	$C_5H_{10}O$	Toluene	C_7H_8	366.55	0.7788	101.20	OX
2-Methyl-3-buten-2-ol	$C_5H_{10}O$	Water	H_2O	359.25	0.4230	101.33	LX
1-Methylcyclobutene	C_5H_8	Acetone	C_3H_6O	307.75	0.7780	101.33	OX
1-Methylcyclobutene	C_5H_8	Dimethoxymethane	$C_3H_8O_2$	309.05	0.7100	101.33	OX
Methylcyclohexane	C_7H_{14}	Methyl propanoate	$C_4H_8O_2$	352.45	0.1044	101.33	OX
Methylcyclohexane	C_7H_{14}	Nitromethane	CH_3NO_2	354.85	0.4877	101.33	EX
Methylcyclohexane	C_7H_{14}	3-Pentanone	$C_5H_{10}O$	366.95	0.5559	99.82	OX
Methylcyclohexane	C_7H_{14}	2-Propanol	C_3H_8O	350.85	0.3470	101.33	OX
Methylcyclohexane	C_7H_{14}	Propyl acetate	$C_5H_{10}O_2$	368.40	0.5254	100.90	OX
Methylcyclopentane	C_6H_{12}	2-Methyl-1-propanol	$C_4H_{10}O$	343.15	0.9433	100.35	OX
Methylcyclopentane	C_6H_{12}	2-Methyl-2-propanol	$C_4H_{10}O$	339.35	0.7441	99.93	OX
Methylcyclopentane	C_6H_{12}	1-Propanol	C_3H_8O	340.85	0.8271	101.19	OX
Methylcyclopentane	C_6H_{12}	2-Propanol	C_3H_8O	336.45	0.7100	98.14	OX
Methylenecyclobutane	C_5H_8	2-Methyl-2-propanol	$C_4H_{10}O$	314.65	0.9850	101.33	OX
Methylenecyclobutane	C_5H_8	Propanal	C_3H_6O	311.30	0.6400	101.33	OX
Methylenecyclobutane	C_5H_8	Water	H_2O	313.15	0.9788	101.30	EX
4-Methylenetetrahydropyran	$C_6H_{10}O$	Toluene	C_7H_8	381.15	0.5253	101.20	OX
Methyl formate	$C_2H_4O_2$	Pentane	C_5H_{12}	294.85	0.5740	101.33	OX
5-Methyl-1-heptanol, (±)-	$C_8H_{18}O$	1,2-Propanediol	$C_3H_8O_2$	456.85	0.4309	101.33	OX
3-Methylisoquinoline	$C_{10}H_9N$	3,4-Xylenol	$C_8H_{10}O$	524.35	0.9189	101.33	ON
Methyl methacrylate	$C_5H_8O_2$	Octane	C_8H_{18}	373.70	0.9651	100.16	OX
Methyl methacrylate	$C_5H_8O_2$	Water	H_2O	354.45	0.5004	101.33	EX
2-Methylnaphthalene	$C_{11}H_{10}$	Quinoline	C_9H_7N	511.05	0.0787	101.33	OX
2-Methyl-1-pentene	C_6H_{12}	Tetrahydrofuran	C_4H_8O	334.65	0.7133	101.29	OX
2-Methyl-1-pentene	C_6H_{12}	Trichloromethane	$CHCl_3$	333.95	0.3765	101.19	OX
Methyl propanoate	$C_4H_8O_2$	Water	H_2O	344.75	0.6950	101.33	EX
2-Methyl-1-propanol	$C_4H_{10}O$	Nitroethane	$C_2H_5NO_2$	375.81	0.5920	101.33	OX
2-Methyl-1-propanol	$C_4H_{10}O$	Octane	C_8H_{18}	376.58	0.6700	101.30	OX
2-Methyl-1-propanol	$C_4H_{10}O$	Tetrachloromethane	CCl_4	348.95	0.0920	101.33	OX
2-Methyl-1-propanol	$C_4H_{10}O$	Toluene	C_7H_8	374.35	0.4941	101.33	OX
2-Methyl-1-propanol	$C_4H_{10}O$	*m*-Xylene	C_8H_{10}	380.35	0.9300	101.33	OX
2-Methyl-1-propanol	$C_4H_{10}O$	*p*-Xylene	C_8H_{10}	380.30	0.9200	101.33	OX
2-Methyl-2-propanol	$C_4H_{10}O$	Octane	C_8H_{18}	343.15	0.9680	61.18	OX
2-Methyl-2-propanol	$C_4H_{10}O$	2-Propanol	C_3H_8O	343.05	0.4449	60.27	ON
2-Methyl-2-propanol	$C_4H_{10}O$	Toluene	C_7H_8	353.44	0.9200	93.61	OX
2-Methyl-2-propanol	$C_4H_{10}O$	1,1,2-Trichloro-1,2,2-trifluoroethane	$C_2Cl_3F_3$	319.95	0.0574	101.09	OX

Component 1	Mol. form. 1	Component 2	Mol. form. 2	T_{AZ}/K	$y_{1,AZ}$	P_{AZ}/kPa	Type
2-Methyl-2-propanol	$C_4H_{10}O$	2,2,4-Trimethylpentane	C_8H_{18}	339.28	0.6040	59.49	OX
2-Methyl-2-propanol	$C_4H_{10}O$	Water	H_2O	353.00	0.5989	101.33	OX
2-Methylpropenal	C_4H_6O	Tetrachloromethane	CCl_4	339.15	0.4000	97.86	OX
1-Methyl-2-propylbenzene	$C_{10}H_{14}$	Phenol	C_6H_6O	447.75	0.4199	101.33	OX
1-Methyl-3-propylbenzene	$C_{10}H_{14}$	Phenol	C_6H_6O	446.35	0.4736	101.33	OX
1-Methyl-4-propylbenzene	$C_{10}H_{14}$	Phenol	C_6H_6O	447.15	0.4425	101.33	OX
2-Methylpyridine	C_6H_7N	Nonane	C_9H_{20}	402.35	0.8790	101.33	OX
2-Methylpyridine	C_6H_7N	Octane	C_8H_{18}	394.27	0.4610	101.33	OX
2-Methylpyridine	C_6H_7N	Phenol	C_6H_6O	458.33	0.2148	101.32	ON
3-Methylpyridine	C_6H_7N	Phenol	C_6H_6O	462.93	0.3082	101.32	ON
4-Methylpyridine	C_6H_7N	Water	H_2O	370.50	0.1028	101.33	OX
2-Methylquinoline	$C_{10}H_9N$	3,4-Xylenol	$C_8H_{10}O$	521.17	0.8353	101.33	ON
3-Methylquinoline	$C_{10}H_9N$	3,4-Xylenol	$C_8H_{10}O$	523.60	0.8848	101.33	ON
7-Methylquinoline	$C_{10}H_9N$	3,4-Xylenol	$C_8H_{10}O$	525.85	0.9534	101.33	ON
2-Methylquinoline	$C_{10}H_9N$	3,5-Xylenol	$C_8H_{10}O$	520.65	0.9906	101.33	ON
Molybdenum(V) chloride	Cl_5Mo	Tungsten(VI) chloride	Cl_6W	274.70	0.9750	101.33	OX
Naphthalene	$C_{10}H_8$	2,4,6-Trimethylphenol	$C_9H_{12}O$	486.70	0.6839	101.33	OX
Naphthalene	$C_{10}H_8$	2,3-Xylenol	$C_8H_{10}O$	485.45	0.5877	101.33	OX
Naphthalene	$C_{10}H_8$	2,4-Xylenol	$C_8H_{10}O$	481.25	0.3565	101.33	OX
Naphthalene	$C_{10}H_8$	2,5-Xylenol	$C_8H_{10}O$	481.25	0.3565	101.33	OX
Naphthalene	$C_{10}H_8$	2,6-Xylenol	$C_8H_{10}O$	475.70	0.0619	101.33	OX
Naphthalene	$C_{10}H_8$	3,4-Xylenol	$C_8H_{10}O$	490.95	0.8842	101.33	OX
Naphthalene	$C_{10}H_8$	3,5-Xylenol	$C_8H_{10}O$	489.33	0.7399	101.33	OX
Niobium(V) chloride	Cl_5Nb	Phosphoryl chloride	Cl_3OP	536.15	0.5980	101.33	ON
Nitric acid	HNO_3	Water	H_2O	393.20	0.3820	101.33	ON
Nitromethane	CH_3NO_2	Nonane	C_9H_{20}	369.29	0.8403	99.73	EX
Nitromethane	CH_3NO_2	Octane	C_8H_{18}	363.38	0.6964	99.73	EX
Nitromethane	CH_3NO_2	Undecane	$C_{11}H_{24}$	373.16	0.9619	99.73	EX
Nitromethane	CH_3NO_2	Water	H_2O	356.90	0.4840	101.33	EX
Nonane	C_9H_{20}	1-Pentanol	$C_5H_{12}O$	404.45	0.3758	101.33	OX
Nonane	C_9H_{20}	Pentyl acetate	$C_7H_{14}O_2$	419.20	0.4620	101.32	OX
Nonane	C_9H_{20}	Phenol	C_6H_6O	419.18	0.7820	101.32	OX
Nonane	C_9H_{20}	1-Propanol	C_3H_8O	369.95	0.0775	101.33	OX
Nonane	C_9H_{20}	Pyridine	C_5H_5N	388.15	0.0650	101.33	OX
Nonane	C_9H_{20}	Water	H_2O	367.95	0.1720	101.33	EX
Nonane	C_9H_{20}	*o*-Xylene	C_8H_{10}	417.40	0.1502	101.33	OX
1-Nonanol	$C_9H_{20}O$	Undecane	$C_{11}H_{24}$	468.45	0.0925	101.33	OX
1-Nonanol	$C_9H_{20}O$	Water	H_2O	373.00	0.0154	101.33	EX
1-Nonene	C_9H_{18}	Phenol	C_6H_6O	413.15	0.8703	86.57	OX
Octane	C_8H_{18}	1-Pentanol	$C_5H_{12}O$	393.15	0.7153	101.33	OX
Octane	C_8H_{18}	Phenol	C_6H_6O	398.17	0.9310	101.32	OX
Octane	C_8H_{18}	1-Propanol	C_3H_8O	366.85	0.2517	101.33	OX
Octane	C_8H_{18}	2-Propanol	C_3H_8O	354.63	0.1010	101.33	OX
Octane	C_8H_{18}	Tetrachloroethene	C_2Cl_4	371.90	0.1219	53.44	OX
Octane	C_8H_{18}	Water	H_2O	362.75	0.3150	101.33	EX
1-Octanol	$C_8H_{18}O$	Undecane	$C_{11}H_{24}$	460.05	0.4772	101.33	OX
1-Octanol	$C_8H_{18}O$	Water	H_2O	372.75	0.0180	101.33	EX
1-Octene	C_8H_{16}	Tetrachloroethene	C_2Cl_4	393.15	0.4100	101.33	OX
cis-4-Octene	C_8H_{16}	Tetrachloroethene	C_2Cl_4	393.65	0.2900	101.33	OX
trans-4-Octene	C_8H_{16}	Tetrachloroethene	C_2Cl_4	393.45	0.3300	101.33	OX
1-Octyne	C_8H_{14}	1-Propanol	C_3H_8O	369.00	0.1400	101.33	OX
Pentane	C_5H_{12}	1-Pentyne	C_5H_8	307.55	0.6950	101.33	OX
1-Pentanol	$C_5H_{12}O$	Water	H_2O	369.08	0.1367	101.33	EX
2-Pentanol	$C_5H_{12}O$	Water	H_2O	363.15	0.2450	92.49	EX
3-Pentanone	$C_5H_{10}O$	Water	H_2O	356.05	0.5250	101.33	EX
Phenol	C_6H_6O	Propylbenzene	C_9H_{12}	428.15	0.1150	91.85	OX
Phenol	C_6H_6O	Tetradecane	$C_{14}H_{30}$	452.48	0.9650	101.32	OX
Phenol	C_6H_6O	1,2,3,4-Tetrahydronaphthalene	$C_{10}H_{12}$	448.15	0.9031	84.25	OX
Phenol	C_6H_6O	1,2,3,5-Tetramethylbenzene	$C_{10}H_{14}$	454.25	0.7957	101.33	OX
Phenol	C_6H_6O	1,2,4,5-Tetramethylbenzene	$C_{10}H_{14}$	453.36	0.7857	101.33	OX
Phenol	C_6H_6O	1,2,3-Trimethylbenzene	C_9H_{12}	443.45	0.3936	101.33	OX
Phenol	C_6H_6O	1,2,4-Trimethylbenzene	C_9H_{12}	440.65	0.2409	101.33	OX
Phenol	C_6H_6O	1,3,5-Trimethylbenzene	C_9H_{12}	436.95	0.1828	101.33	OX

Component 1	Mol. form. 1	Component 2	Mol. form. 2	T_{AZ}/K	$y_{1,AZ}$	P_{AZ}/kPa	Type
Phenol	C_6H_6O	1-Undecene	$C_{11}H_{22}$	443.15	0.6426	92.31	OX
Phosphorus(III) chloride	Cl_3P	Sulfuryl chloride	Cl_2O_2S	364.15	0.5000	101.33	ON
Phosphorus(III) chloride	Cl_3P	Thionyl chloride	Cl_2OS	345.85	0.5800	101.33	OX
Phosphoryl chloride	Cl_3OP	Tantalum (V) chloride	Cl_5Ta	558.85	0.4650	101.33	ON
Propanal	C_3H_6O	Water	H_2O	320.65	0.9400	101.33	LX
1,2-Propanediol diacetate	$C_7H_{12}O_4$	Water	H_2O	358.15	0.0260	59.41	EX
Propanoic acid	$C_3H_6O_2$	Pyridine	C_5H_5N	421.75	0.6860	101.33	ON
Propanoic acid	$C_3H_6O_2$	Water	H_2O	373.05	0.0500	101.33	OX
1-Propanol	C_3H_8O	Propyl acetate	$C_5H_{10}O_2$	367.88	0.6190	101.33	OX
1-Propanol	C_3H_8O	Styrene	C_8H_8	369.08	0.9884	98.13	OX
1-Propanol	C_3H_8O	Tetrachloromethane	CCl_4	346.28	0.1968	101.33	OX
1-Propanol	C_3H_8O	1,1,2,2-Tetrafluoroethyl 1,1,1-trifluoroethyl ether	$C_4H_3F_7O$	329.23	0.0350	101.30	OX
1-Propanol	C_3H_8O	Toluene	C_7H_8	365.35	0.6770	101.33	OX
1-Propanol	C_3H_8O	2,2,4-Trimethylpentane	C_8H_{18}	357.89	0.4580	101.30	OX
1-Propanol	C_3H_8O	Water	H_2O	360.80	0.4320	101.33	OX
1-Propanol	C_3H_8O	*o*-Xylene	C_8H_{10}	369.85	0.9886	98.66	OX
1-Propanol	C_3H_8O	*m*-Xylene	C_8H_{10}	369.90	0.9531	99.06	OX
1-Propanol	C_3H_8O	*p*-Xylene	C_8H_{10}	369.60	0.9531	99.99	OX
2-Propanol	C_3H_8O	Tetrachloromethane	CCl_4	341.83	0.3314	101.33	OX
2-Propanol	C_3H_8O	Toluene	C_7H_8	354.65	0.8370	101.33	OX
2-Propanol	C_3H_8O	Trichloromethane	$CHCl_3$	334.15	0.0500	101.33	OX
2-Propanol	C_3H_8O	1,1,2-Trichloro-1,2,2-trifluoroethane	$C_2Cl_3F_3$	319.35	0.0841	100.95	OX
2-Propanol	C_3H_8O	2,2,3-Trimethyloxirane	$C_5H_{10}O$	346.10	0.1400	101.33	OX
2-Propanol	C_3H_8O	2,2,4-Trimethylpentane	C_8H_{18}	349.58	0.6350	101.30	OX
2-Propanol	C_3H_8O	Water	H_2O	353.70	0.6740	101.33	OX
Propyl acetate	$C_5H_{10}O_2$	Water	H_2O	355.91	0.4772	101.33	EX
Propyl formate	$C_4H_8O_2$	Water	H_2O	344.85	0.6910	101.33	EX
2-Propylphenol	$C_9H_{12}O$	Quinoline	C_9H_7N	513.60	0.2757	101.33	ON
3-Propylphenol	$C_9H_{12}O$	Quinoline	C_9H_7N	514.70	0.3891	101.33	ON
4-Propylphenol	$C_9H_{12}O$	Quinoline	C_9H_7N	515.35	0.4549	101.33	ON
Propyl propanoate	$C_6H_{12}O_2$	Water	H_2O	362.05	0.3400	101.33	EX
Pyridine	C_5H_5N	Toluene	C_7H_8	383.19	0.2250	101.33	OX
Pyridine	C_5H_5N	Water	H_2O	367.30	0.2500	101.33	OX
Pyrrole	C_4H_5N	Water	H_2O	348.15	0.2486	50.13	EX
Quinoline	C_9H_7N	2,3-Xylenol	$C_8H_{10}O$	513.30	0.7316	101.33	ON
Quinoline	C_9H_7N	2,4-Xylenol	$C_8H_{10}O$	512.30	0.8283	101.33	ON
Quinoline	C_9H_7N	2,5-Xylenol	$C_8H_{10}O$	512.30	0.8283	101.33	ON
Quinoline	C_9H_7N	2,6-Xylenol	$C_8H_{10}O$	511.00	0.9110	101.33	ON
Quinoline	C_9H_7N	3,4-Xylenol	$C_8H_{10}O$	514.77	0.6093	101.33	ON
Quinoline	C_9H_7N	3,5-Xylenol	$C_8H_{10}O$	513.58	0.6713	101.33	ON
Styrene	C_8H_8	Water	H_2O	367.15	0.2000	101.33	EX
Tetrachloroethene	C_2Cl_4	1,1,2-Trichloroethane	$C_2H_3Cl_3$	385.95	0.2115	101.33	OX
Tetrahydrofuran	C_4H_8O	Water	H_2O	336.67	0.8172	101.33	OX
1,2,4,5-Tetramethylbenzene	$C_{10}H_{14}$	2,4-Xylenol	$C_8H_{10}O$	474.05	0.8131	101.33	OX
1,2,4,5-Tetramethylbenzene	$C_{10}H_{14}$	2,5-Xylenol	$C_8H_{10}O$	474.35	0.8237	101.33	OX
1,2,4,5-Tetramethylbenzene	$C_{10}H_{14}$	2,6-Xylenol	$C_8H_{10}O$	468.85	0.6520	101.33	OX
Toluene	C_7H_8	Water	H_2O	357.25	0.4770	101.33	EX
Tributylamine	$C_{12}H_{27}N$	Water	H_2O	372.80	0.0238	101.46	EX
Trichloroethene	C_2HCl_3	Water	H_2O	346.55	0.6440	101.33	EX
Trichloromethane	$CHCl_3$	Water	H_2O	329.27	0.8397	101.33	EX
1,1,2-Trichloro-1,2,2-trifluoroethane	$C_2Cl_3F_3$	2,2,2-Trifluoroethanol	$C_2H_3F_3O$	316.58	0.7770	101.33	EX
2,2,4-Trimethylpentane	C_8H_{18}	Water	H_2O	351.95	0.5580	101.33	EX
4-Vinyl-1,3-dioxane	$C_6H_{10}O_2$	Water	H_2O	367.65	0.1045	101.33	EX
Water	H_2O	*m*-Xylene	C_8H_{10}	365.15	0.7667	101.33	EX
Water	H_2O	*p*-Xylene	C_8H_{10}	365.15	0.7450	101.33	EX

VISCOSITY OF GASES

Marcia L. Huber and Allan H. Harvey

The following table gives the viscosity of some common gases as a function of temperature. Since the difference between the viscosity at 100 kPa and the limiting value at zero pressure is generally less than 2%, these values may be used for pressures ranging from very low to normal atmospheric pressure. Uncertainties for the viscosities of gases in this table are usually less than 3%; uncertainty information on specific fluids can be found in the references. Viscosity is given in units of μPa s; note that 1 μPa s = 10^{-5} poise. Substances are listed by name.

References

1. Lemmon, E. W., and Jacobsen, R. T, Viscosity and Thermal Conductivity Equations for Nitrogen, Oxygen, Argon, and Air, *Int. J. Thermophys.* 25, 21, 2004.
2. Vogel, E., Jäger, B., Hellmann, R., and Bich, E., *Ab initio* Pair Potential Energy Curve for the Argon Atom Pair and Thermophysical Properties for the Dilute Argon Gas. II. Thermophysical Properties for Low-Density Argon, *Mol. Phys.* 108, 3335, 2010.
3. May, E. F., Berg, R. F., and Moldover, M. R., Reference Viscosities of H_2, CH_4, Ar, and Xe at Low Densities, *Int. J. Thermophys.* 28, 1085, 2007.
4. Vogel, E., Reference Viscosity of Argon at Low Density in the Temperature Range from 290 K to 680 K, *Int. J. Thermophys.* 31, 447, 2010.
5. Ho, C. Y., Ed., *Properties of Inorganic and Organic Fluids, CINDAS Data Series on Materials Properties,* Vol. V-1, Hemisphere Publishing Corp., New York, 1988.
6. Strehlow, T., and Vogel, E., Temperature Dependence and Initial Density Dependence of the Viscosity of Sulfur Hexafluoride, *Physica A* 161, 101, 1989.
7. Mehl, J. B., Huber, M. L., and Harvey, A. H., *Ab Initio* Transport Coefficients of Gaseous Hydrogen, *Int. J. Thermophys.* 31, 740, 2010.
8. Assael, M. J., Mixafendi, M., and Wakeham, W. A., The Viscosity of Normal Deuterium in the Limit of Zero Density, *J. Phys. Chem. Ref. Data* 16, 189, 1987.
9. Huber, M. L., Perkins, R. A., Laesecke, A., Friend, D. G., Sengers, J. V., Assael, M. J., Metaxa, I. M., Vogel, E., Mares, R., and Miyagawa, K., New International Formulation for the Viscosity of Water, *J. Phys. Chem. Ref. Data* 38, 101, 2009.
10. Matsunaga, N., and Nagashima, A., Transport Properties of Liquid and Gaseous D_2O over a Wide Range of Temperature and Pressure, *J. Phys. Chem. Ref. Dat*, 12, 933, 1983.
11. Schmidt, K. A. G. , Quinones-Cisneros, S. E., Carroll, J. J., and Kvamme, B., Hydrogen Sulfide Viscosity Modeling, *Energy & Fuels* 22, 3424, 2008.
12. Fenghour, A., Wakeham, W. A., Vesovic, V., Watson, J. T. R., Millat, J., and Vogel, E., The Viscosity of Ammonia, *J. Phys. Chem. Ref. Data* 24, 1649, 1995.
13. Cencek, W., Komasa, J., Przybytek, M., Mehl, J. B., Jeziorski, B., and Szalewicz, K., Effects of Aadiabatic, Relativistic, and Quantum Electrodynamics Interactions in Helium Dimer on Thermophysical Properties of Helium, *J. Chem. Phys.*, to be submitted (2011).
14. Bich, E., Millat, J., and Vogel, E., The Viscosity and Thermal Conductivity of Pure Monatomic Gases from Their Normal Boiling Point up to 5000 K in the Limit of Zero Density and at 0.101325 MPa, *J. Phys. Chem. Ref. Data* 19, 1289, 1990.
15. Seibt, D., Herrmann, S., Vogel, E., Bich, E., and Hassel, E., Simultaneous Measurements on Helium and Nitrogen with a Newly Designed Viscometer-Densimeter over a Wide Range of Temperature and Pressure, *J. Chem. Eng. Data* 54, 2626, 2009.
16. Millat, J., Vesovic, V., and Wakeham, W. A., The Viscosity of Nitrous Oxide and Tetrafluoromethane in the Limit of Zero Density, *Int. J. Thermophys.* 12, 265, 1991.
17. Bich, E., Hellmann, R., and Vogel, E., *Ab initio* Potential Energy Curve for the Neon Atom Pair and Thermophysical Properties for the Dilute Neon Gas. II. Thermophysical Properties for Low-Density Neon, *Mol. Phys.* 106, 813, 2008.
18. Fenghour, A., Wakeham, W. A., and Vesovic, V., The Viscosity of Carbon Dioxide, *J. Phys. Chem. Ref. Data* 27, 31, 1998.
19. Hellmann, R., Bich, E., Vogel, E., Dickinson, A. S., and Vesovic, V., Calculation of the Transport and Relaxation Properties of Methane. I. Shear Viscosity, Viscomagnetic Effects, and Self-Diffusion, *J. Chem. Phys.* 129, 064302, 2008.
20. Xiang, H.-W., Huber, M. L., and Laesecke, A., A New Reference Correlation for the Viscosity of Methanol, *J. Phys. Chem. Ref. Data* 35, 1597, 2006.
21. Holland, P. M., Eaton, B. E., and Hanley, H. J. M., A Correlation of the Viscosity and Thermal Conductivity Data of Gaseous and Liquid Ethylene, *J. Phys. Chem. Ref. Data* 12, 917, 1983.
22. Friend, D. G., Ingham, H., and Ely, J. F., Thermophysical Properties of Ethane, *J. Phys. Chem. Ref. Data* 20, 275, 1991.
23. Vogel, E., Küchenmeister, C., Bich, E., and Laesecke, A., Reference Correlation of the Viscosity of Propane, *J. Phys. Chem. Ref. Data* 27, 947, 1998.
24. Vogel, E., Küchenmeister, C., and Bich, E., Viscosity for n-Butane in the Fluid Region, *High Temp. – High Press.* 31, 173, 1999.
25. Vogel, E., Küchenmeister, C., and Bich, E., Viscosity Correlation for Isobutane over Wide Ranges of the Fluid Region, *Int. J. Thermophys.* 21, 343, 2000.

Name	Mol. form.	p/kPa	η(100 K)/ μPa s	η(200 K)/ μPa s	η(300 K)/ μPa s	η(400 K)/ μPa s	η(500 K)/ μPa s	η(600 K)/ μPa s	Ref.
Acetylene	C_2H_2	100			10.4	13.5	16.5		5
Air		100	7.1	13.3	18.5	23.1	27.1	30.8	1
Ammonia	H_3N	100			10.2	14.0	17.9	21.7	12
Argon	Ar	0	8.1	15.9	22.7	28.6	33.9	38.8	2,3*,4*
Boron trifluoride	BF_3	100		12.3	17.1	21.7	26.1	30.2	5
Butane	C_4H_{10}	100			7.5	9.9	12.2	14.5	24
Carbon dioxide	CO_2	100		10.1	15.0	19.7	24.0	28.0	18
Carbon monoxide	CO	100	6.7	12.9	17.8	22.1	25.8	29.1	5
Diethyl ether	$C_4H_{10}O$	100			7.6	10.1	12.4		5
Ethane	C_2H_6	100		6.4	9.4	12.2	14.8	17.1	22
Ethanol	C_2H_6O	100				11.6	14.5	17.0	5
Ethylene	C_2H_4	100		7.0	10.4	13.6	16.5	19.2	21

Name	Mol. form.	*p*/kPa	*η*(100 K)/ μPa s	*η*(200 K)/ μPa s	*η*(300 K)/ μPa s	*η*(400 K)/ μPa s	*η*(500 K)/ μPa s	*η*(600 K)/ μPa s	Ref.
Helium	He	0	9.6	15.1	19.9	24.3	28.3	32.2	13
Hexane	C_6H_{14}	100				8.6	10.8	12.8	5
Hydrogen	H_2	0	4.1	6.8	8.9	10.9	12.8	14.5	3*,7
Hydrogen-*d*$_2$	D_2	0	5.9	9.6	12.6	15.4	17.9	20.3	8
Hydrogen chloride	ClH	100			14.6	19.7	24.3		5
Hydrogen sulfide	H_2S	100			12.5	16.9	21.2	25.4	11
Isobutane	C_4H_{10}	100			7.5	9.9	12.2	14.4	25
Krypton	Kr	0		17.4	25.5	32.9	39.6	45.8	14
Methane	CH_4	0	3.9	7.7	11.1	14.2	17.0	19.5	3*,19
Methanol	CH_4O	0		6.6	9.7	13.0	16.4	19.8	20
Neon	Ne	0	14.4	24.1	31.9	38.6	44.8	50.6	17
Nitric oxide	NO	100		13.8	19.2	23.8	28.0	31.9	5
Nitrogen	N_2	100	7.0	12.9	17.9	22.2	26.1	29.6	1,15*
Nitrous oxide	N_2O	0		10.0	15.0	19.8	24.1	27.9	16
Oxygen	O_2	100	7.7	14.7	20.7	25.8	30.5	34.7	1
Pentane	C_5H_{12}	100			6.7	9.2	11.4	13.4	5
Propane	C_3H_8	100			8.2	10.8	13.3	15.6	23
Sulfur dioxide	O_2S	100		8.6	12.9	17.5	21.7		5
Sulfur hexafluoride	F_6S	0			15.3	19.7	23.8	27.6	6
Trichloromethane	$CHCl_3$	100			10.2	13.7	16.9	20.1	5
Water	H_2O	0			9.8	13.4	17.3	21.4	9
Water-*d*$_2$	D_2O	0			10.2	13.7	17.8	22.0	10
Xenon	Xe	0		15.7	23.2	30.5	37.2	43.5	3*,14

* Reference contains more accurate data covering a restricted temperature range.

VISCOSITY OF LIQUIDS

The absolute viscosity of some common liquids at temperatures between –25 °C and 100 °C is given in this table. Values were derived by fitting experimental data to suitable expressions for the temperature dependence. The substances are arranged by name. All values are given in units of millipascal seconds (mPa s); this unit is identical to centipoise (cp).

Viscosity values correspond to a nominal pressure of 1 atmosphere. However, if a value is given at a temperature above the normal boiling point, the applicable pressure is understood to be the vapor pressure of the liquid at that temperature. A few values are given at a temperature slightly below the normal freezing point; these refer to the supercooled liquid.

The accuracy ranges from 1% in the best cases to 5 to 10% in the worst cases. Additional significant figures are included in the table to facilitate interpolation.

References

1. Viswanath, D. S. and Natarajan, G., *Data Book on the Viscosity of Liquids*, Hemisphere Publishing Corp., New York, 1989.
2. Daubert, T. E., Danner, R. P., Sibul, H. M., and Stebbins, C. C., *Physical and Thermodynamic Properties of Pure Compounds: Data Compilation*, extant 1994 (core with 4 supplements), Taylor & Francis, Bristol, PA (also available as database).
3. Ho, C. Y., Ed., *CINDAS Data Series on Material Properties*, Vol. V-1, *Properties of Inorganic and Organic Fluids*, Hemisphere Publishing Corp., New York, 1988.
4. Stephan, K. and Lucas, K., *Viscosity of Dense Fluids*, Plenum Press, New York, 1979.
5. Vargaftik, N. B., *Tables of Thermophysical Properties of Liquids and Gases*, 2nd ed., John Wiley, New York, 1975.

Viscosity of Liquids as a Function of Temperature

Name	Mol. form.	η(-25 °C)/ mPa s	η(0 °C)/ mPa s	η(25 °C)/ mPa s	η(50 °C)/ mPa s	η(75 °C)/ mPa s	η(100 °C)/ mPa s
Acetic acid	$C_2H_4O_2$			1.056	0.786	0.599	0.464
Acetic anhydride	$C_4H_6O_3$		1.241	0.843	0.614	0.472	0.377
Acetone	C_3H_6O	0.540	0.395	0.306	0.247		
Acetonitrile	C_2H_3N		0.400	0.369	0.284	0.234	
Acetophenone	C_8H_8O			1.681			0.634
Acetyl chloride	C_2H_3ClO			0.368	0.294		
Allyl alcohol	C_3H_6O			1.218	0.759	0.505	
Aniline	C_6H_7N			3.85	2.03	1.247	0.850
Anisole	C_7H_8O			1.056	0.747	0.554	0.427
Benzene	C_6H_6			0.604	0.436	0.335	
Benzonitrile	C_7H_5N			1.267	0.883	0.662	0.524
Benzyl alcohol	C_7H_8O			5.47	2.76	1.618	1.055
Benzylamine	C_7H_9N			1.624	1.080	0.769	0.577
Bromine	Br_2		1.252	0.944	0.746		
Bromobenzene	C_6H_5Br		1.560	1.074	0.798	0.627	0.512
1-Bromobutane	C_4H_9Br		0.815	0.606	0.471	0.379	
Bromoethane	C_2H_5Br	0.635	0.477	0.374			
1-Bromopropane	C_3H_7Br		0.645	0.489	0.387		
2-Bromopropane	C_3H_7Br		0.612	0.458	0.359		
3-Bromopropene	C_3H_5Br		0.620	0.471	0.373		
Butanenitrile	C_4H_7N			0.553	0.418	0.330	0.268
Butanoic acid	$C_4H_8O_2$		2.22	1.426	0.982	0.714	0.542
1-Butanol	$C_4H_{10}O$	12.19	5.18	2.54	1.394	0.833	0.533
2-Butanol	$C_4H_{10}O$			3.10	1.332	0.698	0.419
2-Butanone	C_4H_8O	0.720	0.533	0.405	0.315	0.249	
Butyl acetate	$C_6H_{12}O_2$		1.002	0.685	0.500	0.383	0.305
Butylamine	$C_4H_{11}N$		0.830	0.574	0.409	0.298	
Butylbenzene	$C_{10}H_{14}$			0.950	0.683	0.515	
Butyl formate	$C_5H_{10}O_2$		0.937	0.644	0.472	0.362	0.289
Carbon disulfide	CS_2		0.429	0.352			
2-Chloroaniline	C_6H_6ClN			3.32	1.913	1.248	0.887
Chlorobenzene	C_6H_5Cl	1.703	1.058	0.753	0.575	0.456	0.369
1-Chlorobutane	C_4H_9Cl		0.556	0.422	0.329	0.261	
1-Chloro-1,1-difluoroethane	$C_2H_3ClF_2$	0.477	0.376				
Chloroethane	C_2H_5Cl	0.416	0.319				
2-Chlorophenol	C_6H_5ClO			3.59	1.835	1.131	0.786
3-Chlorophenol	C_6H_5ClO				4.04		
1-Chloropropane	C_3H_7Cl		0.436	0.334			
2-Chloropropane	C_3H_7Cl		0.401	0.303			

Name	Mol. form.	η(-25 °C)/ mPa s	η(0 °C)/ mPa s	η(25 °C)/ mPa s	η(50 °C)/ mPa s	η(75 °C)/ mPa s	η(100 °C)/ mPa s
3-Chloropropene	C_3H_5Cl		0.408	0.314			
2-Chlorotoluene	C_7H_7Cl		1.390	0.964	0.710	0.547	0.437
3-Chlorotoluene	C_7H_7Cl		1.165	0.823	0.616	0.482	0.391
4-Chlorotoluene	C_7H_7Cl			0.837	0.621	0.483	0.390
o-Cresol	C_7H_8O				3.03	1.562	0.961
m-Cresol	C_7H_8O			12.91	4.42	2.09	1.207
Cyclohexane	C_6H_{12}			0.894	0.615	0.447	
Cyclohexanol	$C_6H_{12}O$			57.5	12.29	4.27	1.98
Cyclohexanone	$C_6H_{10}O$			2.02	1.321	0.919	0.671
Cyclohexene	C_6H_{10}		0.882	0.625	0.467	0.364	
Cyclohexylamine	$C_6H_{13}N$			1.944	1.169	0.782	0.565
Cyclopentane	C_5H_{10}		0.555	0.413	0.321		
cis-Decahydronaphthalene	$C_{10}H_{18}$	12.79	5.64	3.04	1.875	1.271	0.924
trans-Decahydronaphthalene	$C_{10}H_{18}$	6.19	3.24	1.948	1.289	0.917	0.689
Decane	$C_{10}H_{22}$	2.19	1.277	0.838	0.598	0.453	0.359
Decanoic acid	$C_{10}H_{20}O_2$				4.33	2.65	
1-Decanol	$C_{10}H_{22}O$			10.91	4.59		
Diacetone alcohol	$C_6H_{12}O_2$	28.7	6.62	2.80	1.829	1.648	
1,2-Dibromoethane	$C_2H_4Br_2$			1.595	1.116	0.837	0.661
Dibromomethane	CH_2Br_2	1.948	1.320	0.980	0.779	0.652	
Dibutylamine	$C_8H_{19}N$		1.509	0.918	0.619	0.449	0.345
Dibutyl ether	$C_8H_{18}O$	1.417	0.918	0.637	0.466	0.356	0.281
Dibutyl phthalate	$C_{16}H_{22}O_4$	483	66.4	16.63	6.47	3.50	2.43
o-Dichlorobenzene	$C_6H_4Cl_2$		1.958	1.324	0.962	0.739	0.593
m-Dichlorobenzene	$C_6H_4Cl_2$		1.492	1.044	0.787	0.628	0.525
1,1-Dichloroethane	$C_2H_4Cl_2$			0.464	0.362		
1,2-Dichloroethane	$C_2H_4Cl_2$		1.125	0.779	0.576	0.447	
cis-1,2-Dichloroethene	$C_2H_2Cl_2$	0.786	0.575	0.445			
trans-1,2-Dichloroethene	$C_2H_2Cl_2$	0.522	0.398	0.317	0.261		
Dichloromethane	CH_2Cl_2	0.727	0.533	0.413			
Diethanolamine	$C_4H_{11}NO_2$				109.5	28.700	9.10
Diethylamine	$C_4H_{11}N$			0.319	0.239		
Diethylene glycol	$C_4H_{10}O_3$			30.2	11.13	4.92	2.51
Diethyl ether	$C_4H_{10}O$		0.283	0.224			
Diethyl sulfide	$C_4H_{10}S$		0.558	0.422	0.331	0.267	
Diisobutylamine	$C_8H_{19}N$		1.115	0.723	0.511	0.384	0.303
Diisopropylamine	$C_6H_{15}N$			0.393	0.300	0.237	
N,N-Dimethylacetamide	C_4H_9NO			1.927			
Dimethylamine	C_2H_7N	0.300	0.232				
N,N-Dimethylaniline	$C_8H_{11}N$		1.996	1.300	0.911	0.675	0.523
N,N-Dimethylformamide	C_3H_7NO		1.176	0.794	0.624		
Dimethyl phthalate	$C_{10}H_{10}O_4$		63.2	14.36	5.31	2.82	1.98
Dimethyl sulfide	C_2H_6S		0.356	0.284			
Dimethyl sulfoxide	C_2H_6OS			1.987	1.290		
1,4-Dioxane	$C_4H_8O_2$			1.177	0.787	0.569	
Diphenyl ether	$C_{12}H_{10}O$				2.13	1.407	1.023
Diphenylmethane	$C_{13}H_{12}$					1.265	0.929
Dipropylamine	$C_6H_{15}N$		0.751	0.517	0.377	0.288	0.228
Dipropyl ether	$C_6H_{14}O$		0.542	0.396	0.304	0.242	
Dodecane	$C_{12}H_{26}$		2.28	1.383	0.930	0.673	0.514
Epichlorohydrin	C_3H_5ClO	2.49	1.570	1.073	0.781	0.597	0.474
1,2-Ethanediol	$C_2H_6O_2$			16.06	6.55	3.34	1.975
Ethanethiol	C_2H_6S		0.364	0.287			
Ethanol	C_2H_6O	3.26	1.786	1.074	0.694	0.476	
Ethanolamine	C_2H_7NO			21.1	8.56	3.93	2.00
Ethoxybenzene	$C_8H_{10}O$			1.197	0.817	0.594	0.453
Ethyl acetate	$C_4H_8O_2$		0.578	0.423	0.325	0.259	
N-Ethylaniline	$C_8H_{11}N$		3.98	2.05	1.231	0.825	0.596
Ethylbenzene	C_8H_{10}		0.872	0.631	0.482	0.380	0.304
Ethyl butanoate	$C_6H_{12}O_2$			0.639	0.453		

Name	Mol. form.	η(-25 °C)/ mPa s	η(0 °C)/ mPa s	η(25 °C)/ mPa s	η(50 °C)/ mPa s	η(75 °C)/ mPa s	η(100 °C)/ mPa s
Ethylcyclohexane	C_8H_{16}		1.139	0.784	0.579		
Ethyl formate	$C_3H_6O_2$		0.506	0.380	0.300		
2-Ethyl-1-hexanol	$C_8H_{18}O$		20.7	6.27	2.63	1.360	0.810
Ethyl propanoate	$C_5H_{10}O_2$		0.691	0.501	0.380	0.299	0.242
Fluorobenzene	C_6H_5F		0.749	0.550	0.423	0.338	
Formamide	CH_3NO		7.11	3.34	1.833		
Formic acid	CH_2O_2			1.607	1.030	0.724	0.545
Furan	C_4H_4O	0.661	0.475	0.361			
Furfural	$C_5H_4O_2$		2.50	1.587	1.143	0.906	0.772
Glycerol	$C_3H_8O_3$			934	152	39.8	14.76
Heptane	C_7H_{16}	0.757	0.523	0.387	0.301	0.243	
Heptanoic acid	$C_7H_{14}O_2$			3.84	2.28	1.488	1.041
1-Heptanol	$C_7H_{16}O$			5.81	2.60	1.389	0.849
2-Heptanol, (±)-	$C_7H_{16}O$			3.95	1.799	0.987	0.615
3-Heptanol, (*S*)-	$C_7H_{16}O$				1.957	0.976	0.584
4-Heptanol	$C_7H_{16}O$			4.21	1.695	0.882	0.539
2-Heptanone	$C_7H_{14}O$			0.714	0.407	0.297	
1-Heptene	C_7H_{14}		0.441	0.340	0.273	0.226	
Heptylamine	$C_7H_{17}N$			1.314	0.865	0.600	0.434
Hexadecane	$C_{16}H_{34}$			3.03	1.879	1.260	0.899
Hexafluorobenzene	C_6F_6			2.79	1.730	1.151	
Hexane	C_6H_{14}		0.405	0.300	0.240		
Hexanenitrile	$C_6H_{11}N$			0.912	0.650	0.488	0.382
1-Hexanol	$C_6H_{14}O$			4.58	2.27	1.270	0.781
2-Hexanone	$C_6H_{12}O$	1.300	0.840	0.583	0.429	0.329	0.262
1-Hexene	C_6H_{12}	0.441	0.326	0.252	0.202		
Hydrazine	H_4N_2			0.876	0.628	0.480	0.384
Hydrogen cyanide	CHN		0.235	0.183			
Indan	C_9H_{10}		2.23	1.357	0.931	0.692	0.545
Iodobenzene	C_6H_5I		2.35	1.554	1.117	0.854	0.683
Iodoethane	C_2H_5I		0.723	0.556	0.444	0.365	
Iodomethane	CH_3I		0.594	0.469			
1-Iodopropane	C_3H_7I		0.970	0.703	0.541	0.436	0.363
2-Iodopropane	C_3H_7I		0.883	0.653	0.506	0.407	
Isobutyl acetate	$C_6H_{12}O_2$			0.676	0.493	0.370	0.286
Isobutylamine	$C_4H_{11}N$		0.770	0.571	0.367		
Isopentane	C_5H_{12}	0.376	0.277	0.214			
Isophorone	$C_9H_{14}O$		4.20	2.33	1.415	0.923	0.638
Isopropylamine	C_3H_9N		0.454	0.325			
Isopropylbenzene	C_9H_{12}		1.075	0.737	0.547		
Mercury	Hg			1.526	1.402	1.312	1.245
Mesityl oxide	$C_6H_{10}O$	1.291	0.838	0.602	0.465	0.381	0.326
Methanol	CH_4O	1.258	0.793	0.544			
Methyl acetate	$C_3H_6O_2$		0.477	0.364	0.284		
Methylamine	CH_5N	0.319	0.231				
2-Methylaniline	C_7H_9N		10.33	3.82	1.936	1.198	0.839
3-Methylaniline	C_7H_9N		8.18	3.31	1.679	1.014	0.699
N-Methylaniline	C_7H_9N		4.12	2.04	1.222	0.825	0.606
Methyl benzoate	$C_8H_8O_2$			1.857			
Methyl butanoate	$C_5H_{10}O_2$		0.759	0.541	0.406	0.318	0.257
2-Methyl-1-butanol, (±)-	$C_5H_{12}O$			4.45	1.963	1.031	0.612
3-Methyl-1-butanol	$C_5H_{12}O$		8.63	3.69	1.842	1.031	0.631
2-Methyl-2-butene	C_5H_{10}		0.255	0.203			
Methylcyclohexane	C_7H_{14}		0.991	0.679	0.501	0.390	0.316
Methylcyclopentane	C_6H_{12}	0.927	0.653	0.479	0.364		
N-Methylformamide	C_2H_5NO		2.55	1.678	1.155	0.824	0.606
Methyl formate	$C_2H_4O_2$		0.424	0.325			
4-Methyl-3-heptanol	$C_8H_{18}O$		1.904	1.085	0.702	0.497	0.375
5-Methyl-3-heptanol	$C_8H_{18}O$		2.05	1.178	0.762	0.536	0.401
3-Methylhexane	C_7H_{16}			0.350			

Name	Mol. form.	η(-25 °C)/ mPa s	η(0 °C)/ mPa s	η(25 °C)/ mPa s	η(50 °C)/ mPa s	η(75 °C)/ mPa s	η(100 °C)/ mPa s
Methyl isobutanoate	$C_5H_{10}O_2$		0.672	0.488	0.373	0.296	
2-Methylpentane	C_6H_{14}		0.372	0.286	0.226		
3-Methylpentane	C_6H_{14}		0.395	0.306			
4-Methyl-2-pentanone	$C_6H_{12}O$			0.545	0.406		
Methyl propanoate	$C_4H_8O_2$		0.581	0.431	0.333	0.266	
2-Methylpropanoic acid	$C_4H_8O_2$		1.857	1.226	0.863	0.639	0.492
2-Methyl-2-propanol	$C_4H_{10}O$			4.31	1.421	0.678	
Methyl salicylate	$C_8H_8O_3$					1.102	0.815
Morpholine	C_4H_9NO			2.02	1.247	0.850	0.627
Nitrobenzene	$C_6H_5NO_2$		3.04	1.863	1.262	0.918	0.704
Nitroethane	$C_2H_5NO_2$	1.354	0.940	0.688	0.526	0.415	0.337
Nitrogen dioxide	NO_2		0.532	0.402			
Nitromethane	CH_3NO_2	1.311	0.875	0.630	0.481	0.383	0.317
1-Nitropropane	$C_3H_7NO_2$	1.851	1.160	0.798	0.589	0.460	0.374
Nonane	C_9H_{20}		0.964	0.665	0.488	0.375	0.300
Nonanoic acid	$C_9H_{18}O_2$			7.01	3.71	2.23	1.475
1-Nonanol	$C_9H_{20}O$			9.12	4.03		
5-Nonanone	$C_9H_{18}O$			1.199	0.834	0.619	0.484
Octadecane	$C_{18}H_{38}$				2.49	1.609	1.132
Octane	C_8H_{18}		0.700	0.508	0.385	0.302	0.243
Octanoic acid	$C_8H_{16}O_2$			5.02	2.66	1.654	1.147
1-Octanol	$C_8H_{18}O$			7.29	3.23	1.681	0.991
Paraldehyde	$C_6H_{12}O_3$			1.079	0.692	0.485	0.362
Pentachloroethane	C_2HCl_5		3.76	2.25	1.491	1.061	
Pentane	C_5H_{12}	0.351	0.274	0.224			
1-Pentanol	$C_5H_{12}O$	25.4	8.51	3.62	1.820	1.035	0.646
2-Pentanol	$C_5H_{12}O$			3.47	1.447	0.761	0.465
3-Pentanol	$C_5H_{12}O$			4.15	1.473	0.727	0.436
2-Pentanone	$C_5H_{10}O$		0.641	0.470	0.362	0.289	0.238
3-Pentanone	$C_5H_{10}O$		0.592	0.444	0.345	0.276	0.227
1-Pentene	C_5H_{10}	0.313	0.241	0.195			
Pentylamine	$C_5H_{13}N$		1.030	0.702	0.493	0.356	
Phenol	C_6H_6O				3.44	1.784	1.099
Phenylhydrazine	$C_6H_8N_2$			13.03	4.55	1.850	0.848
Phosphorus(III) chloride	Cl_3P	0.870	0.662	0.529	0.439		
Piperidine	$C_5H_{11}N$			1.573	0.958	0.649	0.474
Propanal	C_3H_6O			0.321	0.249		
1,2-Propanediol	$C_3H_8O_2$		248	40.4	11.30	4.77	2.75
Propanenitrile	C_3H_5N			0.294	0.240	0.202	
1-Propanethiol	C_3H_8S		0.503	0.385			
2-Propanethiol	C_3H_8S		0.477	0.357	0.280		
Propanoic acid	$C_3H_6O_2$		1.499	1.030	0.749	0.569	0.449
1-Propanol	C_3H_8O	8.64	3.81	1.945	1.107	0.685	
2-Propanol	C_3H_8O		4.62	2.04	1.028	0.576	
Propyl acetate	$C_5H_{10}O_2$		0.768	0.544	0.406	0.316	0.255
Propylamine	C_3H_9N			0.376			
Propyl formate	$C_4H_8O_2$		0.669	0.485	0.370	0.293	
Pyridine	C_5H_5N		1.361	0.879	0.637	0.497	0.409
Pyrrole	C_4H_5N		2.08	1.225	0.828	0.612	
Pyrrolidine	C_4H_9N	1.914	1.071	0.704	0.512		
Quinoline	C_9H_7N			3.34	1.892	1.201	0.833
Styrene	C_8H_8		1.050	0.695	0.507	0.390	0.310
Sulfolane	$C_4H_8O_2S$				6.28	3.82	2.56
1,1,1,2-Tetrachloroethane	$C_2H_2Cl_4$	3.66	2.20	1.437	1.006	0.741	0.570
Tetrachloroethene	C_2Cl_4		1.114	0.844	0.663	0.535	0.442
Tetrachloromethane	CCl_4		1.321	0.908	0.656	0.494	
Tetrachlorosilane	Cl_4Si			99.4	96.2		
Tetradecane	$C_{14}H_{30}$			2.13	1.376	0.953	0.697
Tetrahydrofuran	C_4H_8O	0.849	0.605	0.456	0.359		
Tetrahydrothiophene	C_4H_8S			0.973	0.912		

Name	Mol. form.	η(-25 °C)/ mPa s	η(0 °C)/ mPa s	η(25 °C)/ mPa s	η(50 °C)/ mPa s	η(75 °C)/ mPa s	η(100 °C)/ mPa s
Toluene	C_7H_8	1.165	0.778	0.560	0.424	0.333	0.270
Tribromomethane	$CHBr_3$			1.857	1.367	1.029	
1,1,1-Trichloroethane	$C_2H_3Cl_3$	1.847	1.161	0.793	0.578	0.428	
Trichloroethene	C_2HCl_3		0.703	0.545	0.444	0.376	
Trichlorofluoromethane	CCl_3F	0.740	0.539	0.421			
Trichloromethane	$CHCl_3$	0.988	0.706	0.537	0.427		
Trichlorosilane	Cl_3HSi		0.415	0.326			
1,1,2-Trichloro-1,2,2-trifluoroethane	$C_2Cl_3F_3$	1.465	0.945	0.656	0.481		
Tridecane	$C_{13}H_{28}$		2.91	1.724	1.129	0.796	0.594
Triethanolamine	$C_6H_{15}NO_3$			609	114	31.5	11.70
Triethylamine	$C_6H_{15}N$		0.455	0.347	0.273	0.221	
Trifluoroacetic acid	$C_2HF_3O_2$			0.808	0.571		
Undecane	$C_{11}H_{24}$		1.707	1.098	0.763	0.562	0.433
Water	H_2O		1.793	0.890	0.547	0.378	0.282
o-Xylene	C_8H_{10}		1.084	0.760	0.561	0.432	0.345
m-Xylene	C_8H_{10}		0.795	0.581	0.445	0.353	0.289
p-Xylene	C_8H_{10}			0.603	0.457	0.359	0.290

VISCOSITY OF CARBON DIOXIDE ALONG THE SATURATION LINE

The table below gives the viscosity of gas and liquid CO_2 along the liquid–vapor saturation line.

References

1. Fenghour, A., Wakeham, W. A., and Vesovic, V., *J. Phys. Chem. Ref. Data* 27, 31, 1998.
2. Angus, S. et al., *International Tables for the Fluid State: Carbon Dioxide*, Pergamon Press, Oxford, 1976.

T/K	*P*/kPa	Gas η/µPa s	Liquid η/µPa s
205	227	10.33	
210	327	10.60	
215	465	10.87	
220	600	11.13	241.68
225	735	11.41	221.72
230	894	11.69	203.75
235	1075	11.98	187.48
240	1283	12.27	172.67
245	1519	12.58	159.13
250	1786	12.90	146.69
255	2085	13.24	135.20
260	2419	13.61	124.30
265	2790	14.02	114.63
270	3203	14.47	105.21
275	3658	14.99	96.44
280	4160	15.61	87.89
285	4712	16.37	79.64
290	5315	17.36	71.47
295	5984	18.79	63.01
300	6710	21.29	53.33
302	6997	23.52	48.30

VISCOSITY AND DENSITY OF AQUEOUS HYDROXIDE SOLUTIONS

The viscosity and density of aqueous hydroxide solutions at 25 °C are tabulated here as a function of concentration. Viscosity is given in millipascal second, which is equal to the c.g.s. unit centipoise (cP). The last entry in each column refers to the saturated solution.

Reference

Sipos, P. M., Hefter, G., and May, P. M., *J. Chem. Eng. Data* 45, 613, 2000.

Viscosity in mPa s

c/mol L^{-1}	LiOH	NaOH	KOH	CsOH	$(CH_3)_4NOH$
0.5	1.017	0.997	0.937	0.91	1.017
1.0	1.169	1.116	0.990	0.94	1.186
1.5	1.340	1.248	1.050	0.97	1.430
2.0	1.537	1.396	1.116	1.03	1.762
3.0	2.050	1.754	1.269	1.19	3.031
4.0	2.734	2.228	1.448	1.41	7.238
5.0		2.867	1.657	1.67	
6.0		3.727	1.902	1.98	
7.0		4.869	2.196	2.40	
8.0		6.351	2.554	3.09	
9.0		8.230	3.005	4.31	
10.0		10.554	3.581	6.46	
11.0		13.362	4.328		
12.0		16.677	5.303		
13.0		20.503	6.577		
14.0		24.826	8.235		
15.0		29.604			
16.0		34.767			
17.0		40.212			
18.0		45.800			
19.0		51.354			
Sat.	3.311	51.911	8.526		8.850

Density in g/cm^3

c/mol L^{-1}	LiOH	NaOH	KOH	CsOH	$(CH_3)_4NOH$
0.5	1.012	1.019	1.022	1.063	0.999
1.0	1.025	1.040	1.045	1.128	1.002
1.5	1.038	1.059	1.068	1.193	1.005
2.0	1.050	1.078	1.090	1.257	1.009
3.0	1.072	1.115	1.133	1.383	1.019
4.0	1.093	1.149	1.174	1.508	1.030
5.0		1.182	1.214	1.632	
6.0		1.213	1.253	1.755	
7.0		1.243	1.290	1.876	
8.0		1.271	1.326	1.997	
9.0		1.299	1.362	2.117	
10.0		1.325	1.396	2.236	
11.0		1.350	1.429	2.354	
12.0		1.374	1.462	2.471	
13.0		1.397	1.494	2.587	
14.0		1.419	1.524	2.703	
15.0		1.441			
16.0		1.461			
17.0		1.481			
18.0		1.499			
19.0		1.517			
Sat.	1.109	1.519	1.529	2.800	1.032

VISCOSITY OF LIQUID METALS

This table gives the viscosity of several liquid metals as a function of temperature. Experimental data from some of the references was smoothed to produce the table. Viscosity is given in millipascal second (mPa s), which equals the cgs unit centipoise (cP).

References

1. Shpil'rain, E. E., Yakimovich, K. A., Fomin, V. A., Skovorodjko, S. N., and Mozgovoi, A. G., in *Handbook of Thermodynamic and Transport Properties of the Alkali Metals*, Ohse, R. H., Ed., Blackwell Scientific Publishers, Oxford, 1985. [Li, Na, K, Rb, Cs]
2. Culpin, M. F., *Proc. Phys. Soc.* 70, 1079, 1957. [Ca]
3. *Landolt-Börnstein, Numerical Data and Functional Relationships in Science and Technology,* 6th Ed., II/5a, *Transport Phenomena I (Viscosity and Diffusion)*, Springer-Verlag, Heidelberg, 1961. [Au, Mg]
4. Assael, M. J., Kakosimos, K., Banish, R. M., Brillo, J., Egry, I., Brooks, R., Quested, P. N., Mills, K. C., Nagashima, A., Sato, Y., and Wakeham, W. A., *J. Phys. Chem. Ref. Data* 35, 285, 2006. [Al, Fe]
5. Assael, M. J., Kalyva, A. E., Antoniadis, K. D., Banish, R. M., Egry, I., Wu, J., Kaschnitz, E., and Wakeham, W. A., *J. Phys. Chem. Ref. Data* 39, 033105-1, 2010. [Cu, Sn]
6. Assael, M. J., Kalyva, A. E., Antoniadis, K. D., Banish, R. M., Egry, I., Wu, J., Kaschnitz, E., and Wakeham, W. A., *High Temp.-High Press.* 41, 161, 2012. [Sb, Bi, Pb, Ni, Ag]
7. Assael, M. J., Armyra, I. J., Brillo, J., Stankus, S. V., Wu, J., and Wakeham, W. A., *J. Phys. Chem. Ref. Data* 41, 033101-1, 2010. [Cd, Co, Ga, In, Hg, Si, Tl, Zn]

	Viscosity in mPa s							
t/°C	Lithium	Sodium	Potassium	Rubidium	Cesium	Gallium	Calcium	Magnesium
50				0.542	0.598			
100		0.687	0.441	0.435	0.469			
150		0.542	0.358	0.365	0.389			
200	0.566	0.451	0.303	0.316	0.334			
250	0.503	0.387	0.263	0.280	0.294			
300	0.453	0.341	0.234	0.252	0.264			
350	0.412	0.306	0.211	0.230	0.240	1.369		
400	0.379	0.278	0.193	0.212	0.221	1.158		
450	0.352	0.255	0.178	0.197	0.206	1.016		
500	0.328	0.237	0.166	0.185	0.192	0.915		
550	0.308	0.221	0.155	0.174	0.181	0.840		
600	0.290	0.208	0.146	0.165	0.171	0.783		
650	0.275	0.196	0.138	0.157	0.163	0.737		
700	0.261	0.186	0.132	0.150	0.156	0.700		1.10
750	0.249	0.177	0.126	0.143	0.149	0.669		0.96
800	0.238	0.170	0.120	0.138	0.143	0.643		0.84
850	0.228	0.163	0.115	0.133	0.138		1.107	0.74
900	0.219	0.156	0.111	0.128	0.134		0.959	0.67
950	0.211	0.151	0.107	0.124	0.129			
1000	0.204	0.146	0.104	0.120	0.125			
1050	0.197	0.141	0.101	0.117	0.122			
1100	0.191	0.137	0.098	0.114	0.119			
1150	0.185	0.133	0.095	0.111	0.116			
1200	0.180	0.129	0.092	0.108	0.113			
1250	0.175	0.126	0.090	0.105	0.110			
1300	0.170	0.123	0.088	0.103	0.108			
1350	0.166	0.120	0.086	0.101	0.106			
1400	0.162	0.117	0.084	0.099	0.104			
1450	0.158	0.115	0.082	0.097	0.102			
1500	0.155	0.113	0.081	0.095	0.100			
1550	0.151	0.110	0.079	0.093	0.098			
1600	0.148	0.108	0.078	0.092	0.097			
1650	0.145	0.106	0.076	0.090	0.095			
1700	0.142	0.105	0.075		0.094			
1750	0.139	0.103	0.074		0.092			
1800	0.137	0.101			0.091			
1850	0.135	0.100			0.090			
1900	0.132	0.098			0.089			
1950	0.130	0.097			0.088			
2000	0.128	0.096			0.086			

t/°C	Aluminum	Cobalt	Copper	Gold	Indium	Iron	Nickel	Silver	Tin
	Viscosity in mPa s								
250									1.77
300									1.55
350									1.39
400									1.27
450					1.748				1.17
500					1.521				1.09
550					1.357				1.02
600					1.234				0.97
650					1.139				0.92
700	1.24				1.063				0.88
750	1.13				1.001				0.85
800	1.04				0.951				0.82
850	0.96				0.908				0.79
900	0.90				0.871				0.77
950	0.84				0.840				0.75
1000					0.813				0.73
1050									
1100			3.92	5.130					
1150			3.61	4.874					
1200			3.34	4.640					
1250			3.11	4.429				4.05	
1300			2.91	4.240				3.75	
1350			2.73					3.49	
1400			2.58					3.27	
1450			2.44					3.08	
1500			2.31					2.91	
1550			2.20						
1600			2.10			5.22			
1650			2.01			4.79			
1700			1.92			4.42			
1750						4.09	5.01		
1800		4.543				3.79	4.64		
1850		4.123				3.54	4.31		
1900		3.761				3.31	4.02		
1950		3.446				3.10	3.77		
2000		3.172				2.92	3.54		
2050		2.932				2.75	3.34		
2100		2.719				2.60	3.15		
2150						2.46			
2200						2.34			
2250						2.22			

t/°C	Antimony	Bismuth	Cadmium	Lead	Mercury	Silicon	Thallium	Zinc
	Viscosity in mPa s							
250					1.875			
300					1.531			
350					1.324			
400					1.187			
450					1.091			
500					1.020			
550		1.74			0.965			
600		1.55	2.708		0.921		2.434	
650		1.41	2.326	2.30			2.155	
700		1.30	2.043	2.07			1.941	3.737
750		1.21	1.825	1.88			1.773	3.254
800		1.14	1.654	1.73			1.638	2.883
850		1.08	1.516	1.61				2.591
900	1.43	1.03	1.403	1.51				2.356
950	1.30	0.98		1.43				2.164

	Viscosity in mPa s							
t/°C	Antimony	Bismuth	Cadmium	Lead	Mercury	Silicon	Thallium	Zinc
1000	1.19	0.95		1.36				2.005
1050	1.10			1.29				1.871
1100	1.03			1.24				1.756
1150	0.96			1.19				
1200	0.91			1.15				
1250	0.86			1.11				
1300	0.81			1.08				
1350				1.05				
1400				1.02				
1450								
1500								
1550								
1600								
1650								
1700						0.605		
1750						0.571		
1800						0.541		
1850						0.514		
1900						0/490		

THERMAL CONDUCTIVITY OF GASES

Marcia L. Huber and Allan H. Harvey

The following table gives the thermal conductivity of some common gases as a function of temperature. The difference between the thermal conductivity at 100 kPa and the limiting value at zero pressure is usually less than 1%. Uncertainties for the thermal conductivities of gases in this table are generally less than 3%; uncertainty information on specific fluids can be found in the references. Thermal conductivity is given in units of mW m^{-1} K^{-1}.

Name	Formula	p/kPa	λ(100 K)/ mW m^{-1} K^{-1}	λ(200 K)/ mW m^{-1} K^{-1}	λ(300 K)/ mW m^{-1} K^{-1}	λ(400 K)/ mW m^{-1} K^{-1}	λ(500 K)/ mW m^{-1} K^{-1}	λ(600 K)/ mW m^{-1} K^{-1}	Ref.
Acetone	$(CH_3)_2CO$	100			11.5	20.2	30.6	42.7	4
Acetylene	HC≡CH	100			21.4	33.3	45.4	56.8	4
Air		100	9.5	18.5	26.4	33.5	39.9	46.0	1
Ammonia	NH_3	100			25.1	37.2	53.1	68.6	9
Argon	Ar	0	6.3	12.4	17.7	22.4	26.5	30.3	2,3*
Boron trifluoride	BF_3	100			19.0	24.6			4
Butane	C_4H_{10}	100			16.7	28.3	43.0	60.9	20
Carbon dioxide	CO_2	100		9.6	16.8	25.2	33.5	41.6	15
Carbon monoxide	CO	0			25.0	32.3	39.2	45.7	14
Dichlorodifluoromethane	CF_2Cl_2	100			9.9	15.0	20.1	25.2	13
1,2-Dichloro-1,1,2,2-tetrafluoroethane	$C_2Cl_2F_4$	100			10.3	15.7	21.1		13
Diethyl ether	$(C_2H_5)_2O$	100			15.1	25.0	37.1		4
Ethane	C_2H_6	100		10.7	21.2	36.0	53.8	73.3	18
Ethanol	C_2H_5OH	100			14.4	25.8	38.4	53.2	4
Ethylene	$CH_2{=}CH_2$	100		11.3	20.6	34.7	49.9	68.6	17
Helium	He	0	74.7	118.3	155.7	189.6	221.4	251.6	10
Hexane	C_6H_{14}	100				23.4	35.4	48.7	4
Hydrogen	H_2	0	68.2	132.8	186.6	230.9	270.9	309.1	6
Hydrogen chloride	HCl	100		9.2	14.5	19.5	24.0	28.1	4
Hydrogen sulfide	H_2S	100			14.6	20.5	26.4	32.4	4
Isobutane	C_4H_{10}	100			17.1	28.9	43.2	60.2	21
Krypton	Kr	0		6.5	9.5	12.3	14.8	17.1	11
Methane	CH_4	0	10.4	21.8	34.4	50.0	68.4	88.6	16
Methanol	CH_3OH	100				26.2	38.6	53.0	4
Neon	Ne	0	22.3	37.4	49.4	59.9	69.5	78.5	12
Nitric oxide	NO	100		17.8	25.9	33.1	39.6	46.2	4
Nitrogen	N_2	100	9.4	18.3	26.0	32.8	39.0	44.8	1
Nitrous oxide	N_2O	100		9.8	17.4	26.0	34.1	41.8	4
Oxygen	O_2	100	9.1	18.2	26.5	34.0	41.0	47.7	1
Pentane	C_5H_{12}	100				24.9	37.8	52.7	4
Perfluorocyclobutane	c-C_4F_8	100			12.5	19.5			13
Propane	C_3H_8	100			18.5	31.0	46.4	64.6	19
Sulfur dioxide	SO_2	100			9.6	14.3	20.0	25.6	4
Sulfur hexafluoride	SF_6	0			13.0	20.6	27.5	33.8	5
Tetrafluoromethane	CF_4	0			16.0	24.1	32.2	39.9	5
Trichloromethane	$CHCl_3$	100			7.5	11.1	15.1		4
1,1,2-Trichloro-1,2,2-trifluoroethane	$C_2Cl_3F_3$	100			9.0	13.6	18.3		13
Water	H_2O	0			18.6	26.1	35.6	46.2	7
Water-d_2	D_2O	0			18.2	26.6	36.3	47.6	8
Xenon	Xe	0		3.7	5.5	7.2	8.8	10.3	3*,11

* Reference contains more accurate data covering a restricted temperature range.

References

1. Lemmon, E. W., and Jacobsen, R. T, Viscosity and Thermal Conductivity Equations for Nitrogen, Oxygen, Argon, and Air, *Int. J. Thermophys.* 25, 21, 2004.
2. Vogel, E., Jäger, B., Hellmann, R., and Bich, E., *Ab initio* Pair Potential Energy Curve for the Argon Atom Pair and Thermophysical Properties for the Dilute Argon Gas. II. Thermophysical Properties for Lowdensity Argon, *Mol. Phys.* 108, 3335, 2010.
3. May, E. F., Berg, R. F., and Moldover, M. R., Reference Viscosities of H_2, CH_4, Ar, and Xe at Low Densities, *Int. J. Thermophys.* 28, 1085, 2007.
4. Ho, C. y., Ed., *Properties of Inorganic and Organic Fluids, CINDAS Data Series on Materials Properties,* Vol. V-1, Hemisphere Publishing Corp., New York, 1988.
5. Uribe, F. J., Mason, E. A., and Kestin, J., Thermal Conductivity of Nine Polyatomic Gases at Low Density, *J. Phys. Chem. Ref. Data* 19, 1123, 1990.
6. Mehl, J. B., Huber, M. L., and Harvey, A. H., *Ab Initio* Transport Coefficients of Gaseous Hydrogen, *Int. J. Thermophys.* 31, 740, 2010.
7. Sengers, J. V., and Watson, J. T. R., Improved International Formulations for the Viscosity and Thermal Conductivity of Water Substance, *J. Phys. Chem. Ref. Data* 15, 1291, 1986.
8. Matsunaga, N., and Nagashima, A., Transport Properties of Liquid and Gaseous D_2O over a Wide Range of Temperature and Pressure, *J. Phys. Chem. Ref. Data* 12, 933, 1983.
9. Tufeu, R., Ivanov, D. Y., Garrabos, Y., and Le Neindre, B., Thermal Conductivity of Ammonia in a Large Temperature and Pressure Range Including the Critical Region, *Ber. Bunsenges. Phys. Chem.* 88, 422, 1984.
10. Cencek, W., Komasa, J., Przybytek, M., Mehl, J. B., Jeziorski, B., and Szalewicz, K., Effects of Adiabatic, Relativistic, and Quantum Electrodynamics Interactions in Helium Dimer on Thermophysical Properties of Helium, *J. Chem. Phys.*, to be submitted (2011).
11. Bich, E., Millat, J., and Vogel, E., The Viscosity and Thermal Conductivity of Pure Monatomic Gases from Their Normal Boiling Point up to 5000 K in the Limit of Zero Density and at 0.101325 MPa, *J. Phys. Chem. Ref. Data* 19, 1289, 1990.
12. Bich, E., Hellmann, R., and Vogel, E., *Ab Initio* Potential Energy Curve for the Neon Atom Pair and Thermophysical Properties for the Dilute Neon Gas. II. Thermophysical Properties for Low-Density Neon, *Mol. Phys.* 106, 813, 2008.
13. Krauss, R., and Stephan, K., Thermal Conductivity of Refrigerants in a Wide Range of Temperature and Pressure, *J. Phys. Chem. Ref. Data* 18, 43, 1989.
14. Millat J., and Wakeham, W. A., The Thermal Conductivity of Nitrogen and Carbon Monoxide in the Limit of Zero Density, *J. Phys. Chem. Ref. Data* 18, 565, 1989.
15. Vesovic, v., Wakeham, W. A., Olchowy, G. A., Sengers, J. V., Watson, J. T. R., and Millat, J., The Transport Properties of Carbon Dioxide, *J. Phys. Chem. Ref. Data,* 19, 763, 1990.
16. Hellmann, R., Bich, E., Vogel, E., Dickinson, A. S., and Vesovic, V., Calculation of the Transport and Relaxation Properties of Methane. II. Thermal Conductivity, Thermomagnetic Effects, Volume Viscosity, and Nuclear-Spin Relaxation, *J. Chem. Phys.* 130, 124309, 2009.
17. Holland, P. M., Eaton, B. E., and Hanley, H. J. M., A Correlation of the Viscosity and Thermal Conductivity Data of Gaseous and Liquid Ethylene, *J. Phys. Chem. Ref. Data* 12, 917, 1983.
18. Friend, D. G., Ingham, H., and Ely, J. F., Thermophysical Properties of Ethane, J. Phys. Chem. Ref. Data 20, 275, 1991.
19. Marsh, K., Perkins, R. A., and Ramires, M. L. V., Measurement and Correlation of the Thermal Conductivity of Propane from 86 to 600 K at Pressures to 70 MPa, *J. Chem. Eng. Data* 47, 932, 2002.
20. Perkins, R. A., Ramires, M. L. V., Nieto de Castro, C. A., and Cusco, L., Measurement and Correlation of the Thermal Conductivity of Butane from 135 K to 600 K at Pressures to 70 MPa, *J. Chem. Eng. Data* 47, 1263, 2002.
21. Perkins, R. A., Measurement and Correlation of the Thermal Conductivity of Isobutane from 114 K to 600 K at Pressures to 70 MPa, *J. Chem. Eng. Data* 47, 1272, 2002.

THERMAL CONDUCTIVITY OF LIQUIDS

This table gives the thermal conductivity λ of 275 liquids at temperatures between –25 °C and 100 °C. Values refer to nominal atmospheric pressure; however, when an entry is given for a temperature above the normal boiling point of the liquid, the pressure is understood to be the saturation vapor pressure at that temperature. Reference 1 contains data on many of these liquids at high pressures. Data on halocarbon refrigerants over a wide range of temperature and pressure may be found in Reference 6.

Values given to three decimal places (i.e., to 0.001 W/m K) have an uncertainty of 2% to 5%. Values given to 0.0001 W/m K should be accurate to 1% or better.

References

1. Vargaftik, N. B., Filippov, L. P., Tarzimanov, A. A., and Totskii, E. E., *Handbook of Thermal Conductivity of Liquids and Gases*, CRC Press, Boca Raton FL, 1994.
2. Daubert, T. E., Danner, R. P., Sibul, H. M., and Stebbins, C. C., *Physical and Thermodynamic Properties of Pure Compounds: Data Compilation*, extant 1994 (core with four supplements), Taylor and Francis, Bristol, PA (also available as a database).
3. Watanabe, H., *J. Chem. Eng. Data* 48, 124, 2003.
4. Watanabe, H., and Seong, D. J., *Int. J. Thermophys.* 23, 337, 2002.
5. Nieto de Castro, C. A., Li, S. F. Y., Nagashima, A., Trengove, R. D., and Wakeham, W. A., *J. Phys. Chem. Ref. Data* 15, 1073, 1986.
6. Krauss, R., and Stephan, K., *J. Phys. Chem. Ref. Data* 18, 43, 1989.
7. Assael, M. J., Ramires, M. L. V., Nieto de Castro, C. A., and Wakeham, W. A., *J. Phys. Chem. Ref. Data* 19, 113, 1990.
8. Ramires, M. L. V., Nieto de Castro, C. A., Nagasaka, Y., Nagashima, A., Assael, M. J., and Wakeham, W. A., *J. Phys. Chem. Ref. Data* 24, 1377, 1995.
9. Ramires, M. L. V., Nieto de Castro, C. A., Perkins, R. A., Nagasaka, Y., Nagashima, A., Assael, M. J., and Wakeham, W. A., *J. Phys. Chem. Ref. Data* 29, 133, 2000.
10. Marsh, K. N., Ed., *Recommended Reference Materials for the Realization of Physicochemical Properties*, Blackwell Scientific Publications, Oxford, 1987.
11. Beaton, C. F., and Hewitt, G. F., *Physical Property Data for the Design Engineer*, Hemisphere Publishing Corp., New York, 1989.

Name	Mol. form.	λ(-25 °C)/ W m⁻¹ K⁻¹	λ(0 °C)/ W m⁻¹ K⁻¹	λ(25 °C)/ W m⁻¹ K⁻¹	λ(50 °C)/ W m⁻¹ K⁻¹	λ(75 °C)/ W m⁻¹ K⁻¹	λ(100 °C)/ W m⁻¹ K⁻¹	Ref.
Acetic acid	$C_2H_4O_2$			0.158	0.153	0.149	0.144	2
Acetic anhydride	$C_4H_6O_3$		0.170	0.164	0.158	0.152	0.146	1
Acetone	C_3H_6O		0.169	0.161				2
Acetonitrile	C_2H_3N	0.208	0.198	0.188	0.178	0.168		2
Acetophenone	C_8H_8O			0.147	0.146	0.144	0.142	1
Acrylonitrile	C_3H_3N	0.186	0.176	0.166	0.156	0.146	0.136	
Allyl alcohol	C_3H_6O			0.162				1
Aniline	C_6H_7N		0.175					1
Anisole	C_7H_8O			0.145	0.142	0.139	0.136	1
Benzaldehyde	C_7H_6O			0.153	0.148	0.143	0.139	1
Benzene	C_6H_6			0.1411	0.1329	0.1247		7
Benzonitrile	C_7H_5N			0.148	0.142	0.136	0.130	1
Benzyl alcohol	C_7H_8O			0.159	0.158	0.156	0.154	1
Bromobenzene	C_6H_5Br	0.119	0.115	0.111	0.107	0.103	0.099	1
1-Bromobutane	C_4H_9Br	0.112	0.107	0.103	0.098	0.093	0.088	1
Bromoethane	C_2H_5Br	0.107	0.104	0.101				1
1-Bromohexane	$C_6H_{13}Br$	0.115	0.111	0.108	0.104	0.101	0.097	1
1-Bromonaphthalene	$C_{10}H_7Br$			0.110	0.109	0.108	0.106	1
1-Bromononane	$C_9H_{19}Br$		0.116	0.112	0.109	0.106	0.103	1
1-Bromopentane	$C_5H_{11}Br$	0.113	0.109	0.105	0.101	0.097	0.093	1
1-Bromopropane	C_3H_7Br	0.108	0.104	0.099	0.094			1
1,2-Butadiene	C_4H_6	0.147	0.134					1
Butanal	C_4H_8O		0.155	0.147	0.140	0.132		1
1-Butanol	$C_4H_{10}O$		0.158	0.153	0.147	0.142	0.137	1
2-Butanone	C_4H_8O	0.158	0.151	0.145	0.139	0.133		2
Butyl acetate	$C_6H_{12}O_2$		0.143	0.136	0.130	0.123	0.116	1
Butylbenzene	$C_{10}H_{14}$			0.126	0.121	0.116	0.111	1
sec-Butylbenzene, (±)-	$C_{10}H_{14}$		0.129	0.124	0.119	0.114	0.108	1
tert-Butylbenzene	$C_{10}H_{14}$			0.117	0.114	0.110	0.106	1
Butyl formate	$C_5H_{10}O_2$			0.136	0.130	0.123	0.117	1
Butyl oleate	$C_{22}H_{42}O_2$			0.157	0.153	0.149	0.145	1
Butyl palmitate	$C_{20}H_{40}O_2$			0.151	0.148	0.144	0.140	1
Butyl propanoate	$C_7H_{14}O_2$			0.139	0.133	0.126	0.121	1
2-Butyne	C_4H_6	0.137	0.129	0.121				2
Carbon disulfide	CS_2		0.154	0.149				2
2-Chloroaniline	C_6H_6ClN			0.148				1

Name	Mol. form.	λ(-25 °C)/ W m⁻¹ K⁻¹	λ(0 °C)/ W m⁻¹ K⁻¹	λ(25 °C)/ W m⁻¹ K⁻¹	λ(50 °C)/ W m⁻¹ K⁻¹	λ(75 °C)/ W m⁻¹ K⁻¹	λ(100 °C)/ W m⁻¹ K⁻¹	Ref.
Chlorobenzene	C_6H_5Cl	0.137	0.132	0.127	0.123	0.118	0.113	1
Chloroethane	C_2H_5Cl	0.145	0.132	0.119	0.106	0.093		2
1-Chloronaphthalene	$C_{10}H_7Cl$			0.126				1
1-Chlorononane	$C_9H_{19}Cl$		0.132	0.128	0.124	0.120	0.115	1
1-Chlorooctane	$C_8H_{17}Cl$		0.130	0.127	0.124	0.121	0.119	1
1-Chloropentane	$C_5H_{11}Cl$		0.125	0.120	0.115	0.109		1
1-Chloropropane	C_3H_7Cl	0.129	0.123	0.116	0.110	0.104	0.098	1
o-Cresol	C_7H_8O			0.153				1
m-Cresol	C_7H_8O			0.149	0.147	0.145		1
Cycloheptane	C_7H_{14}			0.123	0.118	0.112	0.108	1
Cyclohexane	C_6H_{12}			0.123	0.117	0.111		2
Cyclohexanol	$C_6H_{12}O$			0.138	0.134	0.130	0.126	1
Cyclohexanone	$C_6H_{10}O$			0.138	0.134	0.130	0.126	1
Cyclohexene	C_6H_{10}	0.142	0.136	0.130	0.124	0.118		2
Cyclohexylbenzene	$C_{12}H_{16}$			0.121	0.119	0.117		1
Cyclopentane	C_5H_{10}	0.140	0.133	0.126				2
Cyclopentene	C_5H_8	0.143	0.136	0.129				2
trans-Decahydronaphthalene	$C_{10}H_{18}$			0.113				1
Decanal	$C_{10}H_{20}O$		0.149	0.144	0.139	0.134	0.129	1
Decane	$C_{10}H_{22}$	0.142	0.1360	0.1296	0.1232	0.1167	0.110	4
Decanoic acid	$C_{10}H_{20}O_2$				0.148	0.144	0.140	1
1-Decanol	$C_{10}H_{22}O$			0.162	0.159	0.155	0.151	1
1-Decene	$C_{10}H_{20}$	0.138	0.132	0.126	0.120	0.114	0.109	1
Decyl acetate	$C_{12}H_{24}O_2$			0.146	0.136	0.126		1
1,2-Dibromoethane	$C_2H_4Br_2$			0.100	0.096	0.092	0.088	1
Dibromomethane	CH_2Br_2	0.120	0.114	0.108	0.103	0.097		2
1,2-Dibromotetrafluoroethane	$C_2Br_2F_4$	0.071	0.066	0.061	0.057	0.053	0.049	1
1,2-Dibutoxyethane	$C_{10}H_{22}O_2$			0.140	0.134	0.127	0.120	1
Dibutyl ether	$C_8H_{18}O$		0.139	0.132	0.125	0.118	0.112	1
Dibutyl phthalate	$C_{16}H_{22}O_4$		0.139	0.136	0.134	0.131	0.129	1
o-Dichlorobenzene	$C_6H_4Cl_2$		0.125	0.121	0.117	0.113	0.109	1
m-Dichlorobenzene	$C_6H_4Cl_2$		0.120	0.116	0.113	0.109		1
p-Dichlorobenzene	$C_6H_4Cl_2$				0.112	0.108	0.105	1
1,2-Dichloroethane	$C_2H_4Cl_2$	0.144	0.139	0.133	0.128	0.122	0.117	1
Dichloromethane	CH_2Cl_2	0.158	0.149	0.140	0.133	0.128	0.127	1
1,2-Diethoxyethane	$C_6H_{14}O_2$				0.140	0.133	0.125	1
o-Diethylbenzene	$C_{10}H_{14}$		0.133	0.127	0.122	0.116	0.111	1
Diethylene glycol dibutyl ether	$C_{12}H_{26}O_3$		0.150	0.146	0.143	0.139	0.135	1
Diethylene glycol monobutyl ether	$C_8H_{18}O_3$			0.163	0.158	0.153	0.148	1
Diethylene glycol monoethyl ether	$C_6H_{14}O_3$				0.188	0.184	0.180	1
Diethylene glycol monomethyl ether	$C_5H_{12}O_3$			0.190	0.185	0.180	0.175	1
Diethyl ether	$C_4H_{10}O$	0.150	0.140	0.130	0.120	0.110	0.100	2
Diethyl oxalate	$C_6H_{10}O_4$			0.157				1
Diethyl phthalate	$C_{12}H_{14}O_4$			0.172	0.169	0.166		1
Diiodomethane	CH_2I_2			0.098	0.093	0.088	0.083	1
N,N-Dimethylacetamide	C_4H_9NO			0.175	0.172	0.168		1
N,N-Dimethylaniline	$C_8H_{11}N$			0.122	0.119	0.115		1
2,2-Dimethylbutane	C_6H_{14}	0.108	0.1006	0.0934	0.0861	0.0788	0.072	3
2,3-Dimethylbutane	C_6H_{14}	0.115	0.1076	0.1003	0.0930	0.0857	0.078	3
N,N-Dimethylformamide	C_3H_7NO			0.183	0.175	0.167	0.159	1
2,2-Dimethylpentane	C_7H_{16}	0.111	0.1046	0.0980	0.0913	0.0847	0.078	3
2,3-Dimethylpentane	C_7H_{16}	0.120	0.1127	0.1059	0.0990	0.0922	0.085	3
2,4-Dimethylpentane	C_7H_{16}	0.116	0.1089	0.1020	0.0951	0.0882	0.081	3
3,3-Dimethylpentane	C_7H_{16}	0.113	0.1068	0.1001	0.0934	0.0867	0.080	3
Dimethyl phthalate	$C_{10}H_{10}O_4$			0.1473	0.1443	0.1409	0.1373	10
Dioctyl hexanedioate	$C_{22}H_{42}O_4$			0.157	0.153	0.149	0.145	1
1,4-Dioxane	$C_4H_8O_2$			0.159	0.147	0.135	0.123	2
Dipentyl ether	$C_{10}H_{22}O$			0.131	0.125	0.121	0.116	1
Diphenyl ether	$C_{12}H_{10}O$				0.139	0.135	0.131	2
Dipropyl ether	$C_6H_{14}O$		0.137	0.130	0.123	0.117		1

Name	Mol. form.	λ(-25 °C)/ W m⁻¹ K⁻¹	λ(0 °C)/ W m⁻¹ K⁻¹	λ(25 °C)/ W m⁻¹ K⁻¹	λ(50 °C)/ W m⁻¹ K⁻¹	λ(75 °C)/ W m⁻¹ K⁻¹	λ(100 °C)/ W m⁻¹ K⁻¹	Ref.
Dodecane	$C_{12}H_{26}$			0.135	0.130	0.124	0.119	1
1-Dodecanol	$C_{12}H_{26}O$				0.167	0.163	0.159	1
Epichlorohydrin	C_3H_5ClO	0.142	0.137	0.131	0.125	0.119	0.114	2
1,2-Ethanediol	$C_2H_6O_2$		0.248	0.254	0.258	0.261	0.261	1
Ethanol	C_2H_6O	0.181	0.174	0.167	0.160	0.153	0.148	1
Ethanolamine	C_2H_7NO			0.240	0.238	0.236		1
Ethoxybenzene	$C_8H_{10}O$	0.151	0.145	0.140	0.135	0.130		1
2-Ethoxyethanol	$C_4H_{10}O_2$			0.190	0.182	0.174	0.165	1
Ethyl acetate	$C_4H_8O_2$		0.151	0.144	0.136			1
Ethyl acetoacetate	$C_6H_{10}O_3$			0.155	0.152	0.148	0.144	1
N-Ethylaniline	$C_8H_{11}N$			0.150				1
Ethylbenzene	C_8H_{10}	0.143	0.137	0.130	0.123	0.116	0.110	1
Ethyl benzoate	$C_9H_{10}O_2$			0.141				1
Ethyl butanoate	$C_6H_{12}O_2$		0.143	0.137	0.131	0.126		1
Ethyl formate	$C_3H_6O_2$	0.181	0.171	0.160	0.149	0.138		1
Ethyl hexanoate	$C_8H_{16}O_2$		0.142	0.137	0.133	0.128	0.123	1
Ethyl hexyl ether	$C_8H_{18}O$		0.131	0.126	0.120	0.114	0.109	1
3-Ethylpentane	C_7H_{16}	0.128	0.1203	0.1128	0.1053	0.0978	0.090	3
Ethyl pentanoate	$C_7H_{14}O_2$			0.132				1
Ethyl propanoate	$C_5H_{10}O_2$				0.133	0.121		1
Fluorobenzene	C_6H_5F			0.136	0.131	0.126		1
Formic acid	CH_2O_2			0.267	0.265	0.263	0.261	1
Furan	C_4H_4O	0.142	0.134	0.126				2
Furfuryl alcohol	$C_5H_6O_2$			0.179				1
Germanium(IV) chloride	Cl_4Ge	0.111	0.105	0.100	0.095	0.090	0.084	1
Glycerol	$C_3H_8O_3$			0.285	0.288	0.292	0.296	1
Heptanal	$C_7H_{14}O$			0.140				1
Heptane	C_7H_{16}	0.1378	0.1303	0.1228	0.1152	0.1077		5
Heptanoic acid	$C_7H_{14}O_2$			0.140	0.137	0.133		1
1-Heptanol	$C_7H_{16}O$	0.160	0.158	0.153	0.149	0.144	0.139	1
3-Heptanone	$C_7H_{14}O$		0.143	0.137	0.131	0.125	0.119	1
4-Heptanone	$C_7H_{14}O$			0.136	0.131	0.125	0.120	1
1-Heptene	C_7H_{14}	0.139	0.132	0.125	0.118	0.111		1
Heptyl acetate	$C_9H_{18}O_2$			0.135	0.128	0.122	0.116	1
Heptyl butanoate	$C_{11}H_{22}O_2$			0.139	0.134	0.129	0.123	1
Heptyl formate	$C_8H_{16}O_2$		0.141	0.137	0.132	0.128	0.123	1
Heptyl propanoate	$C_{10}H_{20}O_2$			0.137	0.132	0.127	0.122	1
Hexadecane	$C_{16}H_{34}$			0.140	0.135	0.130	0.125	2
Hexafluorobenzene	C_6F_6				0.083			1
Hexane	C_6H_{14}	0.133	0.1250	0.1167	0.1083	0.0999	0.092	4
Hexanedinitrile	$C_6H_8N_2$			0.174	0.168			1
Hexanoic acid	$C_6H_{12}O_2$		0.148	0.142	0.137	0.131		1
1-Hexanol	$C_6H_{14}O$	0.161	0.157	0.152	0.147	0.142	0.137	1
2-Hexanone	$C_6H_{12}O$		0.156	0.145	0.134	0.124	0.115	1
1-Hexene	C_6H_{12}	0.138	0.129	0.121	0.113			1
Hexyl acetate	$C_8H_{16}O_2$			0.135	0.129	0.123	0.118	1
Hexylbenzene	$C_{12}H_{18}$		0.141	0.137	0.132	0.128	0.124	1
Hexyl butanoate	$C_{10}H_{20}O_2$			0.137	0.132	0.127	0.121	1
Hexyl formate	$C_7H_{14}O_2$			0.141	0.133	0.126	0.119	1
Indan	C_9H_{10}			0.135				1
Iodobenzene	C_6H_5I	0.106	0.103	0.101	0.098	0.095	0.092	1
1-Iodobutane	C_4H_9I		0.094	0.090	0.085	0.081	0.077	1
Iodoethane	C_2H_5I		0.091	0.087	0.083	0.079		1
1-Iodohexane	$C_6H_{13}I$		0.098	0.095	0.091	0.088	0.084	1
1-Iodononane	$C_9H_{19}I$		0.105	0.102	0.099	0.095	0.092	1
1-Iodopentane	$C_5H_{11}I$		0.096	0.092	0.088	0.084	0.081	1
1-Iodopropane	C_3H_7I	0.096	0.092	0.087	0.083	0.078	0.074	1
2-Iodopropane	C_3H_7I	0.089	0.085	0.082	0.078	0.074	0.071	1
Isopentane	C_5H_{12}			0.111				1
Isopropylbenzene	C_9H_{12}	0.132	0.128	0.123	0.118	0.112	0.107	1

Name	Mol. form.	λ(-25 °C)/ W m⁻¹ K⁻¹	λ(0 °C)/ W m⁻¹ K⁻¹	λ(25 °C)/ W m⁻¹ K⁻¹	λ(50 °C)/ W m⁻¹ K⁻¹	λ(75 °C)/ W m⁻¹ K⁻¹	λ(100 °C)/ W m⁻¹ K⁻¹	Ref.
1-Isopropyl-4-methylbenzene	$C_{10}H_{14}$	0.132	0.127	0.122	0.117	0.112	0.107	2
Mercury	Hg	7.85	8.175	8.514	8.842	9.161	9.475	11
Mesityl oxide	$C_6H_{10}O$	0.170	0.163	0.156	0.149	0.142	0.134	2
Methanol	CH_4O	0.218	0.210	0.202	0.195	0.189	0.182	1
2-Methoxyethanol	$C_3H_8O_2$			0.190	0.180	0.170		1
Methyl acetate	$C_3H_6O_2$	0.174	0.164	0.153	0.143	0.133	0.122	2
2-Methylaniline	C_7H_9N			0.162				1
3-Methylaniline	C_7H_9N			0.161				1
Methyl benzoate	$C_8H_8O_2$			0.147				1
2-Methyl-1,3-butadiene	C_5H_8	0.141	0.130	0.119				1
Methyl butanoate	$C_5H_{10}O_2$			0.140				1
2-Methyl-2-butanol	$C_5H_{12}O$		0.119	0.116	0.113	0.109	0.106	1
N-Methylformamide	C_2H_5NO			0.203	0.201	0.199	0.196	2
Methyl formate	$C_2H_4O_2$		0.194	0.187				1
2-Methylheptane	C_8H_{18}	0.127	0.1206	0.1139	0.1072	0.1005	0.094	3
3-Methylheptane	C_8H_{18}	0.128	0.1216	0.1149	0.1081	0.1014	0.095	3
2-Methylhexane	C_7H_{16}	0.125	0.1177	0.1105	0.1033	0.0961	0.089	3
3-Methylhexane	C_7H_{16}	0.126	0.1184	0.1112	0.1040	0.0968	0.090	3
Methyl hexanoate	$C_7H_{14}O_2$			0.136	0.131	0.126	0.121	1
Methyl methacrylate	$C_5H_8O_2$		0.156	0.147	0.137	0.127	0.117	1
Methyloxirane	C_3H_6O		0.181	0.171				1
2-Methylpentane	C_6H_{14}	0.120	0.1127	0.1050	0.0972	0.0894	0.082	3
3-Methylpentane	C_6H_{14}	0.122	0.1142	0.1064	0.0986	0.0909	0.083	3
Methyl pentanoate	$C_6H_{12}O_2$		0.143	0.138	0.132	0.127		1
Methyl propanoate	$C_4H_8O_2$			0.141	0.137			1
2-Methyl-2-propanol	$C_4H_{10}O$			0.112	0.110	0.109	0.108	1
N-Methyl-2-pyrrolidinone	C_5H_9NO			0.167	0.162	0.157		1
Nitrobenzene	$C_6H_5NO_2$			0.149	0.145	0.142	0.139	1
Nitroethane	$C_2H_5NO_2$			0.173	0.161	0.149		1
Nitromethane	CH_3NO_2	0.226	0.215	0.204	0.193	0.182	0.171	1
1-Nitropropane	$C_3H_7NO_2$			0.152	0.144	0.137		1
Nonane	C_9H_{20}	0.141	0.1337	0.1269	0.1201	0.1133	0.106	4
Nonanoic acid	$C_9H_{18}O_2$			0.150	0.146	0.142	0.138	1
1-Nonanol	$C_9H_{20}O$		0.164	0.159	0.155	0.150	0.145	1
1-Nonene	C_9H_{18}	0.136	0.130	0.123	0.116	0.110	0.104	1
Octadecane	$C_{18}H_{38}$				0.146	0.142	0.137	2
Octane	C_8H_{18}	0.139	0.1317	0.1244	0.1171	0.1097	0.102	4
Octanoic acid	$C_8H_{16}O_2$			0.146	0.143	0.139	0.135	1
1-Octanol	$C_8H_{18}O$		0.162	0.158	0.153	0.148	0.143	1
2-Octanone	$C_8H_{16}O$		0.141	0.135	0.129	0.124	0.118	1
1-Octene	C_8H_{16}	0.139	0.133	0.126	0.120	0.114	0.107	1
Octyl butanoate	$C_{12}H_{24}O_2$			0.139	0.134	0.129	0.125	1
Octyl propanoate	$C_{11}H_{22}O_2$			0.135	0.130	0.125	0.120	1
Paraldehyde	$C_6H_{12}O_3$			0.130				1
Pentanal	$C_5H_{10}O$		0.146	0.139	0.133	0.127	0.121	1
Pentane	C_5H_{12}	0.130	0.1207	0.1113	0.1018	0.0923	0.083	4
1,5-Pentanediol	$C_5H_{12}O_2$		0.221	0.222				
2,4-Pentanedione	$C_5H_8O_2$			0.154	0.150	0.146	0.143	1
Pentanoic acid	$C_5H_{10}O_2$			0.140	0.137	0.133	0.130	1
1-Pentanol	$C_5H_{12}O$	0.159	0.155	0.150	0.145	0.141	0.136	1
2-Pentanone	$C_5H_{10}O$		0.149	0.142	0.135	0.128	0.121	1
3-Pentanone	$C_5H_{10}O$		0.151	0.144	0.137	0.129	0.122	1
1-Pentene	C_5H_{10}	0.131	0.124	0.116				2
Pentyl acetate	$C_7H_{14}O_2$		0.141	0.134	0.126	0.120	0.113	1
Pentylbenzene	$C_{11}H_{16}$		0.135	0.130	0.125	0.120	0.115	
Pentyl propanoate	$C_8H_{16}O_2$			0.138	0.132			1
1-Pentyne	C_5H_8	0.144	0.136	0.127	0.119			1
Perfluorocyclobutane	C_4F_8	0.082	0.072	0.063	0.053	0.044	0.034	1
Perfluoroheptane	C_7F_{16}	0.068	0.064	0.060	0.056	0.053		1
Perfluorohexane	C_6F_{14}		0.067	0.065	0.064			1

Name	Mol. form.	λ(-25 °C)/ W m⁻¹ K⁻¹	λ(0 °C)/ W m⁻¹ K⁻¹	λ(25 °C)/ W m⁻¹ K⁻¹	λ(50 °C)/ W m⁻¹ K⁻¹	λ(75 °C)/ W m⁻¹ K⁻¹	λ(100 °C)/ W m⁻¹ K⁻¹	Ref.
Perfluorooctane	C_8F_{18}		0.066	0.062	0.059	0.055	0.052	1
Perfluoropropane	C_3F_8	0.062	0.056	0.051	0.046	0.041	0.035	1
Phenol	C_6H_6O				0.153	0.149	0.147	1
2-Phenoxyethanol	$C_8H_{10}O_2$			0.169	0.168	0.166	0.165	1
1,2-Propanediol	$C_3H_8O_2$	0.199	0.200	0.200	0.200	0.199	0.197	1
Propanoic acid	$C_3H_6O_2$		0.147	0.144	0.141	0.139	0.136	1
1-Propanol	C_3H_8O	0.162	0.158	0.154	0.149	0.145	0.141	2
2-Propanol	C_3H_8O	0.146	0.141	0.135	0.129	0.124	0.118	2
Propyl acetate	$C_5H_{10}O_2$		0.146	0.140	0.135	0.130	0.124	1
Propylbenzene	C_9H_{12}	0.134	0.130	0.125	0.120	0.115	0.109	1
Propyl formate	$C_4H_8O_2$		0.151	0.144	0.137	0.130		1
Propyl propanoate	$C_6H_{12}O_2$				0.133			1
Pyridine	C_5H_5N		0.171	0.166	0.162	0.157	0.153	1
Quinoline	C_9H_7N			0.147	0.144	0.141	0.138	1
Styrene	C_8H_8	0.148	0.142	0.137	0.131	0.126	0.120	2
1,1,2,2-Tetrachloro-1,2-difluoroethane	$C_2Cl_4F_2$			0.082	0.078	0.074	0.069	1
1,1,2,2-Tetrachloroethane	$C_2H_2Cl_4$	0.124	0.118	0.111	0.104	0.098	0.091	1
Tetrachloroethene	C_2Cl_4		0.117	0.110	0.104	0.098	0.093	1
Tetrachloromethane	CCl_4		0.109	0.103	0.098	0.092	0.087	1
Tetrachlorosilane	Cl_4Si			0.099	0.096			2
Tetradecane	$C_{14}H_{30}$			0.139	0.134	0.129	0.124	1
1-Tetradecanol	$C_{14}H_{30}O$				0.167	0.162	0.157	2
1-Tetradecene	$C_{14}H_{28}$			0.136	0.131	0.126	0.121	1
Tetraethylene glycol	$C_8H_{18}O_5$			0.191	0.192			1
Tetrahydrofuran	C_4H_8O	0.132	0.126	0.120	0.114			2
1,2,3,4-Tetrahydronaphthalene	$C_{10}H_{12}$			0.131	0.129	0.128	0.126	1
Thiophene	C_4H_4S			0.199	0.195	0.191	0.186	2
Tin(IV) chloride	Cl_4Sn	0.123	0.117	0.112	0.106	0.101	0.095	1
Titanium(IV) chloride	Cl_4Ti		0.143	0.138	0.134	0.129	0.124	1
Toluene	C_7H_8	0.1455	0.1385	0.1310	0.1235	0.1162	0.1095	9
Tributylamine	$C_{12}H_{27}N$			0.129				1
1,2,3-Trichlorobenzene	$C_6H_3Cl_3$				0.110	0.108	0.106	1
1,2,4-Trichlorobenzene	$C_6H_3Cl_3$			0.112	0.109	0.106		1
1,1,1-Trichloroethane	$C_2H_3Cl_3$		0.106	0.101	0.096			2
Trichloroethene	C_2HCl_3	0.128	0.121	0.114	0.106	0.098	0.090	1
Trichlorofluoromethane	CCl_3F	0.102	0.096	0.089	0.083	0.076	0.070	1
Trichloromethane	$CHCl_3$	0.127	0.122	0.117	0.112	0.107	0.102	2
1,1,2-Trichloro-1,2,2-trifluoroethane	$C_2Cl_3F_3$	0.0847	0.0790	0.0736	0.0683			6
Tridecane	$C_{13}H_{28}$			0.130	0.125	0.120	0.115	1
1-Tridecene	$C_{13}H_{26}$			0.130	0.125	0.120	0.115	1
Triethylamine	$C_6H_{15}N$	0.146	0.139	0.132	0.125	0.118	0.111	1
Triethylene glycol	$C_6H_{14}O_4$		0.193	0.195	0.196	0.196	0.196	1
Triethylene glycol dimethyl ether	$C_8H_{18}O_4$			0.169	0.158	0.147		1
Trimethylamine	C_3H_9N	0.143	0.133					2
1,2,4-Trimethylbenzene	C_9H_{12}			0.129	0.124	0.118	0.114	1
1,3,5-Trimethylbenzene	C_9H_{12}	0.143	0.139	0.134	0.129	0.123	0.117	1
2,2,3-Trimethylbutane	C_7H_{16}	0.107	0.1011	0.0950	0.0889	0.0828	0.077	3
2,2,4-Trimethylpentane	C_8H_{18}	0.107	0.1007	0.0948	0.0888	0.0829	0.077	3
2,3,4-Trimethylpentane	C_8H_{18}	0.115	0.1093	0.1035	0.0976	0.0918	0.086	3
Undecane	$C_{11}H_{24}$			0.136	0.128	0.122	0.116	1
Undecanoic acid	$C_{11}H_{22}O_2$				0.153	0.149		1
1-Undecanol	$C_{11}H_{24}O$			0.169	0.165	0.161	0.158	1
6-Undecanone	$C_{11}H_{22}O$			0.137	0.132	0.127		1
1-Undecene	$C_{11}H_{22}$			0.126	0.118	0.114	0.108	1
Vinyl acetate	$C_4H_6O_2$			0.151	0.141	0.131	0.120	1
Water	H_2O		0.5562	0.6062	0.6423	0.6643	0.6729	8
o-Xylene	C_8H_{10}			0.131	0.126	0.120	0.114	2
m-Xylene	C_8H_{10}			0.130	0.124	0.118	0.113	2
p-Xylene	C_8H_{10}			0.130	0.124	0.118	0.112	2

DIFFUSION IN GASES

This table gives diffusion coefficients D_{12} for binary gas mixtures as a function of temperature. Values refer to atmospheric pressure. The diffusion coefficient is inversely proportional to pressure as long as the gas is in a regime where binary collisions dominate. See Reference 1 for a discussion of the dependence of D_{12} on temperature and composition.

References

1. Marrero, T. R., and Mason, E. A., *J. Phys. Chem. Ref. Data*, 1, 1, 1972.
2. Kestin, J., et al., *J. Phys. Chem. Ref. Data*, 13, 229, 1984.

D_{12}/cm² s⁻¹ for *p* = 101.325 kPa and the Specified *t*/°C

Mixture	D_{12}(-73 °C)/ cm² s⁻¹	D_{12}(0 °C)/ cm² s⁻¹	D_{12}(20 °C)/ cm² s⁻¹	D_{12}(100 °C)/ cm² s⁻¹	D_{12}(200 °C)/ cm² s⁻¹	D_{12}(300 °C)/ cm² s⁻¹	D_{12}(400 °C)/ cm² s⁻¹
Large Excess of Air							
Ar-air		0.167	0.189	0.289	0.437	0.612	0.810
CH_4-air			0.210	0.321	0.485	0.678	0.899
CO-air			0.208	0.315	0.475	0.662	0.875
CO_2-air			0.160	0.252	0.390	0.549	0.728
H_2-air		0.668	0.756	1.153	1.747	2.444	3.238
H_2O-air			0.242	0.399	0.638	0.873	1.135
He-air		0.617	0.697	1.057	1.594	2.221	2.933
SF_6-air				0.150	0.233	0.329	0.438
Equimolar Mixture							
Ar-CH_4				0.306	0.467	0.657	0.876
Ar-CO		0.168	0.190	0.290	0.439	0.615	0.815
Ar-CO_2		0.129	0.148	0.235	0.365	0.517	0.689
Ar-H_2		0.698	0.794	1.228	1.876	2.634	3.496
Ar-He	0.381	0.645	0.726	1.088	1.617	2.226	2.911
Ar-Kr	0.064	0.117	0.134	0.210	0.323	0.456	0.605
Ar-N_2		0.168	0.190	0.290	0.439	0.615	0.815
Ar-Ne	0.160	0.277	0.313	0.475	0.710	0.979	1.283
Ar-O_2		0.166	0.187	0.285	0.430	0.600	0.793
Ar-SF_6				0.128	0.202	0.290	0.389
Ar-Xe	0.052	0.095	0.108	0.171	0.264	0.374	0.498
CH_4-H_2			0.708	1.084	1.648	2.311	3.070
CH_4-He			0.650	0.992	1.502	2.101	2.784
CH_4-N_2			0.208	0.317	0.480	0.671	0.890
CH_4-O_2			0.220	0.341	0.523	0.736	0.978
CH_4-SF_6				0.167	0.257	0.363	0.482
CO_2-C_3H_8			0.084	0.133	0.209		
CO_2-H_2	0.315	0.552	0.627	0.964	1.470	2.066	2.745
CO_2-H_2O			0.162	0.292	0.496	0.741	1.021
CO_2-He	0.300	0.513	0.580	0.878	1.321		
CO_2-N_2			0.160	0.253	0.392	0.553	0.733
CO_2-N_2O	0.055	0.099	0.113	0.177	0.276		
CO_2-Ne	0.131	0.227	0.258	0.395	0.603	0.847	
CO_2-O_2			0.159	0.248	0.380	0.535	0.710
CO_2-SF_6				0.099	0.155		
CO-CO_2			0.162	0.250	0.384		
CO-H_2	0.408	0.686	0.772	1.162	1.743	2.423	3.196
CO-He	0.365	0.619	0.698	1.052	1.577	2.188	2.882
CO-Kr		0.131	0.149	0.227	0.346	0.485	0.645
CO-N_2	0.133	0.208	0.231	0.336	0.491	0.673	0.878
CO-O_2			0.202	0.307	0.462	0.643	0.849
CO-SF_6				0.144	0.226	0.323	0.432
D_2-H_2	0.631	1.079	1.219	1.846	2.778	3.866	5.103
H_2-He	0.775	1.320	1.490	2.255	3.394	4.726	6.243
H_2-Kr	0.340	0.601	0.682	1.053	1.607	2.258	2.999
H_2-N_2	0.408	0.686	0.772	1.162	1.743	2.423	3.196
H_2-Ne	0.572	0.982	1.109	1.684	2.541	3.541	4.677

Mixture	D_{12}(-73 °C)/ cm² s⁻¹	D_{12}(0 °C)/ cm² s⁻¹	D_{12}(20 °C)/ cm² s⁻¹	D_{12}(100 °C)/ cm² s⁻¹	D_{12}(200 °C)/ cm² s⁻¹	D_{12}(300 °C)/ cm² s⁻¹	D_{12}(400 °C)/ cm² s⁻¹
H_2-O_2		0.692	0.782	1.188	1.792	2.497	3.299
H_2O-N_2			0.242	0.399			
H_2O-O_2			0.244	0.403	0.645	0.882	1.147
H_2-SF_6			0.412	0.649	0.998	1.400	1.851
H_2-Xe		0.513	0.581	0.890	1.349	1.885	2.493
He-Kr	0.330	0.559	0.629	0.942	1.404	1.942	2.550
He-N_2	0.365	0.619	0.698	1.052	1.577	2.188	2.882
He-Ne	0.563	0.948	1.066	1.592	2.362	3.254	4.262
He-O_2		0.641	0.723	1.092	1.640	2.276	2.996
He-SF_6			0.400	0.592	0.871	1.190	1.545
He-Xe	0.282	0.478	0.538	0.807	1.201	1.655	2.168
Kr-N_2		0.131	0.149	0.227	0.346	0.485	0.645
Kr-Ne	0.131	0.228	0.258	0.392	0.587	0.812	1.063
Kr-Xe	0.035	0.064	0.073	0.116	0.181	0.257	0.344
N_2-Ne			0.317	0.483	0.731	1.021	1.351
N_2-O_2			0.202	0.307	0.462	0.643	0.849
N_2-SF_6				0.148	0.231	0.328	0.436
N_2-Xe		0.107	0.122	0.188	0.287	0.404	0.539
Ne-Xe	0.111	0.193	0.219	0.332	0.498	0.688	0.901
O_2-SF_6			0.097	0.154	0.238	0.334	0.441

DIFFUSION OF GASES IN WATER

This table gives values of the diffusion coefficient, D, for diffusion of several common gases in water at various temperatures. For simple one-dimensional transport, the diffusion coefficient describes the time–rate of change of concentration, dc/dt, through the equation

$$dc/dt = D d^2c/dx^2$$

where x is, for example, the perpendicular distance from a gas–liquid interface. The values below have been selected from the references indicated; in some cases data have been refitted to permit interpolation in temperature.

Gas–liquid diffusion coefficients are difficult to measure, and large differences are found between values obtained by different authors and through different experimental methods. See References 1 and 2 for a discussion of measurement techniques.

References

1. Jähne, B., Heinz, G., and Dietrich, W., *J. Geophys. Res.*, 92, 10767, 1987.
2. Himmelblau, D. M., *Chem. Rev.*, 64, 527, 1964.
3. Boerboom, A. J. H., and Kleyn, G., *J. Chem. Phys.*, 50, 1086, 1969.
4. O'Brien, R. N., and Hyslop, W. F., *Can. J. Chem.*, 55, 1415, 1977.
5. Maharajh, D. M., and Walkley, J., *Can. J. Chem.*, 51, 944, 1973.
6. *Landolt-Börnstein, Numerical Data and Functional Relationships in Science and Technology, Sixth Edition*, II/5a, *Transport Phenomena I (Viscosity and Diffusion)*, Springer-Verlag, Heidelberg, 1969.

Diffusion Coefficient in Water at the Indicated Temperature

Name	Formula	D(10 °C)/ 10^{-5} cm² s⁻¹	D(15 °C)/ 10^{-5} cm² s⁻¹	D(20 °C)/ 10^{-5} cm² s⁻¹	D(25 °C)/ 10^{-5} cm² s⁻¹	D(30 °C)/ 10^{-5} cm² s⁻¹	D>(35 °C)/ 10^{-5} cm² s⁻¹	Ref.
Acetylene	HC≡CH	1.43	1.59	1.78	1.99	2.23		2
Ammonia	NH_3		1.3	1.5				2
Argon	Ar				2.5			4
Bromomethane	CH_3Br				1.35			5
Carbon dioxide	CO_2	1.26	1.45	1.67	1.91	2.17	2.47	1
Chlorine	Cl_2		1.13	1.5	1.89			2,6
Chloromethane	CH_3Cl				1.40			5
Dichlorofluoromethane	$CHCl_2F$				1.80			5
Helium	He	5.67	6.18	6.71	7.28	7.87	8.48	1
Hydrogen	H_2	3.62	4.08	4.58	5.11	5.69	6.31	1
Hydrogen bromide	HBr				3.15			6
Hydrogen chloride	HCl				3.07			6
Hydrogen sulfide	H_2S				1.36			2,6
Krypton	Kr	1.20	1.39	1.60	1.84	2.11	2.40	1
Methane	CH_4	1.24	1.43	1.62	1.84	2.08	2.35	1
Neon	Ne	2.93	3.27	3.64	4.03	4.45	4.89	1
Nitrogen	N_2				2.0			2
Nitrogen dioxide	NO_2			1.23	1.4	1.59		2,6
Nitrous oxide	N_2O		1.62	2.11	2.57			2,6
Oxygen	O_2		1.67	2.01	2.42			2,6
Radon	Rn	0.81	0.96	1.13	1.33	1.55	1.80	1
Sulfur dioxide	SO_2			1.62	1.83	2.07	2.32	2
Xenon	Xe	0.93	1.08	1.27	1.47	1.70	1.95	1

DIFFUSION COEFFICIENTS IN LIQUIDS AT INFINITE DILUTION

This table lists diffusion coefficients D_{AB} at infinite dilution for some binary liquid mixtures. Values are given at 25 °C when available; it should be noted that the diffusion coefficient generally increases by 10% to 20% for a 10 °C increase above ambient temperature. The diffusion coefficient is defined by:

$$dc/dt = D_{AB}d^2c/dx^2$$

where x is the perpendicular distance from the point at which the solute is introduced.

Solvents are listed in alphabetical order, as are the solutes within each solvent group.

References

1. Safi, A., Nicolas, C., Neau, E., and Chevalier, J.-L., *J. Chem. Eng. Data* 52, 977, 2007.
2. Safi, A., Nicolas, C., Neau, E., and Chevalier, J.-L., *J. Chem. Eng. Data* 52, 126, 2007.
3. Sanni, S. A., Fell, C. J. D., and Hutchison, H. P., *J. Chem. Eng. Data* 16, 424, 1971.
4. Fan, Y., Qian, R., Shi, M., and Shi, J., *J. Chem. Eng. Data* 40, 1053, 1995.
5. *Landolt-Börnstein Numerical Data and Functional Relationships in Science and Technology*, Sixth Edition, Vol. II/5a, Springer-Verlag, Heidelberg, 1969.

Diffusion Coefficients at Infinite Dilution

Solute	Solvent	t/°C	D_{AB}/ 10^{-5} cm² s⁻¹	Ref.
Acetic acid	Acetone	25	3.31	5
Benzene	Acetone	25	4.25	2
Benzoic acid	Acetone	25	2.62	5
Formic acid	Acetone	25	3.77	5
Nitrobenzene	Acetone	20	2.94	5
Tetrachloromethane	Acetone	25	3.29	5
Trichloromethane	Acetone	25	3.64	5
Water	Acetone	25	4.56	5
Acetic acid	Benzene	25	2.09	5
Aniline	Benzene	25	1.96	5
Benzoic acid	Benzene	25	1.38	5
Bromobenzene	Benzene	8	1.45	5
2-Butanone	Benzene	30	2.09	5
Chloroethene	Benzene	8	1.77	5
Cyclohexane	Benzene	25	2.09	3
Ethanol	Benzene	25	3.02	5
Formic acid	Benzene	25	2.28	5
Heptane	Benzene	25	1.79	3
Methanol	Benzene	25	3.80	5
Toluene	Benzene	25	1.85	3
1,2,4-Trichlorobenzene	Benzene	8	1.34	5
Trichloromethane	Benzene	25	2.26	5
Benzene	1-Butanol	25	1.00	5
Biphenyl	1-Butanol	25	0.63	5
Butanoic acid	1-Butanol	30	0.51	5
p-Dichlorobenzene	1-Butanol	25	0.82	5
1,6-Hexanedioic acid	1-Butanol	30	0.40	5
Methanol	1-Butanol	30	0.59	5
cis-9-Octadecenoic acid	1-Butanol	30	0.25	5
Propane	1-Butanol	25	1.57	5
Water	1-Butanol	25	0.56	5
Benzene	Cyclohexane	25	1.92	1
Chlorobenzene	Cyclohexane	25	1.34	1
4-Chlorotoluene	Cyclohexane	25	1.28	1
Ethylbenzene	Cyclohexane	25	1.36	1
Naphthalene	Cyclohexane	25	1.18	1
Perylene	Cyclohexane	25	0.79	1
Pyrene	Cyclohexane	25	0.95	1
Tetrachloromethane	Cyclohexane	25	1.49	3
Toluene	Cyclohexane	25	1.66	1
Benzene	Decane	25	2.16	1
Chlorobenzene	Decane	25	1.98	1
4-Chlorotoluene	Decane	25	1.80	1
Ethylbenzene	Decane	25	1.79	1
Naphthalene	Decane	25	1.65	1
Perylene	Decane	25	1.08	1
Pyrene	Decane	25	1.23	1
Toluene	Decane	25	1.93	1
Trichloromethane	Diethyl ether	25	4.48	3
Allyl alcohol	Ethanol	20	0.98	5
Benzene	Ethanol	25	1.88	2
Iodine	Ethanol	25	1.32	5
Iodobenzene	Ethanol	20	1.00	5
3-Methyl-1-butanol	Ethanol	20	0.81	5
Pyridine	Ethanol	20	1.10	5
Tetrachloromethane	Ethanol	25	1.50	5
Water	Ethanol	25	1.24	5
Acetic acid	Ethyl acetate	20	2.18	5
Acetone	Ethyl acetate	20	3.18	5
2-Butanone	Ethyl acetate	30	2.93	5
Ethyl benzoate	Ethyl acetate	20	1.85	5
Nitrobenzene	Ethyl acetate	20	2.25	5
Water	Ethyl acetate	25	3.20	5
Benzene	Heptane	25	3.75	1
Chlorobenzene	Heptane	25	3.42	1
4-Chlorotoluene	Heptane	25	3.11	1
Ethylbenzene	Heptane	25	3.15	1
Naphthalene	Heptane	25	2.81	1
Perylene	Heptane	25	1.89	1
Pyrene	Heptane	25	2.16	1
Toluene	Heptane	25	3.42	1
Benzene	Hexane	25	4.70	1
Bromobenzene	Hexane	8	2.60	5
2-Butanone	Hexane	30	3.74	5
Chlorobenzene	Hexane	25	4.16	1
4-Chlorotoluene	Hexane	25	3.74	1
Dodecane	Hexane	25	2.73	5
Ethylbenzene	Hexane	25	3.73	1
Iodine	Hexane	25	4.45	5
Methane	Hexane	25	0.09	5
Naphthalene	Hexane	25	3.55	1
Perylene	Hexane	25	2.30	1
Propane	Hexane	25	4.87	5
Pyrene	Hexane	25	2.58	1

Solute	Solvent	t/°C	D_{AB}/ 10^{-5} cm^2 s^{-1}	Ref.
Tetrachloromethane	Hexane	25	3.70	5
Toluene	Hexane	25	4.12	1
Benzene	Octane	25	3.19	1
Chlorobenzene	Octane	25	2.89	1
4-Chlorotoluene	Octane	25	2.62	1
Ethylbenzene	Octane	25	2.58	1
Naphthalene	Octane	25	2.35	1
Perylene	Octane	25	1.58	1
Pyrene	Octane	25	1.81	1
Toluene	Octane	25	2.83	1
1,3,5-Trimethylbenzene	Octane	30	2.21	4
o-Xylene	Octane	30	2.65	4
p-Xylene	Octane	30	2.82	4
Acetone	Tetrachloromethane	25	1.75	5
Benzene	Tetrachloromethane	25	1.42	5
Cyclohexane	Tetrachloromethane	25	1.30	5
Ethanol	Tetrachloromethane	25	1.90	5
Iodine	Tetrachloromethane	30	1.63	5
Trichloromethane	Tetrachloromethane	25	1.66	5
Acetic acid	Toluene	25	2.26	5
Benzene	Toluene	25	2.54	3
Benzoic acid	Toluene	25	1.49	5
Cyclohexane	Toluene	25	2.42	3
Formic acid	Toluene	25	2.65	5
Water	Toluene	25	6.19	5
Acetone	Trichloromethane	25	2.55	5
Benzene	Trichloromethane	25	2.89	5
2-Butanone	Trichloromethane	25	2.13	5
Butyl acetate	Trichloromethane	25	1.71	5
Cyclohexane	Trichloromethane	25	1.28	3
Diethyl ether	Trichloromethane	25	2.13	3
Ethanol	Trichloromethane	15	2.20	5
Ethyl acetate	Trichloromethane	25	2.02	5
Benzene	2,2,4-Trimethylpentane	30	3.46	4
Ethylbenzene	2,2,4-Trimethylpentane	30	2.63	4
Toluene	2,2,4-Trimethylpentane	30	3.15	4
1,3,5-Trimethylbenzene	2,2,4-Trimethylpentane	30	2.26	4
o-Xylene	2,2,4-Trimethylpentane	30	2.62	4
p-Xylene	2,2,4-Trimethylpentane	30	3.03	4
Acetic acid	Water	25	1.29	5
Acetone	Water	25	1.28	5
Acetonitrile	Water	15	1.26	5
L-Alanine	Water	25	0.91	5
Allyl alcohol	Water	15	0.90	5
Aniline	Water	20	0.92	5
DL-Arabinose	Water	20	0.69	5
Benzene	Water	20	1.02	5
1-Butanol	Water	25	0.56	5
Caprolactam	Water	25	0.87	5
Chloroethene	Water	25	1.34	5
Cyclohexane	Water	20	0.84	5
Diethylamine	Water	20	0.97	5
1,2-Ethanediol	Water	25	1.16	5
Ethanol	Water	25	1.24	5
Ethanolamine	Water	25	1.08	5
Ethyl acetate	Water	20	1.00	5
Ethylbenzene	Water	20	0.81	5
Ethyl carbamate	Water	15	0.80	5
α-D-Glucose	Water	25	0.67	5
Glycerol	Water	25	1.06	5
Glycine	Water	25	1.05	5
β-D-Lactose	Water	15	0.38	5
α-Maltose	Water	15	0.38	5
D-Mannitol	Water	15	0.50	5
Methane	Water	25	1.49	5
Methanol	Water	15	1.28	5
3-Methyl-1-butanol	Water	10	0.69	5
Methylcyclopentane	Water	20	0.85	5
Phenol	Water	20	0.89	5
1-Propanol	Water	15	0.87	5
Propene	Water	25	1.44	5
Pyridine	Water	25	0.58	5
Raffinose	Water	15	0.33	5
Sucrose	Water	25	0.52	5
Toluene	Water	20	0.85	5
Urea	Water	25	1.38	5

Section 7
Biochemistry

PROPERTIES OF AMINO ACIDS

This table gives selected properties of some important amino acids and closely related compounds. Compounds are listed by name in alphabetical order. The three-letter symbol is included for the 20 "standard" amino acids that are the basic constituents of proteins. Structures are given in the following table.

Symbol: Three-letter symbol for standard amino acids
M_r: Molecular weight
t_m: Melting point
pK_a, pK_b, pK_c, pK_d: Negative of the logarithm of the acid dissociation constants for the COOH and NH_2 groups (and, in some cases, other groups) in the molecule (at 25 °C)
pI: pH at the isoelectric point
S: Solubility in water in units of grams of compound per kilogram of water. A temperature of 25 °C is understood unless otherwise stated. When quantitative data are not available, a qualitative indication is given.
$\overline{V}_2$: Partial molar volume in aqueous solution at infinite dilution at 25 °C

Data on the enthalpy of formation of many of these compounds are included in the table "Standard Thermodynamic Properties of Chemical Substances" in Section 5 of this *Handbook*. Absorption spectra and optical rotation data can be found in Ref. 3. Partial molar volume is taken from Ref. 5; other thermodynamic properties, including solubility as a function of temperature, are given in Refs. 3 and 5. Most of the pK values come from Refs. 1, 6, and 7.

References

1. Dawson, R. M. C., Elliott, D. C., Elliott, W. H., and Jones, K. M., *Data for Biochemical Research, Third Edition*, Clarendon Press, Oxford, 1986.
2. O'Neil, Maryadele J., Ed., *The Merck Index, Fifteenth Edition*, Royal Society of Chemistry, Cambridge, 2013.
3. Sober, H. A., Ed., *CRC Handbook of Biochemistry. Selected Data for Molecular Biology*, CRC Press, Boca Raton, FL, 1968.
4. Voet, D., and Voet, J. G., *Biochemistry, Second Edition*, John Wiley & Sons, New York, 1995.
5. Hinz, H. J., Ed., *Thermodynamic Data for Biochemistry and Biotechnology*, Springer-Verlag, Heidelberg, 1986.
6. Fasman, G. D., Ed. *Practical Handbook of Biochemistry and Molecular Biology*, CRC Press, Boca Raton, FL, 1989.
7. Smith, R. M., and Martell, A. E., *NIST Standard Reference Database 46: Critically Selected Stability Constants of Metal Complexes Database*, Version 3.0, National Institute of Standards and Technology, Gaithersburg, MD, 1997.
8. Ramasami, P., *J. Chem. Eng. Data*, 47, 1164, 2002.

Properties of Amino Acids

Symbol	Name	Mol. form.	Mol. wt.	t_m/°C	pK_a	pK_b	pK_c	pK_d	pI	S/g kg^{-1}	$\overline{V}_2$/ cm^3 mol^{-1}
Standard amino acids											
Ala	L-Alanine	$C_3H_7NO_2$	89.094	297 dec	2.33	9.71			6.00	166.9	60.54
Arg	L-Arginine	$C_6H_{14}N_4O_2$	174.201	260	2.03	9.00	12.10		10.76	182.6	127.42
Asn	L-Asparagine	$C_4H_8N_2O_3$	132.118	235	2.16	8.73			5.41	25.1	78.0
Asp	L-Aspartic acid	$C_4H_7NO_4$	133.104	270	1.95	9.66	3.71		2.77	5.04	74.8
Cys	L-Cysteine	$C_3H_7NO_2S$	121.159	240 dec	1.91	10.28	8.14		5.07	vs	73.45
Gln	L-Glutamine	$C_5H_{10}N_2O_3$	146.144	182	2.18	9.00			5.65	42	
Glu	L-Glutamic acid	$C_5H_9NO_4$	147.130	160 dec	2.16	9.58	4.15		3.22	8.6	89.85
Gly	Glycine	$C_2H_5NO_2$	75.067	290 dec	2.34	9.58			5.97	239	43.26
His	L-Histidine	$C_6H_9N_3O_2$	155.154	287 dec	1.70	9.09	6.04		7.59	43.5	98.3
Ile	L-Isoleucine	$C_6H_{13}NO_2$	131.173	284 dec	2.26	9.60			6.02	34.2	105.80
Leu	L-Leucine	$C_6H_{13}NO_2$	131.173	293	2.32	9.58			5.98	23.8	107.77
Lys	L-Lysine	$C_6H_{14}N_2O_2$	146.187	224 dec	2.15	9.16	10.67		9.74	5.8	108.5
Met	L-Methionine	$C_5H_{11}NO_2S$	149.212	281 dec	2.16	9.08			5.74	56	105.57
Phe	L-Phenylalanine	$C_9H_{11}NO_2$	165.189	283 dec	2.18	9.09			5.48	27.9	121.5
Pro	L-Proline	$C_5H_9NO_2$	115.131	221 dec	1.95	10.47			6.30	1625	82.76
Ser	L-Serine	$C_3H_7NO_3$	105.093	228 dec	2.13	9.05			5.68	250	60.62
Thr	L-Threonine	$C_4H_9NO_3$	119.119	256 dec	2.20	8.96			5.60	90.6	76.90
Trp	L-Tryptophan	$C_{11}H_{12}N_2O_2$	204.225	289 dec	2.38	9.34			5.89	13.2	143.8
Tyr	L-Tyrosine	$C_9H_{11}NO_3$	181.188	343 dec	2.24	9.04	10.10		5.66	0.51	
Val	L-Valine	$C_5H_{11}NO_2$	117.147	315	2.27	9.52			5.96	88	90.75
Other important amino acids and related compounds											
	N-Acetylglutamic acid	$C_7H_{11}NO_5$	189.166	199						s	
	N^6-Acetyl-L-lysine	$C_8H_{16}N_2O_3$	188.224	265 dec	2.12	9.51					
	β-Alanine	$C_3H_7NO_2$	89.094	200 dec	3.51	10.08				890	58.28
	2-Aminoadipic acid	$C_6H_{11}NO_4$	161.156	207.0	2.14	4.21	9.77		3.18	2.2^{40}	
	DL-2-Aminobutanoic acid	$C_4H_9NO_2$	103.120	304 dec	2.30	9.63			6.06	210	75.6
	DL-3-Aminobutanoic acid	$C_4H_9NO_2$	103.120	194.3	3.43	10.05			7.30	1250	76.3
	4-Aminobutanoic acid	$C_4H_9NO_2$	103.120	203 dec	4.02	10.35				971	73.2
	10-Aminodecanoic acid	$C_{10}H_{21}NO_2$	187.280								167.3

Symbol	Name	Mol. form.	Mol. wt.	t_m/°C	pK_a	pK_b	pK_c	pK_d	pI	S/g kg^{-1}	$\overline{V}_2$/ cm^3 mol^{-1}
	7-Aminoheptanoic acid	$C_7H_{15}NO_2$	145.200							[vs]	120.0
	6-Aminohexanoic acid	$C_6H_{13}NO_2$	131.173	205					7.29	850	104.2
	2-Amino-2-methylpropanoic acid	$C_4H_9NO_2$	103.120	335	2.36	10.21			5.72	137	77.55
	L-3-Amino-2-methylpropanoic acid	$C_4H_9NO_2$	103.120	185						[s]	
	9-Aminononanoic acid	$C_9H_{19}NO_2$	173.253								151.3
	8-Aminooctanoic acid	$C_8H_{17}NO_2$	159.227								136.1
	5-Amino-4-oxopentanoic acid	$C_5H_9NO_3$	131.130	118	4.05	8.90					
	5-Aminopentanoic acid	$C_5H_{11}NO_2$	117.147	157 dec						[s]	87.6
	Aniline-2-carboxylic acid	$C_7H_7NO_2$	137.137	144.6	2.05	4.95				3.5[14]	
	Azaserine	$C_5H_7N_3O_4$	173.128	150 dec		8.55				[vs]	
	Canavanine	$C_5H_{12}N_4O_3$	176.174	172	2.50	6.60	9.25		7.93	[vs]	
	L-γ-Carboxyglutamic acid	$C_6H_9NO_6$	191.138	167	1.70	9.90	4.75	3.20			
	Carnosine	$C_9H_{14}N_4O_3$	226.232	260	2.51	9.35	6.76			322	
	Citrulline	$C_6H_{13}N_3O_3$	175.185	222	2.32	9.30			5.92	[s]	
	Creatine	$C_4H_9N_3O_2$	131.133	303 dec	2.63	14.30				16	
	L-Cysteic acid	$C_3H_7NO_5S$	169.157	260 dec	1.89	8.70	1.30			[vs]	
	L-Cystine	$C_6H_{12}N_2O_4S_2$	240.300	260 dec	1.50	8.80	2.05	8.03		0.17	
	2,4-Diaminobutanoic acid	$C_4H_{10}N_2O_2$	118.134	118.1	1.85	8.24	10.44		9.27	[s]	
	3,5-Dibromo-L-tyrosine	$C_9H_9Br_2NO_3$	338.980	245						2.72	
	3,5-Dichloro-L-tyrosine	$C_9H_9Cl_2NO_3$	250.078							1.97	
	3,5-Diiodo-L-tyrosine	$C_9H_9I_2NO_3$	432.981	213	2.12	9.10	6.16			0.62	
	Dopamine	$C_8H_{11}NO_2$	153.179			10.36	8.88			[s]	
	L-Ethionine	$C_6H_{13}NO_2S$	163.238	273 dec	2.18	9.05	13.10				
	N-Glycylglycine	$C_4H_8N_2O_3$	132.118	220	3.13	8.10				231	
	Guanidinoacetic acid	$C_3H_7N_3O_2$	117.107	282	2.82					5	
	Histamine	$C_5H_9N_3$	111.145	83		9.83	6.11			[vs]	
	L-Homocysteine	$C_4H_9NO_2S$	135.185	232	2.15	8.57	10.38		5.55	[s]	
	Homocystine	$C_8H_{16}N_2O_4S_2$	268.354	264	1.59	9.44	2.54	8.52		0.20	
	L-Homoserine	$C_4H_9NO_3$	119.119	203 dec	2.27	9.28			6.17	1100	
	3-Hydroxy-DL-glutamic acid	$C_5H_9NO_5$	163.129						3.28		
	5-Hydroxylysine	$C_6H_{14}N_2O_3$	162.186		2.13	8.85	9.83		9.15		
	trans-4-Hydroxy-L-proline	$C_5H_9NO_3$	131.130	274	1.82	9.47			5.74	361	84.49
	L-3-Iodotyrosine	$C_9H_{10}INO_3$	307.084	205 dec	2.20	9.10	8.70			[sls]	
	L-Kynurenine	$C_{10}H_{12}N_2O_3$	208.213	194 dec						[sls]	
	L-Lanthionine	$C_6H_{12}N_2O_4S$	208.235	294 dec						1.5	
	Levodopa	$C_9H_{11}NO_4$	197.188	277	2.32	8.72	9.96	11.79		1.65[20]	
	L-1-Methylhistidine	$C_7H_{11}N_3O_2$	169.181	249	1.69	8.85	6.48			200	
	L-Norleucine	$C_6H_{13}NO_2$	131.173	301 dec	2.31	9.68			6.09	15	107.7
	L-Norvaline	$C_5H_{11}NO_2$	117.147	307	2.31	9.65				107	91.8
	L-Ornithine	$C_5H_{12}N_2O_2$	132.161	140	1.94	8.78	10.52		9.73	[vs]	
	O-Phosphoserine	$C_3H_8NO_6P$	185.073	166 dec	2.14	9.80	5.70				
	L-Pyroglutamic acid	$C_5H_7NO_3$	129.115	162	3.32						
	Sarcosine	$C_3H_7NO_2$	89.094	212 dec	2.18	9.97				428	
	Taurine	$C_2H_7NO_3S$	125.147	328	-0.3	9.06				105	
	L-Thyroxine	$C_{15}H_{11}I_4NO_4$	776.871	235	2.20	10.01	6.45			[sls]	

s lsSlightly soluble
s Soluble
v sVery soluble
1 4At 14 °C
2 0At 20 °C
4 0At 40 °C

Biochem

STRUCTURES OF COMMON AMINO ACIDS

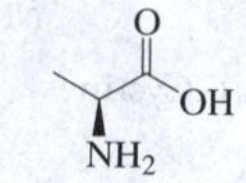
L-Alanine (Ala)

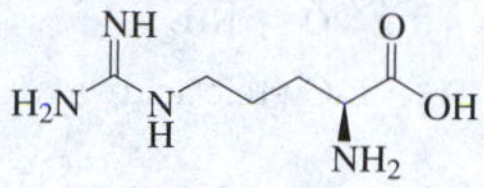
L-Arginine (Arg)

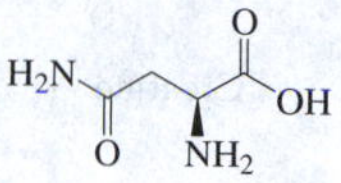
L-Asparagine (Asn)

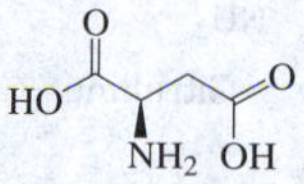
L-Aspartic acid (Asp)

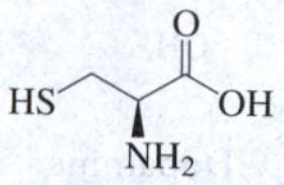
L-Cysteine (Cys)

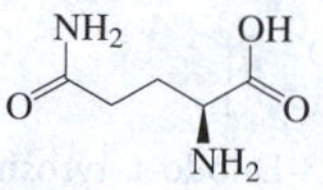
L-Glutamine (Gln)

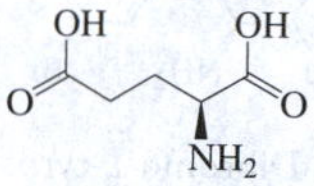
L-Glutamic acid (Glu)

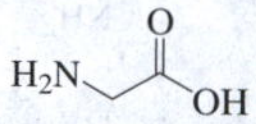
Glycine (Gly)

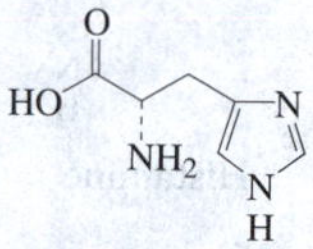
L-Histidine (His)

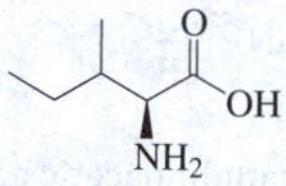
L-Isoleucine (Ile)

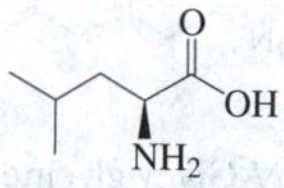
L-Leucine (Leu)

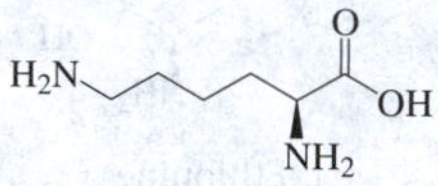
L-Lysine (Lys)

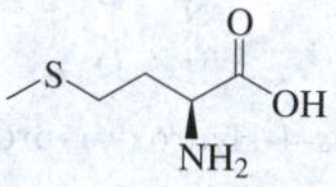
L-Methionine (Met)

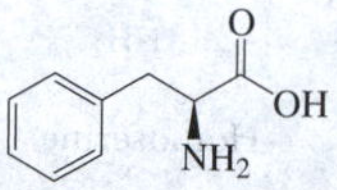
L-Phenylalanine (Phe)

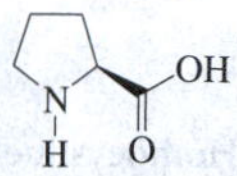
L-Proline (Pro)

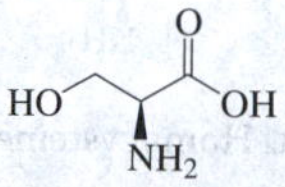
L-Serine (Ser)

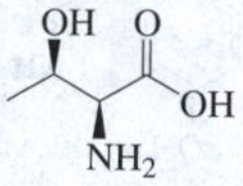
L-Threonine (Thr)

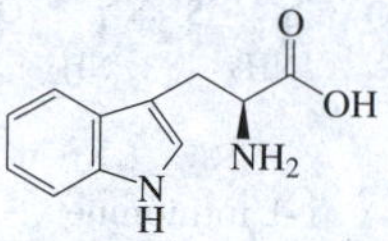
L-Tryptophan (Trp)

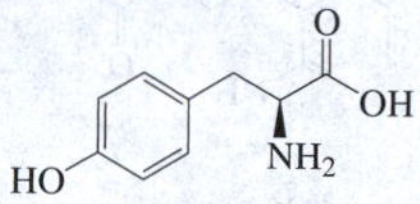
L-Tyrosine (Tyr)

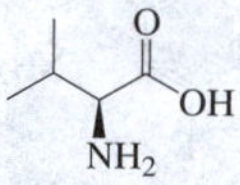
L-Valine (Val)

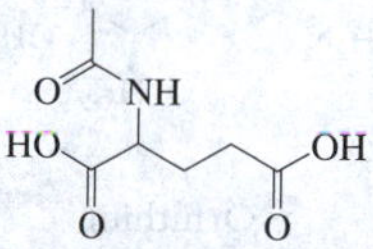
N-Acetylglutamic acid

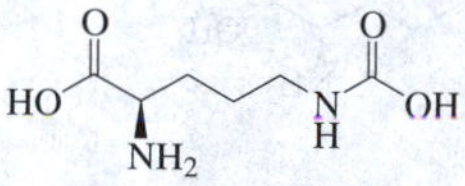
*N*6-Acetyl-L-lysine

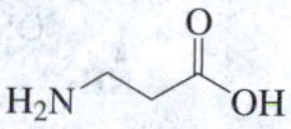
β-Alanine

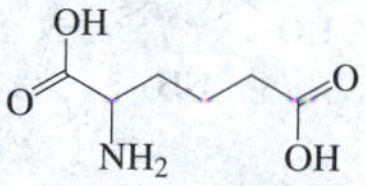
2-Aminoadipic acid

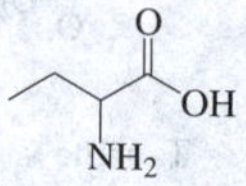
DL-2-Aminobutanoic acid

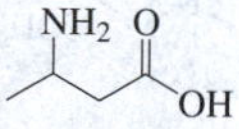
DL-3-Aminobutanoic acid

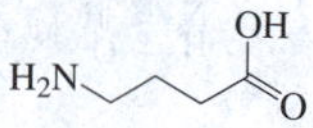
4-Aminobutanoic acid

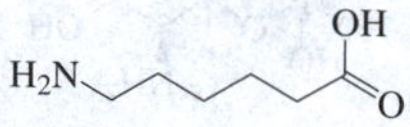
6-Aminohexanoic acid

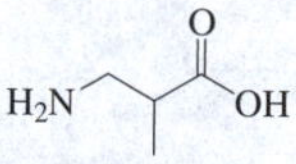
L-3-Amino-2-methylpropanoic acid

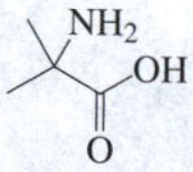
2-Amino-2-methylpropanoic acid

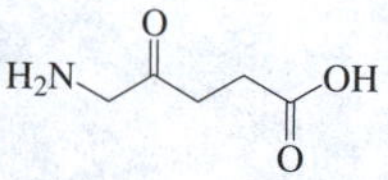
5-Amino-4-oxopentanoic acid

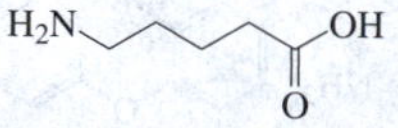
5-Aminopentanoic acid

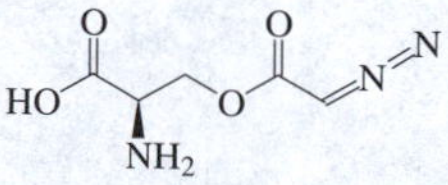
Azaserine

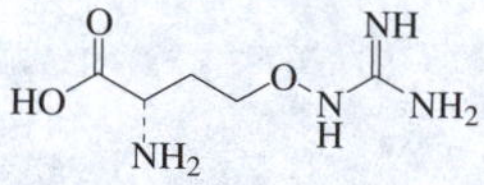
Canavanine

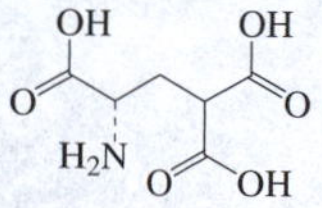
L-γ-Carboxyglutamic acid

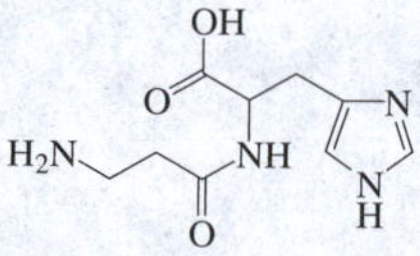
Carnosine

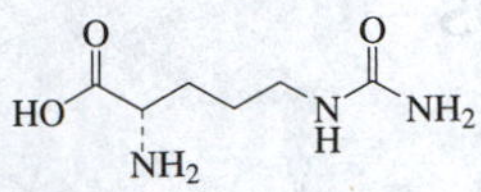
Citrulline

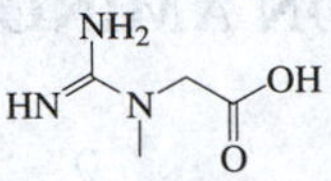
Creatine

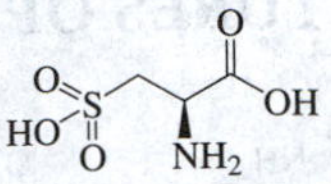
L-Cysteic acid

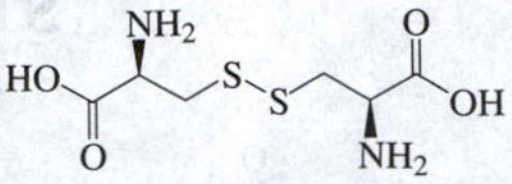
L-Cystine

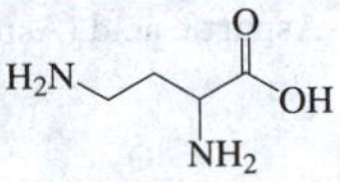
2,4-Diaminobutanoic acid

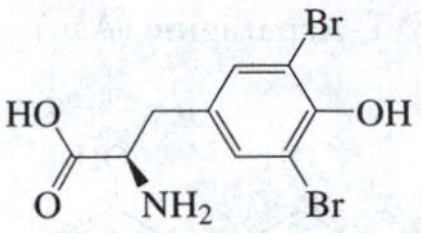
3,5-Dibromo-L-tyrosine

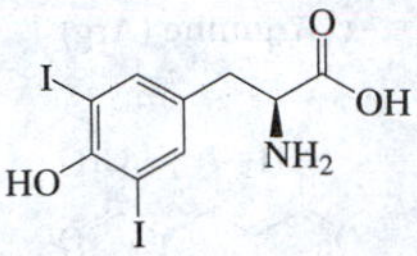
3,5-Diiodo-L-tyrosine

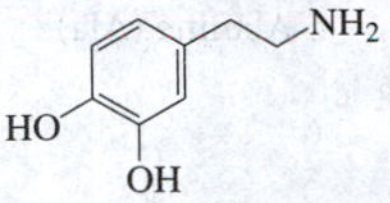
Dopamine

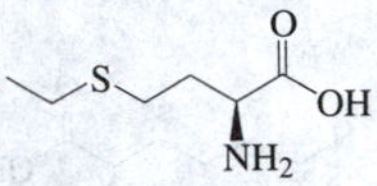
L-Ethionine

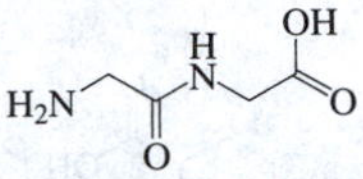
N-Glycylglycine

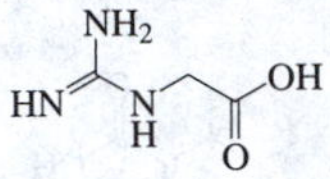
Guanidinoacetic acid

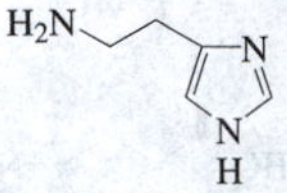
Histamine

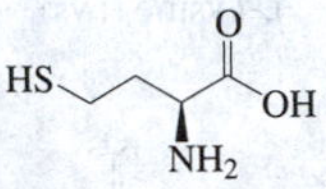
L-Homocysteine

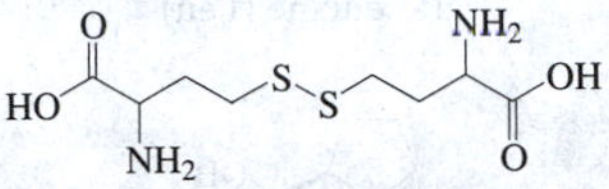
Homocystine

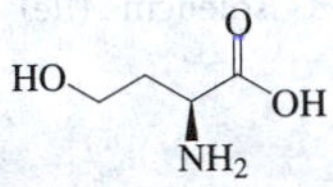
L-Homoserine

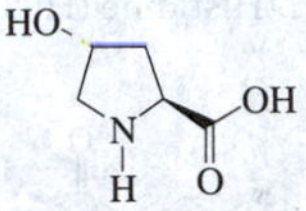
trans-4-Hydroxy-L-proline

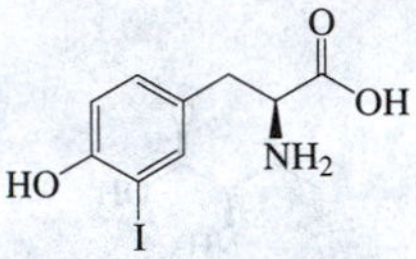
L-3-Iodotyrosine

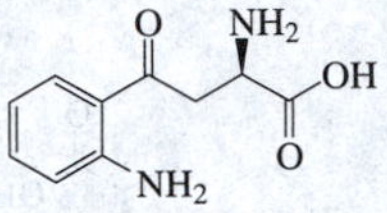
L-Kynurenine

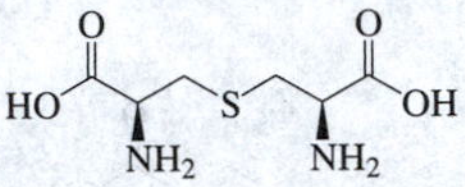
L-Lanthionine

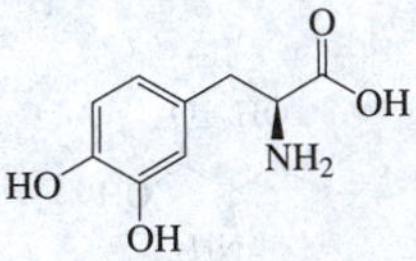
Levodopa

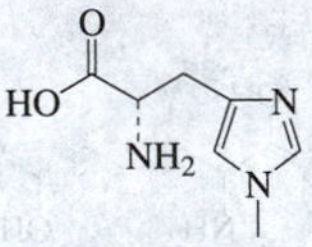
L-1-Methylhistidine

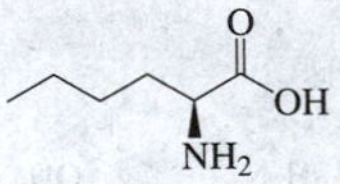
L-Norleucine

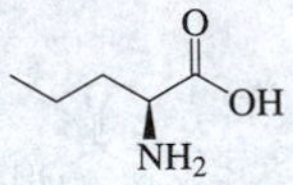
L-Norvaline

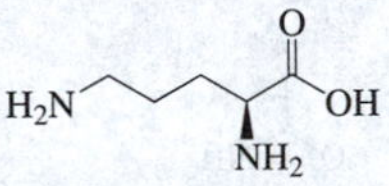
L-Ornithine

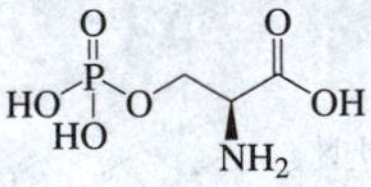
O-Phosphoserine

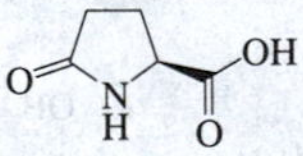
L-Pyroglutamic acid

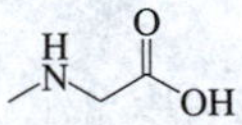
Sarcosine

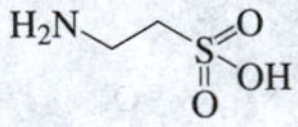
Taurine

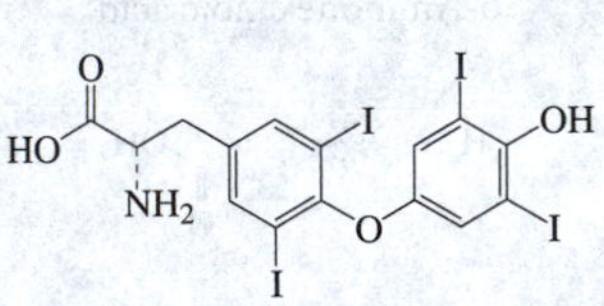
L-Thyroxine

PROPERTIES OF PURINE AND PYRIMIDINE BASES

This table lists some of the important purine and pyrimidine bases that occur in nucleic acids. The pK_a values (negative logarithm of the acid dissociation constant) are given for each ionization stage. The last column gives the aqueous solubility S at the indicated temperature in units of grams per 100 grams of solution.

The numbering system in the rings is:

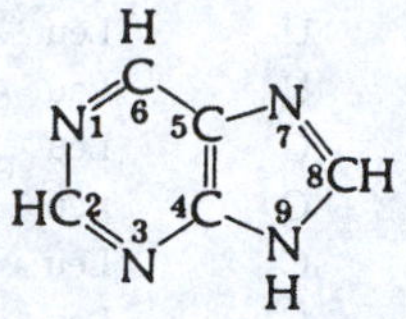

Purine

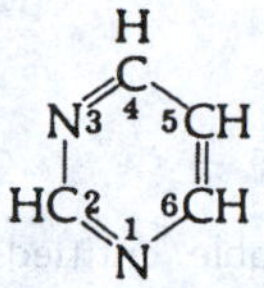

Pyrimidine

References

1. Dawson, R. M. C., et al., *Data for Biochemical Research, 3rd ed.*, Clarendon Press, Oxford, 1986.
2. O'Neil, M. J., Ed., *The Merck Index, 13th ed.*, Merck and Co., Rahway, NJ, 2001.

Common name	Systematic name	Mol. form.	Mol. wt.	pK_a values			S/mass % (temp.)
Pyrimidines							
Cytosine	4-Amino-2-hydroxypyrimidine	$C_4H_5N_3O$	111.10	4.60	12.16		0.73 (25 °C)
5-Methylcytosine	4-Amino-2-hydroxy-5-methylpyrimidine	$C_5H_7N_3O$	125.13	4.6	12.4		0.45 (25 °C)
5-Hydroxymethylcytosine	4-Amino-2-hydroxy-5-hydroxymethylpyrimidine	$C_5H_7N_3O_2$	141.13	4.3	13		
Uracil	2,4-Dihydroxypyrimidine	$C_4H_4N_2O_2$	112.09	0.5	9.5	>13	0.27 (25 °C)
Thymine	5-Methyluracil	$C_5H_6N_2O_2$	126.11	9.94	>13		0.35 (25 °C)
Orotic acid	Uracil-6-carboxylic acid	$C_5H_4N_2O_4$	156.10	2.4	9.5	>13	0.18 (18 °C)
Purines							
Adenine	6-Aminopurine	$C_5H_5N_5$	135.14	<1	4.3	9.83	0.104 (25 °C)
Guanine	2-Amino-6-hydroxypurine	$C_5H_5N_5O$	151.13	3.3	9.2	12.3	0.0068 (40 °C)
7-Methylguanine	7-Methyl-2-amino-6-hydroxypurine	$C_6H_7N_5O$	165.16	3.5	9.9		
Isoguanine	6-Amino-2-hydroxypurine	$C_5H_5N_5O$	151.13	4.5	9.0		0.006 (25 °C)
Xanthine	2,6-Dioxopurine	$C_5H_4N_4O_2$	152.11	0.8	7.4	11.1	0.05 (20 °C)
Hypoxanthine	6-Hydroxypurine	$C_5H_4N_4O$	136.11	2.0	8.9	12.1	0.07 (19 °C)
Uric acid	2,6,8-Trihydroxypurine	$C_5H_4N_4O_3$	168.11	5.4	11.3		0.002 (20 °C)

THE GENETIC CODE

This table gives the correspondence between a messenger RNA codon and the amino acid that it specifies. The symbols for bases in the codon are:

U uracil
C cytosine
A adenine
G guanine

The amino acid symbols are given in the table entitled "Structures of Common Amino Acids." A chain-initiating codon is indicated by **init** and a chain-terminating codon by **term**.

Example: UCA codes for **Ser**, UAC codes for **Tyr**, etc.

1st position	**2nd position = U**	**2nd position = C**	**2nd position = A**	**2nd position = G**	**3rd position**
U	Phe	Ser	Tyr	Cys	U
U	Phe	Ser	Tyr	Cys	C
U	Leu	Ser	**term**	**term**	A
U	Leu	Ser	**term**	Trp	G
C	Leu	Pro	His	Arg	U
C	Leu	Pro	His	Arg	C
C	Leu	Pro	Gln	Arg	A
C	Leu	Pro	Gln	Arg	G
A	Ile	Thr	Asn	Ser	U
A	Ile	Thr	Asn	Ser	C
A	Ile	Thr	Lys	Arg	A
A	Met (**init**)	Thr	Lys	Arg	G
G	Val	Ala	Asp	Gly	U
G	Val	Ala	Asp	Gly	C
G	Val	Ala	Glu	Gly	A
G	Val (**init**)	Ala	Glu	Gly	G

Biochem

PROPERTIES OF FATTY ACIDS AND THEIR METHYL ESTERS

This table gives the names and selected properties of some important fatty acids and their methyl esters. It includes most of the acids that are significant constituents of naturally occurring oils and fats. Compounds are listed first by number of carbon atoms and, secondly, by the degree of unsaturation. Both the systematic name and the common or trivial name are given, as well as the Chemical Abstracts Service Registry Number and the shorthand acid code that is frequently used. The first number in this code gives the number of carbon atoms; the number following the colon is the number of unsaturated centers (mainly double bonds). The location and orientation of the unsaturated centers follow. The symbols used are: c = *cis*; t = *trans*; a = acetylenic center; e = ethylenic center at end of chain; ep = *epoxy*. Thus 9c,11t indicates a double bond with *cis* orientation at the No. 9 carbon and another with *trans* orientation at the No. 11 carbon. More details on the codes can be found in Ref. 1.

The table gives the molecular weight and melting point of the acid and the melting and boiling points of the methyl ester of the acid when available. A superscript on the boiling point indicates the pressure in mmHg (torr); if there is no superscript, the value refers to one atmosphere (760 mmHg). The references cover many other fatty acids beyond those listed here and give additional properties.

We are indebted to Frank D. Gunstone for advice on the content of the table.

References

1. Gunstone, F. D., Harwood, J. L., and Dijkstra, A. J., eds., *The Lipid Handbook, Third Edition*, CRC Press, Boca Raton, FL, 2006.
2. Gunstone, F. D., and Adlof, R. O., *Common (non-systematic) Names for Fatty Acids*, www.aocs.org/member/division/analytic/fanames.asp, 2003.
3. Firestone, D., *Physical and Chemical Characteristics of Oils, Fats, and Waxes, Second Edition*, AOCS Press, Urbana, IL, 2006.
4. Dawson, R. M. C., Elliott, D. C., Elliott, W. H., and Jones, K. M., *Data for Biochemical Research, Third Edition*, Clarendon Press, Oxford, 1986.
5. Altman, P. L., and Dittmer, D. S., eds., *Biology Data Book, Second Edition*, Vol. 1, Federation of American Societies for Experimental Biology, Bethesda, MD, 1972.
6. Fasman, G. D., Ed., *Practical Handbook of Biochemistry and Molecular Biology*, CRC Press, Boca Raton, FL, 1989.

Systematic name	Common name	Mol. form.	Acid code	CAS Reg. No.	Mol. weight	mp/°C	Methyl ester mp/°C	Methyl ester bp/°C
Butanoic acid	Butyric acid	$C_4H_8O_2$	4:0	107-92-6	88.106	–5.1	–85.8	102.8
Pentanoic acid	Valeric acid	$C_5H_{10}O_2$	5:0	109-52-4	102.132	–33.6		127.4
3-Methylbutanoic acid	Isovaleric acid	$C_5H_{10}O_2$	4:0 3-Me	503-74-2	102.132	–29.3		116.5
Hexanoic acid	Caproic acid	$C_6H_{12}O_2$	6:0	142-62-1	116.158	–3	–71	149.5
Heptanoic acid	Enanthic acid	$C_7H_{14}O_2$	7:0	111-14-8	130.185	–7.2	–56	174
Octanoic acid	Caprylic acid	$C_8H_{16}O_2$	8:0	124-07-2	144.212	16.5	–40	192.9
Nonanoic acid	Pelargonic acid	$C_9H_{18}O_2$	9:0	112-05-0	158.238	12.4		213.5
Decanoic acid	Capric acid	$C_{10}H_{20}O_2$	10:0	334-48-5	172.265	31.4	–18	224
9-Decenoic acid	Caproleic acid	$C_{10}H_{18}O_2$	10:1 9e	14436-32-9	170.249	26.5		120^{20}
Undecanoic acid		$C_{11}H_{22}O_2$	11:0	112-37-8	186.292	28.6		123^{10}
Dodecanoic acid	Lauric acid	$C_{12}H_{24}O_2$	12:0	143-07-7	200.318	43.8	5.1	268
cis-9-Dodecenoic acid	Lauroleic acid	$C_{12}H_{22}O_2$	12:1 9c	2382-40-3	198.302			
Tridecanoic acid		$C_{13}H_{26}O_2$	13:0	638-53-9	214.344	41.5	6.5	92^{1}
Tetradecanoic acid	Myristic acid	$C_{14}H_{28}O_2$	14:0	544-63-8	228.371	54.2	19	295
cis-9-Tetradecenoic acid	Myristoleic acid	$C_{14}H_{26}O_2$	14:1 9c	13147-06-3	226.355	–4		
Pentadecanoic acid		$C_{15}H_{30}O_2$	15:0	1002-84-2	242.398	52.3	18.5	153.5
Hexadecanoic acid	Palmitic acid	$C_{16}H_{32}O_2$	16:0	57-10-3	256.424	62.5	29.6	324
cis-9-Hexadecenoic acid	Palmitoleic acid	$C_{16}H_{30}O_2$	16:1 9c	373-49-9	254.408	0.5	–33.7	325
Heptadecanoic acid	Margaric acid	$C_{17}H_{34}O_2$	17:0	506-12-7	270.451	61.3	30	185^{9}
Octadecanoic acid	Stearic acid	$C_{18}H_{36}O_2$	18:0	57-11-4	284.478	69.3	38.7	353
cis-6-Octadecenoic acid	Petroselinic acid	$C_{18}H_{34}O_2$	18:1 6c	593-39-5	282.462	29.8		
cis-9-Octadecenoic acid	Oleic acid	$C_{18}H_{34}O_2$	18:1 9c	112-80-1	282.462	13.4	–19.7	347
trans-9-Octadecenoic acid	Elaidic acid	$C_{18}H_{34}O_2$	18:1 9t	112-79-8	282.462	45	13.5	218^{24}
cis-11-Octadecenoic acid	*cis*-Vaccenic acid	$C_{18}H_{34}O_2$	18:1 11c	506-17-2	282.462	15		$163^{0.1}$
trans-11-Octadecenoic acid	Vaccenic acid	$C_{18}H_{34}O_2$	18:1 11t	693-72-1	282.462	44		172^{3}
cis-12,13-Epoxy-*cis*-9-octadecenoic acid	Vernolic acid	$C_{18}H_{32}O_3$	18:1 12,13-ep,9c	503-07-1	296.445	32.5		
12-Hydroxy-*cis*-9-octadecenoic acid	Ricinoleic acid	$C_{18}H_{34}O_3$	18:1 12-OH,9c	141-22-0	298.461	–8.28		226^{15}
cis,trans-9,11-Octadecadienoic acid	Rumenic (CLA)	$C_{18}H_{32}O_2$	18:2 9c,11t	1839-11-8	280.446	20		
cis,cis-9,12-Octadecadienoic acid	Linoleic acid	$C_{18}H_{32}O_2$	18:2 9c,12c	60-33-3	280.446	–7	–36.6	347

Systematic name	Common name	Mol. form.	Acid code	CAS Reg. No.	Mol. weight	mp/°C	Methyl ester mp/°C	Methyl ester bp/°C
trans,cis-10,12-Octadecadienoic acid	(CLA)	$C_{18}H_{32}O_2$	18:2 10t,12c	22880-03-1	280.446	23	–12	
cis-9-Octadecen-12-ynoic acid	Crepenynic acid	$C_{18}H_{30}O_2$	18:2 9c,12a	2277-31-8	278.430			
cis,cis,cis-5,9,12-Octadecatrienoic acid	Pinolenic acid	$C_{18}H_{30}O_2$	18:3 5c,9c,12c	27213-43-0	278.430			
trans,cis,cis-5,9,12-Octadecatrienoic acid	Columbinic acid	$C_{18}H_{30}O_2$	18:3 5t,9c,12c	2441-53-4	278.430			
cis,cis,cis-6,9,12-Octadecatrienoic acid	γ-Linolenic acid	$C_{18}H_{30}O_2$	18:3 6c,9c,12c	506-26-3	278.430			$162^{0.5}$
trans,trans,cis-8,10,12-Octadecatrienoic acid	Calendic acid	$C_{18}H_{30}O_2$	18:3 8t,10t,12c	28872-28-8	278.430	40		
cis,trans,cis-9,11,13-Octadecatrienoic acid	Punicic acid	$C_{18}H_{30}O_2$	18:3 9c,11t,13c	544-72-9	278.430	45		
cis,trans,trans-9,11,13-Octadecatrienoic acid	α-Eleostearic acid	$C_{18}H_{30}O_2$	18:3 9c,11t,13t	506-23-0	278.430	49		148^{1}
trans,trans,cis-9,11,13-Octadecatrienoic acid	Catalpic acid	$C_{18}H_{30}O_2$	18:3 9t,11t,13c	4337-71-7	278.430	32		
trans,trans,trans-9,11,13-Octadecatrienoic acid	β-Eleostearic acid	$C_{18}H_{30}O_2$	18:3 9t,11t,13t	544-73-0	278.430	71.5	13	162^{1}
cis,cis,cis-9,12,15-Octadecatrienoic acid	α-Linolenic acid	$C_{18}H_{30}O_2$	18:3 9c,12c,15c	463-40-1	278.430	–11.3	–49	348
6,9,12,15-Octadecatetraenoic acid, all *cis*	Stearidonic acid	$C_{18}H_{28}O_2$	18:4 6c,9c,12c,15c	20290-75-9	276.414	–57		
cis,trans,trans,cis-9,11,13,15-Octadecatetraenoic acid	Parinaric acid	$C_{18}H_{28}O_2$	18:4 9c,11t,13t,15c	593-38-4	276.414	86		
Nonadecanoic acid		$C_{19}H_{38}O_2$	19:0	646-30-0	298.504	69.4	41.3	190^{4}
Eicosanoic acid	Arachidic acid	$C_{20}H_{40}O_2$	20:0	506-30-9	312.531	76.5	46.4	371
3,7,11,15-Tetramethylhexadecanoic acid	Phytanic acid	$C_{20}H_{40}O_2$	16:0 3,7,11,15-tetramethyl	14721-66-5	312.531	–65		
cis-5-Eicosenoic acid		$C_{20}H_{38}O_2$	20:1 5c	7050-07-9	310.515	27		
cis-9-Eicosenoic acid	Gadoleic acid	$C_{20}H_{38}O_2$	20:1 9c	29204-02-2	310.515	24.5	–34	378
cis-11-Eicosenoic acid	Gondoic acid	$C_{20}H_{38}O_2$	20:1 11c	2462-94-4	310.515	24		
cis,cis,cis-8,11,14-Eicosatrienoic acid	Dihomo-γ-linolenic acid	$C_{20}H_{34}O_2$	20:3 8c,11c,14c	1783-84-2				
5,8,11,14-Eicosatetraenoic acid, all *cis*	Arachidonic acid	$C_{20}H_{32}O_2$	20:4 5c,8c,11c,14c	506-32-1	304.467	–38		$195^{0.7}$
5,8,11,14,17-Eicosapentaenoic acid, all *cis*	Timnodonic acid, EPA	$C_{20}H_{30}O_2$	20:5 5c,8c,11c,14c,17c	10417-94-4	302.451	–54		
Heneicosanoic acid		$C_{21}H_{42}O_2$	21:0	2363-71-5	326.557	82	49	207^{4}
Docosanoic acid	Behenic acid	$C_{22}H_{44}O_2$	22:0	112-85-6	340.583	81.5	53.3	402
cis-11-Docosenoic acid	Cetolic acid	$C_{22}H_{42}O_2$	22:1 11c	506-36-5	338.567	33		
cis-13-Docosenoic acid	Erucic acid	$C_{22}H_{42}O_2$	22:1 13c	112-86-7	338.567	34.7	–1.1	400
trans-13-Docosenoic acid	Brassidic acid	$C_{22}H_{42}O_2$	22:1 13t	506-33-2	338.567	61.9	35	
cis,cis-5,13-Docosadienoic acid		$C_{22}H_{40}O_2$	22:2 5c,13c	676-39-1	336.552	–4		
7,10,13,16,19-Docosapentaenoic acid, all *cis*		$C_{22}H_{34}O_2$	22:5 7c,10c,13c,16c,19c					
4,7,10,13,16,19-Docosahexaenoic acid, all *cis*	Cervonic acid, DHA	$C_{22}H_{32}O_2$	22:6 4c,7c,10c,13c,16c,19c	2091-24-9		–45		
Tricosanoic acid		$C_{23}H_{46}O_2$	23:0	2433-96-7		79.6	53.4	
Tetracosanoic acid	Lignoceric acid	$C_{24}H_{48}O_2$	24:0	557-59-5	368.637	87.5	60	
cis-15-Tetracosenoic acid	Nervonic acid	$C_{24}H_{46}O_2$	24:1 15c	506-37-6	366.621	43	15	$165^{0.02}$
Pentacosanoic acid		$C_{25}H_{50}O_2$	25:0	506-38-7	382.664	77.5	62	
Hexacosanoic acid	Cerotic acid	$C_{26}H_{52}O_2$	26:0	506-46-7	396.690	88.5	63.8	286^{15}
Heptacosanoic acid		$C_{27}H_{54}O_2$	27:0	7138-40-1		87.6	64	
Octacosanoic acid	Montanic acid	$C_{28}H_{56}O_2$	28:0	506-48-9	424.744	90.9	67	
Nonacosanoic acid		$C_{29}H_{58}O_2$	29:0	4250-38-8	438.770	90.3	69	
Triacontanoic acid	Melissic acid	$C_{30}H_{60}O_2$	30:0	506-50-3	452.796	93.6	72	
Hentriacontanoic acid		$C_{31}H_{62}O_2$	31:0	38232-01-8	466.823	93.1		
Dotriacontanoic acid	Lacceric acid	$C_{32}H_{64}O_2$	32:0	3625-52-3	480.849	96.2		$192^{0.01}$

PROPERTIES OF FATTY ACID METHYL AND ETHYL ESTERS RELATED TO BIOFUELS

Robert D. Chirico and Michael Frenkel

These tables provide critically evaluated properties and sources of experimental data for fatty acid methyl esters and ethyl esters found in biofuels. All listed property values are based on experimental results reported in Refs. 1 to 103. Critical evaluation was carried out dynamically with the NIST ThermoData Engine (TDE) software, version 7.0 released in 2012 (Ref. 104). The software algorithms have been described in the literature (see Refs. 105 to 112). As of March, 2016, the most recent TDE version 10.0 was released by NIST in 2015 (Ref. 113).

The table of properties lists the common name; molecular formula; acid code; normal melting temperature (mp/°C); density of the liquid phase at 25 °C and 50 °C near pressure p = 0.1 MPa (ρ^{25} and ρ^{50} expressed in kg m^{-3}); vapor pressure p_{sat} at the temperatures 100 °C and 150 °C (p_{sat}^{100} and p_{sat}^{150} in Pa); the normal boiling temperature (bp/°C); heat capacity of the liquid phase at 25 °C and 50 °C near pressure p = 0.1 MPa (C_p^{25} and C_p^{50} in J K^{-1} mol^{-1}); and viscosity of the liquid at 25 °C and 50 °C near pressure p = 0.1 MPa (η^{25} and η^{50} in mPa s). In some cases, property values for the liquid cannot be evaluated at 25 °C because the compound is solid at this temperature. In such cases, values are given at 40 °C, as indicated with a superscript "40." All normal boiling temperatures are estimates based on extrapolation of vapor pressures measured at lower temperatures. Digits in parentheses following a property value give the uncertainty in the last digit(s) of the value; these are the estimated expanded uncertainties (0.95 level of confidence; see Refs. 111 to 113) for the property values. A discussion of uncertainty terminology with particular application to thermodynamic and transport properties is given in Ref. 114.

Compounds are listed in order of the acid code. This is a common shorthand in which the first number represents the number of carbon atoms in the parent fatty acid, and the second number indicates the number of double bonds present. See the table "Properties of Fatty Acids and Their Methyl Esters" for further details on the acid code.

The second table provides up to five data sources reporting experimental measurements of the indicated property. Generally, these are the most important sources that were used in the critical evaluation process. The numbers correspond to those in the reference list. The systematic name and CAS Reg. No. are also given here.

Properties of Methyl and Ethyl Esters of Fatty Acids

Common name	Mol. form.	Acid code	mp/°C	bp/°C	ρ^{25}/kg m^{-3}	ρ^{50}/ kg m^{-3}	p_{sat}^{100}/ Pa	p_{sat}^{150}/ Pa	C_p^{25}/ J mol^{-1} K^{-1}	C_p^{50}/ J mol^{-1} K^{-1}	η^{25}/ mPa s	η^{50}/ mPa s
Methyl Esters												
Methyl laurate	$C_{13}H_{26}O_2$	12:0	5.1(2)	268(2)	865.4(5)	845.6(6)	218(6)	2660(50)	437(2)	451(2)	2.80(5)	1.71(4)
Methyl myristate	$C_{15}H_{30}O_2$	14:0	19.0(5)	295(10)	863.2(10)	844.1(6)	53.8(3)	850(60)	501(3)	511(2)	4.07(8)	2.35(5)
Methyl palmitate	$C_{17}H_{34}O_2$	16:0	29.6(5)	324(6)	851(3)^{40}	842(2)	12.2(10)	270(11)	572(3)^{40}	578(3)	3.75(10)^{40}	3.04(10)
Methyl palmitoleate	$C_{17}H_{32}O_2$	16:1	-33.7(6)	325	865.1(7)	846.1(6)	9.9(13)	242(32)			3.75(10)	2.2(1)
Methyl stearate	$C_{19}H_{38}O_2$	18:0	38.7(6)	353(8)	849.7(10)^{40}	842.5(8)	2.7(3)	84(4)	633(5)^{40}	641(4)	5.0(1)^{40}	4.0(1)
Methyl oleate	$C_{19}H_{36}O_2$	18:1	-19.7(5)	347(5)	871.1(6)	853.2(6)	3.3(3)	102(3)			5.6(1)	3.1(1)
Methyl linoleate	$C_{19}H_{34}O_2$	18:2	-36.6(5)	347(5)	883.4(6)	865.1(6)	3.7(2)	106(2)			4.4(1)	2.7(1)
Methyl linolenate	$C_{19}H_{32}O_2$	18:3	-49(4)	348(8)	895.1(15)	877.1(15)	3.4(3)	100(3)			3.9(1)	2.5(1)
Methyl arachidate	$C_{21}H_{42}O_2$	20:0	46.4(3)	371(15)	848.7(10)^{40}	841.9(10)	0.7(2)	32(7)	700(7)^{40}	708(6)	6.1(2)^{40}	4.8(1)
Methyl gondoate	$C_{21}H_{40}O_2$	20:1	-34(2)	378	870.2(15)	852.3(15)	0.7(2)	31(5)			7.4(1)	4.1(1)
Methyl behenate	$C_{23}H_{46}O_2$	22:0	53.3(4)	402(16)	844.4(10)^{40}	838.5(10)	0.3(3)	1.3(6)			7.8(14)^{40}	6.0(5)
Methyl erucate	$C_{23}H_{44}O_2$	22:1	-1.1(5)	400(9)	867.1(7)	849.3(6)	0.2(1)	1.1(2)			9.1(1)	4.8(1)
Ethyl Esters												
Ethyl laurate	$C_{14}H_{28}O_2$	12:0	-1.8(4)	276(3)	858.1(6)	839.2(7)	132(8)	1782(40)			3.01(5)	1.85(4)
Ethyl myristate	$C_{16}H_{32}O_2$	14:0	12.3(8)	308(3)	857.1(6)	838.0(5)	32(2)	572(15)			4.2(1)	2.46(5)
Ethyl palmitate	$C_{18}H_{36}O_2$	16:0	24.2(4)	334(7)	857.2(7)	838.4(4)	5.4(3)	157(30)			5.7(1)	3.20(7)
Ethyl palmitoleate	$C_{18}H_{34}O_2$	16:1	-36(1)								4.5(10)	2.8(10)
Ethyl stearate	$C_{20}H_{40}O_2$	18:0	33.1(7)	356(6)	844.1(9)^{40}	837.2(8)	2.2(3)	65(6)	667(3)	683(2)	7.5(1)	4.05(10)
Ethyl oleate	$C_{20}H_{38}O_2$	18:1	-21(2)	357(9)	865.9(8)	847.6(8)	2.5(2)	78(25)			5.8(1)	3.31(8)
Ethyl linoleate	$C_{20}H_{36}O_2$	18:2	-55(5)	351(10)	877.0(15)	861.1(15)	1.9(15)	64(30)			4.8(1)	2.9(1)

Common name	Mol. form.	Acid code	mp/°C	bp/°C	ρ^{25}/kg m^{-3}	ρ^{50}/ kg m^{-3}	p_{sat}^{100}/ Pa	p_{sat}^{150}/ Pa	C_p^{25}/ J mol^{-1} K^{-1}	C_p^{50}/ J mol^{-1} K^{-1}	η^{25}/ mPa s	η^{50}/ mPa s
Ethyl linolenate	$C_{20}H_{34}O_2$	18:3		357(10)	889.5(15)	870.9(15)	2.4(11)	75(20)			3.95(10)	2.50(7)
Ethyl arachidate	$C_{22}H_{44}O_2$	20:0	41.7(5)	389		838.7(15)	0.4(4)	22(10)				5.0(1)

Sources of Data

Common name	Systematic name	Acid code	CAS Reg. No.	mp source	Density source	Pressure source	Heat capacity source	Viscosity source
Methyl Esters								
Methyl laurate	Methyl dodecanoate	12:0	111-82-0	1,25,81,82,88	14,53,74,91,98	14,16,21,63,81	96	14,19,33,53,74
Methyl myristate	Methyl tetradecanoate	14:0	124-10-7	1,81,82,88	14,32,52,72,74	14,21,64,80,81	30,96	14,19,33,62,74
Methyl palmitate	Methyl hexadecanoate	16:0	112-39-0	7,43	73,99	54,99		44,45,73
Methyl stearate	Methyl octadecanoate	18:0	112-61-8	12,46,48,65,84	12,14,52,68,74	21,34,42,81,88	96	12,14,24,33,74
Methyl oleate	Methyl *cis*-9-octadecenoate	18:1	112-62-9	6,7,17,39,90	3,15,32,68,74	42,54,77,81,87		3,19,44,74,100
Methyl linoleate	Methyl *cis,cis*-9,12-Octadecadienoate	18:2	112-63-0	7	32,68,74	42,47,54,64,81		19,33,44,45,74
Methyl linolenate	Methyl *cis,cis,cis*-9,12,15-octadecatrienoate	18:3	301-00-8	39,75,76	32,68,73	42,47,54,69,81		19,33,44,45,73
Methyl arachidate	Methyl eicosanoate	20:0	1120-28-1	1,4,23,25,46	32,73	21,27,47,95	96	73
Methyl gondoate	Methyl *cis*-11-eicosenoate	20:1	2390-09-2	101	73	54		73
Methyl behenate	Methyl docosanoate	22:0	929-77-1	4,21,23,46,57	73	18,21,47,89		73
Methyl erucate	Methyl *cis*-13-docosenoate	22:1	1120-34-9	39	32,56,73	47,54		33,73
Ethyl Esters								
Ethyl laurate	Ethyl dodecanoate	12:0	106-33-2	41,58,60,60,93	49,52,74,83,98	10,47,70,85,89		36,44,45,51,74
Ethyl myristate	Ethyl tetradecanoate	14:0	124-06-1	22,41,57,60,79	52,67,74,78,83	10,60,78,85		36,37,45,74
Ethyl palmitate	Ethyl hexadecanoate	16:0	628-97-7	29,55,59,60,71	12,28,49,74,83	11,60,85,86,94		12,36,37,45,74
Ethyl palmitoleate	Ethyl *cis*-9-hexadecenoate	16:1	56219-10-4	43				37
Ethyl stearate	Ethyl octadecanoate	18:0	111-61-5	12,26,29,59,82	8,12,49,74,83	20,85,86,88,97	2	12,20,37,45,74
Ethyl oleate	Ethyl *cis*-9-octadecenoate	18:1	111-62-6	6,13,17,43,100	3,20,40,49,74	13,20,85,86		3,40,44,74,100
Ethyl linoleate	Ethyl *cis,cis*-9,12-octadecadienoate	18:2	544-35-4	103	73	85,86		37,44,45,73
Ethyl linolenate	Ethyl *cis,cis,cis*-9,12,15-octadecatrienoate	18:3	1191-41-9		73,102	102		37,44,45,73
Ethyl arachidate	Ethyl eicosanoate	20:0	18281-05-5	4,29,57,66,92	73	5,9,50		73

References

1. Adriaanse, N., Dekker, H., and Coops, J., *Recl. Trav. Chim. Pays-Bas* 83, 557, 1964.
2. Agafonova, L. E., Varushenko, R. M., Druzhinina, A. I., Polyakova, O. I., and Kolesov, Y. S., *Zh. Fiz. Khim.* 85, 1633, 2011.
3. Albert, O., *Z. Phys. Chem. Abt. A* 182, 421, 1938.
4. Aleby, S., and Fischmeister, I., *Ark. Kemi* 31, 267, 1970.
5. Backer, H. J., and Strating, J., *Recl. Trav. Chim. Pays-Bas* 59, 933, 1940.
6. Bailey, A. V., Harris, J. A., and Skau, E. L., *J. Chem. Eng. Data* 16, 473, 1971.
7. Bailey, A. V., Harris, J. A., and Skau, E. L., *J. Chem. Eng. Data* 15, 583, 1970.
8. Baker, W. O., and Smyth, C. P., *J. Am. Chem. Soc.* 60, 1229, 1938.
9. Becker, H. J., and Strating, J., *Recl. Trav. Chim. Pays-Bas* 59, 933, 1940.
10. Benziane, M., Khimeche, K., Mokbel, I., Sawaya, T., Dahmani, A., and Jose, J., *J. Chem. Eng. Data* 56, 4736, 2011.
11. Bhattacharyya, S. C., Chakravarty, K. K., and Kumar, V., *Chem. Ind. (London)* 1352, 1959.
12. Boelhouwer, J. W. M., Nederbragt, G. W., and Verberg, G. W., *Appl. Sci. Res.* 2, 249, 1950.
13. Boeseken, J., and Belinfante, A. H., *Recl. Trav. Chim. Pays-Bas* 45, 914, 1926.
14. Bonhorst, C. W., Althouse, P. M., and Triebold, H. O., *Ind. Eng. Chem.* 40, 2379, 1948.
15. Bridgman, P. W., *Proc. Am. Acad. Arts Sci.* 67, 1, 1932.
16. Bureau, N., Jose, J., Mokbel, I., and de Hemptinne, J. C., *J. Chem. Thermodyn.* 33, 1485, 2001.
17. Candy, L., Vaca-Garcia, C., and Borredon, E., *J. Am. Oil Chem. Soc.* 82, 271, 2005.
18. Cason, J., Allinger, N. L., Sumrell, G., and Williams, D. E., *J. Org. Chem.* 16, 1170, 1951.

19. Ceriani, R., Goncalves, C. B., Rabelo, J., Caruso, M., Cunha, A. C. C., Cavaleri, F. W., Batista, E. A. C., and Meirelles, A. J. A., *J. Chem. Eng. Data* 52, 965, 2007.
20. Cheshko, F. F., and Shvaika, T. N., *Zh. Prikl. Khim.* 44, 1107, 1971.
21. Chickos, J. S., Zhao, H., and Nichols, G., *Thermochim. Acta* 424, 111, 2004.
22. Collinet, E., and Gmehling, J., *Fluid Phase Equilib.* 230, 131, 2005.
23. Craig, B. M., *Can. J. Chem.* 31, 499, 1953.
24. Muñoz Cueto, M. J. M., Galán Vallejo, M. G., and Flores Luque, V., *Grasas Aceites* 42, 14, 1991.
25. Davies, M., and Kybett, B., *Trans. Faraday Soc.* 61, 1893, 1965.
26. Deffet, L., *Bull. Soc. Chim. Belg.* 49, 223, 1940.
27. Ehrenstein, R., and Stuewer, H., *J. Prakt. Chem.* 105, 199, 1923.
28. Ellis, C. P., *J. Chem. Soc.* 1697, 1932.
29. Francis, F., Collins, F. J. E., and Piper, S. H., *Proc. R. Soc. London, Ser. A* 158, 691, 1937.
30. Fuchs, R., *J. Chem. Thermodyn.* 11, 959, 1979.
31. Gaikwad, B. R., and Subrahmanyam, V. V. R., *J. Indian Chem. Soc.* 65, 266, 1988.
32. Gouw, T. H., and Vlugter, J. C., *J. Am. Oil Chem. Soc.* 41, 142, 1964.
33. Gouw, T. H., Vlugter, J. C., and Roelands, C. J. A., *J. Am. Oil Chem. Soc.* 43, 433, 1966.
34. Greaves, W. S., Linstead, R. P., Shephard, B. R., Thomas, S. L. S., and Weedon, B. C. L., *J. Chem. Soc.* 3326, 1950.
35. Grob, C. A., Jenny, E. F., and Utzinger, H., *Helv. Chim. Acta* 34, 2249, 1951.
36. Gros, A. T., and Feuge, R. O., *J. Am. Oil Chem. Soc.* 29, 313, 1952.
37. Hamazaki, T., Kobayashi, S., Urakaze, M., Yano, S., and Fujita, T., *Biorheology* 22, 221, 1985.
38. Hou, J., Xu, S., Ding, H., and Sun, T., *J. Chem. Eng. Data* 57, 2632, 2012.
39. Jantzen, E., and Andreas, H., *Chem. Ber.* 92, 1427, 1959.
40. Keffler, L., and MacLean, J. H., *J. Soc. Chem. Ind. (London)* 54, 178, 1935.
41. King, A. M., and Garner, W. E., *J. Chem. Soc.* 1449, 1934.
42. Klykov, V. N., Serebrennikova, G. A., and Perobrazhenskii, N. A., *Zh. Org. Khim.* 2, 1782, 1966.
43. Knothe, G., *Energy Fuels* 22, 1358, 2008.
44. Knothe, G., and Steidley, K. R., *Fuel* 86, 2560, 2007.
45. Knothe, G., and Steidley, K. R., *Fuel* 84, 1059, 2005.
46. Krewson, C. F., *J. Am. Chem. Soc.* 73, 1365, 1951.
47. Krop, H. B., and Velzen, M. J. M., *J. Am. Oil Chem. Soc.* 74, 309, 1997.
48. Kurtz, F. E., *J. Am. Chem. Soc.* 74, 1902, 1952.
49. Lautsch, W., *Z. Phys. Chem. (Leipzig)* 1, 115, 1928.
50. Levene, P. A., and Taylor, F. A., *J. Biol. Chem.* 59, 905, 1924.
51. Liau, W., Tang, M., and Chen, Y. P., *J. Chem. Eng. Data* 43, 826, 1998.
52. Liew, K. Y., and Seng, C. E., *J. Am. Oil Chem. Soc.* 69, 734, 1992.
53. Liew, K. Y., Seng, C. E., and Oh, L. L., *J. Am. Oil Chem. Soc.* 69, 155, 1992.
54. Lipkind, D., Kapustin, Y., Umnahanant, P., and Chickos, J. S., *Thermochim. Acta* 456, 94, 2007.
55. Loury, M., and Stora, C., *Bull. Soc. Chim. France* 262, 1949.
56. Maksimuk, Yu. V., Antonova, Z. A., and Fes'ko, V. V., *Russ. J. Phys. Chem.* A86, 224, 2012.
57. Marosi, L., and Schlenk, W., *Liebigs Ann. Chem.* 4, 584, 1973.
58. Meyer, J. D., and Reid, E. E., *J. Am. Chem. Soc.* 55, 1574, 1933.
59. Michel, J., *Bull. Soc. Chim. Belg.* 48, 105, 1939.
60. Mumford, S. A., and Phillips, J. W. C., *J. Chem. Soc.* 75, 1950.
61. Mumford, S. A., and Phillips, J. W. C., *Recl. Trav. Chim. Pays-Bas* 52, 181, 1933.
62. Munoz, M. J., Rodriguez, M., Martin, R., and Galan, M., *An. Quim.* 90, 177, 1994.
63. Nevin, C. S., Althouse, P. M., and Triebold, H. O., *J. Am. Oil Chem. Soc.* 28, 325, 1951.
64. Norris, F. A., and Terry, D. E., *Oil and Soap* 22, 41, 1945.
65. O'Connor, R. T., Field, E. T., and Singleton, W. S., *J. Am. Oil Chem. Soc.* 28, 154, 1951.
66. Omar, M. M., *J. Chem. Soc. C* 2038, 1967.
67. Ortega, J., Placido, J., and Vidal, M., *J. Chem. Thermodyn.* 31, 151, 1999.
68. Ott, L. S., Huber, M. L., and Bruno, T. J., *J. Chem. Eng. Data* 53, 2412, 2008.
69. Painter, E. P., and Nesbitt, L. L., *Ind. Eng. Chem. Anal. Ed.* 15, 123, 1943.
70. Palfray, L., and Sabetay, S., *Bull. Soc. Chim. France* 3, 682, 1936.
71. Paquot, C., Sorba, Mme. J., and Wieme, N., *Ind. Chim. Belge,* 20, Spec. No. 727, 1955.
72. Postigo, M. A., Garcia, P. H., Ortega, J., and Tardajos, G., *J. Chem. Eng. Data* 40, 283, 1995.
73. Pratas, M. J., Freitas, S., Oliveira, M. B., Monteiro, S. C., Lima, A. S., and Coutinho, J. A. P., *J. Chem. Eng. Data* 56, 2175, 2011.
74. Pratas, M. J., Freitas, S., Oliveira, M., Monteiro, S., Lima, A. S., and Coutinho, J. A. P., *J. Chem. Eng. Data* 55, 3983, 2010.
75. Rakoff, H., and Emken, E. A., *Chem. Phys. Lipids* 31, 215, 1982.
76. Riemenschneider, R. W., Herb, S. F., and Nichols Jr., P. L., *J. Am. Oil Chem. Soc.* 26, 371, 1949.
77. Rose, A., and Schrodt, V. N., *J. Chem. Eng. Data* 9, 12, 1964.
78. Ruhoff, J. R., and Reid, E. E., *J. Am. Chem. Soc.* 55, 3825, 1933.
79. Sackmann, V. H., and Dorfler, H. D., *Z. Phys. Chem. (Leipzig)* 251, 303, 1972.
80. Sauer, J. C., Hain, B. E., and Boutwell, P. W., *Org. Synth.* 20, 67, 1940.
81. Scott, T. A., Macmillan, D., and Melvin, E. H., *Ind. Eng. Chem.* 44, 172, 1952.
82. Sedgwick, R. S., Hoerr, C. W., and Harwood, H. J., *J. Org. Chem.* 17, 327, 1952.
83. Shigley, J. W., Bonhorst, C. W., Liang, C. C., Althouse, P. M., and Triebold, H. O., *J. Am. Oil Chem. Soc.* 32, 213, 1955.
84. Shreve, O. D., Heether, M. R., Knight, H. B., and Swern, D., *Anal. Chem.* 22, 1498, 1950.
85. Silva, L. Y. A., Falleiro, R. M. M., Meirelles, A. J. A., and Krahenbuhl, M. A., *J. Chem. Thermodyn.* 43, 943, 2011.
86. Silva, L. Y. A., Falleiro, R. M. M., Meirelles, A. J. A., and Krahenbuhl, M. A., *Thermochim. Acta* 512, 178. 2011.
87. Simmons, H. E., and Smith, R. D., *J. Am. Chem. Soc.* 81, 4256, 1959.
88. Spizzichino, C., *J. Rech. C. N. R. S.* 34, 1, 1956.
89. Strating, J., Backer, H. J., Lolkema, J., and Benninga, N., *Recl. Trav. Chim. Pays-Bas* 55, 903, 1936.
90. Suppes, G. J., Goff, M. J., and Lopes, S., *Chem. Eng. Sci.* 58, 1751, 2003.
91. Tat, M. E., and Van Gerpen, J. H., Measurement of Biodiesel Speed of Sound and Its Impact on Injection Timing, NREL/SR-510-31462, National Renewable Energy Laboratory, Golden, CO, 2003.
92. Tiedt, J., and Truter, E. V., *J. Chem. Soc.* 4628, 1952.
93. Timmermans, J., *Bull. Soc. Chim. Belg.* 31, 389, 1922.
94. Torii, S., Inokuchi, T., Kondo, K., and Ito, H., *Bull. Chem. Soc. Jpn.* 58, 1347, 1985.
95. van Genderen, A. C. G., van Miltenburg, J. C., Blok, J. G., Van Bommel, M. J., Van Ekeren, P. J., Van Den Berg, G. J. K., and Oonk, H. A. J., *Fluid Phase Equilib.* 202, 109, 2002.
96. van Miltenburg, J. C., Oonk, H. A. J., and Van Bommel, M. J., *J. Chem. Eng. Data* 49, 1036, 2004.
97. Vasil'eva, T. A., Leont'eva, S. A., and Toropov, A. P., *Izv. Vyssh. Uchebn. Zaved., Khim. Khim. Tekhnol.* 7, 758, 1964.
98. Vogel, A. I., *J. Chem. Soc.* 624, 1948.
99. Watanabe, Y., *Bull. Chem. Soc. Jpn.* 33, 1319, 1960.
100. Yao, L., Hammond, E., and Wang, T., *J. Am. Oil Chem. Soc.* 85, 77, 2008.
101. Murray, S. M., O'Brien, R. A., Mattson, K. M., Ceccarelli, C., Sykora, R. E., West, K. N., and Davis, J. H., Jr., *Angew. Chem. Int. Ed.* 49, 2755, 2010.
102. McCutcheon, J. W., *Can. J. Res.* 18B, 231, 1940.
103. Demirbas, A., *Energy Sources, Part A* 30, 1896, 2008.
104. Frenkel, M., Chirico, R. D., Diky V., Muzny, C. D., Kazakov, A. F., Magee, J. W., Abdulagatov, I. M., Kroenlein, K., Diaz-Tovar, C., Kang, J. W., and Gani, R., *NIST ThermoData Engine, NIST Standard Reference Database 103b-Pure Compounds, Binary Mixtures, and Chemical Reactions,* Version 7.0; Standard Reference Data Program, National Institute of Standards and Technology, Gaithersburg, MD, 2012.
105. Frenkel, M., Chirico, R. D., Diky, V., Yan, X., Dong, Q., and Muzny, C. D., *J. Chem. Inf. Model.* 45, 816, 2005.

106. Diky, V., Muzny, C. D., Lemmon, E. W., Chirico, R. D., and Frenkel, M., *J. Chem. Inf. Model.* 47, 1713, 2007.
107. Diky, V., Chirico, R. D., Kazakov, A. F., Muzny, C. D., and Frenkel, M., *J. Chem. Inf. Model.* 49, 503, 2009.
108. Diky, V., Chirico, R. D., Kazakov, A. F., Muzny, C. D., and Frenkel, M., *J. Chem. Inf. Model.* 49, 2883, 2009.
109. Diky V., Chirico, R. D., Kazakov, A. F., Muzny, C. D., Magee, J. W., Abdulagatov, I., Kang, J. W., Kroenlein, K., and Frenkel, M., *J. Chem. Inf. Model.* 51, 181, 2011.
110. Diky V., Chirico, R. D., Muzny, C. D., Kazakov, A. F., Kroenlein, K., Magee, J. W., Abdulagatov, I., Kang, J. W., Frenkel, M., *J. Chem. Inf. Model.* 52, 260, 2011.
111. Diky, V., Chirico, R. D., Muzny, C. D., Kazakov, A. F., Kroenlein, K., Magee, J. W., Abdulagatov, I., Jeong Won Kang, Gani, R., Frenkel, M. *J. Chem. Inf. Model.* 53, 249, 2013.
112. Diky, V., Chirico, R. D., Muzny, C. D., Kazakov, A. F., Kroenlein, K., Magee, J. W., Abdulagatov, I., Frenkel, M. *J. Chem. Inf. Model.* 53, 3418, 2013.
113. Diky, V., Muzny, C. D., Smolyanitsky, A. Y., Bazyleva, A., Chirico, R. D., Magee, J. W., Paulechka, U., Kazakov, A. F., Townsend, S. A., Lemmon, E. W., Frenkel, M., Kroenlein, K. G., *NIST ThermoData Engine, NIST Standard Reference Database 103b – Pure Compounds, Binary Mixtures, Ternary Mixtures, and Chemical Reactions,* Version 10.0, Standard Reference Data Program, National Institute of Standards and Technology, Gaithersburg, MD, 2015.
114. *Guide to the Expression of Uncertainty in Measurement* (International Organization for Standardization, Geneva, Switzerland, 1993). This *Guide* was prepared by ISO Technical Advisory Group 4 (TAG 4), Working Group 3 (WG 3). ISO/TAG 4 has as its sponsors the BIPM (Bureau International des Poids et Mesures), IEC (International Electrotechnical Commission), IFCC (International Federation of Clinical Chemistry and Laboratory Medicine), ISO, IUPAC (International Union of Pure and Applied Chemistry), IUPAP (International Union of Pure and Applied Physics), and OIML (Organisation Internationale de Métrologie Légale). Although the individual members of WG 3 were nominated by the BIPM, IEC, ISO, or OIML, the *Guide* is published by ISO in the name of all seven organizations.
115. *U.S. Guide to the Expression of Uncertainty in Measurement*, ANSI/NCSL Z540-2-1997, NCSL International, Boulder, CO, 1997.
116. Taylor, B. N., and Kuyatt, C. E., Guidelines for Evaluating and Expressing the Uncertainty of NIST Measurement Results, *NIST Technical Note 1297*, NIST, Gaithersburg, MD, 1994.
117. Chirico, R. D., Frenkel, M., Diky, Marsh, K. N., and Wilhoit, R. C., *J. Chem. Eng. Data* 48, 1344, 2003.

COMPOSITION AND PROPERTIES OF COMMON OILS AND FATS

This table lists some of the most common naturally occurring oils and fats. The oils and fats consist mainly of esters of glycerol (i.e., triglycerides) with fatty acids of 10 to 22 carbon atoms. The four fatty acids with the highest concentration are given for each oil; concentrations are given in weight percent. Because there is often a wide variation in composition depending on the source of the oil sample, a range (or sometimes an average) is generally given. More complete data on composition, including minor fatty acids, sterols, and tocopherols, can be found in the references.

The oils and fats are listed alphabetically by common name. The source — plant, fish and other marine life, or land animal — follows the name. The acids are designated by the codes described in the table "Properties of Fatty Acids and Their Methyl Esters" in this section; that table gives the systematic and common names of the acids. Thus 18:2 9c,12c indicates a C_{18} acid with two double bonds in the 9 and 12 positions, both with a *cis* configuration (*cis,cis*-9,12-octadecadienoic acid, or linoleic acid).

The density and refractive index of the oils are typical values; superscripts indicate the temperature in °C.

Notes:

The composition figure given for oleic acid (18:1 9c) often includes low levels of other 18:1 isomers.

In some oils where a concentration is given for 18:2 9c,12c (linoleic acid), other isomers of 18:2 may be included.

Likewise, where a concentration is given for 18:3 9c,12c,15c (α-linolenic acid), other isomers of 18:3 may be included.

The acid 20:5 6c,9c,12c,15c,17c, which is prevalent in many fish oils, is often abbreviated as 20:5 ω-3 or 20:5 n-3.

Jojoba oil consists primarily of wax esters of the acids listed here and long-chain alcohols.

Kapok oil also contains up to 15% cyclopropene acids.

Stillingia oil also contains 5-10% *trans,cis*-2,4-decadienoic acid (stillingic acid, 10:2 2t,4c).

Capelin oil also contains about 10% 16:1.

The assistance of Frank D. Gunstone in preparing this table is gratefully acknowledged.

References

1. Firestone, D., *Physical and Chemical Characteristics of Oils, Fats, and Waxes, Second Edition*, AOCS Press, Urbana, IL, 2006.
2. Gunstone, F. D., Harwood, J. L., and Dijkstra, A. J., eds., *The Lipid Handbook, Third Edition*, CRC Press, Boca Raton, FL, 2006.
3. Dawson, R. M. C., Elliott, D. C., Elliott, W. H., and Jones, K. M., *Data for Biochemical Research, Third Edition*, Clarendon Press, Oxford, 1986.
4. Altman, P. L., and Dittmer, D. S., eds., *Biology Data Book*, Second Edition, Vol. 1, Federation of American Societies for Experimental Biology, Bethesda, MD, 1972.

Fatty Acid Composition and Physical Properties of Oils and Fats

Oil	Source	Acid	Conc/ wt %	mp/°C	ρ/g cm^{-3}	n_D	Sapon. value	Iodine value
Almond kernel oil	Plants	18:1 9c	43-70%		0.910^{25}	1.467^{26}	188-200	89-101
		24-30%	18:2 9c,12c					
		4-13%	16:0					
		1-10%	18:0					
Anchovy oil	Marine animals	16:0	16-20%				191-194	163-169
		10-20%	20:5 6c,9c,12c,15c,17c					
		9-14%	18:1 9c					
		8-12%	16:1 undefined					
Apricot kernel oil	Plants	18:1 9c	58-66%		0.910^{25}	1.469^{25}	185-199	97-110
		29-33%	18:2 9c,12c					
		4.6-6%	16:0					
		1%	18:0					
Argan seed oil	Plants	18:1 9c	42-55%		0.912^{20}	1.467^{20}	189-195	92-102
		30-34%	18:2 9c,12c					
		12-16%	16:0					
		2-7%	18:0					
Avocado pulp oil	Plants	18:1 9c	56-74%		0.912^{25}	1.466^{25}	177-198	85-90
		10-17%	18:2 9c,12c					
		9-18%	16:0					
		3-9%	16:1 9c					
Babassu palm oil	Plants	12:0	40-55%	24	0.914^{25}	1.450^{40}	245-256	10-18
		11-27%	14:0					
		9-20%	18:1 9c					
		5.2-11%	16:0					

Biochem

Oil	Source	Acid	Conc/ wt %	mp/°C	ρ/g cm^{-3}	n_D	Sapon. value	Iodine value
Beef tallow	Land animals	18:1 undefined	31-50%	47	0.902^{25}	1.454^{40}	190-200	33-47
		25-40%	18:0					
		20-37%	16:0					
		1-6%	14:0					
Blackcurrant oil	Plants	18:2 9c,12c	45-50%		0.923^{20}	1.480^{20}	185-195	173-182
		14-20%	18:3 6c,9c,12c					
		12-15%	18:3 9c,12c,15c					
		9-13%	18:1 9c					
Borage (starflower) oil	Plants	18:2 9c,12c	36-40%				189-192	141-160
		17-25%	18:3 6c,9c,12c					
		14-21%	18:1 9c					
		9.4-12%	16:0					
Borneo tallow	Plants	18:0	39-43%	38	0.855^{100}	1.456^{40}	189-200	29-38
		34-37%	18:1 9c					
		18-21%	16:0					
		1.0%	20:0					
Butterfat	Land animals	16:0	28.1% (av.)	32	0.934^{15}	1.455^{40}	210-232	26-40
		20.8% (av.)	18:1 9c					
		10.8% (av.)	14:0					
		10.6% (av.)	18:0					
Cameline oil	Plants	18:3 9c, 12c, 15c	33-38%		0.924^{15}	1.477^{20}	180-190	127-155
		15-16%	18:2 9c,12c					
		14-16%	20:1 total					
		12-24%	18:1 9c					
Canola (rapeseed) oil (low erucic)	Plants	18:1 9c	52-67%	-10	0.915^{20}	1.466^{40}	182-193	110-126
		16-25%	18:2 9c,12c					
		6-14%	18:3 9c, 12c, 15c					
		3.3-6.0%	16:0					
Canola (rapeseed) oil (low linolenic)	Plants	18:1 9c	59-66%	-10				91
		24-29%	18:2 9c,12c					
		4-5%	16:0					
		2-3%	18:3 9c, 12c, 15c					
Capelin oil	Marine animals	18:1 9c	12-18%			1.463^{50}	185-202	94-164
		9-25%	22:1 undefined					
		9-27%	20:1 undefined					
		8-12%	16:0					
Caraway seed oil	Plants	18:1 9c	40%			1.471^{35}	178	128
		30%	18:2 9c,12c					
		26%	18:1 6c					
		3%	16:0					
Cashew nut oil	Plants	18:1 9c	57-80%		0.914^{15}	1.463^{40}	180-196	79-89
		16-22%	18:2 9c,12c					
		4-17%	16:0					
		2-12%	18:0					
Castor oil	Plants	18:1 12-OH,9c	88%	-18	0.952^{25}	1.475^{25}	176-187	81-91
		3-5%	18:2 9c,12c					
		2.9-6%	18:1 undefined					
		2.1%	22:0					
Cherry kernel oil	Plants	18:2 9c,12c	42-45%		0.918^{25}	1.468^{40}	190-198	110-118
		35-49%	18:1 9c					
		4-9%	16:0					
		3-10	18:3 9c,11t,13t					
Chicken egg lipids, yolk	Land animals	16:0	28%					
		25%	18.1 9c					
		17%	18:0					
		16%	18:2 9c, 12c					
Chicken fat	Land animals	18:1 undefined	37%		0.918^{15}	1.456^{40}		76-80
		22%	16:0					
		20%	18:2					

Oil	Source	Acid	Conc/ wt %	mp/°C	ρ/g cm^{-3}	n_D	Sapon. value	Iodine value
		6%	18:0					
Chinese vegetable tallow	Plants	16:0	58-72%	44	0.887^{25}	1.456^{40}	200-218	16-29
		20-35%	18:1 9c					
		1-8%	18:0					
		0.5-3.7%	14:0					
Cocoa butter	Plants	18:0	31-37%	34	0.974^{25}	1.457^{40}	192-200	32-40
		31-35%	18:1 9c					
		25-27%	16:0					
		2.8-4.0%	18:2 9c,12c					
Coconut oil	Plants	12:0	45-51%	25	0.913^{40}	1.449^{40}	248-265	5-13
		17-21%	14:0					
		7.7-10.2%	16:0					
		5.4-9.9%	18:1 9c					
Cod liver oil	Marine animals	18:1 undefined	19-27%		0.924^{15}	1.482^{25}	180-192	142-176
		14-20%	18:1 9c					
		10-14%	16:0					
		8-14%	20:5 6c,9c,12c,15c,17c					
Cohune nut oil	Plants	12:0	44-48%		0.914^{25}	1.450^{40}	251-260	9-14
		16-17%	14:0					
		8-10%	18:1 9c					
		7-10%	16:0					
Coriander seed oil	Plants	18:1 6c	53%		0.908^{25}	1.464^{25}	182-191	86-100
		32%	18:1 9c					
		7-14%	18:2 9c,12c					
		3-8%	16:0					
Corn oil	Plants	18:2 9c,12c	40-66%	-20	0.919^{20}	1.472^{25}	187-195	107-135
		20-42%	18:1 9c					
		9-16%	16:0					
		0-3%	18:0					
Cottonseed oil	Plants	18:2 9c,12c	47-58%	-1	0.920^{20}	1.462^{40}	189-198	96-115
		18-26%	16:0					
		14-22%	18:1 9c					
		2.1-3.3%	18:0					
Crambe oil	Plants	22:1 13c	55-60%		0.906^{25}	1.470^{25}		87-113
		12-15%	18:1 9c					
		8-10%	18:2 9c,12c					
		6-7%	18:3 9c, 12c, 15c					
Cuphea oil	Plants	18:2 9c,12c	62-67%					
		17-18%	16:0					
		12-14%	18:1 9c					
		2%	18:0					
Euphorbia lagascae seed oil	Plants	18:1 12,13-ep,9c	64%		0.952^{25}	1.473^{25}		102
		19%	18:1 other					
		9%	18:2 9c,12c					
		4%	16:0					
Evening primrose oil	Plants	18:2 9c,12c	65-80%			1.479^{20}	193-198	147-155
		8-14%	18:3 6c,9c,12c					
		6-10%	16:0					
		5-12%	18:1 9c					
Grape seed oil	Plants	18:2 9c,12c	58-78%		0.923^{20}	1.475^{40}	188-194	130-138
		12-28%	18:1 9c					
		5.5-11%	16:0					
		3-6%	18:0					
Hazelnut oil (Chilean)	Plants	18:1 9c	39%					
		22.7%	16:1 11c					
		9.7%	20:1 total					
		9.5%	22:1 total					
Hazelnut oil (Filbert)	Plants	18:1 9c	72-84%		0.909^{25}	1.473^{25}	188-197	83-90
		5.7-22%	18:2 9c,12c					
		4.1-7.2%	16:0					

Oil	Source	Acid	Conc/ wt %	mp/°C	ρ/g cm^{-3}	n_D	Sapon. value	Iodine value
		1.5-2.4%	18:0					
Hempseed oil	Plants	18:2 9c,12c	45-60%		0.921^{25}	1.472^{40}	190-195	145-166
		15-30%	18:3 9c, 12c, 15c					
		11-16%	18:1 9c					
		6-12%	16:0					
Herring oil	Marine animals	16:0	8-25%		0.914^{20}	1.474^{25}	161-192	115-160
		6-20%	20:1					
		5-22%	18:1 undefined					
		4-31%	22:1					
Illipe (mowrah) butter	Plants	18:1 9c	34%	27	0.862^{100}	1.460^{40}	188-207	53-70
		23%	16:0					
		23%	18:0					
		14%	18:2 9c,12c					
Jojoba oil	Plants	20:1 total	66-74%					
		9-19%	22:1 undefined					
		5-12%	18:1 9c					
		1-5%	24:1 15c					
Kapok seed oil	Plants	18:1 9c	45-65%	30	0.926^{15}	1.469^{25}	189-197	86-110
		10-28%	16:0					
		7-35%	18:2 9c,12c					
		2-9%	18:0					
Kokum butter	Plants	18:0	49-56%	41		1.456^{40}	192	33-37
		39-49%	18:1 9c					
		2-5%	16:0					
		1-2%	18:2 9c,12c					
Kusum oil	Plants	18:1 9c	57-62%			1.461^{40}	220-230	48-58
		20-25%	20:0					
		5-8%	16:0					
		2-6%	18:0					
Linola oil	Plants	18:2 9c,12c	72%					142
		16%	18:1 9c					
		5.6%	16:0					
		4.0%	18:0					
Linseed oil	Plants	18:3 9c, 12c, 15c	52-58%	-24	0.924^{25}	1.480^{25}	188-196	170-203
		18-20%	18:1 9c					
		17%	18:2 9c,12c					
		16%	18:2 undefined					
Macadamia nut oil	Plants	18:1 9c	56-59%					
		21-22%	16:1 9c					
		8-9%	16:0					
		2-4%	18:0					
Mango seed oil	Plants	18:1 9c	38-50%		0.912^{15}	1.461^{25}	188-195	39-48
		31-49%	18:0					
		3-6%	18:2 9c,12c					
		2-6%	20:0					
Meadowfoam seed oil	Plants	20:1 5c	58-77%			1.464^{40}	168	86-91
		8-24%	22:1 total					
		7-15%	22:2 5c,13c					
		1-3%	18:1 9c					
Melon oil	Plants	18:2 9c,12c	67% (av.)					
		12% (av.)	18:1 9c					
		11% (av.)	16:0					
		9% (av.)	18:0					
Menhaden oil	Marine animals	16:0	14-23%		0.920^{15}		192-199	150-200
		12-18%	20:5 6c,9c,12c,15c,17c					
		7-15%	16:1					
		6-16%	18:1 undefined					
Milk fats, cow	Land animals	16:0	28.2% (av.)					
		21.4% (av.)	18:1 9c					
		12.6% (av.)	18:0					

Biochem

Oil	Source	Acid	Conc/ wt %	mp/°C	ρ/g cm^{-3}	n_D	Sapon. value	Iodine value
		10.6% (av.)	14:0					
Milk fats, human	Land animals	18:1 9c	31.1% (av.)					
		21.6% (av.)	16:0					
		11.7% (av.)	18:2 9c,12c					
		6.6% (av.)	14:0					
Moringa peregrina seed oil	Plants	18:1 9c	70%		0.903^{24}	1.460^{40}	185	70
		9%	16:0					
		3.8%	18:0					
		2.4%	22:0					
Mustard seed oil	Plants	22:1 13c	43%		0.913^{20}	1.465^{40}	170-184	92-125
		22-50%	22:1 undefined					
		12%	18:3 9c,12c,15c					
		10-24%	18:2 9c,12c					
Mutton tallow	Land animals	18:1 undefined	30-42%	48	0.946^{15}	1.455^{40}		35-46
		22-34%	18:0					
		20-27%	16:0					
		2-4%	14:0					
Neem oil	Plants	18:1 9c	49-62%	-3	0.912^{30}	1.462^{40}	195-205	68-71
		14-24%	18:0					
		13-18%	16:0					
		7-15%	18:2 9c,12c					
Niger seed oil	Plants	18:2 9c,12c	52-78%		0.924^{15}	1.468^{40}	188-193	126-135
		5-12%	16:0					
		4-10%	18:1 9c					
		2-12%	18:0					
Nutmeg butter	Plants	14:0	76-83%	45		1.468^{40}	170-190	48-85
		5-11%	18:1 9c					
		4-10%	16:0					
		3-6%	12:0					
Oat bean oil	Plants	18:2 9c,12c	36%		0.904^{25}	1.4723^{30}	181-187	86-96
		31%	18:1 9c					
		3.8%	16:0					
		2.4%	20:0					
Oat oil	Plants	18:2 9c,12c	24-48%		0.917^{25}	1.467^{40}	190-199	105-116
		18-53%	18:1 9c					
		13-39%	16:0					
		0.5-4%	18:0					
Oiticica oil	Plants	18:3 9c, 11t, 13t	70-80%		0.972^{20}	1.514^{25}	188-193	140-150
		7%	16:0					
		5%	18:0					
		4-7%	18:1 9c					
Olive oil	Plants	18:1 9c	55-83%	-6	0.911^{20}	1.469^{20}	184-196	75-94
		9%	18:2 9c, 12c					
		7.5-20%	16:0					
		3.5-21%	18:2 9c,12c					
Palm kernel oil	Plants	12:0	40-55%	24	0.922^{15}	1.450^{40}	230-250	14-21
		14-18%	14:0					
		12-21%	18:1 9c					
		6.5-10%	16:0					
Palm oil	Plants	16:0	40-48%	35	0.914^{15}	1.455^{40}	190-209	49-55
		36-44%	18:1 9c					
		6.5-12%	18:2 9c,12c					
		3.5-6.5%	18:0					
Palm olein	Plants	18:1 9c	40-44%		0.91^{40}	1.459^{40}	194-202	>56
		38-43%	16:0					
		10-13%	18:2 9c,12c					
		3.7-4.8%	18:0					
Palm stearin	Plants	16:0	48-74%		0.884^{60}	1.449^{40}	193-205	<48
		16-36%	18:1 9c					
		3.9-5.6%	18:0					

Oil	Source	Acid	Conc/ wt %	mp/°C	ρ/g cm^{-3}	n_D	Sapon. value	Iodine value
		3.2-9.8%	18:2 9c,12c					
Parsley seed oil	Plants	18:1 6c	69-76%			1.4800^{40}		110-120
		12-15%	18:1 9c					
		6-14%	18:2 9c,12c					
		2%	16:0					
Peanut oil	Plants	18:1 9c	36-67%	3	0.914^{20}	1.463^{40}	187-196	86-107
		14-43%	18:2 9c,12c					
		8.3-14%	16:0					
		2.1-4.4%	22:0					
Perilla oil	Plants	18:3 9c,12c,15c	59%		0.924^{25}	1.477^{25}	188-197	192-208
		14-18%	18:2 9c,12c					
		11-13%	18:1 9c					
		6-9%	16:0					
Phulwara butter	Plants	16:0	57-61%	43	0.862^{100}	1.458^{40}	188-200	40-51
		30-36%	18:1 9c					
		3-4%	18:2 9c,12c					
		3-4%	18:0					
Pine nut oil	Plants	18:2 9c,12c	47-51%		0.919^{15}		193-197	118-121
		36-39%	18:1 9c					
		6-8%	16:0					
		2-3%	18:0					
Poppy seed oil	Plants	18:2 9c,12c	62-73%	-15	0.916^{25}	1.469^{40}	188-196	132-146
		16-30%	18:1 9c					
		7-11%	16:0					
		1-4%	18:0					
Pork lard	Land animals	18:1 undefined	35-62%	30	0.898^{20}			
		20-32%	16:0					
		5-24%	18:0					
		3-16%	18:2					
Rice bran oil	Plants	18:1 9c	38-48%		0.916^{25}	1.472^{25}	181-189	92-108
		16-36%	18:2 9c,12c					
		16-28%	16:0					
		2-4%	18:0					
Safflower seed oil	Plants	18:2 9c,12c	68-83%		0.924^{15}	1.474^{25}	186-198	136-148
		8.4-30%	18:1 9c					
		5.3-8.0%	16:0					
		1.9-2.9%	18:0					
Safflower seed oil (high oleic)	Plants	18:1 9c	74-80%		0.921^{20}	1.470^{25}		91-95
		13-18%	18:2 9c,12c					
		5-6%	16:0					
		1.5-2.0%	18:0					
Sal fat	Plants	18:0	33-57%	33		1.456^{40}	175-192	31-45
		31-52%	18:1 9c					
		6-23%	16:0					
		1-8%	20:0					
Salmon oil	Marine animals	22:6 4c,7c,10c,13c,16c,19	18%		0.924^{15}	1.475^{25}	183-186	130-160
		13%	20:5 6c,9c,12c,15c,17c					
		9.8%	16:0					
		4.8%	16:1					
Sardine oil	Marine animals	20:5 6c,9c,12c,15c,17c	9-35%		0.915^{25}	1.464^{65}	188-199	159-192
		9-22%	16:0					
		7-17%	18:1 undefined					
		6-13%	16:1					
Seal blubber oil, harp	Marine animals	18:1 9c	21%					
		12%	20:1					
		7.6%	22:6 4c,7c,10c,13c,16c,19c					
		6.4%	20:5 6c,9c,12c,15c,17c					
Sesame seed oil	Plants	18:2 9c,12c	40-51%	-6	0.917^{20}	1.467^{40}	187-195	104-120

Oil	Source	Acid	Conc/ wt %	mp/°C	ρ/g cm^{-3}	n_D	Sapon. value	Iodine value
		33-44%	18:1 9c					
		7.9-10.2%	16:0					
		4.4-6.7%	18:0					
Shark liver oil	Marine animals	18:1 undefined	45%		0.917^{25}	1.476^{25}	170-190	150-300
		21%	16:0					
		12%	20:1					
		9%	22:1					
Sheanut butter	Plants	18:1 9c	45-50%	38	0.863^{100}	1.465^{40}	178-198	52-66
		36-41%	18:0					
		4-8%	16:0					
		4-8%	18:2 9c,12c					
Soybean oil	Plants	18:2 9c,12c	50-57%	-16	0.920^{20}	1.468^{40}	189-195	118-139
		18-28%	18:1 9c					
		9-13%	16:0					
		5.5-9.5%	18:3 9c, 12c, 15c					
Stillingia seed kernel oil	Plants	18:3 9c, 12c, 15c	41-54%		0.937^{25}	1.483^{25}	202-212	169-191
		24-30%	18:2 total					
		7-10%	18:1 9c					
		6-9%	16:0					
Sunflower seed oil	Plants	18:2 9c,12c	48-74%	-17	0.919^{20}	1.474^{25}	188-194	118-145
		13-40%	18:1 9c					
		5-8%	16:0					
		2.5-7.0%	18:0					
Tall oil	Plants	18:2 9c,12c	41-52%		0.969^{25}	1.494^{25}	154-180	140-180
		41-48%	18:1 9c					
		5-6%	16:0					
		2-3%	18:0					
Trout lipids	Marine animals	16:0	21-24%					
		18-31%	18:1 undefined					
		7-16%	18:2					
		4-10%	16:1					
Tung oil	Plants	18:3 9c,11t,13t	71-82%	-2	0.912^{25}	1.517^{25}	189-195	160-175
		8-15%	18:2 9c,12c					
		4-10%	18:1 9c					
		3%	18:0					
Ucuhuba butter oil	Plants	14:0	64-73%		0.870^{100}	1.451^{50}	221-229	11-17
		13-15%	12:0					
		6-8%	18:1 9c					
		3-9%	16:0					
Vernonia seed oil	Plants	18:1 12,13-ep,9c	62-72%		0.901^{30}	1.486^{32}	176	55
		9-17%	18:2 9c,12c					
		3-7%	16:0					
		2-6%	18:0					
Walnut oil	Plants	18:2 9c,12c	56-60%		0.921^{25}	1.474^{25}	189-197	138-162
		17-19%	18:1 9c					
		13-14%	18:3 9c, 12c, 15c					
		6-8%	16:0					
Whale oil, minke	Marine animals	18:1 undefined	18%					
		17%	20:1					
		11%	22:1					
		9%	16:1					
Wheatgerm oil	Plants	18:2 9c,12c	50-59%		0.926^{25}	1.479^{25}	179-217	100-128
		13-23%	18:1 9c					
		12-20%	16:0					
		2-9%	18:3 9c, 12c, 15c					

CARBOHYDRATE NAMES AND SYMBOLS

The following table lists the systematic names and symbols for selected carbohydrates and some of their derivatives. The symbols for monosaccharide residues and derivatives are recommended by IUPAC for use in describing the structures of oligosaccharide chains. A more complete list can be found in the reference.

Reference

McNaught, A. D., *Pure Appl. Chem.*, 68, 1919–2008, 1996.

Carbohydrate Names and Symbols

Common name	Symbol	Systematic name
Abequose	Abe	3,6-Dideoxy-D-*xylo*-hexose
N-Acetyl-2-deoxyneur-2-enaminic acid	Neu2en5Ac	
N-Acetylgalactosamine	GalNAc	
N-Acetylglucosamine	GlcNAc	
N-Acetylneuraminic acid	Neu5Ac	
Allose	All	*allo*-Hexose
Altrose	Alt	*altro*-Hexose
Apiose	Api	3-*C*-(Hydroxymethyl)-*glycero*-tetrose
Arabinitol	Ara-ol	Arabinitol
Arabinose	Ara	*arabino*-Pentose
Arcanose		2,6-Dideoxy-3-*C*-methyl-3-*O*-methyl-*xylo*-hexose
Ascarylose		3,6-Dideoxy-D-*arabino*-hexose
Boivinose		2,6-Dideoxy-D-gulose
Chalcose		4,6-Dideoxy-3-*O*-methyl-D-*xylo*-hexose
Cladinose		2,6-Dideoxy-3-*C*-methyl-3-*O*-methyl-D-*ribo*-hexose
Colitose		3,6-Dideoxy-D-*xylo*-hexose
Cymarose		6-Deoxy-3-*O*-methyl-*ribo*-hexose
3-Deoxy-D-*manno*-oct-2-ulosonic acid	Kdo	
2-Deoxyribose	dRib	2-Deoxy-*erythro*-pentose
2,3-Diamino-2,3-dideoxy-D-glucose	GlcN3N	
Diginose		2,6-Dideoxy-3-*O*-methyl-*lyxo*-hexose
Digitalose		6-Deoxy-3-*O*-methyl-D-galactose
Digitoxose		2,6-Dideoxy-D-*ribo*-hexose
3,4-Di-*O*-methylrhamnose	Rha3,4Me_2	
Ethyl glucopyranuronate	Glc*p*A6Et	
Evalose		6-Deoxy-3-*C*-methyl-D-mannose
Fructose	Fru	*arabino*-Hex-2-ulose
Fucitol	Fuc-ol	6-Deoxy-D-galactitol
Fucose	Fuc	6-Deoxygalactose
β-D-Galactopyranose 4-sulfate	β-D-Gal*p*4*S*	
Galactosamine	GalN	2-Amino-2-deoxygalactose
Galactose	Gal	*galacto*-Hexose
Glucitol	Glc-ol	
Glucosamine	GlcN	2-Amino-2-deoxyglucose
Glucose	Glc	*gluco*-Hexose
Glucuronic acid	GlcA	
N-Glycoloylneuraminic acid	Neu5Gc	
Gulose	Gul	*gulo*-Hexose
Hamamelose		2-*C*-(Hydroxymethyl)-D-ribose
Idose	Ido	*ido*-Hexose
Iduronic acid	IdoA	
Lactose	Lac	β-D-Galactopyranosyl-(1→4)-D-glucose
Lyxose	Lyx	*lyxo*-Pentose
Maltose		α-D-Glucopyranosyl-(1→4)-D-glucose
Mannose	Man	*manno*-Hexose
2-*C*-Methylxylose	Xyl2CMe	
Muramic acid	Mur	2-Amino-3-*O*-[(R)-1-carboxyethyl]-2-deoxy-D-glucose
Mycarose		2,6-Dideoxy-3-*C*-methyl-D-*ribo*-hexose
Mycinose		6-Deoxy-2,3-di-*O*-methyl-D-allose
Neuraminic acid	Neu	5-Amino-3,5-dideoxy-D-*glycero*-D-*galacto*-non-2-ulosonic acid

Biochem

Common name	Symbol	Systematic name
Panose		α-D-Glucopyranosyl-(1→6)-α-D-glucopyranosyl-(1→4)-D-glucose
Paratose		3,6-Dideoxy-D-*ribo*-hexose
Primeverose		β-D-Xylopyranosyl-(1→6)-D-glucose
Psicose	Psi	*ribo*-Hex-2-ulose
Quinovose	Qui	6-Deoxyglucose
Raffinose		β-D-Fructofuranosyl-α-D-galactopyranosyl-(1→6)-α-D-glucopyranoside
Rhamnose	Rha	6-Deoxymannose
Rhodinose		2,3,6-Trideoxy-D-*threo*-hexose
Ribose	Rib	*ribo*-Pentose
Ribose 5-phosphate	Rib5*P*	
Ribulose	Ribulo (Rul)	*erythro*-Pent-2-ulose
Rutinose		α-D-Rhamnopyranosyl-(1→6)-D-glucose
Sarmentose		2,6-Dideoxy-3-*O*-methyl-D-*xylo*-hexose
Sedoheptulose		D-*altro*-Hept-2-ulose
Sorbose	Sor	*xylo*-Hex-2-ulose
Streptose		5-Deoxy-3-*C*-formyl-D-lyxose
Sucrose		β-D-Fructofuranosyl-α-D-glucopyranoside
Tagatose	Tag	*lyxo*-Hex-2-ulose
Talose	Tal	*talo*-Hexose
Turanose		α-D-Glucopyranosyl-(1→3)-D-fructose
Tyvelose	Tyv	3,6-Dideoxy-D-*arabino*-hexose
Xylose	Xyl	*xylo*-Pentose
Xylulose	Xylulo (Xul)	*threo*-Pent-2-ulose

STANDARD TRANSFORMED GIBBS ENERGIES OF FORMATION FOR BIOCHEMICAL REACTANTS

Robert N. Goldberg and Robert A. Alberty

This table contains values of the standard transformed Gibbs energies of formation $\Delta_f G'^\circ$ for 130 biochemical reactants. Values of $\Delta_f G'^\circ$ are given at pH 7.0, the temperature 298.15 K, and the pressure 100 kPa for three ionic strengths: $I = 0$, $I = 0.1$ mol/L and $I = 0.25$ mol/L. The table can be used for calculating apparent equilibrium constants K' and standard apparent reduction potentials E'° for biochemical reactions. Such a listing is more compact than tabulating the actual apparent equilibrium constants or standard apparent reduction potentials, which would require a very large number of reactant–product combinations. In the table, all reactants are in aqueous solution unless indicated otherwise.

A biochemical reactant is a sum of species. For example, ATP consists of an equilibrium mixture of the aqueous species ATP^{4-}, $HATP^{3-}$, H_2ATP^{2-}, $MgATP^{2-}$, etc. Similarly, phosphate refers to the equilibrium mixture of the aqueous species PO_4^{3-}, HPO_4^{2-}, $H_2PO_4^-$, H_3PO_4, $MgHPO_4$, etc. Biochemical reactions are written using biochemical reactants in terms of an apparent equilibrium constant K', which is distinct from the standard equilibrium constant K. This subject is discussed in an IUPAC report (see Ref. 1 below).

The apparent equilibrium constant K' and the standard transformed Gibbs energy change $\Delta_r G'^\circ$ for a biochemical reaction can be calculated from the $\Delta_f G'^\circ$ values by using the relationship

$$-RT \ln K' = \Delta_r G'^\circ = \Sigma \nu'_i \, \Delta_f G'^\circ,$$

where the summation is over all of the biochemical reactants. The quantity ν'_i is the stoichiometric number of reactant i (ν'_i is positive for reactants on the right side of the equation and negative for reactants on the left side); R is the gas constant. As an example, the hydrolysis reaction of ATP is

$$\text{ATP} + H_2O(l) = \text{ADP} + \text{phosphate}.$$

At pH 7.00 and $I = 0.25$ M, $\Delta_r G'^\circ$ and K' are calculated as follows:

$$\Delta_r G'^\circ = \{-1424.70 - 1059.49 - (-2292.50 - 155.66)\} \cdot (\text{kJ mol}^{-1}) = -36.03 \text{ kJ mol}^{-1}$$

$$K' = \exp[-(-36030 \text{ J mol}^{-1})/\{(8.3145 \text{ J mol}^{-1} \text{ K}^{-1}) \cdot (298.15 \text{ K})\} = 2.05 \cdot 10^6$$

An example involving a biochemical half-cell reaction is

$$\text{acetaldehyde(aq)} + 2\, e^- = \text{ethanol(aq)}.$$

At 298.15 K, pH 7.00, and $I = 0$, the standard apparent reduction potential E'° can be calculated as follows

$$E'^\circ = -(1/nF) \cdot \{\Delta_f G'^\circ(\text{ethanol}) - \Delta_f G'^\circ(\text{acetaldehyde})\},$$

where n is the number of electrons in the half-cell reaction and F is the Faraday constant. Then,

$$E'^\circ = [-1/(2 \cdot 9.6485 \cdot 10^4 \text{ C mol}^{-1})] \cdot (58.10 \cdot 10^3 \text{ J mol}^{-1} - 20.83 \cdot 10^3 \text{ J mol}^{-1}) = -0.193 \text{ V}$$

References

1. Alberty, R.A., Cornish-Bowden, A., Gibson, Q.H., Goldberg, R.N., Hammes, G., Jencks, W., Tipton, K.F, Veech, R., Westerhoff, H.V., and Webb, E.C. *Pure Appl. Chem.* 66, 1641-1666, 1994.
2. Alberty, R.A., *Arch. Biochem. Biophys.*, 353, 116-130, 1998; 358, 25-39, 1998.
3. Alberty, R.A., *Thermodynamics of Biochemical Reactions*, Wiley-Interscience, New York, 2003.
4. Alberty, R.A., *BasicBiochemData2: Data and Programs for Biochemical Thermodynamics*, <http://library.wolfram.com/infocenter/MathSource/797>.

Reactant	Note	$\Delta_f G'^\circ(I=0)$/ kJ mol^{-1}	$\Delta_f G'^\circ(I=0.1)$/ kJ mol^{-1}	$\Delta_f G'^\circ(I=0.25)$/ kJ mol^{-1}
Acetaldehyde		20.83	23.27	24.06
Acetate ion [CH_3COO^-]		-249.46	-248.23	-247.83
Acetone		80.04	83.71	84.90
Acetyl Coenzyme A		-60.49	-58.65	-58.06
Acetylphosphate		-1109.34	-1107.57	-1107.02
cis-Aconitate		-797.26	-800.93	-802.12
Adenine		510.45	513.51	514.50
Adenosine		324.93	332.89	335.46
Adenosine 5'-(trihydrogen diphosphate)	ADP	-1428.93	-1425.55	-1424.70
Adenosine 5'-monophosphate	AMP	-562.04	-556.53	-554.83
Adenosine 5'-triphosphate	ATP	-2292.61	-2292.16	-2292.50
D-Alanine		-91.31	-87.02	-85.64
Ammonia		80.50	82.34	82.93
α-D-Arabinose		-342.67	-336.55	-334.57
L-Asparagine		-206.28	-201.38	-199.80
L-Aspartate		-456.14	-453.08	-452.09
1,3-Biphosphoglycerate		-2202.06	-2205.69	-2207.30
Butanoate		-72.94	-69.26	-68.08
1-Butanol		227.72	233.84	235.82
Citrate		-963.46	-965.49	-966.23
Isocitrate		-956.82	-958.84	-959.58
Coenzyme A	CoA	-7.98	-7.43	-7.26
Carbon monoxide	aqueous	-119.90	-119.90	-119.90
Carbon monoxide	gas	-137.17	-137.17	-137.17
Carbon dioxide	aqueous (total)	-547.33	-547.15	-547.10
Carbon dioxide	gas	-394.36	-394.36	-394.36
Creatine		100.41	105.92	107.69
Creatinine		256.55	260.84	262.22
L-Cysteine		-59.23	-55.01	-53.65
L-Cystine		-187.03	-179.69	-177.32
Cytochrome c [oxidized]		0.00	-5.51	-7.29
Cytochrome c [reduced]		-24.51	-26.96	-27.75
Dihydroxyacetone phosphate		-1096.60	-1095.91	-1095.70
Ethanol		58.10	61.77	62.96
Ethyl acetate		-18.00	-13.10	-11.52
Ferredoxin [oxidized]		0.00	-0.61	-0.81
Ferredoxin [reduced]		38.07	38.07	38.07
Flavine adenine dinucleotide	oxidized	1238.65	1255.17	1260.51
Flavine adenine dinucleotide	reduced	1279.68	1297.43	1303.16

Reactant	Note	$\Delta_f G'^\circ(I=0)$/ kJ mol^{-1}	$\Delta_f G'^\circ(I=0.1)$/ kJ mol^{-1}	$\Delta_f G'^\circ(I=0.25)$/ kJ mol^{-1}
Flavin adenine dinucleotide-enzyme (FADenz) [oxidized]		1238.65	1255.17	1260.51
Flavin adenine dinucleotide-enzyme (FADenz) [reduced]		1229.96	1247.71	1253.44
Flavin mononucleotide (FMN) [oxidized]		759.17	768.35	771.32
Flavin mononucleotide (FMN) [reduced]		800.20	810.61	813.97
Formate ion [CHOO$^-$]		-311.04	-311.04	-311.04
D-Fructose		-436.03	-428.69	-426.32
D-Fructose 1,6-diphosphate		-2202.84	-2205.66	-2206.78
D-Fructose 6-phosphate		-1321.71	-1317.16	-1315.74
Fumarate		-521.97	-523.19	-523.58
D-Galactose		-429.45	-422.11	-419.74
Galactose 1-dihydrogen phosphate		-1317.50	-1313.01	-1311.60
α-D-Glucose		-436.42	-429.08	-426.71
α-D-Glucose 1-phosphate		-1318.03	-1313.34	-1311.89
D-Glucose 6-phosphate		-1325.00	-1320.37	-1318.92
Glutamate		-377.82	-373.54	-372.16
D-Glutamine		-128.46	-122.34	-120.36
Glutathione	oxidized	1198.69	1214.60	1219.74
Glutathione	reduced	625.75	634.76	637.62
Glutathione-coenzyme A		563.49	572.06	574.83
D-Glyceraldehyde 3-phosphate		-1088.94	-1088.25	-1088.04
Glycerol		-177.83	-172.93	-171.35
L-Glycerol 1-phosphate		-1080.22	-1077.83	-1077.13
Glycine		-180.13	-177.07	-176.08
Glycolate		-411.08	-409.86	-409.46
N-Glycylglycine		-200.55	-195.65	-194.07
Glyoxylate		-428.64	-428.64	-428.64
Hydrogen	aqueous	97.51	98.74	99.13
Hydrogen	gas	79.91	81.14	81.53
Hydrogen peroxide	aqueous	-54.12	-52.89	-52.50
3-Hydroxypropanoate		-318.62	-316.17	-315.38
Hypoxanthine		249.33	251.77	252.56
1*H*-Indole		503.49	507.78	509.16
Lactate		-316.94	-314.49	-313.70
α-Lactose		-688.29	-674.83	-670.48
L-Leucine		167.18	175.14	177.71
L-Isoleucine		175.53	183.49	186.06
D-Lyxose		-349.58	-343.46	-341.48
Malate		-682.88	-682.85	-682.85
α-Maltose		-695.65	-682.19	-677.84
D-Mannitol		-383.22	-374.65	-371.89
DL-Mannose		-430.52	-423.18	-420.81
Methane	aqueous	125.50	127.94	128.73
Methane	gas	109.11	111.55	112.34
Methanol		-15.48	-13.04	-12.25
L-Methionine		-63.40	-56.67	-54.49
Nicotinamide Adenine Dinucleotide (NAD) [oxidized]		1038.86	1054.17	1059.11
Nicotinamide Adenine Dinucleotide (NAD) [reduced]		1101.47	1115.55	1120.09
Nicotinamide Adenine Dinucleotide Phosphate (NADP) [oxidized]		163.73	173.52	176.68
Nicotinamide Adenine Dinucleotide Phosphate (NADP) [reduced]		229.67	235.79	237.77
Nitrogen	aqueous	18.70	18.70	18.70
Nitrogen	gas	0.00	0.00	0.00
Oxalate ion [$C_2O_4^{-2}$]		-673.90	-676.35	-677.14
Oxaloacetate		-713.38	-714.60	-715.00
Oxalosuccinate		-979.05	-979.05	-979.05
2-Oxoglutarate		-633.58	-633.58	-633.58
Oxygen	aqueous	16.40	16.40	16.40
Oxygen	gas	0.00	0.00	0.00
Palmitate		979.25	997.61	1003.54
L-Phenylalanine		232.42	239.15	241.33
Phosphate ion [PO_4^{-3}]		-1058.56	-1059.17	-1059.49
2-Phospho-D-glycerate		-1340.72	-1341.32	-1341.79
3-Phospho-D-glycerate		-1346.38	-1347.19	-1347.73
Phospho*enol*pyruvate		-1185.46	-1188.53	-1189.73
1-Propanol		143.84	148.74	150.32
2-Propanol		134.42	139.32	140.90
Pyrophosphate ion [$P_2O_7^{-4}$]		-1934.95	-1939.13	-1940.66
Pyruvate		-352.40	-351.18	-350.78
Retinal (all *trans*)		1118.78	1135.91	1141.45
Retinol		1170.78	1189.14	1195.07
D-Ribose		-339.23	-333.11	-331.13
D-Ribose 1-phosphate		-1215.87	-1212.24	-1211.14
D-Ribose 5-phosphate		-1223.95	-1220.32	-1219.22
D-Ribulose		-336.38	-330.26	-328.28
L-Serine		-231.18	-226.89	-225.51
D-Sorbose		-432.47	-425.13	-422.76
Succinate		-530.72	-530.65	-530.64
Succinyl Coenzyme A		-349.90	-348.06	-347.47
Sucrose		-685.66	-672.20	-667.85
Thioredoxin [oxidized]		0.00	0.00	0.00
Thioredoxin [reduced]		54.32	55.41	55.74
DL-Tryptophan		364.78	372.12	374.49
L-Tyrosine		68.82	75.55	77.73
Ubiquinone [oxidized]		3596.07	3651.15	3668.94
Ubiquinone [reduced]		3586.06	3642.37	3660.55
Urate		-206.03	-204.81	-204.41
Urea		-42.97	-40.53	-39.74
Uric acid		-197.07	-194.63	-193.84
L-Valine		80.87	87.60	89.78
Water	liquid	-157.28	-156.05	-155.66
D-Xylose		-350.93	-344.81	-342.83
D-Xylulose		-346.59	-340.47	-338.49

APPARENT EQUILIBRIUM CONSTANTS FOR ENZYME-CATALYZED REACTIONS

Robert N. Goldberg

This table contains values of apparent equilibrium constants K' for selected enzyme-catalyzed reactions at specified temperatures T and pHs. In those cases where the ionic strength I and/or the pMg (pMg = $-\log_{10}[Mg^{2+}]$) have been reported, the values of these quantities are given. The Enzyme Commission numbers (Ref. 1) of the enzymes that were used to catalyze the reactions are also given.

There are two fundamentally different types of equilibrium constants. This is illustrated by the following example for the hydrolysis of adenosine 5′-triphosphate (ATP) to adenosine 5′-diphosphate (ADP) and phosphate:

$$\text{ATP} + H_2O = \text{ADP} + \text{phosphate} \qquad (1)$$

The apparent equilibrium constant for the overall biochemical reaction (1) is

$$K' = [\text{ADP}][\text{phosphate}]/([\text{ATP}]c^o) \qquad (2)$$

The biochemical reactants ATP, ADP, and phosphate each exist in several different ionized and metal bound forms. For example, ATP is an equilibrium mixture of the species ATP^{4-}, $HATP^{3-}$, H_2ATP^{2-}, $MgATP^{2-}$, $MgHATP^{-}$, Mg_2ATP^0. Additional species would also have to be considered if Ca^{2+} were present. Thus, ATP has often been denoted in the literature as ΣATP or as $(ATP)_{tot}$. When it is clear that one is dealing with total amounts of substances, it is not necessary to use either the Σ or "tot." Thus, these designations are not used in this table. In the above equation, c^o=1 mol dm^{-3}; it is included to make K' dimensionless. The standard transformed Gibbs energy of reaction $\Delta_r G'^o$ at specified conditions of temperature T, pressure P, ionic strength I, pH, and pMg can be calculated from K':

$$\Delta_r G'^o = -RT\ln K' \qquad (3)$$

The molar gas constant, R, is equal to 8.314462 J K^{-1} mol^{-1}. $\Delta_r G'^o$ and the apparent equilibrium constant, K', can be used to calculate the position of equilibrium of overall biochemical reactions.

It is also possible to choose a chemical reference reaction that involves selected solute species:

$$ATP^{4-} + H_2O = ADP^{3-} + HPO_4^{2-} + H^+ \qquad (4)$$

The equilibrium constant for this reference reaction is

$$K = [ADP^{3-}][HPO_4^{2-}][H+]/\{[ATP^{4-}](c^o)^2\} \qquad (5)$$

Equations and algorithms that relate these two different types of equilibrium constants have been published in Refs. 2–4. To calculate the equilibrium constant K for the reference reaction from the apparent equilibrium constant K', or vice versa, one needs the equilibrium constants for the binding of H^+ and for the relevant metal ions to ATP^{4-}, ADP^{3-}, and HPO_4^{2-}.

To avoid confusion between the two different types of equilibrium constants (K' and K) and to avoid ambiguity about whether specific species or sums of species are intended, the word "ammonia," for example, rather than NH_3 or NH_4^+, is used for total ammonia, and chemical formulas are used for specific chemical species. Other substances such as carbon dioxide (CO_2, HCO_3^-, and CO_3^{2-}), and phosphate ($H_2PO_4^-$, HPO_4^{2-}, and PO_4^{3-}) are treated in the same manner. Exceptions are made for water, which is always written as H_2O, and for gaseous hydrogen and oxygen, which are written as H_2(g) and O_2(g), respectively.

For symmetrical reactions, there is no concern about the units used to calculate the value of an equilibrium constant. However, care must be exercised for reactions that are not symmetrical. In such cases, the units "mol dm^{-3}" have been used for all concentrations. As stated above, a c^o (1 mol dm^{-3}) is then used to make all equilibrium constants dimensionless.

All substances are assumed to be in aqueous solutions unless specified otherwise.

Values of $\Delta_r G'^o$ and K' can also be calculated for many biochemical reactions by using the table "Standard Transformed Gibbs Energies of Formation for Biochemical Reactants" in this section.

Abbreviations

ADP: adenosine 5′-diphosphate
AMP: adenosine 5′-monophosphate
ATP: adenosine 5′-triphosphate
CoA: coenzyme A
GDP: guanosine 5′-diphosphate
GMP: guanosine 5′-monophosphate
GTP: guanosine 5′-triphosphate
IDP: inosine 5′-diphosphate
IMP: inosine 5′-monophosphate
ITP: inosine 5′-triphosphate
NAD_{ox}: β-nicotinamide-adenine dinucleotide, oxidized form
NAD_{red}: β-nicotinamide-adenine dinucleotide, reduced form
$NADP_{ox}$: β-nicotinamide-adenine dinucleotide phosphate, oxidized form
$NADP_{red}$: β-nicotinamide-adenine dinucleotide phosphate, reduced form
UDP: uridine 5′-diphosphate
UTP: uridine 5′-triphosphate

References

1. Webb, E.C., *Enzyme Nomenclature 1992*, Academic Press, New York, 1992. See also http://www.chem.qmul.ac.uk/iubmb/enzyme/
2. Akers, D.L., and Goldberg, R.N., *Mathematica J.*, 8, 86–113 (2001).
3. Alberty, R.A., *J. Biol. Chem.*, 243, 1337–1343, 1969.
4. Alberty, R.A., *Thermodynamics of Biochemical Reactions*, Wiley-Interscience, Hoboken, NJ, 2003.
5. Goldberg, R.N., Tewari, Y.B., Bell, D., Fazio, K., and Anderson, E., *J. Phys. Chem. Ref. Data*, 22, 515-582 1993.
6. Goldberg, R.N., and Tewari, Y.B., *J. Phys. Chem. Ref. Data*, 23, 547-617, 1994.
7. Goldberg, R.N., and Tewari, Y.B., *J. Phys. Chem. Ref. Data*, 23, 1035-1103, 1994.
8. Goldberg, R.N., and Tewari, Y.B., *J. Phys. Chem. Ref. Data*, 24, 1669-1698, 1995.
9. Goldberg, R.N., and Tewari, Y.B., *J. Phys. Chem. Ref. Data*, 24, 1765-1801, 1995.
10. Goldberg, R.N., *J. Phys. Chem. Ref. Data*, 28, 931–965, 1999.
11. Goldberg, R.N., Tewari, Y.B., and Bhat, T.N., *Bioinformatics*, 20, 2874–2877, 2004; http://xpdb.nist.gov/enzyme_thermodynamics/
12. Goldberg, R.N., Tewari, Y.B., and Bhat, T.N., *J. Phys. Chem. Ref. Data*, 36, 1347–1397, 2007.

Apparent Equilibrium Constants for Enzyme-Catalyzed Reactions

Reaction	K'	Enzyme Commission number	T/K	pH	I mol dm^{-3}	pMg
benzyl alcohol + NAD$_{ox}$ = benzaldehyde + NAD$_{red}$	$9.8\cdot10^{-4}$	1.1.1.1	298.15	7.5		
1-butanol + NAD$_{ox}$ = butanal + NAD$_{red}$	$1.8\cdot10^{-3}$	1.1.1.1	298.15	8.3		
cyclohexanol + NAD$_{ox}$ = cyclohexanone + NADH$_{red}$	0.09	1.1.1.1	298.15	7.2		
1-hexanol + NAD$_{ox}$ = hexanal + NAD$_{red}$	$2.87\cdot10^{-3}$	1.1.1.1	298.15	8.3		
1-octanol + NAD$_{ox}$ = octanal + NAD$_{red}$	1.1·10–3	1.1.1.1	298.15	8.3		
L-homoserine + NADP$_{ox}$ = L-aspartate 4-semialdehyde + NADP$_{red}$	$6.3\cdot10^{-4}$	1.1.1.3	298.15	7.9		
xylitol + NAD$_{ox}$ = L-xylulose + NAD$_{red}$	$2.97\cdot10^{-4}$	1.1.1.10	298.15	7.00		
D-sorbitol + NAD$_{ox}$ = D-fructose + NAD$_{red}$	0.032	1.1.1.14	298.15	7.0		
quinate + NAD$_{ox}$ = 5-dehydroquinate + NAD$_{red}$	$4.61\cdot10^{-3}$	1.1.1.24	305.15	7.2		
shikimate + NADP$_{ox}$ = 5-dehydroshikimate + NADP$_{red}$	0.036	1.1.1.25	303.15	7.0		
2-hydroxybutanoate + NAD$_{ox}$ = 2-oxobutanoate + NAD$_{red}$	$3.0\cdot10^{-3}$	1.1.1.27	298.65	8.0		
(*R*)-3-hydroxybutanoate + NAD$_{ox}$ = 3-oxobutanoate + NAD$_{red}$	$1.9\cdot10^{-3}$	1.1.1.30	298.15	7.0		
D-glucose 6-phosphate + NADP$_{ox}$ = D-glucono-1,5-lactone 6-phosphate + NADP$_{red}$	1.50	1.1.1.49	301.15	6.40		
5α-androstane-3α-ol-17-one + NAD$_{ox}$ = 5α-androstane-3,17-dione + NAD$_{red}$	0.058	1.1.1.50	298.15	7.0		
5α-pregnane-3α,17α,21-triol-20-one + NAD$_{ox}$ = 5α-pregnane-17α,21-diol-3,20-dione + NAD$_{red}$	0.0113	1.1.1.50	298.15	7.0		
5α-androstane-3β,17α-diol + NAD$_{ox}$ = 5α-androstane-17α-ol-3-one + NAD$_{red}$	0.0211	1.1.1.51	298.15	7.0		
4-androstene-17β-ol-3-one + NAD$_{ox}$ = 4-androstene-3,17-dione + NAD$_{red}$	0.378	1.1.1.51	298.15	7.0		
1,2-propanediol + NADP$_{ox}$ =L -lactaldehyde + NADP$_{red}$	$6.0\cdot10^{-5}$	1.1.1.55	298.15	8.4		
ribitol + NAD$_{ox}$ = D-ribulose + NAD$_{red}$	$3.1\cdot10^{-3}$	1.1.1.56	310.15	7.4		
3-hydroxypropanoate + NAD$_{ox}$ = 3-oxopropanoate + NAD$_{red}$	$9.0\cdot10^{-3}$	1.1.1.59	298.15	9.0		
estradiol-17β + NAD$_{ox}$ = estrone + NAD$_{red}$	0.18	1.1.1.62	298.15	7.00		
benzyl alcohol + NAD$_{ox}$ = benzaldehyde + NAD$_{red}$	0.097	1.1.1.90	300.15	9.5		
L-carnitine + NAD$_{ox}$ = 3-dehydrocarnitine + NAD$_{red}$	$1.3\cdot10^{-4}$	1.1.1.108	303.15	7.0		
L-threonate + NAD$_{ox}$ = 3-oxo-L-threonate + NAD$_{red}$	$3.42\cdot10^{-4}$	1.1.1.129	298.15	7.0		
prostaglandin E$_1$ + NAD$_{ox}$ = 15-oxo-prostaglandin E$_1$ + NAD$_{red}$	0.65	1.1.1.141	298.15	7.0		
7,8-dihydrobiopterin + NADP$_{ox}$ = sepiapterin + NADP$_{red}$	0.045	1.1.1.153	298.15	8.0		
glycine + acetaldehyde = L-threonine	56	2.1.2.1	310.15	7.6		
sedoheptulose 7-phosphate + D-glyceraldehyde 3-phosphate = D-ribose 5-phosphate + D-xylulose 5-phosphate	0.48	2.2.1.1	311.15	7.0	0.25	3.0
acetyl-CoA + choline = CoA + *O*-acetylcholine	1.60	2.3.1.7	298.15	7.0	0.25	
acetyl-CoA + acyl-carrier protein = CoA + acetyl-[acyl-carrier protein]	2.09	2.3.1.38	311.15	6.5		
UDPglucose + D-fructose = UDP + sucrose	6.7	2.4.1.13	298.15	7.5		
cellobiose + orthophosphate = D-glucose + α-D-glucose 1-phosphate	0.23	2.4.1.20	310.15	7.0		
laminaritriose + orthophosphate = laminaribiose + α-D-glucose 1-phosphate	0.26	2.4.1.31	310.15	6.5		
α,α-trehalose + orthophosphate = D-glucose + β-D-glucose 1-phosphate	0.24	2.4.1.64	310.15	7.0		
UDPglucose + sinapate = UDP + 1-sinapoyl-D-glucose	0.21	2.4.1.120	303.15	6.0		
inosine + orthophosphate = hypoxanthine + α-D-ribose 1-phosphate	0.0164	2.4.2.1	311.15	7.0	0.25	3.0
xanthosine + orthophosphate = xanthine + α-D-ribose 1-phosphate	0.0156	2.4.2.1	311.15	7.0	0.25	3.0
uridine + orthophosphate = uracil + α-D-ribose 1-phosphate	0.44	2.4.2.2	310.15	7.0		
adenine + 5-phospho-α-D-ribose 1-diphosphate = AMP + pyrophosphate	$2\cdot10^{3}$	2.4.2.7	311.15	7.4	0.25	3.0
GMP + hypoxanthine = IMP + guanine	0.38	2.4.2.8	310.15	7.4		
guanine + 5-phospho-α-D-ribose 1-diphosphate = GMP + pyrophosphate	$1\cdot10^{5}$	2.4.2.8	311.15	7.4	0.25	3.0
hypoxanthine + 5-phospho-α-D-ribose 1-diphosphate = IMP + pyrophosphate	$1\cdot10^{5}$	2.4.2.8	311.15	7.4	0.25	3.0
ATP + ammonium carbamate = ADP + carbamoyl phosphate	0.042	2.7.2.2	283.15	9.4		
ATP + creatine = ADP + phosphocreatine	$5.78\cdot10^{-3}$	2.7.3.2	310.15	7.11	0.25	2.47
ATP + L-arginine = ADP + N$^{\omega}$-phospho-L-arginine	0.10	2.7.3.3	285.15	7.25		
ATP + sulfate = adenosine 5′-phosphosulfate + pyrophosphate	$4\cdot10^{-8}$	2.7.7.4	303.15	7.5		
UTP + α-D-glucose 1-phosphate = pyrophosphate + UDPglucose	0.48	2.7.7.9	310.15	8.0		
succinyl-CoA + acetoacetate = succinate + acetoacetyl-CoA	$2.8\cdot10^{-3}$	2.8.3.5	303.15	7.0		
acetylcholine + H$_2$O = acetate + choline	$5.38\cdot10^{2}$	3.1.1.7	296.15	5.1		
IMP + H$_2$O = inosine + orthophosphate	$1.58\cdot10^{2}$	3.1.3.1	298.15	8.55	1.53	4.44
phosphorylcholine + H$_2$O = choline + orthophosphate	49.9	3.1.3.1	311.15	6.90		
3-*O*-β-D-galactopyranosyl-D-arabinose + H$_2$O = D-galactose + D-arabinose	56	3.1.3.1	308.15	7.0		
cytidine 2′:3′-(cyclic)phosphate + H$_2$O = cytidine 3′-monophosphate	$1.06\cdot10^{3}$	3.1.27.5	298.15	6.0		
isomaltose + H$_2$O = 2 D-glucose	17.2	3.2.1.3	298.15	5.65		
β-gentiobiose + H$_2$O = 2 D-glucose	17.7	3.2.1.21	298.15	5.65		
3-*O*-β-D-galactopyranosyl-D-arabinose + H$_2$O = D-galactose + D-arabinose	$1.04\cdot10^{2}$	3.2.1.23	298.15	5.65		
lactulose + H$_2$O = D-galactose + D-fructose	$1.28\cdot10^{2}$	3.2.1.23	298.15	5.65		
4′,5′-anhydroadenosine + H$_2$O = adenosine	0.48	3.3.1.1	310.15	7.0		

Reaction	K′	Enzyme Commission number	T/K	pH	I mol dm^{-3}	pMg
pteroylglutamate + H_2O = pteroate + L-glutamate	15.6	3.4.19.9	310.15	7.3		
N-acetyl-L-phenylalanine methyl ester + H_2O = *N*-acetyl-L-phenylalanine + methanol	$5.88 \cdot 10^{2}$	3.4.21.1	293.15	5.5		
hippurylanilide + H_2O = hippuric acid + aniline	11	3.4.22.2	312.15	5.0		
ammonium carbamate + H_2O = 2 ammonia + carbon dioxide	$1.92 \cdot 10^{3}$	3.5.1.5	293.15	6.5		
ampicillin + H_2O = 6-aminopenicillanic acid + D(−)-α-aminophenylacetic acid	0.013	3.5.1.11	298.15	5.0		
cephalexin + H_2O = 7-aminodeacetoxycephalosporanic acid + cephem-4-carboxylic acid	0.044	3.5.1.11	298.15	5.8		
cephaloridine + H_2O = 2-thienylacetic acid + 7-amino-3-(1-pyridyl-methyl)-3-cephem-4-carboxylic acid	0.015	3.5.1.11	298.15	5.0		
penicillin G + H_2O = 6-aminopenicillanic acid + phenylacetic acid	0.445	3.5.1.11	298.15	6.71		
N-acetyl-L-alanine + H_2O = acetate + L-alanine	7	3.5.1.14	298.15	6.0		
ampicillin + H_2O = ampicillinoic acid	95	3.5.2.6	282.35	5.55		
penicillin G + H_2O = penicillinoic acid	2.9	3.5.2.6	298.15	6.01		
cytidine + H_2O = uridine + ammonia	$1.03 \cdot 10^{4}$	3.5.4.5	298.15	7.00		
N^4-methylcytidine + H_2O = uridine + methylamine	$4.88 \cdot 10^{2}$	3.5.4.5	298.15	7.50		
5,10-methenyltetrahydrofolate + H_2O = 10-formyltetrahydrofolate	50	3.5.4.9	298.15	7.0		
ITP + oxaloacetate + H_2O = IDP + phospho*enol*pyruvate + carbon dioxide	12	4.1.1.32	303.15	7.6		
2-deoxy-D-ribose 5-phosphate = D-glyceraldehyde 3-phosphate + acetaldehyde	$2.5 \cdot 10^{-4}$	4.1.2.4	295.15	7.5		
6-phospho-2-dehydro-3-deoxy-D-gluconate = pyruvate + D-glyceraldehyde 3-phosphate	$1.2 \cdot 10^{-3}$	4.1.2.14	298.15	8.0	0.37	
L-fuculose 1-phosphate = glycerone phosphate + (*S*)-lactaldehyde	$4.6 \cdot 10^{-4}$	4.1.2.17	310.15	7.2		
L-rhamnulose 1-phosphate = glycerone phosphate + (*S*)-lactaldehyde	0.083	4.1.2.19	310.15	7.5		
isocitrate = succinate + glyoxylate	$2.3 \cdot 10^{-3}$	4.1.3.1	303.15	7.7		
(*S*)-2-methylmalate = acetate + pyruvate	0.151	4.1.3.22	298.15	7.4	0.845	
isocitrate = citrate	14.7	4.2.1.3	298.15	7.4		
3-dehydroquinate = 3-dehydroshikimate + H_2O	15	4.2.1.10	302.15	7.4		
(3*R*)-3-hydroxybutanoyl-CoA = *cis*-but-2-enoyl-CoA + H_2O	0.18	4.2.1.17	298.15	7.5		
indole + D-glyceraldehyde 3-phosphate = 1-(indol-3-yl)glycerol 3-phosphate	$1.2 \cdot 10^{4}$	4.2.1.20	298.15	7.54		
(*R*)-malate = maleate + H_2O	$4.88 \cdot 10^{-4}$	4.2.1.31	298.15	7.00	0.10	
(*R*)-2-methylmalate = 2-methylmaleate + H_2O	0.0962	4.2.1.35	298.15	7.0	0.10	
D-glutamate = 5-oxo-D-proline + H_2O	24.3	4.2.1.48	293.4	7.9		
L-*threo*-3-methylaspartate = 2-methylfumarate + ammonia	0.238	4.3.1.2	298.15	7.9		
L-histidine = urocanate + ammonia	3.01	4.3.1.3	298.25	8.41	0.167	
L-phenylalanine = *trans*-cinnamate + ammonia	2.47	4.3.1.5	298.05	7.69		
ATP = adenosine 3′:5′-(cyclic)phosphate + diphosphate	0.065	4.6.1.1	298.2	7.0		
L,L-2,6-diaminoheptanedioate = *meso*-diaminoheptanedioate	1.9	5.1.1.7	310.15	7.0		
D-ribulose 5-phosphate = D-xylulose 5-phosphate	1.82	5.1.3.1	311.15	7.0	0.25	3.0
UDPglucose = UDPgalactose	0.33	5.1.3.2	298.15	8.7		
GDPmannose = GDP-L-galactose	0.52	5.1.3.18	310.15	8.0		
all-*trans*-retinal = 11-*cis*-retinal	0.05	5.2.1.3	309.15	7.0		
9-*cis*,12-*cis*-octadecadienoate = 9-*cis*,11-*trans*-octadecadienoate	61	5.2.1.5	308.15	7.0		
D-erythrose = D-erythrulose	2.3	5.3.1.2	308.15	5.8		
D-arabinose = D-ribulose	0.146	5.3.1.3	320.25	7.4		
L-fucose = L-fuculose	0.12	5.3.1.3	310.15	8.0		
L-arabinose = L-ribulose	0.11	5.3.1.4	298.15	7.0		
D-psicose = β-D-allose	2.15	5.3.1.4	317.25	7.4		
D-ribose 5-phosphate = D-ribulose 5-phosphate	0.83	5.3.1.6	311.15	7.0	0.25	3.0
D-rhamnose = D-rhamnulose	0.58	5.3.1.7	303.15	7.4		
D-mannose 6-phosphate = D-fructose 6-phosphate	0.99	5.3.1.8	298.15	8.50		
6-amino-D-glucose 6-phosphate = 6-amino-D-fructose 6-phosphate	0.202	5.3.1.9	278.85	8.7		
D-glucosamine 6-phosphate + H_2O = D-fructose 6-phosphate + ammonia	0.15	5.3.1.10	310.15	8.4		
D-lyxose = D-xylulose	0.23	5.3.1.15	298.15	7.0		
D-ribose = D-ribulose	0.391	5.3.1.20	313.15	7.4		
keto-phenylpyruvate = *enol*-phenylpyruvate	0.1	5.3.2.1	298.15	7.8		
L-lysine = (3*S*)-3,6-diaminohexanoate	5.3	5.4.3.2	303.15	7.7		
(*R*)-methylmalonyl-CoA = succinyl-CoA	23.1	5.4.99.2	298.15	7.4		
(−)-4-carboxymethyl-Δ^{α}-but-2-en-4-olide = *cis*,*trans*-hexadienedioate	4.0	5.5.1.1	303.15	8.0		
ATP + heptanoate + CoA = AMP + diphosphate + *n*-heptanoyl-CoA	1.11	6.2.1.2	311.15	8.0		
GTP + succinate + CoA = GDP + phosphate + succinyl-CoA	1.68	6.2.1.4	298.15	7.15	0.25	2.91
GTP + IMP + L-aspartate = GDP + phosphate + adenylosuccinate	2.9	6.3.4.4	310.15	8.0		
ATP + L-citrulline + L-aspartate = AMP + diphosphate + L-arginosuccinate	2.14	6.3.4.5	311.15	6.91		
ATP + propanoyl-CoA + carbon dioxide = ADP + phosphate + (*S*)-methylmalonyl-CoA	$8.1 \cdot 10^{-3}$	6.4.1.3	310.15	8.15		

APPARENT EQUILIBRIUM THERMODYNAMICS OF PROTEIN-LIGAND BINDING REACTIONS

Mark A. Williams

These tables contain values of the observed thermodynamics properties of the binding reactions of selected proteins with chemically related series of small molecule ligands at a specified temperature (T) and pH.

The standard Gibbs free energy change, $\Delta_b G'^{\circ}$, upon a ligand, L, binding a protein, P, determines the relative populations of protein, ligand and protein:ligand complex (P:L) at equilibrium. The apparent association constant (K_b') for the binding equilibrium is defined by

$$K_b' = [\mathrm{P{:}L}]\, c^{\circ} / [\mathrm{P}][\mathrm{L}]$$

and the Gibbs energy determined from K_b' via

$$\Delta_b G'^{\circ} = -RT \ln(K_b')$$

where the molar gas constant R = 8.3144598 J K^{-1} mol^{-1}, and c° is the standard state concentration of unit molarity (1 mol dm^{-3}). The standard enthalpy change of the binding reaction ($\Delta_b H'^{\circ}$) is the heat released (negative) or taken up (positive) upon forming the complex. All reactions reported here are carried out at a constant 1 atmosphere pressure (0.1 MPa). Under these conditions, the standard entropy change, $\Delta_b S'^{\circ}$, associated with a reaction is determined by the Gibbs-Helmholtz relation,

$$\Delta_b G'^{\circ} = \Delta_b H'^{\circ} - T\,\Delta_b S'^{\circ}$$

All experimental data reported here are obtained by isothermal titration microcalorimetry (ITC); a method that takes advantage of the fact that almost all binding reactions are accompanied by a measurable exchange of heat with their environment. An ITC experiment determines both the Gibbs energy and enthalpy change of a reaction by analysis of a single titration. In the most common experimental arrangement, a titration of ligand against protein is performed in a series of small increments. At each increment, the ligand:protein ratio is increased, and the net heat change arising from the shift in the binding equilibrium is measured. As the binding site of the protein becomes saturated, the incremental heat change diminishes to zero. $\Delta_b H'^{\circ}$ is determined from the sum total heat change at saturation and the known concentrations of reactants. $\Delta_b G'^{\circ}$ is determined from the dependence of the incremental heat (which corresponds to the fractional saturation) on the ligand:protein concentration ratio according to the analysis described in Ref. 1. For very high affinity interactions, almost all added ligand binds at each increment and the protein will abruptly saturate at a 1:1 concentration ratio making the data difficult to analyze to obtain accurate values for $\Delta_b G'^{\circ}$. In such circumstances, the binding affinity of the ligand of interest is artefactually diminished in experiments by competition with a saturating excess of a lower affinity ligand that binds to the same site on the protein. The ligand of interest is titrated to displace the competitor, and the observed thermodynamics are then corrected for the known thermodynamics of the competitor binding according to procedures in Ref. 2. Those reactions carried out in this competitive manner are indicated in the tables.

Biochemical binding reactions may be complex events. An idealized scenario of two rigid, chemically distinct, molecular species simply coming into close proximity is rarely the case. The idealized binding event is frequently accompanied by changes in other conformational or chemical equilibria involving the protein or ligand, e.g., due to only one conformation being able to bind or due to altered interactions with other solution components between the bound and free states. The apparent binding thermodynamics reported is for the total of all processes that necessarily accompany the binding. Here we also report the buffer composition in which the reactions were carried out as it is possible that altered interactions with these components affect the observations. Within each selected protein-ligand series, protein and ligands have similar bound conformations and experiments have been carried out under identical or very similar conditions meaning that, in so far as possible, differences in the thermodynamic properties between compounds in a series reflect differences between the ligands and in their interactions with the protein only and are more amenable to structural interpretation (Ref. 20).

Proteins and many ligands contain a variety of titratable acidic and basic groups, and the fractional protonation state of those groups may change upon binding if the protonated state of each group is stabilized or destabilized by interactions between the ligand and protein. Such direct effects of binding are a necessary consequence of the interaction and are appropriately included in the apparent thermodynamic properties. However, any net change in protonation of protein and ligand is coupled to change in protonation of the buffer that introduces an artefactual buffer-dependent heat of ionization (see "Thermodynamic Quantities for the Ionization Reactions of Buffers in Water" in this section). Heats of reaction are consequently corrected for buffer ionization where appropriate.

The amino acid residue sequence of the protein is given by reference to the standard translated gene sequence found in the UNIPROT database, Ref. 3. Any differences from the standard sequence are noted.

References

1. Wiseman, T., Williston, S., Brandts, J. F., and Lin, L. N., *Anal. Biochem.*, 179, 131, 1989.
2. Sigurskjold, B.W., *Anal. Biochem.*, 277, 260, 2000.
3. The Uniprot Consortium, *Nucl. Acids Res.*, 43, D204, 2013.
4. Özen, C. and Serpersu, E. H., *Biochemistry*, 43, 14667, 2004.
5. Martin, M. P., Zhu, J. Y., Lawrence, H. R., Pireddu, R., Luo, Y., Alam, R., Ozcan, S., Sebti, S. M., Lawrence, N. J., and Schonbrunn, E., *ACS Chem. Biol.*, 7, 698, 2012.
6. Gloster, T. M., Meloncelli, P., Stick, R. V., Zechel, D., Vasella, A., and Davies, G. J., *J. Am. Chem. Soc.*, 129, 2345, 2007.
7. Zechel, D. L., Boraston, A. B., Gloster, T., Boraston, C. M., Macdonald, J. M., Tilbrook, D. M. G., Stick, R. V., and Davies, G., *J. Am. Chem. Soc.*, 125, 14313, 2003.
8. Reddy, S. G., Scapin, G., and Blanchard, J. S., *Biochemistry*, 35, 13294, 1996.
9. Sharrow, S. D., Novotny, M. V., and Stone, M. J., *Biochemistry*, 42, 6302, 2003.
10. Bingham, R. J., Findlay, J. B. C., Hsieh, S. -Y., Kalverda, A. P., Kjellberg, A., Perazzolo, C., Phillips, S. E. V., Seshadri, K., Trinh, C. H., Turnbull, W. B., Bodenhausen, G., and Homans, S.W., *J. Am. Chem. Soc.*, 126, 1675, 2004.
11. Malham, R., Johnstone, S., Bingham, R. J., Barratt, E., Phillips, S. E. V., Laughton, C. A., and Homans, S. W., *J. Am. Chem. Soc.*, 127, 17061, 2005.
12. Barratt, E., Bingham, R. J., Warner, D. J., Laughton, C. A., Phillips, S. E. V., and Homans, S. W., *J. Am. Chem. Soc.*, 127, 11827, 2005.
13. Payne, J. W., Grail, B. M., Gupta, S., Ladbury, J. E., Marshall, N. J., O'Brien, R., and Payne, G. M., *Arch. Biochem. Biophys.*, 384, 9, 2000.

14. Sleigh, S. H., Seavers, P. R., Wilkinson, A. J., Ladbury, J. E., and Tame, J. R. H., *J. Mol. Biol.*, 291, 393, 1999.
15. Baum, B., Mohamed, M., Zayed, M., Gerlach, C., Heine, A., Hangauer, D., and Klebe, G. J., *Mol. Biol.*, 390, 56, 2009.
16. Baum, B., Muley, L., Heine, A., Smolinski, M., Hangauer, D., and Klebe, G. J., *Mol. Biol.*, 391, 552, 2009.
17. Beila, A., Sielaff, F., Terwesten, F., Heine, A., Steinmetzer, T., and Klebe, G., *J. Med. Chem.*, 55, 6094, 2012.
18. Talhout, R. and Engberts, J. B., *Eur. J. Biochem.*, 268, 1554, 2001.
19. Talhout, R., Villa, A., Mark, A. E., and Engberts, J. B., *J. Am. Chem. Soc.*, 125, 10570, 2003.
20. Olsson, T. S. G., Williams, M. A., Pitt, W. R., and Ladbury, J. E., *J. Mol. Biol.*, 384, 1002, 2008.

Description of Protein Systems

In the table following this section, thermodynamic data for a set of widely-studied proteins binding to a variety of ligands are presented. Below are given more detailed descriptions of the proteins as well as structural and interaction details.

Enterococcus faecalis Aminoglycoside 3'-phosphotransferase

Uniprot ID: P0A3Y5

Aminoglycoside 3'-phosphtransferase is able to phosphorylate many aminoglycoside antibiotics rendering them inactive. The reaction proceeds by transferring a phosphate from ATP bound in a neighboring site to the antibiotic, e.g., ATP + kanamycin = ADP + kanamycin 3'-phosphate. The reaction is dependent on Mg^{2+}. Replacing magnesium with calcium ions leads to formation of the ternary ATP:aminoglycoside:enzyme complex, but no product is formed. The enzyme is monomeric, and crystallographic analysis shows that it undergoes only local structural change on ATP and/or aminoglycoside binding. The thermodynamics of formation of the binary complex by titration of aminoglycoside to enzyme and the ternary complex by titration of aminoglycoside into enzyme in the presence of Ca^{2+} and ATP have been measured. Data below are corrected for coupled protonation of the buffer according to procedures described in Ref. 4.

Homo sapiens Aurora A kinase

Uniprot ID: O14965

Aurora A kinase transfers phosphate from ATP to serine and threonine residues in other proteins in the context of specific amino acid sequences. Aurora A activity regulates processes in normal cell division. Increased concentrations, and thus activity, of Aurora A in cells leads to aberrant cell division and cancer. Consequently, inhibitors of its action may be useful as anticancer drugs. The series of (bis)anilinopyrimidine inhibitors has a common binding mode that prevents a conformational change required for enzyme activity.

Thermotoga maritima β-glucosidase A

Uniprot ID: Q08638

Glycoside hydrolases are a large family of related enzymes that act to degrade oligosaccharides in a variety of extracellular digestive and intracellular metabolic processes. Several members of the family are implicated in diseases. β-glucosidases attack the β-1-4 linkage of nonreducing (acetal) β-D-glucosyl residues releasing β-D-glucose. β-glucosidase A is an experimentally tractable and stable member of the family that is used to study the binding of a series of inhibitors that mimic the transition state of the reaction. At the pH of these experiments, the binding of 1-deoxynojirimycin and isofagomine are associated with the net release of a single proton upon binding (Ref. 6). This protonation event is thought to be associated with interaction with the glutamic acid, which acts as the general base in the catalytic mechanism. Given the high degree of structural similarity between the ligand complexes (Ref. 7), it is assumed all of the compounds in the series below undergo this same net protonation. Consequently, all of the enthalpies in this series are corrected here by -3.38 kJ mol^{-1} to account for the small heat of ionization of the citrate buffer.

Escherichia coli Dihydrodipicolinate reductase

Uniprot ID: P04036

Dihydrodipicolinate reductase catalyzes the cofactor-dependent reaction, NAD(P)H + dihydropicolinic acid = tetrahydropicolinic acid + $NAD(P)^+$, and is an essential component of the biosynthetic pathway of L-lysine and meso-diaminopimelic acid in bacteria. Given the requirement for these amino acids in, respectively, protein synthesis and the formation of peptidoglycan in the bacterial cell wall, inhibition of this enzyme is considered to be a possible therapeutic strategy. The native enzyme is a homotetramer. There is no evidence for cooperativity of binding of the cofactor and thermodynamic data are analyzed assuming that the four sites are equivalent. The thermodynamics of binding of the cofactor(s) and several analogs have been determined in the absence of the dihydropicolinic acid substrate.

Mus musculus Major urinary protein

Uniprot ID: P02762

Mouse major urinary proteins are a family of monomeric sequence-similar proteins that bind pheromones that are released from the urine of male mice. These pheromones affect the sexual behavior of females. The pheromone binding site is a deep pocket lined with nonpolar amino acids and has consequently been used as an archetypal hydrophobic binding site to investigate factors influencing hydrophobic small-molecule binding.

Escherichia coli Periplasmic dipeptide binding protein DppA

Uniprot ID: P23847

DppA is a component of the dipeptide ATP-binding cassette transporter, a multiprotein complex that is responsible for transport of dipeptides across the cytoplasmic inner membrane of enteric bacteria.

Salmonella typhimurium/enterica Oligopeptide binding protein OppA

Uniprot ID: P06202

OppA is a component of the oligopeptide permease peptide transport system. It is evolutionarily related to the dipeptide binding proteins, but has a larger binding pocket capable of binding peptides up to five amino acids long with high affinity.

Homo sapiens Thrombin (in complex *Hirudo medicinalis* Hirudin)

Uniprot ID: P00734 (thrombin) & P01050 (hirudin)

Thrombin is a protease that is important in the coagulation of blood, in which process the protease specifically cleaves and activates fibrinogen, the clot-forming protein. Thrombin is a serine protease, in which the active site contains a serine-apartate-histidine catalytic triad, arranged such that the serine catalyzes hydrolysis of the peptide bond of a bound substrate protein or peptide. The structure of the active site is complementary to a limited number of peptide sequences. There is specificity for cleavage of peptides to the C-terminal side of a proline-arginine sequence. The series of ligands below contain or structurally mimic some of the features of this dipeptide sequence. The enzyme is inhibited by a peptide derived from the leech anticoagulant, hirudin and stabilized by Na^+ binding. Both these binding events are remote from the active site and there is no direct interaction with the titrated ligands.

Homo sapiens Thrombin (in complex *Hirudo medicinalis* Hirudin 2)

Uniprot ID: P00734 (thrombin) & P09945 (hirudin2)

For the ligands in this set of data, titrations in different buffers show a fractional net change in protonation state of the protein and ligand. Enthalpy changes have been corrected for buffer ionization effects as reported in Ref. 17. The thermodynamics of the very high affinity benzylsulfonyl-containing compounds have been determined via competition with the other two ligands.

Bos taurus Trypsin

Uniprot ID: P00760

Trypsin is a serine protease important in the digestion of proteins and also a useful model system due to it being readily purifiable from animal sources and its stability under laboratory conditions. Trypsin is evolutionarily related to thrombin and cleaves a similar but broader range of peptides, in this case preferentially to the C-terminal side of arginine or lysine residues. Benzamidine-based inhibitors mimic the arginine side-chain interactions. These inhibitors bind as the benzamidinium ion to the carboxylate side-chain of an active-site aspartic acid. The enzyme is stabilized by Ca^{2+} ion binding, but this binding is remote from the active site with no direct interaction with substrates or inhibitors. Experiments with titrations in different buffers show no net change in protonation upon formation of the benzamadine complex at pH 8.0, although fractional protonation changes do occur at lower pH. Because of their chemical and structural similarity it is assumed that no protonation corrections are required for any of this series of compounds.

Thermodynamic Data of Protein-Ligand Reactions

Sequence modification	Ligand	$\Delta_b G°$/ kJ mol^{-1}	$\Delta_b H°$/ kJ mol^{-1}	$-T\Delta_b S°$/ kJ mol^{-1}	*T*/K	Ref.	Buffer composition	Alternative sytematic ligand name
***Enterococcus faecalis* Aminoglycoside 3'-phosphotransferase (P0A3Y5)**								
None	Amikacin	-23.8	-74.9	51.0	310.15	4	50 mM Tris, 100 mM KCl, pH 7.5	
	Kanamycin A	-31.0	-187.9	156.9	310.15	4	50 mM Tris, 100 mM KCl, pH 7.5	
	Kanamycin B	-38.5	-172.8	134.3	310.15	4	50 mM Tris, 100 mM KCl, pH 7.5	
	Lividomycin A	-36.0	-239.7	203.8	310.15	4	50 mM Tris, 100 mM KCl, pH 7.5	
	Neomycin B	-38.9	-233.0	194.1	310.15	4	50 mM Tris, 100 mM KCl, pH 7.5	
	Paramomycin I	-37.7	-183.7	146.0	310.15	4	50 mM Tris, 100 mM KCl, pH 7.5	
	Ribostamycin	-33.9	-107.1	73.2	310.15	4	50 mM Tris, 100 mM KCl, pH 7.5	
	Tobramycin	-36.4	-196.6	160.2	310.15	4	50 mM Tris, 100 mM KCl, pH 7.5	
	Amikacin	-26.4	-117.2	90.8	310.15	4	50 mM Tris, 100 mM KCl, 1.5 mM $CaCl_2$, 1 mM ATP, pH 7.5	
	Kanamycin A	-35.6	-129.7	94.1	310.15	4	50 mM Tris, 100 mM KCl, 1.5 mM $CaCl_2$, 1 mM ATP, pH 7.5	
	Kanamycin B	-39.3	-121.8	82.4	310.15	4	50 mM Tris, 100 mM KCl, 1.5 mM $CaCl_2$, 1 mM ATP, pH 7.5	
	Lividomycin A	-36.4	-153.1	116.7	310.15	4	50 mM Tris, 100 mM KCl, 1.5 mM $CaCl_2$, 1 mM ATP, pH 7.5	
	Neomycin B	-41.0	-151.9	110.9	310.15	4	50 mM Tris, 100 mM KCl, 1.5 mM $CaCl_2$, 1 mM ATP, pH 7.5	
	Paramomycin I	-40.6	-116.7	76.1	310.15	4	50 mM Tris, 100 mM KCl, 1.5 mM $CaCl_2$, 1 mM ATP, pH 7.5	
	Ribostamycin	-38.9	-60.7	21.8	310.15	4	50 mM Tris, 100 mM KCl, 1.5 mM $CaCl_2$, 1 mM ATP, pH 7.5	
	Tobramycin	-39.3	-128.4	89.1	310.15	4	50 mM Tris, 100 mM KCl, 1.5 mM $CaCl_2$, 1 mM ATP, pH 7.5	
***Homo sapiens* Aurora A kinase (O14965)**								
Residues 123-390 of the full-length protein with a T287D mutation	4-{[4-(Phenylamino)pyrimidin-2-yl] amino}benzoic acid	-42.3	-58.6	16.3	298.15	5	100 mM Na/K phosphate, pH 7.4	
	2-({2-[(4-Carboxyphenyl)amino] pyrimidin-4- yl}amino)benzoic acid	-42.6	-56.9	14.3	298.15	5	100 mM Na/K phosphate, pH 7.4	

Biochem

Sequence modification	Ligand	$\Delta_bG°$/ kJ mol-1	$\Delta_bH°$/ kJ mol-1	$-T\Delta_bS°$/ kJ mol-1	T/K	Ref.	Buffer composition	Alternative sytematic ligand name
	4-[[4-(Biphenyl2-ylamino) pyrimidin-2-yl]amino}benzoic acid	-37.2	-13.7	-23.5	298.15	5	100 mM Na/K phosphate, pH 7.4	
	4-[(4-{[2-(Trifluoromethyl)phenyl]amino} pyrimidin- 2-yl)amino]benzoic acid	-41.7	-59.3	17.6	298.15	5	100 mM Na/K phosphate, pH 7.4	
	4-[(4-{[2-(Trifluoromethoxy)phenyl] amino}pyrimidin- 2-yl)amino]benzoic acid	-42.2	-46.0	3.8	298.15	5	100 mM Na/K phosphate, pH 7.4	
	4-({4-[(2-Fluorophenyl)amino] pyrimidin-2- yl}amino)benzoic acid	-44.5	-70.1	25.7	298.15	5	100 mM Na/K phosphate, pH 7.4	
	4-({4-[(2-Chlorophenyl)amino] pyrimidin-2- yl}amino)benzoic acid	-44.7	-71.1	26.5	298.15	5	100 mM Na/K phosphate, pH 7.4	
	4-({4-[(2-Bromophenyl)amino] pyrimidin-2-yl}amino)benzoic acid	-45.0	-62.3	17.3	298.15	5	100 mM Na/K phosphate, pH 7.4	
	4-({4-[(2-Cyanophenyl)amino] pyrimidin-2-yl}amino)benzoic acid	-43.4	-65.3	21.9	298.15	5		
***Thermotoga maritima* β-glucosidase A (Q08638)**								
Residues 2-446 of the full-length protein with a N-terminal hexa-His tag of sequence MGSSHHHHHHSSGLVPRGSHMAS	1-Deoxynojirimycin	-27.9	-22.6	-5.3	298.15	7	100 mM Na citrate, pH 5.8	(2*R*,3*R*,4*R*,5*S*)-2-(Hydroxymethyl) piperidine-3,4,5-triol
	Isofagomine	-41.6	-29.6	-12.0	298.15	7	100 mM Na citrate, pH 5.8	(3*R*,4*R*,5*R*)-5-(Hydroxymethyl) piperidine-3,4-diol
	Noeuromycin	-37.9	-41.4	3.5	298.15	7	100 mM Na citrate, pH 5.8	(2*R*,3*S*,4*R*,5*R*)-5-(Hydroxymethyl) piperidine-2,3,4-triol
	Tetrahydrooxazine	-35.8	-47.4	11.6	298.15	7	100 mM Na citrate, pH 5.8	(4*R*,5*S*,6*R*)-6-(Hydroxymethyl) oxazinane-4,5-diol
	Azafagomine	-41.0	-49.4	8.4	298.15	7	100 mM Na citrate, pH 5.8	(3*R*,4*R*,5*R*)-3-(Hydroxymethyl)-3,4,5,6-tetrahydropyridazine-4,5-diol
	Castanospermine	-32.4	-28.9	-3.5	298.15	7	100 mM Na citrate, pH 5.8	(1*S*,6*S*,7*R*,8*R*,8a*R*)-1,2,3,5,6,7,8,8a-Octahydroindolizine-1,6,7,8-tetrol
	Calystegine B2	-31.3	-15.7	-15.6	298.15	7	100 mM Na citrate, pH 5.8	(1*R*,2*S*,3*R*,4*S*,5*R*)-8-Azabicyclo[3.2.1] octane-2,3,4,5-tetrol
	Isofagomine lactam	-37.3	-60.1	22.8	298.15	7	100 mM Na citrate, pH 5.8	(3*S*,4*R*,5*R*)-3,4-Dihydroxy-5-(hydroxymethyl)piperidin-2-one
	Gluco-hydroximolactam	-36.6	-32.8	-3.9	298.15	7	100 mM Na citrate, pH 5.8	(2*Z*,3*S*,4*S*,5*R*,6*R*)-2-Hydroxyimino-6-(hydroxymethyl)piperidine-3,4,5-triol
	Galacto-hydroximolactam	-34.0	-27.4	-6.6	298.15	7	100 mM Na citrate, pH 5.8	(3*S*,4*S*,5*S*,6*R*)-2-Hydroxyimino-6-(hydroxymethyl)piperidine-3,4,5-triol
	Glucotetrazole	-37.8	-50.7	13.0	298.15	7	100 mM Na citrate, pH 5.8	(5*R*,6*R*,7*S*,8*S*)-5-(Hydroxymethyl)-5,6,7,8-tetrahydro-[1,2,3,4] tetrazolo[5,1-f]pyridine-6,7,8-triol
	Glucoimidazole	-41.4	-40.9	-0.5	298.15	7	100 mM Na citrate, pH 5.8	(5*R*,6*R*,7*S*,8*S*)-5-(Hydroxymethyl)-5,6,7,8-tetrahydro-1*H*-imidazo[2,1-f] pyridin-4-ium-6,7,8-triol
	2-Phenylethyl-glucoimidazole	-45.8	-22.8	-23.0	298.15	7	100 mM Na citrate, pH 5.8	(5*R*,6*R*,7*S*,8*S*)-5-(Hydroxymethyl)-2-phenethyl-5,6,7,8-tetrahydro-1*H*-imidazo[2,1-f]pyridin-4-ium-6,7,8-triol
	2-Phenylaminomethylglucoimidazole	-45.5	-38.7	-6.8	298.15	7	100 mM Na citrate, pH 5.8	(5*R*,6*R*,7*S*,8*S*)-5-(Hydroxymethyl)-2-(phenylazanylmethyl)-5,6,7,8-tetrahydroimidazo[2,1-f]pyridine-6,7,8-triol
	2-Methoxycarbonylglucoimidazole	-40.7	-34.7	-6.0	298.15	7	100 mM Na citrate, pH 5.8	(5*R*,6*R*,7*S*,8*S*)-2-[(S)-Hydroxy-methoxy-methyl]-5-(hydroxymethyl)-5,6,7,8-tetrahydro-1*H*-imidazo[2,1-f] pyridin-4-ium-6,7,8-triol
	2-Methoxycarbonylmethylglucoimidazole	-41.8	-40.3	-1.5	298.15	7	100 mM Na citrate, pH 5.8	Methyl 2-[(5*R*,6*R*,7*S*,8*S*)-6,7,8-trihydroxy-5-(hydroxymethyl)-5,6,7,8-tetrahydro-1*H*-imidazo[1,2-a] pyridin-4-ium-2-yl]ethanoate
	2-Carboxylate-glucosimidazole	-36.3	-41.0	4.7	298.15	7	100 mM Na citrate, pH 5.8	(5*R*,6*R*,7*S*,8*S*)-6,7,8-Trihydroxy-5-(hydroxymethyl)-5,6,7,8-tetrahydro-1*H*-imidazo[1,2-a]pyridin-4-ium-2-carboxylic acid
	2-Carboxymethylglucosimidazole	-40.0	-48.2	8.2	298.15	7	100 mM Na citrate, pH 5.8	2-[(5*R*,6*R*,7*S*,8*S*)-6,7,8-Trihydroxy-5-(hydroxymethyl)-5,6,7,8-tetrahydro-1*H*-imidazo[1,2-a]pyridin-4-ium-2-yl] ethanoic acid
***Escherichia coli* Dihydrodipicolinate reductase (P04036)**								
None	β-Nicotinamide adenine dinucleotide 2'-phosphate, reduced	-32.7	-45.9	13.3	301.15	8	20 mM Hepes/KOH, pH 7.8	
	β-Nicotinamide hypoxanthine dinucleotide 2'-phosphate, reduced	-31.1	-35.7	4.6	301.15	8	20 mM Hepes/KOH, pH 7.8	
	β-Thionicotinamide adenine dinucleotide 2'-phosphate, reduced	-33.0	-53.1	20.1	301.15	8	20 mM Hepes/KOH, pH 7.8	

Sequence modification	Ligand	$\Delta_b G'°$/ kJ mol⁻¹	$\Delta_b H'°$/ kJ mol⁻¹	$-T\Delta_b S'°$/ kJ mol⁻¹	T/K	Ref.	Buffer composition	Alternative sytematic ligand name
	β-Nicotinamide adenine dinucleotide 3'-phosphate, reduced	-30.5	-51.0	20.5	301.15	8	20 mM Hepes/KOH, pH 7.8	
	β-Nicotinamide adenine dinucleotide, reduced	-36.6	-37.4	0.8	301.15	8	20 mM Hepes/KOH, pH 7.8	
	β-Nicotinamide hypoxanthine dinucleotide, reduced	-33.6	-35.0	1.4	301.15	8	20 mM Hepes/KOH, pH 7.8	
	β-Thionicotinamide adenine dinucleotide, reduced	-37.7	-44.1	6.4	301.15	8	20 mM Hepes/KOH, pH 7.8	
	3-Acetylpyridine adenine dinucleotide, reduced	-37.7	-34.4	-3.2	301.15	8	20 mM Hepes/KOH, pH 7.8	
***Mus musculus* Major urinary protein (P02762)**								
Residues 19-180 of the full-length protein only	2-*sec*-Butyl-4,5-dihydrothiazole	-35.0	-46.9	11.9	298.75	9	10 mM Na/K phosphate, pH 6.3	2-(1-Methylpropyl)-4,5-dihydro-1,3-thiazole
	2-Isobutyl-4,5-dihydrothiazole	-37.8	-49.8	12.0	298.85	9	10 mM Na/K phosphate, pH 6.3	2-(2-Methylpropyl)-4,5-dihydro-1,3-thiazole
	2-Propyl-4,5-dihydrothiazole	-34.1	-44.8	10.7	298.65	9	10 mM Na/K phosphate, pH 6.3	2-Propyl-4,5-dihydro-1,3-thiazole
	2-Isopropyl-4,5-dihydrothiazole	-32.7	-45.2	12.5	298.55	9	10 mM Na/K phosphate, pH 6.3	2-(1-Methylethyl)-4,5-dihydro-1,3-thiazole
	2-Ethyl-4,5-dihydrothiazole	-28.9	-41.8	12.9	298.55	9	10 mM Na/K phosphate, pH 6.3	2-Ethyl-4,5-dihydro-1,3-thiazole
	2-Methyl-4,5-dihydrothiazole	-23.4	-40.6	17.1	298.45	9	10 mM Na/K phosphate, pH 6.3	2-Methyl-4,5-dihydro-1,3-thiazole
Residues 19-180 of the full-length protein with a N-terminal hexa-His tag (MRGSHHHHHHGS) and an N68K mutation	2-Methoxy-3-isopropylpyrazine	-33.9	-44.5	10.6	308.00	10	Phosphate buffered saline, pH 7.4	2-Methoxy-3-(1-methylethyl)-pyrazine
	2-Methoxy-3-isobutylpyrazine	-38.5	-47.9	9.4	308.00	10	Phosphate buffered saline, pH 7.4	2-Methoxy-3-(2-methylpropyl)-pyrazine
	Pentan-1-ol	-23.1	-41.0	17.9	300.00	11	Phosphate buffered saline, pH 7.4	
	Hexan-1-ol	-28.3	-47.6	19.3	300.00	11	Phosphate buffered saline, pH 7.4	
	Heptan-1-ol	-32.5	-53.4	20.9	300.00	11	Phosphate buffered saline, pH 7.4	
	Octan-1-ol	-35.6	-58.0	22.4	300.00	11	Phosphate buffered saline, pH 7.4	
	Nonan-1-ol	-38.6	-68.6	30.0	300.00	11	Phosphate buffered saline, pH 7.4	
Residues 19-180 of the full-length protein with a N-terminal hexa-His tag (MRGSHHHHHHGS) and N68K and Y138F mutations	2-Methoxy-3-isobutylpyrazine	-31.4	-35.3	3.9	298.00	12	Phosphate buffered saline, pH 7.4	2-Methoxy-3-(2-methylpropyl)-pyrazine
	2-Methoxy-3-isobutylpyrazine	-33.6	-35.9	2.3	308.00	12	Phosphate buffered saline, pH 7.4	2-Methoxy-3-(2-methylpropyl)-pyrazine
***Escherichia coli* Periplasmic dipeptide binding protein DppA (P23847)**								
None	L-Alanine-L-threonine	-44.4	-53.3	9.0	298.15	13	50 mM Na/K phosphate, pH 7.0	
	L-Alanine-L-valine	-44.0	-41.3	-2.7	298.15	13	50 mM Na/K phosphate, pH 7.0	
	L-Lysine-L-alanine	-43.5	-49.1	5.6	298.15	13	50 mM Na/K phosphate, pH 7.0	
	L-Alanine-L-alanine	-43.4	-56.1	12.7	298.15	13	50 mM Na/K phosphate, pH 7.0	
	L-Alanine-L-serine	-43.3	-52.6	9.3	298.15	13	50 mM Na/K phosphate, pH 7.0	
	L-Aanine-L-lysine	-43.1	-45.1	2.0	298.15	13	50 mM Na/K phosphate, pH 7.0	
	L-Alanine-L-leucine	-42.8	-31.9	-10.8	298.15	13	50 mM Na/K phosphate, pH 7.0	
	L-Alanine-L-isoleucine	-41.3	-37.3	-4.0	298.15	13	50 mM Na/K phosphate, pH 7.0	
	L-Alanine-L-phenylalanine	-41.1	-51.1	10.0	298.15	13	50 mM Na/K phosphate, pH 7.0	
	L-Serine-L-alanine	-41.1	-53.9	12.8	298.15	13	50 mM Na/K phosphate, pH 7.0	
	L-Threonine-L-serine	-40.5	-75.6	35.1	298.15	13	50 mM Na/K phosphate, pH 7.0	
	L-Alanine-L-tyrosine	-38.6	-51.8	13.1	298.15	13	50 mM Na/K phosphate, pH 7.0	
	L-Alanine-L-proline	-38.6	-37.3	-1.3	298.15	13	50 mM Na/K phosphate, pH 7.0	

Sequence modification	Ligand	$\Delta_b G°$/ kJ mol^{-1}	$\Delta_b H°$/ kJ mol^{-1}	$-T\Delta_b S°$/ kJ mol^{-1}	T/K	Ref.	Buffer composition	Alternative sytematic ligand name
	L-Alanine-L-tryptophan	-37.7	-46.4	8.7	298.15	13	50 mM Na/K phosphate, pH 7.0	
	L-Alanine-L-aspartate	-33.9	-34.9	1.0	298.15	13	50 mM Na/K phosphate, pH 7.0	
	L-Glycine-L-leucine	-33.8	-43.6	9.7	298.15	13	50 mM Na/K phosphate, pH 7.0	
	L-Glycine-L-tyrosine	-31.7	-35.5	3.8	298.15	13	50 mM Na/K phosphate, pH 7.0	
	L-Leucine-L-tryptophan	-30.6	-53.1	22.5	298.15	13	50 mM Na/K phosphate, pH 7.0	
	L-Aspartate-L-alanine	-29.7	-43.2	13.4	298.15	13	50 mM Na/K phosphate, pH 7.0	
Salmonella typhimurium/enterica **Oligopeptide binding protein OppA (P06202)**								
Residues 27-543 of the full-length protein only	L-Lysine-L-alanine-L-lysine	-41.4	20.1	-61.5	298.15	14	50 mM Sodium phosphate pH 7.0	
	L-Lysine-L-cysteine-L-lysine	-40.7	7.9	-48.6	298.15	14	50 mM Sodium phosphate pH 7.0	
	L-Lysine-L-aspartate-L-lysine	-29.8	8.1	-37.9	298.15	14	50 mM Sodium phosphate pH 7.0	
	L-Lysine-L-glutamate-L-lysine	-39.0	11.3	-50.3	298.15	14	50 mM Sodium phosphate pH 7.0	
	L-Lysine-L-phenylalanine-L-lysine	-41.5	22.0	-63.5	298.15	14	50 mM Sodium phosphate pH 7.0	
	L-Lysine-L-glycine-L-lysine	-33.6	14.1	-47.7	298.15	14	50 mM Sodium phosphate pH 7.0	
	L-Lysine-L-histidine-L-lysine	-39.3	20.6	-59.9	298.15	14	50 mM Sodium phosphate pH 7.0	
	L-Lysine-L-isoleucine-L-lysine	-38.2	20.5	-58.7	298.15	14	50 mM Sodium phosphate pH 7.0	
	L-Lysine-L-Lysine-L-lysine	-31.6	39.4	-71.0	298.15	14	50 mM Sodium phosphate pH 7.0	
	L-Lysine-L-leucine-L-lysine	-34.0	24.6	-58.6	298.15	14	50 mM Sodium phosphate pH 7.0	
	L-Lysine-L-methionine-L-lysine	-40.5	14.6	-55.1	298.15	14	50 mM Sodium phosphate pH 7.0	
	L-Lysine-L-asparagine-L-lysine	-40.2	7.7	-47.9	298.15	14	50 mM Sodium phosphate pH 7.0	
	L-Lysine-L-proline-L-lysine	-30.2	16.6	-46.8	298.15	14	50 mM Sodium phosphate pH 7.0	
	L-Lysine-L-glutamine-L-lysine	-42.4	11.4	-53.8	298.15	14	50 mM Sodium phosphate pH 7.0	
	L-Lysine-L-arginine-L-lysine	-33.8	36.0	-69.8	298.15	14	50 mM Sodium phosphate pH 7.0	
	L-Lysine-L-serine-L-lysine	-42.0	8.9	-50.9	298.15	14	50 mM Sodium phosphate pH 7.0	
	L-Lysine-L-threonine-L-lysine	-40.6	17.3	-57.9	298.15	14	50 mM Sodium phosphate pH 7.0	
	L-Lysine-L-valine-L-lysine	-41.9	22.4	-64.3	298.15	14	50 mM Sodium phosphate pH 7.0	
	L-Lysine-L-tryprophan-L-lysine	-39.3	29.3	-68.6	298.15	14	50 mM Sodium phosphate pH 7.0	
	L-Lysine-L-tyrosine-L-lysine	-37.6	20.7	-58.3	298.15	14	50 mM Sodium phosphate pH 7.0	
Homo sapiens **Thrombin (P00734) in complex** ***Hirudo medicinalis*** **Hirudin (P01050)**								
Residues 328-363 and 364-622 of full-length thrombin. Residues 54-64 of hirudin where residue 63 is a non-standard amino acid O-sulpho-L-tyrosine.	Titrations of three of the ligands in the series below in different buffers show no difference in observed enthalpy changes, implying that there is no net change in protonation upon formation of the complex . Because of their equivalent complement of ligand titratable groups and crystallographic analyses showing a high degree of structural similarity of the complexes, it is assumed that no buffer ionization correction is required for any of these ligands.							
	D-Phenylalanyl-*N*-(3-fluorobenzyl)-L-prolinamide	-31.2	-13.1	-18.1	298.15	15	50 mM Tris,100 mM NaCl, 2.5% Dimethylsulfoxide, 0.1% Polyethylene glycol 8000, pH 7.8	

Sequence modification	Ligand	$\Delta_b G°$/ kJ mol^{-1}	$\Delta_b H°$/ kJ mol^{-1}	$-T\Delta_b S°$/ kJ mol^{-1}	T/K	Ref.	Buffer composition	Alternative sytematic ligand name
	D-Phenylalanyl-N-(3-chlorobenzyl)-L-prolinamide	-35.4	-37.1	1.7	298.15	15	50 mM Tris,100 mM NaCl, 2.5% Dimethylsulfoxide, 0.1% Polyethylene glycol 8000, pH 7.8	
	D-Phenylalanyl-N-(3-bromobenzyl)-L-prolinamide	-35.8	-34.5	-1.3	298.15	15	50 mM Tris,100 mM NaCl, 2.5% Dimethylsulfoxide, 0.1% Polyethylene glycol 8000, pH 7.8	
	D-Phenylalanyl-N-(3-iodobenzyl)-L-prolinamide	-34.5	-38.0	3.5	298.15	15	50 mM Tris,100 mM NaCl, 2.5% Dimethylsulfoxide, 0.1% Polyethylene glycol 8000, pH 7.8	
	D-Phenylalanyl-N-(3-methylbenzyl)-L-prolinamide	-34.8	-28.5	-6.3	298.15	15	50 mM Tris,100 mM NaCl, 2.5% Dimethylsulfoxide, 0.1% Polyethylene glycol 8000, pH 7.8	
	D-Phenylalanyl-N-(3-ethylbenzyl)-L-prolinamide	-32.9	-16.5	-16.4	298.15	15	50 mM Tris,100 mM NaCl, 2.5% Dimethylsulfoxide, 0.1% Polyethylene glycol 8000, pH 7.8	
	D-Phenylalanyl-N-(4-benzenecarboximidamide)-L-prolinamide	-46.1	-40.1	-6.0	298.15	15	50 mM Tris,100 mM NaCl, 2.5% Dimethylsulfoxide, 0.1% Polyethylene glycol 8000, pH 7.8	
	bis(3-Phenyl)-D-alanyl-N-(4-benzenecarboximidamide)-L-prolinamide	-48.5	-47.5	-1.0	298.15	16	50 mM Tris,100 mM NaCl, 2.5% Dimethylsulfoxide, 0.1% Polyethylene glycol 8000, pH 7.8	
	Methyl-D-alanyl-N-(4-benzenecarboximidamide)-L-prolinamide	-40.1	-38.7	-1.4	298.15	16	50 mM Tris,100 mM NaCl, 2.5% Dimethylsulfoxide, 0.1% Polyethylene glycol 8000, pH 7.8	
	D-Phenylalanyl-N-benzyl-L-prolinamide	-31.7	-13.6	-18.1	298.15	15	50 mM Tris,100 mM NaCl, 2.5% Dimethylsulfoxide, 0.1% Polyethylene glycol 8000, pH 7.8	
	D-Phenylalanyl-N-(3,6-dichloro-benzyl)-L-prolinamide	-38.4	-41.3	2.9	298.15	15	50 mM Tris,100 mM NaCl, 2.5% Dimethylsulfoxide, 0.1% Polyethylene glycol 8000, pH 7.8	
	D-Phenylalanyl-N-((3-chloro-6-methyl)-benzyl)-L-prolinamide	-37.2	-33.5	-3.7	298.15	15	50 mM Tris,100 mM NaCl, 2.5% Dimethylsulfoxide, 0.1% Polyethylene glycol 8000, pH 7.8	
	D-Phenylalanyl-N-((3-chloro-6-fluoro)-benzyl)-L-prolinamide	-37.3	-41.0	3.7	298.15	15	50 mM Tris,100 mM NaCl, 2.5% Dimethylsulfoxide, 0.1% Polyethylene glycol 8000, pH 7.8	
	D-Phenylalanyl-N-(3,6-dimethylbenzyl)-L-prolinamide	-34.4	-31.9	-2.5	298.15	15	50 mM Tris,100 mM NaCl, 2.5% Dimethylsulfoxide, 0.1% Polyethylene glycol 8000, pH 7.8	
	D-Phenylalanyl-N-(3-chlorobenzyl)-L-prolinamide	-35.4	-37.1	1.7	298.15	16	50 mM Tris,100 mM NaCl, 2.5% Dimethylsulfoxide, 0.1% Polyethylene glycol 8000, pH 7.8	
	bis(3-Phenyl)-D-alanyl-N-(3-chlorobenzyl)-L-prolinamide	-39.6	-45.6	6.0	298.15	16	50 mM Tris,100 mM NaCl, 2.5% Dimethylsulfoxide, 0.1% Polyethylene glycol 8000, pH 7.8	

Sequence modification	Ligand	$\Delta_b G°$/ kJ mol⁻¹	$\Delta_b H°$/ kJ mol⁻¹	$-T\Delta_b S°$/ kJ mol⁻¹	T/K	Ref.	Buffer composition	Alternative sytematic ligand name
	3-Cyclohexyl-D-alanyl-*N*-(3-chlorobenzyl)-L-prolinamide	-38.9	-35.6	-3.3	298.15	16	50 mM Tris,100 mM NaCl, 2.5% Dimethylsulfoxide, 0.1% Polyethylene glycol 8000, pH 7.8	
	Methyl-D-alanyl-*N*-(3-chlorobenzyl)-L-prolinamide	-31.3	-33.5	2.2	298.15	16	50 mM Tris,100 mM NaCl, 2.5% Dimethylsulfoxide, 0.1% Polyethylene glycol 8000, pH 7.8	

***Homo sapiens* Thrombin (P00734) in complex *Hirudo medicinalis* Hirudin 2 (P09945)**

Sequence modification	Ligand	$\Delta_b G°$/ kJ mol⁻¹	$\Delta_b H°$/ kJ mol⁻¹	$-T\Delta_b S°$/ kJ mol⁻¹	T/K	Ref.	Buffer composition	Alternative sytematic ligand name
Residues 328-363 and 364-622 of full-length thrombin where residue N416 is gycosylated by N-acetyl-D-glucosamine. Residues 53-65 of hirudin where residue 63 is a non-standard amino acid O-sulpho-L-tyrosine.	For the ligands below titrations in different buffers show a fractional net change in protonation state of the protein and ligand. Enthalpy changes have been corrected for buffer ionization effects as reported in [17]. The thermodynamics of the very high affinity benzylsulfonyl containing compounds have been determined via competition with the other two ligands.							
	N-(Benzylsulfonyl)-glycyl-*N* [2 (aminomethyl)- 5-chlorobenzyl]-L-prolinamide	-47.8	-35.4	-12.4	298.15	17	50 mM Tris,100 mM NaCl, 3% Dimethylsulfoxide, 0.1% Polyethylene glycol 8000, pH 7.8	
	N-(Benzylsulfonyl)-D-alanyl-*N*-[2-(aminomethyl)- 5-chlorobenzyl]-L-prolinamide	-47.2	-32.1	-15.1	298.15	17	50 mM Tris,100 mM NaCl, 3% Dimethylsulfoxide, 0.1% Polyethylene glycol 8000, pH 7.8	
	N-(Benzylsulfonyl)-D-valyl-*N*-[2-(aminomethyl)- 5-chlorobenzyl]-L-prolinamide	-54.2	-36.2	-18.0	298.15	17	50 mM Tris,100 mM NaCl, 3% Dimethylsulfoxide, 0.1% Polyethylene glycol 8000, pH 7.8	
	N-(Benzylsulfonyl)-D-leucyl-*N*-[2-(aminomethyl)- 5-chlorobenzyl]-L-prolinamide	-54.3	-30.1	-24.2	298.15	17	50 mM Tris,100 mM NaCl, 3% Dimethylsulfoxide, 0.1% Polyethylene glycol 8000, pH 7.8	
	N-(Benzylsulfonyl)-3-cyclohexyl-D-alanyl-*N*-[2-(aminomethyl)-5-chlorobenzyl]-L-prolinamide	-54.5	-28.7	-25.8	298.15	17	50 mM Tris,100 mM NaCl, 3% Dimethylsulfoxide, 0.1% Polyethylene glycol 8000, pH 7.8	
	N-(Benzylsulfonyl)-glycyl-*N*-[4-carbamimidoylbenzyl]-L-prolinamide	-44.9	-14.3	-30.6	298.15	17	50 mM Tris,100 mM NaCl, 3% Dimethylsulfoxide, 0.1% Polyethylene glycol 8000, pH 7.8	
	N-(Benzylsulfonyl)-D-alanyl-*N*-[4-carbamimidoylbenzyl]-L-prolinamide	-42.6	-15.9	-26.7	298.15	17	50 mM Tris,100 mM NaCl, 3% Dimethylsulfoxide, 0.1% Polyethylene glycol 8000, pH 7.8	
	N-(Benzylsulfonyl)-D-valyl-*N*-[4-carbamimidoylbenzyl]-L-prolinamide	-47.1	-13.8	-33.3	298.15	17	50 mM Tris,100 mM NaCl, 3% Dimethylsulfoxide, 0.1% Polyethylene glycol 8000, pH 7.8	
	N-(Benzylsulfonyl)-D-leucyl-*N*-[4-carbamimidoylbenzyl]-L-prolinamide	-51.9	-10.4	-41.5	298.15	17	50 mM Tris,100 mM NaCl, 3% Dimethylsulfoxide, 0.1% Polyethylene glycol 8000, pH 7.8	
	N-(Benzylsulfonyl)-3-cyclohexyl-D-alanyl-*N*-[4-carbamimidoylbenzyl]-L-prolinamide	-53.7	-11.4	-42.3	298.15	17	50 mM Tris,100 mM NaCl, 3% Dimethylsulfoxide, 0.1% Polyethylene glycol 8000, pH 7.8	
	D-Phenylalanyl-*N*-(3,6-dichloro-benzyl)-L-prolinamide	-37.6	-37.5	-0.1	298.15	17	50 mM Tris,100 mM NaCl, 3% Dimethylsulfoxide, 0.1% Polyethylene glycol 8000, pH 7.8	

Sequence modification	Ligand	$\Delta_b G°$/ kJ mol⁻¹	$\Delta_b H°$/ kJ mol⁻¹	$-T\Delta_b S°$/ kJ mol⁻¹	*T*/K	Ref.	Buffer composition	Alternative sytematic ligand name
	N-(Benzylethyl)-L-phenylanalyl-*N*-[4-carbamimidoylbenzyl]-L-prolinamide	-37.9	-17.1	-20.8	298.15	17	50 mM Tris,100 mM NaCl, 3% Dimethylsulfoxide, 0.1% Polyethylene glycol 8000, pH 7.8	
***Bos taurus* Trypsin (P00760)**								
Residues 24-246 of the full-length protein only	Benzamdine	-26.6	-18.9	-7.7	298.25	18	50 mM Tris, 10 mM $CaCl_2$, pH 8.0	Benzenecarboximidamide
	4-Aminobenzamidine	-29.1	-26.9	-2.2	298.25	18	50 mM Tris, 10 mM $CaCl_2$, pH 8.0	4-Aminobenzenecarboximidamide
	4-Methoxybenzamidine	-25.3	-15.7	-9.6	298.25	18	50 mM Tris, 10 mM $CaCl_2$, pH 8.0	4-Methoxybenzenecarboximidamide
	4-Carboxamidobenzamidine	-23.9	-12.3	-11.6	298.25	18	50 mM Tris, 10 mM $CaCl_2$, pH 8.0	4-Carboxyamidobenzenecarboximidamide
	4-Methylbenzamidine	-27.6	-18.5	-9.1	298.25	18	50 mM Tris, 10 mM $CaCl_2$, pH 8.0	4-Methylbenzenecarboximidamide
	4-Ethylbenzamidine	-25.4	-13.9	-11.5	298.15	19	50 mM Tris, 10 mM $CaCl_2$, pH 8.0	4-Ethylbenzenecarboximidamide
	4-Propylbenzamidine	-25.7	-12.7	-13.0	298.15	19	50 mM Tris, 10 mM $CaCl_2$, pH 8.0	4-Propylbenzenecarboximidamide
	4-Isopropylbenzamidine	-22.7	-7.0	-15.7	298.15	19	50 mM Tris, 10 mM $CaCl_2$, pH 8.0	4-(1-Methylethyl) benzenecarboximidamide
	4-Butylbenzamidine	-26.2	-9.9	-16.3	298.15	19	50 mM Tris, 10 mM $CaCl_2$, pH 8.0	4-Butylbenzenecarboximidamide
	4-Pentylbenzamidine	-27.2	-9.9	-17.3	298.15	19	50 mM Tris, 10 mM $CaCl_2$, pH 8.0	4-Pentylbenzenecarboximidamide
	4-Hexylbenzamidine	-29.2	-10.6	-18.6	298.15	19	50 mM Tris, 10 mM $CaCl_2$, pH 8.0	4-Hexylbenzenecarboximidamide

THERMODYNAMIC QUANTITIES FOR THE IONIZATION REACTIONS OF BUFFERS IN WATER

Robert N. Goldberg, Nand Kishore, and Rebecca M. Lennen

This table contains selected values for the pK, standard molar enthalpy of reaction $\Delta_r H°$, and standard molar heat-capacity change $\Delta_r C°_p$ for the ionization reactions of 64 buffers, many of which are relevant to biochemistry and to biology (Ref. 1). The values pertain to the temperature T = 298.15 K and the pressure p = 0.1 MPa. The standard state is the hypothetical ideal solution of unit molality. These data permit one to calculate values of the pK and of $\Delta_r H°$ at temperatures in the vicinity $\{T \approx (274 \text{ K to } 350 \text{ K})\}$ of the reference temperature θ = 298.15 K by using the following equations (Ref. 2).

$$\Delta_r G_T^\circ = -RT \ln K_T = \ln(10) \cdot RT \cdot pK_T , \quad (1)$$

$$R \ln K_T = -\left(\Delta_r G_\theta^\circ / \theta\right) + \Delta_r H_\theta^\circ \left\{(1/\theta) - (1/T)\right\} + \Delta_r C_{p\theta}^\circ \left\{(\theta / T) - 1 + \ln(T/\theta)\right\} , \quad (2)$$

$$\Delta_r H_T^\circ = \Delta_r H_\theta^\circ + \Delta_r C_{p\theta}^\circ (T - \theta) , \quad (3)$$

Here, $\Delta_r G°$ is the standard molar Gibbs energy change and K is the equilibrium constant for a reaction; R is the gas constant (8.314 462 J K^{-1} mol^{-1}). The subscripts T and θ denote the temperature to which a quantity pertains, the subscript p denotes constant pressure, and the subscript r denotes that the quantity refers to a reaction. Combination of equations (1) and (2) yields the following equation that gives pK as a function of temperature:

$$pK_T = -\{R \cdot \ln(10)\}^{-1}[-\{\ln(10) \cdot RT \cdot pK_\theta / \theta\} + \Delta_r H_\theta^\circ \{(1/\theta) - 1/T)\} \Delta_r C_{p\theta}^\circ \{(\theta/T) - 1 + \ln(T/\theta)\} , \quad (4)$$

The above equations neglect higher order terms that involve temperature derivatives of $\Delta_r C°_p$. Also, it is important to recognize that the values of pK and $\Delta_r H°$ effectively pertain to ionic strength I = 0. However, the values of pK and $\Delta_r H°$ are almost always dependent on the ionic strength and the actual composition of the solution. These issues are discussed in Reference 1, which also gives an approximate method for making appropriate corrections.

References

1. Goldberg, R. N., Kishore, N., and Lennen, R. M., "Thermodynamic Quantities for the Ionization Reactions of Buffers," *J. Phys. Chem. Ref. Data*, 31, 231, 2002.
2. Clarke, E. C. W., and Glew, D. N., *Trans. Faraday Soc.*, 62, 539-547, 1966.

Selected Values of Thermodynamic Quantities for the Ionization Reactions of Buffers in Water at T = 298.15 K and p = 0.1 MPa

Reaction	pK	$\Delta_r H°$/ kJ mol^{-1}	$\Delta_r C°_p$/ J mol^{-1} K^{-1}
ACES			
$HL^\pm = H^+ + L^-$, ($HL = C_4H_{10}N_2O_4S$)	6.847	30.43	-49
Acetate			
$HL = H^+ + L^-$, ($HL = C_2H_4O_2$)	4.756	-0.41	-142
ADA			
$H_3L^+ = H^+ + H_2L^\pm$, ($H_2L = C_6H_{10}N_2O_5$)	1.59		
$H_2L^\pm = H^+ + HL^-$	2.48	16.7	
$HL^- = H^+ + L^{2-}$	6.844	12.23	-144
2-Amino-2-methyl-1,3-propanediol			
$HL^+ = H^+ + L$, ($L = C_4H_{11}NO_2$)	8.801	49.85	-44
2-Amino-2-methyl-1-propanol			
$HL^+ = H^+ + L$, ($L = C_4H_{11}NO$)	9.694	54.05	-21[a]
3-Amino-1-propanesulfonic acid			
$HL = H^+ + L^-$, ($HL = C_3H_9NO_3S$)	10.2		
Ammonia			
$NH_4^+ = H^+ + NH_3$	9.245	51.95	8
AMPSO			
$HL^\pm = H^+ + L^-$, ($HL = C_7H_{17}NO_5S$)	9.138	43.19	-61

Reaction	pK	$\Delta_r H°$/ kJ mol^{-1}	$\Delta_r C°_p$/ J mol^{-1} K^{-1}
Arsenate			
$H_3AsO_4 = H^+ + H_2AsO_4^-$	2.31	-7.8	
$H_2AsO_4^- = H^+ + HAsO^{2-}$	7.05	1.7	
$HAsO_4^{2-} = H^+ + AsO_4^{3-}$	11.9	15.9	
Barbital			
$H_2L = H^+ + HL^-$, ($H_2L = C_8H_{12}N_2O_3$)	7.980	24.27	-135
$HL^- = H^+ + L^{2-}$	12.8		
BES			
$HL^\pm = H^+ + L^-$, ($HL = C_6H_{15}NO_5S$)	7.187	24.25	-2
Bicine			
$H_2L^+ = H^+ + HL^\pm$, ($HL = C_6H_{13}NO_4$)	2.0		
$HL^\pm = H^+ + L^-$	8.334	26.34	0
Bis-tris			
$H_3L^+ = H^+ + H_2L^\pm$, ($H_2L = C_8H_{19}NO_5$)	6.484	28.4	27
Bis-tris propane			
$H_2L^{2+} = H^+ + HL^+$, ($L = C_{11}H_{26}N_2O_6$)	6.65		
$HL^+ = H^+ + L$	9.10		
Borate			
$H_3BO_3 = H^+ + H_2BO_3^-$	9.237	13.8	-240[a]

Reaction	pK	$\Delta_r H°$/ kJ mol^{-1}	$\Delta_r C°_p$/ J mol^{-1} K^{-1}
Cacodylate			
$H_2L^+ = H^+ + HL$, ($HL = C_2H_6AsO_2$)	1.78	-3.5	
$HL = H^+ + L^-$	6.28	-3.0	-86
CAPS			
$HL^\pm = H^+ + L^-$, ($HL = C_9H_{19}NO_3S$)	10.499	48.1	57
CAPSO			
$HL^\pm = H^+ + L^-$, ($HL = C_9H_{19}NO_4S$)	9.825	46.67	21
Carbonate			
$H_2CO_3 = H^+ + HCO_3^-$	6.351	9.15	-371
$HCO_3^- = H^+ + CO^{2-}$	10.329	14.70	-249
CHES			
$HL^\pm = H^+ + L^-$, ($HL = C_8H_{17}NO_3S$)	9.394	39.55	9
Citrate			
$H_3L = H^+ + H_2L^-$, ($H_3L = C_6H_8O_7$)	3.128	4.07	-131
$H_2L^- = H^+ + HL^{2-}$	4.761	2.23	-178
$HL^{2-} = H^+ + L^{3-}$	6.396	-3.38	-254
L-Cysteine			
$H_3L^+ = H^+ + H_2L$, ($H_2L = C_3H_7NO_2S$)	1.71	-0.6[a]	
$H_2L = H^+ + HL^-$	8.36	36.1	-66[a]
$HL^- = H^+ + L^{2-}$	10.75	34.1	-204[a]
Diethanolamine			
$HL^+ = H^+ + L$, ($L = C_4H_{11}NO_2$)	8.883	42.08	36
Diglycolate			
$H_2L = H^+ + HL^-$, ($H_2L = C_4H_6O_5$)	3.05	-0.1	-142[a]
$HL^- = H^+ + L^{2-}$	4.37	-7.2	-138[a]
3,3-Dimethylglutarate			
$H_2L = H^+ + HL^-$, ($H_2L = C_7H_{12}O_4$)	3.70		
$HL^- = H^+ + L^{2-}$	6.34		
DIPSO			
$HL^\pm = H^+ + L^-$, ($HL = C_7H_{17}NO_6S$)	7.576	30.18	42
Ethanolamine			
$HL^+ = H^+ + L$, ($L = C_2H_7NO$)	9.498	50.52	26
***N*-Ethylmorpholine**			
$HL^+ = H^+ + L$, ($L = C_6H_{13}NO$)	7.77	27.4	
Glycerol 2-phosphate			
$H_2L = H^+ + HL^-$, ($H_2L = C_3H_9NO_6P$)	1.329	-12.2	-330
$HL^- = H^+ + L^{2-}$	6.650	-1.85	-212
Glycine			
$H_2L^+ = H^+ + HL^\pm$, ($HL = C_2H_5NO_2$)	2.351	4.00	-139
$HL^\pm = H^+ + L^-$	9.780	44.2	-57
Glycine amide			
$HL^+ = H^+ + L$, ($L = C_2H_6N_2O$)	8.04	42.9	
Glycylglycine			
$H_2L^+ = H^+ + HL^\pm$, ($HL = C_4H_8N_2O_3$)	3.140	0.11	-128
$HL^\pm = H^+ + L^-$	8.265	43.4	-16
Glycylglycylglycine			
$H_2L^+ = H^+ + HL^\pm$, ($HL = C_6H_{11}N_3O_4$)	3.224	0.84	
$HL^\pm = H^+ + L^-$	8.090	41.7	
HEPES			
$H_2L^+ = H^+ + HL^\pm$, ($HL = C_8H_{18}N_2O_4S$)	3.0[a]		
$HL^\pm = H^+ + L^-$	7.564	20.4	47
HEPPS			
$HL^\pm = H^+ + L^-$, ($HL = C_9H_{20}N_2O_4S$)	7.957	21.3	48
HEPPSO			
$HL^\pm = H^+ + L^-$, ($HL = C_9H_{20}N_2O_5S$)	8.042	23.70	47
L-Histidine			
$H_3L^{2+} = H^+ + H_2L^+$, ($HL = C_6H_9N_3O_2$)	1.54[b]	3.6	
$H_2L^+ = H^+ + HL$	6.07	29.5	176
$HL = H^+ + L^-$	9.34	43.8	-233
Hydrazine			
$H_2L^{2+} = H^+ + HL^+$, ($L = H_4N_2$)	-0.99	38.1	
$HL^+ = H^+ + L$	8.02	41.7	
Imidazole			
$HL^+ = H^+ + L$, ($L = C_3H_4N_2$)	6.993	36.64	-9
Maleate			
$H_2L = H^+ + HL^-$, ($H_2L = C_4H_4O_4$)	1.92	1.1	-21[a]
$HL^- = H^+ + L^{2-}$	6.27	-3.6	-31[a]
2-Mercaptoethanol			
$HL = H^+ + L^-$, ($HL = C_2H_6OS$)	9.75[b]	26.2	
MES			
$HL^\pm = H^+ + L^-$, ($HL = C_6H_{13}NO_4S$)	6.270	14.8	5
Methylamine			
$HL^+ = H^+ + L$, ($L = CH_5N$)	10.645	55.34	33
2-Methylimidazole			
$HL^+ = H^+ + L$, ($L = C_4H_6N_2$)	8.01[b]	36.8	
MOPS			
$HL^\pm = H^+ + L^-$, ($HL = C_7H_{15}NO_4S$)	7.184	21.1	25
MOPSO			
$H_2L^+ = H^+ + HL^\pm$, ($HL = C_7H_{15}NO_5S$)	0.060		
$HL^\pm = H^+ + L^-$	6.90	25.0	38[a]
Oxalate			
$H_2L = H^+ + HL^-$, ($H_2L = C_2H_2O_4$)	1.27	-3.9	-231[a]
$HL^- = H^+ + L^{2-}$	4.266	7.00	-231
Phosphate			
$H_3PO_4 = H^+ + H_2PO_4^-$	2.148	-8.0	-141
$H_2PO_4^- = H^+ + HPO_4^{2-}$	7.198	3.6	-230
$HPO_4^{2-} = H^+ + PO_4^{3-}$	12.35	16.0	-242
Phthalate			
$H_2L = H^+ + HL^-$, ($H_2L = C_8H_6O_4$)	2.950	-2.70	-91
$HL^- = H^+ + L^{2-}$	5.408	-2.17	-295
Piperazine			
$H_2L^{2+} = H^+ + HL^+$, ($L = C_4H_{10}N_2$)	5.333	31.11	86
$HL^+ = H^+ + L$	9.731	42.89	75
PIPES			
$HL^\pm = H^+ + L^-$, ($HL = C_8H_{18}N_2O_6S_2$)	7.141	11.2	22

Reaction	pK	$\Delta_r H°$/ kJ mol^{-1}	$\Delta_r C°_p$/ J mol^{-1} K^{-1}
POPSO			
$HL^{\pm} = H^+ + L^-$, ($HL = C_{10}H_{22}N_2O_8S_2$)	8.0[a]		
Pyrophosphate			
$H_4P_2O_7 = H^+ + H_3P_2O_7^-$	0.83	-9.2	-90[a]
$H_3P_2O_7^- = H^+ + H_2P_2O_7^{2-}$	2.26	-5.0	-130[a]
$H_2P_2O_7^{2-} = H^+ + HP_2O_7^{3-}$	6.72	0.5	-136
$HP_2O_7^{3-} = H^+ + P_2O_7^{4-}$	9.46	1.4	-141
Succinate			
$H_2L = H^+ + HL^-$, ($H_2L = C_4H_6O_4$)	4.207	3.0	-121
$HL^- = H^+ + L^{2-}$	5.636	-0.5	-217
Sulfate			
$HSO_4^- = H^+ + SO_4^{2-}$	1.987	-22.4	-258
Sulfite			
$H_2SO_3 = H^+ + HSO_3^-$	1.857	-17.80	-272
$HSO_3^- = H^+ + SO_3^{2-}$	7.172	-3.65	-262
TAPS			
$HL^{\pm} = H^+ + L^-$, ($HL = C_7H_{17}NO_6S$)	8.44	40.4	15
TAPSO			
$HL^{\pm} = H^+ + L^-$, ($HL = C_7H_{17}NO_7S$)	7.635	39.09	-16
L(+)-Tartaric acid			
$H_2L = H^+ + HL^-$, ($H_2L = C_4H_6O_6$)	3.036	3.19	-147
$HL^- = H^+ + L^{2-}$	4.366	0.93	-218
TES			
$HL^{\pm} = H^+ + L^-$, ($HL = C_6H_{15}NO_6S$)	7.550	32.13	0
Tricine			
$H_2L^+ = H^+ + HL^{\pm}$, ($HL = C_6H_{13}NO_5$)	2.023	5.85	-196
$HL^{\pm} = H^+ + L^-$	8.135	31.37	-53
Triethanolamine			
$HL^+ = H^+ + L$, ($L = C_6H_{15}NO_3$)	7.762	33.6	50
Triethylamine			
$HL^+ = H^+ + L$, ($L = C_6H_{15}N$)	10.72	43.13	151
Tris			
$HL^+ = H^+ + L$, ($L = C_4H_{11}NO_3$)	8.072	47.45	-59

[a] Approximate value
[b] Last digit uncertain

Biochem

BIOLOGICAL BUFFERS

This table of frequently used buffers gives the pK_a value at 25 °C and the useful pH range of each buffer. The buffers are listed in order of increasing pH.

The table is reprinted with permission of Sigma Chemical Company, St. Louis, MO.

Biological Buffers

Acronym	Name	Mol. wt.	pK_a	Useful pH range
MES	2-(*N*-Morpholino)ethanesulfonic acid	195.2	6.1	5.5–6.7
BIS TRIS	Bis(2-hydroxyethyl)iminotris(hydroxymethyl)methane	209.2	6.5	5.8–7.2
ADA	*N*-(2-Acetamido)-2-iminodiacetic acid	190.2	6.6	6.0–7.2
ACES	2-[(2-Amino-2-oxoethyl)amino]ethanesulfonic acid	182.2	6.8	6.1–7.5
PIPES	Piperazine-*N,N′*-bis(2-ethanesulfonic acid)	302.4	6.8	6.1–7.5
MOPSO	3-(*N*-Morpholino)-2-hydroxypropanesulfonic acid	225.3	6.9	6.2–7.6
BIS TRISPROPANE	1,3-Bis[tris(hydroxymethyl)methylamino]propane	282.3	6.8[a]	6.3–9.5
BES	*N,N*-Bis(2-hydroxyethyl)-2-aminoethanesulfonic acid	213.2	7.1	6.4–7.8
MOPS	3-(*N*-Morpholino)propanesulfonic acid	209.3	7.2	6.5–7.9
HEPES	*N*-(2-Hydroxyethyl)piperazine-*N′*-(2-ethanesulfonic acid)	238.3	7.5	6.8–8.2
TES	*N*-Tris(hydroxymethyl)methyl-2-aminoethanesulfonic acid	229.2	7.5	6.8–8.2
DIPSO	3-[*N,N*-Bis(2-hydroxyethyl)amino]-2-hydroxypropanesulfonic acid	243.3	7.6	7.0–8.2
TAPSO	3-[*N*-Tris(hydroxymethyl)methylamino)-2-hydroxypropanesulfonic acid	259.3	7.6	7.0–8.2
TRIZMA	Tris(hydroxymethyl)aminomethane	121.1	8.1	7.0–9.1
HEPPSO	*N*-(2-hydroxyethyl)piperazine-*N′*-(2-hydroxypropanesulfonic acid)	268.3	7.8	7.1–8.5
POPSO	Piperazine-*N,N′*-bis(2-hydroxypropanesulfonic acid)	362.4	7.8	7.2–8.5
EPPS	*N*-(2-Hydroxyethyl)piperazine-*N′*-(3-propanesulfonic acid)	252.3	8.0	7.3–8.7
TEA	Triethanolamine	149.2	7.8	7.3–8.3
TRICINE	*N*-Tris(hydroxymethyl)methylglycine	179.2	8.1	7.4–8.8
BICINE	*N,N*-Bis(2-hydroxyethyl)glycine	163.2	8.3	7.6–9.0
TAPS	*N*-Tris(hydroxymethyl)methyl-3-aminopropanesulfonic acid	243.3	8.4	7.7–9.1
AMPSO	3-[(1,1-Dimethyl-2-hydroxyethyl)amino]-2-hydroxypropanesulfonic acid	227.3	9.0	8.3–9.7
CHES	2-(*N*-Cyclohexylamino)ethanesulfonic acid	207.3	9.3	8.6–10.0
CAPSO	3-(Cyclohexylamino)-2-hydroxy-1-propanesulfonic acid	237.3	9.6	8.9–10.3
AMP	2-Amino-2-methyl-1-propanol	89.1	9.7	9.0–10.5
CAPS	3-(Cyclohexylamino)-1-propanesulfonic acid	221.3	10.4	9.7–11.1

[a] pK_a = 9.0 for the second dissociation stage.

Biochem

TYPICAL pH VALUES OF BIOLOGICAL MATERIALS AND FOODS

This table gives typical pH ranges for various biological fluids and common foods. All values refer to 25 °C.

Typical pH Values of Biological Materials

Biological materials	pH
Blood, human	7.35–7.45
Blood, dog	6.9–7.2
Spinal fluid, human	7.3–7.5
Saliva, human	6.5–7.5
Gastric contents, human	1.0–3.0
Duodenal contents, human	4.8–8.2
Feces, human	4.6–8.4
Urine, human	4.8–8.4
Milk, human	6.6–7.6
Bile, human	6.8–7.0

Typical pH Values of Common Foods

Foods	pH
Apples	2.9–3.3
Apricots	3.6–4.0
Asparagus	5.4–5.8
Bananas	4.5–4.7
Beans	5.0–6.0
Beers	4.0–5.0
Beets	4.9–5.5
Blackberries	3.2–3.6
Bread, white	5.0–6.0
Butter	6.1–6.4
Cabbage	5.2–5.4
Carrots	4.9–5.3
Cheese	4.8–6.4
Cherries	3.2–4.0
Cider	2.9–3.3
Corn	6.0–6.5
Crackers	6.5–8.5
Dates	6.2–6.4
Eggs, fresh white	7.6–8.0
Flour, wheat	5.5–6.5
Gooseberries	2.8–3.0
Grapefruit	3.0–3.3
Grapes	3.5–4.5
Hominy (lye)	6.8–8.0
Jams, fruit	3.5–4.0
Jellies, fruit	2.8–3.4
Lemons	2.2–2.4
Limes	1.8–2.0
Maple syrup	6.5–7.0
Milk, cows	6.3–6.6
Olives	3.6–3.8
Oranges	3.0–4.0
Oysters	6.1–6.6
Peaches	3.4–3.6
Pears	3.6–4.0
Peas	5.8–6.4
Pickles, dill	3.2–3.6
Pickles, sour	3.0–3.4
Pimento	4.6–5.2
Plums	2.8–3.0
Potatoes	5.6–6.0
Pumpkin	4.8–5.2
Raspberries	3.2–3.6
Rhubarb	3.1–3.2
Salmon	6.1–6.3
Sauerkraut	3.4–3.6
Shrimp	6.8–7.0
Soft drinks	2.0–4.0
Spinach	5.1–5.7
Squash	5.0–5.4
Strawberries	3.0–3.5
Sweet potatoes	5.3–5.6
Tomatoes	4.0–4.4
Tuna	5.9–6.1
Turnips	5.2–5.6
Vinegar	2.4–3.4
Water, drinking	6.5–8.0
Wines	2.8–3.8

PROPERTIES AND FUNCTIONS OF COMMON DRUGS

Jessica L. Burger

This table lists the names, properties, categories, and therapeutic uses of selected single-component and multicomponent drugs. The drugs have been chosen to represent a variety of categories and are widely used throughout the world. Table 1 contains the generic (and in some cases chemical) name of single-component drugs with data on their melting point, boiling point, density, and solubility in various solvents, when available. Table 2 classifies the drugs by therapeutic categories and gives additional information about their uses. Table 3 provides an index that can be used to locate an individual single-component drug by its trade name. It should be noted that not all trade names are provided as many drugs are manufactured by more than one organization. Table 4 provides similar information for multicomponent drugs and links the components to the property data given in Table 1 through the ordering number. In all tables, the ordering number is unique to a specific single-component drug as listed in Table 1. It should be noted that most drug products in the marketplace have one or more inactive ingredients. These are not included in these tables.

References

1. *The Combined Chemical Dictionary on DVD, Version 18:1*, CRC Press, Boca Raton, FL, June 2014; also available on the Internet at www.chemnetbase.com.
2. Milne, G. W. A., *Drugs: Synonyms and Properties*, Ashgate Publishing, Aldershot, Hampshire, UK, 2000.
3. Corey, E. J., Czakó, B., and Kürti, L., *Molecules and Medicine*, John Wiley & Sons, Hoboken, NJ, 2007.
4. *Physicians' Desk Reference, 67th Edition*, Thomson PDR, Montvale, NJ, 2013.
5. O'Neil, M. J., Editor, *The Merck Index, 15th Edition*, Royal Society of Chemistry, 2013.
6. Santa Cruz Biotechnology, Inc. <http://www.scbt.com>. Accessed August 2016.
7. Wishart, D. S. et al, *Nucleic Acids Res.* 34, D668, 2006.
8. *ChemACX. Version 16.16.2;* CambridgeSoft Corporation: Cambridge, MA. Accessed August 2016.
9. *PubChem Database*, National Center for Biotechnology Information <http://pubchem.ncbi.nlm.nih.gov>. Accessed August 2016.

Physical Properties of Common Single-Component Drugs

Table 1 contains physical properties of 240 common single-component drugs. The data come from a variety of sources as listed in the reference section. The properties, units, and special notations are given in the table below. A few drugs have no properties listed except their physical state (in some) cases. These include monoclonal antibodies (indicated by an * after the name), drugs produced as recombinant DNA (indicated by a + after the name), and vaccines.

ORDERING: Entry number; also used to link to information in other tables

Common name: Common name of the drug: generic or chemical

t_m/°C: Melting point in °C; superscript p indicates predicted value; lit. indicates literature value

t_b/°C: Boiling point in °C; superscript p indicates predicted value; all values are for 760 mmHg unless otherwise indicated

ρ/g mL^{-1}: Density in g mL^{-1}; all values are for 25 °C unless otherwise indicated by numerical superscript; superscript p indicates predicted value

Phys. state: Indication of whether the drug is a solid or liquid

Storage: Recommendations for normal storage; RT = room temperature

Solubility H_2O: Solubility in water in mg mL^{-1} unless otherwise indicated; temperature values, when available, are given in the superscript

Solubility CH_3OH: Solubility in methanol in mg mL^{-1} unless otherwise indicated; temperature values, when available, are given in the superscript

Solubility C_2H_5OH: Solubility in ethanol in mg mL^{-1} unless otherwise indicated; temperature values, when available, are given in the superscript

Solubility DMSO: Solubility in dimethylsulfoxide (DMSO) in mg mL^{-1} unless otherwise indicated; temperature values, when available, are given in the superscript

Solubility DMF: Solubility in dimethyl formamide (DMF) in mg mL^{-1} unless otherwise indicated; temperature values, when available, are given in the superscript

Table 1. Physical Properties of Common Single-Component Drugs

ORDERING	Common name	t_m/°C	t_b/°C	ρ/ g mL^{-1}	Phys. state	Storage	Solubility H_2O/ mg mL^{-1}	Solubility CH_3OH/ mg mL^{-1}	Solubility C_2H_5OH/ mg mL^{-1}	Solubility DMSO/ mg mL^{-1}	Solubility DMF/ mg mL^{-1}
1	Abatacept										
2	Abiraterone acetate	127-130	506.7^p	~1.1^p	Solid	-20 °C	<l^{25}	sparingly	28^{25}	~2	~16
3	Acetaminophen	168-172	387.83^p	1.29^{20}	Solid	RT	0.014^{20}	50	100 mmol L^{-1}	100 mmol L^{-1}	freely
4	Acetylsalicylic acid				Solid	RT			80	41	
5	Adalimumab: heavy chain*				Liquid						
6	Adalimumab: light chain*				Liquid						
7	Aflibercept			1.081	Liquid	2-8 °C	>100				
8	Albuterol sulfate	153-155		1.4^p	Solid	-20 °C		freely	freely		
9	Alendronate sodium		616.7^p		Solid	RT					
10	Allopurinol				Solid	4 °C				freely	
11	Alprazolam	228-229.5			Solid		40 (pH 7)				

Biochem

ORDERING	Common name	t_m/°C	t_b/°C	ρ/ g mL^{-1}	Phys. state	Storage	Solubility H_2O/ mg mL^{-1}	Solubility CH_3OH/ mg mL^{-1}	Solubility C_2H_5OH/ mg mL^{-1}	Solubility DMSO/ mg mL^{-1}	Solubility DMF/ mg mL^{-1}
12	Amitriptyline hydrochloride	187-189.5			Solid		0.00971^{24}				
13	Amlodipine	133-135	527.2^{p}	1.2^{p}	Solid	RT	75.3	freely	25	25	25
14	Amlodipine besylate	133-135	527.16^{p}	1.2^{p}	Solid	RT	75.3	freely	25	25	25
15	Amoxicillin	329.94^{p}	701.8^{p}	1.6^{p}	Solid	4 °C					
16	Amphetamine aspartate monohydrate				Solid		slightly				
17	Amphetamine sulfate				Solid		slightly				
18	Apixaban	326.53^{p}	770.5^{p}	1.4^{p}	Solid	4 °C	$<1^{25}$		$<1^{25}$	18^{25}	~3
19	Aripiprazole	134-136			Solid	-20 °C			freely		
20	Atazanavir	207-209		1.2^{p}	Solid	4 °C		freely	~2	~16	~25
21	Atenolol				Solid	RT		freely		freely	
22	Atomoxetine				Solid	-20 °C		freely			
23	Atorvastatin calcium	176-178	722.2^{p}		Solid	RT	$<1^{25}$	freely	$<1^{25}$	100^{25}	
24	Azithromycin	114	822.1^{p}	1.2^{p}	Solid	-20 °C	$<1^{25}$		100^{25}	100^{25}	16
25	Bacitracin	221-225			Solid	4 °C	50				
26	Bendamustine hydrochloride	149-151	585.2^{p}		Solid	-20 °C	$\geq 15^{25}$	s	$\geq 17^{25}$	$\geq 79^{25}$	
27	Bevacizumab*				Liquid						
28	Bimatoprost	66-68	629.8^{p}	1.1^{p}	Solid	-20 °C		freely	~50	~25	
29	Bisacodyl	131.1-135	492.33^{p}	1.2^{p}	Solid	4 °C			slightly		
30	Bismuth subsalicylate				Solid		59.7				
31	Bortezomib	139-143	684.30^{p}	1.21^{p}	Solid	-20 °C	5-10 μmol L^{-1}(est. max)	freely	200	200	
32	Brimonidine tartrate	207-208	432.6		Solid	-20 °C	50 mmol L^{-1}	freely		freely	
34	Buprenorphine				Solid		0.0168				
35	Caffeine	238		1.23	Solid		666^{100}				
36	Canagliflozin	68-72	642.9^{p}	1.4^{p}	Solid	-20 °C	$<1^{25}$	freely	$<1^{25}$	88^{25}	
37	Capecitabine	115-120	517^{p}	1.49^{p}	Solid	RT	26	>40%	207	72^{25}	~14
38	Carisoprodol	89-91					0.300^{25}				
39	Carvedilol	117	655.22^{p}	1.25^{p}	Solid	RT	partly miscible	33.8	10	>25	
40	Cefdinir				Solid	-20 °C				freely	freely
41	Celecoxib	156-157	529.01^{p}	1.43^{p}	Solid	-20 °C	$<1^{5}$	freely	33^{25}	76^{25}	
42	Cephalexin				Solid	-20 °C		freely		freely	
43	Certolizumab pegol*										
44	Cetuximab*										
45	Chlorpheniramine maleate	130-135			Solid	RT	freely				
46	Cimetidine		487.97^{p}	1.27^{p}	Solid	4 °C	freely^{37}				
47	Cinacalcet		440.9		Solid	-20 °C	1^{25}	freely	33^{25}	79^{25}	
48	Ciprofloxacin	253-257	581.8^{p}	1.5^{p}	Solid	RT	30^{20}	freely	$<1^{25}$	$<1^{25}$	
49	Clavulanate potassium	117.5-118					300				
50	Clonazepam	238-240			Solid		0.100^{25}				
51	Clonidine hydrochloride	312	319.3^{p}		Solid	4 °C	50	freely	freely	75	
52	Clopidogrel bisulfate				Solid	RT				freely	
53	Cobicistat	87-105			Solid	4 °C					
54	Codeine	157.5					9^{20}		freely		
55	Colesevelam hydrochloride				Solid		freely				
56	Conjugated estrogens	173			Solid		0.0036				
57	Cyclobenzaprine hydrochloride	215-217			Solid	-20 °C	>200	freely	50		
58	Dabigatran etexilate mesylate				Solid	4 °C					
59	Daptomycin				Solid	-20 °C	100^{25}	5	50 mmol L^{-1}	100 mmol L^{-1}	freely
60	Darbepoetin alfa^{+}				Liquid						
61	Darunavir	74-76		1.3^{p}	Solid	-20 °C		freely			
62	Dasatinib	272-274			Solid	-20 °C	very poorly		very poorly	200	
63	Denosumab*										
64	Desvenlafaxine				Solid	4 °C	>20			>20	
65	Dexlansoprazole				Solid		very slightly	freely	freely		
66	Dexmethylphenidate hydrochloride				Solid		0.82				
67	Dextroamphetamine saccharate				Solid		slightly				
68	Dextroamphetamine sulfate				Solid		slightly				
69	Dextromethorphan hydrobromide		427.5^{p}		Solid	RT	40 mmol L-1		1:10		
70	Diazepam	125-126			Solid						
71	Diclofenac sodium	275-277			Solid	RT	50	>24	~35	~35	~35
72	Digoxin	217-221					0.0648^{25}				
73	Dimethyl fumarate	102-106	192-193	1.37^{20}	Solid	RT	1.6^{25}	30-36	10^{25}	29^{25}	~12
74	Diphenhydramine hydrochloride	169-169.9		1.2	Solid	RT	1000	freely	500	58^{25}	~10
75	Dipyridamole	165-166	806.48^{p}		Solid	RT	slightly	freely	10 mmol L^{-1}	100 mmol L^{-1}	

ORDERING	Common name	t_m/°C	t_b/°C	ρ/ g mL^{-1}	Phys. state	Storage	Solubility H_2O/ mg mL^{-1}	Solubility CH_3OH/ mg mL^{-1}	Solubility C_2H_5OH/ mg mL^{-1}	Solubility DMSO/ mg mL^{-1}	Solubility DMF/ mg mL^{-1}
76	Docusate sodium	153 to 157		1.120	Solid		15		freely		
77	Doxycycline	201			Solid	-20 °C	50	1:4	1:60	1	
78	Duloxetine	157^{p}	466.2	1.2	Liquid	-20 °C	slightly				
79	Dutasteride	247-248	620.3^{p}	1.3^{p}	Solid	4 °C	freely	64	44	62	
80	Eculizumab*										
81	Efavirenz	134-136	341^{p}	1.5^{p}	Solid	-20 °C		freely			
82	Elvitegravir	93-96			Solid	4 °C					
83	Emtricitabine	153-155	443.3^{p}	1.8^{p}	Solid	-20 °C	≥ 50		≥ 17	≥ 50	
84	Enalapril maleate	143-144			Solid	RT	25	20	8		
85	Enoxaparin sodium						> 200				
86	Epoetin alfa+				Liquid						
87	Erlotinib hydrochloride	223-225			Solid	-20 °C	<1	sparingly	<1	3	
88	Erythromycin				Solid			freely	freely		
89	Escitalopram oxalate	155			Solid		very slightly				
90	Esomeprazole				Solid	-20 °C	$<1^{25}$	s	143^{25}	143^{25}	
91	Eszopiclone				Solid		Very slightly		slightly		
92	Etanercept+				Liquid/ solid						
93	Ethinyl estradiol	142-146			Solid				1 part in 6		
94	Etonogestrel	182-184			Solid	-20 °C					
95	Everolimus	110-115	998.75^{p}	1.18^{p}	Solid	-20 °C	$<1^{25}$	freely	100	100	
96	Ezetimibe	162-164	655^{p}	1.3^{p}	Solid	4 °C	$<1^{25}$	freely	83^{25}	82^{25}	
97	Febuxostat	208-210			Solid	4 °C				freely	
98	Fenofibrate	79-83	470^{p}	1.2^{p}	Solid	-20 °C	freely	slightly	1	15	30
99	Fexofenadine hydrochloride	193-196			Solid	-20 °C		freely			
100	Filgrastim+										
101	Fingolimod				Solid	-20	freely				
102	Fluconazole	138-142	79.8^{p}	1.5^{p}	Solid	RT	<1(25)		61^{25}	61^{25}	
103	Fluoxetine hydrochloride	158-159			Solid		4^{RT}	freely^{RT}		$>5^{RT}$	
104	Fluticasone propionate	275	568	1.3	Solid	4 °C				≥10	
105	Formoterol fumarate dihydrate	136-142			Solid	4 °C	freely			20	
106	Furosemide		582.1^{p}	1.6^{p}	Solid	RT	$<1^{25}$	50	$<1^{25}$	66^{25}	freely
107	Gabapentin	162-166	314.44^{p}	1.1^{p}	Solid	RT	10	freely	sparingly	sparingly	sparingly
108	Glatiramer acetate										
109	Glimepiride	212-215	750^{p}	1.3^{p}	Solid	RT	$<1^{25}$		$<1^{25}$	>10	
110	Glyburide	173-175		1.4^{p}	Solid	4 °C	freely	1:250	5	25	freely
111	Guaifenesin	79	215 (19 mmHg)	1.2^{p}	Solid	RT	50^{25}	freely	50	40^{25}	freely
112	Human papillomavirus quadrivalent										
113	Hydrochlorothiazide	273-277			Solid	RT		freely			
114	Hydrocodone	198			Solid		freely		freely		
115	Hydrocodone bitartrate	198			Solid		freely		freely		
116	Hydrocortisone	211-214			Solid	RT			15		
117	Ibuprofen	77-78	157 (4 mmHg)	1.0^{p}	Solid	RT	freely		25		
118	Imatinib mesylate	214-224			Solid	4 °C	200	freely	~0.2	200	~10
119	Infliximab*										
120	Influenza vaccine for intramuscular injection										
121	Insulin aspart (rdna origin)				Solid		insoluble				
122	Insulin detemir (rdna origin)				Solid		freely				
123	Insulin glargine	232-234			Solid		insoluble				
124	Insulin lispro				Solid		insoluble				
125	Interferon beta-1a+										
126	Ipilimumab*										
127	Ipratropium bromide	239-240			Solid	4 °C	83^{25}		83^{25}	83^{25}	
128	Isosorbide mononitrate	86-88	364.5^{p}	1.6^{p}	Solid	RT	freely	freely	freely		
129	Lansoprazole	173-175	555.8^{p}	1.5^{p}	Solid	4 °C	freely	freely	14^{25}	74^{25}	~30
130	Ledipasvir				Solid	-20 °C					
131	Lenalidomide	269-271	614^{p}	1.5^{p}	Solid	-20 °C	$<1^{25}$	freely	$<1^{25}$	$≥52^{25}$	
132	Levofloxacin	218-220			Solid	4 °C					
133	Levothyroxine sodium				Solid		0.015%				
134	Linagliptin	190-196	661.2^{p}	1.4^{p}	Solid	-20 °C	$<1^{25}$	freely	1^{25}	17^{25}	
135	Linezolid	171-175	585.5	1.3^{p}	Solid	4 °C	$<1^{25}$	freely	8^{25}	68^{25}	
136	Liraglutide				Solid		freely (pH>7)	68	1.1		
137	Lisdexamfetamine dimesylate				Solid		792				

ORDERING	Common name	t_m/°C	t_b/°C	ρ/ g mL^{-1}	Phys. state	Storage	Solubility H_2O/ mg mL^{-1}	Solubility CH_3OH/ mg mL^{-1}	Solubility C_2H_5OH/ mg mL^{-1}	Solubility DMSO/ mg mL^{-1}	Solubility DMF/ mg mL^{-1}
138	Lisinopril	326^{p}	666.4^{p}	1.3^{p}	Solid	-20 °C			sparingly	sparingly	sparingly
139	Loperamide hydrochloride	223			Solid	RT	160	286	53.7	205	
140	Loratadine	134-136	531.3^{p}	1.3^{p}	Solid	RT	<1^{25}		77^{25}	50	~30
141	Lorazepam	192-194					80				
142	Losartan potassium	265	682^{p}		Solid	RT	>500		≥20	100 mmol L^{-1}	
143	Loteprednol etabonate	226-233			Solid	-20 °C					
144	Lovastatin	175.4	559.2^{p}	1.1^{p}	Solid	-20 °C	0.4 μg mL^{-1}	freely	10^{RT}	10^{RT}	freely
145	Lubiprostone				Solid		freely				
146	Lurasidone hydrochloride	198-205			Solid	4 °C	0.00789				
147	Medroxyprogesterone acetate	214.5			Solid		0.022				
148	Meloxicam	255	588.7	1.6	Solid	4 °C	22	slightly	slightly	25	freely
149	Memantine hydrochloride		239.8^{p}		Solid	RT	100mmol L^{-1}		43^{25}	43^{25}	
150	Metformin hydrochloride	215-218			Solid	-20 °C		freely		freely	
151	Methylprednisolone				Solid	RT					
152	Metoprolol succinate	142.1-148.2	398.6		Solid	RT	miscible				
153	Miconazole nitrate	170-185			Solid	RT	<1^{25}	freely	10 mmol L^{-1}	95^{5}	
154	Mometasone furoate				Solid	-20 °C		freely			
155	Montelukast sodium	135.5	750.5^{p}		Solid	-20 °C	>100^{25}	freely	100 mmol L^{-1}	100 mmol L^{-1}	~30
156	Morphine sulfate				Solid	255	0.149^{20}				
157	Naloxone	184	532.8^{p}	1.4^{p}	Solid	RT					
158	Naproxen sodium	152-154 (lit.)	403.9	1.2	Solid	RT	>3^{25}	freely	freely	freely	freely
159	Nebivolol				Solid		0.0403				
160	Neomycin sulfate						500				
161	Norelgestromin						0.00605				
162	Norethindrone acetate	203.5					0.00704^{25}				
163	Norgestimate	230-232			Solid	-20 °C	freely	freely		5	
164	Octreotide acetate	153-156			Solid	-20 °C		freely		freely	
165	Olmesartan acid	189-192	738.3^{p}	1.3^{p}	Solid	-20 °C		sparingly			20 mmol L^{-1}
166	Olmesartan medoxomil	180-182			Solid	-20 °C			freely		
167	Olopatadine hydrochloride				Solid	4 °C	sparingly				
168	Omalizumab*										
169	Omeprazole magnesium	156			Solid		0.5	freely	freely		
170	Oseltamivir phosphate	197-212			Solid	-20 °C	>500		<1^{25}	2^{25}	
171	Oxycodone hydrochloride	219			Solid		slightly		slightly		
172	Oxymetazoline hydrochloride		431.9^{p}		Solid	RT	50	freely	59	59	
173	Paliperidone palmitate	114-116			Solid						
174	Palivizumab*										
175	Paroxetine hydrochloride	147-150			Solid					25	
176	Pegfilgrastim										
177	Pemetrexed disodium				Solid	-20 °C	freely	freely			
178	Pioglitazone	183-185	75^{p}	1.3^{p}	Solid	-20 °C	<1^{25}		<1^{25}	15^{25}	2.5
179	Pneumococcal 13-valent conjugate vaccine										
180	Polyethylene glycol 3350						freely				
181	Polymyxin b				Solid		25				
182	Pramoxine hydrochloride				Solid						
183	Prasugrel	122	493.5^{p}	1.3^{p}	Solid	4 °C	30^{25}		7^{25}	<1^{25}	
184	Pravastatin sodium	171.2-173			Solid	4 °C	freely				
185	Prednisone	233-235 (d)			Solid		very slightly	slightly	freely		
186	Pregabalin						freely				
187	Promethazine hydrochloride	225-227			Solid	4 °C					
188	Quetiapine fumarate	173-175			Solid	RT	<1		<1	36	
189	Rabeprazole sodium	99-100	538.9-668.9^{p}	1.3^{p}	Solid	-20 °C	freely			freely	
190	Raloxifene hydrochloride	257.8	728.2^{p}		Solid	-20 °C	freely	freely	<1^{25}	50 mmol L^{-1}	
191	Raltegravir				Solid	-20 °C	<1^{25}		<1^{25}	88^{25}	
192	Ramipril	109	616.2^{p}	1.2^{p}	Solid	RT	<1^{25}	freely	83^{25}	83^{25}	
193	Ranibizumab										
194	Ranitidine	69-72	437.1^{p}	1.2^{p}	Solid		freely				
195	Ranolazine	119-120	624.1	1.2	Solid			freely			
196	Rifaximin	218-228			Solid	4 °C	<1		157	47	
197	Rilpivirine	153-155	443.3^{p}	1.8^{p}	Solid	-20 °C	≥ 50		≥ 17	≥ 50	
198	Risedronate sodium	252-262	692.3^{p}		Solid	4 °C	5		<1	1	
199	Ritonavir	120-122	947^{p}	1.2^{p}	Solid	4 °C	partly	freely	~5	100^{25}	~15

ORDERING	Common name	t_m/°C	t_b/°C	ρ/ g mL^{-1}	Phys. state	Storage	Solubility H_2O/ mg mL^{-1}	Solubility CH_3OH/ mg mL^{-1}	Solubility C_2H_5OH/ mg mL^{-1}	Solubility DMSO/ mg mL^{-1}	Solubility DMF/ mg mL^{-1}
200	Rituximab*										
201	Rivaroxaban	228-229	732.6^{p}	1.5^{p}	Solid	4 °C	$<1^{25}$		$<1^{25}$	87^{25}	
202	Rivastigmine tartrate	126					freely		s		
203	Rosuvastatin calcium	151-156			Solid	-20 °C	$<1^{25}$	freely	$<1^{25}$	100^{25}	
204	Salmeterol	243.5^{p}	603.0^{p}	1.1^{p}	Solid	4 °C			50 mmol L^{-1}	100 mmol L^{-1}	
205	Saxagliptin	96-102	548.7^{p}	1.4^{p}	Solid	-20 °C		freely		freely	
206	Sertraline hydrochloride	243-245			Solid	RT	3		10	>25	
207	Sevelamer hydrochloride				Solid		freely				
208	Sildenafil citrate	185-187			Solid	RT	3.5	freely		freely	
209	Simeprevir			1.4^{p}	Solid	4 °C				freely	
210	Simvastatin	127-132	564.9	1.1	Solid	-20 °C	0.03	freely	>25	>25	
211	Sitagliptin	110-120	529.9^{p}	1.6^{p}	Solid	RT					
212	Sofosbuvir	100-106			Solid	-20 °C					
213	Solifenacin succinate				Solid	-20 °C	96^{25}		5^{25}	3^{25}	
214	Spironolactone	207-208	597.0^{p}	1.2^{p}	Solid	RT	0.022^{25}	freely	50 mmol L^{-1}	100 mmol L^{-1}	
215	Sulfamethoxazole	168-170			Solid	4 °C	$<1^{25}$		23^{25}	51^{25}	
216	Tadalafil	276-279	679.1^{p}		Solid	-20 °C	$<1^{25}$	freely	$<1^{25}$	78^{25}	
217	Tamsulosin hydrochloride	226-228	595.5	1.2^{p}	Solid	RT	10 mmol L^{-1}			100 mmol L^{-1}	
218	Telmisartan	261-263	771.9^{p}	1.2^{p}	Solid	RT	freely		$<1^{25}$	13^{25}	~1.6
219	Tenofovir disoproxil fumarate	113-115			Solid	4 °C	$<1^{25}$	freely	44^{25}	128^{25}	
220	Testosterone	155			Solid		0.0234^{25}		1 part in 5		
221	Timolol	202-203	487.2^{p}		Solid		freely		freely	freely	
222	Tiotropium bromide	218-220			Solid	4 °C		freely		freely	
223	Tolterodine tartrate	205-207			Solid	4 °C	0.0212	freely	6	≥20	
224	Tramadol hydrochloride	180-181			Solid		freely				
225	Trastuzumab*										
226	Travoprost				Liquid		$>16^{25}$				
227	Trazodone hydrochloride	87			Solid		sparingly				
228	Triamcinolone acetonide	270			Solid		0.080^{25}				
229	Triamterene				Solid	RT	freely	freely			
230	Trimethoprim	234-236	405.2^{p}	1.3^{p}	Solid	4 °C	0.4	12.1	2.6^{20}		
231	Ustekinumab										
232	Valacyclovir hydrochloride	196-200			Solid	RT	170	freely	$<1^{25}$	14^{25}	
233	Valsartan	108-110	684.9^{p}	1.2^{p}	Solid			freely	freely		
234	Vardenafil hydrochloride	227-230			Solid	-20 °C					
235	Varenicline						51.5^{25}				
236	Venlafaxine hydrochloride	217	397.6^{p}		Solid	RT	>25			50 mmol L^{-1}	
237	Vilazodone hydrochloride	203-205			Solid	-20 °C					
238	Warfarin sodium	161			Solid	RT	freely		freely		
239	Zolpidem tartrate	196			Solid		23				
240	Zoster vaccine live										

Functions of Common Single-Component Drugs

Table 2 contains information on the therapeutic class, mechanism of action indications (use) of the common single-component drugs listed in Table 1. The information comes from a variety of sources as listed in the reference section. While every precaution has been taken to ensure this information is correct, it must not be used to make medical decisions of any type. Therapeutic class and Indication information are based on commonly occurring situations and must not be used in place of professional advice. This general information must not be interpreted as applicable to any specific individual health or medical situation.

ORDERING: Entry number; also used to link to information in other tables

Common name: Common name of the drug: generic or chemical

Therapeutic class: Classification based on chemical characteristics, structure, and functionality; this information must not be used to make medical decisions of any type

Mechanism of action: General description of the mechanism(s) by which the drug commonly works; this information must not be used to make medical decisions of any type

Indications (Use): General description of the result achieved by use of the drug; this information must not be used to make medical decisions of any type

Table 2. Functions of Common Single-Component Drugs

Drug no.	Drug name	Therapeutic class	Mechanism of action	Indications (Use)
Addiction				
34	Buprenorphine	Partial opioid agonist/opioid antagonist	Efficacy is due to its properties as a partial agonist at the mu-opioid receptor and an antagonist at the kappa-opioid receptor.	Used in the management of opioid dependence and should be used as part of a complete treatment plan that includes counseling and psychosocial support.
157	Naloxone		Efficacy is due to its properties as an antagonist at the mu-opioid receptor.	Used to reverse opioid overdoses.
235	Varenicline	Nicotinic acetylcholine receptor agonist	Works by binding to $\alpha4\beta2$ neuronal nicotinic acetylcholine receptors and providing agonist activity as well as limiting nicotine binding at these receptors.	Used to aid in the cessation of smoking.
Antiallergic and Asthma Control				
8	Albuterol sulfate	Short-acting β2-agonist	Use results in the relaxation of smooth muscles in the airways by activating β2-adrenergic receptors causing the activation of adenylcyclase and an increase in intracellular cAMP. Increased cAMP leads to activation of protein kinase A, which inhibits the phosphorylation of myosin and lowers intracellular ionic calcium concentrations.	Used in the treatment of acute bronchospasm caused by bronchial asthma, chronic bronchitis, and other bronchopulmonary disorders.
33	Budesonide	Corticosteroid/long-acting β2-agonist	Corticosteroid	Used in the management of asthma and airflow obstruction in patients with chronic obstructive pulmonary disease, including chronic bronchitis and emphysema.
45	Chlorpheniramine maleate	Antihistamine	Antihistamine	Used to provide temporary relief from allergy symptoms.
74	Diphenhydramine hydrochloride	Antihistamine	Antihistamine	Used to manage allergies.
99	Fexofenadine hydrochloride	Antihistamine	Antihistamine	
104	Fluticasone propionate	Corticosteroid	Corticosteroid	Used as a prophylactic therapy in the management of asthma.
105	Formoterol fumarate dihydrate		Use results in the relaxation of smooth muscles in the airways by activating β2-adrenergic receptors causing the activation of adenylcyclase and an increase in intracellular cAMP. Increased cAMP leads to activation of protein kinase A, which inhibits the phosphorylation of myosin and lowers intracellular ionic calcium concentrations.	
127	Ipratropium bromide	Anticholinergic/β2 agonist	Works by antagonizing the action of acetylcholine, which inhibits vagally mediated reflexes and stops the increase in intracellular concentration of Ca^{2+}, which is caused by interaction of acetylcholine with the muscarinic receptors on bronchial smooth muscle.	For use in the management of bronchial spasms associated with COPD and asthma.
140	Loratadine	Antihistamine	Antihistamine	Used to provide temporary relief of symptoms due upper respiratory allergies.
154	Mometasone furoate	Corticosteroid	Corticosteroid	
155	Montelukast sodium	Leukotriene receptor antagonist	Works by binding to cysteinyl leukotriene (CysLT) receptors found in the airway and on other proinflammatory cells and inhibiting the actions of leukotriene D4 at the CysLT type 1 receptor.	Used in the management of asthma, perennial allergic rhinitis, seasonal allergic rhinitis, and exercise-induced bronchoconstriction.
168	Omalizumab	Monoclonal antibody/IgE blocker	Works by reducing the binding of IgE to the high-affinity IgE receptor on the surface of mast cells and basophils, which reduces the release of mediators of allergic response.	Used in the management of moderate to severe persistent asthma associated with allergies. Also used in the management of chronic idiopathic urticaria in patients who remain symptomatic despite H1 antihistamine management.
172	Oxymetazoline hydrochloride	Decongestant	Vasoconstrictor	Used in the short-term reduction of nasal congestion and sinus pressure caused by the common cold, or upper respiratory allergies.
187	Promethazine hydrochloride	Phenothiazine derivative	Works as an H1 receptor-blocking agent with antihistaminic properties and delivers sedative and antiemetic effects.	Used in the treatment of allergic disorders, and nausea/vomiting.
204	Salmeterol		Works by increasing the activity of intracellular adenyl cyclase, which leads to an increase in the concentration of cAMP. Increased cAMP concentration causes the relaxation of bronchial smooth muscle and inhibition of release of mediators of immediate hypersensitivity from cells, particularly, mast cells.	
222	Tiotropium bromide	Anticholinergic	Works by inhibiting M3-receptors on smooth muscle, which leads to bronchodilation.	Used in the management of bronchospasm.
228	Triamcinolone acetonide	Corticosteroid	Corticosteroid	
Antibiotics				
15	Amoxicillin	Semisynthetic ampicillin derivative	Bactericidal effect is due to the inhibition of the biosynthesis of cell wall.	Used to treat a range of infections caused by Gram-positive and a limited range of Gram-negative organisms, including those of the skin, ear, nose, throat, genitourinary tract, lower respiratory tract infections, acute uncomplicated gonorrhea, and in combination therapy for *Helicobacter pylori* eradication.

Drug no.	Drug name	Therapeutic class	Mechanism of action	Indications (Use)
24	Azithromycin	Macrolide	The mechanism of action is due to the binding of 23S rRNA of the 50S ribosomal subunit, and interfering with bacterial protein synthesis by impeding the assembly of the 50S ribosomal subunit.	Used in the treatment of bacterial infections including acute bacterial sinusitis, chlamydia, gonorrhea, community-acquired pneumonia, pelvic inflammatory disease, pediatric otitis media, and pharyngitis.
25	Bacitracin	Antibiotic	Antibacterial action against a number of Gram-positive and a few Gram-negative organisms.	Polypeptide antibiotic derived from *Bacillus subtilis.*
40	Cefdinir	Cephalosporin (3rd generation)	Works by inhibiting bacterial cell-wall synthesis.	Used in the treatment of a variety of bacterial infections.
42	Cephalexin	Cephalosporin (1st generation)	Efficacy is mediated by the inhibition of bacterial cell-wall synthesis.	Used to treat a variety of bacterial infections including upper respiratory infections, ear infections, skin infections, and urinary tract infections. Mostly used against Gram-positive bacteria.
48	Ciprofloxacin	Fluoroquinolone	Works by inhibiting topoisomerase II, the bacterial enzymes required for DNA replication, transcription, repair, and recombination.	Used in the treatment of infections from a wide range of Gram-positive and Gram-negative microorganisms including: UTIs, chronic bacterial prostatitis, lower respiratory tract infections, acute exacerbations of chronic bronchitis, acute sinusitis, skin and skin structure infections, bone and joint infections, complicated intra-abdominal infections, infectious diarrhea, typhoid fever, and uncomplicated cervical and urethral gonorrhea. Also used to reduce the incidence or progression of postexposure inhalation of anthrax. Used in the treatment of plague, and as prophylaxis for plague.
49	Clavulanate potassium		Bactericidal effect is due to the inactivation of a variety of β-lactamase enzymes frequently found in microorganisms resistant to penicillin and cephalosporin.	
59	Daptomycin	Cyclic lipopeptide	Efficacy due to cell death caused by the loss of membrane potential as a result of daptomcin binding to bacterial cell membranes.	Used in the treatment of complicated skin and skin structure infections caused by susceptible Gram-positive microorganisms, and of *Staphylococcus aureus* bloodstream infections (bacteremia), including those with right-sided infective endocarditis, caused by methicillin-susceptible and methicillin-resistant isolates.
77	Doxycycline	Tetracycline	Efficacy is thought to be due to a reduction in bacterial protein synthesis.	Used in the treatment of a variety of bacterial infections.
88	Erythromycin	Macrolide	Efficacy is due to the blocking of protein synthesis by binding to the 50S ribosomal subunit.	Used in the treatment of mild to moderate upper and lower respiratory tract infections, skin infections, listeriosis, pertussis, diphtheria infections, erythrasma, intestinal amebiasis, acute pelvic inflammatory disease, primary syphilis, and chlamydial infections.
132	Levofloxacin	Fluoroquinolone	Efficacy is due to inhibiting bacterial topoisomerase IV and DNA gyrase, enzymes necessary for DNA replication, transcription, repair, and recombination.	Used in the treatment of a variety of bacterial infections.
135	Linezolid	Oxazolidinone class antibacterial	Works by binding to a site on the bacterial 23S ribosomal RNA of the 50S subunit and inhibiting the formation of a functional 70S initiation complex, which is an essential component of the bacterial translation process.	Used in the treatment of nosocomial pneumonia; community-acquired pneumonia (including patients with concurrent bacteremia); complicated and uncomplicated skin and skin structure infections, and vancomycin-resistant *Enterococcus faecium* infections.
181	Polymyxin B	Antibacterial	Changes outer membrane of bacteria increasing permeability to water and cell death.	Used against Gram-negative infections.
196	Rifaximin	Semisynthetic rifampin analogue	Works by binding to β-subunit of bacterial DNA-dependent RNA polymerase and inhibiting transcription and thereby stemming protein synthesis and growth.	Used in the treatment of traveler's diarrhea, hepatic encephalopathy, and irritable bowel syndrome.
215	Sulfamethoxazole	Sulfonamide/tetrahydrofolic acid inhibitor	Works by reducing the bacterial synthesis of dihydrofolic acid by competing with para-aminobenzoic acid.	Used to treat a broad spectrum of bacterial infections.
230	Trimethoprim		Efficacy is mediated by reducing the production of tetrahydrofolic acid from dihydrofolic acid by inhibiting dihydrofolate reductase.	
Anti-Diabetic				
36	Canagliflozin	Sodium-glucose cotransporter 2 (SGLT2) inhibitor	Efficacy is mediated by decreasing the reabsorption of filtered glucose and lowering the renal threshold for glucose leading to an increase in urinary glucose excretion.	Used to increase glycemic control in adults with type 2 diabetes mellitus.
109	Glimepiride	Sulfonylurea (2nd generation)	Efficacy is due to stimulating insulin release from pancreatic β cells which leads to a lowering of blood glucose.	Used in the treatment of type 2 diabetes mellitus.
110	Glyburide	Sulfonylurea (2nd generation)	Efficacy is due to stimulating insulin release from pancreatic β cells which leads to a lowering of blood glucose.	Used in the treatment of type 2 diabetes mellitus.

Drug no.	Drug name	Therapeutic class	Mechanism of action	Indications (Use)
121	Insulin aspart	Insulin (rapid-acting)	Works by binding insulin receptors on muscle and fat cells increasing the cellular uptake of glucose while concurrently decreasing the output of glucose from the liver.	Used to increase glycemic control in patients with diabetes mellitus (DM).
122	Insulin detemir	Insulin (long-acting)	Long-acting insulin. Works by binding insulin receptors on muscle and fat cells increasing the cellular uptake of glucose while concurrently decreasing the output of glucose from the liver.	Used to increase glycemic control in patients with diabetes mellitus (DM).
123	Insulin glargine	Insulin (long-acting)	Long-acting insulin. Works by binding insulin receptors on muscle and fat cells increasing the cellular uptake of glucose while concurrently decreasing the output of glucose from the liver.	Used to increase glycemic control in patients with diabetes mellitus (DM).
124	Insulin lispro	Insulin (rapid-acting)	Works by binding insulin receptors on muscle and fat cells increasing the cellular uptake of glucose while concurrently decreasing the output of glucose from the liver.	Used to increase glycemic control in patients with diabetes mellitus (DM).
134	Linagliptin	Dipeptidyl peptidase-4 (DPP-4) inhibitor	Efficacy is due to an increase in the concentrations of active incretin hormones, stimulating the release of insulin in a glucose-dependent manner, and reducing glucagon levels.	Used to improve glycemic control in patients with type 2 diabetes mellitus.
136	Liraglutide	Glucagon-like peptide-1 (GLP-1) receptor agonist	Works by increasing intracellular cAMP, which causes insulin to be released in the presence of elevated glucose concentrations. Reduces glucagon secretion in a glucose-dependent manner and delays gastric emptying.	Used to improve glycemic control in adults with type 2 diabetes mellitus.
150	Metformin hydrochloride	Biguanide	Works by decreasing hepatic glucose production, intestinal absorption of glucose, and increasing peripheral glucose uptake and utilization.	Used in the management of type 2 diabetes mellitus.
178	Pioglitazone	Thiazolidinedione (glitazone)	Efficacy is due to a decrease in insulin resistance in the periphery and liver, leading to an increase in insulin-dependent glucose disposal and decreased hepatic glucose output.	Used in the management of type 2 diabetes mellitus.
205	Saxagliptin	Dipeptidyl peptidase-4 (DPP4) inhibitor	Increases the concentration of incretin in the blood by reducing the inactivation of incretin hormones. This reduces fasting and postprandial glucose concentrations in a glucose-dependent manner.	Used to increase glycemic control in patients with type 2 diabetes mellitus.
211	Sitagliptin	Dipeptidyl peptidase-4 (DPP4) inhibitor	Reduces the inactivation of incretin hormones, which promotes insulin release and decreases glucagon levels in the circulation in a glucose-dependent manner.	Used to increase glycemic control in adults with type 2 diabetes mellitus.
Anti-Inflammatory				
4	Acetylsalicylic acid	Nonsteroidal anti-inflammatory drug (NSAID)	Nonsteroidal anti-inflammatory drug (NSAID). The mechanism of action is not completely understood but is believed to involve the inhibition of COX-1 and COX-2.	Temporarily relieves minor aches and pains.
41	Celecoxib	Nonsteroidal anti-inflammatory Drug (NSAID)	Nonsteroidal anti-inflammatory drug (NSAID). The mechanism of action is not completely understood but is believed to involve the inhibition of COX-1 and COX-2.	Used in the management of pain associated with osteoarthritis, rheumatoid arthritis, and ankylosing spondylitis. Also used in the management of acute pain in adults, primary dysmenorrhea and juvenile rheumatoid arthritis.
71	Diclofenac sodium	Nonsteroidal anti-inflammatory drug (NSAID)	Nonsteroidal anti-inflammatory drug (NSAID). The mechanism of action is not completely understood but is believed to involve the inhibition of COX-1 and COX-2.	Used in the management of osteoarthritis and may be used topically.
117	Ibuprofen	Nonsteroidal anti-inflammatory drug (NSAID)	Nonsteroidal anti-inflammatory drug (NSAID). The mechanism of action is not completely understood but is believed to involve the inhibition of COX-1 and COX-2.	Used to reduce pains associated with muscular aches, headaches, backaches, menstrual cramps, toothaches, the common cold, and arthritis.
148	Meloxicam	Nonsteroidal anti-inflammatory drug (NSAID)	Nonsteroidal anti-inflammatory drug (NSAID). The mechanism of action is not completely understood but is believed to involve the inhibition of COX-1 and COX-2.	Used in the treatment of arthritis and osteoarthritis.
151	Methylprednisolone	Glucocorticoid	Works by decreasing the body's immune responses to various stimuli.	Used to treat inflammation.
158	Naproxen sodium	Nonsteroidal anti-inflammatory drug (NSAID)	Nonsteroidal anti-inflammatory drug (NSAID). The mechanism of action is not completely understood but is believed to involve the inhibition of COX-1 and COX-2.	Used to reduce pains associated with muscular aches, headaches, backaches, menstrual cramps, toothaches, the common cold, and arthritis.
185	Prednisone	Glucocorticoid	Works by modifying the body's immune responses to various stimuli.	Used in the management of many undesirable inflammatory conditions.
Antiviral and Antifungal				
102	Fluconazole	Azole antifungal	Works by inhibiting fungal CYP450-dependent enzyme lanosterol 14-α-demethylase. This leads to an increase in concentration of 14-α-methylsterols in fungi and is believed to be responsible for the fungistatic activity.	Used in the treatment of *Candidiasis* and *Crytococcal meningitis*.

Drug no.	Drug name	Therapeutic class	Mechanism of action	Indications (Use)
130	Ledipasvir	HCV NS5A inhibitor/HCV nucleotide analogue NS5B polymerase inhibitor	Works by inhibiting HCV NS5A protein thereby preventing viral replication.	Used in the treatment of hepatitis C.
153	Miconazole nitrate	Azole antifungal	Azole antifungal	Used in the treatment of athlete's foot (*Tinea pedis*), jock itch (*Tinea cruris*), and ringworm (*Tinea corporis*).
168	Oseltamivir phosphate	Neuraminidase inhibitor	Efficacy is mediated by inhibiting influenza virus neuraminidase which limits budding from host cells.	Used in the treatment and prophylaxis of influenza.
174	Palivizumab	Monoclonal antibody/RSV F-protein blocker	Efficacy is mediated by the binding the RSV envelope fusion protein on the surface of the virus and blocking the membrane fusion process.	Used in the prevention of serious lower respiratory tract disease caused by respiratory syncytial virus (RSV) in children at high risk of RSV disease.
209	Simeprevir	HCV NS3/4A protease inhibitor	Acts as an antiviral agent against the virus that causes hepatitis C.	Used in the treatment of hepatitis C.
212	Sofosbuvir	HCV nucleotide analogue NS5B polymerase inhibitor	Works by inhibiting HCV NS5B RNA-dependent RNA polymerase thereby preventing viral replication.	Used in the treatment of hepatitis C.
232	Valacyclovir hydrochloride	Nucleoside analogue	Efficacy is due to valacylovir's conversion to acyclovir which inhibits viral DNA polymerase by being incorporated into the DNA and causing the termination of the DNA growth.	Used in the treatment of cold sores (*Herpes labialis*), shingles (*Herpes zoster*), and genital herpes (*Herpes simplex*).
Autoimmune				
1	Abatacept	Selective costimulation modulator	Efficacy is mediated by inhibiting T-cell activation by binding to CD80 and CD86, thereby blocking interaction with CD28.	Used in the management of moderate to severe active rheumatoid arthritis and polyarticular juvenile idiopathic arthritis in order to limit the progression of structural damage, and to improve physical function.
5, 6	Adalimumab	Monoclonal antibody/TNF blocker	Works by binding to TNF-α and blocking its interaction with the p55 and p75 cell surface TNF receptors.	Used in the management of conditions where the suppression of immune response is desired. Reduces symptoms, inhibits progression of structural damage, and improves physical function in patients with rheumatoid arthritis, polyarticular juvenile idiopathic arthritis, ankylosing spondylitis, psoriatic arthritis, Crohn's disease, ulcerative colitis, and plaque psoriasis.
43	Certolizumab pegol	Tumor necrosis factor (TNF) blocker	Efficacy is mediated by the binding and neutralization of TNF-α, which is important in inflammatory processes.	Used in the management of Crohn's disease, moderate to severe rheumatoid arthritis, psoriatic arthritis, and ankylosing spondylitis.
80	Eculizumab	Monoclonal antibody/protein C5 blocker	Works by binding to the complement protein C5 with high affinity, which inhibits its cleavage to C5a and C5b and inhibits the production of the terminal complement complex C5b-9.	Used in the treatment of paroxysmal nocturnal hemoglobinuria and atypical hemolytic uremic syndrome.
92	Etanercept	Tumor necrosis factor (TNF) blocker	Works by reducing the binding of TNF-α and TNF-β to cell surface TNF-receptors, which results in TNF being biologically inactive.	Used in the management of rheumatoid arthritis, psoriatic arthritis, ankylosing spondylitis, and plaque psoriasis. Also may be used in the management of juvenile idiopathic arthritis.
119	Infliximab	Monoclonal antibody/TNF blocker	Efficacy is mediated by infliximab's binding to the soluble and transmembrane forms of TNF-α which inhibits the binding of TNF-α with its receptors.	Used in the management of autoimmune diseases. Decreases symptoms, inhibits progression of structural damage, and improves physical function with moderately to severely active Chrohn's disease, rheumatoid arthritis, psoriatic arthritis, ankylosing spondylitis, and plaque psoriasis.
231	Ustekinumab	Monoclonal antibody	Works by binding to the p40 protein subunit used by both the IL-12 and IL-23 cytokines.	Used in the management of moderate to severe plaque psoriasis.
Birth Control				
93	Ethinyl estradiol	Estrogen	Used with other birth control drugs to achieve efficacy primarily as mediated by suppressing ovulation.	Used in the prevention of pregnancy.
94	Etonogestrel	Estrogen/progestogen combination	Efficacy is primarily mediated by suppressing ovulation.	Used in the prevention of pregnancy.
161	Norelgestromin	Progestogen	Efficacy is primarily mediated by suppressing ovulation.	Used in combination with estrogen in the prevention of pregnancy.
162	Norethindrone acetate	Progestogen	Works by binding to an intracellular protein, FKBP-12, creating an inhibitory complex formation with mTOR complex 1, which results in the reduction of mTOR kinase activity.	Used in combination with estrogen in the prevention of pregnancy.
163	Norgestimate	Estrogen/progestogen combination	Efficacy is primarily mediated by suppressing ovulation.	Used in the prevention of pregnancy.
Blood Treatment				
86	Epoetin alfa	Erythropoiesis-stimulating agent (ESA)	Erythropoiesis-stimulating glycoprotein.	Used in the management of anemia associated with chronic kidney disease (including patients on dialysis), anemia caused by zidovudine in HIV-infected patients, and anemic patients with nonmyeloid malignancies.

Drug no.	Drug name	Therapeutic class	Mechanism of action	Indications (Use)
100	Filgrastim	Granulocyte colony-stimulating factor (G-CSF)	Works by increasing proliferation, differentiation, commitment, and end cell activation by binding to specific cell surface receptors on hematopoietic cells.	Used to decrease the chance of infection by increasing the concentration of neutrophils, in patients receiving myelosuppressive anticancer treatments, and may also be used to increase survival in patients exposed to radiation.
176	Pegfilgrastim	Granulocyte colony-stimulating factor (G-CSF)	Works by stimulating proliferation, differentiation, commitment, and end cell activation by binding to specific cell surface receptors on hematopoietic cells.	Used to increase the number of neutrophils, thereby decreasing the chance of infection in patients with nonmyeloid malignancies receiving myelosuppressive anticancer treatments.
60	Darbepoetin alfa	Erythropoiesis-stimulating agent	Efficacy is due to increasing erythropoiesis.	Used in the management of anemia caused by chronic kidney disease and where anemia is caused by the effect of concomitant myelosuppressive chemotherapy.
Cancer Therapy				
2	Abiraterone acetate	Antiandrogen	Efficacy is mediated by inhibiting 17 α-hydroxylase/C17,20-lyase (CYP17), an enzyme expressed in testicular, adrenal, and prostatic tumor tissues and required for androgen biosynthesis.	Used in combination with prednisone for the management of patients with metastatic castration-resistant prostate cancer.
26	Bendamustine hydrochloride	Alkylating agent	Efficacy is due to electrophilic alkyl groups that may form covalent bonds with electron-rich nucleophilic moieties. The result is DNA cross-links which leads to cell death. Active against both quiescent and dividing cells.	Used in the management of chronic lymphocytic leukemia. Also used in the management of indolent B-cell non-Hodgkin lymphoma that has progressed during or within 6 months of treatment with rituximab or a rituximab-containing regimen.
27	Bevacizumab	Vascular endothelial growth factor (VEGF) inhibitor	Efficacy is due to the binding of VEGF to the receptors Flt-1 and KDR, which leads to endothelial cell proliferation and new blood vessel formation.	1st- or 2nd-line treatment of some metastatic cancers.
31	Bortezomib	Proteasome inhibitor	Efficacy is mediated by reversibly inhibiting chymotrypsin-like activity of the 26S proteasome in cells.	Used in the management of multiple myeloma and mantle cell lymphoma.
37	Capecitabine	Fluoropyrimidine carbamate	Efficacy is mediated by binding to thymidylate synthase to form a covalently bound ternary complex that prevents the formation of thymidylate from 2′-deoxyuridylate, thereby preventing DNA synthesis, and also interferes with RNA processing and protein synthesis.	Used in the treatment of metastatic breast cancer and metastatic colorectal cancer.
44	Cetuximab	Monoclonal antibody/ epidermal growth factor receptor (EGFR) antagonist	Works by binding to EGFR on normal and tumor cells and decreases the binding of ligands such as epidermal growth factor and transforming growth factor-alpha.	Used in the treatment of squamous cell carcinoma of the head and neck and metastatic colorectal cancer.
62	Dasatinib	Kinase inhibitor	Efficacy is due to inhibiting BCR-ABL, SRC family, c-KIT, EPHA2, and PDGFRβ kinases.	Used in the management of newly diagnosed Philadelphia chromosome-positive chronic myeloid leukemia in chronic phase. Also used in the management of chronic, accelerated, or myeloid or lymphoid blast phase Ph+ CML with resistance or intolerance to prior therapy, including imatinib, and in the management of Ph+ acute lymphoblastic leukemia with resistance or intolerance to prior therapy.
87	Erlotinib hydrochloride	Kinase inhibitor	Efficacy is mediated by reversibly blocking the kinase activity of EGFR, this prevents autophosphorylation of tyrosine residues accompanying the receptor and reduces further downstream signaling.	Used in the treatment of metastatic non-small cell lung cancer where the tumors have epidermal growth factor receptor (EGFR) exon 19 deletions or exon 21 (L858R) substitution mutations.
95	Everolimus	Kinase inhibitor	Works by binding to an intracellular protein, FKBP-12, creating an inhibitory complex formation with mTOR complex 1, which results in the reduction of mTOR kinase activity.	Used in the treatment of advanced hormone receptor-positive, HER2-negative breast cancer, advanced neuroendocrine tumors, advanced renal cell carcinoma, renal angiomyolipoma with tuberous sclerosis complex, and subependymal giant cell astrocytoma with tuberouse sclerosis complex.
118	Imatinib mesylate	Kinase inhibitor	Efficacy is mediated by inhibiting the BCR-ABL tyrosine kinase. This reduces the proliferation and induces apoptosis in BCR-ABL positive cell lines as well as leukemic cells. Inhibits the receptor tyrosine kinases for platelet-derived growth factor and stem cell factor, c-Kit, and inhibits their mediated cellular events.	Used in the treatment of some forms of cancer, including: Philadelphia chromosome positive chronic myeloid leukemia (Ph+ CML), Ph+ acute lymphoblastic leukaemia, myelodysplastic/myeloproliferative diseases, aggressive systemic mastocytosis, hypereosinophilic syndrome and/or chronic eosinophilic leukemia (CEL), dermatofibrosarcoma protuberans, and malignant gastrointestinal stromal tumors (GIST).
126	Ipilimumab	Monoclonal antibody/CTLA-4 blocker	Efficacy is mediated by the binding of ipilimumab to CTLA-4 and blocking interactions with the ligands CD80/CD86. This causes an increase in T-cell activation and proliferation, including antitumor immune response.	Used in the treatment of unresectable or metastatic melanoma.
131	Lenalidomide	Thalidomide analogue	Works by inducing apoptosis and reducing the proliferation and of certain hematopoietic tumor cells. Immunomodulatory properties include activation and increased number of natural killer T cells, and inhibition of proinflammatory cytokines by monocytes.	Used in the treatment of myelodysplastic syndrome, multiple myeloma, and mantle cell lymphoma.

Drug no.	Drug name	Therapeutic class	Mechanism of action	Indications (Use)
164	Octreotide acetate	Somatostatin analogue	Works by mimicking the hormone somatostatin. Octreotide suppresses luteinizing hormone response to gonadotropin-releasing hormone, decreases splanchnic blood flow, and reduces the release of serotonin, gastrin, vasoactive intestinal peptide, secretin, motilin, and pancreatic polypeptide.	Used in the management of acromegaly and in the reduction of diarrhea associated with metastatic carcinoid tumors, and vasoactive intestinal peptide-secreting tumors.
177	Pemetrexed disodium	Antifolate	Works by blocking thymidylate synthase, dihydrofolate reductase, and glycinamide ribonucleotide formyl transferase, which are required for cell replication.	Used in the treatment of metastatic nonsquamous non-small cell lung cancer and malignant pleural mesothelioma.
225	Trastuzumab	Monoclonal antibody/HER2 blocker	Works by inhibiting the proliferation of tumor cells characterized by HER2 overexpression.	Used in the treatment of HER2-overexpressing breast cancers. May be used in combination with cisplatin and capecitabine or 5-fluorouracil for the treatment of HER2-overexpressing metastatic gastric or gastroesophageal junction adenocarcinoma.
200	Rituximab	Monoclonal antibody/CD20 blocker	Works by binding to CD20 antigen expressed on the surface of B-lymphocytes and mediates B-cell lysis.	Used in treatment of diseases, which are characterized by excessive numbers of B cells, overactive B cells, or dysfunctional B cells such as CD20-positive non-Hodgkins lymphoma, chronic lymphocytic leukemia, and rheumatoid arthritis.
Cholesterol Treatment				
23	Atorvastatin calcium	HMG-CoA reductase inhibitor (statin)	Works by preventing the conversion of HMG-CoA to mevalonate (precursor of sterols).	Used in the management of high cholesterol and triglyceride levels.
55	Colesevelam hydrochloride	Bile acid sequestrant	Efficacy is due to LDL clearance from the blood due to the binding of bile acids in the intestine, which hinders their readsorption.	Used in order to decrease LDL levels either as monotherapy or in combination with a statin. Used in order to improve glycemic control in patients with type 2 diabetes mellitus.
96	Ezetimibe	Cholesterol absorption inhibitor	Works to reduce blood cholesterol by reducing the absorption of cholesterol by the sterol transporter, Niemann-Pick C1-Like 1 in the small intestine.	Used in the management of high cholesterol.
144	Lovastatin	HMG-CoA reductase inhibitor (statin)	Works by preventing the conversion of HMG-CoA to mevalonate (precursor of sterols).	Used in the management of high cholesterol and triglyceride levels.
184	Pravastatin sodium	HMG-CoA reductase inhibitor (statin)	Effect is mediated by inhibiting the conversion of HMG-CoA to mevalonate.	Used in the management of high cholesterol and triglyceride levels.
203	Rosuvastatin calcium	HMG-CoA reductase inhibitor (statin)	Works by increasing lipid-modifying effects by increasing the concentration of hepatic LDL receptors on the cell surface and enhancing the uptake and catabolism of LDL and by reducing hepatic synthesis of VLDL, which decreases the total number of VLDL and LDL particles.	Used in the management of high cholesterol and triglyceride levels.
210	Simvastatin	HMG-CoA reductase inhibitor (statin)	Efficacy is believed to be mediated by the inhibition of the conversion of HMG-CoA to mevalonate, which is a rate-limiting step in the biosynthetic pathway for cholesterol. Decreases concentrations of VLDL, TGs, and increases HDL.	Used in the management of high cholesterol and triglyceride levels.
Cold Therapy				
69	Dextromethorphan hydrobromide	Antitussive/expectorant	Cough suppressant	Used to treat cough caused by throat and bronchial irritation, and loosen phlegm and bronchial secretions.
111	Guaifenesin		Expectorant	
172	Oxymetazoline hydrochloride	Decongestant	Vasoconstrictor	Used in the short-term reduction of nasal congestion and sinus pressure caused by the common cold, or upper respiratory allergies.
Dementia Treatment				
149	Memantine hydrochloride	NMDA receptor antagonist	Efficacy is believed to be mediated by its action as a low to moderate affinity uncompetitive (open-channel) NMDA receptor antagonist.	Used in the management of moderate to severe dementia associated with Alzheimer's disease.
202	Rivastigmine tartrate	Acetylcholinesterase (AChE) inhibitor	Efficacy is believed to be mediated by an increase in cholinergic function by increasing concentration of acetylcholine through reversible inhibition of its hydrolysis by cholinesterase.	Used in the management of dementia accompanying Alzheimer's disease and Parkinson's disease.
Depression and Anxiety Therapy				
11	Alprazolam	Benzodiazepine	Efficacy is believed to be mediated by binding at stereo specific receptors at multiple sites within the CNS.	Used in the management of anxiety symptoms and in anxiety disorders. Also used in the treatment of panic disorders, with or without agoraphobia.
12	Amitriptyline hydrochloride	Tricyclic antidepressant (TCA)	Efficacy is believed to be due to inhibiting the membrane pump mechanism responsible for uptake of norepinephrine and serotonin in adrenergic and serotonergic neurons.	Used mainly in the management of depression, but may also be used in the treatment of agitation and insomnia, chronic pain, diabetic neuropathy, irritable bowel syndrome, sleep disorders, and migraine prophylaxis.
16	Amphetamine aspartate monohydrate	CNS stimulant	Efficacy is believed to be mediated by reducing the reuptake of norepinephrine and dopamine into the presynaptic neuron and increasing the release of these monoamines into the extraneuronal space.	Used in the management of attention-deficit hyperactivity disorder and narcolepsy.

Drug no.	Drug name	Therapeutic class	Mechanism of action	Indications (Use)
17	Amphetamine sulfate	CNS stimulant	Efficacy is believed to be mediated by reducing the reuptake of norepinephrine and dopamine into the presynaptic neuron and increasing the release of these monoamines into the extraneuronal space.	Used in the management of attention-deficit hyperactivity disorder and narcolepsy.
19	Aripiprazole	Atypical antipsychotic	Efficacy is mediated by its effect as a partial D2/5HT1A agonist and 5HT2A antagonist.	Used in the management of schizophrenia, irritability accompanying autistic disorder, and Tourette's disorder. Also used in the management of manic and mixed episodes accompanying bipolar I disorder. Used in adjunctive management of major depressive disorder (MDD).
22	Atomoxetine	Selective norepinephrine reuptake inhibitor	Efficacy is believed to be mediated by selectively inhibiting the presynaptic norepinephrine transporter.	Used in the management of attention deficit hyperactivity disorder.
50	Clonazepam	Benzodiazepine	Efficacy is believed to be mediated by its ability to enhance the activity of gamma-aminobutyric acid, the major inhibitory neurotransmitter in the CNS.	Used in the management of Lennox-Gastaut syndrome (petit mal variant), akinetic, and myoclonic seizures. Clonazepam has been used in patients with absence seizures (petit mal) who have failed to respond to succinimides and is also used in the treatment of panic disorder with or without agoraphobia.
64	Desvenlafaxine	Serotonin and norepinephrine reuptake inhibitor (SNRI)	Mechanism of efficacy has not been fully established. Efficacy may be associated with the potentiation of serotonin and norepinephrine in the CNS through inhibition of their reuptake.	Used in the management of major depressive disorder.
66	Dexmethylphenidate hydrochloride	CNS stimulant	Mechanism of efficacy has not been fully established. Efficacy is believed to be due to a reduction in the reuptake of norepinephrine and dopamine into the presynaptic neuron and an increase in the release of monoamines into the extraneuronal space.	Used in the management of attention deficit hyperactivity disorder (ADHD).
67	Dextroamphetamine saccharate	CNS stimulant	Efficacy is believed to be mediated by reducing the reuptake of norepinephrine and dopamine into the presynaptic neuron and increasing the release of these monoamines into the extraneuronal space.	Used in the management of attention deficit hyperactivity disorder and narcolepsy.
68	Dextroamphetamine sulfate	CNS stimulant	Efficacy is believed to be mediated by reducing the reuptake of norepinephrine and dopamine into the presynaptic neuron and increasing the release of these monoamines into the extraneuronal space.	Used in the management of attention deficit hyperactivity disorder and narcolepsy.
70	Diazepam	Benzodiazepine	Efficacy is believed to be mediated by facilitating the actions of gamma aminobutyric acid.	Used in the management of anxiety disorders or for temporary relief of the symptoms of anxiety. Used in the management of acute alcohol withdrawal. Used for relief of skeletal muscle spasm and used in the management of convulsive disorders.
78	Duloxetine	Serotonin and norepinephrine reuptake inhibitor (SNRI)	Efficacy is thought to be due to the potentiation of serotonergic and noradrenergic activity in the central nervous system.	Used in the management of major depressive disorder, generalized anxiety disorder, diabetic peripheral neuropathy, fibromyalgia, and chronic musculoskeletal pain.
89	Escitalopram oxalate	Selective serotonin reuptake inhibitor (SSRI)	Efficacy is believed to be due to a reduction of the CNS neuronal uptake of serotonin.	Used in the management of generalized anxiety disorder and major depressive disorder.
103	Fluoxetine hydrochloride	Selective serotonin reuptake inhibitor (SSRI)	Efficacy is believed to be mediated by inhibiting the CNS neuronal uptake of serotonin.	Used in the management of major depressive disorder, social anxiety disorder, panic disorders, with or without agoraphobia, obsessive-compulsive disorder, premenstrual dysphoric disorder, and post-traumatic stress disorder.
137	Lisdexamfetamine dimesylate	CNS stimulant	Works by reducing the reuptake of dopamine and norepinephrine into the presynaptic neuron and increasing their release into the extraneuronal space.	Used in the management of attention deficit hyperactivity disorder and moderate to severe binge eating disorder.
141	Lorazepam	Benzodiazepine	Efficacy is believed to be mediated by binding at stereo specific receptors at several sites within the CNS.	Used in the treatment of anxiety disorders.
146	Lurasidone hydrochloride	Atypical antipsychotic	Efficacy may be mediated by central dopamine type 2 and serotonin type 2-receptor antagonism.	Used in the management of schizophrenia, and major depressive episodes associated with bipolar I disorder.
173	Paliperidone palmitate	Atypical antipsychotic	Efficacy is believed to be mediated by a combination of central dopamine type 2 and 5HT2A receptor antagonism.	Used in the management of schizophrenia.
175	Paroxetine hydrochloride	Selective serotonin reuptake inhibitor (SSRI)	Efficacy is believed to be mediated by inhibiting the CNS neuronal reuptake of serotonin.	Used in the management of major depressive disorder, panic disorder with or without agoraphobia, obsessive-compulsive disorder, social anxiety disorder, generalized anxiety disorder, post-traumatic stress disorder, and premenstrual dysphoric disorder.
188	Quetiapine fumarate	Atypical antipsychotic	Efficacy is believed to be mediated through a combination of dopamine type 2 and serotonin type 2A antagonism.	Used in the management of schizophrenia, acute manic, depressive, or mixed episodes accompanying bipolar I disorder, and as adjunctive therapy to antidepressants for the management of major depressive disorder in adults.
206	Sertraline hydrochloride	Selective serotonin reuptake inhibitor (SSRI)	Efficacy is believed to be mediated by reducing the CNS neuronal uptake of serotonin.	Used in the management of major depressive disorder, social anxiety disorder, panic disorders, premenstrual dysphoric disorder, post-traumatic stress disorder and obsessive-compulsive disorder.

Drug no.	Drug name	Therapeutic class	Mechanism of action	Indications (Use)
227	Trazodone hydrochloride	Triazolopyridine derivative	Efficacy may be linked to the selective inhibition of neuronal reuptake of serotonin and its activity as an antagonist at 5-HT-2A/2C serotonin receptors.	Used in the treatment of major depressive disorder.
236	Venlafaxine hydrochloride	Serotonin and norepinephrine reuptake inhibitor (SNRI)	Mechanism of efficacy has not been fully established. Efficacy may be associated with the potentiation of serotonin and norepinephrine in the CNS through inhibition of their reuptake.	Used in the management of major depressive disorder, generalized anxiety disorder, social anxiety disorder, and panic disorder.
237	Vilazodone hydrochloride	Selective serotonin reuptake inhibitor (SSRI)/5-HT1A-receptor partial agonist	Efficacy may be associated with the potentiation of serotonin in the CNS through inhibition of its reuptake.	Used in the management of major depressive disorder.
Dermatology				
73	Dimethyl fumarate	Immunomodulatory agent	Efficacy is believed to be due to dimethyl fumarate and the metabolite (monomethyl fumarate [MMF]) activating the nuclear factor (erythroid-derived 2)-like 2 (Nrf2) pathway, which is important in the cellular response to oxidative stress.	
116	Hydrocortisone	Corticosteroid	Corticosteroid	Used in the management of skin conditions such as redness, itching, and swelling.
154	Mometasone furoate	Corticosteroid	Corticosteroid	Used in the management of asthma and the management of nasal symptoms of seasonal and perennial allergic rhinitis.
182	Pramoxine hydrochloride	Local anesthetic	Decreases permeability of membranes to Na influx.	Temporarily relieves itching and other minor skin irritations.
228	Triamcinolone acetonide	Corticosteroid	Corticosteroid	Used in the management of various disorders in which corticosteroids are indicated such as eczema, dermatitis, allergies, and rash.
Digestive Treatment				
29	Bisacodyl	Stimulant laxative	Stimulant laxative.	Used in the treatment of occasional constipation and irregularity.
76	Docusate sodium	Stool softener	Increases concentration of water in the stool.	Used to treat occasional constipation.
139	Loperamide hydrochloride	Antidiarrheal	Works by extending the transit time of intestinal contents.	Controls symptoms of diarrhea.
145	Lubiprostone	Chloride channel activator	Works by increasing chloride-rich intestinal fluid secretion, and motility in the intestine.	Used in the management of chronic idiopathic constipation, opioid-induced constipation, and irritable bowel syndrome with constipation.
180	Polyethylene glycol 3350	Osmotic laxative	Osmotic laxative.	Used in the treatment of occasional constipation.
Heart Therapy				
13	Amlodipine		Works by reducing the transmembrane influx of Ca^{2+} ions into vascular smooth muscle and cardiac muscle. Reduces peripheral vascular resistance and BP by acting directly on vascular smooth muscle.	
14	Amlodipine besylate	Calcium channel blocker (CCB)	Works by reducing the transmembrane influx of Ca^{2+} ions into vascular smooth muscle and cardiac muscle. Reduces peripheral vascular resistance and BP by acting directly on vascular smooth muscle.	Used in the management of hypertension, angina, and coronary artery disease.
18	Apixaban	Selective factor Xa inhibitor	Efficacy is mediated by decreasing thrombin production and thrombus development by inhibiting FXa and does not require antithrombin III for antithrombotic activity.	Used to decrease the chance of stroke and systemic embolism associated with nonvalvlular atrial fibrillation and to decrease the risk of developing venous thrombosis after some orthopedic surgical procedures.
21	Atenolol	Selective β1-blocker	The exact mechanism of action is not fully understood. Believed to work by competitively antagonizing catecholamines at peripheral adrenergic neuron sites. This causes a decreased cardiac output, reduced sympathetic outflow to the periphery, and suppression of renin activity.	Used in the treatment of hypertension and angina pectoris.
39	Carvedilol	α1/β-adrenergic blocker	The mechanism of action has not been fully established. Carvedilol has α1-adrenoreceptor blocking activity and attenuates the pressor effects of phenylephrine, causes vasodilation, and reduces peripheral vascular resistance.	Used in the management of mild to moderate heart failure of ischemic or cardiomyopathic origin.
51	Clonidine hydrochloride	Alpha-adrenergic agonist	Efficacy is due to the stimulation of α-adrenoreceptors in the brain stem, decreasing sympathetic outflow from CNS and decreasing peripheral resistance, renal vascular resistance, HR, and BP.	Used in the treatment of hypertension and prophylaxis of vascular migraine headaches. Used in the management of vasomotor symptoms associated with menopause, rapid detoxification in the management of substance withdrawal, and in the treatment of attention deficit hyperactivity disorder.
52	Clopidogrel bisulfate	Antiplatelet agent	Efficacy is mediated by irreversibly and selectively inhibiting the binding of adenosine diphosphate to its platelet P2Y12 receptor and the consequent ADP-mediated activation of the glycoprotein GPIIb/IIIa complex.	Used in the treatment of myocardial infarction, recent stroke, peripheral arterial disease, and acute coronary syndrome.

Drug no.	Drug name	Therapeutic class	Mechanism of action	Indications (Use)
58	Dabigatran etexilate mesylate	Direct thrombin inhibitor (DTI)	Works by inhibiting free and clot-bound thrombin and thrombin-induced platelet aggregation.	Used to decrease the risk of stroke and systemic embolism in patients with nonvalvular atrial fibrillation. Used in the management of deep venous thrombosis and pulmonary embolism in patients who have been treated with a parenteral anticoagulant for 5-10 days.
72	Digoxin	Cardiac glycoside	Works by inhibiting Na+-K+ ATPase.	Used in the management of congestive heart failure, arrhythmias, and heart failure.
75	Dipyridamole	Antiplatelet agent	Efficacy is due to a reduction in the uptake of adenosine into platelets, endothelial cells, and erythrocytes.	Used to decrease the chance of stroke in patients who have had transient ischemia of the brain or complete ischemic stroke due to thrombosis.
84	Enalapril maleate	ACE inhibitor	Efficacy is due to a decrease in plasma angiotensin II, which leads to reduced vasopressor activity and reduced aldosterone secretion.	Used in the treatment of hypertension and symptomatic congestive heart failure.
85	Enoxaparin sodium	Low molecular weight heparin	Enoxaparin sodium has antithrombotic properties.	Used in the treatment of myocardial infarction, unstable angina, and deep vein thrombosis.
98	Fenofibrate	Fibric acid derivative	Efficacy is due to the activation of the peroxisome proliferator-activated receptor α. This results in an increase in lipolysis and elimination of TG-rich particles from plasma by activating lipoprotein lipase and reducing production of apoprotein C-III.	Used in the management of high cholesterol and triglycerides.
128	Isosorbide mononitrate	Nitrate vasodilator	Works by relaxing vascular smooth muscle, producing dilatation of peripheral arteries and veins.	Used in the treatment of acute and chronic angina pectoris, hypertension, myocardial infarction, and in the prevention of angina pectoris due to coronary artery disease.
138	lisinopril	ACE inhibitor	Mechanism of action is due to a reduction in plasma angiotensin II, which causes a decrease in vasopressor activity and decreased aldosterone secretion.	Used in the treatment of high blood pressure, heart failure, and acute myocardial infarction.
142	Losartan potassium	Angiotensin II receptor blocker (ARB)	Works by blocking the vasoconstrictor effects of angiotensin II by blocking the binding of angiotensin II to the AT1 receptor in vascular smooth muscle.	Used in the management of high blood pressure.
152	Metoprolol succinate	Selective β1-blocker	Efficacy is believed to be due to a decrease in cardiac output, reduced sympathetic outflow to the periphery, and suppressed renin activity.	Used in the management of hypertension. It is also used to prevent further heart problems after myocardial infarction.
159	Nebivolol	Selective β1-blocker.	Selective β1-blocker.	Used in the management of high blood pressure.
165	Olmesartan acid		Works by inhibiting the vasoconstrictor effects of angiotensin II by blocking the binding of angiotensin II to the AT1 receptor in vascular smooth muscle.	
166	Olmesartan medoxomil	Angiotensin II receptor blocker (ARB)	Works by blocking the vasoconstrictor effects of angiotensin II by selectively blocking the binding of angiotensin II to the AT1 receptor in vascular smooth muscle.	Used in the management of high blood pressure.
183	Prasugrel	Platelet aggregation inhibitor	Works by inhibiting platelet activation and aggregation through binding of its active metabolite to the P2Y12 class of ADP receptors on platelets.	Used to decrease the risk of atherothrombotic events in patients with acute coronary symptoms.
192	Ramipril	ACE inhibitor	Mechanism of action is due to a decrease in plasma angiotensin II, which leads to decreased vasopressor activity and decreased aldosterone secretion.	Used in the treatment of high blood pressure and heart failure.
195	Ranolazine	Miscellaneous antianginal	Antianginal	Used in the management of chronic angina.
201	Rivaroxaban	Selective factor Xa inhibitor	Efficacy is mediated by reducing factor Xa. Rivaroxaban decreases thrombin generation and indirectly inhibits platelet aggregation induced by thrombin.	Used to decrease the risk of stroke and systemic embolism in patients with nonvalvular atrial fibrillation. Also used to decrease the risk of the recurrence of deep vein thrombosis and pulmonary embolism, and used in the prophylaxis of deep vein thrombosis, which may lead to pulmonary embolism in patients undergoing knee or hip replacement surgery.
218	Telmisartan	Angiotensin II receptor blocker (ARB)	Works by inhibiting angiotensin II's vasoconstrictor and aldosterone-secreting effects by selectively blocking the binding of angiotensin II to the AT1 receptor.	Used in the management of hypertension and to reduce the risk of MI, stroke, or death from cardiovascular causes.
233	Valsartan	Angiotensin II receptor blocker (ARB)	Works by inhibiting the vasoconstrictor and aldosterone-secreting effects of angiotensin II by blocking the binding of angiotensin II to the AT1 receptor.	Used in the management of high blood pressure.
238	Warfarin sodium	Vitamin K-dependent coagulation factor inhibitor	Efficacy is believed to be mediated by the inhibition of the C1 subunit of the vitamin K epoxide reductase enzyme complex, thus decreasing the regeneration of vitamin K1 epoxide.	Used in the management of venous thrombosis, pulmonary embolism, and thromboembolic complications associated with atrial fibrillation and/or cardiac valve replacement. Used to decrease the risk of death, recurrent myocardial infarction, and thromboembolic events, such as stroke or systemic embolization.
HIV Treatment				
20	Atazanavir	Protease inhibitor	Inhibits the development of mature virions. Inhibits virus-specific processing of viral Gag and Gag-Pol polyproteins in HIV-1 infected cells.	Used in the management of HIV-1 infection in combination with other antiretrovirals.

Drug no.	Drug name	Therapeutic class	Mechanism of action	Indications (Use)
53	Cobicistat	CYP3A inhibitor/HIV integrase strand transfer inhibitor/ nucleoside reverse transcriptase inhibitor (NRTI)	Efficacy is due to an increase in systemic exposure to CYP3A substrates by inhibiting CYP3A-mediated metabolism (e.g., elvitegravir).	For use in the management of HIV-1 infection.
61	Darunavir	Protease inhibitor	Prevents the cleavage of HIV-1 encoded Gag-Pol polyproteins in infected cells, limiting the formation of mature virus particles.	Used in the management of HIV-1 infection in combination with ritonavir and other antiretroviral agents.
81	Efavirenz		Works by inhibiting HIV-1 transcriptase noncompetitively.	Used in the management of HIV-1 infection.
82	Elvitegravir		Limits HIV-1 infection by preventing the formation of HIV-1 provirus and therefore viral propagation by inhibiting the strand transfer activity of HIV-1 integrase.	Used in the management of HIV-1 infection.
83	Emtricitabine		Works by competing with deoxycytidine 5'-triphosphate and by being incorporated into nascent viral DNA, resulting in chain termination.	Used in the management of HIV-1 infection.
191	Raltegravir	HIV-integrase strand transfer inhibitor	Works by preventing the catalytic action of HIV-1 integrase, which inhibits the formation of HIV-1 provirus and limits the propagation of the viral infection.	Used in the management of HIV-1 infection.
197	Rilpivirine		Works by preventing HIV-1 replication by noncompetitive inhibition of HIV-1 RT.	Used in the management of HIV-1 infection.
199	Ritonavir	Protease inhibitor	Works by inhibiting the processing the Gag-Pol polyprotein precursor, which decreases the production of mature HIV-1 particles.	Used in the management of HIV-1 infection.
219	Tenofovir disoproxil fumarate	Nucleotide analogue reverse transcriptase inhibitor	Works by inhibiting HIV-1 reverse transcriptase and HBV reverse transcriptase by competing deoxyadenosine 5'-triphosphate and, following incorporation into DNA, by DNA chain termination.	Used in the management of HIV-1 infection and chronic hepatitis B.
Hormonal Treatment				
56	Conjugated estrogens	Estrogen	Works by increasing concentrations of estrogens, which bind to estrogen receptors in a number of tissues and regulate the pituitary secretions.	Used in the prevention of postmenopausal osteoporosis, and in the treatment of moderate to severe vasomotor symptoms and/or vulvar and vaginal atrophy caused by menopause. Used in the treatment of hypoestrogenism due to hypogonadism, castration, or primary ovarian failure.
133	Levothyroxine sodium	Thyroid replacement hormone	Efficacy is believed to be due to effects exerted through control of DNA transcription and protein synthesis.	Used in the management of congenital or acquired hypothyroidism.
147	Medroxyprogesterone acetate		Synthetic progestin (more potent than progesterone). Works by diffusing into cells in the female reproductive tract, mammary gland, hypothalamus and pituitary and binds progesterone receptor which slows the release of gonadotropin from the hypothalamus.	
Multiple Sclerosis Therapy				
73	Dimethyl fumarate	Immunomodulatory agent	Efficacy is believed to be due to dimethyl fumarate and the metabolite (monomethyl fumarate [MMF]) activating the nuclear factor (erythroid-derived 2)-like 2 (Nrf2) pathway, which is important in the cellular response to oxidative stress.	Used in the management of relapsing forms of multiple sclerosis.
101	Fingolimod	Sphingosine 1-phosphate receptor modulator	The exact mechanism is unknown but efficacy may be due to a reduction of lymphocyte migration into the CNS.	Used in the management of relapsing forms of multiple sclerosis.
108	Glatiramer acetate	Immunomodulatory agent	Immunomodulatory agent.	Used in the management of multiple sclerosis to reduce the frequency of relapses.
125	Interferon beta-1a	Biological response modifier	Biological response modifier.	Used in the management of relapsing forms of multiple sclerosis.
Opthamalogic Treatment				
7	Aflibercept	Vascular endothelial growth factor inhibitor	Aflibercepts binds VEGF-A and placental growth factor, and results in the decreased binding and activation of the cognate VEGF receptors.	Used in the treatment of macular edema, neovascular age-related macular degeneration, diabetic macular edema, and diabetic retinopathy.
28	Bimatoprost	Prostaglandin analogue	Efficacy is due to its ability to mimic prostamides.	Used in the treatment of elevated intraocular pressure in patients with open-angle glaucoma or ocular HTN.
32	Brimonidine tartrate	Selective $\alpha2$ agonist	Works by decreasing aqueous humor production and increasing uveoscleral outflow.	Used in the management of increased pressure in the eye caused by open-angle glaucoma or a condition called hypertension of the eye (ocular hypertension).
143	Loteprednol etabonate	Corticosteroid	Corticosteroid	Used in the management of inflammation and pain caused by several conditions, including allergies, allergic conjunctivitis, uveitis, acne rosacea, superficial punctate keratitis, herpes zoster keratitis, iritis, cyclitis, and selected infective conjunctivitides.
167	Olopatadine hydrochloride	H1 antagonist and mast cell stabilizer	Works by reducing the type 1 immediate hypersensitivity reaction.	Used in the management of ocular itching with accompanying allergic conjunctivitis.

Drug no.	Drug name	Therapeutic class	Mechanism of action	Indications (Use)
193	Ranibizumab	Monoclonal antibody/vascular endothelial growth factor (VEGF)-A blocker	Works by inhibiting the interaction of VEGF-A with its receptors (VEGFR1 and VEGFR2) on the surface of endothelial cells, limiting endothelial cell proliferation, vascular leakage, and new blood vessel formation.	Used in the management of neovascular age-related macular degeneration, macular edema following retinal vein occlusion, diabetic macular edema, and diabetic retinopathy.
221	Timolol		Nonselective β-blocker.	
226	Travoprost	Prostaglandin analogue	Efficacy is believed to be mediated by travoprost being a selective FP prostanoid receptor agonist and is thought to decrease intraocular pressure by increasing uveoscleral outflow.	Used in the treatment of elevated intraocular pressure in patients with open-angle glaucoma or ocular HTN.
Osteoporosis				
9	Alendronate sodium	Bisphosphonate	Works by inhibiting osteoclast-mediated bone resorption by binding to hydroxyapatite found in bone.	Used in the treatment and prevention of osteoporosis and in the treatment of Paget's disease of bone.
63	Denosumab	IgG2 monoclonal antibody	Denosumab decreases bone resorption and increases bone mass and strength by binding to the receptor activator of nuclear factor kappa-B ligand (RANKL) and inhibiting RANKL from activating its receptor.	Used to increase bone mass in patients with a high risk of fracture.
190	Raloxifene hydrochloride	Selective estrogen receptor modulator	Works by binding to estrogen receptors, which results in activation of estrogenic pathways in some tissues and deactivation of estrogenic pathways in others. For example, it works as an estrogen agonist in bone; decreasing bone resorption and bone turnover, while increasing bone mineral density, and decreasing fracture incidence.	Used in management and prevention of osteoporosis in postmenopausal women. Also, reduces risk of invasive breast cancer in postmenopausal women.
198	Risedronate sodium	Bisphosphonate	Functions as an antiresorptive agent and inhibits osteoclasts.	Used in the management and prevention of osteoporosis, glucocorticoid-induced osteoporosis in patients who are either starting or continuing systemic glucocorticoid treatment for chronic diseases, and the treatment of Paget's disease.
Pain Relief				
3	Acetaminophen	Analgesic	Efficacy is believed to be due to effects on the CNS, increasing the pain threshold by inhibiting both isoforms of cyclooxygenase, COX-1, COX-2, and COX-3 enzymes involved in prostaglandin (PG) synthesis. In contrast to NSAIDs, acetaminophen does not inhibit cyclooxygenase in peripheral tissues and exhibits no peripheral anti-inflammatory actions.	Temporarily reduces fever and relieves minor aches and pains.
35	Caffeine		Stimulant; pain reliever aid.	
38	Carisoprodol	Skeletal muscle relaxant (centrally acting)	The mechanism of action is not completely understood. Efficacy may be due to altered interneuronal activity in the spinal cord and the descending reticular formation of the brain.	Used in the treatment of skeletal muscle spasms.
54	Codeine	Opioid analgesic	Opioid analgesic.	Used in the management of moderate to severe pain.
57	Cyclobenzaprine hydrochloride	Skeletal muscle relaxant (centrally acting)	Works by reducing tonic somatic motor activity, affecting both gamma and alpha motor systems.	Used in the treatment of skeletal muscle spasms.
107	Gabapentin	GABA analogue	While the mechanism of action is not fully known, efficacy is believed to be mediated by gabapentin's high-affinity binding to the α2-delta subunit of voltage-activated Ca^{2+} channels.	Used as an anticonvulsant and to treat peripheral neuropathic pain associated with diabetes and shingles.
114	Hydrocodone	Opioid analgesic and antitussive	The mechanism of action has not been fully established. Efficacy is believed to be related to the existence of opiate receptors in CNS.	Used in the management of severe pain and as an antitussive.
115	Hydrocodone bitartrate	Opioid analgesic and antitussive	The mechanism of action has not been fully established. Efficacy is believed to be related to the existence of opiate receptors in CNS.	Used in the management of severe pain and as an antitussive.
156	Morphine sulfate	Opioid analgesic	Opioid analgesic.	Used in the management of severe pain.
171	Oxycodone hydrochloride	Opioid analgesic	Opioid agonist.	Used in the management of severe pain.
186	Pregabalin	GABA analogue	While the mechanism of action is not fully established, efficacy is believed to be due to its binding with high affinity to the α2-delta site in CNS tissues.	Used in the management of neuropathic pain accompanying diabetic peripheral neuropathy or spinal cord injury. Also used in the management of postherpetic neuralgia and fibromyalgia and as an adjunctive therapy for patients with partial onset seizures.
224	Tramadol hydrochloride	Centrally acting analgesic	Centrally acting analgesic.	Used in the treatment of moderate to severe pain.
Sleep Aid				
91	Eszopiclone	Nonbenzodiazepine hypnotic agent	Efficacy is believed to be mediated by its interaction with GABA-receptor complexes at binding domains located close to or allosterically coupled to benzodiazepine receptors.	Used in the management of insomnia.
239	Zolpidem tartrate	GABAA agonist	The mechanism of action is due to interactions with a gamma-aminobutyric acid-BZ receptor complex. Zolpidem binds the BZ1 receptor preferentially with a high-affinity ratio of the α1/α5 subunits.	Used in the management of insomnia.

Drug no.	Drug name	Therapeutic class	Mechanism of action	Indications (Use)
Ulcer Treatment				
30	Bismuth subsalicylate	Salicylate	Salicylate.	Used in the treatment of diarrhea and upset stomach.
46	Cimetidine	H2 blocker	Works by competitively blocking the action of histamine at the histamine H2 receptors of the parietal cells.	Used in the management of heartburn, other symptoms accompanying GERD, and for the treatment of duodenal ulcers.
65	Dexlansoprazole	Proton pump inhibitor (PPI)	Works by reducing gastric acid secretion by specific inhibition of the (H^+/K^+)-ATPase in the gastric parietal cell, which is the last step in acid production.	Used in the management of gastroesophageal reflux disease (GERD), and in the management of heartburn. Used in the management of duodenal ulcers and used in combination with amoxicillin and clarithromycin for the treatment of *Helicobacter pylori* infection and DU disease for *H. pylori* eradication. Also used in long-term management of pathological hypersecretory conditions.
90	Esomeprazole	Proton pump inhibitor (PPI)	Works by reducing gastric acid secretion by specific inhibition of the (H^+/K^+)-ATPase in the gastric parietal cell.	Used in the treatment of erosive or ulcerative gastroesophageal reflux disease (GERD) and GERD symptoms. Also used in the treatment of duodenal ulcers (DUs) and in combination with amoxicillin and clarithromycin for the treatment of *Helicobacter pylori* infection.
129	Lansoprazole	Proton pump inhibitor (PPI)	Works to decrease gastric acid secretion by inhibition of the (H^+/K^+)-ATPase enzyme system at the secretory surface of the gastric parietal cell.	Used in the management of acid-reflux disorders, peptic ulcer disease, and *Helicobacter pylori* eradication.
169	Omeprazole magnesium	Proton pump inhibitor (PPI)	Limits the last step of gastric acid secretion. Works by reducing gastric acid secretion by specific inhibition of the (H^+/K^+)-ATPase in the gastric parietal cell.	Used in the treatment of erosive or ulcerative gastroesophageal reflux disease (GERD) and GERD symptoms. Also used in the treatment of duodenal ulcers (DUs) and in combination with amoxicillin and clarithromycin for the treatment of *Helicobacter pylori* infection.
189	Rabeprazole sodium	Proton pump inhibitor (PPI)	Impedes the last step of gastric acid secretion. Substituted benzimidazole limits acid secretion by reducing the gastric H^+/K^+-ATPase at the secretory surface of the gastric parietal cell.	Used in the treatment of erosive or ulcerative gastroesophageal reflux disease (GERD) and GERD symptoms. Also used in the treatment of duodenal ulcers (DUs) and in combination with amoxicillin and clarithromycin for the treatment of *Helicobacter pylori* infection.
194	Ranitidine	H2 blocker	Efficacy is mediated by blocking the action of histamine on parietal cells in the stomach, which limits the production of acid.	Used in the treatment of erosive or ulcerative gastroesophageal reflux disease (GERD) and GERD symptoms.
Urinary Tract and Reproductive Therapy				
10	Allopurinol	Xanthine oxidase inhibitor	Efficacy is mediated by decreasing the production of uric acid by acting on purine catabolism.	Used in the management of hyperuricemia connected to primary or secondary gout, or hyperuricemia and recurrent renal calculi associated with chemotherapy.
79	Dutasteride		Works by preventing the conversion of testosterone to dihydrotestosterone (the androgen mostly responsible for the enlargement of the prostate gland).	Used in the treatment of difficulties with urination caused by an enlarged prostate.
97	Febuxostat	Xanthine oxidase inhibitor	Efficacy is mediated by decreasing the production of uric acid.	Used in the management of hyperuricemia in patients with gout.
106	Furosemide	Loop diuretic	Works by limiting the absorption of Na^+ nd Cl^- not only in the proximal and distal tubules but also in the loop of Henle.	Used in the management of edema caused by congestive heart failure, kidney or liver disease, and other medical conditions.
113	Hydrochlorothiazide	Thiazide diuretic	Limits the reabsorption of Na^+ and Cl^- ions, which leads to an increase in the concentration of Na+ traversing the distal tubule and increases the volume of water excreted.	Used in the management of hypertension and edema.
207	Sevelamer hydrochloride	Phosphate binder	Lowers serum concentrations of phosphate by interacting with phosphate molecules through ionic and hydrogen bonding, and limiting absorption.	Used in the regulation of serum phosphorus in patients with chronic kidney disease.
208	Sildenafil citrate	Phosphodiesterase 5 (PDE5) inhibitor	Increases the effect of nitric oxide by inhibiting PDE-5, which leads to an increase in cGMP in the corpus cavernosum, resulting in smooth muscle relaxation and the inflow of blood.	Used in the management of erectile dysfunction.
213	Solifenacin succinate	Muscarinic antagonist	Works by inhibiting muscarinic receptors, which mediate contractions of the bladder smooth muscle.	Used in the management of overactive bladder with symptoms including urinary incontinence, urgency, and urinary frequency.
214	Spironolactone	Aldosterone blocker	Efficacy is due to competitively binding receptors at aldosterone-dependent Na^+ - K^+ exchange sites in distal convoluted renal tubule, this leads to an increase in Na^+ and water excretion and K^+ retention.	Used in the management of hypertension, edema, hypokalemia, severe heart failure, and primary hyperaldosteronism.
216	Tadalafil	Phosphodiesterase 5 (PDE5) inhibitor	Works by increasing the concentrations of cGMP that leads to smooth muscle relaxation and increased blood flow into the corpus cavernosum.	Used in the management of erectile dysfunction and benign prostatic hyperplasia.
217	Tamsulosin hydrochloride	$\alpha 1$ Antagonist	Efficacy is due to the selective blockade of $\alpha 1$ receptors in the prostate. This causes a relaxation of the smooth muscles of the bladder neck and prostate, which improves urine flow.	Used in the treatment of urinary obstruction with symptoms such as hesitancy, terminal dribbling of urine, interrupted, and weak stream.

Drug no.	Drug name	Therapeutic class	Mechanism of action	Indications (Use)
220	Testosterone	Androgen	Steroid hormone which can activate the androgen receptor or be converted to estradiol and activation of certain estrogen receptors.	Used to treat conditions resulting from a deficiency or absence of endogenous testosterone.
223	Tolterodine tartrate	Muscarinic antagonist	Works by being a competitive antagonist of acetylcholine at postganglionic muscarinic receptors mediating urinary bladder contraction.	Used in the treatment of overactive bladder.
229	Triamterene	Potassium-sparing diuretic	Works by inhibiting the reabsorption of Na^+ ions in exchange for K^+ and H^+ ions at segment of distal tubule.	Used in the management of edema caused by congestive heart failure, kidney or liver disease, and other medical conditions.
234	Vardenafil hydrochloride	Phosphodiesterase-5 (PDE-5) inhibitor	Works by increasing the concentrations of cGMP that leads to smooth muscle relaxation and increased blood flow into the corpus cavernosum.	Used in the management of erectile dysfunction.
47	Cinacalcet	Calcimimetic agent	Decreases parathyroid hormone concentrations by increasing the sensitivity of the Ca^{2+}-sensing receptor to extracellular Ca^{2+}.	Used in the management of hyperparathyroidism (a consequence of chronic kidney disease on dialysis). Also used in the management of hypercalcemia in patients with parathyroid carcinoma or primary hyperparathyroidism.
Vaccine				
112	Human papillomavirus quadrivalent	Vaccine	Vaccine	Used in the vaccination of patients 9-26 yrs. of age for the prevention human papillomavirus (HPV). This may lead to the prevention of cervical, vulvar, vaginal, and anal cancer in women and in males may lead to the prevention of anal cancer.
120	Influenza vaccine for intramuscular injection	Vaccine	Vaccine	Used in the immunization of patients ages 6 months and older against infection by influenza virus subtypes A and type B present in the vaccine
179	Pneumococcal 13-valent conjugate vaccine	Vaccine	Vaccine	Streptococcus pneumoniae immunization.
240	Zoster vaccine live	Vaccine	Vaccine	Used in the prevention of herpes zoster (shingles) in individuals ≥50 yrs. of age.

Index of Trade Names of Common Single-Component Drugs

Table 3 gives the trade names of common single-component drugs and links them to the physical properties (Table 1) and function information (Table 2) through the ordering number. Note that many trade name single-component drugs not only have a single active component, but also can contain one or more inactive ingredients. Solid drugs are often dissolved or suspended in water for ease of administration.

Table 3. Index of Trade Names of Common Drugs

Trade name	Ordering number	Drug name
Abilify	19	Aripiprazole
Aciphex	188	Rabeprazole sodium
Actonel	197	Risedronate sodium
Actos	177	Pioglitazone
Adcirca	215	Tadalafil
Adoxa	77	Doxycycline
Advil	117	Ibuprofen
Afinitor	95	Everolimus
Afluria	120	Influenza vaccine for intramuscular injection
Afrin	171	Oxymetazoline hydrochloride
Aggrenox	75	Dipyridamole
Aldactone	213	Spironolactone
Aleve	158	Naproxen sodium
Alimta	176	Pemetrexed disodium
Allegra	99	Fexofenadine hydrochloride
Alphagan P	32	Brimonidine tartrate
Altace	191	Ramipril
Amaryl	109	Glimepiride
Ambien	238	Zolpidem tartrate
Amitiza	145	Lubiprostone
Amitriptyline	12	Amitriptyline hydrochloride
Amoxicillin	15	Amoxicillin
Androderm	219	Testosterone
Antara	98	Fenofibrate
Aranesp	60	Darbepoetin alfa
Asmanex	154	Mometasone furoate
Aspirin	4	Acetylsalicylic acid
Atelvia	197	Risedronate sodium
Ativan	141	Lorazepam
Atripla	81	Efavirenz
Avastin	27	Bevacizumab
Avodart	79	Dutasteride
Avonex	125	Interferon beta-1a
Aygestin	161	Norethindrone acetate
Bacitracin	25	Bacitracin
Bactrim	214	Sulfamethoxazole
Benadryl	74	Diphenhydramine hydrochloride
Benicar	165	Olmesartan medoxomil
Bystolic	159	Nebivolol
Caldolor	117	Ibuprofen
Catapres	51	Clonidine hydrochloride
Cefdinir	40	Cefdinir
Celebrex	41	Celecoxib
Chantix	234	Varenicline
Chlor-Trimeton	45	Chlorpheniramine maleate

Trade name	Ordering number	Drug name
Cialis	215	Tadalafil
Cimetidine	46	Cimetidine
Cimzia	43	Certolizumab pegol
Cipro	48	Ciprofloxacin
Claritin	140	Loratadine
Clonazepam	50	Clonazepam
Colace	76	Docusate sodium
Combivent Respimat	127	Ipratropium bromide
Copaxone	108	Glatiramer acetate
Coreg	39	Carvedilol
Cortaid	116	Hydrocortisone
Coumadin	237	Warfarin sodium
Cozaar	142	Losartan potassium
Crestor	202	Rosuvastatin calcium
Cubicin	59	Daptomycin
Cyclobenzaprine	57	Cyclobenzaprine hydrochloride
Cymbalta	78	Duloxetine
Depo-Provera	147	Medroxyprogesterone acetate
Detrol	222	Tolterodine tartrate
Dexilant	65	Dexlansoprazole
DiaBeta	110	Glyburide
Diflucan	102	Fluconazole
Diovan	232	Valsartan
Dulcolax	29	Bisacodyl
Duramorph	156	Morphine sulfate
Dyrenium	228	Triamterene
Edurant	196	Rilpivirine
Effient	182	Prasugrel
Eliquis	18	Apixaban
Elvitegravir	82	Elvitegravir
Embeda	156	Morphine sulfate
Emtriva	83	Emtricitabine
Enbrel	92	Etanercept
Epogen	86	Epoetin alfa
Erbitux	44	Cetuximab
Ery-Tab	88	Erythromycin
Estinyl	93	Ethinyl estradiol
Evista	189	Raloxifene hydrochloride
Exelon	201	Rivastigmine tartrate
Eylea	7	Aflibercept
Ezvio	157	Naloxone
Fenoglide	98	Fenofibrate
Flomax	216	Tamsulosin hydrochloride
Flovent HFA	104	Fluticasone propionate
Focalin	66	Dexmethylphenidate hydrochloride
Formoterol fumarate dihydrate	105	Formoterol fumarate dihydrate
Fortamet	150	Metformin hydrochloride
Fosamax	9	Alendronate sodium
Gardasil	112	Human papillomavirus quadrivalent
Gilenya	101	Fingolimod
Gleevec	118	Imatinib mesylate
Gralise	107	Gabapentin
Harvoni	130	Ledipasvir
Herceptin	224	Trastuzumab
Humalog	124	Insulin lispro
Humira	5, 6	Adalimumab: heavy chain
Hydrocodone	114	Hydrocodone
Hysingla	115	Hydrocodone bitartrate
Imdur	128	Isosorbide mononitrate

Trade name	Ordering number	Drug name
Imodium	139	Loperamide hydrochloride
Invega Sustenna	172	Paliperidone palmitate
Invokana	36	Canagliflozin
Isentress	190	Raltegravir
Januvia	210	Sitagliptin
Kadian	156	Morphine sulfate
Keflex	42	Cephalexin
Khedezla	64	Desvenlafaxine
Klonopin	50	Clonazepam
Lanoxin	72	Digoxin
Lantus	123	Insulin glargine
Lasix	106	Furosemide
Latuda	146	Lurasidone hydrochloride
Levaquin	132	Levofloxacin
Levemir	122	Insulin detemir
Levitra	233	Vardenafil hydrochloride
Lexapro	89	Escitalopram oxalate
Lipitor	23	Atorvastatin calcium
Lipofen	98	Fenofibrate
Lofibra	98	Fenofibrate
Lotemax	143	Loteprednol etabonate
Lotrimin AF	153	Miconazole nitrate
Lovastin	144	Lovastatin
Lovenox	85	Enoxaparin sodium
Lucentis	192	Ranibizumab
Lumigan	28	Bimatoprost
Lunesta	91	Eszopiclone
Lyrica	185	Pregabalin
Medrol	151	Methylprednisolone
Micardis	217	Telmisartan
Microzide	113	Hydrochlorothiazide
Midol	117	Ibuprofen
Miralax	179	Polyethylene glycol 3350
Mobic	148	Meloxicam
Morphine	156	Morphine sulfate
Motrin	117	Ibuprofen
Namenda	149	Memantine hydrochloride
Nasacort	227	Triamcinolone acetonide
Nasonex	154	Mometasone furoate
NeoProfen	117	Ibuprofen
Neulasta	175	Pegfilgrastim
Neupogen	100	Filgrastim
Neurotin	107	Gabapentin
Nexium	90	Esomeprazole
Norvasc	14	Amlodipine besylate
Norvir	198	Ritonavir
Novolog	121	Insulin aspart
Nuvaring	94	Etonogestrel
Oleptro	226	Trazodone hydrochloride
Olysio	208	Simeprevir
Olmesartan acid	164	Olmesartan acid
Onglyza	204	Saxagliptin
Oracea	77	Doxycycline
Orencia	1	Abatacept
Ortho Tri-Cyclen	162	Norgestimate
OxyContin	170	Oxycodone hydrochloride
Pataday	166	Olopatadine hydrochloride
Patanol	166	Olopatadine hydrochloride
Paxil	174	Paroxetine hydrochloride
Pepto-Bismol Caplets	30	Bismuth subsalicylate

Biochem

Trade name	Ordering number	Drug name
Phenadoz	186	Promethazine hydrochloride
Plavix	52	Clopidogrel bisulfate
Polymyxin b	180	Polymyxin b
Pradaxa	58	Dabigatran etexilate mesylate
Pramocaine	181	Pramoxine hydrochloride
Pravachol	183	Pravastatin sodium
Premarin Tablets	56	Conjugated estrogens
Premarin Vaginal Cream	56	Conjugated estrogens
Prevacid	129	Lansoprazole
Prevnar 13	178	Pneumococcal 13-valent conjugate vaccine
Prezista	61	Darunavir
Prilosec	168	Omeprazole magnesium
Primolut-Nor	161	Norethindrone acetate
Primsol	229	Trimethoprim
Prinivil	138	lisinopril
Pristiq	64	Desvenlafaxine
Procrit	86	Epoetin alfa
Prolia	63	Denosumab
Promethazine	186	Promethazine hydrochloride
Promethegan	186	Promethazine hydrochloride
Proventil HFA	8	Albuterol sulfate
Provera	147	Medroxyprogesterone acetate
Prozac	103	Fluoxetine hydrochloride
Ranexa	194	Ranolazine
Rayos	184	Prednisone
Rebif	125	Interferon beta-1a
Remicade	119	Infliximab
Renagel	206	Sevelamer hydrochloride
Revlimid	131	Lenalidomide
Reyataz	20	Atazanavir
Rituxan	199	Rituximab
Robitussin	111	Guaifenesin
Sandostatin	163	Octreotide acetate
Sensipar	47	Cinacalcet
Serevent	203	Salmeterol
Seroquel	187	Quetiapine fumarate
Singulair	155	Montelukast sodium
Soliris	80	Eculizumab
Soma	38	Carisoprodol
Sovaldi	211	Sofosbuvir
Spiriva	221	Tiotropium bromide
Sprycel	62	Dasatinib
Stelara	230	Ustekinumab
Strattera	22	Atomoxetine
Stribild	53	Cobicistat
Suboxone	34	Buprenorphine
Sulfatrim	214	Sulfamethoxazole
Sustiva	81	Efavirenz
Symbicort	33	Budesonide
Synagis	173	Palivizumab vh region medi-498

Trade name	Ordering number	Drug name
Synthroid	133	Levothyroxine sodium
Tamiflu	169	Oseltamivir phosphate
Tarceva	87	Erlotinib hydrochloride
Tecfidera	73	Dimethyl fumarate
Tenormin	21	Atenolol
Timoptic	220	Timolol
Toprol-XL	152	Metoprolol succinate
Tradjenta	134	Linagliptin
Travatan Z	225	Travoprost
Treanda	26	Bendamustine hydrochloride
Triamcinolone Acetonide Cream	227	Triamcinolone acetonide
Tricor	98	Fenofibrate
Triglide	98	Fenofibrate
Truvada	83	Emtricitabine
Tylenol	3	Acetaminophen
Uloric	97	Febuxostat
Ultram	223	Tramadol hydrochloride
Valium	70	Diazepam
Valtrex	231	Valacyclovir hydrochloride
Vasotec	84	Enalapril maleate
Velcade	31	Bortezomib
Venlafaxine Tablets	235	Venlafaxine hydrochloride
Ventolin HFA	8	Albuterol sulfate
Vesicare	212	Solifenacin succinate
Viagra	207	Sildenafil citrate
Vibramycin	77	Doxycycline
Victoza	136	Liraglutide
Viibryd	236	Vilazodone hydrochloride
Viread	218	Tenofovir disoproxil fumarate
Voltaren Gel	71	Diclofenac sodium
Vyvanse	137	Lisdexamfetamine dimesylate
WelChol	55	Colesevelam hydrochloride
Xanax	11	Alprazolam
Xarelto	200	Rivaroxaban
Xeloda	37	Capecitabine
Xgeva	63	Denosumab
Xifaxan	195	Rifaximin
Xolair	167	Omalizumab
Yervoy	126	Ipilimumab
Zantac	193	Ranitidine
Zetia	96	Ezetimibe
Zmax	24	Azithromycin
Zocor	209	Simvastatin
Zohydro	115	Hydrocodone bitartrate
Zoloft	205	Sertraline hydrochloride
Zortress	95	Everolimus
Zostavax	239	Zoster vaccine live
Zyloprim	10	Allopurinol
Zytiga	2	Abiraterone acetate
Zyvox	135	Linezolid

Components of Common Trade Name Multicomponent Drugs

Many common drug products are multicomponent, designed to provide a combination of therapeutic effects. Table 4 lists the trade name of a number of multicomponent drugs, indexed to the individual components listed in Table 1 through the drug number (ordering number in Table 1). Note that multicomponent trade name drugs not only have two or more active components, but also can contain one or more inactive ingredients. Solid drugs are often dissolved or suspended in water for ease of administration.

Table 4. Components of Common Multicomponent Drugs

Trade name	Drug no.	Components
Advair Diskus	104	Fluticasone propionate
	204	Salmeterol
Aggrenox	75	Dipyridamole
	4	Acetylsalicylate acid
Asasantin	75	Dipyridamole
	4	Acetylsalicylate acid
Atripla	219	Tenofovir disoproxil fumarate
	81	Efavirenz
	83	Emtricitabine
Augmentin	15	Amoxicillin
	49	Clavulanate potassium
Azor	13	Amlodipine
	166	Olmesartan medoxomil
Bactrim	215	Sulfamethoxazole
	229	Trimethoprim
Benicar HCT	166	Olmesartan medoxomil
	113	Hydrochlorothiazide
Combigan	32	Brimonidine tartrate
	221	Timolol maleate
Combivent	127	Ipratropium bromide
	8	Albuterol sulfate
Complera	83	Emtricitabine
	197	Rilpivirine
	219	Tenofovir disoproxil fumarate
DouNeb	127	Ipratropium bromide
	8	Albuterol sulfate
Dulera	105	Formoterol fumarate dihydrate
	154	Mometasone furoate
Excedrin	3	Acetaminophen
	4	Acetylsalicylic acid
	35	Caffeine
Exforge	233	Valsartan
	13	Amlodipine
Harvoni	130	Ledipasvir
	212	Sofosbuvir
Janumet	150	Metformin hydrochloride
	211	Sitagliptin
Lo Loestrin Fe	93	Ethinyl estradiol
	162	Norethindrone acetate
		Ferrous fumarate (no therapeutic purpose)
Minastrin 24 Fe	93	Ethinyl estradiol
	162	Norethindrone acetate
		Ferrous fumarate (no therapeutic purpose)
Mucinex DM	69	Dextromethorphan hydrobromide
	111	Guaifenesin
Neosporin +	160	Neomycin sulfate
	181	Polymyxin B
	182	Pramoxine hydrochloride
NuvaRing	93	Ethinyl estradiol
	94	Etonogestrel
Ortho Tri-Cyclin	93	Ethinyl estradiol
	163	Norgestimate
Percocet	3	Acetaminophen
	171	Oxycodone hydrochloride
Premphase/Prempro	56	Conjugated estrogens
	147	Medroxyprogesterone acetate
Robitussin DM	69	Dextromethorphan hydrobromide
	111	Guaifenesin
Septra	215	Sulfamethoxazole
	229	Trimethoprim
Seretide	104	Fluticasone propionate
	204	Salmeterol
Stribild	53	Cobicistat
	82	Elvitegravir
	83	Emtricitabine
	219	Tenofovir disoproxil fumarate
Suboxone	219	Tenofovir disoproxil fumarate
	157	Naloxone
Symbicort	33	Budesonide
	105	Formoterol fumarate dihydrate
Truvada	83	Emtricitabine
	219	Tenofovir disoproxil fumarate
Tylenol #3	3	Acetaminophen
	54	Codeine phosphate
Tylenol with codeine	3	Acetaminophen
	54	Codeine phosphate
Vicodin	3	Acetaminophen
	114	Hydrocodone
Vytorin	96	Ezetimibe
	210	Simvastatin
Xulane	93	Ethinyl estradiol
	161	Norelgestromin
ZenHale	105	Formoterol fumarate dihydrate
	154	Mometasone furoate
Zubsolv	34	Buprenorphine
	157	Naloxone

CHEMICAL CONSTITUENTS OF HUMAN BLOOD

This table lists typical concentrations of some of the chemical constituents of human blood. The table covers elements and compounds of relatively low molecular weight. Refs. 1 and 4 give extensive information on enzymes, hormones, vitamins, and other blood constituents.

The values given for the normal range refer to healthy adults who have not been exposed to unusual environmental agents. In keeping with IUPAC practice, all values refer to a volume of one liter, and thus are stated in units of g/L, mg/L, μg/L, or mmol/L. Many clinical test results, especially in the United States, are reported on a deciliter (dL) rather than a liter basis; thus the values in this table should be divided by 10 to place them on a dL basis. The symbols S (for serum), P (plasma), and WB (whole blood) in the second column indicate the nature of the blood sample to which the values apply. In some cases only a single mean value has been reported, rather than a range; these are given in italics.

The total volume of blood in a 100 kg (220 lb) adult is 7.5 L for a male and 6.7 L for a female. The corresponding volume of plasma is 4.4 L and 4.3 L, respectively (Ref. 1).

Values from Ref. 1 are so-called "reference values" against which clinical tests of blood chemistry are compared.The Lower limit and Upper limit define the "normal range", which is understood to include about 95% of the healthy population. The remaining 5% may show values outside the normal range without necessarily implying a medical problem. Note that these reference values may vary slightly from one testing laboratory to another, depending on the detailed test procedure.

Accurate measurements on trace elements are very difficult to make, and wide variations can be found in the literature. Preferred measurement methods are discussed in Refs. 2 and 6. Values for the trace elements can also vary from one country to another, depending on dietary or environmental factors. Thus cadmium levels tend to be higher in Japan because of the prevalence of seafood in the diet, and lead levels are higher in regions where lead additives are still used in gasoline. Variations with gender, age, geography, and occurrence of diseases are reviewed in Ref. 6.

The Critical values column gives levels that deviate far enough from the normal range to suggest a probable medical issue. Such values from Ref. 3 are the Biological Exposure Indexes (BEI) that are specified by the American Council of Government Industrial Hygienists (ACGIH) as danger signals for the levels of pollutants in the workplace.

References

1. Wallach, J., *Interpretation of Diagnostic Tests, Eight Edition*, Wolters Kluwer, Philadelphia, 2007.
2. IUPAC Commission on Toxicology, "Sample Collection Guidelines for Trace Elements in Blood and Urine," *Pure & Appl. Chem.*, 67, 1575, 1995.
3. *2008 TLV's and BEI's*, American Conference of Governmental Industrial Hygienists, 1330 Kemper Meadow Drive, Cincinnati, OH 45240–1634, 2008 (www.acgih.org).
4. Altman, P. L., and Dittmer, D. S., Eds., *Biology Data Book, Second Edition, Vol. III*, Federation of American Societies for Experimental Biology, Bethesda, MD, 1974.
5. Bowen, H. J. M., *Trace Elements in Biochemistry*, Academic Press, New York, 1966.
6. Versieck, J., and Cornelis, R., *Trace Elements in Human Plasma or Serum*, CRC Press, Boca Raton, FL, 1989.

Component	Source	Unit	Lower limit	Upper limit	Critical values	Ref.
Inorganic						
Aluminum	S	μg/L	1	10	>60	6,2
Ammonia	P	μg/L	190	600	>700	1
Antimony	S,P	μg/L		*1*		6
Arsenic	S	μg/L	0.5	5		6,2
Barium	S,P	μg/L		*79*		4,5
Beryllium	S,P	μg/L		*<4*		4,5
Bicarbonate (HCO_3^-)	WB	mmol/L	22	28	<10 or >40	1
Bromine	S,P	mg/L	2	11		6,4
Cadmium	S	μg/L	0.1	1	>5	6,2,3
Calcium, total	S	mg/L	90	105	<65 or >140	1
Calcium ion (Ca^{++})	WB	mg/L	30	45		1
Carbon dioxide	P	mmol/L	21	30	<11 or >40	1
Carbon monoxide*	WB	%CO-Hb	0	5%	30%	1
Cesium	S,P	μg/L	0.5	2.0		6
Chloride (Cl^-)	S	mmol/L	98	106	<80 or >115	1
Chromium	S	μg/L	0.1	0.4		6,2
Cobalt	S	μg/L	0.05	0.35	>1	6,2,3
Copper	S	mg/L	0.7	1.4		1,2,6
Fluorine	S,P	μg/L	33	236		6
Hydrogen ion (H^+)	WB	pH	7.38	7.44	<7.10 or >7.59	1
Iodine (total)	S,P	μg/L	59	76		4
Iron	S	mg/L	0.5	1.7		1
Lead	S	μg/L	5	100	>300	1,3,6
Lithium	S,P	μg/L		*8*		6
Magnesium	S	mg/L	18	30	<10 or >47	1
Manganese	S	μg/L	0.3	1.0		6,2
Mercury	S	μg/L	0.5	3	>15	2,3

Biochem

Component	Source	Unit	Lower limit	Upper limit	Critical values	Ref.
Molybdenum	S,P	µg/L	0.3	1.3		6
Nickel	S	µg/L	0.1	1.3		6,2
Oxygen (arterial)	WB	% saturation	96%	100%		1
Oxygen (venous)	WB	% saturation	60%	85%		1
Phosphorus (inorganic)	S	mg/L	30	45	<11	1
Potassium	S	mmol/L	3.5	5.0	<2.8 or >6.2	1
Potassium	S	mg/L	137	196		
Rubidium	S,P	µg/L	100	300		6
Selenium	S,P	µg/L	40	160		2,6
Silver	S,P	µg/L		*1*		6
Sodium	S	mmol/L	135	145	<120 or >160	1
Sodium	S	g/L	3.11	3.34		
Strontium	S,P	µg/L		*57*		4,5
Sulfur (total)	S,P	mg/L		*780*		4
Tellurium	S,P	µg/L		*30*		4,5
Titanium	S,P	µg/L		*33*		4,5
Tin	S,P	µg/L		*1*		4,5
Vanadium	S,P	µg/L	0.02	1.0		6
Zinc	S,P	mg/L	0.5	1.2		6,2,4
Zirconium	S,P	µg/L		*400*		4,5
Organic						
Acetoacetate ion	P	mg/L		*<10*		1
Acetone	S,P	mg/L	3	20		1
Alanine	S,P	mg/L	30	37		4
Arginine	S,P	mg/L	12	19		4
Asparagine	S,P	mg/L	5.4	6.5		4
Cholesterol, total	P	mg/L	1000	2000**	>2400	1,4
HDL Cholesterol	P	mg/L	400	600		1
LDL Cholesterol	P	mg/L	0	1000	>1900	1
Citrulline	S,P	mg/L	2.1	9.7		4
Creatine	S,P	mg/L	2.8	6.2		4
Creatinine	S	mg/L	5	15	>50	1
Fructose	WB	mg/L	5	50		4
Glucosamine	S,P	mg/L	760	1110		4
Glucose (fasting)	S	mg/L	600	1000	<450 or >1300	1
Glutamic acid	S,P	mg/L	4.3	11.5		4
Glutamine	S,P	mg/L	61	102		4
Glycine	S,P	mg/L	13.4	17.3		4
Histidine	S,P	mg/L	7.9	14.8		4
Homocysteine	P	mg/L	0.54	1.62		1
Isoleucine	S,P	mg/L	6.9	12.8		4
Lactate (venous)	P	mg/L	50	150		1
Leucine	S,P	mg/L	14	23		4
Lysine	S,P	mg/L	25	30		4
Methionine	S,P	mg/L	3.3	4.3		4
Ornithine	S,P	mg/L	6.2	8.0		4
Phenylalanine	S,P	mg/L	5.8	14.0		1
Proline	S,P	mg/L	20	33		4
Serine	S,P	mg/L	10.1	12.5		4
Taurine	S,P	mg/L	4.1	8.2		4
Threonine	S,P	mg/L	12	17		4
Triglyceride	S	mg/L	250	1750		1
Tyrosine	S,P	mg/L	8.1	14.5		4
Urea	S	mmol/L	3.5	7.0	<0.7 or >28	1
Urea nitrogen (BUN)	S	mg/L (of N)	100	200	<20 or >800	1
Uric acid (males)	S	mg/L	25	80		1
Uric acid (females)	S	mg/L	13	60		1
Valine	S,P	mg/L	24	37		4

* Measured as the percent of hemoglobin bound to CO. Typical value for heavy smokers is 5%–10%. Major symptoms begin around 30%, and respiratory failure sets in at >60%.

** This is the desirable upper limit. Values between 2000 and 2400 mg/L are considered borderline high.

CHEMICAL COMPOSITION OF THE HUMAN BODY

The elemental composition of the "standard man" of mass 70 kg is given below.

References

1. Padikal, T. N., and Fivozinsky, S. P., *Medical Physics Data Book, National Bureau of Standards Handbook 138*, U. S. Government Printing Office, Washington, DC, 1981.
2. Snyde, W. S., et al., *Reference Man: Anatomical, Physiological, and Metabolic Characteristics*, Pergamon, New York, 1975.

Chemical Composition of the Human Body

Element	Amount in grams	Percent of total body mass
Oxygen	43,000	61
Carbon	16,000	23
Hydrogen	7000	10
Nitrogen	1800	2.6
Calcium	1000	1.4
Phosphorus	780	1.1
Sulfur	140	0.20
Potassium	140	0.20
Sodium	100	0.14
Chlorine	95	0.12
Magnesium	19	0.027
Silicon	18	0.026
Iron	4.2	0.006
Fluorine	2.6	0.0037
Zinc	2.3	0.0033
Rubidium	0.32	0.00046
Strontium	0.32	0.00046
Bromine	0.20	0.00029
Lead	0.12	0.00017
Copper	0.072	0.00010
Aluminum	0.061	0.00009
Cadmium	0.050	0.00007
Boron	<0.048	0.00007
Barium	0.022	0.00003
Tin	<0.017	0.00002
Manganese	0.012	0.00002
Iodine	0.013	0.00002
Nickel	0.010	0.00001
Gold	<0.010	0.00001
Molybdenum	<0.0093	0.00001
Chromium	<0.0018	0.000003
Cesium	0.0015	0.000002
Cobalt	0.0015	0.000002
Uranium	0.00009	0.0000001
Beryllium	0.000036	
Radium	$3.1 \cdot 10^{-11}$	

Biochem

NUTRIENT VALUES OF FOODS

The U.S. Department of Agriculture maintains the USDA National Nutrient Database for Standard Reference, which contains over 7000 food items with data on the energy content, minerals, vitamins, and other properties of nutritional interest. The table here includes about 600 common foods extracted from that database. The properties listed are the energy content (in effect, the enthalpy of combustion); the content of carbohydrates, proteins, and lipids (fats); the cholesterol content; and the amount of sodium, potassium, calcium, magnesium, iron, copper, zinc, manganese, phosphorus, and selenium. All values are given for a 100 gram sample of the food.

To conform with common practice in nutritional science, the energy content is given in kilocalories. For conversion to kilojoules, this number should be multiplied by 4.184. For conversion to the avoirdupois units frequently used in the United States, note that:

1 oz = 28.35 g (thus the 100 g basis in this table is approximately 3.5 oz or 02.18 lb, slightly less than a quarter pound)

1 lb = 453.6 g

The full USDA database covers specific proprietary brands of many processed foods, and it includes vitamin content as well as the minerals given in this table.

Reference

U.S. Department of Agriculture, Agricultural Research Service. 2005. USDA National Nutrient Database for Standard Reference, Release 18. Nutrient Data Laboratory Home Page, <www.ars.usda.gov/ba/bhnrc/ndl>.

Nutrient Values of Common Foods

Food description	Energy [kcal/ 100 g]	Carb. [g/ 100 g]	Protein [g/ 100 g]	Fat [g/ 100 g]	Chol. [mg/ 100 g]	Na [mg/ 100 g]	K [mg/ 100 g]	Ca [mg/ 100 g]	Mg [mg/ 100 g]	Fe [mg/ 100 g]	Cu [mg/ 100 g]	Zn [mg/ 100 g]	Mn [mg/ 100 g]	P [mg/ 100 g]	Se [μg/ 100 g]
Alcoholic beverage, beer, light	29	1.6	0.2			4	21	4	5	0.03	0.006	0.01	0.006	12	0.4
Alcoholic beverage, beer, regular	43	3.6	0.5			4	27	4	6	0.02	0.005	0.01	0.008	14	0.6
Alcoholic beverage, dessert wine, sweet	160	13.7	0.2			9	92	8	9	0.24	0.045	0.07	0.119	9	0.5
Alcoholic beverage, distilled (gin, rum, vodka, whiskey), 80 proof	231					1	2			0.04	0.021	0.04	0.018	4	
Alcoholic beverage, distilled (gin, rum, vodka, whiskey), 90 proof	263					1	2			0.04	0.021	0.04	0.018	4	
Alcoholic beverage, distilled, all, 100 proof	295					1	2			0.04	0.021	0.04	0.018	4	
Alcoholic beverage, liqueur, coffee, 63 proof	308	32.2	0.1	0.3		8	30	1	3	0.06	0.040	0.03	0.017	6	0.3
Alcoholic beverage, table wine, red	85	2.6	0.1			4	127	8	12	0.46	0.011	0.14	0.132	23	0.2
Alcoholic beverage, table wine, white	83	2.6	0.1			5	71	9	10	0.27	0.004	0.12	0.117	18	0.1
Almonds, dry roasted, w/o salt	597	19.3	22.1	52.8		1	746	266	286	4.51	1.170	3.54	2.620	489	2.8
Almonds, oil roasted, w/o salt	607	17.7	21.2	55.2		1	699	291	274	3.68	0.955	3.07	2.460	466	2.8
Apple juice, canned or bottled, unsweetened, w/o vitamin C	47	11.7	0.1	0.1		3	119	7	3	0.37	0.022	0.03	0.113	7	0.1
Apples, raw, w/o skin	48	12.8	0.3	0.1			90	5	4	0.07	0.031	0.05	0.038	11	
Apples, raw, with skin	52	13.8	0.3	0.2		1	107	6	5	0.12	0.027	0.04	0.035	11	
Apricots, dried, sulfured, stewed, w. sugar	113	29.3	1.2	0.2		3	443	15	15	1.52	0.138	0.24	0.088	38	
Apricots, dried, sulfured, stewed, w/o sugar	85	22.2		0.2		4	411	19	11	0.94	0.121	0.14	0.083	25	0.8
Artichokes, (globe or French), boiled, w/o salt	50	11.2	3.5	0.2		95	354	45	60	1.29	0.233	0.49	0.259	86	0.2
Arugula, raw	25	3.7	2.6	0.7		27	369	160	47	1.46	0.076	0.47	0.321	52	0.3
Asparagus, boiled	22	4.1	2.4	0.2		14	224	23	14	0.91	0.165	0.60	0.154	54	6.1
Asparagus, raw	20	3.9	2.2	0.1		2	202	24	14	2.14	0.189	0.54	0.158	52	2.3
Avocados, raw, all commercial varieties	160	8.5	2.0	14.7		7	485	12	29	0.55	0.190	0.64	0.142	52	0.4
Bagel, plain, toasted, enriched w. calcium propanoate	257	50.5	10.0	1.6		448	75	89	22	6.05	0.130	1.90	0.515	87	22.8
Bagels, cinnamon-raisin, toasted	294	59.3	10.6	1.8		346	163	20	23	4.09	0.160	0.81	0.328	83	33.3
Baked beans, canned, no salt	105	20.6	4.8	0.4		1	296	50	32	0.29	0.206	1.40		104	4.5
Bamboo shoots, boiled, w/o salt	12	1.9	1.5	0.2		4	533	12	3	0.24	0.082	0.47	0.113	20	0.4
Bananas, raw	89	22.8	1.1	0.3		1	358	5	27	0.26	0.078	0.15	0.270	22	1.0
Bass, freshwater, mixed species, cooked, dry heat	146		24.2	4.7	87	90	456	103	38	1.91	0.119	0.83	1.140	256	16.2
Bass, striped, cooked, dry heat	124		22.7	3.0	103	88	328	19	51	1.08	0.040	0.51	0.019	254	46.8
Beans, French, mature seeds, boiled, w/o salt	129	24.0	7.1	0.8		6	370	63	56	1.08	0.115	0.64	0.382	102	1.2
Beans, kidney, all types, mature seeds, boiled, w/o salt	127	22.8	8.7	0.5		1	405	35	42	2.22	0.216	1.00	0.430	138	1.1
Beans, lima, immature seeds, canned	71	13.3	4.1	0.3		252	285	28	34	1.61	0.162	0.64	0.700	71	1.1
Beans, navy, mature seeds, canned	113	20.5	7.5	0.4		448	288	47	47	1.85	0.208	0.77	0.375	134	5.8
Beans, pinto, mature seeds, boiled, w/o salt	143	26.2	9.0	0.7		1	436	46	50	2.09	0.219	0.98	0.453	147	6.2
Beans, snap, green, boiled, w/o salt	35	7.9	1.9	0.3		1	146	44	18	0.65	0.057	0.25	0.285	29	0.2
Beef, bottom sirloin, tri-tip roast, lean & fat, 0" fat, choice, roasted	218		25.7	12.4	94	50	308	17	20	1.70	0.083	4.52	0.009	189	27.3

Food description	Energy [kcal/ 100 g]	Carb. [g/ 100 g]	Protein [g/ 100 g]	Fat [g/ 100 g]	Chol. [mg/ 100 g]	Na [mg/ 100 g]	K [mg/ 100 g]	Ca [mg/ 100 g]	Mg [mg/ 100 g]	Fe [mg/ 100 g]	Cu [mg/ 100 g]	Zn [mg/ 100 g]	Mn [mg/ 100 g]	P [mg/ 100 g]	Se [µg/ 100 g]
Beef, brisket, flat half, lean & fat, 1/8" fat, select, braised	280		29.0	17.4	71	49	237	17	19	2.40	0.090	6.92	0.010	180	28.1
Beef, chuck, arm pot roast, lean, 1/8" fat, choice, braised	224		34.7	8.4	81	56	275	15	23	3.04	0.129	8.20	0.012	213	34.7
Beef, ground, 75% lean meat / 25% fat, patty, broiled	278		25.6	18.7	89	78	289	30	20	2.37	0.075	6.19	0.010	189	21.4
Beef, ground, 95% lean meat / 5% fat, patty, broiled	171		26.3	6.6	76	65	348	7	22	2.83	0.096	6.43	0.014	206	21.7
Beef, loin, porterhouse steak, lean & fat, 1/4" fat, all grades, broiled	329		22.5	25.8	72	62	255	8	20	2.68	0.118	4.13	0.014	177	19.3
Beef, loin, T-bone steak, lean & fat, 1/4" fat, all grades, broiled	306		23.5	22.8	65	67	282	7	22	3.09	0.122	4.30	0.014	185	11.6
Beef, rib, eye, small end (ribs 10-12), lean, 0" fat, choice, broiled	205		28.9	9.0	91	60	363	16	25	1.98	0.092	5.49	0.011	227	33.6
Beef, rib, large end (ribs 6-9), lean & fat, 1/4" fat, choice, roasted	383		22.3	32.0	85	63	283	10	19	2.27	0.086	5.58	0.013	168	21.9
Beef, round, bottom round roast, lean, 0" fat, select, roasted	169		28.3	5.3	72	38	238	6	19	2.35	0.079	4.92	0.010	183	35.7
Beef, round, bottom round, lean & fat, 1/8" fat, all grades, roasted	218		26.4	11.6	75	35	214	6	17	2.16	0.063	4.43	0.009	164	27.0
Beef, round, eye of round, lean & fat, 1/8" fat, all grades, roasted	208		28.3	9.7	62	37	227	7	18	2.29	0.067	4.70	0.010	174	28.7
Beef, tenderloin, lean & fat, 1/8" fat, all grades, broiled	267		26.5	17.1	90	54	329	19	22	1.69	0.079	4.76	0.009	205	28.6
Beef, top sirloin, lean & fat, 1/8" fat, select, broiled	230		27.1	12.7	67	57	345	22	23	1.66	0.074	4.93	0.009	217	29.3
Beets, boiled	44	10.0	1.7	0.2		77	305	16	23	0.79	0.074	0.35	0.326	38	0.7
Blackberries, raw	43	9.6	1.4	0.5		1	162	29	20	0.62	0.165	0.53	0.646	22	0.4
Blueberries, raw	57	14.5	0.7	0.3		1	77	6	6	0.28	0.057	0.16	0.336	12	0.1
Bluefish, cooked, dry heat	159		25.7	5.4	76	77	477	9	42	0.62	0.068	1.04	0.027	291	46.8
Bologna, beef	314	4.0	10.3	28.2	56	1080	172	31	14	1.10	0.068	9.10	0.044	172	
Bologna, pork	247	0.7	15.3	19.9	59	1184	281	11	14	0.77	0.080	2.03	0.036	139	12.7
Bratwurst, beef & pork, smoked	297	2.0	12.2	26.3	78	848	283	7	15	1.00	0.080	2.47	0.041	130	14.1
Bratwurst, pork, cooked	333	2.9	13.7	29.2	74	846	259	28	21	0.53	0.079	2.49	0.012	225	26.9
Brazil nuts, dried, unblanched	656	12.3	14.3	66.4		3	659	160	376	2.43	1.743	4.06	1.223	725	1917.0
Bread, cracked-wheat	260	49.5	8.7	3.9		538	177	43	52	2.81	0.222	1.24	1.371	153	25.3
Bread, Italian	271	50.0	8.8	3.5		584	110	78	27	2.94	0.191	0.86	0.464	103	27.2
Bread, mixed-grain, toasted (includes whole-grain, 7-grain)	272	50.4	10.9	4.1		530	222	99	58	3.77	0.276	1.38	1.615	191	32.1
Bread, oatmeal	269	48.5	8.4	4.4		599	142	66	37	2.70	0.209	1.02	0.940	126	24.6
Bread, pita, white, enriched	275	55.7	9.1	1.2		536	120	86	26	2.62	0.168	0.84	0.481	97	27.1
Bread, pumpernickel	250	47.5	8.7	3.1		671	208	68	54	2.87	0.287	1.48	1.305	178	24.5
Bread, rye	259	48.3	8.5	3.3		660	166	73	40	2.83	0.186	1.14	0.824	125	30.9
Bread, white, commercially prepared	266	50.6	7.6	3.3		681	100	151	23	3.74	0.253	0.74	0.478	99	17.3
Bread, whole-wheat, commercially prepared	246	46.1	9.7	4.2		527	252	72	86	3.30	0.284	1.94	2.324	229	36.6
Broccoli, boiled, w/o salt	35	7.2	2.4	0.4		41	293	40	21	0.67	0.061	0.45	0.194	67	1.6
Broccoli, raw	34	6.6	2.8	0.4		33	316	47	21	0.73	0.049	0.41	0.210	66	2.5
Brussels sprouts, boiled, w/o salt	36	7.1	2.6	0.5		21	317	36	20	1.20	0.083	0.33	0.227	56	1.5
Butter, w. salt	717	0.1	0.9	81.1	215	576	24	24	2	0.02		0.09		24	1.0
Butter, w/o salt	717	0.1	0.9	81.1	215	11	24	24	2	0.02	0.016	0.09	0.004	24	1.0
Cabbage, boiled, w/o salt	22	4.5	1.0	0.4		8	97	31	8	0.17	0.012	0.09	0.117	15	0.6
Cabbage, raw	24	5.6	1.4	0.1		18	246	47	15	0.59	0.023	0.18	0.159	23	0.9
Cake, angel food, commercially prepared	258	57.8	5.9	0.8		749	93	140	12	0.52	0.078	0.07	0.085	32	7.3
Cake, carrot, dry mix, pudding-type	415	79.2	5.1	9.8		567	169	172	8	1.80	0.050	0.20	0.528	247	14.9
Cake, fruitcake, commercially prepared	324	61.6	2.9	9.1	5	270	153	33	16	2.07	0.050	0.27	0.220	52	2.0
Cake, pound, commercially prepared, butter	388	48.8	5.5	19.9	221	398	119	35	11	1.38	0.035	0.46	0.090	137	8.8
Cake, white, dry mix	426	78.0	4.5	10.9		664	117	192	11	1.39	0.081	0.46	0.205	337	8.6
Cake, yellow, dry mix, enriched	432	78.1	4.4	11.6	2	657	82	135	10	1.50	0.072	0.27	0.191	310	3.0
Candies, fudge, chocolate, prepared-from-recipe	411	76.5	2.4	10.4	14	47	131	45	36	1.77	0.333	1.10	0.423	69	2.5
Candies, gumdrops, starch jelly pieces	396	98.9				44	5	3	1	0.40	0.012		0.010	1	0.8
Candies, hard	394	98.0		0.2		38	5	3	3	0.30	0.029	0.01	0.010	3	0.6
Candies, marshmallows	318	81.3	1.8	0.2		80	5	3	2	0.23	0.097	0.04	0.008	8	1.7
Candies, milk chocolate	535	59.4	7.7	29.7	23	79	372	189	63	2.35	0.491	2.01	0.471	208	4.5
Candies, milk chocolate w. almond bites	550	51.0	9.8	35.7	19	74	471	220	59	1.50	0.200	1.34	0.010	227	3.4
Candies, peanut brittle, prepared-from-recipe	486	71.2	7.6	19.0	12	445	168	27	42	1.22	0.254	0.87	0.593	106	2.6
Candies, semisweet chocolate	479	63.1	4.2	30.0		11	365	32	115	3.13	0.700	1.62	0.800	132	4.2
Candies, sweet chocolate	505	59.6	3.9	34.2		16	290	24	113	2.76	0.574	1.50	0.494	147	2.8
Candies, white chocolate	539	59.2	5.9	32.1	14	90	286	199	12	0.24	0.060	0.74	0.008	176	4.5

Biochem

Food description	Energy [kcal/ 100 g]	Carb. [g/ 100 g]	Protein [g/ 100 g]	Fat [g/ 100 g]	Chol. [mg/ 100 g]	Na [mg/ 100 g]	K [mg/ 100 g]	Ca [mg/ 100 g]	Mg [mg/ 100 g]	Fe [mg/ 100 g]	Cu [mg/ 100 g]	Zn [mg/ 100 g]	Mn [mg/ 100 g]	P [mg/ 100 g]	Se [µg/ 100 g]
Capers, canned	23	4.9	2.4	0.9		2964	40	40	33	1.67	0.374	0.32	0.078	10	1.2
Carbonated beverage, cola, contains caffeine	37	9.6	0.1			4	2	2		0.11	0.001	0.02	0.002	10	0.1
Carbonated beverage, ginger ale	34	8.8				7	1	3	1	0.18	0.018	0.05	0.013		0.1
Carbonated beverage, orange	48	12.3				12	2	5	1	0.06	0.015	0.10	0.013	1	
Carbonated beverage, root beer	41	10.6				13	1	5	1	0.05	0.007	0.07	0.013		0.1
Carbonated beverage, tonic water	34	8.8				12		1		0.01	0.006	0.10	0.001		
Carrots, boiled, w/o salt	35	8.2	0.8	0.2		58	235	30	10	0.34	0.017	0.20	0.155	30	0.7
Carrots, raw	41	9.6	0.9	0.2		69	320	33	12	0.30	0.045	0.24	0.143	35	0.1
Cashew nuts, dry roasted, w. salt	574	32.7	15.3	46.4		640	565	45	260	6.00	2.220	5.60	0.826	490	11.7
Cashew nuts, oil roasted, w. salt	581	30.2	16.8	47.8		308	632	43	273	6.05	2.043	5.35	1.668	531	20.3
Cashew nuts, raw	553	30.2	18.2	43.9		12	660	37	292	6.68	2.195	5.78	1.655	593	19.9
Catfish, channel, farmed, cooked, dry heat	152		18.7	8.0	64	80	321	9	26	0.82	0.122	1.05	0.020	245	14.5
Catsup	97	25.1	1.7	0.4		1114	382	18	19	0.51	0.181	0.26	0.128	33	0.3
Cauliflower, boiled, w/o salt	23	4.1	1.8	0.5		15	142	16	9	0.33	0.027	0.18	0.138	32	0.5
Cauliflower, raw	25	5.3	2.0	0.1		30	303	22	15	0.44	0.042	0.28	0.156	44	0.6
Celery, raw	14	3.0	0.7	0.2		80	260	40	11	0.20	0.035	0.13	0.103	24	0.4
Cereals, corn grits, white, enriched, cooked w. water, w/o salt	59	12.9	1.4	0.2		2	21	3	5	0.60	0.018	0.07	0.018	11	3.1
Chard, Swiss, boiled, w/o salt	20	4.1	1.9	0.1		179	549	58	86	2.26	0.163	0.33	0.334	33	0.9
Chard, Swiss, raw	19	3.7	1.8	0.2		213	379	51	81	1.80	0.179	0.36	0.366	46	0.9
Cheese, American cheddar, imitation	239	11.6	16.7	14.0	36	1345	242	562	29	0.33	0.033	2.59		712	15.2
Cheese, blue	353	2.3	21.4	28.7	75	1395	256	528	23	0.31	0.040	2.66	0.009	387	14.5
Cheese, brie	334	0.5	20.8	27.7	100	629	152	184	20	0.50	0.019	2.38	0.034	188	14.5
Cheese, camembert	300	0.5	19.8	24.3	72	842	187	388	20	0.33	0.021	2.38	0.038	347	14.5
Cheese, cheddar	403	1.3	24.9	33.1	105	621	98	721	28	0.68	0.031	3.11	0.010	512	13.9
Cheese, Cheshire	387	4.8	23.4	30.6	103	700	95	643	21	0.21	0.042	2.79	0.012	464	14.5
Cheese, Colby	394	2.6	23.8	32.1	95	604	127	685	26	0.76	0.042	3.07	0.012	457	14.5
Cheese, cottage, creamed, large or small curd	103	2.7	12.5	4.5	15	405	84	60	5	0.14	0.028	0.37	0.003	132	9.0
Cheese, cottage, low fat, 1% milk fat	72	2.7	12.4	1.0	4	406	86	61	5	0.14	0.028	0.38	0.003	134	9.0
Cheese, cream	349	2.7	7.6	34.9	110	296	119	80	6	1.20	0.016	0.54	0.004	104	2.4
Cheese, cream, low fat	231	7.0	10.6	17.6	56	296	167	112	8	1.68	0.022	0.76		146	4.0
Cheese, edam	357	1.4	25.0	27.8	89	965	188	731	30	0.44	0.036	3.75	0.011	536	14.5
Cheese, feta	264	4.1	14.2	21.3	89	1116	62	493	19	0.65	0.032	2.88	0.028	337	15.0
Cheese, gouda	356	2.2	24.9	27.4	114	819	121	700	29	0.24	0.036	3.90	0.011	546	14.5
Cheese, gruyere	413	0.4	29.8	32.3	110	336	81	1011	36	0.17	0.032	3.90	0.017	605	14.5
Cheese, mozzarella, whole milk	300	2.2	22.2	22.4	79	627	76	505	20	0.44	0.011	2.92	0.030	354	17.0
Cheese, muenster	368	1.1	23.4	30.0	96	628	134	717	27	0.41	0.031	2.81	0.008	468	14.5
Cheese, parmesan, grated	431	4.1	38.5	28.6	88	1529	125	1109	38	0.90	0.238	3.87	0.085	729	17.7
Cheese, pimento	375	1.7	22.1	31.2	94	1428	162	614	22	0.42	0.033	2.98	0.016	744	14.5
Cheese, provolone	351	2.1	25.6	26.6	69	876	138	756	28	0.52	0.026	3.23	0.010	496	14.5
Cheese, Swiss	380	5.4	26.9	27.8	92	192	77	791	38	0.20	0.043	4.36	0.005	567	18.2
Cheese, tilsit	340	1.9	24.4	26.0	102	753	65	700	13	0.23	0.026	3.50	0.013	500	14.5
Cheesecake, commercially prepared	321	25.5	5.5	22.5	55	207	90	51	11	0.63	0.020	0.51	0.140	93	5.2
Cherries, sour, red, raw	50	12.2	1.0	0.3		3	173	16	9	0.32	0.104	0.10	0.112	15	
Cherries, sweet, raw	63	16.0	1.1	0.2			222	13	11	0.36	0.060	0.07	0.070	21	
Chestnuts, European, boiled & steamed	131	27.8	2.0	1.4		27	715	46	54	1.73	0.472	0.25	0.854	99	
Chestnuts, European, roasted	245	53.0	3.2	2.2		2	592	29	33	0.91	0.507	0.57	1.180	107	1.2
Chicken, broilers or fryers, breast, meat & skin, fried	222	1.6	31.8	8.9	89	76	259	16	30	1.19	0.057	1.10	0.026	233	23.9
Chicken, broilers or fryers, breast, meat & skin, roasted	197		29.8	7.8	84	71	245	14	27	1.07	0.050	1.02	0.018	214	24.7
Chicken, broilers or fryers, breast, meat only, fried	187	0.5	33.4	4.7	91	79	276	16	31	1.14	0.054	1.08	0.021	246	26.2
Chicken, broilers or fryers, breast, meat only, roasted	165		31.0	3.6	85	74	256	15	29	1.04	0.049	1.00	0.017	228	27.6
Chicken, broilers or fryers, dark meat, meat only, roasted	205		27.4	9.7	93	93	240	15	23	1.33	0.080	2.80	0.021	179	18.0
Chicken, broilers or fryers, drumstick, meat & skin, fried	245	1.6	27.0	13.7	90	89	229	12	23	1.34	0.080	2.89	0.028	176	18.4
Chicken, broilers or fryers, thigh, meat & skin, roasted	247		25.1	15.5	93	84	222	12	22	1.34	0.078	2.36	0.021	174	19.5
Chicken, broilers or fryers, thigh, meat only, roasted	209		25.9	10.9	95	88	238	12	24	1.31	0.081	2.57	0.021	183	29.0
Chicken, broilers or fryers, wing, meat & skin, fried	321	2.4	26.1	22.2	81	77	177	15	19	1.25	0.061	1.76	0.028	150	21.3
Chicken, Cornish game hens, meat & skin, roasted	260		22.3	18.2	131	64	245	13	18	0.91	0.061	1.49	0.015	146	15.5

Food description	Energy [kcal/ 100 g]	Carb. [g/ 100 g]	Protein [g/ 100 g]	Fat [g/ 100 g]	Chol. [mg/ 100 g]	Na [mg/ 100 g]	K [mg/ 100 g]	Ca [mg/ 100 g]	Mg [mg/ 100 g]	Fe [mg/ 100 g]	Cu [mg/ 100 g]	Zn [mg/ 100 g]	Mn [mg/ 100 g]	P [mg/ 100 g]	Se [µg/ 100 g]
Chicken, Cornish game hens, meat only, roasted	134		23.3	3.9	106	63	250	13	19	0.77	0.059	1.53	0.015	149	20.8
Chickpeas, mature seeds, boiled, w/o salt	164	27.4	8.9	2.6		7	291	49	48	2.89	0.352	1.53	1.030	168	3.7
Chicory greens, raw	23	4.7	1.7	0.3		45	420	100	30	0.90	0.295	0.42	0.429	47	0.3
Chocolate syrup	279	65.1	2.1	1.1		72	224	14	65	2.11	0.512	0.73	0.382	129	2.7
Clam, mixed species, breaded & fried	202	10.3	14.2	11.2	61	364	326	63	14	13.91	0.356	1.46	0.540	188	28.9
Clam, mixed species, raw	74	2.6	12.8	1.0	34	56	314	46	9	13.98	0.344	1.37	0.500	169	24.3
Cocoa, dry powder, unsweetened	229	54.3	19.6	13.7		21	1524	128	499	13.86	3.788	6.81	3.837	734	14.3
Coconut meat, raw	354	15.2	3.3	33.5		20	356	14	32	2.43	0.435	1.10	1.500	113	10.1
Cod, Atlantic, cooked, dry heat	105		22.8	0.9	55	78	244	14	42	0.49	0.036	0.58	0.020	138	37.6
Cod, Atlantic, dried & salted	290		62.8	2.4	152	7027	1458	160	133	2.50	0.176	1.59	0.050	950	147.8
Coffee, brewed from grounds, prep w. tap h2o	1		0.1			2	49	2	3	0.01	0.002	0.02	0.023	3	
Coffee, brewed, espresso, rest-prep	2		0.1	0.2		14	115	2	80	0.13	0.050	0.05	0.050	7	
Coleslaw, home-prepared	69	12.4	1.3	2.6	8	23	181	45	10	0.59	0.023	0.20	0.097	32	0.7
Collards, boiled, w/o salt	26	4.9	2.1	0.4		16	116	140	20	1.16	0.038	0.23	0.436	30	0.5
Cookies, brownies, commercially prepared	405	63.9	4.8	16.3	17	312	149	29	31	2.25	0.224	0.72	0.128	101	6.3
Cookies, chocolate chip, commercially prepared, higher fat, unenriched	481	66.8	5.4	22.6		315	135	25	31	1.00	0.212	0.64	0.449	108	
Cookies, chocolate sandwich, w/creme filling	466	71.6	5.3	19.1		483	187	21	48	3.93	0.431	0.98	0.629	92	8.1
Cookies, fig bars	348	70.9	3.7	7.3		350	207	64	27	2.90	0.147	0.39	0.343	62	3.3
Cookies, gingersnaps	416	76.9	5.6	9.8		654	346	77	49	6.40	0.305	0.55	1.555	83	5.1
Cookies, graham crackers, plain or honey	423	76.8	6.9	10.1		605	135	24	30	3.73	0.202	0.81	0.804	104	10.2
Cookies, oatmeal, commercially prepared	450	68.7	6.2	18.1		383	142	37	33	2.58	0.134	0.79	0.839	138	9.8
Cookies, peanut butter sandwich, regular	478	65.6	8.8	21.1		368	192	53	49	2.60	0.237	1.06	0.912	188	7.7
Cookies, shortbread, commercially prepared, plain	502	64.5	6.1	24.1	20	455	100	35	17	2.74	0.144	0.53	0.428	108	7.3
Cookies, vanilla wafers, higher fat	473	71.1	4.3	19.4		306	107	25	12	2.21	0.124	0.33	0.384	64	11.3
Cookies, vanilla wafers, lower fat	441	73.6	5.0	15.2	51	312	97	48	14	2.38	0.100	0.36	0.262	104	11.3
Corn, sweet, white, boiled, w. salt	108	25.1	3.3	1.3		253	249	2	32	0.61	0.053	0.48	0.194	103	0.8
Corn, sweet, yellow, boiled, w. salt	108	25.1	3.3	1.3		253	249	2	32	0.61	0.053	0.48	0.194	103	0.2
Couscous, cooked	112	23.2	3.8	0.2		5	58	8	8	0.38	0.041	0.26	0.084	22	27.5
Cowpeas (black-eyed), immature seeds, boiled, w. salt	97	20.3	3.2	0.4		240	418	128	52	1.12	0.133	1.03	0.572	51	2.5
Crab, Alaska king, cooked, moist heat	97		19.4	1.5	53	1072	262	59	63	0.76	1.182	7.62	0.040	280	40.0
Crab, Alaska king, raw	84		18.3	0.6	42	836	204	46	49	0.59	0.922	5.95	0.035	219	36.4
Crab, blue, cooked, moist heat	102		20.2	1.8	100	279	324	104	33	0.91	0.645	4.22	0.190	206	40.2
Crab, blue, raw	87		18.1	1.1	78	293	329	89	34	0.74	0.669	3.54	0.150	229	37.4
Crab, Dungeness, cooked, moist heat	110	1.0	22.3	1.2	76	378	408	59	58	0.43	0.734	5.47	0.097	175	47.6
Crackers, cheese, regular	503	58.2	10.1	25.3	13	995	145	151	36	4.77	0.210	1.13	0.629	218	8.6
Crackers, matzo, plain	395	83.7	10.0	1.4		2	112	13	25	3.16	0.060	0.68	0.650	89	36.9
Crackers, saltines (includes oyster, soda, soup)	428	70.9	9.2	11.4		1072	154	68	22	5.64	0.298	0.83	0.653	101	10.3
Crackers, wheat, regular	473	64.9	8.6	20.6		795	183	49	62	4.40	0.318	1.60	1.781	220	6.3
Cranberries, raw	46	12.2	0.4	0.1		2	85	8	6	0.25	0.061	0.10	0.360	13	0.1
Cranberry juice, unsweetened	46	12.2	0.4	0.1		2	77	8	6	0.25	0.055	0.10		13	0.1
Crayfish, mixed species, farmed, cooked, moist heat	87		17.5	1.3	137	97	238	51	33	1.11	0.580	1.48	0.217	241	34.2
Cream, fluid, half and half	130	4.3	3.0	11.5	37	41	130	105	10	0.07	0.010	0.51	0.001	95	1.8
Cream, half & half, fat free	59	9.0	2.6	1.4	5	144	206	96	16		0.016	0.81	0.002	151	2.9
Cream, sour, cultured	214	4.3	3.2	21.0	44	53	144	116	11	0.06	0.019	0.27	0.003	85	2.2
Cream, whipped, cream topping, pressurized	257	12.5	3.2	22.2	76	130	147	101	11	0.05	0.010	0.37	0.001	89	1.4
Croissants, butter	406	45.8	8.2	21.0	67	744	118	37	16	2.03	0.080	0.75	0.330	105	22.7
Croutons, plain	407	73.5	11.9	6.6		698	124	76	31	4.08	0.163	0.89	0.500	115	37.5
Cucumber, peeled, raw	12	2.2	0.6	0.2		2	136	14	12	0.22	0.071	0.17	0.073	21	0.1
Cucumber, with peel, raw	15	3.6	0.7	0.1		2	147	16	13	0.28	0.041	0.20	0.079	24	0.3
Curry powder	325	58.2	12.7	13.8		52	1543	478	254	29.59	0.815	4.05	4.289	349	17.1
Dandelion greens, boiled, w/o salt	33	6.4	2.0	0.6		44	232	140	24	1.80	0.115	0.28	0.230	42	0.3
Dandelion greens, raw	45	9.2	2.7	0.7		76	397	187	36	3.10	0.171	0.41	0.342	66	0.5
Duck, domesticated, meat & skin, roasted	337		19.0	28.4	84	59	204	11	16	2.70	0.227	1.86	0.019	156	20.0
Duck, domesticated, meat only, roasted	201		23.5	11.2	89	65	252	12	20	2.70	0.231	2.60	0.019	203	22.4
Egg, white, raw, fresh	52	0.7	10.9	0.2		166	163	7	11	0.08	0.023	0.03	0.011	15	20.0
Egg, whole, fried	201	0.9	13.6	15.3	457	204	147	59	13	1.98	0.111	1.20	0.041	208	34.2
Egg, whole, hard-boiled	155	1.1	12.6	10.6	424	124	126	50	10	1.19	0.013	1.05	0.026	172	30.8
Egg, whole, poached	147	0.8	12.5	9.9	422	294	133	53	12	1.83	0.102	1.10	0.039	190	31.6
Egg, whole, scrambled	166	2.2	11.1	12.2	352	280	138	71	12	1.20	0.014	1.00	0.022	170	22.5
Egg, yolk, raw, fresh	322	3.6	15.9	26.5	1234	48	109	129	5	2.73	0.077	2.30	0.055	390	56.0
Eggplant, boiled, w/o salt	35	8.7	0.8	0.2		1	123	6	11	0.25	0.059	0.12	0.113	15	0.1

Food description	Energy [kcal/ 100 g]	Carb. [g/ 100 g]	Protein [g/ 100 g]	Fat [g/ 100 g]	Chol. [mg/ 100 g]	Na [mg/ 100 g]	K [mg/ 100 g]	Ca [mg/ 100 g]	Mg [mg/ 100 g]	Fe [mg/ 100 g]	Cu [mg/ 100 g]	Zn [mg/ 100 g]	Mn [mg/ 100 g]	P [mg/ 100 g]	Se [μg/ 100 g]
Eggplant, raw	24	5.7	1.0	0.2		2	230	9	14	0.24	0.082	0.16	0.250	25	0.3
Endive, raw	17	3.4	1.3	0.2		22	314	52	15	0.83	0.099	0.79	0.420	28	0.2
English muffins, whole-wheat, toasted	221	44.1	9.6	2.3		692	228	288	77	2.66	0.225	1.74	1.946	307	43.8
English muffins, plain, toasted, enriched w. calcium propanoate (includes sourdough)	270	52.7	10.3	2.0		477	129	197	28	4.65	0.160	1.40	0.630	107	26.3
Fat, beef tallow	902			100	109										0.2
Fat, chicken	900			99.8	85										0.2
Fat, duck	900			99.8	100										0.2
Fat, goose	900			99.8	100										0.2
Fat, turkey	900			99.8	102										0.2
Fennel, bulb, raw	31	7.3	1.2	0.2		52	414	49	17	0.73	0.066	0.20	0.191	50	0.7
Figs, dried, stewed	107	27.6	1.4	0.4		4	294	70	29	0.88	0.124	0.24	0.220	29	0.2
Figs, raw	74	19.2	0.8	0.3		1	232	35	17	0.37	0.070	0.15	0.128	14	0.2
Fish oil, cod liver	902			100	570										
Frankfurter beef	330	4.1	11.2	29.6	53	1140	156	14	14	1.51	0.184	2.46	0.082	160	8.2
Frankfurter, chick	257	6.8	12.9	19.5	101	1370	84	95	10	2.00	0.050	1.04	0.015	107	18.4
Frankfurter, pork	269	0.3	12.8	23.7	66	816	264	267	15	3.70	0.074	2.09	0.016	171	27.8
Frankfurter, turkey	226	1.5	14.3	17.7	107	1426	179	106	14	1.84	0.100	3.11	0.016	134	15.4
Game meat, beaver, roasted	212		34.9	7.0	117	59	403	22	29	10.00	0.189	2.27		292	43.1
Game meat, beefalo, roasted	188		30.7	6.3	58	82	459	24		3.05		6.40		250	13.1
Game meat, bison, lean, roasted	143		28.4	2.4	82	57	361	8	26	3.42	0.107	3.68	0.008	209	35.5
Game meat, boar, wild, roasted	160		28.3	4.4	77	60	396	16	27	1.12	0.056	3.01		134	13.0
Game meat, deer, roasted	158		30.2	3.2	112	54	335	7	24	4.47	0.300	2.75	0.046	226	12.9
Game meat, elk, roasted	146		30.2	1.9	73	61	328	5	24	3.63	0.142	3.16	0.013	180	13.0
Game meat, moose, roasted	134		29.3	1.0	78	69	334	6	24	4.22	0.079	3.68	0.009	176	12.8
Game meat, rabbit, domesticated, roasted	197		29.1	8.1	82	47	383	19	21	2.27	0.189	2.27	0.032	263	38.5
Game meat, raccoon, roasted	255		29.2	14.5	97	79	398	14	30	7.10	0.189	2.27		261	18.0
Game meat, squirrel, roasted	173		30.8	4.7	121	119	352	3	28	6.81	0.148	1.78	0.032	211	15.1
Goose, domesticated, meat & skin, roasted	305		25.2	21.9	91	70	329	13	22	2.83	0.264	2.62	0.023	270	21.8
Goose, domesticated, meat only, roasted	238		29.0	12.7	96	76	388	14	25	2.87	0.276	3.17	0.024	309	25.5
Goose, liver, raw	133	6.3	16.4	4.3	515	140	230	43	24	30.53	7.522	3.07		261	68.1
Gooseberries, raw	44	10.2	0.9	0.6		1	198	25	10	0.31	0.070	0.12	0.144	27	0.6
Grape juice, canned or bottled, unsweetened, w/o vitamin C	61	15.0	0.6	0.1		3	132	9	10	0.24	0.028	0.05	0.360	11	0.1
Grapefruit juice, white, canned, unsweetened	38	9.0	0.5	0.1		1	153	7	10	0.20	0.038	0.09	0.020	11	0.1
Grapefruit, raw, pink & red, all areas	42	10.7	0.8	0.1			135	22	9	0.08	0.032	0.07	0.022	18	0.1
Grapefruit, raw, white, all areas	33	8.4	0.7	0.1			148	12	9	0.06	0.050	0.07	0.013	8	1.4
Grapes, American type (slip skin), raw	67	17.2	0.6	0.4		2	191	14	5	0.29	0.040	0.04	0.718	10	0.1
Grouper, mixed species, cooked, dry heat	118		24.8	1.3	47	53	475	21	37	1.14	0.045	0.51	0.012	143	46.8
Haddock, cooked, dry heat	112		24.2	0.9	74	87	399	42	50	1.35	0.033	0.48	0.030	241	40.5
Haddock, smoked	116		25.2	1.0	77	763	415	49	54	1.40	0.042	0.50	0.030	251	42.9
Halibut, Atlantic & Pacific, cooked, dry heat	140		26.7	2.9	41	69	576	60	107	1.07	0.035	0.53	0.020	285	46.8
Ham, sliced, regular (approx 11% fat)	163	3.8	16.6	8.6	57	1304	287	24	22	1.02	0.089	1.35	0.557	153	20.7
Herring, Atlantic, kippered	217		24.6	12.4	82	918	447	84	46	1.51	0.135	1.36	0.050	325	52.6
Hominy, canned, white or yellow	72	14.3	1.5	0.9		210	9	10	16	0.62	0.030	1.05	0.070	35	3.0
Hominy, canned, yellow	72	14.3	1.5	0.9		210	9	10	16	0.62	0.030	1.05	0.070	35	3.0
Hummus, commercial	166	14.3	7.9	9.6		379	228	38	71	2.44	0.527	1.83	0.773	176	2.6
Ice creams, chocolate	216	28.2	3.8	11.0	34	76	249	109	29	0.93	0.135	0.58	0.140	107	2.5
Ice creams, French vanilla, soft-serve	222	22.2	4.1	13.0	91	61	177	131	12	0.21	0.030	0.52	0.005	116	3.0
Ice creams, strawberry	192	27.6	3.2	8.4	29	60	188	120	14	0.21	0.037	0.34	0.078	100	1.9
Ice creams, vanilla	201	23.6	3.5	11.0	44	80	199	128	14	0.09	0.023	0.69	0.008	105	1.8
Jams and preserves	278	68.9	0.4	0.1		32	77	20	4	0.49	0.100	0.06	0.040	19	2.0
Jellies	266	70.0	0.2			30	54	7	6	0.19	0.011	0.03	0.132	6	0.4
Kale, boiled, w/o salt	28	5.6	1.9	0.4		23	228	72	18	0.90	0.156	0.24	0.416	28	0.9
Kale, raw	50	10.0	3.3	0.7		43	447	135	34	1.70	0.290	0.44	0.774	56	0.9
Kiwi fruit, (Chinese gooseberries), fresh, raw	61	14.7	1.1	0.5		3	312	34	17	0.31	0.130	0.14	0.098	34	0.2
Knockwurst, pork or beef	307	3.2	11.1	27.7	60	930	199	11	11	0.66	0.060	1.66	0.021	98	13.5
Kumquats, raw	71	15.9	1.9	0.9		10	186	62	20	0.86	0.095	0.17	0.135	19	
Lamb, domestic, composite of retail cuts, lean & fat, 1/4" fat, choice, cooked	294		24.5	20.9	97	72	310	17	23	1.88	0.119	4.46	0.022	188	26.4
Lamb, domestic, composite of retail cuts, lean & fat, 1/8" fat, choice, cooked	271		25.5	18.0	96	72	318	16	24	1.93	0.121	4.74	0.024	193	27.2
Lamb, domestic, composite of retail cuts, lean, 1/4" fat, choice, cooked	206		28.2	9.5	92	76	344	15	26	2.05	0.128	5.27	0.028	210	26.1
Lamb, domestic, leg, shank half, lean & fat, 1/8" fat, choice, roasted	217		26.7	11.4	90	65	329	9	25	1.99	0.118	4.72	0.026	200	29.7

Food description	Energy [kcal/ 100 g]	Carb. [g/ 100 g]	Protein [g/ 100 g]	Fat [g/ 100 g]	Chol. [mg/ 100 g]	Na [mg/ 100 g]	K [mg/ 100 g]	Ca [mg/ 100 g]	Mg [mg/ 100 g]	Fe [mg/ 100 g]	Cu [mg/ 100 g]	Zn [mg/ 100 g]	Mn [mg/ 100 g]	P [mg/ 100 g]	Se [μg/ 100 g]
Lamb, domestic, loin, lean & fat, 1/4" fat, choice, roasted	309		22.6	23.6	95	64	246	18	23	2.12	0.119	3.41	0.020	180	24.6
Lamb, domestic, rib, lean & fat, 1/8" fat, choice, roasted	341		21.8	27.5	96	74	277	22	20	1.62	0.117	3.62	0.021	170	22.3
Lamb, ground, broiled	283		24.8	19.7	97	81	339	22	24	1.79	0.128	4.67	0.024	201	27.7
Lard	902			100	95							0.11			0.2
Leeks, (bulb & lower leaf-portion), boiled, w/o salt	31	7.6	0.8	0.2		10	87	30	14	1.10	0.062	0.06	0.247	17	0.5
Lemon juice, canned or bottled	21	6.5	0.4	0.3		21	102	11	8	0.13	0.037	0.06	0.020	9	0.1
Lemons, raw, with peel	20	10.7	1.2	0.3		3	145	61	12	0.70	0.260	0.10		15	
Lemons, raw, without peel	29	9.3	1.1	0.3		2	138	26	8	0.60	0.037	0.06	0.030	16	0.4
Lentils, mature seeds, boiled, w. salt	116	20.1	9.0	0.4		238	369	19	36	3.33	0.251	1.27	0.494	180	2.8
Lentils, sprouted, stir-fried, w. salt	101	21.3	8.8	0.5		246	284	14	35	3.10	0.337	1.60	0.502	153	0.6
Lettuce, iceberg (includes crisp head types), raw	14	3.0	0.9	0.1		10	141	18	7	0.41	0.025	0.15	0.125	20	0.1
Lettuce, romaine, raw	17	3.3	1.2	0.3		8	247	33	14	0.97	0.048	0.23	0.155	30	0.4
Lima beans, large, mature seeds, boiled, w/o salt	115	20.9	7.8	0.4		2	508	17	43	2.39	0.235	0.95	0.516	111	4.5
Lima beans, thin seeded (baby), mature seeds, boiled, w/o salt	126	23.3	8.0	0.4		3	401	29	53	2.40	0.215	1.03	0.585	127	4.9
Limes, raw	30	10.5	0.7	0.2		2	102	33	6	0.60	0.065	0.11	0.008	18	0.4
Liverwurst spread	305	5.9	12.4	25.5	118	700	170	22	12	8.85	0.240	2.30	0.155	230	58.0
Lobster, northern, cooked, moist heat	98	1.3	20.5	0.6	72	380	352	61	35	0.39	1.940	2.92	0.061	185	42.7
Macadamia nuts, dry roasted, w/o salt	718	13.4	7.8	76.1		4	363	70	118	2.65	0.570	1.29	3.036	198	3.6
Macaroni, cooked, enriched	158	30.9	5.8	0.9		1	45	7	18	1.33	0.103	0.50	0.317	58	26.4
Mackerel, Atlantic, cooked, dry heat	262		23.9	17.8	75	83	401	15	97	1.57	0.094	0.94	0.020	278	51.6
Mackerel, king, cooked, dry heat	134		26.0	2.6	68	203	558	40	41	2.28	0.033	0.72	0.006	318	46.8
Mackerel, Pacific & jack, mixed species, cooked, dry heat	201		25.7	10.1	60	110	521	29	36	1.49	0.119	0.86	0.019	160	46.8
Mackerel, salted	305		18.5	25.1	95	4450	520	66	60	1.40	0.100	1.10		254	73.4
Mackerel, Spanish, cooked, dry heat	158		23.6	6.3	73	66	554	13	38	0.74	0.065	0.62	0.012	271	40.6
Mangos, raw	65	17.0	0.5	0.3		2	156	10	9	0.13	0.110	0.04	0.027	11	0.6
Margarine, regular, hard, corn (hydrogenated)	719	0.9	0.9	80.5		943	42	30	3					23	
Margarine, regular, hard, safflower & soybean (hydrogenated)	719	0.9	0.9	80.5		943	42	30	3					23	
Margarine, regular, hard, soybean (hydrogenated)	719	0.9	0.9	80.5		943	42	30	3					23	
Margarine, regular, hard, soybean (hydrogenated) & palm (hydrogenated)	719	0.9	0.9	80.5		943	42	30	3					23	
Margarine, regular, hard, soybean (hydrogenated), cottonseed (hydrogenated), & soybean	719	0.9	0.9	80.5		943	42	30	3					23	
Margarine, regular, stick, unsalted, 80% fat	719	0.9	0.9	80.5		2	25	17	2					13	
Margarine, regular, tub, unsalted, 80% fat	716	0.5	0.8	80.4		28	38	26	2					20	
Marmalade, orange	246	66.3	0.3			56	37	38	2	0.15	0.090	0.04	0.020	4	0.6
Mayonnaise dressing, no cholesterol	688	0.3		77.8		486	14	7	1	0.23		0.13		25	1.6
Mayonnaise, low sodium, diet	231	16.0	0.3	19.2	24	110	10					0.11			1.6
Melons, honeydew, raw	36	9.1	0.5	0.1		18	228	6	10	0.17	0.024	0.09	0.027	11	0.7
Milk, buttermilk, fluid, cultured, reduced fat	56	5.3	4.1	2.0	8	86	180	143	13	0.06	0.008	0.24		82	2.3
Milk, goat, fluid	69	4.5	3.6	4.1	11	50	204	134	14	0.05	0.046	0.30	0.018	111	1.4
Milk, low fat, fluid, 1% milk fat, w. vitamin A	42	5.0	3.4	1.0	5	44	150	119	11	0.03	0.010	0.42	0.003	95	3.3
Milk, low sodium, fluid	61	4.5	3.1	3.5	14	3	253	101	5	0.05	0.010	0.38	0.004	86	2.0
Milk, reduced fat, fluid, 2% milk fat, w. added vitamin A	50	4.7	3.3	2.0	8	41	150	117	11	0.03	0.012	0.43	0.003	94	2.5
Muffins, blueberry, commercially prepared	277	48.0	5.5	6.5	30	447	123	57	16	1.61	0.074	0.49	0.440	197	11.2
Muffins, corn, commercially prepared	305	50.9	5.9	8.4	26	521	69	74	32	2.81	0.299	0.54	0.355	284	15.2
Muffins, oat bran	270	48.3	7.0	7.4		393	507	63	157	4.20	0.330	1.84	2.630	376	11.0
Mullet, striped, cooked, dry heat	150		24.8	4.9	63	71	458	31	33	1.41	0.141	0.88	0.022	244	46.8
Mushrooms, boiled, w/o salt	28	5.3	2.2	0.5		2	356	6	12	1.74	0.504	0.87	0.115	87	11.9
Mushrooms, oyster, raw	35	6.4	3.3	0.4		18	420	3	18	1.33	0.244	0.77	0.113	120	2.6
Mushrooms, portabella, grilled	35	5.1	4.1	0.8		10	521	4	15	0.56	0.499	0.73	0.075	150	17.7
Mushrooms, raw	22	3.3	3.1	0.3		5	318	3	9	0.50	0.318	0.52	0.047	86	9.3
Mushrooms, shiitake, cooked, w/o salt	56	14.4	1.6	0.2		4	117	3	14	0.44	0.896	1.33	0.204	29	24.8
Mussels, blue, cooked, moist heat	172	7.4	23.8	4.5	56	369	268	33	37	6.72	0.149	2.67	6.800	285	89.6
Mussels, blue, raw	86	3.7	11.9	2.2	28	286	320	26	34	3.95	0.094	1.60	3.400	197	44.8
Mustard greens, boiled, w/o salt	15	2.1	2.3	0.2		16	202	74	15	0.70	0.084	0.11	0.274	41	0.6
Nectarines, raw	44	10.6	1.1	0.3			201	6	9	0.28	0.086	0.17	0.054	26	
Noodles, egg, cooked, enriched, w. salt	138	25.2	4.5	2.1	29	165	38	12	21	1.47	0.098	0.65	0.315	76	23.9

Food description	Energy [kcal/ 100 g]	Carb. [g/ 100 g]	Protein [g/ 100 g]	Fat [g/ 100 g]	Chol. [mg/ 100 g]	Na [mg/ 100 g]	K [mg/ 100 g]	Ca [mg/ 100 g]	Mg [mg/ 100 g]	Fe [mg/ 100 g]	Cu [mg/ 100 g]	Zn [mg/ 100 g]	Mn [mg/ 100 g]	P [mg/ 100 g]	Se [µg/ 100 g]
Noodles, Japanese, soba, cooked	99	21.4	5.1	0.1		60	35	4	9	0.48	0.008	0.12	0.374	25	
Nutmeg, ground	525	49.3	5.8	36.3		16	350	184	183	3.04	1.027	2.15	2.900	213	1.6
Ocean perch, Atlantic, cooked, dry heat	121		23.9	2.1	54	96	350	137	39	1.18	0.033	0.61	0.020	277	55.5
Oil, avocado	884			100											
Oil, canola	884			100											
Oil, canola and soybean	884			100											
Oil, coconut	862			100						0.04					
Oil, corn and canola	884			100											
Oil, mustard	884			100											
Oil, olive, salad or cooking	884			100		2	1	1		0.56					
Oil, palm	884			100						0.01					
Oil, palm kernel	862			100											
Oil, peanut, salad or cooking	884			100						0.03		0.01			
Oil, safflower, salad or cooking, linoleic >70%	884			100											
Oil, safflower, salad or cooking, oleic >70%	884			100											
Oil, sesame, salad or cooking	884			100											
Oil, soybean lecithin	763			100											
Oil, soybean, salad or cooking	884			100						0.02					
Oil, sunflower, linoleic, approx. 65%	884			100											
Okra, boiled, w/o salt	22	4.6	1.9	0.2		6	135	77	36	0.28	0.085	0.43	0.294	32	0.4
Olives, pickled, canned or bottled, green	145	3.8	1.0	15.3		1556	42	52	11	0.49	0.120	0.04		4	0.9
Olives, ripe, canned (small-extra large)	115	6.3	0.8	10.7		872	8	88	4	3.30	0.251	0.22	0.020	3	0.9
Onions, boiled, w/o salt	44	10.2	1.4	0.2		3	166	22	11	0.24	0.067	0.21	0.153	35	0.6
Onions, raw	42	10.1	0.9	0.1		3	144	22	10	0.19	0.038	0.16	0.132	27	0.5
Orange juice, includes from concentrate	44	10.1	0.8	0.3		1	190	10	11	0.17	0.040	0.04	0.023	11	0.1
Oranges, raw, all commercial varieties	47	11.8	0.9	0.1			181	40	10	0.10	0.045	0.07	0.025	14	0.5
Oranges, raw, California, Valencias	49	11.9	1.0	0.3			179	40	10	0.09	0.037	0.06	0.023	17	
Oranges, raw, Florida	46	11.5	0.7	0.2			169	43	10	0.09	0.039	0.08	0.024	12	0.5
Oranges, raw, navels	49	12.5	0.9	0.2		1	166	43	11	0.13	0.039	0.08	0.029	23	
Ostrich, ground, cooked, pan-broiled	175		26.2	7.1	83	80	323	8	23	3.43	0.136	4.33	0.017	224	33.5
Ostrich, inside leg, cooked	141		29.0	1.9	73	83	352	6	25	3.12	0.148	4.71	0.018	244	36.5
Ostrich, top loin, cooked	155		28.1	3.9	93	77	353	6	25	3.31	0.148	4.72	0.018	245	36.6
Oyster, eastern, breaded & fried	197	11.6	8.8	12.6	81	417	244	62	58	6.95	4.294	87.13	0.490	159	66.5
Oyster, eastern, farmed, raw	59	5.5	5.2	1.6	25	178	124	44	33	5.78	0.738	37.92	0.394	93	63.7
Oyster, eastern, wild, raw	68	3.9	7.1	2.5	53	211	156	45	47	6.66	4.452	90.81	0.367	135	63.7
Oyster, Pacific, cooked, moist heat	163	9.9	18.9	4.6	100	212	302	16	44	9.20	2.679	33.24	1.222	243	154.0
Oyster, Pacific, raw	81	5.0	9.5	2.3	50	106	168	8	22	5.11	1.576	16.62	0.643	162	77.0
Palm hearts, raw	115	25.6	2.7	0.2		14	1806	18	10	1.69	0.644	3.73		140	0.7
Papayas, raw	39	9.8	0.6	0.1		3	257	24	10	0.10	0.016	0.07	0.011	5	0.6
Parsley, raw	36	6.3	3.0	0.8		56	554	138	50	6.20	0.149	1.07	0.160	58	0.1
Pastrami beef 98% fat-free	95	1.5	19.6	1.2	47	1010	228	9	18	2.78	0.079	4.26	0.013	150	10.4
Pate de foie gras, canned, smoked	462	4.7	11.4	43.8	150	697	138	70	13	5.50	0.400	0.92	0.120	200	44.0
Pate, liver, not specified, canned	319	1.5	14.2	28.0	255	697	138	70	13	5.50	0.400	2.85	0.120	200	41.6
Peaches, raw	39	9.5	0.9	0.3			190	6	9	0.25	0.068	0.17	0.061	20	0.1
Peanut butter, chunk style, w/o salt	589	21.6	24.1	49.9		17	745	45	160	1.90	0.578	2.79	1.800	319	8.2
Peanut butter, smooth style, w/o salt	588	19.6	25.1	50.4		17	649	43	154	1.87	0.473	2.91	1.466	358	5.6
Peanuts, all types, dry-roasted, w. salt	585	21.5	23.7	49.7		813	658	54	176	2.26	0.671	3.31	2.083	358	7.5
Peanuts, all types, oil-roasted, w. salt	599	15.3	28.0	52.5		320	726	61	176	1.52	0.533	3.28	1.845	397	3.3
Pears, raw	58	15.5	0.4	0.1		1	119	9	7	0.17	0.082	0.10	0.049	11	0.1
Peas, edible-podded, boiled, w. salt	42	7.1	3.3	0.2		240	240	42	26	1.97	0.077	0.37	0.168	55	0.7
Peas, edible-podded, raw	42	7.6	2.8	0.2		4	200	43	24	2.08	0.079	0.27	0.244	53	0.7
Peas, green, boiled, w/o salt	84	15.6	5.4	0.2		3	271	27	39	1.54	0.173	1.19	0.525	117	1.9
Pecans, dry roasted, w. salt	710	13.6	9.5	74.3		383	424	72	132	2.80	1.167	5.07	3.933	293	4.0
Pecans, oil roasted, w. salt	715	13.0	9.2	75.2		393	392	67	121	2.47	1.200	4.47	3.700	263	6.0
Peppers, hot chili, green, raw	40	9.5	2.0	0.2		7	340	18	25	1.20	0.174	0.30	0.237	46	0.5
Peppers, hot chili, red, raw	40	8.8	1.9	0.4		9	322	14	23	1.03	0.129	0.26	0.187	43	0.5
Peppers, jalapeno, raw	30	5.9	1.4	0.6		1	215	10	19	0.70	0.133	0.23	0.250	31	0.3
Peppers, sweet, green, boiled, w/o salt	28	6.7	0.9	0.2		2	166	9	10	0.46	0.065	0.12	0.115	18	0.3
Peppers, sweet, green, raw	20	4.6	0.9	0.2		3	175	10	10	0.34	0.066	0.13	0.122	20	
Peppers, sweet, red, boiled, w/o salt	28	6.7	0.9	0.2		2	166	9	10	0.46	0.065	0.12	0.115	18	0.3
Peppers, sweet, red, raw	26	6.0	1.0	0.3		2	211	7	12	0.43	0.017	0.25	0.112	26	0.1
Peppers, sweet, yellow, raw	27	6.3	1.0	0.2		2	212	11	12	0.46	0.107	0.17	0.117	24	0.3
Persimmons, native, raw	127	33.5	0.8	0.4		1	310	27		2.50				26	
Pheasant, cooked, total edible	247		32.4	12.1	89	43	271	16	22	1.43	0.084	1.37		242	20.7
Pickle, cucumber, sour	11	2.3	0.3	0.2		1208	23		4	0.40	0.085	0.02	0.011	14	
Pickle, cucumber, sweet	117	31.8	0.4	0.3		939	32	4	4	0.59	0.105	0.08	0.015	12	

Food description	Energy [kcal/ 100 g]	Carb. [g/ 100 g]	Protein [g/ 100 g]	Fat [g/ 100 g]	Chol. [mg/ 100 g]	Na [mg/ 100 g]	K [mg/ 100 g]	Ca [mg/ 100 g]	Mg [mg/ 100 g]	Fe [mg/ 100 g]	Cu [mg/ 100 g]	Zn [mg/ 100 g]	Mn [mg/ 100 g]	P [mg/ 100 g]	Se [µg/ 100 g]
Pickles, cucumber, dill	18	4.1	0.6	0.2		1282	116	9	11	0.53	0.079	0.14	0.015	21	
Pie, apple, commercially prepared, enriched flour	237	34.0	1.9	11.0		266	65	11	7	0.45	0.046	0.16	0.182	24	1.0
Pie, blueberry, commercially prepared	232	34.9	1.8	10.0		325	50	8	5	0.30	0.046	0.16	0.176	23	1.4
Pie, cherry, commercially prepared	260	39.8	2.0	11.0		246	81	12	8	0.48	0.040	0.18	0.140	29	1.2
Pie, chocolate creme, commercially prepared	304	33.6	2.6	19.4	5	136	127	36	21	1.07	0.050	0.23	0.200	68	7.5
Pie, coconut creme, commercially prepared	298	37.2	2.1	16.6		255	65	29	20	0.80	0.068	0.47	0.438	85	5.3
Pie, egg custard, commercially prepared	210	20.8	5.5	11.6	33	240	106	80	11	0.58	0.024	0.52	0.060	112	7.1
Pie, lemon meringue, commercially prepared	268	47.2	1.5	8.7	45	146	89	56	15	0.61	0.001	0.49	0.060	105	3.0
Pie, mince, prepared from recipe	289	48.0	2.6	10.8		254	203	22	14	1.49	0.113	0.22	0.263	42	6.6
Pie, peach	223	32.9	1.9	10.0		270	125	8	6	0.50	0.053	0.09	0.152	22	1.3
Pie, pecan, commercially prepared	400	57.2	4.0	18.5	32	424	74	17	18	1.04	0.195	0.57	0.789	77	5.0
Pie, pumpkin, commercially prepared	210	27.3	3.9	9.5	20	282	154	60	15	0.79	0.048	0.45	0.240	71	2.6
Pike, northern, cooked, dry heat	113		24.7	0.9	50	49	331	73	40	0.71	0.065	0.86	0.310	282	16.2
Pike, walleye, raw	93		19.1	1.2	86	51	389	110	30	1.30	0.178	0.62	0.800	210	12.6
Pimento, canned	23	5.1	1.1	0.3		14	158	6	6	1.68	0.049	0.19	0.092	17	0.2
Pine nuts, pinyon, dried	629	19.3	11.6	61.0		72	628	8	234	3.06	1.035	4.28	4.333	35	
Pineapple juice, canned, unsweetened, w/o vitamin C	53	12.9	0.4	0.1		2	130	13	12	0.31	0.069	0.11	0.504	8	0.1
Pineapple, raw, all variety	48	12.6	0.5	0.1		1	115	13	12	0.28	0.099	0.10	1.177	8	0.1
Pineapple, raw, traditional variety	45	11.8	0.6	0.1		1	125	13	12	0.25	0.081	0.08	1.593	9	
Pistachio nuts, dry roasted, w/o salt	571	27.7	21.4	46.0		10	1042	110	120	4.20	1.325	2.30	1.275	485	9.3
Plantains, cooked	116	31.2	0.8	0.2		5	465	2	32	0.58	0.066	0.13		28	1.4
Plums, dried (prunes), uncooked	240	63.9	2.2	0.4		2	732	43	41	0.93	0.281	0.44	0.299	69	0.3
Plums, raw	46	11.4	0.7	0.3			157	6	7	0.17	0.057	0.10	0.052	16	
Pollock, Atlantic, cooked, dry heat	118		24.9	1.3	91	110	456	77	86	0.59	0.064	0.60	0.019	283	46.8
Pollock, walleye, cooked, dry heat	113		23.5	1.1	96	116	387	6	73	0.28	0.055	0.60	0.020	482	43.4
Pomegranates, raw	68	17.2	1.0	0.3		3	259	3	3	0.30	0.070	0.12		8	0.6
Pompano, Florida, cooked, dry heat	211		23.7	12.1	64	76	636	43	31	0.67	0.078	0.69	0.025	341	46.8
Popcorn, air-popped	387	77.8	12.9	4.5		8	329	7	144	3.19	0.262	3.08	1.113	358	
Pork sausage, fresh, cooked	339		19.4	28.4	84	749	294	13	17	1.36	0.086	2.08	0.005	163	
Pork, cured, bacon, broiled, pan-fried or roasted	541	1.4	37.0	41.8	110	2310	565	11	33	1.44	0.164	3.50	0.022	533	62.0
Pork, cured, breakfast strips, cooked	459	1.1	29.0	36.7	105	2099	466	14	26	1.97	0.153	3.68	0.044	265	24.7
Pork, cured, Canadian-style bacon, grilled	185	1.4	24.2	8.4	58	1546	390	10	21	0.82	0.054	1.70	0.027	296	24.7
Pork, cured, ham, boneless, ex lean (approx 5% fat), roasted	145	1.5	20.9	5.5	53	1203	287	8	14	1.48	0.079	2.88	0.054	196	19.5
Pork, cured, ham, boneless, regular (approx 11% fat), roasted	178		22.6	9.0	59	1500	409	8	22	1.34	0.145	2.47	0.041	281	19.8
Pork, cured, ham, whole, lean & fat, roasted	243		21.6	16.8	62	1187	286	7	19	0.87	0.083	2.32	0.014	214	22.7
Pork, cured, salt pork, raw	748		5.1	80.5	86	1424	66	6	7	0.44	0.050	0.90	0.005	52	5.8
Pork, fresh, composite of retail cuts (leg, loin, & shoulder), lean, cooked	212		29.3	9.7	86	59	375	21	26	1.10	0.061	2.97	0.018	237	45.0
Pork, fresh, leg (ham), rump half, lean & fat, roasted	252		28.9	14.3	96	62	374	12	27	1.05	0.103	2.82	0.023	272	46.8
Pork, fresh, leg (ham), shank half, lean & fat, roasted	289		25.3	20.1	92	59	338	15	22	0.98	0.098	3.06	0.028	257	43.3
Pork, fresh, loin, blade (chops), bone-in, lean & fat, broiled	320		22.5	24.9	86	70	344	29	22	0.93	0.083	3.37	0.008	212	37.1
Pork, fresh, loin, center loin (chops), bone-in, lean & fat, broiled	240		28.7	13.1	82	58	358	33	25	0.80	0.046	2.26	0.003	232	44.3
Pork, fresh, loin, center rib (chops), bone-in, lean, broiled	219		30.8	9.7	81	65	420	31	28	0.82	0.070	2.38	0.020	245	47.3
Pork, fresh, loin, center rib (chops), boneless, lean & fat, broiled	260		27.6	15.8	82	62	401	28	26	0.77	0.068	2.26	0.018	237	44.0
Pork, fresh, loin, center rib (roasts), boneless, lean & fat, roasted	252		27.0	15.2	81	48	346	6	22	0.93	0.016	2.64	0.010	214	40.3
Pork, fresh, loin, country-style ribs, lean & fat, braised	296		23.9	21.5	87	59	328	29	17	1.22	0.093	3.56	0.012	167	39.7
Pork, fresh, loin, sirloin (chops), bone-in, lean & fat, broiled	259		26.7	16.1	86	68	383	17	29	0.99	0.058	2.57	0.010	246	47.7
Pork, fresh, loin, sirloin (chops), boneless, lean & fat, broiled	208		30.5	8.6	91	56	372	18	27	1.21	0.053	2.62	0.003	243	50.5
Pork, fresh, loin, tenderloin, lean & fat, broiled	201		29.9	8.1	94	64	444	5	35	1.39	0.067	2.89	0.012	290	47.7
Pork, fresh, spareribs, lean & fat, braised	397		29.1	30.3	121	93	320	47	24	1.85	0.142	4.60	0.014	261	37.4
Potato chips, plain, salted	547	49.7	6.6	37.5		525	1642	24	70	1.61	0.398	2.39	0.664	155	8.1
Potatoes, baked, flesh & skin, w. salt	93	21.2	2.5	0.1		244	535	15	28	1.08	0.118	0.36	0.219	70	0.4
Potatoes, boiled in skin, flesh, w. salt	87	20.1	1.9	0.1		240	379	5	22	0.31	0.188	0.30	0.138	44	0.3
Potatoes, boiled w/o skin, flesh, w. salt	86	20.0	1.7	0.1		241	328	8	20	0.31	0.167	0.27	0.140	40	0.3

Food description	Energy [kcal/ 100 g]	Carb. [g/ 100 g]	Protein [g/ 100 g]	Fat [g/ 100 g]	Chol. [mg/ 100 g]	Na [mg/ 100 g]	K [mg/ 100 g]	Ca [mg/ 100 g]	Mg [mg/ 100 g]	Fe [mg/ 100 g]	Cu [mg/ 100 g]	Zn [mg/ 100 g]	Mn [mg/ 100 g]	P [mg/ 100 g]	Se [µg/ 100 g]
Potatoes, French fried, all types, frozen, oven-heated	172	28.7	2.7	5.2		32	451	12	26	0.74	0.135	0.38	0.210	97	0.2
Potatoes, hashed brown, home-prepared	265	35.1	3.0	12.5		342	576	14	35	0.55	0.293	0.47	0.247	70	0.5
Potatoes, mashed, home-prepared, whole milk & margarine added	113	16.9	2.0	4.2	1	333	328	22	19	0.26	0.153	0.30	0.112	49	0.8
Potatoes, red, flesh & skin, baked	89	19.6	2.3	0.2		8	545	9	28	0.70	0.174	0.40	0.173	72	0.5
Potatoes, white, flesh & skin, baked	94	21.1	2.1	0.2		7	544	10	27	0.64	0.127	0.35	0.189	75	0.5
Pretzels, hard, plain, salted	380	79.8	10.3	2.6		1357	146	18	29	5.20	0.264	1.43	1.789	113	6.0
Prunes, dehydrated (low-moisture), stewed	113	29.7	1.2	0.2		2	353	24	21	1.17	0.204	0.25	0.104	37	
Pumpkin, boiled, w/o salt	20	4.9	0.7	0.1		1	230	15	9	0.57	0.091	0.23	0.089	30	0.2
Pumpkin, raw	26	6.5	1.0	0.1		1	340	21	12	0.80	0.127	0.32	0.125	44	0.3
Quail, cooked, total edible	234		25.1	14.1	86	52	216	15	22	4.43	0.592	3.10		279	21.8
Quinces, raw	57	15.3	0.4	0.1		4	197	11	8	0.70	0.130	0.04		17	0.6
Radicchio, raw	23	4.5	1.4	0.3		22	302	19	13	0.57	0.341	0.62	0.138	40	0.9
Radishes, raw	16	3.4	0.7	0.1		39	233	25	10	0.34	0.050	0.28	0.069	20	0.6
Raisins, golden seedless	302	79.5	3.4	0.5		12	746	53	35	1.79	0.363	0.32	0.308	115	0.7
Raisins, seeded	296	78.5	2.5	0.5		28	825	28	30	2.59	0.302	0.18	0.267	75	0.6
Raisins, seedless	299	79.2	3.1	0.5		11	749	50	32	1.88	0.318	0.22	0.299	101	0.6
Raspberries, raw	52	11.9	1.2	0.7		1	151	25	22	0.69	0.090	0.42	0.670	29	0.2
Rhubarb, frozen, cooked, w/sugar	116	31.2	0.4	0.1		1	96	145	12	0.21	0.027	0.08	0.073	8	0.9
Rhubarb, raw	21	4.5	0.9	0.2		4	288	86	12	0.22	0.021	0.10	0.196	14	1.1
Rice, brown, long-grain, cooked	111	23.0	2.6	0.9		5	43	10	43	0.42	0.100	0.63	0.905	83	9.8
Rice, brown, medium-grain, cooked	112	23.5	2.3	0.8		1	79	10	44	0.53	0.081	0.62	1.097	77	
Rice, white, glutinous, cooked	97	21.1	2.0	0.2		5	10	2	5	0.14	0.049	0.41	0.262	8	5.6
Rice, white, long-grain, regular, cooked	130	28.2	2.7	0.3		1	35	10	12	1.20	0.069	0.49	0.472	43	7.5
Rice, white, medium-grain, cooked	130	28.6	2.4	0.2			29	3	13	1.49	0.038	0.42	0.377	37	7.5
Rice, white, short-grain, cooked	130	28.7	2.4	0.2			26	1	8	1.46	0.072	0.40	0.357	33	7.5
Rockfish, Pacific, mixed species, cooked, dry heat	121		24.0	2.0	44	77	520	12	34	0.53	0.037	0.53	0.020	228	46.8
Rolls, hamburger or hotdog, plain	279	49.5	9.5	4.3		479	94	138	21	3.32	0.220	0.66	0.272	62	19.5
Rutabagas, boiled, w/o salt	39	8.7	1.3	0.2		20	326	48	23	0.53	0.041	0.35	0.174	56	0.7
Sablefish, cooked, dry heat	250		17.2	19.6	63	72	459	45	71	1.64	0.028	0.41	0.019	215	46.8
Sablefish, smoked	257		17.7	20.1	64	737	471	50	74	1.69	0.036	0.43	0.020	222	50.2
Salami, cooked, turkey	152	0.4	15.3	9.4	76	1004	216	40	22	1.25	0.190	2.32	0.020	266	26.4
Salami, dry or hard, pork	407	1.6	22.6	33.7	79	2260	378	13	22	1.30	0.160	4.20	0.070	229	25.4
Salami, dry or hard, pork, beef	385	3.8	23.2	30.1	100	2010	378	8	17	1.51	0.080	3.23		142	26.1
Salami, Italian, pork	425	1.2	21.7	37.0	80	1890	340	10	22	1.52	0.160	4.20	0.070	229	25.4
Salmon, Atlantic, farmed, cooked, dry heat	206		22.1	12.4	63	61	384	15	30	0.34	0.049	0.43	0.016	252	41.4
Salmon, Atlantic, farmed, raw	183		19.9	10.9	59	59	362	12	28	0.36	0.049	0.40	0.015	233	36.5
Salmon, Atlantic, wild, cooked, dry heat	182		25.4	8.1	71	56	628	15	37	1.03	0.321	0.82	0.021	256	46.8
Salmon, Atlantic, wild, raw	142		19.8	6.3	55	44	490	12	29	0.80	0.250	0.64	0.016	200	36.5
Salmon, Chinook, cooked, dry heat	231		25.7	13.4	85	60	505	28	122	0.91	0.053	0.56	0.019	371	46.8
Salmon, Chinook, raw	179		19.9	10.4	50	47	394	26	95	0.25	0.041	0.44	0.015	289	36.5
Salmon, Chinook, smoked	117		18.3	4.3	23	784	175	11	18	0.85	0.230	0.31	0.017	164	32.4
Salmon, Chinook, smoked (lox)	117		18.3	4.3	23	2000	175	11	18	0.85	0.230	0.31	0.017	164	38.1
Salmon, chum, cooked, dry heat	154		25.8	4.8	95	64	550	14	28	0.71	0.071	0.60	0.019	363	46.8
Salmon, coho, farmed, cooked, dry heat	178		24.3	8.2	63	52	460	12	34	0.39	0.089	0.47	0.021	332	14.1
Salmon, coho, farmed, raw	160		21.3	7.7	51	47	450	12	31	0.34	0.048	0.43	0.012	292	12.6
Salmon, coho, wild, cooked, dry heat	139		23.5	4.3	55	58	434	45	33	0.61	0.071	0.56	0.019	322	38.0
Salmon, coho, wild, raw	146		21.6	5.9	45	46	423	36	31	0.56	0.051	0.41	0.014	262	36.5
Salmon, pink, cooked, dry heat	149		25.6	4.4	67	86	414	17	33	0.99	0.099	0.71	0.019	295	57.2
Salmon, pink, raw	116		19.9	3.5	52	67	323	13	26	0.77	0.077	0.55	0.015	230	44.6
Salmon, sockeye, cooked, dry heat	216		27.3	11.0	87	66	375	7	31	0.55	0.067	0.51	0.020	276	37.8
Salmon, sockeye, raw	168		21.3	8.6	62	47	391	6	24	0.47	0.052	0.54	0.014	215	33.7
Sausage, chicken, beef, pork, skinless, smoked	216	8.1	13.6	14.3	120	1034	246	100	14	4.80	0.063	2.68	0.015	132	20.2
Sausage, Italian, pork, cooked	344	4.3	19.1	27.3	57	1207	304	21	18	1.43	0.080	2.39		170	22.0
Sausage, Italian, sweet, links	149	2.1	16.1	8.4	30	570	194	25	12	1.19	0.040	1.52	0.007	103	10.8
Sausage, Polish, beef w. chicken, hot	259	3.6	17.6	19.4	66	1540	237	12	14	0.88	0.090	1.93	0.049	136	17.7
Sausage, smoked link sausage, pork & beef	320	2.4	12.0	28.7	58	911	179	12	13	0.75	0.077	1.26	0.048	121	
Sausage, turkey, breakfast links, mild	235	1.6	15.4	18.1	60	585	197	32	25	1.07	0.111	2.13	0.066	185	22.2
Scallops, (bay & sea), steamed	112		23.2	1.4	53	265	476	115	55	3.00	0.300	3.00		338	27.9
Scrapple, pork	213	14.1	8.1	13.9	49	659	158	7	13	1.89	0.212	1.06		76	17.4
Sea bass, mixed species, cooked, dry heat	124		23.6	2.6	53	87	328	13	53	0.37	0.024	0.52	0.020	248	46.8
Shad, American, cooked, dry heat	252		21.7	17.7	96	65	492	60	38	1.24	0.082	0.47	0.054	349	46.8
Shark, mixed species, batter-dipped & fried	228	6.4	18.6	13.8	59	122	155	50	43	1.11	0.042	0.48	0.050	194	34.0

Food description	Energy [kcal/ 100 g]	Carb. [g/ 100 g]	Protein [g/ 100 g]	Fat [g/ 100 g]	Chol. [mg/ 100 g]	Na [mg/ 100 g]	K [mg/ 100 g]	Ca [mg/ 100 g]	Mg [mg/ 100 g]	Fe [mg/ 100 g]	Cu [mg/ 100 g]	Zn [mg/ 100 g]	Mn [mg/ 100 g]	P [mg/ 100 g]	Se [µg/ 100 g]
Shortening, frying (heavy duty), beef tallow & cottonseed	900			100	100										
Shortening, frying (heavy duty), palm (hydrogenated)	884			100											
Shortening, frying (reg), soybean (hydrogenated) & cottonseed (hydrogenated)	884			100											
Shortening, frying heavy duty, soybean hydrogenated, linoleic 30% w/silicon	884			100											
Shortening, frying heavy duty, soybean, hydrogenated, linoleic <1%	884			100											
Shortening, household, soybean (hydrogenated) & palm	884			100											
Shrimp, mixed species, cooked, moist heat	99		20.9	1.1	195	224	182	39	34	3.09	0.193	1.56	0.034	137	39.6
Shrimp, mixed species, imitation, made from surimi	101	9.1	12.4	1.5	36	705	89	19	43	0.60	0.032	0.33	0.011	282	22.9
Shrimp, mixed species, raw	106	0.9	20.3	1.7	152	148	185	52	37	2.41	0.264	1.11	0.050	205	38.0
Smelt, rainbow, cooked, dry heat	124		22.6	3.1	90	77	372	77	38	1.15	0.178	2.12	0.900	295	46.8
Snails, raw	90	2.0	16.1	1.4	50	70	382	10	250	3.50	0.400	1.00		272	27.4
Snapper, mixed species, cooked, dry heat	128		26.3	1.7	47	57	522	40	37	0.24	0.046	0.44	0.017	201	49.0
Sour cream, light	136	7.1	3.5	10.6	35	71	212	141	10	0.07	0.016	0.50		71	3.1
Sour cream, reduced fat	181	7.0	7.0	14.1	35	70	211	141	11	0.06	0.010	0.27		85	4.1
Soybeans, green, boiled, w/o salt	141	11.1	12.4	6.4		14	539	145	60	2.50	0.117	0.91	0.502	158	1.4
Soybeans, green, raw	147	11.1	13.0	6.8		15	620	197	65	3.55	0.128	0.99	0.547	194	1.5
Soybeans, mature cooked, boiled, w/o salt	173	9.9	16.6	9.0		1	515	102	86	5.14	0.407	1.15	0.824	245	7.3
Spaghetti, cooked, enriched, w. salt	157	30.6	5.8	0.9		128	45	7	18	1.33	0.103	0.50	0.317	58	26.4
Spaghetti, whole-wheat, cooked	124	26.5	5.3	0.5		3	44	15	30	1.06	0.167	0.81	1.379	89	25.9
Spinach, boiled, w/o salt	23	3.8	3.0	0.3		70	466	136	87	3.57	0.174	0.76	0.935	56	1.5
Spinach, raw	23	3.6	2.9	0.4		79	558	99	79	2.71	0.130	0.53	0.897	49	1.0
Squash, summer, all varieties, boiled, w/o salt	20	4.3	0.9	0.3		1	192	27	24	0.36	0.103	0.39	0.213	39	0.2
Squash, summer, all varieties, raw	16	3.4	1.2	0.2		2	262	15	17	0.35	0.051	0.29	0.175	38	0.2
Squash, summer, zucchini, includes skin, boiled, w/o salt	16	3.9	0.6	0.1		3	253	13	22	0.35	0.086	0.18	0.178	40	0.2
Squash, summer, zucchini, includes skin, raw	16	3.4	1.2	0.2		10	262	15	17	0.35	0.051	0.29	0.175	38	0.2
Squash, winter, acorn, baked, w/o salt	56	14.6	1.1	0.1		4	437	44	43	0.93	0.086	0.17	0.242	45	0.7
Squash, winter, butternut, baked, w/o salt	40	10.5	0.9	0.1		4	284	41	29	0.60	0.065	0.13	0.172	27	0.5
Squash, winter, Hubbard, baked, w/o salt	50	10.8	2.5	0.6		8	358	17	22	0.47	0.045	0.15	0.170	23	0.6
Squash, zucchini, baby, raw	21	3.1	2.7	0.4		3	459	21	33	0.79	0.097	0.83	0.196	93	0.3
Strawberries, raw	32	7.7	0.7	0.3		1	153	16	13	0.42	0.048	0.14	0.386	24	0.4
Sturgeon, mixed species, cooked, dry heat	135		20.7	5.2	77	69	364	17	45	0.90	0.053	0.54	0.030	271	16.2
Sturgeon, mixed species, raw	105		16.1	4.0	60	54	284	13	35	0.70	0.041	0.42	0.025	211	12.6
Sturgeon, mixed species, smoked	173		31.2	4.4	80	739	379	17	47	0.93	0.050	0.56	0.030	281	20.1
Sugars, brown	377	97.3				39	346	85	29	1.91	0.298	0.18	0.320	22	1.2
Sugars, granulated	387	100					2	1		0.01					0.6
Sugars, maple	354	90.9	0.1	0.2		11	274	90	19	1.61	0.099	6.06	4.422	3	0.8
Sweet potato, baked in skin, w/o salt	90	20.7	2.0	0.2		36	475	38	27	0.69	0.161	0.32	0.497	54	0.2
Sweet potato, boiled, w/o skin	76	17.7	1.4	0.1		27	230	27	18	0.72	0.094	0.20	0.266	32	0.2
Swordfish, cooked, dry heat	155		25.4	5.1	50	115	369	6	34	1.04	0.162	1.47	0.020	337	61.7
Swordfish, raw	121		19.8	4.0	39	90	288	4	27	0.81	0.126	1.15	0.019	263	48.1
Syrups, corn, dark	286	77.6				155	44	18	8	0.37	0.053	0.04	0.100	11	2.9
Syrups, corn, high-fructose	281	76.0				2				0.03	0.029	0.02	0.094		0.7
Syrups, maple	261	67.1		0.2		9	204	67	14	1.20	0.074	4.16	3.298	2	0.6
Syrups, table blends, cane & 15% maple	278	69.5				104	53	12	2	0.19	0.017	0.63	0.495		0.5
Tangerines, (mandarin oranges), raw	53	13.3	0.8	0.3		2	166	37	12	0.15	0.042	0.07	0.039	20	0.1
Tempeh, cooked	196	9.4	18.2	11.4		14	401	96	77	2.13	0.540	1.57	1.285	253	
Tilefish, cooked, dry heat	147		24.5	4.7	64	59	512	26	33	0.31	0.052	0.53	0.015	236	51.5
Tilefish, raw	96		17.5	2.3	50	53	433	26	28	0.25	0.041	0.37	0.010	187	36.5
Tofu, fried	271	10.5	17.2	20.2		16	146	372	60	4.87	0.398	1.99	1.495	287	28.5
Tomato juice, canned, w. salt	17	4.2	0.8	0.1		269	229	10	11	0.43	0.061	0.15	0.070	18	0.3
Tomatoes, green, raw	23	5.1	1.2	0.2		13	204	13	10	0.51	0.090	0.07	0.100	28	0.4
Tomatoes, orange, raw	16	3.2	1.2	0.2		42	212	5	8	0.47	0.062	0.14	0.088	29	0.4
Tomatoes, red, ripe, cooked	18	4.0	1.0	0.1		11	218	11	9	0.68	0.075	0.14	0.105	28	0.5
Tomatoes, red, ripe, raw, year-round average	18	3.9	0.9	0.2		5	237	10	11	0.27	0.059	0.17	0.114	24	
Tomatoes, sun-dried	258	55.8	14.1	3.0		2095	3427	110	194	9.09	1.423	1.99	1.846	356	5.5
Tomatoes, yellow, raw	15	3.0	1.0	0.3		23	258	11	12	0.49	0.101	0.28	0.120	36	0.4
Trout, mixed species, cooked, dry heat	190		26.6	8.5	74	67	463	55	28	1.92	0.241	0.85	1.091	314	16.2
Trout, rainbow, farmed, cooked, dry heat	169		24.3	7.2	68	42	441	86	32	0.33	0.061	0.49	0.020	266	15.0

Food description	Energy [kcal/ 100 g]	Carb. [g/ 100 g]	Protein [g/ 100 g]	Fat [g/ 100 g]	Chol. [mg/ 100 g]	Na [mg/ 100 g]	K [mg/ 100 g]	Ca [mg/ 100 g]	Mg [mg/ 100 g]	Fe [mg/ 100 g]	Cu [mg/ 100 g]	Zn [mg/ 100 g]	Mn [mg/ 100 g]	P [mg/ 100 g]	Se [μg/ 100 g]
Trout, rainbow, wild, cooked, dry heat	150		22.9	5.8	69	56	448	86	31	0.38	0.058	0.51	0.021	269	13.2
Tuna, fresh, bluefin, cooked, dry heat	184		29.9	6.3	49	50	323	10	64	1.31	0.110	0.77	0.020	326	46.8
Tuna, fresh, bluefin, raw	144		23.3	4.9	38	39	252	8	50	1.02	0.086	0.60	0.015	254	36.5
Tuna, fresh, skipjack, raw	103		22.0	1.0	47	37	407	29	34	1.25	0.086	0.82	0.015	222	36.5
Tuna, fresh, yellowfin, raw	108		23.4	1.0	45	37	444	16	50	0.73	0.064	0.52	0.015	191	36.5
Tuna, skipjack, fresh, cooked, dry heat	132		28.2	1.3	60	47	522	37	44	1.60	0.110	1.05	0.019	285	46.8
Tuna, white, canned in oil	186		26.5	8.1	31	396	333	4	34	0.65	0.130	0.47	0.016	267	60.1
Tuna, white, canned in water	128		23.6	3.0	42	377	237	14	33	0.97	0.039	0.48	0.019	217	65.7
Tuna, yellowfin, fresh, cooked, dry heat	139		30.0	1.2	58	47	569	21	64	0.94	0.082	0.67	0.019	245	46.8
Turbot, European, cooked, dry heat	122		20.6	3.8	62	192	305	23	65	0.46	0.047	0.28	0.022	165	46.8
Turkey breast meat	104	4.2	17.1	1.7	43	1015	302	8	21	1.44	0.057	1.33	0.018	162	22.8
Turkey, all classes, breast, meat & skin, roasted	189		28.7	7.4	74	63	288	21	27	1.40	0.047	2.03	0.020	210	29.1
Turkey, all classes, dark meat, meat & skin, raw	160		18.9	8.8	72	71	261	17	20	1.69	0.137	2.95	0.021	170	26.4
Turkey, all classes, leg, meat & skin, roasted	208		27.9	9.8	85	77	280	32	23	2.30	0.154	4.27	0.023	199	37.8
Turkey, all classes, meat only, roasted	170		29.3	5.0	76	70	298	25	26	1.78	0.094	3.10	0.021	213	36.8
Turkey, all classes, wing, meat & skin, roasted	229		27.4	12.4	81	61	266	24	25	1.46	0.056	2.10	0.020	197	29.9
Turnip greens, boiled, w/o salt	20	4.4	1.1	0.2		29	203	137	22	0.80	0.253	0.14	0.337	29	0.9
Turnips, boiled, w/o salt	22	5.1	0.7	0.1		16	177	33	9	0.18	0.002	0.12	0.071	26	0.2
Turnips, raw	28	6.4	0.9	0.1		67	191	30	11	0.30	0.085	0.27	0.134	27	0.7
Veal, composite of retail cuts, fat, cooked	642		9.4	66.7	73	57	173	4	10	1.00	0.044	0.87	0.012	116	5.5
Veal, composite of retail cuts, lean & fat, cooked	231		30.1	11.4	114	87	325	22	26	1.15	0.114	4.76	0.036	239	12.3
Veal, composite of retail cuts, lean, cooked	196		31.9	6.6	118	89	338	24	28	1.16	0.120	5.10	0.038	250	13.0
Veal, ground, broiled	172		24.4	7.6	103	83	337	17	24	0.99	0.103	3.87	0.035	217	13.7
Veal, leg (top round), lean & fat, roasted	160		27.7	4.7	103	68	389	6	28	0.91	0.129	3.04	0.030	234	11.2
Veal, loin, lean & fat, roasted	217		24.8	12.3	103	93	325	19	25	0.87	0.110	3.03	0.029	212	11.0
Veal, rib, lean & fat, roasted	228		24.0	14.0	110	92	295	11	22	0.97	0.099	4.09	0.030	197	10.5
Veal, shoulder, arm, lean & fat, roasted	183		25.5	8.3	108	90	348	26	26	1.15	0.141	4.18	0.030	221	11.1
Veal, sirloin, lean & fat, roasted	202		25.1	10.5	102	83	351	13	26	0.92	0.129	3.35	0.029	223	11.1
Veal, sirloin, lean, roasted	168		26.3	6.2	104	85	365	14	27	0.91	0.136	3.54	0.030	231	11.5
Vinegar, distilled	18					2	2	6	1	0.03	0.006	0.01	0.055	4	0.5
Watercress, raw	11	1.3	2.3	0.1		41	330	120	21	0.20	0.077	0.11	0.244	60	0.9
Watermelon, raw	30	7.6	0.6	0.2		1	112	7	10	0.24	0.042	0.10	0.038	11	0.4
Wheat flour, whole-grain	339	72.6	13.7	1.9		5	405	34	138	3.88	0.382	2.93	3.799	346	70.7
Wheat flours, bread, unenriched	361	72.5	12.0	1.7		2	100	15	25	0.90	0.182	0.85	0.792	97	39.7
Wolffish, Atlantic, cooked, dry heat	123		22.4	3.1	59	109	385	8	38	0.12	0.037	1.00	0.019	256	46.8
Wolffish, Atlantic, raw	96		17.5	2.4	46	85	300	6	30	0.09	0.029	0.78	0.015	200	36.5
Yam, boiled, or baked, w/o salt	116	27.6	1.5	0.1		8	670	14	18	0.52	0.152	0.20	0.371	49	0.7
Yellowtail, mixed species, cooked, dry heat	187		29.7	6.7	71	50	538	29	38	0.63	0.058	0.67	0.019	201	46.8
Yogurt, fruit varieties, non-fat	94	19.0	4.4	0.2	2	58	194	152	15	0.07	0.011	0.74	0.035	119	6.0
Yogurt, plain, low fat, 12 grams protein per 8 oz.	63	7.0	5.3	1.6	6	70	234	183	17	0.08	0.013	0.89	0.004	144	3.3
Yogurt, plain, skim milk, 13 grams protein per 8 oz.	56	7.7	5.7	0.2	2	77	255	199	19	0.09	0.015	0.97	0.005	157	3.6
Yogurt, plain, whole milk, 8 grams protein per 8 oz.	61	4.7	3.5	3.3	13	46	155	121	12	0.05	0.009	0.59	0.004	95	2.2

Section 8
Analytical Chemistry

ABBREVIATIONS AND SYMBOLS USED IN ANALYTICAL CHEMISTRY

Abbreviations Used in Analytical Chemistry

A: peak area; surface area of solid granular adsorbent; eddy diffusion term in the Van Deemter equation
AAA: absolute activation analysis
AAD: atomic absorption detector
AAS: atomic absorption spectroscopy
AC: alternating current, affinity chromatography
ACP: alternating current plasma
ADXPS: angular dependent x-ray photoelectron spectroscopy
AED: atomic emission detector
AEM: analytical electron microscope (microscopy)
AES: Auger electron spectroscopy, atomic emission spectroscopy
AFID: alkali flame ionization detector
AFM: atomic force microscopy
AFS: atomic force spectroscopy
AIS: average of individual samples
AL: action level
AM: amplitude modulation
AMS: accelerator mass spectrometry
AN: area normalization
ANRF: area normalization with response factors
AOTF: acousto-optical tunable filter
AP: analytical pyrolysis
APCI: atmospheric pressure chemical ionization
APD: azimuthal photoelectron diffraction
API: atmospheric pressure ionization
APSTM: analytical photon scanning tunneling microscope
APT: attached proton test
ARAES: angle resolved Auger electron spectroscopy
ARF: absolute response factor
ARM: atomic resolution microscopy
ARPES: angle resolved photoelectron spectroscopy
ARUPS: angle resolved ultraviolet photoelectron spectroscopy
ASE: accelerated solvent extraction
AsFIFFF (AF4): asymmetrical flow field flow fractionation
ATD: above-threshold dissociation
ATI: above-threshold ionization
ATR: attenuated total reflection
BB: band broadening
BE: magnetic sector – electric sector tandem mass spectrometer (note: also called a MIKE spectrometer)
BEE: magnetic sector – electric sector – electric sector mass spectrometer
BET: Brunauer-Emmett-Teller (adsorption isotherm)
BIFL: burst integrated fluorescence lifetime
BIS: Bremsstrahlung isochromat spectroscopy
BJH: Barrett-Joyner-Halenda (method)
BL: bioluminescence
BLRF: bispectral luminescence radiance factor
BQQ: magnetic sector-double quadrupole mass spectrometer BTOF
BTOF: magnetic sector – time of flight tandem mass spectrometer
CAD: collision activated dissociation
CAR: continuous addition of reagent
CARS: coherent anti-Stokes Raman spectroscopy
CCC: counter-current chromatography
CCD: charge-coupled device
CCT: constant current topography
CD: circular dichroism
CE: capillary electrophoresis, counter electrode
CEC: capillary electrokinetic chromatography, capillary electrochromatography
CED: cohesive energy density
CFA: continuous flow analysis
CF-FAB: continuous flow-fast atom bombardment
CFM: chemical force microscopy
CGE: capillary gel electrophoresis
CHEMFET: chemical-sensing field effect transistor
CI: chemical ionization
CID: collision induced dissociation
CIEF: capillary isoelectric focusing
CITP: capillary isotachophoresis
CL: chemiluminescence
CLLE: continuous liquid-liquid extraction
CMA: cylindrical mirror analyzer
COSY: correlation spectroscopy
CPAA: charged particle activation analysis
CP/MAS: cross polarization/magic angle spinning
CRDS: cavity ring-down spectroscopy
CRF: chromatographic response function
CRM: certified reference material
CS: carbon strip (adsorbent)
CT: cryogenic trapping
CTD: charge transfer device
CV: cyclic voltammetry
CV-ASS: cold vapor atomic absorption spectrometry
CVD: chemical vapor deposition
CW: continuous wave
CZE: capillary zone electrophoresis
DA: diode array
DAD: diode array detector (UV-Vis)
DADI: direct analysis of daughter ions
dB: de Boar t-plot
DBE: double bond equivalent
DC: direct current
DCI: desorption chemical ionization
DCP: direct-current plasma
DEP: differential electrolytic potentiometry
DEPT: distortionless enhancement by polarization transfer
DETA: dielectric thermal analysis
DIN: direct injection nebulizer
DLI: direct liquid introduction
DLS: dynamic light scattering
DMA: dynamic mechanical analysis
DME: dropping mercury electrode
DNMR: dynamic nuclear magnetic resonance
DPP: differential pulse polarography
DRIFT: diffuse-reflectance infrared Fourier transform
DSC: differential scanning calorimetry
DTA: differential thermal analysis
DTC: differential thermal calorimetry
EAES: electron-excited Auger electron spectroscopy
EB: electric sector – magnetic sector tandem mass spectrometer
EBE: electric sector – magnetic sector-electric sector tandem mass spectrometer

Analytical

EC: electrochemical
ECD: electron capture detector
ECMS: electron capture mass spectrometry
ECNIMS: electron capture negative ionization mass spectrometry
EDL: electrodeless discharge lamp
EDS: energy-dispersive spectrometer
EDXRF: energy-dispersive x-ray fluorescence
EELS: electron energy-loss spectroscopy
EFFF: electric field flow fractionation
EG: electrogravimetry
EGA: evolved gas analysis
EIA: enzyme-linked immunoassay
EI(I): electron impact (ionization)
EIMS: electron impact mass spectrometry
ELCD: electrolytic conductivity detector
ELISA: enzyme-linked immunosorbent assay
ELSD: evaporative light scattering detector
EM: electron microscopy
EMIRS: electrochemically modulated IR spectroscopy
EOF: electro-osmotic flow
EPL: enhanced photoactivated luminescence
EPMA: electron-probe microanalysis
EPR: electron paramagnetic resonance
EPXMA: electron-probe x-ray microanalysis
EQL: estimated quantitation limit
ERD: elastic recoil detection
ESA: electrostatic analyzer
ESCA: electron spectroscopy for chemical analysis
ESEM: environmental scanning electron microscope
ESI: electro-spray ionization
ESP: electrospray
ESR: electron spin resonance
ET: electrometric titration
ETA: electrothermal analyzer, emanation thermal analysis
EXAFS: extended x-ray absorption fine structure
FAA: flame atomic absorption
FAAS: flame atomic absorption spectroscopy
FABMS: fast-atom bombardment mass spectrometry
FAES: flame atomic emission spectroscopy
FAFS: flame atomic fluoroescence spectroscopy
FAM: field analytical method
FAS: flame absorption spectroscopy
FD: field desorption
FD/FI: field desorption/field ionization
FES: flame emission spectroscopy
FFEM: freeze-fracture electron microscopy
FFF: field flow fractionation
FFFF: flow field flow fractionation
FFM: friction force microscopy
FFS: flame fluorescence spectroscopy
FFT: fast Fourier transform
FGC: fast gas chromatography
FI: flow injection, field ionization
FIA: flow injection analysis
FIB: focused ion beam
FID: flame ionization detector, free-induction decay
FIM: field ion microscopy
FNAA: fast neutron activation analysis
FOCS: fiber optic chemical sensor
FPD: flame photometric detector
FSOT: fused silica open tubular (column)
FT: Fourier transform
FT-ICR: Fourier transform ion cyclotron resonance
FT-IR: Fourier transform infrared (often "FT/IR," "FTIR," "FT IR")
FT-IRRAS: FT-IR reflection-absorption spectroscopy
FT-MS: Fourier transform mass spectrometry
FWHM: full-width half-maximum
GC: gas chromatography
GC-IR: gas chromatography–infrared spectrometry
GCMS: gas chromatography mass spectrometry
GDL: glow discharge lamp
GDMS: glow discharge mass spectrometry
GE: gel electrophoresis, gradient elution
GEMBE: gradient elution moving boundary electrophoresis
GFAAS: graphite furnace atomic absorption spectroscopy
GLC: gas-liquid chromatography
GPC: gel permeation chromatography
GS: Gram-Schmidt (algorithm)
GSC: gas-solid chromatography
GSED: gaseous secondary electron detector
HCL: hollow cathode lamp
HCOT: helically coiled open tubular (column)
HDC: hydrodynamic chromatography
HETCOR: heteronuclear correlation
HETP: height equivalent of (a) theoretical plate(s)
HG: hydride generation
HIC: hydrophoric interaction chromatography
HMBC: heteronuclear multiple-bond correlation
HPAC: high-performance affinity chromatography
HPIAC: high-performance immunoaffinity chromatography
HPLC: high-performance liquid chromatography, high-pressure liquid chromatography
HPTLC: high-performance thin-layer chromatography
HRCGC: high-resolution capillary gas chromatography
HRGC: high-resolution gas(-liquid) chromatography
HS: head space
HSA: hemispherical analyzer
HSC: heteronuclear shift correlation
HSQC: heteronuclear single quantum coherence
HTC: high temperature combustion
IA: isocratic analysis
IAC: immunoaffinity chromatography
IAES: ion-excited Auger electron spectroscopy
IC: ion chromatography
ICMS: ion chromatography mass spectrometry
ICP: inductively coupled plasma
ICP-OES: ICP optical emission spectrometry
ICR: ion cyclotron resonance
IDMS: isotope dilution mass spectrometry
IEC: ion-exchange chromatography
IEF: isoelectric focusing
IF: intermediate frequency
IGC: inverse gas chromatography
IGF: inert gas fusion
ILDA: intensified linear diode array
IMAC: immobilized metal-ion affinity chromatography
INADEQUATE: incredible natural abundance double-quantum transfer experiment
INEPT: insensitive nuclei enhancement by polarization transfer
INAA: instrumental neutron activation analysis
IP: ion pairing
IPC: ion-pair chromatography
IPG: immobilized pH gradient
IPMA: ion probe microanalysis
IR: infrared (spectrophotometry)
IRN: indicator radionuclide(s)

Analytical

IRS: internal reflection spectroscopy
ISCA: ionization spectroscopy for chemical analysis
ISE: ion selective electrode
ISP: ion spray
ISS: ion scattering spectrometry
LAMMS: laser micro mass spectrometry
LARIMS: laser atomization resonance ionization mass spectrometry
LARIS: laser atomization resonance ionization spectroscopy
LASER: light amplification by stimulated emission of radiation
LBB: Lambert-Beer-Bouguer law
LC: liquid chromatography
LC-LS: multidimensional liquid chromatography
LDMS: laser desorption mass spectrometry
LDR: linear dynamic range
LEAFS: laser-excited atomic fluorescence spectrometry
LED: light-emitting diode
LEED: low-energy electron diffraction
LEEM: low-energy electron microscopy
LEI: laser-enhanced ionization
LEISS: low energy ion scattering spectrometry
LESS: laser-excited Shpol'skii spectroscopy
LFM: lateral force microscopy
LIDAR: light detection and ranging
LIFD: laser-induced fluorescence detection
LIMS: laboratory information management system
LLC: liquid-liquid chromatography
LLD: lower-limit detection
LLE: liquid-liquid extraction
LNRI: laser non-resonant ionization
LO: local oscillator
LOC: lab on a chip
LOD: limit of detection
LPDA: linear photodiode array
LPSIRS: linear potential-sweep IR reflectance spectroscopy
LRI: laser resonance ionization
LRMA: laser Raman microanalysis
LSC: liquid-solid chromatography
LSE: liquid-solid extraction
LTP: low-temperature phosphorescence
MAE: microwave assisted extraction
MALDI: matrix-assisted laser desorption/ionization
MAS: magic angle spinning
MCD: magnetic circular dichroism
MCP: microchannel plate
MDGC: multidimensional gas chromatography
MDL: method detection limit
MDM: minimum detectable mass
MDQ: minimum detectable quantity
MEIS: medium-energy ion scattering
MEKC: micellar electrokinetic chromatography
MFM: magnetic force microscopy
MID: multiple ion detection
MIKE: mass analyzed ion kinetic energy mass spectrometry
MIP: microwave-induced plasma, mercury intrusion porosimetry
MIRS: multiple internal reflection spectroscopy
MLC: micellar liquid chromatography
MLLSQ: multiple linear least squares
MMF: minimum mass fraction
MMLLE: microporous membrane liquid-liquid extraction
MPD: microwave plasma detector
MPI: multiphoton ionization
MRDL: maximum residual disinfectant level (in water analysis)
MRI: magnetic resonance imaging
MS: mass spectrometry
MS-MS: tandem mass spectrometry
MSPD: matric solid-phase dispersion
MSRTP: micelle-stabilized room-temperature phosphorescence
MWD: microwave (assisted) digestion
NAA: neutron activation analysis
NCIMS: negative chemical ionization mass spectrometry
NDP: neutron depth profiling
NEXAFS: near edge x-ray absorption fine structure
NHE: normal-hydrogen electrode
NICI: negative ion chemical ionization
NIR: near-infrared, near-IR
NIRA: near-infrared reflectance analysis
nm: nanometer
NMR: nuclear magnetic resonance
NOE (nOe): nuclear Overhauser effect
NOESY: nuclear Overhauser effect spectroscopy
NPD: nitrogen-phosphorus detector, normal photoelectron diffraction
NPLC: normal phase liquid chromatography
ODMR: optically detected magnetic resonance
ODS: octadecylsilane
OES: optical emission spectrometry, optical emission spectroscopy
OID: optoelectronic imaging device
OMA: optical multichannel analyzer
OPO: optical parametric oscillator
OPTLC: over-pressured thin-layer chromatography
ORD: optical rotary dispersion
OTE: optically transparent electrodes
PA: proton affinity
PAA: photon activation analysis
PAGE: polyacrylamide gel electrophoresis
PAH: polycyclic aromatic hydrocarbon
PAS: photoacoustic spectroscopy
PB: particle beam
PC: paper chromatography
PCA: principal component analysis
PCR: polymerase chain reaction
PCS: photon correlation spectroscopy
PCSE: partially coherent solvent evaporation
PD: plasma desorption
PDA: photodiode array
PDHID: pulsed discharge helium ionization detector
PDMS: plasma desorption mass spectrometry, polydimethyl siloxane
PED: pulsed electrochemical detection, plasma emission detector, photoelectron diffraction
PES: photoelectron spectroscopy
PET: positron emission tomography
PFIA: process flow injection analysis
PGC: packed-column gas chromatography
pH: negative logarithm of hydrogen ion concentration
PICI: positive ion chemical ionization
PID: photoionization detector
PIXE: particle-induced x-ray emission
p*K*: negative logarithm of an equilibrium constant
PLE: pressurized liquid extraction
PLOT: porous-layer open tubular
PLOT-cryo: porous-layer open tubular (column) cryoadsorption
PMT: photomultiplier tube
ppb: parts per billion

ppm: parts per million
ppt: parts per thousand, parts per trillion
PSD: position sensitive detector
PTFE: polytetrafluoroethylene
PTR: proton transfer reaction (in mass spectrometry)
PTV: programmable temperature vaporizer
PVD: pulsed voltammetric detection, physical vapor deposition
QCL: quantum cascade laser
QCM: quartz-crystal microbalance
QFAA: quartz furnace atomic adsorption
QIT: quadrupole ion trap
qNMR: quantitative nuclear magnetic resonance
QQQ, QqQ, Q_1qQ_3: triple quadrupole mass spectrometer
QTH: quartz tungsten halogen
QTOF: tandem quadrupole time of flight mass spectrometer
RAA: running annual average
RBS: Rutherford backscattering spectrometry
REELS: reflection electron energy loss spectrometry
RES: reflection electron spectrometry
RF: radio frequency
RHEED: reflection high energy electron diffraction
RIC: reconstructed ion chromatogram
RI: refractive index, retention index
RID: refractive-index detector
RIMS: resonance ionization mass spectrometry
RIS: resonance ionization spectroscopy, range of individual samples
RM: reference material
RNAA: radiochemical neutron activation analysis
ROA: Raman optical activity
ROESY: rotating frame nuclear Overhauser effect spectroscopy
RPLC: reversed-phase liquid chromatography
RRDE: rotating ring-disk electrode
RS: Raman spectroscopy
RSF: relative sensitivity factor
RTIL: room-temperature ionic liquid
RTP: room-temperature phosphorescence
S/N: signal-to-noise ratio
SAE: sonication assisted extraction
SAM: scanning Auger microscopy, self-assembly monolayers
SANS: small-angle neutron scattering
SAW: surface acoustic wave
SAXS: small-angle x-ray scattering
SBSE: stirbar sorptive extraction
SCE: standard calomel electrode, saturated calomel electrode
SCF: supercritical fluid
SCOT: support-coated open tubular
SDD: silicon drift detector
SdFFF: sedimentation field flow fractionation
SEC: size-exclusion chromatography
SEM: scanning electron microscope
SERS: surface-enhanced Raman spectroscopy
SFC: supercritical-fluid chromatography
SFE: supercritical-fluid extraction
SFFF: sedimentation field flow fractionation
SF-MS: sector field mass spectrometry
SFS: synchronous fluorescence spectroscopy
SHE: standard hydrogen electrode
SIA: sequential injection analysis
SIDA: stable isotope dilution assay
SIMS: secondary ion mass spectrometry
SIRIS: sputter initiated resonance ionization spectroscopy
SMCL: secondary maximum contaminant level (in water analysis)
SMDE: static mercury drop electrode
SMSS: spark source mass spectrometry
SNIFTIRS: subtractively normalized interfacial FT-IR spectroscopy
SNMS: sputtered neutral mass spectrometry
SPAES: spin-polarized Auger electron spectroscopy
SPE: solid-phase extraction
SPME: solid-phase microextraction
SPR: surface plasmon resonance
SRE: stray radiant energy
SRM: standard reference material
SSMS: spark source mass spectrometry
SSRTF: solid-surface room-temperature fluorescence
SSRTP: solid-surface room-temperature phosphorescence
STEM: scanning transmission electron microscope
STM: scanning tunneling microscope
SVE: solvent vapor exit
SWE: supercritical-water extraction
TCA: thermochemical analysis
TCD: thermal-conductivity detector
TCT-GC-MS: thermal cold trap gas chromatography mass spectrometry thermodilatometry
TD: thermodilatometry
TDL: tunable diode laser
TEA: thermal energy analyzer
TED: thermionic emission detector
TEELS: transmission electron energy loss spectrometry
TEM: transmission electron microscope
TET: thermometric enthalpimetric titration
TFFF: thermal field flow fractionation
TGA: thermogravimetric analysis
TGA-IR: thermogravimetric analysis – infrared
THEED: transmission high energy electron diffraction
ThFFF: thermal field flow fractionation
TIC: total ion current chromatogram, tentatively identified compound
TIMS: thermoionization mass spectrometry
TLC: thin-layer chromatography
TLE: thin-layer electrode
TLM: thermal lens microscopy
TLV: threshold limit value
TMA: thermomechanical analysis
TMS: tetramethylsilane
TOCSY: total correlation spectroscopy
TOF: time-of-flight
TOF-MS: time-of-flight mass spectrometry
TSP: thermospray
UHV: ultrahigh vacuum
USE: ultrasonic extraction
UV: ultraviolet
UVPES, UPS: ultraviolet photoelectron spectroscopy
UV-VIS, UV-Vis: ultraviolet-visible
VAR: variable angle reflectance
Vis: visible (radiation)
VOC: volatile organic compound(s)
VOX: volatile organic halogens
VUV: vacuum ultraviolet
W: Wein filter (used in mass spectrometry)
WCOT: wall-coated open tubular
WDS: wavelength dispersive spectrometer
WWCOT: whisker-wall coated open tubular (column)
WWPLOT: whisker-wall coated porous layer open tubular (column)

WWSCOT: whisker-wall support-coated open tubular (column)
XAES: X-ray excited Auger electron spectroscopy, X-ray adsorption edge spectrometry
XANES: X-ray absorption near-edge spectroscopy
XPS: X-ray photoelectron spectroscopy
XRD: X-ray diffraction
XRF: X-ray fluorescence
XRFS: X-ray fluorescence spectroscopy
XRS: X-ray spectroscopy
ZAF: Z (element number) absorption fluorescence

Symbols Used in Analytical Chemistry

α: Auger yield, fine structure constant
a_0: Bohr radius
A: absorbance
A: peak asymmetry factor
B: magnetic field strength
[c]: concentration of component c
d_p: particle diameter (HPLC stationary phase)
D_{ab}: diffusion coefficient
e: electron elementary charge
ε: extinction coefficient
E: energy
E: electrode potential
E_b: binding energy
E_{ea}: electron affinity
E_i: ionization energy
ν: frequency
γ: gyromagnetic ratio
n: refractive index
h: Planck constant
H: enthalpy, plate height
I_0: incident intensity
J: coupling constant
k: coverage factor
k': capacity factor
λ: wavelength, thermal conductivity
m/z: mass-to-elementary-charge ratio (mass spectrometry)
Q_{crit}: Q value (outlier test)
q: quadrupole parameter (mass spectrometry)
r: correlation coefficient
R: resolution
R_∞: Rydberg constant
ρ: density
s: standard deviation
s^2: variance
δ: chemical shift
δ^*: solubility parameter
τ: true value of a measured quantity
τ_{crit}: Chauvanet's criterion (outlier test)
$t_{1/2}$: half-life
t_M: mobile-phase hold up
t_R: retention time
$t_R{}^0$: specific retention time
T: transmittance
$T1$: spin-lattice relaxation time
$T2$: spin-spin relaxation time
$\bar{u}$: carrier phase velocity
V_M: carrier hold-up volume
V_R: retention volume
$V_R{}^0$: specific retention volume

Analytical

BASIC INSTRUMENTAL TECHNIQUES OF ANALYTICAL CHEMISTRY

Thomas J. Bruno

The following section provides brief descriptions of the major instrumental methods of chemical analysis. Please note that these paragraphs are general descriptions and are not meant to convey a comprehensive knowledge on these topics. The reader is referred to one of many excellent texts on instrumental methods of chemical analysis for additional details.

Suggested Reading

1. Skoog, D. A., Holler, F. J., and Crouch, S. R., *Principles of Instrumental Analysis, 6th Edition*, Thomson Brooks/Cole Publishing, Belmont, CA, 2007.
2. Robinson, J. W., Skelly Frame, E. M., and Frame II, G. M., *Undergraduate Instrumental Analysis, 6th Edition*, CRC Press, Boca Raton, FL, 2004.
3. Pungor, E., *A Practical Guide to Instrumental Analysis*, CRC Press, Boca Raton, FL, 1994.

Separation Methods

Gas Chromatography (GC): A separation method in which the sample or solute is vaporized (usually in a solvent, but sometimes neat or free of solvent) and passed through a medium under the influence of a carrier gas. The medium is called the stationary phase, in contrast to the carrier gas, which is mobile. The most common modern stationary phases are based on open tubular or capillary columns, in which the separation medium coats the inside periphery of a tube (typically tenths of millimeters in inside diameter) that is between 25 m and 60 m long. Older media are packed columns, consisting of packed beds, which are still used for gas analysis. In these applications, a solid sorbent is very common, and this is called gas-solid chromatography. Some open tubular columns are available with solid sorbents as well. Interactions of the solute with the separation medium affect the separation of the components of the mixture. A wide variety of detectors are available for general or specific applications. One of the most useful combinations is gas chromatography coupled with mass spectrometry (GC-MS). Solutes amenable to analysis by GC are usually of moderate volatility and a relative molecular mass usually not exceeding 400. The most common stationary phases are cross-linked polymers based on dimethyl polysiloxane, the backbone of which can be derivatized with ligands to provide specific interactions. It is also possible to incorporate stereogenic (chiral) stationary phases as well.

Liquid Chromatography (LC, HPLC): A separation method in which a sample or solute (usually in a solvent, but sometimes neat or free of solvent) is passed through a medium under the influence of a carrier liquid. The medium is called the stationary phase, in contrast to the carrier, which is the mobile phase. Unlike gas chromatography, where the carrier gas plays little role other than mass transfer, the mobile phase in liquid chromatography is a controllable variable whose polarity and other properties are varied, in addition to the interactions with the stationary phase, to affect separation. The stationary phase in liquid chromatography is usually a micrometer-size particle packed bed that requires a high-pressure solvent system to cause mass flow. Liquid chromatographic systems have therefore been called high-pressure liquid chromatography (HPLC), although the acronym is usually taken to mean high-performance liquid chromatography. Many variations of the method have been developed for specific analysis. For example, gel permeation chromatography is an adaptation used for the separation of polymers. Affinity chromatography is similar in concept to gel permeation chromatography, but uses the specific interactions between an antibody and an antigen. The use of stereogenic (or chiral) stationary phases is also an important development, especially in the analysis of pharmaceuticals.

Thin-Layer Chromatography (TLC): A separation method in which a stationary phase (typically a polar adsorbent such as alumina or silica gel) coated on a sheet of plastic, aluminum, or glass is used with a mobile phase usually consisting of a solvent or mixture of solvents in a beaker. The solute is applied as a blotted spot just above the end of the adsorbent-coated plate, and then the end is immersed into the solvent (but not so far as to immerse the solute spot). Commercial plates are robust, plastic sheets that can be cut to the desired size. In earlier applications, filter paper has been used in TLC, giving rise to the term "paper chromatography." This term is rarely used today. Solvent is then drawn up through the adsorbent coating by capillary action. Separation results from a combination of interactions with the adsorbent and solvent. Commonly, the separated components are rendered visible by spraying stains or reactants on the plate after separation has been completed. It is also common to view the "developed" plate under an ultraviolet lamp, to visualize spots that may be fluorescent.

Supercritical Fluid Chromatography (SFC) and Supercritical Fluid Extraction (SFE): A separation technology similar to other extraction and chromatographic methods, but in which the mobile phase is actually a fluid in its supercritical fluid state. A supercritical fluid is a fluid that is held above its critical temperature and pressure, and for which no application of additional pressure can result in the development of a liquid phase. Supercritical fluids are unique in that while they possess liquid-like densities, the mass transfer behavior is superior to that of liquids. Supercritical fluid chromatography remains a niche method that is applicable to pharmaceuticals and other high relative molecular mass solutes. Supercritical extraction, on the other hand, is more widely used as a sample preparation method, especially in pharmaceutical analyses, polymers, and environmental analyses.

Electrophoresis: A family of separation methods based on the motion induced in particles by an applied, uniform electric field. The most common application of electrophoresis is gel electrophoresis, in which sample (typically proteins, amino acids, nucleic acids, etc.) is applied to a channel that is formed in a cross-linked polymer, usually polyacrylamide or agarose (the gel). The speed at which the individual species move through the gel under the influence of the field is determined largely by the size of the species, as expressed by the mass-to-charge ratio. After separation, the individual species usually appear as discrete bands that may be better visualized by staining with ethidium bromide, silver, or Coomassie Brilliant Blue dye. Other related and more specific techniques include isoelectric focusing, pulsed-field gel electrophoresis, immunoelectrophoresis, and isotachophoresis.

Purge-and-Trap Sampling: Purge and trap sampling is a family of methods that are used to capture the headspace above a condensed phase for subsequent analysis, most often for complex mixtures, environmental samples, etc. The headspace is the vapor space that develops above any condensed (solid or liquid) phase. Thermodynamics assures us that the concentration of a particular analyte found in the headspace will be different than that found in the condensed phase, but often the relationship is predictable. The value in the method comes from the simplicity; sample preparation is usually far simpler than the cleanup that is typically required for many complex mixtures. Purge and trap methods fall into two general categories: static and dynamic. The dynamic method typically uses a sweep gas to continuously purge vapor analytes into a cold (cryogenic) trap or an adsorbent. Modern dynamic purge and trap methods include PLOT column cryoadsorption, which uses a combined adsorbent and cryotrapping on a high efficiency platform. Static methods typically employ a syringe to pressurize the headspace above a condensed phase, followed by uptake of the pressurized headspace into a trap. A modern static method (that usually is done without pressurization) is solid-phase microextraction (SPME). This method utilizes a fiber coated with a stationary phase (similar to stationary phases used for gas and liquid chromatography) at the end of a wire mounted in a syringe needle.

Spectroscopic Methods

Mass Spectrometry (MS): An analytical technique in which charged particles or radical ions are produced from a sample by either electron impact (bombardment with a stream of electrons) or chemical ionization (interaction with a small charged ion). Analysis on the basis of the mass-to-charge ratio is performed on fragments of the molecule that develop after the initial ionization. The method is very useful for mass determination and structure determination on the basis of the induced fragmentation pattern. The charged fragments can be separated or analyzed by a magnetic sector, a quadrupole, an ion trap, by time of flight, or by cyclotroning. When coupled to separation techniques such as gas or liquid chromatography, a nearly universal qualitative detection capability is provided. Structure determination can be performed by comparison to well known fragmentation patterns or characteristics. When a mass spectrometer is capable of high mass resolution (0.0001 RMM units), a nearly unequivocal identification of a compound formula is possible. While direct interfaces and gas chromatographs are the most common sample introduction techniques, many others are available for specific applications. Matrix-assisted laser desorption/ionization (MALDI) is often used in time-of-flight instruments, and is especially useful for analysis of biopolymers. Inductively coupled plasma ionization is capable of producing mass spectra of high sensitivity for many metals and some nonmetals. When coupled to liquid chromatography instrumentation, thermospray and electrospray methods have been used with HPLC for analytes that are not amenable to gas chromatography separation. In addition to the interfaces discussed above, tandem mass spectral methods are of great value, especially for confirmatory analysis. In this respect, a fragment is selected for further fragmentation and analysis. The most common types of tandem mass spectrometry involve quadrupoles coupled with a time of flight (Q-TOF) and multiple (usually three) quadrupoles in series, called a triple quad (QQQ or QqQ). More details can be found in the entry "Varieties of Hyphenated Gas Chromatography and Mass Spectrometry."

Ultraviolet Spectrophotometry (UV, UV-Vis): A spectroscopic technique that focuses on electronic transitions in the visible and ultraviolet regions of the electromagnetic spectrum used for excitation and detection. The practical ultraviolet region extends from 190 nm to 400 nm in wavelength. The UV region can be divided into subranges: near-UV: 300 nm to 400 nm, mid-UV: 300 nm to 200 nm, far-UV: 200 nm to 122 nm, and vacuum-UV: 200 nm to 100 nm. Other divisions are possible, but these are less important for analytical chemistry. The visible region, so called because of the response of human vision, extends from about 390 nm to 750 nm in wavelength. Although it is possible to use ultraviolet visible spectrophotometry for structure determination, it is most often used as a quantitative tool. The wavelength-structure correlations are not as detailed, nor the spectra as sharp, as with other spectroscopic methods such as infrared spectrophotometry. The utility of UV-Vis absorptions for many organic compounds has led to this instrument being adapted as a detector for liquid chromatography. Related to ultraviolet spectrophotometry are fluorescence spectroscopic methods. In these methods, the energy emitted is at a different wavelength (usually longer, of lower energy) than the incident radiation, and this is typically detected perpendicular to the incident excitation beam. This makes fluorescence spectroscopy more sensitive than UV-Vis spectroscopy. This type of instrument is also incorporated as a detector for liquid chromatography.

Infrared Spectrophotometry (IR, FTIR): A spectroscopic technique that focuses on molecular vibrations (with a concurrent change in dipole moment) in the infrared (IR) region of the electromagnetic spectrum for excitation and detection. This region is further divided into three separate but overlapping ranges. The near-infrared (high energy IR, approximately 12 500 cm^{-1} to 4000 cm^{-1}, 0.8 μm to 2.5 μm wavelength) is used to study overtone or harmonic vibrations. The mid-infrared (mid-range energy IR, approximately 4000 cm^{-1} to 400 $^{-1}$, 2.5 μm to 25 μm wavelength) is used to study the fundamental vibrations and associated rotational-vibrational combinations. The far-infrared (low energy IR, approximately 400 cm^{-1} to 10 cm^{-1}, 25 μm to 1000 μm wavelength) is adjacent to the microwave region and is used to study rotational transitions. Most modern instruments use the Fourier transform (FT) technique to record the spectrum over all wavelengths, rather than by scanning through the wavelengths. The absorbances of the IR radiation are associated with specific chemical moieties, and a study of the spectra can be used to aid in structure determination. One often uses structure correlation charts to aid in assignment of absorbance bands. An analysis of the intensity of the absorptions can also be used for quantitative analysis.

Nuclear Magnetic Resonance Spectrometry (NMR): A spectroscopic method that takes advantage of the fact that magnetic nuclei (nuclei with an odd number of protons and/or neutrons, having an intrinsic magnetic moment and angular momentum), when placed in a magnetic field, will absorb pulses of electromagnetic radiation and then radiate this energy back out for detection. For these nuclei, the energy and signal intensity are proportional to the applied magnetic field. The power of NMR results from the ability to probe the molecular environment around a particular nucleus, thus making it a favorite tool of the organic chemist. This is done by measurement of the chemical shift (or frequency) of an absorption, and by the analysis of splitting patterns, which are caused by the influence of adjacent nuclei. New high-field, high-sensitivity instruments have given this technique more applications in analytical chemistry. The most commonly studied nuclei are ^{1}H (proton, the most NMR-sensitive isotope after the radioactive ^{3}H) and ^{13}C. With high-field instruments, additional nuclei are accessible: ^{2}H, ^{10}B, ^{11}B, ^{14}N, ^{15}N, ^{17}O, ^{19}F, ^{23}Na, ^{29}Si, ^{31}P, ^{35}Cl, ^{113}Cd, ^{129}Xe, and ^{195}Pt.

Raman Spectroscopy: A vibrational spectroscopic method that arises from the inelastic scattering of monochromatic radiation by molecules that undergo a change in polarizability during the vibration. This is in contrast to infrared spectrophotometry, in which a change in the dipole moment occurs during the vibration. When radiation (typically light from a laser in the visible, near-infrared, or near-ultraviolet range) is scattered, a small fraction of the scattered radiation is observed to have a different frequency (the Raman effect). The variations of Raman spectroscopy are used to locate functional groups or chemical bonds in molecules. There are several variations in the approach to Raman spectroscopy. In resonance Raman spectroscopy, the excitation wavelength is matched to an electronic transition of the molecule, enhancing the vibrational modes. In coherent anti-Stokes Raman spectroscopy (CARS), two laser beams are used to generate a coherent anti-Stokes frequency beam. In surface-enhanced Raman spectroscopy (SERS), surface plasmons (a quantum of plasma oscillation) on a silver or gold colloid on a surface (such as a mirror) are excited by the laser, resulting in an increase in the electric fields surrounding the metal.

Atomic Absorption Spectroscopy (AAS) and Atomic Emission Spectroscopy (AES): Two related spectroscopic methods applied primarily to the analysis of inorganic compounds. Atomic absorption procedures use the absorption of optical radiation (light) by free atoms in the gaseous state. The light can be produced by a hollow cathode lamp, an electrodeless discharge lamp, or a deuterium lamp. The light is absorbed by the analyte during an electronic transition, the wavelength of which corresponds to only one element in the analyte, and the width of an absorption line is of the order of only a few picometers. This method can be used for the quantitative determination (on the basis of a calibration curve) of approximately 70 different elements in solution or directly in solid samples. Atomic emission spectroscopy (AES) uses the light emitted by a vaporized sample in a flame, plasma, arc, spark, or laser, at a particular wavelength, to determine the atomic spectrum (for determination of the elemental composition) and to determine the quantity of an element in a sample. The wavelength of the atomic spectral line gives the identity of the element while the intensity of the emitted light is proportional to the number of atoms of the element. No single source, as described above, is optimal for a given sample, and it is the choice of source that distinguishes the various techniques.

Miscellaneous Methods

Colorimetry, Spot Tests, Presumptive Tests: Rapid, simple tests based on color change are frequently used as the basis of preliminary conclusions or approximate concentration measurement. Colorimetric methods use simple comparative instruments to determine the concentration of colored compounds in solution. These devices, called colorimeters, can be manual or automatic, and use filtered light in the visible region (between 400 nm and 700 nm). In both cases, the operation depends upon having multiple solutions, including a blank, for comparison with a solution of unknown concentration. Manual colorimeters function by measuring the variable light path through the unknown solution as compared to a known solution until a match is achieved visually. The product of concentration and path length matches when the concentrations are the same, so an unknown concentration may be obtained by a proportion. Automated devices function similarly but with photocells. Colorimetric tests are usually done following four basic protocols: (1) an unknown can be treated/reacted with a reagent to form a new compound which is colored, (2) a chelate complex is formed having a different color than the starting compound, (3) a colored compound is oxidized or "bleached" by another compound, resulting is less color, or (4) an intermediate is formed that can be oxidized or reduced later to a colored compound. Related to colorimetric methods are spot tests, often called presumptive tests in forensic science (since they are used to establish probable cause, or for ordering and performing more sophisticated tests), which are the observations of color changes upon the addition of one or more reagents to an unknown. Spot tests are inherently simple, and are done with a drop or two of reagent(s), on a filter paper or other suitable medium, and generate minimal waste. Usually instrumentation is not used, one merely notes a color change, although simple devices such as handheld color space analyzers (used in the paint industry) can be helpful. Other approaches include the recording of a digital photograph of the spot and the determination of the L*, a*, and b* axes in LAB color space. An example of spot testing for presumptive purposes includes the addition of cobalt thiocyanate ($CoC_2N_2S_2$) to cocaine, producing a blue color (the Scott test). These tests are presumptive rather than confirmatory because of the potential of false positives. Thus, the Scott test will also produce a blue coloration in the presence of lidocaine and diphenylhydramine.

Refractometry: Refractometry is the measurement of the degree to which the path of electromagnetic radiation, specifically light, is bent upon traversing from one medium to another. Indirectly, a refractometer measures the speed of light since the refractive index, the measurand provided by a refractometer, is defined as the speed of light in vacuo divided by the speed of light in a particular medium (which can be a gas, liquid, or solid). Refractive index typically varies between 1.3000 and 1.7000 for most compounds, and being a ratio, it is dimensionless. The refractive index of vacuum is by definition unity. Refraction is the result of differing densities of media; light passes through dense media slower than it passes through less dense media. Refractive index is dependent on temperature and the wavelength of light used for the measurement, and when both are known for a media, it can be used as an indication of composition or concentration. Thus, the refractive index is typically measured with light from the sodium D-line (589 nm), and often at a temperature that is near ambient (20 °C). The measurand is thus indicated as n_D^{20}. As an analytical tool, refractometry is used in four primary ways. First, it is used to help identify compounds by comparison with known values. Second, it is used to assess purity by comparison with the refractive index of a pure material. Third, it is used as a measure of solute concentration of a solution, by reference to a calibration curve. Finally, it is used as the basis of universal detection in HPLC. Refractometers are commercially available as small handheld instruments, both manual and digital (that can be carried in a pocket), Abbe refractometers (the typical lab bench device usually equipped with a thermostat), and in-line refractometers such as those used for HPLC detectors.

Thermal Analysis (TA) Methods: A family of analytical techniques in which various properties of a sample are examined as a function of changing temperature at a particular rate of change. For chemical analysis, the most common thermal analysis techniques include differential thermal analysis (DTA) and thermogravimetric analysis (TGA). In chemical analysis, there are many other thermal analysis methods available, perhaps the common of which is differential thermal analysis, used for determination of phase transitions. In chemical analysis, TGA is used to determine a mass change as a function of temperature. This is used to determine decomposition or degradation temperatures, moisture content of materials (although Karl Fisher coulombic titrimetry is also used for this), the level of inorganic and organic components in materials, decomposition points of explosives, and solvent residues in materials. DTA, on the other hand, monitors a temperature change rather than a mass change and is used for determination of phase transitions. In this respect, it is useful as a complementary technique to probe the energetics of decomposition, moisture loss, etc. Other thermal analysis methods, such as differential scanning calorimetry (DSC), are not primarily analytical tools but rather thermophysical property measurement tools.

X-Ray Methods: A family of techniques that utilize radiation in the x-ray region, with wavelengths between 0.01 nm to 10 nm (corresponding to frequencies in the 3×10^{16} Hz to 3×10^{19} Hz range) to analyze for the presence of elements in a sample. In x-ray fluorescence spectroscopy (XRFS), short wavelength x-rays are used to excite secondary or fluorescent x-rays from a sample. The wavelength of x-rays used to excite the sample must be shorter than the expected fluorescence wavelength. The sample is typically presented as a powdered solid on a glass plate. Spectra are generated by changing the incident angle between the source and the sample with a device called a goniometer. The fluorescent signals obtained are very precise and specific, and can rival wet chemical methods for the identification of elements. Auger electron spectroscopy (AES) is a surface analysis method that measures the electrons emitted from a surface by electron bombardment of the surface. This method is considered to be in the same family as x-ray methods because during electron bombardment, the surface can lose energy either by electron emission (the Auger effect, via Auger electron emission) or x-ray emission. These methods are in contrast to x-ray diffraction methods, in which a crystal structure measurement is desired.

Flow Injection Analysis (FIA): An analytical protocol that seeks to replace the manual "test tube and beaker" aspects of wet chemical analysis by injecting an analyte in a flowing stream of a carrier reactant. As the analyte flows with the reactant stream, it diffuses into the reactant and product forms. Ultimately, the product zone, under the influence of the moving reactant, is passed into a detector section. The detection devices can consist of the same wide variety as is used in HPLC. The major advantage of flow injection analysis is the automation and decreased uncertainty associated with sampling and reagent addition. Strict control of reagent concentration, flow rate, and analyte volume is possible. Modern applications of FIA include the sequential addition of analyte and reactant in a stream so that the two are "stacked" in an inert carrier. They then mix by the parabolic flow profile of a laminar flowing stream in a tube. This arrangement can be miniaturized within a sampling valve, forming the so-called lab-in-a-valve approach.

ANALYTICAL STANDARDIZATION AND CALIBRATION

Thomas J. Bruno

Overview

Most modern instrumental techniques used in analytical chemistry produce an output or signal that is not absolute; the signal or peak is not a direct quantitative measure of concentration or target analyte quantity. Thus, to perform quantitative analysis, one must convert the raw output from an instrument (information) into a quantity (knowledge). This is done by standardizing or calibrating the raw response from an instrument (Refs. 1-4). Here, we briefly summarize the most common methods applied in analytical chemistry, recognizing that this is a very large field. We note that the common use of the term "standardization" is not to be confused with the application of standard methods as specified by regulatory or consensus standard organizations.

Samples

In all of the discussion to follow, we assume that the sample has been properly drawn from the parent population material, and properly prepared. Clearly, the most precise analytical methods and the most painstaking calibration methods are useless if applied to a sample that does not represent reality. Nevertheless, the term "sampling," which describes the process of obtaining the sample (from the population material), implies the existence of a sampling uncertainty (arising mainly from population material heterogeneity) (Refs. 5 and 6). Thus, the analytical result is an estimate of what would be obtained from the parent population material. The theory, concepts, and nomenclature regarding samples and sampling constitute a complex, statistically based, subspecialty of analytical chemistry well beyond the scope presented here. We begin with some simplified definitions (Ref. 7):

Amount of substance: The amount of substance is the fundamental quantity of material measured in the number of moles.

Analyte: The analyte is the target component or compound in the sample for which one desires a measurement.

Aliquot: An aliquot is a known fraction of a homogeneous mass or volume.

Bias: As applied to sampling, bias refers to a systematic displacement, error, uncertainty. or mistake caused by a flaw in the sampling procedure.

Convenience sample: A sample chosen on the basis of accessibility, expediency, cost, efficiency, or other reason not directly concerned with sampling parameters.

Determination: The determination is the entire analytical procedure or method performed on a test portion.

Matrix: The matrix is the background or carrier material of the sample that includes all components except the analyte(s) of interest.

Phase: The phase describes the physical state of a substance: primarily solid, liquid, gas, but this term might include more detailed descriptions to include supercritical fluid, plasma, etc. Note that the term "vapor" typically refers to a gas phase above and in equilibrium with a condensed phase (often called the headspace in chemical analysis).

Population material: The population material is the entirety of the bulk material from which the sample is drawn. This might be a plot of earth, a warehouse full of sugar, or a tank of jet fuel.

Quantity: The quantity refers to the mass or volume of a substance.

Random sample: A sample selected so that any portion of the sample has an equal or known chance of being chosen for measurement.

Representative sample: A sample resulting from a sampling process that can be expected to adequately reflect the properties of interest of the parent population. A representative sample may be a random sample. The degree of representativeness of the sample is limited by cost or convenience.

Selective sample: A sample that is deliberately chosen by using a sampling plan that eliminates materials with certain characteristics or selects only a material with other relevant characteristics desired for an analysis.

Test portion: The test portion is the actual material removed from a sample for analysis.

Umpire sample: A sample taken, prepared, and stored in an agreed upon manner for the purpose of settling a dispute, arrived at by agreement that will include the test method and procedure, serving as the basis for acceptance, rejection, or economic adjustment. This is sometimes called a referee sample.

Test portion: The test portion is the actual material removed from a sample for analysis.

Unknown: The unknown is a term that describes the target measurement or unknown quantity that is desired for the analyte, or the analyte itself.

Sampling uncertainty is that part of the total uncertainty in an analytical procedure or determination that results from using only a fraction of the population material. In this respect, sampling by any method is an extrapolation process. Since the sampling uncertainty is usually ignored for an individual analysis on an individual test portion, the sampling uncertainty is considered as being due entirely to the variability of the test portion. It is therefore assessed, when necessary, by replication of the sampling from the parent population material, and statistically isolating the uncertainty thus introduced by analysis of the variance. Typically, the problems associated with liquid population material are less complex but must not be ignored. Sample stratification, concentration and thermal gradients, poor mixing, and gradients associated with flow are all real effects that must be considered. Sampling uncertainty is often minimized by field and laboratory processing, with procedures that can include mixing, reduction, coning and quartering, riffling, milling, and grinding.

Another aspect that must be considered subsequent to sampling is sample preservation and handling. The integrity of the sample must be preserved during the inevitable delay between sampling and analysis. Sample preservation may include the addition of preservatives or buffer solutions, pH adjustment, use of an inert gas "blanket," and cold storage or freezing.

Calibration and Standardization

External Standard Methods

The external standard method can be applied to nearly all instrumental techniques, within the general limits discussed here, and the specific limitations that may be applicable with individual techniques. This method is conceptually simple; the user constructs a calibration curve of instrumental response with prepared mixtures containing the analyte(s) over a range of concentrations, an example of which is shown in Figure 1a. Thus, the curve represents the raw instrumental response of the analyte as a function of analyte concentration or amount. Each point on this plot must be measured several times so that the repeatability can be assessed. Only random uncertainty should be observed in the replicates; trends of increasing or decreasing response (hysteresis) must be remedied by identifying the source and adjusting the method accordingly. The calibration solutions should be randomized (that is, measured in random order). Although called a calibration "curve," ideally the signal versus concentration plot is linear, or substantially linear (that is, areas of nonlinearity are unimportant, otherwise they are localized, minor, and properly treated by the measurement technique). In some cases, the response may be linearizable (for example, by calculating the logarithm of the raw response). If a curve shows nonlinearity in an area that is important for the analysis, one must measure more concentrations (data points) in the region of curvature.

In practice, the line that results from the calibration is fit with an appropriate model, and the desired value for the unknown concentration is calculated. The curve can be used graphically if approximation suffices. Mixtures prepared for external standard calibration can contain one or many analytes. Once a calibration curve is prepared, it can often be used for some time period, provided such a procedure has been previously validated (that is, the stability of the standards and the instrument over the time of use has been assessed). Otherwise, it is best to measure the unknown and the standards within a short period of time. Moreover, if any major change is made to the instrumentation (changing a detector or detector parameters, changing a chromatographic column, etc.), the standards must be re-measured.

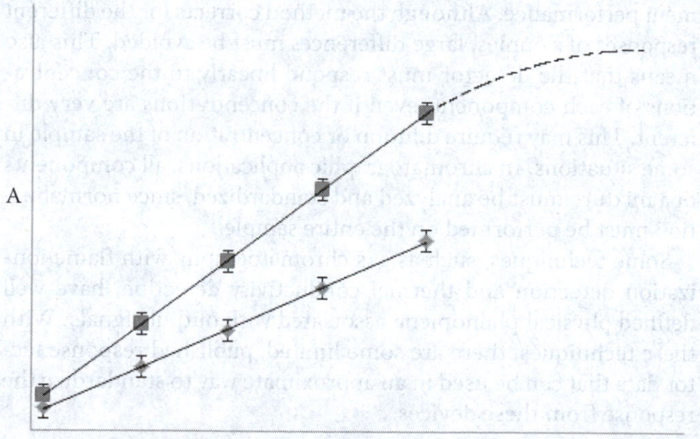

FIGURE 1a. An example calibration curve prepared by use of the external standard method. The instrument response is represented by A, and the concentration resulting in that response is [A]. While curves for two analytes are shown, in principle, one can plot as many analytes as desired. While five points per analyte have been shown, one can measure as many as required. Note that a region of nonlinearity is shown in the latter part of the curve for one of the components. One would require a larger number of points to adequately represent and fit any nonlinear areas.

To successfully use the method, the standard mixtures must be in a concentration range that is comparable to that of the unknown analyte, and ideally should bracket the unknown. Multiple measurements of each standard mixture should be made to establish repeatability of points on the curve. Many instrumental methods have operation ranges (frequency, temperature, etc.) in which the uncertainty is minimized, so components and concentrations for standard mixtures must respect this. The standard mixtures should be in the same matrix as the unknown, and the matrix must not interfere with the unknown or other standard mixture components. Any pretreatment of the unknown must also be reflected in the standard mixtures. As with any calibration method, components in the standard mixtures must be available at a high (or at least known) purity, they must be stable during preparation, and must be soluble in the required matrix. Unless the physical phenomenon of a measurement is well understood, extrapolation beyond the curve is not recommended (and indeed is usually strongly discouraged); nevertheless, extrapolation is occasionally done in practice. In those cases, one must be cautious, report exactly how the extrapolation was done, and assess any increase in uncertainty that may result. Note that the curve might not extrapolate through the origin. This is usually the result of adsorption (of components on container walls), carryover hysteresis, absorption (of components in seals or septa), or component degradation or evaporation.

A major consideration with external standardization is that typically, the sample size (for example, the injection volume in chromatography) must be maintained constant for standard mixtures and the unknowns. If the sample size varies slightly, it is often possible to apply a correction to the raw signal. One should not attempt to generate a calibration curve by varying the sample size (that is, for example, injecting increasing volumes into a gas chromatograph). This caution does not preclude serial dilution methods (see below), in which multiple solutions are generated for separate measurement. Other issues that can hinder successful application of the external standards method include instrumental aspects that might not be readily apparent. In chromatographic methods, for example, one can overload the column or detector. In older instruments, settings of signal attenuation were typically made manually, while in newer instruments, this may occur through software, sometimes without operator interaction or knowledge.

Note, *inter alia,* that in Figure 1a (and indeed all the examples presented here), the uncertainty is only indicated for the variable on the y-axis. In reality, we must recognize that there is uncertainty for the values plotted on the x-axis as well, but we often only treat the largest uncertainty, or the uncertainty that is most important for our application. Note, also, that it is critical to maintain the integrity of standards; decomposition, degradation, moisture uptake, etc., will adversely affect the validity of the calibration.

Abbreviated External Standard Methods

In many situations in chemical analysis, a full calibration curve is not prepared because of the complexity, time, or cost. In such situations, abbreviated external standard methods are often used. Under no circumstances can an abbreviated method be used if the raw signal response is nonlinear. Moreover, these methods are not generally appropriate for analyses in regulatory, forensic, or health care environments where the consequences can be far-reaching.

Single Standard

This method uses a simple proportion approach to standardize an instrument response. It can be used only when the system

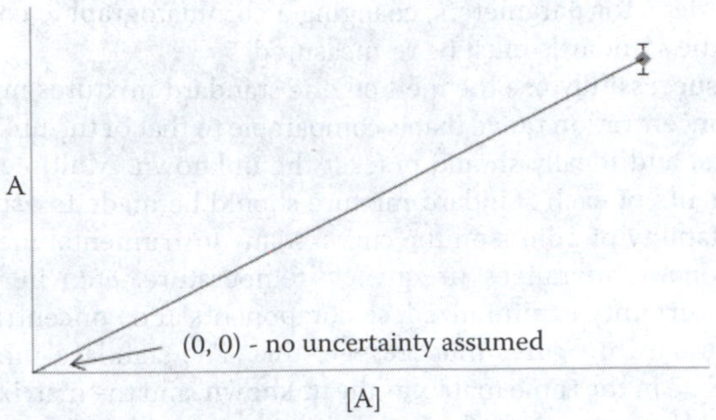

FIGURE 1b. An example of a single-point calibration curve. The instrument response is represented by *A*, and the concentration resulting in that response is [A]. The origin (0,0) is assigned as part of the curve, and is assumed to have no uncertainty.

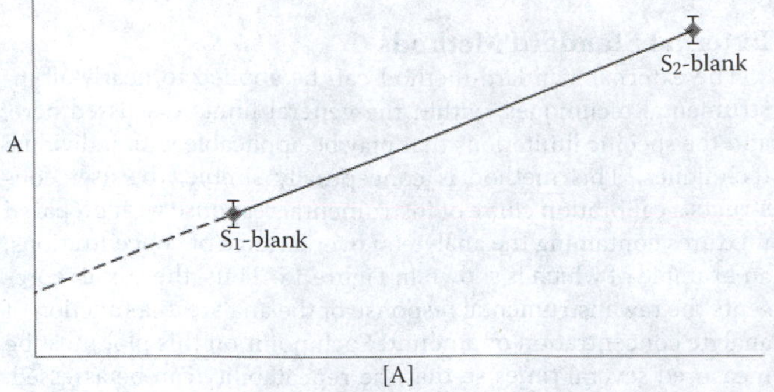

FIGURE 1c. An example of two standards plus a blank calibration curve. The blank is subtracted from each of the standards. The instrument response is represented by *A*, and the concentration resulting in that response is [A].

has no constant, determinate error or bias*, and when the reagents used give a zero blank response (that is, the instrument response from the matrix and measurement system only, without the analyte). A standard should be prepared such that the concentration is close to that of the unknown. One then calculates the concentration of the unknown, [X], as:

$$[X] = (A_x/A_s)\,[S] \qquad (1)$$

where A_x is the instrument response of the unknown, A_s is the instrument response of the standard, and [S] is the concentration of the standard.

Single Standard Plus Assumed Zero

This method, illustrated schematically in Figure 1b, assumes that the blank reading will be zero. One uses a two-point calibration in which the origin is included as the first point. It is important to ensure, by experiment or experience, that such a method is adequate to the task.

Single Standard Plus Blank

If the analytical method has no determinate error or bias, but does produce a finite blank value, then one must also perform a blank measurement, which is subtracted from the instrument response of the standard and the unknown. Then the same procedure (Eq. 1) is used as for the single standard. If multiple samples are to be measured, it is important to measure the blank between each measurement.

Two Standards Plus Blank

When the analytical method has both a determinate error (or bias), and a finite blank value, at least three calibrations must be made: two standards and one blank. The standard concentrations are typically prepared widely spaced in concentration, and the higher concentration should be chosen to represent the limit of linearity of the instrument or method. If this is not practical, the higher concentration should simply be the highest expected concentration of the analyte (unknown). This method is illustrated schematically in Figure 1c. If multiple samples are to be measured, it is important to measure the blank between each measurement.

Internal Normalization Method

As mentioned above, the raw signal from an analytical instrument is typically not an absolute measure of concentration of the analyte(s), because the instrument may respond differently to each component. In some cases, such as with chromatographic methods, it is possible to apply response factors, determined from a standard mixture containing all constituents of the unknown sample, for standardization (Ref. 8). The standard mixture is gravimetrically prepared (with known mass percents for each component), and the instrument response is measured, for example as chromatographic areas. The total mass percent and the total area percent each sum to 100. One calculates the ratio of each mass percent to each area, choosing one component as the reference, which is assigned a response factor of unity. To obtain the response factors of all the other components, one divides its (mass%-to-area ratio) with that of the reference. This is done for all components, producing a response factor for all components, except of course for the reference, defined as unity. When the unknown sample is measured, the response factor is multiplied by each raw area, and the resulting area percent provides the normalized mass percent of each component in the unknown.

This method corrects for minor variations in sample size (earlier defined as the test portion), although large differences in sample size must be avoided so that one is assured of consistent instrument performance. Although the method corrects for the different responses of samples, large differences must be avoided. This also means that the detector must respond linearly to the concentrations of each component, even if the concentrations are very different. This may require dilution or concentration of the sample in some situations. In chromatographic applications, all components of a mixture must be analyzed and standardized, since normalization must be performed on the entire sample.

Some techniques, such as gas chromatography with flame ionization detection and thermal conductivity detection, have well defined physical phenomena associated with output signals. With these techniques, there are some limited, published response factor data that can be used in an approximate way to standardize the response from these devices.

In Situ Standardization

While it is rare that an analytical method can be calibrated by use of a single solution, some instances of spectrophotometry and electroanalytical methods can qualify. To use this method one sequentially and incrementally adds known masses of standard analytes to a solution, with an instrument response being measured after each addition. This procedure can only be used if the analytical method itself does not change the analyte concentration

* Determinate error and bias are related terms that describe uncertainty that arises from a fixed cause, and that can, in principle, be eliminated if recognized. Determinate error (or systematic error) is most often associated with a measurement, while bias can be associated with either a measurement or with the sampling procedure.

(nondestructive) and does not lead to a loss of solution volume. A solid crystalline analyte is an example. One must also minimize changes in solution volume over the course of the standardization.

Standard Addition Methods

Samples presented for analysis often are contained in complex matrices with many impurities that may interact with the analyte, potentially enhancing or diminishing a signal from an instrumental technique. In such cases, the preparation of an external standard calibration curve will be impossible, because it might be very difficult to reproduce the matrix. In these cases, the standard addition method may be used. A standard solution containing the target analyte is prepared and added to the sample, thus accounting for the unknown impurities and their effects. While the quantity of target analyte in the target sample is unknown, the added quantity is known, and its incremental additive effect on the instrument signal can be measured. Then, the quantity of the unknown analyte is determined by what is effectively an extrapolation. In practice, the volume of the standard solution added is kept small to avoid dilution of the unknown impurities by no more than 1% of the total signal. This method can only be used if there is a verified linear relationship between the signal and quantity of analyte. If a determinate error is present, then the slope of the line must be known. Moreover, the sample cannot contain any components that can respond as the analyte (that is, masquerade).

Single Standard Addition

In the simplest case, one addition of analyte is made after first measuring the response of the analyte in the unknown sample. Thus, two measurements are required:

$$A_{xo} = m[X_o] \quad (2)$$

$$A_{xi} = m([X_o] + [S]), \quad (3)$$

where A_{xo} is the instrument response of the analyte in the unknown sample, $[X_0]$ is the concentration in the unknown sample, and A_{xi} is the instrument response upon the addition of the standard, [S] (additive in equation because X and S are the same compound). The assumed slope is the proportionality constant, *m*. The two equations are solved simultaneously for $[X_0]$. This technique is very rapid and economical, but there are serious drawbacks. There is no built in check for mistakes on the part of the analyst, there is no means to average random uncertainties, and there is no way to detect interference (mentioned above as masquerade).

Multiple Standard Addition

This standard addition method alleviates some of the problems inherent in single standard addition. Here, the unknown sample is first measured in the instrument. Then that sample is "spiked" with incrementally increasing concentrations of the analyte, generating a curve such as that shown in Figure 1d. The curve should extrapolate to zero signal at zero concentration. The concentration of the analyte in the unknown is read or calculated from the abscissa (*x*-axis).

Internal Standard Methods

An internal standard is a compound added to a sample at a known concentration, the purpose of which is to exhibit a similar signal when measured in an instrument, but be distinguishable from the signal of the desired analyte. It provides the highest level of reliability in quantitation by chromatographic methods, and is not affected by large differences in sample size (Ref. 8). Unlike

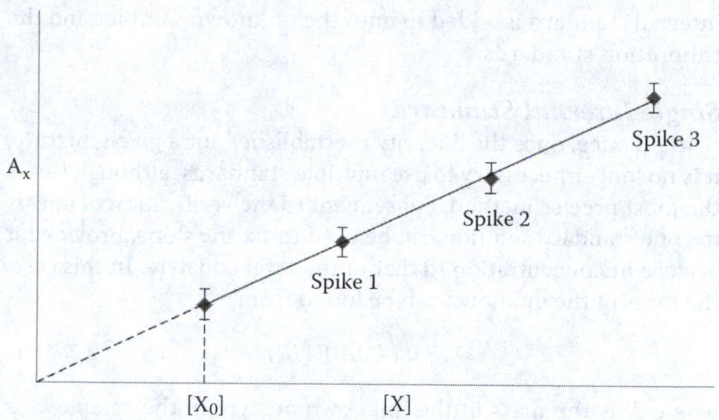

FIGURE 1d. An example of calibration by multiple standard addition. Three additions (spikes) of the analyte X are shown, as is the extrapolation to the unknown concentration, X_0.

the internal normalization method, it is not necessary to elute or measure all the components of the sample, one need focus only on the component(s) of interest. In atomic spectrometry, this method is not affected by changes in gas flow rates, sample aspiration rates, and flame suppression or enhancement. Another situation in which this method is valuable is when the sample matrix is either unknown or very complex, precluding the preparation of external standards.

Multiple Internal Standards

A set of calibration solutions is prepared by mass, containing the target analyte, X, and a standard that is not present in the unknown sample, A. The instrument response (for example, a chromatographic area) is measured for each calibration solution, and a plot is made to establish linearity as in Figure 1a. The ordinate axis is the ratio of the response of the unknown analyte component, A_x, to the response of the chosen standard, A_s. The abscissa is the ratio of mass of X to the mass of S for that standard mixture. Once the linearity is confirmed in the concentration range of interest, the unknown is spiked with a known mass of S, the instrument response is measured, and the area ratio A_x/A_s is calculated. Either the graph or a fit of the data on Figure 1e is then used to determine the corresponding mass fraction, from which the mass of X may be determined. Note that the calculations could be simplified if the same mass per volume of the

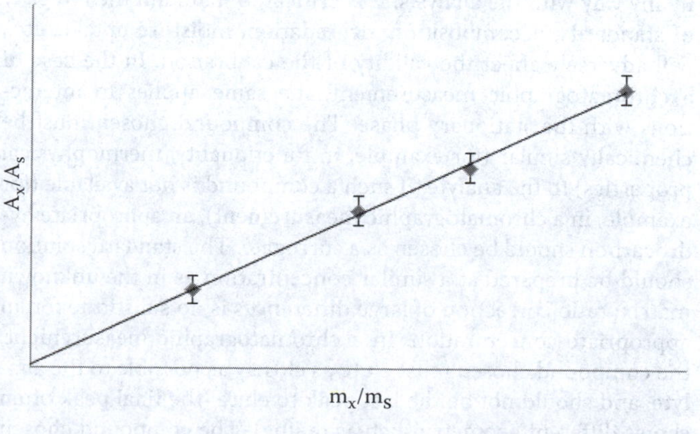

FIGURE 1e. An example of the multiple internal standard method. The ordinate (*y*) axis is the ratio of the response of the unknown analyte component, A_x, to the response of the chosen standard, A_s. The abscissa (*x*) axis is the ratio of mass of X to the mass of S for that standard mixture.

internal standard is added to both the unknown samples and the calibration standards.

Single Internal Standard

In practice, once the linearity is established for a given mixture, it is no longer necessary to use multiple standards, although this is the most precise method. Subsequent to the verification of linearity, one standard solution can be used to fix the slope, provided it is close in concentration to that of the target analyte. In this case, the mass of the unknown can be found from:

$$X/S = (A_x/A_s)(1/R), \tag{4}$$

where X is the mass of the unknown analyte in the sample, S is the mass of the added internal standard in the sample, A_x and A_s are the instrument responses (areas) of the unknown and internal standard, respectively. R is a ratio determined from the standard solution prepared with both X and S: (mass, unknown analyte/mass, internal standard)/(signal, unknown analyte/signal, internal standard) = R.

$$\left(\frac{\left(\dfrac{\text{mass, unknown analyte}}{\text{mass, internal standard}}\right)}{\left(\dfrac{\text{signal, unknown analyte}}{\text{signal, internal standard}}\right)}\right) = R \tag{5}$$

Since R is the slope of the calibration curve discussed above, once linearity is established, one solution suffices. There are many conditions that must be fulfilled in order to use the internal standard method, and it is rare that all of them can actually be met. Indeed, in practice, one tries to meet as many as possible, but those that are mandatory are italicized. The compound chosen *must not be present* already in the unknown. The compound chosen *must be separable from the analyte* present in the unknown. An exception occurs when an isotopically labeled standard is used, in conjunction with mass discrimination or radioactive counting detection. In a chromatographic measurement, this is typically at least baseline resolution, although this would be a minimally acceptable degree of separation. On the other hand, the unknown analyte peak and the internal standard peak should be close to each other (temporally) on the chromatogram. The compound chosen *must be miscible* with the solvent at the temperature of reagent preparation and measurement. The compound chosen must not react chemically with the sample or solvent, or interfere in any way with the analysis. It is critical to maintain the integrity of standards; decomposition, degradation, moisture uptake, etc., will adversely affect the validity of the calibration. In the case of a chromatographic measurement, the same applies to interactions with the stationary phase. The compound chosen must be chemically similar (for example, in functionality, thermophysical properties) to the analyte. If such a compound is not available (for example, in a chromatographic measurement), an appropriate hydrocarbon should be chosen as a surrogate. The standard solution should be prepared at a similar concentration as in the unknown matrix; ratio correction of large differences is no substitute for an appropriate concentration. In a chromatographic measurement, the compound chosen must elute as closely as possible to the analyte, and should not be the last peak to elute (the final peak often shows different geometry such as tailing). The compound chosen must be sufficiently nonvolatile to allow for storage as needed. When there is the potential for the unknown analyte to be lost by adsorption, absorption, or some other interaction with the matrix or container, a compound called a carrier is sometimes added in large excess. The carrier is similar, chemically and physically, to the unknown analyte, but easily separated from it. Its purpose is to saturate or season the matrix and prevent analyte loss.

Serial Dilution

Serial dilution is less a standardization method as it is a method of generating solutions to be used for standardizations. Nevertheless, its importance and utility, as well as the popularity of its application, warrants mention in this section. A serial dilution is the stepwise dilution of a substance, observant of a specified, constant progression, usually geometric (or logarithmic). One first prepares a known volume of stock solution of a known concentration, followed by withdrawing some small fraction of it to another container or vial. This subsequent container is then filled to the same volume as the stock solution with the same solvent or buffer. The process is then repeated for as many standard solutions as are desired. A ten-fold serial dilution could be 1 M, 0.1 M, 0.01 M, 0.001 M, etc. A ten-fold dilution for each step is called a logarithmic dilution or log-dilution, a 3.16 fold ($10^{0.5}$ fold) dilution is called a half-logarithmic dilution or half-log dilution, and a 1.78 fold ($10^{0.25}$ fold) dilution is called a quarter-logarithmic dilution or quarter-log dilution. In practice, the ten-fold dilution is the most common. The serial dilution procedure is not only used in chemical analysis but also in serological preparations in which cellular materials such as bacteria are diluted. A critical aspect of serial dilution is that the initial solution concentration must be prepared and determined with great care, since any mistake here will be propagated into all resulting solutions.

Traceability

Analytical measurements and certifications often contain a statement of traceability. Traceability describes the "result or measurement whereby it can be related to appropriate standards, generally international or national standards, through an unbroken chain of comparisons" (Ref. 9). Traceability typically includes the application of a reference material (RM) or a standard reference material (SRM) for instrument calibration before standardization for the analytes of interest. The **true value** of a measured quantity (τ) cannot typically be determined. The true value is defined as characterizing a quantity that is perfectly defined. It is an ideal value which could be arrived at only if all causes of measurement uncertainty were eliminated, and the entire population was sampled.

Uncertainty

As stated in Section 2, the result of a measurement is only an approximation or estimate of the true value of the measurand or quantity subject to measurement. In the determination of the combined standard uncertainty and ultimately the expanded uncertainty, it is critical to include the uncertainty of calibration in the process, as discussed above. The process of arriving at the uncertainty U_y of a quantity y that is based upon measured quantities $x_i,\ldots,x_z$ is called the propagation of uncertainty. A full discussion of propagation of uncertainty is beyond the scope of this section; a simplified prescription, in the form of general and specific formulae, is provided here. In general, the propagated random uncertainty in y can be determined from:

$$U_y = \sqrt{\left(\frac{\partial y}{\partial x_i}Ux_i\right)^2 + \cdots\left(\frac{\partial y}{\partial x_z}Ux_z\right)^2}$$

This approach can be used when the uncertainties are random (not systematic), are relatively small, and are independent or uncorrelated (that is, in the absence of covariance). Relatively large uncertainties (such as those approaching the magnitude of the measurand itself) cannot be treated with this approach, especially if the measurand is a nonlinear function of the measured quantity. Note that by convention the use of upper case U_y denotes the expanded uncertainty (which is the uncertainty multiplied by a coverage factor *k*; for the 95% confidence level, the coverage factor is 2).

It is possible to reduce this general formulation to more specific formulae in the cases of common mathematical operations. These are provided in Table 1.

References

1. Chalmers, R. A., *Chapter 2: Standards and Standardization in Chemical Analysis*, Vol. 3, Elsevier, Amsterdam, 1975.
2. Danzer, K., and Currie, L. A., *Pure Appl. Chem.* 70, 993, 1998.
3. Danzer, K., Otto, M., and Currie, L. A., *Pure Appl. Chem.* 76, 1215, 2004.
4. Woodget, B. W., and Cooper, D., *Samples and Standards, Analytical Chemistry by Open Learning*, John Wiley & Sons, Chichester, 1987.
5. Gy, P., *Sampling for Analytical Purposes*, John Wiley & Sons, Chichester, 1998.
6. Vitt, J. E., and Engstrom, R. C., *J. Chem. Educ.* 76, 99, 1999.
7. Horowitz, W., *Pure Appl. Chem.* 62, 1193, 1990.
8. Grob, R. L., *Modern Practice of Gas Chromatography*, Wiley Interscience, New York, 1995.
9. Inczedy, J., Lengyel, T., and Ure, A. M., *Compendium of Analytical Nomenclature, Third Edition*, International Union of Pure and Applied Chemistry, 1997

Table 1. Specific Formulae for the Propagation of Random, Independent Uncertainty

Measurand argument	Arithmetic uncertainty formula
y (where y is a counted random event over a time interval)	$U_y = \sqrt{y}$
$y = A \times x$ (where A is a constant with no uncertainty)	$U_y = \lvert A \rvert \times U_x$
$y = x_1 + x_2$	$U_y = \sqrt{U_{x_1}^2 + U_{x_2}^2}$
$y = x_1/x_2$ $y = x_1 \times x_2$	$\frac{U_y}{\lvert y \rvert} = \sqrt{\left(\frac{U_{x_1}}{x_1}\right)^2 + \left(\frac{U_{x_2}}{x_2}\right)^2}$
$y = (x_1 \times x_2)/x_3$	$\frac{U_y}{\lvert y \rvert} = \sqrt{\left(\frac{U_{x_1}}{x_1}\right)^2 + \left(\frac{U_{x_2}}{x_2}\right)^2 + \left(\frac{U_{x_3}}{x_3}\right)^2}$
$y = \log(x)$	$U_y = \left(\frac{1}{2.303}\right)\left(\frac{U_x}{x}\right)$
$y = \ln(x)$	$U_y = \frac{U_x}{x}$
$y = e^x$	$U_y = y \times U_x$
$y = x^a$	$\frac{U_y}{\lvert y \rvert} = \lvert a \rvert \times \left(\frac{U_x}{x}\right)$
$y = 10^x$	$\frac{U_y}{\lvert y \rvert} = 2.303 \times U_x$

Analytical

FIGURES OF MERIT

Thomas J. Bruno

In the interpretation of results from a particular technique, one often must evaluate the performance of the measurement on the basis of objective measures that are called figures of merit. The following table provides a description of the more common figures of merit used in analytical chemistry. Even more important than the evaluation of performance is the validation of metrology and the assessment of validation. The same terms used as figures of merit are critical in the validation process. Because many standardization bodies will issue their own definitions regarding these terms, it is not practical to be inclusive; the reader must be aware that other terms may sometimes be substituted for those used here.

References

1. General Requirements for the Competence of Testing and Calibration Laboratories, ISO/IEC 17025, International Organization for Standardization, 2005.
2. General Chapter 1058 Analytical Instrument Qualification, USP 32 – NF 27, U.S. Pharmacopeial Convention. 2009.

Figures of Merit Used in Analytical Chemistry

Parameter or metric	Definition	Comments
Accuracy	The deviation of a measured value from the true value (τ), which cannot typically be determined, but which is often approached by use of a standard method under standard conditions.	Can be expressed as: $$\text{accuracy} = (x_{av} - \tau)/\tau,$$ where x_{av} is the mean of a series of measurements.
Precision	The reproducibility of a series of replicate measurements, typically under the same operating conditions in a relatively short period of time.	High precision does not imply high accuracy; reproducibility and repeatability often are used interchangeably, but the former typically refers to precision between two instruments or laboratories.
Sensitivity	Typically the ability to distinguish between small differences in concentration between samples at a desired confidence level.	Often specified by response to a *de facto* or consensus standard; a simple measure is the slope of a calibration curve adjusted for recovery in the sampling process.
Limit of Detection (LOD)	The lowest measureable concentration of an analyte by use of a particular metrology at a desired confidence level.	Typically defined as a response that is some multiplicative factor of the noise level; specified signal-to-noise ratios of 2 to 4 are typical; as with sensitivity, must adjust for recovery in sample preparation processes.
Linear Dynamic Range (LDR)	The linear range of a calibration curve obtainable with a particular metrology and analyte.	Within the LDR, the difference in response for two concentrations of a given compound is proportional to the difference in concentration of the two samples.
Selectivity (α)	The ability to distinguish the signal of an analyte from the signal of interferences, the matrix or degradants.	Typically expressed as a ratio, such as the ratio of capacity factors in chromatography; sometimes called specificity.
Analysis Speed	The time required for sample preparation and measurement.	Sometimes necessary to include data processing time.
Throughput	The number of samples that can be run in a given time.	Related to ease of automation, below.
Ease of Automation	The ability of a measurement to be performed without operator attention.	Related to throughput, above, since high throughput typically requires automation and minimal operator attention.
Ruggedness	The response of an instrument or technique to adverse conditions.	Adverse conditions include extremes in temperature, humidity, dust; includes rough handling; closely related to reproducibility, since it carries similar information content; distinguished from robustness, which is the ability of an instrument to remain unaffected by a small but intentional change in operating parameter, and still perform within specification.
Portability	The ability to use an instrument in other than a fixed location or installation.	Can include the ability to use a device in the field in addition to the laboratory.
Environmental Acceptability	The efficiency in terms of low waste generation, low power consumption.	Also described as sustainability, and often focused on the minimization of use of hazardous substances.
Economics	The sum of costs required to operate the sample preparation, analysis and data processing steps of a measurement.	Typically includes the costs of equipment, supplies, labor, utilities, and insurance; labor cost must consider the skill level required of personnel.

MASS- AND VOLUME-BASED CONCENTRATION UNITS

Thomas J. Bruno and Paris D. N. Svoronos

A variety of concentration units are used in analytical chemistry, and the most common are provided in this table (Ref. 1). The reference below provides additional details.

Reference

1. Bruno, T. J., and Svoronos, P. D. N., *CRC Handbook of Basic Tables for Chemical Analysis*, 3rd Edition, CRC Press, Boca Raton, FL, 2011.

Parts per Million

Parts per Million		Percent
1 ppm	=	0.0001 %
10 ppm	=	0.001 %
100 ppm	=	0.01 %
1000 ppm	=	0.1 %
10 000 ppm	=	1.0 %
100 000 ppm	=	10.0 %
1 000 000 ppm	=	100.0 %

Parts per Billion

Parts per Billion		Percent
10	=	0.000 001 %
100	=	0.000 01 %
1000	=	0.0001 %
10 000	=	0.001 %
100 000	=	0.01 %
1 000 000	=	0.1 %

Parts per Trillion

Parts per Trillion		Percent
100	=	1×10^{-8} %
10 000	=	0.000 001 %
1 000 000	=	0.0001 %
100 000 000	=	0.01 %

Because the mass of one liter of water is approximately one kg, mg/L units of dilute aqueous solution are nearly equal to ppm units. The precise equivalence is obtained by dividing by the density, ρ:

$$\text{ppm} = (\text{mg/L})/\rho$$

where the solution density, ρ, is in g cm^{-3}. Some sources will substitute specific gravity for density in the above equation. The specific gravity is the ratio of the solution density to that of the density of pure water at 4 °C. Since the density of pure water at 4 °C is very nearly 1 g cm^{-3}, the specific gravity is numerically equal (within an uncertainty of 25 ppm) to the solution density when the latter is expressed in units of g cm^{-3}.

Concentration Units Nomenclature

The following table provides guidance in the use of base-10 concentration units (presented in the three preceding tables), since there are differences in usage worldwide.

Number	Number of Zeros	Name (Scientific Community)	Name (United Kingdom, France, Germany)
1000.	3	thousand	thousand
1 000 000.	6	million	million
1 000 000 000.	9	billion	milliard, or thousand million
1 000 000 000 000.	12	trillion	billion
1 000 000 000 000 000.	15	quadrillion	thousand billion

Molar-Based Concentration Units

Molarity*, M:	(moles of solute)/(liters of solution)
Molality, *m*:	(moles of solute)/(kilograms of solvent)
Normality, N:	(equivalents* of solute)/(liters of solution)
Formality, F:	(moles of solute)/(kilograms of solution)

* This unit is temperature dependent.

To convert from ppm to formality units:

$F = \text{ppm}/(1000\ RMM)$, where RMM is the relative molecular mass of the solute i

To convert from ppm to molality units:

$$m = [\text{ppm}/(1000\ RMM)]\ [1/(1 - tds/1\,000\,000)]$$

where tds is the total dissolved solids in ppm in the solution

To convert from ppm to molarity units:

$$M = [\text{ppm}/(1000\ RMM)]\ \rho$$

where ρ is the solution density

Analytical

DETECTION OF OUTLIERS IN MEASUREMENTS

Thomas J. Bruno and Paris D. N. Svoronos

The field of outlier detection and treatment is considerable, and a rigorous mathematical discussion is well beyond any treatment that is possible here. Moreover, the practice in the treatment of analytical results is usually simplified, since the number of observations is often not very large. The two most common methods used by analysts to detect outliers in measured data are versions of the Q-test (Refs. 1–3, 6) and Chauvenet's criterion (Refs. 4–6), both of which assume that the data are sampled from a population that is normally distributed.

References

1. Dean, R. B., and Dixon, W. J., *Anal. Chem.* 23, 636, 1951.
2. Day, R. A., and Underwood, A.L., *Quantitative Analysis*, 6th Edition, Prentice Hall, Englewood Cliffs, NJ, 1991.
3. Efstathiou, C. E., *Dixon's Q-test: Detection of a single outlier, http://www.chem.uoa.gr/applets/AppletQtest/Text_Qtest2.htm*, Laboratory of Analytical Chemistry, Department of Chemistry, National and Kapodistrian University of Athens, 2008.
4. Taylor, J. R., *An Introduction to Error Analysis*, 2nd edition, University Science Books, Sausalito, CA, 1997.
5. Benziger, J. B., and Aksay, I.A., *Notes on Data Analysis, http://www.princeton.edu/~che346/Notes/Analysis.pdf*, Department of Chemical Engineering, Princeton University, 1999.
6. Bruno, T. J., and Svoronos, P. D. N., *CRC Handbook of Basic Tables for Chemical Analysis*, 3rd Edition, CRC Press, Boca Raton, FL, 2011.

Q-Test

To perform the Q-test, one calculates the Q value given by:

$$Q = Q_{gap}/R$$

where Q_{gap} is the difference between the suspected outlier and the measured value closest to it, and R is the range of all the measured values in the data set. One then compares the calculated Q value with the critical Q values in the following table:

Number of Observations	Q_{crit}, 90% Confidence Level	Q_{crit}, 95% Confidence Level	Q_{crit}, 99% Confidence Level
3	0.941	0.970	0.994
4	0.765	0.829	0.926
5	0.642	0.710	0.821
6	0.560	0.625	0.740
7	0.507	0.568	0.680
8	0.468	0.526	0.634
9	0.437	0.493	0.598
10	0.412	0.466	0.568

If the calculated value of Q is greater than the appropriate value of Qcrit, then the value is a suspected outlier.

Chauvenet's Criterion

To perform Chauvenet's test on a set of measurements, one first must calculate the mean and standard deviation of the data. Then one calculates:

$$\tau = (x_i - x_{ave})/\sigma$$

where x_i is the suspected outlier, x_{ave} is the mean of all the measurements, and σ is the standard deviation. One then compares the calculated value of τ with τ_{crit} in the following table:

Number of Observations, N	τ_{crit}	Number of Observations, N	τ_{crit}
5	1.65	25	2.33
6	1.73	50	2.57
7	1.81	100	2.81
8	1.86	150	2.93
9	1.91	200	3.02
10	1.96	500	3.29
15	2.12	1000	3.48
20	2.24		

If the calculated value of τ is greater than the value of τ_{crit}, then the value is a suspected outlier.

For numbers of observations between those given in the table, especially for a large number of observations, one may use the following plot to estimate the value of Chauvenet's τ_{crit}:

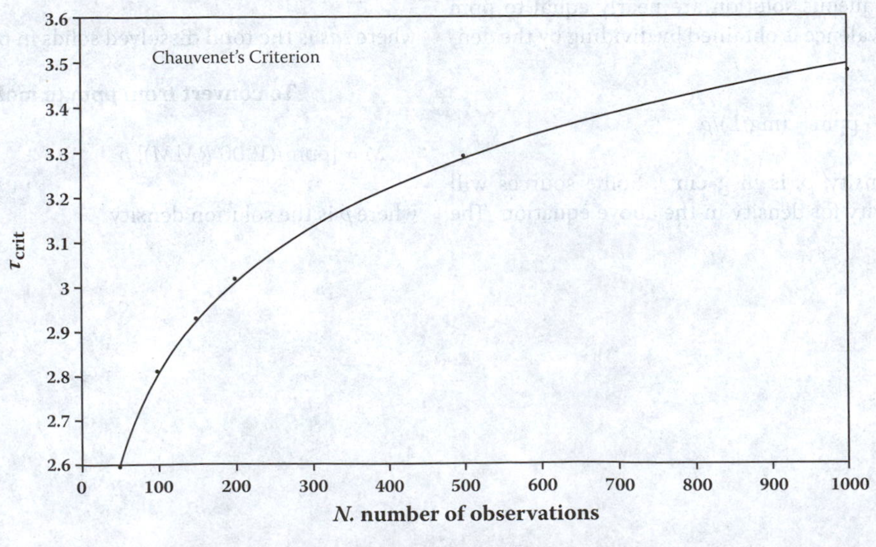

Analytical

PROPERTIES OF CARRIER GASES FOR GAS CHROMATOGRAPHY

The following is a list of carrier gases sometimes used in gas chromatography, with properties relevant to the design of chromatographic systems. All data refer to normal atmospheric pressure (101.325 kPa). Additional properties related to the carrier gas, including split ratio, pressure drop in capillary columns, the Martin-James compressibility factor, and the Giddings plate height correction factor can be found in Ref. 2.

M_r: Molecular weight (relative molar mass)
ρ_{25}: Density at 25 °C in g L^{-1}
c_p: Specific heat at 25 °C in J g^{-1} °C^{-1}
λ: Thermal conductivity in mW m^{-1} °C^{-1}
η: Viscosity in μPa s (equal to 10^{-3} cP)

References

1. Lide, D. R., and Kehiaian, H. V., *CRC Handbook of Thermophysical and Thermochemical Data*, CRC Press, Boca Raton, FL, 1994.
2. Bruno, T. J., and Svoronos, P. D. N., *CRC Handbook of Basic Tables for Chemical Analysis, 3rd Edition*, CRC Press, Boca Raton, FL, 2011.

Properties of Carrier Gases for Gas Chromatography

Gas	M_r	ρ_{25}/g L^{-1}	c_p(25 °C)/ J g^{-1} °C^{-1}	λ(25 °C)/ mW m^{-1} °C^{-1}	η(25 °C)/ μPa s	λ(250 °C)/ mW m^{-1} °C^{-1}	η(250 °C)/ μPa s
Hydrogen	2.016	0.0824	14.3	185.9	8.9	280	13.1
Helium	4.003	0.1636	5.20	154.6	19.9	230	29.5
Argon	39.948	1.6329	0.521	17.8	22.7	27.7	35.3
Nitrogen	28.014	1.1449	1.039	25.9	17.9	39.6	26.8
Oxygen	31.998	1.3080	0.919	26.2	20.7	42.6	31.8
Carbon monoxide	28.010	1.1449	1.039	24.8	17.8	40.7	26.5
Carbon dioxide	44.010	1.7989	0.843	16.7	14.9	35.5	24.9
Sulfur hexafluoride	146.055	5.9696	0.664	13.1	28.1	15.3	24.8
Methane	16.043	0.6556	2.23	34.5	11.1	75.0	17.6
Ethane	30.069	1.2291	1.75	20.9	9.4	57.7	15.5
Ethylene	28.053	1.1465	1.53	20.5	10.3	53.8	17.2
Propane	44.096	1.8025	1.67	17.9	8.3	49.2	14.0

COMMON SYMBOLS USED IN GAS AND LIQUID CHROMATOGRAPHIC SCHEMATIC DIAGRAMS

Thomas J. Bruno

The literature of gas and liquid chromatography frequently contains schematic diagrams that depict analytical apparatus and peripherals. The interpretation of such diagrams can be facilitated by the following graphics that show the most common symbols used to represent chromatographic components.

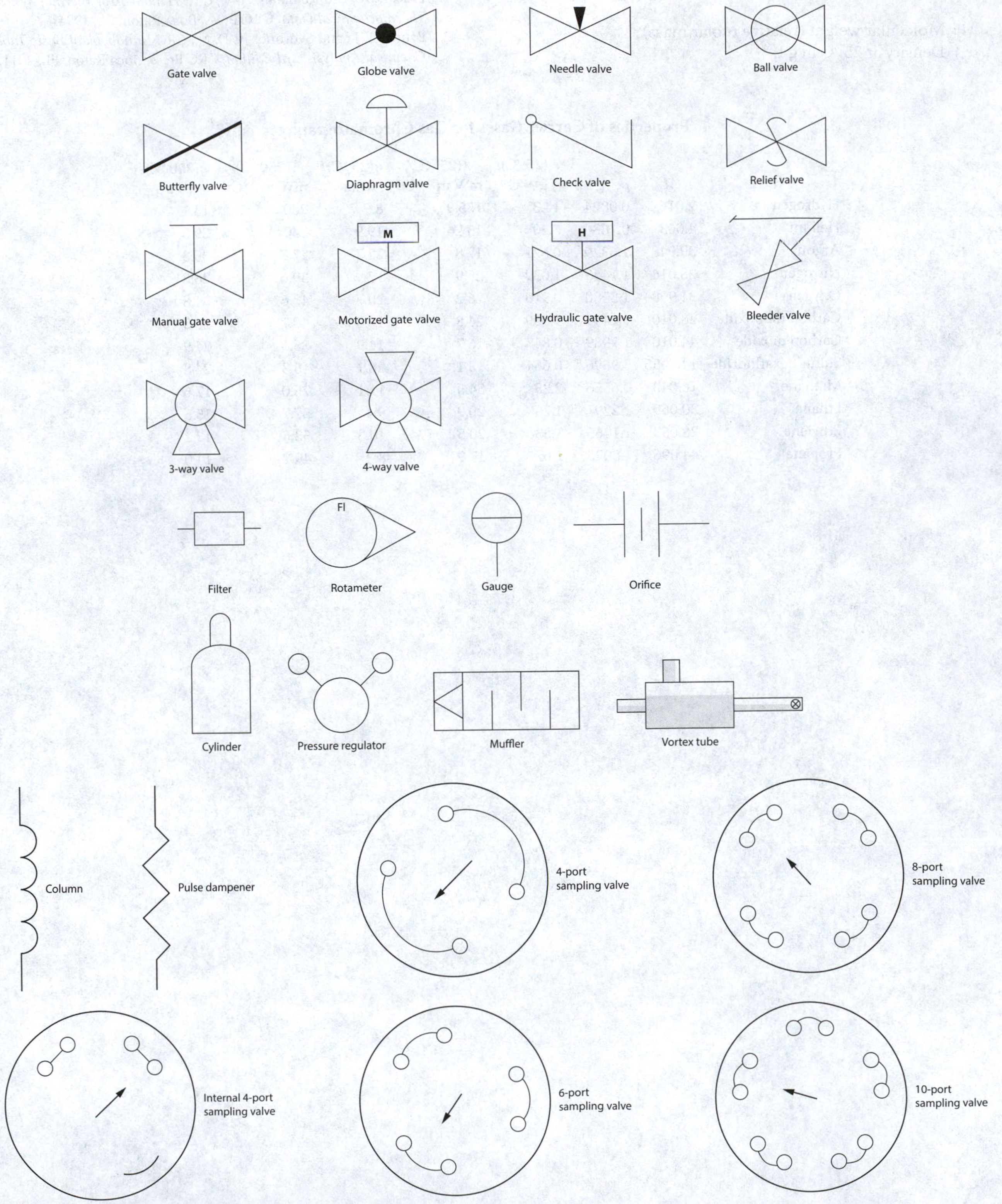

STATIONARY PHASES FOR POROUS-LAYER OPEN TUBULAR COLUMNS

Thomas J. Bruno and Paris D. N. Svoronos

The practical application of solid adsorbents in capillary chromatography typically makes use of porous-layer open tubular (PLOT) columns. Early capillary PLOT columns suffered from detector spiking due to particulates becoming dislodged and entering the detector. This was resolved by placing a short length of a polymeric column (usually polydimethyl siloxane) as a particle trap before the detector; however, recent advances in manufacturing methods have resulted in nearly complete binding of particles on the inside of the column. In this table, several of the more common PLOT column stationary phases are listed, along with the separations that may be affected and some additional information. The maximum temperatures listed represent the point of severe resolution loss. The materials are often chemically stable to much higher temperatures. The user should also be aware that the adsorption of water during use will often change retention characteristics dramatically, sometimes resulting in a reversal of positions of adjacent peaks. This can result from traces of water in the carrier gas. Due to surface adsorption of solutes, some experimentation with temperature may be necessary to prevent tailing or to avoid statistical correlation (or a propagating error) among replicate analyses (Ref. 5). Note that PLOT columns have been used in HPLC for the separation of large molecules such as proteins; this table is limited to smaller molecules.

References

1. Jeffery, P. G., and Kipping, P. J., *Gas Analysis by Gas Chromatography*, Pergamon Press, Oxford, 1972.
2. Cowper, C. J., and DeRose, A. J., *The Analysis of Gases by Chromatography*, Pergamon Press, Oxford, 1983.
3. Breck, D. W., *Zeolite Molecular Sieves*, John Wiley & Sons, New York, 1973.
4. Poole, C. F., *The Essence of Chromatography*, Elsevier, Amsterdam, 2003.
5. Bruno, T. J., and Svoronos, P. D. N., *CRC Handbook of Basic Tables for Chemical Analysis*, 3rd Ed., CRC Press, Boca Raton, FL, 2011.

Phase	Max. temp. /°C	Separation affected	Notes
Silica gel	250	H_2, Air, CO, C_1 to C_4 normal hydrocarbons, alkenes and alkynes, inorganic gases, volatile ethers	Very hydrophilic; requires activation; can be unpredictable; largely replaced by porous polymers; bonded versions are suitable for use with GC-MS, because of the absence of particles.
Alumina, deactivated with KCl	250	C_1 to C_8 hydrocarbons, especially useful for resolution of propadiene and butadiene from ethylene and propylene	Least polar of the alumina phases; lowest retention of olefins relative to the corresponding paraffin; specified in many standard methods.
Alumina, deactivated with Na_2SO_4	250	C_1 to C_8 hydrocarbons, resolves acetylene from butane and propylene from isobutane	Medium and high polarity phases are available among the alumina phases; specified in many standard methods.
Cyclodextrins	260	Light hydrocarbons (C_1 to C_{10}) and halocarbons	α- and β-cyclodextrins have been used; care should be taken with halocarbon analysis, due to the potential of HF contamination of the sample. See the information on cyclodextrin phases in Reference 5.
Styrene – divinylbenzene	250	C_1 to C_3 hydrocarbons; paraffins up to C_{12}; CO from air, ethers, sulfur gases, water	See the information on porous polymers in Reference 5.
Divinylbenzene ethylene glycol dimethacrylate	280	C_1 to C_7 hydrocarbon isomers; CO_2, CH_4, amines, common solvents, alcohols, aldehydes, ketones	More polar than styrene – divinyl benzene phases.
Molecular sieve, 5A	350	Air and light gas analysis; H_2, O_2, N_2, (CH_4, CO, NO, SF_6 coelute); thick film phase can resolve Ar from O_2 at 35 °C; deuterated hydrocarbons	Synthetic calcium alumino-silicate (zeolite) having an effective pore diameter of 0.5 nm CO_2 is adsorbed strongly; 5A usually gives the best results of all synthetic zeolites; thick and thin film columns are available.
Molecular sieve, 13X	350	Same separations as those performed on 5A, but with C_1 to C_4, alkanes, alkenes, and alkynes being separated as well	Sodium alumino-silicate (zeolite), having a larger pore size than 0.5 nm, thus producing lower retention times and less resolution; 28.6% (mass/mass) water capacity.
Monolithic carbon	350	C_1 to C_5 hydrocarbon isomers; acetylene in ethylene; methane	Phase consists of a bonded carbon monolith; historically more suitable for use with GC-MS, because of the absence of particles.

COOLANTS FOR CRYOTRAPPING

Thomas J. Bruno and Paris D. N. Svoronos

The following table provides fluids (in some cases mixtures) that can be used to chill trapping sorbents (or any cold trap) for the purge and trap sampling of solutes (Refs. 1-4). In each case the ratio is mass/mass. Practical use of these methods can require optimization and proper insulation.

References

1. Gordon, A .J., and Ford, R. A., *The Chemists Companion – A Handbook of Practical Data, Techniques and References*, Wiley Interscience, New York, 1972.
2. Bruno, T. J., Chromatographic Cryofocusing and Cryotrapping with the Vortex Tube, *J. Chromatogr. Sci.* 32, 112, 1994.
3. Bruno, T. J., Simple, Quantitative Headspace Analysis by Cryoadsorption on a Short Alumina PLOT Column, *J. Chromatogr. Sci.* 47, 1, 2009.
4. Bruno, T. J., and Svoronos, P. D. N., *CRC Handbook of Basic Tables for Chemical Analysis, 3rd Edition*, CRC Press, Boca Raton, FL, 2011.

Temperature Reached by Various Coolants Used for Cryotrapping

Coolant	Temperature/°C
Crushed ice + sodium chloride (3 : 1)	-21
Crushed ice + calcium chloride (1.2 : 2)	-39
Vortex tube	-40
Liquid nitrogen + butylamine (slush)	-50
Crushed ice + calcium chloride (1.4 : 2)	-55
Liquid nitrogen + chloroform (1 : 1)	-63
Liquid nitrogen + *tert*-butylamine (slush)	-68
Dry ice	-78
Liquid nitrogen + acetone (slush)	-95
Liquid nitrogen + ethanol (slush)	-120
Liquid nitrogen + methylcyclohexane (slush)	-126
Liquid nitrogen + pentane (slush)	-131
Liquid nitrogen + 1,5-hexadiene (slush)	-141
Liquid nitrogen + isopentane (slush)	-160
Liquid argon	-186
Liquid nitrogen	-196

Analytical

PROPERTIES OF COMMON CROSS-LINKED SILICONE STATIONARY PHASES

Thomas J. Bruno and Paris D. N. Svoronos

Modern gas chromatography is most often performed with high efficiency capillary (open tubular) columns coated with a cross-linked polymeric stationary phase. We provide chromatographic data on the two most common cross-linked phases (Ref. 1). These data are useful for the interpretation of chromatograms, and for quality control and assurance. The retention indices presented in these tables were measured at 120 °C isothermally. Retention indices are temperature dependent; the temperature dependence of the Kovats indices have been studied for many compounds (Ref. 2). For more extensive information on other cross-linked phases, other silicone phases, mesogenic phases, and solid sorbents, the reader is advised to consult Ref. 1.

References

1. Bruno, T. J., and Svoronos, P. D. N., *CRC Handbook of Basic Tables for Chemical Analysis*, 3rd Edition, CRC Press, Boca Raton, FL, 2011.
2. Bruno, T. J., Wertz, K. H., and Caciari, M., *Anal. Chem.* 68, 1347, 1996.

Phase: 5% Phenyl Dimethylpolysiloxane

Temperature Ranges:

–60 °C to 325 °C isothermally, –60 °C to 350 °C programmed for < 0.32 mm I.D. columns
–60 °C to 300 °C isothermally, –60 °C to 320 °C programmed for 0.53 mm I.D. columns
–60 °C to 260/280 °C for >2.0 µm films

Similar Phases:

DB-5, Ultra-2, SPB-5, CP-Sil 8 CB, Rtx-5, BP-5, OV-5, 007-2 (MPS-5), SE-52, SE-54, XTI-5, PTE-5, HP-5MS, ZB-5, AT-5, MDN-5

Notes: This phase is probably the most commonly used stationary phase in gas chromatography, because it combines boiling point separation with a minor contribution of a specific interaction; typically used as the first phase in any method development; versatile for hydrocarbons and more polar compounds.

Probe compound	McReynolds constant	McReynolds code	Kovats retention index
Hexane			600
1-Butanol	66	y'	656
Benzene	31	x'	684
2-Pentanone	61	z'	688
Heptane			700
1,4-Dioxane	64	L	718
2-Methyl-2-pentanol	41	H	731
1-Nitropropane	93	u'	745
Pyridine	62	s'	761
Octane			800
Iodobutane	22	J	840
2-Octyne	35	K	876
Nonane			900
Decane			1000

Phase: Dimethylpolysiloxane

Temperature Ranges:

–60 °C to 325 °C for normal operations, periodic operation to 350 °C can be used to facilitate column clean-up
–60 °C to 260/280 °C for >2.0 µm films

Similar Phases:

DB-1, OV-1, HP-1, DB-1ms, HP-1ms, Rtx-1, Rtx-1ms, CP-Sil 5 CB Low Bleed/MS, MDN-1, AT-1

Notes: Useful for the separation of hydrocarbons, pesticides, PCBs, phenols, sulfur compounds, flavors and fragrances, and some amines; columns are typically stable and low bleed; a good all-purpose column used to begin method development protocols.

Probe compound	McReynolds constant	McReynolds code	Kovats retention index
Hexane			600
1-Butanol	54	y'	644
Benzene	16	x'	669
2-Pentanone	44	z'	671
Heptane			700
1,4-Dioxane	46	L	700
2-Methyl-2-pentanol	31	H	721
1-Nitropropane	62	u'	714
Pyridine	44	s'	743
Octane			800
Iodobutane	3	J	821
2-Octyne	23	K	864
Nonane			900
Decane			1000

DETECTORS FOR GAS CHROMATOGRAPHY

Thomas J. Bruno and Paris D. N. Svoronos

The following table provides some comparative data to aid in interpreting results from the more common detectors applied to capillary and packed-column gas chromatography (Refs. 1–8). For more detailed information regarding operation and interpretation of results, see Ref. 8.

References

1. Hill, H.H., and McMinn, D., eds., *Detectors for Capillary Chromatography*, John Wiley & Sons, New York, 1992.
2. Buffington, R., and Wilson, M. K., *Detectors for Gas Chromatography — a Practical Primer*, Hewlett Packard Corp., Avondale, PA, 1987.
3. Buffington, R., *GC-Atomic Emission Spectroscopy using Microwave Plasmas*, Hewlett Packard Corp., Avondale, PA, 1988.
4. Liebrand, R. J., Ed., *Basics of GC/IRD and GC/IRD/MS*, Hewlett Packard Corp., Avondale, PA, 1993.
5. Bruno, T. J., *Sep. Purif. Method* 29, 63, 2000.
6. Bruno, T. J., *Sep. Purif. Method* 29, 27, 2000.
7. Sevcik, J., Detectors in Gas Chromatography, *Journal of Chromatography Library*, Vol. 4, Elsevier, Amsterdam, 1976.
8. Bruno, T. J., and Svoronos, P. D. N., *CRC Handbook of Basic Tables for Chemical Analysis*, 3rd Edition, CRC Press, Boca Raton, FL, 2011.

Detector	Limit of detection	Linearity	Selectivity	Comments
Thermal conductivity detector (TCD, katharometer)	1×10^{-10} g propane (in helium carrier gas)	1×10^{6}	Universal response, concentration detector	Ultimate sensitivity depends on analyte thermal conductivity difference with carrier gas Since thermal conductivity is temperature dependent, response depends on cell temperature Wire selection depends on chemical nature of analyte Helium is recommended as carrier and make-up gas When analyzing mixtures containing hydrogen, one can use a mixture of 8.5% (mass/mass) hydrogen in helium
Gas density balance detector (GADE)	1×10^{-9} g: H_2 with SF_6 as carrier gas	1×10^{6}	Universal response, concentration detector	Response and sensitivity are based on difference in relative molecular mass of analyte with that of the carrier gas; approximate calibration can be done on the basis of relative density The sensing elements (hot wires) never touch sample, thus making GADE suitable for the analysis of corrosive analytes such as acid gases; gold sheathed tungsten wires are most common Best used with SF_6 as a carrier gas, switched between nitrogen when analyses are required Detector can be sensitive to vibrations, and should be isolated on a cushioned base
Flame ionization detector (FID)	1×10^{-11} g to 1×10^{-10} g	1×10^{7}	Organic compounds with C–H bonds	Ultimate sensitivity depends on the number of C–H bonds on analyte Nitrogen is recommended as carrier gas and make-up gas to enhance sensitivity Sensitivity depends on carrier make-up and jet gas flow rates Column must be positioned 1 mm to 2 mm below the base of the flame tip Jet gases must be of high purity
Nitrogen-phosphorus detector (NPD, thermionic detector, alkali flame ionization detector)	4×10^{-13} g to 1×10^{-11} g of nitrogen compounds 1×10^{-13} g to 1×10^{-12} g of phosphorus compounds	1×10^{4}	10^5 to 10^6 by mass selectivity of N or P over carbon	Does not respond to inorganic nitrogen such as N_2 or NH_3 Jet gas flow rates are critical to optimization Response is temperature dependent Used for trace analysis only, and is very sensitive to contamination Avoid use of phosphate detergents or leak detectors Avoid tobacco use nearby Solvent-quenching is often a problem
Electron capture detector (ECD)	5×10^{-14} g to 1×10^{-12} g	1×10^{4}	Selective for compounds with high electron affinity, such as chlorinated organics; concentration detector	Sensitivity depends on number of halogen atoms on analyte Used with nitrogen or argon/methane (95/5, mass/mass) carrier and make-up gases Carrier and make-up gases must be pure and dry The radioactive ^{63}Ni source is subject to regulation and periodic inspection
Flame photometric detector (FPD)	2×10^{-11} g of sulfur compounds 9×10^{-13} g of phosphorous compounds	1×10^{3} for sulfur compounds 1×10^{4} for phosphorous compounds	10^5 to 1 by mass selectivity of S or P over carbon	Hydrocarbon quenching can result from high levels of CO_2 in the flame Self-quenching of S and P analytes can occur with large samples Gas flows are critical to optimization Response is temperature dependent Condensed water can be a source of window fogging and corrosion

Detector	Limit of detection	Linearity	Selectivity	Comments
Photoionization detector (PID)	1×10^{-12} g to 1×10^{-11} g	1×10^{7}	Depends on ionization potentials of analytes	Used with lamps with energies of 10.0 eV to 10.2 eV Detector will have response to ionizable compounds such as aromatics and unsaturated organics, some carboxylic acids, aldehydes, esters, ketones, silanes, iodo- and bromoalkanes, alkylamines and amides, and some thiocyanates
Sulfur chemiluminescence detector (SCD)	1×10^{-12} g of sulfur in sulfur compounds	1×10^{4}	10^{7} by mass selectivity of S over carbon	Equimolar response to all sulfur compounds to within 10% Requires pure hydrogen and oxygen combustion gases Instrument generates ozone *in situ*, which must be catalytically destroyed at detector outlet Catalyst operates at 950 °C to 975 °C Detector operated at reduced pressure (10^{-3} Pa)
Electrolytic conductivity detector (ECD, Hall detector)	10×10^{-13} g to 1×10^{-12} g of chlorinated compounds: 2×10^{-12} g of sulfur compounds: 4×10^{-12} g of nitrogen compounds	1×10^{6} for chlorinated compounds, 10^{4} for sulfur and nitrogen compounds	10^{6} by mass selectivity of Cl over carbon, 10^{5} to 10^{6} by mass selectivity of S and N over carbon	Only high purity solvents should be used Carbon particles in conductivity chamber can be problematic Frequent cleaning and maintenance is required Often used in conjunction with a photoionization detector For chlorine, use hydrogen as the reactant gas and 1-propanol as the electrolyte For nitrogen or sulfur, hydrogen or oxygen can be used as reactant gas, and water or methanol as the electrolyte Ultrahigh purity reactant gases are required
Ion mobility detector (IMD)	1×10^{-12} g	1×10^{3} to 1×10^{4}	10^{3}	Amenable to use in handheld instruments Linear dynamic range of 10^{3} for radioactive sources and 10^{5} for photoionization sources Selectivity depends on mobility differences of ions Has been used for a wide variety of compounds including amino acids, halogenated organics, explosives The radioactive ^{63}Ni source is subject to regulation and periodic inspection
Mass selective detector (MSD, mass spectrometer, MS)	1×10^{-11} g (single ion monitoring); 1×10^{-8} g (scan mode)	1×10^{5}	Universal	Single quadrupole, multiple quadrupole, ion trap, time of flight and magnetic sector instruments available (see separate table entitled: "Varieties of Hyphenated Gas Chromatography with Mass Spectrometry") Must operate under moderate vacuum (1×10^{-4} Pa) Requires a molecular jet separator to operate with packed columns Amenable to library searching for qualitative identification Requires tuning of electronic optics over the entire m/e range of interest
Infrared detector (IRD)	1×10^{-9} g of a strong infrared absorber	1×10^{3}	Universal for compounds with mid-infrared active functionality	A costly and temperamental instrument that requires high purity carrier gas, a nitrogen purge of optical components (purified air will, in general, not be adequate) Must be isolated from vibrations Presence of carbon dioxide is a typical impurity band at 2200 cm^{-1} to 2300 cm^{-1} Requires frequent cleaning and optics maintenance Amenable to library searching for qualitative identification
Atomic emission detector (AED)	1×10^{-13} g to 2×10^{-11} g of each element	1×10^{3} to 1×10^{4}	10^{3} to 10^{5}, element to element	Requires the use of ultra high purity carrier and plasma gases Plasma produced in a microwave cavity operated at 2450 MHz Scavenger gases (H_2, O_2) are used as dopants Photodiode array is used to detect emitted radiation

VARIETIES OF HYPHENATED GAS CHROMATOGRAPHY WITH MASS SPECTROMETRY

Thomas J. Bruno

Because of the very powerful and common application of mass spectrometry with gas chromatography, the various technologies warrant more detailed consideration (Refs. 1 to 4). The following table provides basic information on the capabilities and applicability of the most common approaches. Clearly, the use of the single quadrupole is by far the most prevalent and economical method, but the other techniques are important and advantageous. We exclude methods that are typically not interfaced with gas chromatographic separations and that are highly specific in research settings (such as ion cyclotron mass spectrometry, and combined quadrupoles beyond the triple quad, and the various hybrid sector-quad-time of flight-ion trap combinations) that are used in fundamental ion chemistry research. We also exclude magnetic sector instruments that are seldom used with chromatography.

References

1. March, R. E., and Todd, J. F., *Quadrupole Ion Trap Mass Spectrometry*, Wiley-Intersceince, New York, 2005.
2. Portoles, T., Sancho, J. V., Hernandez, A., Newton, A., and Hancock, P., Potential of Atmospheric Chemical Ionizationsource in GC-QTOF for Pesticide Residue Analysis, *J. Mass Spectrom.* 45, 926, 2010.
3. Busch, K. L., Glish, G. L., and McLuckey, S. A., *Mass Spectrometry Mass Spectrometry: Techniques and Applications of Tandem Mass Spectrometry*, VCH Publishers, New York, 1988.
4. Wong, P. S., and Cooks, R. G., Ion Trap Mass Spectrometry, *Curr. Sep.* 16, 85, 1997.

Method (with accepted acronyms and abbreviations)	Modes	Advantages	Limitations
Gas Chromatography Mass Spectrometry GC-MS (single quadrupole, SQ)	Scan, selected ion monitoring (SIM) and Scan/SIM	Relatively simple and relatively inexpensive; compound identification by library search; dynamic range = 10^5, but typically limited to >10^4; mass/charge range = 10^3 to 10^4, resolution (at m/z = 1000) 10^3 to 10^4 for most ions; well-developed hardware and software; used in standard protocols.	Coelution of compounds compromise library identifications; often user must interpret fragmentation patterns. SIM mode provides higher sensitivity but for target ions only, sensitivity decreases with increasing number of SIM ions. Scan/SIM methods must balance scan and SIM sensitivity requirements.
Gas Chromatography-Mass Spectrometry GC-MS (ion trap, IT)	Scan or selected ion monitoring (SIM); tandem MS-MS; gas or liquid chemical ionization	High sensitivity, compact design, tandem mass spectrometry is possible; dynamic range = 10^4, but typically limited to < 10^4; mass/charge range = 10^4 to 10^5, resolution (at m/z = 1000) 10^4; well-developed hardware and software; used in standard protocols.	Space charge effects can lead to relatively poor dynamic range, however, for "clean" samples, the dynamic range can be as high as single quadrupole units; specific libraries are limited; SIM is for target ions only with a sensitivity lower than that obtainable by a SIM analysis via single quadrupole, above; MS-MS analyses are limited to approximately 120 to 150 compounds per analysis; MS-MS analysis is typically slower than that done with a tandem MS-MS (process is done in time rather than space).
Gas Chromatography (time of flight) Mass Spectrometry, low resolution GC-TOF	Scan or selected ion monitoring	Identification possible with good chromatographic separations, dynamic range = 10^4, mass/charge range = 10^5, resolution (at m/z = 1000) 10^3 to 10^4.	Unit mass resolution limits identification capability; libraries are limited; if chromatographic separation is poor, comprehensive (GCXGC) might be needed; cannot distinguish neutral losses.
Comprehensive Gas Chromatography (time of flight) Mass Spectrometry, low resolution GCXGC-TOF	Scan or selected ion monitoring; normal column configuration (nonpolar – polar) or reversed column configuration (polar – nonpolar)	Identification possible with good chromatographic separations in two dimensions on a nonpolar (long) and a polar (short) column; identification of families with help of principal component analysis tools; dynamic range = 10^4, mass/charge range = 10^5, resolution (at m/z = 1000) 10^3 to 10^4, well-developed hardware and software; used in standard protocols.	Unit mass resolution limits identification capability; libraries are limited; GCXGC may not remedy all aspects of component coelution.

Method (with accepted acronyms and abbreviations)	Modes	Advantages	Limitations
Gas Chromatography Mass Spectrometry (triple quadrupole) GC-MS(QQQ, or QqQ)	Multiple and selected reaction monitoring	Provides product and precursor ion scans; very sensitive for target compounds or functional groups; developed to provide enhanced daughter ion resolution; relatively simple construction with straightforward scanning procedures; no high voltage arcing; dynamic range = 10^5, mass/charge range = 10^3 to 10^4, resolution (at m/z = 1000) 10^3 - 10^4.	Unit mass, making identification ambiguous due to multiple structures as source of breakdown mass; empirical formula determination can be difficult, spectra and fragmentations must often be interpreted manually; full scan data can be acquired (similar to the single quad procedure) by turning off the collision cell.
Gas Chromatography (time of flight) Mass Spectrometry, high resolution GC-TOF	Scan or selected ion monitoring, tandem MS-MS	Might obviate the need for GCXGC separations with accurate mass determinations; deconvolution software can aid in identification of multiple components under peaks; sensitivity intermediate between multiple reaction monitoring and product ion scan of a QQQ; dynamic range = 10^4, mass/charge range = 10^5, resolution (at m/z = 1000) 10^3 to 10^4.	Very large data files (currently approaching 2 Gb), software is currently developmental.
Gas Chromatography Tandem Mass Spectrometry GC-QTOF	Scan or selected ion monitoring, tandem MS-MS	Might obviate the need for GCXGC separations with accurate mass determinations; isolation of parent ions and subsequent fragmentation provides identification; will detect any daughter ion passed into the TOF; sensitivity intermediate between multiple reaction monitoring and product ion scan of a QQQ dynamic range = 10^4, mass/charge range = 10^5, resolution (at m/z = 1000) 10^3 to 10^4.	Large data files are produced; Sophisticated software is needed for processing and deconvolution; requires accurate mass and high resolution.

SOLID-PHASE MICROEXTRACTION SORBENTS

Thomas J. Bruno and Paris D. N. Svoronos

While trapping sorbents have been used for many years in headspace analysis (most commonly with gas chromatography), the modern techniques of solid-phase microextraction (SPME) are particularly applicable to survey analyses (Ref. 1). In the following tables, we provide information for the selection and application of the various fibers, modifications to fiber materials, and data on salting out reagents (Refs. 1–5). For information on other trapping sorbents, chelating agents, resins, and polymeric phases used for headspace analysis, please consult Ref. 1.

References

1. Bruno, T. J., and Svoronos, P. D. N., *CRC Handbook of Basic Tables for Chemical Analysis, 3rd Edition*, CRC Press, Boca Raton, FL, 2011.
2. Haynes, W. M., Ed., *CRC Handbook of Chemistry and Physics, 95th Edition*, CRC Press, Boca Raton, FL, 2011.
3. *NIST Chemistry Web Book*, <www.webbook.nist.gov/chemistry/>, 2009.
4. Machata, G., *Clin. Chem. Newsletter* 4, 29, 1972.
5. Ioffe, B. V., and Vitenberg, A. G., *Head Space Analysis and Related Methods in Gas Chromatography*, Wiley Interscience, New York, 1983.
6. Fritz, J. S., *Analytical Solid-Phase Extraction*, Wiley-VCH, New York, 1999.

Fiber Selection Criteria

The main fiber selection parameters are polarity and relative molecular mass. The table below provides general guidelines on the applicability of available fibers relative to these two parameters. The fibers are characterized by the extraction mechanism, either adsorption or absorption. Adsorbent fibers contain particles suspended in polydimethyl siloxane (PDMS) or polyethylene glycol (PEG, Carbowax).

Fiber	Type of fiber	Polarity	*RMM* Range
7 μm PDMS	Absorbent	Nonpolar	150–700
30 μm PDMS	Absorbent	Nonpolar	80–600
85 μm Polyacrylate	Absorbent	Moderately polar	60–450
100 μm PDMS	Absorbent	Nonpolar	55–400
50 μm Carbowax (PEG)	Adsorbent	Polar	50–400
PDMS-DVB	Adsorbent	Bipolar	50–350
Carbowax-DVB	Adsorbent	Polar	50–350
PDMS-DVB-Carboxen	Adsorbent	Bipolar	40–270
PDMS-Carboxen	Adsorbent	Bipolar	35–180
Carbopak Z-PDMS	Adsorbent	Nonpolar	50–500

PDMS – Polydimethylsiloxane
DVB – Divinylbenzene (3 μm to 5 μm particles)
PEG – Polyethylene glycol
Carboxen – Carboxen 1006 (contains micro-, meso-, and macro-tapered pores) (3 μm to 5 μm particles)
***RMM* Range** – Relative molecular mass range that is the ideal range for optimum extraction. Ranges can be extended by varying extraction times, but results will not be optimized

Phase Material Characteristics

Polydimethylsiloxane (PDMS):
Similar in properties to the OV-1 or SE-30 silicone phases (Ref. 1); nonpolar fluid suitable for nonpolar or slightly polar analytes; thicker coatings extract more analyte, but require longer extraction times; thicker coatings may show carryover effects; useful for hydrocarbons, aldehydes, ketones, and esters.

Polyacrylate:
Rigid solid material; moderate polarity; diffusion of analytes through bulk is relatively slow because of rigidity of material; relatively higher desorption temperatures required because of rigidity of material; can be oxidized easily at higher temperatures; must use oxygen-free carrier gas and ensure gas chromatographic system is leak-free; fibers are very solvent resistant; darkens to a brown color upon exposure to temperatures in excess of 280 °C, but fiber is generally still usable until color becomes black; useful for alcohols.

Carbowax (polyethylene glycol, PEG):
Similar in properties to the PEG coatings used extensively in chromatography; moderately polar; highly cross-linked to counteract water solubility; sensitive to attack by oxygen at temperatures in excess of 220 °C, at which point the fiber will darken and become powdery; requires use of high purity carrier gas (typically He at 99.999 % mass/mass) treated for oxygen contamination.

Divinylbenzene (DVB):
Similar to the properties of divinylbenzene porous polymer phases; higher polarity than Carbowax, and when combined with Carbowax results in a more polar phase; like polyacrylate, it is a solid particle that must be carried in a liquid to coat on a fiber.

Carboxen:
Similar to the material used in Carboxen porous-layer open tubular (PLOT) columns; structure has an approximately even distribution of macro-, meso-, and micropores, making it valuable for smaller analytes; larger analytes can show hysteresis that must be addressed by desorption at 280 °C; useful for hydrocarbons, ethers, aldehydes, and ketones.

Extraction Capability of Solid-Phase Microextraction Sorbents

This table shows the extraction capability of the fibers for acetone, a small, moderately polar analyte, for 4-nitrophenol, a medium size polar analyte, and benzo(GHI)perylene, a large nonpolar analyte. This provides a general guideline for fiber selection.

Approximate Linear Concentration Ranges for Acetone, 4-Nitrophenol, and Benzo[ghi]perylene

Fiber	Acetone 10 min ext[a] (FID)	4-Nitrophenol 20 min ext[b] (GC/MS)	Benzo(GHI) Perylene 20 min ext
7 µm PDMS	100 ppm and up	Not extracted	100 ppt to 500 ppb
30 µm PDMS	10 ppm and up	10 ppm and up	100 ppt to 10 ppm
85 µm Polyacrylate	1 ppm to 1000 ppm	5 ppb to 100 ppm	500 ppt to 10 ppm
100 µm PDMS	500 ppb to 1000 ppm	500 ppb to 500 ppm	500 ppt to 10 ppm
50 µm Carbowax (PEG)	1 ppm to 1000 ppm	5 ppb to 50 ppm	25 ppb to 10 ppm
PDMS-DVB	50 ppb to 100 ppm	25 ppb to 10 ppm	10 ppb to 1 ppm
Carbowax-DVB	100 ppb to 100 ppm	5 ppb to 10 ppm	50 ppb to 5 ppm
PDMS-DVB-Carboxen	25 ppb to 10 ppm	50 ppb to 10 ppm	100 ppb to 1 ppm poorly desorbed
PDMS-Carboxen	5 ppb to 5 ppm	100 ppb to 10 ppm	Not desorbed
Carbopak Z-PDMS	10 ppm to 500 ppm	5 ppm to 100 ppm	500 ppt to 100 ppb

Note: In each case, the concentration is expressed on a mass basis (e.g., ppm mass/mass).

[a] Water sample contains 25% NaCl (mass/mass)

[b] Water sample contains 2% NaCl (mass/mass) acidified to pH = 2 with 0.05 M phosphoric acid

1 ppm = 1 part in 1×10^6

1 ppb = 1 part in 1×10^9

1 ppt = 1 part in 1×10^{12}

Typical Phase Volumes of SPME Fiber Coatings

Fiber coating thickness/type	Type of fiber core	Fiber core diameter/mm	Phase volume/mm³ or µL
PDMS	Fused silica	0.110	0.612
100 µm PDMS	Metal	0.130	0.598
30 µm PDMS	Fused silica	0.110	0.132
30 µm PDMS	Metal	0.130	0.136
7 µm PDMS	Fused silica	0.110	0.028
7 µm PDMS	Metal	0.130	0.030
85 µm PA	Fused silica	0.110	0.543
60 µm PEG	Metal	0.130	0.358
15 µm Carbopack Z/PDMS	Metal	0.130	0.068
65 µm PDMS/DVB	Fused silica	0.120	0.418
65 µm PDMS/DVB	Proprietary	0.130	0.440
65 µm PDMS/DVB	Metal	0.130	0.440
75 µm Carboxen-PDMS	Fused silica	0.120	0.502
85 µm Carboxen-PDMS	Proprietary	0.130	0.528
85 µm Carboxen-PDMS	Metal	0.130	0.528
50/30 µm DVB/Carboxen	Metal		
Carboxen layer		0.130	0.151
DVB layer		0.190	0.377
50/30 µm DVB/Carboxen	Metal		
Carboxen layer		0.130	0.151
DVB layer		0.190	0.377
60 µm PDMS-DVB HPLC	Proprietary	0.160	0.459

Bonded-Phase Modified Silica Substrates for Solid-Phase Extraction

The following table provides the most commonly used bonded-phase modified silica substrates used in solid-phase extraction (Ref. 6). Additional information on many of these materials can be found in the table entitled "More Common HPLC Stationary Phases" in the HPLC chapter of Ref. 2.

Phase	Polarity of phase	Designation
Octadecyl, endcapped	Strongly apolar	C18ec
Octadecyl	Strongly apolar	C18
Octyl	Apolar	C8
Ethyl	Slightly polar	C2
Cyclohexyl	Slightly polar	CH
Phenyl	Slightly polar	PH
Cyanopropyl	Polar	CN
Diol	Polar	2OH
Silica gel	Polar	SiOH
Carboxymethyl	Weak cation exchanger	CBA
Aminopropyl	Weak anion exchanger	NH_2
Propylbenzene sulfonic acid	Strong cation exchanger	SCX
Trimethylaminopropyl	Strong anion exchanger	SAX

Salting-Out Reagents for Headspace Analysis

The following table provides data on the common salts used for salting out in chromatographic headspace analysis, as applied to direct injection methods and to solid-phase microextraction. Data are provided for the most commonly available salts, although others are possible. Sodium citrate, for example, occurs as the dihydrate and the pentahydrate. The pentahydrate is not as stable as the dihydrate, however, and dries out on exposure to air, forming cakes. Potassium carbonate occurs as the dihydrate, trihydrate, and sesquihydrate; however, data are provided only for the anhydrous material. The solubility is provided as the number of grams that can dissolve in 100 mL of water at the indicated temperature. The vapor enhancement cited is the degree of increase of the concentration of vapor over the solution of a 2 % (mass/mass) ethanol solution in water at 60 °C.

Properties of Salting-Out Reagents

Salt	Formula	Mol. wt.	Density/g cm^{-3}	Solubility/g mL^{-1} H_2O Cold water	Solubility/g mL^{-1} H_2O Hot water	Vapor enhancement
Potassium carbonate	K_2CO_3	138.206	2.428 at 14 °C	112[a]	156[b]	8
Ammonium sulfate	$(NH_4)_2SO_4$	132.140	1.769 at 50 °C	70.6[c]	103.8[b]	5
Sodium citrate (dihydrate)	$Na_3C_6H_5O_7 \times 2H_2O$	294.099		72[d]	167[b]	5
Sodium chloride	NaCl	58.443	2.165[e]	37.5[a]	39.12[b]	3
Ammonium chloride	NH_4Cl	53.492	1.527	29.7[c]	75.8[b]	2

[a] 20 °C
[b] 100 °C
[c] 0 °C
[d] 25 °C
[e] Specific gravity, 25 °C/4 °C

Analytical

GAS CHROMATOGRAPHIC RETENTION INDICES

Thomas J. Bruno

The interpretation of results from chromatographic measurements can often be augmented with an appropriate mathematical treatment of the solute retention that is observed. The goal of the treatment is to make the resulting metric as independent of the instrument as possible. A typical situation that arises from analysis by gas chromatography with mass spectrometry is that the library search routine produces "hits" that are ambiguous if not nonsensical (Ref. 1). The correct interpretation of the mass spectrum must then be done manually (Refs. 2, 3), and the mass spectral data should be augmented by additional analytical techniques. The specific techniques that should be used must be determined on a case-by-case basis by a qualified person. One additional datum that is typically already present in gas chromatography with mass spectrometry detection is chromatographic retention. The raw datum from a chromatographic measurement is a retention time, t_r, of each eluted peak, and a corresponding intensity. Here, we will not treat other aspects of the output, such as the width and shape of the chromatographic signal. The retention time (for a given stationary phase) is dependent on the column temperature, column pressure, column geometry (length and inside diameter; phase ratio), and ambient (atmospheric) pressure.

If the volumetric carrier gas flow rate (at the column exit) is measured and multiplied by the retention time, the retention volume, V_R, is obtained. The adjusted retention volume, V_R', is the retention volume corrected for the void volume (or mobile phase holdup) of the column. It is obtained by simply subtracting the retention volume of an unretained solute (V_M):

$$V_R' = V_R - V_M. \tag{1}$$

While it is possible to calculate the corresponding adjusted retention time, t_r', by subtracting the retention time of the unretained peak, t_m, it is better to work with volumes since average flow-rate variations between individual analyses are then accounted for. Note that with each level of refinement beyond the raw retention time, a facet of instrument dependence from the resulting parameter is removed. V_R is independent of the flow rate; V_R' is, further, independent of column geometry. Continuing with this approach, the net retention volume, V_N, is defined, by applying a factor, j, to account for the pressure drop across the column:

$$V_N = jV_R', \tag{2}$$

where j is usually the Martin-James compressibility factor[3]:

$$j = \frac{3}{2}\left[\frac{\left(\frac{P_i}{P_o}\right)^2 - 1}{\left(\frac{P_i}{P_o}\right)^3 - 1}\right] \tag{3}$$

In Eq. 3, P_i is the inlet pressure (absolute) and P_o is the outlet pressure (usually atmospheric pressure). The net retention volume is important because it is independent of the inlet and outlet pressures, as well as being independent of the carrier gas flow rate and column geometry.

The specific retention volume, V_g, corrects the net retention volume for the amount of stationary phase actually on the column, and the column temperature is adjusted or corrected to 0 °C:

$$V_g = (273.15)\frac{V_N}{(W_s T_{col})} \tag{4}$$

where T_{col} is the column temperature (in K), and W_s is the mass of the stationary phase in the column. The V_g value is a characteristic for a particular solute on a particular stationary phase, and is instrument independent. This is a quantity that may be compared from instrument to instrument, and laboratory to laboratory with a high level of confidence provided the stationary phase used is a single, pure compound or a well characterized mixture. If the mass of the stationary phase is not known, or is not meaningful, one may use the net retention volume directly, or one may correct the net retention volume to a column temperature of 0 °C (represented by V_N^0) by simply not including the term for W_s (that is, setting it equal to unity).

It is also extremely valuable to calculate a relative retention, $r_{a/b}$:

$$r_{a/b} = \left(\frac{V_g^b}{V_g^a}\right) = \left(\frac{V_N^b}{V_N^a}\right) \tag{5}$$

where the numerical superscripts refer to the retention volumes of solutes "a" and "b." In this case, solute "a" is a reference compound. The relative retention is dependent only on the column temperature and the type of stationary phase. For reasons of operational simplicity, this parameter is usually one of the best to use for qualitative analysis. It can account for small differences in the column temperature, stationary phase considerations, column history, and minor disturbances in the carrier gas flow rate. It is possible to account for the column temperature by plotting the logarithm of the retention parameters against $1/T$, where T is the thermodynamic temperature. The column pressure is accounted for by variations in the volume measurement; therefore, there is no pressure dependence to these parameters.

We can go beyond the simple retention parameters discussed earlier to incorporate a logarithmic interpolation on a uniform scale by use of the Kovats retention index (Ref. 4). The isothermal Kovats retention index is calculated by use of the following defining equation:

$$I_{sample}(T) = 100\left[\frac{\log X_S - \log X_L}{\log X_H - \log X_L} + n_L\right] \tag{6}$$

Here, I_{sample} is the dimensionless Kovats retention index that is a function of both temperature and the stationary phase employed. The terms represented by X are retention parameters of the sample and standards. Following this convention, X_S is the retention parameter of the sample under consideration. Any retention parameter, such as the adjusted retention time, t', the net retention volume, V_N, the adjusted net retention volume, V_N^0, and the relative retentions, $r_{a/b}$, can be used. X_S is the retention parameter of the sample under consideration, X_L is the retention parameter of a normal alkane (that is, straight chain or unbranched) of carbon number n_L that elutes earlier than the sample, and X_H is the retention parameter of a normal alkane having a carbon number greater than n_L+1 that elutes after the sample. The retention index of a sample is, therefore, 100 multiplied by the carbon number of a hypothetical normal alkane that shows the same retention parameter on the stationary phase at that temperature. Thus, a sample that has a retention index of 785, for example, would coelute with a hypothetical normal alkane that has 7.85 carbon atoms. By definition, the retention indices of the normal alkanes (on any stationary phase) are equal to 100 multiplied by the carbon number. Thus, for *n*-hexane, I = 600, and so on for the other *n*-alkanes in this homologous series. The zero point in the scale is defined for hydrogen, for which I = 0. Kovats retention indices have been determined for selected compounds on the most common stationary phases [Refs. 5-7].

The temperature dependence of I for a given sample is known to follow a hyperbolic form similar to the familiar Antoine equation used to represent vapor pressure:

$$I_{\text{sample}}(T) = A + \frac{B}{(T+C)} \tag{7}$$

In this equation, A, B, and C are empirically determined constants, and T is the temperature (in °C). A nonlinear fitting routine should be used to determine these constants. When retention indices are available for at least three temperatures, initial values for A, B, and C can be determined by use of the following equations:

$$C = \frac{(T_2 - T_1)(I_3T_3 - I_1T_1) + (T_3 - T_1)(I_1T_1 - I_2T_2)}{(T_3 - T_1)(I_2 - I_1) - (T_2 - T_1)(I_3 - I_1)}, \tag{8}$$

$$A = \frac{I_2T_2 - I_1T_1 + C(I_2 - I_1)}{T_2 - T_1}, \tag{9}$$

$$B = (I_2 - A)(T_2 + C)\cdot \tag{10}$$

In these equations, I_1, I_2, and I_3 are retention indices of the sample measured at temperatures T_1, T_2, and T_3. When additional retention indices are available at other temperatures, we advocate the use of minimum deviation estimates from these three equations to furnish the starting values for the nonlinear fit. When retention indices at four temperatures are available, the best starting values are obtained from the I_1, I_2, and I_3 triplet that minimizes the deviation with the experimental value with that produced by Eq. 7. This approach provides the fastest convergence, and also helps avoid converging to local minima. Predictions made by use of Eq. 7 can be used for retention indices within the measured temperature range as well as extrapolation somewhat beyond that range on a case-by-case basis.

It is also of value to report and use the temperature dependence as a slope coefficient, $\delta I_{\text{sample}}/10$ °C, the variation of I_{sample} for a particular stationary phase over a particular temperature range. While not as reliable as the Antoine-type fit, this coefficient is useful for predictions within the range of the measured results.

In all of the above discussion, one must understand that the column temperature is fixed. The Kovats retention indices can be made applicable to temperature programmed analysis by use of

$$I = 100 \times \left[n + (N - n)\frac{t_{r(\text{unknown})} - t_{r(n)}}{t_{r(N)} - t_{r(n)}} \right], \tag{11}$$

where I is the retention index, n is the number of carbon atoms in the smaller *n*-alkane, N is the the number of carbon atoms in the larger *n*-alkane, and t_r is the retention time of the indicated peak.

A useful alternative to the Kovats system is the Lee retention index (isothermal and temperature dependent), based on the polynuclear aromatic hydrocarbon (PAH) standard compounds: naphthalene (I = 200), phenanthrene (I = 300), chrysene (I = 400), and picene or benzo(g,h,i)perylene (I = 500). Isothermal (and temperature-dependent) Kovats and Lee retention indices for many compounds are tabulated in the NIST Chemistry Web Book (Ref. 8).

References

1. NIST/EPA/NIH (NIST 11) Mass Spectral Database, NIST Standard Reference Database 1A, National Institute of Standards and Technology, Standard Reference Data Program, Gaithersburg, MD, 20899, 2011.
2. Bruno, T. J., and Svoronos, P. D. N., *CRC Handbook of Fundamental Spectroscopic Correlation Charts*, CRC Press, Boca Raton, FL, 2006.
3. Bruno, T. J., and Svoronos, P. D. N, *CRC Handbook of Basic Tables for Chemical Analysis*, 3rd. Edition, CRC Press, Boca Raton, FL, 2011.
4. Kovats, E., *Helv. Chim. Acta* 41, 1915, 1958.
5. Lubek, A. J., and Sutton, D. L., *J. High Res. Chromatogr. and Chromatogr. Commun.* 6, 328, 1983.
6. Bruno, T. J., Wertz, K. H., and Cacairi, M., *Anal. Chem.* 68, 1347, 1996.
7. Miller, K. E., and Bruno, T. J., *J. Chromatogr. A* 1007, 117, 2003.
8. NIST Chemistry Web Book, NIST Standard Reference Database Number 69, *Standard Reference Data Program, National Institute of Standards and Technology*, Gaithersburg, MD, 20899, 2012.

ELUOTROPIC VALUES OF SOLVENTS ON OCTADECYLSILANE AND OCTYLSILANE

Thomas J. Bruno and Paris D. N. Svoronos

The following table provides, for comparative and interpretive purposes, eluotropic values on bonded octadecylsilane (ODS) and octylsilane (OS) for common solvents used in HPLC (Refs. 1–3). For additional information on common, specific, and chiral stationary phases for HPLC, and for solvents, derivatizing reagents, and detectors, see Ref. 3.

References

1. Krieger, P. A., *High Purity Solvent Guide*, Burdick and Jackson Laboratories, McGaw Park, IL, 1984.
2. Ahuja, S., *Trace and Ultratrace Analysis by HPLC*, John Wiley and Sons, New York, 1992.
3. Bruno, T. J., and Svoronos, P. D. N., *CRC Handbook of Basic Tables for Chemical Analysis, 3rd Edition*, CRC Press, Boca Raton, FL, 2011.

Eluotropic Values of Solvents on Octadecylsilane and Octylsilane

Solvent	Eluotropic value, ODS	Eluotropic value, OS
Acetic acid		2.7
Acetone	8.8	9.3
Acetonitrile	3.1	3.3
1,4-Dioxane	11.7	13.5
N,N-Dimethylformamide	7.6	9.4
Methanol	1.0	1.0
Ethanol	3.1	3.2
1-Propanol	10.1	10.8
2-Propanol	8.3	8.4
Tetrahydrofuran	3.7	

INSTABILITY OF HPLC SOLVENTS

Thomas J. Bruno and Paris D. N. Svoronos

Solvents that are commonly used in high-performance liquid chromatography frequently have inherent chemical instabilities that must be considered when designing an analysis, or in the interpretation of results (Refs. 1-4). In many cases, such solvents are obtainable with stabilizers added to control the instability or to slow the reaction. Reactive solvents that do not have stabilizers must be used quickly or be given proper treatment. In either case, it is important to understand that the solvents (as they may be used in an analysis) are not necessarily pure materials.

Although not specifically considered below, the preparation of aqueous-organic solvents for HPLC can cause the introduction of impurities. Mixing acetonitrile and water is endothermic and can sometimes exacerbate air dissolution. Buffers are also a special case in that the pH can change due to exposure to air and the development of microbes (Ref. 4). Buffers should be used the same day of preparation and often incorporate the preservative sodium metabisulfite.

References

1. Sadek, P. C., *The HPLC Solvent Guide, 2nd Edition*, Wiley Interscience, New York, 2002.
2. Bruno, T. J., and Straty, G. C., *J. Res. Natl. Bur. Stds. (U.S.)*, 91, 135, 1986.
3. Straty, G. C., Palavra, A. M. F., and Bruno, T. J., *Int. J. Thermophys.* 7, 1077, 1986.
4. Bruno, T. J., and Svoronos, P. D. N., *CRC Handbook of Basic Tables for Chemical Analysis, 3rd Edition*, CRC Press, Boca Raton, FL, 2011.

Solvent	Contaminants, reaction products	Stabilizers
Ethers		
Diethyl ether	Peroxides[1]	2–3% (vol/vol) ethanol[2]
		1–10 ppm (mass/mass) BHT[3]
		(1.5–3.5% ethanol) + (0.2–0.5% water) + (5–10 ppm (mass/mass) BHT[3])
Isopropyl ether	Peroxides[1]	0.01% (mass/mass) hydroquinone
		5–100 ppm (mass/mass) BHT[3]
1,4-Dioxane	Peroxides[1]	25–1500 ppm (mass/mass) BHT[3]
Tetrahydrofuran	Peroxides[1]	25–250 ppm (mass/mass) BHT[3]
Chlorinated Alkanes		
Chloroform	Hydrochloric acid, chlorine, phosgene (CCl_2O)	0.5–1% (vol/vol) ethanol
		50–150 ppm (mass/mass) amylene[4]
		various ethanol amylene blends
Dichloromethane	Hydrochloric acid, chlorine, phosgene (CCl_2O)	25 ppm (mass/mass) amylene
		25 ppm (mass/mass) cyclohexene
		400–600 ppm (mass/mass) methanol
		various amylene methanol blends
Alcohols		
Ethanol	Water, numerous denaturants are commonly added	
Methanol	Water; formal dehydrate (at elevated temperature)	
Ketones		
Acetone	Diacetone alcohol, and higher oligomers	

[1] The peroxide concentration that is usually considered hazardous is 250 ppm (mass/mass).
[2] Ethanol does not actually stabilize diethyl ether, nor is it a peroxide scavenger, although it was thought to be so in the past. It is still available in chromatographic solvents to preserve the utility of retention relationships and analytical methods.
[3] 2,6-Di-*tert*-butyl-*p*-cresol.
[4] Amylene is a generic name for 2-methyl-2-butene.

DETECTORS FOR LIQUID CHROMATOGRAPHY

Thomas J. Bruno and Paris D. N. Svoronos

The following table provides some comparative data for interpretation of results from the more common detectors applied to high-performance liquid chromatography (Refs. 1-6). In general, the operational parameters provided are for optimized systems, and represent the maximum obtainable in terms of sensitivity and linearity. In this table, the molar extinction coefficient is represented by ε.

References

1. Pryde, A., and Gilbert, M. T., *Applications of High Performance Liquid Chromatography*, Chapman and Hall, London, 1979.
2. Hamilton, R. J., and Sewell, P. A., *Introduction to High Performance Liquid Chromatography*, Chapman and Hall, London, 1977.
3. Ahuja, S., *Trace and Ultratrace Analysis by HPLC*, Chemical Analysis Series, John Wiley and Sons, New York, 1991.
4. Snyder, L. R., Kirkland, J. J., and Glajch, J., *Practical HPLC Method Development*, John Wiley and Sons, New York, 1997.
5. Bruno, T. J., *Sep. Purif. Meth.* 29, 63, 2000.
6. Bruno, T. J., and Svoronos, P. D. N, *CRC Handbook of Basic Tables for Chemical Analysis, 3rd Edition*, CRC Press, Boca Raton, FL, 2011.

Detector	Sensitivity	Linearity	Selectivity	Comments
Ultraviolet spectrophotometer	1×10^{-9} g (for compounds of ε = 10 000 to 20 000)	1×10^4	For UV-active functionalities, on the basis of absorptivity	Relatively insensitive to flow and temperature fluctuations; non-destructive, useful with gradient elution; use mercury lamp for 254 nm, and quartz-iodine lamp for 350 nm to 700 nm; often a diode-array instrument is used to obtain entire UV-vis spectrum.
Refractive index detector (RID)	1×10^{-7} g	1×10^4	Universal, dependent on refractive-index difference with mobile phase	Relatively insensitive to flow fluctuations, but sensitive to temperature fluctuations; non-destructive, cannot be used with gradient elution; solvents must be degassed to avoid bubble formations; laser-based RI detectors offer higher sensitivity.
Fluorometric detector	1×10^{-11} g	1×10^5	For fluorescent species with conjugated bonding and/or aromaticity	Relatively insensitive to temperature and flow fluctuations; non-destructive; can be used with gradient elution; often, chemical derivatization is done on analytes to form fluorescent species; uses deuterium lamp for 190 nm to 400 nm, or tungsten lamp for 350 nm to 600 nm.
Amperometric detector	1×10^{-9} g	1×10^4	Responds to –OH functionalities	Used for aliphatic and aromatic —OH compounds, amines, and indoles; pulsed potential units are most sensitive, can be used with gradient elution and organic mobile phases; senses compounds in oxidatitive or reductive modes; mobile phases must be highly pure and purged of O_2.
Conductivity detector	1×10^{-9} g	2×10^4	Specific to ionizable compounds	Uses post-column derivatization to produce ionic species; especially useful for certain halogen, sulfur, and nitrogen compounds.
Mass spectrometers	Interface dependent	Interface dependent	Universal, within limits imposed by interface	Complex, expensive devices highly dependent on an efficient interface; electrospray and thermospray interfaces are most common; linear response is difficult to achieve.

SOLVENTS FOR ULTRAVIOLET SPECTROPHOTOMETRY

Thomas J. Bruno

This table lists some solvents commonly used for sample preparation for ultraviolet spectrophotometry. The properties given are:

- λ_c: cutoff wavelength, below which the solvent absorption becomes excessive.
- ε: dielectric constant (relative permittivity); the temperature in °C is given as a superscript.
- t_b: normal boiling point.

References

1. Bruno, T. J., and Svoronos, P. D. N., *CRC Handbook of Basic Tables for Chemical Analysis*, CRC Press, Boca Raton, FL, 1989.
2. *Landolt-Börnstein, Numerical Data and Functional Relationships in Science and Technology, New Series*, IV/6, *Static Dielectric Constants of Pure Liquids and Binary Liquid Mixtures*, Springer–Verlag, Heidelberg, 1991.

Solvents for Ultraviolet Spectrophotometry

Solvent	λ_c/nm	ε	t_b/°C
Acetic acid	260	6.20^{20}	117.9
Acetone	330	21.01^{20}	56.08
Acetonitrile	190	36.64^{20}	81.6
Benzene	280	2.2825^{20}	80.08
2-Butanol	260	17.26^{20}	99.4
2-Butanone	330	18.56^{20}	79.6
Butyl acetate	254	5.07^{20}	126.0
Carbon disulfide	380	2.6320^{20}	46.2
1-Chlorobutane	220	7.276^{20}	78.4
Cyclohexane	210	2.0243^{20}	80.7
1,2-Dichloroethane	226	10.42^{20}	83.4
Dichloromethane	235	8.93^{25}	39.8
Diethyl ether	218	4.2666^{20}	34.4
1,2-Dimethoxyethane	240	7.30^{24}	85.0
N,N-Dimethylacetamide	268	38.85^{21}	165.9
N,N-Dimethylformamide	270	38.25^{20}	152.8
Dimethyl sulfoxide	265	47.24^{20}	191.9
1,4-Dioxane	215	2.2189^{20}	101.2
Ethanol	210	25.3^{20}	78.24
2-Ethoxyethanol	210	13.38^{25}	134.7
Ethyl acetate	255	6.0814^{20}	77.1
Glycerol	207	46.53^{20}	289
Heptane	197	1.9209^{20}	98.38
Hexadecane	200	2.0460^{20}	286.9
Hexane	210	1.8865^{20}	68.72
Methanol	210	33.0^{20}	64.5
2-Methoxyethanol	210	17.2^{25}	124.3
Methylcyclohexane	210	2.024^{20}	100.9
4-Methyl-2-pentanone	335	13.11^{20}	115.7
2-Methyl-1-propanol	230	17.93^{20}	107.84
N-Methyl-2-pyrrolidinone	285	32.55^{20}	204.2
Nitromethane	380	37.27^{20}	101.19
Pentane	210	1.8371^{20}	36.06
Pentyl acetate	212	4.79^{20}	149.4
1-Propanol	210	20.8^{20}	97.04
2-Propanol	210	20.18^{20}	82.21
Pyridine	330	13.260^{20}	115.2
Tetrachloroethene	290	2.268^{30}	121.2
Tetrachloromethane	265	2.2379^{20}	76.7
Tetrahydrofuran	220	7.52^{22}	66.0
Toluene	286	2.379^{23}	110.60
Trichloromethane	245	4.8069^{20}	61.2
1,1,2-Trichloro-1,2,2-trifluoroethane	231	2.41^{25}	47.6
2,2,4-Trimethylpentane	215	1.943^{20}	99.2
Water	191	80.1^{20}	99.974
o-Xylene	290	2.562^{20}	144.4
m-Xylene	290	2.359^{20}	139.1
p-Xylene	290	2.2735^{20}	138.3

CORRELATION TABLE FOR ULTRAVIOLET ACTIVE FUNCTIONALITIES

Thomas J. Bruno and Paris D. N. Svoronos

The following table presents a correlation between common chromophoric functional groups and the expected absorptions from ultraviolet (UV) spectrophotometry. While not as informative as infrared correlations, UV can often provide valuable qualitative information. In these tables, λ_{max} is the wavelength in nm at which the maximum absorption occurs, and ε_{max} is the extinction coefficient.

References

1. Willard, H. H., Merritt, Jr., L. L., Dean, J. A., and Settle, F. A., *Instrumental Methods of Analysis, 7th Edition*, Wadsworth Publishing Co., Belmont, CA, 1988.
2. Silverstein, R. M., and Webster, F. X., *Spectrometric Identification of Organic Compounds, 6th Edition*, Wiley, New York, 1998.
3. Lambert, J. B., Shurvell, H. F., Lightner D. A., Verbit, L., and Cooks, R. G., *Organic Structural Spectroscopy*, Prentice Hall, Upper Saddle River, NJ, 1998.
4. Bruno, T. J., and Svoronos, P. D. N., *CRC Handbook of Basic Tables for Chemical Analysis, 3rd Edition*, CRC Press, Boca Raton, FL, 2011.
5. Woodward, R. B., *J. Am. Chem. Soc.* 63, 1123, 1941.
6. Woodward, R. B., *J. Am. Chem. Soc.* 64, 72, 1942.
7. Woodward, R. B., *J. Am. Chem. Soc.* 64, 76, 1942.
8. Fieser, L. F., and Fieser, M., *Natural Products Related to Phenanthrene, 3rd Edition*, Reinhold, New York, 1949.

UV Wavelength Maxima and Extinction Coefficients for Different Chromophores

Chromophore	Functional group	λ_{max}/nm	ε_{max}	λ_{max}/nm	ε_{max}	λ_{max}/nm	ε_{max}
Ether	–O–	185	1000				
Thioether	–S–	194	4600	215	1600		
Amine	$–NH_2–$	195	2800				
Amide	$–CONH_2$	<210	—				
Thiol	–SH	195	1400				
Disulfide	–S–S–	194	5500	255	400		
Bromide	–Br	208	300				
Iodide	–I	260	400				
Nitrile	–C≡N	160	—				
Acetylide (alkyne)	–C≡C–	175–180	6000				
Sulfone	$–SO_2–$	180	—				
Oxime	–NOH	190	5000				
Azido	>C=N–	190	5000				
Alkene	–C=C–	190	8000				
Ketone	>C=O	195	1000	270–285	18–30		
Thioketone	>C=S	205	strong				
Esters	–COOR	205	50				
Aldehyde	–CHO	210	strong	280–300	11–18		
Carboxyl	–COOH	200–210	50–70				
Sulfoxide	>S→O	210	1500				
Nitro	$–NO_2$	210	strong				
Nitrite	–ONO	220–230	1000–2000	300–4000	10		
Azo	–N=N–	285–400	3–25				
Nitroso	–N=O	302	100				
Nitrate	$–ONO_2$	270 (shoulder)	12				
Conjugated hydrocarbon	$–(C=C)_2–$ (acyclic)	210–230	21 000				
Conjugated hydrocarbon	$–(C=C)_3–$	260	35 000				
Conjugated hydrocarbon	$–(C=C)_4–$	300	52 000				
Conjugated hydrocarbon	$–(C=C)_5–$	330	118 000				
Conjugated hydrocarbon	$–(C=C)_2–$ (alicyclic)	230–260	3000–8000				
Conjugated hydrocarbon	C=C–C≡C	219	6500				
Conjugated system	C=C–C=N	220	23 000				
Conjugated system	C=C–C=O	210–250	10 000–20 000			300–350	weak
Conjugated system	$C=C–NO_2$	229	9500				
Phenyl	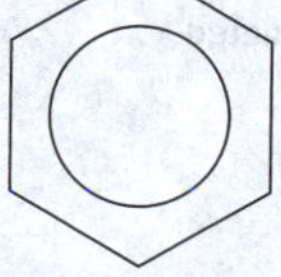	184	46 700	202	6900	255	170
Diphenyl	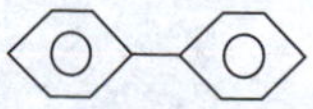			246	20 000		

Chromophore	Functional group	λ_{max}/nm	ε_{max}	λ_{max}/nm	ε_{max}	λ_{max}/nm	ε_{max}
Naphthalene	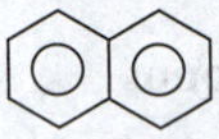	220	112 000	275	5600	312	175
Anthracene	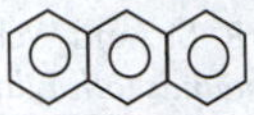	252	199 000	375	7900		
Pyridine	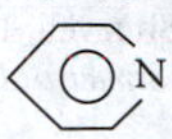	174	80 000	195	6000	251	1700
Quinoline	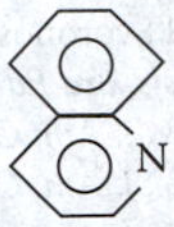	227	37 000	270	3600	314	2750
Isoquinoline	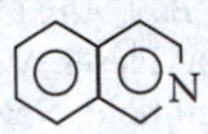	218	80 000	266	4000	317	3500

Wavenumber Adjustments for Bathochromic Shifts (Woodward's Rules)

Conjugated systems show bathochromic shifts in their $\pi \rightarrow \pi^*$ transition bands. Empirical methods for predicting those shifts were originally formulated by Woodward (Refs. 5–7) and Fieser and Fieser (Ref. 8). This section includes the most important conjugated system rules. The reader should consult Refs. 6 and 8 for more details on how to apply the wavelength increment data.

(a) Rules for Diene Absorption

Base value for diene: 214 nm

Increments for each (in nm):

Woodward's Rules for Diene Absorption

Heteroannular diene	+ 0
Homoannular diene	+39
Extra double bond	+30
Alkyl substituent or ring residue	+5
Exocyclic double bond	+5
Polar groups:	
–OOCR	+0
–OR	+6
–S–R	+30
halogen	+5
$-NR_2$	+60
λ Calculated	= Total

(b) Rules for Enone Absorption*

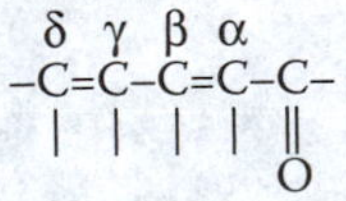

Base value for acyclic (or six-membered) α,β-unsaturated ketone: 215 nm

Base value for five-membered α,β-unsaturated ketone: 202 nm

Base value for α,β-unsaturated aldehydes: 210 nm

Base value for α,β-unsaturated esters or carboxylic acids: 195 nm

Increments for each (in nm):

Woodward's Rules for Enone Absorption

Heteroannular diene	+0
Homoannular diene	+39
Double bond	+30
Alkyl group:	
α–	+10
β–	+12
γ– and higher	+18
Polar groups:	
–OH	
α–	+35
β–	+30
δ–	+50
–OOCR	
α, β, γ, δ	+6
–OR	
α–	+35
β–	+30
γ–	+17
δ–	+31
–SR	
β–	+85
–Cl	
α–	+15
β–	+12
–Br	
α–	+25
β–	+30
$-NR_2$	
β–	+95
Exocyclic double bond	+5
λ Calculated	= Total

(c) Rules for Monosubstituted Benzene Derivatives
Parent Chromophore (benzene): 250 nm

Woodward's Rules for Monosubstituted Benzene Derivatives Absorption

Substituent	Increment (in nm)
–R	-4
–COR	-4
–CHO	0
–OH	-16
–OR	-16
–COOR	-16

where R is an alkyl group, and the substitution is on C_6H_5–.

(d) Rules for Disubstituted Benzene Derivatives
Parent Chromophore (benzene): 250 nm [Increment in nm]

Woodward's Rules for Disubstituted Benzene Derivatives Absorption

Substituent	*o*–	*m*–	*p*–
–R	+3	+3	+10
–COR	+3	+3	+10
–OH	+7	+7	+25
–OR	+7	+7	+25
$–O^-$	+11	+20	+78 (variable)
–Cl	+0	+0	+10
–Br	+2	+2	+15
$–NH_2$	+13	+13	+58
$–NHCOCH_3$	+20	+20	+45
$–NHCH_3$	—	—	+73
$–N(CH_3)_2$	+20	+20	+85

R indicates an alkyl group.

MIDDLE-RANGE INFRARED ABSORPTION CORRELATION CHARTS

Thomas J. Bruno and Paris D. N. Svoronos

The following charts provide characteristic middle-range infrared absorptions obtained from particular functional groups on molecules (Refs. 1 and 2). These include a general mid-range correlation chart, a chart for aromatic absorptions, and a chart for carbonyl moieties. Charts for near infrared absorptions and for inorganic moieties can be found in the cited references.

References

1. Bruno, T. J. and Svoronos, P. D. N., *CRC Handbook of Basic Tables for Chemical Analysis*, 3rd Edition, CRC Press, Boca Raton, FL, 2011.
2. Bruno, T. J., and Svoronos, P. D. N., *CRC Handbook of Fundamental Spectroscopic Correlation Charts*, CRC Press, Boca Raton, FL, 2006.

Notes:

AR = aromatic
b = broad
sd = solid
sn = solution
sp = sharp
? = unreliable

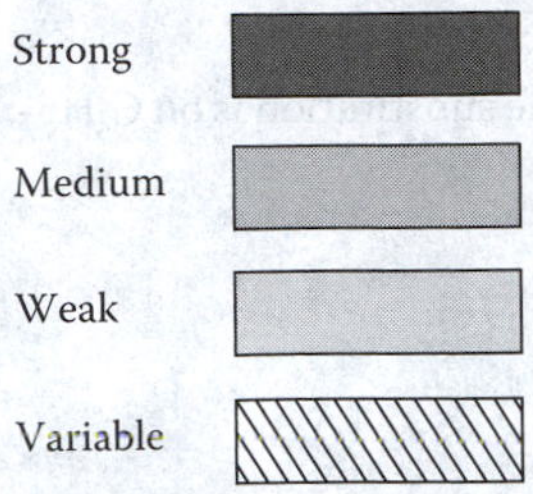

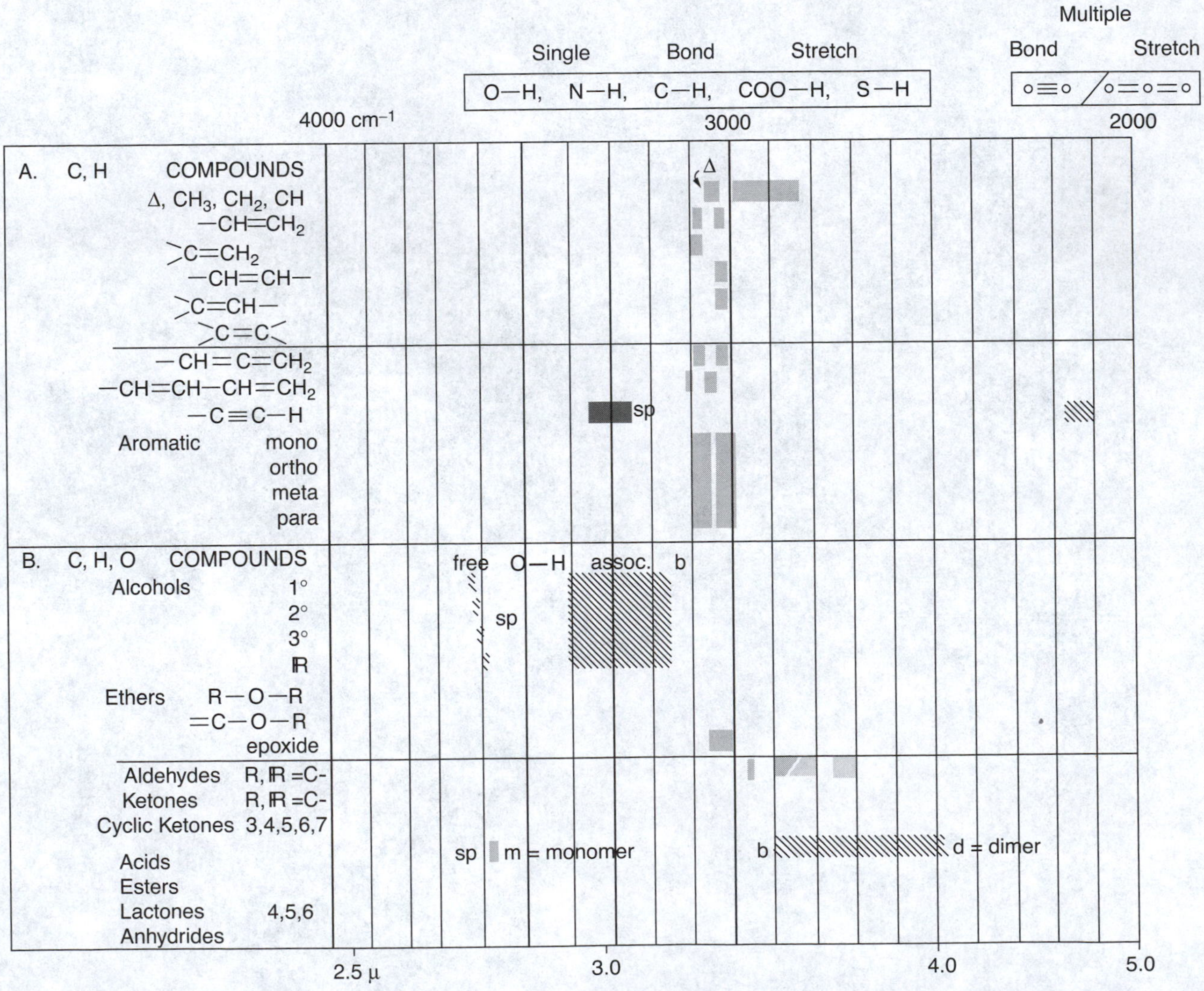

Analytical

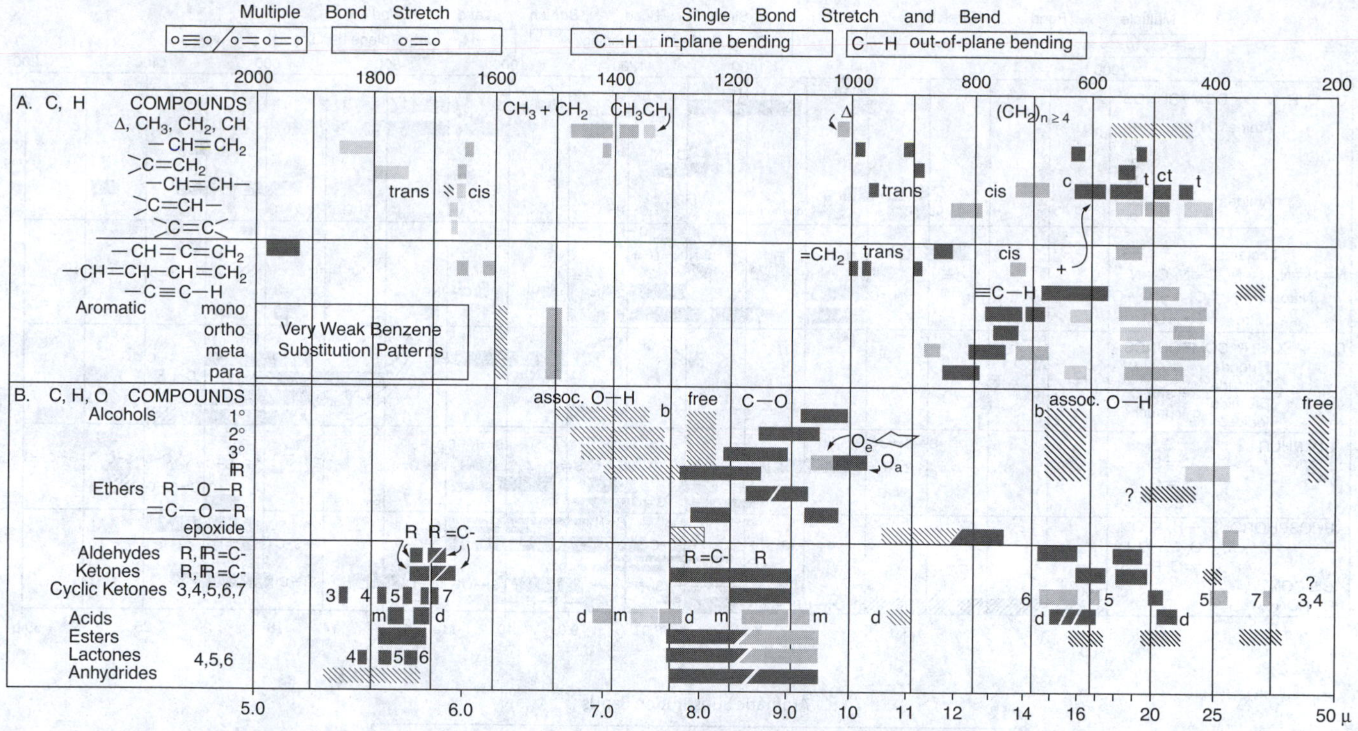
Multiple Bond Stretch
Single Bond Stretch and Bend
C—H in-plane bending
C—H out-of-plane bending
2000 1800 1600 1400 1200 1000 800 600 400 200
A. C, H COMPOUNDS
Δ, CH3, CH2, CH
—CH=CH2
>C=CH2
—CH=CH—
>C=CH—
>C=C<
—CH=C=CH2
—CH=CH—CH=CH2
—C≡C—H
Aromatic mono ortho meta para
Very Weak Benzene Substitution Patterns
CH3 + CH2
CH3CH
(CH2)n ≥ 4
trans cis
=CH2
≡C—H
B. C, H, O COMPOUNDS
Alcohols 1° 2° 3° ΦR
Ethers R—O—R
=C—O—R
epoxide
assoc. O—H
free
C—O
Aldehydes R, ΦR=C-
Ketones R, ΦR=C-
Cyclic Ketones 3,4,5,6,7
Acids
Esters
Lactones 4,5,6
Anhydrides
5.0 6.0 7.0 8.0 9.0 10 11 12 14 16 20 25 50 μ

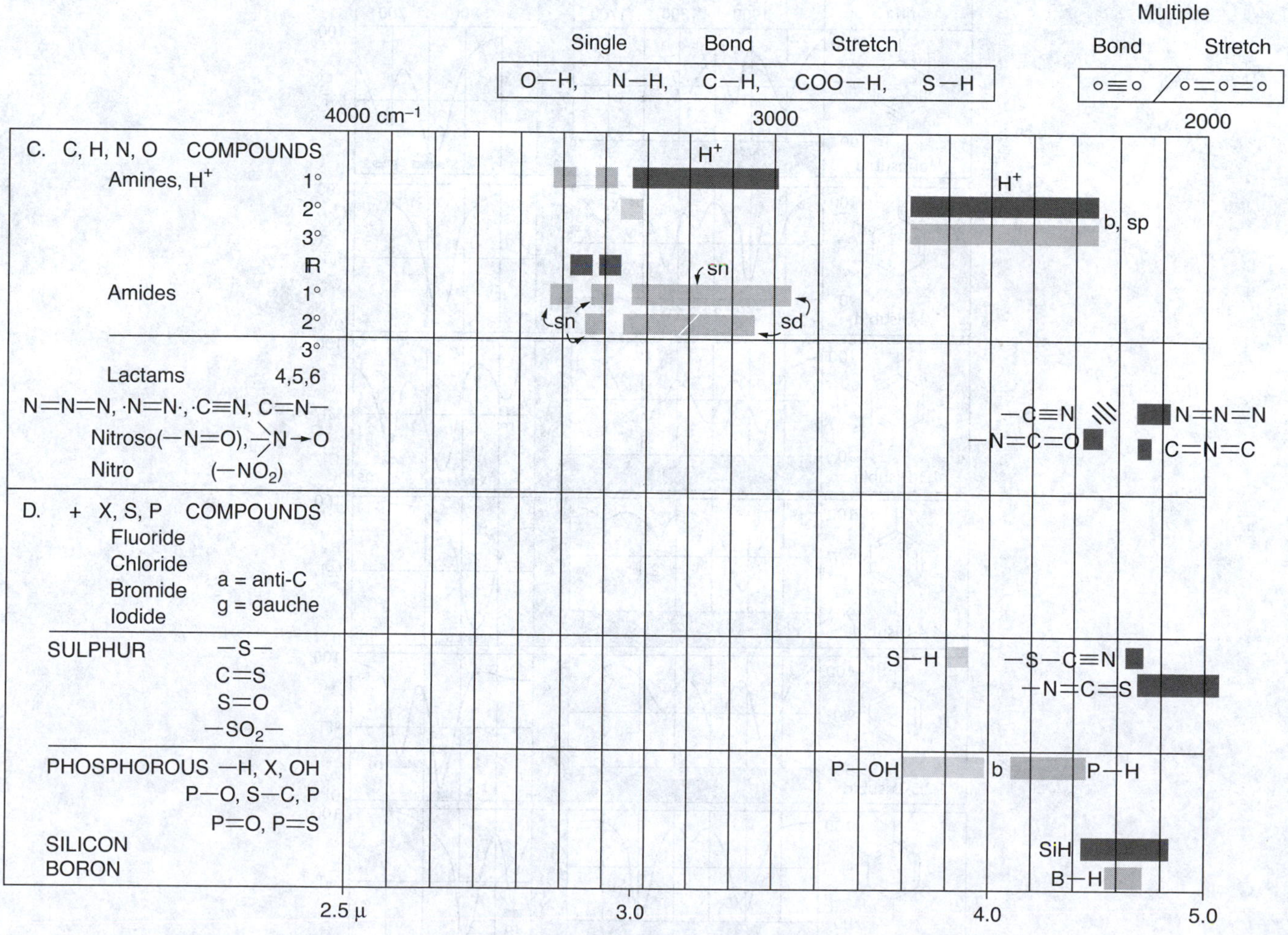
Single Bond Stretch
O—H, N—H, C—H, COO—H, S—H
Multiple Bond Stretch
4000 cm−1 3000 2000
C. C, H, N, O COMPOUNDS
Amines, H+ 1° 2° 3° ΦR
Amides 1° 2° 3°
Lactams 4,5,6
N=N=N, ·N=N·, ·C≡N, C=N—
Nitroso(—N=O), N→O
Nitro (—NO2)
H+
b, sp
sn
sd
—C≡N
—N=C=O
N=N=N
C=N=C
D. + X, S, P COMPOUNDS
Fluoride
Chloride
Bromide
Iodide
a = anti-C
g = gauche
SULPHUR —S— C=S S=O —SO2—
S—H
—S—C≡N
—N=C=S
PHOSPHOROUS —H, X, OH
P—O, S—C, P
P=O, P=S
P—OH
P—H
SILICON
BORON
SiH
B—H
2.5 μ 3.0 4.0 5.0

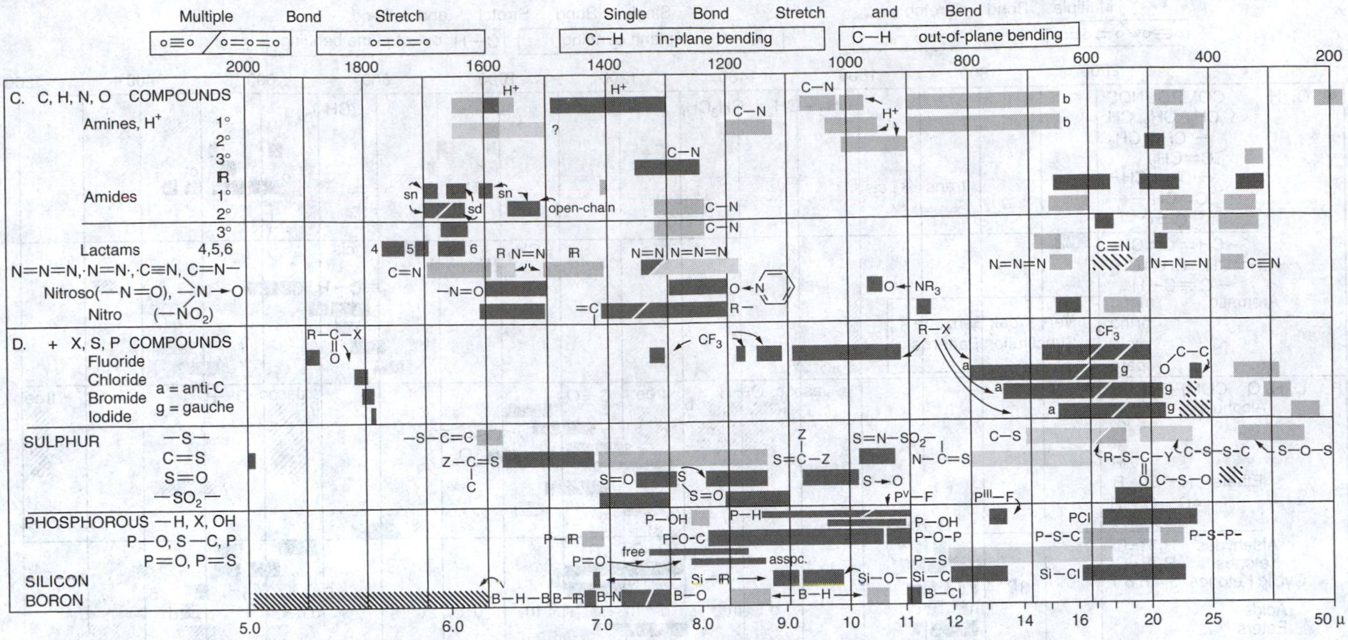

Aromatic Substitution Bands

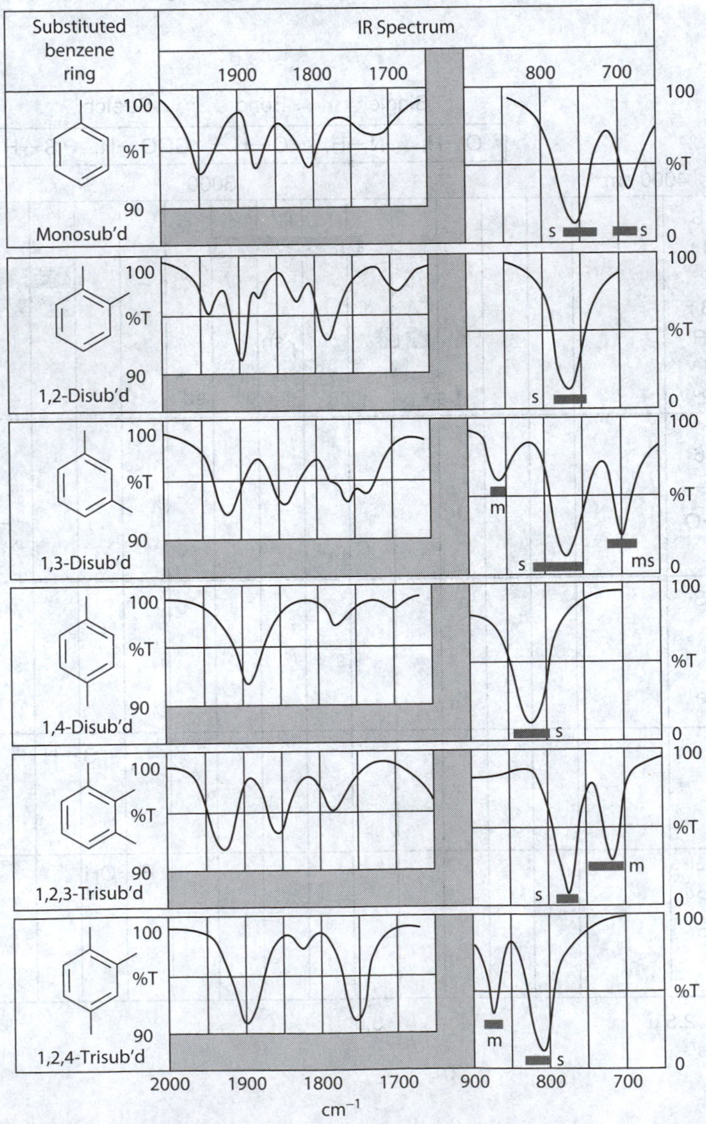

Aromatic Substitution Bands

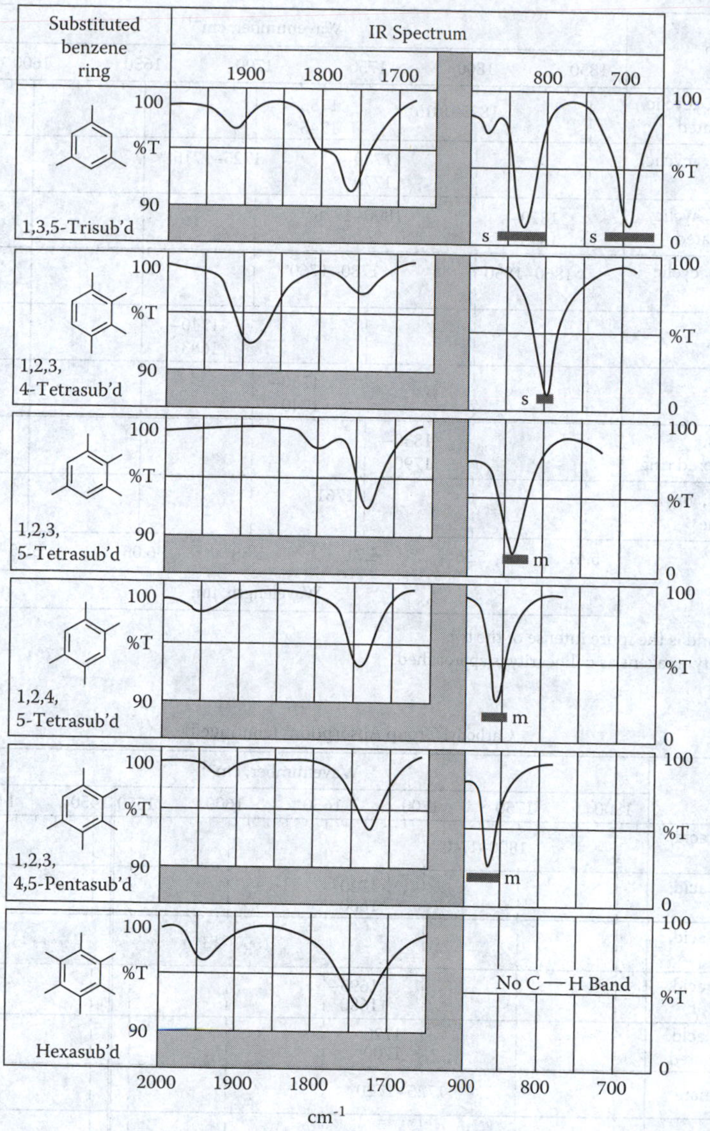

Carbonyl Group Absorptions

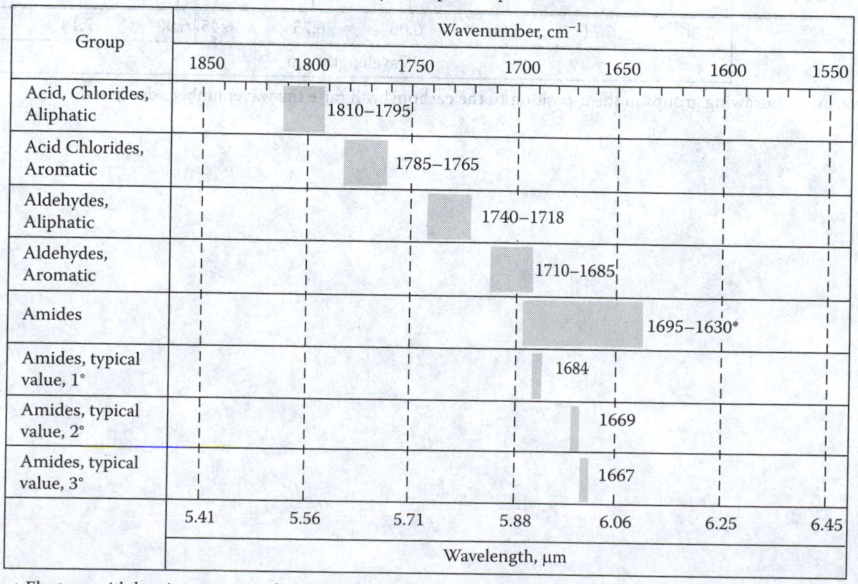

* Electron withdrawing groups at the α-position to the carbonyl will raise the wavenumber of the absorption.

Analytical

Carbonyl Group Absorptions (continued)

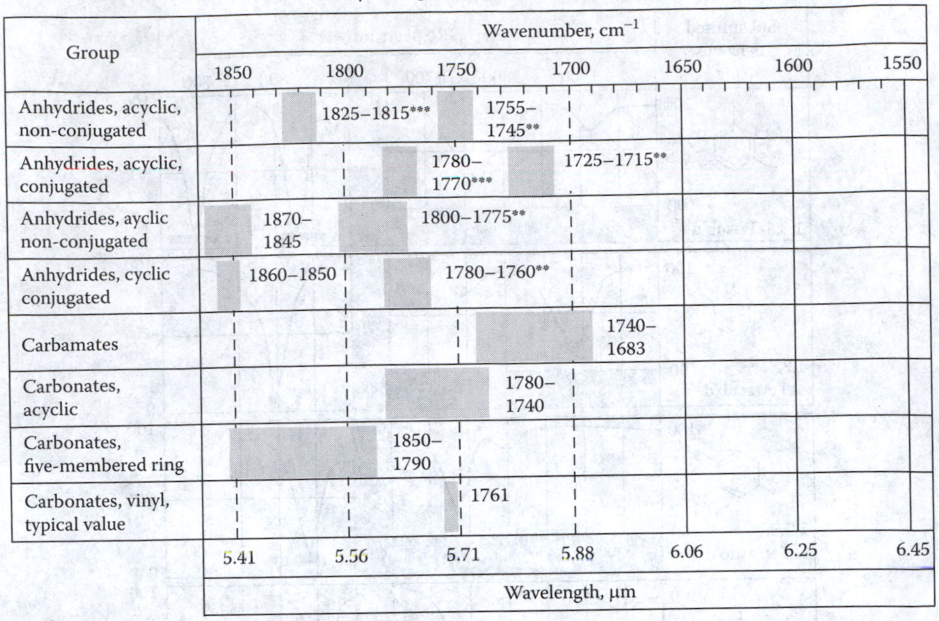

** This band is the more intense of the two.
*** Intensity weakens as colinearity is approached.

Carbonyl Group Absorptions (continued)

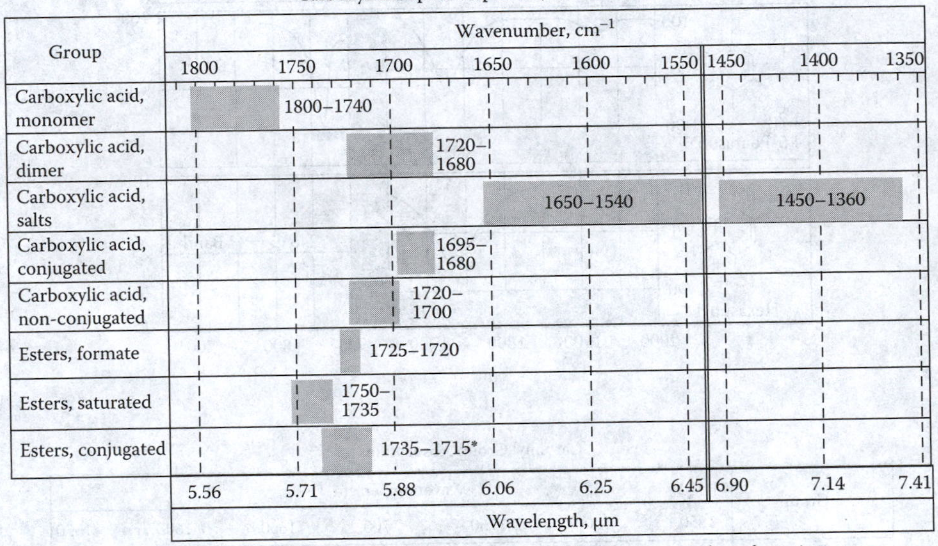

* Electron withdrawing groups in the α-position to the carbonyl will raise the wavenumber adsorption.

Analytical

Carbonyl Group Absorptions (continued)

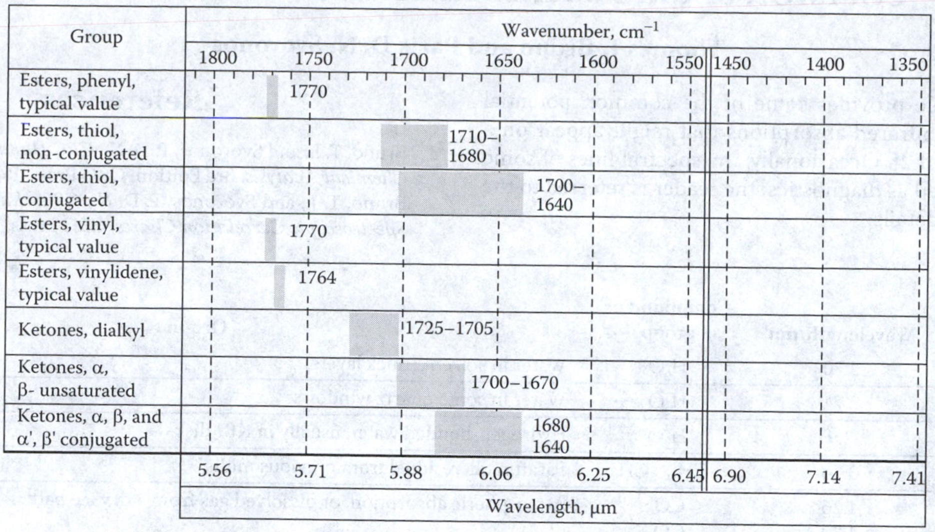

Carbonyl Group Absorptions (continued)

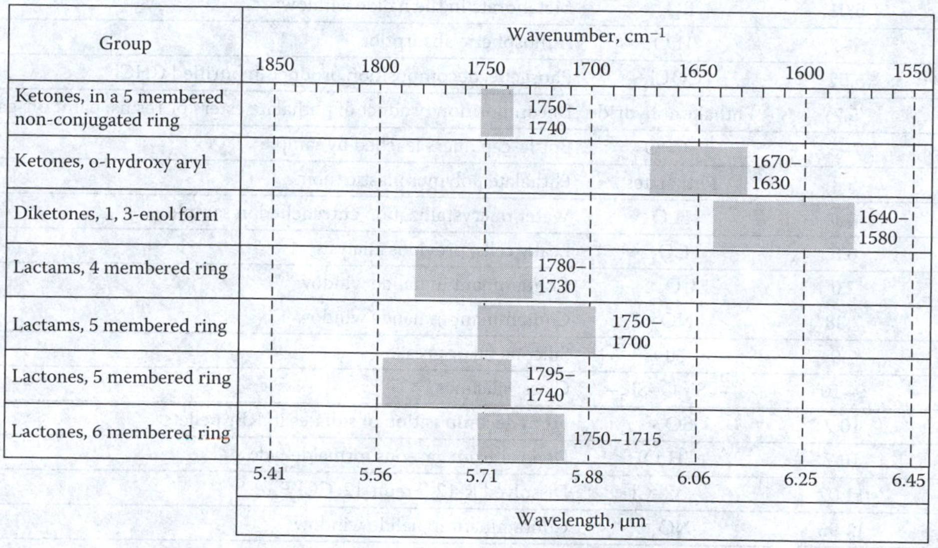

COMMON SPURIOUS INFRARED ABSORPTION BANDS

Thomas J. Bruno and Paris D. N. Svoronos

The following table provides some of the common potential sources of spurious infrared absorptions that might appear on a spectrum (Refs. 1 and 2]. Occasionally, the spectral lines of some impurities can be used as diagnostics; the reader is referred to the references for more details.

References

1. Bruno, T. J., and Svoronos, P. D. N., *CRC Handbook of Basic Tables for Chemical Analysis*, 3rd Edition, CRC Press, Boca Raton, FL, 2011.
2. Bruno, T. J., and Svoronos, P. D. N., *CRC Handbook of Fundamental Spectroscopic Correlation Charts*, CRC Press, Boca Raton, FL, 2006.

Approximate wavenumber/cm^{-1}	Wavelength/μm	Compound or group	Origin
3700	2.70	H_2O	Water in solvent (thick layers)
3650	2.74	H_2O	Water in some quartz windows
3450	2.9	H_2O	Hydrogen-bonded water, usually in KBr disks
2900	3.44	$-CH_3$, $>CH_2$	Paraffin oil, residual from previous mulls
2350 2330	4.26 4.30	CO_2 CO_2	Atmospheric absorption, or dissolved gas from a dry ice bath
2300 and 2150	4.35 and 4.65	CS_2	Leaky cells, previous analysis of samples dissolved in carbon disulfide
1996	5.01	BO_2^-	Metaborate in the halide window
1400–2000	5–7	H_2O	Atmospheric absorption
1820	5.52	$COCl_2$	Phosgene, decomposition product in purified $CHCl_3$
1755	5.7	Phthalic anhydride	Decomposition product of phthalate esters or resins; paint off-gas product
1700–1760	5.7–5.9	>C=O	Bottle-cap liners leached by sample
1720	5.8	Phthalates	Phthalate polymer plastic tubing
1640	6.1	H_2O	Water of crystallization entrenched in sample
1520	6.6	CO_2	Leaky cells, previous analysis
1430	7.0	CO_3^{-2}	Contaminant in halide window
1360	7.38	NO_3^-	Contaminant in halide window
1270	7.9	>SiO–	Silicone oil or grease
1000–1110	9–10	–>Si–O–Si<–	Glass; silicones
980	10.2	SO_4^{-2}	From decomposition of sulfates in KBr pellets
935	10.7	$(CH_2O)_x$	Deposit from gaseous formaldehyde
907	11.02	–>C–Cl	Dissolved R-12 (Freon-12, CCl_2F_2)
837	11.95	NO_3^-	Contaminant in halide window
823	12.15	KNO_3	From decomposition of nitrates in KBr pellets
794	12.6	CCl_4 vapor	Leaky cells, from CCl_4 used as a solvent
788	12.7	CCl_4 liquid	Incomplete drying of cell or contamination, from CCl_4 used as a solvent
720 and 730	13.7 and 13.9	Polyethylene	Various experimental sources
728	13.75	–>Si–F	SiF_4, found in NaCl windows
667	14.98	CO_3^{-2}	Atmospheric carbon dioxide
Any	Any	Fringes	If refractive index of windows is too high, or if the cell is partially empty, or the solid sample is not fully pulverized

NUCLEAR SPINS, MOMENTS, AND OTHER DATA RELATED TO NMR SPECTROSCOPY

David R. Lide

This table presents the following data relevant to nuclear magnetic resonance spectroscopy:

Z: Atomic number
Isotope: Element symbol and mass number
Abundance: Natural abundance of the isotope in percent. An * indicates a radioactive nuclide; if no value is given, the nuclide is not present in nature or its abundance is highly variable.
I: Nuclear spin
ν: Resonant frequency in megahertz for an applied field H_0 of 1 tesla (in cgs units, 10 kilogauss). The resonant frequency scales with H_0.
Relative sensitivity: Sensitivity relative to ^{1}H (=1) assuming an equal number of nuclei and constant temperature. Values were calculated from the expressions:

For constant H_0: $0.0076508(\mu/\mu_N)^3(I+1)/I^2$
For constant ν: $0.23871(\mu/\mu_N)(I+1)$

μ/μ_N: Nuclear magnetic moment in units of the nuclear magneton μ_N
Q: Nuclear quadrupole moment in units of femtometers squared (1 fm^2 = 10^{-2} barn). Because the determination of quadrupole moments requires knowledge of the electron configuration near the nucleus, values of *Q* in the literature tend to scatter considerably. The values quoted here come mainly from the review of Pyykkö (Ref. 3), otherwise from Ref. 1.

The table includes all stable nuclides of non-zero spin for which spin and magnetic moment values have been measured, as well as selected radioactive nuclides of current or potential interest. At least one isotope is included for each element through *Z* = 95 for which data are available. See Reference 1 for a complete listing of spins and moments.

The assistance of P. Pyykkö in providing data on nuclear quadrupole moments is gratefully acknowledged.

References

1. Holden, N. E., Table of the Isotopes, in Haynes, W. M., Ed., *CRC Handbook of Chemistry and Physics*, 96th Ed., CRC Press, Boca Raton, FL, 2015.
2. Raghavan, P., *At. Data Nucl. Data Tables* 42, 189, 1989.
3. Pyykkö, P., *Mol. Phys.* 106, 1965, 2008.
4. Stone, N. J., *At. Data Nucl. Data Tables* 90, 75, 2005.
5. IUPAC Commission on Physiochemical Symbols, Terminology and Units, *Quantities, Units, and Symbols in Physical Chemistry*, 3rd Ed., Royal Society of Chemistry, Cambridge, 2007.

Z	Isotope	Abundance %	I	ν/MHz for H_0 = 1 T	Relative Sensitivity Const. H_0	Const. ν	μ/μ_N	Q/fm^2
1	^{1}n		1/2	29.1647	0.32139	0.6850	−1.91304272	
1	^{1}H	99.9885	1/2	42.5759	1.00000	1.0000	+2.792847337	
1	^{2}H	0.0115	1	6.53566	0.00965	0.4094	+0.857438228	+0.2860
1	^{3}H	*	1/2	45.4129	1.21354	1.0667	+2.9789625	
2	^{3}He	0.000134	1/2	32.4380	0.44220	0.7619	−2.127750	
3	^{6}Li	7.59	1	6.2661	0.00850	0.3925	+0.8220467	−0.0808
3	^{7}Li	92.41	3/2	16.5483	0.29356	1.9434	+3.25644	−4.01
4	^{9}Be	100	3/2	5.9842	0.01388	0.7028	−1.1776	+5.288
5	^{10}B	19.9	3	4.5752	0.01985	1.7193	+1.800645	+8.459
5	^{11}B	80.1	3/2	13.6630	0.16522	1.6045	+2.688649	+4.059
6	^{13}C	1.07	1/2	10.7054	0.01591	0.2515	+0.7024118	
7	^{14}N	99.636	1	3.0756	0.00101	0.1928	+0.4037610	+2.044
7	^{15}N	0.364	1/2	4.3142	0.00104	0.1014	−0.2831888	
8	^{17}O	0.038	5/2	5.772	0.02910	1.5822	−1.89379	−2.558
9	^{19}F	100	1/2	40.0541	0.83400	0.9413	+2.628868	
10	^{21}Ne	0.27	3/2	3.3631	0.00246	0.3949	−0.661797	+10.155
11	^{23}Na	100	3/2	11.2688	0.09270	1.3234	+2.217522	+10.4
12	^{25}Mg	10.00	5/2	2.6083	0.00268	0.7147	−0.85545	+19.94
13	^{27}Al	100	5/2	11.1031	0.20689	3.0424	+3.641507	+14.66
14	^{29}Si	4.685	1/2	8.4578	0.00786	0.1988	−0.55529	
15	^{31}P	100	1/2	17.235	0.06652	0.4052	+1.13160	
16	^{33}S	0.75	3/2	3.2654	0.00227	0.3842	+0.6438212	−6.78
17	^{35}Cl	75.76	3/2	4.1717	0.00472	0.4905	+0.8218743	−8.165
17	^{37}Cl	24.24	3/2	3.4765	0.00272	0.4083	+0.6841236	−6.435
18	^{37}Ar	*	3/2	5.819	0.01276	0.6833	+1.145	+7.6
18	^{39}Ar	*	7/2	3.46	0.01130	1.7080	−1.59	−12
19	^{39}K	93.2581	3/2	1.9893	0.00051	0.2336	+0.3914662	+5.85
19	^{40}K	0.0117	4	2.4737	0.00523	1.5493	−1.298100	−7.3
19	^{41}K	6.7302	3/2	1.0919	0.00008	0.1282	+0.2148701	+7.11

Analytical

Z	Isotope	Abundance %	I	ν/MHz for H_0 = 1 T	Relative Sensitivity Const. H_0	Const. ν	μ/μ_N	*Q*/fm²
20	^{43}Ca	0.135	7/2	2.8697	0.00643	1.4154	−1.317643	−4.08
21	^{45}Sc	100	7/2	10.3591	0.30244	5.1094	+4.756487	−22.0
22	^{47}Ti	7.44	5/2	2.4041	0.00210	0.6588	−0.78848	+30.2
22	^{49}Ti	5.41	7/2	2.4048	0.00378	1.1861	−1.10417	+24.7
23	^{50}V	0.250	6	4.2505	0.05571	5.5905	+3.345689	+21
23	^{51}V	99.750	7/2	11.2133	0.38360	5.5307	+5.1487057	−5.2
24	^{53}Cr	9.501	3/2	2.4115	0.00091	0.2832	−0.47454	−15
25	^{55}Mn	100	5/2	10.5763	0.17881	2.8981	+3.46872	+33
26	^{57}Fe	2.119	1/2	1.3816	0.00003	0.0324	+0.0906230	+16
27	^{59}Co	100	7/2	10.077	0.27841	4.9703	+4.627	+42
28	^{61}Ni	1.1399	3/2	3.8114	0.00359	0.4476	−0.75002	+16.2
29	^{63}Cu	69.15	3/2	11.3188	0.09393	1.3292	+2.2273456	−22.0
29	^{65}Cu	30.85	3/2	12.1027	0.11484	1.4213	+2.38161	−20.4
30	^{67}Zn	4.102	5/2	2.6685	0.00287	0.7312	+0.875205	+15.0
31	^{69}Ga	60.108	3/2	10.2478	0.06971	1.2035	+2.01659	+17.1
31	^{71}Ga	39.892	3/2	13.0208	0.14300	1.5291	+2.56227	+10.7
32	^{73}Ge	7.76	9/2	1.4897	0.00141	1.1547	−0.8794677	−19.6
33	^{75}As	100	3/2	7.3150	0.02536	0.8590	+1.439475	+31.4
34	^{77}Se	7.63	1/2	8.1568	0.00703	0.1916	+0.5350422	
35	^{79}Br	50.69	3/2	10.667	0.07945	1.2570	+2.106400	+31.3
35	^{81}Br	49.31	3/2	11.498	0.09951	1.3550	+2.270562	+26.2
36	^{83}Kr	11.500	9/2	1.6442	0.00190	1.2744	−0.970669	+25.9
37	^{85}Rb	72.17	5/2	4.1253	0.01061	1.1304	+1.35298	+27.6
37	^{87}Rb	27.83	3/2	13.9814	0.17704	1.6419	+2.75131	+13.35
38	^{87}Sr	7.00	9/2	1.8525	0.00272	1.4358	−1.093603	+30.5
39	^{89}Y	100	1/2	2.0949	0.00012	0.0492	−0.1374154	
40	^{91}Zr	11.22	5/2	3.9748	0.00949	1.0892	−1.30362	−17.6
41	^{93}Nb	100	9/2	10.4523	0.48821	8.1013	+6.1705	−32
42	^{95}Mo	15.90	5/2	2.7874	0.00327	0.7638	−0.9142	−2.2
42	^{97}Mo	9.56	5/2	2.8463	0.00349	0.7799	−0.9335	+25.5
43	^{99}Tc	*	9/2	9.6294	0.38174	7.4635	+5.6847	−12.9
44	^{99}Ru	12.76	5/2	1.9553	0.00113	0.5358	−0.6413	+7.9
44	^{101}Ru	17.06	5/2	2.1916	0.00159	0.6005	−0.7188	+45.7
45	^{103}Rh	100	1/2	1.3477	0.00003	0.0317	−0.08840	
46	^{105}Pd	22.33	5/2	1.957	0.00113	0.5364	−0.642	+66.0
47	^{107}Ag	51.839	1/2	1.7331	0.00007	0.0407	−0.1136796	
47	^{109}Ag	48.161	1/2	1.9924	0.00010	0.0468	−0.1306906	
48	^{111}Cd	12.80	1/2	9.0692	0.00966	0.2130	−0.5948861	
48	^{113}Cd	12.22	1/2	9.4871	0.01106	0.2228	−0.6223009	
49	^{113}In	4.29	9/2	9.3655	0.35121	7.2589	+5.5289	+75.9
49	^{115}In	95.71	9/2	9.3856	0.35348	7.2745	+5.5408	+77.0
50	^{115}Sn	0.34	1/2	14.0077	0.03561	0.3290	−0.91883	
50	^{117}Sn	7.68	1/2	15.2610	0.04605	0.3584	−1.00104	
50	^{119}Sn	8.59	1/2	15.9660	0.05273	0.3750	−1.04728	
51	^{121}Sb	57.21	5/2	10.2551	0.16302	2.8101	+3.3634	−54.3
51	^{123}Sb	42.79	7/2	5.5532	0.04659	2.7390	+2.5498	−69.2
52	^{123}Te	0.89	1/2	11.2349	0.01837	0.2639	−0.7369478	
52	^{125}Te	7.07	1/2	13.5446	0.03219	0.3181	−0.8884509	
53	^{127}I	100	5/2	8.5778	0.09540	2.3504	+2.813273	−69.6
54	^{129}Xe	26.4006	1/2	11.8604	0.02162	0.2786	−0.7779763	
54	^{131}Xe	21.2324	3/2	3.5159	0.00282	0.4129	+0.6918619	−11.4
55	^{133}Cs	100	7/2	5.6234	0.04838	2.7736	+2.582025	−0.343
56	^{135}Ba	6.592	3/2	4.2617	0.00501	0.5005	+0.838627	+16.0
56	^{137}Ba	11.232	3/2	4.7634	0.00700	0.5594	+0.937365	+24.5
57	^{138}La	0.090	5	5.6615	0.09404	5.3189	+3.713646	+45
57	^{139}La	99.910	7/2	6.0612	0.06058	2.9895	+2.7830455	+20.0
58	^{137}Ce	*	3/2	4.88	0.00752	0.5729	0.96	
58	^{139}Ce	*	3/2	5.39	0.01012	0.6326	1.06	
58	^{141}Ce	*	7/2	2.37	0.00364	1.1709	1.09	
59	^{141}Pr	100	5/2	13.0359	0.33483	3.5720	+4.2754	−5.9

Z	Isotope	Abundance %	I	ν/MHz for H_0 = 1 T	Relative Sensitivity Const. H_0	Const. ν	μ/μ_N	Q/fm²
60	^{143}Nd	12.2	7/2	2.319	0.00339	1.1440	−1.065	−63
60	^{145}Nd	8.3	7/2	1.429	0.00079	0.7047	−0.656	−33
61	^{143}Pm	*	5/2	11.59	0.23510	3.1748	+3.80	
61	^{147}Pm	*	7/2	5.62	0.04827	2.7714	+2.58	+74
62	^{147}Sm	14.99	7/2	1.7748	0.00152	0.8754	−0.8149	−26
62	^{149}Sm	13.82	7/2	1.4631	0.00085	0.7216	−0.6718	+7.4
63	^{151}Eu	47.81	5/2	10.5856	0.17929	2.9006	+3.4718	+90.3
63	^{153}Eu	52.19	5/2	4.6745	0.01544	1.2809	+1.5331	+241
64	^{155}Gd	14.80	3/2	1.312	0.00015	0.1541	−0.2582	+127
64	^{157}Gd	15.65	3/2	1.720	0.00033	0.2020	−0.3385	+135
65	^{159}Tb	100	3/2	10.23	0.06945	1.2019	+2.014	+143.2
66	^{161}Dy	18.889	5/2	1.4654	0.00048	0.4015	−0.4806	+250.7
66	^{163}Dy	24.896	5/2	2.0508	0.00130	0.5619	+0.6726	+265
67	^{165}Ho	100	7/2	9.0883	0.20423	4.4826	+4.173	+358
68	^{167}Er	22.869	7/2	1.2281	0.00050	0.6057	−0.5639	+356.5
69	^{169}Tm	100	1/2	3.531	0.00057	0.0829	−0.2316	−120
70	^{171}Yb	14.28	1/2	7.5261	0.00552	0.1768	+0.49367	
70	^{173}Yb	16.13	5/2	2.0730	0.00135	0.5680	−0.67989	+280
71	^{175}Lu	97.41	7/2	4.8626	0.03128	2.3984	+2.2327	+349
71	^{176}Lu	2.59	7	3.451	0.03975	6.0518	+3.169	+497
72	^{177}Hf	18.60	7/2	1.7282	0.00140	0.8524	+0.7935	+336.5
72	^{179}Hf	13.62	9/2	1.0856	0.00055	0.8414	−0.6409	+379.3
73	^{181}Ta	99.988	7/2	5.1627	0.03744	2.5464	+2.3705	+317
74	^{183}W	14.31	1/2	1.7716	0.00008	0.0422	+0.1177848	
75	^{185}Re	37.40	5/2	9.7176	0.13870	2.6628	+3.1871	+218
75	^{187}Re	62.60	5/2	9.8170	0.14300	2.6900	+3.2197	+207
76	^{187}Os	1.96	1/2	0.9856	0.00001	0.0231	+0.06465189	
76	^{189}Os	16.15	3/2	3.3536	0.00244	0.3938	+0.659933	+85.6
77	^{191}Ir	37.3	3/2	0.7658	0.00003	0.0899	+0.1507	+81.6
77	^{193}Ir	62.7	3/2	0.8319	0.00004	0.0977	+0.1637	+75.1
78	^{195}Pt	33.832	1/2	9.2922	0.01039	0.2182	+0.60952	
79	^{197}Au	100	3/2	0.7406	0.00003	0.0870	+0.145746	+54.7
80	^{199}Hg	16.87	1/2	7.7123	0.00594	0.1811	+0.5058855	
80	^{201}Hg	13.18	3/2	2.8469	0.00149	0.3343	−0.5602257	+38.7
81	^{203}Tl	29.52	1/2	24.7316	0.19598	0.5809	+1.6222579	
81	^{205}Tl	70.48	1/2	24.9749	0.20182	0.5866	+1.6382146	
82	^{207}Pb	22.1	1/2	9.0340	0.00955	0.2122	+0.59258	
83	^{209}Bi	100	9/2	6.9630	0.14433	5.3968	+4.1106	−51.6
84	^{209}Po	*	1/2	11.7	0.02096	0.2757	+0.77	
86	^{211}Rn	*	1/2	9.16	0.00997	0.2152	+0.601	
87	^{223}Fr	*	3/2	5.95	0.01362	0.6982	+1.17	+117
88	^{223}Ra	*	3/2	1.3746	0.00017	0.1614	+0.2705	+121
88	^{225}Ra	*	1/2	11.187	0.01814	0.2627	−0.7338	
89	^{227}Ac	*	3/2	5.6	0.01131	0.6565	+1.1	+170
90	^{229}Th	*	5/2	1.40	0.00042	0.3843	+0.46	+430
91	^{231}Pa	100	3/2	10.2	0.06903	1.1995	2.01	−172
92	^{235}U	0.7204	7/2	0.83	0.00015	0.4082	−0.38	+493.6
93	^{237}Np	*	5/2	9.57	0.13264	2.6234	+3.14	+388.6
94	^{239}Pu	*	1/2	3.09	0.00038	0.0727	+0.203	
95	^{243}Am	*	5/2	4.6	0.01446	1.2532	+1.5	+421

PROPERTIES OF IMPORTANT NMR NUCLEI

Thomas J. Bruno and Paris D. N. Svoronos

The following table lists the magnetic properties at higher field strengths required for choosing the nuclei to be used in NMR experiments (Refs. 1–15). The table shows NMR frequency at the indicated field strength for a variety of isotopes. The reader is referred to several excellent texts and the literature for guidelines in nucleus selection. For more detailed information on these and other less common nuclei at 10 kG, the reader should consult the table entitled "Nuclear Spins, Moments and Other Data Related to NMR Spectroscopy" in this section.

References

1. Silverstein, R. M., Bassler, G. C., and Morrill, T. C., *Spectrometric Identification of Organic Compounds*, 5th Edition, John Wiley and Sons, New York, 1991.
2. Yoder, C. H., and Shaeffer, C. D., *Introduction to Multinuclear NMR*, Benjamin/Cummings, Menlo Park, CA, 1987.
3. Gordon, A. J., and Ford, R. A., *The Chemist's Companion*, Wiley Interscience, New York, 1971.
4. Silverstein, R. M., and Webster F. X., *Spectrometric Identification of Organic Compounds, 6th Edition*, John Wiley and Sons, New York, 1998.
5. Becker, E. D., *High Resolution NMR, Theory and Chemical Applications, 2nd Edition*, Academic Press, New York, 1980.
6. Gunther, H., *NMR Spectroscopy: Basic Principles, Concepts and Applications in Chemistry*, John Wiley and Sons, New York, 2003.
7. Rahman, A.-u., *Nuclear Magnetic Resonance*, Springer-Verlag, New York, 1986.
8. Harris, R. K., *Chem. Soc. Rev.* 5, 1, 1976.
9. Kitamaru, R., *Nuclear Magnetic Resonance: Principles and Theory*, Elsevier Science, 1990.
10. Lambert, J. B., Holland, L. N., and Mazzola, E. P., *Nuclear Magnetic Resonance Spectroscopy: Introduction to Principles, Applications and Experimental Methods*, Prentice Hall, Englewood Cliffs, NJ, 2003.
11. Bovey, F. A., and Mirau, P. A., *Nuclear Magnetic Resonance Spectroscopy, 2nd Edition*, Academic Press, New York, 1988.
12. Harris, R. K., and Mann, B. E., *NMR and the Periodic Table*, Academic Press, London, 1978.
13. Hore, P. J., *Nuclear Magnetic Resonance*, Oxford University Press, Oxford, 1995.
14. Nelson, J. H., *Nuclear Magnetic Resonance Spectroscopy, 2nd Edition*, John Wiley and Sons, New York, 2003.
15. Bruno, T. J., and Svoronos, P. D. N., *CRC Handbook of Basic Tables for Chemical Analysis, 3rd Edition*, CRC Press, Boca Raton, FL, 2011.

NMR Frequency of Important Nuclei (in MHz) at the Indicated Field Strength in kG**

Isotope	Percent abundance	Spin I	10.000	14.092	21.139	23.487	51.567	93.950	140.925	223.131
$^{1}_{1}H$	99.985	1/2	42.5759	60.0000	90.0000	100.0000	220.0000	400.0000	600.0000	950.0000
$^{2}_{1}H$	0.015	1	6.53566	9.21037	13.81555	15.35061	33.77134	61.40262	92.10380	145.9830
$^{3}_{1}H^{*}$		1/2	45.4129	63.9980	95.9971	106.6634	234.6595	426.6542	639.9813	1013.3024
$^{13}_{6}C$	1.108	1/2	10.7054	15.0866	22.6298	25.1443	55.3174	100.5735	150.8659	2388.5150
$^{14}_{7}N$	99.635	1	3.0756	4.3343	6.5014	7.2238	15.924	28.9104	43.3615	68.6557
$^{15}_{7}N$	0.365	1/2	4.3142	6.0798	9.1197	10.1330	22.2925	40.5306	60.7960	96.2601
$^{17}_{8}O$	0.037	5/2	5.772	8.134	12.201	13.557	29.825	54.1811	81.3186	128.5801
$^{19}_{9}F$	100	1/2	40.0541	42.3537	63.5305	94.0769	206.9692	376.2515	564.3781	893.5963
$^{29}_{14}Si$	4.70	1/2	8.4578	11.9191	17.8787	19.8652	43.7035	79.4638	119.1956	188.72
$^{31}_{15}P$	100	1/2	17.235	24.288	36.433	40.481	89.057	161.9828	242.9741	384.7086
$^{33}_{16}S$	0.76	3/2	3.2654	4.6018	6.9026	7.6696	16.8731	30.6826	46.0238	72.8710
$^{35}_{16}S^{*}$		3/2	5.08	7.16	10.74	11.932	26.250	47.7267	71.5875	113.3508
$^{35}_{17}Cl$	75.53	3/2	4.1717	5.8790	8.8184	9.7983	21.5562	39.1948	58.7902	93.0876
$^{36}_{17}Cl^{*}$		2	4.8931	6.8956	10.3434	11.4927	25.2838	45.9638	68.9432	109.1639
$^{76}_{35}Br^{*}$		1	4.18	5.89	8.84	9.82	21.60	39.2768	58.9130	93.2822
$^{79}_{35}Br$	50.54	3/2	10.667	15.032	22.549	25.054	55.119	100.2133	150.3202	238.0064
$^{81}_{35}Br$	49.46	3/2	11.498	16.204	24.305	27.006	59.413	108.0258	162.0386	256.5608
$^{183}_{74}W$	14.40	1/2	1.7716	2.4966	3.7449	4.1610	9.1543	16.6430	24.9646	39.5272

* Nucleus is radioactive

** 1 kG = 10^{-1} T, the corresponding SI unit

Analytical

PROTON NMR ABSORPTION OF MAJOR CHEMICAL FAMILIES

Thomas J. Bruno and Paris D. N. Svoronos

The following table gives the region of the expected nuclear magnetic resonance absorptions of major chemical families (Refs. 1–12). These absorptions are reported in the dimensionless units of parts per million (ppm) versus the standard compound tetramethylsilane (TMS), which is recorded as 0.0 ppm.

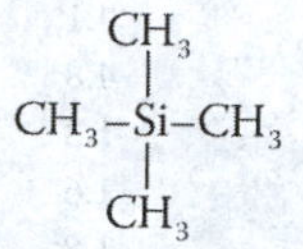

The use of this unit of measure makes the chemical shifts independent of the applied magnetic field strength or the radio frequency. For most proton NMR spectra, the protons in TMS are more shielded than almost all other protons. The chemical shift in this dimensionless unit system is then defined by:

$$\delta = \frac{\nu_s - \nu_r}{\nu_r} \times 10^6$$

where ν_s and ν_r are the absorption frequencies of the sample proton and the reference (TMS) protons (twelve, magnetically equivalent), respectively. In these tables, the proton(s) whose proton NMR shifts are cited are indicated by underscore. Reference 1 provides additional details on the absorptions of other moieties, as well as correlation charts.

References

1. Bruno, T. J., and Svoronos, P. D. N., *CRC Handbook of Basic Tables for Chemical Analysis*, 3rd Edition, CRC Press, Boca Raton, FL, 2011.
2. Silverstein, R. M., and Webster, F. X., *Spectrometric Identification of Organic Compounds*, 6th Edition, Wiley, New York, 1998.
3. Rahman, A.-u., *Nuclear Magnetic Resonance*, Springer Verlag, New York, 1986.
4. Gordon, A. J., and Ford, R. A., *The Chemist's Companion*, Wiley Interscience, New York, 1971.
5. Becker, E. D., *High Resolution NMR, Theory and Chemical Applications*, 2nd Edition, Academic Press, New York, 1980.
6. Gunther, H., *NMR Spectroscopy: Basic Principles, Concepts and Applications in Chemistry, Wiley*, New York, 2003.
7. Kitamaru, R., *Nuclear Magnetic Resonance: Principles and Theory*, Elsevier Science, 1990.
8. Lambert, J. B., Holland, L. N., and Mazzola, E. P., *Nuclear Magnetic Resonance Spectroscopy: Introduction to Principles, Applications and Experimental Methods*, Prentice Hall, Englewood Cliffs, NJ , 2003.
9. Bovey, F. A., and Mirau, P. A., *Nuclear Magnetic Resonance Spectroscopy*, 2nd Edition, Academic Press, New York, 1988.
10. Hore, P. J., *Nuclear Magnetic Resonance*, Oxford University Press, Oxford, 1995.
11. Nelson, J. H., *Nuclear Magnetic Resonance Spectroscopy*, 2nd Edition, Wiley, New York, 2003.
12. Abraham, R. J., Fisher, J., and Loftus, P., *Introduction to NMR Spectroscopy*, Wiley, New York, 1988.

Proton NMR Absorption of Hydrocarbons

Group	δ/ppm of underlined protons
Alkanes	
CH_3-R	~0.8
-CH_2-R	~1.1
>CH-R	~1.4
Cyclopropane	0.2
Alkenes	
CH_3-C=C<	~1.6
-CH_2-C=C<	~2.1
>CH- C=C<	~2.5
CH_3-C-C=C<	~1.0
-CH_2-C-C=C<	~1.4
>CH-C-C=C<	~1.8
>C=C-H	4.2 - 6.2
Alkynes	
CH_3-C≡C-	~1.7
-CH_2-C≡C-	~2.2
>CH -C≡C-	~2.7
CH_3-C-C≡C-	~1.2
>CH-C-C≡C-	~1.5
>CH -C-C≡C-	~1.8
R-C≡C-H	~2.4
Aromatics	
C_6H_5-G	8.5 - 6.9

Proton NMR Absorption of Organic Oxygen Compounds

Group	Approximate δ/ppm of underlined protons
Alcohols	
CH_3-OH	3.2
RCH_2-OH	3.4
R_2CH-OH	3.6
CH_3-C-OH	1.2
RCH_2-C-OH	1,5
R_2CH-C-OH	1.8
R-O-H	1 to 5
Aldehydes	
CH_3-CHO	2.2
RCH_2-CHO	2.4
R_2CH-CHO	2.5
CH_3-C-CHO	1.1
RCH_2-C-CHO	1.6
Amides	
See organic nitrogen compounds	
Anhydrides, acylic	
CH_3-C(=O)O-	1.8
RCH_2-C(=O)O-	2.1
R_2CH-C(=O)O-	2.3
CH_3-C-C(=O)O-	1.2
RCH_2-C-C(=O)O-	1.8
R_2CH-C-C(=O)O-	2.0
Anhydrides, cyclic	
Succinic anhydride	3.0
Maleic anhydride	7.1
Carboxylic acids	
CH_3-COOH	2.1
RCH_2-COOH	2.3
R_2CH-COOH	2.5
CH_3-C-COOH	1.1
R-CH_2-C-COOH	1.6
R_2CH-C-COOH	2.0
R-COO-H	11 to 12
Cyclic ethers	
Oxirane	2.5
Oxetane	4.7 (adjacent to O), 2.7
Tetrahydrofuran	3.8 (adjacent to O), 1.9
Tetrahydropyran	3.6 (adjacent to O), 1.6, 1.6
1,4-Dioxane	3.6
1,3-Dioxane	4.7, 3.8 1.7
Furan	7.4 (adjacent to O), 6.3
Dihydropyran	6.2 (adjacent to O), 4.5, 1.9
Epoxides	
See cyclic ethers	
Esters	
CH_3-COOR (R=alkyl)	1.9
CH_3-COOR (R=aryl)	2.0
RCH_2-COOR (R=alkyl)	2.1
RCH_2-COOR (R=aryl)	2.2
R_2CH-COOR (R=alkyl)	2.3
R_2CH-COOR (R=aryl)	2.4
CH_3-C-COOR	1.1
RCH_2-C-COOR	1.7
R_2CH-C-COOR	1.9
CH_3-OOC-R	3.6
RCH_2-OOC-R	4.1
R_2CH-OOC-R	4.8
CH_3-C-OOC-R	1.3
RCH_2-C-OOC-R	1.6
R_2CH-C-OOC-R	1.8
Cyclic esters	
2H-Pyran-2-one	H_3=2.3, H_4=1.6, H_5=1.6, H_6=4.1
Ethers	
CH_3-O-R (R=alkyl)	3.2
CH_3-O-R (R=aryl)	3.9
RCH_2-O-R (R=alkyl)	3.4
RCH_2-O-R (R=aryl)	4.1
R_2CH-O-R (R=alkyl)	3.6
R_2CH-O-R (R=aryl)	4.5
CH_3-C-O-R (R=alkyl)	1.2
CH_3-C-O-R (R=aryl)	1.3
RCH_2-C-O-R (R=alkyl)	1.5
RCH_2-C-O-R (R=aryl)	1.6
R_2CH-C-O-R (R=alkyl)	1.8
R_2CH-C-O-R (R=aryl)	2.0
Isocyanates	
See nitrogen compounds	
Ketones	
CH_3-C(=O)-	1.9
CH_3-C(=O)-	2.4
RCH_2-C(=O)- (R=alkyl)	2.1
RCH_2-C(=O)- (R=aryl)	2.7
R_2CH-C(=O)- (R=alkyl)	2.3
R_2CH-C(=O)- (R=aryl)	3.4
CH_3-C(=O)-	1.1
CH_3-C(=O)-	1.2
RCH_2-C(=O)- (R=alkyl)	1.6
RCH_2-C(=O)- (R=aryl)	1.6
R_2CH-C(=O)- (R=alkyl)	2.0
R_2CH-C(=O)- (R=aryl)	2.1
Cyclic ketones (*n* = ring carbons):	
α-hydrogens, *n*=3	1.7
α-hydrogens, *n*=4	3.0
α-hydrogens, *n*>5	2.0 to 2.3
β-hydrogens	1.9 to 1.5
Lactones	
See cyclic esters	
Nitro compounds	
See organic nitrogen compounds	
Phenols	
Ar-O-H	9 to 10

Proton NMR Absorption of Organic Nitrogen Compounds

Amides:

δ/ppm of underlined protons	Primary R–C(=O) NH_2 δ/ppm	Secondary R–C(=O) NHR_1 δ/ppm	Tertiary R–C(=O) NR_1R_2 δ/ppm
(i) N-substitution			
$\underline{R–C(=O)N–H}$	5–12	5–12	
(a) alpha			
$–C(=O)N–\underline{CH_3}$		~2.9	~2.9
$–C(=O)N–\underline{CH_2}–$		~3.4	~3.4
$–C(=O)N–\underline{CH}–$		~3.8	~3.8
(b) Beta			
$–C(=O)N–C–\underline{CH_3}$	~1.1	~1.1	~1.1
$–C(=O)N–C–\underline{CH_2}–$	~1.5	~1.5	~1.5
$–C(=O)N–C–\underline{CH}–$	~1.9	~1.9	~1.9
(i) C-substitution			
(a) alpha			
$\underline{CH_3}–C(=O)N$	~1.9	~2.0	~2.1
$\underline{RCN_2}–C(=O)N$	~2.1	~2.1	~2.1
$\underline{R_2CH}–C(=O)N$	~2.2	~2.2	~2.2
(b) Beta			
$\underline{CH_3}–C–C(=O)N$	~1.1	~1.1	~1.1
$\underline{CH_2}–C–C(=O)N$	~1.5	~1.5	~1.5
$–\underline{CH}–C–C(=O)N$	~1.8	~1.8	~1.8

Amines:

δ of underlined protons	Primary $R–NH_2$ δ/ppm	Secondary RN–HR δ/ppm	Tertiary RRRN δ/ppm
(i) Alpha protons			
$>N–\underline{CH_3}$	~2.5	2.3–3.0	~2.2
$>N–\underline{CH_2}–$	~2.7	2.6–3.4	~2.4
$>N–\underline{CH}<$	~3.1	2.9–3.6	~2.8
(ii) Beta protons			
$>N–C–\underline{CH_3}$			~1.1
$>N–C–\underline{CH_2}–$			~1.4
$>N–C–\underline{CH}<$			~1.7

Cyano Compounds

α hydrogens (underlined)	δ/ppm	β hydrogens (underlined)	δ/ppm
Nitriles			
$\underline{CH_3}–C≡N$	~2.1	$\underline{CH_3}–C–C≡N$	~1.2
$–\underline{CH_2}–C≡N$	~2.5	$–\underline{CH_2}–C–C≡N$	~1.6
$\underline{–CH}–C≡N$	~2.9	$\underline{CH}–C–C≡N$	~2.0
Imides			
$\underline{CH_3}–C(=O)NHC(=O)–$	~2.0	$\underline{CH_3}–C(=O)C–NH–C(=O)–$	~1.2
$\underline{CH_2}–C(=O)NHC(=O)–$	~2.1	$\underline{CH_2}–C(=O)C–NH–C(=O)–$	~1.3
$\underline{CH}–C(=O)NHC(=O)–$	~2.2	$–CH–C(=O)C–NH–C(=O)–$	~1.4
Isocyanates			
$\underline{CH_3}–N=C=O$	~3.0		
$–\underline{CH_2}–N=C=O$	~3.3		
$\underline{–CH}–N=C=O$	~3.6		
Isocyanides (isonitriles)			
$\underline{CH_3}–N=C<$	~2.9		
$\underline{CH_2}–N=C<$	~3.3		
$\underline{CH}–N=C<$	~4.9		
Isothiocyanates			
$\underline{CH_3}–N=C=S$	~3.4		
$\underline{CH_2}–N=C=S$	~3.7		
$\underline{>CH}–N=C=S$	~4.0		
Nitroso			
$–\underline{CH_2}–O–N=O$	~4.8		

Nitro Compounds

Nitro compounds (underlined)	δ/ppm
$\underline{CH_3}–NO_2$	~ 4.1
$\underline{CH_3}–C–NO_2$	~1.6
$–\underline{CH_2}–NO_2$	~4.2
$–\underline{CH_2}–C–NO_2$	~2.1
$–\underline{CH}–NO_2$	~4.4
$–\underline{CH}–C–NO_2$	~2.5

Proton NMR Absorption of Organic Sulfur Compounds

Family	δ/ppm of underlined protons
Benzothiopyrans	
2H–1– sp^3 $\underline{C\text{–}H}$	~3.3
4H–1–sp^3 $\underline{C\text{–}H}$	~3.2
2,3,4H–1–sp^3 $\underline{C\text{–}H}$	1.9–2.8
sp^2 $\underline{C\text{–}H}$	5.8–6.4
sp^2 $\underline{C\text{–}H}$	5.9–6.3
$\underline{\text{aromatic}}$	~6.8
$\underline{\text{aromatic}}$	~6.9
$\underline{\text{aromatic}}$	~7.1
Disulfides	
$\underline{CH_3}$–S–S–R	~2.4
$\underline{CH_2}$–S–S–R	~2.7
$\underline{CH}$–S–S–R	~3.0
$\underline{CH_3}$–C–S–S–R	~1.2
$\underline{CH_2}$–C–S–S–R	~1.6
$\underline{CH}$–C–S–S–R	~2.0
Isothiocyanates	
$\underline{CH_3}$–N=C=S	~2.4
–$\underline{CH_2}$–N=C=S	~2.7
–$\underline{CH}$–N=C=S	~3.0
Mercaptans (thiols)	
$\underline{CH_3}$–S–H	~2.1
–$\underline{CH_2}$–S–H	~2.6
–$\underline{CH}$–S–H	~3.1
$\underline{CH_3}$–C–S–H	~1.3
–$\underline{CH_2}$–C–S–H	~1.6
–$\underline{CH}$–C–S–H	~1.7
S-methyl salts	
>S+–$\underline{CH_3}$	~3.2
Sulfates	
$(\underline{CH_3}\text{–O})_2S(=O)_2$	~3.4
Sulfides	
$\underline{CH_3}$–S–	1.8–2.1
R–$\underline{CH_2}$–S–	1.9–2.4
R–$\underline{CHR}$–S–	2.8–3.4
Ar–$\underline{CH_2}$–S–	4.1–4.2
Ar–$\underline{CHR}$–S–	3.6–4.2
Ar_2–$\underline{CH}$–S–	5.1–5.2
$\underline{CH_3}$–CH_2–S–	1.1–1.2
$\underline{CH_3}$–CHR–S–	0.8–1.2
$\underline{CH_3}$–CHAr–S–	1.3–1.4
$\underline{CH_3}$–CR_2–S–	1.0
Ar–$\underline{CH_2}$–CHR–S–	3.0–3.2
>C=C–$\underline{CH_2}$–CHAr–S–	2.4–2.6
>C=C–$\underline{CH_2}$–CAr_2–S–	2.5
$R_2\underline{CH}$–CH_2–S–	2.6–3.0
$Ar_2\underline{CH}$–CH_2–S–	4.0–4.2
>C=C–$\underline{CHR}$–CHAr–S–	2.3–2.4
>C=C–$\underline{CHR}$–CAr_2–S–	2.8–3.2
Sulfilimines	
$\underline{CH_3}$(R)S=N–R^2	~2.5
Sulfonamides	
$\underline{CH_3}$–SO_2NH_2	~3.0
Sulfonates	
$\underline{CH_3}$–SO_2–OR	~3.0
Sulfones	
$\underline{CH_3}$–SO_2–R^2	~2.6
Sulfonic acids	
$\underline{CH_3}$–SO_3H	~3.0
Sulfoxides	
$\underline{CH_3}$–S(=O)R	~2.5
–$\underline{CH_2}$–S(=O)R	~3.1
Thiocyanates	
$\underline{CH_3}$–S–C≡N	~2.7
–$\underline{CH_2}$–S–C≡N	~3.0
–$\underline{CH}$–S–C≡N	~3.3
Thiols	
See mercaptans	

Note: Ar represents aryl.

Analytical

Some Useful ¹H Coupling Constants

The following chart gives the values of some useful proton NMR coupling constants (in Hz). The single numbers indicate a typical or average value, while in some cases, the range is provided.

1. Freely rotating chains.

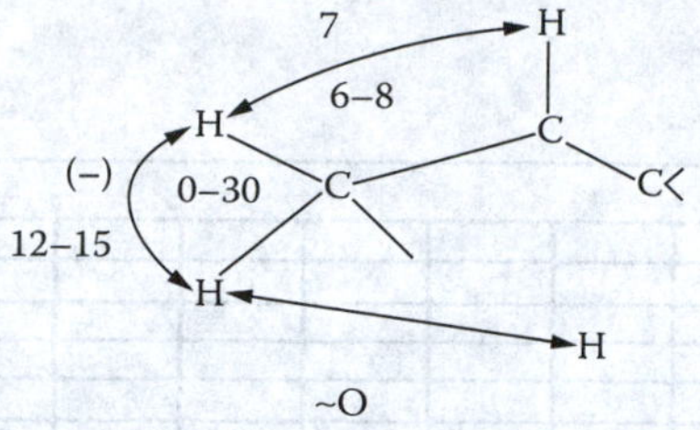

2. Alcohols with no exchange as in DMSO.
1° = triplet
2° = doublet (broad)
3° = singlet
Upon addition of TFA, a sharp singlet results.

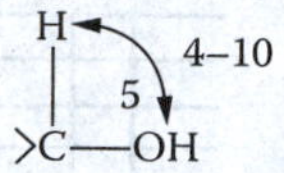

3. Alkenes

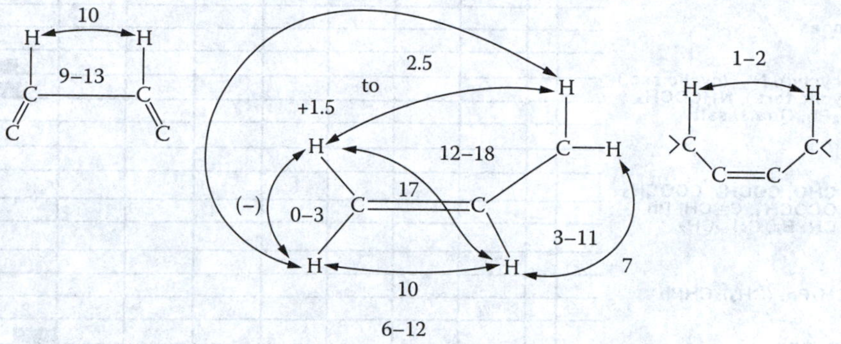

4. Alkynes

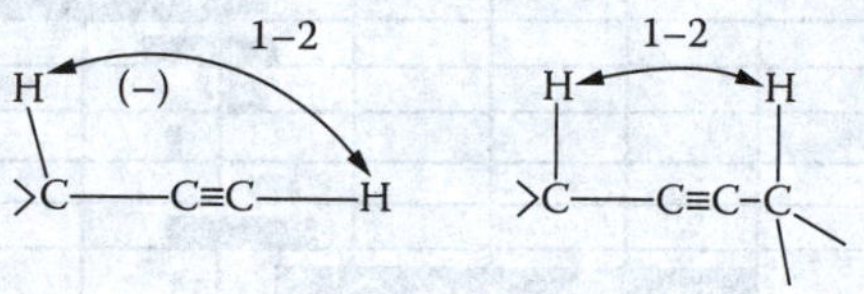

5. Aldehydes

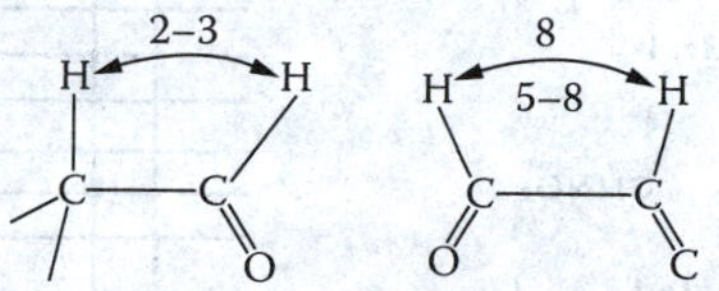

6. Aromatic

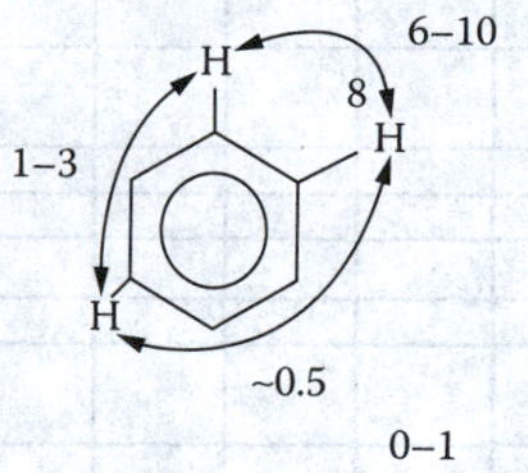

PROTON NMR CORRELATION CHART FOR MAJOR ORGANIC FUNCTIONAL GROUPS

The chart below summarizes the range of chemical shifts for protons in several classes or organic compounds and substituent groups. The chemical shifts δ are given in parts per million relative to tetramethylsilane.

Reference

Mohacsi, E., *J. Chem. Edu.* 41, 38, 1964 (with permission).

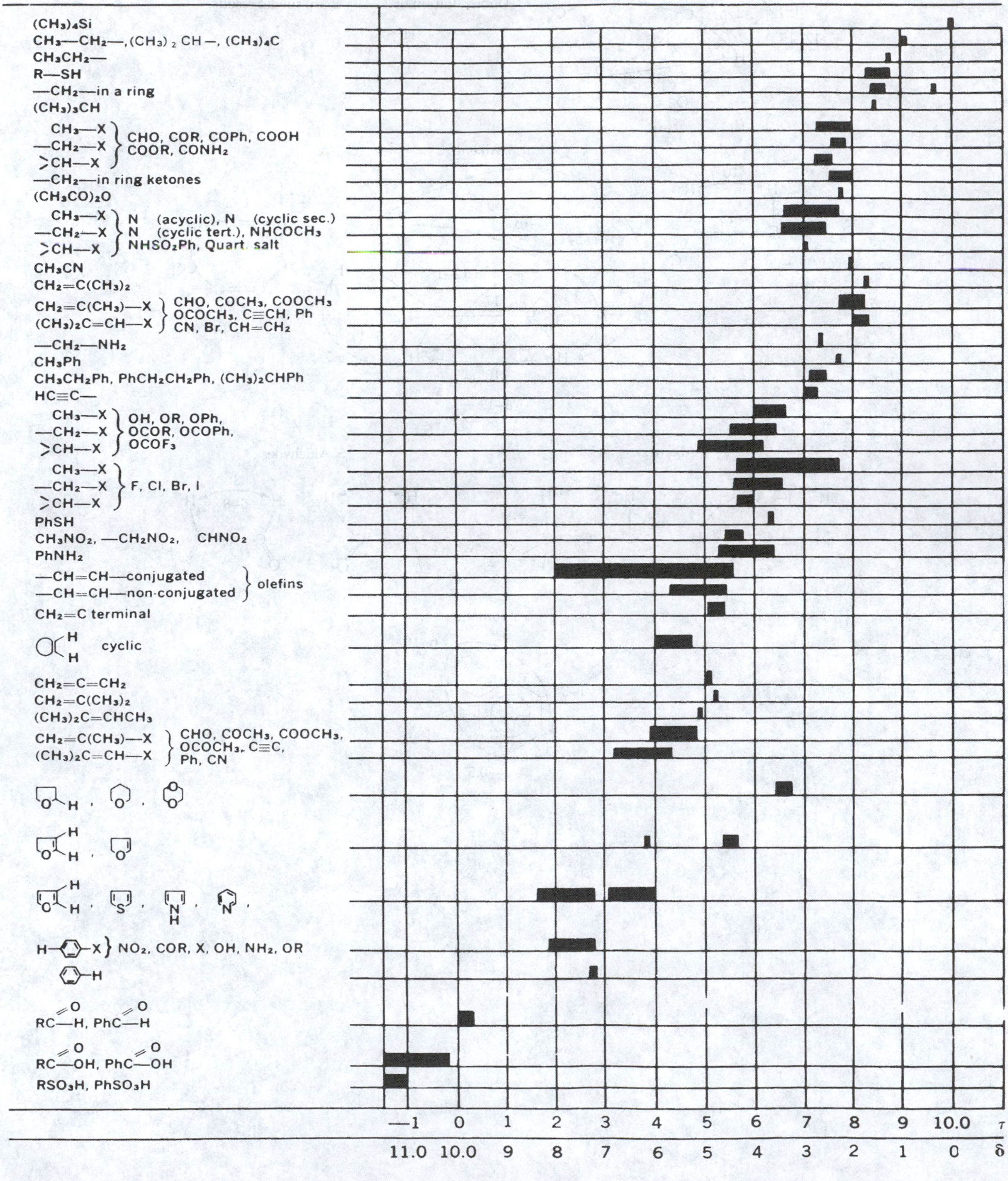

PROTON NMR SHIFTS OF COMMON ORGANIC LIQUIDS

The table below lists the ^{1}H chemical shifts for over 300 organic liquids. The solvent in which the shift was measured is given in the third column. Shifts are given in parts per million relative to tetramethylsilane (TMS) and are listed in order of smallest to largest shift. In many cases the peaks show additional small splittings. Compounds are listed by name.

References

1. Lide, D. R., Editor, *Properties of Organic Compounds*, <www.chemnetbase.com/tours/poc/intro.jsf>; Lide, D. R., and Milne, G. W. A., Editors, *Handbook of Data on Organic Compounds, Third Edition*, CRC Press, Boca Raton, FL, 1993.
2. Spectral Database for Organic Compounds, SDBS, National Institute of Advanced Industrial Science and Technology (AIST), Japan, <sdbs.db.aist.go.jp/>.

^{1}H NMR Shifts (in ppm) of Common Organic Liquids Relative to TMS

Compound	Mol. form.	Solvent						Ref.
Acetic acid	$C_2H_4O_2$	$CDCl_3$	2.10	11.4				2
Acetic anhydride	$C_4H_6O_3$	CCl_4	2.2					1
Acetone	C_3H_6O	$CDCl_3$	2.1					1
Acetonitrile	C_2H_3N	CCl_4	1.9					1
Acrolein	C_3H_4O	$CDCl_3$	6.4	9.5				1
Acrylonitrile	C_3H_3N	$CDCl_3$	6.3					1
Allyl alcohol	C_3H_6O	$CDCl_3$	3.6	4.1	5.1	5.3	6.0	1
Allylamine	C_3H_7N	$CDCl_3$	1.5	3.3	5.0	5.1	5.9	1
2-Amino-2-methyl-1-propanol	$C_4H_{11}NO$	CCl_4	1.1	2.8	3.2			1
Aniline	C_6H_7N	CCl_4	3.3	6.4	6.6	7.0		1
Anisole	C_7H_8O	$CDCl_3$	3.8	7.1				1
Benzaldehyde	C_7H_6O	$CDCl_3$	7.7	10.0				1
Benzene	C_6H_6	$CDCl_3$	7.34					2
Benzeneacetonitrile	C_8H_7N	CCl_4	0.8	0.9	1.6	2.3		1
Benzenethiol	C_6H_6S	CCl_4	3.2	6.9				1
Benzonitrile	C_7H_5N	CCl_4	7.5					1
Benzyl acetate	$C_9H_{10}O_2$	$CDCl_3$	2.1	5.1	7.3			1
Benzyl alcohol	C_7H_8O	$CDCl_3$	2.4	4.6	7.3			1
Bis(2-aminoethyl)amine	$C_4H_{13}N_3$	$CDCl_3$	1.23	2.69	2.79			2
Bis(2-chloroethyl) ether	$C_4H_8Cl_2O$	$CDCl_3$	3.66	3.77				2
Bis(2-ethylhexyl) phthalate	$C_{24}H_{38}O_4$	CCl_4	0.9	1.5	4.2	7.4	7.7	1
Bis(2-hydroxyethyl) sulfide	$C_4H_{10}O_2S$	$CDCl_3$	2.8	3.8	4.2			1
Bromobenzene	C_6H_5Br	CCl_4	7.1	7.4				1
1-Bromobutane	C_4H_9Br	CCl_4	1.0	1.4	1.8	3.4		1
2-Bromobutane, (±)-	C_4H_9Br	$CDCl_3$	1.1	1.7	1.8	4.1		1
1-Bromo-2-chloroethane	C_2H_4BrCl	$CDCl_3$	3.3	4.0				1
Bromochloromethane	CH_2BrCl	CCl_4	5.2					1
1-Bromodecane	$C_{10}H_{21}Br$	CCl_4	0.9	1.8	3.3			1
Bromoethane	C_2H_5Br	$CDCl_3$	1.7	3.4				1
2-Bromo-2-methylpropane	C_4H_9Br	CCl_4	1.8					1
1-Bromonaphthalene	$C_{10}H_7Br$	CCl_4	7.4	8.1				1
1-Bromopentane	$C_5H_{11}Br$	CCl_4	0.9	1.4	1.9	3.3		1
1-Bromopropane	C_3H_7Br	CCl_4	1.0	1.9	3.4			1
2-Bromopropane	C_3H_7Br	$CDCl_3$	1.7	4.3				1
2-Bromopropene	C_3H_5Br	$CDCl_3$	2.3	5.3	5.5			1
Butanal	C_4H_8O	$CDCl_3$	1.0	1.7	2.4	9.7		1
Butanenitrile	C_4H_7N	CCl_4	1.1	1.7	2.3			1
1-Butanethiol	$C_4H_{10}S$	$CDCl_3$	0.9	1.2	1.5	2.5		1
Butanoic acid	$C_4H_8O_2$	CCl_4	0.9	1.7	2.3	12.0		1
Butanoic anhydride	$C_8H_{14}O_3$	CCl_4	1.0	1.7	2.4			1
1-Butanol	$C_4H_{10}O$	$CDCl_3$	0.94	1.39	1.53			2
2-Butanol	$C_4H_{10}O$	$CDCl_3$	0.93	1.17	1.46			2
2-Butanone	C_4H_8O	$CDCl_3$	1.06	2.14	2.45			2
trans-2-Butenal	C_4H_6O	$CDCl_3$	2.0	6.1	6.9	9.5		1
2-Butoxyethanol	$C_6H_{14}O_2$	CCl_4	0.9	1.3	3.3	3.7		1
Butyl acetate	$C_6H_{12}O_2$	$CDCl_3$	0.9	1.4	2.0	4.1		1
Butylamine	$C_4H_{11}N$	$CDCl_3$	0.92	1.33	1.43			2

Compound	Mol. form.	Solvent							Ref.
tert-Butylamine	$C_4H_{11}N$	$CDCl_3$	1.1	1.2					1
Butylbenzene	$C_{10}H_{14}$	CCl_4	0.9	1.4	2.6	7.1			1
sec-Butylbenzene, (±)-	$C_{10}H_{14}$	CCl_4	0.8	1.2	1.6	2.5	7.1		1
tert-Butylbenzene	$C_{10}H_{14}$	CCl_4	1.3	7.2					1
Butyl formate	$C_5H_{10}O_2$	$CDCl_3$	0.9	1.5	4.2	8.1			1
1-*tert*-Butyl-4-methylbenzene	$C_{11}H_{16}$	$CDCl_3$	1.30	2.31	7.11				2
Butyl vinyl ether	$C_6H_{12}O$	CCl_4	0.9	1.4	3.6	3.8	4.0	6.3	1
γ-Butyrolactone	$C_4H_6O_2$	$CDCl_3$	4.4						1
Caprolactam	$C_6H_{11}NO$	$CDCl_3$	1.7	2.4	3.2	7.8			1
2-Chloroaniline	C_6H_6ClN	CCl_4	3.8	6.8					1
Chlorobenzene	C_6H_5Cl	$CDCl_3$	7.3						2
2-Chlorobutane	C_4H_9Cl	CCl_4	1.1	1.5	1.7	3.9			1
Chloroethane	C_2H_5Cl	$CDCl_3$	1.5	3.6					1
2-Chloroethanol	C_2H_5ClO	$CDCl_3$	2.8	3.7	3.8				1
(Chloromethyl)benzene	C_7H_7Cl	CCl_4	4.5	7.3					1
1-Chloro-3-methylbutane	$C_5H_{11}Cl$	$CDCl_3$	0.9	1.7	3.6				1
1-Chloro-2-methylpropane	C_4H_9Cl	CCl_4	1.0	1.9	3.3				1
2-Chloro-2-methylpropane	C_4H_9Cl	CCl_4	1.6						1
1-Chloronaphthalene	$C_{10}H_7Cl$	CCl_4	7.1	7.5	8.2				1
1-Chlorooctane	$C_8H_{17}Cl$	CCl_4	0.9	1.3	1.8	3.5			1
1-Chloropentane	$C_5H_{11}Cl$	CCl_4	0.9	1.6	3.4				1
1-Chloropropane	C_3H_7Cl	CCl_4	1.0	1.8	3.4				1
3-Chloropropene	C_3H_5Cl	CCl_4	4.0	5.2	5.3	5.9			1
2-Chlorotoluene	C_7H_7Cl	$CDCl_3$	2.4	7.2					1
3-Chlorotoluene	C_7H_7Cl	CCl_4	2.3	7.1					1
Cyclohexane	C_6H_{12}	$CDCl_3$	1.43						2
Cyclohexanol	$C_6H_{12}O$	CCl_4	1.6	3.5	4.2				1
Cyclohexanone	$C_6H_{10}O$	CCl_4	1.8	2.3					1
Cyclohexene	C_6H_{10}	CCl_4	1.6	2.0	5.6				1
Cyclohexylamine	$C_6H_{13}N$	CCl_4	1.4	1.5	2.6				1
Cyclopentane	C_5H_{10}	CCl_4	1.5						1
Cyclopentanone	C_5H_8O	CCl_4	2.0						1
cis-Decahydronaphthalene	$C_{10}H_{18}$	$CDCl_3$	1.42	1.62					2
trans-Decahydronaphthalene	$C_{10}H_{18}$	$CDCl_3$	0.87	0.93	1.23				2
Decane	$C_{10}H_{22}$	CCl_4	0.9	1.3					1
Diacetone alcohol	$C_6H_{12}O_2$	$CDCl_3$	1.3	2.2	2.6	3.7			1
1,2-Dibromoethane	$C_2H_4Br_2$	$CDCl_3$	3.65						2
Dibromomethane	CH_2Br_2	CCl_4	4.9						1
1,2-Dibromopropane	$C_3H_6Br_2$	CCl_4	1.8	3.5	3.8	4.2			1
Dibutylamine	$C_8H_{19}N$	CCl_4	0.5	0.9	1.4	2.5			1
Dibutyl ether	$C_8H_{18}O$	CCl_4	0.9	1.4	3.3				1
Dibutyl sebacate	$C_{18}H_{34}O_4$	CCl_4	1.0	1.5	2.2	4.0			1
o-Dichlorobenzene	$C_6H_4Cl_2$	CCl_4	7.2						1
m-Dichlorobenzene	$C_6H_4Cl_2$	CCl_4	7.2	7.4					1
1,1-Dichloroethane	$C_2H_4Cl_2$	$CDCl_3$	2.06	5.90					2
1,2-Dichloroethane	$C_2H_4Cl_2$	CCl_4	3.7						1
1,1-Dichloroethene	$C_2H_2Cl_2$	CCl_4	5.5						1
cis-1,2-Dichloroethene	$C_2H_2Cl_2$	$CDCl_3$	6.28						2
trans-1,2-Dichloroethene	$C_2H_2Cl_2$	$(CH_3)_4Si$	6.24						2
Dichloromethane	CH_2Cl_2	CCl_4	5.3						1
(Dichloromethyl)benzene	$C_7H_6Cl_2$	CCl_4	6.6	7.4					1
1,2-Dichloropropane, (±)-	$C_3H_6Cl_2$	$CDCl_3$	1.61	3.59	3.74				2
2,4-Dichlorotoluene	$C_7H_6Cl_2$	CCl_4	2.3	7.0	7.3				1
3,4-Dichlorotoluene	$C_7H_6Cl_2$	CCl_4	2.3	7.0					1
Diethanolamine	$C_4H_{11}NO_2$	D_2O	2.7	3.7					1
1,1-Diethoxyethane	$C_6H_{14}O_2$	$CDCl_3$	1.2	1.3	3.5	3.7	4.7		2
1,2-Diethoxyethane	$C_6H_{14}O_2$	$CDCl_3$	1.22	3.54	3.58				1
Diethylamine	$C_4H_{11}N$	CCl_4	0.9	1.0	2.6				1
Diethyl carbonate	$C_5H_{10}O_3$	$CDCl_3$	1.3	4.2					1
Diethylene glycol	$C_4H_{10}O_3$	$CDCl_3$	3.7	4.2					1
Diethylene glycol dimethyl ether	$C_6H_{14}O_3$	$CDCl_3$	3.3	3.5					1

Compound	Mol. form.	Solvent							Ref.
Diethylene glycol monoethyl ether	$C_6H_{14}O_3$	CCl_4	1.2	3.1	3.5	3.6			1
Diethylene glycol monoethyl ether acetate	$C_8H_{16}O_4$	$CDCl_3$	1.22	2.08	3.54				2
Diethylene glycol monomethyl ether	$C_5H_{12}O_3$	$CDCl_3$	3.3	3.4	3.6				1
Diethyl ether	$C_4H_{10}O$	$CDCl_3$	1.21	3.47					2
Diethyl sulfide	$C_4H_{10}S$	CCl_4	1.2	2.5					1
Diisopropylamine	$C_6H_{15}N$	CCl_4	0.7	1.0	2.9				1
Diisopropyl ether	$C_6H_{14}O$	CCl_4	1.0	3.5					1
1,2-Dimethoxybenzene	$C_8H_{10}O_2$	CCl_4	3.7	6.8					1
1,2-Dimethoxyethane	$C_4H_{10}O_2$	CCl_4	3.3	3.4					1
Dimethoxymethane	$C_3H_8O_2$	CCl_4	3.2	4.4					1
N,N-Dimethylacetamide	C_4H_9NO	$CDCl_3$	2.1	2.9	3.0				1
2,4-Dimethylaniline	$C_8H_{11}N$	CCl_4	2.0	2.2	3.4	6.4	6.7		1
2,2-Dimethylbutane	C_6H_{14}	CCl_4	0.9	1.1	1.3	1.1	1.3		1
2,3-Dimethylbutane	C_6H_{14}	CCl_4	0.9	1.5					1
N,N-Dimethylformamide	C_3H_7NO	$CDCl_3$	2.9	3.0	8.0				1
Dimethyl glutarate	$C_7H_{12}O_4$	$CDCl_3$	2.0	2.4	3.7				1
2,6-Dimethyl-4-heptanone	$C_9H_{18}O$	CCl_4	0.9	2.1					1
2,5-Dimethylhexane	C_8H_{18}	CCl_4	0.9	1.4					1
Dimethyl maleate	$C_6H_8O_4$	CCl_4	3.7	6.2					1
2,2-Dimethylpentane	C_7H_{16}	$CDCl_3$	0.9	0.9	1.2				1
2,4-Dimethylpentane	C_7H_{16}	CCl_4	0.9	1.1	1.6				1
2,4-Dimethyl-3-pentanone	$C_7H_{14}O$	CCl_4	1.0	2.6					1
2,4-Dimethylpyridine	C_7H_9N	$CDCl_3$	2.3	2.5	7.0	7.0	8.4		1
2,6-Dimethylpyridine	C_7H_9N	$CDCl_3$	2.51	6.93	7.42				2
Dimethyl sulfoxide	C_2H_6OS	$CDCl_3$	2.62						2
1,4-Dioxane	$C_4H_8O_2$	$CDCl_3$	3.69						2
1,3-Dioxolane	$C_3H_6O_2$	$CDCl_3$	3.88	4.90					2
Dipentyl ether	$C_{10}H_{22}O$	$CDCl_3$	0.9	1.4	3.4				1
Dipropylamine	$C_6H_{15}N$	$CDCl_3$	1.5	2.6					1
Dodecane	$C_{12}H_{26}$	CCl_4	0.9	1.3					1
1-Dodecene	$C_{12}H_{24}$	CCl_4	0.9	1.3	2.0	5.4			1
Epichlorohydrin	C_3H_5ClO	CCl_4	2.6	2.8	3.2	3.5	3.6		1
1,2-Epoxybutane	C_4H_8O	CCl_4	1.0	1.5	2.3	2.6	2.7		1
1,2-Ethanediamine	$C_2H_8N_2$	CCl_4	1.2	2.6					1
1,2-Ethanediol	$C_2H_6O_2$	D_2O	3.7						1
1,2-Ethanediol, diacetate	$C_6H_{10}O_4$	CCl_4	2.0	4.2					1
Ethanol	C_2H_6O	$CDCl_3$	1.23	2.61	3.69				2
Ethanolamine	C_2H_7NO	$CDCl_3$	2.7	2.8	3.5				1
Ethoxybenzene	$C_8H_{10}O$	CCl_4	1.3	3.9	6.9				1
2-Ethoxyethanol	$C_4H_{10}O_2$	$CDCl_3$	1.22	2.70	3.55				2
2-Ethoxyethyl acetate	$C_6H_{12}O_3$	CCl_4	1.2	2.0	3.4	3.5	4.1		1
Ethyl acetate	$C_4H_8O_2$	$CDCl_3$	1.26	2.04	4.12				2
Ethyl acetoacetate	$C_6H_{10}O_3$	$CDCl_3$	1.3	1.9	2.2	3.3	4.1	4.9	1
Ethyl acrylate	$C_5H_8O_2$	CCl_4	1.3	4.1	5.7	6.1	6.3		1
Ethylamine	C_2H_7N	D_2O	1.1	2.6					1
Ethylbenzene	C_8H_{10}	$CDCl_3$	1.3	2.7	7.2				1
Ethyl benzoate	$C_9H_{10}O_2$	CCl_4	1.3	4.3	7.4	8.0			1
Ethyl butanoate	$C_6H_{12}O_2$	CCl_4	0.9	1.2	1.7	2.2	4.1		1
Ethyl cyanoacetate	$C_5H_7NO_2$	CCl_4	1.3	3.4	4.3				1
Ethylcyclohexane	C_8H_{16}	$CDCl_3$	0.9	1.9	1.4				1
Ethylene carbonate	$C_3H_4O_3$	$CDCl_3$	4.5						1
Ethyl formate	$C_3H_6O_2$	CCl_4	1.3	4.2	7.9				1
2-Ethyl-1,3-hexanediol	$C_8H_{18}O_2$	$CDCl_3$	1.0	1.4	3.8				1
2-Ethyl-1-hexanol	$C_8H_{18}O$	$CDCl_3$	0.9	1.3	1.8	3.5			1
Ethyl 3-methylbutanoate	$C_7H_{14}O_2$	$CDCl_3$	1.0	1.3	1.9	4.1			1
3-Ethyl-2-methylpentane	C_8H_{18}	CCl_4	0.9	0.9	1.5				1
Fluorobenzene	C_6H_5F	CCl_4	7.0						1
2-Fluorotoluene	C_7H_7F	CCl_4	2.2	6.9					1
3-Fluorotoluene	C_7H_7F	CCl_4	2.3	6.9					1
4-Fluorotoluene	C_7H_7F	CCl_4	2.2	6.8	7.0				1
Furan	C_4H_4O	$CDCl_3$	6.38	7.44					2

Compound	Mol. form.	Solvent							Ref.
Furfural	$C_5H_4O_2$	$CDCl_3$	6.6	7.3	7.7	9.7			1
Furfuryl alcohol	$C_5H_6O_2$	$CDCl_3$	2.8	4.6	6.3	7.4			1
Glycerol	$C_3H_8O_3$	D_2O	3.6						1
Glycerol triacetate	$C_9H_{14}O_6$	$CDCl_3$	2.1	4.2	4.3	5.2			1
Heptane	C_7H_{16}	$CDCl_3$	0.88	1.27	1.30				2
1-Heptanol	$C_7H_{16}O$	CCl_4	0.9	1.4	3.4	3.5			1
3-Heptanol, (*S*)-	$C_7H_{16}O$	CCl_4	0.9	1.4	2.3	3.4			1
2-Heptanone	$C_7H_{14}O$	CCl_4	0.9	1.3	2.0	2.3			1
3-Heptanone	$C_7H_{14}O$	CCl_4	1.0	1.4	2.3				1
1-Heptene	C_7H_{14}	CCl_4	0.9	1.4	2.0	4.9	5.7		1
Hexane	C_6H_{14}	$CDCl_3$	0.89	1.27	1.29				2
Hexanedinitrile	$C_6H_8N_2$	$CDCl_3$	1.8	2.5					1
Hexanenitrile	$C_6H_{11}N$	$CDCl_3$	0.9	1.5	2.3				1
Hexanoic acid	$C_6H_{12}O_2$	$CDCl_3$	0.9	1.4	2.4	11.4			1
1-Hexanol	$C_6H_{14}O$	$CDCl_3$	0.90	1.32	1.56				2
Hexyl acetate	$C_8H_{16}O_2$	CCl_4	0.9	1.4	2.0	4.0			1
3-Hydroxypropanenitrile	C_3H_5NO	$CDCl_3$	2.6	3.4	3.9				1
Iodobenzene	C_6H_5I	CCl_4	6.8	7.5	7.7				1
1-Iodobutane	C_4H_9I	$CDCl_3$	1.0	1.7	1.9	4.2			1
2-Iodobutane, (±)-	C_4H_9I	$CDCl_3$	1.0	1.7	1.9	4.2			1
Iodoethane	C_2H_5I	$CDCl_3$	1.2	2.6	3.7				1
Iodomethane	CH_3I	$CDCl_3$	2.2						1
1-Iodopropane	C_3H_7I	CCl_4	1.0	1.8	3.2				1
2-Iodopropane	C_3H_7I	$CDCl_3$	1.9	4.3					1
Isobutanal	C_4H_8O	CCl_4	1.1	2.4	9.6				1
Isobutyl acetate	$C_6H_{12}O_2$	CCl_4	0.9	1.9	2.0	3.8			1
Isobutylbenzene	$C_{10}H_{14}$	CCl_4	0.9	1.9	2.4	7.1			1
Isobutyl formate	$C_5H_{10}O_2$	$CDCl_3$	1.0	2.0	3.9	8.0			1
Isobutyl isobutanoate	$C_8H_{16}O_2$	CCl_4	0.9	1.2	1.9	2.5	3.8		1
Isopentyl acetate	$C_7H_{14}O_2$	$CDCl_3$	0.9	1.5	2.0	4.0			1
Isophorone	$C_9H_{14}O$	$CDCl_3$	1.04	1.95	2.19				2
Isopropyl acetate	$C_5H_{10}O_2$	$CDCl_3$	0.9	1.4	1.6	2.4			1
Isopropylbenzene	C_9H_{12}	$CDCl_3$	1.3	2.4	2.9	7.3			1
1-Isopropyl-4-methylbenzene	$C_{10}H_{14}$	$CDCl_3$	1.2	2.3	2.9	7.1			1
Isoquinoline	C_9H_7N	$CDCl_3$	8.5	9.3					1
d-Limonene	$C_{10}H_{16}$	CCl_4	1.4	1.7	1.9	4.6	5.3		1
Mesityl oxide	$C_6H_{10}O$	$CDCl_3$	1.89	2.14	2.16				2
Methanol	CH_4O	$CDCl_3$	3.43	3.66					2
2-Methoxyethanol	$C_3H_8O_2$	$CDCl_3$	2.5	3.4	3.5	3.7	3.5	3.7	1
2-Methoxyethyl acetate	$C_5H_{10}O_3$	$CDCl_3$	2.09	3.39	3.59				2
Methyl acetate	$C_3H_6O_2$	CCl_4	2.0	3.7					1
2-Methylacrylonitrile	C_4H_5N	$CDCl_3$	2.0	5.7	5.8				1
2-Methylaniline	C_7H_9N	CCl_4	2.0	3.2	6.7				1
3-Methylaniline	C_7H_9N	CCl_4	2.2	3.3	6.4	6.9			1
N-Methylaniline	C_7H_9N	CCl_4	2.7	3.3	6.4	6.6	7.1		1
Methyl benzoate	$C_8H_8O_2$	CCl_4	3.8	7.4	8.0				1
3-Methylbutanoic acid	$C_5H_{10}O_2$	CCl_4	1.0	2.2	11.0				1
3-Methyl-1-butanol	$C_5H_{12}O$	CCl_4	0.9	1.5	3.5	4.1			1
Methyl cyanoacetate	$C_4H_5NO_2$	$CDCl_3$	3.5	3.8					1
Methylcyclohexane	C_7H_{14}	CCl_4	0.9	1.4					1
cis-4-Methylcyclohexanol	$C_7H_{14}O$	$CDCl_3$	0.9	1.5	2.9	3.9			1
N-Methylformamide	C_2H_5NO	$CDCl_3$	2.82	7.4	8.16				2
Methyl formate	$C_2H_4O_2$	$CDCl_3$	3.76	8.07					2
2-Methylheptane	C_8H_{18}	CCl_4	0.9	1.3					1
4-Methylheptane	C_8H_{18}	$CDCl_3$	0.8	0.9	1.4				1
2-Methylhexane	C_7H_{16}	$CDCl_3$	0.9	0.9	1.4				1
5-Methyl-2-hexanone	$C_7H_{14}O$	CCl_4	0.9	1.4	2.0	2.3			1
Methyl methacrylate	$C_5H_8O_2$	$CDCl_3$	2.0	3.8	5.6	6.1			1
2-Methyloctane	C_9H_{20}	CCl_4	0.9	1.0	1.3				1
2-Methylpentane	C_6H_{14}	CCl_4	0.8	0.9	1.5				1
3-Methylpentane	C_6H_{14}	CCl_4	0.8	1.5					1

Compound	Mol. form.	Solvent							Ref.
2-Methyl-2,4-pentanediol	$C_6H_{14}O_2$	CCl_4	1.1	1.2	1.3	1.4	1.6	4.2	1
4-Methylpentanenitrile	$C_6H_{11}N$	CCl_4	1.0	1.6	2.3				1
4-Methyl-2-pentanol	$C_6H_{14}O$	CCl_4	0.9	1.1	1.3	1.8	3.5	3.7	1
3-Methyl-3-pentanol	$C_6H_{14}O$	CCl_4	0.9	1.1	1.4	1.8			1
4-Methyl-2-pentanone	$C_6H_{12}O$	$CDCl_3$	0.9	2.1	2.3				1
2-Methylpropanenitrile	C_4H_7N	$CDCl_3$	1.3	2.7					1
2-Methylpropanoic acid	$C_4H_8O_2$	CCl_4	1.2	2.6					1
2-Methyl-1-propanol	$C_4H_{10}O$	CCl_4	0.9	1.7	3.3	4.0			1
2-Methyl-2-propanol	$C_4H_{10}O$	$CDCl_3$	1.3	1.4					1
2-Methylpyridine	C_6H_7N	CCl_4	2.5	6.9	7.4	8.4			1
3-Methylpyridine	C_6H_7N	CCl_4	2.3	7.0	7.4	8.4			1
N-Methyl-2-pyrrolidinone	C_5H_9NO	$CDCl_3$	2.1	2.4	2.8	3.4			1
Methyl salicylate	$C_8H_8O_3$	CCl_4	3.9	6.7	6.9	7.3	7.7	10.6	1
2-Methylthiophene	C_5H_6S	$CDCl_3$	2.5	6.7	6.9	7.0			1
Morpholine	C_4H_9NO	$CDCl_3$	2.59	2.86	3.67				2
Nitrobenzene	$C_6H_5NO_2$	CCl_4	7.52	7.65	8.19				2
Nitroethane	$C_2H_5NO_2$	$CDCl_3$	1.6	4.3					1
Nitromethane	CH_3NO_2	CCl_4	4.2						1
1-Nitropropane	$C_3H_7NO_2$	$CDCl_3$	1.0	2.1	4.4				1
Nonane	C_9H_{20}	$CDCl_3$	0.9	1.3					1
Octane	C_8H_{18}	$CDCl_3$	0.88	1.26					1
1-Octanol	$C_8H_{18}O$	$CDCl_3$	0.88	1.29	1.5				2
2-Octanone	$C_8H_{16}O$	CCl_4	0.9	1.4	2.1	2.4			1
1-Octene	C_8H_{16}	CCl_4	0.9	1.3	2.1	4.8	4.9	5.7	1
Pentane	C_5H_{12}	$CDCl_3$	0.88	1.26	1.30				2
1,5-Pentanediol	$C_5H_{12}O_2$	D_2O	1.5	3.6					1
2,4-Pentanedione	$C_5H_8O_2$	CCl_4	2.0	2.2	3.5	5.4	14.7		1
Pentanenitrile	C_5H_9N	CCl_4	1.0	1.6	2.3				1
Pentanoic acid	$C_5H_{10}O_2$	CCl_4	0.9	1.5	2.3	11.7			1
1-Pentanol	$C_5H_{12}O$	CCl_4	0.9	1.4	3.5	4.4			1
2-Pentanol	$C_5H_{12}O$	$CDCl_3$	0.9	1.2	1.3	2.2	3.7		1
3-Pentanol	$C_5H_{12}O$	CCl_4	0.9	1.4	3.3	3.4			1
2-Pentanone	$C_5H_{10}O$	CCl_4	0.9	1.6	2.0	2.3			1
3-Pentanone	$C_5H_{10}O$	CCl_4	1.0	2.4					1
Pentyl acetate	$C_7H_{14}O_2$	CCl_4	0.9	1.4	2.0	4.0			1
Pentylamine	$C_5H_{13}N$	$CDCl_3$	0.9	1.4	2.7				1
α-Pinene	$C_{10}H_{16}$	$CDCl_3$	0.84	1.16	1.27				2
Piperidine	$C_5H_{11}N$	$CDCl_3$	1.53	2.18	2.79				2
Propanal	C_3H_6O	$CDCl_3$	1.1	2.4	9.8				1
1,2-Propanediol	$C_3H_8O_2$	$CDCl_3$	1.1	3.4	3.9	4.3			1
1,3-Propanediol	$C_3H_8O_2$	D_2O	1.8	3.7					1
Propanenitrile	C_3H_5N	CCl_4	1.3	2.3					1
Propanoic acid	$C_3H_6O_2$	$CDCl_3$	1.1	2.4	10.5				1
Propanoic anhydride	$C_6H_{10}O_3$	CCl_4	1.2	2.4					1
1-Propanol	C_3H_8O	$CDCl_3$	0.9	1.6	2.3	3.6			1
2-Propanol	C_3H_8O	$CDCl_3$	1.2	1.6	4.0				1
Propargyl alcohol	C_3H_4O	$CDCl_3$	2.5	2.8	4.3				1
Propyl acetate	$C_5H_{10}O_2$	CCl_4	0.9	1.6	2.0	4.0			1
Propylamine	C_3H_9N	CCl_4	0.9	1.5	2.1	2.7			1
Propylbenzene	C_9H_{12}	$CDCl_3$	0.9	1.6	2.6				1
Propyl formate	$C_4H_8O_2$	CCl_4	1.0	1.7	4.1	7.9			1
Pyridine	C_5H_5N	$CDCl_3$	7.23	7.62	8.59				2
Pyrrole	C_4H_5N	$CDCl_3$	6.2	6.7	8.0				1
Pyrrolidine	C_4H_9N	C_6H_6	1.5	2.4	2.7				1
2-Pyrrolidone	C_4H_7NO	$CDCl_3$	2.2	3.4	7.7				1
Quinoline	C_9H_7N	CCl_4	7.1	7.8	8.8				1
Salicylaldehyde	$C_7H_6O_2$	$CDCl_3$	7.0	7.4	9.8	11.0			1
Styrene	C_8H_8	CCl_4	5.1	5.6	6.6	7.2			1
Sulfolane	$C_4H_8O_2S$	$CDCl_3$	2.2	3.0					1
α-Terpinene	$C_{10}H_{16}$	$CDCl_3$	1.02	1.77	2.09				2
1,1,1,2-Tetrachloroethane	$C_2H_2Cl_4$	$CDCl_3$	4.29						2

Compound	Mol. form.	Solvent							Ref.
1,1,2,2-Tetrachloroethane	$C_2H_2Cl_4$	$CDCl_3$	5.91						2
Tetraethylene glycol	$C_8H_{18}O_5$	CCl_4	3.5	3.6					1
Tetrahydrofuran	C_4H_8O	$CDCl_3$	1.84	3.73					2
1,2,3,4-Tetrahydronaphthalene	$C_{10}H_{12}$	$CDCl_3$	1.8	2.8	7.1				1
Tetrahydropyran	$C_5H_{10}O$	CCl_4	1.6	3.6					1
Tetrahydrothiophene	C_4H_8S	$CDCl_3$	1.9	2.8					1
Tetramethylsilane	$C_4H_{12}Si$	CCl_4	0.0						1
Tetramethylurea	$C_5H_{12}N_2O$	CCl_4	2.8						1
Thiophene	C_4H_4S	$CDCl_3$	7.1	7.3					1
Toluene	C_7H_8	$CDCl_3$	2.34	7.18					2
Tribromomethane	$CHBr_3$	CCl_4	6.8						1
Tributylamine	$C_{12}H_{27}N$	CCl_4	0.9	1.3	2.3				1
1,1,1-Trichloroethane	$C_2H_3Cl_3$	CCl_4	2.7						1
1,1,2-Trichloroethane	$C_2H_3Cl_3$	$CDCl_3$	4.0	5.8					1
Trichloroethene	C_2HCl_3	CCl_4	6.5						1
Trichloroethylsilane	$C_2H_5Cl_3Si$	CCl_4	1.3						1
Trichloromethane	$CHCl_3$	CCl_4	7.2						1
(Trichloromethyl)benzene	$C_7H_5Cl_3$	CCl_4	7.3	7.8					1
Tridecane	$C_{13}H_{28}$	CCl_4	0.9	1.3					1
Triethanolamine	$C_6H_{15}NO_3$	D_2O	2.7	3.6					1
Triethylamine	$C_6H_{15}N$	CCl_4	1.0	2.4					1
Triethylene glycol	$C_6H_{14}O_4$	$CDCl_3$	3.5	3.7					1
Triethyl phosphate	$C_6H_{15}O_4P$	$CDCl_3$	1.4	4.1					1
2,2,2-Trifluoroethanol	$C_2H_3F_3O$	$CDCl_3$	3.4	3.9					1
(Trifluoromethyl)benzene	$C_7H_5F_3$	CCl_4	7.5						1
Trimethylamine	C_3H_9N	CCl_4	2.12						2
1,2,3-Trimethylbenzene	C_9H_{12}	$CDCl_3$	2.2	2.3	7.0				1
1,2,4-Trimethylbenzene	C_9H_{12}	CCl_4	2.2	6.8					1
1,3,5-Trimethylbenzene	C_9H_{12}	$CDCl_3$	2.3	6.8					1
2,2,3-Trimethylbutane	C_7H_{16}	CCl_4	0.8	1.3					1
2,2,5-Trimethylhexane	C_9H_{20}	CCl_4	0.9	1.2					1
2,3,3-Trimethylpentane	C_8H_{18}	CCl_4	0.8	0.8	1.4				1
2,3,4-Trimethylpentane	C_8H_{18}	CCl_4	0.8	1.9					1
Trimethyl phosphate	$C_3H_9O_4P$	$CDCl_3$	3.78						2
2,4,6-Trimethylpyridine	$C_8H_{11}N$	CCl_4	2.2	2.4	6.6				1
1-Undecene	$C_{11}H_{22}$	CCl_4	0.9	1.3	2.0	4.8	4.9	5.6	1
Vinyl acetate	$C_4H_6O_2$	$CDCl_3$	2.1	4.6	4.9	7.3			1
o-Xylene	C_8H_{10}	$CDCl_3$	2.22	7.07					2
m-Xylene	C_8H_{10}	$CDCl_3$	2.28	6.95	7.11				2
p-Xylene	C_8H_{10}	$CDCl_3$	2.30	7.05					2

PROTON CHEMICAL SHIFTS OF CONTAMINANTS IN DEUTERATED SOLVENTS

Thomas J. Bruno

The following table provides the expected chemical shift of solutes that commonly occur as impurities in NMR spectra measured at 298 K. These chemical shifts can vary as a function of (deuterated) solvents. These data are adapted from the compilation of Gottlieb et al. (Ref. 1), as expanded by Fulmer et al. (Ref. 2). Only the more common solvents and impurities are considered here. The reader is referred to these sources for additional impurities and solvents. When the same moiety contains multiple protons, bold print indicates the protons under consideration. When protons differ by ring position, the appropriate protons are indicated parenthetically. Note that the chemical shifts of the -OH protons will vary with concentration as a result of hydrogen bonding, so these have not been included.

References

1. Gottlieb, H. E., Kotlyar, K., and Nudelman, A., *J. Org. Chem.* 62, 7512, 1997.
2. Fulmer, G. R., Miller, A. J. M., Sherdan, N. H., Gottleib, H. E., Nudelman, A., Stoltz, B. M., Bercaw, J. E., and Goldberg, K. I., *Organometallics* 29, 2176, 2010.

^{1}H Chemical Shifts of Contaminants in Deuterated Solvents

Name	Proton	THF-d_8	CD_2Cl_2	$CDCl_3$	Toluene-d_8	C_6D_6	CD_3OD	D_2O
Acetic acid	CH_3	1.89	2.06	2.1	1.57	1.52	4.87	
Acetone	CH_3	2.05	2.12	2.17	1.57	1.55	2.15	2.22
Acetonitrile	CH_3	1.95	1.97	2.1	0.69	0.58	2.03	2.06
Benzene	CH	7.31	7.35	7.36	7.12	7.15	7.33	
Cyclohexane	CH_2	1.44	1.44	1.43	1.4	1.4	1.45	
1,4-Dioxane	CH_2	3.56	3.65	3.71	3.33	3.35	3.66	3.75
Ethanol	CH_3	1.1	1.19	1.25	0.97	0.96	1.19	1.17
Ethanol	CH_2	3.51	3.66	3.72	3.36	3.39	3.6	3.65
Ethyl acetate	CH_3CO	1.94	2	2.05	1.69	1.65	2.01	2.07
Ethyl acetate	CH_2	4.04	4.08	4.12	3.87	3.89	4.09	4.14
Ethyl acetate	CH_3	1.19	1.23	1.26	0.94	0.92	1.24	1.24
Hexane	CH_3	0.89	0.89	0.88	0.88	0.89	0.9	
Hexane	CH_2	1.29	1.27	1.26	1.22	1.24	1.29	
2-Propanol	CH_3	1.08	1.17	1.22	0.95	0.95	1.5	1.17
2-Propanol	CH	3.82	3.97	4.04	3.65	3.67	3.92	4.02
Toluene	CH_3	2.31	2.34	2.36	2.11	2.11	2.32	
Toluene	CH (2,4,6)	7.1	7.15	7.17	6.99(av)	7.02	7.16	
Toluene	CH (3,5)	7.19	7.24	7.25	7.09	7.13	7.16	
Trichloromethane	CH	7.89	7.32	7.26	6.1	6.15	7.9	

^{13}C-NMR ABSORPTIONS OF MAJOR FUNCTIONAL GROUPS

Thomas J. Bruno and Paris D. N. Svoronos

The table below lists the ^{13}C chemical shift ranges (in ppm) with the corresponding functional groups in descending order. Some typical simple compounds for every family are given to illustrate the corresponding range. The shifts for the carbons of interest are given in parentheses, either for each carbon as it appears from left to right in the formula, or by the underscore (Refs. 1–17). In Ref. 1, the reader can find additional details, correlation charts, additivity rules, and the expected ^{13}C peaks attributed to common solvents. All shifts are relative to tetramethylsilane (TMS).

References

1. Bruno, T. J., and Svoronos, P. D. N., *CRC Handbook of Basic Tables for Chemical Analysis, 3rd Edition*, CRC Press, Boca Raton, FL, 2011.
2. Yoder, C. H., and Schaeffer, C. D., Jr., *Introduction to Multinuclear NMR: Theory and Application*, The Benjamin/Cummings Publishing Co., Menlo Park, CA, 1987.
3. Brown, D. W., *J. Chem. Educ.* 62, 209, 1985.
4. Silverstein, R. M., and Webster F. X., *Spectrometric Identification of Organic Compounds, 6th Edition*, John Wiley and Sons, New York, 1998.
5. Becker, E. D., *High Resolution NMR, Theory and Chemical Applications, 2nd Edition*, Academic Press, New York, 1980.
6. Gunther, H., *NMR Spectroscopy: Basic Principles, Concepts and Applications in Chemistry*, Wiley, New York, 2003.
7. Kitamaru, R., *Nuclear Magnetic Resonance: Principles and Theory*, Elsevier Science, 1990.
8. Lambert, J. B., Holland, L. N., and Mazzola, E. P., *Nuclear Magnetic Resonance Spectroscopy: Introduction to Principles, Applications and Experimental Methods*, Prentice Hall, Englewood Cliffs, NJ, 2003.
9. Bovey, F. A., and Mirau, P. A., *Nuclear Magnetic Resonance Spectroscopy, 2nd Edition*, Academic Press, New York, 1988.
10. Harris, R. K., and Mann, B. E., *NMR and the Periodic Table*, Academic Press, London, 1978.
11. Hore, P. J., *Nuclear Magnetic Resonance*, Oxford University Press, Oxford, 1995.
12. Nelson, J. H., *Nuclear Magnetic Resonance Spectroscopy, 2nd Edition*, Wiley, New York, 2003.
13. Levy, G. C., Lichter, R. L., and Nelson, G. L., *Carbon-13 Nuclear Magnetic Resonance Spectroscopy, 2nd Edition*, Wiley, New York, 1980.
14. Pihlaja, K., and Kleinpeter, E., *Carbon-13 NMR Chemical Shifts in Structural and Stereochemical Analysis*, VCH, New York, 1994.
15. *Aldrich Library of ^{1}H and ^{13}C FT-NMR Spectra*, Aldrich Chemical Company, Milwaukee, WI, 1996.
16. Balci, M., *Basic 1H- and 13C-NMR Spectroscopy*, Elsevier, London, 2005.
17. http://www.chem.wisc.edu/areas/reich/handouts/nmr-c13/cdata.htm, 2009.

δ/ppm range	Group	Family	Example	δ/ppm of underlined carbon
220–165	>C=O	Ketones	$(CH_3)_2\underline{C}O$	(206.0)
			$(CH_3)_2CH\underline{C}OCH_3$	(212.1)
		Aldehydes	$CH_3\underline{C}HO$	(199.7)
		α,β-unsaturated carbonyls	$CH_3CH{=}CHCHO$	(192.4)
			$CH_2{=}CH\underline{C}OCH_3$	(169.9)
		Carboxylic acids	$H\underline{C}O_2H$	(166.0)
			$CH_3\underline{C}O_2H$	(178.1)
		Amides	$H\underline{C}ONH_2$	(165.0)
			$CH_3\underline{C}ONH_2$	(172.7)
		Esters	$CH_3\underline{C}O_2CH_2CH_3$	(170.3)
			$CH_2{=}CH\underline{C}O_2CH_3$	(165.5)
140–120	>C=C<	Aromatics, alkenes	C_6H_6	(128.5)
			$CH_2{=}CH_2$	(123.2)
			$CH_2{=}CHCH_3$	(115.9, 136.2)
			$CH_2{=}CHCH_2Cl$	(117.5, 133.7)
			$CH_3CH{=}\underline{C}HCH_2CH_3$	(132.7)
125–115	–C≡N	Nitriles	$CH_3{-}\underline{C}{\equiv}N$	(117.7)
80–70	–C≡C–	Alkynes	$H\underline{C}{\equiv}CH$	(71.9)
			$CH_3\underline{C}{\equiv}CCH_3$	(73.9)
70–45	>C–O–	Esters, alcohols	$\underline{C}H_3OOCCH_2CH_3$	(57.6, 67.9)
			$HO\underline{C}H_3$	(49.0)
	>C–OH			
			$HO\underline{C}H_2CH_3$	(57.0)
40–20	>C–NH$_2$	Amines	$\underline{C}H_3NH_2$	(26.9)
			$CH_3\underline{C}H_2NH_2$	(35.9)
30–15		Sulfides (thioethers)	$C_6H_5{-}S{-}\underline{C}H_3$	15.6
30–(–2.3)		Alkanes, cycloalkanes	$\underline{CH_4}$	(–2.3)
			$\underline{CH_3}CH_3$	(5.7)
			$\underline{CH_3}\underline{CH_2}CH_3$	(15.8, 16.3)
			$CH_3CH_2CH_2\underline{CH_3}$	(13.4, 25.2)
			$\underline{CH_3}\underline{CH_2}\underline{CH_2}CH_2CH_3$	(13.9, 22.8, 34.7)
			Cyclohexane	(26.9)

Analytical

^{13}C-NMR CHEMICAL SHIFTS OF COMMON ORGANIC SOLVENTS

The following table gives the expected carbon-13 chemical shifts, relative to tetramethylsilane, for various useful NMR solvents. In some solvents, slight changes can occur with change of concentration. (Refs. 2-3).

References

1. Bruno, T. J., and Svoronos, P. D. N., *CRC Handbook of Basic Tables for Chemical Analysis, 3rd Edition*, CRC Press, Boca Raton, FL, 2011.
2. Silverstein, R. M., Bassler, G. C., and Morrill, T. C., *Spectrometric Identification of Organic Compounds*, John Wiley & Sons, Now York, 1981.
3. Rahman, A.-U., *Nuclear Magnetic Resonance. Basic Principles*, Springer-Verlag, New York, 1986.
4. Pretsch, E., Clerc, T., Seibl, J., and Simon, W., *Spectral Data for Structure Determination of Organic Compounds, Second Edition*, Springer-Verlag, Heidelberg, 1989.

^{13}C-NMR Chemical Shifts of Common Organic Solvents (in ppm) Relative to Tetramethylsilane

Solvent	Formula	Chemical shift (ppm)
Acetic acid-d_4	CD_3COOD	20.0 (CD_3) 205.8 (C=O)
Acetone	$(CH_3)_2C{=}O$	30.7 (CH_3) 206.7 (C=O)
Acetone-d_6	$(CD_3)_2C{=}O$	29.2 (CD_3) 204.1 (C=O)
Acetonitrile-d_3	$CD_3C{\equiv}N$	1.3 (CD_3) 117.1 (C≡N)
Benzene	C_6H_6	128.5
Benzene-d_6	C_6D_6	128.4
Carbon disulfide	CS_2	192.3
Carbon tetrachloride	CCl_4	96.0
Chloroform	$CHCl_3$	77.2
Chloroform-d_3	$CDC1_3$	77.05
Cyclohexane-d_{12}	C_6D_{12}	27.5
Dichloromethane-d_2	CD_2C1_2	53.6
Dimethylformamide-d_7	$(CD_3)_2NCDO$	31 (CD_3) 36 (CD_3) 162.4 (C=O)
Dimethylsulfoxide-d_6	$(CD_3)_2S{=}O$	39.6
Dioxane-d_8	$C_4D_8O_2$	67.4
Formic acid-d_2	DCOOD	165.5
Methanol-d_4	CD_3OD	49.3
Nitromethane-d_3	CD_3NO_2	57.3
Pyridine	C_5H_5N	123.6 (C_3) 135.7 (C_4) 149.8 (C_2)
Pyridine-d_5	C_5D_5N	123.9 (C_3) 135.9 (C_4) 150.2 (C_2)
1,1,2,2-Tetrachloroethane-d_2	$CDC1_2CDC1_2$	75.5
Tetrahydrofuran-d_8	C_4D_8O	25.8 (C_2) 67.9 (C_1)
Trichlorofluoromethane	$CFC1_3$	117.6

Analytical

^{15}N-NMR CHEMICAL SHIFTS OF MAJOR CHEMICAL FAMILIES

Thomas J. Bruno and Paris D. N. Svoronos

The following table contains ^{15}N-NMR chemical shifts of various organic nitrogen compounds. Chemical shifts are expressed relative to different standards (NH_3, NH_4Cl, CH_3NO_2, NH_4NO_3, HNO_3, etc.) and are interconvertible. Chemical shifts are sensitive to hydrogen bonding and are solvent dependent as seen in the case of pyridine (see footnote a below). Consequently, the reference as well as the solvent should always accompany chemical shift data. All shifts are relative to ammonia unless otherwise specified (Refs. 1–16). A section of "miscellaneous" data gives the chemical shift of special compounds relative to unusual standards. In Ref. 16, the reader can find additional details, correlation charts, and spin-spin coupling ranges.

References

1. Levy, G. C., and Lichter, R. L., *Nitrogen-15 Nuclear Magnetic Resonance Spectroscopy*, John Wiley and Sons, New York, 1979.
2. Yoder, C. H., and Schaeffer, C. D., Jr., *Introduction to Multinuclear NMR*, Benjamin/Cummings, Menlo Park, CA, 1987.
3. Duthaler, R. O., and Roberts, J. D., *J. Am. Chem. Soc.* 100, 4969, 1978.
4. Duthaler, R. O., and Roberts, J. D., *J. Am. Chem. Soc.* 100, 3889, 1978.
5. Kozerski, L., and von Philipsborn, W., *Org. Magn. Res.* 17, 306, 1981.
6. Duthaler, R. O., and Roberts, J. D., *J. Am. Chem. Soc.* 100, 3882, 1978.
7. Duthaler, R. O., and Roberts, J. D., *J. Magn. Res.* 34, 129, 1979.
8. Psota, L., Franzen-Sieveking, M., Turnier, J., and Lichter, R. L., *Org. Magn. Res.* 11, 401, 1978.
9. Subramanian, P. K., Chandra Sekara, N., and Ramalingam, K., *J. Org. Chem.* 47, 1933, 1982.
10. Schuster, I. I., and Roberts, J. D., *J. Org. Chem.* 45, 284, 1980.
11. Kupce, E., Liepins, E., Pudova, O., and Lukevics, E., *J. Chem. Soc., Chem. Commun.*, 581, 1984.
12. Allen, M., and Roberts, J. D., *J. Org. Chem.* 45, 130, 1980.
13. Brownlee, R. T. C., and Sadek, M., *Magn. Res. Chem.* 24, 821, 1986.
14. Dega-Szafran, Z., Szafran, M., Stefaniak, L., Brevard, C., and Bourdonneau, M., *Magn. Res. Chem.* 24, 424, 1986.
15. Lambert, J. B., Shurvell, H. F., Verbit, L., Cooks, R. G., and Stout, G. H., *Organic Structural Analysis*, MacMillan, New York, 1976.
16. Bruno, T. J., and Svoronos, P. D. N., *CRC Handbook of Basic Tables for Chemical Analysis*, 3rd Edition, CRC Press, Boca Raton, FL, 2011.

^{15}N-NMR Chemical Shifts of Major Chemical Families (Relative to NH_3 Unless Otherwise Indicated)

Chemical shift range/ppm	Family	Example (δ)
<930	Nitroso compounds	C_6H_5–NO (913, 930)
608	Sodium nitrite	$NaNO_2$
~500	Azo compounds	C_6H_5–N=N–C_6H_5 (510)
380–350	Nitro compounds	$C_6H_5NO_2$ (370.3); CH_3NO_2 (380.2); 4-F–C_6H_4–NO_2 (368.5); 1,3-$(NO_2)_2C_6H_4$ (365.4)
367	Nitric acid (8.57 M)	HNO_3
360–325	Nitramines	CH_3NHNO_2 (355.6); $CH_3O_2CNHNO_2$ (334.9)
350–300	Pyridines	C_5H_5N (317)[a] (gas); 4-CH_3–C_5H_4N (309.3); 4-NH_2–C_5H_4N (271.5); 4-NC–C_5H_4N (327.9)
~310	Imines (aromatic)	$(C_6H_5)_2C$=NH (308); C_6H_5CH=NCH_3 (318); C_6H_5CH=NC_6H_5 (326)
310.1	Nitrogen (gas)	N_2
250–200	Pyridinium salts	$C_5H_5NH^+$ (215)
260–175	Cyanides (nitriles)	CH_3CN (239.5, 245); C_6H_5CN (258.7); KCN 177.8
~160	Pyrroles, isonitriles	C_4H_4NH (158)CH_3NC (162)
~150	Thioamides	$CH_3C(=S)NH_2$ (150.2)
120–110	Lactams	$HN(CH_2)_3C$=O (5-membered ring; 114.7) $HN(CH_2)_6C$=O (8-membered ring; 117.7)
110–100	Amides	$C_6H_5CONH_2$ (100); CH_3CONH_2 (103.4); $CH_3CONHCH_3$ (105.8); $CH_3CON(CH_3)_2$ (103.8); $HCONH_2$ (108.5)
125–90	Sulfonamides	$CH_3SO_2NH_2$ (95); $C_6H_5SO_2NH_2$ (94.3)
~100	Hydrazines	$C_6H_5NHNHC_6H_5$ (96)
110–60	Ureas	$[H_2N]_2CO$ (75, 82); $[(CH_3)_2N]_2CO$ (63.5); $[C_6H_5NH]_2CO$ (107.7)
100–70	Aminophospines, aminophosphine oxides	$C_6H_5NHP(CH_3)_2$ (71.1) $C_6H_5NHPO(CH_3)_2$ (86.6)
70–50	Aromatic amines	$C_6H_5NH_2$ (55, 59), (–322.3)[b]; $C_6H_5NH_3^+$ (48), (–326.4[b], 26.1[c]; p–O_2N–C_6H_4–NH_2 (70)
40–0	Aliphatic amines	CH_3NH_2 (1.3)[d], (–371)[b]; $(CH_3)_2NH$ (–363.3)[b], (–364.9)[e], 6.7[d]; $(CH_3)_3N$ (–356.9) [b], (–360.7)[e], 13.0[d]
50–10	Isonitriles	CH_3NCO (14.1); C_6H_5NCO (46.5)
65–20	Ammonium salts	NH_4Cl (26.1)[d]; CH_3NH_3Cl (24.5); $(CH_3)_2NH_2Cl$ (26.6); $(CH_3)_3NHCl$ (33.8); $(CH_3)_4NCl$ (44.7)
~15	Isocyanates	CH_3NCO (14.1)
Miscellaneous:		
(–130)–(–110) and ~(–212)	Imidazoles	*N*-methylimdazole (–111.4, pyridine N and -215.7, pyrrole N)[b]
(–345)–(–310)[b]	Piperidine, hydrochloride salts	piperidinium hydrochloride (–344.8); 2-methyl piperdinium hydrochloride (–322.1)[e]
	Decahydroquinolines, hydrochloride salts	*trans*-decahydroquinolinium hydrochloride (–322.5); *cis*-decahydroquinolinium hydrochloride (–328.5)

Chemical shift range/ppm	Family	Example (δ)
(−293)–(−280)[f]	Enaminones	$CH_3C(=O)CH=CHNHCH_3$ [(*E*)-(−294.2); (*Z*)-(−285.9)]
35–15[g]	4-Aminotetrahydropyrans, 4-aminotetrahydro-thiopyrans	2,6-diphenyl 4-aminotetrahydropyran (34.5) 2,6-diphenyl 4-aminotetrahydrothiopyran (33.6)
(−325)–(−310)[c]	1-Naphthylamines	8-nitro-1-naphthylamine (313.9)
(−350)–(−300)[h]	Silylamines	$HN[Si(CH_3)_3]_2$ (−354.2)

[a] Varies with solvent. For instance: cyclohexane (315.5), benzene (312.1), chloroform (304.5), methanol (292.1), water (289), 2,2,2-trifluoroethanol (277.1). All chemical shifts relative to ammonia.

[b] Upfield from external HNO_3 1M (CH_3OH).

[c] In ppm upfield from external 1M $D^{15}NO_3$ in D_2O (DMSO) (Ref. 10).

[d] Downfield from anhydrous liquid ammonia, ±0.2 ppm unless otherwise specified.

[e] Upfield from external HNO_3 1M (cyclohexane).

[f] Relative to external $CH_3{}^{15}NO_2$.

[g] With respect to an external standard of 5M $^{15}NH_4NO_3$ in 2M HNO_3 ($^{15}NH_4NO_3$ = 21.6 ppm relative to anhydrous ammonia) (Ref. 9).

[h] Relative to $N(SiH_3)_3$ (50 % in $CDCl_3$) (Ref. 11).

Analytical

NATURAL ABUNDANCE OF IMPORTANT ISOTOPES

Thomas J. Bruno and Paris D. N. Svoronos

The following table lists the atomic masses and relative percent concentrations of naturally occurring isotopes of importance in mass spectroscopy (Refs. 1–6).

References

1. de Hoffmann, E., and Stroobant, V., *Mass Spectrometry: Principles and Applications*, 3rd Edition, Wiley Interscience, Winchester, United Kingdom, 2007.
2. Johnstone, R. A. W., and Rose, M. E., *Mass Spectrometry for Chemists and Biochemists*, Cambridge University Press, 1996.
3. Lide, D. R., Ed., *CRC Handbook for Chemistry and Physics*, 90th Edition, CRC Press, Boca Raton, FL, 2010.
4. McLafferty, F. W., and Turecek, F., *Interpretation of Mass Spectra*, 4th Edition, University Science Books, Mill Valley, CA, 1993.
5. Watson, J. T., *Introduction to Mass Spectrometry*, 3rd Edition, Lippincott-Raven, Philadelphia, PA, 1997.
6. Bruno, T. J., and Svoronos, P. D. N., *CRC Handbook of Basic Tables for Chemical Analysis*, 3rd Edition, CRC Press, Boca Raton, FL, 2011.

Natural Abundance of More Prominent Isotopes (Mass in u, Percent Abundance)

Element	Primary isotope	Other isotopes	Other isotopes
Hydrogen	^{1}H (1.00783, 99.985)	^{2}H (2.01410, 0.015)	
Boron	^{10}B (10.01294, 19.8)	^{11}B (11.00931, 80.2)	
Carbon	^{12}C (12.00000, 98.9)	^{13}C (13.00335, 1.1)	
Nitrogen	^{14}N (14.00307, 99.6)	^{15}N (15.00011, 0.4)	
Oxygen	^{16}O (15.99491, 99.8)		^{18}O (17.9992, 0.2)
Fluorine	^{19}F (18.99840, ≈100.0)		
Silicon	^{28}Si (27.97693, 92.2)	^{29}Si (28.97649, 4.7)	^{30}Si (29.97376, 3.1)
Phosphorus	^{31}P (30.97376, ≈100.0)		
Sulfur	^{32}S (31.972017, 95.0)	^{33}S (32.97146, 0.7)	^{34}S (33.96786, 4.2)
Chlorine	^{35}Cl (34.96885, 75.5)		^{37}Cl (36.96590, 24.5)
Bromine	^{79}Br (78.9183, 50.5)		^{81}Br (80.91642, 49.5)
Iodine	^{127}I (126.90466, ≈100.0)		

COMMON MASS SPECTRAL FRAGMENTATION PATTERNS OF ORGANIC COMPOUND FAMILIES

Thomas J. Bruno and Paris D. N. Svoronos

The following table provides a guide to the identification and interpretation of commonly observed mass spectral fragmentation patterns for common organic functional groups (Refs. 1–10). It is of course highly desirable to augment mass spectroscopic data with as much other structural information as possible. Especially useful in this regard will be the confirmatory information of infrared and ultraviolet spectrophotometry as well as nuclear magnetic resonance spectrometry.

The reader is referred to other tables in this book for additional information that can be used in assessing or predicting fragmentation patterns, particularly the tables of "Proton Affinities" and "Ionization Energies of Gas-Phase Molecules" in Section 10.

References

1. Bowie, J. H., Williams, D. H., Lawesson, S. O., Madsen, J. O., Nolde, C., and Schroll, G., *Tetrahedron* 22, 3515, 1966.
2. Johnstone, R. A. W., and Rose, M. E., *Mass Spectrometry for Chemical and Biochemists*, Cambridge University Press, New York, 1996.
3. Lee, T. A., *A Beginner's Guide to Mass Spectral Interpretation*, Wiley, New York, 1998
4. McLafferty, F. W., and Turecek, F., *Interpretation of Mass Spectra, 4th Edition*, University Science Books, Mill Valley, CA, 1993.
5. Pasto, D. J., and Johnson, C. R., *Organic Structure Determination*, Prentice Hall, Englewood Cliffs, NJ, 1969.
6. Silverstein, R. M., Bassler, G. C., and Morrill, T. C., *Spectroscopic Identification of Organic Compounds, 6th Edition*, John Wiley and Sons, New York, 1998.
7. Smakman, R., and deBoer, T. J., *Org. Mass Spec.* 3, 1561, 1970.
8. Smith, R. M., *Understanding Mass Spectra: A Basic Approach*, John Wiley and Sons, New York, 1999.
9. Watson, T. J., and Watson, J. T., *Introduction to Mass Spectrometry*, Lippincott, Williams and Wilkins, Philadelphia, PA, 1997.
10. Bruno, T. J., and Svoronos, P. D. N., *CRC Handbook of Basic Tables for Chemical Analysis, 3rd Edition*, CRC Press, Boca Raton, FL, 2011.

Mass Spectral Fragmentation Patterns of Organic Compound Families

Family	Molecular ion peak	Common fragments and characteristic peaks
Acetals		Cleavage of all C–O, C–H, and C–C bonds around the original aldehydic carbon
Alcohols	Weak for 1° and 2°; not detectable for 3°; strong for benzyl alcohols	Loss of 18 (H_2O — usually by cyclic mechanism); loss of H_2O and olefin simultaneously with four (or more) carbon-chain alcohols; prominent peak at m/z = 31($CH_2\ddot{O}H$)$^+$ for 1° alcohols; prominent peak at m/z = ($RCH\ddot{O}H$)$^+$ for 2° alcohols; and m/z = ($R_2C\ddot{O}H$)$^+$ for 3° alcohols
Aldehydes	Low intensity	Loss of aldehydic hydrogen (strong M-1 peak, especially with aromatic aldehydes); strong peak at m/z = 29($HC{\equiv}O^+$); loss of chain attached to alpha carbon (beta cleavage); McLafferty rearrangement via beta cleavage if gamma hydrogen is present
Alkanes		
(a) Chain	Low intensity	Loss of 14 mass units (CH_2)
(b) Branched	Low intensity	Cleavage at the point of branch; low intensity ions from random rearrangements
(c) Alicyclic	Intense	Loss of 28 mass units ($CH_2{=}CH_2$) and side chains
Alkenes (olefins)	Rather high intensity (loss of π-electron) especially in case of cyclic olefins	Loss of units of general formula CnH_{2n-1}; formation of fragments of the composition CnH_{2n} (via McLafferty rearrangement); retro Diels-Alder fragmentation
Alkyl halides	Abundance of molecular ion F<Cl<Br<I; intensity decreases with increase in size and branching	Loss of fragments equal to the mass of the halogen until all halogens are cleaved off
(a) Fluorides	Very low intensity	Loss of 20 (HF); loss of 26 (C_2H_2) in case of fluorobenzenes
(b) Chlorides	Low intensity; characteristic isotope cluster	Loss of 35 (Cl) or 36 (HCl); loss of chain attached to the gamma carbon to the carbon carrying the Cl
(c) Bromides	Low intensity; characteristic isotope cluster	Loss of 79 (Br); loss of chain attached to the gamma carbon to the carbon carrying the Br
(d) Iodides	Higher than other halides	Loss of 127 (I)
Alkynes	Rather high intensity (loss of π-electron)	Fragmentation similar to that of alkenes
Amides	High intensity	Strong peak at m/z = 44 indicative of a 1° amide ($O{=}C{=}NH^+_2$); base peak at m/z = 59 ($CH_2{=}C(OH)N^+H_2$); possibility of McLafferty rearrangement; loss of 42 (C_2H_2O) for amides of the form $RNHCOCH_3$ when R is aromatic ring
Amines	Hardly detectable in case of acyclic aliphatic amines; high intensity for aromatic and cyclic amines	Beta cleavage yielding >C=N$^+$<; base peak for all 1° amines at m/z = 30 ($CH_2{=}N^+H_2$); moderate M-1 peak for aromatic amines; loss of 27 (HCN) in aromatic amines; fragmentation at alpha carbons in cyclic amines
Aromatic hydrocarbons (arenes)	Intense	Loss of side chain; formation of RCH=CHR' (via McLafferty rearrangement); cleavage at the bonds beta to the aromatic ring; peaks at m/z = 77 (benzene ring; especially mono-substituted), 91 (tropyllium); the ring position of alkyl substitution has very little effect on the spectrum

Family	Molecular ion peak	Common fragments and characteristic peaks
Carboxylic acids	Weak for straight-chain monocarboxylic acids; large if aromatic acids	Base peak at *m/z* = 60 (CH_2=C(OH)$_2$) if alpha hydrogen is present; peak at *m/z* = 45 (COOH); loss of 17 (–OH) in case of aromatic acids or short-chain acids
Disulfides	Low intensity	Loss of olefins (*m/z* equal to R–S–S–$H^{+\cdot}$); strong peak at *m/z* = 66 ($HSSH^+$)
Phenols	Highly intense peak (base peak* generally)	Loss of 28 (C=O) and 29 (CHO); strong peak at *m/z* = 65 ($C_5H_5^+$)
Sulfides (thioethers)	Rather low intensity peak but higher than that of corresponding ether	Similar to those of ethers (–O– substituted by –S–); aromatic sulfides show strong peaks at *m/z* = 109 ($C_6H_5S^{+\cdot}$); 65 ($C_5H_5^{+\cdot}$); 91 (tropyllium ion)
Sulfonamides	Intense	Loss of *m/z* = 64 ($SONH_2$) and *m/z* = 27 (HCN) in case of benzenesulfonamide
Esters	Weak intensity	Base peak at *m/z* equal to the mass of R–C≡O^+; peaks at *m/z* equal to the mass of $^+$O≡C–OR', the mass of OR' and R'; McLafferty rearrangement possible in case of: (a) presence of a beta hydrogen in R' (peak at *m/z* equal to the mass of R–C($^+$OH)OH, and (b) presence of a gamma hydrogen in R (peak at *m/z* equal to the mass of (CH_2=C($^+$OH)OR); loss of 42 (CH_2=C=O) in case of benzyl esters; loss of ROH via the ortho effect in case of *o*–substituted benzoates
Ketones	High intensity	Loss of R–groups attached to the >C=O (alpha cleavage); peak at *m/z* = 43 for all methyl ketones (CH_3CO^+); McLafferty rearrangement via beta cleavage if gamma hydrogen is present; loss of *m/z* = 28 (C=O) for cyclic ketones after initial alpha cleavage and McLafferty rearrangement
Mercaptan (thiols)	Rather low intensity but higher than that of corresponding alcohol	Similar to those of alcohols (–OH substituted by –SH); loss of *m/z* = 45 (CHS) and *m/z* = 44 (CS) for aromatic thiols
Nitriles	Unlikely to be detected except in case of acetonitrile (CH_3CN) and propionitrile (C_2H_5CN)	M+1 ion may appear (especially at higher pressures); M-1 peak is weak but detectable (R–CH=C=N^+); base peak at *m/z* = 41 (CH_2=C=N^+H); McLafferty rearrangement possible; loss of HCN is case of cyanobenzenes
Nitrites	Absent (or very weak at best)	Base peak at *m/z* = 30 (NO^+); large peak at *m/z* = 60 (CH_2=O^+NO) in all unbranched nitrites at the alpha carbon; absence of *m/z* = 46 permits differentiation from nitro compounds
Nitro compounds	Seldom observed	Loss of 30 (NO); subsequent loss of CO (in case of aromatic nitro- compounds); loss of NO_2, *m/z* = 46, from molecular ion peak
Sulfones	High intensity	Similar to sulfoxides; loss of mass equal to RSO_2; aromatic heterocycles show peaks at M-32 (sulfur), M-48 (SO), M-64 (SO_2)
Sulfoxides	High intensity	Loss of 17 (OH); loss of alkene (*m/z* equal to $RSOH^{+\cdot}$); peak at *m/z* = 63 (CH_2=SOH)$^+$; aromatic sulfoxides show peak at *m/z* = 125 ($^+$S–CH=CHCH=CHCHC=O), 97($C_5H_5S^+$), 93($C_6H_5O^+$); aromatic heterocycles show peaks at M-16 (oxygen), M-29(COH); M-48(SO)

* The base peak is the most intense peak in the mass spectrum, and is often the molecular ion peak, M^+.

COMMON MASS SPECTRAL FRAGMENTS LOST

Thomas J. Bruno and Paris D. N. Svoronos

The following table gives a list of neutral species that are most commonly lost when measuring the mass spectra of organic compounds. The list is suggestive rather than comprehensive, and should be used in conjunction with other sources [1–5]. The listed fragments include only combinations of carbon, hydrogen, oxygen, nitrogen, sulfur, and the halogens.

References

1. Hamming, M., and Foster, N., *Interpretation of Mass Spectra of Organic Compounds*, Academic Press, New York, 1972.
2. McLafferty, F. W., Turecek, F., *Interpretation of Mass Spectra*, 4th Edition, University Science Books, Mill Valley, CA, 1993.
3. Silverstein, R. M., Bassler, G. C., and Morrill, T. C., *Spectroscopic Identification of Organic Compounds*, 6th Edition, John Wiley and Sons, New York, 1996.
4. Bruno, T. J., *CRC Handbook for the Analysis and Identification of Alternative Refrigerants*, CRC Press, Boca Raton, FL, 1995.
5. Bruno, T. J., and Svoronos, P. D. N., *CRC Handbook of Basic Tables for Chemical Analysis*, 3rd Edition, CRC Press, Boca Raton, FL, 2011.

Common Mass Spectral Fragments Lost for Different Mass Losses

Mass lost	Fragment lost
1	$H^{\bullet}$
15	$CH_3^{\bullet}$
17	$OH^{\bullet}$
18	H_2O
19	$F^{\bullet}$
20	HF
26	$HC{\equiv}CH$; ${}^{\bullet}C{\equiv}N$
27	$CH_2{-}CH^{\bullet}$; $HC{\equiv}N$
28	$CH_2{=}CH_2$; $C{=}O$; (HCN and $H^{\bullet}$)
29	$CH_3CH_2^{\bullet}$; $H{-}{}^{\bullet}C{=}O$
30	${}^{\bullet}CH_2NH_2$; $HCHO$; NO
31	$CH_3O^{\bullet}$; ${}^{\bullet}CH_2OH$; CH_3NH_2
32	CH_3OH; S
33	$HS^{\bullet}$
34	H_2S
35	$Cl^{\bullet}$
36	HCl_2H_2O
37	H_2Cl
38	$C_3H_2^{\bullet}$; C_2N; F_2
39	C_3H_3; HC_2N
40	$CH_3C^{\bullet}CH$
41	$CH_2{=}CHCH_2^{\bullet}$
42	$CH_2{=}CHCH_3$; $CH_2{=}C{=}O$; $(CH_2)_3$; NCO; $NCNH_2$
43	$C_3H_7^{\bullet}$; $CH_3C{=}O^{\bullet}$; $CH_2{=}CH{-}O^{\bullet}$; $HCNO$
44	$CH_2{=}CHOH$; CO_2; N_2O; $CONH_2$; $NHCH_2CH_3$
45	CH_3CHOH; $CH_3CH_2O^{\bullet}$; CO_2H; $CH_3CH_2NH_2$
46	CH_3CH_2OH; ${}^{\bullet}NO_2$
47	$CH_3S^{\bullet}$
48	CH_3SH; SO; O_3
49	${}^{\bullet}CH_2Cl$
51	${}^{\bullet}CHF_2$
52	$C_4H_4^{\bullet}$, C_2N_2
54	$CH_2{=}CHCH{=}CH_2$
55	$CH_2{=}CH{-}CH^{\bullet}CH_3$
56	$CH_2{=}CH{-}CH_2CH_3$; $CH_3CH{=}CHCH_3$; CO (2 moles)
57	$C_4H_9^{\bullet}$
58	${}^{\bullet}NCS$; $(CH_3)_2C{=}O$; (NO and CO)
59	$CH_3OC{=}O^{\bullet}$; CH_3CONH_2; $C_2H_3S^{\bullet}$
60	C_3H_7OH
61	$CH_3CH_2S^{\bullet}$; $(CH_2)_2S^{\bullet}H$
62	[H_2S and $CH_2{=}CH_2$]
63	${}^{\bullet}CH_2CH_2Cl$
64	$S_2^{\bullet}$; $SO_2^{\bullet}$; $C_5H_4^{\bullet}$
68	$CH_2{=}CHC(CH_3){=}CH_2$
69	$CF_3^{\bullet}$; $C_5H_9^{\bullet}$
71	$C_5H_{11}^{\bullet}$
73	$CH_3CH_2OC^{\bullet}{=}O$
74	C_4H_9OH
75	C_6H_3
76	C_6H_4; CS_2
77	C_6H_5; HCS_2
78	$C_6H_6^{\bullet}$; $H_2CS_2^{\bullet}$, C_5H_4N
79	$Br^{\bullet}$; C_5H_5N
80	HBr
85	${}^{\bullet}CClF_2$
100	$CF_2{=}CF_2$
119	$CF_3CF_2^{\bullet}$
122	$C_6H_5CO_2H$
127	$I^{\bullet}$
128	HI

Analytical

MAJOR REFERENCE MASSES IN THE SPECTRUM OF HEPTACOSAFLUOROTRIBUTYLAMINE (PERFLUOROTRIBUTYLAMINE)

Thomas J. Bruno and Paris D. N. Svoronos

The following list tabulates the major reference masses (with their relative intensities and formulas) of the mass spectrum of heptacosafluorotributylamine (CAS No. 311-89-7) (Ref. 1). This is one of the most widely used reference compounds in mass spectrometry. For guidance in the selection of additional reference compounds, the reader is advised to consult Ref. 1.

Reference

1. Bruno, T. J., and Svoronos, P. D. N., *CRC Handbook of Basic Tables for Chemical Analysis, Third Edition*, CRC Press, Boca Raton, FL, 2011.

Mass	Relative intensity	Formula	Mass	Relative intensity	Formula
613.9647	2.6	$C_{12}F_{24}N$	180.9888	1.9	C_4F_7
575.9679	1.7	$C_{12}F_{22}N$	175.9935	1.0	C_4F_6N
537.9711	0.4	$C_{12}F_{20}N$	168.9888	3.6	C_3F_7
501.9711	8.6	$C_9F_{20}N$	163.9935	0.7	C_3F_6N
463.9743	3.8	$C_9F_{18}N$	161.9904	0.3	C_4F_6
425.9775	2.5	$C_9F_{16}N$	149.9904	2.1	C_3F_6
413.9775	5.1	$C_8F_{16}N$	130.9920	31	C_3F_5
375.9807	0.9	$C_8F_{14}N$	118.9920	8.3	C_2F_5
325.9839	0.4	$C_7F_{12}N$	113.9967	3.7	C_2F_4N
313.9839	0.4	$C_6F_{12}N$	111.9936	0.7	C_3F_4
263.9871	10	$C_5F_{10}N$	99.9936	12	C_2F_4
230.9856	0.9	C_5F_9	92.9952	1.1	C_3F_3
225.9903	0.6	C_5F_8N	68.9952	100	CF_3
218.9856	62	C_4F_9	49.9968	1.0	CF_2
213.9903	0.6	C_4F_8N	30.9984	2.3	CF

Analytical

MASS SPECTRAL PEAKS OF COMMON ORGANIC LIQUIDS

David R. Lide

The strongest peaks in the mass spectra of 374 important organic liquids are listed in this table. The *m/z* value for each peak is followed by the relative intensity in parentheses, with the strongest peak assigned an intensity of 100. The peaks for each compound are listed in order of decreasing intensity. Compounds are listed by name.

Data on the physical properties of the same compounds may be found in Section 15 in the table "Laboratory Solvents and Other Liquid Reagents".

References

1. NIST 2011 Mass Spectral Library (NIST 11), National Institute of Standards and Technology,<www.nist.gov/srd/nist1a.cfm>.
2. Lide, D. R., and Milne, G. W. A., Editors, *Handbook of Data on Organic Compounds, Third Edition*, CRC Press, Boca Raton, FL, 1994.
3. Lide, D. R., Editor, *Properties of Organic Compounds*, <www.chemnetbase.com/tours/poc/intro.jsf>

Mass Spectral Peaks of Common Organic Liquids (*m/z*; Relative Intensities in Parentheses)

Compound										
Acetic acid	43(100)	45(87)	60(57)	15(42)	42(14)	29(13)	14(13)	28(7)	18(6)	16(6)
Acetic anhydride	43(100)	42(35)	45(29)	60(18)	29(9)	41(8)	40(2)	26(2)	87(1)	61(1)
Acetone	43(100)	15(34)	58(23)	27(9)	14(9)	42(8)	26(7)	29(5)	28(5)	39(4)
Acetonitrile	41(100)	40(46)	39(13)	14(9)	38(6)	28(4)	26(4)	25(3)	42(2)	27(2)
Acrolein	27(100)	56(74)	28(65)	26(54)	55(52)	29(37)	25(8)	53(5)	38(5)	57(4)
Acrylonitrile	53(100)	26(85)	52(79)	51(34)	27(13)	50(8)	25(7)	38(5)	54(3)	37(3)
Allyl alcohol	57(100)	31(34)	29(32)	28(31)	58(25)	39(22)	27(20)	30(16)	32(14)	26(11)
Allylamine	30(100)	56(80)	28(76)	57(33)	39(21)	29(20)	27(18)	26(13)	41(8)	18(8)
2-Amino-2-methyl-1-propanol	58(100)	41(18)	18(17)	42(13)	28(11)	56(10)	30(10)	29(8)	43(6)	59(5)
Aniline	93(100)	66(32)	65(16)	39(13)	92(10)	94(7)	41(5)	40(5)	67(4)	64(3)
Anisole	108(100)	65(76)	78(60)	39(44)	51(21)	77(20)	93(16)	79(14)	50(13)	63(12)
Benzaldehyde	51(100)	77(81)	50(55)	106(44)	105(43)	52(26)	78(16)	39(13)	27(10)	74(8)
Benzene	78(100)	77(20)	52(19)	51(17)	50(15)	39(12)	79(6)	76(5)	74(4)	38(4)
Benzeneacetonitrile	117(100)	90(43)	116(35)	89(22)	51(13)	39(11)	63(10)	118(9)	91(8)	50(8)
Benzenethiol	110(100)	66(28)	109(23)	39(15)	77(14)	51(14)	84(13)	69(11)	50(11)	45(11)
Benzonitrile	103(100)	76(34)	50(13)	104(9)	75(7)	51(7)	77(5)	52(4)	39(4)	74(3)
Benzyl acetate	108(100)	43(76)	91(60)	90(48)	79(27)	107(17)	77(17)	65(15)	51(15)	89(14)
Benzyl alcohol	79(100)	108(83)	77(74)	107(66)	51(35)	105(23)	106(22)	50(17)	78(16)	39(16)
Bis(2-aminoethyl)amine	44(100)	73(59)	30(35)	19(18)	56(16)	28(16)	27(16)	42(11)	99(8)	43(8)
Bis(2-chloroethyl) ether	93(100)	63(74)	27(38)	95(32)	65(24)	31(9)	49(4)	28(4)	94(3)	62(3)
Bis(2-ethylhexyl) phthalate	149(100)	57(32)	167(29)	71(21)	43(21)	70(18)	150(11)	113(10)	55(10)	41(9)
Bis(2-hydroxyethyl) sulfide	61(100)	45(68)	31(38)	104(36)	91(34)	47(26)	44(26)	27(24)	60(18)	43(17)
Bromobenzene	77(100)	158(64)	156(64)	51(39)	50(17)	78(8)	76(6)	75(6)	28(5)	159(4)
1-Bromobutane	57(100)	41(56)	29(45)	27(29)	56(13)	39(12)	28(12)	55(7)	43(7)	138(6)
2-Bromobutane, (±)-	57(100)	41(60)	29(57)	27(34)	39(20)	28(9)	26(9)	136(1)		
1-Bromo-2-chloroethane	63(100)	27(85)	65(31)	26(23)	144(8)	81(8)	79(8)	28(7)	142(6)	93(6)
Bromochloromethane	49(100)	130(67)	128(52)	51(31)	93(23)	81(20)	79(20)	95(17)	132(16)	47(8)
1-Bromodecane	43(100)	135(94)	137(91)	57(81)	41(58)	55(56)	71(38)	69(36)	85(33)	29(27)
Bromoethane	108(100)	110(97)	29(62)	27(51)	28(35)	26(14)	93(6)	32(6)	95(5)	81(5)
2-Bromo-2-methylpropane	57(100)	41(67)	29(45)	39(30)	27(18)	28(8)	40(5)	38(5)	58(4)	55(4)
1-Bromonaphthalene	44(100)	206(39)	127(39)	208(37)	36(31)	69(29)	131(13)	29(13)	126(12)	63(12)
1-Bromopentane	43(100)	71(80)	41(56)	27(51)	42(37)	29(34)	55(33)	39(30)	28(12)	26(9)
1-Bromopropane	43(100)	41(77)	28(69)	27(60)	39(49)	124(42)	122(41)	42(34)	32(20)	29(15)
2-Bromopropane	43(100)	27(47)	41(43)	39(22)	124(8)	122(8)	26(7)	81(6)	79(6)	38(6)
2-Bromopropene	41(100)	39(58)	122(37)	120(36)	38(12)	37(8)	40(6)	81(5)	79(5)	42(4)
Butanal	44(100)	43(74)	72(57)	41(56)	27(55)	29(48)	57(23)	39(22)	28(15)	42(11)
Butanenitrile	41(100)	29(62)	27(37)	28(10)	39(9)	26(7)	40(5)	42(4)	38(4)	15(4)
1-Butanethiol	56(100)	41(74)	90(66)	47(43)	27(43)	28(36)	29(33)	57(17)	39(16)	61(15)
Butanoic acid	60(100)	27(50)	73(27)	42(25)	41(24)	43(22)	29(21)	45(19)	39(15)	28(11)
Butanoic anhydride	71(100)	43(59)	27(26)	41(19)	42(10)	39(10)	73(9)	28(7)	72(5)	55(5)
1-Butanol	31(100)	56(81)	41(62)	43(60)	27(50)	42(31)	29(31)	28(17)	39(16)	55(12)
2-Butanol	45(100)	31(22)	27(22)	59(20)	29(18)	43(13)	41(12)	44(8)	18(8)	28(5)
2-Butanone	43(100)	72(24)	29(19)	27(12)	57(7)	42(5)	26(4)	28(3)	44(2)	39(2)

Compound										
trans-2-Butenal	41(100)	39(97)	70(82)	69(65)	27(49)	29(39)	42(30)	38(29)	40(27)	37(18)
2-Butoxyethanol	57(100)	45(38)	29(35)	41(31)	87(16)	27(12)	56(11)	31(9)	75(7)	28(7)
2-Butoxyethyl acetate	57(100)	43(86)	56(50)	87(33)	41(26)	29(22)	85(18)	88(11)	44(11)	27(7)
Butyl acetate	43(100)	56(34)	41(17)	27(16)	29(15)	73(11)	61(10)	28(7)	55(6)	39(6)
sec-Butyl acetate	43(100)	56(21)	87(15)	41(14)	29(8)	57(6)	73(4)	61(4)	55(4)	27(4)
Butylamine	30(100)	73(10)	28(5)	41(3)	27(3)	18(3)	44(2)	42(2)	31(2)	29(2)
tert-Butylamine	58(100)	41(21)	42(15)	18(9)	30(8)	15(8)	39(7)	57(6)	28(6)	59(4)
Butylbenzene	91(100)	92(55)	134(20)	65(13)	27(10)	39(9)	105(8)	51(7)	78(6)	41(6)
sec-Butylbenzene, (±)-	105(100)	134(18)	91(14)	77(10)	27(9)	106(9)	51(7)	79(7)		
tert-Butylbenzene	119(100)	91(65)	41(40)	134(24)	39(15)	79(14)	77(13)	51(13)	120(11)	65(7)
Butyl butanoate	43(100)	71(90)	56(80)	89(69)	41(67)	27(52)	29(47)	57(29)	39(22)	60(19)
Butyl formate	56(100)	41(60)	31(58)	29(53)	27(45)	43(34)	28(21)	39(19)	55(11)	42(11)
1-*tert*-Butyl-4-methylbenzene	133(100)	105(38)	41(23)	148(18)	93(16)	91(14)	115(13)	134(11)	39(11)	116(10)
Butyl vinyl ether	29(100)	41(74)	56(56)	57(43)	27(42)	44(26)	15(16)	85(14)	39(14)	43(13)
γ-Butyrolactone	28(100)	42(74)	29(48)	27(33)	41(27)	56(25)	86(24)	26(18)	85(10)	39(10)
Caprolactam	55(100)	113(87)	30(81)	56(66)	84(60)	85(57)	42(51)	41(33)	28(26)	43(17)
Carbon disulfide	76(100)	32(22)	44(17)	78(9)	38(6)	28(5)	77(3)	64(1)	46(1)	39(1)
2-Chloroaniline	127(100)	129(32)	92(17)	65(16)	128(10)	91(9)	64(9)	39(8)	63(7)	99(6)
Chlorobenzene	112(100)	77(63)	114(33)	51(29)	50(14)	75(8)	113(7)	78(5)	76(5)	28(4)
1-Chlorobutane	56(100)	41(65)	27(50)	43(35)	29(24)	39(18)	28(16)	26(11)	55(8)	15(8)
2-Chlorobutane	56(100)	57(100)	41(90)	27(77)	29(57)	63(46)	39(34)	92(1)		
1-Chloro-1,1-difluoroethane	65(100)	45(31)	85(14)	31(10)	64(8)	44(7)	35(6)	26(6)	87(5)	81(4)
Chloroethane	64(100)	28(91)	29(84)	27(75)	66(32)	26(28)	49(25)	51(8)	63(6)	65(4)
2-Chloroethanol	31(100)	15(13)	29(10)	28(10)	27(9)	43(8)	44(7)	26(5)	18(5)	14(5)
(Chloromethyl)benzene	91(100)	126(20)	65(14)	92(9)	39(9)	63(8)	128(6)	45(6)	89(5)	125(3)
1-Chloro-3-methylbutane	43(100)	55(56)	41(55)	27(51)	70(49)	42(37)	29(34)	57(30)	39(30)	56(15)
1-Chloro-2-methylpropane	43(100)	41(67)	42(50)	27(33)	39(26)	15(11)	29(10)	56(7)	49(6)	38(5)
2-Chloro-2-methylpropane	57(100)	41(80)	77(44)	29(26)	39(24)	79(14)	27(14)	59(7)	56(7)	38(6)
1-Chloronaphthalene	162(100)	127(36)	164(30)	126(20)	163(10)	77(8)	101(6)	75(6)	128(5)	28(4)
1-Chlorooctane	91(100)	41(83)	43(76)	27(62)	29(56)	55(55)	39(34)	93(32)	57(32)	69(29)
Chloropentafluoroethane	85(100)	69(61)	31(38)	87(32)	50(17)	35(8)	119(6)	66(4)	100(3)	47(3)
1-Chloropentane	42(100)	41(90)	70(89)	55(87)	27(73)	29(55)	43(40)	39(40)	28(19)	57(18)
1-Chloropropane	42(100)	29(46)	27(37)	41(23)	28(15)	43(14)	39(12)	78(6)	63(6)	49(5)
3-Chloropropene	41(100)	39(73)	76(28)	38(16)	37(13)	40(12)	27(12)	26(11)	78(9)	49(5)
2-Chlorotoluene	91(100)	126(68)	89(23)	128(22)	90(18)	65(17)	125(15)	92(13)	127(10)	63(6)
3-Chlorotoluene	91(100)	126(27)	63(15)	65(12)	89(11)	39(11)	128(9)	125(8)	92(8)	62(6)
Cyclohexane	56(100)	84(71)	41(70)	27(37)	55(36)	39(35)	42(30)	69(23)	28(18)	43(14)
Cyclohexanol	57(100)	44(68)	41(68)	39(51)	32(40)	43(38)	31(32)	42(22)	67(18)	82(16)
Cyclohexanone	55(100)	42(85)	41(34)	27(33)	98(31)	39(27)	69(26)	70(20)	43(14)	28(14)
Cyclohexene	67(100)	54(72)	82(37)	41(35)	39(33)	27(15)	53(12)	81(9)	51(8)	79(6)
Cyclohexylamine	56(100)	43(23)	28(17)	99(10)	70(8)	57(6)	30(6)	93(5)	54(4)	41(4)
Cyclopentane	42(100)	70(30)	55(29)	41(29)	39(22)	27(15)	40(7)	29(5)	28(4)	43(3)
Cyclopentanone	55(100)	28(50)	84(42)	41(38)	56(29)	27(24)	39(19)	42(15)	26(9)	29(7)
cis-Decahydronaphthalene	67(100)	81(87)	41(81)	138(67)	96(62)	82(62)	39(50)	55(45)	27(44)	95(42)
trans-Decahydronaphthalene	41(100)	68(91)	67(88)	82(67)	27(65)	96(61)	95(55)	138(51)	81(51)	29(51)
Decane	43(100)	57(90)	41(41)	71(33)	29(30)	85(24)	27(20)	56(17)	55(14)	42(14)
Diacetone alcohol	43(100)	59(41)	58(17)	101(10)	41(9)	31(9)	83(6)	56(6)	55(6)	29(6)
1,2-Dibromoethane	27(100)	107(77)	109(72)	26(24)	28(10)	81(5)	79(5)	25(5)	95(4)	93(4)
Dibromofluoromethane	111(100)	113(98)	192(29)	43(16)	41(16)	190(15)	194(14)	81(9)	79(9)	122(7)
Dibromomethane	174(100)	93(96)	95(84)	172(53)	176(50)	91(11)	81(9)	79(9)	94(5)	65(5)
1,2-Dibromopropane	41(100)	121(66)	123(65)	39(48)	27(28)	107(11)	38(10)	26(10)	109(9)	42(9)
1,2-Dibromotetrafluoroethane	179(100)	181(97)	129(34)	131(33)	100(17)	31(13)	260(12)	50(8)	69(7)	262(6)
Dibutylamine	86(100)	72(52)	30(48)	44(40)	29(31)	57(24)	41(21)	73(15)	28(15)	43(13)
Dibutyl ether	57(100)	41(34)	29(30)	56(25)	87(21)	27(9)	58(8)	55(6)	39(5)	28(5)
Dibutyl oxalate	57(100)	41(61)	56(24)	44(9)	55(8)	43(7)	58(5)	42(4)	103(2)	73(2)
Dibutyl phthalate	149(100)	86(18)	57(18)	223(17)	205(17)	150(17)	104(17)	56(17)	41(17)	65(16)
Dibutyl sebacate	241(100)	185(71)	41(37)	56(35)	55(32)	57(31)	242(23)	143(21)	98(21)	125(20)
o-Dichlorobenzene	146(100)	148(64)	111(38)	75(23)	113(12)	74(12)	50(11)	150(10)	73(9)	147(7)
m-Dichlorobenzene	146(100)	148(65)	111(36)	75(25)	50(19)	74(16)	150(11)	113(11)	73(11)	147(8)
1,1-Dichloroethane	63(100)	27(71)	65(31)	26(19)	83(11)	85(7)	61(7)	35(6)	98(5)	62(5)
1,2-Dichloroethane	62(100)	27(91)	49(40)	64(32)	26(31)	63(19)	98(14)	51(13)	61(12)	100(9)

Compound										
1,1-Dichloroethene	61(100)	96(61)	98(38)	63(32)	26(16)	60(15)	62(7)	25(7)	100(6)	35(6)
cis-1,2-Dichloroethene	61(100)	96(73)	98(47)	63(32)	26(30)	60(21)	25(13)	35(12)	62(9)	100(8)
trans-1,2-Dichloroethene	61(100)	96(67)	98(43)	26(34)	63(32)	60(24)	25(15)	62(10)	100(7)	47(7)
Dichlorofluoromethane	67(100)	69(32)	47(13)	35(13)	31(9)	32(8)	48(7)	83(5)	102(4)	49(4)
Dichloromethane	49(100)	84(64)	86(39)	51(31)	47(14)	48(8)	88(6)	50(3)	85(2)	83(2)
(Dichloromethyl)benzene	125(100)	127(32)	160(14)	89(13)	162(9)	63(9)	126(8)	62(7)	105(5)	39(5)
1,2-Dichloropropane, (±)-	63(100)	62(71)	27(57)	41(49)	39(32)	65(31)	76(27)	64(25)	49(13)	77(12)
1,2-Dichloro-1,1,2,2-tetrafluoroethane	85(100)	135(52)	87(33)	137(17)	101(9)	31(9)	103(6)	100(6)	50(5)	69(4)
2,4-Dichlorotoluene	125(100)	160(61)	162(40)	127(32)	89(23)	159(16)	161(14)	63(13)	62(11)	126(10)
3,4-Dichlorotoluene	125(100)	160(47)	127(32)	162(31)	89(15)	159(11)	161(10)	63(10)	126(8)	62(8)
Diethanolamine	30(100)	74(82)	28(77)	56(69)	18(50)	42(46)	29(36)	27(34)	45(30)	43(19)
1,1-Diethoxyethane	44(100)	43(92)	29(77)	31(76)	45(74)	27(52)	72(48)	73(23)	28(17)	46(15)
1,2-Diethoxyethane	31(100)	59(71)	29(58)	45(43)	27(33)	74(27)	43(15)	15(14)	28(12)	44(10)
Diethylamine	30(100)	58(81)	44(28)	73(18)	29(18)	28(17)	72(12)	42(11)	27(11)	59(4)
Diethyl carbonate	29(100)	45(70)	31(53)	27(39)	91(24)	28(15)	63(11)	26(10)	30(6)	43(5)
Diethylene glycol	45(100)	75(23)	31(20)	44(16)	27(14)	76(12)	29(12)	43(11)	42(9)	41(4)
Diethylene glycol dimethyl ether	59(100)	58(43)	31(34)	29(32)	45(28)	28(19)	89(15)	43(9)	27(5)	60(4)
Diethylene glycol monobutyl ether acetate	43(100)	87(93)	57(82)	41(28)	45(25)	56(18)	29(18)	72(14)	85(12)	101(10)
Diethylene glycol monoethyl ether	45(100)	59(56)	72(37)	73(22)	60(14)	31(13)	75(11)	44(9)	104(8)	103(7)
Diethylene glycol monoethyl ether acetate	43(100)	29(51)	31(42)	45(40)	59(24)	72(18)	44(10)	73(9)	42(9)	30(6)
Diethylene glycol monomethyl ether	45(100)	31(42)	59(41)	29(38)	28(32)	58(21)	43(14)	27(13)	44(11)	32(10)
Diethyl ether	31(100)	29(63)	59(40)	27(35)	45(33)	74(23)	15(17)	43(9)	28(9)	26(9)
Diethyl oxalate	29(100)	31(16)	45(14)	27(14)	74(11)	28(9)	43(4)	30(4)	73(3)	75(2)
Diethyl sulfide	75(100)	47(81)	90(73)	62(60)	29(55)	61(54)	27(48)	28(23)	46(17)	45(17)
Diisopropylamine	44(100)	86(30)	58(14)	42(13)	28(13)	41(12)	43(11)	27(11)	15(11)	39(6)
Diisopropyl ether	45(100)	43(39)	87(15)	41(12)	59(10)	27(8)	39(4)	69(3)	42(3)	31(3)
1,2-Dimethoxybenzene	138(100)	95(65)	77(48)	123(44)	52(42)	41(33)	65(30)	51(29)	39(19)	63(17)
1,2-Dimethoxyethane	45(100)	60(13)	29(13)	90(7)	58(6)	31(5)	28(5)	43(4)	59(3)	46(2)
Dimethoxymethane	45(100)	75(61)	29(59)	31(13)	30(6)	15(6)	47(5)	76(2)	46(2)	44(2)
N,N-Dimethylacetamide	44(100)	87(69)	43(46)	45(23)	42(19)	72(15)	15(11)	30(8)	28(5)	88(4)
Dimethylamine	44(100)	45(81)	18(32)	28(30)	43(19)	42(15)	15(9)	46(5)	41(5)	27(5)
2,4-Dimethylaniline	121(100)	120(78)	106(57)	77(15)	28(15)	91(12)	122(9)	18(8)	93(6)	118(5)
2,2-Dimethylbutane	43(100)	57(98)	71(73)	41(61)	29(51)	27(36)	56(29)	39(25)	55(15)	15(12)
2,3-Dimethylbutane	43(100)	42(97)	41(27)	71(25)	27(11)	86(10)	39(9)	29(6)	55(5)	44(4)
Dimethyl disulfide	94(100)	45(63)	79(59)	46(38)	47(26)	15(18)	48(14)	61(12)	64(11)	96(9)
N,N-Dimethylformamide	73(100)	44(86)	42(36)	30(22)	28(20)	29(8)	43(7)	72(6)	58(5)	74(4)
Dimethyl glutarate	59(100)	100(51)	55(49)	42(33)	129(32)	101(32)	41(26)	43(22)	128(19)	87(15)
2,6-Dimethyl-4-heptanone	57(100)	85(82)	41(46)	43(39)	58(33)	28(30)	26(30)	39(22)	42(12)	142(11)
2,5-Dimethylhexane	57(100)	43(93)	42(31)	41(26)	99(19)	71(19)	29(11)	70(10)	55(9)	27(9)
Dimethyl maleate	113(100)	59(83)	26(72)	29(40)	85(37)	54(31)	114(25)	53(24)	55(13)	82(9)
2,2-Dimethylpentane	57(100)	43(73)	41(46)	56(40)	85(34)	29(31)	27(23)	39(15)	15(7)	55(6)
2,4-Dimethylpentane	43(100)	57(73)	56(41)	41(35)	42(24)	85(17)	29(16)	27(12)	39(9)	58(3)
2,4-Dimethyl-3-pentanone	43(100)	71(31)	41(13)	27(9)	70(6)	39(6)	114(5)	42(5)	44(3)	72(2)
2,4-Dimethylpyridine	107(100)	106(63)	79(44)	92(22)	65(22)	51(19)	77(17)	80(12)	52(11)	50(11)
2,6-Dimethylpyridine	107(100)	39(39)	106(29)	66(22)	92(18)	65(18)	38(12)	27(11)	79(9)	63(9)
Dimethyl sulfoxide	63(100)	78(70)	15(40)	45(35)	29(16)	61(13)	46(12)	31(11)	48(10)	47(10)
1,4-Dioxane	28(100)	29(37)	88(31)	58(24)	31(17)	15(17)	27(15)	30(13)	43(11)	26(9)
1,3-Dioxolane	73(100)	29(56)	44(53)	45(28)	28(21)	43(20)	27(13)	31(7)	74(5)	42(3)
Dipentene	68(100)	93(50)	67(44)	94(22)	39(22)	107(18)	92(18)	53(18)	136(16)	79(16)
Dipentyl ether	71(100)	43(92)	29(43)	70(40)	41(36)	27(34)	42(23)	69(17)	55(16)	39(16)
Dipropylamine	30(100)	72(79)	44(40)	43(32)	27(25)	28(24)	41(22)	86(11)	58(10)	42(10)
Dodecane	57(100)	43(91)	71(53)	41(45)	85(31)	29(27)	55(19)	56(18)	28(16)	42(14)
1-Dodecene	43(100)	56(83)	55(83)	41(75)	69(62)	70(58)	57(57)	83(47)	29(38)	84(33)
Epichlorohydrin	57(100)	27(39)	29(32)	49(25)	31(22)	62(18)	28(16)	92(1)		
1,2-Epoxybutane	43(100)	44(58)	27(52)	29(50)	45(46)	26(27)	28(27)	72(5)		
1,2-Ethanediamine	30(100)	18(13)	42(6)	43(5)	27(5)	44(4)	29(4)	17(4)	15(4)	41(3)
1,2-Ethanediol	31(100)	33(35)	29(13)	32(11)	43(6)	27(5)	28(4)	62(3)	30(3)	44(2)
1,2-Ethanediol, diacetate	43(100)	86(11)	42(7)	15(7)	116(4)	73(4)	44(3)	29(3)	103(2)	45(2)
Ethanol	31(100)	45(44)	46(18)	27(18)	29(15)	43(14)	30(6)	42(3)	19(3)	14(3)
Ethanolamine	30(100)	18(30)	28(15)	42(7)	31(6)	17(6)	61(5)	15(5)	43(3)	29(3)
Ethoxybenzene	94(100)	122(39)	28(12)	66(11)	39(9)	77(8)	95(7)	65(7)	51(7)	29(6)

Compound										
2-Ethoxyethanol	31(100)	29(52)	59(50)	27(27)	45(26)	72(14)	43(14)	15(14)	28(8)	26(6)
2-Ethoxyethyl acetate	43(100)	31(34)	59(31)	72(28)	44(25)	29(24)	45(12)	27(11)	15(11)	87(7)
Ethyl acetate	43(100)	29(46)	27(33)	45(32)	61(28)	28(25)	42(18)	73(11)	88(10)	70(10)
Ethyl acetoacetate	43(100)	29(24)	88(18)	28(16)	85(14)	27(12)	42(11)	60(9)	130(6)	45(6)
Ethyl acrylate	55(100)	27(32)	29(15)	56(12)	45(9)	73(8)	28(8)	26(6)	99(5)	85(5)
Ethylamine	30(100)	28(32)	44(20)	45(19)	27(13)	15(10)	42(9)	29(8)	41(5)	40(5)
Ethylbenzene	91(100)	106(31)	51(14)	39(10)	77(8)	65(8)	105(7)	92(7)	78(7)	27(6)
Ethyl benzoate	105(100)	77(65)	122(34)	51(34)	27(17)	150(16)	29(14)	50(13)	106(12)	78(6)
Ethyl butanoate	43(100)	71(88)	29(83)	27(43)	88(40)	41(28)	60(22)	45(20)	73(17)	42(17)
Ethyl cyanoacetate	29(100)	68(59)	27(34)	40(21)	28(16)	41(14)	15(13)	45(10)	43(9)	26(9)
Ethylcyclohexane	83(100)	55(65)	82(42)	41(36)	112(23)	56(11)	67(10)	39(10)	42(9)	84(8)
Ethylene carbonate	29(100)	44(62)	43(54)	88(40)	30(16)	28(11)	45(7)	58(6)	42(6)	73(4)
Ethyl formate	31(100)	28(73)	27(51)	29(38)	45(34)	26(17)	74(11)	43(9)	47(8)	56(4)
2-Ethyl-1,3-hexanediol	56(100)	55(71)	41(60)	43(55)	29(46)	27(44)	31(34)	57(33)	73(32)	39(20)
2-Ethyl-1-hexanol	57(100)	43(41)	41(40)	29(29)	55(28)	83(27)	56(23)	70(20)	27(17)	31(13)
N-Ethyl-*N*-isopropyl-2-propanamine	72(100)	114(77)	44(41)	129(22)	42(21)	30(20)	43(18)	41(17)	27(15)	70(11)
Ethyl 3-methylbutanoate	29(100)	41(52)	27(51)	57(43)	43(42)	60(39)	88(38)	85(38)	61(26)	45(24)
3-Ethyl-2-methylpentane	43(100)	70(50)	41(27)	71(25)	29(19)	27(19)	85(18)	55(18)	57(15)	42(14)
Ethyl propanoate	57(100)	29(84)	102(17)	27(17)	75(15)	28(14)	45(13)	74(12)	73(7)	43(6)
Fluorobenzene	96(100)	70(17)	97(7)	95(6)	75(6)	50(6)	51(4)	39(4)	69(3)	57(3)
2-Fluorotoluene	109(100)	110(55)	83(18)	57(11)	63(10)	39(9)	51(6)	50(6)	107(5)	62(5)
3-Fluorotoluene	109(100)	110(54)	83(11)	57(5)	111(4)	107(4)	63(3)	39(3)	108(2)	89(2)
4-Fluorotoluene	109(100)	110(60)	83(13)	57(10)	108(8)	39(8)	63(7)	107(5)	51(5)	50(5)
Furan	68(100)	39(64)	40(9)	38(9)	42(6)	29(6)	37(5)	69(4)	34(2)	67(1)
Furfural	39(100)	96(55)	95(52)	38(38)	29(35)	37(29)	40(11)	97(9)	50(7)	42(7)
Furfuryl alcohol	98(100)	41(65)	39(59)	81(55)	53(53)	97(51)	42(49)	69(39)	70(36)	29(28)
Glycerol	61(100)	43(90)	31(57)	44(54)	29(38)	18(32)	27(12)	42(11)	60(10)	45(10)
Glycerol triacetate	43(100)	103(44)	145(34)	116(17)	115(13)	44(10)	86(9)	28(8)	73(7)	42(7)
Heptane	43(100)	41(56)	29(49)	57(47)	27(46)	71(45)	56(27)	42(26)	39(23)	70(18)
1-Heptanol	41(100)	70(87)	56(86)	31(78)	43(72)	29(70)	55(67)	27(65)	42(54)	69(41)
3-Heptanol, (*S*)-	59(100)	69(73)	87(37)	41(37)	29(33)	55(30)	31(24)	116(0)		
2-Heptanone	43(100)	58(60)	71(14)	41(11)	27(11)	59(9)	39(8)	29(8)	42(5)	114(4)
3-Heptanone	57(100)	29(76)	41(32)	85(29)	72(22)	43(15)	39(11)	114(10)	27(7)	55(5)
1-Heptene	41(100)	56(88)	29(70)	55(59)	42(52)	27(45)	39(41)	70(37)	69(27)	57(27)
Hexane	57(100)	43(78)	41(77)	29(61)	27(57)	56(45)	42(39)	39(27)	28(16)	86(14)
Hexanedinitrile	41(100)	68(50)	54(42)	40(21)	55(20)	27(17)	39(16)	28(13)	52(7)	42(6)
Hexanenitrile	41(100)	54(68)	27(59)	55(55)	29(44)	39(35)	57(30)	43(30)	68(24)	28(22)
Hexanoic acid	60(100)	73(42)	27(36)	41(33)	43(27)	29(26)	45(20)	39(16)	42(15)	55(14)
1-Hexanol	56(100)	43(78)	31(74)	41(71)	27(64)	29(59)	55(58)	42(53)	39(37)	69(27)
2-Hexanone	43(100)	58(60)	57(17)	100(16)	29(15)	41(13)	85(8)	27(8)	71(7)	59(5)
Hexyl acetate	43(100)	56(66)	41(38)	55(37)	61(35)	42(33)	84(31)	69(19)	73(17)	57(7)
3-Hydroxypropanenitrile	41(100)	31(97)	29(20)	42(17)	52(15)	40(13)	53(10)	51(9)	39(7)	26(7)
Iodobenzene	204(100)	77(82)	51(32)	50(20)	127(8)	205(6)	78(6)	74(6)	102(5)	76(5)
1-Iodobutane	57(100)	29(78)	41(56)	27(39)	184(36)	39(17)	28(16)	26(7)	127(6)	55(6)
2-Iodobutane, (±)-	57(100)	29(36)	184(36)	41(36)	27(10)	39(9)	127(6)	58(4)		
Iodoethane	156(100)	29(75)	27(63)	127(31)	26(14)	28(9)	128(8)	141(2)	25(2)	140(1)
Iodomethane	142(100)	127(38)	141(14)	15(13)	139(5)	140(4)	128(3)	14(1)	13(1)	71(0)
1-Iodopropane	43(100)	170(68)	41(35)	27(26)	127(14)	39(11)	44(4)	141(3)	128(3)	42(3)
2-Iodopropane	43(100)	170(46)	41(45)	27(44)	39(19)	127(17)	42(4)	38(4)	128(3)	44(3)
Isobutanal	43(100)	41(84)	27(47)	72(46)	39(30)	29(25)	42(10)	70(5)	38(5)	28(5)
Isobutyl acetate	43(100)	56(26)	73(15)	41(10)	29(5)	71(3)	57(3)	39(3)	27(3)	86(2)
Isobutylamine	30(100)	28(9)	41(6)	73(5)	27(5)	39(4)	29(3)	15(3)	58(2)	56(2)
Isobutylbenzene	91(100)	92(58)	43(22)	134(20)	65(12)	41(12)	39(11)	27(8)	51(7)	93(5)
Isobutyl formate	43(100)	56(82)	41(78)	31(65)	29(64)	27(53)	60(39)	39(36)	42(26)	15(14)
Isobutyl isobutanoate	43(100)	41(94)	56(76)	71(70)	57(68)	27(62)	29(59)	89(37)	39(32)	42(17)
Isopentyl acetate	43(100)	70(49)	55(38)	61(15)	42(15)	41(14)	27(12)	87(11)	29(10)	73(9)
Isophorone	82(100)	39(20)	138(17)	54(13)	27(12)	41(10)	53(8)	83(7)	29(7)	55(6)
Isopropyl acetate	43(100)	61(17)	41(14)	87(9)	59(8)	27(8)	42(7)	39(4)	45(3)	44(2)
Isopropylbenzene	105(100)	120(25)	77(13)	51(12)	79(10)	106(9)	39(9)	27(8)	103(6)	91(5)
1-Isopropyl-4-methylbenzene	119(100)	91(42)	134(33)	39(27)	41(20)	117(18)	65(18)	77(17)	27(16)	120(15)
Isoquinoline	129(100)	102(26)	51(20)	128(18)	50(11)	130(10)	75(10)	76(9)	103(8)	74(7)

Analytical

Compound										
d-Limonene	68(100)	93(50)	67(49)	41(22)	94(21)	79(21)	39(21)	136(20)	53(19)	121(16)
Mesityl oxide	55(100)	83(89)	43(73)	29(42)	98(36)	39(32)	27(28)	53(11)	41(10)	56(5)
Methanol	31(100)	29(72)	32(67)	15(42)	28(12)	14(10)	30(9)	13(6)	12(3)	16(2)
2-Methoxyethanol	45(100)	31(15)	29(14)	28(11)	47(9)	76(6)	43(6)	58(4)	46(4)	27(4)
2-Methoxyethyl acetate	43(100)	45(48)	58(42)	29(10)	42(4)	31(4)	73(3)	27(3)	59(2)	26(2)
Methyl acetate	43(100)	74(52)	28(38)	42(19)	59(17)	44(8)	32(8)	29(6)	31(4)	75(2)
2-Methylacrylonitrile	41(100)	39(54)	67(44)	40(26)	38(24)	52(23)	27(23)	37(20)	66(17)	51(12)
Methylamine	30(100)	31(87)	28(56)	29(19)	32(15)	15(12)	27(9)			
2-Methylaniline	106(100)	107(83)	77(17)	79(13)	39(12)	53(10)	52(10)	54(9)	51(9)	28(9)
3-Methylaniline	106(100)	107(84)	79(17)	77(17)	108(7)	78(6)	80(5)	89(4)	65(4)	53(4)
N-Methylaniline	106(100)	107(79)	77(23)	51(12)	79(11)	65(9)	39(9)	78(8)	108(7)	50(6)
Methyl benzoate	105(100)	77(81)	51(45)	136(24)	50(18)	106(8)	78(6)	28(6)	39(5)	27(5)
Methyl butanoate	43(100)	74(90)	71(66)	41(32)	27(31)	59(28)	87(19)	42(15)	28(15)	39(14)
3-Methylbutanoic acid	60(100)	43(61)	41(54)	27(33)	45(31)	29(27)	74(24)	39(24)	87(21)	57(20)
3-Methyl-1-butanol	55(100)	42(90)	43(82)	41(81)	70(71)	31(61)	29(59)	27(59)	39(44)	57(31)
Methyl *tert*-butyl ether	41(100)	73(78)	57(71)	29(60)	43(48)	39(34)	28(28)	56(23)	55(18)	45(16)
Methyl cyanoacetate	59(100)	15(65)	68(60)	40(38)	28(17)	29(16)	55(11)	54(10)	39(10)	67(8)
Methylcyclohexane	83(100)	55(82)	41(60)	98(44)	42(35)	56(30)	27(29)	39(27)	69(23)	70(22)
cis-4-Methylcyclohexanol	57(100)	58(54)	70(38)	81(36)	96(33)	55(25)	41(23)	114(17)	71(15)	56(12)
N-Methylformamide	59(100)	30(54)	28(34)	29(13)	58(8)	15(7)	60(3)	41(3)	27(3)	31(2)
Methyl formate	31(100)	29(63)	32(34)	60(28)	30(7)	28(7)	44(2)	18(2)	61(1)	59(1)
2-Methylheptane	43(100)	57(91)	42(39)	41(27)	29(16)	70(15)	27(13)	71(12)	99(10)	55(9)
4-Methylheptane	43(100)	71(53)	70(46)	41(27)	29(23)	27(23)	55(15)	57(14)	42(14)	39(10)
6-Methyl-1-heptanol	41(100)	55(94)	43(93)	69(84)	57(81)	56(73)	29(50)	84(48)	70(47)	27(46)
2-Methylhexane	43(100)	42(38)	41(35)	85(32)	57(26)	29(22)	27(22)	56(20)	39(11)	55(5)
5-Methyl-2-hexanone	43(100)	58(34)	27(14)	41(13)	15(13)	57(11)	39(9)	71(8)	59(8)	29(8)
Methyl methacrylate	41(100)	69(66)	39(40)	100(34)	15(20)	40(10)	59(8)	99(6)	38(6)	55(5)
2-Methyloctane	43(100)	57(51)	41(35)	71(28)	42(28)	29(23)	27(21)	85(17)	56(15)	84(13)
2-Methylpentane	43(100)	42(53)	41(35)	27(31)	71(29)	39(20)	29(18)	57(11)	15(10)	70(7)
3-Methylpentane	57(100)	56(76)	41(68)	29(60)	27(40)	43(29)	39(22)	55(9)	15(9)	28(8)
2-Methyl-2,4-pentanediol	59(100)	43(61)	56(25)	45(17)	41(16)	57(13)	42(13)	85(11)	61(10)	31(10)
4-Methylpentanenitrile	55(100)	41(52)	43(46)	27(39)	39(29)	57(27)	54(26)	29(22)	82(13)	28(12)
4-Methyl-2-pentanol	45(100)	43(47)	69(30)	41(27)	27(19)	39(13)	29(12)	87(11)	84(10)	57(10)
3-Methyl-3-pentanol	73(100)	55(38)	43(35)	45(28)	27(25)	29(21)	41(12)	87(11)	31(11)	15(9)
4-Methyl-2-pentanone	43(100)	58(84)	29(65)	41(56)	57(44)	27(42)	39(31)	85(19)	100(14)	42(14)
2-Methylpropanenitrile	42(100)	68(45)	28(45)	54(26)	41(26)	27(26)	29(25)	26(15)	39(13)	15(13)
Methyl propanoate	57(100)	29(72)	59(31)	88(26)	27(18)	28(9)	31(5)	44(4)	26(4)	58(3)
2-Methylpropanoic acid	43(100)	41(42)	27(40)	73(22)	39(15)	45(14)	42(11)	29(9)	88(7)	28(6)
2-Methyl-1-propanol	43(100)	33(73)	31(72)	41(66)	42(60)	27(43)	29(18)	39(17)	28(8)	74(6)
2-Methyl-2-propanol	59(100)	31(33)	41(22)	43(18)	29(13)	27(11)	57(10)	42(4)	60(3)	28(3)
2-Methylpyridine	93(100)	66(41)	39(31)	92(20)	78(19)	51(19)	65(16)	38(13)	50(12)	52(11)
3-Methylpyridine	93(100)	39(51)	66(46)	92(31)	65(29)	40(19)	38(18)	67(13)	63(11)	51(11)
N-Methyl-2-pyrrolidinone	99(100)	44(89)	98(80)	42(60)	41(38)	43(17)	28(17)	71(13)	39(11)	70(10)
Methyl salicylate	120(100)	92(59)	152(47)	121(32)	65(22)	39(22)	93(15)	64(14)	18(14)	63(13)
2-Methylthiophene	97(100)	98(57)	45(22)	39(14)	53(9)	99(8)	27(8)	69(6)	58(6)	59(5)
Morpholine	57(100)	29(100)	87(69)	28(69)	30(38)	56(33)	86(28)	31(28)	27(12)	15(7)
Nitrobenzene	77(100)	51(59)	123(42)	50(25)	30(15)	65(14)	39(10)	93(9)	74(7)	78(6)
Nitroethane	29(100)	30(12)	28(11)	26(9)	27(8)	43(5)	41(5)	14(5)	15(3)	46(2)
Nitromethane	30(100)	61(64)	46(39)	28(30)	45(8)	27(8)	44(7)	29(7)	60(5)	43(4)
1-Nitropropane	43(100)	27(93)	41(90)	39(34)	30(25)	44(20)	42(20)	26(20)	28(13)	54(12)
2-Nitropropane	43(100)	41(73)	27(71)	39(30)	30(18)	15(11)	42(9)	28(8)	26(8)	38(6)
Nonane	43(100)	57(75)	41(29)	84(26)	85(22)	29(22)	71(18)	56(16)	27(13)	42(12)
Octane	43(100)	57(30)	85(25)	41(25)	71(19)	29(17)	56(14)	70(10)	42(10)	27(10)
1-Octanol	41(100)	56(85)	43(82)	55(81)	31(69)	27(69)	29(68)	42(62)	70(53)	69(48)
2-Octanone	43(100)	58(79)	41(56)	59(52)	71(49)	27(46)	29(36)	39(27)	57(18)	55(17)
1-Octene	43(100)	41(82)	55(80)	56(67)	42(67)	70(54)	29(44)	27(31)	69(30)	39(29)
Pentachloroethane	167(100)	165(91)	117(90)	119(89)	83(58)	169(54)	130(43)	132(42)	60(40)	85(37)
Pentane	43(100)	42(55)	41(45)	27(42)	29(26)	39(19)	57(13)	28(9)	15(9)	72(8)
1,5-Pentanediol	31(100)	56(85)	41(67)	57(59)	55(51)	44(45)	29(37)	43(31)	68(29)	27(26)
2,4-Pentanedione	43(100)	85(31)	100(20)	27(12)	42(10)	29(10)	41(7)	39(7)	31(5)	26(5)
Pentanenitrile	41(100)	43(97)	54(54)	27(34)	55(21)	28(19)	29(16)	39(15)	42(5)	26(5)

Compound										
Pentanoic acid	60(100)	73(34)	27(33)	29(28)	41(21)	43(17)	45(16)	28(14)	42(12)	39(12)
1-Pentanol	42(100)	70(72)	55(65)	41(56)	31(47)	29(41)	27(26)	57(22)	28(22)	43(21)
2-Pentanol	45(100)	43(20)	55(18)	27(17)	29(11)	41(9)	31(9)	15(9)	44(8)	39(8)
3-Pentanol	59(100)	31(83)	41(42)	27(35)	29(34)	58(15)	43(15)	57(14)	39(12)	15(9)
2-Pentanone	43(100)	41(17)	86(12)	42(12)	27(11)	39(8)	71(7)	58(7)	45(7)	44(3)
3-Pentanone	57(100)	29(50)	86(26)	27(13)	58(4)	56(4)	28(4)	26(3)	43(2)	42(2)
Pentyl acetate	43(100)	70(90)	42(52)	28(51)	61(50)	55(41)	73(21)	41(20)	29(14)	69(11)
Pentylamine	30(100)	87(8)	41(4)	28(4)	45(3)	42(3)	27(3)	56(2)	44(2)	43(2)
Pentyl lactate	43(100)	70(44)	71(32)	55(28)	41(22)	29(20)	27(15)	45(12)	42(9)	57(7)
α-Pinene	93(100)	92(30)	39(24)	41(23)	77(22)	91(21)	27(21)	79(18)	121(13)	53(10)
β-Pinene	93(100)	41(64)	69(47)	39(33)	27(31)	79(20)	77(18)	53(14)	94(13)	91(13)
Piperidine	84(100)	85(53)	56(46)	57(43)	28(41)	29(37)	44(34)	42(30)	30(30)	43(25)
Propanal	29(100)	58(59)	28(58)	27(39)	57(20)	31(4)	30(4)	42(3)	39(3)	59(2)
1,2-Propanediol	45(100)	18(46)	29(21)	43(19)	31(18)	27(17)	28(11)	19(8)	44(6)	61(5)
1,3-Propanediol	28(100)	58(93)	31(76)	57(70)	29(40)	27(26)	45(24)	43(23)	19(18)	30(17)
Propanenitrile	28(100)	54(63)	26(20)	27(17)	52(11)	55(10)	51(9)	15(9)	53(7)	25(7)
Propanoic acid	28(100)	29(84)	74(79)	27(62)	45(56)	73(48)	57(30)	26(21)	55(17)	56(16)
Propanoic anhydride	57(100)	29(55)	28(20)	27(20)	74(19)	73(12)	45(11)	26(5)	30(4)	58(3)
1-Propanol	31(100)	27(19)	29(18)	59(11)	42(9)	60(7)	41(7)	28(7)	43(3)	32(3)
2-Propanol	45(100)	43(19)	27(17)	29(12)	41(7)	31(6)	19(6)	42(5)	44(4)	59(3)
Propargyl alcohol	55(100)	39(25)	28(20)	27(19)	29(16)	38(14)	26(11)	37(8)	53(6)	56(4)
Propyl acetate	43(100)	61(19)	31(18)	27(15)	42(11)	73(9)	41(9)	29(9)	59(5)	39(5)
Propylamine	30(100)	28(13)	59(8)	27(7)	41(5)	42(3)	39(3)	29(3)	26(3)	18(3)
Propylbenzene	91(100)	120(21)	92(10)	38(10)	65(9)	78(6)	51(6)	27(5)	63(4)	105(3)
Propylene carbonate	28(100)	57(69)	43(66)	29(51)	27(45)	30(42)	26(19)	42(17)	31(16)	58(13)
Propyl formate	27(100)	29(92)	31(81)	42(60)	41(53)	43(29)	39(29)	26(28)	47(22)	30(16)
Pyridine	79(100)	52(62)	51(31)	50(19)	78(11)	53(7)	39(7)	80(6)	27(3)	77(2)
Pyrrole	67(100)	41(58)	39(58)	40(51)	28(42)	38(20)	37(12)	66(7)	68(5)	27(3)
Pyrrolidine	43(100)	28(52)	70(33)	71(26)	42(22)	41(20)	27(16)	39(15)	29(10)	30(9)
2-Pyrrolidone	85(100)	42(43)	41(36)	28(33)	30(29)	56(16)	84(14)	40(12)	27(12)	29(9)
Quinoline	129(100)	51(28)	76(25)	128(24)	44(24)	50(20)	32(19)	75(18)	74(12)	103(11)
Salicylaldehyde	28(100)	122(98)	121(92)	39(79)	65(52)	76(31)	32(31)	44(30)	93(26)	38(23)
Styrene	104(100)	103(41)	78(32)	51(28)	77(23)	105(12)	50(12)	52(11)	39(11)	102(10)
Sulfolane	41(100)	28(94)	56(82)	55(72)	120(37)	27(32)	39(19)	29(17)	26(11)	48(5)
α-Terpinene	121(100)	93(85)	136(43)	91(40)	77(34)	39(33)	27(33)	79(27)	41(26)	43(18)
1,1,1,2-Tetrachloro-2,2-difluoroethane	167(100)	169(96)	117(85)	119(82)	171(31)	85(29)	121(26)	82(14)	47(14)	101(13)
1,1,2,2-Tetrachloro-1,2-difluoroethane	101(100)	103(64)	167(54)	169(52)	117(19)	119(18)	171(17)	105(11)	31(11)	132(9)
1,1,1,2-Tetrachloroethane	131(100)	133(96)	117(76)	119(73)	95(34)	135(31)	121(23)	97(23)	61(19)	60(18)
1,1,2,2-Tetrachloroethane	83(100)	85(63)	95(11)	87(10)	168(8)	133(8)	131(8)	96(8)	61(8)	60(8)
Tetrachloroethene	166(100)	164(82)	131(71)	129(71)	168(45)	94(38)	47(31)	96(24)	133(20)	59(17)
Tetrachloromethane	117(100)	119(98)	121(31)	82(24)	47(23)	84(16)	35(14)	49(8)	28(8)	36(6)
Tetraethylene glycol	45(100)	89(10)	44(8)	43(6)	31(6)	29(6)	27(6)	101(5)	75(5)	28(5)
Tetrahydrofuran	42(100)	41(52)	27(33)	72(29)	71(27)	39(24)	43(22)	29(22)	40(13)	15(10)
1,2,3,4-Tetrahydronaphthalene	104(100)	132(53)	91(43)	51(17)	39(17)	131(15)	117(15)	115(14)	78(13)	77(13)
Tetrahydropyran	41(100)	28(64)	56(57)	45(57)	29(51)	27(49)	85(47)	86(42)	39(28)	55(23)
Tetrahydrothiophene	60(100)	88(54)	45(37)	46(32)	47(26)	27(24)	59(18)	87(16)	39(14)	54(13)
Tetramethylsilane	73(100)	43(14)	45(12)	74(8)	29(7)	15(5)	75(4)	44(4)	42(4)	31(4)
Tetramethylurea	72(100)	44(27)	116(24)	42(13)	15(13)	17(7)	28(5)	73(4)	56(4)	18(4)
1*H*-Tetrazole	42(100)	28(60)	27(25)	29(24)	41(13)	43(11)	70(8)	26(7)	40(2)	38(2)
Thiophene	84(100)	58(65)	45(58)	39(29)	57(13)	38(8)	69(7)	37(7)	83(6)	50(6)
Toluene	91(100)	92(73)	39(20)	65(14)	63(11)	51(11)	50(7)	27(6)	93(5)	90(5)
Tribromomethane	173(100)	171(50)	175(49)	93(22)	91(22)	79(18)	81(17)	94(13)	92(13)	254(11)
Tributylamine	142(100)	100(19)	143(11)	29(8)	185(7)	57(6)	44(6)	41(6)	30(5)	86(4)
1,2,4-Trichlorobenzene	180(100)	182(96)	184(30)	145(30)	109(26)	147(19)	75(11)	74(9)	111(8)	181(7)
1,1,1-Trichloroethane	97(100)	99(64)	61(58)	26(31)	27(24)	117(19)	63(19)	119(18)	35(17)	62(11)
1,1,2-Trichloroethane	97(100)	83(95)	99(62)	85(60)	61(58)	26(23)	96(21)	63(19)	27(17)	98(15)
Trichloroethene	95(100)	130(90)	132(85)	60(65)	97(64)	35(40)	134(27)	47(26)	62(21)	59(13)
Trichloroethylsilane	135(100)	133(100)	126(67)	128(46)	137(37)	98(22)	63(21)	35(16)	100(14)	127(11)
Trichlorofluoromethane	101(100)	103(66)	66(13)	105(11)	35(11)	47(9)	31(8)	82(4)	68(4)	37(4)
Trichloromethane	83(100)	85(64)	47(35)	35(19)	48(16)	49(12)	87(10)	37(6)	50(5)	84(4)
(Trichloromethyl)benzene	159(100)	161(64)	89(14)	163(11)	28(10)	63(9)	160(8)	123(8)	62(8)	124(6)

Compound										
1,2,3-Trichloropropane	75(100)	39(58)	49(42)	110(38)	61(34)	77(33)	112(22)	27(16)	97(15)	38(15)
1,1,2-Trichloro-1,2,2-trifluoroethane	101(100)	151(68)	103(64)	85(45)	31(45)	153(44)	35(20)	66(19)	47(18)	87(14)
Tridecane	57(100)	43(91)	71(51)	41(34)	85(24)	29(23)	56(14)	55(12)	27(11)	42(10)
Triethanolamine	118(100)	56(69)	45(60)	42(56)	44(27)	43(25)	41(14)	116(8)	57(8)	86(7)
Triethylamine	86(100)	30(68)	58(37)	28(24)	29(23)	27(19)	44(18)	101(17)	42(16)	56(8)
Triethylene glycol	45(100)	58(11)	89(9)	31(8)	29(8)	75(7)	44(7)	43(7)	27(7)	28(5)
Triethyl phosphate	99(100)	81(71)	155(56)	82(45)	45(45)	109(44)	127(41)	43(24)	125(16)	111(14)
Trifluoroacetic acid	45(100)	69(70)	51(36)	28(28)	50(15)	44(11)	43(7)	97(5)	31(5)	29(5)
2,2,2-Trifluoroethanol	31(100)	33(24)	61(19)	29(19)	51(16)	69(9)	32(6)	49(5)	83(4)	81(4)
(Trifluoromethyl)benzene	146(100)	145(40)	127(34)	96(28)	77(10)	51(10)	147(8)	75(6)	50(6)	128(3)
Trimethylamine	58(100)	59(47)	30(29)	42(26)	44(17)	15(14)	28(10)	18(10)	43(8)	57(7)
1,2,3-Trimethylbenzene	105(100)	120(47)	39(22)	77(17)	91(14)	51(14)	27(14)	79(12)	119(11)	106(9)
1,2,4-Trimethylbenzene	105(100)	120(56)	119(17)	77(15)	39(15)	51(11)	91(10)	27(10)	106(9)	79(7)
1,3,5-Trimethylbenzene	105(100)	120(64)	119(15)	77(13)	39(11)	106(9)	91(9)	51(8)	27(7)	121(6)
2,2,3-Trimethylbutane	57(100)	43(71)	56(63)	41(53)	85(30)	29(26)	27(18)	39(14)	15(6)	55(5)
2,2,5-Trimethylhexane	57(100)	56(35)	71(18)	41(17)	43(14)	29(8)	70(4)	58(4)	113(3)	55(3)
2,2,4-Trimethylpentane	57(100)	41(31)	56(28)	43(24)	29(16)	27(9)	39(7)	58(4)	55(4)	99(2)
2,3,3-Trimethylpentane	43(100)	71(45)	70(36)	57(36)	41(29)	85(25)	27(18)	55(16)	29(16)	39(11)
2,3,4-Trimethylpentane	43(100)	71(62)	70(41)	41(25)	27(18)	55(17)	57(16)	29(14)	39(10)	42(7)
Trimethyl phosphate	110(100)	109(35)	79(34)	95(25)	80(23)	15(20)	140(18)	47(10)	31(7)	139(5)
2,4,6-Trimethylpyridine	121(100)	39(27)	79(26)	120(24)	106(17)	27(16)	77(13)	51(11)	42(11)	122(10)
1-Undecene	41(100)	43(87)	55(80)	70(67)	56(67)	69(55)	29(55)	83(51)	57(50)	27(46)
Vinyl acetate	43(100)	28(45)	42(26)	44(24)	86(20)	31(10)	32(7)	29(7)	45(2)	41(2)
o-Xylene	91(100)	106(40)	39(21)	105(17)	51(17)	77(15)	27(12)	65(10)	92(8)	79(8)
m-Xylene	91(100)	106(65)	105(29)	39(18)	51(15)	77(14)	27(10)	92(8)	79(8)	78(8)
p-Xylene	91(100)	106(62)	105(30)	51(16)	39(16)	77(13)	27(11)	92(7)	78(7)	65(7)

COMMON SPURIOUS SIGNALS OBSERVED IN MASS SPECTROMETERS

Thomas J. Bruno and Paris D. N. Svoronos

The following table provides guidance in the recognition and interpretation of potentially spurious signals (*m/z* peaks) that will sometimes be observed in measured mass spectra (Ref. 1). Often, the occurrence of these signals can be predicted by the recent history of the instrument or the method being used. This is especially true if the mass spectrometer is interfaced to a gas chromatograph.

Reference

1. Bruno, T. J., and Svoronos, P. D. N., *CRC Handbook of Basic Tables for Chemical Analysis, 3rd Ed.*, CRC Press, Boca Raton, FL, 2011.

Ions observed, *m/z*	Possible compound	Possible source
13, 14, 15, 16	Methane*	Chlorine reagent gas
18	Water*	Residual impurity, outgasing of ferrules, septa, and seals
14, 28	Nitrogen*	Residual impurity, outgasing of ferrules, septa, and seals; leaking seal
16, 32	Oxygen*	Residual impurity, outgasing of ferrules, septa, and seals; leaking seal
44	Carbon dioxide*	Residual impurity, outgasing of ferrules, septa, and seals; leaking seal; note it may be mistaken for propane in a sample
31, 51, 69, 100, 119, 131, 169, 181, 214, 219, 264, 376, 414, 426, 464, 502, 576, 614	Perfluorotributylamine (PFTBA), and related ions**	This is a common tuning compound; may indicate a leaking valve
31	Methanol	Solvent; can be used as a leak detector
43, 58	Acetone	Solvent; can be used as a leak detector
78	Benzene	Solvent; can be used as a leak detector
91, 92	Toluene	Solvent; can be used as a leak detector
105, 106	Xylenes	Solvent; can be used as a leak detector
151, 153	Trichloroethane	Solvent; can be used as a leak detector
69	Rough (fore) pump fluid, PFTBA	Back diffusion of fore pump fluid, possible leaking valve of tuning compound vial
73, 147, 207, 221, 281, 295, 355, 429	Dimethylpolysiloxane	Bleed from a column or septum, often during high temperature program methods in GC-MS
77, 94, 115, 141, 168, 170, 262, 354, 446	Diffusion pump fluid	Back diffusion from diffusion pump, if present
149	Phthalates	Plasticizer in vacuum seals, gloves
X – 14 peaks	Hydrocarbons	Loss of a methylene group indicates a hydrocarbon sample

* It is possible to operate the analyzer to ignore these common background impurities. They will be present to contribute to poor vacuum if these impurities result from a significant leak.

** See the table "Major Reference Masses in the Spectrum of Heptacosafluorotributylamine (Perfluorotributylamine)" in this section for additional details.

Analytical

CHLORINE–BROMINE COMBINATION ISOTOPE INTENSITIES IN MASS SPECTRAL PATTERNS

Thomas J. Bruno and Paris D. N. Svoronos

Due to the distinctive mass spectral patterns caused by the presence of chlorine and bromine in a molecule, interpretation of a mass spectrum can be much easier if the results of the relative isotopic concentrations are known. The following table provides peak intensities (relative to the molecular ion (M^+) at an intensity normalized to 100 percent) for various combinations of chlorine and bromine atoms, assuming the absence of all other elements except carbon and hydrogen (Refs. 1–4). The mass abundance calculations were based upon the most recent atomic mass data (Ref. 1).

References

1. Bruno, T. J., and Svoronos, P. D. N., *CRC Handbook of Basic Tables for Chemical Analysis, 3rd Edition*, CRC Press, Boca Raton, FL, 2011.
2. McLafferty, F. W., and Turecek, F., *Interpretation of Mass Spectra, 4th Edition*, University Science Books, Mill Valley, CA, 1993.
3. Silverstein, R. H., Bassler, G. C., and Morrill, T. C., *Spectroscopic Identification of Organic Compounds, 6th Edition*, John Wiley & Sons, New York, 1998.
4. Williams, D. H., and Fleming, I., *Spectroscopic Methods in Organic Chemistry, 4th Edition*, McGraw-Hill, London, 1989.

Relative Intensities of Isotope Peaks for Combinations of Bromine and Chlorine in Mass Spectral Patterns (M^+ = 100%)

Cl_n	Br_0	Br_1	Br_2	Br_3	Br_4
Cl_0					
P + 2		98.0	196.0	294.0	390.8
P + 4			96.1	288.2	574.7
P + 6				94.1	375.3
P + 8					92.0
Cl_1					
P + 2	32.5	130.6	228.0	326.1	424.6
P + 4		31.9	159.0	383.1	704.2
P + 6			31.2	187.4	564.1
P + 8				30.7	214.8
P + 10					30.3
Cl_2					
P + 2	65.0	163.0	261.1	359.3	456.3
P + 4	10.6	74.4	234.2	490.2	840.3
P + 6		10.4	83.3	312.8	791.6
P + 8			10.2	91.7	397.5
P + 10				9.8	99.2
P + 12					10.1
Cl_3					
P + 2	97.5	195.3	294.0	393.3	489
P + 4	31.7	127.0	99.7	609.8	989
P + 6	3.4	34.4	159.4	473.8	1064
P + 8		3.3	37.1	193.9	654
P + 10			3.2	39.6	229
P + 12				3.0	42
P + 14					3.2
Cl_4					
P + 2	130.0	228.3	326.6	4.2	522
P + 4	63.3	190.9	414.9	735.3	1149
P + 6	13.7	75.8	263.1	670.0	1388
P + 8	1.2	14.4	88.8	347.1	1002
P + 10		1.1	15.4	102.2	443
P + 12			1.3	16.2	117
P + 14				0.7	17
Cl_5					
P + 2	162.6	260.7	358.9		
P + 4	105.7	265.3	520.8		
P + 6	34.3	137.9	397.9		
P + 8	5.5	39.3	174.5		
P + 10	0.3	5.8	44.3		
P + 12		0.3	5.7		
P + 14			0.5		
Cl_6					
P + 2	195.3				
P + 4	158.6				
P + 6	68.8				
P + 8	16.6				
P + 10	2.1				
P + 12	0.1				
Cl_7					
P + 2	227.8				
P + 4	222.1				
P + 6	120.3				
P + 8	39.0				
P + 10	7.5				
P + 12	0.8				
P + 14	0.05				

REDUCTION OF WEIGHINGS IN AIR TO VACUO

When the mass M of a body is determined in air, a correction is necessary for the buoyancy of the air. The corrected mass is given by $M + kM/1000$, where k is a function of the material used for the weights, given by

$$k = 1000\rho_{air}(1/\rho_{body} - 1/\rho_{weight})$$

and ρ is density. The table below is computed for an air density of 0.0012 g/cm³ and for densities of three common weights: platinum-iridium (21.6 g/cm³), brass (8.5 g/cm³), and aluminum or quartz (2.65 g/cm³).

Value of k at Specified Density of Body for Buoyancy of Air Corrections for Masses of Different Materials

Density of body (g cm^{-3})	k (Pt-Ir)	k (Brass)	k (Quartz or Al)
0.5	2.34	2.26	1.95
0.6	1.94	1.86	1.55
0.7	1.66	1.57	1.26
0.8	1.44	1.36	1.05
0.9	1.28	1.19	0.88
1.0	1.14	1.06	0.75
1.1	1.04	0.95	0.64
1.2	0.94	0.86	0.55
1.3	0.87	0.78	0.47
1.4	0.80	0.72	0.40
1.5	0.74	0.66	0.35
1.6	0.69	0.61	0.30
1.7	0.65	0.56	0.25
1.8	0.61	0.53	0.21
1.9	0.58	0.49	0.18
2.0	0.54	0.46	0.15
2.5	0.42	0.34	0.03
3.0	0.34	0.26	-0.05
4.0	0.24	0.16	-0.15
6.0	0.14	0.06	-0.25
8.0	0.09	0.01	-0.30
10.0	0.06	-0.02	-0.33
15.0	0.02	-0.06	-0.37
20.0	0.00	-0.08	-0.39
22.0	0.00	-0.09	-0.40

For a more accurate calculation, use the following values of the density of air at the specified temperatures (assuming 50% relative humidity and 0.04% CO_2).

Density of Air in g/cm³ at Indicated Pressure and Temperature

Name	P/kPa	10 °C	20 °C	30 °C
Air	85	0.001043	0.001005	0.000968
Air	90	0.001105	0.001065	0.001025
Air	95	0.001166	0.001124	0.001083
Air	100	0.001228	0.001184	0.001140
Air	105	0.001290	0.001243	0.001198

Formulas for calculating the density of air over more extended ranges of temperature, pressure, and humidity may be found in the references.

References

1. Kaye, G. W. C., and Laby, T. H., *Tables of Physical and Chemical Constants, 16th Edition*, pp. 25–28, Longman, London, 1995.
2. Giacomo, P., *Metrologia*, 18, 33, 1982.
3. Davis, R. S., *Metrologia*, 29, 67, 1992.

Analytical

STANDARDS FOR LABORATORY WEIGHTS

Thomas J. Bruno and Paris D. N. Svoronos

The following table provides a summary of the requirements for metric weights and mass standards commonly used in chemical analysis (Refs. 1–3). The actual specifications are under the jurisdiction of ASTM Committee E-41 on General Laboratory Apparatus and are the direct responsibility of subcommittee E-41.06 that deals with weighing devices. These standards do not generally refer to instruments used in commerce. Weights are classified according to type (either Type I or Type II), grade (S, O, P, or Q), and class (1, 2, 3, 4, 5, or 6). Information on these mass standards is presented to allow the user to make appropriate choices when using analytical weights for the calibration of electronic analytical balances, for making large-scale mass measurements (such as those involving gas cylinders), and in the use of dead-weight pressure balances. Some historical context can be found in Ref. 4.

References

1. *Annual Book of ASTM Standards, ANSI/ASTM E617-97 Standard Specification for Laboratory Weights and Precision Mass Standards,* Book of Standards Vol. 14.04, 2008.
2. Battino, R., and Williamson, A. G., *J. Chem. Educ.* 61, 51, 1984.
3. Bruno, T. J., and Svoronos, P. D. N., *CRC Handbook of Basic Tables for Chemical Analysis,* 3rd Edition, CRC Press, Boca Raton, FL, 2011.
4. NBS Circular 547, *Precision Laboratory Standards of Mass and Laboratory Weights,* Washington, D.C., 1954 (reprinted as NIST-IR 78-1476, Washington, D.C., 1978)

Type — Classification by Design

Type I — One-piece construction; contains no added adjusting material; used for highest accuracy work.

Type II — Can be of any appropriate and convenient design, incorporating plugs, knobs, rings, etc.; adjusting material can be added if it is contained so that it cannot become separated from the weight.

Grade — Classification by Physical Property

Grade S:	Density:	7.7 g cm^{-3} to 8.1 g cm^{-3} (for 50 mg and larger)
	Surface area:	not to exceed that of a cylinder of equal height and diameter
	Surface finish:	highly polished
	Surface protection:	none permitted
	Magnetic properties:	no more magnetic than 300 series stainless steels
	Corrosion resistance:	same as 303 stainless steel
	Hardness:	at least as hard as brass
Grade O:	Density:	7.7 g cm^{-3} to 9.1 g cm^{-3} (for 1 g and larger)
	Surface area:	same as grade S
	Surface finish:	same as grade S
	Surface protection:	may be plated with suitable material such as platinum or rhodium
	Magnetic properties:	same as grade S
	Corrosion resistance:	same as grade S
	Hardness:	at least as hard as brass when coated; smaller weights at least as hard as aluminum
Grade P:	Density:	7.2 g cm^{-3} to 10 g cm^{-3} (for 1 g or larger)
	Surface area:	no restriction
	Surface finish:	smooth, no irregularities
	Surface protection:	may be plated or lacquered
	Magnetic properties:	same as grades S and O
	Corrosion resistance:	surface must resist corrosion and oxidation
	Hardness:	same as grade O
Grade Q:	Density:	7.2 g cm^{-3} to 10 g cm^{-3} (for 1 g or larger)
	Surface area:	same as grade P
	Surface finish:	same as grade P
	Surface protection:	may be plated, lacquered, or painted
	Magnetic properties:	no more magnetic than unhardened unmagnetized steel
	Corrosion resistance:	same as grade P
	Hardness:	same as grades O and P

Tolerance — Classification by Deviation[1]

CLASS 1			CLASS 2		
Grams	Individual Tolerance/mg	Group Tolerance/mg	Grams	Individual Tolerance/mg	Group Tolerance/mg
500	1.2		500	2.5	
300	0.75		300	1.5	
200	0.50		200	1.0	
100	0.25	1.35	100	0.5	2.7
50	0.12		50	0.25	
30	0.074		30	0.15	
20	0.074		20	0.10	
10	0.050	0.16	10	0.074	0.29
5	0.034		5	0.054	
3	0.034		3	0.054	
2	0.034		2	0.054	
1	0.034	0.065	1	0.054	0.105

CLASS 3		CLASS 4		CLASS 5		CLASS 6	
Grams	Tolerance/mg	Grams	Tolerance/mg	Grams	Tolerance/mg	Grams	Tolerance/mg
500	5.0	500	10	500	30	500	50
300	3.0	300	6.0	300	20	300	30
200	2.0	200	4.0	200	15	200	20
100	1.0	100	2.0	100	9	100	10
50	0.6	50	1.2	50	5.6	50	7
30	0.45	30	0.9	30	4.0	30	5
20	0.35	20	0.7	20	3.0	20	3
10	0.25	10	0.5	10	2.0	10	2
				5	1.3	5	2
				3	0.95	3	2
				2	0.75	2	2
				1	0.50	1	2

[1] In simple terms, the permitted deviation between the assigned nominal mass value of the weight and the actual mass of the weight. Verification of tolerance should be possible on reasonably precise equipment, without using a buoyancy correction, within the political jurisdiction or organizational bounds of a given weight specification.

Applications for Weights and Mass Standards[1]

Application	Type	Grade	Class
Reference standards used for calibrating other weights	I	S	1,2,3, or 4[1]
High-precision standards for calibration of weights and precision balances	I or II[2]	S or O[2]	1 or 2[3]
Working standards for calibration and precision analytical work, dead-weight pressure balances	I or II[2]	S or O	2
Laboratory weights for routine analytical work	II	O	2 or 3
Built-in weights, high-quality analytical balances	I or II	S	2
Moderate precision laboratory balances	II	P	3 or 4
Dial scales and trip balances	II	Q	4 or 5
Platform scales	II	Q	5 or 6

[1] Primary standards are for reference use only and should be calibrated. Since the actual values for each weight are stated, close tolerances are neither required nor desirable.
[2] Type I and Grade S will have a higher constancy but will probably be higher priced.
[3] Since working standards are used for the calibration of measuring instruments, the choice of tolerance depends upon the requirements of the instrument. The weights are usually used at the assumed nominal values and appropriate tolerances should be chosen.

Source: Reprinted (with modification) with permission of the ASTM International (formerly American Society for Testing and Materials), 100 Barr Harbor Drive, West Conshohocken, Pennsylvania, USA.

INDICATORS FOR ACIDS AND BASES

Thomas J. Bruno and Paris D. N. Svoronos

The following table lists the most common indicators together with their pH range and colors in acidic and basic media. Since the color change is not instantaneous at the pK_a value, a pH range is given where a combination of colors is present. This pH range, which varies between indicators, generally falls between the pK_a with a spread or uncertainty of 1 pH unit. All solutions are either aqueous or ethanol/aqueous (% ethanol, vol/vol) (Refs. 1–4). Reference 4 provides additional solution properties that are related to buffer and solvent properties, and Ref. 5 lists the exact quantities needed for the indicator solutions.

References

1. Lange, N. A., *Lange's Handbook of Chemistry*, 8th Edition, Handbook Publishers, New York, 1952.
2. Kolthoff, I. M., and V. A. Stenger (translated in English by N. H. Furman), *Volumetric Analysis*, 2nd Edition, Interscience Publishers, New York, 1942.
3. Sabnis, R.W., *Handbook of Acid-Base Indicators*, CRC Press, Taylor and Francis, Boca Raton, FL, 2008.
4. Bruno, T. J., and Svoronos, P. D. N., *CRC Handbook of Basic Tables for Chemical Analysis*, 3rd Edition, CRC Press, Boca Raton, FL, 2011.
5. http://www.csudh.edu/oliver/chemdata/ind-prep.htm, accessed June 2011.

Name	Synonym	pH Range	Solvent	Color change (acidic to basic)
Crystal Violet	Gentian Violet	0.0-2.0	Aqueous	yellow to blue-violet
Pentamethoxy Red	2,4,2',4',2"-Pentamethoxytriphenylcarbinol	1.2-2.3	70% ethanol	red-violet to colorless
Thymol Blue		1.2-2.8	Aqueous	red to yellow
4'-Anilinoazobenzene-4-sulfonic acid, sodium salt	Tropaeolin OO	1.3-3.2	Aqueous	red to yellow
2,4-Dinitrophenol		2.4-4.0	50% ethanol	colorless to yellow
4-(Dimethylamino)azobenzene	Methyl Yellow	2.9-4.0	90% ethanol	red to yellow
Methyl Orange	Sodium *p*-dimethylaminoazobenzenesulfonate	3.1-4.4	Aqueous	red to orange
Bromophenol Blue	Bromphenol Blue	3.0-4.6	Aqueous	yellow to blue-violet
Tetrabromophenol Blue		3.0-4.6	Aqueous	yellow to blue
Congo Red		3.0-5.0	Aqueous	blue-violet to red
4-(Phenylazo)-1-naphthalenamine	α-Naphthyl Red	3.7-5.0	70% ethanol	red to yellow
Alizarin Red S	Sodium alizarinesulfonate	3.7-5.2	Aqueous	yellow to violet
Etoxazene	*p*-Ethoxychrysoidine	3.5-5.5	Aqueous	red to yellow
Bromocresol Green	3,3',5,5'-Tetrabromo-*m*-cresolsulfonephthalein	4.0-5.6	Aqueous	yellow to blue
Methyl Red	Benzoic acid, 2-[[4-(dimethylamino)phenyl]azo]-	4.4-6.2	Aqueous	red to yellow
Bromocresol Purple	Bromcresol Purple	5.2-6.8	Aqueous	yellow to purple
4-Nitrophenol		5.0-7.0	Aqueous	colorless to yellow
Chlorophenol Red		5.4-6.8	Aqueous	yellow to red
Azolitmin		5.0-8.0	Aqueous	red to blue
Bromothymol Blue	Bromthymol Blue	6.0-7.6	70% ethanol	yellow to blue
Bromophenol Blue	Bromphenol Blue	6.2-7.6	Aqueous	yellow to blue
Phenol Red	Phenolsulfonphthalein	6.4-8.0	Aqueous	yellow to red
Neutral Red		6.8-8.0	70% ethanol	red to yellow
Aurin	Rosolic acid	6.8-8.0	90% ethanol	yellow to red
Cresol Red	*o*-Cresolsulfonphthalein	7.2-8.8	Aqueous	yellow to red
p-Naphtholbenzein	α-Naphtholbenzein	7.3-8.7	70% ethanol	rose to green
Orange I	Tropaeolin 000	7.6-8.9	Aqueous	yellow to rose-red
Thymol Blue		8.0-9.6	Aqueous	yellow to blue
Phenolphthalein	3,3-Bis(4-hydroxyphenyl)-1(3*H*)-isobenzofuranone	8.0-10.0	70% ethanol	colorless to red
p-Naphtholbenzein	α-Naphtholbenzein	9.0-11.0	70% ethanol	yellow to blue
Thymolphthalein		9.4-10.6	90% ethanol	colorless to blue
Nile Blue	C.I. Basic Blue 12	10.1-11.1	Aqueous	blue to red
Alizarin Yellow R		10.0-12.0	Aqueous	yellow to lilac
4-Hydroxy-3'-nitroazobenzene-3-carboxylic acid, sodium salt	Salicylic Yellow	10.0-12.0	90% ethanol	yellow to orange-brown
4-[(4-Nitrophenyl)azo]-1,3-benzenediol	Diazo Violet	10.1-12.0	Aqueous	yellow to violet
4-[(2,4-Dihydroxyphenyl)azo] benzenesulfonic acid, sodium salt	Tropaeolin O	11.0-13.0	Aqueous	yellow to orange-brown
N-Methyl-*N*,2,4,6-tetranitroaniline	Nitramine	11.0-13.0	70% ethanol	colorless to orange-brown
Poirrier Blue C4B	C.I. Acid Blue 22	11.0-13.0	Aqueous	blue to violet-pink
2,4,6-Trinitrobenzoic acid		12.0-13.4	Aqueous	colorless to orange-red

PREPARATION OF SPECIAL ANALYTICAL REAGENTS

Paris D. N. Svoronos and Thomas J. Bruno

This listing of analytical reagents has been updated to include formulations based on more recent research, and to also include safety notes (Refs. 1–3). In the absence of specific cautions, the user must observe sound laboratory practice and housekeeping to avoid exposure and environmental contamination. When less familiar reagents are described, we include a chemical formula and a CAS registry number to avoid ambiguity. When a reagent calls for 95% ethanol, the azeotrope with water is specified unless noted. Additional details can be found in the references listed below.

References

1. Lide, D. R., Ed., *CRC Handbook for Chemistry and Physics*, 90th Edition, CRC Press, Boca Raton, FL, 2010.
2. Bruno, T. J., and Svoronos, P. D. N., *CRC Handbook of Basic Tables for Chemical Analysis*, 3rd Edition, CRC Press, Boca Raton, FL, 2011.
3. Svoronos, P. D. N., Sarlo, E., and Kulawiec, R., *Organic Chemistry Laboratory Manual*, WCB McGraw Hill, New York, 1996.

Aluminon (qualitative test for aluminum). Aluminon is the name for the triammonium salt of aurintricarboxylic acid (5-[(3-carboxy-4-hydroxyphenyl)(3-carboxy-4-oxo-2,5-cyclohexadienylidene)methyl]-2-hydroxybenzoic acid triammonium salt, $C_{22}H_{23}N_3O_9$, CAS No. 569-58-4). Dissolve 1 g of the salt in 1 L of distilled water. Shake the solution well to ensure thorough mixing.

Bang's reagent (for glucose estimation). Dissolve (in the exact order) 100 g of potassium carbonate, 66 g of potassium chloride and 160 g of anhydrous potassium bicarbonate in approximately 700 mL of distilled water at 30 °C. Add (while stirring) 4.4 g of copper (II) sulfate and dilute to 1 L with distilled water after all CO_2 has been released. This solution should be shaken and saved in an air-tight flask. After 24 hours, 300 mL of it are diluted to 1 L with a saturated aqueous potassium chloride solution, shaken gently and used after 24 hours. 50 mL of this solution is equivalent to 10 mg glucose.

Barfoed's reagent (test for glucose). See Cupric acetate.

Baudisch's reagent. See Cupferron.

Benedict's solution (qualitative reagent for glucose). Dissolve 173 g of sodium citrate and 100 g of sodium carbonate in 800 mL of distilled water. Filter, if necessary, and dilute to 850 mL with distilled water. Dissolve 17.3 g of copper (II) sulfate pentahydrate in 100 mL of distilled water. Pour the latter solution, with constant stirring, into the carbonate-citrate solution and dilute to 1 L with distilled water.

Benzidine hydrochloride solution (for sulfate determination). Prepare a paste of 8 g of benzidine hydrochloride ($C_{12}H_8(NH_2)_2{\cdot}2HCl$) and 20 mL of distilled water, add 20 mL of 20% (mass/mass) HCl and dilute to 1 L with distilled water. Each mL of this solution is equivalent to 0.00357 g of H_2SO_4. Note that the reagent is often called benzidine dihydrochloride.

Bertrand's reagent (glucose estimation). Consists of the following solutions:

1. Dissolve 200 g of Rochelle salt (potassium sodium tartrate, $NaKC_4H_4O_6$) and 150 g of NaOH in sufficient distilled water to make 1.0 L of solution.
2. Dissolve 40 g of copper (II) sulfate pentahydrate in sufficient distilled water to a total of 1.0 L of solution.
3. Dissolve 50 g of iron (III) sulfate and 200 g of H_2SO_4 (sp. gr. 1.84) in sufficient distilled water to a total of 1.0 L of solution.
4. Dissolve 5 g of potassium permanganate in sufficient distilled water to a total of 1.0 L solution.

Bial's reagent (for pentose). Dissolve 1 g of orcinol (5-methyl-1,3-benzenediol, $C_7H_8O_2$, CAS No. 504-15-4) in 500 mL of 30% (mass/mass) HCl to which 30 drops of a 10% aqueous solution of iron (III) chloride have been added.

Boutron — Boudet soap solution: Consists of the following solutions:

1. Dissolve 100 g of pure castile soap (olive oil based) in about 2.5 L of 56% (vol/vol) aqueous ethanol.
2. Dissolve 0.59 g of barium nitrate in 1.0 L of water. Adjust the castile soap solution (1) so that 2.4 mL of it will produce a permanent lather with 40 mL of solution (2). When adjusted, 2.4 mL of this soap solution is equivalent to 220 parts per million of hardness (as calcium carbonate) for a 40 mL sample. See also Soap solution.

Brucke's reagent (protein precipitation). See Potassium iodide mercuric iodide.

Cobaltic cyanide paper (Rinnmann's test for zinc detection). Dissolve 4 g of potassium cobalt (III) hexacyanide ($K_3Co(CN)_6$) and 1 g of potassium chlorate in 100 mL of water. Soak filter paper in solution and dry at 100 °C. Apply a drop of the zinc solution and heat in an evaporating dish. A green disk on the filter paper is obtained if zinc is present.

Congo red. Dissolve 0.5 g of Congo red (3,3'-([1,1'-biphenyl]-4,4'-diyl)bis(4-amino naphthalene-1-sulfonic acid, $C_{32}H_{22}N_6Na_2O_6S_2$, CAS No. 573-58-0) in 90 mL of distilled water and 10 mL of ethanol.

Cupferron (Baudisch's reagent for iron analysis). Dissolve 6 g of Cupferron, the ammonium salt of *N*-hydroxy-*N*-nitrosoaniline ($NH_4[C_6H_5N(O)NO]$) in 100 mL of distilled water. The reagent is good only for one week and must be kept in the dark.

Cupric acetate (Barfoed's reagent for reducing monosaccharides). Dissolve 66 g of copper (II) acetate and 10 mL of glacial acetic acid in water and dilute to 1.0 L.

Cupric oxide, ammoniacal; Schweitzer's reagent (dissolves cotton, linen, and silk, but not wool). Dissolve 5 g of cupric sulfate in 100 mL of boiling water, and add sodium hydroxide until precipitation is complete. Wash the precipitate well, and dissolve it in a minimal quantity of ammonium hydroxide. Bubble a slow stream of air through 300 mL of concentrated ammonium hydroxide solution containing 50 g of fine copper turnings. Continue stirring for 1 hour.

Cupric sulfate in glycerin-potassium hydroxide (reagent for silk). Dissolve 10 g of copper (II) sulfate pentahydrate in 100 mL of water and add 5 g of glycerol. Add 6N KOH solution slowly until a deep blue solution is obtained.

Cupron (precipitates copper). Dissolve 5 g of benzoinoxime in 100 mL 95% ethanol.

Cuprous chloride, acidic (reagent for CO in gas analysis).

1. Cover the bottom of a 2 L flask with a layer of copper (II) oxide about 1.5 cm deep, suspend a coil of copper wire so as to reach from the bottom to the top of the solution, and fill the flask with hydrochloric acid 20% (vol/vol) HCl

(aq) with continuous stirring. When the solution becomes nearly colorless, transfer to reagent bottles, which should also contain copper wire. The stock bottle may be refilled with dilute (6N) HCl until either the cupric oxide or the copper wire is used up. Copper (II) sulfate may be substituted for copper oxide in the above procedure.

2. Dissolve 340 g of copper (II) chloride dihydrate in 600 mL of concentrated HCl and reduce the cupric chloride by adding 190 mL of a saturated solution of stannous chloride or until the solution is colorless. The stannous chloride is prepared by treating 300 g of metallic tin in a 500 mL flask with concentrated HCl until no more tin is dissolved.
3. (Winkler method). Add a mixture of 86 g of copper (II) chloride and 17 g of finely divided metallic copper, prepared by the reduction of 8 g to12 g of cupper (II) oxide with hydrogen gas, to a solution of HCl, made by diluting 650 mL of concentrated HCl with 325 mL of distilled water. After the mixture has been added slowly and with frequent stirring, a spiral of copper wire is suspended in the bottle, reaching all the way to the bottom. The solution is ready to use when the solution becomes colorless.

Cuprous chloride, ammoniacal (reagent for CO in gas analysis).

1. The acid solution of copper (II) chloride as prepared above is neutralized with 6N ammonium hydroxide until the ammonia odor persists. An excess of metallic copper must be kept in the solution.
2. Pour 800 mL of acidic copper (II) chloride, prepared by the Winkler method, into approximately 4 L of water. Transfer the precipitate to a 250 mL graduated cylinder. After several hours, siphon off the liquid above the 50 mL mark and refill with 7.5% ammonium hydroxide solution which may be prepared by diluting 50 mL of concentrated ammonium hydroxide with 150 mL of distilled water. The solution, which should have a faint odor of ammonia should be well stirred and allowed to stand for several hours.

Note: Proper precautions must be taken for the safe handling of gaseous samples that contain CO.

Dichlorofluorescein indicator. Dissolve 1 g dichlorofluorescein ($C_{20}H_{10}Cl_2O_5$, CAS No. 76-54-0) in 1 L 70% (vol/vol) ethanol or 1 g of the sodium salt in 1 L of distilled water.

Dimethyglyoxime, 0.01 N. Dissolve 0.6 g of dimethylglyoxime (2,3-butanedione oxime) in 500 mL of 95% ethanol. This is an especially sensitive test for nickel (II) and produces a very characteristic crimson color.

Diphenylamine (reagent for rayon). Dissolve 0.2 g diphenylamine in 100 mL of concentrated sulfuric acid.

Diphenylamine sulfonate (for titration of iron with $K_2Cr_2O_7$). Dissolve 0.32 g of the barium salt of diphenylamine sulfonic acid in 100 mL of distilled water, add 0.5 g of sodium sulfate and filter off the precipitated barium sulfate.

Diphenylcarbazide. Dissolve 0.2 g of diphenylcarbazide[$(C_6H_5NHNH)_2CO$] in 10 mL of glacial acetic acid and dilute to 100 mL with 95% ethanol.

2,4-Dinitrophenylhydrazine (2,4-DNP) Reagent. Dissolve 3 g of 2,4-dinitrophenylhydrazine in 15 mL to 20 mL of concentrated sulfuric acid. Add this solution with stirring to a solution of 10 mL distilled water in 70 mL of 95% hydrous ethanol. Use the filtrate to test for the presence of aldehydes and ketones as follows: To 2 mL of the solution add 10 drops of the unknown. A precipitate (which may be recrystallized from 95% hydrous ethanol) is indicative of a positive test and its melting point can be matched to the corresponding melting points of various aldehydes and ketones.

Esbach's reagent (estimation of protein). Dissolve 10 g of picric acid and 20 g of citric acid in sufficient distilled water to make a 1.0 L of solution.

Eschka's compound. Two parts of calcinated ("light") magnesia are thoroughly mixed with 1 part of anhydrous sodium carbonate.

Fehling's solution (reagent for reducing sugars).

1. Copper (II) sulfate solution: Dissolve 34.66 g of copper (II) sulfate pentahydrate in distilled water and dilute to 500 mL.
2. Alkaline tartrate solution: Dissolve 173 g of potassium sodium tartrate (Rochelle salt, $KNaC_4H_4O_6{\cdot}4H_2O$) and 50 g of NaOH in distilled water, cool and dilute to 500 mL. Mix equal volumes of solutions (1) and (2) just before using.

Ferric-alum indicator. Dissolve 140 g of iron (III) ammonium sulfate crystals in 400 mL of hot distilled water. Cool, filter, and add enough 6N nitric acid until the 500 mL mark.

Folin's mixture (for uric acid). To 650 mL of distilled water add 500 g of ammonium sulfate, 5 g of uranium (VI) acetate, and 6 g of glacial acetic acid and then dilute to 1.0 L with distilled water.

Formaldehyde — sulfuric acid (Marquis' reagent for alkaloids). Add 100 mL of concentrated sulfuric acid to 5 mL of 37% (mass/mass) aqueous formaldehyde solution.

Froehde's reagent. See Sulfomolybdic acid.

Fuchsin (reagent for linen). Dissolve 1 g of fuchsin (an aniline dye) in 100 mL of ethanol.

Fuchsin - sulfurous acid (Schiff's reagent for aldehydes). Dissolve 0.5 g of fuchsin and 9 g of sodium bisulfite in 500 mL of distilled water, and add 10 mL of concentrated HCl. Keep in well-stoppered bottles and protect from light. The fuchsin used to make the Schiff reagent should incorporate a high content of pararosanilin.

Gunzberg's reagent (detection of HCl in gastric juice). Dissolve 4 g of phloroglucinol (1,3,5-benzenetriol, $C_6H_3(OH)_3$) and 2 g of vanillin ($C_8H_8O_3$) in 100 mL of absolute ethanol.

Hager's reagent. See Picric acid.

Hanus solution (for iodine number). Dissolve 13.2 g of resublimed iodine in 1 L of glacial acetic acid, which should pass the dichromate test for reducible matter. Add sufficient bromine to double the halogen content, as determined by titration (approximately 3 mL). The iodine may be dissolved by heating, but the solution should be cold when the bromine is added.

Hopkins-Cole Reagent (test for presence of the indole ring in the tryptophan moiety of a protein or peptide). Add 5 mL cold water to 10 g of magnesium in a conical flask and add 250 mL of a cold saturated aqueous oxalic acid solution with vigorous stirring. Filter, add 25 mL of glacial acetic acid and dilute with distilled water to the 1.0 L mark. Place 10 drops of the saturated protein or peptide solution in a test tube, add 10 drops of concentrated nitric acid and heat in a water bath for three minutes. Allow the mixture to cool and add 4 mL 6N NaOH. A positive test is confirmed by the formation of a yellow color.

Iodine, tincture of. To 50 mL of distilled water add 70 g of iodine and 50 g of potassium iodide. Dilute to 1 L with absolute ethanol.

Iodo-potassium iodide (Wagner's reagent for alkaloids). Dissolve 2 g of iodine and 6 g of KI in 100 mL of distilled water.

Jones Reagent (test is positive for 1° and 2° alcohols, aldehydes, mercaptans, sulfides and sulfoxides). To 1 mL of 5% (mass/mass) sodium (or potassium) dichromate add 10 drops of the unknown compound. The formation of a green color (sometimes appearing as brown due to the mixture of the green chromium (III) and the unreacted orange chromium (VI) colors) indicates a positive test.

Litmus (indicator). Extract litmus powder three times with boiling alcohol (ethanol or isopropanol), each treatment lasting for an hour. Discard the alcoholic extract. Treat the residue with an equal mass of cold distilled water and filter; then add five times its weight of boiling distilled water. Cool, filter, and combine all aqueous extracts.

Magnesia mixture (reagent for phosphates and arsenates). Dissolve 50 g of magnesium chloride ($MgCl_2{\cdot}6H2O$) and 100 g of ammonium chloride (NH_4Cl) in 500 mL of distilled water, add a slight excess of NH_4OH, and allow to stand overnight. Filter, make the solution just acidic with HCl, and dilute with distilled water to 1 L. Let stand for several days and separate the precipitated solid by decanting. The decantate should be used just prior to testing; otherwise, if stored for any period of time it becomes turbid.

Magnesium uranyl acetate. Dissolve 100 g of uranyl (VI) acetate dehydrate ($UO_2(C_2H_3O_2)_2{\cdot}2H_2O$) in 60 mL of glacial acetic acid and dilute to 500 mL with distilled water. Dissolve 330 g magnesium acetate tetrahydrate ($Mg(C_2H_3O_2)_2{\cdot}4H_2O$) in 60 mL of glacial acetic acid and dilute to 200 mL. Heat both solutions to boiling until clear; pour the magnesium acetate solution into the uranyl acetate solution, cool and dilute with distilled water to 1 L. Let stand overnight and filter if necessary.

Marme's reagent. See Potassium-cadmium iodide.

Marquis' reagent. See Formaldehyde-sulfuric acid.

Mayer's reagent (white precipitate with most alkaloids in slightly acid solutions). Dissolve 1.358 g of mercury (II) chloride in 60 mL of distilled water and pour into a solution of 5 g of potassium iodide in 10 mL of distilled water. Add sufficient distilled water to a final volume of 100 mL.

Methyl orange indicator. Dissolve 1 g of methyl orange in 1 L of distilled water. Filter, if necessary.

Methyl orange, modified. Dissolve 2 g of methyl orange and 2.8 g of xylene cyanole FF ($C_{25}H_{27}N_2NaO_6S_2$, CAS No: 2650-17-1) in 1 L of 50% (vol/vol) ethanol.

Methyl red indicator. Dissolve 1 g of methyl red in 600 mL of ethanol and dilute with 400 mL of distilled water.

Methyl red, modified. Dissolve 0.50 g of methyl red and 1.25 g of xylene cyanole FF ($C_{25}H_{27}N_2NaO_6S_2$) in 1 L of 90% (vol/vol) ethanol. Alternatively dissolve 1.25 g of methyl red and 0.825 g of methylene blue in 1 L of 90% (vol/vol) ethanol.

Millon's reagent (for albumins and phenols). Dissolve 1 part of mercury in 1 part of cold fuming nitric acid. Dilute with twice the volume of distilled water and decant the clear solution after several hours. All appropriate precautions must be observed when handling mercury.

Molisch's reagent. See 1-Naphthol.

1-Naphthol (Molisch's reagent for wool). Dissolve 15 g of 1-naphthol in 100 mL of 95% (vol/vol) ethanol or chloroform.

Nessler's reagent (for ammonia). Dissolve 50 g of potassium iodide in the smallest possible quantity of cold distilled water (50 mL). Add a saturated solution of mercury (II) chloride (about 22 g in 350 mL of distilled water will be needed) until an excess is indicated by the formation of a precipitate. Then add 200 mL of 5 N NaOH and dilute to 1 L with distilled water. Allow to settle and draw off the clear liquid.

Nickel oxide, ammoniacal (reagent for silk). Dissolve 5 g of nickel (II) sulfate in 100 mL of distilled water, and add 6N NaOH solution until nickel hydroxide is completely precipitated. Wash the precipitate well and dissolve in 25 mL of concentrated ammonium hydroxide and 25 mL of distilled water.

Nitron (detection of the nitrate radical). Dissolve 10 g of nitron (1,4-diphenyl-3-(phenylamino)-1,2,4-triazolium hydroxide, $C_{20}H_{18}N_4O$) in 5 mL of glacial acetic acid and 95 mL of distilled water. The solution may be filtered with slight suction through an Alundum thimble crucible and kept in a dark bottle.

1-Nitroso-2-naphthol. Make a saturated solution 1-nitroso-2-naphthol in 50% (vol/vol) aqueous acetic acid. The reagent should be used as soon as it is prepared.

Nylander's solution (carbohydrates). Dissolve 20 g of bismuth subnitrate ($Bi_5O(OH)_9(NO_3)_4$) and 40 g of Rochelle salt (potassium sodium tartrate, $KNaC_4H_4O_6{\cdot}4H_2O$)in 1 L of 8% (mass/mass) aqueous NaOH solution. Cool and filter.

Obermayer's reagent (for indoxyl in urine). Dissolve 4 g of iron (III) chloride in 1 L of aqueous (vol/vol) 40% HCl.

Oxine. Dissolve 14 g of 8-hydroxyquinoline (C_9H_7NO) in 30 mL of glacial acetic acid. Warm slightly, if necessary to dissolve. Dilute to 1 L with distilled water.

Oxygen absorbent. Dissolve 300 g of ammonium chloride in 1 L distilled water and add 1 L of concentrated ammonium hydroxide solution. Shake the solution thoroughly. For use as an oxygen absorbent, the gas to be tested is passed through a bottle that is half full of copper turnings filled nearly to the top with this ammonium chloride - ammonium hydroxide solution.

Pasteur's salt solution. To 1 L of distilled water add 2.5 g of potassium phosphate, 0.25 g of calcium phosphate, 0.25 g of magnesium sulfate, and 12.00 g of ammonium tartrate.

Pavy's solution (glucose reagent). To 120 mL of Fehling's solution, add 300 mL of 6N ammonium hydroxide (sp. gr. 0.88) and dilute to 1 L with distilled water.

Phenanthroline ferrous ion indicator. Dissolve 1.485 g of 1,10-phenanthroline monohydrate ($C_{12}H_8N_2{\cdot}H_2O$) in 100 mL of 0.025 M aqueous iron (II) sulfate solution.

Phenolphthalein. Dissolve 1 g of phenolphthalein in 50 mL of ethanol and add 50 mL of distilled water.

Phenolsulfonic acid (determination of nitrogen as nitrate). Dissolve 25 g of phenol in 150 mL of concentrated H_2SO_4, add 75 mL of fuming sulfuric acid (approximately 15% SO_3), stir well and heat for 2 hours at 100 °C.

Phloroglucinol solution (pentosans). Make a 3% phloroglucinol (1,3,5-benzenetriol) solution in alcohol. Store in a dark bottle.

Phosphomolybdic acid (Sonnenschein's reagent for alkaloids).

1. Prepare ammonium phosphomolybdate and after washing with distilled water, boil with nitric acid to expel all ammonia. Evaporate to dryness and dissolve in 2 M HNO_3.
2. Dissolve ammonium molybdate in HNO_3 and treat with phosphoric acid. Filter, wash the precipitate, and boil with aqua regia (a mixture of concentrated nitric acid and concentrated hydrochloric acid, 1:3 vol/vol) until the ammonium salt is decomposed. Evaporate to dryness. The residue is dissolved in 10% (vol/vol) HNO_3. The solution constitutes Sonnenschein's reagent.

Phosphoric acid — sulfuric acid mixture. Dilute 150 mL of concentrated sulfuric acid and 100 mL of concentrated phosphoric acid with distilled water to a final volume of 1 L.

Phosphotungstic acid (Scheibler's reagent for alkaloids).

1. Dissolve 20 g of sodium tungstate and 15 g of sodium phosphate in 100 mL of distilled water slightly acidified with dilute nitric acid.
2. The reagent is a 10% (mass/mass) solution of phosphotungstic acid in distilled water. The phosphotungstic acid is prepared by evaporating a mixture of 10 g of sodium tungstate dissolved in 5 g of phosphoric acid (sp. gr. 1.13) and enough boiling water to a complete solution. Crystals of phosphotungstic acid will separate.

Picric acid (Hager's reagent for alkaloids, wool and silk). Dissolve 1 g of picric acid in 100 mL of distilled water.

Potassium antimonate (reagent for sodium). Boil 22 g of potassium antimonate with 1 L of distilled water until nearly all of the salt has dissolved, cool quickly, and add 35 mL of 10% (mass/mass) potassium hydroxide. Filter after standing overnight.

Potassium-cadmium iodide (Marme's reagent for alkaloids). Add 2 g of cadmium (II) chloride to a boiling solution of 4 g of potassium iodide in 12 mL of distilled water, and then mix with 12 mL of saturated aqueous potassium iodide solution.

Potassium hydroxide (for CO_2 absorption). Dissolve 360 g of potassium hydroxide in distilled water and dilute to 1 L.

Potassium iodide — mercuric iodide (Brucke's reagent for proteins). Dissolve 50 g of potassium iodide in 500 mL of distilled water, and saturate with mercury (II) iodide (about 120 g). Dilute to 1 L with distilled water.

Potassium pyrogallate (for oxygen absorption). For mixtures of gases containing less than 28% (mass/mass) oxygen, add 100 mL of potassium hydroxide solution (50 g of KOH to 100 mL of distilled water) to 5 g of pyrogallol. For mixtures containing more than 28% (mass/mass) oxygen, the KOH solution should contain instead 120 g of KOH in 100 mL of distilled water.

Pyrogallol, alkaline. The reagent is a mixture of two solutions

1. Dissolve 75 g of pyrogallic acid in 75 mL of distilled water.
2. Dissolve 500 g of KOH in 250 mL distilled water. When cooled, adjust the solution with distilled water until the concentration is 50% (vol/vol).

For use, add 270 mL of solution (2) to 30 mL of solution (1).

Rosolic acid (indicator). Dissolve 1 g of rosolic acid ($C_{19}H_{14}O_3$) in 10 mL of ethanol and add 100 mL of distilled water.

Sakaguchi Reagent (for the presence of arginine in proteins or peptides). To 10 drops of the saturated protein or peptide solution, add 5 drops of 6 N NaOH followed by 5 drops of 0.05% (mass/mass) ethanolic 1-naphthol solution and 10 drops of 0.5% aqueous sodium hypochlorite. A positive test is indicated by the formation of a red color that fades away upon standing.

Scheibler's reagent. See Phosphotungstic acid.

Schiff's reagent. See Fuchsin-sulfurous acid.

Schweitzer's reagent. See Cupric oxide, ammoniacal.

Soap solution (reagent for hardness in water). Dissolve 100 g of dry castile soap in 1 L of 80% (vol/vol) ethanol (4 parts ethanol to 1 part distilled water). Allow to stand for several days and dilute with 70% ethanol until 6.4 mL of this solution produces a permanent lather with 20 mL of standardized calcium solution. The latter solution is made by dissolving 0.2 g of calcium carbonate in a small amount of dilute HCl, evaporating to dryness and then dissolving the precipitate with distilled water to a final volume of 1 L.

Sodium bismuthate (for the oxidation of manganese). Heat 20 parts of sodium hydroxide nearly to redness in an iron or nickel crucible and add slowly 10 parts of basic bismuth (III) nitrate which has been previously dried. Add 2 parts of sodium peroxide, and pour the brownish-yellow fused mass onto an iron plate to cool. When cooled, break up in a mortar, extract with distilled water, and collect on an asbestos filter.

Sodium hydroxide (for CO_2 absorption). Dissolve 330 g of sodium hydroxide in distilled water and dilute to 1 L.

Sodium nitroprusside (reagent for hydrogen sulfide and wool). Use a freshly prepared solution of 1 g of sodium nitroferricyanide in 10 mL of distilled water.

Sodium oxalate (primary standard). Dissolve 30 g of the commercial salt of sodium oxalate in 1 L of distilled water, make slightly alkaline with sodium hydroxide, and let stand until clear. Filter and evaporate the filtrate to 100 mL. Cool and filter. Pulverize the residue and wash it several times with small volumes of distilled water. The procedure is repeated until the mother liquor is sulfate-free and is neutral to phenolphthalein.

Sodium plumbite (reagent for wool). Dissolve 5 g of sodium hydroxide in 100 mL distilled water. Add 5 g of litharge (lead (II) oxide) and boil until dissolved.

Sodium polysulfide. Dissolve 480 g of sodium sulfide nonahydrate in 500 mL of distilled water, add 40 g of NaOH and 18 g of sulfur. Stir thoroughly and dilute to 1 L with distilled water.

Sonnenschein's reagent. See Phosphomolybdic acid.

Starch solution.

1. Make a paste with 2 g of soluble starch and 0.01 g of mercury (II) iodide with a small amount of distilled water. Add the mixture slowly to 1 L of boiling distilled water and boil further for a few minutes. Keep in a glass stoppered bottle. If other than soluble starch is used, the solution will not be clear on boiling; it should then be allowed to stand and the clear liquid decanted.
2. A solution of starch that keeps stable indefinitely is made as follows: Mix 500 mL of aqueous saturated NaCl solution (filtered), 80 mL of glacial acetic acid, 20 mL of distilled water and 3 g of starch. Bring slowly to a boil and further heat for 2 minutes.
3. Make a paste with 1 g of soluble starch and 5 mg of mercury (II) iodide using as little cold distilled water as possible. Then pour about 200 mL of boiling distilled water on the paste and stir immediately. This will give a clear solution if the paste is prepared correctly and the water is actually boiling. Cool and add 4 g of potassium iodide. Starch solution decomposes on standing due to bacterial action, but this solution will be stable if stored under a layer of toluene.

Stoke's reagent. Dissolve 30 g of iron (II) sulfate and 20 g of tartaric acid in distilled water and dilute to 1 L. Just before using, add concentrated ammonium hydroxide until the precipitate that is initially formed is redissolved.

Sulfanilic acid (reagent for nitrites). Dissolve 0.5 g of sulfanilic acid in a mixture of 15 mL of glacial acetic acid and 135 mL of recently boiled distilled water.

Sulfomolybdic acid (Froehde's reagent for alkaloids and glucosides). Dissolve 10 g of molybdic acid or sodium molybdate in 100 mL of concentrated sulfuric acid.

Tannic acid (reagent for albumin, alkaloids, and gelatin). Dissolve 10 g of tannic acid in 10 mL of ethanol and dilute with distilled water to 100 mL.

Titration mixture (residual chlorine in water analysis). Prepare 1 L of dilute HCl (100 mL of HCl (sp. gr. 1.19) in sufficient distilled water to make 1 L). Dissolve 1 g of *o*-tolidine (3,3′-dimethylbenzidine , CAS No. 119-93-7) in 100 mL dilute hydrochloride, stir well and dilute to 1 L using dilute HCl solution.

Tollen's Reagent (confirming the presence of aldehydes). 5 drops of 5% NaOH is added to 2 mL 10% aqueous silver nitrate in a test tube. The insoluble silver (I) oxide dissolved by the drop wise addition of 10% aqueous ammonia (ammonium hydroxide) yields a clear solution. Excess ammonium hydroxide should be avoided as it may give false positive result. Approximately 10 drops of the aldehyde will yield a silver mirror coating on the test tube inner wall, especially if the mixture is warmed up to 50 °C.

Trinitrophenol solution. See Picric acid.

Turmeric tincture (reagent for borates). Digest ground turmeric root with several quantities of distilled water which are discarded. Dry the residue and digest it several days with six times its weight of ethanol. Filter.

Uffelmann's reagent (turns yellow in presence of lactic acid). To a 2% solution of pure phenol in distilled water, add an aqueous solution of iron (III) chloride until the phenol solution becomes violet in color.

Wagner's reagent. See Iodo-potassium iodide.

Wagner's solution (used in phosphate rock analysis to prevent precipitation of iron and aluminum). Dissolve 25 g of citric acid and 1 g of salicylic acid in distilled water and dilute to 1 L. Use 50 mL of the reagent.

Wij's iodine monochloride solution (for iodine number). Dissolve 13 g of resublimed iodine in 1 L of glacial acetic acid that will pass the dichromate test for reducible matter. Set aside 25 mL of this solution. Bubble into the remainder of the solution dry chlorine gas (dried and washed by passing through concentrated sulfuric acid) until the characteristic color of free iodine has been dissipated. Add the 25 mL of the iodine solution that was set aside, until all free chlorine has been dissipated. A slight excess of iodine does little or no harm, but an excess of chlorine must be avoided. Preserve in well stoppered, amber colored bottles. Avoid the use of solutions that have been stored for more than 30 days.

Wij's special solution (for iodine number). To 200 mL of glacial acetic acid that will pass the dichromate test for reducible matter, add 12 g of dichloramine T (*N,N*-dichloro-4-methylbenzenesulfonamide, CAS No. 473-34-7), and 16.6 g of dry potassium crystals (in small quantities with continual shaking until all the potassium iodide has dissolved). Dilute to 1 L with the same quality of acetic acid used above and preserve in a dark colored bottle.

Zimmermann-Reinhardt reagent (determination of iron). Dissolve 70 g of manganese (II) sulfate tetrahydrate in 500 mL of water, add slowly 125 mL of concentrated sulfuric acid and 125 mL of 85% phosphoric acid, and dilute to 1 L with water.

Zinc chloride solution, basic (reagent for silk). Dissolve 1000 g of zinc chloride in 850 mL of distilled water, and add 40 g of zinc oxide. Heat until complete dissolving is complete.

Zinc uranyl acetate (reagent for sodium). Dissolve 10 g of $UO_2(C_2H_3O_2)2{\cdot}2H_2O$ in 6 g of 30% acetic acid, heat, if necessary, and dilute to 50 mL. Dissolve 30 g of zinc acetate dehydrate in 3 g of 30% acetic acid and dilute to 50 mL. Mix the two solutions, add 50 mg of sodium chloride, allow to stand overnight and filter. Use distilled water in all dilutions.

ORGANIC ANALYTICAL REAGENTS FOR THE DETERMINATION OF INORGANIC IONS

Paris D. N. Svoronos and Thomas J. Bruno

In this table of organic reagents used for determination of inorganic elements and compounds we present the major tests, reagents, and some guidance as to the expected result or method of observation. The abbreviations used here are defined in the abbreviations table in this section. In addition, for brevity, when a determination calls for a spectrophotometric measurement at a particular wavelength, for example at 500 nm, we denote this as: "spec λ = 500 nm." No wavelength is specified if it is variable, for example, with pH; here we indicate: "spec determination." When a determination calls for spectrofluorimetric determination, the excitation and emission wavelengths are provided, thus: specf λ_{ex} = 303.5 nm, λ_{em} = 353 nm. Note that a common surfactant used in many of these tests is cetyltrimethylammonium bromide, abbreviated (CTAB). Information and data on this and other surfactants commonly used in chemical analysis are provided in Section 6. When relevant, we provide an approximate limit of detection (LOD), and when the uncertainty can vary, this is expressed as a relative standard deviation (RSD). Some of the procedures listed here require the use of hazardous chemicals (carcinogens such as benzene, strong acids such as HF). Appropriate precautions must be observed.

While a great deal of the information presented here is from the recent literature, the reader is referred to several excellent reviews and monographs for additional information (Refs. 1–10).

References

1. Marczenko, Z., *Separation and Spectrophotometric Determination of Elements*, Ellis Horwood, Chichester, 1986.
2. Sandell, E. B., and Onishi, H., *Photometric Determination of Traces of Metals. General Aspects, Part I,4th Edition*, John Wiley & Sons, New York, 1986.
3. Onishi, H., *Photometric Determination of Traces of Metals. Part IIa: Individual Metals, Aluminium to Lithium,4th Edition*, John Wiley & Sons, New York, 1986.
4. Onishi, H., *Photometric Determination of Traces of Metals. Part IIb: Individual Metals, Magnesium to Zinc*, 4th Edition, John Wiley & Sons, New York, 1986.
5. Townshend, A., Burns, D. T., Guilbault, G. G., Lobinski, R., Marczenko, Z., Newman, E., and Onishi, H., *Dictionary of Analytical Reagents*, Chapman & Hall, London, 1993.
6. West, T. S., and Nürnberg, H. W., *The Determination of Trace Metals in Natural Waters*, Blackwell, Oxford, 1988.
7. Savvin, S. B., Shtykov, S. N., and Mikhailova, A. V., Russ. *Chem. Rev.* 75, 341, 2006.
8. Ueno, K., Imamura, T., and Cheng, K. L., *Handbook of Organic Analytical Reagents*, CRC Press, Boca Raton, FL, 1992.
9. American Chemical Society, *Reagent Chemicals: Specifications and Procedures: American Chemical Society Specifications, Official from January 1, 2006*, American Chemical Society, Washington, D.C., 2006.
10. Crompton, T. R., *Determination of Anions: A Guide for the Analytical Chemist*, Springer-Verlag, Berlin, Heidelberg, 1996.

I. Organic Analytical Reagents for the Determination of Cations

Reagents	Results
Aluminum	
Alizarin red S	Red color develops; spectrophotometric determination preferred
Aluminon	Lake pigment stabilized with CTAB
Chrome azurol S	Spec λ; = 500 nm, stabilized with CTAB
Chromazol KS	Spec λ = 625 nm, stabilized with cetylpyridinium bromide
Eriochrome cyanine R (also known as mordant blue 3) + CTAB	Red dye lake (pH = 6) stabilized with CTAB
Eriochrome cyanine R (also known as mordant blue 3) + *N,N*-dodecyltrimethylammonium bromide (DTAB)	Spec determination by use of cationic surfactants
8-Hydroxyquinoline	Produces tris(8-hydroxyquinolinato) aluminum (Alq3), found in organic light emitting diodes (OLEDs)
Bromopyrogallol red + cetyltrimethyl ammonium bromide	Spec λ = 627 nm
Bromopyrogallol red + nonylphenol tetradecaethylene glycol ether	Spec λ = 612 nm
Pyrocatechol violet	Spec λ = 578 nm after separation of aluminum from the matrix materials by chloroform extraction of its acetylacetone complex (pH = 6.5), from an ammonium acetate-hydrogen peroxide medium
2,2′,3,4-Tetrahydroxy-3′,5′-disulfoazobenzene	Spec determination of the binary system (pH = 5)
Antimony	
Brilliant green	Isolated as the hexachloroantimonate (V) salt extracted by either toluene or benzene
Bromopyrogallol red	Used for the determination of antimony (III) with EDTA, cyanide or fluoride ions as masking agents
Catechol violet	Ternary complex stabilized with CTAB
Malachite green (basic green 4)	Isolated as the hexachloantimonate (V) complex after benzene extraction
Phenyl fluorene	Sensitive color reaction with antimony (III) by use of cationic surfactants
Potassium iodide	Antimony (III)-iodide complex formation in the presence of ascorbic acid

Analytical

Reagents	Results
Rhodamine B	Ion pair or ion association extraction by use of toluene or benzene as solvents
Silver diethyldithiocarbamate	Spec λ = 504 nm to avoid arsenic interference
Thiourea	Determination by hydride generation inductively coupled plasma atomic emission spectrometry after reduction of antimony (V) to antimony (III) by thiourea
Arsenic	
Michler's ketone	Spec λ = 640 nm trophotometric determination of trace arsenic (V) in water
Silver diethyldithiocarbamate	Spec λ = 600 nm to avoid antimony interference
Thiourea	Determination by hydride generation inductively coupled plasma atomic emission spectrometry after reduction of arsenic (V) to arsenic (III) by thiourea
Barium	
Dimethylsulfonazo-III (DMSA-III)	Spec λ = 662 nm of the chelate complex
Alizarin S red	Red color used spectrophotometrically as a stain to determine the amount of barium in bone
Beryllium	
Aluminon	Lake pigment derivative
Ammonium bifluoride	Derivative detected by fluorescence
Beryllon II	A resin phase spectrophotometric method that detects a change in absorbance of the resin phase immobilized with beryllon II
Beryllon III	Determination achieved via third-derivative spectrophotometry and decolorization of excess reagent
Chrome azurol S	Spec determination with chrome azurol S in the presence of EDTA or CTAB
Eriochrome cyanine R + (CTAB)	Spec λ = 590 nm of ternary complex
Sulfon black F	Derivative has a long color-development time
Bismuth	
Amberlite XAD-7	Determination system implemented with (HG-ICP-AES) associated with flow injection (FI)
Sodium azide	Azidodimethylbismuthine precipitate formed by the reaction of the corresponding bismuthine with sodium azide
Diethyldithiocarbamate	Heterometric micro-determination of lead with sodium diethyldithiocarbamate by use of a mixture of EDTA, cyanide and ammonium hydroxide (λ = 400 nm)
Dithizone	Orange-red derivative that is extracted in carbon tetrachloride
Pyrocatechol violet	Spec determination of bismuth (III) with pyrocatechol violet in the presence of septonex CTAB
Quinolin-8-ol	Determination system implemented with HG-ICP-AES associated with flow injection (FI)
Thiourea	Determination by HG-ICP-AES
Xylenol orange	Derivative used for sol-gel thin films that serve as bismuth (III) sensors
Boron	
Azomethine H	Spec determination
Carminic acid	Spec determination
Curcumin	Spec determination of rosocyanine and rubrocurcumin formed by the reaction between borates and curcumin
Methylene blue	Spec determination of complex formed between fluoroborate ions and methylene blue after treatment with hydrofluoric and sulfuric acids and extraction with ethylene chloride
Cadmium	
2-(5-Bromo-2-pyridylazo)-5-diethylaminophenol (PAR)	Spec determination in the presence of cationic surfactant cetylpyridinium chloride
Cadion	Determination by β-correction spectrophotometry with cadion and surfactant Triton-X
Dithizone	Determination of dithizonate derivative by the extraction – spectrophotometric method
1-(2-Pyridylazo)-2-naphthol (PAN)	Two-dimensional absorption spec determination of complex in aqueous micellar solutions
4–(2-Pyridylazo)resorcinol	Preconcentration by cloud point extraction of the complex; determination by ICP optic emission spectrometry. Simultaneous spectrophotometric determination of cadmium and mercury
Calcium	
Alizarin S	Red dye used in staining bones for calcium determination
Chlorophosphonazo III	Spec λ = 667.5 nm (pH = 2.2)
Eriochrome black T + EDTA	Complexometric titration where the initially formed red calcium-eriochrome black T color is replaced with the blue calcium-EDTA color at the end point
Glyoxal-bis(2-hydroxyanil)	Spectrophotometric titration of the complex without preliminary extraction
Murexide	Spec λ = 506 nm (pH = 11.3)
Phthalein purple	High-performance chelation ion chromatography involving dye-coated resins
Cerium	
Butaperazine dimaleate propericiazine	Spec determination of the colored complex in a phosphoric acid medium
Persulfate oxidation to cerium (IV)	UV spec λ = 320 nm
Propericiazine	Spec determination of the colored complex in a phosphoric acid medium

Reagents	Results
Propionyl promazine phosphate (PPP)	Spec λ = 513 nm of the red-colored radical cation formed upon the reaction of PPP with cerium(IV) in a phosphoric acid medium
N-Benzoyl-*N*-phenylhydroxylamine	Spec titration of the cerium (IV) complex (pH = 8 to 10)
Sodium triphosphate	Specf λ_{ex} = 303.5 nm, λ_{em} = 353 nm of the cerium(III) complex
8-Hydroxyquinoline	Spec determination of the metal-ligand complex
Chromium	
Alizarin S	Lake pigment complex formation
1,5-Diphenylcarbazide	Spec determination during sonication in carbonated aqueous solutions saturated with CCl_4 that produces chlorine radicals
3-(2-Pyridyl)-5,6 bis(5-(2 furyl disulfonic acid))-1,2,4-triazine disodium salt (ferene-TM)	Indirect spec λ = 593 nm in aqueous samples with a chromogen ferene-TM
4-(2-Pyridylazo)resorcinol (PAR)	Spec determination of the ternary chromium-peroxo-PAR ternary complex
4-(2-Pyridylazo)resorcinol (PAR) + hydrogen peroxide	Spec determination of the ternary chromo-peroxo-PAR mixture after ethyl acetate extraction in 0.1 M sulfuric acid
4-(2-Pyridylazo)resorcinol (PAR) + xylometazolonium (XMH) chloride	Spec determination of the orange-red anionic complex formed in a heated acetate buffer medium (pH = 4.0 to 5.5) and extracted with the xylometazolonium (XMH) chloride
Sulfanilic acid	Spec λ = 360 nm on the catalytic effect of chromium (VI) in the oxidation of sulfanilic acid by hydrogen peroxide with *p*-aminobenzoic acid as an activator
Cobalt	
8-Hydroxyquinoline	Spec determination of the metal-ligand complex
p-Nitroso-*N*,*N*-dimethylaniline	Spec determination of the binary complex
Nitroso-R salt	AAS by use of a continuous on-line precipitation-dissolution procedure
1-Nitroso-2-naphthol	Spec determination by use of non-ionic surfactant Triton X-100
1-Nitroso-2-naphthol	Spec Tween 80 micellar determination
1-Nitroso-2-naphthol	AAS by use of a continuous on-line precipitation-dissolution procedure based on 1-nitroso-2-naphthol
2-Nitroso-1-naphthol	Spec λ = 530 nm after isoamyl acetate extraction
1-(2-Pyridylazo)-2-naphthol + surfactants	Spec λ = 620 nm of the cobalt complex in the presence of surfactants (Triton X-100 combined with sodium dodecylbenzene sulfonate (DBS)) and trace of ammonium persulfate (pH = 5.0)
4-(2-Pyridylazo)resorcinol (PAR)	Spec determination of complex at both pH = 7.2 to 7.9 and in 1 M H_2SO_4
4-(2-Pyridylazo)resorcinol (PAR) + triethanolamine	Ion-pair reversed-phase high-performance liquid chromatography of the complex
Copper	
Bathocuproine disulfonic acid	Spec λ = 470 nm to 550 nm of the bathocuproine-disulfonic acid complex after extraction with chloroform and methanol by use of (PLS_2)
Dithizone	Spec determination of the dithizone complex at pH = 2.3
Neocuproine	Spec determination of the deep orange-red $Cu(neocuproine)_2^+$ complex color
Cuprizone	Spec determination of the highly chromogenic copper (III) cuprizone complex
p-Nitroso-*N*,*N*-dimethylaniline	Spec determination of the binary complex
1-Nitroso-2-naphthol	Spec determination by use of non-ionic surfactant Triton X-100
4-(2-Pyridylazo)resorcinol	Preconcentration of copper by cloud point extraction of the complex and determination by ICP optic emission spectrometry
Europium	
1-Nitroso-2-naphthol	Spec Tween 80 micellar determination
ChromAsurol S	Spec determination of the binary complex
4-(2-Pyridylazo)resorcinol (PAR) + tetradecyldimethylbenzylammonium chloride (TDBA)	Spec λ = 510 nm of the ion-associate complex extracted with chloroform at pH = 9.7
Gallium	
Chrome azurol S + (CTAB)	Spec λ = 640 nm of the ternary complex
Haematoxylin or its oxidized form + (CTAB)	Spec determination of the ternary complex of indium with haematoxylin or its oxidized form in the presence of cationic, anionic and non-ionic surfactants such as (CTAB)
Pyrocatechol violet + diphenylguanidine	Spec determination of the ternary complex
8-hydroxyquinoline	Spec determination of the gallium complex
1-(2-Pyridylazo)-2-naphthol	Spectrofluorimetric determination of Ga(III) with 1-(2-pyridylazo)-2-naphthol in sodium dodecyl sulfate micellar medium
4-(2-Pyridylazo)resorcinol (PAR)	Extraction and spec λ = 510 nm of gallium with 4-(2-pyridylazo)resorcinol
Rhodamine B	Comparison of the determination of gallium by a rhodamine B spectrophotometric method and by an AA method based on preliminary solvent extraction
Xylenol orange + 8-hydroxyquinoline	Spectrophotometric determination of the ternary complex
Germanium	
Brilliant Green + Molybdate	Spec λ = 430 nm of the yellow germanomolybdic acid
Phenylfluorone	Spec determination of the complex previously extracted with carbon tetrachloride at pH = 3.1

Reagents	Results
Gold	
5-(4-Diethylaminobenzylidene) rhodanine	Immobilized 5-(4-dimethylamino-benzylidene) rhodanine serves as a stable solid sorbent for trace amounts of gold (III) ions (pH = 2 to 4)
Di(methylheptyl)methyl phosphonate (DMHMP)	Trace gold determination by on-line preconcentration with flow injection atomic absorption spectrometry, by use of di(methylheptyl)methyl phosphonate (DMHMP) as the immobilized phase loaded onto a macroporous resin
2-Mercaptobenzothiazole	Radiochemical separation and determination of gold complex matrices employing substoichiometric thermal neutron activation analysis
Molybdate + nile blue (NB)	A spectrophotometric method based on the reaction of gold(III) with molybdate and nile blue (NB) to form an ion-association complex in the presence of poly(vinyl alcohol)
Rhodamine B	Aqueous spec determination of gold with rhodamine B and surfactant
Hafnium	
Arsenazo III	Spec determination of the arsenazo III complex in 10 M HCl or H_2SO_4
Indium	
Bromopyrogallol red	Spec determination of indium after an ether extraction from hydrobromic acid, and benzyl alcohol extraction of its complex with bromopyrogallol red (pH 9.0)
5-Bromine-salicylaldehyde salicyloylhydrazone (5-Br-SASH)	Specf λ_{ex} = 395 nm, λ_{em} = 461 nm of the indium: 5-Br-SASH chelate in a water–ethanol (63%) medium (pH = 4.6)
Chrome azurol S	Spec determination of the binary complex
Chrome azurol S + benzyldodecyldimethylammonium bromide (BDDMAB)	Spec determination of the mixed complex with chrome azurol S and (BDDMAB)
Chrome azurol S + (CTAB)	Spec λ = 630 nm of the ternary complex after *n*-butyl acetate extraction from hydrobromic acid
Chrome azurol S + cationic surfactants	Spec determination by use of chrome azurol S and surfactants such as CTA, CP, or zephiramine
Dithizone	Spec determination of the binary complex
Haematoxylin or its oxidized form + (CTAB)	Spec determination of the ternary complex of indium with haematoxylin or its oxidized form in the presence of cationic, anionic and non-ionic surfactants such as (CTAB)
8-Hydroxyquinoline	Spec determination of the binary complex
Methylthymol blue + zepheramine	Spec determination of the ternary complex
1-(2-Pyridylazo)-2-naphthol (PAN)	Spec determination of the chelate complex after chloroform extraction (pH = 6)
4-(2-Pyridylazo)resorcinol	Spec λ = 520 nm after indium extraction from the aqueous phase (pH 5.0 to 5.5) into chloroform with *N-p*-chlorophenyl-2-furohydroxamic acid and formation of the 4-(2-pyridylazo)resorcinol red chelate
Pyrocatechol violet	Spec determination of the binary complex
Pyrocatechol violet + tridodecylammonium bromide	Spec determination of the ternary complex
Quinalizarin (1,2,5,8-tetrahydroxyanthraquinone)	Spec λ = 565 nm of the binary 3:1 Quinalizarin:In(III) colored complex in dimethylformamide-water solution
2,2′,3,4-Tetrahydroxy-3′,5′-disulfoazobenzene	Spec determination of the binary system (pH = 5)
Xylenol orange	Spec determination of the ternary complex
Iridium	
1,5-Diphenylcarbazide	Spec determination of the complex (pH = 5.0)
1-(2-Pyridylazo)-2-naphthol	Spec λ = 550 nm of the red complex (pH = 5.1) after chloroform extraction
Iron	
Bathophenanthroline	Determination of iron(II) in the presence of thousand-to-one ratio of iron(III) by use of bathophenanthroline
Bathophenantroline-disulfonic acid	Spec λ = 470 nm to 550 nm of the bathophenantroline-disulfonic acid complex after extraction with chloroform and methanol
2,2′-Bipyridine	Spec determination of the iron (II)-2,2′-bipyridine dark red complex
FerroZine	Spec λ = 562 nm of iron (II)-ferrozine complex after all iron(III) has been reduced by ascorbic acid
Hematoxylin + (CTAB)	Spec of the ternary complex. Addition of CTAB shifts λ_{max} from 630 nm to 640 nm
1-Nitroso-2-naphthol	Spec determination by use of non-ionic surfactant Triton X-100
1-Nitroso-2-naphthol	Spec determination Tween 80 micellar determination
1,10-Phenanthroline (o-Phen)	Spec λ = 508 nm of the $Fe(o\text{-}Phen)_3^{+2}$ complex
1,10-Phenanthroline + bromothymol blue	Spec determination of the $Fe(o\text{-}Phen)_3^{+2}$ complex in the presence of bromothymol blue
Phenylfluorone	Spec λ = 530 nm of the binary complex (pH = 9.0)
Phenylfluorone + Triton X	Spec λ = 555 nm of the binary complex sensitized with Triton X-100 (pH = 9.0)
Lanthanum	
Ammonium purpurate	Spec determination of lanthanum with ammonium purpurate as a chromogenic reagent
Arsenazo III	Spec determination of the lanthanum complexation with reagents of the arsenazo III group on the solid phase of fibrous ion exchangers
Eriochrome cyanine R (ECR)	Spec λ = 540 nm of the binary complex

Reagents	Results
N-Phenylbenzohydroxamic acid + xylenol orange	Solvent extraction followed by the spec λ = 600 nm of the ternary complex (pH = 8.8 to 9.5)
Lead	
Dithizone	Orange-red derivative that is extracted in carbon tetrachloride
Sodium diethyldithiocarbamate	Electrothermal AAS determination of the diethyldithiocarbamate derivative extracted in carbon tetrachloride
Sodium diethyldithiocarbamate	Heterometric micro-determination of lead diethyldithiocarbamate derivative
4-(2-Pyridylazo)resorcinol	Spec λ = 520 nm of 4-(2-pyridylazo)-resorcinol:lead (1:1) complex in an ammonia-ammonium chloride medium at pH = 10 after extracting the lead in isobutyl methyl ketone
Lithium	
1-(*o*-Arsenophenylazo)-2-naphthol-3, 6-disulfonate (Thoron)	Spec λ = 480 nm of Thoron-lithium complex in an alkaline acetone medium is measured against the reagent as reference
Magnesium	
Chlorophosphonazo III	Spec λ = 669 nm (pH = 7.0)
Eriochrome black T + EDTA	Complexometric titration where the initially formed red magnesium-eriochrome black T color is replaced with the blue magnesium-EDTA color at the end point
8-Hydroxyquinoline	Volumetric, titrimetric, and colorimetric determination
8-Hydroxyquinoline + butylamine	Spec λ = 380 nm of the complex after chloroform extraction
Titan yellow	Spec determination of magnesium by titan yellow in biological fluids
Xylidyl blue	Spec determination of magnesium in biological fluids
Manganese	
Formaldoxime	Spec determination of the formaldoxime/ammonia complex (pH = 8.8 to 8.9)
1-(2-Pyridylazo)-2-naphthol (PAN)	Preconcentration determination by use of (PAN) anchored SiO_2 nanoparticles
Mercury	
Dithizone	Spec λ = 500 nm of dithizone complex after chloroform extraction at pH = 0.3
Michler's thioketone	Spec λ = 560 nm of the binary complex in acetate buffer
4-(2-Pyridylazo)resorcinol	Simultaneous spec determination of cadmium and mercury with 4-(2-pyridylazo) resorcinol
Rhodamine 6G	Photoelectrochemical determination of mercury (II) in aqueous solutions by use of a rhodamine 6G derivative (RS) and polyaniline (PANI) coated optical probe in a photoelectrochemical cell
Xylenol orange + amine buffer	Spec λ = 590 nm of Hg(II)/xylenol orange complex display a sharp hyperchromic effect in the presence of amine buffers (pH= 7.5)
Xylenol orange + citric acid-phosphate buffer	Spec λ = 580 nm of Hg(II)/xylenol orange complex display a sharp hypochromic effect upon substituting amine buffers with a citric acid-phosphate (pH = 7.5)
Molybdenum	
Bromopyrogallol red + cetylpyridium chloride	Sequential injection analysis to the determination of CPC based on the sensitized molybdenum-bromopyrogallol red reaction
Phenylfluorone	Spec λ = 560 nm of molybdenum with phenylfluorone (pH of 1.5 to 3)
8-Hydroxyquinoline-5-sulfonic acid and phenylfluorone	Diffuse reflection spectrometry with 8-hydroxyquinoline-5-sulfonic acid and phenylfluorone after sorbing on a disk of an anion exchange fibrous material
Pyrocatechol violet	Preconcentration (by use of basic anion exchanger AV-17-10P) and determination via diffuse reflection spectroscopy. The colored surface compound to be determined involves Mo(VI) sorption on the resin and subsequent treatment of the concentrate obtained with pyrocatechol violet
Neodymium	
Semi-xylenol orange + cetylpyridinium chloride	Fourth-order derivative spectrophotometric determination of the ternary complex
Neptunium	
Arsenazo III	Spec λ after separation by use of thenoyltrifluoroacetone extraction method and determination in 5M HNO_3
Nickel	
2-(5-Bromo-2-pyridylazo)-5-diethylaminophenol	Spec λ = 520 nm and 560 nm of the red-violet complex in water-ethanol (pH = 5.5)
Dimethylglyoxime + ammonia	Spec λ = 543 nm of the complex
Dimethylglyoxime, voltammetry	Nickel voltammetric determination at a chemically modified electrode based on dimethylglyoxime-containing carbon paste
2,2′-Furildioxime	Spec λ = 438 nm after separation by adsorption of its α-furildioxime complex on naphthalene
Hematoxylin	Spec λ = 595 nm of the binary system (pH = 7.8 to 8.3)
Hematoxylin + (CTAB)	Spec λ = 608 nm of the ternary system (pH = 7.4 to 8.1)
1-Nitroso-2-naphthol	Spec Tween 80 micellar determination
p-Nitroso-*N*,*N*-dimethylaniline	Spec determination of the binary complex
2-(2-Pyridylazo)-2-naphthol	Derivative spectrophotometry λ= 569 nm determination of nickel complex in Tween 80 micellar solutions

Analytical

Reagents	Results
4-(2-Pyridylazo)resorcinol	Preconcentration of nickel by cloud point extraction of the complex and determination by ICP optic emission spectrometry
1-(2-Thiazolylazo)-2-naphthol (TAN)	Flow-injection solid phase spectrophotometry by use of TAN immobilized on C_{18}-bonded silica (λ = 595 nm)
Xylenol orange	Spec determination by mean centering of ratio kinetic profile (pH = 5.3)
Niobium	
N-Benzoyl-*N*-phenylhydroxylamine	Separation and determination of the complex from a tartrate solution at pH > 2
Bromopyrogallol red	Spec λ = 610 nm of complex extracted into isopentyl acetate containing di-*n*-octylmethylamine
O-Hydroxyhydroquinonephthalein (Qnph) + hexadecyltrimethylammonium chloride (HTAC)	Spec λ = 520 nm of the complex in strong acidic media
4-(2-Pyridylazo)resorcinol + citrate	Determination of the ternary complex by ion-interaction reversed-phase HPLC on a C_{18} column with a 5 mM citrate buffer (pH = 6.5, λ = 540 nm)
Pyrocatechol violet	Extraction and spec determination of niobium complex in the presence of pyridine and trichloroacetic acid
Sulfochlorophenol S	Spec determination of the complex after extraction into amyl alcohol
Xylenol orange	Spec λ = 530 nm of chelate (pH = 5.0)
Osmium	
1,5-Diphenylcarbazide	Spec λ = 560 nm of the bluish-violet complex after extraction with isobutyl methyl ketone
Thiourea	Spec λ = 540 nm of the red-rose complex in acid medium
Palladium	
Ammonia + iodide	Thermogravimetric determination of palladium as $Pd(NH_3)_2I_2$
2,2- *bis*-[3-(2-Thiazolylazo)-4-4-hydroxyphenyl-propane], (TAPHP)	Application of TAPHP immobilized on silica beads to determine the palladium concentration in a trans-luminance configuration (pH = 2) by use of flow-through spectrophotometric sensing phase
2-(5-Bromo-2-pyridylazo)-5-diethylaminophenol	Spec determination of the complex in a sulfuric acid medium in the presence of ethanol
Dithizone	Volumetric determination of palladium with dithizone in acid medium after extraction of its dimethylglyoxime complex with chloroform; PAS determination of the dithizone extraction solution into a thermally thin solid film
Dithiazone + iodide	Spec determination of palladium with dithizone, by use of an iodide medium in the presence of sulfite at pH = 3 to 5, to separate platinum
Dithizone + stannous chloride	Extractive separation and spec determination of palladium in the presence of stannous chloride to separate platinum
Isonitrosobenzoylacetone	Radiochemical separation and determination of palladium in complex matrices employing substoichiometric thermal NAA
2-Nitroso-1-naphthol	Spec λ = 370 nm of the violet complex
4-(2-Pyridylazo)resorcinol	Spec determination of complex at both pH = 7.2 to 7.9 and in 1 M H_2SO_4
4-(2-Pyridylazo)resorcinol + diphenylguanidine	Spec determination after extraction of the red Pd(II) chelate with 4-(2-pyridylazo) resorcinol in the presence of *N,N′*-diphenylguanidine into *n*-butanol
Platinum	
N-Phenylbenzimidoylthiourea (PBITU)	Flow injection analysis spec λ = 345 nm of the Pd:PBITU complex in (0.2 to 2.0) M HCl in 10% (vol/vol) ethanol solution
Dithizone + iodide	Spec determination of platinum with dithizone, by use of an iodide medium in the presence of sulfite at pH = 3 to 5, to separate palladium
Dithizone + stannous chloride	Extractive separation and spectrophotometric determination of platinum in the presence of stannous chloride to separate palladium
2-Mercaptobenzothiazole	Radiochemical separation and determination of platinum complex matrices employing substoichiometric thermal NAA
Protactinium	
Arsenazo III	Spec λ = 680 nm of the complex after extraction with 7 N H_2SO_4 and isoamyl alcohol
EDTA + tannic acid	Gravimetric analysis of the tannic acid precipitate (pH = 5.0)
Rhenium	
2,2′-Furildioxime	Spec λ = 532 nm of the complex
N,N-Diethyl-*N′*-benzoylthiourea	Spec λ = 383 nm of the green complex in hydrochloric acid medium in the presence of tin(II) chloride
Tin(II)	Spectrophotometric titration for the determination of rhenium by use of tin(II) as the titrant
Rhodium	
1,5-Diphenylcarbazide	Spec determination of the complex after isobutyl alcohol extraction (pH = 5.0)
p-Nitrosodimethylaniline	Spec λ = 510 nm of the cherry-red binary complex (pH = 4.4)
1-(2-Pyridylazo)-2-naphthol	Spec λ = 598 nm of the green complex (pH = 5.1) after chloroform extraction
Ruthenium	
4-Benzylideneamino-3-mercapto-6-methyl-1,2,4-triazine (4H)-5-one	Spec λ = 620 nm

Reagents	Results
1,10-Phenanthroline	Chemiluminescence determination of chlorpheniramine by use of tris(1,10-phenanthroline)-ruthenium(II) peroxydisulfate system and sequential injection analysis; spec determination of the complex after reducing Ru(IV) to Ru(II)
Thiourea	Spec λ = 640 nm of ruthenium complex in carbon supported Pt-Ru-Ge catalyst in 5 M HCl
1,4-Diphenylthiosemicarbazide	Spec determination of the bright red complex
Scandium	
Alizarin red S	Cathodic adsorptive stripping of the scandium-alizarin red S complex onto a carbon paste electrode
Ammonium purpurate	Spec determination of scandium with ammonium purpurate as a chromogenic reagent
Chrome azurol S	Spec determination of the binary complex
Xylenol orange	Spec λ = 553 nm of the binary mixture
Selenium	
3,3′-Diaminobenzidine	Spec λ = 350 nm of the yellow complex (pH = 3.0)
2,3-Diaminonaphthaline	Fluorometric determination of the binary complex in water
Thiourea	Determination of selenium by hydride generation inductively coupled plasma atomic emission spectrometry
Silver	
Dithizone	Graphite-furnace atomic-absorption spectrophotometric determination of silver on suspended dithizone particles from acidic sample solutions with ultrasonics to facilitate the separation
Eosin + 1,10-phenanthroline	Spec λ = 540 nm to 555 nm of the 1,10-phenanthroline (PHEN) and eosin (2,4,5,7-tetrabromofluorescein) association complexes
Strontium	
Phthalein purple	High-performance chelation ion chromatography involving dye-coated resins
Tantalum	
N-Benzoyl-*N*-phenylhydroxylamine	Separation and determination of the complex from a tartrate solution at pH < 1.5
Crystal violet (CV) + *N,N′*-diphenylbenzamidine (DPBA)	Spec λ = 600 nm of Ta(V)-F-CV+ cation complex with a benzene solution of DPBA from sulfuric acid solution
O-Hydroxyhydroquinonephthalein (Qnph) + hexadecyltrimethylammonium chloride (HTAC)	Spec λ = 510 nm of the complex in strong acidic media
Malachite green	Spec λ = 623 nm of the complex in HF after extraction with benzene or toluene
Methyl violet	Spec determination of the complex
4-(2-Pyridylazo)resorcinol + citrate	Determination of the ternary complex by ion-interaction reversed-phase HPLC on a C18 column in 5 mM citrate buffer (pH = 6.5, λ = 540 nm)
Phenylfluorone	Spectrophotometric determination from HF-HCl solution with methyl isobutyl ketone
Victoria blue	Spectrophotometric determination of the complex after benzene extraction
Tellurium	
Ammonium pyrrolidinedithiocarbamate	Differential determination of tellurium(IV) and tellurium(VI) by AAS with a carbon-tube atomizer
Bismuthiol II	Spec λ = 330 mm of the yellow complex in acidic medium (pH = 3.5) after chloroform extraction
Dithizone	Differential determination of tellurium(IV) and tellurium(VI) by AAS with a carbon-tube atomizer
Sodium diethyldithiocarbamate	Differential determination of tellurium(IV) and tellurium(VI) by AAS with a carbon-tube atomizer
Thiourea	Determination of selenium by hydride generation inductively coupled plasma atomic emission spectrometry
Thallium	
Brilliant green	Spec determination of the binary complex after toluene extraction
Dithizone	Spec λ = 505 mm of the complex after chloroform extraction from a citrate-sulfite-cyanide medium at pH = 10.6
8-Hydroxyquinoline	Spectrophotometric determination of the metal-ligand complex
Rhodamine B	Fluorimetric determination of thallium (in silicate rocks) with rhodamine B after separation by adsorption on a crown ether polymer
Rhodamine B hydroxide	Spec λ = 565 nm via oxidation of rhodamine B hydrazide by thallium(I) in acidic medium to give a pinkish-violet radical cation
Thorium	
Arsenazo III	Spec determination of the arsenazo III complex in 10 M HCl or H_2SO_4
Eriochrome cyanine R (ECR)	Spec λ = 540 nm of the binary complex
Thoron	Spectrophotometric determination of binary complex in tartaric acid
Xylenol orange	Spec λ = 570 nm of the binary complex (pH = 4.0)
Xylenol orange + CTAB bromide	Spec λ = 600 nm of the complex sensitized by CTMAB (pH = 2.5)

Reagents	Results
Tin	
Catechol violet + CTAB bromide	Spectrophotometric determination of the green complex
Pyrocatechol violet (and + CTAB bromide)	Spec λ = 660 nm of the complex (pH = 2.0)
Gallein	Spectrophotometric determination of complex in acid medium
Phenylfluorone	Spec determination of the complex (pH = 1) in 36% aqueous ethanol after a preliminary solvent extraction of the tin as the iodide
Toluene-3,4-dithiol + dispersant	Spec determination of the complex
Titanium	
Chromotropic acid	Spec λ = 443 nm of the complex by use of a flow injection manifold
Diantipyrinylmethane	Spec λ = 390 nm of the yellow complex
3-Hydroxy-2-methyl-1-(4-tolyl)-4-pyridone (HY)	Spec λ = 355 nm of the ternary complex formed in perchloric acid and is extracted by chloroform
Methylene blue-ascorbic acid redox reaction	Spec λ = 665 nm of the titanium-methylene blue-ascorbic acid redox reaction
Tiron	Spec λ= 420 nm of the yellow Tiron derivative (pH = 5.2 to 5.6)
Tungsten	
Cyanate	Spectrophotometric determination of tungsten with thiocyanate after both tungsten(VI) and molybdenum(VI) are extracted into chloroform as benzoin α-oxime complexes
Pyrocatechol violet	Determination of the 2:1 green complex in acid medium which is fixed on a dextran-type anion-exchange resin (Sephadex QAEA-25) by first-derivative solid-phase spectrophotometry (λ = 674 nm)
Tetraphenylarsonium chloride + thiocyanate	Gravimetric determination of tungsten with tetraphenylarsonium chloride after its extraction as thiocyanate (pH = 2 to 4)
Toluene-3,5-dithiol	Spec λ = 630 nm of the toluene-3,4-dithiol derivative after extraction with isoamyl acetate
Uranium	
Arsenazo III (1,8-dihydroxynaphthalene-3,6-disulfonic acid-2,7-bis[(azo-2)-phenylarsonic acid])	Spec determination of the binary complex
2-(5-Bromo-2-pyridylazo)diethylaminophenol	Spec λ = 578 nm of the complex (pH = 7.6)
Chlorophosphonazo III	Spec λ = 673 nm of the complex after extraction into 3-methyl-1-butanol from (1.5 to 3.0) M HCl
8-Hydroxyquinoline	Spectrophotometric determination of the metal (UO_2 (II))-ligand complex
2-(2-Pyridylazo)-5-diethylaminophenol (PADAP)	Spec λ = 564 nm of the complex after extraction into methyl isobutyl ketone (pH = 8.2)
1-(2-Pyridylazo)-2-naphthol (PAN)	Spec λ = 560 nm of the deep red precipitate in ammoniacal solutions extracted with chloroform
4-(2-Pyridylazo)-resorsinol (PAR)	Spec λ = 530 nm of the intensely deep red complex after extraction into methyl isobutyl ketone (pH = 8.0)
2-(2-Thiazolylazo)-*p*-cresol + surfactants	Spec λ = 588 nm of the complex (pH = 6.5) with surfactants such as Triton X-100 or N-cetyl-*N,N,N*-trimethyl ammonium bromide (CTAB)
Vanadium	
Chrome azurol S	Spectrophotometric determination of the binary complex
N-Benzoyl-*N*-phenylhydroxylamine	Spec determination of the violet complex in acidic medium
3,5-Dinitrocatechol (DNC) + brilliant green	Extraction-spectrophotometric determination of the system V(V)- 3,5-dinitrocatechol (DNC)-brilliant green chelate complex
8-Hydroxyquinoline (oxine)	Extraction with *n*-butanol and spec λ = 390 nm of a ternary complex (vanadium:oxine:*n*-butanol = 1:2:2)
8-Hydroxyquinoline-5-sulfonic acid and phenylfluorone	Diffuse reflection spectrometry with 8-hydroxyquinoline-5-sulfonic acid and phenylfluorone after sorbing on a disk of an anion exchange fibrous material
4-(2-Pyridylazo)resorcinol (PAR)	Spec determination of the binary complex based on the extraction tetraphenylphosphonium or tetraphenylarsonium chloride
Xylenol orange	Spec λ = 490 nm of chelate (pH = 5.0)
Yttrium	
Alizarin red S	Spec λ = 550 nm of the complex
Ammonium purpurate	Spec determination of yttrium with ammonium purpurate as a chromogenic reagent
Arsenazo III	Spec λ = 660 nm of the blue-colored complex after preliminary purification with hydroxide and subsequent acidification
Eriochrome cyanine R (ECR)	Spec λ = 540 nm of the binary complex
Pyrocatechol violet	Spec λ = 665 nm of the complex
Zinc	
2-(5-Bromo-2-pyridylazo)-5-diethylaminophenol	Spec determination by use of 2-(5-bromo-2-pyridylazo)-5-diethyl aminophenol in the presence of cationic surfactant cetylpyridinium chloride
Carbonic anhydrase	Enzymatic determination by use of carbonic anhydrase after removing zinc by dialysis against dipicolinic acid
Dithizone	Spec λ = 530 nm of binary complex after chloroform extraction
Eriochrome black T + EDTA	Complexometric titration where the initially formed red zinc-eriochrome black T color is replaced with the blue zinc-EDTA color at the end point (pH = 10)

Analytical

Reagents	Results
7-(4-Nitrophenylazo)-8-hydroxyquinoline-5-sulfonic acid (*p*-NIAZOXS)	Spec λ = 520 nm of the zinc derivative (pH = 9.2, borax buffer)
Phenylglyoxal mono(2-pyridyl) hydrazone (PGMPH)	Spec λ = 464 nm to 470 nm of the yellow-orange complex (pH = 7.2 to 8.5) in 40% (v/v) ethanol
1-(2-Pyridylazo)-2-naphthol (PAN)	Two-dimensional absorption spec determination of complex in aqueous micellar solutions
1-(2-Pyridylazo)-2-naphthol (PAN)	Preconcentration determination by use of 1-(2-pyridylazo)-2-naphthol (PAN) anchored. SiO_2 nanoparticles
4-(2-Pyridylazo)resorcinol	Spec determination at pH = 7.0; preconcentration of zinc by cloud point extraction of the complex and determination by ICP optic emission spectrometry
1-(2-Thiazolylazo)-2-naphthol (TAN)	Flow-injection solid phase spec λ = 595 nm with TAN immobilized on C18-bonded silica
Xylenol orange	Spec determination of 1:1 zinc-xylenol orange red-violet complex (pH= 5.8 to 6.2)
Xylenol orange	Sequential injection analysis (SIA) based on the spec λ = 568 nm of zinc by use of xylenol orange as a color reagent
Xylenol orange	Spec determination by mean centering of ratio kinetic profile (pH = 5.3)
Xylenol orange + cetylpyridinium chloride	Spec λ = 580 nm of the 1:2:4 ratio for the metal:ligand:surfactant ternary complex (pH = 5.0 to 6.0)
Zirconium	
Alizarin red S	Spec determination of the binary complex after phosphate extraction
Arsenazo III (1,8-dihydroxynaphthalene-3,6-disulfonic acid-2,7-bis[(azo-2)-phenylarsonic acid])	Spec determination of the binary complex
N-p-Chlorophenylbenzohydroxamic acid (*N-p*-Cl-BHA) + morin	Spec λ = 420 nm and fluorimetric determination of the *N-p*-Cl-BHA greenish-yellow complex after extraction with isoamyl alcohol from (0.2 to 0.5) N H_2SO_4 followed by morin addition; it is critical that the morin be very pure
Pyrocatechol violet	Spec λ = 650 nm of the blue complex in sulfuric acid solution
Pyrocatechol violet + tri-*n*-octylphosphine oxide (TOPO)	Spec λ = 655 nm of the blue complex after extraction with tri-*n*-octylphosphine oxide (TOPO) in cyclohexane
Morin	Fluorimetric determination of the complex after EDTA addition
Xylenol orange	Flow injection spectrophotometric determination in sulfuric acid medium

II. Organic Analytical Reagents for the Determination of Anions

Method	Results/Comments
Arsenate, AsO_4^{-3}	
X-ray absorption near edge structure (XANES)	Use of self-consistent multiple-scattering methods that measure the AsO_4^{-3} adsorption on TiO_2
Dual detection ion chromatography	Simultaneous determination of AsO_4^{-3} and AsO_3^{-3} (LOD = 0.044 mg/L)
Ion chromatography/inductively-coupled plasma/MS	Validation established after ultrasonic-assisted enzymatic extraction
Ion chromatography	Post-column generation and detection of the AsO_4^{-3} molybdate heteropoly ion
Surface-enhanced Raman scattering (SERS)	Use of Ag nanostructured multilayer films; LOD = 5 μg/L
Liquid chromatography-atomic fluorescence detection	Simultaneous determination of six As ions; LOD < 4 μg/L
LC-ICP/MS	Simultaneous determination of five As ions; LOD < 4 μg/L
HPLC/hydride generation atomic absorption	Simultaneous determination of six As ions; LOD = 3 ng to 9 ng
Ion exchange determination	Separation of AsO_4^{-3} and AsO_3^{-3} using the weakly basic anion exchanger A63-X4A by difference of respective pK_a values
Titration	As^{+5} is reduced by NH_2NH_2 and titration with Br^-/BrO_3^-
Bromide, Br^-	
Ion chromatography using electrochemical detection	Use of Ag working electrode; LOD = 10 ppb (mass/mass)
Carbonate, CO_3^{-2}	
GC	Carbonation of salt and use of a GC with methanation reformer and a flame ionization detector
GC-MS	Determination of pentafluorobenzyl derivative
FTIR	CO_3^{-2} measurement in organic matter (875 cm^{-1} and 712 cm^{-1})
Gas measurement	Conversion to CO_2 measured by a barometer or manometer
Ion chromatography	Measurement in the presence of formyl ions
Photoacoustic IR spectroscopy	Use of several multi-calibration methods typically in sedimentary rocks
Thermal optical analysis	Use of semi-continuous carbon elemental analyzer
Liquid chromatography	Use of a C18 HPLC column, MeOH/AcOH mobile phase and UV detection (λ = 290 nm)
Raman spectroscopy	Use of OH stretch band of H_2O as internal standard
Thermogravimetry	CO_3^{-2} measurement in organic matter after identification of evolved gaseous species
Chlorate, ClO_3^-	
Flow injection analysis	Reduction of methylene blue's leuco form to methylene blue followed by optical detection in the visible region
Spectrophotometry	Reduction to Cl_2 via Tl^{+3} which oxidizes the leuco form of methylene

Analytical

Method	Results/Comments
FTIR	Absorption band at 973 cm^{-1} (λ = 10 277 nm)
Anodic electrochemical dissociation	Determination made during the anodic dissociation of Ti grade VT-1
Potentiometric titration	Reduction to Cl^- (pH = 7) using As^{+3} (OsO_4 as catalyst)
Titrimetric analysis	ClO_3^- converted to Cl^- which is titrated against $Hg(NO_3)_2$ (diphenylcarbazone as indicator)
Titrimetric analysis	Iodometric titration
Titrimetric analysis	Use of methyl orange (LOD = 10^{-2} μg, mass/mass)
Chromate, CrO_4^{-2}	
Potentiometric titration	Selective electrode based on the N,N-butylenebis (salicylideneiminato)copper(II); LOD = 3.0×10^{-6} mol/L
Spectrophotometry	Determination by oxidation of ascorbic acid (λ = 548 nm)
Ion exchange chromatography	Diaminetetraacetic acid (DCTA)/CH_3CN used as eluent
Ion pair chromatography	Tetrabutylammonium hydroxide (TBAH)/aq. CH_3CN used as eluent
Capillary electrophoresis	KH_2PO_4 buffer (pH = 7)
Titration	Iodometric titration
X-ray absorption near edge structure (XANES)	LOD = 10 ppm using a 150 μm synchrotron X-ray beam
Cyanide, CN^-	
Headspace GC/MS	Use of gas chromatography with a silica PLOT column coupled with mass selective detection
Headspace GC/MS	Use of gas chromatography with a nitrogen-phosphorus detector; LOD = 0.05 μg/mL to 5 μg/mL
Headspace GC/MS	Use of headspace solid-phase microextraction (SPME); LOD = 0.006 μg/mL
Headspace GC/MS	Determined as the pentafluorobenzylbromide derivative in the presence of TDMBC
Headspace GC	Use of GC with thermoionic detection (NPD)
Headspace GC/MS	Use of GC with the flame ionization detector (FID)
Headspace GC/MS	Determination after reaction with chloramine-T
	Derivatization by the pyridine-pyazolone method
Ion chromatography using electrochemical detection	Use of Ag working electrode; LOD = 1 μg/L
Cyclic voltammetry	Au electrode; LOD = 10^{-5} mol/L to 10^{-6} mol/L
Differential cathodic stripping voltammetry	LOD = 2×10^{-7} M in the presence of 5×10^{-5} M [S^{-2}] and 1×10^{-3} M [Cu^{+2}]
Differential electrolytic potentiometry (DEP)	DEP coupled with the sequential injection/flow analysis (SIA/FIA) of aqueous solutions; con. range 6 μg/mL to 39 μg/mL; LOD = 0.31 μg/mL; RSD = 2.55
Isotachophoresis	Determination as [$Au(CN)_2^-$], [$Ag(CN)_2^-$], or [$Cu(CN)_3^{-2}$] complexes
Potentiometric determination	Use of PVC sensors responsive to $Ag(CN)_2^-$
Spectrofluorometry	Use of naphthoquinone imidazole boronic-based sensors (M-NQB and p-NQB) with CTAB (spec λ = 460 nm)
Fluorescence	Derivatization with naphthalene 2,3-dicarboxyaldehyde and taurine (specf λ_{ex} = 418 nm, λ_{em} = 460 nm); LOD = 0.002 μg/mL to 0.025 μg/mL
Argentometric method	Argentometric titration/quantification using KCNS ($FeNH_4SO_4$) as indicator
Capillary electrophoresis/UV spectrophotometry	Conversion of CN^- to SCN^- via rhodanese, separated by capillary electrophoresis and measured by UV spectrophotometry (spec λ = 200 nm); LOD = 3 μmol
Piezoelectric quartz crystal (PQC) sensors	Method makes use of silver coated sensors; LOD = 1.25×10^{-8} mol/L
Visible spectrophotometry colorimetric assay	Picrate/resorcinol derivative in bicarbonate (spec λ = 488 nm)
Visible spectrophotometry colorimetric assay	Lithium/sodium picrate derivative (spec λ = 500 nm to 520 nm); LOD = 2.5 μg/100 mg sample
Visible spectrophotometry colorimetric assay	Isonicotinic acid-barbituric acid derivative after treatment with chloramine-T (spec λ = 600 nm)
Visible spectrophotometry colorimetric assay	4-hydroxy-3-(2-oxoindoline-3-ylideneamino)-2-thioxo-2H-1,3-thiazin-5(3H)-one) HOTT derivative (spec λ = 466 nm)
Visible spectrophotometry colorimetric assay	Ethanol extraction of the red complex formed on picric acid test strips (spec λ = 500 nm); LOD = 0.5 μg to 1.0 μg
Visible spectrophotometry colorimetric assay	Derivatization with *p*-nitrobenzaldehyde followed by reaction with o-dinitrobenzene to yield a purple color; LOD = 10 μg/L
Visible spectrophotometry colorimetric assay	Barbituric or N,N-dimethylbarbituric acid derivative
Visible spectrophotometry colorimetric assay	Spectrophotometric determination by use of a modified Konig reaction with NaOCl as the chlorinating agent
Visible spectrophotometry colorimetric assay	Complexation with pyridine or 4-pyridinecarboxylic acid
Visible spectrophotometry colorimetric assay	Reaction of CN^- with Pd-dimethylglyoximate complex that releases dimethylglyoxime, isolated as the Ni derivative
Visible spectrophotometry colorimetric assay	Determination (spec λ = 570 nm) of pyridine/barbituric acid derivative after chloramine-T derivative
Visible spectrophotometry colorimetric assay	Measurement of the pyridine-pyrazolone derivative

Method	Results/Comments
Visible spectrophotometry colorimetric assay	Measurement of the isocotinic acid/1-phenyl-3-methyl-5-pyrazolone derivative after reaction with chloramine-T (spec λ = 638 nm)
Visible spectrophotometry colorimetric assay	Measurement of the derivative with the Hg(II)-*p*-dimethylaminobenzylidenerhodanine complex (spec λ = 452 nm)
Visible spectrophotometry colorimetric assay	Measurement of the Hg(II)-diphenylcarbazone derivative (spec λ = 562 nm)
Visible spectrophotometry colorimetric assay	Measurement of the pyridine-benzidine complex; LOD = 0.02 μg
Visible spectrophotometry colorimetric assay	Measurement of the pentacyano(N-methylpyrazinium) ferrate (II) complex (spec λ = 658 nm)
Flow injection analysis	Use of Ag and Pt electrodes to determine CN^- as Cu complexes
Flow injection analysis	Analyzed as the teracyanonickelate (II) complex
Fluorescent sensors	Fluresecein derivative emits a green light
Fluorescent sensors	Naphthoquinone-imidazole boronic-based sensors (spec λ = 460 nm); LOD = 1.4 μM
FTIR	Use of a horizontal attenuated total reflectance (HATR) accessory; detection range = 0.3 M to 1.0 M
Chemiluminescence	Measurement of intensity of the reaction with luminal, hemin, and peroxide
Amperometry/flow injection analysis	Use of a multimeter as a detection device; LOD = 200 μg/L
Photoacoustic IR detection	Applied to the determination of HCN and NCCl in air
X-ray diffraction	LOD = 0.1% in waste-water analyses
Coulometric determination	LOD = 10^{-9} mg/100 μL
Cavity ringdown spectroscopy	Use of external cavity tunable diode laser
CN^-/SCN^- (only)	
Chemical ionization GC/MS	Derivatization with pentafluorobenzyl bromide (PFBBr) via extraction with ethyl acetate and tetrabutylammonium sulfate (TBAS); LOD = 1 μM (CN^-) and LOD < 50 nM (SCN^-)
Dichromate, $Cr_2O_7^{-2}$	
FTIR	Absorption band at 950 cm^{-1}
Paper electrophoresis-photo density scanning	Photo density scanning determination after development with diphenyl carbazide
Cell voltage variation	Applied to the electrochemical synthesis of sodium dichromate from sodium chromate
Conductimetric determination	Performed in the presence of CrO_4^{-2}
Adsorptive stripping voltamemtry	Commonly applied to body fluids
Potentiometric titration	Determination of Pu via the $Fe^{+2}/Cr_2O_7^{-2}$ redox titration
Sequential injection analysis	$Cr_2O_7^{-2}$ determination based on oxidation of pindolol (spec λ = 640 nm)
Spectrophotometric analysis	Applied to the determination of pyridine derivatives
	Reduction to Cr^{+3} in organic carbon soil analysis
Titrimetric analysis	Reduction by Fe^{+3} titration
Thin-layer chromatography (TLC)	Applied to determination of unsaturated fats and further spectrophotometric analysis (spec λ = 350 nm)
X-ray absorption fine structure spectroscopy	Applied to the speciation of $CrO_4^{-2}/Cr_2O_7^{-2}$ equilibrium
Hypochlorite, ClO^-	
Titrimetric determination	Iodometric titration
Polarography	Determination by conversion of Br^- to BrO_3^-
Spectrophotometry	Oxidation of Ir^{+3} to Ir^{+4} (spec λ = 488 nm)
Spectrophotometry	Measurement of ClO^- absorbance at pH > 7
Spectrophotometry	Measurement of chloramine formed by addition of aq. NH_3
Voltammetry	Technique compared to iodometric titration
Potentiometric titration	ClO^- measured by subtraction of BrO^- in a ClO/BrO^-
Potentiometric titration	Reduction to Cl^- (pH < 7, OsO_4 as catalyst)
Iodide, I^-	
Ion chromatography using electrochemical detection	Use of Ag working electrode; LOD = 100 ppb
Isocyanates, NCO^-	
Mass spectrometry	Use of a proton transfer reaction (PTR), time-of-flight mass spectrometer and high sensitivity PTR quadruple mass spectrometer
FTIR	Use of sealed liquid cell for polymer analysis
HPLC	Analysis of ethanol derivative (spec λ = 245 nm)
Capillary electrophoresis	Use of tris(2,2'-bipyridine)ruthenium (II) by electrochemiluminescence
Potentiometric titration	Titration of carboxylic acid hydrazide derivative with $NaNO_2$
Spectrophotometric analysis	Conversion of CH_3NCO to CH_3NH_2 and reaction with 1 fluoro-2,4-dinitrobenzene to yield $CH_3NHC_6H_3(NO_2)_2$ (spec λ = 352 nm)
Manganate, MnO_4^{-2}	
Titrimetric analysis	Oxidation of As^{+3}, Cr^{+3}, Sb^{+3}, Te^{+4}, PO_3^{-3}, various carboxylic acids, trimethylammoniated compounds
Gravimetric analysis	Precipitation of $BaMnO_4$
Potentiometric titration	Use of a Pt electrode

Method	Results/Comments
Voltammetry	Use of glassy carbon working electrode
Spectrophotometry	LOD = 0.185 μg/mL; (spec λ = 635 nm)
Anodic oxidation	Rotating disk electrode
Gravimetric analysis	Precipitation of $BaMnO_4$
Nitrite, NO_2^-	
Ion chromatography	Use of high capacity, OH^--selective, anion exchange column with a guard column) in simultaneous determination of BrO_3^-; LOD = 0.05 μg/L; RSD = 3.61 % to 4.77 %
HPLC	Simultaneous determination of NO_3^-, SO_4^{-2}, and SCN^-; samples are ultrasonically extracted with $EtOH/CCl_3COOH$ and purified by C18 SPE column (highly retentive alkyl-bonded phase); LOD = 0.01
HPLC	KOH or NaOH aq. solution used as mobile phase for 10 μL to 25 μL sample size
Quantitative sequential injection analysis (SIA)	LOD = 0.002 mg/L
Diffuse UV-visible reflectance method	Reaction of nitrite with sulphadiazine and *o*-naphthol; LOD = 0.02 mg/L; RSD = 5% to 8%
Cyclic voltammetry (CV)/pulse voltammetry	Electropolymerization of Nile Blue at pre-polarized glassy carbon electrode (phosphate buffer, pH = 7.1) leading to a NO_2^- sensor
Cyclic voltammetry (CV)/pulse voltammetry	Use of a solid-state microelectrode formed through the electrochemical co-deposition of Pt-Fe nanoparticles on Au microelectrode
Cyclic voltammetry (CV)/pulse voltammetry	Use of a sol-gel electrode using 3-aminopropyltrimethoxy silane for covalent immobilization of toluidine blue
Capillary electrophoresis	Simultaneous determination of NO_3^- and $C_2O_4^{-2}$
Capillary zone electrophoresis	Simultaneous determination of NO_3^- and Cl^- by use of a photoiodie-array detector (PDA)
UV-visible spectrophotometry	Derivatization with $Zn(OH)_2$/basic fuchsin
UV-visible spectrophotometry	Derivatization with Griess reagent (spec λ = 537 nm); RSD ~1.5 %
Microfluidic gradient elution moving boundary electrophoresis (GEMBE)	Simultaneous determination of NO_3^- and NH_4^+
Microfluidic gradient elution moving boundary electrophoresis (GEMBE)	Simultaneous determination of NO_3^-, Cl^-, $C_2O_4^{-2}$, NO_3^-, and SO_4^{-2}
Gasometric analysis	The NO_2^- concentration is quantified by the formation of N_2(g) produced *vis a vis* its reaction of with sulfamic acid
Seeping layer test paper	Derivatization with *p*-aminobenzenesulfonic acid/*N*-(1-naphthyl) ethylenediamine dihydrochloride
NO_2^- simultaneous NO_2^-/NO_3^- (only)	
HPLC	Pre-column derivatization with Griess reagents (spec λ = 540 nm)
UV-visible spectrophotometry	Following derivatization with HCl/naphthyl ethylenediamine
	LOD = 0.5 mg to 18 mg nitrogen per L (NO_3^-) and 0.5 mg to 5 mg nitrogen per L (NO_2^-)
HPLC-UV	LOD = 2.5 mg/kg (NO_3^-) and 5.0 mg/kg (NO_2^-)
GC-mass spectrometry	Quantification of ^{15}N-labeled nitrate and nitrite biological fluids
Chemiluminescence	LOD = 10 nmol/L; RSD = 3 % (200 nmol/L) to 5% (400 nmol/L)
Oxalate, $C_2O_4^{-2}$	
Capillary electrophoresis	Use of acid run buffer to revert electroosmotic flow
Capillary gas chromatography	LOD = 500 μg by use of a flame ionization detector
Chromogenic strip	Test strip equipped with sorghum leaf oxalate oxidase cross-linked with glutaraldehyde and *o*-toluidine
FT-IR (attenuated)	LOD = 0.07% of the matrix
Capillary electrophoresis	Use of acid run buffer to revert electroosmotic flow
Capillary gas chromatography	LOD = 500 μg by use of a flame ionization detector
Flow injection analysis	Oxalate oxidase and peroxidase are immobilized on Au electrode
Gas chromatography	Analysis of methylated product
Ion exchange chromatography method	LOD = 6 μg/L to 9 μg/L
Isotachophoresis	Detection at λ = 254 nm; LOD = 20 nmol/L
LC-tandem MS	System attached to online weak anion exchange chromatography
Raman spectroscopy	Intensity of peak at 1462 cm^{-1} is used
Spectrophotometric analysis	Analysis of catalytic effect of reaction products of methylene blue/$Cr_2O_7^{-2}$
Spectrophotometric analysis	Analysis of the NaDH ⊠ NAD^+ reaction with $C_2O_4^{-2}$
Titrimetric analysis	Titration against MnO_4^-
Titrimetric analysis	Titration against Ce^{+4}
Perchlorate, ClO_4^-	
Spectrophotometric analysis	Analysis of catalytic effect of reaction products of methylene blue/$Cr_2O_7^{-2}$
Spectrophotometric analysis	Analysis of the NaDH ⊠ NAD^+ reaction with $C_2O_4^{-2}$
Titrimetric analysis	Titration against MnO_4^-
Titrimetric analysis	Titration against Ce^{+4}
Spectrophotometry	Yb-pyrocatechol violet complex

Method	Results/Comments
Spectrophotometry	Determination of reddish-violet color produced by adding methylene blue/$ZnSO_4$ in presence of NO_3^-
HPLC/MS	LOD = 0.2 ppb
Capillary electrophoresis	Analysis typically performed on tobacco plants
MS	Determination of organic base-phosphate complex
Ion pair extraction-MS ESI	Use of cationic surfactants
Titrimetric analysis	Titration against H-tetramethylammonium sulfate with various indicators
Titrimetric analysis	Back titration of excess $TiCl_3$ against $Fe_2(SO_3)_2$
Phosphate, PO_4^{-3}	
FTIR	Use of 1037 cm^{-1} ($Ca_3(PO_4)_2$) and 1010 cm^{-1} ($MgNH_4PO_4$)
Phosphide, P^{-3}	
Gravimetric	Derivatization as Ag_3PO_3 after formation of Ag_3P and further oxidation with HNO_3
ESR	Derivatization to GaP
X-ray electron spectrometry	Derivatization to AlP
X-ray fluorescence	Derivatization to AlP
X-ray diffractometry	Derivatization to AlP
X-ray photoelectron spectroscopy	Derivatization to functionalized GaP
Fluorescence	Derivatization to Ca_3P_2
Phosphite, PO_3^{-3}	
Fluorescence	Enzymatic oxidation to phosphate with NAD^+ as co-substrate to produce fluorescence
^{31}P NMR	Estherification of OH^- in organic polymers
Laser Raman spectrometry	Ar-ion laser allows quantitative analysis of $H_2PO_3^-$ and HPO_3^{-2}
Ion exchange chromatography	Method makes use of conductivity detection
Visible spectrophotometry	Measurement of 3,5-$(O_2N)_2C_6H_3COOH$ derivative
Silcate, SiO_4^{-4}	
Raman spectroscopy	Use of *ab initio* quantum chemical simulation
Electron probe microanalysis (EPMA)	Analysis of Na, Mg, Al, Si, K, Ca, Ti, Mn, and Fe salts
Ion-exclusion chromatography	Use of ascorbate solution as both an eluent and a reducing agent following by molybdate derivatization
Laser-induced plasma spectroscopy	Use of a Q switched Nd:YAG laser and time-resolved spectroscopy
Time-of-flight secondary ion mass spectrometry	Use of C60 primary ions
Laser ablation-inductively coupled plasma mass spectrometry (LA-ICPMS)	No need for internal standard
Sulfide, S^{-2}	
Ion chromatography using electrochemical detection	Use of Ag working electrode; LOD = 30 ppb
Sulfite, SO_3^{-2}	
Amperometric titration	Use of iodine solution
Potentiometric titration	Pt wire coated with Hg is balanced against calomel electrode
Spectrophotometric analysis	Product of reaction with *p*-rosaniline acid bleached dye and HCHO (spec λ = 560 nm)
Ion chromatography	
Titration analysis	Titration against BrO_3^{-2}
Coulometric determination	LOD = 10^{-11} mg/100 μL
Thiocyanate, CNO^-	
Flow injection analysis	Formation of the 2-(5-bromo-2-pyridylazo)-5-diethylaminophenol derivative in $Cr_2O_7^{-2}/H^+$
Flow injection analysis	Analyzed as the teracyanonickelate (II) complex

PRECIPITATION OF SULFIDES

Robert Joyce and Blaine C. McKusick

TABLE 1. Precipitation of Sulfides

Precipitated at pH 7	Not precipitated at low pH	Forms a soluble complex at high pH
Ag^{+}		
As^{3+}[a]		X
Au^{+}[a]		X
Bi^{3+}		
Cd^{2+}		
Co^{2+}	X	
Cr^{3+}[a]		
Cu^{2+}		
Fe^{2+}[a]	X	
Ge^{2+}		X
Hg^{2+}		X
In^{3+}	X	
Ir^{4+}		X
Mn^{2+}[a]	X	
Mo^{3+}		X
Ni^{2+}	X	
Os^{4+}		
Pb^{2+}		
Pd^{2+}[a]		
Pt^{2+}[a]		X
Re^{4+}		
Rh^{2+}[a]		
Ru^{4+}		
Sb^{3+}[a]		X
Se^{2+}		X
Sn^{2+}		X
Te^{4+}		X
Tl^{+}[a]	X	
V^{4+}[a]		
Zn^{2+}	X	

[a] Higher oxidation states of this ion are reduced by sulfide ion and precipitated as this sulfide.

pH RANGE FOR PRECIPITATION OF METAL HYDROXIDES AND OXIDES

Robert Joyce and Blaine C. McKusick

TABLE 1. pH Range for Precipitation of Metal Hydroxides and Oxides

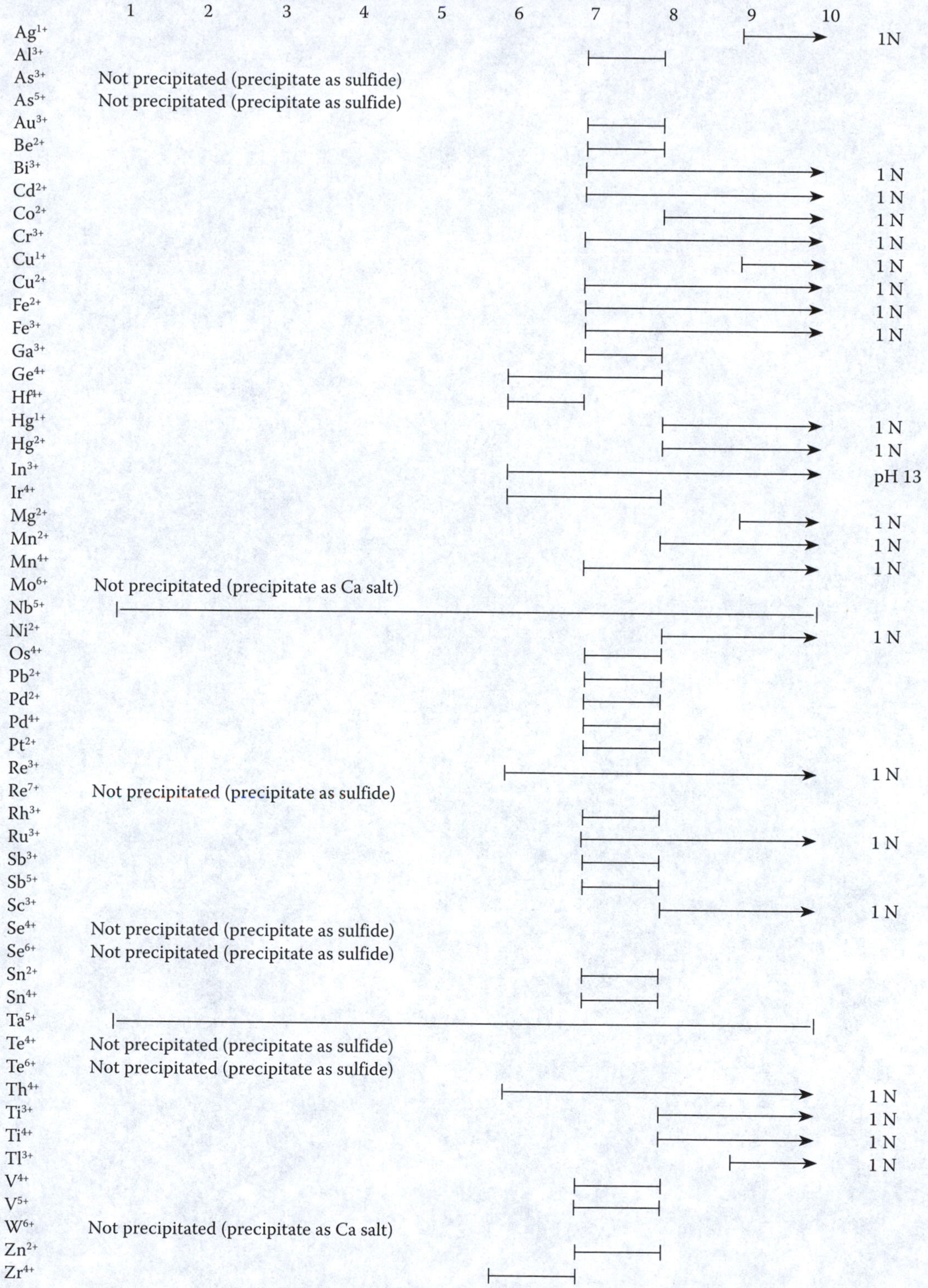

References

L. Erdey, *Gravimetric Analysis*, Part II, Pergamon Press, New York, 1965.
D. T. Burns, A. Towsend, and A. H. Carter, *Inorganic Reaction Chemistry*, Vol. 2, Ellis Horwood, New York, 1981.

Section 9
Molecular Structure and Spectroscopy

BOND LENGTHS IN CRYSTALLINE ORGANIC COMPOUNDS

The following table gives average interatomic distances for bonds between the elements H, B, C, N, O, F, Si, P, S, Cl, As, Se, Br, Te, and I as determined from x-ray and neutron diffraction measurements on organic crystals. The table has been derived from an analysis of high-precision structure data on about 10,000 crystals contained in the 1985 version of the Cambridge Structural Database, which is maintained by the Cambridge Crystallographic Data Center. The explanation of the columns is:

Column 1: Specification of elements in the bond, with coordination number given in parentheses, and bond type (single, double, etc.). For carbon, the hybridization state is given.
Column 2: Substructure in which the bond is found. The target bond is set in boldface. Where X is not specified, it denotes any element type. C# indicates any sp^3 carbon atom, and C* denotes an sp^3 carbon whose bonds, in addition to those specified in the linear formulation, are to C and H atoms only.
Column 3: d is the unweighted mean in Å units of all the values for that bond length found in the sample.
Column 4: m is the median in Å units of all values.
Column 5: σ is the standard deviation in the sample.
Column 6: q_1 is the lower quartile for the sample (i.e., 25% of values are less than q_1 and 75% exceed it).
Column 7: q_u is the upper quartile for the sample.
Column 8: n is number of observations in the sample.
Column 9: Notes refer to the footnotes in Appendix 1.

References to special cases are given in a shorthand form and listed in Appendix 2. Further information on the method of analysis of the data may be found in the reference cited below.

The table is reprinted with permission of the authors, the Royal Society of Chemistry, and the International Union of Crystallography.

Reference

Frank H. Allen, Olga Kennard, David G. Watson, Lee Brammer, A. Guy Orpen, and Robin Taylor, *J. Chem. Soc. Perkin Trans.* II, S1–S19, 1987.

Bond	Substructure	d	m	σ	q_1	q_u	n	Note
As(3)–As(3)	X$_2$–**As–As**–X$_2$	2.459	2.457	0.011	2.456	2.466	8	
As–B	see CUDLOC (2.065), CUDLUI (2.041)							
As–BR	see CODDEE, CODDII (2.346–3.203)							
As(4)–C	**X$_3$–As–CH$_3$**	1.903	1.907	0.016	1.893	1.916	12	
	(X)$_2$(C,O,S=)**As–C*sp*^{3}**	1.927	1.929	0.017	1.921	1.937	16	
	As–C*ar* in Ph$_4$As$^+$	1.905	1.909	0.012	1.897	1.912	108	
	(X)$_2$(C,O,S=)**As–C*ar***	1.922	1.927	0.016	1.908	1.934	36	
As(3)–C	X$_2$–**As–C*sp*^{3}**	1.963	1.965	0.017	1.948	1.978	6	
	X$_2$–**As–C*ar***	1.956	1.956	0.015	1.944	1.964	41	
As(3)–Cl	X$_2$–**As–Cl**	2.268	2.256	0.039	2.247	2.281	10	
As(6)–F	in **AsF$_6^-$**	1.678	1.676	0.020	1.659	1.695	36	
As(3)–I	see OPIMAS (2.579, 2.590)							
As(3)–N(3)	X$_2$–**As–N**–X$_2$	1.858	1.858	0.029	1.839	1.873	19	
As(4)=N(2)	see TPASSN (1.837)							
As(4)–O	(X)$_2$(O=)**As–OH**	1.710	1.712	0.017	1.695	1.726	6	
As(3)–O	see ASAZOC, PHASOC01 (1.787–1.845)							
As(4)=O	X$_3$–**As=O**	1.661	1.661	0.016	1.652	1.667	9	
As(3)=P(3)	see BELNIP (2.350, 2.362)							†
As(3)–P(3)	see BUTHAZ10 (2.124)							†
As(3)–S	X$_2$–**As–S**	2.275	2.266	0.032	2.247	2.298	14	
As(4)=S	X$_3$–**As=S**	2.083	2.082	0.004	2.080	2.086	9	
As(3)–Se(2)	see COSDIX, ESEARS (2.355–2.401)							†
As(3)–Si(4)	see BICGEZ, MESIAD (2.351–2.365)							†
As(3)–Te(2)	see ETEARS (2.571, 2.576)							†
B(n)–B(n)	n = 5–7 in boron cages	1.775	1.773	0.031	1.763	1.786	688	
B(4)–B(4)	see CETTAW (2.041)							
B(4)–B(3)	see COFVOI (1.698)							
B(3)–B(3)	X$_2$–**B–B**–X$_2$	1.701	1.700	0.014	1.691	1.712	8	
B(6)–BR		1.967	1.971	0.014	1.954	1.979	7	†
B(4)–BR		2.017	2.008	0.031	1.990	2.044	15	†
B(n)–C	n = 5–7: **B–C** in cages	1.716	1.717	0.020	1.707	1.728	96	
	n = 3–4: **B–C*sp*^{3}** not cages	1.597	1.599	0.022	1.585	1.611	29	1
	n = 4: **B–C*ar***	1.606	1.607	0.012	1.596	1.615	41	
	n = 4: **B–C*ar*** in Ph$_4$B$^-$	1.643	1.643	0.006	1.641	1.645	16	
B(n)–C	n = 3: **B–C*ar***	1.556	1.552	0.015	1.546	1.566	24	
B(n)–Cl	B(5)–**Cl** and B(3)–**Cl**	1.751	1.751	0.011	1.743	1.761	14	

Molecular

Bond	Substructure	d	m	σ	q_1	q_u	n	Note
	B(4)–Cl	1.833	1.833	0.013	1.821	1.843	22	
B(4)–F	**B–F** (B neutral)	1.366	1.368	0.017	1.356	1.375	25	
	B$^-$–F in BF_4^-	1.365	1.372	0.029	1.352	1.390	84	
B(4)–I	see TMPBTI (2.220, 2.253)							
B(4)–N(3)	X_3–**B–N**(=C)(X)	1.611	1.617	0.013	1.601	1.625	8	
	in pyrazaboles	1.549	1.552	0.015	1.536	1.560	10	
B(3)–N(3)	X_2–**B–N**–C_2: all coplanar	1.404	1.404	0.014	1.389	1.408	40	2
	for τ(BN) > 30° see BOGSUL, BUSHAY, CILRUK (1.434–1.530)							
	S_2–**B–N**–X_2	1.447	1.443	0.013	1.435	1.470	14	
B(4)–O	**B$^-$–O** in BO_4^-	1.468	1.468	0.022	1.453	1.479	24	
	for neutral B–O see Note 3							3
B(3)–O(2)	X_2–**B–O**–X	1.367	1.367	0.024	1.349	1.382	35	
B(*n*)–P	*n* = 4: **B–P**	1.922	1.927	0.027	1.900	1.954	10	
	n = 3: see BUPSIB10 (1.892, 1.893)							
B(4)–S	**B(4)–S(3)**	1.930	1.927	0.009	1.925	1.934	10	
	B(4)–S(2)	1.896	1.896	0.004	1.893	1.899	6	
B(3)–S	N–B–S_2	1.806	1.806	0.010	1.799	1.816	28	
	(=X–)(N–)**B–S**	1.851	1.854	0.013	1.842	1.859	10	
Br–Br	see BEPZEB, TPASTB	2.542	2.548	0.015	2.526	2.551	4	
Br–C	**Br–C***	1.966	1.967	0.029	1.951	1.983	100	4
	Br–C*sp*^{3} (cyclopropane)	1.910	1.910	0.010	1.900	1.914	8	
	Br–C*sp*^{2}	1.883	1.881	0.015	1.874	1.894	31	4
	Br–C*ar* (mono-Br +*m*.*p*-Br_2)	1.899	1.899	0.012	1.892	1.906	119	4
	Br–C*ar* (o-Br_2)	1.875	1.872	0.011	1.864	1.884	8	4
$^-$Br(2)–Cl	see TEACBR (2.362–2.402)							†
Br–I	see DTHIBR10 (2.646), TPHOSI (2.695)							
Br–N	see NBBZAM (1.843)							
Br–O	see CIYFOF	1.581	1.581	0.007	1.574	1.587	4	
Br–P	see CISTED (2.366)							
Br–S(2)	see BEMLIO (2.206)							†
Br–S(3)	see CIWYIQ (2.435, 2.453)							†
Br–S(3)$^+$	see THINBR (2.321)							†
Br–Se	see CIFZUM (2.508, 2.619)							
Br–Si	see BIZJAV (2.284)							
Br–Te	in **Br_6Te^{2-}** see CUGBAH (2.692–2.716)							
	Br–Te(4) see BETUTE10 (3.079, 3.015)							
	Br–Te(3) see BTUPTE (2.835)							
C*sp*^{3}–C*sp*^{3}	C#–**CH_2–CH_3**	1.513	1.514	0.014	1.507	1.523	192	
	(C#)$_2$–**CH–CH_3**	1.524	1.526	0.015	1.518	1.534	226	
	(C#)$_3$–**C–CH_3**	1.534	1.534	0.011	1.527	1.541	825	
	C#–**CH_2–CH_2**–C#	1.524	1.524	0.014	1.516	1.532	2459	
	(C#)$_2$–**CH–CH_2**–C#	1.531	1.531	0.012	1.524	1.538	1217	
	(C#)$_3$–**C–CH_2**–C#	1.538	1.539	0.010	1.533	1.544	330	
	(C#)$_2$–**CH–CH**–(C#)$_2$	1.542	1.542	0.011	1.536	1.549	321	
	(C#)$_3$–**C–CH**–(C#)$_2$	1.556	1.556	0.011	1.549	1.562	215	
	(C#)$_3$–**C–C**–(C#)$_3$	1.588	1.580	0.025	1.566	1.610	21	
	C*–C* (overall)	1.530	1.530	0.015	1.521	1.539	5777	5,6
	in cyclopropane (any subst.)	1.510	1.509	0.026	1.497	1.523	888	7
	in cyclobutane (any subst.)	1.554	1.553	0.021	1.540	1.567	679	8
	in cyclopentane (C,H-subst.)	1.543	1.543	0.018	1.532	1.554	1641	
	in cyclohexane (C,H-subst.)	1.535	1.535	0.016	1.525	1.545	2814	
	cyclopropyl-C* (exocyclic)	1.518	1.518	0.019	1.505	1.531	366	7
	cyclobutyl-C* (exocyclic)	1.529	1.529	0.016	1.519	1.539	376	8
	cyclopentyl-C* (exocyclic)	1.540	1.541	0.017	1.527	1.549	956	
	cyclohexyl-C* (exocyclic)	1.539	1.538	0.016	1.529	1.549	2682	
	in cyclobutene (any subst.)	1.573	1.574	0.017	1.566	1.586	25	8
	in cyclopentene (C,H-subst.)	1.541	1.539	0.015	1.532	1.549	208	
	in cyclohexene (C,H-subst.)	1.541	1.541	0.020	1.528	1.554	586	
	in oxirane (epoxide)	1.466	1.466	0.015	1.458	1.474	249	9
	in aziridine	1.480	1.481	0.021	1.465	1.496	67	9
	in oxetane	1.541	1.541	0.019	1.527	1.557	16	

Bond	Substructure	d	m	σ	q_l	q_u	n	Note
	in azetidine	1.548	1.543	0.018	1.536	1.558	22	
	oxiranyl-C* (exocyclic)	1.509	1.507	0.018	1.497	1.519	333	9
	aziridinyl-C* (exocyclic)	1.512	1.512	0.018	1.496	1.526	13	9
Csp^3–Csp^2	**CH$_3$–C**=C	1.503	1.504	0.011	1.497	1.509	215	
	C#–**CH$_2$–C**=C	1.502	1.502	0.013	1.494	1.510	483	
	(C#)$_2$–**CH–C**=C	1.510	1.510	0.014	1.501	1.518	564	
	(C#)$_3$–**C–C**=C	1.522	1.522	0.016	1.511	1.533	193	
Csp^3–Csp^2	**C*–C**=C (overall)	1.507	1.507	0.015	1.499	1.517	1456	5
	C*–C=C (endocyclic)							
	in cyclopropene	1.509	1.508	0.016	1.500	1.516	20	10
	in cyclobutene	1.513	1.512	0.018	1.500	1.525	50	8
	in cyclopentene	1.512	1.512	0.014	1.502	1.521	208	
	in cyclohexene	1.506	1.505	0.016	1.495	1.516	391	
	in cyclopentadiene	1.502	1.503	0.019	1.490	1.515	18	
	in cyclohexa-1,3-diene	1.504	1.504	0.017	1.491	1.517	56	
	C*–C=C (exocyclic):							
	cyclopropenyl-C*	1.478	1.475	0.012	1.470	1.485	7	10
	cyclobutenyl-C*	1.489	1.483	0.015	1.479	1.496	11	8
	cyclopentenyl-C*	1.504	1.506	0.012	1.495	1.512	115	
	cyclohexenyl-C*	1.511	1.511	0.013	1.502	1.519	292	
	C*CH=O in aldehydes	1.510	1.510	0.008	1.501	1.518	7	
	(C*)$_2$–C=O							
	in ketones	1.511	1.511	0.015	1.501	1.521	952	11
	in cyclobutanone	1.529	1.530	0.016	1.514	1.545	18	
	in cyclopentanone	1.514	1.514	0.016	1.505	1.523	312	
	acyclic and 6 + rings	1.509	1.509	0.016	1.499	1.519	626	
	C*–COOH in carboxylic acids	1.502	1.502	0.014	1.495	1.510	176	
	C*–COO$^-$ in carboxylate anions	1.520	1.521	0.011	1.516	1.528	57	
	C*–C(=O)(–OC*)							
	in acyclic esters	1.497	1.496	0.018	1.484	1.509	553	12
	in β-lactones	1.519	1.519	0.020	1.500	1.538	4	13
	in γ-lactones	1.512	1.512	0.015	1.501	1.521	110	12
	in δ-lactones	1.504	1.502	0.013	1.495	1.517	27	12
	cyclopropyl (**C**)–**C**=O in ketones, acids and esters	1.486	1.485	0.018	1.474	1.497	105	7
	C*–C(=O)(–NH$_2$) in acyclic amides	1.514	1.512	0.016	1.506	1.526	32	14
	C*–C(=O)(–NHC*) in acyclic amides	1.506	1.505	0.012	1.498	1.515	78	14
	C*–C(=O)[–N(C*)$_2$] in acyclic amides	1.505	1.505	0.011	1.496	1.517	15	14
Csp^3–C*ar*	**CH$_3$–C*ar***	1.506	1.507	0.011	1.501	1.513	454	
	C#–**CH$_2$–C*ar***	1.510	1.510	0.009	1.505	1.516	674	
	(C#)$_2$–**CH–C*ar***	1.515	1.515	0.011	1.508	1.522	363	
	(C#)$_3$–**C–C*ar***	1.527	1.530	0.016	1.517	1.539	308	
	C*–C*ar* (overall)	1.513	1.513	0.014	1.505	1.521	1813	
	cyclopropyl (**C**)–**C*ar***	1.490	1.490	0.015	1.479	1.503	90	7
Csp^3–Csp^1	**C*–C**≡C	1.466	1.465	0.010	1.460	1.469	21	15
	C#–C≡C	1.472	1.472	0.012	1.464	1.481	88	15
	C*–C≡N	1.470	1.469	0.013	1.463	1.479	106	7b
	cyclopropyl (**C**)–**C**≡N	1.444	1.447	0.010	1.436	1.451	38	7
Csp^2–Csp^2	C=**C–C**=C							
	(conjugated)	1.455	1.455	0.011	1.447	1.463	30	16,18
	(unconjugated)	1.478	1.476	0.012	1.470	1.479	8	17,18
	(overall)	1.460	1.460	0.015	1.450	1.470	38	
	C=**C–C**=C–C=C	1.443	1.445	0.013	1.431	1.454	29	18
	C=**C–C**=C (endocyclic in TCNQ)	1.432	1.433	0.012	1.424	1.441	280	19
	C=**C–C**(=O)(–C*)							
	(conjugated)	1.464	1.462	0.018	1.453	1.476	211	16,18
	(unconjugated)	1.484	1.486	0.017	1.475	1.497	14	17,18
	(overall)	1.465	1.462	0.018	1.453	1.478	226	
	C=**C–C**(=O)–C=C							
	in benzoquinone (C,H-subst. only)	1.478	1.476	0.011	1.469	1.488	28	
	in benzoquinone (any subst.)	1.478	1.478	0.031	1.464	1.498	172	
	non-quinonoid	1.456	1.455	0.012	1.447	1.464	28	

Bond	Substructure	*d*	*m*	*σ*	q_1	q_u	*n*	Note
	C=**C–C**OOH	1.475	1.476	0.015	1.461	1.488	22	
	C=**C–C**OOC*	1.488	1.489	0.014	1.478	1.497	113	
	C=**C–C**OO$^-$	1.502	1.499	0.017	1.488	1.510	11	
	HOO**C–C**OOH	1.538	1.537	0.007	1.535	1.541	9	
	HOO**C–C**OO$^-$	1.549	1.552	0.009	1.546	1.553	13	
	$^-$OO**C–C**OO$^-$	1.564	1.559	0.022	1.554	1.568	9	
	formal C*sp*^{2}–C*sp*^{2} single bond in selected non-fused heterocycles:							
	in 1*H*-pyrrole (C3–C4)	1.412	1.410	0.016	1.401	1.427	29	
	in furan (C3–C4)	1.423	1.423	0.016	1.412	1.433	62	
	in thiophene (C3–C4)	1.424	1.425	0.015	1.415	1.433	40	
	in pyrazole (C3–C4)	1.410	1.412	0.016	1.400	1.418	20	
	in isoxazole (C3–C4)	1.425	1.425	0.016	1.413	1.438	9	
	in furazan (C3–C4)	1.428	1.427	0.007	1.422	1.435	6	
	in furoxan (C3–C4)	1.417	1.417	0.006	1.412	1.422	14	
C*sp*^{2}–C*ar*	C=**C–C*ar***							
	(conjugated)	1.470	1.470	0.015	1.463	1.480	37	16,18
C*sp*^{2}–C*ar*		1.488	1.490	0.012	1.480	1.496	87	17,18
	(overall)	1.483	1.483	0.015	1.472	1.494	124	
	cyclopropenyl (C=**C)–C*ar***	1.447	1.448	0.006	1.441	1.452	8	10
	C*ar*–C(=O)–C*	1.488	1.489	0.016	1.478	1.500	84	
	C*ar*–C(=O)–C*ar*	1.480	1.481	0.017	1.468	1.494	58	
	C*ar*–COOH	1.484	1.485	0.014	1.474	1.491	75	
	C*ar*–C(=O)(–OC*)	1.487	1.487	0.012	1.480	1.494	218	
	C*ar*–COO$^-$	1.504	1.509	0.014	1.495	1.512	26	
	C*ar*–C(–O)–NH$_2$	1.500	1.503	0.020	1.498	1.510	19	
	C*ar*–C=N–C#							
	(conjugated)	1.476	1.478	0.014	1.466	1.486	27	16
	(unconjugated)	1.491	1.490	0.008	1.485	1.496	48	17
	(overall)	1.485	1.487	0.013	1.481	1.493	75	
	in indole (C3–C3a)	1.434	1.434	0.011	1.428	1.439	40	
C*sp*^{2}–C*sp*^{1}	C=**C–C**≡C	1.431	1.427	0.014	1.425	1.441	11	7b
	C=**C–C**≡N in TCNQ	1.427	1.427	0.010	1.420	1.433	280	19
C*ar*–C*ar*	in biphenyls (*ortho* subst. all **H**)	1.487	1.488	0.007	1.484	1.493	30	
	(≥1 non-**H***ortho*-subst.)	1.490	1.491	0.010	1.486	1.495	212	
C*ar*–C*sp*^{1}	**C*ar*–C**≡C	1.434	1.436	0.006	1.430	1.437	37	
	C*ar*–C≡N	1.443	1.444	0.008	1.436	1.448	31	
C*sp*^{1}–C*sp*^{1}	C≡**C–C**≡C	1.377	1.378	0.012	1.374	1.384	21	
C*sp*^{2}=C*sp*^{2}	C*–**CH=CH**$_2$	1.299	1.300	0.027	1.280	1.311	42	
	(C*)$_2$–**C=CH**$_2$	1.321	1.321	0.013	1.313	1.328	77	
	C*–**CH=CH**–C*							
	(*cis*)	1.317	1.318	0.013	1.310	1.323	106	
	(*trans*)	1.312	1.311	0.011	1.304	1.320	19	
	(overall)	1.316	1.317	0.015	1.309	1.323	127	
	(C*)$_2$–**C=CH**–C*	1.326	1.328	0.011	1.319	1.334	168	
	(C*$_2$–**C=C**–(C*)$_2$	1.331	1.330	0.009	1.326	1.334	89	
	(C*,H)$_2$–**C=C**–(C*,H)$_2$ (overall)	1.322	1.323	0.014	1.315	1.331	493	5
	in cyclopropene (any subst.)	1.294	1.288	0.017	1.284	1.302	10	10
	in cyclobutene (any subst.)	1.335	1.335	0.019	1.324	1.347	25	8
	in cyclopentene (C,H-subst.)	1.323	1.324	0.013	1.314	1.331	104	
	in cyclohexene (C,H-subst.)	1.326	1.325	0.012	1.318	1.334	196	
	C=C=C (allenes, any subst.)	1.307	1.307	0.005	1.303	1.310	18	
	C=C–**C=C** (C,**H** subst., conjugated)	1.330	1.330	0.014	1.322	1.338	76	16
	C=C–**C=C**–**C=C** (C,**H** subst., conjugated)	1.345	1.345	0.012	1.337	1.350	58	16
	C=C–C*ar* (C,H subst., conjugated)	1.339	1.340	0.011	1.334	1.346	124	16
	C=C in cyclopenta-1,3-diene (any subst.)	1.341	1.341	0.017	1.328	1.356	18	
	C=C in cyclohexa-1,3-diene (any subst.)	1.332	1.332	0.013	1.323	1.341	56	
	in **C=C**–C=O							
	(C,**H** subst., conjugated)	1.340	1.340	0.013	1.332	1.348	211	16,18
	(C,**H** subst., unconjugated)	1.331	1.330	0.008	1.326	1.339	14	17,18
	(C,**H** subst., overall)	1.340	1.339	0.013	1.332	1.348	226	
	in cyclohexa-2,5-dien-1-ones	1.329	1.327	0.011	1.321	1.335	28	

Bond	Substructure	*d*	*m*	*σ*	q_l	q_u	*n*	Note
	in *p*-benzoquinones							
	(C*,**H** subst.)	1.333	1.337	0.011	1.325	1.338	14	
	(any subst.)	1.349	1.339	0.030	1.330	1.364	86	
	in TCNQ							
	(endocyclic)	1.352	1.353	0.010	1.345	1.358	142	19
	(exocyclic)	1.392	1.391	0.017	1.379	1.405	139	19
	C=**C**–OH in enol tautomers	1.362	1.360	0.020	1.349	1.370	54	
	in heterocycles (any subst.):							
	1*H*-pyrrole (C2–C3, C4–C5)	1.375	1.377	0.018	1.361	1.388	58	
	furan (C2–C3, C4–C5)	1.341	1.342	0.021	1.329	1.351	125	
	thiophene (C2–C3, C4–C5)	1.362	1.359	0.025	1.346	1.377	60	
	pyrazole (C4–C5)	1.369	1.372	0.019	1.362	1.383	20	
	imidazole (C4–C5)	1.360	1.361	0.014	1.352	1.367	44	
	isoxazole (C4–C5)	1.341	1.336	0.012	1.331	1.355	9	
	indole (C2–C3)	1.364	1.363	0.012	1.355	1.371	40	
Car≃ *Car*	in phenyl rings with C*,**H** subst. only							
	H–**C**≃**C**–H	1.380	1.381	0.013	1.372	1.388	2191	
	C*–**C**≃**C**–H	1.387	1.388	0.010	1.382	1.393	891	
	C*–**C**≃**C**–C*	1.397	1.397	0.009	1.392	1.403	182	
	C≃**C** (overall)	1.384	1.384	0.013	1.375	1.391	3264	
	F–**C**≃**C**–F	1.372	1.374	0.011	1.366	1.380	84	4
	Cl–**C**≃**C**–Cl	1.388	1.389	0.014	1.380	1.398	152	4
	in naphthalene (D_{2h}, any subst.)							
	C1–C2	1.364	1.364	0.014	1.356	1.373	440	
	C2–C3	1.406	1.406	0.014	1.397	1.415	218	
	C1–C8a	1.420	1.419	0.012	1.412	1.426	440	
	C4a–C8a	1.422	1.424	0.011	1.417	1.429	109	
Car≃ *Car*	in anthracene (D_{2h}, any subst.)							
	C1–C2	1.356	1.356	0.009	1.350	1.360	56	
	C2–C3	1.410	1.410	0.010	1.401	1.416	34	
	C1–C9a	1.430	1.430	0.006	1.426	1.434	56	
	C4a–C9a	1.435	1.436	0.007	1.429	1.440	34	
	C9–C9a	1.400	1.402	0.009	1.395	1.406	68	
	in pyridine (C,H subst.)	1.379	1.381	0.012	1.371	1.387	276	20
	(any subst.)	1.380	1.380	0.015	1.371	1.389	537	20
	in pyridinium cation							
	(N^+–H; C,H subst. on C)							
	C2–C3	1.373	1.375	0.012	1.368	1.380	30	
	C3–C4	1.379	1.380	0.011	1.371	1.388	30	
	(N^+–X; C,H subst. on C)							
	C2–C3	1.373	1.372	0.019	1.362	1.382	151	
	C3–C4	1.383	1.385	0.019	1.372	1.394	151	
	in pyrazine (H subst. on C)	1.379	1.377	0.010	1.370	1.388	10	
	(any subst. on C)	1.405	1.405	0.024	1.388	1.420	60	
	in pyrimidine (C,H subst. on C)	1.387	1.389	0.018	1.379	1.400	28	
Csp^1≡Csp^1	X–**C**≡**C**–X	1.183	1.183	0.014	1.174	1.193	119	15
	C,H–**C**≡**C**–C,H	1.181	1.181	0.014	1.173	1.192	104	15
	in **C**≡**C**–C(sp^2,*ar*)	1.189	1.193	0.010	1.181	1.195	38	15
	in **C**≡**C**–C≡**C**	1.192	1.192	0.010	1.187	1.197	42	15
	in **CH**≡**C**–C#	1.174	1.174	0.011	1.167	1.180	42	15
Csp^3–Cl	Omitting 1,2-dichlorides:							
	C–**CH$_2$–Cl**	1.790	1.790	0.007	1.783	1.795	13	4
	C_2–**CH–Cl**	1.803	1.802	0.003	1.800	1.807	8	4
	C_3–**C–Cl**	1.849	1.856	0.011	1.837	1.858	5	4
	X–CH_2–Cl (X = C,H,N,O)	1.790	1.791	0.011	1.783	1.797	37	4
	X_2–CH–Cl (X = C,H,N,O)	1.805	1.803	0.014	1.800	1.812	26	4
	X_3–**C–Cl** (X = C,H,N,O)	1.843	1.838	0.014	1.835	1.858	7	4
	X_2–**C–Cl$_2$** (X = C,H,N,O)	1.779	1.776	0.015	1.769	1.790	18	4
	X–**C–Cl$_3$** (X = C,H,N,O)	1.768	1.765	0.011	1.761	1.776	33	4
	Cl–CH(–C)–**CH**(–C)–**Cl**	1.793	1.793	0.013	1.786	1.800	66	4
	Cl–C(–C_2)–**C**(–C_2)–**Cl**	1.762	1.760	0.010	1.757	1.765	54	4

Bond	Substructure	d	m	σ	q_1	q_u	n	Note
	cyclopropyl–Cl	1.755	1.756	0.011	1.749	1.763	64	
Csp^2–Cl	C=C–Cl (C,H,N,O subst. on C)	1.734	1.729	0.019	1.719	1.748	63	4
	C=C–Cl_2 (C,H,N,O subst. on C)	1.720	1.716	0.013	1.708	1.729	20	4
	Cl–C=C–Cl	1.713	1.711	0.011	1.705	1.720	80	4
Car–Cl	*Car*–Cl (mono–Cl + m,p-Cl_2)	1.739	1.741	0.010	1.734	1.745	340	4
	Car–Cl (*o*–Cl_2)	1.720	1.720	0.010	1.713	1.717	364	4
Csp^1Cl	see HCLENE10 (1.634, 1.646)							
Csp^3–F	Omitting 1,2-difluorides							
	C–CH_2–Fand C_2–CH–F	1.399	1.399	0.017	1.389	1.408	25	4
	C_3–C–F	1.428	1.431	0.009	1.421	1.435	11	4
	$(C^*,H)_2$–C–F_2	1.349	1.347	0.012	1.342	1.356	58	4
	C*–C–F_3	1.336	1.334	0.007	1.330	1.344	12	4
	F–C*–C*–F	1.371	1.374	0.007	1.362	1.375	26	4
	X_3–C–F (X = C,H,N,O)	1.386	1.389	0.033	1.373	1.408	70	4
	X_2–C–F_2 (X = C,H,N,O)	1.351	1.349	0.013	1.342	1.356	58	4
	X–C–F_3 (X = C,H,N,O)	1.322	1.323	0.015	1.314	1.332	309	4
	F–C(–X)$_2$–C(–X)$_2$–F (X = C,H,N,O)	1.373	1.374	0.009	1.362	1.377	30	4
	F–C(–X)$_2$–NO_2 (X = any subst.)	1.320	1.319	0.009	1.312	1.327	18	
Csp^2–F	C=C–F (C,H,N,O subst. on C)	1.340	1.340	0.013	1.334	1.346	34	4
Car–F	*Car*–F (mono-F +*m,p*-F_2)	1.363	1.362	0.008	1.357	1.368	38	4
	Car–F (*o*-F_2)	1.340	1.340	0.009	1.336	1.344	167	4
Csp^3–H	C–C–H_3 (methyl)	1.059	1.061	0.030	1.039	1.083	83	21
	C_2–C–H_2 (primary)	1.092	1.095	0.013	1.088	1.099	100	21
	C_3–C–H (secondary)	1.099	1.097	0.004	1.095	1.103	14	21
	$C_{2,3}$–C–H (primary and secondary)	1.093	1.095	0.012	1.089	1.100	118	21
	X–C–H_3 (methyl)	1.066	1.074	0.028	1.049	1.087	160	21
	X_2–C–H_2 (primary)	1.092	1.095	0.012	1.088	1.099	230	21
	X_3–C–H (secondary)	1.099	1.099	0.007	1.095	1.103	117	21
	$X_{2,3}$–C–H (primary and secondary)	1.094	1.096	0.011	1.091	1.100	348	21
Csp^2–H	C–C=C–H	1.077	1.079	0.012	1.074	1.085	14	21
Car–H	*Car*–H	1.083	1.083	0.011	1.080	1.087	218	21
Csp^3–I	C*–I	2.162	2.159	0.015	2.149	2.179	15	4
Car–I	*Car*–I	2.095	2.095	0.015	2.089	2.104	51	4
Csp^3–N(4)	C*–NH_3^+	1.488	1.488	0.013	1.482	1.495	298	
	(C*)$_2$–NH_2^+	1.494	1.493	0.016	1.484	1.503	249	
	(C*)$_3$–NH^+	1.502	1.502	0.015	1.491	1.512	509	
	(C*)$_4$–N^+	1.510	1.509	0.020	1.496	1.523	319	
	C*–N^+ (overall)	1.499	1.498	0.018	1.488	1.510	1370	
Csp^3–N(3)	C*–N^+ in N-subst. pyridinium	1.485	1.484	0.009	1.477	1.490	32	
	C*–NH_2 (Nsp^3: pyramidal)	1.469	1.470	0.010	1.462	1.474	19	22
	(C*)$_2$–NH (Nsp^3: pyramidal)	1.469	1.467	0.012	1.461	1.477	152	5,22
	(C*)$_3$–N (Nsp^3: pyramidal)	1.469	1.468	0.014	1.460	1.476	1042	5,22
	C*–Nsp^3 (overall)	1.469	1.468	0.014	1.460	1.476	1201	
	Csp^3–Nsp^3							
	in aziridine	1.472	1.471	0.016	1.464	1.482	134	
	in azetidine	1.484	1.481	0.018	1.472	1.495	21	
	in tetrahydropyrrole	1.475	1.473	0.016	1.464	1.483	66	
	in piperidine	1.473	1.473	0.013	1.460	1.479	240	
	Csp^3–Nsp^2 (N planar) in:							23
	acyclic amides C*–NH–C=O	1.454	1.451	0.011	1.446	1.461	78	14
	β-lactams C*–N(–X)–C=O (endo)	1.464	1.465	0.012	1.458	1.475	23	13
	γ-lactams							
	C*–NH–C=O (endo)	1.457	1.458	0.011	1.449	1.465	20	13
	C*–N(–C*)–C=O (endo)	1.462	1.461	0.010	1.453	1.466	15	13
	C*–N(–C*)–C=O (exo)	1.458	1.456	0.014	1.448	1.465	15	13
	δ-lactams							
	C*–NH–C=O (endo)	1.478	1.472	0.016	1.467	1.491	6	14
	C*–N(–C*)–C=O (endo)	1.479	1.476	0.007	1.475	1.482	15	14
	C*–N(–C*)–C=O (exo)	1.468	1.471	0.009	1.462	1.477	15	14
	nitro compounds (1,2-dinitro omitted):							
	C–CH_2–NO_2	1.485	1.483	0.020	1.478	1.502	8	

Bond	Substructure	d	m	σ	q_l	q_u	n	Note
	C_2–**CH**–**NO$_2$**	1.509	1.509	0.011	1.502	1.511	12	
	C_3–**C**–**NO$_2$**	1.533	1.533	0.013	1.530	1.539	17	
	C_2–**C**–(**NO$_2$**)$_2$	1.537	1.536	0.016	1.525	1.550	19	
	1,2-dinitro: NO_2–C*–C*–NO_2	1.552	1.550	0.023	1.536	1.572	32	
C*sp*³–N(2)	**C#**–**N**=N	1.493	1.493	0.020	1.477	1.506	54	
	C*–**N**=C–*Car*	1.465	1.468	0.011	1.461	1.472	75	
C*sp*²–N(3)	C=**C**–**NH$_2$**N*sp*² planar	1.336	1.344	0.017	1.317	1.348	10	23
	C=**C**–**NH**–C# N*sp*² planar	1.339	1.340	0.016	1.327	1.351	17	23
	C=**C**–**N**–(C#)$_2$							
	N*sp*² planar	1.355	1.358	0.014	1.341	1.363	22	23
	N*sp*³ pyramidal	1.416	1.418	0.018	1.397	1.432	18	22
	C*sp*²–N*sp*² (N planar) in:							23
	acyclic amides							
	NH$_2$–**C**=O	1.325	1.323	0.009	1.318	1.331	32	14
	C*–**NH**–**C**=O	1.334	1.333	0.011	1.326	1.343	78	14
	(C*)$_2$–**N**–**C**=O	1.346	1.342	0.011	1.339	1.356	5	14
	β-lactams C*–**NH**–**C**=O	1.385	1.388	0.019	1.374	1.396	23	13
	γ-lactams							
	C*–**NH**–**C**=O	1.331	1.331	0.011	1.326	1.337	20	13
	C*–**N**(–C*)–**C**=O	1.347	1.344	0.014	1.335	1.359	15	13
	δ-lactams							
	C*–**NH**–**C**=O	1.334	1.334	0.006	1.330	1.339	6	14
	(C*)–**N**(–C*)–**C**=O	1.352	1.353	0.010	1.344	1.356	15	14
	peptides C#–**N**(–X)–**C**(–C#)(=O)	1.333	1.334	0.013	1.326	1.340	380	24
	ureas							
	(**NH$_2$**)$_2$–**C**=O	1.334	1.334	0.008	1.329	1.339	48	25,26
	(C#–**NH**)$_2$–**C**=O	1.347	1.345	0.010	1.341	1.354	26	25
	[(C#)$_n$–**N**]$_2$–**C**=O	1.363	1.359	0.014	1.354	1.370	40	25,27
	thioureas	1.346	1.343	0.023	1.328	1.361	192	
	(X$_2$**N**)$_2$–**C**=S							
	imides							
	[C#–**C**(=O)]$_2$–**NH**	1.376	1.377	0.012	1.369	1.383	64	
	[C#–**C**(=O)]$_2$–**N**–C#	1.389	1.383	0.017	1.376	1.404	38	
	[C*sp*²–**C**(=O)]$_2$–**N**–C#	1.396	1.396	0.010	1.389	1.403	46	
	[C*sp*²–**C**(=O)]$_2$–**N**–C*sp*²	1.409	1.406	0.020	1.391	1.419	28	
	guanidinium [**C**–(**NH$_2$**)$_3$]$^+$ (unsubst.)	1.321	1.320	0.008	1.314	1.327	39	
	(any subst.)	1.328	1.325	0.015	1.317	1.333	140	
	in heterocyclic systems (any subst.)							
	1*H*-pyrrole (N1–C2, N1–C5)	1.372	1.374	0.016	1.363	1.384	58	
	indole (N1–C2)	1.370	1.370	0.012	1.364	1.377	40	
	pyrazole (N1–C5)	1.357	1.359	0.012	1.347	1.365	20	
	imidazole (N1–C2)	1.349	1.349	0.018	1.338	1.358	44	
	imidazole (N1–C5)	1.370	1.370	0.010	1.365	1.377	44	
C*sp*²–N(2)	in imidazole (N3–C4)	1.376	1.377	0.011	1.369	1.384	44	
Car–N(4)	*Car*–**N**$^+$–(C,H)$_3$	1.465	1.466	0.007	1.461	1.470	23	
Car–N(3)	*Car*–**NH$_2$**							
	(N*sp*²: planar)	1.355	1.360	0.020	1.340	1.372	33	23
	(N*sp*³: pyramidal)	1.394	1.396	0.011	1.385	1.403	25	22
	(overall)	1.375	1.377	0.025	1.363	1.394	98	28
Car–N(3)	*Car*–**NH**–C#							
	(N*sp*²: planar)	1.353	1.353	0.007	1.347	1.359	16	23
	(N*sp*³: pyramidal)	1.419	1.423	0.017	1.412	1.432	8	22
	(overall)	1.380	1.364	0.032	1.353	1.412	31	28
	Car–**N**–(C#)$_2$							
	(N*sp*²: planar)	1.371	1.370	0.016	1.363	1.382	41	23
	(N*sp*³: pyramidal)	1.426	1.425	0.011	1.421	1.431	22	22
	(overall)	1.390	1.385	0.030	1.366	1.420	69	28
	in indole (N1–C7a)	1.372	1.372	0.007	1.367	1.376	40	
	Car–**NO$_2$**	1.468	1.469	0.014	1.460	1.476	556	
Car–N(2)	*Car*–**N**=N	1.431	1.435	0.020	1.422	1.442	26	
C*sp*²=N(3)	in furoxan ($^+$N2=C3)	1.316	1.316	0.009	1.311	1.324	14	

Molecular

Bond	Substructure	d	m	σ	q_1	q_u	n	Note
Csp^2=N(2)	*Car*–**C=N**–C#	1.279	1.279	0.008	1.275	1.285	75	
	(C,H)$_2$–**C=N**–OH in oximes	1.281	1.280	0.013	1.273	1.288	67	
	S–**C=N**–X	1.302	1.302	0.021	1.285	1.319	36	
	in pyrazole (N2=C3)	1.329	1.331	0.014	1.315	1.339	20	
	in imidazole (C2=N3)	1.313	1.314	0.011	1.307	1.319	44	
	in isoxazole (N2=C3)	1.314	1.315	0.009	1.305	1.320	9	
	in furazan (N2=C3, C4=N5)	1.298	1.299	0.006	1.294	1.303	12	
	in furoxan (C4=N5)	1.304	1.306	0.008	1.300	1.308	14	
Car≃ N(3)	**C≃N**$^+$–H (pyrimidinium)	1.335	1.334	0.015	1.325	1.342	30	
	C≃N$^+$–C* (pyrimidinium)	1.346	1.346	0.010	1.340	1.352	64	
	C≃N$^+$–O$^-$ (pyrimidinium)	1.362	1.359	0.013	1.353	1.369	56	
Car≃ N(2)	**C≃N** (pyridine)	1.337	1.338	0.012	1.330	1.344	269	
	C≃N (pyrazine)	1.336	1.335	0.022	1.319	1.347	120	
	C≃N≃C (pyrimidine)	1.339	1.338	0.015	1.333	1.342	28	
	N≃C≃N (pyrimidine)	1.333	1.335	0.013	1.326	1.337	28	
	C≃N (pyrimidine) (overall) in any 6-membered N-containing aromatic ring:	1.336	1.337	0.014	1.331	1.339	56	
	H–C≃N≃C–H	1.334	1.334	0.014	1.327	1.341	146	
	H–C≃N≃C–C*	1.339	1.341	0.013	1.336	1.345	38	
	C*–C≃N≃C–C*	1.345	1.345	0.008	1.342	1.348	24	
	C≃N≃C (overall)	1.336	1.337	0.014	1.329	1.344	204	
Csp^1≡N(2)	X–S–**N≡C**$^-$ (isothiocyanide)	1.144	1.147	0.006	1.140	1.148	6	
Csp^1≡N(1)	C*–**C≡N**	1.136	1.137	0.010	1.131	1.142	140	
	C=C–**C≡N** in TCNQ	1.144	1.144	0.008	1.139	1.149	284	19
	Car–**C≡N**	1.138	1.138	0.007	1.133	1.143	31	
	X–**C≡N**	1.144	1.141	0.012	1.138	1.151	10	
	(S–**C≡N**)$^-$	1.155	1.156	0.012	1.147	1.165	14	
Csp^3–O(2)	in alcohols							
	CH$_3$–OH	1.413	1.414	0.018	1.395	1.425	17	
	C–**CH$_2$–OH**	1.426	1.426	0.011	1.420	1.431	75	
	C$_2$–**CH-OH**	1.432	1.431	0.011	1.425	1.439	266	
	C$_3$–**C–OH**	1.440	1.440	0.012	1.432	1.449	106	
	C*–OH (overall)	1.432	1.431	0.013	1.424	1.441	464	
	in dialkyl ethers							29
	CH$_3$–O–C*	1.416	1.418	0.016	1.405	1.426	110	
	C–**CH$_2$–O**–C*	1.426	1.424	0.011	1.418	1.435	34	
	C$_2$–**CH–O**–C*	1.429	1.430	0.010	1.420	1.437	53	
	C$_3$–**C–O**–C*	1.452	1.450	0.011	1.445	1.458	39	
	C*–O–C* (overall)	1.426	1.425	0.019	1.414	1.437	236	5
	in aryl alkyl ethers							29
	CH$_3$–O–*Car*	1.424	1.424	0.012	1.417	1.431	616	
	C–**CH$_2$–O**–*Car*	1.431	1.430	0.013	1.422	1.438	188	
	C$_2$–**CH–O**–*Car*	1.447	1.446	0.020	1.435	1.466	58	
	C$_3$–**C–O**–*Car*	1.470	1.469	0.018	1.456	1.483	55	
	C*–O–*Car* (overall)	1.429	1.427	0.018	1.419	1.436	917	
	in alkyl esters of carboxylic acids							12,29
	CH$_3$–O–C(=O)–C*	1.448	1.449	0.010	1.442	1.455	200	
	C–CH$_2$–O–C(=O)–C*	1.452	1.453	0.009	1.445	1.458	32	
	C$_2$–**CH–O**–C(=O)–C*	1.460	1.460	0.010	1.454	1.465	78	
	C$_3$–**C–O**–C(=O)–C*	1.477	1.475	0.008	1.472	1.484	6	
	C*–O–C(=O)–C* (overall)	1.450	1.451	0.014	1.442	1.459	314	
	in alkyl esters of α,β-unsaturated acids:							
	C*–O–C(=O)–C=C (overall)	1.453	1.452	0.013	1.444	1.459	112	
	in alkyl esters of benzoic acid							
	C*–O–C(=O)–C (phenyl) (overall)	1.454	1.454	0.012	1.446	1.463	219	
	in ring systems							
	oxirane (epoxides) (any subst.)	1.446	1.446	0.014	1.438	1.456	498	9
	oxetane (any subst.)	1.463	1.460	0.015	1.451	1.474	16	
	tetrahydrofuran (C,H subst.)	1.442	1.441	0.017	1.430	1.451	154	
Csp^3–O(2)	tetrahydropyran (C,H subst.)	1.441	1.442	0.015	1.431	1.451	22	
	β-lactones: **C*–O**–C(=O)	1.492	1.494	0.010	1.481	1.501	4	16

Bond	Substructure	*d*	*m*	*σ*	q_l	q_u	*n*	Note
	γ-lactones: **C*–O**–C(=O)	1.464	1.464	0.012	1.455	1.473	110	12
	δ-lactones: **C*–O**–C(=O)	1.461	1.464	0.017	1.452	1.473	27	12
	O–C–O system in *gem*-diols, and pyranose and furanose sugars:							30,31
	HO–C*–OH	1.397	1.401	0.012	1.388	1.405	18	
	C_5–O_5–C_1–O_1H in pyranoses							
	O_1 axial (α):							
	C_5–O_5	1.439	1.440	0.008	1.432	1.445	29	
	O_5–C_1	1.427	1.426	0.012	1.421	1.432	29	
	C_1–O_1	1.403	1.400	0.012	1.391	1.412	29	
	O_1 equatorial (β):							
	C_5–O_5	1.435	1.436	0.008	1.429	1.440	17	
	O_5–C_1	1.430	1.431	0.010	1.424	1.436	17	
	C_1–O_1	1.393	1.393	0.007	1.386	1.399	17	
	α + β (overall):							
	C_5–O_5	1.439	1.440	0.008	1.432	1.446	60	
	O_5–C_1	1.430	1.429	0.012	1.421	1.436	60	
	C_1–O_1	1.401	1.399	0.011	1.392	1.407	60	
	C_4–O_4–C_1–O_1H in furanoses (overall values)							
	C_4–O_4	1.442	1.446	0.012	1.436	1.449	18	
	O_4–C_1	1.432	1.432	0.012	1.421	1.443	18	
	C_1–O_1	1.404	1.405	0.013	1.397	1.409	18	
	C_5–O_5–C_1–O_1–C* in pyranoses							
	O_1 axial (α):							
	C_5–O_5	1.439	1.438	0.010	1.433	1.446	67	
	O_5–C_1	1.417	1.417	0.009	1.410	1.424	67	
	C_1–O_1	1.409	1.409	0.014	1.401	1.417	67	
	O_1–C*	1.435	1.435	0.013	1.427	1.443	67	
	O_1 equatorial (β):							
	C_5–O_5	1.434	1.435	0.006	1.429	1.439	39	
	O_5–C_1	1.424	1.424	0.008	1.418	1.431	39	
	C_1–O_1	1.390	1.390	0.011	1.381	1.400	39	
	O_1–C*	1.437	1.438	0.013	1.428	1.445	39	
	α + β (overall):							
	C_5–O_5	1.436	1.436	0.009	1.431	1.442	126	
	O_5–C_1	1.419	1.419	0.011	1.412	1.426	126	
	C_1–O_1	1.402	1.403	0.016	1.391	1.413	126	
	O_1–C*	1.436	1.436	0.013	1.428	1.445	126	
	C_4–O_4–C_1–O_1–C* in furanoses (overall values)							
	C_4–O_4	1.443	1.445	0.013	1.429	1.453	23	
	O_4–C_1	1.421	1.418	0.012	1.413	1.431	23	
	C_1–O_1	1.410	1.409	0.014	1.401	1.420	23	
	O_1–C*	1.439	1.437	0.014	1.429	1.449	23	
	Miscellaneous:							
	C#–O–SiX_3	1.416	1.416	0.017	1.405	1.428	29	
	C*–O–SO_2–C	1.465	1.461	0.014	1.454	1.475	33	
Csp^2–O(2)	in enols: C=**C–OH**	1.333	1.331	0.017	1.324	1.342	53	
	in enol esters: C=**C–O–C***	1.354	1.353	0.016	1.341	1.363	40	
	in acids:							
	C*–**C**(=O)–**OH**	1.308	1.311	0.019	1.298	1.320	174	
	C=C–**C**(=O)–**OH**	1.293	1.295	0.019	1.279	1.307	22	
	C*ar*–**C**(=O)–**OH**	1.305	1.311	0.020	1.291	1.317	75	
	in esters:							
	C*–**C**(=O)–**O**–C*	1.336	1.337	0.014	1.328	1.346	551	12,29
	C=C–**C**(=O)–**O**–C*	1.332	1.331	0.011	1.324	1.339	112	
	C*ar*–**C**(=O)–**O**–C*	1.337	1.335	0.013	1.329	1.344	219	12
	C*–**C**(=O)–**O**–C=C	1.362	1.359	0.018	1.351	1.374	26	
	C*–C(=O)–**O**–**C**=C	1.407	1.405	0.017	1.394	1.420	26	
	C*–C(=O)–**O**–C*ar*	1.360	1.359	0.011	1.355	1.367	40	12
	in anhydrides: O=**C–O**–C=O	1.386	1.386	0.011	1.379	1.393	70	
	in ring systems:							
	furan (O1–C2, O1–C5)	1.368	1.369	0.015	1.359	1.377	125	

Bond	Substructure	d	m	σ	q_1	q_u	n	Note
	isoxazole (O1–C5)	1.354	1.354	0.010	1.345	1.360	9	
	β-lactones: C*–**C**(=O)–**O**–C*	1.359	1.359	0.013	1.348	1.371	4	13
	γ-lactones: C*–**C**(=O)–**O**–C*	1.350	1.349	0.012	1.342	1.359	110	12
	δ-lactones: C*–**C**(=O)–**O**–C*	1.339	1.339	0.016	1.332	1.347	27	12
Car–O(2)	in phenols: ***Car*–OH**	1.362	1.364	0.015	1.353	1.373	551	
	in aryl alkyl ethers: ***Car*–O**–C*	1.370	1.370	0.011	1.363	1.377	920	29,32
Car–O(2)	in diaryl ethers: ***Car*–O–*Car***	1.384	1.381	0.014	1.375	1.391	132	
	in esters: ***Car*–O**–C(=O)–C*	1.401	1.401	0.010	1.394	1.408	40	12
C*sp*²=O(1)	in aldehydes and ketones:							
	C*–**CH=O**	1.192	1.192	0.005	1.188	1.197	7	
	$(C^*)_2$–**C=O**	1.210	1.210	0.008	1.206	1.215	474	5
	$(C\#)_2$–**C=O**							
	in cyclobutanones	1.198	1.198	0.007	1.194	1.204	12	
	in cyclopentanones	1.208	1.208	0.007	1.203	1.212	155	
	in cyclohexanones	1.211	1.211	0.009	1.207	1.216	312	
	C=C–**C=O**	1.222	1.222	0.010	1.216	1.229	225	
	$(C{=}C)_2$–**C=O**	1.233	1.229	0.010	1.226	1.242	28	
	Car–**C=O**	1.221	1.218	0.014	1.212	1.229	85	
	$(Car)_2$–**C=O**	1.230	1.226	0.015	1.220	1.238	66	
	C=O in benzoquinones	1.222	1.220	0.013	1.211	1.231	86	
	delocalized double bonds in carboxylate anions:							
	H–**C≃O**$_2^-$ (formate)	1.242	1.243	0.012	1.234	1.252	24	
	C*–**C≃O**$_2^-$	1.254	1.253	0.010	1.247	1.261	114	
	C=C–**C≃O**$_2^-$	1.250	1.248	0.017	1.238	1.261	52	
	Car–**C≃O**$_2^-$	1.255	1.253	0.010	1.249	1.262	22	
	HOOC–C≃O$_2^-$ (hydrogen oxalate)	1.243	1.247	0.015	1.232	1.256	26	
	$^-$**O**$_2$**≃C–C≃O**$_2^-$ (oxalate)	1.251	1.251	0.007	1.248	1.254	18	
	in carboxylic acids (X–COOH)							
	C*–**C**(=**O**)–OH	1.214	1.214	0.019	1.203	1.224	175	
	C=C–**C**(=**O**)–OH	1.229	1.226	0.017	1.218	1.237	22	
	Car–**C**(=**O**)–OH	1.226	1.223	0.020	1.211	1.241	75	
	in esters:							
	C*–**C**(=**O**)–O–C*	1.196	1.196	0.010	1.190	1.202	551	12
	C=C–**C**(=**O**)–O–C*	1.199	1.198	0.009	1.193	1.203	113	
	Car–**C**(=**O**)–O–C*	1.202	1.201	0.009	1.196	1.207	218	12
	C*–**C**(=**O**)–O–C=C	1.190	1.190	0.014	1.184	1.198	26	
	C*–**C**(=**O**)–O–*Car*	1.187	1.188	0.011	1.181	1.195	40	12
	in anhydrides: **O=C**–O–**C=O**	1.187	1.187	0.010	1.184	1.193	70	
	in β-lactones: C*–**C**(=**O**)–O–C*	1.193	1.193	0.006	1.187	1.198	4	13
	γ-lactones: C*–**C**(=**O**)–O–C*	1.201	1.202	0.009	1.196	1.206	109	12
	δ-lactones: C*–**C**(=**O**)–O–C*	1.205	1.207	0.008	1.201	1.209	27	12
	in amides:							
	NH_2–**C**(–C*)=**O**	1.234	1.233	0.012	1.225	1.243	32	14
	(C*–)(C*,H–)N–**C**(–C*)=**O**	1.231	1.231	0.012	1.224	1.238	378	14
	β-lactams: C*–NH–**C=O**	1.198	1.200	0.012	1.193	1.204	23	13
	γ-lactams:							
	C*–NH–**C=O**	1.235	1.235	0.008	1.232	1.240	20	13
	C*–N(–C*)–C=O	1.225	1.226	0.011	1.217	1.233	15	13
	δ-lactams:							
	C*–NH–**C=O**	1.240	1.241	0.003	1.237	1.243	6	14
	C*–N(–C*)–**C=O**	1.233	1.233	0.007	1.229	1.239	15	14
	in ureas:							
	$(NH)_2)_2$–**C=O**	1.256	1.256	0.007	1.249	1.261	24	25,26
	$(C\#{-}NH)_2$–**C=O**	1.241	1.237	0.011	1.235	1.245	13	25
	$[(C\#)_n{-}N]_2$–**C=O**	1.230	1.230	0.007	1.224	1.234	20	25,27
C*sp*³–P(4)	C_3–**P**⁺–**C***	1.800	1.802	0.015	1.790	1.812	35	33
	C_2–**P**(=O)–**CH**$_3$	1.791	1.790	0.006	1.786	1.795	10	
	C_2–**P**(=O)–**CH**$_2$–C	1.806	1.806	0.009	1.801	1.813	45	
	C_2–**P**(=O)–**CH**–C_2	1.821	1.821	0.009	1.815	1.828	15	
	C_2–**P**(=O)–**C**–C_3	1.841	1.842	0.008	1.835	1.847	14	
	C_2–**P**(=O)–**C*** (overall)	1.813	1.811	0.017	1.800	1.822	84	

Bond	Substructure	*d*	*m*	*σ*	q_l	q_u	*n*	Note
Csp^3–P(3)	C_2–P–C*	1.855	1.857	0.019	1.840	1.870	23	
Car–P(4)	C_3–P^+–*Car*	1.793	1.792	0.011	1.786	1.800	276	
	C_2–P(=O)–*Car*	1.801	1.802	0.011	1.796	1.807	98	
	Ph_3–P=N^+=P–Ph_3	1.795	1.795	0.008	1.789	1.800	197	
Car–P(3)	C_2–P–*Car*	1.836	1.837	0.010	1.830	1.844	102	
	$(N\simeq)_2$P–*Car* (P ≃ N aromatic)	1.795	1.793	0.011	1.788	1.803	43	
Csp^3–S(4)	C*–SO_2–C (C* = CH_3 excluded)	1.786	1.782	0.018	1.774	1.797	75	
	C*–SO_2–C (overall)	1.779	1.778	0.020	1.764	1.790	94	
	C*–SO_2–O–X	1.745	1.744	0.009	1.738	1.754	7	34
	C*–SO_2–N–X_2	1.758	1.756	0.018	1.746	1.773	17	34
Csp^3–S(3)	C*–S(=O)–C (C* = CH_3 excluded)	1.818	1.814	0.024	1.802	1.829	69	
	C*–S(=O)–C (overall)	1.809	1.806	0.025	1.793	1.820	88	
	CH_3–S^+–X_2	1.786	1.787	0.007	1.779	1.792	21	
	C*–S^+–X_2 (C* = CH_3 excluded)	1.823	1.820	0.016	1.812	1.834	18	
	C*–S^+–X_2 (overall)	1.804	1.794	0.025	1.788	1.820	41	
Csp^3–S(2)	C*–SH	1.808	1.805	0.010	1.800	1.819	6	
	CH_3–S–C*	1.789	1.787	0.008	1.784	1.794	9	
Csp^3–S(2)	C–CH_2–S–C*	1.817	1.816	0.013	1.808	1.824	92	
	C_2–CH–S–C*	1.819	1.819	0.011	1.811	1.825	32	
	C_3–C–S–C*	1.856	1.860	0.011	1.854	1.863	26	
	C*–S–C* (overall)	1.819	1.817	0.019	1.809	1.827	242	
	in thiirane	1.834	1.835	0.025	1.810	1.858	4	9
	in thiirane: see ZCMXSP (1.817, 1.844)							
	in tetrahydrothiophene	1.827	1.826	0.018	1.811	1.837	20	
	in tetrahydrothiopyran	1.823	1.821	0.014	1.812	1.832	24	
	C–CH_2–S–S–X	1.823	1.820	0.014	1.813	1.832	41	
	C_3–C–S–S–X	1.863	1.865	0.015	1.848	1.878	11	
	C*–S–S–X (overall)	1.833	1.828	0.022	1.818	1.848	59	
Csp^2–S(2)	C=C–S–C*	1.751	1.755	0.017	1.740	1.764	61	
	C=C–S–C=C (in tetrathiafulvalene)	1.741	1.741	0.011	1.733	1.750	88	
	C=C–S–C=C (in thiophene)	1.712	1.712	0.013	1.703	1.722	60	
	O=C–S–C#	1.762	1.759	0.018	1.747	1.778	20	
Car–S(4)	*Car*–SO_2–C	1.763	1.764	0.009	1.756	1.769	96	
	Car–SO_2–O–X	1.752	1.750	0.008	1.749	1.756	27	
	Car–SO_2–N–X_2	1.758	1.759	0.013	1.749	1.765	106	35
Car–S(3)	*Car*–S(=O)–C	1.790	1.790	0.010	1.783	1.798	41	
	Car–S^+–X_2	1.778	1.779	0.010	1.771	1.787	10	
Car–S(2)	*Car*–S–C*	1.773	1.774	0.009	1.765	1.779	44	
	Car–S–*Car*	1.768	1.767	0.010	1.762	1.774	158	
	Car–S–*Car* (in phenothiazine)	1.764	1.764	0.008	1.760	1.769	48	
	Car–S–S–X	1.777	1.777	0.012	1.767	1.785	47	
Csp^1–S(2)	N≡C–S–X	1.679	1.683	0.026	1.645	1.698	10	
Csp^1–S(1)	$(N≡C–S)^-$	1.630	1.630	0.014	1.619	1.641	14	
Csp^2=S(1)	$(C^*)_2$–C=S: see IPMUDS (1.599)							
	$(Car)_2$–C=S: see CELDOM (1.611)							
	$(X)_2$–C=S (X = C,N,O,S)	1.671	1.675	0.024	1.656	1.689	245	
	X_2N–C(=S)–S–X	1.660	1.660	0.016	1.648	1.674	38	
	$(X_2N)_2$–C=S (thioureas)	1.681	1.684	0.020	1.669	1.693	96	
	N–C$(\simeq S)_2$	1.720	1.721	0.012	1.709	1.731	20	
Csp^3–Se	C#–Se	1.970	1.967	0.032	1.948	1.998	21	
Csp^2–Se(2)	C=C–Se–C=C (in tetraselenafulvalene)	1.893	1.895	0.013	1.882	1.902	32	
Car–Se(3)	Ph_3–Se^+	1.930	1.929	0.006	1.924	1.936	13	
Csp^3–Si(5)	C#–Si^-–X_4	1.874	1.876	0.015	1.859	1.884	9	
Csp^3–Si(4)	CH_3–Si–X_3	1.857	1.857	0.018	1.848	1.869	552	
	C*–Si–X_3 (C* = CH_3 excluded)	1.888	1.887	0.023	1.872	1.905	124	
	C*–Si–X_3 (overall)	1.863	1.861	0.024	1.850	1.875	681	
Car–Si(4)	*Car*–Si–X_3	1.868	1.868	0.014	1.857	1.878	178	
Csp^1–Si(4)	C≡C–Si–X_3	1.837	1.840	0.012	1.824	1.849	8	
Csp^3–Te	C#–Te	2.158	2.159	0.030	2.128	2.177	13	
Car–Te	*Car*–Te	2.116	2.115	0.020	2.104	2.130	72	
Csp^2=Te	see CEDCUJ (2.044)							

Bond	Substructure	*d*	*m*	σ	q_1	q_u	*n*	Note
Cl–Cl	see PHASCL (2.306, 2.227)							
Cl–I	see CMBIDZ (2.563), HXPASC (2.541, 2.513), METAMM (2.552), BQUINI (2.416, 2.718)							
Cl–N	see BECTAE (1.743–1.757), BOGPOC (1.705)							
Cl–O(1)	in ClO^-_4	1.414	1.419	0.026	1.403	1.431	252	
Cl–P	$(N{\simeq})_2$**P–Cl** (N ≃ P aromatic)	1.997	1.994	0.015	1.989	2.004	46	
	Cl–P (overall)	2.008	2.001	0.035	1.986	2.028	111	
Cl–S	**Cl–S** (overall)	2.072	2.079	0.023	2.047	2.091	6	
	see also longer bonds in CILSAR (2.283), BIHXIZ (2.357), CANLUY (2.749)							
Cl–Se	see BIRGUE10, BIRHAL10, CTCNSE(2.234–2.851)							
Cl–Si(4)	**Cl–Si**–X_3 (monochloro)	2.072	2.075	0.009	2.066	2.078	5	
	$\mathbf{Cl_2}$–**Si**–X_2 and $\mathbf{Cl_3}$–**Si**–X	2.020	2.012	0.015	2.007	2.036	5	
Cl–Te	Cl–Te in range 2.34–2.60	2.520	2.515	0.034	2.493	2.537	22	36
	see also longer bonds in BARRIV, BOJPUL, CETUTE, EPHTEA, OPNTEC10 (2.73–2.94)							
F–N(3)	**F–N**–C_2 and $\mathbf{F_2}$–**N**–C	1.406	1.404	0.016	1.395	1.416	9	
F–P(6)	in hexafluorophosphate, PF^-_6	1.579	1.587	0.025	1.563	1.598	72	
F–P(3)	$(N{\simeq})_2$**P–F**(N ≃ P aromatic)	1.495	1.497	0.016	1.481	1.510	10	
F–S	43 observations in range 1.409–1.770 in a wide variety of environments; F–S(6) in F_2–SO_2–C_2 (see FPSULF10, BETJOZ)	1.640	1.646	0.011	1.626	1.649	6	
	F–S(4) in $\mathbf{F_2}$–**S**(=O)–N (see BUDTEZ)	1.527	1.528	0.004	1.524	1.530	24	37
F–Si(6)	in SiF^{2-}_6	1.694	1.701	0.013	1.677	1.703	6	
F–Si(5)	**F–Si**$^-$–X_4	1.636	1.639	0.035	1.602	1.657	10	
F–Si(4)	**F–Si**–X_3	1.588	1.587	0.014	1.581	1.599	24	
F–Te	see CUCPIZ (F–Te(6) = 1.942, 1.937), FPHTEL(F–Te(4) = 2.006)							
H–N(4)	X_3–**N**$^+$–**H**	1.033	1.036	0.022	1.026	1.045	87	21
H–N(3)	X_3–**N–H**	1.009	1.010	0.019	0.997	1.023	95	21
H–O(2)	in alcohols C*–**O–H**	0.967	0.969	0.010	0.959	0.974	63	21
	C#–**O–H**	0.967	0.970	0.010	0.959	0.974	73	21
	in acids O=C–**O–H**	1.015	1.017	0.017	1.001	1.031	16	21,38
I–I	in I^-_3	2.917	2.918	0.011	2.907	2.927	6	
I–N	see BZPRIB, CMBIDZ, HMTITI, HMTNTI, IFORAM, IODMAM (2.042–2.475)							
I–O	X–**I–O** (see BZPRIB, CAJMAB, IBZDAC11)	2.144	2.144	0.028	2.127	2.164	6	
	for IO^-_6 see BOVMEE (1.829–1.912)							
I–P(3)	see CEHKAB (2.490–2.493)							†
I–S	sec DTHIBR10 (2.687), ISUREA10 (2.629), BZTPPI (3.251)							
I–Te(4)	**I–Te**–X_3	2.926	2.928	0.026	2.902	2.944	8	
N(4)–N(3)	X_3–**N**$^+$–**N**^{0}–X_2 (N^0 planar)	1.414	1.414	0.005	1.412	1.418	13	
N(3)–N(3)	(C)(C,H)–$\mathbf{N_a}$–$\mathbf{N_b}$ (C)(C,H)							5,39
	$\mathbf{N_a}$, $\mathbf{N_b}$ pyramidal	1.454	1.452	0.021	1.444	1.457	44	40
	$\mathbf{N_a}$ pyramidal, $\mathbf{N_b}$ planar	1.420	1.420	0.015	1.407	1.433	68	40
	$\mathbf{N_a}$, $\mathbf{N_b}$ planar	1.401	1.401	0.018	1.384	1.418	40	40
	overall	1.425	1.425	0.027	1.407	1.443	139	
N(3)–N(2)	in pyrazole (N1–N2)	1.366	1.366	0.019	1.350	1.375	20	
	in pyridaznium (Nl$^+$≃N2)	1.350	1.349	0.010	1.345	1.361	7	
N(2)≃N(2)	**N≃N** (aromatic) in pyridazine							
	with C,H as *ortho* substituents	1.304	1.300	0.019	1.287	1.326	6	
	with N,Cl as *ortho* substituents	1.368	1.373	0.011	1.362	1.375	9	
N(2)=N(2)	C#–N=N–C#							
	cis	1.245	1.244	0.009	1.239	1.252	21	
	trans	1.222	1.222	0.006	1.218	1.227	6	
	(overall)	1.240	1.241	0.012	1.230	1.251	27	
	Car–N=N–*Car*	1.255	1.253	0.016	1.247	1.262	13	
	X–**N=N**=N (azides)	1.216	1.226	0.028	1.202	1.237	19	
N(2)=N(1)	X–N=**N=N** (azides)	1.124	1.128	0.015	1.114	1.137	19	
N(3)–O(2)	$(C,H)_2$–**N–OH** (Nsp^2: planar)	1.396	1.394	0.012	1.390	1.401	28	
	C_2–**N–O**–C							
	(Nsp^3: pyramidal)	1.463	1.465	0.012	1.457	1.468	22	
	(Nsp^2: planar)	1.397	1.394	0.011	1.388	1.409	12	
	in furoxan (N2–O1)	1.438	1.436	0.009	1.430	1.447	14	

Bond	Substructure	*d*	*m*	σ	q_1	q_u	*n*	Note
N(3)–O(1)	$(C\simeq)_2N^+$–**O**$^-$ in pyridine *N*-oxides	1.304	1.299	0.015	1.291	1.316	11	
	in furoxan ($^+$N2–O6$^-$)	1.234	1.234	0.008	1.228	1.240	14	
N(2)–O(2)	in oximes							
	$(C\#)_2$–C=**N–OH**	1.416	1.418	0.006	1.416	1.420	7	
	(H)(C*sp*^{2})–C=**N–OH**	1.390	1.390	0.011	1.380	1.401	20	
	(C#)(C*sp*^{2})–C=**N–OH**	1.402	1.403	0.010	1.393	1.410	18	
	$(Csp^2)_2$–C=**N–OH**	1.378	1.377	0.017	1.365	1.393	16	
	$(C,H)_2$–C=**N–OH** (overall)	1.394	1.395	0.018	1.379	1.408	67	
	in furazan (O1–N2, O1–N5)	1.385	1.383	0.013	1.378	1.392	12	
	in furoxan (O1–N5)	1.380	1.380	0.011	1.370	1.388	14	
	in isoxazole (O1–N2)	1.425	1.425	0.010	1.417	1.434	9	
N(3)=O(1)	in nitrate ions NO_3^-	1.239	1.240	0.020	1.227	1.251	105	
	in nitro groups							
	C*–NO_2	1.212	1.214	0.012	1.206	1.221	84	
	C#–NO_2	1.210	1.210	0.011	1.203	1.218	251	
	Car–NO_2	1.217	1.218	0.011	1.211	1.215	1116	
	C–NO_2 (overall)	1.218	1.219	0.013	1.210	1.226	1733	
N(3)–P(4)	X_2–P(=X)–NX_2							
	N*sp*^{2}: planar	1.652	1.651	0.024	1.634	1.670	205	
	N*sp*^{3}: pyramidal	1.683	1.683	0.005	1.680	1.686	6	
	(overall)	1.662	1.662	0.029	1.639	1.682	358	
	subsets of this group are:							
	O_2–**P**(=S)–**NX_2**	1.628	1.624	0.015	1.615	1.634	9	
	C–**P**(=S)–$(NX_2)_2$	1.691	1.694	0.018	1.678	1.703	28	
	O–**P**(=S)–$(NX_2)_2$	1.652	1.654	0.014	1.642	1.664	28	
	P(=O)–$(NX_2)_3$	1.663	1.668	0.026	1.640	1.679	78	
N(3)–P(3)	–**NX**–**P**(–X)–NX–**P**(–X)–(P_2N_2ring)	1.730	1.721	0.017	1.716	1.748	20	
	–**NX**–**P**(=S)–NX–**P**(=S)–(P_2N_2ring)	1.697	1.697	0.015	1.690	1.703	44	
	in P-substituted phosphazenes:							
	$(N\simeq)_2$**P–N** (amino) (aziridinyl)	1.637	1.638	0.014	1.625	1.651	16	
		1.672	1.674	0.010	1.665	1.676	15	
N(2)=P(4)	Ph_3–**P=N**$^+$=P–Ph_3	1.571	1.573	0.013	1.563	1.580	66	
N(2)=P(3)	Ph_3–**P=N**–C,S	1.599	1.597	0.018	1.580	1.615	7	
N(2) ≃ P(3)	**N≃P**aromatic							
	in phosphazenes	1.582	1.582	0.019	1.571	1.594	126	
	in P ≃ N ≃ S	1.604	1.606	0.009	1.594	1.612	36	
N(3)–S(4)	C–SO_2–**NH_2**	1.600	1.601	0.012	1.591	1.610	14	35
	C–SO_2–**NH**–C#	1.633	1.633	0.019	1.615	1.652	47	35
	C–SO_2–**N**–$C(\#)_2$	1.642	1.641	0.024	1.623	1.659	38	35
N(3)–S(2)	C–**S–N**X_2N*sp*^{2}: planar	1.710	1.707	0.019	1.698	1.722	22	23
	(for N*sp*^{3} pyramidal see MODIAZ: 1.765)							
	X–S–NX_2N*sp*^{2}: planar	1.707	1.705	0.012	1.699	1.715	30	23
N(2)–S(2)	C=**N–S**–X	1.656	1.663	0.027	1.632	1.677	36	
N(2) ≃ S(2)	**N≃S** aromatic in **P≃N≃S**	1.560	1.558	0.011	1.554	1.563	37	
N(2)=S(2)	**N=S** in **N=S=N** and **N=S=S**	1.541	1.546	0.022	1.521	1.558	37	
N(3)–Se	see COJCUZ (1.830), DSEMOR10 (1.846, 1.852), MORTRS10 (1.841)							
N(2)–Se	see SEBZQI (1.805), NAPSEZ10 (1.809, 1.820)							
N(2)=Se	see CISMUM (1.790, 1.791)							
N(3)–Si(5)	see DMESIP01, BOJLER, CASSAQ, CASYOK, CECXEN, CINTEY, CIPBUY, FMESIB, MNPSIL, PNPOSI (1.973–2.344)							
N(3)–Si(4)	X_3–**Si–N**X_2 (overall)	1.748	1.746	0.022	1.735	1.757	170	
	subsets of this group are:							
	X_3–**Si–N**HX	1.714	1.719	0.014	1.702	1.727	16	
	X_3–**Si–N**X–Si–X_3 acyclic	1.743	1.744	0.016	1.731	1.755	45	
	N–Si–N in 4-membered rings	1.742	1.742	0.009	1.735	1.748	53	
	N–Si–N in 5-membered rings	1.741	1.742	0.019	1.726	1.749	33	
N(2)–Si(4)	X_3–**Si–N**$^-$–Si–X_3	1.711	1.712	0.019	1.693	1.729	15	
N–Te	see ACLTEP (2.402), BIBLAZ (1.980), CESSAU (2.023)							
O(2)–O(2)	C*–**O–O**–C*,H							
	τ(OO) = 70–85°	1.464	1.464	0.009	1.458	1.472	12	
	τ(OO)*ca.*180°	1.482	1.480	0.005	1.478	1.486	5	

Bond	Substructure	*d*	*m*	*σ*	q_1	q_u	*n*	Note
	overall	1.469	1.471	0.012	1.461	1.478	17	
	O=C–**O–O**–C=O see ACBZPO01 (1.446), CEYLUN (1.452), CIMHIP (1.454)							
	Si–**O–O**–Si	1.496	1.499	0.005	1.490	1.499	10	
O(2)–P(5)	X–**P**–(**OX**)$_4$							41
	trigonal bipyramidal:							
	axial	1.689	1.685	0.024	1.675	1.712	20	
	equatorial	1.619	1.622	0.024	1.604	1.628	20	
	square pyramidal	1.662	1.661	0.020	1.649	1.673	28	
O(2)–P(4)	C–**O–P**(≃ O)$_3^{2-}$	1.621	1.622	0.007	1.615	1.628	12	
	(H–**O**)$_2$–**P**(≃ O)$_2^-$	1.560	1.561	0.009	1.555	1.566	16	
	(C–**O**)$_2$–**P**(≃ O)$_2^-$	1.608	1.607	0.013	1.599	1.615	16	
	(C#–**O**)$_3$–**P**=O	1.558	1.554	0.011	1.550	1.564	30	
	(*Car*–**O**)$_3$–**P**=O	1.587	1.588	0.014	1.572	1.599	19	
	X–**O–P**(=O)–(C,N)$_2$	1.590	1.585	0.016	1.577	1.601	33	
	(X–**O**)$_2$–**P**(=O)–(C,N)	1.571	1.572	0.013	1.563	1.579	70	
O(2)–P(3)	(N≃)$_2$**P–O**–C (N ≃ P aromatic)	1.573	1.573	0.011	1.563	1.584	16	
O(1)=P(4)	C–O–**P**(≃**O**)$_3^{2-}$ (delocalized)	1.513	1.512	0.008	1.508	1.518	42	
	(H–O)$_2$–**P**(≃**O**)$_2^-$ (delocalized)	1.503	1.503	0.005	1.499	1.508	16	
	(C–O)$_2$–**P**(≃**O**)$_2^-$ (delocalized)	1.483	1.485	0.008	1.474	1.490	16	
	(C–O)$_3$–**P=O**	1.449	1.448	0.007	1.446	1.452	18	
	C$_3$–**P=O**	1.489	1.486	0.010	1.481	1.496	72	
	N$_3$–**P=O**	1.461	1.462	0.014	1.449	1.470	26	
	(C)$_2$(N)–**P=O**	1.487	1.489	0.007	1.479	1.493	5	
	(C,N)$_2$(O)–**P=O**	1.467	1.462	0.007	1.462	1.472	33	
	(C,N)(O)$_2$–**P=O**	1.457	1.458	0.009	1.454	1.462	35	
O(2)–S(4)	C–**O–SO**$_2$–C	1.577	1.576	0.015	1.566	1.584	41	
	C–**O–SO**$_2$–CH$_3$	1.569	1.569	0.013	1.556	1.582	7	
	C–**O–SO**$_2$–*Car*	1.580	1.578	0.015	1.571	1.588	27	
O(1)=S(4)	C–**SO**$_2$–C	1.436	1.437	0.010	1.431	1.442	316	42
	X–**SO**$_2$–NX$_2$	1.428	1.428	0.010	1.422	1.434	326	
	C–**SO**$_2$–N–(C,H)$_2$	1.430	1.430	0.009	1.425	1.435	206	
	C–**SO**$_2$–O–C	1.423	1.423	0.008	1.418	1.428	82	
	in **SO**$_4^{2-}$	1.472	1.473	0.013	1.463	1.481	104	
O(1)=S(3)	C–**S**(=**O**)–C	1.497	1.498	0.013	1.489	1.505	90	5
O–Se	see BAPPAJ, BIRGUE10, BIRHAL10, CXMSEO, DGLYSE, SPSEBU (1.597 for **O=Se** to 1.974 for **O–Se**)							
O(2)–Si(5)	(X–**O**)$_3$–**Si**–(N)(C)	1.663	1.658	0.023	1.650	1.665	21	
O(2)–Si(4)	X$_3$–**Si–O**–X (overall)	1.631	1.630	0.022	1.617	1.646	191	
O(2)–Si(4)	subsets of this group are:							
	X$_3$–**Si–O**–C#	1.645	1.647	0.012	1.634	1.652	29	
	X$_3$–**Si–O**–Si–X$_3$	1.622	1.625	0.014	1.614	1.631	70	
	X$_3$–**Si–O**–**O**–**Si**–X$_3$	1.680	1.676	0.008	1.673	1.688	10	
O(2)–Te(6)	(X–**O**)$_6$–**Te**	1.927	1.927	0.020	1.908	1.942	16	
O(2)–Te(4)	(X–**O**)$_2$–**Te**–X$_2$	2.133	2.136	0.054	2.078	2.177	12	
P(4)–P(4)	X$_3$–**P–P**–X$_3$	2.256	2.259	0.025	2.243	2.277	6	
P(4)–P(3)	see CECHEX (2.197), COZPIQ (2.249)							
P(3)–P(3)	X$_2$–**P–P**–X$_2$	2.214	2.210	0.022	2.200	2.224	41	
P(4)=P(4)	see BUTSUE (2.054)							
P(3)=P(3)	see BALXOB (2.034)							
P(4)=S(1)	C$_3$–**P=S**	1.954	1.952	0.005	1.950	1.957	13	
	(N,O)$_2$(C)–**P=S**	1.922	1.924	0.014	1.913	1.927	26	
	(N,O)$_3$–**P=S**	1.913	1.914	0.014	1.906	1.921	50	
P(4)=Se(1)	X$_3$–**P=Se**	2.093	2.099	0.019	2.075	2.108	12	
P(3)–Si(4)	X$_2$–**P–Si**–X$_3$: 3- and 4-rings excluded (see BOPFER, BOPFIV, CASTOF10, COZVIW: 2.201–2.317)	2.264	2.260	0.019	2.249	2.283	22	
P(4)=Te(1)	see MOPHTE (2.356), TTEBPZ (2.327)							
S(2)–S(2)	C–**S–S**–C							
	τ(SS) = 75–105°	2.031	2.029	0.015	2.021	2.038	46	
	τ(SS) = 0–20°	2.070	2.068	0.022	2.057	2.077	28	
	(overall)	2.048	2.045	0.026	2.028	2.068	99	
	in polysulphide chain **–S–S–S–**	2.051	2.050	0.022	2.037	2.065	126	

Bond	Substructure	d	m	σ	q_l	q_u	n	Note
S(2)–S(1)	X–N=S–S	1.897	1.896	0.012	1.887	1.908	5	
S–Se(4)	see BUWZUO (2.264, 2.269)							
S–Se(2)	X–**Se–S** (any)	2.193	2.195	0.015	2.174	2.207	9	
S(2)–Si(4)	X_3–**Si–S**–X	2.145	2.138	0.020	2.130	2.158	19	
S(2)–Te	X–**S–Te** (any)	2.405	2.406	0.022	2.383	2.424	10	
	X=**S–Te** (any)	2.682	2.686	0.035	2.673	2.694	28	
Se(2)–Se(2)	X–**Se–Se**–X	2.340	2.340	0.024	2.315	2.361	15	
Se(2)–Te(2)	see BAWFUA, BAWGAH (2.524–2.561)							†
Si(4)–Si(4)	X_3–**Si–Si**–X_3 3-membered rings excluded: see CIHRAM (2.511)	2.359	2.359	0.012	2.349	2.366	42	
Te–Te	see CAHJOK (2.751, 2.704)							

† The standard deviation in the sample for the bond type is greater than for the other entries.

Appendix 1. Footnotes to Table

1. Sample dominated by B–CH_3. For longer bonds in B^-–CH_3 see LIT-MEB10 [B(4)–CH_3 = 1.621–1.644].
2. p(π)–p(π) Bonding with Bsp^2 and Nsp^2 coplanar (τBN = 0 ± 15°) predominates. See G. Schmidt, R. Boese, and D. Bläser, *Z. Naturforsch.*, 1982, **37b**, 1230.
3. 84 observations range from 1.38 to 1.61 Å and individual values depend on substituents on B and O. For a discussion of borinic acid adducts see S. J. Rettig and J. Trotter, *Can. J. Chem.*, 1982, **60**, 2957.
4. See M. Kaftory in *The Chemistry of Functional Groups. Supplement D: The Chemistry of Halides, Pseudohalides, and Azides*, S. Patai and Z. Rappoport, Eds., Wiley: New York, 1983, Part 2, ch. 24.
5. Bonds which are endocyclic or exocyclic to any 3- or 4-membered rings have been omitted from all averages in this section.
6. The overall average given here is for Csp^3–Csp^3 bonds which carry only C or H substituents. The value cited reflects the relative abundance of each 'substitution' group. The 'mean of means' for the 9 subgroups is 1.538 (σ = 0.022) Å
7. See F. H. Allen, (a) *Acta Crystallogr.*, 1980, **B36**, 81; (b) 1981, **B37**, 890.
8. See F. H. Allen, *Acta Crystallogr.*, 1984, **B40**, 64.
9. See F. H. Allen, *Tetrahedron*, 1982, **38**, 2843.
10. See F. H. Allen, *Tetrahedron*, 1982, **38**, 645.
11. Cyclopropanones and cyclobutanones excluded.
12. See W. B. Schweizer and J. D. Dunitz, *Helv. Chim. Acta*, 1982, **65**, 1547.
13. See L. Norskov-Lauritsen, H.-B. Bürgi, P. Hoffmann, and H. R. Schmidt, *Helv. Chim. Acta*, 1985, **68**, 76.
14. See P. Chakrabarti and J. D. Dunitz, *Helv. Chim. Acta*, 1982, **65**, 1555.
15. See J. L. Hencher in *The Chemistry of the C≡C Triple Bond*, S. Patai, Ed., Wiley, New York, 1978, ch. 2.
16. Conjugated: torsion angle about central C–C single bond is 0 ± 20° (*cis*) or 180 ± 20° (*trans*).
17. Unconjugated: torsion angle about central C–C single bond is 20–160°.
18. Other conjugative substituents excluded.
19. TCNQ is tetracyanoquinodimethane.
20. No difference detected between C2 ≃ C3 and C3 ≃ C4 bonds.
21. Derived from neutron diffraction results only.
22. Nsp^3: pyramidal; mean valence angle at N is in range 108–114°.
23. Nsp^2: planar; mean valence angle at N is ≥ 117.5°.
24. Cyclic and acyclic peptides.
25. See R. H. Blessing, *J. Am. Chem. Soc.*, 1983, **105**, 2776.
26. See L. Lebioda, *Acta Crystallogr.*, 1980, **B36**, 271.
27. n = 3 or 4,i.e. tri- or tetra-substituted ureas.
28. Overall value also includes structures with mean valence angle at N in the range 115–118°.
29. See F. H. Allen and A. J. Kirby, *J. Am. Chem. Soc.*, 1984, **106**, 6197.
30. See A. J. Kirby, 'The Anomeric Effect and Related Stereoelectronic Effects at Oxygen,' Springer, Berlin, 1983.
31. See B. Fuchs, L. Schleifer, and E. Tartakovsky, *Nouv. J. Chim.*, 1984, **8**, 275.
32. See S. C. Nyburg and C. H. Faerman, *J. Mol. Struct.*, 1986, **140**, 347.
33. Sample dominated by P–CH_3 and P–CH_2–C.
34. Sample dominated by C* = methyl.
35. See A. Kalman, M. Czugler, and G. Argay, *Acta Crystallogr.*, 1981, **B37**, 868.
36. Bimodal distribution resolved into 22 'short' bonds and 5 longer outliers.
37. All 24 observations come from BUDTEZ.
38. 'Long' O–H bonds in centrosymmetric O---H---O H–bonded dimers are excluded.
39. N–N bond length also dependent on torsion angle about N–N bond and on nature of substituent C atoms; these effects are ignored here.
40. N pyramidal has average angle at N in range 100–113.5°; N planar has average angle of ≥ 117.5°.
41. See R. R. Holmes and J. A. Deiters, *J. Amer. Chem. Soc.*, 1977, **99**, 3318.
42. No detectable variation in S=O bond length with type of C-substituent.

Appendix 2. Reference Codes

The table below gives short-form codes to references cited in the main table. A full list of CSD bibliographic entries is given in SUP 56701.

Code	Reference
ACBZPO01	*J. Am. Chem. Soc.*, 1975, **97,** 6729.
ACLTEP	*J. Organomet. Chem.*, 1980, **184,** 417.
ASAZOC	*Dokl. Akad. Nauk SSSR*, 1979, **249,** 120.
BALXOB	*J. Am. Chem. Soc.*, 1981, **103,** 4587.
BAPPAJ	*Inorg. Chem.*, 1981, **20,** 3071.
BARRIV	*Acta Chem. Scand., Ser. A*, 1981, **35,** 443.
BAWFUA	*Cryst. Struct. Commun.*, 1981, **10,** 1345.
BAWGAH	*Cryst. Struct. Commun.*, 1981, **10,** 1353.
BECTAE	*J. Org. Chem.*, 1981, **46,** 5048, 1981.
BELNIP	*Z. Naturforsch., Teil B*, 1982, **37,** 299.
BEMLIO	*Chem. Ber.*, 1982, **115,** 1126.
BEPZEB	*Cryst. Struct. Commun.*, 1982, **11,** 175.
BETJOZ	*J. Am. Chem. Soc.*, 1982, **104,** 1683.
BETUTE10	*Acta Chem. Scand., Ser. A*, 1976, **30,** 719.
BIBLAZ	*Zh. Strukt. Khim.*, 1981, **22,** 118.
BICGEZ	*Z. Anorg. Allg. Chem.*, 1982, **486,** 90.

Code	Reference
BIHXIZ	*J. Chem. Soc., Chem. Commun.*, 1982, 982.
BIRGUE10	*Z. Naturforsch., Teil B*, 1983, **38,** 20.
BIRHAL10	*Z. Naturforsch., Teil B*, 1982, **37,** 1410.
BIZJAV	*J. Organomet. Chem.*, 1982, **238,** C1.
BOGPOC	*Z. Naturforsch., Teil B*, 1982, **37,** 1402.
BOGSUL	*Z. Naturforsch., Teil B*, 1982, **37,** 1230.
BOJLER	*Z. Anorg. Allg. Chem.*, 1982, **493,** 53.
BOJPUL	*Acta Chem. Scand., Ser. A*, 1982, **36,** 829.
BOPFER	*Chem. Ber.*, 1983, **116,** 146.
BOPFIV	*Chem. Ber.*, 1983, **116,** 146.
BOVMEE	*Acta Crystallogr.,Sect. B* , 1982, **38,** 1048.
BQUINI	*Acta Crystallogr., Sect. B*, 1979, **35,** 1930.
BTUPTE	*Acta Chem. Scand., Ser. A*, 1975, **29,** 738.
BUDTEZ	*Z. Naturforsch., Teil B*, 1983, **38,** 454.
BUPSIB10	*Z. Anorg. Allg. Chem.*, 1981, **474,** 31.
BUSHAY	*Z. Naturforsch., Teil. B*, 1983, **38,** 692.
BUTHAZ10	*Inorg. Chem.*, 1984, **23,** 2582.
BUTSUE	*J. Chem. Soc., Chem. Commun.*, 1983, 862.
BUWZUO	*Acta Chem. Scand., Ser A*, 1983, **37,** 219.
BZPRIB	*Z. Naturforsch., Teil B*, 1981, **36,** 922.
BZTPPI	*Inorg. Chem.*, 1978, **17,** 894.
CAHJOK	*Inorg. Chem.*, 1983, **22,** 1809.
CAJMAB	*Chem. Z*, 1983, **107,** 169.
CANLUY	*Tetrahedron Lett.*, 1983, **24,** 4337.
CASSAQ	*J. Struct. Chem.*, 1983, **2,** 101.
CASTOF10	*Acta Crystallogr., Sect. C*, 1984, **40,** 1879.
CASYOK	*J. Struct. Chem.*, 1983, **2,** 107.
CECHEX	*Z. Anorg. Allg. Chem.*, 1984, **508,** 61.
CECXEN	*J. Struct. Chem.*, 1983, **2,** 207.
CEDCUJ	*J. Org. Chem.*, 1983, **48,** 5149.
CEHKAB	*Z. Naturforsch., Teil B*, 1984, **39,** 139.
CELDOM	*Acta Crystallogr., Sect. C*, 1984, **40,** 556.
CESSAU	*Acta Crystallogr., Sect. C*, 1984, **40,** 653.
CETTAW	*Chem. Ber.*, 1984, **117,** 1089.
CETUTE	*Acta Chem. Scand., Ser A*, 1975, **29,** 763.
CEYLUN	*Izv. Akad. Nauk SSSR, Ser. Khim.*, 1983, 2744.
CIFZUM	*Acta Chem. Scand., Ser A*, 1984, **38,** 289.
CIHRAM	*Angew. Chem., Int. Ed. Engl.*, 1984, **23,** 302.
CILRUK	*J. Chem. Soc., Chem. Commun.*, 1984, 1023.
CILSAR	*J. Chem. Soc., Chem. Commun.*, 1984, 1021.
CIMHIP	*Acta Crystallogr., C*, 1984, **40,** 1458.
CINTEY	*Dokl. Akad. Nauk SSSR*, 1984, **274,** 615.
CIPBUY	*J. Struct. Chem.*, 1983, **2,** 281.
CISMUM	*Z. Naturforsch., Teil B*, 1984, **39,** 485.
CISTED	*Z. Anorg. Allg. Chem.*, 1984, **511,** 95.
CIWYIQ	*Inorg. Chem.*, 1984, **23,** 1946.
CIYFOF	*Inorg. Chem.*, 1984, **23,** 1790.
CMBIDZ	*J. Org. Chem.*, 1979, **44,** 1447.
CODDEE	*Z. Naturforsch., Teil B*, 1984, **39,** 1257.
CODDII	*Z. Naturforsch., Teil B*, 1984, **39,** 1257.
COFVOI	*Z. Naturforsch., Teil B*, 1984, **39,** 1027.
COJCUZ	*Chem. Ber.*, 1984, **117,** 2686.
COSDIX	*Z. Naturforsch., Teil B*, 1984, **39,** 1344.
COZPIQ	*Chem. Ber.*, 1984, **117,** 2063.
COZVIW	*Z. Anorg. Allg. Chem.*, 1984, **515,** 7.
CTCNSE	*J. Am. Chem. Soc.*, 1980, **102,** 5430.
CUCPIZ	*J. Am. Chem. Soc.*, 1984, **106,** 7529.
CUDLOC	*J. Cryst. Spectrosc.*, 1985, **15,** 53.
CUDLUI	*J. Cryst. Spectrosc.*, 1985, **15,** 53.
CUGBAH	*Acta Crystallogr., Sect. C*, 1985, **41,** 476.
CXMSEO	*Acta Crystallogr., Sect. B*, 1973, **29,** 595.
DGLYSE	*Acta Crystallogr., Sect. B*, 1975, **31,** 1785.
DMESIP01	*Acta Crystallogr., Sect. C*, 1984, **40,** 895.
DSEMOR10	*J. Chem. Soc., Dalton Trans.*, 1980, 628.
DTHIBR10	*Inorg. Chem.*, 1971, **10,** 697.
EPHTEA	*Inorg. Chem.*, 1980, **19,** 2487.
ESEARS	*J. Chem. Soc. C*, 1971, 1511.
ETEARS	*J. Chem. Soc. C*, 1971, 1511.
FMESIB	*J. Organomet. Chem.*, 1980, **197,** 275.
FPHTEL	*J. Chem. Soc., Dalton Trans.*, 1980, 2306.
FPSULF10	*J. Am. Chem. Soc.*, 1982, **104,** 1683.
HCLENE10	*Acta Crystallogr., Sect. B*, 1982, **38,** 3139.
HMTITI	*Acta Crystallogr., Sect. B*, 1975, **31,** 1505.
HMTNTI	*Z. Anorg. Allg. Chem.*, 1974, **409,** 237.
HXPASC	*J. Chem. Soc., Dalton Trans.*, 1975, 1381.
IBZDAC11	*J. Chem. Soc., Dalton Trans.*, 1979, 854.
IFORAM	*Monatsh. Chem.*, 1974, **105,** 621.
IODMAM	*Acta Crystallogr., Sect. B*, 1977, **33,** 3209.
IPMUDS	*Acta Crystallogr., Sect. B*, 1973, **29,** 2128.
ISUREA10	*Acta Crystallogr., Sect. B*, 1972, **28,** 643.
LITMEB10	*J. Am. Chem. Soc.*, 1975, **97,** 6401.
MESIAD	*Z. Naturforsch., Teil B*, 1980, **35,** 789.
METAMM	*Acta Crystallogr.*, 1964, **17,** 1336.
MNPSIL	*J. Am. Chem. Soc.*, 1969, **91,** 4134.
MODIAZ	*J. Heterocycl. Chem.*, 1980, **17,** 1217.
MOPHTE	*Acta Chem. Scand., Ser. A*, 1980, **34,** 333.
MORTRS10	*J. Chem. Soc., Dalton Trans.*, 1980, 628.
NAPSEZ10	*J. Am. Chem. Soc.*, 1980, **102,** 5070.
NBBZAM	*Z. Naturforsch., Teil B*, 1977, **32,** 1416.
OPIMAS	*Aust. J. Chem.*, 1977, **30,** 2417.
OPNTEC10	*J. Chem. Soc., Dalton Trans.*, 1982, 251.
PHASCL	*Acta Crystallogr., Sect. B*, 1981, **37,** 1357.
PHASOC01	*Aust. J. Chem.*, 1975, **28,** 15.
PNPOSI	*J. Am. Chem. Soc.*, 1968, **90,** 5102.
SEBZQI	*J. Chem. Soc., Chem. Commun.*, 1977, 325.
SPSEBU	*Acta Chem. Scand., Ser. A*, 1979, **33,** 403.
TEACBR	*Cryst. Struct. Commun.*, 1974, **3,** 753.
THINBR	*J. Am. Chem. Soc.*, 1970, **92,** 4002.
TMPBTI	*Acta Crystallogr., Sect. B*, 1975, **31,** 1116.
TPASSN	*J. Chem. Soc., Dalton Trans.*, 1977, 514.
TPASTB	*Cryst. Struct. Commun.*, 1976, **5,** 39.
TPHOSI	*Z. Naturforsch., Teil B*, 1979, **34,** 1064.
TTEBPZ	*Z. Naturforsch., Teil B*, 1979, **34,** 256.
ZCMXSP	*Cryst. Struct. Commun.*, 1977, **6,** 93.

BOND LENGTHS IN ORGANOMETALLIC COMPOUNDS

This table summarizes the average values of interatomic distances of representative metal–ligand bonds. Sigma bonds between *d*- and *f*-block metals and the elements C, N, O, P, S, and As are included. The values are extracted from a much larger list in Reference 1. The tabulated values are the unweighted means of reported measurements on compounds in each category. If four or more measurements are available, the standard deviation is given in parentheses. All values are in Ångström units (10^{-10} m).

The first part of the table covers metal–carbon bonds in different ligand categories, while the second part covers metal bonds to other elements. R stands for any alkyl group; Me for a CH_3 group; C_6R_5 indicates an aryl group; and C(=O)R an acyl group. Metals are listed in atomic number order.

Reference

1. Orpen, A. G., Brammer, L., Allen, F. H., Kennard, O., Watson, D. G., and Taylor, R., *J. Chem. Soc. Dalton Trans.*, 1989, S1-S83.

Metal-Carbon Bond Lengths in Å

M	M–CH_3	M–CH_2R	M–CR=CR_2	M–C_6R_5	M–C(=O)R
Ti		2.167	2.215(0.042)	2.148	
V				2.114(0.012)	
Cr	2.168		2.035(0.009)	2.075(0.019)	
Mn	2.095(0.030)	2.176(0.024)	2.007	2.064(0.021)	2.044
Fe	2.074	2.091(0.030)	1.991(0.039)	2.031(0.062)	1.997(0.033)
Co	2.014(0.023)	2.039(0.032)	1.934(0.019)	1.974	1.990
Ni	2.029	1.964	1.892(0.017)	1.917(0.038)	1.850(0.059)
Cu				2.020	
Zn		1.964			
Zr	2.292(0.049)		2.257		
Nb	2.336	1.319			
Mo	2.254(0.065)	2.250(0.061)	2.204(0.049)	2.193(0.054)	2.109
Ru	2.179(0.045)	2.036(0.010)	2.063	2.092(0.057)	2.091
Rh	2.092(0.027)	2.100	2.040(0.054)	2.011(0.026)	1.995(0.031)
Pd		2.028	2.000(0.024)	1.981(0.032)	1.982(0.029)
Hf	2.275(0.049)		2.205		
Ta	2.217(0.035)	2.225(0.056)		2.199(0.073)	
W	2.189(0.039)	2.175	2.224		
Re	2.173(0.051)	2.290		2.027	2.190(0.027)
Os		2.221	2.052	2.090(0.032)	2.161
Ir	2.175		2.071(0.044)	2.070(0.038)	2.019
Pt	2.083(0.045)	2.062(0.031)	2.024(0.037)	2.049(0.046)	1.991(0.025)
Au	2.066(0.045)		2.042	2.059(0.024)	
Hg	2.072(0.026)	2.125		2.086(0.040)	
Th	2.567				

Bond Lengths of Metals to Other Elements

M	M–NH_3	M–OH_2	M–PMe_3	M–SR	M–AsR_3
Ti		2.066(0.052)		2.369	2.686
V		2.129(0.131)	2.510(0.010)	2.378(0.007)	
Cr	2.069(0.008)	1.997(0.070)	2.389(0.069)	2.362	2.460(0.040)
Mn		2.189(0.040)	2.455(0.164)	2.366(0.054)	2.400(0.013)
Fe		2.085(0.066)	2.246(0.042)	2.271(0.028)	2.352(0.043)
Co	1.965(0.021)	2.085(0.064)	2.217(0.043)	2.254(0.025)	2.323(0.021)
Ni	2.074(0.093)	2.079(0.038)	2.204(0.031)	2.187(0.007)	2.333(0.035)
Cu	1.987(0.017)	2.186(0.215)			2.367(0.016)
Zn	2.044	2.090(0.061)		2.295	
Y		2.398(0.068)			
Zr			2.692		
Nb		2.248(0.137)			2.741(0.008)
Mo	2.217	2.201(0.094)	2.462(0.046)	2.401(0.050)	2.582(0.036)
Ru	2.126(0.024)	2.074(0.051)	2.307(0.050)		2.446(0.031)
Rh	2.114(0.018)	2.190(0.096)	2.266(0.036)		2.416(0.039)
Pd	2.032	2.200	2.287(0.018)		2.386(0.052)

M	M–NH_3	M–OH_2	M–PMe_3	M–SR	M–AsR_3
Ag		2.350			
Cd		2.318(0.065)		2.444	
La		2.556(0.062)			
Ce		2.565(0.063)			
Pr		2.518(0.038)			
Nd		2.533(0.058)			
Sm		2.459(0.050)			
Eu		2.441(0.055)			
Gd		2.443(0.074)			
Tb		2.455			
Dy		2.409(0.074)			
Ho		2.407(0.069)			
Er		2.404(0.083)			
Yb		2.353(0.066)			
Lu		2.404(0.116)			
Ta			2.589(0.044)		
W		2.115(0.065)	2.485(0.039)		
Re	2.253	2.199(0.091)	2.369(0.065)		2.575(0.006)
Os	2.136	2.166	2.328(0.029)		
Ir	2.050(0.021)		2.323(0.028)	2.461	
Pt			2.295(0.036)	2.320(0.015)	2.366(0.058)
Au		2.157		2.293	
Hg		2.690(0.083)		2.402(0.065)	
Th		2.483(0.032)			
U		2.455(0.047)			

STRUCTURE OF FREE MOLECULES IN THE GAS PHASE

David R. Lide

This table gives information on the geometric structure of selected molecules in the gas phase, including the overall geometry, interatomic distances, and bond angles. The molecules have been chosen to provide data on a wide variety of chemical bonds and to illustrate the influence of molecular environment on bond distances and angles. The table is restricted to molecules with conventional covalent or ionic bonds, but it should be pointed out that structure data on many loosely bonded complexes of the van der Waals type have recently become available. The references below contain data on many molecules that are not included here and give additional information such as uncertainties and isotopic variations.

The two techniques for gas phase structure determination are spectroscopy and electron diffraction. The following codes are used to indicate the method used for each set of data:

ED – Gas phase electron diffraction
MW – Microwave spectroscopy, including both measurements in bulk gases and molecular beams
IR – Infrared spectroscopy
R – Raman spectroscopy
UV – Electronic spectroscopy in the ultraviolet and visible regions, including fluorescence measurements
ESR – Electron spin resonance

In some cases, data from two sources have been combined to derive the structure; these are labeled by "ED, MW", for example.

Because of the internal vibrations that are present in all molecules, even in their lowest energy state, the definition of interatomic distance is not a simple matter. The ideal measure is the equilibrium distance in the hypothetical non-vibrating state, designated by r_e. This is the value of the separation of the atoms at the minimum of the potential function that describes the forces between the two atoms. All other measures represent some form of average, generally complex, over the vibrational motions. Since the potential function is asymmetric and less steep at distances beyond the potential minimum, the average distance is normally greater than r_e. Distances determined by electron diffraction (ED) represent an average over all vibrational states that are populated at the temperature of the measurement; the most common measure is designated r_g. Distances determined by spectroscopy (MW, IR, R, or UV) through measurements on the ground vibrational state of the molecule, designated by r_0, describe some form of average, not easily defined, over the zero-point vibrations. Another measure that is frequently used in microwave spectroscopy is the "substitution" distance r_s, which is operationally defined through a series of measurements on different isotopic species. In simple cases, r_s often lies between r_0 and r_e and is therefore a closer approximation to r_e. Several other types of averages have been used; good discussions can be found in Volumes II/25 and II/28 of the *Landolt-Börnstein* series (Ref. 1) and in Refs. 4 and 5.

Unless otherwise specified, distances and angles given in this table are r_0 values if the method is spectroscopic and r_g values if the method is electron diffraction. When given, equilibrium and substitution distances are designated by r_e and r_s, respectively.

Many interatomic distances and angles calculated by *ab initio* techniques have been reported in the recent literature. However, it should be emphasized that all data in this table are obtained from direct experimental measurements. In a few cases, *ab initio* calculations of vibration-rotation interaction constants have been combined with the primary experimental measurements to derive r_e values in the table.

The number of significant figures in the values is an indication of the precision of the measurement; thus a distance quoted to three decimal places is probably reliable to about 0.005 Å or better. However, discrepancies between r_e, r_0, and r_g values for the same bond are often the order of 0.01 Å because of vibrational averaging considerations, so care must be taken in comparing bond distances in different molecules. Some distances in simple molecules are given here to four or five decimal places, but little chemical significance can be attached to differences beyond the third decimal place.

The table is presented in two parts: Part 1 covers molecules that do not contain carbon while Part 2 lists carbon-containing molecules. Because many of the entries in Part 1 are free radicals or other transient species whose systematic chemical names are unfamiliar, the listing in Part 1 is in order of chemical formula. Part 2 is ordered by name. In both parts, information on the overall configuration of the molecule is given through structure diagrams and/or indication of the point group of the equilibrium geometry. Subsequent columns give the values of the bond distances and angles; distances are given in Å units, where 1 Å = 10^{-10} m or 0.1 nm, and angles are given in degrees. The last column indicates the experimental method.

The contributions of Kozo Kuchitsu in preparing an earlier version of this table and in giving advice on the new version are gratefully acknowledged.

References

1. *Landolt-Börnstein Numerical Data and Functional Relationships in Science and Technology*, Springer-Verlag, Berlin. The following volumes are in the series *Structure Data of Free Polyatomic Molecules*: II/7, 1976 II/15, 1987 II/21, *Supplement to II/7 and II/15*, 1992 II/23, *Supplement to II/7, II/15, and II/21*, 1995 II/25A, *Inorganic Molecules*, 1998 II/25B, *Molecules Containing One or Two Carbon Atoms*, 1999 II/25C, *Molecules Containing Three or Four Carbon Atoms*, 2000 II/25D, *Molecules Containing Five or More Carbon Atoms*, 2003 II/28A, *Inorganic Molecules*, 2006 II/28B, *Molecules Containing One or Two Carbon Atoms*, 2006 II/28C, *Molecules Containing Three or Four Carbon Atoms*, 2007 II/28D, *Molecules Containing Five or More Carbon Atoms*, 2007.
2. Harmony, M. D., Laurie, V. W., Kuczkowski, R. L., Schwendeman, R. H., Ramsay, D. A., Lovas, F. J., Lafferty, W. J., and Maki, A. G., "Molecular Structure of Gas-Phase Polyatomic Molecules Determined by Spectroscopic Methods", *J. Phys. Chem. Ref. Data* 8, 619, 1979.
3. Huber, K. P., and Herzberg, G., *Molecular Spectra and Molecular Structure IV. Constants of Diatomic Molecules*, Van Nostrand Reinhold, London, 1979.
4. Hargittai, M., "Molecular Structure of Metal Halides," *Chem. Rev.* 100, 2233-2301, 2000.
5. Harmony, M. D., and Berry, R. J., *Struct. Chem.* 1, 49, 1989.

Part 1. Molecules Not Containing Carbon

Structures of Non-Carbon Molecules: Bond Distances in Å and Angles in Degrees

Compound	Structure							Method
AgBr		Ag—Br (r_e)	2.3931					MW
AgCl		Ag—Cl (r_e)	2.2808					MW
AgF		Ag—F (r_e)	1.9832					MW
AgH		Ag—H (r_e)	1.617					UV
AgI		Ag—I (r_e)	2.5446					MW
AgLi		Ag—Li	2.41					UV
AgO		Ag—O (r_e)	2.0030					UV
AgOH	bent	Ag—O	2.016	O—H	0.952	∠HOAg	108.3[a]	MW
AlBr		Al—Br (r_e)	2.295					UV
$AlBr_3$	D_{3h}	Al—Br	2.221					ED
AlCa		Al—Ca	3.148					UV
AlCl		Al—Cl (r_e)	2.1301					MW
$AlCl_3$	D_{3h}	Al—Cl	2.063					ED
AlCo		Al—Co	2.283					UV
AlCu		Al—Cu	2.339					UV
AlF		Al—F (r_e)	1.6544					MW
AlF_3	D_{3h}	Al—F	1.633					ED
AlH		Al—H (r_e)	1.6482					UV
AlI		Al—I (r_e)	2.5371					MW
AlI_3	D_{3h}	Al—I	2.461					ED
AlK		Al—K	3.88					UV
AlMn		Al—Mn	2.638					UV
AlNi		Al—Ni	2.321					UV
AlO		Al—O (r_e)	1.6176					UV
AlS		Al—S (r_e)	2.029					UV
AlV		Al—V	2.620					UV
AlZn		Al—Zn	2.696					UV
Al_2		Al—Al (r_e)	2.701					UV
Al_2Br_6	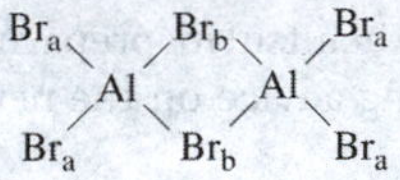	Al—Br_a	2.234	Al—Br_b	2.433			ED
	D_{2h}	∠Br_bAlBr_b	91.6	∠Br_aAlBr_a	122			
Al_2Cl_6	See Al_2Br_6	Al—Cl_a	2.061	Al—Cl_b	2.250			ED
	D_{2h}	∠Cl_bAlCl_b	90.0	∠Cl_aAlCl_a	122			
$AsBr_3$	C_{3v}	As—Br	2.324	∠BrAsBr	99.6			ED
$AsCl_3$	C_{3v}	As—Cl	2.165	∠ClAsCl	98.6			ED, MW
AsF_3	C_{3v}	As—F	1.710	∠FAsF	95.9			ED, MW
AsF_5	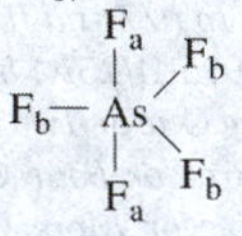 D_{3h}	As—F_a	1.711	As—F_b	1.656			ED
AsH		As—H (r_e)	1.5232					UV
AsH_3	C_{3v}	As—H (r_e)	1.511	∠HAsH (θ_e)	92.1			MW, IR
AsI_3	C_{3v}	As—I	2.557	∠IAsI	100.2			ED
AsN		As—N (r_e)	1.6184					UV
AsO		As—O (r_e)	1.6236					UV
AsP		As—P (r_e)	1.99954					MW
As_2		As—As (r_e)	2.1026					UV
AuH		Au—H (r_e)	1.5237					UV
Au_2		Au—Au (r_e)	2.4719					UV
BBr		B—Br (r_e)	1.888					UV
BBr_3	D_{3h}	B—Br	1.893					ED
BCl		B—Cl (r_e)	1.7153					UV
$BClF_2$	C_{2v}	B—Cl (r_s)	1.728	B—F	1.315	∠FBF	118.1	MW
BCl_3	D_{3h}	B—Cl	1.742					ED
BF		B—F (r_e)	1.2626					UV
BF_2H		B—H	1.189	B—F	1.311	∠FBF	118.3	MW

Compound	Structure							Method
BF_2OH	F_aF_bBOH	B—F_a (r_e)	1.3229	B—F_b (r_e)	1.3129	B—O (r_e)	1.3448	MW
	planar	∠FBF (θ_e)	118.36	∠F_aBO (θ_e)	122.25	∠BOH (θ_e)	113.14	
	F_a *cis* to OH	O—H (r_e)	0.9574					
BF_3	D_{3h}	B—F	1.313					ED, IR
BH		B—H (r_e)	1.2325					UV
BH_2NH_2	planar	B—N	1.391	B—H	1.195	N—H	1.004	MW
		∠HBH	122.2	∠HNH	114.2			
BH_3	planar	B—H	1.1900					IR
BH_3PH_3	staggered form	B—P	1.937	B—H	1.212	P—H	1.399	MW
		∠PBH	103.6	∠BPH	116.9	∠HBH	114.6	
		∠HPH	101.3					
BI_3	D_{3h}	B—I	2.118					ED
BN		B—N (r_e)	1.281					UV
BO		B—O (r_e)	1.2045					EPR
BO_2	linear	B—O	1.265					UV
BS		B—S	1.6091					UV
B_2		B—B (r_e)	1.590					UV
B_2H_6	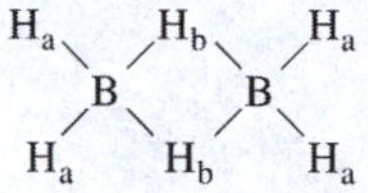	B—H_a	1.19	B—H_b	1.33	B···B	1.77	IR, ED
		∠H_aBH_a	122	∠H_bBH_b	97			
$B_3H_3O_3$		B—O	1.376	∠BOB	120	∠OBO	120	ED
$B_3H_6N_3$	C_2	B—N	1.435	B—H	1.26	N—H	1.05	ED
		∠BNB	121	∠NBN	118			
BaBr		Ba—Br (r_e)	2.8445					UV
$BaBr_2$		Ba—Br	2.912	∠BrBaBr	137.0			ED
BaCl		Ba—Cl (r_e)	2.6828					UV
BaF		Ba—F (r_e)	2.163					UV
BaH		Ba—H (r_e)	2.2318					UV
BaI		Ba—I (r_e)	3.0848					UV
BaI_2		Ba—I	3.150	∠IBaI	137.6			ED
BaO		Ba—O (r_e)	1.9397					MW
BaOH	linear	Ba—O	2.200	O—H	0.927			UV
BaS		Ba—S (r_e)	2.5074					MBE
$BeCl_2$	linear	Be—Cl (r_e)	1.791					ED,IR
BeF		Be—F (r_e)	1.3609					UV
BeF_2	linear	Be—F (r_e)	1.3730					IR
BeH		Be—H (r_e)	1.3431					UV
BeH_2	linear	Be—H (r_e)	1.3264					IR
BeO		Be—O (r_e)	1.3308					UV
BeS		Be—S (r_e)	1.7415					UV
BiBr		Bi—Br (r_e)	2.6095					MW
$BiBr_3$	C_{3v}	Bi—Br	2.577	∠BrBiBr	98.6			ED
BiCl		Bi—Cl (r_e)	2.4716					MW
$BiCl_3$	C_{3v}	Bi—Cl	2.424	∠ClBiCl	97.5			ED
BiF		Bi—F (r_e)	2.0516					MW
BiF_3	C_{3v}	Bi—F	1.987	∠FBiF	96.1			ED
BiH		Bi—H (r_e)	1.805					UV
BiI		Bi—I (r_e)	2.8005					MW
BiI_3	C_{3v}	Bi—I	2.807	∠IBiI	99.5			ED
BiO		Bi—O (r_e)	1.934					UV
BiP		Bi—P (r_e)	2.29345					IR
Bi_2		Bi—Bi (r_e)	2.6596					UV
BrCl		Br—Cl (r_e)	2.1361					MW
BrF		Br—F (r_e)	1.7590					MW
BrF_3	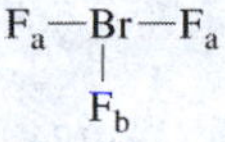	Br—F_a	1.810	∠F_{ax}BrF_{eq}	85.1	∠F_aBrF_b	86.2	MW
	C_{2v}	Br—F_b	1.721					
BrF_5	C_{4v}	Br—F (av.)	1.753	(Br—F_{eq}) – (Br—F_{ax})	0.069	∠F_{ax}BrF_{eq}	85.1	ED, MW
BrN_3	$BrN_aN_bN_c$	N_a—N_b	1.113[a]	N_b—N_c	1.247	N_a—Br	1.899	ED
	planar	∠NNN	170.7	∠BrNN	109.7			

Compound	Structure							Method
BrO		Br—O (r_e)	1.7172					MW
BrO_2	C_{2v}	Br—O (r_e)	1.644	∠OBrO (θ_e)	114.3			MW
Br_2		Br—Br (r_e)	2.2811					R
$CaBr_2$	linear	Ca—Br	2.62					ED
CaCl		Ca—Cl (r_e)	2.43676					UV
$CaCl_2$	linear	Ca—Cl	2.483					ED
CaF		Ca—F (r_e)	1.967					UV
CaH		Ca—H (r_e)	2.002					UV
CaI		Ca—I (r_e)	2.8286					UV
CaI_2	linear	Ca—I	2.840					ED
CaO		Ca—O (r_e)	1.8221					UV
CaOH	linear	Ca—O	1.985	O—H	0.921			UV
CaS		Ca—S (r_e)	2.3178					UV
CdH		Cd—H (r_e)	1.781					EPR
CdH_2	linear	Cd—H	1.6792					IR
$CdBr_2$	linear	Cd—Br	2.394					ED
$CdCl_2$	linear	Cd—Cl	2.284					ED
CdI_2	linear	Cd—I	2.582					ED
CeF_4	T_d	Ce—F	2.036					ED
CeI_3	quasiplanar	Ce—I	2.948					ED
ClBS	linear	B—Cl	1.681	B—S	1.606			MW
ClF		Cl—F (r_e)	1.6283					MW
ClF_3	F_a—Cl—F_a | F_b	Cl—F_a	1.698	Cl—F_b	1.598	∠F_aClF_b	87.5	MW
ClN_3	$ClN_aN_bN_c$ planar	N_a—N_b ∠NNN	1.253 171.0	N_b—N_c ∠ClNN	1.113 108.7	N_a—Cl	1.746	MW
ClO		Cl—O (r_e)	1.5696					MW, UV
ClO_2	C_{2v}	Cl—O	1.470	∠OClO	117.38			MW
Cl_2		Cl—Cl (r_e)	1.9878					UV
Cl_2O	C_{2v}	Cl—O	1.6959	∠ClOCl	110.89			MW
$CoBr_2$	linear	Co—Br	2.241					ED
$CoCl_2$	linear	Co—Cl	2.113					ED
CoF_2	linear	Co—F	1.754	[Co—F (r_e)]	1.738			ED
CoF_3	D_{3h}	Co—F	1.732					ED
CoH		Co—H (r_e)	1.542					UV
CrF_2	linear	Cr—F	1.795					ED
CrF_3	D_{3h}	Cr—F	1.732					ED
CrF_4	T_d	Cr—F	1.706					ED
CrH		Cr—H (r_e)	1.656					UV
CrO		Cr—O (r_e)	1.615					UV
CsBr		Cs—Br (r_e)	3.0723					MW
CsCl		Cs—Cl (r_e)	2.9063					MW
CsF		Cs—F (r_e)	2.3454					MW
CsH		Cs—H (r_e)	2.4938					UV
CsI		Cs—I (r_e)	3.3152					MW
CsO		Cs—O (r_e)	2.3007					MW
CsOH	linear; large amplitude bending mode	Cs—O (r_e)	2.395	O—H (r_e)	0.97			MW
Cs_2		Cs—Cs (r_e)	4.47					UV
CuBr		Cu—Br (r_e)	2.1734					MW
CuCl		Cu—Cl (r_e)	2.0512					MW
CuF		Cu—F (r_e)	1.7449					MW
CuF_2	linear	Cu—F	1.713					ED
CuH		Cu—H (r_e)	1.4626					UV
CuI		Cu—I (r_e)	2.3383					MW
CuLi		Cu—Li	2.26					UV
CuO		Cu—O (r_e)	1.7244					UV
CuOH	bent	Cu—O (r_s)	1.769	O—H	0.952	∠HOCu (θ_s)	110.24	MW
CuS		Cu—S	2.051					UV
Cu_2		Cu—Cu (r_e)	2.2197					UV
$DyBr_3$	quasiplanar	Dy—Br	2.609					ED

Compound	Structure							Method
$DyCl_3$	quasiplanar	Dy—Cl	2.461					ED
FN_3	$FN_aN_bN_c$	N_a—N_b	1.253	N_b—N_c	1.132	N_a—F	1.439	MW
	planar	∠NNN	170.3	∠FNN	103.8			
F_2		F—F (r_e)	1.4119					R
$FeBr_2$	linear	Fe—Br	2.294					ED
$FeCl_2$	linear	Fe—Cl	2.132					UV,ED
FeF_2	linear	Fe—F	1.769	[Fe—F (r_e)]	1.755			ED
FeF_3	D_{3h}	Fe—F	1.763					ED
FeH		Fe—H	1.620					IR
FeO		Fe—O	1.444					UV
FeS		Fe—S	2.017					MW
GaBr		Ga—Br (r_e)	2.3525					MW
$GaBr_3$	D_{3h}	Ga—Br	2.249					ED
GaCl		Ga—Cl (r_e)	2.2017					MW
$GaCl_3$	D_{3h}	Ga—Cl	2.110					ED
GaF		Ga—F (r_e)	1.7744					MW
GaF_3	D_{3h}	Ga—F	1.725					ED
GaH		Ga—H (r_e)	1.663					UV
GaI		Ga—I (r_e)	2.5747					MW
GaI_3	D_{3h}	Ga—I	2.458					ED
GaO		Ga—O	1.744					UV
Ga_2Br_6	See Al_2Br_6	Ga—Br_a	2.250	Ga—Br_b	2.453			ED
	D_{2h}	∠Br_aGaBr_a	92.7	∠Br_bGaBr_b	123			
Ga_2Cl_6	See Al_2Br_6	Ga—Cl_a	2.116	Ga—Cl_b	2.305			ED
	D_{2h}	∠Cl_aGaCl_a	90	∠Cl_bGaCl_b	124.5			
$GdBr_3$	C_{3v}	Gd—Br	2.641					ED
$GdCl_3$	C_{3v}	Gd—Cl	2.488					ED
GdF_3	C_{3v}	Gd—F	2.053					ED
GdI_3	C_{3v}	Gd—I	2.840	∠IGdI	108			ED
$GeBrH_3$	C_{3v}	Ge—H	1.526	Ge—Br	2.299	∠HGeH	106.2	MW, IR
$GeBr_2$		Ge—Br (r_e)	2.359	∠BrGeBr	101.0			ED
$GeBr_4$	T_d	Ge—Br	2.272					ED
$GeClH_3$	C_{3v}	Ge—H	1.537	Ge—Cl	2.150	∠HGeH	111.0	IR, MW
$GeCl_2$		Ge—Cl (r_e)	2.186	∠ClGeCl	100.3			ED
$GeCl_4$	T_d	Ge—Cl	2.113					ED
$GeFH_3$	C_{3v}	Ge—H	1.522	Ge—F	1.732	∠HGeH	113.0	MW, IR
GeF_2		Ge—F (r_e)	1.7321	∠FGeF (θ_e)	97.15			MW
GeH		Ge—H (r_e)	1.5880					UV
GeHI		Ge—I	2.525	Ge—H	1.593	∠HGeI	93.5	UV
GeH_4	T_d	Ge—H	1.5251					IR, R
GeI_2		Ge—I	2.540	∠IGeI	102.1			ED
GeI_4	T_d	Ge—I	2.515					ED
GeO		Ge—O (r_e)	1.6246					MW
GeS		Ge—S (r_e)	2.0121					MW
GeSe		Ge—Se (r_e)	2.1346					MW
GeTe		Ge—Te (r_e)	2.3402					MW
Ge_2H_6		Ge—Ge	2.403	Ge—H	1.541			ED
		∠HGeH	106.4	∠GeGeH	112.5			
HBr		H—Br (r_e)	1.4145					MW
HCl		H—Cl (r_e)	1.2746					MW
HClO	ClOH (bent)	Cl—O	1.690	O—H	0.975	∠HOCl	102.5	MW, IR
$HClO_4$	/H	Cl—O_a	1.407	Cl—O_b	1.639			ED
		∠O_aClO_a	114.3	∠O_aClO_b	104.1			
HF		H—F (r_e)	0.9169					MW
HFO	FOH (bent)	F—O	1.442	O—H	0.96	∠HOF	97.2	MW
HI		H—I (r_e)	1.6090					MW
HIO	IOH (bent)	I—O	1.9941	O—H	0.967	∠HOI	103.9	MW
HNO	bent	N—O	1.212	N—H	1.063	∠HNO	108.6	UV
HNO_2	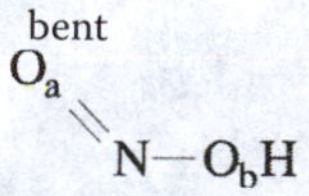	*s-trans* conformer		*s-cis* conformer				MW
		O_b—H	0.958	O_b—H	0.98			

Compound	Structure							Method
		N—O_b	1.432	N—O_b	1.39			
		N—O_a	1.170	N—O_a	1.19			
		∠O_aNO_b	110.7	∠O_aNO_b	114			
		∠NO_bH	102.1	∠NO_bH	104			
HNO_3	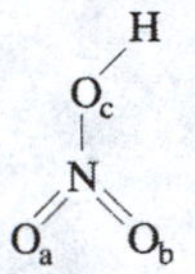	N—O_a	1.20	N—O_b	1.21	N—O_c	1.41	MW
		O_c—H	0.96	∠O_cNO_b	115.9	∠HO_cN	102.2	
	planar	∠O_cNO_a	113.9					
HNSO	planar *cis*	N—S	1.512	S—O	1.451	N—H	1.029	MW
		∠NSO	120.4	∠HNS	115.8			
HN_3	$HN_aNM_bN_c$	N_a—N_b	1.245	N_b—N_c	1.134	N_a—H	1.015	MW
	planar	∠NNN	171.8	∠HNN	109.2			
HPO		P—O	1.4843	P—H	1.473	∠HPO	104.57	MW
H_2		H—H (r_e)	0.74144					UV
H_2O	C_{2v}	O—H (r_e)	0.9575	∠HOH (θ_e)	104.51			MW, IR
H_2O_2	C_2	O—O	1.475	∠OOH	94.8	dihedral angle	119.8	IR
H_2S	C_{2v}	H—S (r_e)	1.3356	∠HSH (θ_e)	92.12			MW, IR
H_2SO_4	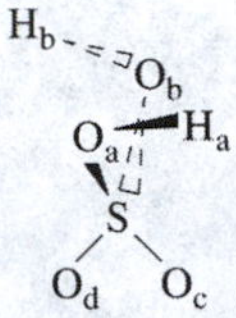 C_2	O—H	0.97	S—O_a	1.574	S—O_c	1.422	MW
		∠O_aSO_b	101.3	∠O_cSO_d	123.3	∠O_aSO_c	108.6	
		∠O_aSO_d	106.4	∠H_aO_aS	108.5	dihedral angle between the H_aO_aS and O_aSO_c planes	20.8	
	C_2	dihedral angle between the H_aO_aS and O_aSO_b planes	90.9	dihedral angle between the H_aSO_b and O_cSO_d planes	88.4			
H_2S_2	C_2	S—S	2.055	S—H	1.327	∠SSH	91.3	ED, MW
		dihedral angle	90.6					
$HfBr_4$	T_d	Hf—Br	2.450					ED
$HfCl_4$	T_d	Hf—Cl	2.316					ED
HfF		Hf—F	1.8596					UV
HfF_4	T_d	Hf—F	1.909					ED
HfI_4	T_d	Hf—I	2.662					ED
$HgBr_2$	linear	Hg—Br	2.384					ED
$HgCl_2$	linear	Hg—Cl	2.252					ED
HgH		Hg—H (r_e)	1.7404					UV
HgI_2	linear	Hg—I	2.568					ED
$HoCl_3$		Ho—Cl	2.462					ED
HoF_3		Ho—F	2.007					ED
HoO		Ho—O	1.797					UV
IBr		I—Br (r_e)	2.4691					MW
ICl		I—Cl (r_e)	2.3210					MW
IF		I—F (r_e)	1.9098					UV
IF_5	C_{4v}	I—F (av.)	1.860	(I—F_{eq}) – (I—F_{ax})	0.03	∠$F_{ax}IF_{eq}$	82.1	ED, MW
IO		I—O (r_e)	1.8676					MW
I_2		I—I (r_e)	2.6663					R
InBr		In—Br (r_e)	2.5432					MW
InCl		In—Cl (r_e)	2.4012					MW
$InCl_3$		In—Cl	2.291					ED
InF		In—F (r_e)	1.9854					MW
InH		In—H (r_e)	1.8376					UV
InI		In—I (r_e)	2.7537					MW
IrF_6	O_h	Ir—F	1.831					ED

Compound	Structure							Method
KBH_4	$H_a(BH_3)K$ (C_{3v})	B—H (BH_3)	1.272	B—H_a	1.233	K—B	2.656	MW
KBr		K—Br (r_e)	2.8208					MW
KCl		K—Cl (r_e)	2.6667					MW
KF		K—F (r_e)	2.1716					MW
KH		K—H (r_e)	2.244					UV
KI		K—I (r_e)	3.0478					MW
KOH	linear; large amplitude bending mode	K—O	2.212	O—H	0.91			MW
K_2		K—K (r_e)	3.9051					UV
KrF_2	linear	Kr—F	1.89					ED
LaBr		La—Br (r_e)	2.65208					MW
$LaBr_3$	C_{3v}	La—Br	2.742					ED
LaCl		La—Cl (r_e)	2.49804					MW
$LaCl_3$	C_{3v}	La—Cl	2.589					ED
LaF		La—F (r_e)	2.02338					MW
LaI		La—I (r_e)	2.87885					MW
LaO		La—O (r_e)	1.82591					UV
$LiBH_4$	$H_a(BH_3)Li$ (C_{3v})	B—H (H_3)	1.257	B—H_a	1.218	Li—B	1.939	MW
LiBr		Li—Br (r_e)	2.1704					MW
LiCl		Li—Cl (r_e)	2.0207					MW
LiF		Li—F (r_e)	1.5639					MW
LiH		Li—H (r_e)	1.5949					MW
LiI		Li—I (r_e)	2.3919					MW
LiO		Li—O (r_e)	1.68822					UV
LiOH	linear	Li—O (r_e)	1.5776	O—H (r_e)	0.949			MW
Li_2		Li—Li (r_e)	2.6729					UV
Li_2Cl_2	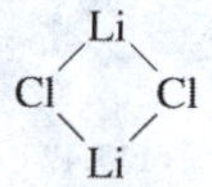	Li—Cl	2.23	Cl—Cl	3.61	∠ClLiCl	108	ED
Li_2O	linear	Li—O	1.606					UV
$LuBr_3$	C_{3v}	Lu—Br	2.557					ED
$LuCl_3$	C_{3v}	Lu—Cl	2.417	∠ClLuCl	112			ED
LuI_3	C_{3v}	Lu—I	2.768					ED
MgBr		Mg—Br (r_e)	2.34742					MW
MgCl		Mg—Cl (r_e)	2.1964					UV
$MgCl_2$	linear	Mg—Cl	2.179					ED
MgF		Mg—F (r_e)	1.7500					UV
MgF_2	linear	Mg—F	1.771					ED
MgH		Mg—H (r_e)	1.7297					UV
MgO		Mg—O (r_e)	1.749					UV
MgOH	linear	Mg—O	1.770	O—H	0.912			UV
Mg_2		Mg—Mg (r_e)	3.891					UV
$MnBr_2$	linear	Mn—Br	2.344					ED
$MnCl_2$	linear	Mn—Cl	2.202					ED
MnF_2	linear	Mn—F	1.811	[Mn—F (r_e)]	1.797			ED
MnH		Mn—H (r_e)	1.7308					UV
MnI_2	linear	Mn—I	2.538					ED
$MoCl_4O$	C_{4v}	Mo—Cl	2.279	Mo—O	1.658			ED
		∠ClMoCl	87.2					
MoF_4		Mo—F	1.851					ED
MoF_6	O_h	Mo—F	1.821					ED
NBr		N—Br (r_e)	1.79					UV
NCl		N—Cl (r_e)	1.6107					UV
$NClH_2$		N—H	1.017	N—Cl	1.748			MW, IR
		∠HNCl	103.7	∠HNH	107			
NCl_3		N—Cl	1.759	∠ClNCl	107.1			ED
NF		N—F (r_e)	1.3170					UV
NF_2		N—F	1.3528	∠FNF	103.18			MW
NH_2		N—H	1.024	∠HNH	103.3			UV
NH_2NO_2		N—N	1.427	N—H	1.005			MW

Compound	Structure							Method
		dihedral angle between NH_2 and NNO_2 planes	128.2	∠HNH	115.2	∠ONO	130.1	
NH_3	C_{3v}	N—H (r_e)	1.012	∠HNH (θ_e)	106.7			IR
NH_4Cl	H_3N····HCl (C_{3v})	N—Cl	3.136					MW
NH		N—H (r_e)	1.0362					LMR
NH_2OH	bisector of HNH angle is *trans* to OH bond	N—O	1.453	N—H	1.02	O—H	0.962	MW
		∠HNO	103.3	∠HNH	107	∠NOH	101.4	
NO		N—O (r_e)	1.1506					IR
NOCl		N—O	1.14	N—Cl	1.975	∠ONCl	113	MW
NOF		N—O	1.136	N—F	1.512	∠FNO	110.1	MW
NO_2		N—O	1.193	∠ONO	134.1			MW
NO_2Cl	C_{2v}	N—O	1.202	N—Cl	1.840	∠ONO	130.6	MW
NO_2F	C_{2v}	N—O	1.1798	N—F	1.467	∠ONO	136	MW
NS		N—S (r_e)	1.4940					IR
N_2		N—N (r_e)	1.0977					UV
N_2H_4	H_a atom is closer to the C_2 axis, H_b is farther from the C_2 axis	N—N	1.449	N—H	1.021	$∠NNH_b$	106	ED, MW
		∠HNH	106.6[a]	$∠NNH_a$	112			
		dihedral angle of internal rotation	91					
N_2O		N—N (r_e)	1.1284	N—O (r_e)	1.1841			MW, IR
N_2O_3	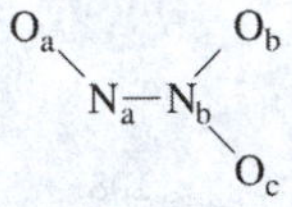	N_a—N_b	1.864	N_a—O_a	1.142			MW
		N_b—O_b	1.202	N_b—O_c	1.217			
		$∠O_aN_aN_b$	105.05	$∠N_aN_bO_b$	112.72	$∠N_aN_bO_c$	117.47	
N_2O_4	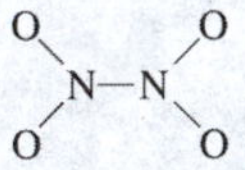 D_{2h}	N—N	1.782	N—O	1.190	∠ONO	135.4	ED
$NaBH_4$	$H_a(BH_3)Na$ (C_{3v})	B—H (BH_3)	1.278	B—H_a	1.238	Na—B	2.308	MW
NaBr		Na—Br (r_e)	2.5020					MW
NaCl		Na—Cl (r_e)	2.3609					MW
NaF		Na—F (r_e)	1.9260					MW
NaH		Na—H (r_e)	1.8873					UV
NaI		Na—I (r_e)	2.7115					MW
NaO		Na—O (r_e)	2.05155					UV
Na_2		Na—Na (r_e)	3.0789					UV
$NbCl_4$	T_d	Nb—Cl	2.279					ED
$NbCl_5$	D_{3h}	Nb—Cl_{ax}	2.307	Nb—Cl_{eq}	2.276			ED
NbO		Nb—O (r_e)	1.691					UV
NdI_3	C_{3v}	Nd—I	2.879					ED
NiBr		Ni—Br	2.1963					UV
$NiBr_2$	linear	Ni—Br	2.201					ED
$NiCl_2$	linear	Ni—Cl	2.076					ED
NiF_2	linear	Ni—F	1.729	[Ni—F (r_e)]	1.715			ED
NiH		Ni—H (r_e)	1.476					UV
NiI		Ni—I	2.348					UV
NpF_6	O_h	Np—F	1.982					ED
OF		O—F (r_e)	1.3579					LMR
OF_2	C_{2v}	O—F (r_e)	1.4053	∠FOF (θ_e)	103.07			MW
OH		O—H (r_e)	0.96966					UV
$O(SiH_3)_2$		Si—H	1.486	Si—O	1.634	∠SiOSi	144.1	ED
O_2		O—O (r_e)	1.2074					MW
O_2F_2	C_2	O—O	1.217	F—O	1.575	∠OOF	109.5	MW
		dihedral angle of internal rotation	87.5					
O_3	C_{2v}	O—O (r_e)	1.2716	∠OOO (θ_e)	117.47			MW
OsF_6	O_h	Os—F	1.832					ED

Compound	Structure							Method
OsO_4	T_d	Os—O	1.712					ED
PBr_3	C_{3v}	P—Br	2.220	∠BrPBr	101.0			ED
PCl		P—Cl (r_e)	2.01461					UV
PCl_3	C_{3v}	P—Cl	2.039	∠ClPCl	100.27			ED
PCl_5	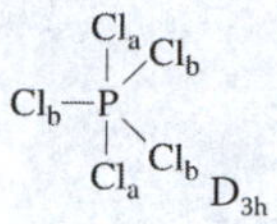 D_{3h}	P—Cl_a	2.124	P—Cl_b	2.020			ED
PF		P—F (r_e)	1.5896					UV
PF_3	C_{3v}	P—F	1.570	∠FPF	97.8			ED, MW
PF_5	D_{3h}	P—F_{eq}	1.534	P—F_{ax}	1.577			ED
PH		P—H (r_e)	1.4223					LMR
PH_2		P—H	1.418	∠HPH	91.70			UV
PH_3	C_{3v}	P—H	1.4200	∠HPH	93.345			MW
PN		N—P (r_e)	1.49087					MW
PO		O—P (r_e)	1.4759					UV
$POCl_3$	C_{3v}	P—O	1.449	P—Cl	1.993	∠ClPCl	103.3	ED
POF_3	C_{3v}	P—O	1.436	P—F	1.524	∠FPF	101.3	ED, MW
P_2		P—P (r_e)	1.8931					UV
P_2F_4	*trans* conformer	P—F	1.587	P—P	2.281	∠FPF	99.1	P_2F_4
		∠PPF	95.4					
P_4	T_d	P—P	2.21					ED
P_4O_6	T_d	P—O	1.638	∠POP	126.4			ED
$PbBr_2$	bent	Pb—Br (r_e)	2.598					ED
$PbCl_2$	bent	Pb—Cl (r_e)	2.444					ED
$PbCl_4$	T_d	Pb—Cl	2.369					ED
PbF		Pb—F (r_e)	2.0575					UV
PbF_2	bent	Pb—F (r_e)	2.041					ED
PbH		Pb—H (r_e)	1.839					UV
PbI_2	bent	Pb—I (r_e)	2.807					ED
PbO		Pb—O (r_e)	1.9218					MW
PbS		Pb—S (r_e)	2.2869					MW
PbSe		Pb—Se (r_e)	2.4022					MW
PbTe		Pb—Te (r_e)	2.5950					MW
$PrCl_3$	C_{3v}	Pr—Cl	2.554					ED
PrF_3	C_{3v}	Pr—F	2.091					ED
PrI_3	C_{3v}	Pr—I	2.901	∠IPrI	113			ED
PtC		Pt—C (r_e)	1.6767					UV
PtH		Pt—H (r_e)	1.52852					UV
PtN		Pt—N (r_e)	1.682					MW
PtO		Pt—O (r_e)	1.7273					UV
PtS		Pt—S (r_e)	2.03983					MW
PtSi		Pt—Si (r_e)	2.0612					MW
PuF_6	O_h	Pu—F	1.972					ED
RbBr		Rb—Br (r_e)	2.9447					MW
RbCl		Rb—Cl (r_e)	2.7869					MW
RbF		Rb—F (r_e)	2.2703					MW
RbH		Rb—H (r_e)	2.367					UV
RbI		Rb—I (r_e)	3.1768					MW
RbO		Rb—O (r_e)	2.25420					UV
RbOH	linear; large amplitude bending mode	Rb—O	2.301	O—H	0.957			MW
$ReClO_3$	C_{3v}	Re—O	1.702	Re—Cl	2.229	∠ClReO	109.4	MW
$ReCl_5$	D_{3h}	Re—Cl_{eq}	2.238	Re—Cl_{ax}	2.263			ED
ReF_6	O_h	Re—F	1.832					ED
ReF_7	pseudorotation	Re—F	1.835					ED
RhB		Rh—B	1.691					UV
RhC		Rh—C	1.614					UV
RhS		Rh—S	2.059					UV
RuO_4	T_d	Ru—O	1.706					ED
SCl_2	C_{2v}	S—Cl	2.006	∠ClSCl	103.0			ED

Molecular

Compound	Structure							Method
SF		S—F (r_e)	1.6006					MW
SF_2		S—F	1.5921	∠FSF	98.20			MW
SF_6	O_h	S—F	1.561					ED
SH		S—H (r_e)	1.34066					UV
SO		S—O (r_e)	1.4811					MW
$SOCl_2$		S—O	1.44	S—Cl	2.072			MW
		∠ClSCl	97.2	∠OSCl	108.0			
SOF_2		S—O	1.420	S—F	1.583			ED
		∠FSF	92.2	∠OSF	106.2			
SOF_4	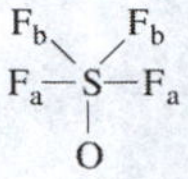 C_{2v}	S—O	1.403	S—F_a	1.575	S—F_b	1.552	ED
		∠OSF_a	90.7	∠OSF_b	124.9			
		∠F_aSF_b	89.6	∠F_bSF_b	110.2			
SO_2		S—O (r_e)	1.4308	∠OSO (θ_e)	119.329			MW
SO_2Cl_2	C_{2v}	S—Cl	2.011	S—O	1.404			ED
		∠ClSCl	100.0	∠OSO	123.5			
SO_2F_2	C_{2v}	S—F	1.530	S—O	1.397			ED
		∠FSF	97	∠OSO	123			
SO_3	D_{3h}	S—O	1.4198					IR
$S(SiH_3)_2$		Si—S	2.136	Si—H	1.494	∠SiSSi	97.4	ED
S_2		S—S (r_e)	1.8892					R
S_2Br_2	C_2	S—Br	2.24	S—S	1.98	∠SSBr	105	ED
		dihedral angle of internal rotation	83.5					
S_2Cl_2	C_2	S—Cl	2.057	S—S	1.931	∠SSCl	108.2	ED
		dihedral angle of internal rotation	84.1					
S_2O_2	planar *cis* form	S—S	2.025	S—O	1.458	∠OSS	112.8	MW
S_8	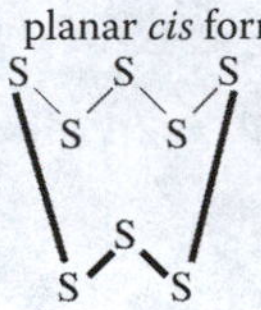	S—S	2.07	∠SSS (D_{4d})	105			ED
$SbBr_3$	C_{3v}	Sb—Br	2.490	∠BrSbBr	98.2			ED
$SbCl_3$	C_{3v}	Sb—Cl	2.334	∠ClSbCl	97.1			ED
$SbCl_5$	D_{3h}	Sb—Cl*eq*	2.277	Sb—Cl_{ax}	2.338			ED
SbF		Sb—F (r_e)	1.918					UV
SbF_3	C_{3v}	Sb—F	1.880	∠FSbF	94.9			ED
SbH		Sb—H	1.723					UV
SbH_3	C_{3v}	Sb—H	1.704	∠HSbH	91.6			MW
SbI_3	C_{3v}	Sb—I	2.721	∠ISbI	99.0			ED
SbO		Sb—O (r_e)	1.826					UV
SbP		Sb—P (r_e)	2.20544					MW
$ScCl_3$	D_{3h}	Sc—Cl	2.291					ED
ScF		Sc—F (r_e)	1.788					UV
ScF_3	D_{3h}	Sc—F	1.847					ED
SeF		Se—F	1.742					MW
SeF_6	O_h	Se—F	1.69					ED
SeH		Se—H (r_e)	1.48					UV
SeO		Se—O (r_e)	1.6393					MW
$SeOF_2$		Se—O	1.576	Se—F	1.730			MW
		∠OSeF	104.82	∠FSeF	92.22			
SeO_2		Se—O (r_e)	1.6076	∠OSeO (θ_e)	113.83			MW
SeO_3	D_{3h}	Se—O	1.69					ED
Se_2		Se—Se (r_e)	2.1660					UV
Se_6	six-membered ring with chair conformation	Se—Se	2.34	∠SeSeSe	102			ED
$SiBrF_3$	C_{3v}	Si—F	1.559	Si—Br	2.156	∠FSiBr	108.5	MW
$SiBrH_3$	C_{3v}	Si—Br	2.210	Si—H	1.486	∠HSiBr	107.8	MW

Compound	Structure							Method
SiCl		Si—Cl (r_e)	2.058					UV
$SiClH_3$	C_{3v}	Si—Cl	2.049	Si—H	1.486	∠HSiCl	107.9	MW
$SiCl_4$	T_4	Si—Cl	2.019					ED
SiF		Si—F	1.6008					UV
$SiFH_3$	C_{3v}	Si—F	1.593	Si—H	1.486	∠HSiH	110.63	MW, IR
SiF_2		Si—F (r_e)	1.590	∠FSiF (θ_e)	100.8			MW
SiF_3H	C_{3v}	Si—H ((r_e)	1.4468	Si—F (r_e)	1.5624	∠HSiF (θ_e)	110.64	MW
SiF_4	T_d	Si—F	1.553					ED
SiH		Si—H (r_e)	1.5201					UV
SiH_3I	C_{3v}	Si—I	2.437	Si—H	1.486	∠HSH	107.8	MW
SiH_4	T_d	Si—H	1.4798					IR
SiN		Si—N (r_e)	1.572					UV
SiO		Si—O (r_e)	1.5097					MW
SiS		Si—S (r_e)	1.9293					MW
SiSe		Si—Se (r_e)	2.0583					MW
Si_2		Si—Si (r_e)	2.246					UV
Si_2Cl_6		Si—Si	2.32	Si—Cl	2.009	∠ClSiCl	109.7	ED
Si_2F_6		Si—Si	2.317	Si—F	1.564	∠FSiF	108.6	ED
Si_2H_6		Si—Si	2.331	Si—H	1.492			ED
		∠SiSiH	110.3	∠HSiH	108.6			
$SnBr_2$		Sn—Br (r_e)	2.501	∠BrSnBr	100.0			ED
SnCl		Sn—Cl (r_e)	2.361					UV
$SnCl_2$		Sn—Cl (r_e)	2.335	∠ClSnCl	99.1			ED
$SnCl_4$	T_d	Sn—Cl	2.281					ED
SnF		Sn—F (r_e)	1.944					UV
SnH		Sn—H (r_e)	1.7815					UV
SnH_4	T_d	Sn—H	1.711					R, IR
SnI_2		Sn—I (r_e)	2.688					ED
SnO		Sn—O (r_e)	1.8325					MW,UV
SnS		Sn—S (r_e)	2.2090					MW
SnSe		Sn—Se (r_e)	2.3256					MW
SnTe		Sn—Te (r_e)	2.5228					MW
SrBr		Sr—Br (r_e)	2.7352					UV
$SrBr_2$	quasilinear	Sr—Br	2.783					
$SrCl_2$		Sr—Cl	2.630	∠ClSrCl	155			ED
SrF		Sr—F (r_e)	2.0754					UV
SrH		Sr—H (r_e)	2.1456					UV
SrI		Sr—I (r_e)	2.9436					UV
SrI_2	linear	Sr—I	3.01					ED
SrO		Sr—O (r_e)	1.9198					MW
SrOH		Sr—O	2.111	O—H	0.922			UV
SrS		Sr—S (r_e)	2.4405					UV
$TaBr_5$	D_{3h}	Ta—Br_{eq}	2.412	Ta—Br_{ax}	2.473			ED
$TaCl_5$	D_{3h}	Ta—Cl_{eq}	2.268	Ta—Cl_{ax}	2.315			ED
TaO		Ta—O (r_e)	1.6875					UV
$TbCl_3$	C_{3v}	Tb—Cl	2.476					ED
TeF_6	O_h	Te—F	1.815					ED
TeH		Te—H	1.74					UV
TeO		Te—O (r_e)	1.825					UV
Te_2		Te—Te (r_e)	2.5574					UV
$ThCl_4$	T_d	Th—Cl	2.567					ED
ThF_4	T_d	Th—F	2.124					ED
ThO		Th—O (r_e)	1.84032					UV
$TiBr_4$	T_d	Ti—Br	2.339					ED
$TiCl_3$	D_{3h}	Ti—Cl	2.208					ED
$TiCl_4$	T_d	Ti—Cl	2.170					ED
TiF		Ti—F	1.8342					MW
TiF_4	T_d	Ti—F	1.756					ED
TiI_3	D_{3h}	Ti—I	2.568					ED
TiI_4	T_d	Ti—I	2.546					ED
TiO		Ti—O (r_e)	1.620					UV

Compound	Structure							Method
TiS		Ti—S (r_e)	2.0825					UV
TlBr		Tl—Br (r_e)	2.6182					MW
TlCl		Tl—Cl (r_e)	2.4848					MW
TlF		Tl—F (r_e)	2.0844					MW
TlH		Tl—H (r_e)	1.870					UV
TlI		Tl—I (r_e)	2.8137					MW
UCl_4	T_d	U—Cl	2.506					ED
UCl_6	O_h	U—F	2.46					ED
UF_4	T_d	U—F	2.059					ED
UF_6	O_h	U—F	2.000					ED
UI_3	C_{3v}	U—I	2.88					ED
VCl_3O	C_{3v}	V—O	1.570	V—Cl	2.142	∠ClVCl	111.3	ED, MW
VBr_4	T_d (Jahn-Teller effect)	V—Br	2.276					ED
VCl_4	T_d (Jahn-Teller effect)	V—Cl	2.138					ED
VF_3	D_{3h}	V—F	1.751					ED
VF_5		V—F_{eq}	1.709	V—F_{ax}	1.736			ED
VMo		V—Mo	1.876					UV
VO		V—O (r_e)	1.5893					UV
$WClF_5$	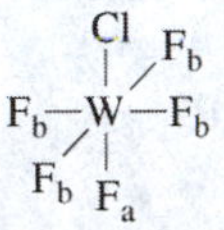	W—F (av.)	1.836	W—Cl	2.251	$∠F_aWF_b$	88.7	MW
WCl_5	D_{3h}	W—Cl_{eq}	2.243	W—Cl_{ax}	2.293			ED
WCl_6	O_h	W—Cl	2.290					ED
WF_4O	C_{4v}	W—O	1.666	W—F	1.847	∠FWF	86.2	ED
WF_6	O_h	W—F	1.833					ED
XeF_2	linear	Xe—F	1.977					IR
XeF_4	D_{4h}	Xe—F	1.94					ED
XeF_6	O_h	Xe—F	1.890					ED
XeO_4	T_d	Xe—O	1.736					ED
YCl		Y—Cl	2.385					UV
YCl_3		Y—Cl	2.437					ED
YF		Y—F (r_e)	1.9257					UV
YI_3		Y—I	2.817					ED
YO		Y—O (r_e)	1.790					UV
YbBr		Yb—Br (r_e)	2.6454					UV
YbH		Yb—H (r_e)	2.0526					UV
$ZnBr_2$	linear	Zn—Br	2.204					ED
$ZnCl_2$	linear	Zn—Cl	2.072					ED
ZnF		Zn—F (r_e)	1.7677					MW
ZnF_2	linear	Zn—F	1.742	[Zn—F (r_e)]	1.729			ED
ZnH		Zn—H (r_e)	1.5949					UV
ZnI_2	linear	Zn—I	2.401					ED
$ZrBr_4$	T_d	Zr—Br	2.465					ED
$ZrCl_4$	T_d	Zr—Cl	2.328					ED
ZrF_4	T_d	Zr—F	1.902					ED
ZrI_4	T_d	Zr—I	2.660					ED
ZrO		Zr—O (r_e)	1.7116					UV

[a] Assumed value

Part 2. Molecules Containing Carbon

Structures of Carbon-Containing Molecules: Bond Distances in Å and Angles in Degrees

Structure							Method
Acetaldehyde							
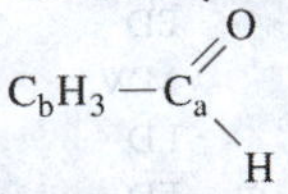	C_a—O	1.210	C_a—C_b	1.515			ED, MW
	C_a—H	1.128	C_b—H	1.107			
	$∠C_bC_aO$	124.1	$∠C_bC_aH$	115.3	$∠HC_bH$	109.8	

Molecular

Structure							Method
Acetamide							
CH_3CONH_2	C—O	1.220	C—N	1.380			ED
	C—C	1.519	N—H	1.022	C—H	1.124	
	∠CCN	115.1	∠NCO	122.0			
Acetic acid							
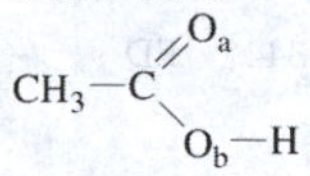	C—C	1.520	C—O_a	1.214	C—O_b	1.364	ED
	C—H	1.10	∠CCO_a	126.6	∠CCO_b	110.6	
Acetone							
$(CH_3)_2CO$	C—C	1.520	C—O	1.213	C—H	1.103	ED, MW
Symmetry axis of each CH_3 is tilted 2° from the C—C bond	∠CCC	116.0	∠HCH	108.5			
Acetonitrile							
CH_3CN (C_{3v})	C—N	1.159	C—C	1.468	C—H	1.107	ED, MW
	∠CCH	109.7					
Acetonitrile-*N*-oxide							
CH_3CNO (C_{3v})	C—C	1.442	C—N	1.169	N—O	1.217	MW
Acetyl chloride							
CH_3COCl	C—C	1.506	C—O	1.187	C—H	1.105	ED, MW
	C—Cl	1.798	∠HCH	108.6	∠OCCl	121.2	
	∠CCCl	111.6					
Acetylene							
HC≡CH	C—C (r_e)	1.203	C—H (r_e)	1.060			IR
Acrolein							
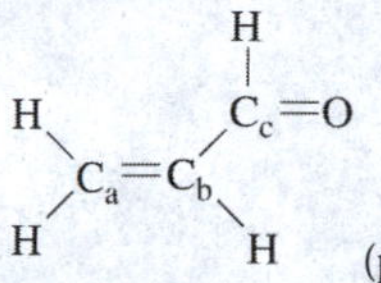 (planar *s-trans* form)	C_a—C_b	1.345	C_b—C_c	1.484	C_c—O	1.217	ED, MW
	C_a—H	1.10	C_c—H	1.13	∠H$C_c$$C_b$	114	
	∠$C_a$$C_b$$C_c$	120.3	∠$C_b$$C_c$O	123.3	Other CCH (av.)	122	
Acrylonitrile							
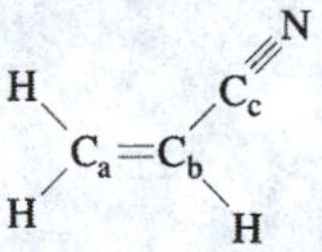	C_a—C_b	1.343	C_b—C_c	1.438	C_c—N	1.167	ED, MW
	C_a—H	1.114	∠$C_b$$C_c$N	178	∠$C_a$$C_b$$C_c$	121.7	
	∠HCC	120					
Allene							
CH_2=C=CH_2	C—C	1.3084	C—H	1.087	∠HCH	118.2	IR
Aniline							
$C_6H_5NH_2$	C—C	1.392	C—N	1.431	N—H	0.998	MW
	∠HNH	113.9	dihedral angle between NH_2 plane and N— C bond	140.6			
Azetidine							
CH_2—CH_2 | | CH_2—NH	C—N	1.482	C—C	1.553			ED
	C—H	1.107	N—H	1.03			
	∠CCC	86.9	∠CCN	85.8	∠CNC	92.2	
	dihedral angle between CCC and CNC planes	147					
Benzamide							
C_6H_5—$C_a$$ONH_2$	C—C (ring)	1.401	C (ring)—C_a	1.511	C_a—O	1.225	ED

Structure							Method
	C—H	1.112	C—N	1.380			
	∠CCN	117.8	∠CCC (ring) (ass.)	120	∠CCO	121.2	
Benzene							
C_6H_6	C—C	1.399	C—H	1.101			ED, IR
***p*-Benzoquinone** 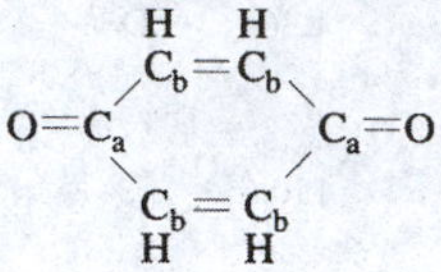	C_a—O	1.225	C_a—C_b	1.481	C_b—C_b	1.344	ED
	$\angle C_bC_aC_b$	118.1					
Bicyclo[1.1.0]butane H_a	C_a—C_a	1.497	C_a—C_b	1.498	C_a—H_a	1.071	MW
	C_b—H_b	1.093	C_b—H_c	1.093	$\angle H_bC_bH_c$	115.6	
	$\angle C_bC_aH_a$	130.4	$\angle C_aC_aH_a$	128.4	$\angle C_aC_bC_a$	60.0	
	dihedral angle between the two $C_aC_aC_b$ planes	121.7					
Bicyclo[2.2.1]heptane							
See preceeding structure	C_a—C_b	1.54	C_b—C_b	1.56	C_a—C_c	1.56	ED
C_7H_{12}	C—C (av.)	1.549	$\angle C_aC_cC_a$	93.1			
	dihedral angle between the two $C_aC_bC_bC_a$ planes	113.1					
Bicyclo[2.2.0]hexa-2,5-diene 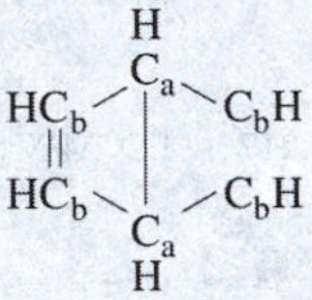	C_b—C_b	1.345	C_a—C_a	1.574	C_a—C_b	1.524	ED
	dihedral angle between the two $C_aC_bC_bC_a$ planes	117.3					
Bicyclo[2.2.2]octane							
$HC_a(C_bH_2C_bH_2)_3C_aH$	C_a—C_b	1.54	C_b—C_b	1.55	C—C (av.)	1.542	ED
large-amplitude torsional motion about D_{3h} symmetry axis	$\angle C_aC_bC_b$	109.7					
Bicyclo[1.1.1]pentane							
C_5H_8	C—C	1.557	∠CCC	74.2			ED
Bicyclo[2.1.0]pentane 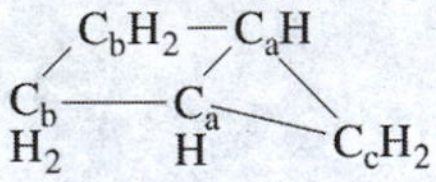	C_a—C_a	1.536	C_b—C_b	1.565	C_a—C_c	1.507	MW
	C_a—C_b	1.528	dihedral angle between the $C_aC_aC_bC_b$ and $C_aC_aC_c$ planes	112.7			
Biphenyl 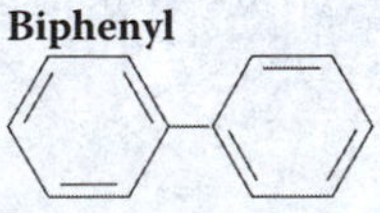	C—C (intraring)	1.396	C—C (interring)	1.49			ED
	torsional dihedral angle between the two rings	40[a]					
4,4′-Bipyridine 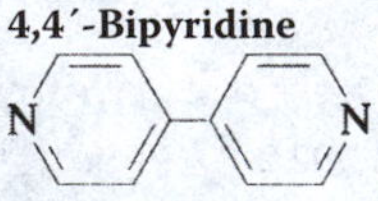	C—C (interring)	1.465	C—C (intraring)	1.375	C—N (intraring)	1.375	ED
	torsional dihedral angle between the two rings	37[a]					
Bis(cyclopentadienyl) beryllium							

Structure							Method
$(C_5H_5)_2Be$ (C_{5v})	Be—(cyclopentadienyl plane)	c	C—C	1.423			ED
Bis(cyclopentadienyl) iron [Ferrocene]							
$(C_5H_5)_2Fe$ (D_{5h})	Fe—C	2.064	C—C	1.440	C—H	1.104	ED
Bis(cyclopentadienyl) lead							
$(C_5H_5)_2Pb$ (D_{5h})	Pb—C	2.79	C—C	1.430			ED
	dihedral angle between the two C_5H_5 planes	d					
Bis(cyclopentadienyl) manganese [Manganocene]							
$(C_5H_5)_2Mn$ (D_{5h})	Mn—C	2.383	C—C	1.429			ED
Bis(cyclopentadienyl) nickel [Nickelocene]							
$(C_5H_5)_2Ni$ (D_{5h})	Ni—C	2.196	C—C	1.430			ED
Bis(cyclopentadienyl) ruthenium [Ruthenocene]							
$(C_5H_5)_2Ru$ (D_{5h})	Ru—C	2.196	C—C	1.439			ED
Bis(cyclopentadienyl) tin							
$(C_5H_5)_2Sn$ (D_{5h})	Sn—C	2.71	C—C	1.431	C—H	1.14	ED
Borane carbonyl							
BH_3CO (C_{3v})	C—O	1.131	B—C	1.540	B—H	1.194	MW
	∠BCO	180	∠HBH	113.9			
Bromobenzene							
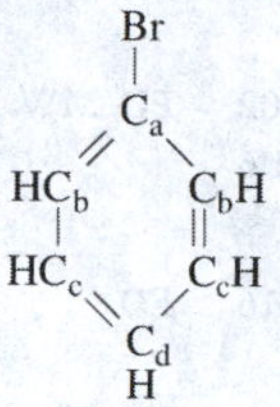	C_a—C_b	1.42	C_b—C_c	1.375	C_c—C_d	1.401	MW
	C—Br	1.85	C—H	1.072	$∠C_bC_aC_b$	117.4	
Bromochloroacetylene							
ClC≡CBr	C—Cl	1.636	C—Br	1.784	C—C	1.206	ED
Bromoiodoacetylene							
IC≡CBr	C—I	1.972	C—Br	1.795	C—C	1.206	ED
Bromomethane [Methyl bromide]							
CH_3Br	C—Br (r_e)	1.933	C—H (r_e)	1.086	∠HCH (θ_e)	111.2	MW, IR
Bromomethyl							
CH_2Br (planar)	C—Br	1.848	C—H	1.084	∠HCH (ass.)	124.5	MW
Bromomethylene							
CHBr (bent)	C—Br	1.857	C—H	1.110	∠HCH	101.0	UV
Bromomethylmercury							
CH_3HgBr (C_{3v})	C—Hg	2.07	Hg—Br	2.406			MW
1,3-Butadiene							
C_aH_2	C_a—C_b	1.349	C_b—C_b	1.467	C—H (av.)	1.108	ED
	∠CCC	124.4	$∠C_bC_aH$	120.9			
(C_{2h})							
1,3-Butadiyne							
HC_a≡C_bC_b≡C_aH (linear)	C_a—C_b	1.218	C_b—C_b	1.384	C—H	1.09	ED
Butane							
$CH_3CH_2CH_2CH_3$	C—C	1.531	C—H	1.117	∠CCC	113.8	ED
	∠CCH	111.0	dihedral angle for the *gauche* conformer	65			
2,3-Butanedione							
$CH_3COCOCH_3$	C—O	1.215	C—C (av.)	1.524	C—H	1.108	ED
trans conformer	∠CCC	116.2	∠CCO	119.5			
2-Butanone							

Structure							Method
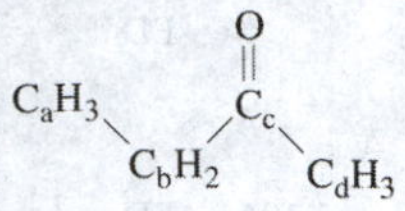	C—C (av.)	1.518	C_c—O	1.219	C—H (av.)	1.102	ED
trans conformer	∠$C_aC_bC_c$	113.5	∠C_bC_cO	121.9	∠C_dC_cO	121.9	
1,2,3-Butatriene							
$H_2C_a{=}C_b{=}C_b{=}C_aH_2$ (D_{2h})	C_a—C_b	1.32	C_b—C_b	1.28	C—H	1.08	ED
***cis*-2-Butene**							
$C_aH_3C_bH{=}C_bHC_aH_3$	C_a—C_b	1.506	C_b—C_b	1.346	∠$C_aC_bC_b$	125.4	ED
***trans*-2-Butene**							
$C_aH_3C_bH{=}C_bHC_aH_3$	C_a—C_b	1.508	C_b—C_b	1.347	∠$C_aC_bC_b$	123.8	ED
1-Buten-3-yne							
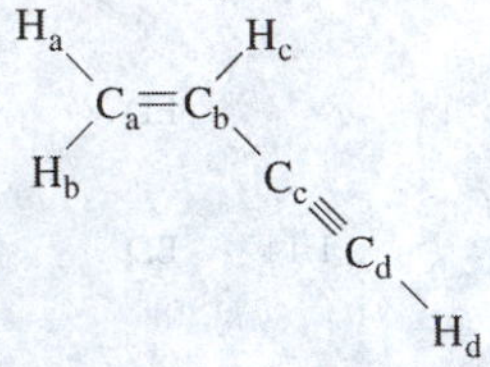	C_a—C_b	1.344	C_b—C_c	1.434	C_c—C_d	1.215	ED, MW
	C_a—H_a	1.11	C_d—H_d	1.09			
	∠$C_aC_bC_c$	123.1	∠$C_bC_cC_d$	178	∠$H_aC_aC_b$	119	
	∠$H_bC_aC_b$	122	∠$H_cC_bC_a$	122	∠$C_cC_dH_d$	182	
***tert*-Butyl chloride**							
$(CH_3)_3CCl$	C—C	1.528	C—Cl	1.828	C—H	1.102	ED, MW
	∠CCCl	107.3	∠CCH	110.8	∠CCC	111.6	
2-Butyne							
C_aH_3—$C_b{\equiv}C_b$—C_aH_3	C_b—C_b	1.214	C_a—C_b	1.468	C—H	1.116	ED
	∠C_bC_aH	110.7					
Carbon dimer							
C_2	C—C (r_e)	1.2425					UV
Carbon trimer							
C_3 (linear)	C—C	1.277					UV
Carbon dioxide							
CO_2 (linear)	C—O (r_e)	1.1600					IR
Carbon disulfide							
CS_2 (linear)	C—S (r_e)	1.5526					IR
Carbon monobromide							
CBr	C—Br	1.8209					UV
Carbon monoselenide							
CSe	C—Se (r_e)	1.67609					UV
Carbon monosulfide							
CS	C—S (r_e)	1.5349					MW
Carbon monoxide							
CO	C—O (r_e)	1.1283					MW
Carbon oxyselenide							
OCSe (linear)	C—O	1.159	C—Se	1.709			MW
Carbon oxysulfide							
OCS (linear)	C—O (r_e)	1.1578	C—S (r_e)	1.5601			MW
Carbon phosphide							
CP	C—P (r_e)	1.562					UV
Carbon sulfide selenide							
SCSe (linear)	C—S	1.553	C—Se	1.693			MW
Carbon sulfide telluride							

Molecular

Structure							Method
SCTe (linear)	C—S	1.557	C—Te	1.904			MW
Carbon suboxide							
OCCCO (linear)	C—C	1.289	C—O	1.163			ED
Carbonyl bromide							
$COBr_2$	C—O	1.178	C—Br	1.923	∠BrCBr	112.3	ED, MW
Carbonyl chloride							
$COCl_2$	C—O	1.179	C—Cl	1.742	∠ClCCl	111.8	ED, MW
Carbonyl chloride fluoride							
COClF	C—O	1.173	C—F	1.334	C—Cl	1.725	ED, MW
	∠ClCO	127.5	∠FCCl	108.8			
Carbonyl dicyanide							
$CO(CN)_2$	C—O	1.209	C—C	1.466	C—N	1.153	ED, MW
	∠CCC	115	∠CCN	180			
Carbonyl fluoride							
COF_2	C—O	1.172	C—F	1.3157	∠FCF	107.71	ED, MW
Chloroacetylene							
HC≡CCl	C—Cl	1.6368	C—C	1.2033	C—H	1.0550	MW
Chlorobenzene							
C_6H_5Cl	C—C	1.400	C—Cl	1.737	C—H	1.083	ED
Chlorocyanoacetylene							
ClC≡C—CN	C—Cl	1.624	C—N	1.160	C—C	1.205	ED
	C—CN	1.362					
Chloroethane [Ethyl chloride]							
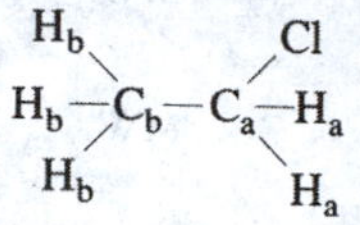	C—C	1.528	C—Cl	1.802	C—H	1.103	ED, MW
	∠CCCl	110.7	$∠H_bC_bH_b$	109.8	$∠H_aC_aH_a$	109.2	
	$∠C_bC_aH_a$	110.6	$C_a—H_a = C_b—_Hb$ (ass.)				
2-Chloroethanol							
$ClCH_2CH_2OH$ (*gauche*)	C—O	1.413	C—C	1.519	C—Cl	1.801	ED
	O—H	1.033	C—H	1.093			
	∠CCCl	110.7	∠CCO	113.8	dihedral angle of internal rotation	62.4	
Chloroiodoacetylene							
ClC≡CI	C—Cl	1.63	C—I	1.99	C—C (ass.)	1.209	MW
Chloromethane [Methyl chloride]							
CH_3Cl	C—Cl	1.785	C—H	1.090	∠HCH	110.8	MW, IR
Chloromethylidyne							
CCl	C—Cl	1.6512					UV
Chloromethylmercury							
CH_3HgCl (C_{3v})	C—Hg	2.06	Hg—Cl	2.282			MW
***trans*-1-Chloropropene**							
$CH_3CH{=}CHCl$	C—Cl	1.728	∠CCCl	121.9			MW
3-Chloropropene							
$CH_2ClCH{=}CH_2$ *cis* conformer	C—Cl	1.811	∠CCCl	115.2			MW
skew conformer	C—Cl	1.809	∠CCCl	109.6	dihedral angle of internal rotation	122.4	
Chlorotrifluoromethane							
$CClF_3$(C_{3v})	C—Cl	1.752	C—F	1.325	∠FCF	108.6	ED, MW
Chromium carbonyl							
$Cr(CO)_6$	Cr—C	1.92	C—O	1.16	∠CrCO	180	ED
Cobalt cyanide							
CoC≡N	Co—C	1.883	C—N	1.131			MW

Molecular

Structure							Method
Copper cyanide							
CuC≡N	Cu—C	1.832	C—N	1.158			MW
Cyanamide							
$H_2N_aCN_b$	N_a—C	1.346	C—N_b	1.160	N—H	1.00	MW
	∠HNH	114	dihedral angle between NH_2 plane and N—C bond	142			
Cyanide							
CN	C—N(r_e)	1.1718					MW
Cyanoacetylene							
$HC_a≡C_b—C_cN$	C_a—C_b	1.205	C_b—C_c	1.378	C—H	1.058	MW
	C_c—N	1.159					
Cyanocyclopropane [Cyclopropanecarbonitrile]							
$C_3H_5C_aN$	C—C (ring)	1.513	C—C_a	1.472	C_a—N	1.157	MW
	C—H	1.107	∠C_aCH	119.6	∠HCH	114.6	
Cyanogen							
N≡C—C≡N (linear)	C—N	1.163	C—C	1.393			ED
Cyanogen azide							
N≡C—N=N≡N (planar)	C—N	1.312	N=N	1.252	N≡N	1.133	MW
	C≡N	1.164	∠CNN	120.2	∠NCN	176.0	
Cyanogen bromide							
BrCN (linear)	C—N(r_e)	1.157	C—Br(r_e)	1.790			MW
Cyanogen chloride							
ClCN (linear)	C—Cl(r_e)	1.629	C—N(r_e)	1.160			MW
Cyanogen fluoride							
FCN (linear)	C—F	1.262	C—N	1.159			MW
Cyanogen iodide							
ICN (linear)	C—I	1.995	C—N	1.159			MW
1-Cyano-2-propyne							
$HC_a≡C_bC_cH_2C_d≡N$	C_a—C_b (ass)	1.207	C_b—C_c (ass.)	1.465	C_c—C_d	1.454	MW
	C_d—N (ass.)	1.159	C_a—H (ass.)	1.057	C_c—H (ass.)	1.090	
	∠$C_bC_cC_d$	113.4	∠HC_cH (ass.)	109.4	∠C_bC_cH	111.3	
Cyclobutane							
$(CH_2)_4$	C—C	1.555	C—H	1.113			ED
	dihedral angle between the two CCC planes	145					
Cyclobutanone							
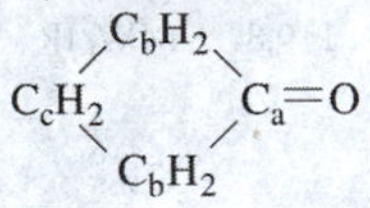	C_a—C_b	1.527	C_b—C_c	1.556			MW
	∠$C_bC_aC_b$	93.1	∠$C_aC_bC_c$	88.0			
Cyclobutene							
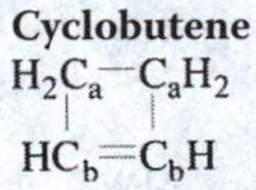	C_a—C_a	1.566	C_b—C_b	1.342	C_a—C_b	1.517	MW
	C_a—H	1.094	C_b—H	1.083			
	∠$C_aC_bC_b$	94.2	∠C_bC_bH	133.5	∠HC_aH	109.2	
	∠C_aC_aH	114.5	∠$C_aC_aC_b$	85.8	dihedral angle between CH_2 plane and C_a—C_a bond	135.8	

Structure							Method
2,4,6-Cycloheptatrien-1-one 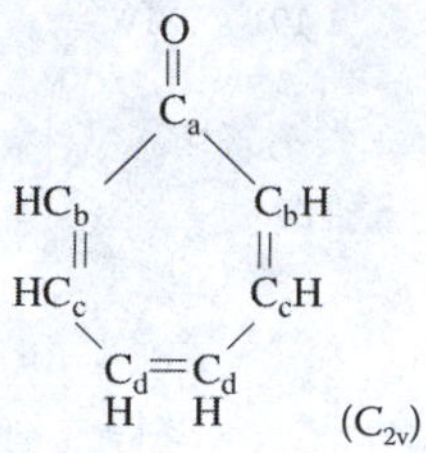 (C_{2v})	C_a—C_b	1.45	C_b—C_c	1.36	C_c—C_d	1.46	ED
	C_d—C_d	1.34	C_a—O	1.23	∠$C_bC_aC_b$	122	
	∠$C_aC_bC_c$	133	∠$C_bC_cC_d$	126	∠$C_cC_dC_d$	130	
Cyclohexane C_6H_{12} (chair form)	C—C	1.536	C—H	1.119	∠CCC	111.3	ED
Cyclohexene 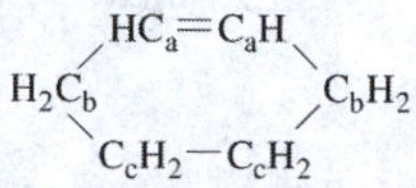 half-chair form (C_2)	C_a—C_a	1.334	C_a—C_b	1.50	C_b—C_c	1.52	ED
	C_c—C_c	1.54	∠$C_aC_aC_b$	123.4	∠C_aC_bCc	112.0	
	∠$C_bC_cC_c$	110.9					
Cyclooctatetraene 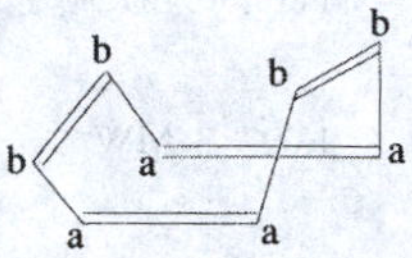 tub form (D_{2d})	C_a—C_b	1.476	C_a—C_a	1.340	C_b—C_b	1.340	ED
	C—H	1.100	∠$C_bC_aC_a$	126.1	∠$C_aC_bC_b$	126.1	
	dihedral angle between $C_aC_aC_aC_a$ and $C_aC_bC_bC_a$ planes	136.9					
1,3-Cyclopentadiene 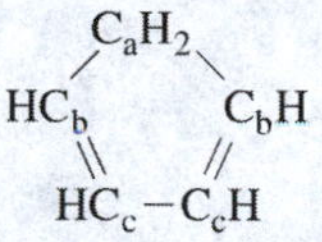	C_a—C_b	1.509	C_b—C_c	1.342	C_c—C_c	1.469	MW
	∠$C_aC_bC_c$	109.3	∠$C_bC_cC_c$	109.4	∠$C_bC_aC_b$	102.8	
Cyclopentadienylindium 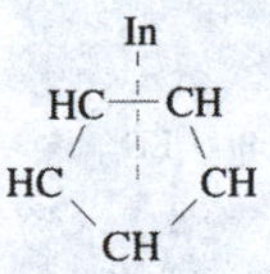	C—In	2.621	C—C	1.426	(C_{5v})		ED
Cyclopentane $(CH_2)_5$	C—C	1.546	C—H	1.114	∠CCH	111.7	ED
Cyclopentene 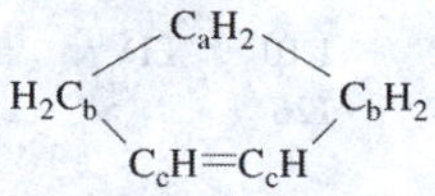	C_a—C_b	1.546	C_b—C_c	1.519	C_c—C_c	1.342	ED
	∠$C_aC_bC_c$	103.0	∠$C_bC_cC_c$	110.0	∠$C_bC_aC_b$	104.0	
	dihedral angle between $C_bC_aC_b$ and $C_bC_cC_cC_b$ planes	151.2					
Cyclopropane $(CH_2)_3$	C—C	1.512	C—H	1.083	∠HCH	114.0	R

Molecular

Structure							Method
Cyclopropanone							
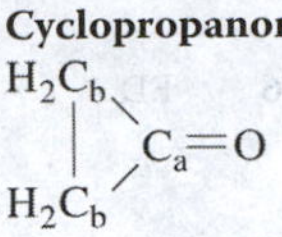	C_a—C_b	1.475	C_b—C_b	1.575	C_a—O	1.191	MW
	C—H	1.086	∠$C_aC_bC_b$	57.7	∠HC_bH	114	
	dihedral angle between CH_2 plane and C_b—C_b bond	151					
Cyclopropene							
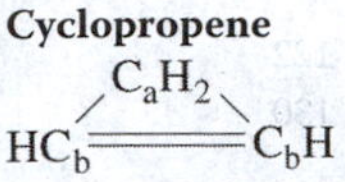	C_a—C_b	1.505	C_b—C_b	1.293	C_a—H	1.085	MW
	C_b—H	1.072	∠C_bC_bH	150	∠HC_aH	114.3	
Cyclopropenone							
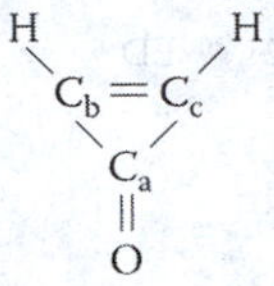	C_a—C_b (r_s)	1.423	C_b—C_c (r_s)	1.349	C_a—O (r_s)	1.212	MW
C_{2v}	C—H (r_s)	1.079	∠HC_bC_c (θ_s)	144.3	$C_bC_aC_c$ (θ_s)	56.6	
Decalin							
$C_{10}H_{18}$	C—C (av.)	1.530	C—H (av.)	1.113	∠CCC (av.)	111.4	ED
Diazirine							
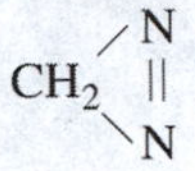	C—N	1.482	N—N	1.228	C—H	1.09	MW
	∠HCH	117					
Diazoacetonitrile							
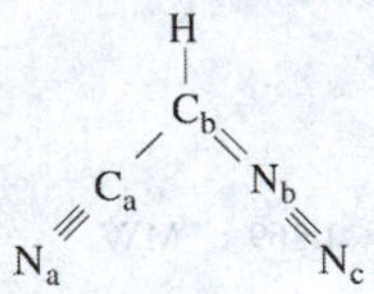	C_a—C_b	1.424	C_a—N_a	1.165	C_b—N_b	1.280	MW
	N_b—N_c	1.132	C—H	1.082			
	∠C_aC_bH	117	∠$C_aC_bN_b$	119.5			
Diazomethane							
CH_2N_2	C—N	1.32	N—N	1.12	C—H	1.075	MW, IR
	∠HCH	126.0					
1,2-Dibromoethane							
CH_2BrCH_2Br	C—C	1.506	C—Br	1.950	C—H	1.108	ED
	∠CCBr	109.5	∠CCH	110			
Dibromomethane							
CH_2Br_2	C—Br	1.924	C—H	1.08	∠HCBr	109	ED
	∠BrCBr	113.2					
2,2'-Dichlorobiphenyl							
C_6H_4Cl—C_6H_4Cl	C—C (rings)	1.398	C—C (interring)	1.495	C—H	1.10	ED
	C—Cl	1.732	∠CCCl	121.4	∠CCH	126	
	dihedral angle between the two rings (defined as 0 for *cis* conformer)	74					
***trans*-1,4-Dichlorocyclohexane**							
$C_6H_{10}Cl_2$	C—C	1.530	C—Cl	1.810	C—H	1.102	ED
	∠CCC	111.5					
equatorial:	∠CCCl	108.6	∠HCCl	111.5			
axial:	∠CCCl	110.6	∠HCCl	107.6			
1,1-Dichloroethane							
$CHCl_2CH_3$	C—C	1.540	C—Cl	1.766			MW

Structure							Method
	∠ClCCl	112.0	∠CCCl	111.0			
1,2-Dichloroethane CH_2ClCH_2Cl	C—C	1.531	C—Cl	1.790	C—H	1.11	ED
	∠CCCl	109.0	∠CCH	113			
1,1-Dichloroethene $CH_2{=}CCl_2$ (C_{2v})	C—C (ass.)	1.32	C—Cl	1.73			MW
	∠ClCC	123					
***cis*-1,2-Dichloroethene** CHCl=CHCl	C—C	1.354	C—Cl	1.718			ED
	∠ClCC	123.8					
Dichloromethane CH_2Cl_2	C—Cl (r_e)	1.765	C—H (r_e)	1.087			MW, IR
	∠ClCCl (θ_e)	112.0	∠HCH (θ_e)	111.5			
1,2-Dicyanocyclobutene 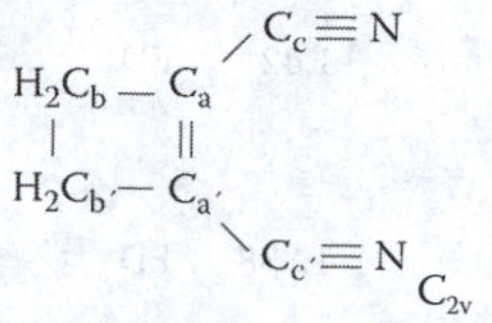	$C_a—C_a'$	1.361	$C_a—C_b$	1.515	$C_b—C_b'$	1.567	MW
	$C_a—C_c$	1.420	Cc—N	1.157	$C_b—H$	1.088	
	$∠C_a'C_aC_b$	93.9	$∠C_aC_bC_b'$	86.1	$∠C_aC_cN$	178.2	
	$∠C_bC_aC_c$	133.3	$∠C_aC_bH$	114.7	$∠C_a'C_aC_bH$	115.8	
Difluorocyanamide $F_2N_b—C≡N_a$	C—Na	1.158	C—Nb	1.386	$N_b—F$	1.399	MW
	$∠N_aCN_b$	174	$∠CN_bF$	105.4	$∠FN_bF$	102.8	
Difluorocyclopropenone 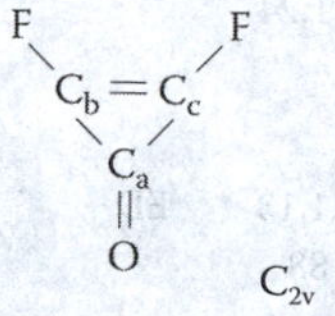	$C_a—C_b$	1.453	$C_b—Cc$	1.324	$C_a—O$	1.192	MW
	C—F	1.314	$∠FC_bC_c$	145.7			
Difluorodimethylsilane $(CH_3)_2SiF_2$	C—Si	1.844	Si—F	1.585	C—H (ass.)	1.093	MW
	∠CSiC	115.2	∠FSiF	106.1	∠SiCH (ass.)	110.8	
1,1-Difluoroethane CH_3CHF_2	C—C	1.498	C—F	1.364	C—H (av.)	1.081	ED
	∠CCF	110.7	∠CCH (av.)	111.0	dihedral angle between CCF planes	118.9	
1,2-Difluoroethane CH_2FCH_2F	C—C	1.503	C—F	1.389	C—H	1.103	ED
	∠CCF	110.3	∠CCH	111	dihedral angle of internal rotation	109	
1,1-Difluoroethene $CH_2{=}CF_2$	C—C	1.340	C—F	1.315	C—H	1.091	ED, MW
	∠CCF	124.7	∠CCH	119.0			
***cis*-1,2-Difluoroethene** CHF=CHF	C—C	1.33	C—F	1.342	C—H	1.099	ED, MW
	∠CCF	122.0	∠CCH	124.1			
Difluoromethane CH_2F_2	C—F	1.357	C—H	1.093			MW
	∠FCF	108.3	∠HCH	113.7			

Structure							Method
Dimethoxymethane 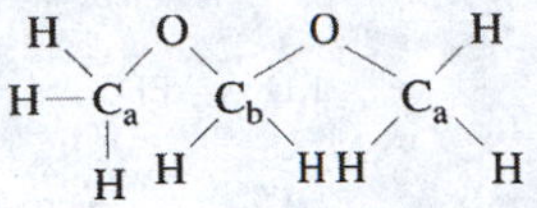	C_a—O	1.432	C_b—O	1.382	C—H (av.)	1.108	ED
	∠COC	114.6	∠OCO	114.3	∠OCH	110.3	
Dimethylamine							
$(CH)_2NH$	C—N	1.455	N—H	1.00	C—H	1.106	ED
	∠CNC	111.8	∠CNH	107	∠NCH	112	
	∠HCH	107					
Dimethylberyllium							
$(CH_3)_2Be$ (CBeC linear)	C—Be	1.698	C—H	1.127	∠BeCH	113.9	ED
Dimethyl cadmium							
$(CH_3)_2Cd$	C—Cd	2.112	∠HCH	108.4			R
Dimethyl carbonate							
$(C_aH_3O_a)_2C_b{=}O_b$	C_b—O_b	1.209	C_b—O_a	1.34	C_a—O_a	1.42	ED
	∠$O_aC_bO_a$	107	∠$C_bO_aC_a$	114.5			
Dimethylcyanamide							
$(C_aH_3)_2N_a$—$C_b{\equiv}N_b$	C_b—N_b	1.161	C_a—N_a	1.463	C_b—N_a	1.338	ED
***trans*-Dimethyldiazene**							
$CH_3N{=}NCH_3$	C—N	1.482	N—N	1.247	∠CNN	112.3	ED
	∠C_aNC_a	115.5	∠C_aNC_b	116.0			
1,2-Dimethyldiborane 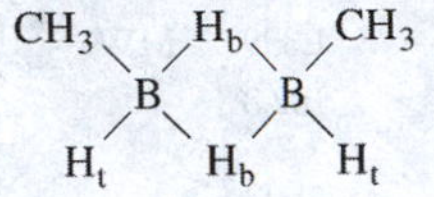	B—B	1.799	B—C	1.580			ED
	B—H_b (*cis*)	1.358	B—H_b (*trans*)	1.365	B—H_t	1.24	
	∠BBC (*cis*)	122.6	∠BBC (*trans*)	121.8			
Dimethyl diselenide							
$(CH_3)_2Se_2$	C—Se	1.95	Se—Se	2.326	C—H	1.13	ED
	∠CSeSe	98.9	∠HCSe	108	CSeSeC dihedral angle	88	
Dimethyl disulfide							
$(CH_3)_2S_2$	C—S	1.816	S—S	2.029	C—H	1.105	ED
	∠SSC	103.2	∠SCH	111.3	CSSC dihedral angle	85	
***S,S′*-Dimethyl dithiocarbonate**							
$C_aH_3SC_bSC_aH_3$ (with =O on C_b) *syn-syn* conformer	C_a—S	1.802	C_b—S	1.777	Cb—O	1.206	ED
	∠OCS	124.9	∠CSC	99.3			
Dimethyl ether							
$(CH_3)_2O$	C—O	1.416	C—H	1.121			ED
	∠COC	112	∠HCH	108			
***N,N′*-Dimethylhydrazine**							
CH_3NH—$NHCH_3$	C—N	1.46	N—N	1.42	N—H	1.03	ED
	C—H	1.12	∠NNC	112	CNNC dihedral angle	90	
Dimethyl mercury							
$(CH_3)_2Hg$	C—Hg	2.083	C—H (ass.)	1.160	Hg···H	2.71	ED
Dimethylphosphine							
$(CH_3)_2PH$	C—P	1.848	P—H	1.419			MW
	∠CPC	99.7	∠CPH	97.0			
2,2-Dimethylpropanenitrile							
$(C_cH_3)_3C_b$—$C_a{\equiv}N$	C_a—C_b	1.495	C_b—C_c	1.536	C_a—N	1.159	MW
	∠$C_cC_bC_c$	110.5					

Structure							Method
Dimethyl selenide							
$(CH_3)_2Se$	C—Se	1.943	C—H	1.093			MW
	∠CSeC	96.2	∠SeCH	108.7	∠HCH	110.3	
Dimethylsilane							
$(CH_3)_2SiH_2$	C—Si	1.868	C—H	1.089	Si—H	1.482	MW
	∠CSiC	110.9	∠CSiH	109.5	∠SiCH	110.9	
	∠HSiH	107.8					
Dimethyl sulfide							
$(CH_3)_2S$	C—S	1.802	C—H	1.090			ED, MW
	∠CSC	98.80	∠HCH	109.3			
Dimethyl sulfone							
$(CH_3)_2SO_2$	C—S	1.771	S—O	1.435	C—H	1.114	ED
	∠CSC	102	∠OSO	121			
Dimethyl sulfoxide							
$(CH_3)_2SO$	C—S	1.799	S—O	1.485	C—H	1.081	MW
	∠CSC	96.6	∠CSO	106.7	∠HCH	110.3	
	dihedral angle between SCC plane and S—O bond	115.5					
Dimethyl zinc							
$(CH_3)_2Zn$	C—Zn	1.929	∠HCH	107.7			R
1,4-Dioxane							
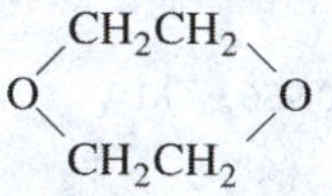	C—C	1.523	C—O	1.423	C—H	1.112	ED
	∠CCO	109.2	∠COC	112.45			
Ethane							
C_2H_6	C—C	1.5351	C—H	1.0940	∠CCH	111.17	MW
staggered conformation	C—C (r_e)	1.522					
1,2-Ethanediamine							
$H_2NCH_2CH_2NH_2$	C—C	1.545	C—N	1.469	C—H	1.11	ED
gauche conformer	∠CCN	110.2	dihedral angle between NCC and CCN planes	64			
Ethanethiol							
$C_bH_3—C_aH_2—SH$	$C_a—C_b$	1.530	$C_a—S$	1.829	S—H	1.350	MW
	$C_a—H$	1.090	$C_b—H$	1.093	$∠C_aSH$	96.4	
	$∠C_bC_aS$	108.3	$∠C_bC_aH$	109.6	$∠C_aC_bH$	109.7	
Ethanol							
$C_bH_3C_aH_2OH$	C—C	1.512	C—O	1.431	O—H	0.971	MW
staggered conformation	$C_a—H$	1.10	$C_b—H$	1.09	∠COH	105	
	∠CCO	107.8	$∠C_bC_aH$	111	$∠C_aC_bH$	110	
Ethylene							
$CH_2=CH_2$	C—C (r_s)	1.329	C—H (r_s)	1.082	∠HCH (θ_s)	117.2	MW, IR
Ethyleneimine							
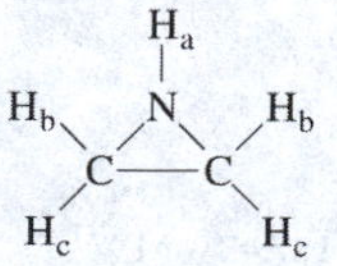	C—C	1.481	N—C	1.475			MW
	C—H	1.084	N—H	1.016			
	∠CNC	60.3	$∠H_aNC$	109.3	$∠H_bCH_c$	115.7	
	$∠H_bCC$	117.8	$∠H_bCN$	118.3	$∠H_cCC$	119.3	
	$∠H_cCN$	114.3					
Ethyl methyl ether							
$C_2H_5OCH_3$	C—C	1.520	C—O (av.)	1.418	C—H (av.)	1.118	ED
	∠COC	111.9	∠OCC	109.4	∠HCH	109.0	

Molecular

Structure							Method
Ethyl methyl sulfide							
$C_2H_5SCH_3$	C—C	1.536	C—S (av.)	1.813	C—H	1.111	ED
gauche conformer	∠CSC	97	∠SCC	114.0	∠HCH	110	
Fluoroketene							
HFC=C=O	C—C	1.317	C—O	1.167	C—F	1.360	MW
	C—H	1.102	∠CCO	178.0	∠CCF	119.5	
	∠CCH	122.3					
Fluoromethane [Methyl fluoride]							
CH_3F	C—F (r_e)	1.382	C—H (r_e)	1.095	∠HCH (θ_e)	110.45	MW, IR
Fluoromethylidyne							
CF	C—F (r_e)	1.2718					UV
(Fluoromethylidyne) phosphine							
FC≡P	C—F	1.285	C—P	1.541			MW
2-Fluoropropane							
CH_3CHFCH_3	C—C	1.522	C—F	1.398			MW
	∠CCC	113.4	∠CCF	108.2			
Formaldehyde							
H_2CO	C—O	1.208	C—H	1.116	∠HCH	116.5	MW
Formaldehyde azine							
H_2C=N—N=CH_2	C—N	1.277	N—N	1.418	C—H	1.094	ED
trans conformer	∠CNN	111.4	∠HCN	120.7			
Formaldehyde oxime							
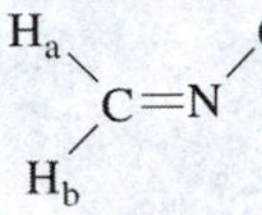	C—N	1.276	N—O	1.408	O—H$_c$	0.956	MW
	C—H$_a$	1.085	C—H$_b$	1.086	∠CNO	110.2	
	∠H$_a$CN	121.8	∠H$_b$CN	115.6	∠NOH$_c$	102.7	
Formamide							
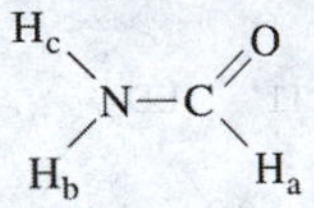	C—N	1.368	C—O	1.212	C—H$_a$	1.125	ED, MW
	N—H	1.027	∠CNH (av.)	119.2	∠NCO	125.0	
Formic acid							
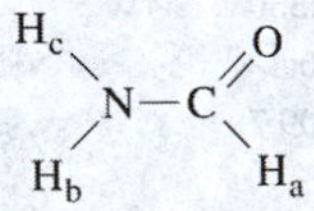	C—O$_a$	1.202	C—O$_b$	1.343	O$_b$—H	0.972	MW
	C—H	1.097					
	∠O$_a$CO$_b$	124.9	∠HCO$_a$	124.1	∠CO$_b$H	106.3	
Formic acid dimer							
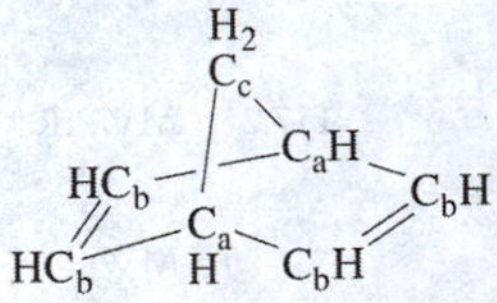	C—O$_a$	1.220	C—O$_b$	1.323	O$_a$···O$_b$	2.703	ED
	∠O$_a$CO$_b$	126.2	∠CO$_a$O$_b$	108.5			
Formyl radical [Oxomethyl]							
HC=O	C—O	1.1712	C—H	1.110	∠HCO	127.43	MW
Fulvene							
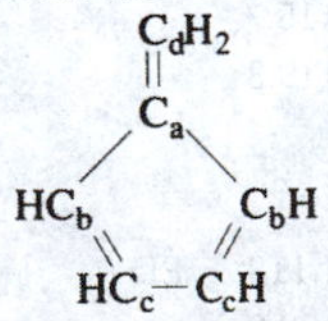	C$_a$—C$_d$	1.349	C$_a$—C$_b$	1.470	C$_b$—C$_c$	1.355	MW
	C$_c$—C$_c$	1.476	C$_b$—H	1.078	C$_c$—H	1.080	

Structure									Method
	C_d—H	1.13	∠$C_bC_aC_b$	106.6	∠$C_bC_cC_c$	109			
	∠$C_aC_bC_c$	107.7	∠C_aC_bH	124.7	∠C_bC_cH	126.4			
	∠HC_dH	117							
Furan 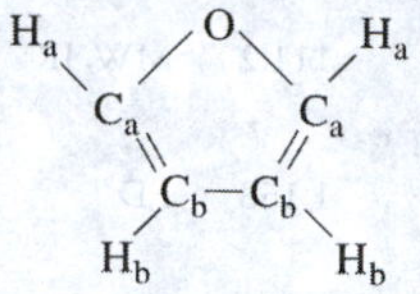	C_a—C_b	1.361	C_b—C_b	1.431	C_a—O	1.362			MW
	C_a—H_a	1.075	C_b—H_b	1.077					
	∠$C_aC_bC_b$	106.1	∠C_bC_aO	110.7	∠C_aOC_a	106.6			
	∠$C_bC_bH_b$	128.0	∠OC_aH_a	115.9					
Furfural 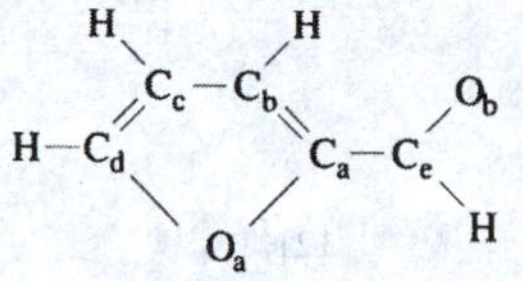	C_a—C_e	1.458	C_e—O_b	1.250	C_e—H	1.088			MW
	∠C_aC_eO	121.6	∠$C_eC_aC_b$	133.9	∠C_aC_eH	116.9			
	trans conformer (with respect to O_a and O_b atoms)								
Glycolaldehyde 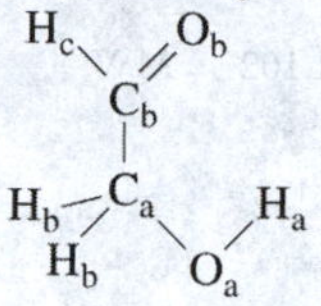	C_a—C_b	1.499	C_a—O_a	1.437	C_b—O_b	1.209			MW
	C_a—H_b	1.093	C_b—H_c	1.102	O_a—H_a	1.051			
	∠$C_aC_bO_b$	122.7	∠$C_bC_aO_a$	111.5					
	∠$C_aC_bH_c$	115.3	∠$C_bC_aH_b$	109.2	∠$H_bC_aH_b$	107.6			
	∠$C_aO_aH_a$	101.6	∠$H_bC_aO_a$	109.7					
Glyoxal									
CHOCHO	C—C	1.526	C—O	1.212	C—H	1.132			ED, UV
trans conformer	∠CCO	121.2	∠HCO	112					
Hexachloroethane									
Cl_3CCCl_3	C—C	1.56	C—C_l	1.769	∠CCCl	110.0			ED
2,4-Hexadiyne									
$C_aH_3C_b{\equiv}C_cC_c{\equiv}C_bC_aH_3$	C_a—C_b	1.450	C_b—C_c	1.208	C_c—C_c	1.377			ED
	C_a—H	1.09							
Hexafluoroethane									
F_3CCF_3	C—C	1.545	CF	1.326	∠CCF	109.8			ED
Hexafluoropropene									
$CF_2{=}CFCF_3$	C—C	1.513	C=C (ass.)	1.329	C—F (ass.)	1.329			ED
	∠CCC	127.8	∠FCC (CF)	120	∠FCC(CF_2)	124			
	∠FCC(CF3)	110							
***trans*-1,3,5-Hexatriene**									
$H_2C_a{=}C_bHC_cH{=}C_cHC_bH{=}C_aH_2$	C_a—C_b	1.337	C_b—C_c	1.458	C_c—C_c	1.368			ED
	∠$C_aC_bC_c$	121.7	∠$C_bC_cC_c$	124.4					
Hydrogen cyanide									
HCN (linear)	C—H (r_e)	1.0655	C—N (r_e)	1.1532					MW, IR
Iminocyanide radical									
HNCN	N—H	1.034	N···N	2.470					UV
	∠HNC	116.5	∠NCN	180[a]					
Iodoacetylene									
IC≡CH	C—C	1.218	C—I	1.980	C—H	1.059			IR

Structure							Method
Iodocyanoacetylene							
$IC_a \equiv C_bC_c \equiv N$ (linear)	$C_a—C_b$	1.207	$C_b—C_c$	1.370	$C_c—N$	1.160	MW
	$C_a—I$	1.985					
Iodomethane [Methyl iodide]							
CH_3I	C—I (r_e)	2.132	C—H (r_e)	1.084	∠HCH(θ_e)	111.2	MW, IR
Iron pentacarbonyl							
$Fe(CO)_5$ (D_{3h})	F_e—C (av.)	1.821	$(F_e—C)_{eq} - (Fe—C)_{ax}$	0.020	C—O (av.)	1.153	ED
Isobutane							
$(C_bH_3)_3C_aH$	$C_a—C_b$	1.535	$C_a—H$	1.122	$C_b—H$	1.113	ED, MW
	$∠C_bC_aC_b$	110.8	$∠C_aC_bH$	111.4			
Isobutene							
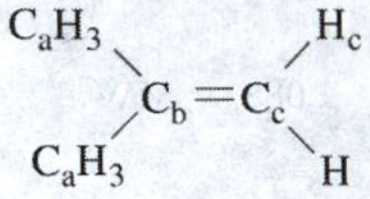	$C_a—C_b$	1.508	$C_b—C_c$	1.342	$C_a—H$	1.119	ED, MW
	$C_c—H_c$	1.10					
	$∠C_aC_bC_a$	115.6	$∠C_aC_bC_c$	122.2	$∠C_bC_cH$	121	
	$∠HC_aC_b$ (av.)	111.4	$∠HC_aH$	107.9	$∠H_cC_cH_c$	118.5	
Isocyanic acid							
HNCO (bent)	N—C	1.209	C—O	1.166	N—H	0.986	MW
	∠NCO	180	∠HNC	128.0			
Isocyanomethane							
$C_aH_3—N \equiv C_b$	$C_a—N$	1.424	$N—C_b$	1.166	$C_a—H$	1.102	MW
	$∠NC_aH$	109.12					
	∠HCH	123.0					
Isofulminic acid							
HCNO (linear)	C—N	1.161	N—O	1.207	H—C	1.027	MW
Isothiocyanic acid							
HNCS	N—C	1.216	C—S	1.561	N—H	0.989	MW
	∠NCS	180	∠HNC	135.0			
Ketene							
$H_2C=C=O$	C—C	1.315	C—O	1.163			MW
	C—H	1.090	∠HCH	123.5			
Malononitrile							
$CH_2(CN)_2$	C—C	1.480	C—N	1.147	C—H	1.091	MW
	∠CCC	110.4	∠CCN	176.6	∠HCH	108.4	
Methane							
CH_4	C—H (r_e)	1.0870					IR
Methanethioamide [Thioformamide]							
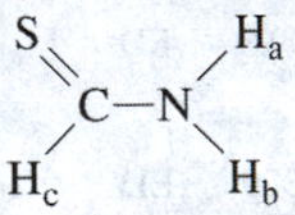	C—S	1.626	C—N	1.358	$C—H_c$	1.10	MW
	$N—H_a$	1.002	$N—H_b$	1.007			
	∠NCS	125.3	$∠H_aN_C$	117.9	$∠H_bN_C$	120.4	
	$∠SCH_c$	127	$∠H_aNH_b$	121.7	∠NCHc	108	
Methanethiol							
CH_3SH	C—S	1.819	S—H	1.34	C—H	1.09	MW
	∠HSC	96.5	∠HCH	109.8	angle between CH_3 symmetry axis and C—S bond	2.2	
Methanol							
CH_3OH	C—O	1.4246	C—H	1.0936	O—H	0.9451	MW
	∠COH	108.53	∠HCH	108.63	angle between CH_3 symmetry axis and C—O bond	3.27	

Molecular

Structure							Method
Methyl							
$\cdot CH_3$ planar (D_{3h})	C—H	1.076					R
***N*-Methylacetamide**							
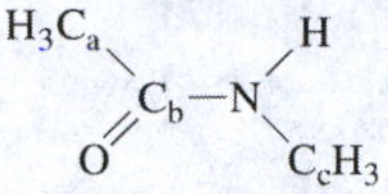	C_a—C_b	1.520	C_b—N	1.386	C_c—N	1.469	ED
	C_b—O	1.225	C—H	1.107			
	$\angle C_bNC_c$	119.7	$\angle NC_bO$	121.8	$\angle C_aC_bN$	114.1	
Methylamine							
CH_3NH_2	C—N	1.471	N—H	1.019	C—H	1.095	MW
	∠HNC	110.3	∠HNH	106.6	∠HCH	108.1	
	angle between CH_3 symmetry axis and C—N bond	4.3					
Methyl azide							
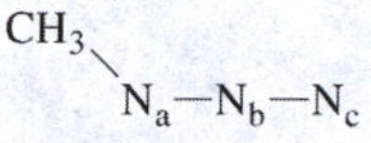	C—N_a	1.468	N_a—N_b	1.216	N_b—N_c	1.113	ED
	C—H	1.09	$\angle C_{N}a_{N}b$	116.8			
3-Methyl-3*H*-diazirine							
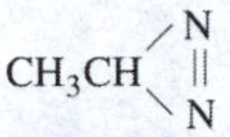	C—C	1.501	C—N	1.481	N—N	1.235	MW
	∠NCN	49.3	dihedral angle between CNN plane and C—C bond	122.3			
Methylene							
$:CH_2$	C—H (r_e)	1.0748	∠HCH(θe)	133.84			IR,MW
Methylenecyclopropane							
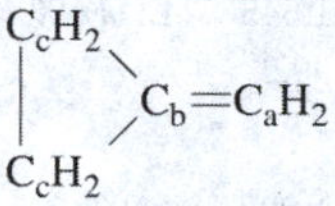	C_a—C_b	1.332	C_b—C_c	1.457	C_c—C_c	1.542	MW
	C_c—H	1.09	$\angle C_cC_bC_c$	63.9	$\angle HC_aH$	114.3	
	$\angle HC_cH$	113.5	dihedral angle between C_cH_2 plane and C_c—C_c bond	150.8			
3-Methyleneoxetane							
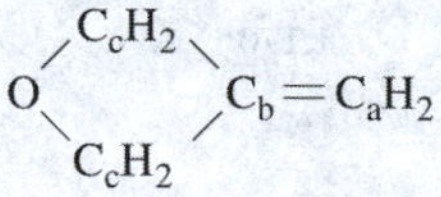	C_a—C_b	1.33	C_b—C_c	1.52	C_c—O	1.45	MW
	C—H (ass)	1.09	$\angle HC_cH$ (ass)	114	$\angle HC_aH$ (ass)	120	
	$\angle C_cC_bC_c$	87					
Methylenephosphine							
CH_cH_t=PH	C—P	1.673	C—H_c	1.09	C—H_t	1.09	MW
planar	P—H	1.420	∠CPH	97.4			
	∠HCH	117.2	$\angle PCH_c$	124.4	$\angle PCH_t$	118.4	
Methyl formate							
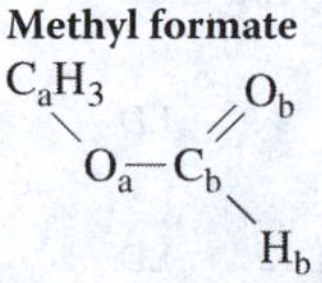	C_b—O_b	1.206	C—O (av.)	1.393	C_a—H	1.08	ED
	C_b—H (ass)	1.101					
	∠COC	114	$\angle O_aC_bO_b$	127	$\angle O_aC_aH$	110	
Methylgermane							
CH_3GeH_3	C—Ge	1.945	Ge—H	1.529	C—H	1.083	MW
	∠HGeH	109.3	∠HCH	108.4			
Methyl hypochlorite							
CH_3OCl	C—O	1.389	O—Cl	1.674	C—H	1.103	MW

Molecular

Structure							Method
	∠COC l	112.8	∠HCH	109.6			
Methylidyne							
:CH	C—H (r_e)	1.1198					UV
Methylidynephosphine							
HCP	C—P (r_e)	1.5398	C—H (r_e)	1.0692			MW
Methylketene [1-Propen-1-one]							
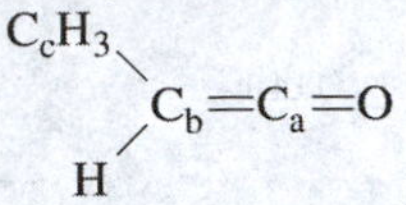	C_a—C_b	1.306	C_b—C_c	1.518	C_a—O	1.171	MW
	C_b—H	1.083	C_c—H	1.10			
	∠OC_aC_b	180.5	∠$C_aC_bC_c$	122.6	∠C_aC_bH	113.7	
	∠C_cC_bH	123.7	∠HCH	109.2			
Methyl nitrate							
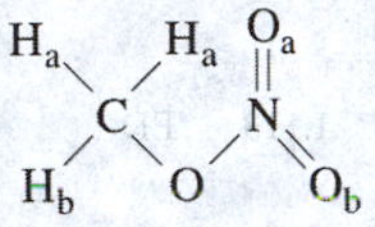	C—O	1.437	C—H_a	1.10	C—H_b	1.09	MW
	O—N	1.402	N—O_a	1.205	N—O_b	1.208	
	∠CON	112.7	∠ONO_a	118.1	∠ONO_b	112.4	
	∠OCH_a	110	∠OCH_b	103			
Methyloxirane							
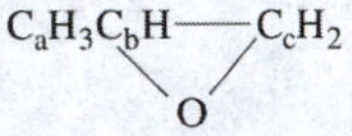	C_a—C_b	1.51	∠$C_aC_bC_c$	121.0	dihedral angle between C_bC_cO plane and $C_aC_b$bond	123.8	MW
Methylphosphine							
CH_3PH_2	C—P	1.858	C—H	1.094			ED
Methylphosphonic difluoride							
CH_3POF_2	C—P	1.770	P—O	1.444	P—F	1.545	ED,MW
	∠OPC	117.8	∠FPC	103.7	∠FPF	99.2	
Methylsilane							
CH_3SiH_3	C—Si	1.867	Si—H	1.485	C—H	1.093	MW
	∠HCH	107.7	∠HSiH	108.3			
Methylstannane							
CH_3SnH_3	C—Sn	2.143	Sn—H	1.700			MW
Methyl thiocyanate							
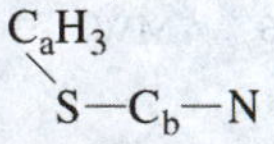	S—C_a	1.824	S—C_b	1.684	C_b—N	1.170	MW
	C—H	1.081					
	∠C_aSC_b	99.0	∠HCH	110.6	∠HCS	108.3	
Methyltrioxorhenium							
CH_3ReO_3	Re—C	2.074	Re—O	1.703	C—H	1.088	MW
	∠ReCH	108.9	∠CReO	106.4			
Molybdenum carbide							
MoC	Mo—C	1.676					UV
Molybdenum carbonyl							
$Mo(CO)_6$ (O_h)	Mo—C	2.063	C—O	1.145			ED
Naphthalene							
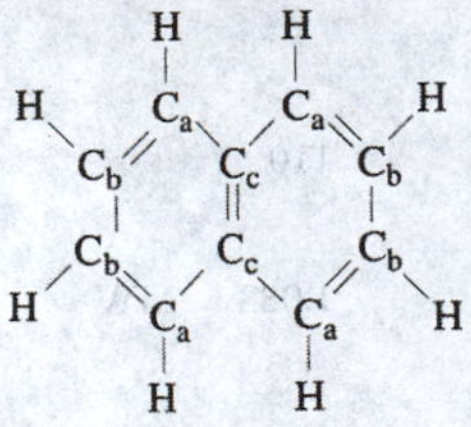	C_a—C_b	1.37	C_b—C_b	1.41	C_a—C_c	1.42	ED
	C_c—C_c	1.42	C—C (av.)	1.40	∠$C_aC_cC_c$	119.4	

Molecular

Structure									Method
Neopentane									
$C(CH_3)_4$	C—C	1.537	C—H	1.114	∠CCH	112			ED
Nickel carbonyl									
$Ni(CO)_4$ (T_d)	Ni—C	1.839	C—O	1.121					IR
Nickel monocarbonyl									
NiCO (linear)	Ni—C	1.64	C—O	1.19					IR
Nickel cyanide									
NiC≡N (linear)	Ni—C	1.828	C—N	1.158					MW
Nitromethane									
CH_3NO_2	C—N	1.489	N—O	1.224	C—H (ass.)	1.088			MW
	∠ONO	125.3	∠NCH	107					
***N*-Nitrosodimethylamine**									
$(CH_3)_2NNO$	C—N	1.461	N—O	1.235	N—N	1.344			ED
	∠CNC	123.2	∠CNN	116.4	∠ONN	113.6			
Nitrosomethane									
CH_3NO	C—N	1.49	N—O	1.22	C—H	1.084			MW
	∠CNO	112.6	∠NCH	109.0					
2,5-Norbornadiene 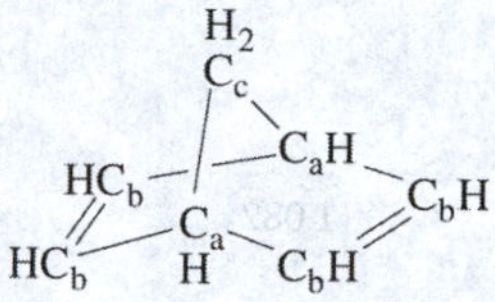 (C_{2v})	C_a—C_b	1.535	C_b—C_b	1.343	C_a—C_c	1.573			ED
	C—H	1.12	$∠C_aC_cC_a$	94					
	dihedral angle between the two $C_aC_bC_bC_a$ planes	115.6							
1,2,5-Oxadiazole 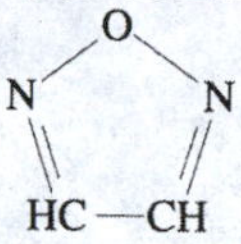	C—C	1.421	C—N	1.300	O—N	1.380			MW
	C—H	1.076	∠CCH	130.2	∠NCH	120.9			
	∠CCN	109.0	∠NON	110.4	∠ONC	105.8			
1,3,4-Oxadiazole 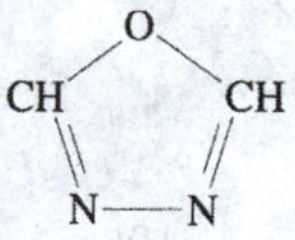	C—O	1.348	C—N	1.297	N—N	1.399			MW
	C—H	1.075	∠OCH	118.1	∠NCH	128.5			
	∠CNN	105.6	∠COC	102.0	∠OCN	113.4			
Oxalic acid 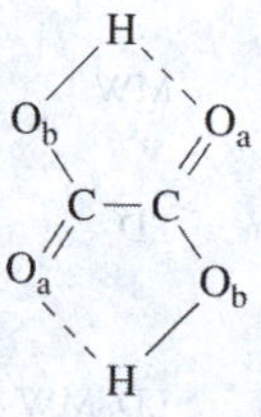	C—C	1.544	C—O_a	1.205	C—O_b	1.336			ED
	O_b—H	1.05							
	$∠CCO_a$	123.1	$∠O_aCO_b$	125.0	$∠CO_bH$	104			
Oxalyl chloride 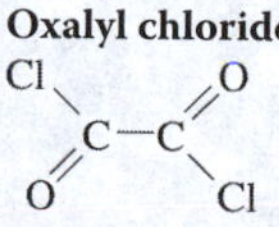	C—C	1.534	C—O	1.182	C—Cl	1.744			ED
	∠CCO	124.2	∠CCCl	111.7	68% *trans*, 32% *gauche* at 0 °C				

Molecular

Structure							Method
Oxetane 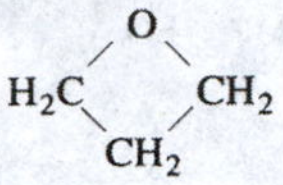	C—C	1.546	C—O	1.448	C—H (av.)	1.090	MW
	∠CCC	85	∠COC	92	∠OCC	92	
	∠HCH (av.)	109.9					
Oxirane 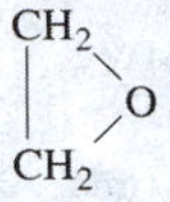	C—C	1.466	C—O	1.431	C—H	1.085	MW
	∠HCH	116.6	dihedral angle between NH_2 plane and N— C bond	158.0			
Phenol 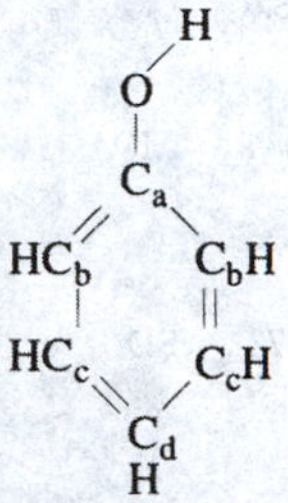	C—C (av.)	1.397	C_a—O	1.364	O—H	0.956	MW
	C_b—H	1.084	C_c—H	1.076	C_d—H	1.082	
	∠COH	109.0					
Phosphirane 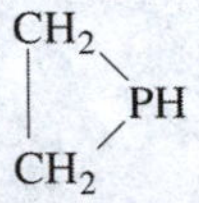	C—C	1.502	C—P	1.867	P—H	1.43	MW
	C—H	1.09	∠CPC	47.4	∠HPC	95.2	
	∠HCH	114.4	∠CCH	118	dihedral angle between PCC plane and PH bond	95.7	
Piperazine 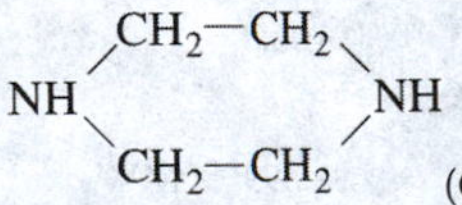 (C_{2h})	C—C	1.540	C—N	1.467	C—H	1.110	ED
	∠CNC	109.0	∠CCN	110.4			
Palladium carbide PdC	Pd—C	1.712					UV
Platinum carbide PtC	Pt—C (*re*)	1.6767					UV
Potassium carbide KC	K—C	2.528					MW
Propane C_3H_8	C—C	1.532	C—H	1.107			ED
	∠CCC	112	∠HCH	107			
Propene 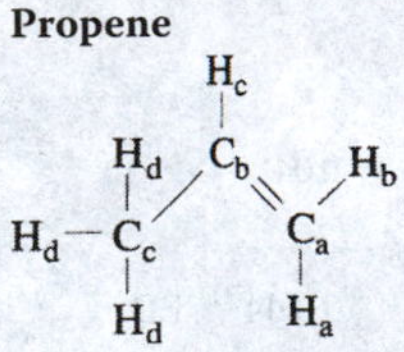	C_a—C_b	1.341	C_b—C_c	1.506			ED, MW
	C_a—H_a	1.104	C_c—H_d	1.117			
	$∠C_aC_bC_c$	124.3	$∠C_bC_aH_{a,b,c}$	121.3	$∠C_bC_cH_d$	110.7	

Structure							Method
2-Propenoyl chloride							
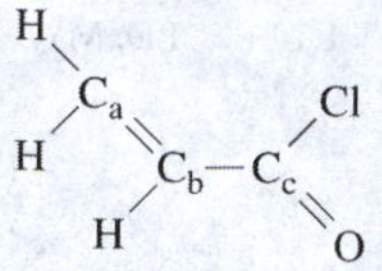	C_a—C_b	1.35	C_b—C_c	1.48	C_c—Cl	1.82	MW
	C_c—O	1.19	C—H (ass.)	1.086			
	∠$C_aC_bC_c$	123	∠C_bC_cCl	116	∠C_bC_cO	127	
	∠C_aC_bH (ass.)	120	∠C_bC_aH (ass.)	121.5			
2-Propynal							
H_aC_a≡C_b—C_cH_cO (planar)	C_a—C_b	1.211	C_b—C_c	1.453	C_c—O	1.214	ED, MW
	C_a—H_a	1.085	C_c—H_c	1.130			
	∠$C_aC_bC_c$	178.6	∠C_bC_cO	124.2	∠$C_bC_cH_c$	113.7	
Propyne							
H_3C_c—C_b≡C_aH	C_c—C_b	1.459	C_b—C_a	1.206			MW
	C_a—H	1.056	C_c—H	1.105	∠HC_cC_b	110.2	
Propynal isocyanide							
H_3C_c—C_b≡C_a—N≡C	C_c—C_b (r_s)	1.456	C_b—C_a (r_s)	1.206	C_a—N (r_s)	1.316	MW
	N—C (r_s)	1.175	C_c—H (r_s)	1.090	∠HC_cC_b (θ_s)	110.7	
Pyrazine							
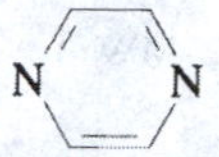	C—C	1.339	C—N	1.403	C—H	1.115	ED
	∠CCH	123.9	∠CCN	115.6			
Pyridazine							
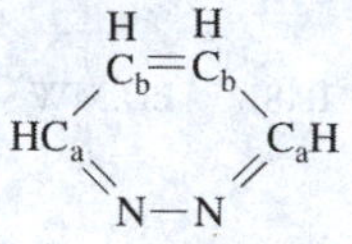	C_a—C_b	1.393	C_b—C_b	1.375	C_a—N	1.341	ED, MW
	N—N	1.330	∠NCC	123.7	∠NNC	119.3	
Pyridine							
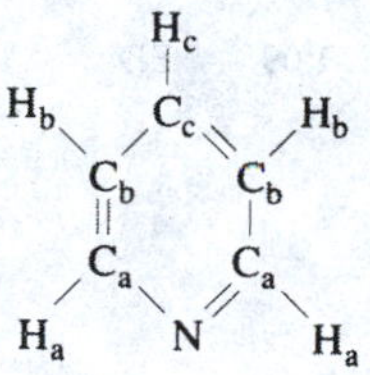	C_a—C_b	1.395	C_b—C_c	1.394	C_a—N	1.340	MW
	C_a—H_a	1.084	C_b—H_b	1.081	C_c—H_c	1.077	
	∠$C_aC_bC_c$	118.5	∠$C_bC_cC_b$	118.3	∠$C_cC_bH_b$	121.3	
	∠C_aNC_a	116.8	∠NC_aC_b	123.9	∠NC_aH_a	115.9	
Pyrimidine							
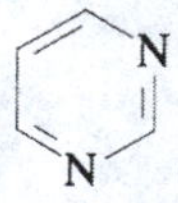 (C_{2v} assumed)	C—C	1.393	C—N	1.340			ED
	∠NCN	127.6	∠CNC	115.5			
Pyrrole							
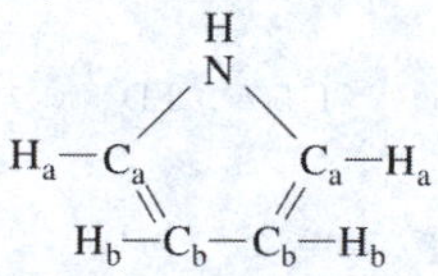	C_a—C_b	1.382	C_b—C_b	1.417	C_a—N	1.370	MW
	C_a—H_a	1.076	C_b—H_b	1.077	N—H	0.996	
	∠$C_aC_bC_b$	107.4	∠C_aNC_a	109.8	∠NC_aC_b	107.7	
	∠C_bC_bH	127.1	∠NC_aH_a	121.5			

Structure							Method
Pyruvonitrile 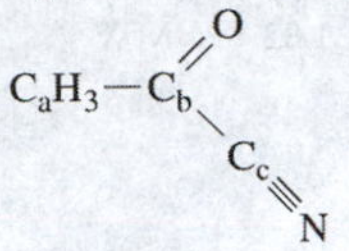	C_a—C_b	1.518	C_b—C_c	1.477	C—H	1.12	ED, MW
	C—N	1.17	C—O	1.208	∠HCH	109.2	
	∠$C_aC_bC_c$	114.2	∠C_aC_bO	124.5	∠CCN	179	
Ruthenium carbide RuC	Ru—C	1.607					UV
Silacyclobutane CH_2—CH_2 / CH_2—SiH_2	C—C	1.571	C—S_i	1.885	C—H	1.100	ED
	Si—H	1.47	∠CCC	99.8	∠CSiC	77.2	
	∠SiCC	84.8	dihedral angle between CCC and CSiC planes	146			
Silaethene $H_2Si{=}CH_2$	Si—C (r_e)	1.704	Si—H (r_e)	1.467	C—H (r_e)	1.082	MW
	∠HCSi	122.0	∠HSiC	122.4			
Silicon dicarbide CSiC (ring)	C—C(r_s)	1.269	Si—C (r_s)	1.832	∠CSiC (θ_s)	40.5	MW
Silylchloroacetylene $SiH_3C{\equiv}CCl$	C—C	1.234	Si—C	1.812	C—Cl	1.620	ED
	Si—H	1.488	∠HSiC	109.4			
Silyl cyanide $SiH_3C{\equiv}N$	Si—C	1.850	C—N	1.156	Si—H	1.487	ED,MW
	∠HSiC	107.25					
Sodium carbide NaC	Na—C	2.232					MW
Spiro[2.2]pentane 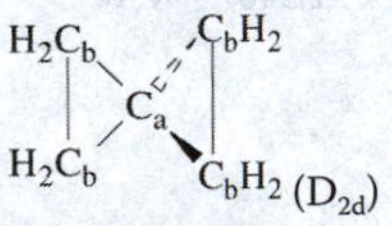	C_b—C_b	1.52	C_a—C_b	1.47	C—H	1.09	ED
	∠$C_bC_aC_b$	62	∠HCH	118			
Strontium methyl $SrCH_3$	Sr—C	2.487	C—H (ass.)	1.104	∠HCH	105.8	UV
Succinonitrile CH_2CN | CH_2CN	C—C	1.561	C—C(N)	1.465	C—N	1.161	ED
	C—H	1.09	∠CCC	110.4	dihedral angle of CCCC for *gauche* conformer	75	
Tetrabromomethane CBr_4 (T_d)	C—Br	1.935					ED
Tetrachloroethene $CCl_2{=}CCl_2$	C—C	1.354	C—C_l	1.718	∠ClCCl	115.7	ED
Tetrachloromethane CCl_4 (T_d)	C—Cl	1.767					ED
Tetracyanoethene $(CN)_2C{=}C(CN)_2$	C—C	1.435	C=C	1.357	C—N	1.162	ED
	∠CC =C	121.1					

Structure							Method
2,2,4,4-Tetrafluoro-1,3-dithietane							
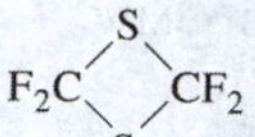 (D$_{2h}$ assumed)	C—S	1.785	C—F	1.314	∠CSC	83.2	ED
	∠FCS	113.7					
Tetrafluoroethene							
CF_2=CF_2	C—C	1.31	C—F	1.319	∠CCF	123.8	ED
Tetrafluoromethane							
CF_4 (T_d)	C—F	1.323					ED
Tetrahydrofuran							
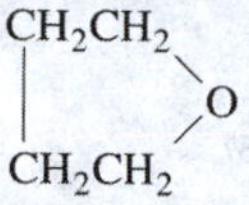	C—C	1.536	C—O	1.428	C—H	1.115	ED
Tetrahydropyran							
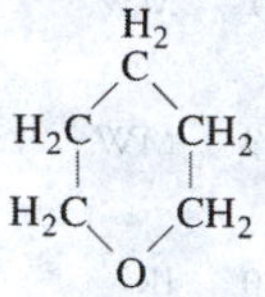	C—C	1.531	C—O	1.420	C—H	1.116	ED
	∠COC	111.5	∠OCC	111.8	∠CCC (C)	108	
	∠CCC (O)	111					
Tetrahydrothiophene							
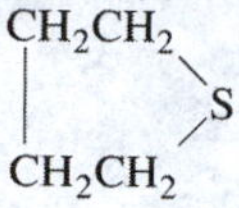	C—C	1.536	C—S	1.839	C—H	1.120	ED
	∠CCC	105.0	∠CSC	93.4	∠SCC	106.1	
Tetraiodomethane							
CI_4 (T_d)	C—I	2.15					ED
Tetramethylgermane							
$(CH_3)_4Ge$	C—G$_e$	1.945	C—H	1.12	∠GeCH	1.08	ED
Tetramethyl lead							
$(CH_3)_4Pb$	C—P$_b$	2.238					ED
Tetramethylsilane							
$(CH_3)_4Si$	C—Si	1.875	C—H	1.115	∠HCH	109.8	ED
Tetramethylstannane							
$(CH_3)_4Sn$	C—Sn	2.144	C—H	1.12			ED
1,2,5-Thiadiazole							
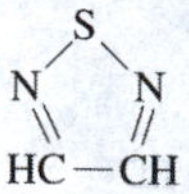	C—C	1.420	C—N	1.328	S—N	1.631	MW
	C—H	1.079					
	∠CCN	113.8	∠NSN	99.6	∠CCH	126.2	
1,3,4-Thiadiazole							
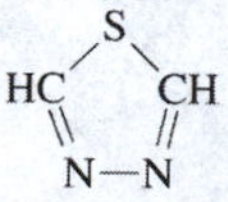	C—S	1.721	C—N	1.302	N—N	1.371	MW
	C—H	1.08	∠CSC	86.4	∠SCN	114.6	
	∠CCN	112.2	∠NCH	123.5	∠SCH	121.9	
Thietane							
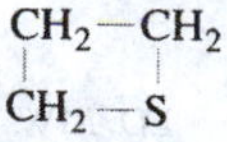	C—C	1.549	C—S	1.847	C—H (av.)	1.100	ED, MW
	∠CSC	76.8	∠HCH (av.)	112	dihedral angle between CCC and CSC planes	154	

Molecular

Structure							Method
Thiirane							
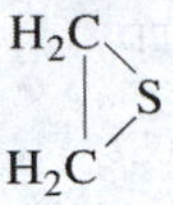	C—C	1.484	C—S	1.815	C—H	1.083	MW
	∠CSC	48.3	∠CCS	65.9	∠HCH	116	
	dihedral angle between CH_2 plane and C—C bond	152					
Thioacetaldehyde							
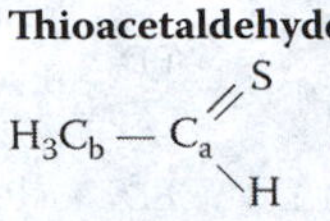	C_a—S (r_s)	1.610	C_a—C_b (r_s)	1.506			MW
	C_a—H (r_s)	1.089	C_b—H (r_s)	1.094[b]			
	∠C_bC_aS (θ_s)	125.3	∠C_bC_aH (θ_s)	119.4	∠HC_bC_a (θ_s)	110.6[b]	
Thiocarbonyl fluoride							
F_2CS	C—S	1.589	C—F	1.315	∠FCF	107.1	MW
Thioformaldehyde							
CH_2S	C—S	1.611	C—H	1.093	∠HCH	116.9	MW
Thioketene							
H_2C=C=S	C—C (r_s)	1.314	C—S (r_s)	1.554	C—H (r_s)	1.080	IR
C_{2v}	∠HCH (θ_s)	119.8					
Thiophene							
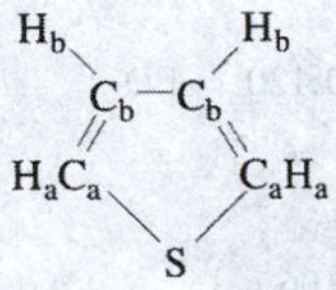	C_a—C_b	1.370	C_b—C_b	1.423	C_a—S	1.714	MW
	C_a—H_a	1.078	C_b—H_b	1.081			
	∠$C_aC_bC_b$	112.5	∠C_aSC_a	92.2	∠$SC_{aC}b$	115.5	
	∠S_Ca_Ha	119.9	∠$C_bC_bH_b$	124.3			
Toluene							
C_6H_5—CH_3	C—C (ring)	1.399	C—CH_3	1.524	C—H (av.)	1.11	ED
1,1,1-Tribromoethane							
CH_3CBr_3	C—C (ass.)	1.51	C—Br	1.93	C—H (ass.)	1.095	MW
	∠BrCBr	111	∠CC Br	1.8	∠CCH (ass.)	109.0	
Tribromomethane							
$CHBr_3$ (C_{3v})	C—Br	1.924	C—H	1.11	∠BrCBr	111.7	ED, MW
Tri-*tert*-butyl methane							
$HC_a[C_b(C_cH_3)_3]_3$	C_a—C_b	1.611	C_b—C_c	1.548	C—H	1.111	ED
	∠$C_aC_bC_c$	113.0					
Trichloroacetonitrile							
CCl_3CN	C—C	1.460	C—N	1.165	C—Cl	1.763	ED
	∠ClCCl	110.0					
1,1,1-Trichloroethane							
CH_3CCl_3	C—C	1.541	C—Cl	1.771	C—H	1.090	MW
	∠CCCl	109.6	∠ClCCl	109.4	∠HCH	110.0	
	∠CCH	108.9					
Trichlorofluoromethane							
CCl_3F	C—Cl	1.754	C—F	1.362	∠ClCCl	111	MW
Trichloromethane							
$CHCl_3$	C—Cl	1.758	C—H	1.100	∠ClCCl	111.3	MW
Trichloromethylgermane							
CH_3GeCl_3	C—Ge	1.89	Ge—Cl	2.132	C—H (ass.)	1.103	ED, MW
	∠ClGeCl	106.4	∠GeCH (ass.)	110.5			

Molecular

Structure							Method
Trichloromethylsilane							
CH_3SiCl_3	C—Si	1.876	Si—Cl	2.021			MW
Trichloromethylstannane							
CH_3SnCl_3	C—Sn	2.10	Sn—Cl	2.304	C—H	1.100	ED
1,1,1-Trichloro-2,2,2-trifluoroethane							
CF_3CCl_3	C—C	1.54	C—F	1.33	C—Cl	1.77	MW
(staggered configuration)	∠CCF	110	∠CCCl	109.6			
	∠CSnCl	113.9	∠ClSnCl	104.7	∠SnCH	108	
Triethylenediamine							
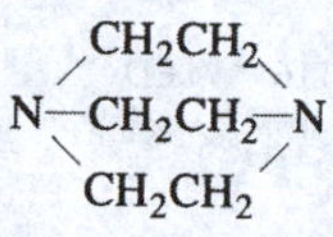 (D_{3h})	C—C	1.562	C—N	1.472	∠CNC	108.7	ED
	∠NCC	110.2					
Trifluoroacetic acid							
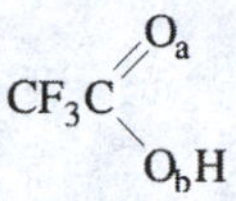	C—C	1.546	C—O_a	1.192	C—O_b	1.35	ED
	C—F	1.325	O—H (ass.)	0.96			
	∠CCO_a	126.8	∠CCO_b	111.1	∠CCF	109.5	
1,1,1-Trifluoroethane							
CH_3CF_3	C—C	1.494	C—F	1.340	C—H	1.081	ED
Trifluoroiodomethane							
CF_3I (C_{3v})	C—F	1.330	C—I	2.138	∠FCF	108.1	ED, MW
Trifluoromethane							
CHF_3 (C_{3v})	C—F	1.332	C—H	1.098	∠FCF	108.8	MW
Trifluoromethanesulfonyl fluoride							
$CF_3SO_2F_a$	C—S	1.835	C—F (av.)	1.325	S—O	1.410	ED
	S—F_a	1.543	∠CSF_a	95.4	∠CSO	108.5	
	∠OSO	124.1	∠FCF	109.8			
Trifluoromethyliminosulfur difluoride							
$CF_3N{=}SF_2$	C—N	1.409	S—N	1.477	S—F	1.594	ED,MW
	C—F	1.331	∠CNS	127.2	∠NSF	112.7	
	∠FSF	92.8	∠FCF	108.1			
Trifluoromethyl peroxide							
CF_3OOCF_3	O—O	1.42	C—O	1.399	C—F	1.320	ED
	∠COO	107	∠FCF	109.0	COOC dihedral angle of internal rotation	123	
	∠CCF	119.2	∠CCH	112			
Trimethyl aluminium							
$(CH_3)_3Al$	C—Al	1.957	C—H	1.113			ED
	∠CAlC	120	∠AlCH	111.7			
Trimethylamine							
$(CH_3)_3N$	C—N	1.458	C—H	1.100			ED
	∠CNC	110.9	∠HCH	110			
Trimethylarsine							
$(CH_3)_3As$	C—As	1.979	∠CAsC	98.8	∠AsCH	111.4	ED
Trimethyl bismuth							
$(CH_3)_3Bi$	C—Bi	2.263	C—H	1.07	∠CBiC	97.1	ED
Trimethylborane							
$(CH_3)_3B$	C—B	1.578	C—H	1.114			ED
	∠CBC	120	∠BCH	112.5			

Structure							Method
Trimethylphosphine							
$(CH_3)_3P$	C—P	1.847	C—H	1.091			ED
	∠CPC	98.6	∠PCH	110.7			
1,3,5-Trioxane							
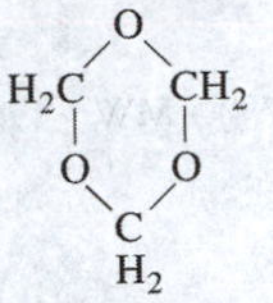	C—O	1.422	∠OCO	112.2	∠COC	110.3	MW
Triphenylamine							
$(C_6H_5)_3N$ (C_3)	C—C	1.392	C—N	1.42	∠CNC	116	ED
	torsional dihedral angle of phenyl rings	47					
Tungsten carbide							
WC	W—C	1.7135					UV
Tungsten carbonyl							
$W(CO)_6$ (O_h)	W—C	2.059	C—O	1.149			ED
Vanadium carbonyl							
$V(CO)_6$ (O_h, involving dynamic Jahn-Teller effect)	V—C	2.015	C—O	1.138			ED
Vinyl bromide [Bromoethene]							
See Vinyl chloride	C—C	1.3256	C—Br	1.8835	C—H_a	1.0780	MW
	C—H_b	1.0804	C—H_c	1.0794	∠CCBr	122.62	
	∠CCH_a	124.34	∠CCH_b	119.28	∠CCH_c	122.03	
Vinyl chloride [Chloroethene]							
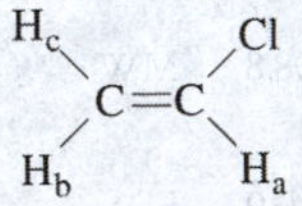	C—C	1.3262	C—Cl	1.7263	C—H_a	1.0783	MW
	C—H_b	1.0796	C—H_c	1.0796	∠CCCl	122.75	
	∠CCH_a	123.91	∠CCH_b	119.28	∠CCH_c	121.77	
Vinyl fluoride [Fluoroethene]							
See Vinyl chloride	C—C	1.3210	C—F	1.3428	C—H_a	1.0796	MW
	C—H_b	1.0774	C—H_c	1.0789	∠CCF	121.70	
	∠CCH_a	125.95	∠CCH_b	118.97	∠CCH_c	121.34	
Vinyl iodide [Iodoethene]							
See Vinyl chloride	C—C	1.3276	C—I	2.0830	C—H_a	1.0787	MW
	C—H_b	1.0823	C—H_c	1.0799	∠CCI	122.97	
	∠CCH_a	123.54	∠CCH_b	119.36	∠CCH_c	122.30	
Zinc cyanide							
ZnC≡N (linear)	Zn—C	1.955	C—N	1.146			MW

[a] Approximate
[b] Average
[c] Distances from Be to planes are 1.47 and 1.92 Å
[d] 40°~50° (The two rings are not parallel)

CHARACTERISTIC BOND LENGTHS IN FREE MOLECULES

David R. Lide

This is a summary of typical bond lengths in gas-phase molecules. The value given for each bond is near the mid-range of values found in simple molecules. Bond lengths usually vary by 1% or 2%, and often by more, depending on the nature of the other bonds attached to the two atoms in question. References 1 and 2 give measured bond lengths in individual gas-phase molecules, as determined by spectroscopic and electron diffraction methods. All bond lengths are given in Å (1 Å = 10^{-10} m).

References

1. Structure of Free Molecules in the Gas Phase, *CRC Handbook of Chemistry and Physics, 97th Edition*, 2016.
2. Harmony, M. D., Laurie, V. W., Kuczkowski, R. L., Schwendeman, R. H., Ramsay, D. A., Lovas, F. J., Lafferty, W. J., and Maki, A. G., *J. Phys. Chem. Ref. Data* 8, 619, 1979.
3. Lide, D. R., *Tetrahedron* 17, 125, 1962.

A. Characteristic Lengths of Single Bonds in Å

	As	B	Br	C	Cl	F	Ge	H	I	N	O	P	S	Se	Si
As	2.10														
Br	2.32	1.89	2.28												
C	1.98	1.58	1.93	1.53											
Cl	2.17	1.74	2.14	1.77	1.99										
F	1.71	1.31	1.76	1.35	1.63	1.41									
Ge			2.30	1.95	2.15	1.73	2.40								
H	1.51	1.19	1.41	1.09	1.28	0.92	1.53	0.74							
I		2.12	2.47	2.13	2.32	1.91	2.51	1.61	2.67						
N		1.39		1.46	1.90	1.37		1.02		1.45					
O		1.35		1.42	1.70	1.42		0.96		1.43	1.48				
P		1.94	2.22	1.85	2.04	1.57		1.42		1.65		2.25			
S			2.24	1.82	2.05	1.56		1.34					2.00		
Sb			2.49		2.33	1.88		1.70	2.27						
Se				1.95		1.71		1.47						2.33	
Si			2.21	1.87	2.05	1.58		1.48	2.44		1.63		2.14		2.33
Sn			2.50	2.14	2.30	1.94		1.71	2.67						
Te						1.82		1.66							

B. Lengths of Multiple Bonds (Non-Ring Molecules) in Å

C=C	1.34	N≡N	1.13
C≡C	1.20	N=O	1.18
C=N	1.21	O=O	1.21
C≡N	1.16	S=S	1.89
C=O	1.21	S=O	1.48
C=S	1.61	Se=Se	2.17
N=N	1.24	Se=O	1.65

C. Effect of Environment on Carbon-Carbon Single-Bond Lengths in Å (Other Single Bonds Not Shown) from Reference 3

Configuration	C–C Length	Examples of Molecules
C–C	1.526	$H_3C–CH_3$
C–C=	1.501	$H_3C–CH=CH_2$
C–C≡	1.459	$H_3C–C≡CH$
=C–C=	1.467	$H_2C=CH–CH=CH_2$
≡C–C=	1.445	$HC≡C–CH=CH_2$
≡C–C≡	1.378	HC≡C–C≡CH

D. Some Metal-Carbon Bond Lengths in Gas-Phase Molecules in Å

Al–C	1.96
Be–C	1.70
Bi–C	2.26
Cd–C	2.11
Co–C	1.88
Cr–C	1.92*
Cu–C	1.83
Fe–C	1.82*
Hg–C	2.08
Mo–C	2.06*
Ni–C	1.83
Pb–C	2.24
Sn–C	2.14
V–C	2.02*
W–C	2.06*
Zn–C	1.93

* In carbonyl molecules

Molecular

ATOMIC RADII OF THE ELEMENTS

Manjeera Mantina, Rosendo Valero, Christopher J. Cramer, and Donald G. Truhlar

Atomic radii are not precisely defined, but are nevertheless very widely used parameters in modeling and understanding molecular structure and interactions. Three main classes of radii may be identified: van der Waals radii (which include radii used to characterize steric interactions), covalent radii, and ionic radii. This table is concerned with the first two; ionic radii are covered in a table in Section 12 of the *Handbook* called "Ionic Radii in Crystals."

There are many scales of van der Waals radii, but they are not fully consistent with one another. The van der Waals radii determined by Bondi (Ref. 1) from x-ray diffraction data, crystal densities, gas kinetic collision cross sections, critical densities, and liquid-state properties are the most widely used values, but Bondi recommended radius values for only 28 of the 44 main-group elements in the periodic table plus 9 transition metals and one actinide. Rowland and Taylor (Ref. 2) redetermined nine of the main-group radii from crystal structure data and recommended that Bondi's values be accepted except for H, for which they recommended a new value; we accepted their recommendation for H and adopted Bondi's 27 other values for main-group elements. Radii for the 16 remaining main-group elements were determined from electronic structure calculations on selected van der Waals molecules by analyzing the results in a way designed to yield radii compatible with Bondi's scale for main-group elements (Ref. 3). Bondi's values for the transition metals and actinide are smaller than expected based on ionization potentials and covalent radii, so we do not adopt them, but defer to later recommendations based on analysis of a more extensive set of crystal data. Van der Waals radii for the remaining elements through atomic number $Z = 93$ were taken from Hu et al. (Ref. 4), who determined them from bond valence parameters (this gives radii usually within 0.1 Å to 0.15 Å of the radii from average atomic volumes in crystals as obtained by statistical analysis of the Cambridge Structural Database (Ref. 5), but the results show a smoother variation with atomic number). Van der Waals radii for the elements with $Z = 94$ to 103 were based on the work of Guzei and Wendt (Ref. 6), who modeled the zero-potential distance for the interaction potentials involved in nonbonded steric interactions; we increased their values by 6%, which is the value they found is needed to make their steric radii scale most consistent with Bondi's van der Waals radii.

Covalent radii are more straightforward, especially for elements that tend to form only single bonds, although some researchers distinguish metallic bonds from covalent bonds. The covalent radii tabulated here are recommendations for single covalent bonds, and they are based on a comprehensive evaluation of experimental data by Cordero et al. (Ref. 7), who recommended covalent radii for all elements up to $Z = 96$, and on an analysis combining experimental data and theoretical calculations by Pyykkö and Atsumi (Ref. 8), who recommended single-bond covalent radii for all elements up to $Z = 118$. If one is interested in a specific coordination number, oxidation state, or type of ligand, one might find a more appropriate radius in the specialized literature or in the "Characteristic Bond Lengths in Free Molecules" table in this *Handbook* since the values in the table below are generic average values. In particular, we give the Pyykkö-Atsumi values for carbon and for $Z = 97$ to 118 and an average of the values from the two sources for all other elements. For Mn, Fe, and Co, Cordero et al. give two values, and we used the lower of these on the average because it corresponds to a smoother periodic trend.

Please note that the van der Waals radii of bromine and lithium have typographical errors in Ref. 3; the correct values (1.85 for Br and 1.82 for Li) in the table below are from Ref. 1.

All values are rounded to the nearest 0.01 Å, but in most cases the uncertainty in the value is of the order of 0.1 Å.

References

1. Bondi, A., *J. Phys. Chem.* 68, 441, 1964.
2. Rowland, R. S., and Taylor, R., *J. Phys. Chem.* 100, 7384, 1996.
3. Mantina, M., Chamberlin, A. C., Valero, R., Cramer, C. J., and Truhlar, D. G., *J. Phys. Chem., A* 113, 5806, 2009.
4. Hu, S.-Z., Zhou, Z.-H., and Robertson, B. E., *Z. Kristallogr.* 224, 375, 2009.
5. Hu, S.-Z., Zhou, Z.-H., and Tsai, K.-R., *Acta Physico-Chimica Sinica* 19, 1073, 2003.
6. Guzei, I. A. and Wendt, M., *Dalton Trans.*, 2006, 3991, 2006.
7. Cordero, B., Gómez, V., Platero-Prats, A. E., Revés, M., Echeverría, J., Cremades, E., Barragán, F., and Alvarez, S., *Dalton Trans.* 2008, 2832, 2008.
8. Pyykkö, P., and Atsumi, M., *Chem. Eur. J.* 15, 186, 2009.

Van der Waals and Covalent Radii of the Elements

Element	Symbol	R_{vdw}/Å	R_{cov}/Å
Actinium	Ac	2.47	2.01
Aluminum	Al	1.84	1.24
Americium	Am	2.44	1.73
Antimony	Sb	2.06	1.40
Argon	Ar	1.88	1.01
Arsenic	As	1.85	1.20
Astatine	At	2.02	1.48
Barium	Ba	2.68	2.06
Berkelium	Bk	2.44	1.68
Beryllium	Be	1.53	0.99
Bismuth	Bi	2.07	1.50
Bohrium	Bh		1.41
Boron	B	1.92	0.84
Bromine	Br	1.85	1.17
Cadmium	Cd	2.18	1.40
Calcium	Ca	2.31	1.74
Californium	Cf	2.45	1.68
Carbon	C	1.70	0.75
Cerium	Ce	2.42	1.84
Cesium	Cs	3.43	2.38
Chlorine	Cl	1.75	1.00
Chromium	Cr	2.06	1.30
Cobalt	Co	2.00	1.18
Copernicium	Cn		1.22
Copper	Cu	1.96	1.22
Curium	Cm	2.45	1.68
Darmstadtium	Ds		1.28
Dubnium	Db		1.49
Dysprosium	Dy	2.31	1.80
Einsteinium	Es	2.45	1.65
Erbium	Er	2.29	1.77
Europium	Eu	2.35	1.83
Fermium	Fm	2.45	1.67
Flerovium	Fl		1.43
Fluorine	F	1.47	0.60
Francium	Fr	3.48	2.42
Gadolinium	Gd	2.34	1.82
Gallium	Ga	1.87	1.23
Germanium	Ge	2.11	1.20
Gold	Au	2.14	1.30
Hafnium	Hf	2.23	1.64
Hassium	Hs		1.34

Molecular

Element	Symbol	R_{vdw}/Å	R_{cov}/Å
Helium	He	1.40	0.37
Holmium	Ho	2.30	1.79
Hydrogen	H	1.10	0.32
Indium	In	1.93	1.42
Iodine	I	1.98	1.36
Iridium	Ir	2.13	1.32
Iron	Fe	2.04	1.24
Krypton	Kr	2.02	1.16
Lanthanum	La	2.43	1.94
Lawrencium	Lr	2.46	1.61
Lead	Pb	2.02	1.45
Lithium	Li	1.82	1.30
Livermorium	Lv		1.75
Lutetium	Lu	2.24	1.74
Magnesium	Mg	1.73	1.40
Manganese	Mn	2.05	1.29
Meitnerium	Mt		1.29
Mendelevium	Md	2.46	1.73
Mercury	Hg	2.23	1.32
Molybdenum	Mo	2.17	1.46
Moscovium	Mc		1.62
Neodymium	Nd	2.39	1.88
Neon	Ne	1.54	0.62
Neptunium	Np	2.39	1.80
Nickel	Ni	1.97	1.17
Nihonium	Nh		1.36
Niobium	Nb	2.18	1.56
Nitrogen	N	1.55	0.71
Nobelium	No	2.46	1.76
Oganesson	Og		1.57
Osmium	Os	2.16	1.36
Oxygen	O	1.52	0.64
Palladium	Pd	2.10	1.30
Phosphorus	P	1.80	1.09
Platinum	Pt	2.13	1.30
Plutonium	Pu	2.43	1.80
Polonium	Po	1.97	1.42
Potassium	K	2.75	2.00
Praseodymium	Pr	2.40	1.90
Promethium	Pm	2.38	1.86
Protactinium	Pa	2.43	1.84
Radium	Ra	2.83	2.11
Radon	Rn	2.20	1.46
Rhenium	Re	2.16	1.41
Rhodium	Rh	2.10	1.34
Roentgenium	Rg		1.21
Rubidium	Rb	3.03	2.15
Ruthenium	Ru	2.13	1.36
Rutherfordium	Rf		1.57
Samarium	Sm	2.36	1.85
Scandium	Sc	2.15	1.59
Seaborgium	Sg		1.43
Selenium	Se	1.90	1.18
Silicon	Si	2.10	1.14
Silver	Ag	2.11	1.36
Sodium	Na	2.27	1.60
Strontium	Sr	2.49	1.90
Sulfur	S	1.80	1.04
Tantalum	Ta	2.22	1.58
Technetium	Tc	2.16	1.38
Tellurium	Te	2.06	1.37
Tennessine	Ts		1.65
Terbium	Tb	2.33	1.81
Thallium	Tl	1.96	1.44
Thorium	Th	2.45	1.90
Thulium	Tm	2.27	1.77
Tin	Sn	2.17	1.40
Titanium	Ti	2.11	1.48
Tungsten	W	2.18	1.50
Uranium	U	2.41	1.83
Vanadium	V	2.07	1.44
Xenon	Xe	2.16	1.36
Ytterbium	Yb	2.26	1.78
Yttrium	Y	2.32	1.76
Zinc	Zn	2.01	1.20
Zirconium	Zr	2.23	1.64

DIPOLE MOMENTS

David R. Lide

These tables give selected values of the electric dipole moment μ for over 800 molecules. When available, values determined by microwave spectroscopy, molecular beam electric resonance, and other high-resolution spectroscopic techniques were selected. Otherwise, the values come from measurements of the dielectric constant in the gas phase or, if these do not exist, in the liquid phase. Entries are listed alphabetically by compound name.

The dipole moment is given in debye units (D). The conversion factor to SI units is 1 D = 3.33564×10^{-30} C m.

When the accuracy of a value is explicitly stated, e.g., 1.234(12), where digit(s) in parentheses represent the uncertainty in the last digit(s) of the value, the stated uncertainty generally indicates two standard deviations. When no uncertainty is given, the value may be assumed to be precise to a few units in the last decimal place. However, if more than three decimal places are given, the exact interpretation of the final digits may require analysis of the vibrational averaging. Other information on molecules that have been studied by spectroscopy, such as the components of the dipole moment in the molecular framework and the variation with vibrational state and isotopic species, is given in the references.

Values measured in the gas phase that are questionable because of undetermined error sources are preceded by ≈. Values obtained by liquid phase measurements, which sometimes have large errors because of association effects, are followed by liq.

Table 1 gives dipole moments of individual molecules in the gas or liquid phase. Table 2 lists dipole moments of conformers whose configurations are related by internal rotation around single bonds. These conformers exist long enough to be measured by gas-phase microwave spectroscopic techniques but interconvert too rapidly to be isolated as bulk samples. The conformers are designated as *gauche*, *trans*, *axial*, etc. The meaning of these terms can be found in the references.

References

1. Nelson, R. D., Lide, D. R., and Maryott, A. A., *Selected Values of Electric Dipole Moments for Molecules in the Gas Phase*, Natl. Stand. Ref. Data Ser. — Nat. Bur. Stnds. 10, 1967.
2. *Landolt-Börnstein, Numerical Data and Functional Relationships in Science and Technology, New Series*, II/6 (1974), Springer-Verlag, Heidelberg.
3. *Landolt-Börnstein, Numerical Data and Functional Relationships in Science and Technology, New Series*, II/14a (1982), Springer-Verlag, Heidelberg.
4. *Landolt-Börnstein, Numerical Data and Functional Relationships in Science and Technology, New Series*, II/14b (1983), Springer-Verlag, Heidelberg.
5. *Landolt-Börnstein, Numerical Data and Functional Relationships in Science and Technology, New Series*, II/19c (1992), Springer-Verlag, Heidelberg.
6. *Landolt-Börnstein, Numerical Data and Functional Relationships in Science and Technology, New Series*, II/24c (2002), Springer-Verlag, Heidelberg.
7. Riddick, J. A., Bunger, W. B., and Sakano, T. K., *Organic Solvents, Fourth Edition*, John Wiley & Sons, New York, 1986.
8. Kasuya, T., Lafferty, W. J., and Lide, D. R., *J. Chem. Phys.* 48, 1, 1968.
9. Kirchhoff, W. H., and Lide, D. R., *J. Chem. Phys.* 51, 467, 1969.
10. Durig, J. R., Li, Y. S., and Rizzolo, J. J., *J. Chem. Phys.* 77, 5885, 1982.
11. Ogata, T., Mochizuki, A. and Yamashita, E., *J. Chem. Phys.* 87, 2531, 1987.
12. Rego, A., and Cox, A. P., *J. Chem. Phys.* 89, 124, 1988.
13. Tyblewski, M., et al., *J. Chem. Phys.* 97, 6168, 1992.
14. Kawashima, Y., et al., *J. Chem. Phys.* 99, 820, 1993.
15. Caminati, W., Melandri, S., and Favero, L., *J. Chem. Phys.* 100, 8569, 1994.
16. Cederberg, J., et al., *J. Chem. Phys.* 105, 3361, 1996.
17. Bauder, A., et al., *J. Chem. Phys.* 106, 7558, 1997.
18. Muller, H. S. P., Miller, C. E., and Cohen, E. A., *J. Chem. Phys.* 107, 8292, 1997.
19. Burgh, D. J., Suenram, R. D., and Stevens, W. J., *J. Chem. Phys.* 111, 3526, 1999.
20. Blake, T. A., et al., *J. Chem. Phys.* 98, 6031, 1993.
21. Ruoff, R. S., et al., *J. Chem. Phys.* 89, 138, 1988.
22. Muenter, J. S., *J. Chem. Phys.* 90, 4048, 1989.
23. Peterson, J. I., Suenram, R. D., and Lovas, F. J., *J. Chem. Phys.* 90, 5964, 1989.
24. Suenram, R. D., Lovas, F. J., and Matsumura, K., *Astrophys. J. Lett.* 342, 103, 1989.
25. Groner, P., et al., *J. Chem. Phys.* 91, 1434, 1989.
26. Suenram, R. D., Lovas, F. J., Fraser, G. T., and Matsumura, K., *J. Chem. Phys.* 92, 4724, 1990.
27. Andrews, A. M., et al., *J. Chem. Phys.* 93, 7030, 1990.
28. Peterson, K. I., Suenram, R. D., and Lovas, F. J., *J. Chem. Phys.* 94, 106, 1991.
29. Iida, M., Ohshima, Y., and Endo, Y., *J. Chem. Phys.* 95, 4772, 1991.
30. Andrews, A. M., Hillig, K. W., and Kuczkowski, R. L., *J. Chem. Phys.* 96, 1784, 1992.
31. Ruoff, R. S., et al., *J. Chem. Phys.* 96, 3441, 1992.
32. Germann, T. C., Tschopp, S. L., and Gutowsky, H. S., *J. Chem. Phys.* 97, 1619, 1992.
33. Taleb-Bendiab, A., Hillig, K. W., and Kuczkowski, R. L., *J. Chem. Phys.* 97, 2996, 1992.
34. Taleb-Bendiab, A., Hillig, K. W., and Kuczkowski, R. L., *J. Chem. Phys.* 98, 3627, 1993.
35. Xu, L-W., and Kuczkowski, R. L., *J. Chem. Phys.* 100, 15, 1994.
36. Peterson, K. I., Suenram, R. D., and Lovas, F. J., *J. Chem. Phys.* 102, 7807, 1995.
37. Tatamitani, Y., and Ogata, T., *J. Chem. Phys.* 121, 9885, 2004.
38. Medvedev, I., et al., *Astrophys. J. Suppl.* 148, 593, 2003.
39. Lesarri, A., Suenram, R. D., and Brugh, D., *J. Chem. Phys.* 117, 9651, 2002.
40. Arunan, E., et al., *J. Chem. Phys.* 117, 9766, 2002.
41. Smith, T. C., Clouthier, D. J., and Steimle, T. C., *J. Chem. Phys.* 115, 817, 2001.
42. Peebles, S. A., Sun, L., and Kuczkowski, R. L., *J. Chem. Phys.* 110, 6804, 1999.
43. Namiki, K. C., Robinson, J. S., and Steimle, T. C., *J. Chem. Phys.* 109, 5283, 1998.
44. Peebles, S. A., and Kuczkowski, R. L., *J. Chem. Phys.* 109, 5276, 1998.
45. Sauer, B. E., Wang, J., and Hinds, E. A., *J. Chem. Phys.* 105, 7412, 1996.
46. Fry, J. L., Drouin, B. J., and Miller, C. E., *J. Chem. Phys.* 124, 084304, 2006.
47. Christiansen, J. J., *J. Mol. Spectrosc.* 231, 131, 2005.
48. Kisiel, Z., et al., *Chem. Phys. Lett.* 325, 523, 2000.
49. Lovas, F. J., et al., *Astrophys. J. Lett.* 455, 201, 1995.
50. Suenram, R. D., and Lovas, F. J., *Astrophys. J. Lett.* 429, 89, 1994.
51. Biermann, S., et al., *J. Chem. Phys.* 105, 9754, 1996.
52. McGlone, S., and Bauder, A., *J. Chem. Phys.* 109, 5383, 1998.
53. Peebles, S. A., and Kuczkowski, R. L., *J. Chem. Phys.* 111, 10511, 1999.
54. Plusquellic, D. F., et al., *J. Chem. Phys.* 115, 3057, 2001.
55. Muller, H. S. P., and Cohen, E. A., *J. Chem. Phys.* 116, 2407, 2002.
56. Andrews, A. M., and Kuczkowski, R. L., *J. Chem. Phys.* 98, 791, 1993.
57. Lovas, F. J., et al., *J. Chem. Phys.* 92, 891, 1990.

58. Klots, T. D., Emilsson, T., and Gutowsky, H. S., *J. Chem. Phys.* 97, 5335, 1992.
59. Careless, A. J., Kroto, H. W., and Landsberg, B. M., *Chem. Phys.* 1, 371, 1973.
60. Costain, C. C., and Kroto, H. W., *Can. J. Phys.* 50, 1453, 1972.
61. Kroto, H. W., Nixon, J. F., and Ohno, K., *J. Mol. Spectrosc.* 90, 367, 1981.
62. Kroto, H. W., Nixon, J. F., and Simmons, N. P. C., *J. Mol. Spectrosc.* 82, 185, 1980.
63. Kroto, H. W., Nixon, J. F., and Ohno, K., *J. Mol. Spectrosc.* 77, 270, 1979.
64. Cox, A. P., Ewart, I. C., and Gayton, T. R., *J. Mol. Spectrosc.* 125, 76, 1987.
65. Cohen, E. A., and Pickett, H. M., *J. Mol. Spectrosc.* 87, 582, 1981.
66. Peebles, S. A., and Peebles, R. A., *J. Mol. Struct.* 607, 19, 2002.
67. Suenram, R. D., Lovas, F. J., and Pickett, H. M., *J. Mol. Spectrosc.* 116, 406, 1986.
68. Kroto, H. W., and Landsberg, B. M., *J. Mol. Spectrosc.* 62, 346, 1976.

Table 1. Dipole Moments of Gases and Liquids

Name	Mol. form.	μ/D	Ref.
Acenaphthene	$C_{12}H_{10}$	≈0.85	1
Acetaldehyde	C_2H_4O	2.750(6)	3
Acetamide	C_2H_5NO	3.68(3)	5
Acetic acid	$C_2H_4O_2$	1.70(3)	2
Acetic anhydride	$C_4H_6O_3$	≈2.8	1
Acetone	C_3H_6O	2.88(3)	1
Acetonitrile	C_2H_3N	3.92519	5
Acetophenone	C_8H_8O	3.02(6)	1
Acetyl chloride	C_2H_3ClO	2.72(14)	1
Acetylene-carbon dioxide complex	$C_2H_2{\cdot}CO_2$	0.161(1)	22
Acetylene-carbon monoxide complex	$C_2H_2{\cdot}CO$	0.311(1)	32
Acetylene-carbon oxysulfide trimer complex	$C_2H_2{\cdot}C_3O_3S_3$	1.23(2)	53
Acetylene-hydrogen cyanide complex	$C_2H_2{\cdot}CHN$	3.29(3)	32
Acetyl fluoride	C_2H_3FO	2.96(3)	1
Acrylonitrile	C_3H_3N	3.92(7)	5
Allyl alcohol	C_3H_6O	1.60(8)	1
Allylamine	C_3H_7N	≈1.2	1
Aluminum monofluoride	AlF	1.53(15)	1
Ammonia	H_3N	1.4718(2)	5
Aniline	C_6H_7N	1.13(2)	3
Anisole	C_7H_8O	1.38(7)	1
Arsenic(III) chloride	$AsCl_3$	1.59(8)	1
Arsenic(III) fluoride	AsF_3	2.59(5)	1
Arsine	AsH_3	0.217(3)	5
Azulene	$C_{10}H_8$	0.80(2)	1
Barium oxide	BaO	7.954(3)	5
Barium sulfide	BaS	10.86(2)	3
Benzaldehyde	C_7H_6O	3.0 liq	7
Benzeneacetonitrile	C_8H_7N	3.5 liq	7
Benzene-hydrogen sulfide complex	$C_6H_6{\cdot}H_2S$	1.14(2)	40
Benzene-krypton complex	$C_6H_6{\cdot}Kr$	0.136(2)	58
Benzene-sulfur dioxide complex	$C_6H_6{\cdot}O_2S$	2.061(2)	33
Benzenethiol	C_6H_6S	1.23 liq	7
Benzonitrile	C_7H_5N	4.18(8)	1
Benzyl acetate	$C_9H_{10}O_2$	1.22 liq	7
Benzyl alcohol	C_7H_8O	1.71(9)	1
Benzyl benzoate	$C_{14}H_{12}O_2$	2.06 liq	7
Bis(2-aminoethyl)amine	$C_4H_{13}N_3$	1.89 liq	7
Bis(2-chloroethyl) ether	$C_4H_8Cl_2O$	2.58 liq	7
Bis(2-ethylhexyl) phthalate	$C_{24}H_{38}O_4$	2.84 liq	7
Borane carbonyl	CH_3BO	1.698(20)	3
Bromine chloride	BrCl	0.519(4)	3
Bromine dioxide	BrO_2	2.8(1)	18
Bromine fluoride	BrF	1.422(16)	3
Bromine monoxide	BrO	1.76(4)	2
Bromine pentafluoride	BrF_5	1.51(15)	1
Bromoacetylene	C_2HBr	0.22962	5
Bromobenzene	C_6H_5Br	1.70(3)	1
1-Bromobutane	C_4H_9Br	2.08(10)	1
2-Bromobutane, (±)-	C_4H_9Br	2.23(11)	1
1-Bromo-2-chloroethane	C_2H_4BrCl	1.2 liq	7
Bromochlorofluoromethane	CHBrClF	1.5(3)	17
Bromochloromethane	CH_2BrCl	1.66 liq	7
1-Bromodecane	$C_{10}H_{21}Br$	1.93 liq	7
Bromoethane	C_2H_5Br	2.04(2)	5
Bromoethene	C_2H_3Br	1.42(3)	1
Bromofluoroacetylene	C_2BrF	0.448(2)	5
1-Bromoheptane	$C_7H_{15}Br$	2.16(11)	1
Bromomethane	CH_3Br	1.8203(4)	5
2-Bromo-2-methylpropane	C_4H_9Br	2.17 liq	7
1-Bromonaphthalene	$C_{10}H_7Br$	1.55 liq	7
1-Bromopentane	$C_5H_{11}Br$	2.20(11)	1
1-Bromopropane	C_3H_7Br	2.18(11)	1
2-Bromopropane	C_3H_7Br	2.21(11)	1
2-Bromopropene	C_3H_5Br	1.51 liq	7
3-Bromopropene	C_3H_5Br	≈1.9	1
Bromosilane	BrH_3Si	1.319	3
Bromotrifluoromethane	$CBrF_3$	0.65(5)	1
Bromotrifluorosilane	BrF_3Si	0.835(7)	64
1,2-Butadiene	C_4H_6	0.403(2)	1
Butanal	C_4H_8O	2.72(5)	1
1,4-Butanediol	$C_4H_{10}O_2$	2.58 liq	7
1-Butanethiol	$C_4H_{10}S$	1.53 liq	7
Butanoic acid	$C_4H_8O_2$	1.65 liq	7
1-Butanol	$C_4H_{10}O$	1.66(3)	1
2-Butanone	C_4H_8O	2.779(15)	2
trans-2-Butenal	C_4H_6O	3.67(7)	1
cis-2-Butene	C_4H_8	0.253(5)	2
cis-2-Butene-1,4-diol	$C_4H_8O_2$	2.48 liq	7
trans-2-Butene-1,4-diol	$C_4H_8O_2$	2.45 liq	7
trans-2-Butenoic acid	$C_4H_6O_2$	2.13 liq	7
cis-2-Buten-1-ol	C_4H_8O	1.96(3)	5
trans-2-Buten-1-ol	C_4H_8O	1.90(2)	5
1-Buten-3-yne	C_4H_4	0.22(2)	3
2-Butoxyethanol	$C_6H_{14}O_2$	2.08 liq	7
Butyl acetate	$C_6H_{12}O_2$	1.87 liq	7
sec-Butyl acetate	$C_6H_{12}O_2$	1.87 liq	7
Butylamine	$C_4H_{11}N$	≈1.0	1
sec-Butylamine	$C_4H_{11}N$	1.28 liq	7
tert-Butylamine	$C_4H_{11}N$	1.29 liq	7
tert-Butylbenzene	$C_{10}H_{14}$	≈0.83	1
Butyl ethyl ether	$C_6H_{14}O$	1.24 liq	7
Butyl formate	$C_5H_{10}O_2$	2.03 liq	7
Butyl stearate	$C_{22}H_{44}O_2$	1.88 liq	7
Butyl vinyl ether	$C_6H_{12}O$	1.25 liq	7
1-Butyne	C_4H_6	0.782(4)	5
γ-Butyrolactone	$C_4H_6O_2$	4.27(3)	3
Calcium methoxide	CH_3CaO	1.58(8)	43
Calcium monochloride	CaCl	≈3.6	4
Camphor, (+)	$C_{10}H_{16}O$	3.1 liq	7
Caprolactam	$C_6H_{11}NO$	3.9 liq	7

Name	Mol. form.	μ/D	Ref.
Carboimidic difluoride	CHF_2N	1.393(1)	5
Carbon dioxide dimer-water complex	$C_2O_4{\cdot}H_2O$	1.989(2)	23
Carbon dioxide-mercury complex	$CO_2{\cdot}Hg$	0.107(3)	29
Carbon dioxide-water dimer complex	$CO_2{\cdot}H_4O_2$	1.746(10)	28
Carbon disulfide-sulfur dioxide complex	$CO_2{\cdot}O_2S$	1.096(1)	42
Carbon monoselenide	CSe	1.99(4)	3
Carbon monosulfide	CS	1.958(5)	2
Carbon monoxide	CO	0.10980	3
Carbon monoxide dimer-water complex	$C_2O_2{\cdot}H_2O$	1.57(5)	36
Carbon oxyselenide	COSe	0.73(2)	1
Carbon oxysulfide	COS	0.715189	5
Carbon oxysulfide-carbon dioxide dimer complex	$COS{\cdot}C_2O_4$	0.69(5)	44
Carbon oxysulfide-water complex	$COS{\cdot}H_2O$	2.668(3)	37
Carbonyl chloride	CCl_2O	1.17(1)	1
Carbonyl fluoride	CF_2O	0.95(1)	1
Cesium chloride	ClCs	10.387(4)	2
Cesium fluoride	CsF	7.884(1)	2
Cesium sodium	CsNa	4.75(20)	2
Chlorine fluoride	ClF	0.888061	5
Chlorine oxide (ClO)	ClO	1.297(1)	5
Chlorine trifluoride	ClF_3	0.6(1)	1
Chloroacetyl chloride	$C_2H_2Cl_2O$	2.23(11)	1
Chloroacetylene	C_2HCl	0.44408	5
2-Chloroaniline	C_6H_6ClN	1.77 liq	7
Chlorobenzene	C_6H_5Cl	1.69(3)	1
Chloroborane	$BClH_2$	0.75(5)	14
1-Chlorobutane	C_4H_9Cl	2.05(4)	1
2-Chlorobutane	C_4H_9Cl	2.04(10)	1
1-Chloro-1,1-difluoroethane	$C_2H_3ClF_2$	2.14(4)	1
Chlorodifluoromethane	$CHClF_2$	1.42(3)	1
Chloroethane	C_2H_5Cl	2.05(2)	1
2-Chloroethanol	C_2H_5ClO	1.78(9)	1
Chloroethene	C_2H_3Cl	1.45(3)	1
1-Chloro-4-fluorobenzene	C_6H_4ClF	0.12(1)	66
1-Chloro-1-fluoroethane	C_2H_4ClF	2.068(14)	3
Chlorofluoromethane	CH_2ClF	1.82(4)	1
Chlorogermane	$ClGeH_3$	2.13(2)	1
Chloromethane	CH_3Cl	1.8963(2)	5
(Chloromethyl)benzene	C_7H_7Cl	1.82 liq	7
1-Chloro-3-methylbutane	$C_5H_{11}Cl$	1.92 liq	7
1-Chloro-2-methylpropane	C_4H_9Cl	2.00(10)	1
2-Chloro-2-methylpropane	C_4H_9Cl	2.13(4)	1
1-Chloronaphthalene	$C_{10}H_7Cl$	1.57 liq	7
1-Chloro-2-nitrobenzene	$C_6H_4ClNO_2$	4.64(9)	1
1-Chloro-3-nitrobenzene	$C_6H_4ClNO_2$	3.73(7)	1
1-Chloro-4-nitrobenzene	$C_6H_4ClNO_2$	2.83(6)	1
1-Chlorooctane	$C_8H_{17}Cl$	2.00 liq	7
Chloropentafluoroethane	C_2ClF_5	0.52(5)	1
1-Chloropentane	$C_5H_{11}Cl$	2.16(11)	1
4-Chlorophenol	C_6H_5ClO	2.11(11)	1
1-Chloropropane	C_3H_7Cl	2.05(4)	1
2-Chloropropane	C_3H_7Cl	2.17(11)	1
cis-1-Chloropropene	C_3H_5Cl	1.67(8)	1
trans-1-Chloropropene	C_3H_5Cl	1.97(10)	1
2-Chloropropene	C_3H_5Cl	1.647(10)	3
3-Chloropropene	C_3H_5Cl	1.94(10)	1
4-Chloropyridine	C_5H_4ClN	0.756(5)	3
Chlorosilane	ClH_3Si	1.31(1)	1
Chlorosyl fluoride	ClFO	1.93(2)	55
2-Chlorotoluene	C_7H_7Cl	1.56(8)	1
3-Chlorotoluene	C_7H_7Cl	1.82 liq	7
4-Chlorotoluene	C_7H_7Cl	2.21(4)	1
Chlorotrifluoroethene	C_2ClF_3	0.40(10)	1
Chlorotrifluoromethane	$CClF_3$	0.50(1)	1
Chlorotrifluorosilane	ClF_3Si	0.636(4)	5
Chromium monoxide	CrO	3.88(13)	5
Copper(I) fluoride	CuF	5.77(29)	2
Copper(II) oxide	CuO	4.5(5)	5
o-Cresol	C_7H_8O	1.45 liq	7
m-Cresol	C_7H_8O	1.48 liq	7
p-Cresol	C_7H_8O	1.48 liq	7
Cyanamide	CH_2N_2	4.28(10)	5
Cyanoacetylene	C_3HN	3.73172	5
Cyanoformamide	$C_2H_2N_2O$	4.10(12)	47
Cyanogen azide	CN_4	2.96(7)	60
Cyanogen chloride	CClN	2.8331(2)	3
Cyanogen fluoride	CFN	2.120(1)	3
Cyanogen iodide	CIN	3.67(2)	5
Cyanomethylmercury	C_2H_3HgN	4.7(1)	12
Cyclobutanecarbonitrile	C_5H_7N	4.04(4)	5
Cyclobutanone	C_4H_6O	2.89(3)	2
Cyclobutene	C_4H_6	0.132(1)	1
1,3-Cycloheptadiene	C_7H_{10}	0.740	3
2,4,6-Cycloheptatrien-1-one	C_7H_6O	4.1(3)	3
3,5-Cyclohexadiene-1,2-dione	$C_6H_4O_2$	4.23(2)	3
Cyclohexanone	$C_6H_{10}O$	3.246(6)	5
Cyclohexylamine	$C_6H_{13}N$	1.26 liq	7
1,3-Cyclopentadiene	C_5H_6	0.419(4)	1
2,4-Cyclopentadien-1-one	C_5H_4O	3.132(7)	3
Cyclopentanone	C_5H_8O	≈3.3	1
Cyclopentene	C_5H_8	0.20(2)	1
3-Cyclopenten-1-one	C_5H_6O	2.79(3)	3
Cyclopropane-sulfur dioxide complex	$C_3H_6{\cdot}O_2S$	1.681(1)	30
Cyclopropanone	C_3H_4O	2.67(13)	2
Cyclopropene	C_3H_4	0.454(10)	1
Cyclopropylamine	C_3H_7N	1.19(1)	2
Cyclopropyl methyl ketone	C_5H_8O	2.62(25)	2
Diacetone alcohol	$C_6H_{12}O_2$	3.24 liq	7
Diazomethane	CH_2N_2	1.50(1)	1
Dibromodifluoromethane	CBr_2F_2	0.66(5)	1
1,2-Dibromoethane	$C_2H_4Br_2$	1.19 liq	7
Dibromomethane	CH_2Br_2	1.43(3)	1
1,2-Dibromopropane	$C_3H_6Br_2$	1.2 liq	7
Dibutylamine	$C_8H_{19}N$	0.98 liq	7
Dibutyl ether	$C_8H_{18}O$	1.17(6)	1
Dibutyl phthalate	$C_{16}H_{22}O_4$	2.82 liq	7
Dibutyl sebacate	$C_{18}H_{34}O_4$	2.48 liq	7
Dibutyl sulfide	$C_8H_{18}S$	1.61 liq	7
o-Dichlorobenzene	$C_6H_4Cl_2$	2.50(5)	1
m-Dichlorobenzene	$C_6H_4Cl_2$	1.72(9)	1
1,4-Dichlorobutane	$C_4H_8Cl_2$	2.22(11)	1
1,1-Dichloro-2,2-difluoroethene	$C_2Cl_2F_2$	0.50	7
Dichlorodifluoromethane	CCl_2F_2	0.51(5)	1
1,1-Dichloroethane	$C_2H_4Cl_2$	2.06(4)	1
1,2-Dichloroethane	$C_2H_4Cl_2$	1.83 liq	7
1,1-Dichloroethene	$C_2H_2Cl_2$	1.34(1)	1
cis-1,2-Dichloroethene	$C_2H_2Cl_2$	1.90(4)	1
Dichlorofluoromethane	$CHCl_2F$	1.29(3)	1
1,1-Dichloro-2-fluoropropene	$C_3H_3Cl_2F$	2.43(2)	3
Dichloromethane	CH_2Cl_2	1.60(3)	1

Name	Mol. form.	μ/D	Ref.
(Dichloromethyl)benzene	$C_7H_6Cl_2$	2.07 liq	7
Dichloromethylborane	CH_3BCl_2	1.419(13)	5
1,2-Dichloropropane, (±)-	$C_3H_6Cl_2$	1.85 liq	7
1,3-Dichloropropane	$C_3H_6Cl_2$	2.08(4)	1
Dichlorosilane	Cl_2H_2Si	1.17(2)	1
1,2-Dichloro-1,1,2,2-tetrafluoroethane	$C_2Cl_2F_4$	≈0.5	1
2,4-Dichlorotoluene	$C_7H_6Cl_2$	1.70 liq	7
3,4-Dichlorotoluene	$C_7H_6Cl_2$	2.95 liq	7
Diethanolamine	$C_4H_{11}NO_2$	2.8 liq	7
1,1-Diethoxyethane	$C_6H_{14}O_2$	1.38 liq	7
Diethylamine	$C_4H_{11}N$	0.92(5)	1
Diethyl carbonate	$C_5H_{10}O_3$	1.10(6)	1
Diethylene glycol	$C_4H_{10}O_3$	2.31 liq	7
Diethylene glycol dimethyl ether	$C_6H_{14}O_3$	1.97 liq	7
Diethylene glycol monoethyl ether	$C_6H_{14}O_3$	1.6 liq	7
Diethylene glycol monoethyl ether acetate	$C_8H_{16}O_4$	1.8 liq	7
Diethylene glycol monomethyl ether	$C_5H_{12}O_3$	1.6 liq	7
Diethyl ether	$C_4H_{10}O$	1.098(1)	38
Diethyl malonate	$C_7H_{12}O_4$	2.54 liq	7
Diethyl oxalate	$C_6H_{10}O_4$	2.49 liq	7
Difluoramine	F_2HN	1.92(2)	1
Difluorine dioxide	F_2O_2	1.44(7)	1
o-Difluorobenzene	$C_6H_4F_2$	2.46(5)	2
m-Difluorobenzene	$C_6H_4F_2$	1.51(2)	2
Difluoroborane	BF_2H	0.971(10)	8
1,1-Difluorocyclohexane	$C_6H_{10}F_2$	2.556(10)	3
3,3-Difluorocyclopropene	$C_3H_2F_2$	2.98(2)	3
cis-Difluorodiazine	F_2N_2	0.16(1)	1
1,1-Difluoroethane	$C_2H_4F_2$	2.27(5)	1
1,1-Difluoroethene	$C_2H_2F_2$	1.3893(2)	5
cis-1,2-Difluoroethene	$C_2H_2F_2$	2.42(2)	1
Difluoromethane	CH_2F_2	1.9785(21)	3
Difluoromethylborane	CH_3BF_2	1.668(3)	3
Difluoromethylene	CF_2	0.47(2)	3
1,1-Difluoro-1-propene	$C_3H_4F_2$	0.889(7)	2
Difluorosilane	F_2H_2Si	1.55(2)	1
Difluorosilylene	F_2Si	1.23(2)	2
3,6-Dihydro-1,2-dioxin	$C_4H_6O_2$	2.329(1)	3
2,3-Dihydro-1,4-dioxin	$C_4H_6O_2$	0.939(8)	3
2,3-Dihydrofuran	C_4H_6O	1.32(3)	2
2,5-Dihydrofuran	C_4H_6O	1.63(1)	5
Dihydro-3-methyl-2(3*H*)-furanone	$C_5H_8O_2$	4.56(2)	5
Dihydro-5-methyl-2(3*H*)-furanone,(±)-	$C_5H_8O_2$	4.71(5)	5
3,4-Dihydro-2*H*-pyran	C_5H_8O	1.400(8)	5
3,6-Dihydro-2*H*-pyran	C_5H_8O	1.283(5)	3
2,3-Dihydrothiophene	C_4H_6S	1.61(20)	5
2,5-Dihydrothiophene	C_4H_6S	1.75(1)	3
Diiodomethane	CH_2I_2	1.08 liq	7
Diisopentyl ether	$C_{10}H_{22}O$	1.23 liq	7
Diisopropylamine	$C_6H_{15}N$	1.15 liq	7
Diisopropyl ether	$C_6H_{14}O$	1.13(10)	1
Diketene	$C_4H_4O_2$	3.53(7)	1
1,2-Dimethoxybenzene	$C_8H_{10}O_2$	1.29 liq	7
Dimethoxymethane	$C_3H_8O_2$	0.74 liq	7
N,N-Dimethylacetamide	C_4H_9NO	3.7 liq	7
Dimethylamine	C_2H_7N	1.01(2)	2
2,4-Dimethylaniline	$C_8H_{11}N$	1.40 liq	7
2,6-Dimethylaniline	$C_8H_{11}N$	1.63 liq	7
N,N-Dimethylaniline	$C_8H_{11}N$	1.68(17)	1
3,3-Dimethyl-1-butyne	C_6H_{10}	0.661(4)	1

Name	Mol. form.	μ/D	Ref.
1,1-Dimethylcyclopropane	C_5H_{10}	0.142(1)	3
3,3-Dimethylcyclopropene	C_5H_8	0.287(3)	3
Dimethyl disulfide	$C_2H_6S_2$	1.85 liq	7
Dimethyl ether	C_2H_6O	1.30(1)	1
N,N-Dimethylformamide	C_3H_7NO	3.82(8)	1
2,6-Dimethyl-4-heptanone	$C_9H_{18}O$	2.66 liq	7
Dimethyl maleate	$C_6H_8O_4$	2.48 liq	7
2,4-Dimethyl-3-pentanone	$C_7H_{14}O$	2.74 liq	7
2,2-Dimethylpropanal	$C_5H_{10}O$	2.66(5)	2
2,2-Dimethylpropanenitrile	C_5H_9N	3.95(4)	1
2,4-Dimethylpyridine	C_7H_9N	2.30 liq	7
2,6-Dimethylpyridine	C_7H_9N	1.66 liq	7
Dimethyl sulfide	C_2H_6S	1.554(4)	3
Dimethyl sulfoxide	C_2H_6OS	3.96(4)	1
1,3-Dioxane	$C_4H_8O_2$	2.06(4)	2
1,3-Dioxolane	$C_3H_6O_2$	1.19(6)	3
Dipentyl ether	$C_{10}H_{22}O$	1.20 liq	7
Diphenyl ether	$C_{12}H_{10}O$	≈1.3	1
Dipropylamine	$C_6H_{15}N$	1.03 liq	7
Dipropyl ether	$C_6H_{14}O$	1.21(6)	1
Disiloxane	H_6OSi_2	0.24(2)	1
1,3-Dithiane	$C_4H_8S_2$	2.14(4)	5
Divinyl ether	C_4H_6O	0.78(5)	2
Epichlorohydrin	C_3H_5ClO	1.8 liq	7
1,2-Epoxybutane	C_4H_8O	1.891(11)	3
1,2-Ethanediamine	$C_2H_8N_2$	1.99(10)	1
1,2-Ethanediol	$C_2H_6O_2$	2.36(10)	5
1,2-Ethanediol, diacetate	$C_6H_{10}O_4$	2.34 liq	7
1,2-Ethanedithiol	$C_2H_6S_2$	2.03(8)	5
Ethanol	C_2H_6O	1.69(3)	1
Ethanolamine	C_2H_7NO	2.27 liq	7
Ethoxybenzene	$C_8H_{10}O$	1.45(15)	1
2-Ethoxyethanol	$C_4H_{10}O_2$	2.08 liq	7
2-Ethoxyethyl acetate	$C_6H_{12}O_3$	2.25 liq	7
Ethyl acetate	$C_4H_8O_2$	1.78(9)	1
Ethyl acrylate	$C_5H_8O_2$	1.96 liq	7
Ethylamine	C_2H_7N	1.22(10)	1
Ethylbenzene	C_8H_{10}	0.59(5)	1
Ethyl benzoate	$C_9H_{10}O_2$	2.00(10)	1
Ethyl butanoate	$C_6H_{12}O_2$	1.74 liq	7
Ethyl *trans*-cinnamate	$C_{11}H_{12}O_2$	1.84 liq	7
Ethyl cyanate	C_3H_5NO	4.72(9)	5
Ethyl cyanoacetate	$C_5H_7NO_2$	2.17 liq	7
Ethylene carbonate	$C_3H_4O_3$	4.9 liq	7
Ethyleneimine	C_2H_5N	1.90(1)	1
Ethylene-sulfur dioxide complex	$C_2H_4{\cdot}O_2S$	1.650(3)	27
Ethylene-water complex	$C_2H_4{\cdot}H_2O$	1.10(1)	56
Ethyl formate	$C_3H_6O_2$	1.93	1
2-Ethyl-1-hexanol	$C_8H_{18}O$	1.74 liq	7
2-Ethylhexyl acetate	$C_{10}H_{20}O_2$	1.8 liq	7
Ethyl lactate	$C_5H_{10}O_3$	2.4 liq	7
Ethyl propanoate	$C_5H_{10}O_2$	1.74 liq	7
Ethyl vinyl ether	C_4H_8O	1.26 liq	7
Fluoramine	FH_2N	2.27(18)	5
Fluorine azide	FN_3	≈1.3	5
Fluorine monoxide	F_2O	0.308180	5
Fluorine oxide	FO	0.0043(4)	5
Fluoroacetylene	C_2HF	0.7207(3)	3
Fluorobenzene	C_6H_5F	1.60(8)	1
Fluoroborane(1)	BF	≈0.5	2
1-Fluorocyclohexene	C_6H_9F	1.942(10)	5

Name	Mol. form.	μ/D	Ref.
Fluoroethane	C_2H_5F	1.937(7)	5
Fluoroethene	C_2H_3F	1.468(3)	5
Fluorogermane	$FGeH_3$	2.33(12)	2
Fluoromethane	CH_3F	1.858(2)	3
Fluoromethylidyne	CF	0.645(5)	3
(Fluoromethylidyne)phosphine	CFP	0.279(1)	62
Fluoromethylsilane	CH_5FSi	1.700(8)	5
1-Fluoro-4-nitrobenzene	$C_6H_4FNO_2$	2.87(6)	1
2-Fluoropropane	C_3H_7F	1.958(1)	5
cis-1-Fluoropropene	C_3H_5F	1.46(3)	1
trans-1-Fluoropropene	C_3H_5F	≈1.9	1
2-Fluoropropene	C_3H_5F	1.61(3)	1
3-Fluoropropyne	C_3H_3F	1.73(2)	5
3-Fluoropyridine	C_5H_4FN	2.09(26)	3
Fluorosilane	FH_3Si	1.2969(6)	5
2-Fluorotoluene	C_7H_7F	1.37(7)	1
3-Fluorotoluene	C_7H_7F	1.82(4)	2
4-Fluorotoluene	C_7H_7F	2.00(10)	1
Formaldehyde	CH_2O	2.332(2)	3
Formaldehyde dimer	$C_2H_4O_2$	0.858(5)	57
Formamide	CH_3NO	3.73(7)	1
Formic acid	CH_2O_2	1.425(2)	5
Formyl fluoride	CHFO	2.081(1)	5
Fulminic acid	CHNO	3.09934	5
Fulvene	C_6H_6	0.4236(13)	2
Furan	C_4H_4O	0.66(1)	1
Furfural	$C_5H_4O_2$	3.54 liq	7
Furfuryl alcohol	$C_5H_6O_2$	1.92 liq	7
Gallium monofluoride	FGa	2.45(5)	2
Germanium(II) fluoride	F_2Ge	2.61(2)	2
Germanium(II) oxide	GeO	3.2823(1)	2
Germanium(II) selenide	GeSe	1.65(5)	2
Germanium(II) sulfide	GeS	2.00(6)	2
Germanium(II) telluride	GeTe	1.06(7)	2
Germylazide	GeH_3N_3	2.58(2)	25
Glycerol	$C_3H_8O_3$	2.56 liq	7
Glycolaldehyde	$C_2H_4O_2$	2.73(5)	2
Hafnium monoxide	HfO	3.431(5)	26
Hafnium(IV) oxide	HfO_2	7.92(1)	39
2-Heptanol, (±)-	$C_7H_{16}O$	1.71 liq	7
3-Heptanol, (*S*)-	$C_7H_{16}O$	1.71 liq	7
2-Heptanone	$C_7H_{14}O$	2.59 liq	7
3-Heptanone	$C_7H_{14}O$	2.78 liq	7
Hexaborane(10)	B_6H_{10}	2.50(5)	3
Hexamethylphosphoric triamide	$C_6H_{18}N_3OP$	5.5 liq	7
Hexanoic acid	$C_6H_{12}O_2$	1.13 liq	7
2-Hexanone	$C_6H_{12}O$	2.66 liq	7
sec-Hexyl acetate	$C_8H_{16}O_2$	1.9 liq	7
1-Hexyne	C_6H_{10}	0.83(5)	1
Hydrazine	H_4N_2	1.75(9)	1
Hydrazoic acid	HN_3	1.70(9)	3
Hydrogen bromide	BrH	0.8272(3)	3
Hydrogen chloride	ClH	1.1086(3)	3
Hydrogen cyanide	CHN	2.985188	5
Hydrogen cyanide trimer	$C_3H_3N_3$	10.6	21
Hydrogen fluoride	FH	1.826178	2
Hydrogen iodide	HI	0.448(1)	2
Hydrogen isocyanide	CHN	3.05(15)	3
Hydrogen peroxide	H_2O_2	1.573(1)	65
Hydrogen sulfide	H_2S	0.97833	5
p-Hydroquinone	$C_6H_6O_2$	2.38(5)	15

Name	Mol. form.	μ/D	Ref.
Hydroxyl	HO	1.655(1)	5
Hydroxylamine	H_3NO	0.59(5)	2
Hypochlorous acid	ClHO	≈1.3	2
Hypofluorous acid	FHO	2.23(11)	3
Imidazole	$C_3H_4N_2$	3.8(4)	2
Imidogen	HN	1.39(7)	3
Indium(I) chloride	ClIn	3.79(19)	2
Indium(I) fluoride	FIn	3.40(7)	2
Iodine bromide	BrI	0.726(3)	5
Iodine chloride	ClI	1.24(2)	2
Iodine fluoride	FI	1.948(20)	3
Iodine monoxide	IO	2.45(5)	2
Iodine pentafluoride	F_5I	2.18(11)	1
Iodoacetylene	C_2HI	0.02525	5
Iodobenzene	C_6H_5I	1.70(9)	1
1-Iodobutane	C_4H_9I	1.93 liq	7
2-Iodobutane, (±)-	C_4H_9I	2.12(11)	1
Iodoethane	C_2H_5I	1.976(2)	5
Iodoethene	C_2H_3I	1.311(5)	5
Iodomethane	CH_3I	1.6406(4)	5
1-Iodo-2-methylpropane	C_4H_9I	1.87 liq	7
Iodomethylsilane	CH_5ISi	1.862(5)	5
1-Iodopropane	C_3H_7I	2.04(10)	1
2-Iodopropane	C_3H_7I	1.95 liq	7
Isobutane	C_4H_{10}	0.132(2)	1
Isobutene	C_4H_8	0.503(10)	1
Isobutyl acetate	$C_6H_{12}O_2$	1.86 liq	7
Isobutylamine	$C_4H_{11}N$	1.27 liq	7
Isobutyl formate	$C_5H_{10}O_2$	1.88 liq	7
Isobutyl isobutanoate	$C_8H_{16}O_2$	1.9 liq	7
Isocyanic acid (HNCO)	CHNO	≈1.6	2
Isocyanobenzene	C_7H_5N	4.018(3)	5
Isocyanocyclopropane	C_4H_5N	4.03(10)	3
2-Isocyanopropane	C_4H_7N	4.055(1)	5
Isopentane	C_5H_{12}	0.13(5)	1
Isopentyl acetate	$C_7H_{14}O_2$	1.86 liq	7
Isopropylamine	C_3H_9N	1.19(6)	3
Isopropylbenzene	C_9H_{12}	≈0.79	1
Isopropyl methyl ether	$C_4H_{10}O$	1.247(3)	5
Isoquinoline	C_9H_7N	2.73(14)	1
Isoxazole	C_3H_3NO	2.95(4)	3
Isoxazole-carbon monoxide complex	$C_3H_3NO{\cdot}CO$	2.873(4)	52
Ketene	C_2H_2O	1.42215	3
Lanthanum monoxide	LaO	3.207(11)	26
Lead(II) oxide (litharge)	OPb	4.64(50)	2
Lead(II) sulfide	PbS	3.59(18)	2
Lithium bromide	BrLi	7.268(1)	2
Lithium chloride	ClLi	7.12887	2
Lithium fluoride	FLi	6.3274(2)	3
Lithium fluoride-sodium fluoride complex	FLi·FNa	2.62(2)	51
Lithium hydride	HLi	5.884(1)	2
Lithium hydroxide	HLiO	4.754(2)	3
Lithium iodide	ILi	7.428(1)	2
Lithium monoxide	LiO	6.84(3)	2
Lithium potassium	KLi	3.45(20)	2
Lithium rubidium	LiRb	4.0(1)	2
Lithium sodium	LiNa	0.463(2)	2
Magnesium oxide	MgO	6.2(6)	5
Mercapto	HS	0.7580(1)	3
Mesityl oxide	$C_6H_{10}O$	2.79 liq	7

Name	Mol. form.	μ/D	Ref.
Methacrylic acid	$C_4H_6O_2$	1.65 liq	7
Methanethiol	CH_4S	1.52(8)	1
Methanol	CH_4O	1.70(2)	1
2-Methoxyethyl acetate	$C_5H_{10}O_3$	2.13 liq	7
1-Methoxy-1,2-propadiene	C_4H_6O	0.963(20)	5
N-Methylacetamide	C_3H_7NO	4.3 liq	7
Methyl acetate	$C_3H_6O_2$	1.72(9)	1
Methyl acrylate	$C_4H_6O_2$	1.77 liq	7
2-Methylacrylonitrile	C_4H_5N	3.69(18)	1
Methylamine	CH_5N	1.31(3)	1
2-Methylaniline	C_7H_9N	1.60 liq	7
3-Methylaniline	C_7H_9N	1.45 liq	7
4-Methylaniline	C_7H_9N	1.52 liq	7
Methyl azide	CH_3N_3	2.17(4)	2
Methyl benzoate	$C_8H_8O_2$	1.94 liq	7
2-Methyl-1,3-butadiene	C_5H_8	0.25(1)	1
3-Methylbutanoic acid	$C_5H_{10}O_2$	0.63 liq	7
2-Methyl-1-butanol, (±)-	$C_5H_{12}O$	1.88 liq	7
2-Methyl-2-butanol	$C_5H_{12}O$	1.82 liq	7
3-Methyl-2-butenenitrile	C_5H_7N	4.61(13)	10
2-Methyl-1-buten-3-yne	C_5H_6	0.513(20)	2
Methyl cyanate	C_2H_3NO	4.26(18)	5
cis-3-Methylcyclohexanol, (±)-	$C_7H_{14}O$	1.91 liq	7
trans-3-Methylcyclohexanol, (±)-	$C_7H_{14}O$	1.75 liq	7
3-Methylcyclopentanone, (±)-	$C_6H_{10}O$	3.14(3)	5
3-Methyl-2-cyclopenten-1-one	C_6H_8O	4.33(2)	5
Methylcyclopropane	C_4H_8	0.139(4)	2
Methyldiborane(6)	CH_8B_2	0.566(6)	3
Methyldifluorophosphine	CH_3F_2P	2.056(6)	3
Methylenecyclohexane	C_7H_{12}	0.62(1)	5
Methylenecyclopropene	C_4H_4	1.90(1)	5
Methylenephosphine	CH_3P	0.869(3)	61
N-Methylformamide	C_2H_5NO	3.83(8)	1
Methyl formate	$C_2H_4O_2$	1.77(4)	1
2-Methylfuran	C_5H_6O	0.65(5)	2
3-Methylfuran	C_5H_6O	1.03(2)	2
5-Methyl-2(3*H*)-furanone	$C_5H_6O_2$	4.08(2)	5
Methyl hydroperoxide	CH_4O_2	≈0.65	13
Methylidyne	CH	≈1.46	2
Methyl isocyanate	C_2H_3NO	≈2.8	1
Methyl isothiocyanate	C_2H_3NS	3.453(3)	5
4-Methylisoxazole	C_4H_5NO	3.583(5)	5
Methyl methacrylate	$C_5H_8O_2$	1.67 liq	7
2-Methyloxazole	C_4H_5NO	1.37(7)	5
4-Methyloxazole	C_4H_5NO	1.08(5)	5
5-Methyloxazole	C_4H_5NO	2.16(4)	5
Methyloxirane	C_3H_6O	2.01(2)	1
2-Methyl-2,4-pentanediol	$C_6H_{14}O_2$	2.9 liq	7
4-Methylpentanenitrile	$C_6H_{11}N$	3.5 liq	7
Methylphosphonic difluoride	CH_3F_2OP	3.69(26)	3
N-Methylpropanamide	C_4H_9NO	3.61	7
2-Methylpropanenitrile	C_4H_7N	4.29(9)	3
2-Methyl-2-propanethiol	$C_4H_{10}S$	1.66(3)	3
2-Methylpropanoic acid	$C_4H_8O_2$	1.08 liq	7
2-Methyl-1-propanol	$C_4H_{10}O$	1.64(8)	1
2-Methyl-2-propanol	$C_4H_{10}O$	1.66 liq	7
2-Methylpropenal	C_4H_6O	2.68(13)	1
2-Methylpyridine	C_6H_7N	1.85(4)	2
3-Methylpyridine	C_6H_7N	2.40 liq	7
4-Methylpyridine	C_6H_7N	2.70(2)	2
2-Methylpyrimidine	$C_5H_6N_2$	1.676(10)	3
5-Methylpyrimidine	$C_5H_6N_2$	2.881(6)	3
N-Methylpyrrolidine	$C_5H_{11}N$	0.572(3)	5
N-Methyl-2-pyrrolidinone	C_5H_9NO	4.1 liq	7
Methyl salicylate	$C_8H_8O_3$	2.47 liq	7
Methylsilane	CH_6Si	0.73456	5
Methyl silyl ether	CH_6OSi	1.15(2)	2
Methylstannane	CH_6Sn	0.68(5)	1
3-Methylthietane	C_4H_8S	2.046(9)	5
2-Methylthiophene	C_5H_6S	0.674(5)	2
3-Methylthiophene	C_5H_6S	0.914(15)	3
Methyl vinyl ether	C_3H_6O	0.965(2)	5
Morpholine	C_4H_9NO	1.55(3)	3
Nitric acid	HNO_3	2.17(2)	1
Nitric oxide	NO	0.15872	2
2-Nitroanisole	$C_7H_7NO_3$	5.0 liq	7
Nitrobenzene	$C_6H_5NO_2$	4.22(8)	1
Nitroethane	$C_2H_5NO_2$	3.23(3)	2
Nitrogen dioxide	NO_2	0.316(10)	1
Nitrogen sulfide	NS	1.81(2)	2
Nitrogen trichloride	Cl_3N	0.39(1)	3
Nitrogen trifluoride	F_3N	0.235(4)	1
Nitrogen trioxide	N_2O_3	2.122(10)	2
Nitromethane	CH_3NO_2	3.46(2)	1
1-Nitropropane	$C_3H_7NO_2$	3.66(7)	1
2-Nitropropane	$C_3H_7NO_2$	3.73(7)	1
Nitrosyl bromide	BrNO	≈1.8	1
Nitrosyl fluoride	FNO	1.730(3)	3
Nitrosyl hydride	HNO	1.62(3)	3
Nitrous oxide	N_2O	0.16083	1
Nitryl chloride	$ClNO_2$	0.53	1
Nitryl fluoride	FNO_2	0.466(5)	2
Nonanoic acid	$C_9H_{18}O_2$	0.79 liq	7
2,5-Norbornadiene	C_7H_8	0.0587(1)	5
cis-9-Octadecenoic acid	$C_{18}H_{34}O_2$	1.18 liq	7
Octanoic acid	$C_8H_{16}O_2$	1.15 liq	7
1-Octanol	$C_8H_{18}O$	1.76 liq	7
2-Octanol	$C_8H_{18}O$	1.71 liq	7
2-Octanone	$C_8H_{16}O$	2.70 liq	7
1,4-Oxathiane	C_4H_8OS	0.295(3)	3
Oxazole	C_3H_3NO	1.503(30)	3
Oxetane	C_3H_6O	1.94(1)	1
2-Oxetanone	$C_3H_4O_2$	4.18(3)	1
3-Oxetanone	$C_3H_4O_2$	0.887(5)	2
Oxirane	C_2H_4O	1.89(1)	1
Ozone	O_3	0.53373	3
Paraldehyde	$C_6H_{12}O_3$	1.43(7)	1
Pentaborane(9)	B_5H_9	2.13(4)	1
Pentachloroethane	C_2HCl_5	0.92(5)	1
cis-1,3-Pentadiene	C_5H_8	0.500(15)	2
trans-1,3-Pentadiene	C_5H_8	0.585(10)	2
1,3-Pentadiyne	C_5H_4	1.207(1)	5
1,5-Pentanediol	$C_5H_{12}O_2$	2.5 liq	7
2,4-Pentanedione	$C_5H_8O_2$	2.78 liq	7
Pentanenitrile	C_5H_9N	4.12(8)	1
Pentanoic acid	$C_5H_{10}O_2$	1.61 liq	7
1-Pentanol	$C_5H_{12}O$	1.7 liq	7
2-Pentanol	$C_5H_{12}O$	1.66 liq	7
3-Pentanol	$C_5H_{12}O$	1.64 liq	7
2-Pentanone	$C_5H_{10}O$	2.70 liq	7
3-Pentanone	$C_5H_{10}O$	2.82 liq	7
1,2,3-Pentatriene	C_5H_6	0.51(5)	11

Name	Mol. form.	μ/D	Ref.
1-Pentene	C_5H_{10}	≈0.5	1
1-Penten-3-yne	C_5H_6	0.66(2)	2
cis-3-Penten-1-yne	C_5H_6	0.78(2)	2
trans-3-Penten-1-yne	C_5H_6	1.06(5)	2
Pentyl acetate	$C_7H_{14}O_2$	1.75(10)	1
Pentyl formate	$C_6H_{12}O_2$	1.90(10)	1
Perchloryl fluoride	$ClFO_3$	0.023(1)	3
Perfluoropyridine	C_5F_5N	0.98(8)	3
Peroxynitric acid	HNO_4	1.99(2)	67
Peroxynitrous acid	HNO_3	1.07(2)	46
Phenol	C_6H_6O	1.224(8)	3
Phenylacetylene	C_8H_6	0.656(5)	3
Phenylsilane	C_6H_8Si	0.845(12)	3
1-Phosphapropyne	C_2H_3P	1.499(1)	63
Phosphine	H_3P	0.5740(3)	3
Phosphorothioc trifluoride	F_3PS	0.64(2)	1
Phosphorus(III) chloride	Cl_3P	0.56(2)	2
Phosphorus(III) fluoride	F_3P	1.03(1)	1
Phosphorus monoxide	OP	1.88(7)	5
Phosphorus nitride	NP	2.7470(1)	2
Phosphoryl chloride	Cl_3OP	2.54(5)	2
Phosphoryl fluoride	F_3OP	1.8685(1)	3
Piperidine	$C_5H_{11}N$	1.19 liq	3
Potassium bromide	BrK	10.628(1)	2
Potassium chloride	ClK	10.269(1)	2
Potassium fluoride	FK	8.585(3)	2
Potassium hydroxide	HKO	7.415(2)	16
Potassium iodide	IK	≈10.8	2
Potassium sodium	KNa	2.693(14)	3
Propanal	C_3H_6O	2.72	1
Propane	C_3H_8	0.084(1)	1
1,2-Propanediol	$C_3H_8O_2$	2.25 liq	7
1,3-Propanediol	$C_3H_8O_2$	2.55 liq	7
Propanenitrile	C_3H_5N	4.05(3)	3
Propanoic acid	$C_3H_6O_2$	1.75(9)	1
Propargyl alcohol	C_3H_4O	1.13(6)	2
Propene	C_3H_6	0.366(1)	1
Propene-sulfur dioxide complex	$C_3H_6{\cdot}O_2S$	1.34(3)	35
Propyl acetate	$C_5H_{10}O_2$	1.78 liq	7
Propylamine	C_3H_9N	1.17(6)	1
Propylene carbonate	$C_4H_6O_3$	4.9 liq	7
Propyl formate	$C_4H_8O_2$	1.89 liq	7
2-Propynal	C_3H_2O	2.78(2)	5
Propyne	C_3H_4	0.784(1)	3
Propyne-argon complex	$C_3H_4{\cdot}Ar$	0.730(5)	20
4*H*-Pyran-4-one	$C_5H_4O_2$	3.79(2)	5
4*H*-Pyran-4-thione	C_5H_4OS	3.95(5)	5
1*H*-Pyrazole	$C_3H_4N_2$	2.20(1)	3
Pyridazine	$C_4H_4N_2$	4.22(2)	2
Pyridine	C_5H_5N	2.215(10)	3
2-Pyridinecarbonitrile	$C_6H_4N_2$	5.78(11)	3
3-Pyridinecarbonitrile	$C_6H_4N_2$	3.66(11)	3
4-Pyridinecarbonitrile	$C_6H_4N_2$	1.96(3)	3
2-Pyridinecarboxaldehyde	C_6H_5NO	3.56(7)	3
3-Pyridinecarboxaldehyde	C_6H_5NO	1.44	3
4-Pyridinecarboxaldehyde	C_6H_5NO	1.66	3
Pyrimidine	$C_4H_4N_2$	2.334(10)	2
Pyrrole	C_4H_5N	1.767(1)	5
Pyrrolidine	C_4H_9N	1.57 liq	7
2-Pyrrolidone	C_4H_7NO	3.5 liq	7
Quinoline	C_9H_7N	2.29(11)	1

Name	Mol. form.	μ/D	Ref.
Rubidium bromide	BrRb	≈10.9	2
Rubidium chloride	ClRb	10.510(5)	2
Rubidium fluoride	FRb	8.5465(5)	2
Rubidium iodide	IRb	≈11.5	2
Rubidium sodium	NaRb	3.1(3)	2
Salicylaldehyde	$C_7H_6O_2$	2.86 liq	7
Selenium dioxide	O_2Se	2.62(5)	2
Selenium tetrafluoride	F_4Se	1.78(9)	2
Selenoformaldehyde	CH_2Se	1.41(1)	5
Silicon dicarbide	C_2Si	2.393(6)	24
Silicon methylidyne	CHSi	0.066(2)	41
Silicon monosulfide	SSi	1.73(9)	2
Silicon monoxide	OSi	3.0982	2
Silver(I) bromide	AgBr	5.62(3)	5
Silver(I) chloride	AgCl	6.08(6)	5
Silver(I) fluoride	AgF	6.22(30)	2
Silver(I) iodide	AgI	4.55(5)	5
Sodium bromide	BrNa	9.1183(6)	2
Sodium chloride	ClNa	9.00117	2
Sodium fluoride	FNa	8.156(1)	2
Sodium iodide	INa	9.236(3)	2
Stibine	H_3Sb	0.12(5)	1
Strontium oxide	OSr	8.900(3)	2
Styrene	C_8H_8	0.123(3)	5
Succinonitrile	$C_4H_4N_2$	3.7 liq	7
Sulfolane	$C_4H_8O_2S$	4.8 liq	7
Sulfur dichloride	Cl_2S	0.36(1)	3
Sulfur difluoride	F_2S	1.05(5)	2
Sulfur dioxide	O_2S	1.63305	3
Sulfur fluoride [SSF_2]	F_2S_2	1.03(5)	1
Sulfur fluoride [FSSF]	F_2S_2	1.45(2)	1
Sulfur monofluoride	FS	0.794(12)	3
Sulfur monoxide	OS	1.55(2)	1
Sulfur oxide (SSO)	OS_2	1.47(3)	1
Sulfur tetrafluoride	F_4S	0.632(3)	1
Sulfuryl chloride	Cl_2O_2S	1.81(4)	1
Sulfuryl fluoride	F_2O_2S	1.12(2)	1
Tetraborane(10)	B_4H_{10}	0.486(2)	3
1,1,2,2-Tetrabromoethane	$C_2H_2Br_4$	1.38 liq	7
1,1,2,2-Tetrachloroethane	$C_2H_2Cl_4$	1.32(7)	1
1,2,3,4-Tetrafluorobenzene	$C_6H_2F_4$	2.42(5)	3
1,2,3,5-Tetrafluorobenzene	$C_6H_2F_4$	1.46(6)	3
1,1,1,2-Tetrafluoroethane	$C_2H_2F_4$	1.80(22)	5
Tetrafluorosilane-ammonia complex	$F_4Si{\cdot}H_3N$	5.61(2)	31
Tetrahydrofuran	C_4H_8O	1.75(4)	2
Tetrahydrofurfuryl alcohol	$C_5H_{10}O_2$	2.1 liq	7
Tetrahydro-4*H*-pyran-4-one	$C_5H_8O_2$	1.720(3)	3
1,2,5,6-Tetrahydropyridine	C_5H_9N	1.007(3)	3
Tetrahydrothiophene	C_4H_8S	1.90 liq	7
Tetramethylurea	$C_5H_{12}N_2O$	3.5 liq	7
1*H*-Tetrazole	CH_2N_4	2.19(5)	3
Thallium(I) bromide	BrTl	4.49(5)	2
Thallium(I) chloride	ClTl	4.54299	2
Thallium(I) fluoride	FTl	4.2282(8)	2
Thallium(I) iodide	ITl	4.61(7)	2
Thiacyclohexane	$C_5H_{10}S$	1.781(10)	3
1,2,5-Thiadiazole	$C_2H_2N_2S$	1.579(7)	3
Thietane	C_3H_6S	1.85(9)	1
Thietane 1,1-dioxide	$C_3H_6O_2S$	4.8(1)	5
Thioacetaldehyde	C_2H_4S	2.33(2)	68
Thiocarbonyl fluoride	CF_2S	0.080	59

Name	Mol. form.	μ/D	Ref.
Thioformaldehyde	CH_2S	1.6491(4)	3
Thionitrosyl chloride	ClNS	1.87(2)	2
Thionitrosyl fluoride (NSF)	FNS	1.902(12)	2
Thionyl chloride	Cl_2OS	1.45(3)	1
Thionyl fluoride	F_2OS	1.63(1)	1
Thiophene	C_4H_4S	0.55(1)	2
2-Thiophenecarbonitrile	C_5H_3NS	4.59(2)	3
3-Thiophenecarbonitrile	C_5H_3NS	4.13(2)	3
4*H*-Thiopyran-4-thione	$C_5H_4S_2$	3.9(2)	5
Tin(II) oxide	OSn	4.32(22)	2
Tin(II) sulfide	SSn	3.18(16)	2
Titanium(II) oxide	OTi	2.96(5)	5
Toluene	C_7H_8	0.375(10)	3
Toluene-sulfur dioxide complex	$C_7H_8{\cdot}O_2S$	1.87(3)	34
1*H*-1,2,4-Triazole	$C_2H_3N_3$	2.7(1)	3
Tribromomethane	$CHBr_3$	0.99(2)	1
Tributylamine	$C_{12}H_{27}N$	0.78 liq	7
Tributyl borate	$C_{12}H_{27}BO_3$	0.77 liq	7
Tributyl phosphate	$C_{12}H_{27}O_4P$	3.07 liq	7
3-Thioxo-1,2-propadienylidene	C_3S	3.704(9)	50
1,1,1-Trichloroethane	$C_2H_3Cl_3$	1.755(15)	2
1,1,2-Trichloroethane	$C_2H_3Cl_3$	1.4 liq	7
Trichloroethene	C_2HCl_3	0.8 liq	7
Trichloroethylsilane	$C_2H_5Cl_3Si$	2.04 liq	7
Trichlorofluoromethane	CCl_3F	0.46(2)	2
Trichlorofluorosilane	Cl_3FSi	0.49(1)	2
Trichloromethane	$CHCl_3$	1.04(2)	2
(Trichloromethyl)benzene	$C_7H_5Cl_3$	2.03 liq	7
Trichloromethylsilane	CH_3Cl_3Si	1.91(1)	2
Trichlorosilane	Cl_3HSi	0.86(1)	2
Tri-*o*-cresyl phosphate	$C_{21}H_{21}O_4P$	2.87 liq	7
Tri-*m*-cresyl phosphate	$C_{21}H_{21}O_4P$	3.05 liq	7
Tri-*p*-cresyl phosphate	$C_{21}H_{21}O_4P$	3.18 liq	7
Triethanolamine	$C_6H_{15}NO_3$	3.57 liq	7
Triethylamine	$C_6H_{15}N$	0.66(5)	1
Triethyl phosphate	$C_6H_{15}O_4P$	3.12 liq	7
Trifluoramine oxide	F_3NO	0.0390(4)	9
Trifluoroacetic acid	$C_2HF_3O_2$	2.28(25)	1
Trifluoroacetonitrile	C_2F_3N	1.262(10)	3
1,2,4-Trifluorobenzene	$C_6H_3F_3$	1.402(9)	5
1,1,1-Trifluorodisilane	$F_3H_3Si_2$	2.03(10)	3
1,1,1-Trifluoroethane	$C_2H_3F_3$	2.347(5)	3
Trifluoroethene	C_2HF_3	1.32(3)	2
Trifluoroiodomethane	CF_3I	1.048(3)	3
Trifluoroiodosilane	F_3ISi	1.11(3)	5
Trifluoroisocyanomethane	C_2F_3N	1.153(10)	5
Trifluoromethane	CHF_3	1.65150	3
(Trifluoromethyl)benzene	$C_7H_5F_3$	2.86(6)	1
Trifluoromethylsilane	CH_3F_3Si	2.3394(2)	5
(Trifluoromethyl)silane	CH_3F_3Si	2.32(2)	5
3,3,3-Trifluoro-1-propene	$C_3H_3F_3$	2.45(5)	1
3,3,3-Trifluoro-1-propyne	C_3HF_3	2.317(13)	5
Trifluorosilane	F_3HSi	1.27(3)	1
Trimethylamine	C_3H_9N	0.612(3)	1
Trimethyl phosphate	$C_3H_9O_4P$	3.18 liq	7
2,4,6-Trimethylpyridine	$C_8H_{11}N$	2.05 liq	7
1,3,5-Trioxane	$C_3H_6O_3$	2.08(2)	1
Vinyl acetate	$C_4H_6O_2$	1.79 liq	7
Vinyl formate	$C_3H_4O_2$	1.49(1)	1
2-Vinylfuran	C_6H_6O	0.69(7)	5
Vinylsilane	C_2H_6Si	0.657(2)	5
Water	H_2O	1.8546(40)	3
Water dimer-hydrogen bromide complex	$BrH{\cdot}H_4O_2$	2.281(3)	48
Water dimer-hydrogen chloride complex	$ClH{\cdot}H_4O_2$	2.328(3)	48
o-Xylene	C_8H_{10}	0.640(5)	2
2,4-Xylenol	$C_8H_{10}O$	1.4 liq	7
2,5-Xylenol	$C_8H_{10}O$	1.45 liq	7
2,6-Xylenol	$C_8H_{10}O$	1.40 liq	7
3,4-Xylenol	$C_8H_{10}O$	1.56 liq	7
3,5-Xylenol	$C_8H_{10}O$	1.55 liq	7
Ytterbium monofluoride	FYb	3.91(4)	45
Yttrium monoxide	OY	4.524(7)	26
Zirconium(II) oxide	OZr	2.55(1)	26
Zirconium(IV) oxide	O_2Zr	7.80(2)	19

Table 2. Dipole Moments of Conformers

Name	Mol. form.	Conformer	μ/D	Ref.
Acrolein	C_3H_4O	*trans*	3.117(4)	5
Acrolein	C_3H_4O	*cis*	2.552(3)	5
Allyl alcohol	C_3H_6O	*gauche*	1.55(8)	3
Butanenitrile	C_4H_7N	*gauche*	3.91(4)	5
Butanenitrile	C_4H_7N	*anti*	3.73(6)	5
1-Butene	C_4H_8	*cis*	0.438(7)	2
1-Butene	C_4H_8	*skew*	0.359(11)	2
Chlorocyclohexane	$C_6H_{11}Cl$	*equitorial*	2.44(7)	5
Chlorocyclohexane	$C_6H_{11}Cl$	*axial*	1.91(2)	5
1-Chloropropane	C_3H_7Cl	*gauche*	2.02(3)	5
1-Chloropropane	C_3H_7Cl	*trans*	1.95(2)	5
Cyclohexene	C_6H_{10}	*half-chair*	0.332(12)	2
Diethyl sulfide	$C_4H_{10}S$	*trans-trans*	1.556(4)	54
Diethyl sulfide	$C_4H_{10}S$	*trans-gauche*	1.591(9)	54
Diethyl sulfide	$C_4H_{10}S$	*gauche-gauche*	1.645(1)	54
1,2-Difluoroethane	$C_2H_4F_2$	*gauche*	2.67(13)	2
Ethanethiol	C_2H_6S	*gauche*	1.61(8)	3
Ethanethiol	C_2H_6S	*trans*	1.58(8)	3
Ethanol	C_2H_6O	*gauche*	1.68(3)	3
Ethanol	C_2H_6O	*trans*	1.44(3)	2
Ethylamine	C_2H_7N	*gauche*	1.210(15)	5
Ethylamine	C_2H_7N	*trans*	1.304(11)	5
Ethyl formate	$C_3H_6O_2$	*gauche*	1.81(2)	2
Ethyl formate	$C_3H_6O_2$	*trans*	1.98(2)	2
Ethyl methyl ether	C_3H_8O	*trans*	1.17(2)	3
Ethyl methyl sulfide	C_3H_8S	*gauche*	1.593(4)	5
Ethyl methyl sulfide	C_3H_8S	*trans*	1.56(3)	3
Fluorocyclohexane	$C_6H_{11}F$	*equitorial*	2.11(4)	2
Fluorocyclohexane	$C_6H_{11}F$	*axial*	1.81(4)	2
1-Fluoropropane	C_3H_7F	*gauche*	1.90(10)	1
1-Fluoropropane	C_3H_7F	*trans*	2.05(4)	1
3-Fluoropropene	C_3H_5F	*gauche*	1.939(15)	1
3-Fluoropropene	C_3H_5F	*cis*	1.765(14)	1
Glycine	$C_2H_5NO_2$	*Conformer I*	1.147(5)	49
Glycine	$C_2H_5NO_2$	*Conformer II*	5.45(5)	49

Name	Mol. form.	Conformer	μ/D	Ref.
Glyoxal	$C_2H_2O_2$	*cis*	4.8(2)	2
3-Hydroxypropanenitrile	C_3H_5NO	*gauche*	3.17(2)	5
Isobutanal	C_4H_8O	*gauche*	2.69(1)	5
Isobutanal	C_4H_8O	*trans*	2.86(1)	5
2-Methoxyethanol	$C_3H_8O_2$	*gauche*	2.36(5)	2
3-Methyl-1-butene	C_5H_{10}	*gauche*	0.398(4)	3
3-Methyl-1-butene	C_5H_{10}	*trans*	0.320(10)	3
2-Methyl-2-propenol	C_4H_8O	*skew*	1.295(22)	5
Methyl propyl ether	$C_4H_{10}O$	*trans-trans*	1.107(13)	3
Nitrous acid	HNO_2	*cis*	1.423(5)	2
Nitrous acid	HNO_2	*trans*	1.855(16)	2
1-Pentyne	C_5H_8	*gauche*	0.769(28)	2
1-Pentyne	C_5H_8	*trans*	0.842(10)	2
Piperidine	$C_5H_{11}N$	*equitorial*	0.82(2)	3
Piperidine	$C_5H_{11}N$	*axial*	1.19(2)	3
Propanal	C_3H_6O	*gauche*	2.86(1)	5
Propanal	C_3H_6O	*cis*	2.52(5)	1
1-Propanethiol	C_3H_8S	*gauche*	1.683(10)	3
1-Propanethiol	C_3H_8S	*trans*	1.60(8)	3
2-Propanethiol	C_3H_8S	*gauche*	1.53(3)	3
2-Propanethiol	C_3H_8S	*trans*	1.61(3)	3
Propanoic acid	$C_3H_6O_2$	*cis*	1.46(7)	2
1-Propanol	C_3H_8O	*gauche*	1.58(3)	2
1-Propanol	C_3H_8O	*trans*	1.55(3)	2
2-Propanol	C_3H_8O	*trans*	1.58(3)	2
Propyleneimine	C_3H_7N	*cis*	1.77(9)	2
Propyleneimine	C_3H_7N	*trans*	1.57(3)	2
Tetrafluorohydrazine	F_4N_2	*gauche*	0.257(2)	5
Tetrahydropyran	$C_5H_{10}O$	*chair*	1.58(3)	3

HINDERED INTERNAL ROTATION

I. Ozier and N. Moazzen-Ahmadi

In asymmetric rotors like methyl alcohol, CH_3OH, and symmetric rotors like CH_3SiH_3, the methyl group can undergo internal rotation relative to the rest of the molecule, traditionally called the frame (Refs. 1 and 2). Although various rotating groups are considered here, all have three-fold symmetry. In such cases, the potential V hindering the internal rotation can be written:

$$V(\alpha)= V_3(½)(1 - \cos 3\alpha) + V_6(½)(1 - \cos 6\alpha) + V_9(½)(1 - \cos 9\alpha) + ...,$$

where α is the deviation from equilibrium of the angle between the top and frame that measures the torsional motion. If only the first two terms are retained, then V_3 is the height of the hindering potential and V_6 is the shape parameter. For symmetric tops like CH_3CH_3 where the top and frame are identical, α is replaced by 2γ and the origin for γ is often taken as the eclipsed configuration. In the expansion, $-\cos 6n\gamma$ is then replaced by $(-1)n^{+1}\cos 6n\gamma$, where n = 1,2,... In cases where different forms of the expansion have been used in the original works, the values of the parameters published there have been converted to the conventions defined here.

In Tables 1 and 2, values are given for V_3 for a selection of asymmetric and symmetric rotors, respectively. In cases where the higher order parameters have been determined, these are given in the Comments column. Where appropriate, this column also indicates the specific top, isomer, state, and/or isotopomer that has been studied. For ethane, three symmetric top isotopomers are listed to illustrate the isotopic dependence of V_3 and V_6. In all other cases, only one isotopomer is listed, even if several have been studied. In all but one of these cases, the isotopomer reported is the one with the highest natural abundance. However, CH_3OCDO is listed because the results obtained are more precise than for CH_3OCHO. The molecules are listed alphabetically in Hill order according to the molecular formula.

The determinations listed for the potential parameters are effective values that incorporate to varying degrees effects from other molecular parameters. For example, the apparent value of V_3 can be changed significantly if the reduced rotational constant F is calculated from the structure, rather than being determined independently (Ref. 1). Other examples include such mechanisms as coupling to excited skeletal vibrations (Ref. 2) and redundancies connecting some of the torsional parameters (Refs. 3 and 4). The experimental uncertainties quoted are taken from the original works; no attempt has been made to standardize the definitions. All the potential parameters are given in cm^{-1}. Where the original work has reported these values in other units, the conversion to cm^{-1} has been carried out using standard factors: 1 calorie = 4.1868 joules and 1 calorie/mole = 0.34998915 cm^{-1} (Ref. 5).

A variety of different methods have been used to measure V_3, V_6, and V_9 (Refs. 1 and 2); only a few of the more important will be discussed here. *For asymmetric rotors*, both the pure rotational spectrum and its torsion-rotation counterpart are electric dipole allowed and are affected in lowest order by the leading terms in the torsional Hamiltonian. Both types of spectra have been used extensively to determine V_3 (Ref. 1). *For symmetric tops* with a single torsional degree of freedom, either the permanent electric dipole moment vanishes, as in CH_3CH_3, or the normal rotational spectrum is independent of V_3 in lowest order, as in CH_3SiH_3. In the latter case, the molecular beam avoided crossing method can often be used (Ref. 2). The torsion-rotation spectrum is forbidden in lowest order, but becomes weakly allowed through interactions with the infrared active skeletal vibrations (Ref. 2). By employing long absorption path lengths, this spectrum has been used to determine V_3 in a number of molecules. For both asymmetric and symmetric tops, the most precise determinations of the molecular parameters have been made in cases where both rotational and torsion-rotation spectra have been investigated.

References

1. Lin, C. C., and Swalen, J. D., *Rev. Mod. Phys.* 31, 841 1959.
2. Ozier, I., and Moazzen-Ahmadi, N., Internal rotation in symmetric tops, in: Arimondo, E., Berman, P. R., and Lin, C. C., (Eds.), *Advances in Atomic, Molecular and Optical Physics*, vol. 54, p. 423, Elsevier, Amsterdam, 2007.
3. Lees, R. M., and Baker, J. G., *J. Chem. Phys.* 48, 5299, 1968.
4. Moazzen-Ahmadi, N., and Ozier, I., *J. Mol. Spectrosc.* 126, 99, 1987.
5. Demaison, J., and Wlodarczak, G., Hindered rotation-Asymmetric top molecules, in: Hüttner, W. (Ed), *Landolt-Börnstein Numerical Data and Functional Relationships in Science and Technology, New Series: Group II: Molecules and Radicals, volume 24, subvolume C, Molecular Constants mostly from Microwave, Molecular Beam, and Sub-Doppler Laser Spectroscopy*, Springer-Verlag, Heidelberg, 2002.
6. Suenram, R. D., Lovas, F. J., Plusquellic, D. F., Ellzy, M. W., Lochner, J. M., Jensen, J. O., and Samuels, A. C., *J. Mol. Spectrosc.* 235, 18, 2006.
7. Sastry, K. V. L. N., Herbst, E., Booker, R. A., and De Lucia, F. C., *J. Mol. Spectrosc.* 116, 120, 1986.
8. Tyblewksi, M., Ha, T.-K., and Bauder, A., *J. Mol. Spectrosc.* 115, 353, 1986.
9. Durig, J. R., Guirgis, G. A., and Van Der Veken, B. J., *J. Raman Spectrosc.* 18, 549, 1987.
10. Eltayeb, S., Guirgis, G. A., Fanning, A. R., and Durig, J. R., *J. Raman Spectrosc.* 27, 111, 1996.
11. Krisher, L. C., *J. Chem. Phys.* 33, 1237, 1960.
12. Alonso, J. L., López, J. C., Blanco, S., and Guarnieri, A., *J. Mol. Spectrosc.* 182, 148, 1997.
13. Pierce, L., and Krisher, L. C., *J. Chem. Phys.* 31, 875, 1959.
14. Moloney, M. J., and Krisher, L. C., *J. Chem. Phys.* 45, 3277, 1966.
15. Kleiner, I., Hougen, J. T., Grabow, J.-U., Belov, S. P., Tretyakov, M. Yu., and Cosléou, J., *J. Mol. Spectrosc.* 179, 41, 1996.
16. Ilyushin, V. V., Alekseev, E. A., Dyubko, S. F., and Kleiner, I., *J. Mol. Spectrosc.* 220, 170, 2003.
17. Fliege, E., Dreizler, H., Demaison, J., Boucher, D., Burie, J., and Dubrulle, A., *J. Chem. Phys.* 78, 3541, 1983
18. Schnell, M., Grabow, J.-U., Hartwig, H., Heineking, N., Meyer, M., Stahl, W., and Caminati, W., *J. Mol. Spectrosc.* 229, 1, 2005.
19. Niide, Y., and Hayashi, M., *J. Mol. Spectrosc.* 223, 152, 2004.
20. Shiki, Y., Hasegawa, A., and Hayashi, M., *J. Mol. Structure* 78, 185, 1982.
21. Groner, P., Gillies, C. W., Gillies, J. Z., Zhang, Y., and Block, E., *J. Mol. Spectrosc.* 226, 169, 2004.
22. Alonso, J. L., Lesarri, A., López, J. C., Blanco, S., Kleiner, I., and Demaison, J., *Molec. Phys.* 91, 731, 1997.
23. Demaison, J., Maes, H., van Eijck, B. P., Wlodarczak, G. and Lasne, M. C., *J. Mol. Spectrosc.* 125, 214, 1987.
24. Antolínez, S., López, J. C., and Alonso, J. L., *J. Chem. Soc., Faraday Trans.* 93, 1291, 1997.
25. Butcher, S. S., and Wilson, E. B., *J. Chem. Phys.* 40, 1671, 1964.
26. Groner, P., *J. Mol. Structure* 550–551, 473, 2000.

27. Marstokk, K.-M., Mollendal, H., Samdal, S., and Steinborn, D., *J. Mol. Structure* 567, 41, 2001.
28. Stiefvater, O. L., *J. Chem. Phys.* 62, 233, 1975.
29. de Luis, A., Eugenia Sanz, M., Lorenzo, F. J., López, J. C., and Alonso, J. L., *J. Mol. Spectrosc.* 184, 60, 1997.
30. Kasten, W., and Dreizler, H., *Z. Naturforsch.* A 41, 944, 1986.
31. Vormann, K., and Dreizler, H., *Z. Naturforsch.* A 43, 338, 1988.
32. Marstokk, K.-M., Møllendal, H., and Samdal, S., *J. Mol. Structure* 376, 11, 1996.
33. Bestmann, G., Lalowski, W., and Dreizler, H., *Z. Naturforschung* A 40, 271, 1985.
34. Dreizler, H., and Scappini, F., *Z. Naturforsch.* A 36, 1187, 1981.
35. Typke, V., Botskor, I., and Wiedenmann, K.-H., *J. Mol. Spectrosc.* 120, 435, 1986.
36. Suenram, R. D., Lovas, F. J., Plusquellic, D. F., Lesarri, A., Kawashima, Y., Jensen, J. O., and Samuels, A. C., *J. Mol. Spectrosc.* 211, 110, 2002.
37. Ohashi, N., and Hougen, J. T., *J. Mol. Spectrosc.* 243, 162, 2007.
38. Grabow, J.-U., Hartwig, H., Heineking, N., Jäger, W., Mäder, H., Nicolaisen, H. W., and Stahl, W., *J. Mol. Structure* 612, 349, 2002.
39. Charro, M. E., and Alonso, J. L., *J. Mol. Spectrosc.* 176, 251, 1996.
40. Nair, K. P. R., Demaison, J., Wlodarczak, G., and Merke, I., *J. Mol. Spectrosc.* 237, 137, 2006.
41. Suenram, R. D., DaBell, R. S., Hight Walker, A. R., Lavrich, R. J., Plusquellic, D. F., Ellzy, M. W., Lochner, J. M., Cash, L., Jensen, J. O., and Samuels, A. C., *J. Mol. Spectrosc.* 224, 176, 2004.
42. Groner, P., Attia, G. M., Mohamad, A. B., Sullivan, J. F., Li, Y. S., and Durig, J. R., *J. Chem. Phys.* 91, 1434, 1989.
43. Varma, R., Ramaprasad, K. R., and Nelson, J. F., *J. Chem. Phys.* 63, 915, 1975.
44. Odom, J. D., Kalasinsky, V. F., and Durig, J. R., *Inorg. Chem.* 14, 2837, 1975.
45. Kuczkowski, R. L., and Lide, D. R., *J. Chem. Phys.* 46, 357, 1967.
46. Durig, J. R., Li, Y. S., Carreira, L. A., and Odom, J. D., *J. Amer. Chem. Soc.* 95, 2491, 1973.
47. Lide, D. R., Johnson, D. R., Sharp, K. G., and Coyle, T. D., *J. Chem. Phys.* 57, 3699, 1972.
48. Krisher, L. C., Watson, W. A., and Morrison, J. A., *J. Chem. Phys.* 61, 3429, 1974.
49. Styger, C., Ozier, I., Wang, S.-X., and Bauder, A., *J. Mol. Spectrosc.* 239, 115, 2006.
50. Laurie, V. W., *J. Chem. Phys.* 30, 1210, 1959.
51. Cahill, P., and Butcher, S., *J. Chem. Phys.* 35, 2255, 1961.
52. Wang, S.-X., Schroderus, J., Ozier, I., Moazzen-Ahmadi, N., Horneman, V.-M., Ilyushyn, V. V., Alekseev, E. A., Katrich, A. A., and Dyubko, S. F., *J. Mol. Spectrosc.* 214, 69, 2002.
53. Nakagawa, J., Yamada, K., Bester, M., and Winnewisser, G., *J. Mol. Spectrosc.* 110, 74, 1985.
54. Merke, I., Stahl, W., Kassi, S., Petitprez, D., and Wlodarczak, G., *J. Mol. Spectrosc.* 216, 437, 2002.
55. di Lauro, C., Bunker, P. R., Johns, J. W. C., and McKellar, A. R. W., *J. Mol. Spectrosc.* 184, 177, 1997.
56. Voges, K., Gripp, J., Hartwig, H., and Dreizler, H., *Z. Naturforsch.* A 51, 299, 1996.
57. Borvayeh, L., Moazzen-Ahmadi, N., and Horneman, V.-M., *J. Mol. Spectrosc.* 242, 77, 2007.

TABLE 1. Asymmetric Rotor Potential Parameters

Mol. form.	Name	Line form.	Ref.	V_3/cm^{-1}	Comments
CHF_3S	Trifluoromethanethiol	CF_3SH	5	500.83 ± 0.03	
CH_3F_2OP	Methylphosphonic difluoride	$CH_3P(=O)F_2$	6	676 ± 25	
CH_4O	Methanol	CH_3OH	5	373.594 ± 0.007	V_6 = −1.597 ± 0.051; V_9 = 1.04 ± 0.20
CH_4S	Methanethiol	CH_3SH	7	443.029 ± 0.070	V_6 = −1.6451 ± 0.0144
CH_4S_2	Methyldisulfane	CH_3SSH	8	609.0 ± 14.0	
C_2F_3NO	Trifluoromethyl isocyanate	$CF_3N{=}C{=}O$	5	47.8769 ± 0.0051	
C_2HF_3O	Trifluoroacetaldehyde	$CF_3C(H){=}O$	9	298 ± 10	
C_2HF_5	Pentafluoroethane	CF_3CHF_2	10	1190 ± 4	
C_2H_3BrO	Acetyl bromide	$CH_3C(Br){=}O$	11	456.7 ± 10.5	
$C_2H_3ClF_2$	1-Chloro-1,1-difluoroethane	CH_3CClF_2	12	1311.8 ± 1.4	
C_2H_3ClO	Acetyl chloride	$CH_3C(Cl){=}O$	5	442.74 ± 1.05	^{35}Cl
C_2H_3FO	Acetyl fluoride	$CH_3C(F){=}O$	13	364.3 ± 2.1	
$C_2H_3FO_2$	Methyl fluoroformate	$CH_3OC(F){=}O$	5	374.1 ± 0.2	
$C_2H_3F_3O$	Methyl trifluoromethyl ether	CH_3OCF_3	5	382 ± 10	CH_3
C_2H_3IO	Acetyl iodide	$CH_3C({=}O)I$	14	455.3 ± 10.5	
C_2H_3NO	Methyl cyanate	$CH_3OC{\equiv}N$	5	399.0 ± 17.5	
C_2H_4ClF	1-Chloro-1-fluoroethane	CH_3CHClF	5	1334.9 ± 3.8	
$C_2H_4F_2$	1,1-Difluoroethane	CH_3CHF_2	5	1163.0 ± 2.5	
C_2H_4O	Acetaldehyde	$CH_3C(H){=}O$	15	407.716 ± 0.010	V_6 = −12.068 ± 0.037
C_2H_4OS	Thioacetaldehyde *S*-oxide	$CH_3C(H){=}S{=}O$	5	285.6 ± 0.3	*Z* isomer
$C_2H_4O_2$	Acetic acid	CH_3COOH	16	170.1742 ± 0.0002	V_6 = −6.4725 ± 0.0001
$C_2H_3DO_2$	Methyl formate	$CH_3OC(D){=}O$	5	400.60 ± 0.03	deuterated
C_2H_5F	Fluoroethane	CH_3CH_2F	17	1172.1 ± 1.4	
C_2H_5NO	Nitrosoethane	$CH_3CH_2N{=}O$	5	903 ± 25	*gauche* conformer
C_2H_5NO		$CH_3CH_2N{=}O$	5	911 ± 25	*cis* conformer
C_2H_5NO	Acetamide	$CH_3C(NH_2){=}O$	5	24.949 ± 0.008	
$C_2H_6F_2Si$	Difluorodimethylsilane	$(CH_3)_2SiF_2$	18	439.4 ± 2.5	
$C_2H_6N_2O$	*N*-Nitrosodimethylamine	$(CH_3)_2NN{=}O$	5	145.8 ± 0.25	*cis* CH_3
$C_2H_6N_2O$		$(CH_3)_2NN{=}O$	5	737.4 ± 13.3	*trans* CH_3
C_2H_6O	Ethanol	CH_3CH_2OH	5	1173.76 ± 2.20	*trans* isomer
C_2H_6O	Dimethyl ether	$(CH_3)_2O$	19	926.0 ± 3.5	

Mol. form.	Name	Line form.	Ref.	V_3/cm^{-1}	Comments
C_2H_6S	Dimethyl sulfide	$(CH_3)_2S$	19	751.1 ± 4.8	
C_2H_6Si	Vinylsilane	$SiH_3C(H){=}CH_2$	20	520.1 ± 1.8	
$C_2H_6S_2$	Dimethyl disulfide	CH_3SSCH_3	5	535.1 ± 1.8	
$C_2H_6Se_2$	Dimethyl diselenide	$CH_3SeSeCH_3$	21	395 ± 2	
C_2H_8Si	Dimethylsilane	$(CH_3)_2SiH_2$	19	578.0 ± 3.5	
$C_3H_3F_3$	3,3,3-Trifluoropropene	$CF_3C(H){=}CH_2$	22	653.06 ± 0.83	
$C_3H_3NO_2$	Methyl cyanoformate	$CH_3OC(C{\equiv}N){=}O$	5	406.6 ± 1.1	*s-trans* conformer
C_3H_4S	(Methylthio)acetylene	$CH_3SC{\equiv}CH$	23	592.0 ± 3.3	
$C_3H_5F_3$	1,1,1-Trifluoropropane	$CH_3CH_2CF_3$	24	922.2 ± 1.4	
C_3H_5I	2-Iodopropene	$CH_3C(I){=}CH_2$	5	905.8 ± 4.2	
C_3H_5N	Ethyl isocyanide	$CH_3CH_2N{\equiv}C$	5	1167.6 ± 18.2	
C_3H_6	Propene	$CH_3C(H){=}CH_2$	5	697.499 ± 0.048	$V_6 = -13.0$ (fixed)
C_3H_6O	Propanal	$CH_3CH_2C(H){=}O$	25	798 ± 39	*cis* conformer
C_3H_6O	Acetone	$(CH_3)_2C{=}O$	26	251.4 ± 2.6	$V_6 = -6.92 \pm 0.65$
C_3H_6S	(Methylthio)ethene	$CH_3SC(H){=}CH_2$	27	1138 ± 13	
$C_3H_6O_2$	Propanoic acid	CH_3CH_2COOH	28	819.0 ± 10.5	*cis* conformer
$C_3H_6O_2S$	Methyl mercaptoacetate	$CH_3OC({=}O)C(H_2)SH$	5	411 ± 8	state 0$^+$
$C_3H_6O_2S$		$CH_3OC({=}O)C(H_2)SH$	5	412 ± 9	state 0$^-$
C_3H_7Br	2-Bromopropane	$(CH_3)_2CHBr$	5	1437.0 ± 2.5	^{79}Br
C_3H_7Cl	1-Chloropropane	$CH_3C(H_2)C(H_2)Cl$	29	1017.8 ± 1.4	*gauche* conformer
C_3H_7Cl		$CH_3C(H_2)C(H_2)Cl$	29	966.0 ± 7.0	*trans* conformer
C_3H_7Cl	2-Chloropropane	$(CH_3)_2CHCl$	5	1374.03 ± 1.00	^{35}Cl
C_3H_7F	1-Fluoropropane	$CH_3C(H_2)C(H_2)F$	30	965.3 ± 12.2	*gauche* conformer
C_3H_7F		$CH_3C(H_2)C(H_2)F$	30	948.5 ± 2.8	*trans* conformer
C_3H_7F	2-Fluoropropane	$(CH_3)_2CHF$	5	1162.79 ± 0.84	
C_4H_7N	Butanenitrile	$CH_3C(H_2)C(H_2)C{\equiv}N$	31	1087.4 ± 8.4	*gauche* conformer
C_4H_7N		$CH_3C(H_2)C(H_2)C{\equiv}N$	31	1088.5 ± 13.3	*trans* conformer
C_3H_7NO	Propanamide	$CH_3CH_2C({=}O)NH_2$	32	761 ± 42	*syn* conformer
C_3H_7NO	*N,N*-Dimethylformamide	$(CH_3)_2NC(H){=}O$	5	366.04 ± 0.26	*cis* CH_3
C_3H_7NO		$(CH_3)_2NC(H){=}O$	5	772.4 ± 7.4	*trans* CH_3
C_3H_8	Propane	$(CH_3)_2CH_2$	33	1108.1 ± 9.5	
C_3H_8Ge	Cyclopropylgermane	$C(H_2)C(H_2)C(H)(GeH_3)$	5	466.6 ± 16.7	GeH_3
$C_3H_8N_2O$	*N*-Nitrosoethylmethylamine	$CH_3CH_2N(CH_3)N{=}O$	5	310 ± 30	*N*-methyl top, OGM conformer
C_3H_8O	1-Propanol	$CH_3C(H_2)C(H_2)OH$	34	956 ± 21	*trans* conformer
C_3H_8Si	Cyclopropylsilane	$C(H_2)C(H_2)C(H)(SiH_3)$	35	670.9 ± 1.5	
C_3H_8Si	Dimethyl(methylene)silane	$(CH_3)_2Si{=}CH_2$	5	351.4 ± 5.9	
$C_3H_9O_3P$	Dimethyl methylphosphonate	$(OCH_3)_2P({=}O)CH_3$	36	662 ± 6	*P*-methyl top
$C_3H_9O_3P$		$(OCH_3)_2P({=}O)CH_3$	37	278.82 ± 0.06	*O*-methyl top #1
$C_3H_9O_3P$		$(OCH_3)_2P({=}O)CH_3$	37	181.82 ± 0.01	*O*-methyl top #2
C_4H_3FO	But-2-ynoyl fluoride	$CH_3C{\equiv}CC(F){=}O$	5	2.20 ± 0.12	
C_4H_5N	*cis*-2-Butenenitrile	$CH_3C(H){=}C(H)C{\equiv}N$	5	485.50 ± 0.25	
C_4H_5N	2-Methylacrylonitrile	$CH_2{=}C(CH_3)C{\equiv}N$	5	695.2 ± 2.1	
C_4H_5NO	2-Methyloxazole	$N{=}C(CH_3)OC(H){=}C(H)$	5	251.70 ± 1.17	
C_4H_5NO	4-Methyloxazole	$N{=}C(H)OC(H){=}C(CH_3)$	5	429.44 ± 0.33	
C_4H_5NO	5-Methyloxazole	$N{=}C(H)OC(CH_3){=}C(H)$	5	477.90 ± 1.34	
C_4H_5NO	5-Methylisoxazole	$C(H){=}NOC(CH_3){=}C(H)$	5	272.05 ± 1.00	
C_4H_5NS	2-Methylthiazole	$N{=}C(CH_3){=}SC(H){=}C(H)$	38	34.938 ± 0.020	
C_4H_5NS	4-Methylisothiazole	$N{=}C(H)C(CH_3){=}C(H)S$	5	105.767 ± 0.043	
$C_4H_6O_2$	4-Methyl-2-oxetanone	$OC({=}O)C(H_2)C(H)(CH_3)$	5	1256.5 ± 10.5	

Mol. form.	Name	Line form.	Ref.	V_3/cm^{-1}	Comments
C_4H_7F	*trans*-1-Fluoro-2-butene	$CH_3C(H)=C(H)CH_2F$	5	596 ± 7	anticlinal conformer
C_4H_7N	1-Isocyanopropane	$CH_3C(H_2)C(H_2)N\equiv C$	5	1012.3 ± 8.4	*gauche* conformer
C_4H_7N		$CH_3C(H_2)C(H_2)N\equiv C$	5	1033.8 ± 7.7	*trans* conformer
C_4H_8	Isobutene	$(CH_3)_2C=CH_2$	5	761.58 ± 1.05	
C_4H_8	*cis*-2-Butene	$CH_3CH=CHCH_3$	5	259.89 ± 0.42	
C_4H_8O	3-Methoxy-1-propene	$CH_3OC(H_2)C(H)=CH_2$	5	728.0 ± 10.5	*skew-gauche* conformer
C_4H_8O		$CH_3OC(H_2)C(H)=CH_2$	5	829.5 ± 10.5	*syn-trans* conformer
C_4H_8O	2,2-Dimethyloxirane	$OC(CH_3)(CH_3)C(H_2)$	5	945.61 ± 0.75	
C_4H_8O	*cis*-2,3-Dimethyloxirane	$OC(H)(CH_3)C(H)(CH_3)$	5	577.80 ± 1.84	*cis* conformer
C_4H_8O		$OC(H)(CH_3)C(H)(CH_3)$	5	862.52 ± 1.84	*trans* conformer
C_4H_8O	2-Methyloxetane	$OC(H_2)C(H_2)C(H)(CH_3)$	5	1166.5 ± 4.9	
C_4H_8O	3-Methyloxetane	$OC(H_2)C(H)(CH_3)C(H_2)$	5	1149.4 ± 4.2	
C_4H_8OS	3-Methoxythietane	$SC(H_2)C(H)(OCH_3)C(H_2)$	5	1071.0 ± 10.5	
C_4H_8S	3-(Methylthio)-1-propene	$CH_3SC(H_2)C(H)=CH_2$	5	619 ± 28	
C_4H_8S	2,2-Dimethylthiirane	$SC(CH_3)(CH_3)C(H_2)$	5	1268.3 ± 3.0	
C_4H_{10}	Butane	$CH_3C(H_2)C(H_2)CH_3$	5	948 ± 24	
$C_4H_{10}N_2O$	*N*-Methyl-*N*-nitrosopropylamine	$CH_3C(H_2)C(H_2)N(CH_3)N=O$	5	320 ± 30	*N*-methyl top, conformer OMGA
$C_5H_8O_2$	Dihydro-3-methyl-2(3*H*)-furanone	$OC(=O)C(H)(CH_3)C(H_2)C(H_2)$	5	913.8 ± 2.5	
$C_5H_8O_2$	Dihydro-4-methyl-2(3*H*)-furanone	$OC(=O)C(H_2)C(H)(CH_3)C(H_2)$	39	1437.8 ± 8.4	
$C_5H_8O_2$	Dihydro-5-methyl-2(3*H*)-furanone	$OC(=O)C(H_2)C(H_2)C(H)(CH_3)$	39	1233.0 ± 4.2	
C_5H_9NO	*tert*-Butyl isocyanate	$(CH_3)_3C\equiv N=C=O$	5	41.510 ± 0.015	$(CH_3)_3$ C group
$C_5H_{12}O$	Methyl *tert*-butyl ether	$(CH_3)_3COCH_3$	5	498.6 ± 1.5	*O*-methyl top
$C_6H_{10}O$	2-Methylcyclopentanone	$C(=O)C(H)(CH_3)C(H_2)C(H_2)C(H_2)$	5	844.2 ± 2.4	
$C_6H_{10}O$	3-Methylcyclopentanone	$C(=O)C(H_2)C(H)(CH_3)C(H_2)C(H_2)C(H_2)$	5	1233.8 ± 1.7	
$C_6H_{14}O$	*tert*-Butyl ethyl ether	$(CH_3)_3COC(H_2)CH_3$	5	1025 ± 3	ethyl CH_3
$C_7H_6F_2$	2,4-Difluorotoluene	$C(H)=C(CH_3)C(F)=C(H)C(F)=C(H)$	5	204.04 ± 0.23	
C_7H_7Cl	2-Chlorotoluene	$C(H)=C(H)C(Cl)=C(CH_3)C(H)=C(H)$	40	513.8 ± 2.7	^{35}Cl
C_7H_9N	2,6-Dimethylpyridine	$C(H)=C(H)C(CH_3)=NC(CH_3)=C(H)$	5	98.24 ± 0.27	
$C_7H_{16}FO_2P$	1,2,2-Trimethylpropyl methylphosphonofluoridate	$(CH_3)_3CC(H)(CH_3)OP(O)(F)CH_3$	41	821 ± 5	*P*-methyl top, conformer GD-I
$C_7H_{16}FO_2P$		$(CH_3)_3CC(H)(CH_3)OP(O)(F)CH_3$	41	738 ± 5	*P*-methyl top, conformer GD-II
GeH_3N_3	Germyl azide	$GeH_3\text{-}N=N\equiv N$	42	86.598 ± 0.062	
H_5PSi	Silylphospine	SiH_3PH_2	43	537.2 ± 14.0	

TABLE 2. Symmetric Rotor Potential Parameters

Mol. form.	Name	Line form.	Ref.	V_3/cm^{-1}	Comments
BF_3H_3P	Phosphine-trifluoroborane	H_3PBF_3	44	1169 ± 123	
BF_3H_3P	Trihydro(phosphorus trifluoride)boron	F_3PBH_3	45	1134 ± 53	
BH_6P	Trihydro(phosphine)boron	H_3PBH_3	46	864.5 ± 17.5	

Mol. form.	Name	Line form.	Ref.	V_3/cm^{-1}	Comments
CF_6Si	Trifluoro(trifluoromethyl)silane	CF_3SiF_3	47	489 ± 50	
CH_3F_3Ge	Trifluoromethylgermane	CF_3GeH_3	48	448 ± 53	
CH_3F_3Si	Trifluoromethylsilane	CH_3SiF_3	49	414.147 ± 0.030	
CH_6Ge	Methylgermane	CH_3GeH_3	50	433.6 ± 8.8	
CH_6Si	Methylsilane	CH_3SiH_3	2	603.3878 ± 0.0037	
CH_6Sn	Methylstannane	CH_3SnH_3	51	227 ± 10	
$C_2H_3F_3$	1,1,1-Trifluoroethane	CH_3CF_3	52	1112.24 ± 0.16	
C_2H_6	Ethane	CH_3CH_3	2	1013.28 ± 0.10	V_6 = 8.798 ± 0.041
$C_2H_3D_3$	Ethane-1,1,1-d_3	CH_3CD_3	2	1001.876 ± 0.023	V_6 = 9.328 ± 0.018
C_2D_6	Ethane-d_6	CD_3CD_3	2	989.946 ± 0.090	V_6 = 9.51 ± 0.10
C_3H_6Si	1-Silylpropyne	$CH_3C{\equiv}CSiH_3$	53	3.77 ± 0.70	
C_3H_9ClSi	Trimethylchlorosilane	$(CH_3)_3SiCl$	54	576.9 ± 0.9	
C_4H_6	2-Butyne	$CH_3C{\equiv}CCH_3$	55	6.067 ± 0.040	V_6 = 0.1240 ± 0.0144; V_9 = −0.0916 ± 0.0180
$C_5H_{10}Ge$	Ethynyltrimethylgermane	$(CH_3)_3GeC{\equiv}CH$	56	376.2 ± 16.7	
H_6Si_2	Disilane	SiH_3SiH_3	57	412.033 ± 0.010	

BOND DISSOCIATION ENERGIES

Yu-Ran Luo and Jin-Pei Cheng

The bond dissociation energy (enthalpy) is also referred to as bond disruption energy, bond energy, bond strength, or binding energy (abbreviation: BDE, BE, or D). It is defined as the standard enthalpy change of the following fission: R–X → R + X. The BDE, denoted by $D°(R–X)$, is usually derived by the thermochemical equation, $D°(R–X) = \Delta_f H°(R) + \Delta_f H°(X) – \Delta_f H°(RX)$. The enthalpy of formation $\Delta_f H°$ of a large number of atoms, free radicals, ions, clusters and compounds is available from the Web sites of NIST, NASA, CODATA, and IUPAC. Most authors prefer to use the BDE values at 298.15 K.

The following seven tables provide essential information of experimental BDE values of R–X and R^+–X bonds.

Table 1: Bond Dissociation Energies in Diatomic Molecules
Table 2: Enthalpy of Formation of Gaseous Atoms
Table 3: Bond Dissociation Energies in Polyatomic Molecules
Table 4: Enthalpies of Formation of Free Radicals and Other Transient Species
Table 5: Bond Dissociation Energies of Some Common Organic Molecules
Table 6: Bond Dissociation Energies in Diatomic Cations
Table 7: Bond Dissociation Energies in Polyatomic Cations

The data in these tables have been revised through September 2016. The Editors of the *Handbook* wish to express our sorrow at the death of Dr. Yu-Ran Luo, who so ably constructed these tables in the past.

Bond Dissociation Energies in Diatomic Molecules

The BDEs in diatomic species have usually been measured by spectroscopy or mass spectrometry. In the absence of data on the enthalpy function, the values at 0 K, $D°(A–B)$, are converted to $D°_{298}$ by the approximate equation:

$$D°_{298}(A–B) \approx D°(A–B) + (3/2)RT = D°(A–B) + 3.7181 \text{ kJ mol}^{-1}$$

This table has been arranged in an alphabetical order of the atoms A in the diatomics A–B.

Table 1. Bond Dissociation Energies in Diatomic Molecules

A–B	$D°_{298}$/kJ mol^{-1}	Ref.
Ac–O	794	1
Ac–S	505 ± 68	22
Ag–Ag	162.9 ± 2.9	1
Ag–Al	183.7 ± 9.2	1
Ag–Au	202.5 ± 9.6	1
Ag–Bi	192 ± 42	1
Ag–Br	280.3 ± 1.3	1
Ag–Cl	279.1 ± 8.4	1
Ag–Cu	171.5 ± 9.6	1
Ag–D	226.8	1
Ag–Dy	130 ± 19	1
Ag–Eu	127 ± 13	1
Ag–F	356.9 ± 5.8	1
Ag–Ga	159 ± 17	1
Ag–Ge	174.5 ± 21	1
Ag–H	202.4 ± 9.1	1
Ag–Ho	124 ± 19	1
Ag–I	234 ± 29	1
Ag–In	166.5 ± 4.9	1
Ag–Li	186.1	1
Ag–Mn	99.2 ± 21	1
Ag–Na	133.1 ± 12.6	1
Ag–Nd	<213	1
Ag–O	221 ± 21	1
Ag–S	216.7 ± 14.6	1
Ag–Sb	156.3 ± 4.9	30
Ag–Se	210.0 ± 14.6	1
Ag–Si	185.1 ± 9.6	1
Ag–Sn	136 ± 21	1
Ag–Te	195.8 ± 14.6	1
Al–Al	264.3 ± 0.5	1
Al–Ar	5.69	1
Al–As	202.7 ± 7.1	1
Al–Au	325.9 ± 6.3	1
Al–Br	429.2 ± 5.8	1
Al–C	267.7	1
Al–Ca	52.7	1
Al–Cl	502	1
Al–Co	181.6 ± 0.2	1
Al–Cr	222.9 ± 0.9	1
Al–Cu	227.1 ± 1.2	1
Al–D	290.4	1
Al–F	675	1
Al–H	288 ± 13	1
Al–I	369.9 ± 2.1	1
Al–Kr	6.05	1
Al–Li	76.1	1
Al–N	≤368 ± 15	1
Al–Ne	3.9	1
Al–Ni	224.7 ± 4.8	1
Al–O	501.9 ± 10.6	1
Al–P	216.7 ± 12.6	1
Al–Pd	254.4 ± 12.1	1
Al–S	332 ± 10	1
Al–Sb	216.3 ± 6	1
Al–Se	318 ± 13	1
Al–Si	246.9 ± 12.6	1
Al–Te	268 ± 13	1
Al–Ti	263.4	1
Al–U	326 ± 29	1
Al–V	147.4 ± 1.0	1
Al–Xe	7.39	1
Am–O	582 ± 34	18
Am–S	375 ± 33	22
Ar–Ar	4.91	1
Ar–Au	5.50 ± 0.16	19
Ar–B	4.62	1
Ar–Br	~5.0	1
Ar–C	5.158	1
Ar–Ca	4.44 ± 0.60	1
Ar–Cd	5.57 ± 0.05	1
Ar–Ga	3.96	1
Ar–Ge	<5.4	1
Ar–He	3.96	1
Ar–Hg	5.32	1
Ar–I	~5.3	1
Ar–In	4.18	1
Ar–Kr	5.11	1
Ar–Li	~7.82	1
Ar–Mg	~3.7	1
Ar–Na	~4.2	1
Ar–Ne	4.27	1
Ar–Si	5.86	1
Ar–Sn	<5.1	1
Ar–Tl	4.09	1
Ar–Xe	5.28	1
Ar–Zn	5.0	1
As–As	385.8 ± 10.5	1
As–Cl	448	1
As–D	270.3	1
As–F	410	1
As–Ga	202.5 ± 4.8	1
As–H	274.0 ± 2.9	1
As–I	296.6 ± 24	1
As–In	201 ± 10	1
As–N	489 ± 2.1	1
As–O	484 ± 8	1
As–P	433.5 ± 12.6	1
As–S	379.5 ± 6.3	1
As–Sb	330.5 ± 5.4	1
As–Se	96	1
As–Tl	198.3 ± 14.6	1
Au–Au	226.2 ± 0.5	1
Au–B	367.8 ± 10.5	1
Au–Ba	275.2 ± 6.3	25
Au–Be	237.7 ± 4.0	1
Au–Bi	293 ± 8.4	1
Au–Br	213 ± 21	1
Au–Ca	250.4 ± 4.0	1
Au–Ce	322 ± 18	1
Au–Cl	280 ± 13	1
Au–Co	218.0 ± 16.4	1
Au–Cr	223.7 ± 28.9	1
Au–Cs	253 ± 3.5	1
Au–Cu	227.1 ± 1.2	1
Au–D	322.2	1
Au–Dy	259 ± 24	1
Au–Eu	245 ± 12	1
Au–F	294.1	1
Au–Fe	187.0 ± 19.3	1
Au–Ga	290 ± 15	1
Au–Ge	273.2 ± 14.6	1
Au–H	300.5 ± 2.6	4
Au–Ho	267 ± 35	1
Au–I	276	1
Au–In	286.0 ± 5.7	1
Au–K	271.5 ± 19.3	28
Au–Kr	6.59 ± 0.23	19

A–B	D°_{298}/kJ mol^{-1}	Ref.
Au–La	457 ± 28	1
Au–Li	284.5 ± 6.7	1
Au–Lu	332 ± 19	1
Au–Mg	179.1 ± 2.7	1
Au–Mn	197.7 ± 21	1
Au–Na	247.4 ± 7.8	27
Au–Nd	294 ± 29	1
Au–Ni	247 ± 16.4	1
Au–O	223 ± 21	1
Au–Pb	133 ± 42	1
Au–Pd	142.7 ± 21	1
Au–Pr	311 ± 25	1
Au–Rb	243 ± 3.5	1
Au–Rh	232.6 ± 29	1
Au–S	253.6 ± 14.6	1
Au–Sb	241.3 ± 5.8	30
Au–Sc	280 ± 40	1
Au–Se	251.0 ± 14.6	1
Au–Si	304.6 ± 6.0	1
Au–Sn	256.5 ± 7.2	1
Au–Sr	246.4 ± 4.8	25
Au–Tb	285 ± 33	1
Au–Te	237.2 ± 14.6	1
Au–U	318 ± 29	1
Au–V	246.0 ± 8.7	1
Au–Xe	11.33 ± 0.23	19
Au–Y	310 ± 12	1
B–B	290	1
B–Br	390.9 ± 0.5	1
B–C	448 ± 29	1
B–Cd	301.0	1
B–Ce	305 ± 21	1
B–Cl	427	1
B–D	341.0 ± 6.3	1
B–F	732	1
B–H	345.2 ± 2.5	1
B–I	361	1
B–Ir	512.2 ± 17	1
B–La	335 ± 63	1
B–N	377.9 ± 8.7	1
B–Ne	3.97	1
B–O	809	1
B–P	347 ± 16.7	1
B–Pd	351.5 ± 16.7	1
B–Pt	477.8 ± 16.7	1
B–Rh	475.8 ± 21	1
B–Ru	446.9 ± 21	1
B–S	577 ± 9.2	1
B–Sc	272 ± 63	1
B–Se	462 ± 14.6	1
B–Si	317 ± 12	1
B–Te	354 ± 20	1
B–Th	297 ± 33	1
B–Ti	272 ± 63	1
B–U	322 ± 33	1
B–Y	289 ± 63	1
Ba–Br	359.9	23
Ba–Cl	439.3	23
Ba–D	≤193.7	1
Ba–F	580.0	23
Ba–H	192.0	1
Ba–I	321.0	23
Ba–O	562 ± 13.4	1
Ba–Pd	221.8 ± 5.0	1
Ba–Rh	259.4 ± 25	1
Ba–S	418 ± 21	1
Be–Be	59	1
Be–Br	316	1
Be–Cl	384	1
Be–D	203.1	1
Be–F	573	1
Be–H	221	1
Be–I	261	1
Be–O	437	1
Be–S	372 ± 59	1
Be–T	204.4	1
Bi–Bi	204.4	1
Bi–Br	240.2	1
Bi–Cl	300.4 ± 4.2	1
Bi–D	283.7	1
Bi–F	366.5 ± 12.5	1
Bi–Ga	158.6 ± 16.7	1
Bi–H	≤283.3	1
Bi–I	186.1 ± 5.8	1
Bi–In	153.6 ± 1.7	1
Bi–Li	149.4	1
Bi–O	337.2 ± 12.6	1
Bi–P	281.7 ± 13	1
Bi–Pb	142.4 ± 3.0	1
Bi–S	315.5 ± 4.6	1
Bi–Sb	252.7 ± 3.9	1
Bi–Se	280.3 ± 5.9	1
Bi–Sn	193 ± 13	1
Bi–Te	232.2 ± 11.3	1
Bi–Tl	120.9 ± 12.6	1
Bk–O	598	1
Br–Br	193.859 ± 0.120	1
Br–C	318.0 ± 8.4	1
Br–Ca	339	1
Br–Cd	159 ± 96	1
Br–Ce	375.2	23
Br–Cl	219.32 ± 0.05	1
Br–Co	326 ± 42	1
Br–Cr	328.0 ± 24.3	1
Br–Cs	389.1 ± 4.2	1
Br–Cu	331 ± 25	1
Br–D	370.74	1
Br–Dy	315.7	23
Br–Er	363.2	23
Br–Eu	328.8	23
Br–F	280 ± 12	1
Br–Fe	243 ± 84	1
Br–Ga	402 ± 13	1
Br–Gd	374.5	23
Br–Ge	347 ± 8	1
Br–H	366.16 ± 0.20	1
Br–Hg	74.9	1
Br–Ho	323.9	23
Br–I	179.1 ± 0.4	1
Br–In	409 ± 10	1
Br–K	379.1 ± 4.2	1
Br–La	448.6	23
Br–Li	418.8 ± 4.2	1
Br–Lu	303.3	23
Br–Mg	317.96	1
Br–Mn	314.2 ± 9.6	1
Br–Mo	313.4	1
Br–N	280.8 ± 21	1
Br–Na	363.1 ± 4.2	1
Br–Nd	341.8	23
Br–Ni	360 ± 13	1
Br–O	237.6 ± 0.4	1
Br–P	≤329	1
Br–Pb	248.5 ± 14.6	1
Br–Pm	337.6	23
Br–Pr	346.3	23
Br–Rb	380.7 ± 4.2	1
Br–S	218 ± 17	1
Br–Sb	314 ± 59	1
Br–Sc	444 ± 63	1
Br–Se	297 ± 84	1
Br–Si	358.2 ± 8.4	1
Br–Sm	334.3	23
Br–Sn	337 ± 13	1
Br–Sr	365	1
Br–T	372.77	1
Br–Tb	386.4	23
Br–Th	364	1
Br–Ti	373	1
Br–Tl	331 ± 21	1
Br–Tm	300.9	23
Br–U	377 ± 15	1
Br–V	439 ± 42	1
Br–W	329.3	1
Br–Xe	5.94 ± 0.02	1
Br–Y	481 ± 84	1
Br–Yb	297.7	23
Br–Zn	138 ± 29	1
Br–Zr	420	1
C–C	605.03 ± 0.28	33
C–Ce	443 ± 30	1
C–Cl	394.9 ± 13.4	1
C–D	341.4	1
C–F	513.8 ± 10.0	1
C–Fe	367.7 ± 4.2	20
C–Ge	455.7 ± 11	1
C–H	338.72 ± 0.11	33
C–Hf	540 ± 25	1
C–I	253.1 ± 35.6	1
C–Ir	631 ± 5	1
C–La	463 ± 20	1
C–Mo	482 ± 16	1
C–N	749.31 ± 0.14	33
C–Nb	523.8 ± 14.5	1
C–Ni	337.0	1
C–O	1076.63 ± 0.06	33
C–Os	608 ± 25	1
C–P	507.5 ± 8.8	1
C–Pd	436 ± 20	1
C–Pt	577.8 ± 6.8	13
C–Rh	580 ± 4	1
C–Ru	648 ± 13	1
C–S	713.3 ± 1.2	1
C–Sc	444 ± 21	1
C–Se	590.4 ± 5.9	1
C–Si	447	1
C–Tc	564 ± 29	1
C–Th	453 ± 17	1
C–Ti	423 ± 30	1
C–U	455 ± 15	1
C–V	396.42±0.24	36
C–Y	418 ± 14	1
C–Zr	495.8 ± 38.6	1
Ca–Ca	16.52 ± 0.11	1
Ca–Cl	409 ± 8.7	1
Ca–D	≤169.9	1
Ca–F	529	1
Ca–H	223.8	1
Ca–I	284.7 ± 8.4	1
Ca–Kr	5.15 ± 0.72	1
Ca–Li	84.9 ± 8.4	1
Ca–O	383.3 ± 5.0	1
Ca–Pd	347 - 360	1
Ca–S	335 ± 21	1
Ca–Xe	7.31 ± 0.96	1
Cd–Cd	7.36	1
Cd–Cl	208.4	1
Cd–F	305 ± 21	1
Cd–H	69.0 ± 0.4	1
Cd–I	97.2 ± 2.1	1
Cd–In	134	1
Cd–K	7.3	1
Cd–Kr	5.17	1
Cd–Na	10.2	1
Cd–Ne	3.97	1
Cd–O	236 ± 84	1
Cd–S	208.5 ± 20.9	1
Cd–Se	127.6 ± 25.1	1
Cd–Te	100.0 ± 15.1	1
Cd–Xe	6.54	1
Ce–Ce	251.7	1
Ce–Cl	460.0	23
Ce–F	621.6	23
Ce–I	335.5	23
Ce–Ir	575 ± 9	1
Ce–N	519 ± 21	1
Ce–O	790	1
Ce–Os	524 ± 20	1
Ce–Pd	319 ± 21	1
Ce–Pt	550 ± 5	1
Ce–Rh	545 ± 7	1
Ce–Ru	494 ± 12	1
Ce–S	569	1
Ce–Se	494.5 ± 14.6	1
Ce–Te	189.4 ± 12.6	1
Cf–O	498	1
Cl–Cl	242.851 ± 0.096	8
Cl–Co	343.9	11
Cl–Cr	380.3	11
Cl–Cs	445.7 ± 7.7	1
Cl–Cu	377.8 ± 7.5	1
Cl–D	436.303 ± 0.011	1
Cl–Dy	395.1	23
Cl–Er	451.0	23

A–B	$D°_{298}$/kJ mol^{-1}	Ref.
Cl–Eu	408.4	23
Cl–F	260.83	1
Cl–Fe	335.5	11
Cl–Ga	463 ± 13	1
Cl–Gd	453.9	23
Cl–Ge	390.8 ± 9.6	1
Cl–H	431.361 ± 0.013	1
Cl–Hg	92.0 ± 9.2	1
Cl–Ho	411.6	23
Cl–I	211.3 ± 0.4	1
Cl–In	436 ± 8	1
Cl–K	433.0 ± 8.4	1
Cl–La	524.4	23
Cl–Li	469 ± 13	1
Cl–Lu	383.3	23
Cl–Mg	312	1
Cl–Mn	337.6	11
Cl–N	333.9 ± 9.6	1
Cl–Na	412.1 ± 8.4	1
Cl–Nd	421.1	23
Cl–Ni	372.3	11
Cl–O	267.47 ± 0.08	1
Cl–P	≤376	1
Cl–Pb	301 ± 50	1
Cl–Pm	417.0	23
Cl–Pr	425.7	23
Cl–Ra	343 ± 75	1
Cl–Rb	427.6 ± 8.4	1
Cl–S	241.8	1
Cl–Sb	360 ± 50	1
Cl–Sc	331	1
Cl–Se	322	1
Cl–Si	416.7 ± 6.3	1
Cl–Sm	422.1	23
Cl–Sn	350 ± 8	1
Cl–Sr	409	1
Cl–T	438.64	1
Cl–Ta	544	1
Cl–Tb	474.2	23
Cl–Th	489	1
Cl–Ti	405.4 ± 10.5	1
Cl–Tl	372.8 ± 2.1	1
Cl–Tm	380.3	23
Cl–U	439	1
Cl–V	477 ± 63	1
Cl–W	419	1
Cl–Xe	7.08	1
Cl–Y	523 ± 84	1
Cl–Yb	377.1	23
Cl–Zn	229 ± 8	1
Cl–Zr	530	1
Cm–O	709 ± 43	18
Cm–S	504 ± 25	22
Co–Co	<127	1
Co–Cu	161.1 ± 16.4	1
Co–D	270.2 ± 5.8	1
Co–F	431 ± 63	1
Co–Ge	230 ± 21	1
Co–H	244.9 ± 4.8	1
Co–I	280 ± 21	1
Co–Mn	50 ± 8	1
Co–Nb	267.02 ± 0.10	1
Co–O	397.4 ± 8.7	1
Co–S	331	1
Co–Sc	240.1	7
Co–Si	274.4 ± 17	1
Co–Ti	235.37 ± 0.10	1
Co–Y	253.71 ± 0.10	1
Co–Zr	306.39 ± 0.10	1
Cr–Cr	152.0 ± 6	1
Cr–Cu	154.4 ± 14.5	1
Cr–F	523 ± 19	1
Cr–Fe	~75	1
Cr–Ge	154 ± 7	1
Cr–H	189.9 ± 6.7	1
Cr–I	287.0 ± 24.3	1
Cr–N	377.8 ± 18.8	1
Cr–Nb	295.72 ± 0.06	1
Cr–O	461 ± 8.7	1
Cr–Pb	105 ± 2	1
Cr–S	331	1
Cr–Sn	141 ± 3	1
Cr–W	280.32±0.10	35
Cs–Cs	43.919 ± 0.010	1
Cs–F	517.1 ± 7.7	1
Cs–H	175.364	1
Cs–Hg	8	1
Cs–I	338.5 ± 2.1	1
Cs–Li	72.9 ± 1.2	5
Cs–Na	63.2 ± 1.3	1
Cs–O	293 ± 25	1
Cs–Rb	49.57 ± 0.01	1
Cu–Cu	182.0 ± 2.5	32
Cu–D	270.3	1
Cu–Dy	144 ± 18	1
Cu–F	414	1
Cu–Ga	215.9 ± 15	1
Cu–Ge	208.8 ± 21	1
Cu–H	254.8 ± 6	1
Cu–Ho	144 ± 19	1
Cu–I	289 ± 63	1
Cu–In	187.4 ± 7.9	1
Cu–Li	191.9	1
Cu–Na	176.1 ± 16.7	1
Cu–Ni	201.7 ± 9.6	1
Cu–O	293.4 ± 2.9	32
Cu–S	274.5 ± 14.6	1
Cu–Sb	186.7 ± 5.1	30
Cu–Se	255.2 ± 14.6	1
Cu–Si	221.3 ± 6.3	1
Cu–Sn	170 ± 10	1
Cu–Tb	191 ± 18	1
Cu–Te	230.5 ± 14.6	1
D–D	443.3197 ± 0.0003	1
D–F	576.236 ± 0.011	1
D–Ga	<276.5	1
D–Ge	≤322	1
D–H	439.2223 ± 0.0002	1
D–Hg	42.05	1
D–I	302.33	1
D–In	246	1
D–K	182.4	1
D–Li	240.24	1
D–Lu	302	1
D–Mg	161.33 ± 0.32	1
D–Mn	312 ± 6	1
D–N	341.6	1
D–Ni	≤302.9	1
D–O	429.64	1
D–P	299.0	1
D–Pt	≤350.2	1
D–S	350.62 ± 1.20	1
D–Si	302.5	1
D–Sr	167.7	1
D–T	444.91	1
D–Tl	193.0	1
D–Zn	88.7	1
Dy–Dy	70.3	1
Dy–F	531.1	23
Dy–I	277.2	23
Dy–O	615	1
Dy–S	414 ± 42	1
Dy–Se	322 ± 20	1
Dy–Te	234 ± 20	1
Er–Er	75 ± 29	1
Er–F	572.6	23
Er–I	317.6	23
Er–O	606	1
Er–S	418 ± 21	1
Er–Se	326 ± 20	1
Er–Te	238 ± 20	1
Es–O	460	1
Eu–Eu	45.2	1
Eu–F	543.0	23
Eu–I	290.4	23
Eu–Li	268.1 ± 12.6	1
Eu–O	473	1
Eu–Rh	238 ± 34	1
Eu–S	365.7 ± 13.4	1
Eu–Se	302.9 ± 14.6	1
Eu–Te	251.0 ± 14.6	1
F–F	158.670 ± 0.096	1
F–Fe	447	1
F–Ga	584 ± 13	1
F–Gd	594.6	23
F–Ge	523 ± 13	1
F–H	569.680 ± 0.011	1
F–Hf	650 ± 15	1
F–Hg	~180	1
F–Ho	517.7	23
F–I	≤271.5	1
F–In	516 ± 13	1
F–K	489.2	1
F–Kr	6.6	1
F–La	665.1	23
F–Li	577 ± 21	1
F–Lu	523.4	23
F–Mg	445.6	1
F–Mn	445.2 ± 7.5	1
F–Mo	464	1
F–N	≤349	1
F–Na	477.3	1
F–Nd	548.7	23
F–Ni	439.7 ± 5.9	2
F–Np	430 ± 50	1
F–O	220	1
F–P	≤405	1
F–Pb	355 ± 13	1
F–Pm	561.3	23
F–Pr	582.0	23
F–Pu	538 ± 29	1
F–Rb	494 ± 21	1
F–Ru	402	1
F–S	343.5 ± 6.7	1
F–Sb	439 ± 96	1
F–Sc	599.1 ± 13.4	1
F–Se	339 ± 42	1
F–Si	576.4 ± 17	1
F–Sm	565.2	23
F–Sn	476 ± 8	1
F–Sr	538	1
F–T	579.009 ± 0.108	1
F–Ta	573 ± 13	1
F–Tb	647.3	23
F–Th	652	1
F–Ti	569 ± 33	1
F–Tl	439 ± 21	1
F–Tm	509.1	23
F–U	648	1
F–V	590 ± 63	1
F–W	≤544	1
F–Xe	14.18	1
F–Y	685.3 ± 13.4	1
F–Yb	525.1	23
F–Zn	364 ± 63	1
F–Zr	627.2 ± 10.5	1
Fe–Fe	118	1
Fe–Ge	210.9 ± 29	1
Fe–H	148 ± 3	1
Fe–I	123	1
Fe–O	407.0 ± 1.0	1
Fe–S	328.9 ± 14.6	1
Fe–Si	297 ± 25	1
Fm–O	443	1
Ga–Ga	<106.4	1
Ga–H	265.9 ± 5.9	4
Ga–I	334 ± 13	1
Ga–In	94.0 ± 3	1
Ga–Kr	4.08	1
Ga–Li	133.1 ± 14.6	1
Ga–O	374 ± 21	1
Ga–P	229.7 ± 12.6	1
Ga–Sb	192.0 ± 12.6	1
Ga–Te	265 ± 21	1
Ga–Xe	5.27	1
Gd–Gd	206.3 ± 67.5	1
Gd–I	336.0	23
Gd–O	715	1
Gd–S	526.8 ± 10.5	1
Gd–Se	430 ± 15	1
Gd–Te	341 ± 15	1
Ge–Ge	264.4 ± 6.8	1
Ge–H	263.2 ± 4.8	1
Ge–I	268 ± 25	1

A–B	D°_{298}/kJ mol^{-1}	Ref.
Ge–Ni	290.3 ± 10.9	1
Ge–O	657.5 ± 4.6	4
Ge–Pb	142.2 ± 6.8	31
Ge–Pd	254.7 ± 10.5	1
Ge–S	534 ± 3	1
Ge–Sc	270 ± 11	1
Ge–Se	484.7 ± 1.7	1
Ge–Si	297	1
Ge–Sn	230.1 ± 13	1
Ge–Te	396.7 ± 3.3	1
Ge–Y	279 ± 11	1
H–H	435.7799 ± 0.0001	1
H–Hg	39.844	1
H–I	298.26 ± 0.10	1
H–In	243.1	1
H–K	174.576	1
H–Li	238.039 ± 0.006	1
H–Mg	127.18 ± 0.006	10
H–Mn	251 ± 5	1
H–Mo	202.5 ± 18.3	9
H–N	358.8 ± 0.2	26
H–Na	192.71 ± 0.01	34
H–Nb	>221.9 ± 9.6	1
H–Ni	240 ± 8	1
H–O	429.74 ± 0.03	33
H–P	297.0 ± 2.1	1
H–Pb	≤157	1
H–Pd	234 ± 25	1
H–Pt	330	1
H–Rb	172.6	1
H–Rh	241.0 ± 5.9	1
H–Ru	223 ± 15	1
H–S	353.57 ± 0.30	1
H–Sb	239.7 ± 4.2	1
H–Sc	205 ± 17	1
H–Se	312.5	1
H–Si	293.3 ± 1.9	1
H–Sn	264 ± 17	1
H–Sr	164 ± 8	1
H–T	440.49	1
H–Te	270.7 ± 1.7	1
H–Ti	204.6 ± 8.8	1
H–Tl	195.4 ± 4	1
H–V	209.3 ± 6.8	1
H–Yb	183.1 ± 2.0	1
H–Zn	85.8 ± 2	1
He–He	3.809	1
He–Hg	3.8	1
He–Xe	3.8	1
Hf–Hf	328 ± 58	1
Hf–N	535 ± 30	1
Hf–O	801 ± 13	1
Hg–Hg	8.10 ± 0.18	1
Hg–I	34.69 ± 0.96	1
Hg–K	8.8	1
Hg–Kr	5.75	1
Hg–Li	13.16 ± 0.38	1
Hg–Na	10.8	1
Hg–Ne	4.14	1
Hg–O	269	1
Hg–Rb	8.4	1
Hg–S	217.3 ± 22.2	1
Hg–Se	144.3 ± 30.1	1
Hg–T	43.14	1
Hg–Te	<142	1
Hg–Tl	2.9	1
Hg–Xe	6.65	1
Hg–Zn	7.3	1
Ho–Ho	70.3	1
Ho–I	277.0	23
Ho–O	606	1
Ho–S	428.4 ± 14.6	1
Ho–Se	333 ± 15	1
Ho–Te	≤259 ± 15	1
I–I	152.25 ± 0.57	1
I–In	306.9 ± 1.1	1
I–K	322.5 ± 2.1	1
I–Kr	5.67	1
I–La	414.8	23
I–Li	345.2 ± 4.2	1
I–Lu	264.8	23
I–Mg	229	1
I–Mn	282.8 ± 9.6	1
I–Mo	266.9	1
I–N	159 ± 17	1
I–Na	304.2 ± 2.1	1
I–Nd	303.3	23
I–Ni	293 ± 21	1
I–O	233.4 ± 1.3	12
I–Pb	194 ± 38	1
I–Pm	299.1	23
I–Pr	307.8	23
I–Rb	318.8 ± 2.1	1
I–Si	243.1 ± 8.4	1
I–Sm	295.8	23
I–Sn	235 ± 3	1
I–Sr	301	1
I–Tb	339.6	23
I–Te	192 ± 42	1
I–Th	361 ± 25	1
I–Ti	306	1
I–Tl	285 ± 21	1
I–Tm	262.4	23
I–U	299 ± 27	1
I–Xe	~6.9	1
I–Y	422.6 ± 12.5	1
I–Yb	259.3	23
I–Zn	153.1 ± 6.3	1
I–Zr	127	1
In–In	82.0 ± 5.7	1
In–Kr	4.85	1
In–Li	92.5 ± 14.6	1
In–O	346 ± 30	1
In–P	197.9 ± 8.4	1
In–S	287.9 ± 14.6	1
In–Sb	151.9 ± 10.5	1
In–Se	245.2 ± 14.6	1
In–Te	215.5 ± 14.6	1
In–Xe	6.48	1
In–Zn	32.2	1
Ir–Ir	361 ± 68	1
Ir–La	577 ± 12	1
Ir–Nb	465 ± 25	1
Ir–O	410.1 ± 42.4	29
Ir–Si	462.8 ± 21	1
Ir–Th	574 ± 42	1
Ir–Ti	422 ± 13	1
Ir–Y	457 ± 15	1
K–K	56.96	1
K–Kr	4.6	1
K–Li	82.0 ± 4.2	1
K–Na	65.994 ± 0.008	1
K–Zn	6.5	1
K–O	271.5 ± 12.6	1
K–Rb	53.723 ± 0.005	1
K–Xe	5.0	1
Kr–Kr	5.39	1
Kr–Li	~12.1	1
Kr–Mg	6.71 ± 0.96	1
Kr–Na	~4.53	1
Kr–Ne	4.31	1
Kr–O	<8	1
Kr–Tl	4.14	1
Kr–Xe	5.66	1
Kr–Zn	5.0	1
La–La	244.9	1
La–N	519 ± 42	1
La–O	798	1
La–Pt	505 ± 12	1
La–Rh	550 ± 12	1
La–S	573.4 ± 1.7	1
La–Se	485.7 ± 14.6	1
La–Te	385.6 ± 15	1
La–Y	197 ± 21	1
Li–Li	105.0	1
Li–Mg	67.4 ± 6.3	1
Li–Na	87.181 ± 0.001	1
Li–O	340.5 ± 6.3	1
Li–Pb	78.7 ± 8	1
Li–S	312.5 ± 7.5	1
Li–Sb	169.0 ± 10.0	1
Li–Si	149	1
Li–Sm	193.3 ± 18.8	1
Li–Tm	276.1 ± 14.6	1
Li–Xe	~12.1	1
Li–Yb	143.5 ± 12.6	1
Lr–O	665	1
Lu–Lu	142 ± 33	1
Lu–O	669	1
Lu–Pt	402 ± 34	1
Lu–S	508.4 ± 14.4	1
Lu–Se	418 ± 15	1
Lu–Te	325 ± 15	1
Md–O	418	1
Mg–Mg	11.3	1
Mg–Ne	~4.1	1
Mg–O	358.2 ± 7.2	1
Mg–S	234	1
Mg–Xe	9.70 ± 1.79	1
Mn–Mn	61.6 ± 9.6	1
Mn–O	362 ± 25	1
Mn–S	301 ± 17	1
Mn–Se	239.3 ± 9.2	1
Mo–Mo	435.5 ± 1.0	1
Mo–Nb	452 ± 25	1
Mo–O	502	1
N–N	944.87 ± 0.05	33
N–O	630.57 ± 0.06	33
N–P	617.1 ± 20.9	1
N–Pt	374.2 ± 9.6	1
N–Pu	469 ± 63	1
N–S	467 ± 24	1
N–Sb	460 ± 84	1
N–Sc	464 ± 84	1
N–Si	437.1 ± 9.9	1
N–Ta	607 ± 84	1
N–Th	577 ± 33	1
N–Ti	476 ± 33	1
N–U	531 ± 21	1
N–V	485.97 ± 36	36
N–Xe	26.9	1
N–Y	477 ± 63	1
N–Zr	565 ± 25	1
Na–Na	74.805 ± 0.586	1
Na–Ne	~3.8	1
Na–O	270 ± 4	1
Na–Rb	63.887 ± 0.024	1
Na–Xe	~5.12	1
Nb–Nb	513	1
Nb–Ni	271.9 ± 0.1	1
Nb–O	726.5 ± 10.6	1
Nb–Ti	302.0 ± 0.1	1
Nb–V	369.3 ± 0.1	1
Nd–Nd	82.8	1
Nd–O	703	1
Nd–S	471.5 ± 14.6	1
Nd–Se	393.9	1
Nd–Te	305 ± 15	1
Ne–Ne	4.070	1
Ne–Xe	4.31	1
Ne–Zn	3.92	1
Ni–Ni	204	1
Ni–O	366 ± 30	1
Ni–Pd	140.9	1
Ni–Pt	273.7 ± 0.3	1
Ni–S	356 ± 21	1
Ni–Si	318 ± 17	1
Ni–V	206.3 ± 0.2	1
Ni–Y	283.92 ± 0.10	1
Ni–Zr	279.8 ± 0.1	1
No–O	268	1
Np–O	744 ± 21	18
Np–S	495 ± 55	22
O–O	498.458 ± 0.004	33
O–Os	575	1
O–P	589	1
O–Pa	801 ± 59	18
O–Pb	382.4 ± 3.3	4
O–Pd	238.1 ± 12.6	1
O–Pr	740	1
O–Pt	418.6 ± 11.6	13
O–Pu	656.1	1
O–Rb	276 ± 12.6	1

A–B	D^o_{298}/kJ mol^{-1}	Ref.
O–Re	627 ± 84	1
O–Rh	405 ± 42	1
O–Ru	528 ± 42	1
O–S	517.90 ± 0.05	1
O–Sb	434 ± 42	1
O–Sc	671.4 ± 1.0	1
O–Se	429.7 ± 6.3	1
O–Si	799.6 ± 13.4	1
O–Sm	573	1
O–Sn	528	1
O–Sr	426.3 ± 6.3	1
O–Ta	839	1
O–Tb	694	1
O–Tc	548	1
O–Te	377 ± 21	1
O–Th	871 ± 25	18
O–Ti	666.5 ± 5.6	1
O–Tl	213 ± 84	1
O–Tm	514	1
O–U	758 ± 13	18
O–V	629.7 ± 13.0	24
O–W	720 ± 71	1
O–Xe	36.4	1
O–Y	714.1 ± 10.2	1
O–Yb	387.7 ± 10	1
O–Zn	≤250	1
O–Zr	766.1 ± 10.6	1
Os–Os	415 ± 77	1
P–P	489.1	1
P–Pt	≤416.7 ± 16.7	1
P–Rh	353.1 ± 16.7	1
P–S	442 ± 10	1
P–Sb	356.9 ± 4.2	1
P–Se	363.7 ± 10.0	1
P–Si	363.6	1
P–Te	297.9 ± 10.0	1
P–Th	372 ± 29	1
P–Tl	209 ± 13	1
P–U	293 ± 21	1
P–W	305 ± 4	1
Pa–S	545 ± 91	22
Pb–Pb	86.6 ± 0.8	1
Pb–S	398	1
Pb–Sb	161.5 ± 10.5	1
Pb–Se	302.9 ± 4.2	1
Pb–Si	166.6 ± 10.4	31
Pb–Sn	124.8 ± 2.8	31
Pb–Te	249.8 ± 10.5	1
Pd–Pd	>136	1
Pd–Pt	191.0	1
Pd–Si	261 ± 12	1
Pd–Y	241 ± 15	1
Po–Po	187	1
Pr–Pr	129.1	1
Pr–S	492.5 ± 4.6	1
Pr–Se	446.4 ± 23.0	1
Pr–Te	326 ± 20	1
Pt–Pt	306.7 ± 1.9	1
Pt–Si	501 ± 18	1
Pt–Th	551 ± 42	1
Pt–Ti	397.5 ± 10.6	1
Pt–Y	474 ± 12	1
Pu–S	446 ± 30	22
Rb–Rb	48.898 ± 0.005	1
Re–Re	432 ± 30	1
Rh–Rh	235.85 ± 0.05	1
Rh–Sc	444 ± 11	1
Rh–Si	395.0 ± 18.0	1
Rh–Th	513 ± 21	1
Rh–Ti	390.8 ± 14.6	1
Rh–U	519 ± 17	1
Rh–V	364 ± 29	1
Rh–Y	446 ± 11	1
Ru–Ru	193.0 ± 19.3	1
Ru–Si	397.1 ± 21	1
Ru–Th	592 ± 42	1
Ru–V	414 ± 29	1
S–S	430.03 ± 0.03	17
S–Sb	378.7	1
S–Sc	478.2 ± 12.6	1
S–Se	371.1 ± 6.7	1
S–Si	617 ± 5	1
S–Sm	389	1
S–Sn	467	1
S–Sr	338.5 ± 16.7	1
S–Ta	669.5 ± 13.5	1
S–Tb	515 ± 42	1
S–Te	335 ± 42	1
S–Th	608 ± 77	22
S–Ti	418 ± 3	1
S–Tm	368 ± 21	1
S–U	510.4 ± 63	22
S–V	441.32 ± 0.24	36
S–Y	528.4 ± 10.5	1
S–Yb	167	1
S–Zn	224.8 ± 12.6	1
S–Zr	572.0 ± 11.6	1
Sb–Sb	301.7 ± 6.3	1
Sb–Te	277.4 ± 3.8	1
Sb–Tl	126.7 ± 10.5	1
Sc–Sc	163 ± 21	1
Sc–Se	385 ± 17	1
Sc–Si	227.2 ± 14	1
Sc–Te	289 ± 17	1
Se–Se	330.5	1
Se–Si	538 ± 13	1
Se–Sm	331.0 ± 14.6	1
Se–Sn	401.2 ± 5.9	1
Se–Sr	251.0 ± 12.6	1
Se–Tb	423 ± 20	1
Se–Te	293.3	1
Se–Ti	381 ± 42	1
Se–Tm	274 ± 40	1
Se–V	347 ± 21	1
Se–Y	435 ± 13	1
Se–Zn	170.7 ± 25.9	1
Si–Si	310	1
Si–Sn	238.4 ± 3.6	31
Si–Te	429.2	3
Si–Y	258 ± 17	1
Sm–Sm	54 ± 21	1
Sm–Te	272.4 ± 14.6	1
Sn–Sn	187.1 ± 0.3	1
Sn–Te	338.1 ± 6.3	1
Sr–Sr	16.64 ± 1.12	1
T–T	446.67	1
Ta–Ta	390 ± 96	1
Tb–Tb	138.8	1
Tb–Te	339 ± 42	1
Tc–Tc	330	1
Te–Te	257.6 ± 4.1	1
Te–Ti	289 ± 17	1
Te–Tm	182 ± 40	1
Te–Y	339 ± 13	1
Te–Zn	117.6 ± 18.0	1
Th–Th	≤289 ± 33	1
Ti–Ti	117.6	1
Ti–V	203.2 ± 0.1	1
Ti–Zr	214.3 ± 0.1	1
Tl–Tl	59.4	1
Tl–Xe	4.18	1
Tm–Tm	54 ± 17	1
U–U	222 ± 21	1
V–V	269.3 ± 0.1	1
V–Zr	260.6 ± 0.3	1
W–W	666	1
Xe–Xe	6.023	1
Y–Y	~270 ± 39	1
Yb–Yb	16.3	1
Zn–Zn	22.2 ± 6.3	1
Zr–Zr	298.2 ± 0.1	1

References

1. Luo, Y. R., *Comprehensive Handbook of Chemical Bond Energies*, CRC Press, Boca Raton, FL, 2007.
2. Hildenbrand, D. L., and Lau, K.H., *J. Phys. Chem. A* 110, 11886, 2006.
3. Chattopadhyaya, S., Pramanik, A., Banerjee, A., and Das, K. K., *J. Phys. Chem. A* 110, 12303, 2006.
4. Brutti, S., Balducci, G., and Gigli, G., *Rapid Commun. Mass Spectrom.* 21, 89, 2007.
5. Staanum, P., Pashov, A., Knöckel, H., and Tiemann, E., *Phys. Rev. A* 75, 042513, 2007.
6. Ciccioli, A., Gigli, G., Meloni, G., and Testani, E., *J. Chem. Phys.* 127, 054303/1, 2007.
7. Nagarajan, R., and Morse, M. D., *J. Chem. Phys.* 127, 074304/1, 2007.
8. Li, J., Hao, Y., Yang, J., Zhou, C., and Mo, Y., *J. Chem. Phys.* 127, 104307/1, 2007.
9. Armentrout, P. B., *Organometallics* 26, 5473, 2007.
10. Shayesteh, A., Henderson, R. D. E., Le Roy, R. J., and Bernath, P. F., *J. Phys. Chem. A* 111, 12495, 2007.
11. Hildenbrand, D. H., *J. Phys. Chem. A* 112, 3813, 2008.
12. Dooley, K. S., Geidosch, J. N., and North, S. W., *Chem. Phys. Lett.* 457, 303, 2008.
13. Citir, M., Metz, R. B., Belau, L., and Ahmed, M., *J. Phys. Chem. A* 112, 9584, 2008.
14. Hildenbrand, D. L., Lau, K. H., Perez-Mariano, J., and Sanjurjo, A., *J. Phys. Chem. A* 112, 9978, 2008.
15. Gibson, J. K., Haire, R. G., Santos, M., Pires de Matos, A., and Marçalo, J., *J. Phys. Chem. A* 112, 11373, 2008.
16. Ciccioli, A., Gigli, G., and Meloni, G., *Chem. Eur. J.* 15, 9543, 2009.
17. Frederix, P. W. J. M., Yang, C.-H., Groenenboom, G. C., Parker, D. H., Alnama, K., Western, C. M., and Orr-Ewing, A. J., *J. Phys. Chem. A* 113, 14995, 2009.
18. Marçalo, J., and Gibson, J. K., *J. Phys. Chem. A* 113, 12599, 2009.
19. Hopkins, W.S., Woodham, A. P., Plowright, R. J., Wright, T. G., and Mackenzie, S. R., *J. Chem. Phys.*, 132, 21403, 2010.
20. Tzeli, D., and Mavridis, A., *J. Chem. Phys.*, 132, 194312, 2010.
21. Huang, H.-Y., Lu, T.-L., Whang, T.-J., Chang, Y.-Y., and Tsai, C.-C., *J. Chem. Phys.* 133, 044301, 2010.
22. Pereira, C. C. L., Marsden, C. J., Marçalo, J., and Gibson, J. K., *Phys. Chem. Chem. Phys.* 13, 12940, 2011.
23. Mucklejohn, S. A., *J. Phys. D: Appl. Phys.* 44, 224010, 2011.
24. Guina, A. O., Lopatin, S. I., and Shugurov, S. M., *Inorg. Chem.*, 51, 4918–4924, 2012.
25. Ciccioli, A., Gigli, G., and Lauricella, M., *J. Chem. Phys.*, 136, 184306, 2012.
26. Ruscic, B., Active Thermochemical Table, http://atct.anl.gov/inde.html, Dec. 16, 2010.
27. Ciccioli, A., and Gigli, G., *J. Phys. Chem.* A 117, 4956, 2013.

28. Stangassinger. A., Knight, A. M., and Duncan, M. A., *J. Phys. Chem.* A 103, 1547, 1999.
29. Armentrout, P. B., and Li, F.-X., *J. Phys. Chem.* A 117, 7754, 2013.
30. Carta, V., Ciccioli, A., and Gigli, G., *J. Chem. Phys.* 140, 064305, 2014.
31. Ciccioli, A., and Gigli, G., *ECS Trans.* 46, 99, 2013.
32. Parry, I. S., Hermes, A. C., Kartouzian, A., and Mackenzie, S. R., *Phys. Chem. Chem. Phys.* 16, 458, 2014.
33. Ruscic, B., Feller, D., and Peterson, K. A., *Theor. Chem. Acc.*. 133, 1415, 2014.
34. Walji, S.-D., Sentjens, K. M., and Le Roy, R. J., *J. Chem. Phys.* 142, 044305, 2015.
35. Matthew, D.J., et al., *J. Chem. Phys.*, 144, 214306, 2016.
36. Johnson, E. L. et al., *J. Chem. Phys.* 144, 234306, 2016.

Enthalpy of Formation of Gaseous Atoms

Table 2. Enthalpy of Formation of Gaseous Atoms

Atom	$\Delta_f H°_{298}$/kJ mol^{-1}	Ref.	Atom	$\Delta_f H°_{298}$/kJ mol^{-1}	Ref.	Atom	$\Delta_f H°_{298}$/kJ mol^{-1}	Ref.	Atom	$\Delta_f H°_{298}$/kJ mol^{-1}	Ref.
Ac	406	5	Dy	290.4 ± 2.1	4	N	472.44 ± 0.03	9	Se	227.2 ± 4	1
Ag	284.9 ± 0.8	2	Er	316.4 ± 2.1	4	Na	107.5 ± 0.7	3	Si	450.0 ± 8	2
Al	330.9 ± 4.0	2	Es	133	6	Nb	733.0 ± 8	3	Sm	206.7 ± 2.1	4
Am	284	6	Eu	177.4 ± 2.1	4	Nd	326.9 ± 2.1	4	Sn	301.2 ± 1.5	2
As	302.5 ± 13	1	F	79.335 ± 0.068	8	Ni	430.1 ± 8.4	3	Sr	164.0 ± 1.7	3
Au	368.2 ± 2.1	1	Fe	415.5 ± 1.3	3	Np	464.8	6	Ta	782.0 ± 2.5	1
B	565 ± 5	2	Ga	271.96 ± 2.1	3	O	249.229 ± 0.002	7	Tb	388.7 ± 2.1	4
Ba	179.1 ± 5.0	3	Gd	397.5 ± 2.1	4	Os	787 ± 6.3	1	Tc	678	5
Be	324 ± 5	2	Ge	372 ± 3	2	P	316.5 ± 1.0	2	Te	196.6 ± 2.1	1
Bi	209.6 ± 2.1	1	H	217.998 ± 0.006	2	Pa	563	5	Th	602 ± 6	2
Bk	310	6	Hf	618.4 ± 6.3	3	Pb	195.2 ± 0.8	2	Ti	473 ± 3	2
Br	111.85 ± 0.06	9	Hg	61.38 ± 0.04	2	Pd	376.6 ± 2.1	1	Tl	182.2 ± 0.4	1
C	716.87 ± 0.06	9	Ho	300.6 ± 2.1	4	Pr	356.9 ± 2.1	4	Tm	232.2 ± 2.1	4
Ca	177.8 ± 0.8	2	I	106.757 ± 0.002	9	Pt	565.7 ± 1.3	1	U	533 ± 8	2
Cd	111.80 ± 0.20	2	In	243 ± 4	1	Pu	345	6	V	515.5 ± 8	3
Ce	420.1 ± 2.1	4	Ir	669 ± 4	1	Ra	159	5	W	851.0 ± 6.3	3
Cf	196	6	K	89.0 ± 0.8	2	Rb	80.9 ± 0.8	2	Y	424.7 ± 2.1	4
Cl	121.302 ± 0.002	9	La	431.0 ± 2.1	4	Re	774 ± 6.3	1	Yb	155.6 ± 2.1	4
Cm	386	6	Li	159.3 ± 1.0	2	Rh	556 ± 4	1	Zn	130.40 ± 0.40	2
Co	426.7	3	Lu	427.6 ± 2.1	4	Ru	650.6 ± 6.3	1	Zr	610.0 ± 8.4	3
Cr	397.48 ± 4.2	3	Mg	147.1 ± 0.8	2	S	277.17 ± 0.15	2			
Cs	76.5 ± 1.0	2	Mn	283.3 ± 4.2	3	Sb	264.4 ± 2.5	1			
Cu	337.4 ± 1.2	2	Mo	658.98 ± 3.8	3	Sc	377.8 ± 4	1			

References

1. Brewer, L., and Rosenblatt, G. M., *Adv. High Temp. Chem.* 2, 1, 1969.
2. Cox, J. D., Wagman, D. D., and Medvedev, V. A., Eds., *CODATA Key Values for Thermodynamics*, Hemisphere Publishing Corporation, New York, 1989; updated e-version: http://www.codata/org/codata.
3. NIST Chemistry WebBook, http://webbook.nist.gov, *NIST-JANAF Thermochemical Table*, 4th Edn., Chase, Jr., M. W., Ed., ACS, AIP, New York, 1998.
4. Chandrasekharaiah, M.S., and Gingerich, K.A., Thermodynamic properties of gaseous species, in *Handbook on the Chemistry and Physics of Rare Earths*, Gschneidner, Jr., K.A., and Ering, L., Eds., Elsevier, Amsterdam, 1989, Vol. 12, Chap. 86, pp. 409–431.
5. Lias, S.G., Bartmess, J. E., Liebman, J. F., Holmes, J. L., Levin, R. D., and Mallard, W. G., *J. Phys. Chem. Ref. Data* 17, Suppl. 1, 1988.
6. Kleinschmidt, P.D., Ward, J. W., Matlack, G. M., and Haire, R. G., *High Temp. Sci.* 19, 267, 1985.
7. Ruscic, B., Pinzon, R.E., Morton, M. E., Srinivasan, N. K., Su, M.-C., Sutherland, J. W., and Michael, J. V., *J. Phys. Chem. A* 110, 6592, 2006.
8. Luo, Y. R., University Chem., 27, 80, 2012.
9. Ruscic, B., Active Thermochemical Table, http://atct.anl.gov/index.html, Dec. 16, 2010.

Bond Dissociation Energies in Polyatomic Molecules

The $D°_{298}$ values in polyatomic molecules are notoriously difficult to measure accurately since the mechanism of the kinetic systems involved in many of the measurements are seldom straightforward. Thus, much lively controversy has taken place in the literature and is likely to continue for some time to come. We will continue updating and presenting our assessment of the most reliable BDE data every year.

The references relating to each of the $D°_{298}$ values listed in Table 3 are contained in the *Comprehensive Handbook of Chemical Bond Energies*, by Yu-Ran Luo, CRC Press, 2007. Many $D°_{298}$ in Table 3 are derived from the equation

$$D°_{298}(\text{R-X}) = \Delta_f H°(\text{R}) + \Delta_f H°(\text{X}) - \Delta_f H°(\text{RX})$$

Here, the enthalpies of formation of the atoms and radicals are taken from Tables 2 and 4, respectively, and the enthalpies of formation of the molecules are from reference sources listed in the above *Comprehensive Handbook of Chemical Bond Energies*.

Table 3 presents **H**-C, **C**-C, C-**halogen**, **O**-, **N**-, **S**-, **Si**-, **Ge**-, **Sn**-, **Pb**-, **P**-, **As**-, **Sb**-, **Bi**-, **Se**-, **Te**-, and **metal**-X BDEs. The **boldface** in the species indicates the dissociated fragment. The **metal**-X BDEs are arranged on the basis of the Periodic Table with the new IUPAC notation for Groups 1 to 18, see inside front cover of this *Handbook*.

Table 3. Bond Dissociation Energies in Polyatomic Molecules

Bond	D°_{298}/kJ mol^{-1}	Ref.
(1) C–H BDEs		
CH_3–H	439.3 ± 0.4	1
CH_3CH_2–H	420.5 ± 1.3	1
$CH_3CH_2CH_2$–H	422.2 ± 2.1	1
$CH_3CH_2CH_3$	410.5 ± 2.9	1
$CH_3CH_2CH_2CH_2$–H	421.3	1
$CH_3CH_2CH_2CH_3$	411.1 ± 2.2	1
$(CH_3)_2CHCH_2$–H	419.2 ± 4.2	1
$(CH_3)_3C$–H	400.4 ± 2.9	1
$(CH_3)_3CCH_2$–H	419.7 ± 4.2	1
$(CH_3CH_2)CH(CH_3)_2$	400.8	1
$CH_3CH_2(CH_2)_2CH_3$	415.1	1
$(C_3H_7)CH(CH_3)_2$	396.2 ± 8.4	1
$CH_3CH(CH_3)CH(CH_3)_2$	399.2 ± 13.0	1
$CH_3CH_2(CH_2)_3CH_3$	410	1
$CH_3CH_2(CH_2)_4CH_3$	410	1
HCC–H	557.81 ± 0.30	1
HCCCC–H	539 ± 12	1
$CHCCH_2$–H	384.1 ± 4.2	1
CH_3CCCH_2–H	379.5	1
$HCCCH_2CH_3$	373.0	1
CH_2=$CHCCCH_2$–H	363.3	1
$CH_3CCCH_2CH_3$	365.3 ± 9.6	1
$HCCCH_2CH_2CH_3$	349.8 ± 8.4	1
$HCCCH(CH_3)_2$	345.2 ± 8.4	1
$CH_3CCCH(CH_3)_2$	344.3 ± 11.3	1
HCCCCCC–H	~543 ± 13	1
H_2C=CH–H	464.2 ± 2.5	1
CH_2=C=CH–H	371.1 ± 12.6	1
CH_3CH=CH–H	464.8	1
CH_2=$CHCH_2$–H	369 ± 3	1
CH_2=CH-CH_2CH_2–H	410.5	1
CH_2=$CHCH_2CH_3$	350.6	1
CH_2=$C(CH_3)CH_2$–H	372.8	1
CH_2=CHCH=$CHCH_2$–H	347.3 ± 12.6	1
$(CH_2$=$CH)_2CH$–H	320.5 ± 4.2	1
CH_2=$CHCH_2CH_2CH_3$	348.8	1
CH_2=$CHCH(CH_3)_2$	332.6 ± 7.1	1
CH_2=$C(CH_3CH_2)CH_2$–H	356.1 ± 8.4	1
$(CH_2$=$CH)_2C(CH_3)$–H	322.2	1
H-*cyclo*-C_3H_5	444.8 ± 1.0	1
H-CH_2-*cyclo*-C_3H_5	407.5 ± 6.7	1
H-*cyclo*-C_4H_7	409.2 ± 1.3	1
H-*cyclo*-C_5H_9	400.0 ± 4.2	1
H-*cyclo*-C_6H_{11}	416.3	1
H-C_6H_5	465.9 ± 0.6	12
H-$CH_2C_6H_5$	375.5 ± 5.0	1
H-$CH(CH_3)C_6H_5$	357.3 ± 6.3	1
H-$CH(C_6H_5)_2$	353.5 ± 2.1	1
H-$CH(C_6H_4$-*p*-$OH)_2$	375.8 ± 4.7	1
H-$C(CH_3)_2C_6H_5$	348.1 ± 4.2	1
H-$C(C_6H_5)_3$	338.9 ± 8.4	1
1-H-$C_{10}H_7$	469.4 ± 5.4	1
2-H-$C_{10}H_7$	468.2 ± 5.9	1
H-CF_3	445.2 ± 2.9	1
H-CHF_2	431.8 ± 4.2	1
H-CH_2F	423.8 ± 4.2	1
H-$CClF_2$	421.3 ± 8.4	1
H-CCl_2F	410.9 ± 8.4	1
H-$CBrF_2$	415.5 ± 12.6	1
H-CHClF	421.7 ± 10.0	1
H-CCl_3	392.5 ± 2.5	1
H-$CHCl_2$	400.6 ± 2.0	1
H-CH_2Cl	419.0 ± 2.3	1
H-CFClBr	413 ± 21	1
H-CHClBr	406.0 ± 2.4	1
H-CCl_2Br	387 ± 21	1
H-$CClBr_2$	371 ± 21	1
H-CBr_3	399.2 ± 8.4	1
H-$CHBr_2$	412.6 ± 2.7	3
H-CH_2Br	427.2 ± 2.4	1
H-CI_3	423 ± 29	1
H-CHI_2	431.0 ± 8.4	1
H-CH_2I	431.6 ± 2.8	1
CF_3CF_2–H	429.7 ± 2.1	1
CHF_2CF_2–H	431.0 ± 18.8	1
CH_2FCF_2–H	433.0 ± 14.6	1
CHF_2CFH–H	426.8 ± 14.6	1
CF_3CH_2–H	446.4 ± 4.5	1
CH_3CF_2–H	416.3 ± 4.2	1
CH_2FCHF–H	413.4 ± 12.6	1
CHF_2CH_2–H	433.0 ± 14.6	1
CH_2FCH_2–H	433.5 ± 8.4	1
CH_3CHF–H	410.9 ± 8.4	1
CF_3CHCl–H	425.9 ± 6.3	1
CF_3CClBr–H	404.2 ± 6.3	1
$CClF_2CHF$–H	412.1 ± 2.1	1
CCl_3CCl_2–H	397.5 ± 8.4	1
$CHCl_2CCl_2$–H	393.3 ± 8.4	1
CH_3CCl_2–H	397.9 ± 5.0	1
CH_3CHCl–H	406.6 ± 1.5	1
CH_2ClCH_2–H	423.1 ± 2.4	1
CH_3CBr_2–H	397.1 ± 5.0	1
CH_2BrCH_2–H	415.1 ± 8.4	1
CH_3CHBr–H	415.0 ± 2.7	3
CF_2=CF–H	464.4 ± 8.4	1
$CF_3CF_2CF_2$–H	432.2	1
CH_3CH_2CHCl–H	407.0 ± 3.5	1
CH_2=CH-CHF–H	370.7 ± 4.6	1
CH_2=CHCHCl–H	370.7 ± 4.6	1
CH_2=CHCHBr–H	374.0 ± 4.6	1
H–C_6F_5	487.4	1
H–CH_2OH	401.92 ± 0.63	1
CH_2CHOH	467 ± 11	1
CH_3CH_2OH	401.2 ± 4.2	1
CH_3CH_2OH	421.7 ± 8	1
$CH_3CH_2CH_2OH$	392	1
$CH_3CH_2CH_2OH$	394.6 ± 8.4	1
$CH_3CH_2CH_2OH$	406.3 ± 8.4	1
$(CH_3)_2CHOH$	383.7 ± 8.4	1
$(CH_3)_2CHOH$	394.6 ± 8.4	1
CH_2=$CHCH_2OH$	341.4 ± 7.5	1
$(CH_3)_3COH$	418.4 ± 8.4	1
$(CH_2$=$CH)_2CHOH$	288.7	1
Ph_2CHOH	326	1
$CH_3CH(OH)_2$	~385	1
$(CH_2OH)_2$	385.3	1
$HOCH_2(CH_2)_2(OH)$	399.2	1
CH–H		
CH_3OCH_3	402.1	1
CHF_2OCF_3	443.5 ± 4.2	1
CHF_2OCHF_2	435.1 ± 4.2	1
CH_3OCF_3	426.8 ± 4.2	1
$CH_3OCH_2CH_3$	389.1	1
$(CH_3)_3COC(CH_3)_3$	402.1	1
$CH_3CH_2OCH_2CH_3$	389.1	1
CH_3CH_2O*t*-$C(CH_3)_3$	405.4	1
CH_3OPh	385.0	1
H-2-oxiran-2-yl	420.5 ± 6.5	1
H-tetrahydrofuran-2-yl	385.3 ± 6.7	1
HC(O)–H	368.40 ± 0.67	1
FC(O)–H	423.0	1
$CH_3C(O)$–H	374.0 ± 1.3	1
$CF_3C(O)$–H	390.4	1
$C_2H_5C(O)$–H	374.5	1
CH_2=CHC(O)–H	372.8	1
$C_3H_7C(O)$–H	371.2	1
iso-$C_3H_7C(O)$–H	364.5	1
$C_4H_9C(O)$–H	372.0	1
$(CH_3)_2CHCH_2C(O)$–H	362.5	1
$C_2H_5CH(CH_3)C(O)$–H	360.8	1
tert-BuC(O)–H	375.1	1
$Et_2CHC(O)$–H	367.2	1
$CH_3(CH_2)_8C(O)$–H	373.3	1
$C_6H_5C(O)$–H	371.1 ± 10.9	1
$PhCH_2C(O)$–H	362.0	1
$PhC(CH_3)_2C(O)$–H	362.9	1
H–CH=C=O	448.1	1
$CH_3C(O)H$	394.5 ± 9.2	1
$CH_3C(O)Cl$	≤423.4	1
$CH_3CH_2C(O)H$	383.7	1
CH_3COCH_3	401.2 ± 2.9	1
$CF_3C(O)CH_3$	465.6	1
$CH_3COCH_2CH_3$	403.8	1
$MeCOCH_2Me$	386.2 ± 7.1	1
$EtCOCH_2Me$	396.5 ± 2.8	1
$CH_3CH_2COC_6H_5$	402.8 ± 3.6	1
$MeCH_2COPh$	388.7	1
H–C(O)OH	404.2	1
$CH_3C(O)OH$	398.7 ± 12.1	1
$ClCH_2C(O)OH$	398.9	1
H–$C(O)OCH_3$	399.2 ± 8.4	1
$CH_3C(O)OCH_3$	406.3 ± 10.5	1
$CH_3C(O)OCH_3$	404.6	1
$CH_3C(O)OCH_2CH_3$	401.7	1
$CH_3C(O)OPh$	419.2 ± 5.4	1
$CH_3CH_2C(O)OEt$	400	1
$PhCH_2C(O)OEt$	370.7	1
$Me_2CHC(O)OEt$	387.4	1
PhCHMe(C(O)OEt)	358.2	1
H-furaylmethyl	361.9 ± 8.4	1
CH_3NH_2	392.9 ± 8.4	1
CH_3N=CH_2	407.9 ± 14.6	1
$CH_3CH_2NH_2$	377.0 ± 8.4	1
$C_2H_5CH_2NH_2$	380.7 ± 8.4	1
$C_3H_7CH_2NH_2$	393.3 ± 8.4	1
$C_4H_9CH_2NH_2$	387.7 ± 8.4	1
$HOCH_2CH_2NH_2$	379.5 ± 8.4	1
$(CH_3CH_2)_2NH$	370.7 ± 8.4	1

Bond	D°_{298}/kJ mol^{-1}	Ref.
$(C_3H_7CH_2)_2NH$	379.9 ± 8.4	1
$(C_4H_9CH_2)_2NH$	384.5 ± 8.4	1
$(C_2H_5)_2NCH_2CH_3$	379.5 ± 1.7	1
$(C_2H_5CH_2)_3N$	376.6 ± 8.4	1
$((CH_3)_2CCH_2)_3N$	388.3 ± 8.4	1
$(Bu)_2NCH_2$(*n*Pr)	381 ± 10.0	1
$((CH_3)_2CH)_3N$	387.0 ± 8.4	1
$(CH_3)_2CHNH_2$	372.0 ± 8.4	1
CH_3NHCH_3	364.0 ± 8.4	1
$(CH_3)_3N$	380.7 ± 8.4	1
tert-$BuN(CH_3)_2$	376.6 ± 8.4	1
$((HOCH_2CH_2)_2(CH_3))N$	364.4 ± 8.4	1
$(HOCH_2CH_2)_3N$	379.9 ± 8.4	1
$((HOCH_2)CH(CH_3))_3N$	379.9 ± 8.4	1
$PhCH_2NH_2$	368.2	1
$PhN(CH_2CH_3)_2$	383.3 ± 4.2	1
Ph_2NCH_3	379.5 ± 1.7	1
$PhN(CH_2Ph)_2$	357.3 ± 8.8	1
$N(CH_2Ph)_3$	372.8 ± 2.5	1
$PhN(CH_2CH{=}CH_2)_2$	339.3 ± 2.9	1
$N(CH_2CH{=}CH_2)_3$	345.6 ± 3.3	1
$H_2NNH(CH_3)$	410	1
$HNN(CH_3)_2$	410	1
$(CH_3)_2NC_6H_5$	383.7 ± 5.4	1
H–CN	528.5 ± 0.8	1
CH_3CN	405.8 ± 4.2	1
CH_3CH_2CN	393.3 ± 12.6	1
$PhCH_2CN$	338.5±9.6	18
$C_6F_5CH_2CN$	350.6	1
FCH_2CN	379.5 ± 11.7	19
$CH_2(CN)_2$	366.5	1
$CH_2(CN)(NH_2)$	355.2	1
$(CH_3)_2CHCN$	384.5	1
CH_3NC	389.1 ± 12.6	1
H–HCNN	405.8 ± 8.4	1
H–CNN	331 ± 17	1
CH_3NO_2	415.4	1
$CH_3CH_2NO_2$	410.5	1
$C_2H_5CH_2NO_2$	410.5	1
Me_2CHNO_2	394.9	1
$C_6H_5C(NO_2)CHCH_3$	357.3	1
H–C(S)H	399.6 ± 5.0	1
CH_3SH	392.9 ± 8.4	1
CH_3SCH_3	392.0 ± 5.9	1
$PhSCH_3$	389.1	1
$PhCH_2SPh$	352.3	1
$(PhS)_2CHPh$	341.0	1
$PhSCHPh_2$	344.8	1
CH_3SOCH_3	393.3	1
$CH_3SO_2CH_3$	414.2	1
$CH_3SO_2CF_3$	431.0	1
CH_3SO_2Ph	414.2	1
$PhCH_2SO_2Me$	380.7	1
$PhCH_2SO_2CF_3$	372.4	1
$PhCH_2SO_2$*t*Bu	376.6	1
Ph_2CHSO_2Ph	365.3	1
$CH_2(SPh)_2$	372.4	1
$H{-}CH_2SiMe_3$	418 ± 6.3	1
$H{-}CH_2C(CH_3)_2SiMe_3$	409 ± 5	1
$H{-}CH_2SiMe_2Ph$	410.1	1
$H{-}CH((CH_3)_3Si)_2$	397 ± 13	1

Bond	D°_{298}/kJ mol^{-1}	Ref.
$H{-}CH_2B(RO)_2$	412.5	1
$H{-}CH((CH_3)_2P)_2$	385 ± 13	1
(2) C–C BDEs		
$CH_3{-}CH_3$	377.4 ± 0.8	1
$CH_3{-}C_2H_5$	370.3 ± 2.1	1
$CH_3{-}C_3H_7$	372.0 ± 2.9	1
CH_3-*iso*-C_3H_7	369.0 ± 3.8	1
$CH_3{-}C_4H_9$	371.5 ± 2.9	1
CH_3-*iso*-C_4H_9	370.3 ± 4.6	1
CH_3-*sec*-C_4H_9	368.2 ± 2.9	1
CH_3-*tert*-C_4H_9	363.6 ± 2.9	1
$CH_3{-}C_5H_{11}$	368.4 ± 6.3	1
$CH_3{-}CH(C_2H_5)_2$	365.7 ± 4.2	1
$CH_3{-}C(CH_3)_2(CH_2CH_3)$	360.9 ± 6.3	1
$CH_3{-}C_6H_{13}$	368.2 ± 6.3	1
$C_2H_5{-}C_2H_5$	363.2 ± 2.5	1
$C_3H_7{-}C_3H_7$	366.1 ± 3.3	1
iso-C_3H_7-*iso*-C_3H_7	353.5 ± 4.6	1
$C_4H_9{-}C_4H_9$	364.0 ± 3.8	1
iso-C_4H_9-*iso*-C_4H_9	362.3 ± 6.3	1
sec-C_4H_9-*sec*-C_4H_9	348.5 ± 3.3	1
tert-C_4H_9-*tert*-C_4H_9	322.6 ± 4.2	1
CH_3-*cyclo*-C_5H_9	358.2 ± 5.0	1
CH_3-*cyclo*-C_6H_{11}	377.0 ± 7.5	1
cyclo-C_6H_{11}-*cyclo*-C_6H_{11}	369.0 ± 8.4	1
$CH_3{-}CH_2C{\equiv}CH$	320.5 ± 5.0	1
$CH_3{-}CH_2C{\equiv}CCH_3$	308.4 ± 6.3	1
$CH_3{-}CH(CH_3)C{\equiv}CH$	305.4 ± 8.4	1
$CH_3{-}CH(CH_3)C{\equiv}CCH_3$	320.9 ± 6.3	1
$CH_3{-}C(CH_3)_2C{\equiv}CH$	295.8 ± 6.3	1
$CH_3{-}C(CH_3)_2C{\equiv}CCH_3$	303.3 ± 6.3	1
$CH_3{-}CHCH_2$	426.3 ± 6.3	1
$CH_3{-}CH{=}CCH_2$	359.8 ± 5.9	1
CH_3-cyclopro-en-1-yl	340.6 ± 20.9	1
$CH_3{-}CH_2CH{=}CH_2$	317.6 ± 3.8	1
$CH_3{-}CH_2C(CH_3){=}CH_2$	310.0 ± 4.2	1
$CH_3{-}CH(CH_3)CH{=}CH_2$	302.5 ± 6.3	1
$CH_3{-}C(CH_3)_2CH{=}CH_2$	282.4 ± 6.3	1
CH_3-*cyclo*-C_5H_7	299.2 ± 8.4	1
$CH_3{-}C_6H_5$	433.9 ± 4.2	20
$HCC{-}C_6H_5$	597.9 ± 5.9	20
$C_2H_3{-}C_6H_5$	489.1 ± 5.4	20
$CH_3{-}CH_2C_6H_5$	325.1 ± 4.2	1
$CH_3{-}CH(CH_3)C_6H_5$	318.8 ± 8.4	1
$CH_3{-}C(CH_3)_2C_6H_5$	303.3 ± 8.4	1
$CH_3{-}CH_2CHCHPh$	295.4	1
$CH_3{-}CH(C_6H_5)_2$	315.9 ± 6.3	1
$CH_3{-}C(CH_3)(C_6H_5)_2$	290.8 ± 8.4	1
$C_6H_5{-}C_6H_5$	492.9 ± 6.3	20
$C_6H_5{-}CH_2C_6H_5$	390.8 ± 8.4	20
$C_6H_5CH_2{-}CH_2C_6H_5$	272.8 ± 9.2	1
$C_6H_5{-}CH(C_6H_5)_2$	368.2 ± 8.4	20
$C_6H_5{-}C(C_6H_5)_3$	331.4 ± 12.6	20
$Ph_2CH{-}CHPh_2$	247.3 ± 8.4	1
$PhCH_2{-}CPh_3$	234.7 ± 14.6	1
R-R, π-dimer, R = phenalenyl	42	1
R-R, σ-dimer, R = phenalenyl	42.7	1
R-R, R = 9-phenylfluorenyl	63.6	1

Bond	D°_{298}/kJ mol^{-1}	Ref.
$CF_3{-}CF_3$	413.0 ± 5.0	1
$CF_3{-}CHF_2$	399.6 ± 8.4	1
$CF_3{-}CClF_2$	373.6 ± 12.5	1
$CF_3{-}CH_2F$	397.5 ± 8.4	1
$CF_3{-}CCl_3$	332.2 ± 5.4	1
$CF_3{-}CHBrCl$	377.0 ± 10.5	1
$CF_3{-}CH_2Br$	399.6 ± 8.4	1
$CF_3{-}CH_2I$	408.4 ± 10.5	1
$CF_3{-}CH_3$	429.3 ± 5.0	1
$CHF_2{-}CHF_2$	382.4 ± 15.5	1
$CClF_2{-}CClF_2$	378.7 ± 12.6	1
$CF_2Cl{-}CFCl_2$	358.6 ± 12.6	1
$CHF_2{-}CH_2F$	394.1 ± 16.7	1
$CH_2F{-}CH_2F$	368.2 ± 8.4	1
$CHF_2{-}CH_3$	405.0 ± 8.4	1
$CH_2F{-}CH_3$	388.3 ± 8.4	1
$CHClF{-}CH_3$	399.6 ± 12.6	1
$CF_2Br{-}CHClF$	369.4	1
$CF_2Br{-}CH_3$	396.6 ± 15.1	1
$CCl_3{-}CCl_3$	285.8 ± 6.3	1
$CCl_3{-}CClF_2$	282.0 ± 12.6	1
$CCl_3{-}CHCl_2$	303.3 ± 6.3	1
$CCl_3{-}CH_2Cl$	323.8 ± 8.4	1
$CCl_3{-}CH_3$	362.3 ± 6.3	1
$CHCl_2{-}CHCl_2$	326.9 ± 4.1	1
$CHCl_2{-}CH_2Cl$	352.2 ± 5.9	1
$CHCl_2{-}CH_3$	361.3 ± 2.5	1
$CHBrCl{-}CH_3$	384.5	1
$CHClBr{-}CHClBr$	317.1 ± 12.6	1
$CH_2Cl{-}CH_2Cl$	360.7 ± 8.4	1
$CH_2Cl{-}CH_3$	375.7 ± 9.2	1
$Br_3C{-}CH_3$	356.9 ± 12.6	1
$Br_3C{-}CBr_3$	278.7 ± 16.7	1
$CHBr_2{-}CH_3$	372.8	1
$CH_2Br{-}CH_2Cl$	378.2	1
$CH_2Br{-}CH_2Br$	379.9 ± 8.4	1
$CH_2I{-}CH_2I$	387.0 ± 10.5	1
$CH_3{-}CH_2Br$	381.6 ± 8.4	1
$CH_3{-}CH_2I$	384.5 ± 8.4	1
$CF_3{-}CF_2CF_3$	424.3 ± 13.6	1
$CF_3{-}CF{=}CF_2$	420.5	1
$CH_3{-}CH_2CH_2Cl$	371.4 ± 2.8	1
$CH_3{-}CHClCH_3$	367.5 ± 2.0	1
$CH_2Cl{-}CHClCH_3$	356.5 ± 8.4	1
$CH_2Cl{-}CH_2CClH_2$	369.0 ± 8.4	1
$CH_3{-}CCl_2CH_3$	362.8 ± 8.4	1
$CH_2Br{-}CHBrCH_3$	369.4 ± 8.4	1
$CH_2ClCH_2{-}CHClCH_3$	364.4 ± 8.4	1
$CH_2ClCH_2{-}CH_2CClH_2$	369.0 ± 8.4	1
$CH_3CHBr{-}CHBrCH_3$	355.6 ± 8.4	1
$CF_3{-}C_6H_5$	470.3 ± 12.6	20
$CCl_3{-}C_6H_5$	395.8 ± 8.4	20
$CH_3{-}C_6F_5$	439.3	1
$CF_3{-}C_6F_5$	435.1	1
$CF_3{-}CH_2C_6H_5$	365.7 ± 12.6	1
$C_6F_5{-}C_6F_5$	488.3	1
$CF_3{-}CHPh_2$	352.3 ± 16.7	1
$CF_3{-}CPh_3$	290.8 ± 16.7	1
$CF_2CF{-}CFCF_2$	558.1 ± 12.6	1
$CH_2FCH_2{-}CPh_3$	274.9 ± 16.7	1
$CHF_2CH_2{-}CPh_3$	264.0 ± 16.7	1

Bond	D^o_{298}/kJ mol^{-1}	Ref.
CH_3-CH_2OH	364.8 ± 4.2	1
CF_3-CH_2OH	405.4 ± 6.3	1
C_2H_5-CH_2OH	356.9 ± 5.0	1
C_3H_7-CH_2OH	357.3 ± 3.3	1
iso-**C_3H_7**-CH_2OH	354.8 ± 4.2	1
C_4H_9-CH_2OH	355.6 ± 4.2	1
sec-**C_4H_9**-CH_2OH	352.7 ± 4.2	1
iso-**C_4H_9**-CH_2OH	354.0 ± 5.4	1
C_6H_5-CH_2OH	409.0 ± 5.4	20
HOH_2C-CH_2OH	358.2 ± 6.3	1
NH_2CH_2-CH_2OH	335.6 ± 10.5	1
CH_3-CH_2OCH_3	363.2 ± 5.0	1
CH_3OCH_2-CH_2OCH_3	338.9 ± 10.5	1
CH_3-$C(O)H$	354.8 ± 1.7	1
CCl_3-$C(O)H$	309.2 ± 5.0	1
CH_3-$C(O)F$	417.6 ± 6.3	1
CH_3-$C(O)Cl$	367.8 ± 6.3	1
CCl_3-$C(O)Cl$	289.1 ± 6.3	1
$CHCl_2$-$C(O)Cl$	312.5 ± 8.4	1
$CClH_2$-$C(O)Cl$	340.2 ± 8.4	1
C_6H_5-$C(O)H$	415.5 ± 4.2	20
C_6H_5-$C(O)Cl$	424.5 ± 6.3	20
CH_3-$C(O)CH_3$	351.9 ± 2.1	1
C_2H_5-$C(O)CH_3$	347.3 ± 2.9	1
C_3H_7-$C(O)CH_3$	348.5 ± 2.9	1
iso-**C_3H_7**-$C(O)CH_3$	340.2 ± 3.8	1
C_4H_7-$C(O)CH_3$	346.9 ± 5.4	1
tert-**C_4H_9**-$C(O)CH_3$	329.3 ± 4.2	1
C_6H_5-$C(O)CH_3$	413.8 ± 4.6	20
$C_6H_5CH_2$-$C(O)CH_3$	299.7 ± 8.4	1
$HC(O)$–$C(O)H$	295.8 ± 6.3	1
$ClC(O)$–$C(O)Cl$	292.5 ± 8.4	1
$CH_3C(O)$–$C(O)H$	302.5 ± 8.4	1
$CH_3C(O)$–$C(O)CH_3$	307.1 ± 4.2	1
$C_6H_5C(O)$–$C(O)C_6H_5$	288.3 ± 16.7	1
CH_3–$C(O)OH$	384.9 ± 8.4	1
CF_3–$C(O)OH$	370.7 ± 8.4	1
CCl_3–$C(O)OH$	310.5 ± 12.6	1
$CClH_2$–$C(O)OH$	357.7 ± 8.4	1
CH_2Br–$C(O)OH$	358.2 ± 8.4	1
NH_2CH_2–$C(O)OH$	349.4 ± 8.4	1
CH_3NHCH_2–$C(O)OH$	300.4 ± 8.4	1
C_6H_5–$C(O)OH$	436.8 ± 8.4	20
C_6F_5–$C(O)OH$	470.0 ± 10.5	1
$HOCH_2$–$C(O)OH$	371.5 ± 5.4	1
$HOC(O)$–$C(O)OH$	334.7 ± 6.3	1
CH_3NHCH_2–$C(O)OH$	301.2 ± 16.7	1
$CH_3CH(NH_2)$–$C(O)OH$	331.4 ± 16.7	1
NH_2CH_2–$CH_2C(O)OH$	325.5 ± 16.7	1
CN–**CN**	571.9 ± 6.7	1
$HC(O)$–**CN**	455.2 ± 8.4	1
$HC(S)$–**CN**	530.1 ± 8.4	1
CF_3–**CN**	469.0 ± 4.2	1
CH_3–**CN**	521.7 ± 9.2	1
NCC –**CN**	462.3	1
C_2H_5–**CN**	506.7 ± 7.5	1
CH_3–CH_2CN	348.1 ± 12.6	1
C_6H_5–CH_2CN	400.8 ± 8.4	20
CH_3–$CH(CH_3)CN$	332.6 ± 8.4	1
CH_3–$C(CH_3)_2CN$	340.6 ± 16.7	1
CH_3–$C(CH_3)(CN)C_6H_5$	250.6	1

Bond	D^o_{298}/kJ mol^{-1}	Ref.
$(Ph)_2(CN)C$–$C(CN)(Ph)_2$	109.6	1
$(NO_2)_3C$–$C(NO_2)_3$	308.8	1
C_{58}–C_2	955.2 ± 14.5	1
(3) C–halogen BDEs		
F–CN	482.8	1
F–CF_3	546.8 ± 2.1	1
F–CHF_2	533.9 ± 5.9	1
F–CH_2F	496.2 ± 8.8	1
F–CF_2Cl	507.4 ± 9.0	17
F–$CFCl_2$	482.3 ± 9.0	17
F–$CHFCl$	462.3 ± 10.0	1
F–CCl_3	439.2 ± 4.2	17
F–$CFClBr$	501.6 ± 6.8	17
F–CH_2Cl	465.3 ± 9.6	1
F–CH_3	460.2 ± 8.4	1
F–C≡CH	521.3	1
F–C≡CF	519 ± 21	1
F–$CF=CF_2$	546.4 ± 12.6	1
F–CF_2CF_3	532.2 ± 6.3	1
F–CH_2CF_3	457.7	1
F–CF_2CH_3	522.2 ± 8.4	1
F–C_2H_3	517.6 ± 12.6	1
F–C_2H_5	467.4 ± 8.4	1
F–C_3H_7	474.9 ± 8.4	1
F–*iso*-C_3H_7	483.8 ± 8.4	1
F–*tert*-C_4H_9	495.8 ± 8.4	1
F–C_6H_5	532.0 ± 1.2	12
F–C_6F_5	485 ± 25	1
F–$CH_2C_6H_5$	412.8 ± 4.2	1
F–COH	497.9 ± 10.5	1
F–COF	510.3	1
F–COCl	484.5	1
F–$C(O)CH_3$	511.7 ± 12.6	1
Cl–CN	422.6 ± 8.4	1
Cl–CF_3	362.7 ± 3.9	17
Cl–CHF_2	364 ± 8	1
Cl–CH_2F	354.4 ± 11.7	1
Cl–CF_2Cl	334.3 ± 9.0	17
Cl–CF_2Br	354.4 ± 12.9	17
Cl–$CFCl_2$	320.9 ± 8.4	1
Cl–$CHFCl$	346.0 ± 13.4	1
Cl–CCl_3	288.7 ± 4.2	17
Cl–$CHCl_2$	311.1 ± 2.0	1
Cl–CH_2Cl	338.0 ± 3.3	1
Cl–$CBrCl_2$	287 ± 10.5	1
Cl–CH_2Br	332.8 ± 4.6	1
Cl–CH_2I	328.2 ± 6.9	1
Cl–CH_3	350.2 ± 1.7	1
Cl–C≡CCl	443 ± 50	1
Cl–C≡CH	435.6 ± 8.4	1
Cl–CH_2CN	267.4	1
Cl–$CCl=CCl_2$	383.7	1
Cl–$CH=CH_2$	394.1 ± 3.1	2
Cl–$CF=CF_2$	434.7 ± 8.4	1
Cl–CF_2CF_3	346.0 ± 7.1	1
Cl–CF_2CF_2Cl	331.4 ± 20.9	1
Cl–CCl_2CF_3	307.9	1
Cl–CCl_2CCl_3	303.8	1
Cl–$CHClCCl_3$	330.5 ± 4.2	1
Cl–CCl_2CHCl_2	311.7	1
Cl–$CHClCH_3$	327.9 ± 1.8	1

Bond	D^o_{298}/kJ mol^{-1}	Ref.
Cl–CH_2CH_2Cl	345.1 ± 5.0	1
Cl–$CHBrCH_3$	331.8 ± 8.4	1
Cl–CH_2CH_3	352.3 ± 3.3	1
Cl–$CH_2CH=CH_2$	298.3 ± 5.0	1
Cl–C_3H_7	352.7 ± 4.2	1
Cl–$CH_2CH_2CH_2Cl$	348.9	1
Cl–*iso*-C_3H_7	354.8 ± 3.3	10
Cl–$CH_2CHCH=CH_2$	342.7	1
Cl–C_4H_9	350.6 ± 6.3	1
Cl–*sec*-C_4H_9	350.2 ± 6.3	1
Cl–*tert*-C_4H_9	351.9 ± 6.3	1
CH_2CHCH**Cl**(CH_3)	300.0 ± 6.3	1
Cl–C_5H_{11}	350.6 ± 6.3	1
Cl–$C(CH_3)_2(C_2H_5)$	352.7 ± 6.3	1
Cl–*cyclo*-C_6H_{11}	360.2 ± 6.5	1
Cl–C_6H_5	406.4 ± 0.8	12
Cl–C_6F_5	383.3 ± 8.4	1
Cl–$CH_2C_6H_5$	299.9 ± 4.3	1
Cl–C(O)Cl	318.8 ± 8.4	1
Cl–COF	376.6	1
Cl–$C(O)CH_3$	354.0 ± 8.4	1
Cl–$C(O)CH_2CH_3$	353.3 ± 6.3	1
Cl–$C(O)C_6H_5$	341.0 ± 8.4	1
Cl–$CH_2C(O)C_6H_5$	309	1
Cl–$CH_2C(O)OH$	310.9 ± 2.2	1
Cl–$C(O)OC_6H_5$	364	1
Cl–$C(NO_2)_3$	302.1	1
Br–CN	364.8 ± 4.2	1
Br–CF_3	296.2 ± 1.3	1
Br–CHF_2	288.7 ± 8.4	1
Br–CF_2Cl	290.5 ± 8.7	17
Br–CCl_3	231.4 ± 4.2	1
Br–CH_2Cl	277.3 ± 3.6	1
Br–CBr_3	242.3 ± 8.4	1
Br–$CHBr_2$	274.9 ± 13.0	1
Br–CH_2Br	276.1 ± 5.3	1
Br–CH_2I	274.5 ± 7.5	1
Br–CH_3	294.1 ± 2.1	1
Br–C≡CH	410.5	1
Br–$CH=CH_2$	338.3 ± 3.1	1
Br–CF_2CF_3	283.3 ± 6.3	1
Br–$CClBrCF_3$	251.0 ± 6.3	1
Br–CF_2CF_2Br	282.8 ± 6.7	1
Br–$CHClCF_3$	274.9 ± 6.3	1
Br–CF_2CH_3	287.0 ± 5.4	1
Br–CH_2CH_2Cl	292.5 ± 8.4	1
Br–$CHClCH_3$	272.0 ± 8.4	1
Br–C_2H_5	292.9 ± 4.2	1
Br–$CH_2CH=CH_2$	233.3 ± 5.3	17
Br–C_3H_7	298.3 ± 4.2	1
Br–*iso*-C_3H_7	295.1 ± 3.3	10
Br–$CH_2CH_2CH_2Br$	324.7	1
Br–$CF_2CF_2CF_3$	278.2 ± 10.5	1
CF_3CF**Br**CF_3	274.2 ± 4.6	1
Br–C_4H_9	296.6 ± 4.2	1
Br–*sec*-C_4H_9	300.0 ± 4.2	1
Br–*tert*-C_4H_9	292.9 ± 6.3	1
Br–C_6H_5	344.2 ± 1.3	12
Br–C_6F_5	~328	1
Br–$CH_2C_6H_5$	239.3 ± 6.3	1
Br–$CH_2C_6F_5$	225.1 ± 6.3	1

Bond	$D°_{298}$/kJ mol^{-1}	Ref.
Br–1–$C_{10}H_7$	339.7	1
Br–2–$C_{10}H_7$	341.8	1
Br–anthracenyl	322.6	1
Br–C(O)CH_3	292.0 ± 8.4	1
Br–C(O)C_6H_5	276.6 ± 8.4	1
Br–$CH_2C(O)CH_3$	257.9 ± 10.5	1
Br–$CH_2C(O)C_6H_5$	271	1
Br–$CH_2C(O)OH$	257.4 ± 3.7	1
Br–$C(NO_2)_3$	218.4	1
I–CN	320.1	1
I–CF_3	227.2 ± 1.3	1
I–CCl_3	168 ± 42	1
I–CH_2Cl	221.8 ± 4.2	1
I–CH_2Br	219.2 ± 5.4	1
I–CH_2I	216.9 ± 7.9	1
I–CH_3	232.4 ± 0.4	13
I–CH_2CN	187.0 ± 6.3	1
I–CF_2CF_3	219.2 ± 2.1	1
I–CF_2CF_2I	217.6 ± 6.7	1
I–CH_2CF_3	235.6 ± 4.2	1
I–$CHFCClF_2$	202 ± 2	1
I–CF_2CH_3	217.6 ± 4.2	1
I-$CFICH_3$	218.0 ± 4.2	1
CF_3CFICF_3	215.1	1
I–CH=CH_2	259.0 ± 4.2	1
I–C_2H_5	233.5 ± 6.3	1
I–CH_2CH=CH_2	185.8 ± 6.3	1
I–C_3H_7	236.8 ± 4.2	1
I–*iso*-C_3H_7	233.1 ± 3.3	10
I–C_4F_9	205.8	1
I–*tert*-C_4H_9	227.2 ± 6.3	1
I–C_6H_5	272.0 ± 4.2	1
I–C_6F_5	<301.7	1
I–$CH_2C_6H_5$	187.8 ± 4.8	1
I–1-naphthyl	274.5 ± 10.5	1
I–2-naphthyl	272.0 ± 10.5	1
I–CH_2CN	187.0 ± 8.4	1
I–CH_2OCH_3	229.4 ± 8.4	1
I–CH_2SCH_3	216.8 ± 6.3	1
I–C(O)CH_3	223.0 ± 8.4	1
I–C(O)C_6H_5	212.1 ± 8.4	1
I–$CH_2C(O)OH$	197.5 ± 2.7	1
I–$C(NO_2)_3$	144.8	1

(4) O–X BDEs

Bond	$D°_{298}$/kJ mol^{-1}	Ref.
HO–H	497.321 ± 0.02	16
FO–H	425.1	1
ClO–H	393.7	1
BrO–H	405	1
IO–H	403.3	1
CH_3O–H	440.2 ± 3	1
CF_3O–H	497.1	1
HC≡CO–H	443.1	1
C_2H_5O–H	441.0 ± 5.9	1
CH_2=CHO–H	355.6	1
CF_3CH_2O–H	447.7 ± 10.5	1
C_3H_7O–H	≤433 ± 2	1
iso-C_3H_7O–H	442.3 ± 2.8	1
C_4H_9O–H	432.3	1
sec-C_4H_9O–H	441.4 ± 4.2	1
tert-C_4H_9O–H	444.9 ± 2.8	1
tert-BuCH_2O–H	436.1	1
$C_6H_5CH_2O$–H	442.7 ± 8.8	1
$CH_3C(OH)O$–H	446.9 ± 6.3	1
$(CH_3)_2C(OH)O$–H	450.6 ± 6.3	1
HC(O)O–H	468.6 ± 12.6	1
$CH_3C(O)O$–H	468.6 ± 12.6	1
$C_2H_5C(O)O$–H	472.8	1
iso-$C_3H_7C(O)O$–H	472.8	1
$C_6H_5C(O)O$–H	464.4 ± 16.7	1
OO–H	205.732 ± 0.16	16
HOO–H	365.71 ± 0.17	16
CH_3OO–H	370.3 ± 2.1	1
CF_3OO–H	383	1
CH_2FOO–H	379	1
CCl_3OO–H	386	1
$CHCl_2OO$–H	383	1
CH_2ClOO–H	379	1
CBr_3OO–H	383	1
CH_2BrOO–H	379	1
C_2H_5OO–H	354.8 ± 9.2	1
$CH_3CHClOO$–H	377	1
CH_3CCl_2OO–H	383	1
$CF_3CHClOO$–H	384	1
C_2Cl_5OO–H	383	1
iso-C_3H_7OO–H	356	1
CH_2=$CHCH_2OO$–H	372.4	1
tert-C_4H_9OO–H	352.3 ± 8.8	1
C_6H_5OO–H	384	1
$C_6H_5CH_2OO$–H	363	1
$(C_6H_5)_2CHOO$–H	370	1
trans-HC(O)OO–H	393 ± 13.8	6
cis-HC(O)OO–H	406.3 ± 13.8	6
trans-$CH_3C(O)OO$–H	381 ± 4	8
cis-$CH_3C(O)OO$–H	403 ± 14	8
$CCl_2(CN)OO$–H	384	1
$OHCH_2OO$–H	368	1
H–ONO	330.7	1
H–OONO	299.2	1
H–ONH_2	318	1
H–ONO_2	426.8	1
H–ONNOH	189	1
H–OPO_2	465.7 ± 12.6	1
H–OSO_2OH	441.4 ± 14.6	1
H–OSiMe3	495	1
$(CH_3)CHNO$-H	354.4	1
$(CH_3)_2CNO$-H	354.0	1
$(C_6H_5)CHNO$-H	368.6	1
PhO–H	362.8 ± 2.9	1
α-tocopherol RO-H	323.4	1
β-tocopherol RO-H	335.6	1
γ-tocopherol RO-H	335.1	1
δ-tocopherol RO-H	342.8	1
p-$C_6H_5CH_2$-C_6H_4O-H	356.2	1
HO–O	274.47 ± 1.6	16
O–O_2	107.483 ± 0.039	16
HO–OH	210.45 ± 0.04	16
HO–OF	199.7 ± 8.4	1
HO–OCl	~146	1
HO–OBr	138.5 ± 8.4	1
FO–OF	199.6	1
ClO–OCl	72.4 ± 2.8	1
IO–OI	74.9 ± 17	1
trans-perp-HO-ONO	≤67.8 ± 0.4	1
cis-cis-HO-ONO	83.3 ± 2.1	1
HO–ONO_2	163.2 ± 8.4	1
HO–OCH_3	189.1 ± 4.2	1
HO–OCF_3	201.3 ± 20.9	1
HO–OC_2H_5	178.7 ± 6.3	1
HO–O-*iso*-C_3H_7	185.8 ± 6.3	1
HO–O-*tert*-C_4H_9	186.2 ± 4.2	1
HO–OC(O)CH_3	169.9 ± 2.1	1
HO–OC(O)C_2H_5	169.9 ± 2.1	1
CH_3O–OCH_3	167.4 ± 6.3	1
CF_3O–OCF_3	198.7 ± 2.1	1
C_2H_5O–OC_2H_5	166.1	1
C_3H_7O–OC_3H_7	155.2 ± 4.2	1
iso-C_3H_7O-O-*iso*-C_3H_7	157.7	1
sec-C_4H_9O-O-*sec*-C_4H_9	152.3 ± 4.2	1
tert-BuO-O-*tert*-Bu	162.8 ± 2.1	1
tert-BuCH_2O-OCH_2-*tert*-Bu	152.3	1
EtC(Me)$_2$O–OC(Me)$_2$Et	164.4 ± 4.2	1
$(CF_3)_3CO$–$OC(CF_3)_3$	148.5 ± 4.6	1
Ph_3CO–$OCPh_3$	131.4	1
SF_5O–OSF_5	155.6	1
SF_5O–$OOSF_5$	126.8	1
$(CH_3)_3CO$–$OSi(CH_3)_3$	196.6	1
tert-BuO-$OGeEt_3$	192.5	1
tert-BuO-$OSnEt_3$	192.5	1
CF_3OO–OCF_3	126.8 ± 8.4	1
HC(O)O–OH	199.2 ± 8.4	1
FC(O)O–OC(O)F	96.2	1
$CH_3C(O)O$–ONO_2	131.4 ± 8.4	1
$CH_3C(O)O$–OC(O)CH_3	140.2 ± 21	1
$CF_3C(O)O$–OC(O)CF_3	125.5	1
$CF_3OC(O)O$–OC(O)F	121.3 ± 4.2	1
$CF_3OC(O)O$–OCF_3	142.3 ± 2.9	1
$CF_3OC(O)O$–OC(O)OCF_3	119.2	1
$C_2H_5C(O)O$–OC(O)C_2H_5	150.6	1
$C_3H_7C(O)O$–OC(O)C_3H_7	150.6	1
$FS(O)_2O$-$OS(O)_2F$	92–100	1
O–CO	532.18 ± 0.03	16
HO–CF_3	≤482.0 ± 1.3	1
FO–CF_3	408 ± 17	1
HO–CH_3	384.93 ± 0.71	1
HO–C_2H_5	391.2 ± 2.9	1
HO–CH_2CF_3	408.4 ± 8.4	1
HO–CH_2CH=CH_2	332.6 ± 4.2	1
HO–C_3H_7	392.0 ± 2.9	1
HO–*iso*-C_3H_7	397.9 ± 4.2	1
HO–C_4H_9	389.9 ± 4.2	1
HO–*sec*-C_4H_9	396.1 ± 4.2	1
HO–*iso*-C_4H_9	394.1 ± 4.2	1
HO–*tert*-C_4H_9	398.3 ± 4.2	1
HO-$CH(CH_3)(nC_3H_7)$	398.3 ± 4.2	1
HO–$CH(C_2H_5)_2$	399.2 ± 4.2	1
HO–$C(CH_3)_2(C_2H_5)$	395.8 ± 6.3	1
HO–C_6H_5	470.7 ± 4.2	20
HO–C_6F_5	446.9 ± 9.2	1
HO–$CH_2C_6H_5$	334.1 ± 2.6	1

Bond	D^o_{298}/kJ mol^{-1}	Ref.
HO–$C(CH_3)_2C_6H_5$	339.3 ± 6.3	1
cyclo-C_5H_9–OH	385.8 ± 6.3	1
1-$C_{10}H_7$–OH	468.6 ± 6.3	1
2-$C_{10}H_7$–OH	467.8 ± 6.3	1
$(CH_3)_2(NH_2)C$–OH	310.4 ± 6.3	1
$CH_3C(O)$-OH	459.4 ± 4.2	1
$HOCH_2$-OH	411.3	1
CH_3–OCH_3	351.9 ± 4.2	1
ICH_2–OCH_3	373.2 ± 12.6	1
CH_3O–C_2H_5	355.2 ± 5.4	1
CH_3O–$CHClCH_3$	370.3 ± 8.4	1
CH_3O–C_3H_7	358.6 ± 6.3	1
CH_3O-*iso*-C_3H_7	360.7 ± 4.2	1
CH_3O-C_4H_9	346.0 ± 6.3	1
CH_3O-*tert*-C_4H_9	353.1 ± 6.3	1
C_6H_5–OCH_3	425.9 ± 5.9	20
$C_6H_5CH(CH_3)$–OCH_3	313.4 ± 9.6	1
C_6H_5–OC_6H_5	333.9 ± 4.2	20
CH_3–OC(O)H	383.7 ± 12.6	1
HC(O)–OH	457.7 ± 2.1	1
$CH_3C(O)$–OH	459.4 ± 4.2	1
$C_6H_5C(O)$–OH	447.7 ± 10.5	1
HO–$CH_2C(O)OH$	368.2 ± 10.5	1
CH_3–$OC(O)CH_3$	380.3 ± 12.6	1
HC(O)–OCH_3	423.8 ± 4.2	1
$CH_3C(O)$–OCH_3	424.3 ± 6.3	1
$C_6H_5C(O)$–OCH_3	421.3 ± 12.6	1
$C_6H_5C(O)$–OC_6H_5	307.5 ± 8.4	1
CH_3OCH_2–OCH_3	367.5 ± 8.4	1
$CH_3C(O)$–$OC(O)CH_3$	382.4 ± 12.6	1
$C_6H_5C(O)$–$OC(O)C_6H_5$	384.9 ± 16.7	1
CH_3–OOH	300.4 ± 12.6	1
C_2H_5–OOH	332.2 ± 20.9	1
C_3H_7–OOH	364.4	1
iso-C_3H_7-OOH	298.3	1
tert-C_4H_9-OOH	309.2 ± 4.2	1
CH_3–$OOCH_3$	292.5 ± 8.4	1
CF_3–$OOCF_3$	361.5 ± 8.4	1
CH_3–OO	137.0 ± 3.8	1
CF_3–OO	169.0	1
$CClF_2$–OO	127.6	1
CCl_2F-OO	124.7	1
CH_2Cl-OO	122.4 ± 10.5	1
$CHCl_2$–OO	108.2 ± 8.2	1
CCl_3–OO	92.0 ± 6.4	1
HC(O)–OOH	290.0	1
$CH_3C(O)$–$OOC(O)CH_3$	315.1	1
ClO–CF_3	≤369.9 ± 1.3	1
CH_3–ONO	245.2	1
C_2H_5–ONO	260.2	1
C_3H_7–ONO	249.4 ± 6.3	1
iso-C_3H_7–ONO	254.4 ± 6.3	1
C_4H_9–ONO	256.5 ± 6.3	1
iso-C_4H_9–ONO	254.0 ± 6.3	1
sec-C_4H_9–ONO	253.6 ± 6.3	1
tert-C_4H_9–ONO	252.7 ± 6.3	1
$(C_2H_5)(CH_3)_2C$–ONO	254.0 ± 8.4	1
CH_3–ONO_2	340.2	1
C_2H_5–ONO_2	344.8	1
CH_3O–CH_2CN	393.3	1
O–N_2	167.4 ± 0.4	1
O–NO	306.301	16
O–NO_2	206.3	1
NO–NO	40.6 ± 2.1	1
O_2N–ONO_2	95.4 ± 1.5	1
cis-HO–NO	207.0	1
trans-HO-NO	200.64 ± 0.19	1
FO–NO	132.5 ± 17	1
cis-ClO–NO	127.6 ± 8.4	1
trans-ClO–NO	116.6 ± 8.4	1
cis-BrO–NO	138.1 ± 8.4	1
trans-BrO–NO	121.6 ± 8.4	1
trans-perp-HOO–NO	114.2 ± 4	1
CH_3O–NO	176.6 ± 3.3	1
C_2H_5O–NO	185.4 ± 4.2	1
C_3H_7O–NO	179.1 ± 6.3	1
iso-C_3H_7O–NO	175.3 ± 4.2	1
C_4H_9O–NO	177.8 ± 6.5	1
iso-C_4H_9O–NO	175.7 ± 6.5	1
sec-C_4H_9O–NO	173.6 ± 3.3	1
tert-C_4H_9O–NO	176.1 ± 5.9	1
tert-AmO–NO	171.1 ± 0.4	1
C_6H_5O–NO	87.0	1
HO–NO_2	205.4	1
FO–NO_2	131.8 ± 12.6	1
ClO–NO_2	110.9	4
BrO–NO_2	118.0 ± 6.3	1
IO–NO_2	~100	1
CH_3O–NO_2	176.1 ± 4.2	1
C_2H_5O–NO_2	174.5 ± 4.2	1
C_3H_7O–NO_2	177.0 ± 4.2	1
iso-C_3H_7O–NO_2	175.7 ± 4.2	1
HOO–NO_2	99.2 ± 4.6	1
CH_3OO–NO_2	93.5 ± 0.3	21
C_4H_9OO–NO_2	93.5 ± 0.6	21
CF_3OO–NO_2	105	1
CF_2ClOO–NO_2	106.7	1
$CFCl_2OO$–NO_2	106.7	1
CCl_3OO–NO_2	95.8	1
$CH_3N(O)$-O	305.3 ± 4.4	1
$C_6H_5N(O)$-O	392 ± 8	1
C_5H_5N-O	264.9 ± 2.0	1
$C_6H_5N{=}N(O)(C_6H_5)$-O	309.4 ± 3.5	1
$C_6H_5(O)N{=}N(O)$ (C_6H_5)-O	309.4 ± 3.6	1
O–SO	551.1	1
O–SOF_2	513.3	1
O–$SOCl_2$	398.5	1
O–$S(OH)_2$	493.7 ± 25	1
HO–SH	293.3 ± 16.7	1
HO–SOH	313.4 ± 12.6	1
HO–$S(OH)O_2$	384.9 ± 8.4	1
HO–SCH_3	303.8 ± 12.6	1
HO–SO_2CH_3	360.2 ± 12.6	1
F–OH	215.1	1
F–OF	164.1	1
F–OCF_3	200.8 ± 4.2	1
F–OCH_3	>196.6	1
F–ONO_2	143.1	1
Cl–OH	233.5	1
Cl–OCl	142	1
Cl–OCF_3	≤220.9 ± 8.4	1
Cl–OCH_3	200.8	1
Cl–O-*tert*-C_4H_9	198.3	1
Cl–OOCl	91.2	1
Cl–ONO_2	172.0	1
Br–OH	209.6 ± 4.2	1
Br–OBr	125	1
Br–O-*tert*-C_4H_9	183.3	1
Br–ONO_2	143.1 ± 6.3	1
I–OH	213.4	1
I–OI	130.1	1
I-ONO_2	>140.6	1
(5) N–X BDEs		
H–NH	390.7 ± 0.2	16
H–NH_2	449.6 ± 0.2	16
H–NF_2	316.7 ± 10.5	1
H–NNH	254.4	1
H–N_3	≤389	1
H–$N{=}CH_2$	364 ± 25	1
H–NO	199.5	1
H–NHOH	341	1
H–NCO	460.7 ± 2.1	1
H–NCS	≤396.6 ± 4.6	1
H–NCS	347.3 ± 8.4	1
CH_3NH_2	425.1 ± 8.4	1
tert-$BuNH_2$	397.5 ± 8.4	1
$C_6H_5CH_2NH_2$	418.4	1
$(CH_3)_2NH$	395.8 ± 8.4	1
H–$NHNH(CH_3)$	276 ± 21	1
H–$NHN(CH_3)_2$	356 ± 21	1
NH_2CN	414.2	1
$(NH_2)_2C{=}O$	464.4	1
$(NH_2)_2C{=}S$	389.1	1
CH_3CSNH_2	380.7	1
$PhCSNH_2$	380.7	1
$(PhNH)_2C{=}S$	364.0	1
$(NH_2)_2C{=}NH$	435.1	1
$Ph_2C{=}NH$	489.5	1
H–$N(SiMe_3)_2$	464	1
H–NHPh	375.3	1
C_6H_5NHOH	292	1
$C_6H_5NH(CONMe_2)$	387.9	1
H–NPh_2	364.8	1
HN–N_2	63	1
ON–N	480.7 ± 0.4	1
ON–NO	8.49 ± 0.12	1
ON–NO_2	42.5	1
O_2N–NO_2	57.3 ± 1	1
H_2N–NH_2	277.0 ± 1.3	1
F_2N–NF_2	92.9 ± 12.6	1
H_2N–$NHCH_3$	275.8 ± 8.4	1
H_2N–$N(CH_3)_2$	259.8 ± 8.4	1
H_2N–NHC_6H_5	227.6 ± 8.4	1
H_2N–NO_2	230	1
$H_2NN(CH_3)$–NO	179.6	1
$(C_6H_5)_2N$-NO	94.6	1
N_3–CH_3	335.1 ± 20.5	1
N_3–C_6H_5	389.9 ± 20.9	20
N_3–$CH_2C_6H_5$	211.3 ± 14.2	1
CH_3–NC	413.0 ± 3.3	1
C_2H_5–NC	413.4 ± 8.4	1
iso-C_3H_7–NC	423.0 ± 8.4	1

Bond	$D°_{298}$/kJ mol^{-1}	Ref.
$tert\text{-}C_4H_9\text{–}\mathbf{NC}$	399.6 ± 5.4	1
$NC\text{–}\mathbf{NO}$	204.4	1
$CH_3\text{–}\mathbf{NO}$	172	1
$CF_3\text{–}\mathbf{NO}$	167	1
$CCl_3\text{–}\mathbf{NO}$	125	1
$C_2H_5\text{–}\mathbf{NO}$	171.5	1
$CH_2CHCH_2\text{–}\mathbf{NO}$	102.4 ± 3.2	15
$iso\text{-}C_3H_7\text{–}\mathbf{NO}$	152.7 ± 12.6	1
$tert\text{-}C_4H_9\text{–}\mathbf{NO}$	167	1
$C_6H_5\text{–}\mathbf{NO}$	229.7 ± 1.4	12
$C_6F_5\text{–}\mathbf{NO}$	211.3 ± 4.2	1
$C_6H_5CH_2\text{–}\mathbf{NO}$	123	1
$CH_3\text{–}\mathbf{NO_2}$	260.7 ± 2.1	1
$C_2H_5\text{–}\mathbf{NO_2}$	254.4	1
$C_3H_7\text{–}\mathbf{NO_2}$	256.5	1
$iso\text{-}C_3H_7\text{–}\mathbf{NO_2}$	259.8	1
$C_4H_9\text{–}\mathbf{NO_2}$	254.8	1
$sec\text{-}C_4H_9\text{–}\mathbf{NO_2}$	263.2	1
$tert\text{-}C_4H_9\text{–}\mathbf{NO_2}$	258.6	1
$C_6H_5\text{–}\mathbf{NO_2}$	302.9 ± 4.2	20
$C_6H_5CH_2\text{–}\mathbf{NO_2}$	210.3 ± 6.3	1
$(NO_2)CH_2\text{–}\mathbf{NO_2}$	207.1	1
$(NO_2)_3C\text{–}\mathbf{NO_2}$	176.1	1
$CF_3\text{–}\mathbf{NF_2}$	280.7	1
$C_6H_5CH_2\text{–}\mathbf{NF_2}$	237.2 ± 14.6	1
$CH_3\text{–}\mathbf{NH_2}$	356.1 ± 2.1	1
$C_2H_5\text{–}\mathbf{NH_2}$	352.3 ± 6.3	1
$C_3H_7\text{–}\mathbf{NH_2}$	356.1 ± 2.9	1
$iso\text{-}C_3H_7\text{–}\mathbf{NH_2}$	357.7 ± 3.8	1
$C_4H_9\text{–}\mathbf{NH_2}$	356.1 ± 2.9	1
$sec\text{-}C_4H_9\text{–}\mathbf{NH_2}$	359.0 ± 2.9	1
$iso\text{-}C_4H_9\text{–}\mathbf{NH_2}$	254.8 ± 5.0	1
$tert\text{-}C_4H_9\text{–}\mathbf{NH_2}$	355.6 ± 6.3	1
$pyridin\text{-}2\text{-}yl\text{–}\mathbf{NH_2}$	431	1
$C_6H_5\text{–}\mathbf{NH_2}$	436.4 ± 4.2	20
$C_6H_5CH_2\text{–}\mathbf{NH_2}$	306.7 ± 6.3	1
$C_6H_5CH(CH)_3\text{–}\mathbf{NH_2}$	307.5 ± 9.6	1
$HC(O)\text{–}\mathbf{NH_2}$	421.7 ± 8.4	1
$CH_3C(O)\text{–}\mathbf{NH_2}$	414.6 ± 8.4	1
$HS\text{–}\mathbf{NO}$	138.9	1
$CH_3S\text{–}\mathbf{NO}$	104.6 ± 4.2	1
$tert\text{-}BuS\text{–}\mathbf{NO}$	115.1	1
$PhCH_2S\text{–}\mathbf{NO}$	120.5	1
$C_6H_5S\text{–}\mathbf{NO}$	81.2 ± 5.4	1
$\mathbf{SCN}\text{–}SCN$	255.6	1
$\mathbf{FSO_2}\text{–}NF_2$	163	1
$\mathbf{F}\text{–}NO$	235.26	1
$\mathbf{F}\text{–}NO_2$	221.3	1
$\mathbf{F}\text{–}NF_2$	254.0	1
$\mathbf{F}\text{–}NH_2$	286.6	1
$\mathbf{Cl}\text{–}NO$	158.8 ± 0.8	1
$\mathbf{Cl}\text{–}NO_2$	141.8 ± 1.3	1
$\mathbf{Cl}\text{–}NF_2$	~134	1
$\mathbf{Cl}\text{–}NH_2$	253.1	1
$\mathbf{Br}\text{–}NO$	120.1 ± 0.8	1
$\mathbf{Br}\text{–}NO_2$	82.0 ± 7.1	1
$\mathbf{Br}\text{–}NF_2$	<227.2	1
$\mathbf{I}\text{–}NO$	75.6 ± 4	1
$\mathbf{I}\text{–}NO_2$	79.6 ± 4	1

(6) S–X BDEs

Bond	$D°_{298}$/kJ mol^{-1}	Ref.
$\mathbf{H}\text{–}SH$	381.18 ± 0.05	1
$\mathbf{H}\text{–}SCH_3$	365.7 ± 2.1	1
$\mathbf{H}\text{–}SCHCH_2$	351.5 ± 8.4	1
$\mathbf{H}\text{–}SC_2H_5$	365.3	1
$\mathbf{H}\text{–}SC_3H_7$	365.7	1
$\mathbf{H}\text{–}S\text{-}iso\text{-}C_3H_7$	369.9 ± 8.4	1
$\mathbf{H}\text{–}S\text{-}tert\text{-}C_4H_9$	362.3 ± 9.2	1
$\mathbf{H}\text{–}SOH$	330.5 ± 14.6	1
$\mathbf{H}\text{–}SCOCH_3$	370.7	1
$\mathbf{H}\text{–}SCO\ Ph$	364	1
$\mathbf{H}\text{–}SO_2CH_3$	≤397	1
$\mathbf{H}\text{–}SSCH_3$	330.5 ± 14.6	1
$\mathbf{H}\text{–}SPh$	349.4 ± 4.5	1
$\mathbf{H}\text{–}SSH$	318.0 ± 14.6	1
$\mathbf{H}\text{–}SSSH$	292.9 ± 6.5	1
$\mathbf{HS}\text{–}SH$	270.7 ± 8.4	1
$\mathbf{FS}\text{–}SF$	362.3	1
$\mathbf{ClS}\text{–}SCl$	329.7	1
$\mathbf{HS}\text{–}SCH_3$	272.0	1
$\mathbf{HS}\text{–}SPh$	255.2 ± 6.3	1
$\mathbf{CH_3S}\text{–}SCH_3$	272.8 ± 3.8	1
$\mathbf{C_2H_5S}\text{–}SC_2H_5$	276.6	1
$\mathbf{MeS}\text{–}SPh$	272.0 ± 6.3	1
$\mathbf{C_6H_5S}\text{–}SC_6H_5$	214.2 ± 12.6	1
$\mathbf{F_5S}\text{–}SF_5$	305 ± 21	1
$\mathbf{HS}\text{–}CH_3$	312.5 ± 4.2	1
$\mathbf{HS}\text{–}C_2H_5$	307.9 ± 2.1	1
$\mathbf{HS}\text{–}C_3H_7$	310.5 ± 2.9	1
$\mathbf{HS}\text{–}iso\text{-}C_3H_7$	307.1 ± 3.8	1
$\mathbf{HS}\text{–}C_4H_9$	309.2 ± 2.9	1
$\mathbf{HS}\text{–}sec\text{-}C_4H_9$	307.5 ± 2.9	1
$\mathbf{HS}\text{–}iso\text{-}C_4H_9$	310.0 ± 4.6	1
$\mathbf{HS}\text{–}tert\text{-}C_4H_9$	301.2 ± 3.8	1
$\mathbf{HS}\text{–}C_6H_5$	367.8 ± 6.3	20
$\mathbf{HS}\text{–}CH_2C_6H_5$	258.2 ± 6.3	1
$\mathbf{HS}\text{–}C(O)H$	309.6 ± 8.4	1
$\mathbf{HS}\text{–}C(O)CH_3$	307.9 ± 6.3	1
$CH_3S\text{–}\mathbf{CH_3}$	307.9 ± 3.3	1
$HOS\text{–}\mathbf{CH_3}$	284.9 ± 12.6	1
$CH_3SO\text{–}\mathbf{CH_3}$	221.8 ± 8.4	1
$HOSO_2\text{–}\mathbf{CH_3}$	324.3 ± 12.6	1
$CH_3SO_2\text{–}\mathbf{CH_3}$	279.5	1
$\mathbf{F_5S}\text{–}CF_3$	392 ± 43	1
$\mathbf{F}\text{–}SF_5$	391.6	1
$\mathbf{F}\text{–}SO_2(F)$	379	1
$\mathbf{Cl}\text{–}SF_5$	<272	1
$\mathbf{Cl}\text{–}SO_2CH_3$	293	1
$\mathbf{Cl}\text{-}SO_2Ph$	297	1
$\mathbf{Br}\text{-}SBr$	259 ± 17	1
$\mathbf{Br}\text{–}SF_5$	<230	1
$\mathbf{I}\text{-}SH$	206.7 ± 8.4	1
$\mathbf{I}\text{-}SCH_3$	206.3 ± 7.1	1

(7) Si-, Ge-, Sn-, and Pb–X BDEs

Bond	$D°_{298}$/kJ mol^{-1}	Ref.
$SiH_3\text{–}\mathbf{H}$	383.7 ± 2.1	1
$Me_3Si\text{–}\mathbf{H}$	396 ± 7	1
$H_5Si_2\text{–}\mathbf{H}$	373 ± 8	1
$(C_2H_5)_3Si\text{–}\mathbf{H}$	396 ± 4	1
$C_6H_5SiH_2\text{–}\mathbf{H}$	382 ± 5	1
$(CH_3S)_3Si\text{–}\mathbf{H}$	364.0	1
$(iPrS)_3Si\text{–}\mathbf{H}$	376.6	1
$PhMe_2Si\text{–}\mathbf{H}$	377 ± 7	1
$Ph_2SiH\text{–}\mathbf{H}$	379 ± 7	1
$Ph_2MeSi\text{–}\mathbf{H}$	361 ± 0	1
$SiF_3\text{–}\mathbf{H}$	432 ± 5	1
$SiCl_3\text{–}\mathbf{H}$	391	5
$SiBr_3\text{–}\mathbf{H}$	334 ± 8	1
$SiH_3\text{–}\mathbf{SiH_3}$	321 ± 4	1
$\mathbf{SiH_3}\text{–}Si_2H_5$	313 ± 8	1
$\mathbf{Ph_3Si}\text{–}SiPh_3$	368.2	1
$\mathbf{F_3Si}\text{–}SiF_3$	453.1 ± 25	1
$SiH_3\text{–}\mathbf{CH_3}$	375 ± 5	1
$SiF_3\text{–}\mathbf{CH_3}$	355.6	1
$H_3Si\text{–}\mathbf{NO}$	158.2 ± 5.7	1
$H_3Si\text{–}\mathbf{PH_2}$	331.4	1
$SiH_3\text{–}\mathbf{F}$	638 ± 5	1
$SiH_3\text{–}\mathbf{Cl}$	458 ± 7	1
$SiH_3\text{–}\mathbf{Br}$	376 ± 9	1
$SiH_3\text{–}\mathbf{I}$	299 ± 8	1
$GeH_3\text{–}\mathbf{H}$	348.9 ± 8.4	1
$Me_3Ge\text{–}\mathbf{H}$	364.0	1
$Ph_3Ge\text{–}\mathbf{H}$	359.8	1
$\mathbf{(CH_3)_3Ge}\text{-}Ge(CH_3)_3$	280.3	1
$\mathbf{(CH_3)_3Ge}\text{-}CH_3$	288.7	1
$\mathbf{Me_3Sn\text{–}H}$	307.5 ± 10.7	11
$Ph_3Sn\text{–}\mathbf{H}$	294.6	1
$\mathbf{(CH_3)_3Sn}\text{-}Sn(CH_3)_3$	252.6 ± 14.8	11
$\mathbf{(CH_3)_3Sn}\text{-}Cl$	425 ± 7	1
$\mathbf{(CH_3)_3Sn}\text{–}CH_3$	284.1 ± 9.9	11
$\mathbf{(CH_3)_3Pb}\text{-}Pb(CH_3)_3$	228.4	
$\mathbf{Cl_3Pb}\text{-}Cl$	271 ± 4	1
$\mathbf{(CH_3)_3Pb}\text{-}CH_3$	238 ± 21	1

(8) P-, As-, Sb-, Bi–X BDEs

Bond	$D°_{298}$/kJ mol^{-1}	Ref.
$H_2P\text{–}\mathbf{H}$	351.0 ± 2.1	1
$CH_3PH\text{–}\mathbf{H}$	322.2 ± 12.6	1
$\mathbf{H_2P}\text{–}PH_2$	256.1	1
$\mathbf{(C_2H_5)_2P}\text{–}P(C_2H_5)_2$	359.8	1
$\mathbf{F_2P}\text{–}F$	549	1
$\mathbf{Cl_2P}\text{–}Cl$	356 ± 8	1
$\mathbf{Br_2P}\text{–}Br$	<259	1
$\mathbf{I_2P}\text{–}I$	217	1
$\mathbf{H_2P}\text{–}SiH_3$	331.4	1
$\mathbf{H_2As}\text{–}H$	319.2 ± 0.8	1
$\mathbf{H_2Sb}\text{–}H$	288.3 ± 2.1	1
$\mathbf{F_2Bi}\text{–}F$	435 ± 19	1
$\mathbf{Br_2Bi}\text{–}Br$	>297.1	1

(9) Se- and Te–X BDEs

Bond	$D°_{298}$/kJ mol^{-1}	Ref.
$\mathbf{H}\text{–}SeH$	334.93 ± 0.75	1
$\mathbf{H}\text{–}SeC_6H_5$	326.4 ± 16.7	1
$\mathbf{PhSe}\text{–}SePh$	280 ± 9	1
$\mathbf{H}\text{–}TeH$	277.0 ± 5.0	1
$\mathbf{H}\text{–}TeC_6H_5$	≤264	1
$\mathbf{PhTe}\text{–}TePh$	138.1 ± 12.6	1

(10) Metal-Centered BDEs

Arranged by the Periodic Table

(10.1) Group 1

Bond	$D°_{298}$/kJ mol^{-1}	Ref.
$\mathbf{Li}\text{–}OH$	431.0	
$\mathbf{Li}\text{–}C_2H_5$	214.6 ± 8.4	1
$\mathbf{Li}\text{–}nC_4H_9$	197.9 ± 16.3	1
$\mathbf{Na}\text{–}OH$	342.3	1
$\mathbf{Na}\text{–}O_2$	<200	1
$\mathbf{K}\text{–}OH$	359	1

Bond	D°_{298}/kJ mol^{-1}	Ref.
Rb–OH	356.2 ± 4.2	1
Cs–OH	373	1

(10.2) Group 2

Bond	D°_{298}/kJ mol^{-1}	Ref.
BeO–H	469	1
Be(OH)–OH	476	1
MgO–H	441	1
Mg(OH)–OH	349	1
BrMg–CH_3	253	1
BrMg–CH_2CH_3	205	1
BrMg–i-C_3H_7	184	1
BrMg–t-C_4H_9	174	1
BrMg–C_6H_5	289	1
BrMg–$CH_2C_6H_5$	201	1
BrMg–$C(C_6H_5)_3$	180	1
Ca(OH)–OH	409	1
Sr(OH)–OH	407	1
Ba(OH)–OH	443	1

(10.3) Group 3

Bond	D°_{298}/kJ mol^{-1}	Ref.
Sc–CH_3	116 ± 29	1
Sc–C_6H_6	60.8	1
$La(\eta^5\text{-}C_5Me_5)_2$–$CH(SiMe_3)_2$	278.7 ± 10.5	1
$Nd(\eta^5\text{-}C_5Me_5)_2$–$CH(SiMe_3)_2$	236.8 ± 10.5	1
$(\eta^5\text{-}C_5Me_5)_2Sm$–H	226.8 ± 12.6	1
$(\eta^5\text{-}C_5Me_5)_2Sm$–$OCH_3$	343.1	1
$(\eta^5\text{-}C_5Me_5)_2Sm$–$(\eta^3\text{-}C_3H_5)$	188.3 ± 6.3	1
$(\eta^5\text{-}C_5Me_5)_2Sm$–S-$nC_3H_7$	295.4 ± 10.0	1
$(\eta^5\text{-}C_5Me_5)_2Sm$–$N(CH_3)_2$	201.7 ± 7.5	1
$(\eta^5\text{-}C_5Me_5)_2Sm$–$SiH(SiMe_3)_2$	179.9 ± 21	1
$(\eta^5\text{-}C5Me_5)_2Sm$–$P(Et)_2$	136.4 ± 8.4	1
$(\eta^5\text{-}C_5Me_5)_2Eu$–I	238.9 ± 8.4	1
$(\eta^5\text{-}C_5Me_5)_2Yb$–I	256.1 ± 6.3	1
$Lu(\eta^5\text{-}C_5Me_5)_2$–$CH(SiMe_3)_2$	279.1 ± 10.5	1
$(\eta^5\text{-}C_5H_4SiMe_3)_3Th$–H	277 ± 6	1
$(\eta^5\text{-}C_5H_4SiMe_3)_3Th$–O	371 ± 24	1
$(\eta^5\text{-}C_5H_5)_3Th$–$CH_3$	375 ± 9	1
$(\eta^5\text{-}C_5H_5)_3Th$–$CH_2Si(CH_3)_3$	369 ± 12	1
$(C_9H_7)_3Th$–$CH_2C_6H_5$	342 ± 9	1
$(\eta^5\text{-}C_5H_4tBu)_3U$–H	249.7 ± 5.7	1
$(\eta^5\text{-}C_5H_4SiMe_3)_3U$–H	253.7 ± 5.1	1
$[HB(3,5\text{-}Me_2Pz)_3]U(Cl)_2$–Cl	422.6	1
$(\eta^5\text{-}C_5H_4SiMe_3)_3U$–I	265.6 ± 4.3	1
$(\eta^5\text{-}C_5H_4tBu)_3U$–O	307 ± 9	1
$(\eta^5\text{-}C_5H_4SiMe_3)_3U$–CO	43.1 ± 0.8	1
$(C_9H_7)_3U$–CH_3	196.3 ± 6.6	1
$(\eta^5\text{-}C_5Me_5)_2U(Cl)$–$C_6H_5$	358 ± 11	1
$(\eta^5\text{-}C_5H_4SiMe_3)_3U$–THF	41.0 ± 0.8	1

(10.4) Group 4

Bond	D°_{298}/kJ mol^{-1}	Ref.
$Ti(\eta^5\text{-}C_5H_5)_2$–Cl	471	1
$Ti(Cl)(\eta^5\text{-}C_5H_5)_2$–Cl	390	1
$Ti(\eta^5\text{-}C_5Me_5)_2$–I	219	1
$Ti(\eta^5\text{-}C_5H_5)_2$–CO	174	1
$Ti(CO)(\eta5\text{-}C_5H_5)_2$–CO	170	1
Ti–CH_3	174 ± 29	1
$Ti(Cl)(\eta^5\text{-}C_5H_5)_2$–$CH_3$	276	1
$Ti(Cl)((\eta^5\text{-}C_5H_5)_2$–$C_6H_5$	292	1
$Ti(C_6H_6)$–C_6H_5	308.7	1
$Zr(\eta^5\text{-}C_5Me_5)_2$–H	351.0 ± 7.5	1
$Zr(H)(\eta^5\text{-}C_5Me_5)_2$–H	326.4 ± 4	1
$Zr(\eta^5\text{-}C_5Me_5)_2$–Cl	481.2	1
$Zr(\eta^5\text{-}C_5Me_5)_2$–Br	410.0	1
$Zr(I)(\eta^5\text{-}C_5Me_5)_2$–I	336.4 ± 2.1	1
$Zr(\eta^5\text{-}C_5Me_5)_2(Ph)$–OH	482.4 ± 6.3	1
$Zr(\eta^5\text{-}C_5Me_5)_2(Ph)(OH)$–OH	482.8 ± 10.5	1
$Zr(\eta^5\text{-}C_5Me_5)_2(NH_2)H$–$NH_2$	421.3 ± 15.1	1
$Zr(\eta^5\text{-}C_5Me_5)_3$–$CH_3$	276 ± 10	1
$Zr(\eta^5\text{-}C_5H_5)_2(C_6H_5)$–$C_6H_5$	300 ± 10	1
$Zr(\eta^5\text{-}C_5H_5)2(Si(SiMe_3)_3)$–$SiMe_3$	188 ± 30	1
$Hf(H)(\eta^5\text{-}C_5Me_5)_2$–H	346.0 ± 7.9	1
$Hf(\eta^5\text{-}C_5Me_5)(C_4H_9)$–$C_4H_9$	274 ± 10	1

(10.5) Group 5

Bond	D°_{298}/kJ mol^{-1}	Ref.
$(\eta^5\text{-}C_5H_5)(CO)_3V$–$\eta^2H_2$	90 ± 20	1
$(\eta^5\text{-}C_5H_5)(CO)_3V$–CO	146 ± 21	1
V–CH_3	169 ± 18	1
V–C_6H_6	76.2	1
$V(C_6H_6)$–C_6H_6	307.8	1
$Nb(\eta^5\text{-}C_5H_5)_2H_3$–TFE	18.8 ± 1.3	1
$Ta(CH_3)_5$–CH_3	261 ± 5	1
$(Me_3SiCH_2)_4Ta$–(CH_2SiMe_3)	184.1 ± 8.4	1

(10.6) Group 6

Bond	D°_{298}/kJ mol^{-1}	Ref.
$[Cr(CO)_3(\eta^5\text{-}C_5Me_5)]_2$–Hg	61.5	1
$[Cr(CO)_3(\eta^5\text{-}C_5Me_5)]$–Hg	111.3	1
$Cr(CO)_5$–Xe	37.7 ± 3.8	1
$(CO)_2(PPh_3)(\eta^5\text{-}C_5H_5)Cr$–H	250.2 ± 4.2	1
$(\eta^5\text{-}C_5H_5)Cr(CO)_3$–H	257	1
$Cr(CO)_5$–H_2	78 ± 4	1
$(P(C_6H_{11})_3)_2(CO)_3Cr$–$H_2$	30.5 ± 0.4	1
$(\eta^6\text{-}C_6H_6)(CO)_3Cr$–$H_2$	251 ± 17	1
$Cr(CO)_5$–N_2	81 ± 4	1
$(P(C_6H_{11})_3)_2(CO)_3Cr$–$N_2$	38.9 ± 0.8	1
$(\eta^5\text{-}C_5Me_5)(CO)_3Cr$–SH	193	1
$Cr(CO)_5$–CO	154.0 ± 8.4	1
$Cr(CO)_5$–CH_4	~33.5 ± 8	1
Cr–C_6H_6	9.6 ± 5.8	1
$Cr(C_6H_6)$–C_6H_6	268.2 ± 15.4	1
$Cr(CO)_5$–C_6H_6	57.3 ± 3.3	1
$(P(C_6H_{11})_3)_2(CO)_3Cr$–$P(OMe_3)_3$	68.6 ± 2.5	1
$(\eta5\text{-}C_5H_5)Mo(CO)_3$–H	290	1
$Mo(\eta^5\text{-}C_5H_5)_2$–H	246	1
$Mo(H)(\eta^5\text{-}C_5H_5)_2$–H	256.9 ± 8.4	1
$Mo(CO)_3(\eta^5\text{-}C_5H_5)$–I	216.7 ± 4.2	1
$(\eta^5\text{-}C_5Me_5)_2Mo$–O	272	1
$(P(C_6H_{11})_3)_2(CO)_3Mo$–$H_2$	27.2 ± 0.8	1
$(P(C_6H_{11})_3)_2(CO)_3Mo$–$N_2$	37.7 ± 2.5	1
$Mo(CO)_5$–CO	169.5 ± 8.4	1
$Mo(CO)_3(\eta^5\text{-}C_5H_5)$–$CH_3$	203 ± 8	1
$W(CO)_5$–Xe	35.1 ± 0.8	1
$W(CO)_3(\eta^5\text{-}C_5H_5)$–H	303	1
$W(H)(\eta^5\text{-}C_5H_5)_2$–H	310.9 ± 4.2	1
$W(I)(\eta^5\text{-}C_5H_5)_2$–H	273 ± 14	1
$(CO)_5W$–H_2	≥67	1
$(P(C_6H_{11})_3)(CO)_3W$–$(\eta^2\text{-}H_2)$	28.5 ± 2.1	1
$W(CO)_5$–CO	192.5 ± 8.4	1
$W(CH_3)(\eta^5\text{-}C_5H_5)_2$–$CH_3$	220.9 ± 4	1

(10.7) Group 7

Bond	D°_{298}/kJ mol^{-1}	Ref.
F_3Mn–MnF_3	210.9 ± 2.5	1
$(CO)_5Mn$–$Mn(CO)_5$	185 ± 8	1
$(CO)_5Mn$–H	284.5	1
$(PPh_3)Mn(CO)_4$–H	286.2	1
$MnBr(CO)_4$–CO	184	1
$(\eta^5\text{-}C_5H_5)(CO)_2Mn$–CO	195.8 ± 9.2	1
Mn–CH_3	>35 ± 12	1
$Mn(CO)_5$–CH_3	187.0 ± 3.8	1
$Mn(CO)_5$–C_6H_5	207 ± 11	1
$(CO)_5Mn$–$Re(CO)_5$	149 ± 11	1
$(\eta^5\text{-}C_5H_5)Mn(CO)_2$–PhMe	59.4 ± 3.3	1
$(CO)_5Tc$–$Tc(CO)_5$	177.5 ± 1.9	1
$(CO)_5Re$–$Re(CO)_5$	187 ± 4.8	1
$(CO)_5Re$–H	313	1
$(CO)_5Re$–CH_3	220 ± 8	1

(10.8) Group 8

Bond	D°_{298}/kJ mol^{-1}	Ref.
$(CO)_4Fe$–$Fe(CO)_5$	171.5	1
$(CO)_4Fe(H)_x$–H	259.4 ± 8.4	1
$(\eta^5\text{-}C_5H_5)(CO)_2Fe$–H	239	1
$Fe(CO)_3(N_2)$–N_2	37.7 ± 19.2	1
$Fe(C_2H_2)(CO)_4$–CO	88 ± 2.3	1
$Fe(CO)_2(PMe_3)$–CO	>125	1
$Fe(CO)_3(PPh_3)$–CO	<177.8 ± 5	1
Fe–NH_3	31.4 ± 4.2	1
Fe–CH_2	364 ± 29	1
Fe–CH_3	135 ± 29	1
$Fe(C_2H_4)(CO)_3$–C_2H_4	89.1 ± 8	1
Fe–C_3H_5	218	1
Fe–C_3H_6	79	1
$Fe(CO)_5$–$Ni(CO)_4$	37.7	1
$Fe(CO)_5$–$(\eta^3\text{-}C_3H_5)$	176	1
$Fe(C_3H_6)(CO)_3$–C_3H_6	~79.5	1
$(CO_2)(\eta^5\text{-}C_5H_5)Ru$–H	272	1
$(PMe_3)_2(\eta^5\text{-}C_5Me_5)Ru$–H	167.4	1
$(CO)_2(\eta^5\text{-}C_5Me_5)Ru$–Cl	337.6	1
$(\eta^5\text{-}C_5Me_5)(PMe_3)_2Ru$–Cl	<138	1
$(\eta^5\text{-}C_5Me_5)(PMe_3)_2Ru$–OH	204.6	1
$(CO)_4Ru$–CO	115 ± 1.7	1
$(\eta^5\text{-}C_5Me_5)(PMe_3)_2Ru$–$CH_3$	142.3	1
$Os(H)(CO)_4$–H	326.4	1
$(CO)_4Os$–CO	133 ± 2.6	1
$Os(C_2H_2)(CO)_4$–CO	99.5 ± 0.8	1

Bond	D^o_{298}/kJ mol^{-1}	Ref.
(10.9) Group 9		
(CO)$_4$Co–Co(CO)$_4$	83 ± 29	1
(CO)$_4$Co–Mn(CO)$_5$	96 ± 12	1
(CO)$_4$Co–Re(CO)$_5$	113 ± 15	1
Co(CO)$_4$–H	278	1
Co(CO)$_3$(PPh$_3$)–H	272	1
(CO)$_3$HCo–CO	~54	1
(η^5-C$_5$H$_5$)Co(CO)–CO	184.3 ± 4.8	1
Co–CH$_2$	331 ± 38	1
Co–CH$_3$	178 ± 8	1
cobalamin–CH$_3$	150.6	1
cobinamide–iC$_4$H$_9$	104	1
Co–C bonds in B$_{12}$	123.8 ± 6.3	1
Cl(CO)$_2$Rh–Rh(CO)$_2$Cl	94.6	1
HRh(m-xylyl)Rh–H	255.6 ± 1.7	1
(PiPr$_3$)$_2$(Cl)Rh–H$_2$	136.0	1
(PiPr$_3$)$_2$(Cl)Rh–N$_2$	69.0	1
(PiPr$_3$)$_2$(Cl)Rh–CO	201.7	1
HRh(m-xylyl)Rh–CH$_2$OH	195.4 ± 7.5	1
Ir(Cl)(CO)(PMe$_3$)$_2$–H	251	1
Ir(H)(η^5-C$_5$Me$_5$)(PMe$_3$)– H	310.5 ± 21	1
Ir(Cl)(H)(CO)(PEt$_3$)$_2$–H	243.1	1
Ir(Cl)(H)(CO)(PPh$_3$)$_2$–H	246.9	1
(Cl)(CO)(PPh$_3$)$_2$Ir–H$_2$	62.8	1
(Cl)(CO)(PPh$_3$)$_2$Ir–CO	45.2	1
Ir(H)(η^5-C$_5$Me$_5$)(PMe$_3$)– C$_6$H$_5$	321	1
(10.10) Group 10		
Ni–H$_2$O	~29	1
Ni(CO)$_3$–N$_2$	~42	1
Ni(CO)$_3$–CO	104.6 ± 8.4	1
Ni–CH$_3$	208 ± 8	1
Ni–C$_2$H$_2$	193 ± 25	1
Ni–C$_2$H$_4$	147.3 ± 17.6	1
Ni–propyne	155 ± 21	1
Ni–2-butyne	121 ± 21	1
Pd–OH	213	1
***trans*-Pt(PPh$_3$)$_2$(Cl)**–H	307 ± 37	1
[Ph$_2$PCH$_2$]$_2$MePt–H	104.6	1
[Ph$_2$PCH$_2$]$_2$MePt –OH	167.4	1
[Ph$_2$PCH$_2$]$_2$MePt –SH	90.0	1
Pt(η^5-C$_5$H$_5$)(CH$_3$)$_2$–CH$_3$	163 ± 21	1
cis-Pt(PEt$_3$)$_2$(CH$_3$)–CH$_3$	269 ± 13	1
(10.11) Group 11		
Cu–OH	>406	1
Cu–CO	25 ± 5	1
Cu–CH$_3$	223 ± 5	1
Cu–NH$_3$	47 ± 15	1
Cu(NH$_3$)–NH$_3$	83.7 ± 4.2	1
Cu–C$_6$H$_6$	16.4 ± 12.5	1
Cu(C$_6$H$_6$)–C$_6$H$_6$	27.0 ± 19.3	1
Ag–CH$_3$	134.1 ± 6.8	1
Ag–NH$_3$	8 ± 13	1
Ag(NH$_3$)–NH$_3$	62.8 ± 4.2	1
Au–OH	>262	1
Au–NH$_3$	76 ± 6	1
Au–CH$_3$	≥191.6	1
Au–C$_6$H$_6$	8.4	1
(10.12) Group 12		
Zn–CH$_3$	70 ± 10	1
Zn(CH$_3$)–CH$_3$	266.5 ± 6.3	1
Zn–C$_2$H$_5$	92.0 ± 17.6	1
Zn(C$_2$H$_5$)–C$_2$H$_5$	219.2 ± 8.4	1
Cd–CH$_3$	63.6 ± 10.0	1
Cd(CH$_3$)–CH$_3$	234.3 ± 6.3	1
Hg–CH$_3$	22.6 ± 12.6	1
Hg(CH$_3$)–CH$_3$	239.3 ± 6.3	1
ClHg–CH$_3$	280.0 ± 12.6	1
BrHg–CH$_3$	270 ± 38	1
IHg–CH$_3$	258.6 ± 12.6	1
(10.13) Group 13		
H$_3$B–BH$_3$	172	1
H$_3$B–NH$_3$	130.1 ± 4.2	1
(CH$_3$)$_3$B–NH$_3$	57.7 ± 1.3	1
F$_3$B–N(CH$_3$)$_3$	130 ± 4.6	1
Cl$_3$B–N(CH$_3$)$_3$	127.6	1
F$_2$B–CH$_3$	397 - 418	1
Al–OH	547 ± 13	1
Al–C$_2$H$_2$	>54	1
Cl$_3$Al–N(CH$_3$)$_3$	198.7 ± 8.4	1
(CH$_3$)$_3$Al–N(CH$_3$)$_3$	130	1
(CH$_3$)$_3$Al–O(CH$_3$)$_2$	92	1
CH$_3$Ga–GaH$_3$	59 ± 16	1
(CH$_3$)$_3$Ga–O(C$_2$H$_5$)$_2$	50.6 ± 0.8	1
Cl$_3$Ga-S(C$_2$H$_5$)$_2$	235.1	1
In–CH$_3$	216.3	1
In(CH$_3$)–CH$_3$	318.8	1
In(CH$_3$)$_2$–CH$_3$	587.4	1
In(CH$_3$)$_3$–N(CH$_3$)$_3$	83.3 ± 2.1	1
Tl–OH	330 ± 30	1

References

1. Luo, Y. R. *Comprehensive Handbook of Chemical Bond Energies*, CRC Press, Boca Raton, FL, 2007.
2. Shuman, N. S., Ochieng, M. A., Sztáray, B., and Baer, T., *J. Phys. Chem. A* 112, 5647, 2008.
3. Seetula, J. A., and Eskola, A. J., *Chem. Phys.* 351, 141, 2008.
4. Golden, D. M., *Int. J. Chem. Kinet.* 41, 573, 2009.
5. Shuman, N. S., Spencer, A. P., and Baer, T., *J. Phys. Chem. A* 113, 9458, 2009.
6. Villano, S. M., Eyet, N., Wren, S. W., Ellison, G. B., Bierbaum, V. M., and Lineberger, W. C., *J. Phys. Chem. A* 114, 191, 2010.
7. Downs, A.J., Greene, T. M., Johnsen, E., Pulham, C. R., Robertson, H. E., and Wann, D. A., *Dalton Trans.* 39, 5637, 2010.
8. Villano, S. M., Eyet, N., Wren, S. W., Ellison, G. B., Bierbaum, V. M., and Lineberger, W. C., *Eur. J. Mass Spectrom.* 16, 255, 2010.
9. Bodi, A., Kercher, J. P., Bond, C., Meteesatien, P., Sztáray, B., and Baer, T., *J. Phys. Chem. A* 110, 13425, 2006. Derived from heats of formation of compounds.
10. Stevens, W. R., Bodi, A., and Baer, T., *J. Phys. Chem. A* 114, 11285, 2010. Derived from heats of formation of compounds.
11. Da'valos, J. Z., Herrero, R., Shuman, N. S., and Baer, T., *J. Phys. Chem. A* 115, 402, 2011.
12. Stevens, W. R., Ruscic, B., and Baer, T., *J. Phys. Chem. A*, 114, 13134, 2010.
13. Bodi, A., Shuman, N. S., and Baer, T., *Phys. Chem. Chem. Phys.*, 11, 11013, 2009.
14. Rissanen, M. P., Amedro, D., Krasnoperov, L., Marshall, P., and Timonen, R. S., *J. Phys. Chem. A*, 117, 793, 2013.
15. Borkar, S., Sztáray, B., and Bodi, A., *Int. J. Mass Spectrom.*, 330-332, 100, 2012. Derived from heats of formation of compounds.
16. Ruscic, B., Active Thermochemical Table, http://atct.anl.gov/index.html, Dec. 16, 2010.
17. Harvey, J., Tuckett, R. P., and Bodi, A., *J. Phys. Chem. A*, 116, 9696, 2012. Derived from heats of formation of compounds.
18. Dixon, A. R., Khuseynov, D., and Sanov, A., *Chem. Phys. Lett.* 614, 72-77, 2014.
19. Dixon, A. R., Khuseynov, D., and Sanov, A., *J. Phys. Chem. A* 118, 8533-8541, 2014.
20. Re-derived based on updated $\Delta_f H^0(C_6H_5)$ = 337.3±0.6 kJ/mol.
21. McKee, K., et al., *J. Phys. Chem. A*, 120, 1408-1420, 2016.

Enthalpies of Formation of Free Radicals and Other Transient Species

Table 4. Enthalpies of Formation of Free Radicals and Other Transient Species

Radical	$\Delta_f H^\circ_{298}$/kJ mol^{-1}	Ref.
(1) Carbon-Centered Species		
CH	595.8 ± 0.6	1
CH_2 (triplet)	391.2 ± 1.6	1
CH_2 (singlet)	428.8 ± 1.6	1
$\cdot CH_3$, methyl	146.7 ± 0.3	1
$\cdot C_2H$, acetenyl, $CH\equiv C\cdot$	567.4 ± 2.1	1
$\cdot C_2H_2$, vinylidene $CH_2{=}C\cdot\cdot$	419.7 ± 16.7	1
$\cdot C_2H_3$, vinyl, $CH_2{=}C\cdot H$	299.6 ± 3.3	1
$\cdot C_2H_5$, ethyl, $CH_3C\cdot H_2$	118.8 ± 1.3	1
$\cdot C_3H_3$, propargyl, $CH\equiv CC\cdot H_2$	351.9	2
$\cdot C_3H_3$, $CH_3C\equiv C\cdot$	515 ± 13	1
$\cdot C_3H_3$, $CH_2{=}C{=}CH\cdot \leftrightarrow CH\equiv CC\cdot H_2$	351.9	2
$\cdot C_3H_3$, cyclopro-2-en-1-yl	497.5 ± 16.7	15
$\cdot C_3H_5$, allyl, $CH_2{=}CHC\cdot H_2$	171.0 ± 3.0	1
$\cdot C_3H_5$, $CH_3CH{=}C\cdot H$	267 ± 6	1
$\cdot C_3H_5$, $CH_3C\cdot{=}CH_2$	242.8±7.1	16
$\cdot C_3H_5$, cyclopropyl	279.9 ± 10.5	1
n-$C_3H_7\bullet$, n-propyl, $CH_3CH_2C\bullet H_2$	100 ± 2	1
i-$C_3H_7\bullet$, i-propyl, $CH_3C\bullet HCH_3$	88 ± 3	1
$\bullet n$-C_4H_3, $CH\equiv CCH{=}C\bullet H$	547.3	1
$\bullet i$-C_4H_3, $CH_2{=}C\bullet C\equiv CH$	499.2	1
$\cdot C_4H_5$, $CH_3C\equiv CC\cdot H_2$	304.5	1
$\cdot C_4H_5$, $CH\equiv CC\cdot HCH_3$	316.5	1
$\cdot C_4H_5$, $\cdot CH{=}CHCHCH_2$	364.4	1
$\cdot C_4H_5$, $CH_2{=}CHC\cdot CH_2$	313.3	1
$\cdot C_4H_7$, $CH_3CH{=}CHC\cdot H_2$	138.4±5.8	16
$\cdot C_4H_7$, $CH_2{=}CHCH_2C\cdot H_2$	209.2±4.2	16
$\cdot C_4H_7$, $CH_2{=}C(CH_3)C\cdot H_2$	137.9	1
$\cdot C_4H_7$, $CH_2{=}CHC\cdot HCH_3$	136.2	1
C_4H_7, $\cdot CH{=}CHCH_2CH_3$	247.1±0.9	16
$\cdot C_4H_7$, cyclopropylmethyl	213.8 ± 6.7	1
$\cdot C_4H_7$, cyclobutyl	219.2 ± 4.2	1
n-$C_4H_9\bullet$, n-butyl, $CH_3CH_2CH_2C\bullet H_2$	77.8 ± 2.1	1
i-$C_4H_9\bullet$, i-butyl, $(CH_3)_2CHC\bullet H_2$	70 ± 4	1
s-$C_4H_9\bullet$, s-butyl, $CH_3C\bullet HCH_2CH_3$	67.8 ± 2.1	1
t-$C_4H_9\bullet$, t-butyl, $(CH_3)_3C\bullet$	48 ± 3	1
$\cdot C_5H_3$, $CH\equiv C\text{-}C\equiv CC\cdot H_2$	579.1	1
$\cdot C_5H_3$, $(CH\equiv C)_2C\cdot H$	573.2	1
$\cdot C_5H_5$, $CH_2{=}CHC\equiv CC\cdot H_2$	351.5	1
$\cdot C_5H_5$, $CH_2{=}CH\text{-}C\cdot H\text{-}C\equiv CH$	372.4	1
$\cdot C_5H_5$, cyclopenta-1,3-dien-5-yl	274.1 ± 7.3	1
$\cdot C_5H_7$, $CH_3C\equiv CC\cdot HCH_3$	272.8 ± 9.2	1
$\cdot C_5H_7$, $CH\equiv CC\cdot HC_2H_5$	277.0 ± 8.4	1
$\cdot C_5H_7$, $CH\equiv CC\cdot(CH_3)_2$	257.3 ± 9.2	1
$\cdot C_5H_7$, $CH_2{=}CHCH{=}CHC\cdot H_2$	205.0 ± 12.6	1
$\cdot C_5H_7$, $(CH_2{=}CH)_2C\cdot H$	208.0 ± 4.2	1
$\cdot C_5H_7$, $CH_3CH{=}C{=}CHC\cdot H_2$	278.0	1
$\cdot C_5H_7$, spiropentyl	380.7 ± 4.2	1
$\cdot C_5H_7$, cyclopent-1-en-3-yl	160.7 ± 4.2	1
$\cdot C_5H_9$, cyclopentyl	105.9 ± 4.2	1
$\cdot C_5H_9$, $CH_2{=}CHC\cdot HCH_2CH_3$	109.6 ± 8.4	1
$\cdot C_5H_9$, $CH_3CH{=}CHC\cdot H(CH_3)$	92	1
$\cdot C_5H_9$, $CH_3CH{=}C(CH_3)C\cdot H_2$	92.0	1
$\cdot C_5H_9$, $CH_2{=}CHC\cdot(CH_3)_2$	87.0 ± 8.4	1
$\cdot C_5H_9$, $CH_2{=}C(CH_3)C\cdot H(CH_3)$	93.7	1
$\cdot C_5H_9$, $CH_2{=}C(C\cdot H_2)CH_2CH_3$	114.2	1
$\cdot C_5H_9$, $CH_2{=}CH(CH_2)_2C\cdot H_2$	179.5	1
$nC_5H_{11}\bullet$, $CH_3CH_2CH_2CH_2C\bullet H_2$	54.4	1
$\cdot C_5H_{11}$, $(C_2H_5)_2C\cdot H$	47.0	1
$\cdot C_5H_{11}$, $(nC_3H_7)(CH_3)C\cdot H$	50.2	1
$\cdot C_5H_{11}$, $(CH_3)_3C\cdot CH_2$	36.4 ± 8.4	1
$\cdot C_5H_{11}$, $(C_2H_5)(CH_3)_2C\cdot$	29	1
$\cdot C_6H_5$, phenyl	337.3 ± 30.6	11
$\cdot C_6H_7$, cyclohexa-1,3-dien-5-yl	199.2	1
$\cdot C_6H_7$, cyclohexa-1,4-dien-3-yl	208.0 ± 3.9	5
$\cdot C_6H_9$, $CH_3C\equiv CC\cdot(CH_3)_2$	221.8 ± 9.2	1
$\cdot C_6H_9$, $(CH_2{=}CH)_2C\cdot(CH_3)$	193.7	1
$\cdot C_6H_9$, cyclohexa-1-en-3-yl	119.7	1
$\cdot C_6H_{11}$, $CH_2{=}CH(CH_2)_3C\cdot H_2$	158.6	1
$\cdot C_6H_{11}$, $CH_2{=}CHC\cdot H(CH_2)_2CH_3$	89.0	1
$\cdot C_6H_{11}$, $CH_2{=}C(CH_3)C\cdot(CH_3)_2$	37.7 ± 6.3	1
$\cdot C_6H_{11}$, $(CH_3)_2C{=}C(CH_3)C\cdot H_2$	39.7 ± 6.3	1
$\cdot C_6H_{11}$, $(CH_3)_2C{=}CHC\cdot H(CH_3)$	47.3	1
$\bullet C_6H_{11}$, (Z)-$CH_3CH{=}CHC\bullet(CH_3)_2$	54.4	1
$\cdot C_6H_{11}$, cyclohexyl	75.3 ± 6.3	1
$nC_6H_{13}\bullet$, $CH_3CH_2CH_2CH_2CH_2C\bullet H_2$	33.5	1
$\bullet C_6H_{13}$, $(nC_4H_9)(CH_3)C\bullet H$	29.3	1
$\cdot C_6H_{13}$, 2-methyl-2-pentyl	3.3 ± 8.4	1
$\cdot C_6H_{13}$, 3-methyl-3-pentyl	14.2	1
$\cdot C_6H_{13}$, 2,3-dimethyl-2-butyl	3.1 ± 10	1
$\cdot C_7H_3$, $(CH\equiv C)_3C\cdot$	784.5	1
$\cdot C_7H_7$, benzyl, $C_6H_5C\cdot H_2$	208.0 ± 1.7	1
$\cdot C_7H_7$, quadricyclolan-5-yl	578.6 ± 5.4	1
$\cdot C_7H_7$, quadricyclolan-4-yl	587.4 ± 5.4	1
$\cdot C_7H_7$, norborna-2,5-dien-7-yl	511.7 ± 7.9	1
$\cdot C_7H_7$, cyclohepta-1,3,5-trien-7-yl	285.3 ± 12.6	1
$\cdot C_7H_9$, $CH_2{=}CH(CH{=}CH)_2CC\cdot H_2$	251.0	1
$\cdot C_7H_9$, $(CH_2{=}CH)_3C\cdot$	274.0	1
$\cdot C_7H_{11}$, norborn-1-yl	136.4 ± 10.5	1
$\cdot C_7H_{11}$, cycloheptenyl	119.2	1
$\cdot C_7H_{13}$, cycloheptyl	50.6 ± 4.2	1
$\cdot C_7H_{13}$, cyclo-$[C\cdot(CH_3)(CH_2)_5]$	22.6	1
$\cdot C_7H_{13}$, cyclo-$[C\cdot(CH_2CH_3)(CH_2)_4]$	47.0	1
$\bullet C_7H_{15}$, $(nC_5H_{11})(CH_3)CH\bullet$	8.4	1
$\cdot C_7H_{15}$, $(CH_3)_2CHCHC\cdot(CH_3)_2$	−21.8 ± 5.2	1
$\cdot C_8H_7$, cubyl	831.0 ± 16.7	1
$\cdot C_8H_7$, $C_6H_5C\cdot{=}CH_2$	309.6	1
$\cdot C_8H_7$, $C_6H_5CH{=}CH\cdot$	387.0	1
$\cdot C_8H_9$, $C_6H_5C\cdot H(CH_3)$	175.7 ± 7.5	1
$\cdot C_8H_9$, $C_6H_5CH_2C\cdot H_2$	236.0 ± 7.5	1
$\bullet C_8H_9$, p-$CH_3C_6H_4C\bullet H_2$	167.4	1
$\bullet C_8H_9$, m-$CH_3C_6H_4C\bullet H_2$	167.4	1
$\bullet C_8H_9$, o-$CH_3C_6H_4C\bullet H_2$	167.4	1
$\cdot C_8H_9$, 1-vinyl-cyclohexa-2,4-dienyl	247.7 ± 14.2	1
$\cdot C_8H_9$, 2-vinyl-cyclohexa-2,4-dienyl	249.8 ± 14.2	1
$\cdot C_8H_9$, 3-vinyl-cyclohexa-2,4-dienyl	269.4 ± 14.2	1
$\cdot C_8H_9$, 6-vinyl-cyclohexa-2,4-dienyl	284.5 ± 14.2	1
$\cdot C_8H_{13}$, $CH_2{=}CHCH{=}CHC\cdot H(CH_2)_2CH_3$	130.5	1
$\cdot C_8H_{13}$, $CH_2{=}CHC\cdot H(CH_2)_3CH{=}CH_2$	130.5	1
$\cdot C_8H_{13}$, bicyclooct-1-yl	92.0	1
$\cdot C_8H_{15}$, $CH_2{=}CHC\cdot H(CH_2)_4CH_3$	49.8	1
$\bullet C_8H_{15}$, (E)-$CH_3CH{=}C\bullet(CH_2)_4CH_3$	29.7	1
$\bullet C_8H_{15}$, (Z)-$(CH_3)_2C\bullet CH{=}CHCH(CH_3)_2$	9.2	1
$\cdot C_8H_{15}$, cyclooctanyl	59.4	1

Radical	$\Delta_f H^\circ_{298}$/kJ mol^{-1}	Ref.
$^{\bullet}C_8H_{15}$, cyclo-[$C^{\bullet}(CH_2CH_3)(CH_2)_5$]	10.0	1
$^{\bullet}C_9H_7$, indenyl	297.1	1
$^{\bullet}C_9H_9$, indanyl-1	204.2 ± 8.4	1
$^{\bullet}C_9H_{11}$, 2,6-dimethylbenzyl	124.7	1
$^{\bullet}C_9H_{11}$, 3,6-dimethylbenzyl	124.7	1
$^{\bullet}C_9H_{11}$, 3,5-dimethylbenzyl	124.7	1
$^{\bullet}C_9H_{11}$, $C_6H_5C^{\bullet}(CH_3)_2$	133.9 ± 4.2	1
$\bullet C_9H_{11}$, *o*-$\bullet C_6H_4C_2H_5$	279.5 ± 7.5	1
$^{\bullet}C_9H_{17}$, cyclononanyl	52.3	1
$^{\bullet}C_{10}H_7$, naphth-1-yl	401.7 ± 5.4	1
$^{\bullet}C_{10}H_7$, naphth-2-yl	400.4 ± 5.9	1
$^{\bullet}C_{10}H_{11}$, tetralin-1-yl	154.8 ± 5.0	1
$^{\bullet}C_{10}H_{13}$, 1-phenyl-but-4-yl	192.0	1
$^{\bullet}C_{10}H_{13}$, $(C_6H_5CH_2)(C_2H_5)C^{\bullet}H$	184.5	1
$^{\bullet}C_{10}H_{13}$, $(C_6H_5CH_2CH_2)(CH_3)C^{\bullet}H$	184.5	1
$^{\bullet}C_{10}H_{13}$, $(C_6H_5C^{\bullet}HCH_2CH_2CH_3$	134.7	1
$^{\bullet}C_{10}H_{15}$, 1-adamantyl	51.5	1
$^{\bullet}C_{10}H_{15}$, 2-adamantyl	61.9	1
$^{\bullet}C_{10}H_{19}$, cyclodecanyl	32.2	1
$^{\bullet}C_{11}H_9$, 1-naphthylmethyl	252.7	1
$^{\bullet}C_{11}H_{21}$, cycloundecanyl	7.5	1
$^{\bullet}C_{12}H_{23}$, cyclododecanyl	−38.5	1
$^{\bullet}C_{13}H_9$, 9-fluorenyl	297.5	1
$^{\bullet}C_{13}H_9$, 1-phenalenyl	264.3	17
$^{\bullet}C_{13}H_{11}$, $(C_6H_5)_2C^{\bullet}H$	302.1 ± 4.2	1
$^{\bullet}C_{13}H_{11}$, 9-methyl-9-fluorenyl	268.2	1
$^{\bullet}C_{14}H_{11}$, 9,10-dihydroanthracen-9-yl	261.0	1
$^{\bullet}C_{15}H_{11}$, 9-anthracenylmethyl	337.6	1
$^{\bullet}C_{15}H_{11}$, 9-phenanthrenylmethyl	311.3	1
$^{\bullet}C_{16}H_{31}$, $CH_2{=}CHC^{\bullet}H(CH_2)_{12}CH_3$	−118.8	1
$^{\bullet}C_{19}H_{15}$, trityl, $(C_6H_5)_3C^{\bullet}$	392.0 ± 8.4	1
$^{\bullet}C_{35}H_{25}$, pentamethylcyclopentadienyl	67.4	1
CF	243.9 ± 3.9	13
CF_2	−199.2 ± 5.6	12
$FC^{\bullet}(O)$	−161.2 ± 8.4	1
CHF	143.0 ± 12.6	1
CClF	31.0 ± 13.4	1
CCl	443.1 ± 13.0	1
CCl_2	226	1
$ClC^{\bullet}(O)$	−21.8 ± 2.5	1
CHCl	326.4 ± 8.4	1
CClBr	267	1
CBr	510 ± 63	1
CHBr	373 ± 18	1
CBr_2	343.5	1
CI	570 ± 35	1
CI_2	468 ± 60	1
$^{\bullet}CF_3$	−465.7 ± 2.1	1
$^{\bullet}CHF_2$	−249.8 ± 6.6	13
$^{\bullet}CH_2F$	−31.8 ± 4.2	1
$^{\bullet}CClF_2$	−279.0 ± 8.4	1
$^{\bullet}CCl_2F$	−89.0 ± 8.4	1
$^{\bullet}CBrClF$	−35.5 ± 6.3	1
$^{\bullet}CHClF$	−60.7 ± 10.0	1
$^{\bullet}CBrF_2$	−224.7 ± 12.6	1
$^{\bullet}CCl_3$	71.1 ± 2.5	1
$^{\bullet}CHCl_2$	87.1 ± 1.6	1
$^{\bullet}CH_2Cl$	117.2 ± 2.9	1
$^{\bullet}CHBrCl$	140 ± 4	1
$^{\bullet}CHBr_2$	199.1 ± 2.7	3
$^{\bullet}CBr_2Cl$	163 ± 8	1

Radical	$\Delta_f H^\circ_{298}$/kJ mol^{-1}	Ref.
$^{\bullet}CBrCl_2$	124 ± 8	1
$^{\bullet}CBr_3$	214.8	1
$^{\bullet}CH_2Br$	171.1 ± 2.7	1
$^{\bullet}CI_3$	424.9 ± 2.8	1
$^{\bullet}CHI_2$	314.4 ± 3.3	1
$^{\bullet}CH_2I$	229.7 ± 8.4	1
$^{\bullet}C_2F$, $FC{\equiv}C^{\bullet}$	460.0 ± 21.0	1
$^{\bullet}C_2Cl$, $ClC{\equiv}C^{\bullet}$	568 ± 26	1
$^{\bullet}C_2F_3$, $CF_2{=}C^{\bullet}F$	−192.0 ± 8.4	1
$^{\bullet}C_2F_2H$, $CF_2{=}C^{\bullet}H$	−92.9 ± 8.4	1
$^{\bullet}C_2F_2H$, $CHF{=}C^{\bullet}F$	−50.6 ± 8.4	1
$^{\bullet}CCl_2H$, $CHCl{=}C^{\bullet}Cl$	234.7 ± 8.4	1
$^{\bullet}CClH_2$, $CH_2{=}C^{\bullet}Cl$	>251	1
$^{\bullet}C_2F_5$, $CF_3C^{\bullet}F_2$	−892.9 ± 4.2	1
$^{\bullet}C_2HF_4$, $CF_3C^{\bullet}HF$	−680.8 ± 9.6	1
$^{\bullet}C_2HF_4$, $CHF_2C^{\bullet}F_2$	−664.8	1
$^{\bullet}C_2H_2F_3$, $CF_3C^{\bullet}H_2$	−517.1 ± 8.4	1
$^{\bullet}C_2H_2F_3$, $CHF_2C^{\bullet}HF$	-456.0	1
$^{\bullet}C_2H_2F_3$, $CH_2FC^{\bullet}F_2$	-449.8	1
$^{\bullet}C_2H_2F_2Cl$, $CF_2ClC^{\bullet}H_2$	−310.9 ± 7.0	1
$^{\bullet}C_2H_3F_2$, $CH_3C^{\bullet}F_2$	−302.5 ± 8.4	1
$^{\bullet}C_2H_3F_2$, $CHF_2C^{\bullet}H_2$	-285.8	1
$^{\bullet}C_2H_3F_2$, $CH_2FC^{\bullet}HF$	-238.5	1
$^{\bullet}C_2H_4F$, $CH_3C^{\bullet}HF$	−70.3 ± 8.4	1
$^{\bullet}C_2H_4F$, $CH_2FC^{\bullet}H_2$	−59.4 ± 8.4	1
$^{\bullet}C_2H_2F_2Cl$, $CF_2ClC^{\bullet}H_2$	−315.2 ± 6	1
$^{\bullet}C_2F_4Cl$, $CF_2ClC^{\bullet}F_2$	−686.0	1
$^{\bullet}C_2HF_3Cl$, $CClF_2C^{\bullet}HF$	−450.6 ± 12.6	1
$^{\bullet}C_2F_4Cl$, $CF_3C^{\bullet}FCl$	−728.0	1
$^{\bullet}C_2F_3Cl_2$, $CF_3C^{\bullet}Cl_2$	-564.0	1
$^{\bullet}C_2F_3ClBr$, $CF_3C^{\bullet}ClBr$	−504.2 ± 8.4	1
$^{\bullet}C_2Cl$, $ClC{\equiv}C^{\bullet}$	534 ± 50	1
$^{\bullet}C_2Cl_3$, $CCl_2{=}C^{\bullet}Cl$	190 ± 50	1
$^{\bullet}C_2Cl_5$, $CCl_3C^{\bullet}Cl_2$	35.1 ± 5.4	1
$^{\bullet}C_2HCl_4$, $CHCl_2C^{\bullet}Cl_2$	23.4 ± 8.4	1
$^{\bullet}C_2HCl_4$, $CCl_3C^{\bullet}HCl$	51.0	1
$^{\bullet}C_2H_2Cl_3$, $CH_2ClC^{\bullet}Cl_2$	26.4	1
$^{\bullet}C_2H_2Cl_3$, $CHCl_2C^{\bullet}HCl$	46.4	1
$^{\bullet}C_2H_2Cl_3$, $CCl_3C^{\bullet}H_2$	71.5 ± 8	1
$^{\bullet}C_2H_3Cl_2$, $CH_3C^{\bullet}Cl_2$	42.5 ± 1.7	1
$^{\bullet}C_2H_3Cl_2$, $CH_2ClC^{\bullet}ClH$	65.3	1
$^{\bullet}C_2H_3Cl_2$, $CHCl_2C^{\bullet}H_2$	90.1 ± 0.8	1
$^{\bullet}C_2H_4Cl$, $CH_3C^{\bullet}HCl$	76.5 ± 1.6	1
$^{\bullet}C_2H_4Cl$, $CH_2ClC^{\bullet}H_2$	93.0 ± 2.4	1
$^{\bullet}C_2H_3Br_2$, $CH_3C^{\bullet}Br_2$	140.2 ± 5.4	1
$^{\bullet}C_2H_4Br$, $BrCH_2C^{\bullet}H_2$	135.1	1
$^{\bullet}C_2H_4Br$, $CH_3C^{\bullet}HBr$	133.4 ± 3.4	3
$^{\bullet}C_2Br$, $CBrC^{\bullet}$	623.8	1
$^{\bullet}C_2Br_3$, $CBr_2C^{\bullet}Br$	385.3	1
$^{\bullet}C_2Br_5$, $CBr_3C^{\bullet}Br_2$	283.3	1
$^{\bullet}C_3H_6Cl$, $CH_3CH_2C^{\bullet}HCl$	56.6	1
$^{\bullet}C_3H_6Cl$, $CH_3C^{\bullet}ClCH_3$	29.9 ± 0.6	1
$^{\bullet}C_3H_6Br$, $C^{\bullet}H_2CH_2CH_2Br$	120.1 ± 1.3	1
$^{\bullet}C_3H_6Br$, $CH_3C^{\bullet}HCH_2Br$	96.7 ± 5.9	1
$^{\bullet}C_3H_6Br$, $CH_3CH_2C^{\bullet}HBr$	107.5 ± 2.5	1
$^{\bullet}C_6F_5$	−547.7 ± 8.4	1
$^{\bullet}CH_3O$, $HOC^{\bullet}H_2$	−17.0 ± 0.7	1
$^{\bullet}CH_2ClO$, $HOC^{\bullet}ClH$	−60.7 ± 7.5	1
$^{\bullet}CHCl_2O$, $HOC^{\bullet}Cl_2$	−94.1 ± 7.5	1
$^{\bullet}CH_2ClO$, $ClOC^{\bullet}H_2$	135.6 ± 9.2	1
$^{\bullet}CH_2BrO$, $BrOC^{\bullet}H_2$	151 ± 16	1

Radical	$\Delta_f H^o_{298}$/kJ mol^{-1}	Ref.
$\cdot C_2H_3O$, $C\cdot H{=}CHOH$	121 ± 11	1
$\cdot C_2H_3O$, $C\cdot H_2CHO$	13.0 ± 2	1
$\cdot C_2H_5O$, $CH_3C\cdot HOH$	-54.7±3.6	16
$\cdot C_2H_4ClO$, $CH_3C\cdot ClOH$	−108.4 ± 8.8	1
$\cdot C_2H_4ClO$, $C\cdot H_2CHClOH$	−73.2 ± 8.8	1
$\cdot C_2H_3Cl_2O$, $C\cdot H_2CCl_2OH$	−99.6 ± 8.8	1
$\cdot C_2H_5O$, $C\cdot H_2CH_2OH$	−28.9 ± 9.3	16
$\cdot C_2H_3O$, oxiran-2-yl	149.8 ± 6.3	1
$\cdot C_3H_5O$,$CH_2{=}CHC\cdot HOH$	0 ± 8.4	1
$\cdot C_3H_7O$, $CH_3CH_2C\cdot HOH$	−81 ± 4	1
$\cdot C_3H_7O$, $(CH_3)C\cdot HCH_2OH$	−78.7 ± 8.4	1
$\cdot C_3H_7O$, $HOCH_2CH_2C\cdot H_2$	−66.9 ± 8.4	1
$\cdot C_3H_7O$, $(CH_3)_2C\cdot OH$	-96.4	1
$\cdot C_3H_7O$, $\cdot CH_2CH(OH)CH_3$	−62.8 ± 11.7	1
$\cdot C_4H_9O$, $\cdot CH_2C(OH)(CH_3)_2$	−147.3 ± 8.4	1
$\cdot C_2H_5O_3$, $C\cdot H_2OCH_2OOH$	109.6 ± 4.2	1
$PhCH\cdot OH$	29.3 ± 8.4	1
$Ph_2C\cdot OH$	152.3 ± 6.3	1
$\cdot C_2H_5O$, $CH_3OC\cdot H_2$	0 ± 4.2	1
$\cdot C_3H_7O$, $CH_3OC\cdot HCH_3$	−57.7 ± 8.4	1
$\cdot C_3H_7O$, $CH_3CH_2OC\cdot H_2$	−45.2 ± 8.4	1
$\cdot C_3H_7O$, $C\cdot H_2CH_2OCH_3$	−7.1 ± 4.2	1
$\cdot C_4H_9O$, $(CH_3)_2CHOC\cdot H_2$	−70.3 ± 7.1	1
$\cdot C_4H_9O$, $CH_3CH_2OC\cdot HCH_3$	−81.2 ± 4.2	1
$\cdot C_4H_9O$, $C\cdot H_2CH(CH_3)OCH_3$	−42.3 ± 3.8	1
$\cdot C_4H_9O$, $(CH_3)_2C\cdot OCH_3$	−72.4 ± 10	1
$\cdot C_5H_{11}O$, $(CH_3)_3COC\cdot H_2$	−102.5 ± 8.4	1
$\cdot C_2H_5O_2$, $HOCH_2C\cdot HOH$	−220.1 ± 8.4	1
$C\cdot H{=}C{=}O$, ketenyl	177.5 ± 8.8	1
$HC\cdot(O)$	42.5 ± 0.5	1
$C\cdot CO$	381.2 ± 2.1	1
$CH_3C\cdot(O)$	−10.3 ± 1.8	1
$CF_3C\cdot(O)$	-608.7	1
$CH_2ClC\cdot(O)$	−21 ± 12.6	1
$CHCl_2C\cdot(O)$	−17.6 ± 23	1
$CCl_3C\cdot(O)$	-19.7	1
$CH_3CH_2C\cdot(O)$	−31.7 ± 3.4	1
$CH_2CHC\cdot(O)$	88.5	1
$CH_2C(CH_3)C\cdot(O)$	58.6 ± 16.7	1
$CH_3CH_2CH_2C\cdot(O)$	54.4 ± 4.2	1
$(CH_3)_2CHC\cdot(O)$	−64.0 ± 3.8	1
$(CH_3)_3CC\cdot(O)$	−102.9 ± 6.3	1
$C_6H_5C\cdot(O)$	116.3 ± 10.9	1
$HC(O)CH_2\cdot$	10.5 ± 9.2	1
$ClC(O)CH_2\cdot$	−52.7 ± 13	1
E-$C\bullet HClC(O)H$	−27.2 ± 10.5	1
Z-$C\bullet HClC(O)H$	−23.4 ± 10.5	1
$C\cdot Cl_2C(O)H$	−55.6 ± 14.2	1
E-$C\bullet HClC(O)Cl$	−88.7 ± 15.1	1
$C\cdot H_2C(O)F$	−273.0 ± 5.8	1
Z-$C\bullet HClC(O)Cl$	−84.9 ± 13.8	1
$C\cdot Cl_2C(O)Cl$	−101.7 ± 15.5	1
$CH_3C(O)CH_2\cdot$	−34 ± 3	1
$CH_3C(O)C\cdot HCH_3$	−70.3 ± 7.1	1
$CH_3C(O)C\cdot{=}CH_2$	113.4	1
$C_2H_5C(O)C\cdot HCH_3$	−107.5 ± 20.9	1
i$PrC(O)C\bullet(CH_3)_2$	−173.6 ± 20.9	1
t$C_4H_9C(O)C\bullet H_2$	−115.5 ± 12.6	1
$PhC(O)C\cdot H_2$	84.5 ± 12.6	1
$PhC(O)C\cdot HCH_3$	41.4 ± 20.9	1
$PhC\cdot HC(O)CH_2Ph$	134.3 ± 20.9	1
$PhC(O)OC\cdot H_2$	-69.9	1
$\bullet C(O)OH$-*trans*	≥−194.6 ± 2.9	1
$\bullet C(O)OH$-*cis*	-219.7	1
$\cdot C(O)OCH_3$	-161.5	1
$C\cdot H_2C(O)OH$	−248.9 ± 12.0	1
$C\cdot H(CH_3)C(O)OH$	−293 ± 3	1
$C\cdot H_2C(O)OCH_3$	−236.8 ± 8.4	1
$C\cdot H_2C(O)OCH_2CH_3$	−260.2 ± 12.6	1
$C\cdot H_2C(O)OPh$	−28.0	1
$\cdot C_4H_7O$, tetrahydrofuran-2-yl	−18.0 ± 6.3	1
$\cdot C_4H_8O$, cyclopentanon-2-yl	−41.8 ± 12.6	1
$\cdot C_4H_7O_2$, 1,4-dioxan-2-yl	−131.8 ± 12.6	1
$\cdot C_7H_5O_2$, 2-$C(O)OH$-$\cdot C_6H_4$	-33.0	1
$\cdot C_7H_5O_2$, 3-$C(O)OH$-$\cdot C_6H_4$	-35.0	1
$\cdot C_7H_5O_2$, 4-$C(O)OH$-$\cdot C_6H_4$	-36.0	1
$\cdot CH_3O_2$, $C\cdot H_2OOH$	66.1	1
$\cdot C_2H_5O_2$, $C\cdot H_2CH_2OOH$	46.0 ± 4.6	1
$\cdot C_2H_5O_2$, $CH_3CH\cdot OOH$	26.9	1
$\cdot C_3H_7O_2$, $CH_3CH\cdot CH_2OOH$	10.9 ± 5.4	1
$\cdot C_3H_7O_2$, $C\cdot H_2CH(OOH)CH_3$	2.9 ± 6.3	1
$\cdot C_4H_9O_2$, $(CH_3)_2C\cdot CH_2OOH$	−30.1 ± 5.4	1
$\cdot C_4H_9O_2$, $C\cdot H_2C(CH_3)_2OOH$	−26.8 ± 5.4	1
$\cdot C_2H_3O_3$, $C\cdot H_2C(O)OOH$	-137.9	1
$\cdot CHN_2$	494.5	1
$\cdot CH_2N{=}CH_2$	263.6 ± 12.6	1
$\cdot CH_2NH_2$	151.9 ± 8.4	1
$CH_3C\cdot HNH_2$	111.7 ± 8.4	1
$(CH_3)_2C\cdot NH_2$	69.9 ± 8.4	1
$\cdot CH_2NHCH_3$	156.6	1
$\cdot CH_2N(CH_3)_2$	148.0	1
$(C_2H_5)_2NC\cdot HCH_3$	68.6 ± 2.1	1
$\cdot CH_2N(CH_3)Ph$	266.0 ± 12.6	1
$\cdot CN$	439.3 ± 2.9	1
$\cdot CH_2CN$	252.6 ± 4	1
$CH_3C\cdot HCN$	226.7 ± 12.6	1
$\cdot CH_2CH_2CN$	245.4 ± 12.6	1
$(CH_3)_2C\cdot CN$	190.4 ± 12.6	1
$Ph(CH_3)C\cdot CN$	248.5 ± 8.4	1
$NCC\cdot HCH_2CN$	381.8 ± 12.6	1
$\cdot CH_2NC$	334.7 ± 16.7	1
$\cdot C(O)NC$	210.0 ± 10	1
$\cdot C(O)NH_2$	−15.1 ± 4	1
$C\cdot NN$	569 ± 21	1
$HC\cdot NN$	460 ± 8	1
$H_2C\cdot NN$	292.5 ± 2.1	1
$\cdot CH_2NO$	157 ± 4	1
$\cdot CH_2NO_2$	115.1 ± 12.6	1
$CH_3C\cdot HNO_2$	61.9 ± 12.6	1
$(CH_3)_2C\cdot NO_2$	6.3 ± 12.6	1
$PhC\cdot HNO_2$	169.0 ± 12.6	1
$\cdot C_6H_6N$, 3-NH_2-C_6H_4	320.1	1
$\cdot C_6H_6N$, 4-NH_2-C_6H_4	327.8	1
$\cdot C_6H_4NO_2$, 3-NO_2-C_6H_4	340.6 ± 10.0	1
$\cdot C_6H_4NO_2$, 4-NO_2-C_6H_4	302.7	1
$\cdot C_6H_4CH_3$, 2-Me-C_6H_4	315.1 ± 10.5	1
$\cdot C_6H_4CH_3$, 4-Me-C_6H_4	296.6 ± 9.6	1
$\cdot C_6H_3N_2O_4$, 3,5-$(NO_2)_2$-C_6H_3	305.4	1
$\cdot C_7H_6NO_2$, 2-Me-4-NO_2-C_6H_3	295.4 ± 8.4	1
$\cdot C_4H_3N$, pyrrol-2-yl	385.8	1
$\cdot C_4H_3N$, pyrrol-3-yl	385.8	1
$\cdot C_4H_8N$, pyrrolidin-2-yl	142.7 ± 12.6	1

Radical	$\Delta_f H°_{298}$/kJ mol^{-1}	Ref.
$\cdot C_5H_4N$, pyrid-2-yl	362.0	1
$\cdot C_5H_4N$, pyrid-3-yl	391.0	1
$\cdot C_5H_4N$, pyrid-4-yl	384.1 ± 8.4	12
$\cdot C_4H_7N_2$, piperad-2-yl	119.7	1
$\cdot C_4H_3N_2$, pyrazin-2-yl	428.0 ± 2.5	12
$\cdot C_4H_3N_2$, pyrimid-2-yl	388.0 ± 12.6	1
$\cdot C_4H_3N_2$, pyrimid-3-yl	529.5 ± 3.3	12
$\cdot C_4H_3N_2$, pyrimid-4-yl	409.0 ± 12.6	1
$\cdot C_4H_3N_2$, pyrimid-5-yl	452.3 ± 3.3	12
$\cdot C_3H_2N_3$, pyrimid-5-yl	459.0 ± 3.3	12
$\cdot CH(NO_2)_2$	139.1	1
$\cdot C(NO_2)_3$	201.2	1
$\cdot CH_2C(NO_2)_3$	150.6	1
$\cdot CH_2CH(NO_2)_2$	103.3	1
$\cdot CH_2CH_2C(NO_2)_3$	133.9	1
$\cdot CH_2N(NO_2)CH_2C(NO_2)_3$	173.6	1
$\cdot CH_2N(NO_2)CH_2CH(NO_2)_2$	126.4	1
$\cdot CH_2CH_2N(NO_2)CH_2C(NO_2)_3$	168.6	1
$\cdot CH_2CH_2ONO_2$	37.7	1
$\cdot CH_2(ONO_2)CHCH_2ONO_2$	-25.5	1
$\cdot CH(CH_2ONO_2)_2$	-57.3	1
$\cdot CH_2C(CH_2ONO_2)_3$	-158.2	1
$\cdot CH_2NHNO_2$	164.8	1
$\cdot CH_2N(NO_2)CH_3$	149.4	1
$\cdot CH_2N(NO_2)_2$	210.5	1
$\cdot CH_2CH_2N(NO_2)CH_3$	144.3	1
$\cdot CH_2N(NO_2)CH_2N(NO_2)CH_3$	202.1	1
$\cdot CH_2N(NO_2)(CH_2)N(NO_2)CH_3$	173.2	1
$C\cdot(S)H$	300.4 ± 8.4	1
$\cdot CH_2SH$	151.9 ± 8.4	1
$\cdot CH_2SCH_3$	136.8 ± 5.9	1
$\cdot CH_2SPh$	268.6 ± 12.6	1
$\cdot CH_2SOCH_3$	23.8 ± 12.6	1
$HOC\cdot(S)S$	110.5	1
$\cdot CH_2SO_2CH_3$	−177.0 ± 12.6	1
$\cdot CH_2SO_2Ph$	−57.3 ± 12.6	1
$PhC\cdot HSO_2CH_3$	−109.2 ± 12.6	1
$PhC\cdot HSO_2Ph$	7 ± 12.6	1
$Ph_2C\cdot SO_2Ph$	102 ± 12.6	1
$Ph_2C\cdot SPh$	435.6 ± 12.6	1
$NC\cdot(O)$	127.2	1
$\cdot CNH$	207.9 ± 12.1	1
$\cdot CNO$	323 ± 30	1
$\cdot CH_2SiMe_3$	−32 ± 6	1
$\cdot CH_2C(CH_3)_2SiMe_3$	-125	1
$\cdot CP$	450 ± 9	1
(2) Oxygen-Centered Species		
$HO\cdot$	37.50 ± 0.03	14
$FO\cdot$	109 ± 10	1
$ClO\cdot$	101.63 ± 0.1	1
$BrO\cdot$	126.2 ± 1.7	1
$IO\cdot$	115.9 ± 5.0	1
$HOO\cdot$	12.27 ± 0.16	14
$FOO\cdot$	25.4 ± 2	1
$ClOO\cdot$	98.0 ± 4	1
$BrOO\cdot$	108 ± 40	1
$IOO\cdot$	96.6 ± 15	1
$OFO\cdot$	378.6 ± 20	1
$OClO\cdot$	95.4	1
$ClOOClO\cdot$	142 ± 12	1
$ClClO\cdot$	90 ± 30	1

Radical	$\Delta_f H°_{298}$/kJ mol^{-1}	Ref.
$NCO\cdot$	184.1	1
$CNO\cdot$	386.6	1
$HONNO\cdot$	172	1
sym-ClO_3	217.2 ± 21	1
$HSO\cdot$	−21.8 ± 2.1	1
$HSOO\cdot$	112	1
$CH_3SOO\cdot$	76	1
$CF_3SO_2O\cdot$	-912	1
$NCO\cdot$	184.0	1
$O_2NO\cdot$	73.7 ± 1.4	1
$ONOO\cdot$	82.8	1
$HOS(O)_2O\cdot$	-511.7	1
$CH_3O\cdot$	21.0 ± 2.1	1
$CF_3O\cdot$	−635.1 ± 7.1	1
$CCl_3O\cdot$	−38.1 ± 9.2	1
$CH_2ClO\cdot$	−21.3 ± 9.2	1
$CHCl_2O\cdot$	−32.2 ± 9.2	1
$CH_2{=}CH\text{-}O\cdot$	18.4 ± 1.3	1
$CF_3CHFO\cdot$	-851.0	1
$C_2H_5O\cdot$	−13.6 ± 3.3	1
$CH_3CHClO\cdot$	−61.9 ± 12.1	1
$CH_3CCl_2O\cdot$	−91.6 ± 11.7	1
$nC_3H_7O\cdot$	−38.3 ± 9.6	16
$iC_3H_7O\cdot$	−48.5 ± 3.3	1
$(CH_3)_2CClO\cdot$	−108.4 ± 8.4	1
$nC_4H_9O\cdot$	-62.8	1
$sC_4H_9O\cdot$	-69.5	1
$tC_4H_9O\cdot$	−85.8 ± 3.8	1
$CH_2{=}CHCH_2O\cdot$	87.0	1
$C_6H_5O\cdot$	48.5 ± 2.9	1
o-Cl-$C_6H_4O\cdot$	30.6	1
$C_6Cl_5O\cdot$	~63	1
p-Cl-$C_6H_4O\cdot$	~9	1
o-OH-$C_6H_4O\cdot$	-186.3	1
p-OH-$C_6H_4O\cdot$	-143.6	1
o-CH_3O-$C_6H_4O\cdot$	-125.5	1
p-CH_3O-$C_6H_4O\cdot$	-81.1	1
$C_6H_5CH_2O\cdot$	136.0 ± 12.6	1
$C_{10}H_7O\cdot$, naphthoxy-1	165.3	1
$C_{10}H_7O\cdot$, naphthoxy-2	174.1	1
$HC(O)O\cdot$	−129.7 ± 12.6	1
$FC(O)O\cdot$	368.0	1
$CH_3C(O)O\cdot$	−179.9 ± 12.6	1
$CF_3C(O)O\cdot$	-797.0	1
$CF_3OC(O)O\cdot$	−958.1 ± 16.7	1
$C_6H_5C(O)O\cdot$	−50.2 ± 16.7	1
$CH_3OO\cdot$	20.1 ± 5.1	1
$C_2H_3OO\cdot$, $CH_2{=}CHOO\cdot$	101.7 ± 1.7	1
$C_2H_5OO\cdot$	−25.5 ± 5.6	16
$C_3H_5OO\cdot$, $CH_2{=}CHCH_2OO\cdot$	88.7	1
$nC_3H_7OO\cdot$	-43.6±2.1	16
$iC_3H_7OO\cdot$	−65.4 ± 11.3	1
$C_4H_7OO\cdot$, $CH_3CH{=}CHCH_2OO\cdot$	82.6 ± 5.3	1
$nC_4H_9OO\cdot$	-65.7±3.6	16
$tC_4H_9OO\cdot$	−109.7 ± 3.9	8
neo-$C_5H_{11}OO\cdot$	-115.5	1
$HOCH_2OO\cdot$	-162.1	1
$HOOCH_2CH_2OO\cdot$	100	1
$C_6H_5CH_2OO\cdot$	114.6 ± 4.2	1
c-$C_6H_{11}OO\cdot$	−25.0 ± 10.5	1
$(C_2H_5)N(CH3)CHOO\cdot$	−36.0 ± 12.6	1

Radical	$\Delta_f H^\circ_{298}$/kJ mol^{-1}	Ref.
$CF_3OO^\cdot$	-635.0	1
$CF_2ClOO^\cdot$	–406.7 ± 14.6	1
$CFCl_2OO^\cdot$	-213.7	1
$CH_2ClOO^\cdot$	–5.1 ± 13.6	1
$CHCl_2OO^\cdot$	–19.2 ± 11.2	1
$CCl_3OO^\cdot$	–20.9 ± 8.9	1
$CH_3CHClOO^\cdot$	–54.7 ± 3.4	1
$CH_3CCl_2OO^\cdot$	–63.8 ± 9.8	1
$CH_3OCH_2OO^\cdot$	–142.2 ± 4.2	1
$CH_3C(O)CH_2OO^\cdot$	–142.1 ± 4	1
cis-HC(O)OO*	85.8 ± 14.6	7
trans-HC(O)OO*	95.4 ± 14.6	7
$CH_3C(O)OO^\cdot$	–154.4 ± 5.8	1
$HOOO^\cdot$	19.3 ± 0.5	9
$CH_3OOO^\cdot$	33.4 ± 12.6	1
$C_2H_5OOO^\cdot$	5.4 ± 12.6	1
(3) Nitrogen-Centered Species		
ON	91.09 ± 0.06	14
NO_2	34.02 ± 0.07	14
N_2O	82.05 ± 0.4	1
NH	358.8 ± 0.2	14
$^\cdot NH_2$	186.1 ± 0.2	14
$^\cdot NNH$	249.5	1
$^\cdot NCO$	131.8	1
$^\cdot N_3$	414.2 ± 20.9	1
$^\cdot N_2H_3$	243.5	1
(*Z*)-N_2H_2	213.0 ± 10.9	1
NF	209.2	1
$^\cdot NF_2$	42.3 ± 8	1
$^\cdot NHF$	112 ± 15	1
NBr	301 ± 21	1
HNO	107.1 ± 2.5	1
FNO	–65.7 ± 1.7	1
ClNO	51.71 ± 0.42	1
BrNO	82.13 ± 0.8	1
INO	112.1 ± 20.9	1
NCO	120.9	1
NCN	464.8 ± 2.9	1
NSi	372 ± 63	1
$NH_2C(O)N^\cdot H$	0.8 ± 12.6	1
$CH_3C(O)N^\cdot H$	–6.7 ± 12.6	1
$NH_2C(S)N^\cdot H$	194 ± 12.6	1
$CH_3C(S)N^\cdot H$	173 ± 12.6	1
$PhC(S)N^\cdot H$	307 ± 12.6	1
$HCON^\cdot H$	49.8 ± 12.6	1
$NH_2C(NH)N^\cdot H$	250.6 ± 12.6	1
$^\cdot NHCN$	319.2 ± 2.9	1
$CH_2N^\cdot H$	104.6 ± 12.6	1
$CH_3N^\cdot H$	184.1 ± 8.4	1
*t*BuN•H	95.4 ± 12.6	1
$C_6H_5CH_2N^\cdot H$	288.3 ± 12.6	1
$C_6H_5N^\cdot H$	244.3 ± 4.2	1
$(CH_3)_2N^\cdot$	158.2 ± 4.2	1
$(C_6H_5)(CH_3)N^\cdot$	241.0 ± 6.3	1
$(C_6H_5)_2N^\cdot$	366.0 ± 6.3	1
1-pyrrolyl	269.2 ± 12.6	1
1-pyrazolyl	413.0 ± 2.1	1
carbazol-9-yl	383.3 ± 8.4	1
$CH_3N_2^\cdot$	215.5 ± 7.5	1
$C_2H_5N_2^\cdot$	187.4 ± 10.5	1
i$C_3H_7N_2$•	146.0 ± 8.4	1
n$C_4H_9N_2$•	140.6 ± 8.4	1
t$C_4H_9N_2$•	97.5 ± 4.2	1
$(NO_2)HN^\cdot$	162.3	1
$(CH_3)(NO_2)N^\cdot$	139.0	1
$(NO_2)_2N^\cdot$	200.0	1
$CH_3N^\cdot CH_2N(NO_2)CH_3$	185.4	1
(4) Sulfur-Centered Species		
$HOS^\cdot$	–6.7 ± 2.1	1
$HC(O)S^\cdot$	56.5	1
$HS^\cdot O_2$	-221.8	1
$HOS^\cdot O_2$	-384.9	1
$NCS^\cdot$	300 ± 8	1
$HS^\cdot$	143.0 ± 0.8	1
$CH_3S^\cdot$	124.7 ± 1.7	1
$C_2H_5S^\cdot$	101	1
nC_3H_7S•	80	1
iC_3H_7S•	74.9 ± 8.4	1
tC_4H_9S•	43.9 ± 8.4	1
$C_6H_5S^\cdot$	242.7 ± 4.6	1
$C_6Cl_5S^\cdot$	~184	1
$C_6H_5CH_2S^\cdot$	246	1
$CH_3S^\cdot O$	–67 ± 10	1
$CH_3S^\cdot O_2$	-239.3	1
$HSS^\cdot$	115.5 ± 14.6	1
$CH_3SS^\cdot$	68.6 ± 8.4	1
$C_2H_5SS^\cdot$	43.5 ± 8.4	1
iC_3H_7SS•	13.8 ± 8.4	1
tC_4H_9SS•	–19.2 ± 8.4	1
$HOC(S)S^\cdot$	110.5 ± 4.6	1
$HC(O)S^\cdot$	56.5	1
SF	13.0 ± 6.3	1
SF_2	–296.7 ± 16.7	1
SF_3	–503.0 ± 33.5	1
SF_4	–763.2 ± 20.9	1
SF_5	–879.9 ± 15.1	1
$ClS^\cdot$	156.5 ± 16.7	1
SN	263.6 ± 105	1
SCl	156.5 ± 16.7	1
(5) Si-, Ge-, Sn-, Pb-Centered Species		
SiF	–20.1 ± 12.6	1
SiF_2	–638 ± 6	1
$^\cdot SiF_3$	–987 ± 20	1
SiCl	198.3 ± 6.7	1
$SiCl_2$	–169 ± 3	1
$^\cdot SiCl_3$	321 ± 8	6
SiBr	235 ± 46	1
$SiBr_2$	46 ± 8	1
$^\cdot SiBr_3$	–201.7 ± 63	1
SiI	313.8 ± 42	1
SiI_2	92.5 ± 8.4	1
$^\cdot SiI_3$	35.3 ± 63	1
SiH	376.6 ± 8.4	1
$SiH_2(^1A_1)$	273 ± 2	1
$SiH_2(^3B_1)$	360.7	1
$^\cdot SiH_3$	200.4 ± 2.5	1
$MeSi^\cdot H_2$	141 ± 6	1
$Me_2Si^\cdot H$	78 ± 6	1
$Me_3Si^\cdot$	15 ± 7	1
$^\cdot Si_2H_3$	~402	1
$H_3SiSi^\cdot H_2$	234 ± 6	1

Radical	$\Delta_f H°_{298}$/kJ mol^{-1}	Ref.
$C_6H_5Si{\cdot}H_2$	274	1
$H_3SiSi{\cdot}H$	312 ± 8	1
$MeSi{\cdot}$	302.2	1
$MeSi{\cdot}H$	202 ± 6	1
$Me_2Si{\cdot\cdot}$	135 ± 8	1
SiN	313.8 ± 42	1
${\cdot}GeH_3$	221.8 ± 8.4	1
GeF	−71 ± 10	1
GeF_2	−574 ± 20	1
${\cdot}GeF_3$	−807 ± 50	1
GeCl	69 ± 18	1
$GeCl_2$	−171 ± 5	1
${\cdot}GeCl_3$	−268 ± 50	1
GeBr	137 ± 5	1
$GeBr_2$	−61 ± 5	1
${\cdot}GeBr_3$	−119 ± 50	1
GeI	211 ± 25	1
GeI_2	50.2 ± 4	1
${\cdot}GeI_3$	42 ± 50	1
SnF	−95 ± 7.2	1
SnF_2	−511 ± 9.2	1
${\cdot}SnF_3$	−647 ± 50	1
SnCl	35 ± 12	1
$SnCl_2$	−202.6 ± 7.1	1
${\cdot}SnCl_3$	−292 ± 50	1
SnBr	76 ± 12	1
$SnBr_2$	−119 ± 2.8	1
${\cdot}SnBr_3$	−159 ± 50	1
SnI	173 ± 12	1
SnI_2	−8.1 ± 4.2	1
${\cdot}SnI_3$	−8 ± 50	1
${\cdot}Sn(CH_3)_3$	116.6 ± 9.7	10
${\cdot}Sn(C_6H_5)_3$	518.8 ± 21	1
PbH	236.2 ± 19.2	1
PbF	−80.3 ± 10.5	1
PbF_2	−435.1 ± 8.4	1
${\cdot}PbF_3$	−490 ± 60	1
PbCl	15.1 ± 50	1
$PbCl_2$	−174.1 ± 1.3	1
${\cdot}PbCl_3$	−178 ± 80	1
PbBr	70.9 ± 42	1
$PbBr_2$	−104.4 ± 6.3	1
${\cdot}PbBr_3$	−104 ± 80	1
PbI	107.4 ± 37.7	1
PbI_2	−3.2 ± 4.2	1
${\cdot}PbI_3$	22 ± 80	1

References

1. Luo, Y. R., *Comprehensive Handbook of Chemical Bond Energies*, CRC Press, Boca, Raton, FL, 2007.
2. Wheeler, S. E., Robertson, K. A., Allen, W. D., Schaefer III, H. F., Bomble, Y. J., and Stanton, J. F., *J. Phys. Chem. A* 111, 3819, 2007.
3. Seetula, J. A., and Eskola, A. J., *Chem. Phys.* 351, 141, 2008.
4. Denis, P. A., and Ornellas, F. R., *J. Phys. Chem. A* 113, 499, 2009.
5. Gao, Y., DeYonker, N. J., Garrett III, E. C., Wilson, A. K., Cundari, T. R., and Marshall, P., *J. Phys. Chem. A* 113, 6955, 2009.
6. Shuman, N. S., Spencer, A. P., and Baer, T., *J. Phys. Chem. A* 113, 9458, 2009.
7. Villano, S. M., Eyet, N., Wren, S. W., Ellison, G. B., Bierbaum, V. M., and Lineberger, W. C., *J. Phys. Chem. A* 114, 191, 2010.
8. Shuman, N. S., Bodi, A., and Baer, T., *J. Phys. Chem. A* 114, 232, 2010.
9. Le Picard, S. D., Tizniti, M., Canosa, A., Sims, I. R., and Smith, I. W. M., *Science* 328, 1258, 2020.
10. Da'valos, J. Z., Herrero, R., Shuman, N. S., and Baer, T., *J. Phys. Chem. A* 115, 402, 2011.
11. Stevens, W. R., Ruscic, B., and Baer, T., *J. Phys. Chem. A*, 114, 13134, 2010.
12. Wren, S. W., Vogelhuber, K. M., Garver, J. M., Kato, S., Sheps, L., Bierbaum, V. M., and Lineberger, W. C., *J. Am. Chem. Soc.*, 134, 6584, 2012.
13. Harvey, J., Bodi, A., Tuckett, R. P., and Sztáray, B., *Phys. Chem. Chem. Phys.*, 14, 3935, 2012.
14. Ruscic, B., Active Thermochemical Table, http://atct.anl.gov/index.html, Dec. 16, 2010.
15. Tian, Z., Lis, L., and Kass, S. R., *J. Org. Chem.* 78, 12650, 2013.
16. Bukke, S. M., Simmie, J. M., and Curran, H. J., *J. Phys. Chem. Ref. Data* 44, 013101, 2015.
17. Rossi, M. J., *Int. J. Chem. Kinet.* 40, 395–415, 2008.

Bond Dissociation Energies of Some Organic Molecules

$D°_{298}$(R-X)/kJ mol^{-1} of some organic compounds are listed below. All data are from Tables 1 and 3.

Table 5. Bond Dissociation Energies of Some Organic Molecules

	X=H	F	Cl	Br	I	OH	OCH_3	NH_2	NO	CH_3	$COCH_3$	CF_3	CCl_3
R=H	435.7799	569.658	431.361	366.16	298.26	497.10	440.2	450.08	199.5	439.3	374.0	445.2	392.5
CH_3	439.3	460.2	350.2	294.1	238.9	384.93	351.9	356.1	172.0	377.4	351.9	429.3	362.3
C_2H_5	420.5	447.4	352.3	292.9	233.5	391.2	355.2	355.2	171.5	370.3	347.3		
i-C_3H_7	410.5	483.8	354.8	295.1	233.1	397.9	360.7	357.7	152.7	369.0	340.2		
t-C_4H_9	400.4	495.8	351.9	292.9	227.2	398.3	353.1	355.6	167	363.6	329.3		
C_6H_5	433.9	525.5	399.6	336.4	272.0	470.7	425.9	436.4	226.8	433.9	413.8	470.3	395.8
$C_6H_5CH_2$	375.5	412.8	299.9	239.3	187.8	334.1		306.7	123	325.1	299.7	365.7	
CCl_3	392.5	439.2	288.7	231.4	168				125	362.3		332.2	285.8
CF_3	445.2	546.8	362.7	296.2	227.2	482.0[a]			167	429.3		413.0	332.2
C_2F_5	429.7	532.2	346.0	283.3	219.2							424.3	
CH_3CO	374.0	511.7	354.0	292.0	223.0	459.4	424.3	414.6		351.9	307.1		
CN	528.5	482.8	422.6	364.8	320.1				204.4	521.7		469.0	
C_6F_5	487.4	485	383.3	328[b]	301.7[a]	446.9			211.3	439.3		435.1	

[a] Upper limit
[b] Approximate

Molecular

Bond Dissociation Energies in Diatomic Cations

From thermochemistry, we have

$$D^\circ_{298}(A^+\text{–}B) = \Delta_f H^\circ(A^+) + \Delta_f H^\circ(B) - \Delta_f H^\circ(AB^+) = D^\circ_{298}(A\text{–}B) + IP(A) - IP(AB)$$

Thus, $D^\circ_{298}(A^+\text{-}B)$ may be derived using Table 1 and the ionization potentials of species A and AB. The following table has been arranged in an alphabetical order of the atoms. The **boldface** in the species indicates the dissociated fragment.

Table 6. Bond Dissociation Energies in Diatomic Cations

A^+–B	D°_{298}/kJ mol^{-1}	Ref.
Ac^+–S	465 ± 48	20
Ag^+–Ag	167.9 ± 8.7	1
Ag^+–Cl	32 ± 30	1
Ag^+–F	24 ± 27	1
Ag^+–H	43.5 ± 5.9	1
Ag^+–O	123 ± 5	1
Ag^+–S	138.8 ± 11.6	21
Al^+–Al	121	1
Al^+–Ar	15.47	1
Al^+–Ca	148.5	1
Al^+–Cl	173 ± 42	1
Al^+–F	314 ± 21	1
Al^+–Kr	5.54	1
Al^+–O	166.7 ± 12.0	1
Al^+–Se	114 ± 49	1
Am^+–O	560 ± 28	14
Am^+–S	334 ± 27	20
Ar^+–Ar	130.323 ± 0.087	1
Ar^+–He	2.9 ± 0.8	1
Ar^+–Ne	7.5 ± 0.8	1
As^+–As	364 ± 22	1
As^+–H	290.8 ± 3.0	1
As^+–O	495	1
As^+–P	367 ± 59	1
As^+–S	433.2 ± 12.5	1
Au^+–Al	170 ± 30	1
Au^+–Au	234.5	1
Au^+–B	329 ± 50	1
Au^+–Be	401 ± 29	1
Au^+–C	311.5 ± 7.7	4
Au^+–F	79	1
Au^+–Ge	292 ± 24	1
Au^+–H	209.2 ± 10.6	18
Au^+–I	230~280	1
Au^+–O	111.8 ± 7.7	19
Au^+–Xe	130 ± 13	1
B^+–Ar	32.7	1
B^+–B	187	1
B^+–Br	164 ± 21	1
B^+–C	284 ± 58	1
B^+–Cl	308 ± 21	1
B^+–F	460 ± 10	1
B^+–H	198 ± 5	1
B^+–O	326 ± 48	1
B^+–Pt	314 ± 98	1
B^+–Se	298 ± 98	1
B^+–Si	365 ± 15	1
Ba^+–Ar	11.85	1
Ba^+–Br	418 ± 10	1
Ba^+–Cl	468.2 ± 10	1
Ba^+–D	245.2 ± 9.6	1
Ba^+–F	640 ± 29	1
Ba^+–I	335 ± 10	1
Ba^+–O	441.4 ± 15	1
Be^+–Ar	49.0 ± 2.4	1
Be^+–Au	410 ± 29	1
Be^+–Be	196 ± 0.5	8
Be^+–Cl	417 ± 50	1
Be^+–F	575 ± 98	1
Be^+–H	307.3 ± 5.0	1
Be^+–O	362.0 ± 6.2	1
Bi^+–Bi	199 ± 10	1
Bi^+–O	174	1
Bi^+–S	179 ± 50	1
Bi^+–Se	184 ± 29	1
Bi^+–Te	125 ± 50	1
Bi^+–Tl	100 ± 42	1
Bk^+–O	610	1
Br^+–Br	318.858 ± 0.024	1
Br^+–C	451.5 ± 8.6	1
Br^+–Cl	303.000 ± 0.048	1
Br^+–F	251.5 ± 12.6	1
Br^+–H	379.26 ± 2.89	1
Br^+–O	365.7 ± 3.1	1
C^+–Ar	72.3	1
C^+–Br	398 ± 8.6	1
C^+–C	601.9 ± 19.3	1
C^+–Cl	614	1
C^+–F	721 ± 40	1
C^+–H	397.848 ± 0.013	1
C^+–N	524.5 ± 4.2	1
C^+–O	810.7 ± 0.8	1
C^+–P	587 ± 50	1
C^+–S	706.6 ± 2.1	1
C^+–Se	587 ± 50	1
Ca^+–Al	144.7	1
Ca^+–Ar	12.99 ± 0.60	1
Ca^+–Au	306 ± 29	1
Ca^+–Br	417.6 ± 10	1
Ca^+–Ca	104.1	1
Ca^+–Cl	433.4 ± 12	1
Ca^+–F	556.5 ± 8.4	1
Ca^+–H	284.2 ± 10	1
Ca^+–I	293.7 ± 10.8	1
Ca^+–Kr	18.60 ± 0.72	1
Ca^+–Ne	4.95 ± 0.06	1
Ca^+–O	348 ± 5	1
Ca^+–Xe	25.38 ± 0.96	1
Cd^+–Cd	122.5 ± 10	1
Cd^+–H	179.5	1
Ce^+–Au	278 ± 34	1
Ce^+–Br	341.0	1
Ce^+–C	254 ± 96	1
Ce^+–Ce	207 ± 42	1
Ce^+–Cl	429.5	1
Ce^+–F	586 ± 63	1
Ce^+–I	295.5	1
Ce^+–Ir	530 ± 96	1
Ce^+–N	494 ± 63	1
Ce^+–O	852 ± 15	1
Ce^+–Pd	255 ± 53	1
Ce^+–Pt	467 ± 96	1
Ce^+–Rh	423 ± 96	1
Ce^+–S	524 ± 59	1
Cl^+–Ar	169	1
Cl^+–Cl	385.746 ± 0.096	6
Cl^+–D	457.284 ± 0.017	1
Cl^+–F	291 ± 10	1
Cl^+–H	452.714 ± 0.018	1
Cl^+–N	650 ± 10	1
Cl^+–O	468.0 ± 2.1	1
Cm^+–O	670 ± 38	14
Cm^+–S	455 ± 16	20
Co^+–Ar	52.89 ± 0.06	1
Co^+–Br	>289	1
Co^+–C	351 ± 29	1
Co^+–Cl	285 ± 12	1
Co^+–Co	269	1
Co^+–D	199.6 ± 5.8	1
Co^+–H	195 ± 6	1
Co^+–He	16.4 ± 0.4	1
Co^+–I	211.7 ± 8.4	1
Co^+–Kr	68.37 ± 0.18	1
Co^+–Ne	12.8 ± 0.4	1
Co^+–O	317.3 ± 4.8	1
Co^+–S	288.3 ± 8.7	1
Co^+–Si	317.1 ± 6.7	1
Co^+–Xe	85.7 ± 6.8	1
Cr^+–Ar	31.7 ± 3.9	1
Cr^+–C	277 ± 24	1
Cr^+–Cl	>211	1
Cr^+–Cr	129	1
Cr^+–D	135 ± 9	1
Cr^+–F	279 ± 42	1
Cr^+–H	136 ± 9	1
Cr^+–He	7.8 ± 0.4	1
Cr^+–Ne	9.5 ± 0.4	1
Cr^+–O	359	1
Cr^+–S	258.6 ± 16.4	1
Cr^+–Si	203 ± 15	1
Cr^+–Xe	71.9 ± 10.0	1
Cs^+–Ar	8.2	1
Cs^+–Br	60.5 ± 10	1
Cs^+–Cl	107.4 ± 10	1
Cs^+–Cs	62.6 ± 9.6	1
Cs^+–F	43.7 ± 10	1

A^+–B	D^o_{298}/kJ mol^{-1}	Ref.
Cs$^+$–He	5.1	1
Cs$^+$–I	29.3 ± 10	1
Cs$^+$–Kr	15.1	1
Cs$^+$–Na	48.1 ± 4.2	1
Cs$^+$–Ne	6.11	1
Cs$^+$–O	59	1
Cs$^+$–Rb	68.3 ± 10	1
Cs$^+$–Xe	14.7	1
Cu$^+$–Ar	51.9 ± 6.8	1
Cu$^+$–Cl	91 ± 10	1
Cu$^+$–Cu	155.2 ± 7.7	1
Cu$^+$–F	117 ± 21	1
Cu$^+$–Ge	231 ± 23	1
Cu$^+$–H	93 ± 13	1
Cu$^+$–Kr	24.3 ± 0.8	1
Cu$^+$–O	133.9 ± 11.6	1
Cu$^+$–S	203.3 ± 14.5	1
Cu$^+$–Si	260 ± 8	1
Cu$^+$–Xe	102.1 ± 5.8	1
D$^+$–D	263.4405 ± 0.0003	1
Dy$^+$–Br	324.2	1
Dy$^+$–Cl	407.9	1
Dy$^+$–Cu	196 ± 42	1
Dy$^+$–F	535 ± 24	1
Dy$^+$–I	279.9	1
Dy$^+$–O	597 ± 15	1
Er$^+$–Br	315.8	1
Er$^+$–Cl	406.7	1
Er$^+$–F	546 ± 34	1
Er$^+$–I	271.6	1
Er$^+$–O	583 ± 15	1
Es$^+$–O	470 ± 60	1
Eu$^+$–Ag	85 ± 50	1
Eu$^+$–Au	252 ± 97	1
Eu$^+$–Br	333.8	1
Eu$^+$–Cl	430.7	1
Eu$^+$–F	543 ± 29	1
Eu$^+$–I	290.7	1
Eu$^+$–O	393 ± 15	1
Eu$^+$–S	257 ± 32	1
F$^+$–Ar	161.1	1
F$^+$–F	325.393 ± 0.096	1
F$^+$–He	181.62 ± 0.08	1
F$^+$–Kr	152.4	1
F$^+$–Xe	188	1
Fe$^+$–Ar	14.2 ± 7.7	1
Fe$^+$–Br	>293	1
Fe$^+$–C	402.0 ± 4.2	16
Fe$^+$–Cl	>343	1
Fe$^+$–Co	259 ± 21	1
Fe$^+$–Cr	209 ± 29	1
Fe$^+$–Cu	222 ± 29	1
Fe$^+$–D	227	1
Fe$^+$–F	360–423	1
Fe$^+$–Fe	272	1
Fe$^+$–H	211.2 ± 9.6	1
Fe$^+$–I	>239	1
Fe$^+$–Kr	33.5 ± 6.7	1
Fe$^+$–N	485	1
Fe$^+$–Nb	285 ± 21	1
Fe$^+$–Ni	268 ± 21	1
Fe$^+$–O	343.3 ± 1.9	13
Fe$^+$–S	295.2 ± 5.8	1
Fe$^+$–Sc	200 ± 21	1
Fe$^+$–Si	277 ± 9	1
Fe$^+$–Ta	301 ± 21	1
Fe$^+$–Ti	251 ± 25	1
Fe$^+$–V	314 ± 21	1
Fe$^+$–Xe	46.0 ± 5.8	1
Ga$^+$–Bi	62 ± 98	1
Ga$^+$–Br	56.5 ± 16	1
Ga$^+$–Cl	86 ± 21	1
Ga$^+$–F	136 ± 15	1
Ga$^+$–Ga	126.3	1
Ga$^+$–I	41.6 ± 15	1
Ga$^+$–O	46 ± 50	1
Ga$^+$–Sb	38 ± 96	1
Ga$^+$–Te	19 ± 29	1
Gd$^+$–Cd	122.5 ± 10	1
Gd$^+$–H	179.5	1
Ge$^+$–Br	398 ± 42	1
Ge$^+$–C	223 ± 31	1
Ge$^+$–Cl	473 ± 50	1
Ge$^+$–F	565 ± 21	1
Ge$^+$–Ge	274 ± 10	1
Ge$^+$–H	377 ± 84	1
Ge$^+$–O	344 ± 21	1
Ge$^+$–S	283 ± 21	1
Ge$^+$–Se	234 ± 10	1
Ge$^+$–Si	268 ± 21	1
Ge$^+$–Te	233 ± 19	1
H$^+$–D	261.1021 ± 0.0002	1
H$^+$–H	259.4659 ± 0.0002	1
He$^+$–H	123.9	1
He$^+$–He	229.687 ± 0.019	1
Hf$^+$–C	311.5 ± 2.9	10
Hf$^+$–H	207.3 ± 7.7	17
Hf$^+$–O	670.4 ± 10.6	10
Hg$^+$–Ar	22.2 ± 1.2	1
Hg$^+$–H	207	1
Hg$^+$–Hg	134	1
Hg$^+$–Kr	37.9 ± 1.3	1
Hg$^+$–Xe	72.2 ± 1.3	1
Ho$^+$–Ag	155 ± 61	1
Ho$^+$–Au	250 ± 60	1
Ho$^+$–Br	320.6	1
Ho$^+$–Cl	410.3	1
Ho$^+$–Cu	214 ± 35	1
Ho$^+$–F	542 ± 50	1
Ho$^+$–Ho	88 ± 96	1
Ho$^+$–I	270.4	1
Ho$^+$–O	551 ± 25	1
I$^+$–Br	184.90 ± 0.02	1
I$^+$–Cl	247.5 ± 0.4	1
I$^+$–F	262.9 ± 2.1	1
I$^+$–H	304.70 ± 0.10	1
I$^+$–I	262.90 ± 0.04	1
I$^+$–O	316.3 ± 10.5	1
In$^+$–Br	65.2 ± 12.6	1
In$^+$–Cl	193 ± 21	1
In$^+$–F	148 ± 50	1
In$^+$–I	51.5 ± 21	1
In$^+$–In	81 ± 30	1
In$^+$–S	171 ± 50	1
In$^+$–Sb	73 ± 50	1
In$^+$–Se	118 ± 50	1
In$^+$–Te	41 ± 50	1
Ir$^+$–C	635.8 ± 4.8	3
Ir$^+$–D	302.8 ± 5.8	1
Ir$^+$–H	305.7 ± 5.8	1
Ir$^+$–O	411.0 ± 8.7	24
K$^+$–Ar	14 ± 7	1
K$^+$–Br	35.7 ± 10.5	1
K$^+$–Cl	51 ± 19	1
K$^+$–He	6.00	1
K$^+$–I	18 ± 45	1
K$^+$–K	83.86 ± 0.15	1
K$^+$–Kr	15.8	1
K$^+$–Li	59.9 ± 5.9	1
K$^+$–Na	58.69 ± 0.08	1
K$^+$–Ne	7.79	1
K$^+$–O	13	1
K$^+$–Xe	19.5	1
Kr$^+$–Ar	55.31 ± 0.14	1
Kr$^+$–H	464	1
Kr$^+$–He	2.1 ± 0.8	1
Kr$^+$–Kr	110.967 ± 0.033	1
Kr$^+$–N	136.9 ± 13	1
Kr$^+$–Ne	3.8 ± 0.8	1
La$^+$–Au	436 ± 97	1
La$^+$–Br	425.9	1
La$^+$–C	427 ± 33	1
La$^+$–Cl	503.6	1
La$^+$–F	589 ± 34	1
La$^+$–H	243 ± 9	1
La$^+$–I	392.4	1
La$^+$–Ir	356 ± 97	1
La$^+$–O	875 ± 25	1
La$^+$–Pt	522 ± 78	1
La$^+$–Rh	345 ± 97	1
La$^+$–S	629 ± 96	1
La$^+$–Si	277.0 ± 9.6	1
Li$^+$–Ar	33 ± 14	1
Li$^+$–Bi	91 ± 50	1
Li$^+$–Br	41.8 ± 10.6	1
Li$^+$–Cl	66 ± 15	1
Li$^+$–F	7 ± 21	1
Li$^+$–He	10.66	1
Li$^+$–I	51.1 ± 6.3	1
Li$^+$–Kr	48.1	1
Li$^+$–Li	137.3 ± 6.3	1
Li$^+$–Ne	15.32	1
Li$^+$–O	38.9 ± 9.6	1
Li$^+$–Sb	129.6 ± 13.9	1
Li$^+$–Xe	56.4	1
Lu$^+$–Br	86.1	1
Lu$^+$–Cl	180.6	1
Lu$^+$–F	376.8	1
Lu$^+$–H	204 ± 15	1
Lu$^+$–I	40.7	1
Lu$^+$–O	524 ± 15	1
Lu$^+$–Si	107 ± 13	1
Mg$^+$–Ar	19.20	1

A^+–B	D^o_{298}/kJ mol^{-1}	Ref.
Mg^+–Au	267 ± 29	1
Mg^+–Cl	327 ± 6.5	1
Mg^+–D	203.6 ± 0.8	1
Mg^+–F	477 ± 50	1
Mg^+–H	190.8 ± 5.8	1
Mg^+–Kr	25.39	1
Mg^+–Mg	125	1
Mg^+–Ne	4.9 ± 0.6	1
Mg^+–O	245.2 ± 10	1
Mg^+–Xe	53.74	1
Mn^+–Cl	>211	1
Mn^+–F	321 ± 24	1
Mn^+–H	202.5 ± 5.9	1
Mn^+–I	>211	1
Mn^+–Mn	129	1
Mn^+–O	285 ± 13	1
Mn^+–S	247 ± 23	1
Mn^+–Se	165 ± 50	1
Mo^+–C	442.7 ± 13.5	1
Mo^+–F	376 ± 29	1
Mo^+–H	170 ± 6	1
Mo^+–Mo	449.4 ± 1.0	1
Mo^+–O	488.2 ± 1.9	1
Mo^+–S	355.1 ± 5.8	1
Mo^+–Xe	>53.1 ± 6.8	1
N^+–Ar	208.4 ± 9.6	1
N^+–F	584 ± 42	1
N^+–H	≥435.67 ± 0.77	1
N^+–N	843.85 ± 0.10	1
N^+–O	115	1
Na^+–Ar	19 ± 8	1
Na^+–Br	58.2 ± 10.6	1
Na^+–Cl	20.3 ± 10	1
Na^+–He	7.55	1
Na^+–I	64.9 ± 3.0	1
Na^+–Kr	~24.9	1
Na^+–Li	95.8 ± 3.9	1
Na^+–Na	98.64 ± 0.29	1
Na^+–Na	6.4	1
Na^+–Ne	~9.04	1
Na^+–O	37 ± 19	1
Na^+–Xe	~28.6	1
Nb^+–Ar	40.87 ± 0.13	1
Nb^+–C	509 ± 15	1
Nb^+–Fe	>251	1
Nb^+–H	220 ± 7	1
Nb^+–Nb	576.8 ± 9.6	1
Nb^+–O	688 ± 11	1
Nb^+–S	501.7 ± 20.3	1
Nb^+–V	404.7 ± 0.2	1
Nb^+–Xe	73.28 ± 0.12	1
Nd^+–Au	267 ± 84	1
Nd^+–Br	352.9	1
Nd^+–Cl	441.4	1
Nd^+–F	309.6	1
Nd^+–I	596 ± 32	1
Nd^+–O	753 ± 15	1
Ne^+–H	1239	1
Ne^+–He	13.0 ± 0.8	1
Ne^+–Ne	125.29 ± 1.93	1
Ni^+–Ar	53.9	1
Ni^+–Br	>289	1
Ni^+–C	418	1
Ni^+–Cl	192 ± 4	1
Ni^+–D	166.0 ± 7.7	1
Ni^+–F	≥456	1
Ni^+–H	158.1 ± 7.7	1
Ni^+–He	12.4 ± 0.4	1
Ni^+–I	>297	1
Ni^+–Ne	9.9 ± 0.4	1
Ni^+–Ni	208	1
Ni^+–O	275.9 ± 7.7	1
Ni^+–S	241.0 ± 3.9	1
Ni^+–Si	326 ± 6.7	1
Np^+–F	730 ± 100	1
Np^+–O	760 ± 10	14
Np^+–S	491 ± 52	20
O^+–Ar	33.8	1
O^+–F	301.8 ± 8.4	1
O^+–H	487.9 ± 0.34	1
O^+–N	1050.64 ± 0.13	1
O^+–O	647.75 ± 0.17	1
Os^+–C	398.2 ± 20.2	28
Os^+–H	223.8 ± 6.6	28
Os^+–O	478 6 ± 1.9	26
P^+–C	512 ± 42	1
P^+–Cl	289	1
P^+–F	490.6 ± 8.4	1
P^+–H	329.6 ± 2.1	1
P^+–N	483 ± 21	1
P^+–O	791.3 ± 8.4	1
P^+–P	481 ± 50	1
P^+–S	606 ± 34	1
Pa^+–O	800 ± 50	14
Pa^+–S	525 ± 86	20
Pb^+–Br	260 ± 63	1
Pb^+–Cl	285 ± 63	1
Pb^+–F	347 ± 32	1
Pb^+–O	247 ± 8.4	1
Pb^+–Pb	214 ± 29	1
Pb^+–S	227.7 ± 10.6	12
Pb^+–Se	169.4 ± 6.3	1
Pb^+–Te	163 ± 63	1
Pd^+–C	528 ± 5	1
Pd^+–H	208.4 ± 8.7	1
Pd^+–O	145 ± 11	1
Pd^+–Pd	197 ± 29	1
Pd^+–S	197 ± 6	1
Pd^+–Si	289 ± 50	1
Pr^+–Au	317 ± 81	1
Pr^+–Br	357.7	1
Pr^+–Cl	445.0	1
Pr^+–F	557 ± 63	1
Pr^+–I	317.0	1
Pr^+–O	796 ± 15	1
Pt^+–Ar	36.4 ± 8.7	1
Pt^+–B	398 ± 105	1
Pt^+–C	530.5 ± 4.8	1
Pt^+–Cl	249.8 ± 14.5	1
Pt^+–H	275 ± 5	1
Pt^+–N	326.9 ± 9.6	1
Pt^+–O	318.4 ± 6.7	1
Pt^+–Pt	318 ± 23	1
Pt^+–Si	515 ± 50	1
Pt^+–Xe	86.6 ± 28.9	1
Pu^+–F	562 ± 50	1
Pu^+–O	651 ± 19	14
Pu^+–S	420 ± 23	20
Rb^+–Ar	12.0	1
Rb^+–Br	17.6v5.1	1
Rb^+–Cl	10.5 ± 10.5	1
Rb^+–I	27 ± 42	1
Rb^+–Kr	14.9	1
Rb^+–Na	50.1 ± 3.9	1
Rb^+–Ne	6.95	1
Rb^+–O	29	1
Rb^+–Rb	75.6 ± 9.6	1
Rb^+–Xe	21.5	1
Re^+–C	497.7 ± 3.9	1
Re^+–H	224.7 ± 6.7	1
Re^+–O	465.1 ± 4.8	25
Rh^+–C	414 ± 17	1
Rh^+–H	164.8 ± 3.8	1
Rh^+–O	295.0 ± 5.8	1
Rh^+–S	251.8 ± 11.6	11
Ru^+–C	594.3 ± 6.8	15
Ru^+–H	160.2 ± 5.0	1
Ru^+–O	372 ± 5	1
Ru^+–S	293.3 ± 9.6	15
S^+–C	620.8 ± 1.3	1
S^+–F	343.5 ± 4.8	1
S^+–H	348.2 ± 1.7	1
S^+–N	516 ± 34	1
S^+–O	524.3 ± 0.4	1
S^+–P	573 ± 21	1
S^+–S	522.4 ± 0.5	1
Sc^+–C	326 ± 6	1
Sc^+–Cl	410 ± 42	1
Sc^+–F	605 ± 32	1
Sc^+–Fe	201 ± 21	1
Sc^+–H	235 ± 8	1
Sc^+–O	689 ± 5	1
Sc^+–S	529.7 ± 17.4	1
Sc^+–Se	475.8 ± 8.4	1
Sc^+–Si	242.3 ± 10.5	1
Se^+–F	364 ± 42	1
Se^+–H	304	1
Se^+–P	514 ± 25	1
Se^+–S	392 ± 19	1
Se^+–Se	413 ± 19	1
Si^+–Au	175 ± 50	1
Si^+–B	351 ± 15	1
Si^+–Br	276 ± 96	1
Si^+–C	365 ± 50	1
Si^+–Cl	591.0 ± 0.6	1
Si^+–F	684.1 ± 5.4	1
Si^+–H	316.6 ± 2.1	1
Si^+–O	478 ± 13.4	1
Si^+–P	272 ± 50	1
Si^+–Pd	237 ± 50	1
Si^+–Pt	525 ± 50	1
Si^+–S	387.5 ± 6.0	1
Si^+–Si	334 ± 19	1

A^+–B	D^o_{298}/kJ mol^{-1}	Ref.
Si^+–Te	347 ± 50	1
Sm^+–Br	343.3	1
Sm^+–Cl	435.4	1
Sm^+–F	620.9	1
Sm^+–I	299.1	1
Sm^+–O	569 ± 15	1
Sn^+–Br	335 ± 50	1
Sn^+–Cu	184 ± 96	1
Sn^+–F	364 ± 29	1
Sn^+–O	281 ± 10	1
Sn^+–S	240 ± 19	1
Sn^+–Se	174 ± 6.3	1
Sn^+–Sn	193	1
Sn^+–Te	168.7 ± 8.4	1
Sr^+–Ar	13.32 ± 2.92	1
Sr^+–Br	378.1 ± 8.4	1
Sr^+–Cl	427 ± 8.4	1
Sr^+–F	615 ± 50	1
Sr^+–H	209 ± 5	1
Sr^+–I	308.2	1
Sr^+–Kr	18.13 ± 6.94	1
Sr^+–Ne	4.52 ± 9.6	1
Sr^+–O	298.7	1
Sr^+–Sr	108.5 ± 1.6	1
Ta^+–C	369.4 ± 3.9	10
Ta^+–H	230 ± 6	1
Ta^+–O	688.7 ± 11.6	10
Ta^+–Ta	666	1
Tb^+–Cu	245 ± 34	1
Tb^+–O	722 ± 15	1
Tc^+–H	197.5	1
Tc^+–O	>167	1
Te^+–H	305 ± 12	1
Te^+–O	339 ± 50	1
Te^+–P	415 ± 97	1
Te^+–Se	342 ± 19	1
Te^+–Si	339.6	5
Te^+–Te	278 ± 29	1
Th^+–C	468.98±28.0	31
Th^+–Cl	499 ± 29	1
Th^+–F	682 ± 29	1
Th^+–H	240.1±6.8	29
Th^+–O	830.60 ± 13.5	31
Th^+–Pt	388 ± 193	1
Th^+–Rh	504 ± 67	1
Th^+–S	570 ± 75	20
Ti^+–C	395 ± 23	1
Ti^+–Cl	426.8	1
Ti^+–F	≈456	1
Ti^+–H	226.6 ± 10.6	1
Ti^+–N	501 ± 13	1
Ti^+–O	667 ± 7	1
Ti^+–Pt	82 ± 96	1
Ti^+–S	461.1 ± 6.8	1
Ti^+–Si	249 ± 16	1
Ti^+–Ti	229	1
Tl^+–Br	52 ± 50	1
Tl^+–Cl	26 ± 4	1
Tl^+–F	13 ± 21	1
Tl^+–I	133 ± 21	1
Tl^+–Tl	22 ± 50	1
Tm^+–Br	312.2	1
Tm^+–Cl	407.9	1
Tm^+–F	537 ± 16	1
Tm^+–I	266.8	1
Tm^+–O	482 ± 15	1
U^+–Br	345 ± 29	1
U^+–C	300 ± 96	1
U^+–Cl	431 ± 34	1
U^+–D	283.4 ± 9.6	1
U^+–F	668 ± 29	1
U^+–H	284 ± 8	1
U^+–N	~485	1
U^+–O	774 ± 13	14
U^+–P	186	1
U^+–S	500 ± 60	20
V^+–Ar	39.39 ± 0.12	1
V^+–C	363.05 ± 0.24	32
V^+–D	202 ± 6	1
V^+–Fe	314 ± 21	1
V^+–H	202 ± 6	1
V^+–Kr	49.46 ± 0.18	1
V^+–N	468.98 ±0.19	32
V^+–Nb	403.5 ± 0.2	1
V^+–O	581.6 ± 9.6	1
V^+–S	358.9 ± 8.7	1
V^+–Si	229 ± 15	1
V^+–V	302	1
V^+–Xe	66.4 ± 0.6	1
W^+–C	464.9 ± 17.4	23
W^+–F	444 ± 96	1
W^+–H	222.5 ± 5	1
W^+–O	656.9 ± 6.8	10
Xe^+–Ar	13.4	1
Xe^+–F	188	1
Xe^+–H	355	1
Xe^+–Kr	41.65 ± 0.08	1
Xe^+–N	66.4 ± 9.6	1
Xe^+–Ne	2.1 ± 0.8	1
Xe^+–Xe	99.6	1
Y^+–C	281 ± 12	1
Y^+–F	677 ± 21	1
Y^+–H	260.5 ± 5.8	1
Y^+–O	718 ± 25	1
Y^+–Pt	466 ± 192	1
Y^+–S	533.9 ± 8	1
Y^+–Si	243 ± 13	1
Y^+–Te	360 ± 96	1
Y^+–Y	281 ± 21	1
Yb^+–Br	307.4	1
Yb^+–Cl	399.6	1
Yb^+–F	557.5 ± 14.4	1
Yb^+–I	271.9±17.4	30
Yb^+–O	376 ± 15	1
Yb^+–Yb	238 ± 96	1
Zn^+–Ar	28.7 ± 1.2	1
Zn^+–H	216 ± 15	1
Zn^+–O	161.1 ± 4.8	1
Zn^+–S	198 ± 12	1
Zn^+–Si	274.1 ± 9.6	1
Zn^+–Zn	60 ± 19	1
Zr^+–Ar	36.09 ± 0.24	1
Zr^+–C	445.8 ± 15.4	1
Zr^+–H	218.8 ± 9.6	1
Zr^+–N	443 ± 46	1
Zr^+–O	753 ± 11	1
Zr^+–S	549.0 ± 9.6	1
Zr^+–Zr	407.0 ± 9.6	1

References

1. Luo, Y. R., *Comprehensive Handbook of Chemical Bond Energies*, CRC Press, Boca Raton, 2007.
2. Parke, L. G., Hinton, C. S., and Armentrout, P. B., *Int. J. Mass Spectrom.* 254, 168, 2006.
3. Li, F.-X., Zhang, X.-G., and Armentrout, P. B., *Int. J. Mass Spectrom.* 255/256, 279, 2006.
4. Li, F.-X., and Armentrout, P. B., *J. Chem. Phys.* 125, 133114/1, 2006.
5. Chattopadhyaya, S., Pramanik, A., Banerjee, A., and Das, K. K., *J. Phys. Chem. A* 110, 12303, 2006.
6. Li, J., Hao, Y., Yang, J., Zhou, C., and Mo, Y., *J. Chem. Phys.* 127, 104307/1, 2007.
7. Gibson, J. K., Haire, R. G., Santos, M., Pires de Matos, A., and Marçalo, J., *J. Phys. Chem. A* 112, 11373, 2008.
8. Merritt, J. M., Kaledin, A. L., Bondybey, V. E., and Heaven, M. C., *Phys. Chem. Chem. Phys.* 10, 4006, 2008.
9. Schröder, D., *J. Phys. Chem. A* 112, 13215, 2008.
10. Hinton, C. S., Li, F.-X., and Armentrout, P. B., *Int. J. Mass Spectrom.* 280, 226, 2009.
11. Armentrout, P. B., and Kretzschmar, I., *J. Phys. Chem. A* 113, 10955, 2009.
12. Armentrout, P. B., and Kretzschmar, I., *Inorg.. Chem.* 48, 10371, 2009.
13. Li, M., Liu, S.-R., and Armentrout, P. B., *J. Chem. Phys.* 131, 144310, 2009.
14. Marçalo, J., and Gibson, J. K., *J. Phys. Chem. A* 113, 12599, 2009.
15. Armentrout, P. B., and Kretzschmar, I., *Phys. Chem. Chem. Phys.* 12, 4078, 2010.
16. Tzeli, D., and Mavridis, A., *J. Chem. Phys.* 132, 194312, 2010.
17. Hinton, C. S., and Armentrout, P. B., *J. Chem. Phys.* 133, 124307, 2010.
18. Li, F.-X., Hinton, C. S., Citir, M., Liu, F., and Armentrout, P. B., *J. Chem. Phys.* 134, 024310, 2011.
19. Li, F.-X., Gorham, K., and Armentrout, P. B., *J. Phys. Chem. A* 114, 11043, 2010.

20. Pereira, C. C. L., Marsden, C. J., Marçalo, J., and Gibson, J. K., *Phys. Chem. Chem. Phys.* 13, 12940, 2011.
21. Armentrout, P. B., and Kretzschmar, I., *J. Chem. Phys.*, 132, 024306, 2010.
22. Hinton, C. S., Citir, M., and Armentrout, P. B., *J. Chem. Phys.*, 135, 234302, 2011.
23. Wnorowski, K., Stano, M., Barszczewska, W., Jówko, A., and Matejcik, Š., *Int. J. Mass Spectrom.* 314, 42, 2012.
24. Armentrout, P. B., and Li, F.-X., *J. Phys. Chem.* A 117, 7754, 2013.
25. Armentrout, P. B., *J. Chem. Phys.* 139, 084305, 2013.
26. Hinton, C. S., Citir, M., and Armentrout. P. B., *Int. J. Mass Spectrosc.* 354, 87, 2013.
27. Parry, I. S., Hermes, A. C., Kartouzian, A., and Mackenzie, S. R., *Phys. Chem. Chem. Phys.* 16, 458, 2014.
28. Armentrout, P. B., Parke, L., Hinton, C., and Citir, M., *ChemPlusChem.* 78, 1157, 2013.
29. Cox, R. M., et al., *J. Phys. Chem. B* 120, 1601-1614, 2016.
30. Sergeev, D. N., et al., *Int. J. Mass Spectrom.*, 374, 1-3, 2014.
31. Cox, R. M., et al., *J. Chem. Phys.*, 184309, 2016.
32. Johnson, E. L., et al., *J. Chem. Phys.*, 144, 234306, 2016.

Bond Dissociation Energies in Polyatomic Cations

This table has been arranged on the basis of the Periodic Table with the IUPAC notation for Groups 1 to 18, see inside front cover of this *Handbook*. The **boldface** in the species indicates the dissociated fragment.

Table 7. Bond Dissociation Energies in Polyatomic Cations

Bond	D°_{298} /kJ mol^{-1}	Ref.
(1) Group 1		
Li^+–H_2	27.2	1
Li^+–CO	57 ± 13	1
Li^+–H_2O	139 ± 8	1
Li^+–NH_3	156 ± 8	1
Li^+–CH_4	130	1
Li^+–CH_3OH	156 ± 8	1
Li^+–CH_3OCH_3	167 ± 10	1
Li^+–pyridine	183.0 ± 14.5	1
Li^+–Gly (glycine)	220 ± 9	1
Na^+–H_2	10.4 ± 0.8	1
Na^+–N_2	33.5	1
Na^+–CO	31 ± 8	1
Na^+–CO_2	66.5	1
Na^+–SO_2	79.1	1
Na^+–O_3	52.3	1
Na^+–H_2O	91.2 ± 6.3	1
$Na^+(H_2O)$-H_2O	82.0 ± 5.8	1
$Na^+(H_2O)_2$–H_2O	66.1	1
$Na^+(H_2O)_3$–H_2O	52.7 ± 0.8	1
Na^+(glycine)–H_2O	75.1 ± 5.3	1
Na^+(glutamine)–H_2O	52 ± 1	1
Na^+–NH_3	106.2 ± 5.4	1
Na^+–HNO_3	86.2	1
Na^+–CH_4	30.1	1
Na^+–CH_3OH	98.8 ± 5.7	1
Na^+–CH_3CN	125.5 ± 9.6	1
Na^+–C_2H_4	44.6 ± 4.4	1
Na^+–CH_3OCH_3	101.4 ± 5.7	1
Na^+–$CH_3C(O)H$	114.4 ± 3.4	1
Na^+–MeCOMe	131.3 ± 4.1	1
Na^+–C_6H_6	97.0 ± 5.9	1
Na^+–pyrrole	103.7 ± 4.8	1
Na^+–Gly (glycine)	166.7 ± 5.1	1
Na^+–Ala (alanine)	167 ± 4	1
Na^+–GlyGly (glycylglycine)	203 ± 8	1
Na^+–(15-crown-5)	299.1 ± 15.5	12
Na^+–(18-crown-6)	300±19	1
K^+–H_2	6.1 ± 0.8	1
K^+–CO_2	35.6	1
K^+–H_2O	74.9	1
$K^+(H_2O)_2$-H_2O	67.4	1
$K^+(H_2O)_3$-H_2O	55.2	1
$K^+(H_2O)_4$-H_2O	11.8	1
$K^+(H_2O)_5$-H_2O	44.8	1
$K^+(H_2O)_6$-H_2O	41.8	1
K^+–NH_3	79 ± 7	1
K^+–C_6H_6	80.3	1
K^+–adenine	95.1 ± 3.2	1
K^+–indole	104.6 ± 12.6	1
K^+–Phe (phenylalanine)	150.5 ± 5.8	1
K^+–Tyr (tyrosine)	165.0 ± 5.8	1
K^+–(15-crown-5)	217.9±10.6	12
K^+–(18-crown-6)	235±13	1
Rb^+–H_2O	66.9 ± 12.6	1
Rb^+–NH_3	78.2	1
Rb^+–CH_3CN	86.6 ± 1.3	1
Rb^+–C_6H_5OH	70.2 ± 3.7	1
Rb^+–(15-crown-5)	176.8±9.7	12
Rb+–(18-crown-6)	192±13	1
Cs^+–H_2O	57.3	1
Cs^+–$C_6H_5NH_2$	70.8 ± 4.5	1
Cs^+–(15-crown-5)	160.5±9.6	12
Cs^+–(18-crown-6)	170±9	1
(2) Group 2		
CH_3Be^+–CH_3	192.9 ± 13.4	1
tert-**$C(CH_3)_3Be^+$**–*tert*-$C(CH_3)_3$	121.8 ± 13.4	1
Mg^+-OH	314 ± 33	1
Mg^+-CO	43.1 ± 5.8	1
Mg^+-CO_2	58.4 ± 5.8	1
Mg^+-H_2O	122.5 ± 12.5	1
Mg^+–NH_3	158.9 ± 11.6	1
Mg^+–CH_4	29.8 ± 6.8	1
Mg^+–MeOH	147.6 ± 6.8	1
Mg^+–C_6H_6	155.2	1
Mg^+–pyridine	200.0 ± 6.4	1
Mg^+–imidazole	243.9 ± 10.4	1
$Mg^{2+}(H_2O)_5$-H_2O	101.3	1
$Mg^{2+}(Me_2CO)_5$-Me_2CO	93.3	1
Ca^+–OH	435.1 ± 14.5	1
Ca^+–H_2O	117.2	1
Ca^+–C_6H_6	134	1
Ca^+–imidazole	186.3 ± 3.9	1
$Ca^{2+}(H_2O)_4$-H_2O	110.0 ± 5.9	1
$Ca^{2+}(Me_2CO)_5$-Me_2CO	101.3	1
Sr^+–CO	20.3	1
Sr^+–CO_2	41.9	1
Sr^+–H_2O	144.3	1

Bond	D°_{298} /kJ mol^{-1}	Ref.
Sr^+–C_6H_6	117	1
$Sr^{2+}(H_2O)_5$–H_2O	87.4	1
Ba^+–OH	530.7 ± 19.3	1
$Ba^{2+}(H_2O)_4$–H_2O	90.8	1
(3) Group 3		
Sc^+–H_2	23.0 ± 1.3	1
Sc^+–CH_2	412 ± 22	1
Sc^+–CH_3	233 ± 10	1
Sc^+–C_2H_2	240 ± 20	1
Sc^+–C_2H_4	≥131	1
Sc^+–C_6H_6	222 ± 21	1
Sc^+–H_2O	131	1
Sc^+–NH	483 ± 10	1
Sc^+–NH_2	347 ± 5	1
Sc^+–pyridine	231.5 ± 10.3	1
Y^+–CH_2	398 ± 13	1
Y^+–CH_3	249 ± 5.0	1
Y^+–C_2H_2	218 ± 13	1
Y^+–C_2H_4	>138	1
Y^+–CO	29.9 ± 10.6	1
Y^+–CS	137.0 ± 7.7	1
$Y^+(O)$–CO_2	86 ± 5	1
La^+–CH	523 ± 33	1
La^+–CH_2	401 ± 7	1
La^+–CH_3	217 ± 15	1
La^+–C_2H_2	262 ± 30	1
La^+–C_2H_4	192.5	1
Lu^+–CH_2	>230 ± 6	1
Lu^+–CH_3	176 ± 20	1
$U^+(F)$–F	552 ± 44	1
$U^+(F)_2$–F	523 ± 38	1
$U^+(F)_3$–F	381 ± 19	1
$U^+(F)_4$–F	243 ± 17	1
$U^+(F)_5$–F	26 ± 11	1
(4) Group 4		
Ti^+–CH	478 ± 5	1
Ti^+–CH_2	391 ± 15	1
Ti^+–CH_3	213.8 ± 3	1
Ti^+–CH_4	70.3 ± 2.5	1
Ti^+–C_2H_2	213 ± 13	1
Ti^+–C_2H_4	146 ± 11	1
Ti^+–C_6H_6	259 ± 9	1
Ti^+–CO	117.7 ± 5.8	1
Ti^+–H_2O	157.7 ± 5.9	1
Ti^+–NH	466 ± 12	1
Ti^+–NH_2	356 ± 13	1
Ti^+–NH_3	197 ± 7	1
Ti^+–pyridine	217.2 ± 9.3	1
Ti^+–imidazole	≤232.4 ± 8.2	1
Zr^+–CH	568 ± 13	1
Zr^+–CH_2	444.8 ± 5	1
Zr^+–CH_3	227.7 ± 9.6	1
Zr^+–C_2H_2	273 ± 14	1
Zr^+–CO	77 ± 10	1
Zr^+–CS	257.6 ± 10.6	1
Hf^+–CH	492.1 ± 14.5	2
Hf^+–CH_2	421.6 ± 6.8	2
Hf^+–CH_2	204.5 ± 25.1	2
Hf^+–C_2H_2	150.6	1
(5) Group 5		
$(CO)_6V^+$–H	220 ± 14	1
V^+–H_2	42.7 ± 2.1	1
V^+–CH	479.5 ± 8.8	10
V^+–CH_2	326 ± 6	1
V^+–CH_3	193 ± 7	1
V^+–C_2H_2	172 ± 8	1
V^+–C_2H_4	124 ± 8	1
V^+–(η^5-C_5H_5)	530.7	1
V^+–C_6H_6	234 ± 10	1
V^+–CO	114.8 ± 2.9	1
V^+–CO_2	72.4 ± 3.8	1
V^+–H_2O	149.8 ± 5.0	1
V^+–NH	423 ± 29	1
V^+–NH_2	293 ± 6	1
V^+–NH_3	192 ± 11	1
V^+–pyridine	218.7 ± 13.5	1
V^+–imidazole	≤243.4 ± 8.0	1
Nb^+–H_2	61.9	1
Nb^+–CH	581 ± 19	1
Nb^+–CH_2	428.4 ± 8.7	1
Nb^+–CH_3	198.8 ± 10.6	1
Nb^+–CH_3NH_2	134	1
Nb^+–C_3H_6	117.7	1
$(NbFe)^+$–C_3H_4	>163	1
Nb^+–CO	95.5 ± 4.8	1
Nb^+–CS	242.2 ± 10.6	1
$Nb_7{}^+$–N_2	<215	1
Ta^+–CH	561.5 ± 15.4	6
Ta^+–CH_2	464.1 ± 2.9	6
Ta^+–CH_3	259.5 ± 13.5	6
Ta^+–C_6H_6	251~301	1
(6) Group 6		
$(CO)_6Cr^+$–H	230 ± 10	1
(η^5-C_5H_5)(NO)$(CO)_2Cr^+$–H	207.1 ± 14	1
Cr^+–H_2	31.8 ± 2.1	1
Cr^+–CH	294 ± 29	1
Cr^+–CH_2	216 ± 4	1
Cr^+–CH_3	110 ± 4	1
Cr^+–C_6H_6	170 ± 10	1
Cr^+–indole	196.6 ± 16.7	1
Cr^+–CO	89.7 ± 5.8	1
Cr^+–OH	298 ± 14	1
Cr^+–H_2O	132.6 ± 8.8	1
Cr^+–N_2	59 ± 4	1
Cr^+–NH_3	183 ± 10	1
$(CO)_6Mo^+$–H	260 ± 9	1
Mo^+–CH	513.3 ± 13.5	1
Mo^+–CH_2	344.4 ± 10	1
Mo^+–CH_3	151.5 ± 8.7	1
Mo^+–CO	193.9 ± 9.6	1
Mo^+–CO_2	49.2 ± 7	1
Mo^+–CS	162 ± 18	1
Mo^+–CS_2	67.5 ± 12.5	1
Mo^+–NH	<385	1
Mo^+–pyrrole	>289	1
$(CO)_6W^+$–H	257 ± 9	1
W^+–CH	580 ± 27	1
W^+–CH_2	456.4 ± 5.8	1
W^+–CH_3	~222.9 ± 9.6	1

Bond	D^o_{298} /kJ mol^{-1}	Ref.
$(PMe_3)_3(CO)_3W^+$–H	259.4	1
W^+–pyrrole	>209	1
(7) Group 7		
$(CO)_5Mn^+$–H	172 ± 10	1
Mn^+–H_2	7.9 ± 1.7	1
Mn^+–CH_2	295 ± 13	1
Mn^+–CH_3	215 ± 10	1
$Mn^+(CO)_5$–CH_3	132 ± 15	1
$Mn^+(CO)_5$–CH_4	>30	1
Mn^+–(η^5-C_5H_5)	326.1 ± 9.6	1
Mn^+–C_6H_6	145 ± 10	1
Mn^+–OH	332 ± 24	1
Mn^+–CO	25 ± 10	1
Mn^+–H_2O	121.8 ± 5.9	1
Mn^+–CH_3OH	134 ± 29	1
Mn^+–$OC(CH_3)_2$	159 ± 14	1
Mn^+–CS	80.0 ± 21	1
Mn^+–NH_2	254 ± 20	1
Mn^+–NH_3	147 ± 8	1
Tc^+–CH_2	<464	1
Tc^+–C_2H_2	<320	1
$Re^+(CH_3)(CO)_5$–H	294 ± 13	1
$(PMe_3)(CO)_2Re^+$–H	300.4	1
(8) Group 8		
$Fe^+(O)$–H	444 ± 17	1
$Fe^+(CO)$–H	120 ± 23	1
$Fe^+(H_2O)$–H	215 ± 14	1
$Fe^+(\eta^5$-$C_5H_5)$–H	193 ± 21	1
$(CO_2)_5Fe^+$–H	299 ± 15	1
Fe^+–H_2	45.2 ± 2.5	1
Fe^+–CH	423 ± 29	1
Fe^+–CH_2	≤342 ± 2	1
Fe^+–CH_3	229 ± 5	1
Fe^+–CH_4	73.2	1
Fe^+–C_2H_2	159.0 ± 2.1	1
Fe^+–C_2H_3	238 ± 10	1
Fe^+–C_2H_4	145 ± 11	1
Fe^+–C_2H_5	233 ± 9	1
Fe^+–C_2H_6	64 ± 6	1
Fe^+–OH	366 ± 12	1
Fe^+–CO	129.3 ± 3.9	1
Fe^+D–CO	53 ± 13	1
Fe^+–CO_2	74.3 ± 7.7	1
Fe^+–H_2O	128.9 ± 0.8	1
Fe^+–N_2	53 ± 4	1
Fe^+–NH_3	184 ± 12	1
Fe^+–CS_2	166.1 ± 4.6	1
Fe^+–imidazole	246.1 ± 13.8	1
Fe^+–SiH	254 ± 13	1
Fe^+–SiH_2	181 ± 9	1
Fe^+–SiH_3	183 ± 9	1
$Ru^+(\eta^5$-$C_5H_5)_2$–H	292 ± 16	1
$(\eta^5$-$C_5Me_5)_2Ru^+$–H	284.5	1
Ru^+–CH	501.7 ± 11.6	1
Ru^+–CH_2	344.4 ± 4.8	1
Ru^+–CH_3	160.2 ± 5.8	1
Ru^+–CS	244.9 ± 17.4	9
Os^+–CH	653.2 ± 14.5	11
Os^+–CH_3	276.7 ± 14.5	11
OsO_4^+–H	552 ± 13	1

Bond	D^o_{298} /kJ mol^{-1}	Ref.
(9) Group 9		
$(\eta^5$-$C_5H_5)(CO)_2Co^+$–H	245 ± 12	1
$(CH_3OD)Co^+$–H	147.6 ± 7.7	1
Co^+–H_2	76.1 ± 4.2	1
$(\eta^5$-$C_5H_5)Co^+$–H_2	67.8	1
Co^+–CH	420 ± 37	1
Co^+–CH_2	317 ± 5	1
Co^+–CH_3	203 ± 4	1
Co^+–CH_4	96.7	1
Co^+–C_{60}	243 ± 67	1
Co^+–CO	173.7 ± 6.7	1
Co^+–H_2O	164.4 ± 5.9	1
Co^+–CS	259 ± 33	1
Co^+–N_2	96.2 ± 7.1	1
Co^+–NH_2	247 ± 7	1
Co^+–NH_3	219 ± 16	1
Co^+–CH_3CN	>255 ± 17	1
Co^+–$P(CH_3)_3$	278 ± 11	1
Co^+–$P(C_2H_5)_3$	339 ± 16	1
$(CH)Rh^+$–H	372 ± 21	1
$(\eta^5$-$C_5H_5)(CO)_2Rh^+$–H	287 ± 12	1
Rh^+–CH	444 ± 12	1
Rh^+–CH_2	356 ± 8	1
Rh^+–CH_3	142 ± 6	1
Rh^+–NO	167 ± 21	1
Rh^+–CS	256.6 ± 18.3	8
$(CO)(\eta^5$-$C_5H_5)(PPh_3)Ir^+$–H	313.4	1
$(CO)_2(\eta^5$-$C_5Me_5)Ir^+$–H	298.3	1
Ir^+–CH	666.7 ± 22.2	3
Ir^+–CH_2	474.7 ± 2.9	3
Ir^+–CH_3	313.6 ± 17.4	3
Ir^+–C_2H_4	234.3	1
(10) Group 10		
$(CO)_4Ni^+$–H	248 ± 9	1
$(\eta^5$-$C_5H_5)(NO)Ni^+$–H	315 ± 14	1
$(\eta^5$-$C_5H_5)(\eta^5$-$C_5H_5)Ni^+$–H	215 ± 13	1
Ni^+–H_2	72.4 ± 1.3	1
Ni^+–CH	301.0 ± 11.6	1
Ni^+–CH_2	306 ± 4	1
Ni^+–CH_3	169.8 ± 6.8	1
Ni^+–CH_4	96.5 ± 4	1
Ni^+–OH	235 ± 19	1
Ni^+–CO	175 ± 11	1
Ni^+–CO_2	104 ± 1	1
Ni^+–H_2O	183.7 ± 3.3	1
Ni^+–CS	234.5 ± 9.6	1
Ni^+–N_2	110.9 ± 10.5	1
Ni^+–NO	227.6 ± 7.5	1
Ni^+–NH_2	232.5 ± 7.7	1
Ni^+–NH_3	238 ± 19	1
Pd^+–CH	536 ± 10	1
Pd^+–CH_2	463 ± 3	1
Pd^+–CH_3	258 ± 8	1
Pd^+–CH_4	170.8 ± 7.7	1
Pd^+–CS	200 ± 14	1
Pd^+–C_2H_2	>28.9 ± 4.8	1
Pt^+–H_2	146.7 ± 11.6	1
Pt^+–CH	536.4 ± 9.6	1
Pt^+–CH_2	471	1
Pt^+–CH_3	257.6 ± 7.7	1

Bond	D°_{298} /kJ mol^{-1}	Ref.
Pt^+-CH_4	170.8 ± 7.7	1
Pt^+-O_2	64.6 ± 4.8	1
Pt^+-CO	218.1 ± 8.7	1
Pt^+-CO_2	59.8 ± 4.8	1
Pt^+-NH_3	274 ± 12	1
$Pt^+-C_2H_4$	229.7	1
(11) Group 11		
Cu^+-H_2	51.9 ± 0.4	1
Cu^+-CH_2	267.3 ± 6.8	1
Cu^+-CH_3	111 ± 7	1
$Cu^+-C_2H_2$	>21.2 ± 9.6	1
$Cu^+-C_2H_4$	176 ± 14	1
$Cu^+-C_6H_6$	218.0 ± 9.6	1
Cu^+-CO	149 ± 7	1
Cu^+-N_2	89 ± 30	1
Cu^+-NO	109.0 ± 4.8	1
Cu^+-H_2O	160.7 ± 7.5	1
Cu^+-NH_2	192 ± 13	1
Cu^+-NH_3	237 ± 15	1
Cu^+-CS	238.3 ± 11.6	1
Cu^+-SiH	246 ± 27	1
Cu^+-SiH_2	≥231 ± 7	1
Cu^+-SiH_3	97 ± 25	1
Ag^+-CH_2	≥107 ± 4	1
Ag^+-CH_3	66.6 ± 4.8	1
$Ag^+-C_2H_5$	65.7 ± 7.5	1
$Ag^+-C_6H_6$	167 ± 19	1
Ag^+-O_2	29.7 ± 0.8	1
Ag^+-CO	89 ± 5	1
Ag^+-H_2O	134 ± 8	1
Ag^+-CS	152 ± 20	1
Ag^+-NH_3	170 ± 13	1
Au^+-CH_2	357.0 ± 6.8	5
Au^+-CH_3	209.4 ± 23.2	5
$Au^+-C_2H_4$	344.5	1
$Au^+-C_6H_6$	289 ± 29	1
Au^+-CO	201 ± 8	1
Au^+-H_2O	164.0 ± 9.6	1
Au^+-H_2S	230 ± 25	1
Au^+-NH_3	297 ± 29	1
Au^+-PH_3	402 ± 33	1
(12) Group 12		
Zn^+-H_2	15.7 ± 1.7	1
Zn^+-CH_3	280 ± 7	1
Zn^+-OH	127.2	1
Zn^+-H_2O	163	1
Zn^+-NO	76.2 ± 9.6	1
Zn^+–pyrimidine	209.6 ± 7.7	1
Zn^+-CS	149 ± 23	1
Cd^+-CH_3	228 ± 3	1
$Cd^+(CH_3)-CH_3$	109 ± 3	1
$Cd^+-C_6H_6$	136 ± 19	1
Hg^+-CH_3	285 ± 3	1
$Hg^+(CH_3)-CH_3$	96 ± 3	1
(13) Group 13		
B^+-H_2	15.9 ± 0.8	1
HB^+-H_2	61.5 ± 2.1	1
$(CH_3)_2B^+-CH_3$	32.6 ± 4.2	1
Al^+-H_2	5.6 ± 0.6	1

Bond	D°_{298} /kJ mol^{-1}	Ref.
Al^+-N_2	5.6	1
Al^+-CO_2	≥29.3	1
Al^+-H_2O	104 ± 15	1
Al^+–MeOH	139.7	1
Al^+–EtC(O)Et	191.2	1
$Al^+-C_6H_6$	147.3 ± 8.4	1
Al^+–pyridine	190.3 ± 10.3	1
Al^+–phenol	154.8 ± 16.7	1
Al^+–imidazole	232.4 ± 8.2	1
Ga^+-NH_3	122.5	1
In^+-NH_3	111.0	1
(14) Group 14		
$C_{58}{}^+-C_2$	955 ± 15	1
$C_{60}{}^+-C_2$	822.0 ± 12.5	1
$C_{62}{}^+-C_2$	846.2 ± 10.6	1
$C_{78}{}^+-C_2$	938.8 ± 10.6	1
$HC_2{}^+-H$	574.749	1
$C_6H_5{}^+-H$	376.3 ± 4.8	1
$C_2H_3{}^+-Cl$	249 ± 1.0	7
$C_2H_5{}^+-Br$	206.3 ± 1.0	7
$C_6H_5{}^+-Br$	266.3	1
$C_2H_3{}^+-I$	196.2 ± 1.4	7
$CH_3{}^+-H_2$	186	1
$CH_5{}^+-H_2$	7.9 ± 0.4	1
$C_2H_5{}^+-H_2$	17	1
$CH_3{}^+-O_2$	80 ± 7	4
CO^+-N_2	67.5 ± 19.3	1
$H_2CH^+-N_2$	31.8	1
CO^+-CO	173.7 ± 14.6	1
$CO^+(CO)-CO$	52.3	1
$CO^+(CO)_2-CO$	30.2	1
$CO^+(CO)_3-CO$	18.4	1
$(CO_2)^+-CO_2$	70.3	1
$(CO_2)^+(CO_2)-CO_2$	34.7	1
$(CO_2)^+(CO_2)_2-CO_2$	21.3	1
$(CO_2)^+(CO_2)_3-CO_2$	20.1 ± 1.3	1
$CH_3{}^+-N_2O$	221.3	1
$CH_3{}^+-SO_2$	253.6	1
$CH_3{}^+-OCS$	239.3	1
$CH_3{}^+-CS_2$	251.9	1
$CH_3{}^+-H_2O$	279	1
$CH_3{}^+(H_2O)-H_2O$	106.3	1
$CH_3{}^+(H_2O)_2-H_2O$	87.9	1
$CH_3{}^+(H_2O)_3-H_2O$	61.9	1
$CH_3{}^+(H_2O)_4-H_2O$	48.5	1
$CH_3{}^+-H_2S$	344.8	1
$CH_2{}^+-CH_2O$	303.0 ± 2.9	1
$CH_3{}^+-NH_3$	431.4	1
$(CH_3)^+-CH_3$	209.2 ± 4.2	1
$CH_3{}^+-CH_4$	166.5	1
$CF_3{}^+-CH_4$	19.0	1
$(CH_5)^+-CH_4$	28.7 ± 1.3	1
$C_6H_6{}^+-CH_4$	12.0	1
$CH_3{}^+-CH_3F$	230	1
$CH_3{}^+-CF_3Cl$	221	1
$CH_3{}^+-CH_3Cl$	259	1
***tert*-$C_4H_9{}^+$**–CH_3OH	63	1
***tert*-$C_4H_9{}^+$**–CH_3CN	85	1
***tert*-$C_4H_9{}^+$**–SO_2F_2	43.5	1
$CH_3{}^+-C_2H_3O$	338.7 ± 2.9	1
$CH_3{}^+-CF_3ClOCl$	252	1

Bond	D^o_{298} /kJ mol^{-1}	Ref.
***tert*-$C_4H_9^+$–$(CH_3)_2S$**	185	1
***tert*-$C_4H_9^+$–C_2H_5OH**	85	1
***tert*-$C_4H_9^+$–C_3H_8**	27.6	1
***tert*-$C_4H_9^+$–t-C_4H_9Cl**	339	1
***tert*-$C_4H_9^+$–$(CH_3)_3CH$**	30.1	1
***tert*-$C_4H_9^+$–C_6H_6**	92	1
$(C_6H_6)^+$–C_6H_6	73.6	1
$(C_6H_6)^+$–indole	54.8	1
$C_6F_6^+$–C_6F_6	30.1 ± 4	1
C_{60}^+–C_{60}	35.89 ± 7.72	1
$PhSiH_2^+$–H	159	1
$Si^+(CH_3)_3$–Cl	178.5 ± 1.9	1
SiH_3^+–CO	≥151	1
SiF_3^+–CO	174.1 ± 1.3	1
$(CH_3)_3Si^+$–H_2O	125.9 ± 7.9	1
$(CH_3)_3Si^+$–NH_3	194.6	1
$Si^+(CH_3)(Cl)_2$–CH_3	60.8 ± 2.9	1
$Si^+(CH_3)_2(Cl)$–CH_3	41.5 ± 1.9	1
Si^+–CH_3	413.9 ± 5.8	1
$Si^+(CH_3)$–CH_3	123 ± 48	1
$Si^+(CH_3)_2$–CH_3	513 ± 27	1
$Si^+(CH_3)_3$–CH_3	66.6 ± 5.8	1
$(CH_3)_3Si^+$–CH_3OH	164.0	1
$(CH_3)_3Si^+$–$(C_2H_5)_2O$	184.9	1
$(CH_3)_3Si^+$–C_6H_6	100.0	1
$(CH_3)_3Si^+$–CH_3NH_2	231.8	1
$(CH_3)_3Ge^+$–H_2O	119.7 ± 2.1	1
$(C_2H_5)_3Ge^+$–H_2O	104.2 ± 2.1	1
$(CH_3)_3Sn^+$–NH_3	154	1
$(CH_3)_3Sn^+$–H_2O	108	1
$(CH_3)_3Sn^+$–$(CH_3)_2CO$	157	1
$(CH_3)_3Sn^+$–C_3H_7SH	143	1
Pb^+–H_2O	93.7	1
Pb^+–NH_3	118.4 ± 0.8	1
Pb^+–CH_3OH	97.5 ± 0.8	1
Pb^+–CH_3NH_2	148.1 ± 1.3	1
Pb^+–C_6H_6	110 ± 2	1
(15) Group 15		
H_2N^+–H	544.43 ± 0.10	1
H_3N^+–H	515.1	1
Me_3N^+–H	376	1
Et_3N^+–H	362	1
$(imidazole)^+$ –Zn	216.1 ± 3.9	1
N_2H^+–H_2	24.7 ± 0.8	1
ON^+–O_2	14.2	1
N^+–N_2	303.8	1
ON^+–N_2	21.3	1
N_2^+–N_2	102.3 ± 14.6	1
HN_2^+–N_2	60.7	1
N_3^+–N_2	18.8 ± 1.3	1
O_2N^+–N_2	19.2 ± 1.3	1
H_4N^+–N_2	54 ± 21	1
ON^+–NO	59.4 ± 0.8	1
ON^+–CO	27.2 ± 1.3	1
ON^+–O_3	<58	1
ON^+–CO_2	32.2	1
N_2O^+–ON_2	72.8 ± 6.3	1
NO^+–ON_2	36.4 ± 0.8	1
$(HON_2)^+$–ON_2	69.9 ± 4	1
ON^+–H_2O	95	1
$ON^+(H_2O)$–H_2O	67.4	1
$ON^+(H_2O)_2$–H_2O	56.5	1
H_4N^+–H_2O	86.2 ± 4.2	1
$H_4N^+(H_2O)$–H_2O	72.8 ± 4.2	1
$H_4N^+(H_2O)_2$–H_2O	57.3 ± 4.2	1
$H_4N^+(H_2O)_3$–H_2O	51.0	1
$H_4N^+(H_2O)_4$–H_2O	44.4	1
$(glycine)H^+$–H_2O	77.2 ± 11.0	1
$(tryptophan)H^+$–H_2O	31.2 ± 2.5	1
$(tryptophanylglicine)H^+$–H_2O	56.0 ± 5.3	1
H_4N^+–H_2S	47.7	1
$H^+(NH_3)$–NH_3	108.8	1
$H^+(NH_3)_2$–NH_3	69.5	1
$H^+(NH_3)_3$–NH_3	57.3	1
$H^+(NH_3)_4$–NH_3	49.0	1
$H^+(NH_3)_5$–NH_3	29.3	1
$H^+(NH_3)_6$–NH_3	27.2	1
NH_4^+–CH_4	15.0	1
ON^+–CH_3OH	97.6	1
O_2N^+–CH_3OH	80.3 ± 9.6	1
$(CH_3CNH)^+$–CH_3CN	130.1 ± 9.6	1
$(pyridineH)^+$–pyridine	105.4 ± 4	1
$(valine\ H)^+$–valine	86.6 ± 8.4	1
$(betainH)^+$–betaine	139.9 ± 4.8	1
H_4P^+–H_2O	54.4	1
$(H_4P)^+$–PH_3	48.1	1
AsH_2^+–H	257	1
I_2As^+–acetone	106 ± 17	1
I_2As^+–benzene	77 ± 17	1
Bi^+–H_2O	95.4	1
Bi^+–NH_3	149	1
Bi^+–C_6H_6	≤149	1
(16) Group 16		
$(H_3O)^+$–H_2	14.6 ± 2.1	1
O^+–O_2	179.5	1
$O^+(O_2)_1$–O_2	28.9	1
$O^+(O_2)_2$–O_2	3.9	1
O_2^+–O_2	38.3 ± 2.1	1
$O_2^+(O_2)$–O_2	24.6 ± 1.3	1
$O_2^+(O_2)_2$–O_2	10.4 ± 0.8	1
$O_2^+(O_2)_3$–O_2	9.0 ± 0.8	1
$O_2^+(O_2)_4$–O_2	8.0 ± 0.8	1
$O_2^+(O_2)_5$–O_2	7.9 ± 1.3	1
O^+–N_2	231.4	1
O_2^+–N_2	22.6	1
$(H_3O)^+$–N_2	22.2 ± 2.1	1
O_4^+–N_2	12.3	1
O_2^+–CO	31.8	1
O_2^+–CO_2	41.0 ± 2.1	1
CO_2^+–CO_2	65.3 ± 4	1
$(H_3O)^+$–CO_2	64.0	1
$(H_3O)^+(CO_2)$–CO_2	51.9	1
$(H_3O)^+(CO_2)_2$–CO_2	43.9	1
$(H_3O)^+(CO_2)_3$–CO_2	18.0	1
O_2^+–ON_2	56.1 ± 4	1
$(H_3O)^+$–ON_2	70.7 ± 6.5	1
$(H_3O)^+(H_2O)$–ON_2	50.6 ± 2.1	1
$(H_3O)^+(H_2O)_2$–ON_2	42.7 ± 2.1	1
O_3^+–O_3	67.5 ± 39	1
$OClO^+$–OClO	246 ± 48	1
O_2^+–H_2O	>67	1
$(OH)^+(H_2O)_2$–H_2O	87.4	1

Bond	$D°_{298}$ /kJ mol^{-1}	Ref.
(OH)$^+$(H$_2$SO$_4$)(H$_2$O)$_4$–H$_2$O	56.9	1
(OH)$^+$(H$_2$SO$_4$)(H$_2$O)$_5$–H$_2$O	49.8	1
(OH)$^+$(H$_2$SO$_4$)(H$_2$O)$_6$–H$_2$O	44.8	1
(H$_2$O)$^+$–H$_2$O	164.0	1
(H$_3$O)$^+$–H$_2$O	140.2	1
(H$_3$O)$^+$(H$_2$O)–H$_2$O	93.3	1
(H$_3$O)$^+$(H$_2$O)$_2$–H$_2$O	71.1	1
(H$_3$O)$^+$(H$_2$O)$_3$–H$_2$O	64.0	1
(H$_3$O)$^+$(H$_2$O)$_4$–H$_2$O	54.4	1
(H$_3$O)$^+$(H$_2$O)$_5$–H$_2$O	49.0	1
(H$_3$O)$^+$(H$_2$O)$_6$–H$_2$O	43.1	1
(HCOOH)H$^+$–H$_2$O	100.8	1
CH$_3$OH$_2$$^+$–H$_2$O	115.6	1
CH$_3$CHOH$^+$–H$_2$O	104.6	1
(CH$_3$)$_2$OH$^+$–H$_2$O	100.4	1
(tetrahydrofuranH)$^+$–H$_2$O	82.8	1
(furanH)$^+$–H$_2$O	43.5	1
furane$^+$–H$_2$O	41.0	1
(phenol)$^+$–H$_2$O	78.0	1
(1-naphthol)$^+$–H$_2$O	66.4	1
H$_3$O$^+$–HC(O)H	137.7	1
H$_3$O$^+$–NH$_3$	229.3	1
H$_3$O$^+$(NH$_3$)–NH$_3$	77.0	1
H$_3$O$^+$(NH$_3$)$_2$–NH$_3$	71.5	1
H$_3$O$^+$(NH$_3$)$_3$–NH$_3$	62.8	1
H$_3$O$^+$–PH$_3$	144	1
H$_3$O$^+$–SO$_3$	74	1
(HCOOH)$^+$–HCOOH	96.5 ± 9.6	1
H$_3$O$^+$–CH$_4$	33.5	1
(CH$_3$OH)$^+$–CH$_3$OH	115.8 ± 19.3	1
CH$_3$OH$_2$$^+$–CH$_3$OH	136.4	1
H$_3$O$^+$–CH$_3$CN	195.4	1
furan$^+$–furan	94.1	1
BH$^+$–B, B = tetrahydofuran	125.1	1
S$^+$–CS$_2$	166	1
CS$^+$–CS$_2$	150.6	1
CS$_2$$^+$–CS$_2$	104.2	1
HCS$_2$$^+$–CS$_2$	46.4	1
OS$^+$–SO$_2$	57.7	1
O$_2$S$^+$–SO$_2$	63.6	1
OCS$^+$–OCS	100.0	1
OCS$^+$–CO$_2$	72.0	1
SO$_2$$^+$–CO$_2$	42.7	1
H$_3$S$^+$–H$_2$O	91.6	1
thiopheneH$^+$–H$_2$O	42.7	1
H$_3$S$^+$–H$_2$S	53.6 ± 6.3	1
H$_3$S$^+$–CH$_4$	16.3	1
(CH$_3$)$_2$Se$^{\cdot+}$–Se(CH$_3$)$_2$	~95 ± 3	1
(CH$_3$)$_2$Te$^{\cdot+}$–Te(CH$_3$)$_2$	97 ± 2	1
(17) Group 17		
HF$^+$–HF	≥138	1
(H$_2$Cl)$^+$ –Cl	39.6	1
HCl$^+$–HCl	83.9	1
Cl$^+$–CCl$_3$	446.7 ± 9.6	1
Cl$^+$–C$_2$H$_3$	685.0 ± 4.8	1
HBr$^+$–HBr	96	1
I$^+$–CH$_3$	330.0	1
I$^+$(CH$_3$I)–CH$_3$	51.1	1
I$^+$(CH$_3$I)$_2$–CH$_3$	112.9	1
(18) Group 18		
He$^+$(He)$_1$–He	17.6	1
He$^+$(He)$_2$–He	2.7 ± 0.6	1
Ne$^+$(Ne)–Ne	10.3 ± 0.6	1
Ne$^+$(Ne)$_2$–Ne	3.3 ± 0.6	1
Ar$^+$(Ar)–Ar	20.4 ± 0.6	1
Ar$^+$(Ar)$_2$–Ar	7.0 ± 0.6	1
Ar$^+$(N$_2$)–Ar	25.1	1
Ar$^+$(N$_2$)(Ar)–Ar	7.1	1
Ar$^+$(N$_2$)(Ar)$_2$–Ar	7.1	1
Kr$^+$(Kr)–Kr	23.3 ± 0.6	1
Kr$^+$(Kr)$_2$–Kr	9.0 ± 0.6	1
Xe$^+$(Xe)–Xe	25.2 ± 0.6	1
Xe$^+$(Xe)$_2$–Xe	11.0 ± 0.6	1
Ar$^+$–H$_2$	93.7	1
Ar$^+$–N$_2$	127.6	1
Ar$^+$(N$_2$)–N$_2$	31.0	1
Ar$^+$(N$_2$)$_2$–N$_2$	10.9	1
Ar$^+$–CO	75 ± 17	1
Ar$^+$(CO)–CO	13	1
Kr$^+$–CO	103.3 ± 7.5	1
Kr$^+$–CO$_2$	79.1 ± 2.9	1
K$^+$–(18-crown-6)	235 ± 13	1

References

1. Luo, Y. R., *Comprehensive Handbook of Chemical Bond Energies*, CRC Press, Boca Raton, FL, 2007.
2. Parke, L. G., Hinton, C. S., and Armentrout, P. B., *Int. J. Mass Spectrom.* 254, 168, 2006.
3. Li, F.-X., Zhang, X.-G., and Armentrout, P. B., *Int. J. Mass Spectrom.* 255/256, 279, 2006.
4. Meloni, G., Zou, P., Klippenstein, S. J., Ahmed, M., Leone, S. R., Taatjes, C. A., and Osborn, D. L., *J. Am. Chem. Soc.* 128, 13559, 2006.
5. Li, F.-X., and Armentrout, P. B., *J. Chem. Phys.* 125, 133114/1, 2006.
6. Parke, L. G., Hinton, C. S., and Armentrout, P. B., *J. Phys. Chem. C* 111, 17773, 2007.
7. Shuman, N. S., Ochieng, M. A., Sztáray, B., and Baer, T., *J. Phys. Chem. A* 112, 5647, 2008.
8. Armentrout, P. B., and Kretzschmar, I., *J. Phys. Chem. A* 113, 10955, 2009.
9. Armentrout, P. B., and Kretzschmar, I., *Phys. Chem. Chem. Phys.* 12, 4078, 2010.
10. Luo, Z., Zhang, Z., Huang, H., Chang, Y.-C., and Ng, C. Y., *J. Chem. Phys.*, 140, 181101, 2014.
11. Armentrout, P. B., Parke, L., Hinton, C., and Citir, M., *ChemPlusChem*, 78, 1157, 2013.
12. Armentrout, P. B., Austin, C. A., and Rodgers, M. T., *J. Phys. Chem. A* 118, 8088-8097, 2014.

ELECTRONEGATIVITY

Electronegativity is a parameter originally introduced by Pauling which describes, on a relative basis, the tendency of an atom in a molecule to attract bonding electrons. While electronegativity is not a precisely defined molecular property, the electronegativity difference between two atoms provides a useful measure of the polarity and ionic character of the bond between them. This table gives the electronegativity χ, on the Pauling scale, for the most common oxidation state. Other scales are described in the references.

References

1. Pauling, L., *The Nature of the Chemical Bond, Third Edition*, Cornell University Press, Ithaca, NY, 1960.
2. Allen, L. C., *J. Am. Chem. Soc.*, 111, 9003, 1989.
3. Allred, A. L., *J. Inorg. Nucl. Chem.*, 17, 215, 1961.

Z	Element	*χ*	*Z*	Element	*χ*	*Z*	Element	*χ*
1	H	2.20	33	As	2.01	65	Tb	
2	He		34	Se	2.55	66	Dy	1.22
3	Li	0.98	35	Br	2.96	67	Ho	1.23
4	Be	1.57	36	Kr		68	Er	1.24
5	B	2.04	37	Rb	0.82	69	Tm	1.25
6	C	2.55	38	Sr	0.95	70	Yb	
7	N	3.04	39	Y	1.22	71	Lu	1.0
8	O	3.44	40	Zr	1.33	72	Hf	1.3
9	F	3.98	41	Nb	1.6	73	Ta	1.5
10	Ne		42	Mo	2.16	74	W	1.7
11	Na	0.93	43	Tc	2.10	75	Re	1.9
12	Mg	1.31	44	Ru	2.2	76	Os	2.2
13	Al	1.61	45	Rh	2.28	77	Ir	2.2
14	Si	1.90	46	Pd	2.20	78	Pt	2.2
15	P	2.19	47	Ag	1.93	79	Au	2.4
16	S	2.58	48	Cd	1.69	80	Hg	1.9
17	Cl	3.16	49	In	1.78	81	Tl	1.8
18	Ar		50	Sn	1.96	82	Pb	1.8
19	K	0.82	51	Sb	2.05	83	Bi	1.9
20	Ca	1.00	52	Te	2.10	84	Po	2.0
21	Sc	1.36	53	I	2.66	85	At	2.2
22	Ti	1.54	54	Xe	2.60	86	Rn	
23	V	1.63	55	Cs	0.79	87	Fr	0.7
24	Cr	1.66	56	Ba	0.89	88	Ra	0.9
25	Mn	1.55	57	La	1.10	89	Ac	1.1
26	Fe	1.83	58	Ce	1.12	90	Th	1.3
27	Co	1.88	59	Pr	1.13	91	Pa	1.5
28	Ni	1.91	60	Nd	1.14	92	U	1.7
29	Cu	1.90	61	Pm		93	Np	1.3
30	Zn	1.65	62	Sm	1.17	94	Pu	1.3
31	Ga	1.81	63	Eu				
32	Ge	2.01	64	Gd	1.20			

FORCE CONSTANTS FOR BOND STRETCHING

David R. Lide

Representative force constants (f) for stretching of chemical bonds are listed in this table. Except where noted, all force constants are derived from values of the harmonic vibrational frequencies ω_e. Values derived from the observed vibrational fundamentals ν, which are noted by footnote [a], are lower than the harmonic force constants, typically by 2 to 3% in the case of heavy atoms (often by 5 to 10% if one of the atoms is hydrogen). Values are given in the SI unit newton per centimeter (N/cm), which is identical to the commonly used cgs unit mdyn/Å.

References

1. Huber, K. P., and Herzberg, G., *Molecular Spectra and Molecular Structure. IV. Constants of Diatomic Molecules*, Van Nostrand Reinhold, New York, 1979.
2. Shimanouchi, T., The Molecular Force Field, in Eyring, H., Henderson, D., and Yost, W., Eds., *Physical Chemistry: An Advanced Treatise*, Vol. IV, Academic Press, New York, 1970.
3. Tasumi, M., and Nakata, M., *Pure and Appl. Chem.*, 57, 121–147, 1985.

Bond	Molecule	f/N cm^{-1}
H-H	H_2	5.75
Be-H	BeH	2.27
B-H	BH	3.05
C-H	CH	4.48
C-H	CH_4	5.44[b]
C-H	C_2H_6	4.83[a, b, c]
C-H	CH_3CN	5.33[b]
C-H	CH_3Cl	5.02[a, b, c]
C-H	$CCl_2=CH_2$	5.57[b]
C-H	HCN	6.22
N-H	NH	5.97
O-H	OH	7.80
O-H	H_2O	8.45
P-H	PH	3.22
S-H	SH	4.23
S-H	H_2S	4.28
F-H	HF	9.66
Cl-H	HCl	5.16
Br-H	HBr	4.12
I-H	HI	3.14
Li-H	LiH	1.03
Na-H	NaH	0.78
K-H	KH	0.56
Rb-H	RbH	0.52
Cs-H	CsH	0.47
C-C	C_2	12.16
C-C	$CCl_2=CH_2$	8.43
C-C	C_2H_6	4.50[a, c]
C-C	CH_3CN	5.16
C-F	CF	7.42
C-F	CH_3F	5.71[a, c]

Bond	Molecule	f/N cm^{-1}
C-Cl	CCl	3.95
C-Cl	CH_3Cl	3.44[a, c]
C-Cl	$CCl_2=CH_2$	4.02[b]
C-Br	CH_3Br	2.89[a, c]
C-I	CH_3I	2.34[a, c]
C-O	CO	19.02
C-O	CO_2	16.00
C-O	OCS	16.14
C-O	CH_3OH	5.42[a, c]
C-S	CS	8.49
C-S	CS_2	7.88
C-S	OCS	7.44
C-N	CN	16.29
C-N	HCN	18.78
C-N	CH_3CN	18.33
C-N	CH_3NH_2	5.12[a, c]
C-P	CP	7.83
Si-Si	Si_2	2.15
Si-O	SiO	9.24
Si-F	SiF	4.90
Si-Cl	SiCl	2.63
N-N	N_2	22.95
N-N	N_2O	18.72
N-O	NO	15.95
N-O	N_2O	11.70
P-P	P_2	5.56
P-O	PO	9.45
O-O	O_2	11.77
O-O	O_3	5.74[a]
S-O	SO	8.30
S-O	SO_2	10.33[a]

Bond	Molecule	f/N cm^{-1}
S-S	S_2	4.96
F-F	F_2	4.70
Cl-F	ClF	4.48
Br-F	BrF	4.06
Cl-Cl	Cl_2	3.23
Br-Cl	BrCl	2.82
Br-Br	Br_2	2.46
I-I	I_2	1.72
Li-Li	Li_2	0.26
Li-Na	LiNa	0.21
Na-Na	Na_2	0.17
Li-F	LiF	2.50
Li-Cl	LiCl	1.43
Li-Br	LiBr	1.20
Li-I	LiI	0.97
Na-F	NaF	1.76
Na-Cl	NaCl	1.09
Na-Br	NaBr	0.94
Na-I	NaI	0.76
Be-O	BeO	7.51
Mg-O	MgO	3.48
Ca-O	CaO	3.61

[a] Derived from fundamental frequency, without anharmonicity correction.

[b] Average of symmetric and antisymmetric (or degenerate) modes.

[c] Calculated from Local Symmetry Force Field (see Reference 2).

FUNDAMENTAL VIBRATIONAL FREQUENCIES OF SMALL MOLECULES

David R. Lide

This table lists the fundamental vibrational frequencies of selected three-, four-, and five-atom molecules. Both stable molecules and transient free radicals are included. The data have been taken from evaluated sources. In general, the selected values are based on gas-phase infrared, Raman, or ultraviolet spectra; when these were not available, liquid-phase or matrix-isolation spectra were used.

Molecules are grouped by structural type. Within each group, related molecules appear together for convenient comparison.

The vibrational modes are described by their approximate character in terms of stretching, bending, deformation, etc. However, it should be emphasized that most such descriptions are only approximate, and that the true normal mode usually involves a mixture of motions. Abbreviations are:

sym.: symmetric
antisym.: antisymmetric
str.: stretch
deform.: deformation
scis.: scissors
rock.: rocking
deg.: degenerate

In the case of free radicals, strong interactions may exist between the electronic and bending vibrational motions. Details can be found in Refs. 3 and 4. The references should be consulted for information on the accuracy of the data and for data on other molecules not listed here.

All fundamental frequencies (more precisely, wavenumbers) are given in units of cm^{-1}.

XY_2 Molecules: Point Groups $D_{\infty h}$ (Linear) and C_{2v} (Bent)

Molecule	Structure	Sym. str.	Bend	Antisym. str.
CO_2	Linear	1333	667	2349
CS_2	Linear	658	397	1535
C_3	Linear	1224	63	2040
CNC	Linear		321	1453
NCN	Linear	1197	423	1476
BO_2	Linear	1056	447	1278
BS_2	Linear	510	120	1015
KrF_2	Linear	449	233	590
XeF_2	Linear	515	213	555
$XeCl_2$	Linear	316		481
H_2O	Bent	3657	1595	3756
D_2O	Bent	2671	1178	2788
F_2O	Bent	928	461	831
Cl_2O	Bent	639	296	686
O_3	Bent	1103	701	1042
H_2S	Bent	2615	1183	2626
D_2S	Bent	1896	855	1999
SF_2	Bent	838	357	813
SCl_2	Bent	525	208	535
SO_2	Bent	1151	518	1362
H_2Se	Bent	2345	1034	2358
D_2Se	Bent	1630	745	1696
NH_2	Bent	3219	1497	3301
NO_2	Bent	1318	750	1618
NF_2	Bent	1075	573	942
ClO_2	Bent	945	445	1111
CH_2	Bent		963	
CD_2	Bent		752	
CF_2	Bent	1225	667	1114
CCl_2	Bent	721	333	748
CBr_2	Bent	595	196	641
SiH_2	Bent	2032	990	2022
SiD_2	Bent	1472	729	1468
SiF_2	Bent	855	345	870
$SiCl_2$	Bent	515		505
$SiBr_2$	Bent	403		400
GeH_2	Bent	1887	920	1864
$GeCl_2$	Bent	399	159	374
SnF_2	Bent	593	197	571
$SnCl_2$	Bent	352	120	334
$SnBr_2$	Bent	244	80	231
PbF_2	Bent	531	165	507
$PbCl_2$	Bent	314	99	299
ClF_2	Bent	500		576

XYZ Molecules: Point Groups $C_{\infty v}$ (Linear) and C_s (Bent)

Molecule	Structure	XY str.	Bend	YZ str.
HCN	Linear	3311	712	2097
DCN	Linear	2630	569	1925
FCN	Linear	1077	451	2323
ClCN	Linear	744	378	2216
BrCN	Linear	575	342	2198
ICN	Linear	486	305	2188
CCN	Linear	1060	230	1917
CCO	Linear	1063	379	1967
HCO	Bent	2485	1081	1868
HCC	Linear	3612		1848
OCS	Linear	2062	520	859
NCO	Linear	1270	535	1921
NNO	Linear	2224	589	1285
HNB	Linear	3675		2035
HNC	Linear	3653		2032
HNSi	Linear	3583	523	1198
HBO	Linear		754	1817
FBO	Linear		500	2075
ClBO	Linear	676	404	1958
BrBO	Linear	535	374	1937
FNO	Bent	766	520	1844
ClNO	Bent	596	332	1800
BrNO	Bent	542	266	1799
HNF	Bent		1419	1000
HNO	Bent	2684	1501	1565
HPO	Bent	2095	983	1179
HOF	Bent	3537	886	1393
HOCl	Bent	3609	1242	725
HOO	Bent	3436	1392	1098
FOO	Bent	579	376	1490

Molecule	Structure	XY str.	Bend	YZ str.
ClOO	Bent	407	373	1443
BrOO	Bent			1487
HSO	Bent		1063	1009
NSF	Bent	1372	366	640
NSCl	Bent	1325	273	414
HCF	Bent		1407	1181
HCCl	Bent		1201	815
HSiF	Bent	1913	860	834
HSiCl	Bent		808	522
HSiBr	Bent	1548	774	408

Symmetric XY_3 Molecules: Point Groups D_{3h} (Planar) and C_{3v} (Pyramidal)

Molecule	Structure	Sym. str.	Sym. deform.	Deg. str.	Deg. deform.
NH_3	Pyramidal	3337	950	3444	1627
ND_3	Pyramidal	2420	748	2564	1191
PH_3	Pyramidal	2323	992	2328	1118
AsH_3	Pyramidal	2116	906	2123	1003
SbH_3	Pyramidal	1891	782	1894	831
NF_3	Pyramidal	1032	647	907	492
PF_3	Pyramidal	892	487	860	344
AsF_3	Pyramidal	741	337	702	262
PCl_3	Pyramidal	504	252	482	198
PI_3	Pyramidal	303	111	325	79
AsI_3	Pyramidal	219	94	224	71
$AlCl_3$	Pyramidal	375	183	595	150
SO_3	Planar	1065	498	1391	530
BF_3	Planar	888	691	1449	480

Linear XYYX Molecules: Point Group $D_{\infty h}$

Molecule	Sym. XY str.	Antisym. XY str.	YY str.	Bend	Bend
C_2H_2	3374	3289	1974	612	730
C_2D_2	2701	2439	1762	505	537
C_2N_2	2330	2158	851	507	233

Planar X_2YZ Molecules: Point Group C_{2v}

Molecule	Sym. XY str.	YZ str.	YX_2 scis.	Antisym. XY str.	YX_2 rock	YX_2 wag
H_2CO	2783	1746	1500	2843	1249	1167
D_2CO	2056	1700	1106	2160	990	938
F_2CO	965	1928	584	1249	626	774
Cl_2CO	567	1827	285	849	440	580
O_2NF	1310	822	568	1792	560	742
O_2NCl	1286	793	370	1685	408	652

Tetrahedral XY_4 Molecules: Point Group T_d

Molecule	Sym. str.	Deg. deform.(e)	Deg. str.(f)	Deg. deform.(f)
CH_4	2917	1534	3019	1306
CD_4	2109	1092	2259	996
CF_4	909	435	1281	632
CCl_4	459	217	776	314
CBr_4	267	122	672	182
CI_4	178	90	555	125
SiH_4	2187	975	2191	914
SiD_4	1558	700	1597	681
SiF_4	800	268	1032	389
$SiCl_4$	424	150	621	221
GeH_4	2106	931	2114	819
GeD_4	1504	665	1522	596
$GeCl_4$	396	134	453	172
$SnCl_4$	366	104	403	134
$TiCl_4$	389	114	498	136
$ZrCl_4$	377	98	418	113
$HfCl_4$	382	102	390	112
RuO_4	885	322	921	336
OsO_4	965	333	960	329

References

1. T. Shimanouchi, Tables of Molecular Vibrational Frequencies, Consolidated Volume I, Natl. Stand. Ref. Data Ser., Natl. Bur. Stand. (U.S.), 39, 1972.
2. T. Shimanouchi, Tables of Molecular Vibrational Frequencies, Consolidated Volume II, *J. Phys. Chem. Ref. Data*, 6, 993, 1977.
3. G. Herzberg, *Electronic Spectra and Electronic Structure of Polyatomic Molecules*, D. Van Nostrand Co., Princeton, NJ, 1966.
4. M. E. Jacox, Ground state vibrational energy levels of polyatomic transient molecules, *J. Phys. Chem. Ref. Data*, 13, 945, 1984.

SPECTROSCOPIC CONSTANTS OF DIATOMIC MOLECULES

David R. Lide

This table lists the leading spectroscopic constants and equilibrium internuclear distance r_e in the ground electronic state for selected diatomic molecules. The constants are those describing the vibrational and rotational energy through the expressions:

$$E_{vib}/hc = \omega_e(v + ½) - \omega_e x_e(v + ½)^2 + \cdots$$

$$E_{rot}/hc = B_v J(J + 1) - D_v[J(J + 1)]^2 + \cdots$$

where

$$B_v = B_e - \alpha_e(v + ½) + \cdots$$

$$D_v = D_e + \cdots$$

Here v and J are the vibrational and rotational quantum numbers, respectively, h is Planck's constant, and c is the speed of light. In this customary formulation the constants ω_e, B_e, etc., have dimensions of inverse length; in this table they are given in units of cm^{-1}.

Users should note that higher order terms in the above energy expressions are required for very precise calculations. The references contain constants for many of these higher terms, as well as more precise values of the lower constants. Also, if the ground electronic state is not $^1\Sigma$, additional terms are needed to account for the interaction between electronic and pure rotational angular momentum. For some molecules in the table the data have been analyzed in terms of the Dunham series expansion:

$$E/hc = \Sigma_{lm} Y_{lm}(v + ½)^l J^m(J + 1)^m$$

In such cases it has been assumed that $Y_{10} = \omega_e$, $Y_{01} = B_e$, etc., although in the highest approximations these identities are not precisely correct. Some of the values of r_e in the table have been corrected for breakdown of the Born-Oppenheimer approximation, which can affect the last decimal place. Because of differences in the method of data analysis and limitations in the model, care should be taken in comparing r_e values for different molecules to a precision beyond 0.001 Å.

Entries in the ω_e column that are marked by * give the interval between $v = 0$ and $v = 1$ states instead of a value of ω_e.

Molecules are listed in alphabetical order by formula as most commonly written. In most cases this form places the more electropositive element first, but there are exceptions such as OH, NH, CH, etc. Refs. 1–5 are evaluated compilations covering many molecules and giving references to the original literature.

References

1. Huber, K. P., and Herzberg, G., *Molecular Spectra and Molecular Structure IV. Constants of Diatomic Molecules*, Van Nostrand Reinhold, New York, 1979.
2. *Landolt-Börnstein, Numerical Data and Functional Relationships in Science and Technology, New Series*, II/6 (1974), II/14a (1982), II/14b (1983), II/19a (1992), II/19d-1 (1995), II/24a (1998), *Molecular Constants*, Springer-Verlag, Heidelberg.
3. Lovas, F. J., and Tiemann, E., *J. Phys. Chem. Ref. Data* 3, 609, 1974.
4. Irikura, K. K., *J. Phys. Chem. Ref. Data* 35, 389, 2007.
5. Lovas, F. J., Tiemann, E., Coursey, J. S., Kotochigova, S. A., Chang, J., Olsen, K., and Dragoset, R. A., *Diatomic Spectral Database* (version 2.0). Available: http://physics.nist.gov/Diatomic, National Institute of Standards and Technology, Gaithersburg, MD, November 2009.
6. Wormsbecher, R. F., Hessel, M. M., and Lovas, F. J., *J. Chem. Phys.* 74, 6893, 1981.
7. Hedderich, H. G., Dulick, M., and Bernath, P. F., *J. Chem. Phys.* 99, 8363, 1993.
8. Ram, R. S., and Bernath, P. F., *J. Mol. Spectrosc.* 176, 320, 1996.
9. Miller, C. E., and Drouin, B. J., *J. Mol. Spectrosc.*. 205, 312, 2001.
10. Tellinghuisen, P. C., Tellinghuisen, J., Coxon, J. A., Velazco, J. E., and Setser, D. W., *J. Chem. Phys.* 68, 5187, 1978.
11. Muntianu, A., Guo, B., and Bernath, P. F., *J. Mol. Spectrosc.* 176, 274, 1996.
12. Yamada, C., Chang, M. C., and Hirota, E., *J. Chem. Phys.* 86, 3804, 1987.
13. Wang, X., Magnes, J., Marjatta Lyyra, A., Ross, A. J., Martin, F., Dove, P. M., and Le Roy, R. J., *J. Chem. Phys.* 117, 9339, 2002.
14. Yamada, C., and Hirota, E., *J. Chem. Phys.* 99, 8489, 1993.
15. James, A. M., Kowalczyk, P., Fournier, R., and Simard, B., *J. Chem. Phys.* 99, 8504, 1993.
16. Campbell, J. M., Dulick, M., Klapstein, D., White, J. B., and Bernath, P. F., *J. Chem. Phys.* 99, 8379, 1993.
17. White, J. B., Dulick, M., and Bernath, P. F., *J. Chem. Phys.* 99, 8371, 1993.
18. Shayesteh, A., Appadoo, D. R. T., Gordon, I., Le Roy, R. J., and Bernath, P. F., *J. Chem. Phys.* 120, 10002, 2004.
19. Sanz, M. E., McCarthy, M. C., and Thaddeus, P., *J. Chem. Phys.* 119, 11715, 2003.
20. Müller, W., and Meyer, W., *J. Chem. Phys.* 80, 3311, 1984.
21. Staanum, P., Pashov, A., Knöckel, H., and Tiemann, E., *Phys. Rev. A* 75, 042513, 2007.
22. Lovas, F. J., Maki, A. G., and Olson, W. B., *J. Mol. Spectrosc.* 87, 449, 1981.
23. Bogey, M., Demuynck, C., and Destombes, J. L., *Chem. Phys.* 66, 99, 1982.
24. Babou, Y., Rivière, Ph., Perrin, M. Y., and Soufiani, A., *Int. J. Thermophys.* 30, 416, 2009.
25. Skatrud, D. D., DeLucia, F. C., Blake, G. A., and Sastry, K. V. L. N., *J. Mol. Spectrosc.* 99, 35, 1983.
26. Le Floch, A., *Mol. Phys.* 72, 133, 1991.
27. George, T., Urban, W., and Le Floch, A., *J. Mol. Spectrosc.* 165, 500, 1994.
28. Mürtz, P., Thümmel, H., Pfelzer, C., and Urban, W., *Mol. Phys.* 86, 1362, 1995.
29. Maki, A. G., Lovas, F. J., and Suenram, R. D., *J. Mol. Spectrosc.* 91, 424, 1982.
30. Engelke, F., Ennen, G., and Meiwes, K. H., *Chem. Phys.* 66, 391, 1982.
31. Tanaka, T., Tamura, M., and Tanaka, K., *J. Mol. Struct.* 413, 153, 1997.

Spectroscopic Constants of Diatomic Molecules

Formula	Molecule	State	ω_e/cm^{-1}	$\omega_e x_e$/cm^{-1}	B_e/cm^{-1}	α_e/cm^{-1}	D_e/10^{-6} cm^{-1}	r_e/Å	Ref.
AgBr	$^{107}Ag^{79}Br$	$^1\Sigma^+$	249.57	0.63	0.064833	0.0002361	0.0175	2.39311	1,2,3
AgCl	$^{107}Ag^{35}Cl$	$^1\Sigma^+$	343.49	1.17	0.12298388	0.00059541	0.06305	2.28079	1,2,3
AgD	$^{107}Ag^{2}H$	$^1\Sigma^+$	1250.70	17.17	3.2572	0.0722	85.9	1.6180	1,2
AgF	$^{107}Ag^{19}F$	$^1\Sigma^+$	513.45	2.59	0.2657020	0.0019206	0.284	1.98318	1,2,3
AgH	$^{107}Ag^{1}H$	$^1\Sigma^+$	1759.9	34.06	6.449	0.201	344	1.618	1,2
AgI	$^{107}Ag^{127}I$	$^1\Sigma^+$	206.50	0.46	0.04486821	0.0001414	0.00847	2.54463	1,2,3
AgO	$^{107}Ag^{16}O$	$^2\Pi_{1/2}$	490.2	3.1	0.3020	0.0025	0.45	2.003	1,2
AlBr	$^{27}Al^{79}Br$	$^1\Sigma^+$	378.0	1.28	0.15919713	0.00086045	0.11285	2.29481	1,2
AlCl	$^{27}Al^{35}Cl$	$^1\Sigma^+$	481.77	2.10	0.24393007	0.00161108	0.25017	2.13014	7
AlD	$^{27}Al^{2}H$	$^1\Sigma^+$	1211.77	15.06	3.3183929	0.0698773	99.42	1.64637	17
AlF	$^{27}Al^{19}F$	$^1\Sigma^+$	802.32	4.85	0.55248021	0.00498426	1.0464	1.65437	4,5
AlH	$^{27}Al^{1}H$	$^1\Sigma^+$	1682.37	29.05	6.3937842	0.1870527	368.53	1.64736	17
AlI	$^{27}Al^{127}I$	$^1\Sigma^+$	316.1	1.0	0.11769985	0.00055859		2.53710	1,2
AlO	$^{27}Al^{16}O$	$^2\Sigma^+$	979.49	7.01	0.6413856	0.0057796	1.08	1.61782	4
AlS	$^{27}Al^{32}S$	$^2\Sigma^+$	617.11	3.33	0.2800368	0.0017823	0.22	2.02828	4
Al_2	$^{27}Al_2$	$^3\Pi_u$	285.8	0.9	0.17127	0.0008		2.701	1,2
AsD	$^{75}As^{2}H$	$^3\Sigma^-$	1484*		3.6688		90	1.5306	1,2
AsH	$^{75}As^{1}H$	$^3\Sigma^-$	2130*		7.3067	0.2117	327	1.52315	1,2
AsN	$^{75}As^{14}N$	$^1\Sigma^+$	1068.54	5.41	0.54551	0.003366	0.53	1.6184	1,2
AsO	$^{75}As^{16}O$	$^2\Pi_{1/2}$	967.08	4.85	0.48482	0.003299	0.49	1.6236	1,2
As_2	$^{75}As_2$	$^1\Sigma_g^+$	429.55	1.12	0.10179	0.000333		2.1026	1,2
AuD	$^{197}Au^{2}H$	$^1\Sigma^+$	1634.98	21.65	3.6415	0.07614	70.9	1.5238	1,2
AuH	$^{197}Au^{1}H$	$^1\Sigma^+$	2305.01	43.12	7.2401	0.2136	279	1.5239	1,2
Au_2	$^{197}Au_2$	$^1\Sigma_g^+$	190.9	0.42	0.028013	0.0000723	0.00250	2.4719	1,2
BBr	$^{11}B^{79}Br$	$^1\Sigma^+$	684.31	3.52	0.4894	0.0035	1.00	1.888	1,2
BCl	$^{11}B^{35}Cl$	$^1\Sigma^+$	840.29	5.49	0.684282	0.006812	1.80	1.71528	1,29
BD	$^{11}B^{2}H$	$^1\Sigma^+$	1703.3	28	6.54	0.17	400	1.2324	1,2
BF	$^{11}B^{19}F$	$^1\Sigma^+$	1402.16	11.82	1.51674399	0.01904848	7.11	1.26267	4
BH	$^{11}B^{1}H$	$^1\Sigma^+$	2366.73	49.34	12.025755	0.421565	1242	1.23217	4
BN	$^{11}B^{14}N$	$^3\Pi$	1514.6	12.3	1.666	0.025	8.1	1.281	1,2
BO	$^{11}B^{16}O$	$^2\Sigma^+$	1885.29	11.69	1.781110	0.016516	6.32	1.20475	4
BS	$^{11}B^{32}S$	$^2\Sigma^+$	1179.91	6.25	0.79478	0.00578	1.4	1.60935	4
B_2	$^{11}B_2$	$^3\Sigma_g^-$	1051.3	9.35	1.212	0.014		1.590	1,2
BaBr	$^{138}Ba^{79}Br$	$^2\Sigma^+$	193.77	0.41	0.0415082	0.0001219	0.00762	2.84449	1,2
BaCl	$^{138}Ba^{35}Cl$	$^2\Sigma^+$	279.92	0.82	0.08396717	0.00033429	0.03022	2.68276	1,2
BaD	$^{138}Ba^{2}H$	$^2\Sigma^+$	829.77	7.32	1.7071	0.02363	28.77	2.2304	1,2
BaF	$^{138}Ba^{19}F$	$^2\Sigma^+$	468.9	1.79	0.2159	0.0012	0.175	2.163	1,2,3
BaH	$^{138}Ba^{1}H$	$^2\Sigma^+$	1168.31	14.50	3.38285	0.06599	112.67	2.23175	1,2
BaI	$^{138}Ba^{127}I$	$^2\Sigma^+$	152.14	0.27	0.02680587	0.00006634	0.00333	3.08476	1,2
BaO	$^{138}Ba^{16}O$	$^1\Sigma^+$	669.76	2.03	0.3126140	0.0013921	0.2724	1.93969	1,2,3
BaS	$^{138}Ba^{32}S$	$^1\Sigma^+$	379.42	0.88	0.10331	0.0003188	0.0306	2.5074	1,2
BeD	$^{9}Be^{2}H$	$^2\Sigma^+$	1530.32	20.71	5.6872	0.1225	313.8	1.3419	1,2
BeF	$^{9}Be^{19}F$	$^2\Sigma^+$	1247.36	9.12	1.4889	0.0176	8.28	1.3610	1,2
BeH	$^{9}Be^{1}H$	$^2\Sigma^+$	2061.24	37.33	10.31992	0.3084	1022	1.34241	4
BeO	$^{9}Be^{16}O$	$^1\Sigma^+$	1487.32	11.83	1.6510	0.0190	8.20	1.3309	1,2
BeS	$^{9}Be^{32}S$	$^1\Sigma^+$	997.94	6.14	0.79059	0.00664	2.00	1.7415	1
BiBr	$^{209}Bi^{79}Br$	O^+	209.62	0.52	0.04321526	0.00013269	0.007347	2.60950	5
BiCl	$^{209}Bi^{35}Cl$	O^+	308.18	1.09	0.9212553	0.0004020	0.0329	2.47152	5
BiD	$^{209}Bi^{2}H$	$^3\Sigma^-$	1173.32	16.1	2.592	0.054	50.6	1.804	1,2
BiF	$^{209}Bi^{19}F$	O^+	513.0	2.35	0.22998897	0.00150262	0.185	2.05154	5
BiH	$^{209}Bi^{1}H$	$^3\Sigma^-$	1635.73	31.6	5.137	0.148	183	1.805	1,2
BiI	$^{209}Bi^{127}I$	O^+	164.12	0.32	0.02722281	0.00006979	0.00300	2.80050	5
Bi_2	$^{209}Bi_2$	$^1\Sigma_g^+$	172.71	0.34	0.022781	0.000055	0.00150	2.6596	1,2
BrCl	$^{79}Br^{35}Cl$	$^1\Sigma^+$	444.28	1.84	0.152470	0.000770	0.07183	2.13607	1,2,3
BrF	$^{79}Br^{19}F$	$^1\Sigma^+$	670.75	4.05	0.35584	0.00261	0.401	1.75894	1,2,3
BrO	$^{79}Br^{16}O$	$^2\Pi_{3/2}$	779	6.8	0.429598	0.003639	0.523	1.717	1,2,3
Br_2	$^{79}Br_2$	$^1\Sigma_g^+$	325.32	1.08	0.082107	0.0003187	0.02092	2.2811	1,2
CCl	$^{12}C^{35}Cl$	$^2\Pi_{1/2}$	876.90	5.45	0.697137	0.006853	1.9	1.64518	4
CD	$^{12}C^{2}H$	$^2\Pi_{1/2}$	2101.05	34.73	7.8079823	0.212240	420	1.11887	4
CF	$^{12}C^{19}F$	$^2\Pi_{1/2}$	1307.93	11.08	1.41626	0.01844	6.6	1.27218	4

Formula	Molecule	State	ω_e/cm^{-1}	$\omega_e x_e$/cm^{-1}	B_e/cm^{-1}	α_e/cm^{-1}	D_e/10^{-6} cm^{-1}	r_e/Å	Ref.
CH	$^{12}C^{1}H$	$^2\Pi_{1/2}$	2860.75	64.44	14.45988	0.53654	1450	1.1199	4
CN	$^{12}C^{14}N$	$^2\Sigma^+$	2068.65	13.10	1.8997830	0.0173717	6.4034	1.17181	24,25
CO	$^{12}C^{16}O$	$^1\Sigma^+$	2169.81	13.29	1.931280985	0.01750439	6.1216	1.12832	24,26,27
CP	$^{12}C^{31}P$	$^2\Sigma^+$	1239.79	6.83	0.79886775	0.00596933	1.33	1.56198	4
CS	$^{12}C^{32}S$	$^1\Sigma^+$	1285.15	6.50	0.82004356	0.00591835	1.336	1.53482	1,4
CSe	$^{12}C^{80}Se$	$^1\Sigma^+$	1035.36	4.86	0.5750	0.00379	0.71	1.67609	1,2,3
C_2	$^{12}C_2$	$^1\Sigma_g^+$	1855.01	13.56	1.82010	0.01801	6.96	1.24244	24
CaBr	$^{40}Ca^{79}Br$	$^2\Sigma^+$	285.3	0.86	0.09446622	0.00040360	0.0413	2.59358	1,5
CaCl	$^{40}Ca^{35}Cl$	$^2\Sigma^+$	367.53	1.31	0.1522302	0.0007990	0.1029	2.43676	1,2
CaD	$^{40}Ca^{2}H$	$^2\Sigma^+$	910*		2.1769	0.035	47.9	2.002	1,2
CaF	$^{40}Ca^{19}F$	$^2\Sigma^+$	581.1	2.74	0.339	0.0026	0.45	1.967	1,2
CaH	$^{40}Ca^{1}H$	$^2\Sigma^+$	1298.34	19.10	4.2766	0.0970	183.7	2.0025	1,2
CaI	$^{40}Ca^{127}I$	$^2\Sigma^+$	238.70	0.63	0.0693263	0.0002634	0.0234	2.82859	1,2
CaO	$^{40}Ca^{16}O$	$^1\Sigma^+$	732.03	4.83	0.444441	0.003282	0.6541	1.8221	1,2
CaS	$^{40}Ca^{32}S$	$^1\Sigma^+$	462.23	1.78	0.1766757	0.0008270	0.1032	2.31775	1,2
CdD	$^{114}Cd^{2}H$	$^2\Sigma^+$			2.704		76	1.775	1,2
CdH	$^{114}Cd^{1}H$	$^2\Sigma^+$	1337.1*		5.323		314	1.781	1,2
ClF	$^{35}Cl^{19}F$	$^1\Sigma^+$	783.45	4.95	0.5164805	0.0043585	0.88	1.62831	4
ClO	$^{35}Cl^{16}O$	$^2\Pi_{3/2}$	853.64	5.52	0.62345797	0.0059357	1.33	1.56962	4
Cl_2	$^{35}Cl_2$	$^1\Sigma_g^+$	559.75	2.69	0.24415	0.00152	0.186	1.9872	4
CrD	$^{52}Cr^{2}H$	$^6\Sigma^+$	1182*		3.14		88.8	1.664	1,2
CrH	$^{52}Cr^{1}H$	$^6\Sigma^+$	1581*		6.220	0.179	347	1.656	1,2
CrO	$^{52}Cr^{16}O$	$^5\Pi$	898.4	6.8	0.5231	0.0070		1.615	1,2
CsBr	$^{133}Cs^{79}Br$	$^1\Sigma^+$	149.66	0.37	0.03606925	0.00012401	0.00838	3.07225	1,2,3
CsCl	$^{133}Cs^{35}Cl$	$^1\Sigma^+$	214.17	0.73	0.07209149	0.00033756	0.03268	2.90627	1,2,3
CsD	$^{133}Cs^{2}H$	$^1\Sigma^+$	619.1*		1.354		20	2.505	1,2
CsF	$^{133}Cs^{19}F$	$^1\Sigma^+$	352.56	1.62	0.18436969	0.0011756	0.20168	2.34535	1,2,3
CsH	$^{133}Cs^{1}H$	$^1\Sigma^+$	891.0	12.9	2.7099	0.0579	113	2.4938	1,2
CsI	$^{133}Cs^{127}I$	$^1\Sigma^+$	119.18	0.25	0.02362736	0.00006826	0.00371	3.31519	1,2,3
CsO	$^{133}Cs^{16}O$	$^2\Sigma^+$	357.5*		0.223073	0.001303	0.348	2.3007	1,2
Cs_2	$^{133}Cs_2$	$^1\Sigma_g^+$	42.02	0.08	0.0127	0.0000264	0.00464	4.47	1,2
CuBr	$^{63}Cu^{79}Br$	$^1\Sigma^+$	314.8	0.96	0.10192625	0.00045214	0.04274	2.17344	1,2
CuCl	$^{65}Cu^{35}Cl$	$^1\Sigma^+$	415.29	1.58	0.17628802	0.00099647	0.12706	2.05118	1,2
CuD	$^{63}Cu^{2}H$	$^1\Sigma^+$	1384.14	18.97	4.0381	0.0917	136.2	1.4626	1,2
CuF	$^{63}Cu^{19}F$	$^1\Sigma^+$	622.7	3.95	0.3794029	0.0032298	0.563	1.74493	1,2,3
CuH	$^{63}Cu^{1}H$	$^1\Sigma^+$	1941.26	37.51	7.9441	0.2563	520	1.46263	1,2
CuI	$^{63}Cu^{127}I$	$^1\Sigma^+$	264.5	0.60	0.07328742	0.00028390	0.02244	2.33832	1,2
CuO	$^{63}Cu^{16}O$	$^2\Pi_{3/2}$	640.17	4.43	0.44454	0.00456	0.85	1.7244	1,2
CuS	$^{63}Cu^{32}S$	$^2\Pi_{3/2}$	415.0	1.75	0.1891		0.18	2.051	1,2
Cu_2	$^{63}Cu_2$	$^1\Sigma_g^+$	264.55	1.02	0.10874	0.000614	0.0716	2.2197	1,2
DBr	$^{2}H^{81}Br$	$^1\Sigma^+$	1884.75	22.72	4.245596	0.084	88.32	1.4145	1,2
DCl	$^{2}H^{35}Cl$	$^1\Sigma^+$	2145.16	27.18	5.448796	0.113292	140	1.27458	1,2
DF	$^{2}H^{19}F$	$^1\Sigma^+$	2998.19	45.76	11.0102	0.3017	594	0.91694	1,2
D_2	$^{2}H_2$	$^1\Sigma_g^+$	3115.50	61.82	30.444	1.0786	11410	0.74152	1,2
F_2	$^{19}F_2$	$^1\Sigma_g^+$	916.93	11.32	0.889294	0.0125952	3.3	1.41264	4
FeO	$^{56}Fe^{16}O$	$^5\Delta$	965*		0.650		0.72	1.444	1,2
GaBr	$^{69}Ga^{81}Br$	$^1\Sigma^+$	263.0	0.81	0.081839	0.0003207	0.032	2.35248	1,2,3
GaCl	$^{69}Ga^{35}Cl$	$^1\Sigma^+$	365.67	1.25	0.1499046	0.0007936	0.1008	2.20169	1,2,3
GaD	$^{69}Ga^{2}H$	$^1\Sigma^+$	1143.23	14.43	3.1218854	0.0689978	93.021	1.66113	16
GaF	$^{69}Ga^{19}F$	$^1\Sigma^+$	622.2	3.2	0.3595161	0.0028642	0.50	1.77437	1,2,3
GaH	$^{69}Ga^{1}H$	$^1\Sigma^+$	1603.94	28.41	6.1434095	0.1906376	359.70	1.66208	16
GaI	$^{69}Ga^{127}I$	$^1\Sigma^+$	216.38	0.47	0.0569359	0.0001897	0.015770	2.57464	1,2,3
GaO	$^{69}Ga^{16}O$	$^2\Sigma$	767.5	6.24	0.4271		0.37	1.744	1,2
GeBr	$^{74}Ge^{79}Br$	$^2\Pi_{1/2}$	295	0.7					1,2
GeCl	$^{74}Ge^{35}Cl$	$^2\Pi_{1/2}$	407.6	1.36					1,2
GeD	$^{72}Ge^{2}H$	$^2\Pi_{1/2}$	1320.09	19	3.415	0.070	83.2	1.5874	1,2
GeH	$^{72}Ge^{1}H$	$^2\Pi_{1/2}$	1833.77	37	6.726	0.192	326	1.5880	1,2
GeO	$^{74}Ge^{16}O$	$^1\Sigma^+$	986.49	4.47	0.4856981	0.0030787	0.4709	1.62464	1,2
GeS	$^{74}Ge^{32}S$	$^1\Sigma^+$	575.8	1.80	0.18656576	0.00074910	0.07883	2.01209	1,2
GeSe	$^{74}Ge^{80}Se$	$^1\Sigma^+$	408.7	1.36	0.09634051	0.00028904	0.02207	2.13463	1,2
GeTe	$^{74}Ge^{130}Te$	$^1\Sigma^+$	323.9	0.75	0.06533821	0.00017246	0.012	2.34017	1,2

Formula	Molecule	State	ω_e/cm^{-1}	$\omega_e x_e$/cm^{-1}	B_e/cm^{-1}	α_e/cm^{-1}	D_e/10^{-6} cm^{-1}	r_e/Å	Ref.
HBr	$^{1}H^{81}Br$	$^1\Sigma^+$	2648.97	45.22	8.46488	0.23328	345.8	1.41444	1,2,3
HCl	$^{1}H^{35}Cl$	$^1\Sigma^+$	2990.92	52.80	10.5933002	0.3069985	531.94	1.27456	4
HF	$^{1}H^{19}F$	$^1\Sigma^+$	4138.39	89.94	20.953712	0.7933704	2150	0.91685	4
HI	$^{1}H^{127}I$	$^1\Sigma^+$	2309.01	39.64	6.4263650	0.1689	206.9	1.60916	1,2,3
H_2	$^{1}H_2$	$^1\Sigma_g^+$	4401.21	121.34	60.853	3.062	47100	0.74144	1,2
HgD	$^{202}Hg^{2}H$	$^2\Sigma^+$	896.12*		2.739		91	1.757	1,2
HgH	$^{202}Hg^{1}H$	$^2\Sigma^+$	1203.24*		5.3888		395.3	1.7662	1,2
IBr	$^{127}I^{79}Br$	$^1\Sigma^+$	268.64	0.81	0.0568325	0.0001969	0.0102	2.46899	1,2,3
ICl	$^{127}I^{35}Cl$	$^1\Sigma^+$	384.29	1.50	0.1141587	0.0005354	0.0403	2.32088	1,2,3
IF	$^{127}I^{19}F$	$^1\Sigma^+$	610.24	3.12	0.2797111	0.0018738	0.2356	1.90976	1,2,3
IO	$^{127}I^{16}O$	$^2\Pi_{3/2}$	681.5	4.3	0.34026	0.00270	0.36	1.8676	1,2
I_2	$^{127}I_2$	$^1\Sigma_g^+$	214.50	0.61	0.03737	0.000114	0.0043	2.666	1,2
InBr	$^{115}In^{81}Br$	$^1\Sigma^+$	221.0	0.65	0.05489468	0.00018672	0.01350	2.54315	1,2,3
InCl	$^{115}In^{35}Cl$	$^1\Sigma^+$	317.39	1.03	0.1090583	0.0005177	0.0515	2.40117	1,2,3
InD	$^{115}In^{2}H$	$^1\Sigma^+$	1048.2	12.4	2.523	0.051	58	1.837	1,2
InF	$^{115}In^{19}F$	$^1\Sigma^+$	535.4	2.6	0.2623241	0.0018798	0.252	1.98540	1,2,3
InH	$^{115}In^{1}H$	$^1\Sigma^+$	1476.0	25.61	4.995	0.143	223	1.8380	1,2
InI	$^{115}In^{127}I$	$^1\Sigma^+$	177.08	0.34	0.03686702	0.00010411	0.00639	2.75364	1,2,3
KBr	$^{39}K^{79}Br$	$^1\Sigma^+$	213	0.80	0.08122109	0.00040481	0.04462	2.82078	1,2,3
KCl	$^{39}K^{35}Cl$	$^1\Sigma^+$	281	1.30	0.1286348	0.0007899	0.1087	2.66665	1,2,3
KD	$^{39}K^{2}H$	$^1\Sigma^+$	707	7.7	1.754	0.0318	50	2.240	1,2
KF	$^{39}K^{19}F$	$^1\Sigma^+$	426.26	2.45	0.27993741	0.00233492	0.4829	2.17146	1,2,3
KH	$^{39}K^{1}H$	$^1\Sigma^+$	983.6	14.3	3.416400	0.085313	163.55	2.243	1,2
KI	$^{39}K^{127}I$	$^1\Sigma^+$	186.53	0.57	0.06087473	0.00026776	0.02593	3.04784	1,2,3
K_2	$^{39}K_2$	$^1\Sigma_g^+$	92.02	0.28	0.056743	0.000165	0.0863	3.9051	1
LaO	$^{139}La^{16}O$	$^2\Sigma^+$	812.8	2.22	0.35252001	0.00142365	0.2626	1.82591	1,2
LiBr	$^{7}Li^{79}Br$	$^1\Sigma^+$	563.2	3.5	0.555399	0.005644	2.159	2.17043	1,2,3
LiCl	$^{7}Li^{35}Cl$	$^1\Sigma^+$	642.95	4.47	0.7065225	0.0080102	3.409	2.02067	1,5
LiCs	$^{7}Li^{133}Cs$	$^1\Sigma^+$	184.70	1.00	0.188003	0.001248	0.7784	3.6681	21
LiD	$^{7}Li^{2}H$	$^1\Sigma^+$	1054.94	13.06	4.23308131	0.09149428	272	1.59526	4,5
LiF	$^{7}Li^{19}F$	$^1\Sigma^+$	910.57	8.21	1.34525715	0.02028749	11.75	1.56386	4,5
LiH	$^{7}Li^{1}H$	$^1\Sigma^+$	1405.50	21.17	7.5137315	0.2163911	859	1.59490	4,5
LiI	$^{7}Li^{127}I$	$^1\Sigma^+$	496.85	2.85	0.4431766	0.0040862	1.4104	2.39192	1,2,3
LiK	$^{7}Li^{39}K$	$^1\Sigma^+$	207		0.265			3.27	1,20
LiO	$^{7}Li^{16}O$	$^2\Pi$	814.62	7.78	1.212830	0.017899	0.1079	1.68822	3,14
Li_2	$^{7}Li_2$	$^1\Sigma_g^+$	351.41	2.58	0.672530	0.007046	9.79	2.6733	13
MgCl	$^{24}Mg^{35}Cl$	$^2\Sigma^+$	462.12*		0.2456154	0.0016204	0.2723	2.19639	1,2
MgD	$^{24}Mg^{2}H$	$^2\Sigma^+$	1077.30	15.52	3.034344	0.066607	96.25	1.72916	18
MgF	$^{24}Mg^{19}F$	$^2\Sigma^+$	711.69*		0.51922	0.00470	1.080	1.7500	1,2
MgH	$^{24}Mg^{1}H$	$^2\Sigma^+$	1492.78	29.85	5.825523	0.177298	354.56	1.72972	18
MgO	$^{24}Mg^{16}O$	$^1\Sigma^+$	785.21	5.13	0.5748414	0.0053223	1.233	1.74817	28,4
MgS	$^{24}Mg^{32}S$	$^1\Sigma^+$	528.74	2.70	0.26797	0.00176	0.276	2.1425	1
Mg_2	$^{24}Mg_2$	$^1\Sigma_g^+$	51.12	1.64	0.09287	0.00378	1.22	3.891	1
MnD	$^{55}Mn^{2}H$	$^7\Sigma$	1103	13.9	2.8957	0.051	79.5	1.7310	1,2
MnH	$^{55}Mn^{1}H$	$^7\Sigma$	1548.0	28.8	5.6841	0.1570	303.9	1.7311	1,2
NBr	$^{14}N^{79}Br$	$^3\Sigma^-$	691.75	4.72	0.444	0.0040		1.79	1,2
NCl	$^{14}N^{35}Cl$	$^3\Sigma^-$	827.96	5.30	0.64976739	0.00641432	1.596	1.61071	1,5
ND	$^{14}N^{2}H$	$^3\Sigma^-$	2399.13	42.11	8.9087	0.2546	491.7	1.03665	8
NF	$^{14}N^{19}F$	$^3\Sigma^-$	1141.37	8.99	1.205679	0.014889	5.4	1.31698	4
NH	$^{14}N^{1}H$	$^3\Sigma^-$	3282.72	79.04	16.66792	0.65038	1710	1.03719	4
NO	$^{14}N^{16}O$	$^2\Pi_{1/2}$	1904.20	14.07	1.67195	0.0171	0.5	1.15077	1,24
NS	$^{14}N^{32}S$	$^2\Pi_{1/2}$	1218.7	7.28	0.769602	0.0064	1.2	1.4940	1,2,3
N_2	$^{14}N_2$	$^1\Sigma_g^+$	2358.56	14.32	1.998236	0.017310	5.737	1.09769	24
NaBr	$^{23}Na^{79}Br$	$^1\Sigma^+$	302	1.5	0.1512533	0.0009410	0.1554	2.50204	1,2,3
NaCl	$^{23}Na^{35}Cl$	$^1\Sigma^+$	364.68	1.78	0.21806302	0.00162479	0.31202	2.36080	4,5
NaD	$^{23}Na^{2}H$	$^1\Sigma^+$	826.1*		2.557089	0.051600	93.46	1.88654	1,2
NaF	$^{23}Na^{19}F$	$^1\Sigma^+$	535.66	3.58	0.43690153	0.0045592	1.16296	1.92595	11
NaH	$^{23}Na^{1}H$	$^1\Sigma^+$	1171.97	19.70	4.90327	0.1370	343.8	1.8870	4,5
NaI	$^{23}Na^{127}I$	$^1\Sigma^+$	258	1.1	0.1178056	0.0006478	0.0973	2.71145	1,2,3
NaK	$^{39}K^{23}Na$	$^1\Sigma^+$	124.03	0.50	0.09519989	0.00044966	0.2206	3.49958	6
NaLi	$^{7}Li^{23}Na$	$^1\Sigma^+$	256.99	1.66	0.376833	0.003810	3.340	2.88851	4,30

Formula	Molecule	State	ω_e/cm^{-1}	$\omega_e x_e$/cm^{-1}	B_e/cm^{-1}	α_e/cm^{-1}	D_e/10^{-6} cm^{-1}	r_e/Å	Ref.
NaO	$^{23}Na^{16}O$	$^2\Pi$	492.3		0.424630	0.004506	1.2638	2.05155	1,2
Na_2	$^{23}Na_2$	$^1\Sigma_g^+$	159.09	0.71	0.15473537	0.0086375	0.58	3.07858	1,4,20
NbO	$^{93}Nb^{16}O$	$^4\Sigma^-$	989.0	3.8	0.4321	0.0021	0.22	1.691	1,2
Nb_2	$^{93}Nb_2$	$^3\Sigma_g^-$	424.89	0.94	0.084054	0.000242	0.016	2.0778	15
NiD	$^{58}Ni^2H$	$^2\Delta_{5/2}$	1390.1	19	3.992	0.092	130	1.465	1,2
NiH	$^{58}Ni^1H$	$^2\Delta_{5/2}$	1926.6	38	7.700	0.23	481	1.476	1,2
OD	$^{16}O^2H$	$^2\Pi_{3/2}$	2720.24	44.05	10.021	0.276	537.4	0.9698	1,2
OF	$^{19}F^{16}O$	$^2\Pi_{3/2}$	1053.01	9.92	1.0587076	0.013295	4.2823	1.35411	9
OH	$^{16}O^1H$	$^2\Pi_{3/2}$	3737.76	84.88	18.911	0.7242	1938	0.96966	1,2,3
O_2	$^{16}O_2$	$^3\Sigma_g^-$	1580.19	11.98	1.445622	0.015933	4.839	1.20752	24
PCl	$^{31}P^{35}Cl$	$^3\Sigma^-$	551.38	2.23	0.2528748	0.0015119	0.2124	2.01461	1,2
PD	$^{31}P^2H$	$^3\Sigma^-$	1699.2	23.0	4.4081	0.0928	116	1.4220	1,2
PF	$^{31}P^{19}F$	$^3\Sigma^-$	846.73	4.49	0.5667427	0.004639	1.0156	1.58933	12
PH	$^{31}P^1H$	$^3\Sigma^-$	2363.77	43.91	8.53904	0.2534	4.462	1.42218	8
PN	$^{31}P^{14}N$	$^1\Sigma^+$	1336.95	6.90	0.7864844	0.0055337	1.091	1.49087	4
PO	$^{31}P^{16}O$	$^2\Pi_{1/2}$	1233.34	6.56	0.733223657	0.005466162	1.3	1.47637	1,4
P_2	$^{31}P_2$	$^1\Sigma_g^+$	780.77	2.84	0.30362	0.00149	0.188	1.8934	1
PbBr	$^{208}Pb^{79}Br$	$^2\Pi_{1/2}$	207.5	0.50					1,2
PbCl	$^{208}Pb^{35}Cl$	$^2\Pi_{1/2}$	303.9	0.88					1,2
PbF	$^{208}Pb^{19}F$	$^2\Pi_{1/2}$	502.73	2.28	0.22875	0.001473	0.183	2.0575	1,2
PbH	$^{208}Pb^1H$	$^2\Pi_{1/2}$	1564.1	29.75	4.971	0.144	201	1.839	1,2
PbO	$^{208}Pb^{16}O$	$^1\Sigma^+$	720.96	3.52	0.30730373	0.00190977	0.2138	1.92181	1,2,3
PbS	$^{208}Pb^{32}S$	$^1\Sigma^+$	429.17	1.26	0.11632307	0.00043510	0.03418	2.28678	1,2,3
PbSe	$^{208}Pb^{80}Se$	$^1\Sigma^+$	277.6	0.51	0.05059953	0.00012993	0.0070	2.40218	1,2,3
PbTe	$^{208}Pb^{130}Te$	$^1\Sigma^+$	212.0	0.43	0.03130774	0.00006743	0.0027	2.59492	1,2,3
Pb_2	$^{208}Pb_2$		110.5	0.35					1,2
PtC	$^{195}Pt^{12}C$	$^1\Sigma^+$	1051.13	4.86	0.53044	0.003273	0.546	1.6767	1,2
PtD	$^{195}Pt^2H$	$^2\Delta_{5/2}$	1644.3*		3.640	0.071	66	1.524	1,2
PtH	$^{195}Pt^1H$	$^2\Delta_{5/2}$	2294.68*		7.1963	0.1996	261	1.52852	1,2
RbBr	$^{85}Rb^{79}Br$	$^1\Sigma^+$	169.46	0.46	0.04752798	0.00018596	0.01496	2.94474	1,2,3
RbCl	$^{85}Rb^{35}Cl$	$^1\Sigma^+$	228	0.92	0.0876404	0.0004537	0.04947	2.78673	1,2,3
RbF	$^{85}Rb^{19}F$	$^1\Sigma^+$	376	1.9	0.2106640	0.0015228	0.2684	2.27033	1,2,3
RbH	$^{85}Rb^1H$	$^1\Sigma^+$	936.9	14.21	3.020	0.072	123	2.367	1,2
RbI	$^{85}Rb^{127}I$	$^1\Sigma^+$	138.51	0.33	0.03283293	0.00010946	0.00738	3.17688	1,2,3
RbO	$^{85}Rb^{16}O$	$^2\Sigma^+$	388.4*		0.246481	0.002174	0.397	2.25420	1,2
SD	$^{32}S^2H$	$^2\Pi_{3/2}$	1885	31	4.95130	0.10308	130	1.34049	1,2
SF	$^{32}S^{19}F$	$^2\Pi_{3/2}$	837.64	4.47	0.555173	0.004459	0.975	1.59624	4,5
SH	$^{32}S^1H$	$^2\Pi_{3/2}$	2696.25	48.74	9.60025	0.27990	480	1.34061	4
SO	$^{32}S^{16}O$	$^3\Sigma^-$	1149.2	5.6	0.7208171	0.005737	1.134	1.48109	1,19,23
S_2	$^{32}S_2$	$^3\Sigma_g^-$	725.71	2.86	0.29539516	0.00159754	0.19	1.88941	4
SbCl	$^{121}Sb^{35}Cl$	$^3\Sigma^-$	374.7	0.6					1,2
SbD	$^{121}Sb^2H$	$^3\Sigma^-$			2.8782		45	1.7194	1,2
SbF	$^{121}Sb^{19}F$	$^3\Sigma^-$	605.0	2.6	0.2792	0.0020	0.23	1.918	1,2
SbH	$^{121}Sb^1H$	$^3\Sigma^-$			5.684		240	1.723	1,2
SbN	$^{121}Sb^{14}N$	$^1\Sigma^+$	942.0	5.6					1,2
SbO	$^{121}Sb^{16}O$	$^2\Pi_{1/2}$	816	4.2	0.3580	0.0022	0.270	1.826	1,2
ScF	$^{45}Sc^{19}F$	$^1\Sigma^+$	735.6	3.8	0.3950	0.00266		1.788	1,2
SeD	$^{80}Se^2H$	$^2\Pi_{3/2}$	1708*		3.94			1.48	1,2
SeH	$^{80}Se^1H$	$^2\Pi_{3/2}$	2400*		8.02	0.23	330	1.48	1,2
SeO	$^{80}Se^{16}O$	$^3\Sigma^-$	914.69	4.52	0.4655	0.00323	0.5	1.648	1,2
Se_2	$^{80}Se_2$	$^3\Sigma_g^-$	385.30	0.96	0.08992	0.000288	0.024	2.166	1,2
SiCl	$^{28}Si^{35}Cl$	$^2\Pi_{1/2}$	535.59	2.18	0.256103	0.001582	0.25	2.05794	4
SiD	$^{28}Si^2H$	$^2\Pi_{1/2}$	1469.32	18.23	3.8840	0.0781	105.4	1.5199	1,2
SiF	$^{28}Si^{19}F$	$^2\Pi_{1/2}$	857.33	4.83	0.58125735	0.00503859	1.065	1.60100	31
SiH	$^{28}Si^1H$	$^2\Pi_{1/2}$	2042.52	36.06	7.503898	0.21814	400	1.51966	4
SiN	$^{28}Si^{14}N$	$^2\Sigma^+$	1151.28	6.46	0.730927	0.005685	1.2	1.57207	4
SiO	$^{28}Si^{16}O$	$^1\Sigma^+$	1241.54	5.97	0.7267521	0.0050379	0.9923	1.50975	1,19,22
SiS	$^{28}Si^{32}S$	$^1\Sigma^+$	749.64	2.58	0.30352788	0.00147308	0.201	1.92926	1,19
SiSe	$^{28}Si^{80}Se$	$^1\Sigma^+$	580.0	1.78	0.1920117	0.0007767	0.0842	2.05832	1,2,3
Si_2	$^{28}Si_2$	$^3\Sigma_g^-$	510.98	2.02	0.2390	0.0014	0.21	2.246	1
SnBr	$^{120}Sn^{79}Br$	$^2\Pi_{1/2}$	247.2	0.6					1,2

Molecular

Formula	Molecule	State	ω_e/cm^{-1}	$\omega_e x_e$/cm^{-1}	B_e/cm^{-1}	α_e/cm^{-1}	D_e/10^{-6} cm^{-1}	r_e/Å	Ref.
SnCl	$^{120}Sn^{35}Cl$	$^2\Pi_{1/2}$	351.1	1.06	0.1117	0.0004		2.361	1,2
SnD	$^{120}Sn^{2}H$	$^2\Pi_{1/2}$	1188.0*		2.6950	0.049	53.4	1.7770	1,2
SnF	$^{118}Sn^{19}F$	$^2\Pi_{1/2}$	577.6	2.69	0.2727	0.0014	0.26	1.944	1,2
SnH	$^{120}Sn^{1}H$	$^2\Pi_{1/2}$			5.31488		207.5	1.78146	1,2
SnI	$^{120}Sn^{127}I$	$^2\Pi_{1/2}$	199.0	0.6					1,2
SnO	$^{120}Sn^{16}O$	$^1\Sigma^+$	822.13	3.72	0.35571998	0.00214432	0.26638	1.83251	1,2,3
SnS	$^{120}Sn^{32}S$	$^1\Sigma^+$	487.26	1.36	0.13686139	0.00050563	0.0424	2.20898	1,2,3
SnSe	$^{120}Sn^{80}Se$	$^1\Sigma^+$	331.2	0.74	0.0649978	0.0001705	0.011	2.32557	1,2,3
SnTe	$^{120}Sn^{130}Te$	$^1\Sigma^+$	259.5	0.50	0.04247917	0.00009543	0.0055	2.52280	1,2,3
SrBr	$^{88}Sr^{79}Br$	$^2\Sigma^+$	216.60	0.52	0.0541847	0.0001827	0.01356	2.73522	1,2
SrCl	$^{88}Sr^{35}Cl$	$^2\Sigma^+$	302.3	0.95					1,2
SrD	$^{88}Sr^{2}H$	$^2\Sigma^+$	841	8.6	1.8609	0.0292	34.7	2.1449	1,2
SrF	$^{88}Sr^{19}F$	$^2\Sigma^+$	502.4	2.3	0.2505346	0.0015513	0.2498	2.07537	1,2
SrH	$^{88}Sr^{1}H$	$^2\Sigma^+$	1206.2	17.0	3.6751	0.0814	135	2.1456	1,2
SrI	$^{88}Sr^{127}I$	$^2\Sigma^+$	173.77	0.35	0.0367097	0.0001060	0.00655	2.94364	1,2
SrO	$^{88}Sr^{16}O$	$^1\Sigma^+$	653.5	3.96	0.33798	0.00219	0.36	1.91983	1,2
T_2	$^{3}H_2$	$^1\Sigma_g^+$	2546.5	41.23	20.335	0.5887		0.74142	1,2
TaO	$^{181}Ta^{16}O$	$^2\Delta_{3/2}$	1028.69	3.51	0.40284	0.00182	0.2450	1.68746	1,2
TeH	$^{(130)}Te^{1}H$	$^2\Pi_{3/2}$			5.56			1.74	1,2
TeO	$^{130}Te^{16}O$	O^+	797.11	4.00	0.3554	0.00237	0.27	1.825	1,2
Te_2	$^{130}Te_2$	$^3\Sigma_g^-$	247.07	0.51	0.039681	0.000106	0.0044	2.5574	1,2
ThO	$^{232}Th^{16}O$	$^1\Sigma^+$	895.77	2.39	0.332644	0.001302	0.1833	1.84032	1,2
TiO	$^{48}Ti^{16}O$	$^3\Delta_1$	1009.02	4.50	0.53541	0.00301	0.603	1.6202	1,2
TlBr	$^{205}Tl^{81}Br$	$^1\Sigma^+$	192.10	0.39	0.0423899	0.0001276	0.0083	2.61817	1,2,3
TlCl	$^{205}Tl^{35}Cl$	$^1\Sigma^+$	284.71	0.86	0.09139702	0.00039784	0.0377	2.48483	1,2,3
TlD	$^{205}Tl^{2}H$	$^1\Sigma^+$	987.7	12.04	2.419	0.057	60	1.869	1,2
TlF	$^{205}Tl^{19}F$	$^1\Sigma^+$	476.86	2.24	0.22315014	0.00150380	0.1955	2.08439	1,2,3
TlH	$^{205}Tl^{1}H$	$^1\Sigma^+$	1390.7	22.7	4.806	0.154	254	1.870	1,2
TlI	$^{205}Tl^{127}I$	$^1\Sigma^+$	150*		0.0271676	0.0000664	0.0036	2.81361	1,2,3
VO	$^{51}V^{16}O$	$^4\Sigma^-$	1011.3	4.86	0.54825	0.00352	0.6	1.5893	1,2
XeF	$^{(132)}Xe^{19}F$	$^2\Sigma$	225.4	10.9	0.19326	0.00699	0.536	2.293	10
YCl	$^{89}Y^{35}Cl$	$^1\Sigma$	380.7	1.3	0.1160	0.0003	0.09	2.41	1,2
YF	$^{89}Y^{19}F$	$^1\Sigma^+$	631.29	2.50	0.29042	0.00163	0.237	1.9257	1,2
YO	$^{89}Y^{16}O$	$^2\Sigma^+$	861.0	2.9	0.3881	0.0018	0.32	1.790	1,2
YbD	$^{174}Yb^{2}H$	$^2\Sigma^+$	886.6	10.57	2.01162	0.03425	41.60	2.0516	1,2
YbH	$^{174}Yb^{1}H$	$^2\Sigma^+$	1249.54	21.06	3.9931	0.0957	161.8	2.0526	1,2
ZnCl	$^{64}Zn^{35}Cl$	$^2\Sigma$	390.5	1.6					1,2
ZnD	$^{64}Zn^{2}H$	$^2\Sigma^+$	1072	28	3.350		124	1.6054	1,2
ZnF	$^{64}Zn^{19}F$	$^2\Sigma$	628	3.5					1,2
ZnH	$^{64}Zn^{1}H$	$^2\Sigma^+$	1607.6	55.14	6.6794	0.2500	466	1.5949	1,2
ZnI	$^{64}Zn^{127}I$	$^2\Sigma$	223.4	0.6					1,2
ZrO	$^{90}Zr^{16}O$	$^1\Sigma^+$	969.8	4.9	0.42263	0.0023	0.319	1.7116	1,2

* Indicates a value for the interval between $v = 0$ and $v = 1$ states instead of a value of ω_e.

Section 10
Atomic, Molecular, and Optical Physics

LINE SPECTRA OF THE ELEMENTS

Joseph Reader and Charles H. Corliss

The original tables from which this table was derived were prepared under the auspices of the Committee on Line Spectra of the Elements of the National Academy of Sciences National Research Council. The table contains the outstanding spectral lines of neutral (I) and singly ionized (II) atoms of the elements from hydrogen through plutonium (Z = 1–94); selected strong lines from doubly ionized (III), triply ionized (IV), and quadruply ionized (V) atoms are also included. Listed are lines that appear in emission from the vacuum ultraviolet to the far infrared. These lines were selected from much larger lists in such a way as to include the stronger observed lines in each spectral region. A more extensive list may be found in Reference 1.

The data were compiled by the following contributors.

J. G. Conway — Lawrence Berkeley Laboratory
C. H. Corliss — National Bureau of Standards
R. D. Cowan — Los Alamos Scientific Laboratory
C. R. Cowley — University of Michigan
Henry M. and Hannah Crosswhite — Argonne National Laboratory
S. P. Davis — University of California, Berkeley
V. Kaufman — National Bureau of Standards
R. L. Kelly — Naval Postgraduate School
J. F. Kielkopf — University of Louisville
W. C. Martin — National Bureau of Standards
T. K. McCubbin — Pennsylvania State University
L. J. Radziemski — Los Alamos Scientific Laboratory
J. Reader — National Bureau of Standards
C. J. Sansonetti — National Bureau of Standards
G. V. Shalimoff — Lawrence Berkeley Laboratory
R. W. Stanley — Purdue University
J. O. Stoner, Jr. — University of Arizona
H. H. Stroke — New York University
D. R. Wood — Wright State University
E. F. Worden — Lawrence Livermore Laboratory
J. J. Wynne — International Business Machines Corporation
R. Zalubas — National Bureau of Standards

All wavelengths are given in Ångstrom units (10^{-10} m). Below 2000 Å the wavelengths are in vacuum (except for the Cu II line at 1999.698 Å, which is in air); above 2000 Å the wavelengths are in air. Wavelengths given to three decimal places have an uncertainty of less than 0.001 Å and are therefore suitable for calibration purposes. In the air region, the elements used most commonly for calibration are Ne, Ar, Kr, Fe, Th, and Hg; in the vacuum region, the most common are C, N, O, Si, Cu.

All data refer to natural isotopic abundance of the elements except that Kr I and Kr II lines below 11,000 Å given to three decimal places are for ^{86}Kr. A separate table for ^{198}Hg contains accurately known wavelengths that are frequently used for calibration.

A large number of the lines for neutral and singly ionized atoms were extracted from the National Bureau of Standards (NBS) *Tables of Spectral Line Intensities* (Reference 2). The intensities of these lines represent quantitative estimates of relative line strengths that take into account varying detection sensitivity at different wavelengths. They are on a linear scale. For nearly all of the other lines the intensities represent qualitative estimates of the relative strengths of lines not greatly separated in wavelength. Because different observers frequently use different scales for their intensity estimates, these intensities are useful only as a rough indication of the appearance of a spectrum. In some cases the intensity scale is not intended to be linear. In the first and second spectra the intensities of the lines of the singly ionized atom (II) relative to those of the neutral atom (I) should be used with caution, inasmuch as the concentration of ions in a light source depends greatly on the excitation conditions.

Descriptive symbols that follow the wavelength have the following meanings:

c — complex
d — line consists of two unresolved lines
h — hazy
l — shaded to longer wavelengths
s — shaded to shorter wavelengths
p — perturbed by a close line
r — easily reversed
w — wide

The table is arranged alphabetically by element name (not symbol); for each element the lines are listed by wavelength. References to the sources of data for each element are given at the end of the table, starting on page 10-88.

General References

1. Reader, J., Corliss, C. H., Wiese, W. L., and Martin, G. A., *Tables of Line Spectra of the Elements, Part 1. Wavelengths and Intensities*, Nat. Stand. Ref. Data Sys. Nat. Bur. Standards (U.S.), No. 68, 1980.
2. Meggers, W. F., Corliss, C. H., and Scribner, B. F., *Tables of Spectral Line Intensities, Part 1. Arranged by Elements*, Nat. Bur. Stand. (U.S.), Monograph 145, 1975.
3. Fuhr, J. R., Martin, W. C., Musgrove, A., Sugar, J., and Wiese, W. L., "NIST Atomic Spectroscopic Database" ver. 1.1, January 1996. *NIST Physical Reference Data*, National Institute of Standards and Technology, Gaithersburg, MD. Available at http://physics.nist.gov/PhysRefData/contents.html

Intensity	Wavelength/Å	
Actinium Ac Z = 89		
2000 h	2952.55	III
2000 h	3392.78	III
3000	3487.59	III
2000 s	3863.12	II
3000 s	4088.44	II
3000 s	4168.40	II
100	4179.98	I
20	4183.12	I
20	4194.40	I
20 l	4384.53	I
20	4396.71	I
2000 h	4413.09	III
20	4462.73	I
3000 h	4569.87	III
1000	5910.85	II
20	6359.86	I
20 l	6691.27	I
Aluminum Al Z = 13		
900	125.53	V
800	126.07	V
800	130.41	V
1000	130.85	V
900	131.00	V
900	131.44	V
800	160.07	IV
1000	278.69	V
900	281.39	V
70	486.884	III
30	486.912	III
250	511.138	III
150	511.191	III
500	560.317	III
200	560.433	III
100	670.068	III
200	671.118	III
500	695.829	III
400	696.217	III
200	725.683	III
300	726.915	III
400	855.034	III
500	856.746	III
400	892.024	III
50	893.887	III
450	893.897	III
800	1042.17	IV
50	1191.812	II
900	1237.19	IV
900	1257.62	IV
800	1264.18	IV
1000	1272.76	IV
150	1350.18	II
800	1384.13	III
800	1447.51	IV
800	1494.79	IV
1000	1526.14	V
800	1537.54	IV
800	1539.830	II
1000	1557.25	IV
100	1569.385	II
900	1582.04	IV
800	1584.46	IV
125	1596.059	II
700	1605.766	III
100	1611.814	III
800	1611.874	III
150	1625.627	II
800	1639.06	IV
100	1644.235	II
100	1644.809	II
1000	1670.787	II
100	1686.250	II
800	1719.440	II
500	1721.244	II
900	1721.271	II
500	1724.952	II
900	1724.984	II
350	1760.104	II
300	1761.975	II
290	1763.00	I
500	1763.869	II
700	1763.952	II
450	1765.64	I
300	1765.815	II
450	1766.38	I
400	1767.731	II
450	1769.14	I
1000	1818.56	IV
600	1828.588	II
400	1832.837	II
250	1834.808	II
1000	1854.716	III
300	1855.929	II
700	1858.026	II
120	1859.980	II
1000	1862.311	II
600	1862.790	III
200	1929.978	II
150	1931.048	II
200	1932.377	II
400	1934.503	II
150	1934.713	II
300	1935.840	III
200	1935.949	III
150	1936.907	II
220	1939.261	II
700	1990.531	II
150	2016.052	II
150	2016.234	II
100	2016.368	II
200	2074.008	II
700	2094.264	II
150	2094.744	II
300	2094.791	II
100	2095.104	II
200	2095.141	II
400	2269.10	I
120	2269.22	I
140	2321.56	I
460	2367.05	I
110	2367.61	I
110	2368.11	I
180	2369.30	I
140	2370.22	I
160	2372.07	I
850	2373.12	I
170	2373.35	I
110	2373.57	I
240	2567.98	I
480	2575.10	I
110	2637.70	II
150	2652.48	I
200	2660.39	I
160	2669.17	II
650	2816.19	II
150	3041.28	II
360	3050.07	I
450	3057.14	I
150	3074.64	II
4500 r	3082.153	I
7200 r	3092.710	I
1800 r	3092.839	I
150	3428.92	II
150	3443.64	I
900	3492.23	IV
800	3508.46	IV
450	3586.56	II
360	3587.07	II
290	3587.45	II
870	3601.63	III
220	3651.06	II
110	3651.10	II
150	3654.98	II
290	3655.00	II
450	3900.68	II
4500 r	3944.006	I
9000 r	3961.520	I
110	3995.86	II
290	4226.81	II
870	4529.19	III
150	4585.82	II
110	4588.19	II
550	4666.80	II
110	4898.76	II
110	4902.77	II
150	5280.21	II
290	5283.77	II
150	5285.85	II
110	5312.32	II
220	5316.07	II
150	5371.84	II
180	5557.06	I
110	5557.95	I
450	5593.23	II
1200	5696.60	III
1000	5722.73	III
110	5853.62	II
220	5971.94	II
290	6001.76	II
220	6001.88	II
450	6006.42	II
150	6061.11	II
290	6068.43	II
110	6068.53	II
450	6073.23	II
110	6181.57	II
150	6181.68	II
290	6182.28	II
220	6182.45	II
450 h	6183.42	II
450	6201.52	II
360	6201.70	II
290	6226.18	II
360	6231.78	II
450	6243.36	II
450	6335.74	II
360	6696.02	I
230	6698.67	I
110	7361.57	I
140	7362.30	I
230	7835.31	I
290	7836.13	I
110	8075.35	I
290	8640.70	II
360	8772.87	I
450	8773.90	I
110	8828.91	I
180	8841.28	I
140	8923.56	I
150	9290.65	II
110	9290.75	II
150	10076.29	II
110	10768.36	I
140	10782.04	I
110	10872.98	I
230	10891.73	I
450	11253.19	I
570	11254.88	I
570	13123.41	I
450	13150.76	I
230	16718.96	I
300	16750.56	I
140	16763.36	I
300	21093.04	I
360	21163.75	I
Antimony Sb Z = 51		
15	722.86	III
15	732.33	III
	861.5	IV
4	876.84	II
4	921.07	II
6	983.57	II
15	999.62	III
6	1001.13	II
6	1009.43	II
40	1011.94	III
6	1052.21	II
8	1056.27	II
8	1057.32	II
40	1065.90	III
6	1073.81	II
30	1075.82	III
	1087.6	IV

Intensity	Wavelength/Å	
8	1104.32	V
30	1151.49	III
40	1157.74	III
	1199.1	IV
50	1205.20	III
50	1210.64	III
12	1226.00	V
6	1230.30	II
8	1274.98	II
20	1306.69	III
8	1327.40	II
6	1358.04	II
8	1384.70	II
20	1404.18	III
6	1407.83	II
8	1436.49	II
20 r	1486.57	I
40 h	1491.36	I
	1499.2	IV
12	1505.70	V
50 r	1512.57	I
12	1524.47	V
120 r	1532.74	I
80 r	1535.06	I
6	1565.51	II
8	1576.11	II
7	1581.36	II
80 r	1599.96	I
10	1606.98	II
200 w	1612.8	I
100 w	1623.3	I
20	1657.04	II
100 w	1662.6	I
15	1673.89	III
15	1711.84	III
80 r	1716.93	I
150 r	1717.45	I
150 r	1723.43	I
15	1725.33	III
100 r	1736.19	I
100 h	1765.76	I
100 r	1780.87	I
100 r	1788.24	I
150	1800.18	I
50 r	1810.50	I
80 r	1814.20	I
100	1829.50	I
50 r	1868.17	I
300 r	1871.15	I
150 r	1882.56	I
100	1927.08	I
200 r	1950.39	I
60 r	2029.49	I
70 r	2039.77	I
150 r	2049.57	I
1000 r	2068.33	I
100	2079.56	I
50 r	2098.41	I
80 r	2118.48	I
100 r	2127.39	I
50 r	2137.05	I

Intensity	Wavelength/Å	
100 r	2139.69	I
10	2141.80	II
50 r	2141.83	I
100 r	2144.86	I
1500 r	2175.81	I
250 r	2179.19	I
200 r	2201.32	I
300 r	2208.45	I
150 r	2220.73	I
100	2221.98	I
120 r	2224.93	I
300 r	2262.51	I
120	2288.98	I
150 r	2293.44	I
300 r	2306.46	I
2500 r	2311.47	I
150	2315.89	I
400 h	2373.67	I
300 h	2383.64	I
100	2395.22	I
150	2422.13	I
250	2426.35	I
400 r	2445.51	I
400	2478.32	I
150	2480.44	I
100	2510.54	I
2000 r	2528.52	I
15	2528.54	II
10	2567.75	II
150	2574.06	I
15	2590.13	III
1500 r	2598.05	I
500 r	2598.09	I
300 r	2612.31	I
12	2617.17	III
200 r	2652.60	I
20	2669.39	III
300 r	2670.64	I
200 r	2682.76	I
120	2692.25	I
150 r	2718.90	I
400 r	2769.95	I
1000 r	2877.92	I
15	2980.96	II
500 r	3029.83	I
600 r	3232.52	I
20	3241.28	II
700 r	3267.51	I
15	3498.46	II
25	3637.80	II
250	3637.83	I
20	3722.78	II
200 r	3722.79	I
20	3850.22	II
200	4033.55	I
20	4033.56	II
20	4133.63	II
15	4140.54	II
15	4195.17	II
20	4219.07	II
20	4314.32	II

Intensity	Wavelength/Å	
15	4514.50	II
30	4596.90	II
20	4599.09	II
15	4604.77	II
30	4647.32	II
20	4675.74	II
40	4711.26	II
20	4757.81	II
20	4765.36	II
30	4784.03	II
20	4802.01	II
20	4832.82	II
20	4877.24	II
15	4947.40	II
15	5044.56	II
20	5238.94	II
20	5354.24	II
40 h	5556.10	I
100 l	5632.02	I
30	5639.75	II
60 h	5830.34	I
100	6005.21	II
20	6053.41	II
30	6079.80	II
50	6130.04	II
20	6154.94	II
20	6611.49	I
30	6647.44	II
30 h	7648.28	I
80	7844.44	I
200	7924.65	I
60	8411.69	I
150	8572.64	I
100	8619.55	I
400	9518.68	I
400	9949.14	I
200	10078.49	I
300	10261.01	I
200	10585.60	I
1000	10677.41	I
800	10741.94	I
80	10794.11	I
600	10839.73	I
200	10868.58	I
400	10879.55	I
300	11012.79	I
150	11266.23	I
5	12116.06	I

Argon Ar Z = 18

Intensity	Wavelength/Å	
3	336.56	V
3	337.56	V
6	338.00	V
2	338.43	V
2	339.01	V
3	339.89	V
3	350.88	V
4	396.87	IV
4	398.55	IV
2	436.67	V
5	446.00	V
8	446.95	V
4	447.53	V
18	449.06	V
4	449.49	V
3	458.12	V
2	458.98	V
6 p	461.23	V
3	462.42	V
7	463.94	V
30	487.227	II
50	490.650	II
30	490.701	II
30	519.327	II
3	522.09	V
5	524.19	V
6	527.69	V
30	542.912	II
200	543.203	II
70	547.461	II
2	554.50	V
70	556.817	II
5	558.48	V
70	573.362	II
30	576.736	II
70	580.263	II
30	583.437	II
70	597.700	II
30	602.858	II
30	612.372	II
6	623.77	IV
3	635.12	V
500	661.867	II
30	664.562	II
200	666.011	II
1000	670.946	II
3000	671.851	II
70	676.242	II
30	677.952	II
30	679.218	II
200	679.401	II
10	683.28	IV
7	688.39	IV
12 p	689.01	IV
6	699.41	IV
8	700.28	IV
3	705.35	V
5	709.20	V
4	715.60	V
3	715.65	V
200	718.090	II
3000	723.361	II
2	725.11	V
500	725.548	II
70	730.930	II
200	740.269	II
200	744.925	II
70	745.322	II
4	754.20	IV
5	761.47	IV
12	769.15	III
5	800.57	IV
10	801.09	IV

Intensity	Wavelength/Å	
10	801.41	IV
5	801.91	IV
20	802.859	I
100	806.471	I
60	806.869	I
30	807.218	I
40	807.653	I
50	809.927	I
120	816.232	I
70	816.464	I
80	820.124	I
4	822.16	V
120	825.346	I
120	826.365	I
5	827.05	V
3	827.35	V
150	834.392	I
4 p	834.88	V
100	835.002	I
2	836.13	V
15	840.03	IV
100	842.805	I
20	843.77	IV
25	850.60	IV
180	866.800	I
150	869.754	I
10	871.10	III
9	875.53	III
180 r	876.058	I
12	878.73	III
8	879.62	III
180 r	879.947	I
9	883.18	III
10	887.40	III
150	894.310	I
5	900.36	IV
9	901.17	IV
1000	919.781	II
1000	932.054	II
1000 r	1048.220	I
500 r	1066.660	I
7	1669.67	III
7	1673.42	III
7	1675.48	III
9	1914.40	III
7	1915.56	III
10	2125.16	III
15	2133.87	III
10	2138.59	III
10	2148.73	III
15	2166.19	III
10	2168.26	III
20	2170.23	III
25	2177.22	III
8	2184.06	III
10	2188.22	III
15	2192.06	III
7	2248.73	III
10	2279.10	III
7	2281.22	III
7	2282.21	III

Intensity	Wavelength/Å	
12	2293.03	III
4	2299.72	IV
10	2300.85	III
15	2302.17	III
9	2317.00	III
15	2317.47	III
12	2318.04	III
10	2319.13	III
10	2319.37	III
9	2345.17	III
7	2351.67	III
9	2360.26	III
10	2395.63	III
12	2399.15	III
10	2413.20	III
7	2415.61	III
10	2418.82	III
5	2420.456	II
12	2423.52	III
12	2423.93	III
7	2443.69	III
8	2447.71	IV
8	2472.95	III
7	2476.10	III
12	2488.86	III
12	2513.28	IV
10	2516.789	II
6	2518.40	IV
9	2525.69	IV
10	2534.709	II
15	2562.087	II
12	2562.17	IV
10	2568.07	IV
7	2569.53	IV
12	2599.47	IV
10	2608.06	IV
7	2608.44	IV
12	2615.68	IV
6	2619.98	IV
12	2621.36	IV
12	2624.92	IV
7	2631.90	III
15	2640.34	IV
10	2654.63	III
8	2674.02	III
9	2678.38	III
9	2682.63	IV
10	2724.84	III
14	2757.92	IV
7	2762.23	III
10	2776.26	IV
12	2784.47	IV
14	2788.96	IV
7	2797.11	IV
16	2809.44	IV
10	2830.25	IV
7	2842.88	III
8	2855.29	III
6	2874.40	IV
9	2884.12	III
25	2891.612	II

Intensity	Wavelength/Å	
12	2913.00	IV
11	2926.33	IV
200	2942.893	II
100	2979.050	II
10	3010.02	III
12	3024.05	III
50	3033.508	II
6	3037.98	IV
12	3054.82	III
10	3064.77	III
8	3077.40	IV
10	3078.15	III
50	3093.402	II
7	3110.41	III
7	3127.90	III
8	3200.37	I
20	3243.689	II
25	3285.85	III
25	3293.640	II
20	3301.88	III
20	3307.228	II
15	3311.25	III
7	3319.34	I
7	3323.59	III
25	3336.13	III
20	3344.72	III
25	3350.924	II
15	3358.49	III
7	3361.28	III
7	3373.47	I
25	3376.436	II
25	3388.531	II
15	3391.85	III
7	3393.73	I
7	3417.49	III
9	3424.25	III
8	3438.04	III
7	3461.07	I
9	3471.32	III
70	3476.747	II
20	3478.232	II
20	3480.55	III
50	3491.244	II
100	3491.536	II
12	3499.67	III
15	3503.58	III
70	3509.778	II
8	3511.12	III
70	3514.388	II
70	3545.596	II
70	3545.845	II
7	3554.306	I
100	3559.508	II
100	3561.030	II
70	3576.616	II
25	3581.608	II
50	3582.355	II
70	3588.441	II
7	3606.522	I
25	3622.138	II
20	3639.833	II

Intensity	Wavelength/Å	
35	3718.206	II
70	3729.309	II
50	3737.889	II
150	3765.270	II
50	3766.119	II
20	3770.369	I
20	3770.520	II
25	3780.840	II
20	3795.37	III
25	3803.172	II
50	3809.456	II
7	3834.679	I
70	3850.581	II
10	3858.32	III
35	3868.528	II
7	3907.84	III
35	3925.719	II
50	3928.623	II
25	3932.547	II
70	3946.097	II
7	3947.505	I
35	3948.979	I
8	3960.53	III
20	3979.356	II
35	3994.792	II
50	4013.857	II
6	4023.60	III
50	4033.809	II
20	4035.460	II
150	4042.894	II
50	4044.418	I
100	4052.921	II
200	4072.005	II
70	4072.385	II
25	4076.628	II
35	4079.574	II
25	4082.387	II
150	4103.912	II
300	4131.724	II
5	4146.70	III
35	4156.086	II
400	4158.590	I
50	4164.180	I
35	4179.297	II
50	4181.884	I
100	4190.713	I
150	4191.029	I
200	4198.317	I
400	4200.674	I
25	4218.665	II
25	4222.637	II
25	4226.988	II
100	4228.158	II
100	4237.220	II
25	4251.185	I
200	4259.362	I
100	4266.286	I
70	4266.527	II
150	4272.169	I
550	4277.528	II
20	4282.898	II

Intensity	Wavelength/Å	
100	4300.101	I
25	4300.650	II
70	4309.239	II
200	4331.200	II
50	4332.030	II
100	4333.561	I
50	4335.338	I
25	4345.168	I
800	4348.064	II
50	4352.205	II
25	4362.066	II
50	4367.832	II
200	4370.753	II
70	4371.329	II
50	4375.954	II
150	4379.667	II
50	4385.057	II
70	4400.097	II
200	4400.986	II
400	4426.001	II
150	4430.189	II
50	4430.996	II
50	4433.838	II
20	4439.461	II
35	4448.879	II
100	4474.759	II
200	4481.811	II
100	4510.733	I
20	4522.323	I
20	4530.552	II
400	4545.052	II
20	4564.405	II
400	4579.350	II
400	4589.898	II
15	4596.097	I
550	4609.567	II
7	4628.441	I
35	4637.233	II
400	4657.901	II
15	4702.316	I
20	4721.591	II
550	4726.868	II
50	4732.053	II
300	4735.906	II
800	4764.865	II
550	4806.020	II
150	4847.810	II
50	4865.910	II
800	4879.864	II
70	4889.042	II
20	4904.752	II
35	4933.209	II
200	4965.080	II
50	5009.334	II
70	5017.163	II
70	5062.037	II
20	5090.495	II
100	5141.783	II
70	5145.308	II
5	5151.391	I
15	5162.285	I
25	5165.773	II
20	5187.746	I
20	5216.814	II
7	5221.271	I
5	5421.352	I
10	5451.652	I
25	5495.874	I
5	5506.113	I
25	5558.702	I
10	5572.541	I
35	5606.733	I
20	5650.704	I
10	5739.520	I
5	5834.263	I
10	5860.310	I
15	5882.624	I
25	5888.584	I
50	5912.085	I
15	5928.813	I
5	5942.669	I
7	5987.302	I
5	5998.999	I
5	6025.150	I
70	6032.127	I
35	6043.223	I
10	6052.723	I
20	6059.372	I
7	6098.803	I
10	6105.635	I
100	6114.923	II
10	6145.441	I
7	6170.174	I
150	6172.278	II
10	6173.096	I
10	6212.503	I
5	6215.938	I
25	6243.120	II
7	6296.872	I
15	6307.657	I
7	6369.575	I
20	6384.717	I
70	6416.307	I
25	6483.082	II
15	6538.112	I
15	6604.853	I
25	6638.221	II
20	6639.740	II
50	6643.698	II
5	6660.676	I
5	6664.051	I
25	6666.359	II
100	6677.282	I
35	6684.293	II
150	6752.834	I
5	6756.163	I
15	6766.612	I
20	6861.269	II
150	6871.289	I
5	6879.582	I
10	6888.174	I
150	6937.664	I
7	6951.478	I
7	6960.250	I
10000	6965.431	I
150	7030.251	I
10000	7067.218	I
100	7068.736	I
25	7107.478	I
25	7125.820	I
1000	7147.042	I
15	7158.839	I
70	7206.980	I
15	7265.172	I
7	7270.664	I
2000	7272.936	I
35	7311.716	I
25	7316.005	I
5	7350.814	I
70	7353.293	I
200	7372.118	I
20	7380.426	II
10000	7383.980	I
20	7392.980	I
15	7412.337	I
10	7425.294	I
25	7435.368	I
10	7436.297	I
20000	7503.869	I
15000	7514.652	I
25000	7635.106	I
15000	7723.761	I
10000	7724.207	I
10	7891.075	I
20000	7948.176	I
20000	8006.157	I
25000	8014.786	I
7	8053.308	I
20000	8103.693	I
35000	8115.311	I
10000	8264.522	I
20	8392.27	I
15000	8408.210	I
20000	8424.648	I
15000	8521.442	I
7	8605.776	I
4500	8667.944	I
20	8771.860	II
180	8849.91	I
20	9075.394	I
35000	9122.967	I
550	9194.638	I
15000	9224.499	I
400	9291.531	I
1600	9354.220	I
25000	9657.786	I
4500	9784.503	I
180	10052.06	I
30	10332.72	I
100	10467.177	II
1600	10470.054	I
13	10478.034	I
180	10506.50	I
200	10673.565	I
11	10681.773	I
7	10683.034	II
30	10733.87	I
30	10759.16	I
7	10812.896	II
11	11078.869	I
30	11106.46	I
12	11441.832	I
400	11488.109	I
200	11668.710	I
12	11719.488	I
200	12112.326	I
50	12139.738	I
50	12343.393	I
200	12402.827	I
200	12439.321	I
100	12456.12	I
200	12487.663	I
150	12702.281	I
30	12733.418	I
12	12746.232	I
200	12802.739	I
50	12933.195	I
500	12956.659	I
200	13008.264	I
200	13213.99	I
200	13228.107	I
100	13230.90	I
500	13272.64	I
1000	13313.210	I
1000	13367.111	I
30	13499.41	I
1000	13504.191	I
11	13573.617	I
30	13599.333	I
400	13622.659	I
200	13678.550	I
1000	13718.577	I
10	13825.715	I
10	13907.478	I
200	14093.640	I
100	15046.50	I
25	15172.69	I
10	15329.34	I
30	15989.49	I
30	16519.86	I
500	16940.58	I
12	18427.76	I
50	20616.23	I
30	20986.11	I
20	23133.20	I
20	23966.52	I
Arsenic As $Z = 33$		
510	871.7	III
325	889.0	III
325	927.5	III
325	937.2	III
325	953.6	III
325	963.8	III
250	987.7	V

Intensity	Wavelength/Å	
340	1021.96	II
250	1029.5	V
340	1082.35	II
500	1139.40	II
615	1149.31	II
555	1181.51	II
555	1189.87	II
615	1196.38	II
615	1196.56	II
340	1207.44	II
800	1211.17	II
800	1218.10	II
340	1223.15	II
760	1241.31	II
965	1243.08	II
870	1245.67	II
800	1258.58	II
965	1263.77	II
800	1266.34	II
800	1267.59	II
715	1280.99	II
715	1287.54	II
715	1305.70	II
340	1307.74	II
760	1333.15	II
965	1341.55	II
760	1355.93	II
965	1369.77	II
800	1373.65	II
1000	1375.07	II
760	1375.78	II
800	1394.64	II
800	1400.31	II
500	1448.59	II
500	1558.88	II
500	1570.99	II
100 r	1593.60	I
500	1660.55	II
340	1860.34	II
1000 r	1890.42	I
500	1912.94	II
800 r	1937.59	I
585 r	1972.62	I
170 r	1990.35	I
100 r	1991.13	I
100 r	1995.43	I
230 r	2003.34	I
100 r	2009.19	I
200	2263.2	IV
350 r	2288.12	I
200	2301.0	IV
350 r	2349.84	I
100 r	2370.77	I
135 r	2381.18	I
250	2417.5	IV
250	2454.0	IV
170 r	2456.53	I
200	2461.4	IV
340	2602.00	II
170 r	2780.22	I
300	2830.359	II
300	2831.164	II
100 r	2860.44	I
300	2884.406	II
80	2926.3	III
615	2959.572	II
300	3003.819	II
300	3116.516	II
340	3842.60	II
325	3922.6	III
715	4190.082	II
615	4197.40	II
615	4242.982	II
500	4315.657	II
500	4323.867	II
500	4336.64	II
500	4352.145	II
425	4352.864	II
375	4371.17	II
615	4427.106	II
615	4431.562	II
715	4458.469	II
340	4461.075	II
715	4466.348	II
500	4474.46	II
800	4494.230	II
850	4507.659	II
615	4539.74	II
715	4543.483	II
615	4602.427	II
340	4629.787	II
340	4707.586	II
340	4730.67	II
340	4888.557	II
340	5105.58	II
500	5107.55	II
425	5231.38	II
500	5331.23	II
340	5497.727	II
425	5558.09	II
425	5651.32	II
425	6110.07	II
500	6170.27	II
300	6511.74	II
300	7092.27	II
300	7102.72	II
340	7990.53	II
300	8174.51	II
200	9300.61	I
230	9597.95	I
290	9626.70	I
230	9833.76	I
170	9915.71	I
290	9923.05	I
290	10024.04	I
170	10614.07	I

***Astatine* At Z = 85**

Intensity	Wavelength/Å	
8	2162.25	I
10	2244.01	I

***Barium* Ba Z = 56**

Intensity	Wavelength/Å	
14	555.48	III
14	587.57	III
18	647.27	III
300	719.86	V
150	721.85	V
1000	766.87	V
40000	794.89	IV
300	877.41	V
50000	923.74	IV
200	946.26	V
200	1486.72	II
400	1504.01	II
300	1554.38	II
200	1572.73	II
	1573.92	II
	1630.40	II
100	1674.51	II
400	1694.37	II
	1697.16	II
	1761.75	II
	1771.03	II
	1786.93	II
100	1904.15	II
500	1924.70	II
	1985.60	II
300	1999.54	II
10	2001.30	III
	2009.20	II
400	2023.95	II
	2052.68	II
	2054.57	II
500	2214.7	II
800	2245.61	II
1000	2254.73	II
1400	2304.24	II
60	2331.10	III
2000	2335.27	II
190	2347.58	II
40	2512.28	III
40	2523.83	III
60	2528.51	II
50	2559.54	III
8 h	2596.64	I
100	2634.78	II
40	2681.89	III
8	2702.63	I
18	2771.36	II
15	2785.28	I
100 r	3071.58	I
40	3079.14	III
10 h	3108.21	I
8	3132.60	I
8 h	3135.72	I
10	3137.70	I
10	3155.34	I
10	3155.67	I
12	3158.05	I
12 h	3158.54	I
25	3165.60	I
15 h	3173.69	I
30	3183.16	I
15	3183.96	I
10	3193.91	I
25 h	3203.70	I
30	3221.63	I
40	3222.19	I
50	3261.96	I
60 r	3262.34	I
40	3281.50	I
15	3281.77	I
50	3322.80	I
80 h	3356.80	I
50	3368.18	III
60 r	3377.08	I
20	3377.39	I
70 r	3420.32	I
25	3421.01	I
30 h	3421.48	I
40	3463.74	I
200 r	3501.11	I
80 h	3524.97	I
30 h	3531.35	I
80 h	3544.66	I
20 h	3547.68	I
100	3552.45	II
200	3567.73	II
100	3576.28	II
30	3577.62	I
80 h	3579.67	I
200	3596.57	II
40	3630.64	I
40 h	3636.83	I
20 h	3688.47	I
400	3735.75	II
200	3816.69	II
200	3842.80	II
100	3854.76	II
20	3889.33	I
1400 l	3891.78	II
20	3892.65	I
40	3909.91	I
500	3914.73	II
25	3926.85	III
50	3935.72	I
20	3937.87	I
200	3939.67	II
500	3949.51	II
25	3993.06	III
80	3993.40	I
30	3995.66	I
300	4036.26	II
200	4083.77	II
30 h	4084.86	I
1500 h	4130.66	II
20	4132.43	I
200	4166.00	II
500	4216.04	II
800	4267.95	II
100	4283.10	I
300	4287.80	II
200	4297.60	II
800	4309.32	II
20 h	4323.00	I

Intensity	Wavelength/Å	
600	4325.73	II
200	4326.74	II
300	4329.62	II
80	4350.33	I
60	4402.54	I
400	4405.23	II
40	4431.89	I
60 h	4488.98	I
50 h	4493.64	I
40	4505.92	I
200	4509.63	II
60 h	4523.17	I
130	4524.93	II
65000	4554.03	II
40	4573.85	I
80	4579.64	I
30	4599.75	I
20 h	4619.92	I
25 h	4628.33	I
300	4644.10	II
30	4673.62	I
35	4691.62	I
20	4700.43	I
800	4708.94	II
40	4726.44	I
800	4843.46	II
300	4847.14	II
200	4850.84	II
30 h	4877.65	I
400	4899.97	II
15	4902.90	I
20000	4934.09	II
8	4947.35	I
1000	4957.15	II
300	4997.81	II
1000	5013.00	II
20 h	5159.94	I
20	5267.03	I
800	5361.35	II
1000	5391.60	II
200	5421.05	II
100	5424.55	I
200	5428.79	II
300	5480.30	II
200	5519.05	I
1000 r	5535.48	I
20 h	5620.40	I
10	5680.18	I
400	5777.62	I
800	5784.18	II
100	5800.23	I
20	5805.69	I
150	5826.28	I
2800	5853.68	II
15	5907.64	I
100	5971.70	I
800	5981.25	II
100	5997.09	I
300	5999.85	II
100	6019.47	I
200	6063.12	I
300	6110.78	I
400	6135.83	II
20000	6141.72	II
150	6341.68	I
500	6378.91	II
10	6383.76	III
90	6450.85	I
150	6482.91	I
12000	6496.90	II
300	6498.76	I
150	6527.31	I
3000	6595.33	I
150	6654.10	I
1500	6675.27	I
1800	6693.84	I
1000	6769.62	II
600	6865.69	I
300 h	6867.85	I
1000	6874.09	II
6000	7059.94	I
2400 hs	7120.33	I
600	7195.24	I
600 hl	7228.84	I
3000	7280.30	I
1200	7392.41	I
300	7417.53	I
900 hl	7459.78	I
600	7488.08	I
450 hl	7636.90	I
600 hl	7642.91	I
1800	7672.09	I
1200	7780.48	I
180 h	7839.57	I
1500	7905.75	I
600	7911.34	I
900 h	8210.24	I
8	8308.69	III
1800 h	8559.97	I
100	8710.74	II
100	8737.71	II
300 h	8799.76	I
300	8860.98	I
450	8914.99	I
300	9219.69	I
300	9308.08	I
300 h	9324.58	I
1500	9370.06	I
300	9455.92	I
8	9521.76	III
450	9589.37	I
900	9608.88	I
300 h	9645.72	I
1500 hl	9830.37	I
900	10001.08	I
600	10032.10	I
1200 h	10233.23	I
300	10471.26	I
120 hl	10791.25	I
180 hl	11012.69	I
150 h	11114.42	I
240	11303.04	I
120 h	11697.45	I
120	13207.30	I
120	13810.50	I
120	14077.90	I
120	15000.40	I
120	20712.00	I
150	25515.70	I
150	29223.90	I

Beryllium Be $Z = 4$

Intensity	Wavelength/Å	
	58.13	IV
	58.57	IV
	59.32	IV
	60.74	IV
	64.06	IV
	75.93	IV
1 h	76.10	III
2	76.48	III
3	78.53	III
4	78.66	III
1 h	78.92	III
5	81.89	III
10	82.38	III
	82.58	II
20	83.20	III
	83.66	II
30	84.76	III
50	88.31	III
	89.16	I
	89.80	II
	90.04	II
	90.21	I
	90.67	I
	91.06	II
	91.36	II
	91.74	II
	92.19	I
	92.61	II
	93.14	II
	93.42	II
	93.93	II
	94.78	II
	95.76	II
	96.29	I
	97.24	I
	97.44	I
	97.86	I
	97.97	I
	98.12	I
	98.37	I
	98.66	I
	98.94	I
	99.19	I
100	100.25	III
	100.86	I
	101.20	I
	102.13	I
	102.49	II
	104.40	II
	104.67	I
	105.80	I
	107.26	I
	107.38	I
3	509.99	III
2	549.31	III
6	582.08	III
4	661.32	III
8	675.59	III
	714.0	II
4	725.59	III
5	725.71	II
5	743.58	II
7	746.23	III
2	767.75	III
8	775.37	II
20	842.06	II
	865.3	II
2	925.25	II
10	943.56	II
10	973.27	II
	981.4	II
	1020.1	II
8	1026.93	II
5	1036.32	II
15	1048.23	II
1	1114.69	III
20	1143.03	II
	1155.9	II
60	1197.19	II
2	1213.12	III
1	1214.32	III
2	1362.25	III
1	1401.52	III
10	1421.26	III
5	1422.86	III
	1426.12	I
1	1435.17	III
2	1440.77	III
	1491.76	I
20	1512.30	II
60	1512.43	II
100	1661.49	I
2 h	1754.69	III
15	1776.12	II
20	1776.34	II
	1907.	I
	1909.0	II
	1912.	I
3	1917.03	III
	1919.	I
5	1929.67	I
10	1943.68	I
60 h	1954.97	III
	1956.	I
50	1964.59	I
5	1985.13	I
	1997.95	I
	1997.98	I
60	1998.01	I
	2033.25	I
	2033.28	I
	2033.38	I
50	2055.90	I

Intensity	Wavelength/Å	
100	2056.01	I
75 h	2076.94	III
60 h	2080.38	III
25	2118.56	III
15 h	2122.27	III
10	2125.57	I
20	2125.68	I
15 h	2127.20	III
5	2137.25	III
25	2145.	I
55	2174.99	I
55	2175.10	I
5	2191.57	III
	2273.5	II
	2324.6	II
	2337.0	I
950	2348.61	I
20	2350.66	I
60	2350.71	I
200	2350.83	I
2	2413.34	II
16	2413.46	II
20	2453.84	II
	2480.6	I
35	2494.54	I
35	2494.58	I
100	2494.73	I
16	2507.43	II
5	2617.99	II
20	2618.13	II
100	2650.45	I
60	2650.55	I
200	2650.62	I
60	2650.69	I
100	2650.76	I
5	2697.46	II
20	2697.58	II
20	2728.88	II
30	2738.05	I
	2764.2	II
20	2898.13	I
10	2898.19	I
20	2898.25	I
30	2986.06	I
10	2986.42	I
60	3019.33	I
30	3019.49	I
30	3019.53	I
20	3019.60	I
10	3046.52	II
30	3046.69	II
	3090.3	I
10	3110.81	I
10	3110.92	I
20	3110.99	I
	3120.	I
480	3130.42	II
320	3131.07	II
	3136.	I
	3150.	I
	3160.6	I
	3163.	I
	3168.	I
	3180.7	II
	3187.	I
20	3193.81	I
20	3197.10	II
30	3197.15	II
20	3208.60	I
	3220.	I
60	3229.63	I
2	3233.52	II
10	3241.62	II
30	3241.83	II
15	3269.02	I
100	3274.58	II
30	3274.67	II
30	3282.91	I
30	3321.01	I
30	3321.09	I
220	3321.34	I
20	3345.43	I
60	3367.63	I
	3405.6	II
5	3451.37	I
300	3455.18	I
20	3476.56	I
300	3515.54	I
10	3555.	I
100	3720.36	III
	3720.92	III
	3722.98	III
100	3736.30	I
700	3813.45	I
40	3865.13	I
80	3865.42	I
1	3865.51	I
6	3865.72	I
100	3866.03	I
90 h	4249.14	III
100	4253.05	I
60	4253.76	I
300	4360.66	II
500	4360.99	II
400	4407.94	I
2	4485.52	III
100 h	4487.30	III
1	4495.09	III
140 h	4497.8	III
	4526.6	I
	4548.	I
12	4572.66	I
700	4673.33	II
1000	4673.42	II
6	4709.37	I
200	4828.16	II
40	4849.16	I
2 h	4858.22	II
80	5087.75	I
8	5218.12	II
20	5218.33	II
3	5255.86	II
64	5270.28	II
500	5270.81	II
20	5403.04	II
20	5410.21	II
	5558.	I
140 h	6142.01	III
10	6229.11	I
16	6279.43	II
30	6279.73	II
30	6473.54	I
60	6547.89	II
60	6558.36	II
30	6564.52	I
2 h	6636.44	II
1	6756.72	II
2	6757.13	II
30	6786.56	I
1 h	6884.22	I
6 h	6884.44	I
100	6982.75	I
6 h	7154.40	I
40 h	7154.65	I
100	7209.13	I
3	7401.20	II
2	7401.43	II
10	7551.90	I
10 h	7618.68	I
20 h	7618.88	I
60	8090.06	I
5 h	8158.99	I
10 h	8159.24	I
4	8254.07	I
10 h	8287.07	I
30	8547.36	I
60	8547.67	I
300	8801.37	I
6	8882.18	I
40	9190.45	I
20 h	9243.92	I
1 h	9343.89	II
40	9392.74	I
2	9476.43	II
16	9477.03	II
20	9847.32	I
10 h	9895.63	I
20 h	9895.96	I
80	9939.78	I
16	10095.52	II
20	10095.73	II
60	10119.92	II
80	10331.03	I
30	11066.46	I
	11173.	II
1	11173.73	II
120	11496.39	I
2 h	11625.16	II
	11659.	II
2	11660.25	II
100	12095.36	II
30	12098.18	II
100	14643.92	I
60	14644.75	I
200	16157.72	I
80	17855.38	I
120	17856.63	I
100	18143.54	I
160	31775.05	I
200	31778.70	I
Bismuth Bi Z = 83		
6	420.7	IV
6	431.2	IV
2	488.39	V
3	563.62	V
5	670.76	III
6	686.88	V
5	730.71	V
10	738.17	V
4	775.16	III
6	790.5	IV
6	790.6	IV
8	792.5	IV
10	820.3	IV
9	822.9	IV
12	824.9	IV
15 d	864.45	V
15	872.6	IV
12	923.9	IV
15	943.3	IV
25	1039.99	III
50 h	1045.76	III
30	1051.81	III
15	1058.88	II
20	1085.47	II
10	1099.20	II
24	1103.4	IV
20	1139.01	III
50	1224.64	III
10	1225.43	II
15	1232.78	II
10	1241.05	II
10	1265.35	II
15	1283.73	II
10	1306.18	II
60	1317.0	IV
20	1325.46	II
40	1326.84	III
20	1329.47	II
60	1346.12	III
20	1350.07	II
25	1372.61	II
15	1376.02	II
20	1393.92	II
35	1423.33	III
35	1423.52	III
45	1436.83	II
25	1447.94	II
50	1455.11	II
60 h	1461.00	III
25	1462.14	II
35	1486.93	II
20	1502.50	II
40	1520.57	II

Intensity	Wavelength/Å	
40	1533.17	II
30	1536.77	II
35	1538.06	II
20	1563.67	II
40	1573.70	II
60	1591.79	II
25	1601.58	II
60 h	1606.40	III
40	1609.70	II
40	1611.38	II
20	1652.81	II
20	1749.29	II
80	1777.11	II
60	1787.47	II
70	1791.93	II
70	1823.80	II
100	1902.41	II
9000	1954.53	I
7000	1960.13	I
25	1989.35	II
7000	2021.21	I
9000	2061.70	I
45 h	2068.9	II
4600	2110.26	I
2500	2133.63	I
15	2143.40	II
15	2143.46	II
60	2186.9	II
40 h	2214.0	II
360	2228.25	I
1700	2230.61	I
340	2276.58	I
100	2311.	IV
100	2326.	IV
16	2368.12	II
12	2368.25	II
100	2376.	IV
190	2400.88	I
75 h	2414.6	III
10	2501.0	II
25	2515.69	I
70	2524.49	I
20 h	2544.5	II
700	2627.91	I
100	2629.	IV
100	2677.	IV
12	2693.0	II
280 c	2696.76	I
20	2713.3	II
140 d	2730.50	I
100	2767.	IV
100	2772.	IV
360	2780.52	I
100	2786.	IV
15	2803.42	II
11	2803.70	II
12	2805.3	II
140 c	2809.62	I
100	2842.	IV
80 h	2855.6	III
4000	2897.98	I

Intensity	Wavelength/Å	
100	2924.	IV
100	2933.	IV
100	2936.	IV
15	2936.7	II
3200	2938.30	I
20	2950.4	II
12	2963.4	II
2800	2989.03	I
700	2993.34	I
100	3012.	IV
2400	3024.64	I
60	3034.87	I
100	3042.	IV
9000 c	3067.72	I
140	3076.66	I
35	3115.0	III
100	3239.	IV
550 c	3397.21	I
10	3430.83	II
12	3431.23	II
40 h	3451.0	III
40	3473.8	III
35	3485.5	III
500 c	3510.85	I
380 c	3596.11	I
45	3613.4	III
100	3643.	IV
12	3654.2	II
100	3682.	IV
50	3695.32	III
50	3695.68	III
100	3734.	IV
70 h	3792.5	II
12	3811.1	II
20	3815.8	II
10	3845.8	II
30	3863.9	II
100	3868.	IV
40 h	4079.1	II
10	4097.2	II
140	4121.53	I
140	4121.86	I
75 h	4259.4	II
25	4272.0	II
70 h	4301.7	II
12 h	4339.8	II
25 h	4340.5	II
12 h	4379.4	II
25 h	4476.8	II
60 h	4705.3	II
600 c	4722.52	I
30	4730.3	II
20	4749.7	II
40 h	4797.4	III
12	4908.2	II
10	4916.6	II
12	4969.7	II
20	4993.6	II
45 h	5079.3	III
10	5091.6	II
50 h	5124.3	II

Intensity	Wavelength/Å	
60 h	5144.3	II
20	5201.5	II
75 h	5209.2	II
40 h	5270.3	II
10	5397.8	II
10 c	5552.35	I
3	5599.41	I
20	5655.2	II
40 h	5719.2	II
6	5742.55	I
12	5818.3	II
20	5860.2	II
20	5973.0	II
15	6059.1	II
15	6128.0	II
6	6134.82	I
3	6475.73	I
3	6476.24	I
15	6497.7	II
10	6577.2	II
40 h	6600.2	II
50 h	6808.6	II
4 h	6991.12	I
12	7033.	II
2	7036.15	I
10 h	7381.	II
2	7502.33	I
10 h	7637.	II
10	7750.	II
3	7838.70	I
2	7840.33	I
20	7965.	II
40	8008.	III
12 h	8050.	II
50	8070.	III
15	8328.	II
15	8388.	II
30	8532.	II
2	8544.54	I
1	8579.74	I
25	8653.	II
2	8754.88	I
3	8761.54	I
25	8863.	II
2	8907.81	I
2000 d	9657.04	I
40	9827.78	I
20	10104.5	I
15	10138.8	I
20	10300.6	I
20	10536.19	I
50	11072.44	I
1500 d	11710.37	I
40	11999.49	I
200	12165.08	I
200	12690.04	I
100	12817.8	I
200	14330.5	I
50	16001.5	I
60	22551.6	I

Intensity	Wavelength/Å	
Boron B Z = 5		
	41.00	V
30	48.59	V
10	52.68	IV
30	60.31	IV
	194.37	V
	262.37	V
160	344.0	IV
450	385.0	IV
40	411.80	III
285	418.7	IV
20	510.77	III
40	510.85	III
	512.53	V
150	518.24	III
75	518.27	III
110	677.00	III
160	677.14	III
40	693.95	II
40	731.36	II
40	731.44	II
	749.74	V
40	758.48	III
70	758.67	III
110	882.54	II
110	882.68	II
40	984.67	II
110	1081.88	II
110	1082.07	II
70	1112.2	IV
450	1168.9	IV
70	1170.9	IV
110	1230.16	II
220	1362.46	II
70	1600.46	I
120	1600.73	I
160	1623.58	II
110	1623.77	II
220	1624.02	II
70	1624.16	II
160	1624.34	II
100	1663.04	I
150	1666.87	I
200	1667.29	I
150	1817.86	I
200	1818.37	I
300	1825.91	I
300	1826.41	I
110	1842.81	II
20	1953.83	III
550	2065.78	III
250	2066.38	I
250	2066.65	I
100	2066.93	I
300	2067.19	I
450	2067.23	III
160	2077.09	III
500	2088.91	I
500	2089.57	I
70	2220.30	II
40	2234.09	III

Intensity	Wavelength/Å	
70	2234.59	III
40	2323.03	II
40	2328.67	II
40	2393.20	II
220	2395.05	II
40	2459.69	II
40	2459.90	II
1000	2496.77	I
1000	2497.73	I
70	2524.7	IV
160	2530.3	IV
450	2821.68	IV
70	2824.57	IV
285	2825.85	IV
160	2918.08	II
110	3032.26	II
70	3179.33	II
110	3323.18	II
110	3323.60	II
450	3451.29	II
285	4121.93	II
110	4194.79	II
40	4242.98	III
70	4243.61	III
110	4472.10	II
110	4472.85	II
220	4487.05	III
360	4497.73	III
70	4784.21	II
110	4940.38	II
110	6080.44	II
70	6285.47	II
70	7030.20	II
40	7031.90	II
110	7835.25	III
70	7841.41	III
20	8667.22	I
70	8668.57	I
800	11660.04	I
570	11662.47	I
125	15629.08	I
200	16240.38	I
250	16244.67	I
235	18994.33	I
***Bromine* Br**	***Z*** **= 35**	
700	379.73	IV
700	400.37	IV
800	482.11	V
900	531.97	V
1000	545.43	IV
1000	547.90	V
1000	559.76	IV
1000	569.19	IV
1000	576.59	IV
1000	585.10	IV
1000	586.71	IV
1000	597.51	IV
1000	600.09	IV
1000	601.27	IV
1000	607.03	IV
1000	617.85	IV
1000	619.87	IV
1000	630.14	IV
1000	642.23	IV
1000	661.53	IV
1000	683.51	IV
1000	697.72	IV
1000	715.39	IV
1000	731.00	IV
1000	800.12	IV
700	812.95	V
1000	813.66	IV
1000	850.81	V
1000	889.23	II
1000	948.97	II
1000	1015.54	II
1000	1049.00	II
1000	1069.15	V
900	1112.13	V
1000	1143.56	V
1000	1189.28	I
1000	1189.50	I
1000	1210.73	I
1000	1221.13	I
1000	1223.24	I
1200	1224.41	I
1200	1226.90	I
7500	1232.43	I
1200	1243.90	I
1500	1251.66	I
1000	1255.80	I
1500	1259.20	I
1200	1261.66	I
1200	1266.20	I
1000	1279.48	I
1000	1286.26	I
3000	1309.91	I
3000	1316.74	I
1000	1317.37	I
2000	1317.70	I
12000	1384.60	I
3000	1449.90	I
50000	1488.45	I
30000	1531.74	I
25000	1540.65	I
30000	1574.84	I
20000	1576.39	I
25000	1582.31	I
75000	1633.40	I
1000	2133.79	IV
1000	2145.02	IV
1000	2257.21	IV
1000	2272.73	IV
1000	2307.40	IV
1000	2408.16	IV
1000	2411.58	IV
700	2491.14	IV
1000	2581.19	IV
600	2661.40	IV
1000	2842.88	IV
1100 h	2907.71	IV
500 h	2972.26	II
500	3041.18	IV
500	3074.42	III
500	3349.64	III
500	3380.56	IV
500	3540.16	III
500	3562.43	III
1200	3815.65	I
1500	3992.36	I
1000	4223.89	II
2000	4365.14	I
1000	4365.60	II
1500	4425.14	I
10000	4441.74	I
10000	4472.61	I
20000	4477.72	I
1000	4490.42	I
3000	4513.44	I
15000	4525.59	I
3000	4575.74	I
2500	4614.58	I
2500	4752.28	I
4000	4780.31	I
1600	4785.19	I
4000	4979.76	I
1200	5395.48	I
1200	5466.22	I
1800	5852.08	I
1600	5940.48	I
2400	6122.14	I
40000	6148.60	I
2000	6177.39	I
1500	6335.48	I
60000	6350.73	I
2500	6410.32	I
1800	6483.56	I
1000	6514.62	I
20000	6544.57	I
1500	6548.09	I
50000 c	6559.80	I
1000	6571.31	I
1800	6579.14	I
20000	6582.17	I
1500	6620.47	I
50000 c	6631.62	I
20000	6682.28	I
10000	6692.13	I
8000	6728.28	I
2000	6760.06	I
2000	6779.48	I
2200	6786.74	I
6500	6790.04	I
1600 c	6791.48	I
1800	6861.15	I
10000	7005.19	I
2000	7260.45	I
10000	7348.51	I
40000	7512.96	I
1600	7591.61	I
1800	7595.07	I
2000	7616.41	I
30000	7803.02	I
1200	7827.23	I
2500 s	7881.45	I
2500	7881.57	I
2500	7925.81	I
30000 c	7938.68	I
3000	7947.94	I
3000	7950.18	I
8000	7978.44	I
10000	7978.57	I
30000	7989.94	I
2000	8026.35	I
2500	8026.54	I
30000	8131.52	I
1000 c	8152.65	I
10000	8153.75	I
25000	8154.00	I
5000	8246.86	I
15000	8264.96	I
75000 c	8272.44	I
20000	8334.70	I
10000	8343.70	I
1200	8384.04	I
40000	8446.55	I
4000	8477.45	I
1500	8513.38	I
1000	8557.73	I
1000	8566.28	I
20000	8638.66	I
4000	8698.53	I
10000 c	8793.47	I
15000	8819.96	I
25000	8825.22	I
4000	8888.98	I
30000	8897.62	I
6000	8932.40	I
1800	8949.39	I
9000	8964.00	I
30000	9166.06	I
15000	9173.63	I
20000	9178.16	I
40000	9265.42	I
15000	9320.86	I
6000	9793.48	I
10000	9896.40	I
3000	10140.08	I
6000	10237.74	I
1000	10299.62	I
1500	10377.65	I
30000	10457.96	I
1000	10742.14	I
3000	10755.92	I
1700	13217.17	I
1800	14354.57	I
1250	14888.70	I
1800	16731.19	I
1200	18568.31	I
3500	19733.62	I
1000	20281.73	I
1000	20624.67	I
1200	21787.24	I
4000	22865.65	I

Intensity	Wavelength/Å	
1000	23513.15	I
500	28346.50	I
500	30380.85	I
600	31630.13	I
150	38345.75	I
120	39964.36	I

Cadmium Cd Z = 48

Intensity	Wavelength/Å	
50	427.01	IV
50	447.85	IV
60	480.90	IV
70	493.00	IV
70	495.13	IV
70	498.14	IV
70	498.53	IV
80	504.09	IV
70	504.20	IV
70	504.50	IV
80	506.31	IV
60	508.01	IV
50	508.95	IV
70	509.55	IV
70	511.40	IV
80	513.00	IV
70	514.50	IV
60	519.42	IV
80	524.41	IV
70	524.47	IV
70	525.10	IV
60	525.19	IV
70	527.07	IV
80	531.09	IV
80	531.51	IV
70	534.29	IV
70	536.77	IV
60	540.90	IV
70	541.74	IV
80	542.60	IV
80	546.55	IV
60	553.06	IV
80	554.05	IV
60	567.01	IV
150	1118.16	IV
100	1164.65	IV
100	1183.40	IV
100	1256.00	II
150	1296.43	II
100	1326.50	II
60	1370.48	IV
150	1370.91	II
60	1418.89	IV
200	1514.26	II
50	1545.17	III
200	1571.58	II
100	1668.60	II
50	1702.47	II
40	1707.16	III
40	1722.95	III
50	1724.41	II
40	1747.67	III
40	1773.06	III
100	1785.84	II
75	1793.40	III
40	1823.41	III
100	1827.70	II
50	1844.66	III
40	1851.13	III
40	1855.85	III
200	1856.67	III
150	1874.08	III
300	1922.23	II
100	1943.54	II
40	1965.54	II
30	1986.89	II
200	1995.43	II
100	2007.49	II
50	2032.45	II
75	2036.23	II
40	2039.83	III
50	2045.61	III
75	2087.91	III
150	2096.00	II
50	2111.60	III
1000 r	2144.41	II
50	2155.06	II
100	2187.79	II
1000	2194.56	II
1000	2265.02	II
1500 r	2288.022	I
1000	2312.77	II
200	2321.07	II
40	2376.82	II
50	2418.69	II
50	2469.73	II
40	2487.93	II
40	2495.58	II
50	2509.11	II
30	2516.22	II
25 h	2525.196	I
50	2544.613	I
50	2551.98	II
25	2553.465	I
3	2565.789	I
500	2572.93	II
50	2580.106	I
30	2592.026	I
25 h	2602.048	I
50	2628.979	I
40	2632.190	I
75	2639.420	I
40	2659.23	II
50 h	2660.325	I
25	2668.20	II
50	2672.62	II
100	2677.540	I
25	2677.748	I
50	2707.00	II
75	2712.505	I
50	2733.820	I
1000	2748.54	II
100 h	2763.894	I
50 h	2764.230	I
50	2774.958	I
30	2823.19	II
200	2836.900	I
25	2856.46	II
100	2868.180	I
200 r	2880.767	I
50 r	2881.224	I
200	2914.67	II
50	2927.87	II
200	2929.27	II
1000 r	2980.620	I
200 r	2981.362	I
50	2981.845	I
50	3030.60	II
150	3080.822	I
25	3081.48	II
30	3082.593	I
100	3092.34	II
200	3133.167	I
50	3146.79	II
150	3250.33	II
300	3252.524	I
300	3261.055	I
50	3343.21	II
50	3385.49	II
30	3388.88	II
800	3403.652	I
50	3417.49	II
50	3442.42	II
100	3464.43	II
1000	3466.200	I
800	3467.655	I
25	3483.08	II
150	3495.44	II
25	3499.952	I
100	3524.11	II
100	3535.69	II
1000	3610.508	I
800	3612.873	I
60	3614.453	I
20	3649.558	I
10	3981.926	I
100	4029.12	II
200	4134.77	II
50	4141.49	II
100	4285.08	II
8	4306.672	I
100	4412.41	II
3	4412.989	I
1000	4415.63	II
30	4440.45	II
8	4662.352	I
200	4678.149	I
30	4744.69	II
300	4799.912	I
50	4881.72	II
50	5025.50	II
1000 h	5085.822	I
6	5154.660	I
100	5268.01	II
100	5271.60	II
1000	5337.48	II
1000	5378.13	II
200	5381.89	II
40	5843.30	II
50	5880.22	II
300	6099.142	I
100	6111.49	I
100	6325.166	I
30	6330.013	I
400	6354.72	II
500	6359.98	II
2000	6438.470	I
400	6464.94	II
25	6567.65	II
500	6725.78	II
100	6759.19	II
30	6778.116	I
50	7237.01	II
100	7284.38	II
1000	7345.670	I
50	8066.99	II
5	8200.309	I
20	9289.	I
15	11652.	I
35	14487.	I
80	15708.	I
55 d	19120.	I
25	24371.	I
35	25448.	I

Calcium Ca Z = 20

Intensity	Wavelength/Å	
250	190.46	V
250	196.97	V
300	199.55	V
250	200.51	V
265	257.98	V
400	267.77	V
300	270.31	V
400	280.99	V
300	284.98	V
450 c	286.96	V
500	322.17	V
300	323.22	V
300	330.94	V
300	334.55	V
250 c	342.45	IV
250	343.93	IV
450	352.92	V
250	377.18	V
200	387.08	V
750	425.00	V
600	434.57	IV
250	437.77	IV
750	443.82	IV
500	450.57	IV
500	558.60	V
400	637.93	V
300	643.12	V
400	646.57	V
750	656.00	IV
300	656.76	V
500	669.70	IV
24	1341.89	II

Intensity	Wavelength/Å	
12	1342.54	II
20	1433.75	II
20	1545.29	III
60	1649.86	II
20	1807.34	II
40	1814.50	II
40	1838.01	II
60	1840.06	II
20	1843.09	II
40	1850.69	II
17	2123.03	III
16	2152.43	III
16	2687.76	III
19	2881.78	III
21	2899.79	III
19	2924.33	III
20	2988.63	III
10	3006.86	I
15	3028.59	III
3	3055.32	I
19	3119.67	III
170	3158.87	II
180	3179.33	II
150	3181.28	II
20	3316.51	II
12	3361.92	I
19	3372.67	III
20	3461.87	II
13	3487.60	I
18	3537.77	III
20	3644.41	I
30	3683.70	II
40	3694.11	II
170	3706.03	II
180	3736.90	II
20	3755.67	II
30	3758.39	II
230	3933.66	II
220	3968.47	II
50	4097.10	II
60	4109.82	II
30	4110.28	II
40	4206.18	II
50	4220.07	II
50	4226.73	I
24	4283.01	I
22	4289.36	I
22	4298.99	I
25	4302.53	I
20	4302.81	III
23	4307.74	I
22	4318.65	I
20	4355.08	I
19	4399.59	III
25	4425.44	I
26	4434.96	I
25	4435.69	I
30	4454.78	I
28	4455.89	I
20	4456.61	I
20	4472.04	II
20	4489.18	II
19	4499.88	III
23	4526.94	I
22	4578.55	I
23	4581.40	I
23	4581.47	I
24	4585.87	I
24	4585.96	I
20	4685.27	I
30	4716.74	II
40	4721.03	II
40	4799.97	II
25	4878.13	I
70	5001.48	II
80	5019.97	II
40	5021.14	II
23	5041.62	I
25	5188.85	I
22	5261.71	I
23	5262.24	I
22	5264.24	I
24	5265.56	I
25	5270.27	I
60	5285.27	II
70	5307.22	II
50	5339.19	II
27	5349.47	I
23	5512.98	I
25	5581.97	I
27	5588.76	I
24	5590.12	I
26	5594.47	I
25	5598.49	I
24	5601.29	I
24	5602.85	I
30	5857.45	I
27	6102.72	I
29	6122.22	I
22	6161.29	I
30	6162.17	I
22	6163.76	I
24	6166.44	I
26	6169.06	I
28	6169.56	I
35	6439.07	I
30	6449.81	I
22	6455.60	I
80	6456.87	II
34	6462.57	I
29	6471.66	I
32	6493.78	I
28	6499.65	I
23	6572.78	I
30	6717.69	I
33	7148.15	I
31	7202.19	I
33	7326.15	I
30	7575.81	II
60	7581.11	II
80	7601.30	II
20	7602.32	II
40	7820.78	II
60	7843.38	II
20	8017.50	II
20	8020.50	II
70	8133.05	II
100	8201.72	II
110	8248.80	II
70	8254.73	II
130	8498.02	II
170	8542.09	II
160	8662.14	II
100	8912.07	II
110	8927.36	II
110	9213.90	II
90	9312.00	II
100	9319.56	II
110	9320.65	II
25	9416.97	I
100	9567.97	II
110	9599.24	II
80	9601.82	II
80	9854.74	II
110	9890.63	II
90	9931.39	II
100	10223.04	II
20	10343.81	I
20	11838.99	II
25	12816.04	I
24	12823.86	I
25	12909.10	I
30	13033.57	I
21	13086.44	I
24	13134.95	I
20	16150.77	I
22	16157.36	I
21	16197.04	I
20	18925.47	I
24	18970.14	I
30	19046.14	I
48	19309.20	I
49	19452.99	I
47	19505.72	I
50	19776.79	I
35	19853.10	I
34	19862.22	I
23	19917.19	I
24	19933.70	I
25	22624.93	I
30	22651.23	I

Carbon C Z = 6

Intensity	Wavelength/Å	
110	34.973	V
450	40.268	V
110	227.19	V
250	244.91	IV
160	248.66	V
160	248.74	V
200	289.14	IV
250	289.23	IV
570	312.42	IV
500	312.46	IV
250	371.69	III
250	371.75	III
150	371.78	III
650	384.03	IV
700	384.18	IV
500	386.203	III
400	419.52	IV
500	419.71	IV
200	450.734	III
400	459.46	III
500	459.52	III
570	459.63	III
250	511.522	III
250	535.288	III
300	538.080	III
350	538.149	III
400	538.312	III
350	574.281	III
9	595.022	II
30	687.053	II
50	687.345	II
10	858.092	II
20	858.559	II
30	903.624	II
60	903.962	II
150	904.142	II
30	904.480	II
800	977.03	III
9	1009.86	II
10	1010.08	II
10	1010.37	II
80	1036.337	II
150	1037.018	II
150	1157.910	I
150	1158.019	I
150	1158.035	I
370	1174.93	III
350	1175.26	III
330	1175.59	III
500	1175.71	III
350	1175.99	III
370	1176.37	III
150	1188.992	I
150	1189.447	I
200	1189.631	I
300	1193.009	I
300	1193.031	I
300	1193.240	I
300	1193.264	I
100	1193.393	I
150	1193.649	I
150	1193.679	I
100	1194.064	I
100	1194.488	I
100	1261.552	I
250	1277.245	I
250	1277.282	I
300	1277.513	I
300	1277.550	I
200	1280.333	I
100	1311.363	I
9	1323.951	II

Intensity	Wavelength/Å	
120	1329.578	I
120	1329.600	I
150	1334.532	II
300	1335.708	II
100	1354.288	I
150	1355.84	I
120	1364.164	I
100	1459.032	I
200	1463.336	I
120	1467.402	I
150	1481.764	I
1000	1548.202	IV
900	1550.774	IV
150	1560.310	I
400	1560.683	I
400	1560.708	I
100	1561.341	I
400	1561.438	I
150	1656.266	I
120	1656.928	I
300	1657.008	I
120	1657.380	I
120	1657.907	I
150	1658.122	I
500	1751.823	I
1000	1930.905	I
250	2162.94	III
40	2270.91	V
5	2277.25	V
20	2277.92	V
800	2296.87	III
800	2478.56	I
250	2509.12	II
350	2512.06	II
200 l	2524.41	IV
300 s	2529.98	IV
250 h	2574.83	II
150	2697.75	III
110 l	2724.85	III
150 l	2725.30	III
150 l	2725.90	III
350 l	2741.28	II
250	2746.49	II
1000	2836.71	II
800	2837.60	II
200	2982.11	III
800 h	2992.62	II
350	3876.19	II
350	3876.41	II
350	3876.66	II
570	3918.98	II
800	3920.69	II
150	4056.06	III
200	4067.94	III
250	4068.91	III
250	4070.26	III
250	4074.52	II
350 l	4075.85	II
150	4162.86	III
250 h	4186.90	III
800	4267.00	II

Intensity	Wavelength/Å	
1000	4267.26	II
200	4325.56	III
600	4647.42	III
520	4650.25	III
375	4651.47	III
200 w	4658.30	IV
200	4665.86	III
200	4771.75	I
200	4932.05	I
5	4943.88	V
5	4944.56	V
200	5052.17	I
350	5132.94	II
350	5133.28	II
350	5143.49	II
570	5145.16	II
400	5151.09	II
300	5380.34	I
250	5648.07	II
350	5662.47	II
450	5695.92	III
250	5801.33	IV
200	5811.98	IV
150	5826.42	III
570	5889.77	II
350	5891.59	II
200	6001.13	I
250	6006.03	I
110	6007.18	I
150	6010.68	I
300	6013.22	I
250	6014.84	I
800	6578.05	II
570	6582.88	II
200	6587.61	I
150	6744.38	III
250	6783.90	II
150 h	7037.25	III
250	7113.18	I
250	7115.19	I
250	7115.63	II
200	7116.99	I
350	7119.90	II
800	7231.32	II
1000	7236.42	II
150	7612.65	III
90 w	7726.2	IV
200	7860.89	I
200	8058.62	I
300 h	8196.48	III
150	8332.99	III
520	8335.15	I
300	8500.32	III
250	9061.43	I
200	9062.47	I
200	9078.28	I
250	9088.51	I
450	9094.83	I
300	9111.80	I
800	9405.73	I
150	9603.03	I

Intensity	Wavelength/Å	
250	9620.80	I
300	9658.44	I
200	10683.08	I
300	10691.25	I
12	11619.29	I
23	11628.83	I
13	11658.85	I
47	11659.68	I
24	11669.63	I
85	11748.22	I
142	11753.32	I
114	11754.76	I
11	11777.54	I
17	11892.91	I
30	11895.75	I
26	12614.10	I
20	13502.27	I
38	14399.65	I
16	14403.25	I
61	14420.12	I
12	14429.03	I
13	14442.24	I
12	16559.66	I
50	16890.38	I
10	17338.56	I
11	17448.60	I
13	18139.80	I
23	19721.99	I

Cerium Ce Z = 58

Intensity	Wavelength/Å	
300	399.36	V
200	482.96	V
40	741.79	IV
30	754.60	IV
75	1332.16	IV
75	1372.72	IV
100	2000.42	IV
100	2009.94	IV
10000	2318.64	III
10000	2372.34	III
10000	2380.12	III
10000	2431.45	III
15000	2439.80	III
10000	2454.32	III
10000	2469.95	III
10000	2483.82	III
10000	2497.50	III
20000	2531.99	III
10000	2603.59	III
340	2651.01	II
270	2830.90	II
250	2874.14	II
10000	2923.81	III
10000	2931.54	III
400	2976.91	II
10000	3022.75	III
50000	3031.58	III
95000	3055.59	III
20000	3056.56	III
40000	3057.23	III
20000	3057.58	III
680	3063.01	II
40000	3085.10	III
20000	3106.98	III
30000	3110.53	III
30000	3121.56	III
20000	3141.29	III
20000	3143.96	III
20000	3147.06	III
710	3194.83	II
990	3201.71	II
710	3218.94	II
880	3221.17	II
710	3227.11	II
20000	3228.57	III
710	3234.16	II
990	3272.25	II
20000	3353.29	III
10000	3395.77	III
30000	3427.36	III
40000	3443.63	III
30000	3454.39	III
40000	3459.39	III
60000	3470.92	III
710	3485.05	II
50000	3497.81	III
60000	3504.64	III
770	3539.08	II
50000	3544.07	III
1200	3560.80	II
1000	3577.45	II
1800	3655.85	II
880	3660.64	II
880	3667.98	II
1000	3709.29	II
1000	3709.93	II
1400	3716.37	II
800	3728.42	II
860	3786.63	II
2500	3801.52	II
800	3803.09	II
1000	3808.11	II
1100	3838.54	II
860	3848.59	II
860	3853.15	II
1200	3854.18	II
1200	3854.31	II
1100	3878.36	II
1500	3882.45	II
1000	3889.98	II
770	3907.29	II
980	3912.44	II
770	3918.28	II
770	3931.09	II
770	3940.34	II
2000	3942.15	II
2700	3942.75	II
770	3943.89	II
3100	3952.54	II
980	3956.28	II
770	3960.91	II
770	3967.05	II
770	3978.65	II

Intensity	Wavelength/Å	
770	3984.68	II
700	3992.39	II
910	3993.82	II
2800	3999.24	II
910	4003.77	II
2700	4012.39	II
910	4014.90	II
840	4024.49	II
840	4028.41	II
840	4031.34	II
2100	4040.76	II
910	4042.58	II
700	4053.51	II
1100	4071.81	II
1800	4073.48	II
1500	4075.71	II
1500	4075.85	II
910	4083.23	II
770	4118.14	II
980	4123.87	II
980	4127.37	II
2700	4133.80	II
2000	4137.65	II
770	4142.40	II
980	4149.94	II
1400	4151.97	II
1300	4165.61	II
3500	4186.60	II
840	4198.72	II
910	4202.94	II
1500	4222.60	II
770	4227.75	II
980	4239.92	II
1100	4248.68	II
2000	4289.94	II
1500	4296.67	II
770	4300.33	II
770	4306.72	II
980	4337.77	II
700	4349.79	II
910	4364.66	II
910	4382.17	II
700	4386.84	II
1700	4391.66	II
980	4418.78	II
770	4449.34	II
2400	4460.21	II
1400	4471.24	II
700	4479.36	II
700	4483.90	II
840	4486.91	II
770	4523.08	II
840	4527.35	II
840	4528.47	II
840	4539.75	II
2100	4562.36	II
1100	4572.28	II
840	4593.93	II
1700	4628.16	II
310	4737.28	II
470	5079.68	II
280	5159.69	I
280	5161.48	I
370	5187.46	II
260	5223.46	I
260	5245.92	I
340	5274.23	II
450	5353.53	II
300	5393.40	II
280	5409.23	II
260	5512.08	II
300	5696.99	I
370	5699.23	I
240	5719.03	I
230	5940.86	I
55	6001.90	I
55	6005.86	I
55	6006.82	I
75	6013.42	I
110	6024.20	I
10000	6032.54	III
110	6043.39	II
55	6047.40	I
10000	6060.91	III
45	6098.34	II
45	6123.67	I
35	6143.36	II
35	6186.17	I
35	6208.98	I
35	6228.94	I
23	6232.45	II
28	6237.45	I
45	6272.05	II
35	6295.58	I
28	6299.51	II
23	6300.21	I
35	6310.01	I
35	6343.95	II
35	6371.11	II
28	6386.84	I
23	6393.02	II
35	6430.07	I
23	6436.40	I
35	6458.03	I
28	6467.39	I
35	6473.72	I
23	6513.59	II
45	6555.65	I
23	6579.10	I
22	6612.06	I
30	6628.93	I
22	6652.72	II
26	6700.66	I
35	6704.27	I
30	6774.28	II
35	6775.59	I
30	6924.81	I
30	6986.02	I
35	7061.75	II
35	7086.35	II
22	7238.36	II
25	7252.75	I
25	7329.91	I
25	7397.77	I
25	7616.11	II
25	7689.17	II
22	7844.94	II
22	7857.54	II
30	8025.56	II
25	8772.14	II
30	8891.20	II

Cesium Ce Z = 55

Intensity	Wavelength/Å	
10000	614.01	III
2000	638.17	III
2500	666.25	III
5000	691.60	III
3500	703.89	III
15000	718.14	II
20000	721.79	III
20000	722.20	III
5000	731.56	III
12000	740.29	III
15000	808.76	II
15000	813.84	II
7500	830.39	III
35000	901.27	II
15000	920.35	III
40000	926.66	II
25000 c	1054.79	III
17 c	1673.99	III
12	1705.25	III
10	1801.83	III
20 c	1822.40	III
11	1823.93	III
12	1824.70	III
12	1841.80	III
25	1915.50	III
25 c	1923.29	III
12	1961.33	III
17	1996.56	III
710	2035.11	III
120	2056.43	III
330	2076.43	III
540	2077.30	III
410	2088.68	III
210	2101.63	III
200	2141.47	III
1000	2316.88	III
230	2325.95	III
390	2340.49	III
1600	2455.81	III
1600	2477.57	III
890	2485.45	III
410	2495.07	III
1400	2525.67	III
430	2573.05	III
16000	2596.86	III
390	2610.12	III
6200	2630.51	III
370	2700.32	III
710	2701.20	III
390	2776.44	III
270	2810.87	III
630	2845.70	III
3100	2859.32	III
200	2893.85	III
180	2921.13	III
3200	2976.86	III
210	3001.28	III
1700	3066.59	III
1100 c	3149.36	III
1400	3152.36	III
8400	3268.32	III
1300	3315.51	III
550	3340.60	III
430	3344.02	III
1200	3349.46	III
400	3463.45	III
580	3476.83	III
480	3559.82	III
7200	3597.45	III
1300	3608.31	III
2300	3618.19	III
300 c	3641.34	III
520	3651.08	III
4800	3661.40	III
640	3699.50	III
430	3837.46	III
2100 c	3876.15	I
2900	3888.37	III
600 c	3888.61	I
2700	3925.60	III
680 c	4001.70	III
3100	4006.55	III
420	4006.78	III
520	4043.42	III
14000	4264.70	II
18000 w	4277.13	II
370	4403.86	III
1200	4410.22	III
940	4425.68	III
530	4471.48	III
12000	4501.55	II
1200	4506.72	III
590	4522.86	III
20000	4526.74	II
1000 c	4555.28	I
460 c	4593.17	I
99900	4603.79	II
420 h	4620.61	III
210	4665.52	III
25000	4830.19	II
140	4851.59	III
19000	4870.04	II
37000	4952.85	II
370	5035.72	III
27000	5043.80	II
75000	5227.04	II
29000	5249.38	II
11000	5274.05	II
10000 c	5349.13	II
22000	5370.99	II
230	5380.79	III
60 c	5465.94	I

Intensity	Wavelength/Å	
37	5502.88	I
39000	5563.02	II
100	5635.21	I
210 c	5664.02	I
27	5745.72	I
24000	5831.14	II
59 c	5838.83	I
300	5845.14	I
51000	5925.63	II
140	5950.14	III
110	5979.97	III
640 c	6010.49	I
86	6034.09	I
150	6043.99	III
870	6079.86	III
9800	6128.61	II
330	6150.42	III
1000	6213.10	I
170	6217.60	I
450	6242.96	III
320 c	6354.55	I
510	6456.33	III
8300	6495.53	II
10000 w	6536.44	II
490	6586.51	I
97	6628.66	I
8800	6646.57	II
3300 c	6723.28	I
9600	6724.47	II
400	6753.12	III
200	6824.65	I
300	6870.45	I
37000	6955.50	II
4800	6973.30	I
16000	6979.67	II
980	6983.49	I
13000 w	7149.54	II
1900 c	7219.60	III
790	7228.53	I
130	7279.90	I
1100	7279.96	I
2600 c	7608.90	I
3300	7943.88	I
22000	7997.44	II
3500	8015.73	I
510	8078.94	I
4500	8079.04	I
59000 c	8521.13	I
15000 c	8761.41	I
61000 c	8943.47	I
18000	9172.32	I
5200	9208.53	I
19000	10024.36	I
4800	10123.41	I
26000	10123.60	I
2900	13424.31	I
38000 c	13588.29	I
8400	13602.56	I
5700	13758.81	I
55000 c	14694.91	I
820	16535.63	I
1500	17012.32	I
760	20138.47	I
880	22811.86	I
1100	23037.98	I
3900	23344.47	I
4400	24251.21	I
850	24374.96	I
890 d	25763.51	I
500	25764.73	I
680 c	29310.06	I
2800	30103.27	I
610 c	30953.06	I
1100	34900.13	I
190	36131.00	I
2 c	39177.28	I
2 d	39421.25	I
1	39424.11	I

Chlorine Cl ***Z = 17***

Intensity	Wavelength/Å	
500	392.43	V
800	486.17	IV
800	534.73	IV
700	535.67	IV
600	536.15	IV
900	537.61	IV
500	538.03	V
600	538.12	IV
800	542.23	V
600	542.30	V
1000	545.11	V
600	546.33	V
1000	547.63	V
500	549.22	IV
700	552.02	IV
600	553.30	IV
700	554.62	IV
600	556.23	III
700	556.61	III
700	557.12	III
350	559.305	II
700	561.53	III
700	561.68	III
700	561.74	III
400	571.904	II
800	574.406	II
500	601.50	IV
500	604.59	IV
500	606.35	III
700	618.057	II
600	619.982	II
800	620.298	II
700	626.735	II
800	635.881	II
1000	636.626	II
1000	650.894	II
1000	659.811	II
1300	661.841	II
2000	663.074	II
1500	682.053	II
1500	687.656	II
1500	693.594	II
2000	725.271	II
2500	728.951	II
2000	777.562	II
5000	787.580	II
5000	788.740	II
5000	793.342	II
500	834.84	IV
500	834.97	IV
6000	839.297	II
8000	839.599	II
600	840.93	IV
5000	851.691	II
2000	888.026	II
2000	893.549	II
2000	961.499	II
500	973.21	IV
600	977.56	IV
40	978.284	I
700	984.95	IV
25	998.372	I
25	998.432	I
75	1002.346	I
500	1005.28	III
600	1008.78	III
150	1013.664	I
700	1015.02	III
90	1025.553	I
6000	1063.831	II
3000	1067.945	II
9000	1071.036	II
6000	1071.767	II
5000	1075.230	II
5000	1079.080	II
200	1084.667	I
200	1085.171	I
250	1085.304	I
400	1088.06	I
350	1090.271	I
250	1090.982	I
250	1092.437	I
400	1094.769	I
350	1095.148	I
350	1095.662	I
400	1095.797	I
250	1096.810	I
300	1097.369	I
200	1098.068	I
200	1099.523	I
500	1107.528	I
800	1139.214	II
800	1167.148	I
3000	1179.293	I
1200	1188.774	I
900	1201.353	I
3000	1335.726	I
10000	1347.240	I
5000	1351.657	I
12000	1363.447	I
2500	1373.116	I
20000	1379.528	I
25000	1389.693	I
20000	1389.957	I
12000	1396.527	I
500	1441.470	II
500	1528.569	II
500	1542.942	II
500	1558.144	II
500	1565.050	II
600	1822.50	III
500	1828.40	III
500	1857.488	II
500	1901.61	III
500	1983.61	III
450 h	1997.370	II
450	2032.116	II
350 h	2088.583	II
350 h	2091.458	II
700	2253.07	III
500	2268.95	III
500	2278.34	III
700	2283.93	III
600	2323.50	III
500	2336.45	III
600	2340.64	III
600	2359.67	III
600	2370.37	III
700	2416.42	III
600	2447.14	III
600	2448.58	III
500	2486.91	III
500	2532.48	III
600	2580.67	III
500	2603.59	III
500	2632.67	III
500	2633.18	III
600	2665.54	III
700	2710.37	III
500	2724.03	IV
500	2751.23	IV
700	2782.47	IV
600	2965.56	III
500	3063.13	IV
600	3076.68	IV
600	3104.46	III
800	3139.34	III
900	3191.45	III
700	3289.80	III
700	3320.57	III
800	3329.06	III
900	3340.42	III
800	3392.89	III
800	3393.45	III
900	3530.03	III
800	3560.68	III
900	3602.10	III
800	3612.85	III
700	3622.69	III
700	3656.95	III
700	3670.28	III
700	3682.05	III
600	3705.45	III
600	3707.34	III
800	3720.45	III

Intensity	Wavelength/Å	
800	3748.81	III
500	3779.35	III
10000	3850.99	II
25000	3860.83	II
500	3925.87	III
700	3991.50	III
600	4018.50	III
600	4059.07	III
500	4104.23	III
500	4106.83	III
10000 h	4132.50	II
500	4608.21	III
40	4623.938	I
50	4654.040	I
80	4661.208	I
45	4691.523	I
40	4721.255	I
45	4740.729	I
13000	4781.32	II
99000	4794.55	II
29000	4810.06	II
16000	4819.47	II
81000	4896.77	II
47000	4904.78	II
26000	4917.73	II
10000	4995.48	II
26000	5078.26	II
30	5099.789	I
56000	5217.94	II
23000	5221.36	II
15000	5392.12	II
99000	5423.23	II
10000	5423.51	II
19000	5443.37	II
10000	5444.21	II
40	5532.162	I
50 d	5796.305	I
45	5799.914	I
30	5856.742	I
50	6019.812	I
200	6140.245	I
160	6194.757	I
150	6434.833	I
300	6932.903	I
300	6981.886	I
600	7086.814	I
7500	7256.62	I
5000	7414.11	I
550	7462.370	I
550	7489.47	I
700	7492.118	I
11000	7547.072	I
2300	7672.42	I
450	7702.828	I
7000	7717.581	I
10000	7744.97	I
2200	7769.16	I
650	7771.09	I
2200	7821.36	I
1700	7830.75	I
3000	7878.22	I
2300	7899.31	I
1800	7915.08	I
3000	7924.645	I
2100	7933.89	I
1700	7935.012	I
650	7952.52	I
1500	7974.72	I
1300	7976.97	I
600	7980.60	I
2900	7997.85	I
2200	8015.61	I
1100	8023.33	I
400	8051.07	I
1700	8084.51	I
2200	8085.56	I
3000	8086.67	I
1300	8087.73	I
2500	8194.42	I
2200	8199.13	I
2200	8200.21	I
800	8203.78	I
18000	8212.04	I
3000	8220.45	I
20000	8221.74	I
18000	8333.31	I
99900	8375.94	I
400	8406.199	I
15000	8428.25	I
2200	8467.34	I
2200	8550.44	I
20000	8575.24	I
750	8578.02	I
75000	8585.97	I
450	8628.54	I
300	8641.71	I
3500	8686.26	I
2200	8912.92	I
3000	8948.06	I
2000	9038.982	I
2500	9045.43	I
1000	9069.656	I
2000	9073.17	I
7500	9121.15	I
3000	9191.731	I
500	9197.596	I
4000	9288.86	I
1500	9393.862	I
3500	9452.10	I
500	9486.964	I
1000	9584.801	I
3500	9592.22	I
250	9632.509	I
1000	9702.439	I
250	9744.426	I
200	9807.057	I
400	9875.970	I
331	10392.549	I
300	11123.05	I
269	11409.69	I
1000	11436.33	I
350	13243.8	I
310	13296.0	I
550	13346.8	I
525	13821.7	I
294	14931.7	I
269	15108.0	I
381	15465.1	I
1094	15520.3	I
1487	15730.1	I
2780	15869.7	I
277	15883.3	I
342	15928.9	I
735	15960.0	I
283	15970.5	I
259	16198.5	I
717	19755.3	I
100	24470.0	I
	39716.0	I
	40085.5	I
	40089.5	I
	40532.2	I

Chromium Cr Z = 24

Intensity	Wavelength/Å	
100	438.62	V
100	464.02	V
100	620.66	IV
100	629.26	IV
80	630.30	IV
100	666.55	IV
100	693.92	IV
60	1030.47	III
100	1033.69	III
100	1036.03	III
80	1055.89	IV
80	1068.41	III
100	1116.48	V
150	1121.07	V
150	1127.63	V
100	1263.50	V
100	1417.42	IV
150	1465.86	V
150	1497.97	V
170	1519.03	V
220	1579.70	V
170	1591.72	V
150	1603.19	V
120	1672.66	IV
120	1758.51	IV
140	1802.72	IV
130	1812.41	IV
200	1837.44	V
140	1873.89	IV
140	1967.18	IV
120	1972.07	IV
19000	2055.52	II
14000	2061.49	II
8900	2065.42	II
200	2226.72	III
200	2235.91	III
150	2237.59	III
150	2244.10	III
150	2284.44	III
150	2324.88	III
130	2383.33	I
140	2408.62	I
170	2496.31	I
110	2502.53	I
190	2504.31	I
110	2516.92	I
390	2519.52	I
190	2527.12	I
160	2549.54	I
130	2560.69	I
150	2571.74	I
100	2577.65	I
380	2591.85	I
250	2653.59	II
250	2658.59	II
320	2663.42	II
440	2666.02	II
280	2668.71	II
350	2671.81	II
280	2672.83	II
1800	2677.16	II
320	2678.79	II
230	2687.09	II
280	2691.04	II
180	2698.41	II
180	2698.69	II
110	2701.99	I
140	2712.31	II
170	2722.75	II
420 h	2726.51	I
280 h	2731.91	I
170 h	2736.47	I
250	2743.64	II
110 h	2748.29	I
330	2748.98	II
390	2750.73	II
280	2751.87	II
110 h	2752.88	I
150	2757.10	I
350	2757.72	II
750	2762.59	II
750	2766.54	II
250 h	2769.92	I
610	2780.70	I
180	2822.37	II
180	2830.47	II
2500	2835.63	II
110	2840.02	II
1700	2843.25	II
1200	2849.84	II
120	2851.36	II
880	2855.68	II
610	2858.91	II
440	2860.93	II
790	2862.57	II
750	2865.11	II
610	2866.74	II
480	2867.65	II
210	2870.44	II
110	2871.63	I
160	2873.48	II

Intensity	Wavelength/Å	
320	2875.99	II
230	2876.24	II
180	2877.98	II
120	2879.27	I
170	2887.00	I
700	2889.29	I
370	2893.25	I
190	2894.17	I
210	2896.75	I
180	2905.49	I
260	2909.05	I
260	2910.90	I
250	2911.14	I
480	2967.64	I
480	2971.11	I
210	2971.91	II
480	2975.48	I
190	2979.74	II
350	2980.79	I
110	2985.32	II
480	2985.85	I
1500	2986.00	I
2100	2986.47	I
660	2988.65	I
160	2989.19	II
480	2991.89	I
230	2994.07	I
300	2995.10	I
700	2996.58	I
210	2998.79	I
1100	3000.89	I
750	3005.06	I
140	3013.03	I
710	3013.71	I
710	3014.76	I
1400	3014.92	I
710	3015.19	I
2800	3017.57	I
430	3018.50	I
240	3018.82	I
430	3020.67	I
2800	3021.56	I
1100	3024.35	I
170	3029.16	I
710	3030.24	I
140	3031.35	I
390	3034.19	I
550	3037.04	I
550	3040.85	I
110	3050.14	II
710	3053.88	I
240	3118.65	II
430	3120.37	II
470	3124.94	II
120	3128.70	II
590	3132.06	II
140	3136.68	II
140	3147.23	II
100	3155.15	I
100	3163.76	I
240	3180.70	II

Intensity	Wavelength/Å	
220	3197.08	II
170	3209.18	II
140	3217.40	II
120	3245.54	I
130	3251.84	I
130	3257.82	I
130	3339.80	II
110	3342.59	II
170	3358.50	II
160	3360.30	II
430	3368.05	II
140	3382.68	II
170	3403.32	II
360	3408.76	II
210	3421.21	II
270	3422.74	II
140	3433.31	II
270	3433.60	I
160	3436.19	I
140	3441.44	I
170	3445.62	I
170	3447.43	I
190	3453.33	I
130	3455.60	I
100	3460.43	I
120	3550.64	I
130	3566.16	I
130	3573.64	I
330 h	3574.80	I
19000	3578.69	I
160 h	3584.33	I
130	3585.30	II
17000	3593.49	I
350	3601.67	I
13000	3605.33	I
130	3632.84	I
350	3636.59	I
630	3639.80	I
220	3641.83	I
220	3649.00	I
170	3653.91	I
220	3656.26	I
130	3663.21	I
120	3685.55	I
130	3686.80	I
130	3687.25	I
130	3730.81	I
150	3732.03	I
480	3743.58	I
570	3743.88	I
340	3749.00	I
230	3757.66	I
260	3768.24	I
130	3791.38	I
130	3792.14	I
120	3793.29	I
130	3793.88	I
140	3797.13	I
200	3797.72	I
530	3804.80	I
110	3806.83	I

Intensity	Wavelength/Å	
110	3807.93	I
180	3815.43	I
180	3819.56	I
130	3826.42	I
130	3830.03	I
380	3841.28	I
190	3848.98	I
140	3849.36	I
290	3850.04	I
140	3852.22	I
190	3854.22	I
110	3855.29	I
140	3855.57	I
260	3857.63	I
660	3883.29	I
570	3885.22	I
380	3886.79	I
260	3894.04	I
360	3902.92	I
960	3908.76	I
120 hd	3911.82	I
120	3915.84	I
190	3916.24	I
1900	3919.16	I
600	3921.02	I
600	3928.64	I
410	3941.49	I
1900	3963.69	I
120	3969.06	I
1600	3969.75	I
1600	3976.66	I
960	3983.91	I
190	3984.34	I
160	3989.99	I
960	3991.12	I
160	3991.67	I
190	3992.84	I
160	4001.44	I
120	4012.47	II
120	4026.17	I
190	4039.10	I
160	4048.78	I
120	4058.77	I
140	4126.52	I
120	4153.82	I
140	4163.62	I
170	4174.80	I
170	4179.26	I
110	4209.37	I
20000	4254.35	I
110	4263.14	I
16000	4274.80	I
10000	4289.72	I
780	4337.57	I
1100	4339.45	I
380	4339.72	I
1900	4344.51	I
380	4351.05	I
2300	4351.77	I
570	4359.63	I
530	4371.28	I

Intensity	Wavelength/Å	
110	4374.16	I
530	4384.98	I
110	4458.54	I
660	4496.86	I
380	4526.47	I
380	4530.74	I
240	4535.72	I
240	4540.50	I
240	4540.72	I
140	4544.62	I
600	4545.96	I
120	4565.51	I
120	4571.68	I
360	4580.06	I
360	4591.39	I
480	4600.75	I
240	4613.37	I
600	4616.14	I
550	4626.19	I
1600	4646.17	I
570	4651.28	I
840	4652.16	I
240 d	4698.46	I
190	4708.04	I
240	4718.43	I
120	4730.71	I
140	4737.35	I
340	4756.11	I
190	4789.32	I
120	4801.03	I
110	4829.38	I
140	4870.80	I
130	4887.01	I
260	4922.27	I
110	4936.33	I
70	4942.50	I
110	4954.81	I
60	5013.32	I
70	5166.23	I
70	5184.59	I
70	5192.00	I
85	5196.44	I
5300	5204.52	I
8400	5206.04	I
11000	5208.44	I
85	5224.94	I
290	5247.56	I
530	5264.15	I
180	5265.72	I
95 h	5275.17	I
70 h	5276.03	I
340	5296.69	I
70 h	5297.36	I
660	5298.27	I
85	5300.75	I
340 h	5328.34	I
70 h	5329.17	I
780	5345.81	I
380	5348.32	I
40	5400.61	I
1400	5409.79	I

Intensity	Wavelength/Å	
24	5628.64	I
7	5642.36	I
24	5664.04	I
24	5694.73	I
40	5698.33	I
24	5702.31	I
24	5712.78	I
24 h	5783.11	I
30 h	5783.93	I
24 h	5785.00	I
19 h	5785.82	I
60 h	5787.99	I
180 h	5791.00	I
35	6330.10	I
22	6362.87	I
19	6661.08	I
21 h	6883.03	I
27 h	6924.13	I
30 h	6978.48	I
85	7355.90	I
130	7400.21	I
150	7462.31	I
40	8947.15	I
19	8976.83	I
Cobalt Co Z = 27		
20	355.52	V
18	355.88	V
12	356.06	V
66	609.16	IV
70	609.21	IV
64	609.28	IV
10	1018.36	V
10	1021.14	V
15	1231.73	V
50	1277.01	V
80	1299.58	II
80	1306.95	II
50	1345.67	V
1000	1696.01	III
800	1697.99	III
1000	1707.35	III
5000	1760.35	III
5000	1773.57	III
2000	1780.05	III
3000	1782.97	III
1000	1787.08	III
1000	1789.07	III
1000	1823.08	III
2000	1830.09	III
2000	1831.44	III
5000	1835.00	III
1500	1842.34	I
1800	1847.89	I
1800	1852.71	I
2400	1855.05	I
2000	1863.83	III
1500	1878.28	I
1800	1936.58	I
1500	1946.79	I
1500	1951.90	I
1800	1954.22	I
1800	1955.17	I
1500	1958.55	I
1500	1961.59	I
1500 h	1968.69	I
1500 h	1968.93	I
3000	1970.71	I
1800 h	1971.16	I
1800 h	1972.52	I
1500	1973.85	I
1800	1976.97	I
2400 h	1980.89	I
1500	1989.80	I
1800	1990.34	I
1500 l	1998.49	I
1500	2002.32	I
900	2008.04	I
50	2011.51	II
1200 h	2014.58	I
900	2016.17	I
50	2022.35	II
50	2027.04	II
900	2031.96	I
1500	2039.95	I
1200	2041.11	I
50	2065.54	II
1500 h	2077.76	I
900	2085.67	I
900	2087.55	I
900	2089.35	I
900	2093.40	I
900	2094.86	I
900	2095.77	I
1200	2097.51	I
1500	2104.73	I
1500	2106.80	I
900	2108.98	I
900 s	2117.68	I
900	2137.78	I
900	2138.97	I
900	2163.03	I
1100	2174.60	I
200	2193.60	II
200	2256.73	II
150	2260.00	II
200	2283.52	II
1000	2286.15	II
200	2291.98	II
300 d	2293.38	II
300	2301.40	II
800 d	2307.85	II
2600	2309.02	I
500	2311.60	II
500	2314.05	II
300	2314.96	II
200 p	2317.06	II
2400	2323.14	I
300 p	2324.31	II
200 d	2326.11	II
500	2326.47	II
1400	2335.99	I
1600	2338.67	I
200	2347.39	II
1600	2352.85	I
200 d	2353.41	II
2000	2353.42	I
500	2363.80	II
400	2378.62	II
1400	2380.48	I
200	2381.76	II
300 p	2383.45	II
1400	2384.86	I
200	2386.36	II
500	2388.92	II
200	2397.38	II
1100 d	2402.06	I
200 p	2404.16	II
5300	2407.25	I
5300	2411.62	I
1600	2412.76	I
4800	2414.46	I
4800	2415.30	I
300	2417.65	II
4100	2424.93	I
3300	2432.21	I
2900	2436.66	I
2400	2439.05	I
200	2442.63	II
200 d	2446.03	II
200 p	2447.69	II
200	2450.00	II
200	2464.20	II
200	2486.44	II
200	2498.82	II
570	2504.52	I
500	2506.46	II
360	2506.88	I
200	2511.16	II
860	2517.87	I
500	2519.82	II
4300	2521.36	I
200 h	2524.65	II
300	2524.97	II
500	2528.62	II
2900	2528.97	I
200 p	2530.09	II
720	2530.13	I
860	2532.18	I
200 d	2533.82	II
2900	2535.96	I
860	2536.49	I
300	2541.94	II
1700	2544.25	I
200	2546.74	II
340	2548.34	I
310	2553.37	I
310	2555.07	I
300	2559.41	II
200	2560.03	II
960	2562.15	I
500	2564.04	II
1100	2567.35	I
960	2574.35	I
800	2580.32	II
300 d	2582.22	II
500	2587.22	II
500	2587.52	II
200	2588.91	II
100 p	2605.71	II
100	2612.50	II
100	2614.36	II
100 p	2628.77	II
100	2632.26	II
100	2636.07	II
310	2646.42	I
770	2648.64	I
100	2653.72	II
100	2663.53	II
200	2666.73	II
100	2675.85	II
100	2684.42	II
100	2702.02	II
200	2706.62	II
200	2707.35	II
190	2715.99	I
100	2727.78	II
80	2734.54	II
190	2745.10	I
100	2753.22	II
190	2764.19	I
100	2766.70	II
100	2774.97	II
100	2791.00	II
100	2793.73	II
150	2815.56	I
80	2835.63	II
80	2847.35	II
80	2871.22	II
190	2886.44	I
100	2918.38	II
100	2930.24	II
100	2954.73	II
690	2987.16	I
690	2989.59	I
60	3022.59	II
3100	3044.00	I
1700	3061.82	I
80	3387.70	II
1100	3388.17	I
2200	3395.38	I
11000	3405.12	I
4500	3409.18	I
6700	3412.34	I
2200	3412.63	I
2700	3417.16	I
50	3423.84	II
2500	3431.58	I
4500	3433.04	I
1600	3442.93	I
8800	3443.64	I
50	3446.39	II
4100	3449.17	I
2100	3449.44	I
21000	3453.50	I

Intensity	Wavelength/Å	
1000	3455.23	I
5100	3462.80	I
5100	3465.80	I
8000	3474.02	I
1900	3483.41	I
4800	3489.40	I
2400	3495.69	I
50	3501.72	II
9600	3502.28	I
7000	3506.32	I
50	3507.77	II
2900	3509.84	I
1400	3510.43	I
4800	3512.64	I
3800	3513.48	I
4800	3518.35	I
1300	3520.08	I
2700	3521.57	I
3800	3523.43	I
60	3523.51	II
6400	3526.85	I
2700	3529.03	I
7300	3529.81	I
1900	3533.36	I
50	3545.03	II
1100	3560.89	I
80	3561.07	II
8800	3569.38	I
50	3574.95	II
1600	3574.96	I
60	3575.32	II
2500	3575.36	I
60	3577.96	II
1000	3585.16	I
6700	3587.19	I
1900	3594.87	I
1600	3602.08	I
100	3621.21	II
1000	3627.81	I
80	3643.61	II
60	3681.35	II
1100	3745.50	I
1400	3842.05	I
6900	3845.47	I
5500	3873.12	I
2800	3873.96	I
7900	3894.08	I
1500	3935.97	I
80 h	3963.10	II
6000	3995.31	I
970	3997.91	I
350	4020.90	I
370	4045.39	I
350	4066.37	I
830	4092.39	I
550	4110.54	I
2800	4118.77	I
4400	4121.32	I
90	4190.71	I
90	4469.56	I
690	4530.96	I
90	4549.66	I
140	4565.59	I
190	4581.60	I
120	4629.38	I
85	4663.41	I
110	4792.86	I
100	4840.27	I
150	4867.88	I
80 h	4964.18	II
50	5212.71	I
50	5230.22	I
50	5247.93	I
50	5342.71	I
50	5352.05	I

Copper Cu Z = 29

Intensity	Wavelength/Å	
80	685.141	II
100	709.313	II
100	718.179	II
150	724.489	II
200	735.520	II
250	736.032	II
80	779.295	II
100	797.455	II
150	810.998	II
200	813.883	II
300	826.996	II
150	848.808	II
250	851.303	II
250	858.487	II
400	861.994	II
400	865.390	II
250	869.336	II
150	873.263	II
200	876.723	II
250	877.012	II
200	877.555	II
500	878.699	II
100	884.133	II
250	885.847	II
600	886.943	II
600	890.567	II
500	892.414	II
800	893.678	II
400	894.227	II
600	896.759	II
400	896.976	II
600	901.073	II
400	906.113	II
800	914.213	II
600	922.019	II
500	924.239	II
400	935.232	II
600	935.898	II
600	943.335	II
600	945.525	II
500	945.965	II
200	954.383	II
250	956.290	II
400	958.154	II
200	960.414	II
250	968.042	II
200	974.759	II
250	977.567	II
100	987.657	II
250	992.953	II
300	1004.055	II
300	1008.569	II
300	1008.728	II
300	1010.269	II
250	1012.597	II
500	1018.707	II
500	1027.831	II
250	1028.328	II
200	1030.263	II
600	1036.470	II
600	1039.348	II
600	1039.582	II
800	1044.519	II
800	1044.744	II
500	1049.755	II
600	1054.690	II
400	1055.797	II
600	1056.955	II
400	1058.799	II
600	1059.096	II
600	1060.634	II
600	1063.005	II
200	1065.782	II
200	1066.134	II
500	1069.195	II
300	1073.745	II
200	1088.395	II
300	1094.402	II
250	1097.053	II
150	1119.947	II
200	1142.640	II
300	1144.856	II
100	1250.048	II
150	1265.506	II
300	1275.572	II
150	1282.455	II
150	1287.468	II
150	1298.395	II
300	1308.297	II
300	1314.337	II
100	1320.686	II
100	1326.395	II
150	1350.594	II
250	1351.837	II
150	1355.305	II
300	1358.773	II
200	1359.009	II
200	1362.600	II
250	1367.951	II
200	1371.840	II
300 r	1376.79	III
200 r	1377.49	III
100	1393.128	II
100	1398.642	II
150	1402.777	II
150	1407.169	II
100	1414.898	II
250	1418.426	II
250	1421.759	II
200	1427.829	II
400	1430.243	II
250	1434.904	II
150	1436.236	II
150	1442.139	II
200	1445.984	II
200	1449.058	II
250	1450.304	II
200	1452.294	II
300	1458.002	II
250	1459.412	II
200	1463.752	II
400	1463.838	II
200	1466.070	II
400	1470.697	II
200	1472.395	II
250	1473.978	II
200	1474.935	II
150	1476.059	II
300 r	1481.23	III
200	1481.544	II
200	1485.328	II
750	1488.831	II
300	1492.834	II
250	1493.366	II
250	1495.430	II
350	1496.687	II
150	1503.368	II
250	1504.757	II
200	1505.388	II
300	1508.632	II
350	1510.506	II
200	1512.465	II
200	1513.366	II
500	1514.492	II
200	1517.631	II
500	1519.492	II
600	1519.837	II
200	1520.540	II
200	1524.860	II
150	1525.764	II
500	1531.856	II
300	1532.131	II
250	1533.986	II
250	1535.002	II
500	1537.559	II
200	1540.239	II
300	1540.389	II
300	1540.588	II
750	1541.703	II
400	1544.677	II
100	1547.958	II
300	1550.653	II
300	1551.389	II
500	1552.646	II
250	1553.896	II
400	1555.134	II
500	1555.703	II
300	1558.345	II

Intensity	Wavelength/Å	
400	1565.924	II
400	1566.415	II
100	1569.416	II
300	1579.492	II
300	1580.626	II
400	1581.995	II
500	1583.682	II
400	1590.165	II
600	1593.556	II
500 r	1593.75	III
400	1598.402	II
400	1602.388	II
200	1604.848	II
300	1605.281	II
400	1606.834	II
250	1608.639	II
150	1610.296	II
200	1617.915	II
600	1621.426	II
400	1622.428	II
250	1630.268	II
100	1636.605	II
1000 r	1642.21	III
250	1649.458	II
30 r	1655.32	I
200	1656.322	II
200	1660.001	II
300	1663.002	II
100	1672.776	II
30	1688.09	I
30	1691.08	I
30 r	1703.84	I
50 r	1713.36	I
150	1717.721	II
50 r	1725.66	I
100	1736.551	II
50 r	1741.57	I
150	1753.281	II
200 r	1774.82	I
100 r	1825.35	I
250	1929.751	II
250	1944.597	II
100	1946.493	II
200	1957.518	II
150	1970.495	II
150	1977.027	II
500	1979.956	II
300	1989.855	II
250	1999.698	II
270	2035.854	II
250	2037.127	II
350	2043.802	II
300	2054.980	II
100	2078.663	II
110	2098.398	II
320	2104.797	II
300	2112.100	II
320	2117.310	II
350	2122.980	II
350	2126.044	II
420	2134.341	II
900	2135.981	II
400	2148.984	II
150	2161.320	II
1300 r	2165.09	I
250	2174.982	II
1600 r	2178.94	I
700	2179.410	II
1700 r	2181.72	I
700	2189.630	II
900	2192.268	II
400	2195.683	II
1700 r	2199.58	I
1300 r	2199.75	I
100	2200.509	II
200	2209.806	II
750	2210.268	II
1600 r	2214.58	I
250	2215.106	II
1000 r	2215.65	I
750	2218.108	II
2100 r	2225.70	I
150	2226.780	II
1600 r	2227.78	I
350	2228.868	II
2500 r	2230.08	I
1100 r	2238.45	I
900	2242.618	II
2300 r	2244.26	I
1000	2247.002	II
1300 r	2260.53	I
2200 r	2263.08	I
150	2263.786	II
200	2276.258	II
100	2286.645	II
2500 r	2293.84	I
170	2294.368	II
1000	2303.12	I
150	2369.890	II
2500 r	2392.63	I
120	2403.337	II
1500	2406.66	I
1000 r	2441.64	I
100	2485.792	II
2000 r	2492.15	I
150	2506.273	II
120	2526.593	II
300	2544.805	II
100	2571.756	II
150	2590.529	II
200	2600.270	II
2500 r	2618.37	I
200	2666.291	II
750	2689.300	II
700	2700.962	II
650	2703.184	II
700	2713.508	II
650	2718.778	II
300	2721.677	II
120	2737.342	II
270	2745.271	II
2500 r	2766.37	I
800	2769.669	II
200	2791.795	II
170	2799.528	II
100	2810.804	II
1250 r	2824.37	I
350	2837.368	II
100	2857.748	II
600	2877.100	II
270	2884.196	II
2500 r	2961.16	I
100	2986.335	II
2000	2997.36	I
2000	3010.84	I
2500	3036.10	I
2500	3063.41	I
1400	3073.80	I
1500	3093.99	I
1250	3099.93	I
2000	3108.60	I
1400 h	3126.11	I
1500	3194.10	I
1400	3208.23	I
1500 h	3243.16	I
10000 r	3247.54	I
10000 r	3273.96	I
1400 h	3282.72	I
400	3290.418	II
1500 h	3290.54	I
110	3300.881	II
250	3301.229	II
2500 h	3307.95	I
200	3316.276	II
1500	3337.84	I
150	3338.648	II
200	3365.648	II
450	3370.454	II
300	3374.952	II
200	3380.712	II
100	3384.945	II
1250 h	3483.76	I
1250	3524.23	I
2000	3530.38	I
1400	3599.13	I
1400	3602.03	I
1000	3686.555	II
150	3786.270	II
170	3797.849	II
100	3818.879	II
140	3826.921	II
160	3864.137	II
280	3884.131	II
150	3892.924	II
170	3903.177	II
140	3920.654	II
120	3933.268	II
120	3987.024	II
150	3993.302	II
140	4003.476	II
1250	4022.63	I
100	4032.647	II
600	4043.484	II
500	4043.751	II
2000	4062.64	I
120	4068.106	II
500	4131.363	II
200	4143.017	II
300	4153.623	II
500	4161.140	II
370	4164.284	II
400	4171.851	II
500	4179.512	II
500	4211.866	II
320	4230.449	II
200	4255.635	II
950	4275.11	I
300	4279.962	II
500	4292.470	II
400	4365.370	II
100	4444.831	II
400	4506.002	II
150	4516.049	II
150	4541.032	II
500	4555.920	II
100	4596.906	II
120	4649.271	II
2000	4651.12	I
120	4661.363	II
320	4671.702	II
300	4673.577	II
450	4681.994	II
100	4758.433	II
400	4812.948	II
120	4851.262	II
300	4854.988	II
100	4873.304	II
150	4901.427	II
1000	4909.734	II
500	4918.376	II
200	4926.424	II
900	4931.698	II
120	4943.026	II
700	4953.724	II
500	4985.506	II
400	5006.801	II
350	5009.851	II
400	5012.620	II
350	5021.279	II
200	5039.016	II
300	5047.348	II
900	5051.793	II
400	5058.910	II
500	5065.459	II
450	5067.094	II
350	5072.302	II
450	5088.277	II
420	5093.816	II
350	5100.067	II
1500	5105.54	I
250	5124.476	II
2000	5153.24	I
100	5158.093	II
100	5183.367	II

Intensity	Wavelength/Å	
2500	5218.20	I
100	5269.991	II
100	5276.525	II
1650	5292.52	I
100	5368.383	II
1500	5700.24	I
1500	5782.13	I
150	5805.989	II
100	5833.515	II
200	5897.971	II
120	5937.577	II
400	5941.196	II
100	5993.260	II
650	6000.120	II
100	6023.264	II
250	6072.218	II
150	6080.343	II
150	6099.990	II
160	6107.412	II
300	6114.493	II
600	6150.384	II
750	6154.222	II
500	6172.037	II
550	6186.884	II
400	6188.676	II
300	6198.092	II
470	6204.261	II
450	6208.457	II
750	6216.939	II
700	6219.844	II
500	6261.848	II
1000	6273.349	II
350	6288.696	II
900	6301.009	II
550	6305.972	II
400	6312.492	II
120	6326.466	II
400	6373.268	II
750	6377.840	II
400	6403.384	II
850	6423.884	II
200	6442.965	II
750	6448.559	II
170	6466.246	II
950	6470.168	II
750	6481.437	II
400	6484.421	II
220	6517.317	II
400	6530.083	II
120	6551.286	II
200	6577.080	II
750	6624.292	II
800	6641.396	II
450	6660.962	II
100	6770.362	II
300	6806.216	II
400	6809.647	II
320	6823.202	II
250	6844.157	II
320	6868.791	II
270	6872.231	II

Intensity	Wavelength/Å	
270	6879.404	II
220	6937.553	II
150	6952.871	II
150	6977.572	II
200	7022.860	II
300	7194.896	II
400	7326.008	II
300	7331.694	II
250	7382.277	II
1000	7404.354	II
270	7434.156	II
500	7562.015	II
700	7652.333	II
1000	7664.648	II
150	7681.788	II
450	7744.097	II
800	7778.738	II
750	7805.184	II
1500	7807.659	II
1000	7825.654	II
350	7860.577	II
300	7890.567	II
700	7902.553	II
1500	7933.13	I
400	7944.438	II
400	7972.033	II
1200	7988.163	II
2000	8092.63	I
500	8277.560	II
800	8283.160	II
250	8503.396	II
750	8511.061	II
200	8609.134	II
500	9813.213	II
250	9827.978	II
200	9830.798	II
600	9861.280	II
600	9864.137	II
200	9883.969	II
550	9916.419	II
500	9917.954	II
550	9925.594	II
450	9938.998	II
500	9960.354	II
450	10006.588	II
550	10022.969	II
550	10038.093	II
650	10054.938	II
450	10080.354	II

***Dysprosium Dy Z* = 66**

Intensity	Wavelength/Å	
260	2356.91	II
240	2410.01	II
260	2439.84	II
220	2585.30	I
440	2634.80	II
220	2755.75	II
300	2816.39	II
390	2913.95	II
610	3038.28	II
830	3135.38	II
500	3141.14	II
1200	3156.52	II
670	3162.83	II
1000	3169.99	II
470	3215.19	II
830	3216.63	II
490	3235.89	II
490	3245.12	II
1200	3251.27	II
890	3280.09	II
490	3282.77	II
1100	3308.88	II
780	3316.32	II
1000	3319.88	II
780	3341.00	II
510	3353.58	II
510	3368.11	II
5300	3385.02	II
610	3388.85	II
3800	3393.57	II
1300	3396.16	II
5300	3407.80	II
1300	3413.78	II
530	3414.82	II
780	3419.63	II
530	3425.06	II
1900	3434.37	II
560	3440.93	II
1300	3441.45	II
3800	3445.57	II
830	3446.99	II
2700	3454.32	II
1300	3456.56	II
4400	3460.97	II
720	3468.43	II
560	3471.14	II
560 d	3471.53	II
1300	3477.07	II
4400	3494.49	II
560	3496.34	II
830	3498.71	II
830	3504.53	II
830	3505.45	II
1300	3506.81	II
560	3517.26	II
4400	3523.98	II
22000	3531.70	II
4400	3534.96	II
5500	3536.02	II
4400	3538.52	II
1700	3542.33	II
1400	3546.83	II
4400	3550.22	II
2200	3551.62	II
440 h	3558.23	II
440	3559.30	II
2200	3563.15	II
560	3563.69	II
780	3573.83	II
1400	3574.15	II
4400	3576.24	II
1700	3576.87	II

Intensity	Wavelength/Å	
830	3577.98	II
440	3580.04	II
3300	3585.06	II
1400	3585.78	II
560	3586.11	II
1100	3591.41	II
560	3591.81	II
560	3592.11	II
1800	3595.04	II
560	3600.38	II
1800	3606.12	II
440	3618.51	II
560	3620.16	II
470	3624.27	II
1100	3629.42	II
4000	3630.24	II
440	3632.78	II
1100	3640.25	II
11000	3645.40	II
1000	3648.78	II
700	3664.62	II
990	3672.30	II
420	3672.70	II
1400	3674.08	II
2200	3676.59	II
640	3678.51	I
820	3684.85	I
1300	3685.78	I
4700	3694.81	II
990	3698.21	II
540	3701.63	II
440	3707.57	II
440	3708.22	II
420	3710.07	II
1600	3724.45	II
930	3739.34	I
1200	3747.82	II
1400	3753.51	II
1400	3753.75	II
1200	3757.05	I
4700	3757.37	II
640	3767.63	I
640	3773.05	I
420	3781.47	I
3300	3786.18	II
1600	3788.44	II
700	3791.87	II
510	3804.14	II
580	3806.27	II
470	3812.27	I
470	3813.67	II
1400	3816.76	II
700	3825.68	II
2300	3836.50	II
1400	3841.31	II
420	3846.34	II
420	3847.02	I
1200	3853.03	II
420	3858.40	I
560	3868.45	II
1600	3868.81	I

Intensity	Wavelength/Å	
820	3869.86	II
7000	3872.11	II
1200	3873.99	II
470	3879.11	II
5800	3898.53	II
540	3914.87	II
540	3915.59	II
540 d	3917.29	I
420	3927.86	I
540	3930.14	I
2100	3931.52	II
10000	3944.68	II
800	3957.79	II
14000	3968.39	II
2700	3978.57	II
1400	3981.92	II
1600	3983.65	II
800	3984.21	II
540	3991.32	II
1600	3996.69	II
8000	4000.45	II
420	4005.84	I
540	4011.29	II
540	4013.82	I
540	4014.70	II
420	4027.78	II
520 d	4028.32	II
520	4032.47	II
420	4033.65	II
420	4036.32	II
12000	4045.97	I
1600	4050.56	II
520	4055.14	II
2500	4073.12	II
7400	4077.96	II
3900	4103.30	II
860	4103.87	I
1500	4111.34	II
490	4124.63	II
990	4129.42	II
1200	4143.10	II
990	4146.06	I
5700	4167.97	I
930	4183.72	I
12000	4186.82	I
2200	4191.64	I
6800	4194.84	I
800	4198.02	I
680	4201.30	I
680	4202.24	I
16000	4211.72	I
1800	4213.18	I
3700	4215.16	I
4400	4218.09	I
4400	4221.11	I
2700	4225.16	I
1000	4308.63	II
540	4409.38	II
740	4449.70	II
420	4577.78	I
2100	4589.36	I
990	4612.26	I
170	4731.84	II
120 h	4775.79	I
480	4957.34	II
70	5022.12	I
160	5042.63	I
95	5070.68	I
120	5077.67	I
80	5090.38	II
80	5110.32	I
130 h	5120.04	I
190	5139.60	II
110	5169.69	II
80	5185.30	I
290	5192.86	II
95	5197.66	II
70	5259.88	I
130	5260.56	I
65	5267.11	I
55	5282.07	I
160	5301.58	I
65	5340.30	I
85	5389.58	II
80	5419.13	I
70	5423.32	I
95	5451.11	I
65	5547.27	I
100	5639.50	I
55 h	5645.99	I
80	5652.01	I
70 h	5718.46	I
55	5745.53	I
55 h	5868.11	II
70	5945.80	I
120	5974.49	I
140	5988.56	I
140	6088.26	I
100	6168.43	I
270	6259.09	I
160	6579.37	I
75	6667.86	I
180	6835.42	I
80	6852.96	I
65	6899.32	II
55	7426.86	II
55	7543.73	I
80	7662.36	I
100	8201.57	II
45	8791.39	II

Erbium Er Z = 68

Intensity	Wavelength/Å	
600	2277.65	III
290	2586.73	II
490	2670.26	II
500	2739.27	III
610	2755.63	II
1000	2904.47	II
1500	2910.36	II
1500	2964.52	II
1200	3002.41	II
1000	3055.10	III
1000	3070.40	III
610	3073.34	II
720	3082.08	II
610	3084.02	II
770	3122.72	II
1500	3166.25	III
870	3181.92	II
870	3220.73	II
610	3223.31	II
2300	3230.58	II
2700	3264.78	II
720	3279.33	II
720	3280.22	II
2000	3301.23	III
2300	3312.42	II
770	3323.19	II
770	3332.70	II
1300	3346.04	II
1400	3364.08	II
1400 d	3368.02	II
7700	3372.71	II
970	3374.17	II
1700	3385.08	II
2300	3392.00	II
770	3441.13	II
970	3471.71	II
610	3479.41	II
970	3485.85	II
6700	3499.10	II
610	3502.78	I
610	3524.91	II
820	3549.84	II
1500	3558.02	I
1000	3559.90	II
920	3570.75	II
1000	3580.52	II
610	3590.76	I
610	3599.50	II
1000	3599.83	II
3100	3616.56	II
720	3628.04	I
1000	3633.54	II
1600	3638.68	I
900	3645.94	II
7900	3692.65	II
1300	3729.52	II
900	3742.64	II
900	3747.43	I
1800	3786.84	II
1600	3810.33	I
4000	3816.78	III
3600	3830.48	II
680	3855.90	I
7500	3862.85	I
1500	3880.61	II
1200	3882.89	II
4200	3892.68	I
5200	3896.23	II
11000	3906.31	II
3200	3937.01	I
2100	3938.63	II
3200	3944.42	I
2700	3973.04	I
3200	3973.58	I
1400	3974.72	II
810	3977.02	I
1100	3982.33	I
810	3987.66	I
14000	4007.96	I
1100	4012.58	I
3000	4020.51	I
1000	4046.96	I
940	4055.47	II
690	4059.78	II
3500	4087.63	I
1100	4098.10	I
6900	4151.11	I
1000	4190.70	I
1400	4218.43	I
690	4286.56	I
40000	4290.06	III
20000	4386.86	III
810	4409.34	I
1000	4606.61	I
570	4675.62	II
15000	4735.56	III
2000	4783.12	III
250	5007.25	I
200	5035.94	I
210	5042.05	II
120	5124.56	I
130	5127.41	II
120	5131.53	I
130	5133.83	II
170	5164.77	II
130	5172.78	I
160	5188.90	II
150	5206.52	I
140	5255.93	II
80	5272.91	I
90	5348.06	I
60	5414.63	II
180	5456.62	I
90	5468.32	I
80	5485.97	II
80	5593.46	I
60	5611.82	I
70	5622.01	I
80	5626.53	II
90	5640.36	I
70	5664.95	I
70	5719.55	I
100	5739.19	I
290	5762.80	I
70	5784.66	I
70	5800.79	I
430	5826.79	I
100	5850.07	I
120	5855.31	I
140	5872.35	I
120	5881.14	I
8000	5903.30	III

Intensity	Wavelength/Å	
70	6022.56	I
70	6061.25	I
60	6076.45	II
360	6221.02	I
55	6262.56	I
60	6268.87	I
130	6308.77	I
55	6326.13	I
55	6492.35	I
60	6583.48	I
70	6601.11	I
70	6759.87	I
35	6790.92	I
70	6848.10	I
55	6865.13	I
55	7459.55	I
120	7469.51	I
35	7680.01	I
35	7797.47	I
35	7921.85	I
30	7937.84	I
35	8312.82	I
55	8409.90	I
9	8866.84	II

Europium Eu Z = 63

Intensity	Wavelength/Å	
30	2124.69	III
200	2350.51	III
4000	2375.46	III
100 d	2435.14	III
1000	2444.38	III
4000	2445.99	III
2000	2513.76	III
200	2522.14	III
160	2564.17	II
110	2568.17	II
230	2577.14	II
1000	2638.77	II
380	2641.27	II
640	2668.34	II
110	2673.42	II
250	2678.29	II
250	2685.66	II
550	2692.03	II
700	2701.14	II
800	2701.90	II
240	2705.28	II
180	2709.99	I
700	2716.98	II
4200	2727.78	II
160	2740.62	II
120	2744.26	II
480	2781.89	II
1900	2802.84	II
220	2811.75	II
3400	2813.94	II
550	2816.18	II
2000	2820.78	II
400 cw	2828.72	II
260	2859.67	II
280	2862.57	II
200	2892.54	I
140	2893.03	I
360	2893.83	I
3200	2906.68	II
160	2908.99	I
850	2925.04	II
200 cw	2952.68	II
260	2960.21	II
300	2991.33	II
100 c	3023.93	III
200 c	3026.79	III
320 cw	3054.94	II
120	3058.98	I
220	3077.36	II
120	3097.45	II
320	3106.18	I
950	3111.43	I
120	3130.73	II
50 c	3171.00	III
50 c	3183.78	III
420	3210.57	I
1000	3212.81	I
420	3213.75	I
150	3272.77	II
210	3277.78	II
150	3301.95	II
140	3308.02	II
140	3313.33	II
950	3334.33	I
110	3350.40	I
140	3369.06	II
190	3391.99	II
280	3396.58	II
150	3425.02	II
150	3441.00	II
130	3461.38	II
470 cw	3521.09	II
150	3542.15	II
180	3552.52	II
150	3603.20	II
6400	3688.42	II
20000 cw	3724.94	II
350	3741.31	II
260	3761.12	II
39000 cw	3819.67	II
140	3844.23	II
190	3865.57	I
150	3884.75	I
28000 cw	3907.10	II
32000 cw	3930.48	II
30000 cw	3971.96	II
180	4011.69	II
150	4017.58	II
120	4039.19	I
120	4085.38	II
33000 cw	4129.70	II
60000 cw	4205.05	II
150	4298.73	I
240	4355.09	II
14000 cw	4435.56	II
3000	4522.57	II
11000	4594.03	I
9800	4627.22	I
8300	4661.88	I
110	4867.62	I
150	4907.18	I
180	4911.40	I
180	5013.17	I
170	5022.91	I
110	5029.54	I
170	5114.37	I
170	5129.10	I
210	5133.52	I
270	5160.07	I
210	5166.70	I
200	5199.85	I
110	5200.96	I
120	5206.44	I
750	5215.10	I
300	5223.49	I
120	5239.24	I
200	5266.40	I
390	5271.96	I
110	5272.48	I
150	5282.82	I
120	5291.26	I
120	5294.64	I
540	5357.61	I
120	5361.61	I
110	5376.94	I
120	5392.94	I
450	5402.77	I
380	5451.51	I
260	5452.94	I
120	5488.65	I
120	5510.52	I
200	5547.44	I
150	5570.33	I
200	5577.14	I
120	5580.03	I
210	5645.80	I
330	5765.20	I
180	5783.69	I
170	5818.74	II
600 cw	5830.98	I
330	5966.07	II
480 cw	5967.10	I
170	5972.75	I
240	5992.83	I
110	6012.56	I
420	6018.15	I
170	6029.00	I
420	6049.51	II
140	6057.36	I
240	6083.84	I
240	6099.35	I
120	6118.78	I
330	6173.05	II
110	6178.76	I
260 cw	6188.13	I
140	6195.07	I
240	6262.25	I
170	6299.77	I
230	6303.41	II
120 cw	6350.04	I
120 cw	6400.93	I
180	6410.04	I
140	6411.32	I
830	6437.64	II
120	6457.96	I
1400	6645.11	II
50	6666.35	III
140	6802.72	I
360	6864.54	I
120	7040.20	I
330	7077.10	II
570	7194.81	II
570	7217.55	II
540	7301.17	II
720	7370.22	II
300	7426.57	II
160	7583.91	I
60 cw	7742.57	I
70	7746.19	I
35	7887.99	I
24 cw	8209.80	I
21 cw	8642.67	I
18	8870.30	I

Fluorine F Z = 9

Intensity	Wavelength/Å	
50	148.00	V
50	163.56	V
90	165.98	V
100	166.18	V
50	186.84	V
60	190.57	V
70	190.84	V
50	196.39	IV
60	196.45	IV
70	200.09	IV
80	201.16	IV
90	208.25	IV
90	240.08	IV
100	251.03	IV
140	419.65	IV
150	420.05	IV
160	420.73	IV
100	429.51	III
110	430.15	III
150	430.76	IV
90	464.29	III
120	465.98	V
130	490.57	IV
160	491.00	IV
50	497.38	IV
60	497.83	IV
70	498.80	IV
90	506.16	V
100	508.08	V
120	508.39	III
60	514.08	V
90	525.29	V
100	526.30	V
120	567.69	III
110	567.75	III

Intensity	Wavelength/Å	
140	570.64	IV
140	571.30	IV
150	571.39	IV
160	572.66	IV
90	605.67	II
100	606.80	II
90	630.20	III
100	647.77	V
110	647.87	V
130	654.03	V
120	656.12	III
130	656.87	III
110	657.23	V
140	657.33	V
140	658.33	III
140	676.12	IV
130	677.15	IV
150	677.22	IV
130	678.99	IV
160	679.21	IV
60	757.04	V
150	806.96	I
125	809.60	I
500	951.87	I
1000	954.83	I
750	955.55	I
500	958.52	I
20	972.40	I
350	973.90	I
100	976.22	I
40	976.51	I
100	977.75	I
60	1082.31	V
70	1088.39	V
80	1219.03	III
80	1266.87	III
90	1267.71	III
70	1297.54	III
70	1359.92	III
110	1498.93	III
120	1502.01	III
110	1504.18	III
140	1504.79	III
130	1506.30	III
110	1506.77	III
100	1553.02	III
110	1557.59	III
100	1563.73	III
100	1565.54	III
100	1623.40	III
100	1650.76	III
130	1670.39	III
140	1677.40	III
100	1716.99	III
120	1770.09	III
150	1770.67	III
110	1772.93	III
140	1773.36	III
160	1791.65	III
110	1803.03	III
170	1805.90	III
110	1839.30	III
120	1839.97	III
110	1840.14	III
100	2027.44	III
120	2030.32	III
120	2217.17	III
50	2298.29	IV
40	2451.58	IV
120	2452.07	III
50	2456.92	IV
130	2464.85	III
130	2470.29	III
120	2478.73	III
150	2484.37	III
120	2542.77	III
120	2580.04	III
130	2583.81	III
120	2593.23	III
130	2595.53	III
140	2599.28	III
130	2625.01	III
140	2629.70	III
120	2656.44	III
130	2755.55	III
160	2759.63	III
120	2788.15	III
160	2811.45	III
40	2820.74	IV
50	2826.13	IV
140	2833.99	III
150	2835.63	III
150	2860.33	III
120	2862.86	III
140	2887.58	III
150	2889.45	III
120	2905.30	III
140	2913.29	III
160	2916.34	III
140	2932.49	III
140	2994.28	III
120 h	2997.21	III
130	2997.53	III
120	2999.47	III
130	3039.25	III
120	3039.75	III
160	3042.80	III
150	3049.14	III
140	3113.62	III
160	3115.70	III
180	3121.54	III
140	3124.79	III
140	3134.23	III
140	3146.99	III
180	3174.17	III
170	3174.76	III
120	3214.00	III
200	3501.45	II
200	3501.57	II
200	3502.96	II
6	3594.10	I
12	3668.17	I
270	3847.09	II
260	3849.99	II
250	3851.67	II
5	3898.48	I
8	3930.69	I
5	3934.26	I
5	3948.56	I
240	4024.73	II
220	4025.01	II
230	4025.49	II
200	4103.51	II
200	4246.23	II
200	4299.17	II
140 h	4420.30	III
120 h	4427.35	III
120 h	4432.32	III
140 h	4479.99	III
6	4960.65	I
150	5012.54	III
160	5110.99	III
15	5230.41	I
12	5279.01	I
18	5540.52	I
12	5552.43	I
10	5577.33	I
20	5624.06	I
12	5626.93	I
15	5659.15	I
40	5667.53	I
90	5671.67	I
18	5689.14	I
25	5700.82	I
25	5707.31	I
140	5753.17	III
120	5761.20	III
12	5950.15	I
25	5959.19	I
70	5965.28	I
50	5994.43	I
150	6015.83	I
80	6038.04	I
900	6047.54	I
100	6080.11	I
150	6091.82	III
140	6125.50	III
800	6149.76	I
400	6210.87	I
130	6233.57	III
13000	6239.65	I
10000	6348.51	I
140	6363.05	III
8000	6413.65	I
450	6569.69	I
300	6580.39	I
400	6650.41	I
1800	6690.48	I
400	6708.28	I
7000	6773.98	I
1500	6795.53	I
9000	6834.26	I
50000	6856.03	I
8000	6870.22	I
15000	6902.48	I
6000	6909.82	I
4000	6966.35	I
45000	7037.47	I
30000	7127.89	I
15000	7202.36	I
1000	7309.03	I
15000	7311.02	I
700	7314.30	I
5000	7331.96	I
120	7336.77	III
130	7354.94	III
10000	7398.69	I
4000	7425.65	I
2200	7482.72	I
2500	7489.16	I
900	7514.92	I
5000	7552.24	I
5000	7573.38	I
7000	7607.17	I
18000	7754.70	I
15000	7800.21	I
300	7879.18	I
500	7898.59	I
350	7936.31	I
300	7956.32	I
80	8016.01	II
1000	8040.93	I
900	8075.52	I
350	8077.52	I
350	8126.56	I
600	8129.26	I
300	8159.51	I
600	8179.34	I
300	8191.24	I
350	8208.63	I
2500	8214.73	I
3000	8230.77	I
500	8232.19	I
1500	8274.62	I
2000	8298.58	I
600	8302.40	I
900	8807.58	I
1000	8900.92	I
300	8912.78	I
350	9025.49	I
400	9042.10	I
350	9178.68	I
200	9433.67	I
25	9505.30	I
12	9662.04	I
25	9734.34	I
15	9822.11	I
12	9902.65	I
80 h	10047.98	II
15	10285.45	I
20	10862.31	I

Francium Fr Z = 87

Intensity	Wavelength/Å	
	7177.	I

Intensity	Wavelength/Å	
Gadolinium Gd Z = 64		
1200	1007.24	IV
1200	1063.84	IV
1600	1476.98	IV
1500	1705.03	IV
1600	1706.01	IV
2000	1736.24	IV
1500	1815.32	IV
2200	1975.24	III
3400	2018.07	III
2800	2359.31	III
1400	2397.87	IV
2800	2697.39	III
2800	2703.28	III
2700	2727.89	III
9000	2904.73	III
9500	2955.53	III
1200	2999.04	II
2100	3010.13	II
1900	3027.60	II
2100	3032.84	II
1600	3034.05	II
2100	3081.99	II
3500	3100.50	II
930	3145.00	II
980	3156.53	II
980	3161.37	II
4000	3176.66	III
1400	3331.38	II
1100	3336.18	II
5400	3350.47	II
4300	3358.62	II
5400	3362.23	II
1100	3392.53	II
1100 d	3407.56	II
6900	3422.47	II
1700	3439.21	II
2700	3439.99	II
1400	3450.38	II
1100	3451.23	II
2700	3463.98	II
1700	3467.27	II
1700	3468.99	II
1400	3473.22	II
2200	3481.28	II
1700	3481.80	II
1700	3494.40	II
1400	3505.51	II
1100	3512.50	II
4300	3545.80	II
3900	3549.36	II
1400	3557.05	II
5400	3584.96	II
1100	3592.71	II
1100	3604.87	I
6100	3646.19	II
3900	3654.62	II
3100	3656.15	II
1400	3662.26	II
2700	3664.60	II
2000	3671.20	II
2000	3684.13	I
3100	3687.74	II
2000	3697.73	II
1300	3699.73	II
2700	3712.70	II
2000	3713.57	I
1400	3716.36	II
2000	3717.48	I
1800 d	3719.45	II
1500	3730.84	II
4500	3743.47	II
1400	3758.31	II
8700	3768.39	II
1400	3770.69	II
2900	3783.05	I
5100	3796.37	II
3700	3813.97	II
3300	3850.69	II
5100	3850.97	II
4300	3852.45	II
1600	3866.99	I
1500	3894.70	II
2200	3916.51	II
1200	3934.79	I
1400	3945.54	I
1200	3957.67	II
1100	4023.14	I
1100	4028.15	I
1400	4037.33	II
1600	4045.01	I
1300	4049.43	II
2200	4049.86	II
2600	4053.64	I
2600	4058.22	I
1900	4063.39	II
1300	4078.44	II
2800	4078.70	I
1500	4085.56	II
1100	4092.71	I
2600	4098.61	II
2200	4130.37	II
1100	4132.28	II
2400	4175.54	I
2400	4184.25	II
2200	4190.78	I
1300	4212.00	II
4800	4225.85	I
1700	4251.73	II
1600	4262.09	I
1100	4306.34	I
1800	4313.84	I
2600 d	4325.57	II
1900	4327.12	I
1000	4344.30	II
2200	4346.46	I
1400	4401.86	I
1400	4422.41	I
1100	4430.63	I
1100	4519.66	I
910	4537.81	I
520	4614.50	I
700	4694.33	I
410	4743.65	I
470	4767.24	I
300	4784.62	I
320	4821.69	I
280	4934.12	I
750	5015.04	I
75	5039.09	I
5000	5091.70	III
130	5098.38	II
910	5103.45	I
180	5108.91	II
120	5125.56	II
860	5155.84	I
190	5176.28	II
410	5197.77	I
280	5219.40	I
130	5233.93	I
320	5251.18	I
120	5252.14	II
140	5255.80	I
280	5283.08	I
280	5301.67	I
220	5302.76	I
280	5307.30	I
130	5321.50	I
280	5321.78	I
110	5327.32	I
170	5333.30	I
300	5343.00	I
200	5348.67	I
300	5350.38	I
240	5353.26	I
3000	5365.96	III
150	5370.63	I
4000	5553.30	III
3000	5587.88	III
190	5617.91	I
110	5632.25	I
260	5643.24	I
3000	5658.98	III
390	5696.22	I
120	5733.86	II
240	5791.38	I
220	5851.63	I
280	5856.22	I
110	5904.56	I
170	5911.45	II
85	5930.29	I
85	5936.84	I
65	5937.71	I
110 h	5988.02	I
430	6114.07	I
75	6305.15	II
40	6331.35	I
40	6380.95	II
40 h	6538.15	I
55	6564.78	I
50	6634.36	II
35	6681.23	II
85	6730.73	I
50	6752.67	II
26	6786.33	II
100	6828.25	I
100	6916.57	I
50	6985.89	II
75	6991.92	I
60	6996.76	II
45	7006.16	II
35	7122.57	I
170	7168.37	I
28	7189.57	II
28	7262.66	I
35	7441.85	I
40	7464.36	I
55	7562.97	I
80	7733.50	I
35	7846.35	II
35	7856.93	I
25	7930.25	II
18	8146.15	I
21	8668.63	I
21 h	8832.06	II
14 h	8849.14	I
18 h	8867.31	I
5000	14332.88	III
Gallium Ga Z = 31		
14	294.53	IV
61	295.67	IV
30	298.44	V
30	300.01	V
30	302.86	V
41	304.99	IV
30	307.03	V
30	308.26	V
30	311.79	V
30	313.68	V
40	319.41	V
40	322.31	V
50	322.99	V
30	323.10	V
40	324.25	V
40	324.95	V
40	326.14	V
30	326.77	V
30	328.65	V
25	423.18	IV
16	439.92	IV
50	620.00	III
40	622.01	III
90	806.51	III
90	817.30	III
50	828.70	III
20	878.17	V
40	973.21	V
40	989.75	V
90	1014.47	V
90	1019.71	V
120	1050.48	V
80	1054.56	V
90	1058.12	V
80	1066.69	V

Intensity	Wavelength/Å	
60	1069.60	V
80	1073.77	V
90	1078.83	V
110	1079.60	V
60	1080.99	V
80	1085.00	III
250	1085.01	V
80	1087.37	V
90	1091.71	V
100	1094.36	V
80	1095.10	V
160	1102.83	V
140	1103.03	V
60	1105.61	III
75	1105.62	V
70	1106.17	V
80	1118.34	V
120	1126.40	V
130	1128.10	V
120	1128.53	V
100	1129.94	V
130	1136.07	V
67	1137.06	IV
130	1150.23	V
90	1150.27	III
70	1156.10	IV
120	1156.51	V
35	1157.74	V
70	1163.60	IV
25	1169.40	V
75	1170.58	IV
48	1171.71	IV
40	1178.95	V
68	1185.23	IV
40	1186.06	IV
73	1190.89	IV
73	1193.02	IV
75	1195.02	IV
69	1201.54	IV
72	1206.89	IV
80	1213.17	V
63	1216.15	IV
50	1228.03	IV
60	1236.38	IV
60	1238.59	IV
75	1245.53	IV
83	1258.77	IV
81	1264.66	IV
82	1267.15	IV
90	1267.16	III
81	1279.24	IV
15	1283.64	V
80	1285.33	IV
80	1293.46	III
60	1295.36	III
82	1295.86	IV
83	1299.46	IV
82	1303.53	IV
80	1309.68	IV
80	1314.82	IV
60	1323.15	III
85	1338.09	IV
77	1347.03	IV
76	1351.06	IV
70	1353.92	III
74	1364.63	IV
60	1395.54	IV
77	1402.55	IV
70	1405.32	IV
73	1465.87	IV
90	1495.07	III
50	1534.46	III
10	1813.98	II
15	1845.30	II
20	2091.34	II
90	2417.70	III
90	2423.98	III
15	2424.36	III
10	2632.66	I
10	2665.05	I
20	2700.47	II
15	2780.15	II
50	3521.77	III
80	3581.19	III
100	3589.34	III
10	3731.10	III
10	3806.60	III
10	4032.99	I
10	4172.04	I
15	4251.16	II
10 h	4254.04	II
10	4255.77	II
40	4262.00	II
100	4380.69	III
150	4381.76	III
100	4863.00	III
150	4993.78	III
10	5808.28	III
20	5848.25	III
15	5993.51	III
10	6334.2	II
2000	6396.56	I
1000	6413.44	I
10 h	7251.4	I
20 h	7403.0	I
30 h	7464.0	I
10 h	7620.5	I
50 h	7734.77	I
100 h	7800.01	I
15 h	8002.55	I
20 h	8074.25	I
100 h	8311.86	I
200 h	8386.49	I
200 h	9492.92	I
200 h	9493.12	I
300 h	9589.36	I
100 h	10905.95	I
400	11949.12	I
200	12109.78	I
60	14996.64	I
60	22016.81	I
70	22568.71	I

Germanium Ge Z = 32

Intensity	Wavelength/Å	
700	294.51	V
1000	295.64	V
200	304.98	V
20	621.52	V
50	724.21	V
60	746.88	V
60	760.05	V
10	862.234	II
15	875.493	II
15	905.977	II
20	920.554	II
300	971.35	V
300	990.66	V
50	999.101	II
300	1004.38	V
100	1016.638	II
900	1045.71	V
700	1072.66	V
100	1075.072	II
300	1085.51	II
40	1088.45	III
800	1089.49	V
200	1098.71	II
500	1106.74	II
1000	1116.94	V
500	1120.46	II
700	1163.39	V
200	1164.27	II
500	1181.19	II
500	1181.65	II
200	1188.73	II
20	1188.99	IV
300	1191.26	II
700	1222.30	V
20	1229.81	IV
500	1237.059	II
500	1261.905	II
100	1264.710	II
200	1401.24	II
200	1538.091	II
500	1576.855	II
75	1581.070	II
100	1602.486	II
3 r	1615.57	I
2 r	1624.130	I
2 r	1630.173	I
3 r	1636.31	I
4 r	1639.730	I
2	1647.531	I
200	1649.194	II
2	1651.528	I
4 r	1651.955	I
3	1661.345	I
4 r	1663.539	I
10 h	1665.275	I
4	1667.802	I
3 r	1670.608	I
100 r	1691.090	I
200 r	1716.784	I
100 h	1739.102	I
100	1742.195	I
50	1746.065	I
200	1750.043	I
100	1758.279	I
100 h	1764.185	I
100 h	1765.284	I
50 h	1766.433	I
200	1774.176	I
200	1785.046	I
100 h	1793.071	I
75 h	1801.432	I
200 h	1841.328	I
200 h	1842.410	I
100 h	1844.410	I
100 h	1845.872	I
100 h	1846.958	I
200	1853.134	I
500 r	1860.086	I
100	1865.052	I
300 r	1874.256	I
100	1895.197	I
500 r	1904.702	I
50 h	1908.434	I
30	1912.409	I
300 r	1917.592	I
100 h	1923.467	I
500 r	1929.826	I
10 h	1934.048	I
100 r	1937.483	I
500	1938.008	II
100 r	1938.300	I
500	1938.891	II
30 s	1944.116	I
200	1944.731	I
200	1955.115	I
500	1962.013	I
30 h	1963.373	I
30	1965.383	I
200	1970.880	I
200	1979.274	II
300 h	1987.849	I
300	1988.267	I
500 r	1998.887	I
200	2011.29	I
1700	2019.068	I
2400 r	2041.712	I
1600 r	2043.770	I
420	2054.461	I
220 h	2057.238	I
750 r	2065.215	I
2600 r	2068.656	I
420	2086.021	I
2000 r	2094.258	I
25	2104.45	III
240	2105.824	I
95 h	2124.744	I
50 h	2186.451	I
340 r	2198.714	I
15	2220.375	I
18	2256.001	I
18	2314.201	I

Intensity	Wavelength/Å	
24	2327.918	I
15	2359.233	I
20	2379.144	I
10	2389.472	I
15	2397.885	I
130	2417.367	I
30	2436.412	I
30	2488.25	IV
90	2497.962	I
500	2500.54	II
70	2533.230	I
20	2542.44	IV
3	2556.298	I
28	2589.188	I
500	2592.534	I
8	2644.184	I
1200	2651.172	I
500	2651.568	I
500	2691.341	I
850	2709.624	I
400	2729.78	II
40	2740.426	I
650	2754.588	I
70	2793.925	I
80	2829.008	I
1000	2831.843	II
1000	2845.527	II
750	3039.067	I
600	3067.021	I
20	3124.816	I
35	3211.86	III
40	3255.05	III
110	3269.489	I
40	3434.03	III
300	3499.21	II
60	3554.19	IV
50	3676.65	IV
200	4178.96	III
70	4226.562	I
200	4260.85	III
150	4291.71	III
10	4685.829	I
1000	4741.806	II
1000	4814.608	II
50	4824.097	II
100	5131.752	II
200	5178.648	II
3	5194.583	I
6	5265.892	I
6	5513.263	I
8	5564.741	I
8	5607.010	I
6	5616.135	I
7	5621.426	I
6	5664.226	I
5	5664.842	I
9	5691.954	I
6	5701.776	I
5	5717.877	I
6	5801.029	I
9	5802.093	I
1000	5893.389	II
500	6021.041	II
75	6283.452	II
100	6336.377	II
100	6484.181	II
6	6557.488	I
50	7049.369	II
30	7145.390	II
5	7353.334	I
7	7384.208	I
10	7833.575	I
10	8031.039	I
6	8044.165	I
10	8256.013	I
10	8482.21	I
9	8700.60	I
5	9068.785	I
5	9095.957	I
6	9398.868	I
20	9474.993	II
20	9475.645	II
4	9492.559	I
7	9625.664	I
5	10039.436	I
4	10200.952	I
10	10382.427	I
10	10404.913	I
8	10734.068	I
8	10947.416	I
10	11125.130	I
230	11252.83	I
600	11714.76	I
1300	12069.20	I
1050	12391.58	I
235	13107.61	I
470	14822.38	I
150	16759.79	I
135	17214.34	I
70	18811.86	I
62	19279.24	I
28	20673.64	I

Gold Au Z = 79

Intensity	Wavelength/Å	
100	843.44	III
100	845.14	III
200	945.10	III
100	1040.63	III
80	1044.49	III
80	1046.81	III
100 h	1239.96	III
100	1278.51	III
100	1314.84	III
100 h	1328.37	I
200	1336.72	III
180	1341.68	III
100	1348.89	III
150	1350.32	III
150	1355.61	III
150	1356.13	III
50	1363.98	I
500	1365.40	III
200	1367.17	III
60	1368.62	I
70	1374.82	I
50	1375.76	I
180	1377.73	III
150	1378.69	III
150	1379.98	III
125	1380.53	III
200	1381.36	III
80	1382.75	I
50	1385.33	I
300	1385.79	III
100	1389.41	III
180	1391.46	III
60	1392.27	I
180	1396.00	III
50	1402.12	I
100	1402.91	III
70	1407.38	I
100	1408.45	I
225	1409.50	III
250	1413.80	III
100	1414.27	III
100	1417.09	III
125	1417.39	III
150	1427.42	III
300	1428.93	III
80	1429.19	I
250	1430.06	III
275	1433.37	III
50	1435.79	I
250	1435.81	III
300	1439.12	III
200	1441.21	III
150	1446.37	III
250	1448.42	III
250	1454.95	III
100	1464.72	III
150	1471.28	III
100	1474.73	III
150	1481.10	III
100	1481.76	I
300	1487.15	III
250	1487.91	III
200	1489.47	III
250	1500.37	III
200	1502.47	III
200	1503.74	III
100	1542.00	III
100	1548.50	III
200	1567.54	III
200	1574.85	III
200	1579.44	III
150	1584.10	III
200	1587.16	I
200	1589.56	III
150	1593.41	III
70	1598.24	I
200	1600.51	III
250	1617.16	III
100	1617.78	III
500	1621.93	III
100	1624.34	I
300 d	1629.13	III
250	1638.88	III
50	1639.90	I
100	1644.17	III
150	1646.67	I
250	1652.74	III
250	1664.77	III
100	1665.76	I
100	1668.11	III
125	1673.93	III
1000	1693.94	III
150	1697.09	III
200	1698.98	III
200	1699.34	I
200	1700.00	III
200	1702.25	III
100	1707.53	III
250	1710.16	III
200	1715.69	III
100	1716.71	III
300	1717.83	III
500	1727.31	III
100 d	1733.17	III
300	1738.48	III
150	1744.39	III
500	1746.10	III
500	1756.92	III
500	1761.95	III
300	1767.42	III
100	1774.42	III
800	1775.17	III
200	1776.40	III
100	1780.57	III
60	1783.22	II
300	1786.11	III
150	1792.65	III
500	1793.76	III
200	1801.98	III
400	1805.24	III
100	1809.81	III
400	1821.17	III
400	1844.89	III
150	1848.83	III
500	1861.80	III
150	1871.92	III
100	1879.83	I
150	1918.28	III
100	1932.04	III
100	1935.42	III
200	1948.79	III
100	1958.47	III
400	1989.63	III
150	1996.85	III
11000	2012.00	I
2600	2021.38	I
150	2082.09	II
300	2083.09	III
60	2110.68	II
100	2159.08	III
80	2167.33	III

Intensity	Wavelength/Å	
200	2172.20	III
100	2184.11	III
500	2188.97	III
70	2248.56	II
80	2263.62	II
300	2322.27	III
180	2352.65	I
100	2382.40	III
120	2387.75	I
150	2402.71	III
150	2405.12	III
2600	2427.95	I
60	2533.52	II
250	2641.48	I
3400	2675.95	I
1100	2748.25	I
100	2780.82	I
1000	2802.04	II
300	2819.79	II
100	2822.55	II
100 h	2825.44	II
300	2837.85	II
100	2846.92	II
100	2856.74	II
300	2883.45	I
300	2891.96	I
100	2893.25	II
100	2907.04	II
300	2913.52	II
300	2918.24	II
100	2954.22	II
100 h	2990.27	II
300	2994.80	II
320	3029.20	I
300	3065.42	I
100	3122.50	II
1600	3122.78	I
100	3194.72	I
100	3227.99	III
300	3230.63	I
300	3308.30	I
300	3309.64	I
100	3309.86	III
100	3320.12	I
100	3355.15	I
100 h	3391.31	I
100	3395.40	I
100	3467.21	I
300	3557.36	I
300	3586.73	I
100	3611.57	I
100 h	3631.31	I
300	3637.90	I
100 h	3645.02	I
100	3650.74	I
100	3709.62	I
100	3796.01	I
100	3874.73	I
100 h	3892.26	I
400	3897.86	I
300	3909.38	I
100	3927.69	I
400	4040.93	I
700	4065.07	I
100	4084.10	I
100	4241.80	I
200	4315.11	I
120 h	4437.27	I
250	4488.25	I
900 h	4607.51	I
100 h	4620.56	I
500	4792.58	I
100	4811.60	I
100	5147.44	I
300	5230.26	I
100 h	5261.76	I
100	5655.77	I
100 h	5721.36	I
300	5837.37	I
100 h	5862.93	I
300 h	5956.96	I
600	6278.17	I
100	6562.68	I
600	7510.73	I
10	8145.06	I
10	9254.28	I

Hafnium Hf ***Z*** **= 72**

Intensity	Wavelength/Å	
220	545.41	V
180	600.00	V
200	618.27	IV
200	644.54	IV
400	647.39	IV
600	665.65	IV
200	673.49	IV
270	867.25	V
180	875.88	V
180	885.58	V
180	896.14	V
180	901.54	V
180	919.10	V
270	921.67	V
245	951.62	V
180	960.12	V
180	964.74	V
160	971.51	V
160	1092.76	V
160	1201.76	V
270	1232.03	V
200	1233.59	V
160	1237.42	V
160	1239.53	V
160	1244.46	V
440	1396.66	V
270	1400.09	V
160	1401.70	V
370	1407.17	V
370	1408.38	V
270	1412.28	V
270	1413.51	V
160	1421.96	V
220	1422.53	V
370	1433.43	V
370	1437.27	V
500	1437.73	V
370	1445.40	V
270	1457.91	V
100	1717.21	IV
270	1719.32	V
550	1729.08	V
750	1731.83	V
750	1733.96	V
440	1741.74	V
1000	1749.11	V
1000	1750.19	V
500	1760.89	V
370	1765.62	V
270	1774.02	V
6200	2012.78	II
8500	2028.18	II
300	2070.94	III
1200	2096.18	II
200 h	2099.30	III
200 h	2110.31	III
200	2155.66	III
200	2183.50	III
200	2195.44	III
540	2210.82	II
200	2234.59	III
320	2254.01	II
160	2255.15	II
250	2266.83	II
620	2277.16	II
200 h	2313.44	III
230	2321.14	II
580	2322.47	II
300	2323.25	II
300	2324.89	II
200	2332.97	II
300	2336.47	III
200	2337.33	II
230	2343.32	II
320	2347.44	II
540	2351.22	II
250	2380.30	II
250	2383.540	III
170	2393.18	II
450	2393.36	II
670	2393.83	II
540	2405.42	II
370	2410.14	II
320	2417.69	II
390	2447.25	II
450	2460.49	II
400	2461.74	III
430	2464.19	II
210	2469.18	II
2000	2495.16	III
290	2496.99	II
580	2512.69	II
580	2513.03	II
1000	2515.16	III
890	2516.88	II
340	2531.19	II
200	2537.33	II
320	2551.40	II
400 h	2560.74	III
250	2563.61	II
300 h	2567.46	III
890	2571.67	II
320	2573.90	II
320	2576.82	II
300	2578.14	II
320	2582.54	II
390	2606.37	II
450	2607.03	II
230	2613.60	II
450	2622.74	II
160	2637.00	I
1100	2638.71	II
1100	2641.41	II
160	2642.75	I
670	2647.29	II
160	2657.84	II
210	2661.88	II
290	2683.35	II
200	2687.22	III
670	2705.61	I
210	2712.42	II
250	2718.59	I
710	2738.76	II
200	2743.64	I
360	2751.81	II
500	2753.60	III
450	2761.63	I
160	2766.96	I
170	2773.02	I
980	2773.36	II
180	2774.02	II
390	2779.37	I
230	2808.00	II
230	2813.86	II
170	2814.48	II
230	2817.68	I
200	2819.74	I
1200	2820.22	II
490	2822.68	II
180	2833.28	I
410	2845.83	I
270	2849.21	II
270	2850.96	I
180	2851.21	II
180	2860.56	I
760	2861.01	II
760	2861.70	II
2100	2866.37	I
210	2887.14	I
800	2889.62	I
1800	2898.26	I
1200	2904.41	I
890	2904.75	I
2000	2916.48	I
580	2918.58	I
320	2919.59	II
180	2924.62	I

Intensity	Wavelength/Å	
490	2929.63	II
450	2929.90	I
710	2937.80	II
2000	2940.77	I
160	2944.71	I
1200	2950.68	I
1100	2954.20	I
540	2958.02	I
1400	2964.88	I
620	2966.93	I
710	2968.81	II
890	2975.88	II
1100	2980.81	I
210	2982.72	I
170	3000.10	II
800	3005.56	I
1100	3012.90	II
540	3016.78	I
1100	3016.94	II
980	3018.31	I
1200	3020.53	I
410	3031.16	II
710	3050.76	I
1100	3057.02	I
850	3067.41	I
2100	3072.88	I
170	3074.10	I
250	3074.79	I
430	3080.84	I
200	3096.76	I
340	3101.40	II
710	3109.12	II
710	3131.81	I
850	3134.72	II
170	3139.65	II
220	3145.32	II
220	3148.41	I
450	3156.63	I
270	3159.82	I
710	3162.61	II
450	3168.39	I
890	3172.94	I
450	3176.86	II
220	3181.01	I
360	3193.53	II
670	3194.19	II
200	3196.93	I
310	3206.11	I
180	3210.98	I
180	3217.30	II
180	3220.61	II
360	3247.66	I
220	3249.53	I
890	3253.70	II
270	3255.28	II
180	3273.66	II
200 h	3279.67	III
270	3279.98	II
160	3291.05	I
210	3306.12	I
340	3310.27	I

Intensity	Wavelength/Å	
670	3312.86	I
180	3317.99	II
890	3332.73	I
370	3352.06	II
230	3358.91	I
180	3360.06	I
180	3378.93	I
230	3384.70	II
170	3386.21	I
800	3389.83	II
230	3392.81	I
230	3394.59	II
230	3397.26	I
230	3397.60	I
2300	3399.80	II
170	3400.21	I
180	3402.51	I
230	3410.17	II
230	3417.34	I
410	3419.18	I
200	3428.37	II
250	3438.24	II
710	3472.40	I
200	3478.99	II
480	3479.28	II
250	3495.75	II
250	3497.16	I
980	3497.49	I
1200	3505.23	II
980	3523.02	I
980	3535.54	II
760	3536.62	I
180	3548.81	I
540	3552.70	II
1300	3561.66	II
270	3567.36	I
1100	3569.04	II
210	3597.42	II
540	3599.87	I
800	3616.89	I
320	3630.87	II
800	3644.36	II
320	3649.10	I
200	3651.84	I
220	3665.35	II
200	3672.27	II
480	3675.74	I
2200	3682.24	I
280	3696.51	I
240	3699.72	II
340	3701.15	II
1000	3717.80	I
650	3719.28	II
160	3729.10	I
460	3733.79	I
160	3737.88	II
400	3746.80	I
170	3766.92	II
200	3768.25	I
1400	3777.64	I
1400	3785.46	I

Intensity	Wavelength/Å	
650	3793.37	II
850 d	3800.38	I
320	3811.78	I
1300	3820.73	I
280	3830.02	I
800	3849.18	I
600	3858.31	I
230	3860.91	I
200	3872.55	II
160	3877.10	II
380	3880.82	II
200	3882.52	I
200	3889.23	I
200	3889.33	I
620	3899.94	I
620	3918.09	II
200	3923.90	II
320	3931.38	I
410	3951.83	I
160	3968.01	I
200	3973.48	I
180	4032.27	I
230	4062.84	I
180	4083.35	I
540	4093.16	II
1100	4174.34	I
160	4206.58	II
190	4209.70	I
170	4228.08	I
170	4232.44	II
170	4260.98	I
200	4263.39	I
170	4272.85	II
320	4294.79	I
160	4330.27	I
180	4336.66	II
250	4356.33	I
180	4370.97	II
160	4417.91	I
200	4438.04	I
250	4565.94	I
500 d	4598.80	I
230	4620.86	I
210	4655.19	I
120	4699.01	I
160	4782.74	I
310	4800.50	I
130	4859.24	I
120	4975.25	I
95	5018.20	I
95	5047.45	I
75	5170.18	I
230	5181.86	I
110	5243.99	I
120	5294.87	I
110	5354.73	I
110	5373.86	I
75	5452.92	I
230	5550.60	I
230	5552.12	I
95	5613.27	I

Intensity	Wavelength/Å	
160	5719.18	I
95	6098.67	I
95	6185.13	I
85	6789.27	I
160	6818.94	I
160	7063.83	I
570	7131.81	I
650	7237.10	I
410	7240.87	I
360	7624.40	I
110	7740.17	I
310	7845.35	I
130	7920.71	I
250	7994.73	I
130	8204.58	I
150	8546.48	I
160	8640.06	I
40	8711.24	I
65	9004.73	I

Helium He $Z = 2$

Intensity	Wavelength/Å	
15	231.454	II
20	232.584	II
30	234.347	II
50	237.331	II
100	243.027	II
300	256.317	II
1000	303.780	II
500	303.786	II
10	320.293	I
2	505.500	I
3	505.684	I
4	505.912	I
5	506.200	I
7	506.570	I
10	507.058	I
15	507.718	I
20	508.643	I
25	509.998	I
35	512.098	I
50	515.616	I
100	522.213	I
400	537.030	I
1000	584.334	I
50	591.412	I
5	958.70	II
6	972.11	II
8	992.36	II
15	1025.27	II
30	1084.94	II
35	1215.09	II
50	1215.17	II
120	1640.34	II
180	1640.47	II
7	2385.40	II
9	2511.20	II
50	2577.6	I
1	2723.19	I
12	2733.30	II
2	2763.80	I
10	2818.2	I
4	2829.08	I

Intensity	Wavelength/Å	
10	2945.11	I
40	3013.7	I
20	3187.74	I
3	3202.96	II
15	3203.10	II
1	3354.55	I
2	3447.59	I
1	3587.27	I
3	3613.64	I
2	3634.23	I
3	3705.00	I
1	3732.86	I
10	3819.607	I
1	3819.76	I
500	3888.65	I
20	3964.729	I
1	4009.27	I
50	4026.191	I
5	4026.36	I
12	4120.82	I
2	4120.99	I
3	4143.76	I
10	4387.929	I
3	4437.55	I
200	4471.479	I
25	4471.68	I
6	4685.4	II
30	4685.7	II
30	4713.146	I
4	4713.38	I
20	4921.931	I
100	5015.678	I
10	5047.74	I
5	5411.52	II
500	5875.62	I
100	5875.97	I
8	6560.10	II
100	6678.15	I
3	6867.48	I
200	7065.19	I
30	7065.71	I
50	7281.35	I
1	7816.15	I
2	8361.69	I
2	9063.27	I
2	9210.34	I
10	9463.61	I
4	9516.60	I
3	9526.17	I
1	9529.27	I
1	9603.42	I
3	9702.60	I
6	10027.73	I
2	10031.16	I
15	10123.6	II
1	10138.50	I
10	10311.23	I
2	10311.54	I
3	10667.65	I
300	10829.09	I
1000	10830.25	I
2000	10830.34	I
9	10913.05	I
3	10917.10	I
4	11626.4	II
30	11969.12	I
20	12527.52	I
50	12784.99	I
20	12790.57	I
7	12845.96	I
10	12968.45	I
2	12984.89	I
12	15083.64	I
200	17002.47	I
1	18555.55	I
6	18636.8	II
500	18685.34	I
200	18697.23	I
100	19089.38	I
20	19543.08	I
1000	20581.30	I
80	21120.07	I
10	21121.43	I
20	21132.03	I
3	30908.5	II
4	40478.90	I

Holmium Ho Z = 67

Intensity	Wavelength/Å	
170	2502.91	II
170	2533.80	I
190	2605.86	II
270	2750.35	II
270	2769.89	II
300	2824.20	II
270 c	2831.69	II
270	2849.10	II
360	2880.26	II
460	2880.98	II
570 c	2909.41	II
410 c	2979.63	II
410	2987.64	II
480 c	3049.38	II
410 c	3054.00	II
500 c	3057.45	II
500 c	3082.34	II
910	3084.36	II
430 c	3086.54	II
760	3118.50	II
580 c	3166.62	II
810	3173.78	II
810 c	3181.50	II
980 c	3281.97	II
630 c	3337.23	II
980 c	3343.58	II
8100 c	3398.98	II
810 c	3410.26	II
1400 c	3414.90	II
5400	3416.46	II
1200	3421.63	II
2000 c	3425.34	II
2000 c	3428.13	II
3200	3453.14	II
16000 c	3456.00	II
1600	3461.97	II
810 c	3473.91	II
5400 c	3474.26	II
6300	3484.84	II
2500 c	3494.76	II
810 c	3498.88	II
810	3510.73	I
4100 c	3515.59	II
410 c	3519.94	II
630	3540.76	II
1600	3546.05	II
1100 c	3556.78	II
410	3560.15	II
410 c	3573.24	II
630 c	3574.80	II
810	3579.12	I
410	3580.75	II
410	3581.83	II
630 c	3592.23	II
1100 cw	3598.77	II
540 c	3600.95	II
410	3618.43	I
430 c	3626.69	II
490	3627.25	II
430 c	3631.76	II
430 c	3638.30	II
1600 c	3662.29	I
1400	3667.97	I
720	3679.19	I
670	3679.70	I
720	3682.65	I
580	3690.65	I
410	3700.04	I
490 c	3702.35	II
430	3712.88	I
450	3720.72	I
1100	3731.40	I
810	3736.35	I
3200 cw	3748.17	II
8900 c	3796.75	II
8900 c	3810.73	II
410 c	3835.35	II
1300 cw	3837.51	II
410 c	3842.05	II
1100	3843.86	II
490 c	3846.73	II
1800 c	3854.07	II
2700 c	3861.68	II
3000 c	3888.96	II
13000 c	3891.02	II
1300 cw	3905.68	II
580	3955.73	I
490	3959.68	I
2700	4040.81	I
5400 c	4045.44	II
8100	4053.93	I
1700	4065.09	II
720	4068.05	I
8900	4103.84	I
2900	4108.62	I
1500	4120.20	I
1300	4125.65	I
4300	4127.16	I
1500	4136.22	I
980 cw	4152.61	II
8100	4163.03	I
2500	4173.23	I
540	4194.35	I
2000	4227.04	I
1300 cw	4254.43	I
490	4264.05	I
1300	4350.73	I
300	4477.64	II
290	4629.10	II
80	4701.69	II
130	4709.84	II
130 c	4717.52	I
290	4742.04	II
100 c	4757.01	I
65	4782.92	I
290	4939.01	I
250 c	4967.21	II
220	4979.97	I
90	4995.05	I
130	5042.37	I
80	5093.07	I
140	5127.81	I
130	5142.59	II
110	5143.22	II
160	5149.59	II
90 c	5167.88	I
130 c	5182.11	I
90	5190.11	II
65	5251.82	I
90	5301.25	I
80	5330.11	I
90	5359.99	I
100	5407.08	I
70	5566.52	I
65	5627.60	I
140	5659.58	I
140 c	5691.47	I
140 c	5696.57	I
140 c	5860.28	I
70 c	5882.99	I
70	5921.76	I
70 cw	5948.03	I
70	5972.76	I
90	5973.52	I
230 c	5982.90	I
120	6081.79	I
70	6208.65	I
70 c	6305.36	I
70	6550.97	I
260	6604.94	I
120	6628.99	I
55 cw	6694.32	I
55 c	6785.43	I
40 cw	6939.49	I
45 cw	6950.39	I
140	7555.09	I
40 c	7628.42	I

Intensity	Wavelength/Å	
50 c	7693.15	I
60 cw	7815.48	I
60	7894.64	I
50	8512.94	I
40	8670.19	I
90	8915.98	II
Hydrogen H Z = 1		
15	926.226	I
20	930.748	I
30	937.803	I
50	949.743	I
100	972.537	I
300	1025.722	I
1000	1215.668	I
500	1215.674	I
5	3835.384	I
6	3889.049	I
8	3970.072	I
15	4101.74	I
30	4340.47	I
80	4861.33	I
120	6562.72	I
180	6562.852	I
5	9545.97	I
7	10049.4	I
12	10938.1	I
20	12818.1	I
40	18751.0	I
5	21655.3	I
8	26251.5	I
15	40511.6	I
4	46525.1	I
6	74578.	I
Indium In Z = 49		
17	378.61	V
17	386.21	V
14	388.91	V
25	393.89	V
25	400.57	V
25	402.39	V
622	472.71	IV
689	479.39	IV
709	498.62	IV
10	882.24	III
10	890.84	III
10	915.87	III
85	954.67	IV
87	973.50	IV
86	991.60	IV
89	1024.68	IV
85	1024.79	IV
88	1031.45	IV
82	1031.98	IV
80	1054.43	IV
84	1063.03	IV
83	1068.25	IV
82	1069.82	IV
86	1077.64	IV
90	1082.10	IV
83	1086.33	IV
82	1096.81	IV
84	1097.18	IV
85	1116.10	IV
80	1124.06	IV
90	1131.46	IV
85	1144.43	IV
80	1145.41	IV
89	1146.62	IV
83	1154.11	IV
84	1154.60	IV
90	1157.71	IV
90	1157.82	IV
85	1159.78	IV
88	1176.50	IV
85	1191.58	IV
83	1204.87	IV
90	1206.55	IV
88	1221.50	IV
85	1221.90	IV
85	1233.58	IV
87	1235.84	IV
90	1373.20	IV
88	1398.77	IV
81	1412.09	IV
100	1625.42	III
100	1748.83	III
30	1842.41	III
40	1850.30	III
30	2154.08	III
50	2205.28	II
40	2281.64	II
100 c	2306.05	II
90 d	2313.21	II
70 d	2327.95	II
80 h	2334.57	II
110 d	2382.63	II
70 h	2427.20	II
100	2447.90	II
110 d	2488.62	II
90	2488.95	II
80	2498.59	II
100	2499.60	II
90 d	2500.99	II
110 d	2512.31	II
100	2521.37	I
100	2527.41	III
160 d	2554.44	II
1100	2560.15	I
200	2601.76	I
90 d	2654.70	II
100 d	2662.63	II
140 d	2668.65	II
140 d	2674.56	II
80	2683.12	II
1600	2710.26	I
300	2713.94	I
80	2726.15	III
130 d	2749.75	II
700	2753.88	I
90 d	2818.97	II
180 c	2836.92	I
80	2865.68	II
120 d	2890.18	II
1100	2932.63	I
100	2941.05	II
100	2982.80	III
110 c	2999.40	II
100	3008.08	III
8000	3039.36	I
110 d	3099.80	II
180 c	3101.8	II
130 c	3138.60	II
80 c	3142.75	II
130 d	3146.70	II
150	3155.77	II
100 c	3158.40	II
90 c	3176.30	II
90 d	3198.11	II
13000	3256.09	I
3000	3258.56	I
90 c	3338.50	II
100 c	3404.28	II
110 d	3438.40	II
180 c	3693.91	II
95 c	3708.13	II
380 w	3716.14	II
120 c	3718.30	II
160 c	3718.72	II
160 c	3723.40	II
170 w	3795.21	II
230 c	3799.21	II
250 c	3834.65	II
200 c	3842.18	II
100	3852.82	III
100	3889.78	II
100 c	3902.07	II
250 w	3962.35	II
120 c	4004.66	II
140 d	4013.92	II
100	4023.77	III
150	4032.32	III
410 w	4056.94	II
100	4071.57	III
100	4072.93	III
17000	4101.76	I
140 c	4205.14	II
100 d	4213.04	II
110 d	4219.66	II
100	4252.68	III
150 d	4372.87	II
150 c	4500.78	II
18000	4511.31	I
110 c	4549.01	II
140 c	4570.85	II
180 w	4578.02	II
180 w	4578.40	II
140 c	4616.08	II
170 c	4617.17	II
250 c	4620.14	II
150 w	4620.70	II
170 c	4627.30	II
140 c	4637.04	II
380 c	4638.16	II
220 c	4644.58	II
360 c	4655.62	II
320 w	4656.74	II
190 c	4681.11	II
450 w	4684.8	II
90 d	4907.06	II
150 c	4973.77	II
80 h	5109.36	II
100 w	5115.14	II
140 c	5117.40	II
270 c	5120.80	II
200 w	5121.75	II
80 d	5129.85	II
240 c	5175.42	II
140 c	5184.44	II
200	5248.77	III
150 c	5309.45	II
80	5411.41	II
140 c	5418.45	II
220 w	5436.70	II
130 c	5497.50	II
140 c	5507.08	II
320 c	5513.00	II
250 w	5523.28	II
130 c	5536.50	II
190 w	5555.45	II
240 c	5576.90	II
200 w	5636.70	II
100	5645.15	III
160 c	5708.50	II
100 c	5721.80	II
100	5819.50	III
210 c	5853.15	II
490 w	5903.4	II
260 w	5915.4	II
120 c	5918.78	II
130 c	6062.9	II
250 c	6095.95	II
210 c	6108.66	II
180 w	6115.9	II
230 w	6128.7	II
240 w	6129.4	II
320 w	6132.1	II
150 c	6140.0	II
90	6143.23	II
140 c	6148.10	II
190 w	6149.5	II
80	6161.15	II
180 w	6162.45	II
200	6197.72	III
100 c	6224.28	II
280 w	6228.3	II
140 w	6231.1	II
270 w	6304.8	II
290 w	6362.3	II
300 w	6469.0	II
210 c	6541.20	II
190 c	6751.88	II
180 c	6765.9	II
100 c	6783.72	II

Intensity	Wavelength/Å	
320 w	6891.5	II
380 w	7182.9	II
180 c	7255.0	II
210 c	7276.5	II
180 c	7303.4	II
320 c	7350.6	II
100 c	7632.7	II
100 c	7682.9	II
210 c	7740.7	II
100 c	7776.96	II
180 c	7789.0	II
90 c	7840.9	II
240 c	8227.0	II
100 w	8813.5	II
80 c	8832.6	II
120 c	9197.7	II
120 c	9202.0	II
220 w	9213.0	II
160 d	9241.1	II
100	9977.86	I
200	10257.03	I
100 h	10744.31	I
20	11334.72	I
20	11731.48	I
10	12912.59	I
9	13429.96	I
7	14719.08	I
6	22291.06	I
7	23879.13	I

Iodine I Z = 53

Intensity	Wavelength/Å	
30	363.78	V
36	380.74	V
45	565.53	V
50	607.57	V
6	612.46	IV
6	666.81	III
8	705.11	III
7	784.64	III
7	784.80	III
8	795.52	III
7	919.28	IV
10000	1034.66	II
8	1094.20	III
10000	1139.80	II
10000	1160.56	II
20000	1166.48	II
10000	1178.65	II
15000	1187.34	II
10000	1190.85	II
200	1218.41	I
20000	1220.89	II
600	1224.05	I
600	1224.08	I
500	1228.89	I
20000	1234.06	II
600	1251.34	I
8	1252.35	III
2500	1259.15	I
3000	1259.51	I
800	1261.27	I
600	1267.57	I
600	1267.60	I
1500	1275.26	I
3000	1289.40	I
10000	1300.34	I
3000	1302.98	I
3000	1313.95	I
3000	1317.54	I
2000	1330.19	I
20000	1336.52	II
5000	1355.10	I
3000	1357.97	I
5000	1360.97	I
3000	1361.11	I
2500	1367.71	I
2500	1368.22	I
4000	1383.23	I
3000	1390.75	I
2000	1392.90	I
2000	1400.01	I
8000	1425.49	I
5000	1446.26	I
5000	1453.18	I
5000	1457.39	I
5000	1457.47	I
10000	1457.98	I
2500	1458.79	I
4000	1459.15	I
2500	1465.83	I
1000	1485.92	I
5000	1492.89	I
5000	1507.04	I
5000	1514.68	I
15000	1518.05	I
2500	1526.45	I
5000	1593.58	I
5000	1617.60	I
2500	1640.78	I
15000	1702.07	I
12000	1782.76	I
5000	1799.09	I
75000	1830.38	I
15000	1844.45	I
2000	2061.63	I
7	2361.13	IV
6	2372.45	IV
7	2376.46	IV
8	2387.11	IV
9	2426.10	IV
8	2475.35	IV
8	2519.74	IV
8	2545.67	IV
7	2545.71	III
8	2652.23	IV
5000	3078.75	II
200	4102.23	I
200	4129.21	I
100 d	4134.15	I
500 d	4321.84	I
250	4763.31	I
1000	4862.32	I
200	4916.94	I
10000	5119.29	I
3000 c	5161.20	II
1000	5234.57	I
3000 c	5245.71	II
10000	5338.22	II
5000 c	5345.15	II
600 c	5427.06	I
3000	5435.83	II
2000 c	5464.62	II
10000	5625.69	II
2000 c	5690.91	II
4000 c	5710.53	II
1000 d	5764.33	I
2000	5894.03	I
5000	5950.25	II
300	5984.86	I
2000 d	6024.08	I
2000 c	6074.98	II
1000	6082.43	I
2000 c	6127.49	II
800	6191.88	I
500	6213.10	I
800	6244.48	I
1000	6293.98	I
500	6313.13	I
800	6330.37	I
400	6333.50	I
2000	6337.85	I
1000	6339.44	I
500	6359.16	I
1000	6566.49	I
2000	6583.75	I
1000	6585.27	I
5000	6619.66	I
500	6661.11	I
500 c	6697.29	I
400	6732.03	I
4000	6812.57	II
500	6989.78	I
500	7120.05	I
1200	7122.05	I
2000	7142.06	I
1000	7164.79	I
400 d	7191.66	I
700	7227.30	I
1000	7236.78	I
500	7237.84	I
5000	7402.06	I
1000	7410.50	I
500	7416.48	I
5000	7468.99	I
500 c	7490.52	I
2000	7554.18	I
500 d	7556.65	I
2000 c	7700.20	I
600	7897.98	I
500	7969.48	I
1000	8003.63	I
99000	8043.74	I
300 d	8065.70	I
1000	8090.76	I
800 c	8169.38	I
500 d	8222.57	I
4000	8240.05	I
10000 c	8393.30	I
1000	8486.11	I
1500 c	8664.95	I
500 c	8700.80	I
250 d	8748.22	I
1000	8853.24	I
2000	8853.80	I
3000	8857.50	I
1000 d	8898.50	I
400	8964.69	I
400	8993.13	I
5000	9022.40	I
15000	9058.33	I
1000	9098.86	I
12000	9113.91	I
600	9128.03	I
600	9227.74	I
1000	9335.05	I
4000	9426.71	I
3000	9427.15	I
2000	9598.22	I
2000	9649.61	I
3000 d	9653.06	I
5000	9731.73	I
500	10003.05	I
750	10131.16	I
1000	10238.82	I
400	10375.20	I
400	10391.74	I
5000	10466.54	I
400	11236.56	I
350	11558.46	I
320	11778.34	I
450	11996.86	I
300	12033.69	I
150	12304.58	I
60	13149.16	I
140	13958.27	I
200	14287.02	I
100	14460.00	I
225	15032.57	I
105	15528.65	I
150	16037.33	I
15	18275.71	I
20	18348.52	I
15	18982.41	I
35	19070.17	I
110	19105.12	I
50	19370.02	I
10	20648.69	I
220	22183.03	I
150	22226.53	I
30	22309.21	I
32	24420.82	I
12	27365.42	I
9	27573.05	I
10	30361.93	I
8	30383.88	I

Intensity	Wavelength/Å	
10	34295.73	I
9	34513.11	I
3	40228.54	I
2	41633.80	I

Iridium Ir *Z* = 77

Intensity	Wavelength/Å	
9900	2010.65	I
8700	2022.35	I
15000	2033.57	I
6200	2052.22	I
5000	2060.64	I
3700	2083.22	I
3100	2085.74	I
17000	2088.82	I
14000	2092.63	I
2700	2112.68	I
1800	2119.54	I
2000	2125.44	I
4500	2126.81	II
2000	2127.52	I
4500	2127.94	I
3700	2148.22	I
2500	2150.54	I
3500	2152.68	II
2900	2155.81	I
7900	2158.05	I
2100	2162.88	I
5800	2169.42	II
4500	2175.24	I
2700	2178.17	I
1600	2187.43	II
1100	2190.38	II
740	2191.64	I
910	2208.09	II
1300	2220.37	I
790	2221.07	II
2500	2242.68	II
620	2245.76	II
2100	2253.38	I
2100	2255.10	I
1400	2255.81	I
350	2258.51	I
1400	2258.86	I
830	2264.61	I
1100	2266.33	I
1000	2268.90	I
660	2280.00	I
950	2281.02	II
660	2281.91	I
330	2284.60	I
330	2295.08	I
790	2298.05	I
460	2299.53	I
910	2300.50	I
2700	2304.22	I
410	2305.47	I
210	2307.27	I
910	2308.93	I
460	2315.38	I
410	2321.45	I
410	2321.58	I
210	2327.98	I
540	2333.30	I
740	2333.84	I
580	2334.50	I
1600	2343.18	I
740	2343.61	I
580	2355.00	I
230	2357.53	II
410	2358.16	I
500	2360.73	I
2500	2363.04	I
370	2368.04	II
3500	2372.77	I
290	2375.09	II
250	2377.28	I
250	2377.98	I
500	2379.38	I
540	2381.62	I
210	2383.17	I
1300	2386.89	I
2500	2390.62	I
2700	2391.18	I
230	2407.59	I
290	2409.37	I
290	2410.17	I
290	2410.73	I
540	2413.31	I
370	2415.86	I
620	2418.11	I
210	2424.89	I
370	2424.99	I
290	2425.66	I
540	2427.61	I
540	2431.24	I
1300	2431.94	I
270	2435.14	I
250	2445.34	I
250	2447.76	I
910	2452.81	I
1300	2455.61	I
230	2455.87	I
210	2457.03	I
210	2457.23	I
870	2467.30	I
3300	2475.12	I
210	2478.11	I
2100	2481.18	I
620	2493.08	I
210	2496.27	I
250	2502.63	I
4100	2502.98	I
210	2513.71	I
990	2533.13	I
1100	2534.46	I
580	2537.22	I
580	2542.02	I
7900	2543.97	I
790	2546.03	I
210	2551.40	I
210	2555.35	I
910	2564.18	I
210	2569.88	I
230	2572.70	I
740	2577.26	I
740	2592.06	I
740	2599.04	I
700	2608.25	I
1800	2611.30	I
210	2614.98	I
330	2617.78	I
210	2619.88	I
250	2625.32	I
700	2634.17	I
250	2639.42	I
3500	2639.71	I
210	2644.19	I
1800	2661.98	I
350	2662.63	I
2700	2664.79	I
520	2669.91	I
520	2671.84	I
330	2673.61	I
270	2692.34	I
3000	2694.23	I
330	2772.46	I
250	2775.55	I
520	2781.29	I
330	2785.22	I
540	2797.35	I
1600	2797.70	I
380	2798.18	I
410	2800.82	I
680	2823.18	I
1200	2824.45	I
820	2836.40	I
1100	2839.16	I
820	2840.22	I
3800	2849.72	I
380	2875.60	I
380	2875.98	I
270	2877.68	I
820	2882.64	I
650	2897.15	I
260	2901.95	I
260	2904.80	I
200	2907.24	I
440	2916.36	I
230	2918.57	I
4400	2924.79	I
1200	2934.64	I
880	2936.68	I
250	2938.47	I
2700	2943.15	I
230	2946.97	I
200	2949.76	I
1200	2951.22	I
200	2974.95	I
440	2980.65	I
300	2996.08	I
220	3002.25	I
600	3003.63	I
270	3017.31	I
380	3029.36	I
330	3039.26	I
300	3047.16	I
300	3049.44	I
300	3057.28	I
1600	3068.89	I
320	3083.22	I
240	3086.44	I
390	3088.04	I
510	3100.29	I
510	3100.45	I
340	3120.76	I
200	3121.78	I
3400	3133.32	I
490	3168.88	I
370	3177.58	I
370	3198.92	I
610	3212.12	I
370	3219.51	I
5100	3220.78	I
300	3229.28	I
470	3241.52	I
200	3262.01	I
390	3266.44	I
200	3322.60	I
560	3368.48	I
660	3437.02	I
410	3448.97	I
3200	3513.64	I
220	3515.95	I
410	3522.03	I
320	3558.99	I
1200	3573.72	I
320	3594.39	I
220	3609.77	I
660	3628.67	I
220	3636.20	I
300	3661.71	I
300	3664.62	I
320	3674.98	I
200	3687.08	I
200	3731.36	II
530	3747.20	I
3100	3800.12	I
230	3817.24	I
480	3902.51	I
480	3915.38	I
400	3934.84	I
590	3976.31	I
460	3992.12	I
350	4033.76	I
370	4069.92	I
150	4070.68	I
100	4092.61	I
140	4115.78	I
90	4172.56	I
260	4268.10	I
220	4311.50	I
160	4399.47	I
65	4403.78	I
110	4426.27	I
55	4478.48	I

Intensity	Wavelength/Å	
55	4545.68	I
30	4548.48	I
35	4568.09	I
75	4616.39	I
26	4656.18	I
50	4728.86	I
26	4756.46	I
65	4778.16	I
30	4795.67	I
50	4938.09	I
26	4970.48	I
25	4999.74	I
25	5002.74	I
30	5014.98	I
30	5123.66	I
35	5364.32	I
75	5449.50	I
45	5625.55	I
35	5894.06	I
20	6110.67	I
12	6288.28	I
10	6686.08	I
6	7834.32	I

Iron Fe *Z* = 26

Intensity	Wavelength/Å	
350	386.16	V
350	386.88	V
400	387.20	V
400	387.50	V
400	387.76	V
400	387.78	V
400	395.90	V
400	404.62	V
400	405.50	V
800	407.42	V
600	407.44	V
400	407.49	V
500	407.75	V
400	409.71	V
400	410.20	V
600	411.55	V
700	417.39	V
700	418.04	V
500	418.47	V
700	421.06	V
500	421.78	V
500	422.31	V
500	426.06	V
500	426.11	V
350	426.97	V
17	525.69	IV
15	526.29	IV
13	526.63	IV
14	536.61	IV
15	537.10	IV
13	537.26	IV
14	537.79	IV
13	537.94	IV
13	552.14	IV
14	607.53	IV
13	608.80	IV
10	813.38	III
10	844.28	III
10 p	861.83	III
10	891.17	III
10	950.33	III
10	981.37	III
10 w	983.88	III
12	1055.27	II
15	1068.36	II
15	1071.60	II
15	1096.89	II
12	1099.12	II
18	1112.09	II
12	1121.99	II
12	1122.86	II
12	1128.07	II
12	1130.43	II
15	1133.41	II
12	1133.68	II
12	1138.64	II
12	1142.33	II
12	1143.23	II
18	1144.95	II
12	1147.41	II
15	1148.29	II
12	1151.16	II
12	1267.44	II
12	1272.00	II
400	1317.86	V
400	1323.27	V
400	1330.40	V
400	1359.01	V
600	1361.82	V
700	1373.59	V
600	1373.67	V
500	1376.34	V
500	1378.56	V
800	1387.94	V
400	1397.97	V
600	1400.24	V
800	1402.39	V
400	1406.67	V
500	1406.82	V
400	1407.25	V
600	1409.45	V
400	1415.20	V
600	1420.46	V
800	1430.57	V
13	1431.43	IV
800	1440.53	V
400	1442.22	V
800	1446.62	V
700	1448.85	V
400	1449.93	V
700	1456.16	V
500	1459.83	V
400	1460.73	V
500	1462.63	V
700	1464.68	V
500	1465.38	V
400	1466.65	V
500	1469.00	V
500	1479.47	V
10 h	1505.17	III
13	1526.60	IV
13	1530.26	IV
14	1532.63	IV
13	1532.91	IV
15	1533.86	IV
13	1533.95	IV
14	1536.58	IV
10 h	1538.63	III
13	1542.16	IV
14	1542.70	IV
12 h	1550.20	III
13	1566.26	IV
14	1568.27	IV
13	1591.51	IV
13	1592.05	IV
13	1598.01	IV
13	1600.58	IV
10 h	1601.21	III
13	1601.67	IV
13	1603.18	IV
13	1603.73	IV
13	1604.88	IV
13	1605.68	IV
15	1605.97	IV
13	1606.98	IV
17	1609.10	IV
14	1609.83	IV
13	1610.47	IV
13	1611.20	IV
13	1613.64	IV
15	1614.02	IV
13	1614.64	IV
13	1615.00	IV
16	1616.68	IV
14	1617.68	IV
12	1618.47	II
14	1619.02	IV
13	1621.16	IV
14	1621.57	IV
13	1623.38	IV
13	1623.53	IV
15	1626.47	IV
14	1626.90	IV
13	1628.54	IV
13	1630.18	IV
17	1631.08	IV
15	1631.12	II
14	1632.40	IV
13	1634.01	IV
18	1635.40	II
15	1636.32	II
15	1639.40	II
14	1639.40	IV
16	1640.04	IV
14	1640.16	IV
12	1641.76	II
15	1641.87	IV
15	1647.09	IV
12	1647.16	II
15	1651.58	IV
15	1652.90	IV
13	1653.41	IV
13	1656.11	IV
15	1656.65	IV
14	1660.10	IV
13	1662.32	IV
13	1662.52	IV
13	1663.54	IV
13	1668.09	IV
12	1670.74	II
14	1671.04	IV
13	1673.68	IV
14	1675.66	IV
13	1681.36	IV
15	1687.69	IV
15	1698.88	IV
12	1702.04	II
13	1709.81	IV
15	1711.41	IV
14	1712.76	IV
14	1717.90	IV
14	1718.16	IV
14	1719.46	IV
14	1722.71	IV
14	1724.06	IV
16	1725.63	IV
13	1761.08	IV
12	1761.38	II
20	1785.26	II
20	1786.74	II
18	1788.07	II
13	1792.10	IV
13	1796.93	IV
13	1827.98	IV
10	1869.83	III
12	1877.99	III
10	1882.05	III
12	1886.76	III
13	1890.67	III
11	1893.98	III
20	1895.46	III
10 s	1907.58	III
19	1914.06	III
15	1915.08	III
15	1922.79	III
10 p	1926.01	III
18	1926.30	III
15	1930.39	III
14	1931.51	III
30	1934.538	I
25	1937.269	I
14	1937.34	III
10 l	1938.90	III
14 s	1943.48	III
12	1945.34	III
50	1946.988	I
10	1950.33	III
12	1951.01	III
25	1951.571	I
30	1952.59	I

Intensity	Wavelength/Å	
11	1952.65	III
30	1953.005	I
13	1953.32	III
10	1953.49	III
10 w	1954.22	III
60	1957.823	I
11	1958.58	III
60	1960.144	I
13	1960.32	III
30	1961.25	I
50	1962.111	I
12	1963.11	II
15	1987.50	III
14	1991.61	III
13	1994.07	III
12	1995.56	III
12	1996.42	III
10	2061.55	III
12	2068.24	III
14	2078.99	III
100	2084.122	I
10	2084.35	III
12	2090.14	III
15	2097.48	III
12	2097.69	III
12	2103.80	III
10	2107.32	III
15	2151.78	III
12	2157.71	III
50	2157.794	I
12	2158.47	III
10	2161.27	III
40	2166.773	I
12	2166.95	III
12	2171.04	III
15	2174.66	III
300	2178.118	I
12	2180.41	III
250	2186.486	I
60	2186.892	I
120	2187.195	I
250	2191.839	I
150	2196.043	I
80	2200.390	I
80	2200.724	I
15	2208.41	II
10 p	2208.85	III
20	2213.65	II
12	2218.26	II
20	2220.38	II
10	2221.83	III
10	2229.27	III
10	2232.43	III
10	2232.69	III
10	2235.91	III
10	2238.16	III
12 p	2241.54	III
50	2250.790	I
60	2251.874	I
300	2259.511	I
12	2261.59	III

Intensity	Wavelength/Å	
60	2264.389	I
80	2267.085	I
10	2267.42	III
80	2267.469	I
50	2270.862	I
150	2272.070	I
150	2276.026	I
80	2279.937	I
150	2284.086	I
150	2287.250	I
300	2292.524	I
10	2293.06	III
15	2295.86	III
200	2297.787	I
600	2298.169	I
80	2299.220	I
300	2300.142	I
50	2301.684	I
100	2303.424	I
150	2303.581	I
120	2308.999	I
150	2313.104	I
10 p	2317.70	III
10	2319.22	III
200	2320.358	I
10 p	2321.71	III
10	2326.95	III
100	2327.40	II
100	2331.31	II
300	2332.80	II
10 p	2336.77	III
200	2338.01	II
10	2338.96	III
600	2343.49	II
80	2343.96	II
150	2344.28	II
200	2348.11	II
250	2348.30	II
200	2359.12	II
150	2360.00	II
120	2360.29	II
200	2364.83	II
80	2365.76	II
80	2368.59	II
80	2369.456	I
80	2369.95	II
120	2371.430	I
300	2373.624	I
150	2373.74	II
120	2374.518	I
120	2376.43	II
80	2379.27	II
120	2380.76	II
150	2381.835	I
1000	2382.04	II
300	2388.63	II
200	2389.973	I
1000	2395.62	II
300	2399.24	II
800	2404.88	II
250	2406.66	II

Intensity	Wavelength/Å	
80	2406.97	II
300	2410.52	II
200	2411.07	II
150	2413.31	II
80	2417.87	II
60	2420.396	I
60	2423.089	I
150	2424.14	II
120	2428.36	II
120	2430.08	II
80	2432.26	II
60	2438.182	I
150	2439.30	II
150	2439.74	I
100	2442.57	I
250	2443.872	I
100	2444.51	II
50	2445.212	I
100	2445.57	II
60	2447.709	I
100	2453.476	I
1500	2457.598	I
150	2458.78	II
80	2461.28	II
100	2461.86	II
100	2462.181	I
1500	2462.647	I
50	2463.730	I
800	2465.149	I
60	2467.732	I
600	2468.879	I
80	2470.67	II
80	2470.965	I
800	2472.336	I
1000	2472.895	I
200	2473.16	I
600	2474.814	I
60	2476.657	I
120	2479.480	I
1200	2479.776	I
100	2480.16	II
80	2482.12	II
100	2482.66	II
10000	2483.271	I
300	2483.533	I
1000	2484.185	I
50	2485.990	I
800	2486.373	I
100	2486.691	I
100	2487.066	I
120	2487.370	I
4000	2488.143	I
100	2488.945	I
80	2489.48	II
1000	2489.750	I
50	2489.913	I
3000	2490.644	I
100	2490.71	II
2000	2491.155	I
100	2491.40	II
100	2493.18	II

Intensity	Wavelength/Å	
500	2493.26	II
60	2494.000	I
50	2494.251	I
100	2495.87	I
600	2496.533	I
150	2498.90	I
1000	2501.132	I
50	2501.693	I
80	2506.09	II
500	2507.900	I
50	2508.753	I
1000	2510.835	I
120	2511.76	II
80	2512.275	I
400	2512.365	I
80	2516.570	I
300	2517.661	I
800	2518.102	I
150	2519.629	I
50	2522.480	I
4000	2522.849	I
200	2523.66	I
500	2524.293	I
100	2525.02	I
200	2525.39	II
300	2526.29	II
2000	2527.435	I
800	2529.135	I
250	2529.55	II
150	2529.836	I
200	2530.687	I
120	2533.63	II
100	2534.42	II
120	2535.49	II
400	2535.607	I
200	2536.792	I
200	2536.80	II
100	2538.80	II
100	2538.91	II
150	2538.99	II
50	2539.357	I
200	2540.66	II
600	2540.972	I
80	2541.10	II
300	2542.10	I
250	2543.92	I
150	2544.70	I
800	2545.978	I
80	2546.67	II
100	2548.74	II
80	2549.08	II
80	2549.39	II
600	2549.613	I
400	2562.53	II
200	2563.48	II
150	2574.36	II
300	2576.691	I
100	2582.58	II
1500	2584.54	I
650	2585.88	II
90	2591.54	II

Intensity	Wavelength/Å	
90	2593.73	II
650	2598.37	II
2000	2599.40	II
300	2599.57	I
60	2605.657	I
300	2606.51	II
800	2606.827	I
650	2607.09	II
600	2611.87	II
320	2613.82	II
320	2617.62	II
250	2618.018	I
90	2620.41	II
400	2623.53	I
200	2625.67	II
150	2628.29	II
250	2631.05	II
250	2631.32	II
100	2632.237	I
300	2635.809	I
50	2641.646	I
200	2643.998	I
300	2666.812	I
60	2666.965	I
600	2679.062	I
500	2684.75	II
400	2689.212	I
10 h	2695.13	III
200	2699.106	I
80	2706.012	I
400	2706.582	I
60	2708.571	I
200	2711.655	I
80	2714.41	II
50	2716.257	I
50	2717.786	I
250	2718.436	I
4000	2719.027	I
100	2719.420	I
50	2720.197	I
1500	2720.903	I
400	2723.578	I
150	2724.953	I
50	2726.235	I
80	2727.54	II
200	2728.020	I
50	2728.820	I
80	2728.90	II
1000	2733.581	I
60	2734.005	I
50	2734.268	I
500	2735.475	I
50	2735.612	I
500	2737.310	I
120	2737.83	I
400	2739.55	II
250	2742.254	I
800	2742.405	I
200	2743.20	II
150	2743.565	I
200	2744.068	I
80	2744.527	I
300	2746.48	II
100	2749.32	II
500	2749.48	II
1200	2750.140	I
80	2753.29	II
150	2754.032	I
100	2754.426	I
800	2755.73	II
250	2756.328	I
100	2757.316	I
120	2761.780	I
150	2761.81	II
150	2762.026	I
120	2762.772	I
120	2763.109	I
80	2766.910	I
250	2767.522	I
300	2772.07	I
600	2778.220	I
3000	2788.10	I
200	2797.78	I
400	2804.521	I
1500	2806.98	I
10 p	2813.24	III
2500	2813.287	I
300	2823.276	I
600	2825.56	I
50	2825.687	I
120	2828.808	I
1500	2832.436	I
120	2835.950	I
200	2838.119	I
200	2843.631	I
1000	2843.977	I
100	2845.594	I
800	2851.797	I
50	2869.307	I
50	2872.334	I
80	2874.172	I
50	2894.504	I
12	2904.43	III
10	2907.50	III
12	2907.70	III
120	2912.157	I
120	2929.007	I
1200	2936.903	I
60	2941.343	I
1000	2947.876	I
600	2953.940	I
250	2957.364	I
150	2965.254	I
1500	2966.898	I
120	2969.36	I
800	2970.099	I
1200	2973.132	I
500	2973.235	I
600	2981.445	I
1000	2983.570	I
1000	2994.427	I
250	2994.502	I
500	2999.512	I
120	3000.451	I
800	3000.948	I
12	3001.62	III
60	3001.655	I
12 h	3007.28	III
200	3007.282	I
500	3008.14	I
120	3009.569	I
15	3013.17	III
60	3017.627	I
60	3018.983	I
500	3020.491	I
1500	3020.639	I
600	3021.073	I
500	3024.032	I
150	3025.638	I
500	3025.842	I
80	3030.148	I
60	3031.214	I
60	3034.484	I
800	3037.389	I
80	3041.637	I
800	3047.604	I
600	3057.446	I
1000	3059.086	I
250	3067.244	I
120	3075.719	I
120	3091.577	I
80	3098.189	I
100	3099.895	I
100	3099.968	I
60	3100.303	I
100	3100.665	I
10 p	3136.43	III
10	3174.09	III
80	3175.445	I
10	3175.99	III
10	3178.01	III
150	3184.895	I
250	3191.659	I
500	3193.226	I
800	3193.299	I
200	3196.928	I
80	3199.500	I
50	3205.398	I
100	3211.88	I
200	3214.011	I
200	3214.396	I
60	3215.938	I
50	3217.377	I
80	3219.583	I
60	3219.766	I
300	3222.045	I
600	3225.78	I
80	3227.796	I
50	3233.967	I
120	3234.613	I
300	3236.222	I
100	3239.433	I
80	3244.187	I
80	3246.005	I
80	3265.046	I
50	3265.617	I
13	3266.88	III
50	3271.000	I
11	3276.08	III
150	3286.75	I
10	3288.81	III
120	3305.97	I
200	3306.343	I
400	3355.227	I
80	3355.517	I
60	3369.546	I
120	3370.783	I
50	3378.678	I
50	3380.110	I
60	3383.978	I
50	3392.304	I
150	3392.651	I
150	3399.333	I
80	3404.353	I
500	3407.458	I
250	3413.131	I
60	3424.284	I
500	3427.119	I
60	3428.748	I
6000	3440.606	I
2500	3440.989	I
1000	3443.876	I
200	3445.149	I
1200	3465.860	I
2000	3475.450	I
500	3476.702	I
2500	3490.574	I
500	3497.840	I
10	3501.76	III
250	3513.817	I
300	3521.261	I
400	3526.040	I
100	3526.166	I
60	3526.237	I
60	3526.381	I
60	3526.467	I
100	3533.199	I
200	3536.556	I
300	3541.083	I
250	3542.075	I
80	3553.739	I
400	3554.925	I
200	3556.878	I
400	3558.515	I
1000	3565.379	I
1200	3570.097	I
800	3570.25	I
120	3571.996	I
100	3573.393	I
60	3573.829	I
60	3573.888	I
4000	3581.19	I
150	3582.199	I
150	3584.660	I

Intensity	Wavelength/Å	
120	3584.929	I
300	3585.319	I
150	3585.705	I
10	3586.04	III
200	3586.103	I
400	3586.984	I
100	3594.633	I
11	3600.94	III
150	3603.204	I
11	3603.88	III
200	3605.454	I
500	3606.680	I
1500	3608.859	I
250	3610.16	I
60	3612.068	I
150	3617.788	I
1500	3618.768	I
200	3621.462	I
150	3622.004	I
150	3623.19	I
100	3631.096	I
1200	3631.463	I
60	3632.041	I
100	3638.298	I
200	3640.389	I
80	3643.717	I
1500	3647.842	I
250	3649.506	I
80	3650.279	I
200	3651.467	I
120	3670.024	I
150	3670.089	I
100	3676.311	I
150	3677.629	I
1500	3679.913	I
200	3682.242	I
120	3683.054	I
150	3684.107	I
120	3685.998	I
500	3687.456	I
120	3689.477	I
150	3694.008	I
120	3695.051	I
150	3701.086	I
80	3704.462	I
1200	3705.566	I
60	3707.041	I
150	3707.821	I
300	3707.919	I
600	3709.246	I
120	3716.442	I
8000	3719.935	I
1500	3722.563	I
120	3724.377	I
60	3725.491	I
60	3727.093	I
500	3727.619	I
150	3732.396	I
1200	3733.317	I
5000	3734.864	I
120	3735.324	I
6000	3737.131	I
100	3738.306	I
400	3743.362	I
6000	3745.561	I
1200	3745.899	I
3000	3748.262	I
80	3748.964	I
3000	3749.485	I
1500	3758.232	I
400	3760.05	I
1500	3763.788	I
400	3765.54	I
600	3767.191	I
60	3776.452	I
250	3785.95	I
100	3786.68	I
250	3787.880	I
250	3790.092	I
150	3794.34	I
400	3795.002	I
120	3797.518	I
250	3798.511	I
400	3799.547	I
200	3805.345	I
80	3806.696	I
600	3812.964	I
60	3813.059	I
1500	3815.840	I
2500	3820.425	I
150	3821.179	I
80	3824.306	I
2500	3824.444	I
1500	3825.880	I
1200	3827.823	I
1000	3834.222	I
120	3839.257	I
500	3840.437	I
800	3841.047	I
120	3843.256	I
80	3846.800	I
200	3849.96	I
120	3850.817	I
2500	3856.372	I
150	3859.212	I
10000	3859.911	I
150	3865.523	I
60	3867.215	I
250	3872.501	I
150	3873.761	I
250	3878.018	I
2000	3878.573	I
4000	3886.282	I
200	3887.048	I
300	3888.513	I
800	3895.656	I
1200	3899.707	I
400	3902.945	I
250	3906.479	I
80	3916.731	I
600	3920.258	I
1200	3922.911	I
1200	3927.920	I
2000	3930.296	I
60	3948.774	I
60	3949.953	I
50	3951.164	I
50	3952.601	I
16	3954.33	III
60	3956.454	I
250	3956.68	I
60	3966.614	I
11	3968.72	III
100	3969.257	I
80	3977.741	I
10 w	3979.42	III
40	3981.771	I
50	3983.956	I
60	3994.114	I
200	3997.392	I
40	3998.053	I
400	4005.241	I
60	4009.713	I
100	4021.867	I
10	4035.42	III
50	4040.638	I
4000	4045.813	I
11	4053.11	III
1500	4063.594	I
50	4066.975	I
50	4067.977	I
1200	4071.737	I
40	4076.629	I
12	4081.00	III
40	4100.737	I
40	4107.489	I
150	4118.544	I
10	4120.90	III
11	4122.02	III
11	4122.78	III
40	4127.608	I
400	4132.058	I
80	4134.676	I
40	4136.997	I
15	4137.76	III
13	4139.35	III
200	4143.415	I
800	4143.869	I
40	4153.898	I
50	4154.500	I
60	4156.799	I
18	4164.73	III
13	4166.84	III
50	4172.744	I
13	4174.26	III
60	4174.912	I
50	4175.635	I
50	4177.593	I
120	4181.754	I
50	4184.891	I
120	4187.038	I
120	4187.795	I
80	4191.430	I
40	4195.329	I
150	4198.304	I
40	4199.095	I
300	4202.029	I
40	4203.984	I
80	4206.696	I
80	4210.343	I
400	4216.183	I
100	4219.360	I
50	4222.212	I
11	4222.27	III
50	4225.956	I
200	4227.423	I
100	4233.602	I
13	4235.56	III
250	4235.936	I
50	4238.809	I
12	4243.75	III
50	4247.425	I
200	4250.118	I
300	4250.787	I
40	4258.315	I
800	4260.473	I
250	4271.153	I
1200	4271.759	I
12 h	4273.40	III
12	4279.72	III
1200	4282.402	I
14 h	4286.16	III
80	4291.462	I
16 h	4296.85	III
250	4299.234	I
18 h	4304.78	III
1200	4307.901	I
20 h	4310.36	III
150	4315.084	I
1500	4325.761	I
80	4352.734	I
80	4369.771	I
11 h	4372.31	III
14 h	4372.53	III
18 h	4372.81	III
800	4375.929	I
3000	4383.544	I
1200	4404.750	I
300	4415.122	I
12	4419.60	III
600	4427.299	I
400	4461.652	I
120	4466.551	I
80	4476.017	I
80	4482.169	I
200	4482.252	I
50	4489.739	I
50	4528.613	I
30	4647.433	I
30	4736.771	I
50	4859.741	I
120	4871.317	I
60	4872.136	I
30	4878.208	I

Intensity	Wavelength/Å	
100	4890.754	I
250	4891.492	I
30	4903.309	I
150	4918.992	I
500	4920.502	I
1500	4957.597	I
80	5001.862	I
30	5005.711	I
100	5006.117	I
60	5012.067	I
30	5014.941	I
150	5041.755	I
30	5049.819	I
30	5051.634	I
25	5074.748	I
150	5110.357	I
40	5139.251	I
100	5139.462	I
25	5151.910	I
12	5156.12	III
80	5166.281	I
2500	5167.487	I
80	5168.897	I
500	5171.595	I
50	5191.454	I
80	5192.343	I
200	5194.941	I
10	5199.08	III
30	5204.582	I
25	5215.179	I
150	5216.274	I
60	5226.862	I
1000	5227.150	I
250	5232.939	I
10	5235.66	III
18	5243.31	III
13 l	5260.34	III
100	5266.555	I
1200	5269.537	I
800	5270.357	I
14	5272.98	III
15	5276.48	III
30	5281.789	I
16	5282.30	III
60	5283.621	I
12	5284.83	III
11	5298.12	III
12	5299.93	III
25	5302.299	I
14 w	5302.60	III
10	5306.76	III
10	5322.74	III
150	5324.178	I
800	5328.038	I
300	5328.531	I
100	5332.899	I
80	5339.928	I
500	5341.023	I
11	5346.88	III
12	5353.77	III
12	5363.76	III
10	5368.06	III
400	5371.489	I
11 l	5375.47	III
40	5393.167	I
300	5397.127	I
250	5405.774	I
250	5429.695	I
100	5434.523	I
200	5446.871	I
120	5455.609	I
25	5497.516	I
20	5501.464	I
30	5506.778	I
30	5569.618	I
60	5572.841	I
120	5586.755	I
200	5615.644	I
20	5624.541	I
50	5662.515	I
11	5719.88	III
10	5756.38	III
20	5762.990	I
18	5833.93	III
10	5854.62	III
30	5862.353	I
15	5891.91	III
30	5914.114	I
10 p	5920.13	III
18 p	5929.69	III
10	5952.31	III
14	5953.62	III
12	5979.32	III
30	5986.956	I
12 h	5989.08	III
18	5999.54	III
16	6032.59	III
13	6036.56	III
11	6048.72	III
11	6054.18	III
40	6065.482	I
30	6102.159	I
40	6136.614	I
40	6137.694	I
40	6191.558	I
30	6213.429	I
30	6219.279	I
40	6230.726	I
20	6246.317	I
80	6247.56	II
30	6252.554	I
20	6393.602	I
30	6399.999	I
20	6411.647	I
20	6421.349	I
30	6430.844	I
200	6456.38	II
60	6494.981	I
20	6546.239	I
20	6592.913	I
40	6677.989	I
25	7164.443	I
80	7187.313	I
30	7207.381	I
30	7445.746	I
40	7495.059	I
60	7511.045	I
80	7937.131	I
60	7945.984	I
80	7998.939	I
60	8046.047	I
50	8085.176	I
150	8220.41	I
120	8327.053	I
20	8331.908	I
120	8387.770	I
30	8468.404	I
15	8514.069	I
60	8661.898	I
150	8688.621	I
52	11422.32	I
87	11439.12	I
91	11593.59	I
255	11607.57	I
160	11638.26	I
230	11689.98	I
160	11783.26	I
580	11882.84	I
225	11884.08	I
1030	11973.05	I
96	14400.56	I
72	14512.23	I
50	14555.06	I
40	14826.43	I
94	15294.58	I
41	15769.42	I
105	18856.65	I

***Krypton Kr** *Z* = 36*

Intensity	Wavelength/Å	
30	467.35	III
150	472.16	V
100	484.39	V
250	496.25	V
120	500.77	V
200	507.20	V
30	540.86	III
60	548.04	V
30	565.64	III
30	569.16	III
30	571.98	III
30	579.83	III
30	585.14	III
30	585.96	III
30	593.70	III
30	594.10	III
30	596.41	III
40	600.17	III
30	603.67	III
50	605.86	III
35	606.47	III
50	611.12	III
35	616.72	III
40	621.45	III
45	622.80	III
50	625.02	III
30	625.76	III
45	628.59	III
50	630.04	III
35	633.09	III
120	637.87	V
50	639.98	III
60	646.41	III
50	651.20	III
50	659.72	III
30	664.86	III
40	672.34	III
35	672.85	III
35	676.57	III
35	680.13	III
35	683.68	III
45	686.25	III
45	687.98	III
	690.86	V
	691.75	V
45	691.93	III
50	695.61	III
30	698.05	III
50	708.36	III
600	708.85	V
50	714.00	III
100 p	722.04	III
60	729.40	II
30	746.70	III
200	761.18	II
100	763.98	II
60	766.20	II
200	771.03	II
60 p	773.69	II
200	782.10	II
100	783.72	II
60	785.97	III
	793.44	IV
	794.11	IV
7	805.76	IV
	810.70	V
18	816.82	IV
60	818.15	II
60	830.38	II
50	837.66	III
22	842.04	IV
100	844.06	II
50	854.73	III
60	862.58	III
60	864.82	II
60	868.87	II
40	870.84	III
50	876.08	III
200	884.14	II
1000	886.30	II
400	891.01	II
75	897.81	III
200	911.39	II
2000	917.43	II
50	945.44	I
50	946.54	I

Intensity	Wavelength/Å	
20	951.06	I
50	953.40	I
50	963.37	I
2000	964.97	II
50	987.29	III
100	1001.06	I
100	1003.55	I
100	1030.02	I
30	1158.74	III
200	1164.87	I
650	1235.84	I
6	1638.82	III
6	1914.09	III
3	2237.34	IV
6	2291.26	IV
3	2329.3	IV
4	2336.75	IV
4	2348.27	IV
3	2358.5	IV
3	2388.05	IV
40	2393.94	III
4	2416.9	IV
3	2428.04	IV
5	2442.68	IV
4	2451.7	IV
6	2459.74	IV
100 h	2464.77	II
5	2474.06	IV
60	2492.48	II
40	2494.01	III
4	2517.0	IV
5	2518.02	IV
6	2519.38	IV
5	2524.5	IV
5	2546.0	IV
6	2547.0	IV
4	2558.08	IV
30	2563.25	III
3	2586.9	IV
5	2606.17	IV
10	2609.5	IV
8	2615.3	IV
7	2621.11	IV
60	2639.76	III
30	2680.32	III
40	2681.19	III
80 h	2712.40	II
3	2730.55	IV
8	2748.18	IV
6	2774.70	IV
3	2829.60	IV
100	2833.00	II
3	2836.08	IV
30	2841.00	III
30	2851.16	III
5	2853.0	IV
3	2859.3	IV
50	2870.61	III
100	2892.18	III
30	2909.17	III
50	2952.56	III
60	2992.22	III
50	3022.30	III
80	3024.45	III
50	3046.93	III
30	3056.72	III
60	3063.13	III
40	3097.16	III
60	3112.25	III
30	3120.61	III
100	3124.39	III
60	3141.35	III
3	3142.01	IV
100	3189.11	III
80	3191.21	III
6	3224.99	IV
40	3239.52	III
40	3240.44	III
300	3245.69	III
3	3261.70	IV
150	3264.81	III
100	3268.48	III
30	3271.65	III
30	3285.89	III
30	3304.75	III
50	3311.47	III
200	3325.75	III
60	3330.76	III
50	3342.48	III
100	3351.93	III
40	3374.96	III
100	3439.46	III
70	3474.65	III
100	3488.59	III
200	3507.42	III
100	3564.23	III
100 h	3607.88	II
200	3631.889	II
30	3641.34	III
250	3653.928	II
80	3665.324	I
150	3669.01	II
100	3679.559	I
80	3686.182	II
30	3690.65	III
300 h	3718.02	II
200	3718.595	II
150	3721.350	II
200	3741.638	II
150	3744.80	II
80	3754.245	II
500	3778.089	II
500	3783.095	II
3	3809.30	IV
5	3860.58	IV
40 h	3868.70	III
150 h	3875.44	II
150	3906.177	II
200	3920.081	II
5	3934.29	IV
100	3994.840	II
100 h	3997.793	II
300	4057.037	II
300	4065.128	II
50	4067.37	III
500	4088.337	II
250	4098.729	II
100	4109.248	II
40	4131.33	III
250	4145.122	II
40	4154.46	III
150	4250.580	II
1000	4273.969	I
100	4282.967	I
600	4292.923	II
200	4300.49	II
500 h	4317.81	II
400	4318.551	I
1000	4319.579	I
150 h	4322.98	II
100	4351.359	I
3000	4355.477	II
500	4362.641	I
200	4369.69	II
800	4376.121	I
300 h	4386.54	II
200	4399.965	I
100	4425.189	I
500	4431.685	II
600	4436.812	II
600	4453.917	I
800	4463.689	I
800	4475.014	II
400 h	4489.88	II
600	4502.353	I
400 h	4523.14	II
200 h	4556.61	II
800	4577.209	II
300	4582.978	II
150 h	4592.80	II
500	4615.292	II
1000	4619.166	II
800	4633.885	II
2000	4658.876	II
500	4680.406	II
100	4691.301	II
200	4694.360	II
3000	4739.002	II
300	4762.435	II
1000	4765.744	II
300	4811.76	II
300	4825.18	II
800	4832.077	II
700	4846.612	II
150	4857.20	II
300	4945.59	II
20 h	5016.45	III
200	5022.40	II
250	5086.52	II
400 h	5125.73	II
500	5208.32	II
200	5308.66	II
500	5333.41	II
200	5468.17	II
10	5501.43	III
500	5562.224	I
2000	5570.288	I
80	5580.386	I
100	5649.561	I
400	5681.89	II
200 h	5690.35	II
100	5832.855	I
3000	5870.914	I
200	5992.22	II
60	5993.849	I
10 h	6037.17	III
60	6056.125	I
10 h	6078.38	III
10	6310.22	III
300	6420.18	II
100	6421.026	I
200	6456.288	I
150	6570.07	II
60	6699.228	I
100	6904.678	I
250	7213.13	II
100	7224.104	I
80	7287.258	I
400	7289.78	II
400	7407.02	II
60	7425.541	I
200	7435.78	II
100	7486.862	I
300	7524.46	II
1000	7587.411	I
2000	7601.544	I
150	7641.16	II
1000	7685.244	I
1200	7694.538	I
250	7735.69	II
150	7746.827	I
800	7854.821	I
200	7913.423	I
180	7928.597	I
200	7933.22	II
120	7973.62	II
100	7982.401	I
1500	8059.503	I
4000	8104.364	I
6000	8112.899	I
60	8132.967	I
3000	8190.054	I
200	8202.72	II
80	8218.365	I
3000	8263.240	I
100	8272.353	I
1500	8281.050	I
5000	8298.107	I
100	8412.430	I
3000	8508.870	I
150	8764.110	I
6000	8776.748	I
2000	8928.692	I
500	9238.48	II

Intensity	Wavelength/Å	
500 hl	9293.82	II
200 h	9320.99	II
300	9361.95	II
100	9362.082	I
200 h	9402.82	II
200 h	9470.93	II
500	9577.52	II
500 h	9605.80	II
400 h	9619.61	II
200	9663.34	II
200 h	9711.60	II
2000	9751.758	I
500	9803.14	II
500	9856.314	I
1000	10221.46	II
100	11187.108	I
200	11257.711	I
150	11259.126	I
500	11457.481	I
150	11792.425	I
1500	11819.377	I
600	11997.105	I
160	12077.224	I
100	12861.892	I
1100	13177.412	I
1000	13622.415	I
2400	13634.220	I
800	13658.394	I
200	13711.036	I
600	13738.851	I
150	13974.027	I
550	14045.657	I
140	14104.298	I
180	14402.22	I
2000	14426.793	I
100	14517.84	I
1600	14734.436	I
550	14762.672	I
450	14765.472	I
400	14961.894	I
120	15005.307	I
140	15209.526	I
1700	15239.615	I
130	15326.480	I
1500	15334.958	I
700	15372.037	I
200	15474.026	I
180	15681.02	I
120	15820.09	I
200	16726.513	I
2000	16785.128	I
1000	16853.488	I
2400	16890.441	I
1600	16896.753	I
1800	16935.806	I
600	17098.771	I
700	17367.606	I
120	17404.443	I
150	17616.854	I
650	17842.737	I
700	18002.229	I
2600	18167.315	I
100	18399.786	I
150	18580.896	I
300	18696.294	I
170	18785.460	I
200	18797.703	I
140	20209.878	I
300	20423.964	I
140	20446.971	I
600	21165.471	I
1800	21902.513	I
120	22485.775	I
180	23340.416	I
120	24260.506	I
180	24292.221	I
600	25233.820	I
180	28610.55	I
1000	28655.72	I
150	28769.71	I
140	28822.49	I
300	29236.69	I
300	30663.54	I
300	30979.16	I
500	39300.6	I
1100	39486.52	I
220	39557.25	I
100	39572.60	I
1400	39588.4	I
1100	39589.6	I
500	39954.8	I
300	39966.6	I
1300	40306.1	I
250	40685.16	I

Lanthanum La Z = 57

Intensity	Wavelength/Å	
100	344.12	IV
400	390.72	V
1000	432.11	V
2500	435.28	V
10000	463.14	IV
5000	482.16	V
7000	498.08	V
15000	499.54	IV
10000	503.58	V
12000	526.76	V
10000	531.07	V
15000	533.23	V
8000	547.44	V
40000	552.02	IV
5000	600.24	V
30000	631.26	IV
400	796.99	III
2000	870.40	III
1000	882.34	III
400	942.86	III
50000	1081.61	III
95000	1099.73	III
2000	1255.63	III
10000	1349.18	III
25000	1368.04	IV
20000	1463.47	IV
15000	1507.87	IV
10000	1523.79	III
4000	1808.66	IV
5000	1902.97	IV
4000 c	2197.45	IV
770	2256.76	II
25000 w	2417.58	IV
50000	2532.75	IV
45000	2582.05	IV
95000 w	2597.50	IV
70000 w	2662.75	IV
420	2808.39	II
50000 w	2848.30	IV
30000 c	2962.58	IV
70000 w	3009.51	IV
90000 c	3056.68	IV
1000	3171.63	III
1500	3171.74	III
510	3245.13	II
550	3265.67	II
800	3303.11	II
1500	3337.49	II
870	3344.56	II
1500	3380.91	II
320	3628.83	II
1000	3645.42	II
550	3713.54	II
2400	3759.08	II
3700	3790.83	II
3900	3794.78	II
600	3840.72	II
1600	3849.02	II
3400	3871.64	II
1700	3886.37	II
1300	3916.05	II
1100	3921.54	II
2200	3929.22	II
9000	3949.10	II
4400	3988.52	II
3600	3995.75	II
2800	4031.69	II
3000	4042.91	II
850	4067.39	II
2800	4077.35	II
5500	4086.72	II
4400	4123.23	II
550	4141.74	II
1100	4151.97	II
1500	4196.55	II
1600	4238.38	II
480	4269.50	II
600	4286.97	II
600	4296.05	II
440	4322.51	II
4600	4333.74	II
550	4354.40	II
2000	4429.90	II
850	4522.37	II
420	4526.12	II
400	4558.46	II
400	4574.88	II
410	4613.39	II
410	4619.88	II
540	4655.50	II
360	4662.51	II
230	4692.50	II
230	4728.42	II
500	4740.28	II
390	4743.09	II
320	4748.73	II
320	4860.91	II
850	4899.92	II
1000	4920.98	II
1000	4921.79	II
370	4949.77	I
340	4970.39	II
370	4986.83	II
720	4999.47	II
210	5050.57	I
470	5114.56	II
470	5122.99	II
450	5145.42	I
290	5158.69	I
580	5177.31	I
850	5183.42	II
260	5188.22	II
720	5211.86	I
520	5234.27	I
340	5253.46	I
370	5271.19	I
370	5301.98	II
180	5303.55	II
500	5455.15	I
470	5501.34	I
240	5648.25	I
180	5740.66	I
370	5769.34	I
320	5789.24	I
450	5791.34	I
140	5821.99	I
320	5930.62	I
720	6249.93	I
260 d	6262.30	II
450	6394.23	I
250	6455.99	I
180	6709.50	I
110	7045.96	I
160	7066.23	II
50	7161.25	I
110 w	7282.34	II
110 w	7334.18	I
75 cw	7483.50	II
50	7498.83	I
85	7539.23	I
40	7964.83	I
75	8086.05	I
85	8324.69	I
95	8346.53	I
65	8545.44	I
300	8583.45	III
40	8674.43	I
35	8825.82	I
120	9184.38	III

Intensity	Wavelength/Å	
100	9212.63	III
140	10284.79	III

Lead Pb Z = 82

Intensity	Wavelength/Å	
10	496.38	IV
12	499.94	IV
14	529.78	IV
20	570.16	IV
10	648.50	IV
20	703.73	V
12	749.46	V
10	752.52	V
10	761.09	IV
18	767.45	V
18	769.49	V
14	771.42	V
14	782.79	V
15	797.02	V
18	802.07	IV
12	802.82	IV
18	809.63	V
10	812.59	IV
10	827.41	IV
12	832.60	IV
12	845.94	IV
18	857.64	IV
16	862.33	IV
20	863.97	V
14	870.44	IV
6	873.71	II
12	879.96	IV
18	883.90	V
14	884.96	IV
14	884.99	IV
14	888.37	V
8	889.68	II
16	890.72	IV
14	894.40	V
12	896.08	V
12	908.51	IV
14	915.71	V
12	917.90	IV
12	918.09	V
12	920.28	V
12	920.66	V
10	922.12	IV
12	922.49	IV
10	927.64	IV
14	932.20	IV
12	954.35	V
10	967.23	II
10	986.71	II
10	995.89	II
10	1016.61	II
14	1028.61	IV
20	1032.05	IV
16	1041.24	IV
18	1044.14	IV
12	1048.9	III
10	1049.82	II
10	1050.77	II
10	1051.26	V
15	1056.53	IV
10	1060.66	II
12	1072.09	IV
18	1080.81	IV
20	1084.17	IV
10	1088.86	V
10	1103.94	II
10	1108.43	II
10	1109.84	II
20	1116.08	IV
10	1119.57	II
10	1121.36	II
10	1133.14	II
18	1137.84	IV
14	1144.93	IV
12	1157.88	V
14	1185.43	V
20	1189.95	IV
10	1203.63	II
10	1231.20	II
11	1233.50	V
10	1291.10	IV
20	1313.05	IV
10	1331.65	II
10	1335.20	II
12	1343.06	IV
10	1348.37	II
16	1388.94	IV
18	1400.26	IV
10	1404.34	IV
10	1433.96	II
10	1512.42	II
14	1535.71	IV
20	1553.1	III
10	1671.53	II
10	1682.15	II
20	1726.75	II
10	1796.670	II
10	1822.050	II
10	1904.77	I
7	1921.471	II
12	1959.34	IV
16	1973.16	IV
10	1998.83	V
5 r	2022.02	I
10	2042.58	IV
12	2049.34	IV
8 r	2053.28	I
12	2079.22	IV
6	2111.758	I
10	2115.066	I
15	2154.01	IV
500 r	2170.00	I
7	2175.580	I
12	2177.46	IV
7	2187.888	I
8	2189.603	I
10	2203.534	II
20	2237.425	I
20	2246.86	I
25	2246.89	I
20	2259.01	V
150	2332.418	I
16	2359.53	IV
180	2388.797	I
550 r	2393.792	I
140	2399.597	I
320 r	2401.940	I
320 r	2411.734	I
16	2417.61	IV
15	2424.81	V
150 r	2443.829	I
160 r	2446.181	I
130 r	2476.378	I
80 r	2577.260	I
500 r	2613.655	I
900 r	2614.175	I
160	2628.262	I
4	2634.256	II
10	2657.094	I
700	2663.154	I
10	2697.541	I
25000 r	2801.995	I
100	2822.58	I
14000 r	2823.189	I
35000 r	2833.053	I
6	2840.557	II
14000 r	2873.311	I
3	2914.442	II
15	2966.460	I
15	2972.991	I
15	2980.157	I
4	2986.876	II
10	3043.85	III
150	3118.894	I
10	3137.81	III
10	3176.50	III
600	3220.528	I
100	3229.613	I
400	3240.186	I
200	3262.355	I
35000	3572.729	I
50000 r	3639.568	I
20000	3671.491	I
70000 r	3683.462	I
10	3713.982	II
25000	3739.935	I
12	3854.08	III
15000	4019.632	I
95000	4057.807	I
14000	4062.136	I
10	4157.814	I
10000	4168.033	I
8	4272.66	III
200	4340.413	I
10	4496.15	IV
6	4499.34	III
16	4534.60	IV
7	4571.21	III
10	4579.051	II
6	4761.12	III
1000	5005.416	I
100	5006.572	I
50	5089.484	I
10	5107.242	I
2000	5201.437	I
10	5372.099	II
40	5692.346	I
200	5895.624	I
2000	6001.862	I
500	6011.667	I
500	6059.356	I
40	6081.409	II
50	6110.520	I
100	6235.266	I
50 c	6660.20	II
20000	7228.965	I
10	7346.676	I
20	7809.259	I
5	7896.737	I
10	8168.001	I
6	8191.886	I
5	8217.711	I
40	8272.690	I
20	8409.384	I
10	8478.492	I
5	8722.810	I
10	8857.457	I
8	9293.476	I
15	9438.05	I
15	9604.297	I
15	9674.351	I
200	10290.458	I
100	10498.965	I
50	10649.249	I
15	10886.688	I
40	10969.53	I
	13512.6	I
	14743.0	I
	15349.6	I
	39039.4	I

Lithium Li Z = 3

Intensity	Wavelength/Å	
	102.9	III
	103.4	III
	104.1	III
	105.5	III
	108.0	III
	113.9	III
	125.5	II
	135.0	III
	136.5	II
	140.5	II
	167.21	II
	168.74	II
	171.58	II
	178.02	II
	199.28	II
	207.5	II
	456.	II
	483.	II
	540.	II
	540.0	III
	729.	II

Intensity	Wavelength/Å	
	729.1	III
	800.	II
	820.	II
	861.	II
	905.5	II
	917.5	II
	936.	II
	945.	II
	965.	II
	972.	II
	988.	II
	1018.	II
	1032.	II
	1036.	II
	1093.	II
	1103.	II
	1109.	II
	1116.	II
	1132.1	II
	1141.	II
	1166.4	II
	1198.09	II
	1215.	II
	1238.	II
	1253.8	II
	1420.89	II
	1424.	II
3	1492.93	II
5	1492.97	II
1	1493.04	II
	1555.	II
3	1653.08	II
5	1653.13	II
1	1653.21	II
	1681.66	II
	1755.33	II
	2009.	II
	2039.	I
	2068.	II
	2131.	II
	2164.	II
	2173.4	I
	2183.	II
	2214.	II
	2222.	II
	2237.	II
h	2249.21	II
	2286.82	II
	2302.57	II
	2303.33	II
	2304.59	I
	2304.92	I
	2305.36	I
	2305.83	I
	2306.29	I
	2306.82	I
	2307.44	I
	2308.97	I
	2309.88	I
	2310.94	I
	2312.11	I

Intensity	Wavelength/Å	
	2313.49	I
	2315.08	I
	2316.95	I
	2319.18	I
	2321.88	I
	2325.11	I
	2329.02	I
	2329.84	II
	2333.94	I
3	2336.88	II
5	2336.91	II
2	2337.00	II
	2340.15	I
	2348.22	I
	2358.93	I
	2373.54	I
	2381.54	II
	2383.20	II
1	2394.39	I
	2402.33	II
	2410.84	II
3	2425.43	I
	2429.81	II
	2460.2	I
10	2475.06	I
	2506.94	II
	2508.78	II
	2518.	I
	2539.49	II
24	2551.7	II
	2559.	II
15	2562.31	I
	2605.08	II
2	2657.29	II
3	2657.30	II
	2674.46	II
	2728.24	II
5	2728.29	II
2	2728.32	II
3	2730.47	II
1	2730.55	II
5	2741.20	I
	2766.99	II
	2790.31	II
	2801.	I
	2846.	I
	2868.	I
	2895.	I
2	2934.02	II
2	2934.07	II
5	2934.12	II
1	2934.25	II
	2968.	I
3	3029.12	II
3	3029.14	II
	3144.	I
3	3155.31	II
4	3155.33	II
1	3196.26	II
9	3196.33	II
4	3196.36	II

Intensity	Wavelength/Å	
5	3199.33	II
2	3199.43	II
17	3232.66	I
	3249.87	II
	3306.28	II
	3488.	I
	3579.8	I
	3618.	I
	3662.	I
	3684.32	II
1	3714.00	II
5	3714.16	II
6 d	3714.27	II
8	3714.29	II
7 d	3714.40	II
10	3714.41	II
1	3714.51	II
	3714.58	II
3	3718.7	I
6	3794.72	I
20	3915.30	I
20	3915.35	I
10	3985.48	I
10	3985.54	I
40	4132.56	I
40	4132.62	I
	4196.	I
20	4273.07	I
20	4273.13	I
5	4325.42	II
5	4325.47	II
1	4325.54	II
	4516.45	II
13	4602.83	I
13	4602.89	I
	4607.34	II
	4671.51	II
6	4671.65	II
2	4671.70	II
3	4678.06	II
1	4678.29	II
	4760.	I
	4763.	II
	4788.36	II
	4843.0	II
4	4881.32	II
4	4881.39	II
1	4881.49	II
8	4971.66	I
8	4971.75	I
	5037.92	II
	5271.	I
	5315.	I
	5395.	I
	5440.	I
600 c	5483.55	II
600 c	5485.65	II
320	6103.54	I
320	6103.65	I
3600	6707.76	I
3600	6707.91	I

Intensity	Wavelength/Å	
48	8126.23	I
48	8126.45	I
	8517.37	II
	9581.42	II
	10120.	II
	12232.	I
	12782.	I
	13566.	I
	17552.	I
	18697.	I
	19290.	I
	24467.	I
	40475.	I
Lutetium Lu Z = 71		
100	563.72	V
500	810.73	III
2000	832.28	III
100	861.92	V
400	876.80	IV
100	880.32	V
100	891.81	V
100	914.72	V
400	1001.18	III
100	1272.42	IV
800	1333.79	IV
400	1429.38	IV
200	1441.76	V
200	1453.35	V
200	1468.99	V
400	1472.12	V
200	1473.71	V
200	1485.58	V
400	1511.26	IV
600	1772.57	IV
100 c	1786.25	V
1000	1854.57	III
1500	2065.35	III
1500 c	2070.56	III
600 c	2086.47	IV
1000 c	2104.41	IV
1000 c	2108.31	IV
1700 h	2195.54	II
1000	2236.14	III
2000	2236.22	III
95	2276.94	II
190	2297.41	II
1300	2392.19	II
120	2399.14	II
80	2419.21	II
130	2459.64	II
370	2536.95	II
930	2571.23	II
1700	2578.79	II
4500 c	2603.35	III
1800	2613.40	II
18000	2615.42	II
1800	2619.26	II
2700	2657.80	II
570 h	2685.08	I
4200	2701.71	II
180 d	2719.09	I

Intensity	Wavelength/Å	
480 h	2728.95	I
3600	2754.17	II
750 h	2765.74	I
2000	2772.55	III
2700	2796.63	II
270 c	2834.35	II
330 h	2845.13	I
3000	2847.51	II
570 h	2885.14	I
6300	2894.84	II
4500	2900.30	II
300	2903.05	I
9000	2911.39	II
270 h	2949.73	I
1200	2951.69	II
4200	2963.32	II
2400	2969.82	II
1800	2989.27	I
3000	3020.54	II
2100	3056.72	II
1000	3057.86	III
7500	3077.60	II
390	3080.11	I
5100 h	3081.47	I
3000	3118.43	I
2400	3171.36	I
260	3191.80	II
1400	3198.12	II
4800	3254.31	II
3800	3278.97	I
7600	3281.74	I
6200	3312.11	I
7600	3359.56	I
6200	3376.50	I
950	3385.50	I
160 h	3391.55	I
1400	3396.82	I
4100	3397.07	II
4800	3472.48	II
8300 c	3507.39	II
1600	3508.42	I
4800	3554.43	II
4800	3567.84	I
340	3596.34	I
800	3623.99	II
680	3636.25	I
2600	3647.77	I
110	3756.70	I
110	3756.79	I
150	3800.67	I
2700	3841.18	I
530	3876.65	II
50	3918.86	I
480	3968.46	I
670	4054.45	I
310	4122.49	I
3100	4124.73	I
150 c	4131.79	I
460	4154.08	I
1600	4184.25	II
150	4277.50	I
250	4281.03	I
330 d	4295.97	I
150	4309.57	I
190 c	4430.48	I
190	4450.81	I
3300	4518.57	I
100 h	4648.21	I
1000	4658.02	I
85 h	4659.03	I
150	4785.42	II
85	4815.05	I
460	4904.88	I
180	4942.34	I
800	4994.13	II
800	5001.14	I
140	5134.05	I
2700	5135.09	I
170	5196.61	I
500	5402.57	I
140 c	5421.90	I
100	5437.88	I
2100	5476.69	II
550	5736.55	I
80	5800.59	I
690 cw	5983.9	II
140	5997.13	I
1400	6004.52	I
440	6055.03	I
150	6159.94	II
600	6198.13	III
160	6199.66	II
2100	6221.87	II
80	6235.36	II
160	6242.34	II
70 h	6345.35	I
1100	6463.12	II
29	6477.67	I
55 c	6523.18	I
35 cw	6611.28	II
23 c	6677.14	I
30 c	6793.77	I
45	6917.31	I
23	7031.24	I
45	7125.84	II
14 ch	7237.98	I
11 c	7441.52	I
9 c	8178.16	I
17	8382.08	I
35	8459.19	II
10 d	8478.50	I
29 c	8508.08	I
35 c	8610.98	I

Magnesium Mg Z = 12

Intensity	Wavelength/Å	
400	146.95	IV
20	186.51	III
20	187.20	III
10	188.53	III
100	231.73	III
80	234.26	III
35	276.58	V
4000	320.99	IV
3000	323.31	IV
30	353.09	V
150	857.29	IV
50	919.03	IV
250	1037.41	IV
300	1210.99	IV
300	1342.19	IV
800	1346.57	IV
300	1346.68	IV
600	1352.05	IV
900	1384.46	IV
500	1385.77	IV
800	1387.53	IV
300	1404.68	IV
1000	1409.36	IV
500	1437.53	IV
1000	1437.64	IV
300	1447.42	IV
300	1459.54	IV
400	1459.62	IV
400	1481.51	IV
350	1490.45	IV
300	1495.50	IV
300	1607.11	IV
5	1668.43	I
500	1683.02	IV
10	1683.41	I
400	1698.81	IV
15	1707.06	I
40	1734.84	II
50	1737.62	II
20	1747.80	I
40	1750.65	II
50	1753.46	II
30	1827.93	I
300	1844.17	IV
9	2025.82	I
25	2064.90	III
20	2091.96	III
20	2177.70	III
3	2329.58	II
20	2395.15	III
6	2449.57	II
1	2557.23	I
1	2560.94	I
1	2562.26	I
1	2564.94	I
1	2570.91	I
1	2572.25	I
2	2574.94	I
1	2577.89	I
1	2580.59	I
1	2584.22	I
2	2585.56	I
3	2588.28	I
1	2591.89	I
1	2593.23	I
2	2595.97	I
2	2602.50	I
4	2603.85	I
5	2606.62	I
1	2613.36	I
2	2614.73	I
3	2617.51	I
3	2628.66	I
6	2630.05	I
8	2632.87	I
2	2644.80	I
3	2646.21	I
4	2649.06	I
8	2660.76	II
8	2660.82	II
6	2668.12	I
8	2669.55	I
10	2672.46	I
3	2693.72	I
5	2695.18	I
6	2698.14	I
8	2731.99	I
10	2733.49	I
12	2736.53	I
5	2765.22	I
7	2768.34	I
38	2776.69	I
32	2778.27	I
90	2779.83	I
8	2781.29	I
32	2781.42	I
36	2782.97	I
1000	2795.53	II
600	2802.70	II
3	2809.76	I
2	2811.11	I
1	2811.78	I
12	2846.72	I
12	2846.75	I
14	2848.34	I
14	2848.42	I
16	2851.65	I
16	2851.66	I
6000	2852.13	I
2	2902.92	I
4	2906.36	I
3	2915.45	I
10	2936.74	I
12	2938.47	I
2	2942.00	I
13	2942.00	I
20	3091.08	I
22	3092.99	I
14	3096.90	I
9	3104.71	II
8	3104.81	II
6	3168.98	II
6	3172.71	II
7	3175.78	II
2	3197.62	I
17	3329.93	I
6	3332.15	I
9	3336.68	I
7	3535.04	II
8	3538.86	II

Atomic

Intensity	Wavelength/Å	
7	3549.52	II
8	3553.37	II
140	3829.30	I
300	3832.30	I
500	3838.29	I
8	3848.24	II
7	3850.40	II
3	3878.31	I
3	3895.57	I
4	3903.86	I
6	3938.40	I
8	3986.75	I
10	4057.50	I
15	4167.27	I
20	4351.91	I
9	4384.64	II
10	4390.59	II
8	4428.00	II
9	4433.99	II
14	4481.16	II
13	4481.33	II
28	4571.10	I
10	4730.03	I
7	4851.10	II
75	5167.33	I
220	5172.68	I
400	5183.61	I
8	5264.21	II
7	5264.37	II
9	5401.54	II
6	5528.41	I
30	5711.09	I
10	6318.72	I
9	6319.24	I
7	6319.49	I
10	6346.74	II
9	6346.96	II
11	6545.97	II
7	6781.45	II
8	6787.85	II
7	6812.86	II
8	6819.27	II
10	7193.17	I
10	7291.06	I
12	7387.69	I
20	7657.60	I
19	7659.15	I
17	7659.90	I
15	7691.55	I
12	7877.05	II
13	7896.37	II
10	8098.72	I
9	8115.22	II
8	8120.43	II
10	8209.84	I
20	8213.03	I
10	8213.99	II
11	8234.64	II
10	8310.26	I
15	8346.12	I
10	8710.18	I
12	8712.69	I
13	8717.83	I
10	8734.99	II
17	8736.02	I
11	8745.66	II
14	8806.76	I
10	8824.32	II
11	8835.08	II
20	8923.57	I
10	8997.16	I
14	9218.25	II
13	9244.27	II
12	9246.50	I
30	9255.78	I
10	9327.54	II
10	9340.54	II
25	9414.96	I
17	9429.81	I
19	9432.76	I
20	9438.78	I
12	9631.89	II
11	9632.43	II
15	9953.20	I
15	9983.20	I
17	9986.47	I
18	9993.21	I
14	10092.16	II
35	10811.08	I
11	10914.23	II
10	10951.78	II
25	10953.32	I
27	10957.30	I
28	10965.45	I
15	11032.10	I
14	11033.66	I
45	11828.18	I
30	12083.66	I
28	14877.62	I
35	15024.99	I
30	15040.24	I
25	15047.70	I
10	15765.84	I
30	17108.66	I
5	26392.90	I

Manganese Mn Z = 25

Intensity	Wavelength/Å	
600	410.30	V
600	410.60	V
600	415.62	V
650	415.98	V
600	428.59	V
600	435.67	V
1000	441.72	V
850	442.49	V
60	579.79	IV
60	581.44	IV
60	581.65	IV
60	585.21	IV
90	1242.25	IV
90	1244.50	IV
95	1251.93	IV
95	1257.28	IV
90	1264.41	IV
500	1283.58	III
400	1287.59	III
300	1291.62	III
1000	1360.72	III
800	1365.20	III
500 h	1609.17	III
1000	1614.14	III
2000	1620.60	III
500	1633.80	III
80	1667.00	IV
80	1698.30	IV
20	1726.47	II
30	1732.70	II
50	1733.55	II
40	1734.49	II
30	1737.93	II
20	1740.16	II
20	1742.00	II
85	1742.10	IV
85	1766.27	IV
80	1795.65	IV
80	1795.79	IV
30	1853.27	II
20	1857.92	II
50	1902.95	II
20	1907.84	II
75	1910.25	IV
30	1911.41	II
20 d	1914.68	II
100	1915.10	II
20	1918.64	II
30	1919.64	II
80	1921.25	II
20	1923.07	II
20	1923.34	II
30	1925.52	II
50	1926.59	II
30	1931.40	II
500	1941.28	III
800	1943.21	III
20	1945.15	II
20	1947.93	II
20	1950.14	II
500	1952.36	III
1000	1952.52	III
30	1953.23	II
20 d	1954.81	II
30	1959.25	II
20	1969.24	II
500	1978.95	III
30	1994.23	II
9700	1996.06	I
14000	1999.51	I
18000	2003.85	I
1000 w	2027.83	III
500 w	2028.14	III
50	2037.31	II
40	2037.64	II
40	2039.97	II
500	2049.68	III
500	2066.38	III
1000	2069.02	III
30	2076.21	II
900	2077.38	III
800	2084.23	III
600	2090.05	III
1500	2092.16	I
500	2094.78	III
20	2097.46	II
500	2097.93	III
500	2099.97	III
20	2102.50	II
1700	2109.58	I
30	2113.96	II
1000	2169.78	III
700	2174.15	III
900	2176.87	III
800	2181.86	III
800	2184.87	III
290	2208.81	I
540	2213.85	I
900	2220.55	III
770	2221.84	I
1000	2227.42	III
20	2373.36	II
20	2427.38	II
50	2427.72	II
30	2427.94	II
30	2437.37	II
20	2437.84	II
30	2452.49	II
50	2499.00	II
30	2507.60	II
20	2516.60	II
30	2516.74	II
20	2521.66	II
20	2530.72	II
20	2531.80	II
50	2532.78	II
50	2533.33	II
30	2534.10	II
80	2534.22	II
100	2535.66	II
30	2535.98	II
100	2537.92	II
50	2541.11	II
80	2542.92	II
50	2543.45	II
100	2548.75	II
50	2551.85	II
30	2553.27	II
75	2556.57	II
30	2556.89	II
95	2558.59	II
30	2559.41	II
150	2563.65	II
30	2565.22	II
580	2572.76	I
480	2575.51	I
12000	2576.10	II
550	2584.31	I

Intensity	Wavelength/Å	
30	2588.97	II
45	2589.71	II
250	2592.94	I
6200	2593.73	II
250	2595.76	I
95	2598.90	II
30	2602.72	II
45	2603.72	II
4300	2605.69	II
190	2610.20	II
500	2618.14	II
140	2622.90	I
150	2624.04	I
40	2624.80	II
200	2625.58	II
190	2632.35	II
130	2638.17	II
80	2639.84	II
27	2650.99	II
60	2655.91	II
30	2666.77	II
30	2667.03	II
110	2672.59	II
55	2673.37	II
55	2674.43	II
45	2680.34	II
30	2680.68	II
30	2681.25	II
55	2684.55	II
55	2685.94	II
110	2688.25	II
27	2693.19	II
55	2695.36	II
27	2698.97	II
85	2701.00	II
50	2701.17	II
160	2701.70	II
100	2703.98	II
130	2705.74	II
80	2707.53	II
110	2708.45	II
45	2709.96	II
80	2710.33	II
110	2711.58	II
30	2716.80	II
30	2717.53	II
30	2719.01	II
50	2719.74	II
30	2722.10	II
30	2724.46	II
55	2728.61	II
6200	2794.82	I
5100	2798.27	I
220	2799.84	I
3700	2801.06	I
110	2809.11	I
60	2815.02	II
30	2816.33	II
60	2870.08	II
30	2872.94	II
80	2879.49	II
70	2886.68	II
160	2889.58	II
55	2892.39	II
50	2898.70	II
80	2900.16	II
140 h	2914.60	I
190 h	2925.57	I
1100	2933.06	II
1500	2939.30	II
250 h	2940.39	I
1900	2949.20	II
30	3019.92	II
55	3031.06	II
30	3035.35	II
330	3044.57	I
120	3045.59	I
200	3047.04	I
30	3050.65	II
250	3054.36	I
140	3062.12	I
170	3066.02	I
170	3070.27	I
160	3073.13	I
140 h	3178.50	I
220	3212.88	I
1000	3228.09	I
300	3230.72	I
850	3236.78	I
330	3243.78	I
650	3248.52	I
100	3251.14	I
310	3252.95	I
310	3256.14	I
220	3258.41	I
180	3260.23	I
180	3264.71	I
200	3330.78	II
720	3441.99	II
50	3460.03	II
360	3460.33	II
360 h	3474.04	II
	3474.13	II
290	3482.91	II
180	3488.68	II
140	3495.84	II
50	3496.81	II
100	3497.54	II
360	3531.85	I
1100	3532.12	I
1300	3547.80	I
1100	3548.03	I
390	3548.20	I
2200	3569.49	I
720	3569.80	I
1400	3577.88	I
720	3586.54	I
290	3595.12	I
150	3601.72	III
420	3607.54	I
420	3608.49	I
360	3610.30	I
290	3619.28	I
220	3623.79	I
140	3629.74	I
100	3660.40	I
280	3693.67	I
180	3696.57	I
210	3706.08	I
130	3718.93	I
130	3731.93	I
260	3790.22	I
110	3800.55	I
3200	3806.72	I
700	3809.59	I
2100	3823.51	I
390	3823.89	I
200	3829.68	I
480	3833.86	I
1300	3834.36	I
350	3839.78	I
670	3841.08	I
350	3843.98	I
120	3926.47	I
130	3982.58	I
150	3985.24	I
190	3986.83	I
150	3987.10	I
1500	4018.10	I
150	4026.44	I
27000	4030.76	I
19000	4033.07	I
11000	4034.49	I
1500	4035.73	I
5600	4041.36	I
210 d	4045.13	I
1100	4048.76	I
150	4055.21	I
1900	4055.54	I
210	4057.95	I
1100	4058.93	I
150	4059.39	I
730	4061.74	I
730	4063.53	I
290	4070.28	I
730	4079.24	I
730	4079.42	I
1100	4082.94	I
1100	4083.63	I
200	4110.90	I
150	4131.12	I
120	4135.04	I
150	4176.60	I
120	4189.99	I
370	4235.14	I
510	4235.29	I
190	4239.72	I
290	4257.66	I
290	4265.92	I
270	4281.10	I
50	4323.63	II
350	4414.88	I
210	4436.35	I
800	4451.59	I
160	4453.00	I
130	4455.01	I
160	4455.32	I
110	4455.82	I
210	4457.55	I
270	4458.26	I
150	4461.08	I
510	4462.02	I
290	4464.68	I
200	4470.14	I
130	4472.79	I
170	4490.08	I
240	4498.90	I
240	4502.22	I
160	4709.72	I
180	4727.48	I
130	4739.11	I
1000	4754.04	I
180	4761.53	I
750	4762.38	I
300	4765.86	I
500	4766.43	I
940	4783.42	I
1000	4823.52	I
19	5004.91	I
30	5074.79	I
200	5079.20	III
150	5100.03	III
60	5117.94	I
50	5150.89	I
50	5196.59	I
85	5255.32	I
160	5341.06	I
19	5349.88	I
95	5377.63	I
95	5394.67	I
50	5399.49	I
95	5407.42	I
35	5413.69	I
85	5420.36	I
35	5432.55	I
150	5454.07	III
12	5457.47	I
60	5470.64	I
200	5474.68	III
40	5481.40	I
30	5505.87	I
50	5516.77	I
40	5537.76	I
21	5551.98	I
200	5946.65	III
140	6013.50	I
200	6016.64	I
290	6021.80	I
200	6231.21	III
17	6440.97	I
24	6491.71	I
14 h	6942.52	I
12	6989.96	I
14	7069.84	I

Intensity	Wavelength/Å	
12	7184.25	I
24 h	7283.82	I
35 h	7302.89	I
50	7326.51	I
12	7680.20	I
12 h	8672.06	I
12 h	8701.05	I
17 h	8703.76	I
30 h	8740.93	I

Mercury 198 Hg Z = 80

Intensity	Wavelength/Å	
80	1250.564	I
8	1259.242	I
100	1268.825	I
5	1307.751	I
20	1402.619	I
10	1435.503	I
1000	1849.492	I
60	2262.210	II
20	2302.065	I
20	2345.440	I
100	2378.325	I
20	2380.004	I
40	2399.349	I
20	2399.729	I
20	2446.900	I
15	2464.064	I
40	2481.999	I
30	2482.713	I
40	2483.821	I
90	2534.769	I
15000	2536.506	I
25	2563.861	I
25	2576.290	I
250	2652.043	I
400	2653.683	I
100	2655.130	I
50	2698.831	I
80	2752.783	I
20	2759.710	I
40	2803.471	I
30	2804.438	I
750	2847.675	II
50	2856.939	I
150	2893.598	I
150	2916.227	II
60	2925.413	I
1200	2967.283	I
300	3021.500	I
120	3023.476	I
30	3025.608	I
50	3027.490	I
400	3125.670	I
320	3131.551	I
320	3131.842	I
80	3341.481	I
2800	3650.157	I
300	3654.839	I
80	3662.883	I
240	3663.281	I
30	3701.432	I
35	3704.170	I
30	3801.660	I
20	3901.867	I
60	3906.372	I
200	3983.839	II
1800	4046.572	I
150	4077.838	I
40	4108.057	I
250	4339.224	I
400	4347.496	I
4000	4358.337	I
80	4916.068	I
1100	5460.753	I
160	5675.922	I
240	5769.598	I
280	5790.663	I
20	6072.713	I
30	6234.402	I
160	6716.429	I
250	6907.461	I
240	11287.407	I

Mercury Hg Z = 80

Intensity	Wavelength/Å	
3	621.44	III
2	679.68	III
2	878.59	III
1	886.48	III
400	893.08	II
300	915.83	II
150	923.39	II
200	940.80	II
100	962.74	II
50	969.13	II
1	988.89	III
2	1009.29	III
5	1068.03	III
800	1099.26	II
2	1161.95	III
80	1250.58	I
8	1259.24	I
100	1268.82	I
5	1307.75	I
300	1307.93	II
400	1321.71	II
400	1331.74	II
80	1350.07	II
200	1361.27	II
20	1402.62	I
200	1414.43	II
10	1435.51	I
15	1619.46	II
120	1623.95	II
20	1628.25	II
150	1649.94	II
50	1653.64	II
200	1672.41	II
9	1681.40	III
100	1702.73	II
100	1707.40	II
120	1727.18	II
250	1732.14	II
15	1759.75	III
20	1775.68	I
40	1783.70	II
30	1796.22	II
200	1796.90	II
60	1798.74	II
30	1803.89	II
40	1808.29	II
400	1820.34	II
5	1832.74	I
1000	1849.50	I
160	1869.23	II
300	1870.55	II
200	1875.54	II
1	1894.77	III
20	1900.28	II
30	1927.60	II
300	1942.27	II
100	1972.94	II
200	1973.89	II
150	1987.98	II
90	2026.97	II
90	2052.93	II
70	2148.00	II
5	2247.55	I
60	2262.23	II
20	2302.06	I
7	2314.15	III
15	2323.20	I
5	2340.57	I
20	2345.43	I
20	2352.48	I
100	2378.32	I
20	2380.00	I
4	2380.55	III
40	2399.38	I
20	2399.73	I
10	2400.49	I
60	2407.35	II
50	2414.13	II
8	2431.65	III
5	2441.06	I
20	2446.90	I
15	2464.06	I
5	2480.56	III
40	2482.00	I
30	2482.72	I
40	2483.82	I
7	2484.50	III
90	2534.77	I
15000	2536.52	I
25	2563.86	I
25	2576.29	I
5	2578.91	I
2	2612.92	III
4	2617.97	III
15	2625.19	I
5	2639.78	I
250	2652.04	I
400	2653.69	I
100	2655.13	I
3	2670.49	III
5	2674.91	I
50	2698.83	I
50	2699.38	I
80	2705.36	II
70	2724.43	III
80	2752.78	I
20	2759.71	I
6	2769.22	III
40	2803.46	I
30	2804.43	I
2	2805.34	I
2	2806.77	I
150	2814.93	II
3	2844.76	III
750	2847.68	II
50	2856.94	I
150	2893.60	I
150	2916.27	II
60	2925.41	I
150	2935.94	II
400	2947.08	II
1200	2967.28	I
300	3021.50	I
120	3023.47	I
30	3025.61	I
50	3027.49	I
15	3090.05	III
400	3125.67	I
320	3131.55	I
320	3131.84	I
400	3208.20	II
400	3264.06	II
5	3283.02	III
12	3312.28	III
80	3341.48	I
100	3385.25	II
8	3389.01	III
5	3450.77	III
400	3451.69	II
3	3500.35	III
4	3538.88	III
200	3549.42	II
5	3557.24	III
2800	3650.15	I
300	3654.84	I
80	3662.88	I
240	3663.28	I
30	3701.44	I
35	3704.17	I
30	3801.66	I
15	3803.51	III
100	3806.38	II
20	3901.87	I
60	3906.37	I
100	3918.92	II
200	3983.96	II
1800	4046.56	I
150	4077.83	I
40	4108.05	I
70	4122.07	III
10	4140.34	III
100	4216.74	III

Intensity	Wavelength/Å	
250	4339.22	I
400	4347.49	I
4000	4358.33	I
100	4398.62	II
15	4470.58	III
12	4552.84	III
90	4660.28	II
50	4797.01	III
80	4855.72	II
10	4869.85	III
5	4883.00	I
5	4889.91	I
80	4916.07	I
5	4970.37	I
80	4973.57	III
5	4980.64	I
20	5102.70	I
40	5120.64	I
100	5128.45	II
20	5137.94	I
30	5210.82	III
20	5290.74	I
5	5316.78	I
60	5354.05	I
30	5384.63	I
1100	5460.74	I
30	5549.63	I
160	5675.86	I
6	5695.71	III
240	5769.60	I
100	5789.66	I
280	5790.66	I
140	5803.78	I
60	5859.25	I
60	5871.73	II
20	5871.98	I
20	6072.72	I
1000	6149.50	II
25	6220.35	III
30	6234.40	I
35	6418.98	III
40	6501.38	III
80	6521.13	II
10	6584.26	III
6	6610.12	III
30	6709.29	III
160	6716.43	I
250	6907.52	I
250	7081.90	I
200	7091.86	I
40	7346.37	II
100	7485.87	II
12	7517.46	III
20	7728.82	I
7	7808.10	III
100	7944.66	II
25	7946.75	III
50	7984.51	III
5	8151.64	III
2000	10139.75	I
240	11287.40	I
120	13209.95	I
140	13426.57	I
60	13468.38	I
80	13505.58	I
500	13570.21	I
450	13673.51	I
200	13950.55	I
500	15295.82	I
100	16881.48	I
400	16920.16	I
300	16942.00	I
500	17072.79	I
400	17109.93	I
20	17116.75	I
20	17198.67	I
20	17213.20	I
70	17329.41	I
30	17436.18	I
50	18130.38	I
40	19700.17	I
	22493.28	I
250	23253.07	I
	32148.06	I
	36303.03	I

Molybdenum Mo Z = 42

Intensity	Wavelength/Å	
50	867.92	IV
100	884.19	IV
60	886.05	IV
50	891.74	IV
100	1169.33	III
100	1254.93	III
100	1258.52	III
100	1262.21	III
100	1263.74	III
100	1274.37	III
100	1276.40	III
200	1277.40	III
200	1277.58	III
200	1278.40	III
150	1281.90	III
150	1283.60	III
100	1854.73	III
80	1926.26	IV
100	1929.24	IV
80	1971.06	IV
70	2010.92	IV
19000	2015.11	II
40000	2020.30	II
21000	2038.44	II
17000	2045.98	II
50	2060.38	IV
4800	2081.68	II
2400	2089.52	II
2200	2092.50	II
4000	2093.11	II
2700	2100.84	II
1500	2104.29	II
1400	2108.02	II
100	2184.37	III
100	2211.02	III
400	2269.69	II
150	2269.71	III
200	2294.97	III
160	2304.25	II
160	2306.97	II
150	2330.93	III
110	2332.12	II
190	2341.59	II
100	2359.76	III
110	2389.20	II
140	2403.61	II
120	2413.01	II
85	2498.28	II
200	2506.19	III
440	2538.46	II
330	2542.67	II
80	2558.88	II
85	2564.34	II
250	2593.70	II
250	2602.80	II
400	2616.78	I
440	2629.85	I
330	2636.67	II
720	2638.76	II
410	2640.99	I
600	2644.35	II
370	2646.49	II
640	2649.46	I
480	2653.35	II
560 h	2655.03	I
640	2660.58	II
720	2672.84	II
250	2673.27	II
1000	2679.85	I
95	2681.36	II
640	2683.23	II
880	2684.14	II
560	2687.99	II
480	2701.42	II
190	2713.51	II
290	2717.35	II
85	2726.97	II
140	2729.68	II
80	2730.20	II
330	2732.88	II
160	2736.96	II
80 h	2737.88	II
290	2746.30	II
110	2756.07	II
220	2763.62	II
240	2769.76	II
160	2773.78	II
190	2774.39	II
1700	2775.40	II
65	2777.86	II
880	2780.04	II
400	2784.99	II
100	2807.74	III
400	2807.76	II
1700	2816.15	II
220	2817.44	II
80	2827.74	II
80	2834.39	II
80	2835.33	II
160	2842.15	II
1700	2848.23	II
370	2853.23	II
370	2863.81	II
220	2866.69	II
1700	2871.51	II
85	2872.88	II
220	2879.05	II
65	2888.15	II
1300	2890.99	II
95	2891.28	II
190	2892.81	II
950	2894.45	II
140	2897.63	II
70	2900.80	II
290	2903.07	II
80	2907.12	II
600	2909.12	II
1100	2911.92	II
120	2918.83	II
1300	2923.39	II
140	2924.32	II
1100	2930.50	II
800	2934.30	II
95	2940.10	II
110	2941.22	II
150	2944.82	II
140	2946.69	II
95	2947.28	II
125	2947.32	III
95	2955.84	II
240	2956.06	II
70	2956.90	II
95	2960.24	II
250	2963.79	II
210	2965.27	II
70	2971.91	II
250	2972.61	II
80	2975.40	II
95	2992.84	II
95	3027.77	II
100	3060.78	II
800	3064.28	I
250	3065.04	II
800	3074.37	I
85	3077.66	II
800	3085.62	I
270	3087.62	II
190	3092.07	II
560	3094.66	I
560	3101.34	I
1400	3112.12	I
290	3122.00	II
14000	3132.59	I
110	3138.72	II
220	3152.82	II
55	3155.64	II
6000	3158.16	I
8700	3170.35	I

Intensity	Wavelength/Å	
95	3172.03	II
160	3172.74	II
120 d	3187.59	II
7600	3193.97	I
880	3205.88	I
3000	3208.83	I
560	3215.07	I
880	3228.22	I
600	3229.79	I
1100	3233.14	I
950	3237.08	I
65	3240.71	II
950	3256.21	I
480	3264.40	I
800	3270.90	I
200	3271.69	III
1100	3289.02	I
950	3290.82	I
190	3292.31	II
100	3313.62	II
190	3320.90	II
640	3323.95	I
1300	3344.75	I
95	3346.40	II
1600	3358.12	I
950	3363.78	I
950	3379.97	I
1900	3384.62	I
130	3395.36	II
640	3404.34	I
1300	3405.94	I
640	3437.22	I
130	3446.08	II
3200	3447.12	I
640	3449.07	I
950	3456.39	I
640	3460.78	I
800	3504.41	I
560	3508.12	I
480	3521.41	I
640	3537.28	I
520	3558.10	I
400	3563.14	I
1400	3581.89	I
1400	3624.46	I
1000	3635.43	I
400	3657.35	I
540	3664.81	I
590	3672.82	I
1300	3680.60	I
65	3688.31	II
180	3692.64	II
1400	3694.94	I
500	3727.69	I
80	3744.37	II
29000	3798.25	I
520	3826.70	I
940	3828.87	I
1700	3833.75	I
29000	3864.11	I
580	3869.08	I
580	3886.82	I
19000	3902.96	I
65	3941.48	II
1400	4062.08	I
2300	4069.88	I
1300	4081.44	I
940	4084.38	I
730	4107.47	I
630	4120.10	I
2900	4143.55	I
480	4185.82	I
2500	4188.32	I
1500	4232.59	I
890	4276.91	I
1200	4277.24	I
1400	4288.64	I
680	4292.13	I
890	4293.21	I
840	4326.14	I
1900	4381.64	I
2500	4411.57	I
990	4434.95	I
480	4457.36	I
630	4474.56	I
400	4536.80	I
460	4626.47	I
640	4707.26	I
700	4731.44	I
770	4760.19	I
410	4819.25	I
410	4830.51	I
180	5014.60	I
80	5029.00	I
65	5030.78	I
100	5047.71	I
50	5055.00	I
200	5059.88	I
100	5080.02	I
100	5096.65	I
130	5097.52	I
130	5109.71	I
80	5114.97	I
150	5145.38	I
110	5147.39	I
80	5163.19	I
100	5167.76	I
160 d	5171.08	I
230 h	5172.94	I
160 h	5174.18	I
110	5200.17	I
50	5200.74	I
50	5211.86	I
80	5219.40	I
65	5231.06	I
100	5234.26	I
460 h	5238.20	I
230 h	5240.88	I
110 h	5242.81	I
100	5245.51	I
150	5259.04	I
65	5261.14	I
65	5279.65	I
210	5280.86	I
55	5292.08	I
55	5295.47	I
55	5313.89	I
80	5354.88	I
65	5356.48	I
560 hl	5360.56	I
110 hl	5364.28	I
65	5394.52	I
50	5400.47	I
55	5435.68	I
65	5437.75	I
50	5501.54	I
7800	5506.49	I
5200	5533.05	I
50	5543.12	I
55	5556.28	I
2500	5570.45	I
100	5610.93	I
330	5632.47	I
50	5634.86	I
230	5650.13	I
55	5674.47	I
460	5689.14	I
80	5705.72	I
210	5722.74	I
620	5751.40	I
520	5791.85	I
55 h	5849.73	I
50 h	5851.52	I
520	5858.27	I
50	5869.33	I
820	5888.33	I
50 h	5893.38	I
160 h	5928.88	I
35	6025.49	I
1300	6030.66	I
40	6101.87	I
40	6357.22	I
35	6401.07	I
100	6424.37	I
230	6619.13	I
50	6650.38	I
110	6733.98	I
50	6746.27	I
35	6753.97	I
40	6838.88	I
35	6914.01	I
110	7109.87	I
150	7242.50	I
40	7245.85	I
40	7391.36	I
140	7485.74	I
27	7720.77	I
40 h	8328.44	I
45 h	8389.32	I
45 h	8483.39	I

Neodymium Nd Z = 60

Intensity	Wavelength/Å	
75	2764.98	I
80	2993.20	II
95	3007.97	II
95	3014.19	II
95	3018.35	II
140	3056.71	II
130	3069.73	II
160	3075.38	II
240	3092.92	II
260	3115.18	II
290	3133.60	II
220	3134.90	II
170	3141.46	II
170	3142.44	II
150	3203.47	II
220	3259.24	II
220	3265.12	II
320	3275.22	II
290	3285.10	II
410	3328.28	II
320	3353.59	II
410	3560.75	II
470	3587.51	II
370	3615.82	II
410	3653.15	II
470	3662.26	II
540	3665.18	II
540	3672.36	II
580	3673.54	II
1200	3685.80	II
440	3687.30	II
410	3689.69	II
410	3697.56	II
470	3713.70	II
640 d	3714.73	II
470	3715.68	II
410	3718.54	II
410	3721.35	II
780	3723.50	II
410	3724.87	II
710	3728.13	II
470	3730.58	II
1000 d	3735.54	II
440	3737.10	II
1000	3738.06	II
580	3752.49	II
510	3757.82	II
930	3758.95	II
930	3763.47	II
510	3769.65	II
1400	3775.50	II
710	3779.47	II
580	3780.40	II
510	3781.32	II
2400	3784.25	II
370	3801.12	II
1200	3803.47	II
2500	3805.36	II
470	3807.23	II
540	3808.77	II
440	3809.06	II
580	3810.49	II
710	3814.73	II

Intensity	Wavelength/Å	
410	3822.47	II
1200	3826.42	II
540	3828.85	II
440	3829.16	II
510	3830.47	II
740	3836.54	II
1700	3838.98	II
410 d	3841.82	II
1700 d	3848.24	II
1500	3848.52	II
470	3850.22	II
2400 d	3851.66	II
3700 d	3863.33	II
850	3869.07	II
470	3875.87	II
1100	3878.58	II
1000	3879.55	II
780	3880.38	II
1200	3880.78	II
540	3887.87	II
1300	3889.93	II
1300	3890.58	II
1300	3890.94	II
580	3891.51	II
470	3892.06	II
810	3894.63	II
440	3897.63	II
2000	3900.21	II
1300	3901.84	II
1700	3905.89	II
510	3907.84	II
2000	3911.16	II
850	3912.23	II
440	3915.13	II
610	3915.95	II
1100	3920.96	II
510	3927.10	II
610	3934.82	II
410	3936.11	II
510	3938.86	II
2000	3941.51	II
2000	3951.16	II
810	3952.20	II
590	3958.00	II
510	3962.21	II
1400	3963.12	II
1100	3973.30	II
740	3973.69	II
740	3976.85	II
740	3979.49	II
470	3986.25	II
1400	3990.10	II
1000	3991.74	II
1100	3994.68	II
410	4000.50	II
540	4004.02	II
410	4007.43	II
3700	4012.25	II
540	4012.70	II
1000	4020.87	II
1000	4021.34	II
1000	4021.78	II
1200	4023.00	II
410	4030.47	II
1200	4031.82	II
3000	4040.80	II
410	4043.59	II
410	4048.81	II
850	4051.15	II
850	4059.96	II
4700	4061.09	II
1100	4069.28	II
710	4075.12	II
470	4075.28	II
470	4080.23	II
1400	4109.08	II
2500	4109.46	II
510	4110.48	II
410	4123.88	II
470	4133.36	II
510	4135.33	II
3000	4156.08	II
510	4156.26	II
410	4168.00	II
810	4175.61	II
2400	4177.32	II
640	4179.59	II
470	4205.60	II
470	4211.29	II
440	4227.73	II
1300	4232.38	II
2000	4247.38	II
850	4252.44	II
410	4261.84	II
470	4282.44	II
710	4284.52	II
5400	4303.58	II
470	4314.52	II
1100	4325.76	II
510	4327.93	II
540	4338.70	II
680	4351.29	II
850	4358.17	II
470 d	4374.93	II
710	4385.66	II
540	4400.83	II
510	4411.06	II
580	4446.39	II
1400	4451.57	II
740	4462.99	II
410	4501.82	II
250	4516.36	II
340	4541.27	II
340	4542.61	II
340	4563.22	II
300	4621.94	I
510	4634.24	I
340	4641.10	I
250	4645.77	II
300	4649.67	I
310	4683.45	I
470	4706.54	II
240	4719.02	I
240	4811.34	II
350	4825.48	II
280	4859.02	II
350	4883.81	I
220	4890.70	II
240	4891.07	I
280	4896.93	I
210	4901.84	I
330	4920.68	II
470	4924.53	I
260	4944.83	I
290	4954.78	I
290	4959.13	II
250	4989.94	II
360	5076.59	II
360	5092.80	II
360	5107.59	II
340	5123.79	II
680	5130.60	II
500	5191.45	II
630	5192.62	II
330	5200.12	II
310	5212.37	II
450	5234.20	II
250	5239.79	II
720	5249.59	II
360	5255.51	II
590	5273.43	II
680	5293.17	II
220	5311.46	II
500	5319.82	II
290	5361.47	II
160	5431.53	II
240	5594.43	II
220	5620.54	I
140 d	5675.97	I
220	5688.53	II
130	5702.24	II
160	5708.28	II
100	5729.29	I
160	5804.02	II
80	5811.57	II
70	5825.87	II
80	5842.39	II
55	5858.91	I
45	6007.67	I
45	6034.24	II
55	6066.03	I
45	6178.59	I
45	6223.39	I
55	6310.49	I
65	6385.20	II
45	6630.14	I
45	6650.57	II
40	6740.11	II
40	6900.43	II
35	7037.30	II
40	7066.89	II
29	7129.35	II
24	7189.42	II
20	7192.01	II
15	7236.54	II
12	7316.81	II
10	7406.62	II
10	7418.18	II
12	7511.16	II
17	7513.73	II
12	7528.99	II
10	7538.26	II
12	7696.56	II
10	7750.95	II
10	7808.47	II
12	7863.04	II
12	7917.01	II
12	7958.95	I
12	7965.73	II
15	7982.09	II
12	7982.68	II
12	8000.76	II
10	8120.93	II
12	8122.07	II
12	8141.75	II
12	8143.27	II
10	8231.52	II
10	8307.72	II
12	8346.36	II
17	8839.10	II
Neon Ne Z = 10		
66	119.01	V
200	122.52	V
66	125.12	V
45	131.99	V
50	132.04	V
150	140.76	V
150	140.79	V
100	142.44	V
100	142.50	V
150	142.72	V
100	143.27	V
150	143.34	V
150	147.13	V
66	151.23	V
120	151.42	V
15	151.82	IV
15	152.23	IV
45	154.50	V
15	158.65	IV
15	158.82	IV
100	164.02	V
100	164.14	V
80	172.62	IV
500	173.93	V
80	177.16	IV
150	186.58	IV
100	194.28	IV
100	208.48	IV
100	208.73	IV
80	208.90	IV
150	212.56	IV
140	223.24	IV
120	223.60	IV

Intensity	Wavelength/Å	
140	234.32	IV
120	234.70	IV
20	251.14	III
20	251.56	III
20	251.73	III
40	267.06	III
40	267.52	III
20	267.71	III
40	283.18	III
160	283.21	III
110	283.69	III
40	283.89	III
220	301.12	III
220	313.05	III
220	313.68	III
40	313.95	III
90	352.956	I
60	354.962	I
50	357.83	IV
400	357.96	V
500	358.47	V
200	358.72	IV
500	359.38	V
90	361.433	II
60	362.455	II
1000	365.59	V
220	379.31	III
125	387.14	IV
100	388.22	IV
150	405.854	II
120	407.138	II
800	416.20	V
150	421.61	IV
200	445.040	II
300	446.256	II
250	446.590	II
180	447.815	II
150	454.654	II
200	455.274	II
10	456.275	II
120	456.348	II
90	456.896	II
1000	460.728	II
500	462.391	II
140	469.77	IV
200	469.82	IV
180	469.87	IV
140	469.92	IV
250	480.41	V
150	481.28	V
250	481.36	V
500	482.99	V
285	488.10	III
220	488.87	III
450	489.50	III
70	489.64	III
220	490.31	III
360	491.05	III
120	521.74	IV
140	521.82	IV
80	541.13	IV
100	542.07	IV
150	543.89	IV
400	568.42	V
250	569.76	V
500	569.83	V
250	572.11	V
800	572.34	V
35	587.213	I
35	589.179	I
35	589.911	I
70	591.830	I
100	595.920	I
75	598.706	I
35	598.891	I
70	600.036	I
170	602.726	I
170	615.628	I
170	618.672	I
120	619.102	I
200	626.823	I
200	629.739	I
1000	735.896	I
400	743.720	I
60	993.88	I
70	1068.65	I
90	1131.72	I
100	1131.85	II
90	1229.83	I
20	1255.03	III
110	1255.68	III
160	1257.19	III
90	1418.38	I
90	1428.58	I
90	1436.09	I
120	1681.68	II
180	1688.36	II
100	1888.11	II
100	1889.71	II
200	1907.49	II
500	1916.08	II
300	1930.03	II
200	1938.83	II
100 c	1945.46	II
80	2007.01	II
65	2018.44	IV
110	2022.19	IV
80	2025.56	II
150	2085.47	II
200	2086.96	III
300	2089.43	III
240	2092.44	III
400	2095.54	III
180	2096.11	II
120	2096.25	II
200	2161.22	III
300	2163.77	III
200	2180.89	III
30	2203.88	IV
200	2209.35	III
200	2211.85	III
240	2213.76	III
300	2216.07	III
10	2220.81	IV
75	2227.42	V
110	2232.41	V
65	2245.48	V
250	2258.02	IV
65	2259.57	V
175	2262.08	IV
240	2263.21	III
65	2263.39	V
110	2264.54	IV
200	2264.91	III
250	2265.71	V
550	2285.79	IV
30	2293.14	IV
250	2293.49	IV
250	2350.84	IV
450	2352.52	IV
700	2357.96	IV
250	2362.68	IV
250	2363.28	IV
110	2365.49	IV
350	2372.16	IV
65	2384.20	IV
350	2384.95	IV
300	2412.73	III
240	2412.94	III
200	2413.78	III
200	2473.40	III
80 p	2562.12	II
90 w	2567.12	II
800	2590.04	III
600	2593.60	III
400	2595.68	III
300	2610.03	III
240	2613.41	III
200	2615.87	III
80	2623.11	II
80	2629.89	II
90 w	2636.07	II
80	2638.29	II
200	2638.70	III
200	2641.07	III
80	2644.10	II
600	2677.90	III
500	2678.64	III
80	2762.92	II
90	2792.02	II
80	2794.22	II
100	2809.48	II
80	2906.59	II
80	2906.82	II
90	2910.06	II
90	2910.41	II
80	2911.14	II
80	2915.12	II
80	2925.62	II
80 w	2932.10	II
80	2940.65	II
90	2946.04	II
150	2955.72	II
150	2963.24	II
150	2967.18	II
100	2973.10	II
15	2974.72	I
100	2979.46	II
12	2982.67	I
150	3001.67	II
120 p	3017.31	II
300	3027.02	II
300	3028.86	II
100	3030.79	II
120	3034.46	II
100	3035.92	II
100	3037.72	II
100	3039.59	II
100	3044.09	II
100	3045.56	II
120	3047.56	II
100	3054.34	II
100	3054.68	II
100	3059.11	II
100	3062.49	II
100	3063.30	II
100	3070.89	II
100	3071.53	II
100	3075.73	II
120	3088.17	II
100	3092.09	II
120	3092.90	II
100	3094.01	II
100	3095.10	II
100	3097.13	II
100	3117.98	II
120	3118.16	II
10	3126.199	I
300	3141.33	II
100	3143.72	II
100 p	3148.68	II
100	3164.43	II
100	3165.65	II
100	3188.74	II
120	3194.58	II
500	3198.59	II
60	3208.96	II
120	3209.36	II
120	3213.74	II
150	3214.33	II
150	3218.19	II
120	3224.82	II
120	3229.57	II
200	3230.07	II
120	3230.42	II
120	3232.02	II
150	3232.37	II
100	3243.40	II
100	3244.10	II
100	3248.34	II
100	3250.36	II
150	3297.73	II
150	3309.74	II
300	3319.72	II

Intensity	Wavelength/Å	
1000	3323.74	II
150	3327.15	II
100	3329.16	II
200	3334.84	II
150	3344.40	II
300	3345.45	II
150	3345.83	II
200	3355.02	II
120	3357.82	II
200	3360.60	II
120	3362.16	II
100	3362.71	II
120	3367.22	II
12	3369.808	I
40	3369.908	I
100	3371.80	II
500	3378.22	II
150	3388.42	II
120	3388.94	II
300	3392.80	II
100	3404.82	II
120	3406.95	II
100	3413.15	II
120	3416.91	II
120	3417.69	II
50	3417.904	I
15	3418.006	I
120	3428.69	II
60	3447.703	I
50	3454.195	I
100	3456.61	II
100	3459.32	II
25	3460.524	I
30	3464.339	I
30	3466.579	I
60	3472.571	I
150	3479.52	II
200	3480.72	II
200	3481.93	II
25	3498.064	I
30	3501.216	I
25	3515.191	I
150	3520.472	I
120	3542.85	II
120	3557.80	II
100	3561.20	II
250	3568.50	II
100	3574.18	II
200	3574.61	II
50	3593.526	I
30	3593.640	I
15	3600.169	I
20	3633.665	I
150	3643.93	II
200	3664.07	II
20	3682.243	I
12	3685.736	I
200	3694.21	II
10	3701.225	I
150	3709.62	II
250	3713.08	II
250	3727.11	II
800	3766.26	II
1000	3777.13	II
100	3818.43	II
120	3829.75	II
150	4219.74	II
100	4233.85	II
120	4250.65	II
120	4369.86	II
70	4379.40	II
150	4379.55	II
100	4385.06	II
200	4391.99	II
150	4397.99	II
150	4409.30	II
100	4413.22	II
100	4421.39	II
100 p	4428.52	II
100 p	4428.63	II
150 p	4430.90	II
150 p	4430.94	II
120	4457.05	II
100	4522.72	II
10	4537.754	I
10	4540.380	I
100	4569.06	II
15	4704.395	I
12	4708.862	I
10	4710.067	I
10	4712.066	I
15	4715.347	I
10	4752.732	I
12	4788.927	I
10	4790.22	I
10	4827.344	I
10	4884.917	I
4	5005.159	I
10	5037.751	I
10	5144.938	I
25	5330.778	I
20	5341.094	I
8	5343.283	I
60	5400.562	I
5	5562.766	I
10	5656.659	I
5	5719.225	I
12	5748.298	I
80	5764.419	I
12	5804.450	I
40	5820.156	I
500	5852.488	I
100	5872.828	I
100	5881.895	I
60	5902.462	I
60	5906.429	I
100	5944.834	I
100	5965.471	I
100	5974.627	I
120	5975.534	I
80	5987.907	I
100	6029.997	I
100	6074.338	I
80	6096.163	I
60	6128.450	I
100	6143.063	I
120	6163.594	I
250	6182.146	I
150	6217.281	I
150	6266.495	I
60	6304.789	I
100	6334.428	I
120	6382.992	I
200	6402.246	I
150	6506.528	I
60	6532.882	I
150	6598.953	I
70	6652.093	I
90	6678.276	I
20	6717.043	I
100	6929.467	I
90	7024.050	I
100	7032.413	I
50	7051.292	I
80	7059.107	I
100	7173.938	I
150	7213.20	II
150	7235.19	II
100	7245.167	I
150	7343.94	II
40	7472.439	I
90	7488.871	I
100	7492.10	II
150	7522.82	II
80	7535.774	I
60	7544.044	I
100	7724.628	I
120	7740.74	II
300	7839.055	I
120	7926.20	II
400	7927.118	I
700	7936.996	I
2000	7943.181	I
2000	8082.458	I
100	8084.34	II
1000	8118.549	I
600	8128.911	I
3000	8136.406	I
2500	8259.379	I
100	8264.81	II
2500	8266.077	I
800	8267.117	I
6000	8300.326	I
100	8315.00	II
1500	8365.749	I
100	8372.11	II
8000	8377.606	I
1000	8417.159	I
4000	8418.427	I
1500	8463.358	I
800	8484.444	I
5000	8495.360	I
600	8544.696	I
1000	8571.352	I
4000	8591.259	I
6000	8634.647	I
3000	8647.041	I
15000	8654.383	I
4000	8655.522	I
100	8668.26	II
5000	8679.492	I
5000	8681.921	I
2000	8704.112	I
4000	8771.656	I
12000	8780.621	I
10000	8783.753	I
500	8830.907	I
7000	8853.867	I
1000	8865.306	I
1000	8865.755	I
3000	8919.501	I
2000	8988.57	I
100	9079.46	II
6000	9148.67	I
6000	9201.76	I
4000	9220.06	I
2000	9221.58	I
2000	9226.69	I
1000	9275.52	I
200	9287.56	II
6000	9300.85	I
1500	9310.58	I
3000	9313.97	I
6000	9326.51	I
2000	9373.31	I
5000	9425.38	I
3000	9459.21	I
5000	9486.68	I
5000	9534.16	I
3000	9547.40	I
120	9577.01	II
1000	9665.42	I
100	9808.86	II
800	10295.42	I
2000	10562.41	I
1500	10798.07	I
2000	10844.48	I
3000	11143.020	I
3500	11177.528	I
1600	11390.434	I
1100	11409.134	I
3000	11522.746	I
1500	11525.020	I
950	11536.344	I
500	11601.537	I
1200	11614.081	I
300	11688.002	I
2000	11766.792	I
1500	11789.044	I
500	11789.889	I
1000	11984.912	I
3000	12066.334	I
800	12459.389	I
1000	12689.201	I

Intensity	Wavelength/Å	
1100	12912.014	I
700	13219.241	I
800	15230.714	I
400	17161.930	I
400	18035.80	I
1000	18083.21	I
350	18221.11	I
250	18227.02	I
2500	18276.68	I
2000	18282.62	I
1200	18303.97	I
250	18359.12	I
1200	18384.85	I
2000	18389.95	I
1000	18402.84	I
1200	18422.39	I
300	18458.65	I
400	18475.79	I
900	18591.55	I
1600	18597.70	I
350	18618.96	I
550	18625.16	I
1200	21041.295	I
750	21708.145	I
300	22247.35	I
350	22428.13	I
2250	22530.40	I
400	22661.81	I
600	23100.51	I
1000	23260.30	I
1050	23373.00	I
850	23565.36	I
3500	23636.52	I
300	23701.64	I
1100	23709.2	I
1800	23951.42	I
600	23956.46	I
1000	23978.12	I
200	24098.54	I
500	24161.42	I
600	24249.64	I
1500	24365.05	I
800	24371.60	I
400	24447.85	I
700	24459.4	I
300	24776.46	I
550	24928.88	I
250	25161.69	I
650	25524.37	I
125	28386.21	I
150	30200.	I
250	33173.09	I
450	33352.35	I
1300	33901.	I
2200	33912.10	I
600	34131.31	I
100	34471.44	I
120	35834.78	I
Neptunium Np Z = 93		
300	3481.93	I
300 h	3501.50	I
300 l	3986.89	I
300 s	5044.66	I
300 l	5601.70	I
300 l	5652.75	I
300 l	5784.39	I
300 l	5878.04	I
300 s	6011.22	I
300	6056.09	I
300 s	6073.90	I
300 s	6080.05	I
300 l	6120.49	I
300	6188.59	I
300 l	6200.00	I
300 s	6215.90	I
300 s	6317.84	I
300 l	6341.38	I
300 l	6566.11	I
300 l	6720.68	I
300 s	6751.32	I
300 s	6795.21	I
300 l	6802.62	I
300 l	6805.81	I
300 s	6816.44	I
300 l	6865.45	I
300 s	6907.13	I
300 h	6912.91	I
1000 s	6930.31	I
300 l	6963.63	I
3000 s	6972.09	I
300	7014.02	I
300 l	7018.91	I
300 s	7039.14	I
300 s	7080.01	I
300 l	7174.83	I
300 l	7184.93	I
300 l	7284.28	I
300 l	7292.29	I
300 l	7332.52	I
300 s	7370.60	I
300 l	7381.03	I
300 l	7381.65	I
300 l	7402.70	I
300 s	7512.22	I
300 l	7515.15	I
300 l	7546.05	I
300 l	7624.83	I
300	7626.85	I
300 s	7681.01	I
300 s	7685.25	I
1000 l	7735.14	I
300 l	7761.61	I
1000 l	7765.75	I
300 s	7776.07	I
300	7787.46	I
1000 l	7791.38	I
300 l	7851.44	I
300 l	7887.88	I
300 l	7901.71	I
300 l	7975.98	I
300 h	8080.32	I
300 s	8124.59	I
300	8155.11	I
300 l	8167.42	I
300 l	8183.06	I
300 l	8188.61	I
300 l	8247.82	I
300 l	8287.11	I
300 s	8287.75	I
300 l	8306.22	I
300 s	8313.66	I
1000 l	8339.12	I
300	8356.79	I
300 l	8367.11	I
3000	8372.88	I
3000	8529.96	I
1000 s	8696.23	I
1000 s	8906.02	I
1000	8942.70	I
1000 s	9004.75	I
1000 l	9006.31	I
10000 l	9016.18	I
3000 l	9141.30	I
3000 s	9379.33	I
3000 l	9468.66	I
3000 s	9679.13	I
3000 l	9930.55	I
10000 l	10091.99	I
10000 s	10817.45	I
10000 l	11695.15	I
10000 l	11776.64	I
10000 s	12148.18	I
10000 s	12377.42	I
10000 l	12407.99	I
10000 l	13834.33	I
Nickel Ni Z = 28		
55	315.24	V
56	315.71	V
72	354.18	V
76	354.42	V
68	354.49	V
500	630.71	III
500	676.94	III
300	713.33	III
300	713.38	III
500	718.48	III
300	722.09	III
500	729.82	III
400	731.70	III
300	732.16	III
300	747.99	III
300	750.05	III
300	757.80	III
400	770.22	III
500	778.81	III
300	788.04	III
500	811.57	III
500	826.14	III
500	842.14	III
400	845.24	III
300	847.43	III
300	860.64	III
300	862.88	III
300	863.22	III
300	867.51	III
300	973.79	III
400	979.59	III
500	1317.22	II
76	1398.19	IV
74	1411.45	IV
70	1438.82	IV
73	1449.01	IV
76	1452.22	IV
73	1482.25	IV
72	1489.83	IV
75	1525.31	IV
74	1527.68	IV
74	1527.80	IV
76	1534.71	IV
73	1537.25	IV
75	1543.41	IV
74	1546.23	IV
300	1604.54	III
300	1652.87	III
400	1687.90	III
1000	1692.51	III
800	1709.90	III
650	1715.30	III
500	1719.46	III
400	1722.28	III
500	1738.25	III
300	1739.78	III
1000	1741.55	II
300	1741.96	III
550	1747.01	III
300	1752.43	III
400	1753.01	III
800	1764.69	III
500	1767.94	III
2000	1769.64	III
400	1776.07	III
300	1807.24	III
300	1819.28	III
800	1823.06	III
400	1830.01	III
650	1847.28	III
800	1854.15	III
300	1858.75	III
1000	2165.55	II
2000	2169.10	II
2000	2174.67	II
1500	2175.15	II
2500	2185.50	II
3000	2192.09	II
5000	2205.55	II
4000	2206.72	II
6000	2216.48	II
1000	2264.46	II
2000	2270.21	II
1600	2289.98	I
630	2300.78	I
1000	2303.00	II
2000	2310.96	I
1700	2312.34	I

Intensity	Wavelength/Å	
1400	2313.66	I
1400	2313.98	I
1000	2316.04	II
1400	2317.16	I
2600	2320.03	I
1900	2321.38	I
1400	2325.79	I
940	2329.96	I
1200	2345.54	I
400	2347.52	I
1000	2375.42	II
240	2386.58	I
1000	2394.52	II
2000	2416.13	II
240	2419.31	I
160	2472.06	I
150	2798.65	I
250	2821.29	I
500	2943.91	I
570	2981.65	I
500	2992.60	I
1000	2994.46	I
4000	3002.49	I
2200	3003.63	I
3700	3012.00	I
1700	3037.94	I
3500	3050.82	I
1500	3054.32	I
1900	3057.64	I
500	3064.62	I
2600	3101.55	I
1300	3101.88	I
2900	3134.11	I
1100	3232.96	I
600	3243.06	I
660	3315.66	I
2000	3331.88	II
2900	3369.57	I
3300	3380.57	I
1300	3391.05	I
3300	3392.99	I
8200	3414.76	I
1600	3423.71	I
2600	3433.56	I
990	3437.28	I
4800	3446.26	I
1300	3452.89	I
5000	3458.47	I
5000	3461.65	I
1600	3472.54	I
550	3483.77	I
5500	3492.96	I
660	3500.85	I
2600	3510.34	I
6600	3515.05	I
660	3519.77	I
8200	3524.54	I
5000	3566.37	I
990	3571.87	I
1300	3597.70	I
1300	3610.46	I

Intensity	Wavelength/Å	
530	3612.74	I
6600	3619.39	I
200	3664.10	I
130	3669.24	I
180	3670.43	I
260	3674.15	I
160	3688.42	I
80	3693.93	I
120	3722.48	I
150	3736.81	I
60	3739.23	I
600	3775.57	I
700	3783.53	I
700	3807.14	I
110	3831.69	I
1200	3858.30	I
110	3973.56	I
110	4401.55	I
85	4459.04	I
55	4470.48	I
65	4605.00	I
75	4648.66	I
110	4714.42	I
45	4786.54	I
45	4855.41	I
40	4904.41	I
45	4980.16	I
45	4984.13	I
50	5017.59	I
100	5035.37	I
100	5080.52	I
65	5081.11	I
40 h	5146.48	I
40 h	5155.76	I
180	5476.91	I
23	5709.56	I
16	5754.68	I
10	5857.76	I
10	5892.88	I
10	6108.12	I
10	6176.81	I
10	6191.18	I
13	6256.36	I
16	6643.64	I
22	6767.77	I
10	6914.56	I
26	7122.20	I
16	7393.60	I
16	7409.35	I
23	7422.28	I
13	7522.76	I
19	7555.60	I
23	7617.00	I
16	7714.32	I
19	7727.61	I
19	7748.89	I
10	7788.94	I
13	7797.59	I
1000	8096.75	II
700	8121.48	II
9	8862.55	I

Intensity	Wavelength/Å	
500 w	9900.92	II

***Niobium* Nb *Z* = 41**

Intensity	Wavelength/Å	
80	464.55	V
80	468.32	V
80	763.77	V
80	774.02	V
60	993.54	IV
400	1005.72	IV
500	1007.05	IV
500	1010.19	IV
100	1116.08	IV
150	1120.02	IV
100	1258.87	V
60	1314.56	III
80	1445.43	III
80	1445.98	III
80	1447.09	III
100	1456.68	III
80	1484.73	III
100	1495.94	III
80	1498.02	III
80	1499.45	III
100	1501.99	III
60	1502.30	IV
80	1513.81	III
60	1524.36	IV
100	1524.91	III
100	1590.21	III
80	1598.86	III
80	1604.72	III
80	1639.51	III
100	1682.77	III
100	1705.44	III
100	1707.14	III
100	1758.33	V
100	1877.34	V
100	1892.92	III
60	1922.41	IV
100	1938.84	III
60	1978.22	IV
3300	2029.32	II
65	2032.53	IV
3000	2032.99	II
2000	2109.42	II
1700	2125.21	II
1100	2126.54	II
80 h	2130.24	III
1500	2131.18	II
80	2273.92	III
100	2275.23	III
80	2279.36	III
100	2281.51	III
80	2284.40	III
100	2290.36	III
370	2295.68	II
280	2302.08	II
100	2313.30	III
100	2338.09	III
80	2344.12	III
90	2349.21	III
80	2355.54	III

Intensity	Wavelength/Å	
100	2362.06	III
80	2362.50	III
80	2365.70	III
100	2372.73	III
170	2376.40	II
110	2387.09	II
100	2387.41	III
140	2387.52	II
80	2388.23	III
45	2388.27	II
160	2398.48	II
80	2404.89	III
55	2405.34	II
55	2405.85	II
140	2412.46	II
100	2413.94	III
160	2416.99	II
140	2418.69	II
100	2421.91	III
75	2433.80	II
40	2435.95	II
45	2437.42	II
40	2442.14	II
28	2442.68	II
65	2451.87	II
65	2453.95	II
100	2456.99	III
55	2458.09	II
65	2462.89	I
80	2468.72	III
80	2475.87	III
110	2477.38	II
65	2478.29	II
65	2479.94	II
35	2483.88	II
100	2499.73	III
110	2511.00	II
110	2521.40	II
390	2544.80	II
100	2545.64	III
110	2551.38	II
130	2556.94	II
80	2557.94	III
130	2562.41	II
110	2571.33	II
390	2583.99	II
390	2590.94	II
80	2598.86	III
80	2633.17	III
200	2642.24	II
320	2646.26	II
330	2647.50	I
330	2654.45	I
310	2656.08	II
80	2657.99	III
110	2665.25	II
110	2666.59	II
110	2667.30	II
400	2671.93	II
200	2673.57	II
200	2675.94	II

Intensity	Wavelength/Å	
160	2691.77	II
1000	2697.06	II
320	2698.86	II
320	2702.20	II
150	2702.52	II
470	2716.62	II
470	2721.98	II
310	2733.26	II
110	2737.09	II
240	2768.13	II
310	2773.20	I
270	2780.24	II
110	2793.05	II
190	2827.08	II
250	2841.15	II
280	2842.65	II
160	2846.28	II
240	2861.09	II
100	2865.61	II
500	2868.52	II
800	2875.39	II
270	2876.95	II
530	2877.03	II
100	2880.72	II
570	2883.18	II
280	2888.83	II
470	2897.81	II
400	2899.24	II
470	2908.24	II
670	2910.59	II
470	2911.74	II
1100	2927.81	II
110	2931.47	II
870	2941.54	II
110 h	2945.88	II
110	2946.12	II
110	2946.90	II
1100	2950.88	II
400	2972.57	II
320	2974.10	II
210	2977.68	II
200	2982.11	II
330	2990.26	II
470	2994.73	II
80	3001.84	III
140	3024.74	II
350	3028.44	II
300	3032.77	II
100	3044.76	II
100	3055.52	II
220	3064.53	II
110	3069.68	II
100	3070.90	II
110	3071.56	II
100	3073.24	II
400	3076.87	II
110	3080.35	II
1800	3094.18	II
140	3099.19	II
270	3127.53	II
1500	3130.79	II
80	3142.26	III
390	3145.40	II
1200	3163.40	II
150	3175.78	II
390	3180.29	II
300	3191.10	II
150	3191.43	II
1000	3194.98	II
120	3203.35	II
300	3206.34	II
390	3215.60	II
800	3225.48	II
140	3229.56	II
400	3236.40	II
200	3247.47	II
120	3248.94	II
320	3254.07	II
230	3260.56	II
160	3263.37	II
200	3283.46	II
160	3292.02	II
320	3296.01	I
400	3312.60	I
120	3319.58	II
130	3341.60	II
1300	3341.97	I
1300	3343.71	I
1700	3349.06	I
420	3349.52	I
340	3354.74	I
1700	3358.42	I
130	3365.58	II
340	3366.96	I
130	3369.16	II
350	3374.92	I
170	3386.24	II
350	3392.34	I
230	3408.68	II
180	3409.19	II
230	3412.94	II
230	3425.42	II
230	3426.57	II
180	3432.70	II
180	3440.59	II
200	3479.56	II
100	3484.05	II
500	3498.63	I
460	3507.96	I
200	3510.26	II
200	3515.42	II
200	3517.67	II
2000	3535.30	I
1300	3537.48	I
250	3540.96	II
500	3544.02	I
300	3550.45	I
1000	3554.66	I
630	3563.50	I
630	3563.62	I
1500	3575.85	I
5000	3580.27	I
500	3584.97	I
750	3589.11	I
500	3589.36	I
500	3593.97	I
500	3602.56	I
300	3619.51	II
420	3649.85	I
400	3651.19	II
200	3659.61	II
630	3660.37	I
900	3664.70	I
1500	3697.85	I
330	3711.34	I
3300	3713.01	I
480	3716.99	I
2700	3726.24	I
2700	3739.80	I
670	3740.73	II
1700	3742.39	I
530	3763.49	I
350	3765.08	I
530	3771.85	I
870	3781.01	I
1700	3787.06	I
1300	3790.15	I
3500	3791.21	I
2700	3798.12	I
2700	3802.92	I
670	3803.88	I
530	3804.74	I
670	3810.49	I
530	3811.03	I
530	3815.51	I
210	3818.86	II
670	3824.88	I
350	3835.18	I
350	3863.38	I
530	3877.56	I
870	3878.82	I
670	3883.14	I
1100	3885.44	I
670	3885.68	I
580	3891.30	I
670	3914.70	I
530	3920.20	I
670	3937.44	I
520	3943.67	I
910 d	3966.09	I
1100	4032.52	I
16000 c	4058.94	I
350	4060.79	I
12000	4079.73	I
440	4100.40	I
6700	4100.92	I
310	4116.90	I
5300	4123.81	I
670	4129.43	I
770	4129.93	I
2300	4137.10	I
440	4139.44	I
2700	4139.71	I
350	4143.21	I
870	4150.12	I
4400	4152.58	I
870	4163.47	I
4400	4163.66	I
4000	4164.66	I
3500	4168.13	I
310	4184.44	I
1200	4190.88	I
870	4192.07	I
870	4195.09	I
1300	4195.66	I
310	4198.51	I
350	4201.52	I
870	4205.31	I
350	4214.73	I
420	4217.94	I
420	4229.15	I
770	4262.05	I
420	4266.02	I
400	4286.99	I
580	4299.60	I
580	4300.99	I
390	4311.27	I
350	4326.33	I
390	4331.37	I
330	4410.21	I
150	4503.04	I
530	4523.41	I
480	4546.82	I
370	4564.53	I
720	4573.08	I
480	4581.62	I
1200	4606.77	I
170	4616.17	I
450	4630.11	I
450	4648.95	I
450	4663.83	I
340	4666.24	I
240	4667.22	I
580	4672.09	I
530	4675.37	I
320	4685.14	I
130 c	4706.14	I
260	4708.29	I
150	4713.50	I
220 c	4749.70	I
130 c	4967.78	I
190	4988.97	I
230	5017.75	I
150	5026.36	I
210	5039.04	I
170	5058.01	I
130	5065.25	I
750	5078.96	I
420	5095.30	I
170	5100.16	I
170	5120.30	I
210	5134.75	I
250	5160.33	I
250	5164.38	I

Intensity	Wavelength/Å	
230	5180.31	I
190	5189.20	I
170	5193.08	I
150	5195.84	I
150	5232.81	I
150 d	5251.62	I
270	5271.53	I
130 c	5276.20	I
250	5318.60	I
460	5344.17	I
340	5350.74	I
110	5437.27	I
85	5551.35	I
170	5642.11	I
130	5664.71	I
170	5665.63	I
130	5729.19	I
110	5760.34	I
110	5819.43	I
130 d	5838.64	I
190 cw	5900.62	I
150	5983.22	I
75	6221.96	I
85 c	6430.46	I
65	6544.61	I
210 cw	6660.84	I
150 cw	6677.33	I
130 c	6723.62	I
85	6828.11	I
85	6990.32	I
190 c	7046.81	I
130	7159.43	I
190 cw	7372.50	I
65	7515.93	I
170 c	7574.58	I
75 c	7726.68	I
35	7885.31	I
40	8135.20	I
29 cw	8320.93	I
29	8346.08	I
35	8905.78	I
Nitrogen N $Z = 7$		
400	181.75	IV
52	186.069	V
62	186.153	V
400	191.7	IV
400	192.9	IV
500	196.87	IV
500	197.23	IV
500	202.60	IV
500	205.94	IV
500	205.97	IV
500	206.03	IV
90	209.303	V
500	217.20	IV
500 d	217.90	IV
500 d	223.4	IV
800 w	225.12	IV
800	225.21	IV
600 w	234.12	IV
600 w	234.20	IV
600 w	234.25	IV
550	236.07	IV
500	237.99	IV
500 w	238.7	IV
600	238.80	IV
500 w	239.62	IV
900	247.20	IV
90	247.561	V
120	247.706	V
500 w	248.43	IV
500 w	248.46	IV
500 w	248.48	IV
500	257.95	III
650	258.50	III
700	259.19	III
800	260.09	III
600	260.45	IV
800	261.28	III
500	262.91	III
500	265.23	III
500	265.27	III
150	266.196	V
200	266.379	V
500	268.70	III
650	270.99	IV
250	283.42	IV
300	283.48	IV
350	283.58	IV
600	285.56	IV
600 w	297.7	IV
700	297.82	IV
650	300.32	IV
90	303.123	IV
500	303.28	IV
150	314.715	III
200	314.850	III
90	314.877	III
150	315.053	IV
120	322.503	IV
150	322.570	IV
200	322.724	IV
120	323.175	IV
600	323.26	III
300	335.050	IV
500	338.35	III
500	340.20	III
500 w	351.93	IV
500	351.98	III
700	353.06	IV
120	362.833	III
150	362.881	III
150	362.946	III
90	362.985	III
300	374.204	III
350	374.441	III
500	387.48	III
500	420.77	IV
250	451.869	III
300	452.226	III
650	463.74	IV
285	644.634	II
360	644.837	II
450	645.178	II
140	647.50	I
360	660.286	II
170	671.016	II
285	671.386	II
150	671.630	II
160	671.773	II
170	672.001	II
500	684.996	III
570	685.513	III
650	685.816	III
500	686.335	III
350	692.70	I
90	713.518	V
150	713.860	V
285	746.984	II
150	748.195	V
200	748.291	V
500	763.336	III
570	764.359	III
570	765.148	IV
250	771.544	III
300	771.901	III
350	772.385	III
200	772.891	III
150	772.975	III
650	775.965	II
90	885.67	I
90	909.697	I
80	910.278	I
40	910.645	I
450	915.612	II
450	915.962	II
550	916.012	II
650	916.701	II
520	921.992	IV
500	922.519	IV
480	923.057	IV
520	924.283	IV
90	953.415	I
100	953.655	I
130	953.970	I
1000	955.335	IV
130	963.990	I
115	964.626	I
70	965.041	I
650	979.842	III
700	979.919	III
900	989.790	III
700	991.514	III
1000	991.579	III
150 w	1036.16	IV
90	1067.614	I
60	1068.612	I
90	1078.71	IV
450	1083.990	II
600	1084.580	II
430	1085.546	II
650	1085.701	II
175	1097.237	I
115	1098.095	I
115	1098.260	I
105	1100.360	I
40	1100.465	I
90	1101.291	I
360	1134.165	I
385	1134.415	I
410	1134.980	I
105	1143.65	I
130	1163.884	I
60	1164.206	I
105	1164.325	I
270	1167.448	I
105	1168.334	I
60	1168.417	I
195	1168.536	I
230	1176.510	I
105	1176.630	I
195	1177.695	I
500	1183.031	III
570	1184.550	III
90	1188.01	IV
410	1199.550	I
385	1200.223	I
360	1200.710	I
175	1225.026	I
160	1225.37	I
130	1228.41	I
160	1228.79	I
1000	1238.821	V
900	1242.804	V
360	1243.179	I
315	1243.306	I
290	1310.540	I
250	1310.95	I
230	1319.00	I
315	1319.68	I
115	1326.57	I
115	1327.92	I
150	1387.371	III
360	1411.94	I
700	1492.625	I
490	1492.820	I
640	1494.675	I
90	1549.336	V
200 l	1616.33	V
350 l	1619.69	V
1000	1718.55	IV
250	1729.945	III
775	1742.729	I
700	1745.252	I
570	1747.848	III
350	1751.218	III
650	1751.657	III
150	1804.486	III
200	1805.669	III
150	1846.42	III
90 w	1860.37	V
350	1885.06	III
400	1885.22	III
200	1907.99	III

Intensity	Wavelength/Å	
150	1919.55	III
150	1919.77	III
300	1920.65	III
150	1920.84	III
200	1921.30	III
200	2064.01	III
250	2064.42	III
120	2068.68	III
90	2071.09	III
90	2080.34	IV
160	2095.53	II
70	2096.20	II
110	2096.86	II
90	2117.59	III
90	2121.50	III
110	2130.18	II
160	2142.78	II
90	2147.31	III
200	2188.20	III
150	2188.38	III
160	2206.09	II
160	2286.69	II
110	2288.44	II
220	2316.49	II
160	2316.69	II
285	2317.05	II
90 w	2318.09	IV
160	2461.27	II
150	2477.69	IV
110	2496.83	II
70	2496.97	II
110	2520.22	II
160	2520.79	II
220	2522.23	II
110	2590.94	II
250	2645.65	IV
300	2646.18	IV
350	2646.96	IV
250 w	2682.18	III
90	2689.20	III
160	2709.84	II
110	2799.22	II
110	2823.64	II
60 l	2859.16	V
160	2885.27	II
90 l	2974.52	V
150 w	2980.78	V
250 w	2981.31	V
60 w	2998.43	V
220	3006.83	II
90	3078.25	IV
120	3367.34	III
360	3437.15	II
90	3463.37	IV
570	3478.71	IV
500	3482.99	IV
400	3484.96	IV
90	3747.54	IV
90	3754.67	III
120	3771.05	III
285	3838.37	II
360	3919.00	II
90	3938.52	III
450	3955.85	II
1000	3995.00	II
150	3998.63	III
200	4003.58	III
360	4035.08	II
550	4041.31	II
360	4043.53	II
150	4057.76	IV
250	4097.33	III
140	4099.94	I
200	4103.43	III
185	4109.95	I
285	4176.16	II
120	4195.76	III
150	4200.10	III
285	4227.74	II
285	4236.91	II
220	4237.05	II
450	4241.78	II
90	4332.91	III
120	4345.68	III
300	4379.11	III
285	4432.74	II
650	4447.03	II
90	4510.91	III
120	4514.86	III
360	4530.41	II
550	4601.48	II
350	4603.73	V
90	4606.33	IV
450	4607.16	II
360	4613.87	II
250	4619.98	V
450	4621.39	II
870	4630.54	II
90	4634.14	III
120	4640.64	III
550	4643.08	II
285	4788.13	II
450	4803.29	II
180	4847.38	I
90	4858.82	III
150	4867.15	III
285	4895.11	II
160	4914.94	I
210	4935.12	I
200 w	4944.56	V
160	4950.23	I
350	4963.98	I
285	4987.37	II
450	4994.36	II
650	5001.48	II
360	5002.70	II
870	5005.15	II
550	5007.32	II
450	5010.62	II
360	5016.39	II
360	5025.66	II
550	5045.10	II
185	5281.20	I
140	5292.68	I
90	5314.35	III
200	5320.82	III
150	5327.18	III
450	5495.67	II
285	5535.36	II
650	5666.63	II
550	5676.02	II
870	5679.56	II
450	5686.21	II
450	5710.77	II
285	5747.30	II
700	5752.50	I
240	5764.75	I
265	5829.54	I
235	5854.04	I
360	5927.81	II
550	5931.78	II
285	5940.24	II
650	5941.65	II
285	5952.39	II
160	5999.43	I
210	6008.47	I
285	6167.76	II
360	6379.62	II
150	6380.77	IV
185	6411.65	I
210	6420.64	I
210	6423.02	I
210	6428.32	I
185	6437.68	I
235	6440.94	I
90	6454.11	III
185	6457.90	I
120	6467.02	III
300	6468.44	I
265	6481.71	I
750	6482.05	II
360	6482.70	I
300	6483.75	I
325	6484.80	I
160	6491.22	I
210	6499.54	I
185	6506.31	I
750	6610.56	II
185	6622.54	I
185	6636.94	I
235	6644.96	I
185	6646.50	I
235	6653.46	I
210	6656.51	I
185	6722.62	I
210	7398.64	I
160	7406.12	I
265	7406.24	I
685	7423.64	I
785	7442.29	I
900	7468.31	I
185	7608.80	I
60 w	7618.46	V
450	7762.24	II
400	8184.87	I
400	8188.02	I
250	8200.36	I
300	8210.72	I
570	8216.34	I
400	8223.14	I
400	8242.39	I
550	8438.74	II
500	8567.74	I
570	8594.00	I
650	8629.24	I
500	8655.89	I
220	8676.08	II
700	8680.28	I
650	8683.40	I
500	8686.15	I
110	8687.43	II
110 h	8699.00	II
500	8703.25	I
160 h	8710.54	II
570	8711.70	I
500	8718.83	I
250	8728.89	I
200	8747.36	I
500	9386.80	I
570	9392.79	I
250	9460.68	I
200	9863.33	I
160 h	9865.41	II
110 h	9868.21	II
160 h	9887.39	II
220 h	9891.09	II
160 h	9961.86	II
220 h	9969.34	II
285 h	10023.27	II
220 h	10035.45	II
220 h	10065.15	II
160 h	10070.12	II
250	10105.13	I
300	10108.89	I
350	10112.48	I
400	10114.64	I
110 h	10126.27	II
250	10539.57	I
200	12074.51	I
380	12186.82	I
225	12288.97	I
290	12328.76	I
310	12381.65	I
180	12438.40	I
510	12461.25	I
920	12469.62	I
500	13429.61	I
840	13581.33	I
180	13587.73	I
180	13602.27	I
290	13624.18	I
250	14757.07	I
100	14868.87	I
160	14966.60	I

Intensity	Wavelength/Å	
180	15582.27	I
120 s	17516.58	I
100 l	17584.86	I
100	17878.26	I

Osmium Os Z = 76

Intensity	Wavelength/Å	
9600	2001.45	I
13000	2003.73	I
17000	2010.15	I
29000	2018.14	I
14000	2022.76	I
14000	2028.23	I
18000	2034.44	I
26000	2045.36	I
8600	2058.69	I
13000	2061.69	I
7800	2067.21	II
4200	2070.67	II
7200	2076.95	I
14000	2079.97	I
2900	2082.54	I
2900	2089.03	I
2900	2089.21	I
6000	2097.60	I
5300	2100.63	I
2100	2117.66	I
4800	2117.96	I
5300	2137.11	I
2600	2154.59	I
1300	2157.84	I
1200	2158.53	I
3100	2166.90	I
1100	2167.75	I
2100	2171.65	I
1100	2234.61	I
1300	2252.15	I
2000	2255.85	II
1400	2264.60	I
1400	2282.26	II
500	2367.35	II
2600	2377.03	I
1700	2387.29	I
1100	2395.88	I
200	2423.07	II
1400	2424.97	I
110	2454.91	II
1800	2461.42	I
110	2468.90	II
530	2486.24	II
4500	2488.55	I
2600	2498.41	I
2400	2513.25	I
780	2538.00	II
1000	2542.51	I
1000	2590.76	I
1800	2613.06	I
3800	2637.13	I
1900	2644.11	I
1900	2658.60	I
2100	2689.82	I
3000	2714.64	I
1300	2720.04	I
960	2770.71	I
2800	2806.91	I
5100	2838.63	I
2300	2844.40	I
1500	2850.76	I
1500	2860.96	I
9600	2909.06	I
2100	2912.33	I
2100	2919.79	I
1100 h	2948.23	I
1400	2949.53	I
4400	3018.04	I
1100	3030.70	I
2900	3040.90	I
120	3042.74	II
8600	3058.66	I
1100	3077.72	I
3100	3156.25	I
180	3173.93	II
150	3213.31	II
1900	3232.06	I
3100	3262.29	I
3100	3267.94	I
1200	3290.26	I
7600	3301.56	I
960	3336.15	I
960	3370.59	I
620	3387.84	I
620	3401.86	I
620	3504.66	I
1200	3528.60	I
1200	3560.86	I
620	3598.11	I
95	3604.48	II
480	3670.89	I
3700	3752.52	I
2100	3782.20	I
730	3876.77	I
1000	3963.63	I
730	3977.23	I
960	4066.69	I
1200	4112.02	I
2500	4135.78	I
1200	4173.23	I
1200	4211.86	I
4900	4260.85	I
560	4293.95	I
560	4311.40	I
4900	4420.47	I
540	4550.41	I
670	4793.99	I
55	5031.83	I
45	5039.12	I
35	5072.88	I
35	5074.77	I
35	5079.09	I
90	5103.50	I
55	5110.81	I
140	5149.74	I
40	5193.52	I
270	5202.63	I
35	5203.23	I
45	5255.82	I
55	5265.15	I
40	5298.78	I
110	5376.79	I
120	5416.34	I
45	5416.69	I
28	5417.51	I
55	5443.31	I
22	5446.93	I
22	5457.30	I
28	5470.00	I
22	5509.33	I
270	5523.53	I
22	5546.82	I
80	5584.44	I
35	5620.08	I
22	5642.56	I
28	5645.25	I
28	5680.88	I
170	5721.93	I
22	5765.05	I
170	5780.82	I
40	5800.60	I
110	5857.76	I
28	5860.64	I
65	5996.00	I
35	6227.70	I
22	6269.41	I
22	6403.15	I
27	6729.56	I
22	7145.54	I
26	7602.95	I
7	8041.29	I

Oxygen O Z = 8

Intensity	Wavelength/Å	
80	124.616	V
110	135.523	V
80	138.109	V
110	139.029	V
80	151.447	V
110	151.477	V
150	151.546	V
80	164.574	V
110	164.657	V
80	164.709	V
80	166.235	V
150	167.99	V
110	170.219	V
450	172.169	V
250	185.745	V
375	192.751	V
450	192.799	V
520	192.906	V
80	193.003	V
200	194.593	V
150	195.86	IV
200	196.01	IV
80	202.161	V
80	202.224	V
80	202.283	V
80	202.334	V
150	202.393	V
110	203.78	V
150	203.82	V
100	203.85	V
200	203.89	V
100	203.94	V
110	207.18	IV
150	207.24	IV
300	207.794	V
150	215.040	V
200	215.103	V
250	215.245	V
250	216.018	V
520	220.352	V
80	227.372	V
80	227.469	V
150	227.511	V
80	227.549	V
80	227.634	V
80	227.689	V
150	231.823	V
140	233.46	IV
150	233.50	IV
110	233.52	IV
200	233.56	IV
110	233.60	IV
90	238.36	IV
180	238.57	IV
110	248.459	V
110	252.56	IV
110	252.95	IV
150	253.08	IV
300	260.39	IV
250	260.56	IV
80 d	264.34	III
110	264.48	III
110	266.97	III
150	266.98	III
150	267.03	III
150	277.38	III
300	279.63	IV
375	279.94	IV
110	285.71	IV
150	285.84	IV
110	286.448	V
80	295.62	III
110	295.66	III
120	295.72	III
150	303.41	III
150	303.46	III
140	303.52	III
160	303.62	III
160	303.69	III
250	303.80	III
200	305.60	III
250	305.66	III
190	305.70	III
300	305.77	III
190	305.84	III
200	306.62	IV
150	306.88	IV

Intensity	Wavelength/Å	
450	320.979	III
300	328.45	III
250	328.74	III
300	345.31	III
110	355.14	III
90	355.33	III
80	355.47	III
200	359.02	III
190	359.22	III
150	359.38	III
210	373.80	III
200	374.00	III
300	374.08	III
190	374.16	III
200	374.33	III
210	374.44	III
450	395.558	III
300	434.98	III
800	507.391	III
900	507.683	III
1000	508.182	III
1000	525.795	III
250	537.83	II
300	538.26	II
220	539.09	II
200	539.55	II
150	539.85	II
700	553.330	IV
775	554.075	IV
850	554.514	IV
700	555.261	IV
700	597.818	III
1000	599.598	III
580	608.398	IV
110	609.70	III
640	609.829	IV
160	610.04	III
200	610.75	III
100	610.85	III
270	616.952	IV
150	617.005	IV
200	617.036	IV
520	624.617	IV
580	625.130	IV
640	625.852	IV
1000	629.730	V
150	644.148	II
200	672.95	II
150	673.77	II
230	681.272	V
70	685.544	I
800	702.332	III
800	702.822	III
900	702.899	III
1000	703.850	III
900	718.484	II
600	718.562	II
70	744.794	I
700	758.678	V
640	759.441	V
580	760.228	V
775	760.445	V
640	761.128	V
700	762.003	V
70	770.793	I
90	771.056	I
520	774.518	V
70	775.321	I
200	779.734	IV
315	779.821	IV
360	779.912	IV
200	779.997	IV
640	787.711	IV
520	790.109	IV
700	790.199	IV
70	791.973	I
300	796.66	II
200	802.200	IV
160	802.255	IV
90	804.267	I
70	804.848	I
70	805.295	I
80	805.810	I
240	832.762	II
600	832.927	III
450	833.332	II
780	833.742	III
600	834.467	II
600	835.096	III
800	835.292	III
40	877.879	I
130	921.296	IV
160	921.366	IV
80	922.008	I
200	923.367	IV
130	923.433	IV
90	935.193	I
40	948.686	I
90	971.738	I
40	976.448	I
160	988.773	I
40	990.204	I
250	1025.762	I
90	1027.431	I
160	1039.230	I
60	1040.942	I
40	1152.152	I
900	1302.168	I
600	1304.858	I
300	1306.029	I
200	1338.612	IV
130	1342.992	IV
230	1343.512	IV
640	1371.292	V
160	1476.89	III
160 w	1506.72	V
285	1590.01	III
160	1591.33	III
315 w	1643.68	V
160	1707.996	V
220	1760.12	III
110	1760.42	III
220	1763.22	III
220	1764.48	III
750	1767.78	III
550	1768.24	III
360	1771.67	III
110	1773.00	III
110	1773.85	III
220	1779.16	III
160	1781.03	III
160	1784.85	III
220	1789.66	III
110	1848.26	III
110	1856.62	III
285	1872.78	III
285	1872.87	III
285	1874.94	III
160	1920.04	III
110	1920.75	III
110	1921.52	III
220	1923.49	III
110	1923.82	III
110	1926.94	III
360	2013.27	III
160	2026.96	III
220	2045.67	III
160	2052.74	III
30 d	2283.42	II
30 d	2284.89	II
110	2293.32	II
200	2300.35	II
30 d	2313.05	II
30 d	2316.12	II
30 d	2316.79	II
50 d	2319.68	II
30 d	2322.15	II
30 d	2339.31	II
200 d	2390.44	III
80	2394.33	III
110	2411.60	II
80	2422.84	III
80	2425.55	II
250	2433.56	II
80 d	2436.06	II
80 d	2438.83	III
80	2444.26	II
300	2445.55	II
200	2449.372	IV
200	2450.040	IV
200	2454.99	III
200	2493.44	IV
200	2493.77	IV
200	2507.73	IV
230	2509.19	IV
200	2517.2	IV
200	2558.06	III
80	2687.53	III
110	2695.49	III
300	2733.34	II
110	2747.46	II
1000	2781.01	V
920	2786.99	V
775	2789.85	V
160	2836.26	IV
160	2921.45	IV
200	2941.33	V
210	2941.65	V
80	2959.68	III
265	2972.29	I
250	2983.78	III
80	3017.63	III
80	3023.45	III
80	3043.02	III
200	3047.13	III
110	3059.30	III
460	3063.42	IV
410	3071.61	IV
80	3121.71	III
160	3122.62	II
220	3129.44	II
110	3132.86	III
450	3134.82	II
285	3138.44	II
160	3144.66	V
160	3209.66	IV
80	3238.57	III
200	3260.98	III
300	3265.46	III
80	3267.31	III
220	3270.98	II
220	3273.52	II
220	3277.69	II
360	3287.59	II
160	3305.15	II
160	3306.60	II
80	3312.30	III
110	3340.74	III
230	3348.08	IV
270	3349.11	IV
160	3354.27	IV
200	3375.40	IV
220	3377.20	II
130	3378.06	IV
360	3381.20	IV
360	3385.52	IV
285	3390.25	II
270	3396.79	IV
360	3403.52	IV
220	3407.38	II
230	3409.66	IV
160	3409.84	II
410	3411.69	IV
230	3413.64	IV
80	3444.10	III
80	3455.12	III
285	3470.81	II
200	3489.83	IV
160	3492.24	IV
230	3560.39	IV
270	3563.33	IV
80	3698.70	III
80	3702.75	III
80	3703.37	III

Intensity	Wavelength/Å	
110	3707.24	III
220	3712.75	II
110	3715.08	III
315 w	3725.93	IV
285	3727.33	II
360	3729.03	IV
410	3736.85	IV
160	3739.92	II
110	3744.00	III
230	3744.89	IV
360	3749.49	II
150	3754.67	III
80	3757.21	III
250	3759.87	III
110	3791.26	III
160	3803.14	II
120	3823.41	I
450	3911.96	II
160	3919.29	II
185	3947.29	I
160	3947.48	I
140	3947.59	I
220	3954.37	II
100	3954.61	I
200	3961.59	III
450	3973.26	II
220	3982.20	II
160	4069.90	II
285	4072.16	II
450	4075.87	II
80 d	4083.91	II
50 d	4087.14	II
150 d	4089.27	II
110	4097.24	II
220	4105.00	II
285	4119.22	II
100	4123.99	V
160	4132.81	II
50	4146.06	II
220	4153.30	II
285	4185.46	II
450	4189.79	II
80	4233.27	I
50 d	4253.74	II
50 d	4253.98	II
50 d	4275.47	II
50 d	4303.78	II
285	4317.14	II
160	4336.86	II
220	4345.56	II
285	4349.43	II
220	4366.90	II
100	4368.25	I
220	4395.95	II
450	4414.91	II
285	4416.98	II
160	4448.21	II
160	4452.38	II
50	4465.45	II
50 d	4466.28	II
50	4467.83	II

Intensity	Wavelength/Å	
50	4469.41	II
360	4590.97	II
285	4596.17	II
80 d	4609.39	II
160	4638.85	II
360	4641.81	II
450	4649.14	II
160	4650.84	II
360	4661.64	II
285	4676.23	II
220	4699.21	II
285	4705.36	II
160	4924.60	II
230 w	4930.27	V
220	4943.06	II
135	5329.10	I
160	5329.68	I
190	5330.74	I
90	5435.18	I
110	5435.78	I
135	5436.86	I
120	5577.34	I
110	5592.37	III
130	5597.91	V
160	5958.39	I
190	5958.58	I
80	5995.28	I
160	6046.23	I
190	6046.44	I
110	6046.49	I
100	6106.27	I
400	6155.98	I
450	6156.77	I
490	6158.18	I
80	6256.83	I
100	6261.55	I
100	6366.34	I
100	6374.32	I
320	6453.60	I
360	6454.44	I
400	6455.98	I
130	6500.24	V
80	6604.91	I
100	6653.83	I
360	7001.92	I
450	7002.23	I
210	7156.70	I
400	7254.15	I
450	7254.45	I
320	7254.53	I
210	7476.44	I
100	7477.24	I
120	7479.08	I
120	7480.67	I
100	7706.75	I
870	7771.94	I
810	7774.17	I
750	7775.39	I
80	7886.27	I
100	7943.15	I
100	7947.17	I

Intensity	Wavelength/Å	
235	7947.55	I
210	7950.80	I
185	7952.16	I
110	7981.94	I
135	7982.40	I
190	7986.98	I
135	7987.33	I
250	7995.07	I
400	8221.82	I
265	8227.65	I
265	8230.02	I
325	8233.00	I
120	8235.35	I
120	8426.16	I
810	8446.25	I
1000	8446.36	I
935	8446.76	I
325	8820.43	I
160 d	9057.01	I
120	9118.29	I
80	9134.71	I
80	9150.14	I
80	9151.48	I
235	9156.01	I
450	9260.81	I
490	9260.84	I
450	9260.94	I
400	9262.58	I
540	9262.67	I
590	9262.77	I
490	9265.94	I
640	9266.01	I
185	9399.19	I
120	9481.16	I
120 d	9482.88	I
235	9487.43	I
140	9492.71	I
265	9497.97	I
160	9499.30	I
235	9505.59	I
210	9521.96	I
120	9523.36	I
120	9523.96	I
100	9528.28	I
100	9622.13	I
120	9625.29	I
160	9677.38	I
80	9694.66	I
65	9694.91	I
235	9741.50	I
235	9760.65	I
120	9909.05	I
140	9936.98	I
120	9940.41	I
160	9995.31	I
120 d	10421.18	I
590	11286.34	I
640	11286.91	I
490	11287.02	I
490	11287.32	I
490	11295.10	I

Intensity	Wavelength/Å	
540	11297.68	I
590	11302.38	I
265	11358.69	I
490	12464.02	I
450	12570.04	I
120	12990.77	I
160	13076.91	I
700	13163.89	I
750	13164.85	I
640	13165.11	I
160	16212.06	I
120	17966.70	I
590	18021.21	I
120	18041.48	I
120	18042.19	I
120	18046.23	I
140	18229.23	I
540	18243.63	I
140	26173.56	I

Palladium Pd Z = 46

Intensity	Wavelength/Å	
200	705.49	III
200	727.72	III
500	763.06	III
500	766.42	III
2000	781.02	III
500	794.08	III
500	797.52	III
500	800.03	III
500	800.10	III
500	803.67	III
500	825.35	III
500	840.58	III
500	856.47	III
500	864.04	III
500	880.59	III
500	888.84	III
1000	889.29	III
300	1596.89	III
500	1741.62	III
4000	1782.55	III
400	1843.49	III
1500	1851.59	III
2000	1852.27	III
1000	1859.21	III
1500	1874.63	III
2000	1885.83	III
1000	1887.40	III
1500	1891.34	III
4000	1914.62	III
1000	1930.33	III
2000	1941.64	III
800	2002.16	III
1000	2004.47	III
500	2055.11	III
500	2149.82	III
500	2177.55	III
500	2177.63	III
100 r	2231.59	II
200 r	2296.53	II
100	2426.87	II
100	2430.94	II

Intensity	Wavelength/Å	
100	2433.11	II
100	2435.32	II
150	2446.17	II
1100	2447.91	I
100	2457.29	II
150	2469.29	II
100	2471.18	II
1700	2476.42	I
250	2486.52	II
300	2488.92	II
200	2498.81	II
150	2505.73	II
150	2551.84	II
150	2565.51	II
100	2569.56	II
150	2658.75	II
1900	2763.09	I
150 h	2776.85	II
100 h	2787.92	II
200	2854.59	II
100 h	2871.37	II
100 h	2878.01	II
520	2922.49	I
650	3002.65	I
1500	3027.91	I
1100	3065.31	I
2600	3114.04	I
11000	3242.70	I
2700	3251.64	I
3500	3258.78	I
3600	3302.13	I
5000	3373.00	I
24000	3404.58	I
13000	3421.24	I
5000	3433.45	I
6400	3441.40	I
7700	3460.77	I
10000	3481.15	I
2000	3489.77	I
12000	3516.94	I
12000	3553.08	I
4500	3571.16	I
20000	3609.55	I
20000	3634.70	I
5500	3690.34	I
1400	3718.91	I
1500	3799.19	I
1500	3832.29	I
2200	3894.20	I
1500	3958.64	I
290	4087.34	I
2500	4212.95	I
180	4473.59	I
160	5163.84	I
120	5295.63	I
55	5542.80	I
75	5670.07	I
55 h	5695.09	I
65	6784.52	I
75	7368.12	I
120	7764.03	I

Intensity	Wavelength/Å	
45	7915.80	I
55	8132.82	I
45	8300.83	I
65	8761.35	I

Phosphorus P Z = 15

Intensity	Wavelength/Å	
250	328.78	V
150	359.899	IV
500	388.318	IV
250	389.50	V
300	390.70	V
300	445.158	IV
375	475.60	V
120	498.180	III
520	542.57	V
600	544.92	V
200	569.853	III
200	581.831	III
350	629.008	IV
400	629.914	IV
500	631.779	IV
450	673.90	V
10	810.24	II
650	823.179	IV
700	824.730	IV
800	827.932	IV
300	847.669	III
350	855.624	III
500	859.652	III
10	865.44	II
450	865.45	V
600	871.39	V
700	877.476	IV
300	913.971	III
300	917.120	III
350	918.665	III
1000	950.655	IV
250	1003.598	III
570	1025.563	IV
500	1028.096	IV
570	1030.517	IV
500	1033.111	IV
500	1035.517	IV
900	1117.98	V
570	1118.551	IV
700	1128.01	V
20	1249.82	II
20	1301.87	II
20	1304.47	II
15	1304.68	II
35	1305.48	II
60	1310.70	II
500	1334.808	III
650	1344.327	III
300	1344.845	III
500	1366.695	IV
15	1372.033	I
400	1372.674	IV
15	1373.500	I
10	1374.732	I
15	1377.080	I
15	1377.937	I

Intensity	Wavelength/Å	
25	1379.429	I
25	1381.469	I
15	1381.637	I
500	1484.507	IV
400	1487.788	IV
350	1502.228	III
80	1532.51	II
120	1535.90	II
450	1610.50	V
150	1618.632	III
200	1618.907	III
140	1671.070	I
100	1671.510	I
180	1671.680	I
140	1672.035	I
140	1672.474	I
600	1674.591	I
600	1679.695	I
140	1685.976	I
100	1694.028	I
100	1694.486	I
100	1706.376	I
100	1707.553	I
600	1774.951	I
500	1782.838	I
400	1787.656	I
140	1834.801	I
140	1847.165	I
100	1849.820	I
140	1851.194	I
100	1852.069	I
500	1858.886	I
400	1859.393	I
140	1864.348	I
650	1888.523	IV
180	1905.481	I
140	1906.403	I
280	1907.665	I
280	2023.489	I
180	2024.516	I
400	2032.432	I
400	2033.477	I
400	2135.465	I
400	2136.182	I
400	2149.145	I
280	2152.940	I
500	2154.080	I
180	2235.732	I
450	2440.93	V
250	2478.256	IV
750	2533.976	I
950	2535.603	I
750	2553.262	I
500	2554.915	I
250	2605.506	IV
300	2632.713	III
400	2644.295	IV
400	2728.770	IV
500	2739.309	IV
250	2739.872	IV
450	2978.55	V

Intensity	Wavelength/Å	
700	3175.09	V
520	3204.04	V
300	3219.307	III
400	3233.602	III
650	3347.736	IV
570	3364.467	IV
400	3371.122	IV
300	3957.641	III
350	3978.307	III
400	4059.312	III
300	4080.084	III
500	4222.195	III
350	4246.720	III
400	4420.71	II
250	4479.776	III
250	4540.288	IV
250	4541.112	IV
500	4588.04	II
500	4589.86	II
600	4602.08	II
300	4626.70	II
300	4658.31	II
500	4943.53	II
300	4954.39	II
300	4969.71	II
100	5079.381	I
100	5098.221	I
100	5100.974	I
140	5109.628	I
140	5154.844	I
180	5162.290	I
300	5253.52	II
140	5293.539	I
400	5296.13	II
250	5316.07	II
300	5344.75	II
180	5345.851	I
100	5364.631	I
250	5378.20	II
300	5386.88	II
400	5425.91	II
100	5428.094	I
400	5450.74	II
140	5458.305	I
180	5477.672	I
140	5477.860	I
140	5478.267	I
100	5514.774	I
100	5516.997	I
250	5588.34	II
500	6024.18	II
400	6034.04	II
500	6043.12	II
250	6055.50	II
150	6083.409	III
350	6087.82	II
180	6097.690	I
350	6165.59	II
500	6199.024	I
180	6210.499	I
140	6375.681	I

Intensity	Wavelength/Å	
100	6388.579	I
250	6435.32	II
600	6459.99	II
600	6503.46	II
600	6507.97	II
100	6717.411	I
150	6992.690	III
100	7102.200	I
100	7158.367	I
180	7165.465	I
180	7175.102	I
180	7176.660	I
200	7443.657	IV
250	7845.63	II
100	8046.801	I
150	8113.528	III
140	8278.058	I
100	8367.856	I
140	8531.475	I
140	8613.835	I
180	8637.578	I
400	8741.529	I
100	8872.174	I
180	9175.819	I
950	9193.85	I
600	9278.88	I
1250	9304.94	I
500	9323.50	I
950	9435.069	I
950	9441.86	I
600	9452.83	I
1250	9493.56	I
1700	9525.73	I
1500	9545.18	I
280	9556.81	I
1700	9563.439	I
280	9593.50	I
750	9609.04	I
400	9638.939	I
500	9676.24	I
180	9706.533	I
1500	9734.750	I
280	9736.680	I
1500	9750.77	I
600	9790.21	I
1700	9796.85	I
280	9834.80	I
400	9903.68	I
280	9976.67	I
229	10084.27	I
458	10511.58	I
962	10529.52	I
1235	10581.57	I
415	10596.90	I
435	10681.40	I
265	10813.13	I
764	11183.23	I
402	11186.75	I
479	14241.64	I
256	14307.83	I
714	15711.52	I
228	15962.53	I
296	16254.77	I
203	16292.97	I
1627	16482.92	I
588	16590.07	I
225	16613.05	I
221	16738.68	I
419	16803.39	I
471	17112.48	I
289	17286.91	I
299	17423.67	I
287	23844.97	I
311	29097.16	I

Platinum Pt Z = 78

Intensity	Wavelength/Å	
30	1621.66	II
30	1723.13	II
30	1751.70	II
50 r	1777.09	II
30	1781.86	II
30	1879.09	II
40	1883.05	II
50	1889.52	II
50	1911.70	II
30	1929.25	II
30	1929.68	II
30	1939.80	II
30	1949.90	II
30	1983.74	II
40	2014.93	II
3200	2030.63	I
4400	2032.41	I
100	2036.46	II
40	2041.57	II
5500	2049.37	I
1500	2067.50	I
3000	2084.59	I
1000	2103.33	I
30	2115.57	II
950	2128.61	I
30	2130.69	II
1900	2144.23	I
100	2144.24	II
600	2165.17	I
1500	2174.67	I
30	2190.32	II
400	2202.22	I
50 h	2202.58	II
320	2222.61	I
50 h	2233.11	II
30 h	2240.99	II
100	2245.52	II
150	2249.30	I
30	2251.52	II
30 h	2251.92	II
190	2268.84	I
30 h	2271.72	II
280	2274.38	I
50 h	2287.50	II
30	2288.20	II
150	2289.27	I
150	2292.40	I
240	2308.04	I
50	2310.96	II
90	2315.50	I
220	2318.29	I
100	2326.10	I
170	2340.18	I
280	2357.10	I
180	2368.28	I
50	2377.28	II
130	2383.64	I
40	2386.81	I
120	2389.53	I
35	2396.17	I
70	2401.87	I
200	2403.09	I
100	2418.06	I
50	2424.87	II
80	2428.04	I
50	2428.20	I
25	2429.10	I
180	2436.69	I
650	2440.06	I
60	2450.97	I
440	2467.44	I
35	2471.01	I
1000	2487.17	I
25	2488.74	II
200	2490.12	I
160	2495.82	I
240	2498.50	I
50	2505.93	I
120	2508.50	I
50	2514.07	I
60	2515.03	I
240	2515.58	I
140	2524.30	I
40	2529.41	I
50	2536.49	I
160	2539.20	I
18	2549.46	I
50	2552.25	I
50	2596.00	I
70	2603.14	I
30	2616.76	II
50	2619.57	I
30	2625.34	II
1100	2628.03	I
130	2639.35	I
1000	2646.89	I
500	2650.86	I
20	2658.17	I
2800	2659.45	I
40	2674.57	I
440	2677.15	I
200	2698.43	I
2000	2702.40	I
1600	2705.89	I
60	2713.13	I
1300	2719.04	I
130	2729.92	I
1800	2733.96	I
70	2738.48	I
70	2747.61	I
80	2753.86	I
200	2754.92	I
30	2769.84	I
500	2771.67	I
40	2773.24	I
20	2774.00	I
50	2774.77	II
50	2793.27	I
100	2794.21	II
40 h	2799.98	II
140	2803.24	I
10	2808.51	I
50	2818.25	I
30 h	2822.27	II
1400	2830.30	I
70	2834.71	I
16	2853.11	I
80 h	2860.68	II
40 h	2865.05	II
40 h	2875.85	II
100 h	2877.52	II
25	2888.20	I
25	2893.22	I
600	2893.86	I
300	2897.87	I
60	2905.90	I
120	2912.26	I
120	2913.54	I
70	2919.34	I
30	2921.38	I
1700	2929.79	I
30	2942.76	I
30	2944.75	I
25	2959.10	I
60	2960.75	I
1800	2997.97	I
35	3001.17	II
220	3002.27	I
30	3017.88	I
30 h	3031.22	II
130	3036.45	I
800	3042.64	I
3200	3064.71	I
30	3071.94	I
130	3100.04	I
320	3139.39	I
140	3156.56	I
120	3200.71	I
320	3204.04	I
30	3230.29	I
20	3233.42	I
20	3250.36	I
40	3251.98	I
160	3255.92	I
25	3268.42	I
25	3281.97	I
120	3290.22	I
500	3301.86	I
60	3315.05	I

Intensity	Wavelength/Å	
35	3323.80	I
340	3408.13	I
35	3427.93	I
60	3483.43	I
160	3485.27	I
120	3628.11	I
70	3638.79	I
70	3643.17	I
50	3663.10	I
80	3671.99	I
80	3674.04	I
35	3699.91	I
18	3706.53	I
80	3818.69	I
40	3900.73	I
110	3922.96	I
35	3948.40	I
100	3966.36	I
20	3996.57	I
110	4118.69	I
80	4164.56	I
40	4192.43	I
18	4327.06	I
18	4391.83	I
80	4442.55	I
14	4445.55	I
25	4498.76	I
12	4520.90	I
35	4552.42	I
12	4879.53	I
14	5044.04	I
30	5059.48	I
35	5227.66	I
40	5301.02	I
12	5368.99	I
12	5390.79	I
14	5475.77	I
14	5478.50	I
6	5763.57	I
20	5840.12	I
8	5844.84	I
6	6026.04	I
7	6318.37	I
8	6326.58	I
9	6523.45	I
10	6710.42	I
20	6760.02	I
60	6842.60	I
20	7113.73	I
10	8224.74	I

Plutonium Pu Z = 94

Intensity	Wavelength/Å	
10000	2806.11	II
10000	2950.06	II
10000	3000.31	II
10000	3200.23	II
10000	3418.88	II
10000	3805.93	I
10000	4097.12	I
10000	4170.95	I
10000	4367.41	I
10000	5590.54	I
10000	7068.90	I
10000	8691.94	I
3000	9533.07	I
3000	12144.46	I
3000	16897.38	I

Polonium Po Z = 84

Intensity	Wavelength/Å	
1500 w	2450.08	I
1500 w	2558.01	I
2500 w	3003.21	I
1200	4170.52	I
800	4493.21	I
500	8618.26	I

Potassium K Z = 19

Intensity	Wavelength/Å	
100	214.35	V
150	271.82	IV
100	273.06	IV
150	282.35	V
150	293.33	V
300	294.84	V
200	296.17	V
200	297.06	V
200	300.25	V
200	300.50	V
200	311.24	V
250	312.77	V
200	315.18	V
250	327.38	V
25	330.68	III
300	340.46	IV
150	340.74	IV
30	341.92	III
15	348.00	III
200	349.50	V
300	354.93	IV
150	356.26	IV
300	359.73	IV
200	359.91	IV
250	362.08	IV
150	362.15	IV
150	363.02	IV
500	372.15	V
200	372.46	V
200	372.77	V
300	375.96	IV
300	375.96	V
250	377.76	V
30	379.12	III
300	379.12	V
300	379.88	IV
25	380.48	III
250	380.48	IV
200	381.70	IV
30	382.23	III
300	382.23	IV
150	382.49	IV
200	382.65	IV
300	382.91	IV
250	384.10	IV
200	386.61	IV
300	387.80	V
250	388.92	IV
250	389.07	IV
250	389.07	V
250	390.11	V
250	390.42	IV
300	390.57	IV
200	391.46	IV
200	392.47	IV
500	393.14	IV
250	395.40	V
200	398.36	V
15	398.63	III
200	398.88	V
200	399.75	V
400	400.21	IV
20	402.10	III
300	402.91	IV
250	403.97	IV
150	404.41	IV
30	406.48	III
250	408.08	IV
40	408.96	III
50	413.79	III
30	414.87	III
250	415.05	V
200	415.79	V
30	416.00	III
150	417.28	IV
30	417.54	III
30	418.62	III
400	422.18	V
300	425.16	V
500	425.59	V
75	434.72	III
50	435.68	III
250	438.02	V
25	441.81	II
200	442.30	IV
300	443.57	IV
75	444.34	III
200	445.61	IV
250	446.83	IV
75	448.60	III
750	448.60	IV
200	449.71	V
200	452.90	V
250	455.67	V
400	456.33	IV
400	456.33	V
75	466.79	III
100	470.09	III
75	471.57	III
45	474.92	III
10	476.03	II
40	479.18	III
10	482.11	III
10	482.41	III
200	482.71	V
200	483.75	V
30	495.14	II
75	497.10	III
10	514.94	III
50	520.61	III
250	523.00	IV
25	523.79	III
200	526.45	IV
150	527.62	IV
40	529.80	III
15	539.71	III
15	546.12	III
750	580.32	V
250	585.51	V
500	586.32	V
30	600.77	II
250	602.27	V
400	603.43	V
25	607.93	II
30	612.62	II
250	638.67	V
750	646.19	IV
300	687.50	V
20	708.84	III
300	720.43	V
400	724.42	V
600	731.86	V
500	737.14	IV
500	741.95	IV
500	745.26	IV
400	746.35	IV
300	749.99	IV
150	754.19	IV
400	754.67	IV
20	765.31	III
30	765.64	III
150	770.29	V
150	771.46	V
35	778.53	III
20	872.31	III
10	873.86	III
15	874.04	III
6	2550.02	III
5	2635.11	III
5	2689.90	III
5	2938.45	III
5	2986.20	III
6	2992.42	III
6	3052.07	III
5	3056.84	III
5	3062.18	II
4	3101.79	I
3	3102.04	I
7	3217.16	I
6	3217.62	I
11	3446.37	I
10	3447.38	I
3	3648.84	I
4	3648.98	I
18	4044.14	I
17	4047.21	I
10	4641.88	I
11	4642.37	I

Intensity	Wavelength/Å	
4	4740.91	I
6	4744.35	I
5	4753.93	I
7	4757.39	I
5	4786.49	I
7	4791.05	I
6	4799.75	I
8	4804.35	I
7	4849.86	I
8	4856.09	I
8	4863.48	I
9	4869.76	I
8	4942.02	I
9	4950.82	I
9	4956.15	I
10	4965.03	I
10	5084.23	I
11	5097.17	I
11	5099.20	I
12	5112.25	I
12	5323.28	I
13	5339.69	I
12	5342.97	I
14	5359.57	I
16	5782.38	I
17	5801.75	I
15	5812.15	I
17	5831.89	I
8	6120.27	II
7	6307.29	II
19	6911.08	I
12	6936.28	I
20	6938.77	I
7	6964.18	I
12	6964.67	I
25	7664.90	I
24	7698.96	I
5	7955.37	I
4	7956.83	I
7	8078.11	I
6	8079.62	I
9	8250.18	I
8	8251.74	I
3	8390.22	I
11	8503.45	I
10	8505.11	I
4	8763.96	I
3	8767.05	I
13	8902.19	I
12	8904.02	I
5	8923.31	I
4	8925.44	I
7	9347.24	I
3	9349.25	I
6	9351.59	I
15	9595.70	I
14	9597.83	I
6	9949.67	I
5	9954.14	I
9	10479.63	I
5	10482.15	I
8	10487.11	I
17	11019.87	I
16	11022.67	I
17	11690.21	I
16	11769.62	I
17	11772.83	I
	12522.11	I
	13377.86	I
	13397.09	I
	15163.08	I
	15168.40	I
	40158.37	I

***Praseodymium* Pr *Z* = 59**

Intensity	Wavelength/Å	
7000	865.90	V
5000	869.17	V
2000	1228.59	IV
5000	1293.22	IV
5000	1295.28	IV
5000	1321.36	IV
5000	1333.57	IV
5000	1354.66	IV
2000	1360.64	IV
2000	1365.77	IV
5000	1374.41	IV
5000	1435.56	IV
2000	1520.98	IV
5000	1574.55	IV
5000	1575.10	IV
3000	1578.38	IV
2000	1622.30	IV
10000	1884.87	IV
2000	2083.23	IV
3300	2246.20	V
2000 c	2378.98	IV
40 h	2598.04	II
100 h	2707.37	II
60	2760.35	II
270	3168.24	II
200 d	3195.99	II
190	3219.48	II
200	3584.21	II
250	3645.66	II
250	3646.30	II
370	3668.83	II
290	3714.05	II
410	3739.18	II
680	3761.87	II
680	3800.30	II
390	3811.84	II
1300 h	3816.02	II
680	3818.28	II
310	3821.80	II
960	3830.72	II
480	3840.99	II
580	3846.59	II
1200	3850.79	II
720 c	3851.55	II
960	3852.80	II
480 c	3865.45	II
480	3876.19	II
1700 c	3877.18	II
680	3880.47	II
440 c	3885.19	II
440 c	3889.34	II
770 c	3908.05	II
630	3912.90	II
310	3913.55	II
1300 c	3918.85	II
420	3919.63	II
960	3925.47	II
480	3927.46	II
370	3929.29	II
370	3935.82	II
730 c	3947.63	II
900 c	3949.43	II
900 c	3953.51	II
380	3956.75	II
470	3962.45	II
560	3964.26	II
1600 c	3964.81	II
560 c	3966.57	II
500	3971.16	II
320	3971.67	II
620 c	3972.14	II
320	3974.85	II
1300 c	3989.68	II
340	3992.16	II
1600	3994.79	II
560 c	3997.04	II
320	3999.12	II
620 c	4000.17	II
730	4004.70	II
1900	4008.69	II
620	4010.60	II
730	4015.39	II
620	4020.96	II
470	4022.71	II
360	4025.54	II
360 c	4029.72	II
730 c	4031.75	II
960	4033.83	II
730	4038.45	II
470	4039.34	II
1300	4044.81	II
340	4047.08	II
450	4051.13	II
2200	4054.88	II
2200	4056.54	II
450	4058.80	II
3400	4062.81	II
500 c	4079.77	II
500 c	4080.98	II
790	4081.85	II
500	4083.34	II
560	4096.82	II
380	4098.40	II
2900 c	4100.72	II
1700 c	4118.46	II
340	4130.77	II
1500 c	4141.22	II
2700	4143.11	II
1700 c	4164.16	II
620	4171.82	II
730	4172.25	II
5200	4179.39	II
2500	4189.48	II
560 c	4191.60	II
2500 c	4206.72	II
500	4208.32	II
320	4211.86	II
320	4217.81	II
3800	4222.93	II
3800	4225.35	II
320	4233.11	II
320 c	4236.15	II
960	4241.01	II
340	4243.51	II
840 c	4247.63	II
500	4254.40	II
320	4269.09	II
790 c	4272.27	II
470 c	4280.07	II
790 c	4282.42	II
450 c	4298.98	II
1500	4305.76	II
1300	4333.97	II
360	4338.70	II
620 cw	4344.30	II
470 c	4347.49	II
340	4350.40	II
450	4354.91	II
410 c	4359.79	II
1200	4368.33	II
320	4371.62	II
430	4405.83	II
1700	4408.82	II
410	4413.77	II
1200 c	4429.13	II
730	4449.83	II
960	4468.66	II
1100	4496.46	II
790	4510.15	II
340 c	4534.15	II
340	4535.92	II
270 c	4628.74	II
270 c	4672.09	II
290	4695.77	I
250	4736.69	I
200	4924.60	I
320	4939.74	I
380	4951.37	I
270	5034.41	II
320	5045.52	I
360	5110.38	II
560	5110.76	II
410	5129.52	II
620	5173.90	II
360	5206.55	II
360	5219.05	II
560	5220.11	II
680	5259.73	II
340 c	5292.02	II
340	5292.62	II

Intensity	Wavelength/Å	
430	5322.76	II
65	5509.15	II
150	5535.17	II
110	5623.05	II
90	5624.45	II
90	5756.17	II
90	5779.28	I
160 d	5815.17	II
90	5823.72	II
90	5859.68	II
160	5939.90	II
7000 w	5956.05	III
90	5956.60	II
110	5967.82	II
90	6006.33	II
150	6017.80	II
150	6025.72	II
140	6055.13	I
65	6087.52	II
9000 w	6090.02	III
65	6114.38	II
65	6148.23	I
5000	6160.24	III
190	6161.18	II
270	6165.94	II
45	6244.35	II
110	6281.28	II
55 c	6359.03	I
55	6411.23	I
45	6429.63	II
45	6431.84	II
45	6486.55	I
45	6566.77	II
55	6616.67	I
75	6656.83	II
55	6673.41	II
75	6673.78	II
35 c	6747.09	I
55 cw	6798.60	I
35 cw	6827.60	II
7000	6910.14	III
40	7021.51	II
5000	7030.39	III
4500	7076.62	III
20	7114.55	I
24	7227.70	II
16	7407.56	II
20 c	7451.74	II
14	7541.02	II
20	7645.66	II
16	7721.84	I
14	7871.67	I
14	8067.44	I
10 cw	8122.78	II
11	8141.10	I
5000 w	8602.74	III
10	8714.59	II

Promethium Pm Z = 61

Intensity	Wavelength/Å	
1000	3892.15	II
1000	3910.26	II
1000	3919.10	II
1000	3957.74	II
1000 r	3998.96	II
1000	4417.96	II
900 r	4728.36	I
900	6100.21	I
1000 d	6520.45	I

Protactinium Pa Z = 91

Intensity	Wavelength/Å	
3000	2599.16	II
3000	2699.22	II
3000	2822.79	II
3000 h	2871.42	II
3000 h	2891.14	II
3000 l	3011.10	II
3000 s	3033.59	II
3000 l	3071.24	II
3000 l	3093.23	II
3000 l	3126.23	II
3000 l	3146.28	II
3000 l	3170.89	II
3000 l	3171.54	II
3000 l	3240.58	II
3000	3274.46	II
3000 l	3332.69	II
3000 s	3346.66	II
3000 l	3452.82	II
3000	3504.97	I
3000 s	3530.65	II
3000	3570.56	I
3000	3571.82	I
3000	3618.07	I
10000	3636.52	I
3000	3702.74	I
3000	3752.67	I
3000	3873.35	I
3000	3931.83	I
3000 s	3952.62	II
10000 l	3957.85	II
3000 s	3970.07	II
3000	3981.82	I
10000	3982.23	I
3000 l	4012.96	II
3000 s	4018.21	II
3000	4030.16	II
3000 s	4046.93	II
10000 s	4056.20	II
10000 s	4070.40	II
3000 l	4176.18	II
10000 l	4217.23	II
10000 s	4248.08	II
3000 s	4291.34	II
3000 s	4601.43	II
3000 l	6035.78	I
3000	6162.56	I
3000 l	6358.61	I
3000	6379.25	I
3000 l	6438.97	I
3000 h	6792.75	I
10000	6945.72	I
3000	6960.09	I
3000 h	6961.78	I
3000 s	6992.73	I
3000	7076.27	I
3000 h	7100.94	I
10000 s	7114.89	I
3000 h	7171.55	I
3000	7227.13	I
3000	7318.79	I
10000 l	7368.25	I
3000 h	7471.89	I
10000 h	7493.15	I
3000 h	7558.26	I
10000 h	7608.20	I
10000	7626.79	I
10000 s	7635.18	I
10000	7669.34	I
3000	7679.20	I
10000 h	7749.19	I
3000	7872.95	I
3000 l	7945.56	I
10000	8039.34	I
10000 h	8099.84	I
10000	8199.04	I
10000	8271.87	I
3000 s	8358.98	I
3000 s	8369.60	I
3000 h	8441.04	I
10000 h	8532.66	I
10000 s	8572.96	I
3000 h	8639.91	I
3000 h	8653.51	I
10000	8735.27	I
3000	10923.32	I
10000	11791.73	I
10000	14344.76	I
3000	18478.61	I

Radium Ra Z = 88

Intensity	Wavelength/Å	
100	3649.55	II
200	3814.42	II
100	4340.64	II
100	4682.28	II
100	4825.91	I
50	5660.81	I
50	7141.21	I
50	8019.70	II

Radon Rn Z = 86

Intensity	Wavelength/Å	
100	4349.60	I
200	7055.42	I
100	7268.11	I
300	7450.00	I
100	7809.82	I
100	8099.51	I
100	8270.96	I
100	8600.07	I

Rhenium Re Z = 75

Intensity	Wavelength/Å	
25000	2003.53	I
16000	2017.87	I
27000	2049.08	I
10000	2085.59	I
9800	2097.12	I
3400	2139.04	II
3700	2156.67	I
4900	2167.94	I
3400	2176.21	I
4200 c	2214.26	II
5200 c	2275.25	II
2900	2287.51	I
2700	2294.49	I
390	2298.09	II
610	2302.99	I
680	2306.54	I
800	2322.49	I
300	2328.66	I
860	2344.78	I
230	2349.39	I
680	2352.07	I
250	2356.50	I
1200	2365.90	I
570	2367.68	I
520	2369.27	I
220	2370.76	II
320	2375.07	I
370	2379.77	I
340	2388.57	I
230	2393.65	I
320	2394.37	I
320	2396.79	I
210 d	2400.72	I
210	2401.68	I
1500	2405.06	I
740	2405.60	I
320	2406.70	I
270	2410.37	I
1200	2419.81	I
300	2421.73	I
300	2421.88	I
2500	2428.58	I
490	2431.54	I
420	2432.18	I
340 c	2441.47	I
230	2442.51	I
250	2444.94	I
610	2446.98	I
610	2449.71	I
390	2461.20	I
800 c	2461.84	II
1200	2483.92	I
390	2485.81	I
980	2487.33	I
370	2496.04	I
370	2501.72	I
570	2502.35	II
230	2504.60	II
270	2505.94	I
1800 c	2508.99	I
570	2520.01	I
540	2521.50	I
370	2534.80	I
570	2540.51	I
740 d	2544.74	I
370	2545.48	I
300	2552.02	I
370	2554.63	II

Intensity	Wavelength/Å	
1000	2556.51	I
250	2559.08	I
340	2564.19	I
540	2568.64	II
370	2571.81	II
380	2586.79	I
290	2599.86	I
290	2603.89	I
660	2608.50	II
610 d	2611.54	I
310	2635.83	II
550	2636.64	I
270	2642.75	I
270	2649.05	I
660	2651.90	I
400	2654.12	I
220	2663.63	I
940	2674.34	I
220	2688.53	I
1300	2715.47	I
220	2732.21	I
610	2733.04	II
220	2758.00	I
210	2763.79	I
310	2767.74	I
220	2768.85	I
220	2769.32	I
350	2770.42	I
550	2783.57	I
220	2791.29	I
220	2814.68	I
880	2819.95	I
310	2834.08	I
220	2843.00	I
270	2850.98	I
240	2867.19	I
2900	2887.68	I
490	2896.01	I
830 c	2902.48	I
210	2905.58	I
550	2909.82	I
830 c	2927.42	I
270	2930.61	I
440	2943.14	I
270	2962.27	I
720	2965.11	I
1500	2965.76	I
310	2976.29	I
210	2978.15	I
220	2980.82	I
220	2982.19	I
220	2988.47	I
1800	2992.36	I
5500	2999.60	I
350	3001.14	I
220	3004.14	I
500	3016.02	I
300	3016.49	I
380	3030.45	I
240	3047.25	I
1600	3067.40	I
320	3069.94	I
260	3071.16	I
550	3082.43	I
340	3088.76	I
700	3100.67	I
700	3108.81	I
340	3110.86	I
340 c	3118.19	I
340	3121.36	I
420	3128.94	I
260	3134.02	I
250	3141.38	I
440	3151.64	I
330	3153.79	I
360 c	3158.31	I
220	3164.52	I
700	3168.37	I
220	3174.61	I
440	3177.71	I
260	3178.61	I
600	3182.87	I
1100	3184.76	I
1100	3185.57	I
260	3190.78	I
260	3192.36	I
220	3198.58	I
1100 c	3204.25	I
380	3235.94	I
600	3258.85	I
600	3259.55	I
300	3268.89	I
280	3296.70	I
280	3296.99	I
280	3301.60	I
240	3302.23	I
320	3303.21	II
280	3303.75	I
240	3313.95	I
600	3322.48	I
2000	3338.18	I
1600	3342.24	I
810	3344.32	I
320	3346.20	I
240 d	3356.33	I
240	3377.74	I
320	3379.06	II
320	3379.70	I
240	3389.43	I
4000	3399.30	I
650	3404.72	I
650	3405.89	I
240	3408.67	I
320	3409.83	I
320	3417.77	I
810	3419.41	I
8000	3424.62	I
400	3426.19	I
300	3427.61	I
320	3437.71	I
400	3449.37	I
16000 c	3451.88	I
240	3453.50	I
55000 c	3460.46	I
40000 c	3464.73	I
400	3467.96	I
240	3476.44	I
400	3480.38	I
320	3480.85	I
240	3482.23	I
560	3503.06	I
320	3516.65	I
320	3517.33	I
320	3537.46	I
240	3549.89	I
240	3570.26	I
360	3579.12	I
810 c	3580.15	II
650	3580.97	I
810	3583.02	I
320	3617.08	I
810	3637.84	I
440	3651.97	I
320	3670.53	I
860 c	3689.50	I
1500 c	3691.48	I
520	3703.24	I
240	3709.93	I
360 c	3717.28	I
4000	3725.76	I
240 c	3735.01	I
810	3735.31	I
910	3740.10	I
300 cw	3745.44	I
700	3787.52	I
240	3869.94	I
240	3875.26	I
240	3876.86	I
380 c	3917.27	I
550	3929.85	I
280	3961.04	I
350 c	3962.48	I
220	4033.31	I
240	4081.43	I
240 c	4110.89	I
240 cw	4133.42	I
1800	4136.45	I
700	4144.36	I
220	4182.90	I
220	4183.06	I
650	4221.08	I
3600 c	4227.46	I
260 c	4257.60	I
380	4358.69	I
360 cw	4394.38	I
2600	4513.31	I
260	4516.64	I
500	4522.73	I
2200 cw	4889.14	I
220	4923.90	I
1300	5270.95	I
1600 cw	5275.56	I
100	5667.88	I
110 c	5752.93	I
110 cw	5776.83	I
550	5834.31	I
200	6307.70	I
200	6321.90	I
100 cw	6605.19	I
180 c	6813.41	I
260	6829.90	I
50 cw	7640.94	I
65 cw	7912.94	I

Rhodium Rh *Z* **= 45**

Intensity	Wavelength/Å	
50	813.44	III
80	882.51	III
100	925.75	III
150	937.28	III
500	991.62	III
400	992.48	III
500 d	1009.60	III
200	1012.22	III
200	1015.17	III
200	1073.87	III
150	1784.24	III
200	1784.94	III
150	1796.50	III
200	1816.03	III
1000	1832.05	III
500	1859.85	III
800	1880.66	III
500	1884.91	III
500	1887.36	III
700	1888.62	III
800	1901.32	III
500	1910.16	III
600	1919.37	III
500	1927.07	III
700	1931.79	III
500	1954.25	III
500	1994.26	III
800	2013.71	III
500	2017.47	III
500	2028.53	III
800	2036.72	III
600	2037.61	III
1000	2040.18	III
3000	2048.67	III
2000	2064.11	III
800	2076.84	III
1000	2118.53	III
1000	2118.63	III
1000	2139.44	III
1000	2152.23	III
3000	2158.17	III
3000	2163.19	III
3000	2167.33	III
150	2276.21	II
140	2288.57	I
110	2309.82	I
350	2322.58	I
140	2326.47	I
190	2334.77	II
300	2361.92	I

Intensity	Wavelength/Å	
110	2368.34	I
270	2382.89	I
230	2383.40	I
270	2386.14	II
80	2415.84	II
130	2427.68	I
230	2429.52	I
110	2437.90	I
330	2440.34	I
90	2461.04	II
130	2473.09	I
150	2487.47	I
100	2490.77	II
130	2502.46	I
300	2504.29	II
150	2505.67	I
350	2509.70	I
300	2511.03	II
200	2515.75	I
130	2520.53	II
110	2537.04	II
350	2545.70	I
550	2555.36	I
150	2622.58	I
230	2625.88	I
100	2630.42	I
110	2647.28	I
400	2652.66	I
100	2680.63	I
400	2703.73	I
100	2715.31	II
180	2718.54	I
160	2728.94	I
100	2771.51	I
130	2783.03	I
150	2826.43	I
180	2826.68	I
280	2862.94	I
110	2878.66	I
140	2882.37	I
160	2907.21	I
65	2910.17	II
180	2924.02	I
130	2929.11	I
130	2931.94	I
230	2968.66	I
160	2977.68	I
450	2986.20	I
110	3004.46	I
50	3006.43	III
130	3023.91	I
50	3052.44	III
180	3083.96	I
140	3121.76	I
240	3123.70	I
130	3155.78	I
140	3189.05	I
470	3191.19	I
190	3197.13	I
520	3263.14	I
520	3271.61	I
2300	3280.55	I
2300	3283.57	I
280	3289.14	I
210	3294.28	I
260	3300.46	I
50	3310.69	III
4200	3323.09	I
330	3338.54	I
280	3360.80	I
420	3368.38	I
1100	3372.25	I
110	3377.14	I
110	3385.78	I
5600	3396.82	I
820	3399.70	I
160	3406.55	I
820	3412.27	I
330	3421.22	I
120 d	3424.38	I
8200	3434.89	I
1400	3440.53	I
120	3447.74	I
120	3450.29	I
400	3455.22	I
180	3457.07	I
220	3457.93	I
5900	3462.04	I
180	3469.62	I
4700	3470.66	I
120	3472.25	I
4700	3474.78	I
2100	3478.91	I
110	3494.44	I
1200	3498.73	I
5900	3502.52	I
2800	3507.32	I
8800	3528.02	I
880 d	3538.14	I
280	3541.91	I
1200	3543.95	I
1800	3549.54	I
1200	3570.18	I
4700	3583.10	I
4700	3596.19	I
5900	3597.15	I
3100	3612.47	I
1800	3626.59	I
8200	3657.99	I
1300	3666.22	I
560	3681.04	I
1900	3690.70	I
9400	3692.36	I
940	3695.52	I
280	3698.26	I
380	3698.60	I
7600	3700.91	I
940	3713.02	I
650	3735.28	I
420	3737.27	I
420	3744.17	I
1200	3748.22	I
240	3754.12	I
380	3754.27	I
490	3755.58	I
1000	3760.40	I
2300	3765.08	I
490	3769.97	I
380	3778.13	I
1000	3788.47	I
1300	3792.18	I
3800	3793.22	I
4900	3799.31	I
760	3805.92	I
1300	3806.76	I
470	3815.01	I
760	3816.47	I
1300	3818.19	I
3800	3822.26	I
2300	3828.48	I
2000	3833.89	I
5900	3856.52	I
490	3870.01	I
380	3877.34	I
120	3913.51	I
240	3922.19	I
2000	3934.23	I
590	3942.72	I
3800	3958.86	I
45	3964.54	II
380	3975.31	I
240	3984.40	I
240	3995.61	I
380	3996.15	I
120	4023.14	I
560	4082.78	I
140	4097.52	I
120	4119.68	I
1100	4121.68	I
1500	4128.87	I
2100	4135.27	I
240	4154.37	I
330	4196.50	I
3300	4211.14	I
820	4288.71	I
4200	4374.80	I
130	4569.00	I
150	4675.03	I
70	4745.11	I
70	5090.63	I
60	5155.54	I
60	5175.97	I
95	5193.14	I
130	5354.40	I
95	5390.44	I
160	5599.42	I
40	5686.38	I
29	5792.66	I
40	5806.91	I
35	5831.58	I
130	5983.60	I
35	6102.72	I
14	6199.99	I
16	6253.72	I
29	6319.53	I
40	6752.35	I
13	6827.33	I
11	6857.68	I
20	6879.94	I
65	6965.67	I
16	6979.15	I
16	7001.58	I
18	7101.64	I
15	7104.45	I
18	7268.18	I
35	7270.82	I
18 h	7442.39	I
12	7475.74	I
12	7495.24	I
11	7557.67	I
29	7791.61	I
55	7824.91	I
21	8029.91	I
29	8045.36	I
15	8136.20	I
8	8425.59	I

Rubidium Rb Z = 37

Intensity	Wavelength/Å	
30	465.85	III
40	481.118	II
500	482.83	III
500	489.66	III
600	493.48	III
90	497.430	II
20	508.434	II
150	513.266	II
300	530.173	II
75	533.801	II
1200	535.86	III
40	542.887	II
200	555.036	II
1200	556.19	III
1500	566.71	III
1000	572.82	III
1500	576.65	III
2500	579.63	III
1500	581.26	III
2500	589.419	II
1000	594.94	III
1300	595.88	III
1200	598.49	III
1500	643.878	II
25	663.76	IV
3000	697.049	II
6000	711.187	II
25	716.24	IV
50	740.85	IV
10000	741.456	II
5000	769.04	III
25	776.89	IV
2500	815.28	III
15	850.18	IV
1000	1604.12	II
5000	1760.50	II
2000	2068.92	II

Intensity	Wavelength/Å	
10000	2075.95	II
30000	2143.83	II
10000	2217.08	II
5000	2291.71	II
50000	2472.20	II
1000	2631.75	III
2000	2956.07	III
500	3086.84	III
500	3111.36	III
5000 c	3148.90	II
25	3157.54	I
50	3227.98	I
500	3286.41	III
60	3348.72	I
75	3350.82	I
100	3587.05	I
40	3591.57	I
5000	3600.60	II
10000	3600.64	II
25000	3940.51	II
1000	4201.80	I
500	4215.53	I
90000	4244.40	II
15000	4273.14	II
20000	4571.77	II
10000	4648.57	II
30000	4775.95	II
2	5087.987	I
2	5132.471	I
10	5150.134	I
10000	5152.08	II
1	5165.023	I
2	5165.142	I
15	5195.278	I
2	5233.968	I
20	5260.034	I
1	5260.228	I
3	5322.380	I
40	5362.601	I
4	5390.568	I
75	5431.532	I
3	5431.830	I
6	5578.788	I
40	5647.774	I
20	5653.750	I
60	5724.121	I
3	5724.614	I
75	6070.755	I
30 c	6159.626	I
75 c	6206.309	I
120 c	6298.325	I
5	6299.224	I
10000	6458.33	II
5000	6560.81	II
100 l	7279.997	I
150	7408.173	I
200 l	7618.933	I
300	7757.651	I
60	7759.436	I
90000 c	7800.27	I
45000 c	7947.60	I
40 l	8271.41	I
30	8271.71	I
40 l	8868.512	I
30	8868.852	I
30 l	9522.65	I
20 l	9540.18	I
2000 c	9689.05	II
35 l	10075.282	I
30 l	10075.708	I
100	13235.17	I
20	13442.81	I
30	13443.57	I
75	13665.01	I
1000	14752.41	I
800	15288.43	I
150	15289.48	I
20	22529.65	I
4	27314.31	I
Ruthenium Ru Z = 44		
250	850.09	III
200	850.30	III
250	919.74	III
500	940.09	III
500	966.54	III
750	974.14	III
900	979.43	III
500	981.35	III
900	986.84	III
900	994.56	III
300	1001.65	III
500	1009.13	III
900	1009.87	III
500	1014.68	III
800	1190.51	III
500	1200.07	III
500	1207.17	III
500	1209.77	III
300	1211.31	III
500	1941.35	III
500	2009.28	III
2400	2076.43	I
2600	2083.77	I
2400	2090.89	I
690	2255.52	I
780	2272.09	I
780	2279.57	I
480	2317.80	I
120	2334.96	II
190 h	2342.85	II
310	2351.33	I
170	2357.91	II
780	2402.72	II
150	2407.92	II
180	2455.53	II
150	2456.44	II
370	2456.57	II
280	2478.93	II
140	2498.42	II
140	2498.57	II
260	2507.01	II
110	2513.32	II
110	2517.32	II
150	2535.59	II
550	2549.58	I
370	2609.06	I
830	2612.07	I
460	2642.96	I
330	2651.84	I
400	2659.62	I
330	2661.61	II
690	2678.76	II
330	2692.06	II
200	2712.41	II
690	2719.52	I
140	2725.47	II
310	2734.35	II
1800	2735.72	I
100	2778.38	II
110	2787.83	II
350	2810.03	I
1700	2810.55	I
350	2818.36	I
400	2829.16	I
640	2854.07	I
420	2861.41	I
550	2866.64	I
1800	2874.98	I
740	2886.54	I
370	2908.88	I
1100	2916.26	I
180	2945.67	II
370	2949.50	I
550	2965.16	I
170	2965.55	II
140	2976.59	II
550	2976.92	I
1400	2988.95	I
460	2994.96	I
440	3006.59	I
330	3017.24	I
310	3020.88	I
390	3064.84	I
330	3096.57	I
830	3099.28	I
740	3100.84	I
490	3294.11	I
370	3301.59	I
930	3339.55	I
3100	3417.35	I
4900	3428.31	I
6400	3436.74	I
8300	3498.94	I
640	3514.49	I
790	3539.37	I
690	3570.59	I
6400	3589.22	I
6900	3593.02	I
6400	3596.18	I
1300	3599.76	I
3100	3634.93	I
6200	3661.35	I
830	3663.37	I
650	3669.49	I
550	3726.10	I
8700	3726.93	I
11000	3728.03	I
7100	3730.43	I
3500	3742.28	I
870	3742.78	I
2800	3745.59	I
760	3753.54	I
870	3755.93	I
1200	3759.84	I
600	3761.51	I
600	3767.35	I
1500	3777.59	I
600	3782.74	I
3900	3786.06	I
6000	3790.51	I
760	3798.05	I
7600	3798.90	I
7600	3799.35	I
600	3812.72	I
760	3817.27	I
760	3819.03	I
650	3822.09	I
550	3824.93	I
760	3831.80	I
930	3839.70	I
760	3850.43	I
1300	3857.55	I
650	3862.69	I
1300	3867.84	I
650	3892.21	I
760	3909.08	I
1500	3923.47	I
3300	3925.92	I
600	3931.76	I
760	3945.57	I
600	3978.44	I
600	3979.42	I
870	3984.86	I
1500	4022.16	I
600	4023.83	I
1400	4051.40	I
710	4054.05	I
760	4068.37	I
980	4076.73	I
6000	4080.60	I
930	4097.79	I
1900	4112.74	I
2000	4144.16	I
650	4145.74	I
870	4167.51	I
550	4197.58	I
550	4198.88	I
7600	4199.90	I
1500	4206.02	I
5400	4212.06	I
760	4214.44	I
930	4217.27	I
550	4230.31	I
760	4241.05	I

Intensity	Wavelength/Å	
760	4243.06	I
760	4284.33	I
550	4295.93	I
3700	4297.71	I
930	4307.60	I
550	4319.87	I
550	4342.07	I
710	4354.13	I
870	4361.21	I
2400	4372.21	I
870	4385.39	I
1300	4385.65	I
1700	4390.44	I
1600	4410.03	I
1100	4460.04	I
5400	4554.51	I
1700	4584.44	I
720	4647.61	I
1400	4709.48	I
500	4757.84	I
550	4869.15	I
160	5011.23	I
450	5057.33	I
120	5076.32	I
200	5093.83	I
530	5136.55	I
170	5142.76	I
250	5147.24	I
110	5151.07	I
500	5155.14	I
920	5171.03	I
180	5195.02	I
130	5284.08	I
260	5309.27	I
110	5335.93	I
130	5361.77	I
110 h	5401.04	I
80	5484.32	I
130	5510.71	I
90	5559.75	I
290	5636.24	I
180	5699.05	I
65	5814.98	I
55	5919.34	I
80	5921.45	I
21 h	5973.38	I
16	5988.67	I
35	5993.65	I
18	6116.77	I
26	6199.42	I
26	6225.20	I
18	6295.22	I
16	6390.23	I
26 h	6444.84	I
21	6663.14	I
55	6690.00	I
21	6766.95	I
30	6775.02	I
21	6824.17	I
26	6911.48	I
110	6923.23	I
26	6982.01	I
26	7027.98	I
35	7238.92	I
16	7393.93	I
18	7468.91	I
26	7485.79	I
70	7499.75	I
26	7559.61	I
18	7621.50	I
18	7722.87	I
22	7791.86	I
30	7847.80	I
80	7881.49	I
18	8264.96	I
22	8710.84	I

Samarium Sm Z = 62

Intensity	Wavelength/Å	
150	2789.38	II
410	3152.52	II
720	3183.92	II
600	3211.73	II
530	3216.85	II
600	3218.61	II
720	3230.56	II
720	3236.64	II
720	3239.66	II
720	3250.37	II
850	3254.38	II
1700	3306.39	II
1200	3321.18	II
1200	3365.86	II
1200	3382.40	II
4200	3568.27	II
4200	3592.60	II
1700	3604.28	II
3400	3609.49	II
1700	3621.23	II
3400	3634.29	II
2200	3661.36	II
2200	3670.84	II
1100	3693.99	II
1600	3728.47	II
2100	3731.26	II
1600	3735.98	II
2900	3739.12	II
1200	3743.87	II
930	3745.46	I
800	3756.41	I
1200	3757.53	II
1900	3760.69	II
1100	3764.37	II
370 d	3773.33	I
1100	3778.14	II
1500	3788.12	II
1600	3793.97	II
1600	3797.73	II
1600	3826.20	II
1100	3831.50	II
560	3834.48	I
1600	3843.50	II
530	3853.30	I
2700	3854.21	II
480	3854.56	I
400	3858.74	I
3700	3885.29	II
1600	3896.98	II
1300	3903.42	II
2500	3922.40	II
1900	3928.28	II
1300	3941.87	II
470	3951.89	I
1500	3963.00	II
1500	3971.40	II
620	3974.66	I
1000	3976.43	II
1500	3990.00	II
1400	4064.58	II
1000	4092.27	II
1900	4118.55	II
1200	4152.21	II
1000	4188.13	II
1100	4203.05	II
1000	4225.33	II
1200	4236.74	II
2100	4256.39	II
1300	4262.68	II
1200	4279.68	II
2200	4280.79	II
710	4282.21	I
470	4282.83	I
1600	4296.74	I
1900	4318.94	II
470	4319.53	I
1800	4329.02	II
440	4330.02	I
1300	4334.15	II
880	4336.14	I
1100	4347.80	II
440	4362.91	I
530	4380.42	I
1600	4390.86	II
410	4401.17	I
470	4419.33	I
1500	4420.53	II
2900	4424.34	II
470	4429.66	I
1600	4433.88	II
1800	4434.32	II
530	4441.81	I
440	4442.28	I
710	4445.15	I
1300	4452.73	II
1200	4454.63	II
1000	4458.52	II
2200	4467.34	II
810	4470.89	I
370	4499.11	I
440	4581.73	I
380	4649.49	I
470 d	4670.75	I
1100	4674.60	II
370	4688.73	I
530	4704.40	II
730	4716.10	I
770	4728.42	I
470	4745.68	II
730	4760.27	I
580	4783.10	I
350	4785.86	I
970	4841.70	I
730	4883.97	I
630	4910.40	I
350	4913.25	II
430	4918.99	I
400	5044.28	I
540	5071.20	I
510	5117.16	I
350	5122.14	I
360	5155.03	II
470	5175.42	I
250	5200.59	I
260	5251.92	I
400	5271.40	I
250	5282.91	I
220	5453.00	I
230	5493.72	I
230	5516.09	I
140	5550.40	I
140	5659.86	I
120	5696.73	I
85	5706.20	I
70	5773.77	I
60	5778.33	I
70 d	5786.98	II
60	5788.38	I
60	5800.52	I
65	5802.84	I
65	5867.79	I
50	5874.21	I
50	5898.96	I
65	5965.71	II
50	6045.00	I
50	6070.06	I
45	6084.12	I
45 h	6159.56	I
45	6256.54	I
100	6267.28	II
140	6569.31	II
110	6589.72	II
50	6671.51	I
120 d	6731.84	II
95	6794.20	II
120	6860.93	I
120	6955.29	II
90	7020.44	II
90	7039.22	II
90	7042.24	II
90	7051.52	II
90	7082.37	II
26	7088.30	I
30	7095.50	I
30	7104.54	I
26	7115.96	I
85 d	7149.60	II

Intensity	Wavelength/Å	
23	7213.82	I
60	7240.90	II
26	7347.30	I
30	7444.56	I
26	7445.41	I
13	7470.76	I
45	7645.09	II
12	7645.82	I
40 w	7835.08	II
16	7895.96	I
90	7928.14	II
40	8048.70	II
16	8065.16	I
45	8068.46	II
40 w	8305.79	II
19	8383.71	I
45 w	8485.99	II
45 w	8708.43	II
95	8913.66	II

Scandium Sc Z = 21

Intensity	Wavelength/Å	
350	180.14	V
500	243.87	V
500	252.85	V
500	253.73	V
900	283.91	V
800	284.45	V
600	288.29	V
900	289.59	V
15	289.85	IV
1000 d	291.93	V
800	293.25	V
15	296.31	IV
15	299.04	IV
700	300.00	V
1000	573.36	V
600	587.94	V
10	785.12	IV
25	1168.61	III
15	1550.80	IV
180	1603.06	III
150	1610.19	III
160	2010.42	III
12	2118.97	IV
11	2185.43	IV
11	2205.46	IV
14	2222.22	IV
11	2271.33	IV
110	2438.62	I
560	2545.22	II
2900	2552.37	II
560	2555.82	II
2300	2560.25	II
1100	2563.21	II
11	2586.93	IV
120	2692.78	I
350	2699.07	III
360	2706.77	I
210	2707.95	I
580	2711.35	I
230	2734.05	III
340	2965.86	I
1200	2974.01	I
1400	2980.75	I
340	2988.95	I
2200	3015.36	I
2700	3019.34	I
360	3030.76	I
120 h	3056.31	I
130	3065.11	II
990	3251.32	II
1500	3255.69	I
4400	3269.91	I
5500	3273.63	I
110 d	3343.28	II
270	3352.05	II
9900	3353.73	II
2000	3359.68	II
1700	3361.27	II
1700	3361.94	II
4000	3368.95	II
6600	3372.15	II
130	3418.51	I
200	3429.21	I
200	3429.48	I
270	3431.36	I
530	3435.56	I
270	3457.45	I
180	3462.19	I
130 d	3469.65	I
110	3471.13	I
200	3498.91	I
2700	3535.73	II
6600	3558.55	II
6100	3567.70	II
13000	3572.53	II
9900	3576.35	II
7700	3580.94	II
4000	3589.64	II
4000	3590.48	II
28000	3613.84	II
110	3617.43	I
20000	3630.75	II
13000	3642.79	II
6600	3645.31	II
110	3646.90	I
5300	3651.80	II
110	3664.25	II
290	3666.54	II
75 h	3717.10	I
270	3833.07	II
610	3843.03	II
20000	3907.49	I
23000	3911.81	I
4400	3933.38	I
5500	3996.61	I
530	4014.49	II
20000	4020.40	I
20000	4023.69	I
220	4030.67	I
140	4031.39	I
220	4043.80	I
200	4046.48	I
2700	4047.79	I
120	4049.95	I
5500	4054.55	I
220	4056.59	I
100	4068.66	III
160 h	4074.97	I
160	4078.57	I
6100	4082.40	I
200	4086.67	I
400	4087.16	I
440 h	4133.00	I
530 h	4140.30	I
720	4152.36	I
1100 h	4165.19	I
110 h	4218.26	I
110 h	4219.73	I
180	4231.93	I
200	4233.61	I
400	4238.05	I
15000	4246.83	II
290	4294.77	II
350	4305.71	II
4200	4314.09	II
3300	4320.74	II
2400	4325.01	II
180	4354.61	II
110	4358.64	I
2000	4374.46	II
130	4384.81	II
1100	4400.37	II
880	4415.56	II
120 h	4557.24	I
160 h	4573.99	I
350	4670.40	II
120	4706.97	I
120	4709.34	I
200	4728.77	I
490	4729.23	I
590	4734.10	I
690	4737.65	I
790	4741.02	I
1200	4743.81	I
200	4753.16	I
220	4779.35	I
170	4839.44	I
90	4909.76	I
90	4922.84	I
90	4934.25	I
170	4954.06	I
120	4973.66	I
150	4980.37	I
140	4991.92	I
530	5031.02	II
250	5064.32	I
530	5070.23	I
250	5075.81	I
2100	5081.56	I
1200	5083.72	I
1100	5085.55	I
750	5086.95	I
390	5087.14	I
270	5089.89	I
390	5096.73	I
620	5099.23	I
370	5101.12	I
180	5109.06	I
150	5112.86	I
320	5116.69	I
390	5210.52	I
280	5219.67	I
350	5239.82	II
280	5258.33	I
210	5285.76	I
120	5341.05	I
350	5349.30	I
120	5349.71	I
210	5355.75	I
530	5356.10	I
270	5375.35	I
370	5392.08	I
270	5446.20	I
120	5451.34	I
750	5481.99	I
530	5484.62	I
570	5514.22	I
660	5520.50	I
660	5526.82	II
70	5564.86	I
110	5591.33	I
80	5640.98	II
250	5657.88	II
1500	5671.81	I
1200	5686.84	I
1100	5700.21	I
10	5706.82	IV
190	5708.61	I
880	5711.75	I
230	5717.28	I
180	5724.08	I
14	5771.63	IV
620	6210.68	I
320	6239.78	I
120	6245.63	II
110	6249.96	I
80	6256.01	III
250	6258.96	I
750	6305.67	I
60	6378.82	I
90	6413.35	I
60	6604.60	II
65	6737.87	I
50	6819.52	I
50	6835.03	I
90	7449.16	III
55 h	7741.17	I
30	7800.44	I
19 h	8761.40	I
50	8829.78	III
30 h	8834.35	I
400	22051.86	I
150	22065.05	I

Intensity	Wavelength/Å	
Selenium Se	***Z* = 34**	
360	613.0	V
360	652.7	IV
450	670.1	IV
360	724.3	III
450	746.4	IV
450	759.1	V
360	808.7	V
360	830.3	V
360	832.7	II
450	839.5	V
360	843.0	III
360	845.8	V
360	912.9	II
360	959.6	IV
360	974.8	III
450	996.7	IV
360	1013.4	II
360	1014.0	II
450	1033.6	II
450	1049.6	II
360	1057.4	II
360	1094.7	V
360	1099.1	III
450	1119.2	III
360	1141.9	II
450	1192.3	II
450	1227.6	V
285	1291.0	II
285	1308.9	II
285	1314.4	IV
120	1435.3	I
120	1435.8	I
150	1449.2	I
150	1500.9	I
250	1530.4	I
150	1531.3	I
200	1531.8	I
150	1575.3	I
150	1577.6	I
150	1577.9	I
150	1579.5	I
200	1580.0	I
150	1587.5	I
150	1593.2	I
250	1606.5	I
200	1617.4	I
150	1621.2	I
150	1643.4	I
250	1671.2	I
250	1675.3	I
250	1690.7	I
250	1793.3	I
300	1795.3	I
300	1855.2	I
250	1858.8	I
400	1898.6	I
350	1913.8	I
300	1919.2	I
500	1960.9	I
500	2039.8	I
285	2057.5	III
500	2074.8	I
285	2136.6	IV
500	2164.2	I
600	2413.5	I
300	2548.0	I
360	2665.5	IV
285	2724.3	IV
285	2767.2	III
220	2773.8	III
160	2951.6	IV
450	3387.2	III
450	3413.9	III
450	3457.8	III
450	3637.6	III
450	3738.7	III
450	3800.9	III
450	4169.1	III
360	4175.3	II
450	4180.9	II
285	4382.9	II
285	4446.0	II
220	4449.2	II
285	4467.6	II
500	4730.8	I
400	4739.0	I
300	4742.2	I
285	4840.6	II
360	4845.0	II
450	5227.5	II
360	5305.4	II
100	5365.5	I
120	5369.9	I
110	5374.1	I
285	5522.4	II
285	5566.9	II
285	5866.3	II
450	6056.0	II
285	6303.8	III
200	6325.6	I
360	6444.2	II
285	6490.5	II
285	6535.0	II
150	6831.3	I
120	6990.690	I
100	6991.792	I
200	7010.809	I
150	7013.875	I
300	7062.065	I
200	7575.1	I
250	7583.4	I
150	7592.2	I
300	8001.0	I
200	8036.4	I
150	8093.2	I
150	8094.7	I
180	8149.3	I
150	8152.0	I
200	8157.7	I
180	8163.1	I
150	8182.9	I
150	8440.47	I
150	8450.38	I
150	8742.33	I
300	8918.86	I
200	9001.97	I
200	9038.61	I
100	9432.50	I
200	10217.25	I
377	10307.45	I
900	10327.26	I
640	10386.36	I
275	11946.87	I
170	11952.64	I
205	11972.93	I
315	14817.93	I
410	14917.47	I
500	15151.44	I
320	15471.00	I
265	15520.97	I
395	15618.40	I
360	16659.44	I
505	16813.78	I
205	16866.54	I
235	21374.24	I
680	21442.56	I
415	21473.48	I
270	21716.36	I
240	21730.60	I
150	23388.85	I
265	24148.18	I
375	24385.99	I
255	25017.51	I
510	25127.43	I
Silicon Si	***Z* = 14**	
10	85.18	V
15	96.44	V
10	97.14	V
20	117.86	V
20	118.97	V
4	457.82	IV
8	566.61	III
8	653.33	III
7	815.05	IV
8	818.13	IV
9	823.41	III
40 h	845.77	II
100	889.72	II
200	892.00	II
9	967.95	III
100	989.87	II
200	992.68	II
10	993.52	III
13	994.79	III
16	997.39	III
50	1023.69	II
8	1066.63	IV
14	1108.37	III
16	1109.97	III
18	1113.23	III
8	1122.49	IV
10	1128.34	IV
100	1190.42	II
200	1193.28	II
250	1194.50	II
100	1197.39	II
30	1206.51	III
30	1206.53	III
9	1207.52	III
10	1210.46	III
50	1226.81	II
100	1227.60	II
150	1228.75	II
200	1229.39	II
100	1246.74	II
150	1248.43	II
100	1250.09	II
150	1250.43	II
200	1251.16	II
40	1256.49	I
50	1258.80	I
1000	1260.42	II
2000	1264.73	II
200	1265.02	II
17	1294.54	III
14	1296.73	III
15	1298.89	III
18	1298.96	III
14	1301.15	III
16	1303.32	III
100	1304.37	II
50 h	1305.59	II
200	1309.27	II
13	1312.59	III
100	1346.87	II
100	1348.54	II
150	1350.06	II
100	1352.64	II
100	1353.72	II
15	1393.76	IV
12	1402.77	IV
13	1417.24	III
90 h	1485.02	II
100 h	1485.51	II
12	1500.24	III
10	1501.19	III
9	1501.87	III
100 h	1509.10	II
50 h	1512.07	II
60 p	1516.91	II
500	1526.72	II
1000	1533.45	II
150	1594.55	I
100	1622.87	I
300	1629.43	I
200	1629.92	I
100	1667.62	I
100	1668.52	I
100	1672.59	I
200	1675.20	I
200	1696.20	I
200	1697.94	I
100 h	1770.92	I

Intensity	Wavelength/Å	
100 h	1776.83	I
100 h	1799.12	I
150	1808.00	II
500 h	1814.07	I
200	1816.92	II
200	1836.51	I
200	1841.44	I
9	1842.55	III
200	1843.77	I
300	1845.51	I
400	1847.47	I
200	1848.14	I
500	1850.67	I
200	1852.46	I
500 h	1874.84	I
200	1881.85	I
200	1887.70	I
200 h	1893.25	I
1000 h	1901.33	I
100 h	1902.46	II
50 h	1910.62	II
50	1941.67	II
100	1949.56	II
100	1954.97	I
50	2058.65	II
50	2059.01	II
200	2072.02	II
200	2072.70	II
100	2124.12	I
50 h	2136.56	II
110	2207.98	I
115	2210.89	I
110	2211.74	I
120	2216.67	I
120	2218.06	I
10	2296.87	III
10	2308.19	III
100 h	2356.30	II
30 h	2357.18	II
50 h	2357.97	II
300	2435.15	I
11	2449.48	III
425	2506.90	I
375	2514.32	I
500	2516.113	I
7	2517.51	IV
350	2519.202	I
425	2524.108	I
450	2528.509	I
110	2532.381	I
25	2541.82	III
10	2546.09	III
14	2559.21	III
30	2563.679	I
85	2568.641	I
45	2577.151	I
190	2631.282	I
11	2640.79	III
14	2655.51	III
9	2817.11	III
1000	2881.579	I
300	2904.28	II
500	2905.69	II
55	2970.355	I
150	2987.645	I
50	3006.739	I
75	3020.004	I
100 h	3030.00	II
9	3040.93	III
100 h	3043.69	II
50 h	3048.30	II
150 h	3053.18	II
25	3086.24	III
20	3093.42	III
16	3096.83	III
9	3165.71	IV
16	3185.13	III
13	3186.02	III
150	3188.97	II
150	3193.09	II
100	3195.41	II
14	3196.50	III
200	3199.51	II
100 h	3203.87	II
200 h	3210.03	II
15	3210.55	III
75	3214.66	II
12	3230.50	III
14	3233.95	III
15	3241.62	III
12	3258.66	III
10	3276.26	III
300	3333.14	II
500	3339.82	II
15	3486.91	III
9	3525.94	III
20	3590.47	III
8	3762.44	IV
20 c	3791.41	III
25	3796.11	III
30	3806.54	III
100 h	3853.66	II
500 h	3856.02	II
200 h	3862.60	II
300	3905.523	I
20	3924.47	III
10	4088.85	IV
70	4102.936	I
9	4116.10	IV
300 h	4128.07	II
500 h	4130.89	II
100 h	4190.72	II
50	4198.13	II
9	4338.50	III
30	4552.62	III
25	4567.82	III
20	4574.76	III
100	4621.42	II
150	4621.72	II
9 h	4631.24	IV
10 h	4654.32	IV
9	4683.02	III
16	4716.65	III
50	4782.991	I
35	4792.212	I
80	4792.324	I
15	4813.33	III
16	4819.72	III
18	4828.97	III
30	4947.607	I
40	5006.061	I
1000	5041.03	II
1000	5055.98	II
10 h	5091.42	III
100	5181.90	II
100 h	5185.25	II
200 h	5192.86	II
500 h	5202.41	II
100 h	5405.34	II
100 h	5438.62	II
100 h	5456.45	II
500 h	5466.43	II
500 h	5466.87	II
100 h	5469.21	II
200 h	5496.45	II
35	5517.535	I
100 h	5540.74	II
150 h	5576.66	II
30	5622.221	I
100 h	5632.97	II
200 h	5639.48	II
90	5645.611	I
150 h	5660.66	II
80	5665.554	I
1000 h	5669.56	II
120	5684.484	I
300 h	5688.81	II
100	5690.425	I
90	5701.105	I
200 h	5701.37	II
100 h	5706.37	II
160	5708.397	I
20	5739.73	III
45	5747.667	I
45	5753.625	I
45	5754.220	I
45	5762.977	I
70	5772.145	I
70	5780.384	I
90	5793.071	I
100	5797.859	I
150 h	5800.47	II
200	5806.74	II
50	5846.13	II
300 h	5868.40	II
40	5873.764	I
10 h	5898.79	III
150	5915.22	II
200	5948.545	I
500	5957.56	II
500	5978.93	II
90	6125.021	I
85	6131.574	I
90	6131.850	I
100	6142.487	I
100	6145.015	I
160	6155.134	I
160	6237.320	I
40	6238.287	I
125	6243.813	I
125	6244.468	I
180	6254.188	I
45	6331.954	I
1000	6347.10	II
1000	6371.36	II
45	6526.609	I
45	6527.199	I
45	6555.462	I
50 h	6660.52	II
100	6671.88	II
7	6701.21	IV
50 h	6717.04	II
100	6721.853	I
50	6829.82	II
30	6848.568	I
80	6976.523	I
180	7003.567	I
180	7005.883	I
90	7017.646	I
250	7034.903	I
6 h	7047.94	IV
200	7165.545	I
100	7226.206	I
100	7235.326	I
180	7250.625	I
160	7275.294	I
400	7289.173	I
375	7405.774	I
200	7409.082	I
275	7415.946	I
425	7423.497	I
9 h	7466.32	III
12 h	7612.36	III
100	7680.267	I
6 h	7723.82	IV
30	7800.008	I
400	7848.80	II
500	7849.72	II
30	7849.967	I
90	7918.386	I
120	7932.349	I
140	7944.001	I
35	7970.306	I
35	8035.619	I
70	8093.241	I
9 h	8102.86	III
11 h	8103.45	III
35	8230.642	I
9 h	8262.57	III
40	8443.982	I
40	8501.547	I
60	8502.221	I
40	8536.165	I
120	8556.780	I

Intensity	Wavelength/Å	
50	8648.462	I
40	8728.011	I
75	8742.451	I
100	8752.009	I
35	8790.389	I
100	9412.72	II
100	9413.506	I
30	10371.269	I
120	10585.141	I
120	10603.431	I
120	10660.975	I
30	10694.251	I
30	10727.408	I
60	10749.384	I
30	10784.550	I
80	10786.856	I
140	10827.091	I
60	10843.854	I
130	10869.541	I
30	10882.802	I
30	10885.336	I
80	10979.308	I
30	10982.061	I
80	11017.965	I
370	11984.19	I
220	11991.57	I
440	12031.51	I
190	15888.39	I
95	16060.03	I
110	19722.50	I

Silver Ag Z = 47

Intensity	Wavelength/Å	
25	730.83	II
30	752.80	II
400	799.41	III
15	1005.32	II
10	1065.49	II
12	1072.23	II
250	1074.22	II
150	1107.03	II
150	1112.46	II
60	1195.83	II
50	1223.33	II
50	1240.80	II
50	1246.87	II
55	1256.81	II
55	1257.55	II
50	1266.63	II
70	1273.67	II
65	1297.51	II
85	1311.20	II
55	1313.81	II
50	1314.61	II
60	1323.84	II
60	1342.09	II
50	1342.57	II
70	1346.62	II
50	1353.54	II
150	1364.50	II
100	1396.00	II
100	1410.93	II
90	1419.72	II
95	1432.60	II
100	1464.72	II
50	1466.23	II
50 r	1507.37	I
100 r	1515.63	I
50 r	1548.58	I
100	1555.16	II
100	1644.50	II
60	1651.52	I
50	1652.10	I
700	1656.18	III
120	1682.82	II
500	1693.51	III
10	1708.11	I
50	1709.27	I
125	1736.44	II
750	1751.03	III
10 h	1766.14	I
75	1790.37	II
600	1917.08	III
700	1957.62	III
100	1967.38	II
600	1975.92	III
500	1977.03	III
600	2000.24	III
150	2015.96	II
150	2033.98	II
200	2061.17	I
100	2069.85	I
80 r	2113.82	II
60	2145.60	II
600	2161.89	III
50	2186.76	II
60	2229.53	II
100 r	2246.43	II
500	2246.51	III
75 r	2248.74	II
75	2280.03	II
30 h	2309.56	I
700	2310.04	III
70 r	2317.05	II
80 r	2320.29	II
70 r	2324.68	II
80 r	2331.40	II
70	2357.92	II
50 h	2375.02	I
75	2411.41	II
90 r	2413.23	II
100 r	2437.81	II
80	2447.93	II
80	2473.84	II
60	2506.63	II
50 h	2575.63	I
60	2660.49	II
60	2721.77	I
75	2767.54	II
100 h	2824.39	I
30 h	3130.02	I
90	3180.70	II
100	3267.35	II
55000 r	3280.68	I
28000 r	3382.89	I
30	3469.16	I
70	3475.82	II
80	3495.28	II
50	3542.61	I
50 h	3624.68	I
75	3682.46	II
30	3682.50	I
80	3683.34	II
50 h	3709.20	I
200	3810.94	I
50	3811.78	I
100 h	3840.74	I
50 h	3907.41	I
50	3909.31	II
50 h	3914.40	I
70	3920.10	II
60	3949.43	II
100 h	3981.58	I
70	3985.19	II
100 h	4055.48	I
80	4085.91	II
100	4185.48	II
90 h	4210.96	I
100	4212.82	I
50	4311.07	I
50 h	4476.04	I
30 h	4615.69	I
80	4620.04	II
50	4620.46	II
60 h	4668.48	I
30 h	4677.60	I
100	4788.40	II
30 h	4847.82	I
100	4874.10	I
80	5027.35	II
1000	5209.08	I
1000	5465.50	I
100	5471.55	I
100	5667.34	I
10 h	6268.50	I
320	7687.78	I
25	8005.4	II
500	8273.52	I
25	8403.8	II
30 h	8645.70	I
10 h	8704.85	I
12	8747.6	II
15	9000.9	II
10	12551.0	I
60	16819.5	I
20	17416.7	I
15	18307.9	I
15	18382.3	I

Sodium Na Z = 11

Intensity	Wavelength/Å	
7	142.232	IV
8	146.064	IV
9	150.298	IV
8	150.687	IV
8	155.510	IV
8	156.537	IV
12	162.448	IV
10	163.190	IV
12	168.411	IV
10	168.546	IV
5	183.95	III
10	190.445	IV
10	199.772	IV
8	202.49	III
8	202.76	III
8 p	203.06	III
8	203.28	III
8	203.33	III
15	229.87	III
50 c	250.52	III
30	251.37	III
25	266.90	III
70	267.65	III
50	267.87	III
50	268.63	III
20 p	272.08	III
20	272.45	III
10	319.644	IV
300	372.08	II
350	376.38	II
100	378.14	III
70	380.10	III
12	408.684	IV
10	409.614	IV
15	410.372	IV
10	411.334	IV
13	412.242	IV
11	1582.18	IV
11 d	1583.98	IV
12	1584.14	IV
12 d	1587.05	IV
11	1615.92	IV
12	1618.57	IV
11	1655.47	IV
15 c	1701.97	IV
20 d	1887.47	III
12	1960.76	IV
11	1965.08	IV
12 d	2106.33	IV
30	2230.33	III
16	2232.19	III
20 h	2246.70	III
300	2315.65	II
18	2386.99	III
17	2394.03	III
300	2420.99	II
300	2424.73	II
25	2459.31	III
18	2468.85	III
20	2474.73	III
1000	2493.15	II
25	2497.03	III
17	2510.26	III
20	2543.84	I
10	2543.87	I
70	2593.87	I
35	2593.92	I

Intensity	Wavelength/Å	
850	2611.81	II
850	2661.00	II
1000	2671.83	II
200	2680.34	I
100	2680.43	I
1000	2841.72	II
400	2852.81	I
200	2853.01	I
2	2893.62	I
1100	2904.92	II
1100	2917.52	II
1100	2919.05	II
1200	2919.85	II
1300	2920.95	II
1000	2923.49	II
1200	2951.24	II
1100	2952.40	II
1000	2977.13	II
1100	2979.66	II
1100	2980.63	II
1300	2984.19	II
1700	3124.42	II
2500	3135.48	II
1700	3137.86	II
2000	3149.28	II
2000	3163.74	II
1000	3179.06	II
1700	3189.79	II
1600	3212.19	II
1500	3257.96	II
1700	3285.60	II
1700	3301.35	II
1200	3302.37	I
600	3302.98	I
1500	3304.96	II
1000	3318.04	II
50	3426.86	I
1500	3533.05	II
1200	3631.27	II
6	4238.99	I
10	4242.08	I
1	4249.41	I
2	4252.52	I
15	4273.64	I
20	4276.79	I
2	4287.84	I
3	4291.01	I
30	4321.40	I
40	4324.62	I
3	4341.49	I
5	4344.74	I
40	4390.03	I
60	4393.34	I
5	4419.88	I
8	4423.25	I
60	4494.18	I
100	4497.66	I
10	4541.63	I
15	4545.19	I
120	4664.811	I
200	4668.560	I
20	4747.941	I
30	4751.822	I
200	4978.541	I
400	4982.813	I
40	5148.838	I
80	5153.402	I
280	5682.633	I
70	5688.193	I
560	5688.205	I
80000	5889.950	I
40000	5895.924	I
120	6154.225	I
240	6160.747	I
130	6530.70	II
130	6544.04	II
130	6545.75	II
20	7373.23	I
10	7373.49	I
50	7809.78	I
25	7810.24	I
4400	8183.256	I
800	8194.790	I
8800	8194.824	I
100	8649.92	I
60	8650.89	I
25	8942.96	I
40	9153.88	I
60	9465.94	I
80	9961.28	I
20	10566.00	I
60	10572.28	I
200	10746.44	I
80	10749.29	I
120	10834.87	I
35	11190.19	I
50	11197.21	I
400	11381.45	I
1000	11403.78	I
400	12679.17	I
60	14767.48	I
100	14779.73	I
60	16373.85	I
100	16388.85	I
400	18465.25	I
50	22056.44	I
25	22083.67	I
60	23348.41	I
100	23379.13	I

Strontium Sr Z = 38

Intensity	Wavelength/Å	
15	298.12	IV
15	300.12	IV
125	330.67	III
500	351.62	III
75	358.80	III
250	363.49	III
150	371.21	III
20	378.53	IV
75	392.44	IV
50	393.00	IV
50	396.22	IV
1000	437.24	III
1875	491.79	III
1250	507.04	III
3750	514.38	III
10	517.28	V
2500	562.75	III
25	578.01	V
30	624.93	V
25	642.23	V
50	649.21	V
25	660.94	V
200	664.43	IV
35	686.23	V
100	710.35	IV
12	747.82	V
50	1025.23	III
35	1125.49	III
50	1236.23	III
1400	2152.84	II
1400	2165.96	II
100	2273.71	III
100	2340.13	III
50	2346.97	IV
160	2428.10	I
100	2486.52	III
40	2555.60	IV
40	2571.04	IV
100	3002.61	III
200	3012.32	III
10	3019.29	IV
100	3021.73	III
50	3061.43	III
50	3182.61	III
100	3235.39	III
400	3351.25	I
650	3380.71	II
50	3430.76	III
950	3464.46	II
600	3969.26	I
1300	4030.38	I
46000	4077.71	II
32000	4215.52	II
9	4298.57	IV
340	4305.45	II
65000	4607.33	I
9	4685.08	IV
3200	4722.28	I
2200	4741.92	I
1400	4784.32	I
4800	4811.88	I
3600	4832.08	I
3000	4872.49	I
2000	4876.32	I
1000	4891.98	I
8000	4962.26	I
1300	4967.94	I
800 h	5156.07	I
1400	5222.20	I
2000	5225.11	I
2000	5229.27	I
2800	5238.55	I
4800	5256.90	I
40	5257.71	III
40	5443.48	III
1500	5450.84	I
7000	5480.84	I
3500	5504.17	I
2600	5521.83	I
2000	5534.81	I
2000	5540.05	I
1000	6380.75	I
900 h	6386.50	I
600 h	6388.24	I
9000	6408.47	I
5500	6504.00	I
1000	6546.79	I
1700	6550.26	I
3000	6617.26	I
1800	6791.05	I
4800	6878.38	I
1200	6892.59	I
5500	7070.10	I
2500	7309.41	I
500	7621.50	I
400 h	7673.06	I
200 h	8422.80	I
120	8505.69	II
200	8688.91	II
100	9294.10	I
400 h	9448.95	I
600	9596.00	I
300	9624.70	I
100	9638.10	I
100 h	9647.70	II
300	10036.66	II
1000	10327.31	II
200	10914.88	II
700	11241.25	I
100	12014.76	II
60	12445.90	II
40	12495.00	I
75	12974.70	II
100	13123.80	II
50	17447.40	I
230	20261.40	I
120	20700.70	I
30	26023.60	I

Sulfur S Z = 16

Intensity	Wavelength/Å	
5	437.4	V
5	438.2	V
5	439.6	V
20	519.3	IV
20	520.1	IV
40	520.8	IV
20	522.0	IV
20	522.5	IV
20	551.2	IV
40	652.5	IV
40	653.0	IV
70	653.6	IV
40	654.0	IV
70	655.6	IV
20	655.9	IV

Intensity	Wavelength/Å	
110	657.3	IV
40	658.3	V
70	659.8	V
40	660.9	IV
160	661.4	IV
110	663.2	V
40	663.7	IV
40	664.8	IV
70	666.1	IV
20	678.1	V
40	680.3	V
110	680.9	V
40	681.6	V
20	693.5	V
70	729.5	III
110	732.42	III
70	735.2	III
70	738.5	III
110	744.9	IV
110	748.4	IV
110	750.2	IV
110	753.8	IV
285	786.5	V
70	789.0	III
70	796.7	III
70	800.5	IV
70	804.0	IV
70	809.7	IV
110	816.0	IV
70	824.9	III
70	836.3	III
160	849.2	V
110	852.2	V
220	854.8	V
110	857.9	V
110	860.5	V
40	906.9	II
40	910.5	II
40	912.7	II
40	937.4	II
40	937.7	II
160	1062.7	IV
160	1073.0	IV
70	1073.5	IV
285	1077.1	III
40	1102.3	II
70	1194.0	III
70	1201.0	III
40	1234.1	II
40	1250.5	II
110	1253.8	II
110	1259.5	II
275	1270.782	I
250	1277.216	I
280	1295.653	I
275	1302.337	I
235	1302.863	I
235	1303.110	I
245	1303.430	I
260	1305.883	I
265	1310.194	I

Intensity	Wavelength/Å	
355	1316.542	I
290	1316.618	I
375	1323.515	I
355	1326.643	I
775	1381.552	I
710	1385.510	I
960	1388.435	I
640	1389.154	I
775	1392.588	I
1000	1396.112	I
300	1409.337	I
510	1425.030	I
425	1433.280	I
300	1436.968	I
300	1448.229	I
425	1472.972	I
550	1473.995	I
300	1474.380	I
355	1481.665	I
485	1483.039	I
300	1483.233	I
330	1485.622	I
390	1487.150	I
20	1624.0	IV
20	1629.2	IV
680	1666.688	I
640	1687.530	I
710	1807.311	I
680	1820.343	I
640	1826.245	I
710	1900.286	I
550	1914.698	I
20	2387.0	IV
40	2398.9	IV
110	2460.5	III
110	2489.6	III
160	2496.2	III
160	2499.1	III
220	2508.2	III
70	2636.9	III
220	2665.4	III
110	2691.8	III
110	2702.8	III
220	2718.9	III
110	2721.4	III
220	2726.8	III
220	2731.1	III
110	2741.0	III
285	2756.9	III
110	2775.2	III
160	2785.5	III
110	2863.5	III
160	2904.3	III
160	2986.0	III
110	3097.5	IV
110	3497.3	III
160	3632.0	III
110	3709.4	III
160	3717.8	III
160	3838.3	III
285	3867.6	I

Intensity	Wavelength/Å	
285	3902.0	I
160	3928.6	III
360	3933.3	II
450	4120.8	I
280	4142.3	II
360	4145.1	II
450	4153.1	II
450	4162.7	II
360	4253.6	III
450	4694.1	I
285	4695.4	I
160	4696.2	I
280	4716.2	II
450	4815.5	II
360	4924.1	II
450	4925.3	II
285	4993.5	I
360	5428.6	II
650	5432.8	II
1000	5453.8	II
1000	5473.6	II
1000	5509.7	II
280	5564.9	II
1000	5606.1	II
450	5640.0	II
450	5640.3	II
280	5647.0	II
650	5659.9	II
450	5664.7	II
160	5706.1	I
450	5819.2	II
450	6052.7	I
280	6286.4	II
450	6287.1	II
450	6305.5	II
450	6312.7	II
280	6384.9	II
280	6397.3	II
280	6398.0	II
360	6413.7	II
160	6743.6	I
285	6748.8	I
450	6757.2	I
450	7579.0	I
450	7629.8	I
285	7686.1	I
450	7696.7	I
1000	7924.0	I
160	7928.8	I
285	7930.3	I
450	7931.7	I
450	7967.4	I
450	7967.4	II
450	8314.7	I
450	8314.7	II
450	8585.6	I
285	8680.5	I
450	8694.7	I
360	8874.5	I
110	8882.5	I
220	8884.2	I

Intensity	Wavelength/Å	
160	9035.9	I
450	9212.9	I
450	9228.1	I
450	9237.5	I
285	9413.5	I
285	9421.9	I
285	9437.1	I
650	9649.9	I
450	9672.3	I
450	9680.8	I
450	9693.7	I
285	9697.3	I
285	9739.7	I
285	9932.3	I
285	9949.8	I
285	9958.9	I
285	10455.5	I
285	10459.5	I

Tantalum Ta Z = 73

Intensity	Wavelength/Å	
60	493.07	V
1000	890.87	V
500	947.30	V
67	999.34	IV
79	1116.10	IV
78	1136.17	IV
85	1175.51	IV
80	1189.28	IV
80	1192.67	IV
85	1213.09	IV
500	1213.42	V
85	1215.53	IV
90	1223.73	IV
88	1238.12	IV
95	1240.06	IV
87	1258.34	IV
94	1264.91	IV
98	1272.42	IV
94	1275.48	IV
86	1275.94	IV
92	1308.51	IV
87	1315.58	IV
92	1332.38	IV
86	1343.30	IV
92	1365.88	IV
5000	1392.56	V
91	1398.78	IV
93	1413.40	IV
91	1454.32	IV
92	1464.41	IV
93	1469.82	IV
90	1495.25	IV
95	1514.19	IV
85	1607.70	IV
7000	1709.10	V
85	1712.16	IV
85	1716.13	IV
85	2055.75	IV
1100	2140.13	II
1500	2146.87	II
1200	2182.71	II
1100	2193.88	II

Intensity	Wavelength/Å	
1500	2196.03	II
90	2199.58	IV
1500	2199.67	II
90	2207.64	IV
1400 d	2210.03	II
1400	2239.48	II
1200	2250.76	II
840	2261.42	II
990	2262.30	II
990	2272.59	II
790	2285.25	II
600	2286.59	II
990	2289.16	II
440	2302.24	II
440	2302.93	II
440	2312.60	II
420	2315.46	II
690	2331.98	II
550	2332.19	II
250	2357.30	I
260	2361.09	I
600	2364.24	II
320	2371.58	I
1400	2387.06	II
2400	2400.63	II
320	2416.89	II
360	2427.64	I
360	2429.71	II
480	2432.70	II
380	2470.90	II
600	2474.62	I
500	2484.95	I
600	2488.70	II
500	2490.46	I
600	2504.45	I
600	2507.45	I
1200 d	2526.35	I
600	2532.12	II
1200	2559.43	I
460	2562.10	I
600	2577.37	II
600	2603.49	II
1400	2608.63	I
1200	2635.58	II
860	2636.90	I
2400	2647.47	I
2600	2653.27	I
1900	2656.61	I
1500	2661.34	I
770	2675.90	II
1500	2685.17	II
470	2694.52	II
1000	2698.30	I
1200	2710.13	I
2600	2714.67	I
470	2727.44	II
1200	2748.78	I
860	2749.83	I
410	2752.49	II
1000	2758.31	I
430	2761.68	II
770	2775.88	I
680	2796.34	I
680	2797.76	II
510	2806.58	I
640	2844.25	I
560	2848.52	I
1500	2850.49	I
1900	2850.98	I
360	2861.98	I
470	2871.42	I
380	2880.02	I
770	2891.84	I
560	2902.05	I
310	2915.49	I
410	2925.19	I
310	2932.70	I
1700	2933.55	I
470	2940.06	I
1200	2940.22	I
510	2951.92	I
340	2953.56	I
1500	2963.32	I
770	2965.13	II
770	2965.54	I
340	2969.47	I
430	2975.56	I
1800	3012.54	II
290 d	3027.48	I
530	3049.56	I
530	3069.24	I
360	3077.24	I
560	3103.25	I
380	3124.97	I
380	3130.58	I
270	3132.64	I
320	3170.29	I
270	3173.59	I
600	3180.95	I
300	3223.83	I
1100	3311.16	I
680	3318.84	I
330 d	3330.99	II
640	3371.54	I
360	3385.05	I
450	3406.94	I
490	3480.52	I
380	3497.85	I
490	3511.04	I
750	3607.41	I
980	3626.62	I
500	3642.06	I
210	3918.51	I
210	3970.10	I
210	3996.17	I
410	4061.40	I
310	4067.91	I
300	4205.88	I
360 c	4510.98	I
340	4574.31	I
260	4619.51	I
450	4681.88	I
200	5037.37	I
100	5067.87	I
110	5115.84	I
100	5141.62	I
100	5143.69	I
330	5156.56	I
110	5212.74	I
110 d	5218.45	I
140	5341.05	I
200	5402.51	I
130	5419.19	I
90	5518.91	I
150	5645.91	I
130	5664.90	I
130	5776.77	I
90	5780.71	I
130	5811.10	I
240	5877.36	I
130	5882.30	I
90	5901.91	I
90	5918.95	I
130	5939.76	I
240	5944.02	I
190 c	5997.23	I
100	6020.72	I
250	6045.39	I
100	6047.25	I
100	6101.58	I
65	6144.56	I
130	6154.50	I
150	6256.68	I
150	6268.70	I
150	6309.58	I
75	6325.08	I
65	6341.17	I
75	6356.16	I
65	6360.84	I
90	6389.45	I
65	6428.60	I
250	6430.79	I
200	6450.36	I
380	6485.37	I
65	6505.52	I
100	6514.39	I
100	6516.10	I
100	6574.84	I
110	6611.95	I
75	6621.30	I
100	6673.73	I
180	6675.53	I
75 c	6740.73	I
75	6771.74	I
160 c	6813.25	I
210	6866.23	I
180	6875.27	I
150	6902.10	I
140	6927.38	I
140	6928.54	I
65	6951.26	I
180	6966.13	I
110 d	6995.22	I
75	7006.96	I
150	7148.63	I
110	7172.90	I
140	7301.74	I
160	7346.41	I
140 c	7352.86	I
100	7356.96	I
90 cw	7369.09	I
160	7407.89	I
100	7882.37	I
75	8026.50	I
75	8281.62	I

Technetium Tc Z = 43

Intensity	Wavelength/Å	
10000 c	3636.07	I
20000 c	4031.63	I
15000	4095.67	I
20000	4262.27	I
30000	4297.06	I
20000	4853.59	I

Tellurium Te Z = 52

Intensity	Wavelength/Å	
8	802.28	II
8	1059.51	II
8	1077.66	II
10	1161.42	II
10	1174.34	II
12	1175.79	II
9	1208.54	II
9	1220.98	II
9	1253.62	II
9	1270.52	II
10	1324.92	II
9	1363.24	II
8	1366.73	II
10	1374.80	II
10	1608.41	II
10	1613.15	II
5	1655.4	I
5	1688.5	I
6	1700.0	I
5	1708.0	I
10	1822.4	I
26000	2002.02	I
6500	2081.16	I
18000	2142.81	I
3200	2147.25	I
500	2259.02	I
1200	2383.26	I
1500	2385.78	I
50	2438.69	II
120	2530.72	I
100	2649.66	II
80	2661.10	II
110	2677.13	I
100	2858.29	II
150	2895.41	II
70	2967.29	II
70	3047.00	II
100	3175.14	I
60	3256.80	II
60	3329.22	II

Intensity	Wavelength/Å	
150	3406.79	II
50	3442.25	II
50	3521.11	II
50	3552.19	II
100	3611.78	II
50	3617.57	II
50	4006.52	II
70	4127.32	II
100	4169.77	II
80	4225.73	II
100	4261.11	II
60	4273.43	II
80	4285.85	II
150	4364.00	II
75	4385.10	II
170	4478.63	II
80	4537.07	II
100	4557.78	II
70	4630.62	II
100	4641.12	II
180	4654.37	II
200	4686.91	II
100	4696.38	II
100	4706.53	II
100	4766.05	II
100	4784.87	II
100	4827.14	II
150	4831.28	II
150	4842.90	II
130	4865.12	II
200	4866.24	II
8	5083.0	I
50	5449.84	II
50	5487.95	II
150	5576.35	II
150	5649.26	II
100	5666.20	II
200	5708.12	II
150	5755.85	II
100	5974.68	II
50	6367.13	II
10 h	6790.0	I
20 h	6837.6	I
20 h	6854.7	I
15 h	7191.1	I
20 h	7263.5	I
12	7460.98	II
15	7468.75	II
15	7921.69	II
15	7943.14	II
10	7950.34	II
30 h	8061.4	I
10	8122.44	II
20	8186.44	II
15	8273.53	II
15	8672.95	II
10	8733.81	II
205	8758.18	I
81	9004.37	I
5660	9722.74	I
532	9868.92	I

Intensity	Wavelength/Å	
689	9956.30	I
325	9977.13	I
5950	10051.41	I
4097	10091.01	I
381	10118.08	I
397	10300.56	I
745	10493.57	I
1880	10918.34	I
10200	11089.56	I
508	11163.74	I
6620	11487.23	I
1580	13247.75	I
1050	14513.51	I
1480	15452.45	I
2430	15546.23	I
3760	16403.90	I
1960	17303.54	I
2780	18291.59	I
1020	21043.73	I
464	21602.50	I
74	22555.29	I
38	26539.17	I
Terbium Tb Z = 65		
1000	1259.40	IV
1000	1327.67	IV
1000	1373.86	IV
5000	1595.39	IV
2000	1633.19	IV
2000	2027.79	IV
1000	2089.98	IV
1000	2332.54	IV
110	2584.61	II
110	2608.57	II
130	2628.69	II
140	2669.29	II
190	2704.07	II
270	2769.53	II
320	2897.44	II
250	2956.21	II
230	3010.59	II
230	3016.18	II
460	3053.55	II
460	3070.05	II
670	3078.86	II
480	3082.36	II
480	3089.58	II
480	3102.96	II
440	3139.64	II
480	3187.26	II
480	3199.56	II
1100	3218.93	II
1200	3219.98	II
480	3252.32	II
760	3280.31	II
760	3281.40	II
1000	3285.04	II
1500	3293.07	II
3800	3324.40	II
760	3349.42	II
760	3364.93	II
810	3454.06	II

Intensity	Wavelength/Å	
810 d	3472.79	II
810	3500.84	II
5700	3509.17	II
1300	3523.66	II
1100	3540.24	II
810	3543.89	II
3200	3561.74	II
810	3567.35	II
4200	3568.52	II
1600	3568.98	II
1100	3579.20	II
710	3585.03	II
810	3596.38	II
1600	3600.44	II
810	3625.54	II
2300	3650.40	II
810	3654.88	II
2000	3658.88	II
3800	3676.35	II
810	3682.26	II
450	3693.58	I
450	3700.12	I
4700	3702.86	II
2400	3703.92	II
1000 d	3711.76	II
650	3745.04	I
870	3747.17	II
870	3747.34	II
1100	3755.24	II
650	3759.35	I
1700	3765.14	I
2100	3776.49	II
600	3783.53	I
410	3789.92	I
760 d	3806.85	II
1500	3830.26	I
540	3833.42	I
920 d	3842.50	II
3700	3848.73	II
3500 w	3874.17	II
480	3888.22	I
490	3894.64	I
2400	3899.20	II
1600	3901.33	I
480	3908.06	I
650	3915.43	I
760	3925.45	II
810 d	3939.52	II
2200 d	3976.84	II
1800	3981.87	II
970	4002.59	II
1900	4005.47	II
760	4012.75	II
870	4032.28	I
2100	4033.03	II
430	4054.12	I
410	4060.37	I
1300	4061.58	I
650	4105.37	I
1100	4144.41	II
350	4158.53	I

Intensity	Wavelength/Å	
390	4196.74	I
650	4203.74	I
600	4206.49	I
480	4215.09	I
480	4232.82	I
650	4266.34	I
760 cw	4278.52	II
450	4310.42	I
2200	4318.83	I
600	4322.23	I
600	4325.83	II
3000	4326.43	I
600	4332.12	I
870	4336.43	I
600	4337.64	I
1700	4338.41	I
700	4340.62	I
870	4356.81	I
330	4382.45	I
300	4388.23	I
260	4390.91	I
350	4423.10	I
240	4436.12	I
240	4448.04	I
430	4493.07	I
75	4514.31	II
110	4549.07	I
110	4550.45	I
110	4556.46	I
110	4563.69	II
210	4578.69	II
65	4584.84	II
65	4591.56	II
75 d	4626.32	II
95	4626.94	II
65	4632.07	I
65 h	4636.59	I
85	4641.00	II
210	4641.98	II
260 cw	4645.31	II
80	4647.23	I
80	4662.79	I
80	4676.90	I
70 c	4681.87	I
80	4688.63	II
80	4693.11	II
200	4702.41	II
110	4707.94	II
80	4739.93	I
70	4747.80	I
410 cw	4752.53	II
180	4786.78	I
100	4813.77	I
80	4875.57	II
80	4881.15	II
95	4915.90	I
65	4931.79	I
85	4993.82	II
110	5078.25	I
75	5089.12	II
85	5186.13	I

Intensity	Wavelength/Å	
120	5228.12	I
75	5248.71	I
75 w	5262.11	II
75	5281.05	I
65	5304.72	I
110	5319.23	I
65 w	5337.90	I
160	5354.88	I
75	5369.72	I
75	5375.98	I
50	5424.10	II
55	5459.81	I
55	5509.61	I
50	5514.54	I
65	5524.12	I
85 c	5747.58	I
75	5795.64	I
75	5803.13	II
65	5815.36	I
65	5851.07	I
65	5870.62	I
65 c	5920.78	I
75	5967.34	II
35	6331.68	II
35 cw	6518.68	I
35	6581.82	I
90	6677.94	II
40 cw	6702.61	I
130	6794.58	II
55	6896.37	II
45 h	6899.95	I
40	6901.98	I
65	7204.28	I
40	7257.73	I
45	7348.88	II
45	7496.12	I
27 h	7582.03	II
45	7590.24	I
65	7596.44	I
30	7627.81	I
30	7737.63	I
30	7855.79	II
27	7927.90	II
30	8025.42	II
30	8085.06	II
65	8194.82	II
95	8212.57	I
40	8450.06	II
30 h	8511.80	I
45	8583.45	II
30	8603.40	I
65	8765.74	II
Thallium Tl Z = 81		
10	570.49	IV
5 r	670.87	II
15 r	696.30	II
5 r	709.23	II
10 r	817.18	II
5 r	836.34	II
8 r	1018.85	II
30	1028.69	IV
20	1034.73	IV
20	1036.61	IV
10 r	1049.73	II
8 r	1050.30	II
5 r	1074.97	II
30	1079.68	IV
10 r	1130.17	II
15 r	1162.55	II
10 r	1167.43	II
10 r	1183.41	II
12 r	1194.84	II
5 r	1246.00	II
10	1266.33	III
15 r	1307.50	II
8 r	1310.20	II
25 r	1321.71	II
8 r	1330.40	II
10 r	1373.52	II
10	1477.14	III
8 r	1489.65	I
10 r	1499.30	II
10 r	1507.82	II
15 r	1561.58	II
10 r	1568.57	II
7 r	1593.26	II
5 h	1616.	I
5	1685.40	I
10 r	1792.76	II
12 r	1814.85	II
25 r	1908.64	II
100 r	2007.56	I
100 r	2210.71	I
30	2298.04	II
140	2315.98	I
900 h	2379.69	I
20	2530.86	II
700	2580.14	I
420	2709.23	I
4400 d	2767.87	I
10	2849.80	II
2800	2918.32	I
20	3091.56	II
15	3185.51	II
15	3186.56	II
15	3187.74	II
1200	3229.75	I
15	3291.01	II
15	3369.15	II
9	3456.34	III
20000	3519.24	I
5000	3529.43	I
8	3540.08	II
9	3560.68	II
12000 w	3775.72	I
10	3832.30	II
10	3887.15	II
7	4109.85	III
6	4269.81	III
20	4274.98	II
40	4306.80	II
20	4737.05	II
15	4981.35	II
25	5078.54	II
25	5152.14	II
18000	5350.46	I
15 d	5384.85	II
10	5410.97	II
25	5949.48	II
10	6179.98	II
10	6378.32	II
16 h	6549.84	I
10	6966.5	II
10	7815.80	I
20	8373.6	I
10	8474.27	I
10	8664.1	II
20	9130.	II
20	9130.5	I
40	9509.4	I
20	9930.4	I
30	10011.9	I
40	10488.80	I
1000	11512.82	I
150	12736.4	I
700	13013.2	I
Thorium Th Z = 90		
150	1707.37	IV
200	1959.02	IV
200	2002.34	IV
200	2413.50	III
200	2427.94	III
200	2431.68	III
200	2441.24	III
500	2565.593	II
480	2692.415	II
520	2747.156	II
410	2752.166	II
800	2832.315	II
1200	2837.295	II
100	2848.084	I
550	2870.406	II
100	2936.086	I
100	2943.729	I
420	3049.092	II
450	3067.729	II
670	3078.828	II
480	3080.217	II
510	3108.296	II
100	3116.263	I
510	3119.526	II
510	3122.963	II
480	3125.507	II
100	3136.216	I
420	3139.306	II
420	3142.835	II
420	3175.726	II
1100	3180.193	II
770	3188.233	II
560	3221.292	II
560	3229.009	II
480	3235.84	II
590	3238.116	II
910	3256.274	II
180	3257.366	I
910	3262.668	II
620	3287.789	II
910	3291.739	II
620	3292.520	II
240	3301.650	I
480	3304.238	I
510	3321.450	II
840	3325.120	II
250	3330.476	I
620	3334.604	II
620	3337.870	II
310	3348.768	I
980	3351.228	II
620	3358.602	II
250	3374.974	I
1300	3392.035	II
200	3396.727	I
250	3398.544	I
200	3405.558	I
250	3413.012	I
390	3421.210	I
270	3423.989	I
980	3433.998	II
770	3435.976	II
1300	3469.920	II
170	3471.218	I
200	3486.552	I
670	3539.587	II
180	3544.018	I
170	3549.595	I
200	3555.013	I
530	3559.451	II
200	3576.557	I
270	3592.780	I
270	3598.120	I
980	3609.445	II
200	3612.427	I
480	3615.133	II
270	3635.943	I
210	3642.248	I
170	3649.735	I
220	3663.202	I
280	3669.968	I
700	3675.567	II
150	3682.486	I
170	3692.566	I
180	3698.105	I
340	3706.767	I
590	3719.435	I
770	3721.825	II
1300	3741.183	II
310	3747.539	I
650	3752.569	II
180	3770.056	I
590	3803.075	I
450	3828.384	I
840	3839.746	II
450	3863.405	II
210	3875.374	I

Intensity	Wavelength/Å	
340	3895.419	I
590	3929.669	II
200	3932.911	I
390	3967.392	I
200	3972.155	I
150	3980.089	I
530	3994.549	II
220	4008.210	I
220	4009.056	I
280	4012.495	I
4200	4019.129	II
250	4030.842	I
250	4036.047	I
250	4063.407	I
700	4086.520	II
700	4094.747	II
150	4100.341	I
840	4108.421	II
240	4112.754	I
280	4115.758	I
1100	4116.713	II
200	4127.411	I
200	4134.067	I
450	4149.986	II
620	4178.060	II
620	4208.890	II
110	4253.538	I
110	4260.333	I
480	4277.313	II
700	4282.042	II
130	4337.277	I
1300	4381.860	II
1100	4391.110	II
110	4498.940	I
280	4510.527	II
90	4723.438	I
50	4840.843	I
280	4863.163	II
260	5017.255	II
110	5067.974	I
120	5148.211	II
95	5216.596	II
110	5231.160	I
95	5247.654	II
60	5343.581	I
60	5587.026	I
95	5707.103	II
70	5760.551	I
85	5989.044	II
60	6169.822	I
50	6182.622	I
50	6274.116	II
50	6274.117	II
50	6355.911	II
60	6457.283	I
50	6462.614	I
50 h	6531.342	I
55	6989.656	I
30	7045.795	II
30	7084.171	I
30	7168.896	I
40	7191.132	II
35	7208.006	I
50	7525.508	II
30	7647.380	I
30	8330.451	I
40	8967.641	I
20	9833.42	I
20	10726.93	I
20	10942.24	II
30	11230.259	I
20	11984.67	II
20	17208.22	II
15	18811.88	I
10	22264.35	II

Thulium Tm ***Z* = 69**

Intensity	Wavelength/Å	
5000	2185.94	III
360	2284.79	II
20000	2296.21	III
5000	2305.03	III
20000	2311.16	III
5000	2312.72	III
5000	2326.19	III
6000	2328.50	III
6000	2329.29	III
3000	2331.80	III
3000	2357.05	III
4000	2406.63	III
450	2409.02	II
450	2426.17	II
770	2480.13	II
30000	2489.44	III
2000	2504.71	III
1300	2509.08	II
3000	2519.78	III
130	2527.02	I
10000	2552.46	III
360	2552.76	I
540	2561.65	II
430	2588.27	II
170 h	2596.49	I
810	2607.06	II
730	2624.33	II
5000	2682.32	III
2000	2707.03	III
3000	2719.47	III
540	2721.19	II
3000	2724.44	III
4000	2727.56	III
680	2794.60	II
730	2797.27	II
2000	2806.77	III
580	2827.92	II
200	2854.17	I
1600	2869.23	II
1000	2947.72	III
490	2973.22	I
1000	2998.28	III
1500	3015.30	II
360	3081.12	I
7400	3131.26	II
2300	3133.89	II
1900	3151.04	II
1500	3157.34	II
450	3172.65	I
2300	3172.83	II
1200	3236.81	II
1600	3240.23	II
2300	3241.54	II
320	3246.96	I
1900	3258.05	II
1600	3266.64	II
1200	3267.40	II
1100	3276.81	II
1200	3283.40	II
1200	3285.61	II
2300	3291.00	II
2000	3302.46	II
1200	3309.80	II
230	3349.99	I
4000	3362.61	II
1700	3397.50	II
850	3410.05	I
340	3412.59	I
340	3416.59	I
6400	3425.08	II
340	3429.33	I
4900	3441.50	II
4900	3453.66	II
8500	3462.20	II
210	3467.51	I
340	3476.69	I
340	3480.98	I
420	3487.38	I
340	3499.95	I
250	3517.60	I
1700	3535.52	II
420	3537.91	I
210	3555.82	I
340	3560.92	I
420	3563.88	I
1300	3566.47	II
420	3567.36	I
280	3586.07	I
2100	3608.77	II
1000	3629.09	III
380	3638.41	I
1100	3668.09	II
4800	3700.26	II
3800	3701.36	II
7700	3717.91	I
2400	3734.12	II
5000	3744.06	I
1700	3751.81	I
6000	3761.33	II
4800	3761.91	II
7100	3795.75	II
770	3798.54	I
600	3807.72	I
290	3826.39	I
1300	3838.20	II
290	3840.87	I
8900	3848.02	II
6800	3883.13	I
1800	3883.44	II
5400	3887.35	I
440	3896.62	I
3500	3916.48	I
1500	3949.27	I
1500	3958.10	II
1800	3996.52	II
220	4024.23	I
380	4044.47	I
10000	4094.19	I
9500	4105.84	I
1100	4138.33	I
8800	4187.62	I
6000	4203.73	I
380	4222.67	I
3000	4242.15	II
270	4271.71	I
150	4298.36	I
2700	4359.93	I
1400	4386.43	I
200	4394.42	I
140	4396.50	I
120	4454.03	I
540	4481.26	II
150	4519.60	I
260	4522.57	II
110	4548.60	I
270	4599.02	I
300	4615.94	II
80	4626.33	II
95	4626.56	II
110	4634.26	II
120	4655.09	I
160	4681.92	I
120	4691.11	I
110	4724.26	I
680	4733.34	I
70	4759.90	I
80	4831.20	II
140	4957.18	I
160	5009.77	II
160	5034.22	II
150	5060.90	I
95	5113.97	I
80	5213.38	I
650	5307.12	I
80	5346.49	II
270	5631.41	I
520	5675.84	I
40	5684.76	II
35	5709.97	II
190	5764.29	I
35	5838.76	II
240	5895.63	I
140	5971.26	I
200	6460.26	I
95	6604.96	I
110	6779.77	I
120	6844.26	I
80	6845.76	I

Intensity	Wavelength/Å	
10	6937.37	I
10	7017.90	I
12	7034.34	I
10	7106.14	I
17	7272.62	I
14	7310.51	I
14	7432.18	I
75	7481.08	I
75	7490.20	I
140	7558.33	I
80	7731.53	I
40	7856.08	I
55	7927.51	I
110	7930.84	I
95	8017.90	I
27	8472.01	II

Tin Sn $Z = 50$

Intensity	Wavelength/Å	
7	169.47	II
150	361.01	V
100	753.01	III
200	910.92	III
500	956.25	IV
7	985.13	II
500	1019.72	IV
1000	1044.49	IV
1000	1073.41	IV
200	1089.35	V
8	1108.19	II
1000	1119.34	IV
1000	1139.29	III
1000	1158.33	III
200	1160.74	V
10	1161.43	II
1000	1184.25	III
2000	1210.52	III
9	1219.07	II
13	1223.70	II
11	1243.00	II
2000	1251.38	V
1000	1259.92	III
20	1290.86	II
200	1294.36	V
1000	1305.97	III
1000	1314.55	IV
20	1316.59	II
1000	1327.34	III
1000	1347.65	III
1000	1386.74	III
25	1400.52	II
1000	1437.52	IV
20	1475.15	II
9	1489.22	II
1000	1570.36	III
10 r	1737.21	I
15 r	1751.46	I
20 r	1764.98	I
30 r	1790.75	I
80 r	1804.60	I
15	1811.34	II
500	1811.71	III
40 r	1815.74	I
120 r	1823.00	I
9	1831.89	II
50 r	1848.75	I
200 r	1860.32	I
80	1886.05	I
100	1891.40	I
12	1899.91	II
50	1909.30	I
80	1925.31	I
500	1941.86	III
150	1952.15	I
50 h	1977.6	I
80	1984.20	I
50	2040.66	I
50	2054.03	I
70	2058.31	I
80	2068.58	I
100	2072.89	I
100	2073.08	I
200	2096.39	I
100	2100.93	I
100 r	2113.93	I
50	2121.26	I
40 r	2148.73	I
20 r	2151.43	I
30	2151.54	II
80	2171.32	I
150 r	2194.49	I
300 r	2199.34	I
400 r	2209.65	I
80 r	2231.72	I
400 r	2246.05	I
60	2251.17	I
400 r	2268.91	I
200 r	2286.68	I
600 r	2317.23	I
300 r	2334.80	I
1000 r	2354.84	I
22	2368.33	II
100	2408.15	I
800 r	2421.70	I
1000 r	2429.49	I
15	2448.98	II
300	2483.39	I
13	2483.48	II
10	2486.99	II
200	2495.70	I
400	2546.55	I
500 r	2571.58	I
200	2594.42	I
200 r	2661.24	I
700 r	2706.51	I
150	2779.81	I
1400 r	2839.99	I
1000 r	2863.32	I
700 r	3009.14	I
850 r	3034.12	I
12	3047.50	II
550 r	3175.05	I
550 r	3262.34	I
50	3283.21	II
110	3330.62	I
60	3351.97	II
10	3472.46	II
11	3575.45	II
280 r	3801.02	I
10	5332.36	II
20	5561.95	II
25	5588.92	II
500	5631.71	I
15	5799.18	II
50	5925.44	I
100	5970.30	I
150	6037.70	I
250	6069.00	I
100	6073.46	I
400	6149.71	I
200	6154.60	I
150	6171.50	I
100	6310.78	I
70	6453.50	II
25	6844.05	II
20	7191.40	II
10	7387.79	II
13	7741.80	II
100	7754.97	I
100 h	8030.5	I
200	8114.09	I
80	8357.04	I
300	8422.72	I
400	8552.60	I
50 h	8681.7	I
50 h	9410.86	I
80 h	9415.37	I
150	9616.40	I
50	9741.1	I
100 h	9742.8	I
300 h	9805.38	I
500	9850.52	I
54	10894.00	I
70	11191.85	I
56	11277.66	I
200	11454.59	I
200	11616.26	I
258	11739.78	I
96	11825.18	I
106	11835.82	I
254	11932.99	I
48	12009.50	I
111	12313.24	I
42	12530.87	I
42	12536.5	I
89	12888.5	I
187	12981.7	I
187	13018.5	I
68	13081.5	I
378	13460.2	I
144	13608.2	I
40	20861.7	I
4	24738.2	I

Titanium Ti $Z = 22$

Intensity	Wavelength/Å	
17	252.96	V
15	498.26	V
14	502.08	V
13	526.57	V
18	779.07	IV
20	1298.66	III
20	1298.97	III
23	1455.19	III
20	1467.34	IV
11	1717.40	V
10	1841.49	V
20	2067.56	IV
18	2103.16	IV
180	2273.28	I
190	2279.96	I
190	2305.67	I
22	2413.99	III
25	2516.05	III
360	2525.60	II
24	2527.84	III
210	2529.85	I
190	2531.25	II
190	2534.62	II
130	2535.87	II
23	2540.06	III
24	2563.44	III
23	2565.42	III
22	2567.56	III
270	2599.92	I
340	2605.15	I
510	2611.28	I
300	2619.94	I
640	2641.10	I
800	2644.26	I
950	2646.64	I
250	2742.32	I
250	2802.50	I
190	2841.94	II
180	2877.44	II
280	2884.11	II
450	2912.08	I
340	2928.34	I
1100	2942.00	I
1300	2948.26	I
1600	2956.13	I
22	2984.75	III
1300 d	3066.22	II
1100	3072.97	II
1600	3075.22	II
2300	3078.64	II
3600	3088.02	II
720	3119.72	I
500	3161.20	II
780	3161.77	II
1000	3162.57	II
1600	3168.52	II
2400	3186.45	I
1000	3190.87	II
3100	3191.99	I
3800	3199.92	I
780	3202.54	II
1100	3217.06	II

Intensity	Wavelength/Å	
1300	3222.84	II
6600	3234.52	II
5200	3236.57	II
4100	3239.04	II
2600	3241.99	II
1200	3248.60	II
1200	3252.91	II
1200	3254.25	II
1200	3261.60	II
840	3314.42	I
2900	3322.94	II
2100	3329.46	II
1800	3335.20	II
1100	3340.34	II
5700	3341.88	I
4300	3349.04	II
12000	3349.41	II
4100	3354.64	I
7200	3361.21	II
1100	3370.44	I
4300	3371.45	I
5700	3372.80	II
2900 d	3377.48	I
1400	3380.28	II
5700	3383.76	II
1400	3385.95	I
1400	3387.84	II
1100	3394.58	II
890	3444.31	II
600	3461.50	II
600	3477.18	II
480	3491.05	II
890	3504.89	II
600	3510.84	II
17	3576.44	IV
600	3610.16	I
4800	3635.46	I
6600	3642.68	I
7200	3653.50	I
600	3671.67	I
3100	3685.20	II
600	3689.91	I
2900	3729.82	I
3300	3741.06	I
330	3741.64	II
5200	3752.86	I
3300	3759.30	II
2900	3761.32	II
840	3786.04	I
500	3882.89	I
530	3900.54	II
2600	3904.78	I
500	3913.46	II
500	3914.34	I
15	3915.47	III
1100	3924.53	I
890	3929.88	I
1100	3947.78	I
4500	3948.67	I
4500	3956.34	I
5200	3958.21	I

Intensity	Wavelength/Å	
950	3962.85	I
950	3964.27	I
4800	3981.76	I
570	3982.48	I
5700	3989.76	I
7800	3998.64	I
950	4008.93	I
1200	4024.57	I
840	4078.47	I
890	4286.01	I
840	4287.40	I
950	4289.07	I
840	4290.94	I
840	4295.76	I
2000	4298.66	I
200	4300.05	II
2900	4300.56	I
4100	4301.09	I
6000	4305.92	I
1200	4314.80	I
330	4395.04	II
890	4427.10	I
230	4443.80	II
840	4449.15	I
550	4450.90	I
840	4453.32	I
950	4455.33	I
1100	4457.43	I
240	4468.50	II
530	4481.26	I
780	4512.74	I
1000	4518.03	I
1000	4522.80	I
780	4527.31	I
6000	4533.24	I
240	4533.97	II
3600	4534.78	I
2400	4535.58	I
1200	4535.92	I
1200	4536.05	I
720	4544.69	I
950	4548.77	I
240	4549.63	II
15	4549.84	III
950	4552.46	I
720	4555.49	I
240	4571.98	II
15 d	4572.20	III
950	4617.27	I
480	4623.09	I
720	4656.47	I
840	4667.59	I
950	4681.92	I
470	4840.87	I
400	4885.08	I
380	4899.91	I
5800	4981.73	I
4600	4991.07	I
4000	4999.51	I
3600	5007.21	I
3200 d	5014.19	I

Intensity	Wavelength/Å	
840	5020.03	I
840	5022.87	I
1200	5035.91	I
840	5036.47	I
740	5038.40	I
1200	5039.95	I
1400	5064.66	I
1100	5173.75	I
1300	5192.98	I
1400	5210.39	I
17	5278.12	III
20	5398.93	IV
340	5512.53	I
270	5514.35	I
320	5514.54	I
250	5644.14	I
130	5675.44	I
95	5689.47	I
95	5715.13	I
85	5739.51	I
400	5866.46	I
230	5899.32	I
120	5918.55	I
150	5922.12	I
120	5941.76	I
300	5953.17	I
200	5965.84	I
270	5978.56	I
340	5999.04	I
110	6064.63	I
120	6085.23	I
120	6091.17	I
120	6126.22	I
17	6246.65	IV
380	6258.10	I
380	6258.70	I
300	6261.10	I
55	6546.28	I
65	6554.23	I
75	6556.07	I
18	6621.58	III
18	6667.99	III
80	6743.12	I
20	7072.64	III
18	7084.57	III
260	7209.44	I
130	7244.86	I
130	7251.72	I
120	7344.72	I
90	7357.74	I
60	7364.11	I
60	7978.88	I
55	8024.84	I
75	8364.24	I
100	8377.85	I
100	8382.54	I
75	8396.87	I
120	8412.36	I
170	8426.52	I
490	8434.94	I
240	8435.70	I

Intensity	Wavelength/Å	
20	8466.87	III
90	8675.39	I

Tungsten W Z = 74

Intensity	Wavelength/Å	
5800	2001.71	II
13000	2008.07	II
5100	2009.98	II
4100	2010.23	II
4100	2014.23	II
7300	2026.08	II
15000	2029.98	II
5300	2049.63	II
9700	2079.11	II
6100	2094.75	II
2100	2118.87	II
2400	2121.59	II
1500	2166.32	II
1300	2204.48	II
460	2249.80	I
510	2277.58	I
530 d	2294.49	I
340	2309.02	I
440	2313.17	I
460	2321.63	I
390 d	2326.56	I
320	2354.61	I
580	2360.44	I
850	2363.07	I
510	2374.47	I
670	2384.82	I
730	2397.09	II
560	2397.73	I
560	2397.98	I
1700 d	2405.58	I
610	2415.68	I
870	2424.21	I
1800	2435.96	I
580	2444.06	I
780	2451.48	II
870	2452.00	I
630	2454.98	I
780	2455.51	I
780	2456.53	I
1100	2459.30	I
1400	2466.85	I
480	2472.51	I
1200	2474.15	I
870	2480.13	I
1500	2481.44	I
480 d	2482.10	I
580	2484.74	I
390	2487.50	I
390	2489.23	II
630	2495.26	I
680	2504.70	I
75	2510.47	II
310	2520.46	I
780	2521.32	I
270	2522.04	II
780	2523.41	I
430	2527.76	I
780	2533.64	I

Intensity	Wavelength/Å	
1200	2547.14	I
780	2550.38	I
2700	2551.35	I
730	2561.97	I
870	2580.49	I
390	2584.39	I
390	2589.17	II
370	2601.96	I
680	2606.39	I
370	2608.32	I
970	2613.08	I
480	2613.82	I
400	2620.25	I
400	2622.21	I
400	2625.22	I
400	2632.48	I
400	2632.70	I
810	2633.13	I
400 d	2638.62	I
650	2646.18	I
400	2646.73	I
1600	2656.54	I
810	2662.84	I
810	2671.47	I
650	2677.28	I
2100	2681.42	I
650	2695.67	I
650	2699.59	I
400	2700.01	I
400	2706.58	I
400	2708.59	I
400 d	2708.80	I
400	2715.50	I
2100	2718.91	I
2600	2724.35	I
400	2725.03	I
650	2748.84	I
400	2762.34	I
400	2764.27	II
400	2769.74	I
810	2770.88	I
810	2774.00	I
810	2774.48	I
810	2792.70	I
400	2799.93	I
810	2818.06	I
1600	2831.38	I
810	2833.63	I
810	2848.02	I
1500	2896.44	I
690	2935.00	I
2400	2944.40	I
2400	2946.99	I
730 d	2979.71	I
360	3013.79	I
520	3016.47	I
770	3017.44	I
210	3024.93	I
310 d	3026.67	I
440 d	3041.73	I
270	3043.80	I

Intensity	Wavelength/Å	
440	3046.44	I
810	3049.69	I
180	3073.28	I
180 d	3084.83	I
370	3093.50	I
240	3107.23	I
240	3108.02	I
230	3117.57	I
260	3120.18	I
290	3163.42	I
320	3176.60	I
190	3181.82	I
390	3191.57	I
390	3198.84	I
520	3207.25	I
1000	3215.56	I
190	3232.49	I
210	3254.36	I
210	3259.66	I
210 d	3266.62	I
730	3300.82	I
440	3311.38	I
440	3326.20	I
440	3331.69	I
390	3373.75	I
230	3429.59	I
240	3443.00	I
400	3495.24	I
650	3545.22	I
240	3570.65	I
1900	3617.52	I
650	3682.08	I
400	3683.30	I
570	3688.06	I
810	3707.92	I
510	3757.92	I
680	3760.13	I
1000	3768.45	I
340	3773.71	I
1000	3780.77	I
290	3809.22	I
190	3810.38	I
260	3810.79	I
1400	3817.48	I
1100	3835.06	I
730	3846.22	I
1800	3867.99	I
730	3881.41	I
8600	4008.75	I
540	4015.22	I
910	4045.59	I
730	4069.95	I
5000	4074.36	I
1000	4102.70	I
540	4137.46	I
450	4171.17	I
220	4207.05	I
250	4219.37	I
540	4244.36	I
1400	4269.38	I
4100	4294.61	I

Intensity	Wavelength/Å	
2200	4302.11	I
200	4378.48	I
180	4384.85	I
200	4408.28	I
640	4484.19	I
170	4588.73	I
640	4659.87	I
640	4680.51	I
790	4843.81	I
380	4886.90	I
220	4982.59	I
820	5053.28	I
770	5224.66	I
220	5514.68	I
65	5648.37	I
55	5735.09	I
45	5804.85	I
40	5902.64	I
55	5947.57	I
55	5965.86	I
55	6012.78	I
40	6021.52	I
45	6292.02	I
35	6404.21	I
40	6445.12	I
17	6611.62	I
13	6678.42	I
15	6693.08	I
13	6984.27	I
15	7140.52	I
9	7162.64	I
11	7200.16	I
10	7278.24	I
15	7285.81	I
15	7296.55	I
10	7509.00	I
17	7569.92	I
17	7614.15	I
13	7688.97	I
11	7784.15	I
22	8017.19	I
22	8055.64	I
13	8123.82	I
10	8338.08	I
27	8585.11	I
10	8594.42	I
13	8865.53	I

Uranium U $Z = 92$

Intensity	Wavelength/Å	
440	2565.41	II
610	2635.53	II
830	2793.94	II
870	2802.56	II
630	2807.05	II
630	2817.96	II
870	2821.12	II
680	2828.90	II
920	2832.06	II
970	2865.68	II
1200	2889.62	II
780	2906.80	II
780	2908.28	II

Intensity	Wavelength/Å	
580	2931.41	II
530 p	2940.37	II
1300	2941.92	II
830	2943.90	II
580	2956.06	II
580	2967.94	II
580	2971.06	II
530	2984.61	II
630	3022.21	II
630	3031.99	II
580	3050.20	II
630	3057.91	II
630	3062.54	II
580	3072.78	II
580	3093.01	II
580	3102.39	II
970	3111.62	II
530	3119.35	II
680	3124.95	II
530	3139.61	II
680	3149.24	II
530	3153.11	II
730	3229.50	II
680	3232.16	II
730	3291.33	II
1100	3305.89	II
730	3390.38	I
580	3424.56	II
580	3435.49	I
630	3466.30	I
680	3482.49	II
1600	3489.37	I
530	3496.41	II
630	3500.08	I
780	3507.34	I
1600	3514.61	I
630	3533.57	II
530	3540.47	II
1200	3550.82	II
680	3555.32	I
1200	3561.80	I
2300	3566.59	I
530	3569.08	I
630	3578.72	II
3200	3584.88	I
840	3638.20	I
2800	3670.07	II
1100	3701.52	II
600	3738.04	II
680	3746.42	II
950	3748.68	II
600	3751.17	I
1900	3782.84	II
570	3793.10	II
1900	3811.99	I
750	3826.51	II
2000	3831.46	II
1200	3839.63	I
2400	3854.64	II
4900	3859.57	II
1900	3865.92	II

Intensity	Wavelength/Å	
1500	3871.03	I
1000	3881.45	II
2200	3890.36	II
2000	3932.02	II
1200	3943.82	I
1200	3985.79	II
1000	4042.75	I
1600	4050.04	II
880	4062.54	II
2200	4090.13	II
810	4116.10	II
880	4153.97	I
1400	4171.59	II
1000	4241.67	II
600	4472.33	II
620	4543.63	II
170	4689.07	II
150	4756.81	I
110	5008.21	II
170	5027.38	I
80	5160.32	II
70	5280.38	I
80	5475.70	II
70	5480.26	II
70	5481.20	II
160	5492.95	II
70	5780.59	I
70	5798.53	II
230	5915.39	I
100	5976.32	I
90	6077.29	I
55	6372.46	I
90	6395.42	I
110	6449.16	I
90	6826.92	I
45	7533.93	I
50	7881.94	I
35	8445.39	I
75	8607.95	I
30	8757.76	I
100	10554.93	I
75	11167.84	I
100	11384.13	I
100	11859.42	I
100	11908.83	I
100	12250.46	I
100	13185.16	I
75	13306.23	I
100	13961.58	I
75	18634.43	I
75	21910.22	I

Vanadium V Z = 23

Intensity	Wavelength/Å	
20	225.46	V
20	251.66	V
20	286.84	V
35	483.01	V
50	633.94	III
200	677.34	IV
500	684.37	IV
400	737.85	IV
100	864.27	III
500	1006.46	III
500	1149.94	III
100	1426.65	IV
1000	1643.03	III
1000	1650.14	III
100	1680.20	V
1000	1694.78	III
1000	1760.07	III
1000	1788.26	III
1000	1794.60	III
1000	1812.19	III
300	1861.56	IV
500	1939.06	IV
400	1951.43	IV
500	1997.72	IV
2100	2092.44	I
500	2268.30	IV
1000	2292.86	III
2500	2330.42	III
2500	2371.06	III
1000	2382.46	III
240	2507.78	I
410	2526.22	I
210	2527.90	II
80 h	2570.72	IV
230	2574.02	I
250	2593.05	III
250	2595.10	III
80 h	2645.54	IV
180	2661.42	I
1100	2687.96	II
680	2700.94	II
530	2706.17	II
640	2715.69	II
180	2731.35	I
240	2864.36	I
900	2891.64	II
900	2892.66	II
1400	2893.32	II
900	2906.46	II
2400	2908.82	II
710	2923.62	I
2400	2924.02	II
1700	2924.64	II
900	2941.37	II
1100	2944.57	II
410	2962.77	I
600	2968.38	II
1200	3056.33	I
1400	3060.46	I
2400	3066.38	I
3800	3093.11	II
3000	3102.30	II
2600	3110.71	II
2000	3118.38	II
1500	3125.28	II
3200	3183.41	I
5300	3183.98	I
3800	3185.40	I
410	3187.71	II
530	3188.51	II
750	3190.68	II
1100	3267.70	II
900	3271.12	II
750	3276.12	II
80 h	3514.25	IV
560	3517.30	II
560	3533.68	I
560	3545.20	II
560	3556.80	II
560	3589.76	II
490	3592.02	II
560	3592.53	I
100	3679.86	III
1300	3688.07	I
1000	3690.28	I
1500	3692.22	I
1000	3695.86	I
3800	3703.58	I
1800	3704.70	I
320	3715.47	II
250	3727.34	II
280	3732.76	II
520	3790.32	I
1100	3794.96	I
570	3799.91	I
570	3803.47	I
1000	3813.49	I
1300	3818.24	I
1700	3828.56	I
2600	3840.75	I
1200	3855.37	I
3000	3855.84	I
1300	3864.86	I
1500	3875.08	I
700	3890.18	I
2400	3902.25	I
700	3909.89	I
540	3990.57	I
430	3998.73	I
170	4005.71	II
1100	4090.58	I
1800	4092.69	I
890	4095.49	I
2800	4099.80	I
590	4102.16	I
2800	4105.17	I
2300	4109.79	I
8900	4111.78	I
4300	4115.18	I
1800	4116.47	I
2000	4123.57	I
3100	4128.07	I
3100	4132.02	I
2300	4134.49	I
20	4200.32	V
360	4232.46	I
560	4268.64	I
460	4271.55	I
460	4276.96	I
430	4284.06	I
460	4330.02	I
510	4332.82	I
760	4341.01	I
1000	4352.87	I
12000	4379.24	I
7000	4384.72	I
4800	4389.97	I
3600	4395.23	I
1400	4400.58	I
2300	4406.64	I
2800	4407.64	I
3600	4408.20	I
4600	4408.51	I
640	4416.47	I
640	4421.57	I
640	4437.84	I
830	4441.68	I
640	4444.21	I
610	4452.01	I
1000	4459.76	I
2000	4460.29	I
610	4462.36	I
510	4577.17	I
640	4580.40	I
830	4586.36	I
1300	4594.11	I
230	4619.77	I
100	4635.18	I
130	4646.40	I
160	4670.49	I
130	4776.36	I
110	4786.51	I
130	4796.92	I
130	4807.53	I
130	4827.45	I
150	4831.64	I
120	4832.43	I
320	4851.48	I
480	4864.74	I
620	4875.48	I
740	4881.56	I
110	5128.53	I
110	5138.42	I
110	5192.99	I
110	5194.83	I
110	5234.07	I
110	5240.87	I
100	5401.93	I
140	5415.26	I
140	5584.50	I
100	5592.42	I
200	5624.60	I
400	5627.64	I
110	5657.44	I
110	5668.36	I
310	5670.85	I
1200	5698.52	I
920	5703.56	I
570	5706.98	I
850	5727.03	I
230	5731.25	I
230	5737.06	I

Intensity	Wavelength/Å	
450	6039.73	I
480	6081.44	I
1300	6090.22	I
600	6119.52	I
450	6199.19	I
450	6216.37	I
430	6230.74	I
710	6243.10	I
280	6251.82	I
130	6268.82	I
170	6274.65	I
200	6285.16	I
200	6292.83	I
170	6296.49	I
110	6531.43	I
65 c	6753.00	I
50 c	6766.49	I
40	6784.98	I
40	7338.92	I
35	7356.54	I
29 c	8027.39	I
120 w	8116.80	I
70 c	8161.07	I
60 c	8919.85	I

Xenon Xe Z = 54

Intensity	Wavelength/Å	
8	657.8	III
8	660.1	III
9	673.8	III
9	674.0	III
9	676.6	III
10	694.0	III
20	698.5	III
12	705.1	III
10	721.2	III
15	731.0	III
10	733.3	III
350	740.41	II
15	742.6	III
10	756.0	III
10	761.5	III
10	769.1	III
25	779.1	III
15	792.9	III
12	796.1	III
15	802.0	III
350	803.07	II
25	823.2	III
30	824.9	III
25	853.0	III
600	880.80	II
350	885.54	II
15	889.3	III
20	894.0	III
20	896.0	III
600	925.87	II
250	935.40	II
10	965.5	III
800	972.77	II
700	976.68	II
35	1003.4	III
35	1017.7	III

Intensity	Wavelength/Å	
500	1032.44	II
700	1037.68	II
1100	1041.31	II
10	1047.8	III
1000	1048.27	II
1200	1051.92	II
12	1066.4	III
2000	1074.48	II
600	1083.86	II
1200	1100.43	II
30	1130.3	III
600	1158.47	II
250	1169.63	II
800 p	1183.05	II
250	1192.04	I
25	1232.1	III
600	1244.76	II
250	1250.20	I
1000	1295.59	I
600	1469.61	I
80	2668.98	III
100	2717.33	III
30	2814.45	III
40	2815.91	III
30	2827.45	III
40	2847.65	III
30	2862.40	III
200	2864.73	II
80 w	2871.10	III
60 w	2871.24	III
30	2871.7	III
150 h	2895.22	II
30	2896.62	III
50	2906.6	III
40	2911.89	III
80 w	2912.36	III
40	2940.2	III
60	2945.2	III
40	2947.5	III
40	2948.1	III
80 w	2970.47	III
400	2979.32	II
40	2992.87	III
30	3004.25	III
100 h	3017.43	II
100	3023.81	III
40	3083.5	III
50	3091.1	III
30	3106.46	III
300	3128.87	II
100 w	3138.3	III
80 c	3150.82	III
40	3185.2	III
100	3242.86	III
80	3268.98	III
30	3287.82	III
80 w	3301.55	III
40	3331.6	III
30	3358.0	III
200 h	3366.72	II
80	3384.12	III

Intensity	Wavelength/Å	
2	3400.07	I
2	3418.37	I
2	3420.00	I
3	3442.66	I
60	3444.2	III
70	3454.2	III
100 w	3458.7	III
100 h	3461.26	II
40	3468.22	III
4	3469.81	I
4	3472.36	I
5	3506.74	I
80	3522.83	III
50	3542.3	III
10	3549.86	I
50	3552.1	III
10	3554.04	I
40	3561.4	III
100	3579.7	III
80	3583.6	III
100 w	3595.4	III
100	3606.06	III
40	3607.0	III
15	3610.32	I
8	3613.06	I
100 w	3615.9	III
40	3623.1	III
600	3624.08	III
6	3633.06	I
10	3669.91	I
50	3676.67	III
40	3685.90	I
40	3693.49	I
40	3776.3	III
300	3781.02	III
100	3841.5	III
200	3877.8	III
60	3880.5	III
100 l	3907.91	II
500	3922.55	III
300	3950.59	III
100	4037.59	II
200	4050.07	III
200 l	4057.46	II
60	4060.4	III
100 h	4098.89	II
100	4109.1	III
100	4145.7	III
200 l	4158.04	II
1000 h	4180.10	II
500 h	4193.15	II
300 h	4208.48	II
100 h	4209.47	II
300 h	4213.72	II
100	4215.60	II
300 h	4223.00	II
400 h	4238.25	II
500 h	4245.38	II
100 l	4251.57	II
30	4285.9	III
500 h	4296.40	II

Intensity	Wavelength/Å	
500 h	4310.51	II
1000 l	4330.52	II
200 h	4369.20	II
100 l	4373.78	II
500 h	4393.20	II
500 l	4395.77	II
200 l	4406.88	II
150 l	4416.07	II
50	4434.2	III
500 h	4448.13	II
100 w	4462.1	III
1000 h	4462.19	II
500 l	4480.86	II
100 l	4521.86	II
100 w	4569.1	III
100 w	4570.1	III
100 w	4641.4	III
30	4673.7	III
60	4683.57	III
30	4723.60	III
600	4734.152	I
100 w	4757.3	III
150	4792.619	I
500	4807.02	I
400	4829.71	I
300	4843.29	I
40	4869.5	III
500	4916.51	I
500	4923.152	I
200 l	4971.71	II
400	4972.71	II
300	4988.77	II
100 l	4991.17	II
200	5028.280	I
200	5044.92	II
1000	5080.62	II
300	5122.42	II
100	5125.70	II
100	5178.82	II
300	5188.04	II
400	5191.37	II
100	5192.10	II
60	5239.0	III
500	5260.44	II
500	5261.95	II
2000	5292.22	II
300	5309.27	II
1000	5313.87	II
2000	5339.33	II
200	5363.20	II
30	5367.1	III
200	5368.07	II
500	5372.39	II
100	5392.80	I
50	5401.0	III
3000	5419.15	II
800	5438.96	II
300	5445.45	II
200	5450.45	II
400	5460.39	II
1000	5472.61	II

Atomic

Intensity	Wavelength/Å	
100 l	5494.86	II
40	5524.4	III
200	5525.53	II
600	5531.07	II
100	5566.62	I
300	5616.67	II
300	5659.38	II
600	5667.56	II
150	5670.91	II
100	5695.75	I
200	5699.61	II
200	5716.10	II
500	5726.91	II
500	5751.03	II
300	5758.65	II
300	5776.39	II
100	5815.96	II
300	5823.89	I
150	5824.80	I
100	5875.02	I
300	5893.29	II
100	5894.99	I
200	5905.13	II
100	5934.17	I
500	5945.53	II
300	5971.13	II
2000	5976.46	II
200	6008.92	II
1000	6036.20	II
2000	6051.15	II
600	6093.50	II
1500	6097.59	II
400	6101.43	II
100	6115.08	II
100	6146.45	II
150	6178.30	I
120	6179.66	I
300	6182.42	I
500	6194.07	II
100	6198.26	I
60	6205.97	III
100	6220.02	II
25	6221.7	III
60	6238.2	III
60	6259.05	III
500	6270.82	II
400	6277.54	II
100	6284.41	II
100	6286.01	I
250	6300.86	II
500	6318.06	I
400	6343.96	II
600	6356.35	II
200	6375.28	II
100	6397.99	II
300	6469.70	I
150	6472.84	I
120	6487.76	I
100	6498.72	I
200 h	6504.18	I
300	6512.83	II

Intensity	Wavelength/Å	
200	6528.65	II
100	6533.16	I
1000	6595.01	II
100	6595.56	I
400	6597.25	II
100	6598.84	II
150	6668.92	I
300	6694.32	II
200	6728.01	I
150	6788.71	II
100	6790.37	II
1000	6805.74	II
200	6827.32	I
100	6872.11	I
300	6882.16	I
80	6910.22	II
100	6925.53	I
800 h	6942.11	II
100	6976.18	I
2000	6990.88	II
150	7082.15	II
500	7119.60	I
50 s	7147.50	II
200	7149.03	II
500	7164.83	II
100	7284.34	II
200	7301.80	II
200	7339.30	II
100	7386.00	I
150	7393.79	I
300	7548.45	II
200	7584.68	I
80	7618.57	II
500	7642.02	I
100	7643.91	I
200	7670.66	II
60	7787.04	II
100	7802.65	I
100	7881.32	I
300	7887.40	I
500	7967.34	I
100	8029.67	I
200	8057.26	I
150	8061.34	I
100	8101.98	I
150 h	8151.80	II
100	8171.02	I
700	8206.34	I
10000	8231.635	I
500	8266.52	I
7000	8280.116	I
2000	8346.82	I
100	8347.24	II
2000	8409.19	I
50 h	8515.19	II
200	8576.01	I
50 h	8604.23	II
250	8648.54	I
100	8692.20	I
200	8696.86	I
50 h	8716.19	II

Intensity	Wavelength/Å	
300	8739.39	I
100	8758.20	I
5000	8819.41	I
300	8862.32	I
200	8908.73	I
200	8930.83	I
1000	8952.25	I
100	8981.05	I
200	8987.57	I
400	9045.45	I
500	9162.65	I
100	9167.52	I
100	9374.76	I
200	9513.38	I
50 h	9591.35	II
150	9685.32	I
50 l	9698.68	II
100	9718.16	I
2000	9799.70	I
3000	9923.19	I
100	10838.37	I
90	11742.01	I
375	12235.24	I
100	12257.76	I
300	12590.20	I
2500	12623.391	I
250	13544.15	I
2000	13657.055	I
1250	14142.444	I
800	14240.96	I
375	14364.99	I
140	14660.81	I
3000	14732.806	I
100	15099.72	I
2500	15418.394	I
150	15557.13	I
250	15979.54	I
100	16039.90	I
1000	16053.28	I
125	16554.49	I
1500	16728.15	I
1500	17325.77	I
350	18788.13	I
150	20187.19	I
3000	20262.242	I
250	21470.09	I
1250	23193.33	I
110	23279.54	I
1800	24824.71	I
175	25145.84	I
2000	26269.08	I
2500	26510.86	I
250	28381.54	I
750	28582.25	I
300	29384.41	I
150	29448.06	I
100	29649.58	I
100	29813.62	I
600	30253.14	I
1500	30475.46	I
100	30504.12	I

Intensity	Wavelength/Å	
500	30794.18	I
6000	31069.23	I
125	31336.01	I
550	31607.91	I
100	32293.08	I
1800	32739.26	I
3500	33666.69	I
150	34014.67	I
450	34335.27	I
170	34744.00	I
5000	35070.25	I
110	35246.92	I
250	36209.21	I
150	36231.74	I
450	36508.36	I
850	36788.83	I
140	38685.98	I
175	38737.82	I
270	38939.60	I
120	39955.14	I

Ytterbium Yb Z = 70

Intensity	Wavelength/Å	
1000	1050.24	IV
1000	1054.46	IV
5000	1134.43	IV
900	1316.04	IV
800	1326.36	IV
900	1350.26	IV
80	1561.42	III
80 h	1765.21	III
800	1791.06	IV
100	1863.32	III
800	1873.91	III
500	1898.25	III
500	1998.82	III
900	2116.65	IV
2500	2116.67	II
800	2123.32	IV
3000	2126.74	II
800	2139.99	IV
20000	2144.77	IV
15000	2154.18	IV
370	2161.60	II
850	2185.71	II
640	2224.46	II
300	2240.11	III
300	2305.32	III
140	2320.81	II
170	2390.74	II
460	2464.50	I
140	2512.06	II
270	2538.67	II
2000	2567.61	III
1000	2579.57	III
800	2599.14	III
600	2621.11	III
1000	2642.56	III
1000	2651.74	III
700	2652.25	III
990	2653.75	II
200	2665.04	II
2000	2666.13	III

Intensity	Wavelength/Å	
2000	2666.99	III
390	2671.96	I
390	2672.66	II
170	2718.35	II
230	2748.66	II
1300	2750.48	II
170	2776.28	II
600	2795.60	III
1000	2803.43	III
600	2816.92	III
1000	2818.72	III
140	2821.15	II
190	2830.99	II
230 h	2847.18	II
360	2851.13	II
430	2859.80	II
140	2861.21	II
200	2867.06	II
45	2873.49	I
200	2888.04	II
3600	2891.38	II
600	2898.30	III
1000	2906.31	III
170	2914.21	II
140	2915.28	II
280	2919.35	II
35	2934.36	I
140	2945.91	II
2000	2970.56	II
200	2983.99	II
170	2994.80	II
800	2998.00	III
310	3005.77	II
160	3017.56	II
160	3026.67	II
2000	3029.49	III
920	3031.11	II
3000	3092.50	III
28	3100.74	I
170	3107.90	II
190	3117.81	II
4000	3126.01	III
1000	3138.58	III
230	3140.94	II
28	3162.29	I
800	3191.35	III
390	3192.88	II
240	3201.16	II
2000	3228.58	III
35	3239.58	I
18000	3289.37	II
130	3305.25	I
140	3305.73	II
80	3319.41	I
2000	3325.51	III
240	3337.17	II
280 d	3342.93	II
240	3375.48	II
2000	3384.01	III
140	3387.50	I
50	3412.45	I
140	3418.39	I
360	3426.04	I
240	3431.11	I
85	3452.40	I
500	3454.08	II
190 d	3458.29	II
360	3460.27	I
2400	3464.37	I
500	3476.30	II
500	3478.84	II
50	3517.00	I
230	3520.29	II
35	3559.03	I
200	3560.33	II
170	3560.70	II
360	3585.47	II
200	3619.80	II
240	3637.76	II
70	3648.15	I
90	3655.73	I
240	3669.69	II
140	3675.08	II
32000	3694.19	II
70	3700.58	I
400	3711.91	III
180	3734.69	I
550	3770.10	I
80	3774.32	I
60 h	3791.74	I
170	3839.91	I
340	3872.85	I
340	3900.85	I
140	3911.27	I
500	3931.23	III
32000	3987.99	I
930	3990.88	I
50	4007.36	I
2000	4028.14	III
70	4052.28	I
440	4089.68	I
470	4149.07	I
120	4174.56	I
340	4180.81	II
300	4213.64	III
150 d	4218.56	II
120	4231.97	I
70	4277.74	I
120	4305.97	I
60 h	4393.69	I
60 h	4430.21	I
440	4439.19	I
85 h	4482.42	I
100	4517.58	III
85 h	4563.95	I
640	4576.21	I
200	4582.36	I
70	4589.21	I
140	4590.83	I
40	4684.27	I
190	4726.08	II
170 h	4781.87	I
170	4786.61	II
35	4816.43	I
40	4837.46	I
40 h	4894.60	I
27	4912.36	I
710	4935.50	I
140	4966.90	I
30	5067.80	I
70	5069.14	I
220	5074.34	I
50	5076.74	I
60	5196.08	I
85	5211.60	I
100	5244.11	I
150 h	5277.04	I
170	5335.15	II
30 h	5351.29	I
150	5352.95	II
30	5363.66	I
40	5449.27	II
60	5481.92	I
40	5505.49	I
17	5524.54	I
85 h	5539.05	I
2400	5556.47	I
60	5651.98	II
220	5719.99	I
27	5771.66	II
35	5833.99	II
35	5837.14	II
27	5854.51	I
17	5989.33	I
40	5991.51	II
60	6152.57	II
60	6274.78	II
200	6328.52	III
35 h	6400.35	I
35 h	6417.91	I
340	6489.06	I
180	6667.82	I
25	6727.61	II
690	6799.60	I
9 h	7244.41	I
8 h	7305.22	I
10 h	7313.05	I
16 h	7350.04	I
25	7448.28	I
30 h	7527.46	I
750	7699.48	I
100	7971.46	III
70 h	8922.56	II
200	10110.60	III
100	10830.36	III

***Yttrium* Y *Z* = 39**

Intensity	Wavelength/Å	
150	264.64	IV
150	273.03	IV
900	333.09	V
500	333.80	V
400	335.14	V
500	336.62	V
500	339.02	V
500	344.59	V
900	355.86	IV
300	370.42	IV
300	372.05	V
400	379.96	V
500	386.82	IV
300	403.45	V
300	420.74	V
600	425.03	IV
300	473.10	IV
4000	584.98	V
2000	630.97	V
5000	805.20	III
7000	809.92	III
15000	989.21	III
25000	996.37	III
5000	1314.51	III
4000	1334.04	III
4000	2068.98	III
10000	2127.98	III
16000	2191.16	III
350	2243.06	II
10000	2284.34	III
10000	2327.31	III
50	2354.20	I
50000	2367.23	III
40000	2414.64	III
560	2422.20	II
60	2694.21	I
95	2723.00	I
70	2742.53	I
140	2760.10	I
90000	2817.04	III
45	2822.56	I
70	2854.43	II
95	2886.48	I
160	2919.05	I
99000	2946.01	III
390	2948.40	I
350	2964.96	I
480	2974.59	I
750	2984.26	I
140	2996.94	I
130	3021.73	I
190	3045.37	I
95	3095.88	II
110	3173.06	II
220	3179.41	II
70	3191.31	I
2300	3195.62	II
2200	3200.27	II
2200	3203.32	II
3900	3216.69	II
6200	3242.28	II
4700	3327.89	II
85	3388.59	I
85	3412.47	I
170	3485.73	I
1700	3496.09	II
3900	3549.01	II
130	3551.80	I

Atomic

Intensity	Wavelength/Å	
540	3552.69	I
170	3558.76	I
190	3571.43	I
260	3576.05	I
3300	3584.52	II
300	3587.75	I
100	3589.69	I
2800	3592.92	I
10000	3600.73	II
6200	3601.92	II
7800	3611.05	II
4300	3620.94	I
1900	3628.71	II
7800	3633.12	II
3000	3664.61	II
170	3692.53	I
13000	3710.30	II
1200	3747.55	II
10000	3774.33	II
1400	3776.56	II
7400	3788.70	II
1300	3818.35	II
4000	3832.88	II
80	3876.82	I
480	3878.28	II
4400	3950.36	II
3600	3982.60	II
940	4039.83	I
2400	4047.64	I
9400	4077.38	I
2000	4083.71	I
9900	4102.38	I
8900	4128.31	I
7500	4142.85	I
100 h	4157.63	I
2400	4167.52	I
2000	4174.14	I
8000	4177.54	II
160	4217.80	I
280 h	4220.63	I
600	4235.73	II
2200	4235.94	I
300	4251.20	I
360 h	4302.30	I
2800	4309.63	II
110	4330.78	I
440 h	4348.79	I
120	4357.73	I
800	4358.73	II
120	4366.03	I
12000	4374.94	II
150 h	4375.61	I
100	4387.74	I
1800	4398.02	II
890	4422.59	II
100	4443.66	I
130	4446.63	I
170	4475.72	I
180	4476.96	I
160	4477.45	I
110	4487.28	I

Intensity	Wavelength/Å	
300	4487.47	I
500	4505.95	I
890	4527.25	I
440	4527.80	I
100	4544.32	I
100	4559.37	I
130	4596.55	I
95	4604.80	I
2000	4643.70	I
200 h	4658.32	I
2000	4674.84	I
180	4696.81	I
170	4728.53	I
160	4752.79	I
410	4760.98	I
120	4781.04	I
170	4786.89	I
180	4799.30	I
140	4819.64	I
120	4822.13	I
770	4839.87	I
550	4845.68	I
410	4852.69	I
120	4854.25	I
890	4854.87	II
330	4859.84	I
1900	4883.69	II
95	4893.44	I
1100	4900.12	II
100	4906.11	I
150	4921.87	I
120	4974.30	I
100	5006.97	I
75	5070.21	I
75	5072.19	I
1100	5087.42	II
180	5135.20	I
960	5200.41	II
1500	5205.72	II
10000	5238.10	III
180	5240.81	I
75	5380.62	I
220	5402.78	II
90	5424.37	I
190	5438.24	I
710	5466.46	I
100	5468.47	I
240	5497.41	II
300	5503.45	I
250	5509.90	II
120	5521.63	I
740	5527.54	I
120	5544.50	I
180	5577.42	I
620	5581.87	I
120	5606.33	I
560	5630.13	I
120	5644.69	I
120	5648.47	I
740	5662.94	II
90	5675.27	I

Intensity	Wavelength/Å	
160	5706.73	I
90	5743.85	I
75	5765.64	I
100	5781.69	II
120	6009.19	I
120	6023.41	I
120	6135.04	I
150	6138.43	I
1200	6191.73	I
300	6222.59	I
1000	6435.00	I
90	6538.60	I
70	6557.39	I
95	6613.75	II
40	6650.61	I
150	6687.58	I
70	6700.71	I
190	6793.71	I
21	6815.16	I
45	6845.24	I
29	6887.22	I
24 h	6950.31	I
24	6979.88	I
29	7052.94	I
35	7191.66	I
35	7264.17	II
50	7346.46	I
29	7450.30	II
9000	7558.71	III
35	7563.13	I
29	7855.52	I
110	7881.90	II
10000	7991.43	III
24	8344.43	I
10000	8796.21	III
95	8800.62	I
19 h	8835.85	II

Zinc Zn Z = 30

Intensity	Wavelength/Å	
200	425.90	IV
200	428.54	IV
200	430.59	IV
1000	677.63	III
750	677.96	III
200	713.90	III
60	1193.23	II
50	1239.12	IV
50	1249.69	IV
500	1265.74	IV
500	1306.66	IV
200	1456.72	III
200	1459.98	IV
300	1499.42	III
300	1500.42	III
300	1505.92	III
300	1515.85	III
300	1552.30	III
90	1572.99	II
200	1629.19	III
200	1639.33	III
200	1673.05	III
80 d	1735.61	II
100	1767.69	III
100	1797.64	II
100 d	1811.05	II
100 d	1833.57	II
100	1864.12	II
100	1866.08	II
100	1872.13	II
100 d	1918.96	II
100 d	1929.67	II
100	1969.40	II
100	1982.11	II
100	1986.99	II
500	2025.48	II
500	2062.00	II
200	2064.23	II
120	2079.08	I
300	2099.94	II
200	2102.18	II
800 r	2138.56	I
1000	2501.99	II
150	2515.81	I
1000	2557.95	II
300	2582.49	I
200	2608.56	I
300	2608.64	I
200	2670.53	I
300	2684.16	I
300	2712.49	I
200	2756.45	I
300	2770.86	I
300	2770.98	I
400	2800.87	I
100	2801.06	I
200	3035.78	I
200	3072.06	I
300	3196.31	II
500 r	3282.33	I
800	3302.58	I
700 r	3302.94	I
800	3345.02	I
500	3345.57	I
50	3883.34	I
300	4680.14	I
400	4722.15	I
400	4810.53	I
800	4911.62	II
500	4924.03	II
200	5181.98	I
500	5894.33	II
500	6021.18	II
500	6102.49	II
500	6214.61	II
1000 h	6362.34	I
300	7588.5	II
300	7732.5	II
100	11054.25	I
100	13053.63	I
100	13150.59	I
100	14038.70	I
20	16483.45	I
20	16491.98	I

Intensity	Wavelength/Å	
20	16505.23	I
10	24375.02	I

Zirconium Zr Z = 40

Intensity	Wavelength/Å	
500	304.01	V
60	480.66	IV
60	497.23	IV
60	500.22	IV
600	628.66	IV
500	633.56	IV
50	690.39	III
2000	740.61	V
10000	800.00	V
10000	806.89	V
10000	812.05	V
3000	841.40	V
300	863.65	IV
500	864.59	IV
9000	1183.97	IV
9000	1201.77	IV
10000	1219.86	IV
500	1303.93	V
500 p	1323.81	V
1000	1469.47	IV
10000	1546.17	IV
10000	1598.95	IV
5000	1607.95	IV
100	1612.38	III
700	1725.02	V
200	1790.19	III
150	1793.56	III
125	1798.13	III
600	1860.86	V
200	1940.25	III
600	2028.54	V
125	2070.43	III
200	2086.78	III
10000	2091.49	IV
10000	2092.36	IV
600	2132.42	V
10000	2163.68	IV
100	2175.80	III
100	2191.15	III
10000	2286.67	IV
100	2301.60	III
90	2539.65	I
570	2567.64	II
1600	2568.87	II
2100	2571.39	II
250	2620.56	III
200	2643.79	III
150	2664.26	III
1800	2678.63	II
90	2687.75	I
750	2700.13	II
1300	2722.61	II
800	2726.49	II
1400	2734.86	II
1100	2742.56	II
660	2745.86	II
660	2752.21	II
530	2758.81	II

Intensity	Wavelength/Å	
620	2814.90	I
390	2818.74	II
530	2825.56	II
710	2837.23	I
660	2844.58	II
350	2848.52	I
350	2851.97	II
340	2869.81	II
490	2875.98	I
300	2915.99	II
270	2918.24	II
320	2926.99	II
320	2948.94	II
320	2955.78	II
320	2960.87	I
320	2962.68	II
320	2968.96	II
320	2978.05	II
820	2985.39	I
320	3003.74	II
820	3011.75	I
350	3020.47	II
500	3028.04	II
880	3029.52	I
350 d	3036.39	II
690	3054.84	II
690	3106.58	II
350	3120.74	I
500	3129.18	II
500	3129.76	II
350	3132.07	I
690	3138.68	II
540	3164.31	II
880	3165.97	II
880	3182.86	II
540	3191.21	I
540	3212.01	I
760	3214.19	II
630	3231.69	II
630	3234.12	I
760	3241.05	II
1000	3273.05	II
1300	3279.26	II
880	3284.71	II
540	3305.15	II
880	3306.28	II
380	3322.99	II
380	3326.80	II
380	3334.25	II
760	3340.56	II
380	3344.79	II
760	3356.09	II
540	3357.26	II
380	3374.73	II
570	3387.87	II
760	3388.30	II
5700	3391.98	II
570	3393.12	II
570	3404.83	II
760	3410.25	II
380	3414.66	I

Intensity	Wavelength/Å	
1000	3430.53	II
4700	3438.23	II
600	3447.36	I
410	3457.56	II
820	3463.02	II
600	3471.19	I
1200	3479.39	II
1300	3481.15	II
4100	3496.21	II
820	3505.67	II
1000	3509.32	I
2000	3519.60	I
440	3525.81	II
440	3533.22	I
630	3542.62	II
1800	3547.68	I
630	3550.46	I
1800	3551.95	II
2100	3556.60	II
1100	3566.10	I
2100	3572.47	II
1100	3575.79	I
1300	3576.85	II
880	3586.29	I
3500	3601.19	I
690	3611.89	II
1100	3613.10	II
1100	3614.77	II
1100	3623.86	I
1100	3663.65	I
390	3671.27	II
800	3674.72	II
390	3697.46	II
960	3698.17	II
720	3709.26	II
560	3745.98	II
880	3751.60	II
480	3764.39	I
480	3766.72	I
340	3766.82	II
720	3780.54	I
560	3791.40	I
560	3822.41	I
2200	3835.96	I
1300	3836.76	II
550	3843.02	II
550	3847.01	I
550	3849.25	I
2900	3863.87	I
770	3864.34	I
990	3877.60	I
1500	3885.42	I
2900	3890.32	I
2000	3891.38	I
610	3921.79	I
1200	3929.53	I
940	3958.22	II
490	3966.66	I
990	3968.26	I
660	3973.50	I
770	3991.13	II

Intensity	Wavelength/Å	
770	3998.97	II
400	4023.98	I
770	4024.92	I
990	4027.20	I
400	4029.68	II
490	4030.04	I
400	4035.89	I
610	4043.58	I
490	4044.56	I
400	4045.61	II
610	4048.67	II
770	4055.03	I
600	4055.71	I
1500	4064.16	I
2000	4072.70	I
240	4078.31	I
2000	4081.22	I
400	4121.46	I
1200	4149.20	II
400	4161.21	II
400	4166.36	I
660	4187.56	I
400	4194.76	I
610	4199.09	I
610	4201.46	I
610	4208.98	II
400	4213.86	I
2000	4227.76	I
2000	4239.31	I
770	4240.34	I
770	4241.20	I
1200	4241.69	I
550	4282.20	I
550	4294.79	I
550	4341.13	I
1000	4347.89	I
290	4359.74	II
310	4360.81	I
350	4366.45	I
550	4507.12	I
610	4535.75	I
490	4542.22	I
490	4575.52	I
350	4602.57	I
700	4633.98	I
2300	4687.80	I
510	4688.45	I
1900	4710.08	I
1400	4739.48	I
870	4772.31	I
700	4815.63	I
250	5046.58	I
360	5064.91	I
470	5078.25	I
300	5155.45	I
200	5158.00	I
100	5191.60	II
270	5385.14	I
160	5664.51	I
160	5797.74	I
340	5879.80	I

Intensity	Wavelength/Å		Intensity	Wavelength/Å		Intensity	Wavelength/Å		Intensity	Wavelength/Å	
170	6045.85	I	300	6313.02	I	170	7103.72	I	790	8070.08	I
170	6121.91	I	150	6953.84	I	590	7169.09	I	390	8132.99	I
680	6127.44	I	150	6990.84	I	160	7944.61	I	280	8212.53	I
340	6134.55	I	540	7097.70	I	160	8005.27	I			
440	6143.20	I	280	7102.91	I	150	8063.09	I			

Sources of Data for Each Element

Numbers following the element name refer to the references on the following pages.

Actinium: 193
Aluminum: 6,8,81,89,127,144,146,227,228,282
Americium: 92
Antimony: 164,167,194,386,406
Argon: 190,203,204,219,367,368,372,373,374,375,414,421
Arsenic: 163,168,197,244,280
Astatine: 188
Barium: 1,78,111,252,259,277,279
Berkelium: 53,339
Beryllium: 15,44,73,102,115,134,135,171,175,198,335
Bismuth: 1,357,358,359,360,361
Boron: 66,69,74,94,104,171,221,222
Bromine: 42,122,124,139,142,240,243,246,248,249,250,316
Cadmium: 44,285,296,353,399
Calcium: 16,25,70,150,270
Californium: 52,331
Carbon: 22,66,211
Cerium: 1,136,166,261,305
Cesium: 78,82,154,155,200,201,259,263,325
Chlorine: 11,28,30,31,85,233,238,239
Chromium: 1,379,380,412
Cobalt: 1,100,125,159,236,276,291
Copper: 199,273,290,295,324
Curium: 51,332
Dysprosium: 1
Einsteinium: 333
Erbium: 1,301
Europium: 1,312
Fluorine: 68,169,224,225,226
Francium: 408
Gadolinium: 1,46,137,151,152
Gallium: 2,19,62,132,140,141,143,195,281
Germanium: 5,119,293,340,341,342
Gold: 38,72,234,393,395
Hafnium: 1,369,404,410,425
Helium: 16,94,173,183,317
Holmium: 1
Hydrogen: 214
Indium: 1,132,348,349,350,351,352,353,435,436
Iodine: 20,21,58,84,124,153,161,176,184
Iridium: 1
Iron: 56,63,71,101,105,138,174,278,381,382
Krypton: 61,121,123,147,208,232,366,390,409,417,421
Lanthanum: 1,78,79,220,309
Lead: 54,64,106,256,274,297,283,329,330
Lithium: 3,15,17,18,37,44,112,284,321,335
Lutetium: 1,148,310,401
Magnesium: 4,7,49,83,103,128,129,177,217,269,315,335
Manganese: 1,126,385,405,433
Mercury (198): 43,50,69,145,229,242
Mercury (Natural): 34,45,90,117,133,189,235,304,327,328,343
Molybdenum: 1,383,420
Neodymium: 1
Neon: 56,58,69,118,150,230,364,365,371,388,389,400,402,413,430
Neptunium: 93
Nickel: 1,294,415,416,422
Niobium: 1,392,407,431
Nitrogen: 66,107,108,212,213,318
Osmium: 1
Oxygen: 23,24,36,66,69,209,210,215
Palladium: 1,287,424
Phosphorus: 179,180,182,336
Platinum: 1,288
Plutonium: 91
Polonium: 47,48
Potassium: 32,59,60,75,76,86,150,160,172,268,314,322
Praseodymium: 1,149,306,308,337,338
Promethium: 196,260
Protactinium: 96
Radium: 253,254
Radon: 251
Rhenium: 1
Rhodium: 1,396
Rubidium: 12,109,130,241,257,258,262,264
Ruthenium: 1,423
Samarium: 1
Scandium: 1,88,150,298,323
Selenium: 9,80,181,216,245,247,275
Silicon: 87,170,237,292,319,320
Silver: 13,99,255,286,289,363,387,398
Sodium: 178,205,206,207,268,299,334
Strontium: 1,109,110,218,231,265,279,313
Sulfur: 29,144,202,209,210,266
Tantalum: 1,411,426
Technetium: 35
Tellurium: 1,344,345,346,347
Terbium: 1,302
Thallium: 1,195,348,354,355,356
Thorium: 1,97,98,156,157,165,434
Thulium: 1,307
Tin: 187,191,399,423
Titanium: 1,378,427,428
Tungsten: 1
Uranium: 1,303
Vanadium: 1,394,397,432
Xenon: 33,116,118,120,232,384,391,429
Ytterbium: 1,40,192,311
Yttrium: 1,77,265,419
Zinc: 39,55,113,131,185,186,370,376,377
Zirconium: 1,362,403,418

References

1. Meggers, W. F., Corliss, C. H., and Scribner, B. F., *Natl. Bur. Stand. (U.S.) Monogr.*, 145, Washington, D.C., 1975.
2. Aksenov, V. P. and Ryabtsev, A. N., *Opt. Spectrosc.*, 37, 860, 1970.
3. Andersen, N., Bickel, W. S., Carriveau, G. W., Jensen, K., and Veje, E., *Phys. Scr.*, 4, 113, 1971.
4. Andersson, E. and Johannesson, G. A., *Phys. Scr.*, 3, 203, 1971.
5. Andrew, K. L. and Meissner, K. W., *J. Opt. Soc. Am.*, 49, 146, 1959.
6. Artru, M. C. and Brillet, W. U. L., *J. Opt. Soc. Am.*, 64, 1063, 1974.
7. Artru, M. C. and Kaufman, V., *J. Opt. Soc. Am.*, 62, 949, 1972.
8. Artru, M. C. and Kaufman, V., *J. Opt. Soc. Am.*, 65, 594, 1975.
9. Badami, J. S. and Rao, K. R., *Proc. R. Soc. London*, 140(A), 387, 1933.
10. Baird, K. M. and Smith, D. S., *J. Opt. Soc. Am.*, 48, 300, 1958.
11. Bashkin, S. and Martinson, I., *J. Opt. Soc. Am.*, 61, 1686, 1971.
12. Beacham, J. R., Ph.D. thesis, Purdue University, 1970.
13. Benschop, H., Joshi, Y. N., and van Kleef, T. A. M., *Can. J. Phys.*, 53, 700, 1975.
14. Berry, H. G., Bromander, J., and Buchta, R., *Phys. Scr.*, 1, 181, 1970.
15. Berry, H. G., Bromander, J., Martinson, I., and Buchta, R., *Phys. Scr.*, 3, 63, 1971.
16. Berry, H. G., Desesquelles, J., and Dufay, M., *Phys. Rev. Sect A.*, 6, 600, 1972.
17. Berry, H. G., Desesquelles, J., and Dufay, M., *Nucl. Instrum. Methods*, 110, 43, 1973.
18. Berry, H. G., Pinnington, E. H., and Subtil, J. L., *J. Opt. Soc. Am.*, 62, 767, 1972.
19. Bidelman, W. P. and Corliss, C. H., *Astrophys. J.*, 135, 968, 1962.
20. Bloch, L. and Bloch, E., *Ann. Phys.* (Paris), 10(11), 141, 1929.
21. Bloch, L., Bloch, E., and Felici, N., *J. Phys. Radium*, 8, 355, 1937.
22. Bockasten, K., *Ark. Fys.*, 9, 457, 1955.
23. Bockasten, K., Hallin, R., Johansson, K. B., and Tsui, P., *Phys. Lett.* (Netherlands), 8, 181, 1964.
24. Bockasten, K. and Johansson, K. B., *Ark. Fys.*, 38, 563, 1969.
25. Borgstrom, A., *Ark. Fys.*, 38, 243, 1968.
26. Borgstrom, A., *Phys. Scr.*, 3, 157, 1971.
27. Bowen, I. S., *Phys. Rev.*, 29, 231, 1927.
28. Bowen, I. S., *Phys. Rev.*, 31, 34, 1928.
29. Bowen, I. S., *Phys. Rev.*, 39, 8, 1932.
30. Bowen, I. S., *Phys. Rev.*, 45, 401, 1934.
31. Bowen, I. S., *Phys. Rev.*, 46, 377, 1934.
32. Bowen, I. S., *Phys. Rev.*, 46, 791, 1934.
33. Boyce, J. C., *Phys. Rev.*, 49, 730, 1936.
34. Boyce, J. C. and Robinson, H. A., *J. Opt. Soc. Am.*, 26, 133, 1936.
35. Bozman, W. R., Meggers, W. F., and Corliss, C. H., *J. Res. Natl. Bur. Stand. Sect. A*, 71, 547, 1967.
36. Bromander, J., *Ark. Fys.*, 40, 257, 1969.
37. Bromander, J. and Buchta, R., *Phys. Scr.*, 1, 184, 1970.
38. Brown, C. M. and Ginter, M. L., *J. Opt. Soc. Am.*, 68, 243, 1978.
39. Brown, C. M., Tilford, S. G., and Ginter, M. L., *J. Opt. Soc. Am.*, 65, 1404, 1975.
40. Bryant, B. W., *Johns Hopkins Spectroscopic Report* No. 21, 1961.
41. Buchet, J. P., Buchet-Poulizac, M. C., Berry, H. G., and Drake, G. W. F., *Phys. Rev. Sect. A*, 7, 922, 1973.
42. Budhiraja, C. J. and Joshi, Y. N., *Can. J. Phys.*, 49, 391, 1971.
43. Burns, K. and Adams, K. B., *J. Opt. Soc. Am.*, 42, 56, 1952.
44. Burns, K. and Adams, K. B., *J. Opt. Soc. Am.*, 46, 94, 1956.
45. Burns, K., Adams, K. B., and Longwell, J., *J. Opt. Soc. Am.*, 40, 339, 1950.
46. Callahan, W. R., Ph.D. thesis, Johns Hopkins University, 1962.
47. Charles, G. W., *J. Opt. Soc. Am.*, 56, 1292, 1966.
48. Charles, G. W., Hunt, D. J., Pish, G., and Timma, D. L., *J. Opt. Soc. Am.*, 45, 869, 1955.
49. Codling, K., *Proc. Phys. Soc.*, 77, 797, 1961.
50. Comite Consulatif Pour La Definition du Metre, *J. Phys. Chem. Ref. Data*, 3, 852, 1974.
51. Conway, J. G., Blaise, J., and Verges, J., *Spectrochim. Acta Part B*, 31, 31, 1976.
52. Conway, J. G., Worden, E. F., Blaise, J., and Verges, J., *Spectrochim. Acta Part B*, 32, 97, 1977.
53. Conway, J. G., Worden, E. F., Blaise, J., Camus, P., and Verges, J., *Spectrochim. Acta Part B*, 32, 101, 1977.
54. Crooker, A. M., *Can. J. Res. Sect. A*, 14, 115, 1936.
55. Crooker, A. M. and Dick, K. A., *Can. J. Phys.*, 46, 1241, 1968.
56. Crosswhite, H. M., *J. Res. Natl. Bur. Stand. Sect. A*, 79, 17, 1975.
58. Crosswhite, H. M. and Dieke, G. H., *American Institute of Physics Handbook*, Section 7, 1972.
59. de Bruin, T. L., *Z. Phys.*, 38, 94, 1926.
60. de Bruin, T. L., *Z. Phys.*, 53, 658, 1929.
61. de Bruin, T. L., Humphreys, C. J., and Meggers, W. F., *J. Res. Natl. Bur. Stand.*, 11, 409, 1933.
62. Dick, K. A., *J. Opt. Soc. Am.*, 64, 702, 1973.
63. Dobbie, J. C., *Ann. Solar Phys. Observ.* (Cambridge), 5, 1, 1938.
64. Earls, L. T. and Sawyer, R. A., *Phys. Rev.*, 47, 115, 1935.
65. Edlen, B., *Z. Phys.*, 85, 85, 1933.
66. Edlen, B., *Nova Acta Reglae Soc. Sci. Ups.*, (IV) 9, No. 6, 1934.
67. Edlen, B., *Z. Phys.*, 93, 726, 1935.
68. Edlen, B., *Z. Phys.*, 94, 47, 1935.
69. Edlen, B., *Rep. Prog. Phys.*, 26, 181, 1963.
70. Edlen, B. and Risberg, P., *Ark. Fys.*, 10, 553, 1956.
71. Edlen, B. and Swings, P., *Astrophys. J.*, 95, 532, 1942.
72. Ehrhardt, J. C. and Davis, S. P., *J. Opt. Soc. Am.*, 61, 1342, 1971.
73. Eidelsberg, M., *J. Phys. B*, 5, 1031, 1972.
74. Eidelsberg, M., *J. Phys. B*, 7, 1476, 1974.
75. Ekberg, J. O. and Svensson, L. A., *Phys. Scr.*, 2, 283, 1970.
76. Ekefors, E., *Z. Phys.*, 71, 53, 1931.
77. Epstein, G. L. and Reader, J., *J. Opt. Soc. Am.*, 65, 310, 1975.
78. Epstein, G. L. and Reader, J., *J. Opt. Soc. Am.*, 66, 590, 1976.
79. Epstein, G. L. and Reader, J., unpublished.
80. Eriksson, K. B. S., *Phys. Lett. A.*, 41, 97, 1972.
81. Eriksson, K. B. S. and Isberg, H. B. S., *Ark. Fys.*, 23, 527, 1963.
82. Eriksson, K. B. S. and Wenaker, I., *Phys. Scr.*, 1, 21, 1970.
83. Esteva, J. M. and Mehlman, G., *Astrophys. J.*, 193, 747, 1974.
84. Even-Zohar, M. and Fraenkel, B. S., *J. Phys. B*, 5, 1596, 1972.
85. Fawcett, B. C., *J. Phys. B*, 3, 1732, 1970.
86. Fawcett, B. C., Culham Laboratory Report ARU-R4, 1971.
87. Ferner, E., *Ark. Mat. Astron. Fys.*, 28(A), 4, 1941.
88. Fischer, R. A., Knopf, W. C., and Kinney, F. E., *Astrophys. J.*, 130, 683, 1959.
89. Fowler, A., *Report on Series in Line Spectra*, Fleetway Press, London, 1922.
90. Fowles, G. R., *J. Opt. Soc. Am.*, 44, 760, 1954.
91. Fred, M., *Argonne Natl. Lab.*, unpublished, 1977.
92. Fred, M. and Tomkins, F. S., *J. Opt. Soc. Am.*, 47, 1076, 1957.
93. Fred, M., Tomkins, F. S., Blaise, J. E., Camus, P., and Verges, J., Argonne National Laboratory Report No. 76–68, 1976.
94. Garcia, J. D. and Mack, J. E., *J. Opt. Soc. Am.*, 55, 654, 1965.
96. Giacchetti, A., *Argonne Natl. Lab.*, unpublished, 1975.
97. Giacchetti, A., Blaise, J., Corliss, C. H., and Zalubas, R., *J. Res. Natl. Bur. Stand. Sect. A*, 78, 247, 1974.
98. Giacchetti, A., Stanley, R. W., and Zalubas, R., *J. Opt. Soc. Am.*, 69, 474, 1970.
99. Gilbert, W. P., *Phys. Rev.*, 47, 847, 1935.
100. Gilroy, H. T., *Phys. Rev.*, 38, 2217, 1931.
101. Glad, S., *Ark. Fys.*, 10, 291, 1956.
102. Goldsmith, S., *J. Phys. B*, 2, 1075, 1969.
103. Goorvitch, D., Mehlmam-Balloffet, G., and Valero, F. P. J., *J. Opt. Soc. Am.*, 60, 1458, 1970.
104. Goorvitch, D. and Valero, F. P. J., *Astrophys.* J., 171, 643, 1972.
105. Green, L. C., *Phys. Rev.*, 55, 1209, 1939.
106. Gutman, F., *Diss. Abstr. Int.* B, 31, 363, 1970.
107. Hallin, R., *Ark. Fys.*, 31, 511, 1966.
108. Hallin, R., *Ark. Fys.*, 32, 201, 1966.
109. Hansen, J. E. and Persson, W., *J. Opt. Soc. Am.*, 64, 696, 1974.
110. Hansen, J. E. and Persson, W., *Phys. Scr.*, 13, 166, 1976.
111. Hellintin, P., *Phys. Scr.*, 13, 155, 1976.
112. Herzberg, G. and Moore, H. R., *Can. J. Phys.*, 37, 1293, 1959.
113. Hetzler, C. W., Boreman, R. W., and Burns, K., *Phys. Rev.*, 48, 656, 1935.
114. Holmstrom, J. E. and Johansson, L., *Ark. Fys.*, 40, 133, 1969.

115. Hontzeas, S., Martinson, I., Erman, P., and Buchta, R., *Nucl. Instrum. Methods*, 110, 51, 1973.
116. Humphreys, C. J., *J. Res. Natl. Bur. Stand.*, 22, 19, 1939.
117. Humphreys, C. J., *J. Opt. Soc. Am.*, 43, 1027, 1953.
118. Humphreys, C. J., *J. Phys. Chem. Ref. Data*, 2, 519, 1973.
119. Humphreys, C. J. and Andrew, K. L., *J. Opt. Soc. Am.*, 54, 1134, 1964.
120. Humphreys, C. J. and Meggers, W. F., *J. Res. Natl. Bur. Stand.*, 10, 139, 1933.
121. Humphreys, C. J. and Paul, E., Jr., *J. Opt. Soc. Am.*, 60, 200, 1970.
122. Humphreys, C. J. and Paul, E., Jr., *J. Opt. Soc. Am.*, 62, 432, 1972.
123. Humphreys, C. J., Paul, E., Jr., Cowan, R. D., and Andrew, K. L., *J. Opt. Soc. Am.*, 57, 855, 1967.
124. Humphreys, C. J., Paul, E., Jr., and Minnhagen, L., *J. Opt. Soc. Am.*, 61, 110, 1971.
125. Iglesias, L., Inst. of Optics, Madrid, unpublished, 1977.
126. Iglesias, L. and Velasco, R., *Publ. Inst. Opt. Madrid*, No. 23, 1964.
127. Isberg, B., *Ark. Fys.*, 35, 551, 1967.
128. Johannesson, G. A., Lundstrom, T., and Minnhagen, L., *Phys. Scr.*, 6, 129, 1972.
129. Johannesson, G. A. and Lundstrom, T., *Phys. Scr.*, 8, 53, 1973.
130. Johansson, I. *Ark. Fys.*, 20, 135, 1961.
131. Johansson, I. and Contreras, R., *Ark. Fys.*, 37, 513, 1968.
132. Johansson, I. and Litzen, U., *Ark. Fys.*, 34, 573, 1967.
133. Johansson, I. and Svensson, K. F., *Ark. Fys.*, 16, 353, 1960.
134. Johansson, L., *Ark. Fys.*, 20, 489, 1961.
135. Johansson, L., *Ark. Fys.*, 23, 119, 1963.
136. Johansson, S. and Litzen, U., *Phys. Scr.*, 6, 139, 1972.
137. Johansson, S. and Litzen, U., *Phys. Scr.*, 8, 43, 1973.
138. Johansson, S. and Litzen, U., *Phys. Scr.*, 10, 121, 1974.
139. Joshi, Y. N., St. Francis Xavier Univ., Nova Scotia, unpublished.
140. Joshi, Y. N., Bhatia, K. S., and Jones, W. E., *Sci. Light Tokyo*, 21, 113, 1972.
141. Joshi, Y. N., Bhatia, K. S., and Jones, W. E., *Spectrochim. Acta Part B*, 28, 149, 1973.
142. Joshi, Y. N. and Budhiraja, C. J., *Can. J. Phys.*, 49, 670, 1971.
143. Joshi, Y. N. and van Kleef, T. A. M., *Can. J. Phys.*, 52, 1891, 1974.
144. Kaufman, V., *Natl. Bur. Stand.*, unpublished.
145. Kaufman, V., *J, Opt. Soc. Am.*, 52, 866, 1962.
146. Kaufman, V., Artru, M. C., and Brillet, W. U. L., *J. Opt. Soc. Am.*, 64, 197, 1974.
147. Kaufman, V. and Humphreys, C. J., *J. Opt. Soc. Am.*, 59, 1614, 1969.
148. Kaufman, V. and Sugar, J., *J. Opt. Soc. Am.*, 61, 1693, 1971.
149. Kaufman, V. and Sugar, J., *J. Res. Natl. Bur. Stand. Sect. A*, 71, 583, 1967.
150. Kelly, R. L. and Palumbo, L. J., *Naval Research Laboratory Report 7599*, Washington, DC., 1973.
151. Kielkopf, J. F., *Univ. of Louisville*, unpublished. 1975.
152. Kielkopf, J. F., *Univ. of Louisville*, unpublished, 1976.
153. Kiess, C. C. and Corliss, C. H., *J. Res. Natl. Bur. Stand. Sect. A*, 63, 1, 1959.
154. Kleiman, H., *J. Opt. Soc. Am.*, 52, 441, 1962.
155. Eriksson, K. B., Johansson, I., and Norlen, G., *Ark. Fys.*, 28, 233, 1964.
156. Klinkenberg, P. F. A., *Physica*, 15, 774, 1949.
157. Klinkenberg, P. F. A., *Physica*, 16, 618, 1950.
158. Krishnamurty, S. G., *Proc. Phys. Soc. London*, 48, 277, 1936.
159. Kruger, P. G. and Gilroy, H. T., *Phys. Rev.*, 48, 720, 1935.
160. Kruger, P. G. and Pattin, H. S., *Phys. Rev.*, 52, 621, 1937.
161. Lacroute, P., *Ann. Phys.* (Paris), 3, 5, 1935.
162. Lang, R. J., *Phys. Rev.*, 30, 762, 1927.
163. Lang, R. J., *Phys. Rev.*, 32, 737, 1928.
164. Lang, R. J., *Phys. Rev.*, 35, 445, 1930.
165. Lang, R. J., *Can. J. Res. Sect. A*, 14, 43, 1936.
166. Lang, R. J., *Can. J. Res. Sect. A*, 14, 127, 1936.
167. Lang, R. J. and Vestine, E. H., *Phys. Rev.*, 42, 233, 1932.
168. Li, H. and Andrew, K. L., *J. Opt. Soc. Am.*, 61, 96, 1971.
169. Liden, K., *Ark. Fys.*, 1, 229, 1949.
170. Litzen, U., *Ark. Fys.*, 28, 239, 1965.
171. Litzen, U., *Phys. Scr.*, 1, 251, 1970.
172. Litzen, U., *Phys. Scr.*, 1, 253, 1970.
173. Litzen, U., *Phys. Scr.*, 2, 103, 1970.
174. Litzen, U. and Verges, I., *Phys. Scr.*, 13, 240, 1976.
175. Lofstrand, B., *Phys. Scr.*, 8, 57, 1973.
176. Luc-Koenig, E., Morillon, C., and Verges, J., *Phys. Scr.*, 12, 199, 1975.
177. Lundstrom, T., *Phys. Scr.*, 7, 62, 1973.
178. Lundstrom, T. and Minnhagen, L., *Phys. Scr.*, 5, 243, 1972.
179. Magnusson, C. E. and Zetterberg, P. O., *Phys. Scr.*, 10, 177, 1974.
180. Magnusson, C. E., and Zetterberg, P. O., *Phys. Scr.*, 15, 237, 1977.
181. Martin, D. C., *Phys. Rev.*, 48. 938, 1935.
182. Svendenius, N., *Phys. Scr.*, 22, 240, 1980.
183. Martin, W. C., *J. Res. Natl. Bur. Stand. Sect. A*, 64, 19, 1960.
184. Martin, W. C. and Corliss, C. H., *J. Res. Natl. Bur. Stand. Sect. A*, 64, 443, 1960.
185. Martin, W. C. and Kaufman, V., *J. Res. Natl. Bur. Stand. Sect. A*, 74, 11, 1970.
186. Martin, W. C. and Kaufman, V., *J. Opt. Soc. Am.*, 60, 1096, 1970.
187. McCormick, W. W. and Sawyer, R. A., *Phys. Rev.*, 54, 71, 1938.
188. McLaughlin, R., *J. Opt. Soc. Am.*, 54, 965, 1964.
189. McLennan, J. C., McLay, A. B., and Crawford, M. F., *Proc. R. Soc. London Ser. A*, 134, 41, 1931.
190. Meissner, K. W., *Z. Phys.*, 39, 172, 1926.
191. Meggers, W. F., *J. Res. Natl. Bur. Stand.*, 24, 153, 1940.
192. Meggers, W. F. and Corliss, C. H., *J. Res. Natl. Bur. Stand. Sect. A*, 70, 63, 1966.
193. Meggers, W. F., Fred, M., and Tomkins, F. S., *J. Res. Natl. Bur. Stand.*, 58, 297, 1957.
194. Meggers, W. F. and Humphreys, C. J., *J. Res. Natl. Bur. Stand.*, 28, 463, 1942.
195. Meggers, W. F. and Murphy, R. J., *J. Res. Natl. Bur. Stand.*, 48, 334, 1952.
196. Meggers, W. F., Scribner, B. F., and Bozman, W. R., *J. Res. Natl. Bur. Stand.*, 46, 85, 1951.
197. Meggers, W. F., Shenstone, A. G., and Moore, C. E., *J. Res. Natl. Bur. Stand.*, 45, 346, 1950.
198. Mehlman, G. and Esteva, J. M., *Astrophys. J.*, 188, 191, 1974.
199. Meinders, E., *Physica*, 84(C), 117, 1976.
200. Sansonetti, C. J., Dissertation, Purdue University, 1981.
201. Sansonetti, C. J., *Natl. Bur. Stand. (U.S.)*, unpublished.
202. Millikan, R. A. and Bowen, I. S., *Phys. Rev.*, 25, 600, 1925.
203. Minnhagen, L., *J. Opt. Soc. Am.*, 61, 1257, 1925.
204. Minnhagen, L., *J. Opt. Soc. Am.*, 63, 1185, 1973.
205. Minnhagen, L., *Phys. Scr.*, 11, 38, 1975.
206. Minnhagen, L., *J. Opt. Soc. Am.*, 66, 659, 1976.
207. Minnhagen, L. and Nietsche, H., *Phys. Scr.*, 5, 237, 1972.
208. Minnhagen. L., Strihed, H., and Petersson, B., *Ark. Fys.*, 39, 471, 1969.
209. Moore, C. E., *Natl. Bur. Stand. (U.S.) Circ.*, 488, 1950.
210. Moore, C. E., *Revised Multiplet Table*, Princeton University Observatory No. 20, 1945.
211. Moore, C. E., National Standard Reference Data Series - National Bureau of Standards 3, Sect. 3, 1970.
212. Moore, C. E., National Standard Reference Data Series - National Bureau of Standards 3, Sect. 4, 1971.
213. Moore, C. E., National Standard Reference Data Series - National Bureau of Standards 3, Sect. 5, 1975.
214. Moore, C. E., National Standard Reference Data Series - National Bureau of Standards 3, Sect. 6, 1972.
215. Moore, C. E., National Standard Reference Data Series - National Bureau of Standards 3, Sect. 7, 1975.
216. Morillon, C. and Verges, J., *Phys. Scr.*, 10, 227, 1974.
217. Newsom, G. H., *Astrophys. J.*, 166, 243, 1971.
218. Newsom, G. H., O'Connor, S., and Learner, R. C. M., *J. Phys. B*, 6, 2162, 1973.
219. Norlen, G., *Phys. Scr.*, 8, 249, 1973.
220. Odabasi, H., *J. Opt. Soc. Am.*, 57, 1459, 1967.
221. Olme, A., *Ark. Fys.*, 40, 35, 1969.
222. Olme, A., *Phys. Scr.*, 1, 256, 1970.
223. Johansson, S., and Litzen, U., *J. Opt. Soc. Am.*, 61, 1427, 1971.
224. Palenius, H. P., *Ark. Fys.*, 39, 15, 1969.
225. Palenius, H. P., *Phys. Scr.*, 1, 113, 1970.
226. Palenius, H. P., *Univ. of Lund, Sweden*, unpublished.
227. Paschen, F., *Ann. Phys.*, Series 5, 12, 509, 1932.
228. Paschen, F. and Ritschl, R., *Ann. Phys.*, Series 5, 18, 867, 1933.

229. Peck, E. R., Khanna, B. N., and Anderholm, N. C., *J. Opt. Soc. Am.*, 52, 53, 1962.
230. Persson, W., *Phys. Scr.*, 3, 133, 1971.
231. Persson, W. and Valind, S., *Phys. Scr.*, 5, 187, 1972.
232. Petersson, B., *Ark. Fys.*, 27, 317, 1964.
233. Phillips, L. W. and Parker, W. L., *Phys. Rev.*, 60, 301, 1941.
234. Platt, J. R. and Sawyer, R. A., *Phys. Rev.*, 60, 866, 1941.
235. Plyer, E. K., Blaine, L. R., and Tidwell, E., *J. Res. Natl. Bur. Stand.*, 55, 279, 1955.
236. Poppe, R., van Kleef, T. A. M., and Raassen, A. J. J., *Physica*, 77, 165, 1974.
237. Radziemski, L. J., Jr. and Andrew, K. L., *J. Opt. Soc. Am.*, 55, 474, 1965.
238. Radziemski, L. J., Jr. and Kaufman, V., *J. Opt. Soc. Am.*, 59, 424, 1969.
239. Radziemski, L. J., Jr. and Kaufman, V., *J. Opt. Soc. Am.*, 64, 366, 1974.
240. Ramanadham, R. and Rao, K. R., *Indian J. Phys.*, 18, 317, 1944.
241. Ramb, R., *Ann. Phys.*, 10, 311, 1931.
242. Rank, D. H., Bennett, J. M., and Bennett, H. E., *J. Opt. Soc. Am.*, 40, 477, 1950.
243. Rao, A. S. and Krishnamurty, S. G., *Proc. Phys. Soc. London*, 46, 531, 1943.
244. Rao, K. R., *Proc. R. Soc. London, Ser. A*, 134, 604, 1932.
245. Rao, K. R. and Badami, J. S., *Proc. R. Soc. London Ser. A*, 131, 154, 1931.
246. Rao, K. R. and Krishnamurty, S. G., *Proc. R. Soc. London Ser. A*, 161, 38, 1937.
247. Rao, K. R. and Murti, S. G. K., *Proc. R. Soc. London* Ser. A, 145, 681, 1934.
248. Rao, Y. B., *Indian J. Phys.*, 32, 497, 1958.
249. Rao, Y. B., *Indian J. Phys.*, 33, 546, 1959.
250. Rao, Y. B., *Indian J. Phys.*, 35, 386, 1961.
251. Rasmussen, E., *Z. Phys.*, 80, 726, 1933.
252. Rasmussen, E., *Z. Phys.*, 83, 404, 1933.
253. Rasmussen, E., *Z. Phys.*, 86, 24, 1934.
254. Rasmussen, E., *Z. Phys.*, 87, 607, 1934.
255. Rasmussen, E., *Phys. Rev.*, 57, 840, 1940.
256. Rau, A. S. and Narayan, A. L., *Z. Phys.*, 59, 687, 1930.
257. Reader, J., *J. Opt. Soc. Am.*, 65, 286, 1975.
258. Reader, J., *J. Opt. Soc. Am.*, 65, 988, 1975.
259. Reader, J., *J. Opt Soc. Am.*, 73, 349, 1983.
260. Reader, J. and Davis, S., *J. Res. Natl. Bur. Stand. Sect. A*, 71, 587, 1967, and unpublished.
261. Reader, J. and Ekberg, J. O., *J. Opt. Soc. Am.*, 62, 464, 1972.
262. Reader, J. and Epstein, G. L., *J. Opt. Soc. Am.*, 62, 1467, 1972.
263. Reader, J. and Epstein, G. L., *J. Opt. Soc. Am.*, 65, 638, 1975.
264. Reader, J. and Epstein, G. L., *Natl. Bur. Stand.*, unpublished.
265. Reader, J., Epstein, G. L., and Ekberg, J. O., *J. Opt. Soc. Am.*, 62, 273, 1972.
266. Kaufman, V., *Phys. Scr.*, 26, 439, 1982.
267. Ricard, R., Givord, M., and George, F., *C. R. Acad. Sci. Paris*, 205, 1229, 1937.
268. Risberg, P., *Ark. Fys.*, 10, 583, 1956.
269. Risberg, G., *Ark. Fys.*, 28, 381, 1965.
270. Risberg, G., *Ark. Fys.*, 37, 231, 1968.
271. Robinson, H. A., *Phys. Rev.*, 49, 297, 1936.
272. Robinson, H. A., *Phys. Rev.*, 50, 99, 1936.
273. Ross, C. B., Jr., Doctoral dissertation, Purdue University, 1969.
274. Ross, C. B., Wood, D. R., and Scholl, P. S., *J. Opt. Soc. Am.*, 66, 36, 1976.
275. Ruedy, J. E. and Gibbs, R. C., *Phys. Rev.*, 46, 880, 1934.
276. Russell, H. N., King, R. B., and Moore, C. E., *Phys. Rev.*, 58, 407, 1940.
277. Russell, H. N. and Moore, C. E., *J. Res. Natl. Bur. Stand.*, 55, 299, 1955.
278. Russell, H. N., Moore, C. E., and Weeks, D. W., *Trans. Am. Philos. Soc.*, 34(2), 111, 1944.
279. Saunders, F., Schneider, E., and Buckingham, E., *Proc. Natl. Acad. Sci.*, 20, 291, 1934.
280. Sawyer, R. A. and Humphreys, C. J., *Phys. Rev.*, 32, 583, 1928.
281. Sawyer, R. A. and Lang, R. J., *Phys. Rev.*, 34, 712, 1929.
282. Sawyer, R. A. and Paschen, F., *Ann. Phys.*, 84(4),1, 1927.
283. Scholl, P. S., M.S. thesis, Wright State Univ., 1975.
284. Schurmann, D., *Z. Phys.*, 17, 4, 1975.
285. Seguier, J., *C. R. Acad. Sci. Paris*, 256, 1703, 1963.
286. Shenstone, A. G., *Phys. Rev.*, 31, 317, 1928.
287. Shenstone, A. G., *Phys. Rev.*, 32, 30, 1928.
288. Shenstone, A. G., *Trans. R. Soc. London*, 237(A), 57, 1938.
289. Shenstone, A. G., *Phys. Rev.*, 57, 894, 1940.
290. Shenstone, A. G. *Philos. Trans. R. Soc. London Ser. A*, 241, 297, 1948.
291. Shenstone, A. G., *Can. J. Phys.*, 38, 677, 1960.
292. Shenstone, A. G., *Proc. R. Soc. London*, 261(A), 153, 1961.
293. Shenstone, A. G., *Proc. R. Soc. London*, 276(A), 293, 1963.
294. Shenstone, A. G., *J. Res. Natl. Bur. Stand. Sect. A*, 74, 801, 1970.
295. Shenstone, A. G., *J. Res. Natl. Bur. Stand. Sect. A*, 79, 497, 1975.
296. Shenstone, A. G. and Pittenger, J. T., *J. Opt. Soc. Am.*, 39, 219, 1949.
297. Smith, S., *Phys. Rev.*, 36, 1, 1930.
298. Smitt, R., *Phys. Scr.*, 8, 292, 1973.
299. Soderqvist, J., *Ark. Mat. Astronom. Fys.*, 32(A), 1, 1946.
300. Sommer, L. A., *Ann. Phys.*, 75, 163, 1924.
301. Spector, N., *J. Opt. Soc. Am.*, 63, 358, 1973.
302. Spector, N. and Sugar, J., *J. Opt. Soc. Am.*, 66, 436, 1976.
303. Steinhaus, D. W., Radziemski, L. J., Jr., and Blaise, J., *Los Alamos Sci. Lab.*, unpublished, 1975.
304. Subbaraya, T. S., *Z. Phys.*, 78, 541, 1932.
305. Sugar, J., *J. Opt. Soc. Am.*, 55, 33, 1965.
306. Sugar, J., *J. Res. Natl. Bur. Stand. Sect. A*, 73, 333, 1969.
307. Sugar, J., *J. Opt. Soc. Am.*, 60, 454, 1970.
308. Sugar, J., *J. Res. Natl. Bur. Stand. Sect. A*, 78, 555, 1974.
309. Sugar, J. and Kaufman, V., *J. Opt. Soc. Am.*, 55, 1283, 1965.
310. Sugar, J. and Kaufman, V., *J. Opt. Soc. Am.*, 62, 562, 1972.
311. Sugar, J., Kaufman, V., and Spector, N., *J. Res. Natl. Bur. Stand., Sect. A*, 83, 233, 1978.
312. Sugar, J. and Spector, N., *J. Opt. Soc. Am.*, 64, 1484, 1974.
313. Sullivan, F. J. *Univ. Pittsburgh Bull.*, 35, 1, 1938.
314. Svensson, L. A. and Ekberg, J. O., *Ark. Fys.*, 37, 65, 1968.
315. Swensson, J. W. and Risberg, G., *Ark. Fys.*, 31, 237, 1966.
316. Tech, J. L., *J. Res. Natl. Bur. Stand. Sect. A*, 67, 505, 1963.
317. Tech, J. L. and Ward, J. F., *Phys. Rev. Lett.*, 27, 367, 1971.
318. Tilford, S. G., *J. Opt. Soc. Am.*, 53, 1051, 1963.
319. Toresson, Y. G., *Ark. Fys.*, 17, 179, 1960.
320. Toresson, Y. G., *Ark. Fys.*, 18, 389, 1960.
321. Toresson, Y. G. and Edlen, B., *Ark. Fys.*, 23, 117, 1963.
322. Tsien, W. Z., *Chin. J. Phys.*, Peiping, 3, 117, 1939.
323. van Deurzen, C. H. H., Conway, J., and Davis, S. P., *J. Opt. Soc. Am.*, 63, 158, 1973.
324. van Kleef, T. A. M., Raassen, A. J. J., and Joshi, Y. N., *Physica*, 84(C), 401, 1976.
325. Sansonetti, C. J., Andrew, K. L., and Verges, J., *J. Opt. Soc. Am.*, 71, 423, 1981.
326. Wheatley, M. A. and Sawyer, R. A., *Phys. Rev.*, 61, 591, 1942.
327. Wilkinson, P. G., *J. Opt. Soc. Am.*, 45, 862, 1955.
328. Wilkinson, P. G and Andrew, K. L., *J. Opt. Soc. Am.*, 53, 710, 1963.
329. Wood, D. and Andrew, K. L., *J. Opt. Soc. Am.*, 58, 818, 1968.
330. Wood, D. R., Ron, C. B., Scholl, P. S., and Hoke, M., *J. Opt. Soc. Am.*, 64, 1159, 1974.
331. Worden, E. F. and Conway, J. G., *Lawrence Livermore Lab.*, unpublished, 1977.
332. Worden, E. F., Hulet, E. K., Gutmacher, R. G., Conway, J. G., *At. Data Nucl. Data Tables*, 18, 459, 1976.
333. Worden, E. F., Lougheed, R. W., Gutmacher, R. G., and Conway, J. G., *J. Opt. Soc. Am.*, 64, 77, 1974.
334. Wu, C. M., Ph.D. thesis, University of British Columbia, 1971.
335. Zaidel, A. N., Prokofev, V. K., Raiskii, S. M., Slavnyi, V. A., and Schreider, E. Y., *Tables of Spectral Lines*, 3rd ed., Plenum, New York, 1970.
336. Zetterberg, P. O. and Magnusson, C. E., *Phys. Scr.*, 15, 189, 1977.
337. Sugar, J., *J. Opt. Soc. Am.*, 55, 1058, 1965.
338. Sugar, J., *J. Opt. Soc. Am.*, 61, 727, 1971.
339. Worden, E. F., and Conway, J. G., *At. Data Nucl. Data Tables*, 22, 329, 1978.
340. Kaufman, V. and Edlen, B., *J. Phys. Chem. Ref. Data*, 3, 825, 1974.
341. Lang, R. J., *Phys. Rev.*, 34, 697, 1929.
342. Ryabtsev, A. N., *Opt. Spectros.*, 39, 455, 1975.

343. Foster, E. W., *Proc. R. Soc. London*, 200(A), 429, 1950.
344. Morillon, C. and Verges, J., *Phys. Scr.*, 12, 129, 1975.
345. Ruedy, J. E., *Phys. Rev.*, 41, 588, 1932.
346. McLennan, J. C., McLay, A. B., and McLeod, J. H., *Philos. Mag.*, 4, 486, 1927.
347. Handrup, M. B. and Mack, J. E., *Physica*, 30, 1245, 1964.
348. Clearman, H. E., *J. Opt. Soc. Am.*, 42, 373, 1952.
349. Paschen, F., *Ann. Physik*, 424, 148, 1938.
350. Paschen, F. and Campbell, J. S., *Ann. Phys.*, 31(5), 29, 1938.
351. Nodwell, R., *Univ. of British Columbia, Vancouver*, unpublished, 1955.
352. Gibbs, R. C. and White, H. E., *Phys. Rev.*, 31, 776, 1928.
353. Green, M., *Phys. Rev.*, 60, 117, 1941.
354. Ellis, C. B. and Sawyer, R. A., *Phys. Rev.*, 49, 145, 1936.
355. McLennan, J. C., McLay, A. B., and Crawford, M. F., *Proc. R. Soc. London Ser. A*, 125, 50, 1929.
356. Mack, J. E. and Fromer, M., *Phys. Rev.*, 48, 346, 1935.
357. Humphreys, C. J. and Paul, E., U.S. Nav. Ord. Lab., Navord Rep. 4589, 25, 1956.
358. Walters, F. M., *Sci. Pap. Bur. Stand.*, 17, 161, 1921.
359. Crawford, M. F. and McLay, A. B., *Proc. R. Soc. London Ser. A*, 143, 540, 1934.
360. McLay, A. D. and Crawford, M. F., *Phys. Rev.*, 44, 986, 1933.
361. Schoepfle, G. K., *Phys. Rev.*, 47, 232, 1935.
362. Acquista, N., and Reader, J., *J. Opt. Soc. Am.*, 70, 789, 1980.
363. Benschop, H., Joshi, Y. N., and van Kleef, T. A. M., *Can. J. Phys.*, 53, 498, 1975.
364. Bockasten, K., Hallin, R., and Hughes, T. P., *Proc. Phys. Soc.*, 81, 522, 1963.
365. Boyce, J. C., *Phys. Rev.*, 46, 378, 1934.
366. Boyce, J. C., *Phys. Rev.*, 47, 718, 1935.
367. Boyce, J. C., *Phys. Rev.*, 48, 396, 1935.
368. Boyce, J. C., *Phys. Rev.*, 49, 351, 1936.
369. Corliss, C. H. and Meggers, W. F., *J. Res. Natl. Bur. Stand.*, 61, 269, 1958.
370. Crooker, A. M. and Dick, K. A., *Can. J. Phys.*, 42, 766, 1964.
371. De Bruin, T. L., *Z. Physik*, 77, 505, 1932.
372. De Bruin, T. L., *Proc. Roy. Acad. Amsterdam*, 36, 727, 1933.
373. De Bruin, T. L., *Zeeman Verhandelingen*, (The Hague), 1935, p. 415.
374. De Bruin, T. L., *Physica*, 3, 809, 1936.
375. De Bruin, T. L., Proc. Roy. Acad. Amsterdam, 40, 339, 1937.
376. Dick, K. A., *Can. J. Phys.*, 46, 1291, 1968.
377. Dick, K. A., unpublished, 1978.
378. Edlen, B. and Swensson, J. W., *Phys. Scr.*, 12, 21, 1975.
379. Ekberg, J. O., *Phys. Scr.*, 7, 55, 1973.
380. Ekberg, J. O., *Phys. Scr.*, 7, 59, 1973.
381. Ekberg, J. O., *Phys. Scr.*, 12, 42, 1975.
382. Ekberg, J. O. and Edlen, B., *Phys. Scr.*, 18, 107, 1978.
383. Eliason, A. Y., *Phys. Rev.*, 43, 745, 1933.
384. Gallardo, M., Massone, C. A., Tagliaferri, A. A., Garavaglia, M., and Persson, W., *Phys. Scr.*, 19, 538, 1979.
385. Garcia-Riquelme, O., *Optica Pura Y Aplicada*, 1, 53, 1968.
386. Gibbs, R. C., Vieweg, A. M., and Gartlein, C. W., *Phys. Rev.*, 34, 406, 1929.
387. Gilbert, W. P., *Phys. Rev.*, 48, 338, 1935.
388. Goldsmith, S. and Kaufman, A. S., *Proc. Phys. Soc.*, 81, 544, 1963.
389. Hermansdorfer, H., *J. Opt. Soc. Am.*, 62, 1149, 1972.
390. Humphreys, C. J., *Phys. Rev.*, 47, 712, 1935.
391. Humphreys, C. J., *J. Res. Natl. Bur. Stand.*, 16, 639, 1936.
392. Iglesias, L., *J. Opt. Soc. Am.*, 45, 856, 1955.
393. Iglesias, L., *J. Res. Natl. Bur. Stand.*, 64A, 481, 1960.
394. Iglesias, L., Anales Fisica Y Quimica, 58A, 191, 1962.
395. Iglesias, L., *J. Res. Natl. Bur. Stand.*, 70A, 465, 1966.
396. Iglesias, L., *Can. J. Phys.*, 44, 895, 1966.
397. Iglesias, L., *J. Res. Natl. Bur. Stand.*, 72A, 295, 1968.
398. Joshi, Y. N., *Can. Spectrosc.*, 15, 96, 1970.
399. Joshi, Y. N. and van Kleef, T. A. M., *Can. J. Phys.*, 55, 714, 1977.
400. Kaufman, A. S., Hughes, T. P., and Williams, R. V., *Proc. Phys. Soc.*, 76, 17, 1960.
401. Kaufman, V. and Sugar, J., *J. Opt. Soc. Am.*, 68, 1529, 1978.
402. Keussler, V., *Z. Physik, 85*, 1, 1933.
403. Kiess, C. C., *J. Res. Natl. Bur. Stand.*, 56, 167, 1956.
404. Klinkenberg, P. F. A., van Kleef, T. A. M., and Noorman, P. E., *Physica*, 27, 1177, 1961.
405. Kovalev, V. I., Romanos, A. A., and Ryabtsev, A. N., *Opt. Spectrosc.*, 43, 10, 1977.
406. Lang, R. J., *Proc. Natl. Acad. Sci.*, 13, 341, 1927.
407. Lang, R. J., *Zeeman Verhandelingen*, (The Hague), 44, 1935.
408. Liberman, S., et al., *C. R. Acad. Sci.* (Paris), 286, 253, 1978.
409. Livingston, A. E., *J. Phys.*, B9, L215, 1976.
410. Meijer, F. G., *Physica*, 72, 431, 1974.
411. Meijer, F. G. and Metsch, B. C., *Physica*, 94C, 259, 1978.
412. Moore, F. L., thesis, Princeton, 1949.
413. Paul, F. W. and Polster, H. D., *Phys. Rev.*, 59, 424, 1941.
414. Phillips, L. W. and Parker, W. L., *Phys. Rev.*, 60, 301, 1941.
415. Poppe, R., *Physica*, 81C, 351, 1976.
416. Raassen, A. J. J., van Kleef, T. A. M., and Metsch, B. C., *Physica*, 84C, 133, 1976.
417. Rao, A. B. and Krishnamurty, S. G., *Proc. Phys. Soc. (London)*, 51, 772, 1939.
418. Reader, J. and Acquista, N., *J. Opt. Soc. Am.*, 69, 239, 1979.
419. Reader, J. and Epstein, G. L., *J. Opt. Soc. Am.*, 62, 619, 1972.
420. Rico, F. R., *Anales, Real Soc. Esp. Fis. Quim.*, 61, 103, 1965.
421. Schonheit, E., *Optik*, 23, 409, 1966.
422. Shenstone, A. G., *J. Opt. Soc. Am.*, 44, 749, 1954.
423. Shenstone, A. G., unpublished, 1958.
424. Shenstone, A. G., *J. Res. Natl. Bur. Stand.*, 67A, 87, 1963.
425. Sugar, J. and Kaufman, V., *J. Opt. Soc. Am.*, 64, 1656, 1974.
426. Sugar, J. and Kaufman, V., *Phys. Rev.*, C12, 1336, 1975.
427. Svensson, L. A., *Phys. Scr.*, 13, 235, 1976.
428. Swensson, J. W. and Edlen, B., *Phys. Scr.*, 9, 335, 1974.
429. Tagliaferri, A. A., Gallego Lluesma, E., Garavaglia, M., Gallardo, M., and Massone, C. A., *Optica Pura Y Aplica*, 7, 89.
430. Tilford, S. G. and Giddings, L. E., *Astrophys. J.*, 141, 1222, 1965.
431. Trawick, M. W., *Phys. Rev.*, 46, 63, 1934.
432. Van Deurzen, C. H. H., *J. Opt. Soc. Am.*, 67, 476, 1977.
433. Yarosewick, S. L. and Moore, F. L., *J. Opt. Soc. Am.*, 57, 1381, 1967.
434. Zalubas, R., unpublished, 1979.
435. Bhatia, K. S., Jones, W. E., and Crooker, A. M., *Can. J. Phys.*, 50, 2421, 1972.
436. van Kleef, T. A. M. and Joshi, Y. N., *Phys. Scr.*, 24, 557, 1981.

ATOMIC TRANSITION PROBABILITIES

J. R. Fuhr, W. L. Wiese, L. I. Podobedova, and D. E. Kelleher

This table contains critically evaluated atomic transition probabilities for about 9000 selected lines of all elements for which reliable data are available on an absolute scale. The values are largely for neutral and singly ionized spectra, but also include some prominent lines of doubly and more highly charged ions of important elements.

Most of the data are obtained from comprehensive compilations of the Data Center on Atomic Transition Probabilities at the National Institute of Standards and Technology. Specifically, data have been taken from recent critical compilations on H, He, and Li (Ref. 1); on Be and B (Ref. 2); on neutral and singly ionized C and N (Ref. 3); and on Na (Ref. 4), Mg (Ref. 4), Al (Ref. 5), and Si (Ref. 6). Material from earlier compilations for the elements H through Ne (Refs. 7 and 8) and Na through Ca (Ref. 9) was supplemented by some more recent material taken directly from the original literature. Most of the original literature is cited in the above tables and in recent bibliographies (Refs. 10 and 11); for lack of space, individual literature references are not cited here.

The wavelength range for the neutral species is normally the visible spectrum or shorter wavelengths; only the very prominent near infrared lines are included. For the higher ions, most of the strong lines are located in the far UV. The tabulation is limited to electric dipole - including intercombination - lines and comprises essentially the fairly strong transitions with estimated uncertainties in the 10% to 50% range. With the exception of hydrogen, helium, and the alkali metals, most transitions are between states with low principal quantum numbers.

The transition probability, A, is given in units of 10^8 s^{-1} and is listed with as many digits as is consistent with the indicated accuracy. Generally, the estimated uncertainties of the A-values are in the range from 25% to 50% for two-digit numbers, 10% to 25% for three-digit numbers and 1% or better for four- and five-digit numbers.

Each transition is identified by the wavelength λ in angstroms and the statistical weights, g_i and g_k, of the lower (i) and upper (k) states [the product g_kA (or g_if) is needed for many applications]. Whenever the wavelengths of individual lines within a multiplet are extremely close, only an average wavelength for the multiplet as well as the multiplet A-value are given, and this is indicated by an asterisk (*) to the left of the wavelength. This also has been done when the transition probability for an entire multiplet has been taken from the literature and values for individual lines cannot be determined because of insufficient knowledge of the coupling of electrons. The wavelength data have been taken either from recent compilations or from the original literature cited in bibliographies published by the Atomic Energy Levels Data Center (Refs. 12 and 13) at the National Institute of Standards and Technology. Wavelength values are consistent with those given in the table "Line Spectra of the Elements," which appears elsewhere in this *Handbook*.

In addition to the transition probability A, the atomic oscillator strength f and the line strength S are often used in the literature. The conversion factors between these quantities are (for electric-dipole transitions):

$$g_if = 1.499{\cdot}10^{-8}\,\lambda^2 g_kA = 303.8\,\lambda^{-1}\,S$$

where λ is in Å, A is in 10^8 s^{-1}, and S is in atomic units, which are $a_0^2e^2 = 7.188{\cdot}10^{-59}$ $m^2\,C^2$.

The table for hydrogen is presented first, followed by the tables for other elements in alphabetical sequence by element name (not symbol). Within each element, the tables are ordered by increasing ionization stage (e.g., Al I, Al II, etc.).

The transition probabilities for hydrogen and hydrogen-like ions are known precisely. Because of the hydrogen degeneracy, a "transition" is actually the sum of all fine-structure transitions between the principal quantum numbers; therefore, the hydrogen table gives weighted average A-values. For hydrogen-like ions of nuclear charge Z, the following scaling laws hold:

$$A_Z = Z^4A_{\text{Hydrogen}}\quad f_Z = f_{\text{Hydrogen}}\quad S_Z = Z^{-2}S_{\text{Hydrogen}}\quad \lambda_Z = Z^{-2}\lambda_{\text{Hydrogen}}$$

For very highly charged hydrogen-like ions, starting at about $Z >$ 25, relativistic values must be used.

A more extensive compilation of over 12,000 lines is available in the web version of the *CRC Handbook of Chemistry and Physics*.

References

1. Wiese, W. L., and Fuhr, J. R., *J. Phys. Chem. Ref. Data* 38, 565, 2009.
2. Fuhr, J. R., and Wiese, W. L., *J. Phys. Chem. Ref. Data* 39, 013101, 2010.
3. Wiese, W. L., and Fuhr, J. R., *J. Phys. Chem. Ref. Data* 36, 1287, 2007.
4. Kelleher, D. E., and Podobedova, L. I., *J. Phys. Chem. Ref. Data* 37, 267, 2008.
5. Kelleher, D. E., and Podobedova, L. I., *J. Phys. Chem. Ref. Data* 37, 709, 2008.
6. Kelleher, D. E., and Podobedova, L. I., *J. Phys. Chem. Ref. Data* 37, 1285, 2008.
7. Wiese, W. L., Smith, M. W., and Glennon, B. M., *Atomic Transition Probabilities* (*H through Ne - A Critical Data Compilation*), National Standard Reference Data Series, National Bureau of Standards 4, Vol. I, U.S. Government Printing Office, Washington, D.C., 1966.
8. Wiese, W. L., Fuhr, J. R., and Deters, T. M., *Atomic Transition Probabilities of Carbon, Nitrogen, and Oxygen, J. Phys. Chem. Ref. Data, Monograph 7*, 1996.
9. Wiese, W. L., Smith, M. W., and Miles, B. M., *Atomic Transition Probabilities* (*Na through Ca - A Critical Data Compilation*), National Standard Reference Data Series, National Bureau of Standards 22, Vol. II, U. S. Government Printing Office, Washington, D.C., 1969.
10. Fuhr, J. R., Miller, B. J., and Martin, G. A., *Bibliography on Atomic Transition Probabilities* (*1914 through October 1997*), National Bureau of Standards Special Publication 505, 1978; Miller, B. J., Fuhr, J. R., and Martin, G. A., *Bibliography on Atomic Transition Probabilities* (*November 1977 through February 1980*), National Bureau of Standards Special Publication 505, Supplement 1, 1980.
11. Wiese, W. L., Reports on Astronomy, *Trans. Int. Astron. Union* 18A, 116, 1982; 19A, 122, 1985; 20A, 117, 1988, Reidel, D., Ed., Kluwer, Dordrecht, The Netherlands.
12. Moore, C. E., *Bibliography on the Analyses of Optical Atomic Spectra*, National Bureau of Standards Special Publication 306 - Section 1, 1968; Sections 2-4, 1969.
13. Hagan, L., and Martin, W. C., *Bibliography on Atomic Energy Levels and Spectra* (*July 1968 through June 1971*), National Bureau of Standards Special Publication 363, 1972; Hagan, L., *Bibliography on Atomic Energy Levels and Spectra* (*July 1971 through June 1975*), National Bureau of Standards Special Publication 363, Supplement 1, 1977; Zalubas, R., and Albright, A., *Bibliography on Atomic Energy Levels and Spectra* (*July 1975 through June 1979*), National Bureau of Standards Special Publication 363, Supplement 2, 1980; Musgrove, A., and Zalubas, R., *Bibliography on Atomic Energy Levels and Spectra* (*July 1979 through December 1983*), National Bureau of Standards Special Publication 363, Supplement 3, 1985.

Hydrogen

H I

λ Å	Weights g_i	Weights g_k	A $10^8\ s^{-1}$
912.765	2	1800	$5.1673 \cdot 10^{-6}$
912.837	2	1682	$6.1221 \cdot 10^{-6}$
912.916	2	1568	$7.2967 \cdot 10^{-6}$
913.004	2	1458	$8.7524 \cdot 10^{-6}$
913.102	2	1352	$1.0571 \cdot 10^{-5}$
913.212	2	1250	$1.2862 \cdot 10^{-5}$
913.337	2	1152	$1.5776 \cdot 10^{-5}$
913.478	2	1058	$1.9519 \cdot 10^{-5}$
913.639	2	968	$2.4380 \cdot 10^{-5}$
913.823	2	882	$3.0769 \cdot 10^{-5}$
914.036	2	800	$3.9276 \cdot 10^{-5}$
914.284	2	722	$5.0767 \cdot 10^{-5}$
914.574	2	648	$6.6540 \cdot 10^{-5}$
914.917	2	578	$8.8574 \cdot 10^{-5}$
915.327	2	512	$1.1997 \cdot 10^{-4}$
915.821	2	450	$1.6572 \cdot 10^{-4}$
916.427	2	392	$2.3409 \cdot 10^{-4}$
917.178	2	338	$3.3927 \cdot 10^{-4}$
918.127	2	288	$5.0659 \cdot 10^{-4}$
919.349	2	242	$7.8340 \cdot 10^{-4}$
920.961	2	200	$1.2631 \cdot 10^{-3}$
923.148	2	162	$2.1425 \cdot 10^{-3}$
926.223	2	128	$3.8694 \cdot 10^{-3}$
930.748	2	98	$7.5684 \cdot 10^{-3}$
937.803	2	72	$1.6440 \cdot 10^{-2}$
949.743	2	50	$4.1250 \cdot 10^{-2}$
972.537	2	32	$1.2785 \cdot 10^{-1}$
1025.72	2	18	$5.5751 \cdot 10^{-1}$
1215.67	2	8	4.6986
3662.23	8	1800	$2.8474 \cdot 10^{-6}$
3663.37	8	1682	$3.3742 \cdot 10^{-6}$
3664.65	8	1568	$4.0224 \cdot 10^{-6}$
3666.07	8	1458	$4.8261 \cdot 10^{-6}$
3667.65	8	1352	$5.8304 \cdot 10^{-6}$
3669.43	8	1250	$7.0963 \cdot 10^{-6}$
3671.45	8	1152	$8.7069 \cdot 10^{-6}$
3673.73	8	1058	$1.0777 \cdot 10^{-5}$
3676.33	8	968	$1.3467 \cdot 10^{-5}$
3679.32	8	882	$1.7005 \cdot 10^{-5}$
3682.78	8	800	$2.1719 \cdot 10^{-5}$
3686.80	8	722	$2.8093 \cdot 10^{-5}$
3691.52	8	648	$3.6851 \cdot 10^{-5}$
3697.12	8	578	$4.9101 \cdot 10^{-5}$
3703.82	8	512	$6.6583 \cdot 10^{-5}$
3711.94	8	450	$9.2102 \cdot 10^{-5}$
3721.91	8	392	$1.3032 \cdot 10^{-4}$
3734.34	8	338	$1.8927 \cdot 10^{-4}$
3750.12	8	288	$2.8337 \cdot 10^{-4}$
3770.60	8	242	$4.3972 \cdot 10^{-4}$
3797.87	8	200	$7.1225 \cdot 10^{-4}$
3835.35	8	162	$1.2156 \cdot 10^{-3}$
3889.02	8	128	$2.2148 \cdot 10^{-3}$
3970.08	8	98	$4.3889 \cdot 10^{-3}$
4101.74	8	72	$9.7320 \cdot 10^{-3}$
4340.47	8	50	$2.5304 \cdot 10^{-2}$
4861.34	8	32	$8.4193 \cdot 10^{-2}$
6562.83	8	18	$4.4101 \cdot 10^{-1}$
8392.19	18	800	$1.5167 \cdot 10^{-5}$
8413.11	18	722	$1.9643 \cdot 10^{-5}$
8437.75	18	648	$2.5804 \cdot 10^{-5}$
8467.04	18	578	$3.4442 \cdot 10^{-5}$
8502.27	18	512	$4.6801 \cdot 10^{-5}$
8545.17	18	450	$6.4901 \cdot 10^{-5}$
8598.18	18	392	$9.2117 \cdot 10^{-5}$
8664.80	18	338	$1.3431 \cdot 10^{-4}$
8750.25	18	288	$2.0207 \cdot 10^{-4}$
8862.55	18	242	$3.1558 \cdot 10^{-4}$
9014.67	18	200	$5.1558 \cdot 10^{-4}$
9228.77	18	162	$8.9050 \cdot 10^{-4}$
9545.70	18	128	$1.6506 \cdot 10^{-3}$
10049.4	18	98	$3.3585 \cdot 10^{-3}$
10938.1	18	72	$7.7829 \cdot 10^{-3}$
12818.1	18	50	$2.2008 \cdot 10^{-2}$
16406.4	32	288	$1.6205 \cdot 10^{-4}$
16805.7	32	242	$2.5565 \cdot 10^{-4}$
17361.2	32	200	$4.2347 \cdot 10^{-4}$
18173.2	32	162	$7.4593 \cdot 10^{-4}$
18751.0	18	32	$8.9860 \cdot 10^{-2}$
19444.5	32	128	$1.4242 \cdot 10^{-3}$
21655.2	32	98	$3.0415 \cdot 10^{-3}$
26251.4	32	72	$7.7110 \cdot 10^{-3}$
27573.0	50	288	$1.4024 \cdot 10^{-4}$
28719.8	50	242	$2.2460 \cdot 10^{-4}$
30381.1	50	200	$3.7999 \cdot 10^{-4}$
32957.8	50	162	$6.9078 \cdot 10^{-4}$
37391.4	50	128	$1.3877 \cdot 10^{-3}$
40511.4	32	50	$2.6993 \cdot 10^{-2}$
43747.2	72	288	$1.2884 \cdot 10^{-4}$
46524.9	50	98	$3.2528 \cdot 10^{-3}$
46706.2	72	242	$2.1096 \cdot 10^{-4}$
51279.2	72	200	$3.6881 \cdot 10^{-4}$
59072.4	72	162	$7.0652 \cdot 10^{-4}$
74598.3	50	72	$1.0254 \cdot 10^{-2}$
75009.1	72	128	$1.5609 \cdot 10^{-3}$
123718.6	72	98	$4.5608 \cdot 10^{-3}$

Aluminum

Al I

λ Å	Weights g_i	Weights g_k	A $10^8\ s^{-1}$
2118.312	2	4	$1.03 \cdot 10^{-1}$
2123.362	4	6	$1.22 \cdot 10^{-1}$
2129.663	2	4	$1.52 \cdot 10^{-1}$
2134.733	4	6	$1.81 \cdot 10^{-1}$
2145.555	2	4	$2.06 \cdot 10^{-1}$
2150.699	4	6	$2.46 \cdot 10^{-1}$
2168.805	2	4	$2.96 \cdot 10^{-1}$
2174.028	4	6	$3.53 \cdot 10^{-1}$
2199.150	2	2	$1.75 \cdot 10^{-2}$
2204.590	4	2	$3.49 \cdot 10^{-2}$
2204.660	2	4	$4.37 \cdot 10^{-1}$
2210.046	4	6	$5.20 \cdot 10^{-1}$
2257.999	2	2	$3.77 \cdot 10^{-2}$
2263.462	2	4	$6.83 \cdot 10^{-1}$
2263.731	4	2	$7.50 \cdot 10^{-2}$
2269.096	4	6	$7.58 \cdot 10^{-1}$
2269.220	4	4	$1.26 \cdot 10^{-1}$
2367.052	2	4	$7.61 \cdot 10^{-1}$
2372.070	2	2	$5.76 \cdot 10^{-2}$
2373.124	4	6	$9.07 \cdot 10^{-1}$
2373.349	4	4	$1.51 \cdot 10^{-1}$
2378.368	4	2	$1.14 \cdot 10^{-1}$
2567.984	2	4	$1.92 \cdot 10^{-1}$
2575.094	4	6	$3.60 \cdot 10^{-1}$
2575.393	4	4	$5.99 \cdot 10^{-2}$
2652.484	2	2	$1.42 \cdot 10^{-1}$
2660.393	4	2	$2.84 \cdot 10^{-1}$
3082.1529	2	4	$5.87 \cdot 10^{-1}$
3092.7099	4	6	$7.29 \cdot 10^{-1}$
3092.8386	4	4	$1.16 \cdot 10^{-1}$
3944.0058	2	2	$4.99 \cdot 10^{-1}$
3961.5200	4	2	$9.85 \cdot 10^{-1}$
5557.063	2	4	$2.30 \cdot 10^{-3}$
5557.948	2	2	$2.29 \cdot 10^{-3}$
6696.015	2	4	$1.00 \cdot 10^{-2}$
6698.673	2	2	$1.00 \cdot 10^{-2}$
7835.309	4	6	$3.71 \cdot 10^{-2}$
7836.134	6	8	$3.97 \cdot 10^{-2}$
8772.866	4	6	$6.47 \cdot 10^{-2}$
8773.896	6	8	$6.95 \cdot 10^{-2}$
8828.909	2	2	$6.72 \cdot 10^{-3}$
8841.277	4	2	$1.34 \cdot 10^{-2}$
8912.900	2	4	$2.28 \cdot 10^{-3}$
8923.555	4	6	$2.73 \cdot 10^{-3}$
8925.504	4	4	$4.54 \cdot 10^{-4}$

Al II

λ Å	Weights g_i	Weights g_k	A $10^8\ s^{-1}$
1047.8893	1	3	$2.33 \cdot 10^{-1}$
1048.5588	3	5	$3.14 \cdot 10^{-1}$
1049.9233	5	7	$4.17 \cdot 10^{-1}$
1189.1854	1	3	$9.30 \cdot 10^{-1}$
1190.0518	3	5	1.12
1191.8111	5	7	1.48
1350.1782	3	5	4.80
1539.8303	3	5	6.70
1670.7867	1	3	$1.41 \cdot 10^{1}$
1719.4400	1	3	6.55
1721.2435	3	3	4.93
1721.2714	3	5	8.82
1724.9519	5	5	2.97
1724.9838	5	7	$1.18 \cdot 10^{1}$
1760.1044	3	5	3.13
1761.9751	1	3	4.12
1763.8692	3	3	3.04
1763.9521	5	5	9.21
1765.8150	3	1	$1.23 \cdot 10^{1}$
1767.7308	5	3	5.13
1772.802	1	3	9.41
1774.002	3	5	$1.26 \cdot 10^{1}$
1774.770	3	3	7.01
1776.975	5	7	$1.67 \cdot 10^{1}$
1777.825	5	5	4.18
1818.352	7	7	5.45
1820.124	3	3	4.57
1855.9286	1	3	$8.38 \cdot 10^{-1}$
1858.0262	3	3	2.49
1862.3111	5	3	4.08

λ Å	Weights g_i	g_k	A 10^8 s^{-1}
1904.326	1	3	2.70
1906.4082	3	5	2.00
1906.596	3	3	2.00
1906.674	3	1	8.20
1910.8252	5	5	5.80
1911.013	5	3	3.40
1931.0481	3	1	$1.04\cdot10^{1}$
1958.77	7	5	5.70
1990.5310	3	5	$1.38\cdot10^{1}$
2192.604	7	9	2.47
2194.189	7	7	$2.74\cdot10^{-1}$
2195.502	3	5	2.07
2816.185	3	1	3.57
2994.277	1	3	$4.04\cdot10^{-2}$
2995.525	3	5	$5.44\cdot10^{-2}$
2998.150	5	7	$7.26\cdot10^{-2}$
3088.516	3	5	$1.08\cdot10^{-1}$
3649.184	1	3	$1.26\cdot10^{-1}$
3649.232	1	3	$1.26\cdot10^{-1}$
3651.065	3	5	$1.70\cdot10^{-1}$
3651.096	3	5	$1.70\cdot10^{-1}$
3654.981	5	7	$2.26\cdot10^{-1}$
3654.998	5	7	$2.26\cdot10^{-1}$
3703.219	3	5	$3.80\cdot10^{-1}$
3731.952	1	3	$4.30\cdot10^{-2}$
3733.908	3	3	$1.30\cdot10^{-1}$
3738.015	5	3	$2.10\cdot10^{-1}$
3866.160	3	1	$3.70\cdot10^{-1}$
3900.675	3	5	$4.80\cdot10^{-3}$
4663.056	5	3	$5.81\cdot10^{-1}$
5388.48	1	3	$1.20\cdot10^{-2}$
5593.302	3	5	$9.26\cdot10^{-1}$
5613.291	5	7	$3.43\cdot10^{-2}$
5853.62	7	9	$1.28\cdot10^{-1}$
5861.53	5	7	$1.14\cdot10^{-1}$
5867.81	3	5	$2.00\cdot10^{-1}$
5971.980	3	5	$4.90\cdot10^{-2}$
5999.70	1	3	$2.07\cdot10^{-2}$
5999.83	1	3	$2.07\cdot10^{-2}$
6001.76	3	5	$2.79\cdot10^{-2}$
6001.88	3	3	$1.55\cdot10^{-2}$
6006.410	5	7	$3.40\cdot10^{-2}$
6061.124	3	1	$7.60\cdot10^{-2}$
6066.32	1	3	$8.88\cdot10^{-3}$
6066.44	1	3	$8.88\cdot10^{-3}$
6068.43	3	3	$2.66\cdot10^{-2}$
6068.53	3	3	$2.66\cdot10^{-2}$
6073.198	5	3	$4.20\cdot10^{-2}$
6226.18	1	3	$6.20\cdot10^{-1}$
6231.745	3	5	$8.40\cdot10^{-1}$
6243.36	5	7	1.11
6335.701	5	3	$1.40\cdot10^{-1}$
6816.69	1	3	$1.10\cdot10^{-1}$
6823.48	3	3	$3.40\cdot10^{-1}$
6837.14	5	3	$5.70\cdot10^{-1}$
6917.93	5	7	$1.60\cdot10^{-1}$
6919.96	3	1	$9.60\cdot10^{-1}$
7042.06	3	5	$5.78\cdot10^{-1}$
7056.60	3	3	$5.74\cdot10^{-1}$
7063.64	3	1	$5.73\cdot10^{-1}$
7449.42	3	5	$1.20\cdot10^{-1}$
7471.41	5	7	$5.57\cdot10^{-1}$
7624.48	1	3	$4.59\cdot10^{-2}$
7627.85	3	5	$6.20\cdot10^{-2}$
7635.33	5	7	$9.00\cdot10^{-2}$
8354.318	7	9	$4.27\cdot10^{-1}$
8359.23	7	7	$4.75\cdot10^{-2}$
8359.492	5	7	$3.79\cdot10^{-1}$
8363.251	5	5	$6.64\cdot10^{-2}$
8363.469	3	5	$4.20\cdot10^{-1}$
8640.705	1	3	$3.00\cdot10^{-1}$

Al III

λ Å	Weights g_i	g_k	A 10^8 s^{-1}
560.390	2	4	$6.10\cdot10^{-1}$
560.390	2	2	$6.31\cdot10^{-1}$
695.817	2	4	$6.83\cdot10^{-1}$
696.212	2	2	$7.17\cdot10^{-1}$
1162.66	6	6	$1.39\cdot10^{-1}$
1162.66	6	8	2.09
1162.67	4	6	1.95
1352.857	6	8	4.26
1352.857	6	6	$2.84\cdot10^{-1}$
1352.857	4	6	3.98
1379.670	2	2	4.61
1384.140	4	2	9.22
1605.7661	2	4	$1.16\cdot10^{1}$
1611.8141	4	4	2.30
1611.8735	4	6	$1.38\cdot10^{1}$
1854.7164	2	4	5.44
1862.7895	2	2	5.36
1935.8404	6	8	$1.19\cdot10^{1}$
1935.8404	6	6	$7.92\cdot10^{-1}$
1935.9489	4	6	$1.11\cdot10^{1}$
3283.316	2	4	$1.99\cdot10^{-2}$
3287.302	4	6	$2.29\cdot10^{-2}$
3601.628	6	4	1.31
3601.926	4	4	$1.46\cdot10^{-1}$
3612.356	4	2	1.45
3702.106	2	2	1.13
3713.123	4	2	2.27
3980.14	6	8	$2.33\cdot10^{-1}$
3980.14	4	6	$2.17\cdot10^{-1}$
4149.915	6	8	2.05
4150.173	4	6	1.91
4357.562	2	4	$7.40\cdot10^{-2}$
4364.642	4	6	$8.67\cdot10^{-2}$
4512.564	2	4	2.09
4528.942	4	4	$4.15\cdot10^{-1}$
4529.194	4	6	2.49
4701.148	6	4	$7.67\cdot10^{-2}$
4701.412	8	6	$7.31\cdot10^{-2}$
4904.10	6	8	$3.51\cdot10^{-1}$
4904.10	4	6	$3.27\cdot10^{-1}$
5696.603	2	4	$8.77\cdot10^{-1}$
5722.728	2	2	$8.65\cdot10^{-1}$

Argon

Ar I

λ Å	Weights g_i	g_k	A 10^8 s^{-1}
1048.22	1	3	5.36
1066.66	1	3	1.29
3948.98	5	3	$4.55\cdot10^{-3}$
4044.42	3	5	$3.33\cdot10^{-3}$
4158.59	5	5	$1.40\cdot10^{-2}$
4181.88	1	3	$5.61\cdot10^{-3}$
4190.71	5	5	$2.80\cdot10^{-3}$
4191.03	1	3	$5.39\cdot10^{-3}$
4198.32	3	1	$2.57\cdot10^{-2}$
4200.67	5	7	$9.67\cdot10^{-3}$
4259.36	3	1	$3.98\cdot10^{-2}$
4266.29	3	5	$3.12\cdot10^{-3}$
4272.17	3	3	$7.97\cdot10^{-3}$
4300.10	3	5	$3.77\cdot10^{-3}$
4333.56	3	5	$5.68\cdot10^{-3}$
4335.34	3	3	$3.87\cdot10^{-3}$
4510.73	3	1	$1.18\cdot10^{-2}$
4752.94	3	3	$4.5\cdot10^{-3}$
4768.68	3	5	$8.6\cdot10^{-3}$
4798.74	7	9	$8.8\cdot10^{-4}$
4835.97	7	9	$9.3\cdot10^{-4}$
4876.26	3	5	$7.8\cdot10^{-3}$
4886.29	7	9	$1.2\cdot10^{-3}$
4887.95	3	3	$1.3\cdot10^{-2}$
4894.69	3	1	$1.8\cdot10^{-2}$
4956.75	7	9	$1.8\cdot10^{-3}$
4989.95	5	7	$1.1\cdot10^{-3}$
5048.81	3	5	$4.6\cdot10^{-3}$
5054.18	3	3	$4.5\cdot10^{-3}$
5060.08	7	9	$3.7\cdot10^{-3}$
5070.99	5	3	$2.6\cdot10^{-3}$
5087.09	5	7	$1.6\cdot10^{-3}$
5118.21	5	7	$2.7\cdot10^{-3}$
5151.39	3	1	$2.39\cdot10^{-2}$
5162.29	3	3	$1.90\cdot10^{-2}$
5177.54	7	5	$2.4\cdot10^{-3}$
5194.02	3	1	$7.8\cdot10^{-3}$
5210.49	7	7	$1.1\cdot10^{-3}$
5214.77	5	3	$2.1\cdot10^{-3}$
5221.27	7	9	$8.8\cdot10^{-3}$
5241.09	5	5	$1.3\cdot10^{-3}$
5246.24	5	7	$1.2\cdot10^{-3}$
5252.79	5	7	$5.4\cdot10^{-3}$
5254.47	3	5	$3.6\cdot10^{-3}$
5286.07	5	7	$9.6\cdot10^{-4}$
5309.52	5	5	$1.2\cdot10^{-3}$
5317.73	5	7	$2.6\cdot10^{-3}$
5373.50	3	5	$2.7\cdot10^{-3}$
5410.48	5	7	$2.0\cdot10^{-3}$
5421.35	7	5	$6.0\cdot10^{-3}$
5439.99	3	3	$1.9\cdot10^{-3}$
5442.24	7	7	$9.3\cdot10^{-4}$
5451.65	3	5	$4.7\cdot10^{-3}$
5457.42	5	3	$3.6\cdot10^{-3}$
5473.46	5	3	$2.0\cdot10^{-3}$
5492.09	3	1	$5.6\cdot10^{-3}$
5495.87	7	9	$1.69\cdot10^{-2}$
5506.11	5	7	$3.6\cdot10^{-3}$
5524.96	7	7	$1.7\cdot10^{-3}$
5534.49	5	3	$2.7\cdot10^{-3}$
5558.70	3	5	$1.42\cdot10^{-2}$

λ Å	Weights g_i	g_k	A $10^8\ s^{-1}$
5559.66	3	5	$2.2 \cdot 10^{-3}$
5572.54	5	7	$6.6 \cdot 10^{-3}$
5588.72	5	5	$1.5 \cdot 10^{-3}$
5597.48	5	7	$4.2 \cdot 10^{-3}$
5606.73	3	3	$2.20 \cdot 10^{-2}$
5618.01	3	3	$2.1 \cdot 10^{-3}$
5623.78	5	5	$1.4 \cdot 10^{-3}$
5635.58	3	5	$9.6 \cdot 10^{-4}$
5639.12	1	3	$2.1 \cdot 10^{-3}$
5650.70	3	1	$3.20 \cdot 10^{-2}$
5659.13	5	5	$2.6 \cdot 10^{-3}$
5681.90	5	7	$2.0 \cdot 10^{-3}$
5683.73	5	5	$2.0 \cdot 10^{-3}$
5700.87	5	7	$5.9 \cdot 10^{-3}$
5739.52	3	5	$8.7 \cdot 10^{-3}$
5772.11	5	7	$2.0 \cdot 10^{-3}$
5773.99	5	5	$1.1 \cdot 10^{-3}$
5802.08	5	3	$4.2 \cdot 10^{-3}$
5882.62	3	1	$1.23 \cdot 10^{-2}$
5888.58	7	5	$1.29 \cdot 10^{-2}$
5928.81	5	3	$1.1 \cdot 10^{-2}$
5942.67	5	5	$1.8 \cdot 10^{-3}$
5949.26	3	3	$1.5 \cdot 10^{-3}$
5968.32	3	3	$1.8 \cdot 10^{-3}$
5971.60	3	1	$1.1 \cdot 10^{-2}$
5987.30	7	7	$1.2 \cdot 10^{-3}$
5999.00	5	5	$1.4 \cdot 10^{-3}$
6005.73	5	3	$1.4 \cdot 10^{-3}$
6013.68	7	5	$1.4 \cdot 10^{-3}$
6025.15	5	3	$9.0 \cdot 10^{-3}$
6043.22	5	7	$1.47 \cdot 10^{-2}$
6052.73	3	5	$1.9 \cdot 10^{-3}$
6064.76	5	7	$5.8 \cdot 10^{-4}$
6090.79	1	3	$3.0 \cdot 10^{-3}$
6098.81	3	3	$5.2 \cdot 10^{-3}$
6101.16	3	3	$3.3 \cdot 10^{-3}$
6105.64	3	5	$1.21 \cdot 10^{-2}$
6128.73	3	5	$8.6 \cdot 10^{-4}$
6145.44	5	7	$7.6 \cdot 10^{-3}$
6155.24	5	3	$5.1 \cdot 10^{-3}$
6165.12	5	5	$9.89 \cdot 10^{-4}$
6170.17	5	5	$5.0 \cdot 10^{-3}$
6173.10	3	5	$6.7 \cdot 10^{-3}$
6212.50	5	7	$3.9 \cdot 10^{-3}$
6215.94	5	5	$5.7 \cdot 10^{-3}$
6248.41	3	5	$6.8 \cdot 10^{-4}$
6296.87	3	5	$9.0 \cdot 10^{-3}$
6307.66	5	5	$6.0 \cdot 10^{-3}$
6364.89	3	1	$5.6 \cdot 10^{-3}$
6369.58	5	3	$4.2 \cdot 10^{-3}$
6384.72	3	3	$4.21 \cdot 10^{-3}$
6416.31	3	5	$1.16 \cdot 10^{-2}$
6466.55	1	3	$1.5 \cdot 10^{-3}$
6538.11	7	7	$1.1 \cdot 10^{-3}$
6604.02	7	5	$2.8 \cdot 10^{-3}$
6660.68	3	1	$7.8 \cdot 10^{-3}$
6664.05	5	5	$1.5 \cdot 10^{-3}$
6698.88	5	3	$1.6 \cdot 10^{-3}$
6719.22	1	3	$2.4 \cdot 10^{-3}$
6752.84	3	5	$1.93 \cdot 10^{-2}$
6754.37	3	3	$2.1 \cdot 10^{-3}$
6756.10	5	5	$3.6 \cdot 10^{-3}$
6766.61	5	3	$4.0 \cdot 10^{-3}$
6779.93	1	3	$1.21 \cdot 10^{-3}$
6827.25	5	3	$2.4 \cdot 10^{-3}$
6851.88	3	5	$6.7 \cdot 10^{-4}$
6871.29	3	3	$2.78 \cdot 10^{-2}$
6879.59	3	5	$1.8 \cdot 10^{-3}$
6887.10	5	7	$1.3 \cdot 10^{-3}$
6888.17	3	5	$2.5 \cdot 10^{-3}$
6925.01	3	3	$1.2 \cdot 10^{-3}$
6937.67	3	1	$3.08 \cdot 10^{-2}$
6951.46	5	5	$2.2 \cdot 10^{-3}$
6960.23	5	5	$2.4 \cdot 10^{-3}$
6965.43	5	3	$6.39 \cdot 10^{-2}$
6992.17	3	1	$7.5 \cdot 10^{-3}$
7030.25	7	5	$2.67 \cdot 10^{-2}$
7067.22	5	5	$3.80 \cdot 10^{-2}$
7068.73	5	3	$2.0 \cdot 10^{-2}$
7086.70	1	3	$1.5 \cdot 10^{-3}$
7107.48	5	5	$4.5 \cdot 10^{-3}$
7125.83	3	3	$6.0 \cdot 10^{-3}$
7147.04	5	3	$6.25 \cdot 10^{-3}$
7158.83	3	1	$2.1 \cdot 10^{-2}$
7206.98	5	3	$2.48 \cdot 10^{-2}$
7229.93	5	5	$6.6 \cdot 10^{-4}$
7265.17	3	3	$1.7 \cdot 10^{-3}$
7270.66	7	7	$1.1 \cdot 10^{-3}$
7272.93	3	3	$1.83 \cdot 10^{-2}$
7285.44	5	3	$1.2 \cdot 10^{-3}$
7311.72	3	3	$1.7 \cdot 10^{-2}$
7316.01	3	3	$9.6 \cdot 10^{-3}$
7350.78	3	1	$1.2 \cdot 10^{-2}$
7353.32	5	7	$9.6 \cdot 10^{-3}$
7372.12	7	9	$1.9 \cdot 10^{-2}$
7383.98	3	5	$8.47 \cdot 10^{-2}$
7392.97	5	3	$7.2 \cdot 10^{-3}$
7412.33	3	5	$3.9 \cdot 10^{-3}$
7422.26	3	5	$6.6 \cdot 10^{-4}$
7425.29	5	7	$3.1 \cdot 10^{-3}$
7435.33	5	5	$9.0 \cdot 10^{-3}$
7436.25	7	5	$2.7 \cdot 10^{-3}$
7484.24	3	5	$3.4 \cdot 10^{-3}$
7503.84	3	1	$4.45 \cdot 10^{-1}$
7510.42	5	5	$4.5 \cdot 10^{-3}$
7514.65	3	1	$4.02 \cdot 10^{-1}$
7618.33	3	5	$2.9 \cdot 10^{-3}$
7628.86	3	5	$2.9 \cdot 10^{-3}$
7635.11	5	5	$2.45 \cdot 10^{-1}$
7670.04	5	3	$2.8 \cdot 10^{-3}$
7704.81	5	7	$6.3 \cdot 10^{-4}$
7723.76	5	3	$5.18 \cdot 10^{-2}$
7724.21	1	3	$1.17 \cdot 10^{-1}$
7798.55	3	5	$8.7 \cdot 10^{-4}$
7868.20	1	3	$3.50 \cdot 10^{-3}$
7891.08	5	5	$9.5 \cdot 10^{-3}$
7916.45	3	3	$1.2 \cdot 10^{-3}$
7948.18	1	3	$1.86 \cdot 10^{-1}$
8006.16	3	5	$4.90 \cdot 10^{-2}$
8014.79	5	5	$9.28 \cdot 10^{-2}$
8037.23	1	3	$3.59 \cdot 10^{-3}$
8046.13	3	1	$1.12 \cdot 10^{-2}$
8053.31	5	3	$8.6 \cdot 10^{-3}$
8066.60	5	5	$1.4 \cdot 10^{-3}$
8103.69	3	3	$2.5 \cdot 10^{-1}$
8115.31	5	7	$3.31 \cdot 10^{-1}$
8264.52	3	3	$1.53 \cdot 10^{-1}$
8384.73	5	7	$2.4 \cdot 10^{-3}$
8408.21	3	5	$2.23 \cdot 10^{-1}$
8424.65	3	5	$2.15 \cdot 10^{-1}$
8490.30	3	5	$9.6 \cdot 10^{-4}$
8521.44	3	3	$1.39 \cdot 10^{-1}$
8605.78	5	5	$1.04 \cdot 10^{-2}$
8620.46	1	3	$9.2 \cdot 10^{-3}$
8667.94	1	3	$2.43 \cdot 10^{-2}$
8761.69	3	5	$9.5 \cdot 10^{-3}$
8784.61	3	1	$2.4 \cdot 10^{-3}$
8799.08	5	3	$4.6 \cdot 10^{-3}$
8962.19	3	3	$1.6 \cdot 10^{-3}$
9075.42	3	1	$1.2 \cdot 10^{-2}$
9122.97	5	3	$1.89 \cdot 10^{-1}$
9194.64	3	3	$1.76 \cdot 10^{-2}$
9224.50	3	5	$5.03 \cdot 10^{-2}$
9291.53	3	1	$3.26 \cdot 10^{-2}$
9354.22	3	3	$1.06 \cdot 10^{-2}$
9657.78	3	3	$5.43 \cdot 10^{-2}$
9784.50	3	5	$1.47 \cdot 10^{-2}$
10470.05	1	3	$9.8 \cdot 10^{-3}$
10478.0	3	3	$2.44 \cdot 10^{-2}$
10950.7	5	3	$3.96 \cdot 10^{-3}$
11078.9	5	5	$8.3 \cdot 10^{-3}$
11393.7	3	1	$2.22 \cdot 10^{-2}$
11441.8	5	3	$1.39 \cdot 10^{-2}$
11467.5	3	5	$3.69 \cdot 10^{-3}$
11488.11	3	3	$1.9 \cdot 10^{-3}$
11668.7	5	5	$3.76 \cdot 10^{-2}$
11719.5	5	3	$9.52 \cdot 10^{-3}$
12026.6	1	3	$4.2 \cdot 10^{-3}$
12112.2	7	7	$3.1 \cdot 10^{-2}$
12139.8	3	3	$4.5 \cdot 10^{-2}$
12343.7	5	7	$2.0 \cdot 10^{-2}$
12402.9	3	3	$1.1 \cdot 10^{-1}$
12439.2	3	5	$4.9 \cdot 10^{-2}$
12456.1	5	3	$8.9 \cdot 10^{-2}$
12487.6	7	5	$1.1 \cdot 10^{-1}$
12554.4	7	5	$1.2 \cdot 10^{-3}$
12702.4	3	3	$7.1 \cdot 10^{-2}$
12733.6	5	5	$1.1 \cdot 10^{-2}$
12746.3	3	3	$2.0 \cdot 10^{-2}$
12802.7	5	5	$5.7 \cdot 10^{-2}$
12933.3	3	1	$1.0 \cdot 10^{-1}$
12956.6	3	3	$7.4 \cdot 10^{-2}$
13008.5	5	3	$8.9 \cdot 10^{-2}$
13214.7	3	1	$8.1 \cdot 10^{-2}$
13273.1	5	7	$1.5 \cdot 10^{-1}$
13313.4	3	5	$1.3 \cdot 10^{-1}$
13504.0	5	7	$1.1 \cdot 10^{-1}$
13599.2	5	5	$2.2 \cdot 10^{-2}$
13622.4	3	5	$7.3 \cdot 10^{-2}$
13678.5	3	5	$6.2 \cdot 10^{-2}$

λ Å	Weights g_i	g_k	A 10^8 s^{-1}
14093.6	1	3	$4.3 \cdot 10^{-2}$
14739.1	5	7	$8.8 \cdot 10^{-4}$
15046.4	1	3	$5.2 \cdot 10^{-2}$
15172.3	1	3	$1.3 \cdot 10^{-2}$
15329.6	5	5	$1.2 \cdot 10^{-3}$
15555.5	5	7	$9.8 \cdot 10^{-5}$
15734.9	5	3	$2.9 \cdot 10^{-4}$
15816.8	5	3	$8.7 \cdot 10^{-4}$
15989.3	1	3	$1.9 \cdot 10^{-2}$
16122.7	5	3	$3.9 \cdot 10^{-4}$
16180.0	5	5	$1.2 \cdot 10^{-3}$
16264.1	3	3	$3.0 \cdot 10^{-4}$
16520.1	3	5	$2.6 \cdot 10^{-3}$
16739.8	3	5	$3.1 \cdot 10^{-3}$
16940.4	5	5	$2.5 \cdot 10^{-2}$
20317.0	1	3	$1.6 \cdot 10^{-3}$
20616.5	5	5	$3.9 \cdot 10^{-3}$
20812.0	5	7	$7.6 \cdot 10^{-4}$
21332.2	3	3	$3.2 \cdot 10^{-4}$
21534.9	3	5	$1.1 \cdot 10^{-3}$
22039.2	3	1	$1.2 \cdot 10^{-3}$
22077.4	5	3	$1.4 \cdot 10^{-3}$
23133.4	3	3	$1.7 \cdot 10^{-3}$
23844.8	9	7	$1.1 \cdot 10^{-2}$
23967.5	3	1	$3.6 \cdot 10^{-3}$

Ar II

λ Å	Weights g_i	g_k	A 10^8 s^{-1}
2942.9	4	4	$5.3 \cdot 10^{-1}$
2979.1	2	2	$4.16 \cdot 10^{-1}$
3139.0	6	6	$5.2 \cdot 10^{-1}$
3169.7	4	6	$4.9 \cdot 10^{-1}$
3181.0	6	4	$3.7 \cdot 10^{-1}$
3243.7	4	2	1.06
3249.8	2	4	$6.3 \cdot 10^{-1}$
3263.6	2	4	$1.55 \cdot 10^{-1}$
3281.7	2	2	$4.2 \cdot 10^{-1}$
3454.1	6	4	$3.14 \cdot 10^{-1}$
3476.7	6	6	1.25
3491.2	4	4	1.79
3491.5	6	8	2.31
3509.8	2	2	2.55
3514.4	4	6	1.36
3520.0	6	6	$5.2 \cdot 10^{-1}$
3521.3	8	8	$2.27 \cdot 10^{-1}$
3535.3	2	4	$5.7 \cdot 10^{-1}$
3548.5	4	4	$8.7 \cdot 10^{-1}$
3559.5	6	8	2.88
3565.0	2	4	$5.5 \cdot 10^{-1}$
3576.6	6	8	2.75
3581.6	2	4	1.76
3582.4	4	6	2.53
3588.4	8	10	3.03
3656.0	6	6	$7.6 \cdot 10^{-2}$
3717.2	6	8	$5.2 \cdot 10^{-2}$
3729.3	6	4	$4.80 \cdot 10^{-1}$
3763.5	8	6	$1.78 \cdot 10^{-1}$
3780.8	8	8	$7.7 \cdot 10^{-1}$
3799.4	6	4	$1.7 \cdot 10^{-1}$
3826.8	6	6	$2.81 \cdot 10^{-1}$
3841.5	4	2	$2.69 \cdot 10^{-1}$
3844.7	6	8	$4.8 \cdot 10^{-2}$
3850.6	4	4	$3.87 \cdot 10^{-1}$
3868.5	4	6	1.4
3872.1	4	4	$1.5 \cdot 10^{-1}$
3880.3	2	2	$2.32 \cdot 10^{-1}$
3900.6	4	6	$7.2 \cdot 10^{-2}$
3928.6	2	4	$2.44 \cdot 10^{-1}$
3932.5	4	4	$9.3 \cdot 10^{-1}$
3952.7	4	4	$2.08 \cdot 10^{-1}$
3979.4	4	2	$9.8 \cdot 10^{-1}$
4013.9	8	8	$1.05 \cdot 10^{-1}$
4042.9	4	4	$4.06 \cdot 10^{-1}$
4052.9	2	4	$6.7 \cdot 10^{-1}$
4072.0	6	6	$5.8 \cdot 10^{-1}$
4079.6	6	4	$1.19 \cdot 10^{-1}$
4131.7	4	2	$8.5 \cdot 10^{-1}$
4228.2	4	6	$1.31 \cdot 10^{-1}$
4237.2	4	4	$1.12 \cdot 10^{-1}$
4266.5	6	6	$1.64 \cdot 10^{-1}$
4277.5	6	4	$8.0 \cdot 10^{-1}$
4282.9	4	2	$1.32 \cdot 10^{-1}$
4300.6	6	6	$5.7 \cdot 10^{-2}$
4331.2	4	4	$5.74 \cdot 10^{-1}$
4332.0	4	2	$1.92 \cdot 10^{-1}$
4348.1	6	8	1.17
4352.2	2	2	$2.12 \cdot 10^{-1}$
4362.1	4	6	$5.5 \cdot 10^{-2}$
4370.8	4	4	$6.6 \cdot 10^{-1}$
4371.3	6	4	$2.21 \cdot 10^{-1}$
4376.0	4	2	$2.05 \cdot 10^{-1}$
4379.7	2	2	1.00
4400.1	4	4	$1.60 \cdot 10^{-1}$
4401.0	8	6	$3.04 \cdot 10^{-1}$
4412.9	6	8	$6.1 \cdot 10^{-2}$
4426.0	4	6	$8.17 \cdot 10^{-1}$
4430.2	2	4	$5.69 \cdot 10^{-1}$
4431.0	6	6	$1.09 \cdot 10^{-1}$
4474.8	4	2	$2.90 \cdot 10^{-1}$
4481.8	6	6	$4.55 \cdot 10^{-1}$
4545.1	4	4	$4.71 \cdot 10^{-1}$
4579.4	2	2	$8.0 \cdot 10^{-1}$
4589.9	4	6	$6.64 \cdot 10^{-1}$
4598.8	4	4	$6.7 \cdot 10^{-2}$
4609.6	6	8	$7.89 \cdot 10^{-1}$
4637.2	6	6	$7.1 \cdot 10^{-2}$
4657.9	4	2	$8.92 \cdot 10^{-1}$
4726.9	4	4	$5.88 \cdot 10^{-1}$
4732.1	6	4	$6.7 \cdot 10^{-2}$
4735.9	6	4	$5.80 \cdot 10^{-1}$
4764.9	2	4	$6.4 \cdot 10^{-1}$
4806.0	6	6	$7.80 \cdot 10^{-1}$
4847.8	4	2	$8.49 \cdot 10^{-1}$
4879.9	4	6	$8.23 \cdot 10^{-1}$
4889.0	2	2	$1.9 \cdot 10^{-1}$
4904.8	6	8	$3.7 \cdot 10^{-2}$
4933.2	4	4	$1.44 \cdot 10^{-1}$
4965.1	2	4	$3.94 \cdot 10^{-1}$
4972.2	2	2	$9.7 \cdot 10^{-2}$
5009.3	4	6	$1.51 \cdot 10^{-1}$
5017.2	4	6	$2.07 \cdot 10^{-1}$
5062.0	2	4	$2.23 \cdot 10^{-1}$
5141.8	6	8	$8.1 \cdot 10^{-2}$
5145.3	4	6	$1.06 \cdot 10^{-1}$
6114.9	10	8	$2.00 \cdot 10^{-1}$
6172.3	8	6	$2.00 \cdot 10^{-1}$
6243.1	8	6	$3.0 \cdot 10^{-2}$
6483.1	4	2	$1.06 \cdot 10^{-1}$
6638.2	6	4	$1.37 \cdot 10^{-1}$
6639.7	4	2	$1.69 \cdot 10^{-1}$
6643.7	10	8	$1.47 \cdot 10^{-1}$
6666.4	2	2	$8.8 \cdot 10^{-2}$
6684.3	8	6	$1.07 \cdot 10^{-1}$
6756.6	4	4	$2.0 \cdot 10^{-2}$
6863.5	6	6	$2.5 \cdot 10^{-2}$
7233.5	2	4	$3.7 \cdot 10^{-2}$
7380.4	4	4	$5.6 \cdot 10^{-2}$
7589.3	6	4	$1.07 \cdot 10^{-1}$

Ar III

λ Å	Weights g_i	g_k	A 10^8 s^{-1}
769.15	5	3	6.0
871.10	5	3	1.59
875.53	3	1	3.74
878.73	5	5	2.79
879.62	3	3	$9.2 \cdot 10^{-1}$
883.18	1	3	1.22
887.40	3	5	$9.0 \cdot 10^{-1}$
3024.1	5	7	2.6
3027.2	5	5	$6.4 \cdot 10^{-1}$
3054.8	3	5	1.9
3064.8	3	3	1.0
3078.2	1	3	1.4
3285.9	5	7	2.0
3301.9	5	5	2.0
3311.3	5	3	2.0
3336.1	7	9	2.0
3344.7	5	7	1.8
3352.1	7	7	$2.2 \cdot 10^{-1}$
3358.5	3	5	1.6
3361.3	5	5	$3.0 \cdot 10^{-1}$
3472.6	5	7	$2.0 \cdot 10^{-1}$
3480.6	7	7	1.6
3499.7	3	3	1.3
3500.6	3	5	$2.6 \cdot 10^{-1}$
3502.7	5	3	$4.3 \cdot 10^{-1}$
3503.6	5	5	1.2
3511.7	7	5	$2.6 \cdot 10^{-1}$

Ar IV

λ Å	Weights g_i	g_k	A 10^8 s^{-1}
840.03	4	2	2.73
843.77	4	4	2.70
850.60	4	6	2.63

Ar VI

λ Å	Weights g_i	g_k	A 10^8 s^{-1}
292.15	2	2	$6.9 \cdot 10^{1}$
294.05	4	2	$1.36 \cdot 10^{2}$

Ar VII

λ Å	Weights g_i	g_k	A 10^8 s^{-1}
*250.41	9	3	$2.78 \cdot 10^{2}$
*477.54	9	15	$9.92 \cdot 10^{1}$
585.75	1	3	$7.83 \cdot 10^{1}$

λ Å	Weights g_i	g_k	A $10^8\ s^{-1}$
*637.30	9	9	$6.7 \cdot 10^{1}$
Ar VIII			
158.92	2	4	$1.1 \cdot 10^{2}$
159.18	2	2	$1.11 \cdot 10^{2}$
229.44	2	2	$1.12 \cdot 10^{2}$
230.88	4	2	$2.21 \cdot 10^{2}$
337.09	4	4	$1.2 \cdot 10^{1}$
337.26	6	4	$1.0 \cdot 10^{2}$
338.22	4	2	$1.1 \cdot 10^{2}$
519.43	2	4	$6.3 \cdot 10^{1}$
526.46	4	6	$7.2 \cdot 10^{1}$
526.87	4	4	$1.2 \cdot 10^{1}$
700.24	2	4	$2.55 \cdot 10^{1}$
713.81	2	2	$2.4 \cdot 10^{1}$
Ar IX			
48.739	1	3	$1.69 \cdot 10^{3}$
Ar XIII			
162.96	5	3	$3.4 \cdot 10^{2}$
*163.08	9	3	$5.3 \cdot 10^{2}$
184.90	5	5	$1.66 \cdot 10^{2}$
186.38	1	3	$8.8 \cdot 10^{1}$
*207.89	9	9	$9.5 \cdot 10^{1}$
*245.10	9	15	$3.7 \cdot 10^{1}$
Ar XIV			
180.29	2	4	$4.5 \cdot 10^{1}$
183.41	2	2	$1.69 \cdot 10^{2}$
187.95	4	4	$1.97 \cdot 10^{2}$
191.35	4	2	$7.5 \cdot 10^{1}$
194.39	2	2	$4.6 \cdot 10^{1}$
203.35	4	2	$7.8 \cdot 10^{1}$
Ar XV			
25.05	1	3	$1.7 \cdot 10^{4}$
221.10	1	3	$9.55 \cdot 10^{1}$
*265.3	9	9	$8.1 \cdot 10^{1}$
Ar XVI			
*23.52	2	6	$1.43 \cdot 10^{4}$
*24.96	6	10	$4.4 \cdot 10^{4}$
353.88	2	4	$1.5 \cdot 10^{1}$
389.11	2	2	$1.1 \cdot 10^{1}$
1268	2	4	1.9
1401	2	2	1.4
2975	2	4	$9.0 \cdot 10^{-2}$
3514	4	6	$6.5 \cdot 10^{-2}$
Arsenic			
As I			
1890.4	4	6	2.0
1937.6	4	4	2.0
1972.6	4	2	2.0
2288.1	6	4	2.8
2344.0	2	4	$3.5 \cdot 10^{-1}$
2349.8	4	2	3.1
2369.7	4	4	$6.0 \cdot 10^{-1}$
2370.8	4	6	$4.2 \cdot 10^{-1}$
2456.5	6	4	$7.2 \cdot 10^{-2}$
2492.9	4	2	$1.2 \cdot 10^{-1}$
2745.0	2	4	$2.6 \cdot 10^{-1}$
2780.2	4	4	$7.8 \cdot 10^{-1}$
2860.4	2	2	$5.5 \cdot 10^{-1}$
2898.7	4	2	$9.9 \cdot 10^{-2}$
Barium			
Ba I			
2427.41	1	3	$5.60 \cdot 10^{-3}$
2472.74	1	3	$4.60 \cdot 10^{-3}$
3071.58	1	3	$4.20 \cdot 10^{-1}$
3501.11	1	3	$3.50 \cdot 10^{-1}$
3889.33	1	3	$1.10 \cdot 10^{-2}$
4132.43	1	3	$1.50 \cdot 10^{-2}$
4181.09	3	5	$4.99 \cdot 10^{-4}$
4181.66	3	5	$5.42 \cdot 10^{-4}$
4182.27	3	5	$6.11 \cdot 10^{-4}$
4182.94	3	5	$6.65 \cdot 10^{-4}$
4183.64	3	5	$6.70 \cdot 10^{-4}$
4184.40	3	5	$7.93 \cdot 10^{-4}$
4185.25	3	5	$8.43 \cdot 10^{-4}$
4186.16	3	5	$9.24 \cdot 10^{-4}$
4187.15	3	5	$9.90 \cdot 10^{-4}$
4188.25	3	5	$1.03 \cdot 10^{-3}$
4189.44	3	5	$1.13 \cdot 10^{-3}$
4190.76	3	5	$1.28 \cdot 10^{-3}$
4192.20	3	5	$1.36 \cdot 10^{-3}$
4193.81	3	5	$1.58 \cdot 10^{-3}$
4195.59	3	5	$1.78 \cdot 10^{-3}$
4323.00	3	5	$8.80 \cdot 10^{-2}$
4402.54	3	5	$2.70 \cdot 10^{-1}$
4488.98	5	7	$2.80 \cdot 10^{-1}$
4493.64	5	5	$1.95 \cdot 10^{-1}$
4573.85	3	1	1.21
4579.64	5	5	$7.00 \cdot 10^{-1}$
4599.72	3	1	$4.07 \cdot 10^{-1}$
4619.92	1	3	$2.70 \cdot 10^{-2}$
4700.42	3	3	$6.10 \cdot 10^{-2}$
4726.43	5	3	$3.30 \cdot 10^{-1}$
4801.30	9	3	$1.39 \cdot 10^{-1}$
4902.85	5	3	$5.40 \cdot 10^{-2}$
5169.53	5	3	$9.00 \cdot 10^{-4}$
5519.04	3	5	$5.70 \cdot 10^{-1}$
5535.48	1	3	1.19
5777.62	5	7	$8.00 \cdot 10^{-1}$
5784.04	3	5	$2.10 \cdot 10^{-1}$
5800.23	5	5	$2.39 \cdot 10^{-1}$
5826.27	5	3	$4.50 \cdot 10^{-1}$
5971.70	5	5	$1.62 \cdot 10^{-1}$
5997.09	3	3	$2.80 \cdot 10^{-1}$
6019.47	3	1	$8.10 \cdot 10^{-1}$
6063.11	5	3	$5.60 \cdot 10^{-1}$
6083.39	3	1	$1.10 \cdot 10^{-1}$
6129.23	3	1	$6.00 \cdot 10^{-2}$
6341.68	5	7	$1.16 \cdot 10^{-1}$
6450.85	3	5	$1.10 \cdot 10^{-1}$
6498.76	7	7	$5.40 \cdot 10^{-1}$
6527.31	5	5	$3.30 \cdot 10^{-1}$
6527.40	15	15	$6.15 \cdot 10^{-1}$
6595.33	3	3	$3.80 \cdot 10^{-1}$
6675.27	5	3	$1.89 \cdot 10^{-1}$
6693.84	7	5	$1.46 \cdot 10^{-1}$
6986.80	5	3	$5.20 \cdot 10^{-3}$
7059.94	7	9	$5.00 \cdot 10^{-1}$
7120.33	3	5	$1.10 \cdot 10^{-1}$
7195.23	1	3	$5.60 \cdot 10^{-2}$
7213.60	5	5	$6.50 \cdot 10^{-4}$
7280.30	5	7	$3.20 \cdot 10^{-1}$
7392.41	3	3	$1.81 \cdot 10^{-1}$
7417.54	7	5	$7.70 \cdot 10^{-3}$
7488.08	7	7	$7.30 \cdot 10^{-2}$
7528.18	5	5	$2.70 \cdot 10^{-2}$
7610.48	5	5	$1.10 \cdot 10^{-2}$
7644.90	9	3	$5.03 \cdot 10^{-1}$
7672.09	3	5	$1.52 \cdot 10^{-1}$
7780.48	5	5	$7.60 \cdot 10^{-2}$
7877.80	3	5	$1.60 \cdot 10^{-2}$
7905.75	5	3	$2.65 \cdot 10^{-1}$
8147.70	5	5	$6.30 \cdot 10^{-2}$
8560.00	5	5	$2.00 \cdot 10^{-1}$
8654.08	5	7	$3.10 \cdot 10^{-3}$
9370.12	5	5	$7.60 \cdot 10^{-2}$
9645.60	7	5	$1.10 \cdot 10^{-1}$
9704.31	3	1	$1.60 \cdot 10^{-1}$
9821.48	3	1	$5.50 \cdot 10^{-2}$
10370.30	3	5	$1.30 \cdot 10^{-2}$
10540.10	5	3	$1.80 \cdot 10^{-2}$
10649.10	5	5	$2.70 \cdot 10^{-2}$
11303.00	5	3	$1.10 \cdot 10^{-3}$
11373.70	3	1	$1.30 \cdot 10^{-1}$
12342.30	3	3	$9.00 \cdot 10^{-4}$
14723.10	3	5	$8.60 \cdot 10^{-3}$
14999.90	5	3	$2.50 \cdot 10^{-3}$
17186.90	3	1	$2.70 \cdot 10^{-2}$
18202.80	5	3	$1.20 \cdot 10^{-2}$
21567.70	5	3	$2.60 \cdot 10^{-3}$
30685.30	5	3	$6.50 \cdot 10^{-3}$
Ba II			
2528.41	2	4	$6.91 \cdot 10^{-1}$
2634.78	4	6	$7.33 \cdot 10^{-1}$
3891.78	2	4	2.17
4130.65	4	6	2.18
4166.00	4	4	$3.54 \cdot 10^{-1}$
4267.92	6	8	$3.10 \cdot 10^{-1}$
4309.26	8	10	$3.10 \cdot 10^{-1}$
4325.75	4	6	$5.65 \cdot 10^{-2}$
4524.93	2	2	$6.63 \cdot 10^{-1}$
4554.03	2	4	1.11
4708.90	2	4	$8.47 \cdot 10^{-2}$
4843.48	4	6	$9.34 \cdot 10^{-2}$
4899.93	4	2	1.04
4934.08	2	2	$9.53 \cdot 10^{-1}$
4957.09	6	8	$5.13 \cdot 10^{-1}$
5012.95	8	10	$5.15 \cdot 10^{-1}$
5361.35	4	6	$4.01 \cdot 10^{-2}$
5391.59	6	8	$4.22 \cdot 10^{-2}$

λ Å	Weights g_i	g_k	A 10^8 s^{-1}
5784.15	2	4	$1.59 \cdot 10^{-1}$
5853.67	4	4	$6.00 \cdot 10^{-2}$
5981.26	4	6	$1.73 \cdot 10^{-1}$
6135.60	2	2	$6.64 \cdot 10^{-2}$
6141.71	6	4	$4.12 \cdot 10^{-1}$
6378.92	4	2	$1.18 \cdot 10^{-1}$
6496.90	4	2	$3.10 \cdot 10^{-1}$
6769.48	6	8	$9.35 \cdot 10^{-1}$
6874.08	8	10	$9.26 \cdot 10^{-1}$
7115.03	8	10	$8.80 \cdot 10^{-3}$
8496.80	2	4	$3.31 \cdot 10^{-2}$
8661.90	6	4	$1.27 \cdot 10^{-2}$
8703.69	4	6	$3.69 \cdot 10^{-2}$
8710.77	6	8	$7.88 \cdot 10^{-1}$
8737.75	4	6	$7.29 \cdot 10^{-1}$
8760.61	8	6	$1.17 \cdot 10^{-2}$
8897.46	6	6	$4.93 \cdot 10^{-2}$
9603.12	2	4	$4.16 \cdot 10^{-1}$
10115.00	4	6	$4.27 \cdot 10^{-1}$
10212.80	4	4	$6.92 \cdot 10^{-2}$
10768.00	2	4	$5.56 \cdot 10^{-2}$
11088.50	4	6	$6.11 \cdot 10^{-2}$
11127.50	4	4	$1.01 \cdot 10^{-2}$
11519.50	2	2	$2.47 \cdot 10^{-2}$
11577.10	2	2	$1.75 \cdot 10^{-1}$
11931.90	4	2	$4.44 \cdot 10^{-2}$
12475.00	4	2	$2.80 \cdot 10^{-1}$
13057.80	2	4	$2.14 \cdot 10^{-1}$
14211.50	2	2	$1.66 \cdot 10^{-1}$
17738.90	6	8	$2.16 \cdot 10^{-1}$
18530.70	8	10	$1.96 \cdot 10^{-1}$
18729.70	2	4	$1.23 \cdot 10^{-1}$
19642.60	4	6	$1.28 \cdot 10^{-1}$
19845.10	4	4	$2.07 \cdot 10^{-2}$
22994.70	2	2	$6.18 \cdot 10^{-2}$
24612.50	4	4	$4.75 \cdot 10^{-3}$
24699.00	4	2	$9.98 \cdot 10^{-2}$
25923.20	6	4	$3.66 \cdot 10^{-2}$
27687.20	2	4	$6.10 \cdot 10^{-2}$
29058.90	4	2	$2.89 \cdot 10^{-2}$
30196.00	2	2	$4.70 \cdot 10^{-2}$
42934.70	6	8	$4.82 \cdot 10^{-3}$
43294.30	4	6	$4.39 \cdot 10^{-3}$
47520.80	6	6	$2.37 \cdot 10^{-4}$

Beryllium

Be I

λ Å	Weights g_i	g_k	A 10^8 s^{-1}
1661.48	1	3	$7.23 \cdot 10^{-2}$
1929.62	3	5	$5.12 \cdot 10^{-2}$
1929.71	5	7	$6.82 \cdot 10^{-2}$
1943.59	1	3	$5.43 \cdot 10^{-2}$
1943.62	3	5	$7.32 \cdot 10^{-2}$
1943.71	5	7	$9.77 \cdot 10^{-2}$
1964.54	1	3	$8.15 \cdot 10^{-2}$
1964.56	3	3	$6.12 \cdot 10^{-2}$
1964.56	3	5	$1.10 \cdot 10^{-1}$
1964.65	5	7	$1.47 \cdot 10^{-1}$
1964.65	5	5	$3.67 \cdot 10^{-2}$
1997.95	1	3	$1.31 \cdot 10^{-1}$
1997.98	3	5	$1.76 \cdot 10^{-1}$
1997.98	3	3	$9.80 \cdot 10^{-2}$
1998.07	5	5	$5.88 \cdot 10^{-2}$
1998.07	5	7	$2.35 \cdot 10^{-1}$
2032.72	5	3	$5.3 \cdot 10^{-2}$
2055.88	1	3	$2.28 \cdot 10^{-1}$
2055.90	3	3	$1.71 \cdot 10^{-1}$
2055.90	3	5	$3.08 \cdot 10^{-1}$
2056.00	5	5	$1.03 \cdot 10^{-1}$
2056.00	5	7	$4.11 \cdot 10^{-1}$
2125.57	3	3	$6.1 \cdot 10^{-2}$
2125.68	5	3	$1.01 \cdot 10^{-1}$
2174.96	1	3	$4.51 \cdot 10^{-1}$
2174.99	3	3	$3.39 \cdot 10^{-1}$
2174.99	3	5	$6.09 \cdot 10^{-1}$
2175.10	5	5	$2.03 \cdot 10^{-1}$
2175.10	5	7	$8.13 \cdot 10^{-1}$
2348.61	1	3	5.54
2350.66	1	3	$4.68 \cdot 10^{-2}$
2350.70	3	3	$1.41 \cdot 10^{-1}$
2350.83	5	3	$2.35 \cdot 10^{-1}$
2494.54	1	3	1.07
2494.58	3	5	1.44
2494.58	3	3	$8.02 \cdot 10^{-1}$
2494.73	5	3	$5.35 \cdot 10^{-2}$
2494.73	5	7	1.93
2494.73	5	5	$4.81 \cdot 10^{-1}$
2650.45	3	5	1.06
2650.55	1	3	1.41
2650.60	3	3	1.06
2650.62	5	5	3.17
2650.70	3	1	4.23
2650.76	5	3	1.76
3193.83	3	5	$5.08 \cdot 10^{-2}$
3229.62	3	5	$7.17 \cdot 10^{-2}$
3282.91	3	5	$1.06 \cdot 10^{-1}$
3321.01	1	3	$1.70 \cdot 10^{-1}$
3321.08	3	3	$5.10 \cdot 10^{-1}$
3321.34	5	3	$8.50 \cdot 10^{-1}$
3367.63	3	5	$1.64 \cdot 10^{-1}$
3476.56	3	1	$3.11 \cdot 10^{-2}$
3515.54	3	5	$2.73 \cdot 10^{-1}$
3736.30	3	1	$5.09 \cdot 10^{-2}$
3813.45	3	5	$4.87 \cdot 10^{-1}$
4407.94	3	1	$1.01 \cdot 10^{-1}$
4572.66	3	5	$7.61 \cdot 10^{-1}$
5252.81	1	3	$3.12 \cdot 10^{-3}$
5365.49	1	3	$3.65 \cdot 10^{-3}$
5546.45	1	3	$5.4 \cdot 10^{-3}$
5857.01	1	3	$7.0 \cdot 10^{-3}$
6085.75	5	3	$1.4 \cdot 10^{-3}$
6229.11	5	7	$2.26 \cdot 10^{-2}$
6319.62	5	3	$2.2 \cdot 10^{-3}$
6473.54	1	3	$7.7 \cdot 10^{-3}$
6564.52	5	7	$3.49 \cdot 10^{-2}$
6711.74	1	3	$3.62 \cdot 10^{-3}$
6711.74	3	5	$4.89 \cdot 10^{-3}$
6711.74	3	3	$2.71 \cdot 10^{-3}$
6711.91	5	7	$6.52 \cdot 10^{-3}$
6711.91	5	5	$1.63 \cdot 10^{-3}$

λ Å	Weights g_i	g_k	A 10^8 s^{-1}
6725.96	5	3	$3.1 \cdot 10^{-3}$
6786.56	3	5	$1.7 \cdot 10^{-3}$
6786.56	3	3	$1.7 \cdot 10^{-3}$
6884.26	3	3	$3.81 \cdot 10^{-3}$
6884.26	3	5	$6.86 \cdot 10^{-3}$
6884.26	1	3	$5.08 \cdot 10^{-3}$
6884.44	5	7	$9.16 \cdot 10^{-3}$
6884.44	5	5	$2.29 \cdot 10^{-3}$
7049.72	1	3	$9.66 \cdot 10^{-4}$
7049.72	3	3	$2.90 \cdot 10^{-3}$
7049.91	5	3	$4.83 \cdot 10^{-3}$
7154.46	3	3	$5.55 \cdot 10^{-3}$
7154.46	1	3	$7.39 \cdot 10^{-3}$
7154.46	3	5	$9.99 \cdot 10^{-3}$
7154.65	5	7	$1.33 \cdot 10^{-2}$
7154.65	5	5	$3.33 \cdot 10^{-3}$
7209.13	5	7	$5.71 \cdot 10^{-2}$
7308.29	3	5	$3.33 \cdot 10^{-3}$
7434.42	1	3	$1.53 \cdot 10^{-3}$
7434.42	3	3	$4.59 \cdot 10^{-3}$
7434.63	5	3	$7.64 \cdot 10^{-3}$
7448.87	3	1	$5.63 \cdot 10^{-3}$
7498.42	3	5	$4.46 \cdot 10^{-3}$
7551.90	5	3	$4.3 \cdot 10^{-3}$
7618.66	3	5	$1.51 \cdot 10^{-2}$
7618.66	1	3	$1.12 \cdot 10^{-2}$
7618.66	3	3	$8.41 \cdot 10^{-3}$
7618.88	5	7	$2.02 \cdot 10^{-2}$
7618.88	5	5	$5.05 \cdot 10^{-3}$
7714.38	3	1	$8.21 \cdot 10^{-3}$
7792.05	3	5	$6.02 \cdot 10^{-3}$
8090.07	1	3	$3.29 \cdot 10^{-3}$
8153.74	3	1	$1.27 \cdot 10^{-2}$
8158.98	1	3	$2.65 \cdot 10^{-3}$
8158.98	3	3	$7.94 \cdot 10^{-3}$
8159.23	5	3	$1.32 \cdot 10^{-2}$
8254.07	3	1	$3.38 \cdot 10^{-1}$
8286.90	3	5	$8.00 \cdot 10^{-3}$
8547.36	3	3	$1.31 \cdot 10^{-2}$
8547.36	1	3	$1.74 \cdot 10^{-2}$
8547.36	3	5	$2.36 \cdot 10^{-2}$
8547.63	5	7	$3.14 \cdot 10^{-2}$
8547.63	5	5	$7.86 \cdot 10^{-3}$
8547.63	5	3	$8.73 \cdot 10^{-4}$
8801.37	5	7	$9.58 \cdot 10^{-2}$
8882.16	7	5	$1.34 \cdot 10^{-3}$
8882.16	5	3	$1.06 \cdot 10^{-3}$
8882.16	3	1	$1.59 \cdot 10^{-3}$
8979.19	3	1	$2.13 \cdot 10^{-2}$
9190.45	5	7	$1.43 \cdot 10^{-2}$
9190.45	3	5	$1.35 \cdot 10^{-2}$
9190.45	7	7	$1.79 \cdot 10^{-3}$
9190.45	5	5	$2.50 \cdot 10^{-3}$
9243.88	3	5	$9.47 \cdot 10^{-3}$
9392.74	5	3	$1.86 \cdot 10^{-3}$
9392.74	3	3	$6.19 \cdot 10^{-4}$
9392.74	5	5	$3.71 \cdot 10^{-4}$
9392.74	7	5	$2.08 \cdot 10^{-3}$
9392.74	3	1	$2.47 \cdot 10^{-3}$
9847.31	5	3	$5.2 \cdot 10^{-3}$

Atomic

λ Å	Weights g_i	Weights g_k	A 10^8 s^{-1}
9895.58	3	3	$1.59 \cdot 10^{-2}$
9895.58	1	3	$5.30 \cdot 10^{-3}$
9895.95	5	3	$2.65 \cdot 10^{-2}$
9939.78	5	7	$2.51 \cdot 10^{-2}$
9939.78	5	5	$4.40 \cdot 10^{-3}$
9939.78	7	7	$3.14 \cdot 10^{-3}$
9939.78	3	5	$2.38 \cdot 10^{-2}$

Be II

λ Å	Weights g_i	Weights g_k	A 10^8 s^{-1}
973.276	4	6	$9.69 \cdot 10^{-1}$
1026.89	2	4	1.46
1026.96	4	6	1.76
1036.30	2	4	1.720
1142.96	2	4	3.14
1143.04	4	6	3.76
1512.27	2	4	9.217
1512.41	4	6	$1.106 \cdot 10^{1}$
1512.42	4	4	1.843
1776.10	2	2	1.361
1776.31	4	2	2.722
2296.81	2	4	$1.09 \cdot 10^{-1}$
2296.91	4	6	$1.30 \cdot 10^{-1}$
2302.96	4	6	$7.28 \cdot 10^{-2}$
2302.98	6	8	$7.80 \cdot 10^{-2}$
2381.95	4	6	$1.13 \cdot 10^{-1}$
2381.98	6	8	$1.22 \cdot 10^{-1}$
2413.34	2	4	$1.64 \cdot 10^{-1}$
2413.45	4	6	$1.97 \cdot 10^{-1}$
2453.84	2	4	$1.06 \cdot 10^{-1}$
2507.41	4	6	$1.82 \cdot 10^{-1}$
2507.45	6	8	$1.94 \cdot 10^{-1}$
2617.99	2	4	$2.65 \cdot 10^{-1}$
2618.13	4	6	$3.18 \cdot 10^{-1}$
2697.59	4	2	$1.48 \cdot 10^{-1}$
2728.85	4	6	$3.23 \cdot 10^{-1}$
2728.89	6	8	$3.46 \cdot 10^{-1}$
3046.52	2	4	$4.66 \cdot 10^{-1}$
3046.69	4	6	$5.60 \cdot 10^{-1}$
3046.70	4	4	$9.33 \cdot 10^{-2}$
3130.42	2	4	1.1292
3131.07	2	2	1.1285
3197.10	4	6	$6.82 \cdot 10^{-1}$
3197.15	6	8	$7.31 \cdot 10^{-1}$
3197.16	6	6	$4.87 \cdot 10^{-2}$
3233.54	6	4	$3.37 \cdot 10^{-2}$
3241.63	2	2	$1.39 \cdot 10^{-1}$
3241.83	4	2	$2.78 \cdot 10^{-1}$
3274.59	2	4	$1.41 \cdot 10^{-1}$
3274.67	2	2	$1.41 \cdot 10^{-1}$
4360.66	2	4	$9.12 \cdot 10^{-1}$
4360.99	4	6	1.09
4361.03	4	4	$1.82 \cdot 10^{-1}$
4673.33	4	6	2.06
4673.42	6	8	2.21
4673.45	6	6	$1.47 \cdot 10^{-1}$
4702.34	2	4	$4.73 \cdot 10^{-2}$
4702.52	4	6	$5.68 \cdot 10^{-2}$
4828.12	6	4	$7.89 \cdot 10^{-2}$
4828.18	4	2	$8.77 \cdot 10^{-2}$
5218.12	2	4	$7.08 \cdot 10^{-2}$
5218.34	4	6	$8.49 \cdot 10^{-2}$
5218.34	4	4	$1.41 \cdot 10^{-2}$
5270.27	2	2	$3.234 \cdot 10^{-1}$
5270.81	4	2	$6.466 \cdot 10^{-1}$
5416.12	2	2	$2.84 \cdot 10^{-2}$
5416.36	4	2	$5.68 \cdot 10^{-2}$
6279.42	2	4	$1.12 \cdot 10^{-1}$
6279.74	4	4	$2.23 \cdot 10^{-2}$
6279.74	4	6	$1.34 \cdot 10^{-1}$
6756.75	2	2	$4.90 \cdot 10^{-2}$
6757.12	4	2	$9.80 \cdot 10^{-2}$
7401.20	2	4	$2.54 \cdot 10^{-2}$
7401.43	2	2	$2.54 \cdot 10^{-2}$
9048.14	2	4	$2.47 \cdot 10^{-2}$
9048.49	4	6	$2.96 \cdot 10^{-2}$
9048.49	4	4	$4.94 \cdot 10^{-3}$
9476.41	2	4	$1.81 \cdot 10^{-1}$
9477.03	4	6	$2.17 \cdot 10^{-1}$
9477.14	4	4	$3.62 \cdot 10^{-2}$

Bismuth

Bi I

λ Å	Weights g_i	Weights g_k	A 10^8 s^{-1}
1954.5	4	6	1.2
2021.2	4	4	$6.0 \cdot 10^{-2}$
2061.7	4	6	$9.9 \cdot 10^{-1}$
2110.3	4	2	$9.1 \cdot 10^{-1}$
2177.3	4	2	$2.6 \cdot 10^{-2}$
2228.3	4	4	$8.9 \cdot 10^{-1}$
2230.6	4	6	2.6
2276.6	4	4	$2.5 \cdot 10^{-1}$
2515.7	4	6	$4.3 \cdot 10^{-2}$
2627.9	4	4	$4.7 \cdot 10^{-1}$
2696.8	4	6	$6.4 \cdot 10^{-2}$
2780.5	4	2	$3.09 \cdot 10^{-1}$
2798.7	6	6	$3.6 \cdot 10^{-2}$
2898.0	4	2	1.53
2938.3	6	4	1.23
2989.0	4	4	$5.5 \cdot 10^{-1}$
2993.3	4	6	$1.6 \cdot 10^{-1}$
3024.6	6	6	$8.8 \cdot 10^{-1}$
3067.7	4	2	2.07
3076.7	4	4	$3.5 \cdot 10^{-2}$
3397.2	6	4	$1.81 \cdot 10^{-1}$
3402.9	6	6	$1.6 \cdot 10^{-2}$
3510.9	6	4	$6.8 \cdot 10^{-2}$
3596.1	2	4	$1.98 \cdot 10^{-1}$
3888.2	2	2	$6.9 \cdot 10^{-2}$
4121.5	2	2	$1.64 \cdot 10^{-1}$
4308.5	2	4	$1.6 \cdot 10^{-2}$
4493.0	2	4	$1.5 \cdot 10^{-2}$
4722.5	4	2	$1.17 \cdot 10^{-1}$
6134.8	4	4	$1.8 \cdot 10^{-2}$

Boron

B I

λ Å	Weights g_i	Weights g_k	A 10^8 s^{-1}
1151.42	4	4	2.44
1378.65	2	4	3.42
1378.87	2	2	$1.37 \cdot 10^{1}$
1378.94	4	4	$1.71 \cdot 10^{1}$
1379.17	4	2	6.84
1465.56	2	4	3.3
1465.66	4	4	6.6
1465.79	6	4	9.9
1566.66	4	6	$3.36 \cdot 10^{-1}$
1587.38	2	4	$6.9 \cdot 10^{-1}$
1587.45	4	2	1.4
1587.59	4	6	$5.0 \cdot 10^{-1}$
1587.66	6	4	$7.5 \cdot 10^{-1}$
1587.75	6	6	1.2
1600.37	2	4	$4.67 \cdot 10^{-1}$
1600.76	4	6	$5.60 \cdot 10^{-1}$
1666.85	2	4	$8.66 \cdot 10^{-1}$
1667.27	4	6	1.04
1818.35	4	2	$6.22 \cdot 10^{-1}$
1825.89	2	4	1.70
1826.40	4	6	2.04
1826.40	4	4	$3.39 \cdot 10^{-1}$
2066.38	4	6	$6.8 \cdot 10^{-1}$
2066.65	6	6	1.6
2066.73	2	4	$9.4 \cdot 10^{-1}$
2066.93	4	4	$3.0 \cdot 10^{-1}$
2067.20	6	4	1.0
2067.20	4	2	1.9
2088.89	2	4	$3.61 \cdot 10^{-1}$
2089.57	4	6	$4.32 \cdot 10^{-1}$
2496.77	2	2	$8.37 \cdot 10^{-1}$
2497.72	4	2	1.67
5761.90	6	8	$5.7 \cdot 10^{-3}$
5942.62	6	8	$8.4 \cdot 10^{-3}$
5942.73	4	6	$7.8 \cdot 10^{-3}$
6244.56	6	8	$1.3 \cdot 10^{-2}$
6244.68	4	6	$1.2 \cdot 10^{-2}$
6819.52	6	8	$2.2 \cdot 10^{-2}$
6819.66	4	6	$2.0 \cdot 10^{-2}$
7208.59	4	2	$1.72 \cdot 10^{-2}$
8211.79	6	8	$3.42 \cdot 10^{-2}$
8211.79	6	6	$2.28 \cdot 10^{-3}$
8212.00	4	6	$3.19 \cdot 10^{-2}$
8667.23	2	2	$1.83 \cdot 10^{-2}$
8668.57	4	2	$3.67 \cdot 10^{-2}$
9576.21	4	6	$5.01 \cdot 10^{-3}$
9576.34	6	8	$5.37 \cdot 10^{-3}$

B II

λ Å	Weights g_i	Weights g_k	A 10^8 s^{-1}
882.676	5	7	$2.44 \cdot 10^{1}$
984.698	5	7	$2.62 \cdot 10^{1}$
1230.17	3	5	$1.36 \cdot 10^{1}$
1624.02	5	5	6.48
2005.87	7	7	2.22
2123.86	5	7	1.58
2220.30	3	3	3.49
2918.08	7	9	$6.57 \cdot 10^{-1}$
3323.21	3	5	$8.62 \cdot 10^{-1}$
3323.60	5	7	1.15
3451.30	3	5	$5.41 \cdot 10^{-1}$
4121.93	5	5	$3.68 \cdot 10^{-1}$
4121.93	5	7	2.10
4121.93	7	9	2.36
4121.93	7	7	$2.63 \cdot 10^{-1}$

λ Å	Weights g_i	g_k	A 10^8 s^{-1}
4121.93	3	5	1.98
4472.15	3	3	$5.44 \cdot 10^{-1}$
4472.86	5	3	$9.07 \cdot 10^{-1}$
4784.20	7	5	$1.94 \cdot 10^{-1}$
4940.37	5	7	1.88
6148.91	5	7	$1.96 \cdot 10^{-1}$
6148.91	3	5	$1.85 \cdot 10^{-1}$
6148.91	7	9	$2.20 \cdot 10^{-1}$
6285.51	5	3	$3.30 \cdot 10^{-1}$
6571.12	3	5	$1.14 \cdot 10^{-1}$
6717.65	5	7	$1.71 \cdot 10^{-1}$
7030.27	3	5	$3.97 \cdot 10^{-1}$
7032.03	3	3	$3.97 \cdot 10^{-1}$
7032.33	3	1	$3.97 \cdot 10^{-1}$
7159.55	3	5	$1.19 \cdot 10^{-1}$
7160.16	1	3	$1.59 \cdot 10^{-1}$
7165.11	3	3	$1.19 \cdot 10^{-1}$
7168.46	3	1	$4.75 \cdot 10^{-1}$
7170.45	5	5	$3.56 \cdot 10^{-1}$
7176.02	5	3	$1.97 \cdot 10^{-1}$
7228.46	1	3	$1.39 \cdot 10^{-1}$
7638.62	5	7	$1.71 \cdot 10^{-1}$
7638.62	3	5	$1.28 \cdot 10^{-1}$
7638.62	1	3	$9.5 \cdot 10^{-2}$
8655.83	5	3	$6.4 \cdot 10^{-2}$
9121.00	3	1	$3.23 \cdot 10^{-1}$
9226.43	7	9	$4.37 \cdot 10^{-1}$
9226.43	5	7	$3.88 \cdot 10^{-1}$
9226.43	3	5	$3.67 \cdot 10^{-1}$
9226.43	7	7	$4.87 \cdot 10^{-2}$
9226.43	5	5	$6.8 \cdot 10^{-2}$
9446.36	3	5	$2.08 \cdot 10^{-1}$
9659.49	3	5	$6.3 \cdot 10^{-2}$

Bromine

Br I

λ Å	g_i	g_k	A 10^8 s^{-1}
1488.5	4	4	1.2
1540.7	4	4	1.4
1574.8	2	4	$2.0 \cdot 10^{-1}$
1576.4	4	6	$2.1 \cdot 10^{-2}$
1633.4	2	4	$8.1 \cdot 10^{-2}$
4365.1	2	4	$7.5 \cdot 10^{-3}$
4425.1	4	2	$4.2 \cdot 10^{-3}$
4441.7	6	4	$7.5 \cdot 10^{-3}$
4472.6	4	4	$9.3 \cdot 10^{-3}$
4477.7	6	8	$1.3 \cdot 10^{-2}$
4513.4	6	4	$2.8 \cdot 10^{-3}$
4525.6	6	6	$7.2 \cdot 10^{-3}$
4575.7	4	4	$1.6 \cdot 10^{-2}$
4614.6	4	6	$5.4 \cdot 10^{-3}$
4979.8	4	4	$2.6 \cdot 10^{-3}$
5245.1	2	4	$3.1 \cdot 10^{-3}$
5345.4	2	4	$7.6 \cdot 10^{-3}$
7348.5	4	6	$1.2 \cdot 10^{-1}$
7513.0	6	4	$1.2 \cdot 10^{-1}$
7803.0	2	4	$5.3 \cdot 10^{-2}$
7938.7	6	6	$1.9 \cdot 10^{-1}$
8131.5	2	4	$3.8 \cdot 10^{-2}$
8343.7	2	2	$2.2 \cdot 10^{-1}$
8446.6	4	4	$1.2 \cdot 10^{-1}$
8638.7	6	4	$9.7 \cdot 10^{-2}$

Br II

λ Å	g_i	g_k	A 10^8 s^{-1}
4704.9	5	7	1.1
4785.5	5	5	$9.4 \cdot 10^{-1}$
4816.7	5	3	1.1

Cadmium

Cd I

λ Å	g_i	g_k	A 10^8 s^{-1}
2288.0	1	3	5.3
2836.9	1	3	$2.8 \cdot 10^{-1}$
2880.8	3	5	$4.2 \cdot 10^{-1}$
2881.2	3	3	$2.4 \cdot 10^{-1}$
2980.6	5	7	$5.9 \cdot 10^{-1}$
2981.4	5	5	$1.5 \cdot 10^{-1}$
3261.1	1	3	$4.06 \cdot 10^{-3}$
3403.7	1	3	$7.7 \cdot 10^{-1}$
3466.2	3	5	1.2
3467.7	3	3	$6.7 \cdot 10^{-1}$
3610.5	5	7	1.3
3612.9	5	5	$3.5 \cdot 10^{-1}$
4140.5	3	5	$4.7 \cdot 10^{-2}$
4662.4	3	5	$5.5 \cdot 10^{-2}$
4678.1	1	3	$1.3 \cdot 10^{-1}$
4799.9	3	3	$4.1 \cdot 10^{-1}$
5085.8	5	3	$5.6 \cdot 10^{-1}$
6438.5	3	5	$5.9 \cdot 10^{-1}$

Cd II

λ Å	g_i	g_k	A 10^8 s^{-1}
2144.4	2	4	2.8
2265.0	2	2	3.0
2572.9	2	2	1.7
2748.5	4	2	2.8
4415.6	4	6	$1.4 \cdot 10^{-2}$

Calcium

Ca I

λ Å	g_i	g_k	A 10^8 s^{-1}
3006.9	5	5	$7.5 \cdot 10^{-1}$
3361.9	5	7	$2.23 \cdot 10^{-1}$
3630.8	3	5	$2.97 \cdot 10^{-1}$
3644.4	5	7	$3.55 \cdot 10^{-1}$
4098.5	7	9	$1.3 \cdot 10^{-1}$
4108.5	5	7	$9.0 \cdot 10^{-1}$
4226.7	1	3	2.18
4283.0	3	5	$4.34 \cdot 10^{-1}$
4289.4	1	3	$6.0 \cdot 10^{-1}$
4299.0	3	3	$4.66 \cdot 10^{-1}$
4302.5	5	5	1.36
4307.7	3	1	1.99
4318.7	5	3	$7.4 \cdot 10^{-1}$
4355.1	5	7	$1.9 \cdot 10^{-1}$
4425.4	1	3	$4.98 \cdot 10^{-1}$
4435.0	3	5	$6.7 \cdot 10^{-1}$
4435.7	3	3	$3.42 \cdot 10^{-1}$
4454.8	5	7	$8.7 \cdot 10^{-1}$
4455.9	5	5	$2.0 \cdot 10^{-1}$
4526.9	5	3	$4.1 \cdot 10^{-1}$
4578.6	3	5	$1.76 \cdot 10^{-1}$
4581.4	5	7	$2.09 \cdot 10^{-1}$
4585.9	7	9	$2.29 \cdot 10^{-1}$
4878.1	5	7	$1.88 \cdot 10^{-1}$
5041.6	5	3	$3.3 \cdot 10^{-1}$
5188.9	3	5	$4.0 \cdot 10^{-1}$
5261.7	3	3	$1.5 \cdot 10^{-1}$
5262.2	3	1	$6.0 \cdot 10^{-1}$
5264.2	5	5	$9.1 \cdot 10^{-2}$
5265.6	5	3	$4.4 \cdot 10^{-1}$
5270.3	7	5	$5.0 \cdot 10^{-1}$
5582.0	5	7	$6.0 \cdot 10^{-2}$
5588.8	7	7	$4.9 \cdot 10^{-1}$
5590.1	3	5	$8.3 \cdot 10^{-2}$
5594.5	5	5	$3.8 \cdot 10^{-1}$
5598.5	3	3	$4.3 \cdot 10^{-1}$
5601.3	7	5	$8.6 \cdot 10^{-2}$
5602.9	5	3	$1.4 \cdot 10^{-1}$
5857.5	3	5	$6.6 \cdot 10^{-1}$
6102.7	1	3	$9.6 \cdot 10^{-2}$
6122.2	3	3	$2.87 \cdot 10^{-1}$
6162.2	5	3	$3.54 \cdot 10^{-1}$
6169.1	5	3	$1.7 \cdot 10^{-1}$
6169.6	7	5	$1.9 \cdot 10^{-1}$
6439.1	7	9	$5.3 \cdot 10^{-1}$
6449.8	3	5	$9.0 \cdot 10^{-2}$
6462.6	5	7	$4.7 \cdot 10^{-1}$
6471.7	7	7	$5.9 \cdot 10^{-2}$
6493.8	3	5	$4.4 \cdot 10^{-1}$
6499.7	5	5	$8.1 \cdot 10^{-2}$

Ca II

λ Å	g_i	g_k	A 10^8 s^{-1}
1341.9	2	4	$1.5 \cdot 10^{-2}$
1342.5	2	2	$1.5 \cdot 10^{-2}$
1649.9	2	4	$3.2 \cdot 10^{-3}$
1652.0	2	2	$3.1 \cdot 10^{-3}$
1673.9	2	4	$2.24 \cdot 10^{-1}$
1680.1	4	6	$2.65 \cdot 10^{-1}$
1680.1	4	4	$4.41 \cdot 10^{-2}$
1807.3	2	4	$3.54 \cdot 10^{-1}$
1814.5	4	6	$4.2 \cdot 10^{-1}$
1814.7	4	4	$7.0 \cdot 10^{-2}$
1843.1	2	2	$1.6 \cdot 10^{-1}$
1850.7	4	2	$3.08 \cdot 10^{-1}$
2103.2	2	4	$8.2 \cdot 10^{-1}$
2112.8	4	6	$9.7 \cdot 10^{-1}$
2113.2	4	4	$1.6 \cdot 10^{-1}$
2197.8	2	2	$3.1 \cdot 10^{-1}$
2208.6	4	2	$6.2 \cdot 10^{-1}$
3158.9	2	4	3.1
3179.3	4	6	3.6
3181.3	4	4	$5.8 \cdot 10^{-1}$
3706.0	2	2	$8.8 \cdot 10^{-1}$
3736.9	4	2	1.7
3933.7	2	4	1.47
3968.5	2	2	1.4

Ca III

λ Å	g_i	g_k	A 10^8 s^{-1}
357.97	1	3	$8.8 \cdot 10^{2}$
439.69	1	3	$1.9 \cdot 10^{-1}$
490.55	1	3	$1.6 \cdot 10^{-2}$

λ Å	Weights g_i	Weights g_k	A $10^8\ s^{-1}$
Ca V			
558.60	5	3	$2.2\cdot10^{1}$
637.93	5	3	3.9
643.12	3	1	9.1
646.57	5	5	6.9
647.88	3	3	2.3
651.55	1	3	2.9
656.76	3	5	2.1
Ca VII			
550.20	5	5	$1.8\cdot10^{1}$
624.39	1	3	3.3
630.54	3	5	4.5
630.79	3	3	2.2
639.15	5	7	5.7
640.41	5	5	1.3
Ca VIII			
182.71	2	2	$1.6\cdot10^{2}$
184.16	4	2	$3.2\cdot10^{2}$
Ca IX			
163.23	5	3	$3.76\cdot10^{2}$
371.89	1	3	$8.8\cdot10^{1}$
373.81	3	5	$1.16\cdot10^{2}$
378.08	5	7	$1.5\cdot10^{2}$
395.03	3	5	$2.2\cdot10^{2}$
466.24	1	3	$1.12\cdot10^{2}$
498.01	3	5	$2.49\cdot10^{1}$
506.18	5	5	$7.2\cdot10^{1}$
515.57	5	3	$3.75\cdot10^{1}$
Ca X			
110.96	2	4	$2.9\cdot10^{2}$
111.20	2	2	$2.92\cdot10^{2}$
151.84	2	2	$2.3\cdot10^{2}$
153.02	4	2	$4.5\cdot10^{2}$
206.57	4	4	$2.9\cdot10^{1}$
206.75	6	4	$2.6\cdot10^{2}$
207.39	4	2	$2.8\cdot10^{2}$
411.70	2	4	$8.3\cdot10^{1}$
419.75	4	6	$9.5\cdot10^{1}$
420.47	4	4	$1.6\cdot10^{1}$
557.76	2	4	$3.50\cdot10^{1}$
574.01	2	2	$3.2\cdot10^{1}$
Ca XI			
30.448	1	3	$6.2\cdot10^{3}$
30.867	1	3	$4.9\cdot10^{4}$
35.212	1	3	$2.0\cdot10^{3}$
Ca XII			
140.05	4	2	$3.7\cdot10^{2}$
147.27	2	2	$1.6\cdot10^{2}$
Ca XV			
141.69	5	3	$4.08\cdot10^{2}$
*142.23	9	3	$6.3\cdot10^{2}$
161.00	5	5	$1.9\cdot10^{2}$
Ca XVII			
19.558	1	3	$3.8\cdot10^{4}$
21.198	3	5	$4.9\cdot10^{4}$
192.82	1	3	$1.21\cdot10^{2}$
218.82	3	5	$2.76\cdot10^{1}$
223.02	1	3	$3.44\cdot10^{1}$
228.72	3	3	$2.37\cdot10^{1}$
232.83	5	5	$6.5\cdot10^{1}$
244.06	5	3	$3.28\cdot10^{1}$
Ca XVIII			
*18.71	2	6	$2.31\cdot10^{4}$
*19.74	6	10	$7.0\cdot10^{4}$
302.19	2	4	$2.0\cdot10^{1}$
344.76	2	2	$1.3\cdot10^{1}$
Carbon			
C I			
1121.52	1	3	$1.74\cdot10^{-2}$
1121.64	3	1	$1.10\cdot10^{-1}$
1121.66	3	3	$3.50\cdot10^{-2}$
1121.92	3	3	$3.60\cdot10^{-2}$
1122.00	5	3	$3.96\cdot10^{-2}$
1122.00	3	5	$8.49\cdot10^{-2}$
1122.10	5	5	$6.71\cdot10^{-2}$
1122.18	5	7	$3.38\cdot10^{-2}$
1122.33	5	7	$1.06\cdot10^{-1}$
1122.45	1	3	$9.04\cdot10^{-2}$
1122.52	1	3	$2.03\cdot10^{-2}$
1122.65	3	3	$2.73\cdot10^{-2}$
1122.79	3	5	$7.25\cdot10^{-2}$
1123.11	5	7	$3.58\cdot10^{-2}$
1128.07	1	3	$2.37\cdot10^{-2}$
1128.17	1	3	$1.83\cdot10^{-2}$
1128.25	3	1	$1.55\cdot10^{-1}$
1128.28	3	3	$4.96\cdot10^{-2}$
1128.63	5	3	$5.80\cdot10^{-2}$
1128.69	3	3	$4.79\cdot10^{-2}$
1128.75	5	5	$1.02\cdot10^{-1}$
1128.82	3	5	$1.61\cdot10^{-1}$
1128.90	5	7	$3.98\cdot10^{-2}$
1129.03	5	3	$1.82\cdot10^{-2}$
1129.13	5	7	$1.70\cdot10^{-1}$
1129.20	1	3	$1.35\cdot10^{-1}$
1129.32	1	3	$3.46\cdot10^{-2}$
1129.40	3	3	$4.86\cdot10^{-2}$
1129.42	5	5	$1.19\cdot10^{-1}$
1129.59	3	1	$6.41\cdot10^{-2}$
1129.62	3	5	$7.29\cdot10^{-2}$
1129.87	5	3	$2.22\cdot10^{-2}$
1129.92	5	7	$4.84\cdot10^{-2}$
1138.38	1	3	$4.13\cdot10^{-2}$
1138.56	3	1	$2.31\cdot10^{-1}$
1138.56	1	3	$1.70\cdot10^{-2}$
1138.60	3	3	$7.31\cdot10^{-2}$
1138.95	5	3	$8.98\cdot10^{-2}$
1139.09	5	5	$1.60\cdot10^{-1}$
1139.30	3	3	$6.24\cdot10^{-2}$
1139.43	5	7	$4.31\cdot10^{-2}$
1139.51	3	5	$2.72\cdot10^{-1}$
1139.65	5	3	$2.26\cdot10^{-2}$
1139.77	3	5	$4.63\cdot10^{-2}$
1139.79	1	3	$2.11\cdot10^{-1}$
1139.81	5	7	$2.96\cdot10^{-1}$
1139.86	5	5	$1.30\cdot10^{-2}$
1140.01	3	3	$9.45\cdot10^{-2}$
1140.01	1	3	$6.10\cdot10^{-2}$
1140.12	5	5	$1.84\cdot10^{-1}$
1140.32	3	1	$1.15\cdot10^{-1}$
1140.36	3	5	$6.39\cdot10^{-2}$
1140.57	5	3	$4.17\cdot10^{-2}$
1140.64	5	7	$6.29\cdot10^{-2}$
1155.81	1	3	$8.21\cdot10^{-2}$
1155.98	3	1	$3.74\cdot10^{-1}$
1156.03	3	3	$1.14\cdot10^{-1}$
1156.20	3	5	$2.39\cdot10^{-2}$
1156.39	5	3	$1.51\cdot10^{-1}$
1156.56	5	5	$2.69\cdot10^{-1}$
1157.33	5	7	$3.99\cdot10^{-2}$
1157.41	3	3	$7.50\cdot10^{-2}$
1157.77	5	3	$2.50\cdot10^{-2}$
1157.77	3	5	$2.62\cdot10^{-1}$
1157.91	1	3	$3.52\cdot10^{-1}$
1158.02	5	7	$5.57\cdot10^{-1}$
1158.03	3	5	$3.10\cdot10^{-1}$
1158.13	3	3	$1.98\cdot10^{-1}$
1158.13	5	5	$2.58\cdot10^{-1}$
1158.32	1	3	$1.09\cdot10^{-1}$
1158.40	5	5	$8.20\cdot10^{-2}$
1158.54	3	3	$3.36\cdot10^{-2}$
1158.67	3	1	$2.19\cdot10^{-1}$
1158.73	3	5	$5.11\cdot10^{-2}$
1158.91	5	3	$8.37\cdot10^{-2}$
1158.97	5	7	$7.22\cdot10^{-2}$
1188.83	1	3	$1.95\cdot10^{-1}$
1188.99	3	1	$7.18\cdot10^{-1}$
1189.07	3	3	$2.01\cdot10^{-1}$
1189.25	3	5	$1.09\cdot10^{-1}$
1189.45	5	3	$2.96\cdot10^{-1}$
1189.63	5	5	$5.28\cdot10^{-1}$
1191.84	5	7	$2.91\cdot10^{-2}$
1192.22	1	3	$2.15\cdot10^{-2}$
1192.83	5	3	$2.19\cdot10^{-2}$
1193.01	3	5	$7.91\cdot10^{-1}$
1193.03	1	3	$6.39\cdot10^{-1}$
1193.24	5	7	1.11
1193.26	3	3	$4.29\cdot10^{-1}$
1193.39	5	5	$3.46\cdot10^{-1}$
1193.65	5	3	$2.25\cdot10^{-2}$
1193.68	3	5	$2.54\cdot10^{-1}$
1194.00	1	3	$1.94\cdot10^{-1}$
1194.06	5	5	$2.98\cdot10^{-1}$
1194.23	3	3	$8.34\cdot10^{-2}$
1194.30	3	5	$4.10\cdot10^{-2}$
1194.41	3	1	$4.40\cdot10^{-1}$
1194.49	5	7	$6.77\cdot10^{-2}$
1194.61	5	3	$1.77\cdot10^{-1}$
1260.74	1	3	$5.70\cdot10^{-1}$

λ Å	Weights g_i	g_k	A 10^8 s^{-1}
1260.93	3	1	1.81
1261.00	3	3	$4.68 \cdot 10^{-1}$
1261.12	3	5	$4.00 \cdot 10^{-1}$
1261.43	5	3	$7.51 \cdot 10^{-1}$
1261.55	5	5	1.34
1274.11	5	7	$1.20 \cdot 10^{-2}$
1276.48	1	3	$5.04 \cdot 10^{-2}$
1276.75	3	3	$9.38 \cdot 10^{-2}$
1277.19	5	3	$1.81 \cdot 10^{-2}$
1277.25	1	3	1.26
1277.28	3	5	1.70
1277.51	3	3	$9.24 \cdot 10^{-1}$
1277.55	5	7	2.28
1277.72	5	5	$6.30 \cdot 10^{-1}$
1277.95	5	3	$5.94 \cdot 10^{-2}$
1279.89	3	5	$3.43 \cdot 10^{-1}$
1280.14	1	3	$3.55 \cdot 10^{-1}$
1280.33	5	5	$6.38 \cdot 10^{-1}$
1280.40	3	3	$1.87 \cdot 10^{-1}$
1280.60	3	1	$8.82 \cdot 10^{-1}$
1280.85	5	3	$3.62 \cdot 10^{-1}$
1328.83	1	3	$7.31 \cdot 10^{-1}$
1329.09	3	1	2.22
1329.10	3	5	$5.39 \cdot 10^{-1}$
1329.12	3	3	$5.55 \cdot 10^{-1}$
1329.58	5	5	1.64
1329.60	5	3	$9.19 \cdot 10^{-1}$
1459.03	5	3	$5.45 \cdot 10^{-1}$
1463.34	5	7	1.79
1467.40	5	3	$5.49 \cdot 10^{-1}$
1470.09	5	7	$1.41 \cdot 10^{-2}$
1472.23	5	3	$7.79 \cdot 10^{-3}$
1481.76	5	5	$3.45 \cdot 10^{-1}$
1560.31	1	3	$6.54 \cdot 10^{-1}$
1560.68	3	5	$8.82 \cdot 10^{-1}$
1560.71	3	3	$4.89 \cdot 10^{-1}$
1561.34	5	5	$2.93 \cdot 10^{-1}$
1561.37	5	3	$3.25 \cdot 10^{-2}$
1561.44	5	7	1.17
1656.27	3	5	$8.72 \cdot 10^{-1}$
1656.93	1	3	1.16
1657.01	5	5	2.61
1657.38	3	3	$8.66 \cdot 10^{-1}$
1657.91	3	1	3.47
1658.12	5	3	1.44
1751.83	1	3	$8.38 \cdot 10^{-1}$
1763.91	1	3	$2.44 \cdot 10^{-2}$
1930.90	5	3	3.39
2478.56	1	3	$2.80 \cdot 10^{-1}$

C II

λ Å	Weights g_i	g_k	A 10^8 s^{-1}
687.053	2	4	$2.35 \cdot 10^{1}$
687.345	4	6	$2.82 \cdot 10^{1}$
687.352	4	4	4.70
806.384	4	6	2.53
806.533	2	4	3.51
806.568	6	6	5.89
806.677	4	4	1.12
806.687	2	2	1.40
806.830	4	2	7.01
806.861	6	4	3.79
858.092	2	2	1.49
858.559	4	2	2.93
903.623	2	4	6.78
903.962	2	2	$2.70 \cdot 10^{1}$
904.142	4	4	$3.39 \cdot 10^{1}$
904.480	4	2	$1.36 \cdot 10^{1}$
1009.86	2	4	5.65
1010.08	4	4	$1.13 \cdot 10^{1}$
1010.37	6	4	$1.69 \cdot 10^{1}$
1036.34	2	2	7.38
1037.02	4	2	$1.46 \cdot 10^{1}$
1065.89	6	4	$1.30 \cdot 10^{1}$
1065.92	4	4	1.44
1066.13	4	2	$1.45 \cdot 10^{1}$
1126.99	2	4	$1.01 \cdot 10^{-6}$
1127.13	2	2	$5.55 \cdot 10^{-7}$
1127.27	4	4	$1.91 \cdot 10^{-7}$
1127.41	4	2	$9.06 \cdot 10^{-8}$
1127.63	6	4	$1.78 \cdot 10^{-6}$
1323.86	6	4	$4.94 \cdot 10^{-1}$
1323.91	4	4	4.38
1323.95	6	6	4.55
1324.00	4	6	$3.27 \cdot 10^{-1}$
1334.53	2	4	2.41
1335.66	4	4	$4.76 \cdot 10^{-1}$
1335.71	4	6	2.88
1384.00	2	4	$4.07 \cdot 10^{-1}$
1384.36	2	2	$3.88 \cdot 10^{-1}$
1490.38	6	4	$2.02 \cdot 10^{-6}$
1720.46	2	4	$8.65 \cdot 10^{-1}$
1721.01	2	2	3.46
1721.68	4	4	4.33
1722.24	4	2	1.72
1760.40	6	4	$3.68 \cdot 10^{-1}$
1760.47	4	4	$4.08 \cdot 10^{-2}$
1760.82	4	2	$4.09 \cdot 10^{-1}$
1915.32	2	4	$9.36 \cdot 10^{-2}$
1916.01	2	2	$9.50 \cdot 10^{-2}$
2323.50	2	4	$1.40 \cdot 10^{-8}$
2324.69	2	2	$5.99 \cdot 10^{-7}$
2325.40	4	6	$4.43 \cdot 10^{-7}$
2326.93	4	4	$8.49 \cdot 10^{-8}$
2328.12	4	2	$6.78 \cdot 10^{-7}$
2509.13	2	4	$4.71 \cdot 10^{-1}$
2511.74	4	4	$9.27 \cdot 10^{-2}$
2512.06	4	6	$5.61 \cdot 10^{-1}$
2836.71	2	4	$3.30 \cdot 10^{-1}$
2837.60	2	2	$3.29 \cdot 10^{-1}$
3183.50	2	4	$2.40 \cdot 10^{-7}$
3187.70	4	4	$8.09 \cdot 10^{-7}$
3922.08	2	4	$1.44 \cdot 10^{-7}$
4309.31	4	4	$3.92 \cdot 10^{-3}$
4309.58	6	4	$3.52 \cdot 10^{-2}$
4312.80	4	2	$3.92 \cdot 10^{-2}$
4735.46	2	4	$9.05 \cdot 10^{-5}$
4737.97	2	2	$5.47 \cdot 10^{-4}$
4744.77	4	4	$5.98 \cdot 10^{-4}$
4747.28	4	2	$2.33 \cdot 10^{-4}$
6578.05	2	4	$3.67 \cdot 10^{-1}$
6582.88	2	2	$3.66 \cdot 10^{-1}$
7231.33	2	4	$3.49 \cdot 10^{-1}$
7236.42	4	6	$4.18 \cdot 10^{-1}$
7237.17	4	4	$6.96 \cdot 10^{-2}$

C III

λ Å	Weights g_i	g_k	A 10^8 s^{-1}
310.170	1	3	6.56
386.203	1	3	$3.46 \cdot 10^{1}$
459.466	1	3	$5.91 \cdot 10^{1}$
459.514	3	5	$7.97 \cdot 10^{1}$
459.627	5	7	$1.06 \cdot 10^{2}$
574.281	3	5	$6.24 \cdot 10^{1}$
977.020	1	3	$1.767 \cdot 10^{1}$
1174.93	3	5	3.293
1175.26	1	3	4.385
1175.59	3	3	3.287
1175.71	5	5	9.856
1175.99	3	1	$1.313 \cdot 10^{1}$
1176.37	5	3	5.468
1247.38	3	1	$2.082 \cdot 10^{1}$
2296.87	3	5	1.376
2849.05	3	1	$1.95 \cdot 10^{-1}$
3703.70	3	3	$5.90 \cdot 10^{-1}$
4325.56	3	5	$1.24 \cdot 10^{-1}$
4647.42	3	5	$7.26 \cdot 10^{-1}$
4650.25	3	3	$7.25 \cdot 10^{-1}$
4651.02	3	5	$2.28 \cdot 10^{-1}$
4651.47	3	1	$7.24 \cdot 10^{-1}$
4652.05	1	3	$3.04 \cdot 10^{-1}$
4659.06	3	3	$2.27 \cdot 10^{-1}$
4663.64	3	1	$9.05 \cdot 10^{-1}$
4665.86	5	5	$6.78 \cdot 10^{-1}$
4673.95	5	3	$3.75 \cdot 10^{-1}$
5244.66	1	3	$5.30 \cdot 10^{-2}$
5253.58	3	3	$1.58 \cdot 10^{-1}$
5272.52	5	3	$2.61 \cdot 10^{-1}$
5695.92	3	5	$4.27 \cdot 10^{-1}$
5858.34	3	1	$1.34 \cdot 10^{-1}$
5863.25	3	3	$3.35 \cdot 10^{-2}$
5871.68	5	3	$1.00 \cdot 10^{-1}$
5880.56	5	5	$1.99 \cdot 10^{-2}$
5894.07	7	5	$1.11 \cdot 10^{-1}$
6727.48	1	3	$1.12 \cdot 10^{-1}$
6731.04	3	5	$1.50 \cdot 10^{-1}$
6742.15	3	3	$8.32 \cdot 10^{-2}$
6744.39	5	7	$1.99 \cdot 10^{-1}$
6762.17	5	5	$4.95 \cdot 10^{-2}$
6773.39	5	3	$5.47 \cdot 10^{-3}$
6851.18	3	5	$7.60 \cdot 10^{-3}$
6853.68	5	7	$5.64 \cdot 10^{-3}$
6857.24	3	3	$3.79 \cdot 10^{-2}$
6862.69	5	5	$3.51 \cdot 10^{-2}$
6868.78	5	3	$1.26 \cdot 10^{-2}$
6872.04	7	7	$4.46 \cdot 10^{-2}$
6881.10	7	5	$7.80 \cdot 10^{-3}$
7353.88	5	3	$3.09 \cdot 10^{-2}$
7707.43	3	5	$1.30 \cdot 10^{-1}$
7771.76	3	1	$1.77 \cdot 10^{-1}$
7780.41	3	3	$1.76 \cdot 10^{-1}$
7796.00	3	5	$1.75 \cdot 10^{-1}$

λ Å	Weights g_i	g_k	A 10^8 s^{-1}
8500.32	1	3	$1.01 \cdot 10^{-1}$
9593.32	3	3	$5.32 \cdot 10^{-3}$
9651.47	5	5	$1.57 \cdot 10^{-2}$
9696.48	5	7	$7.53 \cdot 10^{-3}$
9696.54	3	5	$7.12 \cdot 10^{-3}$
9699.57	7	9	$8.47 \cdot 10^{-3}$
9701.10	1	3	$4.40 \cdot 10^{-2}$
9705.41	3	5	$5.93 \cdot 10^{-2}$
9706.44	3	3	$3.29 \cdot 10^{-2}$
9715.09	5	7	$7.88 \cdot 10^{-2}$
9717.75	5	5	$1.97 \cdot 10^{-2}$
9718.79	5	3	$2.19 \cdot 10^{-3}$

C IV

λ Å	g_i	g_k	A 10^8 s^{-1}
*312.43	2	6	$4.63 \cdot 10^{1}$
*384.13	6	10	$1.76 \cdot 10^{2}$
1548.19	2	4	2.65
1550.77	2	2	2.64
5801.31	2	4	$3.17 \cdot 10^{-1}$
5811.97	2	2	$3.16 \cdot 10^{-1}$

C V

λ Å	g_i	g_k	A 10^8 s^{-1}
34.9728	1	3	$2.554 \cdot 10^{3}$
40.2678	1	3	$8.873 \cdot 10^{3}$
*227.19	3	9	$1.363 \cdot 10^{2}$
247.315	1	3	$1.278 \cdot 10^{2}$
*248.71	9	15	$4.247 \cdot 10^{2}$
*260.19	9	3	$6.680 \cdot 10^{1}$
267.267	3	5	$3.947 \cdot 10^{2}$
*2273.9	3	9	$5.646 \cdot 10^{-1}$
3526.66	1	3	$1.663 \cdot 10^{-1}$
8420.72	3	5	$6.898 \cdot 10^{-2}$
*8433.2	3	9	$6.868 \cdot 10^{-2}$
8448.12	3	1	$6.832 \cdot 10^{-2}$
8449.19	3	3	$6.829 \cdot 10^{-2}$

Cesium

Cs I

λ Å	g_i	g_k	A 10^8 s^{-1}
3203.5	2	4	$7.6 \cdot 10^{-6}$
3205.3	2	4	$7.9 \cdot 10^{-6}$
3207.5	2	4	$8.5 \cdot 10^{-6}$
3210.0	2	4	$9.4 \cdot 10^{-6}$
3212.8	2	4	$1.19 \cdot 10^{-5}$
3216.2	2	4	$1.49 \cdot 10^{-5}$
3220.1	2	4	$1.7 \cdot 10^{-5}$
3220.2	2	2	$1.07 \cdot 10^{-7}$
3224.8	2	4	$2.0 \cdot 10^{-5}$
3225.0	2	2	$1.43 \cdot 10^{-7}$
3230.5	2	4	$2.5 \cdot 10^{-5}$
3230.7	2	2	$1.97 \cdot 10^{-7}$
3237.4	2	4	$2.8 \cdot 10^{-5}$
3237.6	2	2	$2.63 \cdot 10^{-7}$
3245.9	2	4	$3.45 \cdot 10^{-5}$
3246.2	2	2	$3.7 \cdot 10^{-7}$
3256.7	2	4	$4.25 \cdot 10^{-5}$
3257.1	2	2	$7.0 \cdot 10^{-7}$
3270.5	2	4	$5.6 \cdot 10^{-5}$
3271.0	2	2	$9.8 \cdot 10^{-7}$
3288.6	2	4	$1.0 \cdot 10^{-4}$
3289.3	2	2	$2.7 \cdot 10^{-6}$
3313.1	2	4	$1.6 \cdot 10^{-4}$
3314.0	2	2	$5.2 \cdot 10^{-6}$
3347.5	2	4	$2.2 \cdot 10^{-4}$
3348.8	2	2	$1.1 \cdot 10^{-5}$
3397.9	2	4	$4.0 \cdot 10^{-4}$
3400.0	2	2	$2.4 \cdot 10^{-5}$
3476.8	2	4	$6.6 \cdot 10^{-4}$
3480.0	2	2	$6.6 \cdot 10^{-5}$
3611.4	2	4	$1.5 \cdot 10^{-3}$
3617.3	2	2	$2.5 \cdot 10^{-4}$
3876.1	2	4	$3.8 \cdot 10^{-3}$
3888.6	2	2	$9.7 \cdot 10^{-4}$
4555.3	2	4	$1.88 \cdot 10^{-2}$
4593.2	2	2	$8.0 \cdot 10^{-3}$
8521.1	2	4	$3.276 \cdot 10^{-1}$
8943.5	2	2	$2.87 \cdot 10^{-1}$

Chlorine

Cl I

λ Å	g_i	g_k	A 10^8 s^{-1}
1188.8	4	6	2.33
1188.8	4	4	$2.71 \cdot 10^{-1}$
1201.4	2	4	2.39
1335.7	4	2	1.74
1347.2	4	4	4.19
1351.7	2	2	3.23
1363.4	2	4	$7.5 \cdot 10^{-1}$
4323.3	4	4	$1.1 \cdot 10^{-2}$
4363.3	4	6	$6.8 \cdot 10^{-3}$
4379.9	4	4	$1.4 \cdot 10^{-2}$
4389.8	6	8	$1.4 \cdot 10^{-2}$
4526.2	4	4	$5.1 \cdot 10^{-2}$
4601.0	2	2	$4.2 \cdot 10^{-2}$
4661.2	2	4	$1.2 \cdot 10^{-2}$
7256.6	6	4	$1.5 \cdot 10^{-1}$
7414.1	6	4	$4.7 \cdot 10^{-2}$
7547.1	4	4	$1.2 \cdot 10^{-1}$
7717.6	4	4	$3.0 \cdot 10^{-2}$
7745.0	2	4	$6.3 \cdot 10^{-2}$
7769.2	6	6	$6.0 \cdot 10^{-2}$
7821.4	6	8	$9.8 \cdot 10^{-2}$
7830.8	4	4	$9.7 \cdot 10^{-2}$
7878.2	6	6	$1.8 \cdot 10^{-2}$
7899.3	4	6	$5.1 \cdot 10^{-2}$
7924.6	2	4	$2.1 \cdot 10^{-2}$
7935.0	6	8	$3.9 \cdot 10^{-2}$
7997.9	4	4	$2.1 \cdot 10^{-2}$

Cl II

λ Å	g_i	g_k	A 10^8 s^{-1}
3329.1	5	7	1.5
3522.1	7	7	1.4
3798.8	5	7	1.6
3805.2	7	9	1.8
3809.5	3	5	1.5
3851.0	5	7	1.8
3851.4	5	5	1.6
3854.7	3	5	2.2
3861.9	5	7	2.4
3868.6	7	9	2.7
3913.9	9	9	$8.2 \cdot 10^{-1}$
3990.2	5	7	$8.4 \cdot 10^{-1}$
4132.5	5	5	1.6
4276.5	9	7	$7.6 \cdot 10^{-1}$
4768.7	3	5	$7.7 \cdot 10^{-1}$
4781.3	5	7	1.0
4794.6	5	7	1.04
4810.1	5	5	$9.9 \cdot 10^{-1}$
4819.5	5	3	1.00
4904.8	5	7	$8.1 \cdot 10^{-1}$
4917.7	3	5	$7.5 \cdot 10^{-1}$
5078.3	7	7	$7.7 \cdot 10^{-1}$
5219.1	3	9	$8.6 \cdot 10^{-1}$
5392.1	5	7	1.0

Cl III

λ Å	g_i	g_k	A 10^8 s^{-1}
2298.5	4	4	4.2
2340.6	6	6	4.2
2370.4	8	6	2.8
2531.8	2	4	4.4
2532.5	4	6	5.3
2577.1	4	6	4.3
2580.7	6	8	4.7
2601.2	2	4	4.6
2603.6	4	6	5.0
2609.5	6	8	5.7
2617.0	8	10	6.6
2661.6	4	6	3.4
2665.5	6	8	4.8
2691.5	4	4	3.5
2710.4	4	6	3.5
3340.4	6	6	1.5
3392.9	4	4	1.9
3393.5	6	6	1.9
3530.0	6	8	1.8
3560.7	4	6	1.7
3602.1	6	8	1.7
3612.9	4	6	1.2
3720.5	4	6	1.7

Chromium

Cr I

λ Å	g_i	g_k	A 10^8 s^{-1}
1999.95	9	9	1.4
2726.50	5	7	$7.5 \cdot 10^{-1}$
2769.90	7	5	1.1
2780.70	9	7	1.4
2889.22	9	9	$6.6 \cdot 10^{-1}$
2893.25	7	7	$5.2 \cdot 10^{-1}$
2967.64	7	9	$3.9 \cdot 10^{-1}$
2971.10	5	7	$7.1 \cdot 10^{-1}$
2975.48	3	5	$8.9 \cdot 10^{-1}$
2988.64	5	7	$5.2 \cdot 10^{-1}$
2996.57	5	3	2.0
3000.88	7	5	1.6
3005.06	9	7	$9.2 \cdot 10^{-1}$
3013.72	3	5	$8.3 \cdot 10^{-1}$
3015.20	1	3	1.63
3020.67	3	3	1.5
3021.58	9	11	2.91

λ Å	Weights g_i	Weights g_k	A 10^8 s^{-1}
3024.36	5	5	1.27
3030.25	7	7	1.1
3037.05	9	9	$5.4 \cdot 10^{-1}$
3040.84	7	5	$7.4 \cdot 10^{-1}$
3053.87	9	7	$7.97 \cdot 10^{-1}$
3148.44	9	11	$5.6 \cdot 10^{-1}$
3155.16	11	13	$5.7 \cdot 10^{-1}$
3163.76	13	15	$6.0 \cdot 10^{-1}$
3237.73	9	9	1.3
3578.68	7	9	1.48
3593.48	7	7	1.50
3605.32	7	5	1.62
3639.80	13	11	1.8
3743.89	13	13	$7.61 \cdot 10^{-1}$
3757.66	7	7	$4.13 \cdot 10^{-1}$
3768.24	5	5	$5.10 \cdot 10^{-1}$
3804.80	9	9	$6.9 \cdot 10^{-1}$
3963.69	13	15	1.3
3969.75	11	13	1.2
3983.90	7	9	1.05
3991.12	5	7	1.07
4001.44	9	11	$6.8 \cdot 10^{-1}$
4039.10	15	15	$6.7 \cdot 10^{-1}$
4048.78	13	13	$6.4 \cdot 10^{-1}$
4058.78	11	11	$6.7 \cdot 10^{-1}$
4065.71	9	11	$3.5 \cdot 10^{-1}$
4165.52	11	13	$7.5 \cdot 10^{-1}$
4204.48	13	11	$3.1 \cdot 10^{-1}$
4254.33	7	9	$3.15 \cdot 10^{-1}$
4263.15	15	17	$6.4 \cdot 10^{-1}$
4274.81	7	7	$3.07 \cdot 10^{-1}$
4275.98	11	11	$2.2 \cdot 10^{-1}$
4280.42	13	15	$4.7 \cdot 10^{-1}$
4289.73	7	5	$3.16 \cdot 10^{-1}$
4291.97	7	5	$2.4 \cdot 10^{-1}$
4297.75	11	13	$4.9 \cdot 10^{-1}$
4298.05	9	9	$2.6 \cdot 10^{-1}$
4300.52	9	7	$1.9 \cdot 10^{-1}$
4301.19	11	9	$2.6 \cdot 10^{-1}$
4302.78	11	11	$2.5 \cdot 10^{-1}$
4337.25	5	7	$2.0 \cdot 10^{-1}$
4373.65	9	9	$2.8 \cdot 10^{-1}$
4376.80	13	13	$3.2 \cdot 10^{-1}$
4413.86	7	5	$2.7 \cdot 10^{-1}$
4422.70	5	5	$2.7 \cdot 10^{-1}$
4424.29	9	7	$2.1 \cdot 10^{-1}$
4432.77	15	15	$4.9 \cdot 10^{-1}$
4490.55	9	7	$3.9 \cdot 10^{-1}$
4492.31	5	3	$4.47 \cdot 10^{-1}$
4495.28	9	7	$2.0 \cdot 10^{-1}$
4500.29	7	7	$2.1 \cdot 10^{-1}$
4506.84	13	11	$2.7 \cdot 10^{-1}$
4540.72	11	11	$3.14 \cdot 10^{-1}$
4564.17	11	13	$5.1 \cdot 10^{-1}$
4595.60	13	13	$4.7 \cdot 10^{-1}$
4622.47	7	7	$4.1 \cdot 10^{-1}$
4665.90	3	3	$3.0 \cdot 10^{-1}$
4689.38	7	5	$2.3 \cdot 10^{-1}$
4698.46	9	7	$2.2 \cdot 10^{-1}$
4708.02	11	9	$4.31 \cdot 10^{-1}$
4718.43	13	11	$3.4 \cdot 10^{-1}$
4730.69	7	5	$3.83 \cdot 10^{-1}$
4737.33	9	7	$3.38 \cdot 10^{-1}$
4741.09	3	5	$2.2 \cdot 10^{-1}$
4752.07	13	13	$6.2 \cdot 10^{-1}$
4756.09	11	9	$4.0 \cdot 10^{-1}$
4792.49	7	5	$2.6 \cdot 10^{-1}$
4801.02	9	7	$3.06 \cdot 10^{-1}$
4816.13	9	9	$1.8 \cdot 10^{-1}$
4870.79	7	9	$3.5 \cdot 10^{-1}$
4887.01	9	11	$3.2 \cdot 10^{-1}$
4922.28	11	13	$4.0 \cdot 10^{-1}$
5204.51	5	3	$5.09 \cdot 10^{-1}$
5206.02	5	5	$5.14 \cdot 10^{-1}$
5208.42	5	7	$5.06 \cdot 10^{-1}$
5243.38	5	3	$2.19 \cdot 10^{-1}$
5297.37	7	9	$3.88 \cdot 10^{-1}$
5297.99	7	7	$3.0 \cdot 10^{-1}$
5328.36	9	11	$6.2 \cdot 10^{-1}$
5329.17	9	9	$2.25 \cdot 10^{-1}$
5783.11	3	3	$2.1 \cdot 10^{-1}$
5783.89	5	5	$2.02 \cdot 10^{-1}$
5787.97	5	7	$2.35 \cdot 10^{-1}$
Cr II			
2653.57	4	6	$3.5 \cdot 10^{-1}$
2658.59	2	4	$5.8 \cdot 10^{-1}$
2666.02	6	8	$5.9 \cdot 10^{-1}$
2668.71	4	2	1.4
2671.80	6	4	1.0
2672.83	8	6	$5.5 \cdot 10^{-1}$
2744.97	4	6	$8.5 \cdot 10^{-1}$
2787.61	6	6	1.5
2822.38	14	16	2.3
2835.63	10	12	2.0
2840.01	10	12	2.7
2843.24	8	10	$6.4 \cdot 10^{-1}$
2849.83	6	8	$9.2 \cdot 10^{-1}$
2851.35	8	10	2.2
2856.77	4	6	$4.3 \cdot 10^{-1}$
2857.40	6	8	$2.8 \cdot 10^{-1}$
2860.92	2	4	$6.9 \cdot 10^{-1}$
2862.57	8	8	$6.3 \cdot 10^{-1}$
2866.72	4	4	1.2
2867.09	4	4	1.1
2867.65	2	2	1.1
2870.43	6	6	1.3
2873.81	4	2	$8.8 \cdot 10^{-1}$
2880.86	6	4	$7.9 \cdot 10^{-1}$
2898.53	10	12	1.2
2921.81	8	10	$9.0 \cdot 10^{-1}$
2930.83	2	4	1.1
2935.12	6	8	1.8
2953.34	2	2	1.8
2966.03	10	8	$5.4 \cdot 10^{-1}$
2971.90	14	14	2.0
2979.73	12	12	1.8
2985.32	10	10	2.2
2989.18	8	8	2.2
3118.64	2	4	1.7
3120.36	4	6	1.5
3122.59	12	12	$4.4 \cdot 10^{-1}$
3128.69	4	4	$8.1 \cdot 10^{-1}$
3136.68	6	6	$6.4 \cdot 10^{-1}$
4588.22	8	6	$1.2 \cdot 10^{-1}$
Cobalt			
Co I			
2407.25	10	12	3.6
2414.46	6	8	3.4
2415.29	4	6	3.6
2424.93	10	10	3.2
2432.21	8	8	2.6
2436.66	6	6	2.6
2439.04	4	4	2.7
2511.02	10	10	$9.2 \cdot 10^{-1}$
2521.36	10	8	3.0
2528.97	8	6	2.8
2535.96	6	4	1.9
3405.12	10	10	1.0
3409.17	8	8	$4.2 \cdot 10^{-1}$
3412.34	8	10	$6.1 \cdot 10^{-1}$
3433.05	4	4	1.0
3443.64	8	8	$6.9 \cdot 10^{-1}$
3449.17	6	6	$7.6 \cdot 10^{-1}$
3453.51	10	12	1.1
3462.80	4	6	$7.9 \cdot 10^{-1}$
3474.02	6	8	$5.6 \cdot 10^{-1}$
3489.40	8	6	1.3
3495.68	4	6	$4.9 \cdot 10^{-1}$
3502.28	10	8	$8.0 \cdot 10^{-1}$
3506.32	8	6	$8.2 \cdot 10^{-1}$
3512.64	6	4	1.0
3518.34	6	4	1.6
3529.82	8	10	$4.6 \cdot 10^{-1}$
3569.37	8	8	1.6
3587.19	6	6	1.4
3845.47	8	10	$4.6 \cdot 10^{-1}$
3894.07	6	8	$6.9 \cdot 10^{-1}$
3995.31	8	10	$2.5 \cdot 10^{-1}$
4121.32	8	10	$1.9 \cdot 10^{-1}$
5146.75	8	8	$1.5 \cdot 10^{-1}$
5212.70	10	10	$1.9 \cdot 10^{-1}$
5280.63	10	8	$2.8 \cdot 10^{-1}$
5352.05	12	10	$2.7 \cdot 10^{-1}$
6082.43	10	10	$5.4 \cdot 10^{-2}$
6455.00	8	10	$9.0 \cdot 10^{-2}$
7838.12	8	10	$5.4 \cdot 10^{-2}$
8093.93	12	10	$2.0 \cdot 10^{-1}$
8372.79	10	10	$8.7 \cdot 10^{-2}$
Co II			
2286.15	11	13	3.3
2307.85	9	11	2.6
2311.61	7	9	2.8
2314.05	5	7	2.8
2314.97	3	5	2.7
2330.36	5	3	1.32
2344.28	3	3	1.5

λ Å	g_i	g_k	A 10^8 s^{-1}
2353.41	7	7	1.9
2363.80	9	9	2.1
2378.62	11	9	1.9
2383.45	9	7	1.8
2388.92	11	11	2.8
2389.54	5	3	1.5
2404.17	3	3	1.5
2417.66	9	9	$8.5 \cdot 10^{-1}$

Copper

Cu I

λ Å	g_i	g_k	A 10^8 s^{-1}
*2024.3	2	6	$9.8 \cdot 10^{-2}$
2165.1	2	4	$5.1 \cdot 10^{-1}$
2178.9	2	4	$9.13 \cdot 10^{-1}$
2181.7	2	2	1.0
2225.7	2	2	$4.6 \cdot 10^{-1}$
2244.3	2	4	$1.19 \cdot 10^{-2}$
2441.6	2	2	$2.0 \cdot 10^{-2}$
2492.2	2	4	$3.11 \cdot 10^{-2}$
2618.4	6	4	$3.07 \cdot 10^{-1}$
2766.4	4	4	$9.6 \cdot 10^{-2}$
2824.4	6	6	$7.8 \cdot 10^{-2}$
2961.2	6	8	$3.76 \cdot 10^{-2}$
3063.4	4	4	$1.55 \cdot 10^{-2}$
3194.1	4	4	$1.55 \cdot 10^{-2}$
3247.5	2	4	1.39
3274.0	2	2	1.37
3337.8	6	8	$3.8 \cdot 10^{-3}$
4022.6	2	4	$1.90 \cdot 10^{-1}$
4062.6	4	6	$2.10 \cdot 10^{-1}$
4249.0	2	2	$1.95 \cdot 10^{-1}$
4275.1	6	8	$3.45 \cdot 10^{-1}$
4480.4	2	2	$3.0 \cdot 10^{-2}$
4509.4	4	2	$2.75 \cdot 10^{-1}$
4530.8	4	2	$8.4 \cdot 10^{-2}$
4539.7	6	4	$2.12 \cdot 10^{-1}$
4587.0	8	6	$3.20 \cdot 10^{-1}$
4651.1	10	8	$3.80 \cdot 10^{-1}$
4704.6	8	8	$5.5 \cdot 10^{-2}$
5105.5	6	4	$2.0 \cdot 10^{-2}$
5153.2	2	4	$6.0 \cdot 10^{-1}$
5218.2	4	6	$7.5 \cdot 10^{-1}$
5220.1	4	4	$1.50 \cdot 10^{-1}$
5292.5	8	8	$1.09 \cdot 10^{-1}$
5700.2	4	4	$2.4 \cdot 10^{-3}$
5782.1	4	2	$1.65 \cdot 10^{-2}$

Cu II

λ Å	g_i	g_k	A 10^8 s^{-1}
2489.7	5	5	$1.5 \cdot 10^{-2}$
2544.8	9	7	1.1
2689.3	7	7	$4.1 \cdot 10^{-1}$
2701.0	5	5	$6.7 \cdot 10^{-1}$
2703.2	3	3	1.2
2713.5	5	5	$6.8 \cdot 10^{-1}$
2769.7	7	7	$6.1 \cdot 10^{-1}$

Dysprosium

Dy I

λ Å	g_i	g_k	A 10^8 s^{-1}
2862.7	17	15	$6.5 \cdot 10^{-2}$
2964.6	17	17	$6.5 \cdot 10^{-2}$
3147.7	15	17	$1.1 \cdot 10^{-1}$
3263.2	15	13	$1.4 \cdot 10^{-1}$
3511.0	15	13	$3.1 \cdot 10^{-1}$
3571.4	15	13	$2.0 \cdot 10^{-1}$
3757.1	17	19	3.0
3868.8	17	17	3.1
3967.5	17	19	$8.7 \cdot 10^{-1}$
4046.0	17	15	1.5
4103.9	13	11	1.7
4186.8	17	17	1.32
4194.8	17	17	$7.2 \cdot 10^{-1}$
4211.7	17	19	2.08
4218.1	15	15	1.85
4221.1	15	17	1.52
4225.2	13	15	4.5
4268.3	15	15	$3.6 \cdot 10^{-2}$
4276.7	13	13	$7.3 \cdot 10^{-1}$
4292.0	15	15	$5.8 \cdot 10^{-2}$
4577.8	17	19	$2.2 \cdot 10^{-2}$
4589.4	17	15	$1.3 \cdot 10^{-1}$
4612.3	17	15	$8.2 \cdot 10^{-2}$
5077.7	17	17	$5.7 \cdot 10^{-3}$
5301.6	17	15	$1.1 \cdot 10^{-2}$
5547.3	17	17	$2.7 \cdot 10^{-3}$
5639.5	17	19	$4.7 \cdot 10^{-3}$
5974.5	17	17	$4.0 \cdot 10^{-3}$
5988.6	17	15	$5.3 \cdot 10^{-3}$
6010.8	15	15	$2.6 \cdot 10^{-2}$
6088.3	15	13	$3.5 \cdot 10^{-2}$
6168.4	15	17	$2.5 \cdot 10^{-2}$
6259.1	17	19	$8.5 \cdot 10^{-3}$
6579.4	17	15	$7.5 \cdot 10^{-3}$

Erbium

Er I

λ Å	g_i	g_k	A 10^8 s^{-1}
3862.9	13	13	2.5
4008.0	13	15	2.6
4151.1	13	11	1.8

Europium

Eu I

λ Å	g_i	g_k	A 10^8 s^{-1}
2372.9	8	6	$1.9 \cdot 10^{-1}$
2375.3	8	8	$2.0 \cdot 10^{-1}$
2379.7	8	10	$2.0 \cdot 10^{-1}$
2710.0	8	10	$1.4 \cdot 10^{-1}$
2724.0	8	8	$1.2 \cdot 10^{-1}$
2892.5	8	8	$1.0 \cdot 10^{-1}$
2893.0	8	6	$1.0 \cdot 10^{-1}$
2909.0	8	10	$6.9 \cdot 10^{-2}$
3106.2	8	10	$5.5 \cdot 10^{-2}$
3111.4	8	10	$3.0 \cdot 10^{-1}$
3168.3	8	10	$6.9 \cdot 10^{-2}$
3210.6	8	8	$1.1 \cdot 10^{-1}$
3212.8	8	8	$2.9 \cdot 10^{-1}$
3213.8	8	6	$1.8 \cdot 10^{-1}$
3334.3	8	6	$3.4 \cdot 10^{-1}$
4594.0	8	10	1.4
4627.2	8	8	1.3
4661.9	8	6	1.3
5765.2	8	8	$1.1 \cdot 10^{-2}$
6018.2	8	10	$8.5 \cdot 10^{-3}$
6864.5	8	10	$5.8 \cdot 10^{-3}$

Fluorine

F I

λ Å	g_i	g_k	A 10^8 s^{-1}
806.96	4	6	3.3
809.60	2	4	2.8
951.87	4	2	2.6
954.83	4	4	5.77
955.55	2	2	5.1
958.52	2	4	1.3
6239.7	6	4	$2.5 \cdot 10^{-1}$
6348.5	4	4	$1.8 \cdot 10^{-1}$
6413.7	2	4	$1.1 \cdot 10^{-1}$
6708.3	6	4	$1.4 \cdot 10^{-2}$
6774.0	6	6	$1.0 \cdot 10^{-1}$
6795.5	4	2	$5.2 \cdot 10^{-2}$
6834.3	4	4	$2.1 \cdot 10^{-1}$
6856.0	6	8	$4.94 \cdot 10^{-1}$
6870.2	2	2	$3.8 \cdot 10^{-1}$
6902.5	4	6	$3.2 \cdot 10^{-1}$
6909.8	2	4	$2.2 \cdot 10^{-1}$
6966.4	4	2	$1.1 \cdot 10^{-1}$
7037.5	4	4	$3.0 \cdot 10^{-1}$
7127.9	2	2	$3.8 \cdot 10^{-1}$
7309.0	6	8	$4.7 \cdot 10^{-1}$
7311.0	4	2	$3.9 \cdot 10^{-1}$
7314.3	4	6	$4.8 \cdot 10^{-1}$
7332.0	6	4	$3.1 \cdot 10^{-1}$
7398.7	6	6	$2.85 \cdot 10^{-1}$
7425.7	4	2	$3.4 \cdot 10^{-1}$
7482.7	4	4	$5.6 \cdot 10^{-2}$
7489.2	2	2	$1.1 \cdot 10^{-1}$
7514.9	2	2	$5.2 \cdot 10^{-2}$
7552.2	4	6	$7.8 \cdot 10^{-2}$
7573.4	2	4	$1.0 \cdot 10^{-1}$
7607.2	4	4	$7.0 \cdot 10^{-2}$
7754.7	4	6	$3.82 \cdot 10^{-1}$
7800.2	2	4	$2.1 \cdot 10^{-1}$

Gallium

Ga I

λ Å	g_i	g_k	A 10^8 s^{-1}
2195.4	2	2	$1.9 \cdot 10^{-2}$
2199.7	4	2	$3.3 \cdot 10^{-2}$
2214.4	4	6	$1.2 \cdot 10^{-2}$
2235.9	4	2	$4.3 \cdot 10^{-2}$
2255.0	2	2	$3.1 \cdot 10^{-2}$
2259.2	4	6	$3.1 \cdot 10^{-2}$
2294.2	2	4	$7.0 \cdot 10^{-2}$
2297.9	4	2	$5.8 \cdot 10^{-2}$
2338.2	4	6	$9.8 \cdot 10^{-2}$
2371.3	2	2	$5.7 \cdot 10^{-2}$
2418.7	4	2	$1.0 \cdot 10^{-1}$
2450.1	2	4	$2.8 \cdot 10^{-1}$
2500.2	4	6	$3.4 \cdot 10^{-1}$
2659.9	2	2	$1.2 \cdot 10^{-1}$
2719.7	4	2	$2.3 \cdot 10^{-1}$

λ Å	Weights g_i	g_k	A 10^8 s^{-1}
2874.2	2	4	1.2
2943.6	4	6	1.4
2944.2	4	4	$2.7 \cdot 10^{-1}$
4033.0	2	2	$4.9 \cdot 10^{-1}$
4172.0	4	2	$9.2 \cdot 10^{-1}$

Ga II

λ Å	g_i	g_k	A 10^8 s^{-1}
829.60	1	3	$2.2 \cdot 10^{-1}$
1414.4	1	3	$1.88 \cdot 10^{1}$

Germanium

Ge I

λ Å	g_i	g_k	A 10^8 s^{-1}
1944.7	3	1	$7.0 \cdot 10^{-1}$
1955.1	3	3	$2.8 \cdot 10^{-1}$
1988.3	5	3	$2.5 \cdot 10^{-1}$
1998.9	5	5	$5.5 \cdot 10^{-1}$
2041.7	1	3	1.1
2065.2	3	3	$8.5 \cdot 10^{-1}$
2068.7	3	5	1.2
2086.0	3	5	$4.0 \cdot 10^{-1}$
2094.3	5	7	$9.7 \cdot 10^{-1}$
2105.8	5	5	$1.7 \cdot 10^{-1}$
2256.0	5	5	$3.2 \cdot 10^{-2}$
2417.4	5	5	$9.6 \cdot 10^{-1}$
2498.0	1	3	$1.3 \cdot 10^{-1}$
2533.2	3	3	$1.0 \cdot 10^{-1}$
2589.2	5	3	$5.1 \cdot 10^{-2}$
2592.5	3	5	$7.1 \cdot 10^{-1}$
2651.2	5	5	2.0
2651.6	1	3	$8.5 \cdot 10^{-1}$
2691.3	3	3	$6.1 \cdot 10^{-1}$
2709.6	3	1	2.8
2754.6	5	3	1.1
3039.1	5	3	2.8
3124.8	5	5	$3.1 \cdot 10^{-2}$
3269.5	5	3	$2.9 \cdot 10^{-1}$
4226.6	1	3	$2.1 \cdot 10^{-1}$
4685.8	1	3	$9.5 \cdot 10^{-2}$

Ge II

λ Å	g_i	g_k	A 10^8 s^{-1}
999.10	2	4	1.9
1016.6	4	6	2.1
1017.1	4	4	$3.5 \cdot 10^{-1}$
1055.0	2	2	$6.9 \cdot 10^{-1}$
1075.1	4	2	1.3
1237.1	2	4	$1.9 \cdot 10^{1}$
1261.9	4	6	$2.2 \cdot 10^{1}$
1264.7	4	4	3.5
1602.5	2	2	3.4
1649.2	4	2	6.5
4741.8	2	4	$4.6 \cdot 10^{-1}$
4814.6	4	6	$5.1 \cdot 10^{-1}$
4824.1	4	4	$8.6 \cdot 10^{-2}$
5131.8	4	6	1.9
5178.5	6	6	$1.3 \cdot 10^{-1}$
5178.6	6	8	2.0
5893.4	2	4	$9.2 \cdot 10^{-1}$
6021.0	2	2	$8.4 \cdot 10^{-1}$
6336.4	2	2	$4.4 \cdot 10^{-1}$

λ Å	Weights g_i	g_k	A 10^8 s^{-1}
6484.2	4	2	$8.5 \cdot 10^{-1}$

Gold

Au I

λ Å	g_i	g_k	A 10^8 s^{-1}
2427.95	2	4	1.99
2675.95	2	2	1.64
3122.78	6	4	$1.90 \cdot 10^{-1}$
6278.30	4	2	$3.4 \cdot 10^{-2}$

Helium

He I

λ Å	g_i	g_k	A 10^8 s^{-1}
507.058	1	3	$1.5929 \cdot 10^{-1}$
507.718	1	3	$2.1826 \cdot 10^{-1}$
508.643	1	3	$3.1031 \cdot 10^{-1}$
509.998	1	3	$4.6224 \cdot 10^{-1}$
512.099	1	3	$7.3174 \cdot 10^{-1}$
515.617	1	3	1.2582
522.213	1	3	2.4356
537.030	1	3	5.6634
584.334	1	3	$1.7989 \cdot 10^{1}$
*2677.13	3	9	$4.4174 \cdot 10^{-3}$
*2696.12	3	9	$6.0234 \cdot 10^{-3}$
*2723.19	3	9	$8.4996 \cdot 10^{-3}$
*2763.80	3	9	$1.2508 \cdot 10^{-2}$
*2829.08	3	9	$1.9389 \cdot 10^{-2}$
*2945.10	3	9	$3.2006 \cdot 10^{-2}$
*3187.74	3	9	$5.6361 \cdot 10^{-2}$
3231.270	1	3	$5.1015 \cdot 10^{-3}$
3258.273	1	3	$6.9627 \cdot 10^{-3}$
3296.773	1	3	$9.8432 \cdot 10^{-3}$
3354.555	1	3	$1.4537 \cdot 10^{-2}$
3447.589	1	3	$2.2691 \cdot 10^{-2}$
*3554.42	9	15	$7.5971 \cdot 10^{-3}$
*3562.99	9	3	$4.8363 \cdot 10^{-3}$
*3587.28	9	15	$1.8107 \cdot 10^{-2}$
*3599.32	9	3	$6.7245 \cdot 10^{-3}$
3613.642	1	3	$3.8022 \cdot 10^{-2}$
*3634.25	9	15	$2.6062 \cdot 10^{-2}$
*3652.00	9	3	$9.7444 \cdot 10^{-3}$
*3705.02	9	15	$3.9528 \cdot 10^{-2}$
*3732.88	9	3	$1.4895 \cdot 10^{-2}$
*3819.62	9	15	$6.4351 \cdot 10^{-2}$
3833.549	3	5	$9.6470 \cdot 10^{-3}$
3838.100	3	1	$3.7425 \cdot 10^{-3}$
*3867.49	9	3	$2.4466 \cdot 10^{-2}$
3871.786	3	5	$1.3386 \cdot 10^{-2}$
3878.177	3	1	$5.1753 \cdot 10^{-3}$
*3888.64	3	9	$9.4746 \cdot 10^{-2}$
3926.544	3	5	$1.9371 \cdot 10^{-2}$
3935.945	3	1	$7.4475 \cdot 10^{-3}$
3964.729	1	3	$6.9507 \cdot 10^{-2}$
4009.256	3	5	$2.9612 \cdot 10^{-2}$
4023.980	3	1	$1.1281 \cdot 10^{-2}$
*4026.21	9	15	$1.1600 \cdot 10^{-1}$
*4120.84	9	3	$4.4529 \cdot 10^{-2}$
4143.759	3	5	$4.8812 \cdot 10^{-2}$
4168.971	3	1	$1.8298 \cdot 10^{-2}$
4387.929	3	5	$8.9889 \cdot 10^{-2}$
4437.553	3	1	$3.2689 \cdot 10^{-2}$

λ Å	Weights g_i	g_k	A 10^8 s^{-1}
*4471.50	9	15	$2.4578 \cdot 10^{-1}$
*4713.17	9	3	$9.5209 \cdot 10^{-2}$
4921.931	3	5	$1.9863 \cdot 10^{-1}$
5015.678	1	3	$1.3372 \cdot 10^{-1}$
5047.738	3	1	$6.7712 \cdot 10^{-2}$
*5875.66	9	15	$7.0703 \cdot 10^{-1}$
6678.152	3	5	$6.3705 \cdot 10^{-1}$
*7065.25	9	3	$2.7853 \cdot 10^{-1}$
7281.350	3	1	$1.8299 \cdot 10^{-1}$
*7298.04	3	9	$1.2913 \cdot 10^{-3}$
*7499.85	3	9	$1.7942 \cdot 10^{-3}$
*7816.14	3	9	$2.5748 \cdot 10^{-3}$
8094.115	1	3	$1.3791 \cdot 10^{-3}$
8265.701	1	3	$1.8722 \cdot 10^{-3}$
*8361.73	3	9	$3.8126 \cdot 10^{-3}$
8518.036	1	3	$2.6252 \cdot 10^{-3}$
*8582.64	9	15	$4.1927 \cdot 10^{-3}$
*8632.74	9	3	$2.7471 \cdot 10^{-3}$
*8776.74	9	15	$5.7758 \cdot 10^{-3}$
*8849.18	9	3	$3.8377 \cdot 10^{-3}$
8914.772	1	3	$3.8260 \cdot 10^{-3}$
8999.736	5	7	$2.8406 \cdot 10^{-3}$
*9063.32	9	15	$8.2702 \cdot 10^{-3}$
9085.421	3	5	$3.6807 \cdot 10^{-3}$
9111.026	3	1	$2.2000 \cdot 10^{-3}$
*9174.52	9	3	$5.5996 \cdot 10^{-3}$
*9210.34	15	21	$4.7381 \cdot 10^{-3}$
9303.163	3	5	$5.1030 \cdot 10^{-3}$
9340.143	3	1	$3.0562 \cdot 10^{-3}$
*9463.58	3	9	$5.6868 \cdot 10^{-3}$
*9516.60	9	15	$1.2439 \cdot 10^{-2}$
9603.441	1	3	$5.8286 \cdot 10^{-3}$
9625.697	3	5	$7.3744 \cdot 10^{-3}$
9682.388	3	1	$4.4271 \cdot 10^{-3}$
*9702.65	9	3	$8.6511 \cdot 10^{-3}$
10138.424	3	5	$1.1248 \cdot 10^{-2}$
10233.102	3	1	$6.7731 \cdot 10^{-3}$
*10311.27	9	15	$1.9945 \cdot 10^{-2}$
*10667.71	9	3	$1.4471 \cdot 10^{-2}$
*10830.17	3	9	$1.0216 \cdot 10^{-1}$
*10913.00	15	21	$1.9801 \cdot 10^{-2}$
*10996.65	15	9	$1.4253 \cdot 10^{-3}$
11013.072	1	3	$9.2496 \cdot 10^{-3}$
11044.983	3	5	$1.8457 \cdot 10^{-2}$
11225.937	3	1	$1.1168 \cdot 10^{-2}$
*10969.11	9	15	$3.4781 \cdot 10^{-2}$
*12527.48	3	9	$7.0932 \cdot 10^{-3}$
12755.688	5	3	$1.2754 \cdot 10^{-3}$
*12846.01	9	3	$2.7317 \cdot 10^{-2}$
12968.430	3	5	$3.3615 \cdot 10^{-2}$
*12984.88	15	9	$2.7292 \cdot 10^{-3}$
13411.683	3	1	$2.0572 \cdot 10^{-2}$

Indium

In I

λ Å	g_i	g_k	A 10^8 s^{-1}
2560.2	2	4	$4.0 \cdot 10^{-1}$
2710.3	4	6	$4.0 \cdot 10^{-1}$
3039.4	2	4	1.3
3256.1	4	6	1.3

Atomic

Atomic

λ Å	Weights g_i	g_k	A 10^8 s^{-1}
4101.8	2	2	$5.6 \cdot 10^{-1}$
4511.3	4	2	1.02

In II

λ Å	Weights g_i	g_k	A 10^8 s^{-1}
2941.1	3	1	1.4

Iodine

I I

λ Å	Weights g_i	g_k	A 10^8 s^{-1}
1782.8	4	4	2.71
1830.4	4	6	$1.6 \cdot 10^{-1}$

Iridium

Ir I

λ Å	Weights g_i	g_k	A 10^8 s^{-1}
2475.12	10	10	$2.1 \cdot 10^{-1}$
2502.98	10	12	$3.2 \cdot 10^{-1}$
2639.71	10	10	$4.7 \cdot 10^{-1}$
2661.98	10	10	$2.5 \cdot 10^{-1}$
2664.79	10	8	$4.0 \cdot 10^{-1}$
2694.23	10	12	$4.8 \cdot 10^{-1}$
2849.72	10	10	$2.2 \cdot 10^{-1}$
2924.79	10	12	$1.42 \cdot 10^{-1}$
2934.64	8	10	$2.0 \cdot 10^{-1}$
3220.78	10	8	$2.4 \cdot 10^{-1}$
3573.72	8	10	$5.4 \cdot 10^{-2}$
3661.71	8	10	$4.0 \cdot 10^{-2}$
4033.76	8	10	$2.7 \cdot 10^{-2}$
4069.92	6	8	$3.6 \cdot 10^{-2}$
4913.35	12	12	$3.3 \cdot 10^{-2}$

Iron

Fe I

λ Å	Weights g_i	g_k	A 10^8 s^{-1}
2166.77	9	7	2.7
2191.84	5	5	1.2
2196.04	3	3	1.2
2200.39	1	3	$8.9 \cdot 10^{-1}$
2298.17	9	9	$3.09 \cdot 10^{-1}$
2439.74	13	13	3.46
2442.57	11	11	3.12
2457.60	11	11	$4.81 \cdot 10^{-1}$
2462.65	9	9	$5.85 \cdot 10^{-1}$
2465.15	9	9	$4.35 \cdot 10^{-1}$
2468.88	11	11	$2.40 \cdot 10^{-1}$
2472.89	7	7	1.30
2474.81	7	7	$6.13 \cdot 10^{-1}$
2479.78	5	5	1.74
2483.27	9	11	4.80
2484.19	3	3	2.26
2487.07	3	3	$6.40 \cdot 10^{-1}$
2488.14	7	9	4.20
2489.75	1	3	2.31
2490.64	5	7	3.44
2491.16	3	5	2.91
2491.99	9	9	$3.25 \cdot 10^{-1}$
2496.53	9	11	$2.15 \cdot 10^{-1}$
2501.13	9	7	$6.75 \cdot 10^{-1}$
2505.01	9	11	$2.56 \cdot 10^{-1}$
2506.57	7	9	$2.04 \cdot 10^{-1}$
2507.90	7	9	$1.93 \cdot 10^{-1}$
2510.83	7	5	1.29
2518.10	5	3	1.93
2522.85	9	9	2.13
2524.29	3	1	3.23
2527.27	13	13	$3.46 \cdot 10^{-1}$
2527.44	7	7	1.93
2529.14	5	5	$9.91 \cdot 10^{-1}$
2529.31	5	7	4.86
2533.14	11	11	$2.07 \cdot 10^{-1}$
2535.61	1	3	$9.59 \cdot 10^{-1}$
2537.17	13	15	3.70
2540.97	3	5	$9.59 \cdot 10^{-1}$
2542.10	11	13	4.47
2543.92	9	11	4.70
2545.98	5	7	$7.16 \cdot 10^{-1}$
2549.61	7	9	$2.31 \cdot 10^{-1}$
2584.54	11	13	$3.15 \cdot 10^{-1}$
2606.83	9	11	$2.43 \cdot 10^{-1}$
2609.22	7	7	$4.60 \cdot 10^{-1}$
2623.53	7	9	$2.13 \cdot 10^{-1}$
2635.81	5	7	$2.11 \cdot 10^{-1}$
2656.15	13	15	$1.63 \cdot 10^{-1}$
2669.49	11	13	$1.34 \cdot 10^{-1}$
2679.06	11	11	$1.50 \cdot 10^{-1}$
2706.01	13	13	$2.28 \cdot 10^{-1}$
2706.58	7	5	$2.69 \cdot 10^{-1}$
2708.57	9	9	$6.49 \cdot 10^{-1}$
2719.03	9	7	1.42
2719.06	7	7	$7.40 \cdot 10^{-1}$
2719.42	11	11	$3.20 \cdot 10^{-1}$
2720.90	7	5	1.04
2723.58	5	3	$5.69 \cdot 10^{-1}$
2728.82	9	9	$2.98 \cdot 10^{-1}$
2733.58	11	9	$7.10 \cdot 10^{-1}$
2735.48	9	7	$5.03 \cdot 10^{-1}$
2737.31	3	3	$7.25 \cdot 10^{-1}$
2737.64	13	11	$1.14 \cdot 10^{-1}$
2742.25	7	5	$3.41 \cdot 10^{-1}$
2742.41	5	5	$4.70 \cdot 10^{-1}$
2750.14	7	7	$2.74 \cdot 10^{-1}$
2762.03	7	7	$1.76 \cdot 10^{-1}$
2767.52	9	9	$1.48 \cdot 10^{-1}$
2769.30	13	13	$1.80 \cdot 10^{-1}$
2788.10	11	13	$5.92 \cdot 10^{-1}$
2789.80	11	9	$2.36 \cdot 10^{-1}$
2804.86	9	7	$2.40 \cdot 10^{-1}$
2806.98	9	11	$1.15 \cdot 10^{-1}$
2813.29	9	11	$3.42 \cdot 10^{-1}$
2825.56	7	9	$1.32 \cdot 10^{-1}$
2832.44	7	9	$2.38 \cdot 10^{-1}$
2843.98	5	7	$3.17 \cdot 10^{-1}$
2851.80	3	5	$3.37 \cdot 10^{-1}$
2894.50	5	5	$4.83 \cdot 10^{-1}$
2899.41	5	3	$4.68 \cdot 10^{-1}$
2901.91	11	11	$1.78 \cdot 10^{-1}$
2907.52	9	11	$1.61 \cdot 10^{-1}$
2918.02	13	13	1.18
2923.29	11	11	1.39
2923.85	11	11	$2.97 \cdot 10^{-1}$
2925.36	7	9	$1.69 \cdot 10^{-1}$
2929.12	9	9	1.53
2936.90	9	9	$1.40 \cdot 10^{-1}$
2947.88	7	7	$1.83 \cdot 10^{-1}$
2948.43	9	9	$3.32 \cdot 10^{-1}$
2953.49	7	7	$3.64 \cdot 10^{-1}$
2959.99	11	13	$5.02 \cdot 10^{-1}$
2966.90	9	11	$2.72 \cdot 10^{-1}$
2973.24	7	9	$1.83 \cdot 10^{-1}$
2980.53	7	7	$1.66 \cdot 10^{-1}$
2981.85	7	9	$1.86 \cdot 10^{-1}$
2983.57	9	7	$2.79 \cdot 10^{-1}$
2990.39	9	11	$3.5 \cdot 10^{-1}$
2994.43	7	5	$4.39 \cdot 10^{-1}$
2999.51	11	11	$1.70 \cdot 10^{-1}$
3000.95	5	3	$6.42 \cdot 10^{-1}$
3008.14	3	1	1.07
3009.57	9	9	$1.43 \cdot 10^{-1}$
3011.48	7	9	$3.79 \cdot 10^{-1}$
3020.49	5	5	$1.94 \cdot 10^{-1}$
3020.64	9	9	$7.59 \cdot 10^{-1}$
3021.07	7	7	$4.55 \cdot 10^{-1}$
3025.64	13	13	$5.86 \cdot 10^{-1}$
3025.84	1	3	$3.48 \cdot 10^{-1}$
3030.15	11	11	$5.04 \cdot 10^{-1}$
3037.39	3	5	$2.91 \cdot 10^{-1}$
3047.60	5	7	$2.84 \cdot 10^{-1}$
3057.45	11	9	$3.13 \cdot 10^{-1}$
3059.09	7	9	$1.63 \cdot 10^{-1}$
3067.00	11	13	$1.71 \cdot 10^{-1}$
3067.24	9	7	$3.12 \cdot 10^{-1}$
3075.72	7	5	$3.14 \cdot 10^{-1}$
3079.99	9	11	$8.35 \cdot 10^{-2}$
3083.74	5	3	$3.08 \cdot 10^{-1}$
3100.30	5	5	$1.87 \cdot 10^{-1}$
3100.67	7	7	$1.35 \cdot 10^{-1}$
3125.68	13	11	$8.46 \cdot 10^{-2}$
3132.52	9	7	$3.39 \cdot 10^{-1}$
3143.99	9	9	$6.10 \cdot 10^{-1}$
3156.27	7	7	$6.36 \cdot 10^{-1}$
3157.04	9	11	$1.26 \cdot 10^{-1}$
3157.89	5	7	$1.61 \cdot 10^{-1}$
3160.66	9	9	$1.93 \cdot 10^{-1}$
3166.44	9	7	$1.14 \cdot 10^{-1}$
3171.35	9	7	$1.85 \cdot 10^{-1}$
3175.44	11	11	$1.44 \cdot 10^{-1}$
3178.01	11	9	$1.28 \cdot 10^{-1}$
3180.22	7	9	$4.42 \cdot 10^{-1}$
3181.52	7	5	$1.84 \cdot 10^{-1}$
3182.97	5	7	$1.42 \cdot 10^{-1}$
3188.82	3	5	$2.53 \cdot 10^{-1}$
3192.80	3	5	$5.01 \cdot 10^{-1}$
3193.30	5	7	$3.07 \cdot 10^{-1}$
3196.12	11	9	$1.40 \cdot 10^{-1}$
3196.93	9	11	$5.97 \cdot 10^{-1}$
3199.53	9	9	$2.23 \cdot 10^{-1}$
3205.40	3	3	$9.77 \cdot 10^{-1}$
3210.23	9	11	$1.15 \cdot 10^{-1}$
3210.83	5	3	$9.24 \cdot 10^{-1}$
3211.99	11	9	$4.64 \cdot 10^{-1}$
3214.01	7	7	$8.38 \cdot 10^{-1}$

λ Å	Weights g_i	Weights g_k	A $10^8\ s^{-1}$
3214.06	7	5	1.18
3215.94	5	5	$6.19\cdot10^{-1}$
3217.38	11	9	$1.50\cdot10^{-1}$
3219.58	7	9	$4.64\cdot10^{-1}$
3219.80	9	7	$3.61\cdot10^{-1}$
3222.07	11	11	$8.65\cdot10^{-1}$
3225.79	11	13	1.18
3227.80	9	7	$4.96\cdot10^{-1}$
3228.25	5	3	$3.72\cdot10^{-1}$
3229.99	9	11	$1.06\cdot10^{-1}$
3230.21	5	5	$2.06\cdot10^{-1}$
3230.96	7	5	$3.7\cdot10^{-1}$
3233.05	13	15	$4.19\cdot10^{-1}$
3233.97	9	9	$2.08\cdot10^{-1}$
3239.43	9	9	$2.95\cdot10^{-1}$
3244.19	9	11	$3.06\cdot10^{-1}$
3248.20	7	7	$1.92\cdot10^{-1}$
3253.60	7	9	$1.62\cdot10^{-1}$
3254.36	11	13	$4.24\cdot10^{-1}$
3265.62	7	5	$3.06\cdot10^{-1}$
3271.00	5	3	$6.4\cdot10^{-1}$
3280.26	9	11	$4.21\cdot10^{-1}$
3282.89	3	5	$3.42\cdot10^{-1}$
3286.75	7	7	$5.99\cdot10^{-1}$
3292.02	7	9	$5.77\cdot10^{-1}$
3292.59	3	3	$3.0\cdot10^{-1}$
3305.97	5	7	$4.05\cdot10^{-1}$
3306.34	9	9	$5.74\cdot10^{-1}$
3306.35	3	5	$4.84\cdot10^{-1}$
3307.23	13	13	$1.97\cdot10^{-1}$
3314.74	5	7	$7.25\cdot10^{-1}$
3322.47	9	11	$8.21\cdot10^{-2}$
3323.74	5	5	$2.8\cdot10^{-1}$
3328.87	11	11	$2.21\cdot10^{-1}$
3355.23	9	9	$2.59\cdot10^{-1}$
3369.55	9	9	$2.15\cdot10^{-1}$
3370.78	11	11	$2.89\cdot10^{-1}$
3380.11	7	7	$1.66\cdot10^{-1}$
3392.65	7	7	$1.88\cdot10^{-1}$
3399.33	5	5	$2.76\cdot10^{-1}$
3402.26	13	13	$2.19\cdot10^{-1}$
3404.35	5	7	$1.09\cdot10^{-1}$
3406.44	3	5	$2.7\cdot10^{-1}$
3407.46	7	9	$6.09\cdot10^{-1}$
3410.17	3	5	$5.07\cdot10^{-1}$
3413.13	5	7	$3.23\cdot10^{-1}$
3417.84	3	3	$4.01\cdot10^{-1}$
3418.51	3	1	$9.88\cdot10^{-1}$
3422.66	3	5	$1.38\cdot10^{-1}$
3424.28	7	7	$1.61\cdot10^{-1}$
3425.01	9	7	$2.57\cdot10^{-1}$
3426.67	11	11	$1.07\cdot10^{-1}$
3427.12	7	9	$5.04\cdot10^{-1}$
3428.19	5	5	$1.71\cdot10^{-1}$
3440.61	9	7	$1.71\cdot10^{-1}$
3440.99	7	5	$1.24\cdot10^{-1}$
3445.15	5	7	$2.34\cdot10^{-1}$
3450.33	3	3	$2.34\cdot10^{-1}$
3459.91	5	3	$2.17\cdot10^{-1}$
3469.01	9	9	$8.58\cdot10^{-2}$
3476.34	7	7	$2.70\cdot10^{-1}$
3489.67	11	13	$7.47\cdot10^{-2}$
3495.29	9	7	$9.46\cdot10^{-2}$
3497.10	7	7	$9.02\cdot10^{-2}$
3508.47	9	11	$6.46\cdot10^{-2}$
3526.24	7	9	$1.70\cdot10^{-1}$
3526.38	7	7	$4.13\cdot10^{-1}$
3526.47	5	5	$1.29\cdot10^{-1}$
3526.67	5	5	$5.26\cdot10^{-1}$
3527.79	9	9	$2.17\cdot10^{-1}$
3529.82	3	3	$7.75\cdot10^{-1}$
3530.39	13	13	$4.65\cdot10^{-2}$
3533.01	1	3	$8.52\cdot10^{-1}$
3533.20	3	5	$8.25\cdot10^{-1}$
3536.56	5	7	$9.95\cdot10^{-1}$
3537.89	11	11	$8.0\cdot10^{-2}$
3540.12	7	9	$9.48\cdot10^{-2}$
3541.08	9	11	$8.65\cdot10^{-1}$
3542.08	7	9	$9.51\cdot10^{-1}$
3543.67	3	5	$1.6\cdot10^{-1}$
3545.64	9	9	$2.05\cdot10^{-1}$
3547.19	9	9	$7.13\cdot10^{-2}$
3552.83	5	5	$1.74\cdot10^{-1}$
3553.74	11	9	1.09
3554.92	11	13	1.40
3556.88	9	11	$4.1\cdot10^{-1}$
3558.52	5	7	$1.77\cdot10^{-1}$
3559.50	3	3	$2.2\cdot10^{-1}$
3560.70	7	9	$7.4\cdot10^{-2}$
3565.38	7	9	$4.29\cdot10^{-1}$
3567.03	5	7	$8.34\cdot10^{-2}$
3568.82	7	9	$6.72\cdot10^{-2}$
3570.10	9	11	$6.76\cdot10^{-1}$
3572.00	11	11	$2.89\cdot10^{-1}$
3573.39	5	7	$1.05\cdot10^{-1}$
3573.89	9	7	$5.73\cdot10^{-1}$
3575.25	11	9	$7.43\cdot10^{-2}$
3575.37	5	5	$3.06\cdot10^{-1}$
3576.76	11	9	$8.8\cdot10^{-2}$
3581.19	11	13	1.02
3582.20	13	11	$2.35\cdot10^{-1}$
3584.66	11	11	$3.29\cdot10^{-1}$
3584.79	7	5	$1.56\cdot10^{-1}$
3584.96	11	9	$6.74\cdot10^{-1}$
3585.32	7	7	$1.17\cdot10^{-1}$
3586.11	13	11	$7.02\cdot10^{-1}$
3586.98	5	5	$1.66\cdot10^{-1}$
3587.24	7	9	$7.73\cdot10^{-2}$
3588.61	11	11	$1.19\cdot10^{-1}$
3588.92	5	3	$2.15\cdot10^{-1}$
3589.45	9	7	$1.05\cdot10^{-1}$
3594.63	9	9	$3.14\cdot10^{-1}$
3597.02	5	3	$1.8\cdot10^{-1}$
3599.63	11	9	$2.33\cdot10^{-1}$
3602.46	7	7	$1.02\cdot10^{-1}$
3602.53	7	5	$2.12\cdot10^{-1}$
3603.20	11	11	$2.59\cdot10^{-1}$
3605.45	9	9	$4.66\cdot10^{-1}$
3605.50	13	11	$2.12\cdot10^{-1}$
3606.68	11	13	$8.29\cdot10^{-1}$
3608.14	9	11	$6.22\cdot10^{-2}$
3608.86	3	5	$8.13\cdot10^{-1}$
3610.16	13	13	$5.90\cdot10^{-1}$
3612.07	11	13	$1.11\cdot10^{-1}$
3617.79	5	7	$7.09\cdot10^{-1}$
3618.39	9	9	$8.88\cdot10^{-2}$
3618.77	5	7	$7.22\cdot10^{-1}$
3621.46	9	11	$4.45\cdot10^{-1}$
3621.72	11	9	$1.07\cdot10^{-1}$
3622.00	7	7	$5.14\cdot10^{-1}$
3623.19	13	13	$6.68\cdot10^{-2}$
3625.14	11	9	$8.15\cdot10^{-2}$
3630.35	9	7	$1.04\cdot10^{-1}$
3631.10	11	11	$2.15\cdot10^{-1}$
3631.46	7	9	$5.17\cdot10^{-1}$
3632.04	3	5	$6.74\cdot10^{-1}$
3634.33	9	7	$1.05\cdot10^{-1}$
3636.22	5	7	$2.20\cdot10^{-1}$
3637.87	9	9	$5.9\cdot10^{-2}$
3638.30	7	9	$2.36\cdot10^{-1}$
3640.39	9	11	$3.57\cdot10^{-1}$
3645.82	1	3	$4.87\cdot10^{-1}$
3647.42	3	3	$3.38\cdot10^{-1}$
3647.84	9	11	$2.91\cdot10^{-1}$
3649.51	11	9	$3.94\cdot10^{-1}$
3650.03	7	7	$2.26\cdot10^{-1}$
3650.28	11	11	$6.15\cdot10^{-2}$
3651.47	7	9	$5.83\cdot10^{-1}$
3655.46	5	5	$1.18\cdot10^{-1}$
3659.52	9	9	$6.31\cdot10^{-2}$
3667.25	9	7	$1.3\cdot10^{-1}$
3669.15	9	7	$8.03\cdot10^{-2}$
3669.52	9	7	$2.34\cdot10^{-1}$
3670.09	11	13	$7.20\cdot10^{-2}$
3676.31	9	11	$4.63\cdot10^{-2}$
3677.31	5	7	$2.28\cdot10^{-1}$
3677.63	7	5	$6.08\cdot10^{-1}$
3682.17	7	5	$1.04\cdot10^{-1}$
3682.24	5	5	1.5
3684.11	9	7	$2.97\cdot10^{-1}$
3684.14	9	7	$9.29\cdot10^{-2}$
3686.00	9	11	$3.34\cdot10^{-1}$
3687.46	11	9	$8.00\cdot10^{-2}$
3687.66	9	9	$7.38\cdot10^{-2}$
3688.46	7	9	$7.3\cdot10^{-2}$
3689.46	9	9	$3.70\cdot10^{-1}$
3690.73	11	11	$2.99\cdot10^{-1}$
3694.01	5	7	$8.35\cdot10^{-1}$
3695.05	7	9	$2.01\cdot10^{-1}$
3697.43	7	7	$1.94\cdot10^{-1}$
3701.09	7	9	$6.35\cdot10^{-1}$
3703.69	9	11	$6.31\cdot10^{-2}$
3704.46	11	9	$1.42\cdot10^{-1}$
3707.92	7	5	$3.32\cdot10^{-1}$
3709.25	9	7	$1.56\cdot10^{-1}$
3711.41	3	5	$1.28\cdot10^{-1}$
3716.44	9	7	$3.49\cdot10^{-1}$
3719.93	9	11	$1.62\cdot10^{-1}$
3721.50	5	5	$1.94\cdot10^{-1}$
3724.38	5	7	$1.04\cdot10^{-1}$

Atomic

λ Å	Weights g_i	Weights g_k	A 10^8 s^{-1}
3726.93	5	5	$4.57 \cdot 10^{-1}$
3727.09	9	7	$1.71 \cdot 10^{-1}$
3727.62	7	5	$2.24 \cdot 10^{-1}$
3727.81	7	5	$1.91 \cdot 10^{-1}$
3730.39	9	11	$9.73 \cdot 10^{-2}$
3732.40	5	5	$2.69 \cdot 10^{-1}$
3734.86	11	11	$9.01 \cdot 10^{-1}$
3735.32	9	9	$2.70 \cdot 10^{-1}$
3737.13	7	9	$1.41 \cdot 10^{-1}$
3738.31	11	13	$3.44 \cdot 10^{-1}$
3740.24	7	9	$1.3 \cdot 10^{-1}$
3742.62	9	9	$6.75 \cdot 10^{-2}$
3743.36	5	3	$2.60 \cdot 10^{-1}$
3743.47	11	11	$6.05 \cdot 10^{-1}$
3744.10	5	3	$3.17 \cdot 10^{-1}$
3745.56	5	7	$1.15 \cdot 10^{-1}$
3746.93	7	7	$2.33 \cdot 10^{-1}$
3748.96	9	11	$1.48 \cdot 10^{-1}$
3749.49	9	9	$7.63 \cdot 10^{-1}$
3753.61	7	5	$1.22 \cdot 10^{-1}$
3756.94	11	11	$2.2 \cdot 10^{-1}$
3758.23	7	7	$6.34 \cdot 10^{-1}$
3759.15	13	11	$4.55 \cdot 10^{-2}$
3760.05	13	15	$4.47 \cdot 10^{-2}$
3763.79	5	5	$5.44 \cdot 10^{-1}$
3765.54	13	15	$9.51 \cdot 10^{-1}$
3767.19	3	3	$6.39 \cdot 10^{-1}$
3778.51	7	5	$1.17 \cdot 10^{-1}$
3785.95	11	13	$4.14 \cdot 10^{-2}$
3786.19	5	5	$1.3 \cdot 10^{-1}$
3787.16	5	5	$9.9 \cdot 10^{-2}$
3787.88	3	5	$1.29 \cdot 10^{-1}$
3793.48	7	7	$7.92 \cdot 10^{-2}$
3795.00	5	7	$1.15 \cdot 10^{-1}$
3797.51	13	13	$4.57 \cdot 10^{-1}$
3799.55	7	9	$7.31 \cdot 10^{-2}$
3801.98	11	13	$3.7 \cdot 10^{-2}$
3805.34	9	11	$8.60 \cdot 10^{-1}$
3806.22	3	3	$2.5 \cdot 10^{-1}$
3806.70	11	11	$4.35 \cdot 10^{-1}$
3807.54	3	5	$9.37 \cdot 10^{-2}$
3810.76	5	3	$1.94 \cdot 10^{-1}$
3813.88	13	11	$6.62 \cdot 10^{-2}$
3815.84	9	7	1.12
3817.64	11	11	$7.7 \cdot 10^{-2}$
3820.43	11	9	$6.67 \cdot 10^{-1}$
3821.18	11	13	$5.54 \cdot 10^{-1}$
3825.88	9	7	$5.97 \cdot 10^{-1}$
3827.82	7	5	1.05
3834.22	7	5	$4.52 \cdot 10^{-1}$
3836.33	5	5	$3.29 \cdot 10^{-1}$
3839.26	9	9	$2.35 \cdot 10^{-1}$
3840.44	5	3	$4.70 \cdot 10^{-1}$
3841.05	5	3	1.36
3843.26	9	7	$3.70 \cdot 10^{-1}$
3846.41	11	9	$1.68 \cdot 10^{-1}$
3846.80	7	7	$6.20 \cdot 10^{-1}$
3849.97	3	1	$6.05 \cdot 10^{-1}$
3859.21	13	11	$7.25 \cdot 10^{-2}$
3859.91	9	9	$9.69 \cdot 10^{-2}$
3865.52	3	3	$1.55 \cdot 10^{-1}$
3867.22	5	5	$3.16 \cdot 10^{-1}$
3871.75	11	11	$5.83 \cdot 10^{-2}$
3872.50	5	5	$1.05 \cdot 10^{-1}$
3873.76	11	9	$6.57 \cdot 10^{-2}$
3878.02	7	7	$7.72 \cdot 10^{-2}$
3878.67	9	7	$7.02 \cdot 10^{-2}$
3878.73	3	3	$5.34 \cdot 10^{-1}$
3883.28	7	7	$1.28 \cdot 10^{-1}$
3888.51	5	5	$2.50 \cdot 10^{-1}$
3888.82	5	3	$1.95 \cdot 10^{-1}$
3891.93	3	3	$2.71 \cdot 10^{-1}$
3893.39	11	11	$1.00 \cdot 10^{-1}$
3894.01	5	5	$1.03 \cdot 10^{-1}$
3897.89	11	13	$6.20 \cdot 10^{-2}$
3900.52	7	7	$7.9 \cdot 10^{-2}$
3902.95	7	7	$2.14 \cdot 10^{-1}$
3903.90	9	9	$7.61 \cdot 10^{-2}$
3906.75	5	7	$7.05 \cdot 10^{-2}$
3916.73	13	11	$9.83 \cdot 10^{-2}$
3918.42	3	1	$4.22 \cdot 10^{-1}$
3918.64	7	7	$1.17 \cdot 10^{-1}$
3925.94	1	3	$1.67 \cdot 10^{-1}$
3926.01	7	7	$7.26 \cdot 10^{-2}$
3928.08	9	9	$5.64 \cdot 10^{-2}$
3935.81	5	5	$1.14 \cdot 10^{-1}$
3941.28	5	5	$9.1 \cdot 10^{-2}$
3942.44	3	5	$9.62 \cdot 10^{-2}$
3946.99	9	11	$3.91 \cdot 10^{-2}$
3948.10	7	9	$1.31 \cdot 10^{-1}$
3948.77	11	9	$2.08 \cdot 10^{-1}$
3951.16	3	5	$4.29 \cdot 10^{-1}$
3955.34	3	3	$1.5 \cdot 10^{-1}$
3956.46	13	11	$1.76 \cdot 10^{-1}$
3956.68	11	13	$1.22 \cdot 10^{-1}$
3957.02	5	7	$1.67 \cdot 10^{-1}$
3963.10	3	5	$1.5 \cdot 10^{-1}$
3967.42	9	7	$1.52 \cdot 10^{-1}$
3967.96	7	9	$6.09 \cdot 10^{-2}$
3969.26	9	7	$2.26 \cdot 10^{-1}$
3971.32	11	9	$4.97 \cdot 10^{-2}$
3973.65	5	7	$5.81 \cdot 10^{-2}$
3976.61	3	5	$1.20 \cdot 10^{-1}$
3983.96	9	7	$5.72 \cdot 10^{-2}$
3985.39	5	5	$8.53 \cdot 10^{-2}$
3996.96	9	9	$7.95 \cdot 10^{-2}$
3997.39	9	11	$1.26 \cdot 10^{-1}$
3998.05	11	9	$5.70 \cdot 10^{-2}$
4005.24	7	5	$2.04 \cdot 10^{-1}$
4006.31	11	9	$5.1 \cdot 10^{-2}$
4014.53	9	7	$1.53 \cdot 10^{-1}$
4021.87	7	9	$8.55 \cdot 10^{-2}$
4024.73	7	9	$8.09 \cdot 10^{-2}$
4030.49	9	11	$1.04 \cdot 10^{-1}$
4043.90	7	7	$8.69 \cdot 10^{-2}$
4045.59	9	7	$7.39 \cdot 10^{-2}$
4045.81	9	9	$8.62 \cdot 10^{-1}$
4062.44	3	3	$1.85 \cdot 10^{-1}$
4063.59	7	7	$6.65 \cdot 10^{-1}$
4067.98	9	9	$1.51 \cdot 10^{-1}$
4070.77	7	5	$1.1 \cdot 10^{-1}$
4071.74	5	5	$7.64 \cdot 10^{-1}$
4073.76	5	3	$1.68 \cdot 10^{-1}$
4076.63	9	9	$1.32 \cdot 10^{-1}$
4084.49	11	9	$8.66 \cdot 10^{-2}$
4085.30	7	7	$8.92 \cdot 10^{-2}$
4098.18	7	7	$7.49 \cdot 10^{-2}$
4107.49	5	3	$1.74 \cdot 10^{-1}$
4109.80	3	3	$1.51 \cdot 10^{-1}$
4112.96	11	13	$1.1 \cdot 10^{-1}$
4118.55	11	13	$4.96 \cdot 10^{-1}$
4125.62	9	11	$9.9 \cdot 10^{-2}$
4126.18	11	11	$4.2 \cdot 10^{-2}$
4127.61	1	3	$1.43 \cdot 10^{-1}$
4132.06	5	7	$1.18 \cdot 10^{-1}$
4132.90	3	5	$7.70 \cdot 10^{-2}$
4134.68	5	7	$1.25 \cdot 10^{-1}$
4137.00	3	5	$2.75 \cdot 10^{-1}$
4142.59	3	5	$7.5 \cdot 10^{-2}$
4143.41	9	9	$2.70 \cdot 10^{-1}$
4143.87	7	9	$1.33 \cdot 10^{-1}$
4149.37	11	13	$4.23 \cdot 10^{-2}$
4153.90	7	9	$2.05 \cdot 10^{-1}$
4154.50	5	3	$2.64 \cdot 10^{-1}$
4154.81	9	11	$1.40 \cdot 10^{-1}$
4156.80	5	5	$1.20 \cdot 10^{-1}$
4157.78	5	7	$2.18 \cdot 10^{-1}$
4158.79	3	5	$1.6 \cdot 10^{-1}$
4172.12	7	5	$9.80 \cdot 10^{-2}$
4175.64	3	5	$1.14 \cdot 10^{-1}$
4181.75	5	7	$2.32 \cdot 10^{-1}$
4184.89	5	5	$1.03 \cdot 10^{-1}$
4187.04	7	5	$2.15 \cdot 10^{-1}$
4187.80	9	7	$1.52 \cdot 10^{-1}$
4191.43	5	3	$2.73 \cdot 10^{-1}$
4195.33	11	11	$1.11 \cdot 10^{-1}$
4196.21	7	7	$1.09 \cdot 10^{-1}$
4198.25	9	9	$1.47 \cdot 10^{-1}$
4198.30	11	9	$8.03 \cdot 10^{-2}$
4198.63	5	5	$1.25 \cdot 10^{-1}$
4199.10	9	11	$4.92 \cdot 10^{-1}$
4200.92	7	9	$6.25 \cdot 10^{-2}$
4202.03	9	9	$8.22 \cdot 10^{-2}$
4203.94	13	13	$2.97 \cdot 10^{-2}$
4203.98	3	5	$7.37 \cdot 10^{-2}$
4210.34	3	3	$1.48 \cdot 10^{-1}$
4217.55	3	5	$2.46 \cdot 10^{-1}$
4219.36	11	13	$2.88 \cdot 10^{-1}$
4222.21	7	7	$5.76 \cdot 10^{-2}$
4224.17	9	11	$1.06 \cdot 10^{-1}$
4224.51	3	5	$6.81 \cdot 10^{-2}$
4225.45	5	7	$1.65 \cdot 10^{-1}$
4227.43	11	13	$5.29 \cdot 10^{-1}$
4233.60	3	5	$1.85 \cdot 10^{-1}$
4235.94	9	9	$1.88 \cdot 10^{-1}$
4238.81	7	9	$2.41 \cdot 10^{-1}$
4247.43	9	11	$1.94 \cdot 10^{-1}$
4250.12	5	7	$2.07 \cdot 10^{-1}$
4250.79	7	7	$1.02 \cdot 10^{-1}$
4260.47	11	11	$3.99 \cdot 10^{-1}$

λ Å	Weights g_i	g_k	A $10^8\ s^{-1}$
4271.15	7	9	$1.82\cdot10^{-1}$
4271.76	9	11	$2.28\cdot10^{-1}$
4282.40	7	5	$1.21\cdot10^{-1}$
4299.23	9	11	$1.29\cdot10^{-1}$
4307.90	7	9	$3.38\cdot10^{-1}$
4315.08	5	5	$7.76\cdot10^{-2}$
4325.76	5	7	$5.16\cdot10^{-1}$
4327.10	5	5	$1.12\cdot10^{-1}$
4369.77	9	9	$6.09\cdot10^{-2}$
4383.55	9	11	$5.00\cdot10^{-1}$
4388.41	7	7	$1.03\cdot10^{-1}$
4401.29	7	7	$6.4\cdot10^{-2}$
4404.75	7	9	$2.75\cdot10^{-1}$
4415.12	5	7	$1.19\cdot10^{-1}$
4433.22	5	3	$2.1\cdot10^{-1}$
4443.19	1	3	$1.02\cdot10^{-1}$
4455.03	9	7	$4.1\cdot10^{-2}$
4466.55	5	7	$1.20\cdot10^{-1}$
4469.38	5	7	$1.59\cdot10^{-1}$
4476.02	3	5	$1.01\cdot10^{-1}$
4484.22	7	9	$5.04\cdot10^{-2}$
4485.68	3	3	$1.1\cdot10^{-1}$
4528.61	7	9	$5.44\cdot10^{-2}$
4547.85	5	7	$4.48\cdot10^{-2}$
4556.13	7	5	$1.05\cdot10^{-1}$
4619.29	7	5	$5.2\cdot10^{-2}$
4654.61	7	7	$3.68\cdot10^{-2}$
4667.45	7	9	$6.03\cdot10^{-2}$
4673.16	5	7	$3.81\cdot10^{-2}$
4678.85	7	9	$4.97\cdot10^{-2}$
4736.77	9	11	$4.78\cdot10^{-2}$
4789.65	5	7	$4.57\cdot10^{-2}$
4800.65	7	9	$3.01\cdot10^{-2}$
4859.74	5	3	$1.62\cdot10^{-1}$
4871.32	7	5	$2.44\cdot10^{-1}$
4872.14	3	3	$2.54\cdot10^{-1}$
4878.21	1	3	$1.21\cdot10^{-1}$
4890.76	5	5	$2.25\cdot10^{-1}$
4891.49	9	7	$3.08\cdot10^{-1}$
4903.31	3	5	$6.58\cdot10^{-2}$
4918.99	7	7	$1.79\cdot10^{-1}$
4920.50	11	9	$3.58\cdot10^{-1}$
4957.30	9	9	$1.18\cdot10^{-1}$
4957.60	13	11	$4.22\cdot10^{-1}$
4966.09	11	11	$3.31\cdot10^{-2}$
4973.10	3	3	$1.1\cdot10^{-1}$
4978.60	5	3	$1.19\cdot10^{-1}$
4985.25	5	5	$1.48\cdot10^{-1}$
4988.95	7	7	$5.2\cdot10^{-2}$
5001.86	9	7	$3.7\cdot10^{-1}$
5006.12	11	11	$5.87\cdot10^{-2}$
5014.94	7	5	$2.64\cdot10^{-1}$
5021.59	7	9	$6.18\cdot10^{-2}$
5022.24	5	3	$2.4\cdot10^{-1}$
5048.44	3	5	$4.88\cdot10^{-2}$
5068.77	9	7	$3.37\cdot10^{-2}$
5074.75	9	11	$1.4\cdot10^{-1}$
5090.77	7	5	$1.9\cdot10^{-1}$
5121.64	5	5	$7.9\cdot10^{-2}$
5137.38	11	9	$1.0\cdot10^{-1}$
5139.25	7	5	$9.16\cdot10^{-2}$
5139.46	9	9	$8.69\cdot10^{-2}$
5184.27	5	7	$3.8\cdot10^{-2}$
5191.46	5	3	$2.32\cdot10^{-1}$
5192.34	7	7	$1.34\cdot10^{-1}$
5208.59	7	5	$6.23\cdot10^{-2}$
5215.18	5	3	$1.10\cdot10^{-1}$
5226.86	5	5	$1.36\cdot10^{-1}$
5232.94	9	11	$1.94\cdot10^{-1}$
5235.39	9	7	$3.75\cdot10^{-2}$
5242.49	13	11	$2.38\cdot10^{-2}$
5263.31	5	5	$6.36\cdot10^{-2}$
5266.56	7	9	$1.10\cdot10^{-1}$
5273.16	1	3	$8.12\cdot10^{-2}$
5281.79	5	7	$5.00\cdot10^{-2}$
5283.62	7	7	$1.02\cdot10^{-1}$
5302.30	3	5	$9.04\cdot10^{-2}$
5324.18	9	9	$2.06\cdot10^{-1}$
5339.93	5	7	$6.36\cdot10^{-2}$
5364.87	5	7	$5.59\cdot10^{-1}$
5367.47	7	9	$7.13\cdot10^{-1}$
5369.96	9	11	$7.22\cdot10^{-1}$
5373.71	7	9	$3.7\cdot10^{-2}$
5383.37	11	13	$7.81\cdot10^{-1}$
5393.17	7	9	$4.91\cdot10^{-2}$
5398.28	5	5	$9.0\cdot10^{-2}$
5404.15	9	11	$6.92\cdot10^{-1}$
5410.91	7	9	$6.33\cdot10^{-1}$
5415.20	11	13	$7.67\cdot10^{-1}$
5463.28	9	9	$2.9\cdot10^{-1}$
5473.90	7	7	$5.2\cdot10^{-2}$
5476.29	7	9	$2.87\cdot10^{-2}$
5476.56	9	9	$8.70\cdot10^{-2}$
5563.60	5	7	$3.4\cdot10^{-2}$
5569.62	5	3	$2.34\cdot10^{-1}$
5572.84	7	5	$2.28\cdot10^{-1}$
5576.09	3	1	$2.5\cdot10^{-1}$
5586.76	9	7	$2.19\cdot10^{-1}$
5594.66	9	9	$5.20\cdot10^{-2}$
5602.95	3	3	$1.00\cdot10^{-1}$
5615.64	11	9	$2.64\cdot10^{-1}$
5624.54	5	5	$7.41\cdot10^{-2}$
5633.95	11	13	$7.7\cdot10^{-2}$
5638.26	9	7	$4.4\cdot10^{-2}$
5649.99	3	5	$5.1\cdot10^{-2}$
5655.18	7	9	$4.7\cdot10^{-2}$
5658.82	7	7	$4.34\cdot10^{-2}$
5662.52	11	9	$6.18\cdot10^{-2}$
5679.02	5	7	$3.7\cdot10^{-2}$
5686.53	9	11	$6.71\cdot10^{-2}$
5705.99	7	9	$6.1\cdot10^{-2}$
5753.12	3	5	$8.26\cdot10^{-2}$
5762.99	5	7	$9.6\cdot10^{-2}$
5816.37	9	11	$4.49\cdot10^{-2}$
5905.67	5	3	$1.1\cdot10^{-1}$
6301.50	5	5	$6.43\cdot10^{-2}$
6400.00	7	9	$9.27\cdot10^{-2}$
6411.65	5	7	$4.43\cdot10^{-2}$
6419.95	7	7	$1.2\cdot10^{-1}$
6469.19	3	3	$8.3\cdot10^{-2}$
6496.47	5	5	$7.8\cdot10^{-2}$
6569.22	7	9	$6.0\cdot10^{-2}$
6633.75	7	7	$3.44\cdot10^{-2}$
6841.34	5	7	$3.4\cdot10^{-2}$
6855.16	7	9	$2.86\cdot10^{-2}$
7187.32	9	11	$8.36\cdot10^{-2}$
7511.02	11	11	$1.35\cdot10^{-1}$
8220.38	13	11	$1.69\cdot10^{-1}$
8699.45	7	9	$4.08\cdot10^{-2}$
9012.07	11	9	$4.46\cdot10^{-2}$
9401.11	9	11	$2.64\cdot10^{-2}$
9414.04	7	9	$3.98\cdot10^{-2}$
9443.80	5	7	$6.39\cdot10^{-2}$
9569.91	11	11	$2.50\cdot10^{-2}$
9626.50	9	9	$4.51\cdot10^{-2}$
9738.57	11	13	$7.64\cdot10^{-2}$
9763.38	3	5	$5.42\cdot10^{-2}$
9861.74	7	9	$5.49\cdot10^{-2}$
9889.04	9	11	$2.22\cdot10^{-2}$
22473.28	11	11	$3.32\cdot10^{-2}$
23566.67	9	11	$2.21\cdot10^{-2}$
24547.95	11	9	$3.72\cdot10^{-2}$
24729.10	13	11	$5.08\cdot10^{-2}$

Fe II

λ Å	Weights g_i	g_k	A $10^8\ s^{-1}$
1055.26	10	8	$4.6\cdot10^{-1}$
1063.97	10	8	$3.5\cdot10^{-1}$
1068.35	8	8	1.59
1071.58	6	8	1.14
1096.88	10	8	2.26
1112.05	10	12	$2.0\cdot10^{-1}$
1121.97	10	8	1.92
1122.84	8	6	1.81
1125.45	10	8	1.03
1128.05	2	4	1.40
1130.44	6	8	$3.1\cdot10^{-1}$
1133.40	8	10	$2.6\cdot10^{-1}$
1133.67	10	8	$3.1\cdot10^{-1}$
1138.63	8	8	$5.5\cdot10^{-1}$
1142.37	10	8	$2.6\cdot10^{-1}$
1143.23	10	10	$9.8\cdot10^{-1}$
1144.94	10	12	3.52
1147.41	8	8	1.24
1148.28	8	10	3.35
1151.15	6	8	2.23
1267.42	8	6	$9.3\cdot10^{-1}$
1272.61	6	4	$3.3\cdot10^{-1}$
1371.02	14	12	1.74
1563.79	8	8	1.33
1580.63	8	10	$5.8\cdot10^{-1}$
1608.45	10	8	1.91
1610.92	10	10	$1.94\cdot10^{-1}$
1618.47	8	8	$5.53\cdot10^{-1}$
1621.69	8	6	1.32
1623.09	8	8	$1.99\cdot10^{-1}$
1625.52	8	10	$4.04\cdot10^{-1}$
1625.91	6	8	$1.02\cdot10^{-1}$
1629.16	6	6	$8.66\cdot10^{-1}$
1631.13	6	4	$6.93\cdot10^{-1}$
1633.91	6	8	$3.85\cdot10^{-1}$

λ Å	Weights g_i	g_k	A $10^8\ s^{-1}$
1634.35	4	6	$3.21\cdot10^{-1}$
1635.40	8	6	2.28
1636.33	4	4	$9.63\cdot10^{-1}$
1637.40	10	8	$3.57\cdot10^{-1}$
1639.40	2	4	$6.85\cdot10^{-1}$
1641.76	6	4	1.76
1647.16	6	6	$4.98\cdot10^{-1}$
1670.75	10	8	1.06
1676.86	8	8	$6.75\cdot10^{-2}$
1702.04	10	12	1.02
1761.37	8	8	1.42
1785.27	6	8	$1.2\cdot10^{1}$
1786.75	6	6	$1.2\cdot10^{1}$
1788.08	6	4	4.6
1818.52	8	8	$5.70\cdot10^{-2}$
2020.75	6	8	$1.83\cdot10^{-1}$
2078.16	10	10	$2.84\cdot10^{-2}$
2162.02	10	10	$2.54\cdot10^{-1}$
2182.36	10	8	$8.6\cdot10^{-2}$
2191.98	8	8	$7.54\cdot10^{-1}$
2201.59	6	8	$7.77\cdot10^{-1}$
2208.41	10	10	1.59
2209.03	10	8	1.27
2213.66	14	14	$3.26\cdot10^{-1}$
2218.26	8	10	1.57
2220.38	12	12	$4.19\cdot10^{-1}$
2228.73	6	8	1.59
2249.18	10	8	$3.00\cdot10^{-2}$
2253.13	8	8	$4.41\cdot10^{-2}$
2255.77	6	4	$4.75\cdot10^{-1}$
2260.08	10	10	$3.18\cdot10^{-2}$
2267.59	6	8	$3.69\cdot10^{-2}$
2279.92	8	10	$4.49\cdot10^{-2}$
2327.40	6	4	$6.55\cdot10^{-1}$
2327.88	10	12	1.08
2331.31	10	8	$3.17\cdot10^{-1}$
2332.80	8	6	1.31
2338.01	4	4	1.13
2338.54	10	12	$5.6\cdot10^{-2}$
2343.50	10	8	1.73
2343.96	8	6	$3.13\cdot10^{-1}$
2344.28	2	4	$9.27\cdot10^{-1}$
2345.34	14	12	$7.3\cdot10^{-1}$
2348.12	10	8	$6.50\cdot10^{-1}$
2348.30	6	6	1.15
2351.20	12	10	$7.19\cdot10^{-1}$
2351.67	6	6	1.80
2352.31	2	4	4.38
2353.47	12	14	4.98
2353.68	8	8	1.30
2354.48	10	8	$8.13\cdot10^{-1}$
2354.89	6	4	$2.67\cdot10^{-1}$
2359.11	4	6	$5.0\cdot10^{-1}$
2359.60	10	10	$2.25\cdot10^{-1}$
2360.00	10	10	$3.59\cdot10^{-1}$
2360.29	8	6	$6.23\cdot10^{-1}$
2360.53	6	8	$2.22\cdot10^{-1}$
2361.73	8	8	$2.40\cdot10^{-1}$
2362.02	8	8	$1.41\cdot10^{-1}$
2363.86	8	10	5.3
2364.83	8	8	$5.90\cdot10^{-1}$
2365.76	6	6	2.16
2366.59	6	6	$1.01\cdot10^{-1}$
2366.88	8	10	$3.51\cdot10^{-2}$
2368.60	6	4	$6.06\cdot10^{-1}$
2369.95	10	12	5.9
2370.50	4	4	$1.73\cdot10^{-1}$
2373.74	10	10	$4.25\cdot10^{-1}$
2375.19	4	2	$9.81\cdot10^{-1}$
2376.43	12	14	6.4
2378.55	8	8	$1.70\cdot10^{-1}$
2378.70	8	8	$1.49\cdot10^{-1}$
2379.28	8	8	$2.73\cdot10^{-1}$
2379.42	10	10	$3.68\cdot10^{-1}$
2380.76	6	8	$3.10\cdot10^{-1}$
2382.04	10	12	3.13
2382.36	4	6	$3.19\cdot10^{-2}$
2382.90	12	14	$1.62\cdot10^{-1}$
2383.06	8	6	$1.0\cdot10^{-1}$
2383.24	6	6	$3.59\cdot10^{-1}$
2384.39	4	4	$3.22\cdot10^{-1}$
2385.01	6	8	$3.60\cdot10^{-2}$
2388.39	10	12	$2.02\cdot10^{-1}$
2388.63	8	8	1.05
2390.10	14	16	5.5
2390.76	6	6	1.17
2391.48	8	10	$3.77\cdot10^{-2}$
2394.00	12	10	$9.4\cdot10^{-2}$
2395.42	6	4	$2.67\cdot10^{-1}$
2395.63	8	10	2.59
2396.72	10	12	$2.15\cdot10^{-1}$
2399.24	6	6	1.39
2400.05	12	14	4.57
2401.29	6	8	1.89
2402.45	10	10	$5.8\cdot10^{-1}$
2402.63	8	8	$8.19\cdot10^{-1}$
2404.43	4	2	$6.44\cdot10^{-1}$
2404.89	6	8	1.96
2406.66	4	4	1.61
2410.27	8	8	$7.65\cdot10^{-1}$
2410.52	4	6	1.55
2411.07	2	2	2.37
2411.81	10	12	4.33
2412.01	6	8	$1.66\cdot10^{-1}$
2413.31	2	4	1.02
2416.45	8	10	2.38
2417.87	12	12	$9.5\cdot10^{-1}$
2418.44	6	8	2.28
2419.89	10	10	$2.2\cdot10^{-2}$
2422.69	6	8	1.46
2422.93	10	8	$2.94\cdot10^{-2}$
2423.21	4	6	1.40
2424.15	10	12	2.21
2424.39	6	8	$1.61\cdot10^{-1}$
2424.59	6	6	1.24
2424.65	8	8	$6.55\cdot10^{-2}$
2428.36	8	10	2.68
2428.80	4	4	1.38
2429.04	2	4	1.23
2429.39	4	4	$6.9\cdot10^{-1}$
2429.86	8	8	1.51
2430.08	8	10	1.91
2432.26	6	8	1.57
2432.87	14	14	2.86
2433.50	10	12	$1.30\cdot10^{-1}$
2434.06	8	6	$7.2\cdot10^{-1}$
2434.24	8	10	2.01
2434.73	12	12	2.79
2434.95	4	6	1.39
2435.00	8	8	2.02
2436.62	6	8	2.70
2439.30	12	14	2.25
2440.42	6	8	1.18
2441.13	10	10	$8.95\cdot10^{-1}$
2442.38	10	12	2.75
2443.71	8	10	1.44
2444.52	6	8	2.78
2445.11	12	12	2.03
2445.57	4	6	2.07
2445.80	4	6	1.23
2446.11	8	8	1.06
2446.47	12	14	$2.99\cdot10^{-1}$
2447.21	6	6	1.15
2447.33	4	2	2.56
2447.76	12	10	1.97
2449.96	4	4	1.24
2450.21	2	4	1.26
2453.98	8	10	1.31
2454.58	14	12	1.16
2455.71	8	8	1.01
2455.90	4	6	1.73
2457.10	6	4	$4.71\cdot10^{-1}$
2458.78	10	12	2.31
2458.97	6	4	2.51
2460.44	10	12	5.39
2461.28	6	8	2.34
2461.86	8	10	2.43
2463.28	12	10	$7.1\cdot10^{-1}$
2464.01	10	8	1.32
2464.91	6	4	2.22
2465.91	8	6	1.62
2466.50	2	4	2.40
2466.67	4	2	2.64
2466.82	6	4	1.77
2468.30	10	10	$9.8\cdot10^{-2}$
2469.37	10	8	$2.23\cdot10^{-2}$
2469.52	8	6	2.58
2470.41	6	6	$6.0\cdot10^{-1}$
2470.67	8	6	1.54
2471.28	10	8	$4.15\cdot10^{-1}$
2472.61	8	10	3.22
2473.32	2	2	2.74
2475.12	4	6	3.72
2475.54	6	8	3.18
2476.27	8	10	$9.7\cdot10^{-2}$
2477.35	8	8	$1.70\cdot10^{-1}$
2478.57	6	6	$9.1\cdot10^{-1}$
2480.16	10	8	1.55
2481.05	12	12	$1.46\cdot10^{-1}$
2482.12	14	14	$6.5\cdot10^{-1}$

λ Å	Weights g_i	Weights g_k	A 10^8 s^{-1}
2482.33	4	4	2.23
2482.66	12	10	1.25
2482.87	6	4	1.69
2483.72	8	10	$5.4 \cdot 10^{-1}$
2484.24	4	6	$8.3 \cdot 10^{-2}$
2484.44	8	8	2.16
2489.48	12	12	$5.1 \cdot 10^{-1}$
2489.83	12	12	1.94
2490.71	10	12	1.44
2490.86	8	8	$8.8 \cdot 10^{-1}$
2491.40	10	8	1.01
2492.34	10	12	$2.30 \cdot 10^{-1}$
2493.26	14	16	3.04
2493.88	6	6	1.74
2494.12	12	10	$2.97 \cdot 10^{-2}$
2497.82	6	6	1.68
2500.92	6	8	2.41
2501.35	2	2	1.48
2502.39	8	8	1.43
2503.33	12	12	$7.3 \cdot 10^{-1}$
2503.54	8	8	$3.32 \cdot 10^{-1}$
2503.57	10	10	$2.53 \cdot 10^{-1}$
2503.88	10	10	2.23
2506.09	10	10	$9.9 \cdot 10^{-1}$
2506.80	8	10	1.98
2508.34	8	10	$3.79 \cdot 10^{-1}$
2510.57	8	8	$1.54 \cdot 10^{-1}$
2511.76	8	10	2.30
2513.15	10	8	$2.49 \cdot 10^{-1}$
2514.38	8	8	2.11
2517.14	2	4	$9.2 \cdot 10^{-1}$
2519.05	8	6	2.10
2521.09	6	4	2.05
2521.82	8	8	2.36
2525.39	14	14	1.91
2525.92	8	8	$7.4 \cdot 10^{-1}$
2526.08	10	8	$3.52 \cdot 10^{-1}$
2526.29	6	6	2.47
2527.10	12	10	$3.67 \cdot 10^{-1}$
2527.71	10	8	$9.1 \cdot 10^{-1}$
2528.68	10	8	$2.3 \cdot 10^{-2}$
2529.08	4	6	1.80
2529.23	12	10	$3.27 \cdot 10^{-1}$
2529.55	10	10	2.20
2530.10	4	6	$6.6 \cdot 10^{-1}$
2533.63	12	12	1.92
2534.42	8	8	1.83
2535.36	6	4	2.46
2535.49	10	8	$7.47 \cdot 10^{-1}$
2536.67	12	12	$5.7 \cdot 10^{-1}$
2536.81	10	10	1.69
2536.84	12	14	$6.8 \cdot 10^{-1}$
2537.14	10	10	1.44
2538.21	14	12	1.26
2538.40	6	8	$3.7 \cdot 10^{-2}$
2538.50	8	6	$5.9 \cdot 10^{-1}$
2538.68	6	8	$7.4 \cdot 10^{-1}$
2538.91	10	8	1.28
2538.99	14	12	1.93
2539.81	8	8	$5.6 \cdot 10^{-2}$
2540.52	2	2	1.26
2540.66	6	8	1.70
2541.10	8	6	$9.6 \cdot 10^{-1}$
2541.84	8	6	$8.2 \cdot 10^{-1}$
2542.74	2	2	1.61
2543.38	10	12	$6.7 \cdot 10^{-1}$
2543.43	6	4	$8.3 \cdot 10^{-1}$
2544.97	4	6	$3.93 \cdot 10^{-1}$
2545.22	8	10	$5.3 \cdot 10^{-1}$
2545.44	8	10	$1.52 \cdot 10^{-1}$
2546.67	8	8	$7.98 \cdot 10^{-1}$
2547.34	8	8	$2.28 \cdot 10^{-1}$
2548.32	4	6	$2.69 \cdot 10^{-1}$
2548.59	10	10	$2.67 \cdot 10^{-1}$
2548.74	4	2	2.43
2548.92	12	10	$6.0 \cdot 10^{-1}$
2549.08	10	8	1.89
2549.40	4	4	1.65
2549.46	6	6	1.12
2549.77	8	6	$2.35 \cdot 10^{-1}$
2550.03	10	10	1.74
2550.15	8	10	$3.91 \cdot 10^{-1}$
2550.57	12	12	$1.6 \cdot 10^{-2}$
2550.68	12	12	1.07
2551.20	10	8	$2.48 \cdot 10^{-1}$
2554.94	8	8	$2.6 \cdot 10^{-2}$
2555.07	6	8	$1.96 \cdot 10^{-1}$
2555.45	4	6	$2.49 \cdot 10^{-1}$
2557.08	8	10	$2.8 \cdot 10^{-2}$
2557.51	10	8	$1.53 \cdot 10^{-1}$
2559.24	8	8	$6.4 \cdot 10^{-2}$
2559.77	6	8	$2.42 \cdot 10^{-1}$
2559.93	6	8	$2.47 \cdot 10^{-1}$
2560.28	4	4	1.77
2562.09	4	2	1.62
2562.54	8	6	1.79
2563.48	6	4	1.51
2566.22	8	10	2.61
2566.40	8	6	2.29
2566.62	10	12	$7.1 \cdot 10^{-2}$
2566.91	4	2	1.15
2568.41	2	4	$4.77 \cdot 10^{-1}$
2568.89	8	8	$2.8 \cdot 10^{-2}$
2569.78	2	4	1.11
2570.85	8	6	1.84
2571.55	10	10	$2.89 \cdot 10^{-2}$
2572.97	6	8	$7.89 \cdot 10^{-2}$
2573.21	8	10	$1.42 \cdot 10^{-1}$
2573.76	8	8	$2.3 \cdot 10^{-2}$
2574.37	6	4	2.43
2576.86	10	12	1.32
2577.92	2	2	1.24
2581.11	6	6	$7.61 \cdot 10^{-2}$
2582.41	6	8	$2.22 \cdot 10^{-1}$
2582.58	4	4	$8.80 \cdot 10^{-1}$
2583.05	8	10	$2.16 \cdot 10^{-2}$
2585.62	10	10	$3.09 \cdot 10^{-1}$
2585.88	10	8	$8.94 \cdot 10^{-1}$
2586.06	6	4	$5.8 \cdot 10^{-2}$
2587.95	8	10	1.69
2588.19	2	2	$1.5 \cdot 10^{-1}$
2588.80	8	8	$8.4 \cdot 10^{-2}$
2590.55	4	6	$7.9 \cdot 10^{-2}$
2591.54	6	6	$5.72 \cdot 10^{-1}$
2592.79	14	16	2.74
2593.73	2	4	$1.63 \cdot 10^{-1}$
2594.96	8	8	$1.0 \cdot 10^{-1}$
2598.37	8	6	1.43
2599.40	10	10	2.35
2604.05	8	8	$1.49 \cdot 10^{-1}$
2605.04	6	8	2.34
2605.31	4	4	1.99
2605.43	6	6	$3.40 \cdot 10^{-1}$
2605.90	4	2	1.27
2606.52	6	6	2.31
2607.09	6	4	1.73
2608.85	10	8	$5.0 \cdot 10^{-2}$
2609.13	8	10	$2.77 \cdot 10^{-1}$
2609.44	6	8	$6.0 \cdot 10^{-2}$
2609.87	8	8	$1.34 \cdot 10^{-1}$
2611.07	4	6	$7.28 \cdot 10^{-2}$
2611.87	8	8	1.20
2613.57	10	12	$2.0 \cdot 10^{-2}$
2613.82	4	2	2.12
2614.19	8	10	$3.3 \cdot 10^{-2}$
2614.59	10	8	$3.37 \cdot 10^{-2}$
2614.87	8	6	$3.5 \cdot 10^{-2}$
2617.62	6	6	$4.88 \cdot 10^{-1}$
2619.08	10	10	$2.48 \cdot 10^{-1}$
2620.17	6	6	$1.1 \cdot 10^{-1}$
2620.41	4	4	$4.30 \cdot 10^{-2}$
2620.70	8	8	$3.43 \cdot 10^{-1}$
2621.67	2	2	$5.60 \cdot 10^{-1}$
2623.13	14	14	$8.8 \cdot 10^{-2}$
2623.72	6	6	$1.92 \cdot 10^{-1}$
2625.49	12	14	2.55
2625.67	8	10	$3.52 \cdot 10^{-1}$
2626.50	4	6	$3.48 \cdot 10^{-1}$
2626.70	8	8	$1.94 \cdot 10^{-2}$
2628.29	2	4	$8.74 \cdot 10^{-1}$
2628.58	6	6	$3.4 \cdot 10^{-2}$
2629.59	6	8	$7.2 \cdot 10^{-1}$
2630.07	4	6	$5.1 \cdot 10^{-1}$
2631.05	4	6	$8.16 \cdot 10^{-1}$
2631.32	6	8	$6.29 \cdot 10^{-1}$
2631.61	10	12	$6.6 \cdot 10^{-1}$
2633.20	6	4	1.21
2636.70	4	4	$8.8 \cdot 10^{-2}$
2637.50	6	6	$6.2 \cdot 10^{-1}$
2637.64	2	4	$6.6 \cdot 10^{-1}$
2639.57	2	2	$8.0 \cdot 10^{-1}$
2641.12	4	4	$3.7 \cdot 10^{-2}$
2642.01	6	6	$2.29 \cdot 10^{-1}$
2646.21	12	10	$1.44 \cdot 10^{-2}$
2649.47	6	8	1.98
2650.48	6	8	1.60
2652.57	10	8	$4.45 \cdot 10^{-2}$
2654.63	4	4	$8.1 \cdot 10^{-1}$
2657.92	10	10	$3.2 \cdot 10^{-2}$
2658.25	8	8	$2.12 \cdot 10^{-1}$

λ Å	Weights g_i	g_k	A 10^8 s^{-1}
2662.56	2	2	1.33
2664.66	8	10	1.91
2666.64	6	8	1.87
2667.22	4	6	1.02
2669.93	2	4	5.2·10^{-1}
2670.38	6	8	6.0·10^{-2}
2671.39	2	4	6.5·10^{-1}
2680.23	6	8	1.10·10^{-1}
2682.51	8	10	9.2·10^{-1}
2683.00	4	6	7.3·10^{-1}
2684.75	8	10	1.57
2691.74	10	8	5.04·10^{-2}
2692.60	10	12	1.40
2693.86	8	6	4.2·10^{-2}
2697.33	4	4	2.48·10^{-1}
2697.46	4	2	1.65
2697.73	10	8	2.6·10^{-2}
2699.20	4	4	6.2·10^{-1}
2703.99	8	8	1.38
2704.58	8	8	1.66·10^{-2}
2707.13	4	6	8.3·10^{-1}
2709.06	4	6	3.88·10^{-1}
2711.84	12	14	4.36·10^{-1}
2712.39	10	12	1.29·10^{-1}
2714.41	8	6	5.70·10^{-1}
2716.22	6	6	1.15
2716.44	6	6	2.8·10^{-2}
2716.57	14	12	1.35
2717.88	16	14	1.51
2718.64	10	8	1.18
2719.30	6	8	4.44·10^{-1}
2721.81	12	10	5.1·10^{-2}
2722.06	8	8	1.42·10^{-1}
2722.74	6	8	8.2·10^{-1}
2724.88	6	6	9.58·10^{-2}
2726.52	6	8	5.0·10^{-2}
2727.38	12	10	3.12·10^{-1}
2727.54	6	4	9.38·10^{-1}
2728.91	8	10	1.25·10^{-1}
2730.73	4	4	2.79·10^{-1}
2732.01	10	8	7.05·10^{-2}
2732.94	8	6	9.5·10^{-1}
2736.97	4	2	1.22
2739.55	8	8	2.21
2741.39	6	6	2.03·10^{-1}
2743.20	2	4	1.97
2744.90	6	8	3.62·10^{-2}
2746.48	4	6	2.05
2746.98	6	6	1.69
2749.18	4	4	1.21
2749.32	6	8	2.16
2749.49	2	2	1.16
2750.01	10	10	1.8·10^{-2}
2751.13	4	4	2.92·10^{-1}
2752.15	4	4	7.7·10^{-1}
2753.29	10	12	1.89
2754.89	8	6	1.21
2755.74	8	10	2.15
2756.51	6	8	7.3·10^{-2}
2757.03	10	8	8.07·10^{-2}
2761.81	2	4	1.38·10^{-1}
2762.33	6	6	6.0·10^{-1}
2762.45	6	4	3.3·10^{-2}
2763.66	14	12	1.34
2763.91	8	6	2.9·10^{-2}
2764.79	12	12	1.1·10^{-2}
2765.13	10	8	1.47
2767.50	12	14	1.58
2768.93	4	6	4.75·10^{-2}
2769.15	8	10	6.6·10^{-2}
2769.35	12	14	2.07·10^{-1}
2770.50	12	10	4.08·10^{-2}
2771.19	10	12	4.3·10^{-2}
2774.69	2	4	2.73·10^{-1}
2776.18	6	8	2.66·10^{-2}
2776.91	8	8	4.08·10^{-1}
2779.30	10	8	1.00
2779.91	2	4	2.56·10^{-1}
2780.05	2	2	3.3·10^{-1}
2783.69	12	10	1.06
2784.28	2	4	3.4·10^{-2}
2785.19	12	10	1.53
2787.24	8	6	1.83·10^{-1}
2790.56	8	10	2.1·10^{-2}
2793.89	10	12	1.26·10^{-1}
2796.63	10	10	2.0·10^{-1}
2797.92	10	10	3.2·10^{-2}
2799.29	10	8	1.55·10^{-1}
2805.32	4	6	2.5·10^{-2}
2805.79	8	8	3.22·10^{-2}
2809.78	8	8	3.10·10^{-1}
2811.27	12	10	1.2·10^{-2}
2812.49	4	4	2.9·10^{-2}
2813.61	8	10	3.40·10^{-2}
2817.09	6	4	3.37·10^{-1}
2819.34	12	12	9.7·10^{-3}
2826.03	8	6	4.5·10^{-2}
2827.43	12	14	2.4·10^{-2}
2828.63	12	10	6.9·10^{-2}
2831.56	4	6	7.6·10^{-1}
2833.09	6	6	4.55·10^{-1}
2835.71	4	6	5.1·10^{-1}
2836.19	4	4	5.4·10^{-2}
2836.51	2	4	9.8·10^{-2}
2837.30	10	12	1.9·10^{-2}
2838.22	4	2	8.6·10^{-1}
2839.51	10	8	1.47
2839.80	8	10	5.8·10^{-1}
2840.34	12	12	7.7·10^{-2}
2840.65	2	4	7.6·10^{-1}
2840.76	10	12	1.49·10^{-1}
2842.08	8	8	1.5·10^{-2}
2843.32	10	10	1.40·10^{-2}
2843.48	4	6	9.6·10^{-2}
2844.96	2	2	5.5·10^{-1}
2845.60	8	6	1.57
2847.77	4	4	5.1·10^{-1}
2848.11	6	6	9.9·10^{-1}
2848.32	6	4	1.59
2848.91	12	10	5.3·10^{-2}
2849.61	10	12	4.6·10^{-2}
2853.21	6	6	2.3·10^{-2}
2855.67	8	10	9.2·10^{-2}
2856.15	10	10	5.0·10^{-2}
2856.38	6	8	4.42·10^{-1}
2856.91	8	8	1.32
2857.17	6	8	1.22·10^{-1}
2857.42	6	6	2.0·10^{-2}
2858.34	10	12	4.85·10^{-1}
2864.97	8	8	4.3·10^{-2}
2869.16	8	10	1.4·10^{-2}
2869.31	4	6	4.04·10^{-1}
2871.06	10	12	2.2·10^{-2}
2871.13	12	10	3.0·10^{-2}
2872.38	10	8	1.70·10^{-1}
2873.40	8	10	4.56·10^{-1}
2875.35	8	10	1.35·10^{-1}
2876.80	8	8	9.56·10^{-2}
2879.25	10	8	3.6·10^{-2}
2880.76	8	8	2.21·10^{-2}
2883.71	12	14	1.48·10^{-1}
2884.76	6	8	2.46·10^{-1}
2885.93	14	12	3.8·10^{-2}
2888.10	4	6	6.1·10^{-2}
2894.78	10	12	5.7·10^{-2}
2895.22	8	10	1.09·10^{-1}
2897.27	6	4	1.8·10^{-1}
2902.46	10	10	3.2·10^{-2}
2906.12	2	4	4.4·10^{-2}
2910.76	8	8	1.5·10^{-2}
2917.08	6	8	2.7·10^{-2}
2922.02	8	10	3.8·10^{-2}
2926.59	8	10	5.1·10^{-2}
2944.40	4	2	3.5·10^{-1}
2947.65	6	4	2.01·10^{-1}
2949.18	10	8	2.45·10^{-1}
2953.77	6	8	5.2·10^{-2}
2954.05	8	8	1.2·10^{-2}
2959.60	8	6	9.7·10^{-2}
2959.84	8	6	1.36·10^{-1}
2964.13	8	6	4.6·10^{-2}
2964.62	2	2	6.5·10^{-2}
2965.03	4	4	9.43·10^{-2}
2969.94	8	6	2.28·10^{-1}
2970.52	4	6	2.70·10^{-2}
2970.69	10	8	4.15·10^{-2}
2982.06	4	6	2.41·10^{-1}
2984.82	6	6	4.29·10^{-1}
2985.54	2	4	2.39·10^{-1}
2997.30	6	8	8.6·10^{-2}
3000.06	8	6	3.0·10^{-2}
3002.32	6	8	2.0·10^{-2}
3002.65	4	6	1.79·10^{-1}
3036.96	6	6	2.22·10^{-1}
3044.84	8	10	1.2·10^{-2}
3048.99	4	4	3.84·10^{-1}
3056.80	14	12	1.7·10^{-2}
3062.24	12	10	1.36·10^{-1}
3065.32	6	6	2.9·10^{-2}
3070.69	10	8	1.28·10^{-2}

λ Å	Weights g_i	Weights g_k	A 10^8 s^{-1}
3071.12	2	4	$2.59\cdot10^{-1}$
3076.44	4	6	$3.75\cdot10^{-1}$
3077.17	14	12	$1.35\cdot10^{-1}$
3078.68	6	8	$5.5\cdot10^{-1}$
3089.38	6	8	$2.2\cdot10^{-2}$
3096.29	8	8	$1.9\cdot10^{-2}$
3105.17	4	2	$7.5\cdot10^{-2}$
3105.55	2	2	$7.0\cdot10^{-2}$
3106.57	8	8	$1.88\cdot10^{-2}$
3114.30	4	4	$6.4\cdot10^{-2}$
3114.69	2	4	$2.5\cdot10^{-2}$
3116.58	6	4	$5.5\cdot10^{-2}$
3133.05	4	6	$1.5\cdot10^{-2}$
3135.36	6	6	$8.8\cdot10^{-2}$
3144.75	8	6	$2.7\cdot10^{-2}$
3154.20	10	10	$2.06\cdot10^{-1}$
3162.80	8	8	$5.5\cdot10^{-2}$
3167.86	8	8	$1.59\cdot10^{-1}$
3177.53	8	8	$1.74\cdot10^{-1}$
3179.50	6	8	$1.11\cdot10^{-1}$
3180.15	4	6	$7.7\cdot10^{-2}$
3186.74	4	4	$3.85\cdot10^{-2}$
3187.30	10	10	$5.0\cdot10^{-2}$
3192.91	6	6	$1.27\cdot10^{-2}$
3193.80	2	2	$5.4\cdot10^{-2}$
3193.86	8	8	$3.86\cdot10^{-2}$
3196.07	6	8	$1.61\cdot10^{-2}$
3210.44	2	4	$3.63\cdot10^{-2}$
3213.31	4	6	$6.12\cdot10^{-2}$
3227.74	6	8	$8.9\cdot10^{-2}$
3231.71	6	8	$1.4\cdot10^{-2}$
3232.78	8	6	$5.0\cdot10^{-2}$
3237.82	2	4	$6.8\cdot10^{-2}$
3243.72	10	8	$5.1\cdot10^{-2}$
3247.17	4	6	$7.1\cdot10^{-2}$
3258.77	6	8	$9.39\cdot10^{-2}$
3259.05	8	10	$6.7\cdot10^{-2}$
3276.60	6	8	$1.0\cdot10^{-2}$
3289.35	8	8	$2.1\cdot10^{-2}$
3323.06	8	10	$1.4\cdot10^{-2}$
3366.97	8	6	$2.2\cdot10^{-2}$
3381.01	6	4	$3.0\cdot10^{-2}$
3453.62	8	10	$8.5\cdot10^{-3}$
3468.68	8	8	$2.0\cdot10^{-2}$
3493.47	10	10	$3.2\cdot10^{-2}$
3621.27	2	4	$2.2\cdot10^{-2}$
3748.48	6	4	$3.4\cdot10^{-2}$
3759.46	4	2	$3.2\cdot10^{-2}$
3906.04	6	8	$1.1\cdot10^{-2}$
3935.96	8	10	$8.3\cdot10^{-3}$
3938.97	4	6	$8.4\cdot10^{-3}$
4233.17	6	8	$7.22\cdot10^{-3}$
4522.63	6	4	$8.4\cdot10^{-3}$
4549.19	4	6	$9.2\cdot10^{-3}$
4549.47	8	6	$1.00\cdot10^{-2}$
4583.84	10	8	$7.22\cdot10^{-3}$
4635.32	6	8	$1.0\cdot10^{-2}$
4923.93	6	4	$4.28\cdot10^{-2}$
4990.51	6	8	$5.2\cdot10^{-1}$
5001.96	12	14	1.57
5018.44	6	6	$2.0\cdot10^{-2}$
5030.63	10	10	$7.1\cdot10^{-1}$
5035.71	10	12	$9.4\cdot10^{-1}$
5144.35	4	6	$8.5\cdot10^{-1}$
5149.47	8	10	$9.0\cdot10^{-1}$
5169.03	6	8	$4.22\cdot10^{-2}$
5197.58	6	4	$5.4\cdot10^{-3}$
5227.48	12	14	1.22
5247.95	4	6	1.43
5251.23	6	8	$8.0\cdot10^{-1}$
5264.18	8	10	$4.76\cdot10^{-1}$
5272.40	6	6	$3.9\cdot10^{-3}$
5276.00	10	8	$3.76\cdot10^{-3}$
5306.18	6	8	$3.28\cdot10^{-1}$
5316.22	14	14	$3.69\cdot10^{-1}$
5316.62	12	10	$3.89\cdot10^{-3}$
5387.06	12	14	$5.2\cdot10^{-1}$
5395.86	6	8	$5.5\cdot10^{-1}$
5402.06	10	12	$5.6\cdot10^{-1}$
5427.83	12	10	$5.9\cdot10^{-3}$
5429.99	8	10	$6.0\cdot10^{-1}$
5465.93	6	8	$6.2\cdot10^{-1}$
5482.31	10	12	$4.78\cdot10^{-1}$
5493.83	8	10	$4.01\cdot10^{-1}$
5506.19	12	14	1.14
5510.78	10	12	$2.28\cdot10^{-1}$
5529.05	6	6	$2.01\cdot10^{-1}$
5544.76	12	12	$2.49\cdot10^{-1}$
5783.63	8	10	$4.62\cdot10^{-1}$
5885.01	4	6	$6.4\cdot10^{-1}$
5902.83	8	10	$4.98\cdot10^{-1}$
5955.70	6	8	$4.19\cdot10^{-1}$
5961.71	10	12	$7.4\cdot10^{-1}$
5965.62	10	10	$2.19\cdot10^{-1}$
6175.15	8	8	$1.8\cdot10^{-3}$
6305.30	10	10	$1.4\cdot10^{-3}$
6331.95	6	8	$1.8\cdot10^{-3}$
6446.41	8	10	$1.3\cdot10^{-3}$
6456.38	8	6	$1.7\cdot10^{-3}$

Fe III

λ Å	Weights g_i	Weights g_k	A 10^8 s^{-1}
1843.4	9	7	4.8
1844.3	7	5	4.9
1846.9	5	3	5.5
1854.38	3	1	5.7
1865.20	7	7	6.1
1893.98	11	9	5.5
1896.80	13	11	5.0
1904.3	5	5	5.7
1907.58	15	13	5.3
1915.08	13	15	6.0
1922.79	11	13	5.5
1930.39	9	11	5.1
1931.51	9	11	5.3
1937.35	7	9	5.1
1943.48	5	7	5.0
1950.33	13	15	5.5
1951.01	11	11	5.3
1952.65	9	9	4.9
1953.32	7	7	5.1
1987.50	13	13	4.9

Fe VII

λ Å	Weights g_i	Weights g_k	A 10^8 s^{-1}
150.807	5	7	$1.3\cdot10^{3}$
150.852	7	9	$1.3\cdot10^{3}$
151.023	9	11	$1.6\cdot10^{3}$
151.046	7	7	$2.2\cdot10^{2}$
151.145	9	9	$2.1\cdot10^{2}$
151.432	5	7	$2.2\cdot10^{2}$
151.512	5	5	$5.3\cdot10^{2}$
151.675	7	7	$3.9\cdot10^{2}$
151.782	9	9	$2.4\cdot10^{2}$
154.307	3	1	$8.9\cdot10^{2}$
154.335	5	7	$1.2\cdot10^{3}$
154.363	3	3	$4.2\cdot10^{2}$
154.565	5	3	$3.5\cdot10^{2}$
154.650	5	5	$8.8\cdot10^{2}$
154.848	1	3	$7.7\cdot10^{2}$
154.921	3	5	$9.7\cdot10^{2}$
154.941	3	3	$2.4\cdot10^{2}$
154.949	5	7	$1.0\cdot10^{3}$
155.994	9	11	$1.8\cdot10^{3}$
158.481	9	9	$2.3\cdot10^{2}$
165.087	1	3	$6.9\cdot10^{2}$
165.919	7	5	$2.8\cdot10^{3}$
166.365	9	7	$2.9\cdot10^{3}$
173.441	9	9	$3.6\cdot10^{3}$
176.744	9	9	$2.7\cdot10^{3}$
176.928	7	7	$2.4\cdot10^{3}$
177.172	5	5	$1.5\cdot10^{3}$
235.221	5	3	$1.7\cdot10^{2}$
240.053	3	1	$1.3\cdot10^{2}$
243.379	9	7	$2.1\cdot10^{2}$

Fe VIII

λ Å	Weights g_i	Weights g_k	A 10^8 s^{-1}
112.472	4	4	$3.6\cdot10^{2}$
112.486	6	6	$4.3\cdot10^{2}$
116.196	4	6	$4.5\cdot10^{2}$
117.197	6	8	$3.8\cdot10^{2}$
167.486	4	4	$3.0\cdot10^{3}$
168.172	6	6	$3.1\cdot10^{3}$
168.545	6	4	$2.0\cdot10^{3}$
168.929	4	2	$2.1\cdot10^{3}$
185.213	6	8	$1.0\cdot10^{3}$
186.601	4	6	$9.4\cdot10^{2}$

Fe X

λ Å	Weights g_i	Weights g_k	A 10^8 s^{-1}
76.822	2	2	$1.8\cdot10^{3}$
77.865	4	6	$1.6\cdot10^{3}$
100.026	8	10	$2.6\cdot10^{3}$
101.733	6	8	$1.8\cdot10^{3}$
101.846	4	6	$1.7\cdot10^{3}$
102.095	10	12	$2.9\cdot10^{3}$
102.192	10	12	$2.9\cdot10^{3}$
102.829	4	6	$2.1\cdot10^{3}$
103.319	6	8	$2.6\cdot10^{3}$
103.724	6	8	$1.7\cdot10^{3}$
104.638	8	10	$2.1\cdot10^{3}$
174.534	4	6	$1.8\cdot10^{3}$
175.266	2	4	$1.72\cdot10^{3}$

Atomic

λ Å	Weights g_i	Weights g_k	A 10^8 s^{-1}
Fe XI			
72.166	5	7	$2.9{\cdot}10^3$
72.310	5	5	$1.5{\cdot}10^3$
72.635	5	7	$1.6{\cdot}10^3$
91.394	5	7	$2.6{\cdot}10^3$
91.472	7	9	$2.5{\cdot}10^3$
91.63	3	5	$2.3{\cdot}10^3$
91.63	7	9	$3.4{\cdot}10^3$
91.63	5	7	$2.8{\cdot}10^3$
91.733	9	11	$4.1{\cdot}10^3$
92.81	9	11	$3.7{\cdot}10^3$
92.87	11	13	$3.9{\cdot}10^3$
93.433	9	11	$3.2{\cdot}10^3$
179.762	5	7	$1.67{\cdot}10^3$
Fe XII			
65.905	4	4	$2.0{\cdot}10^3$
66.526	6	8	$1.7{\cdot}10^3$
66.960	4	6	$1.6{\cdot}10^3$
67.164	4	2	$1.1{\cdot}10^3$
67.821	4	6	$1.4{\cdot}10^3$
68.382	2	4	$1.7{\cdot}10^3$
80.541	6	6	$8.7{\cdot}10^2$
81.943	6	4	$1.4{\cdot}10^3$
82.226	4	2	$1.9{\cdot}10^3$
84.48	4	6	$4.5{\cdot}10^3$
84.48	8	10	$4.9{\cdot}10^3$
84.52	10	12	$5.2{\cdot}10^3$
84.52	6	8	$4.0{\cdot}10^3$
84.85	6	8	$2.3{\cdot}10^3$
85.14	8	10	$3.4{\cdot}10^3$
85.477	10	12	$4.6{\cdot}10^3$
186.880	6	8	$1.0{\cdot}10^3$
192.394	4	2	$9.0{\cdot}10^2$
193.509	4	4	$9.1{\cdot}10^2$
195.119	4	6	$8.6{\cdot}10^2$
Fe XIII			
62.353	1	3	$2.0{\cdot}10^3$
62.46	5	7	$1.2{\cdot}10^3$
62.699	3	5	$2.3{\cdot}10^3$
63.188	5	7	$3.9{\cdot}10^3$
64.139	1	3	$2.1{\cdot}10^3$
74.845	5	5	$1.0{\cdot}10^3$
75.892	5	3	$7.7{\cdot}10^2$
76.117	5	3	$2.1{\cdot}10^3$
78.452	9	11	$6.3{\cdot}10^3$
84.270	7	9	$5.5{\cdot}10^3$
107.384	7	5	$1.8{\cdot}10^3$
Fe XIV			
72.80	10	12	$7.9{\cdot}10^3$
76.022	4	6	$6.6{\cdot}10^3$
76.152	6	8	$7.0{\cdot}10^3$
190	6	8	$2.8{\cdot}10^2$
211.316	2	4	$3.6{\cdot}10^2$
216	6	8	$1.7{\cdot}10^2$
217	6	8	$4.0{\cdot}10^2$
217	6	6	$2.6{\cdot}10^2$

λ Å	Weights g_i	Weights g_k	A 10^8 s^{-1}
219	2	4	$4.8{\cdot}10^2$
219	4	6	$2.4{\cdot}10^2$
219.123	4	6	$3.9{\cdot}10^2$
220	4	4	$3.2{\cdot}10^2$
221	4	6	$5.9{\cdot}10^2$
226	2	4	$3.9{\cdot}10^2$
264.787	4	4	$3.38{\cdot}10^2$
268	6	6	$2.1{\cdot}10^2$
280	4	6	$2.8{\cdot}10^2$
283	6	8	$2.7{\cdot}10^2$
288.45	6	4	$1.6{\cdot}10^2$
Fe XV			
38.95	1	3	$1.69{\cdot}10^3$
52.911	1	3	$2.94{\cdot}10^3$
59.404	3	5	$3.4{\cdot}10^3$
63.959	5	7	$1.6{\cdot}10^3$
65.370	1	3	$3.2{\cdot}10^2$
65.612	3	3	$9.8{\cdot}10^2$
66.238	5	3	$1.6{\cdot}10^3$
68.860	9	11	$9.2{\cdot}10^3$
69.7	3	1	$1.9{\cdot}10^3$
69.942	3	5	$7.4{\cdot}10^3$
69.989	5	7	$7.9{\cdot}10^3$
70.052	7	9	$8.8{\cdot}10^3$
70.224	1	3	$4.13{\cdot}10^3$
70.53	7	5	$2.6{\cdot}10^2$
70.59	7	7	$1.7{\cdot}10^3$
73.199	7	9	$8.8{\cdot}10^3$
73.473	5	7	$6.2{\cdot}10^3$
233.857	5	7	$2.2{\cdot}10^2$
235	1	3	$2.5{\cdot}10^2$
243	1	3	$2.4{\cdot}10^2$
243	5	7	$2.3{\cdot}10^2$
243.790	3	5	$4.2{\cdot}10^2$
248	3	1	$5.4{\cdot}10^2$
284.160	1	3	$2.28{\cdot}10^2$
Fe XVI			
39.827	2	4	$2.1{\cdot}10^3$
40.153	4	6	$2.5{\cdot}10^3$
40.199	4	6	$1.7{\cdot}10^3$
40.245	6	8	$1.8{\cdot}10^3$
46.661	4	6	$3.46{\cdot}10^3$
46.718	6	8	$3.7{\cdot}10^3$
50.350	2	4	$1.86{\cdot}10^3$
54.142	2	4	$3.41{\cdot}10^3$
54.728	4	6	$4.16{\cdot}10^3$
62.879	2	2	$1.05{\cdot}10^3$
63.719	4	2	$2.18{\cdot}10^3$
66.263	4	6	$9.39{\cdot}10^3$
66.368	6	8	$1.00{\cdot}10^4$
66.392	6	6	$6.69{\cdot}10^2$
76.502	6	4	$6.7{\cdot}10^2$
76.796	4	2	$7.72{\cdot}10^2$
80.192	4	6	$5.2{\cdot}10^2$
80.270	6	8	$5.4{\cdot}10^2$
85.587	2	4	$4.0{\cdot}10^2$
86.133	4	6	$4.8{\cdot}10^2$
96.256	4	6	$8.7{\cdot}10^2$

λ Å	Weights g_i	Weights g_k	A 10^8 s^{-1}
96.348	6	8	$9.3{\cdot}10^2$
117.2	2	4	$3.93{\cdot}10^2$
117.7	2	2	$3.9{\cdot}10^2$
123.4	2	4	$5.9{\cdot}10^2$
124.5	4	6	$7.0{\cdot}10^2$
144.06	4	6	$1.6{\cdot}10^3$
144.25	6	8	$1.6{\cdot}10^3$
148	4	2	$6.5{\cdot}10^2$
266.7	4	6	$3.9{\cdot}10^2$
267.0	6	8	$4.3{\cdot}10^2$
Fe XVII			
11.023	1	3	$2.1{\cdot}10^4$
12.123	1	3	$8.0{\cdot}10^4$
12.264	1	3	$5.9{\cdot}10^4$
12.526	1	3	$3.0{\cdot}10^3$
12.681	1	3	$3.5{\cdot}10^3$
13.823	1	3	$3.3{\cdot}10^4$
13.891	1	3	$3.4{\cdot}10^3$
15.015	1	3	$2.28{\cdot}10^5$
15.262	1	3	$6.0{\cdot}10^4$
16.777	1	3	$8.29{\cdot}10^3$
17.054	1	3	$9.33{\cdot}10^3$
41.37	9	11	$4.8{\cdot}10^3$
49.427	3	3	$4.0{\cdot}10^3$
50.26	7	9	$6.0{\cdot}10^3$
58.76	9	11	$1.2{\cdot}10^4$
Fe XIX			
13.413	5	3	$1.3{\cdot}10^4$
13.426	5	7	$4.8{\cdot}10^4$
13.47	3	1	$1.5{\cdot}10^5$
13.520	5	7	$2.0{\cdot}10^5$
13.56	3	5	$1.0{\cdot}10^4$
13.68	3	1	$8.0{\cdot}10^4$
13.69	5	7	$2.3{\cdot}10^4$
13.700	1	3	$2.7{\cdot}10^5$
13.71	5	5	$2.2{\cdot}10^4$
13.738	5	7	$1.0{\cdot}10^4$
13.796	5	7	$7.0{\cdot}10^4$
13.83	5	5	$1.4{\cdot}10^4$
13.934	1	3	$4.51{\cdot}10^4$
13.961	3	3	$2.0{\cdot}10^4$
14.668	5	7	$1.1{\cdot}10^4$
14.671	5	3	$1.1{\cdot}10^4$
14.929	3	3	$1.2{\cdot}10^4$
14.966	5	3	$2.5{\cdot}10^4$
14.995	5	5	$2.2{\cdot}10^4$
15.015	1	3	$1.4{\cdot}10^4$
16.668	3	1	$1.1{\cdot}10^4$
Fe XX			
12.77	4	4	$2.1{\cdot}10^5$
12.78	4	2	$6.9{\cdot}10^4$
12.78	2	4	$1.4{\cdot}10^5$
12.82	4	4	$1.1{\cdot}10^5$
12.88	6	4	$2.7{\cdot}10^4$
12.89	4	4	$4.4{\cdot}10^4$
12.90	4	6	$1.4{\cdot}10^5$
12.93	4	6	$1.6{\cdot}10^5$

λ Å	Weights g_i	g_k	A 10^8 s^{-1}
12.98	2	2	$6.7{\cdot}10^{4}$
12.99	6	6	$5.1{\cdot}10^{4}$
13.01	2	4	$3.0{\cdot}10^{4}$
13.03	4	2	$8.6{\cdot}10^{4}$
13.13	2	4	$8.9{\cdot}10^{4}$
13.79	6	6	$1.2{\cdot}10^{4}$
Fe XXI			
8.56	5	7	$2.0{\cdot}10^{4}$
8.64	5	7	$1.5{\cdot}10^{4}$
8.65	5	7	$3.9{\cdot}10^{4}$
9.42	3	3	$3.3{\cdot}10^{4}$
9.44	3	5	$1.7{\cdot}10^{4}$
9.45	1	3	$5.2{\cdot}10^{4}$
9.47	5	7	$4.9{\cdot}10^{4}$
9.67	1	3	$5.7{\cdot}10^{4}$
12.02	1	3	$1.3{\cdot}10^{4}$
12.13	3	3	$1.8{\cdot}10^{4}$
12.18	5	7	$2.2{\cdot}10^{4}$
12.21	3	1	$1.5{\cdot}10^{5}$
12.21	3	3	$1.2{\cdot}10^{5}$
12.25	1	3	$2.1{\cdot}10^{5}$
12.28	5	3	$5.2{\cdot}10^{4}$
12.30	5	7	$2.1{\cdot}10^{5}$
12.36	3	3	$3.6{\cdot}10^{4}$
12.37	5	7	$3.1{\cdot}10^{5}$
12.47	5	7	$5.8{\cdot}10^{4}$
12.47	5	3	$1.3{\cdot}10^{4}$
12.49	5	7	$1.3{\cdot}10^{4}$
12.53	5	5	$1.5{\cdot}10^{4}$
12.57	1	3	$7.2{\cdot}10^{4}$
12.73	5	5	$8.2{\cdot}10^{3}$
12.95	3	5	$6.2{\cdot}10^{3}$
13.03	5	5	$1.3{\cdot}10^{4}$
Fe XXII			
9.002	4	6	$5.5{\cdot}10^{4}$
9.006	6	8	$5.7{\cdot}10^{4}$
9.006	6	6	$5.3{\cdot}10^{4}$
9.163	4	6	$6.9{\cdot}10^{4}$
9.183	6	8	$8.3{\cdot}10^{4}$
9.241	4	6	$5.1{\cdot}10^{4}$
11.748	4	4	$1.2{\cdot}10^{5}$
11.748	4	6	$1.6{\cdot}10^{5}$
11.748	4	2	$1.8{\cdot}10^{5}$
11.763	2	4	$1.6{\cdot}10^{5}$
11.789	2	2	$2.6{\cdot}10^{5}$
11.789	6	8	$1.2{\cdot}10^{5}$
11.797	2	4	$1.7{\cdot}10^{5}$
11.823	6	4	$7.9{\cdot}10^{4}$
11.837	6	8	$2.3{\cdot}10^{5}$
11.837	6	6	$1.7{\cdot}10^{5}$
11.886	4	6	$1.3{\cdot}10^{5}$
11.898	2	4	$8.2{\cdot}10^{4}$
11.922	4	6	$1.8{\cdot}10^{5}$
11.976	6	8	$5.9{\cdot}10^{4}$
12.027	2	4	$6.9{\cdot}10^{4}$
12.045	6	8	$2.4{\cdot}10^{5}$
12.045	4	4	$9.7{\cdot}10^{4}$
12.053	4	6	$6.1{\cdot}10^{4}$

λ Å	Weights g_i	g_k	A 10^8 s^{-1}
12.077	2	4	$1.0{\cdot}10^{5}$
12.077	4	6	$2.4{\cdot}10^{5}$
12.095	6	6	$7.8{\cdot}10^{4}$
12.193	2	4	$7.2{\cdot}10^{4}$
12.193	4	6	$9.9{\cdot}10^{4}$
12.325	2	2	$1.5{\cdot}10^{5}$
Fe XXIII			
8.614	5	7	$7.7{\cdot}10^{4}$
8.752	5	7	$1.2{\cdot}10^{5}$
10.927	5	7	$6.0{\cdot}10^{4}$
10.934	3	5	$5.4{\cdot}10^{4}$
11.165	3	5	$6.7{\cdot}10^{4}$
11.298	1	3	$1.3{\cdot}10^{5}$
11.325	3	5	$1.7{\cdot}10^{5}$
11.338	3	3	$9.3{\cdot}10^{4}$
11.433	3	3	$1.2{\cdot}10^{5}$
11.441	5	7	$2.2{\cdot}10^{5}$
11.445	5	5	$5.6{\cdot}10^{4}$
11.485	3	5	$1.40{\cdot}10^{5}$
11.519	5	5	$1.16{\cdot}10^{5}$
11.520	1	3	$2.16{\cdot}10^{5}$
11.524	5	7	$2.3{\cdot}10^{5}$
11.593	5	7	$3.58{\cdot}10^{5}$
11.613	3	5	$1.0{\cdot}10^{5}$
11.691	5	7	$7.7{\cdot}10^{4}$
11.698	5	5	$7.3{\cdot}10^{4}$
11.737	3	5	$1.8{\cdot}10^{5}$
11.898	1	3	$2.03{\cdot}10^{5}$
Fe XXIV			
1.8523	2	2	$1.0{\cdot}10^{5}$
1.8552	2	4	$4.82{\cdot}10^{6}$
1.8563	4	2	$2.43{\cdot}10^{6}$
1.8572	2	2	$3.06{\cdot}10^{6}$
1.858	2	4	$1.2{\cdot}10^{5}$
1.8614	4	4	$6.24{\cdot}10^{6}$
1.8626	2	4	$3.16{\cdot}10^{6}$
1.8627	2	2	$5.47{\cdot}10^{6}$
1.8637	2	2	$1.91{\cdot}10^{6}$
1.8655	4	6	$2.14{\cdot}10^{6}$
1.8672	4	2	$1.63{\cdot}10^{6}$
1.8678	4	4	$3.5{\cdot}10^{5}$
1.8721	4	6	$3.2{\cdot}10^{5}$
1.8721	2	2	$2.0{\cdot}10^{5}$
1.8730	2	4	$1.5{\cdot}10^{5}$
1.8739	4	4	$8.3{\cdot}10^{4}$
1.891	2	2	$9.7{\cdot}10^{4}$
1.897	4	2	$9.8{\cdot}10^{4}$
8.231	2	4	$6.10{\cdot}10^{4}$
8.316	4	6	$7.07{\cdot}10^{4}$
10.619	2	4	$7.28{\cdot}10^{4}$
10.663	2	2	$7.51{\cdot}10^{4}$
11.030	2	4	$1.84{\cdot}10^{5}$
11.171	4	6	$2.18{\cdot}10^{5}$
Fe XXV			
1.4607	1	3	$2.54{\cdot}10^{5}$
1.4945	1	3	$5.05{\cdot}10^{5}$
1.5730	1	3	$1.24{\cdot}10^{6}$

λ Å	Weights g_i	g_k	A 10^8 s^{-1}
1.5749	1	3	$1.5{\cdot}10^{5}$
1.778	3	3	$8.7{\cdot}10^{4}$
1.782	3	1	$4.69{\cdot}10^{6}$
1.787	1	3	$2.57{\cdot}10^{6}$
1.787	5	5	$1.19{\cdot}10^{6}$
1.788	3	5	$2.68{\cdot}10^{6}$
1.788	3	5	$1.63{\cdot}10^{6}$
1.789	1	3	$1.78{\cdot}10^{6}$
1.790	3	3	$1.23{\cdot}10^{6}$
1.791	3	5	$4.10{\cdot}10^{6}$
1.791	3	3	$2.59{\cdot}10^{6}$
1.792	3	1	$4.92{\cdot}10^{6}$
1.792	5	5	$2.81{\cdot}10^{6}$
1.793	3	1	$2.67{\cdot}10^{6}$
1.794	5	3	$2.22{\cdot}10^{6}$
1.797	5	5	$8.8{\cdot}10^{5}$
1.798	3	3	$1.0{\cdot}10^{5}$
1.800	1	3	$8.6{\cdot}10^{4}$
1.802	3	1	$4.1{\cdot}10^{5}$
1.810	3	1	$5.9{\cdot}10^{5}$
1.8502	1	3	$4.57{\cdot}10^{6}$
1.8593	1	3	$4.42{\cdot}10^{5}$
10.038	3	3	$8.08{\cdot}10^{4}$
Krypton			
Kr I			
1164.9	1	3	3.16
1235.8	1	3	3.12
4274.0	5	5	$2.6{\cdot}10^{-2}$
4351.4	3	1	$3.2{\cdot}10^{-2}$
4362.6	5	3	$8.4{\cdot}10^{-3}$
4376.1	3	1	$5.6{\cdot}10^{-2}$
4400.0	5	5	$2.0{\cdot}10^{-2}$
4410.4	3	3	$4.4{\cdot}10^{-3}$
4425.2	3	3	$9.7{\cdot}10^{-3}$
4453.9	3	5	$7.8{\cdot}10^{-3}$
4463.7	3	3	$2.3{\cdot}10^{-2}$
4502.4	3	5	$9.2{\cdot}10^{-3}$
5562.2	5	5	$2.8{\cdot}10^{-3}$
5570.3	5	3	$2.1{\cdot}10^{-2}$
5649.6	1	3	$3.7{\cdot}10^{-3}$
5870.9	3	5	$1.8{\cdot}10^{-2}$
6904.7	3	5	$1.3{\cdot}10^{-2}$
7224.1	3	5	$1.4{\cdot}10^{-2}$
7587.4	3	1	$5.1{\cdot}10^{-1}$
7601.5	5	5	$3.1{\cdot}10^{-1}$
7685.2	3	1	$4.9{\cdot}10^{-1}$
7694.5	5	3	$5.6{\cdot}10^{-2}$
7854.8	1	3	$2.3{\cdot}10^{-1}$
8059.5	1	3	$1.9{\cdot}10^{-1}$
8104.4	5	5	$1.3{\cdot}10^{-1}$
8112.9	5	7	$3.6{\cdot}10^{-1}$
8190.1	3	5	$1.1{\cdot}10^{-1}$
8263.2	3	5	$3.5{\cdot}10^{-1}$
8281.1	3	3	$1.9{\cdot}10^{-1}$
8298.1	3	3	$3.2{\cdot}10^{-1}$
8508.9	3	3	$2.4{\cdot}10^{-1}$
8776.7	3	5	$2.7{\cdot}10^{-1}$
8928.7	5	3	$3.7{\cdot}10^{-1}$

λ Å	Weights g_i	g_k	A 10^8 s^{-1}
Kr II			
4250.6	4	4	$1.2\cdot10^{-1}$
4292.9	4	4	$9.6\cdot10^{-1}$
4355.5	6	8	1.0
4431.7	2	2	1.8
4436.8	2	4	$6.6\cdot10^{-1}$
4577.2	6	8	$9.6\cdot10^{-1}$
4583.0	6	4	$7.6\cdot10^{-1}$
4615.3	4	4	$5.4\cdot10^{-1}$
4619.2	4	6	$8.1\cdot10^{-1}$
4633.9	4	6	$7.1\cdot10^{-1}$
4658.9	6	4	$6.5\cdot10^{-1}$
4739.0	6	6	$7.6\cdot10^{-1}$
4762.4	2	4	$4.2\cdot10^{-1}$
4765.7	4	6	$6.7\cdot10^{-1}$
4811.8	2	4	$1.7\cdot10^{-1}$
4825.2	2	4	$1.9\cdot10^{-1}$
4832.1	4	2	$7.3\cdot10^{-1}$
5208.3	4	4	$1.4\cdot10^{-1}$
5308.7	4	6	$2.4\cdot10^{-2}$
7407.0	6	6	$7.0\cdot10^{-2}$
Lead			
Pb I			
2022.0	1	3	$5.2\cdot10^{-2}$
2053.3	1	3	$1.2\cdot10^{-1}$
2170.0	1	3	1.5
2401.9	3	3	$1.9\cdot10^{-1}$
2446.2	3	3	$2.5\cdot10^{-1}$
2476.4	3	5	$2.8\cdot10^{-1}$
2577.3	5	3	$5.0\cdot10^{-1}$
2613.7	3	3	$2.7\cdot10^{-1}$
2614.2	3	5	1.9
2628.3	5	3	$3.1\cdot10^{-2}$
2657.1	3	5	$9.8\cdot10^{-4}$
2663.2	5	5	$7.1\cdot10^{-1}$
2802.0	5	7	1.6
2823.2	5	5	$2.6\cdot10^{-1}$
2833.1	1	3	$5.8\cdot10^{-1}$
2873.3	5	5	$3.7\cdot10^{-1}$
3572.7	5	3	$9.9\cdot10^{-1}$
3639.6	3	3	$3.4\cdot10^{-1}$
3671.5	5	3	$4.4\cdot10^{-1}$
3683.5	3	1	1.5
3739.9	5	5	$7.3\cdot10^{-1}$
4019.6	5	7	$3.5\cdot10^{-2}$
4057.8	5	3	$8.9\cdot10^{-1}$
4062.1	5	3	$9.2\cdot10^{-1}$
4168.0	5	5	$1.2\cdot10^{-2}$
5005.4	1	3	$2.7\cdot10^{-1}$
5201.4	1	3	$1.9\cdot10^{-1}$
7229.0	5	3	$8.9\cdot10^{-3}$
Lithium			
Li I			
*2394.4	2	6	$2.664\cdot10^{-3}$
*2425.4	2	6	$3.823\cdot10^{-3}$
*2475.1	2	6	$5.735\cdot10^{-3}$
*2562.3	2	6	$8.865\cdot10^{-3}$
*2741.2	2	6	$1.248\cdot10^{-2}$
*3232.7	2	6	$1.002\cdot10^{-2}$
*3671.7	6	10	$1.678\cdot10^{-2}$
*3720.9	6	10	$2.413\cdot10^{-2}$
*3746.6	6	2	$1.01\cdot10^{-2}$
*3795.1	6	10	$3.649\cdot10^{-2}$
*3835.6	6	2	$1.56\cdot10^{-2}$
*3915.3	6	10	$5.957\cdot10^{-2}$
*3985.5	6	2	$2.59\cdot10^{-2}$
*4132.6	6	10	$1.08\cdot10^{-1}$
*4273.1	6	2	$4.76\cdot10^{-2}$
*4602.9	6	10	$2.322\cdot10^{-1}$
*4971.7	6	2	$1.038\cdot10^{-1}$
*6103.6	6	10	$6.8563\cdot10^{-1}$
*6707.8	2	6	$3.6891\cdot10^{-1}$
*8126.4	6	2	$3.3466\cdot10^{-1}$
*10510	6	10	$1.97\cdot10^{-2}$
*11032	6	2	$1.46\cdot10^{-2}$
*12237	6	10	$3.49\cdot10^{-2}$
*12782	10	14	$4.578\cdot10^{-2}$
*12929	10	6	$2.28\cdot10^{-3}$
*13557	6	2	$2.84\cdot10^{-2}$
*17545	6	10	$6.791\cdot10^{-2}$
*18697	10	14	$1.383\cdot10^{-1}$
*19276	10	6	$5.375\cdot10^{-3}$
*24463	6	2	$7.453\cdot10^{-2}$
*26880	2	6	$3.738\cdot10^{-2}$
*38079	6	10	$1.37\cdot10^{-2}$
*41792	10	6	$2.77\cdot10^{-3}$
Li II			
199.279	1	3	$2.5569\cdot10^{2}$
*935.88	9	15	$6.1345\cdot10^{-1}$
*944.72	3	9	1.4329
*965.13	9	15	1.0002
*1017.9	9	15	1.8076
1093.43	1	3	1.3533
1102.46	3	5	1.4070
*1131.9	9	15	3.8492
*1198.1	3	9	2.8969
1237.28	3	5	3.1179
1420.89	1	3	2.8309
*1493.0	9	15	$1.1215\cdot10^{1}$
*1653.1	9	3	2.8585
1681.66	3	5	$1.0069\cdot10^{1}$
1755.33	3	1	2.0499
*2329.8	3	9	$1.1758\cdot10^{-1}$
*2674.4	3	9	$1.9081\cdot10^{-1}$
*3684.7	3	9	$3.0580\cdot10^{-1}$
*5484.5	3	9	$2.2727\cdot10^{-1}$
9581.43	1	3	$5.1423\cdot10^{-2}$
*21061	3	9	$2.5664\cdot10^{-2}$
Lutetium			
Lu I			
3376.5	4	4	2.23
3567.8	4	6	$5.9\cdot10^{-1}$
3620.3	6	4	$1.1\cdot10^{-2}$
3841.2	6	6	$2.5\cdot10^{-1}$
4518.6	4	4	$2.1\cdot10^{-1}$
Magnesium			
Mg I			
1683.412	1	3	$1.88\cdot10^{-2}$
1707.061	1	3	$3.28\cdot10^{-2}$
1747.794	1	3	$6.62\cdot10^{-2}$
1827.934	1	3	$1.60\cdot10^{-1}$
2025.824	1	3	$6.12\cdot10^{-1}$
2731.993	1	3	$6.97\cdot10^{-2}$
2733.493	3	5	$9.37\cdot10^{-2}$
2736.542	5	7	$1.25\cdot10^{-1}$
2776.690	3	5	1.32
2778.270	1	3	1.82
2779.831	3	3	1.36
2779.831	5	5	4.09
2781.416	3	1	5.43
2782.972	5	3	2.14
2809.761	7	7	2.50
2811.112	5	5	1.96
2811.781	3	3	2.11
2846.716	1	3	$1.31\cdot10^{-1}$
2848.342	3	5	$1.77\cdot10^{-1}$
2851.660	5	7	$2.35\cdot10^{-1}$
2852.127	1	3	4.91
2915.453	5	5	4.09
2936.739	1	3	$1.37\cdot10^{-2}$
2938.473	3	3	$4.12\cdot10^{-2}$
2941.995	5	3	$6.83\cdot10^{-2}$
3091.065	1	3	$3.09\cdot10^{-1}$
3092.984	3	5	$3.74\cdot10^{-1}$
3096.890	5	7	$4.96\cdot10^{-1}$
3329.919	1	3	$3.09\cdot10^{-2}$
3332.146	3	3	$1.02\cdot10^{-1}$
3336.674	5	3	$1.70\cdot10^{-1}$
3829.3549	1	3	$8.99\cdot10^{-1}$
3832.2996	3	3	$6.74\cdot10^{-1}$
3832.3037	3	5	1.21
3838.2918	5	7	1.61
3838.2943	5	5	$4.03\cdot10^{-1}$
3890.241	1	3	1.31
3891.906	3	5	1.77
3893.304	3	3	$9.81\cdot10^{-1}$
3895.572	5	7	2.35
3898.120	5	5	$5.88\cdot10^{-1}$
3938.400	3	5	$5.47\cdot10^{-2}$
3986.7533	3	5	$7.30\cdot10^{-2}$
4057.5052	3	5	$1.02\cdot10^{-1}$
4167.2712	3	5	$1.38\cdot10^{-1}$
4351.9056	3	5	$1.84\cdot10^{-1}$
4571.0956	1	3	$2.54\cdot10^{-6}$
4702.9909	3	5	$2.19\cdot10^{-1}$
4730.0285	3	1	$1.34\cdot10^{-2}$
5167.3216	1	3	$1.13\cdot10^{-1}$
5172.6843	3	3	$3.37\cdot10^{-1}$
5183.6042	5	3	$5.61\cdot10^{-1}$
5528.4047	3	5	$1.39\cdot10^{-1}$
5711.0880	3	1	$3.86\cdot10^{-2}$

λ Å	Weights g_i	g_k	A 10^8 s^{-1}
6318.716	3	5	$2.63 \cdot 10^{-3}$
6319.236	3	3	$2.64 \cdot 10^{-3}$
6319.493	3	1	$2.63 \cdot 10^{-3}$
7291.060	1	3	$6.27 \cdot 10^{-4}$
7657.603	3	5	$1.23 \cdot 10^{-2}$
7659.152	3	3	$1.23 \cdot 10^{-2}$
7659.902	3	1	$1.23 \cdot 10^{-2}$
7881.667	5	7	$4.56 \cdot 10^{-3}$
7930.806	7	9	$5.38 \cdot 10^{-3}$
7930.806	3	5	$4.52 \cdot 10^{-3}$
7947.07	1	3	$2.91 \cdot 10^{-4}$
7953.39	5	3	$1.45 \cdot 10^{-3}$
8047.73	1	3	$3.53 \cdot 10^{-3}$
8049.854	3	5	$4.78 \cdot 10^{-3}$
8054.232	5	7	$6.36 \cdot 10^{-3}$
8098.724	7	9	$7.68 \cdot 10^{-3}$
8098.724	3	5	$6.46 \cdot 10^{-3}$
8209.839	5	3	$1.81 \cdot 10^{-3}$
8213.034	5	7	$4.38 \cdot 10^{-2}$
8303.313	1	3	$5.16 \cdot 10^{-3}$
8305.596	3	5	$9.28 \cdot 10^{-3}$
8310.264	5	7	$9.24 \cdot 10^{-3}$
8346.120	7	9	$1.15 \cdot 10^{-2}$
8346.120	3	5	$9.59 \cdot 10^{-3}$
8710.175	1	3	$7.97 \cdot 10^{-3}$
8712.689	3	5	$1.07 \cdot 10^{-2}$
8717.825	5	7	$1.43 \cdot 10^{-2}$
8736.021	7	9	$1.83 \cdot 10^{-2}$
8736.021	3	5	$1.54 \cdot 10^{-2}$
8806.757	3	5	$1.27 \cdot 10^{-1}$
8923.569	1	3	$5.86 \cdot 10^{-3}$
9246.499	5	3	$2.99 \cdot 10^{-3}$
9255.778	5	7	$7.95 \cdot 10^{-2}$
9414.964	5	7	$2.88 \cdot 10^{-2}$
9414.964	7	9	$3.24 \cdot 10^{-2}$
9414.964	3	5	$2.72 \cdot 10^{-2}$
9429.814	1	3	$1.34 \cdot 10^{-2}$
9432.764	3	5	$1.81 \cdot 10^{-2}$
9438.783	5	7	$2.41 \cdot 10^{-2}$
9665.54	3	5	$2.44 \cdot 10^{-3}$
9983.20	1	3	$1.57 \cdot 10^{-3}$
9986.475	3	3	$4.70 \cdot 10^{-3}$
9993.209	5	3	$7.81 \cdot 10^{-3}$
Mg II			
870.2	2	4	$1.04 \cdot 10^{-2}$
870.2	2	2	$1.08 \cdot 10^{-2}$
884.7	2	4	$1.38 \cdot 10^{-2}$
884.7	2	2	$1.44 \cdot 10^{-2}$
907.4	2	4	$1.94 \cdot 10^{-2}$
907.4	2	2	$2.02 \cdot 10^{-2}$
946.703	2	4	$2.69 \cdot 10^{-2}$
946.769	2	2	$2.81 \cdot 10^{-2}$
1025.962	2	4	$3.43 \cdot 10^{-2}$
1026.108	2	2	$3.63 \cdot 10^{-2}$
1239.936	2	4	$1.35 \cdot 10^{-2}$
1240.399	2	2	$1.52 \cdot 10^{-2}$
1248.511	2	2	$3.83 \cdot 10^{-2}$
1249.932	4	2	$7.64 \cdot 10^{-2}$
1271.243	2	4	$1.23 \cdot 10^{-2}$
1271.943	2	2	$4.49 \cdot 10^{-2}$
1272.725	4	6	$1.44 \cdot 10^{-2}$
1273.427	4	2	$8.97 \cdot 10^{-2}$
1306.711	2	4	$2.19 \cdot 10^{-2}$
1307.877	2	2	$6.72 \cdot 10^{-2}$
1308.282	4	6	$2.58 \cdot 10^{-2}$
1309.439	4	2	$1.34 \cdot 10^{-1}$
1365.545	2	4	$4.53 \cdot 10^{-2}$
1367.260	4	6	$5.35 \cdot 10^{-2}$
1367.704	2	2	$1.08 \cdot 10^{-1}$
1369.425	4	2	$2.17 \cdot 10^{-1}$
1476.004	2	4	$1.10 \cdot 10^{-1}$
1478.013	4	6	$1.30 \cdot 10^{-1}$
1480.890	2	2	$1.93 \cdot 10^{-1}$
1482.902	4	2	$3.85 \cdot 10^{-1}$
1734.845	2	4	$4.29 \cdot 10^{-1}$
1737.618	4	6	$5.09 \cdot 10^{-1}$
1750.654	2	2	$4.00 \cdot 10^{-1}$
1753.456	4	2	$7.98 \cdot 10^{-1}$
2329.578	6	8	$1.36 \cdot 10^{-1}$
2329.578	4	6	$1.27 \cdot 10^{-1}$
2449.590	6	8	$2.16 \cdot 10^{-1}$
2449.590	4	6	$2.02 \cdot 10^{-1}$
2660.755	6	8	$3.81 \cdot 10^{-1}$
2660.817	4	6	$3.56 \cdot 10^{-1}$
2790.776	2	4	4.01
2795.528	2	4	2.60
2797.998	4	6	4.79
2802.704	2	2	2.57
2928.634	2	2	1.15
2936.509	4	2	2.30
2968.020	4	6	$1.98 \cdot 10^{-2}$
2969.145	2	2	$2.25 \cdot 10^{-2}$
2971.839	4	2	$4.49 \cdot 10^{-2}$
3104.722	6	8	$7.97 \cdot 10^{-1}$
3104.809	4	6	$7.44 \cdot 10^{-1}$
3165.878	2	4	$2.89 \cdot 10^{-2}$
3168.951	4	6	$3.43 \cdot 10^{-2}$
3172.706	2	2	$3.41 \cdot 10^{-2}$
3175.783	4	2	$6.81 \cdot 10^{-2}$
3534.972	2	4	$5.80 \cdot 10^{-2}$
3538.813	4	6	$6.90 \cdot 10^{-2}$
3549.516	2	2	$5.64 \cdot 10^{-2}$
3553.366	4	2	$1.12 \cdot 10^{-1}$
3613.781	2	4	$1.79 \cdot 10^{-3}$
3615.583	2	2	$1.56 \cdot 10^{-3}$
3848.209	6	4	$2.96 \cdot 10^{-2}$
3848.335	4	4	$3.29 \cdot 10^{-3}$
3850.385	4	2	$3.24 \cdot 10^{-2}$
4384.637	2	4	$1.45 \cdot 10^{-1}$
4390.564	4	6	$1.73 \cdot 10^{-1}$
4427.994	2	2	$1.05 \cdot 10^{-1}$
4433.990	4	2	$2.10 \cdot 10^{-1}$
4436.486	6	8	$6.38 \cdot 10^{-2}$
4436.598	4	6	$5.95 \cdot 10^{-2}$
4481.130	6	8	2.33
4481.327	4	6	2.17
4630.878	6	4	$2.75 \cdot 10^{-3}$
4631.405	4	2	$3.01 \cdot 10^{-3}$
4739.588	6	8	$8.09 \cdot 10^{-2}$
4739.712	4	6	$7.55 \cdot 10^{-2}$
4868.845	6	4	$1.50 \cdot 10^{-3}$
4868.845	8	6	$1.41 \cdot 10^{-3}$
5068.937	6	4	$3.97 \cdot 10^{-3}$
5069.802	4	2	$4.35 \cdot 10^{-3}$
5264.215	6	8	$1.27 \cdot 10^{-1}$
5264.368	4	6	$1.19 \cdot 10^{-1}$
5434.039	6	4	$2.45 \cdot 10^{-3}$
5434.039	8	6	$2.33 \cdot 10^{-3}$
5451.259	2	4	$1.03 \cdot 10^{-2}$
5460.019	2	2	$1.11 \cdot 10^{-2}$
5464.136	4	2	$2.22 \cdot 10^{-2}$
5916.429	6	4	$6.42 \cdot 10^{-3}$
5918.158	4	2	$7.04 \cdot 10^{-3}$
5923.366	2	4	$1.40 \cdot 10^{-2}$
5928.233	4	6	$1.66 \cdot 10^{-2}$
5938.629	2	2	$1.36 \cdot 10^{-2}$
5943.499	4	2	$2.71 \cdot 10^{-2}$
6346.737	6	8	$2.20 \cdot 10^{-1}$
6346.962	4	6	$2.05 \cdot 10^{-1}$
6620.440	6	4	$4.61 \cdot 10^{-3}$
6620.569	8	6	$4.37 \cdot 10^{-3}$
6781.451	2	4	$2.49 \cdot 10^{-2}$
6787.851	4	6	$2.96 \cdot 10^{-2}$
6812.860	2	2	$2.11 \cdot 10^{-2}$
6819.270	4	2	$4.21 \cdot 10^{-2}$
7786.500	2	4	$1.97 \cdot 10^{-3}$
7790.978	2	2	$1.84 \cdot 10^{-3}$
7877.051	2	4	$6.58 \cdot 10^{-1}$
7896.368	4	6	$7.86 \cdot 10^{-1}$
8115.220	6	4	$1.22 \cdot 10^{-2}$
8120.434	4	2	$1.34 \cdot 10^{-2}$
8213.989	2	2	$2.65 \cdot 10^{-1}$
8234.639	4	2	$5.29 \cdot 10^{-1}$
8734.990	2	4	$5.34 \cdot 10^{-2}$
8745.657	4	6	$6.37 \cdot 10^{-2}$
8824.323	2	2	$3.69 \cdot 10^{-2}$
8835.082	4	2	$7.36 \cdot 10^{-2}$
9218.248	2	4	$3.64 \cdot 10^{-1}$
9244.266	2	2	$3.61 \cdot 10^{-1}$
9631.888	6	8	$4.21 \cdot 10^{-1}$
9632.435	4	6	$3.93 \cdot 10^{-1}$
Mg III			
186.5149	1	3	$1.86 \cdot 10^{2}$
187.1977	1	3	$1.26 \cdot 10^{2}$
188.5296	1	3	2.5
231.7333	1	3	$9.12 \cdot 10$
234.2631	1	3	4.98
1229.389	3	3	$5.72 \cdot 10^{-1}$
1239.827	3	1	1.82
1274.831	3	5	2.53
1350.156	5	3	$3.50 \cdot 10^{-1}$
1378.700	3	1	5.32
1393.391	7	5	6.38
1405.170	5	5	1.54
1422.118	3	5	$1.80 \cdot 10^{-1}$
1431.136	5	3	4.30
1435.550	3	1	3.04
1439.770	1	3	$3.54 \cdot 10^{-1}$

λ Å	Weights g_i	g_k	A 10^8 s^{-1}
1443.738	3	3	1.81
1446.254	3	1	$4.15\cdot10^{-1}$
1447.260	5	3	2.22
1458.172	3	1	3.21
1462.305	5	5	1.21
1467.188	3	3	$9.37\cdot10^{-1}$
1482.67	3	5	$1.55\cdot10^{-1}$
1493.097	5	5	1.02
1506.826	3	5	$3.91\cdot10^{-1}$
1550.82	3	1	$1.04\cdot10$
1572.712	3	5	6.92
1586.237	3	3	8.02
1592.360	3	1	8.53
1626.093	5	3	$1.77\cdot10^{-1}$
1635.946	7	5	$2.44\cdot10^{-1}$
1642.826	7	7	$9.17\cdot10^{-1}$
1648.822	3	3	$5.38\cdot10^{-1}$
1652.218	5	5	$9.10\cdot10^{-1}$
1659.244	5	7	$8.48\cdot10^{-1}$
1663.287	5	5	$1.68\cdot10^{-1}$
1675.710	3	5	$4.19\cdot10^{-1}$
1679.470	5	3	$5.88\cdot10^{-1}$
1687.091	3	5	1.41
1697.282	7	7	2.14
1703.108	5	3	$1.71\cdot10^{-1}$
1703.731	3	3	2.97
1704.368	7	5	$1.79\cdot10^{-1}$
1714.783	5	7	$4.76\cdot10^{-1}$
1722.041	5	5	3.23
1730.733	7	7	1.03
1730.778	3	3	1.44
1731.786	5	5	1.60
1738.835	7	9	$1.14\cdot10$
1739.475	5	7	$3.26\cdot10^{-4}$
1743.947	5	5	$3.76\cdot10^{-1}$
1745.009	5	3	$5.37\cdot10^{-2}$
1747.561	3	5	7.38
1748.932	5	7	8.71
1757.176	7	5	$2.04\cdot10^{-1}$
1757.888	1	3	2.10
1761.740	5	3	$3.16\cdot10^{-3}$
1763.805	3	3	3.17
1772.982	3	5	7.67
1775.942	5	5	$1.28\cdot10^{-1}$
1783.253	5	7	9.39
1787.927	5	5	1.46
1791.375	3	3	2.32
1793.207	5	3	$3.36\cdot10^{-1}$
1794.582	3	5	7.73
1800.662	5	7	8.21
1803.087	3	5	$1.03\cdot10^{-1}$
1820.421	1	3	2.98
1820.896	3	3	$3.05\cdot10^{-2}$
1826.750	3	3	$1.32\cdot10^{-1}$
1828.974	3	1	$4.97\cdot10^{-1}$
1838.336	5	7	1.09
1839.878	3	5	$4.55\cdot10^{-1}$
1847.561	5	7	$3.74\cdot10^{-2}$
1858.186	5	3	1.28
1868.225	5	5	1.94
1879.492	5	5	1.76
1887.308	5	3	$7.56\cdot10^{-1}$
1896.304	5	3	$4.20\cdot10^{-1}$
1901.572	3	5	$3.78\cdot10^{-1}$
1901.572	3	3	$5.47\cdot10^{-1}$
1908.500	3	1	5.15
1918.777	5	5	1.98
1921.374	3	3	$2.26\cdot10^{-1}$
1923.896	3	5	1.37
1930.374	3	1	$8.30\cdot10^{-1}$
1930.672	5	5	1.70
1937.843	1	3	1.46
1938.936	5	3	$8.97\cdot10^{-1}$
1941.500	3	5	$1.23\cdot10^{-1}$
1941.500	3	3	$3.27\cdot10^{-1}$
1954.831	1	3	$7.97\cdot10^{-1}$
1962.145	3	3	$7.76\cdot10^{-1}$
1971.514	3	1	2.06
1977.554	3	5	$4.94\cdot10^{-1}$
1979.327	1	3	1.22
1979.43	1	3	$4.10\cdot10^{-1}$
2004.860	5	3	$3.33\cdot10^{-1}$
2039.553	5	5	1.54
2055.491	3	3	2.35
2064.902	5	7	4.21
2085.891	3	3	1.61
2091.963	3	5	2.57
2094.207	3	1	$1.21\cdot10^{-1}$
2097.936	1	3	1.50
2112.773	3	5	1.69
2134.054	3	3	2.36
2177.694	3	5	2.12
2273.414	3	3	$1.27\cdot10^{-2}$
2318.125	3	5	$4.44\cdot10^{-2}$
2395.149	5	3	1.67
2467.751	3	3	$6.91\cdot10^{-1}$
2490.534	1	3	1.61
2529.190	1	3	$1.89\cdot10^{-1}$
2618.011	1	3	$8.72\cdot10^{-1}$
2905.419	1	3	$6.84\cdot10^{-3}$
Mg IV			
129.857	4	6	$5.61\cdot10^{2}$
320.9943	4	2	$1.13\cdot10^{2}$
323.3076	2	2	$5.52\cdot10^{1}$
840.366	6	4	$1.24\cdot10^{1}$
842.087	8	6	$1.19\cdot10^{1}$
1346.543	6	8	$1.07\cdot10^{1}$
1352.020	4	6	6.59
1384.425	8	10	$1.45\cdot10^{1}$
1437.61	6	8	$1.21\cdot10^{1}$
Manganese			
Mn I			
2794.82	6	8	3.7
2798.27	6	6	3.6
2801.08	6	4	3.7
3016.45	10	12	$2.9\cdot10^{-1}$
3043.36	8	8	$5.9\cdot10^{-1}$
3044.57	10	8	$5.7\cdot10^{-1}$
3045.59	10	10	$6.7\cdot10^{-1}$
3047.03	12	12	$6.1\cdot10^{-1}$
3082.71	14	14	$2.9\cdot10^{-1}$
3228.09	10	12	$6.4\cdot10^{-1}$
3230.72	8	8	$3.5\cdot10^{-1}$
3243.78	6	6	$5.3\cdot10^{-1}$
3256.14	4	6	$5.0\cdot10^{-1}$
3267.79	14	14	$3.5\cdot10^{-1}$
3268.72	6	8	$3.3\cdot10^{-1}$
3270.35	12	12	$2.6\cdot10^{-1}$
3273.02	10	10	$2.7\cdot10^{-1}$
3463.66	8	8	$3.2\cdot10^{-1}$
3511.83	12	12	$2.7\cdot10^{-1}$
3577.87	10	8	$9.4\cdot10^{-1}$
3601.27	12	10	$2.3\cdot10^{-1}$
3608.49	6	6	$3.6\cdot10^{-1}$
3660.40	12	14	$9.1\cdot10^{-1}$
3675.67	6	8	$2.2\cdot10^{-1}$
3676.96	10	12	$7.3\cdot10^{-1}$
3680.15	12	10	$1.9\cdot10^{-1}$
3682.09	8	10	$7.6\cdot10^{-1}$
3684.87	6	8	$2.6\cdot10^{-1}$
3706.08	12	14	1.4
3718.92	10	12	$9.6\cdot10^{-1}$
3731.94	8	10	1.0
3771.44	14	14	$1.9\cdot10^{-1}$
3773.86	12	12	$2.5\cdot10^{-1}$
3800.55	6	8	$2.7\cdot10^{-1}$
3806.72	10	12	$5.9\cdot10^{-1}$
3823.51	8	10	$5.21\cdot10^{-1}$
3834.37	6	8	$4.29\cdot10^{-1}$
3841.07	4	6	$3.3\cdot10^{-1}$
3889.46	12	14	$3.1\cdot10^{-1}$
3924.08	2	4	$9.4\cdot10^{-1}$
3926.48	6	8	$5.4\cdot10^{-1}$
3952.84	6	6	$4.1\cdot10^{-1}$
3982.90	6	4	$5.5\cdot10^{-1}$
4011.91	8	8	$2.3\cdot10^{-1}$
4018.11	10	8	$2.54\cdot10^{-1}$
4030.76	6	8	$1.7\cdot10^{-1}$
4041.36	10	10	$7.87\cdot10^{-1}$
4048.75	6	4	$7.5\cdot10^{-1}$
4052.48	6	8	$3.8\cdot10^{-1}$
4055.55	8	8	$4.31\cdot10^{-1}$
4058.94	4	2	$7.25\cdot10^{-1}$
4065.08	12	14	$2.5\cdot10^{-1}$
4066.24	10	8	$2.2\cdot10^{-1}$
4079.42	2	4	$3.8\cdot10^{-1}$
4082.95	4	6	$2.95\cdot10^{-1}$
4083.63	6	8	$2.8\cdot10^{-1}$
4089.94	8	10	$1.7\cdot10^{-1}$
4105.37	10	8	$1.7\cdot10^{-1}$
4135.03	12	12	$3.0\cdot10^{-1}$
4141.06	10	10	$2.6\cdot10^{-1}$
4148.80	8	8	$2.3\cdot10^{-1}$
4176.61	14	12	$2.4\cdot10^{-1}$
4189.99	12	10	$2.0\cdot10^{-1}$
4201.78	10	8	$2.3\cdot10^{-1}$
4235.30	8	6	$9.17\cdot10^{-1}$

λ Å	Weights g_i	g_k	A 10^8 s^{-1}
4265.93	4	4	$4.92 \cdot 10^{-1}$
4281.10	6	6	$2.3 \cdot 10^{-1}$
4411.87	12	10	$2.6 \cdot 10^{-1}$
4414.89	8	6	$2.93 \cdot 10^{-1}$
4419.77	10	8	$2.1 \cdot 10^{-1}$
4436.36	6	4	$4.37 \cdot 10^{-1}$
4451.58	8	8	$7.98 \cdot 10^{-1}$
4453.01	4	2	$5.44 \cdot 10^{-1}$
4455.82	4	6	$1.7 \cdot 10^{-1}$
4457.55	6	6	$4.27 \cdot 10^{-1}$
4458.26	6	8	$4.62 \cdot 10^{-1}$
4461.09	8	8	$1.7 \cdot 10^{-1}$
4462.03	8	10	$7.00 \cdot 10^{-1}$
4464.68	6	6	$4.39 \cdot 10^{-1}$
4470.14	4	4	$3.00 \cdot 10^{-1}$
4479.40	8	10	$3.4 \cdot 10^{-1}$
4490.08	2	4	$2.49 \cdot 10^{-1}$
4498.90	4	6	$2.49 \cdot 10^{-1}$
4502.22	6	8	$1.86 \cdot 10^{-1}$
4605.37	10	12	$3.6 \cdot 10^{-1}$
4626.54	12	14	$3.6 \cdot 10^{-1}$
4709.71	8	8	$1.72 \cdot 10^{-1}$
4727.46	6	6	$1.7 \cdot 10^{-1}$
4739.11	4	4	$2.40 \cdot 10^{-1}$
4754.05	6	8	$3.03 \cdot 10^{-1}$
4761.53	2	4	$5.35 \cdot 10^{-1}$
4762.38	8	10	$7.83 \cdot 10^{-1}$
4765.86	4	6	$4.1 \cdot 10^{-1}$
4766.43	6	8	$4.6 \cdot 10^{-1}$
4783.43	8	8	$4.01 \cdot 10^{-1}$
4823.53	10	8	$4.99 \cdot 10^{-1}$
6013.48	4	6	$1.72 \cdot 10^{-1}$
6021.79	8	6	$3.32 \cdot 10^{-1}$

Mn II

λ Å	g_i	g_k	A 10^8 s^{-1}
2593.72	7	7	2.6
2605.68	7	5	2.7
2933.05	5	3	2.0
2939.31	5	5	1.9
2949.20	5	7	1.9
3441.99	9	7	$4.3 \cdot 10^{-1}$
3460.32	7	5	$3.2 \cdot 10^{-1}$
3474.13	5	3	$1.5 \cdot 10^{-1}$
3482.90	5	5	$2.0 \cdot 10^{-1}$
3488.68	3	3	$2.5 \cdot 10^{-1}$

Mn VI

λ Å	g_i	g_k	A 10^8 s^{-1}
307.999	9	9	$3.7 \cdot 10^{1}$
309.440	9	7	$5.7 \cdot 10^{1}$
309.579	7	5	$4.4 \cdot 10^{1}$
310.058	7	7	$3.4 \cdot 10^{1}$
310.182	5	5	$2.8 \cdot 10^{1}$
311.748	5	3	$5.7 \cdot 10^{1}$
320.598	3	5	$1.5 \cdot 10^{1}$
320.681	1	3	$2.2 \cdot 10^{1}$
320.874	3	1	$7.8 \cdot 10^{1}$
320.979	3	3	$2.2 \cdot 10^{1}$
321.176	5	5	$6.0 \cdot 10^{1}$
321.541	5	3	$2.7 \cdot 10^{1}$
325.146	9	7	$1.3 \cdot 10^{2}$
328.431	5	5	$4.4 \cdot 10^{1}$
328.558	3	5	$1.2 \cdot 10^{1}$
329.043	1	3	$1.1 \cdot 10^{1}$
1236.23	5	3	$1.3 \cdot 10^{1}$
1255.77	3	1	$1.2 \cdot 10^{1}$
1285.10	5	7	$1.1 \cdot 10^{1}$
1333.87	7	9	$1.0 \cdot 10^{1}$

Mercury

Hg I

λ Å	g_i	g_k	A 10^8 s^{-1}
2536.52	1	3	$8.00 \cdot 10^{-2}$
2652.04	3	5	$3.88 \cdot 10^{-1}$
2655.13	3	5	$1.1 \cdot 10^{-1}$
2752.78	1	3	$6.10 \cdot 10^{-2}$
2856.94	3	1	$1.1 \cdot 10^{-2}$
2893.60	3	3	$1.6 \cdot 10^{-1}$
2925.4	5	3	$7.7 \cdot 10^{-2}$
2967.3	1	3	$4.5 \cdot 10^{-1}$
3021.50	5	7	$5.09 \cdot 10^{-1}$
3023.48	5	5	$9.4 \cdot 10^{-2}$
3027.49	5	5	$2.0 \cdot 10^{-2}$
3125.66	3	5	$6.56 \cdot 10^{-1}$
3341.48	5	3	$1.68 \cdot 10^{-1}$
3650.15	5	7	1.3
3654.83	5	5	$1.8 \cdot 10^{-1}$
4046.56	1	3	$2.1 \cdot 10^{-1}$
4077.81	3	1	$4.0 \cdot 10^{-2}$
4108.1	3	1	$3.0 \cdot 10^{-2}$
4339.22	3	5	$2.88 \cdot 10^{-2}$
4347.50	3	5	$8.4 \cdot 10^{-2}$
4358.34	3	3	$5.57 \cdot 10^{-1}$
4916.07	3	1	$5.8 \cdot 10^{-2}$
5025.64	3	3	$2.7 \cdot 10^{-4}$
5460.75	5	3	$4.87 \cdot 10^{-1}$
5769.59	3	5	$2.36 \cdot 10^{-1}$
6234.4	1	3	$5.3 \cdot 10^{-3}$
6716.4	1	3	$4.3 \cdot 10^{-3}$
6907.5	3	5	$2.8 \cdot 10^{-2}$
7728.8	1	3	$9.7 \cdot 10^{-3}$
10139.79	3	1	$2.71 \cdot 10^{-1}$

Molybdenum

Mo I

λ Å	g_i	g_k	A 10^8 s^{-1}
2616.79	3	5	$7.34 \cdot 10^{-1}$
2629.85	5	7	$7.75 \cdot 10^{-1}$
2638.30	5	5	$7.57 \cdot 10^{-1}$
2640.98	7	5	1.20
2649.46	7	9	$9.84 \cdot 10^{-1}$
2655.02	9	7	$4.08 \cdot 10^{-1}$
2658.11	7	7	$6.43 \cdot 10^{-1}$
2679.85	9	11	1.31
2684.16	9	9	$4.18 \cdot 10^{-1}$
2733.39	5	7	$2.95 \cdot 10^{-1}$
2751.47	7	9	$2.54 \cdot 10^{-1}$
2761.53	9	11	$2.06 \cdot 10^{-1}$
2787.83	9	7	$2.85 \cdot 10^{-1}$
2826.75	7	7	$4.23 \cdot 10^{-1}$
2876.54	9	9	$2.84 \cdot 10^{-1}$
2886.60	11	11	$4.74 \cdot 10^{-1}$
2906.06	3	3	$8.04 \cdot 10^{-1}$
2915.38	5	3	$7.31 \cdot 10^{-1}$
2936.50	11	11	$2.33 \cdot 10^{-1}$
2945.43	7	7	$3.66 \cdot 10^{-1}$
2959.48	9	11	$1.75 \cdot 10^{-1}$
2977.27	9	7	$3.28 \cdot 10^{-1}$
2987.92	3	5	$8.43 \cdot 10^{-1}$
2988.23	5	7	$4.28 \cdot 10^{-1}$
2989.80	9	7	$9.27 \cdot 10^{-1}$
3000.85	5	7	$2.58 \cdot 10^{-1}$
3013.39	7	5	$6.06 \cdot 10^{-1}$
3016.78	9	9	$2.75 \cdot 10^{-1}$
3025.00	5	5	$8.49 \cdot 10^{-1}$
3036.31	3	5	$5.81 \cdot 10^{-1}$
3041.70	13	11	$5.94 \cdot 10^{-1}$
3046.80	13	11	$1.63 \cdot 10^{-1}$
3047.31	11	9	$5.01 \cdot 10^{-1}$
3055.32	9	7	$4.29 \cdot 10^{-1}$
3061.59	7	5	$4.41 \cdot 10^{-1}$
3064.27	13	13	$8.46 \cdot 10^{-1}$
3065.04	13	13	$3.08 \cdot 10^{-1}$
3069.96	11	11	$2.72 \cdot 10^{-1}$
3070.89	9	11	$1.87 \cdot 10^{-1}$
3074.37	11	11	1.42
3079.88	9	11	$9.55 \cdot 10^{-1}$
3080.40	7	9	$3.61 \cdot 10^{-1}$
3085.62	9	9	1.63
3089.71	5	7	$2.34 \cdot 10^{-1}$
3094.66	7	7	1.63
3100.88	7	9	1.20
3101.34	5	5	1.92
3117.54	13	13	$1.89 \cdot 10^{-1}$
3132.59	7	9	1.79
3135.90	9	11	$3.68 \cdot 10^{-1}$
3136.75	9	11	$1.57 \cdot 10^{-1}$
3142.75	3	5	$4.10 \cdot 10^{-1}$
3147.35	13	11	$2.41 \cdot 10^{-1}$
3155.19	7	7	$2.75 \cdot 10^{-1}$
3158.17	7	7	$4.63 \cdot 10^{-1}$
3170.34	7	7	1.37
3171.38	5	7	$2.03 \cdot 10^{-1}$
3175.59	13	11	$8.40 \cdot 10^{-1}$
3179.78	11	13	$2.33 \cdot 10^{-1}$
3183.03	11	9	$3.98 \cdot 10^{-1}$
3184.58	7	5	$2.77 \cdot 10^{-1}$
3185.10	7	7	$2.54 \cdot 10^{-1}$
3185.71	5	3	$6.10 \cdot 10^{-1}$
3188.10	7	9	$3.45 \cdot 10^{-1}$
3188.41	5	7	$4.40 \cdot 10^{-1}$
3192.79	9	11	$1.88 \cdot 10^{-1}$
3193.98	7	5	1.53
3194.88	9	11	$1.75 \cdot 10^{-1}$
3195.96	9	7	$4.10 \cdot 10^{-1}$
3198.85	15	13	$7.22 \cdot 10^{-1}$
3205.22	1	3	$4.27 \cdot 10^{-1}$
3205.43	9	11	$2.55 \cdot 10^{-1}$
3205.89	9	9	$5.35 \cdot 10^{-1}$
3208.84	7	5	$2.77 \cdot 10^{-1}$
3210.97	7	5	$6.94 \cdot 10^{-1}$
3214.44	9	7	$2.01 \cdot 10^{-1}$

λ Å	Weights g_i	g_k	A 10^8 s^{-1}
3215.07	3	5	$4.20\cdot10^{-1}$
3216.78	15	13	$2.10\cdot10^{-1}$
3221.73	3	1	1.41
3228.21	5	7	$3.85\cdot10^{-1}$
3229.79	9	11	$1.44\cdot10^{-1}$
3233.14	13	13	$6.33\cdot10^{-1}$
3237.06	7	9	$2.95\cdot10^{-1}$
3251.65	3	5	$3.05\cdot10^{-1}$
3256.21	5	3	$6.89\cdot10^{-1}$
3259.16	11	13	$1.62\cdot10^{-1}$
3262.63	7	9	$3.62\cdot10^{-1}$
3264.40	11	9	$5.42\cdot10^{-1}$
3265.14	5	7	$2.60\cdot10^{-1}$
3266.16	9	11	$1.95\cdot10^{-1}$
3270.90	7	7	$3.59\cdot10^{-1}$
3285.35	9	7	$4.49\cdot10^{-1}$
3289.01	9	9	$5.08\cdot10^{-1}$
3290.82	7	5	$5.44\cdot10^{-1}$
3305.91	7	9	$3.06\cdot10^{-1}$
3323.95	9	7	$2.82\cdot10^{-1}$
3336.56	9	9	$1.64\cdot10^{-1}$
3344.73	3	5	$6.04\cdot10^{-1}$
3346.83	11	11	$1.13\cdot10^{-1}$
3358.12	5	7	$7.59\cdot10^{-1}$
3361.37	9	9	$1.38\cdot10^{-1}$
3363.78	5	7	$2.74\cdot10^{-1}$
3375.65	7	9	$1.56\cdot10^{-1}$
3378.46	13	13	$3.75\cdot10^{-1}$
3379.96	5	5	$4.11\cdot10^{-1}$
3384.61	7	9	$7.32\cdot10^{-1}$
3385.87	9	11	$3.30\cdot10^{-1}$
3389.79	5	7	$1.85\cdot10^{-1}$
3392.17	9	9	$1.97\cdot10^{-1}$
3393.65	11	11	$2.08\cdot10^{-1}$
3404.33	7	7	$2.10\cdot10^{-1}$
3413.37	11	11	$1.25\cdot10^{-1}$
3415.27	9	9	$1.83\cdot10^{-1}$
3415.61	7	9	$1.29\cdot10^{-1}$
3416.14	9	11	$2.45\cdot10^{-1}$
3420.04	5	5	$3.28\cdot10^{-1}$
3422.31	9	9	$2.52\cdot10^{-1}$
3425.13	11	11	$2.29\cdot10^{-1}$
3427.90	11	13	$4.09\cdot10^{-1}$
3434.79	7	7	$1.75\cdot10^{-1}$
3435.45	15	15	1.50
3437.21	11	9	$8.06\cdot10^{-1}$
3445.03	7	9	$1.53\cdot10^{-1}$
3445.26	7	5	$2.96\cdot10^{-1}$
3447.12	9	11	$8.75\cdot10^{-1}$
3449.07	7	9	$1.52\cdot10^{-1}$
3449.85	5	7	$1.65\cdot10^{-1}$
3452.60	7	7	$2.48\cdot10^{-1}$
3456.15	5	5	$3.60\cdot10^{-1}$
3460.78	9	7	$6.03\cdot10^{-1}$
3466.19	9	7	$2.11\cdot10^{-1}$
3466.96	7	7	$1.52\cdot10^{-1}$
3467.85	5	7	$2.63\cdot10^{-1}$
3469.22	5	3	$6.96\cdot10^{-1}$
3469.63	13	15	$1.51\cdot10^{-1}$
3470.92	3	5	$2.91\cdot10^{-1}$
3475.03	3	3	$4.68\cdot10^{-1}$
3479.42	7	5	$2.26\cdot10^{-1}$
3489.43	7	7	$3.27\cdot10^{-1}$
3504.41	7	9	$8.06\cdot10^{-1}$
3505.31	7	9	$2.25\cdot10^{-1}$
3508.11	9	9	$1.59\cdot10^{-1}$
3510.77	13	13	$4.75\cdot10^{-1}$
3517.55	11	11	$5.41\cdot10^{-1}$
3518.21	3	3	$3.64\cdot10^{-1}$
3521.38	9	9	$1.39\cdot10^{-1}$
3521.41	9	11	$6.06\cdot10^{-1}$
3524.98	7	9	$2.25\cdot10^{-1}$
3538.92	11	11	$2.24\cdot10^{-1}$
3540.57	5	3	$4.46\cdot10^{-1}$
3542.17	7	5	$4.93\cdot10^{-1}$
3552.71	9	7	$3.64\cdot10^{-1}$
3555.64	3	3	$3.46\cdot10^{-1}$
3558.09	5	7	$5.43\cdot10^{-1}$
3566.05	9	9	$2.67\cdot10^{-1}$
3566.74	7	7	$1.43\cdot10^{-1}$
3570.64	15	15	$7.18\cdot10^{-1}$
3573.88	3	5	$3.58\cdot10^{-1}$
3580.54	13	11	$5.49\cdot10^{-1}$
3581.88	11	13	$3.81\cdot10^{-1}$
3585.57	7	5	$3.95\cdot10^{-1}$
3590.74	7	9	$2.23\cdot10^{-1}$
3595.55	5	5	$2.32\cdot10^{-1}$
3598.88	13	11	$5.67\cdot10^{-1}$
3600.73	9	9	$2.07\cdot10^{-1}$
3601.88	7	9	$1.15\cdot10^{-1}$
3602.94	5	7	$2.96\cdot10^{-1}$
3604.07	9	7	$3.25\cdot10^{-1}$
3615.16	7	9	$1.96\cdot10^{-1}$
3623.22	11	9	$5.58\cdot10^{-1}$
3624.46	9	11	$5.27\cdot10^{-1}$
3624.62	5	7	$1.37\cdot10^{-1}$
3638.20	5	3	$3.51\cdot10^{-1}$
3638.21	5	3	$3.33\cdot10^{-1}$
3640.62	7	5	$1.94\cdot10^{-1}$
3647.84	7	7	$2.11\cdot10^{-1}$
3657.36	5	7	$2.03\cdot10^{-1}$
3658.13	9	9	$1.86\cdot10^{-1}$
3659.36	7	9	$6.70\cdot10^{-1}$
3662.15	7	9	$1.45\cdot10^{-1}$
3662.99	11	11	$3.48\cdot10^{-1}$
3663.27	7	5	$2.30\cdot10^{-1}$
3664.81	11	13	$9.54\cdot10^{-1}$
3669.34	9	7	$2.16\cdot10^{-1}$
3672.81	9	11	$1.95\cdot10^{-1}$
3672.82	9	9	$1.13\cdot10^{-1}$
3680.68	11	11	$2.96\cdot10^{-1}$
3681.72	9	7	$1.68\cdot10^{-1}$
3687.96	5	7	$2.12\cdot10^{-1}$
3688.97	11	9	$3.26\cdot10^{-1}$
3690.59	11	9	$2.07\cdot10^{-1}$
3694.94	5	7	$6.36\cdot10^{-1}$
3696.04	11	11	$3.59\cdot10^{-1}$
3708.55	7	9	$1.28\cdot10^{-1}$
3715.75	9	7	$2.38\cdot10^{-1}$
3718.48	5	7	$1.34\cdot10^{-1}$
3720.25	7	9	$2.86\cdot10^{-1}$
3725.55	7	7	$1.60\cdot10^{-1}$
3727.68	9	11	$1.51\cdot10^{-1}$
3728.50	7	9	$2.20\cdot10^{-1}$
3733.02	7	7	$1.45\cdot10^{-1}$
3733.41	13	13	$2.80\cdot10^{-1}$
3735.62	11	11	$1.66\cdot10^{-1}$
3742.28	7	7	$1.56\cdot10^{-1}$
3747.19	5	7	$3.07\cdot10^{-1}$
3748.48	9	11	$3.95\cdot10^{-1}$
3755.16	9	9	$2.48\cdot10^{-1}$
3758.52	9	9	$1.22\cdot10^{-1}$
3759.60	9	7	$1.82\cdot10^{-1}$
3760.88	9	9	$2.16\cdot10^{-1}$
3768.73	9	9	$2.88\cdot10^{-1}$
3769.99	7	9	$2.46\cdot10^{-1}$
3777.72	13	11	$1.66\cdot10^{-1}$
3788.25	7	9	$2.87\cdot10^{-1}$
3794.43	9	9	$1.22\cdot10^{-1}$
3798.25	7	9	$6.90\cdot10^{-1}$
3801.84	9	7	$3.16\cdot10^{-1}$
3805.99	5	5	$2.44\cdot10^{-1}$
3819.78	9	11	$1.47\cdot10^{-1}$
3824.78	5	7	$1.40\cdot10^{-1}$
3827.15	7	7	$1.94\cdot10^{-1}$
3828.88	7	7	$1.35\cdot10^{-1}$
3830.81	5	5	$1.83\cdot10^{-1}$
3831.07	7	9	$1.20\cdot10^{-1}$
3832.11	9	9	$3.05\cdot10^{-1}$
3833.75	9	9	$1.70\cdot10^{-1}$
3846.18	7	7	$1.26\cdot10^{-1}$
3848.30	9	9	$1.26\cdot10^{-1}$
3851.99	11	9	$1.78\cdot10^{-1}$
3864.10	7	7	$6.24\cdot10^{-1}$
3866.69	3	5	$1.74\cdot10^{-1}$
3874.15	7	5	$1.67\cdot10^{-1}$
3902.95	7	5	$6.17\cdot10^{-1}$
3909.54	9	7	$1.13\cdot10^{-1}$
3919.55	11	13	$2.24\cdot10^{-1}$
3955.48	13	11	$1.71\cdot10^{-1}$
3973.76	11	13	$4.39\cdot10^{-1}$
3977.90	9	7	$1.35\cdot10^{-1}$
3980.20	5	3	$2.70\cdot10^{-1}$
3991.85	11	9	$1.29\cdot10^{-1}$
4010.13	5	3	$4.38\cdot10^{-1}$
4021.01	9	11	$2.65\cdot10^{-1}$
4051.18	13	11	$1.36\cdot10^{-1}$
4062.08	11	9	$1.96\cdot10^{-1}$
4069.88	13	11	$3.25\cdot10^{-1}$
4076.19	9	9	$1.16\cdot10^{-1}$
4084.37	9	7	$1.94\cdot10^{-1}$
4107.46	7	5	$2.02\cdot10^{-1}$
4120.09	13	15	$6.05\cdot10^{-1}$
4131.92	9	11	$1.56\cdot10^{-1}$
4148.98	9	11	$1.56\cdot10^{-1}$
4157.40	13	11	$2.17\cdot10^{-1}$
4157.90	9	11	$1.60\cdot10^{-1}$
4185.82	11	13	$3.82\cdot10^{-1}$
4188.32	11	13	$3.32\cdot10^{-1}$
4194.56	11	11	$2.70\cdot10^{-1}$

λ Å	Weights g_i	g_k	A $10^8\ s^{-1}$
4232.59	9	11	$3.17{\cdot}10^{-1}$
4240.83	5	5	$1.68{\cdot}10^{-1}$
4246.02	11	13	$2.00{\cdot}10^{-1}$
4251.88	13	11	$1.76{\cdot}10^{-1}$
4254.95	7	9	$2.01{\cdot}10^{-1}$
4269.28	11	11	$1.36{\cdot}10^{-1}$
4276.91	7	9	$2.85{\cdot}10^{-1}$
4277.24	9	11	$1.35{\cdot}10^{-1}$
4317.92	15	15	$1.28{\cdot}10^{-1}$
4325.80	3	3	$1.84{\cdot}10^{-1}$
4326.14	5	7	$2.56{\cdot}10^{-1}$
4340.74	5	7	$1.23{\cdot}10^{-1}$
4381.63	13	13	$2.93{\cdot}10^{-1}$
4382.41	11	13	$3.83{\cdot}10^{-1}$
4409.94	13	13	$1.38{\cdot}10^{-1}$
4411.69	11	11	$2.63{\cdot}10^{-1}$
4434.95	9	9	$2.51{\cdot}10^{-1}$
4446.42	11	11	$1.90{\cdot}10^{-1}$
4457.35	7	7	$1.28{\cdot}10^{-1}$
4474.57	5	5	$2.10{\cdot}10^{-1}$
4491.65	11	11	$2.09{\cdot}10^{-1}$
4536.80	13	15	$5.03{\cdot}10^{-1}$
4598.23	1	3	$1.47{\cdot}10^{-1}$
4624.23	9	9	$1.32{\cdot}10^{-1}$
4633.08	3	5	$2.35{\cdot}10^{-1}$
4652.24	5	7	$1.55{\cdot}10^{-1}$
4686.08	3	3	$1.72{\cdot}10^{-1}$
4688.21	13	15	$1.54{\cdot}10^{-1}$
4707.25	7	9	$3.63{\cdot}10^{-1}$
4718.86	5	5	$2.17{\cdot}10^{-1}$
4723.05	9	9	$1.23{\cdot}10^{-1}$
4731.44	9	11	$4.49{\cdot}10^{-1}$
4758.50	11	9	$3.01{\cdot}10^{-1}$
4760.18	11	13	$4.67{\cdot}10^{-1}$
4764.11	9	7	$2.16{\cdot}10^{-1}$
4811.05	13	11	$4.36{\cdot}10^{-1}$
4819.25	11	9	$2.71{\cdot}10^{-1}$
4830.51	9	7	$4.07{\cdot}10^{-1}$
4858.39	13	11	$1.24{\cdot}10^{-1}$
4868.02	7	5	$3.11{\cdot}10^{-1}$
5037.18	9	7	$1.14{\cdot}10^{-1}$
5044.36	7	5	$1.31{\cdot}10^{-1}$
5163.18	9	11	$2.03{\cdot}10^{-1}$
5171.06	5	7	$1.84{\cdot}10^{-1}$
5172.94	5	5	$4.11{\cdot}10^{-1}$
5174.18	5	3	$5.83{\cdot}10^{-1}$
5191.45	7	9	$1.62{\cdot}10^{-1}$
5238.21	7	9	$3.74{\cdot}10^{-1}$
5240.87	7	7	$3.89{\cdot}10^{-1}$
5242.80	7	5	$2.01{\cdot}10^{-1}$
5261.53	5	7	$1.13{\cdot}10^{-1}$
5280.85	5	5	$1.28{\cdot}10^{-1}$
5355.52	9	9	$1.21{\cdot}10^{-1}$
5356.46	11	11	$2.11{\cdot}10^{-1}$
5360.51	9	11	$6.19{\cdot}10^{-1}$
5364.28	9	9	$2.26{\cdot}10^{-1}$
5460.50	5	3	$3.46{\cdot}10^{-1}$
5493.76	7	5	$2.13{\cdot}10^{-1}$
5506.49	5	7	$3.61{\cdot}10^{-1}$
5533.03	5	5	$3.72{\cdot}10^{-1}$
5570.44	5	3	$3.30{\cdot}10^{-1}$
5849.71	3	3	$3.02{\cdot}10^{-1}$
5851.50	3	5	$1.55{\cdot}10^{-1}$
5893.36	5	5	$2.60{\cdot}10^{-1}$
5895.93	5	7	$3.12{\cdot}10^{-1}$
5926.37	7	7	$1.63{\cdot}10^{-1}$
5928.88	7	9	$5.32{\cdot}10^{-1}$
7154.11	9	9	$3.45{\cdot}10^{-1}$
Neodymium			
Nd II			
3780.4	16	18	$1.4{\cdot}10^{-1}$
3805.4	14	16	$6.9{\cdot}10^{-1}$
3807.2	10	12	$4.9{\cdot}10^{-2}$
3863.3	8	10	$1.5{\cdot}10^{-1}$
3941.5	10	10	$6.1{\cdot}10^{-1}$
3951.2	12	12	$6.0{\cdot}10^{-1}$
3973.3	18	18	$6.3{\cdot}10^{-1}$
3979.5	10	12	$2.7{\cdot}10^{-1}$
3990.1	16	16	$5.2{\cdot}10^{-1}$
4012.3	18	20	$5.5{\cdot}10^{-1}$
4061.1	16	18	$4.4{\cdot}10^{-1}$
4106.6	14	16	$6.8{\cdot}10^{-2}$
4109.5	14	16	$3.7{\cdot}10^{-1}$
4133.4	14	12	$1.5{\cdot}10^{-1}$
4156.1	12	14	$3.4{\cdot}10^{-1}$
4205.6	18	16	$1.8{\cdot}10^{-1}$
4284.5	18	18	$8.5{\cdot}10^{-2}$
4303.6	8	10	$4.7{\cdot}10^{-1}$
4325.8	16	16	$1.6{\cdot}10^{-1}$
4358.2	14	14	$1.5{\cdot}10^{-1}$
4382.7	12	10	$4.0{\cdot}10^{-2}$
4400.8	10	10	$6.8{\cdot}10^{-2}$
4451.6	12	14	$2.5{\cdot}10^{-1}$
4456.4	16	18	$6.4{\cdot}10^{-2}$
4463.0	14	16	$1.8{\cdot}10^{-1}$
4958.1	12	10	$1.2{\cdot}10^{-2}$
5130.6	22	20	$1.6{\cdot}10^{-1}$
5192.6	20	18	$1.7{\cdot}10^{-1}$
5249.6	18	16	$1.8{\cdot}10^{-1}$
5276.9	12	10	$1.2{\cdot}10^{-1}$
5293.2	16	14	$1.2{\cdot}10^{-1}$
5302.3	20	18	$1.1{\cdot}10^{-1}$
5311.5	14	12	$1.1{\cdot}10^{-1}$
5319.8	12	10	$1.6{\cdot}10^{-1}$
5357.0	18	16	$1.8{\cdot}10^{-1}$
5371.9	20	20	$5.1{\cdot}10^{-2}$
5485.7	18	18	$5.7{\cdot}10^{-2}$
5594.4	16	16	$7.0{\cdot}10^{-2}$
5620.6	18	18	$1.3{\cdot}10^{-1}$
5688.5	14	14	$5.9{\cdot}10^{-2}$
5718.1	16	16	$8.7{\cdot}10^{-2}$
5726.8	10	10	$5.6{\cdot}10^{-2}$
5740.9	12	12	$7.2{\cdot}10^{-2}$
5804.0	10	10	$4.6{\cdot}10^{-2}$
5865.1	16	18	$1.3{\cdot}10^{-2}$
6051.9	12	10	$1.1{\cdot}10^{-2}$
Neon			
Ne I			
615.63	1	3	$3.8{\cdot}10^{-1}$
618.67	1	3	$9.3{\cdot}10^{-1}$
619.10	1	3	$3.3{\cdot}10^{-1}$
626.82	1	3	$7.4{\cdot}10^{-1}$
629.74	1	3	$4.8{\cdot}10^{-1}$
735.90	1	3	6.11
743.72	1	3	$4.86{\cdot}10^{-1}$
3369.8	5	5	$1.0{\cdot}10^{-3}$
3369.9	5	3	$7.6{\cdot}10^{-3}$
3375.6	5	3	$2.2{\cdot}10^{-3}$
3417.9	3	5	$9.2{\cdot}10^{-3}$
3418.0	3	3	$2.2{\cdot}10^{-3}$
3423.9	3	3	$1.0{\cdot}10^{-3}$
3447.7	5	5	$2.1{\cdot}10^{-2}$
3450.8	5	3	$4.9{\cdot}10^{-3}$
3454.2	3	1	$3.7{\cdot}10^{-2}$
3460.5	1	3	$7.0{\cdot}10^{-3}$
3464.3	5	5	$6.7{\cdot}10^{-3}$
3466.6	1	3	$1.3{\cdot}10^{-2}$
3472.6	5	7	$1.7{\cdot}10^{-2}$
3498.1	3	5	$5.1{\cdot}10^{-3}$
3501.2	3	3	$1.2{\cdot}10^{-2}$
3510.7	5	3	$2.2{\cdot}10^{-3}$
3515.2	3	5	$6.9{\cdot}10^{-3}$
3520.5	3	1	$9.3{\cdot}10^{-2}$
3593.5	3	5	$9.9{\cdot}10^{-3}$
3593.6	3	3	$6.6{\cdot}10^{-3}$
3600.2	3	3	$4.3{\cdot}10^{-3}$
3633.7	3	1	$1.1{\cdot}10^{-2}$
3682.2	3	5	$1.6{\cdot}10^{-3}$
3685.7	3	3	$3.9{\cdot}10^{-3}$
3701.2	3	5	$2.2{\cdot}10^{-3}$
4536.3	3	3	$5.0{\cdot}10^{-3}$
4702.5	3	3	$2.1{\cdot}10^{-3}$
4708.9	3	3	$4.2{\cdot}10^{-2}$
4955.4	3	3	$3.3{\cdot}10^{-3}$
5113.7	3	3	$1.0{\cdot}10^{-2}$
5120.5	3	3	$5.6{\cdot}10^{-3}$
5154.4	3	3	$1.9{\cdot}10^{-2}$
5191.3	3	3	$1.3{\cdot}10^{-2}$
5326.4	3	3	$6.8{\cdot}10^{-3}$
5333.3	3	3	$5.3{\cdot}10^{-3}$
5341.1	3	3	$1.1{\cdot}10^{-1}$
5400.6	3	1	$9.0{\cdot}10^{-3}$
5418.6	3	3	$5.2{\cdot}10^{-3}$
5433.7	3	3	$2.83{\cdot}10^{-3}$
5652.6	3	3	$8.9{\cdot}10^{-3}$
5662.5	3	3	$6.9{\cdot}10^{-3}$
5852.5	3	1	$6.82{\cdot}10^{-1}$
5868.4	3	3	$1.4{\cdot}10^{-2}$
5881.9	5	3	$1.15{\cdot}10^{-1}$
5913.6	3	3	$4.8{\cdot}10^{-2}$
5939.3	5	3	$2.00{\cdot}10^{-3}$
5944.8	5	5	$1.13{\cdot}10^{-1}$
5961.6	3	3	$3.3{\cdot}10^{-2}$
5975.5	5	3	$3.51{\cdot}10^{-2}$

λ Å	Weights g_i	Weights g_k	A 10^8 s^{-1}
6030.0	3	3	$5.61 \cdot 10^{-2}$
6046.1	3	3	$2.26 \cdot 10^{-3}$
6074.3	3	1	$6.03 \cdot 10^{-1}$
6096.2	3	5	$1.81 \cdot 10^{-1}$
6118.0	5	3	$6.09 \cdot 10^{-3}$
6128.5	3	3	$6.7 \cdot 10^{-3}$
6143.1	5	5	$2.82 \cdot 10^{-1}$
6150.3	3	3	$1.5 \cdot 10^{-2}$
6163.6	1	3	$1.46 \cdot 10^{-1}$
6217.3	5	3	$6.37 \cdot 10^{-2}$
6266.5	1	3	$2.49 \cdot 10^{-1}$
6273.0	3	3	$9.7 \cdot 10^{-3}$
6293.7	3	3	$6.39 \cdot 10^{-3}$
6304.8	3	5	$4.16 \cdot 10^{-2}$
6328.2	5	3	$3.39 \cdot 10^{-2}$
6330.9	3	3	$2.3 \cdot 10^{-2}$
6334.4	5	5	$1.61 \cdot 10^{-1}$
6351.9	1	3	$3.45 \cdot 10^{-3}$
6383.0	3	3	$3.21 \cdot 10^{-1}$
6401.1	3	3	$1.39 \cdot 10^{-2}$
6402.2	5	7	$5.14 \cdot 10^{-1}$
6506.5	3	5	$3.00 \cdot 10^{-1}$
6532.9	1	3	$1.08 \cdot 10^{-1}$
6599.0	3	3	$2.32 \cdot 10^{-1}$
6602.9	3	3	$5.9 \cdot 10^{-3}$
6652.1	3	1	$2.9 \cdot 10^{-3}$
6678.3	3	5	$2.33 \cdot 10^{-1}$
6717.0	3	3	$2.17 \cdot 10^{-1}$
6721.1	3	3	$4.9 \cdot 10^{-4}$
6929.5	3	5	$1.74 \cdot 10^{-1}$
7024.1	3	3	$1.89 \cdot 10^{-2}$
7032.4	5	3	$2.53 \cdot 10^{-1}$
7051.3	3	3	$3.0 \cdot 10^{-2}$
7059.1	3	5	$6.8 \cdot 10^{-2}$
7173.9	3	5	$2.87 \cdot 10^{-2}$
7245.2	3	3	$9.35 \cdot 10^{-2}$
7304.8	1	3	$2.55 \cdot 10^{-3}$
7438.9	1	3	$2.31 \cdot 10^{-2}$
7472.4	3	3	$4.0 \cdot 10^{-2}$
7535.8	3	3	$4.3 \cdot 10^{-1}$
7937.0	5	5	$7.8 \cdot 10^{-3}$
8082.5	3	3	$1.2 \cdot 10^{-3}$
8118.5	3	3	$4.9 \cdot 10^{-2}$
8128.9	3	5	$7.2 \cdot 10^{-3}$
8259.4	5	5	$2.03 \cdot 10^{-2}$
8571.4	3	3	$5.5 \cdot 10^{-2}$
8582.9	3	5	$1.00 \cdot 10^{-2}$
8647.0	5	5	$3.91 \cdot 10^{-2}$
8681.9	3	3	$2.1 \cdot 10^{-1}$
8767.5	3	3	$1.1 \cdot 10^{-3}$
8771.7	3	3	$1.6 \cdot 10^{-1}$
8783.8	3	5	$3.13 \cdot 10^{-1}$
8865.3	3	3	$9.4 \cdot 10^{-3}$
9201.8	3	3	$9.1 \cdot 10^{-2}$
9433.0	3	3	$1.1 \cdot 10^{-3}$
9486.7	3	3	$2.5 \cdot 10^{-2}$
9534.2	3	3	$6.3 \cdot 10^{-2}$
10621	3	3	$2.4 \cdot 10^{-3}$
11409	3	3	$4.2 \cdot 10^{-2}$
11525	3	3	$8.4 \cdot 10^{-2}$
11767	3	3	$6.9 \cdot 10^{-2}$
12459	3	3	$1.5 \cdot 10^{-2}$
Ne II			
*357.03	6	10	$3.8 \cdot 10^{1}$
*361.77	6	2	$1.6 \cdot 10^{1}$
*406.28	6	10	$1.8 \cdot 10^{1}$
*446.37	6	6	$4.07 \cdot 10^{1}$
460.73	4	2	$4.7 \cdot 10^{1}$
462.39	2	2	$2.3 \cdot 10^{1}$
1907.5	4	2	$2.8 \cdot 10^{-1}$
1916.1	4	4	$6.9 \cdot 10^{-1}$
1930.0	2	2	$5.7 \cdot 10^{-1}$
1938.8	2	4	$1.3 \cdot 10^{-1}$
2858.0	6	6	$7.9 \cdot 10^{-1}$
2870.0	6	6	$1.7 \cdot 10^{-1}$
2873.0	6	4	$3.8 \cdot 10^{-1}$
2876.3	4	6	$7.8 \cdot 10^{-1}$
2876.5	6	4	$3.3 \cdot 10^{-1}$
2878.1	2	2	$6.9 \cdot 10^{-2}$
2888.4	4	6	$7.0 \cdot 10^{-2}$
2891.5	4	4	$6.1 \cdot 10^{-2}$
2897.0	6	8	$5.2 \cdot 10^{-2}$
2906.8	2	4	$5.5 \cdot 10^{-1}$
2910.1	4	2	1.7
2910.4	2	4	$5.9 \cdot 10^{-1}$
2916.2	6	4	$9.6 \cdot 10^{-2}$
2925.6	2	2	$5.6 \cdot 10^{-1}$
2933.7	6	6	$6.9 \cdot 10^{-2}$
2955.7	6	4	1.2
3001.7	4	4	$8.7 \cdot 10^{-1}$
3017.3	6	4	$3.5 \cdot 10^{-1}$
3027.0	6	6	1.4
3028.7	4	2	$8.5 \cdot 10^{-1}$
3028.9	2	4	$4.7 \cdot 10^{-1}$
3034.5	6	8	3.1
3037.7	4	4	2.1
3045.6	2	2	2.5
3047.6	4	6	1.8
3054.7	2	4	$9.4 \cdot 10^{-1}$
3092.9	6	6	1.3
3097.1	8	8	1.3
3118.0	8	6	$4.2 \cdot 10^{-2}$
3134.1	6	4	$2.6 \cdot 10^{-1}$
3140.4	8	6	$2.4 \cdot 10^{-1}$
3151.1	6	6	$4.8 \cdot 10^{-2}$
3154.8	8	6	$1.8 \cdot 10^{-2}$
3164.4	8	8	$1.6 \cdot 10^{-1}$
3165.7	6	6	$1.2 \cdot 10^{-1}$
3173.6	6	4	$4.5 \cdot 10^{-2}$
3176.1	4	6	$6.0 \cdot 10^{-2}$
3187.6	4	6	$1.4 \cdot 10^{-2}$
3188.7	6	6	$3.9 \cdot 10^{-1}$
3190.9	4	6	$1.5 \cdot 10^{-1}$
3194.6	4	4	$5.2 \cdot 10^{-1}$
3198.6	6	8	1.7
3198.9	4	4	$2.3 \cdot 10^{-1}$
3209.0	8	8	$1.6 \cdot 10^{-1}$
3209.4	2	4	$6.0 \cdot 10^{-1}$
3213.7	2	4	1.7
3214.3	4	6	2.2
3218.2	8	10	3.6
3224.8	6	8	3.5
3229.5	8	8	$1.3 \cdot 10^{-1}$
3229.6	8	10	3.6
3230.1	6	6	1.8
3230.4	4	6	$1.4 \cdot 10^{-1}$
3232.0	6	4	$2.7 \cdot 10^{-1}$
3232.4	4	4	1.6
3243.4	6	6	$2.3 \cdot 10^{-1}$
3244.1	6	8	1.5
3248.1	4	4	$2.4 \cdot 10^{-1}$
3255.4	6	4	$3.8 \cdot 10^{-2}$
3263.4	2	4	$3.9 \cdot 10^{-1}$
3269.9	4	6	$5.1 \cdot 10^{-1}$
3270.8	6	4	$5.7 \cdot 10^{-2}$
3297.7	6	6	$4.3 \cdot 10^{-1}$
3309.7	4	2	$3.1 \cdot 10^{-1}$
3310.5	4	4	$6.9 \cdot 10^{-2}$
3311.3	4	2	$2.6 \cdot 10^{-1}$
3314.7	6	6	$4.4 \cdot 10^{-2}$
3319.7	4	2	1.6
3320.2	8	6	$2.1 \cdot 10^{-1}$
3323.7	4	4	1.6
3327.2	4	4	$9.1 \cdot 10^{-1}$
3329.2	8	8	$8.8 \cdot 10^{-1}$
3330.7	6	6	$3.9 \cdot 10^{-2}$
3334.8	6	8	1.8
3336.1	4	6	1.1
3344.4	2	2	1.5
3345.5	6	4	1.4
3345.8	4	4	$2.2 \cdot 10^{-1}$
3353.6	4	2	$1.2 \cdot 10^{-1}$
3355.0	4	6	1.3
3356.3	6	6	$2.0 \cdot 10^{-1}$
3357.8	6	6	$5.0 \cdot 10^{-1}$
3360.3	2	4	$8.6 \cdot 10^{-1}$
3360.6	2	4	$8.2 \cdot 10^{-1}$
3362.9	4	2	$3.5 \cdot 10^{-1}$
3371.8	4	6	$2.2 \cdot 10^{-1}$
3374.1	4	4	$3.0 \cdot 10^{-1}$
3378.2	2	2	1.7
3379.3	2	2	$3.0 \cdot 10^{-1}$
3386.2	4	6	$5.5 \cdot 10^{-2}$
3388.4	4	6	2.2
3390.6	2	4	$7.7 \cdot 10^{-2}$
3392.8	2	4	$4.4 \cdot 10^{-1}$
3404.8	4	6	1.9
3407.0	6	8	2.3
3411.4	4	2	$6.1 \cdot 10^{-1}$
3413.2	4	4	1.8
3414.9	4	6	$1.8 \cdot 10^{-2}$
3416.9	6	6	$6.4 \cdot 10^{-1}$
3417.7	6	8	1.6
3438.9	2	2	1.4
3440.7	2	4	$3.5 \cdot 10^{-1}$
3453.1	4	4	$4.6 \cdot 10^{-1}$
3454.8	4	4	1.6
3456.6	2	4	$9.6 \cdot 10^{-1}$
3457.1	4	6	$9.9 \cdot 10^{-2}$

λ Å	Weights g_i	Weights g_k	A $10^8\ s^{-1}$
3459.3	6	6	1.6
3475.2	4	4	$1.2\cdot10^{-2}$
3477.6	4	6	$4.3\cdot10^{-1}$
3481.9	4	2	1.4
3503.6	2	2	2.0
3522.7	4	2	$2.3\cdot10^{-2}$
3538.0	4	2	$7.6\cdot10^{-1}$
3539.9	4	4	$3.6\cdot10^{-2}$
3542.2	6	4	$6.0\cdot10^{-1}$
3542.9	4	6	1.2
3546.2	2	4	$6.3\cdot10^{-2}$
3551.6	2	4	$3.7\cdot10^{-2}$
3557.8	2	2	$1.9\cdot10^{-1}$
3561.2	4	6	$2.1\cdot10^{-1}$
3565.8	4	4	$6.2\cdot10^{-1}$
3568.5	6	8	1.4
3571.2	4	4	$6.3\cdot10^{-1}$
3574.2	6	6	$1.0\cdot10^{-1}$
3574.6	4	6	1.3
3590.4	4	6	$3.6\cdot10^{-2}$
3594.2	4	2	1.3
3612.3	2	4	$2.6\cdot10^{-1}$
3628.0	4	4	$6.0\cdot10^{-1}$
3632.7	4	4	$1.3\cdot10^{-1}$
3643.9	4	4	$3.2\cdot10^{-1}$
3644.9	2	4	$9.9\cdot10^{-1}$
3659.9	4	6	$6.7\cdot10^{-2}$
3664.1	6	4	$7.0\cdot10^{-1}$
3679.8	4	2	$3.2\cdot10^{-1}$
3694.2	6	6	1.0
3697.1	2	2	$2.8\cdot10^{-1}$
3701.8	4	6	$2.7\cdot10^{-1}$
3709.6	4	2	1.1
3713.1	4	6	1.3
3721.8	4	6	$2.0\cdot10^{-1}$
3726.9	4	4	$1.2\cdot10^{-1}$
3727.1	2	4	$9.8\cdot10^{-1}$
3734.9	4	4	$1.9\cdot10^{-1}$
3744.6	2	4	$2.6\cdot10^{-1}$
3751.2	2	2	$1.8\cdot10^{-1}$
3753.8	4	6	$4.5\cdot10^{-1}$
3766.3	4	6	$2.9\cdot10^{-1}$
3777.1	2	4	$4.2\cdot10^{-1}$
3800.0	4	4	$3.7\cdot10^{-1}$
3818.4	2	4	$6.1\cdot10^{-1}$
3829.8	4	6	$8.4\cdot10^{-1}$
3942.3	4	6	$1.0\cdot10^{-2}$

Ne V

λ Å	Weights g_i	Weights g_k	A $10^8\ s^{-1}$
*142.61	9	9	$6.7\cdot10^{2}$
*143.32	9	15	$1.2\cdot10^{3}$
147.13	5	7	$1.5\cdot10^{3}$
151.23	5	5	$3.38\cdot10^{2}$
154.50	1	3	$7.0\cdot10^{2}$
*167.69	9	9	$1.5\cdot10^{2}$
*358.93	9	3	$2.1\cdot10^{2}$
365.59	5	3	$1.35\cdot10^{2}$
*482.15	9	9	$3.01\cdot10^{1}$
*571.04	9	15	$1.0\cdot10^{1}$
2259.6	3	5	1.9
2265.7	5	7	2.4

Ne VII

λ Å	Weights g_i	Weights g_k	A $10^8\ s^{-1}$
97.502	1	3	$1.07\cdot10^{3}$
*115.46	9	3	$4.8\cdot10^{2}$
116.69	3	5	$1.6\cdot10^{3}$
127.66	3	1	$1.9\cdot10^{2}$
465.22	1	3	$4.09\cdot10^{1}$
558.61	3	5	8.11
559.95	1	3	$1.07\cdot10^{1}$
561.38	3	3	7.99
561.73	5	5	$2.39\cdot10^{1}$
562.99	3	1	$3.17\cdot10^{1}$
564.53	5	3	$1.31\cdot10^{1}$

Ne VIII

λ Å	Weights g_i	Weights g_k	A $10^8\ s^{-1}$
*88.09	2	6	$8.4\cdot10^{2}$
*98.208	6	10	$2.77\cdot10^{3}$
770.41	2	4	5.90
780.32	2	2	5.69
2820.7	2	4	$7.20\cdot10^{-1}$
2860.1	2	2	$6.88\cdot10^{-1}$

Nickel

Ni I

λ Å	Weights g_i	Weights g_k	A $10^8\ s^{-1}$
1976.87	7	9	1.1
1990.25	5	7	$8.3\cdot10^{-1}$
2014.25	3	5	$9.3\cdot10^{-1}$
2085.57	5	5	2.6
2158.31	7	5	$6.9\cdot10^{-1}$
2289.98	9	7	2.1
2300.77	7	7	$7.5\cdot10^{-1}$
2312.34	7	7	5.5
2313.98	5	5	5.0
2317.16	7	5	3.8
2320.03	9	11	6.9
2321.38	5	7	5.6
2325.79	7	9	3.5
2329.96	5	3	5.3
2345.54	9	7	2.2
2346.63	7	5	$5.5\cdot10^{-1}$
3002.48	7	7	$8.0\cdot10^{-1}$
3003.62	5	5	$6.9\cdot10^{-1}$
3012.00	5	5	1.3
3037.93	7	7	$2.8\cdot10^{-1}$
3050.82	7	9	$6.0\cdot10^{-1}$
3054.31	5	5	$4.0\cdot10^{-1}$
3057.64	3	3	1.0
3101.56	5	7	$6.3\cdot10^{-1}$
3101.88	5	7	$4.9\cdot10^{-1}$
3134.11	3	5	$7.3\cdot10^{-1}$
3369.56	9	7	$1.8\cdot10^{-1}$
3380.57	5	3	1.3
3392.98	7	7	$2.4\cdot10^{-1}$
3414.76	7	9	$5.5\cdot10^{-1}$
3423.71	3	3	$3.3\cdot10^{-1}$
3433.56	7	7	$1.7\cdot10^{-1}$
3446.26	5	5	$4.4\cdot10^{-1}$
3452.88	5	7	$9.8\cdot10^{-2}$
3458.46	3	5	$6.1\cdot10^{-1}$
3461.66	7	9	$2.7\cdot10^{-1}$
3472.55	5	7	$1.2\cdot10^{-1}$
3492.96	5	3	$9.8\cdot10^{-1}$
3510.33	3	1	1.2
3515.05	5	7	$4.2\cdot10^{-1}$
3524.54	7	5	1.0
3566.37	5	5	$5.6\cdot10^{-1}$
3619.39	5	7	$6.6\cdot10^{-1}$
4027.67	5	7	$1.3\cdot10^{-1}$
4295.88	9	7	$1.7\cdot10^{-1}$
4401.54	9	11	$3.8\cdot10^{-1}$
4462.46	3	5	$1.7\cdot10^{-1}$
4470.48	5	7	$1.9\cdot10^{-1}$
4600.37	5	3	$2.6\cdot10^{-1}$
4604.99	9	7	$2.3\cdot10^{-1}$
4606.23	5	3	$1.0\cdot10^{-1}$
4648.66	11	9	$2.4\cdot10^{-1}$
4686.22	5	5	$1.4\cdot10^{-1}$
4701.54	9	9	$1.4\cdot10^{-1}$
4714.42	13	11	$4.6\cdot10^{-1}$
4715.78	7	7	$2.0\cdot10^{-1}$
4732.47	7	9	$9.3\cdot10^{-2}$
4752.43	3	3	$2.0\cdot10^{-1}$
4756.52	9	9	$1.5\cdot10^{-1}$
4786.54	11	11	$1.8\cdot10^{-1}$
4829.03	5	7	$1.9\cdot10^{-1}$
4831.18	9	7	$1.6\cdot10^{-1}$
4838.64	9	7	$2.2\cdot10^{-1}$
4855.41	5	5	$5.7\cdot10^{-1}$
4904.41	5	3	$6.2\cdot10^{-1}$
4912.03	3	3	$1.5\cdot10^{-1}$
4913.97	1	3	$2.2\cdot10^{-1}$
4918.36	9	7	$2.3\cdot10^{-1}$
4935.83	7	5	$2.4\cdot10^{-1}$
4937.34	9	9	$1.2\cdot10^{-1}$
4953.20	5	5	$1.2\cdot10^{-1}$
4980.17	9	11	$1.9\cdot10^{-1}$
5000.34	7	7	$1.4\cdot10^{-1}$
5012.46	7	7	$1.1\cdot10^{-1}$
5017.58	11	11	$2.0\cdot10^{-1}$
5035.37	7	9	$5.7\cdot10^{-1}$
5042.20	3	5	$1.4\cdot10^{-1}$
5048.85	7	7	$1.6\cdot10^{-1}$
5080.53	9	11	$3.2\cdot10^{-1}$
5081.11	7	9	$5.7\cdot10^{-1}$
5082.35	3	3	$2.5\cdot10^{-1}$
5084.08	7	9	$3.1\cdot10^{-1}$
5099.95	7	7	$2.9\cdot10^{-1}$
5115.40	11	9	$2.2\cdot10^{-1}$
5129.37	7	5	$1.2\cdot10^{-1}$
5155.14	5	5	$1.1\cdot10^{-1}$
5155.76	5	7	$2.9\cdot10^{-1}$
5176.57	5	5	$1.8\cdot10^{-1}$
5371.33	7	7	$1.6\cdot10^{-1}$
5476.91	1	3	$9.5\cdot10^{-2}$
5637.12	3	3	$1.1\cdot10^{-1}$
5664.02	5	7	$1.1\cdot10^{-1}$
5695.00	3	3	$1.7\cdot10^{-1}$
6086.29	3	5	$1.1\cdot10^{-1}$

λ Å	Weights g_i	g_k	A 10^8 s^{-1}
6175.42	3	3	$1.7\cdot10^{-1}$
7122.24	5	7	$2.1\cdot10^{-1}$
7381.94	9	11	$9.7\cdot10^{-2}$
7422.30	7	5	$1.8\cdot10^{-1}$
7727.66	7	7	$1.1\cdot10^{-1}$
Ni II			
2165.55	10	10	2.4
2169.10	8	8	1.58
2174.67	8	10	1.43
2175.15	6	6	1.77
2184.61	4	4	2.90
2201.41	4	6	1.3
2206.72	6	8	1.66
2216.48	10	12	3.4
2220.40	6	8	2.3
2222.96	10	10	$9.8\cdot10^{-1}$
2224.86	8	8	1.55
2226.33	6	6	1.3
2253.85	4	6	1.98
2264.46	6	8	1.43
2270.21	8	10	1.56
2278.77	8	6	2.8
2287.09	6	4	2.8
2296.55	8	8	1.98
2297.14	6	4	2.70
2297.49	4	2	3.0
2298.27	6	6	2.8
2303.00	8	6	2.9
2316.04	10	8	2.88
2334.58	8	8	$8.0\cdot10^{-1}$
2375.42	6	8	$6.6\cdot10^{-1}$
2394.52	8	10	1.70
2416.13	6	8	2.1
2437.89	8	10	$5.4\cdot10^{-1}$
2510.87	8	10	$5.8\cdot10^{-1}$
Ni III			
1692.51	11	13	7.9
1709.90	9	11	6.3
1719.46	5	7	6.0
1722.28	3	5	5.9
1724.52	3	1	6.7
1741.96	9	7	5.7
1752.43	7	5	5.5
1760.56	5	3	6.5
1769.64	11	11	6.2
1823.06	9	9	5.6
Nitrogen			
N I			
951.079	4	6	$8.29\cdot10^{-3}$
951.295	4	4	$1.71\cdot10^{-3}$
952.303	4	6	$1.12\cdot10^{-1}$
952.415	4	4	$1.45\cdot10^{-1}$
952.523	4	2	$7.62\cdot10^{-2}$
953.415	4	2	1.90
953.655	4	4	1.81
953.970	4	6	1.62

λ Å	Weights g_i	g_k	A 10^8 s^{-1}
954.104	4	6	$1.95\cdot10^{-1}$
955.264	4	6	$3.37\cdot10^{-3}$
955.437	4	4	$1.40\cdot10^{-4}$
955.529	4	2	$2.63\cdot10^{-3}$
955.882	4	4	$4.29\cdot10^{-3}$
959.494	4	4	$3.75\cdot10^{-3}$
960.201	4	2	$1.69\cdot10^{-3}$
963.990	4	6	$5.94\cdot10^{-1}$
964.626	4	4	$5.66\cdot10^{-1}$
965.041	4	2	$5.52\cdot10^{-1}$
1003.37	4	4	$1.86\cdot10^{-6}$
1003.38	4	6	$8.40\cdot10^{-6}$
1134.17	4	2	1.51
1134.41	4	4	1.49
1134.98	4	6	1.44
1159.82	4	4	$4.94\cdot10^{-4}$
1160.94	4	2	$2.72\cdot10^{-4}$
1163.88	6	6	$3.25\cdot10^{-1}$
1164.00	4	6	$2.53\cdot10^{-3}$
1164.21	6	4	$2.98\cdot10^{-2}$
1164.32	4	4	$2.82\cdot10^{-1}$
1165.59	6	8	$1.69\cdot10^{-2}$
1165.72	6	6	$8.31\cdot10^{-4}$
1165.84	4	6	$1.99\cdot10^{-3}$
1165.88	6	4	$2.43\cdot10^{-4}$
1166.00	4	4	$6.79\cdot10^{-4}$
1167.45	6	8	1.10
1167.74	6	4	$1.17\cdot10^{-4}$
1167.86	4	4	$7.03\cdot10^{-4}$
1168.22	6	6	$1.30\cdot10^{-2}$
1168.33	4	6	$1.24\cdot10^{-1}$
1168.42	6	6	$3.38\cdot10^{-2}$
1168.54	4	6	$9.32\cdot10^{-1}$
1169.69	6	8	$2.79\cdot10^{-2}$
1170.16	6	6	$3.03\cdot10^{-3}$
1170.28	4	6	$3.06\cdot10^{-2}$
1170.42	6	4	$4.91\cdot10^{-4}$
1170.54	4	4	$1.11\cdot10^{-3}$
1170.67	4	2	$5.86\cdot10^{-2}$
1171.08	6	4	$1.25\cdot10^{-1}$
1171.20	4	4	$3.93\cdot10^{-3}$
1176.51	6	4	$8.52\cdot10^{-1}$
1176.63	4	4	$9.91\cdot10^{-2}$
1177.69	4	2	1.03
1183.28	6	6	$1.88\cdot10^{-5}$
1183.40	4	6	$2.16\cdot10^{-6}$
1184.24	6	4	$2.45\cdot10^{-3}$
1184.36	4	4	$3.34\cdot10^{-4}$
1184.98	4	2	$1.78\cdot10^{-3}$
1199.55	4	6	4.07
1200.22	4	4	4.03
1200.71	4	2	4.00
1243.17	6	4	$3.33\cdot10^{-1}$
1243.18	6	6	3.22
1243.31	4	4	3.10
1243.31	4	6	$2.36\cdot10^{-1}$
1310.54	4	6	$7.68\cdot10^{-1}$
1310.94	2	4	$6.05\cdot10^{-1}$
1310.95	4	4	$1.75\cdot10^{-1}$
1312.87	4	6	$5.53\cdot10^{-4}$

λ Å	Weights g_i	g_k	A 10^8 s^{-1}
1313.07	2	4	$2.79\cdot10^{-3}$
1313.08	4	4	$7.66\cdot10^{-5}$
1313.28	2	2	$2.37\cdot10^{-3}$
1313.28	4	2	$1.15\cdot10^{-3}$
1314.97	2	2	$1.14\cdot10^{-3}$
1314.98	4	2	$6.83\cdot10^{-4}$
1315.43	2	4	$3.65\cdot10^{-3}$
1315.44	4	4	$7.40\cdot10^{-4}$
1316.04	4	6	$1.06\cdot10^{-3}$
1316.29	4	6	$1.12\cdot10^{-2}$
1318.50	4	6	$1.36\cdot10^{-4}$
1318.82	2	4	$1.45\cdot10^{-5}$
1318.83	4	4	$1.00\cdot10^{-2}$
1319.00	2	2	$4.59\cdot10^{-1}$
1319.00	4	2	$2.28\cdot10^{-1}$
1319.67	2	4	$1.70\cdot10^{-1}$
1319.68	4	4	$5.76\cdot10^{-1}$
1326.56	2	4	$9.72\cdot10^{-3}$
1326.57	4	4	$6.71\cdot10^{-2}$
1327.92	2	2	$9.47\cdot10^{-2}$
1327.92	4	2	$4.43\cdot10^{-2}$
1335.18	4	6	$3.25\cdot10^{-6}$
1336.39	2	4	$1.02\cdot10^{-4}$
1336.40	4	4	$5.93\cdot10^{-4}$
1337.19	2	2	$3.16\cdot10^{-4}$
1337.20	4	2	$1.22\cdot10^{-4}$
1411.93	2	4	$4.46\cdot10^{-1}$
1411.94	4	4	$1.01\cdot10^{-1}$
1411.95	4	6	$5.34\cdot10^{-1}$
1492.63	6	4	3.11
1492.82	4	4	$3.26\cdot10^{-1}$
1494.68	4	2	3.46
1742.72	2	4	$2.12\cdot10^{-1}$
1742.73	4	4	1.05
1745.25	2	2	$8.35\cdot10^{-1}$
1745.26	4	2	$4.01\cdot10^{-1}$
6884.31	2	4	$2.90\cdot10^{-6}$
6900.36	4	4	$1.27\cdot10^{-5}$
7380.10	4	6	$1.94\cdot10^{-5}$
7403.39	2	4	$1.14\cdot10^{-4}$
7405.68	6	6	$4.39\cdot10^{-5}$
7421.94	4	4	$1.63\cdot10^{-4}$
7423.64	2	4	$5.64\cdot10^{-2}$
7442.30	4	4	$1.19\cdot10^{-1}$
7447.81	6	4	$1.49\cdot10^{-4}$
7468.31	6	4	$1.96\cdot10^{-1}$
8184.86	4	6	$8.21\cdot10^{-2}$
8188.01	2	4	$1.25\cdot10^{-1}$
8200.36	2	2	$4.68\cdot10^{-2}$
8210.72	4	4	$5.23\cdot10^{-2}$
8216.34	6	6	$2.26\cdot10^{-1}$
8223.13	4	2	$2.62\cdot10^{-1}$
8242.39	6	4	$1.31\cdot10^{-1}$
8567.74	2	4	$4.86\cdot10^{-2}$
8594.00	2	2	$2.09\cdot10^{-1}$
8629.24	4	4	$2.67\cdot10^{-1}$
8655.88	4	2	$1.07\cdot10^{-1}$
8664.39	2	4	$1.63\cdot10^{-3}$
8680.28	6	8	$2.53\cdot10^{-1}$
8683.40	4	6	$1.88\cdot10^{-1}$

λ Å	Weights g_i	Weights g_k	A 10^8 s^{-1}
8686.15	2	4	$1.15 \cdot 10^{-1}$
8703.25	2	2	$2.16 \cdot 10^{-1}$
8711.70	4	4	$1.29 \cdot 10^{-1}$
8718.84	6	6	$6.54 \cdot 10^{-2}$
8728.90	4	2	$3.75 \cdot 10^{-2}$
8747.37	6	4	$9.65 \cdot 10^{-3}$
8758.21	2	4	$2.62 \cdot 10^{-4}$
8767.35	2	2	$1.22 \cdot 10^{-3}$
9020.69	2	4	$3.68 \cdot 10^{-3}$
9028.92	2	2	$3.20 \cdot 10^{-1}$
9060.48	2	4	$3.21 \cdot 10^{-1}$
9386.81	2	4	$2.13 \cdot 10^{-1}$
9392.79	4	6	$2.51 \cdot 10^{-1}$
9419.39	2	4	$2.93 \cdot 10^{-4}$
9460.68	4	4	$3.73 \cdot 10^{-2}$
9464.17	2	2	$3.50 \cdot 10^{-3}$
9493.77	4	4	$4.06 \cdot 10^{-6}$
9660.12	2	4	$2.49 \cdot 10^{-4}$
9708.45	2	2	$3.18 \cdot 10^{-5}$
9716.46	6	4	$1.34 \cdot 10^{-4}$
9740.39	4	2	$9.03 \cdot 10^{-5}$
9742.12	8	6	$4.91 \cdot 10^{-7}$
9776.90	2	4	$1.40 \cdot 10^{-2}$
9786.78	4	6	$1.18 \cdot 10^{-2}$
9788.29	2	2	$3.83 \cdot 10^{-2}$
9798.56	4	4	$3.33 \cdot 10^{-2}$
9810.01	4	2	$5.42 \cdot 10^{-2}$
9814.02	6	8	$7.00 \cdot 10^{-3}$
9822.75	6	6	$5.74 \cdot 10^{-2}$
9834.61	6	4	$4.50 \cdot 10^{-2}$
9863.33	8	8	$1.03 \cdot 10^{-1}$
9872.15	8	6	$2.97 \cdot 10^{-2}$
9883.38	2	2	$1.83 \cdot 10^{-2}$
9905.52	4	2	$1.83 \cdot 10^{-3}$
9909.22	2	4	$4.07 \cdot 10^{-3}$
9931.47	4	4	$2.02 \cdot 10^{-2}$
9965.75	4	6	$3.48 \cdot 10^{-3}$
9968.51	6	4	$2.23 \cdot 10^{-3}$
9980.42	4	6	$4.86 \cdot 10^{-3}$

N II

λ Å	Weights g_i	Weights g_k	A 10^8 s^{-1}
525.983	1	3	$3.94 \cdot 10^{-2}$
526.118	3	3	$8.74 \cdot 10^{-4}$
526.345	5	3	$5.91 \cdot 10^{-3}$
528.529	5	7	$1.86 \cdot 10^{-2}$
529.355	1	3	7.17
529.413	3	1	$2.44 \cdot 10^{1}$
529.491	3	3	6.86
529.637	3	5	4.83
529.722	5	3	$1.04 \cdot 10^{1}$
529.867	5	5	$1.96 \cdot 10^{1}$
533.511	1	3	$2.46 \cdot 10^{1}$
533.581	3	5	$3.29 \cdot 10^{1}$
533.650	3	3	$1.70 \cdot 10^{1}$
533.729	5	7	$4.24 \cdot 10^{1}$
533.815	5	5	9.35
533.884	5	3	$9.54 \cdot 10^{-1}$
534.637	3	5	$4.96 \cdot 10^{-2}$
534.872	5	5	$7.14 \cdot 10^{-4}$
536.300	3	5	$6.40 \cdot 10^{-2}$
536.365	5	7	$1.52 \cdot 10^{-1}$
536.536	5	5	$2.35 \cdot 10^{-2}$
572.069	5	3	$4.24 \cdot 10^{-3}$
574.650	5	7	$3.83 \cdot 10^{1}$
576.060	5	3	$4.22 \cdot 10^{-4}$
576.232	5	5	$3.23 \cdot 10^{-2}$
580.802	5	7	$1.39 \cdot 10^{-2}$
580.904	5	5	$1.38 \cdot 10^{-1}$
582.156	5	5	$2.74 \cdot 10^{1}$
583.925	5	7	$2.32 \cdot 10^{-2}$
584.128	5	5	$2.35 \cdot 10^{-1}$
599.644	1	3	$1.02 \cdot 10^{-4}$
599.819	3	3	$1.69 \cdot 10^{-3}$
600.115	5	3	$3.15 \cdot 10^{-4}$
635.197	1	3	$2.58 \cdot 10^{1}$
640.121	1	3	$2.30 \cdot 10^{-2}$
644.634	1	3	$1.19 \cdot 10^{1}$
644.837	3	3	$3.56 \cdot 10^{1}$
645.178	5	3	$5.94 \cdot 10^{1}$
646.209	1	3	$1.43 \cdot 10^{-2}$
660.286	5	3	$3.17 \cdot 10^{1}$
670.296	1	3	$3.80 \cdot 10^{-1}$
670.515	3	3	$3.02 \cdot 10^{-1}$
670.884	5	3	$3.68 \cdot 10^{-1}$
671.016	3	5	2.84
671.386	5	5	8.52
671.411	1	3	3.40
671.630	3	3	2.53
671.773	3	1	$1.13 \cdot 10^{1}$
672.001	5	3	4.38
693.774	3	5	$5.66 \cdot 10^{-5}$
694.169	5	5	$4.14 \cdot 10^{-4}$
715.254	5	3	$1.06 \cdot 10^{-3}$
745.841	1	3	$1.35 \cdot 10^{1}$
746.984	5	3	$3.69 \cdot 10^{1}$
747.606	5	5	$7.17 \cdot 10^{-5}$
748.369	5	3	3.81
775.965	5	5	$3.14 \cdot 10^{1}$
816.740	1	3	$1.85 \cdot 10^{-4}$
834.070	5	7	$8.13 \cdot 10^{-7}$
834.740	5	5	$2.42 \cdot 10^{-7}$
835.163	5	3	$3.04 \cdot 10^{-8}$
858.376	1	3	$2.56 \cdot 10^{-1}$
860.205	1	3	$2.65 \cdot 10^{-2}$
915.612	1	3	4.23
915.962	3	1	$1.27 \cdot 10^{1}$
916.012	3	5	3.14
916.020	3	3	3.21
916.701	5	5	9.55
916.710	5	3	5.27
1064.95	5	5	$2.30 \cdot 10^{-5}$
1064.96	5	3	$5.16 \cdot 10^{-5}$
1083.99	1	3	2.10
1084.56	3	3	1.54
1084.58	3	5	2.82
1085.53	5	3	$9.96 \cdot 10^{-2}$
1085.55	5	5	$9.10 \cdot 10^{-1}$
1085.70	5	7	3.72
1162.50	3	1	$8.42 \cdot 10^{-5}$
1275.04	7	5	$5.10 \cdot 10^{-1}$
1275.25	5	5	$9.34 \cdot 10^{-2}$
1275.28	3	5	$6.32 \cdot 10^{-3}$
1276.20	5	3	$4.54 \cdot 10^{-1}$
1276.22	3	3	$1.54 \cdot 10^{-1}$
1276.80	3	1	$6.08 \cdot 10^{-1}$
1299.81	5	5	$7.33 \cdot 10^{-6}$
1300.04	5	7	$4.14 \cdot 10^{-5}$
1304.77	5	3	$6.74 \cdot 10^{-4}$
1304.79	3	3	$2.32 \cdot 10^{-4}$
1306.71	1	3	$2.06 \cdot 10^{-5}$
1343.34	7	7	$6.40 \cdot 10^{-2}$
1343.57	5	7	$7.55 \cdot 10^{-3}$
1345.08	7	5	$1.27 \cdot 10^{-2}$
1345.31	5	5	$4.90 \cdot 10^{-2}$
1345.34	3	5	$1.03 \cdot 10^{-2}$
1346.41	5	3	$1.94 \cdot 10^{-2}$
1346.44	3	3	$5.28 \cdot 10^{-2}$
1381.97	5	3	$1.30 \cdot 10^{-7}$
1382.00	3	3	$7.07 \cdot 10^{-5}$
1538.57	3	5	$1.59 \cdot 10^{-5}$
1627.35	3	5	$7.03 \cdot 10^{-3}$
1627.38	5	5	$2.39 \cdot 10^{-2}$
1628.90	3	3	$5.17 \cdot 10^{-3}$
1628.92	5	3	$1.98 \cdot 10^{-2}$
1629.08	1	3	$7.93 \cdot 10^{-3}$
1629.83	3	1	$3.21 \cdot 10^{-2}$
1675.73	3	3	$2.70 \cdot 10^{-1}$
1675.75	5	3	$4.39 \cdot 10^{-1}$
1675.92	1	3	$9.09 \cdot 10^{-2}$
1678.89	1	3	$3.62 \cdot 10^{-6}$
1740.31	5	7	$2.21 \cdot 10^{-1}$
1743.20	3	5	$1.67 \cdot 10^{-1}$
1743.23	5	5	$5.40 \cdot 10^{-2}$
1745.05	3	3	$9.07 \cdot 10^{-2}$
1745.08	5	3	$5.97 \cdot 10^{-3}$
1745.26	1	3	$1.23 \cdot 10^{-1}$
1805.24	3	3	$2.48 \cdot 10^{-4}$
1805.28	5	3	$3.05 \cdot 10^{-5}$
1805.47	1	3	$4.06 \cdot 10^{-5}$
2139.01	3	5	$5.15 \cdot 10^{-7}$
2142.77	5	5	$1.27 \cdot 10^{-6}$
3329.70	5	5	$2.01 \cdot 10^{-2}$
3408.13	3	1	$1.91 \cdot 10^{-1}$
3437.14	3	1	1.91
3775.61	5	5	$6.61 \cdot 10^{-6}$
3919.00	3	3	$7.56 \cdot 10^{-1}$
3955.85	3	5	$1.21 \cdot 10^{-1}$
3977.31	5	5	$7.79 \cdot 10^{-5}$
3995.00	3	5	1.22
4046.74	5	3	$1.44 \cdot 10^{-5}$
4109.59	3	1	$1.24 \cdot 10^{-6}$
4114.33	3	3	$1.65 \cdot 10^{-3}$
4123.12	3	5	$4.37 \cdot 10^{-4}$
4236.36	3	3	$1.93 \cdot 10^{-3}$
4319.06	3	1	$5.59 \cdot 10^{-5}$
4374.99	3	5	$6.31 \cdot 10^{-3}$
4379.58	3	3	$1.76 \cdot 10^{-3}$
4393.85	5	7	$5.21 \cdot 10^{-4}$
4412.50	7	7	$2.93 \cdot 10^{-5}$
4445.03	5	7	$1.20 \cdot 10^{-6}$

λ Å	Weights g_i	g_k	A 10^8 s^{-1}
4447.03	3	5	1.12
4459.94	3	1	$7.99 \cdot 10^{-2}$
4464.13	5	5	$5.85 \cdot 10^{-8}$
4465.53	3	3	$1.59 \cdot 10^{-2}$
4475.89	3	5	$3.96 \cdot 10^{-4}$
4476.27	5	3	$3.40 \cdot 10^{-5}$
4477.68	5	3	$6.44 \cdot 10^{-2}$
4488.09	5	5	$8.62 \cdot 10^{-3}$
4507.56	7	5	$7.39 \cdot 10^{-2}$
4564.76	3	5	$1.65 \cdot 10^{-2}$
4601.48	3	5	$2.22 \cdot 10^{-1}$
4607.15	1	3	$3.15 \cdot 10^{-1}$
4613.87	3	3	$2.12 \cdot 10^{-1}$
4621.39	3	1	$9.04 \cdot 10^{-1}$
4630.54	5	5	$7.48 \cdot 10^{-1}$
4643.09	5	3	$4.39 \cdot 10^{-1}$
4654.53	3	5	$1.92 \cdot 10^{-2}$
4667.21	3	3	$2.31 \cdot 10^{-2}$
4674.91	3	1	$8.54 \cdot 10^{-2}$
4709.44	3	3	$1.49 \cdot 10^{-3}$
4774.24	3	5	$3.07 \cdot 10^{-2}$
4779.72	3	3	$2.49 \cdot 10^{-1}$
4781.19	5	7	$1.92 \cdot 10^{-2}$
4788.14	5	5	$2.50 \cdot 10^{-1}$
4793.65	5	3	$7.73 \cdot 10^{-2}$
4803.29	7	7	$3.17 \cdot 10^{-1}$
4810.30	7	5	$4.75 \cdot 10^{-2}$
4860.17	3	5	$1.87 \cdot 10^{-2}$
4874.57	5	5	$2.17 \cdot 10^{-5}$
4895.12	5	3	$3.04 \cdot 10^{-2}$
4897.54	7	5	$5.61 \cdot 10^{-6}$
4987.38	3	1	$6.98 \cdot 10^{-1}$
4994.37	3	3	$7.11 \cdot 10^{-1}$
5001.13	3	5	$9.65 \cdot 10^{-1}$
5001.47	5	7	1.04
5002.70	1	3	$8.33 \cdot 10^{-2}$
5005.15	7	9	1.14
5007.33	3	5	$7.43 \cdot 10^{-1}$
5010.62	3	3	$2.10 \cdot 10^{-1}$
5016.38	5	5	$1.59 \cdot 10^{-1}$
5025.66	7	7	$1.04 \cdot 10^{-1}$
5040.71	7	5	$3.65 \cdot 10^{-3}$
5045.10	5	3	$3.37 \cdot 10^{-1}$
5073.59	3	3	$2.43 \cdot 10^{-2}$
5114.28	1	3	$3.88 \cdot 10^{-4}$
5123.53	3	3	$5.69 \cdot 10^{-5}$
5138.90	5	3	$3.34 \cdot 10^{-5}$
5238.18	3	5	$3.16 \cdot 10^{-5}$
5355.00	5	7	$7.02 \cdot 10^{-5}$
5383.72	3	5	$3.69 \cdot 10^{-3}$
5390.69	3	3	$2.04 \cdot 10^{-3}$
5452.07	1	3	$9.82 \cdot 10^{-2}$
5454.22	3	1	$3.70 \cdot 10^{-1}$
5462.58	3	3	$1.11 \cdot 10^{-1}$
5478.09	3	5	$5.22 \cdot 10^{-2}$
5480.05	5	3	$1.44 \cdot 10^{-1}$
5493.23	3	5	$3.11 \cdot 10^{-4}$
5495.65	5	5	$2.66 \cdot 10^{-1}$
5666.63	3	5	$3.45 \cdot 10^{-1}$
5674.00	3	5	$4.82 \cdot 10^{-7}$

λ Å	Weights g_i	g_k	A 10^8 s^{-1}
5676.02	1	3	$2.80 \cdot 10^{-1}$
5679.56	5	7	$4.96 \cdot 10^{-1}$
5686.21	3	3	$1.78 \cdot 10^{-1}$
5710.77	5	5	$1.17 \cdot 10^{-1}$
5730.66	5	3	$1.26 \cdot 10^{-2}$
5747.30	3	5	$3.27 \cdot 10^{-2}$
5767.45	3	3	$2.39 \cdot 10^{-2}$
5927.81	1	3	$3.19 \cdot 10^{-1}$
5931.78	3	5	$4.23 \cdot 10^{-1}$
5940.24	3	3	$2.22 \cdot 10^{-1}$
5941.65	5	7	$5.47 \cdot 10^{-1}$
5952.39	5	5	$1.24 \cdot 10^{-1}$
5960.91	5	3	$1.29 \cdot 10^{-2}$
6065.00	3	5	$2.57 \cdot 10^{-3}$
6086.54	5	5	$5.87 \cdot 10^{-5}$
6284.32	5	3	$4.63 \cdot 10^{-2}$
6285.69	5	7	$1.10 \cdot 10^{-3}$
6286.11	3	5	$4.50 \cdot 10^{-4}$
6309.25	5	5	$1.71 \cdot 10^{-4}$
6366.79	1	3	$8.14 \cdot 10^{-5}$
6379.62	3	3	$3.52 \cdot 10^{-2}$
6433.44	3	5	$1.22 \cdot 10^{-4}$
6457.68	3	3	$1.17 \cdot 10^{-4}$
6472.43	3	1	$1.13 \cdot 10^{-4}$
6482.05	3	3	$2.58 \cdot 10^{-1}$
6610.56	5	7	$6.01 \cdot 10^{-1}$
6802.17	5	3	$1.43 \cdot 10^{-5}$
6826.23	5	5	$2.49 \cdot 10^{-4}$
7262.55	3	3	$1.42 \cdot 10^{-7}$
7528.12	5	7	$6.61 \cdot 10^{-5}$
7545.36	5	5	$4.53 \cdot 10^{-4}$
7559.05	5	3	$1.90 \cdot 10^{-5}$
7762.24	5	5	$8.49 \cdot 10^{-2}$
8089.08	5	7	$1.01 \cdot 10^{-4}$
8128.14	5	5	$5.55 \cdot 10^{-4}$
8687.43	3	1	$1.16 \cdot 10^{-2}$
9399.64	1	3	$1.04 \cdot 10^{-4}$
N III			
374.198	2	4	$9.89 \cdot 10^{1}$
451.871	2	2	$1.03 \cdot 10^{1}$
452.227	4	2	$2.05 \cdot 10^{1}$
684.998	2	4	9.63
685.515	2	2	$3.83 \cdot 10^{1}$
685.817	4	4	$4.54 \cdot 10^{1}$
686.336	4	2	$1.95 \cdot 10^{1}$
763.334	2	2	9.58
764.351	4	2	$1.85 \cdot 10^{1}$
771.545	2	4	8.19
771.901	4	4	$1.64 \cdot 10^{1}$
772.384	6	4	$2.45 \cdot 10^{1}$
772.889	6	4	$2.09 \cdot 10^{1}$
772.955	4	2	$2.34 \cdot 10^{1}$
979.832	4	4	8.84
979.905	6	6	9.21
989.799	2	4	4.18
991.511	4	4	$8.17 \cdot 10^{-1}$
991.577	4	6	4.97
1747.85	2	4	1.28
1751.22	4	4	$2.48 \cdot 10^{-1}$

λ Å	Weights g_i	g_k	A 10^8 s^{-1}
1751.66	4	6	1.51
2972.55	2	2	$6.67 \cdot 10^{-1}$
2977.33	4	2	$3.32 \cdot 10^{-1}$
2978.84	2	4	$1.66 \cdot 10^{-1}$
2983.64	4	4	$8.24 \cdot 10^{-1}$
3342.76	2	2	$3.80 \cdot 10^{-1}$
3353.98	2	4	$7.66 \cdot 10^{-1}$
3354.32	4	6	$5.51 \cdot 10^{-1}$
3355.46	4	2	$7.51 \cdot 10^{-1}$
3358.78	2	2	$3.05 \cdot 10^{-1}$
3360.98	4	4	$2.44 \cdot 10^{-1}$
3365.80	4	2	1.52
3367.36	6	6	1.27
3374.07	6	4	$8.13 \cdot 10^{-1}$
3745.95	2	4	$1.90 \cdot 10^{-1}$
3752.63	2	2	$6.67 \cdot 10^{-2}$
3754.69	4	4	$3.78 \cdot 10^{-1}$
3762.60	4	4	$4.24 \cdot 10^{-2}$
3771.03	6	4	$5.59 \cdot 10^{-1}$
3771.36	6	4	$8.28 \cdot 10^{-2}$
3792.97	8	6	$1.03 \cdot 10^{-1}$
3934.50	2	4	$7.49 \cdot 10^{-1}$
3938.51	4	6	$8.96 \cdot 10^{-1}$
3942.88	4	4	$1.49 \cdot 10^{-1}$
4097.36	2	4	$8.70 \cdot 10^{-1}$
4103.39	2	2	$8.67 \cdot 10^{-1}$
4195.74	2	4	$9.37 \cdot 10^{-1}$
4200.07	4	6	1.12
4215.77	4	4	$1.85 \cdot 10^{-1}$
4318.78	2	4	$5.40 \cdot 10^{-2}$
4321.22	2	2	$1.08 \cdot 10^{-1}$
4321.39	4	6	$5.03 \cdot 10^{-2}$
4325.43	4	4	$8.60 \cdot 10^{-2}$
4327.69	6	8	$3.06 \cdot 10^{-2}$
4327.88	4	2	$1.07 \cdot 10^{-1}$
4332.95	6	6	$1.23 \cdot 10^{-1}$
4337.01	6	4	$7.47 \cdot 10^{-2}$
4345.81	8	8	$1.82 \cdot 10^{-1}$
4351.11	8	6	$4.01 \cdot 10^{-2}$
4510.88	2	4	$2.84 \cdot 10^{-1}$
4510.96	4	6	$4.77 \cdot 10^{-1}$
4514.85	6	8	$6.80 \cdot 10^{-1}$
4518.14	2	2	$5.65 \cdot 10^{-1}$
4523.56	4	4	$3.61 \cdot 10^{-1}$
4530.86	4	2	$1.12 \cdot 10^{-1}$
4534.58	6	6	$2.01 \cdot 10^{-1}$
4547.30	6	4	$3.33 \cdot 10^{-2}$
4634.13	2	4	$6.36 \cdot 10^{-1}$
4640.64	4	6	$7.60 \cdot 10^{-1}$
4641.85	4	4	$1.26 \cdot 10^{-1}$
4858.70	2	4	$4.35 \cdot 10^{-1}$
4858.98	4	6	$4.66 \cdot 10^{-1}$
4861.27	6	8	$5.32 \cdot 10^{-1}$
4867.12	4	4	$1.73 \cdot 10^{-1}$
4867.17	8	10	$6.18 \cdot 10^{-1}$
4873.60	6	6	$1.50 \cdot 10^{-1}$
4881.78	6	4	$1.22 \cdot 10^{-2}$
4884.14	8	8	$8.71 \cdot 10^{-2}$
4896.58	8	6	$5.86 \cdot 10^{-3}$
5260.86	2	2	$2.80 \cdot 10^{-2}$

λ Å	Weights g_i	g_k	A 10^8 s^{-1}
5270.57	2	4	$6.95 \cdot 10^{-2}$
5272.68	4	2	$1.39 \cdot 10^{-1}$
5282.43	4	4	$2.21 \cdot 10^{-2}$
5297.75	4	6	$4.93 \cdot 10^{-2}$
5298.95	6	4	$7.38 \cdot 10^{-2}$
5314.36	6	6	$1.14 \cdot 10^{-1}$
5320.87	6	8	$5.68 \cdot 10^{-1}$
5327.19	4	6	$5.29 \cdot 10^{-1}$
5352.46	6	6	$3.72 \cdot 10^{-2}$
6365.84	2	2	$2.18 \cdot 10^{-1}$
6394.75	2	4	$2.15 \cdot 10^{-1}$
6445.34	2	4	$8.89 \cdot 10^{-2}$
6450.79	2	2	$1.77 \cdot 10^{-1}$
6454.08	4	6	$1.49 \cdot 10^{-1}$
6463.09	4	4	$1.13 \cdot 10^{-1}$
6467.02	6	8	$2.11 \cdot 10^{-1}$
6468.57	4	2	$3.52 \cdot 10^{-2}$
6478.76	6	6	$6.31 \cdot 10^{-2}$
6487.84	6	4	$1.05 \cdot 10^{-2}$
7371.51	4	4	$3.53 \cdot 10^{-2}$
7404.54	6	6	$3.61 \cdot 10^{-2}$
8307.51	2	4	$1.65 \cdot 10^{-2}$
8344.95	2	2	$6.52 \cdot 10^{-2}$
8386.39	4	4	$8.03 \cdot 10^{-2}$
8424.56	4	2	$3.17 \cdot 10^{-2}$
N IV			
247.205	1	3	$1.19 \cdot 10^{2}$
*283.52	9	15	$3.05 \cdot 10^{2}$
*322.64	9	3	$8.99 \cdot 10^{1}$
335.047	3	5	$1.845 \cdot 10^{2}$
387.356	3	1	$2.55 \cdot 10^{1}$
765.147	1	3	$2.320 \cdot 10^{1}$
*923.16	9	9	$1.759 \cdot 10^{1}$
955.334	3	1	$2.919 \cdot 10^{1}$
1718.55	3	5	2.321
2649.88	3	3	1.07
3052.20	1	3	$1.33 \cdot 10^{-1}$
3059.60	3	3	$3.95 \cdot 10^{-1}$
3075.19	5	3	$6.48 \cdot 10^{-1}$
3443.61	3	5	$3.46 \cdot 10^{-1}$
3445.22	1	3	$4.60 \cdot 10^{-1}$
3454.65	3	3	$3.42 \cdot 10^{-1}$
3461.36	3	1	1.36
3463.36	5	5	1.02
3474.53	5	3	$5.61 \cdot 10^{-1}$
3478.72	3	5	1.06
*3480.8	3	9	1.06
3483.00	3	3	1.06
3484.93	3	1	1.06
3689.94	3	1	$9.10 \cdot 10^{-2}$
3694.14	3	3	$2.27 \cdot 10^{-2}$
3707.39	5	3	$6.73 \cdot 10^{-2}$
3714.43	5	5	$1.34 \cdot 10^{-2}$
3735.43	7	5	$7.37 \cdot 10^{-2}$
3747.54	3	5	$9.92 \cdot 10^{-1}$
4057.76	3	5	$6.62 \cdot 10^{-1}$
4740.26	3	5	$1.53 \cdot 10^{-2}$
4747.96	3	3	$7.60 \cdot 10^{-2}$
4752.49	5	7	$1.13 \cdot 10^{-2}$

λ Å	Weights g_i	g_k	A 10^8 s^{-1}
4762.09	5	5	$6.99 \cdot 10^{-2}$
4769.86	5	3	$2.50 \cdot 10^{-2}$
4786.92	7	7	$8.79 \cdot 10^{-2}$
4796.66	7	5	$1.53 \cdot 10^{-2}$
5200.41	3	5	$2.67 \cdot 10^{-1}$
5204.28	5	7	$3.55 \cdot 10^{-1}$
5205.15	1	3	$1.97 \cdot 10^{-1}$
5226.70	3	3	$1.46 \cdot 10^{-1}$
5245.60	5	5	$8.66 \cdot 10^{-2}$
5272.35	5	3	$9.48 \cdot 10^{-3}$
5288.25	5	3	$3.22 \cdot 10^{-2}$
5736.93	3	5	$1.84 \cdot 10^{-1}$
5776.31	1	3	$1.85 \cdot 10^{-2}$
5784.76	3	1	$5.51 \cdot 10^{-2}$
5795.09	3	3	$1.37 \cdot 10^{-2}$
5812.31	3	5	$1.36 \cdot 10^{-2}$
5826.43	5	3	$2.25 \cdot 10^{-2}$
5843.84	5	5	$4.01 \cdot 10^{-2}$
6380.75	1	3	$1.42 \cdot 10^{-1}$
7103.24	1	3	$6.28 \cdot 10^{-2}$
7109.35	3	5	$8.46 \cdot 10^{-2}$
7111.28	3	3	$4.70 \cdot 10^{-2}$
*7116.8	9	15	$1.12 \cdot 10^{-1}$
7122.98	5	7	$1.12 \cdot 10^{-1}$
7127.25	5	5	$2.80 \cdot 10^{-2}$
7127.25	5	3	$3.11 \cdot 10^{-3}$
9165.07	3	5	$4.23 \cdot 10^{-2}$
9182.16	5	7	$4.45 \cdot 10^{-2}$
9222.99	7	9	$4.95 \cdot 10^{-2}$
9247.04	5	5	$7.66 \cdot 10^{-3}$
9311.55	7	7	$5.36 \cdot 10^{-3}$
N V			
*209.29	2	6	$1.21 \cdot 10^{2}$
*247.66	6	10	$4.26 \cdot 10^{2}$
1238.82	2	4	3.40
1242.80	2	2	3.37
4603.74	2	4	$4.14 \cdot 10^{-1}$
4619.97	2	2	$4.10 \cdot 10^{-1}$
N VI			
24.8980	1	3	$5.158 \cdot 10^{3}$
28.7870	1	3	$1.809 \cdot 10^{4}$
*161.220	3	9	$2.859 \cdot 10^{2}$
173.275	1	3	$2.697 \cdot 10^{2}$
*173.93	9	15	$8.756 \cdot 10^{2}$
185.192	3	5	$8.205 \cdot 10^{2}$
*1901	3	9	$6.780 \cdot 10^{-1}$
2896.4	1	3	$2.079 \cdot 10^{-1}$
*6991.1	3	9	$8.384 \cdot 10^{-2}$
9622.0	1	3	$3.276 \cdot 10^{-2}$
Oxygen			
O I			
791.973	5	5	4.94
792.938	1	3	2.19
792.967	3	5	1.64
877.798	5	3	2.85
877.879	5	5	5.12

λ Å	Weights g_i	g_k	A 10^8 s^{-1}
922.008	5	7	1.23
935.193	5	5	1.33
1028.16	1	3	$4.22 \cdot 10^{-1}$
1152.15	5	5	5.28
1217.65	1	3	2.06
1302.17	5	3	3.41
1304.86	3	3	2.03
1306.03	1	3	$6.76 \cdot 10^{-1}$
3823.41	7	7	$6.63 \cdot 10^{-3}$
3823.87	5	3	$1.87 \cdot 10^{-3}$
3824.35	5	5	$5.19 \cdot 10^{-3}$
3825.02	3	3	$5.59 \cdot 10^{-3}$
3825.19	5	7	$8.31 \cdot 10^{-4}$
3855.01	5	5	$1.63 \cdot 10^{-2}$
3947.29	5	7	$4.91 \cdot 10^{-3}$
3947.48	5	5	$4.88 \cdot 10^{-3}$
3947.59	5	3	$4.87 \cdot 10^{-3}$
3951.93	3	1	$3.10 \cdot 10^{-3}$
3952.98	5	3	$1.29 \cdot 10^{-3}$
3953.00	1	3	$1.03 \cdot 10^{-3}$
3954.52	3	5	$7.73 \cdot 10^{-4}$
3954.61	5	5	$2.32 \cdot 10^{-3}$
3997.95	5	3	$2.41 \cdot 10^{-2}$
4217.09	3	1	$5.44 \cdot 10^{-3}$
4222.77	5	3	$2.26 \cdot 10^{-3}$
4222.82	1	3	$1.81 \cdot 10^{-3}$
4233.27	5	5	$4.04 \cdot 10^{-3}$
4368.19	3	1	$7.56 \cdot 10^{-3}$
4368.24	3	5	$7.59 \cdot 10^{-3}$
4967.38	3	5	$4.43 \cdot 10^{-3}$
4967.88	5	7	$8.44 \cdot 10^{-3}$
4968.79	7	9	$1.27 \cdot 10^{-2}$
5019.29	5	5	$7.13 \cdot 10^{-3}$
5020.22	7	5	$9.98 \cdot 10^{-3}$
5329.11	3	5	$9.48 \cdot 10^{-3}$
5329.69	5	7	$1.81 \cdot 10^{-2}$
5330.74	7	9	$2.71 \cdot 10^{-2}$
5435.18	3	5	$7.74 \cdot 10^{-3}$
5435.77	5	5	$1.29 \cdot 10^{-2}$
5436.86	7	5	$1.80 \cdot 10^{-2}$
5512.60	3	5	$2.69 \cdot 10^{-3}$
5512.77	5	7	$3.58 \cdot 10^{-3}$
5554.83	3	3	$5.83 \cdot 10^{-3}$
5555.00	5	3	$9.71 \cdot 10^{-3}$
5958.39	3	5	$6.80 \cdot 10^{-3}$
5958.58	5	7	$9.06 \cdot 10^{-3}$
6046.23	3	3	$1.05 \cdot 10^{-2}$
6046.44	5	3	$1.75 \cdot 10^{-2}$
6046.49	1	3	$3.50 \cdot 10^{-3}$
6155.99	3	5	$2.67 \cdot 10^{-2}$
6156.78	5	7	$5.08 \cdot 10^{-2}$
6158.19	7	9	$7.62 \cdot 10^{-2}$
6324.84	7	5	$3.76 \cdot 10^{-5}$
6453.60	3	5	$1.65 \cdot 10^{-2}$
6454.44	5	5	$2.75 \cdot 10^{-2}$
6455.98	7	5	$3.85 \cdot 10^{-2}$
6726.28	5	5	$1.18 \cdot 10^{-5}$
6726.54	5	3	$6.44 \cdot 10^{-6}$
7001.92	3	5	$2.65 \cdot 10^{-2}$
7002.23	5	7	$3.53 \cdot 10^{-2}$

λ Å	Weights g_i	g_k	A $10^8\ s^{-1}$
7254.15	3	3	$2.24\cdot10^{-2}$
7254.45	5	3	$3.73\cdot10^{-2}$
7254.53	1	3	$7.45\cdot10^{-3}$
7771.94	5	7	$3.69\cdot10^{-1}$
7774.17	5	5	$3.69\cdot10^{-1}$
7775.39	5	3	$3.69\cdot10^{-1}$
7981.94	3	3	$2.33\cdot10^{-4}$
7982.40	1	3	$3.09\cdot10^{-4}$
7986.98	3	5	$4.19\cdot10^{-4}$
7987.33	5	5	$1.41\cdot10^{-4}$
7995.07	5	7	$5.63\cdot10^{-4}$
8221.82	7	7	$2.89\cdot10^{-1}$
8227.65	5	3	$8.13\cdot10^{-2}$
8230.00	5	5	$2.26\cdot10^{-1}$
8233.00	3	3	$2.43\cdot10^{-1}$
8235.35	3	5	$4.86\cdot10^{-2}$
8446.25	3	1	$3.22\cdot10^{-1}$
8446.36	3	5	$3.22\cdot10^{-1}$
8446.76	3	3	$3.22\cdot10^{-1}$
8820.42	5	7	$2.93\cdot10^{-1}$
9260.81	3	1	$4.46\cdot10^{-1}$
9260.85	3	3	$3.34\cdot10^{-1}$
9260.94	3	5	$1.56\cdot10^{-1}$
9262.58	5	3	$1.11\cdot10^{-1}$
9262.67	5	5	$2.60\cdot10^{-1}$
9262.78	5	7	$2.97\cdot10^{-1}$
9265.83	7	5	$2.97\cdot10^{-2}$
9265.93	7	7	$1.48\cdot10^{-1}$
9266.01	7	9	$4.45\cdot10^{-1}$
9482.89	5	3	$2.34\cdot10^{-1}$
9622.11	5	3	$5.22\cdot10^{-4}$
9622.16	3	3	$1.57\cdot10^{-3}$
9625.26	7	5	$3.25\cdot10^{-4}$
9625.30	7	7	$1.85\cdot10^{-3}$
9694.66	5	7	$4.54\cdot10^{-4}$
9694.91	5	5	$4.54\cdot10^{-4}$
9695.06	5	3	$4.54\cdot10^{-4}$
O II			
429.918	4	2	$4.25\cdot10^{1}$
430.041	4	4	$4.13\cdot10^{1}$
430.176	4	6	$4.36\cdot10^{1}$
483.760	4	2	$2.05\cdot10^{1}$
483.980	6	4	$1.80\cdot10^{1}$
484.027	4	4	3.22
485.087	6	8	$2.60\cdot10^{1}$
485.470	6	6	1.20
485.518	4	6	$1.93\cdot10^{1}$
2290.85	2	4	$7.41\cdot10^{-2}$
2293.30	2	2	$3.25\cdot10^{-1}$
2300.33	4	4	$4.17\cdot10^{-1}$
2302.81	4	2	$1.67\cdot10^{-1}$
2365.14	4	2	$1.52\cdot10^{-1}$
2375.72	6	4	$1.35\cdot10^{-1}$
2406.38	6	4	$1.85\cdot10^{-1}$
2407.48	4	4	$2.25\cdot10^{-1}$
2411.60	4	2	$2.05\cdot10^{-1}$
2411.64	2	2	$1.10\cdot10^{-1}$
2415.13	4	2	$2.20\cdot10^{-1}$
2418.46	6	6	$2.30\cdot10^{-1}$
2425.57	6	6	$1.77\cdot10^{-1}$
2433.54	2	4	$4.21\cdot10^{-1}$
2436.06	4	4	$1.69\cdot10^{-1}$
2444.25	4	4	$7.56\cdot10^{-2}$
2445.53	4	6	$4.98\cdot10^{-1}$
2517.96	4	6	$7.72\cdot10^{-2}$
2523.21	2	2	$9.63\cdot10^{-2}$
2526.87	4	4	$1.20\cdot10^{-1}$
2530.28	6	8	$8.16\cdot10^{-2}$
2571.46	2	4	$1.15\cdot10^{-1}$
2575.28	4	6	$1.37\cdot10^{-1}$
3134.73	8	6	1.23
3273.43	8	6	$9.99\cdot10^{-1}$
3377.15	2	2	1.27
3390.21	2	4	1.22
3407.28	6	6	1.02
3712.74	2	4	$2.84\cdot10^{-1}$
3727.32	4	4	$5.81\cdot10^{-1}$
3749.48	6	4	$9.31\cdot10^{-1}$
3833.07	6	8	$1.02\cdot10^{-2}$
3842.81	2	4	$7.45\cdot10^{-2}$
3843.58	4	6	$3.55\cdot10^{-2}$
3847.89	2	2	$1.95\cdot10^{-1}$
3850.80	4	6	$6.00\cdot10^{-3}$
3851.03	4	4	$1.59\cdot10^{-1}$
3851.47	8	8	$2.72\cdot10^{-2}$
3856.13	4	2	$2.28\cdot10^{-1}$
3857.16	6	6	$6.59\cdot10^{-2}$
3863.50	6	8	$6.49\cdot10^{-2}$
3864.13	2	2	$9.12\cdot10^{-2}$
3864.43	6	6	$2.15\cdot10^{-1}$
3864.67	6	4	$1.80\cdot10^{-1}$
3874.09	2	4	$3.26\cdot10^{-2}$
3875.80	8	6	$3.38\cdot10^{-2}$
3882.19	8	8	$5.50\cdot10^{-1}$
3882.45	4	4	$8.94\cdot10^{-2}$
3883.14	8	6	$1.13\cdot10^{-1}$
3893.52	4	6	$1.89\cdot10^{-2}$
3907.45	6	6	$8.64\cdot10^{-2}$
3911.96	6	4	1.09
3912.12	4	4	$1.41\cdot10^{-1}$
3919.27	4	2	1.22
3945.04	2	4	$2.05\cdot10^{-1}$
3954.36	2	2	$8.57\cdot10^{-1}$
3973.26	4	4	1.04
3982.71	4	2	$4.27\cdot10^{-1}$
4069.62	2	4	1.52
4069.88	4	6	1.53
4072.15	6	8	1.98
4075.86	8	10	2.11
4078.84	4	4	$5.52\cdot10^{-1}$
4084.65	6	8	$7.28\cdot10^{-2}$
4085.11	6	6	$4.55\cdot10^{-1}$
4092.93	8	8	$2.65\cdot10^{-1}$
4094.14	6	4	$4.70\cdot10^{-2}$
4096.53	4	6	$1.73\cdot10^{-1}$
4097.22	2	4	$3.62\cdot10^{-1}$
4103.00	2	2	$5.09\cdot10^{-1}$
4104.72	4	6	$3.14\cdot10^{-1}$
4104.99	4	4	$9.14\cdot10^{-1}$
4106.02	8	6	$1.70\cdot10^{-2}$
4109.84	6	6	$1.21\cdot10^{-2}$
4110.19	6	4	$2.54\cdot10^{-1}$
4110.79	4	2	$7.70\cdot10^{-1}$
4112.02	6	6	$1.81\cdot10^{-1}$
4113.83	8	6	$2.41\cdot10^{-1}$
4119.22	6	8	1.33
4120.28	6	6	$2.15\cdot10^{-1}$
4120.55	6	4	$2.60\cdot10^{-1}$
4121.46	2	2	$5.60\cdot10^{-1}$
4129.32	4	2	$1.79\cdot10^{-1}$
4132.80	2	4	$9.13\cdot10^{-1}$
4140.70	4	4	$4.09\cdot10^{-2}$
4153.30	4	6	$7.91\cdot10^{-1}$
4156.53	6	4	$2.11\cdot10^{-1}$
4169.22	6	6	$2.71\cdot10^{-1}$
4185.44	6	8	1.91
4189.58	8	8	$7.06\cdot10^{-2}$
4189.79	8	10	1.98
4192.51	6	4	$3.21\cdot10^{-1}$
4196.27	4	4	$3.56\cdot10^{-2}$
4196.70	4	2	$3.56\cdot10^{-1}$
4317.14	2	4	$3.70\cdot10^{-1}$
4319.63	4	6	$2.55\cdot10^{-1}$
4319.87	2	2	$5.62\cdot10^{-1}$
4325.76	2	2	$1.47\cdot10^{-1}$
4327.46	6	6	$6.76\cdot10^{-1}$
4327.85	6	4	$7.24\cdot10^{-2}$
4328.59	4	2	1.12
4331.47	4	6	$4.82\cdot10^{-2}$
4331.86	4	4	$6.50\cdot10^{-1}$
4336.86	4	4	$1.57\cdot10^{-1}$
4345.56	4	2	$8.31\cdot10^{-1}$
4347.22	6	4	$1.19\cdot10^{-1}$
4347.41	4	4	$9.32\cdot10^{-1}$
4349.43	6	6	$6.91\cdot10^{-1}$
4351.26	6	6	$9.89\cdot10^{-1}$
4351.46	4	6	$5.82\cdot10^{-2}$
4359.40	4	6	$1.44\cdot10^{-2}$
4366.89	6	4	$3.98\cdot10^{-1}$
4369.27	4	4	$3.57\cdot10^{-1}$
4395.93	6	6	$3.91\cdot10^{-1}$
4405.98	6	4	$4.30\cdot10^{-2}$
4414.90	4	6	$8.34\cdot10^{-1}$
4416.97	2	4	$7.13\cdot10^{-1}$
4443.01	6	6	$5.05\cdot10^{-1}$
4443.52	6	8	$1.89\cdot10^{-2}$
4447.68	8	6	$2.52\cdot10^{-2}$
4448.19	8	8	$5.10\cdot10^{-1}$
4452.38	4	4	$1.37\cdot10^{-1}$
4466.24	2	4	$9.00\cdot10^{-1}$
4467.46	2	2	$9.00\cdot10^{-1}$
4563.18	4	4	$7.18\cdot10^{-3}$
4590.97	6	8	$8.85\cdot10^{-1}$
4595.96	6	6	$4.87\cdot10^{-2}$
4596.18	4	6	$8.34\cdot10^{-1}$
4638.86	2	4	$3.71\cdot10^{-1}$
4641.81	4	6	$5.96\cdot10^{-1}$
4649.13	6	8	$7.81\cdot10^{-1}$
4650.84	2	2	$6.86\cdot10^{-1}$

λ Å	Weights g_i	g_k	A 10^8 s^{-1}
4661.63	4	4	4.10·10⁻¹
4673.73	4	2	1.35·10⁻¹
4676.23	6	6	2.05·10⁻¹
4690.89	2	4	1.86·10⁻¹
4691.42	2	2	7.43·10⁻¹
4696.35	6	4	3.25·10⁻²
4698.44	6	6	6.59·10⁻²
4699.01	6	8	9.88·10⁻¹
4699.22	4	6	9.36·10⁻¹
4701.18	4	4	9.23·10⁻¹
4701.71	4	2	3.69·10⁻¹
4703.16	4	6	9.20·10⁻¹
4705.35	6	8	1.10
4710.01	4	6	2.98·10⁻¹
4741.70	6	6	4.71·10⁻²
4751.28	6	8	6.39·10⁻²
4752.69	6	6	1.45·10⁻²
4844.92	4	6	1.02·10⁻²
4856.39	4	6	5.58·10⁻²
4856.76	4	4	1.00·10⁻¹
4860.97	2	4	4.70·10⁻¹
4864.88	4	2	8.07·10⁻²
4871.52	4	6	5.60·10⁻¹
4872.02	4	4	9.34·10⁻²
4890.86	4	2	4.80·10⁻¹
4906.83	4	4	4.54·10⁻¹
4924.53	4	6	5.43·10⁻¹
4941.07	2	4	5.87·10⁻¹
4943.01	4	6	7.78·10⁻¹
4955.71	4	4	1.82·10⁻¹
5159.94	2	2	3.29·10⁻¹
5175.90	4	2	1.49·10⁻¹
5190.50	2	4	1.26·10⁻¹
5206.65	4	4	3.58·10⁻¹
5583.22	2	4	2.17·10⁻²
5611.07	2	2	2.14·10⁻²
6627.37	4	4	1.73·10⁻¹
6641.03	2	2	9.88·10⁻²
6666.66	4	2	6.78·10⁻²
6677.87	2	4	3.37·10⁻²
6717.75	2	2	1.33·10⁻¹
6721.39	4	2	1.81·10⁻¹
6810.48	6	8	1.64·10⁻³
6844.10	4	6	2.97·10⁻³
6846.80	8	8	3.17·10⁻²
6869.48	6	6	5.35·10⁻²
6884.88	4	4	6.12·10⁻²
6895.10	10	8	2.72·10⁻¹
6906.44	8	6	2.48·10⁻¹
6907.87	4	2	3.03·10⁻¹
6910.56	6	4	2.43·10⁻¹
O III			
305.656	3	5	1.62·10²
305.767	5	7	2.16·10²
320.978	5	7	2.17·10²
328.448	5	5	1.04·10²
345.312	1	3	1.35·10²
507.680	3	3	4.82·10¹
508.178	5	3	8.04·10¹

λ Å	Weights g_i	g_k	A 10^8 s^{-1}
525.794	5	3	9.60·10¹
599.590	5	5	5.41·10¹
835.289	5	7	5.99
1760.41	3	5	8.38·10⁻¹
1764.46	5	5	2.50
1766.63	1	3	1.11
1772.28	3	1	3.29
1772.97	5	3	1.37
2390.43	3	3	1.62
2454.97	3	1	3.43
2665.68	3	5	6.75·10⁻¹
2674.58	5	5	1.11
2683.66	3	1	1.85
2686.15	7	5	1.54
2687.55	3	3	1.84
2695.48	3	5	1.82
2959.69	3	5	1.83
2983.78	3	5	2.15
2996.48	3	3	4.64·10⁻¹
3004.34	5	5	4.27·10⁻¹
3017.62	7	7	5.38·10⁻¹
3023.43	3	5	4.79·10⁻¹
3024.54	1	3	6.16·10⁻¹
3035.41	3	3	4.59·10⁻¹
3042.07	3	1	1.94
3047.10	5	5	1.49
3059.28	5	3	8.72·10⁻¹
3064.98	1	3	2.17·10⁻¹
3068.13	3	1	6.49·10⁻¹
3068.67	3	5	2.27·10⁻¹
3074.14	5	7	1.84·10⁻¹
3074.72	5	3	3.76·10⁻¹
3075.13	5	5	1.61·10⁻¹
3075.95	7	9	1.07·10⁻¹
3083.65	7	7	3.20·10⁻¹
3084.64	7	5	2.55·10⁻¹
3088.04	9	9	5.30·10⁻¹
3095.79	9	7	1.35·10⁻¹
3115.67	3	1	1.39
3121.63	3	3	1.38
3132.79	3	5	1.37
3201.14	3	3	4.77·10⁻¹
3207.61	5	5	4.40·10⁻¹
3216.07	7	7	5.58·10⁻¹
3260.86	5	7	1.68
3265.33	7	9	1.88
3267.20	3	5	1.58
3281.83	5	5	2.89·10⁻¹
3284.45	7	7	2.06·10⁻¹
3299.39	1	3	1.64·10⁻¹
3312.33	3	3	4.60·10⁻¹
3326.06	3	3	2.65·10⁻¹
3330.30	3	5	6.81·10⁻¹
3330.32	3	5	4.76·10⁻¹
3332.41	5	3	7.92·10⁻¹
3332.93	5	7	5.04·10⁻¹
3336.67	3	3	3.76·10⁻¹
3340.76	5	3	6.57·10⁻¹
3344.20	5	5	1.25·10⁻¹
3344.51	5	7	3.48·10⁻¹

λ Å	Weights g_i	g_k	A 10^8 s^{-1}
3347.98	7	5	4.86·10⁻¹
3350.62	5	3	1.12
3350.92	7	7	9.91·10⁻¹
3355.86	7	7	6.89·10⁻¹
3362.31	7	5	6.87·10⁻¹
3376.61	3	1	1.49
3376.76	3	3	1.12
3377.26	3	5	5.20·10⁻¹
3382.61	5	7	9.86·10⁻¹
3383.31	5	3	3.70·10⁻¹
3383.81	5	5	8.62·10⁻¹
3384.90	7	9	1.48
3394.22	7	7	4.88·10⁻¹
3395.43	7	5	9.75·10⁻²
3406.88	1	3	1.93·10⁻¹
3408.13	3	1	5.79·10⁻¹
3415.26	3	3	1.44·10⁻¹
3428.63	3	5	1.42·10⁻¹
3430.57	5	3	2.37·10⁻¹
3444.05	5	5	4.21·10⁻¹
3446.68	3	5	9.71·10⁻¹
3447.15	1	3	8.09·10⁻¹
3447.97	5	7	1.19
3450.91	7	9	1.44
3451.30	3	3	8.06·10⁻¹
3454.84	5	5	6.89·10⁻¹
3454.99	9	11	1.72
3459.94	7	7	5.14·10⁻¹
3466.13	9	9	2.84·10⁻¹
3520.94	1	3	1.50·10⁻¹
3531.22	3	1	4.45·10⁻¹
3534.90	3	5	1.11·10⁻¹
3555.24	5	3	1.82·10⁻¹
3556.78	5	5	3.26·10⁻¹
3695.38	3	5	4.01·10⁻¹
3698.72	5	7	7.62·10⁻¹
3703.36	7	9	1.14
3704.75	3	3	8.53·10⁻¹
3707.27	3	5	7.34·10⁻¹
3709.54	3	1	1.13
3712.49	5	5	6.59·10⁻¹
3714.03	3	3	4.06·10⁻¹
3715.09	5	7	9.73·10⁻¹
3720.89	7	7	3.74·10⁻¹
3721.95	5	3	2.80·10⁻¹
3725.31	5	5	2.41·10⁻¹
3728.51	5	7	1.29
3728.84	7	9	1.45
3729.80	3	5	1.22
3734.83	7	5	7.40·10⁻²
3742.63	5	5	2.24·10⁻¹
3746.90	7	7	1.59·10⁻¹
3754.70	3	5	7.53·10⁻¹
3757.23	1	3	5.56·10⁻¹
3759.88	5	7	9.79·10⁻¹
3774.03	3	3	3.91·10⁻¹
3791.28	5	5	2.24·10⁻¹
3961.57	5	7	1.25
4072.64	1	3	3.37·10⁻¹
4073.98	3	5	4.54·10⁻¹

λ Å	Weights g_i	Weights g_k	A $10^8\ s^{-1}$
4081.02	5	7	$6.02 \cdot 10^{-1}$
4089.30	3	3	$2.49 \cdot 10^{-1}$
4103.07	5	5	$1.48 \cdot 10^{-1}$
4440.09	5	3	$4.42 \cdot 10^{-1}$
4447.69	5	5	$4.40 \cdot 10^{-1}$
4461.61	5	7	$4.36 \cdot 10^{-1}$
4524.22	3	1	$3.38 \cdot 10^{-1}$
4532.78	5	3	$1.40 \cdot 10^{-1}$
4535.29	3	3	$8.40 \cdot 10^{-2}$
4555.39	5	5	$2.49 \cdot 10^{-1}$
4557.91	3	5	$8.27 \cdot 10^{-2}$
5268.30	1	3	$3.50 \cdot 10^{-1}$
5508.24	5	5	$1.06 \cdot 10^{-1}$
5592.25	3	3	$3.27 \cdot 10^{-1}$
O IV			
238.570	4	6	$3.54 \cdot 10^{2}$
554.513	4	4	$6.06 \cdot 10^{1}$
625.853	6	4	$3.19 \cdot 10^{1}$
779.820	4	4	$1.31 \cdot 10^{1}$
779.912	6	6	$1.36 \cdot 10^{1}$
923.367	4	4	$1.10 \cdot 10^{1}$
1343.51	4	6	2.57
2132.64	4	4	1.29
2493.39	2	4	1.18
2493.75	4	6	$8.48 \cdot 10^{-1}$
2507.73	4	2	2.32
2509.22	6	6	1.94
2510.58	4	2	1.19
2517.37	6	4	1.24
2805.87	2	4	$2.90 \cdot 10^{-1}$
2816.53	4	4	$5.74 \cdot 10^{-1}$
2836.27	6	4	$8.43 \cdot 10^{-1}$
2916.31	2	4	1.06
2921.46	4	6	1.27
3063.43	2	4	1.30
3071.60	2	2	1.29
3194.78	6	6	$1.71 \cdot 10^{-1}$
3209.65	8	8	$2.53 \cdot 10^{-1}$
3348.06	2	4	$8.51 \cdot 10^{-1}$
3349.11	4	6	1.02
3354.27	4	2	$7.71 \cdot 10^{-1}$
3362.55	4	4	$7.65 \cdot 10^{-1}$
3375.40	4	6	$7.56 \cdot 10^{-1}$
3378.02	4	4	$1.66 \cdot 10^{-1}$
3381.21	4	6	$7.19 \cdot 10^{-1}$
3381.30	2	4	$4.28 \cdot 10^{-1}$
3385.52	6	8	1.02
3390.19	2	2	$8.49 \cdot 10^{-1}$
3396.80	4	4	$5.40 \cdot 10^{-1}$
3409.70	6	6	$3.00 \cdot 10^{-1}$
3411.30	4	4	$1.69 \cdot 10^{-1}$
3411.69	4	6	1.02
3489.89	4	6	$7.29 \cdot 10^{-1}$
3492.21	2	4	$6.06 \cdot 10^{-1}$
3560.39	4	6	1.03
3563.33	6	8	1.10
3725.89	2	4	$5.61 \cdot 10^{-1}$
3725.94	4	6	$6.01 \cdot 10^{-1}$
3729.03	6	8	$6.86 \cdot 10^{-1}$
3736.68	4	4	$2.23 \cdot 10^{-1}$
3736.85	8	10	$7.95 \cdot 10^{-1}$
3744.89	6	6	$1.92 \cdot 10^{-1}$
3758.39	8	8	$1.11 \cdot 10^{-1}$
3974.58	4	6	$6.62 \cdot 10^{-2}$
3977.09	6	4	$9.91 \cdot 10^{-2}$
3995.08	6	6	$1.52 \cdot 10^{-1}$
4687.03	2	4	$2.79 \cdot 10^{-1}$
4772.60	2	4	$1.23 \cdot 10^{-1}$
4779.10	2	2	$2.45 \cdot 10^{-1}$
4783.42	4	6	$2.06 \cdot 10^{-1}$
4794.18	4	4	$1.56 \cdot 10^{-1}$
4798.27	6	8	$2.91 \cdot 10^{-1}$
4813.15	6	6	$8.65 \cdot 10^{-2}$
5305.51	4	4	$6.10 \cdot 10^{-2}$
5362.51	6	6	$6.12 \cdot 10^{-2}$
6931.60	2	2	$7.35 \cdot 10^{-2}$
7004.11	4	4	$8.90 \cdot 10^{-2}$
7061.30	4	2	$3.48 \cdot 10^{-2}$
O V			
172.169	1	3	$2.94 \cdot 10^{2}$
*192.85	9	15	$6.90 \cdot 10^{2}$
*215.17	9	3	$1.83 \cdot 10^{2}$
220.353	3	5	$4.292 \cdot 10^{2}$
248.460	3	1	$5.59 \cdot 10^{1}$
629.732	1	3	$2.872 \cdot 10^{1}$
758.677	3	5	5.547
759.442	1	3	7.373
760.227	3	3	5.514
760.446	5	5	$1.652 \cdot 10^{1}$
761.128	3	1	$2.197 \cdot 10^{1}$
762.004	5	3	9.125
774.518	3	1	$3.804 \cdot 10^{1}$
1371.30	3	5	3.336
2729.31	3	5	$4.52 \cdot 10^{-1}$
2731.45	1	3	$5.90 \cdot 10^{-1}$
2743.61	3	3	$4.38 \cdot 10^{-1}$
2752.23	3	1	1.82
2755.13	5	5	1.37
2769.69	5	3	$7.88 \cdot 10^{-1}$
2781.01	3	5	1.40
*2784.0	3	9	1.40
2786.99	3	3	1.39
2789.85	3	1	1.38
3058.68	3	5	1.39
3144.66	3	5	$8.86 \cdot 10^{-1}$
3219.24	3	1	$1.54 \cdot 10^{-1}$
3222.29	1	3	$1.16 \cdot 10^{-1}$
3227.54	3	3	$3.38 \cdot 10^{-2}$
3239.21	3	3	$3.28 \cdot 10^{-1}$
3248.28	5	3	$1.18 \cdot 10^{-1}$
3263.54	5	5	$1.86 \cdot 10^{-2}$
3275.64	5	3	$4.76 \cdot 10^{-1}$
3297.62	7	5	$1.30 \cdot 10^{-1}$
3690.17	3	5	$1.97 \cdot 10^{-2}$
3698.36	3	3	$1.03 \cdot 10^{-1}$
3702.72	5	7	$1.41 \cdot 10^{-2}$
3717.31	5	5	$9.63 \cdot 10^{-2}$
3725.63	5	3	$2.91 \cdot 10^{-2}$
3746.64	7	7	$1.18 \cdot 10^{-1}$
3761.58	7	5	$1.61 \cdot 10^{-2}$
4119.37	3	5	$3.66 \cdot 10^{-1}$
4120.49	3	1	$3.33 \cdot 10^{-1}$
4123.96	5	7	$4.81 \cdot 10^{-1}$
4125.49	1	3	$2.70 \cdot 10^{-1}$
4134.11	3	3	$3.34 \cdot 10^{-1}$
4153.27	3	3	$1.92 \cdot 10^{-1}$
4158.86	3	5	$3.39 \cdot 10^{-1}$
4178.46	5	5	$1.12 \cdot 10^{-1}$
4213.35	5	3	$1.19 \cdot 10^{-2}$
4522.66	5	3	$1.02 \cdot 10^{-2}$
4554.53	3	5	$2.41 \cdot 10^{-1}$
5114.06	1	3	$1.80 \cdot 10^{-1}$
5339.94	1	3	$1.85 \cdot 10^{-2}$
5349.74	3	1	$7.04 \cdot 10^{-2}$
5372.71	3	3	$1.42 \cdot 10^{-2}$
5414.59	3	5	$9.29 \cdot 10^{-3}$
5428.38	5	3	$2.68 \cdot 10^{-2}$
5471.12	5	5	$4.86 \cdot 10^{-2}$
5571.81	1	3	$8.33 \cdot 10^{-2}$
5580.12	3	5	$1.11 \cdot 10^{-1}$
5583.23	3	3	$6.20 \cdot 10^{-2}$
*5589.9	9	15	$1.49 \cdot 10^{-1}$
5597.89	5	7	$1.48 \cdot 10^{-1}$
5604.27	5	5	$3.68 \cdot 10^{-2}$
5607.41	5	3	$4.08 \cdot 10^{-3}$
6330.05	5	7	$1.21 \cdot 10^{-1}$
6460.12	3	5	$9.37 \cdot 10^{-2}$
6466.14	5	7	$1.01 \cdot 10^{-1}$
6500.24	7	9	$1.11 \cdot 10^{-1}$
6543.77	5	5	$1.64 \cdot 10^{-2}$
6601.28	7	7	$1.14 \cdot 10^{-2}$
6764.72	1	3	$4.37 \cdot 10^{-2}$
6789.62	3	5	$5.79 \cdot 10^{-2}$
6817.40	3	3	$3.00 \cdot 10^{-2}$
6828.95	5	7	$7.35 \cdot 10^{-2}$
6878.76	5	5	$1.65 \cdot 10^{-2}$
O VI			
*150.10	2	6	$2.62 \cdot 10^{2}$
*173.03	6	10	$8.78 \cdot 10^{2}$
1031.91	2	4	4.16
1037.61	2	2	4.09
3811.35	2	4	$5.14 \cdot 10^{-1}$
3834.24	2	2	$5.05 \cdot 10^{-1}$
O VII			
18.6270	1	3	$9.365 \cdot 10^{3}$
21.6020	1	3	$3.309 \cdot 10^{4}$
*120.33	3	9	$5.334 \cdot 10^{2}$
128.411	1	3	$8.982 \cdot 10^{2}$
*128.46	9	15	$1.615 \cdot 10^{3}$
135.820	3	5	$1.523 \cdot 10^{3}$
*1630.3	3	9	$7.935 \cdot 10^{-1}$
2448.98	1	3	$2.514 \cdot 10^{-1}$
*5933.1	3	9	$1.002 \cdot 10^{-1}$
8241.76	1	3	$3.864 \cdot 10^{-2}$

λ Å	Weights g_i	g_k	A 10^8 s^{-1}
Phosphorus			
P I			
1671.7	4	2	3.9·10⁻¹
1674.6	4	4	4.0·10⁻¹
1679.7	4	6	3.9·10⁻¹
1775.0	4	6	2.17
1782.9	4	4	2.14
1787.7	4	2	2.13
2135.5	4	4	2.11·10⁻¹
2136.2	6	4	2.83
2149.1	4	2	3.18
2152.9	2	4	4.85·10⁻¹
2154.1	4	4	1.73·10⁻¹
2154.1	4	6	5.8·10⁻¹
2534.0	2	4	2.00·10⁻¹
2535.6	4	4	9.5·10⁻¹
2553.3	2	2	7.1·10⁻¹
2554.9	4	2	3.00·10⁻¹
P II			
1301.9	1	3	5.0·10⁻¹
1304.5	3	1	1.5
1304.7	3	3	3.7·10⁻¹
1305.5	3	5	3.8·10⁻¹
1309.9	5	3	6.2·10⁻¹
1310.7	5	5	1.1
4475.3	5	7	1.3
4499.2	5	7	1.4
4530.8	3	5	1.0
4554.8	3	5	9.6·10⁻¹
4588.0	5	7	1.7
4589.9	3	5	1.6
4602.1	7	9	1.9
4943.5	7	5	6.3·10⁻¹
5253.5	3	5	1.0
5425.9	5	5	6.9·10⁻¹
6024.2	3	5	5.1·10⁻¹
6043.1	5	7	6.8·10⁻¹
P III			
1334.8	2	4	5.5·10⁻¹
1344.3	4	6	6.4·10⁻¹
1344.8	4	4	1.1·10⁻¹
4057.4	4	4	1.0·10⁻¹
4059.3	6	4	9.0·10⁻¹
4080.1	4	2	9.9·10⁻¹
Potassium			
K I			
4044.1	2	4	1.24·10⁻²
4047.2	2	2	1.24·10⁻²
5084.2	2	2	3.50·10⁻³
5099.2	4	2	7.0·10⁻³
5323.3	2	2	6.3·10⁻³
5339.7	4	2	1.26·10⁻²
5343.0	2	4	4.0·10⁻³
5359.6	4	6	4.6·10⁻³
5782.4	2	2	1.23·10⁻²
5801.8	4	2	2.46·10⁻²
5812.2	2	4	2.8·10⁻³
5831.9	4	6	3.2·10⁻³
6911.1	2	2	2.72·10⁻²
6938.8	4	2	5.4·10⁻²
7664.9	2	4	3.87·10⁻¹
7699.0	2	2	3.82·10⁻¹
K II			
607.93	1	3	1.3·10⁻²
K III			
2550.0	6	4	2.0
2635.1	4	4	1.2
2992.4	6	8	2.5
3052.1	4	6	1.7
3202.0	4	4	1.8
3289.1	4	6	2.0
3322.4	6	6	1.3
3421.8	2	4	1.5
K XVI			
206.27	1	3	9.4·10¹
K XVII			
22.020	2	4	4.7·10⁴
22.163	4	6	5.6·10⁴
22.18	4	4	9.3·10³
22.60	2	2	2.5·10³
22.76	4	2	4.7·10³
Praseodymium			
Pr II			
3997.0	15	15	1.87·10⁻¹
4062.8	13	15	1.00
4100.7	17	19	8.4·10⁻¹
4143.1	15	17	5.8·10⁻¹
4179.4	13	15	5.2·10⁻¹
4222.9	11	13	3.91·10⁻¹
4241.0	17	15	2.30·10⁻¹
4359.8	15	15	1.1·10⁻¹
4405.8	17	17	9.0·10⁻²
4429.3	15	15	2.28·10⁻¹
4449.8	13	13	1.24·10⁻¹
4468.7	11	13	1.54·10⁻¹
4510.2	13	15	1.16·10⁻¹
4534.2	15	17	4.9·10⁻²
4734.2	15	13	2.5·10⁻²
4879.1	15	15	1.8·10⁻²
4886.0	15	15	1.3·10⁻²
4912.6	17	15	5.7·10⁻²
5034.4	19	19	1.1·10⁻¹
5110.8	21	19	2.78·10⁻¹
5135.1	17	17	1.25·10⁻¹
5173.9	19	17	3.18·10⁻¹
5219.1	15	15	9.5·10⁻²
5220.1	17	15	2.35·10⁻¹
5251.7	15	13	1.1·10⁻²
5259.7	15	13	2.24·10⁻¹
5292.6	13	13	9.3·10⁻²
5810.6	17	19	2.3·10⁻²
5879.3	15	15	7.6·10⁻²
6200.8	15	17	1.8·10⁻²
6278.7	13	15	2.6·10⁻²
6398.0	11	13	1.9·10⁻²
Rhodium			
Rh I			
3121.76	6	6	1.1·10⁻¹
3189.05	6	6	3.03·10⁻¹
3263.14	6	6	1.3·10⁻¹
3271.61	6	4	2.0·10⁻¹
3280.55	8	8	2.36·10⁻¹
3283.57	6	8	4.4·10⁻¹
3323.09	8	10	6.3·10⁻¹
3396.82	10	10	6.5·10⁻¹
3399.70	6	8	1.2·10⁻¹
3462.04	6	6	6.2·10⁻¹
3470.66	4	4	8.5·10⁻¹
3478.91	6	6	3.32·10⁻¹
3498.73	4	6	2.12·10⁻¹
3502.52	10	10	4.3·10⁻¹
3507.32	6	8	3.4·10⁻¹
3528.02	8	8	8.5·10⁻¹
3543.95	4	4	4.65·10⁻¹
3549.54	6	6	2.22·10⁻¹
3570.18	4	6	1.82·10⁻¹
3583.10	8	10	2.6·10⁻¹
3596.19	6	4	5.5·10⁻¹
3597.15	6	8	5.9·10⁻¹
3612.47	4	2	8.90·10⁻¹
3654.87	8	8	6.0·10⁻²
3657.99	8	6	8.8·10⁻¹
3666.22	6	8	8.4·10⁻²
3690.70	6	4	3.23·10⁻¹
3692.36	10	8	9.1·10⁻¹
3700.91	8	10	3.9·10⁻¹
3788.47	4	6	1.4·10⁻¹
3793.22	8	6	4.2·10⁻¹
3799.31	8	8	5.5·10⁻¹
3806.76	6	6	6.2·10⁻²
3818.19	6	4	5.8·10⁻¹
3822.26	6	6	8.5·10⁻¹
3828.48	6	6	6.2·10⁻¹
3833.89	6	4	5.8·10⁻¹
3856.52	8	10	5.9·10⁻¹
3934.23	8	8	1.58·10⁻¹
3942.72	4	2	7.15·10⁻¹
3958.86	6	8	5.5·10⁻¹
3984.40	4	4	1.1·10⁻¹
4082.78	6	4	1.4·10⁻¹
4121.68	6	6	9.8·10⁻²
4128.87	6	8	1.73·10⁻¹
4135.27	8	8	1.0·10⁻¹
4196.50	6	8	3.9·10⁻²
4211.14	8	10	1.62·10⁻¹
4288.71	6	8	6.1·10⁻²
4374.80	8	10	1.64·10⁻¹

λ Å	Weights g_i	g_k	A 10^8 s^{-1}
5983.60	10	10	$2.1 \cdot 10^{-2}$

Rubidium

Rb I

λ Å	Weights g_i	g_k	A 10^8 s^{-1}
3022.5	2	4	$4.13 \cdot 10^{-5}$
3032.0	2	4	$4.93 \cdot 10^{-5}$
3044.2	2	4	$8.2 \cdot 10^{-5}$
3060.2	2	4	$1.05 \cdot 10^{-4}$
3082.0	2	4	$1.49 \cdot 10^{-4}$
3112.6	2	4	$2.5 \cdot 10^{-4}$
3113.1	2	2	$1.3 \cdot 10^{-4}$
3157.5	2	4	$3.38 \cdot 10^{-4}$
3158.3	2	2	$2.0 \cdot 10^{-4}$
3228.0	2	4	$6.4 \cdot 10^{-4}$
3229.2	2	2	$3.8 \cdot 10^{-4}$
3348.7	2	4	$1.37 \cdot 10^{-3}$
3350.8	2	2	$8.9 \cdot 10^{-4}$
3587.1	2	4	$3.97 \cdot 10^{-3}$
3591.6	2	2	$2.9 \cdot 10^{-3}$
4201.8	2	4	$1.8 \cdot 10^{-2}$
4215.5	2	2	$1.5 \cdot 10^{-2}$
7800.3	2	4	$3.70 \cdot 10^{-1}$
7947.6	2	2	$3.40 \cdot 10^{-1}$

Scandium

Sc I

λ Å	Weights g_i	g_k	A 10^8 s^{-1}
3015.37	4	6	$7.8 \cdot 10^{-1}$
3019.35	6	8	$8.7 \cdot 10^{-1}$
3269.90	4	2	3.13
3273.63	6	4	2.81
3907.48	4	6	1.66
3911.81	6	8	1.79
4020.39	4	4	1.63
4023.68	6	6	1.65
4031.38	6	6	$2.9 \cdot 10^{-1}$
4043.80	8	8	$3.11 \cdot 10^{-1}$
4067.00	6	8	$1.91 \cdot 10^{-1}$
4074.96	4	6	$3.7 \cdot 10^{-1}$
4078.56	2	4	$4.3 \cdot 10^{-1}$
4086.66	6	8	$3.7 \cdot 10^{-1}$
4132.98	4	6	1.19
4140.27	6	8	1.17
4161.85	8	8	$1.77 \cdot 10^{-1}$
4233.59	6	6	$4.0 \cdot 10^{-1}$
4238.05	8	8	$7.1 \cdot 10^{-1}$
4706.94	4	6	$2.81 \cdot 10^{-1}$
4709.31	6	8	$4.0 \cdot 10^{-1}$
4728.77	8	8	$1.16 \cdot 10^{-1}$
4729.24	6	6	$1.93 \cdot 10^{-1}$
4734.11	4	2	1.10
4737.65	6	4	$8.8 \cdot 10^{-1}$
4741.02	8	6	$9.1 \cdot 10^{-1}$
4743.82	10	8	$9.8 \cdot 10^{-1}$
4973.67	4	2	$8.4 \cdot 10^{-1}$
4980.36	6	4	$5.6 \cdot 10^{-1}$
4983.43	4	4	$2.58 \cdot 10^{-1}$
4991.91	6	6	$3.8 \cdot 10^{-1}$
5018.41	6	4	$2.09 \cdot 10^{-1}$
5021.52	4	4	$2.30 \cdot 10^{-1}$
5070.17	6	8	$1.16 \cdot 10^{-1}$
5081.56	10	10	$7.6 \cdot 10^{-1}$
5083.72	8	8	$6.2 \cdot 10^{-1}$
5085.55	6	6	$5.7 \cdot 10^{-1}$
5086.94	4	4	$6.6 \cdot 10^{-1}$
5099.27	4	6	$1.50 \cdot 10^{-1}$
5339.43	6	6	$1.06 \cdot 10^{-1}$
5341.07	4	2	$3.8 \cdot 10^{-1}$
5349.34	6	4	$5.9 \cdot 10^{-1}$
5355.79	6	4	$3.0 \cdot 10^{-1}$
5356.10	8	6	$5.7 \cdot 10^{-1}$
5375.37	8	6	$3.4 \cdot 10^{-1}$
5392.06	10	8	$4.2 \cdot 10^{-1}$
5446.20	8	8	$2.8 \cdot 10^{-1}$
5451.37	6	6	$1.50 \cdot 10^{-1}$
5472.19	8	6	$9.7 \cdot 10^{-2}$
5482.01	8	8	$5.2 \cdot 10^{-1}$
5484.63	6	6	$5.2 \cdot 10^{-1}$
5514.23	6	8	$4.1 \cdot 10^{-1}$
5520.52	8	10	$4.3 \cdot 10^{-1}$
5671.83	10	12	$5.4 \cdot 10^{-1}$
5686.86	8	10	$4.9 \cdot 10^{-1}$
5700.19	6	8	$4.6 \cdot 10^{-1}$
5711.79	4	6	$4.5 \cdot 10^{-1}$
5717.31	8	8	$7.5 \cdot 10^{-2}$
6262.22	4	6	$8.4 \cdot 10^{-2}$
7741.16	10	10	$3.8 \cdot 10^{-2}$
7800.42	8	8	$5.1 \cdot 10^{-2}$

Sc II

λ Å	Weights g_i	g_k	A 10^8 s^{-1}
1880.6	5	3	5.0
2064.3	7	5	2.2
2068.0	5	3	2.0
2273.1	1	3	7.7
2545.20	5	5	$4.0 \cdot 10^{-1}$
2552.35	7	5	2.21
2555.79	3	3	$6.9 \cdot 10^{-1}$
2560.23	5	3	2.01
2563.19	3	1	2.70
2611.19	5	5	2.2
2667.70	3	5	1.5
2746.36	3	1	3.9
2782.31	5	5	1.3
2789.15	7	7	1.3
2801.31	9	9	1.3
2819.49	3	5	2.3
2822.12	5	7	2.5
2826.64	7	9	2.8
2870.85	5	3	1.1
2912.98	5	3	1.1
2979.68	3	5	1.2
2988.92	5	7	2.9
3039.92	7	9	3.5
3045.73	5	7	3.68
3052.92	7	9	3.92
3060.54	7	7	$3.0 \cdot 10^{-1}$
3065.12	9	11	4.00
3075.36	9	9	$2.5 \cdot 10^{-1}$
3128.27	3	3	1.9
3133.07	5	5	1.8
3139.72	7	7	2.1
3190.98	3	3	1.1
3199.33	5	3	1.9
3312.72	5	7	1.2
3320.40	5	3	1.2
3343.23	9	7	1.1
3353.72	5	7	1.51
3359.67	5	5	$2.16 \cdot 10^{-1}$
3361.26	3	3	$3.4 \cdot 10^{-1}$
3361.93	3	1	1.17
3368.94	5	3	$8.3 \cdot 10^{-1}$
3372.15	7	5	$9.9 \cdot 10^{-1}$
3379.16	3	3	2.5
3535.71	5	3	$6.1 \cdot 10^{-1}$
3558.53	5	7	$3.0 \cdot 10^{-1}$
3567.70	3	5	$3.5 \cdot 10^{-1}$
3572.53	7	7	1.38
3576.34	5	5	1.06
3580.93	3	3	1.23
3589.63	5	3	$4.6 \cdot 10^{-1}$
3590.47	7	5	$2.9 \cdot 10^{-1}$
3613.83	7	9	1.48
3630.74	5	7	1.20
3642.78	3	5	1.13
3645.31	7	7	$2.74 \cdot 10^{-1}$
3651.80	5	5	$3.0 \cdot 10^{-1}$
3859.59	7	5	1.1
4246.82	5	5	1.29
4314.08	9	7	$4.1 \cdot 10^{-1}$
4320.75	7	5	$4.0 \cdot 10^{-1}$
4325.00	5	3	$4.3 \cdot 10^{-1}$
4374.46	9	9	$1.48 \cdot 10^{-1}$
4400.39	7	7	$1.43 \cdot 10^{-1}$
4415.54	5	5	$1.47 \cdot 10^{-1}$
4670.41	5	7	$1.16 \cdot 10^{-1}$
5031.01	5	3	$3.5 \cdot 10^{-1}$
5239.81	1	3	$1.39 \cdot 10^{-1}$
5526.79	9	7	$3.3 \cdot 10^{-1}$
5657.91	5	5	$1.04 \cdot 10^{-1}$
5669.06	3	1	$1.31 \cdot 10^{-1}$

Silicon

Si I

λ Å	Weights g_i	g_k	A 10^8 s^{-1}
2207.98	1	3	$2.62 \cdot 10^{-1}$
2210.89	3	5	$3.46 \cdot 10^{-1}$
2211.74	3	3	$1.81 \cdot 10^{-1}$
2216.67	5	7	$4.54 \cdot 10^{-1}$
2218.06	5	5	$1.09 \cdot 10^{-1}$
2218.91	5	3	$1.05 \cdot 10^{-2}$
2438.77	1	3	$7.91 \cdot 10^{-3}$
2443.36	3	3	$1.32 \cdot 10^{-2}$
2452.12	5	3	$1.73 \cdot 10^{-3}$
2506.90	3	5	$5.47 \cdot 10^{-1}$
2514.32	1	3	$7.39 \cdot 10^{-1}$
2516.113	5	5	1.68
2519.202	3	3	$5.49 \cdot 10^{-1}$
2524.108	3	1	2.22
2528.509	5	3	$9.04 \cdot 10^{-1}$
2881.579	5	3	2.17

λ Å	Weights g_i	g_k	A 10^8 s^{-1}
2970.355	5	5	$6.00 \cdot 10^{-4}$
2987.645	5	3	$1.34 \cdot 10^{-2}$
3006.739	3	5	$1.1 \cdot 10^{-5}$
3020.004	5	5	$3.3 \cdot 10^{-5}$
3905.523	1	3	$1.33 \cdot 10^{-1}$
4102.936	1	3	$6.09 \cdot 10^{-4}$
4782.991	5	3	$1.7 \cdot 10^{-2}$
4792.212	3	1	$2.2 \cdot 10^{-2}$
4792.324	5	5	$1.7 \cdot 10^{-2}$
4947.607	3	1	$4.2 \cdot 10^{-2}$
5006.061	3	5	$2.8 \cdot 10^{-2}$
5622.221	3	3	$1.6 \cdot 10^{-2}$
5645.611	3	5	$9.7 \cdot 10^{-3}$
5665.554	1	3	$6.31 \cdot 10^{-3}$
5684.484	5	3	$2.6 \cdot 10^{-2}$
5690.425	3	3	$9.26 \cdot 10^{-3}$
5701.105	3	1	$1.83 \cdot 10^{-2}$
5708.397	5	5	$1.4 \cdot 10^{-2}$
5754.220	5	3	$1.23 \cdot 10^{-2}$
5772.145	3	1	$3.6 \cdot 10^{-2}$
5780.384	1	3	$9.8 \cdot 10^{-3}$
5793.071	3	5	$1.3 \cdot 10^{-2}$
5797.859	5	7	$2.53 \cdot 10^{-3}$
5948.545	3	5	$2.2 \cdot 10^{-2}$
6331.954	3	3	$1.0 \cdot 10^{-4}$
6555.462	7	9	$6.9 \cdot 10^{-3}$
6721.853	3	5	$3.4 \cdot 10^{-2}$
6976.523	3	5	$3.26 \cdot 10^{-2}$
7003.567	5	7	$3.42 \cdot 10^{-2}$
7005.883	7	9	$3.83 \cdot 10^{-2}$
7680.267	3	5	$4.62 \cdot 10^{-2}$
7918.386	3	5	$5.22 \cdot 10^{-2}$
7932.349	5	7	$5.13 \cdot 10^{-2}$
7944.001	7	9	$5.75 \cdot 10^{-2}$
7970.306	5	5	$7.11 \cdot 10^{-3}$
8035.619	7	7	$8.11 \cdot 10^{-3}$
8093.241	3	3	$1.5 \cdot 10^{-2}$
9413.506	3	1	$2.26 \cdot 10^{-1}$

Si II

λ Å	Weights g_i	g_k	A 10^8 s^{-1}
843.72	2	4	$7.05 \cdot 10^{-1}$
845.77	4	4	$1.42 \cdot 10^{-1}$
889.72	2	4	1.43
892.00	4	4	$2.88 \cdot 10^{-1}$
899.41	2	2	$4.63 \cdot 10^{-1}$
901.74	4	2	$9.23 \cdot 10^{-1}$
989.87	2	4	6.81
992.68	4	6	7.11
1020.70	2	2	$8.91 \cdot 10^{-1}$
1023.69	4	2	1.77
1190.42	2	4	6.53
1193.28	2	2	$2.69 \cdot 10^{1}$
1194.50	4	4	$3.45 \cdot 10^{1}$
1197.39	4	2	$1.40 \cdot 10^{1}$
1224.25	4	4	6.72
1224.97	4	6	4.39
1226.81	2	2	$2.77 \cdot 10^{1}$
1226.89	6	4	6.55
1226.99	2	4	$1.39 \cdot 10^{1}$
1227.60	6	6	$1.46 \cdot 10^{1}$
1228.44	4	2	5.53
1228.62	4	4	$1.77 \cdot 10^{1}$
1228.75	4	6	$2.32 \cdot 10^{1}$
1229.39	6	8	$2.25 \cdot 10^{1}$
1246.74	2	4	4.03
1248.43	4	4	8.30
1251.16	6	4	$1.30 \cdot 10^{1}$
1260.42	2	4	$2.57 \cdot 10^{1}$
1264.73	4	6	$3.04 \cdot 10^{1}$
1265.02	4	4	4.73
1304.37	2	2	3.64
1309.27	4	2	6.23
1348.54	2	4	3.36
1350.06	6	6	5.34
1350.52	4	4	1.61
1350.66	2	2	2.02
1352.64	4	2	6.12
1353.72	6	4	3.22
1410.22	6	4	3.47
1509.10	6	4	2.85
1512.07	4	2	3.15
1526.72	2	2	3.81
1533.45	4	2	7.52
1808.00	2	4	$2.54 \cdot 10^{-2}$
1816.92	4	6	$2.65 \cdot 10^{-2}$
1817.45	4	4	$3.23 \cdot 10^{-3}$
2072.02	4	6	$9.6 \cdot 10^{-1}$
2072.70	6	8	1.0
2334.40	2	2	$5.51 \cdot 10^{-5}$
2334.61	4	6	$2.44 \cdot 10^{-5}$
2344.20	4	4	$1.31 \cdot 10^{-5}$
2350.17	4	2	$4.70 \cdot 10^{-5}$
2682.21	4	4	$3.49 \cdot 10^{-2}$
2887.51	6	4	$6.39 \cdot 10^{-2}$
2904.28	4	6	$3.58 \cdot 10^{-1}$
2905.69	6	8	$3.83 \cdot 10^{-1}$
3203.87	2	4	$4.45 \cdot 10^{-1}$
3210.03	4	6	$5.29 \cdot 10^{-1}$
3333.14	2	2	$1.00 \cdot 10^{-1}$
3339.82	4	2	$2.00 \cdot 10^{-1}$
3853.66	4	4	$5.11 \cdot 10^{-2}$
3856.02	6	4	$4.40 \cdot 10^{-1}$
3862.60	4	2	$3.91 \cdot 10^{-1}$
4075.45	6	4	$4.00 \cdot 10^{-2}$
4076.78	4	2	$4.00 \cdot 10^{-2}$
4128.07	4	6	1.49
4130.89	6	8	1.74
4621.42	4	6	$1.28 \cdot 10^{-1}$
4621.72	6	8	$1.37 \cdot 10^{-1}$
5041.03	2	4	$7.00 \cdot 10^{-1}$
5055.98	4	6	1.45
5466.43	4	6	$2.16 \cdot 10^{-1}$
5466.87	6	6	$1.54 \cdot 10^{-2}$
5632.97	8	8	$4.00 \cdot 10^{-2}$
5660.66	6	6	$7.00 \cdot 10^{-2}$
5669.56	10	8	$5.00 \cdot 10^{-1}$
5681.44	4	4	$1.00 \cdot 10^{-1}$
5688.81	8	6	$4.60 \cdot 10^{-1}$
5701.37	6	4	$4.50 \cdot 10^{-1}$
5706.37	4	2	$6.10 \cdot 10^{-1}$
5957.56	2	2	$5.60 \cdot 10^{-1}$
5978.93	4	2	1.13
6347.10	2	4	$5.84 \cdot 10^{-1}$
6371.36	2	2	$6.80 \cdot 10^{-1}$
6660.52	4	6	$3.64 \cdot 10^{-1}$
6665.00	2	4	$2.16 \cdot 10^{-1}$
6671.88	6	8	$4.80 \cdot 10^{-1}$
6699.38	2	2	$4.20 \cdot 10^{-1}$
6750.28	6	6	$1.49 \cdot 10^{-1}$
6818.45	2	4	$1.08 \cdot 10^{-1}$
6829.82	4	4	$2.16 \cdot 10^{-2}$
7848.80	4	6	$3.73 \cdot 10^{-1}$
7849.72	6	8	$3.99 \cdot 10^{-1}$
9412.72	8	8	$4.65 \cdot 10^{-2}$

Si III

λ Å	Weights g_i	g_k	A 10^8 s^{-1}
566.61	1	3	1.21
652.22	3	3	2.67
653.33	5	3	4.40
673.48	5	5	$8.44 \cdot 10^{-2}$
673.48	5	3	$9.33 \cdot 10^{-3}$
800.07	3	1	4.13
823.41	3	5	5.85
883.40	5	7	$5.79 \cdot 10^{2}$
939.09	5	3	$1.11 \cdot 10^{1}$
967.95	5	7	4.68
993.52	1	3	2.68
994.79	3	3	8.05
997.39	5	3	$1.35 \cdot 10^{1}$
1005.37	5	5	$7.06 \cdot 10^{-1}$
1031.16	3	5	2.72
1033.92	5	5	8.11
1037.05	5	3	4.46
1083.22	5	3	3.65
1108.37	1	3	$1.54 \cdot 10^{1}$
1109.97	3	5	$2.07 \cdot 10^{1}$
1113.23	5	7	$2.74 \cdot 10^{1}$
1140.55	1	3	$1.96 \cdot 10^{1}$
1141.58	3	5	$2.67 \cdot 10^{1}$
1142.28	3	3	$1.75 \cdot 10^{1}$
1144.31	5	7	$3.86 \cdot 10^{1}$
1144.96	5	5	$1.19 \cdot 10^{1}$
1145.18	5	5	1.38
1155.00	1	3	9.34
1155.96	3	1	$2.22 \cdot 10^{1}$
1158.10	3	5	7.88
1160.26	5	3	8.65
1161.58	5	5	$1.42 \cdot 10^{1}$
1174.37	3	3	1.00
1174.43	3	5	1.00
1206.51	1	3	$2.55 \cdot 10^{1}$
1206.53	3	5	$4.17 \cdot 10^{1}$
1207.52	5	5	$2.43 \cdot 10^{1}$
1210.46	5	7	$1.60 \cdot 10^{1}$
1235.43	1	3	$2.77 \cdot 10^{1}$
1280.35	5	7	$1.04 \cdot 10^{1}$
1294.54	3	5	5.35
1296.73	1	3	7.10
1298.89	3	3	5.29
1298.96	5	5	$1.59 \cdot 10^{1}$

λ Å	Weights g_i	g_k	A 10^8 s^{-1}
1301.15	3	1	$2.11\cdot10^{1}$
1303.32	5	3	8.71
1312.59	3	1	6.66
1341.47	7	7	8.96
1342.39	5	5	7.54
1343.39	3	3	8.16
1361.60	3	5	$1.06\cdot10^{1}$
1362.37	3	1	$1.06\cdot10^{1}$
1363.47	5	3	8.53
1365.26	7	5	9.40
1367.05	3	3	$1.05\cdot10^{1}$
1369.44	3	1	$1.04\cdot10^{1}$
1373.03	5	5	8.00
1387.99	3	3	$2.09\cdot10^{-1}$
1417.24	3	1	$2.17\cdot10^{1}$
1433.69	5	7	8.67
1435.77	5	7	4.22
1436.17	1	3	7.79
1457.25	5	3	$1.07\cdot10^{1}$
1500.24	7	9	$2.10\cdot10^{1}$
1501.19	5	7	$1.86\cdot10^{1}$
1501.87	3	5	$1.76\cdot10^{1}$
1506.06	3	3	$1.24\cdot10^{1}$
1673.32	5	7	6.70
1842.55	5	3	2.99
2176.89	9	7	1.80
2295.48	5	7	$6.69\cdot10^{-1}$
2300.93	7	7	$5.80\cdot10^{-2}$
2308.19	9	11	$7.16\cdot10^{-1}$
2449.48	5	5	$1.81\cdot10^{-1}$
2483.20	7	9	$1.33\cdot10^{-1}$
2541.82	3	5	$3.22\cdot10^{-1}$
2546.09	5	5	$6.1\cdot10^{-1}$
2559.21	5	7	1.63
2640.79	5	3	1.83
2655.51	7	9	1.35
2817.11	9	7	$2.22\cdot10^{-1}$
2831.49	7	5	$2.12\cdot10^{-1}$
2839.62	5	3	$2.37\cdot10^{-1}$
2959.15	5	3	$1.48\cdot10^{-1}$
2980.52	5	3	$2.37\cdot10^{-3}$
3013.09	5	5	$5.67\cdot10^{-2}$
3034.73	5	7	$4.50\cdot10^{-1}$
3037.29	7	7	$3.93\cdot10^{-2}$
3040.93	9	11	$4.87\cdot10^{-1}$
3043.93	7	5	$2.51\cdot10^{-1}$
3045.08	5	3	$2.24\cdot10^{-1}$
3068.24	5	3	$5.15\cdot10^{-1}$
3086.24	7	5	1.46
3086.46	5	5	$2.61\cdot10^{-1}$
3093.42	5	3	1.30
3093.65	3	3	$4.33\cdot10^{-1}$
3096.83	3	1	1.73
3126.27	5	5	$4.67\cdot10^{-1}$
3147.37	5	3	$2.53\cdot10^{-1}$
3161.61	5	3	$4.91\cdot10^{-1}$
3185.13	3	1	4.04
3186.02	5	7	1.13
3196.50	7	9	1.14
3210.55	9	11	1.20
3216.25	5	7	$1.90\cdot10^{-1}$
3230.50	1	3	$4.75\cdot10^{-1}$
3233.95	3	3	1.42
3241.62	5	3	2.35
3253.40	5	5	$1.82\cdot10^{-1}$
3253.74	7	5	$1.75\cdot10^{-1}$
3254.80	9	7	$1.81\cdot10^{-1}$
3258.66	7	5	1.01
3270.46	3	3	$2.99\cdot10^{-1}$
3276.26	5	3	$8.91\cdot10^{-1}$
3279.26	3	1	1.18
3486.91	5	7	1.54
3525.94	3	3	$2.48\cdot10^{-1}$
3569.67	3	1	$5.78\cdot10^{-1}$
3590.47	3	5	2.53
3622.54	7	7	$1.25\cdot10^{-1}$
3639.45	7	9	$2.24\cdot10^{-1}$
3645.12	5	5	$9.58\cdot10^{-2}$
3681.40	5	3	$2.51\cdot10^{-1}$
3682.15	3	3	$1.50\cdot10^{-1}$
3791.41	1	3	1.76
3796.11	3	5	2.36
3806.54	5	7	3.14
3842.46	3	5	$1.73\cdot10^{-1}$
3924.47	7	9	3.47
3947.49	5	7	$5.59\cdot10^{-2}$
3963.84	7	5	$3.63\cdot10^{-2}$
3981.24	3	3	$2.09\cdot10^{-3}$
4102.42	5	3	$2.46\cdot10^{-1}$
4115.50	7	9	$4.07\cdot10^{-1}$
4338.50	1	3	$1.47\cdot10^{-1}$
4341.40	3	1	$5.25\cdot10^{-3}$
4377.63	7	5	$4.14\cdot10^{-2}$
4405.90	5	5	$4.84\cdot10^{-2}$
4406.72	7	9	$3.10\cdot10^{-1}$
4494.05	3	3	$4.19\cdot10^{-1}$
4552.62	3	5	1.26
4554.00	5	3	$6.75\cdot10^{-1}$
4567.82	3	3	1.25
4574.76	3	1	1.24
4619.66	3	5	$2.89\cdot10^{-1}$
4638.28	1	3	$3.79\cdot10^{-1}$
4665.87	3	3	$2.79\cdot10^{-1}$
4683.02	5	5	$8.30\cdot10^{-1}$
4683.80	3	1	1.10
4716.65	5	7	1.32
4730.52	5	3	$4.47\cdot10^{-1}$
4800.43	1	3	$2.76\cdot10^{-1}$
4813.33	5	7	2.10
4819.72	7	9	2.12
4828.97	9	11	2.25
5113.76	5	7	$3.82\cdot10^{-1}$
5114.12	9	11	$4.17\cdot10^{-1}$
5197.26	3	5	$2.47\cdot10^{-1}$
5451.46	3	5	$4.60\cdot10^{-1}$
5473.05	5	7	$6.07\cdot10^{-1}$
5704.60	7	5	$1.86\cdot10^{-1}$
5716.29	9	7	$1.91\cdot10^{-1}$
5739.73	1	3	$5.41\cdot10^{-1}$
5898.79	7	9	$4.86\cdot10^{-1}$
6314.46	3	1	1.18
6524.36	7	9	$3.82\cdot10^{-1}$
6831.56	5	3	$5.83\cdot10^{-1}$
7461.89	3	1	$4.99\cdot10^{-1}$
7462.62	5	3	$3.75\cdot10^{-1}$
7466.32	7	5	$4.19\cdot10^{-1}$
7612.36	3	5	$9.94\cdot10^{-1}$
8190.43	5	7	$8.09\cdot10^{-1}$
8191.16	7	7	$7.05\cdot10^{-2}$
8191.68	9	11	$8.78\cdot10^{-1}$
8262.57	5	7	$7.18\cdot10^{-1}$
8265.64	5	5	$1.80\cdot10^{-1}$
8269.32	3	5	$5.36\cdot10^{-1}$
8271.38	3	3	$2.98\cdot10^{-1}$
8271.94	1	3	$3.96\cdot10^{-1}$
Si IV			
457.82	2	4	3.90
458.16	2	2	4.05
515.12	2	2	4.66
516.35	4	2	9.32
645.76	6	6	$4.61\cdot10^{-1}$
645.76	6	8	6.92
645.76	4	6	6.46
749.94	6	6	$9.33\cdot10^{-1}$
749.94	6	8	$1.40\cdot10^{1}$
815.05	2	2	$1.18\cdot10^{1}$
818.13	4	2	$2.37\cdot10^{1}$
1066.63	6	6	2.54
1122.49	2	4	$2.14\cdot10^{1}$
1128.34	4	6	$2.53\cdot10^{1}$
1393.76	2	4	8.80
1402.77	2	2	8.63
1722.53	6	4	4.92
1727.38	4	2	5.47
2120.18	2	2	3.06
2127.47	4	2	6.14
2287.04	6	6	$4.27\cdot10^{-1}$
2287.04	6	8	6.41
2287.04	4	6	5.98
2366.76	2	2	$5.18\cdot10^{-1}$
2370.99	4	2	1.04
2482.82	2	4	$6.62\cdot10^{-2}$
2485.38	2	2	$7.06\cdot10^{-2}$
2672.19	2	4	$2.49\cdot10^{-2}$
2675.12	6	4	$2.74\cdot10^{-1}$
2675.25	8	6	$2.61\cdot10^{-1}$
2677.57	4	4	$4.51\cdot10^{-3}$
2723.81	6	8	1.10
2723.81	6	6	$7.34\cdot10^{-2}$
2971.52	6	6	$4.62\cdot10^{-3}$
2971.52	6	4	$9.70\cdot10^{-2}$
2971.52	8	6	$9.24\cdot10^{-2}$
3149.56	2	4	4.02
3165.71	4	6	4.77
3244.19	4	2	$4.21\cdot10^{-1}$
3762.44	4	4	$2.33\cdot10^{-1}$
3773.15	4	2	2.33
4031.39	2	2	$2.29\cdot10^{-1}$
4038.06	4	2	$4.59\cdot10^{-1}$

λ Å	Weights g_i	g_k	A 10^8 s^{-1}
4088.85	2	4	1.56
4116.10	2	2	1.53
4212.41	6	8	1.63
4212.41	6	6	$1.09{\cdot}10^{-1}$
4314.10	2	2	1.06
4328.18	4	2	2.12
4403.73	6	8	$4.09{\cdot}10^{-1}$
4403.73	6	6	$2.73{\cdot}10^{-2}$
4403.73	4	6	$3.82{\cdot}10^{-1}$
4411.65	2	4	$1.74{\cdot}10^{-2}$
4611.27	4	6	$2.23{\cdot}10^{-2}$
4611.27	4	4	$3.71{\cdot}10^{-3}$
4950.11	6	6	$9.46{\cdot}10^{-3}$
4950.11	8	6	$1.89{\cdot}10^{-1}$
4950.11	6	4	$1.99{\cdot}10^{-1}$
5304.97	6	4	$1.90{\cdot}10^{-1}$
5304.97	4	4	$2.11{\cdot}10^{-2}$
5309.49	4	2	$2.10{\cdot}10^{-1}$
6667.56	2	4	1.15
6701.21	4	6	1.36
6998.36	6	6	$3.65{\cdot}10^{-2}$
6998.36	6	8	$5.48{\cdot}10^{-1}$
6998.36	4	6	$5.11{\cdot}10^{-1}$
7047.94	6	4	$9.05{\cdot}10^{-1}$
7068.41	4	2	1.00
7630.50	2	2	$4.40{\cdot}10^{-1}$
7654.56	4	2	$8.82{\cdot}10^{-1}$
8240.61	6	6	$5.85{\cdot}10^{-3}$
8240.61	6	4	$1.23{\cdot}10^{-1}$
8240.61	8	6	$1.17{\cdot}10^{-1}$
8957.25	2	4	$4.26{\cdot}10^{-1}$
9018.16	2	2	$4.17{\cdot}10^{-1}$

Si V

λ Å	g_i	g_k	A 10^8 s^{-1}
96.44	1	3	$2.36{\cdot}10^{3}$
97.14	1	3	$2.42{\cdot}10^{2}$
98.21	1	3	6.88
117.86	1	3	$3.57{\cdot}10^{2}$
118.97	1	3	$3.84{\cdot}10^{1}$

Silver

Ag I

λ Å	g_i	g_k	A 10^8 s^{-1}
2061.2	2	4	$3.1{\cdot}10^{-2}$
2069.9	2	2	$1.5{\cdot}10^{-2}$
3280.7	2	4	1.4
3382.9	2	2	1.3
5209.1	2	4	$7.5{\cdot}10^{-1}$
5465.5	4	6	$8.6{\cdot}10^{-1}$
5471.6	4	4	$1.4{\cdot}10^{-1}$

Sodium

Na I

λ Å	g_i	g_k	A 10^8 s^{-1}
2543.84	2	4	$4.46{\cdot}10^{-4}$
2543.87	2	2	$4.35{\cdot}10^{-4}$
2593.87	2	4	$8.13{\cdot}10^{-4}$
2593.92	2	2	$7.96{\cdot}10^{-4}$
2680.34	2	4	$1.84{\cdot}10^{-3}$
2680.43	2	2	$1.81{\cdot}10^{-3}$
2852.81	2	4	$5.38{\cdot}10^{-3}$
2853.01	2	2	$5.31{\cdot}10^{-3}$
3302.37	2	4	$2.75{\cdot}10^{-2}$
3302.98	2	2	$2.73{\cdot}10^{-2}$
4238.99	2	4	$2.90{\cdot}10^{-3}$
4242.08	4	4	$5.8{\cdot}10^{-4}$
4242.08	4	6	$3.46{\cdot}10^{-3}$
4249.41	2	2	$8.7{\cdot}10^{-4}$
4252.52	4	2	$1.73{\cdot}10^{-3}$
4273.64	2	4	$3.91{\cdot}10^{-3}$
4276.79	4	4	$7.8{\cdot}10^{-4}$
4276.79	4	6	$4.69{\cdot}10^{-3}$
4287.84	2	2	$1.19{\cdot}10^{-3}$
4291.01	4	2	$2.38{\cdot}10^{-3}$
4321.40	2	4	$5.5{\cdot}10^{-3}$
4324.62	4	4	$1.09{\cdot}10^{-3}$
4324.62	4	6	$6.6{\cdot}10^{-3}$
4341.49	2	2	$3.26{\cdot}10^{-3}$
4344.74	4	2	$6.50{\cdot}10^{-3}$
4390.03	2	4	$9.83{\cdot}10^{-3}$
4393.34	4	4	$1.95{\cdot}10^{-3}$
4393.34	4	6	$1.17{\cdot}10^{-2}$
4419.88	2	2	$2.82{\cdot}10^{-3}$
4423.25	4	2	$5.61{\cdot}10^{-3}$
4494.18	2	4	$1.23{\cdot}10^{-2}$
4497.66	4	4	$2.44{\cdot}10^{-3}$
4497.66	4	6	$1.46{\cdot}10^{-2}$
4541.63	2	2	$3.76{\cdot}10^{-3}$
4545.19	4	2	$7.50{\cdot}10^{-3}$
4664.811	2	4	$2.08{\cdot}10^{-2}$
4668.560	4	4	$4.14{\cdot}10^{-3}$
4668.560	4	6	$2.49{\cdot}10^{-2}$
4747.941	2	2	$6.19{\cdot}10^{-3}$
4751.822	4	2	$1.23{\cdot}10^{-2}$
4978.541	2	4	$4.09{\cdot}10^{-2}$
4982.813	4	6	$4.88{\cdot}10^{-2}$
5148.838	2	2	$1.14{\cdot}10^{-2}$
5153.402	4	2	$2.27{\cdot}10^{-2}$
5682.633	2	4	$1.01{\cdot}10^{-1}$
5688.193	4	4	$2.02{\cdot}10^{-2}$
5688.205	4	6	$1.21{\cdot}10^{-1}$
5889.950	2	4	$6.16{\cdot}10^{-1}$
5895.924	2	2	$6.14{\cdot}10^{-1}$
6154.225	2	2	$2.50{\cdot}10^{-2}$
6160.747	4	2	$4.98{\cdot}10^{-2}$
7373.23	2	4	$5.42{\cdot}10^{-4}$
7373.49	2	2	$5.30{\cdot}10^{-4}$
7809.78	2	4	$9.91{\cdot}10^{-4}$
7810.24	2	2	$9.72{\cdot}10^{-4}$
8183.256	2	4	$4.29{\cdot}10^{-1}$
8194.790	4	4	$8.57{\cdot}10^{-2}$
8194.824	4	6	$5.14{\cdot}10^{-1}$
8649.92	2	4	$2.25{\cdot}10^{-3}$
8650.89	2	2	$2.21{\cdot}10^{-3}$
8942.96	6	6	$2.47{\cdot}10^{-4}$
8942.96	6	8	$3.71{\cdot}10^{-3}$
9153.88	6	6	$3.5{\cdot}10^{-4}$
9153.88	6	8	$5.3{\cdot}10^{-3}$
9465.94	6	6	$6.38{\cdot}10^{-4}$
9465.94	6	8	$9.57{\cdot}10^{-3}$
9961.28	6	6	$8.45{\cdot}10^{-4}$
9961.28	6	8	$1.27{\cdot}10^{-2}$

Na II

λ Å	g_i	g_k	A 10^8 s^{-1}
300.15	1	3	$1.18{\cdot}10^{1}$
300.20	1	3	$1.17{\cdot}10^{1}$
301.44	1	3	$3.33{\cdot}10^{1}$
302.45	1	3	1.4
372.08	1	3	$3.13{\cdot}10^{1}$
376.38	1	3	1.70
2315.65	3	1	$1.05{\cdot}10^{-1}$
2493.15	3	1	4.15
2506.30	3	3	$2.73{\cdot}10^{-1}$
2515.46	3	5	$2.25{\cdot}10^{-1}$
2531.54	3	1	$8.44{\cdot}10^{-1}$
2594.96	3	3	$3.71{\cdot}10^{-1}$
2661.00	3	5	1.65
2671.83	3	3	2.64
2678.09	3	1	3.01
2808.71	7	7	$1.93{\cdot}10^{-2}$
2829.87	5	5	$3.36{\cdot}10^{-1}$
2839.56	5	7	1.11
2872.95	3	5	$2.63{\cdot}10^{-1}$
2881.15	3	1	2.50
2886.26	3	5	1.07
2893.95	3	1	1.48
2901.14	7	7	$2.89{\cdot}10^{-1}$
2904.72	7	5	$1.04{\cdot}10^{-1}$
2919.05	7	5	$7.94{\cdot}10^{-2}$
2920.95	1	3	$6.66{\cdot}10^{-1}$
2930.88	5	3	$9.83{\cdot}10^{-1}$
2934.08	5	7	3.17
2937.74	5	5	1.28
2945.70	7	7	$2.50{\cdot}10^{-2}$
2951.24	7	9	4.33
2952.40	5	5	1.29
2960.12	5	5	$5.60{\cdot}10^{-1}$
2970.73	5	7	$1.03{\cdot}10^{-1}$
2974.24	5	5	$2.02{\cdot}10^{-1}$
2974.99	1	3	$6.47{\cdot}10^{-1}$
2977.13	3	3	$1.05{\cdot}10^{-1}$
2979.66	5	7	1.96
2980.63	7	5	$3.30{\cdot}10^{-3}$
3004.15	3	3	$9.44{\cdot}10^{-2}$
3009.14	3	3	$1.73{\cdot}10^{-1}$
3015.40	5	5	$1.51{\cdot}10^{-1}$
3053.67	5	7	2.99
3055.35	1	3	$8.80{\cdot}10^{-2}$
3057.38	5	5	$4.51{\cdot}10^{-1}$
3058.72	5	3	$2.18{\cdot}10^{-1}$
3064.38	3	5	$7.62{\cdot}10^{-2}$
3066.54	3	3	$3.74{\cdot}10^{-1}$
3080.25	3	5	2.81
3087.06	3	1	$1.66{\cdot}10^{-1}$
3094.45	5	5	$1.91{\cdot}10^{-3}$
3095.55	3	5	$3.97{\cdot}10^{-3}$
3104.40	3	1	$5.63{\cdot}10^{-1}$
3124.42	5	7	2.56
3125.21	3	3	$9.72{\cdot}10^{-2}$
3135.48	1	3	$7.42{\cdot}10^{-1}$
3137.86	3	3	1.41

λ Å	Weights g_i	g_k	A 10^8 s^{-1}
3145.71	3	5	$7.23 \cdot 10^{-3}$
3163.74	5	5	1.05
3179.06	5	3	$3.94 \cdot 10^{-1}$
3234.93	3	5	$1.83 \cdot 10^{-1}$
3257.96	5	5	1.03
3260.21	3	1	$4.65 \cdot 10^{-1}$
3274.22	5	3	$4.29 \cdot 10^{-1}$
3301.35	3	5	$4.54 \cdot 10^{-2}$
3304.96	1	3	$3.60 \cdot 10^{-1}$
3318.04	3	3	$4.14 \cdot 10^{-1}$
3327.69	3	1	$9.45 \cdot 10^{-1}$
3711.07	1	3	$1.02 \cdot 10^{-1}$
4123.08	1	3	$3.71 \cdot 10^{-1}$
4344.11	1	3	$4.32 \cdot 10^{-1}$
4368.60	1	3	$3.59 \cdot 10^{-1}$

Na III

λ Å	g_i	g_k	A 10^8 s^{-1}
202.15	4	4	$1.40 \cdot 10^{1}$
202.49	4	2	$1.59 \cdot 10^{2}$
202.71	2	4	$6.40 \cdot 10^{1}$
202.72	4	2	$6.72 \cdot 10^{1}$
202.76	4	4	$1.26 \cdot 10^{2}$
203.28	2	2	$9.24 \cdot 10^{1}$
215.86	2	4	$1.72 \cdot 10^{1}$
216.12	2	4	$1.57 \cdot 10^{1}$
229.87	4	2	$2.91 \cdot 10^{1}$
267.87	2	2	$7.56 \cdot 10^{1}$
378.14	4	2	$8.42 \cdot 10^{1}$
380.10	2	2	$4.13 \cdot 10^{1}$
1336.76	4	2	4.27
1337.36	4	4	1.61
1340.67	6	4	2.29
1342.39	2	2	$8.44 \cdot 10^{-1}$
1342.73	4	2	1.06
1355.28	6	6	3.44
1361.90	2	2	2.24
1372.34	2	4	$5.81 \cdot 10^{-1}$
1420.89	8	6	5.73
1444.19	6	4	4.07
1449.31	8	6	5.12
1562.87	4	4	2.98
1565.29	4	4	$3.95 \cdot 10^{-1}$
1598.18	4	6	$7.96 \cdot 10^{-1}$
1711.12	6	4	1.52
1728.27	4	2	4.10
1731.11	2	4	1.87
1835.22	6	6	3.31
1838.94	8	8	3.31
1844.36	6	6	3.00
1849.56	6	8	6.87
1850.38	4	4	4.37
1855.92	2	2	5.77
1856.71	4	6	4.22
1861.21	2	4	2.34
1887.47	10	12	$1.25 \cdot 10^{1}$
1918.45	4	4	2.29
1926.26	8	10	8.94
1927.24	2	4	5.66
1932.74	4	6	5.37
1933.89	6	8	6.33

λ Å	Weights g_i	g_k	A 10^8 s^{-1}
1946.43	6	8	8.28
1950.91	8	10	8.55
1951.24	6	4	2.38
1977.16	4	6	8.46
1985.57	4	4	1.78
1995.68	6	6	$8.12 \cdot 10^{-1}$
2004.21	2	4	2.07
2005.22	2	4	$9.67 \cdot 10^{-1}$
2008.47	4	6	1.45
2011.87	6	8	6.37
2014.17	4	6	4.88
2017.03	6	8	5.44
2028.56	8	8	1.93
2031.13	4	6	3.08
2035.90	2	2	6.23
2041.66	6	8	1.64
2043.29	4	2	4.02
2044.82	4	4	4.00
2045.44	6	6	1.05
2051.48	6	6	4.36
2060.36	4	6	$8.12 \cdot 10^{-1}$
2066.60	4	4	1.53
2140.72	6	6	3.80
2144.54	4	4	3.88
2202.83	6	6	$8.53 \cdot 10^{-1}$
2225.93	4	4	1.82
2230.33	6	8	3.64
2232.19	4	4	2.34
2246.70	4	6	2.72
2251.47	2	4	1.66
2278.42	2	2	2.98
2285.66	2	4	$7.73 \cdot 10^{-1}$
2309.99	4	2	3.08
2386.99	6	8	3.05
2394.03	4	6	2.80
2406.59	4	4	$5.1 \cdot 10^{-1}$
2459.31	4	6	3.0
2468.85	2	4	2.4
2474.73	6	4	1.38
2497.03	6	6	1.99
2510.26	4	2	2.19
2530.25	4	4	$3.65 \cdot 10^{-1}$
2542.80	2	2	$3.39 \cdot 10^{-1}$

Na IV

λ Å	g_i	g_k	A 10^8 s^{-1}
136.551	5	5	$2.08 \cdot 10^{2}$
136.854	5	7	$1.56 \cdot 10^{2}$
142.232	5	7	$3.06 \cdot 10^{2}$
142.359	5	5	$2.77 \cdot 10^{2}$
146.064	5	5	$6.69 \cdot 10$
146.302	3	5	$2.00 \cdot 10^{2}$
150.298	5	7	$2.38 \cdot 10^{2}$
150.543	3	5	$1.71 \cdot 10^{2}$
150.687	5	5	$9.87 \cdot 10$
151.299	5	5	$6.70 \cdot 10$
155.083	5	3	$1.33 \cdot 10^{2}$
155.240	5	3	$1.88 \cdot 10^{2}$
156.537	5	7	$2.92 \cdot 10^{2}$
162.448	5	7	$6.04 \cdot 10^{2}$
163.190	5	5	$4.32 \cdot 10^{2}$
168.411	3	3	$1.37 \cdot 10^{2}$
168.546	1	3	$1.81 \cdot 10^{2}$
319.644	5	3	$2.52 \cdot 10^{2}$
360.76	1	3	$1.38 \cdot 10^{1}$
408.684	5	3	$2.66 \cdot 10^{1}$
409.614	3	1	$6.34 \cdot 10^{1}$
410.372	5	5	$4.73 \cdot 10^{1}$
411.334	1	3	$2.09 \cdot 10^{1}$
412.242	3	5	$1.55 \cdot 10^{1}$
1580.50	7	9	6.27
1582.18	3	3	7.16
1582.33	3	5	3.34
1583.98	5	5	5.55
1584.14	5	7	6.34
1587.05	7	9	9.46
1613.95	5	7	8.74
1615.92	7	9	8.89
1618.57	9	11	9.23
1655.47	7	9	8.72
1701.97	5	7	7.93
1702.41	3	5	5.95
1960.76	5	7	3.71
1965.08	5	5	3.68
1967.60	5	3	3.67
2018.39	3	5	3.44

Strontium

Sr I

λ Å	g_i	g_k	A 10^8 s^{-1}
2206.2	1	3	$6.6 \cdot 10^{-3}$
2211.3	1	3	$8.5 \cdot 10^{-3}$
2217.8	1	3	$1.2 \cdot 10^{-2}$
2226.3	1	3	$1.6 \cdot 10^{-2}$
2237.7	1	3	$2.3 \cdot 10^{-2}$
2253.3	1	3	$3.7 \cdot 10^{-2}$
2275.3	1	3	$6.7 \cdot 10^{-2}$
2307.3	1	3	$1.2 \cdot 10^{-1}$
2354.3	1	3	$1.8 \cdot 10^{-1}$
2428.1	1	3	$1.7 \cdot 10^{-1}$
2569.5	1	3	$5.3 \cdot 10^{-2}$
2931.8	1	3	$1.9 \cdot 10^{-2}$
4607.3	1	3	2.01

Sr II

λ Å	g_i	g_k	A 10^8 s^{-1}
2018.7	2	2	$1.2 \cdot 10^{-1}$
2051.9	4	2	$2.4 \cdot 10^{-1}$
2282.0	2	4	$8.3 \cdot 10^{-1}$
2322.4	4	6	$9.1 \cdot 10^{-1}$
2324.5	4	4	$1.5 \cdot 10^{-1}$
2423.5	2	2	$2.4 \cdot 10^{-1}$
2471.6	4	2	$4.8 \cdot 10^{-1}$
3464.5	4	6	3.1
3474.9	4	4	$5.1 \cdot 10^{-1}$
4077.7	2	4	1.42
4161.8	2	2	$6.5 \cdot 10^{-1}$
4215.5	2	2	1.27
4305.5	4	2	1.4
4414.8	4	6	$1.1 \cdot 10^{-1}$
4417.5	4	4	$1.8 \cdot 10^{-2}$
4585.9	4	2	$7.0 \cdot 10^{-2}$

λ Å	Weights g_i	g_k	A 10^8 s^{-1}
5303.1	2	4	$1.9\cdot10^{-1}$
5379.1	4	6	$2.2\cdot10^{-1}$
5385.5	4	4	$3.7\cdot10^{-2}$
5723.7	2	2	$7.1\cdot10^{-2}$
5819.0	4	2	$1.4\cdot10^{-1}$
8688.9	4	6	$5.5\cdot10^{-1}$
8719.6	4	4	$9.7\cdot10^{-2}$

Sulfur

S I

λ Å	Weights g_i	g_k	A 10^8 s^{-1}
1295.7	5	5	4.9
1296.2	5	3	2.7
1302.3	3	5	1.8
1302.9	3	3	1.6
1303.1	3	1	6.6
1303.4	5	3	1.9
1305.9	1	3	2.4
1401.5	5	3	$9.1\cdot10^{-1}$
1409.3	3	3	$5.0\cdot10^{-1}$
1412.9	1	3	$1.6\cdot10^{-1}$
1425.0	5	7	4.5
1425.2	5	5	1.2
1433.3	3	5	3.3
1433.3	3	3	1.9
1437.0	1	3	2.4
1448.2	5	3	7.3
1473.0	5	7	$4.2\cdot10^{-1}$
1474.0	5	7	1.6
1474.4	5	5	$5.0\cdot10^{-1}$
1474.6	5	3	$6.2\cdot10^{-2}$
1481.7	3	5	$1.7\cdot10^{-1}$
1483.0	3	5	1.2
1483.2	3	3	$7.5\cdot10^{-1}$
1487.2	1	3	$8.7\cdot10^{-1}$
1666.7	5	5	6.3
1687.5	1	3	$9.4\cdot10^{-1}$
1782.3	1	3	1.9
1807.3	5	3	3.8
1820.3	3	3	2.2
1826.2	1	3	$7.2\cdot10^{-1}$
4694.1	5	7	$6.7\cdot10^{-3}$
4695.4	5	5	$6.7\cdot10^{-3}$
4696.2	5	3	$6.5\cdot10^{-3}$
6403.6	3	5	$5.7\cdot10^{-3}$
6408.1	5	5	$9.5\cdot10^{-3}$
6415.5	7	5	$1.3\cdot10^{-2}$
*6751.2	15	25	$7.9\cdot10^{-2}$
7679.6	3	5	$1.2\cdot10^{-2}$
7686.1	5	5	$2.0\cdot10^{-2}$
7696.7	7	5	$2.8\cdot10^{-2}$

S II

λ Å	Weights g_i	g_k	A 10^8 s^{-1}
1124.4	2	4	1.0
1125.0	4	4	4.6
1131.0	2	2	3.5
1131.6	4	2	1.4
1250.5	4	2	$4.6\cdot10^{-1}$
1253.8	4	4	$4.2\cdot10^{-1}$
1259.5	4	6	$3.4\cdot10^{-1}$
4463.6	8	6	$5.3\cdot10^{-1}$
4483.4	6	4	$3.1\cdot10^{-1}$
4486.7	4	2	$6.6\cdot10^{-1}$
4524.7	4	4	$9.3\cdot10^{-2}$
4525.0	6	4	1.2
4552.4	4	2	1.2
4656.7	2	4	$9.0\cdot10^{-2}$
4716.2	4	4	$2.9\cdot10^{-1}$
4815.5	6	4	$8.8\cdot10^{-1}$
4885.6	2	4	$1.7\cdot10^{-1}$
4917.2	2	2	$6.6\cdot10^{-1}$
4924.1	4	6	$2.2\cdot10^{-1}$
4925.3	2	4	$2.4\cdot10^{-1}$
4942.5	2	2	$1.5\cdot10^{-1}$
4991.9	4	4	$1.5\cdot10^{-1}$
5009.5	4	2	$7.0\cdot10^{-1}$
5014.0	4	4	$8.4\cdot10^{-1}$
5027.2	4	2	$2.6\cdot10^{-1}$
5032.4	6	6	$8.1\cdot10^{-1}$
5047.3	4	2	$3.6\cdot10^{-1}$
5103.3	6	4	$5.0\cdot10^{-1}$
5142.3	2	2	$1.9\cdot10^{-1}$
5201.0	4	4	$7.5\cdot10^{-1}$
5201.3	6	4	$6.5\cdot10^{-2}$
5212.6	4	6	$9.8\cdot10^{-2}$
5212.6	6	6	$8.5\cdot10^{-1}$
5320.7	6	8	$9.2\cdot10^{-1}$
5345.7	4	6	$8.8\cdot10^{-1}$
5345.7	6	6	$1.1\cdot10^{-1}$
5428.6	2	4	$4.2\cdot10^{-1}$
5432.8	4	6	$6.8\cdot10^{-1}$
5453.8	6	8	$8.5\cdot10^{-1}$
5473.6	2	2	$7.3\cdot10^{-1}$
5509.7	4	4	$4.0\cdot10^{-1}$
5526.2	8	8	$8.1\cdot10^{-2}$
5536.8	4	6	$6.6\cdot10^{-2}$
5556.0	4	2	$1.1\cdot10^{-1}$
5564.9	6	6	$1.7\cdot10^{-1}$
5578.8	6	6	$1.1\cdot10^{-1}$
5606.1	10	8	$5.4\cdot10^{-1}$
5616.6	4	4	$1.2\cdot10^{-1}$
5640.0	4	6	$6.6\cdot10^{-1}$
5645.6	6	4	$1.8\cdot10^{-2}$
5647.0	2	4	$5.7\cdot10^{-1}$
5659.9	6	4	$4.6\cdot10^{-1}$
5664.7	4	2	$5.8\cdot10^{-1}$
5819.2	4	4	$8.5\cdot10^{-2}$
6305.5	8	6	$1.8\cdot10^{-1}$
6312.7	6	4	$3.0\cdot10^{-1}$

S III

λ Å	Weights g_i	g_k	A 10^8 s^{-1}
2496.2	7	5	2.5
2508.2	5	3	2.3
2636.9	3	5	$4.5\cdot10^{-1}$
2665.4	5	5	1.4
2680.5	1	3	$6.2\cdot10^{-1}$
2691.8	3	3	$4.6\cdot10^{-1}$
2702.8	3	1	1.9
2718.9	3	3	1.2
2721.4	5	3	$7.7\cdot10^{-1}$
2726.8	3	5	$6.0\cdot10^{-1}$
2731.1	5	5	1.1
2756.9	7	7	1.4
2785.5	3	3	$6.1\cdot10^{-1}$
2856.0	5	7	5.1
2863.5	7	9	5.7
2872.0	3	5	4.7
2950.2	3	5	3.0
2964.8	5	7	4.0
3662.0	3	3	$6.4\cdot10^{-1}$
3717.8	5	3	1.0
3778.9	3	5	$4.4\cdot10^{-1}$
3831.8	1	3	$5.6\cdot10^{-1}$
3837.8	3	3	$4.2\cdot10^{-1}$
3838.3	5	5	1.3
3860.6	3	1	1.6
3899.1	5	3	$6.7\cdot10^{-1}$
4253.6	5	7	1.2
4285.0	3	5	$9.0\cdot10^{-1}$

Tantalum

Ta I

λ Å	Weights g_i	g_k	A 10^8 s^{-1}
3170.3	8	10	$8.5\cdot10^{-2}$
3406.9	4	6	$6.8\cdot10^{-2}$
3419.7	8	8	$1.91\cdot10^{-2}$
3463.8	4	6	$2.62\cdot10^{-2}$
3497.9	6	8	$4.9\cdot10^{-2}$
3505.0	8	6	$2.72\cdot10^{-2}$
3607.4	6	8	$4.6\cdot10^{-2}$
3626.6	8	10	$7.1\cdot10^{-2}$
3642.1	10	12	$5.5\cdot10^{-2}$
3784.3	4	6	$4.3\cdot10^{-2}$
3848.1	10	8	$1.30\cdot10^{-2}$
3922.8	4	4	$3.98\cdot10^{-2}$
3996.2	2	4	$3.35\cdot10^{-2}$
4026.9	4	4	$3.60\cdot10^{-2}$
4029.9	10	10	$2.8\cdot10^{-2}$
4040.9	10	12	$7.3\cdot10^{-3}$
4061.4	2	4	$6.5\cdot10^{-2}$
4064.6	4	4	$3.83\cdot10^{-2}$
4136.2	8	6	$1.82\cdot10^{-2}$
4147.9	10	8	$1.79\cdot10^{-2}$
4175.2	6	8	$2.8\cdot10^{-2}$
4205.9	8	10	$8.9\cdot10^{-3}$
4303.0	6	6	$2.08\cdot10^{-2}$
4386.1	4	6	$1.0\cdot10^{-2}$
4402.5	6	6	$2.28\cdot10^{-2}$
4415.7	2	4	$2.53\cdot10^{-2}$
4441.7	10	8	$9.0\cdot10^{-3}$
4473.5	6	8	$1.36\cdot10^{-2}$
4511.0	10	12	$1.56\cdot10^{-2}$
4530.9	4	6	$2.42\cdot10^{-2}$
4553.7	6	8	$9.5\cdot10^{-3}$
4565.9	8	8	$2.5\cdot10^{-2}$
4619.5	6	4	$5.3\cdot10^{-2}$
4669.1	6	4	$2.85\cdot10^{-2}$
4681.9	6	6	$1.5\cdot10^{-2}$
4691.9	2	4	$4.08\cdot10^{-2}$
4706.1	6	6	$1.4\cdot10^{-2}$

Atomic

λ Å	Weights g_i	g_k	A 10^8 s^{-1}
4740.2	4	4	$5.0 \cdot 10^{-2}$
4758.0	4	6	$7.5 \cdot 10^{-3}$
4769.0	8	8	$2.8 \cdot 10^{-2}$
4780.9	10	8	$2.16 \cdot 10^{-2}$
4812.8	4	4	$1.2 \cdot 10^{-2}$
4825.4	6	6	$2.63 \cdot 10^{-2}$
4832.2	4	4	$1.7 \cdot 10^{-2}$
4852.2	4	4	$1.7 \cdot 10^{-2}$
4884.0	6	8	$1.1 \cdot 10^{-2}$
4904.6	12	10	$1.95 \cdot 10^{-2}$
4921.3	2	4	$1.2 \cdot 10^{-2}$
4926.0	4	4	$1.5 \cdot 10^{-2}$
4936.4	8	6	$4.5 \cdot 10^{-2}$
4969.7	4	4	$1.0 \cdot 10^{-2}$
5012.5	4	4	$1.9 \cdot 10^{-2}$
5037.4	10	8	$4.4 \cdot 10^{-2}$
5043.3	6	4	$2.73 \cdot 10^{-2}$
5067.9	8	6	$2.92 \cdot 10^{-2}$
5087.4	6	4	$1.5 \cdot 10^{-2}$
5090.7	8	6	$9.5 \cdot 10^{-3}$
5136.5	2	2	$4.5 \cdot 10^{-2}$
5143.7	6	4	$1.7 \cdot 10^{-2}$
5147.6	6	4	$9.0 \cdot 10^{-3}$
5161.8	4	6	$6.3 \cdot 10^{-3}$
5218.7	8	6	$8.2 \cdot 10^{-3}$
5295.0	6	6	$7.5 \cdot 10^{-3}$
5336.1	6	8	$5.5 \cdot 10^{-3}$
5349.6	6	4	$2.2 \cdot 10^{-2}$
5435.3	4	6	$1.1 \cdot 10^{-2}$
5499.4	10	10	$6.1 \cdot 10^{-3}$
5518.9	8	10	$3.8 \cdot 10^{-2}$
5620.7	8	10	$6.0 \cdot 10^{-3}$
5640.2	6	8	$4.9 \cdot 10^{-3}$
5645.9	6	8	$1.43 \cdot 10^{-2}$
5811.1	8	6	$5.7 \cdot 10^{-3}$
5877.4	10	12	$2.3 \cdot 10^{-2}$
5939.8	2	4	$1.6 \cdot 10^{-2}$
5944.0	4	6	$2.13 \cdot 10^{-2}$
5997.2	10	10	$2.4 \cdot 10^{-2}$
6020.7	2	4	$1.0 \cdot 10^{-2}$
6045.4	6	8	$2.6 \cdot 10^{-2}$
6047.3	8	10	$9.0 \cdot 10^{-3}$
6249.8	6	6	$3.5 \cdot 10^{-3}$
6258.7	6	8	$3.3 \cdot 10^{-3}$
6309.6	4	6	$1.83 \cdot 10^{-2}$
6360.8	6	8	$4.6 \cdot 10^{-3}$
6428.6	6	6	$6.0 \cdot 10^{-3}$
6430.8	8	8	$2.9 \cdot 10^{-2}$
6450.4	8	10	$2.2 \cdot 10^{-2}$
6485.4	10	10	$5.8 \cdot 10^{-2}$
6514.4	6	4	$2.2 \cdot 10^{-2}$
6516.1	6	8	$1.25 \cdot 10^{-2}$
6612.0	6	4	$1.9 \cdot 10^{-2}$
6673.7	2	4	$9.0 \cdot 10^{-3}$
6771.7	4	4	$5.8 \cdot 10^{-3}$
6866.2	8	6	$2.58 \cdot 10^{-2}$
6927.4	10	12	$1.01 \cdot 10^{-2}$
6928.5	10	8	$1.69 \cdot 10^{-2}$
6951.3	10	10	$3.7 \cdot 10^{-3}$
6953.9	6	8	$8.3 \cdot 10^{-3}$
6966.1	8	8	$1.2 \cdot 10^{-2}$
6969.5	10	10	$2.9 \cdot 10^{-3}$
7407.9	6	4	$2.0 \cdot 10^{-2}$

Thallium

Tl I

λ Å	Weights g_i	g_k	A 10^8 s^{-1}
2104.6	2	4	$4.0 \cdot 10^{-2}$
2118.9	2	2	$2.0 \cdot 10^{-2}$
2129.3	2	4	$5.8 \cdot 10^{-2}$
2151.9	2	2	$3.1 \cdot 10^{-2}$
2168.6	2	4	$9.8 \cdot 10^{-2}$
2237.8	2	4	$1.9 \cdot 10^{-1}$
2316.0	2	2	$7.8 \cdot 10^{-2}$
2379.7	2	4	$4.4 \cdot 10^{-1}$
2507.9	4	2	$1.1 \cdot 10^{-2}$
2538.2	4	2	$1.6 \cdot 10^{-2}$
2580.1	2	2	$1.8 \cdot 10^{-1}$
2609.0	4	6	$1.0 \cdot 10^{-1}$
2609.8	4	4	$1.9 \cdot 10^{-2}$
2665.6	4	2	$5.7 \cdot 10^{-2}$
2709.2	4	6	$1.7 \cdot 10^{-1}$
2710.7	4	4	$3.7 \cdot 10^{-2}$
2767.9	2	4	1.26
2826.2	4	2	$8.0 \cdot 10^{-2}$
2918.3	4	6	$4.2 \cdot 10^{-1}$
2921.5	4	4	$7.6 \cdot 10^{-2}$
3229.8	4	2	$1.73 \cdot 10^{-1}$
3519.2	4	6	1.24
3529.4	4	4	$2.20 \cdot 10^{-1}$
3775.7	2	2	$6.25 \cdot 10^{-1}$
5350.5	4	2	$7.05 \cdot 10^{-1}$

Thulium

Tm I

λ Å	Weights g_i	g_k	A 10^8 s^{-1}
2513.8	8	10	$6.9 \cdot 10^{-2}$
2527.0	8	8	$1.7 \cdot 10^{-1}$
2596.5	8	10	$1.6 \cdot 10^{-1}$
2601.1	8	6	$1.7 \cdot 10^{-1}$
2622.5	8	10	$6.1 \cdot 10^{-2}$
2841.1	6	6	$2.0 \cdot 10^{-1}$
2854.2	8	6	$2.7 \cdot 10^{-1}$
2914.8	8	8	$7.7 \cdot 10^{-2}$
2933.0	8	6	$1.0 \cdot 10^{-1}$
2973.2	8	8	$2.3 \cdot 10^{-1}$
3046.9	8	8	$1.8 \cdot 10^{-1}$
3081.1	8	8	$1.9 \cdot 10^{-1}$
3122.5	6	6	$5.2 \cdot 10^{-1}$
3142.4	6	6	$8.8 \cdot 10^{-2}$
3172.7	8	8	$1.8 \cdot 10^{-1}$
3233.7	8	10	$5.1 \cdot 10^{-2}$
3247.0	6	8	$3.0 \cdot 10^{-1}$
3251.8	6	4	$5.2 \cdot 10^{-1}$
3380.7	6	8	$2.0 \cdot 10^{-1}$
3406.0	6	8	$1.5 \cdot 10^{-1}$
3410.1	8	10	$1.0 \cdot 10^{-1}$
3416.6	8	8	$5.7 \cdot 10^{-2}$
3418.6	6	6	$1.1 \cdot 10^{-1}$
3563.9	8	6	$9.8 \cdot 10^{-2}$
3567.4	8	10	$4.2 \cdot 10^{-2}$
3744.1	8	8	$9.5 \cdot 10^{-1}$
3751.8	8	10	$1.9 \cdot 10^{-1}$
3798.5	6	4	1.2
3807.7	6	6	$3.9 \cdot 10^{-1}$
3883.1	8	6	1.0
3887.4	8	8	$3.8 \cdot 10^{-1}$
3916.5	6	8	1.5
3949.3	6	6	1.0
4022.6	6	8	$4.0 \cdot 10^{-2}$
4044.5	6	4	$2.9 \cdot 10^{-1}$
4094.2	8	6	$9.0 \cdot 10^{-1}$
4105.8	8	10	$6.0 \cdot 10^{-1}$
4138.3	6	4	$7.0 \cdot 10^{-1}$
4158.6	6	8	$5.5 \cdot 10^{-2}$
4187.6	8	8	$6.1 \cdot 10^{-1}$
4203.7	8	10	$2.5 \cdot 10^{-1}$
4222.7	6	8	$1.5 \cdot 10^{-1}$
4271.7	6	6	$1.1 \cdot 10^{-1}$
4359.9	8	6	$1.3 \cdot 10^{-1}$
4386.4	8	8	$4.2 \cdot 10^{-2}$
4394.4	6	4	$1.1 \cdot 10^{-1}$
4643.1	6	6	$3.4 \cdot 10^{-2}$
4681.9	6	8	$3.9 \cdot 10^{-2}$
4691.1	6	6	$3.9 \cdot 10^{-2}$
5307.1	8	10	$2.3 \cdot 10^{-2}$
5658.3	6	8	$1.0 \cdot 10^{-2}$
5675.8	8	10	$1.3 \cdot 10^{-2}$
5760.2	6	6	$1.3 \cdot 10^{-2}$

Tin

Sn I

λ Å	Weights g_i	g_k	A 10^8 s^{-1}
2073.1	1	3	$3.6 \cdot 10^{-2}$
2199.3	3	5	$2.9 \cdot 10^{-1}$
2209.7	5	5	$5.6 \cdot 10^{-1}$
2246.1	1	3	1.6
2268.9	5	7	1.2
2286.7	5	5	$3.1 \cdot 10^{-1}$
2317.2	5	7	2.0
2334.8	3	3	$6.6 \cdot 10^{-1}$
2354.8	3	5	1.7
2380.7	3	5	$3.1 \cdot 10^{-2}$
2408.2	5	3	$1.8 \cdot 10^{-1}$
2421.7	5	7	2.5
2429.5	5	7	1.5
2433.5	5	3	$8.0 \cdot 10^{-3}$
2455.2	5	5	$1.1 \cdot 10^{-2}$
2476.4	5	3	$1.1 \cdot 10^{-2}$
2483.4	5	5	$2.1 \cdot 10^{-1}$
2491.8	1	3	$1.7 \cdot 10^{-1}$
2495.7	5	5	$6.2 \cdot 10^{-1}$
2523.9	5	3	$7.4 \cdot 10^{-2}$
2546.6	1	3	$2.1 \cdot 10^{-1}$
2558.0	1	3	$3.4 \cdot 10^{-1}$
2571.6	5	7	$4.5 \cdot 10^{-1}$
2594.4	5	5	$3.0 \cdot 10^{-1}$
2636.9	1	3	$1.1 \cdot 10^{-1}$
2661.2	3	3	$1.1 \cdot 10^{-1}$
2706.5	3	5	$6.6 \cdot 10^{-1}$
2761.8	5	5	$3.7 \cdot 10^{-3}$

λ Å	Weights g_i	g_k	A 10^8 s^{-1}
2779.8	5	7	$1.8 \cdot 10^{-1}$
2785.0	5	3	$1.4 \cdot 10^{-1}$
2788.0	1	3	$1.4 \cdot 10^{-1}$
2812.6	1	3	$2.3 \cdot 10^{-1}$
2813.6	5	5	$1.2 \cdot 10^{-1}$
2840.0	5	5	1.7
2850.6	5	5	$3.3 \cdot 10^{-1}$
2863.3	1	3	$5.4 \cdot 10^{-1}$
2913.5	1	3	$8.3 \cdot 10^{-1}$
3009.1	3	3	$3.8 \cdot 10^{-1}$
3032.8	1	3	$6.2 \cdot 10^{-1}$
3034.1	3	1	2.0
3141.8	1	3	$1.9 \cdot 10^{-1}$
3175.1	5	3	1.0
3218.7	1	3	$4.7 \cdot 10^{-2}$
3223.6	5	5	$1.2 \cdot 10^{-3}$
3262.3	5	3	2.7
3330.6	5	5	$2.0 \cdot 10^{-1}$
3655.8	1	3	$4.1 \cdot 10^{-2}$
3801.0	5	3	$2.8 \cdot 10^{-1}$
4524.7	1	3	$2.6 \cdot 10^{-1}$
5631.7	1	3	$2.4 \cdot 10^{-2}$
5970.3	5	3	$9.6 \cdot 10^{-2}$
6037.7	5	5	$5.0 \cdot 10^{-2}$
6069.0	1	3	$4.6 \cdot 10^{-2}$
6073.5	3	1	$6.3 \cdot 10^{-2}$
6171.5	3	3	$4.9 \cdot 10^{-2}$

Sn II

λ Å	g_i	g_k	A 10^8 s^{-1}
2368.3	4	2	$4.4 \cdot 10^{-3}$
2449.0	4	6	$3.7 \cdot 10^{-1}$
2487.0	6	8	$5.5 \cdot 10^{-1}$
3283.2	4	6	1.0
3352.0	6	8	1.0
3472.5	2	4	$1.6 \cdot 10^{-1}$
3575.5	4	6	$1.3 \cdot 10^{-1}$
5332.4	2	4	$8.6 \cdot 10^{-1}$
5562.0	4	6	1.2
5588.9	4	6	$8.5 \cdot 10^{-1}$
5596.2	4	4	$1.5 \cdot 10^{-1}$
5797.2	6	6	$2.8 \cdot 10^{-1}$
5799.2	6	8	$8.1 \cdot 10^{-1}$
6453.5	2	4	1.2
6761.5	2	2	$3.2 \cdot 10^{-1}$
6844.1	2	2	$6.6 \cdot 10^{-1}$

Titanium

Ti I

λ Å	g_i	g_k	A 10^8 s^{-1}
2644.28	7	5	1.4
2646.65	9	7	1.5
2733.27	5	5	1.9
2912.07	5	7	1.3
2942.00	5	5	1.0
2948.26	7	7	$9.3 \cdot 10^{-1}$
2956.13	9	9	$9.7 \cdot 10^{-1}$
3186.45	5	7	$8.0 \cdot 10^{-1}$
3191.99	7	9	$8.5 \cdot 10^{-1}$
3199.92	9	11	$9.4 \cdot 10^{-1}$
3341.88	5	7	$6.5 \cdot 10^{-1}$
3354.63	7	9	$6.9 \cdot 10^{-1}$
3371.45	9	11	$7.2 \cdot 10^{-1}$
3377.58	7	5	$6.9 \cdot 10^{-1}$
3385.94	9	7	$5.0 \cdot 10^{-1}$
3635.46	5	7	$8.04 \cdot 10^{-1}$
3642.68	7	9	$7.74 \cdot 10^{-1}$
3653.50	9	11	$7.54 \cdot 10^{-1}$
3724.57	9	9	$9.1 \cdot 10^{-1}$
3741.06	7	7	$4.17 \cdot 10^{-1}$
3752.86	9	9	$5.04 \cdot 10^{-1}$
3786.04	5	3	1.4
3958.21	9	7	$4.05 \cdot 10^{-1}$
3981.76	5	5	$3.76 \cdot 10^{-1}$
3989.76	7	7	$3.79 \cdot 10^{-1}$
3998.64	9	9	$4.08 \cdot 10^{-1}$
4186.12	9	9	$2.10 \cdot 10^{-1}$
4266.23	5	5	$3.1 \cdot 10^{-1}$
4284.99	5	5	$3.2 \cdot 10^{-1}$
4289.07	5	5	$3.0 \cdot 10^{-1}$
4393.93	9	11	$3.3 \cdot 10^{-1}$
4417.27	11	9	$3.6 \cdot 10^{-1}$
4449.14	11	11	$9.7 \cdot 10^{-1}$
4450.90	9	9	$9.6 \cdot 10^{-1}$
4453.31	5	5	$5.98 \cdot 10^{-1}$
4453.71	7	7	$4.7 \cdot 10^{-1}$
4455.32	7	7	$4.8 \cdot 10^{-1}$
4457.43	9	9	$5.6 \cdot 10^{-1}$
4465.81	5	7	$3.28 \cdot 10^{-1}$
4481.26	7	7	$5.7 \cdot 10^{-1}$
4496.15	7	5	$4.4 \cdot 10^{-1}$
4518.02	7	9	$1.72 \cdot 10^{-1}$
4522.80	5	7	$1.9 \cdot 10^{-1}$
4533.24	11	11	$8.83 \cdot 10^{-1}$
4534.78	9	9	$6.87 \cdot 10^{-1}$
4548.76	7	5	$2.85 \cdot 10^{-1}$
4552.45	9	7	$2.1 \cdot 10^{-1}$
4563.43	9	11	$2.1 \cdot 10^{-1}$
4617.27	7	9	$8.51 \cdot 10^{-1}$
4623.10	5	7	$5.74 \cdot 10^{-1}$
4639.94	3	3	$6.64 \cdot 10^{-1}$
4742.79	9	9	$5.3 \cdot 10^{-1}$
4758.12	11	11	$7.13 \cdot 10^{-1}$
4759.27	13	13	$7.40 \cdot 10^{-1}$
4778.26	9	9	$2.0 \cdot 10^{-1}$
4805.42	5	7	$5.8 \cdot 10^{-1}$
4856.01	13	15	$5.2 \cdot 10^{-1}$
4885.08	11	13	$4.90 \cdot 10^{-1}$
4913.62	7	9	$4.44 \cdot 10^{-1}$
4928.34	3	5	$6.2 \cdot 10^{-1}$
4981.73	11	13	$6.60 \cdot 10^{-1}$
4989.14	7	5	$3.25 \cdot 10^{-1}$
4991.07	9	11	$5.84 \cdot 10^{-1}$
4999.50	7	9	$5.27 \cdot 10^{-1}$
5000.99	9	7	$3.52 \cdot 10^{-1}$
5007.21	5	7	$4.92 \cdot 10^{-1}$
5014.28	3	5	$6.8 \cdot 10^{-1}$
5036.47	7	9	$3.94 \cdot 10^{-1}$
5038.40	5	7	$3.87 \cdot 10^{-1}$
5224.30	11	11	$3.6 \cdot 10^{-1}$
5259.98	5	7	$2.3 \cdot 10^{-1}$
5351.07	7	7	$3.4 \cdot 10^{-1}$
5503.90	11	9	$2.6 \cdot 10^{-1}$
5774.04	9	11	$5.5 \cdot 10^{-1}$
5785.98	11	13	$6.1 \cdot 10^{-1}$
5804.27	13	15	$6.8 \cdot 10^{-1}$
6098.66	9	7	$2.5 \cdot 10^{-1}$
6220.46	9	7	$1.8 \cdot 10^{-1}$

Ti II

λ Å	g_i	g_k	A 10^8 s^{-1}
2635.44	4	4	1.9
2638.56	6	6	1.7
2642.02	8	8	1.9
2645.86	10	10	2.7
2746.54	6	8	2.6
2751.59	8	10	3.7
2752.68	8	10	1.1
2757.62	6	8	$7.2 \cdot 10^{-1}$
2758.35	4	6	$9.9 \cdot 10^{-1}$
2804.82	6	8	4.6
2810.30	8	10	5.1
2817.83	10	12	3.8
2819.87	8	8	$6.5 \cdot 10^{-1}$
2821.26	6	8	$7.9 \cdot 10^{-1}$
2827.12	8	10	1.0
2828.06	12	14	4.4
2828.64	6	6	1.2
2828.83	10	10	$9.1 \cdot 10^{-1}$
2834.02	10	12	$7.9 \cdot 10^{-1}$
2836.47	8	8	1.2
2839.64	12	12	$8.3 \cdot 10^{-1}$
2845.93	10	10	1.2
2856.10	12	12	1.5
2877.47	8	8	$5.7 \cdot 10^{-1}$
2884.13	10	10	$5.2 \cdot 10^{-1}$
2926.64	10	8	$8.9 \cdot 10^{-1}$
2931.10	6	6	3.2
2936.02	4	6	2.7
2938.57	6	8	2.4
2941.90	8	10	1.8
2942.97	8	8	1.1
2945.30	10	12	2.7
2954.59	10	12	4.0
2958.80	8	10	4.0
2979.06	4	6	1.2
2990.06	6	8	$5.6 \cdot 10^{-1}$
3017.17	12	12	$3.6 \cdot 10^{-1}$
3022.64	10	10	1.2
3023.67	8	8	1.0
3075.23	6	4	1.13
3078.65	8	6	1.09
3081.52	10	8	1.1
3088.04	10	8	1.25
3089.44	8	6	1.3
3103.81	10	8	1.1
3106.26	6	6	$7.8 \cdot 10^{-1}$
3127.86	6	6	1.6
3128.50	8	8	1.1
3168.55	10	8	$4.1 \cdot 10^{-1}$
3181.73	6	8	$4.6 \cdot 10^{-1}$
3189.49	4	4	$9.2 \cdot 10^{-1}$

λ Å	Weights g_i	g_k	A 10^8 s^{-1}
3190.91	6	8	1.3
3202.56	4	6	1.1
3224.25	12	10	$7.0 \cdot 10^{-1}$
3228.62	4	2	2.0
3232.29	8	6	$6.0 \cdot 10^{-1}$
3234.51	10	10	1.38
3236.58	8	8	1.11
3239.04	6	6	$9.87 \cdot 10^{-1}$
3239.66	6	4	$9.4 \cdot 10^{-1}$
3241.99	4	4	1.16
3278.28	4	4	$9.6 \cdot 10^{-1}$
3278.91	6	4	1.0
3282.32	2	2	1.6
3287.66	8	10	1.4
3321.70	4	4	$7.2 \cdot 10^{-1}$
3322.94	10	10	$3.96 \cdot 10^{-1}$
3332.11	6	4	1.1
3361.23	8	10	1.1
3372.80	6	8	1.11
3383.77	4	6	1.09
3456.40	4	4	$8.2 \cdot 10^{-1}$
3483.63	10	8	$9.7 \cdot 10^{-1}$
3492.37	8	6	$9.8 \cdot 10^{-1}$
3504.90	10	10	$8.2 \cdot 10^{-1}$
3510.86	8	8	$9.3 \cdot 10^{-1}$
3535.41	4	6	$5.5 \cdot 10^{-1}$
3741.64	6	6	$6.2 \cdot 10^{-1}$
3759.30	8	8	$9.4 \cdot 10^{-1}$
3761.33	6	6	$9.9 \cdot 10^{-1}$
4911.18	6	4	$3.2 \cdot 10^{-1}$

Ti III

λ Å	g_i	g_k	A 10^8 s^{-1}
865.79	5	3	$6.6 \cdot 10^{1}$
1002.37	5	5	7.6
1004.67	7	5	$4.3 \cdot 10^{1}$
1005.80	3	3	$1.3 \cdot 10^{1}$
1007.16	5	3	$3.8 \cdot 10^{1}$
1008.12	3	1	$5.1 \cdot 10^{1}$
1286.37	9	9	2.0
1289.30	7	7	2.2
1291.62	5	5	2.4
1293.23	9	7	1.0
1298.97	7	5	4.9
1327.59	5	3	3.2
1420.44	1	3	1.2
1421.63	3	1	4.0
1422.41	5	5	3.0
1424.14	5	3	1.6
1455.19	9	7	6.4
1498.70	5	5	2.8
2007.36	3	3	3.4
2007.60	1	3	1.2
2010.80	5	3	5.4
2097.30	5	7	3.3
2099.86	3	5	2.5
2104.86	3	3	1.1
2105.09	1	3	1.7
2199.22	3	3	5.7
2237.77	7	7	2.4
2331.35	3	1	4.3
2331.66	3	3	1.2
2339.00	5	3	3.0
2346.79	7	5	3.3
2374.99	5	3	4.0
2413.99	5	7	3.8
2516.05	7	9	3.4
2567.56	3	3	2.3
2984.75	5	5	1.9
3066.51	3	3	2.5
3228.89	3	3	1.5
3278.31	7	9	3.4
3320.94	3	5	2.8
3340.20	7	9	3.7
3346.18	9	11	3.7
3354.71	11	13	4.4
3397.24	3	1	1.8
3404.46	3	3	1.8
3417.62	3	5	1.9
3915.47	9	11	2.1
4119.14	5	5	$9.9 \cdot 10^{-1}$
4213.26	9	11	2.2
4215.53	9	11	2.2
4247.62	11	13	1.1
4248.54	5	7	2.3
4250.09	3	5	$9.5 \cdot 10^{-1}$
4259.01	11	13	$9.4 \cdot 10^{-1}$
4269.84	9	11	1.7
4285.61	13	15	3.0
4288.66	11	13	1.1
4296.70	11	13	1.6
4319.56	9	11	1.1
4343.25	3	1	1.0
4378.94	3	5	1.6
4433.91	11	13	1.8
4440.66	1	3	1.2
4533.26	3	5	1.5
4576.53	9	7	1.3
4628.07	3	1	1.5
4652.86	7	9	2.6
4874.00	5	7	1.5
4914.32	3	3	1.1
4971.19	9	11	2.1
5083.80	5	3	$9.7 \cdot 10^{-1}$
5278.33	3	3	$9.4 \cdot 10^{-1}$
7506.87	11	13	1.1

Ti IV

λ Å	g_i	g_k	A 10^8 s^{-1}
423.49	4	6	$4.9 \cdot 10^{1}$
424.16	6	8	$5.3 \cdot 10^{1}$
433.63	4	2	5.5
433.76	6	4	5.0
729.36	4	2	5.7
1183.64	2	2	6.9
1195.21	4	2	$1.4 \cdot 10^{1}$
1451.74	2	4	$1.8 \cdot 10^{1}$
1467.34	4	6	$2.1 \cdot 10^{1}$
2067.56	2	4	5.1
2103.16	2	2	5.0
2541.79	4	6	6.9
2546.88	6	8	7.4
2862.60	4	2	4.1
3576.44	4	6	4.6

Tungsten

W I

λ Å	g_i	g_k	A 10^8 s^{-1}
2879.4	1	3	$2.4 \cdot 10^{-1}$
2911.0	1	3	$7.7 \cdot 10^{-2}$
2923.5	7	9	$1.54 \cdot 10^{-2}$
2935.0	3	5	$1.5 \cdot 10^{-1}$
3013.8	7	9	$6.4 \cdot 10^{-2}$
3016.5	9	11	$9.27 \cdot 10^{-2}$
3017.4	7	9	$1.21 \cdot 10^{-1}$
3024.9	3	3	$1.4 \cdot 10^{-1}$
3046.4	3	5	$5.8 \cdot 10^{-2}$
3049.7	7	5	$1.7 \cdot 10^{-1}$
3064.9	5	7	$1.1 \cdot 10^{-2}$
3084.9	5	5	$1.3 \cdot 10^{-2}$
3093.5	7	9	$4.4 \cdot 10^{-2}$
3107.2	5	7	$2.33 \cdot 10^{-2}$
3108.0	7	9	$1.58 \cdot 10^{-2}$
3145.5	9	9	$4.8 \cdot 10^{-3}$
3170.2	7	5	$6.0 \cdot 10^{-3}$
3176.6	3	5	$2.12 \cdot 10^{-2}$
3183.5	7	7	$2.64 \cdot 10^{-3}$
3184.4	5	3	$2.3 \cdot 10^{-2}$
3191.6	1	3	$3.2 \cdot 10^{-2}$
3198.8	7	9	$4.6 \cdot 10^{-2}$
3207.3	7	9	$3.0 \cdot 10^{-2}$
3208.3	5	5	$4.4 \cdot 10^{-2}$
3215.6	9	11	$2.1 \cdot 10^{-1}$
3221.9	5	7	$1.61 \cdot 10^{-2}$
3223.1	5	3	$3.53 \cdot 10^{-3}$
3232.5	9	9	$2.4 \cdot 10^{-2}$
3235.1	7	5	$2.68 \cdot 10^{-3}$
3259.7	7	7	$1.3 \cdot 10^{-2}$
3300.8	7	9	$8.1 \cdot 10^{-2}$
3311.4	7	5	$5.6 \cdot 10^{-2}$
3363.3	9	7	$6.6 \cdot 10^{-3}$
3371.0	7	5	$1.0 \cdot 10^{-2}$
3371.4	3	3	$6.7 \cdot 10^{-3}$
3386.1	7	7	$2.64 \cdot 10^{-3}$
3413.0	7	9	$9.7 \cdot 10^{-3}$
3459.5	9	9	$2.04 \cdot 10^{-3}$
3510.0	7	9	$5.2 \cdot 10^{-3}$
3545.2	1	3	$3.2 \cdot 10^{-2}$
3570.6	5	3	$6.7 \cdot 10^{-3}$
3606.1	3	5	$9.6 \cdot 10^{-3}$
3617.5	7	7	$1.1 \cdot 10^{-1}$
3631.9	3	5	$1.3 \cdot 10^{-2}$
3675.6	9	11	$1.20 \cdot 10^{-2}$
3682.1	9	11	$2.0 \cdot 10^{-2}$
3707.9	7	7	$2.9 \cdot 10^{-2}$
3757.9	7	9	$1.38 \cdot 10^{-2}$
3760.1	5	7	$1.99 \cdot 10^{-2}$
3768.5	3	3	$3.47 \cdot 10^{-2}$
3780.8	7	5	$4.2 \cdot 10^{-2}$
3809.2	7	5	$9.0 \cdot 10^{-3}$
3817.5	7	7	$3.1 \cdot 10^{-2}$
3829.1	3	3	$3.83 \cdot 10^{-3}$

λ Å	Weights g_i	g_k	A 10^8 s^{-1}
3835.1	5	5	$5.2\cdot10^{-2}$
3846.3	3	5	$2.14\cdot10^{-2}$
3847.5	1	3	$8.3\cdot10^{-3}$
3864.3	5	5	$5.6\cdot10^{-3}$
3868.0	7	9	$4.6\cdot10^{-2}$
3881.4	7	7	$3.6\cdot10^{-2}$
3968.5	1	3	$5.07\cdot10^{-3}$
3975.5	9	11	$4.1\cdot10^{-3}$
4001.4	9	9	$5.6\cdot10^{-3}$
4008.8	7	9	$1.63\cdot10^{-1}$
4019.3	5	3	$6.7\cdot10^{-3}$
4028.8	1	3	$2.0\cdot10^{-2}$
4045.6	7	5	$2.88\cdot10^{-2}$
4055.2	7	9	$1.79\cdot10^{-3}$
4070.0	7	5	$3.60\cdot10^{-2}$
4070.6	3	5	$5.6\cdot10^{-3}$
4074.4	7	7	$1.0\cdot10^{-1}$
4088.3	5	3	$4.13\cdot10^{-3}$
4102.7	9	7	$4.9\cdot10^{-2}$
4115.6	11	11	$4.8\cdot10^{-3}$
4137.5	5	7	$8.4\cdot10^{-3}$
4171.2	7	9	$8.6\cdot10^{-3}$
4203.8	9	7	$4.9\cdot10^{-3}$
4219.4	9	7	$6.1\cdot10^{-3}$
4244.4	9	11	$1.38\cdot10^{-2}$
4269.4	7	5	$3.04\cdot10^{-2}$
4283.8	9	7	$1.69\cdot10^{-3}$
4294.6	7	5	$1.2\cdot10^{-1}$
4302.1	7	7	$3.6\cdot10^{-2}$
4355.2	9	9	$5.1\cdot10^{-3}$
4361.8	9	7	$1.64\cdot10^{-3}$
4378.5	7	5	$3.48\cdot10^{-3}$
4458.1	3	5	$4.2\cdot10^{-3}$
4466.3	7	5	$1.5\cdot10^{-2}$
4472.5	13	11	$1.55\cdot10^{-3}$
4484.2	3	5	$5.6\cdot10^{-3}$
4492.3	9	11	$3.6\cdot10^{-3}$
4495.3	11	11	$3.3\cdot10^{-3}$
4504.8	9	7	$7.0\cdot10^{-3}$
4552.5	9	9	$1.42\cdot10^{-3}$
4586.8	1	3	$4.20\cdot10^{-3}$
4592.6	7	9	$3.4\cdot10^{-3}$
4609.9	7	9	$1.42\cdot10^{-2}$
4613.3	9	9	$2.9\cdot10^{-3}$
4634.8	9	9	$8.8\cdot10^{-3}$
4659.9	1	3	$1.0\cdot10^{-2}$
4680.5	7	7	$1.4\cdot10^{-2}$
4720.4	3	5	$3.22\cdot10^{-3}$
4729.6	7	5	$7.8\cdot10^{-3}$
4752.6	3	3	$5.20\cdot10^{-3}$
4757.5	7	5	$2.72\cdot10^{-3}$
4757.8	11	9	$4.1\cdot10^{-3}$
4788.4	9	11	$2.6\cdot10^{-3}$
4843.8	5	5	$1.9\cdot10^{-2}$
4886.9	9	11	$8.1\cdot10^{-3}$
4924.6	13	11	$1.75\cdot10^{-3}$
4931.6	7	5	$1.0\cdot10^{-2}$
4948.6	9	11	$1.36\cdot10^{-3}$
4972.6	9	11	$3.9\cdot10^{-3}$
4982.6	1	3	$4.17\cdot10^{-3}$
4986.9	11	9	$6.3\cdot10^{-3}$
5006.2	9	7	$1.2\cdot10^{-2}$
5015.3	7	9	$5.4\cdot10^{-3}$
5040.4	3	5	$5.2\cdot10^{-3}$
5053.3	3	3	$1.9\cdot10^{-2}$
5071.5	13	11	$3.4\cdot10^{-3}$
5117.6	11	11	$1.61\cdot10^{-3}$
5124.2	5	5	$4.0\cdot10^{-3}$
5141.2	7	9	$1.12\cdot10^{-3}$
5224.7	7	5	$1.2\cdot10^{-2}$
5243.0	9	7	$1.1\cdot10^{-2}$
5254.5	7	5	$3.86\cdot10^{-3}$
5268.6	9	9	$1.4\cdot10^{-3}$
5500.5	11	9	$6.9\cdot10^{-3}$
5514.7	5	3	$7.3\cdot10^{-3}$
5537.7	9	11	$2.2\cdot10^{-3}$
5617.1	7	7	$1.47\cdot10^{-3}$
5631.9	9	7	$1.43\cdot10^{-3}$
5660.7	13	11	$6.8\cdot10^{-3}$
5675.4	5	5	$2.20\cdot10^{-3}$
5796.5	9	7	$2.21\cdot10^{-3}$
5891.6	7	7	$1.47\cdot10^{-3}$
5947.6	5	7	$2.40\cdot10^{-3}$
5965.9	7	5	$1.0\cdot10^{-2}$
6021.5	5	3	$8.7\cdot10^{-3}$
6081.4	5	3	$4.7\cdot10^{-3}$
6203.5	7	7	$3.0\cdot10^{-3}$
6285.9	7	5	$6.6\cdot10^{-3}$
6292.0	3	5	$2.26\cdot10^{-3}$
6303.2	9	9	$1.84\cdot10^{-3}$
6404.2	5	7	$1.50\cdot10^{-3}$
6439.7	9	9	$1.29\cdot10^{-3}$
6445.1	7	5	$6.4\cdot10^{-3}$
6532.4	3	5	$4.6\cdot10^{-3}$
6538.1	11	9	$2.7\cdot10^{-3}$
6563.2	5	5	$2.04\cdot10^{-3}$
6814.9	9	9	$1.46\cdot10^{-3}$
7285.8	13	11	$1.47\cdot10^{-3}$
7569.9	5	3	$3.73\cdot10^{-3}$
7664.9	5	3	$3.80\cdot10^{-3}$
8017.2	5	7	$1.6\cdot10^{-3}$
8358.7	5	7	$1.89\cdot10^{-3}$
9381.4	9	7	$1.53\cdot10^{-3}$

Uranium

U I

λ Å	Weights g_i	g_k	A 10^8 s^{-1}
3553.0	13	13	$2.0\cdot10^{-2}$
3553.0	9	7	$1.4\cdot10^{-2}$
3553.4	15	13	$2.2\cdot10^{-2}$
3554.5	11	9	$8.4\cdot10^{-3}$
3554.9	15	17	$7.9\cdot10^{-3}$
3555.3	13	15	$2.7\cdot10^{-2}$
3555.8	13	11	$4.1\cdot10^{-3}$
3556.9	13	11	$7.5\cdot10^{-3}$
3557.8	13	13	$2.9\cdot10^{-2}$
3558.0	11	13	$1.6\cdot10^{-2}$
3558.6	9	7	$3.9\cdot10^{-2}$
3559.4	7	9	$1.5\cdot10^{-2}$
3560.3	9	7	$6.4\cdot10^{-2}$
3561.4	15	13	$5.5\cdot10^{-2}$
3561.5	9	9	$2.5\cdot10^{-2}$
3561.8	13	11	$5.7\cdot10^{-2}$
3563.7	13	13	$2.9\cdot10^{-2}$
3563.8	7	7	$1.1\cdot10^{-2}$
3565.0	13	11	$2.9\cdot10^{-2}$
3566.0	13	15	$1.7\cdot10^{-2}$
3566.6	11	11	$2.4\cdot10^{-1}$
3568.8	13	13	$3.8\cdot10^{-2}$
3569.1	17	15	$1.1\cdot10^{-1}$
3569.4	9	9	$1.5\cdot10^{-2}$
3570.1	13	11	$1.3\cdot10^{-2}$
3570.2	11	9	$5.3\cdot10^{-3}$
3570.6	13	15	$2.7\cdot10^{-2}$
3570.7	15	15	$1.2\cdot10^{-2}$
3571.2	11	11	$6.3\cdot10^{-3}$
3571.6	17	15	$1.3\cdot10^{-1}$
3572.9	13	15	$1.5\cdot10^{-2}$
3573.9	13	11	$4.0\cdot10^{-2}$
3574.1	13	15	$3.5\cdot10^{-2}$
3574.8	13	15	$1.9\cdot10^{-2}$
3577.1	17	15	$4.3\cdot10^{-2}$
3577.5	15	13	$7.8\cdot10^{-3}$
3577.8	11	11	$8.3\cdot10^{-3}$
3577.9	13	13	$2.3\cdot10^{-2}$
3578.3	13	11	$2.0\cdot10^{-2}$
3580.0	9	9	$1.2\cdot10^{-2}$
3580.2	11	9	$2.9\cdot10^{-2}$
3580.4	11	13	$7.5\cdot10^{-3}$
3580.9	13	13	$2.1\cdot10^{-2}$
3582.6	13	13	$2.9\cdot10^{-2}$
3584.6	7	5	$2.4\cdot10^{-2}$
3584.9	13	15	$1.8\cdot10^{-1}$
3585.4	11	11	$1.9\cdot10^{-2}$
3585.8	11	9	$2.8\cdot10^{-2}$
3587.8	9	11	$1.3\cdot10^{-2}$
3588.3	7	9	$1.8\cdot10^{-2}$
3589.7	11	13	$2.1\cdot10^{-2}$
3589.8	15	13	$5.9\cdot10^{-2}$
3590.7	9	7	$2.2\cdot10^{-2}$
3591.7	11	9	$5.3\cdot10^{-2}$
3593.0	11	11	$1.4\cdot10^{-2}$
3593.2	13	15	$4.2\cdot10^{-2}$
3593.7	11	11	$7.2\cdot10^{-2}$

Vanadium

V I

λ Å	Weights g_i	g_k	A 10^8 s^{-1}
3053.65	4	4	1.3
3056.33	6	6	1.3
3060.46	8	8	1.4
3066.37	10	10	2.1
3183.41	6	8	2.4
3183.96	8	10	2.5
3183.98	4	6	2.4
3185.38	10	12	2.7
3205.58	8	10	1.3
3212.43	10	12	1.4
3377.62	6	6	$6.0\cdot10^{-1}$
3533.68	6	8	$5.2\cdot10^{-1}$

λ Å	Weights g_i	g_k	A 10^8 s^{-1}
3663.60	4	6	3.1
3667.74	6	8	2.7
3672.41	12	12	$9.2\cdot10^{-1}$
3673.41	8	10	2.7
3676.70	14	14	1.3
3680.12	10	12	2.2
3686.26	10	12	$2.3\cdot10^{-1}$
3687.50	12	14	2.9
3688.07	8	8	$3.5\cdot10^{-1}$
3692.22	6	6	$5.4\cdot10^{-1}$
3695.34	14	16	2.8
3695.86	4	4	$6.6\cdot10^{-1}$
3703.57	10	8	$9.2\cdot10^{-1}$
3704.70	8	6	$6.6\cdot10^{-1}$
3706.03	10	10	$5.2\cdot10^{-1}$
3708.71	12	12	$4.4\cdot10^{-1}$
3794.96	10	10	$2.3\cdot10^{-1}$
3806.79	10	10	$2.5\cdot10^{-1}$
3840.75	8	6	$5.48\cdot10^{-1}$
3855.85	10	8	$5.78\cdot10^{-1}$
3871.07	10	8	$2.8\cdot10^{-1}$
3902.26	10	10	$2.68\cdot10^{-1}$
3930.02	10	10	$3.3\cdot10^{-1}$
3934.01	8	8	$6.2\cdot10^{-1}$
3992.80	12	10	1.2
3998.73	14	12	1.0
4050.96	10	10	1.4
4051.35	12	12	1.3
4090.57	8	10	$8.5\cdot10^{-1}$
4092.68	8	10	$2.30\cdot10^{-1}$
4095.48	6	8	$7.2\cdot10^{-1}$
4099.78	6	8	$4.10\cdot10^{-1}$
4102.15	4	6	$7.1\cdot10^{-1}$
4104.77	10	8	2.1
4105.16	4	6	$4.9\cdot10^{-1}$
4109.78	2	4	$5.00\cdot10^{-1}$
4111.78	10	10	1.01
4115.18	8	8	$5.80\cdot10^{-1}$
4116.47	6	6	$3.2\cdot10^{-1}$
4123.50	4	2	1.00
4128.06	6	4	$7.70\cdot10^{-1}$
4131.99	8	6	$5.5\cdot10^{-1}$
4134.49	10	8	$2.90\cdot10^{-1}$
4232.46	10	10	$9.8\cdot10^{-1}$
4232.95	8	8	$7.7\cdot10^{-1}$
4268.64	14	14	1.2
4271.55	12	12	$9.6\cdot10^{-1}$
4276.95	10	10	$9.4\cdot10^{-1}$
4284.05	8	8	1.2
4291.82	12	14	$8.8\cdot10^{-1}$
4296.10	10	12	$7.7\cdot10^{-1}$
4297.67	8	10	$7.0\cdot10^{-1}$
4298.03	6	8	$7.8\cdot10^{-1}$
4379.23	10	12	1.1
4384.71	8	10	1.1
4389.98	6	8	$6.9\cdot10^{-1}$
4395.22	4	6	$5.5\cdot10^{-1}$
4406.64	10	10	$2.2\cdot10^{-1}$
4407.63	8	8	$4.4\cdot10^{-1}$
4408.20	6	6	$6.0\cdot10^{-1}$
4452.01	14	16	$9.2\cdot10^{-1}$
4457.75	10	12	$2.7\cdot10^{-1}$
4460.33	10	8	$3.0\cdot10^{-1}$
4462.36	12	14	$7.6\cdot10^{-1}$
4468.00	8	10	$2.3\cdot10^{-1}$
4469.71	10	12	$6.2\cdot10^{-1}$
4474.04	10	8	$4.7\cdot10^{-1}$
4496.06	8	6	$4.0\cdot10^{-1}$
4524.21	12	10	$3.0\cdot10^{-1}$
4529.58	10	8	$2.4\cdot10^{-1}$
4545.40	10	12	$7.6\cdot10^{-1}$
4560.72	8	10	$7.0\cdot10^{-1}$
4571.79	6	8	$6.0\cdot10^{-1}$
4578.73	4	6	$6.8\cdot10^{-1}$
4757.47	4	2	$7.6\cdot10^{-1}$
4766.62	6	4	$5.6\cdot10^{-1}$
4776.36	8	6	$5.1\cdot10^{-1}$
4786.50	10	8	$4.7\cdot10^{-1}$
4796.92	12	10	$4.8\cdot10^{-1}$
4807.52	14	12	$5.8\cdot10^{-1}$
5193.00	12	12	$4.0\cdot10^{-1}$
5195.39	8	8	$2.3\cdot10^{-1}$
5234.08	10	10	$4.9\cdot10^{-1}$
5240.87	12	12	$4.3\cdot10^{-1}$
5415.25	12	14	$3.1\cdot10^{-1}$
5487.91	12	10	$2.9\cdot10^{-1}$
5507.75	10	8	$3.5\cdot10^{-1}$
6090.21	8	6	$2.60\cdot10^{-1}$

V II

λ Å	Weights g_i	g_k	A 10^8 s^{-1}
2527.90	13	13	$6.1\cdot10^{-1}$
2528.47	9	9	$5.2\cdot10^{-1}$
2528.83	11	11	$5.3\cdot10^{-1}$
2554.04	9	9	$5.4\cdot10^{-1}$
2589.10	9	9	$7.7\cdot10^{-1}$
2640.86	5	7	1.2
2677.80	3	5	$3.4\cdot10^{-1}$
2679.33	7	7	$3.4\cdot10^{-1}$
2683.09	1	3	$3.4\cdot10^{-1}$
2687.96	9	9	$7.6\cdot10^{-1}$
2689.88	3	1	$9.2\cdot10^{-1}$
2690.25	7	5	$3.4\cdot10^{-1}$
2690.79	5	3	$5.2\cdot10^{-1}$
2700.94	9	11	$3.5\cdot10^{-1}$
2706.17	7	9	$3.4\cdot10^{-1}$
2734.22	9	7	$6.2\cdot10^{-1}$
2753.41	13	11	$4.2\cdot10^{-1}$
2784.20	9	9	1.3
2787.91	7	9	$5.0\cdot10^{-1}$
2825.86	9	7	1.2
2843.82	7	5	$9.9\cdot10^{-1}$
2847.57	9	7	$4.6\cdot10^{-1}$
2854.34	11	9	$5.0\cdot10^{-1}$
2862.31	11	11	$3.6\cdot10^{-1}$
2868.11	5	3	2.1
2869.13	13	11	$4.8\cdot10^{-1}$
2882.49	5	5	$4.2\cdot10^{-1}$
2884.78	3	3	$5.6\cdot10^{-1}$
2889.61	3	1	1.9
2891.64	5	3	1.4
2892.43	9	9	$3.6\cdot10^{-1}$
2892.65	7	5	1.3
2893.31	9	7	1.2
2903.07	3	5	$3.4\cdot10^{-1}$
2906.45	7	7	$7.8\cdot10^{-1}$
2908.81	11	9	1.6
2910.01	5	5	1.1
2910.38	3	3	1.2
2911.05	7	9	$3.7\cdot10^{-1}$
2912.46	11	9	$5.0\cdot10^{-1}$
2915.88	9	7	$4.9\cdot10^{-1}$
2924.02	11	11	1.7
2924.63	9	9	1.2
2930.80	7	7	$5.8\cdot10^{-1}$
2941.37	11	9	$3.5\cdot10^{-1}$
2944.57	9	7	$7.6\cdot10^{-1}$
2948.08	9	11	$4.0\cdot10^{-1}$
2952.07	7	5	$7.2\cdot10^{-1}$
2955.58	7	9	$3.3\cdot10^{-1}$
2968.37	7	9	$7.0\cdot10^{-1}$
2972.26	5	7	$5.2\cdot10^{-1}$
2973.98	9	11	$3.5\cdot10^{-1}$
2985.18	7	9	$4.4\cdot10^{-1}$
3001.20	7	7	$7.5\cdot10^{-1}$
3014.82	5	3	$8.9\cdot10^{-1}$
3016.78	7	5	$5.0\cdot10^{-1}$
3020.21	9	7	$5.0\cdot10^{-1}$
3048.21	11	13	$7.0\cdot10^{-1}$
3063.25	9	11	1.0
3100.94	7	7	$5.8\cdot10^{-1}$
3113.56	11	11	$5.0\cdot10^{-1}$
3122.89	11	13	$7.6\cdot10^{-1}$
3134.93	13	13	$5.9\cdot10^{-1}$
3136.50	11	11	$5.3\cdot10^{-1}$
3139.73	9	9	$5.2\cdot10^{-1}$
3151.32	3	5	$4.4\cdot10^{-1}$
3190.69	9	9	$3.3\cdot10^{-1}$
3250.78	11	9	$5.2\cdot10^{-1}$
3251.87	5	7	$3.5\cdot10^{-1}$
3271.12	7	9	$6.9\cdot10^{-1}$
3276.12	9	11	$5.2\cdot10^{-1}$
3279.84	9	11	$5.8\cdot10^{-1}$
3287.71	5	7	$7.5\cdot10^{-1}$
3337.85	5	7	$5.3\cdot10^{-1}$
3517.30	9	7	$3.8\cdot10^{-1}$
3530.77	5	3	$4.5\cdot10^{-1}$
3545.19	7	5	$4.3\cdot10^{-1}$
3556.80	9	7	$5.1\cdot10^{-1}$
3592.01	7	5	$4.4\cdot10^{-1}$
3618.92	3	5	$3.3\cdot10^{-1}$

V III

λ Å	Weights g_i	g_k	A 10^8 s^{-1}
2318.06	8	10	4.6
2323.82	6	8	3.8
2330.42	10	10	3.2
2331.75	8	8	2.5
2334.21	6	6	2.2
2337.13	4	4	2.7
2343.10	6	8	3.6
2358.73	6	8	4.2

λ Å	Weights g_i	g_k	A 10^8 s^{-1}
2366.31	8	10	4.2
2371.06	10	12	5.2
2373.06	4	6	2.9
2382.46	8	10	5.0
2393.58	6	8	4.3
2404.18	4	6	2.5
2516.14	10	10	3.7
2521.55	8	8	3.5
2548.21	6	4	2.0
2554.22	8	6	1.2
2593.05	6	6	2.8
2595.10	8	8	2.8

V IV

λ Å	Weights g_i	g_k	A 10^8 s^{-1}
677.345	9	9	6.7
680.632	9	7	$1.2 \cdot 10^{1}$
681.145	7	5	$1.1 \cdot 10^{1}$
682.455	7	7	6.5
682.923	5	5	6.9
684.450	7	5	7.7
691.530	5	3	$1.1 \cdot 10^{1}$
723.537	3	1	$1.5 \cdot 10^{1}$
724.068	5	5	$1.1 \cdot 10^{1}$
724.809	5	3	5.6
737.854	9	7	$2.4 \cdot 10^{1}$
750.110	5	5	$1.0 \cdot 10^{1}$
884.146	1	3	4.7
1071.05	5	5	6.1
1110.72	3	3	5.0
1112.20	7	7	6.3
1112.44	5	5	5.0
1127.84	7	5	8.9
1131.26	9	7	9.4
1194.46	7	5	$1.0 \cdot 10^{1}$
1226.52	5	5	$1.5 \cdot 10^{1}$
1243.72	3	1	9.4
1247.07	5	3	4.7
1272.97	3	1	$2.7 \cdot 10^{1}$
1304.17	3	5	$1.5 \cdot 10^{1}$
1305.42	5	7	7.0
1308.06	7	9	7.9
1309.50	5	5	8.7
1312.72	7	7	8.6
1317.57	5	7	8.7
1321.92	7	9	9.9
1326.81	3	5	4.0
1329.29	5	5	$1.5 \cdot 10^{1}$
1329.97	3	3	4.8
1330.36	1	3	6.0
1331.67	3	1	$1.7 \cdot 10^{1}$
1332.46	5	3	7.5
1334.49	9	9	8.3
1355.13	7	9	$2.5 \cdot 10^{1}$
1356.53	5	3	4.9
1395.00	5	7	$1.4 \cdot 10^{1}$
1400.42	5	7	7.5
1403.62	7	9	8.4
1412.69	3	3	$1.1 \cdot 10^{1}$
1414.41	5	7	$1.2 \cdot 10^{1}$
1414.84	5	5	4.6
1418.53	7	7	5.2
1419.58	7	9	$1.3 \cdot 10^{1}$
1423.72	3	5	7.1
1426.65	9	11	$2.2 \cdot 10^{1}$
1429.11	5	5	5.0
1434.84	7	7	5.4
1451.04	3	3	7.0
1454.00	5	3	$1.1 \cdot 10^{1}$
1520.14	5	7	7.2
1522.49	3	5	5.5
1601.92	3	3	$1.2 \cdot 10^{1}$
1611.88	7	7	5.2
1806.18	5	3	7.3
1809.85	3	1	7.2
1817.68	5	3	4.8
1825.84	7	5	5.3
1861.56	5	7	6.6
1939.07	7	9	5.8
1951.43	5	7	5.0
1963.10	3	5	4.8
1997.72	7	7	4.7
2084.43	5	5	4.0
2120.05	7	9	8.1
2141.20	3	5	7.0
2146.83	7	9	6.6
2149.85	5	7	5.1
2151.09	7	9	4.3
2155.34	11	13	$1.2 \cdot 10^{1}$
2446.80	9	11	5.3
2570.72	9	11	7.6
3284.56	7	9	5.3
3496.42	7	9	4.4
3514.25	9	11	4.7

Xenon

Xe I

λ Å	Weights g_i	g_k	A 10^8 s^{-1}
1043.8	1	3	$5.9 \cdot 10^{-1}$
1047.1	1	3	1.3
1050.1	1	3	$8.5 \cdot 10^{-2}$
1056.1	1	3	2.45
1061.2	1	3	$1.9 \cdot 10^{-1}$
1068.2	1	3	3.99
1085.4	1	3	$4.10 \cdot 10^{-1}$
1099.7	1	3	$4.34 \cdot 10^{-1}$
1110.7	1	3	1.5
1129.3	1	3	$4.4 \cdot 10^{-2}$
1170.4	1	3	1.6
1192.0	1	3	6.2
1250.2	1	3	$1.4 \cdot 10^{-1}$
1295.6	1	3	2.46
1469.6	1	3	2.81
4501.0	5	3	$6.2 \cdot 10^{-3}$
4524.7	5	5	$2.1 \cdot 10^{-3}$
4624.3	5	5	$7.2 \cdot 10^{-3}$
4671.2	5	7	$1.0 \cdot 10^{-2}$
4807.0	3	1	$2.4 \cdot 10^{-2}$
7119.6	7	9	$6.6 \cdot 10^{-2}$
7967.3	1	3	$3.0 \cdot 10^{-3}$
8409.2	5	3	$1.0 \cdot 10^{-2}$

Xe II

λ Å	Weights g_i	g_k	A 10^8 s^{-1}
4180.1	4	4	2.2
4330.5	6	8	1.4
4414.8	6	6	1.0
4603.0	4	4	$8.2 \cdot 10^{-1}$
4844.3	6	8	1.1
4876.5	6	8	$6.3 \cdot 10^{-1}$
5260.4	2	4	$2.2 \cdot 10^{-1}$
5262.0	4	4	$8.5 \cdot 10^{-1}$
5292.2	6	6	$8.9 \cdot 10^{-1}$
5372.4	4	2	$7.1 \cdot 10^{-1}$
5419.2	4	6	$6.2 \cdot 10^{-1}$
5439.0	4	2	$7.4 \cdot 10^{-1}$
5472.6	8	8	$9.9 \cdot 10^{-2}$
5531.1	8	6	$8.8 \cdot 10^{-2}$
5719.6	4	6	$6.1 \cdot 10^{-2}$
5976.5	4	4	$2.8 \cdot 10^{-1}$
6036.2	6	6	$7.5 \cdot 10^{-2}$
6051.2	8	6	$1.7 \cdot 10^{-1}$
6097.6	6	4	$2.6 \cdot 10^{-1}$
6270.8	4	6	$1.8 \cdot 10^{-1}$
6277.5	4	6	$3.6 \cdot 10^{-2}$
6805.7	8	6	$6.1 \cdot 10^{-2}$
6990.9	10	8	$2.7 \cdot 10^{-1}$

Ytterbium

Yb I

λ Å	Weights g_i	g_k	A 10^8 s^{-1}
2464.5	1	3	$9.1 \cdot 10^{-1}$
2672.0	1	3	$1.18 \cdot 10^{-1}$
3464.4	1	3	$6.2 \cdot 10^{-1}$
3988.0	1	3	1.76
5556.5	1	3	$1.14 \cdot 10^{-2}$

Yb II

λ Å	Weights g_i	g_k	A 10^8 s^{-1}
3289.4	2	4	1.8
3694.2	2	2	1.4

Yttrium

Y I

λ Å	Weights g_i	g_k	A 10^8 s^{-1}
2984.25	6	8	$4.8 \cdot 10^{-1}$
4077.36	4	6	1.1
4102.36	6	8	1.3
4128.30	6	6	1.6
4142.84	4	4	1.6
4167.51	6	6	$2.38 \cdot 10^{-1}$
4235.93	6	4	$3.0 \cdot 10^{-1}$
4379.33	6	4	$7.83 \cdot 10^{-1}$
4476.95	8	6	$2.8 \cdot 10^{-1}$
4514.01	4	6	$3.34 \cdot 10^{-1}$
4527.78	8	6	$8.33 \cdot 10^{-1}$
4544.31	6	6	$4.10 \cdot 10^{-1}$
4559.36	2	4	$4.0 \cdot 10^{-1}$
4643.70	4	6	$1.8 \cdot 10^{-1}$
4653.78	4	6	$1.6 \cdot 10^{-1}$
4674.85	6	8	$1.3 \cdot 10^{-1}$
4781.03	8	10	$1.0 \cdot 10^{-1}$
4799.30	6	8	$1.6 \cdot 10^{-1}$
4804.31	6	4	$2.6 \cdot 10^{-1}$

λ Å	Weights g_i	g_k	A 10^8 s^{-1}
4804.80	4	4	$3.84 \cdot 10^{-1}$
4845.67	8	8	$6.8 \cdot 10^{-1}$
4852.68	6	6	$6.2 \cdot 10^{-1}$
4856.71	6	6	$2.0 \cdot 10^{-1}$
4859.84	4	4	$7.26 \cdot 10^{-1}$
4893.44	6	4	$2.2 \cdot 10^{-1}$
4900.08	8	6	$2.0 \cdot 10^{-1}$
4906.11	10	8	$1.2 \cdot 10^{-1}$
5380.63	6	4	$3.2 \cdot 10^{-1}$
5424.36	6	4	$3.47 \cdot 10^{-1}$
5466.47	10	12	$6.3 \cdot 10^{-1}$
5513.65	6	6	$2.39 \cdot 10^{-1}$
5527.56	8	10	$5.4 \cdot 10^{-1}$
5606.34	10	10	$5.84 \cdot 10^{-2}$
5630.14	4	6	$4.9 \cdot 10^{-1}$
5675.27	6	6	$9.3 \cdot 10^{-2}$
5732.09	6	6	$7.5 \cdot 10^{-2}$
6087.94	6	4	$1.1 \cdot 10^{-1}$
6437.17	10	8	$4.8 \cdot 10^{-2}$
6538.57	10	10	$1.5 \cdot 10^{-1}$
6815.15	2	4	$7.18 \cdot 10^{-2}$
7035.15	4	4	$6.3 \cdot 10^{-2}$

Y II

λ Å	Weights g_i	g_k	A 10^8 s^{-1}
3112.03	1	3	$1.3 \cdot 10^{-2}$
3179.42	3	5	$3.8 \cdot 10^{-2}$
3195.62	3	3	$8.23 \cdot 10^{-1}$
3200.27	5	5	$4.8 \cdot 10^{-1}$
3203.32	3	1	2.77
3216.69	5	3	2.0
3242.28	7	5	2.0
3448.81	5	5	$4.1 \cdot 10^{-2}$
3467.88	5	3	$2.7 \cdot 10^{-2}$
3496.08	1	3	$3.49 \cdot 10^{-1}$
3549.01	5	7	$3.97 \cdot 10^{-1}$
3584.51	3	5	$4.02 \cdot 10^{-1}$
3600.74	7	7	1.4
3601.91	3	3	1.13
3611.04	5	5	1.04
3628.70	5	3	$3.3 \cdot 10^{-1}$
3664.62	7	5	$3.7 \cdot 10^{-1}$
3710.29	7	9	1.5
3747.55	3	3	$1.9 \cdot 10^{-1}$
3774.34	5	7	1.1
3776.56	5	3	$2.42 \cdot 10^{-1}$
3788.70	3	5	$8.1 \cdot 10^{-1}$
3818.34	5	5	$9.70 \cdot 10^{-2}$
3832.90	7	7	$3.0 \cdot 10^{-1}$
3878.29	7	5	$2.9 \cdot 10^{-2}$
3930.66	5	5	$2.1 \cdot 10^{-2}$
3950.36	3	5	$2.80 \cdot 10^{-1}$
3951.59	5	3	$1.5 \cdot 10^{-2}$
3982.60	5	5	$2.7 \cdot 10^{-1}$
4124.91	5	7	$1.8 \cdot 10^{-2}$
4177.54	5	5	$5.27 \cdot 10^{-1}$
4199.27	3	5	$5.36 \cdot 10^{-3}$
4204.69	1	3	$2.20 \cdot 10^{-2}$
4235.73	5	5	$2.3 \cdot 10^{-2}$
4309.62	7	5	$1.29 \cdot 10^{-1}$
4358.73	3	3	$5.55 \cdot 10^{-2}$
4374.95	5	5	$9.97 \cdot 10^{-1}$
4398.01	5	3	$1.16 \cdot 10^{-1}$
4422.59	3	1	$1.83 \cdot 10^{-1}$
4682.33	5	5	$1.9 \cdot 10^{-2}$
4786.58	7	7	$2.1 \cdot 10^{-2}$
4823.31	5	5	$4.3 \cdot 10^{-2}$
4854.87	5	3	$3.9 \cdot 10^{-1}$
4881.44	5	3	$1.5 \cdot 10^{-3}$
4883.69	9	7	$4.7 \cdot 10^{-1}$
4900.11	7	5	$4.51 \cdot 10^{-1}$
4982.13	7	9	$1.5 \cdot 10^{-2}$
5087.42	9	9	$2.0 \cdot 10^{-1}$
5119.11	5	7	$1.6 \cdot 10^{-2}$
5200.41	5	5	$1.3 \cdot 10^{-1}$
5205.73	7	7	$1.6 \cdot 10^{-1}$
5289.82	7	5	$6.7 \cdot 10^{-3}$
5320.78	9	7	$3.9 \cdot 10^{-3}$
5473.39	3	5	$4.3 \cdot 10^{-2}$
5480.73	1	3	$7.62 \cdot 10^{-2}$
5497.41	5	5	$1.2 \cdot 10^{-1}$
5509.90	5	5	$4.24 \cdot 10^{-2}$
5544.61	3	1	$1.8 \cdot 10^{-1}$
5546.01	5	3	$5.8 \cdot 10^{-2}$
5728.89	5	5	$3.0 \cdot 10^{-2}$
6613.74	5	7	$1.7 \cdot 10^{-2}$
6832.48	5	5	$3.3 \cdot 10^{-3}$
7264.16	5	3	$1.3 \cdot 10^{-2}$

Zinc

Zn I

λ Å	Weights g_i	g_k	A 10^8 s^{-1}
748.29	1	3	$6.0 \cdot 10^{-2}$
765.60	1	3	$7.6 \cdot 10^{-2}$
792.05	1	3	$5.7 \cdot 10^{-2}$
793.85	1	3	$1.8 \cdot 10^{-1}$
809.92	1	3	$2.6 \cdot 10^{-1}$
1109.1	1	3	$3.05 \cdot 10^{-1}$
2138.6	1	3	7.09
3075.9	1	3	$3.29 \cdot 10^{-4}$
3282.3	1	3	$9.0 \cdot 10^{-1}$
3302.6	3	5	1.2
3302.9	3	3	$6.7 \cdot 10^{-1}$
3345.0	5	7	1.7
3345.6	5	5	$4.0 \cdot 10^{-1}$
3345.9	5	3	$4.5 \cdot 10^{-2}$
6362.3	3	5	$4.74 \cdot 10^{-1}$
11054	3	1	$2.43 \cdot 10^{-1}$

Zn II

λ Å	Weights g_i	g_k	A 10^8 s^{-1}
2025.5	2	4	3.3
2064.2	2	4	4.6
2099.9	4	6	5.6
2102.2	4	4	$9.3 \cdot 10^{-1}$
4911.6	4	6	1.6

ELECTRON AFFINITIES

Thomas M. Miller

Electron affinity is defined as the energy difference between the lowest (ground) state of the neutral and the lowest state of the corresponding negative ion. The accuracy of electron affinity measurements has been greatly improved since the advent of laser photodetachment experiments with negative ions. Electron affinities can be determined with optical precision, though a detailed understanding of atomic and molecular states and splittings is required to specify the photodetachment threshold corresponding to the electron affinity.

Atomic and molecular electron affinities are discussed in two excellent articles reviewing photodetachment studies which appear in *Gas Phase Ion Chemistry*, Vol. 3, Bowers, M. T., Ed., Academic Press, Orlando, 1984: Chapter 21 by Drzaic, P. S., Marks, J., and Brauman, J. I., "Electron Photodetachment from Gas Phase Negative Ions," p. 167, and Chapter 22 by Mead, R. D., Stevens, A. E., and Lineberger, W. C., "Photodetachment in Negative Ion Beams," p. 213. Persons interested in photodetachment details should consult these articles and the critical reviews of Andersen, T., Haugen, H. K., and Hotop, H., *J. Phys. Chem. Ref. Data* 28, 1511, 1999, Hotop, H., and Lineberger, W. C., *J. Phys. Chem. Ref. Data* 14, 731, 1985, and Andersen, T., Haugen, H. K., and Hotop, H. *J. Phys. Chem. Ref. Data* 28, 1511, 1999. For simplicity in the tables below, any electron affinity that was discussed in the articles by Drzaic et al. or Hotop and Lineberger is referenced to these sources, where original references are given.

The development of cluster-ion photodetachment apparatuses has brought an explosion of electron affinity estimates for atomic and molecular clusters. The policy in this tabulation is to list (in this order) the electron affinities for atoms, diatoms, triatoms, and selected polyatomic molecules and clusters. If the adiabatic electron affinity has been determined for a simple cluster, the value is given, but the reader is referred to original sources for higher-order clusters. Additional data on molecular electron affinities may be found in Lias, S. G., Bartmess, J. E., Liebman, J. F., Holmes, J. L., Levin, R. D., and Mallard, W. G., Gas Phase Ion and Neutral Thermochemistry, *J. Phys. Chem. Ref. Data* 17, (Supplement No. 1), 1988 and on the NIST WebBook at the Internet address http://webbook.nist.gov/.

For the present tabulation, the 2014 CODATA value $e/(hc)$ = 8065.544005 ± 0.000050 cm^{-1} eV^{-1} (http://physics.nist.gov/constants/) has been used to convert electron affinities from the units used in spectroscopic work, cm^{-1}, into eV for this table. Experimental measurements have improved to the level that the 25 ppb uncertainty in $e/(hc)$ will make a difference in a few cases. For this reason, very accurate electron affinities will be given in cm^{-1} with the relevant references.

Abbreviations used in the tables: calc = calculated value; PT = photodetachment threshold using a lamp as a light source; LPT = laser photodetachment threshold; LPES = laser photoelectron spectroscopy; DA = dissociative attachment; attach = electron attachment/detachment equilibrium; e-scat = electron scattering; kinetic = dissociation kinetics; Knud = Knudsen cell; CT = charge transfer; CD = collisional detachment; and ZEKE = zero electron kinetic energy spectroscopy.

Electron Affinities of Atoms and Molecules

Z	Formula	Name	Elec. aff. (eV)	Uncert. (eV)	Method	Ref.
Atoms (Listed by Atomic Number)						
1	H	Hydrogen (atomic)	0.75420817		calc	205
	H	Hydrogen (atomic)	0.754195	0.000019	LPT	89
	D	Deuterium (atomic)	0.754593	0.000074	LPT	89
	D	Deuterium (atomic)	0.75465629		calc	205
	T	Tritium (atomic)	0.75480545		calc	205
2	He	Helium	not stable		calc	1
3	Li	Lithium	0.618049	0.000020	LPT	185
4	Be	Beryllium	not stable		calc	1
5	B	Boron	0.279723	0.000025	LPES	191
6	C	Carbon	1.262119	0.000020	LPT	28
7	N	Nitrogen (atomic)	not stable		DA	1
8	O	Oxygen (atomic)	1.4611135	0.0000009	LPT	4
9	F	Fluorine (atomic)	3.4011897	0.0000024	LPT	227
10	Ne	Neon	not stable		calc	1
11	Na	Sodium	0.547926	0.000025	LPT	1
12	Mg	Magnesium	not stable		e-scat	1
13	Al	Aluminum	0.43283	0.00005	LPES	208
14	Si	Silicon	1.38952	0.0000007	LPES	4
15	P	Phosphorus	0.746607	0.000010	LPT	377
16	S	Sulfur	2.07710403 (^{32}S)	0.00000051	LPT	334
	S	Sulfur	2.0771043 (^{34}S)	0.0000011	LPT	334
17	Cl	Chlorine (atomic)	3.612725	0.000027	LPT	52
18	Ar	Argon	not stable		calc	1
19	K	Potassium	0.50146	0.000012	LPT	425
20	Ca	Calcium	0.02455	0.00010	LPT	44

Z	Formula	Name	Elec. aff. (eV)	Uncert. (eV)	Method	Ref.
21	Sc	Scandium	0.188	0.020	LPES	1
22	Ti	Titanium	0.079	0.014	LPES	1
23	V	Vanadium	0.525	0.012	LPES	1
24	Cr	Chromium	0.67584	0.00012	LPT	426
25	Mn	Manganese	not stable		calc	1
26	Fe	Iron	0.151	0.003	LPES	27
27	Co	Cobalt	0.66226	0.000046	LPES	417
28	Ni	Nickel	1.15716	0.00012	LPT	116
29	Cu	Copper	1.23578	0.000037	LPT	426
30	Zn	Zinc	not stable		e-scat	1
31	Ga	Gallium	0.43	0.03	LPES	183
32	Ge	Germanium	1.232712	0.000015	LPES	28
33	As	Arsenic	0.804	0.002	LPES	352
34	Se	Selenium	2.020670	0.000025	LPT	1
35	Br	Bromine (atomic)	3.3635882	0.0000019	LPT	74
36	Kr	Krypton	not stable		calc	1
37	Rb	Rubidium	0.48592	0.00002	LPT	1
38	Sr	Strontium	0.05206	0.00006	LPT	122
39	Y	Yttrium	0.307	0.012	LPES	1
40	Zr	Zirconium	0.426	0.014	LPES	1
41	Nb	Niobium	0.916	0.005	LPES	311
42	Mo	Molybdenum	0.74725	0.00025	LPT	426
43	Tc	Technetium	0.55	0.20	calc	1
44	Ru	Ruthenium	1.04638	0.00025	calc	427
45	Rh	Rhodium	1.14289	0.0002	LPT	116
46	Pd	Palladium	0.56214	0.00012	LPT	116
47	Ag	Silver	1.30448	0.000025	LPT	426
48	Cd	Cadmium	not stable		e-scat	1
49	In	Indium	0.404	0.009	LPT	428
50	Sn	Tin	1.112067	0.000015	LPES	28
51	Sb	Antimony	1.0474	0.000019	LPT	429
52	Te	Tellurium	1.970876	0.000007	LPT	261
53	I	Iodine (atomic)	3.0590368	0.0000010	LPES	92
54	Xe	Xenon	not stable		calc	1
55	Cs	Cesium	0.471626	0.000025	LPT	1
56	Ba	Barium	0.14462	0.00006	LPT	195
57	La	Lanthanum	0.55	0.02	LPT	184
58	Ce	Cerium	0.65	0.03	LPT	269
59	Pr	Praseodymium	0.962	0.024	LPES	225
60	Nd	Neodymium	>1.916		LPES	342
63	Eu	Europium	0.864	0.024	LPES	268
65	Tb	Terbium	>1.165		LPES	342
66	Dy	Dysprosium	>0		LPES	342
69	Tm	Thulium	1.029	0.022	LPES	264
70	Yb	Ytterbium	-0.020		calc	196
71	Lu	Lutetium	0.34	0.01	LPT	223
72	Hf	Hafnium	0.014		calc	343
73	Ta	Tantalum	0.322	0.012	LPES	1
74	W	Tungsten	0.81626	0.00007	LPES	360
75	Re	Rhenium	0.15	0.15	calc	1
76	Os	Osmium	1.0778	0.00012	LPT	430
77	Ir	Iridium	1.5638	0.0005	LPT	141
78	Pt	Platinum	2.1251	0.00005	LPT	431
79	Au	Gold	2.30863	0.00003	LPT	1
80	Hg	Mercury	not stable		e-scat	1
81	Tl	Thallium	0.377	0.013	LPES	341
82	Pb	Lead	0.35674	0.000016	LPES	408
83	Bi	Bismuth	0.942362	0.000013	LPT	262
84	Po	Polonium	1.9	0.3	calc	1
85	At	Astatine	2.8	0.2	calc	1
86	Rn	Radon	not stable		calc	1

Z	Formula	Name	Elec. aff. (eV)	Uncert. (eV)	Method	Ref.
87	Fr	Francium	0.46		calc	82
89	Ac	Actinium	0.35		calc	207
114	Fl	Flerovium	<0		calc	354
118	Og	Oganesson	0.056	0.01	calc	140
121	Ubu	Unbiunium	0.57		calc	207
Atomic Families						
57-71		Lanthanides			calc	355
89-103		Actinides			calc	355
Diatomic Molecules (Listed by Formula)						
	Ag_2	Disilver	1.023	0.007	LPES	37
	AgF	Silver(I) fluoride	1.46	0.01	LPES	403
	AgO	Silver(II) oxide	1.654	0.002	LPES	297
	Al_2	Dialuminum	1.10	0.15	LPES	68
	AlO	Aluminum monoxide	2.60	0.02	LPES	143
	AlP	Aluminum phosphide	2.043	0.020	LPES	218
	AlS	Aluminum monosulfide	2.60	0.03	LPES	129
	As_2	Diarsenic	0.739	0.008	LPES	200
	AsH	Arsenic monohydride	1.0	0.1	PT	2
	AsO	Arsenic monoxide	1.286	0.008	LPES	198
	Au_2	Digold	1.9393	0.0003	LPES	37
	AuBi	Gold bismuth	1.38	0.04	LPES	432
	AuH	Gold hydride	0.758	0.020	LPES	276
	AuO	Gold monoxide	2.374	0.007	LPES	282
	AuPd	Gold palladium	1.88		LPES	220
	AuS	Gold monosulfide	2.4734	0.0005	LPES	435
	BN	Boron nitride	3.160	0.005	LPES	189
	BO	Boron monoxide	2.508	0.008	LPES	6
	BeH	Beryllium monohydride	0.7	0.1	PT	2
	Bi_2	Dibismuth	1.271	0.008	LPES	119
	Br_2	Bromine	2.55	0.10	CT	2
	BrO	Bromine monoxide	2.353	0.006	LPES	88
	C_2	Dicarbon	3.269	0.006	LPES	87
	CH	Methylidyne	1.238	0.008	LPES	2
	CN	Cyanide	3.862	0.004	LPES	111
	CRh	Rhodium carbide (RhC)	1.46	0.02	LPES	206
	CS	Carbon monosulfide	0.205	0.021	LPES	2
	CaH	Calcium monohydride	0.93	0.05	PT	2
	Cl_2	Chlorine	2.38	0.10	CT	2
	ClO	Chlorine oxide (ClO)	2.2775	0.0013	LPES	339
	Co_2	Dicobalt	1.110	0.008	LPES	27
	CoH	Cobalt hydride (CoH)	0.671	0.010	LPES	29
	CoD	Cobalt hydride (CoD)	0.680	0.010	LPES	29
	Cr_2	Dichromium	0.505	0.005	LPES	114
	CrH	Chromium monohydride	0.563	0.010	LPES	29
	CrD	Chromium monohydride-*d*	0.568	0.010	LPES	29
	CrO	Chromium monoxide	1.221	0.006	LPES	5
	Cs_2	Dicesium	0.469	0.015	LPES	104
	CsCl	Cesium chloride	0.455	0.010	LPES	30
	CsO	Cesium monoxide	0.273	0.012	LPES	133
	Cu_2	Dicopper	0.836	0.006	LPES	37
	CuH	Copper(I) hydride	0.444	0.006	LPES	308
	CuD	Copper(I) hydride-*d*	0.439	0.006	LPES	308
	CuO	Copper(II) oxide	1.777	0.006	LPES	118
	F_2	Fluorine	3.01	0.07	kinetic	331
	OF	Fluorine oxide	2.272	0.006	LPES	88
	Fe_2	Diiron	0.902	0.008	LPES	27
	FeH	Iron hydride (FeH)	0.934	0.011	LPES	9
	FeD	Iron hydride (FeD)	0.932	0.015	LPES	9
	FeO	Iron(II) oxide	1.493	0.005	LPES	45
	GaAs	Gallium arsenide	1.949	0.020	LPES	218

Z	Formula	Name	Elec. aff. (eV)	Uncert. (eV)	Method	Ref.
	GaO	Gallium monoxide	2.612	0.008	LPES	279
	GaP	Gallium phosphide	1.988	0.020	LPES	218
	Ge_2	Digermanium	2.035	0.001	LPES	123
	HfO	Hafnium monoxide	0.6	0.05	LPES	173
	I_2	Iodine	2.524	0.015	LPES	305
	IBr	Iodine bromide	2.512	0.003	LPES	350
	IO	Iodine monoxide	2.378	0.006	LPES	88
	InBi	Indium bismuthide	1.72		LPES	364
	InP	Indium phosphide	1.845	0.020	LPES	218
	K_2	Dipotassium	0.497	0.012	LPES	104
	KBr	Potassium bromide	0.642	0.010	LPES	30
	KCl	Potassium chloride	0.582	0.010	LPES	30
	KCs	Cesium potassium	0.471	0.020	LPES	104
	KI	Potassium iodide	0.728	0.010	LPES	30
	RbK	Potassium rubidium	0.486	0.020	LPES	104
	LiCl	Lithium chloride	0.593	0.010	LPES	30
	LiH	Lithium hydride	0.342	0.012	LPES	102
	LiD	Lithium hydride-*d*	0.337	0.012	LPES	102
	MgCl	Magnesium monochloride	1.589	0.011	LPES	31
	MgD	Magnesium monohydride-*d*	0.89	0.05	LPES	385
	MgH	Magnesium monohydride	0.9	0.05	LPES	385
	MgI	Magnesium monoiodide	1.899	0.018	LPES	31
	MgO	Magnesium oxide	1.630	0.025	LPES	178
	MnH	Manganese monohydride	0.869	0.010	LPES	9
	MnD	Manganese monohydride-*d*	0.866	0.010	LPES	9
	MnO	Manganese(II) oxide	1.375	0.010	LPES	158
	MoC	Molybdenum carbide (MoC)	1.36	0.003	LPES	386
	MoO	Molybdenum monoxide	1.290	0.006	LPES	127
	NH	Imidogen	0.370	0.004	LPT	32
	NO	Nitric oxide	0.026	0.005	LPES	73
	NRh	Rhodium nitride (RhN)	1.51	0.02	LPES	206
	NS	Nitrogen sulfide	1.194	0.011	LPES	2
	Na_2	Disodium	0.430	0.015	LPES	104
	NaBr	Sodium bromide	0.788	0.010	LPES	30
	NaCl	Sodium chloride	0.727	0.010	LPES	30
	NaF	Sodium fluoride	0.520	0.010	LPES	30
	NaI	Sodium iodide	0.865	0.010	LPES	30
	NbO	Niobium(II) oxide	1.29	0.02	LPES	174
	NaK	Potassium sodium	0.465	0.030	LPES	104
	NbN	Niobium nitride	1.45	0.003	LPES	386
	Ni_2	Dinickel	0.926	0.010	LPES	112
	NiCu	Copper nickel (CuNi)	0.889	0.010	LPES	128
	NiAg	Nickel silver (NiAg)	0.979	0.010	LPES	128
	NiH	Nickel monohydride	0.481	0.007	LPES	29
	NiD	Nickel monohydride-*d*	0.477	0.007	LPES	29
	NiO	Nickel(II) oxide	1.470	0.003	LPES	146
	O_2	Oxygen	0.450	0.002	LPES	222
	OH	Hydroxyl	1.8276488	0.0000009	LPT	226
	OD	Hydroxyl-*d*	1.825533	0.000037	LPT	142
	P_2	Diphosphorus	0.589	0.025	LPES	42
	PH	Phosphorus monohydride	1.027	0.006	LPES	206
	PO	Phosphorus monoxide	1.092	0.010	LPES	2
	Pb_2	Dilead	1.366	0.010	LPES	117
	PbO	Lead(II) oxide (litharge)	0.722	0.006	LPES	105
	PbS	Lead(II) sulfide	1.049	0.010	LPES	228
	Pd_2	Dipalladium	1.685	0.008	LPES	112
	PdCo	Cobalt palladium (CoPd)	0.604	0.010	LPES	160
	PdO	Palladium(II) oxide	1.570	0.006	LPES	290
	PrO	Praseodymium monoxide	0.96	0.01	LPES	415
	Pt_2	Diplatinum	1.898	0.008	LPES	112
	PtN	Platinum nitride (PtN)	1.240	0.010	LPES	46

Z	Formula	Name	Elec. aff. (eV)	Uncert. (eV)	Method	Ref.
	Rb_2	Dirubidium	0.498	0.015	LPES	104
	RbCl	Rubidium chloride	0.544	0.010	LPES	30
	RbCs	Cesium rubidium	0.478	0.020	LPES	104
	Re_2	Dirhenium	1.571	0.008	LPES	33
	RhO	Rhodium monoxide	1.58	0.02	LPES	206
	S_2	Disulfur	1.670	0.015	LPES	53
	SF	Sulfur monofluoride	2.285	0.006	LPES	93
	SH	Mercapto	2.3147282	0.0000015	LPT	47
	SD	Mercapto-*d*	2.315	0.002	LPES	10
	SO	Sulfur monoxide	1.125	0.005	LPES	84
	Sb_2	Diantimony	1.282	0.008	LPES	108
	ScO	Scandium monoxide	1.35	0.02	LPES	171
	Se_2	Diselenium	1.94	0.07	LPES	38
	SeH	Selenium monohydride	2.212519	0.000025	LPT	48
	SeO	Selenium monoxide	1.456	0.020	LPES	41
	Si_2	Disilicon	2.201	0.010	LPES	100
	SiF	Fluorosilylidyne	0.81	0.02	LPES	278
	SiH	Silylidyne	1.277	0.009	LPES	2
	SiN	Silicon nitride	2.949	0.008	LPES	274
	Sn_2	Ditin	1.962	0.010	LPES	117
	SnO	Tin(II) oxide	0.598	0.006	LPES	168
	SnPb	Lead tin (PbSn)	1.569	0.008	LPES	117
	TaC	Tantalum carbide	1.928	0.056	LPES	416
	TaO	Tantalum monoxide	1.07	0.06	LPES	360
	Te_2	Ditellurium	1.92	0.07	LPES	38
	TeH	Tellurium monohydride	2.102	0.015	LPES	39
	TeO	Tellurium monoxide	1.697	0.022	LPES	40
	TiO	Titanium(II) oxide	1.30	0.03	LPES	172
	VO	Vanadium(II) oxide	1.229	0.008	LPES	170
	WC	Tungsten carbide	1.022	0.01	LPES	409
	YO	Yttrium monoxide	1.35	0.02	LPES	171
	ZnF	Zinc monofluoride	1.974	0.008	LPES	179
	ZnH	Zinc monohydride	<0.95		PT	2
	ZnO	Zinc oxide	2.087	0.008	LPES	179
	ZrO	Zirconium(II) oxide	1.26	0.05	LPES	173
Triatomic Molecules (Listed by Formula)						
	Ag_3	Silver trimer	2.32	0.05	LPES	37
	AgCN	Silver(I) cyanide	1.588	0.010	LPES	163
	Al_3	Aluminum trimer	1.916	0.004	LPES	332
	$AlBi_2$	Dibismuth aluminum cluster	1.94	0.08	LPES	399
	AlO_2	Aluminum oxide [AlO_2]	4.23	0.02	LPES	143
	AlP_2	Aluminum phosphide [AlP_2]	1.933	0.007	LPES	217
	AlTiO	Aluminum titanium oxide [AlTiO]	1.70	0.08	LPES	359
	Al_2N	Aluminum nitride (Al_2N)	2.571	0.008	LPES	297
	Al_2P	Aluminum phosphide (Al_2P)	2.513	0.020	LPES	217
	Al_2S	Aluminum sulfide (Al_2S)	0.80	0.12	LPES	129
	As_3	Arsenic trimer	1.45	0.03	LPES	200
	AsH_2	Arsenic hydride [AsH_2]	1.27	0.03	PT	2
	Au_3	Gold trimer	3.7	0.3	LPES	37
	AuBO	Gold-boron oxide complex (AuBO)	1.46	0.02	LPES	336
	$AuBr_2$	Gold dibromide	4.46	0.07	LPES	294
	AuC_2	Gold dicarbide	3.2192	0.0007	LPES	406
	$AuCl_2$	Gold dichloride	4.60	0.07	LPES	294
	AuH_2	Gold dihydride	3.03	0.02	LPES	444
	AuI_2	Gold diiodide	4.226	0.010	LPES	372
	AuOH	Gold hydroxide (AuOH)	1.771	0.015	LPES	336
	AuO_2	Gold dioxide	3.40	0.03	LPES	373
	AuS_2	Gold disulfide (SAuS)	3.42	0.03	LPES	373
	AuS_2	Gold disulfide (AuSS)	2.24	0.03	LPES	373
	Au_2H	Gold hydride (Au_2H)	3.437	0.003	LPES	276
	Au_2Pd	Gold palladium (Au_2Pd)	3.80		LPES	220

Atomic

Z	Formula	Name	Elec. aff. (eV)	Uncert. (eV)	Method	Ref.
	B_3	Boron trimer	2.82	0.02	LPES	221
	BiBO	Bismuth boron oxide cluster	1.84	0.03	LPES	432
	BO_2	Boron dioxide	4.46	0.03	LPES	338
	B_2N	Boron nitride (B_2N)	3.098	0.005	LPES	193
	Bi_3	Bismuth trimer	1.60	0.03	LPES	119
	$BiIn_2$	Bismuth-indium complex ($BiIn_2$)	2.13		LPES	364
	Bi_2In	Bismuth-indium complex (Bi_2In)	2.11		LPES	364
	C_3	Carbon trimer	1.981	0.020	LPES	11
	CBr_2	Dibromomethylene	1.78	0.10	LPES	249
	CCl_2	Dichloromethylene	1.593	0.006	LPES	249
	CF_2	Difluoromethylene	0.180	0.020	LPES	235
	CH_2	Methylene	0.652	0.006	LPES	12
	CD_2	Methylene-d_2	0.645	0.006	LPES	12
	CI_2	Diiodomethylene	2.09	0.07	LPES	235
	CHBr	Bromomethylene	1.454	0.005	LPES	95
	CHCl	Chloromethylene	1.210	0.005	LPES	95
	CHF	Fluoromethylene	0.542	0.005	LPES	95
	CDF	Fluoromethylene-*d*	0.535	0.005	LPES	95
	CHI	Iodomethylene	1.42	0.17	LPES	95
	C_2H	Ethynyl	2.969	0.006	LPES	87
	CNC	Carbon nitride [CNC]	2.7489	0.0010	LPES	346
	C_2Nb	Niobium carbide [NbC_2]	1.380	0.025	LPES	243
	CCO	Dicarbon monoxide	2.3107	0.0006	LPES	323
	CCS	Dicarbon monosulfide	2.7475	0.0006	LPES	323
	CS_2	Carbon disulfide	0.5525	0.0013	LPES	278
	ClO_2	Chlorine dioxide	2.1451	0.0025	LPES	339
	ClO_2	Chlorine superoxide [ClOO]	1.6600	0.0002	LPES	339
	CoH_2	Cobalt hydride [CoH_2]	1.450	0.014	LPES	34
	CoD_2	Cobalt hydride-d_2 [CoD_2]	1.465	0.013	LPES	34
	CoOH	Cobalt hydroxide	1.33	0.04	LPES	400
	CrH_2	Chromium hydride [CrH_2]	>2.5		LPES	34
	Cr_2H	Hydrodichromium	1.474	0.005	LPES	107
	Cr_2D	Hydro-*d*-dichromium	1.464	0.005	LPES	107
	CrO_2	Chromium oxide radical (OCrO)	2.413	0.008	LPES	144
	CrO_2	Chromium oxide radical [$Cr(O_2)$]	1.5	0.06	LPES	241
	Cr_2O	Chromium oxide (Cr_2O)	0.9	0.1	LPES	306
	Cs_3	Cesium trimer	0.864	0.030	LPES	18
	CsI_2	Cesium diiodide	4.52	0.02	LPES	372
	Cu_3	Copper trimer	2.11	0.05	LPES	37
	$CuBr_2$	Copper(II) bromide	4.35	0.05	LPES	177
	CuCN	Copper(I) cyanide	1.466	0.010	LPES	163
	$CuCl_2$	Copper(II) chloride	4.35	0.05	LPES	177
	CuH_2	Copper(II) hydride	2.60	0.05	LPES	308
	CuD_2	Copper(II) hydride-*d*	2.60	0.05	LPES	308
	CuI_2	Copper(II) iodide	4.256	0.010	LPES	372
	Fe_3	Iron trimer	1.43	0.06	LPES	149
	FeC_2	Iron carbide [FeC_2]	1.9782	0.0006	LPES	254
	FeCO	Iron carbonyl (FeCO)	1.157	0.005	LPES	103
	FeH_2	Iron hydride [FeH_2]	1.049	0.014	LPES	34
	FeD_2	Iron hydride-d_2 [FeD_2]	1.038	0.013	LPES	34
	FeO_2	Iron oxide [FeO_2]	2.358	0.030	LPES	130
	Fe_2H	Iron hydride (Fe_2H)	0.564	0.019	LPES	254
	Fe_2O	Iron oxide (Fe_2O)	1.60	0.02	LPES	152
	$GaAs_2$	Gallium arsenide [$GaAs_2$]	1.894	0.033	LPES	192
	GaP_2	Gallium phosphide [GaP_2]	1.666	0.041	LPES	192
	Ga_2As	Gallium arsenide (Ga_2As)	2.428	0.020	LPES	192
	Ga_2N	Gallium nitride (Ga_2N)	2.506	0.008	LPES	302
	Ga_2P	Gallium phosphide (Ga_2P)	2.481	0.015	LPES	192
	Ge_3	Germanium trimer	2.23	0.01	LPES	123
	GeH_2	Germylene	1.097	0.015	LPES	28
	HCO	Oxomethyl [HCO]	0.313	0.005	LPES	35

Z	Formula	Name	Elec. aff. (eV)	Uncert. (eV)	Method	Ref.
	DCO	Oxomethyl-*d* (DCO)	0.301	0.005	LPES	35
	HCl_2	Hydrogen chloride [HCl_2]	4.896	0.005	LPES	69
	HNO	Nitrosyl hydride	0.338	0.015	LPES	14
	DNO	Nitrosyl hydride-*d*	0.330	0.015	LPES	14
	HO_2	Hydroperoxy	1.078	0.006	LPES	15
	DO_2	Hydroperoxy-*d*	1.077	0.005	LPES	15
	HS_2	Thiosulfeno	1.916	0.015	LPES	347
	DS_2	Thiosulfeno-*d*	1.918	0.015	LPES	347
	HfOH	Hafnium hydroxide	1.7	0.05	LPES	173
	HfO_2	Hafnium(IV) oxide	2.14	0.03	LPES	319
	I_3	Iodine trimer	4.226	0.013	LPES	162
	ICN	Cyanogen iodide	1.345	-2	LPES	383
	InP_2	Indium phosphide [InP_2]	1.61	0.05	LPES	137
	In_2P	Indium phosphide (In_2P)	2.36	0.05	LPES	137
	K_3	Potassium trimer	0.956	0.050	LPES	18
	MnH_2	Manganese hydride [MnH_2]	0.444	0.016	LPES	34
	MnD_2	Manganese hydride-d_2 [MnD_2]	0.465	0.014	LPES	34
	MnO_2	Manganese(IV) oxide	2.06	0.03	LPES	158
	N_3	Nitrogen trimer	2.70	0.12	PT	2
	N_3	Nitrogen trimer	2.68	0.01	LPT	255
	NCN	Cyanoimidogen	2.484	0.006	LPES	154
	NCO	Cyanate	3.609	0.005	LPES	111
	NCS	Thiocyanate	3.537	0.005	LPES	111
	NH_2	Amidogen	0.771	0.005	LPES	58
	N_2O	Nitrous oxide	-0.03	0.10	calc	59
	NOAr	Nitric oxide argon cluster	0.084	0.003	LPES	79
	NOKr	Nitric oxide krypton cluster	0.125	0.003	LPES	79
	NO_2	Nitrogen dioxide	2.273	0.005	LPES	63
	(NO)R	Nitric oxide-rare gas cluster	R=Ar,Kr,Xe		LPES	90
	NOXe	Nitric oxide xenon cluster	0.193	0.003	LPES	79
	Na_3	Trisodium	1.019	0.060	LPES	18
	Nb_3	Niobium trimer	1.032	0.010	LPES	175
	Ni_3	Nickel trimer	1.41	0.05	LPES	55
	NiCN	Nickel(I) cyanide	1.771	0.010	LPES	287
	NiCO	Nickel carbonyl (NiCO)	0.804	0.012	LPES	2
	NiH_2	Nickel hydride [NiH_2]	1.934	0.008	LPES	34
	NiD_2	Nickel hydride-d_2 [NiD_2]	1.926	0.007	LPES	34
	ONiO	Nickel oxide [NiO_2]	3.05	0.01	LPES	214
	NiO_2	Nickel peroxide	0.82	0.03	LPES	214
	O_3	Ozone	2.1028	0.0025	LPT	2
	O_2Ar	Argon oxide [ArO_2]	0.52	0.02	LPES	75
	OCS	Carbon oxysulfide	-0.04		LPES	272
	OIO	Iodine dioxide	2.577	0.008	LPES	88
	PH_2	Phosphino	1.263	0.006	LPES	281
	P_2H	Diphosphenyl	1.514	0.010	LPES	281
	PO_2	Phosphorus dioxide	3.42	0.01	LPES	124
	Pb_3	Lead trimer	1.70	0.09	LPES	345
	Pd_3	Palladium trimer	<1.5	0.1	LPES	55
	PdCN	Palladium(I) cyanide	2.543	0.007	LPES	287
	PdCO	Palladium carbonyl	0.606	0.010	LPES	293
	Pt_3	Platinum trimer	1.87	0.02	LPES	55
	PtCN	Platinum(I) cyanide	3.191	0.003	LPES	287
	PtCO	Platinum carbonyl	1.196	0.034	LPES	391
	Rb_3	Rubidium trimer	0.920	0.030	LPES	18
	ReO_2	Rhenium(IV) oxide	2.5	0.1	LPES	216
	S_3	Sulfur trimer	2.093	0.025	LPES	16
	SO_2	Sulfur dioxide	1.107	0.008	LPES	16
	S_2O	Sulfur oxide (SSO)	1.877	0.008	LPES	16
	Sb_3	Antimony trimer	1.85	0.03	LPES	108
	ScO_2	Scandium dioxide	2.32	0.02	LPES	171
	SeO_2	Selenium dioxide	1.823	0.050	LPES	38

Z	Formula	Name	Elec. aff. (eV)	Uncert. (eV)	Method	Ref.
	SiH_2	Silylene	1.124	0.020	LPES	2
	SiF_2	Difluorosilylene	0.10	0.10	LPES	278
	Si_2F	Fluorodisilynylium	1.99	0.28	LPES	17
	Si_2H	Disilynyl	2.31	0.01	LPES	182
	Si_3	Silicon trimer	2.29	0.02	LPES	110
	Sn_3	Tin trimer	2.24	0.01	LPES	289
	SnCN	Tin(I) cyanide	1.922	0.006	LPES	292
	Ta_3	Tantalum trimer	1.36	0.03	LPES	169
	TaO_2	Tantalum(IV) oxide	2.40	0.06	LPES	260
	TiC_2	Titanium carbide [TiC_2]	1.542	0.020	LPES	147
	TiO_2	Titanium(IV) oxide (rutile)	1.59	0.03	LPES	172
	UF_2	Uranium difluoride	1.16	0.03	LPES	412
	UFO	Uranium oxide fluoride	1.27	0.03	LPES	410
	UO_2	Uranium(IV) oxide	1.159	0.02	LPES	413
	V_3	Vanadium trimer	1.107	0.010	LPES	176
	VO_2	Vanadium(IV) oxide	2.3	0.2	CT	101
	WO_2	Tungsten(IV) oxide	1.998	0.050	LPES	299
	YO_2	Yttrium dioxide	2.00	0.03	LPES	171
	ZrO_2	Zirconium(IV) oxide	1.64	0.03	LPES	319
Larger Polyatomic Molecules (Listed by Formula)						
	Ag_n	Silver cluster	n=1-60		LPES	37
	Al_4	Tetraaluminum cluster	2.180	0.02	LPT	421
	Al_5	Aluminum pentamer	2.23	0.05	LPES	238
	Al_n	Aluminum cluster	n=3-32		LPES	68
	$Al_nB_mH_x$ n=1-7 m,x=1-4	Aluminum boride hydride cluster			LPES	382
	Al_nH_m n=3-8 m=1-4	Aluminum hydride cluster			LPES	382
	$AlTiO_2$	Aluminum titanium oxide [$AlTiO_2$]	1.70	0.08	LPES	358
	$AlTiO_3$	Aluminum titanium oxide [$AlTiO_3$]	2.47	0.08	LPES	359
	Al_2C_2	Aluminum carbide [Al_2C_2]	0.64	0.05	LPES	239
	Al_2TiO_2	Aluminum titanium oxide [Al_2TiO_2]	1.17	0.08	LPES	359
	Al_2TiO_3	Aluminum titanium oxide [Al_2TiO_3]	2.2	0.1	LPES	359
	Al_3C	Aluminum carbide (Al_3C)	2.56	0.06	LPES	161
	Al_3C_2	Aluminum carbide [Al_3C_2]	2.19	0.03	LPES	244
	Al_3Ge_2	Aluminum germanium [Al_3Ge_2]	2.43	0.03	LPES	244
	Al_3Si_2	Aluminum silicon [Al_3Si_2]	2.36	0.03	LPES	244
	Al_3O	Oxotrialuminum	1.00	0.15	LPES	68
	Al_5O_4	Aluminum oxide [Al_5O_4]	3.50	0.05	LPES	283
	$Al_5O_4{\cdot}H_2O$	Aluminum oxide (Al_5O_4)-water complex	3.10	0.10	LPES	283
	Al_6N	Aluminum-nitrogen complex (Al_6N)	2.58	0.04	LPES	337
	Al_8N	Aluminum-nitrogen complex (Al_8N)	2.75	0.05	LPES	348
	Al_nO_m	Aluminum oxide cluster	n=1,2	m=1-5	LPES	143
	Al_nO_m	Aluminum oxide cluster	n=3-7	m=2-5	LPES	267
	Al_nP_m	Aluminum phosphide cluster	n=1-4	m=1-4	LPES	217
	Al_nS_m	Aluminum sulfide cluster	n=1-5	m=1-3	LPES	129
	$Ar(H_2O)_n$	Argon-water cluster	n=2,6,7		LPES	77
	Ar_nBr	Argon-bromine cluster	n=2-9		ZEKE	212
	Ar_nI	Argon-iodine cluster	n=2-19		ZEKE	212
	As_4	Arsenic tetramer	<0.8		LPES	200
	As_5	Arsenic pentamer	≈1.7		LPES	200
	As_5	Arsenic pentamer	≈3.5		LPES	253
	Au_n	Gold cluster	n=1-233		LPES	37
	$AuBO_2$	Gold-borate complex ($AuBO_2$)	2.8	0.1	LPES	371
	AuF_6	Gold fluoride [AuF_6]	7.5	estimate	CT	98
	$AuO(BO_2)$	Gold-borate complex ($AuO(BO_2)$)	4.0	0.1	LPES	371
	$Au(BO_2)_2$	Gold-borate complex ($Au(BO_2)_2$)	5.7	0.1	LPES	371
	AuB_3	Gold triboride cluste	2.29	0.02	LPES	440
	AuBrCN	Gold cyanide bromide complex	5.14	0.05	LPES	449
	AuCNCl	Gold cyanide chloride complex	5.38	0.05	LPES	449
	AuCNI	Gold cyanide iodide complex	4.75	0.05	LPES	449
	AuC_2H	Gold ethyne	1.475	0.001	LPES	405
	AuC_4	Gold tetracarbide	3.366	0.001	LPES	405

Z	Formula	Name	Elec. aff. (eV)	Uncert. (eV)	Method	Ref.
	AuC_4H	Gold buta-1,3-diyne	1.778	0.001	LPES	405
	AuC_6H	Gold 1,3,5-hexatriyne	1.962	0.001	LPES	405
	AuC_6	Gold hexacarbide	3.593	0.001	LPES	405
	$Au(SCH_3)_2$	Gold bi(methylthiol)	3.33	0.03	LPES	447
	Au_2Al_2	Gold aluminum cluster	1.4438	0.0008	LPES	436
	Au_2BO	Gold-boron oxide complex (Au_2BO)	4.32	0.02	LPES	336
	Au_2B_3	Bigold triboride cluster	3.17	0.03	LPES	440
	Au_2Br_7	Gold-bromine complex	3.52	0.02	LPES	301
	$Au_2(SCH_3)_3$	Bigold trimethylthiol	4.06	0.03	LPES	447
	Au_3BO	Gold-boron oxide complex (Au_3BO)	3.08	0.02	LPES	336
	Au_3BO_2	Gold-borate complex (Au_3BO_2)	3.1	0.1	LPES	371
	$Au_3O(BO_2)$	Gold-borate complex ($Au_3O(BO_2)$)	4.9	0.1	LPES	371
	Au_3Pd	Gold palladium (Au_3Pd)	2.51		LPES	220
	Au_4	Gold cluster (Au_4)	2.7098	0.0006	LPES	441
	Au_4Pd	Gold palladium (Au_4Pd)	2.69		LPES	220
	Au_4Ti	Gold titanium cluster (Au_4Ti)	2.8	0.05	LPES	438
	Au_6	Gold hexamer	2.06	0.02	LPES	288
	$Au_6(CO)$	Gold hexamer carbonyl	2.04	0.05	LPES	288
	$Au_6(CO)_2$	Gold hexamer dicarbonyl	2.03	0.05	LPES	288
	$Au_6(CO)_3$	Gold hexamer tricarbonyl	1.95	0.05	LPES	288
	$Au_{12}Nb$	Gold-niobium complex	3.70	0.03	LPES	275
	$Au_{12}Ta$	Gold-tantalum complex	3.77	0.03	LPES	275
	$Au_{12}V$	Gold-vanadium complex	3.76	0.03	LPES	275
	B_5	Boron pentamer	2.33	0.02	LPES	245
	BH_3	Borane(3)	0.038	0.015	LPES	62
	BD_3	Borane(3)-d_3	0.027	0.014	LPES	62
	BO_2Fe_n	Boron dioxide-iron cluster	n=1-5		LPES	358
	B_3N	Boron nitride (B_3N)	2.098	0.035	LPES	193
	B_3O_3	Boron oxide complex (B_3O_3)				
	B_3O_3H	Boron oxide hydride complex				
	$B_3(BO)_2$	Boron oxide complex [$B_3(BO)_2$]	4.44	0.02	LPES	440
	B_3BO	Boron oxide complex (B_3BO)	2.71	0.02	LPES	440
	$B_4(BO)$	Boron oxide complex ($B_4(BO)$)				
	$B_4(BO)_n$$n$=1-3	Boron oxide complex $B_4(BO)_n$$n$-1,3				
	LiB_6	Lithium hexaboride	2.3	0.1	LPES	298
	Bi_n	Bismuth cluster	n=2-9		LPES	213
	Bi_2In_2	Bismuth-indium complex (Bi_2In_2)	1.82		LPES	364
	Bi_2In_3	Bismuth-indium complex (Bi_2In_3)	2.36		LPES	364
	Bi_3Ga	Bismuth-gallium complex (Bi_3Ga)	1.87	0.06	LPES	363
	Bi_3Ga_2	Bismuth-gallium complex (Bi_3Ga_2)	2.39	0.05	LPES	363
	Bi_3Ga_3	Bismuth-gallium complex (Bi_3Ga_3)	1.84	0.06	LPES	363
	Bi_3Ga_4	Bismuth-gallium complex (Bi_3Ga_4)	2.29	0.08	LPES	363
	Bi_3In_2	Bismuth-indium complex (Bi_3In_2)	2.42		LPES	364
	Bi_4	Bismuth tetramer	1.05	0.010	LPES	119
	Bi_5	Bismuth pentamer	2.87	0.02	LPES	253
	$Br(CO_2)$	Bromine-carbon dioxide complex	3.582	0.017	LPES	131
	$Br(H_2O)_n$	Bromine-water cluster	n=1-4		LPES	250
	C_n	Carbon cluster	n=2-84		LPES	70
	C_nCr	Carbon-chromium cluster [C_nCr]	n=2-8		LPES	271
	C_nNb	Carbon-niobium cluster	n=2-7		LPES	243
	$(CO_2)_n$	Carbon dioxide cluster	n=1,2		LPES	75
	$(CS)_n$	Carbon monosulfide cluster	n=2		LPES	75
	$(CS_2)_n$	Carbon disulfide cluster	n=1,2		LPES	75
	CAl_3Ge	[μ_3-(Germanetetraylmethanetetrayl)]trialuminum	2.70	0.06	LPES	224
	CAl_3Si	[μ_3-(Methanetetraylsilanetetrayl)]trialuminum	2.77	0.06	LPES	224
	CB_9	Carbon boron complex (CB_9)				
	CCl_4	Tetrachloromethane	¾1.14		CT	266
	CF_3	Trifluoromethyl	1.82	0.05	LPES	187
	CF_3Br	Bromotrifluoromethane	0.91	0.2	CD	2
	CF_3I	Trifluoroiodomethane	1.57	0.2	CD	2
	CFO_2	Fluorooxomethoxy	4.277	0.030	LPES	131

Z	Formula	Name	Elec. aff. (eV)	Uncert. (eV)	Method	Ref.
	$CHBr_2$	Dibromomethyl	1.9	0.2	LPES	367
	$CDBr_2$	Dibromomethyl-*d*	1.9	0.2	LPES	367
	$CHCl_2$	Dichloromethyl	1.3	0.2	LPES	367
	$CDCl_2$	Dichloromethyl-*d*	1.3	0.2	LPES	367
	CHI_2	Diiodomethyl	1.9	0.2	LPES	367
	CHO_2	Formyloxyl	3.4961	0.001	LPES	109
	CHO_2	*cis*-Hydroxyl formyl	1.511	0.005	LPES	395
	CHO_2	*trans*-Hydroxyl formyl	1.381	0.005	LPES	395
	CDO_2	Formyloxyl-*d*	3.5164	0.0010	LPES	109
	CH_2CN	Cyanomethyl radical	1.54583	0.00025	LPES	404
	CD_2CN	Cyanomethyl-d_2 radical	1.53765	0.00025	LPES	404
	CH_2S	Thioformaldehyde	0.465	0.023	LPES	53
	CD_3S	Methyl-d_3-sulfoniumylidene	1.856	0.006	LPT	2
	CH_3	Methyl	0.08	0.03	LPES	2
	CH_3I	Iodomethane	0.11	0.02	LPES	277
	CH_3NO_2	Nitromethane	0.172	0.006	LPES	211
	CH_3O	Methoxy	1.572	0.004	LPES	194
	CD_3O	Methoxy-d_3	1.559	0.004	LPES	194
	CH_3O_2	Methyldioxy	1.161	0.005	LPES	188
	CD_3O_2	Methyl-d_3-dioxy	1.154	0.004	LPES	188
	CH_3S	Methylthio	1.867	0.004	LPES	166
	CH_3S_2	Methyldithio	1.757	0.022	LPES	53
	CD_3S_2	Methyldithio-d_3	1.748	0.022	LPES	53
	CH_3Si	Methylsilylidyne	0.852	0.010	LPES	97
	$CH_2{=}SiH$	Methylenesilyl	2.010	0.010	LPES	97
	CH_3NH	Methylamidogen	0.432	0.015	LPES	215
	CH_5Si	Methylsilyl	1.19	0.04	LPT	65
	CO_3	Carbon oxide [CO_3]	2.69	0.14	LPES	2
	$CO_3(H_2O)$	Carbon oxide (CO_3)-water complex	2.1	0.2	PT	2
	C_2F_2	Difluorovinylidene	2.255	0.006	LPES	106
	C_2B_8	Carbon boron complex (C_2B_8)				
	C_2HO	Oxoethenyl	2.338	0.008	LPES	190
	C_2DO	Oxoethenyl-*d*	2.350	0.020	LPES	13
	C_2HF	Fluorovinylidene	1.718	0.006	LPES	106
	C_2HN	Cyanomethylene	2.003	0.014	LPES	219
	C_2DN	Cyanomethylene-*d*	2.009	0.020	LPES	219
	C_2HN	Isocyanomethylene	1.883	0.013	LPES	219
	C_2DN	Isocyanomethylene-*d*	1.877	0.010	LPES	219
	PdC_2HN	Cyanomethylenepalladium	2.17	0.03	LPES	291
	PdC_2H	Palladium(I) acetylide	1.98	0.03	LPES	287
	PtC_2H	Platinum(I) acetylide	2.650	0.010	LPES	287
	C_2HFe	Ethynyliron	1.4512	0.0025	LPES	254
	C_2HNi	Ethynylnickel	1.063	0.019	LPES	254
	C_2H_2	Vinylidene	0.490	0.006	LPES	83
	C_2HD	Vinylidene-d_1	0.489	0.006	LPES	83
	C_2D_2	Vinylidene-d_2	0.492	0.006	LPES	83
	C_2H_2FO	Acetyl fluoride enolate	2.22	0.09	PT	2
	$C_2H_2BrO_2$	Bromoacetate (CH_2BrCOO)	3.97	0.03	LPES	310
	$C_2H_2ClO_2$	Chloroacetate (CH_2ClCOO)	3.93	0.03	LPES	310
	$C_2H_2FO_2$	Fluoroacetate (CH_2FCOO)	3.80	0.03	LPES	310
	$HFeC_2H$	Ethynylhydroiron	1.328	0.019	LPES	254
	CH_2CN	Cyanomethyl	1.548	0.005	LPES	21
	CD_2CN	Cyanomethyl-d_2	1.538	0.012	LPES	21
	CH_2NC	Isocyanomethyl	1.059	0.024	LPES	21
	CD_2NC	Isocyanomethyl-d_2	1.070	0.024	LPES	21
	$C_2H_2N_3$	1,2,3-Triazolyl	3.447	0.004	LPES	316
	$H_2C_2O_4$	Oxalic acid	0.72	0.05	LPES	393
	$HNiC_2H$	Ethynylhydronickel	2.531	0.005	LPES	287
	C_2H_3	Vinyl	0.667	0.024	LPES	90
	C_2H_3Fe	Ethynyldihydroiron	1.587	0.019	LPES	254
	C_2H_3Ni	Ethynyldihydronickel	1.103	0.019	LPES	254

Z	Formula	Name	Elec. aff. (eV)	Uncert. (eV)	Method	Ref.
	C_2H_3O	Acetaldehyde enolate	1.8249	0.0012	LPT	22
	C_2D_3O	Acetaldehyde-*d*$_3$ enolate	1.8191	0.0012	LPT	22
	CH_3COO	Acetyloxyl radical	3.2528	0.001	LPES	402
	C_2H_5N	Ethylnitrene	0.56	0.01	PT	2
	$C_2H_5NO_2$	Nitroethane	0.3	0.2	LPES	322
	$(C_2H_5NO_2)_n$	Nitroethane cluster	n=0-4		LPES	322
	C_2H_5O	Ethoxy	1.712	0.004	LPES	194
	C_2D_5O	Ethoxy-*d*$_5$	1.699	0.004	LPES	194
	$C_2H_5O_2$	Ethyldioxy	1.186	0.004	LPES	188
	C_2H_5S	Ethylthio	1.953	0.006	LPT	2
	C_2H_5S	(Methylthio)methyl	0.868	0.051	LPES	53
	$C_2H_7O_2$	Methanol-methoxy complex	2.26	0.08	PT	50
	C_3Fe	1,2-Propadiene-1,3-diylideneferrate	1.69	0.08	LPES	132
	C_3H	Cyclopropenylidyne	1.999	0.003	LPES	315
	C_3D	Cyclopropenylidyne-*d*	1.997	0.005	LPES	315
	C_3HFe	2-Propynylidyneferrate	1.58	0.06	LPES	132
	CH_2CC	Propadienylidene	1.794	0.008	LPES	153
	HCCCH	Propargylene	1.156	+0.010, -0	LPES	396
	$C_3H_2F_3O$	1,1,1-Trifluoroacetone enolate	2.625	0.010	LPT	113
	$C_3H_2N_3$	1,3,5-Triazinyl	1.529	0.006	LPES	397
	CH_3CC	1-Propynyl	2.7355	0.0010	LPES	311
	CD_3CC	1-Propynyl-*d*$_3$	2.7300	0.0010	LPES	311
	CH_2CCH	2-Propynyl	0.918	0.008	LPES	153
	C_3H_2D	2-Propynyl-*d*$_1$	0.88	0.15	LPES	24
	C_3D_2H	2-Propynyl-*d*$_2$	0.907	0.023	LPES	24
	C_3H_4N	1-Cyanoethyl	1.247	0.012	LPES	21
	C_3H_5	Allyl	0.481	0.008	LPES	138
	C_3H_4D	Allyl-*d*$_1$	0.373	0.019	LPES	25
	C_3D_5	Allyl-*d*$_5$	0.464	0.006	LPES	138
	C_3H_5	Cyclopropyl	0.397	0.069	kinetic	155
	C_3H_5O	Acetone enolate	1.758	0.019	LPT	113
	C_3H_5O	Propanal enolate	1.621	0.006	LPT	113
	$C_3H_5O_2$	Methyl acetate enolate	1.80	0.06	PT	2
	$C_3H_7NO_2$	1-Nitropropane	0.223	0.006	LPES	392
	C_3H_7O	Propoxy	1.789	0.033	LPES	23
	C_3H_7O	Isopropoxy	1.847	0.004	LPES	194
	C_3H_7S	Propylthio	2.00	0.02	PT	2
	C_3H_7S	Isopropylthio	2.02	0.02	PT	2
	C_3N_2	Dicyanocarbene	<3.25	0.05	LPES	237
	O=C=C=C:	3-Oxo-1,3-propadienylidene	1.237	0.003	LPES	351
	C_3O_2	Carbon suboxide	0.85	0.15	LPES	11
	C_3S	3-Thioxo-1,2-propadienylidene	1.5957	0.0010	LPES	351
	C_3Ti	Titanium carbide [TiC$_3$]	1.561	0.015	LPES	147
	$C_4F_4Cl_2$	1,2-Dichloro-3,3,4,4-tetrafluorocyclobutene	0.87	0.08	attach	258
	$C_4F_4O_3$	3,3,4,4-Tetrafluorodihydro-2,5-furandione	0.5	0.2	CD	2
	c-C_4F_8	Perfluorocyclobutane	0.63	0.05	attach	256
	C_4Fe	1,2,3-Butatriene-1,4-diylideneferrate	<2.2	0.2	LPES	132
	C_4H	1,3-Butadiynyl radical	3.5332	0.0010	LPES	314
	C_4D	1,4-Butadiynyl-*d*	3.5308	0.0012	LPES	314
	HC_4Fe	1,3-Butadiynylferrate	1.67	0.06	LPES	132
	C_4H_2Fe	[(1,2-η)-1,3-Butadiyne]iron	1.633	0.019	LPES	254
	$C_4H_2O_3$	Maleic anhydride	1.44	0.10	CT	61
	C_4H_3Fe	Acetylene-iron complex	1.182	0.019	LPES	254
	$C_4H_3N_2$	1,2-Diazin-4-yl	1.85	0.006	LPES	397
	$C_4H_3N_2$	1,3-Diazin-5-yl	1.84	0.006	LPES	397
	$C_4H_3N_2$	1,4-Diazinyl	1.254	0.006	LPES	397
	C_4H_3Ni	Acetylene-nickel complex	0.824	0.019	LPES	254
	C_4H_4	Vinylvinylidene	0.914	0.015	LPES	125
	C_4D_4	Vinylvinylidene-*d*$_4$	0.909	0.015	LPES	125
	C_4H_4N	1*H*-Pyrrol-1-yl	2.145	0.010	LPES	265
	$NO{\cdot}C_4H_4N_2$	Nitric oxide-pyrimidine complex	0.75		LPT	285

Z	Formula	Name	Elec. aff. (eV)	Uncert. (eV)	Method	Ref.
	$C_4H_4N_3O$	Cytosine, deprotonated	3.037	0.015	LPES	309
	C_4H_5O	Cyclobutanone enolate	1.801	0.008	LPT	113
	C_4H_6	Trimethylenemethane	0.431	0.006	LPES	135
	$C_4H_6O_2$	2,3-Butanedione	0.69	0.10	CT	61
	C_4H_7	2-Methylallyl	0.505	0.006	LPES	138
	C_4H_6D	2-Methylallyl-d_7	0.493	0.008	LPES	138
	C_4H_7O	Butanal enolate	1.67	0.05	PT	2
	C_4H_5DO	2-Butanone-3-d_1 enolate	1.67	0.05	PT	2
	$C_4H_4D_2O$	2-Butanone-3,3-d_2 enolate	1.75	0.06	PT	2
		1-Nitrobutane	0.24	0.006	LPES	392
	C_4H_9O	*tert*-Butoxyl	1.909	0.004	LPES	194
	C_4H_9S	Butylthio	2.03	0.02	PT	2
	C_4H_9S	*tert*-Butylthio	2.07	0.02	PT	2
	C_4N	Carbon nitride (CCCCN)	3.1113	0.0010	LPES	346
	C_4O	4-Oxo-1,2,3-butatrienylidene	2.05	0.15	LPES	11
	C_4O_2	1,2,3-Butatriene-1,4-dione	2.0	0.2	LPES	11
	C_4Ti	Titanium carbide [TiC_4]	1.494	0.020	LPES	147
	C_5	Carbon pentamer	2.853	0.001	LPT	99
	C_5F_5N	Perfluoropyridine	0.70	0.05	attach	259
	$C_5F_6O_3$	Hexafluoroglutaric anhydride	1.5	0.2	CD	2
	C_5H	2,4-Pentadiynylidyne	2.421	0.019	LPES	317
	C_5H	Pentacyclopentyl	2.857	0.028	LPES	317
	C_5HF_4N	Tetrafluoropyridine	0.40	0.08	attach	259
	C_5H_4N	Pyridin-4-yl	1.480	0.006	LPES	397
	C_5H_5	Cyclopentadienyl	1.804	0.007	LPES	11
	C_5D_5	Cyclopentadienyl-d_5	1.790	0.008	LPES	11
	$(C_5H_5N)_nCo_m$	Cobalt-pyridine cluster	n=1-4	m=1-3	LPES	327
	$O_2{\cdot}C_5H_5N$	Oxygen-pyridine complex	1.39		LPES	285
	$NO{\cdot}C_5H_5N$	Nitric oxide-pyridine complex	0.62		LPES	285
	$C_5H_5N_2O_2$	Thymine, deprotonated	3.2635	0.0006	LPES	434
	C_5H_7	2,4-Pentadienyl	0.91	0.03	PT	2
	$O_2{\cdot}C_5H_5N{\cdot}H_2O$	Oxygen-pyridine-water complex	1.87		LPES	285
	C_5H_7O	Cyclopentanone enolate	1.598	0.007	LPT	113
	C_5H_9O	3-Pentanone enolate	1.69	0.05	PT	2
	$C_5H_{11}O$	Neopentoxyl	1.93	0.05	LPT	2
	$C_5H_{11}S$	Pentylthio	2.09	0.02	PT	2
	C_5O_2	1,2,3,4-Pentatetraene-1,5-dione	1.2	0.2	LPES	11
	C_5Ti	Titanium carbide (TiC_5)	1.748	0.050	LPES	147
	C_6	Carbon hexamer	4.180	0.001	LPT	8
	$C_6Br_4O_2$	Tetrabromobenzoquinone	2.44	0.20	CT	2
	$C_6Cl_4O_2$	2,3,5,6-Tetrachloro-2,5-cyclohexadiene-1,4-dione	2.78	0.10	CT	61
	$C_6F_4O_2$	Tetrafluorobenzoquinone	2.70	0.10	CT	61
	C_6F_5Br	Bromopentafluorobenzene	1.15	0.11	CT	67
	C_6F_5Cl	Chloropentafluorobenzene	0.75	0.05	attach	260
	C_6F_5I	Pentafluoroiodobenzene	1.41	0.11	CT	67
	$C_6F_5NO_2$	Perfluoronitrobenzene	1.52	0.11	CT	67
	C_6F_6	Hexafluorobenzene	0.53	0.05	attach	257
	C_6F_{10}	Perfluorocyclohexene	>1.4	0.3	CT	2
	$C_6H_2Cl_2O_2$	2,6-Dichloro-2,5-cyclohexadiene-1,4-dione	2.48	0.10	CT	61
	$C_6H_2O_2$	Benzynequinone	1.859	0.005	LPES	232
	$C_6H_3F_2NO_2$	2,4-Difluoro-1-nitrobenzene	1.17	0.10	CT	61
	$C_6H_3O_2$	Benzoquinonide	<2.18		LPES	232
	C_6H_4	Benzyne	0.560	0.010	LPES	36
	C_6D_4	Benzyne-d_4	0.551	0.010	LPES	36
	$C_6H_4BrNO_2$	1-Bromo-2-nitrobenzene	1.16	0.10	CT	61
	$C_6H_4BrNO_2$	1-Bromo-3-nitrobenzene	1.32	0.10	CT	61
	$C_6H_4BrNO_2$	1-Bromo-4-nitrobenzene	1.29	0.10	CT	61
	$C_6H_4ClNO_2$	1-Chloro-2-nitrobenzene	1.14	0.10	CT	61
	$C_6H_4ClNO_2$	1-Chloro-3-nitrobenzene	1.28	0.10	CT	61
	$C_6H_4ClNO_2$	1-Chloro-4-nitrobenzene	1.26	0.10	CT	61
	C_6H_4ClO	2-Chlorophenoxyl	=2.58	0.08	PT	2

Z	Formula	Name	Elec. aff. (eV)	Uncert. (eV)	Method	Ref.
	$C_6H_4FNO_2$	1-Fluoro-2-nitrobenzene	1.07	0.10	CT	61
	$C_6H_4FNO_2$	1-Fluoro-3-nitrobenzene	1.23	0.10	CT	61
	$C_6H_4FNO_2$	1-Fluoro-4-nitrobenzene	1.12	0.10	CT	61
	$C_6H_4N_2O_4$	1,2-Dinitrobenzene	1.65	0.10	CT	61
	$C_6H_4N_2O_4$	1,3-Dinitrobenzene	1.65	0.10	CT	61
	$C_6H_4N_2O_4$	1,4-Dinitrobenzene	2.00	0.10	CT	61
	$C_6H_4O_2$	*p*-Benzoquinone	1.860	0.005	LPES	284
	C_6H_5	Phenyl	1.096	0.006	LPES	26
	C_6D_5	Phenyl-d_5	1.092	0.020	LPES	26
	C_6H_5N	Phenylimidogen	1.429	0.011	LPT	115
	C_6D_5N	Phenylimidogen-d_5	1.44	0.02	LPES	96
	C_6H_5NH	Anilino	1.607	0.004	LPES	390
	$C_6H_5NO_2$	Nitrobenzene	1.00	0.01	LPES	164
	C_6H_5O	Phenoxy	2.253	0.006	LPES	26
	$HO(C_6H_4)O$	2-Hydroxyphenoxyl radical	2.3289	0.0006	LPES	419
	C_6H_5S	Phenylthio	<2.47	0.06	PT	2
	$(C_6H_6)_nCo_m$	Cobalt-benzene cluster	n=1-4	m=1-5	LPES	326
	$(C_6H_6)_nFe_m$	Iron-benzene cluster	n=1-4	m=1-7	LPES	329
	$NO{\cdot}C_6H_6$	Nitric oxide-benzene complex	0.44		LPES	285
	$O_2{\cdot}C_6H_6$	Oxygen-benzene complex	1.06		LPES	285
	$(C_6H_6)_n$	Benzene cluster	n=53-124		LPES	248
	$(C_7H_8)_n$	Toluene cluster	n=33-139		LPES	248
	C_6H_7	Methylcyclopentadienyl	<1.67	0.04	PT	2
	$C_6H_7NO_2$	Ethyl 2-cyanoacrylate	0.9	0.2	LPES	388
	$(CH_2)_2CC(CH_2)_2$	Tetramethyleneethane	0.855	0.010	LPES	203
	C_6H_8Si	Phenylsilane	1.435	0.004	LPT	65
	$CH_2{=}C(CH_3){-}C(CH_2)_2$	1,1,2-Trimethylene-2-methylethane	0.654	0.010	LPES	203
	C_6H_9O	Cyclohexanone enolate	1.526	0.010	LPT	113
	C_6H_{10}	*tert*-Butylvinylidene	0.645	0.015	LPES	126
	$C_6H_{11}O$	Pinacolone enolate	1.755	+0.05/-0.005	LPT	113
	$C_6H_{11}O$	3,3-Dimethylbutanal enolate	1.82	0.06	PT	2
	C_6N	Carbon nitride (CCCCCCN)	3.3715	0.0010	LPES	346
	C_6N_4	Tetracyanoethene	2.3	0.3	PT	2
	C_6O_6	Carbon monoxide trimer	2.54	0.05	LPES	333
	C_7F_5N	Pentafluorobenzonitrile	1.11	0.11	CT	67
	C_7F_8	Perfluorotoluene	0.86	0.11	CT	67
	C_7F_{14}	Perfluoromethylcyclohexane	1.06	0.10	CT	56
	C_7HF_5O	Pentafluorobenzaldehyde	1.10	0.11	CT	67
	$C_7H_3N_3O_4$	3,5-Dinitrobenzonitrile	2.16	0.10	CT	61
	$C_7H_4F_3NO_2$	1-Nitro-3-(trifluoromethyl)benzene	1.41	0.10	CT	61
	$C_7H_4N_2O_2$	2-Nitrobenzonitrile	1.61	0.10	CT	61
	$C_7H_4N_2O_2$	3-Nitrobenzonitrile	1.56	0.10	CT	61
	$C_7H_4N_2O_2$	4-Nitrobenzonitrile	1.72	0.10	CT	61
	C_7H_6Br	*o*-Bromobenzyl	1.308	0.008	LPES	167
	C_7H_6Br	*m*-Bromobenzyl	1.307	0.008	LPES	167
	C_7H_6Br	*p*-Bromobenzyl	1.229	0.008	LPES	167
	C_7H_6Cl	*o*-Chlorobenzyl	1.257	0.008	LPES	167
	C_7H_6Cl	*m*-Chlorobenzyl	1.272	0.008	LPES	167
	C_7H_6Cl	*p*-Chlorobenzyl	1.174	0.008	LPES	167
	C_6H_5CHCl	Chlorobenzyl	≤1.12		LPES	398
	C_7H_6F	*o*-Fluorobenzyl	1.091	0.008	LPES	167
	C_7H_6F	*m*-Fluorobenzyl	1.173	0.008	LPES	167
	C_7H_6F	*p*-Fluorobenzyl	0.937	0.008	LPES	167
	C_7H_6FO	*m*-Fluoroacetophenone enolate	2.218	0.010	LPT	2
	C_7H_6FO	*p*-Fluoroacetophenone enolate	2.176	0.010	LPT	2
	$C_7H_6FeO_3$	η_4-1,3-Butadiene iron tricarbonyl	0.990	0.10	CT	120
	$C_7H_6N_2O_4$	4-Methyl-1,2-dinitrobenzene	1.77	0.05	PT	60
	$C_7H_6N_2O_4$	1-Methyl-2,3-dinitrobenzene	1.77	0.05	PT	60
	$C_7H_6N_2O_4$	1-Methyl-2,4-dinitrobenzene	1.60	0.05	PT	60
	$C_7H_6N_2O_4$	2-Methyl-1,3-dinitrobenzene	1.55	0.05	PT	60
	C_6H_5CHO	Benzaldehyde	0.35	0.05	LPES	384

Z	Formula	Name	Elec. aff. (eV)	Uncert. (eV)	Method	Ref.
	$C_7H_6O_2$	2-Methyl-2,5-cyclohexadiene-1,4-dione	1.85	0.10	CT	61
	C_7H_7	Benzyl	0.912	0.006	LPES	26
	C_7H_7	1-Quadricyclanide	0.868	0.006	LPES	136
	C_7H_7	2-Quadricyclanide	0.962	0.006	LPES	136
	C_7H_7	Norbornadienide	1.286	0.006	LPES	136
	C_7H_7	Cycloheptatrienide	0.39	0.04	LPES	136
	C_7H_7	1-(1,6-Heptadiynide)	3.046	0.006	LPES	136
	C_7H_7	3-(1,6-Heptadiynide)	>1.140	0.006	LPES	136
	$C_7H_7NO_2$	2-Nitrotoluene	0.92	0.10	CT	61
	$C_7H_7NO_2$	3-Nitrotoluene	0.99	0.10	CT	61
	$C_7H_7NO_2$	4-Nitrotoluene	0.95	0.10	CT	61
	$C_7H_7NO_3$	3-Nitroanisole	1.04	0.10	CT	61
	$C_7H_7NO_3$	4-Nitroanisole	0.91	0.10	CT	61
	C_7H_7O	2-Methylphenoxy	<2.36	0.06	PT	2
	C_7H_7O	Benzyloxy	2.14	0.02	PT	50
	C_7H_8FO	Benzyloxy-hydrogen fluoride complex	<3.05	0.06	PT	50
	C_7H_9	2,4,6-Heptatrienyl	1.27	0.03	PT	2
	C_7H_9O	2-Norbornanone enolate	1.61	0.05	PT	2
	C_7H_9Si	Methylphenylsilyl	1.33	0.04	LPT	65
	$C_7H_{11}O$	Cycloheptanone enolate	1.598	0.007	LPT	113
	$C_7H_{11}O$	2,5-Dimethylcyclopentanone enolate	1.49	0.04	PT	2
	$C_7H_{13}O$	4-Heptanone enolate	1.72	0.06	PT	2
	$C_7H_{13}O$	Diisopropyl ketone enolate	1.46	0.04	PT	2
	$C_8F_{14}N_2$	2,3,5,6-Tetrafluoro-1,4-benzenedicarbonitrile	1.89	0.10	CT	51
	$C_6F_5COCH_3$	Pentafluoroacetophenone	0.88	0.11	CT	67
	$C_8H_3F_6NO_2$	1-Nitro-3,5-bis(trifluoromethyl)benzene	1.79	0.10	CT	61
	$C_8H_4F_3N$	2-(Trifluoromethyl)benzonitrile	0.70	0.05	attach	263
	$C_8H_4F_3N$	3-(Trifluoromethyl)benzonitrile	0.67	0.05	attach	263
	$C_8H_4F_3N$	4-(Trifluoromethyl)benzonitrile	0.83	0.05	attach	263
	$C_8H_4O_3$	Phthalic anhydride	1.21	0.10	CT	61
	C_8H_6	1,3,5-Cyclooctatrien-7-yne	1.044	0.008	LPES	148
	C_6H_5CHCN	Cyanobenzyl	1,9	0.01	LPES	398
	C_8H_7	1,3,5,7-Cyclooctatetraen-1-yl	1.091	0.008	LPES	134
	C_8H_7O	Acetophenone enolate	2.057	0.010	PT	2
	C_8H_7O	Phenylacetaldehyde enolate	2.10	0.08	LPT	2
	C_8H_8	1,3,5,7-Cyclooctatetraene	0.55	0.02	CT	134
	C_8H_8	*m*-Xylylene	0.919	0.008	LPES	139
	$C_8H_9NO_2$	1,3-Dimethyl-2-nitrobenzene	2.61	0.05	PT	60
	$C_8H_9NO_2$	1,2-Dimethyl-3-nitrobenzene	0.86	0.10	CT	61
	$C_8H_9NO_2$	1,3-Dimethyl-5-nitrobenzene	1.21	0.05	PT	60
	$C_8H_{13}O$	Cyclooctanone enolate	1.63	0.06	PT	2
	C_8S_2	Carbon sulfide [C_8S_2]	0.049	0.005	LPES	230
	C_9H_7	Indenyl	1.8016	0.0006	LPES	394
	$C_9H_8FeO_3$	η_4-1,3-Cyclohexadiene iron tricarbonyl	0.76	0.10	CT	120
	C_9H_9O	*m*-Methylacetophenone enolate	2.030	0.010	LPT	2
	$(CH_3)_3SiN$	Trimethylsilylimidogen	1.43	0.10	PT	2
	$C_9H_{11}NO_2$	1,3,5-Trimethyl-2-nitrobenzene	0.70	0.10	CT	61
	$C_9H_{15}O$	Cyclononanone enolate	1.69	0.06	PT	2
	$C_{10}H_4Cl_2O_2$	2,3-Dichloro-1,4-naphthalenedione	2.19	0.10	CT	61
	$C_{10}H_6N_2O_4$	1,3-Dinitronaphthalene	1.78	0.10	CT	61
	$C_{10}H_6N_2O_4$	1,5-Dinitronaphthalene	1.77	0.10	CT	61
	$C_{10}H_6O_2$	1,4-Naphthalenedione	1.81	0.10	CT	61
	$C_{10}H_7$	1-Naphthyl	1.403	0.015	LPES	197
	$C_{10}H_7NO_2$	1-Nitronaphthalene	1.23	0.10	CT	61
	$C_{10}H_7NO_2$	2-Nitronaphthalene	1.18	0.10	CT	61
	$C_{10}H_8$	Azulene	0.790	0.008	LPES	230
	$C_{10}H_8CrO_3$	η_4-1,3,5-Cycloheptatriene chromium tricarbonyl	0.93	0.10	CT	120
	$NO{\cdot}C_{10}H_8$	Nitric oxide-naphthalene complex	0.66		LPES	285
	$O_2{\cdot}C_{10}H_8$	Oxygen-naphthalene complex	1.41		LPES	285
	$O_2{\cdot}C_{10}H_8{\cdot}H_2O$	Oxygen-naphthalene-water complex	2.09		LPES	285
	$O_2{\cdot}C_{10}H_8{\cdot}(H_2O)_2$	Oxygen-naphthalene-water dimer complex	2.72		LPES	285

Z	Formula	Name	Elec. aff. (eV)	Uncert. (eV)	Method	Ref.
	$C_{10}H_8FeO_3$	η4-1,3,5-Cycloheptatriene iron tricarbonyl	0.98	0.10	CT	120
	$C_5(CH_3)_5AlH$	Aluminum hydride pentamethylcyclopentadienyl	1.0	0.2	LPES	424
	$C_5(CH_3)_5Al_2H$	Dialuminum hydride pentamethylcyclopentadienyl	0.8	0.2	LPES	424
	$C_5(CH_3)_5Al_3H$	Trialuminum hydride pentamethylcyclopentadienyl	1.2	0.2	LPES	424
	$C_{10}H_{17}O$	Cyclodecanone enolate	1.83	0.06	PT	2
	$C_{11}H_8FeO_3$	η4-1,3,5,7-Cyclooctatetraene iron tricarbonyl	1.29	0.10	CT	120
	$C_{12}F_{10}$	2,2',3,3',4,4',5,5',6,6'-Decafluoro-1,1'-biphenyl	0.82	0.11	CT	67
	$C_{12}H_4N_4$	Tetracyanoquinodimethane	3.383	0.001	CD	411
	$C_{12}H_{10}OP$	Diphenylphosphide oxide	2.2	0.1	LPES	401
	$C_{12}H_{10}P$	Diphenylphosphide	1.5	0.1	LPES	401
	$C_{13}H_9$	Perinaphthalenyl	1.07	0.10	PT	2
	$NO{\cdot}(C_6H_6)_2$	Nitric oxide-benzene dimer complex	0.79		LPES	285
	$C_{12}H_{15}O$	*tert*-Butylacetophenone enolate	2.032	0.010	LPT	2
	$C_{12}H_{21}O$	Cyclododecanone enolate	1.90	0.07	PT	2
	$C_{13}H_9$	Fluorenyl	1.8751	0.0003	LPES	394
	$C_{13}F_{10}O$	Decafluorobenzophenone	1.52	0.11	CT	67
	$C_{13}H_9FO$	4-Fluorobenzophenone	0.64	0.10	CT	61
	$(C_6H_5)_2CO$	Benzophenone	0.62	0.10	CT	61
	$C_{14}H_9NO_2$	9-Nitroanthracene	1.43	0.10	CT	61
	$C_{14}H_{10}$	Anthracene	0.530	0.005	LPES	286
	$C_{14}H_{10}{\cdot}H_2O$	Anthracene-water complex	0.770	0.005	LPES	286
	$(C_{14}H_{10})_n$	Anthracene cluster	n=1-16		LPES	231
	$C_{16}H_{10}$	Pyrene	0.406	0.010	LPES	270
	$(C_{16}H_{10})_nCo_m$	Cobalt-pyrene cluster	n=1,2	m=1,2	LPES	330
	$C_{18}H_{12}$	Chrysene	0.32	0.01	LPES	303
	$C_{18}H_{12}$	Naphthacene	1.058	0.005	CT	66
	$C_{20}H_{12}$	Benzo[*a*]pyrene	0.79	0.10	CT	66
	$C_{20}H_{12}$	Perylene	0.973	0.005	LPES	236
	$NO{\cdot}(C_{10}H_8)_2$	Nitric oxide-naphthalene dimer complex	1.06		LPES	285
	$C_{22}H_{14}$	Pentacene	1.35	0.10	CT	66
		Coronene	0.47	0.09	LPES	328
	$C_{24}H_{12}Co$	Coronene-cobalt complex	1.15	0.15	LPES	324
	$C_{24}H_{12}Co_2$	Coronene-dicobalt complex	1.15	0.10	LPES	324
	$C_{24}H_{12}Fe$	Coronene-iron complex	1.06		LPES	324
	$C_{24}H_{12}Fe_2$	Coronene-diiron complex	1.59		LPES	324
	$C_{44}Cl_{28}FeN_4$	$FeTPPCl_{28}$	2.59	0.11	CT	186
	$C_{44}Cl_8F_{20}FeN_4$	$FeTPP\beta Cl_8$	3.21	0.03	CT	186
	$C_{44}Cl_9F_{20}FeN_4$	$FeTPPF_{20}\beta Cl_8Cl$	3.35	0.03	CT	186
	$C_{44}H_8F_{20}FeN_4$	$FeTPPF_{20}$	2.15	0.15	CT	186
	$C_{44}H_8ClF_{20}FeN_4$	$FeTPPF_{20}Cl$	3.14	0.03	CT	186
	$C_{44}H_8Cl_{21}FeN_4$	$FeTPPoCl_{20}Cl$	2.93	0.23	CT	186
	$C_{44}H_{12}Cl_{17}FeN_4$	$FeTPPoCl_8\beta Cl_8Cl$	3.14	0.03	CT	186
	$C_{44}H_{20}Cl_8FeN_4$	$FeTPPoCl_8$	1.86	0.03	CT	186
	$C_{44}H_{20}Cl_9FeN_4$	$FeTPPoCl_8Cl$	2.10	0.19	CT	186
	$C_{44}H_{28}FeN_4$	FeTPP	1.87	0.03	CT	186
	$C_{44}H_{28}NiN_4$	NiTPP	1.51	0.01	CT	186
	$C_{44}H_{28}ClFeN_4$	FeTPPCl	2.15	0.15	CT	186
	$C_{44}H_{30}N_4$	H_2TPP	1.69	0.01	CT	186
	$C_{45}H_{29}NiN_4O$	NiTPPCHO	1.74	0.01	CT	186
	$C_{48}H_{24}$	Coronene dimer	0.67	0.09	LPES	328
	$C_{48}H_{24}Fe$	Coronene dimer-iron complex	1.50		LPES	324
	$C_{48}H_{24}Fe_2$	Coronene dimer-diiron complex	1.50		LPES	324
	$C_{52}H_{39}FeN_7O$	FeTPP-val	1.97	0.03	CT	186
	C_{60}	Carbon [fullerene-C_{60}]	2.6835	0.0006	LPES	201
	$C_{60}F_2$	Difluorofullerene-C_{60}	2.74	0.07	Knud	202
	$C_{64}H_{64}FeN_8O_4$	FeTPP-piv	2.07	0.03	CT	186
	C_{70}	Carbon [fullerene-C_{70}]	2.676	0.001	PT	201
	$C_{70}F_2$	Difluorofullerene-C_{70}	2.80	0.07	Knud	202
	CeF_4	Cerium(IV) fluoride	3.8	0.4	CT	98
	$Cl(CO_2)$	Carbonochloridate	3.907	0.010	LPES	131

Z	Formula	Name	Elec. aff. (eV)	Uncert. (eV)	Method	Ref.
	$Cl(H_2O)_n$	Chlorine-water cluster	n=1-4		LPES	250
	ClO_3	Chlorine trioxide	4.25	0.10	LPES	340
	ClO_4	Chlorine tetroxide	5.25	0.10	LPES	340
	Co_n	Cobalt cluster	n=1-108		LPES	251
	CoB_{12}	Cobalt招 boron complex (CoB_{12})	3.23	0.04	LPES	437
	$CoBr_3$	Cobalt bromide [$CoBr_3$]	4.6	0.1	LPES	249
	$CoCO_2NO$	Cobalt-carbon dioxide-nitric oxide complex	1.73	0.03	LPES	199
	$CoCl_3$	Cobalt chloride [$CoCl_3$]	4.7	0.1	LPES	249
	CoF_4	Cobalt fluoride [CoF_4]	6.4	0.3	CT	98
	CoOH	Cobalt hydroxide	1.33	0.04	LPES	400
	Co_2OH	Dicobalt hydroxide	1.14	0.08	LPES	400
	Co_3OH	Tricobalt hydroxide	1.56	0.08	LPES	400
	$Cr(CO)_3$	Chromium carbonyl [$Cr[CO]_3$]	1.349	0.006	LPES	94
	CrO_3	Chromium(VI) oxide	3.66	0.02	LPES	241
	CrO_4	Chromium oxide [CrO_4]	4.98	0.09	LPES	241
	CrO_5	Chromium oxide [CrO_5]	4.4	0.1	LPES	241
	Cr_2O_n	Chromium-oxygen cluster	n=1-7		LPES	306
	CsO_4	Cesium oxide [CsO_4]	2.5	0.2	LPES	252
	Cu_n	Copper cluster	n=1-411		LPES	37
	$Cu_n(CN)_m$	Copper cyanide cluster	n=1-6	m=1-6	LPES	159
	$EuSi_n$	Europium-silicon cluster	n=3-17		LPES	321
	$F(H_2O)_n$	Fluorine-water cluster	n=1-4		LPES	242
	$F(H_2O)_n$	Fluorine-water cluster	n=1-4		LPES	250
	Fe_n	Iron cluster	n=3-34		LPES	149
	$Fe(CO)_2$	Iron carbonyl [$Fe(CO)_2$]	1.22	0.02	LPES	2
	$Fe(CO)_3$	Iron carbonyl [$Fe(CO)_3$]	1.8	0.2	LPES	2
	$Fe(CO)_4$	Iron carbonyl [$Fe(CO)_4$]	2.34	0.1	LPES	389
	$FeBr_3$	Iron(III) bromide	4.26	0.06	LPES	249
	$FeBr_4$	Iron bromide [$FeBr_4$]	5.50	0.08	LPES	249
	$FeCl_3$	Iron(III) chloride	4.22	0.06	LPES	249
	$FeCl_4$	Iron chloride [$FeCl_4$]	6.00	0.08	LPES	249
	FeF_3	Iron(III) fluoride	3.6	0.1	CT	98
	FeF_4	Iron fluoride [FeF_4]	6.0	estimate	CT	98
	Fe_2H_2	Iron hydride [Fe_2H_2]	0.942	0.019	LPES	254
	Fe_4O	Iron oxide cluster (Fe_4O)	1.698	0.0003	LPES	407
	Fe_5O	Iron oxide cluster (Fe_5O)	1.8616	0.0003	LPES	407
	Fe_nO_m	Iron oxide cluster	n=1-4	m=1-6	LPES	152
	Ga_2As_3	Gallium arsenide [Ga_3As_2]	2.783	0.024	LPES	192
	Ga_xAs_y	Gallium arsenide cluster	n=2-50	n=x+y	LPES	229
	Ga_2P_3	Gallium phosphide [Ga_2P_3]	2.991	0.026	LPES	192
	Ge_n	Germanium cluster	n=3-15		LPES	71
	Ge_xAs_y	Germanium-arsenic cluster	n=5-30	n=x+y	LPES	72
	GeH_3	Germyl	<1.74	0.04	PT	2
	$H(NH_3)_n$	Hydrogen-ammonia cluster	n=1,2		LPES	76
	HNO_3	Nitric acid	0.57	0.15	CD	2
	$(H_2O)_n$	Water cluster	n=2-19		LPES	77
	HfOOH	Hafnium oxide hydroxyl radical	1.73	0.05	LPES	173
	$I(CO_2)$	Iodine-carbon dioxide complex	3.225	0.001	LPES	131
	$I(CO_2)_n$	Iodine-carbon dioxide cluster	n=1-3		LPES	350
	$I(H_2O)_n$	Iodine-water cluster	n=1-4		LPES	250
	In_xP_y	Indium phosphide cluster	n=2-8	n=x+y	LPES	137
	IrF_4	Iridium fluoride [IrF_4]	4.7	0.3	CT	98
	IrF_6	Iridium(VI) fluoride	6.5	0.4	CT	98
	K_n	Potassium cluster	n=2-7		LPES	18
	KO_4	Potassium oxide [KO_4]	2.8	0.2	LPES	252
	$LaCl_4$	Lanthanum tetrachloride	7.03	0.01	LPES	145
	LiO_4	Lithium oxide [LiO_4]	3.3	0.2	LPES	252
	$MnBr_3$	Manganese bromide [$MnBr_3$]	5.03	0.06	LPES	249
	$MnCl_3$	Manganese chloride [$MnCl_3$]	5.07	0.06	LPES	249
	MnF_4	Manganese fluoride [MnF_4]	5.5	0.2	CT	98
	MnO_3	Manganese oxide [MnO_3]	3.335	0.010	LPES	158

Z	Formula	Name	Elec. aff. (eV)	Uncert. (eV)	Method	Ref.
	$Mo(CO)_3$	Molybdenum carbonyl [$Mo[CO]_3$]	1.337	0.006	LPES	94
	MoF_5	Molybdenum(V) fluoride	3.5	0.2	CT	98
	MoF_6	Molybdenum(VI) fluoride	3.8	0.2	CT	98
	MoO_3	Molybdenum(VI) oxide	3.17	0.02	LPES	280
	MoO_4	Molybdenum oxide [MoO_4]	5.20	0.07	LPES	86
	MoO_5	Molybdenum oxide [MoO_5]	5.10	0.07	LPES	86
	MoW_2O_6	Molybdenum tungsten oxide [MoW_2O_6]	2.76	0.05	LPES	353
	Mo_2O_2	Molybdenum oxide [Mo_2O_2]	2.24	0.02	LPES	280
	Mo_2O_3	Molybdenum(III) oxide	2.33	0.07	LPES	280
	Mo_2O_4	Molybdenum oxide [Mo_2O_4]	2.13	0.04	LPES	280
	Mo_2WO_6	Molybdenum tungsten oxide [Mo_2WO_6]	2.85	0.05	LPES	353
	Mo_3O_6	Molybdenum oxide [Mo_3O_6]	2.75	0.02	LPES	353
	N_2CH	Cyanoamidogen	2.622	0.005	LPES	154
	N_2CD	Cyanoamidogen-*d*	2.622	0.005	LPES	154
	$(NH_3)_n$	Ammonia cluster	n=41-1100		LPES	77
	$NH_2(NH_3)_n$	Amidogen-ammonia cluster	n=1,2		LPES	78
	$NOAr_n$,n=1,4	Nitric oxide argon cluster ($NOAr_n$,n=1,4)			LPES	79
	$NO(C_2H_6O_2)_n$,n=1-4	Nitric oxide ethylene glycol cluster	$NO(C_2H_6O_2)_n$	n=1-4	LPES	79
	$NO(H_2O)_n$	Nitric oxide-water cluster	n=1,2		LPES	75
	$NO(H_2S)$	Nitric oxide hydrogen sulfide cluster	0.268	0.020	LPES	79
	$NO(NH_3)$	Nitric oxide ammoniua cluster	0.48	0.02	LPES	79
	$NO(N_2O)_n$	Nitric oxide-nitrous oxide cluster	n=1,2		LPES	79
	NO_3	Nitrogen trioxide [NO_3]	3.937	0.014	LPES	85
	$NO_3(H_2O)_n$	Nitrogen trioxide-water cluster	n=0-6		LPES	240
	$(NO)_2$	Nitric oxide dimer	>2.1		LPES	75
	$(N_2O)_n$	Nitrous oxide cluster	n=1,2		LPES	81
	Na_n	Sodium cluster	n=2-5		LPES	18
	$NaCS_2$	Sodium-carbon disulfide complex [$NaCS_2$]	0.80	0.05	LPES	278
	Na_2CS_2	Sodium-carbon disulfide complex [Na_2CS_2]	0.25	0.05	LPES	278
	$(NaF)_n$	Sodium fluoride cluster	n=1-7,12		LPES	64
	$Na(NaF)_n$	Sodium-sodium fluoride cluster	n=5,7-12		LPES	64
	NaO_4	Sodium oxide [NaO_4]	3.1	0.2	LPES	252
	NaO_5	Sodium oxide [NaO_5]	3.2	0.2	LPES	252
	$NaSO_3$	Sodium oxide sulfide (NaO_3S)	2.3	0.2	LPES	252
	Nb_n	Niobium cluster	n=6-17		LPES	181
	Nb_8	Niobium octamer	1.513	0.008	LPES	157
	Nb_2O_5	Niobium(V) oxide	3.33	0.05	LPES	451
	Nb_2O_6	Diniobium hexaoxide	5.35	0.05	LPES	451
	Nb_2O_7	Diniobium heptaoxide	5.25	0.05	LPES	451
	Nb_3O	Niobium oxide (Nb_3O)	1.393	0.006	LPES	169
	Ni_n	Nickel cluster	n=1-100		LPES	247
	$Ni_n(C_6H_6)_m$	Nickel-benzene complex	n=1-3	m=1,2	LPES	295
	$NiBr_3$	Nickel bromide [$NiBr_3$]	4.94	0.08	LPES	249
	$NiCl_3$	Nickel chloride [$NiCl_3$]	5.20	0.08	LPES	249
	$Ni(CO)H$	Nickel-carbon monoxide complex	1.126	0.010	LPES	293
	$Ni(CO)_2$	Nickel carbonyl [$Ni[CO]_2$]	0.643	0.014	LPES	2
	$Ni(CO)_3$	Nickel carbonyl [$Ni[CO]_3$]	1.077	0.013	LPES	2
	$NiAl(CO)_3$	Nickel aluminum tricarbonyl ($NiAl(CO)_3$)	1.064	0.063	LPES	423
	$NICa(CO)_3$	Nickel calcium tricarbonyl ($NiCa(CO)_3$)	1.05	0.064	LPES	423
	$NiMg(CO)_3$	Nickel magnesium tricarbonyl ($NiMg(CO)_3$)	1.541	0.04	LPES	423
	$NiC_8N_4S_4$	Nickel bis(dithiolene)	4.56	0.04	LPES	307
	O_2O_3	Oxygen ozonide	2.16	0.015	LPES	420
	$O_2(O_3)_n$,n=1-4	Oxygen ozonide complex	$O_2(O_3)_n$	n=1-4	LPES	420
	$OH(H_2O)$	Hydroxyl-water complex	<2.95	0.15	PT	2
	$OH(NH_3)$	Hydroxyl-ammonia complex	2.35	0.07	LPES	234
	$OH(N_2O)$	Hydroxyl-nitrous oxide complex	2.14	0.02	LPES	209
	$OH(N_2O)_n$	Hydroxyl-nitrous oxide cluster	n=1-5		LPES	209
	OsF_4	Osmium(IV) fluoride	3.9	0.3	CT	98
	OsF_6	Osmium(VI) fluoride	6.0	0.3	CT	98
	P_5	Phosphorus pentamer	3.88	0.03	LPES	253
	PBr_3	Phosphorus(III) bromide	1.59	0.15	CD	2

Z	Formula	Name	Elec. aff. (eV)	Uncert. (eV)	Method	Ref.
	PBr_2Cl	Phosphorus(III) dibromide chloride	1.63	0.20	CD	2
	$PBrCl_2$	Phosphorus(III) bromide dichloride	1.52	0.20	CD	2
	PCl_3	Phosphorus(III) chloride	0.82	0.10	CD	2
	PF_5	Phosphorus(V) fluoride	0.75	0.15	CT	121
	P_2H_2	*cis*-Diphosphene	1.03	0.01	LPES	281
	P_2H_2	*trans*-Diphosphene	1.00	0.01	LPES	281
	PO_3	Phosphorus oxide [PO_3]	4.95	0.06	LPES	156
	$POCl_2$	Phosphorus oxychloride [$POCl_2$]	3.83	0.25	CD	2
	$POCl_3$	Phosphoryl chloride	1.41	0.20	CD	2
	Pb_4	Lead tetramer	1.55	0.09	LPES	345
	$PdC_8N_4S_4$	Palladium bis(dithiolene)	4.55	0.04	LPES	307
	$PtC_8N_4S_4$	Platinum bis(dithiolene)	4.45	0.04	LPES	307
	$Pt(CO)_2$	Platinum bicarbonyl	0.93	0.042	LPES	391
	$Pt(CO)_3$	Platinum tricarbonyl	1.253	0.032	LPES	391
	PtF_4	Platinum(IV) fluoride	5.5	0.3	CT	98
	PtF_6	Platinum(VI) fluoride	7.0	0.4	CT	98
	ReF_6	Rhenium(VI) fluoride	4.7	estimate	CT	98
	ReO_3	Rhenium(VI) oxide	3.53	0.05	LPES	450
	ReO_4	Rheium tetraoxide	5.58	0.03	LPES	450
	Re_3O_3	Trirhenium trioxide	2.54	0.02	LPES	448
	$RhBr_{12}$	Rhodium dodecabromide	3.45	0.02	LPES	437
	RhF_4	Rhenium fluoride [RhF_4]	5.4	0.3	CT	98
	RuF_4	Ruthenium(IV) fluoride	4.8	0.3	CT	98
	RuF_5	Ruthenium(V) fluoride	5.2	0.4	CT	98
	RuF_6	Ruthenium(VI) fluoride	7.5	0.3	CT	98
	SF_4	Sulfur tetrafluoride	1.5	0.2	CT	91
	SF_5	Sulfur pentafluoride	4.23	0.12	e-scat	204
	SF_6	Sulfur hexafluoride	1.03	0.05	attach	318
	SO_3	Sulfur trioxide (α-form)	1.97	0.10	LPES	165
	$(SO_2)_2$	Sulfur dioxide dimer	0.6	0.2	LPES	80
	Sb_n	Antimony cluster	n=2-9		LPES	213
	Sb_5	Antimony pentamer	3.46	0.03	LPES	253
	$ScBr_4$	Scandium bromide [$ScBr_4$]	6.13	0.08	LPES	249
	$ScCl_4$	Scandium chloride [$ScCl_4$]	6.84	0.01	LPES	145
	Sc_2Sn_n	Scandium-tin cluster	n=2-6		LPES	356
	SeF_6	Selenium hexafluoride	2.9	0.2	CD	2
	Si_4	Tetrasilicon	2.13	0.01	LPES	110
	Si_5	Silicon pentamer	2.59	0.02	LPES	110
	Si_7	Silicon heptamer	1.85	0.02	LPES	110
	Si_n	Silicon cluster	n=3-20		LPES	71
	SiF_3	Trifluorosilyl	2.41	0.22	LPES	17
	SiF_4	Tetrafluorosilane	=0		LPES	17
	SiF_5	Pentafluorosilyl	=4.66		LPES	17
	SiH_3	Silyl	1.406	0.014	LPES	43
	SiD_3	Silyl-d_3	1.386	0.022	LPES	43
	Si_2C_3	Disilicon tricarbide	1.766	0.012	LPES	296
	Si_3H	Silicon hydride (Si_3H)	2.53	0.01	LPES	182
	Si_4H	Silicon hydride (Si_4H)	2.68	0.01	LPES	182
	Si_nNa_m	Silicon-sodium cluster	n=4-11	m=1-3	LPES	210
	Si_nF	Silicon-fluorine cluster	n=2-11		LPES	17
	Sn_n	Tin cluster	n=1-12		LPES	289
	$SnCH_2CN$	Cyanomethylenetin	1.57	0.02	LPES	292
	$Sn(CN)_2$	Tin(II) cyanide	2.622	0.004	LPES	292
	$Sn(CN)(CH_2CN)$	Cyanomethylenetin cyanide	2.29	0.05	LPES	292
	Ta_2B_2	Ditantalum diboride	1.31	0.04	LPES	439
	Ta_2B_3	Ditantalum triboride	1.93	0.03	LPES	439
	Ta_2B_4	Ditantalum tetraboride	1.7	0.07	LPES	439
	Ta_2B_6	Ditantalum hexaboride	1.51	0.03	LPES	439
	Ta_2O_5	Tantalum(V) oxide	3.71	0.05	LPES	451
	Ta_2O_6	Ditantalum hexaoxide	5.28	0.05	LPES	451
	Ta_2O_7	Ditantalum heptaoxide	5.15	0.05	LPES	451

Z	Formula	Name	Elec. aff. (eV)	Uncert. (eV)	Method	Ref.
	Ta_3O	Tantalum oxide (Ta_3O)	1.583	0.010	LPES	169
	TeF_6	Tellurium hexafluoride	3.34	0.17	CD	2
	Ti_n	Titanium cluster	n=1-130		LPES	151
	$TiO_2{\cdot}(H_2O)$	Titanium dioxide hydrate	1.15	0.08	LPES	381
	$TiO_2{\cdot}(H_2O)_2$	Titanium dioxide dihydrate	0.77	0.08	LPES	381
	$TiO_2(H_2O){\cdot}(H_2O)_n$	Titanium dioxide hydrate cluster	n=1-5		LPES	381
	TiO_3	Titanium oxide [TiO_3]	4.2		LPES	172
	UCl_6	Uranium(VI) chloride	5.3	0.2	LPES	414
	UF_3	Uranium(III) fluoride	1.09	0.03	LPES	412
	UF_4	Uranium(IV) fluoride	1.58	0.03	LPES	412
	UF_5	Uranium(V) fluoride	3.885	0.015	LPES	445
	UF_6	Uranium(VI) fluoride	5.1	0.2	CT	98
	UO_2Br_3	Uranyl tribromide	6.27	0.05	LPES	442
	UO_2Cl_3	Uranyl trichloride	6.64	0.05	LPES	442
	UO_2F_3	Uranyl trifluoride	6.25	0.05	LPES	442
	UO_2I_3	Uranyl triiodide	5.6	0.05	LPES	442
	UO_3	Uranium(VI) oxide	1.12	0.03	LPES	433
	UO_4	Uranium tetraoxide	3.60	0.03	LPES	433
	UO_5	Uranium pentaoxide	4.02	0.06	LPES	433
	V_n	Vanadium cluster	n=3-65		LPES	150
	VF_4	Vanadium(IV) fluoride	3.5	0.2	CT	98
	VSi_n	Vanadium-silicon cluster [VSi_n]	n=3-6		LPES	357
	V_2O_n	Vanadium oxide cluster	n=3-7		LPES	246
	V_2Si_n	Vanadium-silicon cluster [V_2Si_n]	n=3-6		LPES	357
	V_3O	Vanadium oxide (V_3O)	1.218	0.008	LPES	169
	V_4O_{10}	Vanadium oxide [V_4O_{10}]	4.2	0.6	CT	101
	$WAlO_n$	Tungsten aluminum oxides	n=2-4		LPES	387
	$W(CO)_3$	Tungsten carbonyl [$W[CO]_3$]	1.859	0.006	LPES	94
	WF_5	Tungsten(V) fluoride	1.25	0.3	CD	18
	WF_6	Tungsten(VI) fluoride	3.5	0.1	CT	19
	WO_3	Tungsten(VI) oxide	3.62	0.05	LPES	86
	$(WO_3)_n$	Tungsten(IV) oxide cluster	n=7-10		LPES	300
	WO_4	Tungsten oxide [WO_4]	5.30	0.05	LPES	86
	WO_5	Tungsten oxide [WO_5]	5.1	0.1	LPES	86
	W_3O_6	Tungsten oxide [W_3O_6]	2.85	0.02	LPES	353
	$YbBr_3$	Ytterbium(III) bromide	4.0	0.2	Knud	379
	YCl_4	Yttrium tetrachloride	7.02	0.01	LPES	145

References

1. Hotop, H., and Lineberger, W. C., *J. Phys. Chem. Ref. Data* 14, 731, 1985.This article was updated by Andersen, T., Haugen, H. K., and Hotop, H., *J. Phys. Chem. Ref. Data*, 28, 1511, 1999.
2. Drzaic, P. S., Marks, J., and Brauman, J. I., in *Gas Phase Ion Chemistry*, Vol. 3, Bowers, M. T., Ed., Academic Press, Orlando, 1984, p. 167. The reference for C_6H_4ClO should read "Richardson et al., 1975c."
3. Schulz, P. A., Mead, R. D., Jones, P. L., and Lineberger, W. C., *J. Chem. Phys.* 77, 1153, 1982.
4. Chaibi, W., Pelàez, R. J., Blondel, C., Drag, C., and Delsart, C., *Eur. Phys. J. D* 58, 29, 2010. EA(^{16}O) = 11784.676 ± 0.007 cm^{-1}, EA(^{32}S) = 16 752.975 3 ± 0.004 1 cm^{-1}, EA(^{34}S) = 16 752.977 6 ± 0.008 5 cm^{-1} (Ref. 334), and EA(^{28}Si) = 11 207.244 ± 0.006 cm^{-1}. EAs given in the paper in eV were obtained using an older CODATA value for hc/e, which affects EA values in the Table 1 in the final decimal place. See also Neumark, D. M., Lykke, K. R., Anderson, T., and Lineberger, W. C., *Phys. Rev. A* 32, 1890, 1985. EA(O) = 11 784.645 ± 0.008 cm^{-1}.
5. Wenthold, P. G., Gunion, R. F., and Lineberger, W. C., *Chem. Phys. Lett.* 258, 101, 1996.
6. Wenthold, P. G., Kim, J. B., Jonas, K. L., and Lineberger, W. C., *J. Phys. Chem.* A 101, 4472, 1997.
7. Klein, R., McGinnis, R. P., and Leone, S. R., *Chem. Phys. Lett.* 100, 475, 1983.
8. Arnold, D. W., Bradforth, S. E., Kitsopoulos, T. N., and Neumark, D. M., *J. Chem. Phys.* 95, 8753, 1991; linear C_n.
9. Stevens, A. E., Fiegerle, C. S., and Lineberger, W. C., *J. Chem. Phys.* 78, 5420, 1983.
10. Breyer, F., Frey, P., and Hotop, H., *Z. Phys.* A 300, 7, 1981.
11. Oakes, J. M., and Ellison, G. B., *Tetrahedron* 42, 6263, 1986.
12. Leopold, D. G., Murray, K. K., Miller, A. E. S., and Lineberger, W. C., *J. Chem. Phys.* 83, 4849, 1985.
13. Oakes, J. M., Jones, M.E., Bierbaum, V. M., and Ellison, G. B., *J. Phys. Chem.* 87, 4810, 1983.
14. Ellis, H. B., Jr., and Ellison, G. B., *J. Chem. Phys.* 78, 6541, 1983.
15. Ramond, T. M., Blanksby, S. J., Kato, S., Bierbaum, V. M., Davico, G. E., Schwartz, R. L., Lineberger, W. C., and Ellison, G. B., *J. Phys. Chem. A* 106, 9641, 2002.
16. Nimlos, M. E., and Ellison, G. B., *J. Chem. Phys.* 90, 2574, 1986.
17. Kawamata, H., Negishi, Y., Kishi, R., Iwata, S., Nakajima, A., and Kaya, K., *J. Chem. Phys.* 105, 5369, 1996.
18. McHugh, K. M., Eaton, J. G., Lee, G. H., Sarkas, H. W., Kidder, L. H., Snodgrass, J. T., Manaa, M. R., and Bowen, K. H., *J. Chem. Phys.* 91, 3792, 1989. See also Ref. 104.
19. George, P. M., and Beauchamp, J. L., *Chem. Phys.* 36, 345, 1979. The lower limit given in this paper (3.4 eV) may be increased to 3.5 eV, as rapid charge transfer from HCO_2– to WF_6 has since been observed (Miller, T. M., and Viggiano, A. A., unpublished).
20. Burnett, S. M., Stevens, A. E., Fiegerle, C. S., and Lineberger, W. C., *Chem. Phys. Lett.* 100, 124, 1983.

21. Moran, S., Ellis, H. B., DeFrees, D. J., McLean, A. D., and Ellison, G. B., *J. Am. Chem. Soc.* 109, 5996, 1987; Moran, S., Ellis, H. B., DeFrees, D. J., McLean, A. D., Paulson, S. E., and Ellison, G. B., *J. Am. Chem. Soc.* 109, 6004, 1987; see also Lykke, K. R., Neumark, D. M., Andersen, T., Trapa, V. J., and Lineberger, W. C., *J. Chem. Phys.* 87, 6842, 1987.
22. Yacovitch, T. I., Garand, E., and Neumark, D. M., *J. Chem. Phys.* 130, 244309, 2009 for C_2H_3O (14,719 ± 9 cm^{-1}). EA(C_2D_3O) was deduced from this and the difference measured by Mead, R. D., Lykke, K. R., Lineberger, W. C., Marks, J., and Brauman, J. I., *J. Chem. Phys.* 81, 4883, 1984.
23. Ellison, G. B., Engelking, P. C., and Lineberger, W. C., *J. Chem. Phys.* 86, 4873, 1982.
24. Oakes, J. M., and Ellison, G. B., *J. Am. Chem. Soc.* 105, 2969, 1983.
25. Ellison, G. B., and Oakes, J. M., *J. Am. Chem. Soc.* 106, 7734, 1984. EA(allyl) and EA(allyl-d_5) are 0.119 and 0.083 eV too low, respectively, in this work, according to Ref. 138. Therefore, EA(allyl-d_1) is likely too low by a similar amount.
26. Gunion, R. F., Gilles, M. K., Polak, M. L., and Lineberger, W. C., *Int. J. Mass Spectrom. Ion Process.* 117, 601, 1992.
27. Leopold, D. G., and Lineberger, W. C., *J. Chem. Phys.* 85, 51, 1986.
28. Scheer, M., Bilodeau, R. C., Brodie, C. A., and Haugen, H. K., *Phys. Rev. A* 58, 2844, 1998.
29. Miller, A. E. S., Fiegerle, C. S., and Lineberger, W. C., *J. Chem. Phys.* 87, 1549, 1987.
30. Miller, T. M., Leopold, D. G., Murray, K. K., and Lineberger, W. C., *J. Chem. Phys.* 85, 2368, 1986.
31. Miller, T. M., and Lineberger, W. C., *Chem. Phys. Lett.* 146, 364, 1988.
32. Neumark, D. M., Lykke, K. R., Andersen, T., and Lineberger, W. C., *J. Chem. Phys.* 83, 4364, 1985.
33. Leopold, D. G., Miller, T. M., and Lineberger, W. C., *J. Am. Chem. Soc.* 108, 178, 1986.
34. Miller, A. E. S., Fiegerle, C. S., and Lineberger, W. C., *J. Chem. Phys.* 84, 4127, 1986.
35. Murray, K. K., Miller, T. M., Leopold, D. G., and Lineberger, W. C., *J. Chem. Phys.* 84, 2520, 1986.
36. Leopold, D. G., Miller, A. E. S., and Lineberger, W. C., *J. Am. Chem. Soc.* 108, 1379, 1986.
37. León, I., Yang, Z., and Wang, L.-S., J. Chem. Phys. 139, 194306, 2013.
38. Snodgrass, J. T., Coe, J. V., McHugh, K. M., Friedhoff, C. B., and Bowen, K. H., *J. Phys. Chem.* 93, 1249, 1989.
39. Friedhoff, C. B., Snodgrass, J. T., Coe, J. V., McHugh, K. M., and Bowen, K. H., *J. Chem. Phys.* 84, 1051, 1986.
40. Friedhoff, C. B., Coe, J. V., Snodgrass, J. T., McHugh, K. M., and Bowen, K. H., *Chem. Phys. Lett.* 124, 268, 1986.
41. Coe, J. V., Snodgrass, J. T., Friedhoff, C. B., McHugh, K. M., and Bowen, K. H., *J. Chem. Phys.* 84, 619, 1986.
42. Snodgrass, J. T., Coe, J. V., Friedhoff, C. B., McHugh, K. M., and Bowen, K. H., *Chem. Phys. Lett.* 122, 352, 1985.
43. Nimlos, M. R., and Ellison, G. B., *J. Am. Chem. Soc.* 108, 6522, 1986.
44. Petrunin, V., Andersen, H., Balling, P., and Andersen, T., *Phys. Rev. Lett.* 76, 744, 1996.
45. Andersen, T., Lykke, K. R., Neumark, D. M., and Lineberger, W. C., *J. Chem. Phys.* 86, 1858, 1987.
46. Murray, K. K., Lykke, K. R., and Lineberger, W. C., *Phys. Rev. A* 36, 699, 1987.
47. Chaibi, W., Delsart, C., Drag, C., and Blondel, C., *J. Mol. Spectrosc.* 239, 11, 2006. EA(^{32}SH) = 18669.543(12) cm^{-1}.
48. Stonemann, R. C., and Larson, D. J., *Phys. Rev. A* 35, 2928, 1987. EA(SeH) = 17,845.17 ± 0.20 cm^{-1}.
49. Nimlos, M. R., Harding, L. B., and Ellison, G. B., *J. Chem. Phys.* 87, 5116, 1987.
50. Moylan, C. R., Dodd, J. A., Han, C.-C., and Braumann, J. I., *J. Chem. Phys.* 86, 5350, 1987.
51. Chowdhury, S., Grimsrud, E. P., Heinis, T., and Kebarle, P., *J. Am. Chem. Soc.* 108, 3630, 1986.
52. Berzinsh, U., Gustafsson, M., Hanstorp, D., Klinkmueller, A. E., Ljungblad, and U., Maartensson-Pendrill, A.-M., *Phys. Rev. A* 51, 231, 1995. EA(Cl) = 29138.59 ± 0.22 cm^{-1}.
53. Moran, S., and Ellison, G. B., *J. Phys. Chem.* 92, 1794, 1988.
54. Murray, K. K., Leopold, D. G., Miller, T. M., and Lineberger, W. C., *J. Chem. Phys.* 89, 5442, 1988.
55. Ervin, K. M., Ho, J., and Lineberger, W. C., *J. Chem. Phys.* 89, 4514, 1988.
56. Grimsrud, E. P., Chowdhury, S., and Kebarle, P., *J. Chem. Phys.* 83, 1059, 1985.
57. Fischer, C. F., *Phys. Rev.* A 39, 963, 1989.
58. Wickham-Jones, C. T., Ervin, K. M., Ellision, G. B., and Lineberger, W. C., *J. Chem. Phys.* 91, 2762, 1989.
59. Kryachko, E. S., Vinckier, C., and Nguyen, M. T., *J. Chem. Phys.* 114, 7911, 2001.
60. Mock, R. S., and Grimsrud, E. P., *J. Am. Chem. Soc.* 111, 2861, 1989.
61. Chowdhury, S., Heinis, T., Grimsrud, E. P., and Kebarle, P., *J. Phys. Chem.* 90, 2747, 1986. The uncertainty and other results are quoted in Ref. 60.
62. Wickham-Jones, C. T., Moran, S., and Ellison, G. B., *J. Chem. Phys.* 90, 795, 1989.
63. Ervin, K. M., Ho, J., and Lineberger, W. C., *J. Phys. Chem.* 92, 5405, 1988.
64. Miller, T. M., and Lineberger, W. C., *Int. J. Mass Spectrom. Ion Process.* 102, 239, 1990.
65. Wetzel, D. M., Salomon, K. E., Berger, S., and Brauman, J. I., *J. Am. Chem. Soc.* 111, 3835, 1989.
66. Crocker, L., Wang, T., and Kebarle, P., *J. Am. Chem. Soc.* 115, 7818, 1993.
67. Dillow, G. W., and Kebarle, P., *J. Am. Chem. Soc.* 111, 5592, 1989.
68. Gantefor, G., Gausa, M., Meiwes-Broer, K. H., and Lutz, H. O., *Z. Phys. D* 9, 253, 1988; Taylor, K. J., Petteitte, C. L., Craycraft, M. J., Chesnovsky, O., and Smalley, R. E., *Chem. Phys. Lett.* 152, 347, 1988.
69. Metz, R. B., Kitsopoulos, T., Weaver, A., and Neumark, D. M., *J. Chem. Phys.* 88, 1463, 1988.
70. Yang, S., Pettiette, C. L., Conceicao, J., Cheshnovsky, O., and Smalley, R. E., *Chem. Phys. Lett.* 139, 233, 1987; Yang, S., Taylor, K. J., Craycraft, M. J., Conceicao, J., Pettiette, C. L., Cheshnovsky, O., and Smalley, R. E., *Chem. Phys. Lett.* 144, 431, 1988; Arnold, D. W., Bradforth, S. E., Kitsopoulos, T. N., and Neumark, D. M., *J. Chem. Phys.* 95, 5479, 1991.
71. Cheshnovsky, O., Yang, S., Pettiette, C. L., Craycraft, M. J., Liu, Y., and Smalley, R. E., *Chem. Phys. Lett.* 138, 119, 1987.
72. Liu, Y., Zhang, Q.-L., Tittel, F. K., Curl, R. F., and Smalley, R. E., *J. Chem. Phys.* 85, 7434, 1986.
73. Travers, M. J., Cowles, D. C., and Ellison, G. B., *Chem. Phys. Lett.* 164, 449, 1989.
74. Blondel, C., Cacciani, P., Delsart, C., and Trainham, R., *Phys. Rev. A* 40, 3698, 1989. EA(Br) = 27,129.170 ± 0.015 cm^{-1} and EA(F) = 27,432.440 ± 0.025 cm^{-1}.
75. Bowen, K. H., and Eaton, J. G., in *The Structure of Small Molecules and Ions*, Naaman, R., and Vager, Z., Eds., Plenum, New York, 1988, pp. 147-169; Arnold, S. T., Eaton, J. G., Patel-Mistra, D., Sarkas, H. W., and Bowen, K. H., in *Ion and Cluster Ion Spectroscopy and Structure*, Maier, J. P., Ed., Elsevier Science, New York, 1989, p. 417.
76. Snodgrass, J. T., Coe, J. V., Friedhoff, C. B., McHugh, K. M., and Bowen, K. H., *Faraday Disc. Chem. Soc.* 88, 1988.
77. Lee, G. H., Arnold, S. T., Eaton, J. G., Sarkas, H. W., Bowen, K. H., Ludewigt, C., and Haberland, H., *Z. Phys. D – At., Mol. Clusters* 20, 9, 1991; Coe, J. V., Lee, G. H., Eaton, J. G., Arnold, S. T., Sarkas, H. W., Bowen, K. H., Ludewigt, C., Haberland, H., and Worsnop, D. R., *J. Chem. Phys.* 92, 3980, 1990.
78. Snodgrass, J. T., Coe, J. V., Freidhoff, C. B., McHugh, K. M., Arnold, S. T., and Bowen, K. H., *J. Phys. Chem.* **99**, 9675, 1995.
79. Hendricks, J. H., de Clercq, H. L., Freidhoff, C. B., Arnold, S. T., Eaton, J. G., Fancher, C., Lyapustina, S. A., Snodgrass, J. T., and Bowen, K. H., *J. Chem. Phys.***116**, 7926, 2002.
80. Snodgrass, J. T., Coe, J. V., Friedhoff, C. B., McHugh, K. M., and Bowen, K. H., *J. Chem. Phys.* 88, 8014, 1988.
81. Coe, J. V., Snodgrass, J. T., Friedhoff, C. B., McHugh, K. M., and Bowen, K. H., *Chem. Phys. Lett.* 124, 274, 1986.
82. Eliav, E., Vilkas, M. J., Ishikawa, Y., and Kaldor, U., *J. Chem. Phys.* 123, 224113(5), 2005.

83. Ervin, K. M., Ho, J., and Lineberger, W. C., *J. Chem. Phys.* 91, 5974, 1989.
84. Polak, M. L., Fiala, B. L., Ervin, K. M., and Lineberger, W. C., *J. Chem. Phys.* 94, 6924, 1991.
85. Weaver, A., Arnold, D. W., Bradforth, S. E., Neumark, D. M., *J. Chem. Phys.* 94, 1740, 1991.
86. Zhai, H.-J., Kirian, B., Cui, L.-F., Li, X., Dixon, D. A., and Wang, L.-S, *J. Am. Chem. Soc.* 126, 16134, 2004.
87. Ervin, K. M., and Lineberger, W. C., *J. Phys. Chem.* 95, 1167, 1991.
88. Gilles, M. K., Polak, M. L., and Lineberger, W. C., *J. Chem. Phys.* 96, 8012, 1992.
89. Lykke, K. R., Murray, K. K., and Lineberger, W. C., *Phys. Rev. A* 43, 6104, 1991. EA(H) = 6082.99 ± 0.15 cm^{-1} and EA(D) = 6086.2 ± 0.6 cm^{-1}.
90. Ervin, K. M., Gronert, S., Barlow, S. E., Gilles, M. K., Harrison, A. G., Bierbaum, V. M., DePuy, C. H., Lineberger, W. C., and Ellison, G. B., *J. Am. Chem. Soc.* 112, 5750, 1990.
91. Viggiano, A. A., Miller, T. M., Miller, A. E. S., Morris, R. A., Van Doren, J. M., and Paulson, J. F., *Int. J. Mass Spectrom. Ion Process.* 109, 327, 1991.
92. Peláez, R. J., Blondel, C., Delsart, C., and Drag, C., *J. Phys. B: At. Mol. Opt. Phys.* 42, 125001, 2009. EA(I) = 24,672.874 ± 0.029 cm^{-1}.
93. Polak, M. L., Gilles, M. K., and Lineberger, W. C., *J. Chem. Phys.* 96, 7191, 1992.
94. Bengali, A. A., Casey, S. M., Cheng, C.-L., Dick, J. P., Fenn, P. T., Villalta, P. W., and Leopold, D. G., *J. Am. Chem. Soc.* 114, 5257, 1992.
95. Gilles, M. K., Ervin, K. M., Ho, J., and Lineberger, W. C., *J. Phys. Chem.* 96, 1130, 1992.
96. Travers, M. J., Cowles, D. C., Clifford, E. P., and Ellison, G. B., *J. Am. Chem. Soc.* 114, 8699, 1992.
97. Bengali, A. A., and Leopold, D. G., *J. Am. Chem. Soc.* 114, 9192, 1992.
98. Rudnyi, E. B., Kaibicheva, E. A., and Sidorov, L. N., *Rapid Commun. Mass Spectrom.* 6, 356, 1992; Sidorov, L. N., *High Temp. Sci.* 29, 153, 1990. See also Srivastava, R. D., Uy, O. M., and Farber, M., *Trans. Faraday Soc.* 67, 2941, 1971.
99. Kitsopoulos, T. N., Chick, C. J., Zhao, Y., and Neumark, D. M., *J. Chem. Phys.* 95, 5479, 1991.
100. Arnold, C. C., Kitsopoulos, T. N., and Neumark, D. M., *J. Chem. Phys.* 99, 766, 1993.
101. Rudnyi, E. B., Kaibicheva, E. A., and Sidorov, L. N., *J. Chem. Thermodyn.* 25, 929, 1993.
102. Sarkas, H. W., Hendricks, J. H., Arnold, S. T., and Bowen, K. H., *J. Chem. Phys.* 100, 1884, 1994.
103. Villalta, P. W., and Leopold, D. G., *J. Chem. Phys.* 98, 7730, 1993.
104. Eaton, J. G., Sarkas, H. W., Arnold, S. T., McHugh, K. M., and Bowen, K. H., *Chem. Phys. Lett.* 193, 141, 1992. See also Ref. 18.
105. Polak, M. L., Gilles, M. K., Gunion, R. F., and Lineberger, W. C., *Chem. Phys. Lett.* 210, 55, 1993.
106. Gilles, M. K., Lineberger, W. C., and Ervin, K. M., *J. Am. Chem. Soc.* 115, 1031, 1993.
107. Casey, S. M., and Leopold, D. G., *Chem. Phys. Lett.* 201, 205, 1993.
108. Polak, M. L., Gerber, G., Ho, J., and Lineberger, W. C., J. *Chem. Phys.* 97, 8990, 1992.
109. Garand, E., Klein, K., Stanton, J. F., Zhou, J., Yacovitch, T. I., and Neumark, D. M., *J. Phys. Chem. A* 114, 1374, 2010.
110. Xu, C., Taylor, T. R., Burton, G. R., and Neumark, D. M., *J. Chem. Phys.* 108, 1395, 1998.
111. Bradforth, S. E., Kim, E. H., Arnold, D. W., and Neumark, D. M., *J. Chem. Phys.* 98, 800, 1993.
112. Ho, J., Polak, M. L., Ervin, K. M., and Lineberger, W. C., *J. Chem. Phys.* 99, 8542, 1993.
113. Brinkman, E. A., Berger, S., Marks, J., and Brauman, J. I., *J. Chem. Phys.* 99, 7586, 1993.
114. Casey, S. M., and Leopold, D. G., *J. Phys. Chem.* 97, 816, 1993.
115. McDonald, R. N., and Davidson, S. J., *J. Am. Chem. Soc.* 115, 10857, 1993.
116. Scheer, M., Brodie, C. A., Bilodeau, R. C., and Haugen, H. K., *Phys. Rev. A*, 58, 2051, 1998
117. Ho, J., Polak, M. L., and Lineberger, W. C., *J. Chem. Phys.* 96, 144, 1992. See also Ref. 289.
118. Polak, M. L., Gilles, M. K., Ho, J., and Lineberger, W. C., *J. Phys. Chem.* 95, 3460, 1991.
119. Polak, M. L., Ho, J., Gerber, G., and Lineberger, W. C., *J. Chem. Phys.* 95, 3053, 1991. See also Ref. 363 confirmation of EA(Bi_3).
120. Sharpe, P., and Kebarle, P., *J. Am. Chem. Soc.* 115, 782, 1993.
121. Miller, T. M., Miller, A. E. S., Viggiano, A. A., Morris, R. A., and Paulson, J. F., *J. Chem. Phys.* 100, 7200, 1994. Accurate calculations have yielded a higher result (0.90 eV); see Lau, J. K.-C. and Li, W.-K., *J. Mol. Struct. (Theochem)* 578, 221, 2002.
122. Andersen, H. H., Petrunin, V. V., Kristensen, P., and Andersen, T., *Phys. Rev. A* 55, 3247, 1997.
123. Arnold, C. C., Xu, C., Burton, G. R., and Neumark, D. M., *J. Chem. Phys.* 102, 6982, 1995; Burton, G. R., Xu, C., Arnold, C. C., and Neumark, D. M., *J. Chem. Phys.* 104, 2757 1996.
124. Xu, C., de Beer, E., and Neumark, D. M., *J. Chem. Phys.* 104, 2749, 1996.
125. Gunion, R. F., Koppel, H., Leach, G. W., and Lineberger, W. C., *J. Chem. Phys.* 103, 1250, 1995.
126. Gunion, R. F., and Lineberger, W. C., *J. Phys. Chem.* 100, 4395, 1996.
127. Gunion, R. F., Dixon-Warren, St. J., and Lineberger, W. C., *J. Chem. Phys.* 104, 1765, 1996.
128. Dixon-Warren, St. J., Gunion, R. F., and Lineberger, W. C., *J. Chem. Phys.* 104, 4902, 1996.
129. Nakajima, A., Zhang, N., Kawamata, H., Hayase, T., Nakao, K., and Kaya, K., *Chem. Phys. Lett.* 241, 295, 1995; Nakajima, A., Taguwa, T., Nakao, K., Hoshino, K., Iwata, S., and Kaya, K., *J. Chem. Phys.* 102, 660, 1995.
130. Fan, J., and Wang, L.-S., *J. Chem. Phys.* 102, 8714, 1995.
131. Arnold, D. W., Bradforth, S. E., Kim, E. H., and Neumark, D. M., *J. Chem. Phys.* 102, 3493, 1995; Zhao, Y., Arnold, C. C., and Neumark, D. M., *J. Chem. Soc. Faraday Trans.* 2 89, 1449, 1992.
132. Fan, J., Lou, L., and Wang, L.-S., *J. Chem. Phys.* 102, 2701, 1995.
133. Sarkas, H. W., Hendricks, J. H., Arnold, S. T., Slager, V. L., and Bowen, K. H., *J. Chem. Phys.* 100, 3358, 1994.
134. Kato, S., Lee, H. S., Gareyev, R., Wenthold, P. G., Lineberger, W. C., DePuy, C. H., and Bierbaum, V. M., *J. Am. Chem. Soc.* 119, 7863, 1997. See also Miller, T. M., Viggiano, A. A., and Miller, A. E. S., *J. Phys. Chem. A* 106, 10200, 2002.
135. Wenthold, P. G., Hu, J., Squires, R. R., and Lineberger, W. C., *J. Am. Chem. Soc.* 118, 475, 1996.
136. Gunion, R. F., Karney, W., Wenthold, P. G., Borden, W. T., and Lineberger, W. C., *J. Am. Chem. Soc.* 118, 5074, 1996. The numbers in the abstract for 1,6-heptadiyne were misprinted. EA(cycloheptatrienide) quoted here derives from the LPES data combined with other thermochemical data in Ref. 136.
137. Xu, C., de Beer, E., Arnold, D. W., Arnold, C. C., and Neumark, D. M., *J. Chem. Phys.* 101, 5406, 1996.
138. Wenthold, P. G., Polak, M. L., and Lineberger, W. C., *J. Phys. Chem.* 100, 6920, 1996.
139. Wenthold, P. G., Kim, J. B., and Lineberger, W. C., *J. Am. Chem. Soc.* 119, 1354, 1997.
140. Eliav, E., Kaldor, U., Ishikawa, Y., and Pyykko, P., *Phys. Rev. Lett.* 77, 5350, 1996.
141. Davies, B. J., Ingram, C. W., Larson, D. J., and Ljungblad, U., *J. Chem. Phys.* 106, 5783, 1997. EA(Ir) = 12,613 ± 4 cm^{-1}.
142. Smith, J. R., Kim, J. B., and Lineberger, W. C., *Phys. Rev. A* 55, 2036, 1997. EA(OH) = 14,741.02 ± 0.03 cm^{-1}. Schulz, P. A., Mead, R. D., Jones, P. L., and Lineberger, W. C., *J. Chem. Phys.* 77, 1153, 1982. EA(OD) = 14,723.92 ± 0.30 cm^{-1}. See also Rudmin, J. D., Ratliff, L. P., Yukich, J. N., and Larson, D. J., *J. Phys. B: At. Mol. Opt. Phys.* 29, L881, 1996.
143. Desai, S. R., Wu, H., Rohlfing, C. M., and Wang, L.-S., *Int. J. Chem. Phys.* 106, 1309, 1997.
144. Wenthold, P. G., Jonas, K.-L., and Lineberger, W. C., *J. Chem. Phys.* 106, 9961, 1997.
145. Yang, J., Wang, X.-B., Xing, X.-P., and Wang, L.-S., *J. Chem. Phys.* 128, 201102, 2008.
146. Moravec, V. D., and Jarrold, C. C., *J. Chem. Phys.* 108, 1804, 1998.
147. Wang, X.-B., Ding, C.-F., and Wang, L.-S., *J. Phys. Chem.* A 101, 7699, 1997.

Atomic

148. Wenthold, P. G., and Lineberger, W. C., *J. Am. Chem. Soc.* 19, 7772, 1997.
149. Wang, L.-S., Li, X., and Zhang, H.-F., *Chem. Phys.* 262, 53, 2000; Wang, L.-S., Cheng, H.-S., and Fan, J., *J. Chem. Phys.* 102, 9480, 1995.
150. Wu, H., Desai, S. R., and Wang, L.-S., *Phys. Rev. Lett.* 77, 2436, 1996.
151. Liu, S.-R., Zhai, H.-J., Castro, M., and Wang, L.-S., *J. Chem. Phys.* 118, 2108, 2003.
152. Gutsev, G. L., Bauschlicher, C. W., Zhai, H.-J., and Wang, L.-S., *J. Chem. Phys.* 119, 11135, 2003; Wang, L.-S., Wu, H., and Desai, S. R., *Phys. Rev. Lett.* 76, 4853, 1996.
153. Robinson, M. S., Polak, M. L., Bierbaum, V. M., DePuy, C. H., and Lineberger, W. C., *J. Am. Chem. Soc.* 117, 6766, 1995.
154. Clifford, E. P., Wenthold, P. G., Lineberger, W. C., Petersson, G. A., and Ellison, G. B., *J. Phys. Chem.* 101, 4338, 1997.
155. Seburg, R. A., and Squires, R. R., *Int. J. Mass Spectrom. Ion Process.* 167/168, 541, 1997.
156. Wang, X.-B., and Wang, L.-S., *Chem. Phys. Lett.* 313, 179, 1999.
157. Marcy, T. P., and Leopold, D. G., *Int. J. Mass Spectrom.* 195/196, 653, 2000.
158. Gutsev, G. L., Rao, B. K., Jena, P., Li, X., and Wang, L.-S., *J. Chem. Phys.* 113, 1473, 2000.
159. Negishi, Y., Yasuike, T., Hayakawa, F., Kizawa, M., Yabushita, S., and Nakajima, A., *J. Chem. Phys.* 113, 1725, 2000.
160. Klopcic, S. A., Moravec, V. D., and Jarrold, C. C., *J. Chem. Phys.* 110, 8986, 1999.
161. Boldyrev, A. I., Simons, J., Li, X., Chen, W., and Wang, L.-S., *J. Chem. Phys.* 110, 8980, 1999.
162. Taylor, T. R., Asmis, K. R., Zanni, M. T., and Neumark, D. M., *J. Chem. Phys.* 110, 7607, 1999.
163. Boldyrev, A., Li, X., and Wang, L.-S., *J. Chem. Phys.* 112, 3627, 2000.
164. Defrançois, C., Périquet, V., Lyapustina, S. A., Lippa, T. P., Robinson, D. W., Bowen, K. H., Nonaka, H., and Compton, R. N., *J. Chem. Phys.* 111, 4569, 1999.
165. Dobrin, S., Boo, B. H., Alconcel, L. S., and Continetti, R. E., *J. Phys. Chem. A* 104, 10695, 2000.
166. Schwartz, R. L., Davico, G. E., and Lineberger, W. C., *J. Electron Spectrosc. Relat. Phenom.*108, 163, 2000.
167. Kim, J. B., Wenthold, P. G., and Lineberger, W. C., *J. Phys. Chem. A* 103, 10833, 1999.
168. Davico, G. E., Ramond, T. M., and Lineberger, W. C., *J. Chem. Phys.* 113, 8852, 2000.
169. Green, S. M. E., Alex, S., Fleischer, N. L., Millam, E. L., Marcy, T. P., and Leopold, D. G., *J. Chem. Phys.* 114, 2653, 2001.
170. Wu, H., and Wang, L.-S., *J. Chem. Phys.* 108, 5310, 1998.
171. Wu, H., and Wang, L.-S., *J. Phys. Chem. A* 102, 9129, 1998.
172. Wu, H., and Wang, L.-S., *J. Chem. Phys.* 107, 8221, 1997.
173. Li, X., Zheng, W., Buonaugurio, A., Buytendyk, A., and Bowen, K., *J. Chem. Phys.* 136, 154306, 2012.
174. Wang, L.-S., private communication quoted in Ref. 169.
175. Marcy, T. P., Ph.D. dissertation, quoted in Ref. 169.
176. Alex, S., Green, M. E., and Leopold, D. G., unpublished, quoted in Ref. 169.
177. Wang, X.-B., Wang, L.-S., Brown, R., Schwerdtfeger, P., Schröder, D., and Schwarz, H., *J. Chem. Phys.* 114, 7388, 2001.
178. Kim, J. H., Li, X., Wang, L.-S., de Clercq, H. L., Fancher, C. A., Thomas, O. C., and Bowen, K. H., *J. Phys. Chem. A* 105, 5709, 2001.
179. Moravec, V. D., Klopcic, S. A., Chatterjee, B., and Jarrold, C. C., *Chem. Phys. Lett.* 341, 313, 2001.
180. Zengin, V., Persson, B. J., Strong, K. M., and Continetti, R. E., *J. Chem. Phys.* 105, 9740, 1996.
181. Kietzmann, H., Morenzin, J., Bechthold, P. S., Ganteför, G., and Eberhardt, W., *J. Chem. Phys.* 109, 2275, 1998.
182. Xu, C., Taylor, T. R., Burton, G. R., and Neumark, D. M., *J. Chem. Phys.* 108, 7645, 1998.
183. Williams, W. W., Carpenter, D. L., Covington, A. M., Koepnick, M. C., Calabrese, D., and Thompson, J. S., *J. Phys. B: At. Mol. Opt. Phys.* 31, L341, 1998.
184. Covington, A. M., Calabrese, D., Thompson, J. S., and Kvale, T. J., *J. Phys. B: At. Mol. Opt. Phys.* 31, L855, 1998. Pan, L., and Beck, D. R., *Phys. Rev. A* 93, 062501, 2016. The experimental uncertainty of ±0.02 eV is assumed applicable to the reinterpretation by Pan and Beck.
185. Haeffler, G., Hanstrorp, D., Kiyan, I., Klinkmueller, A. E., Ljungblad, U., and Pegg, D. J., *Phys. Rev. A* 53, 4127, 1996. EA(Li) = 4984.90 ± 0.17 cm^{-1}.
186. Chen, H. L., Ellis, Jr., P. E., Wijesekera, T., Hagan, T. E., Groh, S. E., Lyons, J. E., and Ridge, D. P., *J. Am. Chem. Soc.* 116, 1086, 1994.
187. Deyerl, H.-J., Alconcel, L. S., and Continetti, R. E., *J. Phys. Chem. A* 105, 552, 2001.
188. Blanksby, S. J., Ramond, T. M., Davico, G. E., Nimlos, M. R., Kato, S., Bierbaum, V. M., Lineberger, W. C., Ellison, G. B., and Okumura, M., *J. Am. Chem. Soc.* 123, 9585, 2001.
189. Asmis, K. R., Taylor, T. R., Xu, C., and Neumark, D. M., *Chem. Phys. Lett.* 295, 75, 1998.
190. Schäfer-Bung, B., Engels, B., Taylor, T. R., Neumark, D. M., Botschwina, P., and Peric, M., *J. Chem. Phys.* 115, 1777, 2001.
191. Scheer, M., Bilodeau, R. C., and Haugen, H. K., *Phys. Rev. Lett.* 80, 2562, 1998.
192. Taylor, T. R., Gómez, H., Asmis, K. R., and Neumark, D. M., *J. Chem. Phys.* 115, 4620, 2001.
193. Asmis, K. R., Taylor, T. R., and Neumark, D. M., *J. Chem. Phys.* 111, 8838, 1999 and 111, 10491, 1999.
194. Ramond, T. M., Davico, G. E., Schwartz, R. L., and Lineberger, W. C., *J. Chem. Phys.* 112, 1158, 2000.
195. Petrunin, V. V., Voldstad, J. D., Balling, P., Kristensen, P., Andersen, T., and Haugen, H. K., *Phys. Rev. Lett.* 75, 1911, 1995.
196. Dzuba, V. A., and Gribakin, G. F., *J. Phys. B: At. Mol. Opt. Phys.* 31, L483, 1998.
197. Ervin, K. M., Ramond, T. M., Davico, G. E., G. E., Schwartz, R. L., Casey, S. M., and Lineberger, W. C., *J. Phys. Chem.* 105, 10822, 2001.
198. Lippa, T. P., Xu, S.-J., Lyapustina, S. A., and Bowen, K. H., *J. Chem. Phys.* 109, 9263, 1998.
199. Turner, N. J., Martel, A. A., and Waller, I. M., *J. Phys. Chem.* 98, 474, 1994.
200. Lippa, T. P., Xu, S.-J., Lyapustina, S. A., Nilles, J. M., and Bowen, K. H., *J. Chem. Phys.* 109, 10727, 1998.
201. Huang, D.-L., Dau, P. D., Liu, H.-T., and Wang, L.-S., *J. Chem. Phys.* 140, 224315, 2014. Brink, C., Andersen, L. H., Hvelplund, P., Mathur, D., Voldstad, J. D., *Chem. Phys. Lett.* 233, 52, 1995.
202. Boltalina, O. V., Sidorov, L. N., Sukhanova, E. V., and Sorokin, I. D., *Chem. Phys. Lett.* 230, 567, 1994.
203. Clifford, E. P., Wenthold, P. G., Lineberger, W. C., Ellison, G. B., Wang, C. X., Grabowski, J. J., Vila, F., and Jordan, K. D., *J. Chem. Soc. Perkin Trans.* 2, 1015, 1998.
204. Spanel, P., Matejcik, S., and Smith, D., *J. Phys. B: At. Mol. Phys.* 28, 2941, 1995. See Miller, A. E. S., Miller, T. M., Viggiano, A. A., Morris, R. A., Van Doren, J. M., Arnold, S. T., and Paulson, J. F., *J. Chem. Phys.* 102, 8865, 1995 for interpretation in terms of EA(SF_5).
205. Kinghom, D. B., and Adamowicz, L., J., *Chem. Phys.* 106, 4589, 1997. EA(H) = 6083.0994 cm^{-1}, EA(D) = 6086.7137 cm^{-1}, and EA(T) = 6087.9168 cm^{-1}.
206. Xi, L., and Wang, L.-S., *J. Chem. Phys.* 109, 5264, 1998.
207. Ephraim, E., Shmulyian, S., Kaldor, U., and Isikawa, Y., *J. Chem. Phys.* 109, 3954, 1998. Also EA(La) = 0.35 eV.
208. Scheer, M., Bilodeau, R. C., Thogersen, J., and Haugen, H. K., *Phys. Rev. A* 57, R1493, 1998.
209. Kim, J. B., Wenthold, P. G., and Lineberger, W. C., *J. Chem. Phys.* 108, 830, 1998.
210. Kishi, R., Kawamata, H., Negishi, Y., Iwata, S., Nakajima, A., and Kaya, K., *J. Chem. Phys.* 107, 10029, 1997.
211. Adams, C. L., Schneider, H., Ervin, K. M. and Weber, J. M., *J. Chem. Phys.* 130, 074307, 2009. Earlier work by Compton, R. N., Carman, Jr., H. S., Desfrançois, C., Abdoul-Carmine, J., Schermann, J. P., Hendricks, J. H., Lyapustina, S. A., and Bowen, K. H., *J. Chem. Phys.* 105, 3472, 1996, found EA(CD_3NO_2) to be 0.02 eV lower than EA(CH_3NO_2). High-level calculations yield 0.172 ± 0.006 eV; see Bull, J. N., Maclagan, R. G. A. R., and Harland, P. W., *J. Phys. Chem. A* 114, 3622, 2010.
212. Yourshaw, I., Zhao, Y., and Neumark, D. M., *J. Chem. Phys.* 105, 351, 1996.
213. Gausa, M., Kaschner, R., Seifert, G., Faehrmann, J. H., Lutz, H. O., and Meiwes-Broer, K., *J. Chem. Phys.* 104, 9719, 1996.
214. Wu, H., and Wang, L.-S., *J. Chem. Phys.* 107, 16, 1997.

215. Radisic, D., Xu, S., and Bowen, K. H., *Chem. Phys. Lett.* 354, 9, 2002.
216. Pramann, A., and Rademann, K., *Chem. Phys. Lett.* 343, 99, 2001.
217. Gómez, H., Taylor, T. R., and Neumark, D. M., *J. Phys. Chem. A* 105, 6886, 2001.
218. Gómez, H., Taylor, T. R., Zhao, Y., and Neumark, D. M., *J. Chem. Phys.* 117, 8644, 2002.
219. Nimlos, M. R., Davico, G., Geise, C. M., Wenthold, P. G., Lineberger, W. C., Blansksby, S. J., Hadad, C. M., Petersson, G. A., and Ellison, G. B., *J. Chem. Phys.* 117, 4323, 2002.
220. Koyasu, K., Mitsui, M., Nakajima, A., and Kaya, K., *Chem. Phys. Lett.* 358, 224, 2002.
221. Zhai, H.-J., Wang, L.-S., Alexandrova, A. N., Boldyrev, A. I., and Zakrzewski, V. G., *J. Phys. Chem.* A 107, 9319, 2003.
222. Schiedt, J., and Weinkauf, *R., Z. Naturforsch.* A 50, 1041, 1995. See also Ervin, K. M., Anusiewicz, I., Skurski, P., Simons, J., and Lineberger, W. C., *J. Phys. Chem.* A 107, 8521, 2003 [EA(O_2) = 0.448 ± 0.006 eV].
223. Davis, V. T., and Thompson, J. S., *J. Phys. B* 34, L433, 2001. Recent high-level calculations give EA(Lu) = 0.029 eV; see Ref. 343.
224. Li, X., Zhai, H.-J., and Wang, L.-S., *Chem. Phys. Lett.* 357, 415, 2002.
225. Davis, V. T., and Thompson, J. S., *J. Phys. B: At. Mol. Opt. Phys.* 35, L11, 2002. High-level caculations imply EA(Pr) ≈ 0.177 eV; see O'Malley, S. M., and Beck, D. R., *Phys. Rev.* A 78, 012510, 2008.
226. Goldfarb, F., Drag, C., Chaibi, W., Kröger, S., Blondel, C., and Delsart, C., *J. Chem. Phys.* 122, 014308, 2005. EA(OH) = 14740.982(7) cm^{-1}.
227. Blondel, C., Delsart, C., and Goldfarb, F., *J. Phys. B* 34, L281, 2001. EA(F) = 27432.446(19) cm^{-1}.
228. Fancher, C. A., de Clercq, H. L., and Bowen, K. H., *Chem. Phys. Lett.* 366, 197, 2002.
229. Jin, C., Taylor, K. J., Conceicao, J., and Smalley, R. E., *Chem. Phys. Lett.* 175, 17, 1990.
230. Schiedt, J., Knott, W. J., Le Barbu, K., Schlag, E. W., and Weinkauf, R., *J. Chem. Phys.* 113, 9470, 2000.
231. Song, J. K., Lee, N. K., Kim, J. H., Han, S. Y., and Kim, S. K., *J. Chem. Phys.* 119, 3071, 2003.
232. Davico, G. E., Schwartz, R. L., Ramond, T. M., and Lineberger, W. C., *J. Amer. Chem. Soc.* 121, 6047, 1999.
233. Andrews, D. H., Gianola, A. J., and Lineberger, W. C., *J. Chem. Phys.* 117, 4074, 2002.
234. Schwartz, R. L., Davico, G. E., Kim, J. B., and Lineberger, W. C., *J. Chem. Phys.* 112, 4966, 2000.
235. Schwartz, R. L., Davico, G. E., Ramond, T. M., and Lineberger, W. C., *J. Phys. Chem.* A 103, 8213, 1999.
236. Schiedt, J., and Weinkauf, R., *Chem. Phys. Lett.* 274, 18, 1997.
237. Goebbert, D. J., Pichugin, K., Khuseynov, D., Wenthold, P. G., and Sanov, A., J. Chem. Phys. 132, 224301, 2010.
238. Geske, G. D., Boldyrev, A. I., Li, X., and Wang, L.-S., *J. Chem. Phys.* 113, 5130, 2000.
239. Cannon, N. A., Boldyrev, A. I., Li, X., and Wang, L.–S., *J. Chem. Phys.* 113, 2671, 2000.
240. Wang, X.-B., Yang, X., and Wang, L.-S., *J. Chem. Phys.* 116, 561, 2002.
241. Gutsev, G. L., Jena, P., Zhai, H.-J., and Wang, L.-S., *J. Chem. Phys.* 115, 7935, 2001.
242. Yang, X., Wang, X.-B., and Wang, L.-S., *J. Chem. Phys.* 115, 2889, 2001.
243. Zhai, H.-J., Liu, S.-R., Li, X., and Wang, L.-S., *J. Chem. Phys.* 115, 5170, 2001.
244. Li, X., Wang, L.-S., Cannon, N. A., and Boldyrev, A. I., *J. Chem. Phys.* 116, 1330, 2002.
245. Zhai, H.-J., Wang, L.-S., Alexandrova, A. N., and Boldyrev, A. I., *J. Chem. Phys.* 117, 7917, 2002.
246. Zhai, H.-J., and Wang, L.-S., *J. Chem. Phys.* 117, 7882, 2002.
247. Liu, S.-R., Zhai, H.-J., and Wang, L.-S., *J. Chem. Phys.* 117, 9758, 2002.
248. Mitsui, M., Nakajima, A., and Kaya, K., *J. Chem. Phys.* 117, 9740, 2002.
249. Yang, X., Wang, X.-B., Wang, L.-S., Niu, S., and Ichiye, T., *J. Chem. Phys.* 119, 8311, 2003.
250. Kim, J., Lee, H. M., Suh, S. B., Majumdar, D., and Kim, K. S., *J. Chem. Phys.* 113, 5259, 2000.
251. Liu, S. R., Zhai, H. J., and Wang, L.-S., *Phys. Rev. B* 64, 153402, 2001.
252. Zhai, H. J., Yang, X., Wang, X. B., Wang, L.-S., Elliott, B., and Boldyrev, A. I., *J. Am. Chem. Soc.* 124, 6742, 2002.
253. Zhai, H. J., Wang, L.-S., Kuznetsov, A. E., and Boldyrev, A. I., *J. Phys. Chem. A* 106, 5600, 2002.
254. Drechsler, G., and Boesl, U., *Int. J. Mass Spectrom.* 228, 1067, 2003.
255. Illenberger, E, Comita, P. B., Brauman, J. I., Fenzlaff, H. P., Heni, M., Heinrich, N., Koch, W., and Fenking, G., *Ber. Bunsen-Ges. Phys. Chem.* 89, 1026, 1985; Jackson, R. L., Pellerite, M. J., and Brauman, J. I., *J. Am. Chem. Soc.* 103, 1802, 1981.
256. Miller, T. M., Friedman, J. F., and Viggiano, A. A., *J. Chem. Phys.* 120, 7024, 2004.
257. Miller, T. M., Van Doren, J. M., and Viggiano, A. A., *Int. J. Mass Spectrom.* 233, 67, 2004.
258. Van Doren, J. M., McSweeney, S. A., Hargus, M. D., Kerr, D. M., Miller, T. M., Arnold, S. T., and Viggiano, A. A., *Int. J. Mass Spectrom.* 228, 541, 2003.
259. Van Doren, J. M., Miller, T. M., and Viggiano, A. A., *J. Chem. Phys.* 123, 114303, 2005.
260. Miller, T. M., and Viggiano, A. A., *Phys. Rev.* A 71, 012702, 2005.
261. Haeffler, G., Klinkmueller, A. E., Rangell, J., Berzinsh, U., and Hanstorp, D., *Z. Phys. D* 38, 211, 1996.
262. Bilodeau, R. C., and Haugen, H. K., *Phys. Rev. A* 64, 024501, 2001.
263. Miller, T. M., Viggiano, A. A., Friedman, J. F., and Van Doren, J. M., *J. Chem. Phys.* 121, 9993, 2004.
264. Davis, V. T., and Thompson, J. S., *Phys. Rev A* 65, 010501, 2001. Theoretical work implies that the measured EA(Tm) was actually for a long-lived excited anion state, and that Tm does not form a stable anion. See O'Malley, S. M., and Beck, D. R., *Phys. Rev. A* 70, 022502, 2004.
265. Gianola, A. J., Ichino, T., Hoenigman, R. L., Kato, S., Bierbaum, V. M., and Lineberger, W. C., *J. Phys. Chem. A* 108, 10326, 2004.
266. Staneke, P. O., Groothuis, G., Ingemann, S., and Nibbering, N. M. M., *Int. J. Mass Spectrom. Ion Process.* 142, 83, 1995. A Gaussian-3 calculation yields EA(CCl_4) = 0.994 eV [Ed.].
267. Meloni, G., Ferguson, M. J., and Neumark, D. M., *Phys. Chem. Chem. Phys.* 5, 4073, 2003.
268. Davis, V. T., and Thompson, J. S., *J. Phys. B: At. Mol. Opt. Phys.* 37, 1961, 2004.
269. Walter, C. W., Gibson, N. D., Janczak, C. M., Starr, K. A., Snedden, A. P., Field III, R. L., and Andersson, P., *Phys. Rev.* A76, 052702, 2007.
270. Ando, N., Kokubo, S., Mitsui, M., and Nakajima, A., *Chem. Phys. Lett.* 389, 279, 2004.
271. Zhai, H.-J., and Wang, L.-S., *J. Chem. Phys.* 120, 8996, 2004.
272. Surber, E., and Sanov, A., *J. Chem. Phys.* 116, 5921, 2002.
273. Dzuba, V. A., and Gribakin, G. F., *Phys. Rev.* A 55, 2443, 1997.
274. Meloni, G., Sheehan, S. M., Ferguson, M. J., and Neumark, D. M., *J. Phys. Chem.* A 108, 9750, 2004.
275. Zhai, H.-J., Li, J., and Wang, L.-S., *J. Chem. Phys.* 121, 8369, 2004.
276. Wu, X., Qin, Z., Xie, H., Cong, R., Wu, X., Tang, Z., and Fan, H., J. Chem. Phys. 133, 044303, 2010.
277. Kim, J., Kelley, J. A., Ayotte, P., Nielsen, S. B., Weddle, G. H., and Johnson, M. A., *J. Am. Soc. Mass Spectrom.* 10, 810, 1999.
278. Cavanagh, S. J., Gibson, S. T., and Lewis, B. R., *J. Chem. Phys.* 137, 144304, 2012.
279. Meloni, G., Sheehan, S. M., and Neumark, D. M., *J. Chem. Phys.* 122, 074317, 2005.
280. Yoder, B. L., Maze, J. T., Raghavachari, K., and Jarrold, C. C., *J. Chem. Phys.* 122, 094313, 2005.
281. Ervin, K. M., and Lineberger, W. C., *J. Chem. Phys.* 122, 194303, 2005.
282. Ichino, T., Gianola, A. J., Andrews, D. H., and Lineberger, W. C., *J. Phys. Chem.* A108, 11307, 2004. See also Ref. 373.
283. Das, U., Raghavachari, K., and Jarrold, C. C., *J. Chem. Phys.* 122, 014313, 2005.
284. Schiedt, J., and Weinkauf, R., *J. Chem. Phys.* 110, 304, 1999.
285. Le Barbu, K., Schiedt, J., Weinkauf, R., Schlag, E. W., Nilles, J. M., Xu, S.-J., Thomas, O. C., and Bowen, J. H., *J. Chem. Phys.* 116, 9663, 2002. Uncertainties not stated.

286. Schiedt, J., and Weinkauf, R., *Chem. Phys. Lett.* 266, 201, 1997. The uncertainty for EA(anthracene) quoted as ±0.008 eV in a later paper (Ref. 230).
287. Chatterjee, B., Akin, F. A., Jarrold, C. C., and Raghavachari, K., *J. Phys. Chem. A* 109, 6880, 2005.
288. Zhai, H.-J., Kiran, B., Dai, B., Li, J., and Wang, L.-S., *J. Am. Chem. Soc.* 127, 12098, 2005.
289. Moravec, V. D., Klopcic, S. A., and Jarrold, C. C., *J. Chem. Phys.* 110, 5079, 1999.
290. Klopcic, S. A., Moravec, V. D., and Jarrold, C. C., *J. Chem. Phys.* 110, 10216, 1999.
291. Moravec, V. D., and Jarrold, C. C., *J. Chem. Phys.* 112, 792, 2000.
292. Moravec, V. D., and Jarrold, C. C., *J. Chem. Phys.* 113, 1035, 2000.
293. Chatterjee, B., Akin, F. A., Jarrold, C. C., and Raghavachari, K., *J. Chem. Phys.* 119, 10591, 2003.
294. Schröder, D., Brown, R., Schwerdtfeger, P., Wang, X. B., Yang, X., Wang, L. S., and Schwarz, H., *Angew. Chem. Int. Ed.* 42, 311, 2003.
295. Zheng, W., Nilles, J. M., Thomas, O. C., and Bowen, K. H., *J. Chem. Phys.* 122, 044306, 2005.
296. Duan, X., Burggraf, L. W., Weeks, D. E., Davico, G. E., Schwartz, R. L., and Lineberger, W. C., *J. Chem. Phys.* 116, 3601, 2002.
297. Meloni, G., Sheehan, S. M., Parsons, B. F., and Neumark, D. M., *J. Phys. Chem. A* 110, 3527, 2006.
298. Alexandrova, A. N., Boldyrev, A. I., Zhai, H.-J., and Wang, L.-S., *J. Chem. Phys.* 122, 054313, 2005.
299. Davico, G. E., Schwartz, R. L., Raymond, T. M., and Lineberger, W. C., *J. Phys. Chem.* A 103, 6167, 1999. EA(WO_2) may be lower than 1.998 eV by either 0.040 eV or 0.080 eV depending on the precise assignment of structures in the photoelectron spectrum.
300. Huang, X., Zhai, H.-J., Li, J., and Wang, L.-S., *J. Phys. Chem. A* 110, 85, 2006.
301. Zhai, H.-J., Wang, L.-S., Zubarev, D. Yu., and Boldyrev, A. I., *J. Phys. Chem.* A 110, 1689, 2006.
302. Sheehan, S. M., Meloni, G., Parsons, B. F., Wehres, N., and Neumark, D. M., *J. Chem. Phys.* 124, 064303, 2006.
303. Tschurl, M., and Boesl, U., *Int. J. Mass Spectrom.* 249/250, 364, 2006.
304. Nee, M. J., Osterwalder, A., Zhou, J., and Neumark, D. M., *J. Chem. Phys.* 125, 014306, 2006.
305. Zanni, M. T., Taylor, T. R., Greenblatt, B. J., Soep, B., and Neumark, D. M., *J. Chem. Phys.* 110, 3748, 1997.
306. Zhai, H.-J., and Wang, L.-S., *J. Chem. Phys.* 125, 164315, 2006.
307. Waters, T., Woo, H.-K., Wang, X.-B., and Wang, L.-S., *J. Am. Chem. Soc.* 128, 4282, 2006.
308. Calvi, R. M. D., Andrews, D. H., and Lineberger, W. C., *Chem. Phys. Lett.* 442, 12, 2007.
309. Parsons, B. F., Sheehan, S. M., Yen, T. A., Neumark, D. M., Wehres, N., and Weinkauf, R., *Phys. Chem. Chem. Phys.* 9, 3291, 2007.
310. Yu, W., Lin, Z., and Ding, C., *J. Chem. Phys.* 126, 114301, 2007.
311. Zhou, J., Garand, E., Eisfeld, W., and Neumark, D. M., *J. Chem. Phys.* 127, 034304, 2007.
312. Zheng, W., Li, X., Eustis, S., Grubisic, A., Thomas, O., de Clercq, H., and Bowen, K., *Chem. Phys. Lett.* 444, 232, 2007.
313. Mitsui, M., Ando, N., and Nakajima, A., *J. Phys. Chem. A* 111, 9644, 2007.
314. Zhou, J. Garand, E., and Neumark, D. M., *J. Chem. Phys.* 127, 154320, 2007. The measured values are EA(C_4D) = 28478(10) cm^{-1} and EA(C_4H) = 28497(8) cm^{-1}.
315. Sheehan, S. M., Parsons, B. F., Zhou, J., Garand, E., Yen, T. A., Moore, D. T., and Neumark, D. M., J., *Chem. Phys.* 128, 034301, 2008.
316. Ichino, T., Andrews, D. H., Rathbone, G. J., Misaizu, F., Calvi, R. M. D., Wren, S. W., Kato, S., Bierbaum, V. M., and Lineberger, W. C., *J. Phys. Chem.* B 112, 545, 2008.
317. Sheehan, S. M., Parsons, B. F., Yen, T. A., Furlanetto, M. R., and Neumark, D. M., *J. Chem. Phys.* 128, 174301, 2008.
318. Troe, J., Miller, T. M., and Viggiano, *J. Chem. Phys.* 136, 121102, 2012. See the calculation by Karton, A., and Martin, J. M. L., *J. Chem. Phys.* 136, 197101, 2012.
319. Zheng, W., Bowen, K. H., Li, J., Dąbkowska, I., Gutowski, M., *J. Phys. Chem. A* 109, 482, 2005. See also Mok, D. K. W., Chau, F.-t., Dyke, J. M., and Lee, E. P. F., *Chem. Phys. Lett.* 458, 11, 2008 and Mok, D. K. W., Lee, E. P. F., Chau, F.-t., and Dyke, J. M., *Phys. Chem. Chem. Phys.* 10, 7270, 2008.
320. Zheng, W., Li, X., Eustis, S., and Bowen, K., *Chem. Phys. Lett.* 460, 68, 2008.
321. Grubisic, A., Wang, H., Ko, Y. J., and Bowen, K. H., *J. Chem. Phys.* 129, 054302, 2008.
322. Stokes, S. T., Bowen, K. H., Sommerfeld, T., Ard, S., Mirsaleh-Kohan, N., Steill, J. D., and Compton, R. N., *J. Chem. Phys.* 129, 064308, 2008.
323. Garand, E., Yacovitch, T. I., and Neumark, D. M., *J. Chem. Phys.* 129, 074312, 2008.
324. Li, X., Eustis, S., Bowen, K. H., Kandalam, A. K., and Jena, P., *J. Chem. Phys.* 129, 074313, 2008. Uncertainties not stated.
325. Kandalam, A. K., Kiran, B., Jena, P., Li, X., Grubisic, A., and Bowen, K. H., *J. Chem. Phys.* 126, 084306, 2007.
326. Gerhards, M., Thomas, O. C., Nilles, J. M., Zhang, W.-J., and Bowen, K. H., *J. Chem. Phys.* 116, 10247, 2002.
327. Edmonds, B. D., Kandalam, A. K., Khanna, S. N., Li, X., Grubisic, A., Khanna, I., and Bowen, K. H., *J. Chem. Phys.* 124, 074316, 2006.
328. Duncan, M. A., Knight, A. M., Negishi, Y., Nagao, S., Nakamura, Y., Kato, A., Nakajima, A., and Kaya, K., *Chem. Phys. Lett.* 309, 49, 1999.
329. Zheng, W., Eustis, S. N., Li, X., Nilles, J. M., Thomas, O. C., Bowen, K. H., and Kandalam, A. K., *Chem. Phys. Lett.* 462, 35, 2008.
330. Kandalam, A. K., Jena, P., Li, X., Eustis, S. N., and Bowen, K. H., *J. Chem. Phys.* 129, 134308, 2008.
331. Wenthold, P. G., and Squires, R. R., *J. Phys. Chem.* 99, 2002, 1995. See also Artau, A., Nizzi, K. E., Hill, B. T., Sunderlin, L. S., and Wenthold, P. G., *J. Am. Chem. Soc.* 122, 10667, 2000.
332. Villalta, P. W., and Leopold, D. G., *J. Chem. Phys.* 130, 024303, 2009.
333. Wyrwas, R. B., and Jarrold, C. C., *J. Am. Chem. Soc.* 128, 13688, 2008.
334. Carette, T., Drag, C., Scharf, O., Blondel, C., Delsart, C., Fischer, C. F., and Godefroid, M., *Phys. Rev. A* 81, 042522, 2010. EA(^{32}S) = 16 752.975 3(41) cm^{-1}, and EA(^{34}S) = 16 752.977 6(85) cm^{-1}.
335. Goebbert, D. J., Khuseynov, D., and Sanov, A., *J. Phys. Chem. A* 114, 2259, 2010.
336. Zubarev, D. Yu., Boldyrev, A. I., Li, J., Zhai, H.-J., and Wang, L.-S., *J. Phys. Chem. A* 111, 1648, 2007.
337. Averkiev, B. B., Boldyrev, A. I., Li, Y., and Wang, L.-S., *J. Phys. Chem. A* 111, 34, 2007.
338. Zhai, H.-J., Wang, L.-M., Li, S.-D., and Wang, L.-S., *J. Phys. Chem. A* 111, 1030, 2007.
339. Distelrath, V., and Boesl, U., *Faraday Discuss.* 115, 161, 2000.
340. Wang, X.-B., and Wang, L.-S., *J. Chem. Phys.* 113, 10928, 2000.
341. Carpenter, D. L., Covington, A. M., and Thompson, J. S., *Phys. Rev. A* 61, 042501, 2000.
342. Davis, V. T., Thompson, J., and Covington, A., *Nucl. Instrum. Methods Phys. Res. B* 241, 118, 2005. High-level calculations give a lower value, EA(Nd) = 0.162 eV (Ref. 343) and 0.169 eV (O'Malley, S. M., and Beck, D. R., *Phys. Rev. A* 77, 012505, 2008).
343. Pan, L., and Beck, D. R., *J. Phys. B* 43, 025002, 2010.
344. Sobhy, M. A., Reveless, J. U., Gupta, U., Khanna, S. N., and Castleman, A. W., *J. Chem. Phys.* 130, 054304, 2009.
345. Ganteför, G., Gausa, M., Meiwes-Broer, K. H., and Lutz, H. O., *Z. Phys. D* 12, 405, 1989.
346. Garand, E., Yacovitch, T. I., and Neumark, D. M., *J. Chem. Phys.* 130, 064304, 2009.
347. Entfellner, M., and Boesl, U., *Phys. Chem. Chem. Phys.* 11, 2657, 2009.
348. Wang, L.-M., Huang, W., Wang, L.-S., Averkiev, B. B., and Boldyrev, A. I., *J. Chem. Phys.* 130, 134303, 2009.
349. Wren, S. W., Vogelhuber, K. M., Ervin, K. M., and Lineberger, W. C., *Phys. Chem. Chem. Phys.* 11, 4745, 2009.
350. Sheps, L., Miller, E. M., and Lineberger, W. C., *J. Chem. Phys.* 131, 064304, 2009.

351. Garand, E., Yacovitch, T. I., and Neumark, D. M., *J. Chem. Phys.* 131, 054312, 2009.
352. Walter, C. W., Gibson, N. D., Field III, R. L., Snedden, A. P., Shapiro, J. Z., Janczak, C. M., and Hanstorp, D., *Phys. Rev. A* 80, 014501, 2009.
353. Rothgeb, D. W., Hossain, E., Kuo, A. T., Troyer, J. L., and Jarrold, C. C., *J. Chem. Phys.* 131, 044310, 2009. See also Ref. 369.
354. Borschevsky, A., Pershina, V., Bliav, E., and Kaldor, U., *Chem. Phys. Lett.* 480, 49, 2009.
355. O'Malley, S. M., and Beck, D. M., *Phys. Rev. A* 79, 012511, 2009 and 80, 032514, 2009.
356. Xu, H.-G., Zhang, Z.-F., Feng, Y., and Zheng, W., *Chem. Phys. Lett.* 498, 22, 2010.
357. Xu, H.-G., Zhang, Z.-F., Feng, Y., Yuan, J., Zhao, Y., and Zheng, W., *Chem. Phys. Lett.* 487, 204, 2010.
358. Feng, Y., Xu, H.-G., Zhang, Z.-G., Gao, Z., and Zheng, W., *J. Chem. Phys.* 132, 074308, 2010.
359. Zhang, Z.-G., Xu, H.-G., Zhao, Y., and Zheng, W., *J. Chem. Phys.* 133, 154314, 2010.
360. Lindahl, A. O., Andersson, P., Diehl, C., Forstner, O., Klason, P., and Hanstorp, D., *Eur. Phys. J. D* 60, 219, 2010. EA(W) = 6583.6 ± 0.6 cm^{-1}.
361. Vogelhuber, K. M., Wren, S. W., Sheps, L., and Lineberger, W. C., *J. Chem. Phys.* 134, 064302, 2011.
362. Feng, Y., Xu, H.-G., Zheng, W., Zhao, H., Kandalam, A. K., and Jena, P., *J. Chem. Phys.* 134, 094309, 2011.
363. Gupta, U., Reveles, J. U., Melko, J. J., Khanna, S. N., and Castleman, A. W., *Chem. Phys. Lett.* 467, 223, 2009.
364. Gupta, U., Reveles, J. U., Melko, J. J., Khanna, S. N., and Castleman, A. W., *J. Phys. Chem. C* 114, 15963, 2010. Uncertainties not given, but are typically ± 0.05 to ± 0.08 eV (See Refs. 360, 363, and 365).
365. Gupta, U., Reber, A. C., Melko, J. J., Khanna, S. N., and Castleman, A. W., *Chem. Phys. Lett.* 505, 92, 2011.
366. Yacovich, T. I., Kim, J. B., Garand, E., van der Poll, D. G., and Neumark, D. M., *J. Chem. Phys.* 134, 134307, 2011.
367. Vogelhuber, K. M., Wren, S. W., McCoy, A. B., Ervin, K. M., and Lineberger, W. C., *J. Chem. Phys.* 134, 184306, 2011.
368. Adams, C. L., and Weber, J. M., *J. Chem. Phys.* 134, 244301, 2011.
369. Rothgeb, D. W., Mann, J. E., Waller, S. E., and Jarrold, C. C., *J. Chem. Phys.* 135, 104312, 2011.
370. Mann, J. E., Waller, S. E., Rothgeb, D. W., and Jarrold, C. C., *J. Chem. Phys.* 135, 104317, 2011.
371. Willis, M., Götz, M., Kandalam, A. K., Ganteför, G. F., and Jena, P., *Angew. Chem. Int. Ed.* 49, 8966, 2010.
372. Wang, Y.-L., Wang, X.-B., Xing, X.-P., Wei, F., Li, J., and Wang, L.-S., *J. Phys. Chem. A* 114, 11244, 2010.
373. Zhai, H.-J., Bürgel, C., Bonacid-Koutecky, V., and Wang, L.-S., *J. Am. Chem. Soc.* 130, 9156 (2008).
374. Chen, W.-J., Zhai, H.-J., Zhang, Y.-F., Huang, X., and Wang, L.-S., *J. Phys. Chem. A* 114, 5958, 2010.
375. Wang, X.-B., Fu, Q., and Yang, J., *J. Phys. Chem. A* 114, 9083, 2010.
376. Gruzinskaya, N. I., Aleshina, V. E., Borshchevskii, A. Ya., and Sidorov, L. N., *J. Anal. Chem.* 65, 1328, 2010.
377. Peláez, R. J., Blondel, C., Vandevraye, M., Drag, C., and Delsart, C., *J. Phys. B* 44, 1950009, 2011. EA(P) = 6021.79 ± 0.08 cm^{-1}.
378. Lenington, M. J., and Wenthold, P. G., *J. Phys. Chem. A* 114, 1334, 2010.
379. Butman, M. F., Sergeev, D. N., Motalov, V. B., Kudin, L. S., Kryuchkov, A. S., and Krämer, K. W., *Russ. J. Phys. Chem. A* 85, 922, 2011.
380. Zheng, W.-J., Thomas, O. C., Nilles, J. M., Bowen, K. H., Reber, A. C., and Khanna, S. N., *J. Chem. Phys.* 134, 224307, 2011.
381. Li, R.-Z., Xu, H.-G., Cao, G.-J., Xu, X.-L., and Zheng, W.-J., *J. Chem. Phys.* 139, 184303, 2013.
382. Wang, H., Zhang, X., Ko, Y. J., Gantefoer, G., Bowen, K. H., Li, X., Kiran, B., and Kandalam, A. K., *J. Chem. Phys.* 140, 164317, 2014. Uncertainty ±0.1 eV (private communication).
383. Miller, E. M., Sheps, L., Lu, Y.-J., Case, A. S., McCoy, A. B., and Lineberger, W. C., *J. Chem. Phys.* 136, 044313, 2012.
384. Buonaugurio, A., Zhang, X., Stokes, S. T., Wang, Y., Ellison, G. B., and Bowen, K. H., *Int. J. Mass Spectrom.* 377, 278, 2015.
385. Buytendyk, A., Graham, J., Wang, H., Zhang, X., Collins, E., Ko, Y.-J., Gantefoer, G., Eichhorn, B., Regmi, A., Kiran, B., and Bowen, K. H., *Int. J. Mass Spectrom.* 365/366, 140, 2014.
386. Liu, Q.-Y., Hu, L., Li, Z.-Y., Ning, C.-G., Ma, J.-B., Chen, H., and He, S. G., *J. Chem. Phys.* 142, 164301, 2015.
387. Mann, J. E., Waller, S. E., and Jarrold, C. C., *J. Chem. Phys.* 137, 044301, 2012.
388. Zhang, X., Tang, X., and Bowen, K., *Chem. Phys. Lett.* 582, 21, 2013.
389. Nakajima, A., Taguwa, T., and Kaya, K., *Chem. Phys. Lett.* 221, 436, 1994.
390. Wren, S. W., Vogelhuber, K. M., Ichimo, T., Stanton, J. F., and Lineberger, W. C., *J. Phys. Chem. A* 116, 3118, 2012.
391. Liu, Z., Xie, H., Qin, Z., Cong, R., Wu, X., Tang, Z., Lu, X., and He, J., *J. Chem. Phys.* 137, 204302, 2012.
392. Adams, C. L., Knurr, B. J., and Weber, M., *J. Chem. Phys.* 136, 064307, 2012.
393. Buonaugurio, A., Graham, J., Buytendyk, A., Bowen, K. H., Ryder, M. R., Keolopile, Z. G., Haranczyk, M., and Gutowski, M., *J. Chem. Phys.* 140, 221103, 2014.
394. Kim, J. B., Weichman, M. L., Yacovitch, T. I., Shih, C., and Neumark, D. M., *J. Chem. Phys.* 139, 104301, 2013.
395. Lu, Y.-J., Lehman, J. H., and Lineberger, W. C., *J. Chem. Phys.* 142, 044201, 2015.
396. Osborn, D. L., Vogelhuber, K. M., Wren, S. W., Miller, E. M., Lu, Y.-J., Case, A. S., Sheps, L., McMahon, R. J., Stanton, J. F., Harding, L. B., Ruscic, B., and Lineberger, W. C., *J. Am. Chem. Soc.* 136, 10361, 2014.
397. Wren, S. W., Vogelhuber, K. M., Garver, J. M., Kato, S., Sheps, L., Bierbaum, V. M., and Lineberger, W. C., *J. Am. Chem. Soc.* 134, 6584, 2012.
398. Dixon, A. R., Khuseynov, D., and Sanov, A., *Chem. Phys. Lett.* 614, 72, 2014.
399. Sun, Z., Xu, H.-G., Feng, G., Xu, X.-L., and Zheng, W.-J., *Chem. Phys. Lett.* 615, 56, 2014.
400. Li, R.-Z., Xu, H.-G., Xu, X.-L., and Zheng, W.-J., *Chem. Phys. Lett.* 607, 105, 2014.
401. Zhang, X., Tang, X., Mayo, D. H., DeCarlo, S., Eichhorn, B., and Bowen, K., *Chem. Phys. Lett.* 597, 110, 2014.
402. Huang, D.-L., Zhu, G.-Z., and Wang, L.-S., *J. Chem. Phys.* 142, 091103, 2015.
403. Dao, D. B., and Mabbs, R., *J. Chem. Phys.* 141, 154304, 2014.
404. Weichman, M. L, Kim, J. B., and Neumark, D. M., *J. Chem. Phys.* 140, 104305, 2014. EA(CH2CN) = 12,468 ± 2 and EA(CH2CN) = 12,402 ± 2 cm-1.
405. León, I., Ruipérez, F., Ugalde, J. M., and Wang, L.-S., *J. Chem. Phys.* 145, 064304, 2016.
406. Le, I., Yang, Z., and Wang, L.-S., *J. Chem. Phys.* 140, 084303, 2014.
407. Weichman, M. L., DeVine, J. A., and Neumark, D. M., *J. Chem. Phys.* 145, 054302, 2016.
408. Chen, X., and Ning, C., *J. Chem. Phys.* 145, 084303, 2016.
409. Li, X., Liu, S. S., Chen, W., and Wang, L.-S., *J. Chem. Phys.* 111, 2464, 1999.
410. Roy, S. K., Jian, T., Lopez, G. V., Li, W.-L., Su, J., Bross, D. H., Peterson, K. A., Wang, L.-S., and Li, J., *J. Chem. Phys.* 144, 084309, 2016.
411. Zhu, G.-Z., and Wang, L.-S., *J. Chem. Phys.* 143, 221102, 2015.
412. Li, W.-L., Hu, H.-S., Jian, T., Lopez, G. V., Su, J., Li, J., and Wang, L.-S., *J. Chem. Phys.* 139, 244303, 2013.
413. Li, W.-L., Su, J., Jian, T., Lopez, G. V., Hu, H.-S., Cao, G.-J., Li, J., and Wang, L.-S., *J. Chem. Phys.* 140, 094306, 2014.
414. Su, J., Dau, P. D., Liu, H.-T., Huang, D. L., Wei, F., Schwarz, W. H. E., Li, J., and Wang, L.-S., *J. Chem. Phys.* 142, 134308, 2015.
415. Kafader, J. O., Ray, M., and Jarrold, C. C., *J. Chem. Phys.* 143, 064305, 2015.
416. Aravind, G., Nrisimhamurty, M., Mane, R. G., Gupta, A. K., and Krishnakumar, E., *Phys. Rev. A* 92, 042503, 2015.
417. Chen, X., and Ning, C., *Phys. Rev. A* 93, 052508, 2016.
418. Tian, W. J., Zhao, L.-J., Chen, Q., Ou, T., Xu, H.-G., Zheng, W. J., Zhai, H.-J., and Li, S.-D., *J. Chem. Phys.* 142, 134305, 2015.
419. Huang, D.-L., Liu, H.-T., Ning, C.-G., and Wang, L.-S., *J. Chem. Phys.* 142, 124309, 2015.

420. Mann, J. E., Troyer, M. E., and Jarrold, C. C., *J. Chem. Phys.* 142, 124305, 2015.
421. Kafle, B., Aviv, O., Chandrasekaran, V., Heber, O., Rappaport, M. L., Rubinstein, H., Schwalm, D., Strasser, D., and Zajfman, D., *Phys. Rev. A* 92, 052503, 2015.
422. Zhao, L.-J., Tian, W.-J., Ou, T., Xu, H.-G., Feng, G., Xu, X.-L., Zhai, H.-J., Li, S.-D., and Zheng, W.-J., *J. Chem . Phys.* 144, 124301, 2016.
423. Xie, H., Zou, J., Yuan, Q., Fan, H., Tang, Z., and Jiang, L., *J. Chem. Phys.* 142, 124303, 2016.
424. Zhang, X., Ganteför, G., Eichorn, B., Mayo, D., Sawyer, W.H., Gill, A. F., Kandalam, A. K., Schockel, H., and Bowen, K.H., *J. Chem. Phys.* 145, 074305, 2016.
425. Anderson, K. T., Sandström, K., Hanstrop, D., and Pegg, D. J., *Phys. Rev. A* 62, 022503, 2000. EA(K) = 4044.54 0.10 cm^{-1}.
426. Bilodeau, R. C., Scheer, M., and Haugen, H. K., *J. Phys. B* 31, 3885, 1998.
427. Norquist, P. L., Beck, D. R., Bilodeau, R. C., Scheer, M., Srawley, R. A., Haugen, H. K., *Phys. Rev. A* 59, 1896, 1999.
428. Williams, W. W., Carpenter, D. L., Covington, A. M., Thompson, J. S., Kvale, T. J., and Seely, D. G., *Phys. Rev. A* 58, 3582, 1998.
429. Scheer, M., Haugen, H. K., and Beck, D. R., *Phys. Rev. Lett.* 79, 4104, 1997.
430. Bilodeau, R. C., and Haugen, H. K., *Phys. Rev. Lett.* 85, 534, 2000.
431. Bilodeau, R. C., Scheer, M., Haugen, H. K., and Brooks, R. L., *Phys. Rev. A* 61, 012505, 2001.
432. Jian, T., Lopez, G. V., and Wang, L.-S., *J. Phys. Chem. B* 120, 1635, 2016.
433. Su, J., Li, W.-Li., Lopez, G. V., Jian, T., Cao, G.-J., Li, W.-L., Schwarz, W. H. E., Wang, L.-S., and Li, J., J. Chem. Phys. A 120, 1084, 2016.
434. Huang, D.-L., Liu, H.-T., Ning, C.-G., Zhu, G.-Z., and Wang, L.-S., *Chem. Sci.* 6, 3129, 2015.
435. Liu, H.-T., Huang, D.-L., Liu, Y., Cheung, L.-F., Dau, P. D., Ning, C.-G., and Wang, L.-S., *J. Phys. Chem. Lett.* 6, 637, 2015.
436. Lopez, G. V., Czekner, J., Jian, T., Li, W.-L., Yang, Z., and Wang, L.-S., J. Chem. Phys. 141, 234309, 2014.
437. Popov, I. A., Li, W.-L., Piazza, Z. A., Boldyrev, A. I., and Wang, L.-S., *J. Phys. Chem. A* 118, 8098, 2014. EA uncertainty assumed to be the same as given for the vertical detachment energies.
438. Erdogdu, Y., Jian, T., Lopez, G. V., Li, W.-L., and Wang, L.-S., *Chem. Phys. Lett.* 610/611, 23, 2014.
439. Li, W.-L., Xie, L., Jian, T., Romanescu, C., Huang, X., and Wang, L.-S., Angew. Chem. 126, 1312, 2014. Xie, L., Li, W.-L., Romanescu, C., Huang, X., and Wang, L.-S., *J. Chem. Phys.* 138, 034308, 2013.
440. Chen, Q., Bai, H., Zhai, H.-J., Li, S.-D., and Wang, L.-S., *J. Chem. Phys.* 139, 044308, 2013.
441. Yang, Z., Leon, I., and Wang, L.-S., *J. Chem. Phys.* 139, 021106, 2013.
442. Su, J., Dau, P. D., Qiu, Y.-H., Liu, H.-T., Xu, C.-F., Huang, D.-l., Wang, L.-S., and Li, J., *Inorg. Chem.* 52, 6617, 2013.
443. Galeev, T. R., Li, W.-L., Romanescu, C., Černuš, I., Wang, L.-S., and Boldyrev, A. I., *J. Chem. Phys.* 137, 234306, 2012. C2B8 has a second isomer with EA = 2.39 ± 0.03 eV.
444. Liu, H.-T., Wang, Y.-L., Xiong, X.-G., Dau, P. D., Piazza, Z. A., Huang, D.-L., Xu, C.-Q., Li, J., and Wang, L.-S., *Chem. Sci.* 3, 3286, 2012.
445. Dau, P. D., Liu, H.-T., Huang, D.-L., and Wang, L.-S., *J. Chem. Phys.* 137, 116101, 2012. Dau, P. D., Su, J., Liu, H.-T., Huang, D.-L., Wei, F., Li, J., and Wang, L.-S., *J. Chem. Phys.* 136, 194304, 2012.
446. Chen, Q., Zhai, H.-J., Li, S.-D., and Wang, L.-S., *J. Chem. Phys.* 137, 044307, 2012.
447. Ning, C.-G., Xiong, X.-G., Wang, Y.-L., Li, J., and Wang, L.-S., *Phys. Chem. Chem. Phys.* 14, 9323, 2012.
448. Zhai, H.-J., Chen, W.-J., Huang, X., and Wang, L.-S., *RSC Adv.* 2, 2707, 2012.
449. Liu, H.-T., Xiong, X.-G., Dau, P. D., Wang, Y.-L., Li, J., and Wang, L.-S., *Chem. Sci.* 2, 2101, 2011.
450. Chen, W.-J., Zhai, H.-J., Huang, X., and Wang, L.-S., *Chem. Phys. Lett.* 512, 49, 2011.
451. Zhai, H.-J., Zhang, X.-H., Chen, W.-J., Huang, X., and Wang, L.-S., *J. Am. Chem. Soc.* 133 3085, 2011.

PROTON AFFINITIES

Proton affinity is a useful parameter for describing gas-phase ion-molecule reactions in fields such as atmospheric chemistry, plasma chemistry, mass spectrometry, and astrophysics. The proton affinity E_{pa} (often designated in the literature as PA) of a molecular species M is defined as the negative of the enthalpy change for the gas-phase reaction

$$M + H^+ \rightarrow MH^+.$$

A closely related quantity is the gas-phase basicity $\Delta_{base}G^\circ$ (often designated as GB), which is the negative of the Gibbs energy change for the same reaction. Thus the two are related by

$$\Delta_{base}G^\circ = E_{pa} + T\Delta S,$$

where T is the temperature and ΔS is the entropy change in the reaction (which can be calculated if the molecular structure of M and M^+ is known).

Direct measurement of the proton affinity is possible for only a few molecules, mainly olefins and carbonyl compounds. However, these measurements have been used to establish a scale of E_{pa} values that permits proton affinities to be determined for many other molecules, including unstable species and reaction intermediates. The basis for this scale is described by Hunter and Lias in Ref. 1.

The E_{pa} and $\Delta_{base}G^\circ$ values at a temperature of 298 K are tabulated below for selected molecules. Many values are given to one decimal place, but the majority are not accurate to better than one or two kilojoules per mole. The methods of measurement are described in Ref. 1, which contains a much more extensive and detailed tabulation.

Compounds are listed by name.

References

1. Hunter, E. P. L., and Lias, S. G., *J. Phys. Chem. Ref. Data* 27, 413, 1998.
2. Hunter, E. P., and Lias, S. G., "Proton Affinity Evaluation", in *NIST Chemistry WebBook*, NIST Standard Reference Database No. 69, Linstrom, P. J., and Mallard, W. G., Eds., March 2003, National Institute of Standards and Technology, Gaithersburg, MD 20899, <http://webbook.nist.gov>.
3. Do, K., Klein, T. P., Pommerening, C. A., Bachrach, S. M., and Sunderlin, L. S., *J. Am. Chem. Soc.* 120, 6093, 1998.
4. Kim, H.-T., Green, R. J., Qian, J., and Anderson, S. L., *J. Chem. Phys.* 112, 5717, 2000.
5. Park, S. T., Kim, S. K., and Kim, M. S., *J. Chem. Phys.* 114, 5568, 2001.
6. Hiraoka, K., Mizuno, T., Eguchi, D., Takao, T., and Ino, S., *J. Chem. Phys.* 116, 7574, 2002.
7. Oresmaa, L. O., Haukka, M., Vainiotalo, P., and Pakkanen, T. A., *J. Org. Chem.* 67, 8216, 2002.
8. Wang, F., Ma, S., Zhang, D., and Cooks, R. G., *J. Phys. Chem. A* 102, 2988, 1998.
9. Bouchoux, G., Gal, J.-F., Szulejko, J. E., McMahon, T. B., Tortajada, J., Luna, A., Yanez, M., and Mo, O., *J. Phys. Chem. A* 102, 9183, 1998.
10. van Beelen, E., Koblenz, T. A., Ingemann, S. and Hammerum, S., *J. Phys. Chem. A* 108, 2728, 2004.

Proton Affinities and Gas-Phase Basicities

Name	Mol. form.	E_{pa}/kJ mol^{-1}	$\Delta_{base}G^\circ$/kJ mol^{-1}	Note
Acenaphthene	$C_{12}H_{10}$	851.7	821.0	
Acetaldehyde	C_2H_4O	768.5	736.5	
Acetamide	C_2H_5NO	863.6	832.6	
Acetic acid	$C_2H_4O_2$	783.7	752.8	
Acetic anhydride	$C_4H_6O_3$	844		Ref. 9
Acetohydroxamic acid	$C_2H_5NO_2$	854.0	823.0	
Acetone	C_3H_6O	812	782.1	
Acetonitrile	C_2H_3N	779.2	748	
Acetophenone	C_8H_8O	861.1	829.3	
4-Acetylanisole	$C_9H_{10}O_2$	895.6	863.7	
4-Acetylbenzonitrile	C_9H_7NO	826.8	795.0	
Acetylene	C_2H_2	641.4	616.7	
4-Acetylphenyl acetate	$C_{10}H_{10}O_3$	853.2	821.3	
4-Acetylthioanisole	$C_9H_{10}OS$	888.2	856.3	
Acridine	$C_{13}H_9N$	972.6	940.7	
Acrolein	C_3H_4O	797.0	765.1	
Acrylamide	C_3H_5NO	870.7	839.8	
Acrylonitrile	C_3H_3N	784.7	753.7	
Adenine	$C_5H_5N_5$	942.8	912.5	
Adenosine	$C_{10}H_{13}N_5O_4$	989.3	956.8	
L-Alanine	$C_3H_7NO_2$	901.6	867.7	
N-L-Alanyl-L-alanine	$C_6H_{12}N_2O_3$		905.6	
N-(*N*-L-Alanyl-L-alanyl)-L-alanine	$C_9H_{17}N_3O_4$		924.1	
Allene	C_3H_4	775.3	745.8	
Allyl	C_3H_5	736	707.4	
Allylamine	C_3H_7N	909.5	875.5	
Allyldimethylamine	$C_5H_{11}N$	957.8	926.8	
Allyl ethyl ether	$C_5H_{10}O$	833.7	804.5	

Name	Mol. form.	E_{pa}/kJ mol^{-1}	$\Delta_{base}G°$/kJ mol^{-1}	Note
N-Allyl-2-propen-1-amine	$C_6H_{11}N$	949.3	916.3	
2-Aminoacetamide	$C_2H_6N_2O$		882.3	
Aminoacetonitrile	$C_2H_4N_2$	824.9	791.0	
4-Aminobenzaldehyde	C_7H_7NO	910.4	878.6	
3-Aminobenzamide	$C_7H_8N_2O$	900.9	869.9	
4-Aminobenzamide	$C_7H_8N_2O$	927.9	896.9	
3-Aminobenzonitrile	$C_7H_6N_2$	842.3	810.4	
4-Amino-1-butanol	$C_4H_{11}NO$	984.5	932.1	
6-Amino-1-hexanol	$C_6H_{15}NO$	969.0	915.7	
2-Aminophenol	C_6H_7NO	898.8	866.9	
3-Aminophenol	C_6H_7NO	898.8	866.9	
1-(4-Aminophenyl)ethanone	C_8H_9NO	908.8	877.0	
3-Aminopropanenitrile	$C_3H_6N_2$	866.4	832.5	
3-Amino-1-propanol	C_3H_9NO	962.5	917.3	
Ammonia	H_3N	853.6	819.0	
Aniline	C_6H_7N	882.5	850.6	
Aniline-2-carboxylic acid	$C_7H_7NO_2$	901.5	869.0	
Aniline-3-carboxylic acid	$C_7H_7NO_2$	864.7	832.3	
Aniline-4-carboxylic acid	$C_7H_7NO_2$	864.7	832.3	
Anilino	C_6H_6N	949.8	917.4	
Anisole	C_7H_8O	839.6	807.2	
Anthracene	$C_{14}H_{10}$	877.3	846.6	
L-Arginine	$C_6H_{14}N_4O_2$	1051.0	1006.6	
Argon	Ar	369.2	346.3	
Arsenic(III) fluoride	AsF_3	636.7	604.2	
Arsine	AsH_3	747.9	712.0	
L-Asparagine	$C_4H_8N_2O_3$	929	891.5	
L-Aspartic acid	$C_4H_7NO_4$	908.9	875	
1-Azabicyclo[2.2.2]octane	$C_7H_{13}N$	983.3	952.5	
Azetidine	C_3H_7N	943.4	908.6	
2-Azetidinone	C_3H_5NO	852.6	821.7	
Azidobenzene	$C_6H_5N_3$	820	787.5	
Azulene	$C_{10}H_8$	925.2	896	
Barium oxide	BaO	1215.4	1187.6	
Benzaldehyde	C_7H_6O	834.0	802.1	
Benzamide	C_7H_7NO	892.1	861.2	
Benzene	C_6H_6	750.4	725.4	
Benzeneacetonitrile	C_8H_7N	805.5	774.8	
1,2-Benzenediamine	$C_6H_8N_2$	896.5	865.8	
1,3-Benzenediamine	$C_6H_8N_2$	929.9	899.2	
1,4-Benzenediamine	$C_6H_8N_2$	905.9	874.0	
Benzeneethanamine	$C_8H_{11}N$	936.2	902.3	
1*H*-Benzimidazole	$C_7H_6N_2$	953.8	920.5	
Benzoic acid	$C_7H_6O_2$	821.1	790.1	
Benzonitrile	C_7H_5N	811.5	780.9	
Benzo[*ghi*]perylene	$C_{22}H_{12}$	876.0	845.2	
Benzophenone	$C_{13}H_{10}O$	882.3	852.5	
p-Benzoquinone	$C_6H_4O_2$	799.1	769.3	
Benzoxazole	C_7H_5NO	891.6	859.8	
Benzyl	C_7H_7	831.4	800.7	
Benzyl alcohol	C_7H_8O	778.3	748.0	
Benzylamine	C_7H_9N	913.3	879.4	
Benzyl methyl ether	$C_8H_{10}O$	816.7	787.5	
Benzyne	C_6H_4	841	808.5	
Bicyclo[2.2.1]heptan-2-one	$C_7H_{10}O$	847.4	815.5	
Bicyclo[2.2.1]hept-2-ene	C_7H_{10}	836.5	804.0	
Biphenyl	$C_{12}H_{10}$	813.6	782.9	
2,2'-Bipyridine	$C_{10}H_8N_2$	965		Ref. 7
Bis(2-cyanoethyl) ether	$C_6H_8N_2O$	813.8	786.4	
Bis(2-propynyl)amine	C_6H_7N	910.0	876.9	
Bis(2-propynyl) ether	C_6H_6O	783.9	756.5	
Bis(2,2,2-trifluoroethyl) ether	$C_4H_4F_6O$	702.3	674.9	
Borazine	$B_3H_6N_3$	802.5	772.8	

Name	Mol. form.	E_{pa}/kJ mol^{-1}	$\Delta_{base}G°$/kJ mol^{-1}	Note
Boric acid	BH_3O_3	728.1	698.4	
Bromine (atomic)	Br	554.4	531.2	
3-Bromoaniline	C_6H_6BrN	873.2	841.4	
Bromobenzene	C_6H_5Br	754.1	725.8	
Bromoethane	C_2H_5Br	696.2	669.7	
2-Bromoethanol	C_2H_5BrO	766.1	735.7	
Bromomethane	CH_3Br	664.2	638.0	
2-Bromopyridine	C_5H_4BrN	904.8	873.0	
3-Bromopyridine	C_5H_4BrN	910.0	878.2	
4-Bromopyridine	C_5H_4BrN	917.8	886.0	
2-Bromotoluene	C_7H_7Br	775.3	745.8	
3-Bromotoluene	C_7H_7Br	782.0	752.5	
4-Bromotoluene	C_7H_7Br	775.3	745.8	
Bromotrifluoromethane	$CBrF_3$	580.0	550.3	
1-Bromo-3-vinylbenzene	C_8H_7Br	822.4	793.5	
1-Bromo-4-vinylbenzene	C_8H_7Br	838.7	809.8	
1,2-Butadiene	C_4H_6	778.9	749.8	
1,3-Butadiene	C_4H_6	783.4	757.6	
1,3-Butadiyne	C_4H_2	737.2	712.8	
Butanal	C_4H_8O	792.7	760.8	
1,4-Butanediamine	$C_4H_{12}N_2$	1005.6	954.3	
1,4-Butanediol	$C_4H_{10}O_2$	915.6	854.9	
2,3-Butanedione	$C_4H_6O_2$	801.9	770.1	
Butanenitrile	C_4H_7N	798.4	767.7	
1-Butanethiol	$C_4H_{10}S$	801.7	770.5	
1,2,4-Butanetriol	$C_4H_{10}O_3$	905.9	841	
1-Butanol	$C_4H_{10}O$	789.2	758.9	
2-Butanol	$C_4H_{10}O$	815.7	784.6	
2-Butanone	C_4H_8O	827.3	795.5	
2-Butenamide	C_4H_7NO	887.1	856.1	
trans-2-Butene	C_4H_8	747	719.9	
trans-2-Butenoic acid	$C_4H_6O_2$	824.0	793	
3-Buten-2-one	C_4H_6O	834.7	802.8	
Butylamine	$C_4H_{11}N$	921.5	886.6	
tert-Butylamine	$C_4H_{11}N$	934.1	899.9	
Butylbenzene	$C_{10}H_{14}$	791.9	764.2	
Butyldimethylamine	$C_6H_{15}N$	969.2	938.2	
tert-Butyl ethyl ether	$C_6H_{14}O$	856.0	826.9	
Butyl formate	$C_5H_{10}O_2$	806.0	775	
tert-Butyl isopropyl ether	$C_7H_{16}O$	870.7	841.5	
Butyl methyl ether	$C_5H_{12}O$	820.3	791.2	
tert-Butyl nitrite	$C_4H_9NO_2$	863.9	831.4	
1-(4-*tert*-Butylphenyl)ethanone	$C_{12}H_{16}O$	882.5	850.6	
2-*tert*-Butylpyridine	$C_9H_{13}N$	961.7	929.8	
4-*tert*-Butylpyridine	$C_9H_{13}N$	957.7	925.8	
Butyl trifluoroacetate	$C_6H_9F_3O_2$	764.8	733.8	
2-Butyne	C_4H_6	775.8	745.1	
γ-Butyrolactone	$C_4H_6O_2$	840.0	808.1	
Calcium oxide	CaO	1190.6	1162.3	
Carbon dioxide	CO_2	540.5	515.8	
Carbon diselenide	CSe_2	725	700.9	
Carbon disulfide	CS_2	681.9	657.7	
Carbon [fullerene-C_{60}]	C_{60}		827.5	
Carbon [fullerene-C_{70}]	C_{70}		827.5	
Carbon monoselenide	CSe	831.8	800.2	Protonation at C
Carbon monosulfide	CS	791.5	760	
Carbon monoxide	CO	594	562.8	Protonation at C
Carbon monoxide	CO	426.3	402.2	Protonation at O
Carbonothioic dichloride	CCl_2S	752.5	721.8	
Carbon oxyselenide	COSe	670	644.1	Protonation at Se
Carbon oxysulfide	COS	628.5	602.6	Protonation at S
Carbon trimer	C_3	767.0	736.3	
Carbonyl fluoride	CF_2O	666.7	637.0	

Name	Mol. form.	E_{pa}/kJ mol^{-1}	$\Delta_{base}G^o$/kJ mol^{-1}	Note
Cesium hydroxide	CsHO	1117.9	1092.2	
Cesium oxide	Cs_2O	1442.9	1412.2	
Chlorine (atomic)	Cl	513.6	490.1	
Chloroacetic acid	$C_2H_3ClO_2$	765.4	734.5	
Chloroacetonitrile	C_2H_2ClN	745.7	715.1	
3-Chloroaniline	C_6H_6ClN	868.1	836.3	
4-Chloroaniline	C_6H_6ClN	873.8	842.0	
3-Chlorobenzaldehyde	C_7H_5ClO	813.0	781.1	
4-Chlorobenzaldehyde	C_7H_5ClO	831.3	799.4	
3-Chlorobenzamide	C_7H_6ClNO	877.2	846.3	
4-Chlorobenzamide	C_7H_6ClNO	877.2	846.3	
Chlorobenzene	C_6H_5Cl	753.1	724.6	
4-Chloro-*N,N*-diethylaniline	$C_{10}H_{14}ClN$	931.0	899.2	
4-Chloro-*N,N*-dimethylaniline	$C_8H_{10}ClN$	922.9	896.4	
Chloroethane	C_2H_5Cl	693.4	666.9	
2-Chloroethanol	C_2H_5ClO	766.1	735.7	
1-Chloro-4-ethynylbenzene	C_8H_5Cl	832.4	801.7	
1-Chloro-4-isopropenylbenzene	C_9H_9Cl	854.3	825.4	
Chloromethane	CH_3Cl	647.3	621.1	
3-Chloro-4-methoxyacetophenone	$C_9H_9ClO_2$	883.7	851.9	
2-Chloro-6-methoxypyridine	C_6H_6ClNO	909.9	878.0	
3-(Chloromethyl)benzonitrile	C_8H_6ClN	811.2	780.6	
4-(Chloromethyl)benzonitrile	C_8H_6ClN	812.8	782.1	
Chloromethylene	CHCl	874.1	839.9	
2-Chloro-4-methylpyridine	C_6H_6ClN	921.2	889.4	
2-Chloro-6-methylpyridine	C_6H_6ClN	908.0	876.2	
1-(3-Chlorophenyl)ethanone	C_8H_7ClO	846.9	815.1	
1-(4-Chlorophenyl)ethanone	C_8H_7ClO	856.6	824.8	
3-Chloropropanenitrile	C_3H_4ClN	773.1	742.4	
6-Chloro-1*H*-purine	$C_5H_3ClN_4$	873.6	841.7	
2-Chloropyridine	C_5H_4ClN	900.9	869	
3-Chloropyridine	C_5H_4ClN	903.4	871.5	
4-Chloropyridine	C_5H_4ClN	916.1	884.2	
2-Chlorotoluene	C_7H_7Cl	790.5	761.1	
3-Chlorotoluene	C_7H_7Cl	783.9	754.5	
4-Chlorotoluene	C_7H_7Cl	762.9	735.2	
Chlorotrifluoromethane	$CClF_3$	571.3	541.5	
Chromium	Cr	791.3	768.4	
Chromium carbonyl	C_6CrO_6	739.2	714.6	
Chrysene	$C_{18}H_{12}$	840.9	810.1	
Cinnoline	$C_8H_6N_2$	936.3	904.4	
Cobalt	Co	742.7	719.8	
Copper	Cu	655.3	632.4	
Coronene	$C_{24}H_{12}$	861.3	835.0	
o-Cresol	C_7H_8O	832	800	Ref. 10
m-Cresol	C_7H_8O	841	809	Ref. 10
p-Cresol	C_7H_8O	814	782	Ref. 10
Cyanamide	CH_2N_2	805.6	774.9	
Cyanide	CN	>595	>564	Protonation at N
Cyanoacetylene	C_3HN	751.2	720.5	
Cyanogen	C_2N_2	674.7	645.8	
Cyanogen bromide	CBrN	749.8	719.2	
Cyanogen chloride	CClN	722.1	691.5	
Cyanogen fluoride	CFN	632	601.3	
Cyclobutanecarboxylic acid	$C_5H_8O_2$	817.4	786.4	
Cyclobutanone	C_4H_6O	802.5	772.7	
Cyclobutene	C_4H_6	784.4	753.6	
Cycloheptanone	$C_7H_{12}O$	845.6	815.9	
2,4,6-Cycloheptatrien-1-one	C_7H_6O	920.8	891.0	
1,3-Cyclohexadiene	C_6H_8	837	804.5	
1,4-Cyclohexadiene	C_6H_8	837	808.0	
Cyclohexane	C_6H_{12}	686.9	666.9	
Cyclohexanecarbonitrile	$C_7H_{11}N$	815.0	784.4	

Name	Mol. form.	E_{pa}/kJ mol^{-1}	$\Delta_{base}G°$/kJ mol^{-1}	Note
Cyclohexanecarboxylic acid	$C_7H_{12}O_2$	823.8	792.8	
cis-1,3-Cyclohexanediol	$C_6H_{12}O_2$	882.2	849.7	
trans-1,3-Cyclohexanediol	$C_6H_{12}O_2$	828.6	797.9	
1,2-Cyclohexanedione	$C_6H_8O_2$	849.6	818.9	
1,3-Cyclohexanedione	$C_6H_8O_2$	881.2	849.4	
1,4-Cyclohexanedione	$C_6H_8O_2$	812.5	782.7	
Cyclohexanemethanamine	$C_7H_{15}N$	926.6	895.8	
Cyclohexanemethanol	$C_7H_{14}O$	802.1	771.7	
Cyclohexanone	$C_6H_{10}O$	841.0	811.2	
Cyclohexene	C_6H_{10}	784.5	752.0	
Cyclohexylamine	$C_6H_{13}N$	934.4	899.6	
Cyclohexyldimethylamine	$C_8H_{17}N$	983.6	952.6	
1-Cyclohexylethanone	$C_8H_{14}O$	841.4	809.5	
Cyclononanone	$C_9H_{16}O$	852.6	822.8	
Cyclooctanone	$C_8H_{14}O$	849.4	819.6	
1,3-Cyclopentadiene	C_5H_6	821.6	798.4	
Cyclopentadienyl	C_5H_5	831.5	799.1	
Cyclopentanecarboxylic acid	$C_6H_{10}O_2$	817.4	786.4	
cis-1,2-Cyclopentanediol	$C_5H_{10}O_2$	885.6	853.1	
Cyclopentanone	C_5H_8O	823.7	794.0	
Cyclopentene	C_5H_8	766.3	733.8	
N-(1-Cyclopenten-1-yl)pyrrolidine	$C_9H_{15}N$	1019.2	988.4	
Cyclopropane	C_3H_6	750.3	722.2	
Cyclopropanecarbonitrile	C_4H_5N	808.2	777.5	
Cyclopropanecarboxylic acid	$C_4H_6O_2$	821.4	790.4	
Cyclopropene	C_3H_4	818.5	787.8	
Cyclopropyl	C_3H_5	738.9	702.0	
Cyclopropylamine	C_3H_7N	904.7	869.9	
Cyclopropylbenzene	C_9H_{10}	834.9	802.4	
Cyclopropyl methyl ketone	C_5H_8O	854.9	823	
L-Cysteine	$C_3H_7NO_2S$	903.2	869.3	
Cytidine	$C_9H_{13}N_3O_5$	982.5	950.0	
Cytosine	$C_4H_5N_3O$	949.9	918	
Decylamine	$C_{10}H_{23}N$	930.4	896.5	
2'-Deoxyadenosine	$C_{10}H_{13}N_5O_3$	991.5	959.1	
Diacetone alcohol	$C_6H_{12}O_2$	822.9	791.1	
1,3-Diacetylbenzene	$C_{10}H_{10}O_2$	852.0	822.3	
1,4-Diacetylbenzene	$C_{10}H_{10}O_2$	850.8	821.0	
Diallyl ether	$C_6H_{10}O$	827.4	800.0	
N,N-Diallyl-2-propen-1-amine	$C_9H_{15}N$	972.3	941.3	
Diazomethane	CH_2N_2	858.9	826.7	
Diborane	B_2H_6	615	586.0	
Dibutylamine	$C_8H_{19}N$	968.5	935.3	
Di-*sec*-butylamine	$C_8H_{19}N$	980.7	947.5	
Dibutyl ether	$C_8H_{18}O$	845.7	818.3	
Di-*sec*-butyl ether	$C_8H_{18}O$	865.9	838.5	
Di-*tert*-butyl ether	$C_8H_{18}O$	887.4	860.0	
2,5-Di-*tert*-butylfuran	$C_{12}H_{20}O$	894.7	863.9	
Di-*tert*-butyl ketone	$C_9H_{18}O$	861.3	831.5	
1,3-Di-*tert*-butyl-5-methylbenzene	$C_{15}H_{24}$	853.7	826.0	
2,4-Di-*tert*-butylpyridine	$C_{13}H_{21}N$	983.8	952.0	
2,6-Di-*tert*-butylpyridine	$C_{13}H_{21}N$	982.9	951	
Dibutyl sulfide	$C_8H_{18}S$	871.8	842.1	
Di-*tert*-butyl sulfide	$C_8H_{18}S$	893.8	864.0	
Dicarbon monoxide	C_2O	774.7	747.0	
Dichloromethylene	CCl_2	861	828.5	
m-Dicyanobenzene	$C_8H_4N_2$	779.3	750.4	
p-Dicyanobenzene	$C_8H_4N_2$	779.0	751.8	
Dicyclopropyl ketone	$C_7H_{10}O$	880.4	850.6	
Diethanolamine	$C_4H_{11}NO_2$	953	920	
N,N-Diethylacetamide	$C_6H_{13}NO$	925.4	894.4	
Diethylamine	$C_4H_{11}N$	952.4	919.4	
N,N-Diethylaniline	$C_{10}H_{15}N$	959.8	927.9	

Atomic

Name	Mol. form.	E_{pa}/kJ mol^{-1}	$\Delta_{base}G°$/kJ mol^{-1}	Note
Diethylene glycol dimethyl ether	$C_6H_{14}O_3$	918.8	870.9	
Diethyl ether	$C_4H_{10}O$	828.4	801	
Diethylmethylamine	$C_5H_{13}N$	971.0	940.0	
N,N-Diethyl-4-methylaniline	$C_{11}H_{17}N$	962.8	931.0	
Diethylpropylamine	$C_7H_{17}N$	978.8	947.9	
2,6-Diethylpyridine	$C_9H_{13}N$	972.3	940.4	
N,N-Diethyl-3-pyridinecarboxamide	$C_{10}H_{14}N_2O$	940.9	909.0	
Diethyl sulfide	$C_4H_{10}S$	856.7	827.0	
o-Difluorobenzene	$C_6H_4F_2$	731.2	703.5	
m-Difluorobenzene	$C_6H_4F_2$	749.7	722	
p-Difluorobenzene	$C_6H_4F_2$	718.7	692.8	
2,2-Difluoroethanol	$C_2H_4F_2O$	727.4	697.0	
1,1-Difluoroethene	$C_2H_2F_2$	734	705.1	
trans-1,2-Difluoroethene	$C_2H_2F_2$	688.6	657.9	
Difluoromethane	CH_2F_2	620.5	589.7	
Difluoromethylene	CF_2	765	732.5	
2,3-Dihydro-1,4-dioxin	$C_4H_6O_2$	823.5	792.8	
2,3-Dihydrofuran	C_4H_6O	866.9	834.4	
2,5-Dihydrofuran	C_4H_6O	823.4	796	
2,3-Dihydro-1*H*-indole	C_8H_9N	957.1	926.3	
3,4-Dihydro-2*H*-pyran	C_5H_8O	865.8	833.4	
Diisobutylamine	$C_8H_{19}N$	958.1	925.1	
Diisopropylamine	$C_6H_{15}N$	971.9	938.6	
Diisopropyl ether	$C_6H_{14}O$	855.5	828.1	
Diisopropyl sulfide	$C_6H_{14}S$	876.4	846.6	
Dilithium	Li_2	1162	1133.1	
Dimagnesium	Mg_2	919	886.5	
1,2-Dimethoxyethane	$C_4H_{10}O_2$	858.0	820.2	
N,N-Dimethylacetamide	C_4H_9NO	908.0	877.0	
Dimethylamine	C_2H_7N	929.5	896.5	
(Dimethylamino)acetonitrile	$C_4H_8N_2$	884.5	853.7	
4'-(Dimethylamino)acetophenone	$C_{10}H_{13}NO$	932.8	906.3	
4-(Dimethylamino)benzaldehyde	$C_9H_{11}NO$	924.8	898.3	
4-(Dimethylamino)benzonitrile	$C_9H_{10}N_2$	889.1	862.6	
1-[3-(Dimethylamino)phenyl]ethanone	$C_{10}H_{13}NO$	928.0	901.5	
3-(Dimethylamino)-1-propyne	C_5H_9N	940.3	909.5	
2,6-Dimethylaniline	$C_8H_{11}N$	901.7	869.8	
N,N-Dimethylaniline	$C_8H_{11}N$	941.1	909.2	
N,N-Dimethylbenzamide	$C_9H_{11}NO$	932.7	901.8	
N,N-Dimethyl-1,4-benzenediamine	$C_8H_{12}N_2$	955.0	928.4	
N,N-Dimethylbenzylamine	$C_9H_{13}N$	968.4	937.4	
2,3-Dimethyl-1,3-butadiene	C_6H_{10}	835.0	807.8	
N,N-Dimethylbutanamide	$C_6H_{13}NO$	921.7	890.8	
3,3-Dimethyl-2-butanone	$C_6H_{12}O$	840.1	808.2	
2,3-Dimethyl-2-butene	C_6H_{12}	813.9	785.9	
Dimethyl carbonate	$C_3H_6O_3$	830.2	799.2	
Dimethylcyanamide	$C_3H_6N_2$	852.1	821.4	
1,2-Dimethylcyclopentene	C_7H_{12}	822.6	791.9	
3,3-Dimethylcyclopropene	C_5H_8	847.8	817.1	
trans-Dimethyldiazene	$C_2H_6N_2$	865.1	834.4	
Dimethyl disulfide	$C_2H_6S_2$	815.3	782.8	
N,N'-Dimethyl-1,2-ethanediamine	$C_4H_{12}N_2$	989.2	946.9	
Dimethyl ether	C_2H_6O	792	764.5	
N,N-Dimethylformamide	C_3H_7NO	887.5	856.6	
2,4-Dimethylfuran	C_6H_8O	894.7	862.3	
2,5-Dimethylfuran	C_6H_8O	865.9	835.2	
3,4-Dimethylfuran	C_6H_8O	869.0	838.3	
1,1-Dimethylhydrazine	$C_2H_8N_2$	927.1	894.7	
Dimethyl hydrogen phosphite	$C_2H_7O_3P$	894.8	862.4	
1,2-Dimethyl-1*H*-imidazole	$C_5H_8N_2$	984.7	952.6	
1,4-Dimethyl-1*H*-imidazole	$C_5H_8N_2$	976.7	944.9	
1,5-Dimethyl-1*H*-imidazole	$C_5H_8N_2$	977.6	945.8	
1,3-Dimethyl-2-imidazolidinone	$C_5H_{10}N_2O$	918.4	886.0	

Name	Mol. form.	E_{pa}/kJ mol^{-1}	$\Delta_{base}G°$/kJ mol^{-1}	Note
Dimethyl isophthalate	$C_{10}H_{10}O_4$	843.5	814.3	
1,4-Dimethyl-7-isopropylazulene	$C_{15}H_{18}$	983.1	950.6	
Dimethyl mercury	C_2H_6Hg	771.6	740.8	
N,N-Dimethyl-3-nitroaniline	$C_8H_{10}N_2O_2$	894.1	867.6	
N,N-Dimethyl-4-nitroaniline	$C_8H_{10}N_2O_2$	896.7	870.2	
2,4-Dimethyl-1-nitrobenzene	$C_8H_9NO_2$	831.0	798.5	
2,4-Dimethyl-1,3-pentadiene	C_7H_{12}	886.5	857.6	
2,4-Dimethyl-3-pentanone	$C_7H_{14}O$	850.3	820.5	
2,4-Dimethyl-2-pentene	C_7H_{14}	812	783.1	
1-(2,4-Dimethylphenyl)ethanone	$C_{10}H_{12}O$	882.6	850.8	
1-(2,5-Dimethylphenyl)ethanone	$C_{10}H_{12}O$	873.5	841.6	
1-(3,4-Dimethylphenyl)ethanone	$C_{10}H_{12}O$	882.8	851.0	
Dimethylphosphine	C_2H_7P	912.0	877.9	
2,2-Dimethylpropanamide	$C_5H_{11}NO$	889.0	857.2	
N,N-Dimethyl-1-propanamine	$C_5H_{13}N$	962.8	931.9	
N,N-Dimethyl-1,3-propanediamine	$C_5H_{14}N_2$	1025.0	975.3	
2,2-Dimethylpropanenitrile	C_5H_9N	810.9	780.2	
2,2-Dimethyl-1-propanethiol	$C_5H_{12}S$	809.5	778.2	
2,2-Dimethyl-1-propanol	$C_5H_{12}O$	795.5	765.2	
N,N-Dimethyl-2-propenamide	C_5H_9NO	904.3	873.4	
2,2-Dimethylpropylamine	$C_5H_{13}N$	928.3	894.0	
2,6-Dimethyl-4*H*-pyran-4-one	$C_7H_8O_2$	941.5	907.3	
1,3-Dimethyl-1*H*-pyrazole	$C_5H_8N_2$	933.9	902.3	
1,5-Dimethyl-1*H*-pyrazole	$C_5H_8N_2$	934.3	902.8	
3,4-Dimethyl-1*H*-pyrazole	$C_5H_8N_2$	927.3	895.4	
3,5-Dimethyl-1*H*-pyrazole	$C_5H_8N_2$	933.5	900.1	
N,N-Dimethyl-2-pyridinamine	$C_7H_{10}N_2$	968.2	941.6	
N,N-Dimethyl-4-pyridinamine	$C_7H_{10}N_2$	997.6	971.1	
2,3-Dimethylpyridine	C_7H_9N	958.9	927.0	
2,4-Dimethylpyridine	C_7H_9N	962.9	930.8	
2,5-Dimethylpyridine	C_7H_9N	958.8	926.9	
2,6-Dimethylpyridine	C_7H_9N	963.0	931.1	
3,4-Dimethylpyridine	C_7H_9N	957.3	925.5	
3,5-Dimethylpyridine	C_7H_9N	955.4	923.5	
2,5-Dimethylpyrrole	C_6H_9N	918.7	887.1	
Dimethyl sulfide	C_2H_6S	830.9	801.2	
Dimethyl sulfoxide	C_2H_6OS	884.4	853.7	
Dimethyl terephthalate	$C_{10}H_{10}O_4$	843.2	812.3	
N,N-Dimethylthioacetamide	C_4H_9NS	925.3	894.4	
N,N'-Dimethylthiourea	$C_3H_8N_2S$	926.0	895.1	
N,N-Dimethyltricyclo[3.3.1.1^{3,7}]decane-1-carboxamide	$C_{13}H_{21}NO$	949.4	917.6	
N,N'-Dimethylurea	$C_3H_8N_2O$	903.3	873.5	
1,3-Dioxane	$C_4H_8O_2$	825.4	796.2	
1,4-Dioxane	$C_4H_8O_2$	797.4	770.0	
Dipentyl ether	$C_{10}H_{22}O$	852.7	825.3	
1,4-Diphenylbutane	$C_{16}H_{18}$	822.0	779.8	
1,2-Diphenylethane	$C_{14}H_{14}$	801.8	774.1	
1,1-Diphenylethene	$C_{14}H_{12}$	885.7	856.9	
1,6-Diphenylhexane	$C_{18}H_{22}$	826.1	783.8	
Diphenylmethane	$C_{13}H_{12}$	802.0	769.5	
1,5-Diphenylpentane	$C_{17}H_{20}$	824.7	782.4	
1,3-Diphenylpropane	$C_{15}H_{16}$	820.1	787.6	
3,5-Diphenyl-1*H*-pyrazole	$C_{15}H_{12}N_2$	946.3	912.7	
Dipropylamine	$C_6H_{15}N$	962.3	929.3	
N,N-Dipropylaniline	$C_{12}H_{19}N$	963.0	931.1	
Dipropyl ether	$C_6H_{14}O$	837.9	810.5	
Dipropyl sulfide	$C_6H_{14}S$	864.7	834.9	
Disiloxane	H_6OSi_2	749	718.3	
Disodium	Na_2	1146.8	1118.2	
Ethane	C_2H_6	596.3	569.9	
1,2-Ethanediamine	$C_2H_8N_2$	951.6	912.5	
1,2-Ethanediol	$C_2H_6O_2$	815.9	773.6	
Ethanethiol	C_2H_6S	789.6	758.4	

Name	Mol. form.	E_{pa}/kJ mol^{-1}	$\Delta_{base}G°$/kJ mol^{-1}	Note
Ethanimidamide	$C_2H_6N_2$	970.7	938.2	
Ethanol	C_2H_6O	776.4	746	
Ethanolamine	C_2H_7NO	930.3	896.8	
Ethenamine	C_2H_5N	898.9	866.5	
3-Ethoxypropanenitrile	C_5H_9NO	807.2	776.5	
N-Ethylacetamide	C_4H_9NO	898.0	867.0	
Ethyl acetate	$C_4H_8O_2$	835.7	804.7	
Ethylamine	C_2H_7N	912.0	878	
3-Ethylaniline	$C_8H_{11}N$	897.9	866.1	
N-Ethylaniline	$C_8H_{11}N$	924.8	892.9	
Ethylbenzene	C_8H_{10}	788.0	760.3	
Ethyl chloroformate	$C_3H_5ClO_2$	764.8	733.8	
Ethyl cyanoformate	$C_4H_5NO_2$	745.7	714.7	
Ethyldimethylamine	$C_4H_{11}N$	960.1	929.1	
Ethyl 1,5-dimethyl-1*H*-pyrazole-3-carboxylate	$C_8H_{12}N_2O_2$	933.4	901.5	
Ethylene	C_2H_4	680.5	651.5	
Ethylene carbonate	$C_3H_4O_3$	814.2	784.4	
Ethyleneimine	C_2H_5N	905.5	872.5	
Ethyl formate	$C_3H_6O_2$	799.4	768.4	
N-Ethyl-*N*-hydroxyethanamine	$C_4H_{11}NO$	914.7	882.2	
Ethyl isocyanide	C_3H_5N	851.3	818.9	
Ethylisopropylamine	$C_5H_{13}N$	960.0	926.7	
Ethyl isopropyl ether	$C_5H_{12}O$	842.7	813.5	
N-Ethyl-*N*-isopropyl-2-propanamine	$C_8H_{19}N$	994.3	963.5	
Ethylmethylamine	C_3H_9N	942.2	909.2	
N-Ethyl-*N*-methylaniline	$C_9H_{13}N$	939.0	912.4	
Ethyl *N*-methylcarbamate	$C_4H_9NO_2$	888.8	857.8	
Ethyl methyl carbonate	$C_4H_8O_3$	842.7	810.8	
3-Ethyl-3-methyldiaziridine	$C_4H_{10}N_2$	903.8	871.3	
Ethyl methyl ether	C_3H_8O	808.6	781.2	
Ethyl methyl sulfide	C_3H_8S	846.5	815.3	
Ethyl nitrite	$C_2H_5NO_2$	818.9	786.4	
2-Ethylpyridine	C_7H_9N	952.4	920.6	
3-Ethylpyridine	C_7H_9N	947.4	915.5	
4-Ethylpyridine	C_7H_9N	951.1	919.2	
Ethyl trichloroacetate	$C_4H_5Cl_3O_2$	790.4	759.4	
Ethyl trifluoroacetate	$C_4H_5F_3O_2$	758.8	727.9	
Ethyl vinyl ether	C_4H_8O	870.1	840.4	
Ethynyl	C_2H	753	720.8	
Ferrocene	$C_{10}H_{10}Fe$	863.6	841.3	
Fluoranthene	$C_{16}H_{10}$	828.6	800.9	
9*H*-Fluorene	$C_{13}H_{10}$	831.5	803.8	
Fluorine	F_2	332	305.5	
Fluorine (atomic)	F	340.1	315.1	
Fluorine oxide	FO	508.7	482.2	
Fluoroacetic acid	$C_2H_3FO_2$	765.4	734.5	
Fluoroacetylene	C_2HF	686	661.3	
3-Fluoroaniline	C_6H_6FN	867.3	835.5	
4-Fluoroaniline	C_6H_6FN	871.5	839.7	
3-Fluorobenzaldehyde	C_7H_5FO	814.3	782.5	
4-Fluorobenzaldehyde	C_7H_5FO	827.1	795.3	
3-Fluorobenzamide	C_7H_6FNO	877.2	846.3	
4-Fluorobenzamide	C_7H_6FNO	877.2	846.3	
Fluorobenzene	C_6H_5F	755.9	726.6	
m-Fluorobenzyl	C_7H_6F	836.5	804	
Fluoroethane	C_2H_5F	683.4	655.8	
2-Fluoroethanol	C_2H_5FO	715.6	685.2	
Fluoroethene	C_2H_3F	729	700.1	
Fluoromethane	CH_3F	598.9	571.5	
Fluoromethylene	CHF	797.9	763.8	
1-(4-Fluorophenyl)ethanone	C_8H_7FO	858.6	826.8	
1-Fluoro-2-propanone	C_3H_5FO	795.4	763.5	
2-Fluoropyridine	C_5H_4FN	884.6	852.7	

Name	Mol. form.	E_{pa}/kJ mol⁻¹	$\Delta_{base}G°$/kJ mol⁻¹	Note
3-Fluoropyridine	C_5H_4FN	902.0	870.1	
2-Fluorotoluene	C_7H_7F	773.3	743.8	
3-Fluorotoluene	C_7H_7F	785.4	756.0	
4-Fluorotoluene	C_7H_7F	763.8	736.1	
Formaldehyde	CH_2O	712.9	683.3	
Formamide	CH_3NO	822.2	791.2	
Formic acid	CH_2O_2	742.0	710.3	
4-Formylbenzonitrile	C_8H_5NO	796.9	766.3	
Formyloxyl	CHO_2	623.4	590.9	
Fulminic acid	$CHNO$	758	725.5	
Furan	C_4H_4O	812	781	Ref. 10
Germane	GeH_4	713.4	687.1	
α-D-Glucose	$C_6H_{12}O_6$		778.9	
β-D-Glucose	$C_6H_{12}O_6$		778.9	
L-Glutamic acid	$C_5H_9NO_4$	913.0	879.1	
L-Glutamine	$C_5H_{10}N_2O_3$	937.8	900	
Glutaric anhydride	$C_5H_6O_3$	816		Ref. 9
Glycerol	$C_3H_8O_3$	874.8	820	
Glycine	$C_2H_5NO_2$	886.5	852.2	
N-Glycylglycine	$C_4H_8N_2O_3$		882	
N-(*N*-Glycylglycyl)glycine	$C_6H_{11}N_3O_4$	966.8	916.8	
Guanidine	CH_5N_3	986.3	949.4	
Guanine	$C_5H_5N_5O$	959.5	927.6	
Guanosine	$C_{10}H_{13}N_5O_5$	993.4	960.9	
Helium	He	177.8	148.5	
1,7-Heptanediamine	$C_7H_{18}N_2$	998.5	944.9	
4-Heptanone	$C_7H_{14}O$	845.0	815.3	
Heptylamine	$C_7H_{17}N$	923.2	889.3	
Hexafluorobenzene	C_6F_6	648.0	624.4	
1,1,1,3,3,3-Hexafluoro-2-propanol	$C_3H_2F_6O$	686.6	656.2	
Hexahydro-1*H*-azepine	$C_6H_{13}N$	956.7	923.5	
2,3,4,6,7,8-Hexahydropyrrolo[1,2-*a*]pyrimidine	$C_7H_{12}N_2$	1038.3	1005.9	
Hexamethylbenzene	$C_{12}H_{18}$	860.6	836.0	
Hexamethyldisiloxane	$C_6H_{18}OSi_2$	846.4	816.2	
Hexamethylphosphoric triamide	$C_6H_{18}N_3OP$	958.6	928.7	
Hexamethylphosphorous triamide	$C_6H_{18}N_3P$	930.1	897.7	
1,6-Hexanediamine	$C_6H_{16}N_2$	999.5	946.2	
2,5-Hexanedione	$C_6H_{10}O_2$	892.0	851.8	
3-Hexanone	$C_6H_{12}O$	843.2	811.3	
1-Hexene	C_6H_{12}	805.2	776.3	
trans-3-Hexen-2-one	$C_6H_{10}O$	865.6	833.8	
Hexylamine	$C_6H_{15}N$	927.5	893.5	
2-Hexylpyridine	$C_{11}H_{17}N$	963.6	931.7	
1-Hexyne	C_6H_{10}	799.8	774.8	
2-Hexyne	C_6H_{10}	806.1	781.1	
Histamine	$C_5H_9N_3$	999.8	961.9	
L-Histidine	$C_6H_9N_3O_2$	988	950.2	
Hydrazine	H_4N_2	853.2	822.4	
Hydrazoic acid	HN_3	756.0	723.5	
Hydrogen	H_2	422.3	394.7	
Hydrogen bromide	BrH	584.2	557.7	
Hydrogen chloride	ClH	556.9	530.1	
Hydrogen cyanide	CHN	712.9	681.6	
Hydrogen fluoride	FH	484	456.7	
Hydrogen iodide	HI	627.5	601.3	
Hydrogen isocyanide	CHN	772.3	739.8	
Hydrogen peroxide	H_2O_2	674.5	643.8	
Hydrogen selenide	H_2Se	707.8	676.4	
Hydrogen sulfide	H_2S	705	673.8	
Hydrogen telluride	H_2Te	735.9	704.5	
Hydroperoxy	HO_2	660	627.5	
Hydroxyl	HO	593.2	564.0	
1-(3-Hydroxyphenyl)ethanone	$C_8H_8O_2$	863.6	831.8	

Name	Mol. form.	E_{pa}/kJ mol^{-1}	$\Delta_{base}G°$/kJ mol^{-1}	Note
1-(4-Hydroxyphenyl)ethanone	$C_8H_8O_2$	883.7	851.9	
Hypoxanthine	$C_5H_4N_4O$	912.3	880.5	
Imidazole	$C_3H_4N_2$	942.8	909.2	
2-Imidazolidinethione	$C_3H_6N_2S$	921.9	891.2	
1*H*-Indazole	$C_7H_6N_2$	900.8	868.9	
Indene	C_9H_8	848.8	819.6	
1*H*-Indole	C_8H_7N	933.4	901.9	
Iodine (atomic)	I	608.2	583.5	
3-Iodoaniline	C_6H_6IN	878.7	846.8	
Iodoethane	C_2H_5I	724.8	698.3	
Iodomethane	CH_3I	691.7	665.5	
1-Iodo-2-methylbenzene	C_7H_7I	780.3	750.8	
Iron	Fe	754	731.1	
Iron(II) oxide	FeO	907	880.5	
Iron pentacarbonyl	C_5FeO_5	833.0	798.5	
Isobutanal	C_4H_8O	797.3	765.5	
Isobutane	C_4H_{10}	677.8	671.3	
Isobutene	C_4H_8	802.1	775.6	
Isobutylamine	$C_4H_{11}N$	924.8	890.8	
Isobutyldimethylamine	$C_6H_{15}N$	968.7	937.8	
1-Isobutylpiperidine	$C_9H_{19}N$	974.5	943.5	
Isocyanic acid (HNCO)	CHNO	753	718.8	
Isocyanobenzene	C_7H_5N	868.4	836.0	
Isocyanomethane	C_2H_3N	839.1	806.6	
2-Isocyano-2-methylpropane	C_5H_9N	870.7	838.3	
1-Isocyanopropane	C_4H_7N	856.8	824.3	
L-Isoleucine	$C_6H_{13}NO_2$	917.4	883.5	
Isophorone	$C_9H_{14}O$	893.5	861.6	
Isopropenylbenzene	C_9H_{10}	864.2	835.3	
Isopropyl acetate	$C_5H_{10}O_2$	836.6	805.6	
Isopropylamine	C_3H_9N	923.8	889.0	
Isopropylbenzene	C_9H_{12}	791.6	763.9	
Isopropyl formate	$C_4H_8O_2$	811.3	780.3	
Isopropylmethylamine	$C_4H_{11}N$	952.4	919.4	
Isopropyl methyl ether	$C_4H_{10}O$	826.3	797.1	
Isopropyl nitrite	$C_3H_7NO_2$	845.5	813.0	
4-Isopropylpyridine	$C_8H_{11}N$	955.7	923.8	
Isoquinoline	C_9H_7N	951.7	919.9	
Isoxazole	C_3H_3NO	848.6	816.8	
Ketene	C_2H_2O	825.3	793.6	
Krypton	Kr	424.6	402.4	
Lanthanum	La	1013	991.9	
L-Leucine	$C_6H_{13}NO_2$	914.6	880.6	
Lithium bromide	BrLi	819	792.5	
Lithium chloride	ClLi	827	800.5	
Lithium hydride	HLi	1021.7	996.4	
Lithium hydroxide	HLiO	1000.1	972.1	
Lithium oxide	Li_2O	1206	1175.3	
Lutetium	Lu	992	970.6	
L-Lysine	$C_6H_{14}N_2O_2$	996	951.0	
Magnesium	Mg	819.6	797.3	
Magnesium oxide	MgO	988	959.4	
Malononitrile	$C_3H_2N_2$	723.0	694.1	
Manganese	Mn	797.3	774.4	
Manganese 2-methylcyclopentadienyl tricarbonyl	$C_9H_7MnO_3$	833.8	801.3	
Mesityl oxide	$C_6H_{10}O$	878.7	846.9	
Metaboric acid (α form)	BHO_2	763.0	730.5	
Methacrylic acid	$C_4H_6O_2$	816.7	785.7	
Methane	CH_4	543.5	520.6	
Methanesulfonic acid	CH_4O_3S	761.3	728.9	
Methanethiol	CH_4S	773.4	742	
Methanol	CH_4O	754.3	724.5	
L-Methionine	$C_5H_{11}NO_2S$	935.4	901.5	

Name	Mol. form.	E_{pa}/kJ mol^{-1}	$\Delta_{base}G°$/kJ mol^{-1}	Note
Methoxyacetonitrile	C_3H_5NO	758.1	727.4	
2-Methoxyaniline	C_7H_9NO	905.2	873.3	
3-Methoxyaniline	C_7H_9NO	913.0	881.1	
4-Methoxyaniline	C_7H_9NO	900.3	868.5	
3-Methoxybenzaldehyde	$C_8H_8O_2$	844.1	812.2	
4-Methoxybenzaldehyde	$C_8H_8O_2$	881.1	849.3	
3-Methoxybenzamide	$C_8H_9NO_2$	900.9	869.9	
4-Methoxybenzamide	$C_8H_9NO_2$	900.3	869.4	
Methoxycyclohexane	$C_7H_{14}O$	840.5	811.3	
4-Methoxy-*N*,*N*-dimethylaniline	$C_9H_{13}NO$	949.1	922.4	
2-Methoxyethanol	$C_3H_8O_2$	768.8	729.8	
2-Methoxyethylamine	C_3H_9NO	928.6	894.6	
1-(3-Methoxyphenyl)ethanone	$C_9H_{10}O_2$	871.2	839.3	
2-Methoxy-1-propene	C_4H_8O	894.9	866.1	
2-Methoxypyridine	C_6H_7NO	934.7	902.8	
3-Methoxypyridine	C_6H_7NO	942.7	910.9	
4-Methoxypyridine	C_6H_7NO	961.7	929.8	
N-Methylacetamide	C_3H_7NO	888.5	857.6	
Methyl acetate	$C_3H_6O_2$	821.6	790.7	
4-Methylacetophenone	$C_9H_{10}O$	875.5	843.6	
Methyl acrylate	$C_4H_6O_2$	825.8	794.8	
2-Methylallyl	C_4H_7	778	747.3	
Methylamidogen	CH_4N	832.8	801.6	
Methylamine	CH_5N	899.0	864.5	
Methyl 4-aminobenzoate	$C_8H_9NO_2$	883.9	853.0	
2-Methylaniline	C_7H_9N	890.9	859.1	
3-Methylaniline	C_7H_9N	895.8	864.0	
4-Methylaniline	C_7H_9N	896.7	864.8	
N-Methylaniline	C_7H_9N	916.6	890.1	
2-Methylanisole	$C_8H_{10}O$	850	818	Ref. 10
3-Methylanisole	$C_8H_{10}O$	860	828	Ref. 10
4-Methylanisole	$C_8H_{10}O$	841	809	Ref. 10
2-Methylanthracene	$C_{15}H_{12}$	887.5	855.1	
9-Methylanthracene	$C_{15}H_{12}$	896.5	865.8	
Methyl azide	CH_3N_3	833	800.5	
1-Methylaziridine	C_3H_7N	934.8	904.1	
3-Methylbenzaldehyde	C_8H_8O	840.0	808.1	
4-Methylbenzaldehyde	C_8H_8O	851.8	820.0	
3-Methylbenzamide	C_8H_9NO	900.9	869.9	
4-Methylbenzamide	C_8H_9NO	900.9	869.9	
1-Methyl-1*H*-benzimidazole	$C_8H_8N_2$	967.0	935.2	
Methyl benzoate	$C_8H_8O_2$	850.5	819.5	
2-Methylbenzofuran	C_9H_8O	859.6	827.2	
1-Methyl-1*H*-benzotriazole	$C_7H_7N_3$	931.2	898.7	
2-Methylbiphenyl	$C_{13}H_{12}$	815.9	783.4	
3-Methylbiphenyl	$C_{13}H_{12}$	828.0	795.5	
4-Methylbiphenyl	$C_{13}H_{12}$	817.9	785.4	
2-Methyl-1,3-butadiene	C_5H_8	826.4	797.6	
2-Methyl-2-butanamine	$C_5H_{13}N$	937.8	903.6	
Methyl butanoate	$C_5H_{10}O_2$	836.4	805.4	
3-Methyl-2-butanone	$C_5H_{10}O$	836.3	804.4	
3-Methyl-2-butenal	C_5H_8O	856.9	825.0	
2-Methyl-2-butene	C_5H_{10}	808.8	779.9	
Methyl *trans*-2-butenoate	$C_5H_8O_2$	851.3	820.4	
cis-2-Methyl-2-butenoic acid	$C_5H_8O_2$	822.5	791.5	
3-Methyl-2-butenoic acid	$C_5H_8O_2$	822.9	791.9	
3-Methyl-3-buten-2-one	C_5H_8O	843.1	811.3	
Methyl *tert*-butyl ether	$C_5H_{12}O$	841.6	812.4	
3-Methyl-1-butyne	C_5H_8	814.9	787.8	
Methyl 3-chlorobenzoate	$C_8H_7ClO_2$	835.4	804.4	
Methyl 4-chlorobenzoate	$C_8H_7ClO_2$	842.1	811.1	
1-Methylcyclobutene	C_5H_8	841.5	807.3	
Methyl cyclohexanecarboxylate	$C_8H_{14}O_2$	846.2	815.3	

Name	Mol. form.	E_{pa}/kJ mol^{-1}	$\Delta_{base}G°$/kJ mol^{-1}	Note
4-Methylcyclohexanone	$C_7H_{12}O$	844.9	813.0	
1-Methylcyclohexene	C_7H_{12}	825.1	792.6	
1-Methylcyclopentene	C_6H_{10}	816.5	787.1	
Methyl cyclopropanecarboxylate	$C_5H_8O_2$	842.1	811.2	
2-Methyl-*N,N*-dimethylaniline	$C_9H_{13}N$	951.8	925.3	
3-Methyl-*N,N*-dimethylaniline	$C_9H_{13}N$	942.1	915.7	
4-Methyl-*N,N*-dimethylaniline	$C_9H_{13}N$	950.0	918.1	
Methyl 2,4-dimethylbenzoate	$C_{10}H_{12}O_2$	868.2	837.2	
Methyl 2,5-dimethylbenzoate	$C_{10}H_{12}O_2$	864.7	833.7	
Methyl 3,5-dimethylbenzoate	$C_{10}H_{12}O_2$	864.3	833.4	
Methyl 2,2-dimethylpropanoate	$C_6H_{12}O_2$	845.2	814.2	
Methyldiphenylphosphine	$C_{13}H_{13}P$	972.1	939.7	
Methyldipropylamine	$C_7H_{17}N$	983.5	950.9	
Methylenecyclopentane	C_6H_{10}	832.4	803.5	
N-Methylformamide	C_2H_5NO	851.3	820.3	
Methyl formate	$C_2H_4O_2$	782.5	751.5	
Methyl 4-formylbenzoate	$C_9H_8O_3$	832.9	801.9	
2-Methylfuran	C_5H_6O	865.9	833.5	
3-Methylfuran	C_5H_6O	854.0	821.5	
4-Methylglutaric anhydride	$C_6H_8O_3$	820		Ref. 9
Methylhydrazine	CH_6N_2	898.8	866.4	
Methyl 3-hydroxybenzoate	$C_8H_8O_3$	850.0	819.1	
Methyl 4-hydroxybenzoate	$C_8H_8O_3$	863.4	832.5	
O-Methylhydroxylamine	CH_5NO	844.8	812.3	
1-Methylimidazol	$C_4H_6N_2$	959.6	927.7	
2-Methyl-1*H*-imidazole	$C_4H_6N_2$	963.4	929.6	
4-Methyl-1*H*-imidazole	$C_4H_6N_2$	952.8	920.9	
1-Methyl-1*H*-indazole	$C_8H_8N_2$	922.4	890.5	
2-Methyl-2*H*-indazole	$C_8H_8N_2$	941.4	909.6	
Methyl isobutanoate	$C_5H_{10}O_2$	836.6	805.7	
Methyl isocyanate	C_2H_3NO	764.4	732.0	
Methyl isothiocyanate	C_2H_3NS	799.2	766.7	
Methyl methacrylate	$C_5H_8O_2$	831.4	800.5	
Methyl 3-methoxybenzoate	$C_9H_{10}O_3$	856.7	825.8	
Methyl 4-methoxybenzoate	$C_9H_{10}O_3$	870.6	839.6	
Methyl 2-methylbenzoate	$C_9H_{10}O_2$	858.3	827.3	
Methyl 3-methylbenzoate	$C_9H_{10}O_2$	857.7	826.8	
Methyl 4-methylbenzoate	$C_9H_{10}O_2$	861.5	830.6	
1-Methyl-2-(1-methylvinyl)benzene	$C_{10}H_{12}$	857.8	828.9	
1-Methyl-3-(1-methylvinyl)benzene	$C_{10}H_{12}$	867.6	838.7	
1-Methyl-4-(1-methylvinyl)benzene	$C_{10}H_{12}$	881.8	852.9	
1-Methylnaphthalene	$C_{11}H_{10}$	834.8	805.3	
2-Methylnaphthalene	$C_{11}H_{10}$	831.9	802.4	
Methyl nitrate	CH_3NO_3	733.6	714.8	
Methyl nitrite	CH_3NO_2	798.9	766.4	
N-Methyl-4-nitroaniline	$C_7H_8N_2O_2$	891.6	865.1	
Methyl 3-nitrobenzoate	$C_8H_7NO_4$	815.7	784.7	
Methyl 4-nitrobenzoate	$C_8H_7NO_4$	813.2	782.3	
N-Methyl-*N*-nitromethanamine	$C_2H_6N_2O_2$	828.3	795.8	
2-Methyl-2-norbornene	C_8H_{12}	845	812.5	
2-Methyl-1,3-pentadiene	C_6H_{10}	864.9	836	
3-Methyl-1,3-pentadiene	C_6H_{10}	852.3	823.4	
2-Methyl-2-pentene	C_6H_{12}	812	783.1	
2-Methylphenoxy	C_7H_7O	874.5	842	
1-(3-Methylphenyl)ethanone	$C_9H_{10}O$	868.2	836.4	
1-Methyl-3-phenyl-1*H*-pyrazole	$C_{10}H_{10}N_2$	932.6	900.8	
1-Methyl-5-phenyl-1*H*-pyrazole	$C_{10}H_{10}N_2$	932.4	900.5	
Methyl phenyl sulfone	$C_7H_8O_2S$	812.7	780.3	
Methylphosphine	CH_5P	851.5	817.6	
1-Methylpiperidine	$C_6H_{13}N$	971.1	940.1	
1-Methyl-2-piperidinone	$C_6H_{11}NO$	924.4	892.6	
2-Methylpropanamide	C_4H_9NO	878.6	846.7	
N-Methylpropanamide	C_4H_9NO	920.4	889.4	

Name	Mol. form.	E_{pa}/kJ mol^{-1}	$\Delta_{base}G°$/kJ mol^{-1}	Note
2-Methylpropanenitrile	C_4H_7N	803.6	772.8	
2-Methyl-1-propanethiol	$C_4H_{10}S$	802.6	771.4	
2-Methyl-2-propanethiol	$C_4H_{10}S$	816.4	785.1	
Methyl propanoate	$C_4H_8O_2$	830.2	799.2	
2-Methyl-1-propanol	$C_4H_{10}O$	793.7	762.2	
2-Methyl-2-propanol	$C_4H_{10}O$	802.6	772.2	
2-Methylpropenal	C_4H_6O	808.7	776.8	
2-Methyl-2-propenamide	C_4H_7NO	880.4	849.4	
Methyl propyl ether	$C_4H_{10}O$	814.9	785.7	
6-Methyl-1*H*-purine	$C_6H_6N_4$	939.2	907.3	
1-Methyl-1*H*-pyrazole	$C_4H_6N_2$	912.0	880.1	
3-Methyl-1*H*-pyrazole	$C_4H_6N_2$	906.0	874.2	
4-Methyl-1*H*-pyrazole	$C_4H_6N_2$	906.8	873.4	
2-Methylpyridine	C_6H_7N	949.1	917.3	
3-Methylpyridine	C_6H_7N	943.4	911.6	
4-Methylpyridine	C_6H_7N	947.2	915.3	
Methyl 3-pyridinecarboxylate	$C_7H_7NO_2$	925.6	893.8	
Methyl 4-pyridinecarboxylate	$C_7H_7NO_2$	926.6	894.7	
3-Methylpyridine-1-oxide	C_6H_7NO	935.2	902.8	
1-Methyl-2(1*H*)-pyridinone	C_6H_7NO	925.8	894.8	
N-Methylpyrrolidine	$C_5H_{11}N$	965.6	934.8	
N-Methyl-2-pyrrolidinone	C_5H_9NO	923.5	891.6	
2-Methylstyrene	C_9H_{10}	855.2	826.3	
3-Methylstyrene	C_9H_{10}	849.4	820.5	
4-Methylstyrene	C_9H_{10}	861.7	832.8	
3-Methylsuccinic anhydride	$C_5H_6O_3$	807		Ref. 9
2-Methyltetrahydrofuran	$C_5H_{10}O$	840.8	811.6	
2-Methylthiazole	C_4H_5NS	930.6	898.7	
Methylthiirane	C_3H_6S	833.3	801.5	
(Methylthio)acetonitrile	C_3H_5NS	784.8	754.1	
(Methylthio)benzene	C_7H_8S	872.6	843.7	
Methyl thiocyanate	C_2H_3NS	796.7	766.1	
(Methylthio)ethene	C_3H_6S	858.2	829.3	
2-Methylthiophene	C_5H_6S	859.0	826.5	
Methyl trifluoroacetate	$C_3H_3F_3O_2$	740.5	709.6	
Methyl trifluoromethyl ether	$C_2H_3F_3O$	719.2	690.0	
Methyl 2,4,6-trimethylbenzoate	$C_{11}H_{14}O_2$	866.3	835.3	
(1-Methylvinyl)cyclopropane	C_6H_{10}	871.6	842.7	
Methyl vinyl ether	C_3H_6O	859.2	830.3	
Molybdenum carbonyl	C_6MoO_6	762.6	738.1	
Morpholine	C_4H_9NO	924.3	891.2	
Naphthacene	$C_{18}H_{12}$	905.5	876.5	
Naphthalene	$C_{10}H_8$	802.9	779.4	
1,8-Naphthalenediamine	$C_{10}H_{10}N_2$	944.5	912.1	
1-Naphthylamine	$C_{10}H_9N$	907.0	875.1	
Neon	Ne	198.8	174.4	
Nickel	Ni	737	714.1	
Nickel carbonyl [Ni(CO)$_4$]	C_4NiO_4	742.3	716.0	
Nickelocene	$C_{10}H_{10}Ni$	935.7	907.3	
L-Nicotine	$C_{10}H_{14}N_2$	963.4	932.6	
Nitramide	$H_2N_2O_2$	757.4	725.0	
Nitric acid	HNO_3	751.4	731.5	
Nitric oxide	NO	531.8	505.3	
4-Nitroaniline	$C_6H_6N_2O_2$	866.0	834.2	
4-Nitrobenzaldehyde	$C_7H_5NO_3$	795.1	763.2	
3-Nitrobenzamide	$C_7H_6N_2O_3$	854.2	823.2	
4-Nitrobenzamide	$C_7H_6N_2O_3$	845.3	814.4	
Nitrobenzene	$C_6H_5NO_2$	800.3	769.5	
4-Nitrobenzenemethanol	$C_7H_7NO_3$	810.5	778.0	
3-Nitrobenzonitrile	$C_7H_4N_2O_2$	781.4	750.7	
4-Nitrobenzonitrile	$C_7H_4N_2O_2$	775.7	745.1	
Nitroethane	$C_2H_5NO_2$	765.7	733.2	
Nitrogen	N_2	493.8	464.5	

Name	Mol. form.	E_{pa}/kJ mol^{-1}	$\Delta_{base}G°$/kJ mol^{-1}	Note
Nitrogen (atomic)	N	342.2	318.7	
Nitrogen dioxide	NO_2	591.0	560.3	
Nitrogen trifluoride	F_3N	568.4	538.6	
Nitromethane	CH_3NO_2	754.6	721.6	
1-(3-Nitrophenyl)ethanone	$C_8H_7NO_3$	826.0	794.1	
1-(4-Nitrophenyl)ethanone	$C_8H_7NO_3$	824.3	792.5	
4-Nitropyridine	$C_5H_4N_2O_2$	874.3	842.5	
4-Nitropyridine 1-oxide	$C_5H_4N_2O_3$	868.0	837.3	
Nitrosobenzene	C_6H_5NO	854.3	823.6	
4-Nitrotoluene	$C_7H_7NO_2$	815.2	782.7	
Nitrous oxide	N_2O	549.8	523.3	Protonation at N
Nitrous oxide	N_2O	575.2	548.7	Protonation at O
5-Nonanone	$C_9H_{18}O$	853.7	821.9	
2,5-Norbornadiene	C_7H_8	849.3	820.3	
1,2,3,4,5,6,7,8-Octahydroanthracene	$C_{14}H_{18}$	845.4	814.7	
1,2,3,4,5,6,7,8-Octahydrophenanthrene	$C_{14}H_{18}$	846.2	815.5	
Octylamine	$C_8H_{19}N$	928.9	895.0	
4-Octylaniline	$C_{14}H_{23}N$	894.5	862	
Osmium(VIII) oxide	O_4Os	676.9	650.6	
7-Oxabicyclo[2.2.1]heptane	$C_6H_{10}O$	844.2	816.8	
7-Oxabicyclo[4.1.0]heptane	$C_6H_{10}O$	848.1	815.6	
Oxazole	C_3H_3NO	876.4	844.5	
Oxepane	$C_6H_{12}O$	834.2	806.8	
Oxetane	C_3H_6O	801.3	773.9	
Oxirane	C_2H_4O	774.2	745.3	
Oxomethyl [HCO]	CHO	636	601.8	
Oxygen	O_2	421	396.3	
Oxygen (atomic)	O	485.2	459.6	
Ozone	O_3	625.5	595.9	
Palladium	Pd	696	673.4	
Pentaborane(9)	B_5H_9	699.4	666.9	
trans-1,3-Pentadiene	C_5H_8	834.1	804.4	
Pentafluorobenzene	C_6HF_5	690.4	662.7	
Pentamethylbenzene	$C_{11}H_{16}$	850.7	823.5	
Pentanal	$C_5H_{10}O$	796.6	764.8	
1,5-Pentanediamine	$C_5H_{14}N_2$	999.6	946.2	
2,4-Pentanedione	$C_5H_8O_2$	873.5	836.8	
Pentanenitrile	C_5H_9N	802.4	771.7	
2-Pentanone	$C_5H_{10}O$	832.7	800.9	
3-Pentanone	$C_5H_{10}O$	836.8	807	
trans-2-Pentenal	C_5H_8O	839.0	807.2	
Pentylamine	$C_5H_{13}N$	923.5	889.5	
2-Pentyne	C_5H_8	810.2	778.0	
Perfluoroacetone	C_3F_6O	670.4	639.7	
Perfluorocyclobutane	C_4F_8	>544		Ref. 6
Perfluoropyridine	C_5F_5N	764.9	733.0	
Perylene	$C_{20}H_{12}$	888.6	859.6	
Phenanthrene	$C_{14}H_{10}$	825.7	795.0	
Phenazine	$C_{12}H_8N_2$	938.4	908.3	
Phenol	C_6H_6O	817.3	786.3	
Phenoxy	C_6H_5O	873.2		Ref. 4
Phenyl	C_6H_5	884	851.5	
Phenylacetylene	C_8H_6	832.0	801.3	
L-Phenylalanine	$C_9H_{11}NO_2$	922.9	888.9	
1-Phenylpiperidine	$C_{11}H_{15}N$	952.9	926.4	
1-Phenyl-1-propanone	$C_9H_{10}O$	867.4	835.6	
1-Phenyl-2-propanone	$C_9H_{10}O$	842.6	810.8	
4-Phenylpyridine	$C_{11}H_9N$	939.7	907.8	
Phenyl-3-pyridinylmethanone	$C_{12}H_9NO$	934.1	902.3	
1-Phenylpyrrolidine	$C_{10}H_{13}N$	941.6	915.1	
Phosphine	H_3P	785	750.9	
Phosphino	H_2P	709.2	675.7	
Phosphorus	P	626.8	604.8	

Name	Mol. form.	E_{pa}/kJ mol⁻¹	$\Delta_{base}G°$/kJ mol⁻¹	Note
Phosphorus(III) fluoride	F_3P	695.3	662.8	
Phosphorus monohydride	HP	670.3	639.6	
Phosphorus monoxide	OP	682	649.5	
Phosphorus nitride	NP	789.4	757.0	
Phosphoryl fluoride	F_3OP	694.0	664.2	
Picene	$C_{22}H_{14}$	851.3	820.6	
Piperazine	$C_4H_{10}N_2$	943.7	914.7	
Piperidine	$C_5H_{11}N$	954.0	921	
Potassium hydroxide	HKO	1101.8	1075.4	
Potassium oxide	K_2O	1342.5	1311.8	
L-Proline	$C_5H_9NO_2$	920.5	886.0	
Propanal	C_3H_6O	786.0	754.0	
Propanamide	C_3H_7NO	876.2	845.3	
Propane	C_3H_8	625.7	607.8	
1,3-Propanediamine	$C_3H_{10}N_2$	987.0	940.0	
1,3-Propanediol	$C_3H_8O_2$	876.2	825.9	
Propanenitrile	C_3H_5N	794.1	763.0	
1-Propanethiol	C_3H_8S	794.9	763.6	
2-Propanethiol	C_3H_8S	803.6	772.3	
Propanoic acid	$C_3H_6O_2$	797.2	766.2	
1-Propanol	C_3H_8O	786.5	756.1	
2-Propanol	C_3H_8O	793.0	762.6	
Propene	C_3H_6	741.6		Ref. 5
1-Propen-1-one	C_3H_4O	834.1	803.4	
cis-1-Propenylbenzene	C_9H_{10}	836.4	807.5	
trans-1-Propenylbenzene	C_9H_{10}	834.2	805.3	
Propyl acetate	$C_5H_{10}O_2$	836.6	805.6	
Propylamine	C_3H_9N	917.8	883.9	
Propylbenzene	C_9H_{12}	790.1	762.4	
Propyleneimine	C_3H_7N	925.1	892.1	
Propyl formate	$C_4H_8O_2$	804.9	773.9	
2-Propylpyridine	$C_8H_{11}N$	955.7	923.8	
Propyl trifluoroacetate	$C_5H_7F_3O_2$	763.9	732.9	
2-Propyn-1-amine	C_3H_5N	887.4	853.5	
Propyne	C_3H_4	748.2	723.0	
2-Propynyl	C_3H_3	741	708.5	
1*H*-Purine	$C_5H_4N_4$	920.1	888.2	
Pyrazine	$C_4H_4N_2$	877.1	847.0	
1*H*-Pyrazol-3-amine	$C_3H_5N_3$	921.5	889.6	
1*H*-Pyrazol-4-amine	$C_3H_5N_3$	907.6	874.0	
1*H*-Pyrazole	$C_3H_4N_2$	894.1	860.5	
Pyrene	$C_{16}H_{10}$	869.2	840.1	
Pyridazine	$C_4H_4N_2$	907.2	877.1	
2-Pyridinamine	$C_5H_6N_2$	947.2	915.3	
3-Pyridinamine	$C_5H_6N_2$	954.4	922.6	
4-Pyridinamine	$C_5H_6N_2$	979.7	947.8	
Pyridine	C_5H_5N	930	898.1	
2-Pyridinecarbonitrile	$C_6H_4N_2$	872.9	841	
3-Pyridinecarbonitrile	$C_6H_4N_2$	877.0	845.1	
4-Pyridinecarbonitrile	$C_6H_4N_2$	880.6	848.8	
4-Pyridinecarboxaldehyde	C_6H_5NO	904.6	872.8	
3-Pyridinecarboxamide	$C_6H_6N_2O$	918.3	886.4	
Pyridine-1-oxide	C_5H_5NO	923.6	892.9	
3-Pyridinol	C_5H_5NO	929.5	897.7	
1-(3-Pyridinyl)ethanone	C_7H_7NO	916.2	884.3	
1-(4-Pyridinyl)ethanone	C_7H_7NO	914.7	882.9	
Pyrimidine	$C_4H_4N_2$	885.8	855.7	
2,4(1*H*,3*H*)-Pyrimidinedithione	$C_4H_4N_2S_2$	911.4	880.5	
Pyrrole	C_4H_5N	875.4	843.8	
Pyrrolidine	C_4H_9N	948.3	915.3	
3-(2-Pyrrolidinyl)pyridine, (*S*)-	$C_9H_{12}N_2$	964.0	931.0	
1*H*-Pyrrolo[2,3-*b*]pyridine	$C_7H_6N_2$	940.2	908.3	
Pyruvonitrile	C_3H_3NO	746.9	716.2	

Name	Mol. form.	E_{pa}/kJ mol^{-1}	$\Delta_{base}G°$/kJ mol^{-1}	Note
Quinoline	C_9H_7N	953.2	921.4	
Quinoxaline	$C_8H_6N_2$	903.8	873.7	
Rhodium	Rh	768	745.4	
Ruthenium	Ru	774	751.4	
Ruthenocene	$C_{10}H_{10}Ru$	899.1	876.8	
Sarcosine	$C_3H_7NO_2$	921.2	888.7	
Scandium	Sc	914	892.0	
Selenoformaldehyde	CH_2Se	764.0	734.9	
L-Serine	$C_3H_7NO_3$	914.6	880.7	
Silane	H_4Si	639.7	613.4	
Silicon	Si	837	814.1	
Silicon monosulfide	SSi	627	596.6	Protonation at Si
Silicon monosulfide	SSi	683	660.2	Protonation at S
Silicon monoxide	OSi	777.8	750.4	Protonation at O
Silicon monoxide	OSi	533	500.5	Protonation at Si
Silylene	H_2Si	839.2	804.1	
Sodium hydride	HNa	1095	1070.6	
Sodium hydroxide	HNaO	1071.8	1044.8	
Sodium oxide	Na_2O	1375.9	1345.2	
Strontium oxide	OSr	1209	1180.7	
Styrene	C_8H_8	839.5	809.2	
Succinic anhydride	$C_4H_4O_3$	797		Ref. 9
Sulfur	S	664.3	640.2	
Sulfur dioxide	O_2S	672.3	643.3	
Sulfur hexafluoride	F_6S	575.3	550.7	
Sulfuric acid	H_2O_4S	717	681	Ref. 3
Sulfur trioxide (α-form)	O_3S	588.3	560.3	
Sulfuryl fluoride	F_2O_2S	605.5	580.5	
Tetraborane(10)	B_4H_{10}	605	572.5	
Tetraethylene glycol	$C_8H_{18}O_5$		>910	
Tetraethylene glycol dimethyl ether	$C_{10}H_{22}O_5$	953.8	897.8	
Tetraethylhydrazine	$C_8H_{20}N_2$	964.3	935.3	
1,2,3,4-Tetrafluorobenzene	$C_6H_2F_4$	700.4	672.7	
1,2,3,5-Tetrafluorobenzene	$C_6H_2F_4$	747.3	719.6	
1,2,4,5-Tetrafluorobenzene	$C_6H_2F_4$	746.5	718.8	
Tetrafluoromethane	CF_4	529.3	503.7	
Tetrafluorosilane	F_4Si	502.9	476.6	
Tetrahydrofuran	C_4H_8O	822.1	794.7	
5,6,7,8-Tetrahydroisoquinoline	$C_9H_{11}N$	966.6	934.7	
1,2,3,4-Tetrahydronaphthalene	$C_{10}H_{12}$	809.7	782.1	
Tetrahydropyran	$C_5H_{10}O$	822.8	795.4	
5,6,7,8-Tetrahydroquinoline	$C_9H_{11}N$	966.0	934.1	
Tetrahydrothiophene	C_4H_8S	849.1	819.3	
N,N,2,6-Tetramethylaniline	$C_{10}H_{15}N$	954.1	923.2	
N,N,3,5-Tetramethylaniline	$C_{10}H_{15}N$	956.1	924.3	
1,2,3,5-Tetramethylbenzene	$C_{10}H_{14}$	845.6	816.5	
N,N,N',N'-Tetramethyl-1,2-benzenediamine	$C_{10}H_{16}N_2$	982.6	950.2	
1,4,7,7-Tetramethylbicyclo[2.2.1]heptan-2-one	$C_{11}H_{18}O$	863.3	831.4	
N,N,N',N'-Tetramethyl-1,4-butanediamine	$C_8H_{20}N_2$	1046.3	992.7	
1,1,3,3-Tetramethyldisiloxane	$C_4H_{14}OSi_2$	845.3	814.6	
N,N,N',N'-Tetramethyl-1,2-ethanediamine	$C_6H_{16}N_2$	1012.8	970.6	
1,1,3,3-Tetramethylguanidine	$C_5H_{13}N_3$	1031.6	997.4	
N,N,N',N'-Tetramethyl-1,6-hexanediamine	$C_{10}H_{24}N_2$	1035.8	982.2	
N,N,N',N'-Tetramethylmethanediamine	$C_5H_{14}N_2$	952.2	919.8	
2,2,6,6-Tetramethylpiperidine	$C_9H_{19}N$	987.0	953.9	
N,N,N',N'-Tetramethyl-1,3-propanediamine	$C_7H_{18}N_2$	1035.2	985.4	
Tetramethylstannane	$C_4H_{12}Sn$	823.7	797.4	
Tetramethylthiourea	$C_5H_{12}N_2S$	947.6	916.6	
Tetramethylurea	$C_5H_{12}N_2O$	930.6	899.6	
2,6,10,14-Tetraoxapentadecane	$C_{11}H_{24}O_4$		895.1	
Thiacyclohexane	$C_5H_{10}S$	855.8	826.0	
2-Thiazolamine	$C_3H_4N_2S$	930.6	898.7	
Thiazole	C_3H_3NS	904	872.1	

Name	Mol. form.	E_{pa}/kJ mol^{-1}	$\Delta_{base}G°$/kJ mol^{-1}	Note
Thietane	C_3H_6S	834.8	805.0	
Thiirane	C_2H_4S	807.4	777.6	
Thioacetamide	C_2H_5NS	884.6	852.8	
Thiocyanate	CNS	751	718.5	
Thioformaldehyde	CH_2S	759.7	730.5	
Thiophene	C_4H_4S	815.0	784.3	
Thiourea	CH_4N_2S	893.7	863.9	
L-Threonine	$C_4H_9NO_3$	922.5	888.5	
Thymidine	$C_{10}H_{14}N_2O_5$	948.3	915.9	
Thymine	$C_5H_6N_2O_2$	880.9	850.0	
Titanium	Ti	876	853.7	
Toluene	C_7H_8	784.0	756.3	
o-Toluic acid	$C_8H_8O_2$	838.8	807.8	
m-Toluic acid	$C_8H_8O_2$	829.8	798.8	
p-Toluic acid	$C_8H_8O_2$	836.7	805.7	
1,3,5-Triazine	$C_3H_3N_3$	848.8	819.6	
1*H*-1,2,3-Triazole	$C_2H_3N_3$	879.3	847.4	
1*H*-1,2,4-Triazole	$C_2H_3N_3$	886.0	855.9	
Tributylamine	$C_{12}H_{27}N$	998.5	967.6	
1,3,5-Tri-*tert*-butylbenzene	$C_{18}H_{30}$	848.8	822.3	
Trichloroacetaldehyde	C_2HCl_3O	722.3	690.5	
Trichloroacetic acid	$C_2HCl_3O_2$	770.0	739.1	
Trichloroacetonitrile	C_2Cl_3N	723.2	692.6	
2,2,2-Trichloroethanol	$C_2H_3Cl_3O$	729.3	698.9	
2,2,2-Trichloro-1-phenylethanone	$C_8H_5Cl_3O$	818.9	787.0	
1,1,1-Trichloro-2-propanone	$C_3H_3Cl_3O$	768.3	736.3	
Tricyclo[3.3.1.1^{3,7}]decan-1-amine	$C_{10}H_{17}N$	948.8	916.3	
Tricyclo[3.3.1.1^{3,7}]decane-1-carbonitrile	$C_{11}H_{15}N$	834.4	803.8	
1-Tricyclo[3.3.1.1^{3,7}]dec-1-ylethanone	$C_{12}H_{18}O$	864.9	833.1	
Triethylamine	$C_6H_{15}N$	981.8	951	
Triethylenediamine	$C_6H_{12}N_2$	963.4	934.6	
Triethylene glycol dimethyl ether	$C_8H_{18}O_4$	946.6	892.4	
Triethyl phosphate	$C_6H_{15}O_4P$	909.3	879.6	
Triethylphosphine	$C_6H_{15}P$	984.5	952.0	
Triethylphosphine oxide	$C_6H_{15}OP$	936.6	906.8	
Triethylsilanol	$C_6H_{16}OSi$	822.1	794.8	
Trifluoroacetic acid	$C_2HF_3O_2$	711.7	680.7	
1,1,1-Trifluoroacetone	$C_3H_3F_3O$	723.9	692.0	
Trifluoroacetonitrile	C_2F_3N	688.4	657.7	
Trifluoroacetyl chloride	C_2ClF_3O	681.6	649.8	
1,2,3-Trifluorobenzene	$C_6H_3F_3$	724.3	696.6	
1,2,4-Trifluorobenzene	$C_6H_3F_3$	729.5	699.4	
1,3,5-Trifluorobenzene	$C_6H_3F_3$	741.9	715.4	
2,2,2-Trifluoro-*N,N*-dimethylacetamide	$C_4H_6F_3NO$	849.0	818.0	
2,2,2-Trifluoroethanol	$C_2H_3F_3O$	700.2	669.9	
Trifluoroethene	C_2HF_3	699.4	666.9	
2,2,2-Trifluoroethylamine	$C_2H_4F_3N$	846.8	812.9	
2,2,2-Trifluoroethyl methyl ether	$C_3H_5F_3O$	747.6	718.4	
Trifluoroiodomethane	CF_3I	628.0	598.2	
Trifluoromethane	CHF_3	619.5	589.7	
Trifluoromethanesulfonic acid	CHF_3O_3S	699.4	666.9	
3-(Trifluoromethyl)aniline	$C_7H_6F_3N$	856.9	825.1	
4-(Trifluoromethyl)benzaldehyde	$C_8H_5F_3O$	805.6	773.8	
3-(Trifluoromethyl)benzonitrile	$C_8H_4F_3N$	791.4	760.8	
4-(Trifluoromethyl)benzonitrile	$C_8H_4F_3N$	787.2	758.3	
Trifluoronitrosomethane	CF_3NO	703.3	670.8	
2,2,2-Trifluoro-1-phenylethanone	$C_8H_5F_3O$	799.2	767.4	
1,3,5-Trimethoxybenzene	$C_9H_{12}O_3$	926.7	898.2	
Trimethylamine	C_3H_9N	948.9	918.1	
Trimethylamine oxide	C_3H_9NO	983.2	953.5	
Trimethylarsine	C_3H_9As	897.3	864.9	
N,N,3-Trimethylbenzamide	$C_{10}H_{13}NO$	927.0	896.0	
N,N,4-Trimethylbenzamide	$C_{10}H_{13}NO$	927.0	896.0	

Name	Mol. form.	E_{pa}/kJ mol^{-1}	$\Delta_{base}G°$/kJ mol^{-1}	Note
1,3,5-Trimethylbenzene	C_9H_{12}	836.2	808.6	
Trimethyl borate	$C_3H_9BO_3$	815.8	783.4	
1,3,5-Trimethyl-2-nitrobenzene	$C_9H_{11}NO_2$	823.8	793.1	
2,2,4-Trimethyl-3-pentanone	$C_8H_{16}O$	856.9	825.0	
Trimethyl phosphate	$C_3H_9O_4P$	890.6	860.8	
Trimethylphosphine	C_3H_9P	958.8	926.3	
Trimethyl phosphite	$C_3H_9O_3P$	929.7	899.9	
N,*N*,2-Trimethylpropenylamine	$C_6H_{13}N$	967.0	934.5	
1,3,5-Trimethyl-1*H*-pyrazole	$C_6H_{10}N_2$	949.3	917.4	
3,4,5-Trimethyl-1*H*-pyrazole	$C_6H_{10}N_2$	949.3	916.0	
Trimethylsilanol	$C_3H_{10}OSi$	814.0	781.5	
Triphenylamine	$C_{18}H_{15}N$	908.9	876.4	
Triphenylarsine	$C_{18}H_{15}As$	908.9	876.4	
Triphenylarsine oxide	$C_{18}H_{15}AsO$	906.2	876.4	
Triphenylene	$C_{18}H_{12}$	819.2	791.2	
Triphenylphosphine	$C_{18}H_{15}P$	972.8	940.4	
Triphenylphosphine oxide	$C_{18}H_{15}OP$	906.2	876.4	
Triphenylphosphine sulfide	$C_{18}H_{15}PS$	906.2	876.4	
Triphenylstibine	$C_{18}H_{15}Sb$	845.5	813.1	
Tripropylamine	$C_9H_{21}N$	991.0	960.1	
L-Tryptophan	$C_{11}H_{12}N_2O_2$	948.9	915	
Tungsten carbonyl	C_6O_6W	758.0	733.4	
L-Tyrosine	$C_9H_{11}NO_3$	926	892.1	
Uracil	$C_4H_4N_2O_2$	872.7	841.7	
Uranium	U	995.2	973.2	
Urea	CH_4N_2O	873.5	841.6	Protonation at O; Ref. 8
Uridine	$C_9H_{12}N_2O_6$	947.6	916.6	
L-Valine	$C_5H_{11}NO_2$	910.6	876.7	
Vanadium	V	859.4	836.8	
Vinyl acetate	$C_4H_6O_2$	813.9	782.9	
Vinylcyclopropane	C_5H_8	816.3	787.5	
4-Vinylpyridine	C_7H_7N	944.1	912.3	
Vinyltrimethylsilane	$C_5H_{12}Si$	833	804.1	
Water	H_2O	691	660.0	
Xenon	Xe	499.6	478.1	
o-Xylene	C_8H_{10}	796.0	768.3	
m-Xylene	C_8H_{10}	812.1	786.2	
p-Xylene	C_8H_{10}	794.4	766.8	
Yttrium	Y	967	945.9	
Zinc	Zn	608.6	586.0	

POLARIZABILITIES OF ATOMS AND MOLECULES

Thomas M. Miller

The *polarizability* of an atom or molecule describes the response of its electron cloud to an external field. The atomic or molecular energy shift ΔW due to an external electric field E is proportional to E^2 for external fields that are weak compared to the internal electric fields between the nucleus and electron cloud. The *electric dipole polarizability* α is the constant of proportionality defined by $\Delta W = -\alpha E^2/2$. The induced electric dipole moment is αE. *Hyperpolarizabilities*, coefficients of higher powers of E, are less often required. Technically, the polarizability is a tensor quantity but for spherically symmetric charge distributions reduces to a single number. In any case, an *average polarizability* is usually adequate in calculations. Frequency-dependent or *dynamic polarizabilities* are needed for electric fields that vary in time, except for frequencies that are much lower than electron orbital frequencies, for which *static polarizabilities* suffice.

Polarizabilities for atoms and molecules in excited states are found to be larger than for ground states and may be positive or negative. Molecular polarizabilities are very slightly temperature dependent because the size of the molecule depends on its rovibrational state. Only in the case of dihydrogen (H_2) has this effect been studied enough to warrant consideration.

Polarizabilities are normally expressed in cgs units of cm³. Ground state polarizabilities are in the range of 10^{-24} cm³ = 1 Å³ and hence are often given in Å³ units. Theorists tend to use atomic units of a_0^3 where a_0 is the Bohr radius. The conversion is $\alpha(\text{cm}^3) = 0.148184 \times 10^{-24} \times \alpha(a_0^3)$. Polarizabilities are only recently encountered in SI units, C m²/ V = J/(V/m)². The conversion from cgs units to SI units is $\alpha(\text{C m}^2/\text{ V}) = 4\pi\varepsilon_0 \times 10^{-6}\alpha(\text{cm}^3)$, where ε_0 is the permittivity of free space in SI units and the factor 10^{-6} simply converts cm³ into m³. Thus, $\alpha(\text{C m}^2/\text{ V}) = 1.11265 \times 10^{-16} \times \alpha(\text{cm3})$. Measurements of excited state polarizabilities by optical methods tend tobe reported using units of MHz/(V/cm)², where the energy shift, ΔW, is expressed in frequency units with a factor of h understood. The polarizability is $-2\ \Delta W/E^2$. The conversion into cgs units is $\alpha(\text{cm}^3) = 5.955214 \times 10^{-16} \times \alpha[\text{MHz/(V/cm)}^2]$.

The polarizability appears in many formulas for low-energy processes involving the valence electrons of atoms or molecules. These formulas are given below in cgs units: the polarizability α is in cm³; masses m or μ are in grams; energies are in ergs; and electric charges are in esu, where $e = 4.8032 \times 10^{-10}$esu. The symbol $\alpha(\nu)$ denotes a frequency (ν) dependent polarizability, where $\alpha(\nu)$ reduces to the static polarizability α for $\nu = 0$. For further information, see Bonin, K. D., and Kresin, V. V., *Electric Dipole Polarizabilities of Atoms, Molecules, and Clusters*, World Scientific, Singapore, 1997; Bonin, K. D., and Kadar-Kallen, M.A., *Int. J. Mod. Phys.* B, 24, 3313, 1994; Miller, T. M., and Bederson, B., *Advances in Atomic and Molecular Physics*, 13, 1, 1977; and Gould, H., and Miller, T. M., *Advances in Atomic, Molecular, and Optical Physics*, 51, 243, 2005. A helpful listing of theoretical results for atomic polarizabilities is given by P. Schwerdtfeger at http://ctcp.massey.ac.nc/dipole-polarizabilities (accessed September 2016). Details on polarizability-related interactions, especially in regard to hyperpolarizabilities and nonlinear optical phenomena, are given by Bogaard, M. P., and Orr, B. J., in *Physical Chemistry, Series Two, Vol. 2, Molecular Structure and Properties*, Buckingham, A. D., Ed., Butterworths, London, 1975, pp. 149–194. A tabulation of tensor and hyperpolarizabilities is included.

An empirical additive formula for molecular polarizabilities at 589 nm frequency has been given in Bosque, R., and Sales, J., *J. Chem. Inf. Comput. Sci.* 42, 1154, 2002:

$$\alpha = 0.32 + 1.51\#C + 0.17\#H + 0.57\#O + 1.05\#N + 2.99\#S + 2.48\#P + 0.22\#F + 2.16\#Cl + 3.29\#Br + 5.45\#I,$$

where #C denotes the number of carbon atoms in the molecule, etc. A helium-elimination additive method has been given by Kassimi, N. E.-B., and Thakkar, A. J., *Chem. Phys. Lett.* 472, 232, 2009.

All polarizabilities in this table are experimental values except those indicated as calculated in the Note column. The experimental polarizabilities are mostly determined by measurements of a dielectric constant or refractive index and are quite precise (0.5% or better). However, one should treat many of the results with several percent of caution because of the age of the data and because some of the results refer to optical frequencies rather than static. Comments given with the references are intended to allow one to judge the degree of caution required. Interested persons should consult these references. In many cases, the reference given is to a theoretical paper in which the experimental results are quoted. These papers, noted in the references, contain valuable information on polarizability calculations and experimental data, which often includes the tensor components of the polarizability.

The table first lists atoms in order of atomic number, followed by molecules in alphabetical order by name.

Formulas Involving Polarizability*

Description	Formula	Remarks
Lorentz-Lorenz relation	$\alpha(\nu) = \frac{3}{4\pi n}\frac{\eta^2(\nu)-1}{\eta^2(\nu)+2}$	For a gas of atoms or nonpolar molecules; the index of refraction is $\eta(\nu)$
Refraction by polar molecules	$\alpha(\nu) + \frac{d^2}{3kT} = \frac{3}{4\pi n}\frac{\eta^2(\nu)-1}{\eta^2(\nu)+2}$	The dipole moment is d, in esu.cm (= 10^{-18} D)
Dielectric constant (dimensionless)	$\kappa(\nu) = 1 + 4\pi n\,\alpha(\nu)$	From the Lorentz-Lorenz relation for the usual case of $\kappa(\nu) \approx 1$
Index of refraction (dimensionless)	$\mu(\nu) = 1 + 2\pi n\alpha(\nu)$	From $\eta^3(\nu) = \kappa(\nu)$
Diamagnetic susceptibility	$\chi_m = e^2(a_0 N\alpha)^{1/2} / 4m_e c^2$	From the approximation that the static polarizability is given by the variational formula $\alpha = (4/9a_0)\Sigma(N_i r_i^2)^2$; N is the number of electrons, m_e is the electron mass; a crude approximation is $\chi_m = (E_i/4m_e c^2)\alpha$, where E_i is the ionization energy

Description	Formula	Remarks
Long-range electron- or ion-molecule interaction energy	$V(r) = -e^2\alpha / 2r^4$	The target molecule polarizability is α
Ion mobility in a gas	$\kappa = 13.87 / (\alpha\mu)^{1/2} \mathrm{cm}^2 / \mathrm{V} \cdot \mathrm{s}$	This one formula is not in c.g.s. units. Enter α in Å^3 or 10^{-24} cm^3 units and the reduced mass μ of the ion-molecule pair in amu. Classical limit; pure polarization potential
Langevin capture cross section	$\sigma(v_0) = (2\pi e / v_0)(\alpha / \mu)^{1/2}$	The relative velocity of apaproach for an ion-molecule pair is v_0; the target molecular polarizability is α and the reduced mass of the ion-molecule pair is μ
Langevin reaction rate coefficient	$k = 2\pi e(\alpha / \mu)^{1/2}$	Collisional rate coefficient for an ion-molecule reaction
Rate coefficient for polar molecules	$kd = 2\pi e[(\alpha / \mu)^{1/2} + cd(2 / \mu\pi kT)^{1/2}]$	The dipole moment of the neutral is d in esu cm; the number c is a "locking factor" that depends on α and d, and is between 0 and 1
Modified effective range cross section for electron-neutral scattering	$\sigma(k) = 4\pi A^2 + 32\pi^4\mu e^2\alpha A k / 3h^2 + \ldots$	Here, k is the electron momentum divided by $h/2\pi$, where h is Planck's constant; A is called the "scattering length"; the reduced mass is μ
van der Waals constant between two systems A, B	$C_6 = \frac{3}{2}\frac{\alpha^A\alpha^B E^A E^B}{E^A + E^B}$	For the interaction potential term $V_6(r) = -C_6 r^6$; $E^{A,B}$ represents average dipole transition energies and $\alpha^{A,B}$ the respective polarizabilities of A, B
Dipole-quadrupole constant between two systems A, B	$C_8 = \frac{15}{4}\frac{\alpha^A\alpha_q^B E^A E_q^B}{E^A + E_q^B} + \frac{15}{4}\frac{\alpha_q^A\alpha^B E_q^A E^B}{E_q^A + E^B}$	For the interaction potential term $V_8(r) = -C_8 r^8$; $E_q^{A,B}$ represents average quadrupole transition energies and $\alpha_q^{A,B}$ are the respective quadrupole polarizabilities of A, B
van der Waals constant between an atom and a surface	$C_3 = \frac{\alpha g E^A E^S}{8\left(E^A + E^S\right)}$	For an interaction potential $V_3(r) = -C_3 r^3$; $E^{A,S}$ are characteristic energies of the atom and surface; $g = 1$ for a free-electron metal and $g = (\varepsilon_\infty - 1)/(\varepsilon_\infty + 1)$ for an ionic crystal
Relationship between α(ν) and oscillator strengths	$\alpha(\nu) = \frac{e^2h^2}{4\pi^2 m_e}\Sigma\frac{f_k}{E_k^2 - (h\nu)^2}$	Here, f_k is the oscillator strength from the ground state to an excited state k, with excitation energy E_k. This formula is often used to estimate static polarizabilities (ν = 0)
Dynamic polarizability	$\alpha(\nu) = \frac{\alpha E_r^2}{E_r^2 - (h\nu)^2}$	Approximate variation of the frequency-dependent polarizability α(ν) from ν = 0 up to the first dipole-allowed electronic transition, of energy E_r; the static dipole polarizability is α(0); infrared contributions ignored
Rayleigh scattering cross section	$\alpha(\nu) = \frac{8\pi}{9c^4}(2\pi\nu)^4 \times 3\alpha^2(\nu) + 2\gamma^2(\nu)/3$	The photon frequency is ν; the polarizability anisotropy (the difference between polarizabilities parallel and perpendicular to the molecular axis) is γ(ν)
Verdet constant	$V(\nu) = \frac{\nu n}{2m_e c^2}\frac{d\alpha(\nu)}{d\nu}$	Defined from $\theta = V(\nu)B$, where θ is the angle of rotation of linearly polarized light through a medium of number density n, per unit length, for a longitudinal magnetic field strength B (Faraday effect)

* The gas number density, n, in the list of formulas is usually taken to be that of 1 atm at 0 °C in reporting experimental data.

Polarizability of Atoms and Molecules

Name	Mol. form.	$\alpha/10^{-24}$ cm^3	Ref.	Note	Z
Atoms					
Hydrogen (atomic)	H	0.666793	1	calculated ("exact")	1
Helium	He	0.2050522	2	calculated ("exact")	2
Helium	He	0.2050519	3	dielectric constant (±0.009%)	2
Lithium	Li	24.33	4	beam (±0.66%)	3
Beryllium	Be	5.60	42	calculated (±1.2%)	4
Boron	B	3.051	31	calculated (±0.07%)	5
Carbon	C	1.67	5	calculated (±2%)	6
Nitrogen (atomic)	N	1.13	2	index of refraction (±5.3%)	7
Nitrogen (atomic)	N	1.1	40	calculated (<2.5%)	
Oxygen (atomic)	O	0.77	2	index of refraction (±7.8%)	8
Fluorine (atomic)	F	0.557	1	calculated (±2%)	9
Neon	Ne	0.39432	6	dielectric constant (±0.003)	10
Sodium	Na	24.11	7	interferom (±0.50%)	11
Sodium	Na	24.11	8	interferom (±0.83%)	11
Magnesium	Mg	8.8	36	beam (±26%)	12
Magnesium	Mg	10.5	25	calculated (<2.9%)	12
Aluminum	Al	6.8	11	beam (±4.4%)	13
Aluminum	Al	8.22	31	calculated (±0.1%)	

Name	Mol. form.	$\alpha/10^{-24}$ cm³	Ref.	Note	Z
Silicon	Si	5.53	5	calculated (±2%)	14
Phosphorus	P	3.69	40	calculated (<2.5%)	15
Sulfur	S	2.87	33	calculated (±0.5%)	16
Chlorine (atomic)	Cl	2.16	33	calculated (±0.5%)	17
Argon	Ar	1.6411	12,13	index/dielectric constant (±0.05%)	18
Potassium	K	42.93	32	interferom (±.16%)	19
Calcium	Ca	23.8	25	calculated (±2.9%)	20
Calcium	Ca	25.1	34	beam (±9%)	20
Scandium	Sc	14.4	36	beam (±9.7%)	21
Scandium	Sc	21.2	39	calculated (<12%)	21
Titanium	Ti	9.4	36	beam (±5.3%)	22
Titanium	Ti	16.9	39	calculated (<12%)	22
Vanadium	V	14.4	39	calculated (<12%)	23
Vanadium	V	10.1	36	beam (±7.96%)	23
Chromium	Cr	11.6	41	calculated (±1%)	24
Chromium	Cr	8.9	36	beam (±39%)	24
Manganese	Mn	9.9	41	calculated (±1%)	25
Iron	Fe	9.47	39	calculated (<12%)	26
Cobalt	Co	8.55	39	calculated (<12%)	27
Nickel	Ni	7.57	39	calculated (<12%)	28
Copper	Cu	6.03	41	calculated (±1%)	29
Copper	Cu	8.7	36	beam (±8%)	29
Zinc	Zn	5.75	15	index of refraction (±2%)	30
Zinc	Zn	5.73	44	calculated	30
Gallium	Ga	7.57	24	calculated (±2.9%)	31
Gallium	Ga	6.9	36	beam (±8.7%)	31
Germanium	Ge	5.84	5	calculated (±2%)	32
Arsenic	As	4.42	40	calculated (<2.5%)	33
Selenium	Se	3.89	35	calculated	34
Bromine (atomic)	Br	3.11	35	calculated	35
Krypton	Kr	2.4844	13	dielectric constant (±0.05%)	36
Rubidium	Rb	47.39	32	interferom (±.17%)	37
Strontium	Sr	27.6	1	beam (±7%)	38
Strontium	Sr	28.3	25	calculated (<2.9%)	38
Yttrium	Y	22.7	14	calculated (±25%)	39
Yttrium	Y	24.1	36	beam (±7.1%)	39
Zirconium	Zr	17.9	14	calculated (±25%)	40
Zirconium	Zr	16.6	36	beam (±11.4%)	40
Niobium	Nb	15.7	14	calculated (±25%)	41
Niobium	Nb	14.5	36	beam (±7.6%)	41
Molybdenum	Mo	10.7	41	calculated (±1%)	42
Molybdenum	Mo	12.9	36	beam (±7%)	42
Technetium	Tc	11.9	41	calculated (±1%)	43
Ruthenium	Ru	9.6	14	calculated (±25%)	44
Rhodium	Rh	8.6	14	calculated (±25%)	45
Rhodium	Rh	1.6	36	beam (±0.5%)	45
Palladium	Pd	4.8	14	calculated (±25%)	46
Silver	Ag	6.8	16	index (±11%)	47
Silver	Ag	6.8	36	beam (±16%)	47
Silver	Ag	5.44	41	calculated (±1%)	47
Cadmium	Cd	7.36	17	index of refraction (±3%)	48
Cadmium	Cd	7.4	10	calculated (±6%)	48
Indium	In	10.2	18	beam (±12%)	49
Indium	In	9.2	36	beam (±13%)	49
Indium	In	9.19	24	calculated (±3.1%)	49
Tin	Sn	10.0	36	beam (±13%)	50
Tin	Sn	6.3	5	beam (±26%)	50
Tin	Sn	7.84	5	calculated	50
Antimony	Sb	6.25	40	calculated (<2.5%)	51
Tellurium	Te	5.5	14	calculated (±25%)	52
Iodine (atomic)	I	5.35	19	index of refraction (±25%)	53

Name	Mol. form.	$\alpha/10^{-24}$ cm³	Ref.	Note	Z
Iodine (atomic)	I	5.11	21	calculated	53
Xenon	Xe	4.044	1	dielectric constant (±0.5%)	54
Cesium	Cs	59.42	20	beam (±0.13%)	55
Cesium	Cs	59.39	32	beam (±0.09%)	55
Barium	Ba	39.7	1	beam (±7%)	56
Barium	Ba	40.7	25	calculated (<2.9%)	56
Lanthanum	La	25.3	36	beam (±4.7	57
Lanthanum	La	31.7	30	calculated (<12%)	57
Cerium	Ce	30.3	30	calculated (<12%)	58
Cerium	Ce	28.4	36	beam (±10.6%)	58
Praseodymium	Pr	35.4	36	beam (±11.6%)	59
Praseodymium	Pr	32	30	calculated (<12%)	59
Neodymium	Nd	27.2	36	beam (±10.7%)	60
Neodymium	Nd	30.9	30	calculated (<12%)	60
Promethium	Pm	29.7	14	calculated	61
Samarium	Sm	23.2	36	beam (±10.3%)	62
Samarium	Sm	28.5	30	calculated (<12%)	62
Europium	Eu	21.6	36	beam (±17.1%)	63
Europium	Eu	27.3	30	calculated (<12%)	63
Gadolinium	Gd	26.1	36	beam (±14.9%)	64
Gadolinium	Gd	23.5	30	calculated (<12%)	64
Terbium	Tb	23.5	36	beam (±6.9%)	65
Terbium	Tb	25.1	30	calculated (<12%)	65
Dysprosium	Dy	23.3	36	beam (±6.9%)	66
Dysprosium	Dy	24.1	30	calculated (<12%)	66
Holmium	Ho	21.5	36	beam (±7.9%)	67
Holmium	Ho	23.2	30	calculated (<12%)	67
Erbium	Er	32.2	36	beam (±17.7%)	68
Erbium	Er	26.3	30	calculated (<12%)	68
Thulium	Tm	19.2	36	beam (±12.5%)	69
Thulium	Tm	21.4	30	calculated (<12%)	69
Ytterbium	Yb	21.8	36	beam (±13.3%)	70
Ytterbium	Yb	20.6	30	calculated (<12%)	70
Lutetium	Lu	18.3	36	beam (±14.8%)	71
Lutetium	Lu	20.3	30	calculated (<12%)	71
Hafnium	Hf	12.4	36	beam (±22.6%)	72
Hafnium	Hf	16.2	14	calculated (±25%)	72
Tantalum	Ta	8.6	36	beam (±20.9%)	73
Tantalum	Ta	13.1	14	calculated (±25%)	73
Tungsten	W	11.1	14	calculated (±25%)	74
Rhenium	Re	9.05	41	calculated (±1%)	75
Osmium	Os	8.5	14	calculated (±25%)	76
Iridium	Ir	7.6	14	calculated (±25%)	77
Platinum	Pt	6.5	14	calculated (±25%)	78
Gold	Au	4.13	41	calculated (±1%)	79
Mercury	Hg	5.02	22	index of refraction (±1%)	80
Mercury	Hg	5.08	45	calculated (±0.5%)	80
Thallium	Tl	7.6	23	atomic beam deflection (±15%)	81
Thallium	Tl	7.72	24	calculated (±3.1%)	81
Lead	Pb	8.3	36	beam (±32.5%)	82
Lead	Pb	6.98	5	beam (±15%)	82
Lead	Pb	7.07	5	calculated	82
Bismuth	Bi	8.1	36	beam (±21%)	83
Bismuth	Bi	7.2	40	calculated (<2.5%)	83
Polonium	Po	6.8	14	calculated (±25%)	84
Astatine	At	6.39	46	calculated	85
Radon	Rn	4.83	40	calculated (<2.5%)	86
Francium	Fr	46.2	9	calculated (±2%)	87
Radium	Ra	36.48	26	calculated (±2%)	88
Actinium	Ac	32.1	14	calculated (±25%)	89
Thorium	Th	32.1	14	calculated (±25%)	90

Name	Mol. form.	$\alpha/10^{-24}$ cm³	Ref.	Note	Z
Protactinium	Pa	25.4	14	calculated (±25%)	91
Uranium	U	24.9	27	atomic beam deflection (±6%)	92
Neptunium	Np	24.8	14	calculated (±25%)	93
Plutonium	Pu	24.5	14	calculated (±25%)	94
Americium	Am	23.3	14	calculated (±25%)	95
Curium	Cm	23.0	14	calculated (±25%)	96
Berkelium	Bk	22.7	14	calculated (±25%)	97
Californium	Cf	20.5	14	calculated (±25%)	98
Einsteinium	Es	19.7	14	calculated (±25%)	99
Fermium	Fm	23.8	14	calculated (±25%)	100
Mendelevium	Md	18.2	14	calculated (±25%)	101
Nobelium	No	15.6	30	calculated (<12%)	102
Dubnium	Db	6.22	29	calculated (±10%)	105
Seaborgium	Sg	5.93	29	calculated (±10%)	106
Bohrium	Bh	5.63	29	calculated (±11%)	107
Hassium	Hs	5.33	29	calculated (±11%)	108
Meitnerium	Mt	5.04	29	calculated (±9%)	109
Darmstadtium	Ds	4.74	29	calculated (±10%)	110
Roentgenium	Rg	4.45	29	calculated (±10%)	111
Copernicium	Cn	4.06	28	calculated (±2%)	112
Nihonium	Nh	4.43	38	calculated	113
Flerovium	Fl	4.72	5	calculated (±2%)	114
Flerovium	Fl	4.37	28	calculated (±2%)	114
Oganesson	Og		37	calculated	118
Ununennium	Uue	25.15	9	calculated (±2%)	119
Unbinilium	Ubn	24.09	9	calculated (±2%)	120
Molecules					
Acenaphthene	$C_{12}H_{10}$	20.61	27		
Acetaldehyde	C_2H_4O	4.6	2	molar refraction; see Ref.2	
Acetaldehyde	C_2H_4O	4.59	18		
Acetamide	C_2H_5NO	5.67	18		
Acetic acid	$C_2H_4O_2$	5.1	2	molar refraction; see Ref.2	
Acetic anhydride	$C_4H_6O_3$	8.9	2	molar refraction; see Ref.2	
Acetone	C_3H_6O	6.33	15		
Acetone	C_3H_6O	6.4	2	molar refraction; see Ref.2	
Acetone	C_3H_6O	6.39	18		
Acetone (1-methylethylidene)hydrazone	$C_6H_{12}N_2$	15.6	2		
Acetonitrile	C_2H_3N	4.40	2	molar refraction; see Ref.2	
Acetonitrile	C_2H_3N	4.48	18		
Acetophenone	C_8H_8O	15.0	2		
Acetyl chloride	C_2H_3ClO	6.62	2		
Acetylene	C_2H_2	3.33	3		
Acetylene	C_2H_2	3.93	2		
Acrolein	C_3H_4O	6.38	2	molar refraction; see Ref.2	
Acrylonitrile	C_3H_3N	8.05	2		
Allyl alcohol	C_3H_6O	7.65	2		
Aluminum 2,4-pentanedioate	$C_{15}H_{21}AlO_6$	51.9	2		
Ammonia	H_3N	2.81	20		
Ammonia	H_3N	2.10	2		
Ammonia	H_3N	2.26	3		
Ammonia	H_3N	2.22	33	calculated	
Ammonia-*d*3	D_3N	1.70	2		
Aniline	C_6H_7N	12.1	2	molar refraction; see Ref.2	
Anisole	C_7H_8O	13.1	2	molar refraction; see Ref.2	
Anthracene	$C_{14}H_{10}$	25.4	17		
Anthracene	$C_{14}H_{10}$	25.93	27		
9,10-Anthracenedione	$C_{14}H_8O_2$	24.46	27		
Arsenic(III) chloride	$AsCl_3$	14.9	2		
Arsenic trinitride	AsN_3	5.75	2		
Benz[*a*]anthracene	$C_{18}H_{12}$	32.86	27		
Benzene	C_6H_6	10.0	25		

Name	Mol. form.	$\alpha/10^{-24}$ cm³	Ref.	Note	Z
Benzene	C_6H_6	10.32	3		
Benzene	C_6H_6	10.74	2		
1,4-Benzenediamine	$C_6H_8N_2$	13.8	2	molar refraction; see Ref.2	
11*H*-Benzo[*b*]fluorene	$C_{17}H_{12}$	30.21	27		
Benzonitrile	C_7H_5N	12.5	2	molar refraction; see Ref.2	
p-Benzoquinone	$C_6H_4O_2$	14.5	2		
Beryllium hydride	BeH_2	4.34	14	calculated	
Beryllium 2,4-pentanedioate	$C_{10}H_{14}BeO_4$	34.1	2		
Bis(4-bromophenyl) ether	$C_{12}H_8Br_2O$	27.8	2	molar refraction; see Ref.2	
Bis(4-methylphenyl) ether	$C_{14}H_{14}O$	24.9	2	molar refraction; see Ref.2	
Borane(1)	BH	3.32	1	calculated	
Borazine	$B_3H_6N_3$	8.0	2	molar refraction; see Ref.2	
Boron trichloride	BCl_3	9.38	20		
Boron trifluoride	BF_3	3.31	2		
Bromine	Br_2	7.02	2		
Bromoacetylene	C_2HBr	7.39	2		
9-Bromoanthracene	$C_{14}H_9Br$	28.32	27		
Bromobenzene	C_6H_5Br	14.7	2		
Bromobenzene	C_6H_5Br	13.62	27		
1-Bromobutane	C_4H_9Br	13.9	2		
1-Bromobutane	C_4H_9Br	10.86	27		
1-Bromo-2-chloroethane	C_2H_4BrCl	9.5	2	molar refraction; see Ref.2	
1-Bromodecane	$C_{10}H_{21}Br$	21.60	27		
Bromodifluoromethane	$CHBrF_2$	5.7	2	molar refraction; see Ref.2	
1-Bromododecane	$C_{12}H_{25}Br$	25.18	27		
Bromoethane	C_2H_5Br	8.05	2		
Bromoethane	C_2H_5Br	7.28	27		
Bromoethene	C_2H_3Br	7.59	2		
1-Bromo-4-fluorobenzene	C_6H_4BrF	13.4	2	molar refraction; see Ref.2	
1-Bromoheptane	$C_7H_{15}Br$	16.8	2	molar refraction; see Ref.2	
1-Bromoheptane	$C_7H_{15}Br$	16.23	27		
1-Bromohexadecane	$C_{16}H_{33}Br$	32.34	27		
1-Bromohexane	$C_6H_{13}Br$	14.44	27		
Bromomethane	CH_3Br	5.87	20		
Bromomethane	CH_3Br	6.03	2		
Bromomethane	CH_3Br	5.55	15		
1-Bromo-4-(4-methylphenoxy)benzene	$C_{13}H_{11}BrO$	26.6	2	molar refraction; see Ref.2	
1-Bromonaphthalene	$C_{10}H_7Br$	20.34	27		
1-Bromononane	$C_9H_{19}Br$	19.81	27		
1-Bromooctadecane	$C_{18}H_{37}Br$	35.92	27		
1-Bromooctane	$C_8H_{17}Br$	18.02	27		
1-Bromopentane	$C_5H_{11}Br$	13.1	2	molar refraction; see Ref.2	
1-Bromo-4-phenoxybenzene	$C_{12}H_9BrO$	24.2	2	molar refraction; see Ref.2	
1-Bromopropane	C_3H_7Br	9.4	2	molar refraction; see Ref.2	
1-Bromopropane	C_3H_7Br	9.07	27		
2-Bromopropane	C_3H_7Br	9.6	2	molar refraction; see Ref.2	
4-Bromotoluene	C_7H_7Br	14.80	27		
1,3-Butadiene	C_4H_6	8.64	2		
Butanal	C_4H_8O	8.2	2	molar refraction; see Ref.2	
Butane	C_4H_{10}	8.20	2		
2,3-Butanedione	$C_4H_6O_2$	8.2	2	molar refraction; see Ref.2	
Butanenitrile	C_4H_7N	8.4	2	molar refraction; see Ref.2	
Butanoic acid	$C_4H_8O_2$	8.58	27		
1-Butanol	$C_4H_{10}O$	8.88	2		
2-Butanone	C_4H_8O	8.13	15		
trans-2-Butenal	C_4H_6O	8.5	2	molar refraction; see Ref.2	
1-Butene	C_4H_8	7.97	2		
1-Butene	C_4H_8	8.52	2		
trans-2-Butene	C_4H_8	8.49	2		
trans-2-Butenedinitrile	$C_4H_2N_2$	11.8	2		
trans-2-Butenenitrile	C_4H_5N	8.2	2	molar refraction; see Ref.2	

Name	Mol. form.	$\alpha/10^{-24}$ cm³	Ref.	Note	Z
Butylamine	$C_4H_{11}N$	13.5	2		
tert-Butylbenzene	$C_{10}H_{14}$	17.2	25		
tert-Butylbenzene	$C_{10}H_{14}$	17.8	2	molar refraction; see Ref.2	
tert-Butylcyclohexane	$C_{10}H_{20}$	19.8	2		
1-Butyne	C_4H_6	7.41	2	molar refraction; see Ref.2	
Carbon dioxide	CO_2	2.911	8		
Carbon disulfide	CS_2	8.74	3		
Carbon disulfide	CS_2	8.86	2		
Carbon [fullerene-C_{60}]	C_{60}	88.9	36		
Carbon [fullerene-C_{60}]	C_{60}	79	31		
Carbon [fullerene-C_{60}]	C_{60}	76.5	24		
Carbon [fullerene-C_{70}]	C_{70}	108.5	36		
Carbon [fullerene-C_{70}]	C_{70}	102	24		
Carbon monoxide	CO	1.95	3		
Carbonothioic dichloride	CCl_2S	10.2	2		
Carbon oxysulfide	COS	5.71	2		
Carbon oxysulfide	COS	5.2	15		
Carbonyl chloride	CCl_2O	7.29	2		
Carbonyl fluoride	CF_2O	1.88	17	calculated	
Cesium bromide dimer	Br_2Cs_2	54.5	16		
Cesium bromide cluster			38	n=3-32	
Cesium chloride dimer	Cl_2Cs_2	42.4	16		
Cesium fluoride dimer	Cs_2F_2	28.4	16		
Cesium iodide dimer	Cs_2I_2	51.8	16		
Cesium potassium	CsK	89	22		
Chlorine	Cl_2	4.61	3		
Chlorine trifluoride	ClF_3	6.32	2		
Chloroacetone	C_3H_5ClO	8.4	2	molar refraction; see Ref.2	
Chloroacetonitrile	C_2H_2ClN	6.10	18		
Chloroacetyl chloride	$C_2H_2Cl_2O$	8.92	2		
Chloroacetylene	C_2HCl	6.07	2		
9-Chloroanthracene	$C_{14}H_9Cl$	27.35	27		
Chlorobenzene	C_6H_5Cl	14.1	2		
Chlorobenzene	C_6H_5Cl	12.3	15		
4-Chloro-1,2-butadiene	C_4H_5Cl	10.0	2	molar refraction; see Ref.2	
1-Chlorobutane	C_4H_9Cl	11.3	2		
2-Chlorobutane	C_4H_9Cl	12.4	2		
2-Chlorobutanoic acid	$C_4H_7ClO_2$	10.87	27		
3-Chlorobutanoic acid	$C_4H_7ClO_2$	10.80	27		
4-Chlorobutanoic acid	$C_4H_7ClO_2$	10.69	27		
2-Chloro-1-butanol	C_4H_9ClO	10.70	27		
3-Chloro-1-butanol	C_4H_9ClO	10.38	27		
1-Chloro-1,1-difluoroethane	$C_2H_3ClF_2$	8.05	2		
Chlorodifluoromethane	$CHClF_2$	6.38	20		
Chlorodifluoromethane	$CHClF_2$	5.91	2		
Chloroethane	C_2H_5Cl	7.27	20		
Chloroethane	C_2H_5Cl	8.29	2		
Chloroethane	C_2H_5Cl	6.4	15		
2-Chloroethanol	C_2H_5ClO	7.1	2	molar refraction; see Ref.2	
2-Chloroethanol	C_2H_5ClO	6.88	27		
Chloroethene	C_2H_3Cl	6.41	2		
2-Chloroethyl ethyl ether	C_4H_9ClO	10.56	27		
1-Chloro-2-fluoroethane	C_2H_4ClF	6.5	2	molar refraction; see Ref.2	
Chlorogermane	$ClGeH_3$	6.7	2	molar refraction; see Ref.2	
Chloromethane	CH_3Cl	5.35	20		
Chloromethane	CH_3Cl	4.72	8		
1-Chloro-2-methoxyethane	C_3H_7ClO	8.71	27		
Chloromethyl methyl ether	C_2H_5ClO	7.1	2	molar refraction; see Ref.2	
1-Chloro-2-methylpropane	C_4H_9Cl	11.1	2		
2-Chloro-2-methylpropane	C_4H_9Cl	12.5	2	molar refraction; see Ref.2	
1-Chloro-2-methylpropene	C_4H_7Cl	10.8	2		

Name	Mol. form.	$\alpha/10^{-24}$ cm³	Ref.	Note	Z
1-Chloronaphthalene	$C_{10}H_7Cl$	19.30	27		
2-Chloronaphthalene	$C_{10}H_7Cl$	19.58	27		
1-Chloro-2-nitrobenzene	$C_6H_4ClNO_2$	14.6	2	molar refraction; see Ref.2	
1-Chloro-3-nitrobenzene	$C_6H_4ClNO_2$	14.6	2	molar refraction; see Ref.2	
1-Chloro-4-nitrobenzene	$C_6H_4ClNO_2$	14.6	2	molar refraction; see Ref.2	
1-Chloro-1-nitroethane	$C_2H_4ClNO_2$	10.9	2		
Chloronitromethane	CH_2ClNO_2	6.9	2	molar refraction; see Ref.2	
1-Chloro-1-nitropropane	$C_3H_6ClNO_2$	10.4	2	molar refraction; see Ref.2	
Chloropentafluoroethane	C_2ClF_5	6.3	2	molar refraction; see Ref.2	
1-Chloropentane	$C_5H_{11}Cl$	12.0	2	molar refraction; see Ref.2	
2-Chloropentanoic acid, (±)-	$C_5H_9ClO_2$	12.69	27		
3-Chloropentanoic acid	$C_5H_9ClO_2$	12.57	27		
4-Chloropentanoic acid	$C_5H_9ClO_2$	12.53	27		
2-Chlorophenol	C_6H_5ClO	13.0	2	molar refraction; see Ref.2	
4-Chlorophenol	C_6H_5ClO	13.0	2	molar refraction; see Ref.2	
1-Chloropropane	C_3H_7Cl	10.0	2		
2-Chloro-1-propanol	C_3H_7ClO	8.89	27		
3-Chloro-1-propanol	C_3H_7ClO	8.84	27		
cis-1-Chloropropene	C_3H_5Cl	8.3	2		
trans-1-Chloropropene	C_3H_5Cl	8.3	2		
2-Chloropropene	C_3H_5Cl	8.3	2		
3-Chloropropene	C_3H_5Cl	8.3	2		
Chlorosilane	ClH_3Si	7.02	2		
4-Chlorotoluene	C_7H_7Cl	13.70	27		
Chlorotrifluoromethane	$CClF_3$	5.72	20		
Chlorotrifluoromethane	$CClF_3$	5.59	8		
Chromium(III) 2,4-pentanedioate	$C_{15}H_{21}CrO_6$	53.7	2		
Chrysene	$C_{18}H_{12}$	33.06	27		
Coronene	$C_{24}H_{12}$	42.50	27		
Cyanogen	C_2N_2	7.99	2		
Cyclohexane	C_6H_{12}	11.0	18		
Cyclohexane	C_6H_{12}	10.87	15		
Cyclohexanol	$C_6H_{12}O$	11.56	18		
Cyclohexene	C_6H_{10}	10.7	2	molar refraction; see Ref.2	
1,3-Cyclopentadiene	C_5H_6	8.64	2		
Cyclopentane	C_5H_{10}	9.15	18		
Cyclopropane	C_3H_6	5.66	2		
Decane	$C_{10}H_{22}$	19.10	27		
Dialuminum	Al_2	19	23		
Diboron hexanitride	B_2N_6	5.73	2		
Dibromodifluoromethane	CBr_2F_2	9.0	2	molar refraction; see Ref.2	
1,2-Dibromoethane	$C_2H_4Br_2$	10.7	2	molar refraction; see Ref.2	
Dibromomethane	CH_2Br_2	9.32	2		
Dibromomethane	CH_2Br_2	8.68	27		
1,4-Di-*tert*-butylbenzene	$C_{14}H_{22}$	24.5	25		
Dibutyl ether	$C_8H_{18}O$	17.2	2		
Dicesium	Cs_2	104	22		
o-Dichlorobenzene	$C_6H_4Cl_2$	14.17	27		
m-Dichlorobenzene	$C_6H_4Cl_2$	14.23	27		
p-Dichlorobenzene	$C_6H_4Cl_2$	14.20	27		
1,4-Dichlorobutane	$C_4H_8Cl_2$	12.0	2	molar refraction; see Ref.2	
2,5-Dichloro-2,5-cyclohexadiene-1,4-dione	$C_6H_2Cl_2O_2$	18.4	2		
1,1-Dichloro-2,2-difluoroethane	$C_2H_2Cl_2F_2$	8.4	2	molar refraction; see Ref.2	
Dichlorodifluoromethane	CCl_2F_2	7.93	20		
Dichlorodifluoromethane	CCl_2F_2	7.81	2		
1,1-Dichloroethane	$C_2H_4Cl_2$	8.64	2		
1,2-Dichloroethane	$C_2H_4Cl_2$	8.0	2	molar refraction; see Ref.2	
1,1-Dichloroethene	$C_2H_2Cl_2$	7.83	27		
cis-1,2-Dichloroethene	$C_2H_2Cl_2$	8.03	27		
trans-1,2-Dichloroethene	$C_2H_2Cl_2$	8.15	27		
Dichlorofluoromethane	$CHCl_2F$	6.82	2		

Name	Mol. form.	$\alpha/10^{-24}$ cm^3	Ref.	Note	Z
Dichloromethane	CH_2Cl_2	6.48	3		
Dichloromethane	CH_2Cl_2	7.93	2		
1,2-Dichloropropane, (±)-	$C_3H_6Cl_2$	10.9	2	molar refraction; see Ref.2	
1,3-Dichloropropene (unspecified isomer)	$C_3H_4Cl_2$	10.1	2	molar refraction; see Ref.2	
Dichlorosilane	Cl_2H_2Si	8.92	2		
1,2-Dichloro-1,1,2,2-tetrafluoroethane	$C_2Cl_2F_4$	8.5	2	molar refraction; see Ref.2	
p-Dicyanobenzene	$C_8H_4N_2$	19.2	2		
1,1-Diethoxyethane	$C_6H_{14}O_2$	13.2	2	molar refraction; see Ref.2	
1,2-Diethoxyethane	$C_6H_{14}O_2$	11.3	2	molar refraction; see Ref.2	
Diethylamine	$C_4H_{11}N$	10.2	2		
Diethylamine	$C_4H_{11}N$	9.61	27		
Diethyl carbonate	$C_5H_{10}O_3$	11.3	2		
Diethyl ether	$C_4H_{10}O$	10.2	2		
Diethyl ether	$C_4H_{10}O$	8.73	15		
Diethyl succinate	$C_8H_{14}O_4$	16.8	2	molar refraction; see Ref.2	
Diethyl sulfide	$C_4H_{10}S$	10.8	2		
o-Difluorobenzene	$C_6H_4F_2$	9.80	27		
m-Difluorobenzene	$C_6H_4F_2$	10.3	2	molar refraction; see Ref.2	
p-Difluorobenzene	$C_6H_4F_2$	9.80	27		
1,1-Difluoroethene	$C_2H_2F_2$	5.01	20		
Diiodomethane	CH_2I_2	12.90	27		
Diketene	$C_4H_4O_2$	8.0	2	molar refraction; see Ref.2	
Dilithium	Li_2	32.8	29		
Dilithium	Li_2	34	22		
Dilithium disodium	Li_2Na_2	60.0	30		
Dilithium sodium	Li_2Na	35.4	30		
Dimethoxymethane	$C_3H_8O_2$	7.7	2	molar refraction; see Ref.2	
Dimethylamine	C_2H_7N	6.37	2		
Dimethylamine	C_2H_7N	5.90	33	calculated	
N,N-Dimethylaniline	$C_8H_{11}N$	16.2	2	molar refraction; see Ref.2	
2,3-Dimethyl-1,3-butadiene	C_6H_{10}	11.8	2	molar refraction; see Ref.2	
Dimethyl carbonate	$C_3H_6O_3$	7.7	2	molar refraction; see Ref.2	
2,5-Dimethyl-2,5-cyclohexadiene-1,4-dione	$C_8H_8O_2$	18.8	2		
Dimethyl ether	C_2H_6O	5.29	20		
Dimethyl ether	C_2H_6O	5.84	2		
Dimethyl ether	C_2H_6O	5.16	15		
N,N-Dimethylformamide	C_3H_7NO	7.81	18		
trans-2,3-Dimethyloxirane	C_4H_8O	8.22	17	calculated	
2,4-Dimethyl-3-pentanone	$C_7H_{14}O$	13.5	15		
2,2-Dimethylpropanenitrile	C_5H_9N	9.59	18		
(1,2-Dimethyl-1-propenyl)benzene	$C_{11}H_{14}$	19.64	27		
1,5-Dimethyl-1*H*-pyrazole	$C_5H_8N_2$	10.72	27		
2,3-Dimethylquinoxaline	$C_{10}H_{10}N_2$	18.70	27		
Dimethyl sulfone	$C_2H_6O_2S$	7.3	2	molar refraction; see Ref.2	
1,4-Dinitrobenzene	$C_6H_4N_2O_4$	18.4	2		
1,4-Dioxane	$C_4H_8O_2$	10.0	2		
1,4-Dioxane	$C_4H_8O_2$	8.60	18		
Dipotassium	K_2	77	22		
Dipotassium	K_2	72	21		
Dipropylamine	$C_6H_{15}N$	13.29	27		
Dipropyl ether	$C_6H_{14}O$	12.8	2		
Dipropyl ether	$C_6H_{14}O$	12.5	15		
Dirubidium	Rb_2	79	22		
Disilane	H_6Si_2	11.1	2		
Disodium	Na_2	40	22		
Disodium	Na_2	38	21		
Divinyl sulfide	C_4H_6S	10.9	2	molar refraction; see Ref.2	
Dodecane	$C_{12}H_{26}$	22.75	27		
Erbium(III) tris(cyclopentadienyl)	$C_{15}H_{15}Er$	28.44	37		
Ethane	C_2H_6	4.47	3		
Ethane	C_2H_6	4.43	2		

Atomic

Name	Mol. form.	$\alpha/10^{-24}$ cm³	Ref.	Note	Z
1,2-Ethanediamine	$C_2H_8N_2$	7.2	2	molar refraction; see Ref.2	
1,2-Ethanediol	$C_2H_6O_2$	5.7	2	molar refraction; see Ref.2	
1,2-Ethanediol	$C_2H_6O_2$	5.61	27		
Ethanethiol	C_2H_6S	7.41	2		
Ethanol	C_2H_6O	5.41	2		
Ethanol	C_2H_6O	5.11	18		
Ethoxybenzene	$C_8H_{10}O$	14.9	2		
2-Ethoxyethanol	$C_4H_{10}O_2$	9.28	27		
Ethyl acetate	$C_4H_8O_2$	9.7	2		
Ethyl acetate	$C_4H_8O_2$	8.62	27		
Ethyl acetoacetate	$C_6H_{10}O_3$	12.9	2	molar refraction; see Ref.2	
Ethylamine	C_2H_7N	7.10	2		
Ethylbenzene	C_8H_{10}	14.2	2		
Ethyl benzoate	$C_9H_{10}O_2$	16.9	2	molar refraction; see Ref.2	
Ethyl 2-chlorobutanoate	$C_6H_{11}ClO_2$	14.16	27		
Ethyl 3-chlorobutanoate	$C_6H_{11}ClO_2$	14.13	27		
Ethyl 4-chlorobutanoate	$C_6H_{11}ClO_2$	14.11	27		
Ethyl chloroformate	$C_3H_5ClO_2$	9.0	2	molar refraction; see Ref.2	
Ethylcyclohexane	C_8H_{16}	15.9	2		
Ethylene	C_2H_4	4.252	8		
Ethyl formate	$C_3H_6O_2$	8.01	2		
Ethyl formate	$C_3H_6O_2$	6.88	27		
Ethyl *trans,trans*-2,4-hexadienoate	$C_8H_{12}O_2$	17.2	2	molar refraction; see Ref.2	
Ethyl methyl ether	C_3H_8O	7.93	2		
1-Ethyl-5-methyl-1*H*-pyrazole	$C_6H_{10}N_2$	12.50	27		
1-Ethylnaphthalene	$C_{12}H_{12}$	21.19	27		
2-Ethylnaphthalene	$C_{12}H_{12}$	21.36	27		
Ethyl nitrite	$C_2H_5NO_2$	7.0	15		
1-Ethyl-1-phenylhydrazine	$C_8H_{12}N_2$	16.62	27		
Ethyl propanoate	$C_5H_{10}O_2$	10.41	27		
Ethyl propyl ether	$C_5H_{12}O$	10.68	27		
Ferrocene	$C_{10}H_{10}Fe$	17.1	26		
9*H*-Fluorene	$C_{13}H_{10}$	21.68	27		
Fluorine	F_2	1.38	7	calculated	
Fluoroanthracene (unspecified isomer)	$C_{14}H_9F$	28.34	27		
Fluorobenzene	C_6H_5F	10.3	2		
1-Fluorodecane	$C_{10}H_{21}F$	19.18	27		
1-Fluorododecane	$C_{12}H_{25}F$	22.83	27		
Fluoroethane	C_2H_5F	4.96	2		
1-Fluoroheptane	$C_7H_{15}F$	13.66	27		
2-Fluorohexane	$C_6H_{13}F$	11.80	27		
1-Fluoro-4-iodobenzene	C_6H_4FI	15.5	2	molar refraction; see Ref.2	
Fluoromethane	CH_3F	2.97	8		
1-Fluoro-4-nitrobenzene	$C_6H_4FNO_2$	12.8	2	molar refraction; see Ref.2	
1-Fluorononane	$C_9H_{19}F$	17.34	27		
1-Fluorooctane	$C_8H_{17}F$	15.46	27		
1-Fluoropentane	$C_5H_{11}F$	9.95	27		
1-Fluorotetradecane	$C_{14}H_{29}F$	26.57	27		
4-Fluorotoluene	C_7H_7F	11.70	27		
1-Fluoroundecane	$C_{11}H_{23}F$	21.00	27		
Formaldehyde	CH_2O	2.8	2	molar refraction; see Ref.2	
Formaldehyde	CH_2O	2.45	18		
Formamide	CH_3NO	4.2	2	molar refraction; see Ref.2	
Formamide	CH_3NO	4.08	18		
Formic acid	CH_2O_2	3.4	2	molar refraction; see Ref.2	
Formic acid dimer	$C_2H_4O_4$	12.7	2		
Formyl fluoride	CHFO	1.76	17	calculated	
Gallium arsenide (variable composition)	GanAsm		28	$n+m$=4-30	
Germanium cluster (Ge_9)		90	35		
Germanium cluster (Ge_{10})		128	35		
Germanium cluster (Ge_{15})		167	35		

Name	Mol. form.	$\alpha/10^{-24}$ cm^3	Ref.	Note	Z
Germanium(IV) chloride	Cl_4Ge	15.1	2		
Heptane	C_7H_{16}	13.61	2		
1-Heptene	C_7H_{14}	13.51	27		
1-Heptyne	C_7H_{12}	12.8	2	molar refraction; see Ref.2	
Hexafluorobenzene	C_6F_6	9.58	27		
Hexafluoroethane	C_2F_6	6.82	2		
Hexamethylbenzene	$C_{12}H_{18}$	20.9	25		
Hexane	C_6H_{14}	11.9	2		
1-Hexene	C_6H_{12}	11.65	27		
1-Hexyne	C_6H_{10}	10.9	2	molar refraction; see Ref.2	
Hydrogen	H_2	0.8023	5	calc v=0,J=0	
Hydrogen	H_2	0.8045	5	calc 293 K	
Hydrogen	H_2	0.8042	6	293 K	
Hydrogen	H_2	0.8059	8	322 K	
Hydrogen-d_2	D_2	0.7921	5	calc v=0,J=0	
Hydrogen-d_2	D_2	0.7954	6	293 K	
Hydrogen-d_1	DH	0.7976	5	calc v=0,J=0	
Hydrogen bromide	BrH	3.61	3		
Hydrogen chloride	ClH	2.63	3		
Hydrogen chloride	ClH	2.77	2		
Hydrogen chloride-d	ClD	2.84	2		
Hydrogen cyanide	CHN	2.59	3		
Hydrogen cyanide	CHN	2.46	2		
Hydrogen fluoride	FH	0.80	27		
Hydrogen iodide	HI	5.44	3		
Hydrogen iodide	HI	5.35	2		
Hydrogen sulfide	H_2S	3.78	3		
Hydrogen sulfide	H_2S	3.95	2		
Iodine chloride	ClI	12.3	2		
Iodobenzene	C_6H_5I	15.5	2	molar refraction; see Ref.2	
1-Iodobutane	C_4H_9I	13.3	2	molar refraction; see Ref.2	
1-Iodobutane	C_4H_9I	12.65	27		
Iodoethane	C_2H_5I	10.0	2		
Iodoethene	C_2H_3I	9.3	2	molar refraction; see Ref.2	
Iodomethane	CH_3I	7.97	2		
1-Iodonaphthalene	$C_{10}H_7I$	22.41	27		
2-Iodonaphthalene	$C_{10}H_7I$	22.95	27		
1-Iodopropane	C_3H_7I	11.5	2	molar refraction; see Ref.2	
4-Iodotoluene	C_7H_7I	17.10	27		
Iron(III) 2,4-pentanedioate	$C_{15}H_{21}FeO_6$	58.1	2		
Isobutane	C_4H_{10}	8.14	27		
Isobutene	C_4H_8	8.29	2		
Isopropenylbenzene	C_9H_{10}	16.05	27		
Isopropylamine	C_3H_9N	7.77	27		
Isopropylbenzene	C_9H_{12}	16.0	2	molar refraction; see Ref.2	
Isopropylcyclohexane	C_9H_{18}	17.2	2		
Isoquinoline	C_9H_7N	16.43	27		
Ketene	C_2H_2O	4.4	2	molar refraction; see Ref.2	
Lithium bromide dimer	Br_2Li_2	18.9	16		
Lithium chloride	ClLi	3.46	10	calculated	
Lithium chloride dimer	Cl_2Li_2	13.1	16		
Lithium cluster	Lin		29	n=2-22	
Lithium disodium	$LiNa_2$	61.2	30		
Lithium fluoride	FLi	10.8	11	calculated	
Lithium fluoride dimer	F_2Li_2	6.9	16		
Lithium hydride	HLi	3.84	12	calculated	
Lithium hydride	HLi	3.68	13	calculated	
Lithium hydride	HLi	3.88	14	calculated	
Lithium iodide dimer	I_2Li_2	23.4	16		
Lithium sodium	LiNa	40	4		
Lithium trisodium	$LiNa_3$	75.6	30		

Name	Mol. form.	$\alpha/10^{-24}$ cm^3	Ref.	Note	Z
Malononitrile	$C_3H_2N_2$	5.79	18		
Mercury(II) bromide	Br_2Hg	14.5	2		
Mercury chloride	$ClHg$	7.4	9	calculated	
Mercury(I) chloride	Cl_2Hg_2	14.7	9		
Mercury(II) chloride	Cl_2Hg	11.6	2		
Mercury(II) iodide (red)	HgI_2	19.1	2		
Methane	CH_4	2.593	8		
Methanol	CH_4O	3.29	2		
Methanol	CH_4O	3.23	15		
Methanol	CH_4O	3.32	18		
2-Methoxyaniline	C_7H_9NO	14.2	2	molar refraction; see Ref.2	
Methoxycyclohexane	$C_7H_{14}O$	13.4	2	molar refraction; see Ref.2	
2-Methoxyethanol	$C_3H_8O_2$	7.44	27		
N-Methylacetamide	C_3H_7NO	7.82	18		
Methyl acetate	$C_3H_6O_2$	6.94	2		
Methyl acetate	$C_3H_6O_2$	6.81	27		
2-Methylacrylonitrile	C_4H_5N	8.0	2	molar refraction; see Ref.2	
Methylamine	CH_5N	4.7	2		
Methylamine	CH_5N	4.01	19		
Methylamine	CH_5N	4.01	33	calculated	
2-Methyl-1,3-butadiene	C_5H_8	9.99	2		
Methyl butanoate	$C_5H_{10}O_2$	10.41	27		
Methyl 2-chlorobutanoate	$C_5H_9ClO_2$	12.33	27		
Methyl 3-chlorobutanoate	$C_5H_9ClO_2$	12.31	27		
Methyl 4-chlorobutanoate	$C_5H_9ClO_2$	12.27	27		
Methyl chloroformate	$C_2H_3ClO_2$	7.1	2	molar refraction; see Ref.2	
Methylcyclohexane	C_7H_{14}	13.1	2		
2-Methyl-1,3-dioxolane	$C_4H_8O_2$	9.44	15		
3-Methylene-1-pentene	C_6H_{10}	11.8	2	molar refraction; see Ref.2	
N-Methylformamide	C_2H_5NO	5.91	18		
Methyl formate	$C_2H_4O_2$	5.05	27		
3-Methylheptane	C_8H_{18}	15.44	27		
1-Methylisoquinoline	$C_{10}H_9N$	18.28	27		
1-Methylnaphthalene	$C_{11}H_{10}$	19.35	27		
2-Methylnaphthalene	$C_{11}H_{10}$	19.52	27		
2-Methyl-2-nitropropane	$C_4H_9NO_2$	10.3	2	molar refraction; see Ref.2	
2-Methyl-1,3-pentadiene	C_6H_{10}	12.1	2	molar refraction; see Ref.2	
3-Methyl-1,3-pentadiene	C_6H_{10}	11.8	2	molar refraction; see Ref.2	
1-Methyl-1-phenylhydrazine	$C_7H_{10}N_2$	14.81	27		
2-Methylpropanenitrile	C_4H_7N	8.05	18		
2-Methylpropanenitrile	C_4H_7N	8.05	32	calculated	
Methyl propanoate	$C_4H_8O_2$	8.97	27		
2-Methyl-1-propanol	$C_4H_{10}O$	8.92	2		
2-Methylpropenal	C_4H_6O	8.3	2	molar refraction; see Ref.2	
Methyl propyl ether	$C_4H_{10}O$	8.86	27		
1-Methyl-1*H*-pyrazole	$C_4H_6N_2$	8.99	27		
2-Methylquinoline	$C_{10}H_9N$	18.65	27		
Naphthacene	$C_{18}H_{12}$	32.27	27		
Naphthalene	$C_{10}H_8$	16.5	17		
Naphthalene	$C_{10}H_8$	17.48	27		
1-Naphthalenecarboxaldehyde	$C_{11}H_8O$	19.75	27		
2-Naphthalenecarboxaldehyde	$C_{11}H_8O$	20.06	27		
1-Naphthylamine	$C_{10}H_9N$	19.50	27		
2-Naphthylamine	$C_{10}H_9N$	19.73	27		
Neodymium(III) tris(cyclopentadienyl)	$C_{15}H_{15}Nd$	33.05	37		
Neopentane	C_5H_{12}	10.20	18		
Nitric oxide	NO	1.70	2		
2-Nitroanisole	$C_7H_7NO_3$	15.7	2	molar refraction; see Ref.2	
3-Nitroanisole	$C_7H_7NO_3$	15.7	2	molar refraction; see Ref.2	
4-Nitroanisole	$C_7H_7NO_3$	15.7	2	molar refraction; see Ref.2	
Nitrobenzene	$C_6H_5NO_2$	14.7	2		

Name	Mol. form.	$\alpha/10^{-24}$ cm^3	Ref.	Note	Z
Nitrobenzene	$C_6H_5NO_2$	12.92	15		
4-Nitrobenzonitrile	$C_7H_4N_2O_2$	19.0	2		
1-Nitrobutane	$C_4H_9NO_2$	10.4	2	molar refraction; see Ref.2	
Nitroethane	$C_2H_5NO_2$	9.63	2		
Nitrogen	N_2	1.7403	6,8		
Nitrogen dioxide	NO_2	3.02	2	molar refraction; see Ref.2	
Nitrogen tetroxide	N_2O_4	6.69	2		
Nitrogen trifluoride	F_3N	3.62	2		
Nitromethane	CH_3NO_2	7.37	2		
1-Nitro-4-phenoxybenzene	$C_{12}H_9NO_3$	24.7	2	molar refraction; see Ref.2	
1-Nitropropane	$C_3H_7NO_2$	8.5	2	molar refraction; see Ref.2	
2-Nitropropane	$C_3H_7NO_2$	8.5	2	molar refraction; see Ref.2	
Nitrous oxide	N_2O	3.03	8		
Nonane	C_9H_{20}	17.36	27		
Octane	C_8H_{18}	15.9	2		
Osmium(VIII) oxide	O_4Os	8.17	2		
Oxirane	C_2H_4O	4.43	18		
Oxygen	O_2	1.5689	34		
Ozone	O_3	3.21	2		
Paraldehyde	$C_6H_{12}O_3$	17.9	2		
Pentachloroethane	C_2HCl_5	14.0	2		
trans-1,3-Pentadiene	C_5H_8	10	2		
2,4-Pentadienenitrile	C_5H_5N	10.5	2	molar refraction; see Ref.2	
Pentafluorobenzene	C_6HF_5	9.63	27		
Pentamethylbenzene	$C_{11}H_{16}$	19.1	25		
Pentane	C_5H_{12}	9.99	2		
2,4-Pentanedione	$C_5H_8O_2$	10.5	2	molar refraction; see Ref.2	
Pentanenitrile	C_5H_9N	10.4	2		
2-Pentanone	$C_5H_{10}O$	9.93	15		
3-Pentanone	$C_5H_{10}O$	9.93	15		
1-Pentene	C_5H_{10}	9.65	27		
2-Pentene (unspecified isomer)	C_5H_{10}	9.84	27		
Pentyl acetate	$C_7H_{14}O_2$	14.9	2		
Pentyl formate	$C_6H_{12}O_2$	14.2	2		
1-Pentyne	C_5H_8	9.12	2		
Perfluoronaphthalene	$C_{10}F_8$	17.64	27		
Phenanthrene	$C_{14}H_{10}$	36.8	17	calculated	
Phenanthrene	$C_{14}H_{10}$	24.70	27		
Phenazine	$C_{12}H_8N_2$	23.43	27		
Phenol	C_6H_6O	11.1	2	molar refraction; see Ref.2	
Phenol	C_6H_6O	9.94	17	calculated	
Phenylhydrazine	$C_6H_8N_2$	12.91	27		
Phosphine	H_3P	4.84	2		
Phosphorus(III) chloride	Cl_3P	12.8	2		
Phosphorus(V) fluoride	F_5P	6.10	2		
Potassium bromide dimer	Br_2K_2	42.0	16		
Potassium chloride dimer	Cl_2K_2	32.1	16		
Potassium cluster			21	n=2,5,7-9,11,	
Potassium fluoride dimer	F_2K_2	21.0	16		
Potassium iodide dimer	I_2K_2	36.3	16		
Potassium sodium	KNa	51	22		
Propanal	C_3H_6O	6.50	2		
Propane	C_3H_8	6.29	3		
Propane	C_3H_8	6.37	2		
Propanenitrile	C_3H_5N	6.70	2		
Propanenitrile	C_3H_5N	6.24	18		
Propanenitrile	C_3H_5N	6.27	32	calculated	
Propanoic acid	$C_3H_6O_2$	6.9	2	molar refraction; see Ref.2	
1-Propanol	C_3H_8O	6.74	2		
2-Propanol	C_3H_8O	7.61	2		
2-Propanol	C_3H_8O	6.97	18		

Atomic

Name	Mol. form.	$\alpha/10^{-24}$ cm³	Ref.	Note
Propene	C_3H_6	6.26	2	
Propylamine	C_3H_9N	7.70	27	
Propylamine	C_3H_9N	9.20	2	
Propyne	C_3H_4	6.18	2	
1*H*-Pyrazole	$C_3H_4N_2$	7.23	27	
Pyrene	$C_{16}H_{10}$	28.22	27	
Pyridazine	$C_4H_4N_2$	9.27	17	calculated
Pyridine	C_5H_5N	9.5	15	
Pyridine	C_5H_5N	9.18	27	
Pyrimidine	$C_4H_4N_2$	8.53	17	calculated
Quinoline	C_9H_7N	15.70	27	
Quinoxaline	$C_8H_6N_2$	15.13	27	
Rubidium bromide dimer	Br_2Rb_2	48.2	16	
Rubidium chloride dimer	Cl_2Rb_2	43.2	16	
Rubidium fluoride dimer	F_2Rb_2	40.7	16	
Rubidium iodide dimer	I_2Rb_2	46.3	16	
Samarium(III) tris(cyclopentadienyl)	$C_{15}H_{15}Sm$	32.01	37	
Selenium hexafluoride	F_6Se	7.33	2	
Silane	H_4Si	5.44	2	
Sodium bromide dimer	Br_2Na_2	26.8	16	
Sodium chloride dimer	Cl_2Na_2	23.4	16	
Sodium cluster			39	*n*=7-93
Sodium cluster			21	*n*=1-40
Sodium fluoride dimer	F_2Na_2	20.7	16	
Sodium iodide dimer	I_2Na_2	26.9	16	
Styrene	C_8H_8	15.0	2	
Styrene	C_8H_8	14.41	27	
Succinonitrile	$C_4H_4N_2$	8.1	2	molar refraction; see Ref.2
Sulfur decafluoride	$F_{10}S_2$	13.2	2	
Sulfur dioxide	O_2S	3.72	3	
Sulfur dioxide	O_2S	4.28	2	
Sulfur hexafluoride	F_6S	6.54	8	
Sulfur trioxide (α-form)	O_3S	4.84	2	
Sulfuryl chloride	Cl_2O_2S	10.5	2	
Tellurium hexafluoride	F_6Te	9.00	2	
1,1,2,2-Tetrachloroethane	$C_2H_2Cl_4$	12.1	2	molar refraction; see Ref.2
Tetrachloromethane	CCl_4	11.2	2	
Tetrachloromethane	CCl_4	10.5	3	
1,2,3,4-Tetrafluorobenzene	$C_6H_2F_4$	9.69	27	
1,2,4,5-Tetrafluorobenzene	$C_6H_2F_4$	9.69	27	
Tetrafluoromethane	CF_4	3.838	8	
Tetrafluorosilane	F_4Si	5.45	2	
Tetramethoxymethane	$C_5H_{12}O_4$	13.0	2	molar refraction; see Ref.2
1,2,4,5-Tetramethylbenzene	$C_{10}H_{14}$	17.3	25	
2,2,4,4-Tetramethyl-1,3-cyclobutanedione	$C_8H_{12}O_2$	18.6	2	
Tetranitromethane	CN_4O_8	15.3	2	
Tetraphosphorus	P_4	13.59	40	
Thiophene	C_4H_4S	9.67	2	
Thorium(IV) 2,4-pentanedioate	$C_{20}H_{28}O_8Th$	79.0	2	
Tin(IV) bromide	Br_4Sn	22.0	2	
Tin(IV) chloride	Cl_4Sn	18	2	
Tin(IV) chloride	Cl_4Sn	13.8	15	
Tin(IV) iodide	I_4Sn	32.3	2	
Titanium(IV) chloride	Cl_4Ti	16.4	2	
Toluene	C_7H_8	11.8	25	
Toluene	C_7H_8	12.26	15	
Toluene	C_7H_8	12.3	2	
Tribromomethane	$CHBr_3$	11.8	27	
1,3,5-Tri-*tert*-butylbenzene	$C_{18}H_{30}$	31.8	25	
Trichloroacetonitrile	C_2Cl_3N	10.42	18	
1,1,1-Trichloroethane	$C_2H_3Cl_3$	10.7	2	

Name	Mol. form.	$\alpha/10^{-24}$ cm^3	Ref.	Note	Z
Trichloroethene	C_2HCl_3	10.03	27		
1,1,2-Trichloro-2-fluoroethane	$C_2H_2Cl_3F$	10.2	2	molar refraction; see Ref.2	
Trichlorofluoromethane	CCl_3F	9.47	2		
Trichloromethane	$CHCl_3$	9.5	8		
Trichloromethane	$CHCl_3$	8.23	27		
Trichloronitromethane	CCl_3NO_2	10.8	2	molar refraction; see Ref.2	
Trichlorosilane	Cl_3HSi	10.7	2		
Triethylamine	$C_6H_{15}N$	13.1	2		
Triethylamine	$C_6H_{15}N$	13.38	27		
1,3,5-Trifluorobenzene	$C_6H_3F_3$	9.74	27		
1,1,1-Trifluoroethane	$C_2H_3F_3$	4.4	2	molar refraction; see Ref.2	
Trifluoromethane	CHF_3	3.52	20		
Trifluoromethane	CHF_3	3.57	8		
Triiodomethane	CHI_3	18.0	27		
Trilithium	Li_3	34.5	29		
Trilithium sodium	Li_3Na	54.8	30		
Trimethylamine	C_3H_9N	8.15	2		
Trimethylamine	C_3H_9N	7.78	33	calculated	
1,3,5-Trimethylbenzene	C_9H_{12}	15.5	25		
1,3,5-Trimethylbenzene	C_9H_{12}	16.14	27		
2,2,4-Trimethylpentane	C_8H_{18}	15.44	27		
Triphenylene	$C_{18}H_{12}$	31.07	27		
Tripropylamine	$C_9H_{21}N$	18.87	27		
Trisodium	Na_3	70	21		
Undecane	$C_{11}H_{24}$	21.03	27		
Uranium(VI) fluoride	F_6U	12.5	2		
Water	H_2O	1.45	2		
Water-*d*$_2$	D_2O	1.26	2		
o-Xylene	C_8H_{10}	14.9	2		
o-Xylene	C_8H_{10}	14.1	15		
m-Xylene	C_8H_{10}	14.2	15		
p-Xylene	C_8H_{10}	13.7	25		
p-Xylene	C_8H_{10}	14.2	15		
p-Xylene	C_8H_{10}	14.9	2		

References to Atoms

1. Miller, T. M., and Bederson, B., *Adv. At. Mol. Phys.* 13, 1, 1977. For simplicity, any value in Table 2 which has not changed since this 1977 review is referenced as 1. Persons interested in original references and further details should consult 1.
2. Alpher, R. A., and White, D. R., *Phys. Fluids* 2, 153, 1959.
3. Schmidt, J. W., Glavioso, R. M., May, E. F., and Moldover, M. R., *Phys. Rev. Lett.* 98, 254504, 2007.
4. Miffre, A., Jacquey, M., Büchner, M., Trénec, G., and Vigué, J., *Phys. Rev. A* 73, 011603(R), 2006.
5. Thierfelder, C., Assadollahzadeh, B., Schwerdtfeger, P., Schäfer, S., and Schäfer, R., *Phys. Rev. A* 78, 052506, 2008. Relativistic calculations for element 6 (C) through element 114 (Flerovium) and measurements for SN and Pb. The measurement uncertainties do not include that of the reference atom Ba. The polarizability of Ba has since been adjusted (see comment with reference 43).
6. Gaiser, C., and Fellmuth, B., *Eur. Phys. Lett.* 90, 63002, 2010.
7. Ekstrom, C. R., Schmiedmayer, J., Chapman, M. S., Hammond, T. D., and Pritchard, D. E., *Phys. Rev. A* 51, 3883, 1995. See theoretical work by Thakkar, A. J., and Lupinetti, C., *Chem. Phys. Lett.* 402, 270, 2005.
8. Holmgren, W. F., Revelle, M. C., Lonij, V. P. A., and Cronin, A. D., *Phys. Rev. A* 81, 053607, 2010.
9. Borschevsky, A., Pershina, V., Eliav, E., and Kaldor, U., *J. Chem. Phys.* 138, 124302, 2013; *Phys. Rev. A* 87, 022502, 2013.
10. Bromley, M. W. J., and Mitroy, J., *Phys. Rev. A* 65, 062505, 2002; 062506, 2002.
11. Milani, P., Moullet, I., and de Heer, W. A., *Phys. Rev. A* 42, 5150, 1990. See theoretical comments on this result, in Fuentealba, P., *Chem. Phys. Lett.* 397, 459, 2004, and in Lupinetti, C., and Thakkar, A. J., *J. Chem. Phys.* 122, 044301, 2005.
12. Newell, A. C., and Baird, R. D., *J. Appl. Phys.* 36, 3751, 1965.
13. Orcutt, R. H., and Cole, R. H., *J. Chem. Phys.* 46, 697, 1967; see also the later references from this group in Reference 8 following the tables for molecular polarizabilities.
14. Doolen, G. D., Los Alamos National Laboratory, unpublished. A relativistic linear response method was used. The method is that described by Zangwill, A., and Soven, P., *Phys. Rev. A* 21, 1561, 1980. Adjustments of less than 10% across the periodic table have been made to these results to bring them into agreement with accurate experimental values when available, for the purpose of presenting "recommended" polarizabilities in Table 2.
15. Goebel, D., Holm, U., and Maroulis, G., *Phys. Rev. A* 54, 1973, 1996.
16. Hu, M., and Kusse, B. R., *Phys. Rev. A* 66, 062506, 2002. Measured $11.4 \pm 1.2 \times 10^{-24}$ cm^3 at 532 nm and extrapolated to a static value.
17. Goebel, D., and Holm, U., *Phys. Rev. A* 52, 3691, 1995.
18. Guella, T. P., Miller, T. M., Bederson, B., Stockdale, J. A. D., and Jaduszliwer, B., *Phys. Rev. A* 29, 2977, 1984.
19. Atoji, M., *J. Chem. Phys.* 25, 174, 1956.
20. Amini, J. M., and Gould, H., *Phys. Rev. Lett.* 91, 153001, 2003.
21. Fleig, T. and Sadlej, A. J., *Phys. Rev. A* 65, 032506, 2002.
22. Goebel, D. and Holm, U., *J. Chem. Phys.* 100, 7710, 1996.
23. Preliminary value from the New York University group. See Guella, T. P., Miller, T. M., Bederson, B., Stockdale, J. A. D., and Jaduszliwer, B., *Phys. Rev. A* 29, 2977, 1984.

24. Borschevsky, A., Zelovich, T., Eliav, E., and Kaldor, U., *Chem. Phys.* 395, 104, 2012.
25. Chattopadhyay, S., Mani, B. K., and Angom, D., *Phys. Rev. A* 89, 022506, 2014.
26. Łach, G., Jezionski, B., and Szalewicz, K., *Phys. Rev. Lett.* 92, 233001, 2004.
27. Kadar-Kallen, M. A., and Bonin, K. D., *Phys. Rev. Lett.* 72, 828, 1994.
28. Pershina, V., Borschevsky, A., Eliav, E., and Kaldor, U., *J. Chem. Phys.* 128, 024707, 2008.
29. Dzuba, V. A., *Phys. Rev. A* 93, 032519,2016.
30. Dzuba, V. A., Koslov, A., and Flambaum, V. V., *Phys. Rev. A* 89, 042507, 2014.
31. Fleig, T., *Phys. Rev. A* 72, 052506, 2005.
32. Gregoire, M. D., Hromada, I., Holmgren, W. F., Trubko, R., and Cronin, A. D., *Phys. Rev. A* 92, 052513, 2015.
33. Lupinetti, C., and Thakkar, A. J., *J. Chem. Phys.* 122, 044301, 2005.
34. Miller, T. M., and Bederson, B., *Phys. Rev. A* 14, 1572, 1976. The polarizability ratio of α(Ca)/ α(Li) was measured to be 1.0316. Because α(Li) is now known more accurately, α(Ca) is reevaluated as 24.9 ± 8% x 10^{-24} cm^3.
35. Medvedď, M., Fowler, P. W., and Hutson, J. M., *Mol. Phys.* 98, 453, 2000.
36. Ma, L., Indergaard, J., Zhang, B., Larkin, I., Moro, R., and deHeer, W. A., *Phys. Rev. A* 91, 010501, 2015. Polarizabilities were normalized to that measured for Al (6.8 ± 0.3 x 10^{-24} cm^3). The uncertainties given in the table do not include that for Al. Note that all theoretical values tend to be 21% greater than the measured value for Al given above (see Reference 33).
37. Pershina, V., Borschevsky, A., Eliav, E., and Kaldor, U., *J. Chem. Phys.* 129, 144106, 2008.
38. Pershina, V., Borschevsky, A., Eliav, E., and Kaldor, U., *J. Phys. Chem. A* 112, 13712, 2008.
39. Pou-Amérigo, R., Merchán, M., Nebot-Gil, I., Widmark, P. -O., and Roos, B. O., *Theor. Chim Acta* (now *Theor. Chem Acc.*) 92, 149, 1995.
40. Roos, B. O., Lindh, R., Malmqvist, P. -Å., Veryazov, V., and Widmark, P. -O., *J. Phys. Chem. A* 108, 2851, 2004.
41. Roos, B. O., Lindh, R., Malmqvist, P. -Å., Veryazov, V., and Widmark, P. -O., *J. Phys. Chem. A* 109, 6575, 2005.
42. Sahoo, B. K.,, and Das, B. P., *Phys. Rev. A* 77, 062516, 2008.
43. Schwartz, H. L., Miller, T. M., and Bederson, B., *Phys. Rev. A* 10, 1924, 1974. The polarizability ratio α(Ba)/α(Li) was measured to be 1.6316 and that for α(Sr)/α(Li) = 1.1350. Because α(Li) is now known more accurately, α(Ba) is reevaluated as 39.3 ± 6% and α(Sr) = 39.3 ± 6%.
44. Singh, Y., and Sahoo, B. K., *Phys. Rev. A* 90, 022511, 2014.
45. Singh, Y., and Sahoo, B. K., *Phys. Rev. A* 91, 030501, 2015.

References to Molecules

46. McCullough, E. A., Jr., *J. Chem. Phys.*, 63, 5050, 1975. This calculation is for the parallel component, not the average polarizability.
47. Maryott, A. A., and Buckley, F., *U. S. National Bureau of Standards Circular No. 537*, 1953. A tabulation of dipole moments, dielectric constants, and molar refractions measured between 1910 and 1952, and used here to determine polarizabilities if no more recent result exists. The polarizability is $3/(4\pi N_A)$ times the molar polarization or molar refraction, where N_A is Avogadro's number. The value $3/(4\pi N_A) = 0.3964308 \times 10^{-24}$ cm^3 was used for this conversion. A dagger (†) following the reference number in the tables indicates that the polarizability was derived from the molar refraction and hence may not include some low-frequency contributions to the static polarizability; these "static" polarizabilities are therefore low by 1 to 30%.
48. Hirschfelder, J. O., Curtis, C. F., and Bird, R. B., *Molecular Theory of Gases and Liquids*, Wiley, New York, 1954, p. 950. Fundamental information on molecular polarizabilities.
49. Miller, T. M., and Bederson, B., *Adv. At. Mol. Phys.*, 13, 1, 1977. Review emphasizing atomic polarizabilities and measurement techniques. The data quoted in Table 3 are accurate to 8 to 12%.
50. Kolos, W., and Wolniewicz, L., *J. Chem. Phys.*, 46, 1426, 1967. Highly accurate molecular hydrogen calculations. See also recent work by Machado, A.M., and Masilli, M., *J. Chem. Phys.* 120, 7505, 2004.
51. Newell, A. C., and Baird, R. C., *J. Appl. Phys.*, 36, 3751, 1965. Highly accurate refractive index measurements at 47.7 GHz (essentially static).
52. Jao, T. C., Beebe, N. H. F., Person, W. B., and Sabin, J. R., *Chem. Phys. Lett.*, 26, 474, 1974. Tensor polarizabilities, derivatives, and other results are reported.
53. Orcutt, R. H., and Cole, R. H., *J. Chem. Phys.*, 46, 697, 1967 (He, Ne, Ar, Kr, H_2, N_2); Sutter, H., and Cole, R. H., *J. Chem. Phys.*, 52, 132, 1970 (CF_3H, CFH_3, $CClF_3$, $CClH_3$); Bose, T. K., and Cole, R. H., *J. Chem. Phys.*, 52, 140, 1970 (CO_2), and 54, 3829, 1971 (C_2H_4); Nelson, R. D., and Cole, R. H., *J. Chem. Phys.*, 54, 4033, 1971 (SF_6, $CClF_3$); Bose, T. K., Sochanski, J. S., and Cole, R. H., *J. Chem. Phys.*, 57, 3592, 1972 (CH_4, CF_4); Kirouac, S., and Bose, T. K., *J. Chem. Phys.*, 59, 3043, 1973 (N_2O), and 64, 1580, 1976 (He). Highly accurate dielectric constant measurements. These modern data give the most accurate polarizabilities available. A criticism of the interpretation of these data in the case of polar molecules is given in Ref. 20, p. 2905.
54. Huestis, D. L., Technical Report #MP 78-25, SRI International (project PYU 6158), Menlo Park, CA 94025. Molar refractions for mercury-chlorine compounds are analyzed.
55. Bounds, D. G., Clarke, J. H. R., and Hinchliffe, A., *Chem. Phys. Lett.*, 45, 367, 1977. Theoretical tensor polarizability for LiCl.
56. Kolker, H. J., and Karplus, M., *J. Chem. Phys.*, 39, 2011, 1963. Theoretical.
57. Cutschick, V. P., and McKoy, V., *J. Chem. Phys.*, 58, 2397, 1973. Theoretical tensor polarizabilities.
58. Gready, J. E., Bacskay, G. B., and Hush, N. S., *Chem. Phys.*, 22, 141, 1977, and 23, 9, 1977. Theoretical.
59. Amos, A. T., and Yoffe, J. A., *J. Chem. Phys.*, 63, 4723, 1975. Theoretical.
60. Stuart, H. A., *Landolt-Börnstein Zahlenwerte and Funktionen*, Vol. 1, Part 3, Eucken, A., and Hellwege, K. H., Eds., Springer-Verlag, Berlin, 1951, p. 511. Tabulation of molecular polarizabilities. Two misprints in the chemical symbols have been corrected.
61. Guella, T., Miller, T. M., Stockdale, J. A. D., Bederson, B., and Vuskovic, L., *J. Chem. Phys.*, 94, 6857, 1991. Beam measurements with accuracies between 12 and 24%.
62. Marchese, F. T., and Jaff, H. H., *Theoret. Chim. Acta* (Berlin), 45, 241, 1977. Theoretical and experimental tensor polarizabilities are tabulated in this paper.
63. Applequist, J., Carl, J. R., and Fung, K.-K., *J. Am. Chem. Soc.*, 94, 2952, 1972. Excellent reference on the calculation of molecular polarizabilities, including extensive tables of tensor polarizabilities, both theoretical and experimental, at 589.3 nm wavelength.
64. Bridge, N. J., and Buckingham, A. D., *Proc. Roy. Soc.* (London), A295, 334, 1966. Measured tensor polarizabilities at 633 nm wavelength.
65. Barnes, A. N. M., Turner, D. J., and Sutton, L. E., *Trans. Faraday Soc.*, 67, 2902, 1971. Dielectric constants yielding polarizabilities accurate from 0.3–8%.
66. Rayane, D., Allouche, A. R., Benichou, E., Antoine, R., Aubert-Frecon, M., Dugourd, Ph., Broyer, M., Ristori, C., Chandezon, F., Huber, B. A., and Guet, C., *Eur. Phys. J.* D 9, 243, 1999. See also Knight, W. D., Clemenger, K., de Heer, W. A., and Saunders, W. A., *Phys. Rev. B* 31, 2539, 1985. These data probably correspond to a very low internal temperature.
67. Tarnovsky, V., Bunimovicz, M., Vuskovic, L., Stumpf, B., and Bederson, B., *J. Chem. Phys.*, 98, 3894, 1993. These data correspond to internal temperatures 480-948 K.
68. Milani, P., Moullet, I., and de Heer, W. A., *Phys. Rev. A*, 42, 5150, 1990. Beam measurements accurate to 11%.
69. Compagnon, I., Antoine, R., Broyer, M., Dugourd, P., Lermé, J., and Rayane, D., *Phys. Rev. A* 64, 025201, 2001. The uncertainties are ±8 $Å^3$ for C_{60} and ±14 $Å^3$ for C_{70}.
70. Aroney, M. J., and Pratten, S. J., *J. Chem. Soc., Faraday Trans.*, 1, 80, 1201, 1984. Uncertainties in the range 1–3%.
71. Le Fevre, R. J. W., Murthy, D. S. N., and Saxby, J. D., *Aust. J. Chem.*, 24, 1057, 1971. Kerr effect.

72. No, K. T., Cho, K. H., Jhon, M. S., and Scheraga, H. A., J. Am. Chem. Soc., 115, 2005, 1993. Theoretical; these results are quoted in numerous valuable papers on calculated polarizabilities, e.g., Miller, K. J., and Savchik, J. A., *J. Am. Chem. Soc.*, 101, 7206, 1979.
73. Schlecht, S., Schäfer, R., Woenckhaus, J., Becker, J. A., *Chem. Phys. Lett.*, 246, 315, 1995.
74. Benichou, E., Antoine, R., Rayane, D., Vezin, B., Dalby, F. W., Dugourd, P., Ristori, C., Chandezon, F., Huber, B. A., Rocco, J. C., Blundell, S. A., and Guet, C., *Phys. Rev. A* 59, R1, 1999. See also Rayane, D., Allouche, A. R., Benichou, E., Antoine, R., Aubert-Frecon, M., Dugourd, Ph., Broyer, M., Ristori, C., Chandezon, F., Huber, B. A., and Guet, C., *Eur. Phys. J.* D 9, 243, 1999.
75. Antoine, R., Rayane, D., Allouche, A. R., Aubert-Frécon, M., Benichou, E., Dalby, F. W., Dugourd, P., Broyer, M., and Guet, C., *J. Chem. Phys.*, 110, 5568, 1999.
76. Ballard, A., Bonin, K., and Louderback, J., *J. Chem. Phys.* 113, 5732, 2000.
77. Ritchie, G.L.D., and Watson, J.N., *J. Phys. Chem. A* 108, 4515, 2004. These measurements are at 632.8 nm frequency, and are stated accurate to 0.4%.
78. Ritchie, G.L.D., and Blanch, E.W., *J. Phys. Chem. A* 107, 2093, 2003. These measurements are at 632.8 nm frequncy, and are stated accurate to better than 1%.
79. May, E. F., Moldover, M. R., and Schmidt, J. W., *Phys. Rev. A* 78, 032522 (2008). The uncertainty is 0.0003 Å;^{3}.
80. Heiles, S., Schäfer, S., and Schäfer, R., *J. Chem. Phys.* 135, 034303, 2011. The uncertainties are ±2.5 Å;*3* for Ge_9 and Ge_{10}, and ±6.1 Å;^{3} for Ge_{15}.
81. Berninger, M., Stefanov, A., Deachapunya, S., and Arndt, M., *Phys. Rev. A* 76, 013607, 2007. The uncertainties (statistical, systematic) are ±(0.9, 5.1) Å;^{3} for C_{60} and ±(2.0, 6.2) Å;^{3} for C_{70}.
82. Hohm, U., and Loose, A., *Chem. Phys. Lett.* 348, 375, 2001. The uncertainties are ±0.25 Å;^{3} for $Nd(C_5H_5)_3$, ±0.13 Å;^{3} for $Sm(C_5H_5)_3$, and ±0.45 Å;^{3} for $Er(C_5H_5)_3$.
83. Rayane, D., Antoine, R., Dugourd, P., and Broyer, M., *J. Chem. Phys.* 113, 4501, 2000. A 10% precision in the measurements is stated.
84. Tikhonov, G., Kasperovich, V., Wong, K., and Kresin, V. V., *Phys. Rev. A* 64, 063202, 2001. The uncertainties increase with cluster size, from ±2.8% (Na_7) to ±12.4% (Na_{93}).
85. Hohm, U., Loose, A., Maroulis, G., and Xenides, D., *Phys. Rev. A* 61, 053202, 2000. The uncertainty is ±0.14 Å;^{3}.

IONIZATION ENERGIES OF ATOMS AND ATOMIC IONS

The ionization energies (often called ionization potentials) of neutral and partially ionized atoms are listed in this table. Data were obtained from the compilations cited below, supplemented by results from the recent research literature. Values for the first and second ionization energies come from Ref. 6. All values are given in electron volts (eV).

Following the traditional spectroscopic notation, columns are headed I, II, III, etc., up to XXX, where I indicates the neutral atom, II the singly ionized atom, III the doubly ionized atom, etc. The first section of the table includes spectra I to VIII of all the elements through rutherfordium; subsequent sections cover higher spectra (ionization stages) for those elements for which data are available.

References

1. Moore, C. E., *Ionization Potentials and Ionization Limits Derived from the Analysis of Optical Spectra*, Natl. Stand. Ref. Data Ser. — Natl. Bur. Stand. (U.S.) No. 34, 1970.
2. Martin, W. C., Zalubas, R., and Hagan, L., *Atomic Energy Levels — The Rare Earth Elements*, Natl. Stand. Ref. Data Ser. — Natl. Bur. Stand. (U.S.), No. 60, 1978.
3. Sugar, J. and Corliss, C., *Atomic Energy Levels of the Iron Period Elements: Potassium through Nickel, J. Phys. Chem. Ref. Data*, Vol.14, Suppl. 2, 1985.
4. References to papers in *J. Phys. Chem. Ref. Data*, in the period 1973–91 covering other elements may be found in the cumulative index to that journal.
5. Martin, W.C., and Wiese, W.L., in *Atomic, Molecular, and Optical Physics Handbook*, Drake, G.W.F., Ed., AIP Press, New York, 1996.
6. Sansonetti, J. E., Martin, W. C., and Young, S. L., *Handbook of Basic Atomic Spectroscopic Data* (version 1.1), NIST Physical Data web site <http://physics.nist.gov/PhysRefData/Handbook> (October 2004); *J. Phys. Chem. Ref. Data*, 34, 1559, 2005.

Neutral Atoms to +7 Ions

Z	Element	I	II	III	IV	V	VI	VII	VIII
1	H	13.598443							
2	He	24.587387	54.417760						
3	Li	5.391719	75.6400	122.45429					
4	Be	9.32270	18.21114	153.89661	217.71865				
5	B	8.29802	25.1548	37.93064	259.37521	340.22580			
6	C	11.26030	24.3833	47.8878	64.4939	392.087	489.99334		
7	N	14.5341	29.6013	47.44924	77.4735	97.8902	552.0718	667.046	
8	O	13.61805	35.1211	54.9355	77.41353	113.8990	138.1197	739.29	871.4101
9	F	17.4228	34.9708	62.7084	87.1398	114.2428	157.1651	185.186	953.9112
10	Ne	21.56454	40.96296	63.45	97.12	126.21	157.93	207.2759	239.0989
11	Na	5.139076	47.2864	71.6200	98.91	138.40	172.18	208.50	264.25
12	Mg	7.646235	15.03527	80.1437	109.2655	141.27	186.76	225.02	265.96
13	Al	5.985768	18.82855	28.44765	119.992	153.825	190.49	241.76	284.66
14	Si	8.15168	16.34584	33.49302	45.14181	166.767	205.27	246.5	303.54
15	P	10.48669	19.7695	30.2027	51.4439	65.0251	220.421	263.57	309.60
16	S	10.36001	23.33788	34.79	47.222	72.5945	88.0530	280.948	328.75
17	Cl	12.96763	23.8136	39.61	53.4652	67.8	97.03	114.1958	348.28
18	Ar	15.759610	27.62966	40.74	59.81	75.02	91.009	124.323	143.460
19	K	4.3406633	31.63	45.806	60.91	82.66	99.4	117.56	154.88
20	Ca	6.11316	11.87172	50.9131	67.27	84.50	108.78	127.2	147.24
21	Sc	6.56149	12.79977	24.75666	73.4894	91.65	110.68	138.0	158.1
22	Ti	6.82812	13.5755	27.4917	43.2672	99.30	119.53	140.8	170.4
23	V	6.74619	14.618	29.311	46.709	65.2817	128.13	150.6	173.4
24	Cr	6.76651	16.4857	30.96	49.16	69.46	90.6349	160.18	184.7
25	Mn	7.43402	15.6400	33.668	51.2	72.4	95.6	119.203	194.5
26	Fe	7.9024	16.1877	30.652	54.8	75.0	99.1	124.98	151.06
27	Co	7.88101	17.084	33.50	51.3	79.5	102.0	128.9	157.8
28	Ni	7.6398	18.16884	35.19	54.9	76.06	108	133	162
29	Cu	7.72638	20.2924	36.841	57.38	79.8	103	139	166
30	Zn	9.394199	17.96439	39.723	59.4	82.6	108	134	174
31	Ga	5.999301	20.51515	30.7258	63.241	86.01	112.7	140.9	169.9
32	Ge	7.89943	15.93461	34.2241	45.7131	93.5			
33	As	9.7886	18.5892	28.351	50.13	62.63	127.6		
34	Se	9.75239	21.19	30.8204	42.9450	68.3	81.7	155.4	
35	Br	11.8138	21.591	36	47.3	59.7	88.6	103.0	192.8
36	Kr	13.99961	24.35984	36.950	52.5	64.7	78.5	111.0	125.802
37	Rb	4.177128	27.2895	40	52.6	71.0	84.4	99.2	136
38	Sr	5.69485	11.0301	42.89	57	71.6	90.8	106	122.3
39	Y	6.2173	12.224	20.52	60.597	77.0	93.0	116	129
40	Zr	6.63390	13.1	22.99	34.34	80.348			
41	Nb	6.75885	14.0	25.04	38.3	50.55	102.057	125	

Neutral Atoms to +7 Ions

Z	Element	I	II	III	IV	V	VI	VII	VIII
42	Mo	7.09243	16.16	27.13	46.4	54.49	68.8276	125.664	143.6
43	Tc	7.28	15.26	29.54					
44	Ru	7.36050	16.76	28.47					
45	Rh	7.45890	18.08	31.06					
46	Pd	8.3369	19.43	32.93					
47	Ag	7.57623	21.47746	34.83					
48	Cd	8.99382	16.90831	37.48					
49	In	5.78636	18.8703	28.03	54				
50	Sn	7.34392	14.6322	30.50260	40.73502	72.28			
51	Sb	8.60839	16.63	25.3	44.2	56	108		
52	Te	9.0096	18.6	27.96	37.41	58.75	70.7	137	
53	I	10.45126	19.1313	33					
54	Xe	12.12984	20.9750	32.1230					
55	Cs	3.893905	23.15744						
56	Ba	5.211664	10.00383						
57	La	5.5769	11.059	19.1773	49.95	61.6			
58	Ce	5.5387	10.85	20.198	36.758	65.55	77.6		
59	Pr	5.473	10.55	21.624	38.98	57.53			
60	Nd	5.5250	10.72	22.1	40.4				
61	Pm	5.582	10.90	22.3	41.1				
62	Sm	5.6437	11.07	23.4	41.4				
63	Eu	5.67038	11.25	24.92	42.7				
64	Gd	6.14980	12.09	20.63	44.0				
65	Tb	5.8638	11.52	21.91	39.79				
66	Dy	5.9389	11.67	22.8	41.47				
67	Ho	6.0215	11.80	22.84	42.5				
68	Er	6.1077	11.93	22.74	42.7				
69	Tm	6.18431	12.05	23.68	42.7				
70	Yb	6.25416	12.176	25.05	43.56				
71	Lu	5.42586	13.9	20.9594	45.25	66.8			
72	Hf	6.82507	15	23.3	33.33				
73	Ta	7.54957							
74	W	7.86403	16.1						
75	Re	7.83352							
76	Os	8.43823							
77	Ir	8.96702							
78	Pt	8.9588	18.563						
79	Au	9.22553	20.20						
80	Hg	10.4375	18.7568	34.2					
81	Tl	6.108194	20.4283	29.83					
82	Pb	7.41663	15.03248	31.9373	42.32	68.8			
83	Bi	7.2855	16.703	25.56	45.3	56.0	88.3		
84	Po	8.414							
85	At								
86	Rn	10.7485							
87	Fr	4.072741							
88	Ra	5.278423	10.14715						
89	Ac	5.17	11.75						
90	Th	6.3067	11.9	20.0	28.8				
91	Pa	5.89							
92	U	6.1941	10.6						
93	Np	6.2657							
94	Pu	6.0260	11.2						
95	Am	5.9738							
96	Cm	5.9914							
97	Bk	6.1979							
98	Cf	6.2817	11.8						
99	Es	6.42	12.0						
100	Fm	6.50							
101	Md	6.58							
102	No	6.65							
103	Lr	4.9							
104	Rf	6.0							

+8 Ions to +15 Ions

Z	Element	IX	X	XI	XII	XIII	XIV	XV	XVI
9	F	1103.1176							
10	Ne	1195.8286	1362.1995						
11	Na	299.864	1465.121	1648.702					
12	Mg	328.06	367.50	1761.805	1962.6650				
13	Al	330.13	398.75	442.00	2085.98	2304.1410			
14	Si	351.12	401.37	476.36	523.42	2437.63	2673.182		
15	P	372.13	424.4	479.46	560.8	611.74	2816.91	3069.842	
16	S	379.55	447.5	504.8	564.44	652.2	707.01	3223.78	3494.1892
17	Cl	400.06	455.63	529.28	591.99	656.71	749.76	809.40	3658.521
18	Ar	422.45	478.69	538.96	618.26	686.10	755.74	854.77	918.03
19	K	175.8174	503.8	564.7	629.4	714.6	786.6	861.1	968
20	Ca	188.54	211.275	591.9	657.2	726.6	817.6	894.5	974
21	Sc	180.03	225.18	249.798	687.36	756.7	830.8	927.5	1009
22	Ti	192.1	215.92	265.07	291.500	787.84	863.1	941.9	1044
23	V	205.8	230.5	255.7	308.1	336.277	896.0	976	1060
24	Cr	209.3	244.4	270.8	298.0	354.8	384.168	1010.6	1097
25	Mn	221.8	248.3	286.0	314.4	343.6	403.0	435.163	1134.7
26	Fe	233.6	262.1	290.2	330.8	361.0	392.2	457	489.256
27	Co	186.13	275.4	305	336	379	411	444	511.96
28	Ni	193	224.6	321.0	352	384	430	464	499
29	Cu	199	232	265.3	369	401	435	484	520
30	Zn	203	238	274	310.8	419.7	454	490	542
31	Ga	210.8	244.0	280.7	319.2	357.2	471.2	508.8	548.3
36	Kr	230.85	268.2	308	350	391	447	492	541
37	Rb	150	277.1						
38	Sr	162	177	324.1					
39	Y	146.2	191	206	374.0				
42	Mo	164.12	186.4	209.3	230.28	279.1	302.60	544.0	570

+16 Ions to +23 Ions

Z	Element	XVII	XVIII	XIX	XX	XXI	XXII	XXIII	XXIV
17	Cl	3946.2960							
18	Ar	4120.8857	4426.2296						
19	K	1033.4	4610.8	4934.046					
20	Ca	1087	1157.8	5128.8	5469.864				
21	Sc	1094	1213	1287.97	5674.8	6033.712			
22	Ti	1131	1221	1346	1425.4	6249.0	6625.82		
23	V	1168	1260	1355	1486	1569.6	6851.3	7246.12	
24	Cr	1185	1299	1396	1496	1634	1721.4	7481.7	7894.81
25	Mn	1224	1317	1437	1539	1644	1788	1879.9	8140.6
26	Fe	1266	1358	1456	1582	1689	1799	1950	2023
27	Co	546.58	1397.2	1504.6	1603	1735	1846	1962	2119
28	Ni	571.08	607.06	1541	1648	1756	1894	2011	2131
29	Cu	557	633	670.588	1697	1804	1916	2060	2182
30	Zn	579	619	698	738	1856			
36	Kr	592	641	786	833	884	937	998	1051
42	Mo	636	702	767	833	902	968	1020	1082

+24 Ions to +29 Ions

Z	Element	XXV	XXVI	XXVII	XXVIII	XXIX	XXX
25	Mn	8571.94					
26	Fe	8828	9277.69				
27	Co	2219.0	9544.1	10012.12			
28	Ni	2295	2399.2	10288.8	10775.40		
29	Cu	2308	2478	2587.5	11062.38	11567.617	
36	Kr	1151	1205.3	2928	3070	3227	3381
42	Mo	1263	1323	1387	1449	1535	1601

IONIZATION ENERGIES OF GAS-PHASE MOLECULES

Sharon G. Lias

This table presents values for the first ionization energies (IP) of approximately 1000 molecules and atoms. Substances are listed by name. Values designated as "approx." are considered not to be well established. Data appearing in the 1988 reference were updated in 1996 for inclusion in the database of ionization energies available at the Internet site of the Standard Reference Data program of the National Institute of Standards and Technology (http://webbook.nist.gov). The list appearing here includes these updates.

The list also includes values for enthalpies of formation of the ions at 298 K, $\Delta_f H_{ion}$, given according to the ion convention used by mass spectrometrists; to convert these values to the electron convention used by thermodynamicists, add 6 kJ/mol. Details on the calculation of $\Delta_f H_{ion}$, as well as data for a much larger number of molecules, may be found in the reference and on the NIST Internet site.

Reference

Lias, S. G., Bartmess, J. E., Liebman, J. F., Holmes, J. L., Levin, R. D., and Mallard, W.G., *Gas-Phase Ion and Neutral Thermochemistry, J. Phys. Chem. Ref. Data*, Vol. 17, Suppl. No. 1, 1988. Additional details at <webbook.nist.gov/chemistry/ion/>

Ionization Energies of Gas-Phase Molecules

Name	Mol. form.	IP/eV	Uncert.	$\Delta_f H_{ion}$/kJ mol^{-1}
Acenaphthene	$C_{12}H_{10}$	7.75	± 0.07	903
Acenaphthylene	$C_{12}H_8$	8.22	approx.	1053
Acetaldehyde	C_2H_4O	10.229	± 0.0007	821
Acetamide	C_2H_5NO	9.65	± 0.03	693
Acetic acid	$C_2H_4O_2$	10.65	± 0.02	595
Acetic anhydride	$C_4H_6O_3$	10.0	approx.	398
Acetone	C_3H_6O	9.703	± 0.006	719
Acetonitrile	C_2H_3N	12.20	± 0.01	1253
Acetophenone	C_8H_8O	9.29	± 0.03	810
Acetyl chloride	C_2H_3ClO	10.82	± 0.04	801
Acetylene	C_2H_2	11.400	± 0.002	1328
Acetyl fluoride	C_2H_3FO	11.5	approx.	667
Acrolein	C_3H_4O	10.103	± 0.006	900
Acrylamide	C_3H_5NO	9.5	approx.	720
Acrylic acid	$C_3H_4O_2$	10.60		701
Acrylonitrile	C_3H_3N	10.91	± 0.01	1237
Actinium	Ac	5.17		905
Allene	C_3H_4	9.692	± 0.004	1126
Allyl alcohol	C_3H_6O	9.67	± 0.05	808
Allylamine	C_3H_7N	8.76	approx.	891
Aluminum	Al	5.98577		905
Aluminum bromide	$AlBr_3$	10.4	approx.	593
Aluminum chloride	$AlCl_3$	12.01	approx.	573
Aluminum fluoride	AlF_3	≤15.45		≤282
Aluminum iodide	AlI_3	9.1	approx.	673
Aluminum monobromide	AlBr	9.3	approx.	913
Aluminum monochloride	AlCl	9.4		855
Aluminum monofluoride	AlF	9.73	± 0.01	673
Aluminum monoiodide	AlI	9.3	± 0.3	965
Americium	Am	5.9738	± 0.0002	860
Amidogen	H_2N	11.14	± 0.01	1264
3-Amino-1-propanol	C_3H_9NO	9.0	approx.	651
Ammonia	H_3N	10.070	± 0.020	925
Aniline	C_6H_7N	7.720	± 0.002	832
Anisole	C_7H_8O	8.22	± 0.03	725
Anthracene	$C_{14}H_{10}$	7.439	± 0.006	948
Antimony	Sb	8.64		1096
Antimony(III) chloride	Cl_3Sb	≤10.7	approx.	≤719
Argon	Ar	15.75962		1521
Arsenic	As	9.8152		1250
Arsenic(III) chloride	$AsCl_3$	10.55	approx.	754
Arsenic(III) fluoride	AsF_3	12.84	approx.	452
Arsine	AsH_3	9.89	approx.	1021

Name	Mol. form.	IP/eV	Uncert.	$\Delta_f H_{ion}$/kJ mol^{-1}
trans-Azoxybenzene	$C_{12}H_{10}N_2O$	8.1	approx.	1123
Azulene	$C_{10}H_8$	7.38	± 0.05	1001
Barium	Ba	5.21170		683
Barium oxide	BaO	6.91	± 0.06	543
Benzaldehyde	C_7H_6O	9.49	± 0.02	878
Benzamide	C_7H_7NO	9.25	approx.	792
Benzene	C_6H_6	9.24378	± 0.00007	975
Benzeneacetic acid	$C_8H_8O_2$	8.26	approx.	479
1,2-Benzenediamine	$C_6H_8N_2$	7.2	approx.	787
1,3-Benzenediamine	$C_6H_8N_2$	7.14	approx.	777
1,4-Benzenediamine	$C_6H_8N_2$	6.87	approx.	759
Benzenethiol	C_6H_6S	8.32	approx.	915
Benzoic acid	$C_7H_6O_2$	9.3	approx.	604
Benzonitrile	C_7H_5N	9.70	± 0.01	1154
Benzophenone	$C_{13}H_{10}O$	9.08	± 0.05	926
p-Benzoquinone	$C_6H_4O_2$	10.01	± 0.06	844
Benzoyl chloride	C_7H_5ClO	9.53	approx.	815
Benzyl alcohol	C_7H_8O	8.3	approx.	701
Benzylamine	C_7H_9N	8.64	approx.	917
Berkelium	Bk	6.23		911
Beryllium	Be	9.32263		1224
Beryllium oxide	BeO	10.1	approx.	1111
Biphenyl	$C_{12}H_{10}$	8.23	± 0.10	977
Bismuth	Bi	7.2855		908
Bismuth trichloride	$BiCl_3$	10.4	approx.	736
Borane(1)	BH	9.77	approx.	1385
Borane(3)	BH_3	12.026	± 0.024	1261
Borane carbonyl	CH_3BO	11.14	± 0.02	962
Boron	B	8.29803		1363
Boron dioxide	BO_2	13.5	approx.	1001
Boron oxide	B_2O_3	13.5	± 0.15	460
Boron tribromide	BBr_3	10.51	approx.	809
Boron trichloride	BCl_3	11.60	± 0.02	718
Boron trifluoride	BF_3	15.7	± 0.3	365
Boron triiodide	BI_3	9.25	approx.	964
Bromine	Br_2	10.516	± 0.005	1046
Bromine (atomic)	Br	11.81381		1252
Bromine chloride	BrCl	11.01		1079
Bromine fluoride	BrF	11.86		1086
Bromine monoxide	BrO	10.46	± 0.02	1135
Bromine pentafluoride	BrF_5	13.172	± 0.002	840
Bromoacetylene	C_2HBr	10.31	± 0.02	1242
Bromobenzene	C_6H_5Br	9.00	± 0.02	971
1-Bromobutane	C_4H_9Br	10.12	approx.	869
2-Bromobutane, (±)-	C_4H_9Br	10.01	± 0.02	845
Bromochlorodifluoromethane	$CBrClF_2$	11.21	approx.	642
Bromochloromethane	CH_2BrCl	10.77	± 0.01	1085
2-Bromo-2-chloro-1,1,1-trifluoroethane	$C_2HBrClF_3$	11.0	approx.	363
Bromodichloromethane	$CHBrCl_2$	10.6		973
Bromoethane	C_2H_5Br	10.29	± 0.01	931
Bromoethene	C_2H_3Br	9.83	± 0.02	1028
Bromomethane	CH_3Br	10.541	± 0.003	979
1-Bromo-2-methylpropane	C_4H_9Br	10.09	± 0.02	861
2-Bromo-2-methylpropane	C_4H_9Br	9.92	± 0.03	823
1-Bromonaphthalene	$C_{10}H_7Br$	8.08	± 0.03	955
Bromopentafluorobenzene	C_6BrF_5	9.67	approx.	222
1-Bromopentane	$C_5H_{11}Br$	10.10	± 0.01	846
1-Bromopropane	C_3H_7Br	10.18	± 0.01	898
2-Bromopropane	C_3H_7Br	10.10	± 0.03	877
3-Bromopropene	C_3H_5Br	9.96	approx.	1008
Bromosilane	BrH_3Si	10.6		943
4-Bromotoluene	C_7H_7Br	8.67	± 0.02	908

Name	Mol. form.	IP/eV	Uncert.	$\Delta_f H_{ion}$/kJ mol^{-1}
Bromotrichloromethane	$CBrCl_3$	10.6	approx.	980
Bromotrifluoromethane	$CBrF_3$	11.40	approx.	451
1,2-Butadiene	C_4H_6	9.03	approx.	1034
1,3-Butadiene	C_4H_6	9.082	± 0.004	986
Butanal	C_4H_8O	9.84	± 0.02	742
Butane	C_4H_{10}	10.53	± 0.10	890
Butanenitrile	C_4H_7N	11.2	approx.	1110
1-Butanethiol	$C_4H_{10}S$	9.14	± 0.01	794
2-Butanethiol	$C_4H_{10}S$	9.10	approx.	781
Butanoic acid	$C_4H_8O_2$	10.17	± 0.05	509
1-Butanol	$C_4H_{10}O$	9.99	± 0.05	689
2-Butanol	$C_4H_{10}O$	9.88	± 0.03	658
2-Butanone	C_4H_8O	9.52	± 0.04	678
trans-2-Butenal	C_4H_6O	9.73	± 0.01	835
1-Butene	C_4H_8	9.55	± 0.06	921
cis-2-Butene	C_4H_8	9.11	± 0.01	871
trans-2-Butene	C_4H_8	9.10	± 0.01	866
cis-2-Butenoic acid	$C_4H_6O_2$	10.08	approx.	625
trans-2-Butenoic acid	$C_4H_6O_2$	9.9	approx.	604
1-Buten-3-yne	C_4H_4	9.58	± 0.02	1230
Butyl acetate	$C_6H_{12}O_2$	9.92	approx.	471
sec-Butyl acetate	$C_6H_{12}O_2$	9.90		453
Butylamine	$C_4H_{11}N$	8.7	± 0.1	748
sec-Butylamine	$C_4H_{11}N$	8.46	± 0.1	711
tert-Butylamine	$C_4H_{11}N$	8.46	± 0.1	695
Butylbenzene	$C_{10}H_{14}$	8.69	± 0.02	826
sec-Butylbenzene, (±)-	$C_{10}H_{14}$	8.68	± 0.02	820
tert-Butylbenzene	$C_{10}H_{14}$	8.68	± 0.05	816
Butylcyclohexane	$C_{10}H_{20}$	9.41	approx.	695
Butylcyclopentane	C_9H_{18}	9.95	approx.	793
Butyl ethyl ether	$C_6H_{14}O$	9.36	approx.	610
Butyl formate	$C_5H_{10}O_2$	10.52	± 0.02	584
1-*tert*-Butyl-4-methylbenzene	$C_{11}H_{16}$	8.12	approx.	730
Butyl methyl ether	$C_5H_{12}O$	9.4	approx.	648
tert-Butyl methyl sulfide	$C_5H_{12}S$	8.38	approx.	687
4-*tert*-Butylphenol	$C_{10}H_{14}O$	7.8	approx.	552
1-Butyne	C_4H_6	10.19	± 0.02	1148
2-Butyne	C_4H_6	9.59	± 0.03	1071
Cadmium	Cd	8.99367		980
Calcium	Ca	6.11316		768
Calcium monochloride	CaCl	5.86	± 0.07	462
Calcium oxide	CaO	6.66	± 0.18	668
Californium	Cf	6.30		805
Camphor, (+)	$C_{10}H_{16}O$	8.76	approx.	577
Caprolactam	$C_6H_{11}NO$	9.07	approx.	629
Carbazole	$C_{12}H_9N$	7.57	approx.	961
Carbon	C	11.26030		1803
Carbon dioxide	CO_2	13.773	± 0.002	935
Carbon disulfide	CS_2	10.0685	± 0.0020	1089
Carbon monosulfide	CS	11.33	± 0.01	1361
Carbon monoxide	CO	14.014	± 0.0003	1242
Carbon oxyselenide	COSe	10.36	± 0.01	929
Carbon oxysulfide	COS	11.18	± 0.01	936
Carbonyl chloride	CCl_2O	11.5	approx.	888
Carbonyl fluoride	CF_2O	13.035	± 0.030	617
Cerium	Ce	5.5387		957
Cesium	Cs	3.89390		452
Cesium chloride	ClCs	7.84	approx.	510
Cesium fluoride	CsF	8.80	approx.	489
Cesium sodium	CsNa	4.05	approx.	535
Chlorine	Cl_2	11.480	± 0.005	1108
Chlorine (atomic)	Cl	12.96764		1373

Name	Mol. form.	IP/eV	Uncert.	$\Delta_f H_{ion}$/kJ mol^{-1}
Chlorine difluoride	ClF_2	12.77	approx.	1128
Chlorine dioxide	ClO_2	10.33	± 0.02	1093
Chlorine fluoride	ClF	12.66	± 0.01	1171
Chlorine monoxide	Cl_2O	10.94		1135
Chlorine oxide (ClO)	ClO	10.95		1159
Chlorine trifluoride	ClF_3	12.65	approx.	1057
Chloroacetaldehyde	C_2H_3ClO	10.48	approx.	815
Chloroacetic acid	$C_2H_3ClO_2$	10.7	approx.	597
Chloroacetyl chloride	$C_2H_2Cl_2O$	≤10.3	approx.	815
Chloroacetylene	C_2HCl	10.58	± 0.02	1276
2-Chloroaniline	C_6H_6ClN	8.50	approx.	883
3-Chloroaniline	C_6H_6ClN	8.09	approx.	835
4-Chloroaniline	C_6H_6ClN	≤8.18	approx.	≤844
Chlorobenzene	C_6H_5Cl	9.07	± 0.02	930
1-Chlorobutane	C_4H_9Cl	10.67	± 0.03	875
2-Chlorobutane	C_4H_9Cl	10.53		857
Chlorodibromomethane	$CHBr_2Cl$	10.59	± 0.01	1030
1-Chloro-1,1-difluoroethane	$C_2H_3ClF_2$	11.98	approx.	626
1-Chloro-2,2-difluoroethene	C_2HClF_2	9.80	± 0.04	628
Chlorodifluoromethane	$CHClF_2$	12.2	approx.	693
Chloroethane	C_2H_5Cl	10.98	± 0.02	947
2-Chloroethanol	C_2H_5ClO	10.5	approx.	756
Chloroethene	C_2H_3Cl	9.99	± 0.02	985
Chlorofluoromethane	CH_2ClF	11.71	± 0.01	870
Chloromethane	CH_3Cl	11.22	± 0.01	1001
(Chloromethyl)benzene	C_7H_7Cl	9.10	± 0.02	897
Chloromethylene	CHCl	9.84		1247
Chloromethylidyne	CCl	8.9	approx.	1244
1-Chloro-2-methylpropane	C_4H_9Cl	10.73	± 0.07	877
2-Chloro-2-methylpropane	C_4H_9Cl	10.61	approx.	842
1-Chloronaphthalene	$C_{10}H_7Cl$	8.13	approx.	906
1-Chloro-3-nitrobenzene	$C_6H_4ClNO_2$	9.92	approx.	995
1-Chloro-4-nitrobenzene	$C_6H_4ClNO_2$	9.96	approx.	999
Chloropentafluorobenzene	C_6ClF_5	9.72	approx.	126
Chloropentafluoroethane	C_2ClF_5	12.6	approx.	99
3-Chlorophenol	C_6H_5ClO	8.655	± 0.001	680
4-Chlorophenol	C_6H_5ClO	≤8.69	approx.	≤692
1-Chloropropane	C_3H_7Cl	10.81	± 0.01	911
2-Chloropropane	C_3H_7Cl	10.79	± 0.02	896
3-Chloropropene	C_3H_5Cl	10.04	± 0.01	965
Chlorosilane	ClH_3Si	11.4		899
2-Chlorotoluene	C_7H_7Cl	8.7	approx.	856
3-Chlorotoluene	C_7H_7Cl	8.83	approx.	869
4-Chlorotoluene	C_7H_7Cl	8.69	approx.	855
Chlorotrifluoroethene	C_2ClF_3	9.81	± 0.03	373
Chlorotrifluoromethane	$CClF_3$	12.6	± 0.2	505
Chromium	Cr	6.76664		1050
Chromium(VI) dichloride dioxide	Cl_2CrO_2	11.6		580
Chrysene	$C_{18}H_{12}$	7.60	± 0.01	1017
Cobalt	Co	7.8810		1187
Copper	Cu	7.72638		1084
Copper(I) fluoride	CuF	10.15	± 0.02	984
Coronene	$C_{24}H_{12}$	7.29	± 0.01	1026
o-Cresol	C_7H_8O	8.24	approx.	670
m-Cresol	C_7H_8O	8.29	± 0.07	668
p-Cresol	C_7H_8O	8.3	approx.	675
Curium	Cm	6.02		966
Cyanamide	CH_2N_2	10.4	approx.	1137
Cyanate	CNO	11.76	± 0.01	1290
Cyanide	CN	13.5984		1748
Cyanoacetylene	C_3HN	11.64	± 0.01	1475
Cyanogen	C_2N_2	13.37	± 0.01	1597

Name	Mol. form.	IP/eV	Uncert.	$\Delta_f H_{ion}$/kJ mol^{-1}
Cyanogen chloride	CClN	12.34	± 0.01	1329
Cyanogen fluoride	CFN	13.34	± 0.02	1325
Cyclobutane	C_4H_8	9.82	approx.	976
Cyclobutanone	C_4H_6O	9.35	approx.	815
Cyclobutene	C_4H_6	9.43	± 0.02	1067
Cycloheptane	C_7H_{14}	9.97		844
Cyclohexane	C_6H_{12}	9.86	± 0.03	828
Cyclohexanol	$C_6H_{12}O$	9.75	approx.	651
Cyclohexanone	$C_6H_{10}O$	9.14	± 0.01	656
Cyclohexene	C_6H_{10}	8.945	± 0.01	859
Cyclohexylamine	$C_6H_{13}N$	8.86	approx.	750
Cyclohexylcyclohexane	$C_{12}H_{22}$	9.41	approx.	690
Cyclooctane	C_8H_{16}	9.75	± 0.05	816
1,3-Cyclopentadiene	C_5H_6	8.55	± 0.02	955
Cyclopentane	C_5H_{10}	10.33	approx.	918
Cyclopentanol	$C_5H_{10}O$	9.72	approx.	695
Cyclopentanone	C_5H_8O	9.26	± 0.01	701
Cyclopentene	C_5H_8	9.01	± 0.01	905
Cyclopropane	C_3H_6	9.86		1005
Cyclopropanecarbonitrile	C_4H_5N	10.25	approx.	1173
Cyclopropanone	C_3H_4O	9.1	approx.	895
Cyclopropene	C_3H_4	9.67	± 0.01	1209
Cyclopropylamine	C_3H_7N	8.8	approx.	926
Cyclopropylbenzene	C_9H_{10}	8.35	approx.	956
Cyclopropyl methyl ketone	C_5H_8O	≤9.46	approx.	796
cis-Decahydronaphthalene	$C_{10}H_{18}$	9.36	± 0.04	734
trans-Decahydronaphthalene	$C_{10}H_{18}$	9.34	± 0.04	720
Decane	$C_{10}H_{22}$	9.65	approx.	682
1-Decene	$C_{10}H_{20}$	9.42	± 0.05	786
Diazomethane	CH_2N_2	8.999	± 0.001	1098
Diborane	B_2H_6	11.38	± 0.05	1134
1,4-Dibromobutane	$C_4H_8Br_2$	10.15	approx.	879
Dibromodifluoromethane	CBr_2F_2	11.03	± 0.04	683
1,2-Dibromoethane	$C_2H_4Br_2$	10.35	± 0.04	961
Dibromomethane	CH_2Br_2	10.50	approx.	1013
1,2-Dibromopropane	$C_3H_6Br_2$	10.1	approx.	903
1,3-Dibromopropane	$C_3H_6Br_2$	≤10.2	approx.	≤919
1,2-Dibromotetrafluoroethane	$C_2Br_2F_4$	11.1	approx.	280
Dibutylamine	$C_8H_{19}N$	7.69	approx.	586
Dibutyl ether	$C_8H_{18}O$	9.28	approx.	≤560
Di-*sec*-butyl ether	$C_8H_{18}O$	9.11	approx.	511
Di-*tert*-butyl ether	$C_8H_{18}O$	8.88	± 0.07	493
Dibutyl sulfide	$C_8H_{18}S$	8.2	approx.	624
Di-*tert*-butyl sulfide	$C_8H_{18}S$	8.0	approx.	583
Dicarbon	C_2	11.4	approx.	2000
Dichloroacetyl chloride	C_2HCl_3O	10.9	approx.	809
Dichloroacetylene	C_2Cl_2	9.9		1165
o-Dichlorobenzene	$C_6H_4Cl_2$	9.06	± 0.02	907
m-Dichlorobenzene	$C_6H_4Cl_2$	9.10	± 0.02	906
p-Dichlorobenzene	$C_6H_4Cl_2$	8.92	± 0.02	885
Dichlorodifluoromethane	CCl_2F_2	12.05	± 0.24	685
Dichlorodimethylsilane	$C_2H_6Cl_2Si$	10.7	approx.	576
1,1-Dichloroethane	$C_2H_4Cl_2$	11.04	± 0.02	935
1,2-Dichloroethane	$C_2H_4Cl_2$	11.04	± 0.02	931
1,1-Dichloroethene	$C_2H_2Cl_2$	9.81	± 0.04	949
cis-1,2-Dichloroethene	$C_2H_2Cl_2$	9.66	± 0.01	936
trans-1,2-Dichloroethene	$C_2H_2Cl_2$	9.64	± 0.02	934
Dichlorofluoromethane	$CHCl_2F$	11.5	approx.	829
Dichloromethane	CH_2Cl_2	11.32	± 0.01	996
Dichloromethylene	CCl_2	9.27	approx.	1058
1,2-Dichloropropane, (±)-	$C_3H_6Cl_2$	10.8	± 0.1	886
1,3-Dichloropropane	$C_3H_6Cl_2$	10.89	± 0.04	892

Name	Mol. form.	IP/eV	Uncert.	$\Delta_f H_{ion}$/kJ mol^{-1}
Dichlorosilane	Cl_2H_2Si	11.4		765
Dichlorosilylene	Cl_2Si	10.93	approx.	887
1,2-Dichloro-1,1,2,2-tetrafluoroethane	$C_2Cl_2F_4$	12.2		252
Dicyclopropyl ketone	$C_7H_{10}O$	9.1	approx.	1041
1,1-Diethoxyethane	$C_6H_{14}O_2$	9.2	approx.	434
Diethylamine	$C_4H_{11}N$	7.85	± 0.1	684
o-Diethylbenzene	$C_{10}H_{14}$	≤8.51	approx.	≤804
m-Diethylbenzene	$C_{10}H_{14}$	8.49	approx.	798
p-Diethylbenzene	$C_{10}H_{14}$	8.40	approx.	790
Diethyl disulfide	$C_4H_{10}S_2$	8.27	approx.	724
Diethylene glycol dimethyl ether	$C_6H_{14}O_3$	≤9.8		≤448
Diethyl ether	$C_4H_{10}O$	9.51	± 0.03	666
Diethyl oxalate	$C_6H_{10}O_4$	9.8	approx.	205
Diethyl sulfide	$C_4H_{10}S$	8.43	approx.	730
Difluoramine	F_2HN	11.53	approx.	1046
Difluoroamidogen	F_2N	11.628	± 0.01	1155
o-Difluorobenzene	$C_6H_4F_2$	9.29	± 0.01	602
m-Difluorobenzene	$C_6H_4F_2$	9.33	± 0.01	591
p-Difluorobenzene	$C_6H_4F_2$	9.1589	± 0.0003	577
Difluoroborane(2)	BF_2	9.4	approx.	317
trans-Difluorodiazine	F_2N_2	12.8	approx.	1315
1,1-Difluoroethane	$C_2H_4F_2$	11.87	approx.	643
1,1-Difluoroethene	$C_2H_2F_2$	10.29	± 0.01	650
cis-1,2-Difluoroethene	$C_2H_2F_2$	10.23	± 0.02	690
Difluoromethane	CH_2F_2	12.71		774
Difluoromethylene	CF_2	11.44	± 0.03	899
Difluorosilane	F_2H_2Si	12.2	approx.	386
Difluorosilylene	F_2Si	10.78	± 0.05	450
3,4-Dihydro-2*H*-pyran	C_5H_8O	8.35	± 0.01	681
2,5-Dihydrothiophene	C_4H_6S	8.4	approx.	898
Diiodomethane	CH_2I_2	9.46	± 0.02	1030
Diisobutylamine	$C_8H_{19}N$	7.8	approx.	574
Diisobutyl sulfide	$C_8H_{18}S$	8.34	approx.	625
Diisopropylamine	$C_6H_{15}N$	7.73	approx.	602
Diisopropyl ether	$C_6H_{14}O$	9.20	± 0.05	569
Diisopropyl sulfide	$C_6H_{14}S$	8.2	approx.	649
Diketene	$C_4H_4O_2$	9.6	approx.	736
Dilithium	Li_2	5.1127	± 0.0003	709
1,2-Dimethoxyethane	$C_4H_{10}O_2$	9.3	approx.	558
Dimethoxymethane	$C_3H_8O_2$	9.7		588
N,N-Dimethylacetamide	C_4H_9NO	8.81	± 0.03	616
Dimethylamine	C_2H_7N	8.24	± 0.08	777
N,N-Dimethylaniline	$C_8H_{11}N$	7.12	± 0.02	787
2,2-Dimethylbutane	C_6H_{14}	10.06	approx.	787
2,3-Dimethylbutane	C_6H_{14}	10.02	approx.	791
3,3-Dimethyl-2-butanone	$C_6H_{12}O$	9.12	± 0.02	589
2,3-Dimethyl-1-butene	C_6H_{12}	9.07	approx.	812
2,3-Dimethyl-2-butene	C_6H_{12}	8.27	± 0.01	729
3,3-Dimethyl-1-butyne	C_6H_{10}	9.90	± 0.04	1060
1,1-Dimethylcyclohexane	C_8H_{16}	9.42	approx.	728
cis-1,2-Dimethylcyclohexane	C_8H_{16}	<9.78	approx.	772
trans-1,2-Dimethylcyclohexane	C_8H_{16}	9.41		728
cis-1,3-Dimethylcyclohexane	C_8H_{16}	<9.98	approx.	778
trans-1,3-Dimethylcyclohexane	C_8H_{16}	9.53		743
cis-1,4-Dimethylcyclohexane	C_8H_{16}	<9.93	approx.	782
trans-1,4-Dimethylcyclohexane	C_8H_{16}	9.56	approx.	738
cis-1,2-Dimethylcyclopentane	C_7H_{14}	9.92	approx.	828
trans-1,2-Dimethylcyclopentane	C_7H_{14}	9.7	± 0.2	799
Dimethyl disulfide	$C_2H_6S_2$	7.4	approx.	690
Dimethyl ether	C_2H_6O	10.025	± 0.025	783
N,N-Dimethylformamide	C_3H_7NO	9.12	approx.	688
2,6-Dimethyl-4-heptanone	$C_9H_{18}O$	9.01	± 0.06	512

Name	Mol. form.	IP/eV	Uncert.	$\Delta_f H_{ion}$/kJ mol^{-1}
1,1-Dimethylhydrazine	$C_2H_8N_2$	7.29	± 0.05	787
Dimethyl oxalate	$C_4H_6O_4$	10.0	approx.	287
2,4-Dimethyl-3-pentanone	$C_7H_{14}O$	8.95	± 0.01	552
Dimethyl phthalate	$C_{10}H_{10}O_4$	9.64	approx.	277
2,2-Dimethylpropanal	$C_5H_{10}O$	9.51	± 0.01	675
2,3-Dimethylpyridine	C_7H_9N	8.85	approx.	922
2,4-Dimethylpyridine	C_7H_9N	8.85	approx.	918
2,5-Dimethylpyridine	C_7H_9N	≤8.80	approx.	≤916
2,6-Dimethylpyridine	C_7H_9N	8.86	± 0.03	913
3,4-Dimethylpyridine	C_7H_9N	≤9.15	approx.	≤953
3,5-Dimethylpyridine	C_7H_9N	≤9.25	approx.	≤965
Dimethyl sulfide	C_2H_6S	8.69	± 0.02	801
Dimethyl sulfoxide	C_2H_6OS	9.10	± 0.03	727
1,3-Dioxane	$C_4H_8O_2$	9.8		607
1,4-Dioxane	$C_4H_8O_2$	9.19	± 0.01	571
1,3-Dioxolane	$C_3H_6O_2$	9.9	approx.	658
Diphenylacetylene	$C_{14}H_{10}$	7.94	± 0.03	1168
Diphenylamine	$C_{12}H_{11}N$	7.16	± 0.04	908
1,2-Diphenylethane	$C_{14}H_{14}$	8.9	± 0.1	1002
Diphenyl ether	$C_{12}H_{10}O$	8.09	approx.	766
Diphenylmethane	$C_{13}H_{12}$	8.55	approx.	963
Diphosphorus	P_2	10.53		1160
Dipotassium	K_2	4.0637	± 0.0002	519
Dipropylamine	$C_6H_{15}N$	7.84	approx.	641
Dipropyl ether	$C_6H_{14}O$	9.27	approx.	602
Dipropyl sulfide	$C_6H_{14}S$	8.30	± 0.02	676
Disilane	H_6Si_2	9.74	± 0.02	1019
Disodium	Na_2	4.894	± 0.003	614
Disulfur	S_2	9.356	± 0.002	1031
Divinyl ether	C_4H_6O	8.7	approx.	827
5,7-Dodecadiyne	$C_{12}H_{18}$	8.67	approx.	1079
Dysprosium	Dy	5.9389		862
Einsteinium	Es	6.42		753
Epichlorohydrin	C_3H_5ClO	10.64	approx.	919
1,2-Epoxybutane	C_4H_8O	≤10.15	approx.	862
Erbium	Er	6.1078		907
Ethane	C_2H_6	11.56	± 0.02	1031
1,2-Ethanediamine	$C_2H_8N_2$	8.6	approx.	812
1,2-Ethanediol	$C_2H_6O_2$	10.16		593
Ethanethiol	C_2H_6S	9.31	± 0.03	851
Ethanol	C_2H_6O	10.43	± 0.05	772
Ethanolamine	C_2H_7NO	8.96		664
Ethoxybenzene	$C_8H_{10}O$	8.13	approx.	683
2-Ethoxyethanol	$C_4H_{10}O_2$	9.6	approx.	529
Ethyl acetate	$C_4H_8O_2$	10.01	± 0.05	522
Ethyl acrylate	$C_5H_8O_2$	≤10.3	approx.	617
Ethylamine	C_2H_7N	8.86	± 0.02	808
N-Ethylaniline	$C_8H_{11}N$	≤7.67	approx.	≤794
Ethylbenzene	C_8H_{10}	8.77	± 0.01	876
Ethyl benzoate	$C_9H_{10}O_2$	8.9	approx.	537
2-Ethyl-1-butene	C_6H_{12}	9.06	approx.	818
Ethylcyclohexane	C_8H_{16}	9.54	approx.	748
Ethylene	C_2H_4	10.5138	± 0.0006	1067
Ethyleneimine	C_2H_5N	9.5	approx.	1044
Ethyl formate	$C_3H_6O_2$	10.61	± 0.01	639
Ethyl isopropyl sulfide	$C_5H_{12}S$	8.35	approx.	689
Ethyl methyl ether	C_3H_8O	9.72	± 0.07	722
Ethyl methyl sulfide	C_3H_8S	8.55	approx.	765
Ethyl pentyl ether	$C_7H_{16}O$	≤9.49	approx.	≤602
4-Ethylphenol	$C_8H_{10}O$	7.84	approx.	613
Ethyl propanoate	$C_5H_{10}O_2$	10.00	approx.	500
Ethyl propyl ether	$C_5H_{12}O$	9.45	approx.	640

Name	Mol. form.	IP/eV	Uncert.	$\Delta_f H_{ion}$/kJ mol^{-1}
Ethyl propyl sulfide	$C_5H_{12}S$	8.50	approx.	716
Ethyl vinyl ether	C_4H_8O	8.98	approx.	709
Ethynyl	C_2H	11.61	approx.	1685
Europium	Eu	5.6704		723
Fermium	Fm	6.50		627
Fluoranthene	$C_{16}H_{10}$	7.9	± 0.1	1052
9*H*-Fluorene	$C_{13}H_{10}$	7.91	± 0.02	952
Fluorine	F_2	15.697	± 0.003	1515
Fluorine (atomic)	F	17.42282		1761
Fluorine monoxide	F_2O	13.11	± 0.01	1290
Fluorine oxide	FO	12.78	± 0.03	1342
Fluorine superoxide [FOO]	FO_2	12.6	approx.	1228
Fluoroacetylene	C_2HF	11.26		1195
Fluorobenzene	C_6H_5F	9.20	± 0.01	772
Fluoroborane(1)	BF	11.12	± 0.01	957
Fluoroethane	C_2H_5F	11.78	approx.	873
Fluoroethene	C_2H_3F	10.36	± 0.01	861
Fluoromethane	CH_3F	12.47	± 0.02	956
Fluoromethylene	CHF	10.06	± 0.05	1121
Fluoromethylidyne	CF	9.11	± 0.01	1134
1-Fluoro-4-nitrobenzene	$C_6H_4FNO_2$	9.90	approx.	826
1-Fluoropropane	C_3H_7F	11.3	approx.	806
2-Fluoropropane	C_3H_7F	11.08	approx.	776
3-Fluoropropene	C_3H_5F	10.11	approx.	821
Fluorosilane	FH_3Si	11.7		752
2-Fluorotoluene	C_7H_7F	8.91	± 0.01	709
3-Fluorotoluene	C_7H_7F	8.91	± 0.01	709
4-Fluorotoluene	C_7H_7F	8.79	± 0.01	701
Formaldehyde	CH_2O	10.88	± 0.01	941
Formamide	CH_3NO	10.16	± 0.06	796
Formic acid	CH_2O_2	11.33	± 0.01	715
Fulminic acid	CHNO	10.83	approx.	1263
Fulvene	C_6H_6	8.36	approx.	1031
Fumaric acid	$C_4H_4O_4$	10.7	approx.	355
Furan	C_4H_4O	8.883	± 0.003	822
Furfural	$C_5H_4O_2$	9.22	± 0.01	739
Gadolinium	Gd	6.1500		991
Gallium	Ga	5.99930		851
Gallium(III) bromide	Br_3Ga	10.40		711
Gallium(III) chloride	Cl_3Ga	11.52		664
Gallium(III) iodide	GaI_3	9.40		765
Gallium monofluoride	FGa	9.6	approx.	700
Germane	GeH_4	≤10.53		≤1108
Germanium	Ge	7.900		1139
Germanium(II) chloride	Cl_2Ge	10.20	approx.	813
Germanium(IV) chloride	Cl_4Ge	11.68	± 0.05	629
Germanium(II) fluoride	F_2Ge	≤11.65	approx.	551
Germanium(IV) fluoride	F_4Ge	15.5	approx.	307
Germanium(IV) iodide	GeI_4	9.42	approx.	850
Germanium(II) oxide	GeO	11.25	± 0.01	1044
Germanium(II) sulfide	GeS	9.98	approx.	1055
Glyoxal	$C_2H_2O_2$	10.2		773
Gold	Au	9.22567		1254
Hafnium	Hf	6.82507	± 0.00004	1278
Hafnium(IV) bromide	Br_4Hf	10.9	approx.	366
Hafnium(IV) chloride	Cl_4Hf	11.7	approx.	246
Helium	He	24.58741		2372
Heptanal	$C_7H_{14}O$	9.65	approx.	668
Heptane	C_7H_{16}	9.93	± 0.10	771
1-Heptanol	$C_7H_{16}O$	9.84	approx.	614
2-Heptanol, (±)-	$C_7H_{16}O$	9.70	approx.	580
3-Heptanol, (*S*)-	$C_7H_{16}O$	9.68	approx.	578

Name	Mol. form.	IP/eV	Uncert.	$\Delta_f H_{ion}$/kJ mol^{-1}
4-Heptanol	$C_7H_{16}O$	9.61	approx.	572
2-Heptanone	$C_7H_{14}O$	9.28	± 0.10	594
3-Heptanone	$C_7H_{14}O$	9.18	± 0.08	589
4-Heptanone	$C_7H_{14}O$	9.10	± 0.06	577
1-Heptene	C_7H_{14}	9.34	± 0.10	839
trans-3-Heptene	C_7H_{14}	8.92	approx.	790
Hexaborane(10)	B_6H_{10}	9.0	approx.	965
Hexachlorobenzene	C_6Cl_6	8.98	approx.	822
Hexachloroethane	C_2Cl_6	11.1	approx.	920
1,5-Hexadiene	C_6H_{10}	9.27	± 0.05	978
Hexafluorobenzene	C_6F_6	9.89	± 0.04	8
Hexafluoroethane	C_2F_6	13.6	approx.	-30
Hexamethylbenzene	$C_{12}H_{18}$	7.85	± 0.01	670
Hexanal	$C_6H_{12}O$	9.72	± 0.05	691
Hexane	C_6H_{14}	10.13		810
Hexanoic acid	$C_6H_{12}O_2$	≤10.12		≤463
1-Hexanol	$C_6H_{14}O$	9.89	approx.	639
2-Hexanol	$C_6H_{14}O$	9.80	approx.	611
3-Hexanol	$C_6H_{14}O$	9.63	approx.	599
2-Hexanone	$C_6H_{12}O$	9.3	± 0.1	626
3-Hexanone	$C_6H_{12}O$	9.12	± 0.02	600
1-Hexene	C_6H_{12}	9.44	± 0.04	869
cis-2-Hexene	C_6H_{12}	8.97	approx.	818
trans-2-Hexene	C_6H_{12}	8.97	approx.	814
Hexylamine	$C_6H_{15}N$	8.63	approx.	699
1-Hexyne	C_6H_{10}	10.03	± 0.05	1089
Holmium	Ho	6.0216		882
Hydrazine	H_4N_2	8.1	± 0.15	877
Hydrazoic acid	HN_3	10.72	± 0.025	1328
Hydrogen (atomic)	H	13.59844		1530
Hydrogen	H_2	15.42593	± 0.00005	1488
Hydrogen bromide	BrH	11.66	± 0.03	1087
Hydrogen chloride	ClH	12.749	± 0.009	1137
Hydrogen cyanide	CHN	13.60	± 0.01	1447
Hydrogen fluoride	FH	16.044	± 0.003	1276
Hydrogen iodide	HI	10.386	± 0.001	1028
Hydrogen isocyanide	CHN	12.5	approx.	1407
Hydrogen peroxide	H_2O_2	10.58	± 0.04	885
Hydrogen selenide	H_2Se	9.892	± 0.005	984
Hydrogen sulfide	H_2S	10.457	± 0.012	989
Hydroperoxy	HO_2	11.35	± 0.01	1106
p-Hydroquinone	$C_6H_6O_2$	7.94	± 0.01	503
Hydroxyl	HO	13.0170	± 0.0002	1294
Hydroxylamine	H_3NO	10.00	approx.	923
Hypochlorous acid	ClHO	11.12	approx.	993
Hypofluorous acid	FHO	12.71	± 0.01	1130
Imidazole	$C_3H_4N_2$	8.81	approx.	997
Imidogen	HN	≤13.49	± 0.01	1678
Indan	C_9H_{10}	8.3	approx.	864
Indene	C_9H_8	8.14	± 0.01	949
Indium	In	5.78636		802
Indium(I) chloride	ClIn	9.51	approx.	842
Indium(I) fluoride	FIn	9.6	approx.	740
1*H*-Indole	C_8H_7N	7.7602	± 0.0006	908
Iodine	I_2	9.3074	± 0.0002	960
Iodine (atomic)	I	10.45126		1115
Iodine bromide	BrI	9.790	± 0.004	986
Iodine chloride	ClI	10.088	± 0.01	991
Iodine fluoride	FI	10.54	± 0.01	922
Iodine pentafluoride	F_5I	12.943	± 0.005	408
Iodobenzene	C_6H_5I	8.685		1003
1-Iodobutane	C_4H_9I	9.23	± 0.01	840

Name	Mol. form.	IP/eV	Uncert.	$\Delta_f H_{ion}$/kJ mol^{-1}
2-Iodobutane, (±)-	C_4H_9I	9.10	± 0.02	815
Iodoethane	C_2H_5I	9.3492	± 0.0006	893
1-Iodohexane	$C_6H_{13}I$	9.179		794
Iodomethane	CH_3I	9.538		936
1-Iodo-2-methylpropane	C_4H_9I	9.19	± 0.01	824
2-Iodo-2-methylpropane	C_4H_9I	9.02	approx.	798
1-Iodopentane	$C_5H_{11}I$	9.20	± 0.01	817
1-Iodopropane	C_3H_7I	9.25	± 0.01	860
2-Iodopropane	C_3H_7I	9.19	± 0.02	845
Iridium	Ir	9.1		1543
Iron	Fe	7.9024		1177
Isobutanal	C_4H_8O	9.71	± 0.01	721
Isobutane	C_4H_{10}	10.57	approx.	886
Isobutene	C_4H_8	9.239	± 0.003	875
Isobutylamine	$C_4H_{11}N$	8.50	± 0.1	721
Isobutylbenzene	$C_{10}H_{14}$	8.69	± 0.02	817
Isocyanic acid (HNCO)	CHNO	11.595	± 0.005	1016
Isopentane	C_5H_{12}	10.32	± 0.05	843
Isophorone	$C_9H_{14}O$	≤9.07	approx.	≤670
Isophthalic acid	$C_8H_6O_4$	9.98	approx.	268
Isopropyl acetate	$C_5H_{10}O_2$	9.99	± 0.03	482
Isopropylamine	C_3H_9N	8.72	approx.	758
Isopropylbenzene	C_9H_{12}	8.73	± 0.01	847
Isopropylcyclohexane	C_9H_{18}	9.33	approx.	704
1-Isopropyl-4-methylbenzene	$C_{10}H_{14}$	8.29	approx.	771
Isopropyl methyl ether	$C_4H_{10}O$	9.45	± 0.04	661
Isopropyl methyl sulfide	$C_4H_{10}S$	8.7	approx.	749
Isoquinoline	C_9H_7N	8.53	± 0.03	1032
Isoxazole	C_3H_3NO	9.93	approx.	1038
Ketene	C_2H_2O	9.617	± 0.003	880
Krypton	Kr	13.99961		1351
Lanthanum	La	5.5770		969
Lead	Pb	7.41666		911
Lead(II) chloride	Cl_2Pb	10.2	approx.	791
Lead(II) fluoride	F_2Pb	11.5	approx.	679
Lead(II) oxide (massicot)	OPb	9.08	± 0.10	939
Lead(II) sulfide	PbS	8.5	approx.	954
Lithium	Li	5.39172		680
Lithium bromide	BrLi	8.7	approx.	685
Lithium chloride	ClLi	9.57		727
Lithium hydride	HLi	7.7		882
Lithium iodide	ILi	7.5	approx.	633
Lithium monoxide	LiO	8.44	approx.	894
Lithium potassium	KLi	4.57	± 0.04	512
Lithium rubidium	LiRb	4.3	± 0.1	486
Lithium sodium	LiNa	5.05	± 0.04	571
Lutetium	Lu	5.42585		950
Magnesium	Mg	7.64624		885
Magnesium fluoride	F_2Mg	13.6	approx.	588
Magnesium oxide	MgO	8.76	approx.	901
Maleic anhydride	$C_4H_2O_3$	10.8	approx.	645
Manganese	Mn	7.43402		998
Mendelevium	Md	6.58		635
Mercapto	HS	10.4219	± 0.0004	1145
Mercury	Hg	10.43750		1069
Mercury(II) bromide	Br_2Hg	10.560	± 0.003	935
Mercury(II) chloride	Cl_2Hg	11.380	± 0.003	952
Mercury(II) iodide (red)	HgI_2	9.5088	± 0.0022	900
Mesityl oxide	$C_6H_{10}O$	9.10	± 0.01	694
Methacrylic acid	$C_4H_6O_2$	10.15	approx.	611
Methane	CH_4	12.61	± 0.01	1143
Methanethiol	CH_4S	9.44	± 0.005	888

Name	Mol. form.	IP/eV	Uncert.	$\Delta_f H_{ion}$/kJ mol^{-1}
Methanol	CH_4O	10.85	± 0.01	845
Methoxy	CH_3O	10.72	approx.	1050
Methyl	CH_3	9.843	± 0.002	1095
Methyl acetate	$C_3H_6O_2$	10.25	± 0.02	579
Methyl acrylate	$C_4H_6O_2$	9.9	approx.	641
2-Methylacrylonitrile	C_4H_5N	10.34		1127
Methylamine	CH_5N	8.80	approx.	826
2-Methylaniline	C_7H_9N	7.44	approx.	772
3-Methylaniline	C_7H_9N	7.50	approx.	778
4-Methylaniline	C_7H_9N	7.24	approx.	753
N-Methylaniline	C_7H_9N	7.34	± 0.04	792
Methyl azide	CH_3N_3	9.81	± 0.02	1227
4-Methylbenzaldehyde	C_8H_8O	9.33	approx.	825
Methyl benzoate	$C_8H_8O_2$	9.32	± 0.03	611
2-Methylbenzonitrile	C_8H_7N	≤9.38	approx.	1085
3-Methylbenzonitrile	C_8H_7N	≤9.34	approx.	1085
4-Methylbenzonitrile	C_8H_7N	9.32	± 0.02	1083
2-Methyl-1,3-butadiene	C_5H_8	8.84	± 0.01	928
Methyl butanoate	$C_5H_{10}O_2$	10.07	approx.	520
3-Methylbutanoic acid	$C_5H_{10}O_2$	≤10.51	approx.	≤499
2-Methyl-1-butanol, (±)-	$C_5H_{12}O$	9.86	approx.	649
2-Methyl-2-butanol	$C_5H_{12}O$	9.8	approx.	615
3-Methyl-2-butanol, (±)-	$C_5H_{12}O$	9.88	approx.	637
3-Methyl-2-butanone	$C_5H_{10}O$	9.30	± 0.01	635
2-Methyl-1-butene	C_5H_{10}	9.12	± 0.01	844
3-Methyl-1-butene	C_5H_{10}	9.52	± 0.01	891
2-Methyl-2-butene	C_5H_{10}	8.69	± 0.01	796
2-Methyl-1-buten-3-yne	C_5H_6	9.25	± 0.02	1152
Methyl *tert*-butyl ether	$C_5H_{12}O$	9.24	approx.	608
Methyl chloroacetate	$C_3H_5ClO_2$	10.3	approx.	575
Methylcyclohexane	C_7H_{14}	9.64		775
1-Methylcyclohexanol	$C_7H_{14}O$	9.8	approx.	586
Methylcyclopentane	C_6H_{12}	9.85	approx.	845
Methylcyclopropane	C_4H_8	9.46	approx.	936
2-Methyldecane	$C_{11}H_{24}$	9.7	approx.	658
2-Methyl-*N,N*-dimethylaniline	$C_9H_{13}N$	7.40	± 0.02	814
Methyl 2,2-dimethylpropanoate	$C_6H_{12}O_2$	9.90	approx.	466
Methylene	CH_2	10.396	± 0.003	1392
N-Methylformamide	C_2H_5NO	9.83	± 0.04	760
Methyl formate	$C_2H_4O_2$	10.835	± 0.005	690
2-Methylfuran	C_5H_6O	8.38	± 0.02	729
3-Methylfuran	C_5H_6O	8.64	approx.	763
2-Methylheptane	C_8H_{18}	9.84	approx.	734
5-Methyl-2-hexanone	$C_7H_{14}O$	9.28	approx.	586
Methylhydrazine	CH_6N_2	7.7	± 0.15	835
Methylidyne	CH	10.64	± 0.01	1622
Methyl isocyanate	C_2H_3NO	10.67	approx.	900
Methyl methacrylate	$C_5H_8O_2$	9.7	approx.	589
1-Methylnaphthalene	$C_{11}H_{10}$	7.97	± 0.03	882
2-Methylnaphthalene	$C_{11}H_{10}$	7.91	± 0.08	877
Methyloxirane	C_3H_6O	10.22	approx.	892
2-Methylpentane	C_6H_{14}	10.12	approx.	802
3-Methylpentane	C_6H_{14}	10.08	approx.	801
3-Methyl-2-pentanone, (±)-	$C_6H_{12}O$	9.21	± 0.01	600
4-Methyl-2-pentanone	$C_6H_{12}O$	9.30	± 0.01	609
2-Methyl-3-pentanone	$C_6H_{12}O$	9.10	± 0.01	592
2-Methyl-1-pentene	C_6H_{12}	9.08	approx.	817
4-Methyl-1-pentene	C_6H_{12}	9.45	± 0.01	862
2-Methyl-2-pentene	C_6H_{12}	8.58	approx.	761
4-Methyl-*cis*-2-pentene	C_6H_{12}	8.98	± 0.01	809
4-Methyl-*trans*-2-pentene	C_6H_{12}	8.97	approx.	804
Methyl pentyl ether	$C_6H_{14}O$	≤9.67	approx.	≤657

Name	Mol. form.	IP/eV	Uncert.	$\Delta_f H_{ion}$/kJ mol^{-1}
2-Methylpropanenitrile	C_4H_7N	11.3	approx.	1115
2-Methyl-1-propanethiol	$C_4H_{10}S$	9.12	approx.	783
2-Methyl-2-propanethiol	$C_4H_{10}S$	9.03	approx.	762
Methyl propanoate	$C_4H_8O_2$	10.15	± 0.03	548
2-Methylpropanoic acid	$C_4H_8O_2$	10.33	± 0.03	516
2-Methyl-1-propanol	$C_4H_{10}O$	10.02	± 0.04	683
2-Methyl-2-propanol	$C_4H_{10}O$	9.90	± 0.02	642
2-Methylpropenal	C_4H_6O	9.92	approx.	834
Methyl propyl ether	$C_4H_{10}O$	9.41	± 0.07	670
Methyl propyl sulfide	$C_4H_{10}S$	8.8	approx.	767
2-Methylpyridine	C_6H_7N	9.02	approx.	970
3-Methylpyridine	C_6H_7N	9.04	approx.	979
4-Methylpyridine	C_6H_7N	9.04	approx.	976
N-Methylpyrrolidine	$C_5H_{11}N$	≤8.41	± 0.02	≤809
N-Methyl-2-pyrrolidinone	C_5H_9NO	≤9.17	approx.	≤676
Methylsilane	CH_6Si	10.7	approx.	1003
2-Methylstyrene	C_9H_{10}	8.20	approx.	908
3-Methylstyrene	C_9H_{10}	8.15	approx.	899
4-Methylstyrene	C_9H_{10}	8.1	approx.	895
2-Methylthiophene	C_5H_6S	8.14	approx.	867
3-Methylthiophene	C_5H_6S	8.40	approx.	893
Methyl vinyl ether	C_3H_6O	8.95	± 0.01	763
Molybdenum	Mo	7.09243		1343
Molybdenum(V) chloride	Cl_5Mo	8.7	approx.	392
Molybdenum(VI) fluoride	F_6Mo	14.5	approx.	-159
Morpholine	C_4H_9NO	8.2	approx.	841
Naphthalene	$C_{10}H_8$	8.1442	± 0.0009	936
1-Naphthol	$C_{10}H_8O$	7.76	± 0.03	719
2-Naphthol	$C_{10}H_8O$	7.87	± 0.06	729
Neodymium	Nd	5.5250		859
Neon	Ne	21.56454		2081
Neopentane	C_5H_{12}	≤10.2	approx.	≤818
Neptunium	Np	6.2657	± 0.0003	1069
Nickel	Ni	7.6398		1167
Nickel carbonyl [Ni(CO)$_4$]	C_4NiO_4	8.27	± 0.04	200
Niobium	Nb	6.75885		1384
Niobium(V) chloride	Cl_5Nb	10.97	approx.	356
Nitric acid	HNO_3	11.95	± 0.01	1019
Nitric oxide	NO	9.26438	± 0.00005	985
2-Nitroaniline	$C_6H_6N_2O_2$	8.27	approx.	861
3-Nitroaniline	$C_6H_6N_2O_2$	8.31	approx.	865
4-Nitroaniline	$C_6H_6N_2O_2$	8.34	approx.	859
Nitrobenzene	$C_6H_5NO_2$	9.86	± 0.02	1019
Nitroethane	$C_2H_5NO_2$	10.88	± 0.05	948
Nitrogen	N_2	15.5808		1503
Nitrogen (atomic)	N	14.53414		1875
Nitrogen dioxide	NO_2	9.586	± 0.002	958
Nitrogen pentoxide	N_2O_5	11.9	approx.	1161
Nitrogen sulfide	NS	8.87	± 0.01	1119
Nitrogen tetroxide	N_2O_4	10.8	approx.	1050
Nitrogen trichloride	Cl_3N	10.12	approx.	1244
Nitrogen trifluoride	F_3N	13.00	± 0.02	1125
Nitromethane	CH_3NO_2	11.08	± 0.07	994
2-Nitrophenol	$C_6H_5NO_3$	9.1	approx.	782
3-Nitrophenol	$C_6H_5NO_3$	9.0	approx.	755
4-Nitrophenol	$C_6H_5NO_3$	9.1	approx.	761
1-Nitropropane	$C_3H_7NO_2$	10.81	approx.	919
2-Nitropropane	$C_3H_7NO_2$	10.71	approx.	894
Nitrosyl bromide	BrNO	10.17	± 0.03	1065
Nitrosyl chloride	ClNO	10.87	± 0.01	1099
Nitrosyl fluoride	FNO	12.63	± 0.03	1152
Nitrosyl hydride	HNO	10.1	approx.	1075

Name	Mol. form.	IP/eV	Uncert.	$\Delta_f H_{ion}$/kJ mol^{-1}
2-Nitrotoluene	$C_7H_7NO_2$	9.24		946
3-Nitrotoluene	$C_7H_7NO_2$	9.45	± 0.1	941
4-Nitrotoluene	$C_7H_7NO_2$	9.46	± 0.05	942
Nitrous acid	HNO_2	≤11.3		≤1011
Nitrous oxide	N_2O	12.886		1325
Nitryl chloride	$ClNO_2$	11.84	approx.	1155
Nitryl fluoride	FNO_2	13.09	approx.	1154
Nobelium	No	6.65		642
Nonane	C_9H_{20}	9.71	± 0.10	709
2-Nonanone	$C_9H_{18}O$	9.16	approx.	545
5-Nonanone	$C_9H_{18}O$	9.07	approx.	530
Octane	C_8H_{18}	9.80	± 0.10	737
1-Octene	C_8H_{16}	9.43	± 0.01	829
1-Octyne	C_8H_{14}	9.95	approx.	1040
2-Octyne	C_8H_{14}	9.31	± 0.01	961
3-Octyne	C_8H_{14}	9.22	± 0.01	952
4-Octyne	C_8H_{14}	9.20	± 0.01	946
Osmium	Os	8.7		1630
Osmium(VIII) oxide	O_4Os	12.32	approx.	850
Oxazole	C_3H_3NO	9.9	approx.	940
Oxetane	C_3H_6O	9.65	± 0.01	851
2-Oxetanone	$C_3H_4O_2$	9.70	approx.	653
Oxirane	C_2H_4O	10.56	± 0.01	966
Oxomethyl [HCO]	CHO	8.55	approx.	826
Oxygen	O_2	12.0697	± 0.0002	1165
Oxygen (atomic)	O	13.61806		1563
Ozone	O_3	12.43		1342
Palladium	Pd	8.3367		1181
Pentaborane(9)	B_5H_9	9.90	± 0.04	1028
Pentachloroethane	C_2HCl_5	11.0	approx.	919
cis-1,3-Pentadiene	C_5H_8	8.63	± 0.03	914
trans-1,3-Pentadiene	C_5H_8	8.59	± 0.02	905
1,4-Pentadiene	C_5H_8	9.60	± 0.02	1032
Pentafluorobenzene	C_6HF_5	9.63	approx.	122
Pentafluorophenol	C_6HF_5O	9.20	approx.	-71
2,3,4,5,6-Pentafluorotoluene	$C_7H_3F_5$	9.4	approx.	64
Pentanal	$C_5H_{10}O$	9.74	± 0.04	709
Pentane	C_5H_{12}	10.28	± 0.10	845
2,4-Pentanedione	$C_5H_8O_2$	8.85	± 0.01	469
Pentanoic acid	$C_5H_{10}O_2$	≤10.53	approx.	≤527
1-Pentanol	$C_5H_{12}O$	10.00	approx.	668
2-Pentanol	$C_5H_{12}O$	9.78	approx.	630
3-Pentanol	$C_5H_{12}O$	9.78		628
2-Pentanone	$C_5H_{10}O$	9.38	± 0.01	646
3-Pentanone	$C_5H_{10}O$	9.31	± 0.01	640
1-Pentene	C_5H_{10}	9.51	± 0.01	896
cis-2-Pentene	C_5H_{10}	9.01	± 0.03	843
trans-2-Pentene	C_5H_{10}	9.04	± 0.01	841
1-Penten-3-yne	C_5H_6	9.00	± 0.01	1119
cis-3-Penten-1-yne	C_5H_6	9.14	± 0.04	1137
trans-3-Penten-1-yne	C_5H_6	9.05	± 0.01	1128
1-Pentyne	C_5H_8	10.10	± 0.01	1119
Perchloryl fluoride	$ClFO_3$	12.945	approx.	1224
Perfluoroacetone	C_3F_6O	11.57	approx.	-282
Perfluorocyclohexane	C_6F_{12}	13.2	approx.	-1095
Perfluoronaphthalene	$C_{10}F_8$	8.85		-368
Perfluoropropane	C_3F_8	13.38	approx.	-491
Perfluoropropene	C_3F_6	10.60	± 0.03	-103
Perylene	$C_{20}H_{12}$	6.960	± 0.001	981
Phenanthrene	$C_{14}H_{10}$	7.8914	± 0.0006	966
Phenol	C_6H_6O	8.49	± 0.02	723
Phosphine	H_3P	9.869	± 0.002	958

Atomic

Name	Mol. form.	IP/eV	Uncert.	$\Delta_f H_{ion}$/kJ mol^{-1}
Phosphorothioc trichloride	Cl_3PS	9.71	± 0.03	573
Phosphorothioc trifluoride	F_3PS	≤11.05	± 0.035	≤58
Phosphorus	P	10.48669		1328
Phosphorus(III) bromide	Br_3P	9.7		798
Phosphorus(III) chloride	Cl_3P	9.91		668
Phosphorus(V) chloride	Cl_5P	10.2	approx.	608
Phosphorus(III) fluoride	F_3P	11.60	± 0.05	161
Phosphorus(V) fluoride	F_5P	15.1	approx.	-137
Phosphorus nitride	NP	11.84	± 0.04	1247
Phosphoryl chloride	Cl_3OP	11.36	± 0.02	540
Phosphoryl fluoride	F_3OP	12.76	± 0.01	-24
Phthalic anhydride	$C_8H_4O_3$	10.1	approx.	603
α-Pinene	$C_{10}H_{16}$	8.07	approx.	808
Piperidine	$C_5H_{11}N$	8.03	± 0.11	726
Platinum	Pt	9.0		1433
Plutonium	Pu	6.025		926
Potassium	K	4.34066		508
Potassium bromide	BrK	7.85	± 0.1	578
Potassium chloride	ClK	8.0	approx.	557
Potassium iodide	IK	7.21	approx.	570
Potassium sodium	KNa	4.41636	± 0.00017	561
Praseodymium	Pr	5.464		883
Promethium	Pm	5.55		536
Propanal	C_3H_6O	9.96	± 0.01	772
Propane	C_3H_8	10.95	± 0.05	952
Propanenitrile	C_3H_5N	11.84	± 0.02	1194
1-Propanethiol	C_3H_8S	9.20	± 0.01	819
2-Propanethiol	C_3H_8S	9.145	± 0.005	806
Propanoic acid	$C_3H_6O_2$	10.525	± 0.003	568
1-Propanol	C_3H_8O	10.18	± 0.06	727
2-Propanol	C_3H_8O	10.17	± 0.02	709
Propargyl alcohol	C_3H_4O	10.49	± 0.02	1060
Propene	C_3H_6	9.73	± 0.02	959
Propyl acetate	$C_5H_{10}O_2$	≤9.92	approx.	501
Propylamine	C_3H_9N	8.78	approx.	777
Propylbenzene	C_9H_{12}	8.713	± 0.010	848
Propylcyclohexane	C_9H_{18}	9.46	approx.	720
Propylcyclopentane	C_8H_{16}	9.34	approx.	753
Propyleneimine	C_3H_7N	9.0	approx.	960
Propyl formate	$C_4H_8O_2$	10.52	± 0.02	555
2-Propynal	C_3H_2O	10.7	approx.	1145
Propyne	C_3H_4	10.37	± 0.01	1187
Protactinium	Pa	5.89		1133
Pyrene	$C_{16}H_{10}$	7.4256	± 0.0006	935
Pyridazine	$C_4H_4N_2$	8.67	± 0.03	1112
Pyridine	C_5H_5N	9.25		1031
Pyrimidine	$C_4H_4N_2$	9.23		1087
Pyrrole	C_4H_5N	8.207	± 0.005	900
Pyrrolidine	C_4H_9N	8.0	approx.	769
2-Pyrrolidone	C_4H_7NO	9.2	approx.	674
Quinoline	C_9H_7N	8.62	± 0.01	1041
Radium	Ra	5.27892		668
Radon	Rn	10.74850		1037
Rhenium	Re	7.88		1530
Rhenium(VII) oxide	O_7Re_2	12.7	approx.	125
Rhodium	Rh	7.45890		1276
Rubidium	Rb	4.17713		484
Rubidium bromide	BrRb	7.94	± 0.03	583
Rubidium chloride	ClRb	8.50	approx.	590
Rubidium sodium	NaRb	4.32	± 0.04	480
Ruthenium	Ru	7.36050		1355
Ruthenium(VIII) oxide	O_4Ru	12.15	± 0.03	988

Name	Mol. form.	IP/eV	Uncert.	$\Delta_f H_{ion}$/kJ mol^{-1}
Samarium	Sm	5.6437		751
Scandium	Sc	6.56144		1010
Selenium	Se	9.75238		1168
Silane	H_4Si	11.00	± 0.02	1095
Silicon	Si	8.15169		1238
Silicon monoxide	OSi	11.49	± 0.20	1008
Silver	Ag	7.57624		1016
Silver(I) chloride	$AgCl$	≤10.08	approx.	≤1065
Silver(I) fluoride	AgF	11.0	approx.	1071
Silylene	H_2Si	8.244	± 0.025	1084
Sodium	Na	5.13908		603
Sodium bromide	$BrNa$	8.31	± 0.1	660
Sodium chloride	$ClNa$	8.92	± 0.06	681
Sodium iodide	INa	7.64	± 0.02	659
Spiro[2.2]pentane	C_5H_8	9.26	approx.	1078
Stannane	H_4Sn	10.75	approx.	1200
Stibine	H_3Sb	9.54	± 0.03	1067
cis-Stilbene	$C_{14}H_{12}$	7.80	approx.	1005
trans-Stilbene	$C_{14}H_{12}$	7.656	± 0.001	973
Strontium	Sr	5.69484		713
Strontium oxide	OSr	6.6	± 0.2	623
Styrene	C_8H_8	8.464	± 0.001	964
Succinic anhydride	$C_4H_4O_3$	10.6	approx.	500
Succinonitrile	$C_4H_4N_2$	12.1	approx.	1377
Sulfolane	$C_4H_8O_2S$	9.8	approx.	577
Sulfur	S	10.36001		1277
Sulfur chloride pentafluoride	ClF_5S	12.335	approx.	144
Sulfur dichloride	Cl_2S	9.45	± 0.03	895
Sulfur difluoride	F_2S	10.08	approx.	676
Sulfur dioxide	O_2S	12.349	± 0.001	894
Sulfur hexafluoride	F_6S	15.32	± 0.02	258
Sulfur monofluoride	FS	10.09		986
Sulfur monoxide	OS	10.294	± 0.004	998
Sulfur oxide (SSO)	OS_2	10.584	± 0.005	971
Sulfur pentafluoride	F_5S	9.60	± 0.05	10
Sulfur tetrafluoride	F_4S	12.0	± 0.3	399
Sulfur trioxide (α-form)	O_3S	12.82	± 0.03	841
Sulfuryl chloride	Cl_2O_2S	12.05		807
Sulfuryl fluoride	F_2O_2S	13.04	± 0.01	501
Tantalum	Ta	7.89		1544
Tantalum(V) chloride	Cl_5Ta	11.08	approx.	303
Technetium	Tc	7.28		1380
Tellurium	Te	9.0096		1066
Terbium	Tb	5.8639		955
Terephthalic acid	$C_8H_6O_4$	9.86	approx.	232
o-Terphenyl	$C_{18}H_{14}$	7.99	approx.	1056
m-Terphenyl	$C_{18}H_{14}$	8.01	approx.	1057
p-Terphenyl	$C_{18}H_{14}$	7.80	± 0.03	1037
Tetraborane(10)	B_4H_{10}	10.76	± 0.04	1105
Tetrabromomethane	CBr_4	10.31	approx.	1079
1,1,2,2-Tetrachloro-1,2-difluoroethane	$C_2Cl_4F_2$	11.3	approx.	563
1,1,1,2-Tetrachloroethane	$C_2H_2Cl_4$	11.1	approx.	920
1,1,2,2-Tetrachloroethane	$C_2H_2Cl_4$	≤11.62	approx.	≤971
Tetrachloroethene	C_2Cl_4	9.326	± 0.001	887
Tetrachloromethane	CCl_4	11.47	± 0.01	1010
Tetrachlorosilane	Cl_4Si	11.79	± 0.01	527
Tetraethylsilane	$C_8H_{20}Si$	8.9	approx.	595
1,2,3,4-Tetrafluorobenzene	$C_6H_2F_4$	9.53	approx.	284
1,2,3,5-Tetrafluorobenzene	$C_6H_2F_4$	9.53	approx.	263
1,2,4,5-Tetrafluorobenzene	$C_6H_2F_4$	9.35	approx.	254
Tetrafluoroethene	C_2F_4	10.12	± 0.02	315
Tetrafluorohydrazine	F_4N_2	11.94	± 0.03	1119

Name	Mol. form.	IP/eV	Uncert.	$\Delta_f H_{ion}$/kJ mol^{-1}
Tetrafluorosilane	F_4Si	15.24	± 0.14	-144
Tetrahydrofuran	C_4H_8O	9.38	± 0.05	721
1,2,3,4-Tetrahydronaphthalene	$C_{10}H_{12}$	8.46	± 0.02	841
Tetrahydropyran	$C_5H_{10}O$	9.25	± 0.01	670
Tetrahydrothiophene	C_4H_8S	8.38		774
1,2,4,5-Tetramethylbenzene	$C_{10}H_{14}$	8.04	± 0.02	730
2,2,3,3-Tetramethylbutane	C_8H_{18}	9.8		720
Tetramethylsilane	$C_4H_{12}Si$	9.80	± 0.04	713
Tetramethylstannane	$C_4H_{12}Sn$	8.89	± 0.05	837
Thallium	Tl	6.10829		771
Thallium(I) bromide	BrTl	9.14	± 0.02	844
Thallium(I) chloride	ClTl	9.70	± 0.03	869
Thallium(I) fluoride	FTl	10.52		835
Thallium(I) iodide	ITl	8.47	± 0.02	826
Thiacyclohexane	$C_5H_{10}S$	8.2	approx.	728
Thietane	C_3H_6S	8.61		892
Thionitrosyl fluoride (NSF)	FNS	11.51	± 0.04	1090
Thionyl chloride	Cl_2OS	10.96		844
Thionyl fluoride	F_2OS	12.25		688
Thiophene	C_4H_4S	8.86	± 0.02	970
Thorium	Th	6.308	± 0.003	1207
Thorium(IV) oxide	O_2Th	8.7	approx.	342
Thulium	Tm	6.18431		827
Tin	Sn	7.34381		1011
Tin(II) bromide	Br_2Sn	9.0		839
Tin(IV) bromide	Br_4Sn	10.6		709
Tin(II) chloride	Cl_2Sn	10.0	approx.	760
Tin(IV) chloride	Cl_4Sn	11.7	± 0.2	656
Tin(II) fluoride	F_2Sn	11.1	approx.	586
Tin(II) oxide	OSn	9.60	± 0.02	944
Tin(II) sulfide	SSn	8.8	approx.	966
Titanium	Ti	6.8282		1127
Titanium(IV) bromide	Br_4Ti	10.3		375
Titanium(IV) chloride	Cl_4Ti	11.5	approx.	349
Titanium(IV) iodide	I_4Ti	9.1	approx.	602
Titanium(IV) oxide (rutile)	O_2Ti	9.54	approx.	623
Toluene	C_7H_8	8.8276	± 0.0006	901
o-Toluic acid	$C_8H_8O_2$	9.1	approx.	558
m-Toluic acid	$C_8H_8O_2$	9.43	approx.	579
p-Toluic acid	$C_8H_8O_2$	9.23	approx.	560
Tribromomethane	$CHBr_3$	10.48	± 0.02	1035
Tributylamine	$C_{12}H_{27}N$	7.4	approx.	492
Trichloroacetyl chloride	C_2Cl_4O	11.0	approx.	827
1,2,4-Trichlorobenzene	$C_6H_3Cl_3$	9.04	approx.	880
1,3,5-Trichlorobenzene	$C_6H_3Cl_3$	9.32	± 0.02	899
1,1,1-Trichloroethane	$C_2H_3Cl_3$	11.0	approx.	917
1,1,2-Trichloroethane	$C_2H_3Cl_3$	11.0	approx.	911
Trichloroethene	C_2HCl_3	9.46	± 0.02	894
Trichlorofluoromethane	CCl_3F	11.77	± 0.02	868
Trichloromethane	$CHCl_3$	11.37	± 0.02	992
(Trichloromethyl)benzene	$C_7H_5Cl_3$	≤9.60	approx.	≤914
Trichloromethylsilane	CH_3Cl_3Si	11.36	approx.	548
Trichlorosilane	Cl_3HSi	11.7	approx.	648
1,1,1-Trichloro-2,2,2-trifluoroethane	$C_2Cl_3F_3$	11.5		386
1,1,2-Trichloro-1,2,2-trifluoroethane	$C_2Cl_3F_3$	11.99	± 0.02	429
Triethanolamine	$C_6H_{15}NO_3$	7.9	approx.	206
Triethylamine	$C_6H_{15}N$	7.50	approx.	631
Trifluoramine oxide	F_3NO	13.31	± 0.06	1121
Trifluoroacetic acid	$C_2HF_3O_2$	11.46		75
Trifluoroacetonitrile	C_2F_3N	13.93	± 0.07	845
1,1,1-Trifluoroethane	$C_2H_3F_3$	13.3	± 0.5	536
Trifluoroethene	C_2HF_3	10.14		489

Name	Mol. form.	IP/eV	Uncert.	$\Delta_f H_{ion}$/kJ mol^{-1}
Trifluoroiodomethane	CF_3I	10.23		397
Trifluoromethane	CHF_3	13.86	approx.	643
Trifluoromethyl	CF_3	8.7	± 0.2	379
(Trifluoromethyl)benzene	$C_7H_5F_3$	9.685	± 0.005	335
3,3,3-Trifluoro-1-propene	$C_3H_3F_3$	10.9	approx.	437
Trifluorosilane	F_3HSi	14.0	approx.	150
Trifluorosilyl	F_3Si	9.99	approx.	-32
Triiodomethane	CHI_3	9.25	± 0.02	1010
Trimethylamine	C_3H_9N	7.82	± 0.06	731
1,2,3-Trimethylbenzene	C_9H_{12}	8.42	± 0.02	803
1,2,4-Trimethylbenzene	C_9H_{12}	8.27	± 0.01	784
1,3,5-Trimethylbenzene	C_9H_{12}	8.41	± 0.01	796
Trimethyl borate	$C_3H_9BO_3$	10.0	approx.	65
Trimethylchlorosilane	C_3H_9ClSi	10.15	approx.	624
2,2,4-Trimethylpentane	C_8H_{18}	9.86	approx.	713
2,2,4-Trimethyl-3-pentanone	$C_8H_{16}O$	8.80	approx.	511
2,4,6-Trimethylpyridine	$C_8H_{11}N$	≤8.9	approx.	≤880
1,3,5-Trioxane	$C_3H_6O_3$	10.3	approx.	528
Trisilane	H_8Si_3	9.2	approx.	1009
Tungsten	W	7.98		1621
Tungsten(VI) chloride	Cl_6W	9.5	approx.	348
Undecane	$C_{11}H_{24}$	9.56	approx.	650
Uranium	U	6.19405		1129
Uranium(VI) fluoride	F_6U	14.00	± 0.10	-796
Uranium(IV) oxide	O_2U	5.4	approx.	57
Uranium(VI) oxide	O_3U	10.5	approx.	214
Urea	CH_4N_2O	9.7		690
Vanadium	V	6.746	± 0.002	1166
Vanadium(IV) chloride	Cl_4V	9.2	approx.	361
Vanadyl trichloride	Cl_3OV	11.6	approx.	425
Vinyl acetate	$C_4H_6O_2$	9.19	± 0.05	572
Water	H_2O	12.6206	± 0.0020	976
Xenon	Xe	12.12987		1170
Xenon difluoride	F_2Xe	12.35	± 0.01	1083
Xenon tetrafluoride	F_4Xe	12.65	± 0.1	1016
o-Xylene	C_8H_{10}	8.56	± 0.01	844
m-Xylene	C_8H_{10}	8.56	± 0.01	843
p-Xylene	C_8H_{10}	8.44	± 0.01	832
2,3-Xylenol	$C_8H_{10}O$	8.26	approx.	640
2,4-Xylenol	$C_8H_{10}O$	8.0	approx.	609
2,6-Xylenol	$C_8H_{10}O$	8.05	approx.	615
3,4-Xylenol	$C_8H_{10}O$	8.09	approx.	624
Ytterbium	Yb	6.25416		754
Yttrium	Y	6.217		1022
Zinc	Zn	9.39405		1037
Zirconium	Zr	6.63390		1251
Zirconium(IV) bromide	Br_4Zr	10.7	approx.	388
Zirconium(IV) chloride	Cl_4Zr	11.2	approx.	210
Zirconium(IV) iodide	I_4Zr	9.3	approx.	500

X-RAY ATOMIC ENERGY LEVELS

The energy levels in this table are the values recommended by Bearden and Burr on the basis of a thorough review of the literature on x-ray wavelengths and related data. All values are in electron volts (eV). Values in parentheses are interpolated, and an asterisk * indicates a level that is not resolved from the level above it. See Ref. 1 for uncertainties in the levels and a complete description of how the recommended values were obtained.

References

1. Bearden, J. A., and Burr, A. F., *Rev. Mod. Phys.*, 39, 125, 1967; also published as *X-Ray Wavelengths and X-Ray Atomic Energy* Levels, Natl. Stand. Ref. Data Sys. — Natl. Bur. Standards (U.S.), No. 14, 1967.
2. Gray, D. E., Ed., *American Institute of Physics Handbook, Third Edition*, pp. 7-158 to 7-167, McGraw-Hill, New York, 1972.

Level	$_1$H	$_2$He	$_3$Li	$_4$Be	$_5$B	$_6$C	$_7$N	$_8$O
K	13.59811	24.58678	54.75	111.0	188.0	283.8	401.6	532.0
L_I								23.7
$L_{II,III}$					4.7	6.4	9.2	7.1

Level	$_9$F	$_{10}$Ne	$_{11}$Na	$_{12}$Mg	$_{13}$Al	$_{14}$Si	$_{15}$P	$_{16}$S
K	685.4	866.9	1072.1	1305.0	1559.6	1838.9	2145.5	2472.0
L_I	(31)	(45)	63.3	89.4	117.7	148.7	189.3	229.2
$L_{II,III}$	8.6	18.3	31.1	51.4	73.1	99.2	132.2	164.8

Level	$_{17}$Cl	$_{18}$Ar	$_{19}$K	$_{20}$Ca	$_{21}$Sc	$_{22}$Ti	$_{23}$V	$_{24}$Cr
K	2822.4	3202.9	3607.4	4038.1	4492.8	4966.4	5465.1	5989.2
L_I	270.2	320	377.1	437.8	500.4	563.7	628.2	694.6
L_{II}	201.6	247.3	296.3	350.0	406.7	461.5	520.5	583.7
L_{III}	200.0	245.2	293.6	346.4	402.2	455.5	512.9	574.5
M_I	17.5	25.3	33.9	43.7	53.8	60.3	66.5	74.1
$M_{II,III}$	6.8	12.4	17.8	25.4	32.3	34.6	37.8	42.5
$M_{IV,V}$					6.6	3.7	2.2	2.3

Level	$_{25}$Mn	$_{26}$Fe	$_{27}$Co	$_{28}$Ni	$_{29}$Cu	$_{30}$Zn	$_{31}$Ga	$_{32}$Ge
K	6539.0	7112.0	7708.9	8332.8	8978.9	9658.6	10367.1	11103.1
L_I	769.0	846.1	925.6	1008.1	1096.1	1193.6	1297.7	1414.3
L_{II}	651.4	721.1	793.6	871.9	951.0	1042.8	1142.3	1247.8
L_{III}	640.3	708.1	778.6	854.7	931.1	1019.7	1115.4	1216.7
M_I	83.9	92.9	100.7	111.8	119.8	135.9	158.1	180.0
M_{II}	48.6	54.0	59.5	68.1	73.6	86.6	106.8	127.9
M_{III}	48.6*	54.0*	59.5*	68.1*	73.6*	86.6*	102.9	120.8
$M_{IV,V}$	3.3	3.6	2.9	3.6	1.6	8.1	17.4	28.7

Level	$_{33}$As	$_{34}$Se	$_{35}$Br	$_{36}$Kr	$_{37}$Rb	$_{38}$Sr	$_{39}$Y	$_{40}$Zr
K	11866.7	12657.8	13473.7	14325.6	15199.7	16104.6	17038.4	17997.6
L_I	1526.5	1653.9	1782.0	1921.0	2065.1	2216.3	2372.5	2531.6
L_{II}	1358.6	1476.2	1596.0	1727.2	1863.9	2006.8	2155.5	2306.7
L_{III}	1323.1	1435.8	1549.9	1674.9	1804.4	1939.6	2080.0	2222.3
M_I	203.5	231.5	256.5		322.1	357.5	393.6	430.3
M_{II}	146.4	168.2	189.3	222.7	247.4	279.8	312.4	344.2
M_{III}	140.5	161.9	181.5	213.8	238.5	269.1	300.3	330.5
M_{IV}	41.2	56.7	70.1	88.9	111.8	135.0	159.6	182.4
M_V	41.2*	56.7*	69.0	88.9*	110.3	133.1	157.4	180.0
N_I			27.3	24.0	29.3	37.7	45.4	51.3
N_{II}	2.5	5.6	5.2	10.6	14.8	19.9	25.6	28.7
N_{III}	2.5*	5.6*	4.6	10.6*	14.0	19.9*	25.6*	28.7*

Level	$_{41}$Nb	$_{42}$Mo	$_{43}$Tc	$_{44}$Ru	$_{45}$Rh	$_{46}$Pd	$_{47}$Ag	$_{48}$Cd
K	18985.6	19999.5	21044.0	22117.2	23219.9	24350.3	25514.0	26711.2
L_I	2697.7	2865.5	3042.5	3224.0	3411.9	3604.3	3805.8	4018.0
L_{II}	2464.7	2625.1	2793.2	2966.9	3146.1	3330.3	3523.7	3727.0
L_{III}	2370.5	2520.2	2676.9	2837.9	3003.8	3173.3	3351.1	3537.5
M_I	468.4	504.6		585.0	627.1	669.9	717.5	770.2

Level	$_{41}$Nb	$_{42}$Mo	$_{43}$Tc	$_{44}$Ru	$_{45}$Rh	$_{46}$Pd	$_{47}$Ag	$_{48}$Cd
M_{II}	378.4	409.7	444.9	482.8	521.0	559.1	602.4	650.7
M_{III}	363.0	392.3	425.0	460.6	496.2	531.5	571.4	616.5
M_{IV}	207.4	230.3	256.4	283.6	311.7	340.0	372.8	410.5
M_{V}	204.6	227.0	252.9	279.4	307.0	334.7	366.7	403.7
N_{I}	58.1	61.8		74.9	81.0	86.4	95.2	107.6
N_{II}	33.9	34.8	38.9	43.1	47.9	51.1	62.6	66.9
N_{III}	33.9*	34.8*	38.9*	43.1*	47.9*	51.1*	55.9	66.9*
$N_{IV,V}$	3.2	1.8		2.0	2.5	1.5	3.3	9.3

Level	$_{49}$In	$_{50}$Sn	$_{51}$Sb	$_{52}$Te	$_{53}$I	$_{54}$Xe	$_{55}$Cs	$_{56}$Ba
K	27939.9	29200.1	30491.2	31813.8	33169.4	34561.4	35984.6	37440.6
L_{I}	4237.5	4464.7	4698.3	4939.2	5188.1	5452.8	5714.3	5988.8
L_{II}	3938.0	4156.1	4380.4	4612.0	4852.1	5103.7	5359.4	5623.6
L_{III}	3730.1	3928.8	4132.2	4341.4	4557.1	4782.2	5011.9	5247.0
M_{I}	825.6	883.8	943.7	1006.0	1072.1		1217.1	1292.8
M_{II}	702.2	756.4	811.9	869.7	930.5	999.0	1065.0	1136.7
M_{III}	664.3	714.4	765.6	818.7	874.6	937.0	997.6	1062.2
M_{IV}	450.8	493.3	536.9	582.5	631.3		739.5	796.1
M_{V}	443.1	484.8	527.5	572.1	619.4	672.3	725.5	780.7
N_{I}	121.9	136.5	152.0	168.3	186.4		230.8	253.0
N_{II}	77.4	88.6	98.4	110.2	122.7	146.7	172.3	191.8
N_{III}	77.4*	88.6*	98.4*	110.2*	122.7*	146.7*	161.6	179.7
N_{IV}	16.2	23.9	31.4	39.8	49.6		78.8	92.5
N_{V}	16.2*	23.9*	31.4*	39.8*	49.6*		76.5	89.9
O_{I}	0.1	0.9	6.7	11.6	13.6		22.7	39.1
O_{II}	0.8	1.1	2.1	2.3	3.3		13.1	16.6
O_{III}	0.8*	1.1*	2.1*	2.3*	3.3*		11.4	14.6

Level	$_{57}$La	$_{58}$Ce	$_{59}$Pr	$_{60}$Nd	$_{61}$Pm	$_{62}$Sm	$_{63}$Eu	$_{64}$Gd
K	38924.6	40443.0	41990.6	43568.9	45184.0	46834.2	48519.0	50239.1
L_{I}	6266.3	6548.8	6834.8	7126.0	7427.9	7736.8	8052.0	8375.6
L_{II}	5890.6	6164.2	6440.4	6721.5	7012.8	7311.8	7617.1	7930.3
L_{III}	5482.7	5723.4	5964.3	6207.9	6459.3	6716.2	6976.9	7242.8
M_{I}	1361.3	1434.6	1511.0	1575.3		1722.8	1800.0	1880.8
M_{II}	1204.4	1272.8	1337.4	1402.8	1471.4	1540.7	1613.9	1688.3
M_{III}	1123.4	1185.4	1242.2	1297.4	1356.9	1419.8	1480.6	1544.0
M_{IV}	848.5	901.3	951.1	999.9	1051.5	1106.0	1160.6	1217.2
M_{V}	831.7	883.3	931.0	977.7	1026.9	1080.2	1130.9	1185.2
N_{I}	270.4	289.6	304.5	315.2		345.7	360.2	375.8
N_{II}	205.8	223.3	236.3	243.3	242	265.6	283.9	288.5
N_{III}	191.4	207.2	217.6	224.6	242*	247.4	256.6	270.9
$N_{IV,V}$	98.9	110.0	113.2	117.5	120.4	129.0	133.2	140.5
$N_{VI,VII}$		0.1	2.0	1.5		5.5	0.0	0.1
O_{I}	32.3	37.8	37.4	37.5		37.4	31.8	36.1
$O_{II,III}$	14.4	19.8	22.3	21.1		21.3	22.0	20.3

Level	$_{65}$Tb	$_{66}$Dy	$_{67}$Ho	$_{68}$Er	$_{69}$Tm	$_{70}$Yb	$_{71}$Lu	$_{72}$Hf
K	51995.7	53788.5	55617.7	57485.5	59389.6	61332.3	63313.8	65350.8
L_{I}	8708.0	9045.8	9394.2	9751.3	10115.7	10486.4	10870.4	11270.7
L_{II}	8251.6	8580.6	8917.8	9264.3	9616.9	9978.2	10348.6	10739.4
L_{III}	7514.0	7790.1	8071.1	8357.9	8648.0	8943.6	9244.1	9560.7
M_{I}	1967.5	2046.8	2128.3	2206.5	2306.8	2398.1	2491.2	2600.9
M_{II}	1767.7	1841.8	1922.8	2005.8	2089.8	2173.0	2263.5	2365.4
M_{III}	1611.3	1675.6	1741.2	1811.8	1884.5	1949.8	2023.6	2107.6
M_{IV}	1275.0	1332.5	1391.5	1453.3	1514.6	1576.3	1639.4	1716.4
M_{V}	1241.2	1294.9	1351.4	1409.3	1467.7	1527.8	1588.5	1661.7
N_{I}	397.9	416.3	435.7	449.1	471.7	487.2	506.2	538.1
N_{II}	310.2	331.8	343.5	366.2	385.9	396.7	410.1	437.0
N_{III}	385.0	292.9	306.6	320.0	336.6	343.5	359.3	380.4
N_{IV}	147.0	154.2	161.0	176.7	179.6	198.1	204.8	223.8
N_{V}	147.0*	154.2*	161.0*	167.6	179.6*	184.9	195.0	213.7

Level	$_{65}$Tb	$_{66}$Dy	$_{67}$Ho	$_{68}$Er	$_{69}$Tm	$_{70}$Yb	$_{71}$Lu	$_{72}$Hf
$N_{VI,VII}$	2.6	4.2	3.7	4.3	5.3	6.3	6.9	17.1
O_{I}	39.0	62.9	51.2	59.8	53.2	54.1	56.8	64.9
O_{II}	25.4	26.3	20.3	29.4	32.3	23.4	28.0	38.1
O_{III}	25.4*	26.3*	20.3*	29.4*	32.3*	23.4*	28.0*	30.6

Level	$_{73}$Ta	$_{74}$W	$_{75}$Re	$_{76}$Os	$_{77}$Ir	$_{78}$Pt	$_{79}$Au	$_{80}$Hg
K	67416.4	69525.0	71676.4	73870.8	76111.0	78394.8	80724.9	83102.3
L_{I}	11681.5	12099.8	12526.7	12968.0	13418.5	13879.9	14352.8	14839.3
L_{II}	11136.1	11544.0	11958.7	12385.0	12824.1	13272.6	13733.6	14208.7
L_{III}	9881.1	10206.8	10535.3	10870.9	11215.2	11563.7	11918.7	12283.9
M_{I}	2708.0	2819.6	2931.7	3048.5	3173.7	3296.0	3424.9	3561.6
M_{II}	2468.7	2574.9	2681.6	2792.2	2908.7	3026.5	3147.8	3278.5
M_{III}	2194.0	2281.0	2367.3	2457.2	2550.7	2645.4	2743.0	2847.1
M_{IV}	1793.2	1871.6	1948.9	2030.8	2116.1	2201.9	2291.1	2384.9
M_{V}	1735.1	1809.2	1882.9	1960.1	2040.4	2121.6	2205.7	2294.9
N_{I}	565.5	595.0	625.0	654.3	690.1	722.0	758.8	800.3
N_{II}	464.8	491.6	517.9	546.5	577.1	609.2	643.7	676.9
N_{III}	404.5	425.3	444.4	468.2	494.3	519.0	545.4	571.0
N_{IV}	241.3	258.8	273.7	289.4	311.4	330.8	352.0	378.3
N_{V}	229.3	245.4	260.2	272.8	294.9	313.3	333.9	359.8
N_{VI}	25.0	36.5	40.6	46.3	63.4	74.3	86.4	102.2
N_{VII}	25.0*	33.6	40.6*	46.3*	60.5	71.1	82.8	98.5
O_{I}	71.1	77.1	82.8	83.7	95.2	101.7	107.8	120.3
O_{II}	44.9	46.8	45.6	58.0	63.0	65.3	71.7	80.5
O_{III}	36.4	35.6	34.6	45.4	50.5	51.7	53.7	57.6
$O_{IV,V}$	5.7	6.1	3.5		3.8	2.2	2.5	6.4

Level	$_{81}$Tl	$_{82}$Pb	$_{83}$Bi	$_{84}$Po	$_{85}$At	$_{86}$Rn	$_{87}$Fr	$_{88}$Ra
K	85530.4	88004.5	90525.9	93105.0	95729.9	98404	101137	103921.9
L_{I}	15346.7	15860.8	16387.5	16939.3	17493	18049	18639	19236.7
L_{II}	14697.9	15200.0	15711.1	16244.3	16784.7	17337.1	17906.5	18484.3
L_{III}	12657.5	13035.2	13418.6	13813.8	14213.5	14619.4	15031.2	15444.4
M_{I}	3704.1	3850.7	3999.1	4149.4	4317[a]	4482[a]	4652[a]	4822.0
M_{II}	3415.7	3554.2	3696.3	3854.1	4008	4159	4327	4489.5
M_{III}	2956.6	3066.4	3176.9	3301.9	3426	3538	3663	3791.8
M_{IV}	2485.1	2585.6	2687.6	2798.0	2908.7	3021.5	3136.2	3248.4
M_{V}	2389.3	2484.0	2579.6	2683.0	2786.7	2892.4	2999.9	3104.9
N_{I}	845.5	893.6	938.2	995.3	1042[a]	1097[a]	1153[a]	1208.4
N_{II}	721.3	763.9	805.3	851	886	929	980	1057.6
N_{III}	609.0	644.5	678.9	705	740	768	810	879.1
N_{IV}	406.6	435.2	463.6	500.2	533.2	566.6	603.3	635.9
N_{V}	386.2	412.9	440.0	473.4			577	602.7
N_{VI}	122.8	142.9	161.9					298.9
N_{VII}	118.5	138.1	157.4					298.9*
O_{I}	136.3	147.3	159.3					254.4
O_{II}	99.6	104.8	116.8					200.4
O_{III}	75.4	86.0	92.8					152.8
O_{IV}	15.3	21.8	26.5	31.4				67.2
O_{V}	13.1	19.2	24.4	31.4*				67.2*
P_{I}		3.1						43.5
$P_{II,III}$		0.7	2.7					18.8

[a] Interpolated

Level	$_{89}$Ac	$_{90}$Th	$_{91}$Pa	$_{92}$U	$_{93}$Np	$_{94}$Pu	$_{95}$Am	$_{96}$Cm
K	106755.3	109650.9	112601.4	115606.1	118678	121818	125027	128220
L_{I}	19840	20472.1	21104.6	21757.4	22426.8	23097.2	23772.9	24460
L_{II}	19083.2	19693.2	20313.7	20947.6	21600.5	22266.2	22944.0	23779
L_{III}	15871.0	16300.3	16733.1	17166.3	17610.0	18056.8	18504.1	18930
M_{I}	5002[a]	5182.3	5366.9	5548.0	5723.2	5932.9	6120.5	6288
M_{II}	4656	4830.4	5000.9	5182.2	5366.2	5541.2	5710.2	5895
M_{III}	3909	4046.1	4173.8	4303.4	4434.7	4556.6	4667.0	4797

Level	$_{89}$Ac	$_{90}$Th	$_{91}$Pa	$_{92}$U	$_{93}$Np	$_{94}$Pu	$_{95}$Am	$_{96}$Cm
M_{IV}	3370.2	3490.8	3611.2	3727.6	3850.3	3972.6	4092.1	4227
M_{V}	3219.0	3332.0	3441.8	3551.7	3665.8	3778.1	3886.9	3971
N_{I}	1269[a]	1329.5	1387.1	1440.8	1500.7	1558.6	1617.1	1643
N_{II}	1080	1168.2	1224.3	1272.6	1327.7	1372.1	1411.8	1440
N_{III}	890	967.3	1006.7	1044.9	1086.8	1114.8	1135.7[a]	1154
N_{IV}	674.9	714.1	743.4	780.4	815.9	848.9	878.7	
N_{V}		676.4	708.2	737.7	770.3	801.4	827.6	
N_{VI}		344.4	371.2	391.3	415.0	445.8		
N_{VII}		335.2	359.5	380.9	404.4	432.4		
O_{I}		290.2	309.6	323.7		351.9		385
O_{II}		229.4	222.9	259.3	283.4	274.1		
O_{III}		181.8	222.9*	195.1	206.1	206.5		
O_{IV}		94.3	94.1	105.0	109.3	116.0	115.8	
O_{V}		87.9	94.1*	96.3	101.3	105.4	103.3	
P_{I}		59.5		70.7				
P_{II}		49.0		42.3				
P_{III}		43.0		32.3				

[a] Interpolavted

Level	$_{97}$Bk	$_{98}$Cf	$_{99}$Es	$_{100}$Fm	$_{101}$Md	$_{102}$No	$_{103}$Lr
K	131590	135960	139490	143090	146780	150540	154380
L_{I}	25275	26110	26900	27700	28530	29380	30240
L_{II}	24385	25250	26020	26810	27610	28440	29280
L_{III}	19452	19930	20410	20900	21390	21880	22360
M_{I}	6556	6754	6977	7205	7441	7675	7900
M_{II}	6147	6359	6574	6793	7019	7245	7460
M_{III}	4977	5109	5252	5397	5546	5688	5710
M_{IV}	4366	4497	4630	4766	4903	5037	5150
M_{V}	4132	4253	4374	4498	4622	4741	4860
N_{I}	1755	1799	1868	1937	2010	2078	2140
N_{II}	1554	1616	1680	1747	1814	1876	1930
N_{III}	1235	1279	1321	1366	1410	1448	1480
O_{I}	398	419	435	454	472	484	490

ELECTRON BINDING ENERGIES OF THE ELEMENTS

Gwyn P. Williams

This table gives the binding energies in electron volts (eV) for selected electronic levels of the elements. For metallic elements the binding energy is referred to the Fermi level; for semiconductors, to the valence band maximum; and for gases and insulators, to the vacuum level. The atomic number is listed after the element name.

References

1. Fluggle, J. C., and Martensson, N., *J. Elect. Spect.*, 21, 275, 1980.
2. Cardona, M. and Ley, L., *Photoemission from Solids*, Springer-Verlag, Heidelberg, 1978.
3. Bearden, J. A. and Burr, A. F., *Rev. Mod. Phys.*, 39, 125, 1967.

Level	Electron	Binding energy/eV
Actinium (89)		
K	1s	106755
L I	2s	19840
L II	$2p_{1/2}$	19083
L III	$2p_{3/2}$	15871
M I	3s	5002
M II	$3p_{1/2}$	4656
M III	$3p_{3/2}$	3909
M IV	$3d_{3/2}$	3370
M V	$3d_{5/2}$	3219
N I	4s	1269[a]
N II	$4p_{1/2}$	1080[a]
N III	$4p_{3/2}$	890[a]
N IV	$4d_{3/2}$	675[a]
N V	$4d_{5/2}$	639[a]
N VI	$4f_{5/2}$	319[a]
N VII	$4f_{7/2}$	319[a]
O I	5s	272[a]
O II	$5p_{1/2}$	215[a]
O III	$5p_{3/2}$	167[a]
O IV	$5d_{3/2}$	80[a]
O V	$5d_{5/2}$	80[a]
P I	6s	
P II	$6p_{1/2}$	
P III	$6p_{3/2}$	
Aluminum (13)		
K	1s	1559.0
L I	2s	117.8[a]
L II	$2p_{1/2}$	72.9[a]
L III	$2p_{3/2}$	72.5[a]
Antimony (51)		
K	1s	30419
L I	2s	4698
L II	$2p_{1/2}$	4380
L III	$2p_{3/2}$	4132
M I	3s	946[b]
M II	$3p_{1/2}$	812.7[b]
M III	$3p_{3/2}$	766.4[b]
M IV	$3d_{3/2}$	537.5[b]
M V	$3d_{5/2}$	528.2[b]
N I	4s	153.2[b]
N II	$4p_{1/2}$	95.6[b, c]
N III	$4p_{3/2}$	95.6[b]
N IV	$4d_{3/2}$	33.3[b]
N V	$4d_{5/2}$	32.1[b]
Argon (18)		
K	1s	3205.9[a]
L I	2s	326.3[a]
L II	$2p_{1/2}$	250.6[a]
L III	$2p_{3/2}$	248.4[a]
M I	3s	29.3[a]
M II	$3p_{1/2}$	15.9[a]
M III	$3p_{3/2}$	15.7[a]
Arsenic (33)		
K	1s	11867
L I	2s	1527.0[a, d]
L II	$2p_{1/2}$	1359.1[a, d]
L III	$2p_{3/2}$	1323.6[a, d]
M I	3s	204.7[a]
M II	$3p_{1/2}$	146.2[a]
M III	$3p_{3/2}$	141.2[a]
M IV	$3d_{3/2}$	41.7[a]
M V	$3d_{5/2}$	41.7[a]
Astatine (85)		
K	1s	95730
L I	2s	17493
L II	$2p_{1/2}$	16785
L III	$2p_{3/2}$	14214
M I	3s	4317
M II	$3p_{1/2}$	4008
M III	$3p_{3/2}$	3426
M IV	$3d_{3/2}$	2909
M V	$3d_{5/2}$	2787
N I	4s	1042[a]
N II	$4p_{1/2}$	886[a]
N III	$4p_{3/2}$	740[a]
N IV	$4d_{3/2}$	533[a]
N V	$4d_{5/2}$	507[a]
N VI	$4f_{5/2}$	210[a]
N VII	$4f_{7/2}$	210[a]
O I	5s	195[a]
O II	$5p_{1/2}$	148[a]
O III	$5p_{3/2}$	115[a]
O IV	$5d_{3/2}$	40[a]
O V	$5d_{5/2}$	40[a]
Barium (56)		
K	1s	37441
L I	2s	5989
L II	$2p_{1/2}$	5624
L III	$2p_{3/2}$	5247
M I	3s	1293[a, d]
M II	$3p_{1/2}$	1137[a, d]
M III	$3p_{3/2}$	1063[a, d]
M IV	$3d_{3/2}$	795.7[a]
M V	$3d_{5/2}$	780.5[a]
N I	4s	253.5[b]
N II	$4p_{1/2}$	192
N III	$4p_{3/2}$	178.6[b]
N IV	$4d_{3/2}$	92.6[b]
N V	$4d_{5/2}$	89.9[b]
N VI	$4f_{5/2}$	
N VII	$4f_{7/2}$	
O I	5s	30.3[b]
O II	$5p_{1/2}$	17.0[b]
O III	$5p_{3/2}$	14.8[b]
Beryllium (4)		
K	1s	111.5[a]
Bismuth (83)		
K	1s	90526
L I	2s	16388
L II	$2p_{1/2}$	15711
L III	$2p_{3/2}$	13419
M I	3s	3999
M II	$3p_{1/2}$	3696
M III	$3p_{3/2}$	3177
M IV	$3d_{3/2}$	2688
M V	$3d_{5/2}$	2580
N I	4s	939[b]
N II	$4p_{1/2}$	805.2[b]
N III	$4p_{3/2}$	678.8[b]
N IV	$4d_{3/2}$	464.0[b]
N V	$4d_{5/2}$	440.1[b]
N VI	$4f_{5/2}$	162.3[b]
N VII	$4f_{7/2}$	157.0[b]
O I	5s	159.3[a, d]
O II	$5p_{1/2}$	119.0[b]
O III	$5p_{3/2}$	92.6[b]
O IV	$5d_{3/2}$	26.9[b]
O V	$5d_{5/2}$	23.8[b]
Boron (5)		
K	1s	188[a]
Bromine (35)		
K	1s	13474
L I	2s	1782[a]
L II	$2p_{1/2}$	1596[a]
L III	$2p_{3/2}$	1550[a]
M I	3s	257[a]
M II	$3p_{1/2}$	189[a]
M III	$3p_{3/2}$	182[a]
M IV	$3d_{3/2}$	70[a]
M V	$3d_{5/2}$	69[a]
Cadmium (48)		
K	1s	26711
L I	2s	4018
L II	$2p_{1/2}$	3727
L III	$2p_{3/2}$	3538
M I	3s	772.0[b]
M II	$3p_{1/2}$	652.6[b]
M III	$3p_{3/2}$	618.4[b]
M IV	$3d_{3/2}$	411.9[b]
M V	$3d_{5/2}$	405.2[b]
N I	4s	109.8[b]
N II	$4p_{1/2}$	63.9[b, c]
N III	$4p_{3/2}$	63.9[b, c]
N IV	$4d_{3/2}$	11.7[b]
N V	$4d_{5/2}$	10.7[b]
Calcium (20)		
K	1s	4038.5[a]
L I	2s	438.4[b]
L II	$2p_{1/2}$	349.7[b]
L III	$2p_{3/2}$	346.2[b]
M I	3s	44.3[b]
M II	$3p_{1/2}$	25.4[b]
M III	$3p_{3/2}$	25.4[b]
Carbon (6)		
K	1s	284.2[a]
Cerium (58)		
K	1s	40443
L I	2s	6548
L II	$2p_{1/2}$	6164
L III	$2p_{3/2}$	5723
M I	3s	1436[a, d]
M II	$3p_{1/2}$	1274[a, d]
M III	$3p_{3/2}$	1187[a, d]
M IV	$3d_{3/2}$	902.4[a]
M V	$3d_{5/2}$	883.8[a]
N I	4s	291.0[a]
N II	$4p_{1/2}$	223.3
N III	$4p_{3/2}$	206.5[a]
N IV	$4d_{3/2}$	109[a]
N V	$4d_{5/2}$	
N VI	$4f_{5/2}$	0.1
N VII	$4f_{7/2}$	0.1

Level	Electron	Binding energy/eV
O I	5s	37.8
O II	$5p_{1/2}$	19.8[a]
O III	$5p_{3/2}$	17.0[a]

Cesium (55)

Level	Electron	Binding energy/eV
K	1s	35985
L I	2s	5714
L II	$2p_{1/2}$	5359
L III	$2p_{3/2}$	5012
M I	3s	1211[a,d]
M II	$3p_{1/2}$	1071[a]
M III	$3p_{3/2}$	1003[a]
M IV	$3d_{3/2}$	740.5[a]
M V	$3d_{5/2}$	726.6[a]
N I	4s	232.3[a]
N II	$4p_{1/2}$	172.4[a]
N III	$4p_{3/2}$	161.3[a]
N IV	$4d_{3/2}$	79.8[a]
N V	$4d_{5/2}$	77.5[a]
N VI	$4f_{5/2}$	
N VII	$4f_{7/2}$	
O I	5s	22.7
O II	$5p_{1/2}$	14.2[a]
O III	$5p_{3/2}$	12.1[a]

Chlorine (17)

Level	Electron	Binding energy/eV
K	1s	2822.0
L I	2s	270[a]
L II	$2p_{1/2}$	202[a]
L III	$2p_{3/2}$	200[a]

Chromium (24)

Level	Electron	Binding energy/eV
K	1s	5989
L I	2s	696.0[b]
L II	$2p_{1/2}$	583.8[b]
L III	$2p_{3/2}$	574.1[b]
M I	3s	74.1[b]
M II	$3p_{1/2}$	42.2[b]
M III	$3p_{3/2}$	42.2[b]

Cobalt (27)

Level	Electron	Binding energy/eV
K	1s	7709
L I	2s	925.1[b]
L II	$2p_{1/2}$	793.2[b]
L III	$2p_{3/2}$	778.1[b]
M I	3s	101.0[b]
M II	$3p_{1/2}$	58.9[b]
M III	$3p_{3/2}$	58.9[b]

Copper (29)

Level	Electron	Binding energy/eV
K	1s	8979
L I	2s	1096.7[b]
L II	$2p_{1/2}$	952.3[b]
L III	$2p_{3/2}$	932.5[b]
M I	3s	122.5[b]
M II	$3p_{1/2}$	77.3[b]
M III	$3p_{3/2}$	75.1[b]

Dysprosium (66)

Level	Electron	Binding energy/eV
K	1s	53789
L I	2s	9046
L II	$2p_{1/2}$	8581
L III	$2p_{3/2}$	7790
M I	3s	2047
M II	$3p_{1/2}$	1842
M III	$3p_{3/2}$	1676
M IV	$3d_{3/2}$	1333
M V	$3d_{5/2}$	1292[a]
N I	4s	414.2[a]
N II	$4p_{1/2}$	333.5[a]
N III	$4p_{3/2}$	293.2[a]
N IV	$4d_{3/2}$	153.6[a]
N V	$4d_{5/2}$	153.6[a]
N VI	$4f_{5/2}$	8.0[a]
N VII	$4f_{7/2}$	4.3[a]
O I	5s	49.9[a]
O II	$5p_{1/2}$	26.3
O III	$5p_{3/2}$	26.3

Erbium (68)

Level	Electron	Binding energy/eV
K	1s	57486
L I	2s	9751
L II	$2p_{1/2}$	9264
L III	$2p_{3/2}$	8358
M I	3s	2206
M II	$3p_{1/2}$	2006
M III	$3p_{3/2}$	1812
M IV	$3d_{3/2}$	1453
M V	$3d_{5/2}$	1409
N I	4s	449.8[a]
N II	$4p_{1/2}$	366.2
N III	$4p_{3/2}$	320.2[a]
N IV	$4d_{3/2}$	167.6[a]
N V	$4d_{5/2}$	167.6[a]
N VI	$4f_{5/2}$	
N VII	$4f_{7/2}$	4.7[a]
O I	5s	50.6[a]
O II	$5p_{1/2}$	31.4[a]
O III	$5p_{3/2}$	24.7[a]

Europium (63)

Level	Electron	Binding energy/eV
K	1s	48519
L I	2s	8052
L II	$2p_{1/2}$	7617
L III	$2p_{3/2}$	6977
M I	3s	1800
M II	$3p_{1/2}$	1614
M III	$3p_{3/2}$	1481
M IV	$3d_{3/2}$	1158.6[a]
M V	$3d_{5/2}$	1127.5[a]
N I	4s	360
N II	$4p_{1/2}$	284
N III	$4p_{3/2}$	257
N IV	$4d_{3/2}$	133
N V	$4d_{5/2}$	1227[a]
N VI	$4f_{5/2}$	0
N VII	$4f_{7/2}$	0
O I	5s	32
O II	$5p_{1/2}$	22
O III	$5p_{3/2}$	22

Fluorine (9)

Level	Electron	Binding energy/eV
K	1s	696.7[a]

Francium (87)

Level	Electron	Binding energy/eV
K	1s	101137
L I	2s	18639
L II	$2p_{1/2}$	17907
L III	$2p_{3/2}$	15031
M I	3s	4652
M II	$3p_{1/2}$	4327
M III	$3p_{3/2}$	3663
M IV	$3d_{3/2}$	3136
M V	$3d_{5/2}$	3000
N I	4s	1153[a]
N II	$4p_{1/2}$	980[a]
N III	$4p_{3/2}$	810[a]
N IV	$4d_{3/2}$	603[a]
N V	$4d_{5/2}$	577[a]
N VI	$4f_{5/2}$	268[a]
N VII	$4f_{7/2}$	268[a]
O I	5s	234[a]
O II	$5p_{1/2}$	182[a]
O III	$5p_{3/2}$	140[a]
O IV	$5d_{3/2}$	58[a]
O V	$5d_{5/2}$	58[a]
P I	6s	34
P II	$6p_{1/2}$	15
P III	$6p_{3/2}$	15

Gadolinium (64)

Level	Electron	Binding energy/eV
K	1s	50239
L I	2s	8376
L II	$2p_{1/2}$	7930
L III	$2p_{3/2}$	7243
M I	3s	1881
M II	$3p_{1/2}$	1688
M III	$3p_{3/2}$	1544
M IV	$3d_{3/2}$	1221.9[a]
M V	$3d_{5/2}$	1189.6[a]
N I	4s	378.6[a]
N II	$4p_{1/2}$	286
N III	$4p_{3/2}$	271
N IV	$4d_{3/2}$	
N V	$4d_{5/2}$	142.6[a]
N VI	$4f_{5/2}$	8.6[a]
N VII	$4f_{7/2}$	8.6[a]
O I	5s	36
O II	$5p_{1/2}$	20
O III	$5p_{3/2}$	20

Gallium (31)

Level	Electron	Binding energy/eV
K	1s	10367
L I	2s	1299.0[a,d]
L II	$2p_{1/2}$	1143.2[b]
L III	$2p_{3/2}$	1116.4[b]
M I	3s	159.5[b]
M II	$3p_{1/2}$	103.5[b]
M III	$3p_{3/2}$	100.0[b]
M IV	$3d_{3/2}$	18.7[b]
M V	$3d_{5/2}$	18.7[b]

Germanium (32)

Level	Electron	Binding energy/eV
K	1s	11103
L I	2s	1414.6[a,d]
L II	$2p_{1/2}$	1248.1[a,d]
L III	$2p_{3/2}$	1217.0[a,d]
M I	3s	180.1[a]
M II	$3p_{1/2}$	124.9[a]
M III	$3p_{3/2}$	120.8[a]
M IV	$3d_{3/2}$	29.8[a]
M V	$3d_{5/2}$	29.2[a]

Gold (79)

Level	Electron	Binding energy/eV
K	1s	80725
L I	2s	14353
L II	$2p_{1/2}$	13734
L III	$2p_{3/2}$	11919
M I	3s	3425
M II	$3p_{1/2}$	3148
M III	$3p_{3/2}$	2743
M IV	$3d_{3/2}$	2291
M V	$3d_{5/2}$	2206
N I	4s	762.1[b]
N II	$4p_{1/2}$	642.7[b]
N III	$4p_{3/2}$	546.3[b]
N IV	$4d_{3/2}$	353.2[b]
N V	$4d_{5/2}$	335.1[b]
N VI	$4f_{5/2}$	87.6[b]
N VII	$4f_{7/2}$	83.9[b]
O I	5s	107.2[a,d]
O II	$5p_{1/2}$	74.2[b]
O III	$5p_{3/2}$	57.2[b]

Hafnium (72)

Level	Electron	Binding energy/eV
K	1s	65351
L I	2s	11271
L II	$2p_{1/2}$	10739
L III	$2p_{3/2}$	9561
M I	3s	2601
M II	$3p_{1/2}$	2365
M III	$3p_{3/2}$	2107
M IV	$3d_{3/2}$	1176
M V	$3d_{5/2}$	1662
N I	4s	538[a]
N II	$4p_{1/2}$	438.2[b]
N III	$4p_{3/2}$	380.7[b]
N IV	$4d_{3/2}$	220.0[b]
N V	$4d_{5/2}$	211.5[b]
N VI	$4f_{5/2}$	15.9[b]
N VII	$4f_{7/2}$	14.2[b]
O I	5s	64.2[b]
O II	$5p_{1/2}$	38[a]
O III	$5p_{3/2}$	29.9[b]

Helium (2)

Level	Electron	Binding energy/eV
K	1s	24.6[a]

Holmium (67)

Level	Electron	Binding energy/eV
K	1s	55618
L I	2s	9394
L II	$2p_{1/2}$	8918
L III	$2p_{3/2}$	8071
M I	3s	2128
M II	$3p_{1/2}$	1923
M III	$3p_{3/2}$	1741
M IV	$3d_{3/2}$	1392

Level	Electron	Binding energy/eV
M V	$3d_{5/2}$	1351
N I	4s	432.4[a]
N II	$4p_{1/2}$	343.5
N III	$4p_{3/2}$	308.2[a]
N IV	$4d_{3/2}$	160[a]
N V	$4d_{5/2}$	160[a]
N VI	$4f_{5/2}$	8.6[a]
N VII	$4f_{7/2}$	5.2[a]
O I	5s	49.3[a]
O II	$5p_{1/2}$	30.8[a]
O III	$5p_{3/2}$	24.1[a]

Hydrogen (1)

Level	Electron	Binding energy/eV
K	1s	13.6

Indium (49)

Level	Electron	Binding energy/eV
K	1s	27940
L I	2s	4238
L II	$2p_{1/2}$	3938
L III	$2p_{3/2}$	3730
M I	3s	827.2[b]
M II	$3p_{1/2}$	703.2[b]
M III	$3p_{3/2}$	665.3[b]
M IV	$3d_{3/2}$	451.4[b]
M V	$3d_{5/2}$	443.9[b]
N I	4s	122.9[b]
N II	$4p_{1/2}$	73.5[b, c]
N III	$4p_{3/2}$	73.5[b, c]
N IV	$4d_{3/2}$	17.7[b]
N V	$4d_{5/2}$	16.9[b]

Iodine (53)

Level	Electron	Binding energy/eV
K	1s	33169
L I	2s	5188
L II	$2p_{1/2}$	4852
L III	$2p_{3/2}$	4557
M I	3s	1072[a]
M II	$3p_{1/2}$	931[a]
M III	$3p_{3/2}$	875[a]
M IV	$3d_{3/2}$	631[a]
M V	$3d_{5/2}$	620[a]
N I	4s	186[a]
N II	$4p_{1/2}$	123[a]
N III	$4p_{3/2}$	123[a]
N IV	$4d_{3/2}$	50[a]
N V	$4d_{5/2}$	50[a]

Iridium (77)

Level	Electron	Binding energy/eV
K	1s	76111
L I	2s	13419
L II	$2p_{1/2}$	12824
L III	$2p_{3/2}$	11215
M I	3s	3174
M II	$3p_{1/2}$	2909
M III	$3p_{3/2}$	2551
M IV	$3d_{3/2}$	2116
M V	$3d_{5/2}$	2040
N I	4s	691.1[b]
N II	$4p_{1/2}$	577.8[b]
N III	$4p_{3/2}$	495.8[b]
N IV	$4d_{3/2}$	311.9[b]
N V	$4d_{5/2}$	296.3[b]
N VI	$4f_{5/2}$	63.8[b]
N VII	$4f_{7/2}$	60.8[b]
O I	52	95.2[a, d]
O II	$5p_{1/2}$	63.0[a, d]
O III	$5p_{3/2}$	48.0[b]

Iron (26)

Level	Electron	Binding energy/eV
K	1s	7112
L I	2s	844.6[b]
L II	$2p_{1/2}$	719.9[b]
L III	$2p_{3/2}$	706.8[b]
M I	3s	91.3[b]
M II	$3p_{1/2}$	52.7[b]
M III	$3p_{3/2}$	52.7[b]

Krypton (36)

Level	Electron	Binding energy/eV
K	1s	14326
L I	2s	1921
L II	$2p_{1/2}$	1730.9[a]
L III	$2p_{3/2}$	1678.4[a]
M I	3s	292.8[a]
M II	$3p_{1/2}$	222.2[a]
M III	$3p_{3/2}$	214.4[a]
M IV	$3d_{3/2}$	95.0[a]
M V	$3d_{5/2}$	93.8[a]
N I	4s	27.5[a]
N II	$4p_{1/2}$	14.1[a]
N III	$4p_{3/2}$	14.1[a]

Lanthanum (57)

Level	Electron	Binding energy/eV
K	1s	38925
L I	2s	6266
L II	$2p_{1/2}$	5891
L III	$2p_{3/2}$	5483
M I	3s	1362[a, d]
M II	$3p_{1/2}$	1209[a, d]
M III	$3p_{3/2}$	1128[a, d]
M IV	$3d_{3/2}$	853[a]
M V	$3d_{5/2}$	836[a]
N I	4s	247.7[a]
N II	$4p_{1/2}$	205.8
N III	$4p_{3/2}$	196.0[a]
N IV	$4d_{3/2}$	105.3[a]
N V	$4d_{5/2}$	102.5[a]
N VI	$4f_{5/2}$	
N VII	$4f_{7/2}$	
O I	5s	34.3[a]
O II	$5p_{1/2}$	19.3[a]
O III	$5p_{3/2}$	16.8[a]

Lead (82)

Level	Electron	Binding energy/eV
K	1s	88005
L I	2s	15861
L II	$2p_{1/2}$	15200
L III	$2p_{3/2}$	13055
M I	3s	3851
M II	$3p_{1/2}$	3554
M III	$3p_{3/2}$	3066
M IV	$3d_{3/2}$	2586
M V	$3d_{5/2}$	2484
N I	4s	891.8[b]
N II	$4p_{1/2}$	761.9[b]
N III	$4p_{3/2}$	643.5[b]
N IV	$4d_{3/2}$	434.3[b]
N V	$4d_{5/2}$	412.2[b]
N VI	$4f_{5/2}$	141.7[b]
N VII	$4f_{7/2}$	136.9[b]
O I	5s	147[a, d]
O II	$5p_{1/2}$	106.4[b]
O III	$5p_{3/2}$	83.3[b]
O IV	$5d_{3/2}$	20.7[b]
O V	$5d_{5/2}$	18.1[b]

Lithium (3)

Level	Electron	Binding energy/eV
K	1s	54.7[a]

Lutetium (71)

Level	Electron	Binding energy/eV
K	1s	63314
L I	2s	10870
L II	$2p_{1/2}$	10349
L III	$2p_{3/2}$	9244
M I	3s	2491
M II	$3p_{1/2}$	2264
M III	$3p_{3/2}$	2024
M IV	$3d_{3/2}$	1639
M V	$3d_{5/2}$	1589
N I	4s	506.8[a]
N II	$4p_{1/2}$	412.4[a]
N III	$4p_{3/2}$	359.2[a]
N IV	$4d_{3/2}$	206.1[a]
N V	$4d_{5/2}$	196.3[a]
N VI	$4f_{5/2}$	8.9[a]
N VII	$4f_{7/2}$	7.5[a]
O I	5s	57.3[a]
O II	$5p_{1/2}$	33.6[a]
O III	$5p_{3/2}$	26.7[a]

Magnesium (12)

Level	Electron	Binding energy/eV
K	1s	1303.0[b]
L I	2s	88.6[a]
L II	$2p_{1/2}$	49.6[b]
L III	$2p_{3/2}$	49.2[a]

Manganese (25)

Level	Electron	Binding energy/eV
K	1s	6539
L I	2s	769.1[b]
L II	$2p_{1/2}$	649.9[b]
L III	$2p_{3/2}$	638.7[b]
M I	3s	82.3[b]
M II	$3p_{1/2}$	47.2[b]
M III	$3p_{3/2}$	47.2[b]

Mercury (80)

Level	Electron	Binding energy/eV
K	1s	83102
L I	2s	14839
L II	$2p_{1/2}$	14209
L III	$2p_{3/2}$	12284
M I	3s	3562
M II	$3p_{1/2}$	3279
M III	$3p_{3/2}$	2847
M IV	$3d_{3/2}$	2385
M V	$3d_{5/2}$	2295
N I	4s	802.2[b]
N II	$4p_{1/2}$	680.2[b]
N III	$4p_{3/2}$	576.6[b]
N IV	$4d_{3/2}$	378.2[b]
N V	$4d_{5/2}$	358.8[b]
N VI	$4f_{5/2}$	104.0[b]
N VII	$4f_{7/2}$	99.9[b]
O I	5s	127[b]
O II	$5p_{3/2}$	83.1[b]
O III	$5p_{3/2}$	64.5[b]
O IV	$5d_{3/2}$	9.6[b]
O V	$5d_{5/2}$	7.8[b]

Molybdenum (42)

Level	Electron	Binding energy/eV
K	1s	20000
L I	2s	2866
L II	$2p_{1/2}$	2625
L III	$2p_{3/2}$	2520
M I	3s	506.3[b]
M II	$3p_{1/2}$	411.6[b]
M III	$3p_{3/2}$	394.0[b]
M IV	$3d_{3/2}$	231.1[b]
M V	$3d_{5/2}$	227.9[b]
N I	4s	63.2[b]
N II	$4p_{1/2}$	37.6[b]
N III	$4p_{3/2}$	35.5[b]

Neodymium (60)

Level	Electron	Binding energy/eV
K	1s	43569
L I	2s	7126
L II	$2p_{1/2}$	6722
L III	$2p_{3/2}$	6208
M I	3s	1575
M II	$3p_{1/2}$	1403
M III	$3p_{3/2}$	1297
M IV	$3d_{3/2}$	1003.3[a]
M V	$3d_{5/2}$	980.4[a]
N I	4s	319.2[a]
N II	$4p_{1/2}$	243.3
N III	$4p_{3/2}$	224.6
N IV	$4d_{3/2}$	120.5[a]
N V	$4d_{5/2}$	120.5[a]
N VI	$4f_{5/2}$	1.5
N VII	$4f_{7/2}$	1.5
O I	5s	37.5
O II	$5p_{1/2}$	21.1
O III	$5p_{3/2}$	21.1

Neon (10)

Level	Electron	Binding energy/eV
K	1s	870.2[a]
L I	2s	48.5[a]
L II	$2p_{1/2}$	21.7[a]
L III	$2p_{3/2}$	21.6[a]

Nickel (28)

Level	Electron	Binding energy/eV
K	1s	8333
L I	2s	1008.6[b]
L II	$2p_{1/2}$	870.0[b]
L III	$2p_{3/2}$	852.7[b]
M I	3s	110.8[b]
M II	$3p_{1/2}$	68.0[b]
M III	$3p_{3/2}$	66.2[b]

Level	Electron	Binding energy/eV
Niobium (41)		
K	1s	18986
L I	2s	2698
L II	$2p_{1/2}$	2465
L III	$2p_{3/2}$	2371
M I	3s	466.6[b]
M II	$3p_{1/2}$	376.1[b]
M III	$3p_{3/2}$	360.6[b]
M IV	$3d_{3/2}$	205.0[b]
M V	$3d_{5/2}$	202.3[b]
N I	4s	56.4[b]
N II	$4p_{1/2}$	32.6[b]
N III	$4p_{3/2}$	30.8[b]
Nitrogen (7)		
K	1s	409.9[a]
L I	2s	37.3[a]
Osmium (76)		
K	1s	73871
L I	2s	12968
L II	$2p_{1/2}$	12385
L III	$2p_{3/2}$	10871
M I	3s	3049
M II	$3p_{1/2}$	2792
M III	$3p_{3/2}$	2457
M IV	$3d_{3/2}$	2031
M V	$3d_{5/2}$	1960
N I	4s	658.2[b]
N II	$4p_{1/2}$	549.1[b]
N III	$4p_{3/2}$	470.7[b]
N IV	$4d_{3/2}$	293.1[b]
N V	$4d_{5/2}$	278.5[b]
N VI	$4f_{5/2}$	53.4[b]
N VII	$4f_{7/2}$	50.7[b]
O I	5s	84[a]
O II	$5p_{1/2}$	58[a]
O III	$5p_{3/2}$	44.5[b]
Oxygen (8)		
K	1s	543.1[a]
L I	2s	41.6[a]
Palladium (46)		
K	1s	24350
L I	2s	3604
L II	$2p_{1/2}$	3330
L III	$2p_{3/2}$	3173
M I	3s	671.6[b]
M II	$3p_{1/2}$	559.9[b]
M III	$3p_{3/2}$	532.3[b]
M IV	$3d_{3/2}$	340.5[b]
M V	$3d_{5/2}$	335.2[b]
N I	4s	87.1[a, d]
N II	$4p_{1/2}$	55.7[b, c]
N III	$4p_{3/2}$	50.9[b, c]
Phosphorus (15)		
K	1s	2145.5
L I	2s	189[a]
L II	$2p_{1/2}$	136[a]
L III	$2p_{3/2}$	135[a]
Platinum (78)		
K	1s	78395
L I	2s	13880
L II	$2p_{1/2}$	13273
L III	$2p_{3/2}$	11564
M I	3s	3296
M II	$3p_{1/2}$	3027
M III	$3p_{3/2}$	2645
M IV	$3d_{3/2}$	2202
M V	$3d_{5/2}$	2122
N I	4s	725.4[b]
N II	$4p_{1/2}$	609.1[b]
N III	$4p_{3/2}$	519.4[b]
N IV	$4d_{3/2}$	331.6[b]
N V	$4d_{5/2}$	314.6[b]
N VI	$4f_{5/2}$	74.5[b]
N VII	$4f_{7/2}$	71.2[b]
O I	5s	101.7[a, d]
O II	$5p_{1/2}$	65.3[a, b]
O III	$5p_{3/2}$	51.7[b]
Polonium (84)		
K	1s	93105
L I	2s	16939
L II	$2p_{1/2}$	16244
L III	$2p_{3/2}$	13814
M I	3s	4149
M II	$3p_{1/2}$	3854
M III	$3p_{3/2}$	3302
M IV	$3d_{3/2}$	2798
M V	$3d_{5/2}$	2683
N I	4s	995[a]
N II	$4p_{1/2}$	851[a]
N III	$4p_{3/2}$	705[a]
N IV	$4d_{3/2}$	500[a]
N V	$4d_{5/2}$	473[a]
N VI	$4f_{5/2}$	184[a]
N VII	$4f_{7/2}$	184[a]
O I	5s	177[a]
O II	$5p_{1/2}$	132[a]
O III	$5p_{3/2}$	104[a]
O IV	$5d_{3/2}$	31[a]
O V	$5d_{5/2}$	31[a]
Potassium (19)		
K	1s	3608.4[a]
L I	2s	378.6[a]
L II	$2p_{1/2}$	297.3[a]
L III	$2p_{3/2}$	294.6[a]
M I	3s	34.8[a]
M II	$3p_{1/2}$	18.3[a]
M III	$3p_{3/2}$	18.3[a]
Praseodymium (59)		
K	1s	41991
L I	2s	6835
L II	$2p_{1/2}$	6440
L III	$2p_{3/2}$	5964
M I	3s	1511
M II	$3p_{1/2}$	1337
M III	$3p_{3/2}$	1242
M IV	$3d_{3/2}$	948.3[a]
M V	$3d_{5/2}$	928.8[a]
N I	4s	304.5
N II	$4p_{1/2}$	236.3
N III	$4p_{3/2}$	217.6
N IV	$4d_{3/2}$	115.1[a]
N V	$4d_{5/2}$	115.1[a]
N VI	$4f_{5/2}$	2.0
N VII	$4f_{7/2}$	2.0
O I	5s	37.4
O II	$5p_{1/2}$	22.3
O III	$5p_{3/2}$	22.3
Promethium (61)		
K	1s	45184
L I	2s	7428
L II	$2p_{1/2}$	7013
L III	$2p_{3/2}$	6459
M I	3s	
M II	$3p_{1/2}$	1471.4
M III	$3p_{3/2}$	1357
M IV	$3d_{3/2}$	1052
M V	$3d_{5/2}$	1027
N I	4s	
N II	$4p_{1/2}$	242
N III	$4p_{3/2}$	242
N IV	$4d_{3/2}$	120
N V	$4d_{5/2}$	120
Protactinium (91)		
K	1s	112601
L I	2s	21105
L II	$2p_{1/2}$	20314
L III	$2p_{3/2}$	16733
M I	3s	5367
M II	$3p_{1/2}$	5001
M III	$3p_{3/2}$	4174
M IV	$3d_{3/2}$	3611
M V	$3d_{5/2}$	3442
N I	4s	1387[a]
N II	$4p_{1/2}$	1224[a]
N III	$4p_{3/2}$	1007[a]
N IV	$4d_{3/2}$	743[a]
N V	$4d_{5/2}$	708[a]
N VI	$4f_{5/2}$	371[a]
N VII	$4f_{7/2}$	360[a]
O I	5s	310[a]
O II	$5p_{1/2}$	232[a]
O III	$5p_{3/2}$	232[a]
O IV	$5d_{3/2}$	94[a]
O V	$5d_{5/2}$	94[a]
P I	6s	
P II	$6p_{1/2}$	
P III	$6p_{3/2}$	
Radium (88)		
K	1s	103922
L I	2s	19237
L II	$2p_{1/2}$	18484
L III	$2p_{3/2}$	15444
M I	3s	4822
M II	$3p_{1/2}$	4490
M III	$3p_{3/2}$	3792
M IV	$3d_{3/2}$	3248
M V	$3d_{5/2}$	3105
N I	4s	1208[a]
N II	$4p_{1/2}$	1058
N III	$4p_{3/2}$	879[a]
N IV	$4d_{3/2}$	636[a]
N V	$4d_{5/2}$	603[a]
N VI	$4f_{5/2}$	299[a]
N VII	$4f_{7/2}$	299[a]
O I	5s	254[a]
O II	$5p_{1/2}$	200[a]
O III	$5p_{3/2}$	153[a]
O IV	$5d_{3/2}$	68[a]
O V	$5d_{5/2}$	68[a]
P I	6s	44
P II	$6p_{1/2}$	19
P III	$6p_{3/2}$	19
Radon (86)		
K	1s	98404
L I	2s	18049
L II	$2p_{1/2}$	17337
L III	$2p_{3/2}$	14619
M I	3s	4482
M II	$3p_{1/2}$	4159
M III	$3p_{3/2}$	3538
M IV	$3d_{3/2}$	3022
M V	$3d_{5/2}$	2892
N I	4s	1097[a]
N II	$4p_{1/2}$	929[a]
N III	$4p_{3/2}$	768[a]
N IV	$4d_{3/2}$	567[a]
N V	$4d_{5/2}$	541[a]
N VI	$4f_{5/2}$	238[a]
N VII	$4f_{7/2}$	238[a]
O I	5s	214[a]
O II	$5p_{1/2}$	164[a]
O III	$5p_{3/2}$	127[a]
O IV	$5d_{3/2}$	48[a]
O V	$5d_{5/2}$	48[a]
P I	6s	26
Rhenium (75)		
K	1s	71676
L I	2s	12527
L II	$2p_{1/2}$	11959
L III	$2p_{3/2}$	10535
M I	3s	2932
M II	$3p_{1/2}$	2682
M III	$3p_{3/2}$	2367
M IV	$3d_{3/2}$	1949
M V	$3d_{5/2}$	1883
N I	4s	625.4[b]
N II	$4p_{1/2}$	518.7[b]
N III	$4p_{3/2}$	446.8[b]
N IV	$4d_{3/2}$	273.9[b]
N V	$4d_{5/2}$	260.5[b]
N VI	$4f_{5/2}$	42.9[a]
N VII	$4f_{7/2}$	40.5[a]
O I	5s	83[b]

Level	Electron	Binding energy/eV
O II	$5p_{1/2}$	45.6[b]
O III	$5p_{3/2}$	34.6[a, d]

Rhodium (45)

Level	Electron	Binding energy/eV
K	1s	23220
L I	2s	3412
L II	$2p_{1/2}$	3146
L III	$2p_{3/2}$	3004
M I	3s	628.1[b]
M II	$3p_{1/2}$	521.3[b]
M III	$3p_{3/2}$	496.5[b]
M IV	$3d_{3/2}$	311.9[b]
M V	$3d_{5/2}$	307.2[b]
N I	4s	81.4[a, d]
N II	$4p_{1/2}$	50.5[b]
N III	$4p_{3/2}$	47.3[b]

Rubidium (37)

Level	Electron	Binding energy/eV
K	1s	15200
L I	2s	2065
L II	$2p_{1/2}$	1864
L III	$2p_{3/2}$	1804
M I	3s	326.7[a]
M II	$3p_{1/2}$	248.7[a]
M III	$3p_{3/2}$	239.1[a]
M IV	$3d_{3/2}$	113.0[a]
M V	$3d_{5/2}$	112[a]
N I	4s	30.5[a]
N II	$4p_{1/2}$	16.3[a]
N III	$4p_{3/2}$	15.3[a]

Ruthenium (44)

Level	Electron	Binding energy/eV
K	1s	22117
L I	2s	3224
L II	$2p_{1/2}$	2967
L III	$2p_{3/2}$	2838
M I	3s	586.2[b]
M II	$3p_{1/2}$	483.3[b]
M III	$3p_{3/2}$	461.5[b]
M IV	$3d_{3/2}$	284.2[b]
M V	$3d_{5/2}$	280.0[b]
N I	4s	75.0[b]
N II	$4p_{1/2}$	46.5[b]
N III	$4p_{3/2}$	43.2[b]

Samarium (62)

Level	Electron	Binding energy/eV
K	1s	46834
L I	2s	7737
L II	$2p_{1/2}$	7312
L III	$2p_{3/2}$	6716
M I	3s	1723
M II	$3p_{1/2}$	1541
M III	$3p_{3/2}$	1419.8
M IV	$3d_{3/2}$	1110.9[a]
M V	$3d_{5/2}$	1083.4[a]
N I	4s	347.2[a]
N II	$4p_{1/2}$	265.6
N III	$4p_{3/2}$	247.4
N IV	$4d_{3/2}$	129.0
N V	$4d_{5/2}$	129.0
N VI	$4f_{5/2}$	5.2
N VII	$4f_{7/2}$	5.2

Level	Electron	Binding energy/eV
O I	5s	37.4
O II	$5p_{1/2}$	21.3
O III	$5p_{3/2}$	21.3

Scandium (21)

Level	Electron	Binding energy/eV
K	1s	4492
L I	2s	498.0[a]
L II	$2p_{1/2}$	403.6[a]
L III	$2p_{3/2}$	389.7[a]
M I	3s	51.1[a]
M II	$3p_{1/2}$	28.3[a]
M III	$3p_{3/2}$	28.3[a]

Selenium (34)

Level	Electron	Binding energy/eV
K	1s	12658
L I	2s	1652.0[a, d]
L II	$2p_{1/2}$	1474.3[a, d]
L III	$2p_{3/2}$	1433.9[a, d]
M I	3s	229.6[a]
M II	$3p_{1/2}$	166.5[a]
M III	$3p_{3/2}$	160.7[a]
M IV	$3d_{3/2}$	55.5[a]
M V	$3d_{5/2}$	54.6[a]

Silicon (14)

Level	Electron	Binding energy/eV
K	1s	1839
L I	2s	149.7[a, d]
L II	$2p_{1/2}$	99.8[a]
L III	$2p_{3/2}$	99.2[a]

Silver (47)

Level	Electron	Binding energy/eV
K	1s	25514
L I	2s	3806
L II	$2p_{1/2}$	3524
L III	$2p_{3/2}$	3351
M I	3s	719.0[b]
M II	$3p_{1/2}$	603.8[b]
M III	$3p_{3/2}$	573.0[b]
M IV	$3d_{3/2}$	374.0[b]
M V	$3d_{5/2}$	368.0[b]
N I	4s	97.0[b]
N II	$4p_{1/2}$	63.7[b]
N III	$4p_{3/2}$	58.3[b]

Sodium (11)

Level	Electron	Binding energy/eV
K	1s	1070.8[b]
L I	2s	63.5[b]
L II	$2p_{1/2}$	30.4[b]
L III	$2p_{3/2}$	30.5[a]

Strontium (38)

Level	Electron	Binding energy/eV
K	1s	16105
L I	2s	2216
L II	$2p_{1/2}$	2007
L III	$2p_{3/2}$	1940
M I	3s	358.7[b]
M II	$3p_{1/2}$	280.3[b]
M III	$3p_{3/2}$	270.0[b]
M IV	$3d_{3/2}$	136.0[b]
M V	$3d_{5/2}$	134.2[b]
N I	4s	38.9[b]
N II	$4p_{1/2}$	21.6[b]
N III	$4p_{3/2}$	20.1[b]

Sulfur (16)

Level	Electron	Binding energy/eV
K	1s	2472
L I	2s	230.9[a, d]
L II	$2p_{1/2}$	163.6[a]
L III	$2p_{3/2}$	162.5[a]

Tantalum (73)

Level	Electron	Binding energy/eV
K	1s	67416
L I	2s	11682
L II	$2p_{1/2}$	11136
L III	$2p_{3/2}$	9881
M I	3s	2708
M II	$3p_{1/2}$	2469
M III	$3p_{3/2}$	2194
M IV	$3d_{3/2}$	1793
M V	$3d_{5/2}$	1735
N I	4s	563.4[b]
N II	$4p_{1/2}$	463.4[b]
N III	$4p_{3/2}$	400.9[b]
N IV	$4d_{3/2}$	237.9[b]
N V	$4d_{5/2}$	226.4[b]
N VI	$4f_{5/2}$	23.5[b]
N VII	$4f_{7/2}$	21.6[b]
O I	5s	69.7[b]
O II	$5p_{1/2}$	42.2[a]
O III	$5p_{3/2}$	32.7[b]

Technetium (43)

Level	Electron	Binding energy/eV
K	1s	21044
L I	2s	3043
L II	$2p_{1/2}$	2793
L III	$2p_{3/2}$	2677
M I	3s	586.1[a]
M II	$3p_{1/2}$	447.6[a]
M III	$3p_{3/2}$	417.7[a]
M IV	$3d_{3/2}$	257.6[a]
M V	$3d_{5/2}$	253.9[a]
N I	4s	69.5[a]
N II	$4p_{1/2}$	42.3[a]
N III	$4p_{3/2}$	39.9[a]

Tellurium (52)

Level	Electron	Binding energy/eV
K	1s	31814
L I	2s	4939
L II	$2p_{1/2}$	4612
L III	$2p_{3/2}$	4341
M I	3s	1006[b]
M II	$3p_{1/2}$	870.8[b]
M III	$3p_{3/2}$	820.0[b]
M IV	$3d_{3/2}$	583.4[b]
M V	$3d_{5/2}$	573.0[b]
N I	4s	169.4[b]
N II	$4p_{1/2}$	103.3[b, c]
N III	$4p_{3/2}$	103.3[b, c]
N IV	$4d_{3/2}$	41.9[b]
N V	$4d_{5/2}$	40.4[b]

Terbium (65)

Level	Electron	Binding energy/eV
K	1s	51996
L I	2s	8708
L II	$2p_{1/2}$	8252
L III	$2p_{3/2}$	7514
M I	3s	1968
M II	$3p_{1/2}$	1768
M III	$3p_{3/2}$	1611
M IV	$3d_{3/2}$	1267.9[a]
M V	$3d_{5/2}$	1241.1[a]
N I	4s	396.0[a]
N II	$4p_{1/2}$	322.4[a]
N III	$4p_{3/2}$	284.1[a]
N IV	$4d_{3/2}$	150.5[a]
N V	$4d_{5/2}$	150.5[a]
N VI	$4f_{5/2}$	7.7[a]
N VII	$4f_{7/2}$	2.4[a]
O I	5s	45.6[a]
O II	$5p_{1/2}$	28.7[a]
O III	$5p_{3/2}$	22.6[a]

Thallium (81)

Level	Electron	Binding energy/eV
K	1s	85530
L I	2s	15347
L II	$2p_{1/2}$	14698
L III	$2p_{3/2}$	12658
M I	3s	3704
M II	$3p_{1/2}$	3416
M III	$3p_{3/2}$	2957
M IV	$3d_{3/2}$	2485
M V	$3d_{5/2}$	2389
N I	4s	846.2[b]
N II	$4p_{1/2}$	720.5[b]
N III	$4p_{3/2}$	609.5[b]
N IV	$4d_{3/2}$	405.7[b]
N V	$4d_{5/2}$	385.0[b]
N VI	$4f_{5/2}$	122.2[b]
N VII	$4f_{7/2}$	117.8[b]
O I	5s	136[a, d]
O II	$5p_{1/2}$	94.6[b]
O III	$5p_{3/2}$	73.5[b]
O IV	$5d_{3/2}$	14.7[b]
O V	$5d_{5/2}$	12.5[b]

Thorium (90)

Level	Electron	Binding energy/eV
K	1s	109651
L I	2s	20472
L II	$2p_{1/2}$	19693
L III	$2p_{3/2}$	16300
M I	3s	5182
M II	$3p_{1/2}$	4830
M III	$3p_{3/2}$	4046
M IV	$3d_{3/2}$	3491
M V	$3d_{5/2}$	3332
N I	4s	1330[a]
N II	$4p_{1/2}$	1168[a]
N III	$4p_{3/2}$	966.4[b]
N IV	$4d_{3/2}$	712.1[b]
N V	$4d_{5/2}$	675.2[b]
N VI	$4f_{5/2}$	342.4[b]
N VII	$4f_{7/2}$	333.1[b]
O I	5s	290[a, c]
O II	$5p_{1/2}$	229[a, c]
O III	$5p_{3/2}$	182[a, c]
O IV	$5d_{3/2}$	92.5[b]

Level	Electron	Binding energy/eV
O V	$5d_{5/2}$	85.4[b]
P I	$6s$	41.4[b]
P II	$6p_{1/2}$	24.5[b]
P III	$6p_{3/2}$	16.6[b]

Thulium (69)

Level	Electron	Binding energy/eV
K	$1s$	59390
L I	$2s$	10116
L II	$2p_{1/2}$	9617
L III	$2p_{3/2}$	8648
M I	$3s$	2307
M II	$3p_{1/2}$	2090
M III	$3p_{3/2}$	1885
M IV	$3d_{3/2}$	1515
M V	$3d_{5/2}$	1468
N I	$4s$	470.9[a]
N II	$4p_{1/2}$	385.9[a]
N III	$4p_{3/2}$	332.6[a]
N IV	$4d_{3/2}$	175.5[a]
N V	$4d_{5/2}$	175.5[a]
N VI	$4f_{5/2}$	
N VII	$4f_{7/2}$	4.6
O I	$5s$	54.7[a]
O II	$5p_{1/2}$	31.8[a]
O III	$5p_{3/2}$	25.0[a]

Tin (50)

Level	Electron	Binding energy/eV
K	$1s$	29200
L I	$2s$	4465
L II	$2p_{1/2}$	4156
L III	$2p_{3/2}$	3929
M I	$3s$	884.7[b]
M II	$3p_{1/2}$	756.5[b]
M III	$3p_{3/2}$	714.6[b]
M IV	$3d_{3/2}$	493.2[b]
M V	$3d_{5/2}$	484.9[b]
N I	$4s$	137.1[b]
N II	$4p_{1/2}$	83.6[b, c]
N III	$4p_{3/2}$	83.6[b, c]
N IV	$4d_{3/2}$	24.9[b]
N V	$4d_{5/2}$	23.9[b]

Titanium (22)

Level	Electron	Binding energy/eV
K	$1s$	4966
L I	$2s$	560.9[b]
L II	$2p_{1/2}$	460.2[b]
L III	$2p_{3/2}$	453.8[b]
M I	$3s$	58.7[b]
M II	$3p_{1/2}$	32.6[b]
M III	$3p_{3/2}$	32.6[b]

Tungsten (74)

Level	Electron	Binding energy/eV
K	$1s$	69525
L I	$2s$	12100
L II	$2p_{1/2}$	11544
L III	$2p_{3/2}$	10207
M I	$3s$	2820
M II	$3p_{1/2}$	2575
M III	$3p_{3/2}$	2281
M IV	$3d_{3/2}$	1949
M V	$3d_{5/2}$	1809
N I	$4s$	594.1[b]
N II	$4p_{1/2}$	490.4[b]
N III	$4p_{3/2}$	423.6[b]
N IV	$4d_{3/2}$	255.9[b]
N V	$4d_{5/2}$	243.5[b]
N VI	$4f_{5/2}$	33.6[a]
N VII	$4f_{7/2}$	31.4[b]
O I	$5s$	75.6[b]
O II	$5p_{1/2}$	453[a, d]
O III	$5p_{3/2}$	36.8[b]

Uranium (92)

Level	Electron	Binding energy/eV
K	$1s$	115606
L I	$2s$	21757
L II	$2p_{1/2}$	20948
L III	$2p_{3/2}$	17166
M I	$3s$	5548
M II	$3p_{1/2}$	5182
M III	$3p_{3/2}$	4303
M IV	$3d_{3/2}$	3728
M V	$3d_{5/2}$	3552
N I	$4s$	1439[a, d]
N II	$4p_{1/2}$	1271[a, d]
N III	$4p_{3/2}$	1043[b]
N IV	$4d_{3/2}$	778.3[b]
N V	$4d_{5/2}$	736.2[b]
N VI	$4f_{5/2}$	388.2[a]
N VII	$4f_{7/2}$	377.4[b]
O I	$5s$	321[a, c, d]
O II	$5p_{1/2}$	257[a, c, d]
O III	$5p_{3/2}$	192[a, c, d]
O IV	$5d_{3/2}$	102.8[b]
O V	$5d_{5/2}$	94.2[b]
P I	$6s$	43.9[b]
P II	$6p_{1/2}$	26.8[b]
P III	$6p_{3/2}$	16.8[b]

Vanadium (23)

Level	Electron	Binding energy/eV
K	$1s$	5465
L I	$2s$	626.7[b]
L II	$2p_{1/2}$	519.8[b]
L III	$2p_{3/2}$	521.1[b]
M I	$3s$	66.3[b]
M II	$3p_{1/2}$	37.2[b]
M III	$3p_{3/2}$	37.2[b]

Xenon (54)

Level	Electron	Binding energy/eV
K	$1s$	34561
L I	$2s$	5453
L II	$2p_{1/2}$	5107
L III	$2p_{3/2}$	4786
M I	$3s$	1148.7[a]
M II	$3p_{1/2}$	1002.1[a]
M III	$3p_{3/2}$	940.6[a]
M IV	$3d_{3/2}$	689.0[a]
M V	$3d_{5/2}$	676.4[a]
N I	$4s$	213.2[a]
N II	$4p_{1/2}$	146.7
N III	$4p_{3/2}$	145.5[a]
N IV	$4d_{3/2}$	69.5[a]
N V	$4d_{5/2}$	67.5[a]
N VI	$4f_{5/2}$	
N VII	$4f_{7/2}$	
O I	$5s$	23.3[a]
O II	$5p_{1/2}$	13.4[a]
O III	$5p_{3/2}$	12.1[a]

Ytterbium (70)

Level	Electron	Binding energy/eV
K	$1s$	61332
L I	$2s$	10486
L II	$2p_{1/2}$	9978
L III	$2p_{3/2}$	8944
M I	$3s$	2398
M II	$3p_{1/2}$	2173
M III	$3p_{3/2}$	1950
M IV	$3d_{3/2}$	1576
M V	$3d_{5/2}$	1528
N I	$4s$	480.5[a]
N II	$4p_{1/2}$	388.7[a]
N III	$4p_{3/2}$	339.7[a]
N IV	$4d_{3/2}$	191.2[a]
N V	$4d_{5/2}$	182.4[a]
N VI	$4f_{5/2}$	2.5[a]
N VII	$4f_{7/2}$	1.3[a]
O I	$5s$	52.0[a]
O II	$5p_{1/2}$	30.3[a]
O III	$5p_{3/2}$	24.1[a]

Yttrium (39)

Level	Electron	Binding energy/eV
K	$1s$	17038
L I	$2s$	2373
L II	$2p_{1/2}$	2156
L III	$2p_{3/2}$	2080
M I	$3s$	392.0[a, d]
M II	$3p_{1/2}$	310.6[a]
M III	$3p_{3/2}$	298.8[a]
M IV	$3d_{3/2}$	157.7[b]
M V	$3d_{5/2}$	155.8[b]
N I	$4s$	43.8[a]
N II	$4p_{1/2}$	24.4[a]
N III	$4p_{3/2}$	23.1[a]

Zinc (30)

Level	Electron	Binding energy/eV
K	$1s$	9659
L I	$2s$	1196.2[a]
L II	$2p_{1/2}$	1044.9[a]
L III	$2p_{3/2}$	1021.8[a]
M I	$3s$	139.8[a]
M II	$3p_{1/2}$	91.4[a]
M III	$3p_{3/2}$	88.6[a]
M IV	$3d_{3/2}$	10.2[a]
M V	$3d_{5/2}$	10.1[a]

Zirconium (40)

Level	Electron	Binding energy/eV
K	$1s$	17998
L I	$2s$	2532
L II	$2p_{1/2}$	2307
L III	$2p_{3/2}$	2223
M I	$3s$	430.3[b]
M II	$3p_{1/2}$	343.5[b]
M III	$3p_{3/2}$	329.8[b]
M IV	$3d_{3/2}$	181.1[b]
M V	$3d_{5/2}$	178.8[b]
N I	$4s$	50.6[b]
N II	$4p_{1/2}$	28.5[b]
N III	$4p_{3/2}$	27.1[b]

[a] Reference 1.
[b] Reference 2 (remaining values from Reference 3).
[c] One-particle approximation not valid.
[d] Derived using energy differences from Reference 3.

NATURAL WIDTH OF X-RAY LINES

Natural Widths of K X-ray Lines in eV

Element	$K\alpha_1$	$K\alpha_2$	$K\beta_1$	$K\beta_3$
Ca	1.00	0.98		
Ti	1.45	2.13		
Cr	2.05	2.64		
Fe	2.45	3.20		
Ni	3.00	3.70		
Zn	3.40	3.96		
Ge	3.75	4.18		
Se	4.10	4.43		
Kr	4.23	4.62		
Sr	5.17	4.97		
Zr	5.70	5.25		
Mo	6.82	6.80		
Ru	7.41	7.96		
Pd	8.80	9.20		
Cd	9.80	10.40		
Sn	11.20	12.40	11.80	11.00
Te	12.80	14.20	13.30	13.10
Xe	14.20	15.10	15.30	14.50
Ba	16.10	16.80	18.15	16.70
Ce	18.60	19.50	20.60	18.60
Nd	21.50	21.50	23.25	21.33
Sm	26.00	24.70	25.65	24.65
Gd	29.50	28.00	29.37	28.00
Dy	33.90	32.20	32.73	32.00
Er	35.00	35.50	36.20	35.70
Yb	38.80	40.60	41.43	41.15
Hf	42.70	44.30	46.00	46.10
W	46.80	48.00	51.83	51.50
Os	49.00	49.40	55.90	55.95
Pt	54.10	54.30	59.98	62.13
Hg	64.75	68.20	65.75	68.95
Pb	67.10	72.30	72.20	73.80
Po	73.20	75.10	78.60	80.10
Rn	80.00	81.50	85.50	86.50
Ra	87.00	88.20	94.20	95.50
Th	94.70	95.00	99.70	101.00
U	103.00	104.30	105.00	107.30

From Salem, S. I. and Lee, P. L., *At. Data Nucl. Data Tables*, 18, 233, 1976.

Natural Widths of L X-ray Lines in eV

Element	$L\alpha_1$	$L\alpha_2$	$L\beta_1$	$L\beta_2$	$L\beta_3$	$L\beta_4$	$L\gamma_1$
Zr	1.68	1.52	1.87	5.13	5.50	5.60	3.34
Mo	1.86	1.80	2.03	5.30	5.90	5.78	3.76
Ru	2.03	1.98	2.18	5.45	6.35	5.96	4.15
Pd	2.21	2.16	2.36	5.63	6.80	6.18	4.50
Gd	2.43	2.40	2.54	5.82	7.23	6.28	4.83
Sn	2.62	2.62	2.75	6.10	7.70	6.60	5.23
Tc	2.88	2.88	2.96	6.25	8.22	6.82	5.60
Xe	3.15	3.15	3.20	6.43	8.70	7.15	5.95
Ba	3.39	3.45	3.45	6.70	9.20	7.42	6.35
Ce	3.70	3.78	3.73	6.86	9.70	7.82	6.75
Nd	3.93	4.08	4.00	7.18	10.30	8.15	7.16
Sm	4.13	4.50	4.33	7.42	10.80	8.60	7.50
Cd	4.46	4.90	4.63	7.70	11.20	9.08	7.83
Dy	4.81	5.35	5.03	7.90	11.50	9.60	8.30
Er	5.17	5.73	5.45	8.28	11.85	10.03	8.75
Yb	5.40	6.22	5.90	8.58	12.20	11.00	9.20
Hf	5.83	6.70	6.36	8.92	12.40	12.80	9.63
W	6.50	7.20	6.90	9.06	13.10	14.60	10.20
Os	7.04	7.70	7.42	9.60	14.60	16.50	10.65
Pt	7.60	8.28	8.00	9.95	16.10	18.00	11.20
Hg	8.10	8.80	8.70	10.40	17.40	19.70	11.80
Pb	8.82	9.35	9.35	10.75	18.65	21.30	12.30
Po	9.50	9.95	10.10	11.25	19.90	22.70	13.05
Rn	10.03	10.50	10.65	11.65	21.00	24.00	13.55
Ra	11.00	11.20	11.60	12.20	22.00	25.20	14.30
Th	11.90	11.80	12.40	12.80	22.85	26.35	15.00
U	12.40	12.40	13.50	13.30	23.70	27.50	15.70
Pu	13.20	13.00	14.10	13.90	24.10	28.30	16.40
Cm	14.80	13.60	15.70	14.60	25.00	29.40	17.10

From Salem, S. I. and Lee, P. L., *At. Data Nucl. Data Tables*, 18, 233, 1976.

PHOTON ATTENUATION COEFFICIENTS

Martin J. Berger and John H. Hubbell

This table gives mass attenuation coefficients for photons for elements of $Z = 1$ to 100 at energies between 1 keV (soft x-rays) and 1 GeV (hard gamma rays). The mass attenuation coefficient μ describes the attenuation of radiation as it passes through matter by the relation

$$I(x)/I_o = e^{-\mu\rho x}$$

where I_o is the initial intensity, $I(x)$ the intensity after path length x, and ρ is the mass density of the element in question. To a high approximation the mass attenuation coefficient is additive for the elements present, independent of the way in which they are bound in chemical compounds.

The first four tables give the attenuation coefficients for elements with atomic number 1–100. The power of ten is indicated beside each number in the table; i.e., 7.41 + 03 means 7.41×10^3. Plots of attenuation coefficient vs. energy (see Ref. 4) show a smooth decline with increasing energy until an absorption edge is reached, where the attenuation coefficient increases sharply and then continues to decline. Thus interpolation in Table 1 is unreliable when the interval includes an absorption edge. Absorption edge energies are listed in the final table.

The attenuation coefficients were calculated with the computer program XCOM (Refs. 1 and 3), which uses a cross-section database compiled at the Photon and Charged Particle Data Center at the National Institute of Standards and Technology. Their accuracy has been confirmed at all energies by extensive comparisons with experimental attenuation coefficients. Such comparisons for X-ray energies up to 100 keV can be found in Ref. 2.

References

1. Berger, M. J. and Hubbell, J. H., *National Bureau of Standards Report* NBSIR-87-3597, 1987.
2. Saloman, E. B., Hubbell, J. H., and Scofield, J. H., *Atomic Data and Nuclear Data Tables*, 38, 1, 1988.
3. XCOM: Photon Cross Sections Database, Version 3.1, NIST Standard Reference Database 8, November 2010, <www.nist.gov/pml/data/xcom/index.cfm>.
4. Tables of X-Ray Mass Attenuation Coefficients and Mass Energy-Absorption Coefficients from 1 keV to 20 MeV for Elements Z = 1 to 92 and 48 Additional Substances of Dosimetric Interest, NIST Standard Reference Database 26, July 2004, <www.nist.gov/pml/data/xraycoef/index.cfm>.

Mass Attenuation Coefficients in cm²/g for H through Sn and Photon Energy from 0.001 to 0.5 MeV

	Atomic no.	0.001 MeV	0.002 MeV	0.005 MeV	0.01 MeV	0.02 MeV	0.05 MeV	0.1 MeV	0.2 MeV	0.5 MeV
H	1	7.21 + 00	1.06 + 00	4.19-01	3.85-01	3.69-01	3.36-01	2.94-01	2.43-01	1.73-01
He	2	6.08 + 01	6.86 + 00	5.77-01	2.48-01	1.96-01	1.70-01	1.49-01	1.22-01	8.71-02
Li	3	2.34 + 02	2.71 + 01	1.62 + 00	3.40-01	1.86-01	1.49-01	1.29-01	1.06-01	7.53-02
Be	4	6.04 + 02	7.47 + 01	4.37 + 00	6.47-01	2.25-01	1.55-01	1.33-01	1.09-01	7.74-02
B	5	1.23 + 03	1.60 + 02	9.68 + 00	1.25 + 00	3.01-01	1.66-01	1.39-01	1.14-01	8.07-02
C	6	2.21 + 03	3.03 + 02	1.91 + 01	2.37 + 00	4.42-01	1.87-01	1.51-01	1.23-01	8.72-02
N	7	3.31 + 03	4.77 + 02	3.14 + 01	3.88 + 00	6.18-01	1.98-01	1.53-01	1.23-01	8.72-02
O	8	4.59 + 03	6.95 + 02	4.79 + 01	5.95 + 00	8.65-01	2.13-01	1.55-01	1.24-01	8.73-02
F	9	5.65 + 03	9.05 + 02	6.51 + 01	8.21 + 00	1.13 + 00	2.21-01	1.50-01	1.18-01	8.27-02
Ne	10	7.41 + 03	1.24 + 03	9.34 + 01	1.20 + 01	1.61 + 00	2.58-01	1.60-01	1.24-01	8.66-02
Na	11	6.54 + 02	1.52 + 03	1.19 + 02	1.56 + 01	2.06 + 00	2.80-01	1.59-01	1.20-01	8.37-02
Mg	12	9.22 + 02	1.93 + 03	1.58 + 02	2.11 + 01	2.76 + 00	3.29-01	1.69-01	1.24-01	8.65-02
Al	13	1.19 + 03	2.26 + 03	1.93 + 02	2.62 + 01	3.44 + 00	3.68-01	1.70-01	1.22-01	8.44-02
Si	14	1.57 + 03	2.78 + 03	2.45 + 02	3.39 + 01	4.46 + 00	4.38-01	1.84-01	1.28-01	8.75-02
P	15	1.91 + 03	3.02 + 02	2.86 + 02	4.04 + 01	5.35 + 00	4.92-01	1.87-01	1.25-01	8.51-02
S	16	2.43 + 03	3.85 + 02	3.49 + 02	5.01 + 01	6.71 + 00	5.85-01	2.02-01	1.30-01	8.78-02
Cl	17	2.83 + 03	4.52 + 02	3.90 + 02	5.73 + 01	7.74 + 00	6.48-01	2.05-01	1.27-01	8.45-02
Ar	18	3.18 + 03	5.12 + 02	4.23 + 02	6.32 + 01	8.63 + 00	7.01-01	2.04-01	1.20-01	7.96-02
K	19	4.06 + 03	6.59 + 02	5.19 + 02	7.91 + 01	1.09 + 01	8.68-01	2.34-01	1.32-01	8.60-02
Ca	20	4.87 + 03	8.00 + 02	6.03 + 02	9.34 + 01	1.31 + 01	1.02 + 00	2.57-01	1.38-01	8.85-02
Sc	21	5.24 + 03	8.70 + 02	6.31 + 02	9.95 + 01	1.41 + 01	1.09 + 00	2.58-01	1.31-01	8.31-02
Ti	22	5.87 + 03	9.86 + 02	6.84 + 02	1.11 + 02	1.59 + 01	1.21 + 00	2.72-01	1.31-01	8.19-02
V	23	6.50 + 03	1.11 + 03	9.29 + 01	1.22 + 02	1.77 + 01	1.35 + 00	2.88-01	1.32-01	8.07-02
Cr	24	7.40 + 03	1.28 + 03	1.08 + 02	1.39 + 02	2.04 + 01	1.55 + 00	3.17-01	1.38-01	8.28-02
Mn	25	8.09 + 03	1.42 + 03	1.21 + 02	1.51 + 02	2.25 + 01	1.71 + 00	3.37-01	1.39-01	8.19-02
Fe	26	9.09 + 03	1.63 + 03	1.40 + 02	1.71 + 02	2.57 + 01	1.96 + 00	3.72-01	1.46-01	8.41-02
Co	27	9.80 + 03	1.78 + 03	1.54 + 02	1.84 + 02	2.80 + 01	2.14 + 00	3.95-01	1.48-01	8.32-02
Ni	28	9.86 + 03	2.05 + 03	1.79 + 02	2.09 + 02	3.22 + 01	2.47 + 00	4.44-01	1.58-01	8.70-02
Cu	29	1.06 + 04	2.15 + 03	1.90 + 02	2.16 + 02	3.38 + 01	2.61 + 00	4.58-01	1.56-01	8.36-02
Zn	30	1.55 + 03	2.37 + 03	2.12 + 02	2.33 + 02	3.72 + 01	2.89 + 00	4.97-01	1.62-01	8.45-02

	Atomic no.	0.001 MeV	0.002 MeV	0.005 MeV	0.01 MeV	0.02 MeV	0.05 MeV	0.1 MeV	0.2 MeV	0.5 MeV
Ga	31	1.70 + 03	2.52 + 03	2.27 + 02	3.42 + 01	3.93 + 01	3.08 + 00	5.20-01	1.62-01	8.24-02
Ge	32	1.89 + 03	2.71 + 03	2.47 + 02	3.74 + 01	4.22 + 01	3.34 + 00	5.55-01	1.66-01	8.21-02
As	33	2.12 + 03	2.93 + 03	2.71 + 02	4.12 + 01	4.56 + 01	3.63 + 00	5.97-01	1.72-01	8.26-02
Se	34	2.32 + 03	3.10 + 03	2.90 + 02	4.41 + 01	4.82 + 01	3.86 + 00	6.28-01	1.74-01	8.13-02
Br	35	2.62 + 03	3.41 + 03	3.21 + 02	4.91 + 01	5.27 + 01	4.26 + 00	6.86-01	1.84-01	8.33-02
Kr	36	2.85 + 03	3.60 + 03	3.43 + 02	5.26 + 01	5.55 + 01	4.52 + 00	7.22-01	1.87-01	8.23-02
Rb	37	3.17 + 03	3.41 + 03	3.74 + 02	5.77 + 01	5.98 + 01	4.92 + 00	7.80-01	1.96-01	8.36-02
Sr	38	3.49 + 03	2.59 + 03	4.06 + 02	6.27 + 01	6.39 + 01	5.31 + 00	8.37-01	2.04-01	8.44-02
Y	39	3.86 + 03	7.42 + 02	4.42 + 02	6.87 + 01	6.86 + 01	5.76 + 00	9.05-01	2.15-01	8.61-02
Zr	40	4.21 + 03	8.12 + 02	4.76 + 02	7.42 + 01	7.24 + 01	6.17 + 00	9.66-01	2.24-01	8.69-02
Nb	41	4.60 + 03	8.89 + 02	5.13 + 02	8.04 + 01	7.71 + 01	6.64 + 00	1.04 + 00	2.34-01	8.83-02
Mo	42	4.94 + 03	9.60 + 02	5.45 + 02	8.58 + 01	1.31 + 01	7.04 + 00	1.10 + 00	2.42-01	8.85-02
Tc	43	5.36 + 03	1.04 + 03	5.84 + 02	9.23 + 01	1.41 + 01	7.52 + 00	1.17 + 00	2.53-01	8.97-02
Ru	44	5.72 + 03	1.12 + 03	6.17 + 02	9.80 + 01	1.50 + 01	7.92 + 00	1.23 + 00	2.62-01	8.99-02
Rh	45	6.17 + 03	1.21 + 03	6.59 + 02	1.05 + 02	1.61 + 01	8.45 + 00	1.31 + 00	2.74-01	9.13-02
Pd	46	6.54 + 03	1.29 + 03	6.91 + 02	1.11 + 02	1.70 + 01	8.85 + 00	1.38 + 00	2.83-01	9.13-02
Ag	47	7.04 + 03	1.40 + 03	7.39 + 02	1.19 + 02	1.84 + 01	9.45 + 00	1.47 + 00	2.97-01	9.32-02
Cd	48	7.35 + 03	1.47 + 03	7.69 + 02	1.24 + 02	1.92 + 01	9.78 + 00	1.52 + 00	3.04-01	9.25-02
In	49	7.81 + 03	1.58 + 03	8.13 + 02	1.32 + 02	2.04 + 01	1.03 + 01	1.61 + 00	3.17-01	9.37-02
Sn	50	8.16 + 03	1.66 + 03	8.47 + 02	1.38 + 02	2.15 + 01	1.07 + 01	1.68 + 00	3.26-01	9.37-02

Mass Attenuation Coefficients in cm²/g for H through Sn and Photon Energy from 1.0 to 1000.0 MeV

	Atomic no.	1.0 MeV	2.0 MeV	5.0 MeV	10.0 MeV	20.0 MeV	50.0 MeV	100.0 MeV	500.0 MeV	1000.0 MeV
H	1	1.26-01	8.77-02	5.05-02	3.25-02	2.15-02	1.42-02	1.19-02	1.14-02	1.16-02
He	2	6.36-02	4.42-02	2.58-02	1.70-02	1.18-02	8.61-03	7.78-03	7.79-03	7.95-03
Li	3	5.50-02	3.83-02	2.26-02	1.53-02	1.11-02	8.68-03	8.21-03	8.61-03	8.87-03
Be	4	5.65-02	3.94-02	2.35-02	1.63-02	1.23-02	1.02-02	9.94-03	1.08-02	1.12-02
B	5	5.89-02	4.11-02	2.48-02	1.76-02	1.37-02	1.19-02	1.19-02	1.32-02	1.37-02
C	6	6.36-02	4.44-02	2.71-02	1.96-02	1.58-02	1.43-02	1.46-02	1.64-02	1.70-02
N	7	6.36-02	4.45-02	2.74-02	2.02-02	1.67-02	1.57-02	1.63-02	1.85-02	1.92-02
O	8	6.37-02	4.46-02	2.78-02	2.09-02	1.77-02	1.71-02	1.79-02	2.06-02	2.13-02
F	9	6.04-02	4.23-02	2.66-02	2.04-02	1.77-02	1.75-02	1.86-02	2.14-02	2.21-02
Ne	10	6.32-02	4.43-02	2.82-02	2.20-02	1.95-02	1.96-02	2.11-02	2.43-02	2.51-02
Na	11	6.10-02	4.28-02	2.75-02	2.18-02	1.97-02	2.03-02	2.19-02	2.53-02	2.62-02
Mg	12	6.30-02	4.43-02	2.87-02	2.31-02	2.13-02	2.23-02	2.42-02	2.81-02	2.90-02
Al	13	6.15-02	4.32-02	2.84-02	2.32-02	2.17-02	2.31-02	2.52-02	2.93-02	3.03-02
Si	14	6.36-02	4.48-02	2.97-02	2.46-02	2.34-02	2.52-02	2.76-02	3.23-02	3.34-02
P	15	6.18-02	4.36-02	2.91-02	2.45-02	2.36-02	2.58-02	2.84-02	3.33-02	3.45-02
S	16	6.37-02	4.50-02	3.04-02	2.59-02	2.53-02	2.79-02	3.08-02	3.62-02	3.75-02
Cl	17	6.13-02	4.33-02	2.95-02	2.55-02	2.52-02	2.81-02	3.11-02	3.67-02	3.80-02
Ar	18	5.76-02	4.07-02	2.80-02	2.45-02	2.45-02	2.76-02	3.07-02	3.62-02	3.75-02
K	19	6.22-02	4.40-02	3.05-02	2.70-02	2.74-02	3.11-02	3.46-02	4.09-02	4.24-02
Ca	20	6.39-02	4.52-02	3.17-02	2.84-02	2.90-02	3.32-02	3.71-02	4.40-02	4.56-02
Sc	21	5.98-02	4.24-02	3.00-02	2.72-02	2.80-02	3.23-02	3.62-02	4.30-02	4.45-02
Ti	22	5.89-02	4.18-02	2.98-02	2.73-02	2.84-02	3.30-02	3.71-02	4.40-02	4.56-02
V	23	5.79-02	4.11-02	2.96-02	2.74-02	2.88-02	3.36-02	3.78-02	4.49-02	4.65-02
Cr	24	5.93-02	4.21-02	3.06-02	2.86-02	3.03-02	3.56-02	4.01-02	4.76-02	4.93-02
Mn	25	5.85-02	4.16-02	3.04-02	2.87-02	3.07-02	3.63-02	4.09-02	4.86-02	5.04-02
Fe	26	5.99-02	4.26-02	3.15-02	2.99-02	3.22-02	3.83-02	4.33-02	5.15-02	5.33-02
Co	27	5.91-02	4.20-02	3.13-02	3.00-02	3.26-02	3.88-02	4.40-02	5.23-02	5.41-02
Ni	28	6.16-02	4.39-02	3.29-02	3.18-02	3.48-02	4.17-02	4.73-02	5.61-02	5.81-02
Cu	29	5.90-02	4.20-02	3.18-02	3.10-02	3.41-02	4.10-02	4.66-02	5.53-02	5.72-02
Zn	30	5.94-02	4.24-02	3.22-02	3.18-02	3.51-02	4.24-02	4.82-02	5.72-02	5.91-02
Ga	31	5.77-02	4.11-02	3.16-02	3.13-02	3.48-02	4.22-02	4.80-02	5.70-02	5.89-02
Ge	32	5.73-02	4.09-02	3.16-02	3.16-02	3.53-02	4.30-02	4.89-02	5.80-02	6.00-02
As	33	5.73-02	4.09-02	3.19-02	3.21-02	3.60-02	4.40-02	5.01-02	5.95-02	6.15-02
Se	34	5.62-02	4.01-02	3.14-02	3.19-02	3.60-02	4.41-02	5.03-02	5.97-02	6.17-02
Br	35	5.73-02	4.09-02	3.23-02	3.29-02	3.74-02	4.60-02	5.24-02	6.22-02	6.43-02

	Atomic no.	1.0 MeV	2.0 MeV	5.0 MeV	10.0 MeV	20.0 MeV	50.0 MeV	100.0 MeV	500.0 MeV	1000.0 MeV
Kr	36	5.63-02	4.02-02	3.20-02	3.28-02	3.74-02	4.61-02	5.26-02	6.25-02	6.46-02
Rb	37	5.69-02	4.06-02	3.25-02	3.36-02	3.85-02	4.75-02	5.43-02	6.45-02	6.67-02
Sr	38	5.71-02	4.08-02	3.29-02	3.41-02	3.93-02	4.87-02	5.56-02	6.61-02	6.83-02
Y	39	5.80-02	4.14-02	3.35-02	3.50-02	4.05-02	5.03-02	5.75-02	6.83-02	7.06-02
Zr	40	5.81-02	4.15-02	3.38-02	3.55-02	4.12-02	5.13-02	5.87-02	6.98-02	7.22-02
Nb	41	5.87-02	4.18-02	3.44-02	3.63-02	4.22-02	5.27-02	6.03-02	7.17-02	7.42-02
Mo	42	5.84-02	4.16-02	3.44-02	3.65-02	4.26-02	5.33-02	6.10-02	7.26-02	7.51-02
Tc	43	5.88-02	4.19-02	3.48-02	3.71-02	4.35-02	5.45-02	6.24-02	7.43-02	7.68-02
Ru	44	5.85-02	4.16-02	3.48-02	3.73-02	4.39-02	5.50-02	6.30-02	7.51-02	7.77-02
Rh	45	5.89-02	4.20-02	3.53-02	3.80-02	4.48-02	5.63-02	6.45-02	7.69-02	7.94-02
Pd	46	5.85-02	4.16-02	3.52-02	3.80-02	4.50-02	5.66-02	6.49-02	7.73-02	8.00-02
Ag	47	5.92-02	4.21-02	3.58-02	3.88-02	4.61-02	5.81-02	6.67-02	7.93-02	8.20-02
Cd	48	5.83-02	4.14-02	3.54-02	3.85-02	4.59-02	5.79-02	6.64-02	7.91-02	8.18-02
In	49	5.85-02	4.15-02	3.56-02	3.90-02	4.65-02	5.88-02	6.75-02	8.04-02	8.32-02
Sn	50	5.80-02	4.11-02	3.55-02	3.90-02	4.66-02	5.90-02	6.78-02	8.07-02	8.35-02

Mass Attenuation Coefficients in cm²/g for Sb through Fm and Photon Energy from 0.001 to 0.5 MeV

	Atomic no.	0.001 MeV	0.002 MeV	0.005 MeV	0.01 MeV	0.02 MeV	0.05 MeV	0.1 MeV	0.2 MeV	0.5 MeV
Sb	51	8.58 + 03	1.77 + 03	8.85 + 02	1.46 + 02	2.27 + 01	1.12 + 01	1.76 + 00	3.38-01	9.45-02
Te	52	8.43 + 03	1.83 + 03	9.01 + 02	1.50 + 02	2.34 + 01	1.14 + 01	1.80 + 00	3.43-01	9.33-02
I	53	9.10 + 03	2.00 + 03	8.43 + 02	1.63 + 02	2.54 + 01	1.23 + 01	1.94 + 00	3.66-01	9.70-02
Xe	54	9.41 + 03	2.09 + 03	6.39 + 02	1.69 + 02	2.65 + 01	1.27 + 01	2.01 + 00	3.76-01	9.70-02
Cs	55	9.37 + 03	2.23 + 03	2.30 + 02	1.79 + 02	2.82 + 01	1.34 + 01	2.12 + 00	3.94-01	9.91-02
Ba	56	8.54 + 03	2.32 + 03	2.41 + 02	1.86 + 02	2.94 + 01	1.38 + 01	2.20 + 00	4.05-01	9.92-02
La	57	9.09 + 03	2.46 + 03	2.58 + 02	1.97 + 02	3.12 + 01	1.45 + 01	2.32 + 00	4.24-01	1.01-01
Ce	58	9.71 + 03	2.61 + 03	2.74 + 02	2.08 + 02	3.31 + 01	1.52 + 01	2.45 + 00	4.45-01	1.04-01
Pr	59	1.06 + 04	2.77 + 03	2.92 + 02	2.21 + 02	3.53 + 01	1.60 + 01	2.59 + 00	4.69-01	1.07-01
Nd	60	6.63 + 03	2.88 + 03	3.06 + 02	2.30 + 02	3.68 + 01	1.65 + 01	2.69 + 00	4.84-01	1.08-01
Pm	61	2.06 + 03	3.05 + 03	3.26 + 02	2.44 + 02	3.92 + 01	1.73 + 01	2.84 + 00	5.10-01	1.12-01
Sm	62	2.11 + 03	3.12 + 03	3.36 + 02	2.50 + 02	4.03 + 01	1.77 + 01	2.90 + 00	5.19-01	1.11-01
Eu	63	2.22 + 03	3.28 + 03	3.54 + 02	2.63 + 02	4.24 + 01	1.85 + 01	3.04 + 00	5.43-01	1.14-01
Gd	64	2.29 + 03	3.36 + 03	3.65 + 02	2.69 + 02	4.36 + 01	3.86 + 00	3.11 + 00	5.54-01	1.14-01
Tb	65	2.40 + 03	3.51 + 03	3.84 + 02	2.82 + 02	4.59 + 01	4.06 + 00	3.25 + 00	5.77-01	1.17-01
Dy	66	2.49 + 03	3.47 + 03	3.99 + 02	2.90 + 02	4.76 + 01	4.23 + 00	3.36 + 00	5.95-01	1.18-01
Ho	67	2.62 + 03	3.59 + 03	4.17 + 02	3.01 + 02	4.98 + 01	4.43 + 00	3.49 + 00	6.18-01	1.20-01
Er	68	2.75 + 03	3.52 + 03	4.36 + 02	3.13 + 02	5.20 + 01	4.63 + 00	3.63 + 00	6.41-01	1.23-01
Tm	69	2.90 + 03	3.69 + 03	4.57 + 02	2.83 + 02	5.45 + 01	4.87 + 00	3.78 + 00	6.68-01	1.26-01
Yb	70	3.02 + 03	3.80 + 03	4.72 + 02	2.94 + 02	5.63 + 01	5.04 + 00	3.88 + 00	6.86-01	1.27-01
Lu	71	3.19 + 03	3.45 + 03	4.94 + 02	2.21 + 02	5.88 + 01	5.28 + 00	4.03 + 00	7.13-01	1.30-01
Hf	72	3.34 + 03	3.60 + 03	5.11 + 02	2.30 + 02	6.09 + 01	5.48 + 00	4.15 + 00	7.34-01	1.32-01
Ta	73	3.51 + 03	3.77 + 03	5.33 + 02	2.38 + 02	6.33 + 01	5.72 + 00	4.30 + 00	7.60-01	1.35-01
W	74	3.68 + 03	3.92 + 03	5.53 + 02	9.69 + 01	6.57 + 01	5.95 + 00	4.44 + 00	7.84-01	1.38-01
Re	75	3.87 + 03	3.77 + 03	5.76 + 02	1.01 + 02	6.84 + 01	6.21 + 00	4.59 + 00	8.12-01	1.41-01
Os	76	4.03 + 03	2.22 + 03	5.93 + 02	1.04 + 02	7.04 + 01	6.41 + 00	4.70 + 00	8.33-01	1.43-01
Ir	77	4.24 + 03	1.03 + 03	6.18 + 02	1.09 + 02	7.32 + 01	6.69 + 00	4.86 + 00	8.63-01	1.46-01
Pt	78	4.43 + 03	1.08 + 03	6.40 + 02	1.13 + 02	7.57 + 01	6.95 + 00	4.99 + 00	8.90-01	1.49-01
Au	79	4.65 + 03	1.14 + 03	6.66 + 02	1.18 + 02	7.88 + 01	7.26 + 00	5.16 + 00	9.22-01	1.53-01
Hg	80	4.83 + 03	1.18 + 03	6.87 + 02	1.22 + 02	8.12 + 01	7.50 + 00	5.28 + 00	9.46-01	1.56-01
Tl	81	5.01 + 03	1.23 + 03	7.07 + 02	1.26 + 02	8.36 + 01	7.75 + 00	5.40 + 00	9.69-01	1.58-01
Pb	82	5.21 + 03	1.29 + 03	7.30 + 02	1.31 + 02	8.64 + 01	8.04 + 00	5.55 + 00	9.99-01	1.61-01
Bi	83	5.44 + 03	1.35 + 03	7.58 + 02	1.36 + 02	8.95 + 01	8.38 + 00	5.74 + 00	1.03 + 00	1.66-01
Po	84	5.72 + 03	1.42 + 03	7.93 + 02	1.43 + 02	9.35 + 01	8.80 + 00	5.99 + 00	1.08 + 00	1.71-01
At	85	5.87 + 03	1.49 + 03	8.25 + 02	1.49 + 02	9.70 + 01	9.19 + 00	6.17 + 00	1.12 + 00	1.77-01
Rn	86	5.83 + 03	1.49 + 03	8.16 + 02	1.48 + 02	9.56 + 01	9.12 + 00	6.09 + 00	1.10 + 00	1.73-01
Fr	87	6.08 + 03	1.56 + 03	8.49 + 02	1.54 + 02	9.93 + 01	9.52 + 00	1.66 + 00	1.14 + 00	1.78-01
Ra	88	6.20 + 03	1.62 + 03	8.74 + 02	1.59 + 02	1.02 + 02	9.85 + 00	1.71 + 00	1.17 + 00	1.82-01
Ac	89	6.47 + 03	1.70 + 03	8.69 + 02	1.65 + 02	1.06 + 02	1.03 + 01	1.79 + 00	1.21 + 00	1.87-01
Th	90	6.61 + 03	1.74 + 03	8.88 + 02	1.69 + 02	9.37 + 01	1.05 + 01	1.83 + 00	1.23 + 00	1.90-01

	Atomic no.	0.001 MeV	0.002 MeV	0.005 MeV	0.01 MeV	0.02 MeV	0.05 MeV	0.1 MeV	0.2 MeV	0.5 MeV
Pa	91	6.53 + 03	1.83 + 03	8.76 + 02	1.77 + 02	7.03 + 01	1.10 + 01	1.92 + 00	1.29 + 00	1.97-01
U	92	6.63 + 03	1.86 + 03	8.89 + 02	1.79 + 02	7.11 + 01	1.12 + 01	1.95 + 00	1.30 + 00	1.98-01
Np	93	6.95 + 03	1.96 + 03	9.32 + 02	1.87 + 02	7.45 + 01	1.18 + 01	2.05 + 00	1.35 + 00	2.05-01
Pu	94	7.19 + 03	2.04 + 03	9.65 + 02	1.94 + 02	7.71 + 01	1.22 + 01	2.13 + 00	1.39 + 00	2.10-01
Am	95	7.37 + 03	2.10 + 03	9.90 + 02	1.98 + 02	7.93 + 01	1.25 + 01	2.19 + 00	1.42 + 00	2.14-01
Cm	96	7.54 + 03	2.15 + 03	1.02 + 03	2.03 + 02	8.14 + 01	1.28 + 01	2.25 + 00	1.44 + 00	2.18-01
Bk	97	7.84 + 03	2.25 + 03	1.06 + 03	2.10 + 02	8.39 + 01	1.34 + 01	2.35 + 00	1.50 + 00	2.25-01
Cf	98	7.89 + 03	2.31 + 03	9.27 + 02	2.15 + 02	8.58 + 01	1.37 + 01	2.41 + 00	1.52 + 00	2.29-01
Es	99	7.79 + 03	2.40 + 03	9.59 + 02	2.22 + 02	4.01 + 01	1.42 + 01	2.51 + 00	1.57 + 00	2.36-01
Fm	100	7.13 + 03	2.46 + 03	9.77 + 02	2.26 + 02	4.09 + 01	1.45 + 01	2.57 + 00	1.59 + 00	2.39-01

Mass Attenuation Coefficients in cm^2/g for Sb through Fm and Photon Energy from 1.0 to 1000.0 MeV

	Atomic no.	1.0 MeV	2.0 MeV	5.0 MeV	10.0 MeV	20.0 MeV	50.0 MeV	100.0 MeV	500.0 MeV	1000.0 MeV
Sb	51	5.80-02	4.10-02	3.56-02	3.92-02	4.70-02	5.96-02	6.85-02	8.16-02	8.44-02
Te	52	5.67-02	4.01-02	3.49-02	3.86-02	4.64-02	5.89-02	6.77-02	8.07-02	8.35-02
I	53	5.84-02	4.12-02	3.61-02	4.00-02	4.82-02	6.13-02	7.04-02	8.40-02	8.69-02
Xe	54	5.78-02	4.08-02	3.58-02	3.99-02	4.82-02	6.12-02	7.04-02	8.40-02	8.69-02
Cs	55	5.85-02	4.12-02	3.64-02	4.06-02	4.91-02	6.25-02	7.19-02	8.58-02	8.88-02
Ba	56	5.80-02	4.08-02	3.61-02	4.04-02	4.90-02	6.25-02	7.19-02	8.58-02	8.88-02
La	57	5.88-02	4.12-02	3.66-02	4.11-02	5.00-02	6.37-02	7.34-02	8.76-02	9.06-02
Ce	58	5.96-02	4.18-02	3.73-02	4.19-02	5.10-02	6.52-02	7.50-02	8.96-02	9.27-02
Pr	59	6.07-02	4.24-02	3.80-02	4.29-02	5.23-02	6.68-02	7.69-02	9.19-02	9.50-02
Nd	60	6.07-02	4.24-02	3.81-02	4.30-02	5.26-02	6.72-02	7.74-02	9.25-02	9.56-02
Pm	61	6.19-02	4.31-02	3.88-02	4.40-02	5.38-02	6.89-02	7.94-02	9.48-02	9.81-02
Sm	62	6.11-02	4.24-02	3.83-02	4.35-02	5.34-02	6.84-02	7.88-02	9.41-02	9.73-02
Eu	63	6.19-02	4.28-02	3.88-02	4.42-02	5.42-02	6.96-02	8.02-02	9.57-02	9.90-02
Gd	64	6.12-02	4.23-02	3.84-02	4.38-02	5.38-02	6.91-02	7.97-02	9.51-02	9.83-02
Tb	65	6.20-02	4.27-02	3.89-02	4.45-02	5.47-02	7.03-02	8.11-02	9.67-02	1.00-01
Dy	66	6.20-02	4.26-02	3.90-02	4.46-02	5.49-02	7.06-02	8.15-02	9.72-02	1.00-01
Ho	67	6.26-02	4.29-02	3.93-02	4.50-02	5.55-02	7.14-02	8.24-02	9.83-02	1.02-01
Er	68	6.32-02	4.32-02	3.96-02	4.55-02	5.61-02	7.23-02	8.34-02	9.95-02	1.03-01
Tm	69	6.40-02	4.36-02	4.01-02	4.61-02	5.70-02	7.35-02	8.48-02	1.01-01	1.04-01
Yb	70	6.40-02	4.35-02	4.00-02	4.61-02	5.70-02	7.35-02	8.49-02	1.01-01	1.04-01
Lu	71	6.48-02	4.39-02	4.05-02	4.66-02	5.77-02	7.45-02	8.60-02	1.02-01	1.06-01
Hf	72	6.50-02	4.39-02	4.05-02	4.68-02	5.80-02	7.48-02	8.64-02	1.03-01	1.06-01
Ta	73	6.57-02	4.41-02	4.08-02	4.72-02	5.85-02	7.56-02	8.73-02	1.04-01	1.07-01
W	74	6.62-02	4.43-02	4.10-02	4.75-02	5.89-02	7.62-02	8.80-02	1.05-01	1.08-01
Re	75	6.69-02	4.46-02	4.14-02	4.79-02	5.95-02	7.70-02	8.89-02	1.06-01	1.09-01
Os	76	6.71-02	4.46-02	4.13-02	4.79-02	5.96-02	7.71-02	8.90-02	1.06-01	1.10-01
Ir	77	6.79-02	4.50-02	4.17-02	4.84-02	6.02-02	7.80-02	9.01-02	1.07-01	1.11-01
Pt	78	6.86-02	4.52-02	4.20-02	4.87-02	6.06-02	7.86-02	9.08-02	1.08-01	1.12-01
Au	79	6.95-02	4.57-02	4.24-02	4.93-02	6.14-02	7.95-02	9.19-02	1.09-01	1.13-01
Hg	80	6.99-02	4.57-02	4.25-02	4.94-02	6.15-02	7.98-02	9.22-02	1.10-01	1.13-01
Tl	81	7.03-02	4.58-02	4.25-02	4.94-02	6.16-02	8.00-02	9.24-02	1.10-01	1.14-01
Pb	82	7.10-02	4.61-02	4.27-02	4.97-02	6.21-02	8.06-02	9.31-02	1.11-01	1.15-01
Bi	83	7.21-02	4.66-02	4.32-02	5.03-02	6.28-02	8.15-02	9.42-02	1.12-01	1.16-01
Po	84	7.39-02	4.75-02	4.40-02	5.12-02	6.40-02	8.32-02	9.61-02	1.15-01	1.18-01
At	85	7.54-02	4.82-02	4.46-02	5.20-02	6.49-02	8.44-02	9.76-02	1.16-01	1.20-01
Rn	86	7.30-02	4.65-02	4.30-02	5.01-02	6.26-02	8.14-02	9.42-02	1.12-01	1.16-01
Fr	87	7.45-02	4.72-02	4.36-02	5.08-02	6.35-02	8.26-02	9.56-02	1.14-01	1.18-01
Ra	88	7.53-02	4.75-02	4.38-02	5.10-02	6.38-02	8.31-02	9.61-02	1.15-01	1.19-01
Ac	89	7.69-02	4.82-02	4.44-02	5.17-02	6.47-02	8.43-02	9.75-02	1.16-01	1.20-01
Th	90	7.71-02	4.81-02	4.42-02	5.15-02	6.45-02	8.40-02	9.72-02	1.16-01	1.20-01
Pa	91	7.94-02	4.93-02	4.52-02	5.26-02	6.59-02	8.60-02	9.95-02	1.19-01	1.23-01
U	92	7.90-02	4.88-02	4.46-02	5.19-02	6.51-02	8.49-02	9.83-02	1.17-01	1.21-01
Np	93	8.13-02	4.99-02	4.56-02	5.30-02	6.65-02	8.68-02	1.01-01	1.20-01	1.24-01
Pu	94	8.26-02	5.05-02	4.60-02	5.34-02	6.71-02	8.76-02	1.01-01	1.21-01	1.25-01
Am	95	8.33-02	5.06-02	4.60-02	5.34-02	6.70-02	8.77-02	1.02-01	1.21-01	1.25-01

	Atomic no.	1.0 MeV	2.0 MeV	5.0 MeV	10.0 MeV	20.0 MeV	50.0 MeV	100.0 MeV	500.0 MeV	1000.0 MeV
Cm	96	8.41-02	5.08-02	4.60-02	5.34-02	6.70-02	8.77-02	1.02-01	1.21-01	1.26-01
Bk	97	8.62-02	5.18-02	4.68-02	5.42-02	6.81-02	8.92-02	1.03-01	1.24-01	1.28-01
Cf	98	8.70-02	5.20-02	4.68-02	5.42-02	6.81-02	8.92-02	1.04-01	1.24-01	1.28-01
Es	99	8.89-02	5.28-02	4.74-02	5.48-02	6.89-02	9.04-02	1.05-01	1.25-01	1.29-01
Fm	100	8.94-02	5.28-02	4.72-02	5.45-02	6.86-02	9.00-02	1.05-01	1.25-01	1.29-01

Absorption Edges in keV

	Atomic no.	K	L1	L2	L3	M1	M2	M3	M4	M5
Na	11	1.07								
Mg	12	1.31								
Al	13	1.56								
Si	14	1.84								
P	15	2.15								
S	16	2.47								
Cl	17	2.82								
Ar	18	3.20								
K	19	3.61								
Ca	20	4.04								
Sc	21	4.49								
Ti	22	4.97								
V	23	5.47								
Cr	24	5.99								
Mn	25	6.54								
Fe	26	7.11								
Co	27	7.71								
Ni	28	8.33	1.01							
Cu	29	8.98	1.10							
Zn	30	9.66	1.19	1.04	1.02					
Ga	31	10.37	1.30	1.14	1.12					
Ge	32	11.10	1.41	1.25	1.22					
As	33	11.87	1.53	1.36	1.32					
Se	34	12.66	1.65	1.48	1.44					
Br	35	13.47	1.78	1.60	1.55					
Kr	36	14.33	1.92	1.73	1.68					
Rb	37	15.20	2.07	1.86	1.80					
Sr	38	16.10	2.22	2.01	1.94					
Y	39	17.04	2.37	2.16	2.08					
Zr	40	18.00	2.53	2.31	2.22					
Nb	41	18.99	2.70	2.47	2.37					
Mo	42	20.00	2.87	2.63	2.52					
Tc	43	21.04	3.04	2.79	2.68					
Ru	44	22.12	3.22	2.97	2.84					
Rh	45	23.22	3.41	3.15	3.00					
Pd	46	24.35	3.60	3.33	3.17					
Ag	47	25.51	3.81	3.52	3.35					
Cd	48	26.71	4.02	3.73	3.54					
In	49	27.94	4.24	3.94	3.73					
Sn	50	29.20	4.47	4.16	3.3					
Sb	51	30.49	4.70	4.38	4.13					
Te	52	31.81	4.94	4.61	4.34	1.01				
I	53	33.17	5.19	4.85	4.56	1.07				
Xe	54	34.56	5.45	5.10	4.78	1.15				
Cs	55	35.98	5.71	5.36	5.01	1.22	1.07			
Ba	56	37.44	5.99	5.62	5.25	1.29	1.14	1.06		
La	57	38.92	6.27	5.89	5.48	1.36	1.20	1.12		
Ce	58	40.44	6.55	6.16	5.72	1.44	1.27	1.19		
Pr	59	41.99	6.84	6.44	5.96	1.51	1.34	1.24		
Nd	60	43.57	7.13	6.72	6.21	1.58	1.40	1.30	1.01	

	Atomic no.	K	L1	L2	L3	M1	M2	M3	M4	M5
Pm	61	45.18	7.43	7.01	6.46	1.65	1.47	1.36	1.05	1.03
Sm	62	46.83	7.74	7.31	6.72	1.72	1.54	1.42	1.11	1.08
Eu	63	48.52	8.05	7.62	6.98	1.80	1.61	1.48	1.16	1.13
Gd	64	50.24	8.38	7.93	7.24	1.88	1.69	1.54	1.22	1.19
Tb	65	52.00	8.71	8.25	7.51	1.97	1.77	1.61	1.28	1.24
Dy	66	53.79	9.05	8.58	7.79	2.05	1.84	1.68	1.33	1.30
Ho	67	55.62	9.39	8.92	8.07	2.13	1.92	1.74	1.39	1.35
Er	68	57.49	9.75	9.26	8.36	2.21	2.01	1.81	1.45	1.41
Tm	69	59.39	10.12	9.62	8.65	2.31	2.09	1.88	1.52	1.47
Yb	70	61.33	10.49	9.98	8.94	2.40	2.17	1.95	1.58	1.53
Lu	71	63.31	10.87	10.35	9.24	2.49	2.26	2.02	1.64	1.59
Hf	72	65.35	11.27	10.74	9.56	2.60	2.37	2.11	1.72	1.66
Ta	73	67.42	11.68	11.14	9.88	2.71	2.47	2.19	1.79	1.74
W	74	69.53	12.10	11.54	10.21	2.82	2.58	2.28	1.87	1.81
Re	75	71.68	12.53	11.96	10.54	2.93	2.68	2.37	1.95	1.82
Os	76	73.87	12.97	12.39	10.87	3.05	2.79	2.46	2.03	1.96
Ir	77	76.11	13.42	12.82	11.22	3.17	2.91	2.55	2.12	2.04
Pt	78	78.39	13.88	13.27	11.56	3.30	3.03	2.65	2.20	2.12
Au	79	80.72	14.35	13.73	11.92	3.43	3.15	2.74	2.29	2.21
Hg	80	83.10	14.84	14.21	12.28	3.56	3.28	2.85	2.39	2.30
Tl	81	85.53	15.35	14.70	12.66	3.70	3.42	2.96	2.49	2.39
Pb	82	88.00	15.86	15.20	13.04	3.85	3.55	3.07	2.59	2.48
Bi	83	90.53	16.39	15.71	13.42	4.00	3.70	3.18	2.69	2.58
Po	84	93.10	16.94	16.24	13.81	4.15	3.85	3.30	2.80	2.68
At[a]	85	95.73	17.49	16.78	14.21	4.32	4.01	3.43	2.91	2.79
Rn[a]	86	98.40	18.05	17.34	14.62	4.48	4.16	3.54	3.02	2.89
Fr[a]	87	101.1	18.64	17.91	15.03	4.65	4.33	3.66	3.14	3.00
Ra[a]	88	103.9	19.24	18.48	15.44	4.82	4.49	3.79	3.25	3.10
Ac[a]	89	106.8	19.84	19.08	15.87	5.00	4.66	3.91	3.37	3.22
Th[a]	90	109.7	20.47	19.69	16.30	5.18	4.83	4.05	3.49	3.33
Pa[a]	91	112.6	21.10	20.31	16.73	5.37	5.00	4.17	3.61	3.44
U[a]	92	115.6	21.76	20.95	17.17	5.55	5.18	4.30	3.73	3.55
Np[a]	93	118.7	22.43	21.60	17.61	5.72	5.37	4.44	3.85	3.67
Pu[a]	94	121.8	23.10	22.27	18.06	5.93	5.54	4.56	3.97	3.78
Am[a]	95	125.0	23.77	22.94	18.50	6.12	5.71	4.67	4.09	3.89
Cm[a]	96	128.2	24.46	23.80	18.93	6.29	5.90	4.80	4.23	3.97
Bk[a]	97	131.6	25.27	24.39	19.45	6.56	6.15	4.98	4.37	4.13
Cf[a]	98	136.0	26.11	25.25	19.93	6.75	6.36	5.11	4.50	4.25
Es[a]	99	139.5	26.90	26.02	20.41	6.98	6.57	5.25	4.63	4.37
Fm[a]	100	143.1	27.70	26.81	20.90	7.21	6.79	5.40	4.77	4.50

[a] Additional absorption edges are present; see Refs. 3, 4.

CLASSIFICATION OF ELECTROMAGNETIC RADIATION

David R. Lide

Through historical usage and, in some cases, action of standardization bodies, the electromagnetic spectrum has been divided into frequency bands or regions where the character of the radiation is similar. Table 1 below shows a division into decade frequency steps running from 3 Hz to 30 ZHz (30×10^{21} Hz), i.e., from low radiofrequencies (RF) to γ-rays. The second column gives the approximate wavelength range corresponding to each frequency range. The wavenumber range and approximate photon energy range are also given. The last column gives the designation of each of the bands. The bands up to 3 THz frequency have been given letter designations by the International Telecommunication Union (ITU), but it should be noted that many other designations are in use. The higher frequency regions are labeled by historical terms like infrared (IR), visible, ultraviolet (UV), etc. There are no sharp boundaries between these regions, and usage varies. The exact relations between these parameters are:

$$\nu = c/\lambda = ck$$
$$\lambda = c/\nu = 1/k$$
$$k = \nu/c = 1/\lambda$$
$$c = \text{speed of light} = 2.99792458 \times 10^8 \text{ m/s}$$

The major regions are illustrated graphically in Figure 1.

The microwave region, which is usually considered to extend from about 1 GHz to 300 GHz, has been subdivided into several bands that are given arbitrary letter designations. These bands, which are important in communications and radar technology, are listed in Table 2.

Table 1. Frequency, Wavelength, Wavenumber, and Energy Ranges of Radiation Bands in Decade Steps

Frequency (ν)	Wavelength (λ)	Wavenumber (k)	Approximate photon energy	Names of bands
3 Hz – 30 Hz	10^8 – 10^7 m	10^{-8} – 10^{-7} m^{-1}		ELF (ELF 1)
30 Hz – 300 Hz	10^7 – 10^6 m	10^{-7} – 10^{-6} m^{-1}		SLF (ELF 2)
300 Hz – 3 kHz	10^6 – 10^5 m	10^{-6} – 10^{-5} m^{-1}		ULF (ELF 3)
3 kHz – 30 kHz	100 km – 10 km	10^{-5} – 10^{-4} m^{-1}		VLF
30 kHz – 300 kHz	10 km – 1 km	10^{-4} – 10^{-3} m^{-1}		LF
300 kHz – 3 MHz	1 km – 100 m	10^{-3} – 10^{-2} m^{-1}		MF
3 MHz – 30 MHz	100 m – 10 m	10^{-2} – 10^{-1} m^{-1}		HF
30 MHz – 300 MHz	10 m – 1 m	10^{-1} – 10^{0} m^{-1}		VHF
300 MHz – 3 GHz	1 m – 100 mm	10^{0} – 10^{1} m^{-1}		UHF
3 GHz – 30 GHz	100 mm – 10 mm	10^{1} – 10^{2} m^{-1}		SHF, microwave
30 GHz – 300 GHz	100 mm –10 mm	1 - 10 cm^{-1}		EHF, microwave
300 GHz – 3 THz	1 mm – 100 μm	10 – 100 cm^{-1}		THF, far IR
3 THz – 30 THz	100 μm – 10 μm	100 – 1000 cm^{-1}		Far IR, mid IR
30 THz – 300 THz	10 μm – 1 μm	1000 – 10,000 cm^{-1}	0.1 eV – 1 eV	Mid IR, near IR
300 THz – 3 PHz	1 μm – 100 nm	1 – 10 μm^{-1}	1 eV – 10 eV	Visible, near UV
3 PHz – 30 PHz	100 nm – 10 nm	10 – 100 μm^{-1}	10 eV – 100 eV	Far (vacuum) UV
30 PHz – 300 PHz	10 nm –1 nm	100 μm^{-1} – 1 nm^{-1}	100 eV – 1 keV	Soft x-ray
300 PHz – 3 EHz	1 nm – 100 pm	1 – 10 nm^{-1}	1 keV – 10 keV	Soft x-ray
3 EHz – 30 EHz	100 pm – 10 pm	10 – 100 nm^{-1}	10 keV – 100 keV	Hard x-ray, soft γ-ray
30 EHz – 300 EHz	10 pm – 1 pm	100 nm^{-1} – 1 pm^{-1}	100 keV – 1 MeV	Soft γ-ray, hard γ-ray
300 EHz – 3 ZHz	1 pm – 100 fm	1 – 10 pm^{-1}	1 MeV – 10 MeV	Hard γ-ray, cosmic γ-ray
3 ZHz – 30 ZHz	100 pm – 10 fm	10 – 100 pm^{-1}	10 Mev – 100 MeV	Cosmic γ-ray

Abbreviations for radio frequency bands: EHF–extremely high frequency; ELF–extremely low frequency; HF–high frequency; ITU–International Telecommunications Union; LF–low frequency; MF–medium frequency; SHF–super high frequency; SLF–super low frequency; UHF–ultra high frequency; ULF–ultra low frequency; VHF–very high frequency; VLF–very low frequency.

Table 2. Designations of Microwave Bands (IEEE Standard 521-1984)

Band	Frequency (GHz)	Wavelength (cm)	Wavenumber (cm^{-1})
L	1—2	30—15	0.033—0.067
S	2—4	15—7.5	0.067—0.133
C	4—8	7.5—3.7	0.133—0.267
X	8—12	3.7—2.5	0.267—0.4
Ku	12—18	2.5—1.7	0.4—0.6
K	18—27	1.7—1.1	0.6—0.9
Ka	27—40	1.1—0.75	0.9—1.33
V	40—75	0.75—0.40	1.33—2.5
W	75—110	0.40—0.27	2.5—3.7

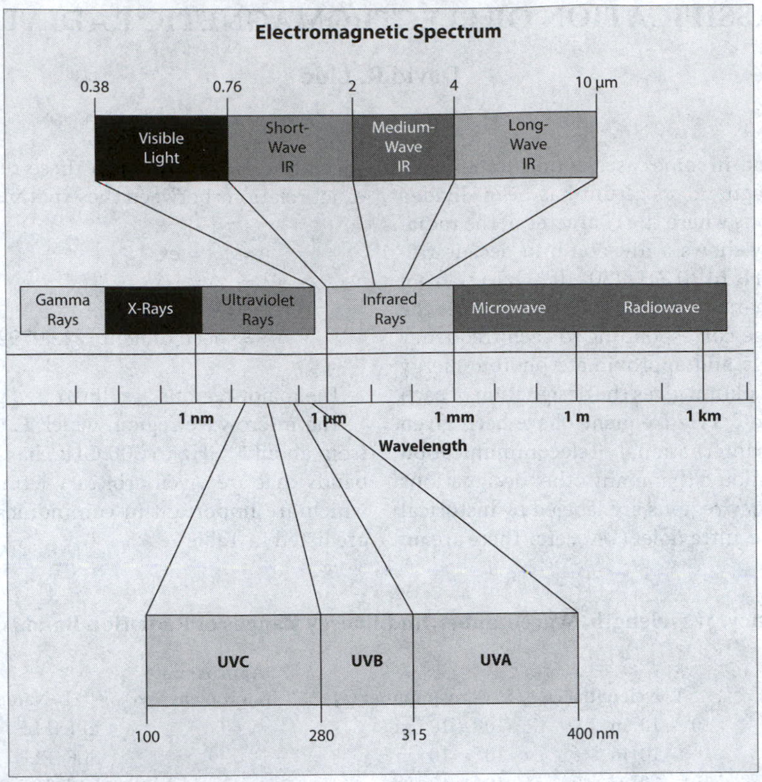

FIGURE 1. Subdivisions of the Electromagnetic Spectrum

SENSITIVITY OF THE HUMAN EYE TO LIGHT OF DIFFERENT WAVELENGTHS

The human eye responds to electromagnetic radiation in the wavelength range from about 360 nm (violet) to 820 nm (red), with a peak sensitivity near 555 nm (green). While the detailed shape of this response curve depends on the individual person, studies on representative samples of human subjects have led to adoption of a standard function relating the perceived brightness (luminous flux) to the actual power of the spectral radiation. This function is referred to as $V(\lambda)$, the photopic spectral luminous efficiency function, and it plays an important role in photometry.

The function $V(\lambda)$, as adopted by the International Commission on Illumination (CIE), is tabulated and plotted below.

References

1. *The Basis for Physical Photometry*, CIE Publication #18.2, 1983.
2. *CIE Standard Colorimetric Observers*, ISO/CIE #10527, 1991.
3. *Kaye and Laby Tables of Physical and Chemical Constants, Sixteenth Edition*, Longman Group Ltd., Harlow, Essex, 1995.

λ/nm	$V(\lambda)$	λ/nm	$V(\lambda)$	λ/nm	$V(\lambda)$
360	0.000004	520	0.710000	670	0.032000
370	0.000012	530	0.862000	680	0.017000
380	0.000039	540	0.954000	690	0.008210
390	0.000120	550	0.994950	700	0.004102
400	0.000396	555	1.000000	710	0.002091
410	0.001210	560	0.995000	720	0.001047
420	0.004000	570	0.952000	730	0.000520
430	0.011600	580	0.870000	740	0.000249
440	0.023000	590	0.757000	750	0.000120
450	0.038000	600	0.631000	760	0.000060
460	0.060000	610	0.503000	770	0.000030
470	0.090980	620	0.381000	780	0.000015
480	0.139020	630	0.265000	790	0.000007
490	0.208020	640	0.175000	800	0.000004
500	0.323000	650	0.107000	810	0.000002
510	0.503000	660	0.061000	820	0.000001

Spectral Luminous Efficiency Function

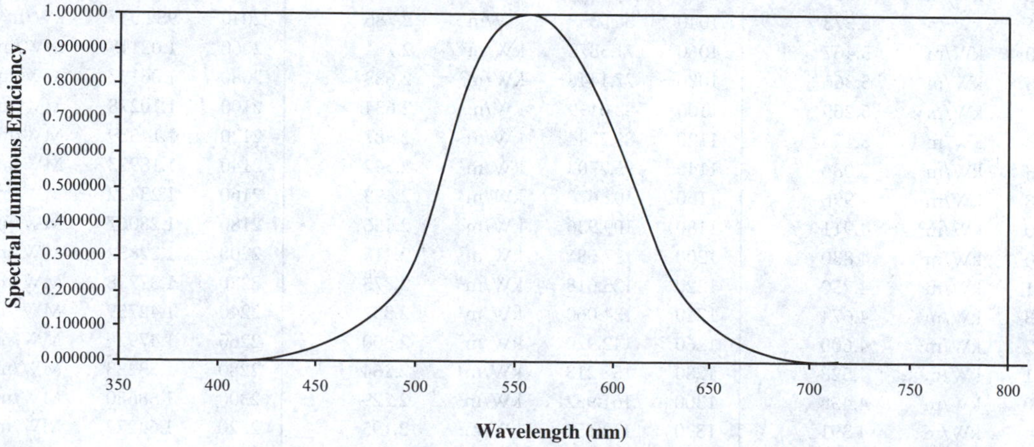

BLACKBODY RADIATION

The total power radiated from an ideal blackbody and the wavelength corresponding to maximum power are given here as a function of absolute temperature. The radiated power in a band $\Delta\lambda$ at λ_{max} may be calculated from:

$$P_{max} = 0.657548\ (\Delta\lambda/\lambda_{max})\ P_{tot}$$

Total Power and Wavelength of Maximum Power as a Function of Temperature

T/K	P_{tot}	Units	λ_{max}/µm	*T*/K	P_{tot}	Units	λ_{max}/µm	*T*/K	P_{tot}	Units	λ_{max}/µm
50	0.354398	W/m²	57.955	720	15.2385	kW/m²	4.025	1440	243.816	kW/m²	2.012
100	5.67037	W/m²	28.978	730	16.1029	kW/m²	3.970	1460	257.646	kW/m²	1.985
150	28.7063	W/m²	19.318	740	17.0035	kW/m²	3.916	1480	272.056	kW/m²	1.958
200	90.7260	W/m²	14.489	750	17.9414	kW/m²	3.864	1500	287.063	kW/m²	1.932
250	221.499	W/m²	11.591	760	18.9176	kW/m²	3.813	1520	302.682	kW/m²	1.906
273	314.965	W/m²	10.614	770	19.9331	kW/m²	3.763	1540	318.929	kW/m²	1.882
280	348.533	W/m²	10.349	780	20.9889	kW/m²	3.715	1560	335.823	kW/m²	1.858
290	401.055	W/m²	9.992	790	22.0861	kW/m²	3.668	1580	353.378	kW/m²	1.834
300	459.300	W/m²	9.659	800	23.2258	kW/m²	3.622	1600	371.614	kW/m²	1.811
310	523.671	W/m²	9.348	810	24.4091	kW/m²	3.577	1620	390.546	kW/m²	1.789
320	594.582	W/m²	9.055	820	25.6370	kW/m²	3.534	1640	410.192	kW/m²	1.767
330	672.461	W/m²	8.781	830	26.9106	kW/m²	3.491	1660	430.570	kW/m²	1.746
340	757.752	W/m²	8.523	840	28.2312	kW/m²	3.450	1680	451.699	kW/m²	1.725
350	850.910	W/m²	8.279	850	29.5997	kW/m²	3.409	1700	473.595	kW/m²	1.705
360	952.405	W/m²	8.049	860	31.0174	kW/m²	3.369	1720	496.278	kW/m²	1.685
370	1.06272	kW/m²	7.832	870	32.4854	kW/m²	3.331	1740	519.767	kW/m²	1.665
380	1.18235	kW/m²	7.626	880	34.0050	kW/m²	3.293	1760	544.079	kW/m²	1.646
390	1.31181	kW/m²	7.430	890	35.5772	kW/m²	3.256	1780	569.235	kW/m²	1.628
400	1.45162	kW/m²	7.244	900	37.2033	kW/m²	3.220	1800	595.253	kW/m²	1.610
410	1.60231	kW/m²	7.068	910	38.8846	kW/m²	3.184	1820	622.153	kW/m²	1.592
420	1.76445	kW/m²	6.899	920	40.6222	kW/m²	3.150	1840	649.954	kW/m²	1.575
430	1.93859	kW/m²	6.739	930	42.4173	kW/m²	3.116	1860	678.677	kW/m²	1.558
440	2.12531	kW/m²	6.586	940	44.2714	kW/m²	3.083	1880	708.342	kW/m²	1.541
450	2.32521	kW/m²	6.439	950	46.1855	kW/m²	3.050	1900	738.969	kW/m²	1.525
460	2.53888	kW/m²	6.299	960	48.1611	kW/m²	3.018	1920	770.578	kW/m²	1.509
470	2.76696	kW/m²	6.165	970	50.1994	kW/m²	2.987	1940	803.190	kW/m²	1.494
480	3.01007	kW/m²	6.037	980	52.3017	kW/m²	2.957	1960	836.827	kW/m²	1.478
490	3.26886	kW/m²	5.914	990	54.4694	kW/m²	2.927	1980	871.510	kW/m²	1.464
500	3.54398	kW/m²	5.796	1000	56.7037	kW/m²	2.898	2000	907.260	kW/m²	1.449
510	3.83612	kW/m²	5.682	1020	61.3779	kW/m²	2.841	2020	944.098	kW/m²	1.435
520	4.14596	kW/m²	5.573	1040	66.3353	kW/m²	2.786	2040	982.047	kW/m²	1.420
530	4.47420	kW/m²	5.467	1060	71.5872	kW/m²	2.734	2060	1.02113	MW/m²	1.407
540	4.82155	kW/m²	5.366	1080	77.1448	kW/m²	2.683	2080	1.06137	MW/m²	1.393
550	5.18875	kW/m²	5.269	1100	83.0199	kW/m²	2.634	2100	1.10278	MW/m²	1.380
560	5.57653	kW/m²	5.175	1120	89.2244	kW/m²	2.587	2120	1.14539	MW/m²	1.367
570	5.98565	kW/m²	5.084	1140	95.7703	kW/m²	2.542	2140	1.18923	MW/m²	1.354
580	6.41688	kW/m²	4.996	1160	102.670	kW/m²	2.498	2160	1.23432	MW/m²	1.342
590	6.87100	kW/m²	4.911	1180	109.936	kW/m²	2.456	2180	1.28067	MW/m²	1.329
600	7.34880	kW/m²	4.830	1200	117.581	kW/m²	2.415	2200	1.32832	MW/m²	1.317
610	7.85111	kW/m²	4.750	1220	125.618	kW/m²	2.375	2220	1.37728	MW/m²	1.305
620	8.37873	kW/m²	4.674	1240	134.060	kW/m²	2.337	2240	1.42759	MW/m²	1.294
630	8.93252	kW/m²	4.600	1260	142.920	kW/m²	2.300	2260	1.47926	MW/m²	1.282
640	9.51331	kW/m²	4.528	1280	152.213	kW/m²	2.264	2280	1.53233	MW/m²	1.271
650	10.1220	kW/m²	4.458	1300	161.952	kW/m²	2.229	2300	1.58680	MW/m²	1.260
660	10.7594	kW/m²	4.391	1320	172.150	kW/m²	2.195	2320	1.64272	MW/m²	1.249
670	11.4264	kW/m²	4.325	1340	182.823	kW/m²	2.163	2340	1.70010	MW/m²	1.238
680	12.1240	kW/m²	4.261	1360	193.985	kW/m²	2.131	2360	1.75897	MW/m²	1.228
690	12.8531	kW/m²	4.200	1380	205.650	kW/m²	2.100	2380	1.81936	MW/m²	1.218
700	13.6146	kW/m²	4.140	1400	217.833	kW/m²	2.070	2400	1.88129	MW/m²	1.207
710	14.4094	kW/m²	4.081	1420	230.550	kW/m²	2.041	2420	1.94479	MW/m²	1.197

T/K	P_{tot}	Units	λ_{max}/μm	*T*/K	P_{tot}	Units	λ_{max}/μm	*T*/K	P_{tot}	Units	λ_{max}/μm
2440	2.00988	MW/m²	1.188	3400	7.57752	MW/m²	0.852	4900	32.6886	MW/m²	0.591
2460	2.07660	MW/m²	1.178	3450	8.03319	MW/m²	0.840	5000	35.4398	MW/m²	0.580
2480	2.14496	MW/m²	1.168	3500	8.50910	MW/m²	0.828	5100	38.3612	MW/m²	0.568
2500	2.21499	MW/m²	1.159	3550	9.00586	MW/m²	0.816	5200	41.4596	MW/m²	0.557
2550	2.39758	MW/m²	1.136	3600	9.52405	MW/m²	0.805	5300	44.7420	MW/m²	0.547
2600	2.59122	MW/m²	1.115	3650	10.0643	MW/m²	0.794	5400	48.2155	MW/m²	0.537
2650	2.79637	MW/m²	1.093	3700	10.6272	MW/m²	0.783	5500	51.8875	MW/m²	0.527
2700	3.01347	MW/m²	1.073	3750	11.2134	MW/m²	0.773	5600	55.7653	MW/m²	0.517
2750	3.24297	MW/m²	1.054	3800	11.8235	MW/m²	0.763	5700	59.8565	MW/m²	0.508
2800	3.48533	MW/m²	1.035	3850	12.4582	MW/m²	0.753	5800	64.1688	MW/m²	0.500
2850	3.74103	MW/m²	1.017	3900	13.1181	MW/m²	0.743	5900	68.7100	MW/m²	0.491
2900	4.01055	MW/m²	0.999	3950	13.8038	MW/m²	0.734	6000	73.4880	MW/m²	0.483
2950	4.29437	MW/m²	0.982	4000	14.5162	MW/m²	0.724	6500	101.220	MW/m²	0.446
3000	4.59300	MW/m²	0.966	4100	16.0231	MW/m²	0.707	7000	136.146	MW/m²	0.414
3050	4.90694	MW/m²	0.950	4200	17.6445	MW/m²	0.690	7500	179.414	MW/m²	0.386
3100	5.23671	MW/m²	0.935	4300	19.3859	MW/m²	0.674	8000	232.258	MW/m²	0.362
3150	5.58282	MW/m²	0.920	4400	21.2531	MW/m²	0.659	8500	295.997	MW/m²	0.341
3200	5.94582	MW/m²	0.906	4500	23.2521	MW/m²	0.644	9000	372.033	MW/m²	0.322
3250	6.32623	MW/m²	0.892	4600	25.3888	MW/m²	0.630	9500	461.855	MW/m²	0.305
3300	6.72461	MW/m²	0.878	4700	27.6696	MW/m²	0.617	10000	567.037	MW/m²	0.290
3350	7.14152	MW/m²	0.865	4800	30.1007	MW/m²	0.604				

The curves below show, for various temperatures, the fraction of radiant power as a function of wavelength. The function plotted is $P_\lambda/\Delta\lambda P_{tot}$, where P_λ is the power at wavelength λ in a small interval $\Delta\lambda$ (in μm), and P_{tot} is the total power.

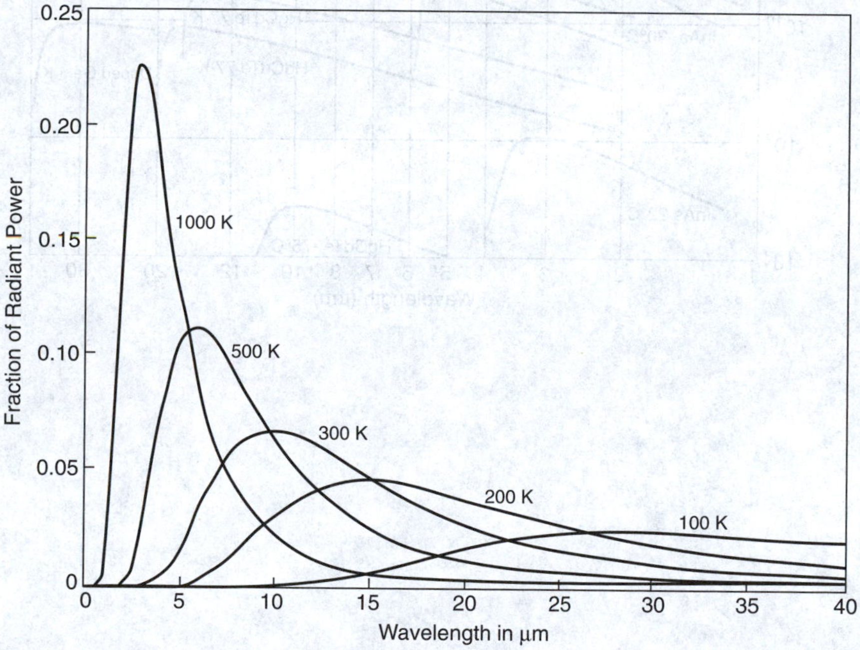

CHARACTERISTICS OF INFRARED DETECTORS

This graph summarizes the wavelength response of some semiconductors used as detectors for infrared radiation. The quantity $D^*(\lambda)$ is the signal-to-noise ratio for an incident radiant power density of 1 W/cm^2 and a bandwidth of 1 Hz (60° field of view). The Ge, InAs, and InSb detectors are photovoltaics, while the HgCdTe series are photoconductive devices. The cutoff wavelength of the latter can be varied by adjusting the relative amounts of Hg, Cd, and Te (three examples are shown at 77 K). The graph also shows the theoretical background limited sensitivity for ideal detectors which introduce no intrinsic noise.

Reference

Infrared Detectors 1995, EG&G Judson, Montgomeryville, PA.

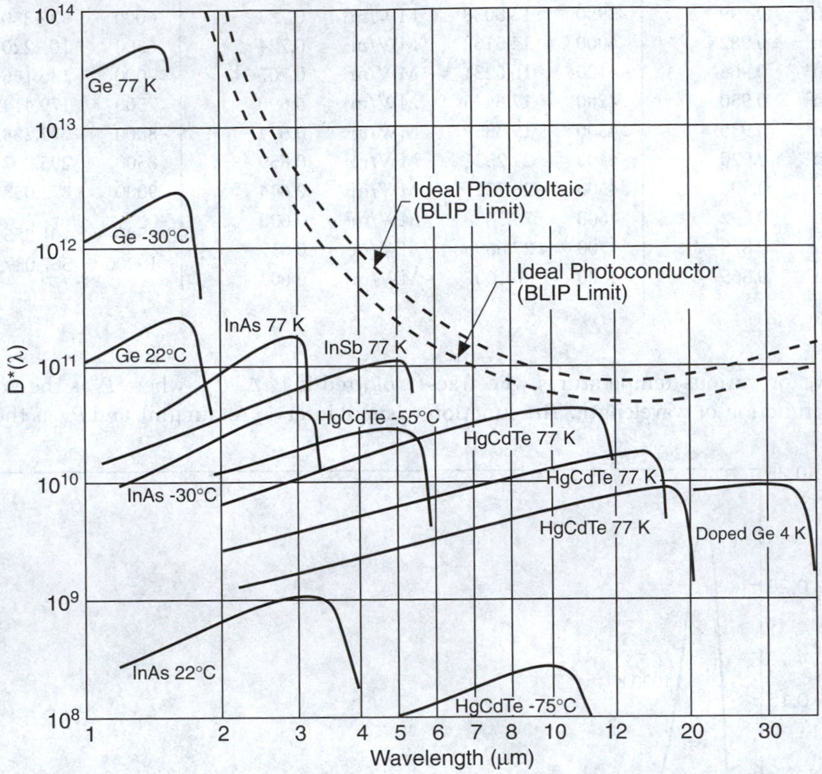

INDEX OF REFRACTION OF INORGANIC CRYSTALS

This table lists the index of refraction of selected crystalline inorganic compounds. When available, values are given as a function of wavelength in the range from the ultraviolet to the far infrared region. For most compounds a value at 589 nm, the wavelength of the principal sodium line, is given. The data have been taken from the references indicated; in many cases, data from a reference have been refitted to generate the index of refraction at the wavelengths used in this table. All values refer to ambient temperature. Entries marked by * are based on extrapolation beyond the range of available experimental data.

Compounds belonging to the cubic crystal system have only a single refractive index value, but other systems are anisotropic, so that the crystal is characterized by two or three unique indexes. Hexagonal, rhombohedral, and tetragonal crystals have two unique indexes that are traditionally labeled n_o and n_e for "ordinary ray" and "extraordinary ray." Orthorhombic, monoclinic, and triclinic crystals are characterized by three indexes that here are called n_x, n_y, and n_z. The table indicates the crystal system for each entry in order to identify the material uniquely.

Compounds are listed in order of their common formula.

The refractive index and other optical properties for metals, semiconductors, and certain other compounds can be found in the tables "Optical Properties of Selected Elements" and "Optical Properties of Selected Inorganic and Organic Solids" in Section 12 of this *Handbook*.

References

1. Li, H. H., "Refractive Index of Alkali Halides and its Wavelength and Temperature Derivatives," *J. Phys. Chem. Ref. Data* 5, 329, 1976.
2. Li, H. H., "Refractive Index of Alkaline Earth Halides and its Wavelength and Temperature Derivatives," *J. Phys. Chem. Ref. Data* 9, 161, 1980.
3. Li, H. H., "Refractive Index of ZnS, ZnSe, and ZnTe and its Wavelength and Temperature Derivatives," *J. Phys. Chem. Ref. Data* 13, 103, 1984.
4. Shannon, R. D., Shannon, R. C., Medenbach, O., and Fischer, R. X., "Refractive Index and Dispersion of Fluorides and Oxides," *J. Phys. Chem. Ref. Data* 31, 931, 2002.
5. Gray, D. E., ed., *American Institute of Physics Handbook*, Sec. 6b, pp. 6–12, McGraw-Hill, New York, 1972.
6. *Landolt-Börnstein Numerical Data and Functional Relationships in Science and Technology, III/11, Elastic, Piezoelectric, Pyroelectric, Piezooptic, Electrooptic Constants, and Nonlinear Dielectric Susceptibilities of Crystals*, Springer-Verlag, Berlin, 1979.
7. *Landolt-Börnstein Numerical Data and Functional Relationships in Science and Technology, III/30A, High Frequency Properties of Dielectric Crystals. Piezooptic and Electrooptic Constants*, Springer-Verlag, Berlin, 1996.
8. Weber, M. J., *CRC Handbook of Laser Science and Technology*, Vol. IV. Optical Materials. Part 2: Properties, CRC Press, Boca Raton, FL, 1986.

Index of Refraction at the Indicated Wavelength

Formula	Cryst. syst.	Ray	*n* (200 nm)	*n* (300 nm)	*n* (589 nm)	*n* (750 nm)	*n* (1 µm)	*n* (2 µm)	*n* (5 µm)	*n* (10 µm)	*n* (20 µm)	Ref.
AgCl	cub	n			2.0668	2.0401	2.0224	2.0062	1.9975	1.9803	1.9069	5
$AlPO_4$	rhomb	n_o			1.5247	1.5203	1.5161	1.5034				6
$AlPO_4$	rhomb	n_e			1.5338	1.5290	1.5245	1.5116				6
Al_2O_3	hex	n_o			1.7673							4
Al_2O_3	hex	n_e			1.7598							4
As_2O_3[a]	cub	n			1.7537							4
BaF_2	cub	n	1.557	1.5010	1.4744	1.4712	1.4686	1.4647	1.4511	1.4014		2
BaO	cub	n			1.9841							4
$BaSO_4$	orth	n_x			1.6362							4
$BaSO_4$	orth	n_y			1.6374							4
$BaSO_4$	orth	n_z			1.6480							4
$BaTiO_3$	tetr	n_o			2.4405							4
$BaTiO_3$	tetr	n_e			2.3831							4
$BaWO_4$	tetr	n_o			1.8426							4
$BaWO_4$	tetr	n_e			1.8405							4
BeO	hex	n_o			1.7184							4
BeO	hex	n_e			1.7342							4
$BeSO_4·4H_2O$	tetr	n_o			1.4713							4
$BeSO_4·4H_2O$	tetr	n_e			1.4328							4
$CaCO_3$[b]	hex	n_o	1.9028	1.7216	1.6584	1.6503	1.6436	1.6249				5
$CaCO_3$[b]	hex	n_e	1.5765	1.5145	1.4864	1.4828	1.4801	1.4753				5
CaF_2	cub	n	1.495	1.4540	1.4338	1.4311	1.4289	1.4239	1.3990	1.299		2
CaO	cub	n			1.8396							4
$CaSO_4$	orth	n_x			1.5698							4
$CaSO_4$	orth	n_y			1.5755							4
$CaSO_4$	orth	n_z			1.6137							4
$CaSO_4·2H_2O$	monocl	n_x			1.5207							4
$CaSO_4·2H_2O$	monocl	n_y			1.5227							4
$CaSO_4·2H_2O$	monocl	n_z			1.5304							4

Formula	Cryst. syst.	Ray	n (200 nm)	n (300 nm)	n (589 nm)	n (750 nm)	n (1 µm)	n (2 µm)	n (5 µm)	n (10 µm)	n (20 µm)	Ref.
$CaWO_4$	tetr	n_o			1.9195							4
$CaWO_4$	tetr	n_e			1.9355							4
CdS	hex	n_o			2.507	2.390	2.334					5
CdS	hex	n_e			2.525	2.409	2.352					5
CdSe	hex	n_o				2.68*	2.5502	2.4682	2.4483	2.4331		7
CdSe	hex	n_e				2.69*	2.5696	2.4873	2.4676	2.4514		7
CdTe	cub	n								2.6724	2.6302	7
CeF_3	hex	n_o			1.6183							4
CeF_3	hex	n_e			1.6113							4
CsBr	cub	n		1.8047	1.6974	1.6861	1.6784	1.6711	1.6678	1.6630	1.6439	1
CsCl	cub	n	1.937	1.712	1.640	1.631	1.626	1.620	1.616	1.606	1.563	1
$CsClO_4$	orth	n_x			1.4752							4
$CsClO_4$	orth	n_y			1.4788							4
$CsClO_4$	orth	n_z			1.4804							4
CsF	cub	n	1.575*	1.506	1.477	1.474	1.472	1.469*	1.461*	1.436*	1.32*	1
CsI	cub	n		1.9790	1.7873	1.7694	1.7576	1.7465	1.7428	1.7396	1.7280	1
Cs_2SO_4	orth	n_x			1.5598							4
Cs_2SO_4	orth	n_y			1.5644							4
Cs_2SO_4	orth	n_z			1.5662							4
CuBr	cub	n			2.117							7
CuCl	cub	n			1.9727	1.9391				1.9245		7
$CuSO_4{\cdot}5H_2O$	tricl	n_x			1.5140							4
$CuSO_4{\cdot}5H_2O$	tricl	n_y			1.5367							4
$CuSO_4{\cdot}5H_2O$	tricl	n_z			1.5436							4
Dy_2O_3	cub	n			1.9757							4
FeF_2	tetr	n_o			1.514							4
FeF_2	tetr	n_e			1.524							4
Gd_2O_3	cub	n			1.96							4
HgS	rhomb	n_o			2.9413	2.7770	2.7120	2.6305		2.6018		6
HgS	rhomb	n_e			3.3072	3.0896	3.0050	2.8776		2.8522		6
KBr	cub	n	2.099*	1.6482	1.5598	1.5498	1.5444	1.5383	1.5345	1.5264	1.4924	1
KCl	cub	n	1.717	1.5455	1.4902	1.4840	1.4798	1.4753	1.4704	1.4564	1.3946	1
$KClO_4$	orth	n_x			1.4730							4
$KClO_4$	orth	n_y			1.4736							4
$KClO_4$	orth	n_z			1.4768							4
KF	cub	n	1.426	1.380	1.362	1.360	1.358	1.355	1.344	1.304*	1.09*	1
KH_2AsO_4	tetr	n_o			1.5674							7
KH_2AsO_4	tetr	n_e			1.5179							7
KH_2PO_4	tetr	n_o	1.617*	1.5450	1.5093	1.5030	1.4957					5
KH_2PO_4	tetr	n_e	1.558*	1.4977	1.4682	1.4641	1.4606					5
KI	cub	n		1.834*	1.665	1.650	1.640	1.631	1.627	1.620	1.593	1
KIO_3	tricl	n_x			1.6959							7
KIO_3	tricl	n_y			1.8317							7
KIO_3	tricl	n_z			1.8343							7
KIO_4	tetr	n_o			1.6205							4
KIO_4	tetr	n_e			1.6476							4
$KNbO_3$	orth	n_x			2.2480	2.3395	2.2612					7
$KNbO_3$	orth	n_y			2.3464	2.2959	2.2622					7
$KNbO_3$	orth	n_x			2.1803	2.1457	2.1288					7
K_2SO_4	orth	n_x			1.4934							4
K_2SO_4	orth	n_y			1.4947							4
K_2SO_4	orth	n_z			1.4973							4
LaF_3	hex	n_o			1.605							4
LaF_3	hex	n_e			1.599							4
LiBr	cub	n	1.92*	1.810	1.783	1.781	1.778	1.774*	1.756*	1.68*	1.33	1
LiCl	cub	n	1.72	1.677	1.662	1.660	1.658	1.654*	1.62*	1.53*		1
$LiClO_4{\cdot}3H_2O$	hex	n_o			1.4832							4
$LiClO_4{\cdot}3H_2O$	hex	n_e			1.4384							4
LiF	cub	n	1.439	1.4087	1.3921	1.3895	1.3871	1.3786	1.3266	1.1005		1
LiI	cub	n		1.979	1.955	1.952	1.950	1.948*	1.940*	1.91*	1.77*	1
$LiIO_3$	hex	n_o			1.8875	1.8713	1.8589	1.8410				6

Formula	Cryst. syst.	Ray	n (200 nm)	n (300 nm)	n (589 nm)	n (750 nm)	n (1 μm)	n (2 μm)	n (5 μm)	n (10 μm)	n (20 μm)	Ref.
$LiIO_3$	hex	n_e			1.7400	1.7268	1.7179	1.7062				6
$LiNbO_3$	rhomb	n_o			2.3007	2.2632	2.2370					7
$LiNbO_3$	rhomb	n_e			2.2116	2.1804	2.1567					7
$LiTaO_3$	rhomb	n_o			2.1864	2.1590	2.1391	2.1066				7
$LiTaO_3$	rhomb	n_e			2.1908	2.1634	2.1432	2.1115				7
$Li_2SO_4{\cdot}H_2O$	monocl	n_x			1.4615							4
$Li_2SO_4{\cdot}H_2O$	monocl	n_y			1.4765							4
$Li_2SO_4{\cdot}H_2O$	monocl	n_z			1.4863							4
Lu_2O_3	cub	n			1.9349							4
MgF_2	tetr	n_o	1.423	1.3930	1.3776	1.375	1.373	1.368	1.34	1.21		2
MgF_2	tetr	n_e	1.436	1.4055	1.3894	1.387	1.385	1.379	1.34	1.21		2
MgO	cub	n			1.7355	1.7283	1.7228	1.7084	1.6361			5
$MgSO_4{\cdot}7H_2O$	orth	n_x			1.4326							4
$MgSO_4{\cdot}7H_2O$	orth	n_y			1.4555							4
$MgSO_4{\cdot}7H_2O$	orth	n_z			1.4607							4
MnF_2	tetr	n_o			1.472							4
MnF_2	tetr	n_e			1.501							4
NaBr	cub	n		1.748	1.642	1.631	1.623	1.616	1.609	1.593*	1.520*	1
$NaBrO_3$	cub	n			1.6168							4
NaCl	cub	n	1.789*	1.6066	1.5441	1.5369	1.5320	1.5265	1.5188	1.4947	1.382*	1
$NaClO_3$	cub	n			1.5151							7
NaF	cub	n	1.383	1.3424	1.3252	1.3231	1.3214	1.3179	1.3017	1.2400		1
$NaH_2PO_4{\cdot}2H_2O$	orth	n_x			1.4400							7
$NaH_2PO_4{\cdot}2H_2O$	orth	n_y			1.4628							7
$NaH_2PO_4{\cdot}2H_2O$	orth	n_z			1.4814							7
NaI	cub	n		1.93*	1.774	1.758	1.74	1.73*	1.73*	1.71*	1.66*	1
$NaNO_2$	orth	n_x			1.6547							7
$NaNO_2$	orth	n_y			1.3455							7
$NaNO_2$	orth	n_z			1.4125							7
$NaNO_3$	rhomb	n_o			1.5840							5
$NaNO_3$	rhomb	n_e			1.3340							5
$Na_2HPO_4{\cdot}7H_2O$	monocl	n_x			1.4411							4
$Na_2HPO_4{\cdot}7H_2O$	monocl	n_y			1.4423							4
$Na_2HPO_4{\cdot}7H_2O$	monocl	n_z			1.4525							4
Na_2SO_4	orth	n_x			1.4669							4
Na_2SO_4	orth	n_y			1.4730							4
Na_2SO_4	orth	n_z			1.4809							4
Nd_2O_3	cub	n			1.92							4
NdF_3	hex	n_o			1.6191							4
NdF_3	hex	n_e			1.6132							4
$NH_4H_2AsO_4$	tetr	n_o		1.6401	1.5777	1.5704	1.5583					7
$NH_4H_2AsO_4$	tetr	n_e		1.5754	1.5232	1.5179	1.5101					7
$NH_4H_2PO_4$	tetr	n_o		1.5668	1.5247	1.5187	1.5084					7
$NH_4H_2PO_4$	tetr	n_e		1.5137	1.4797	1.4754	1.4694					7
NiF_2	tetr	n_o			1.526							4
NiF_2	tetr	n_e			1.561							4
$NiSO_4{\cdot}6H_2O$	tetr	n_o			1.5107							4
$NiSO_4{\cdot}6H_2O$	tetr	n_e			1.4870							4
PbF_2	cub	n		1.94*	1.767	1.754	1.745	1.73	1.70	1.66	1.32	5
$PbSO_4$	orth	n_x			1.8780							4
$PbSO_4$	orth	n_y			1.8834							4
$PbSO_4$	orth	n_z			1.8945							4
PrF_3	hex	n_o			1.6207							4
PrF_3	hex	n_e			1.6146							4
RbBr	cub	n		1.639	1.553	1.544	1.538	1.532	1.530	1.525	1.505*	1
RbCl	cub	n	1.738	1.549	1.493	1.487	1.483	1.479	1.475	1.465	1.424*	1
$RbClO_4$	orth	n_x			1.4691							4
$RbClO_4$	orth	n_y			1.4701							4
$RbClO_4$	orth	n_z			1.4732							4
RbF	cub	n	1.50*	1.428*	1.397	1.394	1.391	1.388	1.379	1.346	1.19*	1
RbH_2AsO_4	tetr	n_o		1.6183	1.5603	1.5538	1.5432					7

Atomic

Formula	Cryst. syst.	Ray	n (200 nm)	n (300 nm)	n (589 nm)	n (750 nm)	n (1 μm)	n (2 μm)	n (5 μm)	n (10 μm)	n (20 μm)	Ref.
RbH_2AsO_4	tetr	n_e		1.5718	1.5232	1.5184	1.5121					7
RbH_2PO_4	tetr	n_o		1.5434	1.5078	1.5021	1.4941					7
RbH_2PO_4	tetr	n_e		1.5106	1.4791	1.4754	1.4704					7
RbI	cub	n		1.808	1.647	1.633	1.623	1.615	1.612	1.608	1.595	1
Rb_2SO_4	orth	n_x			1.5131							4
Rb_2SO_4	orth	n_y			1.5133							4
Rb_2SO_4	orth	n_z			1.5144							4
Sb_2O_3[c]	cub	n			2.8017							4
Sc_2O_3	cub	n			1.9943							4
SiO_2[d]	hex	n_o	1.6493	1.5733	1.5442	1.5394	1.5350	1.5209				5
SiO_2[d]	hex	n_e	1.6605	1.5882	1.5534	1.5484	1.5438	1.5291				5
SnO_2	tetr	n_o			1.993							4
SnO_2	tetr	n_e			2.088							4
SrF_2	cub	n	1.50	1.459	1.4380	1.435	1.433	1.429	1.412	1.35		2
SrO	cub	n			1.8710							4
$SrSO_4$	orth	n_x			1.6214							4
$SrSO_4$	orth	n_y			1.6231							4
$SrSO_4$	orth	n_z			1.6303							4
$SrTiO_3$	cub	n			2.4082	2.3525	2.3160	2.2676	2.1205			5
$SrWO_4$	tetr	n_o			1.8618							4
$SrWO_4$	tetr	n_e			1.8719							4
TbF_3	hex	n_o			1.6034							4
TbF_3	hex	n_e			1.5603							4
TeO_2	tetr	n_o			2.2738		2.2080					7
TeO_2	tetr	n_e			2.4295		2.3520					7
ThO_2	cub	n			2.1113							4
TiO_2[e]	tetr	n_o			2.614	2.533	2.485	2.399	2.290			5
TiO_2[e]	tetr	n_e			2.900	2.805	2.748					5
TiO_2[j]	tetr	n_o			2.562							4
TiO_2[j]	tetr	n_e			2.489							4
Tl_2SO_4	orth	n_x			1.8604							4
Tl_2SO_4	orth	n_y			1.8676							4
Tl_2SO_4	orth	n_z			1.8857							4
TlBr	cub	n			2.418	2.350	2.289	2.103	1.984	2.339	2.322	5
TlCl	cub	n			2.247	2.198	2.145	1.986	1.891	2.193		5
$TlClO_4$	orth	n_x			1.6427							4
$TlClO_4$	orth	n_y			1.6446							4
$TlClO_4$	orth	n_z			1.6542							4
Y_2O_3	cub	n			1.930							4
Yb_2O_3	cub	n			1.9468							4
ZnF_2	tetr	n_o			1.495							4
ZnF_2	tetr	n_e			1.525							4
ZnO	hex	n_o			2.0036	1.9662	1.9435	1.9197				7
ZnO	hex	n_e			2.0199	1.9821	1.9589	1.9330				7
ZnS[f]	cub	n			2.3691	2.3232	2.2932	2.2633				7
ZnS[g]	hex	n_o			2.372	2.331	2.303	2.26	2.25	2.20		3,5
ZnS[g]	hex	n_e			2.368	2.327	2.301					5
ZnSe	cub	n			2.6222	2.5384	2.4888	2.4462	2.4296	2.4065		3
ZnTe	cub	n			3.060	2.880	2.789	2.719	2.698	2.684		3
$ZrSiO_4$[h]	tetr	n_o			1.9255							4
$ZrSiO_4$[h]	tetr	n_e			1.9843							4

* Provisional value based on extrapolation beyond the range of experimental data.
[a] Arsenolite
[b] Calcite
[c] Senarmontite
[d] α-Quartz
[e] Rutile
[f] Sphalerite
[g] Wurtzite
[h] Zircon
[j] Anatase

REFRACTIVE INDEX AND TRANSMITTANCE OF REPRESENTATIVE GLASSES

Typical values of the index of refraction and internal transmittance (fraction of light transmitted through a one centimeter thickness) are tabulated here for selected types of glasses, as well as for synthetic fused (vitreous) silica. Nominal compositions are given in the first part of the table. The second part gives the index of refraction, relative to air, and the internal transmittance for representative samples of each glass at wavelengths in the infrared, visible, and near-ultraviolet regions. It should be emphasized that a wide variation of these parameters may be found among subtypes of each glass. More detailed data may be found in Ref. 3.

Assuming that the Lambert-Beer Law is followed, the transmittance of a glass plate of thickness *d* (in centimeters) can be obtained by raising the transmittance value in the table to the power *d*.

References

1. Weber, M. J., *CRC Handbook of Laser Science and Technology*, Vol. IV, Part 2, CRC Press, Boca Raton, FL, 1988.
2. Gray, D. E., Ed., *American Institute of Physics Handbook, Third Edition*, McGraw Hill, New York, 1972.
3. *Schott Optical Glass*, Schott Glass Technologies, Inc., 400 York Ave., Duryea, PA.
4. Kaye, G. W. C., and Laby, T. H., *Tables of Physical and Chemical Constants, 15th Edition*, Longman, London, 1986.

Glass Composition in Percent by Mass

Type	Name	SiO_2	B_2O_3	Al_2O_3	Na_2O	K_2O	CaO	BaO	ZnO	PbO	P_2O_5
PK	Phosphate crown		3	10		12	5				70
PSK	Dense phosphate crown		3	5			4	28			60
BK	Borosilicate crown	70	10		8	8	1	3			
K	Crown	74			9	11	6				
ZK	Zinc crown	71			17				12		
BaK	Barium crown	60	3		3	10		19	5		
SK	Dense crown	39	15	5				41			
KF	Crown flint	67		2	16				3	12	
BaLF	Barium light flint	51			6	5		20	14	4	
SSK	Extra dense crown	35	10	5				42	8		
LLF	Extra light flint	63			5	8				24	
BaF	Barium flint	46				8		16	8	22	
LF	Light flint	53			5	8				34	
F	Flint	47			2	7				44	
BaSF	Dense barium flint	43			1	7		11	5	33	
SF	Dense flint	33				5				62	
KzFS	Short flint										
SiO_2	Fused silica	100									

Refractive Index and Transmittance of a 1 cm Plate at the Specified Wavelength

Type	Refractive index at 1.060 μm	Refractive index at 546.1 nm	Refractive index at 365.0 nm	Refractive index at 312.6 nm	Transmittance at 1.060 μm	Transmittance at 546.1 nm	Transmittance at 365.0 nm	Transmittance at 310 nm
PK	1.51519	1.52736	1.54503	1.5574	0.997	0.998	0.987	0.46
PSK	1.54154	1.55440	1.57342	1.5868	0.996	0.998	0.984	0.46
BK	1.50669	1.51872	1.53627	1.5486	0.999	0.998	0.987	0.35
K	1.50091	1.51314	1.53189	1.5454	0.998	0.998	0.988	0.40
ZK	1.52220	1.53534	1.55588	1.5708	0.996	0.998	0.976	0.27
BaK	1.55695	1.57124	1.59407	1.6108	0.998	0.997	0.986	0.28
SK	1.59490	1.60994	1.63398		0.998	0.998	0.959	0.28
KF	1.50586	1.51978	1.54251	1.5600	0.998	0.996	0.989	0.49
BaLF	1.57579	1.59166	1.61804		0.996	0.998	0.933	0.010
SSK	1.60402	1.61993	1.64595		0.999	0.998	0.915	0.010
LaK	1.69710	1.71616	1.74573		0.999	0.998	0.882	0.17
LLF	1.52775	1.54344	1.57038		0.998	0.997	0.990	0.32
BaF	1.56873	1.58565	1.61524		0.999	0.997	0.992	0.004
LF	1.56594	1.58482	1.61926		0.999	0.998	0.981	0.008
F	1.58636	1.60718	1.64606		0.997	0.998	0.959	
BaSF	1.60889	1.62987	1.66926		0.999	0.998	0.857	
SF	1.71350	1.74620	1.8145		0.998	0.997	0.650	
KzFS	1.59680	1.61639	1.64849	1.6739		0.998	0.672	0.012
SiO_2	1.44968	1.46008	1.47435[a]	1.53430[b]				

[a] At 366.3 nm.
[b] At 213.9 nm.

INDEX OF REFRACTION OF WATER

This table gives the index of refraction of liquid water at atmospheric pressure, relative to a vacuum, at several temperatures and wavelengths. It is generated from the formulation in Ref. 1, which covers a wide range of temperature, pressure, and wavelength. The wavelengths listed here correspond to prominent lines of cadmium (226.50 and 361.05 nm), potassium (404.41 nm), sodium (589.00 nm), Ne (632.80 nm, from a helium - neon laser), and mercury (1.01398 μm).

References

1. Schiebener, P., Straub, J., Levelt Sengers, J. M. H., and Gallagher, J. S., *J. Phys. Chem. Ref. Data*, 19, 677, 1990; 19, 1617, 1990.
2. Marsh, K. N., Editor, *Recommended Reference Materials for the Realization of Physicochemical Properties*, Blackwell Scientific Publications, Oxford, 1987.

Index of Refraction of Water as a Function of Wavelength

t/°C	226.50 nm	361.05 nm	404.41 nm	589.00 nm	632.80 nm	1.01398 μm
0	1.39450	1.34896	1.34415	1.33432	1.33306	1.32612
10	1.39422	1.34870	1.34389	1.33408	1.33282	1.32591
20	1.39336	1.34795	1.34315	1.33336	1.33211	1.32524
30	1.39208	1.34682	1.34205	1.33230	1.33105	1.32424
40	1.39046	1.34540	1.34065	1.33095	1.32972	1.32296
50	1.38854	1.34373	1.33901	1.32937	1.32814	1.32145
60	1.38636	1.34184	1.33714	1.32757	1.32636	1.31974
70	1.38395	1.33974	1.33508	1.32559	1.32438	1.31784
80	1.38132	1.33746	1.33284	1.32342	1.32223	1.31576
90	1.37849	1.33501	1.33042	1.32109	1.31991	1.31353
100	1.37547	1.33239	1.32784	1.31861	1.31744	1.31114

INDEX OF REFRACTION OF LIQUIDS FOR CALIBRATION PURPOSES

This table gives the index of refraction of six liquids that are available in highly pure form and whose index of refraction has been accurately measured as a function of wavelength and temperature. These values are therefore useful for calibration of refractometers. The estimated uncertainty in the values is:

2,2,4-Trimethylpentane	±0.00003
Hexadecane	±0.00008
trans-Decahydronaphthalene	±0.00008
1-Methylnaphthalene	±0.00008
Toluene	±0.00003
Methylcyclohexane	±0.00003

Full details are given in the references. This table is reprinted from Ref. 1 by permission of the International Union of Pure and Applied Chemistry.

References

1. Marsh, K. N., Editor, *Recommended Reference Materials for the Realization of Physicochemical Properties*, Blackwell Scientific Publications, Oxford, 1987.
2. Tilton, L. W., *J. Opt. Soc. Am.*, 32, 71, 1941.

Index of Refraction at 20 °C

Name	667.81 nm	656.28 nm	589.26 nm	546.07 nm	501.57 nm	486.13 nm	435.83 nm
trans-Decahydronaphthalene	1.46654	1.46688	1.46932	1.47141	1.47420	1.47535	1.48011
Hexadecane	1.43204	1.43235	1.43453	1.43640	1.43888	1.43993	1.44419
Methylcyclohexane	1.42064	1.42094	1.42312	1.42497	1.42744	1.42847	1.43269
1-Methylnaphthalene	1.60828	1.60940	1.61755	1.62488	1.63513	1.63958	
Toluene	1.4918	1.49243	1.49693	1.50086	1.50620	1.50847	1.51800
2,2,4-Trimethylpentane	1.38916	1.38945	1.39145	1.39316	1.39544	1.39639	1.40029

Index of Refraction at 25 °C

Name	667.81 nm	656.28 nm	589.26 nm	546.07 nm	501.57 nm	486.13 nm	435.83 nm
trans-Decahydronaphthalene	1.46438	1.46472	1.46715	1.46923	1.47200	1.47315	1.47789
Hexadecane	1.43001	1.43032	1.43250	1.43436	1.43684	1.43788	1.44213
Methylcyclohexane	1.41812	1.41842	1.42058	1.42243	1.42488	1.42590	1.43010
1-Methylnaphthalene	1.60592	1.60703	1.61512	1.62240	1.63259	1.63701	1.65627
Toluene	1.48903	1.48966	1.49413	1.49803	1.50334	1.50559	1.51506
2,2,4-Trimethylpentane	1.38670	1.38698	1.38898	1.39068	1.39294	1.39389	1.39776

Index of Refraction at 30 °C

Name	667.81 nm	656.28 nm	589.26 nm	546.07 nm	501.57 nm	486.13 nm	435.83 nm
trans-Decahydronaphthalene	1.46222	1.46256	1.46498	1.46705	1.46980	1.47095	1.47567
Hexadecane	1.42798	1.42829	1.43047	1.43232	1.43480	1.43583	1.44007
Methylcyclohexane	1.41560	1.41591	1.41806	1.41989	1.42233	1.42334	1.42752
1-Methylnaphthalene	1.60360	1.60471	1.61278	1.62005	1.63022	1.63463	1.65386
Toluene	1.48619	1.48682	1.49126	1.49514	1.50041	1.50265	1.51206
2,2,4-Trimethylpentane	1.38424	1.38452	1.38650	1.38820	1.39044	1.39138	1.39523

INDEX OF REFRACTION OF AIR

This is a table of the index of refraction n of dry air at 15 °C and a pressure of 101.325 kPa and containing 0.045% by volume of carbon dioxide ("standard air"). The index of refraction is defined by $n = \lambda_{vac}/\lambda_{air}$ where λ is the wavelength of the radiation. The index is calculated from the expression

$$(n-1) \times 10^8 = 8342.54 + 2406147(130 - \sigma^2)^{-1} + 15998(38.9 - \sigma^2)^{-1}$$

where $\sigma = 1/\lambda_{vac}$ and λ_{vac} has units of μm. The equation is valid for λ_{vac} from 200 nm to 2 μm. The table also gives the correction $(n-1)\lambda_{air}$ that must be added to the wavelength in air to obtain λ_{vac}.

If the air is at a temperature t in °C (ITS-90) and a pressure p in Pa, a value of $(n - 1)$ from this table should be multiplied by

$$p[1 + p(60.1 - 0.972t) \times 10^{-10}]/96095.43(1 + 0.003661t)$$

References

1. Birch, K. P., and Downs, M. J., *Metrologia*, 31, 315, 1994.
2. Edlen, B., *Metrologia* 2, 71, 1966.

λ_{vac}	$(n-1) \times 10^8$	$\lambda_{vac} - \lambda_{air}$
200 nm	32409	0.06480 nm
210	31748	0.06665
220	31226	0.06868
230	30801	0.07082
240	30447	0.07305
250	30148	0.07535
260	29892	0.07769
270	29670	0.08009
280	29477	0.08251
290	29307	0.08497
300	29157	0.08745
310	29023	0.08995
320	28904	0.09247
330	28796	0.09500
340	28700	0.09755
350	28612	0.10011
360	28532	0.10269
370	28460	0.10527
380	28393	0.10786
390	28332	0.11046
400	28276	0.11307
410	28224	0.11569
420	28177	0.11831
430	28132	0.12094
440	28091	0.12357
450	28053	0.12620
460	28018	0.12885
470	27985	0.13149
480	27954	0.13414
490	27925	0.13679
500	27897	0.13945
510	27872	0.14211
520	27848	0.14477
530	27825	0.14743
540	27804	0.15010
550	27784	0.15277
560	27765	0.15544
570	27747	0.15811
580	27730	0.16079
590	27714	0.16347
600	27698	0.16614
610	27684	0.16882
620	27670	0.17151
630	27657	0.17419
640	27644	0.17688
650	27632	0.17956
660	27621	0.18225
670	27610	0.18494
680	27600	0.18763
690	27590	0.19032
700	27581	0.19301
710	27572	0.19570
720	27563	0.19840
730	27555	0.20109
740	27547	0.20379
750	27539	0.20649
760	27532	0.20918
770	27525	0.21188
780	27518	0.21458
790	27511	0.21728
800	27505	0.21998
810	27499	0.22268
820	27493	0.22538
830	27488	0.22808
840	27482	0.23079
850	27477	0.23349
860	27472	0.23619
870	27467	0.23890
880	27462	0.24160
890	27458	0.24431
900	27454	0.24701
910	27449	0.24972
920	27445	0.25243
930	27441	0.25513
940	27437	0.25784
950	27434	0.26055
960	27430	0.26326
970	27427	0.26597
980	27423	0.26868
990 nm	27420	0.27138 nm
1.00 μm	27417	0.0002741 μm
1.05	27402	0.0002876
1.10	27390	0.0003012
1.15	27379	0.0003148
1.20	27370	0.0003283
1.25	27361	0.0003419
1.30	27354	0.0003555
1.35	27347	0.0003691
1.40	27341	0.0003827
1.45	27336	0.0003963
1.50	27331	0.0004099
1.55	27327	0.0004234
1.60	27323	0.0004370
1.65	27319	0.0004506
1.70	27316	0.0004642
1.75	27313	0.0004778
1.80	27310	0.0004914
1.85	27307	0.0005050
1.90	27305	0.0005187
1.95	27303	0.0005323
2.00	27301	0.0005459

INDEX OF REFRACTION OF GASES

This table gives the index of refraction of several gases at selected wavelengths ranging from the blue to the red region of the spectrum. The entries at 0.5893 μm correspond to the prominent sodium D line in the yellow region. All values refer to gas at a pressure of one atmosphere (101.325 kPa) and at a temperature of 0 °C.

References

1. Gray, D. E., ed., *American Institute of Physics Handbook, 3rd Edition*, p. 6-110, McGraw-Hill, New York, 1972.
2. Forsythe, W. E., *Smithsonian Physical Tables, Ninth Edition*, p. 533, Smithsonian Institution, Washington, 1954.
3. *Kaye and Laby Tables of Physical and Chemical Constants, Sixteenth Edition*, p. 131, Longman Group Ltd., Harlow, Essex, 1995.

Gas	Formula	*n* (0.4360 μm)	*n* (0.4861 μm)	*n* (0.5461 μm)	*n* (0.5790 μm)	*n* (0.5893 μm)	*n* (0.6563 μm)	*n* (0.6709 μm)
Air		1.0002966	1.0002947	1.0002932	1.0002926	1.0002924	1.0002915	1.0002913
Argon	Ar					1.000281		
Carbon dioxide	CO_2	1.0004563		1.0004506		1.0004493		1.0004471
Chlorine	Cl_2					1.000773		
Helium	He					1.000036		
Hydrogen	H_2	1.0001418	1.0001406	1.0001397	1.0001393	1.0001392	1.0001387	1.0001385
Methane	CH_4					1.000444		
Nitric oxide	NO					1.000297		
Nitrogen	N_2		1.0003012	1.0002998		1.0002990	1.0002982	
Nitrous oxide	N_2O					1.000516		
Oxygen	O_2	1.0002743	1.0002734	1.0002717	1.0002710	1.0002709	1.0002698	1.0002683
Sulfur dioxide	SO_2					1.000686		

CHARACTERISTICS OF LASER SOURCES

William F. Krupke

Light amplification by stimulated emission of radiation was first demonstrated by Maiman in 1960, the result of a population inversion produced between energy levels of chromium ions in a ruby crystal when irradiated with a xenon flashlamp. Since then population inversions and coherent emission have been generated in literally thousands of substances (neutral and ionized gases, liquids, and solids) using a variety of incoherent excitation techniques (optical pumping, electrical discharges, gas-dynamic flow, electron-beams, chemical reactions, nuclear decay).

The extrema of laser output parameters that have been demonstrated to date and the laser media used are summarized in Table 1. Note that the extreme power and energy parameters listed in this table were attained with laser systems rather than with simple laser oscillators.

Laser sources are commonly classified in terms of the state-of-matter of the active medium: gas, liquid, and solid. Each of these classes is further subdivided into one or more types as shown in Table 2. A well-known representative example of each type of laser is also given in Table 2 together with its nominal operation wavelength and the methods by which it is pumped.

The various lasers together cover a wide spectral range from the far ultraviolet to the far infrared. The particular wavelength of emission (usually a narrow line) is presented for some six dozen lasers in Figures 1A and 1B.

By suitably designing the excitation source and/or by controlling the laser resonator structure, laser systems can provide continuous or pulsed radiation as shown in Table 3.

Besides the method of excitation and the temporal behavior of a laser, there are many other parameters that characterize its operation and efficiency, as shown in Tables 4 and 5.

Although many lasers only emit in one or more narrow spectral "lines," an increasing number of lasers can be tuned by changing the composition or the pressure of the medium, or by varying the wavelength of the pump bands. The spectral regions in which these tunable lasers operate are presented in Figure 2.

Reference

Krupke, W. F., in *Handbook of Laser Science and Technology*, Vol. I, Weber, M. J., Ed., CRC Press, Boca Raton, FL, 1986.

TABLE 1. Extrema of Output Parameters of Laser Devices or Systems

Parameter	Value	Laser medium
Peak power	1×10^{14} W (collimated)	Nd:glass
Peak power density	10^{18} W/cm^2 (focused)	Nd:glass
Pulse energy	$>10^5$ J	CO_2, Nd:glass
Average power	10^5 W	CO_2
Pulse duration	3×10^{-15} s continuous wave (cw)	Rh6G dye; various gases, liquids, solids
Wavelength	60 nm ↔ 385 μm	Many required
Efficiency (nonlaser pumped)	70%	CO
Beam quality	Diffraction limited	Various gases, liquids, solids
Spectral linewidth	20 Hz (for 10^{-1} s)	Neon-helium
Spatial coherence	10 m	Ruby

TABLE 2. Classes, Types, and Representative Examples of Laser Sources

Type (characteristic)	Representative example	Nominal operating wavelength (nm)	Method(s) of excitation
Gas			
Atom, neutral (electronic transition)	Neon-Helium (Ne-He)	633	Glow discharge
Atom, ionic (electronic transition)	Argon (Ar^+)	488	Arc discharge
Molecule, neutral (electronic transition)	Krypton fluoride (KrF)	248	Glow discharge; e-beam
Molecule, neutral (vibrational transition)	Carbon dioxide (CO_2)	10600	Glow discharge; gasdynamic flow
Molecule, neutral (rotational transition)	Methyl fluoride (CH_3F)	496000	Laser pumping
Molecule, ionic (electronic transition)	Nitrogen ion (N_2^+)	420	E-beam
Liquid			
Organic solvent (dye-chromophore)	Rhodamine dye (Rh6G)	580–610	Flashlamp; laser pumping
Organic solvent (rare earth chelate)	Europium:TTF	612	Flashlamp
Inorganic solvent (trivalent rare earth ion)	Neodymium:$POCl_4$	1060	Flashlamp
Solid			
Insulator, crystal (impurity)	Neodymium:YAG	1064	Flashlamp, arc lamp
Insulator, crystal (stoichiometric)	Neodymium:UP(NdP_5O_{14})	1052	Flashlamp
Insulator, crystal (color center)	F_2^-:LiF	1120	Laser pumping
Insulator, amorphous (impurity)	Neodymium:glass	1061	Flashlamp
Semiconductor (p-n junction)	GaAs	820	Injection current
Semiconductor (electron-hole plasma)	GaAs	890	E- beam, laser pumping

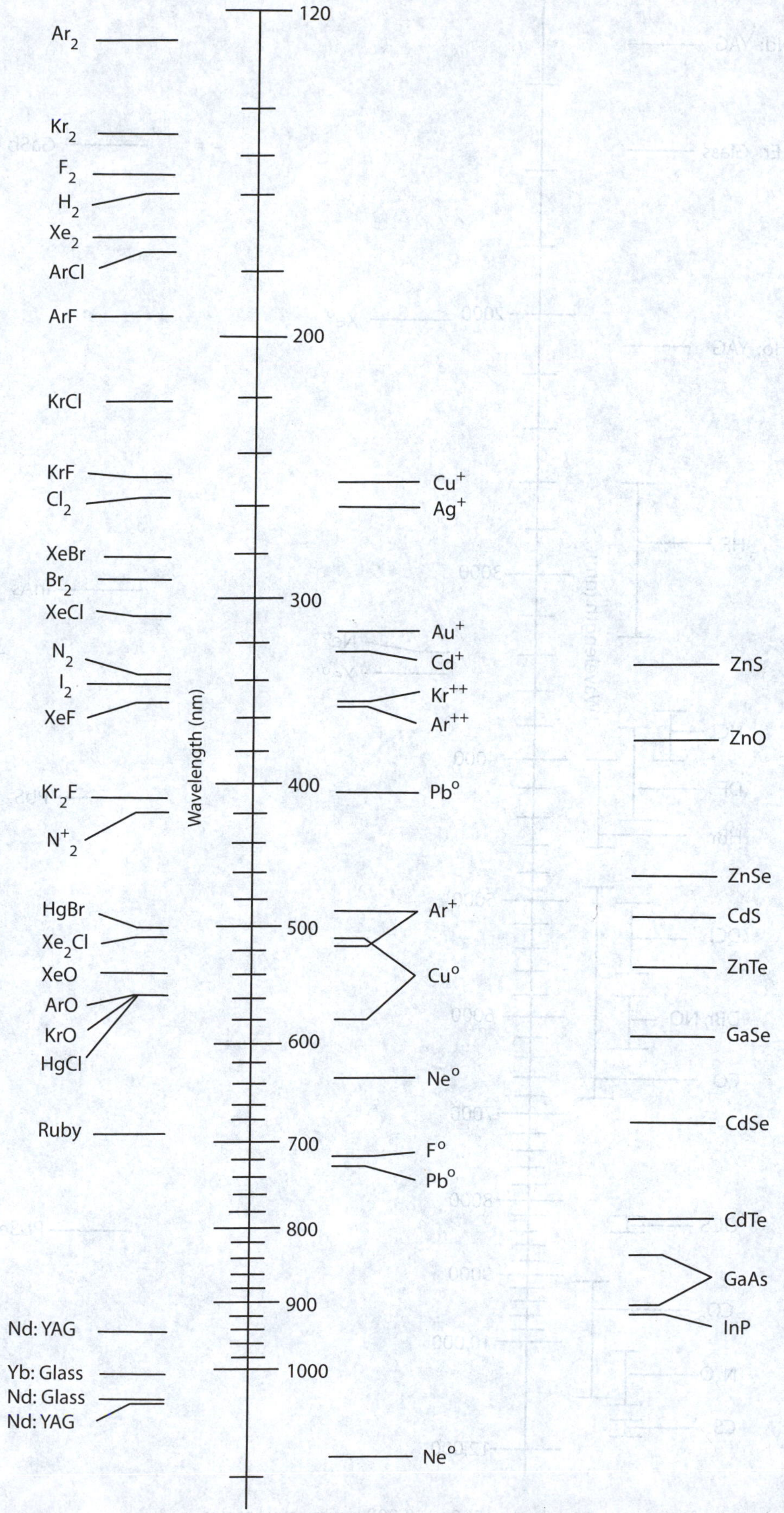

FIGURE 1A. Wavelengths of lasers operating in the 120 to 1200 nm spectral region.

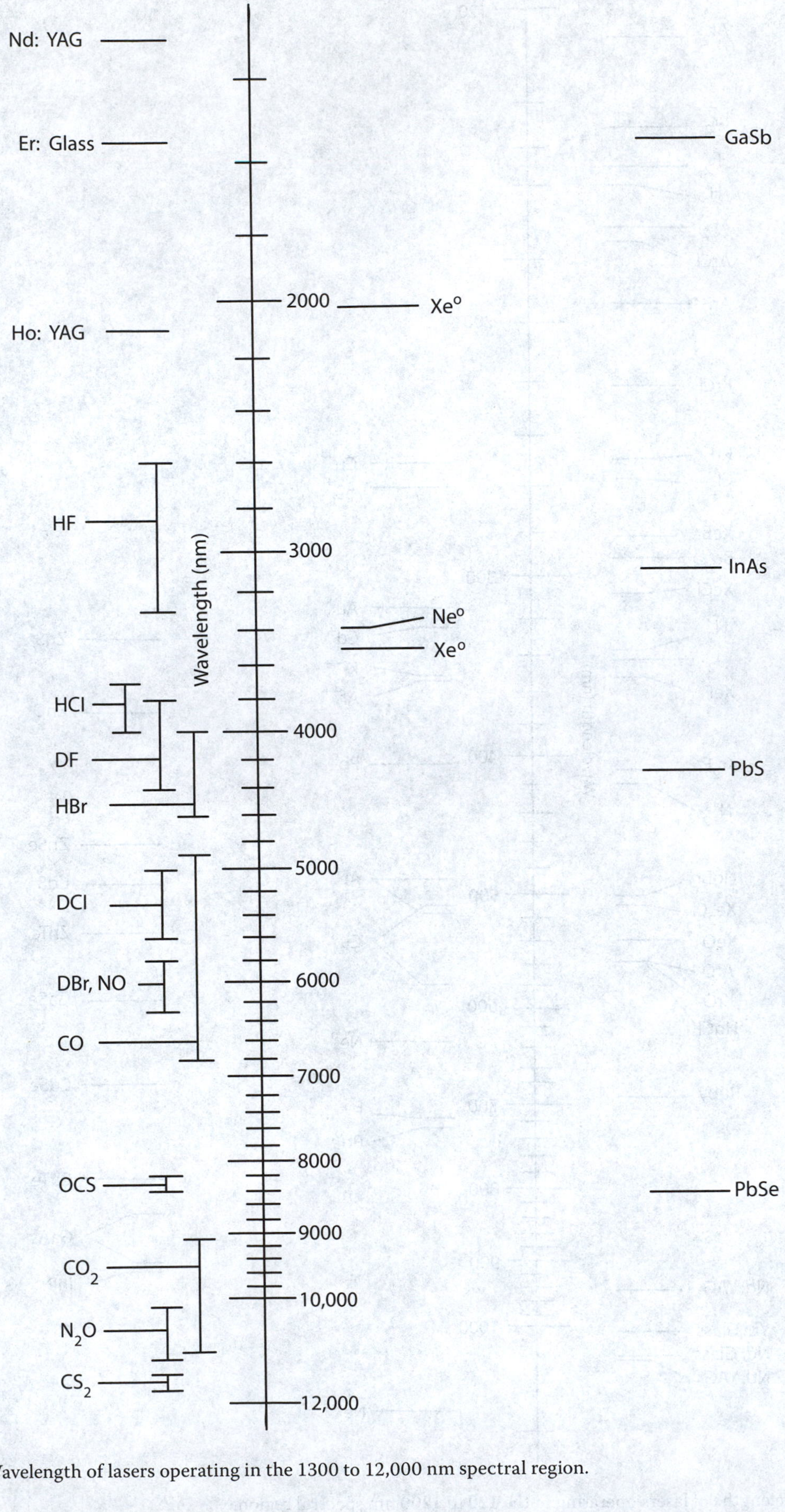

FIGURE 1B. Wavelength of lasers operating in the 1300 to 12,000 nm spectral region.

Table 3. Temporal Characteristics of Lasers and Laser Systems

Form	Technique	Pulse width range(s)
Continuous wave	Excitation is continuous; resonator Q is held constant at some moderate value	∞
Pulsed	Excitation is pulsed; resonator Q is held constant at some moderate value	$10^{-8} - 10^{-3}$
Q-Switched	Excitation is continuous or pulsed; resonator Q is switched from a very low value to a moderate value	$10^{-8} - 10^{-6}$
Cavity dumped	Excitation is continuous or pulsed; resonator Q is switched from a very high value to a low value	$10^{-7} - 10^{-5}$
Mode locked	Excitation is continuous or pulsed; phase or loss of the resonator modes is modulated at a rate related to the resonator transit time	$10^{-12} - 10^{-9}$

Table 4. Properties and Performance of Some Continuous Wave (CW) Lasers

Parameter	Unit	Ne-He (gas)	Ar^+ (gas)	CO_2 (gas)	Rhodamine 6G dye (liquid)	Nd:YAG (solid)	GaAs (solid)
Excitation method		DC discharge	DC discharge	DC discharge	Ar^+ laser pump	Krypton arc lamp	DC injection
Gain medium composition		Neon:helium	Argon	CO_2:N_2:He	Rh 6G:H_2O	Nd:YAG	p:n:GaAs
Gain medium density	Torr	0.1:1.0	0.4	0.4:0.8:5.0			
	ions/cm³				2(18):2(22)	1.5(20):2(22)	2(19):3(18):3(22)
Wavelength	nm	633	488	10600	590	1064	810
Laser cross-section	cm⁻²	3(-13)	1.6(-12)	1.5(-16)	1.8(-16)	7(-19)	~6(-15)
Radiative lifetime (upper level)	s	~1(-7)	7.5(-9)	4(-3)	6.5(-9)	2.6(-4)	~1(-9)
Decay lifetime (upper level)	s	~1(-7)	~5.0(-9)	~4(-3)	6.0(-9)	2.3(-4)	~1(-9)
Gain bandwidth	nm	2(-3)	5(-3)	1.6(-2)	80	0.5	10
Type, gain saturation		Inhomogeneous	Inhomogeneous	Homogeneous	Homogeneous	Homogeneous	Homogeneous
Homogeneous saturation flux	W cm⁻²			~20	3(5)	2.3(3)	~2(4)
Decay lifetime (lower level)	s	~1(-8)	~4(-10)	~5(-6)	<1(-12)	<1(-7)	<1(-12)
Inversion density	cm⁻³	~1(9)	2(10)	2(15)	2(16)	6(16)	1(16)
Small signal gain coefficient	cm⁻¹	~1(-3)	~3(-2)	1(-2)	4	5(-2)	40
Pump power density	W cm⁻³	3	900	0.15	1(6)	150	7(7)
Output power density	W cm⁻³	2.6(-3)	~1	2(-2)	3(5)	95	5(6)
Laser size (diameter:length)	cm:cm	0.5:100	0.3:100	5.0:600	1(-3):0.3	0.6:10	5(-4):7(-3);2(-2)[a]
Excitation current/voltage	A/V	3(-2):2(3)	30:300	0.1:1.5(4)		90:125	1.0/1.7
Excitation current density	A cm⁻²	0.15	600	6(-3)		140	4.5(3)
Excitation power	W	60	9(3)	1.5(3)	4	1.1(4)	1.7
Output power	W	0.06	10	240	0.3	300	0.12
Efficiency[b]	%	0.1	0.1	13	7	2.6	7

[a] Junction thickness:width:length.
[b] Pressure dependent.

Table 5. Properties and Performance of Some Pulsed Lasers

Parameter	Unit	CO_2 (gas)	CO_2 (gas)	KrF (gas)	KrF (gas)	Rhodamine 6G (liquid)	Nd:YAG (solid)	Nd:glass (solid)
Excitation method		TEA-discharge	E- beam/sust.	E-beam	Glow discharge	Xenon flashlamp	Xenon flashlamp	Xenon flashlamp
Gain medium composition		CO_2:N_2:He	CO_2:N_2:He	Ar:Kr:F_2	He:Kr:F_2	Rh6G:alcohol	Nd:YAG	Nd:Glass
Gain medium density	torr	100:50:600	240:240:320	1235:52:3	1070:70:3			–
	ions/cm³					1(18):1.5(22)	1.5(20):1(22)	3(20):2(22)
Wavelength	nm	10600	10600	249	249	590	1064	1061
Laser cross-section	cm⁻²	2(-18)	2(-18)	2(-16)	2(-16)	1.8(-16)	7(-19)	2.8(-20)
Radiative lifetime (upper level)	s	4(-3)	4(-3)	7(-9)	7(-9)	6.5(-9)	2.6(-4)	4.1(-4)
Decay lifetime (upper level)	s	~1(-4)	5(-5)	3(-9)	2(-9)	6.0(-9)	2.3(-4)	3.7(-4)
Gain bandwidth	nm	1	1	2	2	80	0.5	26
Homogeneous saturation fluence	J/cm²	0.2	0.2	4(-3)	4(-3)	2(-3)	0.6	~5
Decay lifetime (lower level)	s	5(-8)[a]	1(-8)[a]	<1(-12)	<1(-12)	<1(-12)	<1(-7)	<1(-8)
Inversion density	cm⁻³	3(17)	6(17)	2(14)	4(14)	2(16)	4(17)	3(18)
Small signal gain coefficient	cm⁻¹	2(-2)	4(-2)	4(-2)	8–92)	4	0.3	8(-2)
Medium excitation energy density	J/cm³	0.1	0.36	0.13	0.15	2.8	0.15	0.6
Output energy density	J/cm³	2(-2)	1.8(-2)	1.2(-2)	1.5(-3)	0.85	5(-2)	2(-2)
Laser dimensions	cm:cm:cm	4.5:4.5:87	10:10:100	8.5:10:100	1.5:4.5:100	1.2:25	0.6:7.5	0.6:8.3
Excitation current/voltage	A/V	6(4)/3.3(3)	2.4(4)/4(4)	1.2(4)/2.5(5)	2.5(4)/1.5(5)	2(5)/2.5(4)		
Excitation current density	A cm⁻²	8.5	22	11.5	170	2.6(3)		
Excitation peak power	W	2(8)	9(8)	3(9)	4(9)	5.4(9)	4(4)	9(4)
Output pulse energy	J	35	180	102	1	32	0.1	1.0
Output pulse length	s	1(-6)	4(-6)	6(-7)	2.5(-8)	3.2(-6)	2(-8)	1(-4)
Output pulse power	W	3.5(7)	4(7)	2(8)	4(7)	1(7)	5(6)	1(4)
Efficiency	%	17	5	10[b]	1	0.2	1.5	3.7

[a] Pressure dependent.
[b] Intrinsic efficiency ≡ energy output/energy deposited in gas.

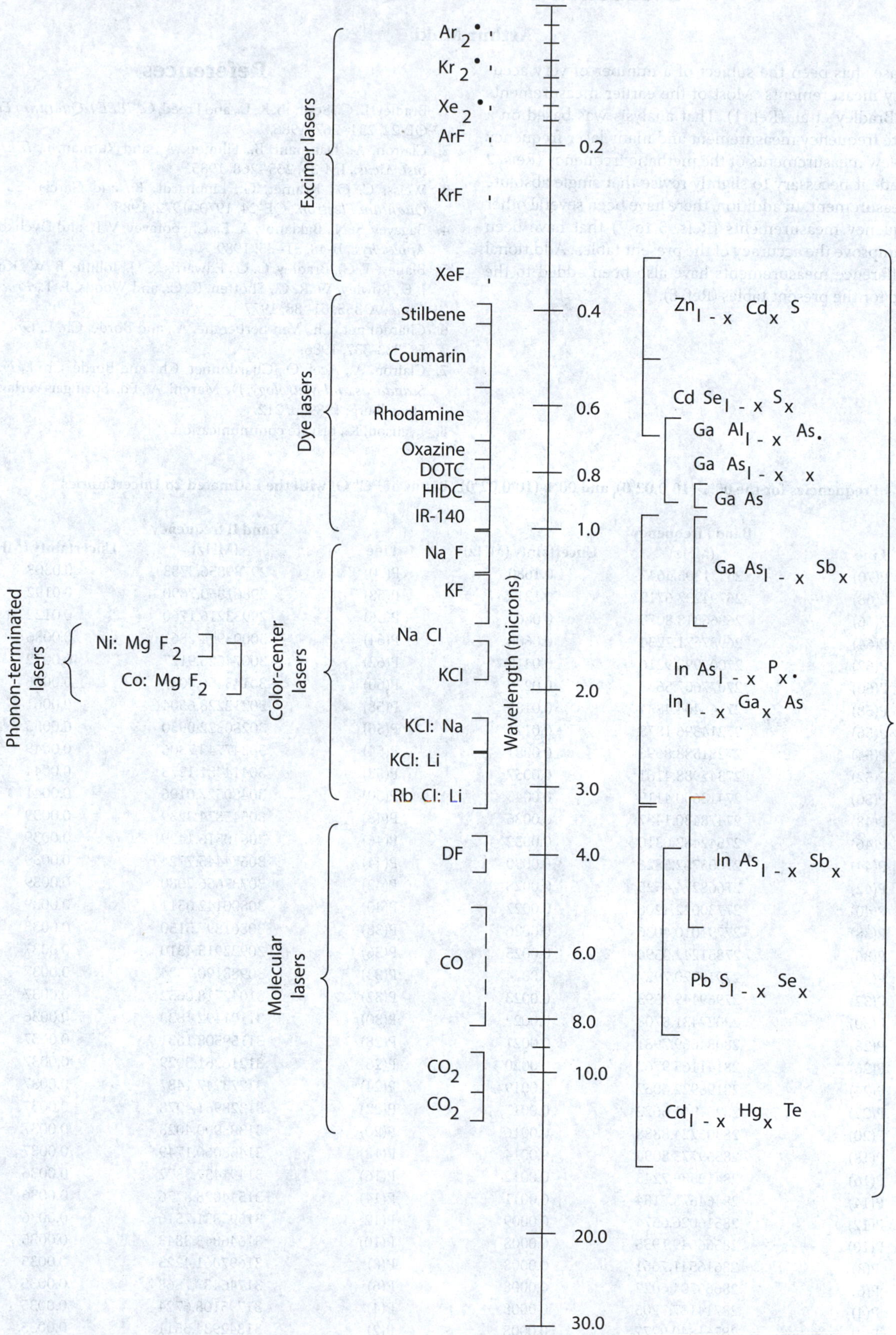

FIGURE 2. Spectral tuning ranges of various types of tunable lasers.

INFRARED LASER FREQUENCIES

Arthur Maki

The CO_2 laser has been the subject of a number of very accurate frequency measurements. Most of the earlier measurements are given by Bradley et al. (Ref. 1). That analysis was based on a single absolute frequency measurement and many laser frequency differences. New measurements of the methane frequency (Refs. 2 to 4) have made it necessary to slightly revise that single absolute frequency measurement. In addition, there have been several other absolute frequency measurements (Refs. 5 to 7) that have been used here to improve the accuracy of the present tables. Additional frequency difference measurements have also been added to the database used for the present tables (Ref. 8).

References

1. Bradley, L. C., Soohoo, K. L., and Freed, C., *IEEE J. Quantum Electron.*, QE-22, 234–267, 1986.
2. Clairon, A., Dahmani, B., Filimon, A., and Rutman, J., *IEEE Trans. Inst. Meas.*, IM-34, 265–268, 1985.
3. Weiss, C. O., Kramer, G., Lipphardt, B., and Garcia, E., *IEEE J. Quantum Electron.*, QE-24, 1970–1972, 1988.
4. Bagayev, S. N., Baklanov, A. E., Chebotayev, V. P., and Dychkov, A. S., *Appl. Phys.*, B-48, 31–35, 1989.
5. Blaney, T. G., Bradley, C. C., Edwards, G. J., Jolliffe, B. W., Knight, D. J. E., Rowley, W. R. C., Shotten, K. C., and Woods, P. T., *Proc. R. Soc. Lond.*, A-355, 61–88, 1977.
6. Chardonnet, Ch., Van Lerberghe, A., and Bordé, Ch. J., *Opt. Comm.*, 58, 333-337, 1986.
7. Clairon, A., Acef, O., Chardonnet, Ch., and Bordé, Ch. J., *Frequency Standards and Metrology*, De Marchi, A., Ed., Springer-Verlag, Berlin, Heidelberg, 1989, p. 212.
8. Evenson, K., private communication.

Frequencies for the 00°1-(10°0,02°0)$_I$ and 00°1-(10°0,02°0)$_{II}$ Bands of $^{12}C^{16}O_2$ with the Estimated 2σ Uncertainties

Line	Band I frequency (MHz)	Uncertainty (MHz)	Line	Band II frequency (MHz)	Uncertainty (MHz)
P(70)	26721305.4647	0.1680	P(70)	29789856.3783	0.0308
P(68)	26794232.6712	0.1217	P(68)	29861850.7690	0.0192
P(66)	26866318.8073	0.0867	P(66)	29933216.1760	0.0122
P(64)	26937571.7234	0.0606	P(64)	30003944.2861	0.0086
P(62)	27007998.9216	0.0415	P(62)	30074026.9127	0.0072
P(60)	27077607.5643	0.0279	P(60)	30143456.0039	0.0066
P(58)	27146404.4834	0.0185	P(58)	30212223.6504	0.0061
P(56)	27214396.1873	0.0121	P(56)	30280322.0930	0.0055
P(54)	27281588.8696	0.0081	P(54)	30347743.7306	0.0049
P(52)	27347988.4161	0.0057	P(52)	30414481.1273	0.0044
P(50)	27413600.4119	0.0043	P(50)	30480527.0196	0.0041
P(48)	27478430.1487	0.0036	P(48)	30545874.3239	0.0039
P(46)	27542482.6310	0.0032	P(46)	30610516.1429	0.0039
P(44)	27605762.5826	0.0030	P(44)	30674445.7724	0.0039
P(42)	27668274.4525	0.0028	P(42)	30737656.7080	0.0039
P(40)	27730022.4206	0.0027	P(40)	30800142.6511	0.0039
P(38)	27791010.4036	0.0026	P(38)	30861897.5150	0.0038
P(36)	27851242.0594	0.0025	P(36)	30922915.4310	0.0037
P(34)	27910720.7927	0.0024	P(34)	30983190.7534	0.0037
P(32)	27969449.7593	0.0023	P(32)	31042718.0652	0.0037
P(30)	28027431.8708	0.0022	P(30)	31101492.1833	0.0036
P(28)	28084669.7981	0.0021	P(28)	31159508.1631	0.0037
P(26)	28141165.9762	0.0020	P(26)	31216761.3029	0.0037
P(24)	28196922.6067	0.0019	P(24)	31273247.1487	0.0037
P(22)	28251941.6622	0.0017	P(22)	31328961.4978	0.0037
P(20)	28306224.8888	0.0016	P(20)	31383900.4028	0.0037
P(18)	28359773.8090	0.0014	P(18)	31438060.1749	0.0037
P(16)	28412589.7245	0.0012	P(16)	31491437.3872	0.0036
P(14)	28464673.7184	0.0011	P(14)	31544028.8776	0.0036
P(12)	28516026.6574	0.0009	P(12)	31595831.7516	0.0036
P(10)	28566649.1935	0.0008	P(10)	31646843.3843	0.0035
P(8)	28616541.7661	0.0008	P(8)	31697061.4225	0.0035
P(6)	28665704.6027	0.0008	P(6)	31746483.7868	0.0035
P(4)	28714137.7205	0.0008	P(4)	31795108.6724	0.0035
P(2)	28761840.9272	0.0008	P(2)	31842934.5511	0.0035
R(0)	28832026.2198	0.0008	R(0)	31913172.5691	0.0035
R(2)	28877902.4382	0.0007	R(2)	31958996.0621	0.0034

Line	Band I frequency (MHz)	Uncertainty (MHz)	Line	Band II frequency (MHz)	Uncertainty (MHz)
R(4)	28923046.4303	0.0006	R(4)	32004017.3822	0.0034
R(6)	28967457.0657	0.0005	R(6)	32048236.2498	0.0034
R(8)	29011133.0054	0.0003	R(8)	32091652.6619	0.0034
R(10)	29054072.7010	0.0001	R(10)	32134266.8917	0.0034
R(12)	29096274.3935	0.0003	R(12)	32176079.4878	0.0034
R(14)	29137736.1129	0.0005	R(14)	32217091.2721	0.0035
R(16)	29178455.6759	0.0007	R(16)	32257303.3386	0.0036
R(18)	29218430.6852	0.0009	R(18)	32296717.0510	0.0037
R(20)	29257658.5269	0.0010	R(20)	32335334.0408	0.0038
R(22)	29296136.3689	0.0011	R(22)	32373156.2044	0.0039
R(24)	29333861.1583	0.0012	R(24)	32410185.7003	0.0041
R(26)	29370829.6191	0.0011	R(26)	32446424.9459	0.0042
R(28)	29407038.2491	0.0011	R(28)	32481876.6140	0.0042
R(30)	29442483.3168	0.0011	R(30)	32516543.6293	0.0042
R(32)	29477160.8582	0.0012	R(32)	32550429.1641	0.0042
R(34)	29511066.6733	0.0013	R(34)	32583536.6340	0.0042
R(36)	29544196.3221	0.0015	R(36)	32615869.6937	0.0041
R(38)	29576545.1205	0.0017	R(38)	32647432.2320	0.0040
R(40)	29608108.1360	0.0019	R(40)	32678228.3665	0.0039
R(42)	29638880.1831	0.0022	R(42)	32708262.4386	0.0038
R(44)	29668855.8183	0.0024	R(44)	32737539.0081	0.0039
R(46)	29698029.3350	0.0027	R(46)	32766062.8469	0.0041
R(48)	29726394.7582	0.0032	R(48)	32793838.9334	0.0045
R(50)	29753945.8385	0.0037	R(50)	32820872.4463	0.0055
R(52)	29780676.0464	0.0042	R(52)	32847168.7576	0.0071
R(54)	29806578.5659	0.0047	R(54)	32872733.4269	0.0099
R(56)	29831646.2878	0.0052	R(56)	32897572.1935	0.0141
R(58)	29855871.8032	0.0058	R(58)	32921690.9701	0.0202
R(60)	29879247.3960	0.0074	R(60)	32945095.8355	0.0288
R(62)	29901765.0357	0.0113	R(62)	32967793.0268	0.0407
R(64)	29923416.3695	0.0186	R(64)	32989788.9322	0.0567
R(66)	29944192.7145	0.0302	R(66)	33011090.0831	0.0780
R(68)	29964085.0488	0.0475	R(68)	33031703.1467	0.1060
R(70)	29983084.0036	0.0720	R(70)	33051634.9172	0.1423

Frequencies for the 00^01-$(10^00,02^00)_I$ and 00^01-$(10^00,02^00)_{II}$ Bands of $^{13}C^{16}O_2$ with the Estimated 2σ Uncertainties

Line	Band I frequency (MHz)	Uncertainty (MHz)	Line	Band II frequency (MHz)	Uncertainty (MHz)
P(66)	25523832.1808	0.7836	P(66)	28512082.5283	1.2894
P(64)	25590013.4703	0.5415	P(64)	28585121.9396	0.9194
P(62)	25655543.6502	0.3629	P(62)	28657449.4180	0.6420
P(60)	25720428.2487	0.2339	P(60)	28729056.6374	0.4375
P(58)	25784672.4840	0.1430	P(58)	28799935.4147	0.2897
P(56)	25848281.2771	0.0810	P(56)	28870077.7187	0.1853
P(54)	25911259.2627	0.0405	P(54)	28939475.6771	0.1135
P(52)	25973610.8005	0.0157	P(52)	29008121.5846	0.0659
P(50)	26035339.9857	0.0045	P(50)	29076007.9109	0.0357
P(48)	26096450.6582	0.0079	P(48)	29143127.3077	0.0180
P(46)	26156946.4123	0.0101	P(46)	29209472.6164	0.0090
P(44)	26216830.6053	0.0101	P(44)	29275036.8754	0.0058
P(42)	26276106.3655	0.0090	P(42)	29339813.3270	0.0050
P(40)	26334776.6003	0.0077	P(40)	29403795.4243	0.0044
P(38)	26392844.0030	0.0068	P(38)	29466976.8383	0.0037
P(36)	26450311.0599	0.0063	P(36)	29529351.4635	0.0032
P(34)	26507180.0565	0.0061	P(34)	29590913.4252	0.0029
P(32)	26563453.0836	0.0060	P(32)	29651657.0844	0.0028
P(30)	26619132.0428	0.0058	P(30)	29711577.0447	0.0028
P(28)	26674218.6515	0.0055	P(28)	29770668.1566	0.0031
P(26)	26728714.4479	0.0054	P(26)	29828925.5239	0.0035
P(24)	26782620.7952	0.0054	P(24)	29886344.5074	0.0041
P(22)	26835938.8858	0.0054	P(22)	29942920.7308	0.0046
P(20)	26888669.7451	0.0055	P(20)	29998650.0838	0.0051
P(18)	26940814.2347	0.0055	P(18)	30053528.7271	0.0054
P(16)	26992373.0555	0.0055	P(16)	30107553.0955	0.0055

Line	Band I frequency (MHz)	Uncertainty (MHz)	Line	Band II frequency (MHz)	Uncertainty (MHz)
P(14)	27043346.7508	0.0054	P(14)	30160719.9016	0.0055
P(12)	27093735.7083	0.0052	P(12)	30213026.1388	0.0054
P(10)	27143540.1624	0.0051	P(10)	30264469.0839	0.0054
P(8)	27192760.1962	0.0049	P(8)	30315046.2994	0.0054
P(6)	27241395.7431	0.0048	P(6)	30364755.6359	0.0055
P(4)	27289446.5880	0.0047	P(4)	30413595.2335	0.0056
P(2)	27336912.3682	0.0046	P(2)	30461563.5231	0.0057
R(0)	27407012.8882	0.0045	P(0)	30531879.5415	0.0057
R(2)	27453013.4589	0.0043	P(2)	30577664.6138	0.0056
R(4)	27498426.5430	0.0040	P(4)	30622575.1885	0.0054
R(6)	27543251.1200	0.0037	P(6)	30666611.0128	0.0051
R(8)	27587486.0225	0.0034	P(8)	30709772.1257	0.0047
R(10)	27631129.9356	0.0031	P(10)	30752058.8571	0.0045
R(12)	27674181.3963	0.0029	P(12)	30793471.8269	0.0044
R(14)	27716638.7917	0.0029	P(14)	30834011.9425	0.0043
R(16)	27758500.3577	0.0029	P(16)	30873680.3976	0.0044
R(18)	27799764.1770	0.0029	P(18)	30912478.6694	0.0044
R(20)	27840428.1773	0.0030	P(20)	30950408.5159	0.0044
R(22)	27880490.1283	0.0029	P(22)	30987471.9732	0.0043
R(24)	27919947.6395	0.0029	P(24)	31023671.3517	0.0042
R(26)	27958798.1567	0.0028	P(26)	31059009.2327	0.0042
R(28)	27997038.9591	0.0028	P(28)	31093488.4642	0.0042
R(30)	28034667.1551	0.0027	P(30)	31127112.1569	0.0043
R(32)	28071679.6785	0.0027	P(32)	31159883.6793	0.0045
R(34)	28108073.2842	0.0026	P(34)	31191806.6529	0.0046
R(36)	28143844.5432	0.0026	P(36)	31222884.9469	0.0048
R(38)	28178989.8377	0.0026	P(38)	31253122.6730	0.0053
R(40)	28213505.3554	0.0028	P(40)	31282524.1795	0.0061
R(42)	28247387.0838	0.0033	P(42)	31311094.0452	0.0077
R(44)	28280630.8035	0.0046	P(44)	31338837.0736	0.0108
R(46)	28313232.0818	0.0083	P(46)	31365758.2858	0.0173
R(48)	28345186.2652	0.0161	P(48)	31391862.9147	0.0295
R(50)	28376488.4720	0.0301	P(50)	31417156.3972	0.0505
R(52)	28407133.5839	0.0531	P(52)	31441644.3679	0.0845
R(54)	28437116.2372	0.0887	P(54)	31465332.6516	0.1366
R(56)	28466430.8141	0.1419	P(56)	31488227.2557	0.2138
R(58)	28495071.4324	0.2188	P(58)	31510334.3631	0.3247
R(60)	28523031.9357	0.3271	P(60)	31531660.3243	0.4800
R(62)	28550305.8819	0.4763	P(62)	31552211.6497	0.6932
R(64)	28576886.5323	0.6781	P(64)	31571995.0017	0.9805
R(66)	28602766.8393	0.9467	P(66)	31591017.1868	1.3619

Frequencies for the 00°1-(10°0,02°0)$_I$ and 00°1-(10°0,02°0)$_{II}$ Bands of $^{12}C^{18}O_2$ with the Estimated 2σ Uncertainties

P(70)	27045326.3119	0.4540	P(70)	30695237.5856	0.0858
P(68)	27114914.0922	0.3324	P(68)	30755520.2231	0.0570
P(66)	27183635.7945	0.2392	P(66)	30815311.4928	0.0364
P(64)	27251496.4118	0.1688	P(64)	30874607.2084	0.0223
P(62)	27318500.7361	0.1165	P(62)	30933403.2309	0.0131
P(60)	27384653.3618	0.0783	P(60)	30991695.4724	0.0075
P(58)	27449958.6881	0.0510	P(58)	31049479.9009	0.0049
P(56)	27514420.9224	0.0319	P(56)	31106752.5446	0.0041
P(54)	27578044.0828	0.0191	P(54)	31163509.4964	0.0040
P(52)	27640832.0010	0.0108	P(52)	31219746.9183	0.0040
P(50)	27702788.3248	0.0059	P(50)	31275461.0455	0.0039
P(48)	27763916.5206	0.0035	P(48)	31330648.1908	0.0039
P(46)	27824219.8762	0.0028	P(46)	31385304.7490	0.0039
P(44)	27883701.5029	0.0026	P(44)	31439427.2006	0.0039
P(42)	27942364.3379	0.0025	P(42)	31493012.1163	0.0038
P(40)	28000211.1464	0.0024	P(40)	31546056.1605	0.0038
P(38)	28057244.5242	0.0022	P(38)	31598556.0954	0.0037
P(36)	28113466.8992	0.0021	P(36)	31650508.7847	0.0037

Line	Band I frequency (MHz)	Uncertainty (MHz)	Line	Band II frequency (MHz)	Uncertainty (MHz)
P(34)	28168880.5335	0.0020	P(34)	31701911.1970	0.0037
P(32)	28223487.5256	0.0019	P(32)	31752760.4093	0.0037
P(30)	28277289.8118	0.0017	P(30)	31803053.6105	0.0037
P(28)	28330289.1679	0.0016	P(28)	31852788.1043	0.0038
P(26)	28382487.2111	0.0015	P(26)	31901961.3125	0.0038
P(24)	28433885.4012	0.0013	P(24)	31950570.7773	0.0038
P(22)	28484485.0420	0.0012	P(22)	31998614.1649	0.0038
P(20)	28534287.2828	0.0011	P(20)	32046089.2669	0.0037
P(18)	28583293.1193	0.0010	P(18)	32092994.0036	0.0037
P(16)	28631503.3952	0.0010	P(16)	32139326.4254	0.0036
P(14)	28678918.8025	0.0009	P(14)	32185084.7154	0.0036
P(12)	28725539.8830	0.0010	P(12)	32230267.1907	0.0036
P(10)	28771367.0288	0.0010	P(10)	32274872.3041	0.0037
P(8)	28816400.4829	0.0010	P(8)	32318898.6455	0.0038
P(6)	28860640.3403	0.0011	P(6)	32362344.9434	0.0039
P(4)	28904086.5477	0.0011	P(4)	32405210.0652	0.0041
P(2)	28946738.9048	0.0011	P(2)	32447493.0185	0.0041
R(0)	29009228.1702	0.0010	P(0)	32509824.0580	0.0042
R(2)	29049894.0586	0.0010	P(2)	32550648.1723	0.0042
R(4)	29089764.2368	0.0009	P(4)	32590887.7542	0.0042
R(6)	29128837.8426	0.0008	P(6)	32630542.4457	0.0041
R(8)	29167113.8668	0.0008	P(8)	32669612.0295	0.0041
R(10)	29204591.1529	0.0009	P(10)	32708096.4282	0.0040
R(12)	29241268.3964	0.0010	P(12)	32745995.7040	0.0040
R(14)	29277144.1444	0.0011	P(14)	32783310.0573	0.0040
R(16)	29312216.7955	0.0012	P(16)	32820039.8258	0.0040
R(18)	29346484.5984	0.0012	P(18)	32856185.4827	0.0040
R(20)	29379945.6517	0.0013	P(20)	32891747.6358	0.0040
R(22)	29412597.9024	0.0013	P(22)	32926727.0254	0.0040
R(24)	29444439.1458	0.0013	P(24)	32961124.5220	0.0040
R(26)	29475467.0236	0.0014	P(26)	32994941.1249	0.0040
R(28)	29505679.0230	0.0015	P(28)	33028177.9594	0.0040
R(30)	29535072.4755	0.0016	P(30)	33060836.2743	0.0040
R(32)	29563644.5557	0.0018	P(32)	33092917.4394	0.0041
R(34)	29591392.2794	0.0020	P(34)	33124422.9429	0.0043
R(36)	29618312.5023	0.0023	P(36)	33155354.3878	0.0046
R(38)	29644401.9182	0.0028	P(38)	33185713.4894	0.0049
R(40)	29669657.0575	0.0036	P(40)	33215502.0716	0.0056
R(42)	29694074.2853	0.0053	P(42)	33244722.0637	0.0068
R(44)	29717649.7992	0.0082	P(44)	33273375.4969	0.0092
R(46)	29740379.6276	0.0128	P(46)	33301464.5003	0.0134
R(48)	29762259.6274	0.0200	P(48)	33328991.2976	0.0199
R(50)	29783285.4820	0.0307	P(50)	33355958.2027	0.0294
R(52)	29803452.6988	0.0461	P(52)	33382367.6161	0.0427
R(54)	29822756.6072	0.0681	P(54)	33408222.0209	0.0607
R(56)	29841192.3558	0.0985	P(56)	33433523.9780	0.0848
R(58)	29858754.9100	0.1401	P(58)	33458276.1228	0.1165
R(60)	29875439.0495	0.1960	P(60)	33482481.1601	0.1576
R(62)	29891239.3658	0.2702	P(62)	33506141.8605	0.2104
R(64)	29906150.2589	0.3673	P(64)	33529261.0556	0.2775
R(66)	29920165.9352	0.4930	P(66)	33551841.6335	0.3621
R(68)	29933280.4042	0.6540	P(68)	33573886.5352	0.4679
R(70)	29945487.4756	0.8581	P(70)	33595398.7493	0.5992

Frequencies for the 00°1-(10°0,02°0)$_I$ and 00°1-(10°0,02°0)$_{II}$ Bands of $^{13}C^{18}O_2$ with the Estimated 2σ Uncertainties

Line	Band I frequency (MHz)	Uncertainty (MHz)	Line	Band II frequency (MHz)	Uncertainty (MHz)
P(70)	25967863.7652	1.1146	P(70)	28960476.2278	0.4069
P(68)	26033448.2798	0.8152	P(68)	29022326.9578	0.2861
P(66)	26098273.9159	0.5860	P(66)	29083661.3546	0.1961
P(64)	26162346.4813	0.4129	P(64)	29144473.5795	0.1303
P(62)	26225671.5466	0.2844	P(62)	29204757.8761	0.0833

Line	Band I frequency (MHz)	Uncertainty (MHz)	Line	Band II frequency (MHz)	Uncertainty (MHz)
P(60)	26288254.4494	0.1906	P(60)	29264508.5768	0.0507
P(58)	26350100.2984	0.1237	P(58)	29323720.1086	0.0290
P(56)	26411213.9778	0.0772	P(56)	29382386.9988	0.0152
P(54)	26471600.1504	0.0459	P(54)	29440503.8809	0.0073
P(52)	26531263.2618	0.0258	P(52)	29498065.4997	0.0038
P(50)	26590207.5442	0.0138	P(50)	29555066.7172	0.0032
P(48)	26648437.0195	0.0077	P(48)	29611502.5178	0.0031
P(46)	26705955.5026	0.0057	P(46)	29667368.0132	0.0031
P(44)	26762766.6051	0.0055	P(44)	29722658.4475	0.0034
P(42)	26818873.7378	0.0055	P(42)	29777369.2022	0.0039
P(40)	26874280.1143	0.0056	P(40)	29831495.8006	0.0044
P(38)	26928988.7531	0.0056	P(38)	29885033.9125	0.0049
P(36)	26983002.4809	0.0056	P(36)	29937979.3584	0.0053
P(34)	27036323.9351	0.0055	P(34)	29990328.1139	0.0054
P(32)	27088955.5657	0.0054	P(32)	30042076.3132	0.0055
P(30)	27140899.6384	0.0051	P(30)	30093220.2534	0.0055
P(28)	27192158.2363	0.0049	P(28)	30143756.3978	0.0054
P(26)	27242733.2620	0.0047	P(26)	30193681.3793	0.0053
P(24)	27292626.4396	0.0044	P(24)	30242992.0038	0.0052
P(22)	27341839.3165	0.0042	P(22)	30291685.2529	0.0051
P(20)	27390373.2651	0.0040	P(20)	30339758.2870	0.0049
P(18)	27438229.4843	0.0037	P(18)	30387208.4477	0.0048
P(16)	27485409.0008	0.0035	P(16)	30434033.2603	0.0046
P(14)	27531912.6704	0.0033	P(14)	30480230.4356	0.0045
P(12)	27577741.1795	0.0031	P(12)	30525797.8725	0.0044
P(10)	27622895.0455	0.0031	P(10)	30570733.6593	0.0043
P(8)	27667374.6182	0.0031	P(8)	30615036.0750	0.0043
P(6)	27711180.0803	0.0033	P(6)	30658703.5912	0.0044
P(4)	27754311.4480	0.0034	P(4)	30701734.8727	0.0045
P(2)	27796768.5718	0.0036	P(2)	30744128.7785	0.0045
R(0)	27859189.3155	0.0036	P(0)	30806522.5414	0.0045
R(2)	27899959.0889	0.0035	P(2)	30847319.2956	0.0044
R(4)	27940052.7921	0.0033	P(4)	30887476.2168	0.0043
R(6)	27979469.5315	0.0031	P(6)	30926993.0424	0.0042
R(8)	28018208.2478	0.0028	P(8)	30965869.7046	0.0041
R(10)	28056267.7161	0.0026	P(10)	31004106.3298	0.0040
R(12)	28093646.5448	0.0025	P(12)	31041703.2379	0.0040
R(14)	28130343.1757	0.0025	P(14)	31078660.9408	0.0040
R(16)	28166355.8825	0.0025	P(16)	31114980.1420	0.0040
R(18)	28201682.7706	0.0025	P(18)	31150661.7340	0.0041
R(20)	28236321.7757	0.0025	P(20)	31185706.7976	0.0042
R(22)	28270270.6628	0.0024	P(22)	31220116.5992	0.0043
R(24)	28303527.0249	0.0024	P(24)	31253892.5891	0.0043
R(26)	28336088.2817	0.0023	P(26)	31287036.3991	0.0044
R(28)	28367951.6781	0.0024	P(28)	31319549.8396	0.0043
R(30)	28399114.2823	0.0025	P(30)	31351434.8973	0.0043
R(32)	28429572.9843	0.0026	P(32)	31382693.7318	0.0042
R(34)	28459324.4940	0.0028	P(34)	31413328.6728	0.0042
R(36)	28488365.3390	0.0029	P(36)	31443342.2165	0.0041
R(38)	28516691.8625	0.0029	P(38)	31472737.0219	0.0040
R(40)	28544300.2211	0.0031	P(40)	31501515.9074	0.0039
R(42)	28571186.3823	0.0032	P(42)	31529681.8467	0.0040
R(44)	28597346.1222	0.0032	P(44)	31557237.9646	0.0042
R(46)	28622775.0223	0.0038	P(46)	31584187.5329	0.0046
R(48)	28647468.4672	0.0071	P(48)	31610533.9656	0.0057
R(50)	28671421.6417	0.0148	P(50)	31636280.8146	0.0088
R(52)	28694629.5272	0.0286	P(52)	31661431.7650	0.0151
R(54)	28717086.8993	0.0510	P(54)	31685990.6298	0.0261
R(56)	28738788.3239	0.0852	P(56)	31709961.3449	0.0434
R(58)	28759728.1540	0.1355	P(58)	31733347.9642	0.0693
R(60)	28779900.5263	0.2075	P(60)	31756154.6537	0.1068

Line	Band I frequency (MHz)	Uncertainty (MHz)	Line	Band II frequency (MHz)	Uncertainty (MHz)
R(62)	28799299.3572	0.3078	P(62)	31778385.6867	0.1594
R(64)	28817918.3393	0.4447	P(64)	31800045.4375	0.2317
R(66)	28835750.9374	0.6283	P(66)	31821138.3761	0.3291
R(68)	28852790.3843	0.8707	P(68)	31841669.0622	0.4581
R(70)	28869029.6768	1.1863	P(70)	31861642.1394	0.6268

Frequencies for the $01^{1e}1$-$(11^{1e}0,03^{1e}0)_I$ and $01^{1e}1$-$(11^{1e}0,03^{1e}0)_{II}$ Bands of $^{12}C^{16}O_2$ with the Estimated 2σ Uncertainties

Line	Band I frequency (MHz)	Uncertainty (MHz)	Line	Band II frequency (MHz)	Uncertainty (MHz)
P(59)	26125213.2723	1.6633	P(59)	30427055.2899	0.1962
P(57)	26191576.6703	1.0880	P(57)	30494640.3229	0.1332
P(55)	26257240.7898	0.6844	P(55)	30561557.5929	0.0865
P(53)	26322208.2302	0.4094	P(53)	30627802.0344	0.0530
P(51)	26386481.4313	0.2286	P(51)	30693368.7014	0.0306
P(49)	26450062.6783	0.1155	P(49)	30758252.7710	0.0175
P(47)	26512954.1076	0.0498	P(47)	30822449.5469	0.0123
P(45)	26575157.7109	0.0191	P(45)	30885954.4624	0.0114
P(43)	26636675.3402	0.0160	P(43)	30948763.0834	0.0109
P(41)	26697508.7115	0.0182	P(41)	31010871.1119	0.0100
P(39)	26757659.4084	0.0177	P(39)	31072274.3882	0.0091
P(37)	26817128.8857	0.0160	P(37)	31132968.8940	0.0091
P(35)	26875918.4726	0.0144	P(35)	31192950.7549	0.0102
P(33)	26934029.3751	0.0131	P(33)	31252216.2430	0.0118
P(31)	26991462.6787	0.0119	P(31)	31310761.7788	0.0134
P(29)	27048219.3509	0.0106	P(29)	31368583.9339	0.0147
P(27)	27104300.2431	0.0096	P(27)	31425679.4328	0.0155
P(25)	27159706.0925	0.0093	P(25)	31482045.1550	0.0157
P(23)	27214437.5237	0.0097	P(23)	31537678.1367	0.0154
P(21)	27268495.0505	0.0104	P(21)	31592575.5725	0.0147
P(19)	27321879.0769	0.0108	P(19)	31646734.8172	0.0137
P(17)	27374589.8987	0.0108	P(17)	31700153.3868	0.0127
P(15)	27426627.7040	0.0104	P(15)	31752828.9602	0.0119
P(13)	27477992.5747	0.0098	P(13)	31804759.3803	0.0113
P(11)	27528684.4867	0.0096	P(11)	31855942.6551	0.0113
P(9)	27578703.3113	0.0101	P(9)	31906376.9582	0.0116
P(7)	27628048.8151	0.0113	P(7)	31956060.6304	0.0122
P(5)	27676720.6609	0.0127	P(5)	32004992.1796	0.0129
P(3)	27724718.4080	0.0141	P(3)	32053170.2819	0.0136
R(1)	27841759.7696	0.0152	P(1)	32170312.0391	0.0149
R(3)	27887393.2105	0.0146	P(3)	32215845.0845	0.0151
R(5)	27932349.2934	0.0135	P(5)	32260620.8121	0.0152
R(7)	27976627.0108	0.0124	P(7)	32304638.8261	0.0152
R(9)	28020225.2521	0.0115	P(9)	32347898.8990	0.0150
R(11)	28063142.8031	0.0110	P(11)	32390400.9714	0.0148
R(13)	28105378.3457	0.0109	P(13)	32432145.1513	0.0145
R(15)	28146930.4576	0.0109	P(15)	32473131.7137	0.0142
R(17)	28187797.6116	0.0107	P(17)	32513361.0997	0.0140
R(19)	28227978.1750	0.0103	P(19)	32552833.9153	0.0140
R(21)	28267470.4088	0.0099	P(21)	32591550.9309	0.0141
R(23)	28306272.4666	0.0099	P(23)	32629513.0796	0.0143
R(25)	28344382.3939	0.0107	P(25)	32666721.4564	0.0144
R(27)	28381798.1267	0.0122	P(27)	32703177.3164	0.0142
R(29)	28418517.4902	0.0141	P(29)	32738882.0732	0.0136
R(31)	28454538.1976	0.0165	P(31)	32773837.2976	0.0136
R(33)	28489857.8477	0.0213	P(33)	32808044.7156	0.0174
R(35)	28524473.9240	0.0312	P(35)	32841506.2063	0.0279
R(37)	28558383.7917	0.0486	P(37)	32874223.8000	0.0462
R(39)	28591584.6963	0.0754	P(39)	32906199.6761	0.0735
R(41)	28624073.7602	0.1131	P(41)	32937436.1606	0.1114
R(43)	28655847.9806	0.1644	P(43)	32967935.7238	0.1624
R(45)	28686904.2261	0.2328	P(45)	32997700.9775	0.2292
R(47)	28717239.2334	0.3239	P(47)	33026734.6728	0.3151
R(49)	28746849.6038	0.4465	P(49)	33055039.6965	0.4238

Line	Band I frequency (MHz)	Uncertainty (MHz)	Line	Band II frequency (MHz)	Uncertainty (MHz)
R(51)	28775731.7988	0.6142	P(51)	33082619.0689	0.5595
R(53)	28803882.1361	0.8465	P(53)	33109475.9403	0.7272

Frequencies for the $01^{1f}1$-$(11^{1f}0,03^{1f}0)_I$ and $01^{1f}1$-$(11^{1f}0,03^{1f}0)_{II}$ Bands of $^{12}C^{16}O_2$ with the Estimated 2σ Uncertainties

P(60)	26051570.0104	4.4521	P(60)	30355115.0204	0.2752
P(58)	26120964.4932	3.0629	P(58)	30425283.5969	0.1926
P(56)	26189552.8496	2.0516	P(56)	30494732.8293	0.1301
P(54)	26257339.6006	1.3305	P(54)	30563455.6325	0.0840
P(52)	26324329.0344	0.8289	P(52)	30631445.1076	0.0512
P(50)	26390525.2136	0.4901	P(50)	30698694.5456	0.0292
P(48)	26455931.9824	0.2698	P(48)	30765197.4310	0.0163
P(46)	26520552.9722	0.1334	P(46)	30830947.4444	0.0111
P(44)	26584391.6075	0.0551	P(44)	30895938.4662	0.0104
P(42)	26647451.1105	0.0181	P(42)	30960164.5794	0.0105
P(40)	26709734.5057	0.0151	P(40)	31023620.0723	0.0105
P(38)	26771244.6242	0.0174	P(38)	31086299.4415	0.0107
P(36)	26831984.1067	0.0157	P(36)	31148197.3941	0.0114
P(34)	26891955.4069	0.0126	P(34)	31209308.8510	0.0126
P(32)	26951160.7945	0.0105	P(32)	31269628.9481	0.0138
P(30)	27009602.3576	0.0096	P(30)	31329153.0395	0.0147
P(28)	27067282.0045	0.0092	P(28)	31387876.6994	0.0151
P(26)	27124201.4662	0.0090	P(26)	31445795.7236	0.0149
P(24)	27180362.2977	0.0089	P(24)	31502906.1318	0.0141
P(22)	27235765.8792	0.0090	P(22)	31559204.1695	0.0128
P(20)	27290413.4182	0.0093	P(20)	31614686.3091	0.0113
P(18)	27344305.9494	0.0096	P(18)	31669349.2515	0.0098
P(16)	27397444.3368	0.0097	P(16)	31723189.9280	0.0086
P(14)	27449829.2733	0.0096	P(14)	31776205.5007	0.0081
P(12)	27501461.2824	0.0096	P(12)	31828393.3642	0.0085
P(10)	27552340.7179	0.0101	P(10)	31879751.1463	0.0095
P(8)	27602467.7649	0.0111	P(8)	31930276.7092	0.0107
P(6)	27651842.4399	0.0125	P(6)	31979968.1497	0.0120
P(4)	27700464.5912	0.0139	P(4)	32028823.8002	0.0131
P(2)	27748333.8988	0.0148	P(2)	32076842.2290	0.0139
R(2)	27864709.8633	0.0146	P(2)	32193218.1935	0.0150
R(4)	27909939.2762	0.0135	P(4)	32238298.4853	0.0151
R(6)	27954412.3294	0.0122	P(6)	32282538.0393	0.0153
R(8)	27998127.7801	0.0112	P(8)	32325936.7244	0.0153
R(10)	28041084.2173	0.0107	P(10)	32368494.6458	0.0154
R(12)	28083280.0620	0.0108	P(12)	32410212.1438	0.0155
R(14)	28124713.5668	0.0110	P(14)	32451089.7941	0.0155
R(16)	28165382.8151	0.0112	P(16)	32491128.4063	0.0154
R(18)	28205285.7213	0.0111	P(18)	32530329.0234	0.0152
R(20)	28244420.0302	0.0110	P(20)	32568692.9211	0.0149
R(22)	28282783.3158	0.0114	P(22)	32606221.6061	0.0147
R(24)	28320372.9812	0.0129	P(24)	32642916.8154	0.0144
R(26)	28357186.2574	0.0149	P(26)	32678780.5147	0.0140
R(28)	28393220.2023	0.0168	P(28)	32713814.8971	0.0133
R(30)	28428471.6994	0.0175	P(30)	32748022.3813	0.0120
R(32)	28462937.4565	0.0165	P(32)	32781405.6101	0.0106
R(34)	28496614.0042	0.0142	P(34)	32813967.4482	0.0122
R(36)	28529497.6934	0.0163	P(36)	32845710.9809	0.0212
R(38)	28561584.6939	0.0309	P(38)	32876639.5111	0.0385
R(40)	28592870.9914	0.0593	P(40)	32906756.5580	0.0646
R(42)	28623352.3850	0.1054	P(42)	32936065.8540	0.1015
R(44)	28653024.4839	0.1779	P(44)	32964571.3426	0.1518
R(46)	28681882.7038	0.2907	P(46)	32992277.1760	0.2184
R(48)	28709922.2632	0.4635	P(48)	33019187.7118	0.3048
R(50)	28737138.1785	0.7231	P(50)	33045307.5105	0.4152

INFRARED AND FAR-INFRARED ABSORPTION FREQUENCY STANDARDS

Arthur Maki

Aside from the CO_2 laser transitions, the absorption spectrum of CO has been more accurately and thoroughly measured than any other spectrum. A bibliography of earlier measurements on CO is given by Maki and Wells (Ref. 1), and the present tables were calculated from the measurements referred to in that work. In addition, some new and very accurate frequency measurements (Refs. 2 and 3) have been made and were incorporated in the present tables. The frequencies of the rotational transitions of HF and HCl were calculated from constants obtained from fitting the measurements of Evenson et al. (Refs. 4 and 5) and Jennings and Wells (Ref. 6).

A new report on infrared wavenumber standards from the International Union of Pure and Applied Chemistry, Commission on Molecular Structure and Spectroscopy, may be found in Ref. 7.

References

1. Maki, A. G. and Wells, J. S., *Wavenumber Calibration Tables from Heterodyne Frequency Measurements*, NIST Special Publication 821, U.S. Dept. of Commerce, Washington, D.C., 1991.
2. Evenson, K. and Stroh, F., private communication.
3. George, T., Urban, W., and co-workers, private communication.
4. Jennings, D. A., Evenson, K. M., Zink, L. R., Demuynck, C., Destombes, J. L., Lemoine, B., and Johns, J. *W. C., J. Mol. Spectrosc.*, 122, 477–480, 1987.
5. Nolt, I. G., Radostitz, J. V., DiLonardo, G., Evenson, K. M., Jennings, D. A., Leopold, K. R., Vanek, M. D., Zink, L. R., Hinz, A., and Chance, K. V., *J. Mol. Spectrosc.*, 125, 274–287, 1987.
6. Jennings, D. A. and Wells, J. S., *J. Mol. Spectrosc.*, 130, 267–268, 1988.
7. High Resolution Wavenumber Standards for the Infrared, *Pure Appl. Chemistry*, 68, 193, 1996.

Wavenumbers for the $v = 1 - 0$ Band of CO

Wavenumber (unc)*/cm⁻¹	Transition	Wavenumber (unc)/cm⁻¹	Transition
		2147.081132(01)	R(0)
2139.426071(01)	P(1)	2150.856006(01)	R(1)
2135.546178(01)	P(2)	2154.595581(01)	R(2)
2131.631574(01)	P(3)	2158.299710(01)	R(3)
2127.682404(01)	P(4)	2161.968245(01)	R(4)
2123.698816(01)	P(5)	2165.601041(01)	R(5)
2119.680957(01)	P(6)	2169.197949(01)	R(6)
2115.628973(01)	P(7)	2172.758824(01)	R(7)
2111.543012(01)	P(8)	2176.283519(01)	R(8)
2107.423221(01)	P(9)	2179.771887(01)	R(9)
2103.269746(01)	P(10)	2183.223782(01)	R(10)
2099.082734(01)	P(11)	2186.639057(01)	R(11)
2094.862333(01)	P(12)	2190.017565(01)	R(12)
2090.608688(01)	P(13)	2193.359161(01)	R(13)
2086.321947(01)	P(14)	2196.663698(01)	R(14)
2082.002256(01)	P(15)	2199.931030(01)	R(15)
2077.649762(01)	P(16)	2203.161010(01)	R(16)
2073.264612(01)	P(17)	2206.353492(01)	R(17)
2068.846952(01)	P(18)	2209.508331(02)	R(18)
2064.396929(01)	P(19)	2212.625379(02)	R(19)
2059.914688(02)	P(20)	2215.704492(02)	R(20)
2055.400377(02)	P(21)	2218.745522(02)	R(21)
2050.854140(02)	P(22)	2221.748326(03)	R(22)
2046.276126(03)	P(23)	2224.712755(03)	R(23)
2041.666479(03)	P(24)	2227.638666(03)	R(24)
2037.025345(03)	P(25)	2230.525912(04)	R(25)
2032.352870(04)	P(26)	2233.374349(04)	R(26)
2027.649200(04)	P(27)	2236.183829(04)	R(27)
2022.914480(04)	P(28)	2238.954210(05)	R(28)
2018.148857(05)	P(29)	2241.685344(05)	R(29)
2013.352474(05)	P(30)	2244.377088(06)	R(30)
2008.525477(06)	P(31)	2247.029296(07)	R(31)
2003.668012(06)	P(32)	2249.641824(08)	R(32)
1998.780224(07)	P(33)	2252.214527(10)	R(33)
1993.862257(09)	P(34)	2254.747262(14)	R(34)
1988.914257(11)	P(35)	2257.239883(18)	R(35)
1983.936367(14)	P(36)	2259.692248(24)	R(36)
1978.928733(18)	P(37)	2262.104213(33)	R(37)
1973.891500(25)	P(38)	2264.475634(45)	R(38)

Wavenumbers for the v = 1 – 0 Band of CO

Wavenumber (unc)*/cm⁻¹	Transition	Wavenumber (unc)/cm⁻¹	Transition
1968.824811(34)	P(39)	2266.806368(61)	R(39)
1963.728813(46)	P(40)	2269.096273(81)	R(40)
1958.603648(61)	P(41)	2271.345206(106)	R(41)
1953.449462(82)	P(42)	2273.553027(139)	R(42)

* The uncertainty in the last digits (twice the standard error) is given in parentheses.

Wavenumbers for the v = 2 – 0 Band of CO

Wavenumber (unc)*/cm⁻¹	Transition	Wavenumber (unc)/cm⁻¹	Transition
		4263.837198(02)	R(0)
4256.217140(02)	P(1)	4267.542066(02)	R(1)
4252.302244(02)	P(2)	4271.176630(02)	R(2)
4248.317633(02)	P(3)	4274.740746(02)	R(3)
4244.263453(02)	P(4)	4278.234264(02)	R(4)
4240.139852(02)	P(5)	4281.657039(02)	R(5)
4235.946975(02)	P(6)	4285.008924(02)	R(6)
4231.684972(02)	P(7)	4288.289772(02)	R(7)
4227.353987(02)	P(8)	4291.499437(02)	R(8)
4222.954169(02)	P(9)	4294.637773(02)	R(9)
4218.485665(02)	P(10)	4297.704631(02)	R(10)
4213.948620(02)	P(11)	4300.699868(02)	R(11)
4209.343182(02)	P(12)	4303.623334(02)	R(12)
4204.669499(02)	P(13)	4306.474886(02)	R(13)
4199.927716(02)	P(14)	4309.254375(02)	R(14)
4195.117980(02)	P(15)	4311.961657(02)	R(15)
4190.240439(02)	P(16)	4314.596584(02)	R(16)
4185.295239(02)	P(17)	4317.159011(02)	R(17)
4180.282526(02)	P(18)	4319.648791(02)	R(18)
4175.202447(02)	P(19)	4322.065779(03)	R(19)
4170.055149(03)	P(20)	4324.409829(03)	R(20)
4164.840777(03)	P(21)	4326.680794(03)	R(21)
4159.559478(03)	P(22)	4328.878530(03)	R(22)
4154.211398(03)	P(23)	4331.002889(04)	R(23)
4148.796683(04)	P(24)	4333.053728(04)	R(24)
4143.315479(04)	P(25)	4335.030899(05)	R(25)
4137.767932(04)	P(26)	4336.934259(06)	R(26)
4132.154187(05)	P(27)	4338.763661(07)	R(27)
4126.474391(06)	P(28)	4340.518961(09)	R(28)
4120.728689(07)	P(29)	4342.200014(11)	R(29)
4114.917226(09)	P(30)	4343.806675(16)	R(30)
4109.040148(12)	P(31)	4345.338799(21)	R(31)
4103.097600(16)	P(32)	4346.796243(29)	R(32)
4097.089728(21)	P(33)	4348.178862(40)	R(33)
4091.016676(29)	P(34)	4349.486513(54)	R(34)
4084.878591(40)	P(35)	4350.719052(73)	R(35)
4078.675618(54)	P(36)	4351.876336(96)	R(36)
4072.407901(73)	P(37)	4352.958224(127)	R(37)
4066.075588(97)	P(38)	4353.964572(166)	R(38)
4059.678822(127)	P(39)	4354.895240(214)	R(39)

* The uncertainty in the last digits (twice the standard error) is given in parentheses.

Wavenumbers for the v = 3 – 0 Band of CO

Wavenumber (unc)*/cm⁻¹	Transition	Wavenumber (unc)/cm⁻¹	Transition
		6354.179057(13)	R(0)
6346.594000(13)	P(1)	6357.813923(13)	R(1)
6342.644103(13)	P(2)	6361.343487(13)	R(2)
6338.589491(13)	P(3)	6364.767599(13)	R(3)
6334.430309(13)	P(4)	6368.086115(13)	R(4)

Wavenumbers for the $v = 3 - 0$ Band of CO

Wavenumber (unc)*/cm^{-1}	Transition	Wavenumber (unc)/cm^{-1}	Transition
6330.166705(13)	P(5)	6371.298887(13)	R(5)
6325.798826(13)	P(6)	6374.405768(12)	R(6)
6321.326819(13)	P(7)	6377.406611(12)	R(7)
6316.750831(12)	P(8)	6380.301271(12)	R(8)
6312.071008(12)	P(9)	6383.089600(12)	R(9)
6307.287498(12)	P(10)	6385.771452(12)	R(10)
6302.400447(12)	P(11)	6388.346680(13)	R(11)
6297.410003(12)	P(12)	6390.815139(13)	R(12)
6292.316311(13)	P(13)	6393.176681(13)	R(13)
6287.119520(13)	P(14)	6395.431160(13)	R(14)
6281.819775(13)	P(15)	6397.578430(13)	R(15)
6276.417224(13)	P(16)	6399.618344(13)	R(16)
6270.912012(13)	P(17)	6401.550757(13)	R(17)
6265.304287(13)	P(18)	6403.375523(13)	R(18)
6259.594194(13)	P(19)	6405.092495(14)	R(19)
6253.781880(13)	P(20)	6406.701527(14)	R(20)
6247.867492(14)	P(21)	6408.202474(14)	R(21)
6241.851176(14)	P(22)	6409.595189(15)	R(22)
6235.733077(14)	P(23)	6410.879527(15)	R(23)
6229.513342(15)	P(24)	6412.055343(16)	R(24)
6223.192117(15)	P(25)	6413.122491(17)	R(25)
6216.769547(16)	P(26)	6414.080825(19)	R(26)
6210.245778(17)	P(27)	6414.930201(23)	R(27)
6203.620957(19)	P(28)	6415.670474(28)	R(28)
6196.895229(23)	P(29)	6416.301500(37)	R(29)
6190.068739(28)	P(30)	6416.823133(50)	R(30)
6183.141633(37)	P(31)	6417.235231(67)	R(31)
6176.114058(50)	P(32)	6417.537649(90)	R(32)
6168.986159(67)	P(33)		
6161.758082(90)	P(34)		

* The uncertainty in the last digits (twice the standard error) is given in parentheses.

Frequencies and Wavenumbers for the Rotational Lines of CO

Frequency/MHz	Uncertainty*/MHz	J'	J''	Wavenumber/cm^{-1}	Uncertainty*/cm^{-1}
115271.2029	0.0004	1	0	3.84503345	0.00000001
230538.0016	0.0008	2	1	7.68991999	0.00000003
345795.9923	0.0012	3	2	11.53451273	0.00000004
461040.7712	0.0016	4	3	15.37866477	0.00000005
576267.9350	0.0019	5	4	19.22222923	0.00000006
691473.0809	0.0021	6	5	23.06505926	0.00000007
806651.8065	0.0023	7	6	26.90700800	0.00000008
921799.7104	0.0025	8	7	30.74792863	0.00000008
1036912.3919	0.0027	9	8	34.58767438	0.00000009
1151985.4515	0.0029	10	9	38.42609848	0.00000010
1267014.4906	0.0031	11	10	42.26305422	0.00000010
1381995.1119	0.0034	12	11	46.09839491	0.00000011
1496922.9195	0.0038	13	12	49.93197392	0.00000013
1611793.5189	0.0042	14	13	53.76364468	0.00000014
1726602.5173	0.0047	15	14	57.59326065	0.00000016
1841345.5237	0.0052	16	15	61.42067535	0.00000017
1956018.1486	0.0057	17	16	65.24574239	0.00000019
2070616.0050	0.0061	18	17	69.06831542	0.00000020
2185134.7075	0.0065	19	18	72.88824816	0.00000022
2299569.8733	0.0069	20	19	76.70539441	0.00000023
2413917.1217	0.0071	21	20	80.51960806	0.00000024
2528172.0747	0.0073	22	21	84.33074306	0.00000024
2642330.3567	0.0074	23	22	88.13865346	0.00000025
2756387.5949	0.0075	24	23	91.94319341	0.00000025

Frequencies and Wavenumbers for the Rotational Lines of CO

Frequency/MHz	Uncertainty*/MHz	J′	J″	Wavenumber/cm⁻¹	Uncertainty*/cm⁻¹
2870339.4194	0.0077	25	24	95.74421713	0.00000026
2984181.4631	0.0080	26	25	99.54157896	0.00000027
3097909.3621	0.0085	27	26	103.33513334	0.00000028
3211518.7558	0.0090	28	27	107.12473480	0.00000030
3325005.2869	0.0096	29	28	110.91023800	0.00000032
3438364.6013	0.0102	30	29	114.69149772	0.00000034
3551592.3489	0.0107	31	30	118.46836884	0.00000036
3664684.1829	0.0111	32	31	122.24070637	0.00000037
3777635.7608	0.0118	33	32	126.00836545	0.00000039
3890442.7435	0.0137	34	33	129.77120137	0.00000046
4003100.7965	0.0179	35	34	133.52906952	0.00000060
4115605.5892	0.0254	36	35	137.28182546	0.00000085
4227952.7954	0.0370	37	36	141.02932487	0.00000123
4340138.0932	0.0531	38	37	144.77142361	0.00000177
4452157.1657	0.0746	39	38	148.50797766	0.00000249
4564005.7001	0.1025	40	39	152.23884318	0.00000342

* The uncertainty given is twice the standard error.

Frequencies and Wavenumbers for the Rotational Lines of HF

Frequency/MHz	Uncertainty*/MHz	J′	J″	Wavenumber/cm⁻¹	Uncertainty*/cm⁻¹
1232476.21	0.12	1	0	41.110981	0.000004
2463428.09	0.19	2	1	82.171116	0.000006
3691334.81	0.25	3	2	123.129676	0.000008
4914682.58	0.51	4	3	163.936165	0.000017
6131968.11	1.10	5	4	204.540439	0.000037
7341702.00	2.00	6	5	244.892818	0.000067
8542412.1	3.21	7	6	284.944197	0.000107
9732646.8	4.72	8	7	324.646153	0.000157
10910978.2	6.51	9	8	363.951056	0.000217
12076004.8	8.55	10	9	402.81216	0.000285
13226355.2	10.81	11	10	441.18372	0.000361
14360689.8	13.25	12	11	479.02105	0.00044
15477704.4	15.86	13	12	516.28065	0.00053
16576131.8	18.61	14	13	552.92024	0.00062
17654744.4	21.48	15	14	588.89888	0.00072
18712356.5	24.44	16	15	624.17703	0.00082
19747825.6	27.43	17	16	658.71656	0.00092
20760054.3	30.32	18	17	692.4809	0.00101
21747991.7	32.91	19	18	725.4349	0.00110
22710634.7	34.94	20	19	757.5452	0.00117
23647028.7	36.08	21	20	788.7800	0.00120
24556268.8	35.93	22	21	819.1090	0.00120
25437499.9	34.12	23	22	848.5037	0.00114
26289917.4	30.32	24	23	876.9373	0.00101
27112767.2	24.41	25	24	904.38457	0.00081
27905345.6	16.88	26	25	930.82214	0.00056
28666999.3	10.80	27	26	956.22817	0.00036
29397124.8	14.65	28	27	980.58253	0.00049
30095168.2	24.62	29	28	1003.86676	0.00082
30760624.2	33.36	30	29	1026.0640	0.00111
31393035.7	36.17	31	30	1047.1590	0.00121

* The uncertainty given is twice the standard error.

Frequencies and Wavenumbers for the Rotational Lines of $H^{35}Cl$

Frequency/MHz	Uncertainty*/MHz	J'	J''	Wavenumber/cm^{-1}	Uncertainty*/cm^{-1}
1876226.517	0.065	3	2	62.584180	0.000002
2499864.439	0.066	4	3	83.386502	0.000002
3121986.563	0.064	5	4	104.138262	0.000002
3742216.601	0.076	6	5	124.826909	0.000003
4360180.042	0.098	7	6	145.439951	0.000003
4975504.51	0.11	8	7	165.964966	0.000004
5587820.10	0.12	9	8	186.389615	0.000004
6196759.76	0.22	10	9	206.701656	0.000007
6801959.63	0.50	11	10	226.888951	0.000017
7403059.41	1.02	12	11	246.939481	0.000034
7999702.7	1.8	13	12	266.841359	0.000062
8591537.3	3.1	14	13	286.582837	0.000103
9178215.8	4.8	15	14	306.152324	0.000161

* The uncertainty given is twice the standard error.

Frequencies and Wavenumbers for the Rotational Lines of $H^{37}Cl$

Frequency/MHz	Uncertainty*/MHz	J'	J''	Wavenumber/cm^{-1}	Uncertainty*/cm^{-1}
1873410.72	0.05	3	2	62.490255	0.000002
2496115.33	0.05	4	3	83.261445	0.000002
3117308.69	0.05	5	4	103.982225	0.000002
3736615.64	0.06	6	5	124.640082	0.000002
4353662.84	0.08	7	6	145.222561	0.000003
4968079.04	0.09	8	7	165.717279	0.000003
5579495.53	0.10	9	8	186.111938	0.000003
6187546.42	0.19	10	9	206.394332	0.000006
6791869.04	0.45	11	10	226.552365	0.000015
7392104.3	0.9	12	11	246.574057	0.000030
7987896.9	1.6	13	12	266.447561	0.000054
8578896.1	2.7	14	13	286.161170	0.000089

* The uncertainty given is twice the standard error.

Section 11
Nuclear and Particle Physics

SUMMARY TABLES OF PARTICLE PROPERTIES

Details of particle properties may be found in the following publication:

Patrignani, C. et al. (Particle Data Group), *Chinese Physics C*, 40, 100001 (2016.)

Abstract

The *Review of Particle Physics* summarizes much of particle physics and cosmology. Using data from previous editions, plus 3,062 new measurements from 721 papers, we list, evaluate, and average measured properties of gauge bosons and the recently discovered Higgs boson, leptons, quarks, mesons, and baryons. We summarize searches for hypothetical particles such as supersymmetric particles, heavy bosons, axions, dark photons, etc. All the particle properties and search limits are listed in the Summary Tables. We also give numerous tables, figures, formulae, and reviews of topics such as Higgs Boson Physics, Supersymmetry, Grand Unified Theories, Neutrino Mixing, Dark Energy, Dark Matter, Cosmology, Particle Detectors, Colliders, Probability and Statistics. Among the 117 reviews are many that are new or heavily revised, including new reviews on Pentaquarks and Inflation.

The complete Review is published online in a journal and on the Web site of the Particle Data Group (http://pdg.lbl.gov). The printed PDG Book contains the Summary Tables and all review articles but no longer includes the detailed tables from the Particle Listings. A Booklet with the Summary Tables and abbreviated versions of some of the review articles is also available.

The full Summary Tables are included in the Internet version of the *CRC Handbook of Chemistry and Physics*.

TABLE OF THE ISOTOPES

Norman E. Holden

This table presents an evaluated set of values for the experimental quantities that characterize the decay of radioactive nuclides. A list of the major references used in this evaluation is given below. When uncertainties are not listed, they are assumed to be five or less in the last digit quoted. If the uncertainty in the value exceeds five in the last digit, the value is preceded by an approximate sign.

For quasi-stable nuclides, the measured width, Γ, of the resonance is given. To estimate the approximate half-life, the Heisenberg relationship may be used: the half-life = 4.56×10^{-22} seconds / Γ (MeV). The effective literature cutoff date for data in this edition of the table is October 2016.

Nuclear

Table Layout

Column number	Column title	Description
1	Isotope or Element	For elements, the atomic number and chemical symbol are listed. For nuclides, the mass number and chemical symbol are listed. Isomers are indicated by the addition of m, m1, or m2.
2	Isotopic Abundance	The abundance of an isotope in normal terrestrial samples of an element, listed in atom percent.
3	Atomic Mass or Atomic Weight	Atomic mass relative to $^{12}C = 12$. Atomic weight of elements is given on the same scale.
4	Half-Life/Resonance Width	Half-life in decimal notation. μs = microseconds; ms = milliseconds; s = seconds; m = minutes; h = hours; d = days; and a = years. For quasi-stable nuclides, the measured width at half-maximum of the energy resonance is given.
5	Decay Mode/Energy	Decay modes are α = alpha particle emission; β = negative beta emission; β+ = positron emission; ec = orbital electron capture; IT = isomeric transition from upper to lower isomeric state; n = neutron emission; p = proton emission; sf = spontaneous fission; ββ, ECEC, 2p, 2n, 3n = double beta, double EC, two proton and multiple neutron decay. Total disintegration energy in MeV units.
6	Particle Energy/Intensity	End point energies of beta transitions and discrete energies of alpha particles, neutrons and protons are given in MeV units and their intensities in percent.
7	Spin and Parity	Nuclear spin or angular momentum of the nuclides in units of $h/2\pi$; parity is positive or negative.
8	Magnetic Dipole Moment	Magnetic dipole moments in nuclear magneton units. An absolute value is indicated in the absence of a positive or a negative sign.
9	Electric Quadrupole Moment	Electric quadrupole moments in barn units (10^{-24} cm^2). An absolute value is indicated in the absence of a positive or a negative sign.
10	Gamma Ray Energy/Intensity	Gamma ray energies are given in MeV units and intensities in percent. Annihilation radiation (ann.rad.) refers to the 511.006 keV photons emitted in the annihilation of positrons in matter.

General Nuclear Data References

The following references represent the major sources of the nuclear data presented, along with subsequent published journal articles and reports:

1. G. Audi, F.G. Kondev, M. Wang, B. Pfeiffer, X. Sun, J. Blachot, M. MacCormick, The NUBASE2012 Evaluation of Nuclear Properties, *Chinese Physics C – High Energy and Nuclear Physics* 36, 1157 (2012).
2. M. Wang, G. Audi, A.H. Wapstra, F.G. Kondev, M. MacCormick, X. Xu, B. Pfeiffer, The AME2012 Atomic Mass Evaluation (II., Tables, graphs and references), *Chinese Physics C – High Energy and Nuclear Physics* 36, 1603 (2012).
3. J. Meija, T.B. Coplen, M. Berglund, W.A. Brand, P. DeBièvre, M. Gröning, N.E. Holden, J. Irrgeher, R.D. Loss, T. Walczyk, T. Prohaska, Atomic Weights of the Elements - 2013, *Pure & Applied Chemistry* 88, 265 (2016).
4. International Union of Pure and Applied Chemistry (IUPAC). Standard Atomic Weight of Ytterbium Revised (press release, 24 August 2015). http://www.iupac.org/news/news-detail/article/standard-atomic-weight-of-ytterbium-revised.html (accessed Dec 2016).
5. J. Meija, T.B. Coplen, M. Berglund, W.A. Brand, P. DeBièvre, M. Gröning, N.E. Holden, J. Irrgeher, R.D. Loss, T. Walczyk, T. Prohaska, Isotopic compositions of the elements 2013 (IUPAC Technical Report). *Pure Applied Chemistry* 88, 293 (2016).
6. E.M. Baum, H.D. Knox, T.R. Miller, *Chart of the Nuclides*, 17th Edition, Knolls Atomic Power Lab (2009).
7. N.E. Holden, Total and Spontaneous Fission Half-Lives for Uranium, Plutonium, Americium and Curium Nuclides, *Pure & Applied Chemistry* 61, 1483 (1989).
8. N.E. Holden, Half-lives of Selected Nuclides, *Pure & Applied Chemistry* 62, 941 (1990).
9. E. Brown, R. Firestone, *Radioactivity Handbook*, Wiley Interscience Press (1986).
10. J.K. Tuli, *Nuclear Wallet Cards*, Brookhaven National Laboratory (October 2011).
11. N.E. Holden, D.C. Hoffman, Spontaneous Fission Half-Lives for Ground State Nuclides, *Pure & Applied Chemistry* 72, 1525 (2000).
12. N.J. Stone, Table of Nuclear Magnetic Dipole and Electrical Quadrupole Moments, *Atomic Data Nuclear Data Tables* 90, 75 (2005).
13. N. J. Stone, Table of Nuclear Electric Quadrupole Moments, *Atomic Data Nuclear Data Tables*, 111-112, 1 (2016).
14. I. M. Villa, P. De Bièvre, N. E. Holden, P. R. Renne, Recommendation on the half-life of ^{87}Rb, *Geochimica Cosmochimica Acta*, 164, 382 (2015).
15. I. M. Villa, M. L. Bonardi, P. De Bièvre, N. E. Holden, P. R Renne, IUPAC-IUGS status report on the half-lives of ^{238}U, ^{235}U and ^{234}U, *Geochimica Cosmochimica Acta* 172, 387 (2016).
16. A. K. Jain, B. Maheshwari, S. Garg, M. Patial, B. Singh, Atlas of Nuclear Isomers, *Nuclear Data Sheets* 128, 1, (2015).

This research was carried out under the auspices of the U.S. Department of Energy Contract No. DE-AC02-98CH10886. The author is at Brookhaven National Laboratory, Upton, NY, and can be contacted at holden@bnl.gov.

Elem. or Isot.	Natural abundance (Atom %)	Atomic mass or weight	Half-life/ Resonance width (MeV)	Decay mode/ Energy (/MeV)	Particle energy/ Intensity (MeV/%)	Spin (h/2π)	Nuclear magnetic mom. (nm)	Elect. quadr. mom. (b)	γ-Energy/ Intensity (MeV/%)
$_{0}$n		1.008664916	610.5 s	β- /0.78235	0.782/100.	1/2+	-1.913043		
				β-, γ	/ 0.0031				
$_{1}$H		**1.00784-1.00811**							
^{1}H	99.972 - 99.999	1.007825032	>2.8 ×10^{23} a			1/2+	+2.7928473		
^{2}H	0.001-0.028	2.014101778				1+	+0.8574382	+2.86 mb	
^{3}H		3.016049278	12.31 a	β-/0.01859	0.01860/100.	1/2+	+2.9789624		
^{4}H		4.026	Γ = 1.2	n/	/100	2-			
^{5}H		5.035	Γ < 0.5	n/	/100	(1/2+)			
^{6}H		6.045	Γ ~ 6.	n/		(2-)			
^{7}H		7.05	Γ ~ 0.1						
$_{2}$He		**4.002602(2)**							
^{3}He	0.0002(2)	3.01602932				1/2+	-2.127750		
^{4}He	99.9998(2)	4.0026032541				0+			
^{5}He		5.0121	Γ = 0.60(2)	n, α		3/2-			
^{6}He		6.0188859	0.8069 s	β-/3.508	3.510/100.	0+			
				β, d	0.35/0.00017				
^{7}He		7.02799	Γ ~ 0.19	n		(3/2)-			
^{8}He		8.033934	0.119 s	β-/10.65	/84.	0+			0.9807/84.
				β- n/	/16.				0.4776/5.
				β, t	/0.82				
^{9}He		9.0440	Γ = 0.10(6)	n	/100	(1/2-)			
^{10}He		10.053	Γ = 0.1	2n	/100	0+			
$_{3}$Li		**6.938-6.997**							
^{4}Li		4.027	Γ = 6.0	p/	/100	2-			
^{5}Li		5.0125	Γ = 1.2	p/α		3/2-			
^{6}Li	(1.9 - 7.8)	6.01512289				1+	+0.822047	-0.81 mb	
^{7}Li	(92.2 - 98.1)	7.01600344				3/2-	+3.25644	-0.0400	
^{8}Li		8.0224863	0.839 s	β-/16.004	12.5/100.	2+	+1.65334	+0.0314	
				α/	α(1.6)				
^{9}Li		9.026790	0.178 s	β-/13.606	13.5/75.	3/2-	3.4368	-0.0304	
				β-/ n	11/25.				
^{10}Li		10.0355	Γ = 0.11(5)	n	/7.	1+			
^{11}Li		11.043724	8.8 ms	β-/20.6	/8.3	3/2(-)	3.671	-0.0339	3.368/33.
				β-, n	/85.7				0.320/7.
				β-, 2n	/4.1				2.590/8.
				β-, 3n	/1.9				5.958/3.
				β-, d	/0.013				2.895/1.5
				β-, t	/0.00009				2.811/1.1
^{12}Li		12.0525	< 0.01 μs						
^{13}Li		13.063	Γ = 1.3						
$_{4}$Be		**9.0121831(5)**							
^{5}Be		5.04		p, 3He		(1/2+)			
^{6}Be		6.01973	Γ = 0.092(6)	2p,α		0+			
^{7}Be		7.0169287	53.28 d	ec/0.8618		3/2-	-1.40		0.4776/10.4
^{8}Be		8.0053051	Γ = 6.8(17) eV	2α/0.046		0+			
^{9}Be	100.	9.0121831				3/2-	-1.1776	+0.0529	
^{10}Be		10.013535	1.39 ×10^{6} a	β-/0.5559	0.555/100.	0+			
^{11}Be		11.021661	13.8 s	β-, β-α/11.51	11.48/61.	1/2+	-1.681		2.125/35.5
				β-, p//10^{-5}					
^{12}Be		12.02692	22.0 ms	β-, (n)/11.71	n//0.5	0+			(0.95 - 4.4)
^{13}Be		13.0361	Γ ~ 1.						
^{14}Be		14.043	4.6 ms	β-/16.2		0+			3.5346/0.9
				β-, n	0.288/94.				3.6845/7.
				β-, 2n	/6.				
				β-, α	/<0.012				
				β-, t	/<0.04				
^{15}Be		15.053	Γ = 0.6	β-					
^{16}Be		16.062	Γ = 0.8	2n		0+			
$_{5}$B		**10.806-10.821**							
^{7}B		7.0297	Γ = 1.4(2)	p, α		(3/2-)			
^{8}B		8.02461	0.770 s	β+, 2α/17.979	13.7(β+)/93.	2+	1.0358	+0.064	ann.rad.
^{9}B		9.01333	Γ = 0.5(2) keV	p, 2α/		3/2-			
^{10}B	(18.9 - 20.4)	10.012937				3+	+1.8006448	+0.0845	
^{11}B	(79.6 - 81.1)	11.009305				3/2-	+2.688649	+0.0406	
^{12}B		12.01435	0.0202 s	β-/13.369		1+	+1.0027	0.0132	4.438/1.3
				β- α/1.6/					3.215/0.00065
^{13}B		13.01778	0.0174 s	β- /13.437	13.4	3/2-	+3.1778	+0.037	3.68/7.6
				β- n/0.25/	2.43(n)/0.09				
					3.55(n)/0.16				
^{14}B		14.0254	14. ms	β-/20.64		2-	1.185	0.030	6.094/90.
^{15}B		15.0311	9.9 ms	β-, (n)/19.09	n//99.7	(3/2-)	2.66	0.038	

Elem. or Isot.	Natural abundance (Atom %)	Atomic mass or weight	Half-life/ Resonance width (MeV)	Decay mode/ Energy (/MeV)	Particle energy/ Intensity (MeV/%)	Spin (h/2π)	Nuclear magnetic mom. (nm)	Elect. quadr. mom. (b)	γ-Energy/ Intensity (MeV/%)
^{16}B		16.0398	Γ < 0.1	n					
^{17}B		17.047	5.1 ms	β-, (n)/22.7			2.55	0.039	
^{18}B		18.056	< 0.026 μs			0-			
^{19}B		19.063	2.9 ms	β-, (n)/26.5	1n//72.	(3/2-)			
					2n//16.				
					3n// < 9.				
$_{6}$**C**		**12.0096-12.0116**							
^{8}C		8.0376	Γ = 0.25(4)	p		0+			
^{9}C		9.03104	127. ms	β+, p, 2α/16.498		(3/2-)	1.391		ann.rad.
^{10}C		10.016853	19.30 s	β+/3.648	1.865	0+			ann.rad.
									0.71829/100.
^{11}C		11.01143	20.4 m	β+, ec/1.982	0.9608/99.	3/2-	-0.964	0.0333	ann.rad.
^{12}C	(98.84-99.04)	12.000000000				0+			
^{13}C	(0.96-1.16)	13.003354835				½-	+0.702412		
^{14}C		14.00324199	5.70 x 103 a	β-/0.15648	0.1565/100.	0+			
^{15}C		15.01060	2.45 s	β-/9.772	4.51/68.	½+	1.72		5.298/68.
					9.82/32.				(7.30–9.05)
^{16}C		16.01470	~ 0.750 s	β-/8.012	β/3.3, 4.3/84, 16	0+			
				β, n	n/0.8, 1.7/84, 16				
^{17}C		17.0226	0.19 s	β-/13.17		3/2+	0.758		1.375
				β-, n	n/1.6–3.7/11.				1.849
									1.906
^{18}C		18.0268	0.092 s	β-/11.81		0+			
				β-, n	n/0.88–4.59/21.				
^{19}C		19.035	0.05 s	n		½+			
^{20}C		20.040	0.02 s	β,n	1n//~ 65.	0+			
					2n// < 19.				
^{21}C		21.049	< 0.03 μs						
^{22}C		22.058	6 ms	β-, n	1n//~ 61.	0+			
					2n// < 37.				
$_{7}$**N**		**14.00643-14.00728**							
^{10}N		10.042	Γ = 2.3(16)						
^{11}N		11.0261	Γ = 0.83			½+			
^{12}N		12.01861	11.00 ms	β+, β+α/17.338	16.38/95.	1+	+0.457	+10. mb	ann.rad.
									4.438/2.
^{13}N		13.005739	9.97 m	β+ /2.2204	1.190/100.	½-	0.3222		
^{14}N	(99.578 -99.663)	14.003074004				1+	+0.403761	+0.02044	
^{15}N	(0.337-0.422)	15.00010899				½-	-0.283189		
^{16m}N			5.3 μs	IT		5+			0.120
^{16}N		16.00610	7.13 s	β- /10.419	4.27/68.	2-			6.129/68.8
					10.44/26.		1.986	-18 mb	7.115/4.7
				β+, α //10^{-3}	1.85/.0012				(0.99–8.87)
					(0.6-2.3)				
^{17}N		17.0084	4.17 s	β-, β- n/8.68	3.7/100.	½-	0.3551		0.871/3.
				0.4–1.7n/95.					2.1842/0.3
				β- α/	8.0, 8.2				
^{18}N		18.0141	0.62 s	β- /13.90	9.4/100.	1-	0.3273	0.012	0.822/48.
				β-, α	(1.08-5.23)/12.				1.65/47.
				β-, n	/14				1.982/77.
									(0.535–7.13)
^{19}N		19.0170	0.336 s	β-/12.53			0.305		(0.096–3.14)
				β-, n	0.45-4.51/42.				
^{20}N		20.0234	0.136 s	β- /17.97					
				β-, n	1.10-3.68/43.				
^{21}N		21.027	0.083 s	β-,n	4.98//90.5				1.222
				β-	//3.5				
^{22}N		22.034	0.02 s	β-,n	1n// ~ 41.				
					2n// < 13.				
^{23}N		23.041	15. ms	β-, n	n//~ 42.				
					2n//~ 8.				
					3n// < 3.4				
^{24}N		24.051	< 0.052 μs						
^{25}N		25.060	< 0.26 μs						
$_{8}$**O**		**15.99903-15.99977**							
^{12}O		12.0343	Γ < 0.07	2p		0+			
^{13}O		13.0248	8.9 ms	β+, p/17.77	1.560 (p)	(3/2-)	1.389	0.011	ann.rad.
					p/(1.00 - 13.5)				4.438/0.56
^{14}O		14.008596	70.62 s	β+ /5.1430	1.81/99.	0+			ann.rad.
									2.312/99.4

Nuclear

Elem. or Isot.	Natural abundance (Atom %)	Atomic mass or weight	Half-life/ Resonance width (MeV)	Decay mode/ Energy (/MeV)	Particle energy/ Intensity (MeV/%)	Spin (h/2π)	Nuclear magnetic mom. (nm)	Elect. quadr. mom. (b)	γ-Energy/ Intensity (MeV/%)
^{15}O		15.003066	122.2 s	β+ /2.754	1.723/100.	1/2-	0.7195		ann.rad.
^{16}O	(99.738 - 99.776)	15.994914620				0+			
^{17}O	(0.0367 - 0.0400)	16.999131757				5/2+	-1.8938	-0.0256	
^{18}O	(0.187-0.222)	17.999159613				0+			
^{19}O		19.00358	26.48 s	β- /4.820	3.25/60.	5/2+	1.5320	3.6 mb	0.197/95.9
					4.60/40.				1.3569/50.4
									(0.11–4.18)
^{20}O		20.00408	13.5 s	β- /3.814		0+			1.057/100.
^{21}O		21.0087	3.4 s	β- /8.11					(0.28–4.6)
^{22}O		22.0100	2.2 s	β- /6.5		0+			0.072/100
									0.638/98
									1.862/63
									(0.918-2.499)
^{23}O		23.016	0.097 s	β -, n	/7.2%				2.234/51.5
									4.066/17.1
									(0.911-3.868)
^{24}O		24.020	~ 65. ms	β-, n	n//18.	0+			1.83/28.
									0.52/14.
									1.31/12.
^{25}O		25.029	Γ ~ 0.02						
^{26}O		26.037	Γ = 0.018	2n		0+			
^{27}O		27.048	< 0.026 μs						
^{28}O		28.056	< 0.10 μs			0+			
$_{9}$F		**18.998403163(6)**							
^{14}F		14.0343	Γ = 0.9						
^{15}F		15.0180	Γ = 0.8(3)	p		(1/2+)			
^{16}F		16.0115	Γ = 0.037(14)	p		0-			
^{17}F		17.002095	64.4 s	β+ /2.761	1.75/	5/2+	+4.721	- 0.076	ann.rad.
^{18}F		18.000937	1.8295 h	β+, ec/1.656	0.635/97.	1+			ann.rad.
^{19}F	100.	18.99840316				½+	+2.62887		
^{20}F		19.9999813	11.00 s	β- /7.0245	5.398/100.	2+	+2.0934	0.056	1.634/100.
									3.33/0.009
^{21}F		20.99995	4.16 s	β- /5.684	3.7/8.	5/2+	3.9	0.011	0.3507/90.
					5.0/63.				1.395/15.
					5.4/29.				(1.746–4.684)
^{22}F		22.00300	4.23 s	β- /10.82	3.48/15.	4+	2.694	0.003	1.2746/100.
					4.67/7.				2.0826/82.
					5.50/62.				(0.82–4.37)
^{23}F		23.0036	2.2 s	β- /8.5		5/2+			1.701/48.
									2.129/34.
									(0.493–3.83)
^{24}F		24.0081	0.3 s	β- /13.5					1.9816/
^{25}F		25.0122	~ 50. ms	β-, (n)	n//14.				1.70/39.
									(0.57–2.19)
^{26m}F			2.2 ms	IT, β-		4+			
^{26}F		26.0200	10. ms	β-, (n)	n//11.				2.02/67.
									1.67/19.
^{27}F		27.026	5.0 ms	β-, (n)	n//90.				2.02/18.
^{28}F		28.035	< 0.04 μs						
^{29}F		29.043	2.5 ms	β-, (n)	n//100.				
^{30}F		30.052	< 0.26 μs						
^{31}F		31.06	> 0.26 μs						
$_{10}$Ne		**20.1797(6)**							
^{15}Ne			Γ = 0.6	2p		(3/2-)			
^{16}Ne		16.0258	Γ = 0.08	2p		0+			
^{17}Ne		17.017714	109. ms	β+, p/14.53	1.4–10.6/6.9	1/2-	0.787		ann.rad./
				β+, α	/0.014				0.495
^{18}Ne		18.005709	1.665 s	β+ /4.446	3.416/92.	0+			ann.rad./
									1.0413/7.8
									(0.658–1.70)
^{19}Ne		19.001881	17.28 s	β+ /3.238	2.24/99.	1/2+	-1.885		ann.rad./
									(0.11–1.55)
^{20}Ne	90.48(3)	19.99244018				0+			
^{21}Ne	0.27(1)	20.9938467				3/2+	-0.66180	+0.102	
^{22}Ne	9.25(3)	21.9913851				0+			
^{23}Ne		22.994467	37.2 s	β- /4.376	3.95/32.	5/2+	-1.08	+0.15	0.440/33.
					4.39/67.				(1.64–2.98)
^{24}Ne		23.993611	3.38 m	β- /2.47	1.10/8.	0+			0.4723/100.
					1.98/92.				0.874/7.9
^{25}Ne		24.9978	0.61 s	β- /7.30	6.3/	1/2+	-1.006		0.0895/96.
					7.3/				(0.98–3.69)

Elem. or Isot.	Natural abundance (Atom %)	Atomic mass or weight	Half-life/ Resonance width (MeV)	Decay mode/ Energy (/MeV)	Particle energy/ Intensity (MeV/%)	Spin (h/2π)	Nuclear magnetic mom. (nm)	Elect. quadr. mom. (b)	γ-Energy/ Intensity (MeV/%)
^{26}Ne		26.0005	197 ms	β-, n/7.3	n //0.13	0+			0.082/100
									1.278/6
									0.233/5
									0.151/3
									1.211/1
									2.489/1
^{27}Ne		27.0076	31. ms	β-, n/12.7	n //3.	(3/2+)			
^{28}Ne		28.012	19. ms	β-, n/12.3	n//12.	0+			2.06/19.
					2n//3.				0.86/3.
^{29}Ne		29.020	15. ms	β-, (n)/15.4	n//29.	(3/2+)			2.92/55.
				β-, 2n	2n//4.				(0.22–1.18)
^{30}Ne		30.025	7. ms	β-, (n)	n//9.	0+			0.151/9.
^{31}Ne		31.03	3. ms						
^{32}Ne		32.040	~ 3.5 ms			0+			
^{33}Ne		33.049	< 0.26 μs						
^{34}Ne		34.057	> 1.5 μs			0+			
$_{11}$**Na**		**22.98976928(2)**							
^{18}Na		18.027	Γ < 0.02	p					
^{19}Na		19.0139	0.03 s	β+, p/11.18					
^{20}Na		20.00735	0.446 s	β+ /13.89		2+	+0.3694	+ 0.10	ann.rad./
				α	2.15/16				1.634/79.
^{21}Na		20.997655	22.42 s	β+ /3.547	2.50/95.	3/2+	+2.3863	+0.14	ann.rad./
									0.351/5.
^{22}Na		21.994437	2.601 a	β+ /90/2.842	0.545/90.	3+	+1.746	+0.18	ann.rad./
				ec/10/					1.2745/99.9
^{23}Na	100.	22.98976928				3/2+	+2.21752	+0.104	
^{24m}Na			20.2 ms	IT, β-		1+	-1.93		0.4723/100.
^{24}Na		23.9909630	14.96 h	β- /5.5158	1.389/>99.	4+	+1.690		1.3686/100.
									2.754/100.
									(0.997–4.238)
^{25}Na		24.98995	59.3 s	β- /3.835	2.6/7.	5/2+	+3.683	-0.0015	0.3897/12.7
					3.15/25.				0.5850/13.
					4.0/65.				0.9747/14.9
									(0.836–2.80)
^{26m}Na			9 μs	IT		1+			0.0825
^{26}Na		25.99264	1.071 s	β- /9.31		3+	+2.851	-5.3 mb	1.809/98.9
									(0.24-7.37)
^{27}Na		26.99408	0.290 s	β- /9.01	7.95/	5/2+	+3.90	-7.1 mb	0.9847/87.4
				β-, n/					1.698/11.9
^{28}Na		27.9989	31. ms	β- /14.0	12.3/	1+	+2.43	+0.039	1.473/37.
				β-, n/					2.389/18.6
^{29}Na		29.00288	44. ms	β-, n/13.3	11.5/	3/2+	+2.45	+0.085	2.560/36.
									(1.04–3.99)
^{30}Na		30.00910	50. ms	β-, n/17.5	n//30.	2+	+2.08		1.483/46.
^{31}Na		31.0132	17.2 ms	β-, n/15.9	n//37.	3/2-	+2.30		1.483/14.
									(0.05–3.54)
^{32}Na		32.020	13.5 ms	β- /19.1					0.240–3.935
				β-, n					0.171-0.894
				β-, 2n					1.483/4.2
^{33}Na		33.026	8.0 ms	β- /20.	/~ 38				0.886/16
				β-, n	0.8,1.02/47(6)				0.546/6.4
				β-, 2n	/13(3)				0.050–2.55
^{34}Na		34.034	5. ms	β- /24.					
^{35}Na		35.041	1.5 ms	β- /24					
^{36}Na		36.05	< 0.26 μs						
^{37}Na		37.06	> 1.5 μs						
$_{12}$**Mg**		**24.304-24.307**							
^{19}Mg		19.0342	4 × 10-12 s	2p					
^{20}Mg		20.0189	91. ms	β+ /10.73	/70	0+			
				β+, p	/30				
^{21}Mg		21.0117	122. ms	β+, p/13.10		5/2+	- 0.98		0.332/51.
				β+, α					
^{22}Mg		21.999571	3.876 s	β+ /4.786	3.05/	0+			0.0729/60.
									0.5820/100.
									(1.28–1.93)
^{23}Mg		22.994124	11.32 s	β+ /4.057	3.09/92.	3/2+	0.536	0.114	0.440/8.2
^{24}Mg	(78.88-79.05)	23.9850417				0+			
^{25}Mg	(9.988 -10.034)	24.9858370				5/2+	- 0.85545	+0.199	
^{26}Mg	(10.96-11.09)	25.9825930				0+			
^{27}Mg		26.9843406	9.45 m	β- /2.6103	1.59/41.	1/2+	- 0.411		0.17068/0.9
					1.75/58.				0.84376/72.
					2.65/0.3				1.01443/28.

Elem. or Isot.	Natural abundance (Atom %)	Atomic mass or weight	Half-life/ Resonance width (MeV)	Decay mode/ Energy (/MeV)	Particle energy/ Intensity (MeV/%)	Spin (h/2π)	Nuclear magnetic mom. (nm)	Elect. quadr. mom. (b)	γ-Energy/ Intensity (MeV/%)
^{28}Mg		27.98388	20.9 h	β- /1.832	0.459/95.	0+			0.0306/95.
									0.4006/36.
									0.9418/36.
									1.342/54.
^{29}Mg		28.9886	1.3 s	β-/7.55	5.4/	3/2+	+ 0.978		0.960/15.
									1.398/16.
									2.224/36.
^{30}Mg		29.99046	0.32 s	β- /7.0		0+			0.224/85.
^{31}Mg		30.99665	0.24 s	β- /11.7	8.4/29.9	1/2+	- 0.8836		1.613/47.
									0.947/37.
									(0.666-4.640)
				β-, n	/1.7				
^{32}Mg		31.99911	0.12 s	β- /10.3		0+			2.765/25.
^{33}Mg		33.00533	91. ms	β- /13.7	/83.	3/2-			1.848/
				β-, n	/17.				
^{34}Mg		34.0089	0.02 s	β- /11.3		0+			
^{35}Mg		35.017	0.07 s			(7/2-)			
^{36}Mg		36.022	4. ms			0+			
^{37}Mg		37.030	> 0.26 μs			(7/2-)			
^{38}Mg		38.037	> 0.26 μs			0+			
^{39}Mg		39.045	< 0.26 μs						
^{40}Mg		40.052	> 0.17 μs			0+			
$_{13}$Al		**26.9815385(7)**							
^{21}Al		21.029	< 0.035 μs						
^{22}Al		22.020	59. ms	β+ /18.6	p/1.3/18.	4+			ann.rad./
				β+, p//55	p/0.48-7.52				0.584/47.
				β+, 2p//11					(1.25-2.15)
				β+, α	α/3.3/0.3				
^{23m}Al			~ 0.35 s	β+, p/0.17					0.554
									0.839
^{23}Al		23.007244	0.47 s	β+ /12.24	p/0.56/46	5/2+	3.89	0.16	ann.rad./
				β+, p/	p/2.01/20				
^{24m}Al			0.131 s	IT/0.4259		1+	3.0		
				β+	13.3				1.3686/5.3
^{24}Al		23.99995	2.07 s	β+ /13.878,p	3.40/48.	4+			1.078(2)/16.
					4.42/41.				1.368(2)/96.
					6.80/3.				2.753(2)/43.
					8.74/8.				4.315(3)/15.
									5.392(3)/20.
									7.0662(2)/41.
^{25}Al		24.990428	7.17 s	β+ /4.277	3.27/	5/2+	3.646	0.24	ann.rad./
									1.6115(2)/100.
									0.975(2)/5.
^{26m}Al			6.345 s	β+ /	3.2/	0+			ann.rad./
^{26}Al		25.9868919	7.1 ×10^{5} a	β+ /82/4.0042	1.16/	5+	+2.804	+0.26	ann.rad./
				ec/18					1.8087/99.8
^{27}Al	100.	26.9815385				5/2+	+3.641507	+0.147	
^{28}Al		27.981910	2.25 m	β- /4.6422	2.865/100.	3+	3.24	0.17	1.7778(6)/100.
^{29}Al		28.98046	6.5 m	β- /3.680	1.4/30.	5/2+			1.2732(8)/89.
					2.5/70.				2.0282(8)/4.
									2.4262(8)/7.
^{30}Al		29.9830	3.68 s	β- /8.56	5.05/	3+	3.01		1.26313(3)/35.
									2.23525(5)/65.
^{31}Al		30.9839	0.64 s	β- /8.00	6.25/	5/2+	3.8	0.135	0.75223(3)/18.
									1.69473(3)/59.
									2.31664(4)/73.
^{32}Al		31.9881	33. ms	β- /13.0		1+	1.96	0.025	
^{33}Al		32.9909	41.7 ms	β- /12.0	/91.5			0.141	1.940/2.5
				β-, n	/8.5				(1.01–4.34)
^{34m}Al			26 ms	β-		1+			
^{34}Al		33.9967	56. ms	β- /17.1	4.255/44	4			0.929/57
				β-, n	/26.				(0.12–4.26)
^{35}Al		34.9998	38. ms	β-/14.3	0.974/48	5/2+			0.064/45.
				β-, n	/ 38.				(0.12–5.63)
^{36}Al		36.006	0.09 s	β-/18.3					
				β-, n	/<31.				
^{37}Al		37.011	11. ms	β-/16.					
^{38}Al		38.017	> 7.6 ms						
^{39}Al		39.023	> 8. ms						
^{40}Al		40.030	> 0.26 μs						
^{41}Al		41.036	> 0.26 μs						
^{42}Al		42.044	> 0.17 μs						
^{43}Al		43.05	> 0.17 μs						

Elem. or Isot.	Natural abundance (Atom %)	Atomic mass or weight	Half-life/ Resonance width (MeV)	Decay mode/ Energy (/MeV)	Particle energy/ Intensity (MeV/%)	Spin (h/2π)	Nuclear magnetic mom. (nm)	Elect. quadr. mom. (b)	γ-Energy/ Intensity (MeV/%)
$_{14}$**Si**		**28.084-28.086**							
^{22}Si		22.036	29. ms	β+, p	1.99/20	0+			
^{23}Si		23.025	40.7 ms	β+, p/5.9/~ 80	1.32,(0.6–11.6)				
				β+, 2p//4					
^{24}Si		24.0115	143 ms	β+, p/10.81	1.44,3.92,1.09	0+			ann.rad./
					(1.66–4.47)				
^{25}Si		25.0041	221 ms	β+, p/12.74	p/4.25/9.5	5/2+			ann.rad./
					p/0.40/4.75				
					p/0.56-6.80				
^{26}Si		25.99234	2.228 s	β+ /5.066	3.282/	0+			ann.rad./
									0.8294(8)/22.
^{27}Si		26.986705	4.14 s	β+ /4.8118	3.85/100.	5/2+	-0.865	0.06	ann.rad./
									2.211(5)/0.2
^{28}Si	(92.191 - 92.318)	27.976926535				0+			
^{29}Si	(4.645-4.699)	28.976494665				1/2+	-0.5553		
^{30}Si	(3.037-3.110)	29.9737701				0+			
^{31}Si		30.9753632	2.62 h	β- /1.4920	1.471/99.9	3/2+			1.2662(5)/0.05
^{32}Si		31.974152	1.5 ×10^{2} a	β- /0.224	0.213/100.	0+			
^{33}Si		32.977977	6.1 s	β- /5.85	3.92	(3/2+)	1.21		1.4313(5)/13.
									1.8477/100.
									2.538(2)/10.
^{34}Si		33.9786	2.8 s	β- /4.60	3.09/	0+			0.42907(5)/60.
									1.17852(2)/64.
									1.60756(5)/36.
^{35}Si		34.9846	0.9 s	β- /10.50					
^{36}Si		35.9867	0.5 s	β- /7.9		0+			
				β-, n	/~ 12.				
^{37}Si		36.993	~ 0.09 s	β- /12.5					
				β-, n	/~ 17.				
^{38}Si		37.9955	> 1 μs	β- /10.7		0+			
				β-, n					
^{39}Si		39.002	48. ms	β- /14.8					
^{40}Si		40.006	33. ms			0+			
^{41}Si		41.013	20. ms						
^{42}Si		42.018	13. ms			0+			
^{43}Si		43.025	> 0.26 μs						
^{44}Si		44.031	> 0.36 μs						
$_{15}$**P**		**30.973761998(5)**							
^{24}P		24.036							
^{25}P		25.021	< 0.03 μs						
^{26}P		26.012	44. ms	β+, p/18.1/37	p/0.41/18.0	3+			
					p/1.98/2.4				
					p/0.78-7.49				
				β+, 2p//2.2					
^{27}P		26.9992	0.3 s	β+, p/11.63	p/0.73, 0.61/0.07	1/2+			
^{28}P		27.99233	270. ms	β+ /14.332	3.94/13.	3+	0.31	0.14	ann.rad./
					5.25/13.				1.779(2)/98.
					6.96/16.				2.839(2)/2.8
					8.8/7.				3.040(2)/3.2
					11.49/52.				4.498(2)/12.
									7.537(2)/9.
^{29}P		28.981801	4.14 s	β+ /4.9431	3.945/98.	1/2+	1.2349		ann.rad./
									1.273/1.32
									2.426/0.39
^{30}P		29.978314	2.50 m	β+ /4.2323	3.245/99.9	1+			ann.rad./
									2.230(3)/0.07
^{31}P	100.	30.973761998				1/2+	+1.13160		
^{32}P		31.9739076	14.27 d	β- /1.7106	1.710/100.	1+	-0.2524		
^{33}P		32.97173	25.3 d	β- /0.249	0.249/100.	1/2+			
^{34}P		33.97365	12.4 s	β- /5.374	3.2/15.	1+			1.78–4.1/
					5.1/85.				2.127(5)/15.
^{35}P		34.97331	47. s	β- /3.989	2.34/100.	1/2+			1.572(1)/100.
^{36}P		35.9783	5.7 s	β- /10.41					0.902/77.
									3.291/100.
^{37}P		36.9796	2.3 s	β- /7.90					0.6462/
									1.5829/
^{38}P		37.9843	0.6 s	β- /12.4					1.2923/
				β-, n	/~ 12.				2.224/
^{39}P		38.986	0.3 s	β- /10.5					
				β-, n	/26				
^{40}P		39.991	0.15 s	β- /14.5					
				β-,n	/~ 30.				

Elem. or Isot.	Natural abundance (Atom %)	Atomic mass or weight	Half-life/ Resonance width (MeV)	Decay mode/ Energy (/MeV)	Particle energy/ Intensity (MeV/%)	Spin (h/2π)	Nuclear magnetic mom. (nm)	Elect. quadr. mom. (b)	γ-Energy/ Intensity (MeV/%)
^{41}P		40.995	0.10 s	β-/~ 13.8					
				β-, n	/~ 30.				
^{42}P		42.001	49. ms	β-/17.					
				β-, n	/~ 50.				
^{43}P		43.005	36. ms	β-/16.					
				β-, n	/100.				
^{44}P		44.011	19. ms						
^{45}P		45.016	> 0.2 μs						
^{46}P		46.024	> 0.2 μs						
$_{16}$S		**32.059-32.076**							
^{26}S		26.029	< 0.08 μs			0+			
^{27}S		27.018	16. ms	β+, p//4	p/2.26, 7.80				
				β+, 2p/18.3/1					
^{28}S		28.004	0.13 s			0+			
^{29}S		28.9966	0.188 s	β+ /13.79		5/2+			ann.rad./
				β+, p/					
^{30}S		29.984907	1.176 s	β+ /6.138	4.42/78.	0+			ann.rad./
					5.08/20.				0.678/79.
^{31}S		30.979557	2.553 s	β+ /5.396	4.39/99.	1/2+	0.48793		ann.rad./
									1.2662(5)/1.2
^{32}S	(94.41-95.29)	31.97207117				0+			
^{33}S	(0.729-0.797)	32.97145891				3/2+	+0.64382	-0.068	
^{34}S	(3.96 -4.77)	33.967867				0+			
^{35}S		34.9690323	87.2 d	β- /0.1672	0.1674/100.	3/2+	+1.00	+0.047	
^{36}S	(0.0129 - 0.0187)	35.967081				0+			
^{37}S		36.971126	5.05 m	β- /4.8653	1.64/94.	7/2-			0.9083(4)/0.06
					4.75/5.6				3.1033(2)/94.2
^{38}S		37.97116	2.84 h	β- /2.94	1.00/	0+			0.1962(4)/0.2
									1.9421(3)/84.
^{39}S		38.9751	11.5 s	β- /6.64					1.301/52.
									1.697/44.
^{40}S		39.97548	9. s	β- /4.7		0+			(0.2116-1.874)
^{41}S		40.97959	~ 2.6 s	β- /8.7					
				β-, n					
^{42}S		41.98107	1.03 s	β- /7.8		0+			(0.118-2.912)
				β-, n	/< 4.				
^{43m}S			~ 0.2 μs	IT					
^{43}S		42.98691	0.26 s	β-/12.					
				β-, n	/~ 40				
^{44m}S			2.62 μs	IT					
^{44}S		43.99012	0.10 s	β-/9.		0+			
				β-, n	/18.				
^{45}S		44.996	68. ms	β-/14.					
				β-, n	/54.				
^{46}S		46.000	0.05 s			0+			
^{47}S		47.008	> 0.2 μs						
^{48}S		48.014	> 0.2 μs			0+			
^{49}S		49.023	< 0.2 μs						
$_{17}$Cl		**35.446-35.457**							
^{28}Cl		28.030							
^{29}Cl		29.015	< 0.02 μs						
^{30}Cl		30.005	< 0.03 μs						
^{31}Cl		30.9924	0.15 s	β+, p/11.98	0.986, 1.52/0.7	3/2+			ann.rad./
									2.234
					0.762-2.75				1.249-4.045
^{32}Cl		31.985685	297. ms	β+ /12.69	9.47/50.	1+	1.11		ann.rad./
				β+, α	/0.05				2.2305/92
				β+, p	/0.026				(1.55–4.77)
^{33}Cl		32.977452	2.506 s	β+ /5.583	4.51/98.	3/2+	+0.752		ann.rad./
									0.8409/0.52
									1.966/0.45
									2.866/0.44
^{34m}Cl			32.2 m	β+/	1.35/24.	3+			ann.rad./
					2.47/28.				
				IT/					0.1457(8)/42.
									2.1276(5)/42.
^{34}Cl		33.9737625	1.528 s	β+ /5.4922	4.50/100.	0+			ann.rad./
^{35}Cl	(75.5 - 76.1)	34.9688527				3/2+	+0.82187	-0.082	
^{36}Cl		35.9683068	3.01 ×10^{5} a	β- /0.7086	0.7093/98.	2+	+1.28547	-0.018	
				β+, ec/1.1421	0.115/0.002				ann.rad./
^{37}Cl	(23.9 - 24.5)	36.9659026				3/2+	+0.68412	-0.0646	

Elem. or Isot.	Natural abundance (Atom %)	Atomic mass or weight	Half-life/ Resonance width (MeV)	Decay mode/ Energy (/MeV)	Particle energy/ Intensity (MeV/%)	Spin (h/2π)	Nuclear magnetic mom. (nm)	Elect. quadr. mom. (b)	γ-Energy/ Intensity (MeV/%)
^{38m}Cl			0.715 s	IT/		5-			0.6714/100
^{38}Cl		37.968010	37.2 m	β- /4.9168	1.11/31.	2-	2.05		1.64267/33.3
					2.77/11.				2.16740/44.8
					4.91/58.				3.80998/0.03
^{39}Cl		38.96801	55.6 m	β- /3.442	1.91/85.	3/2+			0.25026(1)/47.
					2.18/8.				1.26720(5)/54.
					3.45/7.				0.986–1.517
^{40}Cl		39.9704	1.38 m	β- /7.48		2-			0.6431(3)/6.
									1.4608(1)/77.
									2.8402(2)/17.
^{41}Cl		40.9707	34. s	β- /5.7	3.8/				(0.167–1.359)
^{42}Cl		41.973	6.8 s	β- /9.4					
^{43}Cl		42.974	3.3 s	β- /8.0					(0.256-1.338)
^{44}Cl		43.978	~ 0.43 s	β-/12.3					0.518
				β-, n	/ < 8.				(0.107-0.996)
^{45}Cl		44.980	0.40 s	β-/11.					0.930
				β-, n	/24.				(0.130-1.619)
^{46}Cl		45.985	0.23 s	β-/14.9					0.151
				β-, n	/~ 60				(0.118-1.024)
^{47}Cl		46.989	0.10 s	β-/15.					
				β-, n	/ < 3.				
^{48}Cl		47.996	> 0.2 μs						
^{49}Cl		48.001	> 0.17 s						
^{50}Cl		50.009	> 0.62 μs						
^{51}Cl		51.016	> 0.2 μs						
$_{18}$Ar		**39.948(1)**							
^{30}Ar		30.023	< 0.001 μs			0+			
^{31}Ar		31.012	15 ms	β+ //18.4		5/2+			
				β+, p//65	p/2.08				
				β+, 2p//12					
				β+, 3p//4					
^{32}Ar		31.99764	98. ms	β+ /11.2		0+			ann.rad./
				β+, p	3.98-6.40/22.7				
^{33}Ar		32.989926	174. ms	β+ /11.62	3.17,2.10	1/2+	-0.72		ann.rad./
				β+, p/	(1.32–5.72)				0.810(2)/48.
^{34}Ar		33.9802701	0.844 s	β+/6.061	5.0/95.	0+			ann.rad./
									0.6658(1)/2.5
									3.1290(1)/1.3
^{35}Ar		34.975258	1.77 s	β+/5.965	4.94/93.	3/2+	+0.6322	-0.08	ann.rad./
									1.2185(5)/1.22
									1.763(1)/0.25
									2.964(1)/0.2
^{36}Ar	0.3336(210)	35.9675451				0+			
^{37}Ar		36.966776	35.0 d	ec/.813		3/2+	+1.15	+0.08	
^{38}Ar	0.0629(70)	37.962732				0+			
^{39}Ar		38.96431	268. a	β-/0.565	0.565/100.	7/2-	-1.59	-0.12	
^{40}Ar	99.6035(250)	39.96238312				0+			
^{41}Ar		40.964501	1.82 h	β-/2.492	1.198/	7/2-		-0.042	1.29364(5)/99.
									1.6770(3)/0.05
^{42}Ar		41.96305	33. a	β-/0.60	0.60/100.	0+			
^{43}Ar		42.96564	5.4 m	β-/4.6				+0.14	0.4791(2)/10.
									0.7380(1)/43.
									0.9752(1)/100.
									1.4400(3)/39.
^{44}Ar		43.96492	11.87 m	β-/3.55		0+			0.182–1.866
^{45}Ar		44.968040	21.5 s	β-/6.9		7/2-			0.0610/25.
									1.020/35.
									3.707/34.
^{46}Ar		45.9681	8.4 s	β-/5.70		0+			1.944/
^{47}Ar		46.973	1.23 s	β-					0.36/100
									1.66/53
									1.74/41
									(2.02 - 4.01)
^{48}Ar		47.976	0.48 s			0+			
^{49}Ar		48.982	0.17 s	β-, n	n//~ 65.				
^{50}Ar		49.986	~ 0.085 s	β-, n	n//~ 35.	0+			
^{51}Ar		50.994	> 0.2 μs	β-					
^{52}Ar		51.999	10 ms	β-		0+			
^{53}Ar		53.007	> 0.62 μs	β-					
$_{19}$K		**39.0983(1)**							
^{32}K		32.023							
^{33}K		33.008	< 0.025 μs						
^{34}K		33.999	< 0.04 μs						

Elem. or Isot.	Natural abundance (Atom %)	Atomic mass or weight	Half-life/ Resonance width (MeV)	Decay mode/ Energy (/MeV)	Particle energy/ Intensity (MeV/%)	Spin (h/2π)	Nuclear magnetic mom. (nm)	Elect. quadr. mom. (b)	γ-Energy/ Intensity (MeV/%)
^{35}K		34.98805	0.19 s	β+ /11.88		3/2+	0.392		ann.rad./
				β+, p/	/0.37				1.751/14.
									2.5698/26.
									2.9827/51.
^{36}K		35.981302	0.342 s	β+ /12.81	5.3/42.	2+	+0.548		ann.rad./
					9.9/44.				1.97044(5)/82.
				β+, p	/0.048				2.20783(5)/30.
									2.43343(2)/32.
^{37}K		36.973376	1.24 s	β+ /6.149	5.13/	3/2+	+0.2032	0.106	ann.rad./
									2.7944(8)/2.
									3.602(2)/0.05
^{38m2}K			22.0 μs	IT					
^{38m}K			924.5 ms	β+ /6.742	5.02/100.	0+			ann.rad./
				IT	/0.033				
				β-	/<0.001				
^{38}K		37.969081	7.63 m	β+ /5.913	2.60/99.8	3+	+1.37		ann.rad./
									2.1675(3)/99.8
									3.9356(5)/0.2
^{39}K	93.2581(44)	38.96370649				3/2+	+0.39146	+0.059	
^{40}K	0.0117(1)	39.9639982	1.248 ×10^{9} a	β- /1.3111	1.312/89.	4-	-1.29810	-0.073	ann.rad./
				β+, ec/1.505	1.50/10.7				1.4608/10.5
^{41}K	6.7302(44)	40.96182526				3/2+	+0.21487	+0.071	
^{42}K		41.962402	12.36 h	β- /3.525	1.97/19.	2-	-1.1425		0.31260(2)/0.3
					3.523/81.				1.5246(3)/18.1
^{43}K		42.960735	22.3 h	β- /1.82	0.465/8.	3/2+	+0.163		0.2211(2)/4.
					0.825/87.				0.3729(2)/88.
					1.24/3.5				0.3971(2)/11.
					1.814/1.3				0.6178(2)/81.
^{44}K		43.961587	22.1 m	β- /5.66	5.66/34.	2-	-0.856		0.36821/2.2
									1.15700(1)/58.
									2.15079(2)/22.
^{45}K		44.960692	17.8 m	β- /4.20	1.1/23.	3/2+	+0.173		0.1743(5)/80.
					2.1/69.				1.2607(8)/7.
					4.0/8.				1.7056(6)/69.
									2.3542(5)/14.
^{46}K		45.961982	1.8 m	β- /7.72	6.3/	2-	-1.05		1.347(1)/91.
									3.700(5)/28.
^{47}K		46.96166	17.5 s	β- /6.64	4.1/99.	½+	+1.93		0.56474(3)/15.
					6.0/1.				0.58575(3)/85.
									2.0131/100
^{48}K		47.965341	6.8 s	β- /12.09	5.0/	(2-)			0.67122(1)/4.
									0.6723(5)/20.
				β-					0.78016(1)/32.
									3.83153(7)/80.
^{49}K		48.96821	1.26 s	β- /11.0		½	+ 1.34		2.025/
									2.252/
^{50}K		49.97238	0.472 s	β- /14.2					
^{51}K		50.9758	0.365 s	β- /			+ 0.513		1.027/21.7
				β-, n	//68(10)				3.46/3.9
					/2.23/0.218				(1.976-4.035)
^{52}K		51.982	0.105 s	β-	//79(12)				2.563/25.2
				β-, n	/1.040/0.216				2.377/6.9
				β-, 2n					1.027/0.55
^{53}K		52.987	30. ms	β-		3/2+			2.22/15.3
				β-, n	//85(19)				2.56/51.5
^{54}K		53.995	10. ms	β-					
^{55}K		55.01	> 0.62 μs						
^{56}K		56.009	> 0.62 μs						
$_{20}$Ca		**40.078(4)**							
^{34}Ca		34.015	< 0.035 μs			0+			
^{35}Ca		35.005	25.7 ms	β+, p/15.6/96	p/1.43/49				
				β+, 2p//4	1.9–8.8				
^{36}Ca		35.9931	0.100 s	β+, (p)/10.99	p/2.61/32	0+			ann.rad./
				β+, n/	p/1.71/7				
^{37}Ca		36.985898	0.182 s	β+ /11.64	p/3.19/40.7	3/2+			ann.rad./
				β+, n/	p/(0.899-2.00)				1.369
^{38}Ca		37.976319	443.6 ms	β+ /6.74		0+			ann.rad./
									1.5677(5)/25.
									3.210(2)/1.
^{39}Ca		38.970711	0.861 s	β+ /6.531	5.49/100.	3/2+	1.02168	0.04	ann.rad./
^{40}Ca	96.941(156)	39.9625909	> 9.9 ×10^{21} a	ec-ec		0+			
^{41}Ca		40.962278	9.9 ×10^{4} a	ec/0.4214		7/2-	-1.5948	-0.067	
^{42}Ca	0.647(23)	41.958618				0+			

Elem. or Isot.	Natural abundance (Atom %)	Atomic mass or weight	Half-life/ Resonance width (MeV)	Decay mode/ Energy (/MeV)	Particle energy/ Intensity (MeV/%)	Spin (h/2π)	Nuclear magnetic mom. (nm)	Elect. quadr. mom. (b)	γ-Energy/ Intensity (MeV/%)
^{43}Ca	0.135(10)	42.958766				7/2-	-1.31764	-0.055	
^{44}Ca	2.086(110)	43.955482				0+			
^{45}Ca		44.956186	162.7 d	β- /0.257	0.257/100.	7/2-	-1.327	+0.04	
^{46}Ca	0.004(3)	45.95369	>0.4 ×10^{16} a	β-β-		0+			
^{47}Ca		46.95454	4.536 d	β-/1.992	0.684/84.	7/2-	-1.406	+0.08	1.297/75
					1.98/16.				(0.041–1.88)
^{48}Ca	0.187(21)	47.95253	4.4 ×10^{19} a	β-β-		0+			
			>7.1 ×10^{19} a	β-					
^{49}Ca		48.955663	8.72 m	β- /5.262	0.89/7.	3/2-	-1.38	-0.04	3.0844(1)/90.7
					1.95/92.				4.0719(1)/8.12
									(0.143 - 4.738)
^{50}Ca		49.95750	14. s	β- /4.97	3.12/	0+			0.2569/98.
									(0.0715–1.59)
^{51}Ca		50.9610	10. s	β- /7.3		(3/2-)	-1.05	+0.04	
^{52}Ca		51.9632	4.6 s	β- /8.0		0+			
^{53}Ca		52.969	0.4 s	β- /10.9					2.11/56
^{54}Ca		53.973	0.1 s			0+			0.247/65
^{55}Ca		54.980	22 ms						
^{56}Ca		55.985	11 ms			0+			
^{57}Ca		56.993	> 0.62 μs						
^{58}Ca		57.998	> 0.62 μs						
$_{21}$Sc		**44.955908(5)**							
^{36}Sc		36.016	0.102 s						
^{37}Sc		37.004	0.181 s						
^{38}Sc		37.995	< 0.3 μs						
^{39}Sc		38.9848	< 0.3 μs	p					
^{40}Sc		39.97797	0.182 s	β+ /14.320	5.73/50.	4-			ann.rad./
					7.53/15.				0.752/41.
					8.76/15.				3.732/99.5
					9.58/20.				(1.12–3.92)
^{41}Sc		40.969251	0.596 s	β+ /6.4953	5.61/100.	7/2-	+5.431	-0.145	ann.rad./
^{42m}Sc			61.6 s	β+ /	2.82/	7+			ann.rad./
									0.4375(5)/100.
									1.2270(5)/100.
									1.5245(5)/100.
^{42}Sc		41.965517	0.682 s	β+ /6.4259	5.32/100.	0+			ann.rad./
^{43m}Sc			0.44 ms	IT					0.151/100
^{43}Sc		42.96115	3.89 h	β+, ec/2.221	0.82/22.	7/2-	+4.53	-0.27	ann.rad./
					1.22/78.				0.3729(1)/22.
^{44m}Sc			58.2 h	IT/0.27		6+	+3.83	0.21	0.27124(1)/87.
				ec/3.926					(1.00–1.16)
^{44m1}Sc			0.051 ms	IT/0.0783					0.1462
^{44}Sc		43.95940	4.04 h	β+, ec/3.653	1.47/	2+	+2.50	+0.10	ann.rad./
									1.157/100
^{45m}Sc			0.032 s	IT/100				+0.28	0.0124
^{45}Sc	100.	44.955908				7/2-	+4.75649	-0.220	
^{46m2}Sc			18.7 s	IT	0.14253/100	1-			
^{46m1}Sc			~ 9.4 μs	IT	0.05201/100	6+			
^{46}Sc		45.955168	83.81 d	β- /2.367	0.357/100.	4+	+3.04	+0.12	0.8893/100
									1.121/100
^{47}Sc		46.95240	3.349 d	β- /0.600	0.439/69.	7/2-	+5.34	-0.22	0.15938(1)/68.
					0.601/31.				
^{48}Sc		47.95222	43.7 h	β- /3.99	0.655/	6+	3.72		0.9835/100
									1.03750(1)/97.
									1.3121/100
^{49}Sc		48.95001	57.3 m	β- /2.006	2.00/99.9.	7/2-	+ 5.62		1.7619(3)/0.05
^{50m}Sc			0.35 s	IT	0.2569				
^{50}Sc		49.9522	1.71 m	β- /6.89	3.05/76.	(5+)			0.5235(1)/88.
					3.60/24.				1.1210(1)/100.
									1.5537(2)/100.
^{51}Sc		50.9536	12.4 s	β- /6.51	4.4/	7/2-			1.4373(4)/52.
					5.0/				0.718–2.144
^{52}Sc		51.957	8.2 s	β- /9.0		(3+)			
^{53}Sc		52.959	> 3. ms	β- /8.1					
^{54m}Sc			2.8 μs	IT		(5+)			0.110/IT
^{54}Sc		53.964	0.53 s	β- /11.6					0.100/50
									1.70/40
									0.50/40
^{55}Sc		54.968	0.103 s	β- /13					0.593(1)/40
^{56m}Sc			0.08 s	β-					1.161/21
									0.690/19
^{56}Sc		55.973	35. ms	β-		(1+)			1.129/48
^{57}Sc		56.978	13. ms	β-					

Elem. or Isot.	Natural abundance (Atom %)	Atomic mass or weight	Half-life/ Resonance width (MeV)	Decay mode/ Energy (/MeV)	Particle energy/ Intensity (MeV/%)	Spin (h/2π)	Nuclear magnetic mom. (nm)	Elect. quadr. mom. (b)	γ-Energy/ Intensity (MeV/%)
^{58}Sc		57.984	12. ms	β-					
^{59}Sc		58.989	> 0.62 μs						
^{60}Sc		59.996	> 0.62 μs						
^{61}Sc		61.001	> 0.62 μs						
$_{22}$**Ti**		**47.867(1)**							
^{38}Ti		38.011	< 0.12 μs			0+			
^{39}Ti		39.002	29. ms	β+, 15.4/90	p//94				
				β+, 2p//10					
^{40}Ti		39.991	52. ms	β+ /11.7	p/2.16/29	0+			2.467/8.5
				β+, p	3.73/23				
					1.70/22				
					0.242–5.74				
^{41}Ti		40.9832	83. ms	β+, p/12.93	p/4.73/107	3/2+			ann.rad./
					3.10/67				
					3.75/39				
					0.744–6.73				
^{42}Ti		41.973049	0.208 s	β+ /7.000	6.0/48	0+			ann.rad./
									0.6107(5)/56.
^{43m}Ti			0.012 ms	IT	0.3127/100				
^{43}Ti		42.96852	0.50 s	β+ /6.87	5.80/	7/2-	0.85		ann.rad./
^{44}Ti		43.959690	59. a	ec/0.268		0+			0.06787/91
									0.07832/97
^{45m}Ti			3.0 μs	IT	0.037/100				
^{45}Ti		44.95812	3.078 h	β+/86/2.062	1.04	7/2-	0.095	~ 0.015	ann.rad./
				ec/14/					(0.36–1.66)
^{46}Ti	8.25(3)	45.952628				0+			
^{47}Ti	7.44(2)	46.951759				5/2-	-0.78848	+0.30	
^{48}Ti	73.72(3)	47.947942				0+			
^{49}Ti	5.41(2)	48.947866				7/2-	-1.10417	+0.25	
^{50}Ti	5.18(2)	49.944787				0+			
^{51}Ti		50.946611	5.76 m	β- /2.471	1.50/92.	3/2-			0.3197(2)/93.
					2.13/				0.6094–0.9291
^{52}Ti		51.94689	1.7 m	β- /1.97	1.8/100.	0+			0.0170(5)/100.
									0.1245/100
^{53}Ti		52.950	33. s	β- /5.0	(2.2–3)/	3/2-			0.1008(1)/20.
									0.1276(1)/45.
									0.2284(1)/39.
									1.6755(5)/45.
									(1.72–2.8)/
^{54}Ti		53.951	1.5 s	β- /4.3		0+			
^{55}Ti		54.955	1.3 s	β- /7.4					0.672/44
									(0.32 - 1.83)
^{56}Ti		55.958	0.20 s	β- /7.0		0+			
^{57}Ti		56.964	98. ms	β- /11.					
^{58}Ti		57.967	58. ms	β-		0+			0.114
^{59}Ti		58.972	29. ms	β-					
^{60}Ti		59.976	22. ms	β-		0+			
^{61}Ti		60.982	15. ms						
^{62}Ti		61.987	> 0.62 μs						
^{63}Ti		62.994	> 0.62 μs						
^{64}Ti									
$_{23}$**V**		**50.9415(1)**							
^{40}V		40.013							
^{41}V		41.000							
^{42}V		41.992	< 0.055 μs						
^{43}V		42.9808	79. ms	β+ /11.3	p// < 2.5				
^{44m}V			0.15 s						
^{44}V		43.974	0.09 s	β+, α/13.7					ann.rad./
^{45}V		44.9658	0.54 s	β+ /7.13		7/2-			
^{46m}V			1.0 ms	IT	0.8015/100				
^{46}V		45.960199	0.4227 s	β+ /7.051	6.03/100.	0+			ann.rad./
^{47}V		46.954905	32.6 m	β+, ec/2.928	1.90/99.+	3/2-			ann.rad./
									1.7949(8)/0.19
									(0.2–2.16)
^{48}V		47.95225	15.98 d	β+ /4.012	0.698/50.	4+	2.01		ann.rad./
									0.9835/100
									(1.3–2.4)
^{49}V		48.94851	337. d	ec/0.602		7/2-	4.47		
^{50}V	0.250(10)	49.94716	2.0 ×10^{17} a	ec	/ > 93	6+	+3.34569	+0.21	
			> 15 ×10^{17} a	β-	/ < 7				
^{51}V	99.750(10)	50.94396				7/2-	+5.148706	-0.04	
^{52}V		51.94477	3.76 m	β- /3.976	2.47/	3+			1.4341(1)/100.
^{53}V		52.94434	1.56 m	β- /3.436	2.52/	7/2-			1.0060(5)/90.

Elem. or Isot.	Natural abundance (Atom %)	Atomic mass or weight	Half-life/ Resonance width (MeV)	Decay mode/ Energy (/MeV)	Particle energy/ Intensity (MeV/%)	Spin (h/2π)	Nuclear magnetic mom. (nm)	Elect. quadr. mom. (b)	γ-Energy/ Intensity (MeV/%)
									1.2891(3)/10.
^{54m}V			0.9 μs	IT		(5+)			0.108/IT
^{54}V		53.9464	49.8 s	β- /7.04	1.00/5.	3+			0.8348/97.
					2.00/12.				0.9887/80.
					2.95/45.				2.259/46.
					5.20/11.				(0.56–3.38)
^{55}V		54.947	6.5 s	β- /6.0	6.0/	(7/2-)			0.5177/73.
									(0.224–1.21)
^{56}V		55.950	0.22 s	β- /9.1					1.01/30.
									0.688/26.
									(0.82 – 1.32)
^{57}V		56.953	0.35 s	β- /8.1					0.268/52.
									0.692/20.
									(0.25 – 1.31)
^{58}V		57.957	0.19 s	β- /11.6					0.880/62
									1.056/28
									2.217/13
									(1.04 – 1.57)
^{59}V		58.959	97. ms	β- /9.9					0.90/80.
^{60m}V			0.12 s						
^{60}V		59.964	0.07 s	β- /14.					0.102–0.208
^{61}V		60.97	48 ms	β- /12					(0.071-1.144)
				β-, n// > 10					
^{62}V		61.973	34. ms	β-					
^{63}V		62.976	20. ms	β-					
				β-, n//29					
^{64}V		63.983	15 ms	β-					0.081
^{65}V		64.988	> 0.62 μs						
^{66}V		65.994	> 0.62 μs						
^{67}V									
$_{24}$Cr		**51.9961(6)**							
^{42}Cr		42.007	13. ms	β+, p	p/1.90/29	0+			1.623/35
					p/1.50–3.7				
^{43}Cr		42.998	21. ms	β+	/12				1.555/35
				β+ ,p	/81				0.838/6.2
				β+ , 2p	/7.1				1.937/1.8
				β+, 3p	/0.08				
^{44}Cr		43.985	43. ms	β+, (p)/10.3	p/0.95–3.1	0+			0.677/59.
^{45m}Cr			> 0.08 ms	IT	0.0107/100				
^{45}Cr		44.9791	61. ms	β+, p/12.5	p/2.088/19.6	7/2-			ann.rad./
					p/(0.945-1.61)				(1.08-1.37)
^{46}Cr		45.9684	0.3 s	β+ /7.60		0+			ann.rad./
^{47}Cr		46.96290	0.51 s	β+ /7.45		3/2-			ann.rad./
^{48}Cr		47.95403	21.6 h	ec/1.66		0+			ann.rad./
									0.116(2)/95.
									0.305(10)/100.
^{49}Cr		48.95133	42.3 m	β+, ec/2.631	1.39/	5/2-	0.476		ann.rad./
					1.45/				0.09064(1)/51.
					1.54/				0.15293(1)/27.
									(0.062-1.6)
^{50}Cr	4.345(13)	49.94604	>1.3 ×10^{18} a	β+ec		0+			
^{51}Cr		50.94477	27.70 d	ec/0.7527		7/2-	-0.934		0.3201/9.8
									0.00543/2.6
									0.00495/0.02
^{52}Cr	83.789(18)	51.940606				0+			
^{53}Cr	9.501(17)	52.940648				3/2-	-0.47454	-~ 0.15	
^{54}Cr	2.365(7)	53.938879				0+		-0.2	
^{55}Cr		54.940838	3.497 m	β- /2.603	2.5/	3/2-			1.5282(2)/0.04
									(0.13–2.37)
^{56}Cr		55.94065	5.9 m	β- /1.62	1.50/100.	0+			0.026(2)/100.
									0.083(3)/100.
^{57}Cr		56.94361	21. s	β- /5.1	3.3/	3/2-	0.0834		0.850/8.
					3.5/				(0.083-2.62)
^{58}Cr		57.944	7.0 s	β- /4.0		0+			(0.131–0.683)
^{59m}Cr			0.10 ms	IT		(9/2+)			0.208/IT
									0.193
									0.102
^{59}Cr		58.949	1.0 s	β- /7.7					1.236
^{60}Cr		59.950	0.6 s	β- /6.0		0+			
^{61}Cr		60.954	0.23 s	β- /8.8					(0.157–2.378)
^{62}Cr		61.956	0.19 s	β- /7.3		0+			(0.156-1.215)
^{63}Cr		62.962	0.129 s	β-					(0.250-3.454)
^{64}Cr		63.964	0.043 s	β-		0+			0.188

Elem. or Isot.	Natural abundance (Atom %)	Atomic mass or weight	Half-life/ Resonance width (MeV)	Decay mode/ Energy (/MeV)	Particle energy/ Intensity (MeV/%)	Spin (h/2π)	Nuclear magnetic mom. (nm)	Elect. quadr. mom. (b)	γ-Energy/ Intensity (MeV/%)
^{65}Cr		64.970	0.027 s	β-					0.272, 1.368
^{66}Cr		65.974	23. ms	β-		0+			
^{67}Cr		66.980	> 0.3 μs						
^{68}Cr		67.984	> 0.62 μs						
$_{25}$Mn		**54.938044(3)**							
^{44}Mn		44.007	< 0.105 μs						
^{45}Mn		44.994	< 0.07 μs						
^{46}Mn		45.986	36. ms	β+ /17.1	p/3.00/6.5				0.329/11.
				β+, p	//~ 58				(0.0544-1.322)
^{47}Mn		46.9758	88. ms	β+ /12.3	p// <1.7				
^{48}Mn		47.969	0.15 s	β+ /13.5	5.79/58.	4+			
					4.43/10.				
^{49}Mn		48.9596	0.38 s	β+ /7.72	6.69/	5/2-			ann.rad./
^{50m}Mn			1.74 m	β+ /7.887	3.54/	5+	+2.76	+0.8	ann.rad./
									1.0980/94.
									0.783/91.
									(0.66–3.11)
^{50}Mn		49.95424	0.283 s	β+ /7.6330	6.61/	0+			ann.rad./
^{51}Mn		50.94821	46.2 m	β+, ec/3.208	2.2/	5/2-	3.568	0.4	ann.rad./
									0.7491(1)/0.26
									(1.148–1.164)
^{52m}Mn			21.1 m	β+ /98/5.09	2.631/	2+	0.0077		ann.rad./
				IT/2/0.378					0.3778 (I.T.)
									1.43406(1)/98.
									(0.7–4.8)
^{52}Mn		51.94556	5.591 d	β+ /4.712	0.575/	6+	+3.063	+0.5	ann.rad./
				ec/					0.74421(1)/90.
									1.4341/100
^{53}Mn		52.941289	3.7 ×10^{6} a	ec/0.5970		7/2-	5.035	+0.17	
^{54}Mn		53.94036	312.2 d	ec/1.377		3+	+3.306	+0.37	0.8340/100
			6.7 ×10^{8} a	β+	//1.3 ×10^{-7}				
^{55}Mn	100.	54.938044				5/2-	+3.4687	+0.33	
^{56}Mn		55.938904	2.579 h	β- /3.6954	0.718/18.	3+	+3.2266	+0.5	0.84675/98.9
					1.028/34.				1.81072(4)/26.3
									2.113/13.8
									(1.04 – 3.37)
^{57}Mn		56.93829	1.45 m	β- /2.691		5/2-	+3.485	+0.37	
^{58m}Mn			65 s	β- ,IT		4+	+3.064	+0.5	0.0718/~ 10
^{58}Mn		57.94007	3.0 s	β- /6.25	3.8/	1+	+2.624	+ ~ 0.3	0.45916(2)/20.
					5.1/				0.81076(1)/82.
									1.32309(5)/53.
^{59}Mn		58.94039	4.6 s	β- /5.19	4.5/	5/2-	+3.496	+0.34	0.726/
									0.473/
									0.287–2.35
^{60m}Mn			1.77 s	β- /IT	5.7/	4+	+3.402	+0.3	IT/0.272/~ 12
^{60}Mn		59.94314	0.28 s	β- /8.6	8.2/88.	0+	+2.49	~ +0.1	0.8234/12.2
					6.2/5.0				1.150/5.0
					7.4/4.2				1.523/3.0
^{61}Mn		60.94445	0.71 s	β- /7.4		(5/2)	+3.535	+0.36	(0.202-3.146)
^{62m}Mn		61.948	0.67 s	β- /10.4		(4+)	+3.17	+0.6	0.877/
									0.942–1.299
^{62}Mn		61.948	0.09 s			(+1)	+2.39	~ +0.1	
^{63}Mn		62.94967	0.28 s	β- /8.8			+3.441	+0.48	0.356,0.450
^{64m}Mn			0.40 ms						0.135/IT
^{64}Mn		63.95385	0.090s	β- /11.8			+2.086	+0.2	0.746
^{65}Mn		64.95602	0.092 s	β-, n	/7.9				(0.034-4.438)
^{66m}Mn			0.78 ms	IT	0.170/100				
^{66}Mn		65.9605	64 ms	β-					0.471
^{67}Mn		66.964	49 ms						
^{68}Mn		67.970	28 ms						
^{69}Mn		68.974	16 ms						
^{70}Mn		69.979	> 0.62 μs						
^{71}Mn		70.984	> 0.62 μs						
^{72}Mn									
$_{26}$Fe		**55.845(2)**							
^{45}Fe		45.014	2.6 ms	2p /1.14	p//~ 59.				
^{46}Fe		46.001	14. ms	β+, p /13.1/~ 20	p//79.	0+			0.493/23.
				β+, 2p//~ 20					
^{47}Fe		46.992	21.9 ms	β+ /15.6	p//87.				0.892/76.
^{48}Fe		47.980	51. ms	β+ /11.2	p//14	0+			0.313/63.
^{49}Fe		48.973	65. ms	β+ /13.0	p/1.977/34.5	(7/2-)			0.797/23.7
					p/(1.161-1.550)				(0.261-1.279)

Elem. or Isot.	Natural abundance (Atom %)	Atomic mass or weight	Half-life/ Resonance width (MeV)	Decay mode/ Energy (/MeV)	Particle energy/ Intensity (MeV/%)	Spin (h/2π)	Nuclear magnetic mom. (nm)	Elect. quadr. mom. (b)	γ-Energy/ Intensity (MeV/%)
^{50}Fe		49.9630	0.15 s	β+ /8.2		0+			0.651
^{51}Fe		50.9568	0.30 s	β+ /8.02		(5/2-)			ann.rad./
^{52m}Fe			46. s	β+ /4.4		(12+)			ann.rad./
									(0.622–2.286)/
^{52}Fe		51.94811	8.28 h	β+ /57/2.37	0.804/	0+			ann.rad./
				ec/43/					0.16868(1)/99.
				IT/					0.377 (I.T.)/
^{53m}Fe			2.6 m	IT/3.0407		19/2-			0.7011(1)/99.
									1.0115(1)/87.
									1.3281(1)/87.
									2.3396(1)/13.
^{53}Fe		52.94531	8.51 m	β+ /3.743	2.40/42.	7/2-			ann.rad./
					2.80/57.				0.3779(1)/42.
									(1.2–3.2)
^{54}Fe	5.845(105)	53.939610	>3.1 ×10^{22} a	ec-ec		0+			
^{55}Fe		54.938292	2.75 a	ec/0.2314		3/2-			Mn x-ray
^{56}Fe	91.754(106)	55.934936				0+			
^{57}Fe	2.119(29)	56.935393				½-	+0.0906	0.16	
^{58}Fe	0.282(12)	57.933274				0+			
^{59}Fe		58.934874	44.50 d	β- /1.565	0.273/48.	3/2-	-0.336		1.099/57
					0.475/51.				1.292/43.
									(0.14–1.48)
^{60}Fe		59.93407	2.6 ×10^{6} a	β- /0.237	0.184/100.	0+			0.0586/100
^{61m}Fe			0.25 μs	IT		(9/2+)			0.654/IT
									0.207
^{61}Fe		60.93675	6.0 m	β- /3.98	2.5/13.				1.205/44.
					2.63/54.				1.028/43.
					2.80/31.				(0.12–3.37)
^{62}Fe		61.93679	68. s	β- /2.53	2.5/100.	0+			0.5061(1)/100.
^{63}Fe		62.94027	6. s	β- /6.3		5/2-			0.995/
									(1.365–1.427)
^{64}Fe		63.94099	2.0 s	β- /4.9		0+			
^{65m}Fe			1.1 s						(0.413-2.996)
^{65}Fe		64.94501	0.8 s	β- /7.9					(0.128-1.997)
^{66}Fe		65.94625	0.44 s	β- /5.7		0+			0.471–1.425
^{67m}Fe			~ 0.06 ms	IT		(5/2-)			0.388/IT
^{67}Fe		66.951	0.40 s	β- /8.8					0.189/85
									2.089/14
^{68}Fe		67.953	0.19 s	β- /~ 7.6		0+			
^{69}Fe		68.958	0.11 s						
^{70}Fe		69.961	0.06 s			0+			
^{71}Fe		70.967	35 ms						
^{72}Fe		71.970	19 ms			0+			
^{73}Fe		72.976	> 0.4 μs						
^{74}Fe		73.98	> 0.4 μs						
$_{27}$Co		**58.933194(4)**							
^{47}Co		47.01							
^{48}Co		48.00							
^{49}Co		48.989	< 0.035 μs						
^{50}Co		49.981	39. ms	β+ /17.0	p/2.770/41.				0.2614/64.
					p/(1.874-2.296)				(0.482-1.308)
^{51}Co		50.9707	69. ms	β+ /12.8	p// < 3.8				
^{52m}Co			0.10 s	β+		2+			0.849-5.185
^{52}Co		51.964	0.112 s	β+ /14.0		6+			0.849–1.942
^{53m}Co			0.25 s	β+, p/		19/2-			ann.rad./
^{53}Co		52.9542	0.24 s	β+ /8.30		7/2-			ann.rad./
^{54m}Co			1.46 m	β+ /8.44	4.25/100.	7+			ann.rad./
									0.411(1)/99.
									1.130(1)/100.
									1.408(1)/100.
^{54}Co		53.948460	0.1932 s	β+ /8.2430	7.34/100.	0+			ann.rad./
^{55}Co		54.941997	17.53 h	β+ /3.4513	0.53/	7/2-	+4.822		ann.rad./
				ec/	1.03/				0.9312/75.
					1.50/				0.4772/20.
									(0.092–3.11)
^{56}Co		55.939839	77.3 d	β+/4.566	1.459/18.	4+	3.85	~ +0.25	ann.rad./
				ec/					0.8468/99.9
									1.2383/68.
									(0.26–3.61)
^{57}Co		56.936291	271.8 d	ec/0.8361		7/2-	+4.72	+0.5	0.12206/86
									(0.014–0.706)
^{58m2}Co			0.010 ms	IT/100	0.0283				0.053
^{58m}Co			9.1 h	IT/		5+			0.02489/0.035

Elem. or Isot.	Natural abundance (Atom %)	Atomic mass or weight	Half-life/ Resonance width (MeV)	Decay mode/ Energy (/MeV)	Particle energy/ Intensity (MeV/%)	Spin (h/2π)	Nuclear magnetic mom. (nm)	Elect. quadr. mom. (b)	γ-Energy/ Intensity (MeV/%)
^{58}Co		57.93575	70.88 d	β+ /2.307		2+	+4.04	+0.23	ann.rad./
				ec/					0.81076/99
^{59}Co	100.	58.933194				7/2-	+4.63	+0.42	
^{60m}Co			10.47 m	IT/99.8/0.059		2+	+4.4	~ +0.3	0.0586/2.0
				β- /0.2/1.56					
^{60}Co		59.933816	5.271 a	β- /2.824	0.315/99.7	5+	+3.799	+0.5	1.1732/100
									1.3325/100
^{61}Co		60.93248	1.650 h	β- /1.322	1.22/95.	7/2-			0.0674/86.
									0.842–0.909
^{62m}Co			13.9 m	β- /	0.88/25.	5+			1.1635(3)/70.
					2.88/75.				1.1730(3)/98.
									2.0039(3)/19.
^{62}Co		61.9341	1.50 m	β- /5.32	1.03/10.	2+			1.1292(3)/13.
					1.76/5.				1.1730(3)/83.
					2.9/20.				1.9851(1)/3.
					4.05/60.				2.3020(1)/19.
^{63}Co		62.9336	27.5 s	β- /3.67	3.6/	7/2-			0.08713(1)/49.
									0.9817(3)/2.6
									0.156–2.17
^{64}Co		63.9358	0.30 s	β- /7.31	7.0/	1+			
^{65}Co		64.9364	1.14 s	β- /5.96		(7/2)-			(0.063-1.273)
^{66m2}Co			> 0.1 ms	IT		(8-)			0.252/IT
									0.214
									0.175
^{66m1}Co			1.2 μs	IT		(5+)			0.175/IT
^{66}Co		65.9394	0.25 s	β- /10.0					(1.245–1.425)
^{67m}Co			0.05 s	IT/β-					0.4916
^{67}Co		66.94061	0.43 s	β- /8.4					0.694
									(0.189-2.769)
^{68m}Co			1.6 s	β-					
^{68}Co		67.944	0.19 s	β- /11.7					
^{69m}Co			~ 0.8 s	β-					
^{69}Co		68.946	0.23 s	β- /9.3					
^{70m}Co			0.5 s	β-					
^{70}Co		69.950	0.11 s	β- 13.					1.26/102
									0.97/100
									(0.45 – 0.92)
^{71}Co		70.952	0.09 s	β-					0.566/100
				β-,n	// > 3				(0.25 – 0.77)
^{72m}Co			0.2 s	β-					
^{72}Co		71.957	0.056 s	β-					1.096/100
				β-,n	// > 6				0.845
									(0.455 – 1.197)
^{73}Co		72.960	0.040 s	β-					0.524/100
				β-,n	// < 7.9				(0.24 – 0.76)
^{74}Co		73.965	0.031 s	β-					0.739
				β-,n	//~ 18				1.024
^{75}Co		74.969	0.027 s	β-,n	// < 16				
^{76m}Co			3.0 μs						0.192,0.446
^{76}Co		75.97	22 ms	β-					
^{77}Co			13 ms	β-					
$_{28}$Ni		**58.6934(4)**							
^{48}Ni		48.018	~ 2.1 ms	2p//0.7	p //~ 25	0+			
				β, p//0.3					
^{49}Ni		49.01	12. ms		p // ~83				0.965/82
^{50}Ni		49.99	17 ms	2p//14	p //70.	0+			0.063
				β+, p//86	p /1.97/14				0.090
^{51}Ni		50.99	24. ms	β+ /16.0	p/4.66/8.7				0.765/73
					p/1.08-5.66				1.087/29
					p//87.				1.546-1.743
^{52}Ni		51.975	42.8 ms	β+ /11.7	p//31.	0+			2.418/38
					p/1.35/9				0.142
^{53}Ni		52.9682	55. ms	β+, p/13.3	p//23.	7/2-			ann.rad./
					p/1.93/5.4				0.849/13
^{54}Ni		53.9579	0.11 s	β+ /8.80		0+			0.937
^{55}Ni		54.951331	0.20 s	β+ /8.70	7.66/	7/2-	0.98		ann.rad./
^{56}Ni		55.942129	6.08 d	ec/2.14		0+			0.15838/99
				β+ /<10^{-6}					0.81185(3)/87.
									0.2695–0.7500
^{57}Ni		56.939792	35.6 h	β+ /3.264	0.712/10.	3/2-	-0.798		ann.rad./
				ec/	0.849/76.				1.3776/78.
									(0.127–3.177)
^{58}Ni	68.0769(190)	57.935342	>4 ×10^{19} a	ec-ec		0+			

Elem. or Isot.	Natural abundance (Atom %)	Atomic mass or weight	Half-life/ Resonance width (MeV)	Decay mode/ Energy (/MeV)	Particle energy/ Intensity (MeV/%)	Spin (h/2π)	Nuclear magnetic mom. (nm)	Elect. quadr. mom. (b)	γ-Energy/ Intensity (MeV/%)
^{59}Ni		58.934346	~ 7.6 ×10^{4} a	ec/		3/2-			
^{60}Ni	26.2231(150)	59.930786				0+			
^{61}Ni	1.1399(13)	60.931056				3/2-	-0.75002	+0.16	
^{62}Ni	3.6345(40)	61.928345				0+			
^{63m}Ni			1.7 μs	IT	0.0871/100				
^{63}Ni		62.929670	99. a	β- /0.066945	0.065/	½-			
^{64}Ni	0.9256(19)	63.927967				0+			
^{65m}Ni			0.069 ms	IT	0.0634/100				
^{65}Ni		64.930085	2.517 h	β- /2.137	0.65/30.	5/2-	0.69		0.36627(3)/5.
					1.020/11.				1.11553(4)/16.
					2.140/58.				1.48184(5)/23.
^{66}Ni		65.92914	54.6 h	β- /0.23		0+			
^{67m}Ni			13.3 μs	IT		9/2+	0.56		0.313/IT
									0.694
^{67}Ni		66.93157	21. s	β- /3.56	3.8/	½-	+0.601		1.0722/100.
									1.6539/100.
									(0.10–1.98)
^{68m2}Ni			0.34 μs						0.511
^{68m1}Ni			0.86 ms	IT		(5-)			0.814/IT
									2.033
^{68}Ni		67.93187	29. s	β- /2.06		0+			
^{69m2}Ni			0.44 μs	IT		(17/2)			0.148/IT
									0.593
									1.959
^{69m1}Ni			3.5 s						
^{69}Ni		68.93561	11. s	β- /5.4					0.6807(3)/100.
									(0.207–1.213)
^{70m}Ni			0.23 μs	IT		(8+)			0.183/IT
									0.448
									0.970
									1.259
^{70}Ni		69.93643	6.0 s	β- /3.5		0+			
^{71m}Ni			2.3 s	β-					
^{71}Ni		70.94052	2.56 s	β- /6.9					
^{72}Ni		71.94179	1.6 s	β- /5.2		0+			
^{73}Ni		72.94621	0.84 s	β- /9.					
^{74}Ni		73.948	0.51 s	β- /7.		0+			
^{75}Ni		74.953	0.332 s	β-,n	/10				
^{76}Ni		75.955	0.235 s	β-,n	/14	0+			
^{77}Ni		76.961	0.159 s	β-,n	/~ 30				
^{78}Ni		77.96	0.122 s	β-		0+			
^{79}Ni		78.97	~ 0.043 s	β-					
^{80}Ni			~ 0.02 s	β-					
$_{29}$Cu		**63.546(3)**							
^{52}Cu		51.997							
^{53}Cu		52.985	< 0.13 μs						
^{54}Cu		53.977	< 0.075 μs						
^{55}Cu		54.966	~ 27. ms	β+ /13.2	p//15.				
^{56}Cu		55.959	93. ms	β+ /15.3					0.511/233
									2.700/100
									0.9507–3.287
^{57}Cu		56.949213	196. ms	β+ /8.77		3/2-	+ 2.58		0.77–3.01
^{58}Cu		57.944533	3.21 s	β+ /8.563	4.5/15.	1+	+ 0.48	- 0.16	ann.rad./
				ec/	7.439/83.				0.0403(4)/5.
									1.4483(2)/11.
									1.4546(2)/16.
^{59}Cu		58.939498	1.36 m	β+ /4.800	1.9/	3/2-	+ 1.91	- 0.20	ann.rad./
					3.75/				0.3393(1)/8.
									0.8780(1)/12.
									1.3015(1)/15.
									(0.4–2.6)
^{60}Cu		59.93736	23.7 m	β+ /6.127	2.00/69.	2+	+1.219	+ 0.12	ann.rad./
				ec/	3.00/18.				1.3325/88.
					3.92/6.				1.7915/45.
									(0.12–5.048)
^{61}Cu		60.93346	3.32 h	β+ /2.237	0.56/3.	3/2-	+2.109	- 0.22	ann.rad./
					0.94/5.				0.2830/13.
					1.15/2.				0.6560/11.
					1.220/51.				(0.067–2.123)
^{62}Cu		61.932595	9.67 m	β+ /98/3.948	2.93/98.	1+	-0.381	- 0.022	ann.rad./
				ec/					1.17302(1)/0.6
									(0.87–3.37)
^{63}Cu	69.15(15)	62.929598				3/2-	+2.2273	-0.22	

Elem. or Isot.	Natural abundance (Atom %)	Atomic mass or weight	Half-life/ Resonance width (MeV)	Decay mode/ Energy (/MeV)	Particle energy/ Intensity (MeV/%)	Spin (h/2π)	Nuclear magnetic mom. (nm)	Elect. quadr. mom. (b)	γ-Energy/ Intensity (MeV/%)
^{64}Cu		63.929764	12.700 h	β- /0.579/38.5	0.578/	1+	-0.2164	+ 0.08	ann.rad./35.1
				β+ /1.6751/17.6	0.65/				1.3459(3)/0.475
				ec//43.5					
^{65}Cu	30.85(15)	64.927790				3/2-	+2.3817	-0.20	
^{66}Cu		65.928869	5.09 m	β- /2.642	1.65/6.	1+	-0.282	+ 0.06	0.8330(1)/0.22
					2.7/94.				1.0392(2)/9.2
^{67}Cu		66.92773	2.580 d	β- /0.58	0.395/56.	3/2-	+ 2.514	- 0.18	0.09125(1)/7.
					0.484/23.				0.09325(1)/17.
					0.577/20.				0.18453(1)/47.
^{68m}Cu			3.75 m	IT/86/		6-	+1.155	- 0.46	0.0843(5)/70.
				β- /14/1.8					0.1112(5)/18.
									0.5259(5)/74.
									(0.64–1.34)
^{68}Cu		67.92961	31. s	β- /4.46	3.5/40.	1+	+2.393	- 0.09	1.0774(5)/58.
					4.6/31.				1.2613(5)/17.
									(0.15–2.34)
^{69m}Cu			0.36 μs	IT		(13/2+)	+1.5		0.075/IT
									0.190/IT
									0.680
									1.871
^{69}Cu		68.92943	2.8 m	β- /2.68	2.48/80.	3/2-	+2.838	- 0.15	0.5307(3)/3.
									0.8340(5)/6.
									1.0065(8)/10.
^{70m2}Cu			6.6 s	β/93		1+	+1.778	- 0.12	0.8849/100
									1.072/19
				IT/7					0.141/ IT
^{70m1}Cu			33. s	β- /52	2.52/10.	3-	-3.364	- 0.14	0.8848(2)/100.
				IT/48	0.1011				0.9017(2)/90.
									1.2517(5)/60.
									(0.39–3.06)
^{70}Cu		69.93239	44.5 s	β- /6.60	5.42/54.	6-	+1.367	- 0.30	0.8848(2)/100.
					6.09/46.				0.9017/99.7
									(0.438 - 3.062)
^{71m}Cu			0.28 μs	IT		(19/2)			0.133/IT
									0.494
									0.939
									1.189
^{71}Cu		70.93268	20. s	β- /4.56		3/2-	+2.275	- 0.20	0.490/
^{72m}Cu			1.76 μs	IT		(6-)			0.051/IT
									0.082
									0.138
^{72}Cu		71.93582	6.6 s	β- /8.2		2 (-)	- 1.347	+ 0.08	0.652/
^{73}Cu		72.93667	4.2 s	β- /6.3	5.8/43	3/2 (-)	+ 1.743	- 0.21	0.450/100
					6.25/42				0.307–1.559
^{74}Cu		73.93988	1.59 s	β- /9.9		2 (-)	- 1.07	+ 0.27	
^{75}Cu		74.94152	1.2 s	β- /7.9		5/2-	+ 1.006	- 0.28	(0.109-4.895)
^{76m}Cu			1.3 s	β-					
^{76}Cu		75.94528	0.64 s	β- /11.	/88.				
				β-,n	/7.				
^{77}Cu		76.948	0.47 s	β- /~ 10.		5/2-	+ 1.6		0.5057
				β-,n	/30				(.1147-3827)
^{78}Cu		77.952	0.331 s	β- /12.	/35	(6-)			(0.576-0.908)
				β-,n	/65				0.1147
^{79}Cu		78.955	0.241 s	β- //28					
				β-,n//72	/7				
^{80}Cu		79.961	0.113 s						
^{81}Cu		80.966	~ 0.073 s						
^{82}Cu		81.972	> 0.4 μs						
$_{30}$**Zn**		**65.38(2)**							
^{54}Zn		53.992	1.6 ms	2p	p//92	0+			
^{55}Zn		54.966	20. ms		p//91.				
^{56}Zn		55.973	33 s	β+,p/88	(0.83-2.95)	0+			(0.309-1.835)
^{57}Zn		56.965	47. ms	β+, p/14.6		(7/2-)			ann.rad./
^{58}Zn		57.9546	0.09 s	β+		0+			
^{59}Zn		58.94931	183. ms	β+, p/9.09	8.1/	3/2-			ann.rad./
									(0.491–0.914)
^{60}Zn		59.941842	2.40 m	β+ /97/4.16		0+			ann.rad./
				ec/3/					0.669/47.
									(0.062–0.947)
^{61}Zn		60.9395	1.485 m	β+ /5.64	4.38/68.	3/2-			ann.rad./
									0.4748/17.
									(0.15–3.52)
^{62}Zn		61.934334	9.22 h	β+ /3/1.63	0.66/7.	0+			ann.rad./

Nuclear

Elem. or Isot.	Natural abundance (Atom %)	Atomic mass or weight	Half-life/ Resonance width (MeV)	Decay mode/ Energy (/MeV)	Particle energy/ Intensity (MeV/%)	Spin (h/2π)	Nuclear magnetic mom. (nm)	Elect. quadr. mom. (b)	γ-Energy/ Intensity (MeV/%)
				ec/93/					0.0408/25
									0.5967/26.
									(0.20–1.526)/
^{63}Zn		62.93321	38.5 m	β+ /93/3.367	1.02/	3/2-	-0.28164	+0.29	ann.rad./
				ec/7/	1.40/				0.66962(5)/8.4
					1.71/				0.96206(5)/6.6
					2.36/84.				(0.24–3.1)
^{64}Zn	49.17(75)	63.929142	>7. ×10^{20} a	ec-β+		0+			
^{65m}Zn			1.6 μs	IT	0.0539/100	(1/2)-			
^{65}Zn		64.929241	244. d	β+ /98/1.3514	0.325/	5/2-	+0.7690	-0.023	ann.rad./
				ec/1.5/					1.1155/50.2
^{66}Zn	27.73(98)	65.92603				0+			
^{67m}Zn			9.37 μs	IT	0.0933/100	1/2-			
^{67}Zn	4.04(16)	66.92713				5/2-	+0.8753	+0.150	
^{68}Zn	18.45(63)	67.92484				0+			
^{69m}Zn			13.76 h	IT/99+/0.439		9/2+	1.157	-0.5	0.4390(2)/95.
^{69}Zn		68.92655	56. m	β- /0.906	0.905/99.9	½-			0.318/
^{70}Zn	0.61(10)	69.92532	>2.3 ×10^{17} a	β-β-		0+			
^{71m}Zn			4.13 h	β- /	1.45/	9/2+	1.05		0.3864/93.
									0.4874/62.
									0.6203/57.
									(0.099–2.489)
^{71}Zn		70.92772	2.4 m	β- /2.81		½-			0.5116(1)/30.
									0.9103(1)/7.5
									(0.12–2.29)
^{72}Zn		71.92684	46.5 h	β- /0.46	0.25/14.	0+			0.0164(3)/8.
					0.30/86.				0.1447(1)/83.
									0.1915(2)/9.4
^{73m2}Zn			13.0 ms	IT	0.1955/100	(5/2+)			
^{73m}Zn			6. s			(9/2+)			0.042
^{73}Zn		72.92958	24. s	β- /4.29	4.7/	(1/2-)			0.216(1)/100.
									0.496–0.911
^{74}Zn		73.92941	1.60 m	β- /2.3	2.1/	0+			0.0565/
									0.1401/
									(0.05–0.35)
^{75}Zn		74.93284	10.2 s	β- /6.0					0.229/
^{76}Zn		75.93312	5.7 s	β- /4.2	3.6/	0+			0.119/
^{77m}Zn			1.0 s	β- /		(1/2-)			0.772
^{77}Zn		76.93689	2.1 s	β- /7.3	4.8/				0.189/
^{78m}Zn			0.3 μs	IT	0.148/100	(8+)			
^{78}Zn		77.93829	1.5 s	β- /6.4		0+			0.225/
^{79m}Zn			> 0.2 s			1/2+	-1.018		
^{79}Zn		78.94264	1.0 s	β- /8.6		9/2+	-1.187	+0.49	0.702/
				β-,n	/2				
^{80}Zn		79.94455	0.562 s	β- /7.3		0+			0.713/
				β-,n	/ <18				0.2248/
^{81}Zn		80.95040	0.303 s	β- /11.9					(0.351-4.880)
				β-,n//12	/~ 30				
^{82}Zn		81.954	0.178 s	β-,n//67		0+			(0.034-5.5327)
^{83}Zn		82.961	0.12 s						
^{84}Zn		83.965	> 0.4 μs						
^{85}Zn		84.972	> 0.4 μs						
$_{31}$Ga		**69.723(1)**							
^{56}Ga		55.995							
^{57}Ga		56.983							
^{58}Ga		57.975							
^{59}Ga		58.964	< 0.043 μs						
^{60}Ga		59.957	0.07 s	β+					1.004
				β+, p	//~ 1.6				3.848
				β+, α	//~ 0.02				1.555–2.559
^{61}Ga		60.9494	0.17 s	β+ /9.0		3/2-			0.088–1.362
^{62}Ga		61.944190	116.1 ms	β+ /9.17	8.3/99.89	0+			ann.rad./
				ec/					0.954/0.085
									(0.851-5.92)
^{63}Ga		62.93929	32. s	β+ /5.5	4.5/			+0.212	ann.rad./
				ec/					0.6271(2)/10.
									0.6370(2)/11.
									1.0652(4)/45.
^{64m}Ga			0.022 ms	IT	0.0429/100	(2+)			0.0429
^{64}Ga		63.93684	2.63 m	β+ /7.165	2.79/	0+			ann.rad./
					6.05/				0.80785(1)/14.
									0.99152(1)/43.

Elem. or Isot.	Natural abundance (Atom %)	Atomic mass or weight	Half-life/ Resonance width (MeV)	Decay mode/ Energy (/MeV)	Particle energy/ Intensity (MeV/%)	Spin (h/2π)	Nuclear magnetic mom. (nm)	Elect. quadr. mom. (b)	γ-Energy/ Intensity (MeV/%)
									1.38727(1)/12.
									3.3659(1)/13.
^{65}Ga		64.93273	15.2 m	β+ /86/3.255	0.82/10.	3/2-			ann.rad./
				ec/	1.39/19.				0.1151(2)/55.
					2.113/56.				0.1530(2)/96.
					2.237/15.				0.2069(2)/39.
									(0.06–2.4)
^{66}Ga		65.93159	9.31 h	β+ /56/5.175	0.74/1.	0+			ann.rad./
				ec/43/	1.84/54.				1.03935(8)/38.
					4.153/51.				2.7523(1)/23.
									(0.28–5.01)
^{67m}Ga			9.16 μs						(0.093-0.888)
^{67}Ga		66.92820	3.261 d	ec/1.001		3/2-	+1.8507	+ 0.197	0.09332/37.
									0.18459/20.
									0.30024/17.
									(0.091–0.89)
^{68}Ga		67.92798	1.1308 h	β+ /90/2.921	1.83/	1+	0.01175	0.028	ann.rad./
				ec/10/					1.0774(1)/3.
									(0.578-2.338)
^{69}Ga	60.108(50)	68.92557				3/2-	+2.01659	+0.171	
^{70}Ga		69.92602	21.1 m	ec/0.2/0.655		1+		+0.11	0.1755(5)/0.15
				β- /99.8/1.656	1.65/99.				1.042(5)/0.48
^{71}Ga	39.892(50)	70.92470	>2.4 ×10^{26} a	β-		3/2-	+2.56227	+0.107	
^{72m}Ga			39.7 ms	IT	0.1031/100				
^{72}Ga		71.92637	14.10 h	β- /4.001	0.64/40.	3-	-0.13224	+0.53	0.8340/95.53
					1.51/9.				2.202/26.9
					2.52/8.				0.630/26.2
					3.15/11.				(0.113–3.678)
^{73}Ga		72.92517	74.87 h	β- /1.59		1/2-	+ 0.209		0.05344(5)/10.
									0.29732(5)/47.
									(0.01–1.00)/
^{74m}Ga			10. s	IT/		1+			0.0565(1)/75.
^{74}Ga		73.92695	8.1 m	β- /5.4	2.6/	(3-)		+ 0.55	0.5959/92.
									2.354/45.
									(0.23–3.99)
^{75}Ga		74.92650	2.10 m	β- /3.39	3.3/	3/2-	+ 1.836	- 0.29	0.2529/
									0.5746/
									(0.12–2.10)
^{76}Ga		75.92883	29. s	β- /7.0		2-	- 0.95	+ 0.33	0.5629/66.
									0.5455/26.
									(0.34–4.25)
^{77}Ga		76.92915	13.0 s	β- /5.3	5.2/	3/2-	+ 2.020	- 0.21	0.469/
									0.459/
^{78}Ga		77.93161	5.09 s	β- /8.2		2-	- 1.22	+ 0.33	0.619/77.
									1.187/20.
^{79}Ga		78.93285	2.85 s	β- /7.0	4.6/	3/2-	+ 1.047	+ 0.16	0.465/
^{80m}Ga			1.3 s	β-		(3-)		+0.38	
^{80}Ga		79.93642	1.68 s	β- /10.4	10./	(6-)		+0.48	0.659/
				β-,n					
^{81}Ga		80.93813	1.22 s	β-/8.3	5.1/	5/2-	+ 1.747	- 0.05	0.217/
				β-,n	/ <21				
^{82}Ga		81.94318	0.60 s	β- /12.6		2-			1.348/
				β-,n	/22				
^{83}Ga		82.94712	0.308 s	β- /~ 11.5					(0.248-2.910)
^{84}Ga		83.952	~ 0.085 s	β- /14					(0.624-3.502)
				β-,n	/74				(0.248-3.238)
^{85}Ga		84.957	0.092 s						
^{86}Ga		85.963	0.04 s	β-//20					0.527
				β-,n//60					(0.107-0.365)
				β-,2n//20					0.624
^{87}Ga		86.968	> 0.4 μs						
$_{32}$Ge		**72.630(8)**							
^{58}Ge		57.992				0+			
^{59}Ge		58.982	0.013 s	β+					
^{60}Ge		59.970	0.02 s	β+, p		0+			
^{61}Ge		60.964	~ 44. ms	β+ /13.6					
^{62}Ge		61.955	0.08 s			0+			(0.571-2.414)
^{63}Ge		62.9496	0.149 s	β- /9.8					
^{64}Ge		63.94169	1.06 m	β+ /4.4	3.0/	0+			ann.rad./
				ec/					0.1282(2)/11.
				β+, p					0.4270(3)/37.
									0.6671(3)/17.
^{65}Ge		64.93937	31. s	β+ /6.2	0.82/10.				ann.rad./

Elem. or Isot.	Natural abundance (Atom %)	Atomic mass or weight	Half-life/ Resonance width (MeV)	Decay mode/ Energy (/MeV)	Particle energy/ Intensity (MeV/%)	Spin (h/2π)	Nuclear magnetic mom. (nm)	Elect. quadr. mom. (b)	γ-Energy/ Intensity (MeV/%)
				ec/	1.39/19.				0.0620/27.
				ec, p	2.113/56.				0.6497/33.
					2.237/15.				0.8091/21.
				β+, p	//0.011				(0.19–3.28)
^{66}Ge		65.93386	2.26 h	β+ /27/2.10		0+			ann.rad./
				ec/73/					0.0438/29.
									0.3819/28.
									(0.022–1.77)
^{67m}Ge			0.014 ms	IT	0.0182/100	5/2-			
^{67}Ge		66.93273	19.0 m	β+ /96/4.225	1.6/	½-			ann.rad./
				ec/4/	2.3/				0.1670/84.
					3.15/				(0.25–3.73)
^{68}Ge		67.92810	270.8 d	ec/0.11		0+			Ga k x-ray/39.
^{69m2}Ge			2.8 μs	IT	0.3980/100	9/2+		0.8	
^{69m1}Ge			5.1 μs	IT	0.0868/100	1/2-			
^{69}Ge		68.92796	1.63 d	β+ /36/2.2273	0.70/	5/2-	0.735	+0.03	ann.rad./
				ec/64/	1.2/				0.574/13.
									1.1068/36.
									(0.2–2.04)
^{70}Ge	20.52(19)	69.92425				0+			
^{71m}Ge			20.1 ms	IT	I.T./0.0234	9/2+	-1.041	~ 0.34	0.1749
^{71}Ge		70.92495	11.2 d	ec/0.229		½-	+0.547		
^{72}Ge	27.45(15)	71.9220758				0+			
^{73m2}Ge			0.50 s	IT	0.05344/100	1/2-			
^{73m1}Ge			2.92 μs	IT	0.01328/100	5/2+		0.7	
^{73}Ge	7.76(8)	72.9234590	>1.8 ×10^{23} a	β-		9/2+	-0.879468	-0.196	
^{74}Ge	36.52(12)	73.9211778				0+			
^{75m}Ge			48. s	IT/		7/2+			0.13968(3)/39.
^{75}Ge		74.9228584	1.380 h	β- /1.177	1.19/	½-	+0.510		0.26461(5)/11.
									0.41931(5)/0.2
^{76}Ge	7.75(12)	75.9214027	1.5 ×10^{21} a	β-β-		0+			
^{77m}Ge			54. s	IT/19/	0.1597	1/2-			1.605/0.22
				β- /81/2861	2.9/				1.676/0.16
									0.195–1.482
^{77}Ge		76.9235498	11.25 h	β- /2.702	0.71/23.	7/2+			0.2110/29.
					1.38/35.				0.2155/27.
					2.19/42.				0.2644/51.
									(0.15–2.35)
^{78}Ge		77.92285	1.45 h	β- /0.95	0.70/	0+			0.2773(5)/96.
									0.2939(5)/4.
^{79m}Ge			39. s	β- /IT		(7/2+)			
^{79}Ge		78.9254	19.1 s	β- /4.2	4.0/20.	½-			0.1096/21.
					4.3/80.				(0.10–2.59)
									0.5427(4)/15.
^{80}Ge		79.9253	29.5 s	β- /2.67	2.4/	0+			0.1104(4)/6.
									0.2656(4)/25.
^{81m}Ge			~ 7.6 s	β- /	3.75/	(1/2+)			0.3362(4)/
									0.7935(4)/
^{81}Ge		80.92883	~ 7.6 s	β- /6.2	3.44/	9/2+			0.1976(4)/21.
									0.3362(4)/100.
^{82}Ge		81.92977	4.0 s	β- /4.7	1.093/80	0+			(0.093-2.196)
^{83}Ge		82.93454	1.9 s	β- /8.9					
				β-,n	/63				
^{84}Ge		83.93758	0.95 s	β- /7.7		0+			
^{85}Ge		84.94297	0.49 s	β- /10.	n//0.16				
^{86}Ge		85.947	0.23 s	β-		0+			(0.098-0.441)
^{87}Ge		86.953	> 0.3 μs						
^{88}Ge		87.957	> 0.3 μs			0+			
^{89}Ge		88.964	> 0.3 μs						
^{90}Ge		89.969	> 0.4 μs						
$_{33}$**As**		**74.92159(1)**							
^{60}As		59.994							
^{61}As		60.981							
^{62}As		61.974							
^{63}As		62.964	< 0.043 μs						
^{64}As		63.957	~ 0.072 s						
^{65}As		64.959	0.13 s	β+ /9.4					
^{66m2}As			7.9 μs	IT	0.1144/100	9+			
^{66m1}As			1.15 μs	IT	0.1244/100	5+			
^{66}As		65.94415	95.8 ms	β+ /9.55					(0.259-1.946)
^{67}As		66.939251	42. s	β+ /6.0	5.0/	5/2-			0.121/
				ec/					0.123/
									0.244/

Elem. or Isot.	Natural abundance (Atom %)	Atomic mass or weight	Half-life/ Resonance width (MeV)	Decay mode/ Energy (/MeV)	Particle energy/ Intensity (MeV/%)	Spin (h/2π)	Nuclear magnetic mom. (nm)	Elect. quadr. mom. (b)	γ-Energy/ Intensity (MeV/%)
^{68}As		67.93677	2.53 m	β+ /8.1		3+			ann.rad./
									0.652/32.
									0.762/33.
									1.016/77.
									(0.61–3.55)
^{69}As		68.9323	15.2 m	β+ /98/4.01	2.95/	5/2-	+ 1.623		ann.rad./
				ec/2/					0.0868(5)/1.5
									0.1458(3)/2.4
^{70m}As			96 μs	IT	0.03205/100	2+			
^{70}As		69.9309	52.6 m	β+ /84/6.22	1.44/	4+	+2.1061	+0.09	ann.rad./
				ec/16/2.14					1.0395(7)/82.
				/2.89					(0.17–4.4)/
^{71}As		70.92711	2.72 d	β+ /32/2.013		5/2-	+1.6735	-0.02	ann.rad./
				ec/68/					0.1749(2)/84.
									1.0957(2)/4.2
^{72}As		71.92675	26.0 h	β+ /77/4.356	0.669/5.	2-	-2.1566	-0.08	ann.rad./
					1.884/12.				0.83395(5)/80.
					2.498/62.				1.0507(1)/9.6
					3.339/19.				(0.1–4.0)
^{73m}As			5.7 μs	IT		9/2+			
^{73}As		72.92383	80.3 d	ec/0.341		3/2-			0.0133/0.1
									0.0534/10.5
									Se k x-ray/90.
^{74}As		73.92393	17.78 d	β+ /31/2.562	0.94/26.	2-	-1.597		ann.rad./
				ec/37/	1.53/3.				0.59588(1)/60.
				β- /1.353	0.71/16.				0.6084(1)/0.6
					1.35/16.				0.6348(1)/15.
^{75m}As			0.0153 s	IT		9/2+			
^{75}As	100.	74.92159				3/2-	+1.43947	+0.31	
^{76m}As			1.8 μs	IT	0.044425/100	(1)+			
^{76}As		75.92239	26.3 h	β- /2.962	0.54/3.	2-	-0.903		0.5591(1)/45.
				Ec//<0.0002	1.785/8.				0.65703(5)/6.2
					2.410/36.				1.21602(1)/3.4
					2.97/51.				(0.3–2.67)
^{77m}As			0.114 ms	IT		9/2+			
^{77}As		76.92065	38.8 h	β- /0.683	0.70/98.	3/2-	+1.295		0.2391(2)/1.6
									0.2500(3)/0.4
									0.5208/0.43
^{78}As		77.9218	1.512 h	β- /4.21	3.00/12.	2-			0.6136(3)/54.
					3.70/17.				0.6954(3)/18.
					4.42/37.				1.3088(3)/10.
^{79m}As			1.18 μs	IT	0.5423/100	9/2+			0.542/IT
									0.231
^{79}As		78.92095	9.0 m	β- /2.28	1.80/95.	3/2-			0.0955(5)/16.
									0.3645(5)/1.9
^{80}As		79.92248	16. s	β- /5.64	3.38/	1+			0.6662(2)/42.
									(2.5–3.0)
^{81}As		80.92213	33. s	β- /3.856		3/2-			0.4676(2)/20.
									0.4911(2)/8.
^{82m}As			13.6 s	β- /	3.6/	5-			0.6544(1)/77.
									0.344/65.
									(0.561 – 1.894)
^{82}As		81.92474	19. s	β- /7.4	7.2/80.	(2-)			0.6544(1)/54.
									(0.755 – 3.667)
^{83}As		82.92521	13.4 s	β- /5.5					0.7345/100.
									1.1131/34.
									2.0767/28.
^{84m}As			0.6 s	β-					
^{84}As		83.92930	3.2 s	β-, n/7.2		1-			0.6671(2)/21.
									1.4439(5)/49.
									(0.325–5.150)
^{85}As		84.93216	2.03 s	β-, n/8.9		3/2-			0.667(1)/42.
									1.4551(2)/100.
^{86}As		85.93670	0.95 s	β-, n/11.4	n//0.36				(0.051-1.116)
^{87}As		86.94029	0.49 s	β-, n/10.					(0.092-1.399)
^{88}As		87.946	~ 0.2 s						
^{89}As		88.950	> 0.3 μs						
^{90}As		89.956	> 0.3 μs						
^{91}As		90.960	> 0.3 μs						
^{92}As		91.967	> 0.3 μs						
$_{34}$**Se**		**78.971(8)**							
^{63}Se			0.013 s	β+					

Nuclear

Elem. or Isot.	Natural abundance (Atom %)	Atomic mass or weight	Half-life/ Resonance width (MeV)	Decay mode/ Energy (/MeV)	Particle energy/ Intensity (MeV/%)	Spin (h/2π)	Nuclear magnetic mom. (nm)	Elect. quadr. mom. (b)	γ-Energy/ Intensity (MeV/%)
^{64}Se		63.971	> 0.18 μs			0+			
^{65}Se		64.964	0.011 s	β+ /60/14.					
				β+, p	3.55/				
^{66}Se		65.956	0.03 s			0+			
^{67}Se		66.9500	0.13 s	β+ /10.2					ann.rad./
				β+, (p)/					0.352
^{68}Se		67.94183	36. s	β+ /4.7		0+			ann.rad./
									(0.050–0.426)
^{69m}Se			2.0 μs	IT	0.0394/100	5/2-			
^{69}Se		68.93941	27.4 s	β+ /6.78	5.006/				ann.rad./
				ec/					0.0664(4)/27.
				β+, p	// ~ 0.045				0.0982(4)/63.
^{70}Se		69.93352	41.1 m	β+ /2.4		0+			ann.rad
									0.04951(5)/35.
									0.4262(2)/29.
^{71m2}Se			19.0 μs	IT	0.2605/100	9/2+			
^{71m1}Se			~ 5.6 μs	IT	0.0488/100	1/2-			
^{71}Se		70.93221	4.7 m	β+ /4.4	3.4/36.	5/2-			ann.rad
				ec/					0.1472(3)/47.
									0.8309(3)/13.
									1.0960(3)/10.
^{72}Se		71.92714	8.5 d	ec/0.34		0+			0.0460(2)/57.
^{73m}Se			40. m	IT/73/0.0257	0.85	3/2-			ann.rad.
				β+ /27/2.77	1.45/				0.0257(2)/27.
					1.70/				0.2538(1)/2.5
^{73}Se		72.92676	7.1 h	β+ /65/2.74	0.80/	9/2+	0.86		ann.rad
				ec/35/	1.32/95.				0.0670(1)/72.
					1.68/1.				0.3609(1)/97.
									(0.6–1.5)
^{74}Se	0.86(3)	73.9224759	>1.5×10^{19} a	ec-ec		0+			
^{75}Se		74.9225229	119.8 d	ec/0.864		5/2+	0.68	1.1	0.13600/55
									0.26465/58
									(0.024–0.821)
^{76}Se	9.23(7)	75.9192137				0+			
^{77m}Se			17.4 s	IT/		7/2+			0.1619(2)/52.
^{77}Se	7.60(7)	76.9199142				½-	+0.53506		
^{78}Se	23.69(22)	77.917309				0+			
^{79m}Se			3.92 m	IT	0.0957/100	1/2-			0.09573(3)/9.5
^{79}Se		78.918499	3.3 ×10^{5} a	β- /0.151		7/2+	-1.02	+0.8	
^{80}Se	49.80(36)	79.91652				0+			
^{81m}Se			57.3 m	IT	0.10297/99	7/2+			0.10297(1)
									0.2602(2)/0.06
									(0.276-0.767)
^{81}Se		80.91799	18.5 m	β- /1.585	1.6/98.	½-			0.2759/0.85
									0.2901/0.75
									(0.178-0.828)
^{82}Se	8.82(15)	81.91670	9. ×10^{19} a	β-β-		0+			
^{83m}Se			70.1 s	β- /3.96	2.88/	½-			0.35666(6)/17.
					3.92/				0.9879(1)/15.
									1.0305(1)/21.
									2.0514(2)/11.
									(0.19–3.1)
^{83}Se		82.91912	22.3 m	β- /3.668	0.93/	9/2+			0.22516(6)/33.
					1.51/				0.35666(6)/69.
									0.51004(8)/45.
									(0.21–2.42)
^{84}Se		83.91847	3.3 m	β- /1.83	1.41/100.	0+			0.4088(5)/100.
^{85}Se		84.92226	32. s	β- /6.18	5.9/	5/2+			0.3450(1)/22.
									0.6094(1)/41.
^{86}Se		85.92431	15. s	β- /5.10		0+			2.0124(1)/24.
									2.4433(8)/100.
									2.6619(1)/49.
^{87}Se		86.92869	5.4 s	β- /7.28		5/2+			0.468(1)/100.
				n/					1.4979(1)/23.
^{88}Se		87.93142	1.5 s	β-, n/6.85		0+			0.5346/
^{89}Se		88.93667	0.41 s	β-, n/9.0					
^{90}Se		89.940	0.20 s			0+			
^{91}Se		90.946	0.27 s	β-, n/8.					
^{92m}Se			10 μs	IT					(0.503-0.898)
^{92}Se		91.950	> 0.3 μs			0+			
^{93}Se		92.956	> 0.3 μs						
^{94}Se		93.960	> 0.3 μs			0+			
^{95}Se		94.967	> 0.4 μs						

Elem. or Isot.	Natural abundance (Atom %)	Atomic mass or weight	Half-life/ Resonance width (MeV)	Decay mode/ Energy (/MeV)	Particle energy/ Intensity (MeV/%)	Spin (h/2π)	Nuclear magnetic mom. (nm)	Elect. quadr. mom. (b)	γ-Energy/ Intensity (MeV/%)
$_{35}$Br		**79.901-79.907**							
^{67}Br		66.965							
^{68}Br		67.959	< 1.5 μs						
^{69}Br		68.9505	< 0.024 μs	β+ /9.6					
^{70m}Br			2.2 s	ec/β+		9+			
^{70}Br		69.9448	~ 0.08 s	β+ /10.0	/0.75				
^{71}Br		70.93934	21. s	β+ /6.9					
^{72m}Br			10.6 s	IT	0.1013/100				
^{72}Br		71.93659	1.31 m	β+ /8.7		3	0.6		0.4547–1.317
^{73}Br		72.93167	3.4 m	β+ /4.7	3.7/	3/2-			ann.rad
									0.065–0.700
^{74m}Br			46. m	β+ /	4.5/	4(+)	1.82		ann.rad
									0.6348
									0.7285
									(0.2–4.38)
^{74}Br		73.92991	25.4 m	β+ /6.91		4 (+)			ann.rad
									0.6341
									0.6348
									(0.2–4.7)
^{75}Br		74.92581	1.62 h	β+ /76/3.03		3/2-	+0.75		ann.rad
									0.28650
									(0.1–1.56)
^{76m}Br			1.31 s	IT	0.0571/99.4	4+			0.104548
									0.05711
^{76}Br		75.9245	16.0 h	β+ /57/4.96	1.9/	1-	0.54821	+0.255	ann.rad
					3.68/				0.55911
									1.85368
									(0.4–4.6)
^{77m}Br			4.3 m	IT/0.1059		9/2+			0.1059
^{77}Br		76.92138	2.376 d	ec/99/1.365		3/2-	0.973	+0.50	ann.rad.
									0.23898
									0.52069
									(0.08–1.2)
^{78m}Br			0.120 ms	IT	0.1486/100	(4+)			
^{78}Br		77.92115	6.45 m	β+ /92/3.574	1.2/	1+	0.13		ann.rad.
				ec/8/	2.5/				0.61363
									(0.7–3.0)
^{79m}Br			5.1 s	IT	0.2074/100	9/2+			0.2072
^{79}Br	(50.5-50.8)	78.91834				3/2-	+2.106400	+0.31	
^{80m}Br			4.421 h	IT	0.04879/100	5-	+1.3177	+0.71	Br k x-ray
									0.03705/39.1
									0.04885/0.3
^{80}Br		79.91853	17.66 m	β- /92/2.004	1.38 β-/7.6	1+	0.5140	+0.18	ann.rad.
				ec/5.7/1.8706	1.99 β-/82				0.6169/6.7
				β+ /2.6/	0.85 β+ /2.8				(0.64–1.45)
^{81m}Br			0.035 ms	IT	0.2602/100	9/2+			
^{81}Br	(49.2-49.5)	80.91629				3/2-	+2.270562	+0.26	
^{82m}Br			6.1 m	IT	0.045949/100	2-			0.046/0.24
				β- /2 /3.139					(0.62–2.66)
^{82}Br		81.91680	1.471 d	β- /3.093	0.444/	5-	+1.6270	+0.71	0.5544/71
									0.61905/43
									0.77649/84
									(0.013–1.96)
^{83}Br		82.91518	2.37 h	β- /0.972	0.395/1	3/2-			0.52964
					0.925/99				(0.12–0.68)
^{84m}Br			6.0 m	β- /4.97	2.2/100	(6-)			0.4240/100
									0.8817/98
									1.4637/101
^{84}Br		83.9165	31.8 m	β- /4.65	2.70/11	2-	2.		0.8816/41
					3.81/20				1.8976/13
					4.63/34				(0.23–4.12)
^{85}Br		84.91565	2.87 m	β- /2.87	2.57	3/2-			0.80241/2.56
									0.92463/1.6
									(0.09–2.4)
^{86}Br		85.91881	55.5 s	β- /7.63	3.3	(2-)			1.56460/64
					7.4				2.75106/21
									(0.5–6.8)
^{87}Br		86.92067	55.6 s	β- /6.85	6.1/	3/2-			1.41983/55.
				n/					1.578/45.
									(0.173–1.76)
^{88m}Br			5.5 μs	IT	0.111/100				
^{88}Br		87.92408	16.3 s	β- /8.96		1-			0.7649
				n/					0.7753

Elem. or Isot.	Natural abundance (Atom %)	Atomic mass or weight	Half-life/ Resonance width (MeV)	Decay mode/ Energy (/MeV)	Particle energy/ Intensity (MeV/%)	Spin (h/2π)	Nuclear magnetic mom. (nm)	Elect. quadr. mom. (b)	γ-Energy/ Intensity (MeV/%)
									0.8021
									(0.1–6.99)
^{89}Br		88.92671	4.35 s	β- /8.16		3/2-			0.7753
				n/					1.0978
^{90}Br		89.93129	1.91 s	β- /10.4	8.3/	2-			0.6555
				n/	9.8/				0.7071
									1.3626
^{91}Br		90.93440	0.54 s	β- /90 /9.80					0.263
				β- n/10 /					0.803
^{92}Br		91.93963	0.31 s	β- /12.20					0.740
				β- n/					
^{93}Br		92.943	0.15 s	β- /11					0.117-0.593
				β- n	//~ 53				(0.237–3.606)
^{94}Br		93.949	0.07 s	β- n/					
^{95m}Br			~ 7 μs	IT	0.538/100				0.5379
^{95}Br		94.953	> 0.3 μs						
^{96m}Br			~ 3 μs	IT	0.312/100				0.3115
^{96}Br		95.959	> 0.3 μs						
^{97}Br		96.963	> 0.3 μs						
^{98}Br		97.969	> 0.4 μs						
$_{36}$Kr		**83.798(2)**							
^{67}Kr			~ 7.4 ms	2p/1.69	// ~ 0.37				
^{69}Kr		68.965	28 ms	β+, (p)	4.07/				
^{70}Kr		69.956	0.06 s			0+			
^{71}Kr		70.950	100. ms	β+, ec/10.1					(0.198–0.207)
^{72}Kr		71.9421	17.1 s	β+ /5.0		0+			ann.rad
				ec/					0.3099/15.3
									0.4150/12.8
									(0.305 – 3.305)
^{73}Kr		72.93929	28. s	β+ /6.7		5/2-			ann.rad.
				ec/					0.1781/66
				β+, p/	/0.25				(0.06–0.86)
^{74}Kr		73.93308	11.5 m	β+ /3.1		0+			ann.rad.
				ec/					0.08970/31
									0.2030/20
									(0.010–1.06)
^{75}Kr		74.9309	4.3 m	β+ /4.90	3.2/	5/2+	-0.531	+1.1	ann.rad.
				ec/					0.1325/68
									0.1547/21
									(0.02–1.7)
^{76}Kr		75.92591	14.8 h	ec/1.31		0+			Br k x-ray
									0.270/21
									0.3158/39
									(0.03–1.07)
^{77}Kr		76.92467	1.24 h	β+ /80/3.06		5/2+	-0.583	+0.9	ann.rad.
				ec/20/	1.55/				0.1297/80
					1.70/				0.1465/38
					1.87/				(0.02–2.3)
^{78}Kr	0.355(3)	77.920365	9. ×10^{21} a	ec-ec		0+			
^{79m}Kr			50 s	IT	0.1298/100	7/2+	-0.786	+0.40	Kr x-ray
^{79}Kr		78.92008	1.455 d	β+ /7 /1.626		½-	+0.536		ann.rad.
				ec/93 /					0.2613/13
									0.39756/19
									0.6061/8
									(0.04–1.3)
^{80}Kr	2.286(10)	79.916378				0+			
^{81m}Kr			13.1 s	IT	0.1905/100	1/2-	+0.586		0.1904
^{81}Kr		80.91659	2.1 ×10^5 a	ec/0.2807		7/2+	-0.908	+0.644	Br k x-ray
									0.2760
^{82}Kr	11.593(31)	81.91348				0+			
^{83m}Kr			1.85 h	IT	0.03215/100	1/2-	+0.591		Kr k x-ray
									0.00940
									0.03215/0.055
^{83}Kr	11.500(19)	82.914127				9/2+	-0.970669	+0.259	
^{84m}Kr			1.83 μs	IT	0.0635/100	8+		+0.36	
^{84}Kr	56.987(15)	83.91149773				0+			
^{85m2}Kr			1.8 μs	IT	0.0602	(17/2+)			
^{85m}Kr			4.48 h	β- /79 /	0.83/79	½-	+ 0.633		0.30487
				IT/21/	0.30487				0.15118
^{85}Kr		84.91253	10.68 a	β- /0.687	0.15/0.4	9/2+	1.005	+0.443	0.51399
^{86}Kr	17.279(41)	85.91061063				0+			
^{87}Kr		86.913355	1.27 h	β- /3.887	1.33/8	5/2+	-1.023	-0.30	0.40258/49.6
					3.49/43				2.5548/9.2

Elem. or Isot.	Natural abundance (Atom %)	Atomic mass or weight	Half-life/ Resonance width (MeV)	Decay mode/ Energy (/MeV)	Particle energy/ Intensity (MeV/%)	Spin (h/2π)	Nuclear magnetic mom. (nm)	Elect. quadr. mom. (b)	γ-Energy/ Intensity (MeV/%)
					3.89/30				(0.13–3.31)
^{88}Kr		87.91445	2.80 h	β- /2.91		0+			0.19632/26.
									2.392/34.6
									(0.03–2.8)
^{89}Kr		88.91784	3.15 m	β- /4.99	3.8/	3/2 (+)	-0.330	+0.16	0.19746
					4.6/				0.2209/19.9
					4.9/				0.5858/16.4
									1.4728/6.8
									(0.2–4.7)
^{90}Kr		89.91953	32.3 s	β- /4.39	2.6/77	0+			0.12182/32.9
					2.8/6				0.5395/28.6
									1.1187/36.2
									(0.1–4.2)
^{91}Kr		90.92381	8.6 s	β- /6.4	4.33/	5/2 (+)	-0.583	+0.30	0.10878/43.5
					4.59/				0.50658/19.
									(0.2–4.4)
^{92}Kr		91.92617	1.84 s	β- /5.99		0+			0.1424/66.
				n/					(0.14–3.7)
^{93}Kr		92.93115	1.30 s	β- /8.6	7.1/	½+	-0.413		0.1820
				n/	//1.9				0.2534/42.
									0.32309/24.6
									(0.057–4.03)
^{94}Kr		93.9341	0.23 s	β- /7.3		0+			0.2196/67
				n	n//1.0				0.6293/100.
									(0.033-0.984)
^{95m}Kr			1.4 μs	IT	~ 0.0826/100	(7/2+)			
^{95}Kr		94.9397	0.10 s	β- /9.7	n//2.9		-0.410		
^{96}Kr		95.9430	~ 80 ms	β-, n	n//3.7	0+			
^{97}Kr		96.949	0.06 s	β-, n	n//7.				
^{98}Kr		97.952	0.05 s	β-, n	n//7.	0+			
^{99}Kr		98.958	0.04 s	β-, n	n// ~ 11.				
^{100}Kr		99.962	~ 0.007 s			0+			
^{101}Kr		100.969	> 0.4 μs						
$_{37}$**Rb**		**85.4678(3)**							
^{71}Rb		70.965							
^{72}Rb		71.959	< 1.5 μs						
^{73}Rb		72.951	< 0.03 μs						
^{74}Rb		73.94427	64.8 ms	β+ /10.4					0.456/0.0025
									(0.053 – 5.820)
^{75}Rb		74.93857	19. s	β+ /7.02	2.31/	3/2 (-)			ann.rad
									0.179
^{76m}Rb			3.2 μs	IT	0.07055/100	4(+)			
^{76}Rb		75.93507	39. s	β+ /8.50	4.7/	1 (-)	-0.372623	+0.5	ann.rad
									0.4240/92.
									(0.064–1.68)
^{77}Rb		76.93040	3.8 m	β+ /5.34	3.86/	3/2-	+0.654468	+0.8	ann.rad
									0.0665/59
									(0.04–2.82)
^{78m}Rb			5.74 m	IT/9/		4 (-)	+2.549	+1.0	ann.rad
				β+ /91/	3.4				0.4553/81.
				ec/					(0.103–4.01)
^{78}Rb		77.92814	17.7 m	β+ /7.22		0 (+)			ann.rad
				ec/					0.4553/63.
									(0.42–5.57)
^{79}Rb		78.92399	23. m	β+ /84/3.65		5/2+	+0.3358	-0.12	ann.rad.
				ec/16 /					0.68812/23.
									(0.017–3.02)
^{80m}Rb			1.63 μs			6+			
^{80}Rb		79.92252	34. s	β+ /5.72	4.1/22	1+	-0.0836	+0.4	ann.rad.
					4.7/74				0.6167/25.
^{81m}Rb			30.5 m	IT	0.0863/~ 97.6	9/2+	+5.598	-0.9	ann.rad.
				β+, ec/	1.4/~ 2.4				(0.085–1.9)
^{81}Rb		80.91899	4.57 h	β+ /27/2.24	1.05/	3/2-	+2.060	+0.5	ann.rad./
				ec/73					0.19030/64.
									(0.05–1.9)
^{82m}Rb			6.47 h	β+/26/	0.80/	5-	+1.51001	+1.2	ann.rad./
				ec/74/					0.5544/63.
									0.7765/85.
									(0.092–2.3)
^{82}Rb		81.91821	1.258 m	β+/96/4.40	3.3/	1+	+0.554508	+0.2	ann.rad./
				ec/4/					0.7665/13.
									(0.47–3.96)

Elem. or Isot.	Natural abundance (Atom %)	Atomic mass or weight	Half-life/ Resonance width (MeV)	Decay mode/ Energy (/MeV)	Particle energy/ Intensity (MeV/%)	Spin (h/2π)	Nuclear magnetic mom. (nm)	Elect. quadr. mom. (b)	γ-Energy/ Intensity (MeV/%)
^{83}Rb		82.91511	86.2 d	ec/0.91		5/2-	+1.425	+ ~ 0.24	Kr x-ray
									0.5205/46.
									(0.03–0.80)
^{84m}Rb			20.3 m	IT	0.2156/100	6-	+0.21293	+0.7	0.2156/34.
									0.2482/63.
									0.4632/32.
^{84}Rb		83.91438	32.9 d	β+/22/2.681	0.780/11	2-	-1.32412	-0.02	ann.rad./
				ec/75 /	1.658/11				0.8817/68.
				β-/3/0.894	0.893/				(1.02–1.9)
^{85m}Rb			1.015 μs	IT	0.51400/100	9/2+			
^{85}Rb	72.17(2)	84.91178974				5/2-	+1.353	+0.276	
^{86m}Rb			1.017 m	IT	0.5561/100	6-	+1.815	+ ~ 0.45	0.556/98.
^{86}Rb		85.911167	18.68 d	β-/1.775	1.774/8.8	2-	-1.6920	+0.2	1.0768/8.8
^{87}Rb	27.83(2)	86.90918053	4.97 ×10^{10} a	β-/0.283	0.273/100	3/2-	+2.7512	+0.13	
^{88}Rb		87.911316	17.7 m	β-/5.316	5.31	2-	0.508		0.8980/14.4
									1.8360/22.8
									(0.34–4.85)
^{89}Rb		88.91228	15.4 m	β-/4.50	1.26/38	3/2-	+2.38	+0.17	1.032/58.
					1.9/5				1.248/42.
					2.2/34				2.1960/13
					4.49/18				(0.12–4.09)
^{90m}Rb			4.3 m	β-/4.50	1.7/	3 -	+1.616	+0.3	0.1069(IT)
					6.5/				0.8317/94
									(0.20–5.00)
^{90}Rb		89.91480	2.6 m	β-/6.59	6.6	0 -			0.8317/28.
									(0.31–5.60)
^{91}Rb		90.91654	58.0 s	β-/5.861	5.9	3/2 (-)	+2.182	+ ~ 0.19	0.0936/34.
									(0.35–4.70)
^{92}Rb		91.91973	4.48 s	β-/8.11	8.1/94	0 -			0.8148/8.
									(0.1–6.1)
^{93}Rb		92.92204	5.85 s	β-/7.46	7.4/	5/2-	+1.410	+0.2	0.2134/4.8
				n/1					0.4326/12.5
									0.9861/4.9
									(0.16–5.41)
^{94}Rb		93.92639	2.71 s	β-/10.31	9.5/	3 (-)	+1.498	+0.2	0.8369/87.
				n/10					1.5775/32.
									(0.12–6.35)
^{95}Rb		94.92926	0.377 s	β-/9.30	8.6/	5/2-	+1.334	+0.3	0.352/65.
				n/8					0.680/22.
									(0.20–2.27)
^{96m}Rb			1.8 μs	IT	0.040/100	(10-)			0.2999
									0.4612
									0.2400
									0.093–0.369
^{96}Rb		95.93413	0.199 s	β-/11.76	10.8/	2+	+1.466	+0.3	0.815/76.
				n/13/					(0.20–5.42)
^{97m}Rb			5.1 μs	IT	0.0765/100				0.077
^{97}Rb		96.93718	0.169 s	β-/10.42	10.0	3/2+	+1.841	+0.7	0.167/100.
				n/27/					0.585/79.
									0.599/56.
									1.258/52.
									(0.14–2.08)
^{98m}Rb			96 ms	β-		(3)	1.79	+1.43	
^{98}Rb		97.94169	0.107 s	β-/12.34	0.144/	(0)			
				n/13					(0.07–3.68)
^{99}Rb		98.945	59. ms	β-/11.3					
^{100}Rb		99.950	53. ms	β- /13.5					0.129
									(0.058–4.483)
^{101}Rb		100.954	0.03 s	β- /11.8					
^{102}Rb		101.960	0.04 s	β-					(0.126-1.095)
				β-,n/ >46					(0.092-0.363)
^{103}Rb		102.064	0.02 s						
$_{38}$Sr		**87.62(1)**							
^{73}Sr		72.966	> 25 ms						
^{74}Sr		73.956	> 1.5 μs			0+			
^{75}Sr		74.950	88. ms	β+ ,p	p//5.				0.144/4.5
^{76}Sr		75.9418	7.9 s	β+ /6.1		0+			
^{77}Sr		76.9379	9.0 s	β+ /6.9	5.6	5/2 +	-0.348	+1.3	0.147
				β+, p	//0.08				
^{78}Sr		77.93218	155. s	β+ /3.76		0+			(0.025–1.660)
^{79}Sr		78.9297	2.1 m	β+ /5.32	4.1	3/2 (-)	-0.474	+0.66	ann.rad./
									0.039/28.
									0.105/22.

Elem. or Isot.	Natural abundance (Atom %)	Atomic mass or weight	Half-life/ Resonance width (MeV)	Decay mode/ Energy (/MeV)	Particle energy/ Intensity (MeV/%)	Spin (h/2π)	Nuclear magnetic mom. (nm)	Elect. quadr. mom. (b)	γ-Energy/ Intensity (MeV/%)
^{80}Sr		79.92452	1.77 h	β+ /1.87		0+			(0.135–0.612)
									ann.rad./
									0.174/10.
									0.589/39.
^{81m}Sr			~ 6.4 µs	IT	0.07923/100				(0.24–0.55)
^{81}Sr		80.92321	22.3 m	β+ /87/3.93	2.43/	1/2-	+0.544		ann.rad./
				ec/13/	2.68/				0.148/31.
									0.1534/35
^{82}Sr		81.91840	25.36 d	ec/0.18		0+			(0.06–1.7)
^{83m}Sr			5.0 s	IT/0.2591		½-	+0.582		Rb x-ray
^{83}Sr		82.91755	1.350 d	β+/24/2.28	0.465/	7/2+	-0.8298	+0.71	0.2591/87.5
				ec/76/	0.803/				ann.rad./
					1.227/				0.3816/12.
									0.3816
									0.7627/30.
^{84}Sr	0.56(2)	83.91342				0+			(0.094–2.15)
^{85m}Sr			1.127 h	IT/87/0.2387		½-	+0.600		0.2318/84.
				ec/13					(0.15–0.24)
^{85}Sr		84.91293	64.85 d	ec/1.065		9/2+	-1.001	+0.26	0.51399/99.3
^{86}Sr	9.86(20)	85.90926				0+			
^{87m}Sr			2.81 h	IT/0.3884		½-	+0.63		0.3884(IT)
^{87}Sr	7.00(20)	86.90888				9/2+	-1.0936	+0.305	
^{88}Sr	82.58(35)	87.90561				0+			
^{89}Sr		88.90745	50.6 d	β-/1.497	1.492/100	5/2+	-1.148	-0.25	0.9092
^{90}Sr		89.90773	28.9 a	β-/0.546	0.546/100	0+			
^{91}Sr		90.91020	9.5 h	β-/2.70	0.61/7	5/2+	-0.885	+0.04	0.5556/61.
					1.09/33				0.7498/24.
					1.36/29				1.0243/33.
					2.66/26				(0.12–2.4)
^{92}Sr		91.91104	2.64 h	β-/1.91	0.55/96	0+			1.3831/90.
					1.5/3				(0.24–1.1)
^{93}Sr		92.91402	7.4 m	β-/4.08	2.2/10	5/2+	-0.793	+0.24	0.5903/
					2.6/25				0.7104
					3.2/65				0.87573
									0.8883/
^{94}Sr		93.91536	1.25 m	β-/3.511	2.1/	0+			(0.17–3.97)
					3.3/				0.6219
									0.7043
									0.7241
									0.8064
^{95}Sr		94.91935	25.1 s	β-/6.08		½+	-0.537		1.4283
									0.6859
					6.1/50				0.8269
									2.7173
^{96}Sr		95.9217	1.06 s	β-/5.37	4.2/	0+			2.9332
									0.1222
									0.5305
									0.8094
^{97}Sr		96.92637	0.42 s	β-/7.47	5.3	1/2+	-0.498		0.9318
									0.2164
									0.3071
									0.6522
									0.9538
									1.2580
^{98}Sr		97.92869	0.65 s	β-/5.83	5.1	0+			1.9050
									0.0365
									0.1190
									0.4286
									0.4447
									0.5636
^{99}Sr		98.93289	0.27 s	β-/8.0		3/2 +	-0.26	+0.76	
^{100}Sr		99.9358	0.201 s	β-/7.1		0+			
^{101}Sr		100.940	0.115 s	β-/9.5					
^{102}Sr		101.9438	68. ms	β-/8.8		0+			
^{103}Sr		102.949	0.07 s						
^{104}Sr		103.953	0.04 s			0+			
^{105}Sr		104.959	~ 0.04 s						
^{106}Sr		105.963	~ 20 ms						
^{107}Sr		106.969	> 0.4 µs						
$_{39}$Y		**88.90584(2)**							
^{76}Y		75.959	> 0.2 µs						

Elem. or Isot.	Natural abundance (Atom %)	Atomic mass or weight	Half-life/ Resonance width (MeV)	Decay mode/ Energy (/MeV)	Particle energy/ Intensity (MeV/%)	Spin (h/2π)	Nuclear magnetic mom. (nm)	Elect. quadr. mom. (b)	γ-Energy/ Intensity (MeV/%)
^{77}Y		76.9498	~ 57. ms						
^{78m}Y			5.8 s	ec, β+/100		(5+)			
^{78}Y		77.944	53 ms	β+/10.5					0.279/100
									0.504/90
									0.713/40
									(0.152–1.106)
^{79}Y		78.937	15. s	β+/7.1					
^{80m2}Y			4.7 μs	IT	0.084	(2+)			
^{80m}Y			4.8 s	IT	0.2285/81	(1-)			0.2285
				ec, β+	/19				
^{80}Y		79.93436	30. s	β+/7.0	5.5	(4-)			ann.rad./
					5.0/				0.3858/100
									0.5951/42
									0.756–1.396
^{81}Y		80.92946	1.21 m	β+/5.5	3.7/				ann.rad./
					4.2/				0.428
									0.469
^{82}Y		81.92693	9.5 s	β+/7.8	6.3/	1+			ann.rad./
									0.5736
									0.6017
									0.7375
^{83m}Y			2.85 m	β+/95/4.6	2.9	3/2-			ann.rad./
				ec/5 /					0.2591
									0.4218
									0.4945
^{83}Y		82.9225	7.1 m	β+/4.47	3.3	9/2+			ann.rad./
				ec/					0.0355
									0.4899
									0.8821
									(0.03–3.4)
^{84m}Y			4.6 s	β+/		1+			ann.rad./
				ec/					0.7930
^{84}Y		83.92067	40. m	β+/6.4	1.64/47	5-			ann.rad./
				ec/	2.24/25				0.4628
					2.64/21				0.6606
					3.15/7				0.7931
									0.9744
									1.0398
									(0.2–3.3)
^{85m}Y			4.9 h	β+/70/		9/2+	6.2		ann.rad./
				ec/30/					0.2317
									0.5356
									2.1238
									(0.1–3.1)
									0.7673
^{85}Y		84.9164	2.6 h	β+/55/3.26	1.54/	1/2-			ann.rad./
				ec/45/					0.2317
									0.5045
									0.9140
									(0.07–1.4)
^{86m}Y			48. m	IT	0.0102/99.3	(8+)	4.8		ann.rad./
				β+/					0.0102(IT)
				ec/					0.2080
									(0.09–1.1)
^{86}Y		85.9149	14.74 h	β+/5.24		4-	<0.6		ann.rad./
				ec/					0.3070
									0.6277
									1.0766
									1.1531
									1.9207
									(0.1–3.8)
^{87m}Y			13.4 h	IT	0.3808/98	9/2+	+ 6.24	- 0.5	0.3807
				β+/0.7/	1.15/0.7				
				ec/					
^{87}Y		86.91088	3.35 d	ec/99+/1.862	0.78/	1/2-	- 0.19		0.3880
									0.4870
^{88m2}Y			14.0 ms	IT	0.4426/100	8+		+ ~ 0.06	
^{88m1}Y			0.301 ms	IT	0.3929/100				
^{88}Y		87.90950	106.6 d	ec/99+/3.623	0.76/	4-	- 0.42	+ 0.16	ann.rad./
				β+/0.2/					0.89802
									1.83601
									2.73404
									3.2190
^{89m}Y			15.66 s	IT	0.90896/100	9/2+	+ 6.37	~ - 0.43	0.9092(IT)

Elem. or Isot.	Natural abundance (Atom %)	Atomic mass or weight	Half-life/ Resonance width (MeV)	Decay mode/ Energy (/MeV)	Particle energy/ Intensity (MeV/%)	Spin (h/2π)	Nuclear magnetic mom. (nm)	Elect. quadr. mom. (b)	γ-Energy/ Intensity (MeV/%)
^{89}Y	100.	88.90584				1/2-	-0.13742		
^{90m}Y			3.20 h	IT/99+/	0.68204/100	7+	+ 5.28	~ - 0.65	0.2025
				β-/0.002/					0.4794
									0.6820
^{90}Y		89.90714	2.669 d	β-/2.282	2.28/	2-	-1.63	-0.13	(0.203-0.480)
^{91m}Y			49.7 m	IT	0.5556/100	9/2+	5.96		0.5556(IT)
^{91}Y		90.90730	58.5 d	β-/1.544	1.545/	1/2-	0.164		1.208
^{92m}Y			~ 3.3 µs						
^{92}Y		91.9089	3.54 h	β-/3.63	3.64/	2-	- 0.67		0.4485
									0.5611
									0.9345
									1.4054
									(0.4–3.3)
^{93m}Y			0.82 s	IT	0.16850/100	9/2+	+ 6.04	~ - 0.64	0.1686(IT)
									0.5902
^{93}Y		92.9096	10.2 h	β-/2.87	2.88/90	1/2-	- 0.12		0.2669
									0.9471
									1.9178
^{94m}Y			1.30 µs	IT	0.1202/100				0.4322
									0.7699
									1.2024
^{94}Y		93.91159	18.7 m	β-/4.919	4.92/	2-	- 0.24	~ 0.03	0.3816
									0.9188
									1.1389
									(0.3–4.1)
^{95m}Y			0.049 ms	IT					
^{95}Y		94.91282	10.3 m	β-/4.42		1/2-	- 0.16		0.4324
									0.9542
									2.1760
									3.5770
^{96m}Y			9.6 s	β-/		(8+)		-1.0	0.1467
									0.6174
									0.9150
									1.1071
									1.7507
^{96}Y		95.91590	6.2 s	β- /7.09	7.12/	0-			1.594
^{97m2}Y			0.14 s	IT	0.1623/95	(27/2-)		-1/2	
^{97m}Y			1.17 s	β- /7.4	4.8/	9/2+	+ 5.88	~ -0.76	0.1614
					6.0/				0.9700
									1.1030
^{97}Y		96.9183	3.76 s	β- /6.69	6.7	1/2-	- 0.12		0.2969
									1.9960
									3.2876
									3.4013
^{98m2}Y			~ 6.87 µs	IT	0.0498/100	4-			
^{98m}Y			2.0 s	β- /9.8	5.5/		+ 2.98	+ 1.7	0.2415
									0.6205
									0.6473
									1.2228
									1.8016
^{98}Y		97.9224	0.59 s	β- /8.83	8.7/	1+			0.2131
									1.2228
									1.5907
									2.9413
									4.4501
^{99m}Y			0.008 ms	IT	0.0277/100	(17/2+)			(0.273-1.436)
^{99}Y		98.92415	1.47 s	β- /7.57		1/2-			0.1218/43.8
				n	/2.5/				0.5362
									0.7242
									1.0130
^{100m}Y			0.94 s	β-, n /		4+			
^{100}Y		99.9277	0.73 s	β-, n/9.3	n/1.8/	1+			
^{101}Y		100.93015	0.43 s	β-, n/8.6	n/1.5/	(5/2+)	+ 3.22	- 1.5	
^{102}Y		101.93433	0.36 s	β-, n/9.9	n/4.0/				
^{103}Y		102.9372	0.23 s	β-, n	n/8.3/				
^{104}Y		103.942	0.20 s						
^{105}Y		104.945	0.08 s						
^{106}Y		105.951	0.06 s						
^{107}Y		106.955	0.04 s						
^{108}Y		107.960	~ 0.03 s						
^{109}Y		108.964	25 ms						
^{40}Zr		**91.224(2)**							

Elem. or Isot.	Natural abundance (Atom %)	Atomic mass or weight	Half-life/ Resonance width (MeV)	Decay mode/ Energy (/MeV)	Particle energy/ Intensity (MeV/%)	Spin (h/2π)	Nuclear magnetic mom. (nm)	Elect. quadr. mom. (b)	γ-Energy/ Intensity (MeV/%)
^{78}Zr		77.956	> 0.2 μs			0+			
^{79}Zr		78.949	0.06 s						
^{80}Zr		79.94	~ 4.5 s	β+ /8.0		0+			0.290
									0.538
^{81}Zr		80.937	5.3 s	β+ /7.2	6.1	(3/2-)			ann.rad./
^{82}Zr		81.931	32. s	β+ /4.0	3.	0+			ann.rad./
^{83m}Zr			7. s	β+ /7.0		(7/2+)			ann.rad./
^{83}Zr		82.92924	44. s	β+ /5.9	4.8	(1/2-)			0.0556
				ec					0.1050
									0.2560
									0.474
									1.525
^{84}Zr		83.92333	26. m	β+ /2.7		0+			ann.rad./
				ec/					0.0449
									0.1125
									0.3729
									0.667
^{85m}Zr			10.9 s	IT/0.2922		½-			ann.rad./
				β+, ec/					0.2922(IT)
									0.4165
^{85}Zr		84.92144	7.9 m	β+ /4.7	3.1	7/2+			ann.rad./
				ec/					0.2663
									0.4163
									0.4543
^{86}Zr		85.91630	16.5 h	ec/1.47		0+			0.0280
									0.243
									0.612
^{87m}Zr			14.0 s	IT	0.1349/100	1/2	+ 0.64		0.1352(IT)
									0.2010
^{87}Zr		86.91482	1.73 h	β+ /3.67	2.26	9/2+	- 0.895	+ 0.42	ann.rad./
				ec/					0.3811
									1.228
^{88m}Zr			1.32 μs	IT	0.0770/100	8+	- 1.81	+ 0.44	0.077
^{88}Zr		87.91022	83.4 d	ec/0.67		0+			0.3929
^{89m}Zr			4.16 m	IT	0.5878/100	1/2-	+ 0.80		ann.rad./
				β+ /1.5/					0.5877(IT)
				ec/4.7/					1.507
^{89}Zr		88.90888	3.27 d	β+ /23/2.832	0.9/	9/2+	-1.05	+ 0.3	ann.rad./
				ec/77/					0.9092
^{90m}Zr			0.809 s	IT/	0.1327/100	5-	6.3		0.1326
									2.1862
									2.3189(IT)
^{90}Zr	51.45(4)	89.90470				0+			
^{91m}Zr			4.4 μs	IT	0.8792/100	21/2+		0.71	
^{91}Zr	11.22(5)	90.90564				5/2+	-1.30362	-0.176	
^{92}Zr	17.15(3)	91.90503				0+			
^{93}Zr		92.90647	1.5 ×10^{6} a	β- /0.091		5/2+			0.0304
^{94}Zr	17.38(4)	93.90631	>10^{17} a	β-β-		0+			
^{95}Zr		94.90804	64.02 d	β- /1.125	0.366/55	5/2+	1.13	+0.22	0.7242
					0.400/44				0.7567
^{96}Zr	2.80(2)	95.90827	2.3 ×10^{19} a	β-β-		0+			
			>1.7 ×10^{18} a	β-					
^{97}Zr		96.91095	16.75 h	β- /2.658	1.91/	½-	- 0.937		0.7434
^{98m}Zr			1.7 μs	IT	0.0630/100	(17-)			
^{98}Zr		97.9127	30.7 s	β- /2.26	2.2/100	0+			
^{99}Zr		98.9167	2.2 s	β- /4.56	3.9/	½+	- 0.930		0.4692/55.2
					3.5/				0.5459/48
									0.028–1.321
^{100}Zr		99.9180	7.1 s	β- /3.34		0+			0.4006/20
									0.5043/30
^{101}Zr		100.9214	2.1 s	β- /5.49	6.2/	3/2-	- 0.27	+ 0.81	0.1194
									0.2057
									0.2089
^{102}Zr		101.9231	2.9 s	β- /4.61		0+			0.064/15
									0.599/14
									0.535/11
^{103}Zr		102.9272	1.3 s	β- /7.0					
^{104}Zr		103.9294	1.2 s	β- /5.9		0+			
^{105}Zr		104.9340	0.6 s	β- /8.5					
^{106}Zr		105.937	0.19 s			0+			
^{107}Zr		106.942	0.14 s						
^{108}Zr		107.945	73. ms			0+			

Elem. or Isot.	Natural abundance (Atom %)	Atomic mass or weight	Half-life/ Resonance width (MeV)	Decay mode/ Energy (/MeV)	Particle energy/ Intensity (MeV/%)	Spin (h/2π)	Nuclear magnetic mom. (nm)	Elect. quadr. mom. (b)	γ-Energy/ Intensity (MeV/%)
^{109}Zr		108.950	0.06 s						
^{110}Zr		109.954	0.04 s			0+			
^{111}Zr		110.960	24 ms						
^{112}Zr		111.964	~ 0.03 s						
$_{41}$Nb		**92.90637(2)**							
^{81}Nb		80.950	<0.08 μs						
^{82}Nb		81.944	50 ms	β+ /11.					
^{83}Nb		82.937	4.1 s	β+ /7.5					
^{84}Nb		83.934	10. s	β+, ec/9.6		(3+)			0.540
									(0.456-1.427)
^{85m}Nb			3. s						0.069
^{85}Nb		84.92885	21. s	β+ /6.0					
^{86m}Nb			56. s	β+					
^{86}Nb		85.92578	1.46 m	β+ /8.0					ann.rad./
									0.751
									1.003
^{87m}Nb			2.6 m	β+ /		(9/2+)			ann.rad./
				ec/					0.2010
									(0.471-1.884)
^{87}Nb		86.92069	3.7 m	β+ /5.2/		(1/2-)			ann.rad./
				ec/					0.1351
									0.2010
^{88m}Nb			7.7 m	β+ /		(4-)			ann.rad./
				ec/					0.2625
									0.3996
									1.0569
									1.0825
^{88}Nb		87.9182	14.3 m	β+ /7.6	3.2/	8+			ann.rad./
				ec/					1.0570
									1.0828
									(0.07–2.5)
^{89m}Nb			2.0 h	β+ /	3.3/	9/2+			0.5880/10(D)
				ec/					(0.17–4.0)
^{89}Nb		88.9134	1.10 h	β+ /74/4.29	2.8/	(1/2-)	+6.216		ann.rad./
				ec/26 /					0.5074
									0.5880
									0.7696
									1.2775
^{90m3}Nb			6.2 ms	IT	0.2573/100	1+			
^{90m2}Nb			18.8 s	IT/0.1246		4-		-0.26	0.002
^{90m1}Nb			0.063 ms	IT	0.1224/100	6+			0.1225
^{90}Nb		89.91126	14.6 h	β/53 /6.111	0.86/5	8+	4.961	+ 0.01	ann.rad./
				ec/47 /	1.5/92				0.1412
									1.1292
									2.1862
									2.3189
									(0.1–3.3)
^{91m2}Nb			3.8 μs	IT	0.050/100	17/2-			
^{91m}Nb			60.9. d	IT	0.1046/97	1/2-			0.1045(IT)
				ec/3 /					1.2050
^{91}Nb		90.90699	7 ×10^{2} a	ec/1.253		9/2+		-0.25	Mo k x-ray
^{92m2}Nb			5.9 μs	IT	0.0904/100	(2)-			
^{92m}Nb			10.15 d	ec/99+/		2+	6.14		0.9126
									0.9345
									1.8475
^{92}Nb		91.90719	3.7 ×10^{7} a	ec/2.006		7+		-0.35	0.5611
									0.9345
^{93m2}Nb			~ 1.5 μs						
^{93m}Nb			16.1 a	IT	0.03077/100	1/2-			Nb x-ray
^{93}Nb	100.	92.90637				9/2+	+6.1705	-0.32	0.0304
^{94m}Nb			6.26 m	IT/99+	/2.086	3+			Nb k x-ray
				β- /0.5/					0.0409
									0.87109
^{94}Nb		93.90728	2.04 ×10^{4} a	β- /2.045	0.47/	6+			0.70263
									0.87109
^{95m}Nb			3.61 d	IT	0.2357/94.4	1/2-			0.2040
				β- /2.5 /					0.2356
^{95}Nb		94.906832	34.97 d	β- /0.926	0.160/	9/2+	6.14		0.76578
^{96}Nb		95.90810	23.4 h	β- /3.187	0.5/10	6+	4.976		0.7782
					0.75/90				0.2191–1.498
^{97m}Nb			59 s	IT	0.7434/100	1/2-			0.7434

Elem. or Isot.	Natural abundance (Atom %)	Atomic mass or weight	Half-life/ Resonance width (MeV)	Decay mode/ Energy (/MeV)	Particle energy/ Intensity (MeV/%)	Spin (h/2π)	Nuclear magnetic mom. (nm)	Elect. quadr. mom. (b)	γ-Energy/ Intensity (MeV/%)
^{97}Nb		96.90810	1.23 h	β- /1.934	1.27/98	9/2+	6.15		0.4809
									0.6579
^{98m}Nb			51.3 m	β- /4.67		(5+)			0.7874
									0.1726–1.89
^{98}Nb		97.91033	2.9 s	β- /4.59	4.6/	1+			0.6451
									0.7874
									1.0243
^{99m}Nb			2.5 m	β- /	3.2/	1/2-			0.0978/100
									(0.138–3.010)
^{99}Nb		98.9116	15.0 s	β- /3.64	3.5/100	9/2+		-0.4	0.0977
									0.1378/3.1
^{100m2}Nb			0.012 ms	IT	0.028/100	(8-)			
^{100m1}Nb			3.0 s	β- /6.74	5.8	(5+)			Nb k x-ray
									0.159
									0.6364
									1.0637
^{100}Nb		99.9143	1.5 s	β- /6.25	6.2/				0.5354
					5.3/				0.6001–1.566
^{101}Nb		100.91531	7.1 s	β- /4.57	4.3/			+.011	0.1105–0.810
^{102m}Nb			1.3 s	β- /		1+			
^{102}Nb		101.91808	1.3 s	β- /7.21	7.2/				0.2960–2.184
^{103}Nb		102.91946	1.5 s	β- /5.53	5.3/	5/2+			
^{104m}Nb			0.94 s	β-, n/	n/0.06				
^{104}Nb		103.92289	4.9 s	β-, n/8.1	n/0.05				
^{105}Nb		104.92495	2.9 s	β-, n/6.5	n/1.7				
^{106}Nb		105.92893	1.0 s	β-, n/9.3	n/4.5				
^{107}Nb		106.9316	0.30 s	β-, n/7.9	n/6.0				
^{108}Nb		107.9361	0.19 s	β, n/	n/6.2				(0.193–0.590)
^{109}Nb		108.939	0.19 s	β, n/	n/31				
^{110}Nb		109.944	0.17 s	β, n/	n/40				
^{111}Nb		110.948	0.05 s						
^{112}Nb		111.952	~ 3.3 ms						
^{113}Nb		112.957	> 0.15 μs						
^{114}Nb		113.962	17 ms						
^{115}Nb		114.96	0.02 s						
$_{42}$Mo		**95.95(1)**							
^{81}Mo									
^{82}Mo									
^{83}Mo		82.950	~ 6. ms						
^{84}Mo		83.941	~ 2.2 s	β+ /6.		0+			
^{85}Mo		84.9383	3.2 s	β+/8.1		½+			
^{86}Mo		85.93118	19. s	β+ /4.8		0+			
^{87}Mo		86.92820	14. s	ec, β+/6.5					(0.752–1.004)
^{88}Mo		87.92197	8.0 m	β+ /3.4		0+	+0.5		ann.rad./
				ec					0.0800
									0.1399
									0.1707
^{89m}Mo			0.19 s	IT	0.2686/100	(1/2-)			0.118
									0.2686
^{89}Mo		88.91947	2.2 m	β+ /5.58		9/2+			ann.rad./
				ec/					0.659
									0.803
									1.155
									1.272
^{90m}Mo			~ 1.12 μs	IT	0.0632/100	8+	- 1.39	0.61	
^{90}Mo		89.91392	5.7 h	β+ /25/2.489	1.085/	0+			ann.rad./
				ec/75 /					0.04274
									0.12237
									0.25734
^{91m}Mo			1.08 m	IT/50/	0.6529	1/2-			ann.rad./
				β+, ec/50 /	2.5/				0.6529
					2.8/				1.2081
					4.0/				1.5080
									2.2407
^{91}Mo		90.91175	15.5 m	β+ /94/4.43	3.44/94	9/2-			ann.rad./
				ec/6/					1.6373
									2.6321
									3.0286
									(0.1–4.2)
^{92}Mo	14.649(106)	91.906808	> 3 ×10^{17} a	β+-ec		0+			
^{93m2}Mo			1 μs	IT		(39/2-)			
^{93m}Mo			6.9 h	IT/99+ /2.425		21/2+	+9.9		0.26306(IT)
									0.68461

Elem. or Isot.	Natural abundance (Atom %)	Atomic mass or weight	Half-life/ Resonance width (MeV)	Decay mode/ Energy (/MeV)	Particle energy/ Intensity (MeV/%)	Spin (h/2π)	Nuclear magnetic mom. (nm)	Elect. quadr. mom. (b)	γ-Energy/ Intensity (MeV/%)
^{93}Mo		92.906810	3.5 ×10^{3} a	ec/0.405		5/2+			1.47711
^{94}Mo	9.187(33)	93.905085				0+			0.0304
^{95}Mo	15.873(30)	94.905839				5/2+	-0.9142	-0.02	
^{96}Mo	16.673(8)	95.904676				0+			
^{97}Mo	9.582(15)	96.906018				5/2+	-0.9335	+0.26	
^{98}Mo	24.292(80)	97.905405				0+			
^{99m}Mo			15.5 μs	IT	0.097785/100	5/2+			
^{99}Mo		98.907709	2.747 d	β- /1.357	0.45/14	½+	0.375		0.144048
					0.84/2				0.18109
					1.21/84				0.36644
									0.73947
^{100}Mo	9.744(65)	99.90747	7 ×10^{18} a	β-β-		0+			
^{101}Mo		100.91034	14.8 m	β- /2.82	2.23/	1/2+			0.0063
					0.7/				0.19193
									0.5909
									(0.0809–2.405)
^{102}Mo		101.9103	11.3 m	β- /1.01	1.2/	0+			0.1493/89.
									0.2116/100.
									0.2243/32.
^{103}Mo		102.9131	1.13 m	β- /3.8		3/2+			0.1028(2)/
									0.1440(2)
									0.2511(2)
^{104}Mo		103.9137	1.00 m	β- /2.16		0+			0.0686(1)/100.
									0.4239(4)/21.
^{105}Mo		104.9170	36. s	β- /4.95		3/2+			0.0642/
									0.0856/
									0.2495/
^{106}Mo		105.9183	8.4 s	β- /3.52		0+			0.1894(2)/22.
									0.3644(2)/6.
									0.3723(2)/12.
^{107}Mo		106.9221	3.5 s	β- /6.2					
^{108}Mo		107.9240	1.1 s	β- /5.1		0+			(0.028–0.636)
^{109}Mo		108.9284	0.5 s	β- /7.2					
^{110}Mo		109.9307	0.27 s	β- /5.7		0+			Tc k x-ray
									0.142
									(0.039–0.599)
^{111}Mo		110.9357	~ 196 ms						
^{112}Mo		111.938	0.12 s			0+			
^{113}Mo		112.943	0.08 s						
^{114}Mo		113.947	0.06 s			0+			
^{115}Mo		114.952	46 ms						
^{116}Mo		115.955	32 ms			0+			
^{117}Mo		116.961	22 ms						
^{118}Mo			~ 19 ms						
$_{43}$Tc									
^{85}Tc		84.951	< 0.1 ms						
^{86m}Tc			1.1 μs	IT	0.081/100	(6+)			
^{86}Tc		85.945	0.05 s	β+ /11.9					
^{87}Tc		86.93807	2.4 s	β+ /8.6					
^{88m}Tc			~ 6.4 s	ec, β+					
^{88}Tc		87.934	5.8 s	β+ /10.1					
^{89m}Tc			13. s	ec, β+		(1/2-)			
^{89}Tc		88.9265	13. s	β+ /7.5					
^{90m}Tc			49.2 s	β+	5.3/	6+			ann.rad./
									0.9479/
									1.0542/
^{90}Tc		89.92407	8.7 s	β+ /8.9	7.0/15	1+			ann.rad./
					7.9/95.				0.9479/
^{91m}Tc			3.3 m	β+		(1/2)-			ann.rad./170.
				ec					0.8110(5)/5.
									1.6052(1)/7.8
									1.6339(1)/9.1
									1.9023(1)/6.
									2.4509(1)/13.5
^{91}Tc		90.91843	3.14 m	β+ /6.2	5.2	9/2+			ann.rad./200.
^{92m}Tc			1.0 μs	IT	0.05634/100	(4+)			
^{92}Tc		91.91527	4.4 m	β+ /7.87	4.1	8+			ann.rad./200.
				ec					0.0850/
									0.1475
									0.3293
									0.7731
									1.5096

Elem. or Isot.	Natural abundance (Atom %)	Atomic mass or weight	Half-life/ Resonance width (MeV)	Decay mode/ Energy (/MeV)	Particle energy/ Intensity (MeV/%)	Spin (h/2π)	Nuclear magnetic mom. (nm)	Elect. quadr. mom. (b)	γ-Energy/ Intensity (MeV/%)
^{93m2}Tc			10.2 μs	IT		(17/2)-			
^{93m}Tc			43. m	IT	0.3918/77.4	1/2-			0.3924(IT)
				ec/20					0.9437
									2.6445
^{93}Tc		92.91025	2.73 h	β+ /13/3.201	0.81	9/2+	6.3		ann.rad./
				ec/87/					1.3629
									1.4771
									1.5203
									(0.1–3.0)
^{94m}Tc			52. m	β+ /72/4.33		(2+)			ann.rad./
				ec/28/					0.8710
									1.8686
^{94}Tc		93.90965	4.88 h	β+ /11/4.256		7+	5.12		ann.rad./
				ec/89/					0.4491
									0.7026
									0.8496
									0.8710
^{95m}Tc			61 d	IT/4/	0.0389	1/2-			ann.rad./
				β+ /0.3	0.5/				0.0389(IT)
				ec/96	0.7/				0.2041
									0.5821
									0.5821
									0.8351
^{95}Tc		94.90765	20.0 h	ec/100/1.691		9/2+	5.9		0.7657
									1.0738
^{96m}Tc			52. m	IT	0.0342/98	4+			0.0342(IT)
				β+, ec/2/					0.7782
									1.2002
^{96}Tc		95.90787	4.3 d	ec/2.973		7+	+5.1		Mo k x-ray
									0.7782
									0.8125
									0.8498
									1.12168
^{97m}Tc			91. d	IT	0.0956/96	1/2-			Tc k x-ray
				ec	/3.9				0.0965
^{97}Tc		96.90637	4.2 ×10^{6} a	ec/100/0.320		9/2+			Mo k x-ray
^{98m}Tc			14.7 μs			(2)-			
^{98}Tc		97.90721	~ 6.6 ×10^{6} a	β- /1.80	0.40/100	6+			0.65241/100
				ec	//<0.036				0.74535/100
^{99m}Tc			6.007 h	IT/100/0.142		1/2-			Tc k x-ray
									0.14049
									0.14261
^{99}Tc		98.90625	2.13 ×10^{5} a	β- /0.294	0.293/100	9/2+	+5.6847	-0.129	
^{100m2}Tc			3.2 μs	IT	0.043286/100	(6)+			
^{100m1}Tc			8.3 μs	IT	0.02852/100	(4+)			
^{100}Tc		99.90765	15.8 s	β- /3.202	2.2/	1+			0.5396
				ec /2.6(10)-3/	2.9/				0.5908
					3.3				(0.3 79–2.30)
^{101m}Tc			0.64 ms	IT	0.1919/100	1/2-			
^{101}Tc		100.9073	14.0 m	β- /1.61	1.32/	9/2+			0.1272
									0.1841
									0.3068
									0.5451
									(0.073–0.969)
^{102m}Tc			4.4 m	IT/2/4.8	1.8/				0.4184
				β- /98/					0.4752
									0.6281
									0.6302
									1.0464
									1.1033
									1.6163
									2.2447
^{102}Tc		101.9092	5.3 s	β- /4.53	3.4/	1+			0.4686
					4.2				0.4751
					2.2/				1.1055
^{103}Tc		102.9092	54. s	β- /2.66	2.0/	5/2+			0.1361
					2.2/				0.1743
									0.2104
									0.3464
									0.5629
									(0.13–1.0)
^{104m}Tc			0.0035 ms	IT	0.0697/100				
^{104}Tc		103.9114	18.2 m	β- /5.60	5.3/	(3+)			0.3483

Elem. or Isot.	Natural abundance (Atom %)	Atomic mass or weight	Half-life/ Resonance width (MeV)	Decay mode/ Energy (/MeV)	Particle energy/ Intensity (MeV/%)	Spin (h/2π)	Nuclear magnetic mom. (nm)	Elect. quadr. mom. (b)	γ-Energy/ Intensity (MeV/%)
									0.3580
									0.5305
									0.5351
									0.8844
									0.8931
									1.6768
									(0.3–3.7)
^{105}Tc		104.9117	7.6 m	β- /3.6	3.4/	5/2+			0.1079
									0.1432
									0.3215
^{106}Tc		105.9144	36. s	β- /6.55		2+			0.2703
									0.5222
									1.9694
									2.2393
									2.7893
^{107m}Tc			3.9 μs	IT	0.0301/100	(1/2+)			
^{107}Tc		106.9155	21.2 s	β- /4.8					0.1027
									0.1063
									0.1770
									0.4587
^{108}Tc		107.9185	5.1 s	β- /7.72		(3)			0.2422
									0.4656
									0.7078
									0.7326
									1.5835
^{109}Tc		108.9203	1.4 s	β- /6.3	p/0.08				
^{110}Tc		109.9237	0.83 s	β- /8.8	p/0.04				0.2407
^{111}Tc		110.9259	0.30 s	β- .n/7.0	n/0.85				0.150/92.7
									0.063–1.435
^{112}Tc		111.92995	0.26 s	β, n	n/2.6				
^{113}Tc		112.93257	0.15 s	β-, n/8.	/2.1				0.0985/100
									0.0658–1.520
^{114}Tc		113.937	0.15 s	β-, n	/1.3				
^{115}Tc		114.940	0.07 s						
^{116}Tc		115.945	0.06 s						
^{117}Tc		116.948	45 ms						
^{118}Tc		117.953	30 ms						
^{119}Tc		118.957	22 ms						
^{120}Tc		119.962	~ 21 ms						
^{121}Tc			0.02 s						
$_{44}$Ru		**101.07(2)**							
^{85}Ru									
^{86}Ru									
^{87}Ru		86.951	> 1.5 μs						
^{88}Ru		87.942	1.2 s			0+			
^{89}Ru		88.938	1.4 s	β+ .p/8.	/3.				
^{90}Ru		89.93034	12. s	β+ /5.9		0+			ann.rad./
									0.155–1.551
^{91m}Ru			~ 7.6 s			(1/2-)			
^{91}Ru		90.92674	7.9 s	β+, ec/7.4		9/2+			ann.rad./
									(0.205-1.998)
^{92}Ru		91.92023	3.7 m	β+ /53/4.5		0+			ann.rad./
				ec/47/					0.1346
									0.2138
									0.2593
^{93m2}Ru			2.5 μs	IT	0.146/100	(21/2)+		0.04	
^{93m}Ru			10.8 s	IT/22/	0.7344	(1/2)-			ann.rad./
				β+, ec/78/	5.3/				0.7344
									1.1112
									1.3962
									2.0931
^{93}Ru		92.91710	1.0 m	β+ /6.3		9/2+			ann.rad./
				ec/					0.6807
									1.4349
									(0.5–4.2)weak
^{94m}Ru			0.071 ms	IT	0.1461/100	8+			
^{94}Ru		93.91134	52. m	ec/100/1.59		0+			0.3672
									0.5247
									0.8922
^{95}Ru		94.9104	1.64 h	ec/85/2.57	1.20/	5/2+	0.86		ann.rad./
				β+ /15/	0.91/				0.3364/71
									1.097/16.4
									0.6268/15.9

Elem. or Isot.	Natural abundance (Atom %)	Atomic mass or weight	Half-life/ Resonance width (MeV)	Decay mode/ Energy (/MeV)	Particle energy/ Intensity (MeV/%)	Spin (h/2π)	Nuclear magnetic mom. (nm)	Elect. quadr. mom. (b)	γ-Energy/ Intensity (MeV/%)
									0.036–2.424
^{96}Ru	5.54(14)	95.907590	>31.4 ×10^{20} a	β+β+		0+			
^{97}Ru		96.90755	2.84 d	ec/1.12		5/2+	-0.79		Tc k x-ray
									0.2157
									0.3245
									0.4606
^{98}Ru	1.87(3)	97.90529				0+			
^{99}Ru	12.76(14)	98.90593				5/2+	-0.64	+0.079	
^{100}Ru	12.60(7)	99.90421				0+			
^{101m}Ru			17.5 μs	IT	0.22207/100	11/2-			
^{101}Ru	17.06(2)	100.90558				5/2+	-0.72	+0.46	
^{102}Ru	31.55(14)	101.90434				0+			
^{103m}Ru			1.7 ms	IT		11/2-			
^{103}Ru		102.90632	39.26 d	β- /0.763	0.223	3/2+	0.206	+0.62	0.05329
									0.29498
									0.4438
									0.49708
									0.55704
									0.61033
									(0.04–1.6)
^{104}Ru	18.62(27)	103.90543	> 6.6 ×10^{20} a			0+			
^{105}Ru		104.90775	4.44 h	β- /1.917	1.11/22	3/2+	-0.3		0.12968
					1.134/13				0.1491
					1.187/49				0.2629
									0.31664
									0.46943
									0.67634
									0.72420
									(0.1–1.8)
^{106}Ru		105.90733	1.020 a	β- /0.0394	0.0394/100	0+			
^{107}Ru		106.9100	3.8 m	β- /2.9	2.1/				0.1939
					3.2/				0.3741
									0.4625
									0.8488
^{108}Ru		107.9102	4.5 m	β- /1.4	1.2/	0+			0.0923
									0.1651
									0.4339
									0.4975
									0.6189
^{109}Ru		108.9133	34.5 s	β- /4.2					0.1164
									0.3584
^{110}Ru		109.9140	15. s	β- /2.81		0+			0.1121
									0.3737
									0.4397
									0.7967
^{111}Ru		110.9176	1.5 s	β- /5.5					
^{112}Ru		111.9188	4.5 s	β- /4.5		0+			
^{113m}Ru			0.51 s			(7/2-)			
^{113}Ru		112.9228	0.80 s	β- /7.					0.2632
									0.048–2.418
^{114}Ru		113.92462	0.57 s	β- /6.1		0+			0.127/24
									(0.053–0.180)
^{115m}Ru			0.08 s	IT		(9/2-)			0.0617
^{115}Ru		114.9288	0.32 s	β- /8.					0.292
									(0.080-2.249)
^{116}Ru		115.93122	0.20 s			0+			
^{117m}Ru			2.49 μs	IT					(0.058-0.185)
^{117}Ru		116.936	0.14 s						
^{118}Ru		117.939	0.12 s			0+			
^{119}Ru		118.944	0.07 s						
^{120}Ru		119.946	45 ms			0+			
^{121}Ru		120.952	29 ms						
^{122}Ru		121.954	25 ms						
^{123}Ru		122.960	19 ms						
^{124}Ru		123.963	15 ms						
$_{45}$Rh		**102.90550(2)**							
^{89}Rh		88.951	> 0.12 μs	p					
^{90m}Rh			~ 15. ms			(0+)			
^{90}Rh		89.944	1.1 s			(9+)			
^{91m}Rh			1.5 s	IT		(1/2-)			0.387
^{91}Rh		90.937	1.5 s	β +,p	/1.3				(0.438-0.973)
^{92m}Rh			0.5 s	ec,β+		(2+)			0.866
^{92}Rh		91.93237	5.7 s	β+ /11.1					(0.163-0.991)

Elem. or Isot.	Natural abundance (Atom %)	Atomic mass or weight	Half-life/ Resonance width (MeV)	Decay mode/ Energy (/MeV)	Particle energy/ Intensity (MeV/%)	Spin (h/2π)	Nuclear magnetic mom. (nm)	Elect. quadr. mom. (b)	γ-Energy/ Intensity (MeV/%)
				β+,p	/1.9				
^{93}Rh		92.92591	12. s	β+ /8.1					
^{94m}Rh			25.8 s	β+ /		(8+)			(0.138–1.493)
									ann.rad./
									0.1264
									0.3117
									0.7562
									1.0752
									1.4307
^{94}Rh		93.92173	1.18 m	β+ /9.6	6.4/	4+			ann.rad./
									0.1461
									0.3117
									0.7562
									1.4307
^{95m}Rh			1.96 m	IT/88/	0.5433	(1/2)-			ann.rad./
				β+, ec/12/					0.5433(IT)
									0.7837
^{95}Rh		94.91590	5.0 m	β+ /5.1	3.2	9/2+			ann.rad./
									0.2293
									0.4103
									0.6610
									0.9416
									1.3520
^{96m}Rh			1.51 m	IT/60/0.052		3+			(0.2–3.8)
				β+, ec/40/	4.70/				ann.rad./
									Tc,Ru x-rays
									0.8326
									1.0985
									1.6921
^{96}Rh		95.9145	9.6 m	β+/6.45	3.3/	5+			(0.4–3.3)
				ec/					ann.rad./
									0.4299
									0.6315
									0.6853
									0.7418
									0.8326
^{97m}Rh			46. m	IT/5 /	2.6/	1/2-			(0.2–3.4)
				β+, ec/95/					ann.rad./
									0.1886
									0.4215
									2.2452
^{97}Rh		96.9113	31.0 m	β+ /3.52	2.1/	9/2+			ann.rad./
									0.1886
									0.3892
									0.4515
									0.8398
									0.8788
^{98m}Rh			3.6 m	IT/89		(5+)			(0.2–3.5)
				β+/11					ann.rad./
									0.6154
									0.6524
									0.7452
^{98}Rh		97.9107	8.7 m	β+ /90/5.06	3.4/	2+			ann.rad./
									0.6524
									0.7623
^{99m}Rh			4.7 h	β+ /8/	.74/	9/2+	5.67		ann.rad./
				ec/92/					0.2766/
									0.3408
									0.6178
									1.2612
^{99}Rh		98.90813	16. d	β+/4/2.10	0.54/	1/2-			ann.rad./
				ec/97/	0.68/				0.0894/
									0.3530
									0.5277
^{100m}Rh			4.6 m	IT/98.3/		5+			(0.1–2.0)
				ec,β+ /1.7/					ann.rad./
									0.0748/
									0.2647(IT)
^{100}Rh		99.9081	20.8 h	β+ /3.63	2.62/	1-			0.4462
				ec/	2.07/				0.5396
									0.5882
									0.8225
									1.5534
									2.3761

Elem. or Isot.	Natural abundance (Atom %)	Atomic mass or weight	Half-life/ Resonance width (MeV)	Decay mode/ Energy (/MeV)	Particle energy/ Intensity (MeV/%)	Spin (h/2π)	Nuclear magnetic mom. (nm)	Elect. quadr. mom. (b)	γ-Energy/ Intensity (MeV/%)
^{101m}Rh			4.34 d	ec/92.8/		9/2+	+5.47		Rh k x-ray
				IT/7.2/	0.1573				0.1272/
									0.3069
									0.5451
^{101}Rh		100.90616	3.3 a	ec/0.54		1/2-			Ru k x-ray
									0.1272
									0.1980
									0.3252
^{102m}Rh			3.74 a	ec/2.323		6(+)	4.04		0.4751
				IT/0.0419					0.6313
			> 1.2 ×10^6 a	β+	/<0.00025				0.6975
									0.7668
									1.0466
									1.1032
^{102}Rh		101.90684	207. d	ec/62		(2-)	~ 0.5		ann.rad./
				β- /19/					0.4686
				β+ /14/					0.4751
									0.5566
									0.6280
									1.1032
									(0.4–1.6)
^{103m}Rh			56.12 m	IT	0.03976	7/2+	4.54		
^{103}Rh	100.	102.90550				1/2-	-0.0884		
^{104m}Rh			4.34 m	IT/99+ /		5+			Rh k x-ray
				β-	1.3/				0.0514
									0.0971
									0.5558
^{104}Rh		103.90665	42.3 s	β-/99+/2.441	1.88/2	1+			0.3581
				ec/0.4/1.141	2.44/98				0.5558
									1.2370
									(0.35–1.8)
^{105m}Rh			43. s	IT/1.296	0.1298/100	1/2-			Rh k x-ray
									0.1296
^{105}Rh		104.90569	35.3 h	β- /0.567	0.247/30	7/2+	+4.45		0.2801
					0.567/70				0.3061/4.8
									0.3189/17.0
^{106m}Rh			2.18 h	β- /	0.92/	(6)+			0.2217
									0.4510
									0.5119
									0.6162
									0.7173
									0.7484
									1.0458
									1.5277
^{106}Rh		105.90729	29.9 s	β- /3.54	2.4/2	1+	+2.58		0.51186/
					3.0/12				0.61612
					3.54/79				0.62187
									(0.05–3.04)
^{107m}Rh			>0.01 ms	IT	0.2684/100				
^{107}Rh		106.9067	21.7 m	β- /1.51	1.20/65	7/2+			0.2776
					1.5/17				0.3028
						(5+)			0.3925
^{108m}Rh			6.0 m	β- /	1.57/				0.4339
									0.4973
									0.6189
^{108}Rh		107.9087	17. s	β- /4.5		1+			0.4046
									0.4339
									0.4973
									0.5811
									0.6146
									0.9014
									0.9471
^{109m}Rh			1.7 µs	IT	0.2260/100	(3/2)+			
^{109}Rh		108.90875	1.34 m	β- /2.59	2.25/	7/2+			0.1134
									0.1780
									0.2914
									0.3254
									0.3268
									0.4261
									(0.1–1.6)
^{110m}Rh			28 s	β- /	6/	(6+)			0.3737
									0.4397
									0.7967

Elem. or Isot.	Natural abundance (Atom %)	Atomic mass or weight	Half-life/ Resonance width (MeV)	Decay mode/ Energy (/MeV)	Particle energy/ Intensity (MeV/%)	Spin (h/2π)	Nuclear magnetic mom. (nm)	Elect. quadr. mom. (b)	γ-Energy/ Intensity (MeV/%)
^{110}Rh		109.9111	3.1 s	β- /5.4	5.5/	1+			0.3737
									0.4400
									0.5463
									0.6877
									0.8381
									0.9045
^{111}Rh		110.91164	11. s	β- /3.7					0.275
^{112m}Rh			6.8 s	β- /		(6+)			
^{112}Rh		111.9144	3.5 s	β- /6.2		1+			0.3489
^{113}Rh		112.91544	0.9 s	β- /4.9					0.1285
^{114m}Rh			1.9 s	β- /		(7-)			(0.103-1.923)
^{114}Rh		113.9187	1.8 s	β- /6.5		1+			(0.276-0.783)
^{115}Rh		114.92031	0.99 s	β- /6.0					
^{116m}Rh			~ 0.57 s	β- /		(6-)			0.3405
^{116}Rh		115.9241	0.7 s	β- /8.0		1+			0.340
^{117}Rh		116.9260	0.42 s	β- /7.					0.398–1.665
									0.0346
^{118}Rh		117.9303	0.27 s						0.1317
									0.379
									0.575
^{119}Rh		118.9326	0.17 s						0.370–1.037
^{120}Rh		119.937	0.13 s						
^{121}Rh		120.939	0.15 s						
^{122}Rh		121.944	~ 51 ms						
^{123}Rh		122.947	42 ms						
^{124}Rh		123.952	30 ms						
^{125}Rh		124.955	27 ms						
^{126}Rh		125.959	19 ms						
^{127}Rh			~ 0.02 s						
$_{46}$Pd		**106.42(1)**							
^{90}Pd									
^{91}Pd		90.950	> 1.5 μs						
^{92}Pd		91.941	1.0 s			0+			
^{93}Pd		92.937	1.2 s	β+, p	/7.5	9/2+			0.240/81
^{94}Pd		93.92904	9.6 s	ec, β+ /~ 6.6		0+			0.382–0.864
									0.5582
^{95m}Pd		94.92684	13.3 s	ec, β+ /10.2		21/2+			(0.055–0.798)
^{95}Pd		94.92489	7.5 s						
^{96m}Pd			1.8 μs	IT	0.1064/100	8+			
^{96}Pd		95.91822	2.03 m	ec, β+ /3.5	1.15/	0+			0.1248
^{97}Pd		96.91647	3.1 m	β+, ec/4.8	3.5/	5/2+			0.4995
									ann.rad./
									0.2653
									0.4752
									0.7927
^{98}Pd		97.91270	17.7 m	β+ /1.87		0+			(0.2–3.4)
				ec/					ann.rad./
									0.0677
									0.1125
									0.6630
^{99}Pd		98.91178	21.4 m	β+ /49/3.37	2.18/	5/2+			0.8379
				ec/51/					ann.rad./
									0.1360
									0.2636
									0.6734
^{100}Pd		99.9085	3.7 d	ec/0.36		0+			(0.2–2.85)
									0.03271
									0.0748
^{101}Pd		100.90829	8.4 h	β+ /5/1.980	0.776/	5/2+	-0.66		0.0840
				ec/95/					ann.rad./
									0.0244
									0.2963
									0.5904
^{102}Pd	1.02(1)	101.90560	> 6.0 ×10^{18} a	Ec ec		0+			
^{103}Pd		102.90608	16.99 d	ec/0.543		5/2+			Rh k x-ray
									0.03975
									0.3575
									0.4971
^{104}Pd	11.14(8)	103.90403				0+			
^{105m}Pd			36.1 μs	IT	0.1829/100	11/2-			
^{105}Pd	22.33(8)	104.90508				5/2+	-0.642	+0.66	

Elem. or Isot.	Natural abundance (Atom %)	Atomic mass or weight	Half-life/ Resonance width (MeV)	Decay mode/ Energy (/MeV)	Particle energy/ Intensity (MeV/%)	Spin (h/2π)	Nuclear magnetic mom. (nm)	Elect. quadr. mom. (b)	γ-Energy/ Intensity (MeV/%)
^{106}Pd	27.33(3)	105.90348				0+			
^{107m}Pd			21.3 s	IT/0.2149		11/2-			Pd k x-ray
									0.2149(IT)
^{107}Pd		106.90513	6.5 ×10^{6} a	β- /0.033	0.03/	5/2+			
^{108}Pd	26.46(9)	107.90389				0+			
^{109m}Pd			4.70 m	IT	0.18899	11/2-			Pd x-ray
									0.1889(IT)
^{109}Pd		108.90595	13.5 h	β- /1.116	1.028	5/2+			0.0880
									(0.08–1.0)
^{110}Pd	11.72(9)	109.905172	> 2.0 ×10^{20} a	β- β-		0+			
^{111m}Pd			5.57 h	IT/73/0.172		11/2-			0.0704
				β- /27/	0.35				0.1722
					0.77				0.3912
									(0.1–1.97)
^{111}Pd		110.90769	23.4 m	β- /2.19	2.2/95	5/2+			0.0598
									0.2454
									0.5800
									0.6504
									1.3885
									1.4590
^{112}Pd		111.90733	21.04 h	β- /0.29	0.28/	0+			0.018
^{113m}Pd			0.3 s	IT	0.0811/100	(9/2-)			0.0959
^{113}Pd		112.91026	1.55 m	β- /3.34		5/2+			0.0958
									0.4824
									0.6436
									0.7394
^{114}Pd		113.91037	2.48 m	β- /1.45		0+			0.1266
									0.2320
									0.5582
									0.5760
^{115m}Pd			50 s			(7/2-)			0.089
^{115}Pd		114.9137	25. s	β- /4.58		(3/2+)			0.1255
									0.2554
									0.3428
^{116}Pd		115.91430	12.7 s	β- /2.61		0+			0.1015
									0.1147
									0.1778
^{117m}Pd			19 ms	IT					0.203
^{117}Pd		116.91796	4.4 s	β- /5.7		(3/2+)			0.2473
									0.077–0.403
^{118}Pd		117.91907	2.4 s	β- /4.1		0+			0.1254
									0.028–0.596
^{119}Pd		118.9233	0.9 s	β- /6.5					0.2566
									0.070–0.326
^{120}Pd		119.92455	0.49 s	β- /5.0		0+			0.1581
									0.053–0.595
^{121}Pd		120.92895	0.29 s						
^{122}Pd		121.9306	0.18 s			0+			
^{123}Pd		122.935	0.17 s						
^{124m}Pd			> 0.02 ms	IT	0.062/100				0.2548
^{124}Pd		123.937	~ 0.04 s			0+			
^{125}Pd		124.942	0.06 s						
^{126m}Pd			23 ms			(10+)			
^{126}Pd		125.944	49 ms						
^{127}Pd		126.949	38 ms						
^{128m}Pd			6 μs	IT	0.075/100	(8+)			0.0751
^{128}Pd		127.952	35 ms						
^{129}Pd			0.03 s						
$_{47}$**Ag**		**107.8682(2)**							
^{92}Ag									
^{93}Ag		92.950	0.23 μs	p					
^{94m2}Ag			0.40 s	β+	p/1.01/2.2	(21+)			(0.153-1.132)
				β+,p	p/0.79/1.9				
^{94m1}Ag			0.60 s	β+		(7+)			(0.659-0.905)
				β+,p	p//20.				
^{94}Ag		93.944	0.03 s	β+		0+			
				β+,p					
^{95}Ag		94.936	1.9 s	β+, p/	/2.5				(0.089–2.940)
^{96m3}Ag			0.010 ms	IT		(13-)			0.473
^{96m2}Ag			1.5 μs	IT		(15+)			(0.257-1.506)
^{96m}Ag			4.40 s	β+		(2+)			
				β+, p	/7.				
^{96}Ag		95.931	7. s	β+ /11.6		2+			ann.rad./

Elem. or Isot.	Natural abundance (Atom %)	Atomic mass or weight	Half-life/ Resonance width (MeV)	Decay mode/ Energy (/MeV)	Particle energy/ Intensity (MeV/%)	Spin (h/2π)	Nuclear magnetic mom. (nm)	Elect. quadr. mom. (b)	γ-Energy/ Intensity (MeV/%)
				ec/					0.1248
				β+, p	/18.				0.4995
^{97}Ag		96.924	19. s	β+ /7.0			6.1		(0.1066–1.416)
				ec/					ann.rad./
									0.6862
									1.2941
^{98}Ag		97.9216	47.6 s	β+ /8.4		5+	4.6		(0.352–3.294)
				ec/	/36.				ann.rad./
				β+, p	/0.11				0.5711
									0.6786
									0.8631
^{99m}Ag			11. s	IT/100/	0.1636	(1/2-)			(0.153–1.185)
									Ag k x-ray
									0.1636(IT)
^{99}Ag		98.91765	2.07 m	β+ /87 5.4		9/2+	5.8		0.3426
				ec/13/					ann.rad./
									0.2199
									0.2645
									0.8056
									0.8323
^{100m}Ag			2.2 m	β+ /		(2+)			(0.2–3.5)
				ec/					ann.rad./
									0.6657
^{100}Ag		99.91612	2.0 m	β+/7.1	4.7/	5+	4.4		1.6941
				ec/					ann.rad./
									0.2807
									0.4503
									0.6657
									0.7508
^{101m}Ag			3.1 s	IT/0.23	0.176	(1/2)-			0.7732
									Ag k x-ray
									0.0981
^{101}Ag		100.91268	11.1 m	β+/69/4.2	2.7/	9/2+	5.6	+~ 0.35	0.176(IT)
				ec/31/					ann.rad./
									0.2610
					2.18/				0.2747
					2.73/				0.3269
					3.38/				0.4392
									0.6673
									1.1739
^{102m}Ag			~ 7.7 m	β+ /38/	3.4	2+	+4.1		(0.2–3.1)
				ec/13/					ann.rad./
				IT/49/					0.5567
									0.9777
									1.8347
									2.0545
									2.1594
^{102}Ag		101.9117	13.0 m	β+ /78/5.92	2.26/	5(+)	~ 4.6		3.2386
				ec/22/					ann.rad./
									0.5564
									0.7193
^{103m}Ag			5.7 s	IT/0.134		1/2-			0.163–2.242
									Ag k x-ray
^{103}Ag		102.90896	1.10 h	β+ /28/2.69	1.7	7/2+	+4.47	0.8	0.1344
				ec/72/	1.3				ann.rad./
									0.1187
^{104m}Ag			33. m	β+ /64/	2.71/	2+	+3.69		0.1482
				ec/36/					ann.rad./
				IT/0.07/					0.5558
									0.7657
^{104}Ag		103.90862	69. m	β+ /16/4.28	0.99/	5+	3.92	+1.1	(0.5–3.4)
				ec/84/					ann.rad./
									0.5558
									0.9259
									0.9416
^{105m}Ag			7.2 m	IT/98/0.0255		7/2+	+4.41	+~ 0.85	(0.18–2.27)
				ec/2 /					Ag x-ray
									0.3063
									0.3192
^{105}Ag		104.90653	41.3 d	ec/1.35		1/2-	0.101		(0.1–1.0)
									0.0640
									0.2804
									0.3445

Elem. or Isot.	Natural abundance (Atom %)	Atomic mass or weight	Half-life/ Resonance width (MeV)	Decay mode/ Energy (/MeV)	Particle energy/ Intensity (MeV/%)	Spin (h/2π)	Nuclear magnetic mom. (nm)	Elect. quadr. mom. (b)	γ-Energy/ Intensity (MeV/%)
									0.4434
^{106m}Ag			8.3 d	ec/		6+	3.71	+1.1	Pd k x-ray
									0.4510
									0.5118
									0.7173
									1.0458
^{106}Ag		105.90666	24.0 m	β+ /59/2.965	/1.96	1+	+2.9		ann.rad./
				ec/41 /					0.5119
^{107m}Ag			44.3 s	IT/0.093		7/2+	+4.40	1.0	Ag x-ray
									0.0931
^{107}Ag	51.839(8)	106.90509				1/2-	-0.113680		
^{108m}Ag			~ 440 a	ec/92/		6+	3.58	+1.3	Ag k x-ray
				IT/8 /0.079					Pd k x-ray
									0.43392
									0.61427
									0.72290
^{108}Ag		107.90595	2.39 m	β- /97/1.65	1.02/1.7	1+	+2.6884		ann.rad./
				ec/2/	1.65/96				0.43392
				β- /1/1.92	0.88/0.3				0.61885
									0.63298
^{109m}Ag			39.6 s	IT/0.088		7/2+	+4.40	+1.0	Ag k x-ray
									0.0880
^{109}Ag	48.161(8)	108.90476				1/2-	-0.13069		
^{110m}Ag			249.8 d	β- /99/	0.087	6+	+3.61	+1.4	0.65774
				IT/1 /0.1164	0.530				0.76393
									0.88467
									0.93748
									1.38427
									(0.447–1.56)
^{110}Ag		109.90611	24.6 s	β- /2.892	2.22/5	1+	+2.727	0.2	0.65774
					2.89/95				0.8154
									1.1257
^{111m}Ag			1.08 m	IT/99/0.0598		7/2+			Ag k x-ray
				β- /1/					0.0598
									0.2454
^{111}Ag		110.90530	7.45 d	β- /1.037	1.035/	1/2-	-0.146		0.2454
									0.3421
^{112}Ag		111.90705	3.13 h	β- /3.96	3.94/	2(-)	0.0547		0.6067
					3.4				0.6174
									1.3877
									(0.4–2.9)
^{113m}Ag			1.14 m	IT/80 /0.043		7/2+			0.1422
				β- /20 /	1.5				0.2983
									0.3161
									0.3923
^{113}Ag		112.9066	5.3 h	β- /2.02	2.01/	1/2-	0.159		0.2588
									0.2986
^{114m}Ag			1.5 ms	IT					
^{114}Ag		113.90882	4.6 s	β- /5.08	4.9/	1+			0.5582
									0.5760
									1.9946
^{115m}Ag			18 s	β- /79/		7/2+			0.1134
				IT/21/					0.1315
									0.2288
									0.3887
^{115}Ag		114.9088	20. m	β- /3.10		1/2-			0.1316
									0.2128
									0.2291
									0.4727
									(0.13–2.49)
^{116m2}Ag			20. s	β-,IT/7	IT/0.0479	(3+)			
^{116m}Ag			9.3 s	β-/92 /	3.2/	(6-)			0.5134
					2.9				0.7055
				IT/8	IT/.0809				(0.255–2.838)
^{116}Ag		115.91139	2.68 m	β- /6.16	5.3	2-			0.5134
									0.6993
									2.4779
^{117m}Ag			5.3 s	β- /	3.2/	(7/2+)			0.1354
									0.2981
									0.3868
									0.1571
^{117}Ag		116.9118	1.22 m	β- /4.18	2.3	1/2-			0.1354
									0.3377

Elem. or Isot.	Natural abundance (Atom %)	Atomic mass or weight	Half-life/ Resonance width (MeV)	Decay mode/ Energy (/MeV)	Particle energy/ Intensity (MeV/%)	Spin (h/2π)	Nuclear magnetic mom. (nm)	Elect. quadr. mom. (b)	γ-Energy/ Intensity (MeV/%)
^{118m}Ag			2.0 s	β- /59/					0.1277
				IT/41 /0.1277					0.4878
									0.6771
									0.7709
									(0.190-2.778)
^{118}Ag		117.91460	4.0 s	β- /7.1					0.4878
									0.6771
									3.2259
^{119m}Ag			~ 6.0 s	β-		(1/2-)			
^{119}Ag		118.9156	2.1 s	β- /5.35		(7/2+)			0.0674
									0.3662
									0.3991
									0.6264
^{120m2}Ag			~ 0.44 s	IT/68/		7(-)			0.5056
				β-/32/					(0.1145-3.142)
^{120m1}Ag			~ 1.52 s	β-		4(+)			
^{120m}Ag			0.40 s	β- /63.					0.2030
				IT/37.					0.5059
									0.6978
									0.8300
									(0.115-1.644)
^{120}Ag		119.91879	0.9 s	β-					0.5059
				β-,n	n//<0.0030%				0.6978
									0.8171
									(0.442-3.044)
^{121}Ag		120.9201	0.78 s	β- /6.4					0.1150
									0.3148
									0.3537
									0.3696
									0.5007
									1.5105
									(0.11–2.5)
^{122m3}Ag			0.006 ms	IT		(1+)			
^{122m2}Ag			0.2 s	β-		(9-)			
^{122m1}Ag			0.6 s	β-		(1-)			
^{122m}Ag			1. s	β- /					
^{122}Ag		121.9237	0.53 s	β- /9.2		(3+)			0.7617
									(0.351-1.079)
^{123}Ag		122.9253	0.28 s	β- /7.4					
^{124m}Ag			0.14 s	IT		(8-)			(0.0755-0.156)
^{124}Ag		123.929	0.17 s	β- /10.1					
^{125}Ag		124.931	0.16 s	β-		(9/2+)			0.686
									(0.672-0.731)
^{126m}Ag			0.09 s						0.2548
^{126}Ag		125.935	0.05 s	β-					
^{127}Ag		126.937	0.089 s	β-					
^{128}Ag		127.941	58 ms	β-					
^{129m}Ag			~ 0.16 s						
^{129}Ag		128.944	0.052 s	β-, n					
^{130}Ag		129.951	0.042 s						
^{131}Ag			~ 0.035 s						
^{132}Ag			0.03						
$_{48}$Cd		**112.414(4)**							
^{94}Cd									
^{95}Cd		94.950	0.09 s						
^{96m}Cd			0.3 s	β+		(16+)			
^{96}Cd		95.940	1.0 s	β+, p	/ ~ 5.5	0+			
^{97m}Cd			3.8 s	β+, p	/25	(25/2+)			
^{97}Cd		96.935	1.1 s	β+, (p)					
^{98}Cd		97.9274	9.2 s	β+ /5.4		0+			
				(p)	/0.025				
^{99}Cd		98.92493	16. s	β+, ec/6.9					ann.rad./
^{100}Cd		99.92035	1.1 m	β+, ec/3.9		0+			ann.rad./
									(0.090–1.043)
^{101}Cd		100.91859	1.2 m	β+ /83/5.5	4.5	5/2+			In k x-ray
				ec/17/					0.0985
									1.7225
									0.31–2.84)
^{102}Cd		101.91448	5.8 m	β+ /27/2.59		0+			ann.rad./
				ec/73					0.0974
									0.4810
									1.0366
									1.3598

Elem. or Isot.	Natural abundance (Atom %)	Atomic mass or weight	Half-life/ Resonance width (MeV)	Decay mode/ Energy (/MeV)	Particle energy/ Intensity (MeV/%)	Spin (h/2π)	Nuclear magnetic mom. (nm)	Elect. quadr. mom. (b)	γ-Energy/ Intensity (MeV/%)
^{103}Cd		102.91342	7.5 m	β+ /33/4.14		5/2+	-0.81	~ -0.7	ann.rad./
				ec/67/					Ag k x-ray
									1.0799
									1.4487
									1.4618
									(0.1–2.8)
^{104}Cd		103.90986	58. m	ec/1.14		0+			Ag k x-ray
									0.0835
									0.7093
^{105m}Cd			0.005 ms	IT	0.1265/100			+1.0	
^{105}Cd		104.90946	55.5 m	β+ /26/2.739	1.69/	5/2+	-0.7393	+0.37	Ag k x-ray
				ec/74/					0.3469
									0.6072
									0.9618
									1.3025
									(0.25–2.4)
^{106}Cd	1.245(22)	105.90646	≥ 1.1 ×10^{21} a	ec, β+		0+			
^{107}Cd		106.90661	6.52 h	ec/99+/1.417		5/2+	-0.615055	+0.60	Ag k x-ray
				β+ /					0.0931
									0.8289
^{108}Cd	0.888(11)	107.90418	>4.1 ×10^{17} a	ec ec		0+			
^{109m2}Cd			10.9 µs	IT	0.259/100	11/2		-0.9	
^{109m1}Cd			0.012 ms	IT	0.060/100	1/2+			
^{109}Cd		108.90499	462.2 d	ec/0.214		5/2+	-0.827846	+0.60	Ag k x-ray
									0.08804/.0366
^{110}Cd	12.470(61)	109.903007				0+			
^{111m}Cd			48.5 m	IT/	0.1508/100	11/2-	- 1.1052	-0.75	Cd k x-ray
									0.1508(IT)
									0.2454
^{111}Cd	12.795(38)	110.904183				1/2+	-0.594886		
^{112}Cd	24.109(12)	111.902763				0+			
^{113m}Cd			13.9 a	β- /99.9/0.59	0.59/99.9	11/2-	-1.087784	-0.61	0.2637
^{113}Cd	12.227(7)	112.904408	8.04 ×10^{15} a	β-		1/2+	-0.622301		
^{114}Cd	28.754(81)	113.903365	>1.3 ×10^{18} a	β-β-		0+			
^{115m}Cd			44.6 d	β- /1.629	0.68/1.6	11/2	-1.041034	-0.48	0.48450
					1.62/97				0.93381
									1.29064
^{115}Cd		114.905438	2.228 d	β- /1.446	0.593/42	1/2+	-0.648426		0.23141
					1.11/58				0.26085
									0.33624
									0.49227
									0.52780
^{116}Cd	7.512(54)	115.904763	3. ×10^{19} a	β-β-		0+			
^{117m}Cd			3.4 h	β- /2.66	0.72/	11/2-	- 0.998	- 0.32	0.1586
									0.5529
									0.37–2.42
^{117}Cd		116.90723	2.49 h	β- /2.52	0.67/51	1/2+	- 0.7436		0.2209
					2.2/10				0.2733
									0.3445
									1.3033
^{118}Cd		117.9069	50.3 m	β- /0.52		0+			
^{119m}Cd			2.20 m	β- /		11/2(-)	- 0.9642	- 0.13	0.1056
									0.7208
									1.0250
									2.0213
^{119}Cd		118.9099	2.69 m	β- /3.8	~ 3.5/	1/2+	- 0.9201		0.1340
									0.2929
									0.3429
^{120}Cd		119.90987	50.8 s	β- /1.76	1.5/	0+			
^{121m}Cd			8. s	β- /		11/2(-)	- 1.010	~ -0.009	0.1008
									0.9878
									1.0209
									1.1815
									2.0594
^{121}Cd		120.91296	13.5 s	β- /4.9		(3/2+)	0.627	- 0.27	0.2102
									0.3242
									0.3492
									1.0403
^{122}Cd		121.91346	5.3 s	β- /3.0		0+			
^{123m}Cd			1.8 s	β- /		11/2(-)	- 1.0015	~ 0.135	
^{123}Cd		122.91689	2.09 s	β- /6.12		3/2+	0.790	- 0.04	

Elem. or Isot.	Natural abundance (Atom %)	Atomic mass or weight	Half-life/ Resonance width (MeV)	Decay mode/ Energy (/MeV)	Particle energy/ Intensity (MeV/%)	Spin (h/2π)	Nuclear magnetic mom. (nm)	Elect. quadr. mom. (b)	γ-Energy/ Intensity (MeV/%)
^{124}Cd		123.91766	1.24 s	β- /4.17		0+			0.0365
									0.0628
									0.1799
^{125m2}Cd			0.019 ms	IT	0.742/100	(19/2+)			
^{125m}Cd			0.48 s	β- /		11/2(-)	- 0.935	0.27	
^{125}Cd		124.92126	0.68 s	β- /7.16		3/2+	0.860	0.21	
^{126}Cd		125.92243	~ 0.51 s	β- /5.49		0+			0.2601
^{127m}Cd			0.0175 ms	IT		(19/2+)	- 0.870	0.34	(0.110-0.849)
^{127}Cd		126.9265	0.33 s	β- /8.5		3/2+	0.878	0.24	
^{128m}Cd			3.6 μs	IT	0.0687/100	(10+)			(0.0687-1.224)
^{128}Cd		127.92781	0.246 s	β- /7.1		0+			0.247
^{129m2}Cd			3.6 ms	IT		(21/2+)			(0.353-1.587)
^{129m}Cd			0.16 s	β-		3/2+	0.848	+0.13	
^{129}Cd		128.932	0.147 s	β- /5.9		11/2-	0.706	+0.57	0.281
^{130}Cd		129.934	0.127 s	β- /		0+			(0.229-5.583)
				β-, n	/~ 3.5				
^{131}Cd		130.941	68 ms						(0.844-6.039)
^{132}Cd		131.946	0.10 s	β-, n/	/60	0+			
^{133}Cd		132.953	0.06 s						
^{134}Cd			0.07						
$_{49}$In		**114.818(1)**							
^{96}In									
^{97}In		96.949	0.05 s						
^{98m}In			1.3 s	β+, p					
^{98}In		97.942	0.5 s						
^{99}In		98.934	3.1 s	β+ /8.9					
				β+, p	/0.9				
^{100}In		99.931	5.8 s	β+, (p)/10.5	/1.7				(0.297-1.365)
^{101}In		100.926	15. s	β+ /7.3					
^{102}In		101.92411	23. s	ec/8.9		(5)			0.1566
									0.7767
									(0.397–0.923)
^{103m}In			34. s	Ec, β+/67/		(1/2-)			
^{103}In		102.9199	1.1 m	β+, ec6.05	4.2	9/2+			ann.rad./
				ec	/45				0.1879
									(0.157–3.98)
^{104m}In			16. s	IT/0.0935		(3+)			
^{104}In		103.91822	1.84 m	β+, ec/7.9	4.8	(5+)	+4.44	+0.6	ann.rad./
									0.6580
									0.8341
									0.8781
^{105m}In			0.8 m	IT	0.674/100	(1/2-)			In k x-ray
									0.6740
^{105}In		104.9145	5.1 m	β+, ec/4.85	3.7	9/2+	+5.675	~ +0.79	0.1310
									0.2600
									0.6038
^{106m}In			5.2 m	β+ /85/	4.90	(2)+			ann.rad./
				ec/15/					0.6326
									0.8611
									1.7164
^{106}In		105.9135	6.2 m	β+ /65/6.52	2.6	7+	+4.92	+0.9	ann.rad./
				ec/35/					0.2259
									0.6327
									0.8611
									0.9978
									1.0091
^{107m}In			50 s	IT/0.6786		½-			In k x-ray
									0.6785
^{107}In		106.9103	32.4 m	β+ /35/3.43	2.20/	9/2+	+5.59	~ +0.77	ann.rad./
				E.C/65/					Cd k x-ray
									0.2050
									0.3209
									0.5055
									(0.2–2.99)
^{108m}In			57. m	β+ /53/	1.3	6+	+4.94	+0.44	ann.rad./
				ec/47/					Cd k x-ray
									0.6329
									1.9863
									3.4522
^{108}In		107.9097	40. m	β+ /33/5.15	3.49/	7+	+4.561	+0.96	ann.rad./
				ec/67/					Cd k x-ray
									0.2429
									(0.6331-0.8756)

Elem. or Isot.	Natural abundance (Atom %)	Atomic mass or weight	Half-life/ Resonance width (MeV)	Decay mode/ Energy (/MeV)	Particle energy/ Intensity (MeV/%)	Spin (h/2π)	Nuclear magnetic mom. (nm)	Elect. quadr. mom. (b)	γ-Energy/ Intensity (MeV/%)
^{109m2}In			0.21s	IT	0.673	19/2+			
^{109m}In			1.3 m	IT/0.650		½-			In k x-ray
									0.6498
^{109}In		108.90715	4.17 h	β+ /8/2.02	0.79/	9/2+	+5.54	+0.80	ann.rad./
				ec/92/					Cd k x-ray
									0.2035
									0.6235
^{110m}In			4.9 h	ec/		7+	+4.71	+0.95	Cd k x-ray
									0.6577
									0.8847
									0.9375
									(0.1–1.98)
^{110}In		109.9072	1.15 h	β+ /62/3.88	2.22/	2+	+4.37	+0.32	ann.rad./
				ec/38/					Cd k x-ray
									0.6577
									(0.6–3.6)
^{111m}In			7.7 m	IT/0.537		½-	+5.53		In k x-ray
									0.537
^{111}In		110.90511	2.8049 d	ec/0.866		9/2+	+5.50	+0.76	Cd k x-ray
									0.1712
									0.2453
^{112m2}In			2.8 μs	IT	0.263	(8)-		0.092	
^{112m}In			20.7 m	IT/0.155		4+	+ 5.227	+0.68	In k x-ray
									0.1555
^{112}In		111.90554	14.4 m	β+ /22/2.586		1+	+2.82	~ +0.82	ann.rad./
				ec/34/					Cd k x-ray
				β- /0.663					0.6171
^{113m}In			1.658 h	IT/0.3917		½-	-0.21074		In k x-ray
									0.3917
^{113}In	4.281(52)	112.90406				9/2+	+5.5289	0.76	
^{114m2}In			0.043 s	IT	0.31167	8-			
^{114m}In			49.51 d	IT/97/0.190		5+	+4.65	+0.70	In k x-ray
				ec/3 /					0.19027
^{114}In		113.90492	1.198 m	β- /97/1.989	1.984/	1+	+2.82		Cd k x-ray
				ec/3/1.453					0.5584
									0.5727
									1.2998
^{115m}In			4.486 h	IT/95/0.336		½-	-0.2440		In k x-ray
				β- /5 /0.83					0.3362
									0.4974
^{115}In	95.719(52)	114.9038788	4.4 ×10^{14} a	β- /0.495		9/2+	+5.5408	0.77	
^{116m2}In			2.18 s	IT/0.162		8-	+3.22	+0.30	In k x-ray
				ec	/0.023				0.1624
^{116m1}In			54.3 m	β- /	1.0	5+	+4.44	+0.76	0.13792
									0.41688/27
									1.09723/58.5
									1.29349/85
^{116}In		115.905260	14.1 s	β- /3.274	3.3/99	1+	2.788	0.11	0.46313
									1.2526
									1.29349
^{117m}In			1.94 h	β- /53/1.769	1.77/	½-	-0.25174		In k x-ray
				IT/47 /					0.15855
									0.31531
									0.55294
^{117}In		116.90452	44. m	β- /1.455	0.74/	9/2+	+5.519	+0.79	0.15855
									0.3966
									0.55294
^{118m2}In			8.5 s	IT/98/		8-	+3.32	+0.42	In k x-ray
				β- /2/					0.1382
^{118m1}In			4.45 m	β- /	1.3	5+	+4.23	+0.76	0.2086
					2.0				0.6833
									1.2295
^{118}In		117.90636	5.0 s	β- /4.42	4.2/	1+			0.5282
									1.1734
									1.2295
									2.0432
^{119m}In			18.0 m	β- /97/	2.7/	½-	-0.32		0.3114
				IT/3/0.311					0.7631
^{119}In		118.90585	2.3 m	β- /2.36	1.6/	9/2+	+5.52	+0.81	0.0239
									0.6495
									0.7631
									1.2149
^{120m2}In			47 s	β- /6.1		8-	+3.692	+0.50	1.171

Elem. or Isot.	Natural abundance (Atom %)	Atomic mass or weight	Half-life/ Resonance width (MeV)	Decay mode/ Energy (/MeV)	Particle energy/ Intensity (MeV/%)	Spin (h/2π)	Nuclear magnetic mom. (nm)	Elect. quadr. mom. (b)	γ-Energy/ Intensity (MeV/%)
^{120m1}In			46. s	β- /5.8	2.2/	(5)+	+4.30	+0.77	1.023
									1.171
^{120}In		119.9078	3.1 s	β- /5.37	5.6/	(1+)			1.023
					3.1/				0.4146
									0.5924
									0.8637
									1.0232
									1.1714
									(0.4–2.7)
^{121m2}In			0.017 ms			(21/2-)			
^{121m}In			3.9 m	β- /99/ IT/1/0.313	3.7/	1/2-	-0.355		0.0601
									0.3136
									0.9256
									1.0412
									1.1022
									1.1204
^{121}In		120.9079	23. s	β- /3.36	2.5	9/2+	+5.50	+0.77	0.2620
									0.6573
									0.9256
^{122m}In			10. s	β- /	4.4/	8-	+3.78	+0.56	1.0014
									1.1403
^{122}In		121.9103	1.5 s	β- /6.37	5.3/	(1+)			0.2391
									1.0014
									(1.1403-1.1903)
^{123m3}In			> 0.1 ms	IT		(21/2-)			
^{123m2}In			1.4 μs	IT		(17/2-)			
^{123m}In			47. s	β- /	4.6/	(1/2)-	-0.400		0.1258
									1.170
									3.234
^{123}In		122.9104	6.0 s	β- /4.39	3.3/	(9/2)+	+5.49	+0.72	0.6188
									1.0197
									1.1305
^{124m}In			3.7 s	β-		8-	+3.89	+0.63	0.1029
									0.9699
									1.0729
									1.1316
^{124}In		123.9132	3.18 s	β- /7.36	5/	3+	+4.04	+0.6	0.7070
									0.9978
									1.1316
									3.2142
									(0.3–4.6)
^{125m3}In			0.005 s	IT	0.152/100	(23/2-)			
^{126m2}In			0.009 ms	IT	0.056/100	(19/2+)			
^{125m}In			12.2 s	β- /	5.5/	1/2(-)	-0.433		0.1876
^{125}In		124.9136	2.33 s	β- /5.42	4.1/	9/2+	+5.50	+0.68	0.4260
									(1.0318-1.3350)
^{126m2}In			0.022 ms	IT	0.2433	1-			
^{126m}In			1.64 s		4.9/	3(+)	+4.03	+0.47	0.9086
									0.9696
									1.1411
^{126}In		125.9165	1.63 s	β- /8.21	4.2/	8-	+4.061	+0.65	0.1118
									0.9086
									1.1411
^{127m}In			3.73 s	β- /	6.4/	(1/2-)			0.2523
									3.074
^{127}In		126.9174	1.14 s	β- /6.51	4.9/	(9/2+)	+5.52	+0.56	0.4680
									0.6461
									0.8051
									1.5977
^{128m}In			0.7 s	β- /	5.4/	(8-)			1.8670
									1.9739
									(0.1205–2.12)
^{128}In		127.920	0.80 s	β- /8.98	5.0/	3+			0.9352
									(1.1688-4.2970)
^{129m4}In			0.11 s			(29/2+)			
^{129m3}In			0.009 ms			(17/2-)			
^{129m2}In			~ 0.67 s			(23/2-)			
^{129m}In			1.23 s	β- /98/ n/2/	~ 7.5/	(1/2-)			0.3153
									0.9067
									1.2220
^{129}In		128.92181	~ 0.611 s	β- /7.66	5.5/	(9/2+)			0.2853
									0.7693
									1.8650

Nuclear

Elem. or Isot.	Natural abundance (Atom %)	Atomic mass or weight	Half-life/ Resonance width (MeV)	Decay mode/ Energy (/MeV)	Particle energy/ Intensity (MeV/%)	Spin (h/2π)	Nuclear magnetic mom. (nm)	Elect. quadr. mom. (b)	γ-Energy/ Intensity (MeV/%)
									2.1180
^{130m3}In			5.3 µs						0.3888
^{130m2}In			0.53 s	β- /	8.8/	5+			0.0892
									0.7744
									1.2212
^{130m1}In			0.51 s	β- /	6.1/	10-			0.0892
									0.1298
									0.7744
									1.2212
									1.9052
^{130}In		129.9250	0.28 s	β- /10.25	10.0/	1-			
^{131m2}In			0.3 s	β- /		(21/2+)			
^{131m1}In			0.35 s	β- /		(1/2-)			
^{131}In		130.92697	0.261 s	β- /9.18	6.4/	(9/2+)			0.3328
									2.433
^{132}In		131.9330	0.198 s	β- /13.6	6.0/	(7-)			0.1320
					8.8/				0.2992
									(0.3747-0.4041)
^{133m}In			0.18 s	β-, n		(1/2-)			
^{133}In		132.938	0.165 s	β-, (n)		(9/2+)			
^{134}In		133.945	0.13 s						(0.354–2.005)
^{135}In		134.950	0.10 s						
^{136}In			~ 85 ms						
^{137}In			~ 65 ms						
$_{50}$Sn		**118.710(7)**							
^{99}Sn		98.949	> 0.2 µs						
^{100}Sn		99.939	1.1 s	β+ /7.3	3.4/	0+			
^{101}Sn		100.935	2.0 s	β+ /9.		(7/2+)			0.352
				β+, p	/22.				1.065
^{102}Sn		101.930	3.8 s	β+ /5.8		0+			(0.069-1.425)
^{103}Sn		102.9281	7. s	β+ /7.7		(5/2+)			1.3558
				β+,p	p//1.2				(0.351-2.813)
				ec	/ 20.				
^{104}Sn		103.92311	21. s	β+, ec/4.5		0+			(0.913-1.846)
^{105}Sn		104.92127	32.7 s	β+ /6.3	EC//42.				In-x-ray
				β+, p	p//0.11				(0.288–3.819)
^{106}Sn		105.91696	2.0 m	β+ /20/3.18		0+			ann.rad./
				ec/80/					In k x-ray
									0.3865
									0.4772
^{107}Sn		106.91571	2.92 m	ec/5.0	1.2/				0.4218
				β+ /					0.6105
									0.6785
									1.0013
									1.1290
									1.542
^{108}Sn		107.91189	10.3 m	β+ /1/2.09	0.36/	0+			In k x-ray
				ec/99/					0.2724
									0.3965
									(0.105–1.68)
^{109}Sn		108.9113	18.0 m	β+ /9/3.85	1.52/	5/2+	-1.08	+0.3	ann.rad./
				ec/91/					In k x-ray
									0.6498
									1.0992
^{110}Sn		109.90785	4.17 h	ec/0.64		0+			In k x-ray
									0.283
^{111m}Sn			0.013 ms	IT	0.10024/100				
^{111}Sn		110.90774	35. m	β+ /31/2.45	1.5/	7/2+	+0.61	+0.2	In k x-ray
				ec/69/					0.7620
									1.1530
									1.9147
^{112}Sn	0.97(1)	111.904824	> 9.7 ×10^{19} a	β+/ec		0+			
^{113m}Sn			21.4 m	IT/92/0.077		7/2+			Sn k x-ray
				ec/8/					In x-ray
									0.0774
^{113}Sn		112.90518	115.1 d	ec/1.036		½+	-0.879		In k x-ray
									0.25511/3.2
									0.39169/100
^{114}Sn	0.66(1)	113.90278				0+			
^{115m2}Sn			0.159 ms	IT	0.1007	11/2-			
^{115m1}Sn			3.3 µs	IT	0.1155	7/2+			
^{115}Sn	0.34(1)	114.9033447				½+	-0.9188		
^{116}Sn	14.54(9)	115.901743				0+			

Elem. or Isot.	Natural abundance (Atom %)	Atomic mass or weight	Half-life/ Resonance width (MeV)	Decay mode/ Energy (/MeV)	Particle energy/ Intensity (MeV/%)	Spin (h/2π)	Nuclear magnetic mom. (nm)	Elect. quadr. mom. (b)	γ-Energy/ Intensity (MeV/%)
^{117m2}Sn			1.8 μs	IT		(19/2+)			
^{117m}Sn			13.91 d	IT/0.3146		11/2-	-1.396	-0.4	Sn k x-ray
									0.15856
^{117}Sn	7.68(7)	116.902954				½+	-1.0010		
^{118m}Sn			2.5 μs	IT	0.0559/100	10+			
^{118}Sn	24.22(9)	117.901607				0+			
^{119m2}Sn			0.01 ms	IT		(19/2+)			
^{119m}Sn			293. d	IT/0.0896	0.0657/100	11/2-	-1.4	0.21	Sn k x-ray
									0.02387
^{119}Sn	8.59(4)	118.903311				½+	-1.0473		
^{120m2}Sn			6.3 μs	IT	0.0657/100	10+			
^{120m1}Sn			0.012 ms	IT	0.1974/100	7-			
^{120}Sn	32.58(9)	119.90220				0+			
^{121m2}Sn			0.005 ms			19/2+			
^{121m}Sn			44. a	IT/78/0.006		11/2-	-1.388	-0.14	Sn k x-ray
				β- /22/	0.354/				0.03715
^{121}Sn		120.90424	1.128 d	β- /0.388	0.383/100	3/2+	0.698	~ -0.02	
^{122m2}Sn			0.062 ms	IT/2.765	0.075/100	10+			
^{122m1}Sn			8 μs	IT/2.409	0.1632/100	7-			
^{122}Sn	4.63(3)	121.90344				0+			
^{123m4}Sn			0.034 ms	IT/2.713		27/2+			(0.170-0.560)
^{123m3}Sn			0.006 ms	IT/2.153	0.208/100	23/2+			
^{123m2}Sn			0.007 ms	IT/1.945		19/2+			(0.019-0.838)
^{123m}Sn			40.1 m	β- /1.428	1.26/99	3/2+			0.1603
									0.3814
^{123}Sn		122.90573	129.2 d	β- /1.404	1.42/99.4	11/2-	-1.370	~ +0.03	0.1603
									(1.030-1.089)
^{124m2}Sn			0.05 ms	IT/2.657	0.078/100	10+			
^{124m1}Sn			0.003 ms	IT/2.325	0.1204/100	7-			
^{124}Sn	5.79(5)	123.90528	>2.2 ×10^{18} a	β-β-		0+			
^{125m2}Sn			6.2 μs	IT/1.893		19/2+			
^{125m}Sn			9.51 m	β- /2.387	2.03/98	3/2+	+0.764	+0.8	(0.3321-1.404)
^{125}Sn		124.90779	9.63 d	β- /2.364	2.35/82	11/2-	-1.348	~ +0.1	1.0671
									(0.2–2.3)
^{126m2}Sn			0.008 ms	IT/2.564	0.057/100	10+			
^{126m1}Sn			0.007 ms	IT/2.219	0.076/100	7-			
^{126}Sn	1 ×10^{-6}	125.9077	2.0 ×10^{5} a	β- /0.38	0.25/100	0+			0.0643
									0.0876
									(0.4148-0.6950)
^{127m3}Sn			1.2 μs	IT/1.931	0.104/100	(23/2+)			
^{127m2}Sn			4.5 μs	IT/1.827		(19/2+)			(0.017-0.732)
^{127m}Sn			4.15 m	β- /3.21	2.72/	3/2+	+0.757	+0.60	0.4909
									1.3480
									1.5640
^{127}Sn		126.9104	2.12 h	β- /3.20	2.42/	11/2-	-1.33	+0.3	0.8231
					3.2/				(0.120–2.84)
^{128m2}Sn			2.8 μs	IT/2.492	0.079/100	10+			
^{128m}Sn			6.5 s	IT/2.092	0.091/100	(7-)			
^{128}Sn		127.9105	59.1 m	β- /1.27	0.48/	0+			0.4823
					0.63/				(0.5573-0.6805)
^{129m3}Sn			2.2 μs	IT/1.803	0.041/100	(23/2+)			
^{129m2}Sn			3.4 μs	IT/1.762	0.020/100	(19/2+)			
^{129m}Sn			6.9 m	β- /		11/2-	-1.30	~ -0.2	1.1611
^{129}Sn		128.9135	2.23 m	β- /4.0		3/2+	+0.754	~ +0.05	0.6456
^{130m2}Sn			1.50 μs	IT/2.435	0.0965/100	(10+)			(0.0965-0.391)
^{130m}Sn			1.7 m	β- /	/100	(7-)	-0.381	-0.4	0.1449
									0.8992
^{130}Sn		129.91397	3.7 m	β- /2.15	1.10/	0+			0.0700
									0.1925
									0.7798
^{131m}Sn			1.02 m	β- /	3.4/	11/2-	-1.28		0.3043
									0.4500
									0.7985
									1.2260
									(0.08–3.21)
^{131}Sn		130.91705	39. s	β- /4.69	3.8/	3/2+	+0.747	~ -0.04	see 131mSn
^{132m}Sn			2.09 μs	IT/4.849	0.132/100				(0.064-4.416)
^{132}Sn		131.91783	40. s	β- /3.12	1.8/	0+			0.0855
									0.2467
									0.3402
									0.8985
^{133}Sn		132.92391	1.44 s	β- /7.8	7.5/	7/2-			
^{134}Sn		133.92868	1.04 s	β- /6.8		0+			(0.053-2.417)

Elem. or Isot.	Natural abundance (Atom %)	Atomic mass or weight	Half-life/ Resonance width (MeV)	Decay mode/ Energy (/MeV)	Particle energy/ Intensity (MeV/%)	Spin (h/2π)	Nuclear magnetic mom. (nm)	Elect. quadr. mom. (b)	γ-Energy/ Intensity (MeV/%)
^{135}Sn		134.93491	~ 0.515 s	β-					(0.053-0.830)
				β-,n	/21.				0.733–1.855
^{136}Sn		135.940	0.35 s	β-, n	/30.	0+			
^{137}Sn		136.947	0.27 s	β-, n	/~ 50				
^{138}Sn		137.952	0.15 s			0+			
^{139}Sn			~ 0.13 s						
$_{51}$Sb		**121.760(1)**							
^{103}Sb		102.940	> 1.5 μs						
^{104}Sb		103.936	0.5 s						
^{105}Sb		104.9313	1.1 s	β+,p	p//<0.1				
^{106}Sb		105.92864	0.6 s	β+ /10.5					
^{107}Sb		106.92415	4.0 s	β+ /7.9					1.280
									0.1515
									0.6666
									0.553–2.046
									(0.151–1.280)
^{108}Sb		107.92223	7.0 s	β+ /9.5					0.925
^{109}Sb		108.91814	17.3 s	β+ /6.38	4.42/	5/2+			1.062
				ec/	4.67/				0.261–2.127
					4.33/				
^{110}Sb		109.91685	24. s	β+ /9.0	6.8/	3+			ann.rad./
				ec/					0.6365
									0.9847
									1.2117
									1.2433
^{111}Sb		110.9132	1.25 m	β+ /87/4.47	3.3/	5/2+			ann.rad./
				ec/13 /					0.1002
									0.1545
									0.4891
									1.0326
^{112}Sb		111.9124	51.4 s	β+ /90/7.06	4.75/	3+			ann.rad./
				ec/10/					0.6700
									0.9909
									1.2571
^{113}Sb		112.9094	6.7 m	β+ /65/3.91	2.42/	5/2+			ann.rad./
									(0.3–3.6)
				ec/35/					Sn k x-ray
									(0.3324-0.4980)
^{114m}Sb			0.22 ms	IT/0.496	0.322/100	(8-)			
^{114}Sb		113.9093	3.49 m	β+ /78/5.9	3.4/	3+	1.7		ann.rad./
				ec/22/					Sn k x-ray
									0.8876
									1.2999
^{115}Sb		114.9066	32.1 m	β+ /67/3.03	1.51/	5/2+	+3.46	-0.4	ann.rad./
				ec/33/					Sn k x-ray
									0.4973
^{116m}Sb			1.00 h	β+ /78/	1.16/	8-	2.6		ann.rad./
				ec/22/					Sn k x-ray
									0.4073
									0.5429
									0.9725
									1.2935
									(0.100–1.501)
^{116}Sb		115.90679	16. m	β+ /50/4.707	1.3/	3+	2.72		ann.rad./
				ec/50/	2.3/				Sn k x-ray
									0.93180
									(0.138–3.903)
^{117m}Sb			0.36 ms	IT/3.131	0.058/100	(25/2+)			
^{117}Sb		116.9048	2.80 h	β+ /2/1.76	0.57/	5/2+	+3.4		Sn k x-ray
				ec/98/					0.1586
^{118m2}Sb			5.00 h	ec/99/		8-	2.32		Sn k x-ray
									0.25368
									(1.051-1.230)
^{118m1}Sb			0.021 ms	IT/0.0508	0.0508/100	(3)+			
^{118}Sb		117.90553	3.6 m	β+ /74/3.657	2.65/	1+	2.5		ann.rad./
				ec/26/					Sn k x-ray
									1.22964
^{119m}Sb			0.9 s	IT		(27/2+)			
^{119}Sb		118.90395	38.1 h	ec/0.59		5/2+	+3.45	-0.4	Sn k x-ray
									0.0239
^{120m}Sb			5.76 d	ec/		8-	2.34		Sn k x-ray
									0.0898
									0.19730
									1.02301

Elem. or Isot.	Natural abundance (Atom %)	Atomic mass or weight	Half-life/ Resonance width (MeV)	Decay mode/ Energy (/MeV)	Particle energy/ Intensity (MeV/%)	Spin (h/2π)	Nuclear magnetic mom. (nm)	Elect. quadr. mom. (b)	γ-Energy/ Intensity (MeV/%)
^{120}Sb		119.90508	15.89 m	β+ /41/2.68	1.72/	1+	+2.3		1.17121
				ec/59/					ann.rad./
									Sn k x-ray
^{121m}Sb			0.18 ms	IT					(0.7038-1.1712)
^{121}Sb	57.21(5)	120.90381				5/2+	+3.3634	-0.4	
^{122m3}Sb			4.19 m	IT/0.164	0.02609/100	8-			Sb x-ray
^{122m2}Sb			0.53 ms	IT/0.1375	0.761/100	(5)+			(0.0614-0.0761)
^{122m1}Sb			1.9 μs	IT/0.0614	0.0614/100	3+			
^{122}Sb		121.90517	2.72 d	β- /98/1.979	1.414/65	2-	-1.90	+0.9	0.56409
				β+ /2/1.620	1.980/26				0.69277
^{123m}Sb			0.65 ms	IT/2.613		23/2+			(1.1495-1.257)
^{123}Sb	42.79(5)	122.90421				7/2+	+2.5498	-0.5	(0.127-0.376)
^{124m3}Sb			3.2 μs	IT/0.0408					(0.0299-0.0408)
^{124m2}Sb			20.3 m	IT/0.035		8-			
^{124m1}Sb			1.6 m	IT/80/	1.2/	5+			0.4984
				β- /20/	1.7/				0.6027
									0.6458
^{124}Sb		123.90594	60.2 d	β- /2.905	0.61/52	3-	1.20	+1.9	1.1010
					2.301/23				0.60271/97.8
									1.69094/48.2
^{125m2}Sb			0.028 ms	IT/2.112	0.141/100	(19/2-)			(0.027–2.871)
^{125m1}Sb			4.1 μs	IT/1.971		15/2-			(0.882-0.904)
^{125}Sb		124.90525	2.758 a	β- /0.767	0.13/30	7/2+	+2.63		0.0355
					0.302/45				0.17632
					0.62/13				0.38044
									0.42786
									0.46336
									0.60060
									0.63595
^{126m2}Sb			~ 11 s	IT/0.040	0.0227/100	(3-)			L x-ray
^{126m1}Sb			19.2 m	β- /86 /	1.9	(5+)			0.0227
				IT/14 /					0.4148
									0.6663
									0.6950
^{126}Sb		125.9073	12.4 d	β- /3.67	1.9	8-	1.3		(0.222-1.477)
									0.2786
									0.4148
									0.6663
^{127m}Sb			0.011 ms	IT/1.920	/100	(15/2-)			(0.695-0.7205)
^{127}Sb		126.90692	3.84 d	β- /1.581	0.89/	7/2+	2.70		(0.806-0.825)
					1.10/				0.2524
					1.50/				0.2908
									0.4121
									0.4370
									0.6857
^{128m}Sb			10.4 m	β- /96/	2.6/	5+			0.7837
				IT/4/					0.3140
									0.5941
									0.7432
^{128}Sb		127.9091	9.1 h	β- /4.38	2.3/	8-	1.3		0.7539
									0.2148
									0.3141
^{129m3}Sb			1.1 μs	IT/2.137	0.0986/100	(23/2+)			(0.5265-0.7540)
^{129m2}Sb			2.2 μs	IT/1.851		(15/2-)			
^{129m}Sb			17.7 m	β- /1.851		(19/2-)			0.4338
									0.6578
^{129}Sb		128.9091	4.37 h	β- /2.38	0.65/	7/2+	2.79		0.7598
									0.0278
									0.1808
									0.3594
									0.4596
									0.5447
									0.8128
^{130m2}Sb			1.8 μs	IT/1.545	0.037/100	(13+)			(0.9146-1.030)
^{130m}Sb			6.3 m	β- /2.6	2.12/		3.09		0.1023
									0.7934
									0.8394

Elem. or Isot.	Natural abundance (Atom %)	Atomic mass or weight	Half-life/ Resonance width (MeV)	Decay mode/ Energy (/MeV)	Particle energy/ Intensity (MeV/%)	Spin (h/2π)	Nuclear magnetic mom. (nm)	Elect. quadr. mom. (b)	γ-Energy/ Intensity (MeV/%)
^{130}Sb		129.9117	38.4 m	β- /4.96	2.9/	8-			0.1823
									(0.3309-0.8394)
^{131m3}Sb			1.1 μs	IT/2.166	0.096/100	(23/2+)			
^{131m2}Sb			0.004 ms	IT/1.687	0.011/100	(19/2-)			
^{131m1}Sb			0.091 ms	IT/1.676		(15/2-)			
^{131}Sb		130.91199	23.0 m	β- /3.20	1.31/	7/2+	2.89		0.6423
					3.0/				0.6579
									0.9331
									0.9434
^{132m}Sb			2.8 m	β- /	3.9/	4+	3.18		0.1034
									0.3538
									0.6968
									0.9739
									0.9896
^{132}Sb		131.91451	4.1 m	β- /5.49		(8-)			0.1034
									0.1506
									(0.6968-0.9739)
^{133m}Sb			0.017 ms			(21/2+)			
^{133}Sb		132.91527	2.5 m	β- /4.00	1.20/	7/2+	3.00		0.4235
									0.6318
									0.8165
									1.0764
^{134m}Sb			10.1 s	β- /	6.1	(7-)			
^{134}Sb		133.92054	0.8 s	β- /8.4	8.4	0-			0.1152
									0.297
									(0.7063-1.279)
						7/2+			1.127
^{135}Sb		134.92519	1.71 s	β- /8.12					1.279
^{136}Sb		135.93075	0.82 s	β- /9.3					
^{137}Sb		136.936	0.49 s	β-, n	/49				
^{138}Sb		137.941	0.35 s	β-, n	/72	(3-)			
^{139}Sb		138.947	0.09 s	β-, n	/90				
^{140}Sb		139.953	> 0.4 μs						
$_{52}$**Te**		**127.60(3)**							
^{105}Te		104.943	~ 0.63 μs	α	α/4.70	(5/2+)			
^{106}Te		105.938	0.07 ms	α/4.3	/100	0+			
^{107}Te		106.9350	3.1 ms	α/ 70/	3.86(1)/	(5/2+)			(0.090-0.721)
				β+, ec/10.1					
^{108}Te		107.92938	2.1 s	α /68 /	3.314(4)/	0+			
				β+, ec/32 /6.8					
^{109}Te		108.92731	4.6 s	β+ ec/96 /8.7					0.7523
				α/4 /	3.107(4)/				0.287–2.045
^{110}Te		109.92246	19. s	β+, ec/4.5		0+			ann.rad./
									0.2191
									0.6059
^{111}Te		110.92110	19.3 s	β+, ec/8.0		(7/2+)			ann.rad./
									0.267
									0.322
									0.341
^{112}Te		111.9167	2.0 m	β+, ec/4.3		0+			ann.rad./
									0.2962
									0.3727
									0.4187
^{113}Te		112.9159	1.7 s	β+ /85/5.7	4.5/	(7/2+)			ann.rad./
				ec/15/					Sb k x-ray
									0.8144
									1.0181
									1.1812
^{114}Te		113.9121	15. m	β+ /40/3.2		0+			ann.rad./
				ec/60/					Sb k x-ray
									(0.0838-0.0903)
^{115m2}Te			7.5 μs	IT/0.280	0.280/100	11/2-			
^{115m}Te			6.7 m	β+ /45/		(1/2+)			ann.rad./
				ec/55/					Sb k x-ray
									0.7236
									0.7704
^{115}Te		114.9119	5.8 m	β+ /45/4.6	2.7/	7/2+			ann.rad./
				ec/55/					Sb k x-ray
									0.7236
									1.3268
									1.3806
									(0.22–2.7)
^{116}Te		115.9085	2.49 h	ec/1.5		0+			Sb k x-ray

Elem. or Isot.	Natural abundance (Atom %)	Atomic mass or weight	Half-life/ Resonance width (MeV)	Decay mode/ Energy (/MeV)	Particle energy/ Intensity (MeV/%)	Spin (h/2π)	Nuclear magnetic mom. (nm)	Elect. quadr. mom. (b)	γ-Energy/ Intensity (MeV/%)
^{117m}Te			0.103 s	IT/0.296		11/2-			0.0937
^{117}Te		116.9086	1.03 h	ec/75/3.54	1.78/	½+			ann.rad./
				β+ /25/					Sb k x-ray
									0.9197
									1.7164
									2.3000
^{118}Te		117.9059	6.00 d	ec/0.28		0+			Sb k x-ray
^{119m}Te			4.70 d	ec/		11/2-	0.89		Sb k x-ray
									0.15360
									0.2705
									1.21271
^{119}Te		118.9064	16.0 h	β+ /2/2.293	0.627/	½+	0.25		ann.rad.
				ec/98/					Sb k x-ray
									0.6440
									0.6998
^{120}Te	0.09(1)	119.90406	>1.9 ×10^{17} a	β+ ec		0+			
^{121m}Te			164 d	IT (89%)		11/2-	0.90		Te k x-ray
				ec (11%)					0.2122
^{121}Te		120.9049	16.8 d	ec/1.04		½+			Sb k x-ray
									0.5076
									0.5731
^{122}Te	2.55(12)	121.90304				0+			
^{123m}Te			119.7 d	IT/0.247		11/2-	-0.93		Te k x-ray
									0.1590/84.1
^{123}Te	0.89(3)	122.90427	>9.2 ×10^{16} a	ec/0.051		½+	-0.736948		
^{124}Te	4.74(14)	123.90282				0+			
^{125m}Te			57.4 d	IT/0.1448	0.1093/100	11/2-	-0.99	-0.06	Te k x-ray
									0.0355
^{125}Te	7.07(15)	124.90443				½+	-0.8885		
^{126}Te	18.84(25)	125.90331				0+			
^{127m}Te			106 d	IT/98/0.088		11/2-	-1.04	~ 0.2	Te k x-ray
				β- /2/0.77/2.4					0.0883
^{127}Te		126.90523	9.30 h	β- /0.698	0.696/	3/2+	0.635		0.3603
^{128}Te	31.74(8)	127.90446	2.5 ×10^{24} a	β-β-		0+			
^{129m}Te			33.6 d	IT/63/0.105		11/2-	-1.09	0.40	Te k x-ray
				β- /37/	1.60/				0.45984
									0.6959
^{129}Te		128.90660	1.16 h	β- /1.498	0.99/9	3/2+	0.702	0.06	0.0278
					1.45/89				(0.4598-0.4873)
$^{130m1}Te$			1.9 μs	IT/2.665		10+			
^{130}Te	34.08(62)	129.9062227	7. ×10^{20} a	β-β-		0+			
$^{131m2}Te$			0.09 s	IT/1.940	0.360/100	(23/2+)			
^{131m}Te			1.35 d	β- /78/2.4	0.42/	11/2-	-1.04	~ 0.25	0.0811
				IT/22/0.18					0.1021
									0.14973
									0.77369
									0.79375
									0.85225
^{131}Te		130.9085222	25.0 m	β- /2.233	1.35/12	3/2+	0.70		0.14973
$^{132m2}Te$			3.7 μs	IT/2.723		(10+)			(0.022-0.798)
^{132m}Te			0.028 ms	IT/1.925		(7-)			0.151
					1.69/22				0.45327
					2.14/60				0.49269
^{132}Te		131.90855	3.26 d	β- /0.51	0.215	0+			0.049725
									0.11198
									0.22830
^{133m}Te			55.4 m	β- /82/	2.4/30	(11/2-)	-1.13	0.3	Te k x-ray
				IT/18/0.334					0.0949
									0.1689
									0.3121
									0.3341
^{133}Te		132.91097	12.4 m	β- /2.94	2.25/25	3/2+	0.85	0.2	0.3121
					2.65				0.4079
									1.3334
^{134}Te		133.91139	42. m	β- /1.51	0.6/	0+			0.7672/29
					0.7/				(0.079–0.926)
^{135}Te		134.91656	19.0 s	β- /6.0	5.4/		0.7	0.3	0.267
					6.0				0.603
									0.870
^{136}Te		135.92010	17.5 s	β- /5.1	2.5/	0+			2.0779/25
									(0.087–3.235)
^{137}Te		136.92560	2.5 s	β- /98 /6.9	6.8	7/2-			0.2436

Elem. or Isot.	Natural abundance (Atom %)	Atomic mass or weight	Half-life/ Resonance width (MeV)	Decay mode/ Energy (/MeV)	Particle energy/ Intensity (MeV/%)	Spin (h/2π)	Nuclear magnetic mom. (nm)	Elect. quadr. mom. (b)	γ-Energy/ Intensity (MeV/%)
				n/2 /					
^{138}Te		137.92947	1.4 s	β- /6.4		0+			
^{139}Te		138.93537	> 0.15 μs						
^{140}Te		139.9395	> 0.15 μs			0+			
^{141}Te		140.946	> 0.15 μs						
^{142}Te		141.950	> 0.15 μs			0+			
^{143}Te		142.957	> 0.4 μs						
$_{53}$I		**126.90447(3)**							
^{107}I		106.947							
^{108}I		107.943	0.04 s	α/91/4.	3.95				
^{109}I		108.93809	93.5 μs	p					0.593/100
				α	α//0.014				0.717/63
									(0.496–1.057)
^{110}I		109.9351	0.65 s	β+, ec/83/11.4					ann.rad./
				α/17/~ 3.6	3.457(10)/				
				p/11/					
^{111}I		110.93027	2.5 s	β+, E.C./8.5					ann.rad./
									0.2665
									0.3215
									0.3412
^{112}I		111.9280	3.4 s	β+, ec/10.2					ann.rad./
									0.6889
									0.7869
^{113}I		112.9237	5.9 s	β+, ec/7.6					ann.rad./
									0.4625/100
									0.6224/74
									0.0550–1.422
^{114m}I			6.2 s	β+, ec		(7)			
^{114}I		113.922	2.1 s	β+, ec/8.7					ann.rad./
									0.6826
									0.7088
^{115}I		114.9181	1.3 m	β+, ec/6.7		5/2+			ann.rad./
									0.275
									0.284
									(0.460-0.709)
^{116m}I			3.3 μs	IT/0.430	0.0443/100	(7-)			
^{116}I		115.917	2.9 s	β+ /97/7.8	6.7/	1+			ann.rad./
				ec/3/					0.5402
									0.6789
^{117}I		116.9136	2.22 m	β+, ec/4.7	3.2/	(5/2+)	3.1		ann.rad./
									0.2744
									0.3259
^{118m}I			8.5 m	β+, ec/	4.9/	(7-)	4.2		ann.rad./
				IT					0.104
									0.5998
									0.6052
									0.6138
^{118}I		117.9131	14. m	β+, ec/7.0		2-	2.0		ann.rad./
									0.5448
									0.6052
									1.3384
^{119}I		118.9101	19. m	β+ /54/3.5	2.4/	(5/2+)	+2.9		ann.rad./
				ec/46/					Te k x-ray
									0.2575
^{120m}I			53. m	β+ /80/	3.8	(7-)	4.2		ann.rad.
				ec/20/					Te k x-ray
									0.4257
									0.5604
									0.6147
									1.3459
^{120}I		119.9101	1.36 h	β+ /56/5.62	4.03	2-	1.23		ann.rad./
				ec/	4.60				Te k x-ray
									0.5604
									0.6411
									1.5230
									(0.111–3.1)
^{121m}I			0.009 ms	IT/0.1587					
^{121}I		120.90741	2.12 h	β+ /13/2.27	1.2/	5/2+	2.3		ann.rad./
				ec/87/					Te k x-ray
									0.2122
									(0.14–1.1)
^{122m2}I			0.081 ms	IT/0.394	0.051/100	(8+)			
^{122m1}I			0.080 ms	IT/0.380	0.0645/100	(7-)			

Elem. or Isot.	Natural abundance (Atom %)	Atomic mass or weight	Half-life/ Resonance width (MeV)	Decay mode/ Energy (/MeV)	Particle energy/ Intensity (MeV/%)	Spin (h/2π)	Nuclear magnetic mom. (nm)	Elect. quadr. mom. (b)	γ-Energy/ Intensity (MeV/%)
^{122}I		121.90759	3.6 m	β+ /4.234	3.1/	1+	+0.94		ann.rad./
				ec/					Te k x-ray
									0.5641
^{123}I		122.90559	13.22 h	ec/1.242		5/2+	2.82		Te k x-ray
									0.1590
^{124}I		123.90621	4.176 d	β+ /23/3.160	1.54/	2-	1.446		ann.rad./
				ec/77/	2.14/				Te k x-ray
					0.75/				0.6027/62.9
									0.7228/10.3
									1.6910/11.2
									(0.31–1.73)
^{125}I		124.90463	59.4 d	ec/0.1861		5/2+	2.82	-0.78	Te k x-ray
									0.0355
^{126}I		125.90562	13.0 d	ec/		2-	1.438		ann.rad./
				β+ /2.155	1.13/				Te k x-ray
				β- /1.258/47	0.87/				0.3887
					1.25/				0.6622
^{127}I	100.	126.90447				5/2+	+2.8133	-0.71	
^{128}I		127.90581	25.00 m	β- /2.118	2.13/	1+			Te k x-ray
				ec/1.251					0.44287
									0.52658
^{129}I		128.90498	1.7 ×10^{7} a	β- /0.194	0.15/	7/2+	+2.6210	-0.50	Xe k x-ray
									0.0396
^{130m}I			8.8 m	IT/83/0.048		2+			I k x-ray
				β- /17/					0.5361
^{130}I		129.90667	12.36 h	β- /2.949	1.04/	5+	3.35		0.4180
					0.62				0.5361
									(0.6685-0.7395)
^{131m}I			0.024 ms	IT/1.918	0.122	19/2-			
^{131}I		130.906126	8.021 d	β- /0.971	0.606/	7/2+	+2.742	-0.35	0.08017
									0.28431
									0.36446
									0.63699
^{132m}I			1.39 h	IT	0.098/86	(8-)			
^{132}I		131.90799	2.283 h	β- /14/3.58	0.80/	4+	3.09	0.08	I k x-ray
				IT/86/	1.03/				0.0980
					1.2/				0.5059
					1.6/				0.52264
					2.16/				0.63019
									0.6506
									0.66768
									0.77260
									0.95457
^{133m}I			9. s	IT/1.63	0.07441/100	(19/2-)			I kx-ray
									0.0730
									0.6474
									0.9126
^{133}I		132.90780	20.8 h	β- /1.77	1.24/85	7/2+	+2.86	-0.24	0.51056
									0.52989
									0.87537
^{134m}I			3.5 m	IT/98/0.316	0.272	(8-)			I k x-ray
				β- /2/					0.0444
									0.2719
^{134}I		133.90976	52.6 m	β- /4.05	1.2/	4+			0.1354
									0.84702
									0.88409
^{135}I		134.91005	6.57 h	β- /2.63	0.9/	7/2+	2.940		0.2884
					1.3/				0.41768
									0.52658
									1.13156
									1.26046
^{136m}I			47. s	β- /	4.7/	(6-)			0.1973
					5.2/				0.3468
									0.3701
									0.3814
									1.3130
									(0.16–2.36)
^{136}I		135.9146	1.39 m	β- /6.93	4.3/	2-			0.3447
					5.6/				1.3130
									1.3211
									2.2896
									(0.3–6.1)
^{137}I		136.9180	24.5 s	β- /5.88	5.0/	(7/2+)			0.6010

Elem. or Isot.	Natural abundance (Atom %)	Atomic mass or weight	Half-life/ Resonance width (MeV)	Decay mode/ Energy (/MeV)	Particle energy/ Intensity (MeV/%)	Spin (h/2π)	Nuclear magnetic mom. (nm)	Elect. quadr. mom. (b)	γ-Energy/ Intensity (MeV/%)
									1.2180
									1.2201
									1.3026
									(0.25–4.4)
^{138m}I			1.3 μs	IT		(3-)			
^{138}I		137.92273	6.5 s	β- /7.8	6.9/	2-			0.4836
					7.4/				0.5888
									0.8752
									(0.4–5.3)
^{139}I		138.9265	2.30 s	β- /6.81					0.192
				n/					0.198
									0.273
									0.382
									0.386
									0.468
									0.683
									1.313
^{140}I		139.932	0.86 s	β- /8.8		(3)			0.372
				n/					0.377
									0.457
^{141}I		140.936	0.45 s	β- /7.8					
^{142}I		141.941	~ 0.2 s	β-					
^{143}I		142.946	> 0.15 μs						
^{144}I		143.951	> 0.15 μs						
^{145}I		144.956	> 0.4 μs						
$_{54}$Xe		**131.293(6)**							
^{109}Xe		108.950	13. ms	α	α/3.92/31	(7/2+)			
					α/4.06/69				
^{110}Xe		109.944	0.11 s	β+ /9.2		0+			
				α	/~ 64				
^{111m}Xe			0.9 s	ec, β+					
^{111}Xe		110.942	0.7 s	ec, β+ /10.6					
				α/	3.58/63				
				α	3.46/37				
^{112}Xe		111.9356	3. s	ec, β+ /7.2	α/0.8/	0+			
^{113m}Xe			6.9 μs	IT/0.405	0.278/100	11/2			
^{113}Xe		112.93322	2.8 s	ec, β+ /9.1					
^{114}Xe		113.9280	10.0 s	β+, ec/5.9		0+			ann.rad./
									0.1031
									0.1616
									0.3085
									0.6826
									0.7088
^{115}Xe		114.9263	18. s	β+, ec/7.6		(5/2+)			ann.rad./
^{116}Xe		115.9216	56. s	β+, ec/4.3	3.3/	0+			ann.rad./
									0.1042
									0.1916
									0.2477
									0.3107
									0.4127
^{117}Xe		116.9204	1.02 m	β+, ec/6.5		(5/2+)	-0.594	+1.16	ann.rad./
									0.2214
									0.5190
									0.6389
									0.6613
^{118}Xe		117.9162	~ 4. m	β+, ec/3.	2.7/	0+			ann.rad./
									0.0535
									0.0600
									0.1199
^{119}Xe		118.9154	5.8 m	β+, ec/5.0	3.5/	7/2+	-0.654	+1.31	0.0873
									0.1000
									0.2318
									0.4615
^{120}Xe		119.9118	46. m	β+, ec/97/1.96		0+			I k x-ray
				β+ /3/					0.0251
									0.0726
									0.1781
									(0.1–1.03)
^{121}Xe		120.9115	39. m	β+ /44/3.73	2.8/	5/2+	-0.701	+1.33	ann.rad./
				ec/56/					I k x-ray
									0.1328
									0.2527
									0.4452

Elem. or Isot.	Natural abundance (Atom %)	Atomic mass or weight	Half-life/ Resonance width (MeV)	Decay mode/ Energy (/MeV)	Particle energy/ Intensity (MeV/%)	Spin (h/2π)	Nuclear magnetic mom. (nm)	Elect. quadr. mom. (b)	γ-Energy/ Intensity (MeV/%)
									(0.1–3.1)
^{122}Xe		121.9084	20.1 h	ec/0.9		0+			I k x-ray
									0.3501
^{123m}Xe			5.5 μs	IT/0.185		7/2(-)			
^{123}Xe		122.9085	2.00 h	β+ /23/2.68	1.51/	1/2+	-0.150		ann.rad./
				ec/77/					I k x-ray
									0.1489
									0.1781
									(0.1–2.1)
^{124}Xe	0.0952(3)	123.90589	> 10^{17} a	β-β-		0+			
			≥ 2.5 x 10^{21} a	ec, ec					
^{125m}Xe			57. s	IT/0.252		(9/2-)	-0.745	+0.42	γ k x-ray
									0.1111
									0.141
^{125}Xe		124.90639	17.1 h	ec/1.653	0.47/	½+	-0.269		I k x-ray
									0.1884
									0.2434
^{126}Xe	0.0890(2)	125.90430	> 4.3 x 10^{21} a	Ec. ec		0+			
^{127m}Xe			1.15 m	IT/0.297	0.172/100	(9/2-)	-0.884	+0.69	γ k x-ray
									0.1246
									0.1725
^{127}Xe		126.90518	36.34 d	ec/0.662		½+	-0.504		I k x-ray
									0.1721
									0.2029
									0.3750
^{128}Xe	1.9102(8)	127.90353				0+			
^{129m}Xe			8.89 d	IT/0.236	0.1966/100	11/2-	-0.89122	+0.64	γ k x-ray
									0.0396
									0.1966
^{129}Xe	26.4006(82)	128.90478086				½+	-0.77798		
^{130}Xe	4.0710(13)	129.9035094				0+			
^{131m}Xe			11.9 d	IT/0.164		11/2-	-0.99405	+0.73	γ k x-ray
									0.16398
^{131}Xe	21.2324(30)	130.905084				3/2+	+0.69186	-0.12	
^{132m}Xe			8.4 ms	IT/2.752	0.538/100	(10+)			
^{132}Xe	26.9086(33)	131.90415509				0+			
^{133m}Xe			2.20 d	IT/0.233		11/2-	-1.082	+0.77	γ k x-ray
									0.23325
^{133}Xe		132.90591	5.243 d	β- /0.427	0.346/99	3/2+	+0.813	+0.14	Cs k x-ray
									(0.0810-0.1606)
^{134m2}Xe			0.005 ms	IT/3.02	0.028/100	(10+)			
^{134m1}Xe			0.29 s	IT/1.97	0.234/100	7-			
^{134}Xe	10.4357(21)	133.90539	>1.1 ×10^{16} a	β- β-		0+			
^{135m}Xe			15.3 m	IT/	0.5266/99.4	11/2-	1.1030	+0.62	γ k x-ray
									0.52658
^{135}Xe		134.90723	9.10 h	β-/1.15	0.91/	3/2+	0.903	+0.21	0.24975/0.90
									0.60807
^{136m}Xe			0.0030 ms	IT/1.89	0.1973/100	6+			
^{136}Xe	8.8573(44)	135.9072145	2.3 ×10^{21} a	β-β-		0+			
^{137}Xe		136.91158	3.82 m	β- /4.17	4.1/	7/2-	-0.97	-0.48	0.45549
					3.6/				0.8489
									0.9822
									1.2732
									1.7834
									2.8498
^{138}Xe		137.91415	14.1 m	β- /2.77	0.8/	0+			0.1538
					2.4/				0.2426
									0.2583/35
									0.4345/22
									1.76826/19
									2.0158
^{139}Xe		138.91879	39.7 s	β- /5.06	4.5/	3/2-	-0.30	+0.40	0.1750
					5.0/				0.2186
									0.2965
									(0.1–3.37)
^{140}Xe		139.92165	13.6 s	β- /4.1	2.6	0+			0.0801
									0.6220
									0.8055
									1.4137
									(0.04–2.3)
^{141}Xe		140.92679	1.72 s	β- /6.2	6.2/	5/2(-)	+0.010	-0.58	0.1187
									0.9095
									(0.05–2.55)

Nuclear

Elem. or Isot.	Natural abundance (Atom %)	Atomic mass or weight	Half-life/ Resonance width (MeV)	Decay mode/ Energy (/MeV)	Particle energy/ Intensity (MeV/%)	Spin (h/2π)	Nuclear magnetic mom. (nm)	Elect. quadr. mom. (b)	γ-Energy/ Intensity (MeV/%)
^{142}Xe		141.92997	1.22 s	β- /5.0	3.7/	0+			0.0338
					4.2/				0.0729
									0.2038
									0.3091
									0.4145
									0.5382
									0.5718
									0.6181
									0.6448
^{143m}Xe			0.96 s	β-					
^{143}Xe		142.93537	0.30 s	β- /7.3		5/2-	-0.460	+0.93	
^{144}Xe		143.93895	1.2 s	β- /6.1		0+			
^{145}Xe		144.9447	0.9 s	β-, (n)					
^{146}Xe		145.9485	> 0.15 μs			0+			
^{147}Xe		146.954	> 0.15 μs						
^{148}Xe		147.958	> 0.4 μs						
$_{55}$Cs		**132.9054520(1)**							
^{112}Cs		111.950	0.50 ms	p	0.81				
^{113}Cs		112.9444	17. μs	p	0.96				
^{114}Cs		113.9413	0.58 s	β+, ec/11.8		1+			ann.rad./
									0.6826
									0.7088
^{115}Cs		114.936	~ 1.4 s	β+, ec/8.4					ann.rad./
^{116m}Cs			0.7 s	β, ec/					ann.rad./
									0.3935
^{116}Cs		115.933	3.8 s	β+, ec/10.8					ann.rad./
									0.3935
									0.5243
									0.6151
									0.6223
^{117m}Cs			6.5 s	β+, ec/		(3/2+)			
^{117}Cs		116.9286	~ 8.4 s	β+, ec/7.5		(9/2+)			ann.rad./
^{118m}Cs			17. s	β+, ec/			5.		
^{118}Cs		117.9266	14. s	β+, ec/9.		2	+3.88	+1.4	ann.rad./
									0.3372
									0.4727
									0.5865
									0.5906
^{119m}Cs			30.4 s			3/2(+)	+0.84	+0.9	
^{119}Cs		118.9224	43. s	β+, ec/6.3		9/2+	+5.46	+2.8	ann.rad./
									0.169
									0.176
									0.224
									0.257
^{120m}Cs			~ 57 s	β+, ec/		(7-)			
^{120}Cs		119.9207	64. s	β+, ec/7.92		2(+)	+3.87	+1.45	ann.rad./
									0.3224
									0.4735
									0.5534
									(0.3–3.28)
^{121m}Cs			~ 2.03 m	IT/17/		9/2(+)	+5.41	+2.7	ann.rad./
				β+ /83/	4.4				0.1794
									0.1961
^{121}Cs		120.9172	2.3 m	β+, ec/5.40	4.38/	3/2(+)	+0.77	+0.84	ann.rad./
									0.1537
									(0.08–0.56)
^{122m3}Cs			3.7 m	β+, ec		8(-)	+4.77	+3.3	ann.rad./
^{122m2}Cs			0.36 s	IT/0.127/	0.0812/100	(5-)			0.3311
									0.4971
									(0.27-2.22)
^{122m1}Cs			> 1 μs	IT	0.0459/100	(3+)			0.3311
^{122}Cs		121.9161	21. s	β+, ec/7.1	5.8/	1+	-0.133	-0.19	ann.rad./
									0.3311
									0.5120
									0.8179
^{123m}Cs			1.6 s	IT/	0.0617/100	11/2-			Cs k x-ray
									0.0946
^{123}Cs		122.9130	5.87 m	β+ /75/4.20	3.0/	1/2+	+1.38		ann.rad./
				ec/25/					Xe k x-ray
									0.0974
									0.5964
^{124m}Cs			6.4 s	IT		(7+)			
^{124}Cs		123.9123	30. s	β+ /9 /5.92	~ 5.	1+	+0.673	-0.74	ann.rad./

Elem. or Isot.	Natural abundance (Atom %)	Atomic mass or weight	Half-life/ Resonance width (MeV)	Decay mode/ Energy (/MeV)	Particle energy/ Intensity (MeV/%)	Spin (h/2π)	Nuclear magnetic mom. (nm)	Elect. quadr. mom. (b)	γ-Energy/ Intensity (MeV/%)
				ec/8 /					Xe k x-ray
									0.3539
^{125m}Cs			0.90 ms	IT/0.266/		(11/2-)			(0.4925-0.942)
^{125}Cs		124.90973	45. m	β+ /40/3.09	2.06/	1/2(+)	+1.41		ann.rad./
				ec/60/					Xe k x-ray
^{126m2}Cs			0.17 ms	IT/0.596					(0.112-0.526)
^{126m1}Cs			≥ 1 μs	IT/0.272		(4-)			
^{126}Cs		125.9094	1.64 m.	β+ /81/4.83	3.4	1+	+0.78	-0.68	ann.rad./
				ec/19/	3.7/				Xe k x-ray
									0.3886
									(0.491-0.925)
^{127m}Cs			0.055 ms	IT/0.452		(11/2)-			(0.179-0.386)
^{127}Cs		126.90742	6.2 h	β+ /96/2.08	0.65/	1/2+	+1.46		Xe k x-ray
				ec/4/	1.06				0.1247
									0.4119
^{128}Cs		127.90775	3.62 m	β+ /68/3.930	2.44/	1+	+0.97	-0.57	ann.rad./
				ec/32 /	2.88/				Xe k x-ray
									0.4429
^{129}Cs		128.90607	1.336 d	ec/1.195		1/2+	+1.49		Xe k x-ray
									0.3719
									0.4115
^{130m}Cs			3.5 m	IT, β+, ec		5-	+0.629	+1.45	
^{130}Cs		129.9067	29.21 m	β+ /55/2.98	1.98/	1+	+1.46	-0.06	ann.rad./
				ec/43/					Xe k x-ray
				β- /1.6/0.37	0.44/1.6				0.5361
^{131}Cs		130.90547	9.69 d	ec/0.352		5/2+	+3.543	-0.58	Xe k x-ray
^{132}Cs		131.90643	6.48 d	ec/98/		2 +	+2.22	+0.51	Xe k x-ray
				β+ /0.3/2.120					0.4646
				β- / /1.280					0.6302
^{133}Cs	100.	132.9054520				7/2+	+2.58291	-0.00355	0.66769
^{134m}Cs			2.91 h	IT/0.139		8-	+1.0978	+1.0	Cs k x-ray
									0.12749
^{134}Cs		133.9067185	2.065 a	β- /2.059	0.089/27	4+	+2.994	+0.389	0.56327
					0.658/70				0.56935
				ec/1.22					0.60473
									0.79584
^{135m}Cs			53. m	IT/1.627	0.846/100	19/2-	+2.18	+0.9	0.7869
									0.8402
^{135}Cs		134.90598	~ 1.6 ×10^{6} a	β- /0.269	0.205/100	7/2+	+2.7324	+0.050	
^{136m}Cs			17.5 s	IT/		8-	+1.32	+0.7	(0.105-0.518)
^{136}Cs		135.90731	13.16 d	β- /2.548	0.341/	5+	+3.71	+0.23	0.06691
									0.34057
									0.81850
									1.04807
^{137}Cs		136.907089	30.05 a	β- /1.176	0.514/95	7/2+	+2.851	+0.051	Ba k x-ray
									0.66164
^{138m}Cs			2.9 m	IT//81	/0.080	6-	+1.71	-0.40	Cs k x-ray
				β- /19 /	3.3				0.0799
									0.1917
									0.4628
									1.43579
^{138}Cs		137.9110	32.2 m	β- /5.37	2.9/	3-	+0.700	+0.12	0.1381
									0.46269
									1.00969
									1.43579
									2.21788
^{139}Cs		138.91336	9.3 m	β- /4.213	4.21	7/2+	+2.70	-0.07	0.6272
									1.2832
									(0.4–3.66)
^{140}Cs		139.9173	1.06 m	β- /6.22	5.7/	1-	+0.133895	-0.11	0.5283
					6.21/				0.6023
									0.9084
									(0.41–3.94)
^{141}Cs		140.9200	24.9 s	β- /5.26	5.20/	7/2+	+2.44	-0.4	Ba k x-ray
									0.0485
									0.5616
									0.5887
									1.1940
									(0.05–3.33)
^{142}Cs		141.92430	1.8 s	β- /7.31	6.9/	0 -			0.3596

Elem. or Isot.	Natural abundance (Atom %)	Atomic mass or weight	Half-life/ Resonance width (MeV)	Decay mode/ Energy (/MeV)	Particle energy/ Intensity (MeV/%)	Spin (h/2π)	Nuclear magnetic mom. (nm)	Elect. quadr. mom. (b)	γ-Energy/ Intensity (MeV/%)
					7.28				0.9668
									1.1759
									1.3265
^{143}Cs		142.9273	1.78 s	β- /6.24	6.1	3/2+	+0.87	+0.47	0.1955
									0.2324
									(0.17–1.98)
^{144m}Cs			1.1 μs	IT/0.092	0.049/100	(4-)			
^{144}Cs		143.9321	1.01 s	β- /8.47	8.46/	1 (-)	-0.546	+0.30	0.1993
					7.9/				0.5598
									0.6392
									0.7587
^{145}Cs		144.9355	0.59 s	β- /7.89	7.4/	3/2+	+0.784	+0.6	0.1126
					7.9/				0.1755
									0.1990
^{146}Cs		145.9403	0.322 s	β-, (n)/9.38	~ 9.0	1-	-0.515	+0.22	(0.024-2.2798)
^{147}Cs		146.9442	0.227 s	β-, (n)/9.3					
^{148}Cs		147.949	0.15 s	β-, (n)/10.5					
^{149}Cs		148.953	> 50 ms						
^{150}Cs		149.958	> 50 ms						
^{151}Cs		150.963	> 50 ms						
$_{56}$Ba		**137.327(7)**							
^{114}Ba		113.951	0.43 s	β+, (p)	p/20	0+			
				α	/0.9				
^{115}Ba		114.947	0.45 s	β+, (p)	p/<15				
^{116}Ba		115.941	1.3 s	β+, (p)	p/3	0+			
^{117}Ba		116.938	1.8 s	β+, (p), ec/8.4	p/16	(3/2-)			(0.046–0.364)
^{118}Ba		117.933	5.2 s	β+,		0+			(0.040–0.156)
^{119}Ba		118.931	5.4 s	β+, ec/8.					
^{120}Ba		119.926	24. s	β+, ec/5.0		0+			ann.rad./
									0.140
									(0.075–0.146)
^{121}Ba		120.924	30. s	β+, ec/6.8		5/2+	+0.660	+1.8	ann.rad./
^{122}Ba		121.9199	2.0 m	β+, ec/3.8		0+			ann.rad./
^{123}Ba		122.9188	2.7 m	β+, ec/5.5		5/2(+)	-0.680	+1.5	ann.rad./
									0.0306
									0.0927
									0.1161
									0.1235
^{124}Ba		123.9151	12. m	β+, ec/2.65		0+			ann.rad./
									0.1695
									0.1888
									1.2160
^{125m}Ba			8. m	β+, ec/	4.5		0.174		
^{125}Ba		124.9145	3.5 m	β+, ec/4.6	3.4	½(+)	+0.18		ann.rad./
									0.0550
									0.0776
									0.0854
									0.1409
^{126}Ba		125.9113	1.65 h	β+ /2/1.67		0+			Cs k x-ray
				ec/98 /					0.2179
									0.2336
									0.2576
^{127m}Ba			1.9 s	IT/0.080		7/2-	-0.723	+1.6	(0.024-0.802)
^{127}Ba		126.9111	12.9 m	β+ /54/3.5		1/2+	+0.083		ann.rad./
				ec/46/					Cs k x-ray
									0.1148
									0.1808
									(0.07–2.5)
^{128}Ba		127.90834	2.43 d	ec/0.52		0+			Cs k x-ray
									0.27344
^{129m}Ba			2.14 h	ec/98/0.008		7/2+	+0.93	+1.6	Cs k x-ray
				β+ /2/					0.1769
									0.1823
									0.2023
									1.4593
^{129}Ba		128.9087	2.2 h	β+ /20/2.43	1.42/	1/2+	-0.40		ann.rad./
				ec/80/					Cs k x-ray
									0.1291
									0.2143
									0.2208
^{130m}Ba			9.5 ms	IT/2.475	/100.	8-	-0.04	+2.8	0.080–0.882
^{130}Ba	0.106(1)	129.90632	2.2 $\times 10^{21}$ a	β+β+		0+			
^{131m}Ba			14.3 m	IT/0.188	0.0799/100	9/2-	-0.87	+1.5	Ba k x-ray

Elem. or Isot.	Natural abundance (Atom %)	Atomic mass or weight	Half-life/ Resonance width (MeV)	Decay mode/ Energy (/MeV)	Particle energy/ Intensity (MeV/%)	Spin (h/2π)	Nuclear magnetic mom. (nm)	Elect. quadr. mom. (b)	γ-Energy/ Intensity (MeV/%)
^{131}Ba		130.90694	11.7 d	ec/1.37		1/2+	0.70811		0.1085
									Cs k x-ray
									0.12381/28.4
									0.21608/21.3
									0.49636/42.9
									(0.055–1.171)
^{132}Ba	0.101(1)	131.90506	1.3 ×10^{21} a	ec ec		0+			
^{133m}Ba			1.621 d	IT/0.288		11/2-	-0.91	+0.9	Ba k x-ray
									0.2761
^{133}Ba		132.90601	10.53 a	ec/0.517		1/2+	0.77167		Cs k x-ray
									(0.081-0.356)
^{134m}Ba			2.6 μs	IT/2.96	0.121/100	(10+)			
^{134}Ba	2.417(18)	133.904508				0+			
^{135m}Ba			1.20 d	IT/0.2682		11/2-	-1.00	+1.0	Ba k x-ray
									0.2682
^{135}Ba	6.592(12)	134.905688				3/2+	+0.83863	+0.160	
^{136m}Ba			0.308 s	IT/2.0305		7-			Ba k x-ray
									0.8185
									1.0481
^{136}Ba	7.854(24)	135.904576				0+			
^{137m}Ba			2.552 m	IT/0.6617		11/2-	-0.99	+0.8	Ba k x-ray
									0.66164
^{137}Ba	11.232(24)	136.905827				3/2+	+0.93737	+0.245	
^{138}Ba	71.698(42)	137.905247				0+			
^{139}Ba		138.90884	1.396 h	β- /2.317	2.14/27	7/2-	-0.97	-0.57	0.16585
					2.27/72				1.2544
									1.42033
^{140}Ba		139.9106	12.75 d	β- /1.05	0.48	0+			0.16268
					1.0/				0.30485
					1.02/				0.53727
^{141}Ba		140.91440	18.3 m	β- /3.22	2.59/	3/2-	-0.34	+0.45	0.1903
					2.73/				0.2770
									0.3042
									(0.1–2.5)
^{142}Ba		141.91643	10.7 m	β- /2.212	1.0/	0+			0.23152
					1.10/				0.25512
									0.3090
									1.2040
^{143}Ba		142.92063	14.3 s	β- /4.24	4.2/	5/2-	+0.44	-0.88	0.1786
									0.21148
									0.7988
									(0.17–2.4)
^{144}Ba		143.92296	11.4 s	β- /3.1	2.4/	0+			La k x-ray
					2.9/				0.10386
									0.1566
									0.1728
									0.3882
									0.43048
^{145}Ba		144.9275	4.0 s	β- /4.9	4.9/	5/2-	-0.28	+1.22	La k x-ray
									0.0918
^{146}Ba		145.9303	2.5 s	β- /4.12	3.9/	0+			0.09709
									0.0644
									0.2513
									0.3270
									0.3329
									0.3622
^{147}Ba		146.9353	0.892 s	β- /5.75	5.5/				
^{148}Ba		147.9382	0.64 s	β-, n/5.11		0+			
^{149}Ba		148.943	0.36 s	β-, (n)/7.3					
^{150}Ba		149.946	0.3 s			0+			
^{151}Ba		150.951	> 0.15 μs						
^{152}Ba		151.955	> 0.4 μs			0+			
^{153}Ba		152.960							
$_{57}$La		**138.90547(7)**							
^{117}La		116.950	23 ms	p	0.813/	3/2+			(0.189-0.817)
^{118}La		117.947							
^{119}La		118.941							
^{120}La		119.938	2.8 s	ec, β+ /11.					
^{121}La		120.933	5.3 s						
^{122}La		121.931	9. s	ec, β+/~ 9.7					
^{123}La		122.926	17. s	ec/7.					
^{124m}La			29.2 s			(8-)			
^{124}La		123.9246	21 s	ec/ ~ 8.8					

Elem. or Isot.	Natural abundance (Atom %)	Atomic mass or weight	Half-life/ Resonance width (MeV)	Decay mode/ Energy (/MeV)	Particle energy/ Intensity (MeV/%)	Spin (h/2π)	Nuclear magnetic mom. (nm)	Elect. quadr. mom. (b)	γ-Energy/ Intensity (MeV/%)
^{125m}La			0.39 s	IT	0.107/100				
^{125}La		124.9208	1.2 m	β+, ec/5.6		11/2-			ann.rad./
									0.0436
									0.0676
^{126m}La			< 50. s						
^{126}La		125.920	54. s	β+, ec/7.6					ann.rad./
									0.256
									0.455
									0.117–3.853
^{127}La		126.9164	3.7 m	β+, ec/4.7		(3/2+)			ann.rad./
									0.025
									0.0562
^{128}La		127.9156	5.0 m	β+ /80/6.7		(5-)			ann.rad./
				ec/20/					Ba k x-ray
									0.2841/87
									0.4793/54
									(0.315–2.212)
^{129m}La			0.56 s	IT/0.172	0.104/100	(11/2-)			
^{129}La		128.9127	11.6 m	β+ /58/3.72	2.42/	3/2+			ann.rad./
				ec/42/					Ba k x-ray
									0.1105
									0.2786
									(0.1–1.8)
^{130}La		129.9124	8.7 m	β+ /78/5.6		3+			ann.rad./
				ec/22/					Ba k x-ray
									0.3573/81
									0.5506/27
									(0.1965–1.989)
^{131m}La			0.17 ms	IT/0.304	0.109/100	11/2-			
^{131}La		130.9101	59. m	β+ /76/3.0	1.42/	3/2+			ann.rad./
				ec/24/	1.94/				Ba k x-ray
									0.1085
									0.3658
									0.5263
^{132m}La			24. m	IT/76/		6-			La k x-ray
				β+, ec/24/					0.1352
									0.4645
^{132}La		131.9101	4.8 h	β+ /40/4.71	2.6/	2-			ann.rad./
				ec/60/	3.2				Ba k x-ray
					3.7/				0.4645
									0.5671
^{133}La		132.9082	3.91 h	β+ /4/2.2	1.2/	5/2+			Ba k x-ray
				ec/96/					0.2788
									(0.290-0.3024)
^{134m1}La			0.029 ms	IT					
^{134}La		133.9085	6.5 m	β+ /63/3.71	2.67/	1+			ann.rad./
				ec/37/					Ba k x-ray
									0.6047
									(0.5–1.9)
^{135}La		134.9070	19.5 h	ec/1.20		5/2+			Ba k x-ray
									0.4805
^{136m}La			0.114 s	IT	0.088/100				
^{136}La		135.9076	9.87 m	β+ /36/2.9	1.8/	1+	+3.7	~ -0.4	ann.rad./
				ec/64/					Ba k x-ray
									0.8185
^{137}La		136.90645	6 ×10^{4} a	ec/0.60		7/2+	+2.70	+0.21	0.2836
^{138m}La			0.002 ms	IT/0.739		7-			
^{138}La	0.08881(71)	137.90712	1.06 ×10^{11} a			5+	+3.71365	+0.44	1.4358/65
									0.7887/35
^{139}La	99.91119(71)	138.90636				7/2+	+2.783046	+0.20	
^{140}La		139.90948	1.678 d	β- /3.762	1.35	3-	+0.73	+0.09	
					1.24/				
					1.67/				
^{141}La		140.91097	3.90 h	β- /2.502	2.43/	7/2+			
^{142}La		141.91409	1.54 h	β- /4.505	2.11/	2-			
					2.98/				
					4.52/				
^{143}La		142.91608	14.1 m	β- /3.43	3.3/	7/2-			
^{144}La		143.9196	40.7 s	β- /5.5	4.1/				
^{145}La		144.9218	24. s	β- /4.1	4.1/	3/2+			
^{146m}La			10.0 s	β- /6.7	5.5/	(6)			
^{146}La		145.9259	6.3 s	β- /6.6	6.2/	(2-)			
^{147}La		146.9284	4.02 s	β- /5.0	4.6/				

Elem. or Isot.	Natural abundance (Atom %)	Atomic mass or weight	Half-life/ Resonance width (MeV)	Decay mode/ Energy (/MeV)	Particle energy/ Intensity (MeV/%)	Spin (h/2π)	Nuclear magnetic mom. (nm)	Elect. quadr. mom. (b)	γ-Energy/ Intensity (MeV/%)
^{148}La		147.9327	1.3 s	β- /7.26		2-			
^{149}La		148.935	1.10 s	β- /5.5					
									0.1335
									0.009–1.709
^{150}La		149.939	0.51 s						x-ray
									(0.097–0.209)
^{151}La		150.942	> 0.15 μs						
^{152}La		151.947	> 0.15 μs						
^{153}La		152.950	> 0.15 μs						
^{154}La		153.955							
^{155}La		154.959							
$_{58}$**Ce**		**140.116(1)**							
^{119}Ce		118.953							
^{120}Ce		119.947				0+			
^{121}Ce		120.943	1.1 s	β+, p					
^{122}Ce		121.938				0+			
^{123}Ce		122.935	3.8 s	β+, ec/~ 8.6					ann.rad./
^{124}Ce		123.930	6. s	ec/~ 5.6		0+			
^{125m}Ce			4 s	IT		(1/2+)			
^{125}Ce		124.928	9.6 s	β+, ec/7.		(7/2-)			ann.rad./
									0.1346
									0.1666
									0.056–1.329
^{126}Ce		125.9240	50. s	ec/4.		0+			
^{127}Ce		126.9227	29. s	β+, ec/6.1		(5/2+)			ann.rad./
									(0.058–1.961)
^{128}Ce		127.9189	4.1 m	β+, ec/3.2		0+			ann.rad./
									(0.023–0.880)
^{129}Ce		128.9181	3.5 m	β+, ec/5.6					ann.rad./
									(0.0675–1.015)
^{130}Ce		129.9147	26. m	β+, ec/2.2		0+			ann.rad./
									La k x-ray
									0.047–1.431
^{131m}Ce			5.4 m	β+ ec		(1/2+)			ann.rad./
									0.2304
									0.3955
									0.4213
^{131}Ce		130.9144	10. m	β+, ec/4.0	2.8/				ann.rad.
									0.119
									0.169
									0.414
^{132m}Ce			9.4 ms	IT/2.340		(8-)			0.3255
									0.10–0.955
^{132}Ce		131.9115	3.5 h	ec/1.3		0+			La k x-ray
									0.1554
									0.1821
^{133m}Ce			1.6 h	β+, ec/		½+			ann.rad.
									0.0769
									0.0973
									0.5577
^{133}Ce		132.9115	5.4 h	β+/8/2.9	1.3/	9/2-			ann.rad.
				ec/92/					0.0584
									0.1308
									0.4722
									0.5104
^{134}Ce		133.9089	3.16 d	ec/0.5		0+			La k x-ray
									0.1304
									0.1623
									0.6047
^{135m}Ce			20. s	IT/0.446	0.150/100	(11/2-)			Ce k x-ray
									0.0826
									0.1497
									0.2134
^{135}Ce		134.9092	17.7 h	β+/1 /2.026	0.8/	1/2(+)			La k x-ray
				ec/99 /					0.0345
									0.2656
									(0.300-0.6068)
^{136m}Ce			2.2 μs	IT/3.095	1.057/100	10+			
^{136}Ce	0.185(2)	135.907129	> 3. ×10^{16} a	ec ec		0+			
			> 4.2 ×10^{15} a	β- β-					
^{137m}Ce			1.45 d	IT/99 /0.254		11/2-	1.0		Ce k x-ray
				ec/0.8 /					0.1693
									0.2543
^{137}Ce		136.907762	9.0 h	β+/1.222		3/2+	0.96		La k x-ray

Elem. or Isot.	Natural abundance (Atom %)	Atomic mass or weight	Half-life/ Resonance width (MeV)	Decay mode/ Energy (/MeV)	Particle energy/ Intensity (MeV/%)	Spin (h/2π)	Nuclear magnetic mom. (nm)	Elect. quadr. mom. (b)	γ-Energy/ Intensity (MeV/%)
									0.4472
^{138m}Ce			8.7 ms	IT/2.13	0.303/100	7-			
^{138}Ce	0.251(2)	137.9060	>4.4 ×10^{16} a	ec ec		0+			
^{139m}Ce			57.6 s	IT/0.7542		11/2-			Ce k x-ray
									0.7542
^{139}Ce		138.90666	137.6 d	ec/0.28		3/2+	1.06		La k x-ray
									1.320/72.1
									0.255/59.6
									(0.231-2.364)
^{140m}Ce			0.007 ms	IT/2.107	0.246/100	6+			
^{140}Ce	88.450(51)	139.90544				0+			
^{141}Ce		140.90828	32.50 d	β-/0.581	0.436/69	7/2-	1.1		Pr k x-ray
					0.581/31				0.14544/48.0
^{142}Ce	11.114(51)	141.90925	>1.6 ×10^{17} a	β- β-		0+			
^{143}Ce		142.91239	1.38 d	β-/1.462	1.404/	3/2-	0.43		Pr k x-ray
					1.110/47				0.0574
									0.2933
^{144}Ce		143.91365	284.6 d	β-/0.319	0.185/20	0+			Pr k x-ray
					0.318/				0.0801
									0.1335
^{145}Ce		144.9173	3.00 m	β-/2.54	1.7/24	3/2-			Pr k x-ray
					1.3				0.0627
									0.7245
^{146}Ce		145.9188	13.5 m	β-/1.04	0.7/90	0+			Pr k x-ray
									0.0986
									0.2182
									0.3167
									0.0930
^{147}Ce		146.9227	56. s	β-/3.29	3.3/				0.2687
^{148}Ce		147.9244	56. s	β-/2.1	1.66/	0+			0.0904
									0.0985
									0.1212
									0.2918
^{149}Ce		148.9284	5.2 s	β-/4.2					0.0577
									0.0864
									0.3800
									0.1099
^{150}Ce		149.9304	4.4 s	β-/3.0		0+			0.0526
^{151m}Ce				1.0 s	β-				
^{151}Ce		150.9343	1.8 s	β-/5.3		(5/2+)			Pr k x-ray
									(0.035-0.637)
^{152}Ce		151.937	1.4 s	β-/4.4		0+			0.098
									0.115
^{153}Ce		152.941	> 0.15 μs						
^{154}Ce		153.944	> 0.15 μs			0+			
^{155}Ce		154.949	> 0.15 μs						
^{156}Ce		155.952				0+			
^{157}Ce		156.957							
$_{59}$Pr		**140.90766(2)**							
^{121}Pr		120.955	0.01 s	p	p/0.882				
^{122}Pr		121.952							
^{123}Pr		122.946							
^{124}Pr		123.943	1.2 s	β+, ec/12.					ann.rad./
^{125}Pr		124.938	~ 3.3 s	β+					ann.rad./
									0.1358
^{126}Pr		125.935	3.1 s	β+, ec/~ 10.4					ann.rad./
									(0.170–0.985)
^{127}Pr		126.931	4.2 s	β+ /~ 7.5					ann.rad./
									(0.028–0.8949)
^{128}Pr		127.9288	3.0 s	β+, ec/~ 9.3					ann.rad./
									0.207/100
									0.400–1.373
^{129}Pr		128.9251	32 s	β+, ec/5.8					ann.rad./
									(0.0395–1.865)
^{130}Pr		129.9236	40. s	β+, ec/8.1		(5+)			ann.rad./
^{131m}Pr			5.7 s	IT/0.152/96		(11/2-)			(0.06–0.16)
				ec/β+/4					
^{131}Pr		130.9202	1.7 m	β+, ec/5.3				~ 5.5	ann.rad./
									(0.059–0.980)
^{132m}Pr			2.5 μs	IT	0.102/100				
^{132}Pr		131.9193	1.6 m	β+, ec/7.1					ann.rad./
									0.325
									0.496

Elem. or Isot.	Natural abundance (Atom %)	Atomic mass or weight	Half-life/ Resonance width (MeV)	Decay mode/ Energy (/MeV)	Particle energy/ Intensity (MeV/%)	Spin (h/2π)	Nuclear magnetic mom. (nm)	Elect. quadr. mom. (b)	γ-Energy/ Intensity (MeV/%)
^{133m}Pr			1.1 s	IT/0.192	0.130/100	(11/2-)			0.533
									0.1305
^{133}Pr		132.9163	6.5 m	β+, ec/4.3		5/2+			0.0617
									ann.rad./
									0.074
									0.1343
									0.2419
									0.3156
									0.3308
^{134m}Pr			~ 11. m	β+, ec/0.068		(6-)			0.4650
									ann.rad./
									0.294
									0.460
									0.495
^{134}Pr		133.9157	17. m	β+, ec/6.2		2+			0.632
									ann.rad./
^{135m}Pr			0.11 ms	IT		(11/2-)			(0.294-0.495)
^{135}Pr		134.9131	24. m	β+, ec/3.7	2.5/	3/2(+)			(0.113-0.317)
									ann.rad./
									0.0826
									0.2135
									0.2961
^{136}Pr		135.9127	13.1 m	β+ /57 /5.13	2.98/	2+			0.5832
				ec/43					ann.rad./
									Ce k x-ray
^{137m}Pr			2.7 µs	IT/0.561		11/2-			(0.540-0.552)
^{137}Pr		136.9107	1.28 h	β+ /26 /2.70	1.68/	5/2+			(0.331-0.563)
				ec/74 /					ann.rad./
									Ce k x-ray
									0.4339
									0.5140
									0.8367
^{138m}Pr			2.1 h	β+ /24 /	1.65/	7-			(0.16–1.8)
				ec/76 /					ann.rad./
									Ce k x-ray
									0.3027
									0.7887
									1.0378
^{138}Pr		137.9108	1.45 m	β+ /75 /4.44	3.42/	1+			(0.07–2.0)
				ec/25 /					ann.rad./
									Ce k x-ray
^{139}Pr		138.90894	4.41 h	β+ /8 /2.129	1.09/	5/2+			0.7887
				ec/92 /					ann.rad./
									Ce k x-ray
									0.2551
^{140m}Pr			3.1 µs	IT/0.764	0.636/100	(7)-			(1.35-1.63)
^{140}Pr		139.90908	3.39 m	β+ /51 /3.39	2.37/	1+			ann.rad./
				ec/49 /					Ce k x-ray
									0.3069
									1.5965
^{141}Pr	100.	140.90766				5/2+	+4.275	-0.08	
^{142m}Pr			14.6 m	IT/0.004	c.e.	5-	2.2		
^{142}Pr		141.91005	19.12 h	β- /2.162	0.58/4	2-	+0.234	+0.03	0.5088
				ec/0.744	2.16/96				1.57580
^{143}Pr		142.91082	13.57 d	β- /0.934	0.933/	7/2+	+2.70	+0.8	0.7420
^{144m}Pr			7.2 m	IT/99+/0.059		3-			Pr k x-ray
				β- /					0.0590
									0.6965
									0.8142
^{144}Pr		143.91331	17.28 m	β- /2.998	0.807/1	0-			0.69649
					2.30/				1.48912
					2.996/98				2.18562
^{145}Pr		144.91452	5.98 h	β- /1.81	1.80/97	7/2+			0.0725
									0.6758
									0.7483
^{146}Pr		145.9177	24.2 m	β- /4.2	2.2/30	2-			0.4539/48
					3.7/10				1.5247
					4.2/40				
^{147}Pr		146.9190	13.4 m	β- /2.69	1.5/	3/2+			0.3146/24.
					2.1/				0.5779/16
									0.6413/19.
^{148m}Pr			2.0 m	β- /	4.0/	4-			0.3016

Elem. or Isot.	Natural abundance (Atom %)	Atomic mass or weight	Half-life/ Resonance width (MeV)	Decay mode/ Energy (/MeV)	Particle energy/ Intensity (MeV/%)	Spin (h/2π)	Nuclear magnetic mom. (nm)	Elect. quadr. mom. (b)	γ-Energy/ Intensity (MeV/%)
					3.8/				0.4506
				IT	0.77/36				0.6975
^{148}Pr		147.9221	2.27 m	β- /4.9	4.8/	1-			0.3017
					4.5/				
^{149}Pr		148.9237	2.3 m	β- /3.40	3.0	(5/2+)			0.1085
									0.1385
									0.1651
^{150}Pr		149.9267	6.2 s	β- /5.7		1-			0.1302
					~ 5.5				(0.804-0.8527)
^{151m}Pr			0.05 ms	IT	0.035/100				
^{151}Pr		150.9283	22.4 s	β- /4.2					
^{152}Pr		151.9316	3.2 s	β- /6.7		4+			0.0726
									0.164
									0.285
^{153}Pr		152.9339	4.3 s	β- /5.5					
^{154}Pr		153.938	2.3 s	β- /7.9					
^{155}Pr		154.9405	> 0.3 μs						
^{156}Pr		155.945	> 0.3 μs						
^{157}Pr		156.948							
^{158}Pr		157.952							
^{159}Pr		158.956							
$_{60}$Nd		**144.242(3)**							
^{124}Nd		123.952				0+			
^{125}Nd		124.949	0.6 s	β+, p					
^{126}Nd		125.943	> 0.2 μs			0+			
^{127}Nd		126.940	1.8 s	β+, ec/9.		(5/2)			ann.rad./
^{128}Nd		127.935	4. s	β+, ec/6.					ann.rad./
^{129m}Nd			2.6 s			(1/2+)			0.134,0.399
^{129}Nd		128.933	4.9 s	β+, ec/8.		5/2(-)			ann.rad./
									(0.091–0.875)
^{130}Nd		129.9285	28. s	β+, ec/5.		0+			ann.rad./
^{131}Nd		130.9272	0.5 m	β+, ec/6.6					ann.rad./
									(0.09–0.36)
^{132}Nd		131.9233	1.5 m	β+, ec/3.7		0+			ann.rad./
									(0.099–0.567)
^{133m}Nd			~ 70 s			(1/2+)			0.128
^{133}Nd		132.9224	1.2 m	β+, ec/5.6					ann.rad./
									(0.06–0.37)
^{134m}Nd			0.41 ms	IT/2.293		(8-)			(0.167-0.874)
^{134}Nd		133.9188	~ 8.5 m	β+ /17 /2.8		0+			ann.rad./
				ec/83 /					Pr k x-ray
									0.1631/58
									(0.09–1.00)
^{135m}Nd			5.5 m	β+/0.065		(1/2+)			
^{135}Nd		134.9182	12. m	β+ /65 /4.8		9/2(-)	-0.78	+1.9	ann.rad./
				ec/35 /					Pr k x-ray
									0.0415/23.
									0.204/51.
									(0.11–1.8)
^{136}Nd		135.9150	50.6 m	ec/94 /2.21	1.04/	0+			Pr kx-ray
				β+ /6 /					0.0401/21.
									0.1091/35.
									(0.10–0.97)
^{137m}Nd			1.6 s	IT/0.5196	0.234/100	11/2-			Nd k x-ray
									0.1084
									0.1775
									0.2337
^{137}Nd		136.9146	38. m	β+ /40 /3.69	1.7/20	1/2+	-0.63		ann.rad./
				ec/60 /	2.40/20				Pr k x-ray
									0.0755
									0.5806
^{138}Nd		137.9120	5.1 h	ec/1.1		0+			Pr k x-ray
									0.1995
									0.3258
^{139m}Nd			5.5 h	IT/12 /0.231	1.17/	11/2-			Nd k x-ray
				β+ /88 /					Pr k x-ray
									0.1139/34.
									0.7382/30.
^{139}Nd		138.9120	30. m	β+ /25 /2.79	1.77/	3/2+	+0.91	+0.3	ann.rad./
				ec/75 /					Pr k x-ray
									0.4050
^{140m2}Nd			1.2 μs			20+			
^{140m1}Nd			0.6 ms	IT/2.2221	0.420/100	7-			

Elem. or Isot.	Natural abundance (Atom %)	Atomic mass or weight	Half-life/ Resonance width (MeV)	Decay mode/ Energy (/MeV)	Particle energy/ Intensity (MeV/%)	Spin (h/2π)	Nuclear magnetic mom. (nm)	Elect. quadr. mom. (b)	γ-Energy/ Intensity (MeV/%)
^{140}Nd		139.9096	3.37 d	ec /0.22		0+			Pr k x-ray
^{141m}Nd			1.04 m	IT/99+/0.756		11/2-			Nd k x-ray
									0.7565
^{141}Nd		140.90962	2.49 h	ec/98 /1.823	0.802/	3/2+	+1.01	+0.3	Pr k x-ray
				β+ /2 /					(0.15–1.7)
^{142m}Nd			0.017 ms	IT/2.209	0.109/100				
^{142}Nd	27.152(40)	141.90773				0+			
^{143}Nd	12.174(26)	142.90982				7/2-	-1.07	-0.60	
^{144}Nd	23.798(19)	143.91009	2.1 ×10^{15} a	α	1.83	0+			
^{145}Nd	8.293(12)	144.91258				7/2-	-0.66	-0.31	
^{146}Nd	17.189(32)	145.91312				0+			
^{147}Nd		146.91611	10.98 d	β- /0.896	0.805/	5/2-	0.58	0.9	Pr k x-ray
									0.53102
									0.09111–0.686
^{148}Nd	5.756(21)	147.91690	1020 a	β- β-		0+			
^{149}Nd		148.92015	1.73 h	β- /1.691	1.03/25	5/2-	0.35	1.3	Pr k x-ray
					1.13/26				0.1143/19.
					1.42/				0.2113/27.
									(0.026–1.6)
^{150}Nd	5.638(28)	149.92090	1.33 ×10^{20} a	β-β-		0+			
^{151}Nd		150.92384	12.4 m	β- /2.442	1.2/	(3/2+)			Pm k x-ray
									0.1168
									0.2557
									1.1806
									(0.10–1.9)m
^{152}Nd		151.9247	11.4 m	β- /1.1		0+			0.2785/29.
									(0.016–0.66)
^{153m}Nd			1.1 μs	IT/0.192					(0.072-0.1917)
^{153}Nd		152.92772	28.9 s	β- /3.6		(5/2+)			0.418
^{154m2}Nd			0.003 ms	IT/1.298		(4-)			(0.135-0.295)
^{154m1}Nd			0.001 ms						
^{154}Nd		153.929	25.9 s	β- /2.8		0+			0.1519
									0.7998
^{155}Nd		154.9331	8.9 s	β- /5.0					0.1807
^{156}Nd		155.935	5.3 s	β- /4.1		0+			0.0848
^{157}Nd		156.9394	> 0.3 μs						
^{158}Nd		157.942	> 0.3 μs			0+			
^{159}Nd		158.947	> 0.3 μs						
^{160}Nd		159.949	> 0.3 μs			0+			
^{161}Nd		160.954							
$_{61}$**Pm**									
^{128}Pm		127.949	1.0 s	β+, p					ann.rad.
^{129}Pm		128.943	~ 2.4 s						
^{130}Pm		129.941	2.5 s	β+, ec/11.					0.1589
									0.326–1.062
^{131}Pm		130.936	~ 6.3 s	β+					0.185
									0.220
									0.146
^{132}Pm		131.934	6. s	β+, ec/10.					ann.rad./
									(0.213-1.491)
^{133}Pm		132.9298	12. s	β+, ec/~ 7.0					ann.rad./
^{134m}Pm			20. μs	IT	0.0707/100	(7-)			0.071
^{134}Pm		133.9284	24. s	β+, ec/~ 8.9		(5+)			ann.rad./
									0.294/100
									(0.300-1.821)
^{135m}Pm			45 s	β+, ec					
^{135}Pm		134.9248	0.8 m	β+, ec/6.0		11/2-			(0.13–0.47)
^{136m}Pm			1.5 μs	IT	0.0472/100	(8+)			
^{136}Pm		135.9236	1.8 m	β+ /89 /7.9		(3+)			ann.rad./
				ec/11 /					Nd k x-ray
									0.3735
									0.6027
^{137}Pm		136.9205	2.4 m	β+, ec/5.6		(11/2-)			ann.rad./
									0.1086
									0.1775
^{138m}Pm			3.2 m	β+ /50 /~ 7.0	3.9/	3+	3.		ann.rad./
				ec/50 /					Nd k x-ray
									0.5209
									0.7290
^{138}Pm		137.9195	10. s	β+ /6.9	6.1/	1+			ann.rad./
^{139m}Pm			0.18 s	IT/	0.189/99.9	(11/2-)			0.1887
^{139}Pm		138.9168	4.14 m	β+ /68 /4.52	3.52/	(5/2+)			ann.rad./
				ec/32 /					Nd k x-ray

Elem. or Isot.	Natural abundance (Atom %)	Atomic mass or weight	Half-life/ Resonance width (MeV)	Decay mode/ Energy (/MeV)	Particle energy/ Intensity (MeV/%)	Spin (h/2π)	Nuclear magnetic mom. (nm)	Elect. quadr. mom. (b)	γ-Energy/ Intensity (MeV/%)
									0.4028
									(0.27–2.4)
^{140m}Pm			5.87 m	β+ /70 /	3.2	7/2-			ann.rad./
				ec/30 /					Nd k x-ray
									0.4199
									0.7738
									1.0283
^{140}Pm		139.9160	9.2 s	β+ /89 /6.09	5.07/74	1+			ann.rad./
				ec/11 /					Nd k x-ray
									0.7738
									1.4898
^{141}Pm		140.9136	20.9 m	β+ /52 /3.72	2.71	5/2+			ann.rad./
				ec/48 /					Nd k x-ray
									0.8862
									1.2233
^{142m2}Pm			67 μs	IT/2.829		(13-)			(0.208-0.882)
^{142m1}Pm			2.0 ms	IT/0.883	0.434/100	(8-)			
^{142}Pm		141.9129	40.5 s	β+ /86 /4.87	3.8/	1+			ann.rad./
				ec/20 /					Nd k x-ray
									0.6414
									1.5758
^{143}Pm		142.91094	265. d	ec/1.041		5/2+	3.8		Nd k x-ray
				β+ /< 6 ×10^{-6}/					0.7420
^{144}Pm		143.91260	360. d	ec/2.332		5-	1.7		Nd k x-ray
				β+ /7 ×10^{-6}/					0.6180
									0.6965
^{145}Pm		144.91276	17.7 a	ec/0.163		5/2+	+3.8	+0.2	Nd k x-ray
									0.0723
^{146}Pm		145.91470	5.53 a	ec/63 /1.472		3-			Nd k x-ray
				β- /37 /1.542	0.795/				0.4538
									0.7362
									0.7474
^{147}Pm		146.91515	2.623 a	β- /0.224	0.224/	7/2+	+2.6	+0.6	0.1213
									0.1974
^{148m}Pm			41.3 d	β- /95 /2.6	0.4/60	6-	1.8		0.5503/94.
				IT/5 /0.137	0.5/17				0.6300/89.
					0.7/21				0.7257/33
^{148}Pm		147.91748	5.37 d	β- /2.47	1.02/	1-	+2.0	~ +0.2	0.5503
					2.47/				(0.915-1.465)
^{149m}Pm			0.035 ms	IT/0.240		11/2-			
^{149}Pm		148.91834	2.212 d	β- /1.071	0.78/9	7/2+	3.3		0.2859
					1.072/90				0.5909
									0.8594
^{150}Pm		149.9210	2.68 h	β- /3.45	1.6/	(1-)			0.3339/69.
					2.3/				1.1658/16.
					1.8/				1.3245/17.
									(0.25–2.9)
^{151}Pm		150.92122	1.183 d	β- /1.187	0.84/	5/2+	+1.8	1.9	0.1677/8
									0.2751/7
									0.3401/22
^{152m2}Pm			14 m	β-, IT/		(8)			(0.14–1.4)
^{152m1}Pm			7.5 m	β- /		(4-)			0.1218
									0.2447
									0.3404
									1.0971
									1.4375
^{152}Pm		151.9235	4.1 m	β- /3.5	3.5/20	1+			0.1218
					3.50/60				(0.12–2.1)
^{153}Pm		152.9242	5.4 m	β- /1.90	1.7/	(5/2-)			0.0910
									0.1198
									0.1273
^{154m}Pm			1.7 m	β- /	2.0/				0.0820
									0.1848
									1.4403
^{154}Pm		153.9265	1.7 m	β- /4.1	1.9/				0.0820
									0.8396
									1.3940
									2.0589
									(0.08–2.8)
^{155}Pm		154.92814	48. s	β- /3.2		(5/2-)			(0.05–0.78)
^{156}Pm		155.93112	26.7 s	β- /5.16					
^{157}Pm		156.93312	10.9 s	β- /4.6					
^{158}Pm		157.9366	5. s	β- /6.3					

Elem. or Isot.	Natural abundance (Atom %)	Atomic mass or weight	Half-life/ Resonance width (MeV)	Decay mode/ Energy (/MeV)	Particle energy/ Intensity (MeV/%)	Spin (h/2π)	Nuclear magnetic mom. (nm)	Elect. quadr. mom. (b)	γ-Energy/ Intensity (MeV/%)
^{159}Pm		158.9393	1.5 s						(0.072-0.261)
^{160}Pm		159.943	> 0.3 μs						
^{161}Pm		160.946	> 0.3 μs						
^{162}Pm		161.950	> 0.3 μs						
^{163}Pm		162.954	> 0.3 μs						
$_{62}$Sm		**150.36(2)**							
^{129}Sm		128.955	~ 0.55 s	β+, p					
^{130}Sm		129.949				0+			
^{131}Sm		130.946	1.2 s	β+, ec/					ann.rad./
^{132}Sm		131.941	4.0 s	β+		0+			
^{133m}Sm			3.5 s			(1/2-)			
^{133}Sm		132.939	2.9 s	β+, ec/~ 8.4		5/2+			ann.rad./
									0.3696
									0.0845
^{134}Sm		133.934	11. s	β+, ec/5.		0+			ann.rad./
^{135}Sm		134.933	10. s	β+, ec/7.		7/2+			ann.rad./
^{136m}Sm			15. μs	IT/2.265	0.466/100	(8-)			0.466
^{136}Sm		135.9283	42. s	β+, ec/4.5		0+			ann.rad./
^{137}Sm		136.9270	45. s	β+, ec/6.1					ann.rad./
^{138}Sm		137.9232	3.0 m	β+, ec/3.9		0+			ann.rad./
									0.0536
									0.0747
^{139m}Sm			11 s	IT/94 /0.457	0.190	(11/2-)	1.1		Sm k x-ray
				β+ /6 /	4.7				0.1118
									0.1553
									0.1901
									0.2673
^{139}Sm		138.9223	2.6 m	β+ /75 /5.5	4.1/	½+	-0.53		Pm k x-ray
				ec/25 /					0.3678
									0.4028
									(0.27–2.4)
^{140}Sm		139.9190	14.8 m	β+, ec/3.4	1.9/	0+			ann.rad./
									Pm k x-ray
									0.1396
									0.2255
									(0.07–1.7)
^{141m}Sm			22.6 m	β+ /32 /	1.6/	11/2-	-0.84	+1.6	ann.rad./
				ec/68 /	2.19/				Pm k x-ray
				IT/0.3 /0.1758					0.1966
									0.4318
									0.7774
^{141}Sm		140.91848	10.2 m	β+ /52 /4.54	3.2/	½+	-0.74		ann.rad./
				ec/48 /					Pm k x-ray
									0.4382
^{142}Sm		141.91520	1.208 h	β+ /6 /2.10	1.0/	0+			ann.rad./
				ec/94 /					Pm k x-ray
^{143m2}Sm			30 ms	IT/2.794	0.208/100	23/2(-)			
^{143m}Sm			1.10 m	IT/99/0.7540		11/2-			Sm k x-ray
									0.7540
^{143}Sm		142.91464	8.83 m	β+ /46 /3.443	2.47/	3/2+	+1.01	+0.4	ann.rad./
				ec/54 /					Pm k x-ray
									1.0565
^{144}Sm	3.07(7)	143.91201				0+			
^{145}Sm		144.91342	340. d	ec/0.617		7/2-	-1.12	-0.6	Pm k x-ray
									0.0613
									0.4924
^{146}Sm		145.91305	~ 68. ×10^{8} a	α/	2.50/	0+			
^{147}Sm	14.99(18)	146.91490	1.06 ×10^{11} a	α/	2.23/	7/2-	-0.815	-0.26	
^{148}Sm	11.24(10)	147.91483	7 ×10^{15} a	α/	1.96/	0+			
^{149}Sm	13.82(7)	148.91719	10^{16} a	α/		7/2-	-0.672	+0.075	
^{150}Sm	7.38(1)	149.91728				0+			
^{151m}Sm			1.4 μs	IT/0.261		(11/2)-			
^{151}Sm		150.91994	95 a	β- /0.0768	0.076/	5/2-	-0.363	+0.7	0.02154
^{152}Sm	26.75(16)	151.91974				0+			
^{153m}Sm			10.6 ms	IT/0.0984/100	0.0329/100	11/2-			
^{153}Sm		152.92210	1.930 d	β- /0.808	0.64/	3/2+	-0.0216	+1.3	Eu k x-ray
					0.69/				0.0697/4.7
									0.10318/28
									0.075–0.714
^{154}Sm	22.75(29)	153.92222				0+			
^{155m}Sm			~ 2.8 μs	IT/0.0166		5/2+			
^{155}Sm		154.92465	22.2 m	β- /1.627	1.52	3/2-		1.1	Eu k x-ray
									0.1043/75.

Elem. or Isot.	Natural abundance (Atom %)	Atomic mass or weight	Half-life/ Resonance width (MeV)	Decay mode/ Energy (/MeV)	Particle energy/ Intensity (MeV/%)	Spin (h/2π)	Nuclear magnetic mom. (nm)	Elect. quadr. mom. (b)	γ-Energy/ Intensity (MeV/%)
^{156}Sm		155.9255	9.4 h	β- /0.72	0.43/	0+			0.0872
					0.71/				0.1657
									0.2038
^{157}Sm		156.92842	8.0 m	β- /2.7	2.4/	3/2-			Eu k x-ray
									0.1964
									0.1978
									0.3942
^{158}Sm		157.92995	5.5 m	β- /2.0		0+			0.1894/100.
									0.3636/82.
^{159}Sm		158.93322	11.3 s	β- /3.8					0.1898
^{160m}Sm			1.8 μs			(11+)			
^{160}Sm		159.93534	9.6 s	β- /3.6		0+			0.110
^{161}Sm		160.93916	~ 4.8 s						0.264
^{162}Sm		161.941	2.4 s			0+			(0.036-0.741)
^{163}Sm		162.946	> 0.3 μs						
^{164}Sm		163.948	> 0.3 μs			0+			
^{165}Sm		164.953	> 0.3 μs						
$_{63}$**Eu**		**151.964(1)**							
^{130}Eu		129.964	0.9 ms	p	1.027/				
^{131}Eu		130.958	~ 26. ms	β+, p	p/0.95				
^{132}Eu		131.955							
^{133}Eu		132.949							
^{134}Eu		133.946	0.5 s	ec, β+					ann.rad./
^{135}Eu		134.942	1.5 s	ec, β+ /~ 8.7					ann.rad./
^{136m}Eu			~ 3.2 s			7+			0.255
^{136}Eu		135.940	~ 3.9 s	ec, β+ /10.		1+			ann.rad./
^{137}Eu		136.935	11. s	ec/~ 7.5		11/2-			ann.rad./
^{138}Eu		137.9337	12. s	ec, β+ /~ 9.2		7+	5		ann.rad./
^{139m}Eu			10. μs	IT/0.148	0.0264/100	(7/2+)			0.026
^{139}Eu		138.9298	18. s	ec, β+ /6.7			6		ann.rad./
^{140m}Eu			0.125 s	IT		(5-)			ann.rad./
^{140}Eu		139.9281	1.51 s	ec, β+ /8.4		1-	+1.37	+0.31	ann.rad./
^{141m}Eu			2.7 s	β+ /58 /		11/2-			ann.rad./
				ec/9 /					Eu k x-ray
				IT/33 /0.0964					(0.09–1.6)
^{141}Eu		140.9249	40. s	β+ /81 /5.6		5/2+	+3.49	+0.85	ann.rad./
				ec/15 /					Sm k x-ray
									0.3845
									0.3940
^{142m}Eu			1.22 m	β+ /83 /	4.8/	8-	+2.98	+1.4	ann.rad./
				ec/17 /					Sm k x-ray
									0.5566
									0.7680
									1.0233
^{142}Eu		141.9234	2.4 s	β- /94/7.4	7.0/	1+	+1.54	+0.12	ann.rad./
				ec/6 /					0.7680
^{143m}Eu			0.050 ms	IT/0.390		11/2-			(0.118-0.389)
^{143}Eu		142.9203	2.62 m	β+ /72/5.17	4.1/	5/2+	+3.67	+0.51	ann.rad./
				ec/28/	5.1/				Sm k x-ray
									0.1107/7
									1.5368/3.
									1.9127/2.
^{144}Eu		143.9188	10.2 s	β+ /86 /6.33	5.31/	1+	+1.89	+0.10	ann.rad./
				ec/13 /					Sm k x-ray
									1.6601
^{145}Eu		144.91627	5.93 d	β+ /2 /2.660	0.79/	5/2+	+4.00	+0.29	ann.rad./
				ec/98 /1.71					Sm k x-ray
									0.6535
									(0.894-1.659)
^{146m}Eu			0.235 ms	IT/0.666		9+			
^{146}Eu		145.91721	4.57 d	β+ /5 /3.88	1.47/	4-	+1.42	-0.18	ann.rad./
				ec/95 /					Sm k x-ray
									0.6336
									0.6341
									0.7470
									(0.27–2.64)
^{147}Eu		146.91675	24.4 d	ec/99. /1.722		5/2+	+3.73	+0.53	Sm k x-ray
				β+ /0.4 /					0.12113/20.6
									0.19725/24.0
									(0.601-1.077)
^{148}Eu		147.9181	54.5 d	ec/3.11	0.92	5-	+2.34	+0.35	Sm k x-ray
									0.5503/99.
									(0.067–2.17)

Elem. or Isot.	Natural abundance (Atom %)	Atomic mass or weight	Half-life/ Resonance width (MeV)	Decay mode/ Energy (/MeV)	Particle energy/ Intensity (MeV/%)	Spin (h/2π)	Nuclear magnetic mom. (nm)	Elect. quadr. mom. (b)	γ-Energy/ Intensity (MeV/%)
^{149m}Eu			2.5 μs	IT/0.496		11/2-			
^{149}Eu		148.91794	93.1 d	ec/0.692		5/2+	+3.57	+0.75	Sm k x-ray
									0.2770/4.1
									0.3275/4.8
^{150}Eu		149.91971	36. a	ec/2.26		5-	+2.71	+1.13	Sm k x-ray
									0.3340
									0.4394
									0.5843
									(0.25–1.8)
^{150m}Eu			12.8 h	β- /92 /	1.013/	0-			Sm k x-ray
				β+ /0.4 /	1.24/				0.3339
				ec/8 /					0.4065
^{151m}Eu			0.059 ms	IT/0.196		11/2-			(0.175-0.196)
^{151}Eu	47.81(6)	150.91986	~ 5. ×10^{18} a	α		5/2+	+3.472	+0.90	
^{152m2}Eu			1.60 h	IT/0.1478	0.0398/100	8-			Eu k x-ray
									0.0898
^{152m1}Eu			9.31 h	β- /72 /	1.85/	0-			Sm k x-ray
				ec/28 /	0.89/				0.12178
									0.84153
									0.96334
^{152}Eu		151.92175	13.5 a	ec/72 /1.874	0.69/	3-	-1.941	+2.71	Sm k x-ray
				β- /28 /1.818	1.47/				Gd k x-ray
									0.12178
									0.34427
									1.40802
									(0.252–1.528)
^{153}Eu	52.19(6)	152.92124	> 5.5 ×10^{17} a	α		5/2+	+1.533	+2.41	
^{154m2}Eu			46.3 m	IT/0.145		8-			Eu k x-ray
									(0.068-0.101)
^{154m1}Eu			2.2 μs	IT/0.068		2+			
^{154}Eu		153.92299	8.59 a	β- /99.9/1.969	0.27/29	3-	-2.01	+2.8	Gd k x-ray
				ec/0.02/0.717	0.58/38				0.12299/40.
					0.84/17				0.72331/20.
					0.98/4				1.2745/36
					1.87/11				(0.059-1.90)
^{155}Eu		154.92290	4.76 a	β- /0.252	0.15/	5/2+	+1.520	+2.50	Gd k x-ray
									0.0865/30
									0.1053/20
^{156}Eu		155.92476	15.2 d	β- /2.451	0.30/11	0+	≈1.1		0.08899/9.
					0.49/30				0.64623/7.
					1.2/12				0.723441/6.
					2.45/31				0.8118/10.
^{157}Eu		156.92543	15.13 h	β- /1.36	0.98/	5/2+	+1.50	+2.6	Gd k x-ray
					1.30/41				0.0639/100.
									0.3705/48.
									0.4107/76.
^{158}Eu		157.9278	45.9 m	β- /3.5	2.5/	(1-)	+1.44	+0.7	0.0795
									0.8976
									0.9442
									0.9771
^{159}Eu		158.92910	18.1 m	β- /2.51	2.4/	5/2+	+1.38	+2.7	0.0678
					2.57/				0.0786
									0.0957
^{160}Eu		159.9319	38. s	β- /4.1	2.7/	(0-)			0.0753
					4.1/				0.1735
									0.4131
									0.5155
									0.8217
									0.9110
									0.9246
^{161}Eu		160.9337	27. s	β- /3.7					0.0719
^{162}Eu		161.9370	11. s	β- /5.6					
^{163}Eu		162.9392	7.7 s						
^{164}Eu		163.943	4.2 s						
^{165}Eu		164.946	2.7 s						
^{166}Eu		165.950	1.7 s						
^{167}Eu		166.953	> 0.3 μs						
^{168}Eu			> 0.3 μs						
$_{64}$Gd		**157.25(3)**							
^{135}Gd		134.952	1.1 s	β+					(0.163–0.360)
^{136}Gd		135.947	> 0.2 μs			0+			
^{137}Gd		136.945	7. s	ec, β+ /~ 8.8					ann.rad./
^{138m}Gd			6.2 μs	IT/2.233	0.583/100	(8-)			0.583

Nuclear

Elem. or Isot.	Natural abundance (Atom %)	Atomic mass or weight	Half-life/ Resonance width (MeV)	Decay mode/ Energy (/MeV)	Particle energy/ Intensity (MeV/%)	Spin (h/2π)	Nuclear magnetic mom. (nm)	Elect. quadr. mom. (b)	γ-Energy/ Intensity (MeV/%)
^{138}Gd		137.940	~ 4.7 s	ec, β+		0+			0.0647
^{139m}Gd			5.5 s						0.1216
^{139}Gd		138.938	5. s	ec, β+ /~ 7.7					0.104–0.323
^{140}Gd		139.9337	16. s	ec/4.8		0+			0.1748
^{141m}Gd			25. s	ec, β+ /		11/2-			ann.rad./
				IT//11					
^{141}Gd		140.9321	21. s	β+ /7.3		½+			ann.rad./
^{142}Gd		141.9281	1.17 m	ec, β+ /4.2		0+			ann.rad./
^{143m}Gd			1.84 m	β+ /67 /		(11/2-)			ann.rad./
				ec/33 /					Eu k x-ray
				IT/					0.1176
									0.2719
									0.5880
									0.6681
									0.7999
^{143}Gd		142.927	39. s	β+ /82 /6.0		1/2+			ann.rad./
				ec/18 /					Eu k x-ray
									0.2048
									0.2588
^{144}Gd		143.9230	4.5 m	β+ /45 /4.3	3.3/	0+			ann.rad./
				ec/55 /					Eu k x-ray
									0.3332
^{145m}Gd			1.42 m	IT/95 /0.749		11/2-	-1.0		0.0273
				β+ /4 /5.7					0.3295
									0.3866
									0.7214
^{145}Gd		144.9217	23.4 m	β+ /33 /5.05	2.5/	1/2+	-0.74		ann.rad./
				ec/67 /					Eu k x-ray
									1.7579
									1.8806
									(0.32–3.69)
^{146}Gd		145.91832	48.3 d	ec/99.9 /1.03	0.35/	0+			Eu k x-ray
				β+ /0.2					0.1147
									0.1155
									0.1546
^{147}Gd		146.91910	1.588 d	ec/99.8 /2.188	0.93/	7/2-	1.0		Eu k x-ray
				ec/0.2 /					0.2293
									0.3699
									0.3960
									0.9289
									(0.1–1.8)
^{148}Gd		147.91812	71. a	α/3.27	3.1828/	0+			
^{149}Gd		148.91935	9.3 d	ec/1.32		7/2-	0.9		Eu k x-ray
									0.1496
									0.2985
									0.3465
^{150}Gd		149.91866	1.8 ×10^{6} a	α/2.80	2.73/	0+			
^{151}Gd		150.92036	124. d	ec/0.464		7/2-	0.8		Eu k x-ray
									0.1536
									0.2432
^{152}Gd	0.20(1)	151.91980				0+			
^{153m2}Gd			0.076 ms	IT/0.171		(11/2-)			
^{153m1}Gd			3.5 μs	IT/0.095		(9/2+)			
^{153}Gd		152.92176	240. d	ec/0.485		3/2-	0.4		Eu k x-ray
									0.09743
									0.10318
^{154}Gd	2.18(3)	153.92087				0+			
^{155m}Gd			0.032 s	IT/0.121	0.0135/100				
^{155}Gd	14.80(12)	154.92263				3/2-	-0.258	+1.30	
^{156m}Gd			1.3 μs	IT/2.138		7-			
^{156}Gd	20.47(9)	155.92213				0+			
^{157m}Gd			0.019 ms	IT/0.427		11/2-			
^{157}Gd	15.65(2)	156.92397				3/2-	-0.340	+1.36	
^{158}Gd	24.84(7)	157.92411				0+			
^{159}Gd		158.92640	18.6 h	β- 0.971	0.971/58	3/2-	-0.44		Tb k x-ray
					0.913/29				0.36351
					0.607/12				0.058-0.855
^{160}Gd	21.86(19)	159.92706	>1.9 ×10^{19} a	β- β-		0+			
^{161}Gd		160.92968	3.66 m	β- /1.956	1.56/85	5/2-			Tb k x-ray
									0.1023
									0.3149
									0.3609
^{162}Gd		161.93099	8.4 m	β- /1.39	1.0/	0+			0.4030

Elem. or Isot.	Natural abundance (Atom %)	Atomic mass or weight	Half-life/ Resonance width (MeV)	Decay mode/ Energy (/MeV)	Particle energy/ Intensity (MeV/%)	Spin (h/2π)	Nuclear magnetic mom. (nm)	Elect. quadr. mom. (b)	γ-Energy/ Intensity (MeV/%)
^{163}Gd		162.9342	1.13 m	β- /3.1					0.4421
									0.2868
									0.214
									1.685
^{164}Gd		163.936	45. s	β- /2.3		0+			
^{165}Gd		164.939	10 s	β-					
^{166}Gd		165.942	~ 4.8 s			0+			(0.040-1.015)
^{167}Gd		166.945	> 0.3 μs						
^{168}Gd		167.948	> 0.3 μs			0+			
^{169}Gd		168.953	> 0.3 μs						
^{170}Gd			> 0.3 μs						
$_{65}$Tb		**158.92535(2)**							
^{135}Tb		134.965	1.0 ms	p	p/1.179				
^{138}Tb		137.953	> 0.2 μs						
^{139}Tb		138.948	1.6 s						0.109
									0.120
^{140}Tb		139.946	2.4 s	β+, ec/11					0.329
									0.355–0.740
^{141}Tb		140.941	3.5 s	β+, ec/~ 8.3					
^{142m2}Tb			26. μs	IT/0.652		(8+)			0.037
^{142m}Tb			0.30 s	β+, ec/		5-			0.0682
^{142}Tb		141.939	0.60 s	β+, ec/10.		0+			
^{143}Tb		142.9351	12. s	β+, ec/7.4		(11/2-)			
^{144m2}Tb			2.8 μs	IT/0.476	0.079/100	8-			
^{144m}Tb			4.3 s	IT//66		6-			
^{144}Tb		143.9331	< 1.5 s	β+, ec/8.4		1+			
^{145m}Tb			31. s	β+, ec/~ 6.6		(11/2-)			ann.rad./
									0.2577
									0.5370
									0.9876
^{145}Tb		144.929		β+, ec/6.5		½+			
^{146m2}Tb			1.2 ms	IT	0.418/100	(10+)			(0.157-0.417)
^{146m}Tb			23. s	β+ /76 /		(5-)			ann.rad./
				ec/24 /					Gd k x-ray
									1.0789
									1.5795
^{146}Tb		145.9273	~ 8. s	β+ /8.1		1+			
^{147m}Tb			1.8 m	β+ /35 /		(11/2-)			ann.rad./
				ec/65 /					Gd k x-ray
									1.3977
									1.7978
^{147}Tb		146.9241	1.6 h	β+ /42 /4.61		5/2+	+1.70		ann.rad./
				ec/58 /					Gd k x-ray
									0.6944
									(0.120–3.318)
^{148m2}Tb			1.31 μs	IT/8.619		(27+)			
^{148m}Tb			2.2 m	β+ /25 /		(9)+			ann.rad./
				ec/75 /					Gd k x-ray
									0.3945
									0.6319
									0.7845
									0.8824
^{148}Tb		147.9243	1.00 h	β+, ec/5.69		2-	-1.75	-0.3	ann.rad./
									Gd k x-ray
									0.4888
									0.7845
									(0.14–3.8)
^{149m}Tb			4.16 m	ec/88 /		11/2-			ann.rad./
				β+ /12 /					Gd k x-ray
									0.1650
									0.7960
^{149}Tb		148.92325	4.13 h	β+ /4 /3.636	1.8/	½+	+1.35		Gd k x-ray
				α/16/	3.97/				0.1650
									0.3522
									0.3886
									(0.1–3.2)
^{150m}Tb			5.8 m	β+ /17 /		9+			ann.rad./
				ec/83 /					Gd k x-ray
									0.4384
									0.6380
									0.6504
									0.8275
^{150}Tb		149.9237	3.3 h	β+, ec/4.66		2-	-0.90		ann.rad./

Elem. or Isot.	Natural abundance (Atom %)	Atomic mass or weight	Half-life/ Resonance width (MeV)	Decay mode/ Energy (/MeV)	Particle energy/ Intensity (MeV/%)	Spin (h/2π)	Nuclear magnetic mom. (nm)	Elect. quadr. mom. (b)	γ-Energy/ Intensity (MeV/%)
									0.4963
									0.6380
									(0.3–4.29)
^{151m}Tb			25. s	IT/95 /		(11/2-)			0.0229
				β+, ec/7 /					0.0495
									0.3797
									0.8305
^{151}Tb		150.92311	17.61 h	β+/1 /2.565	0.70/	1/2(+)	+0.92		Gd k x-ray
				ec/99 /					0.1083
									0.2517
									0.2870
									(0.1–1.8)
^{152m}Tb			4.2 m	IT/79 /0.5018		8+			Tb k x-ray
				ec/21 /4.35					Gd k x-ray
									0.2833
									0.3443
									0.4111
^{152}Tb		151.9241	17.5 h	β+ /20 /3.99	2.5/	2-	-0.58	+0.3	ann.rad./
				ec/80 /	2.8/				Gd k x-ray
									0.3443
									(0.2–2.88)
^{153m}Tb			0.186 ms	IT/0.163	0.082/100				
^{153}Tb		152.92344	2.34 d	ec/1.570		5/2+	+3.44	+1.1	Gd k x-ray
									0.2119
									(0.05–1.1)
^{154m2}Tb			~ 22.7 h	ec/98 /		(7-)	0.9		Gd k x-ray
				IT/2 /					0.1231
									0.2479
									0.3467
									1.4199
^{154m1}Tb			10.0 h	β+ /78 /		(3-)	+1.7	+3.	Gd k x-ray
				IT/22 /					0.1231
									0.2479
									0.5401
									(0.12–2.57)
^{154}Tb		153.9247	21.5 h	ec/99 /3.56	1.86/	0-			Gd k x-ray
				β+ /1 /	2.45				0.1231
									1.2744
									2.1872
									(0.12–3.14)
^{155}Tb		154.9235	5.3 d	ec/0.82		3/2+	+2.01	+1.41	Gd k x-ray
									0.08654
									0.10530
^{156m2}Tb			1.02 d	IT/		(7-)			Tb k x-ray
									0.0496
^{156m1}Tb			5.3 h	IT/0.0884		(0+)			Tb k x-ray
									0.0884
^{156}Tb		155.92476	5.3 d	ec/2.444		3-	~ 1.7	+2.	Gd k x-ray
									0.08896
									0.19921
									0.53435
									1.22245
^{157}Tb		156.92403	1.1 ×10^{2} a	ec/0.0601		3/2+	+2.01	+1.4	Gd k x-ray
									0.0545
^{158m2}Tb			0.40 ms	IT/0.388		7-			
^{158m}Tb			10.7 s	IT/0.11		0-			Gd k x-ray
									0.0110
^{158}Tb		157.92542	1.8 ×10^{2} a	ec/80 /1.220		3-	+1.76	+2.7	Gd k x-ray
				β- /20 /0.937					0.0795
									0.9442
									0.9621
^{159}Tb	100.	158.92535				3/2+	+2.014	+1.43	
^{160}Tb		159.92716	72.3 d	β- /1.835	0.57/47	3-	+1.79	3.8	Dy k x-ray
					0.86/27				0.08678
									0.29857
									0.87936
									0.96615
^{161}Tb		160.92758	6.91 d	β- /0.593	0.46/23	3/2+	2.2	+1.2	Dy k x-ray
					0.52/66				0.02565
					0.6/10				0.04892
									0.07458
^{162}Tb		161.9295	7.6 m	β- /2.51	1.4	(1/2-)			Dy k x-ray
									0.2600

Elem. or Isot.	Natural abundance (Atom %)	Atomic mass or weight	Half-life/ Resonance width (MeV)	Decay mode/ Energy (/MeV)	Particle energy/ Intensity (MeV/%)	Spin (h/2π)	Nuclear magnetic mom. (nm)	Elect. quadr. mom. (b)	γ-Energy/ Intensity (MeV/%)
									0.8075
									0.8882
^{163}Tb		162.93066	19.5 m	β- /1.785	0.80/	3/2+			Dy k x-ray
									0.3511
									0.3897
									0.4945
^{164}Tb		163.933	3.0 m	β- /3.9	1.7/	(5+)			Dy k x-ray
									0.1689
									0.2157
									0.6110
									0.6885
									0.7548
^{165}Tb		164.935	2.1 m	β- /3.0		3/2+			0.5389
									1.1785
									1.2920
									1.6648
^{166}Tb		165.9379	26 s	β-/					
^{167}Tb		166.940	19 s						0.057
									0.070
^{168}Tb		167.943	8 s						(0.075–0.227)
^{169}Tb		168.946	> 0.3 μs						
^{170}Tb		169.950	> 0.3 μs						
^{171}Tb		170.953	> 0.3 μs						
^{172}Tb			> 0.3 μs						
$_{66}$Dy		**162.500(1)**							
^{139}Dy		138.960	0.6 s	β+, p					
^{140m}Dy			7 μs	IT/2.166		(8-)			0.574
^{140}Dy		139.954				0+			
^{141}Dy		140.951	0.9 s	ec, β+ /9.					
^{142}Dy		141.946	2.3 s	ec, β+ /7.1		0+			
^{143m2}Dy			1.2 μs	IT/0.406					
^{143m1}Dy			3.0 s	ec, β+		(11/2-)			
^{143}Dy		142.9440	3.9 s	ec, β+ /~ 8.8					
^{144}Dy		143.93927	9.1 s	ec, β+ /~ 6.2		0+			
^{145m}Dy		144.9365	14 s	ec, β+		(11/2-)			
^{146m}Dy			0.15 s	IT		(10+)			
^{146}Dy		145.93285	30. s	ec, β+ /5.2		0+			
^{147m}Dy			55 s	IT//31	0.75	(11/2-)	-0.66	+0.7	Dy k x-ray
				β+, ec//69					0.072
									0.6787
^{147}Dy		146.9311	75. s	ec, β+ /6.37		½+	-0.92		ann.rad./
									0.1007
									0.2534
									0.3653
^{148}Dy		147.9272	3.1 m	β+ /4 /2.68	1.2/	0+			ann.rad./
				ec/96 /					Tb k x-ray
									0.6202
^{149m}Dy			0.49	IT/2.661	0.111/99	(27/2-)			
^{149}Dy		148.9273	4.2 m	β+, ec/3.81		(7/2-)	-0.12	-0.62	ann.rad./
									0.1008
									0.1063
									0.2534
									0.6536
									0.7894
									1.7765
									1.8062
^{150}Dy		149.92559	7.18 m	β+, ec/67 /1.79		0+			Tb k x-ray
				α/33 /	4.233/				0.3967
^{151}Dy		150.92619	17. m	β+ /5 /2.871		7/2(-)	-0.95	-0.30	Tb k x-ray
				ec/89 /					0.1764
				α /6 /	4.067/				0.3030
									0.3861
									0.5463
									(0.16–2.09)
^{152}Dy		151.92473	2.37 h	ec/0.60		0+			Tb k x-ray
				α /	3.63/				0.2569
^{153}Dy		152.92577	6.3 h	β+ /1 /2.171	0.89/	7/2(-)	-0.78	~ -0.15	Tb k x-ray
				ec/99 /					0.0807
				α /0.01 /	3.46/				0.0997
									0.2137
									(0.08–1.66)
^{154}Dy		153.92443	3. ×10^6 a	α/2.95	2.87/	0+			
^{155m}Dy			6 μs	IT/0.234		11/2-			

Nuclear

Elem. or Isot.	Natural abundance (Atom %)	Atomic mass or weight	Half-life/ Resonance width (MeV)	Decay mode/ Energy (/MeV)	Particle energy/ Intensity (MeV/%)	Spin (h/2π)	Nuclear magnetic mom. (nm)	Elect. quadr. mom. (b)	γ-Energy/ Intensity (MeV/%)
^{155}Dy		154.9258	9.9 h	β+ /2 /2.095	0.845/	3/2-	-0.385	+1.04	Tb k x-ray
				ec/98 /					0.0655
									0.2269
^{156}Dy	0.056(3)	155.92428	> 6.1 ×10^{14} a	ec-ec		0+			
^{157m2}Dy			22 ms	IT/0.199		11/2-			
^{157m1}Dy			1.3 μs	IT/0.162	0.014/100	9/2+			
^{157}Dy		156.92547	8.1 h	ec/1.34		3/2-	-0.301	+1.30	Tb k x-ray
									(0.061–1.319)
^{158}Dy	0.095(3)	157.92442	> 1. ×10^{15} a	ec-ec		0+			
^{159m}Dy			0.122 ms	IT/0.353		11/2-			
^{159}Dy		158.92575	144. d	ec/0.366		3/2-	-0.354	+1.37	Tb k x-ray
									0.3262
^{160}Dy	2.329(18)	159.92520				0+			
^{161}Dy	18.889(42)	160.92694				5/2+	-0.480	+2.51	
^{162m}Dy			8.3 μs	IT/2.188		8+			
^{162}Dy	25.475(36)	161.92681				0+			
^{163}Dy	24.896(42)	162.92874				5/2-	+0.673	+2.65	
^{164}Dy	28.260(54)	163.92918				0+			
^{165m}Dy			1.26 m	IT/98 /0.108		1/2-			Dy k x-ray
				β- /2 /					0.1082
									0.5155
^{165}Dy		164.93171	2.33 h	β- /1.286	1.29/	7/2+	-0.52	-3.5	Ho k x-ray
									0.09468/3.8
^{166}Dy		165.93281	3.400 d	β- /0.486	0.40/	0+			Ho k x-ray
									0.0282
									0.0825
^{167}Dy		166.9357	6.2 m	β- /~ 2.35	1.78	(1/2-)			Ho k x-ray
									0.2593
									0.3103
									0.5697
									(0.06–1.4)
^{168}Dy		167.937	8.5 m	β- /1.6		0+			Ho k x-ray
									0.1925
									0.4867
^{169}Dy		168.940	~ 39. s	β- /3.2					
^{170}Dy		169.942	> 0.3 μs			0+			
^{171}Dy		170.946	> 0.3 μs						
^{172m}Dy			~ 0.71 s						(0.045-0.758)
^{172}Dy		171.948	3.4 s			0+			
^{173}Dy		172.953	> 0.3 μs						
^{174}Dy			> 0.3 μs						
$_{67}$Ho		**164.93033(2)**							
^{140}Ho		139.969	6 ms	p/	p/1.09				
^{141m}Ho			7.3 μs	p/	p/1.23	1/2+			
^{141}Ho		140.963	4.1 ms	β+, p	p/1.17/99.3				
					p/0.97/0.7				
^{142}Ho		141.960	0.4 s	ec/β+, p					0.307
^{143}Ho		142.955	> 0.2 μs						
^{144}Ho		143.9521	0.7 s	β+, ec/12					
^{145}Ho		144.94727	2.4 s	β+					
^{146}Ho		145.94499	3.3 s	β+, ec/10.7		(10+)			ann.rad./
^{147}Ho		146.94014	5.8 s	β+, ec/8.2		11/2-			ann.rad./
^{148m2}Ho			2.4 ms	IT/0.694	0.373/100	(10+)			(0.033-0.373)
^{148m}Ho			9.6 s	β+, ec/		(5)-			ann.rad./
^{148}Ho		147.9377	2. s	β+, ec/9.4		1+			ann.rad./
									0.6615
									1.6883
^{149m}Ho			21. s	β+, ec/		11/2-			ann.rad./
									1.0733
									1.0911
^{149}Ho		148.9338	56 s	β+, ec/6.01		1/2+			
^{150m}Ho			24 s	β+, ec/		(9)+			ann.rad./
									0.3939
									0.5511
									0.6534
									0.8034
^{150}Ho		149.9335	1.3 m	β+, ec/6.6					ann.rad./
									0.5913
									0.6534
									0.8034
^{151m}Ho			47. s	β+, ec/87 /		(1/2+)			ann.rad./
				α/13	4.605/				0.2102
									0.4889

Elem. or Isot.	Natural abundance (Atom %)	Atomic mass or weight	Half-life/ Resonance width (MeV)	Decay mode/ Energy (/MeV)	Particle energy/ Intensity (MeV/%)	Spin (h/2π)	Nuclear magnetic mom. (nm)	Elect. quadr. mom. (b)	γ-Energy/ Intensity (MeV/%)
									0.6948
									0.7762
^{151}Ho		150.9317	35.2 s	β+, ec/80/5.13		(11/2-)			ann.rad./
				α/20 /	4.519/				0.3522
									0.5274
									(0.968-1.047)
^{152m2}Ho			8.4 μs	IT/3.020	0.734/100	19-			
^{152m}Ho			50. s	β+, ec/90/		9+	+5.9	-1.3	ann.rad./
				α/10/	4.453/				0.4929
									0.6138
									0.6474
									0.6835
^{152}Ho		151.9317	2.4 m	β+, ec/88/6.47		(3+)	-1.02	~ +0.1	ann.rad./
				α/12/	4.387/				0.6140
									0.6476
^{153m}Ho			9.3 m	β+, ec/99+/4.12		1/2+	+1.19		ann.rad./
				α/	4.01/				0.0905
									0.1089
									0.1618
									0.2302
									0.2707
									0.3659
									0.4565
^{153}Ho		152.93020	2.0 m	β+, ec/99+/4.13		11/2-	+6.8	-1.1	ann.rad./
				α/	3.91/				0.2958
									0.3346
									0.4381
									0.6383
^{154m}Ho			3.1 m	β+, ec/		8+	5.7	-1.0	ann.rad./
									0.3346
									0.4124
									0.4771
^{154}Ho		153.9306	12. m	β+, ec/5.75		2-	-0.64	+0.2	ann.rad./
									Dy k x-ray
									0.3346
									(0.570-0.8734)
^{155m}Ho			0.9 ms	IT/0.142	0.0317/100	11/2-			
^{155}Ho		154.9291	48. m	β+/6/3.10		5/2+	+3.51	+1.5	ann.rad./
				ec/94 /					Dy k x-ray
									0.0474
									0.1363
									(0.06–2.24)
^{156m2}Ho			10 s	IT/0.052		1-			
^{156m}Ho			7.6 m	IT/0.035/25		9+	+2.99	+2.3	ann.rad./
				β+ /25 /	1.8/				Dy k x-ray
				ec/75 /	2.9/				0.1378
									0.2666
									(0.28–2.9)
^{156}Ho		155.9298	56. m	β+, ec/4.4		4 -			ann.rad./
									0.1378
									0.2665
^{157}Ho		156.9283	12.6 m	β+/5/2.54	1.18/	7/2-	+4.35	+3.0	ann.rad./
				ec/95/					Dy k x-ray
									0.2800
									0.3411
^{158m2}Ho			28. m	IT/44/		2-	+2.44	+1.6	ann.rad./
				ec/56/					Dy k x-ray
									0.0989
									0.2182
^{158m1}Ho			21. m	β+, ec/		(9+)			ann.rad./
									0.0981
									0.1664
									0.2182
									0.3205
									0.4062
									0.9774
									1.0532
									0.4846
^{158}Ho		157.9289	11.3 m	β+/8/4.24	1.30/	5+	+3.77	+4.1	ann.rad./
				ec/92/					Dy k x-ray
									0.0989
									0.2182
									0.9488

Elem. or Isot.	Natural abundance (Atom %)	Atomic mass or weight	Half-life/ Resonance width (MeV)	Decay mode/ Energy (/MeV)	Particle energy/ Intensity (MeV/%)	Spin (h/2π)	Nuclear magnetic mom. (nm)	Elect. quadr. mom. (b)	γ-Energy/ Intensity (MeV/%)
^{159m}Ho			8.3 s	IT/0.206		1/2+			Ho k x-ray
									0.1660
									0.2059
^{159}Ho		158.92772	33.0 m	ec/1.838		7/2-	+4.28	+3.2	Dy k x-ray
									0.1210
									0.1320
									0.2529
									0.3096
									(0.06–1.2)
^{160m2}Ho			3. s	IT/0.20		(9+)			
^{160m}Ho			5.0 h	IT/67/0.060		2-	+2.52	+1.8	0.0868
				ec/33/3.35					0.1970
									0.6464
									0.7281
									0.8791
									0.9619
									0.9658
^{160}Ho		159.9287	25.6 m	β+, ec/3.29	0.57/	5+	+3.71	+4.0	See Ho[166m]
									0.7282
									0.8794
^{161m}Ho			6.8 s	IT/0.211		1/2+			Ho k x-ray
									0.2112
^{161}Ho		160.92786	2.48 h	ec/0.859		7/2-	+4.25	+3.2	Dy k x-ray
									0.0256
									0.0592
									0.0774
									0.1031
^{162m}Ho			1.12 h	IT/61/		6-	+3.60	+4.	Dy k x-ray
				ec/39/					Ho k x-ray
									0.0807
									0.1850
									0.2828
									0.9372
									1.2200
^{162}Ho		161.92910	15. m	ec/96 /0.295		1+			Dy k x-ray
				β+ /4 /					0.0807
									1.3196
									1.3728
^{163m}Ho			1.09 s	IT/0.298		(1/2+)			Ho k x-ray
									0.2798
^{163}Ho		162.92874	4.57 ×10³ a	ec/0.00258		7/2-	+4.23	+3.6	Dy M x-rays
^{164m}Ho			38 m	IT/0.140	0.046/100	6-			Ho k x-ray
									0.0373
									0.0566
									0.0940
^{164}Ho		163.93024	29. m	ec/58 /0.987		1+			Dy k x-ray
				β- /42 /0.963					0.0734
									0.0914
^{165m}Ho			1.5 μs	IT/0.362		(3/2+)	+ 2.9		
^{165}Ho	100.	164.93033				7/2-	+4.17	+3.6	
^{166m}Ho			0.19 ms	IT/0.190		3+			
^{166m}Ho			1.13 ×10³ a	β- /		7-	3.6	~ -3.	Er k x-ray
									0.18407
									0.71169
									0.81031
^{166}Ho		165.93229	1.117 d	β- /1.855	1.776/48	0-			Er k x-ray
					1.855/51				(0.0806-1.379)
^{167m}Ho			6 μs	IT/0.259		3/2+			
^{167}Ho		166.93314	3.1 h	β- /1.007	0.31/43	(7/2-)			Er k x-ray
					0.61/21				0.0793
					0.96/15				0.0835
					0.97/15				0.2379
									0.3213
									0.3465
^{168m}Ho			2.2 m	IT/0.06		(6+)			
^{168}Ho		167.9355	3.0 m	β- /2.91	2.0/	3+			Er k x-ray
									0.7413
									0.8159
									(0.08–2.34)
^{169m}Ho			0.12 ms	IT/1.386		(19/2+)			
^{169}Ho		168.9369	4.7 m	β- /2.12	1.2/	(7/2-)			
					2.0/				0.1496
									0.7610

Elem. or Isot.	Natural abundance (Atom %)	Atomic mass or weight	Half-life/ Resonance width (MeV)	Decay mode/ Energy (/MeV)	Particle energy/ Intensity (MeV/%)	Spin (h/2π)	Nuclear magnetic mom. (nm)	Elect. quadr. mom. (b)	γ-Energy/ Intensity (MeV/%)
									0.7784
									0.7884
									0.8529
^{170m}Ho			43. s	β- /		1+			0.0787
									0.8123
									1.8940
									1.9726
^{170}Ho		169.9396	2.8 m	β- /3.87		6+			Er k x-ray
									0.1816
									0.2582
									0.8902
									0.9321
									0.9414
									1.1387
^{171}Ho		170.941	53 s	β- /					
^{172}Ho		171.945	25. s	β- /					Er k x-ray
									(0.077–1.186)
^{173}Ho		172.947	> 0.3 μs						
^{174}Ho		173.951	> 0.3 μs						
^{175}Ho		174.954	> 0.3 μs						
^{176}Ho			> 0.3 μs						
$_{68}$Er		**167.259(3)**							
^{144}Er		143.961	> 0.2 μs			0+			
^{145m}Er			0.9 s	β+, ec		(11/2-)			0.067
^{145}Er		144.958	0.9 s	β+		(1/2+)			0.049
^{146}Er		145.95242	~ 1.7 s	β+		0+			
^{147m}Er			1.6 s	ec, β+		(11/2-)			
^{147}Er		146.9500	2.5 s	E.C, β+ /~ 9.1					
^{148m}Er			13 μs	IT/2.913	0.131/100	(10+)			
^{148}Er		147.9447	4.5 s	β+, ec/6.8		0+			
^{149m2}Er			4.8 μs	IT/3.243		(27/2-)			
^{149m}Er			8.9 s	IT/0.742		11/2-			
^{149}Er		148.9423	10.7 s	ecβ+ /8.1		½+			
^{150m}Er			2.6 μs	IT/2.797	0.063/100	10+			
^{150}Er		149.9379	18. s	β+ /36 /4.11		0+			ann.rad./
				ec/64 /					Ho k x-ray
									0.4758
^{151m}Er			0.58 s	IT/95/2.586	0.058/100	(27/2-)			
^{151}Er		150.9374	23. s	β+, ec/5.2		7/2-			ann.rad./
^{152}Er		151.9351	10.2 s	β+, ec/10/3.11		0+			ann.rad./
				α/90/	4.804/				
^{153}Er		152.9351	37.1 s	α/	4.674		-0.934	-0.42	0.351
				β+, ec/47/4.56	4.35/				(0.0945–1.70)
^{154}Er		153.93279	3.7 m	β+, ec/99+/2.03		0+			ann.rad./
				α/0.5/	4.166/				
^{155}Er		154.93322	5.3 m	β+, ec/47/3.84		(7/2-)	-0.669	-0.27	ann.rad./
				ec/53 /					Ho k x-ray
									0.1101
									0.2415
^{156}Er		155.9311	20. m	β+, ec/1.7		0+			ann.rad./
									0.0298
									0.0352
									(0.052-0.1336)
^{157m}Er			0.08 s	IT	0.155/100	(9/2+)			
^{157}Er		156.9319	25. m	β+, ec/3.5		3/2-	-0.412	+0.92	ann.rad./
									0.117
									0.385
									1.320
									1.660
									1.820
									2.000
^{158}Er		157.9299	2.2 h	ec/99.5 /1.78	0.74/	0+			Ho k x-ray
				β+ /0.5 /					0.0719
									0.2486
									0.3868
^{159}Er		158.93069	36. m	β+ /7 /2.769		3/2-	-0.304	+1.17	ann.rad./
				ec/93 /					Ho k x-ray
									0.6245
									0.6493
									(0.07–2.5)
^{160}Er		159.9291	1.191 d	ec/0.33		0+			Ho k x-ray
									(0.05–0.96)
^{161m}Er			0.008 ms	IT/0.396		11/2-			

Nuclear

Elem. or Isot.	Natural abundance (Atom %)	Atomic mass or weight	Half-life/ Resonance width (MeV)	Decay mode/ Energy (/MeV)	Particle energy/ Intensity (MeV/%)	Spin (h/2π)	Nuclear magnetic mom. (nm)	Elect. quadr. mom. (b)	γ-Energy/ Intensity (MeV/%)
^{161}Er		160.9300	3.21 h	ec/2.00		3/2-	-0.37	+1.36	Ho k x-ray
									0.8265
									(0.07–1.74)
^{162}Er	0.139(5)	161.92879				0+			
^{163}Er		162.93004	1.25 h	ec/1.210		5/2-	+0.557	+2.55	Ho k x-ray
									0.4361
									0.4399
									1.1135
^{164}Er	1.601(3)	163.92921				0+			
^{165}Er		164.93073	10.36 h	ec/0.376		5/2-	+0.643	+2.71	Ho k x-ray
^{166}Er	33.503(36)	165.9303				0+			
^{167m}Er			2.27 s	IT/0.208		½-			Er k x-ray
									0.2078
^{167}Er	22.869(9)	166.93205				7/2+	-0.5639	+3.57	
^{168}Er	26.978(18)	167.93238				0+			
^{169}Er		168.93460	9.40 d	β- /0.351	0.35/~ 100	½-	+0.485		Tm k x-ray
									0.1098
									0.1182
^{170}Er	14.910(36)	169.93547				0+			
^{171}Er		170.93804	7.52 h	β- /1.491		5/2-	0.66	2.9	Tm k x-ray
									0.11160
									0.29591
									0.30832
									(0.08–1.4)
^{172}Er		171.93936	2.05 d	β-/0.891	0.28/48	0+			Tm k x-ray
					0.36/46				0.0597
									0.4073
									0.6101
^{173}Er		172.942	1.4 m	β- /2.6		(7/2-)			Tm k x-ray
									0.1928
									(0.1992-0.895)
^{174m}Er			4.0 s	IT/1.111		8-			
^{174}Er		173.944	3.1 m	β- /1.8		0+			Tm k x-ray
									(0.100–0.152)
^{175}Er		174.948	1.2 m	β-					(0.0765–1.17)
^{176}Er		175.950	> 0.3 μs			0+			
^{177}Er		176.954	> 0.3 μs						
^{178}Er			> 0.3 μs						
$_{69}$Tm		**168.93422(2)**							
^{144}Tm		143.976	~ 1.9 μs	p	1.70, 1.43				
^{145}Tm		144.970	3.1 μs	p//~ 10	1.73/91				
					1.40/~ 9.6				
^{146m}Tm			0.198 s	β+, p	p/1.12/99	10+			
					p/0.89/~ 1				
^{146}Tm		145.967	73 ms	β+/14.	p/1.19/68	(5-)			
				p	p/1.01/~ 18				
					p/0.94/~ 14				
^{147m}Tm			0.36 ms	β+, p	p/1.115	3/2+			
^{147}Tm		146.961	0.56 s	ec, β+/85	~ 10.7				
				p/15/	1.052/				
^{148m}Tm		147.958	0.7 s	β+, ec/12.					ann.rad./
^{148}Tm									
^{149}Tm		148.953	0.9 s	β+, ec/~ 9.2		11/2-			
^{150m}Tm			5.2 ms	IT/0.671	0.331/100	(10+)			(0.017-0.340)
^{150}Tm		149.950	2.3 s	β+, ec/~ 11.5		6-			(0.101–2.177)
^{151m}Tm			7 s	ec, β+		1/2+			
^{151}Tm		150.9455	4. s	β+, ec/7.5					ann.rad./
^{152m}Tm			8. s	β+, ec/		9+			
^{152}Tm		151.9444	5. s	β+, ec/8.8					ann.rad./
^{153m}Tm			2.5 s	α/92		(1/2+)			
				ec, β+/8					
^{153}Tm		152.9420	1.6 s	β+,ec/10 /6.46		(11/2-)	6.9	~ +0.5	ann.rad./
				α/90 /	5.109/				
^{154m}Tm			3.3 s	β+, ec/15 /	α/5.031/100	(9+)	+5.9	~ -0.2	ann.rad./
				α/	4.84/0.24				0.4605–0.796
^{154}Tm		153.9416	8.1 s	β+, ec/56 /7.4	α/4.956/100	(2-)	-1.14	~ +0.4	ann.rad./
				α/44 /	4.83/0.45				
^{155m}Tm			45 s	Ec, β+		1/2+			
									(0.1-1.58)
^{155}Tm		154.9392	30. s	β+, ec/5.58					0.0315
				α/	4.46/				0.0638
									0.0881
									0.2268

Elem. or Isot.	Natural abundance (Atom %)	Atomic mass or weight	Half-life/ Resonance width (MeV)	Decay mode/ Energy (/MeV)	Particle energy/ Intensity (MeV/%)	Spin (h/2π)	Nuclear magnetic mom. (nm)	Elect. quadr. mom. (b)	γ-Energy/ Intensity (MeV/%)
									0.5320
									0.6067
^{156m}Tm			19. s	α/	4.46/				
^{156}Tm		155.9390	1.40 m	β+, ec/7.6		2-	+0.40	-0.5	ann.rad./
				α/	4.23/				0.3446
									0.4529
									0.5860
^{157}Tm		156.9369	3.6 m	β+, ec/4.5	2.6	½+	+0.48		ann.rad./
				α/	3.97/				0.1104
									0.3484
									0.3855
									0.4550
									(0.1–1.58)
^{158m}Tm			~ 20 s						
^{158}Tm		157.9370	4.0 m	β+, ec/74 /6.5		2-	+0.04	+0.7	ann.rad./
				ec/26 /					Er k x-ray
									0.1921
									0.3351
									0.6280
									1.1498
									(0.18–2.81)
^{159}Tm		158.9350	9.1 m	β+/23 /3.9		5/2+	+3.42	+1.9	ann.rad./
				ec/77 /					Er k x-ray
									0.0591
									0.0848
									0.2713
									(0.05–1.27)
^{160m}Tm			1.24 m	IT/0.071/85		-5			
				ec//15					
^{160}Tm		159.9353	9.4 m	β+/15 /5.9		1-	+0.16	+0.58	ann.rad./
				ec/85 /					Er k x-ray
									0.1264
									0.2642
									0.7285
									0.8544
									0.8614
									1.3685
^{161}Tm		160.9336	31. m	β+, ec/3.2		7/2+	+2.40	+2.9	ann.rad./
									Er k x-ray
									0.0595
									0.0844
									1.6481
									(0.04–2.15)
^{162m}Tm			24. s	IT/81/		5+			Tm k x-ray
				β+, ec/19/					Er k x-ray
									0.0669
									0.8115
									0.9003
^{162}Tm		161.9340	21.7 m	β+ /8 /4.81		1-	+0.07	+0.69	ann.rad./
				ec/92 /					Er k x-ray
									0.1020
									0.7987
									(0.1–3.75)
^{163}Tm		162.93266	1.81 h	ec/98 /2.439		½+	-0.082		Er k x-ray
				β+ /1 /					0.0692
									0.1043
									0.2414
^{164m}Tm			5.1 m	IT/80 /		6-			0.0914
				β+, ec/20 /					0.1394
									0.2081
									0.2405
									0.3149
^{164}Tm		163.9335	2.0 m	β+ /36 /3.96	2.94/	1+	+2.38	+0.71	ann.rad./
				ec/64 /					Er k x-ray
									0.0914
^{165m2}Tm			0.009 ms	IT/0.160		7/2-			
^{165m1}Tm			0.080 ms	IT/0.080	0.069	7/2+			
^{165}Tm		164.93244	1.253 d	ec/1.593		½+	-0.139		Er k x-ray
									0.0472
									0.0544
									(0.297–0.8064)
^{166m}Tm			0.34 s	IT	0.0344/100	(6-)			
^{166}Tm		165.9336	7.70 h	ec/98 /3.04		2+	+0.092	+2.14	Er k x-ray

Elem. or Isot.	Natural abundance (Atom %)	Atomic mass or weight	Half-life/ Resonance width (MeV)	Decay mode/ Energy (/MeV)	Particle energy/ Intensity (MeV/%)	Spin (h/2π)	Nuclear magnetic mom. (nm)	Elect. quadr. mom. (b)	γ-Energy/ Intensity (MeV/%)
				β+ /2 /					0.0806
									0.1844
									0.7789
									1.2734
									2.0524
^{167}Tm		166.93286	9.24 d	ec/0.748		½+	-0.197		Er k x-ray
									0.0571
									0.20778
^{168}Tm		167.93418	93.1 d	ec/1.679		3+	+0.23	+3.2	Er k x-ray
									0.19825
									0.4475
									0.81595
^{169}Tm	100	168.93422				½+	-0.231	-1.2	
^{170m}Tm			4.1 µs	IT/0.183		(3)+			
^{170}Tm		169.93581	128.6 d	β- /99.8/0.968	0.883/24	1-	+0.246	+0.74	Yb k x-ray
^{171m2}Tm			1.7 µs	IT/1.674					
^{171m1}Tm			2.6 µs	IT/0.424		7/2-			
				ec/0.2 /0.314	0.968/76				0.08425
^{171}Tm		170.93643	1.92 a	β- /0.096	0.03/2	½+	-0.228		0.06674
					0.096/98				
^{172m}Tm			0.13 ms	IT/0.476		6+			
^{172}Tm		171.93841	2.65 d	β- /1.88	1.79/36	2-			Yb k x-ray
					1.88/29				0.07879
									1.38722
									1.46601
									(1.53-1.609)
^{173m}Tm			0.010 ms	IT/0.318		7/2-			
^{173}Tm		172.93961	8.2 h	β- /1.298	0.80/21	½+			Yb k x-ray
					0.86/71				0.3988
									0.4613
^{174m2}Tm			0.11 ms	IT/2.092		14+			
^{174m}Tm			2.29 s			0+			
^{174}Tm		173.9422	5.4 m	β- /3.08	0.70/14	(4-)			Yb k x-ray
					1.20/83				0.07664
									0.17669
									0.27332
									0.3666
									(0.08–1.6)
^{175m}Tm			0.02 ms	IT/1.518		23/2+			
^{175}Tm		174.9438	15.2 m	β- /2.39	0.9/36	(1/2+)			Yb k x-ray
					1.9/23				0.36396
									0.51487
									0.94125
									0.98247
^{176}Tm		175.947	1.9 m	β-/4.2	2.0/	(4+)			Yb k x-ray
					1.2/				0.1898
									0.3819
									1.0691
^{177}Tm		176.949	1.4 m	β-		(7/2-)			
^{178}Tm		177.953	> 0.3 µs						
^{179}Tm		178.955	> 0.3 µs						
^{180}Tm			> 0.3 µs						
^{181}Tm			> 0.3 µs						
$_{70}$Yb		**173.045(10)**							
^{148}Yb		147.968				0+			
^{149}Yb		148.964	0.7 s	β+, p	p/2.5–6.4/				0.647
^{150}Yb		149.959	> 0.2 µs			0+			
^{151m2}Yb			0.020 ms	IT	0.057/100				
^{151m1}Yb			0.003 ms	IT					(0.057-0.259)
^{151}Yb		150.955	1.6 s	β+ /8.5		(1/2+)			
^{152m}Yb			0.030 ms	IT/2.745	0.055/100	(10+)			
^{152}Yb		151.950	3.2 s	β+ ec/5.5		0+			
^{153m}Yb			0.015 ms	IT		27/2-			
^{153}Yb		152.949	4. s	β+ ec/6.7					
^{154}Yb		153.9464	0.40 s	β+ ec/7 /4.49		0+			ann.rad./
				α/93 /	5.32/				
^{155}Yb		154.9458	1.7 s	β+, ec/16 /6.0		(7/2-)	-0.91	-0.5	ann.rad./
				α/84 /	5.19/				
^{156}Yb		155.9428	26. s	β+, ec/21/3.57		0+			ann.rad./
				α/79 /	4.69/				
^{157}Yb		156.9426	39. s	β+, ec/99+/5.5		7/2-	-0.64		ann.rad./
				α/0.5/	4.69/				0.231
									(0.035–0.670)

Elem. or Isot.	Natural abundance (Atom %)	Atomic mass or weight	Half-life/ Resonance width (MeV)	Decay mode/ Energy (/MeV)	Particle energy/ Intensity (MeV/%)	Spin (h/2π)	Nuclear magnetic mom. (nm)	Elect. quadr. mom. (b)	γ-Energy/ Intensity (MeV/%)
^{158}Yb		157.9399	1.5 m	β+, ec/1.9		0+			ann.rad./
									0.0741
									0.2526
^{159}Yb		158.9401	1.4 m	ec, β+/5.1		5/2-	-0.37	-0.22	Tm k x-ray
									0.1661
									0.1772
									0.3297
									0.3903
^{160}Yb		159.9376	4.8 m	β+, ec/2.0		0+			ann.rad./
									0.1404
									0.1737
									0.2158
^{161}Yb		160.9379	4.2 m	β+, ec/3.9		3/2-	-0.33	+1.03	ann.rad./
									Tm k x-ray
									0.0782
									0.5999
									0.6315
^{162}Yb		161.9358	18.9 m	β+, ec/1.7		0+			ann.rad./
									Tm k x-ray
									0.1188
									0.1635
^{163}Yb		162.9363	11.1 m	β+ /26 /3.4	1.4/	3/2-	-0.37	+1.24	ann.rad./
									Tm k x-ray
									0.0636
									0.8603
									(0.06 –1.9)
^{164}Yb		163.9345	1.26 h	ec/1.0		0+			Tm k x-ray
									0.0914
									0.6752
^{165}Yb		164.9352	9.9 m	β+ /10 /2.76 ec/90 /	1.58/	5/2-	+0.48	+2.48	ann.rad./
									Tm k x-ray
									0.0801
									1.0903
^{166}Yb		165.93388	2.363 d	ec/0.30		0+			Tm k x-ray
									0.0828
									0.1844
									0.7789
									1.2734
									2.0524
^{167}Yb		166.93495	17.5 m	β+ /0.5 /1.954 ec/99.5 /	0.639/	5/2-	+0.62	+2.70	Tm k x-ray
									0.06296
									0.10616
									0.11337
									0.17633
^{168}Yb	0.126(1)	167.93389				0+			
^{169m}Yb			46. s	IT/0.0242		1/2-	+0.51		Yb L x-ray
									0.0242
^{169}Yb		168.93518	32.02 d	ec/0.909		7/2+	-0.63	+3.5	0.1979/35.9
									0.3078/10.05
									0.0207–0.261
^{170}Yb	3.023(2)	169.93477				0+			
^{171}Yb	14.216(7)	170.93633				1/2-	+0.49367		
^{172}Yb	21.754(10)	171.93639				0+			
^{173m}Yb			2.9 μs	IT	0.399/100	1/2-			
^{173}Yb	16.098(9)	172.93822				5/2-	-0.67989	+2.80	
^{174m}Yb			0.83 ms	IT		6+			
^{174}Yb	31.896(26)	173.93887				0+			
^{175m}Yb			0.068 s	IT	0.5149/100	1/2-			
^{175}Yb		174.94128	4.19 d	β- /0.470	0.466/73	7/2-	+0.77	+ 3.5	Lu k x-ray
					0.071/21				0.3963/13
					0.353/6.2				(0.114–0.28)
^{176m}Yb			11.4 s	IT/1.051	0.096	(8-)	- 0.15	+ 5.3	Yb k x-ray
									0.0961
									0.1901
									0.2929
									0.3897
^{176}Yb	12.887(30)	175.94258				0+	+ 0.15		
^{177m}Yb			6.41 s	IT/0.3315		1/2-			Yb k x-ray
									0.1131
									0.2084
^{177}Yb		176.94527	1.9 h	β- /1.399	1.40	9/2+	- 0.70	+ 4.0	Lu k x-ray
									0.1504
^{178}Yb		177.9467	1.23 h	β- /0.65	0.25/	0+			0.1415

Elem. or Isot.	Natural abundance (Atom %)	Atomic mass or weight	Half-life/ Resonance width (MeV)	Decay mode/ Energy (/MeV)	Particle energy/ Intensity (MeV/%)	Spin (h/2π)	Nuclear magnetic mom. (nm)	Elect. quadr. mom. (b)	γ-Energy/ Intensity (MeV/%)
									0.3246
									0.3516
									0.3815
									0.6125
^{179}Yb		178.950	8. m	β- /2.4		(1/2-)			
^{180}Yb		179.952	2. m	β-		0+			0.1028–0.442
^{181}Yb		180.956	> 0.3 μs						
^{182}Yb			> 0.3 μs						
^{183}Yb			> 0.3 μs						
^{184}Yb			> 0.3 μs						
^{185}Yb			> 0.3 μs						
$_{71}$Lu		**174.9668(1)**							
^{150m}Lu			0.045 ms	p/1.29/100					
^{150}Lu		149.974	43. ms	p					
^{151m}Lu			16 μs	p/1.31/100		(3/2+)			
^{151}Lu		150.968	0.08 s	p/1.231		(11/2-)			
^{152}Lu		151.964	0.7 s						
^{153m}Lu			0.015 ms	IT/2.633	0.130/100	27/2-			
^{153}Lu		152.959	0.9 s			11/2-			
^{154m2}Lu			0.035 ms	IT	0.130/100	(17+)			
^{154m1}Lu			1.1 s	ec, β+ /98.6		(9+)			
				IT//1.4		(2-)			
^{154}Lu		153.957	1.0 s	β+, ec/10.8					
^{155m2}Lu			2.7 ms	α/7.41/100		(25/2-)			
^{155m1}Lu			0.14 s	α//76		1/2+			
				ec, β+//24					
^{155}Lu		154.9543	0.07 s	ec/8.0		(11/2-)			
				α/	5.66/90				
^{156m}Lu			0.20 s	α/	5.57/	9+			
^{156}Lu		155.9530	~ 0.5 s	β+, ec/9.5		2-			ann.rad./
				α/	5.45/				
^{157m}Lu			4.8 s	β+, ec/94/6.93		(11/2-)			
				α//6	5.00				
^{157}Lu		156.9501	7 s	α	4.925				ann.rad./
^{158}Lu		157.9493	10.4 s	β+, ec/99 /8.0		2-			ann.rad./
				α/	4.67/				0.3682
									0.4770
^{159}Lu		158.9466	12.3 s	β+, ec/6.0					ann.rad./
									0.1505
									(0.1875-0.369)
^{160m}Lu			40 s	ec/β+//100					
^{160}Lu		159.9460	36.1 s	β+, ec/7.3					ann.rad./
									0.2434
									(0.396-0.577)
^{161m}Lu			7.3 ms	IT		(9/2-)			
^{161}Lu		160.9436	1.2 m	β+, ec/5.3		1/2+	+0.223		ann.rad./
									0.0437
									0.0671
									0.1003
									0.1108
									(0.156-0.256)
^{162m2}Lu			1.9 m	ec					
^{162m}Lu			~ 1.5 m	ec/		(4-)			
^{162}Lu		161.9433	1.37 m	β+, ec/6.9		1-	+0.055	+0.52	ann.rad./
									0.1666
									0.6314
^{163}Lu		162.9412	4.1 m	β+, ec/4.6		1/2(+)	+0.077		ann.rad./
									0.0539
									0.0581
									0.1504
									0.1631
									0.3717
^{164}Lu		163.9413	3.14 m	β+, ec/6.3	1.6/	1-	+0.059	+0.61	0.1238
					3.8/				0.2621
									0.7404
									0.8639
									0.8804
^{165}Lu		164.9394	10.7 m	β+, ec/3.9	2.06/	1/2+	-0.0245		ann.rad./
									0.1206
									0.1324
									0.1742
									0.2036
									(0.04–2.0)

Elem. or Isot.	Natural abundance (Atom %)	Atomic mass or weight	Half-life/ Resonance width (MeV)	Decay mode/ Energy (/MeV)	Particle energy/ Intensity (MeV/%)	Spin (h/2π)	Nuclear magnetic mom. (nm)	Elect. quadr. mom. (b)	γ-Energy/ Intensity (MeV/%)
^{166m2}Lu			2.1 m	β+ /35 /		0-			ann.rad./
				ec/65 /					Yb k x-ray
				IT					1.0673
									1.2566
									2.0986
^{166m1}Lu			1.4 m	β+, ec/58 /		3(-)	+0.189	+2.72	ann.rad./
				IT/42 /0.0344					0.1024
									0.2281
									0.2861
									0.8119
									0.8301
^{166}Lu		165.9399	2.8 m	β+ /25 /5.5		6-	+2.91	+4.33	ann.rad./
				ec/75 /					Yb k x-ray
									0.1024
									0.2281
									(0.3375-0.368)
^{167m}Lu			≥ 1 m			1/2+			
^{167}Lu		166.9383	52. m	β+ /2 /3.1	2.1/	7/2+	+2.325	+3.28	Yb k x-ray
				ec/98 /					0.0297
									0.2392
									(0.03–2.0)
^{168m}Lu			6.7 m	β+ /12 /		3+	+1.221	+2.43	ann.rad./
				ec/88 /					Yb k x-ray
				IT/<0.8					0.1988/190
									0.8960/100
									0.9792/128
									0.018–2.65
^{168}Lu		167.9387	5.5 m	β+ /6 /4.5	1.2/	6(-)	+3.02	+4.8	ann.rad./
				ec/94 /					Yb k x-ray
									0.1114
									0.1124
									0.2286
									0.3483
									1.4836
^{169m}Lu			2.7 m	IT/0.0290		1/2-			Lu L x-ray
									0.0290
^{169}Lu		168.93764	1.419 d	ec/2.293	1.271/	7/2+	2.30	3.5	Yb k x-ray
									0.19121
									0.9606
									(0.08–2.1)
^{170m}Lu			0.7 s	IT/0.0929	0.0484/100	(4)-			Lu L x-ray
									0.04449
									0.0484
^{170}Lu		169.9385	2.01 d	ec/3.46	2.44/	0+			Yb k x-ray
									0.58711
									0.5908
									1.28029
									(0.1–3.38)
^{171m}Lu			1.31 m	IT/0.0711		1/2-	+0.59		Lu k x-ray
									0.07119
^{171}Lu		170.93792	8.24 d	ec/1.479	0.362/	7/2+	+2.293	+3.48	Yb k x-ray
									0.01939
									0.66744
									(0.02–1.3)
^{172m2}Lu			0.44 ms	IT/0.067		(1)+			
^{172m}Lu			3.7 m	IT/0.0419		1-	+1.98	+0.76	Lu L x-rays
									0.04186
^{172}Lu		171.93909	6.64 d	ec/2.519		4-	2.90	+3.80	Yb k x-ray
									0.18156
									1.09367
									(0.07–2.2)
^{173m}Lu			0.074 ms	IT/0.1237		5/2-			
^{173}Lu		172.93893	1.37 a	ec/0.671		7/2+	+2.281	+3.53	Yb k x-ray
									0.07860
									0.27198
^{174m}Lu			142. d	IT/99.3/	0.17086	(6)-	+1.49	+4.80	Lu k x-ray
				ec/0.7 /					0.067055
^{174}Lu		173.94034	3.3 a	ec/1.374		(1)-	+1.988	+0.773	Yb k x-ray
									(0.0766-1.242)
^{175m2}Lu			0.98 ms	IT		(19/2+)			
^{175m1}Lu			1.5 μs	IT	0.353/100	5/2-			
^{175}Lu	97.401(13)	174.94078				7/2+	+2.232	+3.49	
^{176m2}Lu			0.040 ms	IT		(14+)			0.073-0.190

Elem. or Isot.	Natural abundance (Atom %)	Atomic mass or weight	Half-life/ Resonance width (MeV)	Decay mode/ Energy (/MeV)	Particle energy/ Intensity (MeV/%)	Spin (h/2π)	Nuclear magnetic mom. (nm)	Elect. quadr. mom. (b)	γ-Energy/ Intensity (MeV/%)
^{176m}Lu			3.66 h	β- /1.315	1.229/	1-	+0.318	-1.47	Hf k x-ray
					1.317/				0.088372
^{176}Lu	2.599(13)	175.94269	3.73 ×10^{10} a	β- /1.192		7-	+3.169	+4.92	Hf k x-ray
				β+/ < 0.9					0.20187
				ec/<0.36					0.30691
^{177m3}Lu			> 6.9 μs	β-/3.530		39/2-			0.089
^{177m2}Lu			160.4 d	IT/22/0.9702		23/2-	+2.32	+5.7	Lu k x-ray
				β- /78					Hf k x-ray
									0.11295/22.7
									0.20836/61.0
									0.37850/29.9
									(0.072-0.466)
^{177m1}Lu			0.16 ms	IT		1/2+			
^{177}Lu		176.94376	6.646 d	β- /0.498	0.497/78.4	7/2+	+2.238	+3.39	0.11295/0.062
					0.177/11.7				0.20836/0.104
					0.385/9.9				
					0.249/0.002				
^{178m}Lu			23.1 m	β- /		(9-)	+4.83	+5.39	0.2166
									0.3317
^{178}Lu		177.94596	28.5 m	β-/2.099	2.03/	1(+)	-1.38	+0.71	Hf k x-ray
									0.0932
									1.3099
									(0.09–1.7)
^{179m}Lu			3 ms	IT	0.592/100	1/2+			
^{179}Lu		178.94733	4.6 h	β- /1.405	1.35/	7/2+	+2.38	+3.32	(0.214-0.3377)
^{180m2}Lu			≥ 1 ms	IT	0.128/100	(9-)			
^{180m1}Lu				~ 1 s		3-			
^{180}Lu		179.9499	5.7 m	β- /3.1	1.49/				0.40795/50.
									(0.07–1.9)
^{181}Lu		180.952	3.5 m	β- /2.5		(7/2+)			0.0458
									0.2059
									0.5749
^{182}Lu		181.955	2.0 m	β- /~ 4.1					0.0978
									0.7208
									0.8182
^{183}Lu		182.957	58. s	β- /		7/2+			
^{184}Lu		183.961	20 s	β-					
^{185}Lu		184.964	> 0.3 μs						
^{186}Lu			> 0.3 μs						
^{187}Lu			> 0.3 μs						
^{188}Lu			> 0.3 μs						
$_{72}$Hf		**178.49(2)**							
^{153}Hf		152.971	> 0.2 μs						
^{154m}Hf			0.009 ms	IT		(10+)			
^{154}Hf		153.965	2. s	ec, β+/~ 6.7		0+			
^{155}Hf		154.963	0.9 s	ec, β+/8.					
^{156m}Hf			0.52 ms	α		8+			
^{156}Hf		155.959	25. ms	α/		0+			
^{157}Hf		156.958	0.11 s	α/					
^{158}Hf		157.9548	2.9 s	ec/54 /5.1		0+			
				α/46 /	5.27/				
^{159}Hf		158.9540	5.6 s	β+, ec/88 /6.9					ann.rad./
				α/12 /	5.09/				
^{160}Hf		159.9507	~ 12. s	β+, ec/97 /4.9		0+			ann.rad./
				α/4.78					
^{161m}Hf			4.8 μs	IT	0.202/100	(13/2+)			
^{161}Hf		160.9503	17. s	α/	4.60/				
^{162}Hf		161.9472	38. s	β+, ec/3.7		0+			ann.rad./
									0.1739
									0.1963
									0.4101
^{163}Hf		162.9471	40. s	β+, ec/5.5					ann.rad./
									0.0454
									0.0621
									0.0710
									0.6882
^{164}Hf		163.9444	2.8 m	ec, β+/3.0		0+			
^{165}Hf		164.9446	1.32 m	ec/4.6		11/2-			
^{166}Hf		165.9422	6.8 m	ec/93 /2.3		0+			ann.rad./
				β+ /7 /					Lu k x-ray
									0.0788
^{167}Hf		166.9426	2.0 m	β+ /40 /4.0		(5/2-)			ann.rad./
				ec/60 /					Lu k x-ray

Elem. or Isot.	Natural abundance (Atom %)	Atomic mass or weight	Half-life/ Resonance width (MeV)	Decay mode/ Energy (/MeV)	Particle energy/ Intensity (MeV/%)	Spin (h/2π)	Nuclear magnetic mom. (nm)	Elect. quadr. mom. (b)	γ-Energy/ Intensity (MeV/%)
									0.1754
									0.3152
^{168}Hf		167.9406	25.9 m	β+, ec/1.8		0+			ann.rad./
									(0.014–1.311)
^{169}Hf		168.9413	3.25 m	ec/85 /3.3		(5/2-)			ann.rad./
				β+ /15 /					Lu k x-ray
									0.3695
									0.4929
^{170}Hf		169.9396	16.0 h	ec/1.1		0+			Lu k x-ray
									0.0985
									0.1202
									0.1647
									0.5729
^{171m}Hf			30. s			(1/2-)	+0.53		0.6207
^{171}Hf		170.9405	12.2 h	ec, β+ /2.4		7/2+	-0.67	+3.46	ann.rad./
									Lu k x-ray
									0.1221
									0.6620
									1.0714
^{172}Hf		171.9395	1.87 a	ec/0.35		0+			Lu k x-ray
									0.02399
									0.12582
									(0.082–0.123)
^{173}Hf		172.9405	23.6 h	ec/1.6		½-	+0.50		Lu k x-ray
									0.12367
									0.13963
									(0.1–2.1)
^{174m2}Hf			3.7 μs	IT		(14+)			
^{174m1}Hf			0.024 ms	IT		(8-)			
^{174}Hf	0.16(1)	173.94005	2.0 ×10^{15} a			0+			
^{175m4}Hf			0.003 ms	IT		(45/2+)			
^{175m3}Hf			1.2 μs	IT		(35/2-)			
^{175m2}Hf			1.1 μs	IT		(19/2)+			
^{175m1}Hf			0.054 ms	IT	0.126/100	1/2-			
^{175}Hf		174.94151	70.7 d	ec/0.686		5/2-	-0.68	+2.72	Lu k x-ray
									(0.0894-0.343)
^{176m4}Hf			0.05 ms	IT		22-			
^{176m3}Hf			0.40 ms	IT		14-			
^{176m2}Hf			9.9 μs	IT		8-			
^{176m1}Hf			9.6 μs	IT		6+			
^{176}Hf	5.26(7)	175.94141				0+			
^{177m3}Hf			51.4 m	IT/2.740		37/2-			Hf k x-ray
									0.2140
									0.2951
									(0.3115-0.327)
^{177m2}Hf			0.056 ms	IT/1.342		(19/2-)			
^{177m1}Hf			1.1 s	IT/		23/2+			Hf k x-ray
									0.20836
									0.22847
									0.37851
^{177}Hf	18.60(9)	176.94323				7/2-	+0.794	+0.337	
^{178m3}Hf			0.068 ms	IT/2.572		14-			
^{178m2}Hf			31. a	IT/		16+	+8.16	+6.0	Hf k x-ray
									0.32555
									0.42635
									0.089–0.574
^{178m1}Hf			4.0 s	IT/1.147	0.0889/100	8-	3		Hf k x-ray
									0.21342
									0.32555
									0.42635
^{178}Hf	27.28(7)	177.94371				0+			
^{179m3}Hf			0.02 ms	IT/3.775	0.336/100	(43/2+)			
^{179m2}Hf			25.1 d	IT/1.1057		25/2-	7.4		Hf k x-ray
									0.1227
									0.1461
									0.3626
									0.4537
^{179m1}Hf			18.7 s	IT/0.375		½-			Hf k x-ray
									0.1607
									0.2141
^{179}Hf	13.62(2)	178.94582				9/2+	-0.641	+3.86	
^{180m5}Hf			0.09 ms	IT/3.559	0.151/100	(18-)			

Elem. or Isot.	Natural abundance (Atom %)	Atomic mass or weight	Half-life/ Resonance width (MeV)	Decay mode/ Energy (/MeV)	Particle energy/ Intensity (MeV/%)	Spin (h/2π)	Nuclear magnetic mom. (nm)	Elect. quadr. mom. (b)	γ-Energy/ Intensity (MeV/%)
^{180m4}Hf			> 0.010 ms	IT/2.538	0.052/100	(14+)			
^{180m3}Hf			0.010 ms	IT/2.486		12+			
^{180m2}Hf			0.02 ms	IT/2.426		(10+)			
^{180m}Hf			5.47 h	IT/1.1416		8-	+9.	+4.6	Hf k x-ray
									0.2152
									0.3323
									0.4432
^{180}Hf	35.08(16)	179.94656				0+			
^{181m3}Hf			0.08 ms	IT//0.595		9/2+			
^{181m2}Hf			~ 0.1 ms	IT		(17/2+)			
^{181m}Hf			1.5 ms	/1.738		25/2-			
^{181}Hf		180.94911	42.4 d	β- /1.027	0.408/	1/2-			Ta k x-ray
									0.482
									(0.133-0.3459)
^{182m2}Hf			0.04 ms	IT	0.264/100	(13+)			
^{182m}Hf			62. m	β- /54 /1.60	0.49/43	(8-)			Hf k x-ray
				IT/46 /1.173	0.95/10				0.0509
									0.2244
									0.3441
									0.4558
									0.5066
									0.9428
^{182}Hf		181.95056	8.9 ×10^{6} a	β- /0.37		0+			Ta k x-ray
									0.2704/79
									(0.098-0.270)
^{183}Hf		182.9535	1.07 h	β- /2.01	1.18/68	3/2-			Ta k x-ray
					1.54/25				0.0732
									0.4591
									0.7837
^{184}Hf		183.9555	4.1 h	β- /1.34	0.74/38	0+			Ta k x-ray
					0.85/16				0.0414
					1.10/46				0.1391
									0.3449
^{185}Hf		184.9589	~ 3.5 m	β- /					0.165
^{186}Hf		185.9609	~ 2.6 m			0+			0.738
^{187}Hf		186.965	> 0.3 μs						
^{188}Hf		187.967	> 0.3 μs			0+			
^{189}Hf		188.971	> 0.3 μs						
^{190}Hf			> 0.3 μs						
$_{73}$Ta		**180.94788(2)**							
^{155}Ta		154.974	3. ms	p/1.44					
^{156m}Ta			0.36 s	ec, β+//96		9+			
				p//4					
^{156}Ta		155.972	0.11 s	β+ /~ 11.6	/29				
				p/	1.01/71				
^{157m2}Ta			1.7 ms	α//100		(25/2-)			
^{157m1}Ta			4.3 ms	α//100		11/2-			
^{157}Ta		156.968	10 ms	α//96.6	6.117	1/2+			
				p/3.4	0.927/3.4				
^{158m2}Ta			6.1 μs	IT//98.6		19-			
^{158m1}Ta			37 ms	α//95		(9+)			
				Ec, β+//5					
^{158}Ta		157.967	0.06 s	α//91	6.05/100	(2-)			
				ec, β+//9	5.97/100				
^{159m1}Ta			0.6 s			11/2-			
^{159}Ta		158.9630	0.8 s	β+, ec/20 /8.5	α/5.52/34	1/2+			ann.rad./
				α/80 /	5.60/55				
^{160m}Ta			1.7 s	α					
^{160}Ta		159.9615	1.55 s	β+, ec/10.1					ann.rad./
				α	5.41/				
^{161}Ta		160.9585	3.16 s	β+, ec/7.5					ann.rad./
				α/	5.15				
^{162}Ta		161.9573	4. s	ec/8.6					
^{163}Ta		162.9543	10.6 s	ec/6.8					
^{164}Ta		163.9535	14.2 s	β+ /8.5		3+			ann.rad./
				α/	4.62/				0.2110
									0.3768
^{165}Ta		164.9508	31. s	ecβ+/5.9					
^{166}Ta		165.9505	34. s	β+ /82 /7.7					ann.rad./
				ec/18 /					Hf k x-ray
									0.1587
									0.3117
									0.8101

Elem. or Isot.	Natural abundance (Atom %)	Atomic mass or weight	Half-life/ Resonance width (MeV)	Decay mode/ Energy (/MeV)	Particle energy/ Intensity (MeV/%)	Spin (h/2π)	Nuclear magnetic mom. (nm)	Elect. quadr. mom. (b)	γ-Energy/ Intensity (MeV/%)
^{167}Ta		166.9481	1.4 m	β+, ec/5.6					ann.rad./
^{168}Ta		167.9481	2.4 m	β+ /77 /6.7		3+			ann.rad./
				ec/23 /					Hf k x-ray
									0.1241
									0.2619
									0.7518
									(0.307-1.985)
^{169}Ta		168.9460	4.9 m	β+, ec/4.4					ann.rad./
									0.0288
									0.1535
									0.1924
^{170}Ta		169.9462	6.8 m	β +/70 /6.0		(3+)			ann.rad./
				ec/35 /					Hf k x-ray
									0.1008
									0.2212
^{171}Ta		170.9445	23.3 m	β+, ec/3.7		(5/2-)			0.0496
									0.5018
									0.5064
									(0.05–1.02)
^{172}Ta		171.9449	36.8 m	β+ /25 /4.9		(3-)			ann.rad./
				ec/75 /					Hf k x-ray
									0.21396
									1.10923
									(0.09 –3.8)
^{173}Ta		172.9438	3.6 h	β+ /24 /3.7		(5/2-)	1.70	-1.9	ann.rad./
				ec/76 /					Hf k x-ray
									0.06972
									0.17219
									(0.06 –2.7)
^{174}Ta		173.9445	1.12 h	β+ /27 /3.8		(3+)			ann.rad./
				ec/73 /					Hf k x-ray
									0.09089
									(0.09–3.64)
^{175m}Ta			2.0 μs	IT		(21/2-)			
^{175}Ta		174.9437	10.5 h	ec/2.0		7/2+	2.27	+3.6	Hf k x-ray
									(0.208-0.339)
^{176m3}Ta			1.0 ms	IT		20-			
^{176m2}Ta			3.8 μs	IT		14-			
^{176m1}Ta			1.1 ms	IT	0.034/100				
^{176}Ta		175.9449	8.1 h	ec/3.1		1-			Hf k x-ray
									0.08837
									1.15735
^{177m3}Ta			0.133 ms	IT/4.656	0.086/100	49/2-			
^{177m2}Ta			5.3 μs	IT/1.355		21/2-			
^{177m1}Ta			3.6 μs	IT/0.186		5/2-			
^{177}Ta		176.94448	2.356 d	ec/1.166		7/2+	2.25		Hf k x-ray
									0.11295
									(0.07–1.06)
^{178m4}Ta			9.3 m	ec		1+			
^{178m3}Ta			0.29 s	IT	0.431	(21-)			
^{178m2}Ta			0.058 s	IT	0.227	15-			
^{178m}Ta			2.4 h	ec/		(7-)			Hf k x-ray
									0.08886
									0.21342
									0.32555
									0.42635
^{178}Ta		177.9457	9.29 m	ec/99 /1.9		1+	+2.74	+0.65	ann.rad./
				β+ /1 /					Hf k x-ray
									0.09316
^{179}Ta		178.94594	1.8 a	ec/0.110		7/2+	+2.29	3.37	Hf k x-ray
^{180m4}Ta			0.017 ms						
^{180m3}Ta			2.0 μs	IT	0.370	(22-)			
^{180m2}Ta			0.031 ms	IT		15-			
^{180m}Ta	0.01201(32)		3.65 ×10^{16} a			(9-)	+4.82	+4.95	
			4.5 ×10^{16} a	β-					
			> 2.0 ×10^{16} a	ec					
^{180}Ta		179.94746	8.15 h	ec/87 /0.854		1+			Hf k x-ray
				β- /13 /0.708	0.61/3				W k x-ray
					0.71/10				(0.093-0.1034)
^{181m4}Ta			0.21 ms	IT//2.228					
^{181m3}Ta			0.025 ms	IT//1.483		21/2-			
^{181m2}Ta			0.018 ms	IT//0.615		1/2+			
^{181m1}Ta			6.0 μs	IT//0.482		9/2-			

Elem. or Isot.	Natural abundance (Atom %)	Atomic mass or weight	Half-life/ Resonance width (MeV)	Decay mode/ Energy (/MeV)	Particle energy/ Intensity (MeV/%)	Spin (h/2π)	Nuclear magnetic mom. (nm)	Elect. quadr. mom. (b)	γ-Energy/ Intensity (MeV/%)
^{181}Ta	99.98799(32)	180.94800				7/2+	+2.370	+3.3	
^{182m2}Ta			15.8 m	IT/0.5198		10-			Ta k x-ray
									(0.1468-0.1716)
^{182m1}Ta			0.283 s	IT		5+			
^{182}Ta		181.95015	114.43 d	β- /1.814	0.25/30	3-	+3.02	+2.6	W k x-ray
					0.44/20				1.12127/100
					0.52/40				1.22138/79
									0.085–1.289
^{183}Ta		182.95137	5.1 d	β- /1.070	0.45/5	7/2+	+2.36		W k x-ray
					0.62/91				0.0847
									0.0991
									0.1079
									0.2461
									0.3540
^{184}Ta		183.9540	8.7 h	β- /2.87	1.11/15	(5-)			W k x-ray
					1.17/81				0.2528/44.
									0.4140/74.
									(0.09–1.4)
^{185m}Ta			12 ms	IT		5/2-			0.280
^{185}Ta		184.9556	49. m	β- /1.99	1.21/5	(7/2+)			W k x-ray
					1.77/81				0.0697
									(0.1739-0.1776)
^{186m}Ta			1.6 m						
^{186}Ta		185.959	10.5 m	β- /3.9	2.2/	(3-)			W k x-ray
									0.1979
									0.2149
									0.5106
									(0.09–1.5)
^{187}Ta		186.960	> 0.3 μs						
^{188m}Ta			3.5 μs						(0.143-0.434)
^{188}Ta		187.964	20 s						0.292
^{189}Ta		188.966	1.3 μs						
^{190}Ta		189.969	5. s						(0.207-0.357)
^{191}Ta		190.972	> 0.3 μs						
^{192}Ta		191.975	2. s						0.219
^{193}Ta			> 0.3 μs						
^{194}Ta			> 0.3 μs						
$_{74}$**W**		**183.84(1)**							
^{157}W		156.979	0.28 s			(7/2-)			
^{158m}W			0.14 ms	α	8.28(3)/	(8+)			
^{158}W		157.975	1.3 ms	α/	6.433/96	0+			
^{159}W		158.973	7. ms	α/					
^{160}W		159.968	0.08 s	α/	5.92/	0+			
^{161}W		160.967	0.41 s	β+, ec/18 /8.1					
				α/82 /	5.78/				
^{162}W		161.9635	1.39 s	β+, ec/54 /5.8		0+			
				α/46 /	5.54/				
^{163}W		162.9625	2.7 s	β+, ec/59 /7.5					
				α/41 /	5.38/				
^{164}W		163.9590	6. s	β+, ec/97 /5.0		0+			ann.rad./
				α/3 /	5.15/				
^{165}W		164.9583	5.1 s	β+, ec/99 /7.0					ann.rad./
				α/1 /	4.91/				
^{166}W		165.9550	16. s	β+, ec/99 /4.2		0+			ann.rad./
				α/1 /	4.74/				
^{167}W		166.9548	20. s	ec/5.6					
^{168}W		167.9518	53. s	ec/3.8		0+			ann.rad./
				α/10^{-5}/	4.40(1)				Ta k x-ray
									0.1755
									(0.037–0.573)
^{169}W		168.9518	1.3 m	ec/5.4					ann.rad./
									Ta k x-ray
									0.123
									(0.097–0.699)
^{170}W		169.9492	2.4 m	ec/2.2		0+			ann.rad./
									Ta k x-ray
									(0.060–0.144)
^{171m}W			1.3 μs			1/2-			
^{171}W		170.9495	2.4 m	ec/4.6					ann.rad./
									Ta k x-ray
									0.1842
									(0.052–0.479)
^{172}W		171.9473	6.6 m	β+, ec/2.5		0+			ann.rad./

Nuclear

Elem. or Isot.	Natural abundance (Atom %)	Atomic mass or weight	Half-life/ Resonance width (MeV)	Decay mode/ Energy (/MeV)	Particle energy/ Intensity (MeV/%)	Spin (h/2π)	Nuclear magnetic mom. (nm)	Elect. quadr. mom. (b)	γ-Energy/ Intensity (MeV/%)
									Ta k x-ray
									0.0389
									(0.034–0.674)
^{173}W		172.9477	6.3 m	ec/4.0					
									ann.rad./
									Ta k x-ray
									0.4576
									(0.035–0.623)
^{174}W		173.9461	35. m	ec/1.9		0+			
									ann.rad./
									Ta k x-ray
									0.3287
									0.4288
									(0.056–0.429)
^{175}W		174.9467	35. m	ec/2.9		½-			
									(0.015–0.27)
^{176}W		175.9456	2.5 h	β+, ec/0.8		0+			
									0.03358
									0.06129
									0.09487
									0.10020
^{177}W		176.9466	2.21 h	ec/2.0		(1/2-)			
									Ta k x-ray
									0.15505
									0.18569
									0.42694
^{178}W		177.9459	21.6 d	ec/0.091		0+			
									Ta k x-ray
^{179m}W			6.4 m	IT/99.7/0.222		(1/2-)			
									W k x-ray
				ec/0.3/					
									0.2220
^{179}W		178.9471	38. m	ec/1.06		(7/2-)			
									Ta k x-ray
									0.0307
^{180m2}W			2.3 μs	IT//3.265		14-			
^{180m1}W			5.5 ms	IT//1.529	0.3906	8-			
^{180}W	0.12(1)	179.94671	~ 1.5 ×10^{13} a	α/		0+			
^{181m}W			14.6 μs	IT//0.366		5/2-			
^{181}W		180.94820	121.1 d	ec/0.188		9/2+			
									Ta k x-ray
									0.13617
									0.15221
^{182m}W			1.3 μs	IT/2.231		(10+)			
^{182}W	26.50(16)	181.94820	>7.7 ×10^{21} a	α/		0+			
^{183m}W			5.15 s	IT/		(11/2+)			
									W k x-ray
									0.0465
									0.0526
									0.0991
									0.1605
^{183}W	14.31(4)	182.95022	>4.1 ×10^{21} a	α/		½-	+0.1177848		
^{184m}W			8.3 μs	IT/1.285		5-			
^{184}W	30.64(2)	183.95093	>8.9 ×10^{21} a	α/		0+			
^{185m}W			1.7 m	IT/0.1974		11/2+			
									W k x-ray
									0.0659
									0.1315
									0.1737
^{185}W		184.95342	74.8 d	β- /0.433	0.433/99.9	3/2-	+ 0.54		0.12536
^{186m}W			0.018 ms	IT//1.517		(7-)			
^{186}W	28.43(19)	185.95436	>8.2 ×10^{21} a	α/		0+			
^{187m}W			1.4 μs	IT	0.411	11/2+			(0.014-0.287)
^{187}W		186.95716	23.9 h	β- /1.311	0.624/66	3/2-	0.62		Re k x-ray
					1.315/16				0.68572/33
					0.081–1.18				0.134–0.773
^{188}W		187.95849	69.78 d	β- /0.349	0.349/99	0+			0.0636
									0.2271
									0.2907
^{189}W		188.9618	9.7 m	β- /2.5	1.4/	(3/2-)			0.2604
					2.5/				(0.1262-1.466)
^{190m}W			~ 0.11 ms	IT//1.839	0.097	10-			(0.0585-0.694)
^{190}W		189.9631	30. m	β- /1.3	0.95/	0+			Re k x-ray
									0.1576
									0.1621
^{191}W		190.9665	> 0.3 μs						
^{192}W		191.968	> 0.3 μs			0+			
^{193}W		192.972	> 0.3 μs						
^{194}W		193.974	> 0.3 μs						
^{195}W			> 0.3 μs						
^{196}W			> 0.3 μs						
^{197}W			> 0.3 μs						
^{200}W									
$_{75}$**Re**		**186.207(1)**							
^{159}Re		158.984	~ 0.02 ms	α	α/6.78/7.5	11/2-			

Elem. or Isot.	Natural abundance (Atom %)	Atomic mass or weight	Half-life/ Resonance width (MeV)	Decay mode/ Energy (/MeV)	Particle energy/ Intensity (MeV/%)	Spin (h/2π)	Nuclear magnetic mom. (nm)	Elect. quadr. mom. (b)	γ-Energy/ Intensity (MeV/%)
				p	p/1.80				
^{160m}Re			2.8 μs	IT	0.096/100	(9+)			(0.038,0.096)
^{160}Re		159.982	0.61 ms	p/	1.26/89				
				α/	6.53/11				
^{161m}Re			15 ms	α/93	6.24	(11/2-)			
				p/7					
^{161}Re		160.978	0.44 ms	p/		1/2+			
^{162m}Re			0.08 s	α/91		(9+)			
				ec/9					
^{162}Re		161.976	0.10 s	α/	6.12/94	(2-)			
					6.09/94				
^{163m}Re			0.21 s	α/66		11/2-			
				ec/34					
^{163}Re		162.9721	0.26 s	β+, ec/9.0	α/5.87/32	1/2+			
				α/	5.92/66				
^{164}Re		163.9705	~ 0.85 s	β+, ec/10.7					
				α/	5.78/				
^{165m}Re			~ 2.37 s	α/	5.502/				
^{165}Re		164.9671	2.6 s	β+, ec/87 /8.1					
				α/	5.49/ < 5.				
^{166}Re		165.9658	2.5 s	β+, ec/9.4					
				α/	5.50/				
^{167m}Re			5.9 s	α, ec/		(9/2-)			
^{167}Re		166.9626	3.4 s	β+, ec/7.4					
				α/	5.015/				
^{168}Re		167.9616	4.4 s	β+, ec/9.1					
				α/	4.833/				0.1117
^{169m}Re			8.1 s	α	4.70/				
					4.87/				
^{169}Re		168.9588	16. s						
^{170}Re		169.9582	9.2 s	β+, ec/9.0					0.1560
									0.3055
									0.4125
^{171}Re		170.9557	15.2 s	ec/~ 5.7					
^{172m}Re			55. s	β+, ec/		(2)			ann.rad./
									0.1234
									0.2537
									0.3504
^{172}Re		171.9554	15. s	β+, ec/7.3					ann.rad./
									0.1234
									0.2537
^{173}Re		172.9532	2.0 m	ec/~ 3.9					ann.rad./
^{174}Re		173.9531	2.4 m	β+, ec/5.6					ann.rad./
									0.1119
									0.2430
^{175}Re		174.9514	5.8 m	β+, ec/4.3					ann.rad./
^{176}Re		175.9516	5.3 m	β+, ec/5.6		(3+)			ann.rad./
									(0.1089-0.2406)
^{177m}Re			0.05 ms	IT	0.0847	5/2+			
^{177}Re		176.9503	14. m	ec/78 /3.4		(5/2-)			ann.rad./
				β+ /22 /					W k x-ray
									0.0797
									0.0843
									0.1968
^{178}Re		177.9510	13.2 m	β+ /11 /4.7	3.3/	(3+)			ann.rad./
				ec/89 /					W k x-ray
									0.1059
									0.2373
									0.9391
^{179m2}Re			0.47 ms	IT/5.408	0.245				
^{179m1}Re			0.10 ms	IT	0.0654	(5/2-)			
^{179}Re		178.9500	19.7 m	ec/99 /2.71	0.95/	(5/2+)	2.8		W k x-ray
				β+ /1 /					0.1199
									0.2900
									0.4154
									(0.430-1.680)
^{180m}Re			0.009 ms	IT		21-			
^{180}Re		179.9508	2.45 m	ec/92 /3.80	1.76/	1-	1.6		ann.rad./
				β+ /8 /					W k x-ray
									0.1036
									(0.07–2.2)
^{181m2}Re			1.2 μs	IT/3.869	0.357	(35/2-)			
^{181m1}Re			0.012 ms	IT/1.881	0.224	(25/2+)			

Elem. or Isot.	Natural abundance (Atom %)	Atomic mass or weight	Half-life/ Resonance width (MeV)	Decay mode/ Energy (/MeV)	Particle energy/ Intensity (MeV/%)	Spin (h/2π)	Nuclear magnetic mom. (nm)	Elect. quadr. mom. (b)	γ-Energy/ Intensity (MeV/%)
^{181}Re		180.9501	20. h	ec /1.74		5/2+	3.19		W k x-ray
									0.3607
									0.3655
									0.6390
^{182m}Re			12.7 h	ec/	0.55/	2(+)	3.3	+1.8	W k x-ray
					1.74/				0.0677
									1.1214
									1.2215
									(0.06–2.2)
^{182}Re		181.951	2.67 d	ec/2.8		7+	2.8	+4.1	W k x-ray
									0.0678
									0.2293
									(1.121-1.221)
^{183m}Re			1.04 ms	IT/1.907		(25/2)+			
^{183}Re		182.9508	70. d	ec/0.56		(5/2+)	+3.16	+2.2	W k x-ray
									0.16232
^{184m}Re			165. d	IT/75 /0.188		8+	+2.9		Re k x-ray
				ec/25 /					0.1047
									0.2165
									0.92093
									(0.10–1.1)
^{184}Re		183.95252	35. d	ec/1.48		3-	+2.53	+3.0	W k x-ray
									0.79207
									08948
									0.90328
^{185}Re	37.40(2)	184.95295				5/2+	+3.1871	+2.19	
^{186m}Re			2.0×10^{5} a	IT/0.150		8+			Re k x-ray
									(0.059-0.148)
^{186}Re		185.95499	3.718 d	β- /92 /1.070	0.973/21	1-	+1.739	+0.62	W k x-ray
				ec/8 /0.582	1.07/71				0.1227/0.6
									0.1372/9.5
									(0.63–0.77)
^{187}Re	62.60(2)	186.95575	4.16×10^{10} a	β- /0.00266	0.0025/	5/2+	+3.2197	+2.07	
^{188m}Re			18.6 m	IT/0.172		(6-)			Re k x-ray
									0.0925
									0.1059
^{188}Re		187.95811	17.00 h	β- /2.120	1.962/20	1-	+1.788	+0.57	Os k x-ray
					2.118/79				0.15502
									0.309–2.022
^{189}Re		188.9592	24. h	β- /1.01	1.01/	(5/2+)			0.1471
									0.2167
									0.2194
									0.2451
^{190m}Re			3.2 h	β- /51 /		(6-)			Re k x-ray
				IT/49 /					0.1191
									0.2238
									0.6731
									(0.1–1.79)
^{190}Re		189.9617	3.0 m	β- /3.2	1.8/	(2-)			Os k x-ray
									0.1867
									0.5580
									0.6051
^{191m}Re			0.08 ms	IT					
^{191}Re		190.9631	9.7 m	β- /2.05	1.8/				
^{192m}Re			0.09 ms	β-					(0.0606-0.146)
^{192}Re		191.9661	16. s	β- /4.2	~ 2.5/				(0.2–0.75)
^{193m}Re			0.074 ms	IT/0.142		(9/2-)			(0.061-0.146)
^{193}Re		192.9675	0.07 ms	β-					
^{194m3}Re			1.7 m	β-					
^{194m2}Re			0.4 m	β-		(11-)			
^{194m1}Re			0.05 ms	IT					
^{194}Re		193.971	1.0 s	β-					
^{195}Re		194.973	6. s						
^{196m}Re			0.004 ms	IT					
^{196}Re		195.976	3. s						
^{197}Re		196.978	> 0.3 μs						
^{198}Re		197.982	> 0.3 μs						
^{199}Re			> 0.3 μs						
^{201}Re									
$_{76}$Os		**190.23(3)**							
^{161}Os		160.989	0.64 ms	α	/5.5	(7/2-)			
^{162}Os		161.984	1.8 ms	α/	6.60	0+			
^{163}Os		162.982	5.5 ms	α/	6.51				

Elem. or Isot.	Natural abundance (Atom %)	Atomic mass or weight	Half-life/ Resonance width (MeV)	Decay mode/ Energy (/MeV)	Particle energy/ Intensity (MeV/%)	Spin (h/2π)	Nuclear magnetic mom. (nm)	Elect. quadr. mom. (b)	γ-Energy/ Intensity (MeV/%)
^{164}Os		163.978	0.04 s	α	/96	0+			
^{165}Os		164.977	0.07 s	α	/88				
^{166}Os		165.9727	0.18 s	β+, ec/28 /6.3	6.27/	0+			ann. rad./
				α/72 /	5.98/84				
^{167}Os		166.9716	0.84 s	β+, ec/76 /8.2					ann.rad./
				α/24 /	5.84/				
^{168}Os		167.9678	2.2 s	β+, ec/51 /5.7		0+			ann. rad./
				α/49 /					
^{169}Os		168.9670	3.3 s	β+, ec/89 /7.7	5.57/80				ann.rad./
				α/13 /	5.51/12				
					5.54/8				
^{170}Os		169.9636	7.1 s	β+, ec/5.0		0+			ann.rad./
				α/	5.40/				(0.162–0.216)
^{171}Os		170.9632	8.4 s	β+, ec/98 /7.1	α/5.24/93.5				ann.rad./
				α/19 /	5.17/6.5				0.190–0.705
^{172}Os		171.9600	19. s	β+, ec/99 /4.5		0+			ann.rad./
				α/1.1/	5.10/				(0.063–1.120)
^{173}Os		172.9598	16. s	β+, ec/6.3					ann.rad./
				α/0.4 /	4.94/				0.142–0.299
^{174}Os		173.9571	44. s	β+, ec/3.9		0+			0.118
				α/0.02 /	4.76/				0.138 / 0.001
									0.158
									0.325
^{175}Os		174.9569	1.4 m	β+, ec/5.3					0.125
									0.181
									0.248
^{176}Os		175.9548	3.6 m	β+, ec/3.2		0+			0.8155
									0.7758
									0.8573
									1.2093
									1.2909
^{177}Os		176.9550	2.8 m	β+, ec/4.5		(1/2-)			0.0848
									0.1958
									0.3002
									1.2686
^{178}Os		177.9533	5.0 m	β+, ec/2.3		0+			ann.rad./
									0.5946
									0.6850
									0.9687
									1.3311
^{179}Os		178.9538	7. m	β+, ec/3.7					ann.rad./
									0.0654
									0.2186
									0.5938
^{180}Os		179.9524	21.5 m	β+, ec/1.5		0+			Re k x-ray
									0.0202–0.7174
^{181m}Os			2.7 m	ec/		7/2-			ann.rad./
^{181}Os			1.75 h	ec/		1/2-			ann.rad./
									0.0489
^{181}Os		180.9532	2.7 m	ec/2.9		(7/2-)			ann.rad./
									0.11794
									0.23868
									0.8267
									(0.07–2.64)
^{182m}Os			0.8 ms	IT/1.831	0.554	(8-)			
^{182}Os		181.9521	21.5 h	ec/0.9		0+			Re k x-ray
									0.1802
									0.5100
^{183m}Os			9.9 h	ec/84 /		½-			Os k x-ray
				IT/16 /					Re k x-ray
									1.1020
									1.1080
^{183}Os		182.9531	13. h	ec/2.1		9/2+	-0.79	+3.1	Re k x-ray
									0.1144
									0.3818
^{184}Os	0.02(1)	183.95249	1.1×10^{13} a	α		0+			
			$> 1.9 \times 10^{14}$ a	ec,ec					
			$> 2.5 \times 10^{16}$ a	ec,β+					
^{185m}Os			3.0 μs			7/2-			
^{185}Os		184.954042	93.6 d	ec/1.013		½-			Re k x-ray
									0.6461
									0.8748
									0.8805

Elem. or Isot.	Natural abundance (Atom %)	Atomic mass or weight	Half-life/ Resonance width (MeV)	Decay mode/ Energy (/MeV)	Particle energy/ Intensity (MeV/%)	Spin (h/2π)	Nuclear magnetic mom. (nm)	Elect. quadr. mom. (b)	γ-Energy/ Intensity (MeV/%)
^{186}Os	1.59(3)	185.95384	2. ×10^{15} a	α/	~ 2.75/	0+			
^{187m}Os			0.231 ms	IT/0.257	0.157	11/2+			
^{187}Os	1.96(2)	186.95575				½-	+0.0646519		
^{188}Os	13.24(8)	187.95584				0+			
^{189m}Os			5.8 h	IT/0.0308		9/2-			Os L x-ray
									0.0308
^{189}Os	16.15(5)	188.95814				3/2-	+0.659933	+0.9	
^{190m}Os			9.9 m	IT/1.705	0.039	10-	-0.6		Os k x-ray
									0.1867
									0.3611
									0.5026
									0.6161
^{190}Os	26.26(2)	189.95844				0+			
^{191m}Os			13.1 h	IT/0.0744		3/2-			Os k x-ray
									0.0744
^{191}Os		190.96093	15.4 d	β- /0.314	0.140/100	9/2-	+0.96	+2.5	Ir k x-ray
									0.1294
^{192m}Os			5.9 s	IT/2.0154	0.302	(10-)			Os k x-ray
									0.2058/65.9
									0.5692/70
									(0.201–1.000)
^{192}Os	40.78(19)	191.96148	> 5.3 ×10^{19} a	β+ β+		0+			
^{193}Os		192.96415	30.0 h	β- /1.141	1.04/20	3/2-	+0.730	+0.47	Ir k x-ray
									0.1389
									0.4605
^{194}Os		193.96518	6.0 a	β- /0.097	0.054/33	0+			Ir L x-ray
					0.096/67				0.0429
^{195}Os		194.9683	~ 6.5 m	β- /2.0	2.0/				
^{196}Os		195.9696	34.9 m	β- /1.16	0.84/	0+			0.1262/5
									0.4079/5.9
^{197}Os		196.973	2.8 m	β-					0.2239
									(0.0412-0.406)
^{198}Os		197.974	> 0.3 μs						
^{199}Os		198.978	5. s						
^{200}Os		199.980	6. s						
^{201}Os		200.984	> 0.3 μs						
^{202}Os		201.986	> 0.3 μs						
^{203}Os			> 0.3 μs						
$_{77}$Ir		**192.217(3)**							
^{164}Ir		163.992	0.06 ms	p	1.78				
^{165}Ir		164.988	0.3 ms	p/87	1.71	11/2-			
				α/13	6.72				
^{166m}Ir			15 ms	α/98.2	6.545	(9+)			
				p/1.8	1.32				
^{166}Ir		165.986	0.010 s	α/93	6.56	(2-)			
				p/6.9	1.15				
^{167m}Ir			26. ms	α/48, β+	6.39/90	11/2-			
				p/32	1.25/0.42				
^{167}Ir		166.9817	32. ms	α/80, β+	6.35/48	1/2+			
				p/0.4	1.06/39.3				
^{168m}Ir			0.16 s						
^{168}Ir		167.9799	0.2 s	α/82					
^{169m}Ir			280. ms	α/	6.12/59				
^{169}Ir		168.9763	353. ms	α/	5.99/42				
^{170}Ir		169.975	0.81 s	α/	6.01/				0.175
					6.05/				
					5.95/				
					6.12/				
^{171}Ir		170.9716	1.3 s	α/	5.73/15				
^{172m}Ir			2.1 s	α	5.811	(7+)			0.288
^{172}Ir		171.9706	4.4 s	ec		(3+)			
									(0.379–0.475)
^{173m}Ir			2.2 s	α//66		(11/2-)			
				ec//34					
^{173}Ir		172.9675	3.0 s	α/	5.665/				0.0493
									(0.092–0.296)
^{174m}Ir			4.9 s	ec//98		(7+)			
^{174}Ir		173.9669	4. s	α/	5.478/				0.1587
									(0.276–1.33)
^{175}Ir		174.9642	~ 4.5 s	α/	5.393/				0.1056
^{176}Ir		175.9637	10.9 s	ec, β+/80					0.260
				α/3.2/	5.118/				(0.135–0.415)
^{177}Ir		176.9613	30. s	ec, β+/5.7					0.184

Elem. or Isot.	Natural abundance (Atom %)	Atomic mass or weight	Half-life/ Resonance width (MeV)	Decay mode/ Energy (/MeV)	Particle energy/ Intensity (MeV/%)	Spin (h/2π)	Nuclear magnetic mom. (nm)	Elect. quadr. mom. (b)	γ-Energy/ Intensity (MeV/%)
				α/0.06/	5.011/				(0.062–0.194)
^{178}Ir		177.9611	12. s	β+, ec/6.3					
									0.1320
									0.2667
									0.3633
^{179}Ir		178.9591	4. m	ec/4.9					0.0975
									(0.045–0.220)
^{180}Ir		179.9592	1.5 m	ec/6.4					0.2765
									((0.132–1.106)
^{181}Ir		180.9576	4.9 m	β+, ec/4.1		(7/2+)			ann.rad./
									0.1076
									(0.020–1.715)
^{182}Ir		181.9581	15. m	β+ /44 /5.6			+2.6	~ -1.7	ann.rad./
				ec/56 /					Os k x-ray
									0.1273
									0.2370
^{183}Ir		182.9568	57. m	β+, ec/3.5			+2.4	~ -1.8	ann.rad./
									0.0877
									0.2285
									0.2824
^{184m}Ir			0.47 ms	IT/0.226		3+			
^{184}Ir		183.9575	3.0 h	β+ /12 /4.6	2.3/	5-	0.70	+2.41	ann.rad./
				ec/88 /	2.9/				Os k x-ray
									0.11968
									0.2640
									0.3904
^{185}Ir		184.9567	14. h	β+ /3 /2.4		(5/2-)	2.60	-2.1	ann.rad./
				ec/97 /					Os k x-ray
									0.2543
									1.8288
^{186m}Ir			1.9 h	ec /		(2-)	0.64	+1.46	Os k x-ray
									0.1371
									0.7675
^{186}Ir		185.9579	15.7 h	ec/98 /3.83		5+	3.9	-2.55	Os k x-ray
				β+ /2 /					0.1372
									0.2968
									0.4348
									(0.13–3.0)
$^{187m2}Ir$			0.002 ms	IT//2.488	0.110	29/2-			
$^{187m1}Ir$			0.030 s	IT//0.186		9/2-			
^{187}Ir		186.9575	10.5 h	ec/1.50		3/2+	+0.17	+0.94	Os k x-ray
									0.0743
									0.4009
									0.4271
									0.6109
									0.9128
^{188m}Ir			4.2 ms			(9-)			
^{188}Ir		187.9588	1.72 d	β+ /2.81	1.13/	1-	0.30	+0.48	Os k x-ray
				ec/99+ /	1.64/				0.1550
									0.4780
									0.6330
									2.2146
$^{189m2}Ir$			3.7 ms	IT//2.333		(25/2)+			
$^{189m1}Ir$			13 ms	IT//0.372		11/2-			
^{189}Ir		188.9587	13.2 d	ec/0.53		3/2+	+0.14	+0.88	Os k x-ray
									0.2449
$^{190m2}Ir$			3.09 h	β+, ec/95 /		(11-)			0.376
				IT/5 /					
$^{190m1}Ir$			1.12 h	IT /0.0263		7+			Ir L x-ray
^{190}Ir		189.96054	11.8 d	ec/2.0		(4+)	0.04	+2.8	Os k x-ray
									0.1867
									0.4072
									0.5186
									0.5580
									0.6051
									(0.2–1.4)
$^{191m2}Ir$			6 s	IT//2.101		31/2(+)			
^{191m}Ir			4.90 s	IT/0.1714	0.042	11/2-	+0.603		Ir k x-ray
									0.1294
^{191}Ir	37.3(2)	190.96059				3/2+	+0.151	+0.82	
$^{192m2}Ir$			~ 241. a	IT/0.168	0.155	11-			Ir k x-ray
$^{192m1}Ir$			1.45 m	IT/0.567		1-			Ir L x-ray
									0.0580

Elem. or Isot.	Natural abundance (Atom %)	Atomic mass or weight	Half-life/ Resonance width (MeV)	Decay mode/ Energy (/MeV)	Particle energy/ Intensity (MeV/%)	Spin (h/2π)	Nuclear magnetic mom. (nm)	Elect. quadr. mom. (b)	γ-Energy/ Intensity (MeV/%)
									0.3165
^{192}Ir		191.96260	73.83 d	β- /1.460		4+	+1.92	+2.15	Pt k x-ray
									0.31649/83.
^{193m2}Ir			0.125 ms	IT/2.277	31/2+				0.46806/48.
^{193m}Ir			10.53 d	IT/0.0802		11/2-			Ir L x-ray
									0.0803
^{193}Ir	62.7(2)	192.96292				3/2+	+0.164	+0.75	
^{194m}Ir			1.7 ×10² d	β- /					Pt k x-ray
									0.3284
									0.4829
									0.5624
^{194m1}Ir			32 ms	IT//0.147		4+			
^{194}Ir		193.96507	19.2 h	β-/2.247	1.92/9	1-	+0.39	+0.34	0.2935
					2.25/86				0.3284
									0.6451
									(0.1–2.2)
^{195m2}Ir			0.004 ms	IT/2.354	0.268	(27/2+)			
^{195m}Ir			3.7 h	β- /	0.41/	11/2-			Pt k x-ray
					0.97/				0.3199/9.6
									0.3649/9.5
									0.4329/9.6
									0.6849/9.6
^{195}Ir		194.9657	2.3 h	β- /1.120	1.0/80	(3/2+)			Pt k x-ray
					1.11/13				0.0989/9.7
^{196m}Ir			1.40 h	β-/	1.16/				Pt k x-ray
									0.3557
									0.3935
									0.4471
									0.5214
									0.6473
^{196}Ir		195.9684	52. s	β- /3.21	2.1/15	0-			0.3329
					3.2/80				0.3557
									0.7796
^{197m2}Ir			8.9 m	β- /		(11/2-)			0.3465
^{197m1}Ir			0.02 ms						see Ir[197]
^{197}Ir		196.9697	5.8 m	β- /2.16	1.5/	(3/2+)			0.0531
					2.0/				0.1351
									0.4306
									0.4697
^{198}Ir		197.972	8. s	β- /4.1					0.4074
									0.5070
^{199}Ir		198.9738	6. s	β-					
^{200}Ir		199.977	~ 44 s	β-					(0.314-1.260)
^{201}Ir		200.979	~ 21 s	β-					(0.333-1.706)
^{202m}Ir			0.003 ms						
^{202}Ir		201.982	11 s	β-					(0.514-2.165)
^{203}Ir		202.984	> 0.3 μs						
^{205}Ir			> 0.3 μs						
$_{78}$Pt		**195.084(9)**							
^{166}Pt		165.995	0.3 ms	α/	7.11/	0+			
^{167}Pt		166.993	0.9 ms	α/	6.98/				
^{168}Pt		167.988	2.1 ms	α	6.82	0+			0.582/69
									0.594/69
									0.725/62
^{169}Pt		168.987	7.0 ms	α	6.69				
^{170}Pt		169.9825	14.0 ms	α	6.55	0+			0.509/100
									0.662/86
									0.214–0.726
^{171}Pt		170.9813	0.048 s	α	6.45				0.4450
									(0.1564-1.208)
^{172}Pt		171.9774	0.10 s	α/	6.31/94	0+			
^{173}Pt		172.9764	0.36 s	β+, ec/8.2	6.23				
				α/	6.20/				
^{174}Pt		173.9728	0.89 s	β+, ec/17 /5.6		0+			
				α/83 /	6.040/				
^{175}Pt		174.9724	2.5 s	β+, ec/65 /7.6					0.0774
				α/35 /	5.831/5				0.1354
					5.96/54				0.2128
					6.038/				
^{176}Pt		175.9689	6.3 s	β+, ec/60 /5.1		0+			ann.rad./
				α/40 /	5.528/0.6				0.2277
					5.750/41				

Elem. or Isot.	Natural abundance (Atom %)	Atomic mass or weight	Half-life/ Resonance width (MeV)	Decay mode/ Energy (/MeV)	Particle energy/ Intensity (MeV/%)	Spin (h/2π)	Nuclear magnetic mom. (nm)	Elect. quadr. mom. (b)	γ-Energy/ Intensity (MeV/%)
^{177m}Pt			2.2 μs	IT	0.147	1/2-			
^{177}Pt		176.9685	11. s	ec/91 /6.8	5.53/				0.0908
				α/9 /	5.485/3				
					5.525/6				
^{178}Pt		177.9657	21. s	ec/93 /4.5		0+			
				α/7 /	5.286/0.2				
					5.442/7				
^{179}Pt		178.9654	33. s	β+, ec/5.7			+0.43		
				α/	5.16/				
^{180}Pt		179.9630	52. s	β+, ec/99.7 /3.7	0+				
				α/0.3 /	5.140/				
^{181}Pt		180.9631	51. s	β+, ec/5.2			+0.48		
^{182}Pt		181.9612	2.7 m	β+, ec/2.9		0+			ann.rad./
									0.1360
									0.1460
									0.2100
^{183m}Pt			43. s	β+, ec/		(7/2-)	+0.78	+3.4	ann.rad./
				IT/					0.3132/26
									0.3164/59
									0.6296/100
									0.058–1.75
^{183}Pt		182.9616	7. m	β+, ec/4.6			+0.50		ann.rad./
									0.119/100
									0.307/93
									0.260/90
									0.058–1.377
^{184m}Pt			1.0 ms	IT/1.840		8-			
^{184}Pt		183.9599	17.3 m	β+, ec/2.3		0+			ann.rad./
									0.1549
									0.1919
									0.5484
^{185m}Pt			33. m	β+, ec/		½-	+0.50		
^{185}Pt		184.9606	1.18 h	β+, ec/3.8		(9/2+)	-0.75	+4.	ann.rad./
									0.1353
									0.1974
									0.2296
									0.2551
^{186}Pt		185.9594	2.0 h	β+, ec/1.38		0+			ann.rad./
									0.6115
									0.6892
^{187m}Pt			0.31 ms	IT/174	0.117	11/2+			
^{187}Pt		186.9606	2.35 h	β+, ec/3.1		3/2-	-0.40	-1.0	ann.rad./
									Ir k x-ray
									0.1064
									0.1100
									0.2015
									0.2849
									0.7092
^{188}Pt		187.95939	10.2 d	ec/0.51		0+			Ir k x-ray
									0.1876
									0.1951
^{189m}Pt			0.14 ms	IT/0.192		(13/2+)			
^{189}Pt		188.9608	10.9 h	β+, ec/1.97		3/2-	-0.43	-1.2	Ir k x-ray
									0.0943
									0.6076
									0.7214
									(0.09–1.47)
^{190}Pt	0.012(2)	189.95993	4.5 ×10^{11} a	α		0+			
			> 8.4 ×10^{14} a	ec,ec					
			> 9.2 ×10^{15} a	ec, β+					
^{191m}Pt			0.10 ms	IT//0.149		(13/2)+			
^{191}Pt		190.96167	2.86 d	ec/1.02		3/2-	-0.50	-0.9	Ir k x-ray
									0.3599
									0.4094
									0.5389
^{192}Pt	0.782(24)	191.96104				0+			
^{193m}Pt			4.33 d	IT/0.1498	0.136	13/2+	-0.75		Pt k x-ray
									0.1355
^{193}Pt		192.96298	60. a	ec/0.0566		(1/2-)	+0.60		Ir k x-rays
^{194}Pt	32.86(40)	193.96268				0+			
^{195m}Pt			4.01 d	IT/0.2952		13/2+	-0.61	~ +1.4	Pt k x-ray
									0.0989
^{195}Pt	33.78(24)	194.96479	7.3 ×10^{18} a	α		1/2-	+0.6095		

Elem. or Isot.	Natural abundance (Atom %)	Atomic mass or weight	Half-life/ Resonance width (MeV)	Decay mode/ Energy (/MeV)	Particle energy/ Intensity (MeV/%)	Spin (h/2π)	Nuclear magnetic mom. (nm)	Elect. quadr. mom. (b)	γ-Energy/ Intensity (MeV/%)
^{196}Pt	25.21(34)	195.96495				0+			
^{197m}Pt			1.590 h	IT/97 /		13/2+			Pt k x-ray
				β- /3 /					0.0530
									0.3465
^{197}Pt		196.96734	19.9 h	β- /0.719		1/2-	0.51		Au k x-ray
									0.1914
									0.2688
^{198}Pt	7.356(130)	197.96789	$> 4.7 \times 10^{17}$ a	α		0+			
			$> 3.5 \times 10^{18}$ a	β- β-					
^{199m}Pt			13.6 s	IT/0.424	0.392	13/2+			Pt k x-ray
									0.3919
^{199}Pt		198.97060	30.8 m	β- /1.70	0.90/18	(5/2-)			0.3170/3.88
					1.14/14				0.49375/4.47
									0.5430/11.7
									(0.055–1.293)
^{200}Pt		199.9714	12.5 h	β- /~ 0.66		0+			Au k x-ray
									0.13590
									0.22747
									0.24371
^{201}Pt		200.9745	2.5 m	β- /2.66		(5/2-)			0.070
									0.152
									0.222
									1.760
^{202m}Pt			0.14 ms	IT/1.788	0.535	(7-)			(0.535-0.719)
^{202}Pt		201.9756	1.8 d			0+			0.440
^{203}Pt		202.979	22. s						
^{204m2}Pt			0.055 ms	IT		(7-)			
^{204m1}Pt			0.006 ms	IT/1.995	1.123	(5-)			
^{204}Pt		203.981	0.055 ms						
^{205}Pt		204.986	> 0.3 μs						
^{206}Pt		205.990	> 0.3 μs						
^{207}Pt			> 0.3 μs						
^{208}Pt			> 0.3 μs						
$_{79}$Au		**196.966569(5)**							
^{170m}Au			0.62 ms	p/58	1.74/				
				α/42	7.11/				
^{170}Au		169.996	0.30 ms	p/89	1.46/				
				α/11	7.00/				
^{171m}Au			1.09 ms	α/66	6.995				
				p/34	1.694				
^{171}Au		170.9919	0.022 ms	p/100	1.437				
^{172m}Au			8. ms		6.80				
^{172}Au		171.9899	22. ms	α/7.02	6.76				
^{173m}Au			12 ms	α/0.21/92	6.732	(11/2-)			
^{173}Au		172.9862	0.02 s	α/94	6.672				
^{174}Au		173.985	0.14 s	α	6.54				
^{175m}Au			0.14 s	α//94	6.43/90	(11/2-)			
^{175}Au		174.9813	0.15 s	α	6.43/~ 90				
^{176m2}Au			1.4 s	IT					
^{176m1}Au			1.0 s	α//75					
^{176}Au		175.9803	0.9 s	β+, ec/10.5					
				α/	6.260/80				
					6.290/20				
^{177m}Au			1.18 s	α	6.12/66				
^{177}Au		176.9769	1.5 s	α/	6.16/40				
					6.150/				
^{178}Au		177.9760	2.6 s	α/	5.920/				
^{179}Au		178.9732	7.5 s	α/	5.85/				
^{180}Au		179.9725	8.1 s	ec/8.6	5.65				0.1522
				α/	5.61				0.2564
					5.50				0.5242
									0.6765
									0.8084
									0.8597
^{181}Au		180.9701	11.4 s	ec/97.5/6.3	5.482/				
				α/2.7/					
^{182}Au		181.9696	21. s	β+, ec/6.9		(2+)	1.3		ann.rad./
				α/0.13/					0.1549
									0.2649
									(0.13–1.4)
^{183}Au		182.9676	42. s	ec/5.5			+1.97		0.1630
				α/0.8/					0.2730
									0.3625

Nuclear

Elem. or Isot.	Natural abundance (Atom %)	Atomic mass or weight	Half-life/ Resonance width (MeV)	Decay mode/ Energy (/MeV)	Particle energy/ Intensity (MeV/%)	Spin (h/2π)	Nuclear magnetic mom. (nm)	Elect. quadr. mom. (b)	γ-Energy/ Intensity (MeV/%)
^{184m}Au			48 s	IT		2+	+1.44	+1.9	0.069(IT)
^{184}Au		183.9675	21. s	ec, β+/7.1		5+	+2.07	+4.7	
				α/0.013/					
^{185m}Au			6.8 m	β+, ec/					
				IT/0.145					
^{185}Au		184.9658	4.3 m	β+, ec/4.71		5/2-	+2.17	-1.1	ann.rad./
				α/0.26/					
^{186m}Au			< 2. m	β+, ec/					0.1915
^{186}Au		185.9660	10.7 m	β+, ec/6.0		3-	-1.28	+3.1	ann.rad./
				α/8(10)-4/					0.1915
									0.2988
^{187m}Au			2.3 s	IT/0.120	0.101/100	9/2-			
^{187}Au		186.9645	8.3 m	β+, ec/3.60		1/2(+)	+0.53		ann.rad./
									0.9152
									1.2668
									1.3321
									1.4081
^{188}Au		187.9653	8.8 m	β+, ec/5.3		1(-)	-0.07		ann.rad./
									0.2660
									0.3404
									0.6061
^{189m}Au			4.6 m	β+, ec/		11/2-	+6.19		0.1667
^{189}Au		188.9639	28.7 m	ec/96 /3.2		1/2+	+0.49		ann.rad./
				β+ /4 /					Pt k x-ray
									0.4478
									(0.713-0.8128)
^{190m}Au			0.13 s	IT		(17-)			
^{190}Au		189.9647	43. m	β+ /2 /4.44		1-	-0.07		ann.rad./
				ec/98 /					Pt k x-ray
									0.2958
									0.3018
									0.5977
^{191m}Au			0.9 s	IT/0.2663	0.014/100	(11/2-)	6.6		Au k x-ray
									0.2414
									0.2526
^{191}Au		190.9637	3.2 h	ec/1.83		3/2+	+0.137	+0.72	Pt k x-ray
									0.5864/16
									(0.088–1.30)
^{192m2}Au			0.16 s	IT/0.432	0.0598/100	11-			
^{192m1}Au			29 ms	IT/0.135		(5)+			
^{192}Au		191.9648	4.9 h	β+ /5 /3.52	2.19/	1-	-0.011	-0.23	ann.rad./
				ec/95 /	2.49/				Pt k x-ray
									0.2959
									0.3165
^{193m}Au			3.9 s	IT/0.2901		11/2-	6.2	+1.98	Au k x-ray
									0.2580
^{193}Au		192.9641	17.6 h	ec/1.07		3/2+	+0.140	+0.66	Pt k x-ray
									0.1862
									0.2556
^{194m2}Au			0.42 s	IT/0.476	0.69	11-			
^{194m1}Au			0.60 s	IT/0.107	0.027	5+			
^{194}Au		193.96542	1.64 d	β+ /3 /2.49	1.49/	1-	+0.076	-0.24	ann.rad./
				ec/97 /					Pt k x-ray
									0.3391
									(0.1015-2.413)
^{195m3}Au			0.013 ms	IT		31/2(-)			
^{195m2}Au			0.02 ms	IT/2.461		29/2+			
^{195m}Au			30.5 s	IT/0.3186		11/2-	6.2	+1.9	Au k x-ray
									0.2617
^{195}Au		194.96504	186.10 d	ec/0.227		3/2+	+0.149	+0.61	Pt k x-ray
^{196m2}Au			9.6 h	IT/0.5954	0.175	12-	5.7		Au k x-ray
									0.1478
									0.1883
^{196m1}Au			8.1 s	IT/0.0846		5+			0.0847
^{196}Au		195.96657	6.17 d	ec/92 /1.506		2-	+0.591	~ 0.81	Pt k x-ray
^{197m}Au			7.7 s	IT/0.4094		11/2-	+6.0	+1.7	Au k x-ray
				β- /8 /0.686					0.1302
									0.2790
^{197}Au	100.	196.966569				3/2+	+0.14575	+0.55	
^{198m}Au			2.27 d	IT/0.812	0.115/100	(12-)	+5.9		Au k x-ray
									0.0972
									0.1803
									0.2419

Elem. or Isot.	Natural abundance (Atom %)	Atomic mass or weight	Half-life/ Resonance width (MeV)	Decay mode/ Energy (/MeV)	Particle energy/ Intensity (MeV/%)	Spin (h/2π)	Nuclear magnetic mom. (nm)	Elect. quadr. mom. (b)	γ-Energy/ Intensity (MeV/%)
^{198}Au		197.968242	2.695 d	β- /1.372	0.290/1	2-	+0.5934	+0.64	Hg k x-ray
					0.961/99				0.411794/95.3
^{199m}Au			0.44 ms	IT/0.549		(11/2-)			
^{199}Au		198.968765	3.14 d	β- /0.453	0.25/22	3/2+	~ +0.2715	+0.51	Hg k x-ray
					0.292/72				0.15837
					0.462/6				0.20820
^{200m}Au			18.7 h	β- /84 /1.0	0.56/	12-	5.9		Au k x-ray
				IT/16 /					0.2559/71
									0.3680/77
									0.4978/73
									0.5793/72
									0.084–0.904)
^{200}Au		199.9708	48.4 m	β- /2.24	0.7/15	1-			0.3679/19
					2.2/77				1.2254/10.6
									(0.077–1.570)
^{201m2}Au			~ 0.3 ms	IT/0.594	0.04				
^{201m1}Au			0.006 ms	IT					
^{201}Au		200.97166	26. m	β- /1.28	1.27/82	3/2+			(0.027–0.732)
^{202}Au		201.9739	29. s	β- /3.0		(1-)			0.4396
^{203m}Au			0.14 ms	IT/0.641		11/2-			
^{203}Au		202.97515	1.0 m	β- /2.14	~ 1.9/	3/2+			(0.04–0.37)
^{204m}Au			2.1 μs	IT	0.977	(16+)			
^{204}Au		203.978	~ 34 s	β- /4.5		(2-)			0.4366
									1.5113
^{205m}Au			6 s			11/2-			
^{205}Au		204.980	0.6 m	β- /					(0.38–1.33)
^{206}Au		205.985	~ 1 m						
^{207}Au		206.988	> 0.3 μs						
^{208}Au		207.993	> 0.3 μs						
^{209}Au		208.997	> 0.3 μs						
^{210}Au		210.003	> 0.3 μs						
$_{80}$Hg		**200.592(3)**							
^{171}Hg		171.004	0.06 ms	α	7.49				
^{172}Hg		171.999	0.3 ms	α	7.36	0+			
^{173}Hg		172.997	0.8 ms	α	7.20				
^{174}Hg		173.9929	1.9 ms	α	7.07	0+			
^{175}Hg		174.9914	0.02 s	α					
^{176}Hg		175.9874	21 ms	α	6.74/94	0+			
^{177m}Hg			1.5 μs	IT/0.323	0.246	(13/2+)			0.246
^{177}Hg		176.9863	0.13 s	α	6.58				
^{178}Hg		177.9825	0.26 s	ec/50 /6.1		0+			
				α/50 /	6.43/89				
^{179m}Hg			~ 0.006 ms	IT/0.171	0.111	(13/2+)			
^{179}Hg		178.9818	1.05 s	ec/8.0					
				α/	6.29/75				
^{180}Hg		179.9783	2.6 s	ec/5.5		0+			0.1250
				α/	6.12/33				0.3005
					5.69/.03				0.3812
^{181m}Hg			0.48 ms			(13/2+)			
^{181}Hg		180.9778	3.6 s	β+ ec/76 /~ 7.3		1/2-	+0.507		0.0663
				α/24 /					0.0811
									0.0924
									0.1474
									0.1587
									0.2142
									0.2398
^{182}Hg		181.9747	10.8 s	β+, ec/85/5.0		0+			0.129/122
				α/15/	5.87/8.6				0.2176/66
					5.45/0.03				0.0256–0.543
^{183}Hg		182.97445	9. s	β+, ec/77/6.3		½-	+0.524		0.0714
				α/	5.83/				0.0874
					5.91/				0.1538
^{184m}Hg			0.8 m						
^{184}Hg		183.9717	30.9 s	β+, ec/99/4.1		0+			0.1565/102
				α/1/	5.54/1.3				0.2367/100
					5.07/0.002				0.2384/18
									(0.018–0.4227)
^{185m}Hg			22 s	β+, ec, IT, α/	5.37/	13/2+	-1.02	~+0.2	0.211
									0.292
^{185}Hg		184.9719	51. s	β+, ec/95/5.8		½-	+0.509		0.02–0.55
^{186m}Hg			0.08 ms	IT/2.217		(8-)			
^{186}Hg		185.9694	1.4 m	β+, ec/3.3		0+			0.1119
				α	5.09/0.02				0.2518

Elem. or Isot.	Natural abundance (Atom %)	Atomic mass or weight	Half-life/ Resonance width (MeV)	Decay mode/ Energy (/MeV)	Particle energy/ Intensity (MeV/%)	Spin (h/2π)	Nuclear magnetic mom. (nm)	Elect. quadr. mom. (b)	γ-Energy/ Intensity (MeV/%)
^{187m}Hg			2.4 m	β+, ec/		13/2(+)	-1.04	+0.5	see Hg187
^{187}Hg		186.9698	2.4 m	β+, ec/4.9		3/2(-)	-0.594	-0.8	0.1034/32
									0.2334/100
									0.2403/33
									0.27151/31
									0.3763/38
									0.5254/30
									0.10–2.18
^{188}Hg		187.9676	3.2 m	β+, ec/2.3		0+			0.0988
				α	4.61				0.1148
									0.1424
									0.1900
^{189m}Hg			8.6 m	ec/		13/2+	-1.06	+0.7	0.0780
									0.3210
									0.4345
									0.5655
									(0.08–2.170)
^{189}Hg		188.9682	7.6 m	ec/4.2		3/2-	-0.609	-0.8	0.2005
									0.2038
									0.2386
									0.2485
^{190}Hg		189.9663	20.0 m	ec/1.5		0+			0.1296
									0.1426
^{191m}Hg			51. m	β+ /6 /		13/2(+)	-1.07	+0.6	ann.rad./
				ec/94 /					Au k x-ray
									0.2741
									0.4203
									0.5787
									(0.07–1.9)
^{191}Hg		190.9672	50. m	β+, ec/3.2		3/2(-)	-0.62	-0.8	0.1963
									0.2247
									0.2524
^{192}Hg		191.9656	5.0 h	ec/~ 0.5		0+			Au k x-ray
									0.1572
									0.2748
									0.3065
^{193m}Hg			11.8 h	β+, ec/91 /		13/2(+)	-1.058430	~ +0.92	Hg k x-ray
				IT/9 /0.2901					0.1866
									0.2580
									0.4076
									0.5733
									0.9324
									(0.1–1.96)
^{193}Hg		192.9666	3.8 h	ec, B+/2.34		3/2(-)	-0.6276	-0.7	0.1866
									0.2580
									0.8611
^{194}Hg		193.96545	520. a	ec/0.04		0+			Au L x-rays
^{195m}Hg			1.73 d	IT/(54)/0.3186		13/2+	-1.04465	+1.1	Hg k x-ray
				ec/(46)/					Au k x-ray
									0.2617
									0.5603
									0.7798
^{195}Hg		194.9667	10.5 h	ec/1.51		1/2-	+0.541475		Au k x-ray
									0.0614
									0.7798
^{196}Hg	0.155(12)	195.96583	>2.5 ×10^{18} a	α		0+			
^{197m}Hg			23.8 h	IT/(93)/0.2989		13/2+	-1.027684	+1.2	Hg k x-ray
									Au k x-ray
									0.13398
^{197}Hg		196.96721	2.69 d	ec/0.600		1/2-	+0.527374		Au k x-ray
									0.07735
^{198}Hg	10.038(16)	197.966769				0+			
^{199m}Hg			42.7 m	IT/0.532		13/2+	-1.014703	+1.2	Hg k x-ray
									0.15841
^{199}Hg	16.938(39)	198.968281				1/2-	+0.505885		
^{200}Hg	23.138(65)	199.968327				0+			
^{201m}Hg			0.094 ms	IT/0.766	0.219	(13/2+)			
^{201}Hg	13.170(66)	200.970303				3/2-	-0.560226	+0.37	
^{202}Hg	29.743(89)	201.970643				0+			
^{203m}Hg			0.22 ms	IT/0.932	0.341	(13/2+)			
^{203}Hg		202.972873	46.61 d	β- /0.492	0.213/100	5/2-	+0.8489	+0.34	Tl k x-ray
									0.279188
^{204}Hg	6.818(35)	203.973494				0+			

Elem. or Isot.	Natural abundance (Atom %)	Atomic mass or weight	Half-life/ Resonance width (MeV)	Decay mode/ Energy (/MeV)	Particle energy/ Intensity (MeV/%)	Spin (h/2π)	Nuclear magnetic mom. (nm)	Elect. quadr. mom. (b)	γ-Energy/ Intensity (MeV/%)
^{205m2}Hg			5.9 μs	IT/3.316		(23/2-)			
^{205m1}Hg			1.09 ms	IT/1.556		13/2+			
^{205}Hg		204.97607	5.2 m	β- /1.531	1.33/4	1/2-	+0.6009		0.20378
									(0.2–1.4)
^{206m}Hg			2.1 μs	IT/2.102	1.034	5-			
^{206}Hg		205.9775	8.2 m	β- /1.31	0.935/34	0+			Tl k x-ray
					1.3/63				0.3052
									0.6502
^{207}Hg		206.9823	2.9 m	β- /4.8		(9/2+)			
^{208}Hg		207.9858	41. m	β-		0+			0.474
^{209}Hg		208.991	6 s	β-					0.324
^{210m}Hg			2.0 μs						(0.170-0.663)
^{210}Hg		209.994	1.1 m	β-		0+			
^{211}Hg		210.999	0.4 m						
^{212}Hg		212.003	> 0.3 μs						
^{213}Hg		213.008	>0.3 μs						
^{214}Hg		214.012	> 0.3 μs						
^{215}Hg		215.017	> 0.3 μs						
^{216}Hg		216.021	>0.3 μs						
$_{81}$Tl		**204.382-204.385**							
^{176}Tl		176.0006	5 ms	p	1.26/~ 100				
^{177m}Tl			0.23 ms	p/51	1.95	(11/2-)			
				α/49	7.48				
^{177}Tl		176.9964	0.017 s	α/73					
				p/27					
^{178}Tl		177.995	0.25 s	α/	6.704				
					6.785				
					6.62				
					6.859				
^{179m}Tl			1.5 ms	α	/7.21/80	(11/2-)			
				α	/7.10/20				
^{179}Tl		178.991	~ 0.27 s	α	6.56/60				
^{180}Tl		179.9901	1.09 s	α//6	6.28/30				
				ec/ β+/94	6.36/30				
				β+,f/0.004	6.21/18				
					6.56/15				
					6.47/7				
^{181m}Tl			1.4 ms	IT/99.6	6.58/96	(9/2-)			
				α/0.4	6.97/2.6				
^{181}Tl		180.9863	3.2 ms	α/ < 10	6.19/100				
^{182}Tl		181.9857	3. s	β+, ec/10.9					0.351
									(0.26–0.41)
^{183m2}Tl			1.5 μs	IT/0.977		(13/2+)			
^{183m}Tl			0.06 s	α//1.5	6.33/80	9/2-			0.0618
					6.38/16				(0.046-0.0894)
					6.46/4				
^{183}Tl		182.9822	5. s	β+, ec/7.7		½+	1.62		0.208
^{184}Tl		183.9819	11. s	β+, ec/(98)/9.2					0.2868
^{184m1}Tl			0.047 s	IT/0.506		10-			
				α/(2)/	6.16/				0.3399
									0.3667
^{185m}Tl			1.9 s	IT/0.453		(9/2-)	3.8		0.1688
				α/5.97	6.01				0.2840
^{185}Tl		184.9788	20. s	ec/β+/6.6		(1/2)	1.61		
^{186m}Tl			2.9 s	IT/0.396		(10-)			0.3738
^{186}Tl		185.9787	28. s	β+, ec/7.5		7+			0.3567
									0.4026
									0.4053
^{187m2}Tl			1.1 μs						
^{187m}Tl			15.6 s	IT/~ 0.33		(9/2+)	+3.71	-2.4	0.2995
^{187}Tl		186.9759	50. s	β+, ec/6.0		½+	1.6		
^{188m2}Tl			0.041 s	IT		(9-)			
^{188m}Tl			1.18 m	β+, ec/		(7+)			Hg k x-ray
									0.4129
									0.5043
									0.5921
^{188}Tl		187.9760	1.2 m	β+, ec/7.8		(2-)	+0.48	+0.129	see Tl[188m]
									0.4129
^{189m}Tl			1.4 m	β+, ec/		(9/2-)	+3.76	-2.29	0.2156
									0.2284
									0.3175
									0.4452

Elem. or Isot.	Natural abundance (Atom %)	Atomic mass or weight	Half-life/ Resonance width (MeV)	Decay mode/ Energy (/MeV)	Particle energy/ Intensity (MeV/%)	Spin (h/2π)	Nuclear magnetic mom. (nm)	Elect. quadr. mom. (b)	γ-Energy/ Intensity (MeV/%)
^{189}Tl		188.9736	2.3 m	β+, ec/5.2		(1/2+)			0.3337
									0.4510
									0.5223
									0.9422
^{190m2}Tl			0.75 ms	IT		(8-)			
^{190m}Tl			3.7 m	β+, ec/	4.2/	(7+)	+0.495	+0.29	0.1968
									0.4164
									0.7311
^{190}Tl		189.9738	2.6 m	β+, ec/7.0	5.7/	(2-)	+0.254	-0.33	0.4164
									0.6254
									0.6838
									1.0999
^{191m}Tl			5.2 m	β+, ec/(98)/		9/2(-)	+3.78	-2.2	0.2157
									0.2647
									0.3256
									0.3359
^{191}Tl		190.97178	2.2 m			(1/2)	+1.588		
^{192m}Tl			10.8 m	β+, ec/		(7+)	+0.518	+0.46	0.1740
									0.4228
									0.6348
									0.7863
									0.7455
^{192}Tl		191.9722	9.6 m	β+, ec/6.4		(2-)	+0.200	-0.33	0.3975
									0.4228
									0.6908
^{193m}Tl			2.1 m	IT/(75)/		(9/2-)	+3.82	-2.20	0.3650
^{193}Tl		192.9705	22. m	β+, ec/3.6		1/2(+)	+1.591		0.2077
									0.3244
									0.3440
									0.6761
									1.0447
									1.5793
^{194m}Tl			32.8 m	β+ /(20)/~ 0.30		7+	+0.540	+0.61	ann.rad./
				ec/(80)/					Hg k x-ray
									0.4282
									0.6363
									0.7490
^{194}Tl		193.9711	33.0 m	β+, ec/5.3		2-	+0.140	-0.28	0.4279/75.2
									0.6452/10.8
									(0.395-1.623)
^{195m}Tl			3.6 s	IT/0.483	0.099	9/2-	3.87		Tl k x-ray
									0.0990
									0.3836
^{195}Tl		194.9698	1.16 h	ec/97/2.8		1/2+	+1.58		ann.rad./
				β+ /(3)/					Hg k x-ray
									0.2422
									0.5635
									0.8845
									1.3639
									(0.13–2.5)
^{196m}Tl			1.41 h	β+, ec/95/4.9		(7+)	+0.55	+0.76	0.0840
									0.4261
									0.6353
									0.6954
									(0.08–1.0)
^{196}Tl		195.9705	1.84 h	β+ /(15)/4.4		2-	+0.072	-0.18	ann.rad./
				ec/(85)/					Hg k x-ray
									0.4257
									0.6105
									(0.03–2.4)
^{197m}Tl			0.54 s	IT/53/0.608	0.222	9/2-	4.03		Tl k x-ray
				β+, ec/47/					0.2262
									0.4118
									0.5872
									0.6367
^{197}Tl		196.9696	2.83 h	β+ /(1)/2.18		1/2+	+1.58		Hg k x-ray
				ec/(99)/					0.1522/8.2
									0.4258
^{198m2}Tl			0.032 s	IT/0.742	0.199				
^{198m}Tl			1.87 h	β+, ec/(53)/		7+	+0.64		Hg k x-ray
				IT/47/0.5347					Tl k x-ray
									0.4118
									0.5872

Elem. or Isot.	Natural abundance (Atom %)	Atomic mass or weight	Half-life/ Resonance width (MeV)	Decay mode/ Energy (/MeV)	Particle energy/ Intensity (MeV/%)	Spin (h/2π)	Nuclear magnetic mom. (nm)	Elect. quadr. mom. (b)	γ-Energy/ Intensity (MeV/%)
									0.6367
^{198}Tl		197.970	5.3 h	ec, β+ /(1)/3.5	1.4/	2-			Hg k x-ray
					2.1/				0.4118
					2.4/				0.6367
									0.6759
									(0.23–2.8)
^{199m1}Tl			0.028 s	IT/0.749		9/2-			
^{199}Tl		198.9699	7.4 h	ec/1.4		1/2+	+1.60		Hg k x-ray
									0.2082
									0.2473
									0.4555
^{200m1}Tl			0.034 s	IT/0.754	0.213	7+			
^{200}Tl		199.97096	1.087 d	ec/2.46	1.07/	2-	0.04		Hg k x-ray
					1.44/				0.36799
									1.2057
									(0.11–2.3)
^{201m1}Tl			2.0 ms	IT/0.919	0.588	9/2-			
^{201}Tl		200.9708	3.038 d	ec/0.48		1/2+	+1.605		Hg k x-ray
									0.13528
									0.16740/10.0
^{202m1}Tl			0.591 ms	IT/0.950	0.46	7+			
^{202}Tl		201.9721	12.47 d	ec/1.36		2-	0.06		Hg k x-ray
									0.43957
^{203m1}Tl			0.008 ms			25/2+			
^{203}Tl	29.52(1)	202.97234				1/2+	+1.622258		
^{204m2}Tl			2.6 μs	IT/2.319	0.669	12-			
^{204m1}Tl			0.062	IT/1.104	0.69	7+			
^{204}Tl		203.97386	3.78 a	β- /97/0.7637	0.763/97	2-	0.09		Hg k x-ray
				ec/(3)/0.347					
^{205m1}Tl			2.6 μs	IT/3.291	0.739	25/2+			
^{205}Tl	70.48(1)	204.97443				1/2+	+1.638215		
^{206m}Tl			3.74 m	IT/2.644		(12-)			Tl k x-ray
									0.2166
									0.2661
									0.4534
									0.6866
									1.0219
^{206}Tl		205.97611	4.20 m	β- /1.533	1.53/99.9	0-			Pb k x-ray
									0.80313
^{207m}Tl			1.3 s	IT/1.350		11/2-			Tl k x-ray
									0.3501
									1.0000
^{207}Tl		206.97742	4.77 m	β- /1.423	1.43/99.8	1/2+	+1.88		0.89723
^{208}Tl		207.98202	3.053 m	β- /5.001	1.28/23	(5+)	+0.29		Pb k x-ray
					1.52/22				0.27728
					1.796/51				0.51061
									0.58302
									2.61448
^{209}Tl		208.9854	2.16 m	β-/3.98	1.8 /100	(1/2+)			Pb k x-ray
									1.5670/100
									0.4651/95
									(0.12–1.33)
^{210}Tl		209.9901	1.30 m	β- /5.48	1.3/25	(5+)			Pb k x-ray
					1.9/56				0.081
									0.2981
									0.79788
^{211}Tl		210.9935	1.3 m	β-					
				β, n	/~ 2				
^{212}Tl		211.998	0.5 m	β-					
				β, n					
^{213}Tl		213.0019	24 s	β-					
				β, n	/ 8				
^{214}Tl		214.007	11 s	β, n	/~ 34				
^{215}Tl		215.011	10 s	β-, n					
^{216}Tl		216.016	6 s	β-, n	/ <11.5				
^{217}Tl		217.020	> 0.3 μs						
$_{82}$Pb		**207.2(1)**							
^{178}Pb		178.0038	~ 0.2 ms			0+			
^{179}Pb		179.0022	~ 3.5 ms	α	7.35				
^{180}Pb		179.9979	4.2 ms	α/	7.25	0+			
^{181}Pb		180.9967	0.036 s	α/	7.02				
^{182}Pb		181.9927	55 ms	α	6.90	0+			

Elem. or Isot.	Natural abundance (Atom %)	Atomic mass or weight	Half-life/ Resonance width (MeV)	Decay mode/ Energy (/MeV)	Particle energy/ Intensity (MeV/%)	Spin (h/2π)	Nuclear magnetic mom. (nm)	Elect. quadr. mom. (b)	γ-Energy/ Intensity (MeV/%)
^{183m}Pb			0.42 s	α	6.70/82.7	13/2+	- 1.25	~ 1.7	
					6.86/1.9				
^{183}Pb		182.9919	0.54 s	α/	6.57/4.3	(3/2-)	- 1.16	~ 0.6	
					6.78/11.0				
^{184}Pb		183.9881	0.48 s	α/~ 80	6.63/	0+			
^{185m}Pb			4.3 s	α	6.41/100	13/2+	-1.19	~ 0.9	
^{185}Pb		184.9876	6.3 s	α/	6.29/56	3/2-	-1.10	~ 0.2	0.205
					6.49/44				0.269
					6.55/<1.4				
^{186}Pb		185.9842	5. s	β+, ec/95/5.5		0+			
				α/(5)/	6.32/				
					6.34/<100				
					6.01/<0.2				
^{187m}Pb			18.3 s	β+, ec/	5.99/	(13/2+)	- 1.21	~ -0.5	0.0674
				α/12	6.19/				0.2080
									0.2755
									0.2995
									0.4487
									0.7477
^{187}Pb		186.9839	18.3 s	ec/7.2		13/2+	- 1.13	~ -0.4	0.1930
				α/7	6.08/				0.3314
									0.3435
									0.3934
^{188}Pb		187.9809	23. s	ec/(78)/4.8		0+			0.1850
				α/(22)/	5.98/<10				0.7582
					5.61/<0.1				
^{189m}Pb			50 s	ec/100		(13/2+)	- 1.19	~ -1.3	
^{189}Pb		188.9808	39 s	ec/6.1		(3/2-)	- 1.08	~ 0.5	
^{190m2}Pb			7 μs	IT/2.658		(11)-			
^{190m1}Pb			25 μs			(12+)			
^{190}Pb		189.9781	1.2 m	β+ (13)/4.1		0+			ann.rad./
				ec/(86)/					Tl k x-ray
				α/(0.9)/	5.58/				0.1415
									0.1512
									0.9422
^{191m}Pb			2.2 m	β+, ec/		13/2+	-1.17	+0.085	ann.rad./
									0.3871
									0.6135
									0.7122
^{191}Pb		190.9783	1.3 m	β+, ec/5.5					ann.rad./
									0.9368
^{192m1}Pb			1.1 μs	IT/2.624		(12+)			
^{192}Pb		191.9758	3.5 m	β+, ec/~ 3.4		0+			ann.rad./
				α/.006/	5.11				0.1675
									0.6082
									1.1954
^{193m}Pb			5.8 m	β+, ec/		13/2+	-1.15	+0.19	ann.rad./
									0.3650
									0.3922
^{193}Pb		192.9762	~ 2. m	ec/5.2		3/2 (-)			
^{194}Pb		193.9740	10. m	β+, ec/2.7		0+			ann.rad./
				α	4.64				0.2036
^{195m2}Pb			0.010 ms	IT/1.759		21/2-			
^{195m}Pb			15. m	β+ /(8)/		13/2+	-1.132	+0.31	ann.rad./
				ec/(92)/					Tl k x-ray
									0.3836
									0.3942
									0.8784
^{195}Pb		194.9745	~ 15. m	β+, ec/5.8					ann.rad./
									0.3836
									0.3937
									0.7776
^{196}Pb		195.9728	37. m	β+, ec/2.1		0+			Tl k x-ray
									0.2531
									0.5021
^{197m2}Pb			1.2 μs	IT/1.914		21/2-			
^{197m}Pb			43. m	ec/79/		13/2+	-1.105	+0.38	Tl k x-ray
				β+ /2/					0.3079
				IT/19/0.3193					0.3877
									0.7743
									(0.2–2.2)
^{197}Pb		196.97343	~ 8. m	ec/97/3.6		(3/2-)	-1.075	~ -0.08	Tl k x-ray
				β+ /3/					0.3755

Elem. or Isot.	Natural abundance (Atom %)	Atomic mass or weight	Half-life/ Resonance width (MeV)	Decay mode/ Energy (/MeV)	Particle energy/ Intensity (MeV/%)	Spin (h/2π)	Nuclear magnetic mom. (nm)	Elect. quadr. mom. (b)	γ-Energy/ Intensity (MeV/%)
									0.3858
									0.7611
^{198m1}Pb			4.2 μs	IT/2.141	0.318	(7)-			
^{198}Pb		197.9720	2.4 h	ec/1.4		0+			Tl k x-ray
									0.1734
									0.2903
									0.3654
^{199m2}Pb			0.010 ms	IT		(29/2-)			
^{199m}Pb			12.2 m	IT/93/0.4248		13/2+			Pb k x-ray
				β+, ec/(7)/					0.4255
^{199}Pb		198.9729	1.5 h	ec/(99)/2.9		5/2-	-1.074	~ +0.08	Tl k x-ray
				β+ /(1)/					0.3534
									0.7202
									1.1350
									(0.22–2.4)
^{200}Pb		199.9718	21.5 h	ec/0.81		0+			Tl k x-ray
									0.14763
^{201m}Pb			1.02 m	IT/0.6291		13/2+			Pb k x-ray
									0.6288
^{201}Pb		200.9729	9.33 h	ec/1.90		5/2-	+0.675	~ -0.009	Tl k x-ray
									0.33120
									0.36131
									(0.11–1.8)
^{202m}Pb			3.53 h	IT/90/2.170		9-	-0.228	~ +0.58	Pb k x-ray
				β+ /10/					Tl k x-ray
									0.42219
									0.78700
									0.96271
^{202}Pb		201.97215	5.3 ×10^{4} a	ec/0.05		0+			Tl L x-ray
^{203m2}Pb			0.48 s	IT/2.949		29/2-			
^{203m}Pb			6.2 s	IT/0.8252		13/2+			Pb k x-ray
									0.8203
									0.8252
^{203}Pb		202.97339	2.163 d	ec/0.98		5/2-	+0.686	+0.10	Tl k x-ray
									0.279188
^{204m}Pb			1.13 h	IT/2.185		9-			Pb k x-ray
									0.37481
									0.89922
									0.91175
^{204}Pb	1.4(1)	203.97304	> 1.4 ×10^{20} a	α		0+			
^{205m2}Pb			5.6 ms	IT/1.014		13/2+			
^{205m1}Pb			0.024 ms	IT		1/2-			
^{205}Pb		204.97448	1.51 ×10^{7} a	ec/0.0512		5/2-	+0.712	+0.23	Tl L x-ray
^{206m1}Pb			0.125 ms	IT/2.200					
^{206}Pb	24.1(1)	205.97447	> 2.5 ×10^{21} a	α		0+			
^{207m}Pb			0.81 s	IT/1.632		13/2+			Pb k x-ray
									0.56915
									1.06310
^{207}Pb	22.1(1)	206.97590	> 1.9 ×10^{21} a	α		1/2-	+0.59258		
^{208}Pb	52.4(1)	207.97665	> 2 ×10^{19} a	sf		0+			
			> 2.6 ×10^{21} a	α					
^{209}Pb		208.98109	3.23 h	β- /0.644	0.645/100	9/2+	-1.474	-0.3	
^{210}Pb		209.98419	22.6 a	β- /0.0635	0.017/81	0+			
					0.061/19				
				α	3.72				
^{211}Pb		210.98874	36.16 m	β- /1.37	0.57/5	9/2+	-1.404	+0.09	0.40486
					1.36/92				0.42700
									0.83186
									(0.314-1.27)
^{212}Pb		211.99190	10.64 h	β- /0.574	0.28/83	0+			Bi k x-ray
					0.57/12				0.23858
^{213}Pb		212.9966	10.2 m	β- /2.1					
^{214m1}Pb			6.2 μs			(8+)			
^{214}Pb		213.99981	27. m	β- /1.0	0.67/48	0+			Bi k x-ray
					0.73/42				0.24192
									0.29509
									0.35187
^{215}Pb		215.005	2 m						
^{216}Pb		216.008	1.7 m						
^{217}Pb		217.013	0.3 m						
^{218}Pb		218.017	~ 15 s						
^{219}Pb		219.022	> 0.3 μs						
^{220}Pb		220.025	> 0.3 μs						

Elem. or Isot.	Natural abundance (Atom %)	Atomic mass or weight	Half-life/ Resonance width (MeV)	Decay mode/ Energy (/MeV)	Particle energy/ Intensity (MeV/%)	Spin (h/2π)	Nuclear magnetic mom. (nm)	Elect. quadr. mom. (b)	γ-Energy/ Intensity (MeV/%)
$_{83}$Bi		208.98040(1)							
^{184m}Bi			0.007 s	α	(7.22-7.85)				0.449
^{184}Bi		184.0013	13. ms	α					0.124
^{185}Bi		184.998	60. μs	p/90	1.55				
				α/10	8.03				
^{186m}Bi			10 ms	α	7.26	(10-)			0.1085
					7.37				
^{186}Bi		185.9966	15 ms	α	7.07-7.23				(0.087-0.520)
				p/<0.5					
^{187m}Bi			0.37 ms	α/12	7.72/100	1/2+			
^{187}Bi		186.9931	40. ms	α/7	7.00/88.	9/2-			
					7.61/9.0				
					7.34/3.0				
^{188}Bi		187.9923	0.271 s	α	6.81				(0.071-0.320)
^{189m}Bi			7.0 ms	α	7.30				
^{189}Bi		188.9892	0.68 s	α					
^{190m}Bi			6.2 s	α/90	6.43	(10-)			(0.105-0.314)
					(6.23-6.72)				
^{190}Bi		189.9886	6.3 s	β+, ec/(30)/8.7	α/6.45	(3+)			(0.089-0.374)
				α/70	(6.39-6.82)				
^{191m}Bi			0.12 ms	α/	6.87/100	(1/2+)			
^{191}Bi		190.98579	12.4 s	β+, ec/(60)/7.3					
				α/(40)/	6.31				
^{192}Bi		191.9855	40. s	β+, ec/(80)/9.0		(10-)			
				α/(20)/	6.06/				
^{193m2}Bi			3 μs			(29/2-)			
^{193m}Bi			3.2 s	β+, ec/		(1/2+)			
				α/	6.48/				
^{193}Bi		192.98296	1.11 m	β+, ec/16/7.1		9/2+			
				α/84	5.91/				
^{194m1}Bi			2.1 m						
^{194}Bi		193.9828	1.9 m	β+, ec/99.9/8.2		(10-)			0.1661
				α/0.1/					0.1740
									0.2802
									0.421
									0.5754
									0.9650
^{195m}Bi			1.45 m	β+, ec/(94)/		(1/2+)			
				α/(6)/	6.11/				
^{195}Bi		194.98065	2.9 m	β+, ec/99.8/5.8		3/2-			
				α/(0.2)	5.45/				
^{196m2}Bi			4.0 m	ec/74		(10-)			
				IT/26					
^{196m1}Bi			~ 0.6 s			(7+)			
^{196}Bi		195.9807	5.1 m	ec/~ 7.4		(3+)			0.1376
									0.3720
									0.6880
									1.0486
^{197}Bi		196.9789	5.0 m	β+, ec/5.2		1/2+			
^{198m}Bi			7.7 s	IT/0.2485		(10-)			0.2485
^{198}Bi		197.9792	11.6 m	β+, ec/6.6		(7+)			0.0900
									0.1976
									0.5624
									1.0635
^{199m}Bi			24.7 m	β+, ec/		(1/2+)			ann.rad./
^{199}Bi		198.9777	27. m	β+, ec/4.3		9/2-	4.6		0.7203
									0.8374
									0.8417
									0.9460
									1.0528
									1.3056
									(0.12–3.2)
^{200m2}Bi			0.4 s			(10-)			
^{200m}Bi			31 m	β+, ec/		(2+)			0.2453
									0.4198
									0.4624
									1.0265
^{200}Bi		199.9781	36. m	ec/(90)/5.9		7+			ann.rad./
				β+ /(10)/					Pb k x-ray
									0.4198
									0.4623
									1.0265
^{201m}Bi			58 m	IT/0.846		(1/2+)			Bi k x-ray

Elem. or Isot.	Natural abundance (Atom %)	Atomic mass or weight	Half-life/ Resonance width (MeV)	Decay mode/ Energy (/MeV)	Particle energy/ Intensity (MeV/%)	Spin (h/2π)	Nuclear magnetic mom. (nm)	Elect. quadr. mom. (b)	γ-Energy/ Intensity (MeV/%)
				β+, ec/					0.8464
^{201}Bi		200.9770	1.8 h	ec/3.84		9/2-	4.8		Pb k x-ray
									0.6288
									0.9357
									1.0138
									(0.13–2.4)
^{202}Bi		201.9777	1.72 h	β+ /(3)/5.16		5+	+4.26	-0.9	ann.rad./
				ec/(97)/					Pb k x-ray
									0.57860
									0.92734
									(0.08–3.5)
^{203m1}Bi			0.31 ms	IT/1.098		1/2+			
^{203}Bi		202.9769	11.8 h	ec/99.8/3.25		9/2-	+4.02	-0.8	Pb k x-ray
				β+ /(0.2)/	1.35/				0.1865
									0.8203
									0.8969
									1.8475
									(0.1–2.9)
^{204m2}Bi			1.07 ms			17+			
^{204m1}Bi			13 ms			10-			
^{204}Bi		203.9778	11.2 h	ec/4.44		6+	+4.32	-0.43	Pb k x-ray
									0.37481
									0.89922
									0.98409
^{205m1}Bi			0.008 ms	IT/1.497		1/2+			
^{205}Bi		204.97739	14.9 d	ec/2.71		9/2-	+4.07	-0.7	Pb k x-ray
									0.70347
									1.76435
									(0.550-1.862)
^{206m2}Bi			0.89 ms	IT/1.045		10-			
^{206m1}Bi			7.7 μs	IT/0.060		4+			
^{206}Bi		205.97850	6.243 d	ec/3.76		6+	+4.36	-0.4	Pb k x-ray
									0.51619
									0.80313
									0.88100
^{207m1}Bi			0.18 ms	IT/2.102		21/2+			
^{207}Bi		206.97847	31.55 a	ec/2.399		9/2-	4.092	-0.6	Pb k x-ray
									0.56915
									1.06310
^{208m}Bi			2.58 ms	IT/1.571	0.921	10-	2.67		0.921
^{208}Bi		207.97974	3.68×10^5 a	ec/2.880		5+	4.60	-0.64	Pb k x-ray
									2.61435
^{209}Bi	100.	208.98040	2.0×10^{19} a	α	3.13/98.8	9/2-	+4.111	-0.6	0.204
^{210m}Bi			3.0×10^6 a	α/	4.420(3)/0.29	9-	+2.73	-0.6	Tl k x-ray
					4.59(3)/3.9				
					4.584(3)/1.4				0.3052
					4.908(4)/39				0.6502
					4.946(3)/55				
^{210}Bi		209.98412	5.01 d	β- /1.163	1.16/99	1-	-0.0445	+0.136	0.2661
									0.3.52
^{211m1}Bi			1.4 μs	IT/1.257		(25/2-)			
^{211}Bi		210.98727	2.14 m	α/(99.7)/	6.279/16	9/2-			Tl k x-ray
				β- /(0.3)/0.58	6.623/84				0.3501
^{212m2}Bi			7.0 m	β- /		(18-)			
^{212m1}Bi			25.0 m	α/(93)/	6.300/40	(9-)			0.120
				β- /(7)/	6.340/53				0.233
									0.275
									0.404
									0.727
^{212}Bi		211.99129	1.009 h	β- /(64)/2.254		(1-)	+0.32	~ +0.1	Tl k x-ray
				α/(36)/	6.051/25				Po k x-ray
					6.090/9.6				0.2881
									0.72725
									0.78551
									1.62066
^{213}Bi		212.99439	45.6 m	β- /(98)/1.43	1.02/31	9/2-	+3.72	-0.71	Po k x-ray
					1.42/66				0.4404
				α/2/	5.549/0.15				(0.15–1.328)
					5.869/1.9				
^{214}Bi		213.9987	19.7 m	β- /3.27					1.10006
									0.60931
									1.12027
									1.76449

Elem. or Isot.	Natural abundance (Atom %)	Atomic mass or weight	Half-life/ Resonance width (MeV)	Decay mode/ Energy (/MeV)	Particle energy/ Intensity (MeV/%)	Spin (h/2π)	Nuclear magnetic mom. (nm)	Elect. quadr. mom. (b)	γ-Energy/ Intensity (MeV/%)
									(0.19–3.2)
^{215m}Bi			37 s	β					(0.158-0.498)
^{215}Bi		215.0018	7.7 m	β- /2.3					0.2937/35.2
									(0.271–1.399)
^{216}Bi		216.0063	2.3 m	β-/4.0					0.5498
									0.4192
^{217}Bi		217.0094	98 s	β/					0.2646/100
									(0.254-1.017)
^{218}Bi		218.0142	33. s	β-					0.5097/134
									0.3857/100
									(0.174-0.703)
^{219}Bi		219.017	~ 15 s						
^{220}Bi		220.022	~ 10 s						
^{221}Bi		221.026	> 0.3 μs						
^{222}Bi		222.031	> 0.3 μs						
^{223}Bi		223.035	> 0.3 μs						
^{224}Bi		224.039	> 0.3 μs						
$_{84}$**Po**									
^{187}Po		187.0030	1.4 ms	α	7.53/100				0.286
^{188}Po		187.9994	0.27 ms	α	7.91/80	0+			
					7.320				
^{189}Po		188.9985	5 ms	α	7.532/8				
					7.259/80				
					7.309/12				
^{190}Po		189.9951	2.4 ms	α/	7.53/96.4	0+			
					7.01/3.3				
^{191m}Po			93. ms	α/94	7.376/50	(13/2+)			
					6.888/46				
^{191}Po		190.99456	22 ms	α/	7.334/77				
					6.97/8				
^{192}Po		191.9913	32. ms	α/8.5	7.17/98.6	0+			
					6.59/1.4				
^{193m}Po			0.37 s	α/	7.00	(3/2-)			
^{193}Po		192.9910	0.25 s	α/	6.95	(13/2+)			
^{194m}Po			0.015 ms	IT		(11-)			
^{194}Po		193.9882	0.2 s	α/	6.84/93	0+			
					6.19/0.22				
^{195m}Po			1.9 s	α/	6.70/	(13/2+)			
^{195}Po		194.9881	~ 3.9 s	α/	6.62/				
^{196}Po		195.9855	~5.75 s	α/(95)/	6.53/94	0+			
				β+, ec/(5)/~ 4.6	5.77/0.02				
^{197m}Po			32 s	α/(84)/	6.385(3)/55	13/2+			
				β+, ec/(16)/					
^{197}Po		196.9857	53. s	α/(44)/	6.282(4)/76	(3/2-)			
				β+, ec/(56)/6.2					
^{198}Po		197.9834	1.76 m	α/(70)/	6.18/57	0+			
				β+, ec/(30)/4.0	5.27/7.6 ×10^{-4}				
^{199m}Po			4.2 m	β+, ec/(51)/		13/2+	~ 0.99		ann.rad./
				α/(39)/	6.059/24				0.2745
									0.4998
									1.0020
^{199}Po		198.9837	5.2 m	β+, ec/(88)/7.		(3/2-)			Bi k x-ray
				α/(12)/	5.952/7.5				0.1877
									0.3616
									1.0214
									1.0344
^{200}Po		199.98178	11.5 m	β+, ec/85/3.4		0+			0.14748
				α/(15)/	5.863/11.1				0.32792
									0.6176
									0.6709
^{201m}Po			9.0 m	β+, ec/(57)/		13/2+	~ 1.00		Bi k x-ray
				IT/40/0.418					Po k x-ray
				α/(3)/	5.786/~ 3.				0.2726
									0.4123
									0.4179
									0.9670
^{201}Po		200.98226	15.3 m	β+, ec/98/4.9		3/2-	~ 0.94		Bi k x-ray
				α/(2)/	5.683(3)/1.1				0.2056
									0.2250
									0.8483
									0.9048
^{202}Po		201.9808	45. m	β+, ec/98/2.8		0+			0.0410
				α/(2)/	5.588/1.9				0.1656

Elem. or Isot.	Natural abundance (Atom %)	Atomic mass or weight	Half-life/ Resonance width (MeV)	Decay mode/ Energy (/MeV)	Particle energy/ Intensity (MeV/%)	Spin (h/2π)	Nuclear magnetic mom. (nm)	Elect. quadr. mom. (b)	γ-Energy/ Intensity (MeV/%)
									0.3158
									0.6884
^{203m}Po			45 s	IT/96/0.6414		13/2+			Bi k x-ray
				β-ec/(4)/					Po k x-ray
									0.6414
^{203}Po		202.9814	35. m	β+, ec/4.2		5/2-	+0.74		0.17516
									0.21477
									0.89350
									0.90863
									1.09095
^{204}Po		203.9803	3.53 h	ec/2.34		0+			Bi k x-ray
				α	5.377/0.66				0.2702
									0.8844
									1.0162
									(0.11–1.9)
^{205m2}Po			57 ms	IT/1.461	0.581	19/2-			
^{205m1}Po			0.65 ms	IT/0.880	0.161	13/2+			
^{205}Po		204.9812	1.7 h	β+, ec/3.53		5/2-	~ +0.76	+0.17	Bi k x-ray
									0.83681
									0.84983
									0.87241
									1.00124
									(0.12–2.7)
^{206}Po		205.98047	8.8 d	ec/(95)/1.85		0+			Bi k x-ray
				α/(5)/	5.223/5.5				0.28644
									0.31156
									0.51134
									0.80737
									1.03228
									(0.11–1.5)
^{207m2}Po			2.8 s	IT/1.383		19/2-			Po k x-ray
									0.2682
									0.30074
									0.81448
^{207m1}Po			0.049 ms	IT/1.115	0.301	13/2+			
^{207}Po		206.98159	5.80 h	ec, β+/2.91		5/2-	~ +0.79	+0.28	Bi k x-ray
									0.74263
									0.91176
									0.99225
^{208}Po		207.98125	2.898 a	α/5.213	4.233/0.0002	0+			
					5.1158/100				
^{209}Po		208.98243	123. a	α/4.976	4.624/0.56	1/2-	+0.7		0.26049
					4.879/99.2				0.8964
^{210}Po		209.98287	138.4 d	α/5.407	4.516/0.001	0+			0.80313
					5.304/100				
^{211m2}Po			0.003 ms	IT/4.87	0.509	(43/2+)			
^{211m1}Po			25.2 s	α/	7.273/91	25/2+			Pb k x-ray
					7.994/1.7				0.32808
					8.316/0.25				0.56915
					8.875/7.0				0.89723
									1.06310
^{211}Po		210.98665	0.516 s	α/7.594	6.570/0.54	9/2+			0.56915
					6.892/0.55				0.89723
					7.450/98.9				
^{212m}Po			45. s	α/	8.514/2.0	(18+)			
					9.086/1.0				
					11.650/97				
^{212}Po		211.98887	0.298 μs	α/8.953	8.784/100	0+			
^{213}Po		212.99286	3.80 μs	α/8.537	7.614/0.003	9/2+			
					8.375/100				
^{214}Po		213.99520	164. μs	α/7.833	6.904/0.01	0+			0.7995
					7.686/99.99				0.298
^{215}Po		214.99942	1.780 ms	α/7.526	6.950/0.02	(9/2+)			0.439
					6.957/0.03				
					7.386/100				
^{216}Po		216.00192	0.145 s	α/6.906	5.895/0.002	0+			
					6.778/99.99				
^{217}Po		217.00632	1.53 s	α/6.662	6.539/				
^{218}Po		218.00897	3.04 m	α/6.114	6.003/99.999	0+			
					5.181/0.11				
^{219}Po		219.0136	~ 2 m						
^{220}Po		220.0163	> 0.3 μs			0+			
^{221}Po		221.0212	~ 1.9 m						

Elem. or Isot.	Natural abundance (Atom %)	Atomic mass or weight	Half-life/ Resonance width (MeV)	Decay mode/ Energy (/MeV)	Particle energy/ Intensity (MeV/%)	Spin (h/2π)	Nuclear magnetic mom. (nm)	Elect. quadr. mom. (b)	γ-Energy/ Intensity (MeV/%)
^{222}Po		222.0241	~ 2.4 m						
^{223}Po		223.029							
^{224}Po		224.032							
^{225}Po		225.037							
^{226}Po		226.040							
^{227}Po		227.045							
$_{85}$At									
^{191m}At			2.1 ms	α	7.65/98	(7/2-)			
					7.72/2				
^{191}At		191.0041	~ 1.7 ms	α	7.55/100				
^{192m}At			0.012 s	α	7.435/56				0.036
					7.47/31				
^{192}At		192.0032	0.09 s	α	7.22/82				0.165
				β, f	7.385/14				0.188
^{193m2}At			27 ms	IT/76		(13/2+)			
				α/24					
^{193m}At			21 ms	α	7.33/98	(7/2-)			
					7.42/2				
^{193}At		192.9999	28 ms	α/	7.24/100	(1/2+)			
^{194m}At			0.31 s	α	7.2				
^{194}At		193.9992	0.25 s	α/					
				β, f					
^{195m}At			0.14 s	α	7.07-7.22	(7/2-)			
^{195}At		194.9963	0.33 s	α/	6.95				
^{196m}At			11 μs	IT	0.158	(5+)			0.158
^{196}At		195.9958	0.371 s	α/97.5	7.05/98				
				β+/2.5	6.75/1.8				
				β+, f / 0.002					
^{197m2}At			1.3 μs	IT/0.311		13/2+			
^{197m}At			2.0 s	α	6.707	(1/2+)			
^{197}At		196.9932	0.39 s	β+, ec/7.8		(9/2-)			
				α/	6.960/				
^{198m}At			1.2 s	β+, ec/(75)/		(10-)			
				α/(25)/	6.856/86				
^{198}At		197.9928	4.1 s	α/	6.755/94				
^{199m}At			0.27 s	IT/0.244	0.103	1/2+			
^{199}At		198.99053	6.9 s	β+, ec/8/5.6		9/2-			
				α/(92)/	6.643/				
^{200m2}At			7 s	β+, ec/90		10-			
				α/10	6.538/12				
^{200m1}At			47 s	α	6.411	(7+)			
^{200}At		199.9904	43. s	β+, ec/65/~ 8.0		5+			
				α/(35)/	6.412/44				
					6.465/57				
^{201m2}At			3.4 μs	IT/2.318		29/2+			
^{200m1}At			45 ms	IT/0.459	0.269	1/2+			
^{201}At		200.9884	1.48 m	β+, ec/29/5.9		9/2-			0.5918/100
			45 ms	α/(71)/6.474	6.344/				(0.3585-0.761)
^{202m}At			0.46 s	IT/0.391		(10-)			
^{202}At		201.9886	3.02 m	β+, ec/88/7.2		(7+)			ann.rad./
				α/(12)/	6.135/7.7				0.4413
					6.225/4.3				0.5697
									0.6753
^{203}At		202.9869	7.4 m	β+, ec/69/5.1		9/2-			0.1458
				α/(31)/6.210	6.088/				0.2459
									0.6414
									1.0020
									1.0340
^{204m}At			0.11 s	IT/0.587		10-			
^{204}At		203.9873	9.1 m	β+, ec/95/6.5		(5+)			Po k x-ray
				α/(5)/	5.951/				0.3271
									0.4254
									0.5156
									0.6837
^{205m1}At			7.8 μs	IT/2.340		(29/2+)			
^{205}At		204.9861	26. m	β+, ec/90/4.54		(9/2-)			Po k x-ray
				α/(10)/6.020	5.902/				0.1543
									0.6696
									0.7194
^{206}At		205.9867	29.4 m	β+, ec/99/5.72		5+			Po k x-ray
				α/(1)/5.881	5.703/				0.20186
									0.39561
									0.47716

Nuclear

Elem. or Isot.	Natural abundance (Atom %)	Atomic mass or weight	Half-life/ Resonance width (MeV)	Decay mode/ Energy (/MeV)	Particle energy/ Intensity (MeV/%)	Spin (h/2π)	Nuclear magnetic mom. (nm)	Elect. quadr. mom. (b)	γ-Energy/ Intensity (MeV/%)
^{207}At		206.9858	1.81 h	β+, ec/90/3.91		9/2-			0.70071
				α/(10)/5.873	5.758/				Po k x-ray
									0.16801
									0.58842
									0.81448
^{208m1}At			1.5 μs	IT/2.276		16-			
^{208}At		207.9866	1.63 h	β+, ec/99/4.97		(6+)			Po k x-ray
				α/(1)/5.752	5.626/0.01				0.1770
					5.641/0.53				0.2060
									0.6601
									0.6852
									0.8450
									1.0281
^{209}At		208.98617	5.4 h	β+, ec/96/3.49		(6+)			Po k x-ray
				α/(4)/5.757	5.647/4.1				0.10422
									0.54503
									0.78189
									0.79020
									(0.1–2.6)
^{210m1}At			5.7 μs	IT/4.028		(19)+			
^{210}At		209.98715	8.1 h	ec/99.8/3.98		5+			Po k x-ray
				α/(0.2)/5.632	5.361/0.05				0.24535
					5.442/0.05				0.52758
									1.18143
									1.43678
									1.48335
									(0.04–2.4)
^{211m1}At			4.2 μs	IT/4.815	0.435	(39/2-)			
^{211}At		210.98750	7.21 h	ec/(58)/0.787		9/2+			Po k x-ray
				α/(42)/5.980	5.211/0.004				0.66956
					5.868/42				0.6870
									0.74263
^{212m2}At			0.15 ms	IT/4.772	0.224	(25-)			
^{212m}At			0.119 s	α/	7.837/65	(9-)			
					7.897/33				
^{212}At		211.99074	0.314 s	α/7.828	7.058/0.4	(1-)			
					7.088/0.6				
					7.618/15				
					7.681/84				
^{213m1}At			0.045 ms	IT	0.306	(49/2+)			
^{213}At		212.99294	0.11 μs	α/9.254	9.080/	9/2-			
^{214m}At			0.76 μs	α/8.762		(9-)			
^{214}At		213.99637	0.56 μs	α/8.987	8.819/100	(1-)			
^{215}At		214.99865	0.10 ms	α/8.178	7.626/0.045	(9/2-)			0.40486
					8.023/99.9				
^{216}At		216.00242	0.30 ms	α/7.947	7.595/0.2	(1-)			
					7.697/2.1				
					7.800/97				
^{217}At		217.00472	32.8 ms	α/7.202	6.812/0.06	(9/2-)	3.8		0.2595
					7.067/99.9				0.3345
									0.5940
^{218}At		218.0087	1.6 s	α/6.883	6.654/6				
					6.695/90				
					6.748/4				
^{219}At		219.01116	50. s	α/6.390	6.275/				
^{220}At		220.0154	3.71 m	β- /3.7					(0.24–0.70)
^{221}At		221.0180	2.3 m	β					
^{222}At		222.0225	0.9 m	β					
^{223}At		223.0252	50. s	β					
^{224}At		224.0297	~ 1.3 m						
^{225}At		225.033							
^{226}At		226.037							
^{227}At		227.040							
^{228}At		228.045							
^{229}At		229.048							
$_{86}$**Rn**									
^{193}Rn		193.0097	1.2 ms	α	7.69/74				0.194
					7.88/26				
^{194}Rn		194.0061	0.8 ms	α	7.70				
^{195m}Rn			5 ms	α	7.56	13/2+			
^{195}Rn		195.0054	6 ms	α	7.54				
^{196}Rn		196.0021	4. ms	α/	7.46	0+			
^{197m}Rn			0.025 s	α	7.36	(13/2+)			

Elem. or Isot.	Natural abundance (Atom %)	Atomic mass or weight	Half-life/ Resonance width (MeV)	Decay mode/ Energy (/MeV)	Particle energy/ Intensity (MeV/%)	Spin (h/2π)	Nuclear magnetic mom. (nm)	Elect. quadr. mom. (b)	γ-Energy/ Intensity (MeV/%)
^{197}Rn		197.0016	0.07 s	α/	7.26				
^{198}Rn		197.9987	64. ms	α	7.205	0+			
^{199m}Rn			0.31 s	α	7.060	(13/2+)			
^{199}Rn		198.9984	0.62 s	α/	6.989	3/2-			
^{200m1}Rn			0.03 ms						
^{200}Rn		199.9957	1.06 s	α/(98)/	6.901/	0+			0.4329
				ec/(2)/5.					0.5043
^{201m}Rn			3.8 s	ec/(10)/		13/2+			
				α/(90)/	6.773/				
^{201}Rn		200.9956	7.0 s	α/(80)/	6.725/	(3/2-)			
				ec/(20)/	α/6.778				
^{202}Rn		201.9933	9.9 s	α/(12)/	6.641/	0+			0.5695
				ec/(88)/					0.288–0.6255
^{203m}Rn			27. s	α/	6.551	13/2+	-0.96	+1.3	
^{203}Rn		202.9934	45. s	α/(66)/6.629	6.499/	3/2-			
				ec/(34)/~ 7.4					
^{204}Rn		203.9914	1.24 m	α/(68)/	6.420/	0+			
				ec/(32)/3.8					
^{205}Rn		204.9917	2.8 m	α/(23)/6.390	6.123(3)/0.02	5/2-	+0.80	+0.06	0.2652
				ec/(77)/5.2	6.262(3)/23				0.3553
									0.4648
									0.6205
									0.6753
									0.7300
^{206}Rn		205.9902	5.7 m	α/(68)/6.384	6.258(3)/	0+			0.06170
				ec/(32)/3.3					0.0968
									0.3245
									0.3862
									0.4822
									0.4973
									0.7728
^{207m}Rn			0.185 ms	IT/0.899	0.234	13/2+			
^{207}Rn		206.9907	9.3 m	β+, ec/77/4.6		5/2-	+0.82	+0.22	At k x-ray
				α/(23)/6.252	5.995(4)/0.02				0.32947
					6.068(3)/0.15				0.34455
					6.126(3)/22.8				0.36767
									0.40267
									0.74723
									(0.18–1.4)
^{208}Rn		207.9896	24.3 m	α/(60)/6.260	5.469(2)/0.003	0+			
				ec/(40)/2.85	6.140(2)/60				
^{209m2}Rn			3.0 μs	IT/3.637	0.479	(35/2+)			
^{209m1}Rn			0.013 ms	IT/1.174	0.376	13/2+			
^{209}Rn		208.9904	29. m	β+ /(83)/3.93	2.16/2.3	5/2-	+0.8388	+0.31	At k x-ray
				α/(17)/	5.887(3)/0.04				0.27933
					5.898(3)/0.02				0.33753
					6.039(2)/16.9				0.40841
									0.68942
									0.74594
									(0.18–3.2)
^{210m1}Rn			1.1 μs	IT/3.812	0.564	(17)-			
^{210}Rn		209.98969	2.4 h	α/(96)/6.157	5.351(2)/0.005	0+			At k x-ray
				ec/(4)/2.37	6.039(2)/96				0.19625
									0.45824
									0.57104
									0.64868
									(0.14–1.7)
^{211}Rn		210.99060	14.6 h	β+, ec/74/2.89		1/2-	+0.60		At k x-ray
				α/(26)/5.964	5.619(1)/0.7				0.16877
					5.784(1)/16.4				0.25022
					5.851(1)/8.8				0.37049
									0.67412
									0.67839
									1.36298
									(0.11–2.7)
^{212}Rn		211.99070	24. m	α/6.385	5.587(4)/0.05	0+			
					6.260(4)/99.95				
^{213m1}Rn			1.4 μs	IT		(31/2-)			
^{213}Rn		212.99388	19. ms	α/8.243	7.552(8)/1.0	9/2+			0.540
					8.087(8)/98.2				
					7.254/0.8				
^{214}Rn		213.9954	0.27 μs	α/9.209	9.037(9)/	0+			
^{215}Rn		214.99875	2.3 μs	α/8.840	8.674(8)/	(9/2+)			

Elem. or Isot.	Natural abundance (Atom %)	Atomic mass or weight	Half-life/ Resonance width (MeV)	Decay mode/ Energy (/MeV)	Particle energy/ Intensity (MeV/%)	Spin (h/2π)	Nuclear magnetic mom. (nm)	Elect. quadr. mom. (b)	γ-Energy/ Intensity (MeV/%)
^{216}Rn		216.00027	45. μs	α		0+			
^{217}Rn		217.00393	0.6 ms	α/7.885	7.500/0.1	9/2+			
					7.742(4)/100				
^{218}Rn		218.00560	33.8 ms	α/7.267	6.534(1)/0.16	0+			0.6093
					7.133(1)/99.8				0.6653
^{219}Rn		219.00948	3.96 s	α/6.946(1)	6.3130(5)/0.05	5/2+	-0.44	+1.0	Po k x-ray
					6.425(3)/7.5				0.13057
					6.5309(4)/0.12				0.27113
					6.5531(3)/12.2				0.40170
					6.8193(3)/81				(0.1–1.075)
^{220}Rn		220.01139	55.6 s	α/6.404	5.7486(5)/0.07	0+			
					6.2883(1)/99.9				
^{221}Rn		221.01554	25. m	α/(22)/6.148	5.778(3)/1.8	7/2+	-0.020	-0.4	Fr L x-ray
				β-/(78)/1.2	5.788(3)/2.2				0.07384
					6.037(3)/18				0.08323
									0.0610
									0.18639
^{222}Rn		222.01758	3.823 d	α/5.590	4.987(1)/0.08	0+			0.510
					5.4897(3)/99.9				
^{223}Rn		223.02189	23. m	β- /		7/2	-0.78	~ +0.80	
^{224}Rn		224.0241	1.8 h	β- /		0+			0.1085
									0.2601
									0.2655
^{225}Rn		225.0285	4.5 m	β- /		7/2-	-0.70	~ +0.84	
^{226}Rn		226.0309	7.4 m	β- /		0+			
^{227}Rn		227.0353	2. s	β- /					
^{228}Rn		228.0378	65. s	β- /		0+			
^{229}Rn		229.0423	12 s						
^{230}Rn		230.045							
^{231}Rn		231.050							
^{232}Rn									
$_{87}$Fr									
^{197}Fr			~ 0.6 ms	α	7.73				
^{198m1}Fr			1. ms	α					
^{198}Fr			15. ms	α					
^{199m}Fr			~ 4.6 ms	α	7.81				
^{199}Fr		199.0073	6. ms	α	7.68				
^{200}Fr		200.0066	49 ms	α	7.47				
^{201m}Fr			~ 8 ms	α/	7.454	(1/2+)			
^{201}Fr		201.0039	~ 60 ms	α/	7.36/	(9/2-)			
^{202m}Fr			0.29 s	α	7.236/	(10-)			
^{202}Fr		202.0033	0.30 s	α/7.590	7.24/100				
^{203m1}Fr			0.043 s			1/2+			
^{203}Fr		203.00094	0.54 s	α/7.280	7.132(5)/	(9/2-)			
^{204m2}Fr			1.7 ms	α	7.01	10(-)			
^{204m1}Fr			2.3 s	α	6.97	7(+)			
^{204}Fr		204.0007	1.8 s	α/	7.03/96	(3+)	+ 4.0	- 0.18	
					6.97/90				
					7.01/74				
^{205m1}Fr			1.15 ms	IT/0.609		1/2+			
^{205}Fr		204.99859	3.9 s	α/7.050	6.914(5)/	(9/2-)	+ 3.8	- 0.35	0.5647/100
									(0.356-0.657)
^{206m}Fr			0.7 s	α/	6.93	10(-)			0.531(IT)
^{206}Fr		205.9987	~ 16 s	α/7.416	6.792(5)/84	7(+)	+ 4.0	- 0.36	
^{207}Fr		206.9969	14.8 s	α/6.900	6.766(5)/	9/2-	+3.9	-0.16	
^{208}Fr		207.9971	59.1 s	α/(77)/6.770	6.636(5)/	7+	+4.8	+ 0.005	
				ec/(23)/6.99					
^{209}Fr		208.9960	50.0 s	α/(89)/5.1	6.646(3)/	9/2-	+3.9	-0.24	0.7978
				ec/(11)/5.16					(0.110–1.384)
^{210}Fr		209.9964	3.2 m	α/6.670/71	6.543(5)/99.87	6+	+4.4	+0.19	0.2030
				ec/6.26	(5.90-6.42)				0.6438
									0.8175
									0.9008
^{211}Fr		210.9956	3.10 m	α/6.660/87	6.534(5)/99.94	9/2-	+4.0	-0.19	0.220
				ec/4.61	(5.87-6.20)				0.2799
									0.5389
									0.9169
^{212m2}Fr			0.024 ms	IT/8.533					
^{212m1}Fr			0.032 ms	IT/1.551	0.162	(11+)			
^{212}Fr		211.9962	20. m	ec/(57)/5.12	6.261(1)/16	5+	+4.6	-0.10	Rn x-ray
				α/(43)/6.529	6.335(1)/4				0.08107
					6.335(1)/4				0.08378
					6.343(1)/1.3				0.2277

Elem. or Isot.	Natural abundance (Atom %)	Atomic mass or weight	Half-life/ Resonance width (MeV)	Decay mode/ Energy (/MeV)	Particle energy/ Intensity (MeV/%)	Spin (h/2π)	Nuclear magnetic mom. (nm)	Elect. quadr. mom. (b)	γ-Energy/ Intensity (MeV/%)
					6.383(1)/10				1.1856
					6.406(1)/9.5				1.2748
					6.08–6.18				0.014–1.178
^{213m1}Fr			3.1 μs	IT/8.095		(65/2-)			
^{213}Fr		212.9962	34.1 s	α/6.905	8.476(4)/51	9/2-	+4.0	-0.14	(0.408-0.577)
^{214m}Fr			3.4 ms	α/	8.547(4)/46	(8-)			
					6.775–8.046				
^{214}Fr		213.9990	5.2 ms	α/8.587	7.409(3)/0.3	(1-)			(0.073-0.966)
					7.605(8)/1.0				
					7.940(3)/1.0				
					8.355(3)/4.7				
					8.427(3)/93				
^{215}Fr		215.00034	0.12 μs	α/9.537	9.360(8)/	(9/2-)			
^{216}Fr		216.00319	0.70 μs	α/9.175	9.005(10)/95				(0.045–0.160)
^{217}Fr		217.00463	0.016 ms	α/8.471	8.315(8)/	(9/2-)			
^{218m}Fr			22. ms	α					
^{218}Fr		218.00758	1. ms	α/8.014	7.384(10)/0.5	(1-)			
					7.542(15)/1.0				
					7.572(10)/5				
					7.732(10)/0.5				
					7.867(2)/93				
^{219}Fr		219.00925	21. ms	α/8.132	6.802(2)/0.25	(9/2-)			
					6.967(2)/0.6				
					7.146(2)/0.25				
					7.313(2)/99				
^{220}Fr		220.01233	27.4 s	α/6.800	6.582(1)/10	1+	-0.67	+0.47	0.0450
					6.630(2)/6				0.061
					6.641(1)/12				0.1060
					6.686(1)/61				0.1539
					6.39–6.58				0.1617
^{221}Fr		221.01426	4.81 m	α/6.457	5.9393(7)/0.17	5/2-	+1.58	-1.0	At k x-ray
					5.9797(7)/0.49				0.0995
					6.0751(7)/0.15				0.21798
					6.1270(7)/				0.4091
					6.2433(3)/1.3				
					6.3410(7)/83.4				
^{222}Fr		222.0176	14.3 m	β- /2.03	1.78/	2-	+0.63	+0.51	
				α/5.850					
^{223}Fr		223.01974	22.0 m	β- /1.149	α/5.291	3/2(-)	+1.17	+1.17	0.1509
				α//0.006	5.314				0.0589
					5.403				0.1453
^{224}Fr		224.0234	3.0 m	β- /2.82		1-	+0.40	+0.517	0.13150
									0.21575
									0.8367
									(0.1–2.21)
^{225}Fr		225.0256	3.9 m	β- /1.87		3/2-	+1.07	+1.3	
^{226}Fr		226.0296	49. s	β- /3.6		1-	+0.071	-1.35	0.18606
									0.25373
^{227}Fr		227.0319	2.48 m	β- /2.5		1/2+	+1.50		
^{228}Fr		228.0358	39. s	β- /~ 3.5		2-	-0.76	+2.4	
^{229}Fr		229.0383	50. s	β- /					
^{230}Fr		230.0424	19. s	β- /		(3)			
^{231}Fr		231.0452	17. s	β- /					
^{232}Fr		232.049	5. s	β- /					(0.0545-0.721)
^{233}Fr		233.053							
$_{88}$Ra									
^{201}Ra		201.013	~ 1.6 ms	α	7.91/				
^{202}Ra		202.0098	~ 0.02 ms	α	7.74	0+			
^{203m}Ra			24 ms	α	7.61	(13/2+)			
^{203}Ra		203.0093	~ 31 ms	α	7.59				
^{204}Ra		204.0065	0.06 s	α	7.48	0+			
^{205m}Ra			~ 0.17 s			(13/2+)			
^{205}Ra		205.0063	0.22 s	α	7.34				
^{206}Ra		206.0038	0.4 s	α/7.416	7.272(5)/	0+			
^{207m1}Ra			0.059 s			(13/2+)			
^{207}Ra		207.0038	1.3 s	α/7.270	7.133(5)/				
^{208}Ra		208.0018	1.1 s	α/7.273/87	7.133(5)/	0+			
^{209m}Ra			0.12 ms	IT/0.882	0.238	(13/2+)			(0.2384-0.644)
^{209}Ra		209.0020	4.6 s	α/7.150	(6.50-7.14)	5/2-	+0.87	+0.40	(0.387-0.634)
^{210m}Ra			2.1 μs	IT/2.051	0.096	8+			(0.0967-0.775)
^{210}Ra		210.0005	3.7 s	α/7.610	7.020(5)/	0+			574.9
^{211m}Ra			9.4 μs	IT/1.198	0.396	(13/2+)			(0.396-0.802)
^{211}Ra		211.0009	13. s	α/7.046	6.907/99.	5/2(-)	+0.878	+0.48	

Elem. or Isot.	Natural abundance (Atom %)	Atomic mass or weight	Half-life/ Resonance width (MeV)	Decay mode/ Energy (/MeV)	Particle energy/ Intensity (MeV/%)	Spin (h/2π)	Nuclear magnetic mom. (nm)	Elect. quadr. mom. (b)	γ-Energy/ Intensity (MeV/%)
				ec/5.0	(6.26-6.79)				(0.120-0.665)
^{212m}Ra			7.1 μs	IT/1.958	0.063	(8)+			(0.440-0.824)
^{212}Ra		211.99979	13.0 s	α/7.033	6.901(2)/	0+			
^{213m}Ra			2.20 ms	IT//99	α/8.47/63	(17/2-)	7.4		0.1612/41
				α//0.6	8.36/33				0.5462/100
					8.27/4				0.1062/97
^{213}Ra		213.0004	2.7 m	ec/(20)/3.88	α/6.225/49	1/2-	+0.613		0.1024
				α/(80)/6.860	6.733/45				0.110
					6.522/5.4				0.215
					6.413/0.22				(0.105-0.511)
^{214m}Ra			0.067 ms	α//0.09	8.95/91	8+			(0.181-1.382)
^{214}Ra		214.00010	2.4 s	α/7.272	7.14/99.8/	0+			0.642
					6.51/0.2				
^{215m2}Ra			1.4 μs	IT/2.247		(29/2-)			
^{215m}Ra			7.3 μs	IT/1.878	0.057	(25/2+)			(0.196-1.048)
^{215}Ra		215.00272	1.64 ms	α/8.864	7.883(6)/2.8	(9/2+)			0.773/100
					8.171(3)/1.4				0.852/74
					8.700(3)/95.9				0.055–1.048
^{216}Ra		216.0035	0.18 μs	α/9.526	9.349(8)/	0+			
^{217}Ra		217.0063	1.6 μs	α/9.161	8.992(8)/	9/2-			
^{218}Ra		218.0071	26. μs	α/8.547	8.390(8)/	0+			
^{219}Ra		219.0101	0.010 s	α/8.132	7.680(10)/65				
					7.982(9)/35				
^{220}Ra		220.0110	18. ms	α/7.593	7.39/5	0+			0.465
					7.45/95				
^{221}Ra		221.01392	29. s	α/6.879	6.254(10)/0.7	5/2+	-0.180	+2.0	
					6.578(5)/3				
					6.585(3)/8				
					6.608(3)/35				
					6.669(3)/21				
					6.758(3)/31				
^{222}Ra		222.01538	33.6 s	α/5.590	6.237(2)/3.0	0+			0.324
					6.556(2)/97				0.145–0.8402
^{223}Ra		223.01850	11.45 d	α/5.979	5.287(1)/0.15	3/2+	+0.271	+1.25	Rn k x-ray
					5.338(1)/0.13				0.12231
					5.365(1)/0.13				0.14418
					5.433(5)/2.3				0.15418
					5.502(1)/1.0				0.15859
					5.540(1)/9.2				0.26939
					5.607(3)/24				0.32388
					5.716(3)/52				0.33328
					5.747(1)/9				0.44494
					5.857(1)/0.32				(0.10–0.7)
					5.872(1)/0.85				
^{224}Ra		224.02021	3.66 d	α/5.789	5.034(10)/0.003	0+			Rn k x-ray
					5.047(1)/0.007				0.2407
					5.164(5)/0.007				0.4093
					5.449(2)/4.9				0.6501
					5.685(2)/95				
^{225}Ra		225.02361	14.9 d	β- /0.36	0.32/100	1/2+	-0.734		Ac k x-ray
				α	5.01 ×10^{-5}				0.0434
					4.98 ×10^{-6}				
^{226}Ra		226.02541	1599. a	α/4.870	4.194(1)/0.001	0+			Rn k x-ray
			> 4 ×10^{18} a	sf/4 ×10^{-14}	4.343(1)/0.006				0.1861/3.64
					4.601(1)/6.16				0.2624
					4.784(1)/93.8				0.053–2.448
^{227}Ra		227.02918	42. m	β- /1.325	1.03/	3/2+	-0.404	+1.5	Ac L x-ray
					1.30/				Ac k x-ray
									0.02739
^{228}Ra		228.03107	5.76 a	β- /0.046	0.039/50	0+			0.0135
				βf//5 ×10^{-12}	0.014/30				(0.006–0.031)
					0.026/20				
^{229}Ra		229.0349	4.0 m	β- /1.76	1.76/	5/2+	+0.503	+3.1	0.0145–0.172
^{230}Ra		230.0371	1.5 h	β- /1.0	0.7/	0+			0.0631
									0.0720
									0.2028
									0.4698
									0.4787
^{231m1}Ra			~ 0.053 ms	IT	0.066	(1/2+)			
^{231}Ra		231.0410	1.7 m	β-					(0.018-1.155)
^{232}Ra		232.0435	4. m	β-		0+			
^{233}Ra		233.0476	30. s	β-					
^{234}Ra		234.0503	~ 30. s	β-/		0+			

Nuclear

Elem. or Isot.	Natural abundance (Atom %)	Atomic mass or weight	Half-life/ Resonance width (MeV)	Decay mode/ Energy (/MeV)	Particle energy/ Intensity (MeV/%)	Spin (h/2π)	Nuclear magnetic mom. (nm)	Elect. quadr. mom. (b)	γ-Energy/ Intensity (MeV/%)
$_{89}$Ac									
^{205}Ac			~ 0.02 s	α	7.94	(9/2-)			
^{206m}Ac			~ 0.03 s	α	7.79	(10-)			
^{206}Ac		206.0145	~ 26 ms	α	7.75				
^{207}Ac		207.0120	27 ms	α/	7.69				
^{208m}Ac			~ 25. ms	α/	7.72	(10-)			
^{208}Ac		208.0116	~ 0.1 s	α/	7.62				
^{209}Ac		209.0095	~ 0.10 s	α/	7.58				
^{210}Ac		210.0094	0.34 s	α/7.610	7.462(8)/				
^{211}Ac		211.0077	0.20 s	α/7.620	7.480(8)/				
^{212}Ac		212.0078	0.9 s	α/7.520	7.379(8)/				
^{213}Ac		213.0066	0.73 s	α/7.500	7.364(8)/	(9/2-)			
^{214}Ac		214.0069	8.2 s	α/(86)/7.350	7.215/54	(5+)			(0.0626-0.754)
				ec/(14)/6.34	7.081/42				
					(6.48-7.15)				
^{215}Ac		215.0065	0.17 s	α/7.750	7.60/99.57	(9/2-)			0.399
					7.21/0.46				0.582
					7.03/0.20				0.654
					6.96/0.14				
^{216m}Ac			0.44 ms	α/	8.198(8)/1.7	(9-)			(0.0826-1.375)
					8.283(8)/2.5				
					9.028(5)/49				
					9.106(5)/46				
^{216}Ac		216.0087	44. ms	α/9.241	8.990(2)/10	(1-)			
					9.070(8)/90				
^{217m}Ac			0.7 μs	α/	10.540/100	(29/2)+			
^{217}Ac		217.0093	0.07 μs	α/9.832	9.650(10)/100	9/2-			
^{218}Ac		218.0116	1.1 μs	α/9.380	9.205(15)/				
^{219}Ac		219.0124	0.012 ms	α/8.830	8.664(10)/	(9/2-)			
^{220}Ac		220.01476	26. ms	α/8.350	7.610(20)/23				
					4.680(20)/21				
					7.790(10)/13				
					7.850(10)/24				
					7.985(10)/4				
					8.005(10)/5				
					8.060(10)/6				
					8.195(10)/3				
^{221}Ac		221.0156	52. ms	α/7.790	7.170(10)/2				
					7.375(10)/10				
					7.440(15)/20				
					7.645(10)/70				
^{222m}Ac			63. s	α/(>89)/	6.710(20)/7				
				ec/(1)/	6.750(20)/13				
				IT/(<10)/	6.810(20)/24				
					6.840(20)/9				
					6.890(20)/13				
					6.970(20)/7				
					7.000(20)/13				
^{222}Ac		222.01784	5. s	α/7.141	6.967(10)/6	1-			
					7.013(2)/94				
^{223}Ac		223.01914	2.1 m	α/(99)/6.783	6.131(2)/0.12	(5/2-)			0.0725
				ec/(1)/0.59	6.177(2)/0.94				0.0839
					6.293(1)/0.47				0.0927
					6.326(1)/0.3				0.0990
					6.332(2)/0.14				0.1917
					6.360(1)/0.22				0.2158
					6.397(1)/0.13				0.3588
					6.448(1)/0.2				0.4768
					6.473(1)/3.1				
					6.523(2)/0.6				
					6.528(1)/3.1				
					6.563(1)/13.6				
					6.582(3)/0.3				
					6.646(1)/44				
					6.661(1)/31				
^{224}Ac		224.02172	2.7 h	ec/(90)/1.403	5.841(1)/0.5	0-			Ra L x-ray
				α/(10)/6.323	5.860(1)/0.75				Ra k x-ray
					5.875(1)/1.7				0.08426
					5.941(1)/4.4				0.13150
					6.000(1)/6.7				0.1571
					6.013(1)/1.4				0.21575
					6.056(1)/22				0.2619
					6.138(1)/26				(0.03–0.3)

Elem. or Isot.	Natural abundance (Atom %)	Atomic mass or weight	Half-life/ Resonance width (MeV)	Decay mode/ Energy (/MeV)	Particle energy/ Intensity (MeV/%)	Spin (h/2π)	Nuclear magnetic mom. (nm)	Elect. quadr. mom. (b)	γ-Energy/ Intensity (MeV/%)
					6.154(1)/1.0				
					6.204(1)/12				
					6.210(1)/20				
^{225}Ac		225.02323	9.920 d	α/5.935	5.286(1)/0.2	3/2			Fr k x-ray
					5.444(3)/0.1				0.06296/0.48
					5.554(1)/0.1				0.09982/1.36
					5.608(1)/1.1				0.1084
					5.636(1)/4.5				0.1116
					5.681(1)/1.4				0.1451
					5.722(1)/2.9				0.150/0.691
					5.731(1)/10				0.15724
					5.791(1)/9				0.18795/0.54
					5.793(1)/18				0.0075–0.809
^{226}Ac		226.02610	1.224 d	ec/(17)/0.640		(1-)			Ra k x-ray
				β- /(83)/1.116					Th k x-ray
				α/(0.006)/5.51	5.399(5)/0.006				0.07218
									0.15816
									0.23034
^{227}Ac		227.02775	21.77 a	β- /98.6/0.045	0.045/54	3/2-	+1.1	+1.7	0.0838/23.
				α/(1.4)/5.043	4.869(1)/0.09				0.0811/14.
					4.938(1)/0.52				0.2696/13.
					4.951(1)/0.65				(0.044–1.27)
^{228}Ac		228.03102	6.15 h	β- /2.127	1.11/32	3+			Th L x-ray
					1.85/12				Th k x-ray
					2.18/11				0.12903
									0.33842
									0.91116
									0.96897
									(0.2–1.96)
^{229}Ac		229.0330	1.04 h	β- /1.10	1.1/	(3/2+)			0.09335/2.43
									0.16451/2.61
									0.56916/2.24
									0.0111–0.898
^{230}Ac		230.0363	2.03 m	β- /2.7	1.4/	1+			Th k x-ray
				β-, sf	/0.000119				0.45497
									0.50820
									(0.12–2.5)
^{231}Ac		231.0384	7.5 m	β- /2.1	2.1/100	(1/2+)			0.14379
									0.18574
									0.22140
									0.28250
									0.3070
^{232}Ac		232.0420	2.0 m	β- /3.7		(2-)			
^{233}Ac		233.0443	2.4 m	β- /		(1/2+)			
^{234}Ac		234.0481	40. s	β- /		(1+)			
^{235}Ac		235.0508							
^{236}Ac		236.0550	~ 1.2 m						
$_{90}$Th		**232.0377(4)**							
^{208}Th		208.0179	~ 1.7 ms	α	8.04				
^{209}Th		209.0178	~ 2.5 ms	α	8.08				
^{210}Th		210.0151	16. ms	α	7.90	0+			
^{211}Th		211.0149	0.04 s	α	7.79				
^{212}Th		212.0130	32. ms	α/	7.80/	0+			
^{213m1}Th			1.4 μs	IT/1.180	0.381	(13/2)+			
^{213}Th		213.0130	0.14 s	α/7.840	7.692(10)/				
^{214m1}Th			1.2 μs	IT/2.181	0.089	(8+)			
^{214}Th		214.0115	0.10 s	α/7.825	7.677(10)/	0+			
^{215}Th		215.0117	1.2 s	α/7.660	7.33(10)/8	(1/2-)			0.134
					7.395(8)/52				0.192
					7.524(8)/40				(0.069-0.295)
^{216m}Th			0.13 ms	α	9.93/74				(0.0905-1.478)
					8.00, 9.31				
^{216}Th		216.0111	27. ms	α/8.071	7.92/99.46	0+			0.628
					7.30/0.54				
^{217m1}Th			0.07 ms			(25/2+)			
^{217}Th		217.0131	0.25 ms	α/9.424	9.27/94.6				(0.546-0.822)
					8.46/3.8				
					8.73/1.6				
^{218}Th		218.0133	0.11 μs	α/9.847	9.665(10)/	0+			
^{219}Th		219.0155	1.01 μs	α/9.510	9.340(20)/				
^{220}Th		220.0157	10. μs	α/8.953	8.790(20)/	0+			
^{221}Th		221.0182	2. ms	α/8.628	7.732/7				
					8.142/72				

Elem. or Isot.	Natural abundance (Atom %)	Atomic mass or weight	Half-life/ Resonance width (MeV)	Decay mode/ Energy (/MeV)	Particle energy/ Intensity (MeV/%)	Spin (h/2π)	Nuclear magnetic mom. (nm)	Elect. quadr. mom. (b)	γ-Energy/ Intensity (MeV/%)
					8.469/21				
^{222}Th		222.0185	2.24 ms	α/8.129	7.980/97.7	0+			
					7.599/2.3				
^{223}Th		223.0208	0.60 s	α/7.454	7.29(1)/41(5)				
					7.32(1)/29(5)				
					7.350(15)/20(5)				
					7.390(15)/10(4)				
^{224}Th		224.0215	1.05 s	α/7.305	6.768(5)/1.2	0+			
					6.997(5)/19				
					7.170(5)/7				
^{225}Th		225.02395	8.72 m	ec/(10)/0.68		(3/2+)			0.3214
				α/(90)/6.920	6.441(2)/15				(0.053-0.486)
					6.479(2)/43				
					6.501(3)/14				
					6.627(3)/3				
					6.650(5)/3				
					6.700(5)/2				
					6.743(3)/7				
					6.796(2)/9				
^{226}Th		226.02490	30.70 m	α/6.454	6.026(1)/0.2	0+			Ra k x-ray
					6.041(1)/0.19				0.1112
					6.098(1)/1.3				0.2421
					6.2283(4)/23				0.1310
					6.3375(4)/75				0.1733–0.9295
^{227}Th		227.02770	18.70 d	α/6.146		(3/2+)			Ra L x-ray
									Ra k x-ray
									0.05014
									0.23597
									0.25624
									(0.02–1.0)
^{228}Th		228.02874	1.913 a	α/5.520	5.1770(2)/0.18	0+			
					5.2114(1)/0.4				
					5.3405(1)/26.7				
					5.4233(1)/73				
^{229m}Th			13.9 h	α	4.83-5.08	(3/2+)			
^{229}Th		229.03176	7.93 $\times 10^{3}$ a	α/5.168	4.814/9.3	5/2+	+0.46	+4.	0.1935/4.3
					4.845(5)/56				0.21089/277
					4.9008(5)/10.2				0.13697/1.21
					4.689–5.077				0.011–0.6036
^{230}Th	0.02(02)	230.03313	7.56 $\times 10^{4}$ a	α/4.771	4.4383(6)/0.03	0+			0.0677/0.46
					4.4798(6)/0.12				0.1439/0.078
			> 2 $\times 10^{18}$ a	sf// < 4 $\times 10^{-12}$	4.6211(6)/23.4				
					4.6876(6)/76.3				
^{231}Th		231.03630	1.063 d	β- /0.390	0.138/22	5/2+			Pa L x-ray
					0.218/20				Pa k x-ray
					0.305/52				0.02564
									0.084203/
									(0.02–0.3)
^{232}Th	99.98(02)	232.03806	1.40 $\times 10^{10}$ a	α/4.081	3.830(10)/0.2	0+			0.0590
			1.2 $\times 10^{21}$ a	sf/1.1 $\times 10^{-9}$	3.952(5)/23				0.124
					4.010(5)/77				
^{233}Th		233.04158	22.3 m	β-/1.245	1.245/	½+			Pa L x-ray
									Pa k x-ray
									0.02938
									0.08653
									0.45930
									(0.02–1.2)
^{234}Th		234.04360	24.10 d	β- /0.273	0.102/20	0+			Pa L x-ray
					0.198/72				0.06329/4.1
									0.09235/2.4
									0.09278/2.4
^{235}Th		235.0473	7.2 m	β- /1.9					0.4162
									0.6594
									0.7272
									0.747
									0.9318
^{236}Th		236.0497	37.5 m	β- /~ 1.0		0+			Pa k x-ray
									0.1107
^{237}Th		237.0536	5.0 m	β-					
^{238}Th		238.057	9.4 m			0+			0.0890
$_{91}$Pa		**231.03588(2)**							
^{212}Pa		212.0232	~ 5 ms	α	8.27				
^{213}Pa		213.0211	7 ms	α	8.24				

Elem. or Isot.	Natural abundance (Atom %)	Atomic mass or weight	Half-life/ Resonance width (MeV)	Decay mode/ Energy (/MeV)	Particle energy/ Intensity (MeV/%)	Spin (h/2π)	Nuclear magnetic mom. (nm)	Elect. quadr. mom. (b)	γ-Energy/ Intensity (MeV/%)
^{214}Pa		214.0209	17 ms	α	8.12				
^{215}Pa		215.0192	15. ms	α	8.08/100				
^{216}Pa		216.0191	0.19 s	α/	7.95/51				0.134
					7.82/45				
					7.79/4				
^{217m}Pa			1.2 ms	α/	10.16/72				0.4504–0.821
					8.306/11				
					9.55/6				
					9.69/2				
^{217}Pa		217.0183	3.8 ms	α/8.490	8.337/99				0.0466–0.634
					7.873/0.4				
					7.728/0.3				
					7.710/0.3				
^{218}Pa		218.0201	0.12 ms	α/	9.54/31				0.092
					9.61/69				
^{219}Pa		219.0199	0.05 μs	α					
^{220}Pa		220.0217	0.8 μs	α					
^{221}Pa		221.0219	6. μs	α	9.08(3)				
^{222}Pa		222.0238	~ 4.3 ms	α/8.700	8.180/50				
					8.330/20				
					8.540/30				
^{223}Pa		223.0240	~ 6.5 ms	α/8.340	8.006(10)/55				
					8.196(10)/45				
^{224}Pa		224.02562	0.84 s	α/7.630	7.555(10)/75(3)				0.1945
					7.46(1)/25(3)				(0.028–0.412)
^{225}Pa		225.0261	1.8 s	α/7.380	7.195(10)/30				
					7.245(10)/70				
^{226}Pa		226.0279	1.8 m	α/(74)/6.987	6.728(10)/0.7				
				ec/(26)/2.83	6.823(10)/35				
					6.863(10)/39				
^{227}Pa		227.02881	38.3 m	α/(85)/6.582	6.357(4)/7	(5/2-)			0.0649
				ec/(15)/1.02	6.376(10)/2.2				0.0669
					6.401(4)/8				0.1100
					6.416(4)/13				
					6.423(10)/10				
					6.465(4)/43				
^{228}Pa		228.03105	22. h	ec/(98)/2.111		(3+)	+3.5		Th k x-ray
				α/(2)	5.779/0.23				0.409/100
					5.805/0.15				0.4631/222
					6.078/0.4				0.91116/242
					6.105/0.25				0.96464/120
					6.118/0.22				0.96897/149
									0.058–1.96
^{229}Pa		229.03210	1.5 d	ec/(99.8)/0.32		(5/2+)			0.04244
				α/(0.2)/5.836	5.536(2)/0.02				(0.024–0.18)
					5.579(2)/0.09				
					5.668(2)/0.05				
^{230}Pa		230.03454	17.4 d	ec/(90)/1.310	0.51/	(2-)	2.0		Th L x-ray
				β-/(10)/0.563					Th k x-ray
									0.4437
									0.45477
									0.89876
									0.91856
									0.95199
									(0.053–1.07)
^{231}Pa	100	231.03588	3.25 ×10^{4} a	α/5.148	4.6781(5)/1.5	3/2-	2.01	-1.7	Ac L x-ray
					4.7102(5)/1.0				Ac k x-ray
			> 2 ×10^{17} a	sf/< 1.6 ×10^{-15}	4.7343(5)/8.4				0.01899
					4.8513(5)/1.4				0.027396
					4.9339(5)/3				0.03823
					4.9505(5)/22.8				0.04639
					4.9858(5)/1.4				0.25586
					5.0131(5)/25.4				0.26029
					5.0292(5)/20				0.28367
					5.0318(5)/2.5				0.30007
					5.0587(5)/11				0.30264
									0.33007
									(0.02–0.61)
^{232m}Pa			0.10 d						
^{232}Pa		232.03859	1.31 d	β- /1.34		(2-)			U k x-ray
									0.10900
									0.15009
									0.89439

Elem. or Isot.	Natural abundance (Atom %)	Atomic mass or weight	Half-life/ Resonance width (MeV)	Decay mode/ Energy (/MeV)	Particle energy/ Intensity (MeV/%)	Spin (h/2π)	Nuclear magnetic mom. (nm)	Elect. quadr. mom. (b)	γ-Energy/ Intensity (MeV/%)
									0.96934
									(0.10–1.17)
^{233}Pa		233.04025	26.97 d	β-/0.571	0.15/40	3/2-	+4.0	-3.0	U L x-ray
					0.256/60				U k x-ray
									0.30017
									0.31201/38.4
									(0.0286-0.456)
^{234m}Pa			1.16 m	β- /99.9/2.29		(0-)			U k x-ray
				IT/0.13/					0.25818/0.07
									0.76641/0.32
									1.0009/0.86
									(0.06–1.96)
^{234}Pa		234.04331	6.69 h	β- /2.197	0.51/	(4+)			U L x-ray
									U k x-ray
									0.1312/0.03
									0.5695/0.02
									0.9256/0.02
									(0.02–1.99)
^{235}Pa		235.0454	24.4 m	β- /1.41	1.4/97	(3/2-)			0.0308–0.659
^{236}Pa		236.0487	9.1 m	β- /2.9	1.1/40	(1-)			U k x-ray
					2.0/50				0.64235
					3.1/10				0.68759
									1.7630
									(0.04–2.18)
^{237}Pa		237.0510	8.7 m	β- /2.3	1.1/60	(1/2+)			0.4986
					1.6/30				0.5293
					2.3/10				0.5407
									0.8536
									0.8650
									(0.04–1.4)
^{238}Pa		238.0546	2.3 m	β- /3.5	1.2/	(3-)			0.10350
					1.7/				0.1785
									0.4484
									0.6350
									0.6800
									1.01446
									(0.04–2.5)
^{239}Pa		239.057	1.8 h						
^{240}Pa		240.061							
$_{92}$U		**238.02891(3)**							
^{215}U			~ 0.7 ms	α	8.43				
^{216m1}U			~ 0.7 ms	α		(8+)			
^{216}U			~ 0.4 ms	α	8.34				
^{217}U		217.025	~ 0.2 ms	α	8.02				
^{218m}U			~ 0.56 ms	α	10.68	(8+)			
^{218}U		218.0235	0.5 ms	α	8.61	0+			
^{219}U		219.0250	~ 0.08 ms	α	9.77				
^{221}U			~ 0.7 µs	α	9.7				
^{222}U		222.026	~ 4.7 µs	α	9.31	0+			
^{223}U		223.0277	0.02 s	α/	8.78(4)/				
^{224}U		224.0276	~ 1. ms	α/	8.46/100	0+			
^{225}U		225.0294	84. ms	α/	7.87/83				
					7.82/15				
					7.63/2				
^{226}U		226.0293	0.26 s	α/7.560	7.56/86	0+			
					7.38/14				
^{227}U		227.0312	1.1 m	α/7.200	6.870/				
^{228}U		228.0314	9.1 m	α/6.803	6.404(6)/0.6	0+			0.095
					6.440(5)/0.7				0.152
					6.589(5)/29				0.187
					6.681(6)/70				0.246
^{229}U		229.03351	58. m	ec/(80)/1.31	6.223/3	(3/2+)			
				α/(20)/6.473	6.297(3)/11				
					6.332(3)/20				
					6.360(3)/64				
^{230}U		230.03394	20.23 d	α/5.992	5.5866(3)/0.01	0+			Th L x-ray
			> 4 $\times 10^{10}$ a	sf/< 10^{-10}	5.6624(3)/0.26				0.07218
					5.6663(3)/0.38				0.15421
					5.8178(3)/32				0.23034
					5.8887(3)/67				(0.081–0.8565)
^{231}U		231.03629	4.2 d	ec/0.36		(5/2-)			Pa L x-ray
				α/(10^{-3})	5.46/1.6 $\times 10^{-3}$				Pa k x-ray
					5.47/1.4 $\times 10^{-3}$				0.02564

Elem. or Isot.	Natural abundance (Atom %)	Atomic mass or weight	Half-life/ Resonance width (MeV)	Decay mode/ Energy (/MeV)	Particle energy/ Intensity (MeV/%)	Spin (h/2π)	Nuclear magnetic mom. (nm)	Elect. quadr. mom. (b)	γ-Energy/ Intensity (MeV/%)
					5.40/1. ×10^{-3}				0.08420
^{232}U		232.03716	70. a	α/5.414	4.9979(1)/0.003	0+			
			2.6 ×10^{15} a	sf/2.7 ×10^{-12}	5.1367(1)/0.3				
					5.2635(1)/31				
					5.3203(1)/69				
^{233}U		233.03964	1.590 ×10^{5} a	α/4.909	4.7830(8)/13.2	5/2+	+0.59	3.66	Th L x-ray
			>2.7 ×10^{17} a	sf/6 ×10^{-11}	4.8247(8)/84.4				0.04244
					4.510–4.804				0.09714
									(0.0252–1.119)
^{234m1}U			0.034 ms	IT/1.421		6-			
^{234}U	0.0054(5)	234.04095	2.453 ×10^{5} a	α/4.856	4.604(1)/0.24	0+			0.05323/0.156
			1.5 ×10^{16} a	sf/1.6 ×10^{-9}	4.7231(1)/27.5				0.12091
					4.776(1)/72.5				
^{235m2}U			4 ms						
^{235m}U			~ 26. m	IT	0.077	1/2+			
^{235}U	0.7204(6)	235.04393	7.03 ×10^{8} a	α/4.6793	4.1525(9)/0.9	7/2-	-0.38	4.94	Th L x-ray
			1.0 ×10^{19} a	sf/7 ×10^{-9}	4.2157(9)/6.				Th k x-ray
					4.3237(9)/4.6				0.10917
					4.3641(9)/19.				0.14378/0.134
					4.370(4)/6				0.16338/0.067
					4.3952(9)/57.				0.18574/0.806
					4.4144(9)/2.1				0.1949/0.009
					4.5025(9)/1.7				0.20533/0.774
					4.5558(9)/4.2				0.2214/0.0014
					4.5970(9)/4.8				(0.03–0.79)
^{236}U		236.04557	2.342 ×10^{7} a	α/4.569	4.332(8)/0.26	0+			Th L x-ray
			2.5 ×10^{16} a	sf// 9 ×10^{-8}	4.445(5)/26				0.04946/100
					4.494(3)/74				0.11279/24.1
									0.17115/0.080
^{237}U		237.04873	6.75 d	β- /0.519	0.24/	1/2+			Np L x-ray
					0.25/				Np k x-ray
									0.05953
									0.20801
^{238}U	99.2742(10)	238.05079	4.47 ×10^{9} a	α	4.0395/0.23	0+			Th L x-ray
			8.2 ×10^{15} a	sf// 5 ×10^{-5}	4.147(5)/23				0.04955/.06
					4.196(5)/77				0.1135/.01
^{239}U		239.05429	23.5 m	β- /1.265	1.2/	5/2+			(0135-1.102)
					1.3/				
^{240}U		240.05659	14.1 h	β- /0.39	0.36/	0+			Np L x-ray
									0.04410
									0.05558
									0.06760
^{242}U		242.063	16.8 m	β- /~ 1.2		0+			
$_{93}$**Np**									
^{219}Np			< 0.005 ms	> 9					
^{225}Np		225.0339	> 2 μs						
^{226}Np		226.035	0.03 s	α/	8.04(2)/				
^{227}Np		227.0350	0.51 s	α/	7.65(2)/				
					7.68(1)/				
^{228}Np		228.036	61. s	ec/60(7)/					
				α/40(7)/, sf					
^{229}Np		229.036	4.0 m	α/7.010	6.890(20)				
^{230}Np		230.0378	4.6 m	ec/97 /3.6					
				α/3	6.660(20)				
^{231}Np		231.0383	48.8 m	ec/98 /1.8		5/2			0.2629
				α/2 /6.368	6.280/2				0.3475
									0.3703
^{232}Np		232.040	14.7 m	ec/99 /2.7		(4-)			U L x-ray
									U k x-ray
									0.3268
									0.81925
									0.86683
^{233}Np		233.0407	36.2 m	ec/1.2		(5/2+)			U L x-ray
									U k x-ray
									0.29887
									0.31201
^{234}Np		234.0429	4.4 d	β+, ec/1.81	0.79/	(0+)			U L x-ray
									U k x-ray
									1.5272
									1.5587
									1.6022
^{235}Np		235.04406	1.085 a	ec/99.9 /0.124		5/2+			U k x-ray
				α/0.001/5.191					

Elem. or Isot.	Natural abundance (Atom %)	Atomic mass or weight	Half-life/ Resonance width (MeV)	Decay mode/ Energy (/MeV)	Particle energy/ Intensity (MeV/%)	Spin (h/2π)	Nuclear magnetic mom. (nm)	Elect. quadr. mom. (b)	γ-Energy/ Intensity (MeV/%)
^{236m}Np			22.5 h	ec/52 /		(1-)			U L x-ray
				β- /48 /					Pu L x-ray
									U k x-ray
									0.64235
									0.68759
^{236}Np		236.0466	1.55 $\times 10^{5}$ a	ec/91 /0.94		(6-)			U L x-ray
				β- /9 /0.49					U k x-ray
									0.10423
									0.16031
^{237}Np		237.04817	2.14 $\times 10^{6}$ a	α/4.957	4.6395(5)/6.5	5/2+	+3.14	+3.87	Pa L x-ray
			1 $\times 10^{18}$ a	sf/2.1 $\times 10^{-10}$	4.766(5)/9.7				Pa k x-ray
					4.7715(5)/22.7				0.029378/15
					4.7884(5)/47.8				0.08653/12
					4.558–4.873				(0.03–0.28)
^{238}Np		238.05095	2.117 d	β- /1.292	1.2/	2+			Pu L x-ray
									Pu k x-ray
									0.98447/25.2
									1.02855/18.3
									(.044–1.026)
^{239}Np		239.05294	2.355 d	β- /0.722	0.341/30	5/2+			Pu L x-ray
					0.438/48				Pu k x-ray
									0.10613
									0.228186/11
									0.27760/15
									(0.04–0.50)
^{240m}Np			7.22 m	β- /99.9 /	2.18/	(1+)			0.25143
				IT/0.1 /					0.26333
									0.55454
									0.59735
^{240}Np		240.0562	1.032 h	β- /2.20	0.89/	5+			0.1471/
									0.5664
									0.6008
^{241}Np		241.0583	13.9 m	β- /1.3	1.3/	5/2+			0.1330/
									0.1740
^{242m}Np			5.5 m	β- /	2.7	(6+)			0.6209
									(0.04-2.37)
^{242}Np		242.062	2.2 m	β-		(1+)			0.1591
									(0.265-0.9448)
^{243}Np		243.0643	1.9 m						
^{244}Np		244.068	2.3 m						
$_{94}$Pu									
^{228}Pu		228.0387	~ 1.1 s	α/	7.81(2)/	0+			
^{229}Pu		229.0401	1.1 m	α/50	7.46/				
				ec/50					
				sf < 7					
^{230}Pu		230.0397	1.7 m	α/	7.06/81	0+			
					7.00/19				
^{231}Pu		231.0411	8.6 m	ec/90					
				α/10	6.72				
^{232}Pu		232.0412	34. m	ec/>80/1.1		0+			
				α/<20/6.716	6.542(10)/38				
					6.600(10)/62				
^{233}Pu		233.0430	20.9 m	ec(99.9)/1.9					0.1503
				α/0.1 /6.416	6.300(20)/0.1				0.1804
									0.2353
									0.5002
									0.5346/
									1.0352/
^{234}Pu		234.04332	8.8 h	ec/94 /0.39		0+			
				α/6 /6.310	6.035(3)/0.024				
					6.149(3)/1.9				
					6.200(3)/4.				
^{235}Pu		235.0453	25.3 m	ec/99+ /1.2		(5/2+)			
				α/0.003/5.957	5.850(20)/0.003				
^{236m}Pu			1.2 µs	IT/1.185		5-			
^{236}Pu		236.04606	2.87 a	α/5.867	5.611/0.21	0+			0.0476/0.07
			1.5 $\times 10^{9}$ a	sf/1.9 $\times 10^{-7}$	5.7210/30.5				0.109/0.02
					5.7677(1)/69.3				(0.17–0.97)
^{237m2}Pu			1.1 µs						
^{237m1}Pu			0.18 s	IT	0.146	1/2+			
^{237}Pu		237.04841	45.7 d	ec/99.9 /0.220		7/2-			Np L x-ray
				α/0.003 /5.747	5.334(4)/0.0015				Np k x-ray
					5.356(4)/0.0006				0.026344

Nuclear

Elem. or Isot.	Natural abundance (Atom %)	Atomic mass or weight	Half-life/ Resonance width (MeV)	Decay mode/ Energy (/MeV)	Particle energy/ Intensity (MeV/%)	Spin (h/2π)	Nuclear magnetic mom. (nm)	Elect. quadr. mom. (b)	γ-Energy/ Intensity (MeV/%)
					5.650(4)/0.0007				0.03319
									0.05954
									(0.03–0.5)
^{238}Pu		238.04956	87.7 a	α/5.593	5.3583(1)/0.10	0+			U k x-ray
			4.75×10^{10} a	sf/1.8×10^{-7}	5.465(1)/28.3				0.04347
					5.4992(1)/71.6				(0.04–1.1)
^{239m1}Pu			0.008 ms			(5/2+)			
^{239}Pu		239.05216	2.410×10^{4} a	α/5.244	5.055/0.047	1/2+	+0.203		U k x-ray
			$8. \times 10^{15}$ a	sf/3×10^{-10}	5.076/0.078				0.05162
					5.106/11.9				0.05682
					5.144/17.1				0.12928
					5.157/70.8				0.37502
					(4.74 –5.03)				0.41369
^{240}Pu		240.05381	6.56×10^{3} a	α/5.255	5.0212(1)/0.07	0+			U L x-ray
			1.14×10^{11} a	sf/5.7×10^{-6}	5.1237(1)/26.4				0.04524
					5.1681(1)/73.5				0.10423
					(4.492-4.863)				(0.04–0.97)
^{241m1}Pu			0.021 ms						
^{241}Pu		241.05685	14.33 a	β-/99+/0.0208	4.853/3×10^{-4}	5/2+	-0.68	+6.	0.14854
				α/0.002 /5.139	4.897/0.002				0.1600
			< 6×10^{16} a	sf/> 2.4×10^{-14}					
^{242}Pu		242.05874	3.75×10^{5} a	α/4.983	4.7546(7)/0.098	0+			U L x-ray
			6.77×10^{10} a	sf/5.5×10^{-4}	4.8564(7)/22.4				0.04491
					4.9006(7)/78				0.10350
^{243}Pu		243.06200	4.956 h	β- /0.582	0.49/21	7/2+			Am L x-ray
					0.58/60				0.0417
									0.0839
^{244}Pu		244.06421	8.12×10^{7} a	α/99.9/4.665	4.546(1)/19.4	0+			U L x-ray
			6.6×10^{10} a	sf/0.12	4.589(1)/80.5				0.0439
^{245}Pu		245.0678	10.5 h	β- /1.21	0.93/57	(9/2-)			Am L x-ray
					1.21/11				Am k x-ray
									0.2804 /
									0.30832
									0.32752
									0.56014
									(0.03–1.2)
^{246}Pu		246.0702	10.85 d	β- /0.40	0.150/85	0+			Am L x-ray
					0.35/10				Am k x-ray
									0.04379
									0.22371
^{247}Pu		247.074	2.3 d						
$_{95}$Am									
^{223}Am			~ 5 ms						
^{229}Am			0.9 s	α	7.99				
^{230}Am		230.046							
^{232}Am		232.046	0.9 m	ec/~ 5.0					
^{233}Am		233.046	~ 3.2 m	α	6.78				
^{234}Am		234.048	2.3 m	ec/4.2					
^{235}Am		235.0479	10.3 m	ec					Pu K x-ray
				α	6.46/0.4				0.291/100
									(0.170-0.828)
^{236m}Am			2.9 m			(1-)			(0.583-0.713)
^{236}Am		236.049	3.6 m	ec		(5-)			(0.158-1.038)
^{237}Am		237.0500	1.22 h	ec/99.98 /1.7		(5/2-)			Pu k x-ray
				α/0.02 /6.20	6.042(5)/0.02				0.14559
									0.28026
									0.43845
^{238m1}Am			0.035 ms						
^{238}Am		238.0520	1.63 h	ec/2.26		1+			Pu L x-ray
				α/0.0001 /6.04	5.940/0.0001				Pu k x-ray
									0.91870
									0.96278
^{239}Am		239.05302	11.9 h	ec/99.99/0.803		5/2-			Pu L x-ray
				α/0.01/5.924	5.734(2)/0.001				Pu k x-ray
					5.776(2)/0.008				0.18172
									0.22818
									0.27760
^{240m1}Am			0.94 ms						
^{240}Am		240.0553	2.12 d	ec/1.38		(3-)			Pu L x-ray
				α/5.592	5.378/16×10^{-4}				Pu k x-ray
									0.88878
									0.98764
									(0.1–1.3)

Elem. or Isot.	Natural abundance (Atom %)	Atomic mass or weight	Half-life/ Resonance width (MeV)	Decay mode/ Energy (/MeV)	Particle energy/ Intensity (MeV/%)	Spin (h/2π)	Nuclear magnetic mom. (nm)	Elect. quadr. mom. (b)	γ-Energy/ Intensity (MeV/%)
^{241m1}Am			1.2 μs						
^{241}Am		241.05683	432.7 a	α/5.637	5.2443(1)/0.002	5/2-	+1.58	+3.1	Np L x-ray
			1.2 ×10^{14} a	sf/3.6 ×10^{-10}	5.3221(1)/0.015				0.02634 /.024
					5.3884(1)/1.4				0.0332/.00126
					5.4431(1)/12.8				0.05954/0.359
					5.4857(1)/85.2				(0.03–1.128)
					5.5116(1)/0.20				
					5.5442(1)/0.34				
^{242m2}Am			14 ms						
^{242m}Am			141. a	IT/99.5/0.048		5-	+1.0	+7.	Am L x-ray
				α/0.5/5.62	5.141(4)/0.026				0.04863
			> 3 ×10^{12} a	sf/< 4.7 ×10^{-9}	5.2070(2)/0.4				0.08648
									0.10944
									0.16304
^{242}Am		242.05955	16.02 h	β- /83 /0.665	0.63/46	1-	+0.388	~ -2.4	Pu L x-ray
				ec/17 /0.750	0.67/37				Cm L x-ray
									Pu k x-ray
									0.0422
									0.04453
^{243m1}Am			0.006 ms						
^{243}Am		243.06138	7.37 ×10^{3} a	α/5.438	5.1798(5)/1.1	5/2-	+1.50	+2.86	0.04354
			2. ×10^{14} a	sf/3.7 ×10^{-9}	5.2343(5)/11				0.07467
					5.2766(5)/88				0.08657
					5.394(5)/0.12				0.11770
					5.3500(5)/0.16				0.14197
^{244m3}Am			~ 0.007 ms						
^{244m2}Am			0.9 ms						
^{244m}Am			26. m	β- /1.498		(1-)			0.0429
^{244}Am		244.06429	10.1 h	β- /1.428					Am L x-ray
									Cm k x-ray
									0.7460
									0.9000
^{245}Am		245.06646	2.05 h	β- /0.894	0.65/19	(5/2+)			Cm L x-ray
					0.90/77				Cm k x-ray
									0.25299
^{246m2}Am			0.07 ms	sf					
^{246m}Am			25.0 m	β- /	1.3/79.	2-			Cm L x-ray
					1.60/14				Cm k x-ray
					2.1/7				0.27002
									0.79881
									1.06201
									1.07885
									(0.04–2.29)
^{246}Am		246.0698	39. m	β- /2.38	1.2/	(7-)			Cm L x-ray
									Cm k x-ray
									0.1529
									0.2046
									0.6786
^{247}Am		247.072	22. m	β- /1.7					Cm L x-ray
									Cm k x-ray
									0.2267 /
									0.2853 /
^{248}Am		248.076							
^{249}Am		249.078							
$_{96}$Cm									
^{232}Cm									
^{233}Cm		233.0508	~ 23 s	α/20	7.34/				
				ec/80					
^{234}Cm		234.0502	~ 51. s	α/27	7.24/	0+			
				ec/71					
				sf/~ 2					
^{235}Cm		235.052							
^{236}Cm		236.0514	6.8 m	ec/1.7/82		0+			
				α/18	6.95				
				sf/< 0.1					
^{237}Cm		237.0529		ec/2.5					
				α/<1	6.66				
^{238}Cm		238.0531	2.4 h	ec/>90 /0.97	α/6.558/86	0+			
				α/<10 /6.632	6.520(50)/14				
^{239}Cm		239.0549	~ 3. h	ec/1.7					
				α/< 0.001					0.0407
									0.1466
									0.1874

Elem. or Isot.	Natural abundance (Atom %)	Atomic mass or weight	Half-life/ Resonance width (MeV)	Decay mode/ Energy (/MeV)	Particle energy/ Intensity (MeV/%)	Spin (h/2π)	Nuclear magnetic mom. (nm)	Elect. quadr. mom. (b)	γ-Energy/ Intensity (MeV/%)
^{240}Cm		240.05553	27. d	α/6.397	5.989/0.014	0+			
					6.147/0.05				
			1.9 ×10^{6} a	sf/3.9 ×10^{-6}	6.2478(6) /28.8				
					6.2906(6) /70.6				
^{241}Cm		241.05765	32.8 d	ec/99 /0.768		1/2+			Am k x-ray
				α/1 /6.184	5.8842(4)/0.12				0.13241
					5.9291(4)/0.18				0.16505
					5.9389(4)/0.69				0.18028
									0.43063
									0.47181
^{242}Cm		242.05884	162.8 d	α/6.216	5.9694(1)/0.035	0+			Pu L x-ray
					6.069(1)/25				0.04408
			7.0 ×10^{6} a	sf/6.4 ×10^{-6}	6.1129(1)/74				0.10189
									(0.04–1.2)
^{243m1}Cm			1.1 μs	IT	0.087	1/2+			
^{243}Cm		243.06139	29.1 a	α/6.167	5.6815(5) /0.2	5/2+	0.40		Pu L x-ray
					5.6856(5)/1.6				Pu k x-ray
			5.5 ×10^{11} a	sf/5.3 ×10^{-9}	5.7420(5)/10.6				0.10612
					5.7859(5)/73.5				0.20975
					5.9922(5)/6.5				0.22819
					6.0103(5)/1.0				0.27760
					6.0589(5)/5				0.28546
					6.0666(5)/1.5				0.33431
									(0.04–0.7)
^{244m1}Cm			0.034 s	IT		6+			
^{244}Cm		244.06275	18.1 a	α/5.902	5.6656/0.02	0+			Pu L x-ray
					5.7528/23				0.04282
			1.32 ×10^{7} a	sf/1.4 ×10^{-4}	5.8050/77				0.09885
					5.515/0.004				0.15262
^{245}Cm		245.06549	8.48 ×10^{3} a	α/5.623	5.235(10)/0.3	7/2+	0.5		Pu L x-ray
					5.3038(10)/5.0				Pu k x-ray
			1.4 ×10^{12} a	sf/6.1 ×10^{-7}	5.3620(7)/93				0.04195
					5.4927(11)/0.8				0.13299
					5.5331(11)/0.6				0.13606
									0.17494
^{246}Cm		246.06722	4.75 ×10^{3} a	α/5.476	5.343(3)/21	0+			Pu L x-ray
			1.8 ×10^{7} a	sf/0.026	5.386(3)/79				0.04453
^{247m1}Cm			0.026 ms	IT/0.277		5/2+			
^{247}Cm		247.07035	1.56 ×10^{7} a	α/5.352	4.818(4)/4.7	9/2-	0.36		Pu k x-ray
					4.8690(20)/71				0.2792
					4.941(4)/1.6				0.2886
					4.9820(20)/2.0				0.3471
					5.1436(20)/1.2				0.4035
					5.2104(20)/5.7				
					5.2659(20)/13.8				
^{248}Cm		248.07235	3.48 ×10^{5} a	α/99.92 /5.162	4.931(5)/0.07	0+			
					5.0349(2)/16.5				
			4.15 ×10^{6} a	sf/8.38	5.0784(2)/(75)/1				
^{249m}Cm			0.023 ms	IT/0.0488		7/2+			
^{249}Cm		249.07596	64.15 m	β- /0.900	0.9/	1/2+			Bk k x-ray
									0.56039/0.84
									0.63431/1.5
									(0.085-0.653)
^{250}Cm		250.0783	~ 9.7 ×10^{3} a	sf/85.8		0+			
				α/5.27					
^{251}Cm		251.0823	16.8 m	β- /1.42	0.90/16	(1/2+)			0.3896 /
									0.5299
									0.5425
^{252}Cm		252.085	< 2 d			0+			
$_{97}$**Bk**									
^{233}Bk			~ 21 s	α	7.77				
^{234}Bk		234.057							
^{236}Bk		236.057							
^{237}Bk		237.057							
^{238}Bk		238.058	2.4 m	ec/5.0					
^{239}Bk		239.058		ec//> 99					
				α//< 1					
				sf//< 1					
^{240}Bk		240.060	~ 4.8 m						
^{241}Bk		241.060	4.6 m	ec					(0.152-0.262)
^{242}Bk		242.062	7.0 m	ec/3.0					
^{243}Bk		243.06301	4.5 h	ec/99.8 /1.508	6.542(4)/0.03	(3/2-)			0.1466
				α/0.15 /6.871	6.5738(2)/0.04				0.1874

Elem. or Isot.	Natural abundance (Atom %)	Atomic mass or weight	Half-life/ Resonance width (MeV)	Decay mode/ Energy (/MeV)	Particle energy/ Intensity (MeV/%)	Spin (h/2π)	Nuclear magnetic mom. (nm)	Elect. quadr. mom. (b)	γ-Energy/ Intensity (MeV/%)
					6.7180(22)/0.02				0.755
					6.7581(20)/0.02				0.840
									0.946
^{244}Bk		244.0652	4.4 h	ec/99.99 /2.26		(4-)			0.1445
				α/0.01 /6.778	6.625(4)/0.003				0.1876
					6.667(4)/0.003				0.2176
									0.9815
									0.9215/
^{245}Bk		245.06636	4.94 d	ec/99.9 /0.810		3/2-			Cm L x-ray
				α/0.1 /6.453	5.8851(5)/0.03				Cm k x-ray
					6.1176(9)/0.01				0.25299
					6.1467(5)/0.02				0.3809
					6.3087(5)/0.014				0.3851
					6.3492(5)/0.018				
^{246}Bk		246.0687	1.80 d	ec/1.35		(2-)			Cm L x-ray
									Cm k x-ray
									0.79881
									1.08142
^{247}Bk		247.07031	1.4 $\times10^3$ a	α/5.889	5.465(5)/1.5	(3/2-)			0.04175
					5.501(5)/7				0.0839
					5.532(5)/45				0.268
					5.6535(20)/5.5				
					5.678(2)/13				
					5.712(2)/17				
					5.753(2)/4.3				
					5.794(2)/5.5				
^{248}Bk		248.0731	23.7 h	β- /70 /0.87	0.86/	(1-)			Cm L x-ray
				ec/30 /0.72					Cf L x-ray
									Cm k x-ray
									Cf k x-ray
									0.5507
^{249m1}Bk			0.3 ms	IT/0.0088		3/2-			
^{249}Bk		249.07499	320. d	β- /0.125	0.125/100	7/2+	2.0		0.327/10^{-5}
				α/0.001 /5.525	5.390(1)/0.0002				0.308/10^{-6}
			1.8 $\times10^9$ a	sf/4.9 $\times10^{-8}$	5.4174(6)/0.001				(0.028-0.280)
^{250m2}Bk			0.21 ms	IT/0.084		7+			
^{250m1}Bk			0.03 ms	IT/0.036		4+			
^{250}Bk		250.07832	3.217 h	β- /1.780	0.74/	2-			Cf L x-ray
									Cf k x-ray
									0.98912
									1.03184
									(0.04–1.6)
^{251}Bk		251.0808	56. m	β- /1.09		(3/2-)			0.02481
									0.1528
									0.1776
^{252}Bk		252.084	1.8 m						
^{253}Bk		253.087							
^{254}Bk		254.091							
$_{98}$Cf									
^{237}Cf		237.062	2.1 s	α//70	8.08				
				sf//30					
^{238}Cf		238.061	21 ms	sf// > 95		0+			
				α// < 5					
^{239}Cf		239.063	~ 0.7 m	α					
^{240}Cf		240.0623	1.1 m	α/7.719/98.5	7.590(10)/	0+			
				sf// 1.5					
^{241}Cf		241.064	4. m	ec/3.3					
				α/7.60	7.335(5)/				
^{242}Cf		242.0638	3.5 m	α/7.509/61	7.351(6)/20	0+			
				sf/<0.014	7.385(4)/80				
				ec/39/					
^{243}Cf		243.065	11. m	ec/86 /2.2	7.060(6)/20	(1/2+)			
				α/14 /7.40	7.170/4				
^{244}Cf		244.06600	20. m	α/7.328	7.168(5)/25	0+			
					7.210(5)/75				
^{245}Cf		245.06805	44. m	α/36 /7.255	7.14/91.7				Cm K x-ray
				ec/64 /1.569	6.983/0.31				0.5709
					7.09/7				0.6014
					7.065/0.68				0.6163
^{246}Cf		246.06881	1.49 d	α/6.869	6.6156(10)/0.18	0+			Cm L x-ray
					6.7086(7)/21.8				0.04221
			1.8 $\times10^3$ a	sf/2.3 $\times10^{-4}$	6.7501(7)/78.0				0.0945
									0.147

Elem. or Isot.	Natural abundance (Atom %)	Atomic mass or weight	Half-life/ Resonance width (MeV)	Decay mode/ Energy (/MeV)	Particle energy/ Intensity (MeV/%)	Spin (h/2π)	Nuclear magnetic mom. (nm)	Elect. quadr. mom. (b)	γ-Energy/ Intensity (MeV/%)
^{247}Cf		247.0710	3.11 h	ec/99.96 /0.65		7/2+			Bk k x-ray
				α/0.04 /6.55	6.301(5)/				0.2941
									0.4778
^{248}Cf		248.07219	334. d	α/6.369	6.220(5)/17	0+			
			3.2 ×10^{4} a	sf/0.0029	6.262(5)/83				
^{249m}Cf			45 μs	IT/0.145		5/2+			
^{249}Cf		249.07485	351. a	α/6.295	5.758/3.7	9/2-			Cm L x-ray
					5.812/85.7				Cm k x-ray
			8. ×10^{10} a	sf/4.4 ×10^{-7}	5.8488(2)/1.0				0.25299/2.5
					5.9029(2)/2.8				0.33351/13.6
					5.9451(2)/4.0				0.38832/63.6
					6.1401(2)/1.1				(0.0376–1.10)
					6.1940(2)/2.2				
^{250}Cf		250.07641	13.1 a	α/6.129	5.8913(4)/0.28	0+			Cm L x-ray
			1.7 ×10^{4} a	sf/0.077	5.9889(4)/17.1				0.04285
					6.0310(4)/82.6				
^{251m}Cf			26.3 μs						
^{251}Cf		251.07959	9.0 ×10^{2} a	α/6.172	5.56448(7)/1.5	1/2+			0.109/19.8
					5.632(1)/4.5				0.1775/17.3
					5.648(1)/3.5				(0.0385-0.354)
					5.6773(6)/35				
					5.762(3)/3.8				
					5.7937(7)/2.0				
					5.8124(8)/4.2				
					5.8514(6)/27				
					6.0140(7)/11.6				
					6.0744(7)/2.7				
^{252}Cf		252.08163	2.65 a	α/96.9 /6.217	5.7977(1)/0.23	0+			Cm L x-ray
			86. a	sf/3.1/	6.0756(4)/15.2				0.04339
					6.1184(4)/81.6				0.1002
^{253}Cf		253.08513	17.8 d	β- /99.7 /0.29	0.27/100	(7/2+)			
				α/0.3 /6.126	5.921(5)/0.02				
^{254}Cf		254.0873	60.5 d	sf/99.7/		0+			
				α/0.3/5.930	5.792(5)/0.05				
					5.834(5)/0.26				
^{255}Cf		255.091	1.4 h	β- /0.7					
^{256}Cf		256.093	12. m	sf		0+			
$_{99}$**Es**									
^{241}Es		241.069	~ 8 s	α	8.11				
^{242}Es		242.070	18 s	α//57	8.03				(0.087-0.122)
				ec//43					
^{243}Es		243.070	22. s	α//61	7.89/79				
				ec/39 /4.0	7.85/16				
				sf// < 1	7.75/4.3				
^{244}Es		244.071	37. s	ec/76 /4.6					
				α/4 /	7.57/4				
^{245}Es		245.071	1. m	α/40 /7.858	7.74				
				ec/60 /3.1					
^{246}Es		246.073	7.7 m	ec/90 /3.9					
				α/10 /	7.35				
^{247}Es		247.0736	4.8 m	ec/93 /2.48					
				α/7 /	7.32				
^{248}Es		248.0755	26. m	ec/99.7 /3.1					
				α/0.3 /	6.87				
^{249}Es		249.0764	1.70 h	ec/99.4 /1.45		(7/2+)			0.3795
				α/0.6 /	6.77				0.8132
^{250m}Es			2.2 h	ec/		1(-)			Cf L x-ray
				β+					Cf k x-ray
									0.9891
									1.0319
^{250}Es		250.079	8.6 h	ec/2.1		(6+)			Cf L x-ray
									Cf k x-ray
									0.30339
									0.34948
									0.82883
^{251}Es		251.07999	1.38 d	ec/99.5 /0.38		(3/2-)			
				α/0.5 /	6.462/0.05				
					6.492/0.4				
^{252}Es		252.0830	1.29 a	α/76 /	6.632/61.0	(5-)			
				ec/24 /1.26	6.562/10.3				
^{253}Es		253.08483	20.47 d	α/	6.633/89.8	7/2+	~ +4.10	7.	0.04180/5.6
			6.3 ×10^{5} a	sf/8.9 ×10^{-6}	6.5916/6.6				0.3892/2.7
									(0.0309-1.106)

Elem. or Isot.	Natural abundance (Atom %)	Atomic mass or weight	Half-life/ Resonance width (MeV)	Decay mode/ Energy (/MeV)	Particle energy/ Intensity (MeV/%)	Spin (h/2π)	Nuclear magnetic mom. (nm)	Elect. quadr. mom. (b)	γ-Energy/ Intensity (MeV/%)
^{254m}Es			1.64 d	β- /99.6 /	0.475	2+	2.9	3.7	Fm L x-ray
				α/0.3 /6.67	6.382	2+			Fm k x-ray
			> 10. a	sf/0.045					0.6488
									0.6938
^{254}Es		254.08802	276. d	α/	6.429	(7+)	4.35		0.064
			> 2.5 ×10^{7} a	sf/< 3 ×10^{-6}					
^{255}Es		255.0903	40. d	β- /92 /0.29		(7/2+)			
				α/8 /	6.26				
			2.6 ×10^{3} a	sf/0.0042	6.300				
^{256m}Es			7.6 h	β- /		(8+)			0.218
									0.232
									0.862
^{256}Es		256.094	25. m	β-/1.7		(1+)			
^{257}Es		257.096	7.7 d	β-					
^{258}Es		258.100							
$_{100}$**Fm**									
^{241}Fm		241.074							
^{242}Fm		242.073	0.8 ms	sf// > 96		0+			
^{243}Fm		243.074	0.22 s	α/	8.55				
				sf/< 0.4					
^{244}Fm		244.074	3.2 ms	sf/> 97		0+			
^{245}Fm		245.075	4. s	α/	8.15/				
				sf/<0.1					
^{246}Fm		246.0754	1.54 s	α//93	8.24/	0+			
				sf//7					
				ec// <1.3					
^{247m}Fm			5.1 s	α// 88	8.78/83	(1/2+)			
				sf//12	8.66/17				
^{247}Fm		247.077	30. s	α/8.20	7.87/70				(0.082-0.167)
				ec/2.9	7.93/30				
^{248}Fm		248.0772	33. s	α/99.9 /8.001	7.83/20	0+			
				sf/0.1/	7.87/80				
^{249}Fm		249.07893	1.6 m	ec/2.4		(7/2+)			
				α/16/	7.57				
				Sf//15.6					
^{250m}Fm			1.9 s	IT/1.199	0.682/100	(8-)			
				sf/<8 ×10^{-5}					
^{250}Fm		250.0795	28. m	α/	7.43/	0+			
				ec/0.8					
				sf/0.007					
^{251m1}Fm			0.021 ms	IT/0.20		5/2+			
^{251}Fm		251.0815	5.3 h	ec/98 /1.47		(9/2-)			
				α/2 /	6.833				
^{252}Fm		252.08247	1.058 d	α/7.154	6.998/15	0+			
				sf/0.0023	7.039/85				
^{253}Fm		253.08519	3.0 d	ec/88/0.333	6.676/	½+			Es k x-ray
				α/12 /	6.943/				0.2719
^{254}Fm		254.08685	3.240 h	α/	7.150	0+			
				sf/0.059	7.192				
^{255}Fm		255.08996	20.1 h	α/	6.9635(5)/5.0	7/2+			0.08148/1.
			1.0 ×10^{4} a	sf/2.3 ×10^{-5}	7.0225(5)/93.4				(0.041-0.900)
^{256m1}Fm			0.07 s	IT/1.426	(7-)				
^{256}Fm		256.09177	2.63 h	sf/91		0+			
				α/19	6.92/				
^{257}Fm		257.09511	100.5 d	α/99.79	6.519	(9/2+)			0.1794
				sf/0.21					0.2410
^{258}Fm		258.097	0.37 ms	sf/		0+			
^{259}Fm		259.101	1.5 s	sf/					
^{260}Fm		260.103	~ 4 ms	sf/		0+			
$_{101}$**Md**									
^{245m}Md			~ 0.4 s	α	8.64, 8.68	(7/2)			
^{245}Md		245.081	0.9 ms	sf					
^{246m}Md			~ 4.4 s	ec/77					
				α/23	8.18				
^{246}Md		246.082	1.3 s	α	8.74				(0.169-0.396)
					8.50–8.56				
^{247m}Md			0.25 s	α// 79	8.78/83	(1/2-)			
				sf// 21	8.66/17				
^{247}Md		247.082	0.9 s	α// > 99.9	8.42/92				0.2096/100
				sf// < 0.1	8.35/6				0.1575/11
					8.62/2				
^{248}Md		248.083	7. s	ec/80 /5.3	8.32/15				
				α/20 /	8.36/5				

Elem. or Isot.	Natural abundance (Atom %)	Atomic mass or weight	Half-life/ Resonance width (MeV)	Decay mode/ Energy (/MeV)	Particle energy/ Intensity (MeV/%)	Spin (h/2π)	Nuclear magnetic mom. (nm)	Elect. quadr. mom. (b)	γ-Energy/ Intensity (MeV/%)
				sf// <0.05					
^{249m1}Md			2 s	α//100		(1/2-)			
^{249}Md		249.083	22. s	α/60/8.46	8.026	(7/2-)			0.2532/100
				ec/40/3.7					0.2004/20
									0.2232/10
^{250}Md		250.084	50. s	ec/94 /4.6	7.75/4				0.1523
				α/6 /8.25	7.83/2				
^{251}Md		251.0848	4.3 m	ec/>94 /3.1	α/7.55/87	7/2-			0.293/100
				α/10	7.59/13				0.243/16
^{252}Md		252.086	2. m	ec/>50 /3.9					
				α/<50 /	7.73/				
^{253}Md		253.0871	~ 6 m	ec/2.0/~ 99	α/7.10				0.3532/100
									0.3042
^{254m}Md			~ 30. m	ec/					
^{254}Md		254.090	10. m	ec/2.7					
^{255}Md		255.09108	27. m	ec/92 /1.04	α/7.33/93	(7/2-)			0.121/100
				α/8 /	7.27/5				0.115/65
				sf// < 0.15	7.75/1				0.136/35
					7.71/1				0.141–0.453
^{256}Md		256.094	1.30 h	ec/89 /2.13	7.21/71				Fm k x-ray
				α/11 /	7.14/22				0.121/409
				sf// < 2.6	7.68/2.5				0.115/266
					7.25/2.5				0.136/143
					7.64/2.1				0.634/119
									0.141–1.37
^{257}Md		257.09554	5.5 h	ec/85 /0.41	7.074	(7/2-)			Fm k x-ray
				α/15, sf// < 1	7.014				(0.181–0.389)
^{258m}Md			57 m	ec// > 70		(1-)			Fm k x-ray
				sf// < 30					
^{258}Md		258.09843	51.5 d	α/7.40	6.718(2)/	(8-)			0.3678
				sf// < 0.003	6.763(4)/				0.057–0.448
^{259}Md		259.101	1.64 h	sf// >98.7		7/2+			
				α// <1.3					
^{260}Md		260.104	~ 27.8 d	sf// 73–100					
^{261}Md		261.106							
$_{102}$**No**									
^{248}No		248.087	< 1.0 μs	sf		0+			
^{249}No		249.088	0.05 ms	sf					
				α/ < 20					
^{250m}No			0.04 ms	sf		6+			
^{250}No		250.088	~ 0.005 ms	sf		0+			
				α/ < 10					
^{251m2}No			~ 2 μs						
^{251m}No			1.0 s	α	8.668/~ 98	(1/2+)			
					8.625/~ 2				
^{251}No		251.089	0.80 s	α// 91	8.612/~ 98				
				sf// 0.26	8.552/~ 1				
^{252m}No			0.11 s	IT/1.253	0.710/100	8-			0.710
^{252}No		252.0890	2.44 s	α/75/8.551	8.42	0+			
				sf/24/	8.37				
				ec, β+/<1.6					
^{253m2}No			0.63 ms						
^{253m1}No			0.031 ms	IT/0.168		5/2+			0.167
^{253}No		253.09056	1.56 m	α/	8.00/96	(9/2-)			0.222/100
				ec/3.2	α/8.08/4				(0.058-0.670)
^{254m3}No			0.184 ms	IT/2.93		16+			(0.133-0.312)
^{254m2}No			0.263 s	IT/1.297		8-			(0.053-0.778)
^{254m1}No			0.28 s	IT/					0.778
				sf// 0.02					0.856
				α// 0.01					
^{254}No		254.0910	49. s	α/	8.09	0+			0.102
				ec/1.1					0.152
				sf// 0.17					
^{255}No		255.0932	3.1 m	α/62 /	8.095/58	½+			0.187
				ec/38/2.01	7.742/19				(0.163-0.358)
					7.903/11				
^{256}No		256.09428	2.9 s	α/	8.43	0+			
				sf// 0.5					
^{257}No		257.09689	24.5 s	α/	8.222/83	(7/2+)			0.0770/100
				sf// <1.5	8.323/17				0.1018/57
					8.19/<4				0.1241/78
^{258}No		258.098	~ 1.2 ms	sf		0+			
^{259}No		259.101	58. m	α/78 /7.794	7.52	(9/2+)			

Elem. or Isot.	Natural abundance (Atom %)	Atomic mass or weight	Half-life/ Resonance width (MeV)	Decay mode/ Energy (/MeV)	Particle energy/ Intensity (MeV/%)	Spin (h/2π)	Nuclear magnetic mom. (nm)	Elect. quadr. mom. (b)	γ-Energy/ Intensity (MeV/%)
				ec/22/0.5	7.55				
				sf// <9.7					
^{260}No		260.103	0.11 s	sf		0+			
^{261}No		261.106							
^{262}No		262.107	~ 8. ms	sf		0+			
^{264}No		264.113							
$_{103}$**Lr**									
^{251}Lr		251.094	39 m	sf					
^{252}Lr		252.095	0.3 s	A	9.02/73				
				sf// <1	8.97/27				
^{253m}Lr			0.7 s	A	8.79				
				sf// 1.3					
^{253}Lr		253.095	1.5 s	α//92	8.72	(1/2-)			
				sf// 8					
^{254}Lr		254.096	18. s	α// 72	8.45				(0.042-0.306)
				ec/5.2	(8.36-8.53)				
				sf// <0.1					
^{255m2}Lr			1.4 ms	IT/1.409		(25/2+)			(0.244-0.301)
^{255m}Lr			2.5 s	α//40	8.46	(7/2-)			
				IT/2.038/60					
^{255}Lr		255.0966	31. s	α/	8.37/67	1/2-			
				ec/3.2	8.43/<3.6				
				sf// < 0.1	8.29/1.2				
^{256}Lr		256.098	27. s	α/99.7 /8.554	8.43/				
				ec/4.2	8.39				
				sf// < 0.03					
^{257}Lr		257.0994	0.65 s	α/	8.80	7/2+			
				ec/2.5					
				sf// < 0.03					
^{258}Lr		258.102	3.9 s	α/	8.60/46				
				ec/3.4	8.62/25				
				sf// < 5	8.56/20				
					8.65/9				
^{259}Lr		259.1029	6.1 s	α/80	8.44(1)				
				sf// 20					
^{260}Lr		260.106	3. m	α/	8.03				
^{261}Lr		261.107	40. m	sf					
^{262}Lr		262.110	3.6 h	ec/2.					
				sf// <10					
^{263}Lr		263.111							
^{264}Lr		264.114							
$_{104}$**Rf**									
^{253m1}Rf			~ 48 μs	sf					
^{253}Rf		253.100	0.011 s			(7/2+)			
^{254m2}Rf			~ 0.25 ms						
^{254m1}Rf			0.005 ms						
^{254}Rf		254.100	23. μs	sf// >98.5		0+			
				α/<1.5					
^{255m1}Rf			0.05 ms	α//100		(5/2+)			
^{255}Rf		255.101	1.6 s	α// ~ 52	8.716/92				0.2036/49
				sf// ~ 48	8.678/3				0.1433/51
					8.906/2.5				
					8.646/1.5				
					8.575/1				
^{256m3}Rf			~ 0.027 ms						
^{256m2}Rf			0.017 ms			(8-)			
^{256m1}Rf			0.025 ms			(5-)			
^{256}Rf		256.1012	6.2 ms	sf// 99.68		0+			
				α/0.32	8.81				
^{257m2}Rf			~ 0.106 ms	IT/100		21/2+			(0.296-0.586)
^{257m}Rf			4.4 s	α/88		(11/2-)			
				ec/11					
^{257}Rf		257.1029	5.5 s	α/9.22	8.77				0.117
				ec/11	9.01				
				sf// <1.4	8.95				
					8.62				
^{258m2}Rf			~ 0.015 ms						
^{258m1}Rf			~ 2 ms						
^{258}Rf		258.1034	6. ms	sf// 87		0+			
				α/13	9.05				
^{259}Rf		259.1056	2.5 s	α/9.09/93	8.77(2)/				
				sf// 7	8.86/				
				ec// 15					

Elem. or Isot.	Natural abundance (Atom %)	Atomic mass or weight	Half-life/ Resonance width (MeV)	Decay mode/ Energy (/MeV)	Particle energy/ Intensity (MeV/%)	Spin (h/2π)	Nuclear magnetic mom. (nm)	Elect. quadr. mom. (b)	γ-Energy/ Intensity (MeV/%)
^{260}Rf		260.106	21. ms	sf		0+			
				α// <35					
^{261m}Rf			1.3 ms	α	8.29				
				sf// < 11	8.34				
^{261}Rf		261.1088	~ 2.6 s	α/18, sf/82	8.52/				
^{262}Rf		262.110	0.05 s	sf// >99.2		0+			
^{263}Rf		263.112	11. m	sf, α					
^{265}Rf		265.116	~ 2. m	α					
^{267}Rf		267.122	~ 1. h	sf					
$_{105}$**Db**									
^{255}Db		255.107	~ 1.5 s	α,					
				sf//~ 20					
^{256}Db		256.108	1.6 s	α/64	9.02/67				
				ec/35	8.89/11				
				sf// 0.05	9.08/11				
					9.12/11				
^{257m}Db			~ 0.67 s	α/87	9.17				
				sf// <13					
^{257}Db		257.107	1.8 s	α/	8.96/33				0.1022
				sf// <6	9.06/38				
					9.12/5.5				
					8.94/9				
					9.02/9				
					8.89/5.5				
^{258m}Db			~ 1.9 s	α	9.20				
^{258}Db		258.109	4.3 s	α/	9.30/				0.1568
				ec/5.3/39	9.17/				0.2215
				sf// <33	9.08/				
^{259}Db		259.1095	~ 0.51 s	sf					
				α/	9.47/				
^{260m}Db			0.3 m						
^{260}Db		260.111	1.5 s	α/	9.05/				
				sf// <9.6	9.08/				
					9.13/				
^{261}Db		261.112	1.8 s	α/	8.93/				
				sf// ~ 73					
^{262}Db		262.114	0.5 m	sf// <33					
				α/	8.45/				
					8.53/				
					8.67/				
^{263}Db		263.115	~ 0.45 m	sf// 57	8.36/				
				α// 41	8.41/				
				ec/3					
^{266}Db		266.112	~ 0.4 h	ec, sf					
^{267}Db		267.122	1.2 h	sf					
^{268}Db		268.126	1.2 d	sf, ec					
^{270}Db		270.131	~ 0.7 d	sf					
$_{106}$**Sg**									
^{258}Sg		258.113	2.6 ms	sf		0+			
				α// <38					
^{259m1}Sg			0.24 s	α/100	9.55/54	(1/2+)			
					9.20/43				
					9.70/3				
^{259}Sg		259.114	0.4 s	α/	9.61	(11/2-)			
				sf// <8.6	9.35				
				ec// <13	9.03				
^{260}Sg		260.1144	5.0 ms	α// 29	9.748	0+			
				sf// 71	9.72				
					9.81				
^{261m}Sg			~ 9 μs	IT/~ 0.2		(11/2-)			
^{261}Sg		261.1159	0.2 s	α	9.56				0.107
				ec// 1.3	9.42				
				sf// 0.6	9.41				
^{262}Sg		262.1163	15. ms	sf		0+			
				α/<16					
^{263m}Sg			0.3 s	α	9.2				
^{263}Sg		263.118	0.8 s	α	9.06				
				sf//~ 13	9.25				
^{264}Sg		264.119	~ 0.04 s	α// <36					
^{265}Sg		265.121	~ 9 s	α/100					
^{265m}Sg			14 s	α// >65	8.84/46				
				sf// <35	8.76/23				
					8.94/23				

Elem. or Isot.	Natural abundance (Atom %)	Atomic mass or weight	Half-life/ Resonance width (MeV)	Decay mode/ Energy (/MeV)	Particle energy/ Intensity (MeV/%)	Spin (h/2π)	Nuclear magnetic mom. (nm)	Elect. quadr. mom. (b)	γ-Energy/ Intensity (MeV/%)
					8.69/8				
^{266}Sg		266.122	0.3 s	sf		0+			
^{267}Sg		267.124	~ 1.3 m	α					
^{269}Sg		269.129	~ 3 m	α	~ 8.50				
^{271}Sg		271.134	2. m	α/70	~ 8.54				
				sf// 30					
$_{107}$**Bh**									
^{260}Bh		260.122	0.04 s	α// 70	10.16				
				sf// < 18					
				ec// < 18					
^{261}Bh		261.121	12. ms	α/, sf <10	10.40				
					10.10				
					10.03				
^{262m}Bh			10 ms	α/	10.37				0.1024
				sf// <24	10.24				
^{262}Bh		262.123	~ 0.08 s	α/	10.1				0.0389
				sf// <11	9.9				0.1565
					9.7				
^{264}Bh		264.125	0.44 s	α/	9.3 – 9.8				
				sf					
^{265}Bh		265.125	0.9 s	α	9.24				
^{266}Bh		266.127	~ 2 s	α	9.08				
^{267}Bh		267.128	~ 17 s	α	8.83				
^{270}Bh		270.133	~ 1. m	α	9.0				
^{271}Bh			1 s	α	~ 9.35				
^{272}Bh		272.138	~ 11. s	α	8.9				
^{274}Bh		274.144	~ 0.9 m	α	8.76				
$_{108}$**Hs**									
^{263}Hs		263.129	0.7 ms	α	10.6				
				sf// < 8.4	10.8				
					10.9				
^{264}Hs		264.1284	0.45 ms	α/, sf//~ 50	11.0	0+			
^{265m}Hs			0.4 ms	α	10.57/63				
					10.73				
					10.52				
					10.34				
^{265}Hs		265.1298	1.9 ms	α/	10.30/90				
				sf// <1	10.43				
					10.37				
					10.25				
^{266}Hs		266.1301	~ 2.3 ms	α	10.2	0+			
^{267m}Hs			~ 94 ms						
^{267}Hs		267.132	0.05 s	α// >88	9.88				
					9.83				
					9.75				
^{269}Hs		269.134	~ 10. s	α	9.07				
					8.92				
^{270m}Hs			0.3 m	α	8.88				
^{270}Hs		270.134	~ 7.6 s	α	9.02	0+			
^{271}Hs		271.137	~ 4. s	α	9.13				
					9.30				
^{273}Hs		273.142	0.8 s	α	9.53				
^{275}Hs		275.147	~ 0.19 s	α	9.3				
^{277m}Hs			~ 0.5 m	Sf					
^{277}Hs		277.152	~ 3. ms	sf					
$_{109}$**Mt**									
^{266m}Mt			~ 1.2 ms	α	10.46–10.81				
^{266}Mt		266.137	~ 3. ms	α	10.48–11.31				
				sf// 0.25					
^{267}Mt		267.137	19 ms	α					
^{268}Mt		268.139	~ 0.02 s	α// >68	10.3 - 10.8				
^{270}Mt		270.140	~ 0.8 s	α	10.0				
^{274}Mt		274.147	~ 0.4 s	α	9.8				
^{275}Mt		275.149	~ 9.7 ms	α	10.3				
^{276m1}Mt			~ 6 s						
^{276}Mt		276.152	0.7 s	α	~ 9.71				
^{277}Mt		277.153	~ 5 ms	sf					
^{278}Mt		278.156	~ 5. s						
$_{110}$**Ds**									
^{267}Ds		267.144	~ 3 μs	α// >32	11.6				
^{269}Ds		269.1448	0.17 ms	α// >75	11.11				
^{270m}Ds			~ 6 ms	α/>70	10.95				
				IT/< 30	11.15				

Elem. or Isot.	Natural abundance (Atom %)	Atomic mass or weight	Half-life/ Resonance width (MeV)	Decay mode/ Energy (/MeV)	Particle energy/ Intensity (MeV/%)	Spin (h/2π)	Nuclear magnetic mom. (nm)	Elect. quadr. mom. (b)	γ-Energy/ Intensity (MeV/%)
					12.15				
^{270}Ds		270.1446	0.1 ms	α	11.03	0+			
^{271m}Ds			0.07 s	α	9.9				
^{271}Ds		271.146	1.6 ms	α	10.8				
^{273m}Ds			0.17 ms	α	11.8				
^{273}Ds		273.149	118 ms	α	9.73				
^{277}Ds		277.156	4. ms	α/	10.6				
^{279}Ds		279.158	0.18 s	sf// 90					
				α// 10	~ 9.70				
^{280}Ds		280.161	~ 7.6 s	sf		0+			
^{281}Ds		281.165	14. s	sf					
^{282}Ds			0.5 ms	sf					
$_{111}$**Rg**									
^{272}Rg		272.153	~ 3.8 ms	α// >68	10.82				
^{274}Rg		274.155	~ 6. ms	α	11.2				
^{278}Rg		278.161	~ 4. ms	α	10.7				
^{279}Rg		279.163	~ 0.17 s	α	10.4				
^{280}Rg		280.165	4.4 s	α	~ 9.75				
^{281}Rg		281.17	17. s	sf//90					
				α//10	9.28				
^{282}Rg		282.169	1. m	α	9.0-9.2				
$_{112}$**Cn**									
^{277}Cn		277.164	~ 0.7 ms	α	11.45				
					11.65				
^{281}Cn		281.170	0.13 s	α	10.3				
^{282}Cn		282.171	0.8 ms	sf					
^{283}Cn		283.173	4. s	sf// < 10					
				α// ~ 100	9.52				
^{284}Cn		284.174	~ 0.9 s	sf					
^{285}Cn		285.177	~ 0.5 m	α	9.15				
$_{113}$**Nh**									
^{278}Nh		278.171	~ 0.24 ms	α	11.7				
^{282}Nh		282.176	~ 0.07 s	α	10.6				
^{283}Nh		283.177	~ 0.1 s	α	10.1				
^{284}Nh		284.179	1.0 s	α	9.97				
^{285}Nh		285.18	4. s	α	9.74				
					9.48				
^{286}Nh		286.182	~ 13. s	α	9.6				
$_{114}$**Fl**									
^{284}Fl			~ 2.5 ms	sf					
^{285}Fl		285.184	0.15 s	α	10.4				
^{286}Fl		286.184	0.12 s	α/50	10.2				
				sf// 50					
^{287}Fl		287.187	0.48 s	α	10.0				
^{288}Fl		288.188	~ 0.58 s	α	9.94				
^{289}Fl		289.190	~ 1.9 s	α	9.82				
$_{115}$**Mc**									
^{287}Mc		287.191	~ 0.03 s	α	10.1				
^{288}Mc		288.193	0.17 s	α	10.5				
^{289}Mc		289.19	~ 0.33 s	α	10.3				
^{290}Mc		290.196	~ 0.2 s	α	9.95				
$_{116}$**Lv**									
^{290}Lv		290.199	8. ms	α	10.8				
^{291}Lv		291.201	0.02 s	α	~ 10.74				
^{292}Lv		292.202	13. ms	α	~ 10.66				
^{293}Lv		293.204	0.06 s	α	10.5				
$_{117}$**Ts**									
^{293}Ts		293.208	~ 22. ms	α	11.0				
^{294}Ts		294.210	~ 0.05 s	α	10.8				
$_{118}$**Og**									
^{294}Og		294.214	0.7 ms	α	11.7				

Nuclear

NEUTRON SCATTERING AND ABSORPTION PROPERTIES

Norman E. Holden

This table presents an evaluated set of values for experimental quantities that characterize the properties for scattering and absorption of neutrons. The neutron cross section is given for room temperature neutrons, 20.43 °C, corresponding to a thermal neutron energy of 0.0253 electron volts (eV) or a neutron velocity of 2200 meters/second. The neutron resonance integral is defined over the energy range from 0.5 eV to 0.1×10^6 eV, or 0.1 MeV.

Bound neutron scattering lengths and neutron cross sections averaged over a Maxwellian spectrum at 30 keV for astrophysical applications are also presented. A list of the major references used is given below. The literature cutoff date is January 2003. Uncertainties are given in parentheses. Parentheses with two or more numbers indicate values to the excited state(s) and to the ground state of the product nucleus.

Table Layout

Column number	Column title	Description
1	Isotope/Element	For elements, atomic number and chemical symbol are listed. For nuclides, mass number and chemical symbol are listed. Isomers are indicated by the addition of m, m1, or m2.
2	Isotopic Abundance	In atom percent.
3	Half-Life	Half-life in decimal notation. μs = microsecond; ms = millisecond; s = second; m = minute; h − hour; d = day; y = year.
4	Thermal Neutron Cross Sections	Cross sections for neutron capture reactions in units of barns (10^{-24} cm^2) or millibarns (mb). Proton, alpha production and fission reactions are designated by σ_p, σ_α, σ_f, respectively. Separate values are listed for isomeric production.
5	Neutron Resonance Integrals	Resonance integrals for neutron capture reactions in barns (10^{-24} cm^2) or millibarns (mb). Proton, alpha production and fission reactions are designated by R.I.$_p$, R.I.$_\alpha$, R.I.$_f$, respectively. Separate values are listed for isomeric production.
6	Neutron Scattering Lengths	Bound coherent scattering lengths for neutron scattering reactions in units of femtometers (fm), which is equal to fermis (10^{-13} cm).
7	Maxwellian Averaged Cross Section	Astrophysical cross sections, averaged over a stellar neutron maxwellian spectrum characterized by a thermal energy of 30 keV, expressed in barns (10^{-24} cm^2), millibarns (mb) or microbarns (μb).

General Nuclear Data References

The following references represent the major sources of the nuclear data presented:

Mughabghab, S.F., Divadeenam, M., Holden, N.E.; Neutron Cross Sections, Vol. 1 *Neutron Resonance Parameters and Thermal Cross Sections*, Part A, *Z* = 1–60. Academic Press Inc., New York, New York (1981); Mughabghab, S.F.; Part B, *Z* = 61–100. Academic Press Inc., Orlando, FL (1984).

Holden, N.E.; *Fifty Years with Nuclear Fission* Conference, Washington, DC, Gaithersburg, MD, April 26–29, 1989, p. 946. American Nuclear Society, LaGrange Park, IL (1989).

Tuli, J.K.; *Nuclear Wallet Cards*, Brookhaven National Laboratory (Jan. 2000).

Holden, N.E.; Half-Lives of Selected Nuclides, *Pure & Applied Chemistry* 62, 941 (1990).

Holden, N.E., Hoffman, D.C.; Spontaneous Fission Half-Lives for Ground State Nuclides, *Pure & Applied Chemistry* 72, 1525 (2000).

Koester, L., Rauch, H., Seymann, E.; Neutron Scattering Lengths: A Survey of Experimental Data and Methods, *Atomic Data Nuclear Data Tables* 49, 65 (1991).

Sears, V.F.; Neutron Scattering Lengths and Cross Sections, *Neutron News* 3, (3), 26 (1992).

Bao, Z.Y., Beer, H., Käpeler, F., Voss, F., Wisshak, K., Raucher, T.; Neutron Cross Sections for Nucleo-Synthesis Studies, *Atomic Data Nuclear Data Tables* 76, 70 (2000).

Elem. or isot.	Natural abundance (%)	Half-life	Thermal neut. cross section (barns)	Resonance integral (barns)	Coh. scat. length (fm)	σ (30 keV) Maxw. avg. (barns)
$_1$H			0.332(2)	0.149(1)	−3.739(1)	
^{1}H	99.9885(70)	>2.8×10^{23} y	0.332(2)	0.149(1)	− 3.741(1)	0.25(2) mb*
^{2}H	0.0115(70)		0.51(1)mb	0.23(2) mb	6.671(4)	2.1(4) μb
^{3}H		12.33 y	< 6. μb		4.79(3)	
$_2$He			< 0.05		3.26(3)	
^{3}He	0.000134(3)		σ_p = 5.33(1)×10^3	RI$_p$ = 2.39(1)×10^3	5.74(7)	
			0.05(1) mb			8.(1) μb*
^{4}He	99.999867(3)				3.26(3)	
$_3$Li			71.(2)	32.(1)	− 1.90(2)	

* Extrapolated value.

Elem. or isot.	Natural abundance (%)	Half-life	Thermal neut. cross section (barns)	Resonance integral (barns)	Coh. scat. length (fm)	σ (30 keV) Maxw. avg. (barns)
^{6}Li	7.59(4)		σ_t = 9.4(1)×10^{2}	RI_t = 422.(4)	2.0(1)	$\sigma_t \approx$ 1.
			39.(5) mb	17.(2) mb		0.06(1) mb*
^{7}Li	92.41(4)		45.(5) mb	20.(2) mb	– 2.22(2)	42.(3) μb
^{8}Li		0.84 s				< ≈ 5.5 μb
$_{4}$Be			8.8(4) mb	3.9(2) mb	7.79(1)	
^{7}Be		53.28 d	σ_p = 3.9(1)×10^{4}	RI_p = 1.75(5)×10^{4}		σ_p = 16(4)*
			$\sigma_\alpha \approx$ 0.1			
^{9}Be	100.		8.8(4) mb	3.9(2) mb	7.79(1)	
^{10}Be		1.52×10^{6} y	<1. mb			
$_{5}$B			7.6(1)×10^{2}	3.4(1)×10^{2}	5.30(4)	
^{10}B	19.9(7)		σ_α = 38.4(1)×10^{2}	RI_α = 17.3(1)×10^{2}	– 0.1(3)	
			0.3(1)	0.13(4)		
			σ_p = 7.(1) mb			
			σ_t = 8.(2) mb			
^{11}B	80.1(7)		5.(3) mb	2.(1) mb	6.65(4)	
$_{6}$C			3.5(1) mb	1.6(1) mb	6.646(1)	
^{12}C	98.93(8)		3.5(1) mb	1.6(1) mb	6.651(2)	16.(1) μb*
^{13}C	1.07(8)		1.4(1) mb	1.7(2) mb	6.19(9)	0.021(4) mb
^{14}C		5715. y	<1.4 μb			3.(1) μb*
$_{7}$N			2.00(6)	0.90(3)	9.36(2)	
^{14}N	99.636(20)		σ_p = 1.93(5)	RI_p = 0.87(3)	9.37(2)	σ_p = 1.8(2) mb*
			0.080(1)	0.034(1)		0.04(1) mb
^{15}N	0.364(20)		0.04(1) mb	0.11(3) mb	6.44(3)	6.(1) μb*
$_{8}$O			0.29(1) mb	0.40(4) mb	5.805(4)	
^{16}O	99.757(16)		0.19(1) mb	0.36(4) mb	5.805(5)	34.(4) μb
^{17}O	0.038(1)		σ_α = 0.257(10)	0.11(1)	5.8(2)	σ_α = 3.9(5) mb*
			0.54(7) mb	0.39(5) mb		
^{18}O	0.205(14)		0.16(1) mb	0.81(4) mb	5.84(7)	9.(1) μb*
$_{9}$F			9.5(1) mb	21.(3) mb	5.65(1)	6.(1) mb
^{19}F	100.		9.5(1) mb	21.(3) mb	5.65(1)	6.(1) mb
$_{10}$Ne			42.(5) mb	19.(3) mb	4.566(6)	
^{20}Ne	90.48(3)		39.(5) mb	18.(3) mb	4.631(6)	0.12(1) mb
^{21}Ne	0.27(1)		0.7(1)	0.31(5)	6.7(2)	≈ 1.5 mb
			σ_α = 0.18(9) mb			
^{22}Ne	9.25(3)		51.(5) mb	23.(3) mb	3.87(1)	58.(4) μb*
$_{11}$Na			0.53(2)	0.32(2)	3.63(2)	2.1(2) mb
^{22}Na		2.605 y	σ_p = 2.8(3)×10^{4}	RI_p < 2.×10^{5}		
			σ_α = 2.6(4)×10^{2}	RI_a = 1.2(2)×10^{2}		
^{23}Na	100.		σ_m = 0.43(3)	RI_m = 0.30(6)	3.63(2)	2.1(2) mb
$_{12}$Mg			66.(6) mb	38.(5) mb	5.375(4)	
^{24}Mg	78.99(4)		0.053(6)	32.(4) mb	5.7(2)	3.3(4) mb
^{25}Mg	10.00(1)		0.20(1)	98.(15) mb	3.6(2)	6.4(4) mb
^{26}Mg	11.01(3)		0.038(1)	25.(2) mb	4.9(2)	0.13(1) mb*
^{27}Mg		9.45 m	0.07(2)	0.03(1)		
$_{13}$Al			0.230(2)	0.17(1)	3.45(1)	
^{26}Al		7.1×10^{5} y	σ_p = 1.97(10)			0.14(2)
			σ_α = 0.34(1)			
^{27}Al	100.		0.230(2)	0.17(1)	3.45(1)	2.9(3) mb
$_{14}$Si			0.166(9)	0.12(2)	4.15(1)	
^{28}Si	92.223(19)		0.17(1)	0.11(2)	4.11(1)	2.9(3) mb
^{29}Si	4.685(8)		0.12(1)	0.08(2)	4.7(1)	7.9(9) mb
^{30}Si	3.092(11)		0.107(3)	0.62(6)	4.61(1)	3.2(3) mb*
^{31}Si		2.62 h	73.(6) mb	33.(3) mb		
^{32}Si		1.6×10^{2} y	< 0.5			
$_{15}$P			0.17(1)	0.08(1)	5.13(1)	
^{31}P	100.		0.17(1)	0.08(1)	5.13(1)	1.7(1) mb
$_{16}$S			0.54(2)	0.24(2)	2.847(1)	

* Extrapolated value.

Nuclear

Elem. or isot.	Natural abundance (%)	Half-life	Thermal neut. cross section (barns)	Resonance integral (barns)	Coh. scat. length (fm)	σ (30 keV) Maxw. avg. (barns)
^{32}S	94.93(31)		0.55(5)	0.25(2)	2.804(2)	4.1(2) mb
			σ_α < 0.5 mb			
^{33}S	0.76(2)		0.46(3)	0.21(2)	4.7(2)	7.4(15) mb
			σ_α = 0.12(1)	RI_α = 0.05(1)		σ_α = 0.18(1)
			σ_p = 2. mb			
^{34}S	4.29(28)		0.25(1)	0.13	3.48(3)	0.23(1) mb
^{36}S	0.02(1)		0.24(2)	0.26(3)		0.17(1) mb*
$_{17}$Cl			33.6(3)	15.(2)	9.58(1)	
^{35}Cl	75.78(4)		43.7(4)	20.(2)	11.7(1)	9.4(3) mb
			σ_p = 0.44(1)	RI_p = 0.2		σ_p = 1.7(2) mb*
			σ_α ≈ 0.08 mb			
^{36}Cl		3.01×10^5 y	σ_p = 46.(2) mb	RI_p = 0.02		σ_p = 91.(8) mb
			<10.			
			σ_α = 0.59(7) mb			σ_α = 0.9(2) mb
^{37}Cl	24.22(4)		(0.05 + 0.38)	(0.04+0.26)	3.1(1)	2.0(2) mb
$_{18}$Ar			0.66(3)	0.42(5)	1.91(1)	
^{36}Ar	0.3365(30)		5.(1)	2.(1)	24.9(1)	
			σ_α = 5.4(3) mb			
			σ_p < 1.5 mb			
^{37}Ar		35.0 d	σ_α = 1.08(8)×10^3	RI_α = 900.		σ_α ≈ 1.3
			σ_p = 37.(4)	RI_p = 31.		σ_p ≈ 0.04
^{38}Ar	0.0632(5)		0.8(2)	0.4(1)	3.5(35)	
^{39}Ar		268. y	6.(2)×10^2			
			σ_α <0.29			
^{40}Ar	99.6003(30)		0.64(3)	0.41(5)	1.83(1)	2.5(3) mb
^{41}Ar		1.82 h	0.5(1)	0.2(1)		
$_{19}$K			2.1(1)	1.0(1)	3.67(2)	
^{39}K	93.2581(44)		2.1(2)	0.9(1)	3.74(2)	11.8(4) mb
			σ_α = 4.3(5) mb			
			σ_p < 0.05 mb			
^{40}K	0.0117(1)	1.26×10^9 y	30.(8)	13.(4)		σ_p = 7.(1) mb
			σ_p = 4.4(4)	2.0(2)		σ_α = 40.(6) mb
			σ_α = 0.42(8)			
^{41}K	6.7302(44)		1.46(3)	1.4(2)	2.69(8)	22.(1) mb
$_{20}$Ca			0.43(2)	0.23(2)	4.70(2)	
^{40}Ca	96.941(156)		0.41(3)	0.22(4)	4.80(2)	6.7(7) mb
			σ_α = 0.13(4) mb			
^{41}Ca		1.02×10^5 y	≈ 4.			
			σ_α = 0.18(3)			
			σ_p = 7.(2) mb			
^{42}Ca	0.647(23)		0.65(10)	0.39(4)	3.4(1)	16.(2) mb
^{43}Ca	0.135(10)		6.(1)	3.9(2)	– 1.56(9)	51.(6) mb
^{44}Ca	2.086(110)		0.8(2)	0.56(1)	1.42(6)	9.(1) mb
^{45}Ca		162.7 d	≈ 15.			
^{46}Ca	0.004(3)	>4×10^{15} y	0.70(3)	0.9(1)	3.6(2)	5.3(5) mb*
^{48}Ca	0.187(21)	4.3×10^{19} y	1.0(1)	0.5(1)	0.39(9)	0.8(1) mb*
$_{21}$Sc			27.2(2)	12.(1)	12.3(1)	
^{45}Sc	100.		(10.+17.)	(5.6+6.4)	12.3(1)	69.(5) mb
^{46}Sc		83.81 d	8.(1)	3.6(5)		
$_{22}$Ti			6.1(1)	2.8(2)	– 3.438(2)	
^{44}Ti		60 y	1.1(2)			
			σ_p < 0.2			
^{46}Ti	8.25(3)		0.6(2)	0.4(1)	4.93(6)	27.(3) mb
^{47}Ti	7.44(2)		1.6(2)	1.6(2)	3.63(1)	64.(8) mb
^{48}Ti	73.72(3)		7.9(9)	3.6(2)	– 6.09(2)	32.(5) mb
^{49}Ti	5.41(2)		1.9(5)	1.2(2)	1.04(5)	22.(2) mb
^{50}Ti	5.18(2)		0.179(3)	0.12(2)	6.18(8)	3.6(4) mb
$_{23}$V			5.0(2)	2.8(1)	– 0.382(1)	

* Extrapolated value.

Elem. or isot.	Natural abundance (%)	Half-life	Thermal neut. cross section (barns)	Resonance integral (barns)	Coh. scat. length (fm)	σ (30 keV) Maxw. avg. (barns)
^{50}V	0.250(4)	1.4×10^{17} y	21.(4)	50.(20)	7.6(6)	
			σ_p = 0.7(4) mb			
^{51}V	99.750(4)		4.9(1)	2.7(2)	– 0.402(2)	38.(4) mb
$_{24}$Cr			3.0(2)	1.7(1)	3.635(7)	
^{50}Cr	4.345(13)	>1.8×10^{17} y	15.(1)	8.(1)	– 4.5(1)	0.05(1)
^{51}Cr		27.70 d	< 10.			
^{52}Cr	83.789(18)		0.8(1)	0.6(2)	4.91(2)	8.8(4) mb
^{53}Cr	9.501(17)		18.(2)	9.(1)	– 4.2(1)	0.06(1)
^{54}Cr	2.365(7)		0.36(4)	0.25(5)	4.6(1)	7.(2) mb
$_{25}$Mn			13.3(1)	14.0(3)	– 3.75(2)	
^{53}Mn		3.7×10^{6} y	70.(10)	32.(5)		
^{54}Mn		312.1 d	< 10.			
^{55}Mn	100.		13.3(1)	14.0(3)	– 3.75(2)	40.(3) mb
$_{26}$Fe			2.7(1)	1.4(2)	9.45(2)	
^{54}Fe	5.845(35)		2.3(2)	1.3(2)	4.2(1)	29.(2) mb
			σ_α = 10. μb	RI_α = 1.1(1) mb		
^{55}Fe		2.73 y	13.(2)	6.(1)		
			σ_α = 0.01			
^{56}Fe	91.754(36)		2.8(3)	1.4(2)	9.93(3)	11.7(5) mb
^{57}Fe	2.119(10)		1.4(2)	0.8(4)	2.3(1)	40.(4) mb
^{58}Fe	0.282(4)		1.3(1)	1.3(2)	15.(7)	12.(1) mb
^{59}Fe		44.51 d	13.(3)	6.(1)		
$_{27}$Co			37.19(8)	74.(2)	2.49(2)	
^{58m}Co		9.1 h	1.4(1)×10^{5}	2.5(10)×10^{5}		
^{58}Co		70.88 d	1.9(2)×10^{3}	7.(1)×10^{3}		
^{59}Co	100.		(20.7+16.5)	(39.+35.)	2.49(2)	38.(4) mb
^{60m}Co		10.47 m	58.(3)	230.(50)		
^{60}Co		5.271 y	2.0(2)	4.3(10)		
$_{28}$Ni			4.5(2)	2.3(2)	10.3(1)	
^{58}Ni	68.0769(89)	>4×10^{19} y	4.6(4)	2.3(2)	14.4(1)	41.(2) mb
			σ_α < 0.03 mb			
^{59}Ni		≈ 7.6×10^{4} y	σ_{abs} = 92.(4)	RI_{abs} = 1.4(1)×10^{2}		
			σ_α = 14.(2)			
			σ_p = 2.(1)			
^{60}Ni	26.2231(77)		2.9(3)	1.5(2)	2.8(1)	25.(1) mb
^{61}Ni	1.1399(6)		2.5(5)	1.5(4)	7.60(6)	82.(8) mb
			σ_α = 0.03 mb			
^{62}Ni	3.6345(17)		15.(1)	6.8(3)	– 8.7(2)	13.(4) mb
^{63}Ni		100. y	20.(5)	9.(2)		
^{64}Ni	0.9256(9)		1.6(1)	1.2(2)	– 0.37(7)	9.(1) mb
^{65}Ni		2.517 h	22.(2)	10.(1)		
$_{29}$Cu			3.8(1)	4.1(4)	7.718(4)	
^{63}Cu	69.15(15)		4.5(2)	5.(1)	6.43(15)	0.09(1)
^{64}Cu		12.701 h	≈ 270.			
^{65}Cu	30.85(15)		2.17(3)	2.2(1)	10.61(19)	41.(5) mb
^{66}Cu		5.09 m	1.4(1)×10^{2}	60.(20)		
$_{30}$Zn			1.1(2)	2.8(4)	5.680(5)	
^{64}Zn	48.27(32)	>2.3×10^{18} y	0.74(5)	1.4(3)	5.23(4)	59.(5) mb
			σ_p < 12. μb			
			σ_α = 11.(3) μb			
^{65}Zn		243.8 d	66.(8)	30.(4)		
			σ_α = 2.0(2)			
^{66}Zn	27.977(77)		0.9(3)	1.8(2)	5.98(5)	35.(3) mb
			σ_α < 0.02 mb			
^{67}Zn	4.102(21)		6.9(1.4)	25.(5)	7.58(8)	0.15(2)
			σ_α = 0.4 mb			
^{68}Zn	19.02(12)		(0.072 + 0.8)	(0.2 + 2.9)	6.04(3)	19.(2) mb
			σ_α <0.02 mb			σ_m = 3.(1) mb

* Extrapolated value.

Elem. or isot.	Natural abundance (%)	Half-life	Thermal neut. cross section (barns)	Resonance integral (barns)	Coh. scat. length (fm)	σ (30 keV) Maxw. avg. (barns)
^{70}Zn	0.631(9)		(8.1+83.) mb	0.9(2)		0.02(1)
$_{31}$Ga			2.9(1)	22.(3)	7.288(2)	
^{69}Ga	60.108(9)		1.68(7)	16.(2)	7.88(4)	0.14(1)
^{71}Ga	39.892(9)	>2.4×10^{26} y	4.7(2)	31.(3)	6.40(3)	0.12(1)
			σ_m = 0.15(5)			
$_{32}$Ge			2.2(1)	6.(2)	8.19(2)	
^{68}Ge		270.8 d	1.0(5)			
^{70}Ge	20.370(89)		(0.3 + 2.7)	2.3(1)	10.0(1)	88.(5) mb
^{72}Ge	27.380(60)		0.9(2)	0.8(3)	8.5(1)	0.07(2)
^{73}Ge	7.759(78)	>1.8×10^{23} y	15.(1)	66.(20)	5.02(4)	0.3(1)
^{74}Ge	36.656(80)		(0.14 + 0.28)	(0.4+0.5)	7.6(1)	53.(7) mb
^{76}Ge	7.835(81)	1.6×10^{21} y	(0.09 + 0.06)	(1.3+0.6)	8.2(15)	0.03(2)
$_{33}$As			4.0(4)	61.(5)	6.58(1)	
^{75}As	100.		4.0(4)	61.(5)	6.58(1)	0.57(4)
$_{34}$Se			12.(1)	14.(3)	7.970(9)	
^{74}Se	0.89(4)		50.(2)	520(50)	0.8(3)	0.2(1)
^{75}Se		119.78 d	3.3(10)×10^{2}			
^{76}Se	9.37(29)		(22. + 63.)	(9.+31.)	12.2(1)	0.16(1)
^{77}Se	7.63(16)		42.(4)	30.(5)	8.25(8)	0.3(1)
			σ_α = 0.97(3) μb			
^{78}Se	23.77(28)		σ_m = 0.38(2)	RI_m = 4.3(4)	8.24(9)	0.1
^{80}Se	49.61(41)		(0.05+0.54)	(0.15+0.85)	7.48(3)	42.(3) mb
^{82}Se	8.73(22)	≈ 1×10^{20} y	(39.+ 5.2) mb	39.(4) mb	6.34(8)	0.04(2)
$_{35}$Br			6.8(2)	92.(8)	6.79(2)	
^{76}Br		16.0 h	224.(42)			
^{79}Br	50.69(7)		(2.5+8.3)	(36.+96.)	6.79(7)	0.63(4)
						σ_m = 0.08(1)
^{81}Br	49.31(7)		(2.4+0.24)	51.(5)	6.78(7)	0.31(2)
$_{36}$Kr			24.(1)	39.(6)	7.81(2)	
^{78}Kr	0.353(3)	>2.3×10^{20} y	(0.17+6.)	20.(1)		(0.11+0.19)
^{80}Kr	2.286(10)		(4.6+7.)	57.(6)		(0.09+0.18)
^{82}Kr	11.593(3)		(14.+7.)	130.(13)		90.(6) mb
^{83}Kr	11.500(19)		183.(30)	183.(20)		0.24(2)
^{84}Kr	56.987(15)		(σ_m+ σ_g) = 0.11	2.4(3)		(16.+33.) mb
			σ_m = 0.09			
^{85}Kr		10.73 y	1.7(2)	1.8(10)		0.07(2)
^{86}Kr	17.279(41)		3.(2) mb	≈ 1. mb	8.1(3)	3.2(4) mb
$_{37}$Rb			0.39(4)	6.(3)	7.08(2)	
^{84}Rb		32.9 d	σ_p = 12.(2)			
^{85}Rb	72.17(2)		(0.06+0.38)	(0.7+7.)	7.0(1)	0.24(1)
^{86}Rb		18.65 d	<20.			
^{87}Rb	27.83(2)	4.88×10^{10} y	0.10(1)	2.3(4)	7.3(1)	16.(1) mb
^{88}Rb		17.7 m	1.2(3)	0.5(1)		
$_{38}$Sr			1.2(1)	10.(1)	7.02(2)	
^{84}Sr	0.56(1)		(0.6+0.2)	(9.+1.)		0.4(1)
^{86}Sr	9.86(1)		σ_m = 0.81(4)	RI_m = 4.(1)	5.68(5)	(48.+22.) mb
^{87}Sr	7.00(1)		16.(3)	118.(30)	7.41(7)	97.(5) mb
^{88}Sr	82.58(1)		5.8(4) mb	0.07(3)	7.16(6)	6.0(2) mb
^{89}Sr		50.52 d	0.42(4)	0.2		
^{90}Sr		29.1 y	10.(1) mb	0.10(2)		
$_{39}$Y			1.25(5)	1.0(1)	7.75(2)	
^{89}Y	100.		(0.001+1.25)	(0.006+1.0)	7.75(2)	19.(1) mb
^{90}Y		2.67 d	<6.5			
^{91}Y		58.5 d	1.4(3)	0.6(1)		
$_{40}$Zr			0.19(1)	0.95(9)	7.16(3)	
			σ_α < 0.1 mb			
^{90}Zr	51.45(40)		≈ 0.014	0.2(1)	6.4(1)	21.(2) mb
^{91}Zr	11.22(5)		1.2(3)	5.(2)	8.8(1)	60.(8) mb

* Extrapolated value.

Elem. or isot.	Natural abundance (%)	Half-life	Thermal neut. cross section (barns)	Resonance integral (barns)	Coh. scat. length (fm)	σ (30 keV) Maxw. avg. (barns)
^{92}Zr	17.15(8)		0.2(1)	0.6(2)	7.5(2)	33.(4) mb
^{93}Zr		1.5×10^{6} y	<4.	16.(5)		0.10(1)
^{94}Zr	17.38(28)	>10^{17} y	0.049(6)	0.25(3)	8.3(2)	26.(1) mb
^{96}Zr	2.80(9)	>1.7×10^{18} y	0.020(3)	5.0(5)	5.5(1)	11.(1) mb
$_{41}$Nb			1.11(1)	8.5(6)	7.14(3)	
			σ_α < 0.1 mb			
^{93}Nb	100.		1.1	(6.3+2.2)	7.14(3)	266.(5) mb
			σ_m = 0.86			
^{94}Nb		2.4×10^{4} y	(σ_m+ σ_g) = 15.(1)	126.(13)		
			σ_m = 0.6(1)			
^{95}Nb		34.97 d	<7.	<200.		
$_{42}$Mo			2.5(1)	26.(5)	6.72(2)	
			σ_α < 0.1 mb			
^{92}Mo	14.77(31)	>3×10^{17} y	0.06	≈ 0.8	6.93(8)	0.07(1)
			σ_m = 0.2 μb			
^{94}Mo	9.226(99)		0.02	≈ 0.8	6.82(7)	0.10(2)
^{95}Mo	15.900(85)		13.4(3)	109.(5)	6.93(6)	0.29(1)
			σ_α = 30.(4) μb			
^{96}Mo	16.674(12)		0.5	17.(3)	6.22(6)	0.11(1)
^{97}Mo	9.560(50)		2.5(2)	14.(3)	7.26(8)	0.34(1)
			s_a = 0.4(2) mb			
^{98}Mo	24.20(25)		0.14(1)	7.2(7)	6.60(7)	0.10(1)
^{100}Mo	9.67(20)	≈ 1×10^{19} y	0.19(1)	3.6(3)	6.75(7)	0.11(1)
$_{43}$Tc						
^{98}Tc		≈ 6.6×10^{6} y	σ_m = 0.9(2)			
^{99}Tc		2.13×10^{5} y	23.(2)	4.0(4)×10^{2}	6.8(3)	0.93(5)
$_{44}$Ru			2.6 (1)	48.(5)	7.03(3)	
^{96}Ru	5.54(14)	>3.1×10^{16} y	0.23(4)	7.(2)		0.21(1)
^{98}Ru	1.87(3)		< 8.			0.3(1)
^{99}Ru	12.76(14)		4.(1)	195.(20)		1.2(3)
^{100}Ru	12.60(7)		5.8(6)	11.(2)		0.21(1)
^{101}Ru	17.06(2)		5.(1)	1.1(3)×10^{2}		1.00(4)
			σ_α <0.15 μb			
^{102}Ru	31.55(14)		1.2(1)	4.3(5)		0.15(1)
^{103}Ru		39.27 d	<20.	≈ 30.		
^{104}Ru	18.62(27)		0.49(2)	6.(2)		0.15(1)
^{105}Ru		4.44 h	0.29(3)	0.13(1)		
^{106}Ru		1.020 y	0.15(4)	2.0(6)		
$_{45}$Rh			145.(2)	1.2(1)×10^{3}	5.88(4)	
^{103}Rh	100.		(11.+ 134.)	(0.08+1.1)×10^{3}	5.88(4)	0.81(1)
^{104m}Rh		4.36 m	800.(100)			
^{104}Rh		42.3 s	40.(30)			
^{105}Rh		35.4 h	1.1(3)×10^{4}	1.7(4)×10^{4}		
$_{46}$Pd			7.(1)	82.(8)	5.91(6)	
^{102}Pd	1.02(1)		3.2(10)	10.(2)		0.3(1)
^{104}Pd	11.14(8)			16.(2)		0.29(3)
^{105}Pd	22.33(8)		22.(2)	60.(20)	5.5(3)	1.20(6)
			σ_α = 0.5(2) μb			
^{106}Pd	27.33(3)		(0.013+0.28)	(0.2+5.5)	6.4(4)	0.25(3)
^{107}Pd		6.5×10^{6} y	1.8(2)	108.(4)		1.34(6)
^{108}Pd	26.46(9)		(0.19+8.5)	(2.+240.)	4.1(3)	0.20(2)
^{110}Pd	11.72(9)		(0.033+0.7)	(0.7+8.)		0.15(2)
$_{47}$Ag			62.(1)	767.(60)	5.922(7)	
^{107}Ag	51.839(8)		(1.+35.)	(3.+105.)	7.56(1)	0.80(3)
^{109}Ag	48.161(8)		(4.1 + 87.)	(0.7+14.1)×10^{2}	4.17(1)	0.79(3)
^{110m}Ag		249.8 d	82.(11)	20.(4)		
^{111}Ag		7.47 d	3.(2)	105.(20)		
$_{48}$Cd			2.52(5)×10^{3}	73.(8)	4.87(5)	

* Extrapolated value.

Elem. or isot.	Natural abundance (%)	Half-life	Thermal neut. cross section (barns)	Resonance integral (barns)	Coh. scat. length (fm)	σ (30 keV) Maxw. avg. (barns)
^{106}Cd	1.25(6)	>2.6×10^{17} y	0.20(3)	4.(1)		0.30(2)
^{108}Cd	0.89(3)	>4.1×10^{17} y	1.	14.(3)	5.4(1)	0.20(1)
^{109}Cd		462.0 d	≈ 180.	6.7(12)×10^3		
			σ_α <0.05			
^{110}Cd	12.49(18)		(0.06+11.)	(6.+34.)	5.9(1)	(0.01+0.22)
^{111}Cd	12.80(12)		3.5(20)	51.(6)	6.5(1)	0.75(1)
^{112}Cd	24.13(21)		(0.012+2.2)	15.	6.4(1)	0.19(1)
^{113}Cd	12.22(12)	7.7×10^{15} y	2.06(4)×10^4	390.(40)	– 8.0(2)	0.67(1)
			s_a <1. mb			
^{114}Cd	28.73(42)		(0.04+0.29)	16.(7)	7.5(1)	(0.01+0.12)
^{116}Cd	7.49(18)	3.8×10^{19} y	(26.+52.) mb	1.2	6.3(1)	(12.+47.) mb
$_{49}$In			197.(4)	3.3(2)×10^3	4.07(2)	
^{113}In	4.29(5)		(3.1+5.0+3.9)	(220.+90.)	5.39(6)	(0.48+0.31)
^{115}In	95.71(5)	4.4×10^{14} y	(88.+73.+44.)	(1.5+1.2+0.7)×10^3	4.01(2)	(0.69+0.02)
$_{50}$Sn			0.61(3)	8.(2)	6.225(2)	
^{112}Sn	0.97(1)		(0.15+0.40)	(8.+19.)		0.21(1)
^{113}Sn		115.1 d	≈ 9.	210.(50)		
^{114}Sn	0.66(1)		≈ 0.12	5.(1)	6.2(3)	134.(3) mb
^{115}Sn	0.34(1)		σ_α = 0.06 mb	29.(6)		0.34(1)
^{116}Sn	14.54(9)		(0.006+0.14)	(0.5+11.)	5.93(5)	91.(2) mb
^{117}Sn	7.68(7)		1.1(1)	16.(5)	6.48(5)	319.(7) mb
^{118}Sn	24.22(9)		σ_m = 4. mb	4.7(5)	6.07(5)	62.(1) mb
^{119}Sn	8.59(4)		2.(1)	2.9(5)	6.12(5)	0.18(1)
^{120}Sn	32.58(9)		(0.001+0.13)	1.2(3)	6.49(5)	(0.5+36.) mb
^{122}Sn	4.63(3)		(0.15+0.001)	0.81(4)	5.74(5)	(18.+4.) mb
^{124}Sn	5.79(5)	>2.2×10^{18} y	(0.13+0.004)	(8.0+0.08)	5.97(5)	12.(2) mb
$_{51}$Sb			5.2(2)	169.(20)	5.57(3)	
^{121}Sb	57.21(5)		(0.4+5.8)	(13.+192.)	5.71(6)	0.53(2)
^{123}Sb	42.79(5)		(0.02+0.04+4.0)	(1.+119.)	5.38(7)	0.30(1)
^{124}Sb		60.20 d	17.(3)	≈ 8.		
$_{52}$Te			4.2(1)	47.(3)	5.80(3)	
^{120}Te	0.09(1)		(1.+5.)	≈ 1.	5.3(5)	0.4(1)
^{122}Te	2.55(12)		(0.4+3.)	(5.+75.)	3.8(2)	295.(3) mb
^{123}Te	0.89(3)	>5.3×10^{16} y	370.(40)	4.5(3)×10^3	– 0.05	0.83(1)
			σ_α = 0.05 mb			
^{124}Te	4.74(14)		(1.+6.)	(1.4+4.)	8.0(1)	155.(2) mb
^{125}Te	7.07(15)		1.1(2)	21.(4)	5.02(8)	431.(4) mb
^{126}Te	18.84(25)		(0.12+0.8)	(0.6+7.4)	5.56(7)	(28.+53.) mb
^{128}Te	31.74(8)	2.2×10^{24} y	(0.03+0.2)	(0.2+1.6)	5.89(7)	(3.+41.) mb
^{130}Te	34.08(62)	8.×10^{20} y	(0.01+0.19)	(0.03+0.3)	6.02(7)	(4.+11.) mb
$_{53}$I			6.2(1)	1.5(1)×10^2	5.28(2)	
^{125}I		59.4 d	900.(100)	1.4(2)×10^4		
^{127}I	100.		6.2(1)	1.5(1)×10^2	5.28(2)	0.64(3)
^{128}I		25.00 m	22.(4)	≈ 10.		
^{129}I		1.7×10^7 y	(20.7+10.3)	36.(4)		0.44(2)
^{130}I		12.36 h	18.(3)	≈ 8.		
^{131}I		8.021 d	≈ 0.7	8.(4)		
$_{54}$Xe			25.(1)	263.(50)	4.92(3)	
^{124}Xe	0.0953(27)	>10^{17} y	(28.+137.)	(0.6+3.0)×10^3		(0.13+0.51)
^{125}Xe		17.1 h	σ_α < 0.03			
^{126}Xe	0.0890(14)		(0.45+3.)	(8.+52.)		(0.04+0.32)
^{127}Xe		36.34 d	σ_α ≈ 0.01			
^{128}Xe	1.910(22)		σ_m = 0.48	RI_m = 38.(10)		0.26(1)
^{129}Xe	26.40(18)		22.(5)	250.(50)		0.62(2)
^{130}Xe	4.071(53)		σ_m = 0.45	RI_m = 16.(4)		0.132(3)
^{131}Xe	21.233(62)		90.(10)	9.(1)×10^2		0.45(8)
^{132}Xe	26.9087(680)		(0.05+0.4)	(0.9+3.7)		(5.+60.) mb
^{133}Xe		5.243 d	190.(90)			

* Extrapolated value.

Elem. or isot.	Natural abundance (%)	Half-life	Thermal neut. cross section (barns)	Resonance integral (barns)	Coh. scat. length (fm)	σ (30 keV) Maxw. avg. (barns)
^{134}Xe	10.436(29)	>1.1×10^{16} y	(0.003 + 0.26)	0.40(4)		20.(2) mb
^{135}Xe		9.10 h	2.65(11)×10^{6}	7.6(5)×10^{3}		
^{136}Xe	8.858(33)	>8×10^{20} y	0.26(2)	0.7(2)		0.9(1) mb
$_{55}$Cs			30.4(8)	422.(50)	5.42(2)	
^{132}Cs		6.48 d	σ_α < 0.15			
^{133}Cs	100.		(2.7+27.3)	(32.+360.)	5.42(2)	(0.04+0.47)
^{134}Cs		2.065 y	140.(10)	54.(9)		
^{135}Cs		2.3×10^{6} y	8.3(3)	38.(3)		
^{137}Cs		30.2 y	(0.20+0.07)	0.36(7)		
$_{56}$Ba			1.3(2)	10.(2)	5.07(3)	
^{130}Ba	0.106(1)	2.2×10^{21} y	(1.+8.)	(25.+200.)	– 3.6(6)	0.76(11)
^{132}Ba	0.101(1)	1.3×10^{21} y	(0.84+9.7)	(4.7+24.)	7.8(3)	0.6(1)
^{133}Ba		10.53 y	4.(1)	85.(30)		
^{134}Ba	2.417(18)		(0.1+1.3)	(5.6+18.)	5.7(1)	0.18(1)
^{135}Ba	6.592(12)		(0.014+5.8)	(0.47+131.)	4.7(1)	0.46(2)
^{136}Ba	7.854(24)		(0.010+0.44)	(0.1+1.5)	4.91(8)	61.(2) mb
^{137}Ba	11.232(24)		5.(1)	4.(1)	6.8(1)	76.(3) mb
^{138}Ba	71.698(42)		0.41(2)	0.4(1)	4.84(8)	4.0(2) mb
^{139}Ba		1.396 h	5.(1)	2.2(5)		
^{140}Ba		12.75 d	1.6(3)	14.(1)		
$_{57}$La			9.2(2)	12.(1)	8.24(4)	
^{138}La	0.090(1)	1.06×10^{11} y	57.(6)	4.1(9)×10^{2}		
^{139}La	99.910(1)		9.2(2)	12.(1)	8.24(4)	38.(3) mb
^{140}La		1.678 d	2.7(3)	69.(4)		
$_{58}$Ce			0.64(4)	0.71(6)	4.84(2)	
^{136}Ce	0.185(2)		(1.0+6.5)	58.(12)	5.80(9)	(0.028+0.3)
^{138}Ce	0.251(2)		(0.025+1.0)	(1.5+5.2)	6.70(9)	179.(5) mb
^{140}Ce	88.450(51)		0.58(4)	0.50(5)	4.84(9)	11.0(4) mb
^{141}Ce		32.50 d	29.(3)	13.(2)		
^{142}Ce	11.114(51)	>1.6×10^{17} y	0.97(3)	1.3(3)	4.75(9)	28.(1) mb
^{143}Ce		1.38 d	6.1(7)	2.7(3)		
^{144}Ce		284.6 d	1.0(1)	2.6(3)		
$_{59}$Pr			11.5(4)	14.(3)	4.58(5)	
^{141}Pr	100.		(4.+7.5)	14.(3)	4.58(5)	111.(2) mb
^{142}Pr		19.12 h	20.(3)	9.(1)		
^{143}Pr		13.57 d	90.(10)	190.(25)		
$_{60}$Nd			51.(2)	49.(5)	7.69(5)	
^{142}Nd	27.2(5)		19.(1)	34.(11)	7.7(3)	35.(1) mb
^{143}Nd	12.2(2)		330.(10) σ_α = 17. mb	128.(30)		0.24(1)
^{144}Nd	23.8(3)	2.1×10^{15} y	3.6(3)	3.9(5)	2.8(3)	81.(2) mb
^{145}Nd	8.3(1)		47.(6) σ_α = 12. μb	260.(40)		0.42(1)
^{146}Nd	17.2(3)		1.5(2)	3.0(4)	8.7(2)	91.(1) mb
^{147}Nd		10.98 d	440.(150)	200.		
^{148}Nd	5.7(1)		2.4(1)	13.(2)	5.7(3)	147.(2) mb
^{150}Nd	5.6(2)	≈ 1×10^{19} y	1.0(1)	14.(2)	5.3(2)	0.16(1)
$_{61}$Pm						
^{146}Pm		5.53 y	8.4(1.7)×10^{3}			
^{147}Pm		2.623 y	(84.+96.)	(1000.+1280.)	12.6(4)	2.(1)
^{148m}Pm		41.3 d	10600.(800)			
^{148}Pm		5.37 d	≈ 10^{3}	2.6(2.4)×10^{3}		
^{149}Pm		2.212 d	1400.(200)			
^{151}Pm		1.183 d	≈ 150.			
$_{62}$Sm			5.6(1)×10^{3}	1.4(2)×10^{3}		
^{144}Sm	3.07(7)		1.6(1)	2.4(3)		92.(6) mb
^{145}Sm		340. d	280.(20)	600.(90)		
^{147}Sm	14.99(18)	1.06×10^{11} y	56.(4), σ_α = 0.6 mb	710.(50)	14.(3)	0.97(1)

* Extrapolated value.

Elem. or isot.	Natural abundance (%)	Half-life	Thermal neut. cross section (barns)	Resonance integral (barns)	Coh. scat. length (fm)	σ (30 keV) Maxw. avg. (barns)
^{148}Sm	11.24(10)	7×10^{15} y	2.4(6)	27.(14)		241.(2) mb
^{149}Sm	13.82(7)	10^{16} y	4.01(6)×10^{4}, σ_α = 31. mb	3.1(5)×10^{3}		1.82(2)
^{150}Sm	7.38(1)		102.(5)	290.(30)	14.(3)	422.(4) mb
^{151}Sm		90. y	1.52(3)×10^{4}	3520.(60)		2.(1)
^{152}Sm	26.75(16)		206.(15)	3.0(3)×10^{3}	– 5.0(6)	473.(4) mb
^{153}Sm		1.929 d	420.(180)			
^{154}Sm	22.75(29)		7.5(3)	32.(6)	9.(1)	0.21(1)
$_{63}$Eu			4570.(100)	3.8(5)×10^{3}	5.3(3)	
^{151}Eu	47.81(6)		(4.+3150.+6000.) σ_α = 8.7(3) μb	(2.+4.)×10^{3}		(1.6+2.2)
^{152m1}Eu		9.30 h	6.8(15)×10^{4}	< 10^{5}		
^{152}Eu		13.5 y	1.1(2)×10^{4}	1.6(2)×10^{3}		5.(2)
^{153}Eu	52.19(6)		300.(20), σ_α <1. mb	1.8(4)×10^{3}	8.2(1)	2.8(1)
^{154}Eu		8.59 y	1.5(3)×10^{3}	1.6(2)×10^{3}		4.4(7)
^{155}Eu		4.76 y	3.9(2)×10^{3}	1.6(2)×10^{4}		1.3(1)
$_{64}$Gd			48.8(6)×10^{3}	400.(10)	9.5(2)	
^{148}Gd		75. y	1.40(14)×10^{4}			
^{152}Gd	0.20(1)	1.1×10^{14} y	700.(200), σ_α <7. mb	700.(200)		1.05(2)
^{153}Gd		240. d	2.(1)×10^{4}, σ_α = 0.03			
^{154}Gd	2.18(3)		(0.035+60.)	230.(50)		1.03(1)
^{155}Gd	14.80(12)		61.(1)×10^{3}, σ_α = .08 mb	1540.(100)		2.65(3)
^{156}Gd	20.47(9)		≈ 2.0	104.(15)	6.3(4)	615.(5) mb
^{157}Gd	15.65(2)		2.54(3)×10^{5}, σ_α <0.05	800.(100)		1.37(2)
^{158}Gd	24.84(7)		2.3(3)	73.(7)	9.(2)	324.(3) mb
^{160}Gd	21.86(19)	>1.9×10^{19} y	1.5(7)	6.(1)	9.15(5)	0.15(2)
^{161}Gd		3.66 m	2.0(6)×10^{4}			
$_{65}$Tb			23.2(5)	420.(50)	7.34(2)	
^{159}Tb	100.		23.2(5)	420.(50)	7.34(2)	1.6(2)
^{160}Tb		72.3 d	570.(110)			
$_{66}$Dy			9.5(2)×10^{2}	1.5(2)×10^{3}	16.9(3)	
^{156}Dy	0.056(3)		33.(3), σ_α < 9. mb	1000.(100)		1.6(2)
^{158}Dy	0.095(3)		43.(6), σ_α < 6. mb	120.(10)	6.1(5)	0.8(2)
^{159}Dy		144. d	8.(2)×10^{3}			
^{160}Dy	2.39(18)		60.(10), σ_α < 0.3 mb	1100.(200)	6.7(4)	0.89(1)
^{161}Dy	18.889(42)		600.(50), σ_α < 1. μb	1100.(100)	10.3(4)	1.96(2)
^{162}Dy	25.475(36)		170.(20)	2755.(300)	– 1.4(5)	446.(4) mb
^{163}Dy	24.896(42)		120.(10), σ_α < 20. μb	1600.(400)	5.0(4)	1.11(1)
^{164}Dy	28.260(54)		(1.7+1.0)×10^{3}	(4.+2.)×10^{2}	49.4(2)	212.(3) mb
^{165m}Dy		1.26 m	2.0(6)×10^{3}			
^{165}Dy		2.33 h	3.5(3)×10^{3}	2.2(3)×10^{4}		
$_{67}$Ho			61.(2)	670.(40)	8.01(8)	
^{163}Ho		4.57×10^{3} y				(0.4+1.7)
^{165}Ho	100.		(3.1+58.), σ_α < 20. μb	(?+670.)	8.01(8)	(0.8+0.5)
^{166m}Ho		1.2×10^{3} y	3.1(8)×10^{3}	10.(3)×10^{3}		
$_{68}$Er			1.5(2)×10^{2}	730.(10)	7.79(2)	
^{162}Er	0.139(5)		19.(3), σ_α < 11. mb	480.(50)	8.8(2)	1.6(1)
^{164}Er	1.601(3)		13.(3), σ_α < 1.2 mb	105.(10)	8.2(2)	1.08(5)
^{166}Er	33.503(36)		(3.+14.), σ_α < 70. μb	96.(12)	10.6(2)	0.56(6)
^{167}Er	22.869(9)		6.5(8)×10^{2}, σ_α = 3. μb	2970.(70)	3.0(3)	1.4(2)
^{168}Er	26.978(18)		2.3(3), σ_α = 0.09 mb	37.(5)	7.4(4)	0.34(4)
^{170}Er	14.910(36)		8.(2)	26.(4)	9.6(5)	0.17(1)
^{171}Er		7.52 h	370.(40)	170.(20)		
$_{69}$Tm			108.(4)	1.5(2)×10^{3}	7.07(3)	
^{169}Tm	100		(8.+100.)	1.5(2)×10^{3}	7.07(3)	1.13(6)
^{170}Tm		128.6 d	100.(20)	460.(50)		
^{171}Tm		1.92 y	≈ 160.	118.(6)		
$_{70}$Yb			52.(10)	1.7(2)×10^{2}	12.43(3)	
^{168}Yb	0.13(1)		2.4(2)×10^{3}, σ_α < 0.1 mb	2.0(5)×10^{4}	–4.07(2)	0.7(4)

* Extrapolated value.

Elem. or isot.	Natural abundance (%)	Half-life	Thermal neut. cross section (barns)	Resonance integral (barns)	Coh. scat. length (fm)	σ (30 keV) Maxw. avg. (barns)
^{169}Yb		32.02 d	3.6(3)×10^3	5200.(500)		
^{170}Yb	3.04(15)		12.(2), σ_α < 10. μb	320.(30)	6.8(1)	0.77(1)
^{171}Yb	14.28(57)		53.(5), σ_α < 1.5 μb	315.(30)	9.7(1)	1.21(1)
^{172}Yb	21.83(67)		≈ 1.3, σ_α < 1. μb	25.(3)	9.4(1)	0.34(1)
^{173}Yb	16.13(27)		16.(2), σ_α < 1. mb	380.(30)	9.56(7)	0.75(1)
^{174}Yb	31.83(92)		(46.+17.), σ_α < 0.02 μb	(13.+16.)	19.3(1)	151.(2) mb
^{176}Yb	12.76(41)		3.1(2), σ_α < 1. μb	8.(2)	8.7(1)	116.(2) mb
$_{71}$Lu			78.(7)	8.3(7)×10^2	7.21(3)	
^{175}Lu	97.41(2)		(16.+8.)	(550.+270.)	7.24(3)	(1.04+0.11)
^{176}Lu	2.59(2)	3.73×10^{10} y	(2.+2100.)	(3.+930.)	6.1(2)	1.53(7)
^{177m}Lu		160.7 d	3.2(3)	1.4(2)		
^{177}Lu		6.65 d	1000.(300)			
$_{72}$Hf			106.(3)	19.7(5)×10^2	7.8(1)	
^{174}Hf	0.16(1)	2.0×10^{15} y	600.(50)	400.(50)	11.(1)	0.8(2)
^{176}Hf	5.26(7)		23.(4)	700.(100)	6.6(2)	0.46(2)
^{177}Hf	18.60(9)		(1.+375.), σ_α < 20. μb	7170.(200)		1.5(1)
^{178m2}Hf		31. y	σ_{m2} = 45.(5)	RI_{m2} = 8(1)×10^2		
^{178}Hf	27.28(7)		(54.+32.)	(0.9+1.0)×10^3	5.9(2)	0.31(1)
^{179}Hf	13.62(2)		(0.43+46.)	(6.8+620.)	7.5(2)	(0.01+0.95)
^{180}Hf	35.08(16)		13.0(5), σ_α < 13. μb	32.(1)	13.2(3)	179.(5) mb
^{181}Hf		42.4 d	30.(25)			
$_{73}$Ta			20.(1)	650(20.)	6.91(7)	
^{179}Ta		1.8 y	9.3(6)×10^2	1.22(7)×10^3		
^{180m}Ta	0.012(2)	> 1.2×10^{15} y	≈ 560.	1350.(100)		
^{181}Ta	99.988(2)		(0.012 + 20.), σ_α <1. μb	(0.4+650.)	6.91(7)	0.77(2)
^{182}Ta		114.43 d	8200.(600)	900.(90)		
$_{74}$W			18.(1)	3.6(3)×10^2	4.86(2)	
^{180}W	0.12(1)	7.4×10^{16} y	≈ 4.	210.(30)		0.54(6)
^{182}W	26.50(16)	8.3×10^{18} y	20.(1)	600.(90)	6.97(4)	274.(8) mb
^{183}W	14.31(4)	1.9×10^{18} y	10.5(3)	340.(50)	6.53(4)	0.52(2)
^{184}W	30.64(2)	4.0×10^{18} y	(0.002 + 2.0)	15.(2)	7.48(6)	0.22(1)
^{185}W		74.8 d	≈ 3.3	300.(50)		
^{186}W	28.43(19)	6.5×10^{18} y	37.(2)	510.(50)	– 0.72(4)	176.(5) mb
^{187}W		23.9 h	70.(10)	2760.(550)		
^{188}W		69.78 h	12.(1)			
$_{75}$Re			90.(4)	8.4(2)×10^2	9.2(3)	
^{185}Re	37.40(2)		(0.33+110.)	1700.(50)	9.0(3)	1.54(6)
^{187}Re	62.60(2)	4.2×10^{10} y	(2.+72.)	(9.+310.)	9.3(3)	1.16(6)
$_{76}$Os			17.(1)	1.5(1)×10^2	10.7(2)	
^{184}Os	0.02(1)	>5.6×10^{13} y	3.3(3)×10^3, σ_α <10. mb	1.4(1)×10^3		0.4(2)
^{186}Os	1.59(3)	2.×10^{15} y	≈ 80., σ_α < 0.1 mb	3.8(9)×10^2	12(2)	0.42(2)
^{187}Os	1.96(2)		2.(1)×10^2, σ_α < 0.1 mb	5.0(7)×10^2		0.90(3)
^{188}Os	13.24(8)		≈ 5., σ_α < 30. μb	1.5(2)×10^2	7.6(3)	0.40(2)
^{189}Os	16.15(5)		(0.00026+40.), σ_α < 10. μb	(0.013+670.)	10.7(3)	1.17(5)
^{190}Os	26.26(2)		(9.+4.), σ_α < 20. μb	(22.+8.)	11.0(3)	0.30(5)
^{191}Os		15.4 d	3.8(6)×10^2	1.7(3)×10^2		
^{192}Os	40.78(19)		3.(1), σ_α < 10. μb	7.(1)	11.5(4)	0.31(5)
^{193}Os		30.5 h	2.5(5)×10^2	1.1(2)×10^2		
$_{77}$Ir			4.2(1)×10^2	2.8(4)×10^3		10.6(3)
^{191}Ir	37.3(2)		(0.14+660.+260.)	(1.0+4.2)×10^3		1.35(4)
^{192}Ir		73.83 d	1.4(3)×10^3	4.8(7)×10^3		
^{193}Ir	62.7(2)		(0.04+6.+109.)	1.4(2)×10^3		0.99(7)
^{194}Ir		19.3 h	1.6(3)×10^3	7.(2)×10^2		
$_{78}$Pt			10.(1)	1.3(1)×10^2	9.60(1)	
^{190}Pt	0.014(1)	4.5×10^{11} y	1.5(1)×10^2, σ_α < 8. mb	70.(10)	9.(1)	0.7(2)
^{192}Pt	0.782(7)		(2.0+6.), σ_α < 0.2 mb	115.(20)	9.9(5)	0.6(1)
^{194}Pt	32.967(99)		(0.1+1.1), σ_α < 5. μb	(4.+?)	10.55(8)	(0.03+0.34)
^{195}Pt	33.832(10)		28.(1), σ_α < 5. μb	365.(50)	8.8(1)	0.9(2)

* Extrapolated value.

Elem. or isot.	Natural abundance (%)	Half-life	Thermal neut. cross section (barns)	Resonance integral (barns)	Coh. scat. length (fm)	σ (30 keV) Maxw. avg. (barns)
^{196}Pt	25.242(41)		(0.045+0.55)	7.(2)	9.89(8)	(0.01+0.19)
^{198}Pt	7.163(55)		(0.3+3.1)	(5.+53.)	7.8(1)	(3.+79.) mb
^{199}Pt		30.8 m	≈ 15.	≈ 7.		
$_{79}$Au			98.7(1)	$1.55(3)\times10^3$	7.63(6)	
^{197}Au	100.		σ_{m+g} = 98.7(1)	RI_{m+g} = $1.55(3)\times10^3$	7.63(6)	582.(9) mb
			σ_m = 8.(2) mb	RI_m = 0.06(2)		
^{198}Au		2.695 d	$26.5(15)\times10^3$	≈ $4.\times10^4$		
^{199}Au		3.14 d	≈ 30.			
$_{80}$Hg			$3.7(1)\times10^2$	87.(5)	12.69(2)	
^{196}Hg	0.15(1)	>2.5×10^{18} y	(105.+3000.)	(53.+410.)	30.(1)	0.4(2)
^{198}Hg	9.97(8)		(0.017+2.)	(1.7+70.)		0.17(2)
^{199}Hg	16.87(10)		$2.1(2)\times10^3$	435(20)	16.9(4)	0.37(2)
^{200}Hg	23.10(16)		≈ 1.	2.1(5)		0.12(1)
^{201}Hg	13.18(8)		≈ 8.	30.(3)		0.26(1)
^{202}Hg	29.86(20)		4.9(5)	4.5(2)	11.(1)	74.(6) mb
^{204}Hg	6.87(4)		0.4(1)	0.8(2)		42.(4) mb
$_{81}$Tl			3.3(1)	12.5(8)	8.776(5)	
^{203}Tl	29.524(14)		11.(1), σ_α < 0.3 mb	41.(2)	7.0(2)	124.(8) mb
^{204}Tl		3.78 y	22.(2)	90.(20)		0.14(5)
^{205}Tl	70.476(14)		0.11(2)	0.6(2)	9.52(7)	54.(4) mb
$_{82}$Pb			0.172(2)	0.14(4)	9.402(2)	
^{204}Pb	1.4(1)		0.68(7)	2.0(2)	10.9(1)	90.(6) mb
^{205}Pb		1.51×10^7 y	≈ 5.	≈ 2.		0.06(1)
^{206}Pb	24.1(1)		0.027(1)	0.10(1)	9.23(5)	16.(1) mb
^{207}Pb	22.1(1)		0.61(3)	0.38(1)	9.28(2)	10.(1) mb
^{208}Pb	52.4(1)	>2×10^{19} y	0.23(1) mb, σ_α < 8. μb	2.0(2) mb	9.50(3)	0.36(4) mb
^{210}Pb		22.6 y	< 0.5			
$_{83}$Bi			0.034(1)	0.19(2)	8.532(2)	
^{209}Bi	100.		(11.+23.) mb, σ_α < 0.3 μb	0.19(2)	8.532(2)	2.7(5) mb
^{210m}Bi		3.0×10^6 y	54.(4) mb	0.20(3)		
$_{84}$Po						
^{210}Po		138.4 d	σ_m < 0.5 mb, σ_α < 2. mb			
			σ_g < 30. mb, σ_f < 0.1			
$_{85}$At						
$_{86}$Rn						
^{220}Rn		55.6 s	<0.2			
^{222}Rn		3.823 d	0.74(5)			
$_{88}$Ra						
^{223}Ra		11.43 d	$1.3(2)\times10^2$, σ_f < 0.7			
^{224}Ra		3.66 d	12.0(5)			
^{226}Ra		1599. y	≈ 13., σ_f < 7. μb	280.(50)		10.(1)
^{228}Ra		5.76 y	36.(5), σ_f < 2.			
$_{89}$Ac						
^{227}Ac		21.77 y	$8.8(7)\times10^2$, σ_f < 0.35 mb	$1.5(4)\times10^3$		
$_{90}$Th			7.4	85.(3)		10.31(3)
^{227}Th		18.72 d	σ_f = $2.0(2)\times10^2$			
^{228}Th		1.913 y	$1.2(2)\times10^2$, σ_f < 0.3	1014.(400)		
^{229}Th		7.9×10^3 y	≈ 60.	$1.0(2)\times10^3$		
			σ_f = 30.(3)	RI_f = 466.(75)		
^{230}Th		7.54×10^4 y	23.4(5)	$1.0(1)\times10^3$		
			σ_f < 0.5 mb			
^{232}Th	100.	1.40×10^{10} y	7.37(4)	85.(3)		10.31(3)
			σ_f = 3.(1) μb			
			σ_α < 1. μb			
^{233}Th		22.3 m	$1.5(1)\times10^3$	$4.(1)\times10^2$		
			σ_f = 15.(2)			

* Extrapolated value.

Elem. or isot.	Natural abundance (%)	Half-life	Thermal neut. cross section (barns)	Resonance integral (barns)	Coh. scat. length (fm)	σ (30 keV) Maxw. avg. (barns)
^{234}Th		24.10 d	1.8(5)			
			σ_f < 0.01			
$_{91}$Pa						
^{230}Pa		17.4 d	1.5(3)×10^3			
^{231}Pa		3.25×10^4 y	2.0(1)×10^2	750.(80)		9.1(3)
			σ_f = 20.(1) mb	RI$_f$ = 0.05(1)		
^{232}Pa		1.31 d	4.6(10)×10^2	300.(70)		
			σ_f = 1.5(5)×10^3	RI$_f$ = 1.0(1)×10^3		
^{233}Pa		27.0 d	39.(2)	(460.+440.)		
			σ_m = 20.(4)			
			σ_g = 19.(3)			
			σ_f < 0.1			
$_{92}$U			3.4(3); σ_f = 4.2(1)	280.(20),RI$_f$ = 2.0		8.417(5)
^{230}U		20.8 d	σ_f ≈ 25.			
^{231}U		4.2 d	σ_f ≈ 250.			
^{232}U		70. y	73.(2)	280.(15)		
			σ_f = 74.(8)	RI$_f$ = 350.(30)		
^{233}U		1.592×10^5 y	47.(2)	137.(6)		10.1(2)
			σ_f = 5.3(1)×10^2	RI$_f$ = 760.(17)		
			σ_α < 0.2 mb			
^{234}U	0.0054(5)	2.455×10^5 y	96.(2)	660.(70)		12.(4)
			σ_f = 0.07(2)	RI$_f$ = 6.5		
^{235}U	0.7204(6)	7.04×10^8 y	95.(5)	144.(6)		10.47(4)
			σ_f = 586.(2)	RI$_f$ = 275(5)		
			σ_α < 0.1 mb			
^{236}U		2.342×10^7 y	5.1(3)	360.(15)		
			σ_f < 1.3 mb	RI$_f$ = 4.38(50)		
^{237}U		6.75 d	≈ 10^2	1200.(200)		
			σ_f < 0.35			
^{238}U	99.2742(10)	4.47×10^9 y	2.7(1)	277.(3)		8.402(5)
			σ_f ≈ 3. μb	1.54(15) mb		
			σ_α = 1.4(5) μb			
^{239}U		23.5 m	22.(2)			
			σ_f = 15.(3)			
$_{93}$Np						
^{234}Np		4.4 d	σ_f = 9.(3)×10^2			
^{235}Np		1.085 y	1.6(1)×10^2			
^{236m}Np		22.5 h	σ_f = 2.7(2)×10^3	7.(4)×10^2		
^{236}Np		1.55×10^5 y	σ_f = 3.0(2)×10^3	1.35(30)×10^3		
^{237}Np		2.14×10^6 y	1.7(1)×10^2	6.5(3)×10^2		10.6(1)
			σ_f = 20.(1) mb	RI$_f$ = 4.7		
^{238}Np		2.117 d	σ_f = 2.6(3)×10^3	1.4(3)×10^3		
^{239}Np		2.355 d	(32.+19.)			
			σ_f < 1.			
$_{94}$Pu						
^{236}Pu		2.87 y	σ_f = 1.6(3)×10^2	1000.(60)		
^{237}Pu		45.7 d	σ_f = 2.3(3)×10^3			
^{238}Pu		87.7 y	5.1(2)×10^2	1.6(2)×10^2		14.1(5)
			σ_f = 17.(1)	RI$_f$ = 26.(2)		
^{239}Pu		2.410×10^4 y	2.7(1)×10^2	2.0(2)×10^2		7.7(1)
			σ_f = 752.(3)	3.0(1)×10^2		
			σ_α ≈ 0.3 mb			
^{240}Pu		6.56×10^3 y	2.9(1)×10^2	8.4(3)×10^3		3.5(1)
			σ_f ≈ 59. mb	RI$_f$ = 3.2		
^{241}Pu		14.4 y	3.7(1)×10^2, σ_α <0.2 mb	1.6(1)×10^2		
			σ_f = 1.01(1)×10^3	5.7(4)×10^2		
^{242}Pu		3.75×10^5 y	19.(1)	1.1(1)×10^3		8.1(1)
			σ_f <0.2	RI$_f$ = 0.23		
^{243}Pu		4.956 h	<100.			

* Extrapolated value.

Elem. or isot.	Natural abundance (%)	Half-life	Thermal neut. cross section (barns)	Resonance integral (barns)	Coh. scat. length (fm)	σ (30 keV) Maxw. avg. (barns)
			$\sigma_f = 2.0(2)\times10^{2}$			
^{244}Pu		8.00×10^{7} y	1.7(1)	41.(3)		
^{245}Pu		10.5 h	$1.5(3)\times10^{2}$	220.(40)		
$_{95}$Am						
^{241}Am		432.7 y	$(0.6+6.4)\times10^{2}$	$(1.+14.)\times10^{2}$		
			$\sigma_f = 3.15(10)$	14.(1)		
^{242m}Am		141. y	$1.7(4)\times10^{3}$	≈ 200.		
			$\sigma_f = 5.9(3)\times10^{3}$	$RI_f = 1.8(1)\times10^{3}$		
^{242}Am		16.02 h	$\sigma_f = 2.1(2)\times10^{3}$	$RI_f = < 300.$		
			$3.3(5)\times10^{2}$	$\approx 1.5\times10^{2}$		
^{243}Am		7.37×10^{3} y	(75.+5.)	$(17.1+1.0)\times10^{2}$		8.3(2)
			$\sigma_f = 79.(2)$ mb	$RI_f = 0.056$		
^{244m}Am		≈ 26. m	$\sigma_f = 1.6(3)\times10^{3}$			
^{244}Am		10.1 h	$\sigma_f = 2.2(3)\times10^{3}$			
$_{96}$Cm						
^{242}Cm		162.8 d	≈ 20.	120.(50)		
			$\sigma_f \approx 5.$			
^{243}Cm		29.1 y	$1.3(1)\times10^{2}$	214.(20)		
			$\sigma_f = 6.2(2)\times10^{2}$	$RI_f = 1.6(1)\times10^{3}$		
^{244}Cm		18.1 y	15.(1)	640.(50)		9.5(3)
			$\sigma_f = 1.1(2)$	$RI_f = 10.8(8)$		
^{245}Cm		8.48×10^{3} y	$3.5(2)\times10^{2}$	110.(10)		
			$\sigma_f = 2.1(1)\times10^{3}$	$RI_f = 8.(1)\times10^{2}$		
^{246}Cm		4.76×10^{3} y	1.2(2)	120.(10)		9.3(2)
			$\sigma_f = 0.16(7)$	13.(2)		
^{247}Cm		1.56×10^{7} y	60.(30)	$5.(1)\times10^{2}$		
			$\sigma_f = 82.(5)$	$7.3(7)\times10^{2}$		
^{248}Cm		3.48×10^{5} y	2.6(3)	270.(30)		7.7(2)
			$\sigma_f = 0.36(7)$	13.(2)		
^{249}Cm		64.15 m	≈ 1.6			
^{250}Cm		$\approx 9.7\times10^{3}$ y	≈ 80.			
$_{97}$Bk						
^{249}Bk		320. d	$7.(1)\times10^{2}$	$9.(1)\times10^{2}$		
			$\sigma_f \approx 0.1$			
^{250}Bk		3.217 h	$\sigma_f = 1.0(2)\times10^{3}$			
$_{98}$Cf						
^{249}Cf		351. y	$5.0(3)\times10^{2}$	$7.7(4)\times10^{2}$		
			$\sigma_f = 1.7(1)\times10^{3}$	$RI_f = 2.1(3)\times10^{3}$		
^{250}Cf		13.1 y	$2.0(2)\times10^{3}$	$12.(2)\times10^{3}$		
			$\sigma_f = 110.(90)$	$RI_f = 160.(40)$		
^{251}Cf		9.0×10^{2} y	$2.9(2)\times10^{3}$	$1.6(1)\times10^{3}$		
			$\sigma_f = 4.5(5)\times10^{3}$	$RI_f = 5.5(3)\times10^{3}$		
^{252}Cf		2.65 y	20.(2)	43.(3)		
			$\sigma_f = 32.(4)$	$RI_f = 1.1(3)\times10^{2}$		
^{253}Cf		17.8 d	18.(2)	8.(1)		
			$\sigma_f = 1.3(2)\times10^{3}$			
^{254}Cf		60.5 d	4.5(10)	2.		
$_{99}$Es						
^{253}Es		20.47 d	(180.+5.8)	$(37.5+1.1)\times10^{2}$		
^{254m}Es		1.64 d	$\sigma_f = 1.8(1)\times10^{3}$			
^{254}Es		276. d	28.(3)	18.(2)		
			$\sigma_f = 1.8(2)\times10^{3}$	$RI_f = 1.2(3)\times10^{3}$		
^{255}Es		40. d	≈ 55.			
$_{100}$Fm						
^{255}Fm		20.1 h	26.(3)	14.(2)		
			$\sigma_f = 3.3(2)\times10^{3}$			
^{257}Fm		100.5 d	$\sigma_f = 3.0(2)\times10^{3}$			

* Extrapolated value.

COSMIC RADIATION

A. G. Gregory and R. W. Clay

The Nature of Cosmic Rays

Primary cosmic radiation, in the form of high energy nuclear particles, electrons and photons from outside the solar system and from the Sun, continually bombards our atmosphere. Secondary radiation, resulting from the interaction of the primary cosmic rays with atmospheric gas, is present at sea level and throughout the atmosphere.

Secondary radiation is collimated by absorption and scattering in the atmosphere and consists of a number of components associated with different particle species. High energy primary particles can produce large numbers of secondary particles forming an extensive air shower. Thus, a number of particles may then be detected simultaneously at sea level.

Primary particle energies accessible in the vicinity of the Earth range from ~10^8 eV to ~10^{20} eV. At the lower energies, the limit is determined by the inability of charged particles to traverse the heliosphere to us through the outward-moving solar wind. The upper energy limit is set by the practicality of building detectors to record particles with the extremely low fluxes found at those energies (O. C. Allkofer, 1975a; J. G. Wilson, 1976).

Primary Cosmic Rays

Primary Particle Energy Spectrum

Figure 1 shows the spectrum of primary particle energies. This includes all particle species. In differential form it is roughly a power law of intensity versus energy with an index of ~ −3. There appears to be a knee (a steepening) at a little above 10^{15} eV and an ankle (a flattening) above ~10^{18} eV. Figure 2 emphasizes the features in the spectrum at the highest energies through multiplying the flux with a strongly rising power law of energy. This figure should be used with caution as errors for the two axes are not now independent.

Data on the high energy cosmic ray spectrum are uncertain largely because of limited event statistics due to the very low flux which might best be measured in particles per square kilometer per century. The highest energy event recorded to 1995 had an energy of 3×10^{20} eV (D. J. Bird et al., 1993).

It is expected that the highest energy cosmic rays will interact with the 2.7 K cosmic microwave background through photoproduction or photodisintegration. These interactions will appreciably reduce the observed flux of cosmic rays with energies above 5×10^{19} eV if they travel further than ~150 million light-years. This process is known as the Greisen-Zatsepin-Kuz'min (GZK) cutoff (P. Sokolsky, 1989).

At energies below ~10^{13} eV, solar system magnetic fields and plasma can modulate the primary component and Figure 3 shows the extent of this modulation between solar maximum and minimum (E. Juliusson, 1975; J. Linsley, 1981).

Primary Particle Energy Density

If the above spectrum is corrected for solar effects, the energy density above a particle energy of 10^9 eV outside the solar system is found to be ~5×10^5 eV m^{-3}. As the threshold energy is increased, the energy density decreases rapidly, being 2×10^4 eV m^{-3} above 10^{12} eV and 10^2 eV m^{-3} above 10^{15} eV. The energy density at lower energies outside the heliosphere is unknown but may be substantially greater if the particle rest mass energy is included together with the kinetic energy (A. W. Wolfendale, 1979).

Primary Particle Isotropy

This is measured as an anisotropy $(I_{max} - I_{min})/(I_{max} + I_{min}) \times 100\%$, where I, the intensity (m^{-2}s^{-1}sr^{-1}), is usually measured with an angular resolution of a few degrees.

The measured anisotropy is small and energy dependent. It is roughly constant in amplitude at between 0.05 and 0.1% (with a phase of 0 to 6 hours in right ascension) for energies between 10^{11} eV and 10^{14} eV and appears to increase at higher energies roughly as $0.4 \times (\text{Energy(eV)}/10^{16})^{0.5}$ % up to ~10^{18} eV. The latter rise may well be an artifact of the progressively more limited statistics as the flux drops rapidly with energy. It appears possible that a real anisotropy has been observed at the highest energies (above a few times 10^{19} eV) with a directional preference for the supergalactic plane (this plane reflects the directions of galaxies within about 100 million light-years) (A. W. Wolfendale, 1979; R. W. Clay, 1987; T. Stanev et al., 1995).

Primary Particle Composition

The composition of low energy cosmic rays is close to universal nuclear abundances except where propagation effects are present. For example, Li, Be, and B which are spallation products, are over-abundant by about six orders of magnitude.

Composition at 10^{11} eV per Nucleus

Charge	1	2	(3–5)	(6–8)	(10–14)	(16–24)	(26–28)	≥30
% Composition	50	25	1	12	7	4	4	0.1

Measurements at higher energies indicate that there is an increase in the relative abundances of nuclei with charges greater than 6 at energies above 50 TeV/nucleus (K. Asakimori et al., 1993) (1 TeV = 10^{12} eV).

Cosmic ray composition at low energies is often quoted at a fixed energy per nucleon. When presented in this way, protons constitute roughly 90% of the flux, helium nuclei about 10%, and the remainder sum to a total of about 1%.

Certain radioactive isotopic ratios show lifetime effects. The ratio of Be^{10}/B^9 abundances is used to measure an "age" of cosmic rays since Be^{10} is unstable with a half-life of about 1.6×10^6 years. A ratio of 0.6 is expected in the absence of Be^{10} decay and a ratio of about 0.2 is found experimentally (E. Juliusson, 1975; P. Meyer, 1981).

At higher energies, composition determinations are indirect and are rather contradictory and controversial. Experiments aim to differentiate between broad composition models. The measurement technique is based on studies of cosmic ray shower development. A rather direct technique for such studies is to use fluorescence observations of the shower development to determine the atmospheric depth of maximum development of the shower. Such observations suggest a heavy composition (large atomic number) at energies ~10^{17} eV which changes with increasing energy to a light composition (perhaps protonic) above ~10^{19} eV (T. K. Gaisser et al., 1993).

Primary Electrons

Primary electrons constitute about 1% of the cosmic ray beam. The positron to negative electron ratio is about 10% (J. M. Clem et al., 1995).

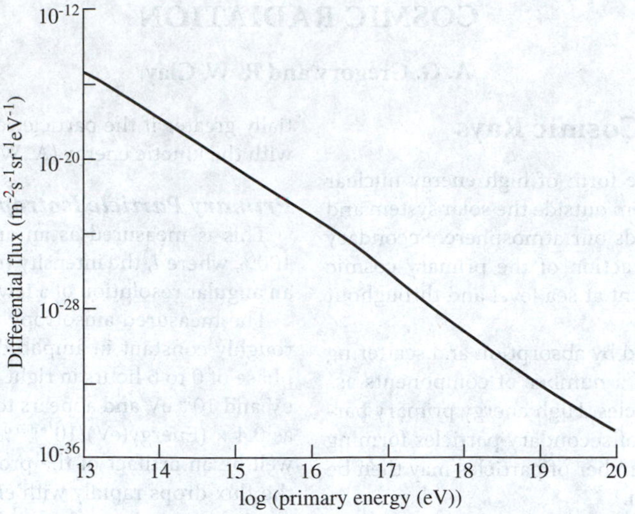

FIGURE 1. The energy spectrum of cosmic ray particles. This spectrum is of a differential form and can be converted to an integral spectrum by integration over all energies above a required threshold (E). Insofar as the spectrum approximates a power law of index −3, a simple conversion to the integral at an energy$E$/1.8 is obtained by multiplying the differential flux by the energy and dividing by 0.62.

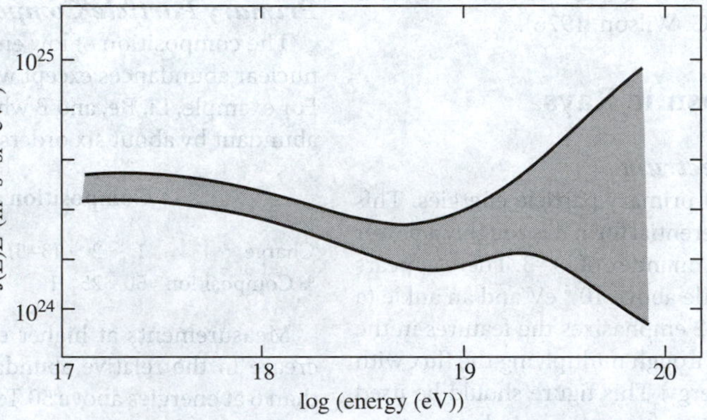

FIGURE 2. Energy spectrum at the highest energies. This spectrum (after Yoshida et al., 1995) has the differential spectrum multiplied by energy cubed. It is from a compilation of a number of measurements and indicates the good general agreement at the lower energies and a spread due to inadequate statistics at the highest energies.

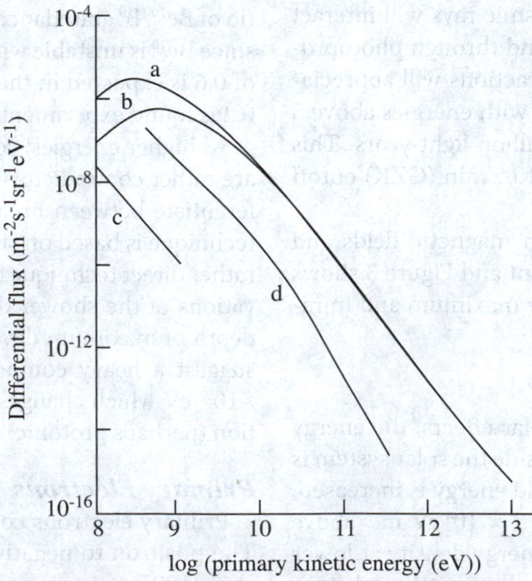

FIGURE 3. Energy spectrum of particles at lower energies. (a) Solar minimum proton energy spectrum. (b) Solar maximum proton energy spectrum. (c) Gamma ray energy spectrum. (d) Local interstellar electron spectrum.

Antimatter in the Primary Beam

The ratio of antiprotons to protons in the primary cosmic ray beam (at about 400 MeV) is about 10^{-5}. At about 10 GeV the ratio is about 10^{-3}. At the highest measured energies (10 TeV), the upper limit to the ratio is about 20% (M. Amenomori et al., 1995; S. Orito et al., 1995).

Primary Gamma Rays

The flux of primary gamma rays is low at high energies. At 1 GeV the ratio of gamma rays to protons is about 10^{-6}. The arrival directions of these gamma rays are strongly concentrated in the plane of the Milky Way although there is a diffuse, near isotropic background flux and some point sources have been detected.

Since the absorption cross section for gamma rays above 100 MeV is approximately 20 mbarn/electron, less than 10% of gamma rays reach mountain altitudes (A. W. Wolfendale, 1979; P. F. Michelson, 1994).

Sea-Level Cosmic Radiation

The sea-level cosmic ray dose is 300 millirad·yr^{-1} and the sea-level ionization is 2.2×10^6 ion pairs m^{-3}s^{-1}. The sea-level flux has a soft component, which can be absorbed in about 100 mm of lead (about 100 g·cm^{-2} of absorber) and a more penetrating (largely muon) hard component. The sea-level radiation is largely produced in the atmosphere and is a secondary component from interactions of the primary particles. The steep primary energy spectrum means that most secondaries at sea level are from rather low energy primaries. Thus the secondary flux is dependent on the solar cycle and the geomagnetic latitude of the observer.

Absolute Flux of the Hard Component

Vertical Integral Intensity $I(0) \sim 100$ m^{-2} s^{-1} sr^{-1}
Angular Dependence $I(\theta) \sim I(0)\cos^2(\theta)$
Integrated Intensity ~200 m^{-2} s^{-1}
(O. C. Allkofer, 1975b)

Flux of the Soft Component

In free air, the soft component comprises about one-third of the total cosmic ray flux.

Latitude Effect

The geomagnetic field influences the trajectories of lower energy cosmic rays approaching the Earth. As a result, the background flux is reduced by about 7% at the geomagnetic equator. The effect decreases toward the poles and is negligible at latitudes above about 40°.

Flux of Protons

The proton component is strongly attenuated by the atmosphere with an attenuation length (reduction by a factor of e) of about 120 g·cm^{-2}. It constitutes about 1% of the total vertical sea-level flux.

Absorption

The soft component is absorbed in about 100 g·cm^{-2} of matter. The hard component is absorbed much more slowly:

Absorption in lead, 6% per 100 g·cm^{-2}
Absorption in rock, 8.5% per 100 g·cm^{-2}
Absorption in water, 10% per 100 g·cm^{-2}
(Absorption for depths less than 100 g·pd cm^{-2} is given by K. Greisen, 1943)

Altitude Dependence

The cosmic ray background in the atmosphere has a maximum intensity of about 15 times that at sea level at a depth of about 150 g·cm^{-2} (15 km altitude). At maximum intensity, the soft and hard components contribute roughly equally but the hard component is then attenuated more slowly (S. Hayakawa, 1969).

Cosmic Ray Showers

High energy cosmic rays produce particle cascades in the atmosphere which can be detected at sea level provided that their energy exceeds about 100 GeV (such low energy cascades may be detected by using the most sensitive atmospheric Cerenkov detectors). The primary particle progressively loses energy which is transferred through the production of successive generations of secondary particles to a cascade of hadrons, an electromagnetic shower component (both positively and negatively charged electrons and gamma rays) and muons. The secondary particles are relativistic and all travel effectively at the speed of light. As a result, they reach sea level at approximately the same time but, due to Coulomb scattering (for the electrons) and production angles (for the pions producing the muons), are spread laterally into a disk-like shower front with a characteristic lateral width of several tens of meters and thickness (near the central shower core) of 2 to 3 m. The number of particles at sea level is roughly proportional to the primary particle energy:

Number of particles at sea level ~10^{-10} × energy (eV).

At altitudes below a few kilometers, the number of particles in a shower attenuates with an *attenuation length* of about 200 g·cm^{-2}, i.e.,

particle number = original number × exp(−(depth increase)/200)

The above applies to an individual shower. The rate of observation of showers of a given size (particle number at the detector) at different depths of absorber attenuates with an *absorption length* of about 100 g·cm^{-2} (J. G. Wilson, 1976).

Atmospheric Background Light from Cosmic Rays

Cosmic ray particles produce Cerenkov light in the atmosphere and produce fluorescent light through the excitation of atmospheric molecules.

Cerenkov Light

High energy charged particles will cause the emission of Cerenkov light in air if their energies are above about 30 MeV (electrons). This threshold is pressure (and hence altitude) dependent. A typical Cerenkov light pulse (at sea level, 100 m from the central shower core) has a time spread of a few nanoseconds. Over this time, the photon flux between 430 and 530 nm would be ~10^{14} m^{-2}s^{-1} for a primary particle energy of 10^{16} eV. For comparison, the night sky background flux is ~6×10^{11} photons m^{-2}s^{-1}sr^{-1} in the same wavelength band (J. V. Jelley, 1967).

Fluorescence Light

Cosmic ray particles in the atmosphere excite atmospheric molecules which then emit fluorescence light. This is weak compared to the highly collimated Cerenkov component when viewed in the direction of the incident cosmic ray particle but is emitted isotropically. Typical pulse widths are longer than 50 ns and may be up to several microseconds for the total pulse from distant large showers (R. M. Baltrusaitis et al., 1985).

Effects of Cosmic Rays

Cerenkov Effects in Transparent Media

Background cosmic ray particles will produce Cerenkov light in transparent material with a photon yield between wavelengths λ_1 and λ_2

$$\sim (2\pi/137)\sin^2(\theta_c)\int_{\lambda_1}^{\lambda_2} d\lambda/\lambda^2 \text{ photons(unit length)}^{-1}$$

where θ_c (the Cerenkov angle) = $\cos^{-1}$ (1/refractive index).

This background light is known to affect light detectors, e.g., photomultipliers, and can be a major source of background noise (R. W. Clay and A. G. Gregory, 1977).

Effects on Electronic Components

If background cosmic ray particles pass through electronic components, they may deposit sufficient energy to affect the state of, e.g., a transistor flip-flop. This effect may be significant where reliability is of great importance or the background flux is high. For instance, it has been estimated that, in communication satellite operation, an error rate of about 2×10^{-3} per transistor per year may be found. Permanent damage may also result. A significant error rate may be found even at sea level in large electronic memories. This error rate is dependent on the sensitivity of the component devices to the deposition of electrons in their sensitive volumes (J. F. Ziegler, 1981).

Biophysical Significance

When cosmic rays interact with living tissue, they produce radiation damage. The amount of the damage depends on the total dose of radiation. At sea level, this dose is small compared with doses from other sources but both the quantity and quality of the radiation change rapidly with altitude. Approximate dose rates under various conditions are:

Dose rates (mrem·yr^{-1})
Sea-level cosmic rays, 30
Cosmic rays at 10 km (subsonic jets), 2000
Cosmic rays at 18 km (supersonic transports), 10,000
(c.f., mean total sea-level dose, 300)

Astronauts would be subject to radiation from galactic (0.05 rads per day) and solar (a few hundred rads per solar flare) cosmic rays as well as large fluxes of low energy radiation when passing through the Van Allen belts (about 0.3 rads per traverse).

Both astronauts and SST travelers would be subject to a small flux of low energy heavy nuclei stopping in the body. Such particles are capable of destroying cell nuclei and could be particularly harmful in the early stages of the development of an embryo. The rates of heavy nuclei stopping in tissue in supersonic transports and spacecraft are approximately as follows:

Stopping nuclei ((cm^3 tissue)$^{-1}$ hr^{-1})
Supersonic transport (16 km), 0.0005
Supersonic transport (20 km), 0.005
Spacecraft, 0.15
(O. C. Allkofer, 1975a; O. C. Allkofer et al., 1974)

Carbon Dating

Radiocarbon is produced in the atmosphere due to the action of cosmic ray slow neutrons. Solar cycle modulation of the very low energy cosmic rays causes an anticorrelation of the atmospheric ^{14}C activity with sunspot number with a mean amplitude of about 0.5%. In the long term, modulation of cosmic rays by a varying magnetic field may be important (A. A. Burchuladze et al., 1979).

Practical Uses of Cosmic Rays

There are few direct practical uses of cosmic rays. Their attenuation in water and snow have, however, enabled automatic monitors of water and snow depth to be constructed. A search for hidden cavities in pyramids has been carried out using a muon "telescope."

Other Effects

Stellar x-rays have been observed to affect the transmission times of radio signals between distant stations by altering the depth of the ionospheric reflecting layer. It has also been suggested that variations in ionization of the atmosphere due to solar modulation may have observable effects on climatic conditions.

References

1. O.C. Allkofer (1975a) *Introduction to Cosmic Radiation*, Verlag Karl Thiemig, Munchen, Germany.
2. O.O. Allkofer (1975b) *J. Phys. G: Nucl. Phys.*, 1, L51.
3. O.C. Allkofer and W. Heinrich (1974) *Health Phys.*, 27, 543.
4. M. Amenomori et al. (1995) *Proc. 24th Int. Cosmic Ray Conf. Rome*, 3, 85. Universita La Sapienza, Roma.
5. K. Asakimori et al. (1993) *Proc. 23rd Int. Cosmic Ray Conf. Calgary*, 2, 25, University of Calgary, Canada.
6. R.M. Baltrusaitis et al. (1985) *Nucl. Inst. Meth.*, A420, 410.
7. D.J. Bird et al. (1993) *Phys. Rev. Lett.*, 71, 3401.
8. A.A. Burchuladze, S.V. Pagava, P. Povinec, G. I. Togonidze, S. Usacev (1979) *Proc. 16th Int. Cosmic Ray Conf. Kyoto*, 3, 201, Univ. of Tokyo, Japan.
9. R.W. Clay (1987) *Aust. J. Phys.*, 40, 423.
10. R.W. Clay and A.G. Gregory (1977) *J. Phys. A: Math. Gen.*, 10, 135.
11. J.M. Clem et al. (1995) *Proc. 24th Int. Cosmic Ray Conf. Rome*, 3, 5, Universita La Sapienza, Roma.
12. T.K. Gaisser et al. (1993) *Phys. Rev. D*, 47, 1919.
13. K. Greisen (1943) *Phys. Rev.*, 63, 323.
14. S. Hayakawa (1969) *Cosmic Ray Physics*, Wiley-Interscience, New York.
15. J.V. Jelley (1967) *Prog. in Elementary Particle and Cosmic Ray Physics*, 9, 41.
16. E. Juliusson (1975) *Proc. 14th Int. Cosmic Ray Conf. Munich*, 8, 2689, Max Planck Institute fur Extraterrestriche Physik, Munchen, Germany.
17. J. Linsley (1981) *Origin of Cosmic Rays*, I.A.U. Symposium 94, 53, D. Reidel Publishing Co. Dordrecht, Holland.
18. P. Meyer (1981) *Origin of Cosmic Rays*, I.A.U. Symposium 94, 7, D. Reidel Publishing Co. Dordrecht, Holland.
19. P.F. Michelson (1994) in *Towards a Major Atmospheric Cerenkov Detector III*, 257, Ed. T. Kifune, Universal Academy Press Inc., Tokyo, Japan.
20. P. Sokolsky (1989) *Introduction to Ultrahigh Energy Cosmic Ray Physics*, Addison Wesley Publishing Company.
21. T. Stanev et al. (1995) *Phys. Rev. Lett.*, 75, 3056.
22. S. Orito et al. (1995) *Proc. 24th Int. Cosmic Ray Conf. Rome*, 3, 76. Universita La Sapienza, Roma.
23. J.G. Wilson (1976) *Cosmic Rays*, Wykeham Pub. (London) Lt., U.K.
24. A.W. Wolfendale (1979) *Pramana*, 12, 631.
25. S. Yoshida et al. (1995) *Astroparticle Phys.*, 3, 105.
26. J.F. Ziegler, (1981) *IEEE Trans. Electron Devices*, ED-28, 560.

Section 12
Properties of Solids

TECHNIQUES FOR MATERIALS CHARACTERIZATION

H. P. R. Frederikse

The many experimental methods, originally designed to study the chemical and physical behavior of solids and liquids, have grown into a new field known as Materials Characterization (or Materials Analysis). During the past 30 years a host of techniques aimed at the study of surfaces and thin films has been added to the many tools for the analysis of bulk samples. The field has benefited particularly from the development of computers and microprocessors, which have vastly increased the speed and accuracy of the measuring devices and the recording of their output. Materials characterization was and is a very important tool in the search for new physical and chemical phenomena. It plays an essential role in new applications of solids and liquids in industry, communications, and medicine. Many of its techniques are used in quality control, in safety regulations, and in the fight against pollution.

In most Materials Characterization experiments the sample is subjected to some kind of radiation: electromagnetic, acoustic, thermal, or particles (electrons, ions, neutrons, etc.). The surface analysis techniques usually require a high vacuum. As a result of interactions between the solid (or liquid) and the incoming radiation a beam of a similar (or a different) nature will emerge from the sample. Measurement of the physical and/or chemical attributes of this emerging radiation will yield qualitative, and often quantitative, information about the composition and the properties of the material being probed.

The modern tendency of describing practically everything in this world by a combination of a few letters (acronyms) has also penetrated the field of Materials Characterization. The table below gives the meaning of the acronym for every technique listed, the form and size of the required sample (bulk, surface, film, liquid, powder, etc.), the nature of the incoming and of the emerging radiation, the depth and the lateral spatial resolution that can be probed, and the information obtained from the experiment. The last column lists one or two major references to the technique described.

Experimental Techniques Used to Determine the Composition, Structure, and Energy States of Solids and Liquids

Technique	Sample	In	Out	Depth	Lateral resolution	Information obtained	Ref.
Optical and Mass Spectroscopies for Chemical Analysis							
1. AAS Atomic Absorption Spectroscopy	Atomize (flame, electro, thermal, etc.)	Light, e.g., glow discharge	Absorption spectrum	–	–	Concentration of atomic species (quantitative, using standards)	1,2
2. ICP-AES Induct. Coupled Plasma – Atomic Emission Spectroscopy	Atomize (flame, electro, thermal, ICP, etc.)	–	Emission spectrum	–	–	Concentration of atomic species (quantitative, using standards)	3
3. Dynamic SIMS Dynamic Secondary Ion Mass Spectroscopy	Surface	Ion beam (1–20 keV)	Secondary ions; analysis with mass spectrometer	2 nm–1 μm (or deeper: ion milling)	0.50 nm	Elemental and isotopic analysis; depth profile (all elements); detection limits: ppb-ppm	4
4. Static SIMS Static Secondary Ion Mass Spectroscopy	Surface	Ion beam (0.5–20 keV)	Secondary ions, analysis with mass spectrometer	0.1–0.5 nm	10 μm	Elemental analysis of surface layers; molecular analysis; detection limits: ppb-ppm	4
5. SNMS Sputtered Neutral Mass Spectroscopy	Surface, bulk	Plasma discharge; noble gases: 0.5–20 keV	Sputtered atoms ionized by atoms or electrons; then mass analyzed	0.1–0.5 nm (or deeper: ion milling)	1 cm	Elemental analysis $Z \geq 3$; depth profile; detection limit: ppm	4,6
6. SALI Surface Analysis by Laser Ionization	Surface	e-beam, ion-beam, or laser for sputtering	Sputtered atoms ionized by laser; then mass analyzed	0.1–0.5 nm up to 3 μm in milling mode	60 nm	Surface analysis; depth profiling	7
7. LIMS Laser Ionization Mass Spectroscopy	Surface, bulk	u.v. laser (ns pulses)	Ionized species; analyzed with mass spectrometer	50–150 nm	5 μm–1 mm	Elemental (micro)analysis; detection limits: 1–100 ppm	8
8. SSMS Spark Source Mass Spectroscopy	Sample in the form of two electrodes	High voltage R.F. spark produces ions	Ions – analyzed in mass spectrometer	1–5 μm	–	Survey of trace elements; detection limit: 0.01–0.05 ppm	9
9. GDMS Glow Discharge Mass Spectroscopy	Sample forms the cathode for a D.C. glow discharge	Sputtered atoms ionized in plasma	Ions – analyzed in mass spectrometer	0.1–100 μm	3–4 mm	(Bulk) trace element analysis; detection limit: sub-ppb	9,10
10. ICPMS Induct. Coupled Plasma Mass Spectroscopy	Liquid-dissolved sample carried by gas stream into R.F. induction coil	Ions produced in argon plasma	Ions – analyzed in quadrupole mass spectrometer	–	–	High sensitivity analysis of trace elements	11
Photons — Absorption, Reflection, and Electron Emission							
11. IRS Infrared Spectroscopy	Thin crystal, glass, liquid	I.R. light (W-filament, globar, Hg-arc)	I.R. spectrum	–	–	Electronic transitions (mainly in semiconductors and superconductors); vibrational modes (in crystals and molecules)	12,13,14

Technique	Sample	In	Out	Depth	Lateral resolution	Information obtained	Ref.
12. FTIR Fourier Transform I.R. Spectroscopy	Solid, liquid; transmission or reflection	White light (all frequencies)	Fourier transform of spectrum (interferometer)	–	–	Spectra obtained at higher speed and resolution	15
13. ATR Attenuated Total Reflection	Surface or thin crystal	–	–	μm's	–	Atomic or molecular spectra of surfaces and films	16
14. (μ)-RS (Micro-) Raman Spectroscopy	Solid, liquid (1 μm–1 cm)	Laser beam, e.g., Ar-line, YAG-line	Raman spectra	0.5 μm	0.5 μm	Molecular and crystal vibrations	12,14,17
15. CARS Coherent Anti-Stokes Raman Spectroscopy	Solid, liquid (50 μm–3 cm)	Pump beam (ω_0)+ probe beam (ω_s)	Anti-Stokes spectrum	–	–	High resolution Raman spectra	14
16. Ellipsometry	Transparent films, crystals, adsorbed layers	Polarized light	Change in polarization	0.05 nm–5 μm	25 μm (or sample thickness)	Refractive index and absorption	18,19
17. UPS Ultraviolet Photoelectron Spectroscopy	Surfaces, adsorbed layers	u.v. light, 10–100 eV; 200 eV (synchrotron)	Electrons	0.2–10 nm	0.1–10 nm	Energies of electronic states of surfaces and free molecules	20,21
18. PSD Photon Stimulated Desorption	Surfaces with adsorbed species	Far u.v. light $E > 10$ eV	Ions – analyzed with mass spectrometer	0.1–2 nm	–	Structure and desorption kinetics of adsorbed atoms and molecules	22
X-Rays							
19. XRD X-Ray Diffraction	Single crystals, powders films	X-rays: $\lambda = 0.05$–0.2 nm (6–17 keV)	Diffracted X-ray beam	1–1000 μm	0.1–10 mm	Identification of crystallographic structures; all elements (low Z difficult)	23,24
20. XRF/EDS X-Ray Fluorescence/Energy Dispersive Spectroscopy	Thin films, single layer	Prim. X-ray beam λ = 0.02–0.1 nm 12–80 keV	Fluorescent X-rays	1–100 μm	10 mm	Elemental analysis; all elements except H, He, Li – (EDS also used in XRD, SEM, TEM and EPMA)	25,26
21. EXAFS Extended X-Ray Absorption Fine Structure	Films, foils	High intensity X-rays (synchrotron)	Spectrum near absorption edge	nm–μm	–	Local atomic structure: order/disorder in vicinity of absorbing atom	27
22. XPS/ESCA X-Ray Photoelectron Spectroscopy/Electron Spect. for Chemical Analysis	Surfaces, thin films (≈20 atomic layers)	Soft X-rays (1–20 keV)	Core electrons; valence electrons	0.5–10 nm	5 nm–50 μm	(Quantitative) identification of all elements in surface layer or film	28,29
Electrons							
23. CL Cathode Luminescence	Insulators, semiconductors	Electrons 5–50 keV	Photons 0.1–5 eV	1 nm–2 μn	1 or 2 μm	Energy levels of impurities and point defects	30
24. APS Appearance Potential Spectroscopy	Surface (≈20 atomic layers)	Electrons (energy scan) 50–2000 eV	X-rays to pinpoint electron energy threshold	–	–	Identification of surface species	21, see also C
25. AES Auger Electron Spectroscopy	Thin films, surfaces	Electrons 3–10 keV	Auger electrons 20–2000 eV	0.3–3 nm	≈30 nm	Elemental composition of surface (except H, He); detection limit 0.1–1%	28,29
26. EELS Electron Energy Loss Spectroscopy	Very thin samples (<200 nm)	Electrons (100–400 keV)	(Retarded) electrons; minus 1–1000 eV	<200 nm	1–100 nm	Local elemental concentration; electronic structure, chem. bonding; interatomic distances	31
27. EXELFS Extended Electron Energy Loss Fine Structure	Thin films	Electrons (100–400 keV)	Electrons energies 0–30 eV above edge	<200 nm	1–100 nm	Density of states of valence electrons (above Fermi level)	27,32
28. ESD Electron Stimulated Desorption	Adsorbed species	Electrons $E > 10$ eV	Ions – analyzed with mass spectrometer	–	–	Structure and desorption properties of adsorbed atoms and molecules	22
29. ESDIAD ESD-Ion Angular Distribution	(See ESD)	(See ESD)	Directional dependence of emitted ions	–	–	Geometries of adsorbed species (atoms or molecules)	22
30. EPMA Electron Probe (X-Ray) Micro Analysis	Solid conductors and insulators <1 cm thick	Electrons 5–30 keV	Characteristic X-ray 0.1–15 keV	100 nm–5 μm	1 μm	Elemental analysis, $Z \leq 4$, major, minor and trace amounts	33,34
31. LEED Low Energy Electron Diffraction	Surface	Mono-energetic electron beam 10–1000 eV	Diffracted electrons	0.4–2 nm	<5 μm	Crystallographic structure of surface; resolution: 0.01 nm	35
32. RHEED Reflection High Energy Electron Diffraction	Surface	Electron beam at grazing angle 5–50 keV	Reflected electrons	0.2–10 nm	<5 μm	Surface symmetry	36,37
33. SEM Scanning Electron Microscopy	Bulk, films (conducting)	High energy electrons usually ~30 keV	Secondary and backscattered electrons	1 nm–5 μm	1–20 nm	Surface image, defect structure; resolution 5–15 nm; magnification 300,000×	33,34

	Technique	Sample	In	Out	Depth	Lateral resolution	Information obtained	Ref.
34.	(S)TEM (Scanning) Transmission Electron Microscopy	Thin specimen – <200 nm	High energy electrons typically 300 keV	Transmitted and diffracted electrons	(Sample thickness)	2–20 nm	(Defect) structure of cryst. solids; microchemistry; high resol.: 0.2 nm	33
35.	FEM Field Emission Microscopy	Metals, alloys (sharp point)	–	Electron emission (with appl. electric field – 50 kV)	≈0.5 nm	10–100 nm	Surface image, crystallographic structure	34
36.	STM Scanning Tunneling Microscopy	Polished or cleaved surface (conducting)	Tunneling current controls distance between sample and very sharp tip	Tunneling current controls distance between sample and very sharp tip	1–5 nm	2–10 nm	Atomic-scale relief map of surface; resolution: vert. 0.002 nm, hor. 0.2 nm	39
37.	SPM Scanned Probe Microscopy	Very flat surface	Any field: e.g., mechan. vibration recorded with laser probe; same with magnetic, electric or thermal field	Any field: e.g., mechan. vibration recorded with laser probe; same with magnetic, electric or thermal field	1–100 nm	1–100 nm	Surface-magnetic field, surface-thermal conductivity, etc.	39a
38.	AFM Atomic Force Microscopy	Very flat surface	Similar to STM; force measured with cantilever spring	Similar to STM; force measured with cantilever spring	0.5–5 nm	0.2–130 nm	Surface topography with atomic resolution; interatomic force	40
Ions and Neutrons								
39.	ISS (or LEIS) Ion Scattering Spectroscopy (Low Energy Ion Scattering)	Surface	Ion beam He^+ or Ne^+ <3 keV	Sputtered ions (energy analysis)	0.1–0.5 nm	1–100 μm	Elemental analysis (better for low *Z*) detection limits: 0.01–1%	41
40.	FIM Field Ion Microscopy	Surface: metals, alloys; very sharp tip	(He gas above sample)	He ions + high electric field produce image	≈0.1 nm	0.1–2 nm	Atomic structure of surface	34,42
41.	RBS Rutherford Back Scattering	Solids, thin films	Mono-energetic ions (H^+ or He^{++}) 0.5–3 MeV	Backscattered ions	10 nm–1 μm	1 mm	Element identification (Li to U) detection limit: 0.01–1%	46
42.	NRA Nuclear Reaction Analysis	Solids, thin films	Mono-energetic ions (Li, Be, B, etc.) 200 keV–6 MeV	Protons, deuterons ^3He, α-particles, γ-rays	0.1–5 μm	10 μm–10 mm	Element identification (all) detection limit: 10-12–10-2	47
43.	PIXE Particle Induced X-ray Emission	Thin films, surface layers	High energy ions (H^+ or He^{++})	Characteristic X-rays	<10 μm	1 μm–2 mm	Trace impurities: *Z*>3 detection limit: 0.1–100 ppm (depending on sample thickness)	48
44.	INS Ion Neutralization Spectroscopy	Surface	He-ions (≈5 eV)	Electrons	–	–	Energies of valence electrons	49
45.	NAA Neutron Activation Analysis	Bulk, >0.5 g	Thermal neutrons	Characteristic γ-rays, (≈1 MeV)	Bulk	–	Trace concentrations (of isotopes) of elements: trans. metals, Pt-group; detection limit: 10^8–10^{14} atoms/cm^3	43
46.	N(P)D Neutron (Powder) Diffraction	Crystalline solids	Thermal neutrons *E* ≈0.0025 eV	Diffracted neutrons	Bulk	–	Crystallographic structure; porosity, particle size	44
47.	SANS Small Angle Neutron Scattering	Inhomogeneous solids; powders; porous samples	Thermal neutrons 2 θ = 10^{-2}–10^{-4}	Scattered neutrons	1–25 mm	–	Average size of inhomogeneities; range: 1 nm^{-1} mm	45
Acoustic								
48.	SLAM Scanning Laser Acoustic Microscopy	Bulk, film	Acoustic wave produced by laser 1 MHz–1 GHz	Reflected acoustic wave	μm–cm	0.1–20 mm	Defect structure; thickness measurement	50
Thermal								
49.	DTA Differential Thermal Analysis	Specimen and reference sample	Uniform heating	Temperature difference	Bulk	–	Phase transitions, crystallization	51
50.	DSC Differential Scanning Calorimetry	Specimen and ref. sample	Controlled heating	Measure heat required for equal temperature	Bulk	–	Phase transitions, crystallization; activation energies	51
51.	TGA Thermo Gravimetric Analysis	Bulk, 1–100 g	Controlled heating	Weight as function of temperature (and time)	Bulk	–	Decomposition, non-stoichiometry, kinetics of reaction	52
Resonance								
52.	EPR (ESR) Electron Paramagnetic (Spin) Resonance	Paramagnetic solids or liquids	Microwave radiation in magnetic field 3–300 GHz; 1–100 kG	Microwave absorption (at resonance)	Bulk	–	Local environment of paramagnetic ion; concentration of paramagnetic, species; detection limit: 10^{11} spins/cm^3	53,54
53.	ECR Electron Cyclotron Resonance	Semiconductors, metals; free electrons (low temperature)	Microwave radiation in magnetic field 10–30 GHz; 5–10 kG	Microwave absorption (at resonance)	Bulk	–	Electronic energy bands, effective masses	55

	Technique	Sample	In	Out	Depth	Lateral resolution	Information obtained	Ref.
54.	Mössbauer Effect	Source and absorber	Mono-energetic γ-rays: 5–100 keV	Mössbauer spectrum (Doppler shifted (lines)	50 m	1 cm	Interaction between nucleus and its environment (local electric, magnetic fields; bonds; valency; diffusion, etc.)	56
55.	NMR (MRI) Nuclear Magnetic Resonance (Magnetic Resonance Imaging)	Solids, liquids	R.F. radiation + magnetic field; e.g., for protons: 60 MHz, 14 kG	R.F. absorption	<1 cm	1 cm	Quant. analysis; local magnetic environment; diffusion; imaging	58
56.	ENDOR Electron Nuclear Double Resonance	Solids, liquids	R.F. + microwave radiation in magn. field.	Microwave absorption	–	–	Hyperfine interaction → local atomic structure	54
57.	NQR Nuclear Quadrupole Resonance	Solids	R.F. radiation 0.5–1000 MHz	R.F. absorption	–	–	Asymmetry of the charge distribution at the nucleus	55,59
Other								
58.	BET Brunauer-Emmett-Teller	(Large) surface area 1–20 m^2/g	Adsorbed gas (e.g., N_2 at low temp.) as function of pressure (monolayer coverage)	Adsorbed gas (e.g., N_2 at low temp.) as function of pressure (monolayer coverage)	–	–	Surface area measurement	60

Solids

General References

1. A. Wachtman, J. B., *Characterization of Materials*, Butterworth-Heinemann, Boston, 1993.
2. B. Brundle, C. R., Evans, C. A., and Wilson, S., Eds., *Encyclopedia of Materials*, Butterworth-Heinemann, Boston, 1992.
3. C. Woodruff, D. P. and Delchar, T. A., *Modern Techniques of Surface Science*, Cambridge University Press, Cambridge, 1986.
4. *D. Metals Handbook*, 9th Edition, Vol. 10, Materials Characterization, Whan, R. E., Coordinator, American Society for Metals, Metals Park, OH, 1986.

Specific References

1. Slavin, M., *Atomic Absorption Spectroscopy*, 2nd Edition, John Wiley & Sons, New York, 1978.
2. Schrenk, W. G., *Analytical Atomic Spectroscopy*, Plenum Press, New York, 1975.
3. Dean, J. A. and Rains, T. E., *Flame Emission and Atomic Absorption Spectroscopy*, Vols. 1–3, Marcel Dekker, New York, 1969.
4. Benninghoven, A., Rudenauer, F. G., and Werner, H. W., *Secondary Ion Mass Spectroscopy*, John Wiley & Sons, New York, 1987.
5. Bird, J. R. and Williams, J. S., Eds., in *Ion Beams for Materials Analysis*, Academic Press, New York, 1989, pp. 515–537.
6. Smith, G. C., *Quantitative Surface Analysis for Materials Science*, The Institute of Metals, London, 1991.
7. Becker, E. H., in *Ion Spectroscopies for Surface Analysis*, Czanderna, A. W. and Hercules, D. M., Eds., Plenum Press, New York, 1991, p. 273.
8. Simons, D. S., *Int. J. Mass Spectrometry and Ion Processes*, 55, 15, 1983.
9. White, F. A. and Wood, G. M., *Mass Spectrometry: Applications in Science and Engineering*, John Wiley & Sons, New York, 1986.
10. Harrison, W. W. and Bentz, B. L., *Prog. Anal. Spectrometry*, 11, 53, 1988.
11. Bowmans, P. W. J. M., *Inductively Coupled Plasma Emission Spectroscopy*, Parts I and II, John Wiley & Sons, New York, 1987.
12. Brame, Jr., E. G. and Grasselli, J., *Infrared and Raman Spectroscopy*, Practical Spectroscopy Series, Vol. I, Marcel Dekker, New York, 1976.
13. Hollas, J. M., *Modern Spectroscopy*, John Wiley & Sons, New York, 1987.
14. Turrell, G., *Infrared and Raman Spectroscopy of Crystals*, Academic Press, New York and London, 1972.
15. Griffith, P. R. and Haseth, J. A., *Fourier Transform Infrared Spectroscopy*, John Wiley & Sons, New York, 1986.
16. Barnowski, M. K., *Fundamentals of Optical Fiber Communications*, Academic Press, New York, 1976.
17. Long, D. A., *Raman Spectroscopy*, McGraw-Hill, New York, 1977.
18. Azzam, R. M. A., *Ellipsometry and Polarized Light*, Elsevier-North Holland, Amsterdam, 1977.
19. Hecht, E., *Optics*, 2nd Edition, Addison-Wesley, Reading MA, 1987.
20. Brundle, C. R., in *Molecular Spectroscopy*, West, A. R., Ed., Heyden, London, 1976.
21. Park, R. L., in *Experimental Methods in Catalytic Research*, Vol. III, Anderson, R. B. and Dawson, P. T., Academic Press, New York, 1976, pp. 1–39.
22. Madey, T. E. and Stockbauer, R., in *Solid State Physics: Surfaces*, Vol. 22 of Methods of Experimental Physics, Park, R.L. and Lagally, M. G., Eds., Academic Press, New York, 1985.
23. Cullity, B. D., *Elements of X-Ray Diffraction*, 2nd Edition, Addison-Wesley, Reading, MA, 1978.
24. Schwartz, L. H. and Cohen, J. B., *Diffraction from Materials*, Springer Verlag, Berlin, 1987.
25. deBoer, D. K. G., in *Advances in X-Ray Analysis*, Vol. 34, Barrett, C. S. et. al., Eds., Plenum Press, New York, 1991.
26. Birks, L. S., *X-Ray Spectrochemical Analysis*, 2nd Edition, John Wiley & Sons, New York, 1969.
27. Bonnelle, C. and Mande, C., *Advances inX-Ray Spectroscopy*, Pergamon Press, Oxford, 1982.
28. *Practical Surface Analysis by Auger and X-Ray Photo-Electric Spectroscopy*, Briggs, D. and Seah, M. P., Eds., John Wiley & Sons, New York, 1983.
29. Powell, C. J. and Seah, M. P., *J. Vac. Sci. Technol. A*, Vol. 8, 735, 1990.
30. Yacobi, G. G. and Holt, D. B., *Cathodeluminescence Microscopy of Inorganic Solids*, Plenum Press, New York, 1990.
31. Egerton, R. F., *Electron Energy Loss Spectroscopy in the Electron Microscope*, Plenum Press, New York, 1986.
32. Disko, M. M., Krivanek, O. L., and Rez, P., *Phys. Rev.*, B25, 4252, 1982.
33. Goldstein, J. I., et. al., *Scanning Electron Microscopy and X-Ray Microanalysis*, 2nd Edition, Plenum Press, New York, 1986.
34. Murr, L. E., *Electron and Ion Microscopy and Microanalysis*, Marcel Dekker, New York, 1982.
35. Armstrong, R. A., in *Experimental Methods in Catalytic Research*, Vol. III, Anderson, R. B., and Dawson, P. T., Eds., Academic Press, New York, 1976.
36. Dobson, P. J. et. al., *Vacuum*, 33, 593, 1983.
37. Rymer, T. B., *Electron Diffraction*, Methuen, London, 1970.
38. Reimer, L., *Transmission Electron Microscopy*, Springer-Verlag, Berlin, 1984.

39. *Scanning Tunneling Microscopy and Related Methods*, Behm, R. J., Garcia, N., and Rohrer, H., Eds., Kluwer Academic Publishers, Norwell, MA, 1990.
39a. Wikramasinghe, H.K., *Scientific American*, Vol. 261, No. 4, pp. 98–105, Oct. 1989.
40. Rugar, D. and Hansma, P., *Physics Today*, 43(10), pp. 23–30, 1990.
41. Feldman, C. C. and Mayer, J. W., *Fundamentals of Surface and Thin Film Analysis*, North-Holland, Amsterdam, 1986.
42. Muller, E. W. and Tsong, T. T., *Field Ion Microscopy*, Elsevier, Amsterdam, 1969.
43. Amiel, S., *Nondestructive Activation Analysis*, Elsevier, Amsterdam, 1981.
44. Bacon, G. E., *Neutron Diffraction*, 3rd Edition, Clarendon Press, Oxford, 1975.
45. Neutron Scattering, Part A., in *Methods of Experimental Physics*, Vol. 23, Skold, K. and Price, D. L., Eds., Academic Press, New York, 1986.
46. Chu, W. K., Mayer, J. W., and Nicolet, M. A., *Backscattering Spectroscopy*, Academic Press, New York, 1987.
47. Rickey, F. A., in *High Energy and Heavy Ion Beams in Materials Analysis*, Tesmer, J. R., et. al., Eds., MRS, 1990, pp. 3–26.
48. Johansson, S. A. E. and Campbell, J. L., *PIXE: A Novel Technique for Elemental Analysis*, John Wiley & Sons, New York, 1988.
49. Hagstrum, H. D., in *Inelastic Ion-Surface Collisions*, Tolk, N. H. et. al., Eds., Academic Press, New York, 1977, pp. 1–46.
50. Nikoonahad, M., in *Research Techniques in Nondestructive Testing*, Vol. VI, Sharpe, R.S., Ed., Academic Press, New York, 1984, pp. 217–257.
51. Gallagher, P. K., *Characterization of Materials by Thermoanalytical Techniques*, MRS - Bulletin, Vol. 13, No. 7, pp. 23–27, 1988.
52. Earnest, C. M., *Compositional Analysis by Thermogravimetry*, ASTM Special Technical Publication 997, 1988.
53. Poole, C. P., *Electron Spin Resonance – A Comprehensive Treatise on Experimental Techniques*, 2nd Edition, John Wiley & Sons, New York, 1983.
54. Atherton, N. M., *Principles of Electron Spin Resonance*, Ellis Horwood Ltd., Chichester, U.K., 1993.
55. Kittel, C., *Introduction to Solid State Physics*, 6th Edition, John Wiley & Sons, New York, 1986, p. 196.
56. Gibb, T. C., *Principles of Mössbauer Spectroscopy*, Chapman & Hall, London, 1976.
57. Slichter, C. P., *Principles of Magnetic Resonance*, 3rd Edition, Springer-Verlag, Berlin, 1990.
58. *NMR Spectroscopy Techniques*, Dybrowski, C. and Lichter, R. L., Eds., Marcel Dekker, New York, 1987.
59. Das, T. P. and Hahn, E. L., *Nuclear Quadrupole Resonance Spectroscopy*, Academic Press, New York, 1958.
60. Somorjai, G. A., *Principles of Surface Chemistry*, Prentice-Hall, Englewood Cliffs, NJ, 1972, p. 216

SYMMETRY OF CRYSTALS

L. I. Berger

The ability of a body to coincide with itself in its different positions regarding a coordinate system is called its symmetry. This property reveals itself in iteration of the parts of the body in space. The iteration may be done by reflection in mirror planes, rotation about certain axes, inversions and translations. These actions are called the symmetry operations. The planes, axes, points, etc., are known as symmetry elements. Essentially, mirror reflection is the only truly primitive symmetry operation. All other operations may be done by a sequence of reflections in certain mirror planes. Hence, the mirror plane is the only true basic symmetry element. But for clarity, it is convenient to use the other symmetry operations, and accordingly, the other aforementioned symmetry elements. The symmetry elements and operations are presented in Table 1.

The entire set of symmetry elements of a body is called its symmetry class. There are thirty-two symmetry classes that describe all crystals that have ever been noted in mineralogy or been synthesized (more than 150,000). The denominations and symbols of the symmetry classes are presented in Table 2.

There are several known approaches to classification of individual crystals in accordance with their symmetry and crystallochemistry. The particles that form a crystal are distributed in certain points in space. These points are separated by certain distances (translations) equal to each other in any chosen direction in the crystal. Crystal lattice is a diagram that describes the location of particles (individual or groups) in a crystal. The lattice parameters are three non-coplanar translations that form the crystal lattice. Three basic translations form the unit cell of a crystal. August Bravais (1848) has shown that all possible crystal lattice structures belong to one or another of fourteen lattice types (Bravais lattices). The Bravais lattices, both primitive and non-primitive, are the contents of Table 3.

Among the three-dimensional figures, there is a group of polyhedrons that are called regular, which have all faces of the same shape and all edges of the same size (regular polygons). It has been shown that there are only five regular polyhedrons. Because of their importance in crystallography and solid state physics, a brief description of these polyhedrons is included in Table 4.

The systematic description of crystal structures is presented primarily in the well-known *Structurbericht*. The classification of crystals by the Structurbericht does not reflect their crystal class, the Bravais lattice, but is based on the crystallochemical type. This makes it inconvenient to use the Structurbericht categories for comparison of some individual crystals. Thus, there have been several attempts to provide a more convenient classification of crystals. Table 5 presents a compilation of different classifications which allows the reader to correlate the Structurbericht type with the international and Schoenflies point and space groups and with Pearson's symbols, based on the Bravais lattice and chemical composition of the class prototype. The information included in Table 5 has been chosen as an introduction to a more detailed crystallophysical and crystallochemical description of solids.

TABLE 1. Symmetry Operations and Elements

	Symmetry element			Presentation on the stereographic projection	
		Symbol			
Symmetry operation	**Name**	**International (Hermann-Mauguin)**	**Schoenflies**	**Parallel**	**Perpendicular**
Reflection in a plane	Plane	m	C_s	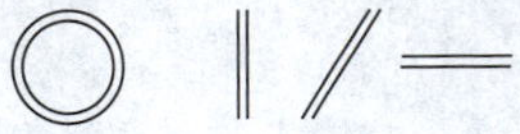	
Rotation by angle α = 360°/n about an axis	Axis	n = 1, 2, 3, 4 or 6	C_n		
		n = 2	C_2		
		n = 3	C_3		
		n = 4	C_4		
		n = 6	C_6		
Rotation about an axis and inversion in a symmetry center lying on the axis	Inversion (improper) axis	$\bar{n} = \bar{3}, \bar{4}, \bar{6}$	C_{ni}		
		$\bar{n} = \bar{3}$	C_{3i}		
		$\bar{n} = \bar{4}$	C_{4i}		
		$\bar{n} = \bar{6}$	C_{6i}		

TABLE 1. Symmetry Operations and Elements

Symmetry operation	Symmetry element: Name	Symbol: International (Hermann-Mauguin)	Symbol: Schoenflies	Presentation on the stereographic projection: Parallel	Presentation on the stereographic projection: Perpendicular
Inversion in a point	Center	$\bar{1}$	C_i	● ○	×
Parallel translation	Translation vector a, b, c				
Reflection in a plane and translation parallel to the plane	Glide–plane	a, b, c, n, d · · ·			
Rotation about an axis and translation parallel to the axis	Screw axis	n_m (m = 1, 2, .., n – 1)			
Rotation about an axis and reflection in a plane perpendicular to the axis	Rotatory-reflection axis	$\tilde{n}$ $\tilde{n} = \tilde{1}, \tilde{2}, \tilde{3}, \tilde{4}, \tilde{6}$	S_n		

TABLE 2. The Thirty-Two Symmetry Classes

Class name[a] and its symbol – International (Int) and Schoenflies (Sch)

Crystal symbol	Primitive Int	Primitive Sch	Central Int	Central Sch	Planal Int	Planal Sch	Axial Int	Axial Sch	Plane-axial Int	Plane-axial Sch	Inversion primitive Int	Inversion primitive Sch	Inversion-planal Int	Inversion-planal Sch
Triclinic	1	C_1	1	C_i										
Monoclinic					m	C_s	2	C_2	2/m	C_{2h}				
Ortho-rhombic			—		mm2	C_{2v}	222	D_2	mmm	D_{2h}				
Trigonal	3	C_3	3	C_{3i}	3m	C_{3v}	32	D_3	$\bar{3}$m	C_{3d}				
Tetragonal	4	C_4	4/m	C_{4h}	4mm	C_{4v}	422	D_4	4/mmm	D_{4h}	$\bar{4}$	S_4	$\bar{4}$2m	D_{2d}
Hexagonal	6	C_6	6/m	C_{6h}	6mm	C_{6v}	622	D_6	6/mmm	D_{6h}	$\bar{6}$	C_{3h}	$\bar{6}$m2	D_{3h}
Cubic	23	T	m$\bar{3}$	T_h	$\bar{4}$3m	T_d	432	O	m3m	O_h				

[a] Per Fedorov Institute of Crystallography, Russian Academy of Sciences, nomenclature.

TABLE 3. The Fourteen Possible Space Lattices (Bravais Lattices)

Crystal system	Metric category of the system	No. of different lattices in the system	Lattice type[a] (marked by +) P	C	I	F	R	No. of identi-points per unit cell	Characteristic parameters (marked by +) a	b	c	α	β	γ	Description of characteristic parameters a⊂X, b⊂Y, c⊂Z α≡(b,c), β≡(a,c), γ≡(a,b)	Symmetry of the lattice Int	Sch
Triclinic	Trimetric	1	+					1	+	+	+	+	+	+	a ≠ b ≠ c, α ≠ β ≠ γ	1	C
Monoclinic	Trimetric	2	+	+				1 or 2	+	+	+		+		a ≠ b ≠ c, α = γ = 90° ≠ β	2/m	C_{2h}
Orthorhombic	Trimetric	4	+	+	+	+		1, 2 or 4	+	+	+				a ≠ b ≠ c, α = β = γ = 90°	mmm	D_{2h}
Trigonal (rhombohedral)	Dimetric	1					+	1	+			+			a = b = c, 120° > α = β = γ ≠ 90°	3m	D_{3d}
Tetragonal	Dimetric	2	+		+			1 or 2	+		+				a = b ≠ c, α = β = γ = 90°	4/mmm	D_{4h}
Hexagonal	Dimetric	1	+					1	+		+				a = b ≠ c, α = β = 90°, γ = 120°	6/mmm	D_{6h}
Isometric (cubic)	Monometric	3	+		+	+		1, 2 or 4	+						a = b = c, α = β = γ = 90°	m3m	O_h

[a] Designations of the space-lattice types: P – primitive, C – side-centered (base-centered), I – body-centered, F – face-centered, R – rhombohedral.

TABLE 4. The Five Possible Regular Polyhedrons

Polyhedron	Symmetry (Schoenflies) Class	Symmetry (Schoenflies) Elements	Form of faces	Number of[a] Faces (F)	Number of[a] Edges (E)	Number of[a] Vertices (V)
Tetrahedron	T	$4C_3 3C_2$	Equilateral triangle	4	6	4
Cube (hexahedron)	O	$3C_4 4C_3 6C_2$	Square	6	12	8

Octahedron	O	$3C_4 4C_3 6C_2$	Equilateral triangle	8	12	6
Pentagonal dodecahedron	J	$6C_5 10C_3 15C_2$	Regular pentagon	12	30	20
Icosahedron	J	$6C_5 10C_3 15C_2$	Equilateral triangle	20	30	12

[a] Per formula by Leonhard Euler: F + V − E = 2

TABLE 5. Classification of Crystals

Strukturbericht symbol	Structure name	Symmetry group: International	Symmetry group: Schoenflies	Pearson symbol[a]	Standard ASTM E157-82a symbol[b]
1	2	3	4	5	6
A1	Cu	Fm3m	O^4_h	cF4	F
A2	W	Im3m	O^9_h	cI2	B
A3	Mg	$P6_3/mmc$	D^4_{6h}	hP2	H
A4	C	Fd3m	O^7_h	cF8	F
A5	Sn	If_1/amd	D^{19}_{4h}	tI4	U
A6	In	I4/mmm	D^{17}_{4h}	tI2	U
A7	As	$R\bar{3}m$	D^5_{3d}	hR2	R
A8	Se	$P3_121$ or $P3_221$	D^4_3 (D^6_3)	hP3	H
A10	Hg	$R\bar{3}m$	D^5_{3d}	hR1	R
A11	Ga	Cmca	D^{18}_{2h}	oC8	Q
A12	α-Mn	$I\bar{4}3m$	T^3_d	cI58	B
A13	β-Mn	$P4_132$	O^7	cP20	C
A15	OW_3	Pm3n	O^3_h	cP8	C
A20	α-U	Cmcm	D^{17}_{2h}	oC4	Q
B1	ClNa	Fm3m	O^5_h	cF8	F
B2	ClCs	Pm3m	O^1_h	cP2	C
B3	SZn	$F\bar{4}3m$	T^2_d	cF8	F
B4	SZn	$P6_3mc$	C^4_{6v}	hP4	H
$B8_1$	AsNi	$P6_3/mmc$	D^4_{6h}	hP4	H
$B8_2$	$InNi_2$	$P6_3/mmc$	D^4_{6h}	hP6	H
B9	HgS	$P3_121$ or $P3_221$	D^4_3 or D^6_3	hP6	H
B10	OPb	P4/nmm	D^7_{4h}	tP4	T
B11	γ-CuTi	P4/nmm	D^7_{4h}	tP4	T
B13	NiS	$R\bar{3}m$	D^5_{3d}	hR6	R
B16	GeS	Pnma	D^{16}_{2h}	oP8	O
B17	PtS	$P4_2/mmc$	D^9_{4h}	tP4	T
B18	CuS	$P6_3/mmc$	D^4_{6h}	hP12	H
B19	AuCd	Pmma	D^5_{2h}	oP4	O
B20	FeSi	$P2_13$	T^4	cP8	C
B27	BFe	Pnma	D^{16}_{2h}	oP8	O
B31	MnP	Pnma	D^{16}_{2h}	oP8	O
B32	NaTl	Fd3m	O^7_h	cF16	F
B34	Pds	$P4_2/m$	C^2_{4h}	tP16	T
B35	CoSn	P6/mmm	D^1_{6h}	hP6	H
B37	SeTl	I4/mcm	D^{18}_{4h}	tI16	U
B_e	CdSb	Pbca	D^{15}_{2h}	oP16	O
B_f (B33)	ξ-BCr	Cmcm	D^{17}_{2h}	oC8	Q
B_g	BMo	$I4_1/amd$	D^{19}_{4h}	tI4	U
B_h	CW	P6m2	D^1_{3h}	hP2	H
B_i	γ′CMo (AsTi)	$P6_3/mmc$	D^4_{6h}	hP8	H
C1	CaF_2	$Fm\bar{3}m$	O^5_h	cF12	F
$C1_b$	AgAsMg	$F\bar{4}3m$	T^2_d	cF12	F
C2	FeS_2	Pa3	T^6_h	cP12	C
C3	Cu_2O	Pn3m	O^4_h	cP6	C
C4	O_2Ti	$P4_2/mnm$	D^{14}_{4h}	tP6	T
C6	CdI_2	P3m1	D^3_{3d}	hP3	H
C7	MoS_2	$P6_3/mmc$	D^4_{6h}	hP6	H
$C11_a$	C_2Ca	I4/mmm	D^{17}_{4h}	tI6	U
$C11_b$	$MoSi_2$	I4/mmm	D^{17}_{4h}	tI6	U
C12	$CaSi_2$	$R\bar{3}m$	D^5_{3d}	hR6	R

TABLE 5. Classification of Crystals

Strukturbericht symbol	Structure name	Symmetry group		Pearson symbol[a]	Standard ASTM E157-82a symbol[b]
		International	Schoenflies		
1	2	3	4	5	6
C14	$MgZn_2$	$P6_3/mmc$	D^4_{6h}	hP12	H
C15	Cu_2Mg	Fd3m	O^7_h	cF24	F
$C15_b$	$AuBe_5$	$F\bar{4}3m$ or F23	T^2_d or T^2	cF24	F
C16	Al_2Cu	I4/mcm	D^{18}_{4h}	tI12	U
C18	FeS_2	Pnnm	D^{12}_{2h}	oP6	O
C19	$CdCl_2$	$R\bar{3}m$	D^5_{3d}	hR3	R
C22	Fe_2P	$P2\bar{6}m$	D^1_{3h}	hP9	H
C23	Cl_2Pb	Pnma	D^{16}_{2h}	oP12	O
C32	AlB_2	P6/mmm	D^1_{6h}	hP3	H
C33	Bi_2STe_2	$R\bar{3}m$	D^5_{3d}	hR5	R
C34	$AuTe_2$	C2/m (P2/m)	C^3_{2h} (C^1_{2h})	mC6	N
C36	$MgNi_2$	$P6_3/mmc$	D^4_{6h}	hP24	H
C38	Cu_2Sb	P4/nmm	D^7_{4h}	tP6	T
C40	$CrSi_2$	$P6_222$	D^4_6	hP9	H
C42	SiS_2	Ibam	D^{26}_{2h}	oI12	P
C44	GeS_2	Fdd2	C^{19}_{2v}	oF72	S
C46	$AuTe_2$	Pma2	C^4_{2v}	oP24	O
C49	Si_2Zr	Cmcm	D^{17}_{2h}	oC12	Q
C54	Si_2Ti	Fddd	D^{24}_{2h}	oF24	S
C_c	Si_2Th	$I4_1/amd$	D^{19}_{4h}	tI12	U
C_e	$CoGe_2$	Aba2	C^{17}_{2v}	oC23	Q
DO_2	As_3Co	Im3	T^5_h	cI32	B
DO_3	BiF_3	Fm3m	O^5_h	cF16	F
DO_9	O_3Re	Pm3m	O^1_h	cP4	C
DO_{11}	CFe_3	Pnma	D^{16}_{2h}	oP16	O
DO_{18}	$AsNa_3$	$P6_3/mmc$	D^4_{6h}	hP8	H
DO_{19}	Ni_3Sn	$P6_3/mmc$	D^4_{6h}	hP8	H
DO_{20}	Al_3Ni	Pnma	D^{16}_{2h}	oP16	O
DO_{21}	Cu_3P	$P\bar{3}c1$	D^4_{3d}	hP24	H
DO_{22}	Cu_3P	I4/mmm	D^{17}_{4h}	tI8	U
DO_{23}	Al_3Zr	I4/mmm	D^{17}_{4h}	tI16	U
DO_{24}	Ni_3Ti	$P6_3/mmc$	D^4_{6h}	hP16	H
DO_c	SiU_3	I4/mcm	D^{18}_{4h}	tI16	U
DO_e	Ni_3P	$I\bar{4}$	S^2_4	tI32	U
$D1_3$	Al_4Ba	I4/mmm	D^{17}_{4h}	tI10	U
$D1_a$	$MoNi_4$	I4/m	C^5_{4h}	tI10	U
$D1_b$	Al_4U	Imma	D^{28}_{2h}	oI20	P
$D1_c$	$PtSn_4$	Aba2	C^{17}_{2v}	oC20	Q
$D1_e$	B_4Th	P4/mbm	D^5_{4h}	tP20	T
$D1_f$	BMn_4	Fddd	D^{24}_{2h}	oF40	S
$D2_1$	B_6Ca	Pm3m	O^1_h	cP7	C
$D2_3$	$NaZn_{13}$	Fm3m	O^5_h	cF112	F
$D2_b$	$Mn_{12}Th$	I4/mmm	D^{17}_{4h}	tI26	U
$D2_c$	MnU_6	I4/mcm	D^{18}_{4h}	tI28	U
$D2_d$	$CaCu_5$	P6/mmm	D^1_{6h}	hP6	H
$D2_f$	$B_{12}U$	Fm3m	O^5_h	cF52	F
$D2_h$	Al_6Mn	Cmcm	D^{17}_{2h}	oC28	Q
$D5_1$	α-Al_2O_3	R3c	D^6_{3d}	hR10	R
$D5_2$	La_2O_3	$P\bar{3}m1$	D^3_{3d}	hP5	H
$D5_3$	Mn_2O_3	Ia3	T^7_h	cI80	B
$D5_8$	S_3Sb_2	Pnma	D^{16}_{2h}	oP20	O
$D5_9$	P_2Zn_3	$P4_2/mmc$	D^9_{4h}	tP40	T
$D5_{10}$	C_2C_3	Pnma	D^{16}_{2h}	oP20	O
$D5_{13}$	Al_3Ni_2	$P\bar{3}m1$	D^3_{3d}	hP5	H
$D5_a$	Si_2U_3	P4/mbm	D^5_{4h}	tP10	T
$D5_c$	C_3Pu_2	$I\bar{4}3d$	T^6_d	cI40	B
$D7_1$	Al_4C_3	$R\bar{3}m$	D^5_{3d}	hR7	R

TABLE 5. Classification of Crystals

Strukturbericht symbol	Structure name	Symmetry group: International	Symmetry group: Schoenflies	Pearson symbol[a]	Standard ASTM E157-82a symbol[b]
1	2	3	4	5	6
$D7_3$	P_4Th_3	$I\bar{4}3d$	T_d^6	cI28	B
$D7_b$	B_4Ta_3	Immm	D_{2h}^{25}	oI14	P
$D8_1$	Fe_3Zn_{10}	Im3m	O_h^9	cI52	B
$D8_2$	Cu_5Zn_8	$I\bar{4}3m$	T_d^3	cI52	B
$D8_3$	Al_4Cu_9	P43m	T_d^1	cP52	C
$D8_4$	C_6Cr23	Fm3m	O_h^5	cF116	F
$D8_5$	Fe_7W_6	$R\bar{3}m$	D_{3d}^5	hR13	R
$D8_6$	$Cu_{15}Si_4$	$I\bar{4}3m$	T_d^3	cI76	B
$D8_8$	Mn_5Si_3	$P6_3/mcm$	D_{6h}^3	hP16	H
$D8_9$	Co_9S_8	Fm3m	O_h^5	cF68	F
$D8_{10}$	Al_8Cr_5	R3m	C_{3v}^5	hR26	R
$D8_{11}$	Al_5Co_2	$P6_3/mcm$	D_{6h}^3	hP28	H
$D8_a$	$Mn_{23}Th_6$	Fm3m	O_h^5	cF116	F
$D8_b$	σ-phase of Cr-Fe	$p\bar{4}_2/mnm$	D_{4h}^{14}	tP30	T
$D8_e$	$(Al,Zn)_{49}Mg_{32}$	Im3	T_h^5	cI162	B
$D8_f$	Ge_7Ir_3	Im3m	O_h^9	cI40	B
$D8_h$	B_5W_2	$P6_3/mmc$	D_{6h}^4	hP14	H
$D8_i$	B_5Mo_2	$R\bar{3}m$	D_{3d}^5	hR7	R
$D8_l$	B_3Cr_5	I4/mcm	D_{4h}^{18}	tI32	U
$D8_m$	Si_3W_5	I4/mcm	D_{4h}^{18}	tI32	U
$D10_1$	C_3Cr_7	P31c	C_{3v}^4	hP80	H
$D10_2$	Fe_3Th_7	$P6_3mc$	C_{6v}^4	hP20	H
$E0_1$	ClFPb	P4/nmm	D_{4h}^7	tP6	T
$E1_1$	$CuFeS_2$	$I\bar{4}2d$	D_{2d}^{12}	tI16	U
$E2_1$	CaO_3Ti	Pm3m	O_h^1	cP5	C
$E2_4$	S_3Sn_2	Pnma	D_{2h}^{16}	oP20	O
E3	Al_2CdS_4	$I\bar{4}$	S_4^2	tI14	U
$E9_3$	$SiFe_3W_3$	Fd3m	O_h^7	cF112	F
$E9_a$	Al_7Cu_2Fe	P4/mnc	D_{4h}^6	tP40	T
$E9_b$	$AlLi_3N_2$	Ia3	T_h^7	cI96	B
$F0_1$	NiSSb	$P2_13$	T^4	cP12	C
$F5_1$	$CrNaS_2$	R3m or R32	D_{3d}^5 or D_3^7	hR4	R
$F5_6$	CuS_2Sb	Pnma	D_{2h}^{16}	oP16	O
$H1_1$	Al_2MgO_4	Fd3m	O_h^7	cF56	F
$H2_4$	Cu_3S_4V	P43m	T_d^1	cP8	C
$H2_5$	$AsCu_3S_4$	$Pmn2_1$	C_{2v}^7	oP16	O
$L1_0$	AuCu	P4/mmm	D_{4h}^1	tP4	T
$L1_2$	$AlCu_3$	Pm3m	O_h^1	cP4	C
$L2_1$	$AlCu_2Mn$	Fm3m	O_h^5	cF16	F
$L2_2$	Sb_2Tl_7	Im3m	O_h^9	cI54	B
$L'2_b$	H_2Th	I4/mmm	D_{4h}^{17}	tI6	U
L′3	Fe_2N	$P6_3/mmc$	D_{6h}^4	hP3	H
$L6_0$	$CuTi_3$	P4/mmm	D_{4h}^1	tP4	T

[a] The first letter denotes the crystal system: triclinic (a), monoclinic (m), orthorhombic (o), tetragonal (t), hexagonal (h) and cubic (c). Trigonal (rhombohedral) system is denoted by combination hR. The second letter of Pearson's symbol denotes lattice type: primitive (P), edge-(base-) centered (C), body-centered (I) or face-centered (F). The following number denotes number of atoms in the crystal unit cell.

[b] Standard ASTM E157-82a has the Bravais lattices designations as following: C – primitive cubic; B – body-centered cubic; F – face-centered cubic; T – primitive tetragonal; U – body-centered tetragonal; R – rhombohedral; H – hexagonal; O – primitive orthorhombic; P – body-centered orthorhombic; Q – base-centered orthorhombic; S – face-centered orthorhombic; M – primitive monoclinic; N – centered monoclinic; A – triclinic.

References

1. A. Schoenflies, *Kristallsysteme und Kristallstructur*, Leipzig, 1891.
2. E. S. Fedorow, Zusammenstellung der kristallographischen Resultate, *Zs. Krist.*, 20, 1892.
3. P. Groth, *Elemente der physikalischen und chemischen Krystallographie*, R. Oldenbourg, München/Berlin, 1921.

4. N. V. Belov, Class Method of Deriving Space Groups of Symmetry, *Trudy Instituta Kristallodraffi imeni Fedorova (Transactions of the Fedorov Inst. of Crystallography)*, 5, 25, 1951, in Russian.
5. W. B. Pearson, *Handbook of Lattice Spacings and Structures of Metals and Alloys*, Vol. 1, Pergamon Press, 1958; Vol. 2, 1967.
6. Ch. Kittel, *Introduction to Solid State Physics*, John Wiley & Sons, 1956.
7. G. S. Zhdanov, *Fizika Tverdogo Tela (Solid State Physics)*, Moscow University Press, 1962, in Russian.
8. M. J. Buerger, *Elementary Crystallography*, John Wiley & Sons, 1963.
9. F. D. Bloss, *Crystallography & Crystal Chemistry*, Holt, Rinehart & Winston, 1971.
10. T. Janssen, *Crystallographic Groups*, North-Holland/American Elsevier, 1973.
11. M. P. Shaskolskaya, *Kristallografiya (Crystallography)*, Vysshaya Shkola, Moscow, 1976, in Russian.
12. T. Hahn, Ed., Internat. *Tables for Crystallography*, Vol. A, D. Reidel Publishing, Boston, 1983.
13. Crystal Data. Determinative Tables, Volumes 1–6, 1966–1983, JCPDS-Intern Centre for Diffraction Data and U.S. Dept. of Commerce.
14. R. W. G. Wyckoff, *Crystal Structures*, 2nd ed., Volumes 1–6, Interscience, New York, 1963.
15. C. J. Bradley and A. P. Cracknell, *The Mathematical Theory of Symmetry in Solids*, Clarendon Press, Oxford, 1972.
16. International Tables for Crystallography. Volume A, *Space–Group Symmetry*, T. Hahn, Ed., 1989; Volume B, *Reciprocal Space*, U. Schmueli, Ed.; Volume C, *Mathematical, Physical and Chemical Tables*, A. J. C. Wilson, Ed., Kluwer Academic Publishers, Dordrecht, 1989.
17. G. R. Desiraju, *Crystal Engineering: The Design of Organic Solids*, Elsevier, Amsterdam, 1989.
18. M. Senechal, *Crystalline Symmetries: An Informal Mathematical Introduction*, Adam Hilger Publ., Bristol, 1990.
19. C. Hammond, *Introduction to Crystallography*, Oxford University Press, 1990.
20. N.W. Alcock, *Bonding and Structure: Structural Principles in Inorganic and Organic Chemistry*, Ellis Norwood Publ., 1990.
21. T. C. W. Mak and G. D. Zhou. *Crystallography in Modern Chemistry: A Resource Book of Crystal Structures*, Wiley–Interscience, New York, 1992.
22. S. C. Abrahams, K. Mirsky, and R. M. Nielson, *Acta Cryst*, B52, 806 (1996); B52, 1057 (1996).
23. C. Marcos, A. Panalague, D. B. Morciras, S. Garcia-Granda and M. R. Dias. *Acta Cryst*, B52, 899 (1996).
24. A. C. Larson, *Crystallographic Computing*, Manksgaard, Copenhagen, 1970.
25. G. M. Sheldrick, SHELXS86. Crystallographic Computing 3, Clarendon Press, Oxford, 1986; SHELXL93. Program for the Refinement of Crystal Structures, University of Göttingen Press, 1993.
26. Inorganic Crystal Structure Database, CD–ROM. Sci. Inf. Service. E-mail: SISI@Delphi.com.

IONIC RADII IN CRYSTALS

Ionic radii are a useful tool for predicting and visualizing crystal structures. This table lists a set of ionic radii R_i in Å units for the most common coordination numbers *CN* of positive and negative ions. The values are based on experimental crystal structure determinations, supplemented by empirical relationships, and theoretical calculations. Anions are listed first, followed by cations.

The advice of Howard T. Evans and Marvin J. Weber in preparing this table is appreciated.

References

1. Shannon, R. D., *Acta Crystallogr.* A32, 751, 1976.
2. Jia, Y. Q., *J. Solid State Chem.* 95, 184, 1991.

Ionic Radii for Ions with Various Coordination Numbers

Ion	*CN*	R_i/Å
Anions		
F^{-1}	6	1.33
Cl^{-1}	6	1.81
Br^{-1}	6	1.96
I^{-1}	6	2.20
OH^{-1}	4	1.35
OH^{-1}	6	1.37
O^{-2}	2	1.21
O^{-2}	6	1.40
O^{-2}	8	1.42
S^{-2}	6	1.84
Se^{-2}	6	1.98
Te^{-2}	6	2.21
Cations		
Ac^{+3}	6	1.12
Ag^{+1}	4	1.00
Ag^{+1}	6	1.15
Ag^{+1}	8	1.28
Ag^{+2}	4 (square)	0.79
Ag^{+2}	6	0.94
Al^{+3}	4	0.39
Al^{+3}	5	0.48
Al^{+3}	6	0.54
Am^{+3}	6	0.98
Am^{+3}	8	1.09
Am^{+4}	6	0.85
Am^{+4}	8	0.95
As^{+3}	6	0.58
As^{+5}	4	0.34
As^{+5}	6	0.46
Au^{+1}	6	1.37
Au^{+3}	4 (square)	0.64
Au^{+3}	6	0.85
Ba^{+2}	6	1.35
Ba^{+2}	8	1.42
Ba^{+2}	12	1.61
Be^{+2}	4	0.27
Be^{+2}	6	0.45
Bi^{+3}	5	0.96
Bi^{+3}	6	1.03
Bi^{+3}	8	1.17
Bi^{+5}	6	0.76
Bk^{+3}	6	0.96
Bk^{+4}	6	0.83
Bk^{+4}	8	0.93
Br^{+5}	3 (pyramidal)	0.31
Br^{+7}	4	0.25
Br^{+7}	6	0.39
C^{+4}	4	0.15
C^{+4}	6	0.16
Ca^{+2}	6	1.00
Ca^{+2}	8	1.12
Ca^{+2}	10	1.23
Ca^{+2}	12	1.34
Cd^{+2}	4	0.78
Cd^{+2}	6	0.95
Cd^{+2}	8	1.10
Cd^{+2}	12	1.31
Ce^{+3}	6	1.01
Ce^{+3}	8	1.14
Ce^{+3}	10	1.25
Ce^{+3}	12	1.34
Ce^{+4}	6	0.87
Ce^{+4}	8	0.97
Ce^{+4}	10	1.07
Ce^{+4}	12	1.14
Cf^{+3}	6	0.95
Cf^{+4}	6	0.82
Cf^{+4}	8	0.92
Cl^{+5}	3 (pyramidal)	0.12
Cl^{+7}	4	0.08
Cm^{+3}	6	0.97
Cm^{+4}	6	0.85
Cm^{+4}	8	0.95
Co^{+2}	4	0.56
Co^{+2}	6	0.65
Co^{+2}	8	0.90
Co^{+3}	6	0.55
Cr^{+2}	6	0.73
Cr^{+3}	6	0.62
Cr^{+4}	4	0.41
Cr^{+4}	6	0.55
Cr^{+6}	4	0.26
Cr^{+6}	6	0.44
Cs^{+1}	6	1.67
Cs^{+1}	8	1.74
Cs^{+1}	10	1.81
Cs^{+1}	12	1.88
Cu^{+1}	2	0.46
Cu^{+1}	4	0.60
Cu^{+1}	6	0.77
Cu^{+2}	4 (square)	0.57
Cu^{+2}	6	0.73
Dy^{+2}	6	1.07
Dy^{+2}	8	1.19
Dy^{+3}	6	0.91
Dy^{+3}	8	1.03
Er^{+3}	6	0.89
Er^{+3}	8	1.00
Eu^{+2}	6	1.17
Eu^{+2}	8	1.25
Eu^{+2}	10	1.35
Eu^{+3}	6	0.95
Eu^{+3}	8	1.07
F^{+7}	6	0.08
Fe^{+2}	4	0.63
Fe^{+2}	6	0.61
Fe^{+2}	8	0.92
Fe^{+3}	4	0.49
Fe^{+3}	6	0.55
Fe^{+3}	8	0.78
Fr^{+1}	6	1.80
Ga^{+3}	4	0.47
Ga^{+3}	6	0.62
Gd^{+3}	6	0.94
Gd^{+3}	8	1.05
Ge^{+2}	6	0.73
Ge^{+4}	4	0.39
Ge^{+4}	6	0.53
Hf^{+4}	4	0.58
Hf^{+4}	6	0.71
Hf^{+4}	8	0.83
Hg^{+1}	6	1.19
Hg^{+2}	2	0.69
Hg^{+2}	4	0.96
Hg^{+2}	6	1.02
Hg^{+2}	8	1.14
I^{+5}	3 (pyramidal)	0.44
I^{+5}	6	0.95
I^{+7}	4	0.42
I^{+7}	6	0.53
In^{+3}	4	0.62
In^{+3}	6	0.80
Ir^{+3}	6	0.68
Ir^{+4}	6	0.63
Ir^{+5}	6	0.57
K^{+1}	4	1.37
K^{+1}	6	1.38
K^{+1}	8	1.51
K^{+1}	12	1.64
La^{+3}	6	1.03
La^{+3}	8	1.16
La^{+3}	10	1.27
La^{+3}	12	1.36
Li^{+1}	4	0.59
Li^{+1}	6	0.76
Li^{+1}	8	0.92
Lu^{+3}	6	0.86
Lu^{+3}	8	0.97
Mg^{+2}	4	0.57
Mg^{+2}	6	0.72
Mg^{+2}	8	0.89
Mn^{+2}	4	0.66
Mn^{+2}	6	0.83
Mn^{+2}	8	0.96

Ion	*CN*	R_i/Å
Mn^{+3}	6	0.58
Mn^{+4}	4	0.39
Mn^{+4}	6	0.53
Mn^{+5}	4	0.33
Mn^{+6}	4	0.26
Mn^{+7}	4	0.25
Mo^{+3}	6	0.69
Mo^{+4}	6	0.65
Mo^{+5}	4	0.46
Mo^{+5}	6	0.61
Mo^{+6}	4	0.41
Mo^{+6}	6	0.59
Mo^{+6}	7	0.73
N^{+3}	6	0.16
N^{+5}	6	0.13
Na^{+1}	4	0.99
Na^{+1}	6	1.02
Na^{+1}	8	1.18
Na^{+1}	9	1.24
Na^{+1}	12	1.39
Nb^{+3}	6	0.72
Nb^{+3}	8	0.79
Nb^{+4}	6	0.68
Nb^{+5}	4	0.48
Nb^{+5}	6	0.64
Nb^{+5}	8	0.74
Nd^{+3}	6	0.98
Nd^{+3}	8	1.12
Nd^{+3}	9	1.16
Nd^{+3}	12	1.27
Ni^{+2}	4 (square)	0.49
Ni^{+2}	6	0.69
Ni^{+3}	6	0.56
Np^{+3}	6	1.01
Np^{+4}	6	0.87
Np^{+5}	6	0.75
Np^{+6}	6	0.72
Os^{+4}	6	0.63
Os^{+5}	6	0.58
Os^{+6}	6	0.55
Os^{+8}	4	0.39
P^{+5}	4	0.17
P^{+5}	6	0.38
Pa^{+3}	6	1.04
Pa^{+4}	6	0.90
Pa^{+5}	6	0.78
Pb^{+2}	6	1.19
Pb^{+2}	8	1.29
Pb^{+2}	10	1.40
Pb^{+2}	12	1.49
Pb^{+4}	4	0.65
Pb^{+4}	6	0.78
Pb^{+4}	8	0.94
Pd^{+2}	4 (square)	0.64
Pd^{+2}	6	0.86
Pd^{+3}	6	0.76
Pd^{+4}	6	0.62
Pm^{+3}	6	0.97
Pm^{+3}	8	1.09
Po^{+4}	6	0.97
Pr^{+3}	6	0.99
Pr^{+3}	8	1.13
Pr^{+4}	6	0.85
Pr^{+4}	8	0.96
Pt^{+2}	4 (square)	0.60
Pt^{+2}	6	0.80
Pt^{+4}	6	0.63
Pu^{+3}	6	1.00
Pu^{+4}	6	0.86
Pu^{+5}	6	0.74
Pu^{+6}	6	0.71
Ra^{+2}	8	1.48
Ra^{+2}	12	1.70
Rb^{+1}	6	1.52
Rb^{+1}	8	1.61
Rb^{+1}	10	1.66
Rb^{+1}	12	1.72
Re^{+4}	6	0.63
Re^{+5}	6	0.58
Re^{+6}	6	0.55
Re^{+7}	4	0.38
Re^{+7}	6	0.53
Rh^{+3}	6	0.67
Rh^{+4}	6	0.60
Rh^{+5}	6	0.55
Ru^{+3}	6	0.68
Ru^{+4}	6	0.62
Ru^{+5}	6	0.57
Ru^{+7}	4	0.38
Ru^{+8}	4	0.36
S^{+4}	6	0.37
S^{+6}	4	0.12
S^{+6}	6	0.29
Sb^{+3}	4 (pyramidal)	0.76
Sb^{+3}	6	0.76
Sb^{+5}	6	0.60
Sc^{+3}	6	0.75
Sc^{+3}	8	0.87
Se^{+4}	6	0.50
Se^{+6}	4	0.28
Se^{+6}	6	0.42
Si^{+4}	4	0.26
Si^{+4}	6	0.40
Sm^{+2}	6	1.19
Sm^{+2}	8	1.27
Sm^{+3}	6	0.96
Sm^{+3}	8	1.08
Sm^{+3}	12	1.24
Sn^{+4}	4	0.55
Sn^{+4}	6	0.69
Sn^{+4}	8	0.81
Sr^{+2}	6	1.18
Sr^{+2}	8	1.26
Sr^{+2}	10	1.36
Sr^{+2}	12	1.44
Ta^{+3}	6	0.72
Ta^{+4}	6	0.68
Ta^{+5}	6	0.64
Tb^{+3}	6	0.92
Tb^{+3}	8	1.04
Tb^{+4}	6	0.76
Tb^{+4}	8	0.88
Tc^{+4}	6	0.65
Te^{+4}	4	0.66
Te^{+4}	6	0.97
Te^{+6}	4	0.43
Te^{+6}	6	0.56
Th^{+4}	6	0.94
Th^{+4}	8	1.05
Th^{+4}	10	1.13
Th^{+4}	12	1.21
Ti^{+2}	6	0.86
Ti^{+3}	6	0.67
Ti^{+4}	4	0.42
Ti^{+4}	6	0.61
Ti^{+4}	8	0.74
Tl^{+1}	6	1.50
Tl^{+1}	8	1.59
Tl^{+1}	12	1.70
Tl^{+3}	4	0.75
Tl^{+3}	6	0.89
Tl^{+3}	8	0.98
Tm^{+2}	6	1.01
Tm^{+2}	7	1.09
Tm^{+3}	6	0.88
Tm^{+3}	8	0.99
U^{+3}	6	1.03
U^{+4}	6	0.89
U^{+4}	8	1.00
U^{+4}	12	1.17
U^{+5}	6	0.76
U^{+6}	2	0.45
U^{+6}	4	0.52
U^{+6}	6	0.73
U^{+6}	8	0.86
V^{+2}	6	0.79
V^{+3}	6	0.64
V^{+4}	5	0.53
V^{+4}	6	0.58
V^{+4}	8	0.72
V^{+5}	4	0.36
V^{+5}	5	0.46
V^{+5}	6	0.54
W^{+4}	6	0.66
W^{+5}	6	0.62
W^{+6}	4	0.42
W^{+6}	5	0.51
W^{+6}	6	0.60
Y^{+3}	6	0.90
Y^{+3}	8	1.02
Y^{+3}	9	1.08
Yb^{+2}	6	1.02
Yb^{+2}	8	1.14
Yb^{+3}	8	0.99
Yb^{+3}	9	1.04
Zn^{+2}	4	0.60
Zn^{+2}	6	0.74
Zn^{+2}	8	0.90
Zr^{+4}	4	0.59
Zr^{+4}	6	0.72
Zr^{+4}	8	0.84
Zr^{+4}	9	0.89

POLARIZABILITIES OF ATOMS AND IONS IN SOLIDS

H. P. R. Frederikse

The polarization of a solid dielectric medium, ***P***, is defined as the dipole moment per unit volume averaged over the volume of a crystal cell. A component of ***P*** can be expanded as a function of the electric field ***E***:

$$P_i = \sum_j a_j E_j + \sum_{jk} b_{jk} E_j E_k$$

For relatively small electric fields in isotropic substances $\boldsymbol{P} = \chi_e \boldsymbol{E}$, where χ_e is the electric susceptibility. If the medium is made up of N atoms (or ions) per unit volume, the polarization is $\boldsymbol{P} = N\boldsymbol{p}_m$ where $\boldsymbol{p}_m$ is the average dipole moment per atom. The polarizability α can be defined as $\boldsymbol{p}_m = \alpha \boldsymbol{E}_0$, where $\boldsymbol{E}_0$ is the local field at the position of the atom. Using the Lorentz method to calculate the local field one finds:

$$\boldsymbol{P} = N\alpha(\boldsymbol{E} + 4\pi\boldsymbol{P}) = \chi_e \boldsymbol{E}$$

Together with the definition of the dielectric constant (relative permittivity), $\varepsilon = 1 + 4\pi\chi_e$, this leads to:

$$\alpha = (3/4\pi N)(\varepsilon - 1)/(\varepsilon + 2)$$

This expression is known as the Clausius-Mossotti equation.

The total polarization associated with atoms, ions, or molecules is due to three different sources:

Electronic polarization arises because the center of the local electronic charge cloud around the nucleus is displaced under the action of the field: $\boldsymbol{P}_e = N\alpha_e \boldsymbol{E}_0$ where α_e is the *electronic polarizability.*

Ionic polarization occurs in ionic materials because the electric field displaces cations and anions in opposite directions: $\boldsymbol{P}_i = N\alpha_i \boldsymbol{E}_0$, where α_i is the *ionic polarizability.*

Orientational polarization can occur in substances composed of molecules that have permanent electric dipoles. The alignment of these dipoles depends on temperature and leads to an *orientational polarizability* per molecule: $\alpha_{or} = p^2/3kT$, where p is the permanent dipole moment per molecule, k is the Boltzmann constant, and T is the temperature.

Because of the different nature of these three polarization processes the response of a dielectric solid to an applied electric field will strongly depend on the frequency of the field. The resonance of the electronic excitation in insulators (dielectrics) takes place in the ultraviolet part of the spectrum; the characteristic frequency of the lattice vibrations is located in the infrared, while the orientation of dipoles requires fields of much lower frequencies (below 10^{10} Hz). This response to electric fields of different frequencies is shown in Figure 1. Values of the electronic polarizabilities for selected atoms and ions are given in Table 1.

References

1. Kittel, C., *Introduction to Solid State Physics, Fourth Edition*, John Wiley & Sons, New York, 1971.
2. Lerner, R.G., and Trigg, G.L., Eds., *Encyclopedia of Physics, Second Edition*, VCH Publishers, New York, 1990.
3. Ralls, K.M., Courtney, T.H., and Wulff, J., *An Introduction to Materials Science and Engineering*, John Wiley & Sons, New York, 1976.

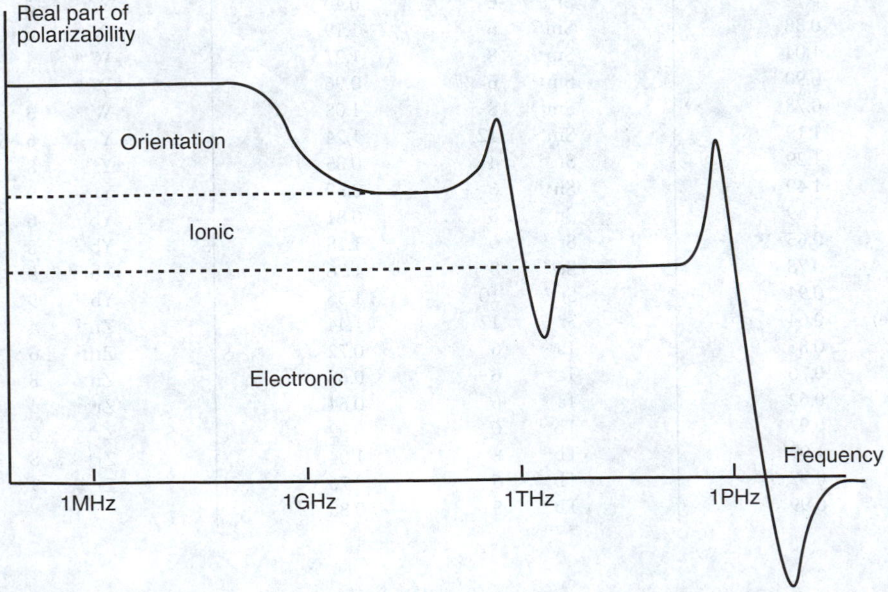

FIGURE 1. Schematic graph of the frequency dependence of the different contributions to polarizability.

TABLE 1. Electronic Polarizabilities in Units of 10^{-24} cm³

I	II	III	IV	VI	VII	VIII
						He 0.205
Li^+ 0.029	**Be^{2+}** 0.008	**B^{3+}** 0.003	**C^{4+}** 0.0013	**O^{2-}** 3.88	**F^-** 1.04	**Ne** 0.394
Na^+ 0.179	**Mg^{2+}** 0.094	**Al^{3+}** 0.052	**Si^{4+}** 0.0165	**S^{2-}** 10.2	**Cl^-** 3.66	**Ar** 1.64
K^+ 0.83	**Ca^{2+}** 0.47	**Sc^{3+}** 0.286	**Ti^{4+}** 0.185	**Se^{2-}** 10.5	**Br^-** 4.77	**Kr** 2.48
Rb^+ 1.40	**Sr^{2+}** 0.86	**Y^{3+}** 0.55	**Zr^{4+}** 0.37	**Te^{2-}** 14.0	**I^-** 7.1	**Xe** 4.04
Cs^+ 2.42	**Ba^{2+}** 1.55	**La^{3+}** 1.04	**Ce^{4+}** 0.73			

Data from Pauling, L., *Proc. R. Soc. London*, A114, 181, 1927. See also Jaswal, S.S. and Sharma, T.P., *J. Phys. Chem. Solids*, 34, 509, 1973.

Values are appropriate for cgs units. To convert to SI, use the relation $\alpha(SI)/C\ m^2 V^{-1} = 1.11265.10^{-16}\alpha(cgs)/cm^3$.

CRYSTAL STRUCTURES AND LATTICE PARAMETERS OF ALLOTROPES OF THE ELEMENTS

H. W. King

The crystal structures of the allotropic forms of the elements are presented in terms of the Pearson symbol, the Strukturbericht designation, and the prototype of the structure. The temperatures of the phase transformations are listed in degrees Celsius and the pressures are in GPa. A consistent nomenclature is used, whereby all allotropes are labeled by Greek letters. The lattice parameters of the units cells are given in nanometers (nm) and are considered to be accurate to ±2 in the last reported digit.

This compilation is restricted to changes in crystal structures that occur as a result of a change in temperature or pressure. Low-temperature structures are included for the diatomic and rare gases, which show many similarities with respect to the metallic elements.

Reprinted with the permission of ASM International from T.B. Massalski, Ed., *Binary Alloy Phase Diagrams*, ASM International, Metals Park, Ohio, 1986; certain data on rare earth elements were provided by K.A. Gschneidner.

Element	t/°C	p/GPa	Pearson symbol	Space group	Strukturbericht designation	Prototype	a/nm	b/nm	c/nm
Ac	25	atm	*cF4*	*Fm3m*	*A1*	Cu	0.5311		
Ag	25	atm	*cF4*	*Fm3m*	*A1*	Cu	0.40857		
αAl	25	atm	*cF4*	*Fm3m*	*A1*	Cu	0.40496		
βAl	25	>20.5	*hP2*	$P6_3/mmc$	*A3*	Mg	0.2693		0.4398
α´Am	25	atm	*hP4*	$P6_3/mmc$	*A3´*	αLa	0.34681		1.1241
αAm	>769	atm	*cF4*	*Fm3m*	*A1*	Cu	0.4894		
βAm	>1074	atm	*cI2*	*Im3m*	*A2*	W	?		
γAm	25	>15	*oC4*	*Cmcm*	*A20*	αU	0.3063	0.5968	0.5169
αAr	<-189.35	atm	*cF4*	*Fm3m*	*A1*	Cu	0.5316		
(βAr)	<-189.40	atm	*hP2*	$P6_3/mmc$	*A3*	Mg	0.3760		0.6141
αAs	25	atm	*hR2*	*R3m*	*A7*	αAs	0.41319		
εAs	>448	atm	*oC8*	*Cmca*	…	P (black)	0.362	1.085	0.448
Au	25	atm	*cF4*	*Fm3m*	*A1*	Cu	0.40782		
βB	25	atm	*hR105*	*R3m*	…	βB	1.017		
αBa	25	atm	*cI2*	*Im3m*	*A2*	W	0.50227		
βBa	25	>5.33	*hP2*	$P6_3/mmc$	*A3*	Mg	0.3901		0.6154
γBa	25	>23	?	?	…	…			
αBe	25	atm	*hP2*	$P6_3/mmc$	*A3*	Mg	0.22859		0.35845
βBe	>1270	atm	*cI2*	*Im3m*	*A2*	W	0.25515		
γBe	25	>9.3	?	…	…	…			
αBi	25	atm	*hR2*	*r3m*	*A7*	αAs	0.47460		
βBi	25	>2.6	*mC4*	*C2/m*	…	βBi	0.6674	0.6117	0.3304
γBi	25	>3.0	*mP3*	?	…	…	0.605	0.42	0.465
σBi	25	>4.3	?	?	…	…			
εBi	25	>6.5	?	?	…	…			
ζBi	25	>9.0	*cI2*	*Im3m*	*A2*	W	0.3800		
αBk	25	atm	*hP4*	$P6_3/mmc$	*A3´*	αLa	0.3416		1.1069
βBk	>977	atm	*cF4*	*Fm3m*	*A1*	Cu	0.4997		
Br	<7.25	atm	*oC8*	*Cmca*	…	Cl	0.668	0.449	0.874
C (graphite)	25	atm	*hP4*	$P6_3/mmc$	*A9*	C (graphite)	0.24612		0.6709
C (diamond)	25	>60	*cF8*	*Fd3m*	*A4*	C (diamond)	0.35669		
C (hd)	25	HP	*hP4*	$P6_3/mmc$	…	C (hd)	0.2522		0.4119
αCa	25	atm	*cF4*	*Fm3m*	*A1*	Cu	0.55884		
βCa	>443	atm	*cI2*	*Im3m*	*A2*	W	0.4480		
γCa	25	>1.5	?	…	…	…			
Cd	25	atm	*hP2*	$P6_3/mmc$	*A3*	Mg	0.29793		0.56196
αCe	<-177	atm	*cF4*	*Fm3m*	*A1*	Cu	0.485		
βCe	25	atm	*hP4*	$P6_3/mmc$	*A3´*	αLa	0.36810		1.1857
γCe	25	atm	*cF4*	*Fm3m*	*A1*	Cu	0.51610		
δ-Ce	>726	atm	*cI2*	*Im3m*	*A2*	W	0.412		
α´Ce	25	>5.4	*oC4*	*Cmcm*	*A20*	αU	0.3049	0.5998	0.5215
αCf	25	atm	*hP4*	$P6_3/mmc$	*A3´*	αLa	0.339		1.1015

Element	t/°C	p/GPa	Pearson symbol	Space group	Strukturbericht designation	Prototype	a/nm	b/nm	c/nm
βCf	>590	atm	cF4	Fm3m	A1	Cu	?		
Cl	<-102	atm	oC8	Cmca	…	Cl	0.624	0.448	0.826
αCm	25	atm	hP4	$P6_3/mmc$	A3′	αLa	0.3496		1.1331
βCm	>1277	atm	cF4	Fm3m	A1	Cu	0.4382		
ϵCo	25	atm	hP2	$P6_3/mmc$	A3	Mg	0.25071		0.40686
αCo	>422	atm	cF4	Fm3m	A1	Cu	0.35447		
αCr	25	atm	cI2	Im3m	A2	W	0.28848		
α′Cr	25	HP	tI2	I4/mmm	…	α′Cr	0.2882		0.2887
aCs	25	atm	cI2	Im3m	A2	W	0.6141		
βCs	25	>2.37	cF4	Fm3m	A1	Cu	0.6465		
β′Cs	25	>4.22	cF4	Fm3m	A1	Cu	0.5800		
γCs	25	>4.27	?	…	…	…	?		
Cu	25	atm	cF4	Fm3m	A1	Cu	0.36146		
α′Dy	<-187	atm	oC4	Cmcm	…	α′Dy	0.3595	0.6184	0.5678
αDy	25	atm	hP2	$P6_3/mmc$	A3	Mg	0.35915		0.56501
βDy	>1381	atm	cI2	Im3m	A2	W	0.403		
γDy	25	>7.5	hR3	R3m	…	αSm	0.3436		2.483
Er	25	atm	hP2	$P6_3/mmc$	A3	Mg	0.35592		0.55850
αEs	25	atm	hP4	$P6_3/mmc$	A3′	αLa	?		
βEs	?	atm	cF4	Fm3m	A1	Cu	?		
Eu	25	atm	cI2	Im3m	A2	W	0.45827		
αF	<-227.6	atm	mC8	C2/c	…	αF	0.550	0.338	0.728
βF	<-219.67	atm	cP16	Pm3n	…	γO	0.667		
αFe	25	atm	cI2	Im3m	A2	W	0.28665		
γFe	>912	atm	cF4	Fm3m	A1	Cu	0.36467		
σFe	>1394	atm	cI2	Im3m	A2	W	0.29315		
ϵFe	25	>13	hP2	$P6_3/mmc$	A3	Mg	0.2468		0.396
αGa	25	atm	oC8	Cmca	A11	αGa	0.45186	0.76570	0.45258
βGa	25	>1.2	tI2	I4/mmm	A6	In	0.2808		0.4458
γGa	-53	>3.0	oC40	Cmcm	…	γGa	1.0593	1.3523	0.5203
αGd	25	atm	hP2	$P6_3/mmc$	A3	Mg	0.36336		0.57810
βGd	>1235	atm	cI2	Im3m	A2	W	0.406		
γGd	25	>3.0	hR3	R3m	…	αSm	0.361		2.603
αGe	25	atm	cF8	Fd3m	A4	C (diamond)	0.56574		
βGe	25	>12	tI4	$I4_1/amd$	A5	βSn	0.4884		0.2692
γGe	25	>12→atm	tP12	$P4_12_12$	…	σGe	0.593		0.698
σGe	LT	>12	cI16	Im3m	…	γSi	0.692		
αH	<-271.9	atm	cF4	Fm3m	A1	Cu	0.5338		
βH	<-259.34	atm	hP2	$P6_3/mmc$	A3	Mg	0.3776		0.6162
αHe	<-268.94	atm	hP2	$P6_3/mmc$	A3	Mg	0.3555		0.5798
βHe	>-258	0.125	cF4	Fm3m	A1	Cu	0.4240		
γHe	<-271.47	0.03	cI2	Im3m	A2	W	0.4110		
αHf	25	atm	hP2	$P6_3/mmc$	A3	Mg	0.31946		0.50510
βHf	>1995	atm	cI2	Im3m	A2	W	0.3610		
αHg	<-38.84	atm	hR1	R3m	A10	αHg	0.3005		
βHg	<-194	HP	tI2	I4/mmm	…	βHg	0.3995		0.2825
γHg	<-194	c.w.	hR1	?	…	…			
αHo	25	atm	hP2	$P6_3/mmc$	A3	Mg	0.35778		0.56178
βHo	25	>7.5	hR3	R3m	…	αSm	0.334		2.45
I	25	atm	oC8	Cmca	…	Cl	0.72697	0.47903	0.97942
In	25	atm	tI2	I4/mmm	A6	In	0.3253		0.49470
Ir	25	atm	cF4	Fm3m	A1	Cu	0.38392		
K	25	atm	cI2	Im3m	A2	W	0.5321		
Kr	<-157.39	atm	cF4	Fm3m	A1	Cu	0.5810		
αLa	25	atm	hP4	$P6_3/mmc$	A3′	αLa	0.37740		1.2171
βLa	>310	atm	cF4	Fm3m	A1	Cu	0.5303		
γLa	>865	atm	cI2	Im3m	A2	W	0.426		
β′La	25	>2.0	cF4	Fm3m	A1	Cu	0.517		
αLi	<-193	atm	hP2	$P6_3/mmc$	A3	Mg	0.3111		0.5093
βLi	25	atm	cI2	Im3m	A2	W	0.35093		

Element	t/°C	p/GPa	Pearson symbol	Space group	Strukturbericht designation	Prototype	a/nm	b/nm	c/nm
γLi	<-201	c.w.	cF4	Fm3m	A1	Cu	0.4388		
Lu	25	atm	hP2	$P6_3/mmc$	A3	Mg	0.35052		0.55494
Mg	25	atm	hP2	$P6_3/mmc$	A3	Mg	0.32094		0.52107
αMn	25	atm	cl58	I43m	A12	αMn	0.89126		
βMn	>710	atm	cP20	$P4_132$	A13	βMn	0.63152		
γMn	>1079	atm	cF4	Fm3m	A1	Cu	0.3860		
σMn	>1143	atm	cI2	Im3m	A2	W	0.3080		
Mo	25	atm	cI2	Im3m	A2	W	0.31470		
αN	<-237.6	atm	cP8	Pa3	…	αN	0.5661		
βN	<-210.00	atm	hP4	$P6_3/mmc$	…	βN	0.4050		0.6604
γN	<-253	>3.3	tP4	$P4_2/mnm$	…	γN	0.3957		0.5109
αNa	<-233	atm	hP2	$P6_3/mmc$	A3	Mg	0.3767		0.6154
βNa	25	atm	cI2	Im3m	A2	W	0.42906		
Nb	25	atm	cI2	Im3m	A2	W	0.33004		
αNd	25	atm	hP4	$P6_3/mmc$	A3′	αLa	0.36582		1.17966
βNd	>863	atm	cI2	Im3m	A2	W	0.413		
γNd	25	>5.0	cF4	Fm3m	A1	Cu	0.480		
Ne	<-243.59	atm	cF4	Fm3m	A1	Cu	0.4462		
Ni	25	atm	cF4	Fm3m	A1	Cu	0.35240		
αNp	25	atm	oP8	Pnma	A_c	αNp	0.6663	0.4723	0.4887
βNp	>280	atm	tP4	$P42_12$	A_d	βNp	0.4883		0.3389
γNp	>576	atm	cI2	Im3m	A2	W	0.352		
αO	<-243.3	atm	mC4	C2m	…	αO	0.5403	0.3429	0.5086
βO	<-229.6	atm	hR2	R3m	…	βO	0.4210		
γO	<-218.79	atm	cP16	Pm3n	…	γO	0.683		
Os	25	atm	hP2	$P6_3/mmc$	A3	Mg	0.27341		0.43918
P (black)	25	atm	oC8	Cmca	…	P (black)	0.33136	1.0478	0.43763
αPa	25	atm	tI2	I4/mmm	A_a	αPa	0.3921		0.3235
βPa	>1170	atm	cI2	Im3m	A2	W	0.381		
αPb	25	atm	cF4	Fm3m	A1	Cu	0.49502		
βPb	25	>10.3	hP2	$P6_3/mmc$	A3	Mg	0.3265		0.5387
Pd	25	atm	cF4	Fm3m	A1	Cu	0.38903		
αPm	25	atm	hP4	$P6_3/mmc$	A3′	αLa	0.365		1.165
βPm	>890	atm	cI2	Im3m	A2	W	0.410?		
αPo	25	atm	cP1	Pm3m	A_h	αPo	0.3366		
βPo	>54	atm	hR1	R3m	…	βPo	0.3373		
αPr	25	atm	hP4	$P6_3/mmc$	A3′	αLa	0.36721		1.18326
βPr	>795	atm	cI2	Im3m	A2	W	0.413		
γPr	25	>4.0	cF4	Fm3m	A1	Cu	0.488		
Pt	25	atm	cF4	Fm3m	A1	Cu	0.39236		
αPu	25	atm	mP16	$P2_1/m$	…	αPu	0.6183	0.4822	1.0963
βPu	>125	atm	mI34	I2/m	…	βPu	0.9284	1.0463	0.7859
γPu	>215	atm	oF8	Fddd	…	γPu	0.31587	0.57682	1.0162
σPu	>320	atm	cF4	Fm3m	A1	Cu	0.46371		
σ′Pu	>463	atm	tI2	I4/mmm	A6	In	0.33261		0.44630
εPu	>483	atm	cI2	Im3m	A2	W	0.36343		
Ra	25	atm	cI2	Im3m	A2	W	0.5148		
αRb	25	atm	cI2	Im3m	A2	W	0.5705		
βRb	25	>1.08	?	…	…	…			
γRb	25	>2.05	?	…	…	…			
Re	25	atm	hP2	$P6_3/mmc$	A3	Mg	0.27609		0.4458
Rh	25	atm	cF4	Fm3m	A1	Cu	0.38032		
Ru	25	atm	hP2	$P6_3/mmc$	A3	Mg	0.27058		0.42816
αS	25	atm	oF128	Fddd	A16	αS	1.0464	1.28660	2.44860
αSb	25	atm	hR2	R3m	A7	αAs	0.45067		
βSb	25	>5.0	cP1	Pm3m	A_h	αPo	0.2992		
γSb	25	>7.5	hP2	$P6_3/mmc$	A3	Mg	0.3376		0.5341
σSb	25	>14.0	mP3	?	…	…	0.556	0.404	0.422
αSc	25	atm	hP2	$P6_3/mm$	A3	Mg	0.33088		0.52680
βSc	>1337	atm	cI2	Im3m	A2	W	0.373?		

Element	t/°C	p/GPa	Pearson symbol	Space group	Strukturbericht designation	Prototype	a/nm	b/nm	c/nm
γSe	25	atm	hP3	$P3_121$	A8	γSe	0.43659		0.49537
αSi	25	atm	cF8	Fd3m	A4	C (diamond)	0.54306		
βSi	25	>9.5	tI4	$I4_1/amd$	A5	βSn	0.4686		0.2585
γSi	25	>16.0	cI16	Im3m	…	γSi	0.6636		
σSi	25	>16→atm	hP4	$P6_3/mmc$	A3′	αLa	0.380		0.628
αSm	25	atm	hR3	R3m	…	αSm	0.36290		2.6207
βSm	>734	atm	hP2	$P6_3/mmc$	A3	Mg	0.36630		0.58448
γ′Sm	>922	atm	cI2	Im3m	A2	W	0.410?		
σSm	25	>4.0	hP4	$P6_3/mmc$	A3′	αLa	0.3618		1.166
αSn	<13	atm	cF8	Fd3m	A4	C (diamond)	0.64892		
βSn	25	atm	tI4	$I4_1/amd$	A5	βSn	0.58318		0.31818
γSn	25	>9.0	tI2	?	…	γSn	0.370		0.337
αSr	25	atm	cF4	Fm3m	A1	Cu	0.6084		
βSr	>547	atm	cI2	Im3m	A2	W	0.487		
β′Sr	25	>3.5	cI2	Im3m	A2	W	0.4437		
Ta	25	atm	cI2	Im3m	A2	W	0.33030		
α′Tb	<-53	atm	oC4	Cmcm	…	α′Dy	0.3605	0.6244	0.5706
aTb	25	atm	hP2	$P6_3/mmc$	A3	Mg	0.36055		0.56966
βTb	>1289	atm	cI2	Im3m	A2	W	0.407?		
γTb	25	>6.0	hR3	R3m	…	αSm	0.341		2.45
Tc	25	atm	hP2	$P6_3/mmc$	A3	Mg	0.2738		0.4393
αTe	25	atm	hP3	$P3_121$	A8	γSe	0.44566		0.59264
βTe	25	>2.0	hR2	R3m	A7	αAs	0.469		
γTe	25	>7.0	hR1	R3m	…	βPo	0.3002		
αTh	25	atm	cF4	Fm3m	A1	Cu	0.50842		
βTh	>1360	atm	cI2	Im3m	A2	W	0.411		
αTi	25	atm	hP2	$P6_3/mmc$	A3	Mg	0.29506		0.46835
βTi	>882	atm	cI2	Im3m	A2	W	0.33065		
ωTi	25	HP→atm	hP3	P6/mmm	…	ωTi	0.4625		0.2813
αTl	25	atm	hP2	$P6_3/mmc$	A3	Mg	0.34566		0.55248
βTl	>230	atm	cI2	Im3m	A2	W	0.3879		
γTl	25	HP	cF4	Fm3m	A1	Cu	?		
Tm	25	atm	hP2	$P6_3/mmc$	A3	Mg	0.35375		0.55540
αU	25	atm	oC4	Cmcm	A20	αU	0.28537	0.58695	0.49548
βU	>668	atm	tP30	$P4_2/mnm$	A_b	βU	1.0759		0.5656
γU	>776	atm	cI2	Im3m	A2	W	0.3524		
V	25	atm	cI2	Im3m	A2	W	0.30240		
W	25	atm	cI2	Im3m	A2	W	0.31652		
Xe	<-111.76	atm	cF4	Fm3m	A1	Cu	0.6350		
αY	25	atm	hP2	$P6_3/mmc$	A3	Mg	0.36482		0.57318
βY	>1478	atm	cI2	Im3m	A2	W	0.410?		
αYb	<-3	atm	hP2	$P6_3/mmc$	A3	Mg	0.38799		0.63859
βYb	25	atm	cF4	Fm3m	A1	Cu	0.54848		
γYb	>795	atm	cI2	Im3m	A2	W	0.444		
Zn	25	atm	hP2	$P6_3/mmc$	A3	Mg	0.26650		0.49470
αZr	25	atm	hP2	$P6_3/mmc$	A3	Mg	0.32316		0.51475
βZr	>863	atm	cI2	Im3m	A2	W	0.36090		
ωZr	25	HP→atm	hP2	P6/mmm	…	ωTi	0.5036		0.3109

? uncertain

PHASE TRANSITIONS IN THE SOLID ELEMENTS AT ATMOSPHERIC PRESSURE

This table gives the phase transition temperatures for the elements that can exist in two or more crystalline forms (allotropes). The crystal phases are labeled by Greek letters in the most common conventions, although some variation is found. All data refer to normal atmospheric pressure.

References

1. Massalski, T. B., Ed., *Binary Alloy Phase Diagrams, Second Edition*, ASM International, Metals Park, OH, 1990.
2. Cordfunke, E. H. P., and Konings, R. J. M., Eds., *Thermochemical Data for Reactor Materials and Fission Products*, North-Holland, Amsterdam, 1990.
3. Greenwood, N. N., and Earnshaw, A., *Chemistry of the Elements, Second Edition*, Butterworth-Heinemann, Oxford, 1997.
4. Rhyne, J. J., "Magnetic Phase Transitions of the Elements", *Bull. Alloy Phase Diag.* 3, 402, 1982.

Solids

Symbol	Transition	t /°C	Comments
Americium			
Am	α→β	769	
Am	β→γ	1077	
Am	γ→liq	1176	
Beryllium			
Be	α→β	1270	
Be	β→liq	1287	
Boron			
B	α→β	1100	
B	β→γ	1500	
B	γ→liq	2075	
Calcium			
Ca	α→β	443	
Ca	β→liq	842	
Californium			
Cf	α→β	590	
Cf	β→liq	900	
Cerium			
Ce	α→β	-177	
Ce	β→γ	61	β-Ce and γ-Ce are magnetic
Ce	γ→δ	726	
Ce	δ→liq	799	
Cobalt			
Co	ε→α	422	magnetic transition at 1115 °C
Co	α→liq	1495	
Curium			
Cm	α→β	1277	magnetic transition at −221 °C
Cm	β→liq	1345	
Dysprosium			
Dy	α'→α	-187	
Dy	α→β	1381	magnetic transitions in α-Dy at −184 °C and −94 °C
Dy	β→liq	1412	
Fluorine			
F_2	α→β	-227.60	
F_2	β→liq	-219.67	
Gadolinium			
Gd	α→β	1235	
Gd	β→liq	1313	
Hafnium			
Hf	α→β	1743	
Hf	β→liq	2233	
Iron			
Fe	α→γ	912	magnetic transition in α-Fe at 771 °C
Fe	γ→δ	1394	
Fe	δ→liq	1538	
Lanthanum			
La	α→β	277	
La	β→γ	860	
La	γ→liq	920	
Lithium			
Li	α→β	-193	
Li	β→liq	180.50	
Manganese			
Mn	α→β	727	magnetic transition in α-Mn at −100 °C
Mn	β→γ	1100	
Mn	γ→δ	1138	
Mn	δ→liq	1246	
Neodymium			
Nd	α→β	855	magnetic transition in α-Nd at −253 °C
Nd	β→liq	1016	
Neptunium			
Np	α→β	280	
Np	β→γ	576	
Np	γ→liq	644	
Nitrogen			
N_2	α→β	-237.54	
N_2	β→liq	-210.0	
Oxygen			
O_2	α→β	-249.29	
O_2	β→γ	-229.35	
O_2	γ→liq	-218.79	
Phosphorus			
P	brown→β-white	-190	several amorphous phases (red, black, gray) exist (Ref. 3)
P	β-white→α-white	-76.9	
P	α-white→liq	44.15	

Symbol	Transition	t /°C	Comments
Plutonium			
Pu	α→β	124.5	
Pu	β→γ	214.8	
Pu	γ→δ	320.0	
Pu	δ→δ'	462.9	
Pu	δ'→ε	482.6	
Pu	ε→liq	640	
Polonium			
Po	α→β	54	
Po	β→liq	254	
Praseodymium			
Pr	α→β	795	
Pr	β→liq	931	
Promethium			
Pm	α→β	890	magnetic transition in α-Pm at −175 °C
Pm	β→liq	1042	
Protactinium			
Pa	α→β	1170	
Pa	β→liq	1572	
Samarium			
Sm	α→β	734	magnetic transition in α-Sm at −167 °C
Sm	β→γ	922	
Sm	γ→liq	1072	
Scandium			
Sc	α→β	1337	
Sc	β→liq	1541	
Selenium			
Se	α-red→gray	180	many allotropes exist (Ref. 3)
Se	gray→liq	220.8	
Sodium			
Na	α→β	-233	
Na	β→liq	97.794	
Strontium			
Sr	α→β	547	
Sr	β→liq	777	
Sulfur			
S	α→β	95.3	many allotropes exist (Ref. 3)
S	β→liq	115.21	
Terbium			
Tb	α'→α	-53	
Tb	α→β	1289	magnetic transition in α-Tb at −230 °C
Tb	β→liq	1359	
Thallium			
Tl	α→β	230	
Tl	β→liq	304	
Thorium			
Th	α→β	1360	
Th	β→liq	1750	
Tin			
Sn	α (gray)→β (white)	13.2	
Sn	β (white)→liq	231.928	defining fixed point on ITS-90
Titanium			
Ti	α→β	882	
Ti	β→liq	1668	
Uranium			
U	α→β	669	
U	β→γ	776	
U	γ→liq	1135	
Ytterbium			
Yb	α→β	3	
Yb	β→γ	795	
Yb	γ→liq	824	
Yttrium			
Y	α→β	1478	
Y	β→liq	1522	
Zirconium			
Zr	α→β	866	
Zr	β→liq	1854.7	

LATTICE ENERGIES

H. D. B. Jenkins and H. K. Roobottom

THERMOCHEMICAL CYCLE AND CALCULATED VALUES

Table 1 contains calculated values of the lattice energies (total lattice potential energies), U_{POT}, of crystalline salts, M_aX_b. U_{POT} is expressed in units of kilojoules per mole, kJ mol^{-1}. M and X can be either simple or complex ions. Substances are arranged by chemical class.

Also listed in the table is the lattice energy, U_{POT}^{BFHC}, obtained from the application of the Born–Fajans–Haber cycle (BHFC) described below, using the "Standard Thermochemical Properties of Chemical Substances" table in Section 5 of this *Handbook*, References 1 through 4, and certain other data that are given in Table 3 below.

The lattice enthalpy, ΔH_L, is given by the cycle:

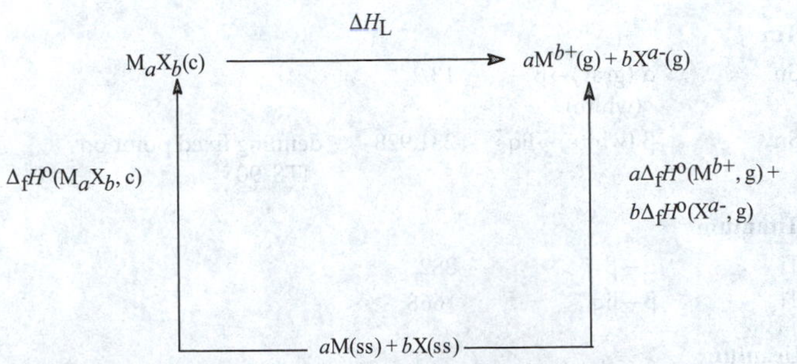

where (ss) is the standard state of the element concerned.

The lattice enthalpy, ΔH_L, is obtained using the equation:

$$\Delta H_L = a\Delta_f H°(M^{b+}, g) + b\Delta_f H°(X^{a-}, g) - \Delta_f H°(M_aX_b, c)$$

and is further related to the total lattice potential energy, U_{POT}, by the relationship:

$$\Delta H_L = U_{POT} + \left[a\left(\frac{n_M}{2} - 2\right) + b\left(\frac{n_X}{2} - 2\right)\right]RT$$

where n_M and n_X equal 3 for monatomic ions, 5 for linear polyatomic ions and 6 for polyatomic non-linear ions.

METHOD OF ESTIMATION OF VALUES NOT TABULATED

In cases where the lattice energy is not tabulated and we want to furnish an estimate, then the Kapustinskii equation (Ref. 5) can be used to obtain a value (in kJ mol^{-1}):

$$U_{POT} = \frac{121.4 z_a z_b \nu}{(r_a + r_b)}\left(1 - \frac{0.0345}{(r_a + r_b)}\right)$$

where z_a and z_b are the moduli of the charges on the ν ions in the lattice and r_a and r_b (in nm) are the thermochemical radii given in Table 2. The r_a for metal ions is taken to be the Goldschmidt radius (Ref. 6).

To cite an example, if we wish to estimate the lattice energy of the salt $[NH^+_4][HF^-_2]$ using the above procedure, we see that Table 2 gives the thermochemical radius (r_a) for NH^+_4 to be 0.136 nm and that for HF^-_2 (r_b) to be 0.172 nm. The lattice potential energy is then estimated to be 700 kJ mol^{-1} compared with the calculated value of 705 kJ mol^{-1} and the Born–Fajans–Haber cycle value of 658 kJ mol^{-1}.

References

1. Wagman, D. D., Evans, W. H., Parker, V. B., Schumm, R. H., Halow, I., Bailey, S. M., Churney, K. L., and Nuttall, R. L., The NBS Tables of Chemical Thermodynamic Properties, *J. Phys. Chem. Ref. Data*, Vol. 11, Suppl. 2, 1982.
2. Chase, M. W., Davies, C. A., Downey, J. R., Frurip, D. J., McDonald, R. A., and Syverud, A. N., JANAF Thermochemical Tables, Third Edition, *J. Phys. Chem. Ref. Data*, Vol. 14, Suppl. 1, 1985.
3. Lias, S. G., Bartmess, J. E., Liebman, J. F., Holmes, J. L., Levin, R. D., and Mallard, W. G., Gas-Phase Ion and Neutral Thermochemistry, *J. Phys. Chem. Ref. Data*, Vol. 17, Suppl. 1, 1988.
4. Jenkins, H. D. B., and Pratt, K. F., *Adv. Inorg. Chem. Radiochem.* 22, 1, 1978.
5. Kapustinskii, A. F., *Quart. Rev.* 10, 283, 1956.
6. Goldschmidt, V. M., *Skrifter Norske Videnskaps-Akad.* Oslo, I, Mat.-Naturn. Kl, 1926. See also Dasent, W. E., *Inorganic Energetics*, 2nd ed., Cambridge University Press, 1982.

Table 1. Lattice Energies (kJ mol^{-1})

Substance	Calc. U_{POT}	U_{POT}^{BHFC}
Acetates		
$Li(CH_3COO)$	–	843
$Na(CH_3COO)$	828	807
$K(CH_3COO)$	749	726
$Rb(CH_3COO)$	715	–
$Cs(CH_3COO)$	682	–
Acetylides		
CaC_2	2911	2902
SrC_2	2788	2782
BaC_2	2647	2652
Azides		
LiN_3	861	875
NaN_3	770	784
KN_3	697	–
RbN_3	674	691
CsN_3	665	674
AgN_3	854	910
TlN_3	689	742
$Ca(N_3)_2$	2186	2316
$Sr(N_3)_2$	2056	2187
$Ba(N_3)_2$	2021	–
$Mn(N_3)_2$	2408	2348
$Cu(N_3)_2$	2730	2738
$Zn(N_3)_2$	2840	2970
$Cd(N_3)_2$	2446	2576
$Pb(N_3)_2$	–	2300
Bihalide Salts		
$LiHF_2$	821	847
$NaHF_2$	755	748
KHF_2	657	660
$RbHF_2$	627	631
$CsHF_2$	607	–
NH_4HF_2	705	658
$CsHCl_2$	601	–
Me_4NHCl_2	427	–
Et_4NHCl_2	346	–
Bu_4NHCl_2	290	–
Bicarbonates		
$NaHCO_3$	820	656
$KHCO_3$	741	573
$RbHCO_3$	707	522
$CsHCO_3$	678	520
NH_4HCO_3	–	577
Borides		
CaB_6	5146	–
SrB_6	5104	–
BaB_6	5021	–
YB_6	7447	–
LaB_6	7406	–
CeB_6	10083	–
PrB_6	7447	–
NdB_6	7447	–
PmB_6	7406	–
SmB_6	7447	–
EuB_6	5104	–
GdB_6	7489	–
TbB_6	7489	–
DyB_6	7489	–
HoB_6	7489	–
ErB_6	7489	–
TmB_6	7489	–
YbB_6	5146	–
LuB_6	7489	–
ThB_6	10167	–
Borohydrides		
$LiBH_4$	778	–
$NaBH_4$	703	694
KBH_4	655	638
$RbBH_4$	648	–
$CsBH_4$	628	–
Borohalides		
$LiBF_4$	699	749
$NaBF_4$	657	674
KBF_4	611	616
$RbBF_4$	577	590
$CsBF_4$	556	565
NH_4BF_4	582	–
$KBCl_4$	506	497
$RbBCl_4$	489	486
$CsBCl_4$	473	–
Carbonates		
Li_2CO_3	2523	2254
Na_2CO_3	2301	2016
K_2CO_3	2084	1846
Rb_2CO_3	2000	1783
Cs_2CO_3	1920	1722
$MgCO_3$	3138	3122
$CaCO_3$	2804	2811
$SrCO_3$	2720	2688
$BaCO_3$	2615	2554
$MnCO_3$	3046	3092
$FeCO_3$	3121	3169
$CoCO_3$	3443	3235
$CuCO_3$	3494	–
$ZnCO_3$	3121	3273
$CdCO_3$	2929	3052
$SnCO_3$	2904	–
$PbCO_3$	2728	2750
Cyanates		
LiNCO	849	–
NaNCO	807	816
KNCO	726	734
RbNCO	692	–
CsNCO	661	–
NH_4NCO	724	–
Cyanides		
LiCN	874	
NaCN	766	759
KCN	692	686
RbCN	638	–
CsCN	601	–
$Ca(CN)_2$	2268	2240
$Sr(CN)_2$	2138	–
$Ba(CN)_2$	2001	2009
NH_4CN	617	691
AgCN	(741)	935

Solids

Substance	Calc. U_{POT}	U_{POT}^{BHFC}
$Zn(CN)_2$	2809	2817
$Cd(CN)_2$	2583	2591
Formates		
$Li(HCO_2)$	865	–
$Na(HCO_2)$	791	804
$K(HCO_2)$	713	722
$Rb(HCO_2)$	685	–
$Cs(HCO_2)$	651	–
$NH_4(HCO_2)$	715	–
Germanates		
Mg_2GeO_4	7991	–
Ca_2GeO_4	7301	7306
Sr_2GeO_4	6987	–
Ba_2GeO_4	6653	6643
Halates		
$LiBrO_3$	883	880
$NaBrO_3$	803	791
$KBrO_3$	740	722
$RbBrO_3$	720	705
$CsBrO_3$	694	681
$NaClO_3$	770	785
$KClO_3$	711	721
$RbClO_3$	690	703
$CsClO_3$	–	679
$LiIO_3$	975	974
$NaIO_3$	883	876
KIO_3	820	780
$RbIO_3$	791	–
$CsIO_3$	761	–
Halides		
LiF	1030	1049
LiCl	834	864
LiBr	788	820
LiI	730	764
NaF	910	930
NaCl	769	790
NaBr	732	754
NaI	682	705
KF	808	829
KCl	701	720
KBr	671	691
KI	632	650
RbF	774	795
RbCl	680	695
RbBr	651	668
RbI	617	632
CsF	744	759
CsCl	657	670
CsBr	632	647
CsI	600	613
FrF	715	–
FrCl	632	–
FrBr	611	–
FrI	582	–
CuCl	992	996
CuBr	969	978
CuI	948	966
AgF	953	974
AgCl	910	918
AgBr	897	905
AgI	881	892
AuCl	1013	1066
AuBr	1029	1059
AuI	1027	1070
InCl	–	764
InBr	–	767
InI	–	733
TlF	–	850
TlCl	738	751
TlBr	720	734
TlI	692	710
Me_4NCl	566	–
Me_4NBr	553	–
Me_4NI	544	–
PH_4Br	616	–
PH_4I	590	–
BeF_2	3464	3526
$BeCl_2$	3004	3033
$BeBr_2$	2950	2914
BeI_2	2780	2813
MgF_2	2926	2978
$MgCl_2$	2477	2540
$MgBr_2$	2406	2451
MgI_2	2293	2340
CaF_2	2640	2651
$CaCl_2$	2268	2271
$CaBr_2$	2132	–
CaI_2	1971	2087
SrF_2	2476	2513
$SrCl_2$	2142	2170
SrI_2	1984	1976
BaF_2	2347	2373
$BaCl_2$	2046	2069
$BaBr_2$	1971	1995
BaI_2	1862	1890
RaF_2	2284	–
$RaCl_2$	2004	–
$RaBr_2$	1929	–
RaI_2	1803	–
$ScCl_2$	2380	–
$ScBr_2$	2291	–
ScI_2	2201	–
TiF_2	2724	–
$TiCl_2$	2439	2514
$TiBr_2$	2360	2430
TiI_2	2259	2342
VCl_2	2607	2593
VBr_2	–	2534
VI_2	–	2470
CrF_2	2778	2939
$CrCl_2$	2540	2601
$CrBr_2$	2377	2536
CrI_2	2269	2440
$MoCl_2$	2737	2746
$MoBr_2$	2742	2753
MoI_2	2630	–
MnF_2	2644	–
$MnCl_2$	2510	2551
$MnBr_2$	2448	2482
MnI_2	2212	–
FeF_2	2849	2967

Substance	Calc. U_{POT}	U_{POT}^{BHFC}
$FeCl_2$	2569	2641
$FeBr_2$	2515	2577
FeI_2	2439	2491
CoF_2	3004	3042
$CoCl_2$	2707	2706
$CoBr_2$	2640	2643
CoI_2	2569	2561
NiF_2	3098	3089
$NiCl_2$	2753	2786
$NiBr_2$	2729	2721
NiI_2	2607	2637
$PdCl_2$	2778	2818
$PdBr_2$	2741	2751
PdI_2	2748	2760
CuF_2	3046	3102
$CuCl_2$	2774	2824
$CuBr_2$	2715	2774
CuI_2	2640	–
AgF_2	2942	2967
ZnF_2	3021	3053
$ZnCl_2$	2703	2748
$ZnBr_2$	2648	2689
ZnI_2	2581	2619
CdF_2	2809	2830
$CdCl_2$	2552	2565
$CdBr_2$	2507	2517
CdI_2	2441	2455
HgF_2	2757	–
$HgCl_2$	2657	2664
$HgBr_2$	2628	2639
HgI_2	2628	2624
SnF_2	2551	–
$SnCl_2$	2297	2310
$SnBr_2$	2251	2256
SnI_2	2193	2206
PbF_2	2535	2543
$PbCl_2$	2270	2282
$PbBr_2$	2219	2230
PbI_2	2163	2177
ScF_3	5492	5540
$ScCl_3$	4874	4901
$ScBr_3$	4729	4761
ScI_3	4640	–
YF_3	4983	–
YCl_3	4506	4524
YI_3	4240	4258
TiF_3	5644	–
$TiCl_3$	5134	5153
$TiBr_3$	5012	5023
TiI_3	4845	–
$ZrCl_3$	–	4791
$ZrBr_3$	–	4758
ZrI_3	–	4591
VF_3	5895	–
VCl_3	5322	5329
VBr_3	5214	5224
VI_3	5121	5136
$NbCl_3$	5062	–
$NbBr_3$	4980	–
NbI_3	4860	–
CrF_3	6033	6065

Substance	Calc. U_{POT}	U_{POT}^{BHFC}
$CrCl_3$	5518	5529
$CrBr_3$	5355	–
CrI_3	5275	5294
MoF_3	6459	–
$MoCl_3$	5246	5253
$MoBr_3$	5156	–
MoI_3	5073	–
MnF_3	6017	–
$MnCl_3$	5544	–
$MnBr_3$	5448	–
MnI_3	5330	–
$TcCl_3$	5270	–
$TcBr_3$	5215	–
TcI_3	5188	–
FeF_3	5870	–
$FeCl_3$	5364	5436
$FeBr_3$	5333	5347
FeI_3	5117	–
$RuCl_3$	5245	5257
$RuBr_3$	5223	5232
RuI_3	5222	5235
CoF_3	5991	–
$RhCl_3$	5641	5665
IrF_3	(6112)	–
$IrBr_3$	(4794)	–
NiF_3	(6111)	–
AuF_3	(5777)	–
$AuCl_3$	(4605)	–
$ZnCl_3$	5832	–
$ZnBr_3$	5732	–
ZnI_3	5636	–
AlF_3	5924	6252
$AlCl_3$	5376	5513
$AlBr_3$	5247	5360
AlI_3	5070	5227
GaF_3	5829	6238
$GaCl_3$	5217	5665
$GaBr_3$	4966	5569
GaI_3	4611	5496
$InCl_3$	4736	5183
$InBr_3$	4535	5117
InI_3	4234	5001
TlF_3	5493	–
$TlCl_3$	5258	5278
$TlBr_3$	5171	–
TlI_3	5088	–
$AsBr_3$	5497	5365
AsI_3	4824	5295
SbF_3	5295	5324
$SbCl_3$	5032	4857
$SbBr_3$	4954	4776
SbI_3	4867	4692
$BiCl_3$	4689	4707
BiI_3	3774	–
LaF_3	4682	–
$LaCl_3$	4263	4242
$LaBr_3$	4209	–
LaI_3	3916	3986
$CeCl_3$	4394	4348
CeI_3	–	4061
$PrCl_3$	4322	4387

Substance	Calc. U_{POT}	U_{POT}^{BHFC}
PrI_3	–	4101
$NdCl_3$	4343	4415
$SmCl_3$	4376	4450
$EuCl_3$	4393	4490
$GdCl_3$	4406	4495
$DyCl_3$	4481	4529
$HoCl_3$	4501	4572
$ErCl_3$	4527	4591
$TmCl_3$	4548	4608
TmI_3	–	4340
$YbCl_3$	–	4651
$AcCl_3$	4096	–
UCl_3	4243	–
$NpCl_3$	4268	–
$PuCl_3$	4289	–
$PuBr_3$	(3959)	–
$AmCl_3$	4293	–
TiF_4	10012	9908
$TiCl_4$	9431	–
$TiBr_4$	9288	9059
TiI_4	9108	8918
ZrF_4	8853	8971
$ZrCl_4$	8021	8144
$ZrBr_4$	7661	7984
ZrI_4	7155	7801
MoF_4	8795	–
$MoCl_4$	8556	9603
$MoBr_4$	8510	9500
MoI_4	8427	–
$SnCl_4$	8355	8930
$SnBr_4$	7970	8852
PbF_4	9519	–
CrF_2Cl	5795	–
CrF_2Br	5753	–
CrF_2I	5669	–
$CrCl_2Br$	5448	–
$CrCl_2I$	5381	5429
$CrBr_2I$	5330	5370
CuFCl	2891	–
CuFBr	2853	–
CuFI	2803	–
CuClBr	2753	–
CuClI	2694	–
CuBrI	2669	–
FeF_2Cl	5711	–
FeF_2Br	5653	–
FeF_2I	5569	–
$FeCl_2Br$	5339	–
$FeCl_2I$	5272	–
$FeBr_2I$	5209	–
$LiIO_2F_2$	845	–
$NaIO_2F_2$	766	756
KIO_2F_2	699	689
$RbIO_2F_2$	674	–
$CsIO_2F_2$	636	–
$NH_4IO_2F_2$	678	–
$AgIO_2F_2$	736	685
Hydrides		
LiH	916	918
NaH	807	807
KH	711	713
RbH	686	684
CsH	648	653
VH	1184	(1344)
NbH	1163	(1633)
PdH	979	1368
CuH	828	1254
TiH	996	1407
ZrH	916	1590
HfH	904	–
LaH	828	–
TaH	1021	–
CrH	1050	–
NiH	929	–
PtH	937	–
AgH	941	–
AuH	1033	1108
TlH	745	–
GeH	950	–
PbH	778	–
BeH_2	3205	3306
MgH_2	2791	2718
CaH_2	2410	2406
SrH_2	2250	2265
BaH_2	2121	2133
ScH_2	2711	2744
YH_2	(2598)	2733
LaH_2	2380	2522
CeH_2	2414	2509
PrH_2	2448	2405
NdH_2	2464	2394
PmH_2	2519	–
SmH_2	2510	2389
GdH_2	2494	2651
AcH_2	2372	–
ThH_2	2711	2738
PuH_2	2519	–
AmH_2	2544	–
TiH_2	2866	2864
ZrH_2	2711	2999
CuH_2	2941	–
ZnH_2	2870	–
HgH_2	2707	–
AlH_3	5924	5969
FeH_3	5724	–
ScH_3	5439	–
YH_3	5063	4910
LaH_3	4895	4493
FeH_3	5724	–
GaH_3	5690	–
InH_3	5092	–
TlH_3	5092	–
Hydroselenides		
NaHSe	703	732
KHSe	644	712
RbHSe	623	689
CsHse	598	669
Hydrosulfides		
LiHS	768	862
NaHS	723	771
RbHS	655	682

Substance	Calc. U_{POT}	U_{POT}^{BHFC}
CsHS	628	657
NH_4HS	661	718
$Ca(HS)_2$	2184	(2171)
$Sr(HS)_2$	2063	–
$Ba(HS)_2$	1979	(1956)
Hydroxides		
LiOH	1021	1028
NaOH	887	892
KOH	789	796
RbOH	766	765
CsOH	721	732
$Be(OH)_2$	3477	3620
$Mg(OH)_2$	2870	2998
$Ca(OH)_2$	2506	2637
$Sr(OH)_2$	2330	2474
$Ba(OH)_2$	2142	2330
$Ti(OH)_2$	–	2953
$Mn(OH)_2$	2909	3008
$Fe(OH)_2$	2653	3044
$Co(OH)_2$	2786	3109
$Ni(OH)_2$	2832	3186
$Pd(OH)_2$	–	3189
$Cu(OH)_2$	2870	3229
CuOH	1006	–
AgOH	918	845
AuOH	1033	–
TlOH	705	874
$Zn(OH)_2$	2795	3151
$Cd(OH)_2$	2607	2909
$Hg(OH)_2$	2669	–
$Sn(OH)_2$	2489	2721
$Pb(OH)_2$	2376	–
$Sc(OH)_3$	5063	5602
$Y(OH)_3$	4707	–
$La(OH)_3$	4443	–
$Cr(OH)_3$	5556	6299
$Mn(OH)_3$	6213	–
$Al(OH)_3$	5627	–
$Ga(OH)_3$	5732	6368
$In(OH)_3$	5280	–
$Tl(OH)_3$	5314	–
$Ti(OH)_4$	9456	–
$Zr(OH)_4$	8619	–
$Mn(OH)_4$	10933	–
$Sn(OH)_4$	9188	9879
Imides		
CaNH	3293	–
SrNH	3146	–
BaNH	2975	–
Metavanadates		
Li_3VO_4	3945	–
Na_3VO_4	3766	–
K_3VO_4	3376	–
Rb_3VO_4	3243	–
Cs_3VO_4	3137	–
Nitrates		
$LiNO_3$	848	854
$NaNO_3$	755	763
KNO_3	685	694
$RbNO_3$	662	671

Substance	Calc. U_{POT}	U_{POT}^{BHFC}
$CsNO_3$	648	650
$AgNO_3$	820	832
$TlNO_3$	690	707
$Mg(NO_3)_2$	2481	2521
$Ca(NO_3)_2$	2268	2247
$Sr(NO_3)_2$	2176	2151
$Ba(NO_3)_2$	2062	2035
$Mn(NO_3)_2$	2318	2478
$Fe(NO_3)_2$	–	(2580)
$Co(NO_3)_2$	2560	2647
$Ni(NO_3)_2$	–	2729
$Cu(NO_3)_2$	–	2739
$Zn(NO_3)_2$	2376	2649
$Cd(NO_3)_2$	2238	2462
$Sn(NO_3)_2$	2155	2254
$Pb(NO_3)_2$	2067	2208
Nitrides		
ScN	7547	7506
LaN	6876	6793
TiN	8130	8033
ZrN	7633	7723
VN	8283	8233
NbN	7939	8022
CrN	8269	8358
Nitrites		
$NaNO_2$	774	772
KNO_2	699	687
$RbNO_2$	724	765
$CsNO_2$	690	–
Oxides		
Li_2O	2799	2814
Na_2O	2481	2478
K_2O	2238	2232
Rb_2O	2163	2161
Cs_2O	2131	2063
Cu_2O	3273	3189
Ag_2O	3002	2910
Tl_2O	2659	2575
LiO_2	(878)	(872)
NaO_2	799	821
KO_2	741	751
RbO_2	706	721
CsO_2	679	696
Li_2O_2	2592	2557
Na_2O_2	2309	22717
K_2O_2	2114	2064
Rb_2O_2	2025	1994
Cs_2O_2	1948	1512
MgO_2	3356	3526
CaO_2	3144	3132
SrO_2	3037	2977
KO_3	697	707
BeO	4514	4443
MgO	3795	3791
CaO	3414	3401
SrO	3217	3223
BaO	3029	3054
TiO	3832	3811
VO	3932	3863
MnO	3724	3745

Substance	Calc. U_{POT}	U_{POT}^{BHFC}
FeO	3795	3865
CoO	3837	3910
NiO	3908	4010
PdO	3736	–
CuO	4135	4050
ZnO	4142	3971
CdO	3806	–
HgO	3907	–
GeO	3919	–
SnO	3652	–
PbO	3520	–
Sc_2O_3	13557	13708
Y_2O_3	12705	–
La_2O_3	12452	–
Ce_2O_3	12661	–
Pr_2O_3	12703	–
Nd_2O_3	12736	–
Pm_2O_3	12811	–
Sm_2O_3	12878	–
Eu_2O_3	12945	–
Gd_2O_3	12996	–
Tb_2O_3	13071	–
Dy_2O_3	13138	–
Ho_2O_3	13180	–
Er_2O_3	13263	–
Tm_2O_3	13322	–
Yb_2O_3	13380	–
Lu_2O_3	13665	–
Ac_2O_3	12573	–
Ti_2O_3	–	14149
V_2O_3	15096	14520
Cr_2O_3	15276	14957
Mn_2O_3	15146	15035
Fe_2O_3	14309	14774
Al_2O_3	15916	–
Ga_2O_3	15590	15220
In_2O_3	13928	–
Pb_2O_3	(14841)	–
CeO_2	9627	–
ThO_2	10397	–
PaO_2	10573	–
VO_2(g)	10644	–
NpO_2	10707	–
PuO_2	10786	–
AmO_2	10799	–
CmO_2	10832	–
TiO_2	12150	–
ZrO_2	11188	–
MoO_2	11648	–
MnO_2	12970	–
SiO_2	13125	–
GeO_2	12828	–
SnO_2	11807	–
PbO_2	11217	–
Perchlorates		
$LiClO_4$	709	715
$NaClO_4$	643	641
$KClO_4$	599	595
$RbClO_4$	564	576
$CsClO_4$	636	550
NH_4ClO_4	583	580

Substance	Calc. U_{POT}	U_{POT}^{BHFC}
$Ca(ClO_4)_2$	1958	1971
$Sr(ClO_4)_2$	1862	1862
$Ba(ClO_4)_2$	1795	1769
Permanganates		
$NaMnO_4$	661	–
$KMnO_4$	607	–
$RbMnO_4$	586	–
$CsMnO_4$	565	–
$Ca(MnO_4)_2$	1937	–
$Sr(MnO_4)_2$	1845	–
$Ba(MnO_4)_2$	1778	–
Phosphates		
$Mg_3(PO_4)_2$	11632	11407
$Ca_3(PO_4)_2$	10602	10479
$Sr_3(PO_4)_2$	10125	10075
$Ba_3(PO_4)_2$	9652	9654
$MnPO_4$	7397	–
$FePO_4$	7251	7300
BPO_4	8201	–
$AlPO_4$	7427	7507
$GaPO_4$	7381	–
Selenides		
Li_2Se	2364	–
Na_2Se	2130	–
K_2Se	1933	–
Rb_2Se	1837	–
Cs_2Se	1745	–
Ag_2Se	2686	–
Tl_2Se	2209	–
BeSe	3431	–
MgSe	3071	–
CaSe	2858	2862
SrSe	2736	–
BaSe	2611	–
MnSe	3176	–
Selenites		
Li_2SeO_3	2171	–
Na_2SeO_3	1950	1916
K_2SeO_3	1774	1749
Rb_2SeO_3	1715	1675
Cs_2SeO_3	1640	–
Tl_2SeO_3	1879	–
Ag_2SeO_3	2113	2148
$BeSeO_3$	3322	–
$MgSeO_3$	3012	2998
$CaSeO_3$	2732	–
$SrSeO_3$	2586	2588
$BaSeO_3$	2460	2451
Selenates		
Li_2SeO_4	2054	–
Na_2SeO_4	1879	–
K_2SeO_4	1732	–
Rb_2SeO_4	1686	–
Cs_2SeO_4	1615	–
Cu_2SeO_4	2201	–
Ag_2SeO_4	2033	–
Tl_2SeO_4	1766	–
Hg_2SeO_4	2163	–
$BeSeO_4$	3448	–

Substance	Calc. U_{POT}	U_{POT}^{BHFC}
$MgSeO_4$	2895	–
$CaSeO_4$	2632	–
$SrSeO_4$	2489	–
Sulfates		
Li_2SO_4	2229	2142
Na_2SO_4	1827	1938
K_2SO_4	1700	1796
Rb_2SO_4	1636	1748
Cs_2SO_4	1596	1658
$(NH_4)_2SO_4$	1766	1777
Cu_2SO_4	2276	2166
Ag_2SO_4	2104	1989
Tl_2SO_4	1828	1722
Hg_2SO_4	–	2127
$CaSO_4$	2489	2480
$SrSO_4$	2577	2484
$BaSO_4$	2469	2374
$MnSO_4$	2920	2825
Sulfides		
Li_2S	2464	2472
Na_2S	2192	2203
K_2S	1979	(2052)
Rb_2S	1929	1949
Cs_2S	1892	1850
$(NH_4)_2S$	2008	(2026)
Cu_2S	2786	2865
Ag_2S	2606	2677
Au_2S	2908	
Tl_2S	2298	2258
Ternary Salts		
Cs_2CuCl_4	1393	–
Rb_2ZnCl_4	1529	–
Cs_2ZnCl_4	1492	–
Rb_2ZnBr_4	1498	–
Cs_2ZnBr_4	1454	–
Cs_2ZnI_4	1386	–
$CsGaCl_4$	494	–
$NaAlCl_4$	556	–
$CsAlCl_4$	486	–
$NaFeCl_4$	492	–
Rb_2CoCl_4	1447	–
Cs_2CoCl_4	1391	–
K_2PtCl_4	1574	1550
Cs_2GeF_6	1573	–
$(NH_4)_2GeF_6$	1657	–
Cs_2GeCl_6	1404	1419
K_2HfCl_6	1345	1461
K_2IrCl_6	1442	1440
Na_2MoCl_6	1526	1504
K_2MoCl_6	1418	1412
Rb_2MoCl_6	1399	1399
Cs_2MoCl_6	1347	1347
K_2NbCl_6	1375	1398
Rb_2NbCl_6	1371	1385
Cs_2NbCl_6	1381	1344
K_2OsCl_6	1447	1447
Cs_2OsCl_6	1409	–
K_2OsBr_6	1396	–
K_2PdCl_6	1481	1493
Rb_2PdCl_6	1449	–

Substance	Calc. U_{POT}	U_{POT}^{BHFC}
Cs_2PdCl_6	1426	–
Rb_2PbCl_6	1343	1343
Cs_2PbCl_6	1344	–
$(NH_4)_2PbCl_6$	1355	–
K_2PtCl_6	1468	1471
Rb_2PtCl_6	1464	–
Cs_2PtCl_6	1444	–
$(NH_4)_2PtCl_6$	1468	–
Tl_2PtCl_6	1546	–
Ag_2PtCl_6	1773	1881
$BaPtCl_6$	2047	2070
K_2PtBr_6	1423	1392
Ag_2PtBr_6	1791	2276
K_2PtI_6	1421	
K_2ReCl_6	1416	1442
Rb_2ReCl_6	1414	–
Cs_2ReCl_6	1398	–
K_2ReBr_6	1375	1375
K_2SiF_6	1670	1765
Rb_2SiF_6	1639	1673
Cs_2SiF_6	1604	1498
Tl_2SiF_6	1675	–
K_2SnCl_6	1363	1390
Rb_2SnCl_6	1361	1363
Cs_2SnCl_6	1358	–
Tl_2SnCl_6	1437	–
$(NH_4)_2SnCl_6$	1370	1344
Rb_2SnBr_6	1309	–
Cs_2SnBr_6	1306	–
Rb_2SnI_6	1226	–
Cs_2SnBr_6	1243	
K_2TeCl_6	1318	1320
Rb_2TeCl_6	1321	–
Cs_2TeCl_6	1323	–
Tl_2TeCl_6	1392	–
$(NH_4)_2TeCl_6$	1318	–
K_2RuCl_6	1451	–
Rb_2CoF_6	1688	–
Cs_2CoF_6	1632	–
K_2NiF_6	1721	–
Rb_2NiF_6	1688	–
Rb_2SbCl_6	1357	–
Rb_2SeCl_6	1409	–
Cs_2SeCl_6	1397	–
$(NH_4)_2SeCl_6$	1420	–
$(NH_4)_2PoCl_6$	1338	–
Cs_2PoBr_6	1286	–
Cs_2CrF_6	1603	–
Rb_2MnF_6	1688	–
Cs_2MnF_6	1620	–
K_2MnCl_6	1462	–
Rb_2MnCl_6	1451	–
$(NH_4)_2MnCl_6$	1464	–
Cs_2TeBr_6	1306	–
Cs_2TeI_6	1246	–
K_2TiCl_6	1412	1447
Rb_2TiCl_6	1415	1416
Cs_2TiCl_6	1402	1384
Tl_2TiCl_6	1560	1553
K_2TiBr_6	1379	1379
Rb_2TiBr_6	1341	1331

Substance	Calc. U_{POT}	U_{POT}^{BHFC}
Cs_2TiBr_6	1339	1306
Na_2UBr_6	1504	–
K_2UBr_6	1484	–
Rb_2UBr_6	1473	–
Cs_2UBr_6	1459	–
K_2WCl_6	1398	1423
Rb_2WCl_6	1397	1434
Cs_2WCl_6	1392	1366
K_2WBr_6	1408	1408
Rb_2WBr_6	1361	1391
Cs_2WBr_6	1362	1332
K_2ZrCl_6	1339	1371
Rb_2ZrCl_6	1341	–
Cs_2ZrCl_6	1339	1307
Tellurides		
Li_2Te	2212	–
Na_2Te	1997	2095
K_2Te	1830	–
Rb_2Te	1837	–
Cs_2Te	1745	–
Cu_2Te	2706	2683
Ag_2Te	2607	2600
Tl_2Te	2084	2172
BeTe	3319	–
MgTe	2878	3081
CaTe	2721	–
Thiocyanates		
LiCNS	764	(765)
NaCNS	682	682
KCNS	623	616
RbCNS	623	619
CsCNS	623	568
NH4CNS	605	611
Ca(CNS)2	2184	2118
Sr(CNS)2	2063	1957
Ba(CNS)2	1979	1852
Mn(CNS)2	2280	2351
Zn(CNS)2	2335	2560
Cd(CNS)2	2201	2374
Hg(CNS)2	2146	2492
Sn(CNS)2	2117	2142
Pb(CNS)2	2058	–
Vanadates		
LiVO3	810	–
NaVO3	761	–
KVO3	686	–
RbVO3	657	–
CsVO3	628	–

TABLE 2. Thermochemical Radii (nm)

Ion	Radius	
Singly Charged Anions		
AgF_4^-	0.231	± 0.019
$AlBr_4^-$	0.321	± 0.023
$AlCl_4^-$	0.317	± 0.019
AlF_4^-	0.214	± 0.023
AlH_4^-	0.226	± 0.019
AlI_4^-	0.374	± 0.019
AsF_6^-	0.243	± 0.019
AsO_2^-	0.211	± 0.019
$Au(CN)_2^-$	0.266	± 0.019
$AuCl_4^-$	0.288	± 0.019
AuF_4^-	0.240	± 0.019
AuF_6^-	0.235	± 0.038
$B(OH)_4^-$	0.229	± 0.019
BF_4^-	0.205	± 0.019
BH_4^-	0.205	± 0.019
Br^-	0.190	± 0.019
BrF_4^-	0.231	± 0.019
BrO_3^-	0.214	± 0.019
$CF_3SO_3^-$	0.230	± 0.049
$CH_3CO_2^-$	0.194	± 0.019
Cl^-	0.168	± 0.019
ClO_2^-	0.195	± 0.019
ClO_3^-	0.208	± 0.019
ClO_4^-	0.225	± 0.019
$ClS_2O_6^-$	0.260	± 0.049
CN^-	0.187	± 0.023
$Cr_3O_8^-$	0.276	± 0.019
$CuBr_4^-$	0.315	± 0.019
F^-	0.126	± 0.019
$FeCl_4^-$	0.317	± 0.019
$GaCl_4^-$	0.328	± 0.019
H^-	0.148	± 0.019
$H_2AsO_4^-$	0.227	± 0.019
$H_2PO_4^-$	0.213	± 0.019
HCO_2^-	0.200	± 0.019
HCO_3^-	0.207	± 0.019
HF_2^-	0.172	± 0.019
HSO_4^-	0.221	± 0.019
I^-	0.211	± 0.019
I_2Br^-	0.261	± 0.019
I_3^-	0.272	± 0.019
I_4^-	0.300	± 0.019
IBr_2^-	0.251	± 0.019
ICl_2^-	0.235	± 0.019
ICl_4^-	0.307	± 0.019
$IO_2F_2^-$	0.233	± 0.019
IO_3^-	0.218	± 0.019
IO_4^-	0.231	± 0.019
IrF_6^-	0.242	± 0.019
MnO_4^-	0.220	± 0.019
MoF_6^-	0.241	± 0.019
$MoOF_5^-$	0.241	± 0.019
N_3^-	0.180	± 0.019
NCO^-	0.193	± 0.019
$NbCl_6^-$	0.338	± 0.049
NbF_6^-	0.254	± 0.019
$Nb_2F_{11}^-$	0.311	± 0.038
NbO_3^-	0.194	± 0.019
NH_2^-	0.168	± 0.019
$NH_2CH_2COO^-$	0.252	± 0.019
NO_2^-	0.187	± 0.019

Ion	Radius	
NO_3^-	0.200	± 0.019
O_2^-	0.165	± 0.019
O_3^-	0.199	± 0.034
OH^-	0.152	± 0.019
OsF_6^-	0.252	± 0.020
PaF_6^-	0.249	± 0.019
PdF_6^-	0.252	± 0.019
PF_6^-	0.242	± 0.019
PO_3^-	0.204	± 0.019
PtF_6^-	0.247	± 0.019
PuF_5^-	0.239	± 0.019
ReF_6^-	0.240	± 0.019
ReO_4^-	0.227	± 0.019
RuF_6^-	0.242	± 0.019
S_6^-	0.305	± 0.019
SCN^-	0.209	± 0.019
$SbCl_6^-$	0.320	± 0.019
SbF_6^-	0.252	± 0.019
$Sb_2F_{11}^-$	0.312	± 0.038
$Sb_3F_{14}^-$	0.374	± 0.038
$SeCl_5^-$	0.258	± 0.038
$SeCN^-$	0.230	± 0.019
SeH^-	0.195	± 0.019
SH^-	0.191	± 0.019
SO_3F^-	0.214	± 0.019
$S_3N_3^-$	0.231	± 0.038
$S_3N_3O_4^-$	0.252	± 0.038
$TaCl_6^-$	0.352	± 0.019
TaF_6^-	0.250	± 0.019
TaO_3^-	0.192	± 0.019
UF_6^-	0.301	± 0.019
VF_6^-	0.235	± 0.019
VO_3^-	0.201	± 0.019
WCl_6^-	0.337	± 0.019
WF_6^-	0.246	± 0.019
WOF_5^-	0.241	± 0.019

Doubly Charged Anions

Ion	Radius	
AmF_6^{2-}	0.255	± 0.019
$Bi_2Br_8^{2-}$	0.392	± 0.055
$Bi_6Cl_{20}^{2-}$	0.501	± 0.073
$CdCl_4^{2-}$	0.307	± 0.019
$CeCl_6^{2-}$	0.352	± 0.019
CeF_6^{2-}	0.249	± 0.019
CO_3^{2-}	0.189	± 0.019
$CoCl_4^{2-}$	0.306	± 0.019
CoF_4^{2-}	0.209	± 0.019
CoF_6^{2-}	0.256	± 0.019
$Cr_2O_7^{2-}$	0.292	± 0.019
CrF_6^{2-}	0.253	± 0.019
CrO_4^{2-}	0.229	± 0.019
$CuCl_4^{2-}$	0.304	± 0.019
CuF_4^{2-}	0.213	± 0.019
$GeCl_6^{2-}$	0.335	± 0.019
GeF_6^{2-}	0.244	± 0.019
HfF_6^{2-}	0.248	± 0.019
HgI_4^{2-}	0.377	± 0.019
$IrCl_6^{2-}$	0.332	± 0.019
$MnCl_6^{2-}$	0.314	± 0.031
MnF_4^{2-}	0.219	± 0.019
MnF_6^{2-}	0.241	± 0.019
$MoBr_6^{2-}$	0.364	± 0.019
$MoCl_6^{2-}$	0.338	± 0.019
MoF_6^{2-}	0.274	± 0.019
MoO_4^{2-}	0.231	± 0.019
$NbCl_6^{2-}$	0.343	± 0.019
NH^{2-}	0.128	± 0.019
$Ni(CN)_4^{2-}$	0.322	± 0.019
NiF_4^{2-}	0.211	± 0.019
NiF_6^{2-}	0.249	± 0.019
O^{2-}	0.141	± 0.019
O_2^{2-}	0.167	± 0.019
$OsBr_6^{2-}$	0.365	± 0.019
$OsCl_6^{2-}$	0.336	± 0.019
OsF_6^{2-}	0.276	± 0.019
$PbCl_4^{2-}$	0.279	± 0.019
$PbCl_6^{2-}$	0.347	± 0.019
PbF_6^{2-}	0.268	± 0.019
$PdBr_6^{2-}$	0.354	± 0.019
$PdCl_4^{2-}$	0.313	± 0.019
$PdCl_6^{2-}$	0.333	± 0.019
PdF_6^{2-}	0.252	± 0.019
$PoBr_6^{2-}$	0.380	± 0.019
PoI_6^{2-}	0.428	± 0.019
$Pt(NO_2)_3Cl_3^{2-}$	0.364	± 0.019
$Pt(NO_2)_4Cl_2^{2-}$	0.383	± 0.019
$Pt(OH)_2^{2-}$	0.333	± 0.019
$Pt(SCN)_6^{2-}$	0.451	± 0.019
$PtBr_4^{2-}$	0.324	± 0.019
$PtBr_6^{2-}$	0.363	± 0.019
$PtCl_4^{2-}$	0.307	± 0.019
$PtCl_6^{2-}$	0.333	± 0.019
PtF_6^{2-}	0.245	± 0.019
$PuCl_6^{2-}$	0.349	± 0.019
$ReBr_6^{2-}$	0.371	± 0.019
$ReCl_6^{2-}$	0.337	± 0.019
ReF_6^{2-}	0.256	± 0.019
ReF_8^{2-}	0.276	± 0.019
ReH_9^{2-}	0.257	± 0.019
ReI_6^{2-}	0.421	± 0.026
RhF_6^{2-}	0.240	± 0.019
$RuCl_6^{2-}$	0.336	± 0.019
RuF_6^{2-}	0.248	± 0.019
S^{2-}	0.189	± 0.019
$S_2O_3^{2-}$	0.251	± 0.019
$S_2O_4^{2-}$	0.262	± 0.019
$S_2O_5^{2-}$	0.270	± 0.019
$S_2O_6^{2-}$	0.283	± 0.019
$S_2O_7^{2-}$	0.275	± 0.019
$S_2O_8^{2-}$	0.291	± 0.019
$S_3O_6^{2-}$	0.302	± 0.019
$S_4O_6^{2-}$	0.325	± 0.019
$S_6O_6^{2-}$	0.382	± 0.019
ScF_6^{2-}	0.276	± 0.019
Se^{2-}	0.181	± 0.019
$SeBr_6^{2-}$	0.363	± 0.019
$SeCl_6^{2-}$	0.336	± 0.019
SeO_4^{2-}	0.229	± 0.019
SiF_6^{2-}	0.248	± 0.019
SiO_3^{2-}	0.195	± 0.019
SmF_4^{2-}	0.218	± 0.019
$Sn(OH)_6^{2-}$	0.279	± 0.020
$SnBr_6^{2-}$	0.374	± 0.019

Ion	Radius	
$SnCl_6^{2-}$	0.345	± 0.019
SnF_6^{2-}	0.265	± 0.019
SnI_6^{2-}	0.427	± 0.019
SO_3^{2-}	0.204	± 0.019
SO_4^{2-}	0.218	± 0.019
$TcBr_6^{2-}$	0.363	± 0.019
$TcCl_6^{2-}$	0.337	± 0.019
TcF_6^{2-}	0.244	± 0.019
TcH_9^{2-}	0.260	± 0.019
TcI_6^{2-}	0.419	± 0.019
Te^{2-}	0.220	± 0.019
$TeBr_6^{2-}$	0.383	± 0.019
$TeCl_6^{2-}$	0.353	± 0.019
TeI_6^{2-}	0.430	± 0.019
TeO_4^{2-}	0.238	± 0.019
$Th(NO_3)_6^{2-}$	0.424	± 0.019
$ThCl_6^{2-}$	0.360	± 0.019
ThF_6^{2-}	0.263	± 0.019
$TiBr_6^{2-}$	0.356	± 0.019
$TiCl_6^{2-}$	0.335	± 0.019
TiF_6^{2-}	0.252	± 0.019
UCl_6^{2-}	0.354	± 0.019
UF_6^{2-}	0.256	± 0.019
VO_3^{2-}	0.204	± 0.019
WBr_6^{2-}	0.363	± 0.019
WCl_6^{2-}	0.339	± 0.019
WO_4^{2-}	0.237	± 0.019
$WOCl_5^{2-}$	0.334	± 0.019
$ZnBr_4^{2-}$	0.335	± 0.019
$ZnCl_4^{2-}$	0.306	± 0.019
ZnF_4^{2-}	0.219	± 0.019
ZnI_4^{2-}	0.384	± 0.019
$ZrBr_4^{2-}$	0.334	± 0.019
$ZrCl_4^{2-}$	0.306	± 0.019
$ZrCl_6^{2-}$	0.348	± 0.019
ZrF_6^{2-}	0.258	± 0.019
Multi-Charged Anions		
AlH_6^{3-}	0.256	± 0.042
AsO_4^{3-}	0.237	± 0.042
$CdBr_6^{4-}$	0.374	± 0.038
$CdCl_6^{4-}$	0.352	± 0.038
CeF_6^{3-}	0.278	± 0.038
CeF_7^{3-}	0.282	± 0.038
$Co(CN)_6^{3-}$	0.349	± 0.038
$Co(NO_2)_6^{3-}$	0.343	± 0.038
$CoCl_5^{3-}$	0.320	± 0.038
CoF_6^{3-}	0.258	± 0.042
$Cr(CN)_6^{3-}$	0.351	± 0.038
CrF_6^{3-}	0.254	± 0.042
$Cu(CN)_4^{3-}$	0.312	± 0.038
$Fe(CN)_6^{3-}$	0.347	± 0.038
FeF_6^{3-}	0.298	± 0.042
HfF_7^{3-}	0.277	± 0.042
InF_6^{3-}	0.268	± 0.038
$Ir(CN)_6^{3-}$	0.347	± 0.038
$Ir(NO_2)_6^{3-}$	0.338	± 0.038
$Mn(CN)_6^{3-}$	0.350	± 0.038
$Mn(CN)_6^{5-}$	0.401	± 0.042
$MnCl_6^{4-}$	0.349	± 0.038
N^{3-}	0.180	± 0.042
$Ni(NO_2)_6^{3-}$	0.342	± 0.038

Ion	Radius	
$Ni(NO_2)_6^{4-}$	0.383	± 0.038
NiF_6^{3-}	0.250	± 0.042
O^{3-}	0.288	± 0.038
P^{3-}	0.224	± 0.042
PaF_8^{3-}	0.299	± 0.042
PO_4^{3-}	0.230	± 0.042
PrF_6^{3-}	0.281	± 0.038
$Rh(NO_2)_6^{3-}$	0.345	± 0.038
$Rh(SCN)_6^{3-}$	0.428	± 0.042
TaF_8^{3-}	0.284	± 0.042
TbF_7^{3-}	0.290	± 0.038
$Tc(CN)_6^{5-}$	0.410	± 0.042
ThF_7^{3-}	0.282	± 0.042
$TiBr_6^{3-}$	0.315	± 0.038
TlF_6^{3-}	0.271	± 0.038
UF_7^{3-}	0.285	± 0.042
YF_6^{3-}	0.275	± 0.038
ZrF_7^{3-}	0.273	± 0.038
Singly Charged Cations		
$N(CH_3)_4^+$	0.234	± 0.019
$N_2H_5^+$	0.158	± 0.019
$N_2H_6^{2+}$	0.158	± 0.029
$NH(C_2H_5)_3^+$	0.274	± 0.019
$NH_3C_2H_5^+$	0.193	± 0.019
$NH_3C_3H_7^+$	0.225	± 0.019
$NH_3CH_3^+$	0.177	± 0.019
NH_3OH^+	0.147	± 0.019
NH_4^+	0.136	± 0.019
$NH_3C_2H_4OH^+$	0.203	± 0.019
$As_3S_4^+$	0.244	± 0.027
$As_3Se_4^+$	0.253	± 0.027
$AsCl_4^+$	0.221	± 0.027
Br_2^+	0.155	± 0.027
Br_3^+	0.204	± 0.027
Br_3^-	0.238	± 0.027
Br_5^+	0.229	± 0.027
$BrClCNH_2^+$	0.175	± 0.027
BrF_2^+	0.183	± 0.027
BrF_4^+	0.172	± 0.027
$C_{10}F_8^+$	0.265	± 0.027
$C_6F_6^+$	0.228	± 0.027
$Cl(SNSCN)_2^+$	0.347	± 0.027
$Cl_2C{=}NH_2^+$	0.173	± 0.027
Cl_2F^+	0.165	± 0.027
Cl_3^+	0.182	± 0.027
ClF_2^+	0.147	± 0.027
ClO_2^+	0.118	± 0.027
$GaBr_4^-$	0.317	± 0.038
I_2^+	0.185	± 0.027
I_3^+	0.225	± 0.027
I_5^+	0.263	± 0.027
IBr_2^+	0.196	± 0.027
ICl_2^+	0.175	± 0.036
IF_6^+	0.209	± 0.027
$N(S_3N_2)_2^+$	0.258	± 0.027
$N(SCl)_2^+$	0.186	± 0.027
$N(SeCl)_2^+$	0.246	± 0.027
$N(SF_2)_2^+$	0.214	± 0.027
N_2F^+	0.156	± 0.027
NO^+	0.145	± 0.027
NO_2^+	0.153	± 0.027

Ion	Radius	
O_2^+	0.140	± 0.027
$O_2(SCCF_3Cl)_2^+$	0.275	± 0.027
$ONCH_3CF_3^+$	0.200	± 0.027
$OsOF_5^-$	0.246	± 0.038
$P(CH_3)_3Cl^+$	0.197	± 0.027
$P(CH_3)_3D^+$	0.196	± 0.027
PCl_4^+	0.235	± 0.027
$ReOF_5^-$	0.245	± 0.038
$S(CH_3)_2Cl^+$	0.207	± 0.027
$S(N(C_2H_5)_3)_3^+$	0.439	± 0.027
$S_2(CH_3)_2Cl^+$	0.265	± 0.027
$S_2(CH_3)_2CN^+$	0.223	± 0.027
$S_2(CH_3)_3^+$	0.233	± 0.027
$S_2Br_5^+$	0.267	± 0.027
S_2N^+	0.159	± 0.034
$S_2N_2C_2H_3^+$	0.211	± 0.027
$S_2NC_2(PhCH_3)_2^+$	0.310	± 0.027
$S_2NC_3H_4^+$	0.218	± 0.027
$S_2NC_4H_8^+$	0.225	± 0.027
$S_3(CH_3)_3^+$	0.239	± 0.027
$S_3Br_3^+$	0.245	± 0.027
$S_3C_3H_7^+$	0.199	± 0.027
$S_3C_4F_6^+$	0.261	± 0.027
$S_3CF_3CN^+$	0.263	± 0.027
$S_3Cl_3^+$	0.233	± 0.027
$S_3N_2^+$	0.201	± 0.027
$S_3N_2Cl^+$	0.232	± 0.027
$S_4N_3^+$	0.231	± 0.027
$S_4N_3(Ph)_2^+$	0.316	± 0.027
$S_4N_4H^+$	0.178	± 0.027
$S_5N_5^+$	0.257	± 0.027
S_7I^+	0.262	± 0.027
$Sb(NPPh_3)_4^+$	0.518	± 0.027
SBr_3^+	0.220	± 0.027
$SCH_3O_2^+$	0.183	± 0.027
$SCH_3P(CH_3)_3^+$	0.248	± 0.027
$SCH_3PCH_3Cl_2^+$	0.205	± 0.027
$SCl(C_2H_5)_2^+$	0.207	± 0.027
$SCl_2CF_3^+$	0.207	± 0.027
$SCl_2CH_3^+$	0.204	± 0.027
SCl_3^+	0.185	± 0.027
$Se_3Br_3^+$	0.253	± 0.027
$Se_3Cl_3^+$	0.245	± 0.027
$Se_3N_2^+$	0.288	± 0.042
$Se_3NC_{12}^+$	0.163	± 0.027
Se_6I^+	0.260	± 0.027
$SeBr_3^+$	0.182	± 0.027
$SeCl_3^+$	0.192	± 0.027
SeF_3^+	0.179	± 0.027
SeI_3^+	0.238	± 0.027
SeN_2Cl^+	0.196	± 0.027
$SeNCl_2^+$	0.157	± 0.027
$(SeNMe_3)_3^+$	0.406	± 0.027
$SeS_2N_2^+$	0.282	± 0.042
$SF(C_6F_5)_2^+$	0.294	± 0.027
$SF_2CF_3^+$	0.198	± 0.027
$SF_2N(CH_3)_2^+$	0.210	± 0.027
SF_3^+	0.172	± 0.027
$SFS(C(CF_3)_2)_2^+$	0.275	± 0.027
$SH_2C_3H_7^+$	0.210	± 0.027
SN^+	0.158	± 0.027
$SNCl_5(CH_3CN)^-$	0.290	± 0.038

Ion	Radius	
$(SNPMe_3)_3^+$	0.308	± 0.027
$SNSC(CH_3)N^+$	0.225	± 0.027
$SNSC(CN)CH^+$	0.209	± 0.027
$SNSC(Ph)N^+$	0.251	± 0.027
$SNSC(Ph)NS_3N_2^+$	0.327	± 0.027
$SNSC(PhCH_3)N^+$	0.264	± 0.027
$(Te(N(SiMe_3)_2)_2^+$	0.371	± 0.027
$Te(N_3)_3^+$	0.226	± 0.027
$Te_4Nb_3OTe_2I_6^+$	0.407	± 0.027
$TeBr_3^+$	0.235	± 0.027
$TeCl_3^+$	0.216	± 0.027
$TeCl_3$(15-crown-5)$^+$	0.282	± 0.027
TeI_3^+	0.243	± 0.027
$Xe_2F_{11}^+$	0.266	± 0.027
$Xe_2F_3^+$	0.221	± 0.027
XeF^+	0.174	± 0.027
XeF_3^+	0.183	± 0.027
XeF_5^+	0.186	± 0.027
$XeOF_3^+$	0.186	± 0.027
Doubly Charged Cations		
$Co_2S_2(CO)_6^{2+}$	0.263	± 0.035
$FeW(Se)_2(CO)^{2+}$	0.260	± 0.035
I_4^{2+}	0.207	± 0.035
$Mo(Te_3)(CO)_4^{2+}$	0.234	± 0.035
S_{19}^{2+}	0.292	± 0.035
$S_2(S(CH_3)_2)_2^{2+}$	0.230	± 0.035
$S_2I_4^{2+}$	0.231	± 0.035
$S_3N_2^{2+}$	0.184	± 0.035
$S_3NCCNS_3^{2+}$	0.220	± 0.035
S_3Se^{2+}	0.326	± 0.035
$S_4N_4^{2+}$	0.186	± 0.035
$S_6N_4^{2+}$	0.232	± 0.035
S_8^{2+}	0.182	± 0.035
Se_{10}^{2+}	0.253	± 0.035
Se_{17}^{2+}	0.236	± 0.035
Se_{19}^{2+}	0.296	± 0.035
$Se_2I_4^{2+}$	0.218	± 0.035
$Se_3N_2^{2+}$	0.182	± 0.035
Se_4^{2+}	0.152	± 0.035
$Se_4S_2N_4^{2+}$	0.224	± 0.035
Se_8^{2+}	0.186	± 0.035
$SeN_2S_2^{2+}$	0.182	± 0.035
$(SNP(C_2H_5)_3)_2^{2+}$	0.312	± 0.035
$TaBr_6^-$	0.351	± 0.049
$Te(trtu)_4^{2+}$	0.328	± 0.035
$Te(tu)_4^{2+}$	0.296	± 0.035
$Te_2(esu)_4Br_2^{2+}$	0.356	± 0.035
$Te_2(esu)_4Cl_2^{2+}$	0.361	± 0.035
$Te_2(esu)_4I_2^{2+}$	0.342	± 0.035
$Te_2Se_2^{2+}$	0.192	± 0.035
$Te_2Se_4^{2+}$	0.222	± 0.035
$Te_2Se_8^{2+}$	0.252	± 0.035
$Te_3S_3^{2+}$	0.217	± 0.035
Te_3Se^{2+}	0.193	± 0.035
Te_4^{2+}	0.169	± 0.035
Te_8^{2+}	0.187	± 0.035
$W(CO)_4$(h3-Te)$^{2+}$	0.234	± 0.035
$W_2(CO)_{10}Se_4^{2+}$	0.290	± 0.035
Multi-Charged Cations		
I_{15}^{3+}	0.442	± 0.051
$Te_2(su)_6^{4+}$	0.453	± 0.034

Ligand abbreviations: su = selenourea; esu = ethyleneselenourea; tu = thiourea; ph = phenyl.

TABLE 3. Ancillary Thermochemical Data (kJ mol⁻¹)

Species	State	$\Delta_f H^o$
AsO_4^{3-}	g	(289)
BrO_3^-	g	−145
ClO_4^-	g	−344
CN^-	g	66
CO_3^{2-}	g	−321
$Fe(NO_3)_2$	c	(−448)
HF_2^-	g	−774
$HfCl_6^{2-}$	g	−1640
$IO_2F_2^-$	g	−693
IO_3^-	g	−208
$IrCl_6^{2-}$	g	−785
$LiCH_3O_2$	c	(−745)
$NbCl_6^{2-}$	g	−1224
$NH_2CH_2CO_2^-$	g	−564
O_2^{2-}	g	553
$PdCl_6^{2-}$	g	−749
PO_4^{3-}	g	291
$PtCl_6^{2-}$	g	−774
$ReBr_6^{2-}$	g	−689
$ReCl_6^{2-}$	g	−919
$Ti(OH)_2$	c	−778

Solids

THE MADELUNG CONSTANT AND CRYSTAL LATTICE ENERGY

If U is the crystal lattice energy and M is the Madelung constant, then[a]

$$U = \frac{NMz_i z_j e^2}{r}(1 - 1/n)$$

Substance	Ion type	Crystal form[b]	M
Sodium chloride, NaCl	M^+, X^-	FCC	1.74756
Cesium chloride, CsCl	M^+, X^-	BCC	1.76267
Calcium chloride, $CaCl_2$	$M^{++}, 2X^-$	Cubic	2.365
Calcium fluoride (fluorite), CaF_2	$M^{++}, 2X^-$	Cubic	2.51939
Cadmium chloride, $CdCl_2$	$M^{++}, 2X^-$	Hexagonal	2.244[c]
Cadmium iodide (α), CdI_2	$M^{++}, 2X^-$	Hexagonal	2.355[c]
Magnesium fluoride, MgF_2	$M^{++}, 2X^-$	Tetragonal	2.381[c]
Cuprous oxide (cuprite), Cu_2O	$2M^+, X^{--}$	Cubic	2.22124
Zinc oxide, ZnO	M^{++}, X^{--}	Hexagonal	1.4985[c]
Sphalerite (zinc blende), ZnS	M^{++}, X^{--}	FCC	1.63806
Wurtzite, ZnS	M^{++}, X^{--}	Hexagonal	1.64132[c]
Titanium dioxide (anatase), TiO_2	$M^{4+}, 2X^{--}$	Tetragonal	2.400[c]
Titanium dioxide (rutile), TiO_2	$M^{4+}, 2X^{--}$	Tetragonal	2.408[c]
β-Quartz, SiO_2	$M^{4+}, 2X^{--}$	Hexagonal	2.2197[c]
Corundum, Al_2O_3	$2M^{3+}, 3X^{--}$	Rhombohedral	4.1719

[a] N is Avogadro's number, z_i and z_j are the integral charges on the ions (in units of e), and e is the charge on the electron in electrostatic units ($e = 4.803 \times 10^{-10}$ esu). r is the shortest distance between cation–anion pairs in centimeters. Then U is in ergs (1 erg = 10^{-7} J).

[b] FCC = face centered cubic; BCC = body centered cubic.

[c] For tetragonal and hexagonal crystals the value of M depends on the details of the lattice parameters.

The Born Exponent, n is:

Ion type	n
He, Li^+	5
Ne, Na^+, F^-	7
Ar, K^+, Cu^+, Cl^-	9
Kr, Rb^+, Ag^+, Br^-	10
Xe, Cs^+, Au^+, I^-	12

For a crystal with a mixed-ion type, an average of the values of n in this table is to be used (6 for LiF, for example).

ELASTIC CONSTANTS OF SINGLE CRYSTALS

H. P. R. Frederikse

This table gives selected values of elastic constants for single crystals. The values believed most reliable were selected from the original literature. The substances are arranged by crystal system and, within each system, alphabetically by name. A reference to the original literature is given for each value; a useful compilation of published values from many sources may be found in Ref. 1.

Data are given for the single-crystal density and for the elastic constants C_{ij}, in units of 10^{11} N/m², which is equivalent to 10^{12} dyn/cm².

General References

1. Simmons, G., and Wang, H., *Single Crystal Elastic Constants and Calculated Aggregate Properties: A Handbook, Second Edition*, The MIT Press, Cambridge, MA, 1971.
2. Gray, D. E., Ed., *American Institute of Physics Handbook, Third Edition*, McGraw-Hill, New York, 1972.

Elastic Constants of Single Crystals: Cubic Crystals

Name	Formula	ρ/g cm^{-3}	*T*/K	C_{11}	C_{12}	C_{44}	Ref.
Aluminum	Al	2.6970	298	1.0675	0.6041	0.2834	1
Aluminum antimonide	AlSb	4.3600	300	0.8939	0.4427	0.4155	2
Ammonium bromide	NH_4Br	2.4314	300	0.3414	0.0782	0.0722	3
Ammonium chloride	NH_4Cl	1.5279	290	0.3814	0.0866	0.0903	4
Argon	Ar	1.7710	4.2	0.0529	0.0135	0.0159	5
Barium fluoride	BaF_2	4.8860	298	0.9199	0.4157	0.2568	6
Barium nitrate	$Ba(NO_3)_2$	3.2560	293	0.2925	0.2065	0.1277	7
Calcium fluoride	CaF_2	3.810	298	1.6420	0.4398	0.8406	8
Calcium telluride	CaTe	5.8544	298	0.5351	0.3681	0.1994	9
Cesium	Cs	1.9800	78	0.0247	0.0206	0.0148	10
Cesium bromide	CsBr	4.4560	298	0.3063	0.0807	0.0750	11
Cesium chloride	CsCl	3.9880	298	0.3644	0.0882	0.0804	11
Cesium iodide	CsI	4.5250	298	0.2446	0.0661	0.0629	11
Chromite	$FeCr_2O_4$	4.4500	[a]	3.2250	1.4370	1.1670	12
Chromium	Cr	7.20	298	3.398	0.586	0.990	13
Cobalt(II) oxide	CoO	6.44	298	2.6123	1.4699	0.8300	14
Cobalt zinc ferrite	$CoZnFeO_2$	5.43	303	2.660	1.530	0.780	12
Copper	Cu	8.932	298	1.683	1.221	0.757	15
Gallium antimonide	GaSb	5.6137	298	0.8839	0.4033	0.4316	16
Gallium arsenide	GaAs	5.3169	298	1.1877	0.5372	0.5944	17
Gallium phosphide	GaP	4.1297	300	1.4120	0.6253	0.7047	18
Germanium	Ge	5.313	298	1.2835	0.4823	0.6666	20
Gold	Au	19.283	296.5	1.9244	1.6298	0.4200	21
Indium antimonide	InSb	5.7890	298	0.6720	0.3670	0.3020	22
Indium arsenide	InAs	5.6720	293	0.8329	0.4526	0.3959	23
Indium phosphide	InP	4.78	[a]	1.0220	0.5760	0.4600	24
Iridium	Ir	22.52	300	5.80	2.42	2.56	25
Iron	Fe	7.8672	298	2.26	1.40	1.16	26
Lead	Pb	11.34	296	0.4966	0.4231	0.1498	27
Lead(II) fluoride	PbF_2	7.79	300	0.8880	0.4720	0.2454	28
Lead(II) nitrate	$Pb(NO_3)_2$	4.547	293	0.3729	0.2765	0.1347	29
Lead(II) telluride	PbTe	8.2379	303.2	1.0795	0.0764	0.1343	30
Lithium	Li	0.5326	298	0.1350	0.1144	0.0878	31
Lithium bromide	LiBr	3.47	[a]	0.3940	0.1880	0.1910	32
Lithium chloride	LiCl	2.068	295	0.4927	0.2310	0.2495	33
Lithium fluoride	LiF	2.638	[a]	1.1397	0.4767	0.6364	34
Lithium iodide	LiI	4.061	[a]	0.2850	0.1400	0.1350	32
Magnesium oxide	MgO	3.579	298	2.9708	0.9536	1.5613	20
Magnetite	Fe_3O_4	5.18	[a]	2.730	1.060	0.971	32
Manganese(II) oxide	MnO	5.39	298	2.23	1.20	0.79	35
Mercury(II) telluride	HgTe	8.079	290	0.548	0.381	0.204	36
Molybdenum	Mo	10.2284	273	4.637	1.578	1.092	37
Nickel	Ni	8.91	298	2.481	1.549	1.242	15
Niobium	Nb	8.578	300	2.4650	1.3450	0.2873	38
Palladium	Pd	12.038	300	2.2710	1.7604	0.7173	39
Platinum	Pt	21.50	300	3.4670	2.5070	0.7650	40

Name	Formula	ρ/g cm⁻³	T/K	C_{11}	C_{12}	C_{44}	Ref.
Potassium	K	0.851	295	0.0370	0.0314	0.0188	41
Potassium bromide	KBr	2.740	298	0.3468	0.0580	0.0507	11
Potassium chloride	KCl	1.984	298	0.4069	0.0711	0.0631	11
Potassium cyanide	KCN	1.553	[a]	0.1940	0.1180	0.0150	32
Potassium fluoride	KF	2.480	295	0.6490	0.1520	0.1232	33
Potassium iodide	KI	3.128	300	0.2710	0.0450	0.0364	42
Pyrite	FeS_2	5.016	[a]	3.818	0.310	1.094	43
Rubidium	Rb	1.58	170	0.0296	0.0250	0.0171	44
Rubidium bromide	RbBr	3.350	300	0.3152	0.0500	0.0380	45
Rubidium chloride	RbCl	2.797	300	0.3624	0.0612	0.0468	45
Rubidium iodide	RbI	3.551	300	0.2556	0.0382	0.0278	45
Silicon	Si	2.331	298	1.6578	0.6394	0.7962	46
Silver	Ag	10.50	300	1.2399	0.9367	0.4612	47
Silver(I) bromide	AgBr	5.585	300	0.5920	0.3640	0.0616	48
Sodium	Na	0.971	299	0.0739	0.0622	0.0419	49
Sodium bromate	$NaBrO_3$	3.339	[a]	0.5450	0.1910	0.1500	32
Sodium bromide	NaBr	3.202	300	0.3970	0.1001	0.0998	33
Sodium chlorate	$NaClO_3$	2.485	[a]	0.4920	0.1420	0.1160	50
Sodium chloride	NaCl	2.163	298	0.4947	0.1288	0.1287	11
Sodium fluoride	NaF	2.804	300	0.9700	0.2380	0.2822	51
Sodium iodide	NaI	3.6689	300	0.3007	0.0912	0.0733	52
Spinel	$MgAl_2O_4$	3.6193	298	2.9857	1.5372	1.5758	53
Strontium fluoride	SrF_2	4.277	300	1.2350	0.4305	0.3128	54
Strontium nitrate	$Sr(NO_3)_2$	2.989	293	0.4255	0.2921	0.1590	29
Strontium oxide	SrO	4.99	300	1.601	0.435	0.590	55
Strontium titanate	$SrTiO_3$	5.123	[a]	3.4817	1.0064	4.5455	56
Tantalum	Ta	16.626	298	2.6023	1.5446	0.8255	57
Tantalum carbide	TaC	14.65	[a]	5.05	0.73	0.79	58
Thallium(I) bromide	TlBr	7.4529	298	0.3760	0.1458	0.0757	59
Thorium	Th	11.694	300	0.7530	0.4890	0.4780	60
Thorium(IV) oxide	ThO_2	9.991	298	3.670	1.060	0.797	61
Tin(II) telluride	SnTe	6.445	300	1.1250	0.0750	0.1172	62
Titanium carbide	TiC	4.940	[a]	5.00	1.13	1.75	107
Tungsten	W	19.257	297	5.2239	2.0437	1.6083	64
Uranium carbide	UC	13.63	300	3.200	0.850	0.647	65
Uranium(IV) oxide	UO_2	10.97	298	3.960	1.210	0.641	66
Vanadium	V	6.022	300	2.287	1.190	0.432	67
Yttrium iron oxide	$Y_3Fe_5O_{12}$	5.17	298	2.680	1.106	0.766	19
Zinc selenide	ZnSe	5.262	298	0.8096	0.4881	0.4405	68
Zinc sulfide (wurtzite)	ZnS	4.088	298	1.0462	0.6534	0.4613	68
Zinc telluride	ZnTe	5.636	298	0.7134	0.4078	0.3115	68
Zirconium carbide	ZrC	6.606	298	4.720	0.987	1.593	63

[a] Room temperature

Elastic Constants of Single Crystals: Tetragonal Crystals

Name	Formula	ρ/g cm⁻³	T/K	C_{11}	C_{12}	C_{13}	C_{16}	C_{33}	C_{44}	C_{66}	Ref.
Ammonium dihydrogen arsenate	$NH_4H_2AsO_4$	2.3110	298	0.6747	-0.106	0.1652		0.3022	0.0685	0.0639	69
Ammonium dihydrogen phosphate	$NH_4H_2PO_4$	1.8030	293	0.6200	-0.050	0.1400		0.3000	0.0910	0.0610	69
Barium titanate	$BaTiO_3$	5.9988	298	2.7512	1.7897	1.5156		1.6486	0.5435	1.1312	70
Calcium molybdate	$CaMoO_4$	4.255	298	1.447	0.664	0.466	0.134	1.265	0.369	0.451	79
Indium	In	7.300	[a]	0.4450	0.3950	0.4050		0.4440	0.0655	0.1220	71
Magnesium fluoride	MgF_2	3.177	[a]	1.237	0.732	0.536		1.770	0.552	0.978	72
Nickel(II) sulfate hexahydrate	$NiSO_4{\cdot}6H_2O$	2.070	[a]	0.3209	0.2315	0.0209		0.2931	0.1156	0.1779	73
Potassium dihydrogen arsenate	KH_2AsO_4	2.867	[a]	0.530	-0.060	-0.020		0.370	0.120	0.070	12
Potassium dihydrogen phosphate	KH_2PO_4	2.388	[a]	0.7140	-0.049	0.1290		0.5620	0.1270	0.0628	71
Rubidium dihydrogen phosphate	RbH_2PO_4	2.800	298	0.5562	-0.064	0.0279		0.4398	0.1142	0.0350	74
Tellurium dioxide	TeO_2	5.99	[a]	0.5320	0.4860	0.2120		1.0850	0.2440	0.5520	76
Tin (white)	Sn	7.29	288	0.7529	0.6156	0.4400		0.9552	0.2193	0.2336	77
Titanium(IV) oxide (rutile)	TiO_2	4.260	298	2.7143	1.7796	1.4957		4.8395	1.2443	1.9477	75
Zirconium(IV) orthosilicate	$ZrSiO_4$	4.70	[a]	2.585	1.791	1.542		3.805	0.733	1.113	78

[a] Room temperature

Elastic Constants of Single Crystals: Orthorhombic Crystals

Name	Formula	ρ/g cm^{-3}	T/K	C_{11}	C_{12}	C_{13}	C_{22}	C_{23}	C_{33}	C_{44}	C_{55}	C_{66}	Ref.
Acenaphthene	$C_{12}H_{10}$	1.220	293	0.1380	0.0210	0.0410	0.1262	0.0460	0.1117	0.0265	0.0290	0.0185	80
Ammonium sulfate	$(NH_4)_2SO_4$	1.774	293	0.3607	0.1651	0.1580	0.2981	0.1456	0.3534	0.1025	0.0717	0.0974	81
Aragonite	$CaCO_3$	2.93	[a]	1.5958	0.3663	0.0197	0.8697	0.1597	0.8503	0.4132	0.2564	0.4274	82
Barite	$BaSO_4$	4.40	[a]	0.8941	0.4614	0.2691	0.7842	0.2676	1.0548	0.1190	0.2874	0.2778	82
Benzene	C_6H_6	1.061	250	0.0614	0.0352	0.0401	0.0656	0.0390	0.0583	0.0197	0.0378	0.0153	83
Benzophenone	$(C_6H_5)_2CO$	1.219	[a]	0.1070	0.0550	0.0169	0.1000	0.0321	0.0710	0.0203	0.0155	0.0353	32
Bronzite	$(Mg,Fe)SiO_4$	3.38	[a]	1.876	0.686	0.605	1.578	0.561	2.085	0.700	0.592	0.544	78
Calcium sulfate	$CaSO_4$	2.962	[a]	0.9382	0.1650	0.1520	1.845	0.3173	1.1180	0.3247	0.2653	0.0926	84
Celestite	$SrSO_4$	3.96	[a]	1.044	0.773	0.605	1.061	0.619	1.286	0.135	0.279	0.266	12
Cesium sulfate	Cs_2SO_4	4.243	293	0.4490	0.1958	0.1815	0.4283	0.1800	0.3785	0.1326	0.1319	0.1323	81
Forsterite	Mg_2SiO_4	3.224	298	3.2848	0.6390	0.6880	1.9980	0.7380	2.3530	0.6515	0.8120	0.8088	85
Iodic acid	HIO_3	4.630	[a]	0.3030	0.1194	0.1169	0.5448	0.0548	0.4359	0.1835	0.2193	0.1736	73
Lithium ammonium tartrate monohydrate	$LiNH_4C_4H_4O_6{\cdot}H_2O$	1.71	[a]	0.3864	0.1655	0.0875	0.5393	0.2007	0.3624	0.1190	0.0667	0.2326	12
Magnesium sulfate heptahydrate	$MgSO_4{\cdot}7H_2O$	1.68	[a]	0.325	0.174	0.182	0.288	0.182	0.315	0.078	0.156	0.090	86
Natrolite	$Na_2Al_2Si_3O_{10}{\cdot}2H_2O$	2.25	[a]	0.716	0.261	0.297	0.632	0.297	1.378	0.196	0.248	0.423	78
Nickel(II) sulfate heptahydrate	$NiSO_4{\cdot}7H_2O$	1.948	[a]	0.353	0.198	0.201	0.311	0.201	0.335	0.091	0.172	0.099	86
Olivine	$(Mg,Fe)SiO_4$	3.324	[a]	3.240	0.590	0.790	1.980	0.780	2.490	0.667	0.810	0.793	87
Potassium pentaborate tetrahydrate	$KB_5O_8{\cdot}4H_2O$	1.74	[a]	0.582	0.229	0.174	0.359	0.231	0.255	0.164	0.046	0.057	71
Potassium sulfate	K_2SO_4	2.665	293	0.5357	0.1999	0.2095	0.5653	0.1990	0.5523	0.195	0.1879	0.1424	81
Potassium sodium tartrate tetrahydrate	$KNaC_4H_4O_6{\cdot}4H_2O$	1.79	[a]	0.255	0.141	0.116	0.381	0.146	0.371	0.134	0.032	0.098	71
Rubidium sulfate	Rb_2SO_4	3.621	293	0.5029	0.1965	0.1999	0.5098	0.1925	0.4761	0.1626	0.1589	0.1407	81
Sodium ammonium tartrate tetrahydrate	$NaNH_4C_4H_4O_6{\cdot}4H_2O$	1.587	[a]	0.3685	0.2725	0.3083	0.5092	0.3472	0.5541	0.1058	0.0303	0.0870	12
Sodium tartrate	$Na_2C_4H_4O_6{\cdot}2H_2O$	1.794	[a]	0.461	0.286	0.320	0.547	0.352	0.665	0.124	0.031	0.098	12
Strontium formate dihydrate	$Sr(CHO_2)_2{\cdot}2H_2O$	2.25	[a]	0.4391	0.1037	-0.149	0.3484	-0.014	0.3746	0.1538	0.1075	0.1724	12
Sulfur (orthorhombic)	S_8	2.07	[a]	0.240	0.133	0.171	0.205	0.159	0.483	0.043	0.087	0.076	12
Thallium(I) sulfate	Tl_2SO_4	6.776	293	0.4106	0.2573	0.2288	0.3885	0.2174	0.4268	0.1125	0.1068	0.0751	81
Topaz	$Al_2(SiO_4)(F,OH)_2$	3.52	[a]	2.8136	1.2582	0.8464	3.8495	0.8815	2.9452	1.0811	1.3298	1.3089	82
Uranium	U	19.0453	293	2.1486	0.4622	0.2176	1.9983	1.0764	2.6763	1.2479	0.7379	0.7454	88
Zinc sulfate heptahydrate	$ZnSO_4{\cdot}7H_2O$	1.970	[a]	0.3320	0.1720	0.2000	0.2930	0.1980	0.3200	0.0780	0.1530	0.0830	86

[a] Room temperature

Elastic Constants of Single Crystals: Monoclinic Crystals

Name	Formula	ρ/g cm^{-3}	T/K	C_{11}	C_{12}	C_{13}	C_{15}	C_{22}	Ref.
Acmite	$NaFe(SiO_3)_2$	3.50	RT	1.858	0.685	0.707	0.098	1.813	89
Anthracene	$C_{14}H_{10}$	1.28	RT	0.0852	0.0672	0.0590	-0.0192	0.1170	90
Cobalt(II) sulfate heptahydrate	$CoSO_4{\cdot}7H_2O$	2.03	RT	0.335	0.205	0.158	0.016	0.378	86
Diopside	$CaMg(SiO_3)_2$	3.30	RT	2.040	0.884	0.0883	-0.193	1.750	91
Feldspar	$K_2O{\cdot}Al_2O{\cdot}6SiO_2$	≈2.6	RT	0.664	0.438	0.259	-0.033	1.710	92
Iron(II) sulfate heptahydrate	$FeSO_4{\cdot}7H_2O$	1.895	RT	0.349	0.208	0.174	-0.020	0.376	86
Lithium sulfate monohydrate	$Li_2SO_4{\cdot}H_2O$	2.06	RT	0.5250	0.1715	0.1730	-0.0196	0.5060	32
Naphthalene	$C_{10}H_8$	1.0253	RT	0.0780	0.0445	0.0340	-0.006	0.0990	93
Potassium hydrogen tartrate	$KHC_4H_4O_6$	1.98	RT	0.4294	0.1399	0.3129	-0.0105	0.3460	12
Potassium tartrate	$K_2C_4H_4O_6$	1.99	RT	0.3110	0.1720	0.1690	0.0287	0.3900	32
Sodium thiosulfate	$Na_2S_2O_3$	1.69	RT	0.3323	0.1814	0.1875	0.0225	0.2953	12
trans-Stilbene	$(C_6H_5CH)_2$	0.9707	RT	0.0930	0.0570	0.0670	-0.003	0.0920	94
Triglycine sulfate	$(NH_2CH_2COOH)_3{\cdot}H_2SO_4$	1.68	RT	0.4550	0.1720	0.1980	-0.030	0.3210	32

Name	Formula	C_{23}	C_{25}	C_{33}	C_{35}	C_{44}	C_{46}	C_{55}	C_{66}	Ref.
Acmite	$NaFe(SiO_3)_2$	0.626	0.094	2.344	0.214	0.692	0.077	0.510	0.474	89
Anthracene	$C_{14}H_{10}$	0.0375	-0.0170	0.1522	-0.0187	0.0272	0.0138	0.0242	0.0399	90
Cobalt(II) sulfate heptahydrate	$CoSO_4{\cdot}7H_2O$	0.158	-0.018	0.371	-0.047	0.060	0.016	0.058	0.101	86
Diopside	$CaMg(SiO_3)_2$	0.482	-0.196	2.380	-0.336	0.675	-0.113	0.588	0.705	91
Feldspar	$K_2O{\cdot}Al_2O{\cdot}6SiO_2$	0.192	-0.148	1.215	-0.131	0.143	-0.015	0.238	0.361	92

Name	Formula	C_{23}	C_{25}	C_{33}	C_{35}	C_{44}	C_{46}	C_{55}	C_{66}	Ref.
Iron(II) sulfate heptahydrate	$FeSO_4 \cdot 7H_2O$	0.172	-0.019	0.360	-0.014	0.064	0.001	0.056	0.096	86
Lithium sulfate monohydrate	$Li_2SO_4 \cdot H_2O$	0.0368	0.0571	0.5400	-0.0254	0.1400	-0.0054	0.1565	0.2770	32
Naphthalene	$C_{10}H_8$	0.0230	-0.0270	0.1190	0.0290	0.0330	-0.0050	0.0210	0.0415	93
Potassium hydrogen tartrate	$KHC_4H_4O_6$	0.1173	0.0176	0.6816	0.0294	0.0961	-0.0044	0.1270	0.0841	12
Potassium tartrate	$K_2C_4H_4O_6$	0.1330	0.0182	0.5540	0.0710	0.0870	0.0072	0.1040	0.0826	32
Sodium thiosulfate	$Na_2S_2O_3$	0.1713	0.0983	0.4590	-0.0678	0.0569	-0.0268	0.1070	0.0598	12
trans-Stilbene	$(C_6H_5CH)_2$	0.0485	-0.005	0.0790	-0.005	0.0325	0.0050	0.0640	0.0245	94
Triglycine sulfate	$(NH_2CH_2COOH)_3 \cdot H_2SO_4$	0.2080	-0.0036	0.2630	-0.0500	0.0950	-0.0026	0.1110	0.0620	32

Elastic Constants of Single Crystals: Hexagonal Crystals

Name	Formula	ρ/g cm^{-3}	*T*/K	C_{11}	C_{12}	C_{13}	C_{33}	C_{55}	Ref.
Apatite	$Ca_5(PO_4)_3(OH,F,Cl)$	3.2	[a]	1.667	0.131	0.655	1.396	0.663	12
Beryl	$Be_3Al_2(SiO_3)_6$	2.64	[a]	2.800	0.990	0.670	2.480	0.658	12
Beryllium	Be	1.85	300	2.923	0.267	0.140	3.364	1.625	95
Beryllium oxide	BeO	3.01	[a]	4.70	1.68	1.19	4.94	1.53	96
Cadmium	Cd	8.69	300	1.1450	0.3950	0.3990	0.5085	0.1985	97
Cadmium selenide	CdSe	5.81	298	0.7046	0.4516	0.3930	0.8355	0.1317	68
Cadmium sulfide	CdS	4.826	298	0.8431	0.5208	0.4567	0.9183	0.1458	98
Cobalt	Co	8.86	298	3.071	1.650	1.027	3.581	0.755	99
Dysprosium	Dy	8.55	298	0.7466	0.2616	0.2233	0.7871	0.2427	100
Erbium	Er	9.07	298	0.8634	0.3050	0.2270	0.8554	0.2809	100
Gadolinium	Gd	7.90	298	0.6667	0.2499	0.2132	0.7191	0.2089	101
Hafnium	Hf	13.3	298	1.881	0.772	0.661	1.969	0.557	102
Ice	H_2O	0.9167[0]	250	0.1410	0.0660	0.0624	0.1515	0.0288	103
Indium	In	7.31	300	0.4535	0.4006	0.4151	0.4515	0.0651	104
Magnesium	Mg	1.74	298	0.5950	0.2612	0.2180	0.6155	0.1635	105
Rhenium	Re	20.8	298	6.1820	2.7530	2.0780	6.8350	1.6060	100
Ruthenium	Ru	12.1	298	5.6260	1.8780	1.6820	6.2420	1.8060	100
Thallium	Tl	11.8	300	0.4080	0.3540	0.2900	0.5280	0.0726	106
Titanium	Ti	4.506	298	1.6240	0.9200	0.6900	1.8070	0.4670	102
Titanium boride	TiB_2	4.38	[a]	6.90	4.10	3.20	4.40	2.50	107
Yttrium	Y	4.47	300	0.7790	0.2850	0.2100	0.7690	0.2431	108
Zinc	Zn	7.134	295	1.6368	0.3640	0.5300	0.6347	0.3879	109
Zinc oxide	ZnO	5.6	298	2.0970	1.2110	1.0510	2.1090	0.4247	110
Zinc sulfide (wurtzite)	ZnS	4.09	298	1.2420	0.6015	0.4554	1.4000	0.2864	96
Zirconium	Zr	6.52	298	1.434	0.728	0.653	1.648	0.320	102

[a] Room temperature

Elastic Constants of Single Crystals: Trigonal Crystals

Name	Formula	Density	*T*/K	C_{11}	C_{12}	C_{13}	C_{14}	C_{33}	C_{44}	Ref.
Aluminum oxide (α)	Al_2O_3	3.99	300	4.9735	1.6397	1.1220	-0.2358	4.9911	1.4739	111
Aluminum phosphate	$AlPO_4$	2.56	[a]	1.0503	0.2934	0.6927	-0.1271	1.3353	0.2314	73
Antimony	Sb	6.70	295	1.0130	0.3450	0.2920	0.2090	0.4500	0.3930	112
Bismuth	Bi	9.79	295	0.6370	0.2490	0.2470	0.0717	0.3820	0.1123	112
Calcium carbonate (calcite)	$CaCO_3$	2.710	300	1.4806	0.5578	0.5464	-0.2058	0.8557	0.3269	113
Iron(III) oxide	Fe_2O_3	5.25	[a]	2.4243	0.5464	0.1542	-0.1247	2.2734	0.8569	82
Lithium niobate	$LiNbO_3$	4.30	[a]	2.030	0.530	0.750	0.090	2.450	0.600	114
Lithium tantalate	$LiTaO_3$	7.45	[a]	2.330	0.470	0.800	-0.110	2.750	0.940	114
Quartz (α)	SiO_2	2.65	298	0.8680	0.0704	0.1191	-0.1804	1.0575	0.5820	115
Selenium	Se	4.838	300	0.1870	0.0710	0.2620	0.0620	0.7410	0.1490	116
Sodium nitrate	$NaNO_3$	2.261		0.8670	0.1630	0.1600	0.0820	0.3740	0.2130	12
Tourmaline	$Na(Mg,Fe,Mn,Li,Al)_3Al_6Si_6O_{18}(BO_3)_3$	3.14	[a]	2.7066	0.6927	0.0872	-0.0774	1.6070	0.6682	82

[a] Room temperature

References

3. Thomas, J. F., *Phys. Rev.*, 175, 955–962, 1968.
4. Bolef, D. I. and M. Menes, *J. Appl. Phys.*, 31, 1426–1427, 1960.
5. Garland, C. W. and C. F. Yarnell, *J. Chem. Phys.*, 44, 1112–1120, 1966.
6. Garland, C. W. and R. Renard, *J. Chem. Phys.*, 44, 1130–1139, 1966.
7. Gsänger, M., H. Egger and E. Lüscher, *Phys. Letters*, 27A, 695–696, 1968.
8. Wong, C. and D. E. Schuele, *J. Phys. Chem. Solids*, 29, 1309–1330, 1968.
9. Haussühl, S., *Phys. Stat. Sol.*, 3, 1072–1076, 1963.
10. Wong, C. and D. E. Schuele, *J. Phys. Chem. Solids*, 28, 1225–1231, 1967.

11. McSkimin, H. J. and D. G. Thomas, *J. Appl. Phys.*, 33, 56–59, 1962.
12. Kollarits, F. J. and J. Trivisonno, *J. Phys. Chem. Solids*, 29, 2133–2139, 1968.
13. Slagle, D. D. and H. A. McKinstry, *J. Appl. Phys.*, 38, 446–458, 1967.
14. Hearmon, R. F. S., *Adv. Phys.*, 5, 323–382, 1956.
15. Sumer, A. and J. F. Smith, *J. Appl. Phys.*, 34, 2691–2694, 1963.
16. Alexandrov, K. S. et al., *Sov. Phys. Sol. State*, 10, 1316–1321, 1968.
17. Epstein, S. G. and O. N. Carlson, *Acta Metal.*, 13, 487–491, 1965.
18. McSkimin, H. J., et al., *J. Appl. Phys.*, 39, 4127–4128, 1968.
19. McSkimin, H. J., et al., *J. Appl. Phys.*, 38, 2362–2364, 1967.
20. Weil, R. and W. O. Groves, *J. Appl. Phys.*, 39, 4049–4051, 1968.
21. Bateman, T. B., *J. Appl. Phys.*, 37, 2194–2195, 1966.
22. Bogardus, E. H., *J. Appl. Phys.*, 36, 2504–2513, 1965.
23. Golding, B., S. C. Moss and B. L. Averbach, *Phys. Rev.*, 158, 637–645, 1967.
24. Bateman, T. B., H. J. McSkimin and J. M. Whelan, *J. Appl. Phys.*, 30, 544–545, 1959.
25. Gerlich, D., *J. Appl. Phys.*, 35, 3062, 1964.
26. Hickernell, F. S. and W. R. Gayton, *J. Appl. Phys.*, 37, 462, 1966.
27. MacFarlane, R. E., et al., *Phys. Letters*, 20, 234–235, 1966.
28. Leese, J. and A. E. Lord Jr., *J. Appl. Phys.*, 39, 3986–3988, 1968.
29. Miller, R. A. and D. E. Schuele, *J. Phys. Chem. Solids*, 30, 589–600, 1969.
30. Wasilik, J. H. and M. L. Wheat, *J. Appl. Phys.*, 36, 791–793, 1965.
31. Haussühl, S., *Phys. Stat. Sol.*, 3, 1072–1076, 1963.
32. Houston, B., et al., *J. Appl. Phys.*, 39, 3913–3916, 1968.
33. Trivisonno, J. and C. S. Smith, *Acta Metal.*, 9, 1064–1071, 1961.
34. Alexandrov, K. S. and T. V. Ryzhova, *Sov. Phys. Cryst.*, 6, 228–252, 1961.
35. Lewis, J. T., A. Lehoczky and C. V. Briscoe, *Phys. Rev.*, 161, 877–887, 1967.
36. Drabble, J. R. and R. E. B. Strathen, *Proc. Phys. Soc.*, 92, 1090–1995, 1967.
37. Oliver, D. W., *J. Appl. Phys.*, 40, 893, 1969.
38. Alper, T., and G. A. Saunders, *J. Phys. Chem. Solids*, 28, 1637–1642, 1967.
39. Dickinson, J. M. and P. E. Armstrong, *J. Appl. Phys.*, 38, 602–606, 1967.
40. Bolef, D. I., *J. Appl. Phys.*, 32, 100–105, 1961.
41. Rayne, J. A., *Phys. Rev.*, 112, 1125–1130, 1958.
42. MacFarlane, R. E., et al., *Phys. Letters*, 18, 91–92, 1965.
43. Smith, P. A. and C. S. Smith, *J. Phys. Chem. Solids*, 26, 279–289, 1965.
44. Norwood, M. H. and C. V. Briscoe, *Phys. Rev.*, 112, 45–48, 1958.
45. Simmons, G. and F. Birch, *J. Appl. Phys.*, 34, 2736–2738, 1963.
46. Gutman, E. J. and J. Trivisonno, *J. Phys. Chem. Sol.*, 28, 805–809, 1967.
47. Ghafelehbashi, M., et al., *J. Appl. Phys.*, 41, 652–666, 1970.
48. McSkimin, H. J. and P. Andreatch, Jr., *J. Appl. Phys.*, 35, 2161–2165, 1964.
49. Neighbours, J. R. and G. A. Alers, *Phys. Rev.*, 111, 707–712, 1958.
50. Hidshaw, W., J. T. Lewis, and C. V. Briscoe, *Phys. Rev.*, 163, 876–881, 1967.
51. Daniels, W. B., *Phys. Rev.*, 119, 1246–1252, 1960.
52. Viswanathan, R., *J. Appl. Phys.*, 37, 884–886, 1966.
53. Miller, R. A. and C. S. Smith, *J. Phys. Chem. Sol.*, 25, 1279–1292, 1964.
54. Claytor, R. N. and B. J. Marshall, *Phys. Rev.*, 120, 332–334, 1960.
55. Schreiber, E., *J. Appl. Phys.*, 38, 2508–2511, 1967.
56. Gerlich, D., *Phys. Rev.*, 136, A1366–A1368, 1964.
57. Johnston, D. L., P. H. Thrasher and R. J. Kearney, *J. Appl. Phys.*, 41, 427–428, 1970.
58. Poindexter, E. and A. A. Giardini, *Phys. Rev.*, 110, 1069, 1958.
59. Soga, N., *J. Appl. Phys.*, 37, 3416–3420, 1966.
60. Bartlett, R. W. and C. W. Smith, *J. Appl. Phys.*, 38, 5428–5429, 1967.
61. Morse, G. E. and A. W. Lawson, *J. Phys. Chem. Sol.*, 28, 939–950, 1967.
62. Armstrong, P. E., O. N. Carlson and J. F. Smith, *J. Appl. Phys.*, 30, 36–41, 1959.
63. Macedo, P. M., W. Capps and J. B. Wachtman, *J. Am. Cer. Soc.*, 47, 651, 1964.
64. Beattie, A. G., *J. Appl. Phys.*, 40, 4818–4821, 1969.
65. Chang, R. and L. J. Graham, *J. Appl Phys.*, 37, 3778–3783, 1966.
66. Lowrie, R. and A. M. Gonas, *J. Appl. Phys.*, 38, 4505–4509, 1967.
67. Graham, L. J., H. Nadler and R. Chang, *J. Appl. Phys.*, 34, 1572–1573, 1963.
68. Wachtman, J. B., Jr., et al., *J. Nucl. Mat.*, 16, 39–41, 1965.
69. Bolef, D. I., *J. Appl. Phys.*, 32, 100–105, 1961.
70. Berlincourt, D., H. Jaffe and L. R. Shiozawa, *Phys. Rev.*, 129, 1009–1017, 1963.
71. Adhav. R. S. J. *Acoust. Soc. Am.*, 43, 835–838, 1968.
72. Berlincourt, D. and H. Jaffe, *Phys. Rev.*, 111, 143–148, 1958.
73. Huntington, H. B., in *Solid State Pysics, Vol. 7*, Seitz, F., and Turnbull, D., Ed., pp. 213–285, Academic Press, New York 1958.
74. Cutler, H. R., J. J. Gibson and K. A. McCarthy, *Sol. State Comm.*, 6, 431–433, 1968.
75. Mason, W. P., *Piezoelectric Crystals and Their Application to Ultrasonics*, D. Van Nostrand Co., Inc., New York, 1950.
76. Adhav, R. S., *J. Appl. Phys.*, 40, 2725–2727, 1969.
77. Manghnani, M. H., *J. Geophys. Res.*, 74, 4317–4328, 1969.
78. Uchida, N. and Y. Ohmachi, *J. Appl Phys.*, 40, 4692–4695, 1969.
79. House, D. G. and E. Y. Vernon, *Br. J. Appl. Phys.*, 11, 254–259, 1960.
80. Ryzhova, T. V., et al., *Bull. Acad. Sci. USSR, Earth Phys. Ser.*, English Transl., no. 2, 111–113, 1966.
81. Alton, W. J. and A. J. Barlow, *J. Appl. Phys.*, 38, 3817–3820, 1967.
82. Michard, F., et al., *C. R. Acad. Sci., Paris*, 265, 565–567, 1967.
83. Haussühl, S., *Acta Cryst.*, 18, 839–842, 1965.
84. Hearmon, R. F. S., *Rev. Mod. Phys.*, 18, 409–440, 1946.
85. Heseltine, J. C. W., D. W. Elliott and O. B. Wilson, *J. Chem. Phys.*, 40, 2584–2587, 1964.
86. Schwerdtner, W. M., et al., *Canad. J. Earth Sci.*, 2, 673–683, 1965.
87. Kumazawa, M. and O. L. Anderson, *J. Geophys. Res.*, 74, 5961–5972, 1969 .
88. Alexandrov, K. S., et al., *Sov. Phys. Cryst.*, 7, 753–755, 1963.
89. Verma, R. K., *J. Geophys. Soc.*, 65, 757–766, 1960.
90. McSkimin, H. J. and E. S. Fisher, *J. Appl. Phys.*, 31, 1627–1639, 1960.
91. Alexandrov, K. S. and T.V. Ryzhova, *Bull. Acad. Sci. USSR, Geophys. Ser.*, English Transl., no. 8, 871–875, 1961.
92. Afanaseva, G. K., et al, *Phys. Stat. Sol.*, 24, K61–K63, 1967.
93. Alexandrov, K. S., et al., *Sov. Phys. Cryst.*, 8, 589–591, 1964.
94. Alexandrov, K. S. and T. V Ryzhova, *Bull Acad. Sci. USSR, Geophys. Ser.*, English Transl., no. 2, 129–131, 1962.
95. Alexandrov, K. S., et al., *Sov. Phys. Cryst.*, 8, 164–166, 1963.
96. Teslenko, V. F., et al., *Sov. Phys. Cryst.*, 10, 744–747, 1966.
97. Smith, J. F. and C. L. Arbogast, *J. Appl. Phys.*, 31, 99–102, 1960.
98. Cline, C. F., H. L. Dunegan and G. M. Henderson, *J. Appl. Phys.*, 38, 1944–1948, 1967.
99. Chang, Y. A. and L. Himmel, *J. Appl. Phys.*, 37, 3787–3790, 1966.
100. Gerlich, D., *J. Phys. Chem. Solids*, 28, 2575–2579, 1967.
101. McSkimin, H. J., *J. Appl. Phys.*, 26, 406–409, 1955.
102. Fisher, E. S. and D. Dever, *Trans. Met. Soc. AIME*, 239, 48–57, 1967.
103. Fisher, E. S. and D. Dever, *Proc. Conf. Rare Earth Res.*, 6th, Gatlinburg, TN, 522–533, 1967.
104. Fisher, E. S. and C. J. Renken, *Phys. Rev.*, 135, A482–A494, 1964.
105. Proctor, T. M., Jr., *J. Acoust. Soc. Am.*, 39, 972–977, 1966.
106. Chandrasekhar, B. S. and J. A. Rayne, *Phys. Rev.*, 124, 1011–1041, 1961.
107. Wazzan, A. R. and L. B. Robinson, *Phys. Rev.*, 155, 586–594, 1967.
108. Ferris, R. W., et al., *J. Appl. Phys.*, 34, 768–770, 1963.
109. Gilman, J. J. and B. W. Roberts, *J. Appl. Phys.*, 32, 1405, 1961.
110. Smith, J. F. and J. A. Gjevre, *J. Appl. Phys.*, 31, 645–647, 1960.
111. Alers, G. A. and J. R. Neighbours, *J. Phys. Chem. Solids*, 7, 58–64, 1908.
112. Bateman, T. B., *J. Appl. Phys.*, 33, 3309–3312, 1962.
113. Tefft, W. E., *J. Res. Natl. Bur. Stand.*, 70A, 277–280, 1966.
114. DeBretteville, Jr., A. et al., *Phys. Rev.*, 148, 575–579, 1966.
115. Dandekar, D. P. and A. L. Ruoff, *J. Appl. Phys.*, 39, 6004–6009, 1968.
116. Warner, A. W., M. Onoe and G. A. Coquin, *J. Acoust. Soc. Am.*, 42, 1223–1231, 1967.
117. McSkimin, H. J., P. Andreatch and R. N. Thurston, *J. Appl. Phys.*, 36, 1624–1632, 1965.
118. Mort, J., *J. Appl. Phys.*, 38, 3414–3415, 1967.

ELECTRICAL RESISTIVITY OF PURE METALS

The first part of this table gives the electrical resistivity, in units of $10^{-8}\ \Omega$ m, for 28 common metallic elements as a function of temperature. The data refer to polycrystalline samples. The number of significant figures indicates the accuracy of the values. However, at low temperatures (especially below 50 K) the electrical resistivity is extremely sensitive to sample purity. Thus the low-temperature values refer to samples of specified purity and treatment. The references should be consulted for further information on this point, as well as for values at additional temperatures.

The second part of the table gives resistivity values in the neighborhood of room temperature for other metallic elements that have not been studied over an extended temperature range.

References

1. C. Y. Ho, et al., *J. Phys. Chem. Ref. Data*, 12, 183–322, 1983; 13, 1069–1096, 1984; 13, 1097–1130, 1984, 13, 1131–1172, 1984.
2. R. A. Matula, *J. Phys Chem. Ref. Data*, 8, 1147–1298, 1979.
3. T. C. Chi, *J. Phys. Chem. Ref. Data*, 8, 339–438, 1979; 8, 439–498, 1979.
4. K. H. Hellwege, Ed., *Landolt-Börnstein Numerical Data and Functional Relationships in Science and Technology*, Group III, Vol. 15, Subvolume a, Springer-Verlag, Heidelberg, 1982.
5. L. A. Hall, *Survey of Electrical Resistivity Measurements on 16 Pure Metals in the Temperature Range 0 to 273 K*, NBS Technical Note 365, U.S. Superintendent of Documents, 1968.

Electrical Resistivity in $10^{-8}\ \Omega$ m

T/K	Aluminum	Barium	Beryllium	Calcium	Cesium	Chromium	Copper
1	0.000100	0.081	0.0332	0.045	0.0026		0.00200
10	0.000193	0.189	0.0332	0.047	0.243		0.00202
20	0.000755	0.94	0.0336	0.060	0.86		0.00280
40	0.0181	2.91	0.0367	0.175	1.99		0.0239
60	0.0959	4.86	0.067	0.40	3.07		0.0971
80	0.245	6.83	0.075	0.65	4.16		0.215
100	0.442	8.85	0.133	0.91	5.28	1.6	0.348
150	1.006	14.3	0.510	1.56	8.43	4.5	0.699
200	1.587	20.2	1.29	2.19	12.2	7.7	1.046
273	2.417	30.2	3.02	3.11	18.7	11.8	1.543
293	2.650	33.2	3.56	3.36	20.5	12.5	1.678
298	2.709	34.0	3.70	3.42	20.8	12.6	1.712
300	2.733	34.3	3.76	3.45	21.0	12.7	1.725
400	3.87	51.4	6.76	4.7		15.8	2.402
500	4.99	72.4	9.9	6.0		20.1	3.090
600	6.13	98.2	13.2	7.3		24.7	3.792
700	7.35	130	16.5	8.7		29.5	4.514
800	8.70	168	20.0	10.0		34.6	5.262
900	10.18	216	23.7	11.4		39.9	6.041

T/K	Gold	Hafnium	Iron	Lead	Lithium	Magnesium	Manganese
1	0.0220	1.00	0.0225		0.007	0.0062	7.02
10	0.0226	1.00	0.0238		0.008	0.0069	18.9
20	0.035	1.11	0.0287		0.012	0.0123	54
40	0.141	2.52	0.0758		0.074	0.074	116
60	0.308	4.53	0.271		0.345	0.261	131
80	0.481	6.75	0.693	4.9	1.00	0.557	132
100	0.650	9.12	1.28	6.4	1.73	0.91	132
150	1.061	15.0	3.15	9.9	3.72	1.84	136
200	1.462	21.0	5.20	13.6	5.71	2.75	139
273	2.051	30.4	8.57	19.2	8.53	4.05	143
293	2.214	33.1	9.61	20.8	9.28	4.39	144
298	2.255	33.7	9.87	21.1	9.47	4.48	144
300	2.271	34.0	9.98	21.3	9.55	4.51	144
400	3.107	48.1	16.1	29.6	13.4	6.19	147
500	3.97	63.1	23.7	38.3		7.86	149
600	4.87	78.5	32.9			9.52	151
700	5.82		44.0			11.2	152
800	6.81		57.1			12.8	
900	7.86					14.4	

T/K	Molybdenum	Nickel	Palladium	Platinum	Potassium	Rubidium	Silver
1	0.00070	0.0032	0.0200	0.002	0.0008	0.0131	0.00100
10	0.00089	0.0057	0.0242	0.0154	0.0160	0.109	0.00115
20	0.00261	0.0140	0.0563	0.0484	0.117	0.444	0.0042
40	0.0457	0.068	0.334	0.409	0.480	1.21	0.0539
60	0.206	0.242	0.938	1.107	0.90	1.94	0.162
80	0.482	0.545	1.75	1.922	1.34	2.65	0.289
100	0.858	0.96	2.62	2.755	1.79	3.36	0.418
150	1.99	2.21	4.80	4.76	2.99	5.27	0.726
200	3.13	3.67	6.88	6.77	4.26	7.49	1.029
273	4.85	6.16	9.78	9.6	6.49	11.5	1.467
293	5.34	6.93	10.54	10.5	7.20	12.8	1.587
298	5.47	7.12	10.73	10.7	7.39	13.1	1.617
300	5.52	7.20	10.80	10.8	7.47	13.3	1.629
400	8.02	11.8	14.48	14.6			2.241
500	10.6	17.7	17.94	18.3			2.87
600	13.1	25.5	21.2	21.9			3.53
700	15.8	32.1	24.2	25.4			4.21
800	18.4	35.5	27.1	28.7			4.91
900	21.2	38.6	29.4	32.0			5.64

T/K	Sodium	Strontium	Tantalum	Tungsten	Vanadium	Zinc	Zirconium
1	0.0009	0.80	0.10	0.000016		0.0100	0.250
10	0.0015	0.80	0.102	0.000137	0.0145	0.0112	0.253
20	0.016	0.92	0.146	0.00196	0.039	0.0387	0.357
40	0.172	1.70	0.751	0.0544	0.304	0.306	1.44
60	0.447	2.68	1.65	0.266	1.11	0.715	3.75
80	0.80	3.64	2.62	0.606	2.41	1.15	6.64
100	1.16	4.58	3.64	1.02	4.01	1.60	9.79
150	2.03	6.84	6.19	2.09	8.2	2.71	17.8
200	2.89	9.04	8.66	3.18	12.4	3.83	26.3
273	4.33	12.3	12.2	4.82	18.1	5.46	38.8
293	4.77	13.2	13.1	5.28	19.7	5.90	42.1
298	4.88	13.4	13.4	5.39	20.1	6.01	42.9
300	4.93	13.5	13.5	5.44	20.2	6.06	43.3
400		17.8	18.2	7.83	28.0	8.37	60.3
500		22.2	22.9	10.3	34.8	10.82	76.5
600		26.7	27.4	13.0	41.1	13.49	91.5
700		31.2	31.8	15.7	47.2		104.2
800		35.6	35.9	18.6	53.1		114.9
900			40.1	21.5	58.7		123.1

Element	T/K	Electrical resistivity $10^{-8}\ \Omega$ m	Element	T/K	Electrical resistivity $10^{-8}\ \Omega$ m
Antimony	273	39	Osmium	273	8.1
Bismuth	273	107	Polonium	273	40
Cadmium	273	6.8	Praseodymium	290–300	70.0
Cerium (β, hex)	290–300	82.8	Promethium	290–300	75 est.
Cerium (γ, cub)	298	74.4	Protactinium	273	17.7
Cobalt	273	5.6	Rhenium	273	17.2
Dysprosium	290–300	92.6	Rhodium	273	4.3
Erbium	290–300	86.0	Ruthenium	273	7.1
Europium	290–300	90.0	Samarium	290–300	94.0
Gadolinium	290–300	131	Scandium	290–300	56.2
Gallium	273	13.6	Terbium	290–300	115
Holmium	290–300	81.4	Thallium	273	15
Indium	273	8.0	Thorium	273	14.7
Iridium	273	4.7	Thulium	290–300	67.6
Lanthanum	290–300	61.5	Tin	273	11.5
Lutetium	290–300	58.2	Titanium	273	39
Mercury	298	96.1	Uranium	273	28
Neodymium	290–300	64.3	Ytterbium	290–300	25.0
Niobium	273	15.2	Yttrium	290–300	59.6

ELECTRICAL RESISTIVITY OF SELECTED ALLOYS

These values were obtained by fitting all available measurements to a theoretical formulation describing the temperature and composition dependence of the electrical resistivity of metals. Some of the values listed here fall in regions of temperature and composition where no actual measurements exist. Details of the procedure may be found in the reference.

Values of the resistivity are given in units of 10^{-8} Ω m. General comments in the preceding table for pure metals also apply here.

Reference

Ho, C. Y., et al., *J. Phys. Chem. Ref. Data*, 12, 183–322, 1983.

Electrical Resistivity in Units of 10^{-8} Ω m at the Indicated Temperature

Wt %	100 K	273 K	293 K	300 K	350 K	400 K
Aluminum-Copper						
Wt % Al						
99[a]	0.531	2.51	2.74	2.82	3.38	3.95
95[a]	0.895	2.88	3.10	3.18	3.75	4.33
90[b]	1.38	3.36	3.59	3.67	4.25	4.86
85[b]	1.88	3.87	4.10	4.19	4.79	5.42
80[b]	2.34	4.33	4.58	4.67	5.31	5.99
70[b]	3.02	5.03	5.31	5.41	6.16	6.94
60[b]	3.49	5.56	5.88	5.99	6.77	7.63
50[b]	4.00	6.22	6.55	6.67	7.55	8.52
40[c]		7.57	7.96	8.10	9.12	10.2
30[c]		11.2	11.8	12.0	13.5	15.2
25[f]		16.3	17.2	17.6	19.8	22.2
15[h]			12.3			
10[g]	8.71	10.8	11.0	11.1	11.7	12.3
5[c]	7.92	9.43	9.61	9.68	10.2	10.7
1[b]	3.22	4.46	4.60	4.65	5.00	5.37
Aluminum-Magnesium						
Wt % Al						
99[c]	0.958	2.96	3.18	3.26	3.82	4.39
95[c]	3.01	5.05	5.28	5.36	5.93	6.51
90[c]	5.42	7.52	7.76	7.85	8.43	9.02
10[b]	14.0	17.1	17.4	17.6	18.4	19.2
5[b]	9.93	13.1	13.4	13.5	14.3	15.2
1[a]	2.78	5.92	6.25	6.37	7.20	8.03
Copper-Gold						
Wt % Cu						
99[c]	0.520	1.73	1.86	1.91	2.24	2.58
95[c]	1.21	2.41	2.54	2.59	2.92	3.26
90[c]	2.11	3.29	3.42	3.46	3.79	4.12
85[c]	3.01	4.20	4.33	4.38	4.71	5.05
80[c]	3.95	5.15	5.28	5.32	5.65	5.99
70[c]	5.91	7.12	7.25	7.30	7.64	7.99
60[c]	8.04	9.18	9.13	9.36	9.70	10.05
50[c]	9.88	11.07	11.20	11.25	11.60	11.94
40[c]	11.44	12.70	12.85	12.90	13.27	13.65
30[c]	12.43	13.77	13.93	13.99	14.38	14.78
25[c]	12.59	13.93	14.09	14.14	14.54	14.94
15[c]	11.38	12.75	12.91	12.96	13.36	13.77
10[c]	9.33	10.70	10.86	10.91	11.31	11.72
5[c]	5.91	7.25	7.41	7.46	7.87	8.28
1[c]	2.00	3.40	3.57	3.62	4.03	4.45
Copper-Nickel						
Wt % Cu						
99[c]	1.45	2.71	2.85	2.91	3.27	3.62
95[c]	6.19	7.60	7.71	7.82	8.22	8.62
90[c]	12.08	13.69	13.89	13.96	14.40	14.81
85[c]	18.01	19.63	19.83	19.90	20.32	20.70
80[c]	23.89	25.46	25.66	25.72	26.12	26.44
70[i]	35.73	36.67	36.72	36.76	36.85	36.89
60[i]	45.76	45.43	45.38	43.35	45.20	45.01
50[i]	50.22	50.19	50.05	50.01	49.73	49.50
40[c]	36.77	47.42	47.73	47.82	48.28	48.49
30[i]	26.73	40.19	41.79	42.34	44.51	45.40
25[c]	22.22	33.46	35.11	35.69	39.67	42.81
15[c]	13.49	22.00	23.35	23.85	27.60	31.38
10[c]	9.28	16.65	17.82	18.26	21.51	25.19
5[c]	5.20	11.49	12.50	12.90	15.69	18.78
1[c]	1.81	7.23	8.08	8.37	10.63	13.18
Copper-Palladium						
Wt % Cu						
99[c]	0.91	2.10	2.23	2.27	2.59	2.92
95[c]	2.99	4.21	4.35	4.40	4.74	5.08
90[c]	5.69	6.89	7.03	7.08	7.41	7.74
85[c]	8.30	9.48	9.61	9.66	10.01	10.36
80[c]	10.74	11.99	12.12	12.16	12.51	12.87
70[c]	15.67	16.87	17.01	17.06	17.41	17.78
60[c]	20.45	21.73	21.87	21.92	22.30	22.69
50[c]	26.07	27.62	27.79	27.86	28.25	28.64
40[c]	33.53	35.31	35.51	35.57	36.03	36.47
30[c]	45.03	46.50	46.66	46.71	47.11	47.47
25[c]	44.12	46.25	46.45	46.52	46.99	47.43
15[c]	31.79	36.52	36.99	37.16	38.28	39.35
10[c]	23.00	28.90	29.51	29.73	31.19	32.56
5[c]	13.09	20.00	20.75	21.02	22.84	24.54
1[c]	8.97	11.90	12.67	12.93	14.82	16.68
Copper-Zinc						
Wt % Cu						
99[b]	0.671	1.84	1.97	2.02	2.36	2.71
95[b]	1.54	2.78	2.92	2.97	3.33	3.69
90[b]	2.33	3.66	3.81	3.86	4.25	4.63
85[b]	2.93	4.37	4.54	4.60	5.02	5.44
80[b]	3.44	5.01	5.19	5.26	5.71	6.17
70[b]	4.08	5.87	6.08	6.15	6.67	7.19
Gold-Palladium						
Wt % Au						
99[c]	1.31	2.69	2.86	2.91	3.32	3.73
95[c]	3.88	5.21	5.35	5.41	5.79	6.17
90[i]	6.70	8.01	8.17	8.22	8.56	8.93
85[b]	9.14	10.50	10.66	10.72	11.10	11.48
80[b]	11.23	12.75	12.93	12.99	13.45	13.93
70[c]	16.44	18.23	18.46	18.54	19.10	19.67
60[b]	24.64	26.70	26.94	27.02	27.63	28.23
50[a]	23.09	27.23	27.63	27.76	28.64	29.42

Wt %	100 K	273 K	293 K	300 K	350 K	400 K
40[a]	19.40	24.65	25.23	25.42	26.74	27.95
30[b]	14.94	20.82	21.49	21.72	23.35	24.92
25[b]	12.72	18.86	19.53	19.77	21.51	23.19
15[a]	8.54	15.08	15.77	16.01	17.80	19.61
10[a]	6.54	13.25	13.95	14.20	16.00	17.81
5[a]	4.58	11.49	12.21	12.46	14.26	16.07
1[a]	3.01	10.07	10.85	11.12	12.99	14.80

Gold-Silver
Wt % Au

Wt % Au	100 K	273 K	293 K	300 K	350 K	400 K
99[b]	1.20	2.58	2.75	2.80	3.22	3.63
95[a]	3.16	4.58	4.74	4.79	5.19	5.59
90[j]	5.16	6.57	6.73	6.78	7.19	7.58
85[j]	6.75	8.14	8.30	8.36	8.75	9.15
80[j]	7.96	9.34	9.50	9.55	9.94	10.33
70[j]	9.36	10.70	10.86	10.91	11.29	11.68
60[j]	9.61	10.92	11.07	11.12	11.50	11.87
50[j]	8.96	10.23	10.37	10.42	10.78	11.14
40[j]	7.69	8.92	9.06	9.11	9.46	9.81
30[a]	6.15	7.34	7.47	7.52	7.85	8.19
25[a]	5.29	6.46	6.59	6.63	6.96	7.30
15[a]	3.42	4.55	4.67	4.72	5.03	5.34
10[a]	2.44	3.54	3.66	3.71	4.00	4.31
5[i]	1.44	2.52	2.64	2.68	2.96	3.25
1[b]	0.627	1.69	1.80	1.84	2.12	2.42

Iron-Nickel
Wt % Fe

Wt % Fe	100 K	273 K	293 K	300 K	350 K	400 K
99[a]	3.32	10.9	12.0	12.4		18.7
95[c]	10.0	18.7	19.9	20.2		26.8
90[c]	14.5	24.2	25.5	25.9		33.2
85[c]	17.5	27.8	29.2	29.7		37.3
80[c]	19.3	30.1	31.6	32.2		40.0
70[b]	20.9	32.3	33.9	34.4		42.4
60[c]	28.6	53.8	57.1	58.2		73.9
50[d]	12.3	28.4	30.6	31.4		43.7
40[d]	7.73	19.6	21.6	22.5		34.0

Wt %	100 K	273 K	293 K	300 K	350 K	400 K
30[c]	5.97	15.3	17.1	17.7		27.4
25[b]	5.62	14.3	15.9	16.4		25.1
15[c]	4.97	12.6	13.8	14.2		21.1
10[c]	4.20	11.4	12.5	12.9		18.9
5[c]	3.34	9.66	10.6	10.9		16.1
1[b]	1.66	7.17	7.94	8.12		12.8

Silver-Palladium
Wt % Ag

Wt % Ag	100 K	273 K	293 K	300 K	350 K	400 K
99[b]	0.839	1.891	2.007	2.049	2.35	2.66
95[b]	2.528	3.58	3.70	3.74	4.04	4.34
90[b]	4.72	5.82	5.94	5.98	6.28	6.59
85[k]	6.82	7.92	8.04	8.08	8.38	8.68
80[k]	8.91	10.01	10.13	10.17	10.47	10.78
70[k]	13.43	14.53	14.65	14.69	14.99	15.30
60[i]	19.4	20.9	21.1	21.2	21.6	22.0
50[k]	29.3	31.2	31.4	31.5	32.0	32.4
40[m]	40.8	42.2	42.2	42.2	42.3	42.3
30[b]	37.1	40.4	40.6	40.7	41.3	41.7
25[k]	32.4	36.67	37.06	37.19	38.1	38.8
15[i]	21.0	27.08	26.68	27.89	29.3	30.6
10[i]	14.95	21.69	22.39	22.63	24.3	25.9
5[b]	8.91	15.98	16.72	16.98	18.8	20.5
1[a]	3.97	11.06	11.82	12.08	13.92	15.70

[a] Uncertainty in resistivity is ± 2%.
[b] Uncertainty in resistivity is ± 3%.
[c] Uncertainty in resistivity is ± 5%.
[d] Uncertainty in resistivity is ± 7% below 300 K and ± 5% at 300 and 400 K.
[e] Uncertainty in resistivity is ± 7%.
[f] Uncertainty in resistivity is ± 8%.
[g] Uncertainty in resistivity is ± 10%.
[h] Uncertainty in resistivity is ± 12%.
[i] Uncertainty in resistivity is ± 4%.
[j] Uncertainty in resistivity is ± 1%.
[k] Uncertainty in resistivity is ± 3% up to 300 K and ± 4% above 300.
[m] Uncertainty in resistivity is ± 2% up to 300 K and ± 4% above 300.

Solids

ELECTRICAL RESISTIVITY OF GRAPHITE MATERIALS

L. I. Berger

At normal conditions, the only stable crystallographic modification of carbon is graphite. The quasi-stable diamond turns into graphite starting from about 1000 °C in air. In industry, a graphitic material is commonly called either *carbon*, if it consists of small and low-oriented crystallites, or *graphite*, the material with highly ordered structure. In the 1970s, the first carbon filaments of about 7 nm in diameter were grown by Morinobu Endo at the University of Orleans, France, by the vapor-growth technique. In 1985, Sir Harold Walter Kroto of Sussex University, UK, and Richard E. Smalley and co-workers at Rice University discovered spherical carbon molecules, C_{60} (or C60), consisting of combinations of carbon atoms organized into hexagons and pentagons, named *buckminsterfullerenes* or *fullerenes* and possessing very promising mechanical and electrical properties. In 1991, Sumio Iijima, NEC Labs, Japan, and David S. Bethune, IBM Almaden Labs, observed the carbon atomic groups in the form of tubes capped by halves of the fullerene molecules and formed on the cathodes of carbon arc devices. The length of the tubes could be up to tens of micrometers and the diameter, naturally, is equal to that of the fullerene molecule. These tubes, called *nanotubes*, may be single wall (SWNT) or consist of several concentric tubes with a common axis (multi-walled nanotubes, MWNT). Two-dimensional *graphene* is another crystallographic modification of graphite (Saroj Nayak, Rensselaer U., 2004) that is a flat hexagonal network of carbon atoms with a thickness equal to the carbon atom size. The nanotube may be considered as formed by strips of graphenes turned into a cylinder. The character of the electrical conductivity (metallic or semiconductive) of a SWNT depends on orientation of the carbon hexagons of the nanotube surface regarding its axis (the chiral angle [Ref. 1]). The following table contains some typical data on electrical and electronic properties of graphite materials.

References

1. M. S. Dresselhaus, G. Dresselhaus, and Ph. Avouris (Eds.), *Carbon Nanotubes. Synthesis, Structure, Properties, and Applications*, Springer-Verlag, 2001.
2. ESPI Metals Catalog, 2007.
3. SPI Supplies Catalog, 2007.
4. F. L. Vogel, *J. Mater. Sci.* 12, 982–986, 1977.
5. K. S. Novoselov et al., *Nature* 438, 197–200, 2005.
6. Y. Zhang et al., *Nature* 438, 201–204, 2005.
7. N. Tombros et al., *Nature* 448, 571–574, 2007.
8. H. Dai, in Ref. 1, pp. 29–53.
9. CTI Carbon Nanotube Cat., 2007.
10. L. Matija et al., *Sci. Forum* 413, 49–52, 2003.

Material	Electrical resistivity ρ at R. T. mΩ cm [μΩ inch]	Energy gap at R. T. eV	Electron mobility $cm^2/V\ s$	(1/ρ)dρ/d*t* near R. T. 10^{-4}°C^{-1}	Ref.
Bulk graphite					
Electromet graphite	1.90 [750]			–5	2
Electro graphite	1.60 [630]			–5	2
Aeromet graphite	1.47 [580]			–5	2
ESPI Superconductive	1.75 [690]			–5	2
Radioelectronics data	30 [11,800]			–5.6	3
Highly ordered pyrolytic graphite	Parallel 0.04 [15.7]				3
	Across 150 [59000]				
Single crystal graphite, normal to *c*-axis	$1 \cdot 10^{-6}$				4
Graphenes					
n-Graphene		≈5 (M); ≈10 (Γ)[c]	10^6		5,6
p-Graphene			10^4		7
Carbon nanotubes					
Metallic SWNT	12 kΩ[a]				1
Semiconducting SWNT		0.7 – 0.9[b]	128[d]		1
MWNT	10^2				9
Carbon fullerenes					
Fullerene (C_{60})	10^{12}	1.95			10

[a] Minimum resistance of individual nanotubes [Ref. 8]
[b] Est. from Ref. 1, p. 47
[c] Est. from Ref. 1, p. 116
[d] Est. from Ref. 1, p. 179.

PERMITTIVITY (DIELECTRIC CONSTANT) OF INORGANIC SOLIDS

H. P. R. Frederikse

This table lists the permittivity ε, frequently called the dielectric constant, of a number of inorganic solids. When the material is not isotropic, the individual components of the permittivity are given. A superscript S indicates a measurement made under constant strain ("clamped" dielectric constant). If the constraint is removed, the measurement yields ε^T, the "unclamped" or free dielectric constant.

The temperature of the measurement is given when available; the symbol r.t. indicates a value at nominal room temperature. The frequency of the measurement is given in the last column (i.r. indicates a measurement in the infrared).

Substances are listed in alphabetical order by chemical formula.

Reference

Young, K. F., and Frederikse, H. P. R., *J. Phys. Chem. Ref. Data* 2, 313, 1973.

Permittivity of Inorganic Solids

Formula	Name	ε_{ijk}	*T*/K	ν/Hz
Ag_3AsS_3	Silver thioarsenate (Proustite)	$\varepsilon^T_{11} = 16.5, \varepsilon^S_{11} = 14.5$	r.t.	2×10^7
		$\varepsilon^T_{33} = 20.0, \varepsilon^S_{11} = 18.0$	r.t.	2×10^7
AgBr	Silver bromide	12.50	r.t.	
AgCN	Silver cyanide	5.6	r.t.	10^6
AgCl	Silver chloride	11.15	r.t.	
$AgNO_3$	Silver nitrate	9.0	293	5×10^5
$AgNa(NO_2)_2$	Silver sodium nitrite	4.5 ± 0.5	r.t.	9.4×10^9
Ag_2O	Silver oxide	8.8	r.t.	
$(AlF)_2SiO_4$	Aluminum fluorosilicate (Topaz)	$\varepsilon_{11} = 6.62$	297	7×10^3
		$\varepsilon_{22} = 6.58$	297	7×10^3
		$\varepsilon_{33} = 6.95$	297	7×10^3
Al_2O_3	Aluminum oxide (Alumina)	$\varepsilon_{11} = \varepsilon_{22} = 9.34$	298	10^2–8×10^9
		$\varepsilon_{33} = 11.54$	298	10^2–8×10^9
$AlPO_4$	Aluminum phosphate	$\varepsilon^T_{11} = 6.05$	r.t.	10^3
AlSb	Aluminum antimonide	11.21	300	i.r.
AsF_3	Arsenic trifluoride	5.7	r.t.	
BN	Boron nitride	7.1	r.t.	i.r.
$BaCO_3$	Barium carbonate	8.53	291	2×10^5
$Ba(COOH)_2$	Barium formate	$\varepsilon_{11} = 7.9$	r.t.	10^3
		$\varepsilon_{22} = 5.9$	r.t.	10^3
		$\varepsilon_{33} = 7.5$	r.t.	10^3
$BaCl_2$	Barium chloride	9.81	r.t.	
$BaCl_2 \cdot 2H_2O$	Barium chloride dihydrate	9.00	r.t.	10^3
BaF_2	Barium fluoride	7.32	292	5×10^2–10^{11}
$Ba(NO_3)_2$	Barium nitrate	4.95	292	2×10^5
$Ba_2NaNb_5O_{15}$	Barium sodium niobate ("Bananas")	$\varepsilon^S_{11} = 222, \varepsilon^T_{11} = 235$	296	10^4
		$\varepsilon^S_{22} = 227, \varepsilon^T_{22} = 247$	296	
		$\varepsilon^S_{33} = 32, \varepsilon^T_{33} = 51$	296	
BaO	Barium oxide (Baria)	34 ± 1	248, 333	60×10^7
BaO_2	Barium peroxide	10.7	r.t.	2×10^6
BaS	Barium sulfide	19.23	r.t.	7.25×10^6
$BaSO_4$	Barium sulfate	11.4	288	10^8
$BaSnO_3$	Barium stannate	18	298	25×10^5
$BaTiO_3$	Barium titanate	$\varepsilon^T_{11} = 3600$	298	10^5
		$\varepsilon^S_{11} = 2300$	298	2.5×10^8
		$\varepsilon^T_{33} = 150$	298	10^5
		$\varepsilon^S_{33} = 80$	298	2.5×10^8
$Ba_6Ti_2Nb_8O_{30}$	Barium titanium niobate	$\varepsilon_{11} = \varepsilon_{22} \approx 190$	298	
		$\varepsilon_{33} \approx 220$	298	
$BaWO_4$	Barium tungstate	$\varepsilon_{11} = \varepsilon_{22} = 35.5 \pm 0.2$	297.5	1.6×10^3
		$\varepsilon_{33} = 37.2 \pm 0.2$	297.5	1.6×10^3
$BaZrO_3$	Barium zirconate	43	r.t.	

Formula	Name	ε_{ijk}	*T*/K	*ν*/Hz
$Be_3Al_2Si_6O_{18}$	Beryllium aluminum silicate (Beryl)	$\varepsilon_{33} = 5.95$	297	7×10^3
		$\varepsilon_{11} = \varepsilon_{22} = 6.86$	297	7×10^3
$BeCO_3$	Beryllium carbonate	9.7	291	2×10^5
BeO	Beryllium oxide (beryllia)	7.35 ± 0.2	293	2×10^6
$BiFeO_3$	Bismuth iron oxide	40 ± 3	300	9.4×10^9
$Bi_{12}GeO_{20}$	Bismuth germanite	$\varepsilon^S_{11} = 38$	r.t.	
$Bi(GeO_4)_3$	Bismuth germanate	16	293	
Bi_2O_3	Bismuth sesquioxide	18.2	r.t.	2×10^6
$Bi_4Ti_3O_{12}$	Bismuth titanate	112	r.t.	10^3
C	Diamond, Type I	5.87 ± 0.19	300	10^3
	Diamond, Type IIa	5.66 ± 0.04	300	10^3
$C_4H_4O_6$	Tartaric acid	$\varepsilon_{11} = \varepsilon_{22} = 4.3$	298	
		$\varepsilon_{33} = 4.5$	298	
		$\varepsilon_{13} = 0.55$	298	
$C_6H_{12}N_4$	Hexamethylene tetramine (HMTA)	2.6 ± 0.2	r.t.	10^9–10^{10}
$C_6H_{14}N_2O_6$	Ethylene diamine tartrate (EDT)	$\varepsilon^T_{11} = 5.0$	293	
		$\varepsilon^T_{22} = 8.3$	293	
		$\varepsilon^T_{33} = 6.0$	293	
		$\varepsilon^T_{13} = 0.7$	293	
$C_6H_{12}O_6NaBr$	Dextrose sodium bromide	$\varepsilon^T_{11} = 4.0$	r.t.	10^3
$(CH_3NH_3)Al(SO_4)_2 \cdot 12H_2O$	Methyl ammonium alum (MASD)	19	197	
$Ca_2B_6O_{11} \cdot 5H_2O$	Calcium hexaborate pentahydrate	$\varepsilon_{11} = 20$	293	10^3
		$\varepsilon_{33} = 25$	293	10^3
$CaCO_3$	Calcium carbonate	$\varepsilon_{11} = 8.67$	r.t.	9.4×10^{10}
		$\varepsilon_{22} = 8.69$	r.t.	9.4×10^{10}
		$\varepsilon_{33} = 8.31$	r.t.	9.4×10^{10}
$CaCeO_3$	Calcium cerate	21	r.t.	
CaF_2	Calcium fluoride	6.81	300	5×10^2–10^{11}
$CaMoO_4$	Calcium molybdate	$\varepsilon_{11} = \varepsilon_{22} = 24.0 \pm 0.2$	297.5	<10
		$\varepsilon_{33} = 20.0 \pm 0.2$	297.5	<10
$Ca(NO_3)_2$	Calcium nitrate	6.54	292	2×10^5
$CaNb_2O_6$	Calcium niobate	$\varepsilon_{11} = 22.8 \pm 1.9$	r.t.	$(5–500) \times 10^3$
$Ca_2Nb_2O_7$	Calcium pyroniobate	~45	r.t.	5×10^7
CaO	Calcium oxide	11.8 ± 0.3	283	2×10^6
CaS	Calcium sulfide	6.699	r.t.	7.25×10^6
$CaSO_4 \cdot 2H_2O$	Calcium sulfate dihydrate	$\varepsilon_{11} = 5.10$	r.t.	
		$\varepsilon_{22} = 5.24$	r.t.	
		$\varepsilon_{33} = 10.30$	r.t.	
$CaTiO_3$	Calcium titanate	165	r.t.	
$CaWO_4$	Calcium tungstate	$\varepsilon_{11} = \varepsilon_{22} = 11.7 \pm 0.1$	297.5	1.59×10^3
		$\varepsilon_{33} = 9.5 \pm 0.2$	297.5	1.59×10^3
Cd_3As_2	Cadmium arsenide	$\varepsilon_{33} = 18.5$	4	
$CdBr_2$	Cadmium bromide	8.6	293	5×10^5
CdF_2	Cadmium fluoride	8.33 ± 0.08	300	10^5–10^7
CdS	Cadmium sulfide	$\varepsilon_{11} = \varepsilon_{22} = 8.7$	300	i.r.
		$\varepsilon_{33} = 9.25$	300	i.r.
		$\varepsilon_{11} = \varepsilon_{22} = 8.37$	8	i.r.
		$\varepsilon_{33} = 9.00$	8	i.r.
		$\varepsilon^T_{11} = 8.48$	77	10^4
		$\varepsilon^T_{33} = 9.48$	77	10^4
		$\varepsilon^S_{11} = 9.02, \varepsilon^T_{11} = 9.35$	298	10^4
		$\varepsilon^S_{33} = 9.53, \varepsilon^T_{33} = 10.33$	298	10^4
CdSe	Cadmium selenide	$\varepsilon^S_{11} = 9.53, \varepsilon^T_{11} = 9.70$	298	10^4
		$\varepsilon^S_{33} = 10.2, \varepsilon^T_{33} = 10.65$	298	10^4
CdTe	Cadmium telluride	$\varepsilon_{11} = \varepsilon_{22} = 10.60 \pm 0.15$	297	i.r.
		$\varepsilon_{33} = 7.05 \pm 0.05$	297	i.r.
$Cd_2Nb_2O_7$	Cadmium pyroniobate	500-580	293	10^3
CeO_2	Cerium oxide	7.0	r.t.	2×10^6
$CoNb_2O_6$	Cobalt niobate	$\varepsilon_{11} = 18.4 \pm 1.1$	r.t.	$(5–500) \times 10^3$
		$\varepsilon_{22} = 21.4 \pm 1.1$	r.t.	$(5–500) \times 10^3$

Formula	Name	ε_{ijk}	T/K	ν/Hz
		$\varepsilon_{33} = 33.0 \pm 0.7$	r.t.	$(5–500) \times 10^3$
CoO	Cobalt oxide	12.9	298	$10^2–10^{10}$
Cr_2O_3	Chromic sesquioxide	$\varepsilon_{11} = \varepsilon_{22} = 13.3$	298.5	10^3
		$\varepsilon_{33} = 11.9$	298.5	10^3
		8	315 (T_N)	6×10^{10}
$CsAl(SO_4)_2 \cdot 12H_2O$	Cesium alum	5.0	r.t.	$20–20 \times 10^3$
CsBr	Cesium bromide	6.38	298	1.6×10^3
Cs_2CO_3	Cesium carbonate	6.53	291	2×10^5
CsCl	Cesium chloride	7.2	298	
CsH_2AsO_4	Cesium dihydrogen arsenate (CDA)	4.8	273	9.5×10^9
CsH_2PO_4	Cesium dihydrogen phosphate (CDP)	6.15	285	9.5×10^9
$CsH_3(SeO_3)_2$	Cesium trihydrogen selenite	$\varepsilon_{11} = 80$	273	10^5
		$\varepsilon_{22} = 63$	273	10^5
		$\varepsilon_{33} = 12$	273	10^5
CsI	Cesium iodide	6.31	298	1.6×10^3
$CsNO_3$	Cesium nitrate	$\varepsilon_{11} = \varepsilon_{22} = 9.4$	r.t.	5×10^5
		$\varepsilon_{33} = 8.3$	r.t.	5×10^5
$CsPbCl_3$	Cesium lead chloride	14.37	300	$10^5–10^6$
CuBr	Cuprous bromide	8.0	293	5×10^5
CuCl	Cuprous chloride	9.8 ± 0.5	r.t.	10^3
CuO	Cupric oxide	18.1	r.t.	2×10^6
Cu_2O	Cuprous oxide (Cuprite)	7.60 ± 0.06	r.t.	10^5
$CuSO_4 \cdot 5H_2O$	Cupric sulfate pentahydrate	6.60	r.t.	
EuF_2	Europium fluoride	7.7 ± 0.2	298	$(1–300) \times 10^3$
$Eu_2(MoO_4)_3$	Europium molybdate	9.5	298	
EuS	Europium sulfide	13.10 ± 0.04	80	$5 \times 10^2–10^5$
FeO	Ferrous oxide	14.2	r.t.	2×10^6
Fe_2O_3	Ferric oxide	4.5	r.t.	$10^5–10^7$
Fe_2O_3-α	Ferric oxide (Hematite)	12		6×10^{10}
Fe_3O_4	Ferrosoferric oxide (Magnetite)	20	r.t.	$10^5–10^7$
GaAs	Gallium arsenide	13.13	300	
		12.90	4	i.r.
GaP	Gallium phosphide	11.1	r.t.	
		10.75 ± 0.1	1.6	i.r.
GaSb	Gallium antimonide	15.69	r.t.	
		15.7	4	i.r.
$Gd_2(MoO_4)_3$	Gadolinium molybdate	$\varepsilon^T = 10$	298	
		$\varepsilon^S = 9.5$	298	10^3
Ge	Germanium	16.0 ± 0.3	4	9.2×10^9
		15.8 ± 0.2	r.t.	$500–3 \times 10^{10}$
GeO_2	Germanium dioxide	$\varepsilon_{11} = \varepsilon_{22} = 7.44$	r.t.	i.r.
HIO_3	Iodic acid	$\varepsilon_{11} = 7.5$	r.t.	10^3
		$\varepsilon_{22} = 12.4$	r.t.	10^3
		$\varepsilon_{33} = 8.1$	r.t.	10^3
H_2O	Ice I (P = 0 kbar)	99	243	
	Ice III (P = 3 kbar)	117	243	
	Ice V (P = 5 kbar)	114	243	
	Ice VI (P = 8 kbar)	193	243	
HgCl	Mercury(I) chloride (Calomel)	$\varepsilon_{11} = \varepsilon_{22} = 14.0$	r.t.	10^{12}
$HgCl_2$	Mercuric chloride	6.5	r.t.	10^{12}
HgS	Mercurous sulfide (Cinnabar)	$\varepsilon_{11} = \varepsilon_{22} = 18.0$	r.t.	i.r.
		$\varepsilon_{33} = 32.5$	r.t.	i.r.
HgSe	Mercurous selenide	25.6	r.t.	$10^4–10^6$
I_2	Iodine	$\varepsilon_{11} = 6$	r.t.	$5 \times 10^4–10^7$
		$\varepsilon_{22} = 3$	r.t.	$5 \times 10^4–10^7$
		$\varepsilon_{33} = 40$	r.t.	$5 \times 10^4–10^7$
InAs	Indium arsenide	14.55 ± 0.3	r.t.	i.r.
		15.15	4	i.r.
InP	Indium phosphide	12.61	r.t.	i.r.
InSb	Indium antimonide	17.88	4	i.r.
$KAl(SO_4)_2 \cdot 12H_2O$	Potassium alum	6.5	r.t.	$20–20 \times 10^3$

Solids

Formula	Name	ε_{ijk}	*T*/K	*ν*/Hz
KBr	Potassium bromide	4.88	300	
		4.53	4.2	
$KBrO_3$	Potassium bromate	7.3	r.t.	2×10^6
KCN	Potassium cyanide	6.15	r.t.	2×10^6
K_2CO_3	Potassium carbonate	4.96	291	2×10^5
$K_2C_4H_4O_6 \cdot ½H_2O$	Dipotassium tartrate (DKT)	$\varepsilon_{11} = 6.44$	r.t.	
		$\varepsilon_{22} = 5.80$	r.t.	
		$\varepsilon_{33} = 6.49$	r.t.	
		$\varepsilon_{13} = 0.005$	r.t.	
KCl	Potassium chloride	4.86 ± 0.02	r.t.	5×10^3
		4.50	4.2	
$KClO_3$	Potassium chlorate	5.1	r.t.	2×10^6
$KClO_4$	Potassium perchlorate	5.9	r.t.	2×10^6
K_2CrO_4	Potassium chromate	7.3	r.t.	6×10^7
$KCr(SO_4)_2 \cdot 12H_2O$	Potassium chrome alum	6.5	100–240	175×10^3
KD_2AsO_4	Potassium dideuterium arsenate (KDDA)	$\varepsilon_{11} = 70$	298	
		$\varepsilon_{33} = 31$	298	
KD_2PO_4	Potassium dideuterium phosphate (KDDP)	50 ± 2	297	10^3
KF	Potassium fluoride	6.05		2×10^6
KH_2AsO_4	Potassium dihydrogen arsenate (KDA)	$\varepsilon_{11} = 60$	298	
		$\varepsilon_{33} = 24$	298	
KH_2PO_4	Potassium dihydrogen phosphate (KDP)	46	298	10^3
		$\varepsilon_{11} = 42$	r.t.	
		$\varepsilon_{33} = 21$	r.t.	
K_2HPO_4	Dipotassium monohydrogen orthophosphate	9.05	r.t.	2×10^6
KI	Potassium iodide	5.00	r.t.	9.4×10^{10}
KIO_3	Potassium iodate	170	255	10^5
		10	293	10^5
		$\varepsilon_{[101]} \approx 40{,}70$	r.t.	10^5
		16.85	r.t.	2×10^6
$(K,H)Al_3(SiO_4)_3$	Mica (muscovite)	5.4	299	10^2–3×10^9
$(K,H)Mg_3Al(SiO_4)_3$	Mica (Canadian)	$\varepsilon_{11} = \varepsilon_{22} = 6.9$	298	10^2–10^4
		$\varepsilon_{33} = 7.3$	298	10^4
KNO_2	Potassium nitrite	25	305	
KNO_3	Potassium nitrate	4.37	293	2×10^5
$KNa(C_4H_4O_6) \cdot 4H_2O$	Sodium potassium tartrate tetrahydrate (Rochelle salt)	$\varepsilon_{11} = 170$	273	10^3
		$\varepsilon_{22} = 9.1$	273	10^3
$KNa(C_4H_2D_2O_6) \cdot 4D_2O$	Sodium potassium tartrate tetradeutrate (double deuterated Rochelle salt)	$\varepsilon_{11} = 70$	273	10^3
		$\varepsilon_{22} = 8.9$	273	10^3
$KNbO_3$	Potassium niobate	700	r.t.	
K_3PO_4	Potassium orthophosphate	7.75	r.t.	2×10^6
KSCN	Potassium thiocyanate	7.9	r.t.	2×10^6
K_2SO_4	Potassium sulfate	6.4	r.t.	2×10^6
$K_2S_3O_6$	Potassium trithionate	5.7	293	1.8×10^6
$K_2S_4O_6$	Potassium tetrathionate	5.5	293	1.8×10^6
$K_2S_5O_6 \cdot H_2O$	Potassium pentathionate	7.8	293	1.8×10^6
$K_2S_6O_6$	Potassium hexathionate	7.8	293	1.8×10^6
K_2SeO_4	Potassium selenate	$\varepsilon_{11} = 5.9$	r.t.	10^3
		$\varepsilon_{22} = 7.7$	r.t.	10^3
$KSr_2Nb_5O_{15}$	Potassium strontium niobate	$\varepsilon_{11} = \varepsilon_{11} \approx 1200$	298	
		$\varepsilon_{33} \approx 800$	298	
$KTaNbO_3$	Potassium tantalate niobate (KTN)	34,000	273	10^4
		6,000	293	10^4
$KTaO_3$	Potassium tantalate	242	298	2×10^5
$LaScO_3$	Lanthanum scandate	30	r.t.	
LiBr	Lithium bromide	12.1	r.t.	2×10^6
Li_2CO_3	Lithium carbonate	4.9	291	2×10^5
LiCl	Lithium chloride	11.05	r.t.	2×10^6
LiD	Lithium deuteride	14.0 ± 0.5	r.t.	i.r.
LiF	Lithium fluoride	9.00	298	10^2–10^7

Formula	Name	ε_{ijk}	*T*/K	*ν*/Hz
		9.11	353	10^2–10^7
$LiGaO_2$	Lithium metagallate	$\varepsilon_{11}^T = 7.0, \varepsilon_{22}^T = 6.0$	r.t.	
		$\varepsilon_{33}^T = 9.5$	r.t.	
		$\varepsilon_{11}^S = 6.8, \varepsilon_{22}^S = 5.8$	r.t.	
Li^6H	Lithium-6 hydride	13.2 ± 0.5	r.t.	
Li^7H	Lithium-7 hydride	12.9 ± 0.5	r.t.	
$LiH_3(SeO_3)_2$	Lithium trihydrogen selenite	29	298	10^4
		$\varepsilon_{11} = 13.0$	r.t.	
		$\varepsilon_{22} = 12.9$	r.t.	
		$\varepsilon_{33} = 46$	r.t.	
LiI	Lithium iodide	11.03	r.t.	2×10^6
$LiIO_3$	Lithium iodate	$\varepsilon_{11} = \varepsilon_{22} = 65$	294.5	10^3
		$\varepsilon_{33} = 554$	298	
$LiNH_4C_4O_6 \cdot H_2O$	Lithium ammonium tartrate (LAT)	$\varepsilon_{11}^T = 7.2$	298	
		$\varepsilon_{22}^T = 8.0$	298	
		$\varepsilon_{33}^T = 6.9$	298	
$LiNa_3CrO_4 \cdot 6H_2O$	Lithium trisodium chromate	8.0	r.t.	10^3
$LiNa_3MoO_4 \cdot 6H_2O$	Lithium trisodium molybdate	$\varepsilon_{11} = 6.7$	r.t.	10^3
		$\varepsilon_{33} = 5.3$	r.t.	10^3
$LiNbO_3$	Lithium niobate	$\varepsilon_{11} = \varepsilon_{22} = 82$	298	10^5
		$\varepsilon_{33} = 30$	298	10^5
$Li_2SO_4 \cdot H_2O$	Lithium sulfate monohydrate	$\varepsilon_{11} = 5.6$	298	
		$\varepsilon_{22} = 10.3$	298	
		$\varepsilon_{33} = 6.5$	298	
		$\varepsilon_{13} = 0.07$	298	
$LiTaO_3$	Lithium tantalate	$\varepsilon_{11} = \varepsilon_{22} = 53$	r.t.	10^5
		$\varepsilon_{33} = 46$	r.t.	10^5
		$\varepsilon_{11}^S = \varepsilon_{22}^S = 41$	r.t.	
		$\varepsilon_{33}^S = 43$	r.t.	
		$\varepsilon_{11}^T = \varepsilon_{22}^T = 51$	r.t.	
		$\varepsilon_{33}^T = 45$	r.t.	
$LiTlC_4H_4O_6 \cdot H_2O$	Lithium thallium tartrate (LTT)	$\varepsilon_{11} \approx 20$	80	
$Mg_3B_7O_{13}Cl$	Magnesium borate monochloride (boracite)	$\varepsilon_{11} = 14.1$	r.t.	5×10^5
$MgCO_3$	Magnesium carbonate	8.1	291	2×10^5
$MgNb_2O_6$	Magnesium niobate	$\varepsilon_{11} = 16.4 \pm 0.5$	r.t.	$(5–500) \times 10^3$
		$\varepsilon_{22} = 20.9 \pm 0.5$	r.t.	$(5–500) \times 10^3$
		$\varepsilon_{33} = 32.4 \pm 0.5$	r.t.	$(5–500) \times 10^3$
MgO	Magnesium oxide (Periclase)	9.65	298	10^2–10^8
$(MgO)_xAl_2O_3$	Spinel	8.6	r.t.	
$MgSO_4$	Magnesium sulfate	8.2	r.t.	
$MgSO_4 \cdot 7H_2O$	Magnesium sulfate heptahydrate	5.46	r.t.	
$MgTiO_3$	Magnesium titanate	13.5	r.t.	
$MgWO_4$	Magnesium tungstate	$\varepsilon_{11} = 18.0 \pm 1$	r.t.	$(5–500) \times 10^3$
		$\varepsilon_{22} = 18.0 \pm 1$	r.t.	$(5–500) \times 10^3$
$MnNb_2O_6$	Manganese niobate	$\varepsilon_{11} = 17.4 \pm 2$	r.t.	$(5–500) \times 10^3$
		$\varepsilon_{22} = 16.1 \pm 0.5$	r.t.	$(5–500) \times 10^3$
		$\varepsilon_{33} = 30.7 \pm 1$	r.t.	$(5–500) \times 10^3$
MnO	Manganese oxide (Pyrolusite)	12.8	r.t.	6×10^{10}
MnO_2	Manganese dioxide	$\sim 10^4$	298	10^4
Mn_2O_3	Manganese sesquioxide	8	r.t.	6×10^{10}
$MnWO_4$	Manganese tungstate	$\varepsilon_{11} = 19.3 \pm 1.3$	r.t.	$(5–500) \times 10^3$
		$\varepsilon_{22} = 14.3 \pm 0.5$	r.t.	$(5–500) \times 10^3$
		$\varepsilon_{33} = 16.5 \pm 1.1$	r.t.	$(5–500) \times 10^3$
$N(CH_3)_4HgBr_3$	Tetramethylammonium tribromomercurate (TTM)	~10	233–373	
$N(CH_3)_4HgI_3$	Tetramethylammonium triiodo mercurate (TTM)	~10	233–373	
$(ND_4)_2BeF_4$	Deuteroammonium tetrafluoroberyllate	$\varepsilon_{11} = 10$	r.t.	
		$\varepsilon_{22} = 9$	r.t.	
		$\varepsilon_{33} = 9$	r.t.	
$(ND_4)_2SO_4$	Deuteroammonium sulfate	$\varepsilon_{11} = 9$	r.t.	
		$\varepsilon_{22} = 10$	r.t.	
		$\varepsilon_{33} = 9$	r.t.	

Formula	Name	ε_{ijk}	*T*/K	*ν*/Hz
$(NH_2 \cdot CH_2COOH)_3 \cdot H_2SO_4$	Triglycine sulfate (TGS)	$\varepsilon_{11} = 9$	273	10^4
		$\varepsilon_{22} = 30$	273	10^4
		$\varepsilon_{33} = 6.5$	273	10^4
$(NH_2 \cdot CH_2COOH)_3 \cdot H_2SeO_4$	Triglycine selenate (TGSe)	200	293	1.6×10^3
$(NH_2 \cdot CH_2COOH)_3 \cdot H_2BeF_4$	Triglycine tetrafluoroberyllate (TGFB)	$\varepsilon_{22} = 12$	273	10^4
$NH_4Al(SO_4)_2 \cdot 12H_2O$	Ammonium alum	6	r.t.	10^{12}
$(NH_4)_2BeF_4$	Ammonium tetrafluoroberyllate	$\varepsilon_{11} = \varepsilon_{22} = 7.8$	123	10^5
		$\varepsilon_{33} = 7.1$	123	10^5
		$\varepsilon_{11} = \varepsilon_{22} = 8.8$	293	10^5
		$\varepsilon_{33} = 9.2$	293	10^5
NH_4Br	Ammonium bromide	7.1	r.t.	7×10^5
NH_4I	Ammonium iodide	9.8	r.t.	
$(NH_4)_2C_2H_6O_6$	Ammonium tartrate	$\varepsilon_{11} = 6.45$	r.t.	10^3
		$\varepsilon_{22} = 6.8$	r.t.	10^3
		$\varepsilon_{33} = 6.0$	r.t.	10^3
$(NH_4)_2Cd_2(SO_4)_3$	Ammonium cadmium sulfate	10.0	r.t.	10^4
NH_4Cl	Ammonium chloride	6.9	r.t.	7×10^5
$NH_4(ClCH_2COO)$	Ammonium monochloroacetate	5	r.t.	2×10^6
$NH_4H(ClCH_2COO)_2$	Hydrogen ammonium chloroacetate	$\varepsilon_{[102]} = 5.9$	r.t.	10^5
$NH_4Cr(SO_4)_2 \cdot 12H_2O$	Ammonium chrome alum	6.5	r.t.	175×10^3
NH_4HSO_4	Ammonium hydrogen sulfate	165	273	5×10^4
$NH_4H_2AsO_4$	Ammonium dihydrogen arsenate (ADA)	5.1	265	9.5×10^9
		$\varepsilon_{11} = \varepsilon_{22} = 85$	298	10^3
		$\varepsilon_{33} = 22$	298	
$NH_4H_2PO_4$	Ammonium dihydrogen phosphate (ADP)	$\varepsilon_{11} = \varepsilon_{22} = 57.1 \pm 0.6$	294.5	10^5–35×10^9
		$\varepsilon_{33} = 14.0 \pm 0.3$	294	10^5–36×10^9
$ND_4D_2PO_4$	Ammonium dideuterium phosphate (ADDP)	$\varepsilon_{11} = \varepsilon_{22} = 74, \varepsilon_{33} = 24$	300	
NH_4NO_3	Ammonium nitrate	10.7	322	$(5–50) \times 10^3$
$(NH_4)_2SO_4$	Ammonium sulfate	$\varepsilon_{11} = \varepsilon_{22} = 8.0$	123	10^5
		$\varepsilon_{33} = 6.3$	123	10^5
		$\varepsilon_{11} = \varepsilon_{22} = 10.0$	293	10^5
		$\varepsilon_{33} = 9.3$	293	10^5
$(NH_4)_2UO_2(C_2O_4)_2$	Ammonium uranyl oxalate	8.03	r.t.	10^4–3.3×10^9
$(NH_4)_2UO_2(C_2O_4)_2 \cdot 3H_2O$	Ammonium uranyl oxalate trihydrate	6.06	r.t.	10^4–3.3×10^9
NaBr	Sodium bromide	6.44	298	1.6×10^3
$NaBrO_3$	Sodium bromate	$\varepsilon_{11}^T = 5.70$	298	10^3
NaCN	Sodium cyanide	7.55	293	10^5
$NaCO_3$	Sodium carbonate	8.75	291	2×10^5
$NaCO_3 \cdot 10H_2O$	Sodium carbonate decahydrate	5.3	r.t.	6×10^7
NaCl	Sodium chloride	5.9	298	10^2–10^7
		5.45	4.2	
$NaClO_3$	Sodium chlorate	$\varepsilon_{11}^T = 5.76$	301	10^3
		5.28	r.t.	10^3
$NaClO_4$	Sodium perchlorate	5.76	r.t.	10^3
NaF	Sodium fluoride	5.08 ± 0.02	r.t.	5×10^3
$NaH_3(SeO_3)_2$	Sodium trihydrogen selenite	$\varepsilon_{11} \approx 75$	273	2×10^5
$NaD_3(SeO_3)_2$	Sodium trideuterium selenite	$\varepsilon_{11} \approx 220$	273	2×10^5
NaI	Sodium iodide	7.28 ± 0.03	r.t.	
$NaNH_4(C_4H_4O_6) \cdot 4H_2O$	Sodium ammonium tartrate (Ammonium Rochelle salt)	$\varepsilon_{11} = 8.4$	298	
		$\varepsilon_{22} = 9.2$	298	
		$\varepsilon_{33} = 9.5$	298	
$NaNbO_3$	Sodium niobate	$\varepsilon_{33} = 670 \pm 13$	r.t.	
		$\varepsilon_{11} = \varepsilon_{22} = 76 \pm 2$	r.t.	
$NaNO_2$	Sodium nitrite	$\varepsilon_{11} = 7.4$	r.t.	5×10^5
		$\varepsilon_{22} = 5.5$	r.t.	5×10^5
		$\varepsilon_{33} = 5.0$	r.t.	5×10^5
$NaNO_3$	Sodium nitrate	6.85	292	2×10^5
$NaSO_4$	Sodium sulfate	7.90	r.t.	
$NaSO_4 \cdot 10H_2O$	Sodium sulfate decahydrate	5.0	r.t.	
$Na_2SO_4 \cdot 5H_2O$	Sodium thiosulfate pentahydrate	7	250–290	300–10^4
$Na_2UO_2(C_2O_4)_2$	Sodium uranyl oxalate	5.18	r.t.	

Formula	Name	ε_{ijk}	T/K	ν/Hz
$NdAlO_3$	Neodymium aluminate	17.5	r.t.	
$NdScO_3$	Neodymium scandate	27	r.t.	
$Ni_3B_7O_{13}I$	Nickel iodine boracite	ε_{11} = 14	260	
$NiNb_2O_6$	Nickel niobate	ε_{11} = 16.0 ± 0.5	r.t.	(5–500) ×10³
		ε_{22} = 23.8 ± 1.8	r.t.	(5–500) × 10³
		ε_{33} = 31.3 ± 2.5	r.t.	(5–500) × 10³
NiO	Nickel oxide	11.9	298	10^5
$NiSO_4 \cdot 6H_2O$	Nickel sulfate hexahydrate	ε_{11} = 6.2	r.t.	
		ε_{33} = 6.8	r.t.	
$NiWO_4$	Nickel tungstate	ε_{11} = 17.4 ± 2.4	r.t.	(5–500) × 10³
		ε_{22} = 13.6 ± 1.0	r.t.	(5–500) × 10³
		ε_{33} = 19.7 ± 0.6	r.t.	(5–500) × 10³
P	Phosphorous (red)	4.1	r.t.	10^8
	Phosphorous (yellow)	3.6	r.t.	10^8
$[P(CH_3)_4]HgBr_3$	Tetramethylphosphonium tribromomercurate (TTM)	~10	233–373	
$PbBr_2$	Lead bromide	>30	293	(0.5–3) × 10^6
$PbCO_3$	Lead carbonate	18.6	288	10.8
$Pb(C_2H_3O_2)_2$	Lead acetate	2.6	290–295	10^6
$PbCl_2$	Lead chloride	33.5	273	(0.5–3) × 10^6
Pb_2CoWO_6	Lead cobalt tungstate	~250	r.t.	
PbF_2	Lead fluoride	26.3	r.t.	
$PbHfO_3$	Lead hafnate	390	300	10^5
		185	400	
PbI_2	Lead iodide	20.8	293	(0.5–3) × 10^6
$Pb_3MgNb_2O_9$	Lead magnesium niobate	10,000	297	
$PbMoO_4$	Lead molybdate	ε_{11} = 34.0 ± 0.4	297.5	1.6 ± 10³
		ε_{33} = 40.6 ± 0.2	297.5	1.6 ± 10³
$Pb(NO_3)_2$	Lead nitrate	16.8	r.t.	(0.5–3) × 10^6
$PbNb_2O_6$	Lead niobate	ε_{33}^{T} = 180	298	
PbO	Lead oxide	25.9	r.t.	2 × 10^6
PbS	Lead sulfide (Galena)	190	77	i.r.
		200 ± 35	r.t.	i.r.
$PbSO_4$	Lead sulfate	14.3	290— 295	10^6
PbSe	Lead selenide	280	r.t.	i.r.
$PbTa_2O_6$	Lead tantalate	ε_{11} = ε_{22} ≈ 300	r.t.	10^4
		ε_{33} = 150	r.t.	10^4
PbTe	Lead telluride	450	r.t.	i.r.
		40	77	10^4–15 × 10^4
		430	4.2	10^4–15 × 10^4
$PbTiO_3$	Lead titanate	~200	r.t.	10^3
$PbWO_4$	Lead tungstate	ε_{11} = ε_{22} = 23.6 ± 0.3	297.5	1.59 × 10³
		ε_{33} = 31.0 ± 0.4	297.5	1.59 × 10³
$Pb(Zn_{1/3}Nb_{2/3})O_3$	Lead zinc niobate	7	300	10^3, 300 × 10^3
$PbZrO_3$	Lead zirconate	200	400	
$RbAl(SO_4)_2 \cdot 12H_2O$	Rubidium alum	5.1	r.t.	10^{12}
RbBr	Rubidium bromide	4.83	300	
Rb_2CO_3	Rubidium carbonate	4.87 ± 0.02	r.t.	5 × 10³
RbCl	Rubidium chloride	4.91 ± 0.02	r.t.	5 × 10³
$RbCr(SO_4)_2 \cdot 12H_2O$	Rubidium chrome alum	5.0	r.t.	10^{12}
RbF	Rubidium fluoride	5.91	r.t.	2 × 10^6
$RbHSO_4$	Rubidium bisulfate	ε_{11} = 7	r.t.	10^5
		ε_{22} = 8	r.t.	10^5
		ε_{33} = 10	r.t.	10^5
RbH_2AsO_4	Rubidium dihydrogen arsenate (RDA)	3.90	273	9.5 × 10^9
RbH_2PO_4	Rubidium dihydrogen phosphate (RDP)	6.15	285	9.5 × 10^9
RbI	Rubidium iodide	4.94 ± 0.02	r.t.	5 × 10³
$RbInSO_4$	Rubidium indium sulfate	6.85	r.t.	
$RbNO_3$	Rubidium nitrate	20—380	433–488	10^6
		30	488–538	10^6
S	Sulfur	ε_{11} = 3.75	298	10^2–10^3
		ε_{22} = 3.95	298	10^2–10^3

Formula	Name	ε_{ijk}	T/K	ν/Hz
		$\varepsilon_{33} = 4.44$	298	10^2–10^3
	(sublimed)	3.69	298	10^2–10^3
$SC(NH_2)_2$	Thiourea	$\varepsilon_{11} = \varepsilon_{22} \approx 3$	77–300	10^3
		$\varepsilon_{22} = 35$	300	10^3
Sb_2O_3	Antimonous sesquioxide	12.8	r.t.	$(1.5–2) \times 10^3$
Sb_2S_3	Antimonous sulfide (stibnite)	$\varepsilon_{11} = \varepsilon_{33} = 15$	r.t.	10^3
		$\varepsilon_{33} = 180$	r.t.	10^3
Sb_2Se_3	Antimonous selenide	~110	r.t.	$(10–16.5) \times 10^9$
SbSI	Antimonous sulfide iodide	2000	273	10^5
		$\varepsilon_{11} = \varepsilon_{22} \approx 25$	r.t.	10^3–10^5
		$\varepsilon_{33} \approx 5 \times 10^4$	295	10^3–10^5
Se	Selenium (monocrystal)	$\varepsilon_{11} = \varepsilon_{22} = 11$	300	24×10^9
		$\varepsilon_{33} = 21$	300	24×10^9
	(amorphous)	6.0	298	10^2—10^{10}
Si	Silicon	12.1	4.2	10^7—10^9
SiC	Silicon carbide, cubic	9.72	r.t.	i.r.
	Silicon carbide, 6H	$\varepsilon_{11} = \varepsilon_{22} = 9.66$	r.t.	i.r.
		$\varepsilon_{33} = 10.03$	r.t.	i.r.
		9.7 ± 0.1	1.8	i.r.
Si_3N_4	Silicon nitride	4.2 (film)	r.t.	10^3
SiO	Silicon monoxide	5.8	r.t.	10^3
SiO_2	Silicon dioxide	$\varepsilon_{11} = 4.42$	r.t.	9.4×10^{10}
		$\varepsilon_{22} = 4.41$	r.t.	9.4×10^{10}
		$\varepsilon_{33} = 4.60$	r.t.	9.4×10^{10}
$Sm_2(MoO_4)_3$	Samarium molybdate	12	298	
SnO_2	Stannic dioxide	$\varepsilon_{11} = \varepsilon_{22} = 14 \pm 2$	r.t.	10^4–10^{10}
		$\varepsilon_{33} = 9.0 \pm 0.5$	r.t.	10^4–10^{10}
SnSb	Tin antimonide	147	r.t.	10^4–10^6
SnTe	Tin telluride	1770 ± 300	r.t.	i.r.
$Sr(COOH)_2 \cdot 2H_2O$	Strontium formate dihydrate	6.1	r.t.	10^3
$SrCO_3$	Strontium carbonate	8.85	298	2×10^5
$SrCl_2$	Strontium chloride	9.19	r.t.	
$SrCl_2 \cdot 6H_2O$	Strontium chloride hexahydrate	8.52	r.t.	
SrF_2	Strontium fluoride	6.50	300	5×10^2–10^{11}
$SrMoO_4$	Strontium molybdate	$\varepsilon_{11} = \varepsilon_{22} = 31.7 \pm 0.2$	297.5	1.59×10^3
		$\varepsilon_{33} = 41.7 \pm 0.2$	297.5	1.59×10^3
$Sr(NO_3)_2$	Strontium nitrate	5.33	292	2×10^5
$Sr_2Nb_2O_7$	Strontium niobate	$\varepsilon_{11} = 75$	r.t.	10^3
		$\varepsilon_{22} = 46$	r.t.	10^3
		$\varepsilon_{33} = 43$	r.t.	10^3
SrO	Strontium oxide	13.3 ± 0.3	273	2×10^6
SrS	Strontium sulfide	11.3	r.t.	7.25×10^6
$SrSO_4$	Strontium sulfate	11.5	r.t.	
$SrTiO_3$	Strontium titanate	332	298	10^3
		2080	78	10^3
$SrWO_4$	Strontium tungstate	$\varepsilon_{11} = \varepsilon_{22} = 25.7 \pm 0.2$	297.5	1.6×10^3
		$\varepsilon_{33} = 34.1 \pm 0.2$	297.5	1.6×10^3
	Tantalum pentoxide, α phase	$\varepsilon_{11} = \varepsilon_{22} = 30$	77	10^3
		$\varepsilon_{33} = 65$	77	10^3
	Tantalum pentoxide, β phase	24	292	10^3
$Tb(MoO_4)_3$	Terbium molybdate	11	298	
		$\varepsilon_{11} = \varepsilon_{22} = 33$	100–200	9.4×10^9
		$\varepsilon_{33} = 53$	100–200	9.4×10^9
Te	Tellurium	$\varepsilon_{11} = \varepsilon_{22} = 33$	r.t.	
		$\varepsilon_{33} = 54$	r.t.	
	polycrystalline	27.5	r.t.	i.r.
	monocrystalline	28.0	r.t.	i.r.
ThO_2	Thorium dioxide	18.9 ± 0.4	r.t.	3×10^5
TiO_2	Titanium dioxide (Rutile)	$\varepsilon_{11} = \varepsilon_{22} = 86$	300	10^4–10^6
		$\varepsilon_{33} = 170$	300	10^4–10^6
Ti_2O_3	Titanium sesquioxide	30	77	6×10^{10}

Formula	Name	ε_{ijk}	T/K	ν/Hz
TlBr	Thallium bromide	30	293	10^3–10^7
TlCl	Thallous chloride	32.2 ± 0.2	293	10^3–10^5
TlI	Thallous iodide (orthorhombic)	20.7 ± 0.2	293	10^4
		37.3	193	10^7
$TlNO_3$	Thallous nitrate	16.5	293	5×10^5
$TlSO_4$	Thallous sulfate	25.5	293	5×10^5
UO_2	Uranium dioxide	24	r.t.	3×10^5
WO_3	Tungsten trioxide	300		
$YMnO_3$	Yttrium manganate	20	r.t.	2×10^7
Y_2O_3	Yttrium oxide	10	r.t.	10^6
$YbMnO_3$	Ytterbium manganate	20	r.t.	2×10^7
Yb_2O_3	Ytterbium oxide	5.0 (film)	r.t.	10^3
ZnO	Zinc oxide	$\varepsilon_{11}^S = 8.33$	r.t.	
		$\varepsilon_{33}^S = 8.84$	r.t.	
		$\varepsilon_{11}^T = 9.26$	r.t.	
		$\varepsilon_{33}^T = 11.0$	r.t.	
		$\varepsilon_{11} = 9.26$	r.t.	
		$\varepsilon_{33} = 8.2$	r.t.	
		8.15	r.t.	i.r.
ZnS	Zinc sulfide	$\varepsilon_{11}^S = 8.08 \pm 2\%$	77	10^4
		$\varepsilon_{11}^S = 8.32 \pm 2\%$	298	10^4
		$\varepsilon_{11}^T = 8.14 \pm 2\%$	77	10^4
		$\varepsilon_{11}^T = 8.37 \pm 2\%$	298	10^4
ZnSe	Zinc selenide	$\varepsilon_{11}^T = \varepsilon_{11}^S = 9.12 \pm 2\%$	298	10^4
ZnTe	Zinc telluride	$\varepsilon_{11}^T = \varepsilon_{11}^S = 10.10 \pm 2\%$	r.t.	
$ZnWO_4$	Zinc tungstate	$\varepsilon_{22} = 16.1 \pm 0.5$	r.t.	$(5–500) \times 10^3$
ZrO_2	Zirconium dioxide (zirconia)	12.5	r.t.	2×10^6

[a] Type I diamond
[b] Type IIa diamond
[c] γ phase
[d] Ice I (P = 0 kbar)
[e] Ice III (P = 3 kbar)
[f] Ice V (P = 5 kbar)
[g] Ice VI (P = 8 kbar)
[h] Muscovite
[i] Canadian mica
[j] ^{6}LiH
[k] ^{7}LiH
[l] α phase
[m] β phase

CURIE TEMPERATURE OF SELECTED FERROELECTRIC CRYSTALS

H. P. R. Frederikse

The following table lists the major ferroelectric crystals and their Curie temperatures, T_C.

Reference

Young, K. F. and Frederikse, H. P. R., *J. Phys. Chem. Ref. Data*, 2, 313, 1973.

Curie Temperature of Ferroelectric Crystals

Systematic name	Acronym or common name	Formula	T_C/K
Potassium dihydrogen phosphate group			
Potassium dihydrogen phosphate	KDP	KH_2PO_4	123
Potassium dihydrogen arsenate	KDA	KH_2AsO_4	97
Potassium dideuterium phosphate	KDDP	KD_2PO_4	213
Potassium dideuterium arsenate	KDDA	KD_2AsO_4	162
Rubidium dihydrogen phosphate	RDP	RbH_2PO_4	146
Rubidium dihydrogen arsenate	RDA	RbH_2AsO_4	111
Rubidium dideuterium phosphate	RDDP	RbD_2PO_4	218
Rubidium dideuterium arsenate	RDDA	RbD_2AsO_4	178
Cesium dihydrogen phosphate	CDP	CsH_2PO_4	159
Cesium dihydrogen arsenate	CDA	CsH_2AsO_4	143
Cesium dideuterium arsenate	CDDA	CsD_2AsO_4	212
Rochelle salt group			
Potassium sodium tartrate tetrahydrate	Rochelle salt	$KNaC_4H_4O_6{\cdot}4H_2O$	255-297
Potassium sodium tartrate-d_2 tetrahydrate	Deuterated Rochelle salt	$KNaC_4H_2D_2O_6{\cdot}4H_2O$	251-308
Sodium ammonium tartrate tetrahydrate	Ammonium Rochelle salt	$NaNH_4C_4H_4O_6{\cdot}4H_2O$	109
Lithium ammonium tartrate monohydrate	LAT	$LiNH_4C_4H_4O_6{\cdot}H_2O$	106
Triglycine sulfate group			
Triglycine sulfate	TGS	$(NH_2CH_2COOH)_3{\cdot}H_2SO_4$	322
Triglycine selenate	TGSe	$(NH_2CH_2COOH)_3{\cdot}H_2SeO_4$	295
Triglycine tetrafluoroberyllate	TGFB	$(NH_2CH_2COOH)_3{\cdot}H_2BeF_4$	346
Ammonium tetrafluoroberyllate	AFB	$(NH_4)_2BeF_4$	176
Hydrogen ammonium chloroacetate		$NH_4H(CH_2ClCOO)_2$	128
Perovskites and related compounds			
Barium titanate	Barium metatitanate	$BaTiO_3$	406, 278, 193
Lead(II) titanate	Lead metatitanate	$PbTiO_3$	765
Potassium niobate		$KNbO_3$	712
Potassium tantalate niobate		$K_3Ta_2NbO_9$	241, 220, 170
Lithium niobate		$LiNbO_3$	1483
Lithium tantalate	Lithium tantalum trioxide	$LiTaO_3$	891
Barium titanium niobate		$Ba_6Ti_2Nb_8O_{30}$	521
Barium sodium niobate	BSN	$Ba_2Na(NbO_3)_5$	833
Potassium iodate		KIO_3	485, 343, 257–263, 83
Lithium iodate		$LiIO_3$	529
Potassium nitrate	Saltpeter	KNO_3	397
Sodium nitrate	Chile saltpeter	$NaNO_3$	548
Rubidium nitrate		$RbNO_3$	437-487
Miscellaneous compounds			
Cesium trihydrogen selenite		$CsH_3(SeO_3)_2$	143
Lithium trihydrogen selenite		$LiH_3(SeO_3)_2$	$T_C > T_{mp}$
Potassium selenate		K_2SeO_4	93
Methyl ammonium aluminum alum	MASD	$(CH_3NH_3)Al(SO_4)_2{\cdot}12H_2O$	177
Ammonium cadmium sulfate	Ammonium cadmium langbeinite	$(NH_4)_2Cd_2(SO_4)_3$	95
Ammonium hydrogen sulfate	Ammonium bisulfate	NH_4HSO_4	271
Ammonium sulfate	Mascagnite	$(NH_4)_2SO_4$	224
Ammonium nitrate		NH_4NO_3	398, 357, 305, 255
Calcium hexaborate pentahydrate	Colemanite	$2CaO{\cdot}3B_2O_3{\cdot}5H_2O$	266
Cadmium niobate		$Cd_2Nb_2O_7$	185
Gadolinium(III) molybdate		$Gd_2(MoO_4)_3$	432

PROPERTIES OF ANTIFERROELECTRIC CRYSTALS

H. P. R. Frederikse

Some important antiferroelectric crystals are listed here with their Curie temperatures T_C. The last column gives the constant T_0 which appears in the Curie–Weiss law describing the dielectric constant of these materials above the Curie temperature:

$$\varepsilon = \text{const.}/(T - T_0)$$

Name	Common name or acronym	Formula	T_C/K	T_0/K
Ammonium dihydrogen phosphate	ADP	$NH_4H_2PO_4$	148	
Ammonium dihydrogen arsenate	ADA	$NH_4H_2AsO_4$	216	
Ammonium dideuterium phosphate	ADDP	$NH_4D_2PO_4$	242[a]	
Ammonium dideuterium arsenate	ADDA	$NH_4D_2AsO_4$	299	
Ammonium-d_4 dideuterium phosphate	A_dDDP	$ND_4D_2PO_4$	243	
Ammonium-d_4 dideuterium arsenate	A_dDDA	$ND_4D_2AsO_4$	304	
Sodium niobate	Sodium niobate	$NaNbO_3$	911[b]	
Lead(II) hafnate	Lead hafnate	$PbHfO_3$	476	378
Lead(II) zirconate	Lead zirconate	$PbZrO_3$	503	475
Lead(II) niobate	Lead metaniobate	$Pb(NbO_3)_2$	843	530
Lead(II) tantalate	Lead metatantalate	$Pb(TaO_3)_2$	543	533
Tungsten(VI) oxide	Tungsten trioxide	WO_3	1010	
Potassium strontium niobate	Potassium strontium niobate	$KSr_2Nb_5O_{15}$	427	413
Sodium nitrite	Sodium nitrite	$NaNO_2$	437	437
Sodium trihydrogen selenite	Sodium trihydrogen selenite	$NaH_3(SeO_3)_2$	193	192
Sodium trideuterium selenite	Sodium trideuterium selenite	$NaD_3(SeO_3)_2$	271	245
Ammonium trihydrogen periodate	Ammonium trihydrogen periodate	$(NH_4)_2H_3IO_6$	245	

[a] Also reported as 245 K
[b] Also reported as 793 K

DIELECTRIC CONSTANTS OF GLASSES

Type	Dielectric constant at 100 MHz (20 °C)	Volume resistivity in MΩ cm (at 350 °C)	Loss factor[a]
Corning 0010	6.32	10	0.015
Corning 0080	6.75	0.13	0.058
Corning 0120	6.65	100	0.012
Pyrex 1710	6.00	2500	0.025
Pyrex 3320	4.71		0.019
Pyrex 7040	4.65	80	0.013
Pyrex 7050	4.77	16	0.017
Pyrex 7052	5.07	25	0.019
Pyrex 7060	4.70	13	0.018
Pyrex 7070	4.00	1300	0.0048
Vycor 7230	3.83		0.0061
Pyrex 7720	4.50	16	0.014
Pyrex 7740	5.00	4	0.040
Pyrex 7750	4.28	50	0.011
Pyrex 7760	4.50	50	0.0081
Vycor 7900	3.9	130	0.0023
Vycor 7910	3.8	1600	0.00091
Vycor 7911	3.8	4000	0.00072
Corning 8870	9.5	5000	0.0085
G. E. Clear (silica glass)	3.81	[d]	0.00038
Quartz (fused)	3.75[b]		0.0002[c]

[a] Power factor × dielectric constant equals loss factor.
[b] 4.1 at 1 MHz
[c] At 1 MHz
[d] Ranges from 4000 MΩ to 30000 MΩ

PROPERTIES OF SUPERCONDUCTORS

L. I. Berger and B. W. Roberts

The following tables include superconductive properties of selected elements, compounds, and alloys. Individual tables are given for thin films, elements at high pressures, superconductors with high critical magnetic fields, and high critical temperature superconductors.

The historically first observed and most distinctive property of a superconductive body is the near total loss of resistance at a critical temperature (T_c) that is characteristic of each material. Figure 1(a) below illustrates schematically two types of possible transitions. The sharp vertical discontinuity in resistance is indicative of that found for a single crystal of a very pure element or one of a few well-annealed alloy compositions. The broad transition, illustrated by broken lines, suggests the transition shape seen for materials that are not homogeneous and contain unusual strain distributions. Careful testing of the resistivity limit for superconductors shows that it is less than 4×10^{-23} ohm cm, while the lowest resistivity observed in metals is of the order of 10^{-13} ohm cm. If one compares the resistivity of a superconductive body to that of copper at room temperature, the superconductive body is at least 10^{17} times less resistive.

The temperature interval ΔT_c, over which the transition between the normal and superconductive states takes place, may be of the order of as little as 2×10^{-5} K or several K in width, depending on the material state. The narrow transition width was attained in 99.9999% pure gallium single crystals.

A Type I superconductor below T_c, as exemplified by a pure metal, exhibits perfect diamagnetism and excludes a magnetic field up to some critical field H_c, whereupon it reverts to the normal state as shown in the *H-T* diagram of Figure 1(b).

The magnetization of a typical high-field superconductor is shown in Figure 1(c). The discovery of the large current-carrying capability of Nb_3Sn and other similar alloys has led to an extensive study of the physical properties of these alloys. In brief, a high-field superconductor, or Type II superconductor, passes from the perfect diamagnetic state at low magnetic fields to a mixed state and finally to a sheathed state before attaining the normal resistive state of the metal. The magnetic field values separating the four stages are given as H_{c1}, H_{c2}, and H_{c3}. The superconductive state below H_{c1} is perfectly diamagnetic, identical to the state of most pure metals of the "soft" or Type I superconductor. Between H_{c1} and H_{c2} a "mixed superconductive state" is found in which fluxons (a minimal unit of magnetic flux) create lines of normal flux in a superconductive matrix. The volume of the normal state is proportional to $-4\pi M$ in the "mixed state" region. Thus at H_{c2} the fluxon density has become so great as to drive the interior volume of the superconductive body completely normal. Between H_{c2} and H_{c3} the superconductor has a sheath of current-carrying superconductive material at the body surface, and above H_{c3} the normal state exists. With several types of careful measurement, it is possible to determine H_{c1}, H_{c2}, and H_{c3}. Table 6 contains some of the available data on high-field superconductive materials.

High-field superconductive phenomena are also related to specimen dimension and configuration. For example, the Type I superconductor, Hg, has entirely different magnetization behavior in high magnetic fields when contained in the very fine sets of filamentary tunnels found in an unprocessed Vycor glass. The great majority of superconductive materials are Type II. The elements in very pure form and a very few precisely stoichiometric and well annealed compounds are Type I with the possible exceptions of vanadium and niobium.

Metallurgical Aspects. The sensitivity of superconductive properties to the material state is most pronounced and has been used in a reverse sense to study and specify the detailed state of alloys. The mechanical state, the homogeneity, and the presence of impurity atoms and other electron-scattering centers are all capable of controlling the critical temperature and the current-carrying capabilities in high-magnetic fields. Well-annealed specimens tend to show sharper transitions than those that are strained or inhomogeneous. This sensitivity to mechanical state underlines a general problem in the tabulation of properties for superconductive materials. The occasional divergent values of the critical temperature and of the critical fields quoted for a Type II superconductor may lie in the variation in sample preparation. Critical temperatures of materials studied early in the history of superconductivity must be evaluated in light of the probable metallurgical state of the material, as well as the availability of less pure starting elements. It has been noted that recent work has given extended consideration to the metallurgical aspects of sample preparation.

Symbols in tables: T_c: Critical temperature; H_o: Critical magnetic field in the $T = 0$ limit; θ_D: Debye temperature; and γ: Electronic specific heat.

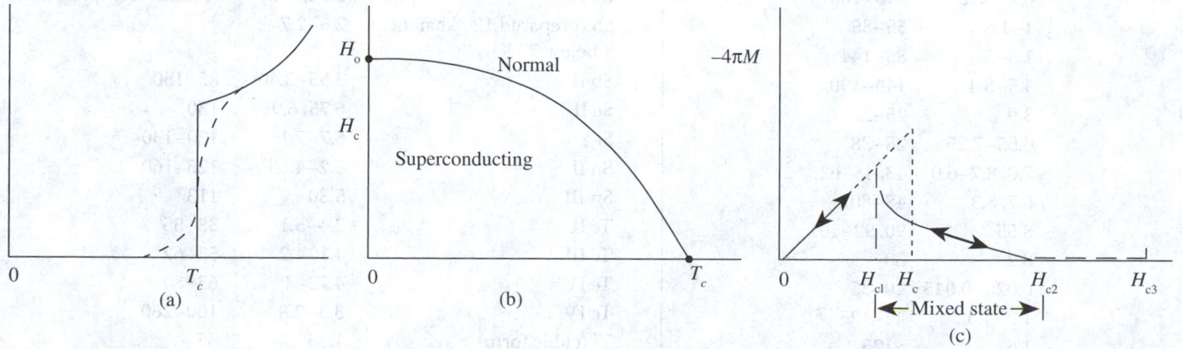

FIGURE 1. Physical properties of superconductors. (a) Resistivity vs. temperature for a pure and perfect lattice (solid line); impure and/or imperfect lattice (broken line). (b) Magnetic-field temperature dependence for Type I or "soft" superconductors. (c) Schematic magnetization curve for "hard" or Type II superconductors.

TABLE 1. Selective Properties of Superconductive Elements

Element	T_c/K	H_o/Oe	θ_D/K	γ/mJ mol^{-1}K^{-1}
Al	1.175 ± 0.002	104.9 ± 0.3	420	1.35
Am* (α,?)	0.6			
Am* (β,?)	1.0			
Be	0.026			0.21
Cd	0.517 ± 0.002	28 ± 1	209	0.69
Ga	1.083 ± 0.001	58.3 ± 0.2	325	0.60
Ga (β)	5.9, 6.2	560		
Ga (γ)	7	950, HF[a]		
Ga (Δ)	7.85	815, HF		
Hf	0.128	12.7		2.21
Hg (α)	4.154 ± 0.001	411 ± 2	87, 71.9	1.81
Hg (β)	3.949	339	93	1.37
In	3.408 ± 0.001	281.5 ± 2	109	1.672
Ir	0.1125 ± 0.001	16 ± 0.05	425	3.19
La (α)	4.88 ± 0.02	800 ± 10	151	9.8
La (β)	6.00 ± 0.1	1096, 1600	139	11.3
Lu	0.1 ± 0.03	350 + 50		
Mo	0.915 ± 0.005	96 ± 3	460	1.83
Nb	9.25 ± 0.02	2060 ± 50, HF	276	7.80
Os	0.66 ± 0.03	70	500	2.35
Pa	1.4			
Pb	7.196 ± 0.006	803 ± 1	96	3.1
Re	1.697 ± 0.006	200 ± 5	4.5	2.35
Ru	0.49 ± 0.015	69 ± 2	580	2.8
Sn	3.722 ± 0.001	305 ± 2	195	1.78
Ta	4.47 ± 0.04	829 ± 6	258	6.15
Tc	7.8 ± 0.1	1410, HF	411	6.28
Th	1.38 ± 0.02	1.60 ± 3	165	4.32
Ti	0.40 ± 0.04	56	415	3.3
Tl	2.38 ± 0.02	178 ± 2	78.5	1.47
U	0.2			
V	5.40 ± 0.05	1408	383	9.82
W	0.0154 ± 0.0005	1.15 ± 0.03	383	0.90
Zn	0.85 ± 0.01	54 ± 0.3	310	0.66
Zr	0.61 ± 0.15	47	290	2.77
Zr (ω)	0.65, 0.95			

[a] HF denotes high magnetic field superconductive properties.

TABLE 2. Range of Critical Temperatures Observed for Superconductive Elements in Thin Films Condensed Usually at Low Temperatures

Element	T_c/K range	Comments
Al	1.15–5.7	HF[a]
Be	5–9.75	HF
Bi	6.17–6.6	
Cd (Disordered)	0.79–0.91	
Cd (Ordered)	0.53–0.59	
Ga	2.5–8.5	HF
Hg	3.87–4.5	
In	2.2–5.6	HF
La	3.55–6.74	
Mo	3.3–8.0	
Nb	2.0–10.1	
Pb	1.8–7.5	
Re	1.7–7	
Sn	3.5–6	
Ta	<1.7–4.51	HF
Tc	4.6–7.7	
Ti	1.3 Max	
Tl	2.33–2.96	
V	1.8–6.02	
W	<1.0–4.1	
Zn	0.77–1.9	

[a] HF denotes high magnetic field superconductive properties.

TABLE 3. Elements Exhibiting Superconductivity Under or After Application of High Pressure

Element	T_c/K Range	p/kbar
Al	1.98–0.075	0–62
As	0.31–0.5	220–140
As	0.2–0.25	140–100
Ba II	1–1.8	55–85
Ba III	1.8–5	85–144
Ba IV	4.5–5.4	144–190
Bi II	3.9	25–27
Bi III	6.55–7.25	28–38
Bi IV	7.0, 8.7–6.0	43, 43–62
Bi V	6.7, 8.3	48–80
Bi VI	8.55	90, 92–101
Bi VII(?)	8.2	30
Ce (α)	0.020–0.045	20–35
Ce (α′)	1.9–1.3	45–125
Cs V	1.5	>125
Ga II	6.38	≥35
Ga II′	7.5	≥35 then P removed
Ge	5.35	115
Lu	0.022–1.0	45–190
P	5.8	170
Pb II	3.55	160
Re II	2.3 Max.	"Plastic" compression
Sb (prepared 120 kbar, held below 77 K)	2.6–2.7	
Sb II	3.55–3.40	85–150
Se II	6.75, 6.95	130
Si	6.7–7.1	120–130
Sn II	5.2–4.85	125–160
Sn III	5.30	113
Te II	2.4–5.1	38–55
Te III	4.1–4.2	53–62
Te IV	4.72–4	63–80
Te IV	3.3–2.8	100–260
Tl (cubic form)	1.45	35
Tl (hexagonal form)	1.95	35
U	2.4–0.4	10–85
Y	1.7–2.5	110–160
Zr (omega form, metastable)	1–1.7	60–130

Properties of Superconductive Compounds and Alloys

In Tables 4 to 6 all compositions are denoted on an atomic basis, i.e., AB, AB_2, or AB_3 for compounds, unless noted. Solid solutions or odd compositions may be denoted as A_zB_{1-z} or A_zB. A series of three or more alloys is indicated as A_xB_{1-x} or by actual indication of the atomic fraction range, such as $A_{0-0.6}B_{1-0.4}$. The critical temperature of such a series of alloys is denoted by a range of values or possibly the maximum value.

The selection of the critical temperature from a transition in the effective permeability, or the change in resistance, or possibly the incremental changes in frequency observed by certain techniques is not often obvious from the literature. Most authors choose the mid-point of such curves as the probable critical temperature of the idealized material, while others will choose the highest temperature at which a deviation from the normal state property is observed. In view of the previous discussion concerning the variability of the superconductive properties as a function of purity and other metallurgical aspects, it is recommended that appropriate literature be checked to determine the most probable critical temperature or critical field of a given alloy.

A very limited amount of data on critical fields, H_o, is available for these compounds and alloys; these values are given in Table 5.

TABLE 4. Superconductive Compounds and Alloys

Substance	T_c/K	Crystal structure type
A. Superconductors with T_c < 10 K		
$Ag_{3.3}Al$	0.34	A12-cI58 (Mn)
$Ag_xAl_yZn_{1-x-y}$	0.15	Cubic
$AgBi_2$	2.87–3.0	
$Ag_7F_{0.25}N_{0.75}O_{10.25}$	0.85–0.90	
Ag_2F	0.0.066	
Ag_7FO_8	0.3	Cubic
$Ag_{0.8-0.3}Ga_{0.2-0.7}$	6.5–8	
Ag_4Ge	0.85	Hex., c.p.
$Ag_{0.438}Hg_{0.562}$	0.64	$D8_2$
$AgIn_2$	~2.4	C16
$Ag_{0.1}In_{0.9}Te$ ($n = 1.4 \times 10^{22}$)*	1.2–1.89	B1
$Ag_{0.2}In_{0.8}Te$ ($n = 1.07 \times 10^{22}$)	0.77–1.00	B1
AgLa	0.94	B2-cP2 (CsCl)
AgLa (9.5 kbar)	1.2	B2
AgLu	0.33	B2-cP2
$AgMo_4S_5$	9.1	hR15 (Mo_6PbS_8)
$Ag_{1.2}Mo_6Se_8$	5.9	Same
Ag_7NO_{11}	1.04	Cubic
Ag_xPb_{1-x}	7.2 max.	
Ag_4Sn	0.1	h**
Ag_xSn_{1-x}	1.5–3.7	
Ag_xSn_{1-x} (film)	2.0–3.8	
$AgTe_3$	2.6	Cubic
AgTh	2.2	C16-tI12 (Al_2Cu)
$AgTh_2$	2.26	C16
$Ag_{0.03}Tl_{0.97}$	2.67	
$Ag_{0.94}Tl_{0.06}$	2.32	
AgY	0.33	B2-cP2 (CsCl)
Ag_xZn_{1-x}	0.5–0.845	
$AlAu_4$	0.4–0.7	Like A13
Al_2Au	0.1	C1-cF12 (CaF_2)
Al_2CMo_3	9.8–10.2	A13+trace 2nd. phase
Al_2CaSi	5.8	
$Al_{0.131}Cr_{0.088}V_{0.781}$	1.46	Cubic
$AlGe_2$	1.75	
Al_2Ge_2U	1.6	LI_2-cP4 (Cu_3Au)
$AlLa_3$	5.57	DO_{19}
Al_2La	3.23	C15
Al_2Lu	1.02	C15-cF24 (Cu_2Mg)
Al_3Mg_2	0.84	F.C.C.
$AlMo_3$	0.58	A15
$AlMo_6Pd$	2.1	
AlN	1.55	B4
Al_2NNb_3	1.3	A13
Al_3Nb	0.64	tI8 (Al_3Ti)
AlOs	0.39	B2
Al_3Os	5.90	
AlPb (film)	1.2–7	
Al_2Pt	0.48–0.55	C1
Al_5Re_{24}	3.35	A12
AlSb	2.8	B4-tI4 (Sn)
Al_2Sc	1.02	C15-cF24 (Cu_2Mg)
Al_2Si_2U	1.34	LI_2-cP4 (Cu_3Au)
$AlTh_2$	0.1	C16-tI12 (Al_2Cu)
Al_3Th	0.75	DO_{19}
$Al_xTi_yV_{1-x-y}$	2.05–3.62	Cubic
$Al_{0.108}V_{0.892}$	1.82	Cubic
Al_2Y	0.35	C15-cF24 (Cu_2Mg)
Al_3Yb	0.94	LI_2-cP4 (Cu_3Au)
Al_xZn_{1-x}	0.5–0.845	
$AlZr_3$	0.73	LI_2
AsBiPb	9.0	
AsBiPbSb	9.0	
AsHfOs	3.2	C22-hP9 (Fe_2P)
AsHfRu	4.9	same
$As_{0.33}InTe_{0.67}$ ($n = 1.24 \times 10^{22}$)	0.85–1.15	B1
$As_{0.5}InTe_{0.5}$ ($n = 0.97 \times 10^{22}$)	0.44–0.62	B1
As_4La_3	0.6	cI28 (Th_3P_4)
$AsNb_3$	0.3	$L1_2$-tP32
$As_{0.50}Ni_{0.06}Pd_{0.44}$	1.39	C2
$AsNi_{0.25}Pd_{0.75}$	1.6	$B8_1$-hP4 (NiAs)
AsOsZr	8.0	C22-hP9 (Fe_2P)
AsPb	8.4	
$AsPd_2$ (low-temp. phase)	0.60	Hexagonal
$AsPd_2$ (high-temp. phase)	1.70	C22
$AsPd_5$	0.46	Complex
As_3Pd_5	1.9	
AsRh	0.58	B31
$AsRh_{1.4-1.6}$	< 0.03–0.56	Hexagonal
AsSn	4.10	
AsSn ($n = 2.14 \times 10^{22}$)	3.41–3.65	B1
$As_{\sim2}Sn_{\sim3}$	3.5–3.6; 1.21–1.17	
As_3Sn_4 ($n = 0.56 \times 10^{22}$)	1.16–1.19	Rhombohedral
AsV_3	0.20	A15-cP8 (Cr_3Si)
Au_5Ba	0.4–0.7	$D2_d$
AuBe	2.64	B20
Au_2Bi	1.80	C15
Au_5Ca	0.34–0.38	$C15_b$
$AuGa_2$	1.6	C1-cF12 (CaF_2)
AuGa	1.2	B31
$Au_{0.40-0.92}Ge_{0.60-0.08}$	<0.32–1.63	Complex

Solids

Substance	T_c/K	Crystal structure type
$AuIn_2$	0.2	C1-cF12
$AuIn$	0.4–0.6	Complex
$AuLu$	<0.35	B2
$AuNb_3$	1.2	A2
$AuPb_2$	3.15	
$AuPb_2$ (film)	4.3	
$AuPb_3$	4.40	
$AuPb_3$ (film)	4.25	
Au_2Pb	1.18; 6–7	C15
$AuSb_2$	0.58	C2
$AuSn$	1.25	$B8_1$
Au_xSn_{1-x} (film)	2.0–3.8	
Au_5Sn	0.7–1.1	A3
$AuTa_{4.3}$	0.55	A15-cP8 (Cr_3Si)
Au_3Te_5	1.62	Cubic
$AuTh_2$	3.08	C16
$AuTl$	1.92	
AuV_3	0.74	A15
Au_xZn_{1-x}	0.50–0.845	
$AuZn_3$	1.21	Cubic
Au_xZr_y	1.7–2.8	A3
$AuZr_3$	0.92	A15
$B_2Ba_{0.67}Pt_3$	5.60	hP12 (B_2BaPt_3)
$BCMo_2$	5.4	Orthorhombic
$BCMo_2$	5.3–7.0	Same
$B_2Ca_{0.67}Pt_3$	1.57	hP12
B_4ErIr_4	2.1	tP18 (B_4CeCo_4)
B_4ErRh_4	4.3	oC108 (B_4LuRh_4)
B_4ErRh_4	8.7	tP18 (B_4CeCo_4)
BHf	3.1	Cubic
B_4HoIr_4	2.0	tP18
B_4HoRh_4	1.4	oC108
B_2Ir_3La	1.65	hP6 ($CaCu_5$)
B_2Ir_3Th	2.09	Same
B_4Ir_4Tm	1.6	tP18
B_6La	5.7	
B_2LaRh_3	2.82	hP6
$B_{12}Lu$	0.48	
B_2LuOs	2.66	oP16 (B_2LuRu)
B_2LuOs_3	4.62	hP6
B_4LuRh_4	6.2	oC108
B_2LuRu	9.86	oP16
B_4LuRu_4	2.0	tI72 (B_4LuRu_4)
BMo	0.5[a]	
BMo_2	4.74	C16
BNb	8.25	B_f
B_4NdRh_4	5.3	tP18
B_2OsSc	1.34	oP16
B_2OsY	2.22	oP16
$B_2Pt_3Sr_{0.67}$	2.78	hP12 (B_2BaPt_3)
BRe_2	2.80; 4.6	
$B_4Rh_{3.4}Ru_{0.6}$	8.38	tI72
B_4Rh_4Sm	2.7	tP18
B_4Rh_4Th	4.3	Same
B_4Rh_4Tm	9.8	Same
B_4Rh_4Tm	5.4	oC108
$B_{0.3}Ru_{0.7}$	2.58	$D10_2$
B_4Ru_4Sc	7.2	tI72
B_2Ru_3Th	1.79	hP6
B_2Ru_3Y	2.85	Same
$B_2Ru Y$	7.80	oP16

Substance	T_c/K	Crystal structure type
B_4Ru_4Y	1.4	tI72
$B_{12}Sc$	0.39	
BTa	4.0	B_f
BTa_2	3.12	C16-tI12 (Al_2Cu)
B_6Th	0.74	
BW_2	3.1	C16
B_6Y	6.5–7.1	
$B_{12}Y$	4.7	
BZr	3.4	Cubic
$B_{12}Zr$	5.82	
$BaBi_3$	5.69	Tetragonal
$Ba_2Mo_{15}Se_{19}$	2.75	hP15 (Mo_6PbS_8)
$Ba_xO_3Sr_{1-x}Ti$ ($n = 4.2 \times 10^{19}$)	<0.1–0.55	
$Ba_{0.13}O_3W$	1.9	Tetragonal
$Ba_{0.14}O_3W$	<1.25–2.2	Hexagonal
$BaRh_2$	6.0	C15
$Be_{22}Mo$	2.51	Cubic ($Be_{22}Re$)
$Be_8Nb_5Zr_2$	5.2	
$Be_{0.98-0.92}Re_{0.02-0.08}$ (quenched)	9.5–9.75	Cubic
$Be_{0.957}Re_{0.043}$	9.62	Cubic ($Be_{22}Re$)
$BeTc$	5.21	Cubic
$Be_{22}W$	4.12	Cubic ($Be_{22}Re$)
$Be_{13}W$	4.1	Tetragonal
Bi_3Ca	2.0	
$Bi_{0.5}Cd_{0.13}Pb_{0.25}Sn_{0.12}$ (weight fractions)	8.2	
$BiCo$	0.42–0.49	
Bi_2Cs	4.75	C15
Bi_xCu_{1-x} (electrodeposited)	2.2	
$BiCu$	1.33–1.40	
Bi_3Fe	1.0	m**
$Bi_{0.019}In_{0.981}$	3.86	
$Bi_{0.05}In_{0.95}$	4.65	α-phase
$Bi_{0.10}In_{0.90}$	5.05	Same
$Bi_{0.15-0.30}In_{0.85-0.70}$	5.3–5.4	α- and β-phases
$Bi_{0.34-0.48}In_{0.66-0.52}$	4.0–4.1	
Bi_3In_5	4.1	
$BiIn_2$	5.65	β-phase
Bi_2Ir	1.7–2.3	
Bi_2Ir (quenched)	3.0–3.96	
BiK	3.6	
Bi_2K	3.58	C15
$BiLi$	2.47	$L1_o$, α-phase
$Bi_{4-9}Mg$	0.7–~1.0	
Bi_3Mo	3–3.7	
$BiNa$	2.25	$L1_o$
$BiNb_3$	4.5	A15-cP8 (Cr_3Si)
$BiNb_3$ (high pressure and temperature)	3.05	A15
$BiNi$	4.25	$B8_1$
Bi_3Ni	4.06	Orthorhombic
$BiNi_{0.5}Rh_{0.5}$	3.0	$B8_1$-hP4 (AsNi)
$Bi_{0.5}NiSb_{0.5}$	2.0	Same
$Bi_{1-0}Pb_{0-1}$	7.26–9.14	
$Bi_{1-0}Pb_{0-1}$ (film)	7.25–8.67	
$Bi_{0.05-0.40}Pb_{0.95-0.60}$	7.35–8.4	H.C.P. to ε-phase
Bi_2Pb	4.25	t**
$BiPbSb$	8.9	
$Bi_{0.5}Pb_{0.31}Sn_{0.19}$ (weight fractions)	8.5	
$Bi_{0.5}Pb_{0.25}Sn_{0.25}$	8.5	
$BiPd_2$	4.0	
$Bi_{0.4}Pd_{0.6}$	3.7–4	Hexagonal, ordered

Solids

Substance	T_c/K	Crystal structure type
BiPd	3.7	Orthorhombic
Bi_2Pd	1.70	Monoclinic, α-phase
Bi_2Pd	4.25	Tetragonal, β-phase
$BiPd_{0.45}Pt_{0.55}$	3.7	$B8_1$-hP4 (NiAs)
BiPdSe	1.0	C2
BiPdTe	1.2	C2
BiPt	1.21	$B8_1$
$Bi_{0.1}PtSb_{0.9}$	2.05; 1.5	$B8_1$-hP4 (NiAs)
BiPtSe	1.45	C2
BiPtTe	1.15	C2
Bi_2Pt	0.155	Hexagonal
Bi_2Rb	4.25	C15
$BiRe_2$	1.9–2.2	
BiRh	2.06	$B8_1$
Bi_3Rh	3.2	Orthorhombic (NiB_3)
Bi_4Rh	2.7	Hexagonal
BiRu	5.7	m**
Bi_3Sn	3.6–3.8	
BiSn	3.8	
Bi_xSn_y	3.85–4.18	
Bi_3Sr	5.62	$L1_2$
Bi_3Te	0.75–1.0	
Bi_5Tl_3	6.4	
$Bi_{0.26}Tl_{0.74}$	4.4	Cubic, disordered
$Bi_{0.26}Tl_{0.74}$	4.15	$L1_2$, ordered (?)
Bi_2Y_3	2.25	
Bi_3Zn	0.8–0.9	
$Bi_{0.3}Zr_{0.7}$	1.51	
$BiZr_3$	2.4–2.8	
$BrMo_6Se_7$	7.1	hP15 (Mo_6PbS_8)
$Br_3Mo_6Se_5$	7.1	Same
CCs_x	0.020–0.135	Hexagonal
CFe_3	1.30	DO_{11}-oP16 (Fe_3C)
$CGaMo_2$	3.7–4.1	Hexagonal
$CHf_{0.5}Mo_{0.5}$	3.4	B1
$CHf_{0.3}Mo_{0.7}$	5.5	B1
$CHf_{0.25}Mo_{0.75}$	6.6	B1
$CHf_{0.7}Nb_{0.3}$	6.1	B1
$CHf_{0.6}Nb_{0.4}$	4.5	B1
$CHf_{0.5}Nb_{0.5}$	4.8	B1
$CHf_{0.4}Nb_{0.6}$	5.6	B1
$CHf_{0.25}Nb_{0.75}$	7.0	B1
$CHf_{0.2}Nb_{0.8}$	7.8	B1
$CHf_{0.9-0.1}Ta_{0.1-0.9}$	5.0–9.0	B1
CK (excess K)	0.55	Hexagonal
C_8K	0.39	Hexagonal
C_2La	1.66	tI6 (CaC_2)
C_2Lu	3.33	Same
$C_{0.40-0.44}Mo_{0.60-0.56}$	9–13	
C_3MoRe	3.8	B1-cF8
$C_{0.6}Mo_{4.8}Si_3$	7.6	$D8_8$
$CMo_{0.2}Ta_{0.8}$	7.5	B1
$CMo_{0.5}Ta_{0.5}$	7.7	B1
$CMo_{0.75}Ta_{0.25}$	8.5	B1
$CMo_{0.8}Ta_{0.2}$	8.7	B1
$CMo_{0.85}Ta_{0.15}$	8.9	B1
CMo_xV_{1-x}	2.9–9.3	B1
CMo_xZr_{1-x}	9.8	B1
$C_{0.984}Nb$	9.8	B1
CNb_2	9.1	
CNb_xTi_{1-x}	<4.2–8.8	B1
$CNb_{0.1-0.9}Zr_{0.9-0.1}$	4.2–8.4	B1
CRb_x (Au)	0.023–0.151	Hexagonal
$CRe_{0.06}W$	5.0	
CRu	2.00	hP2 (CW)
$C_{0.98}7Ta$	9.7	
$C_{0.848-0.987}$	2.04–9.7	
CTa (film)	5.09	B1
CTa_2	3.26	L'_3
$CTa_{0.4}Ti_{0.6}$	4.8	B1
$Cta_{1-0.4}W_{0-0.6}$	8.5–10.5	B1
$CTa_{0.2-0.9}Zr_{0.8-0.1}$	4.6–8.3	B1
CTc (excess C)	3.85	Cubic
$CTi_{0.5-0.7}W_{0.5-0.3}$	6.7–2.1	B1
CW	1.0	
CW_2	2.74	L'_3
CW_2	5.2	F.C.C.
C_2Y	3.88	tI6 (CaC_2)
$Ca_3Co_4Sn_{13}$	5.9	cP40 ($Pr_3Rh_2Sn_{13}$)
$Ca_3Ge_{13}Rh_4$	2.1	Same
CaHg	1.6	B2-cP2 (CsCl)
$CaHg_3$	1.6	hP8 (Ni_3Sn)
$CaIr_2$	6.15	C15
$Ca_3Ir_4Sn_{13}$	7.1	cP40
$Ca_xO_3Sr_{1-x}Ti$ ($n = 3.7–11 \times 10^{19}$)	< 0.1–0.55	
$Ca_{0.1}O_3W$	1.4–3.4	Hexagonal
CaPb	7.0	
$CaRh_2$	6.40	C15
$CaRh_{1.2}Sn_{4.5}$	8.7	cP40
$CaTl_3$	2.0	B2-cP2
$Cd_{0.3-0.5}Hg_{0.7-0.5}$	1.70–1.92	
CdHg	1.77; 2.15	Tetragonal
$Cd_{0.0075-0.05}In_{0.9925-0.95}$	3.24–3.36	Tetragonal
$Cd_{0.97}Pb_{0.03}$	4.2	
CdSn	3.65	
$Cd_{0.17}Tl_{0.83}$	2.3	
$Cd_{0.18}Tl_{0.82}$	2.54	
$CeCo_2$	0.84	C15
$CeCo_{1.67}Ni_{0.33}$	0.46	C15
$CeCo_{1.67}Rh_{0.33}$	0.47	C15
$Ce_xGd_{1-x}Ru_2$	3.2–5.2	C15
$CeIr_3$	3.34	
$CeIr_5$	1.82	
$Ce_{0.005}La_{0.995}$	4.6	
Ce_xLa_{1-x}	1.3–6.3	
$Ce_xPr_{1-x}Ru_2$	1.4–5.3	C15
Ce_xPt_{1-x}	0.7–1.55	
$CeRu_2$	6.0	C15
$Ce_3Mo_6Se_5$	5.7	hR15 (Mo_6PbS_8)
$Ce_2Mo_6Te_6$	1.7	Same
$Co_xFe_{1-x}Si_2$	1.4 (max.)	C1
$CoHf_2$	0.56	$E9_3$
$CoLa_3$	4.28	
$Co_4La_3Sn_{13}$	2.8	cP40
$CoLu_3$	~0.35	
Co_xLuSn_y	1.5	cP40
$Co_{0-0.01}Mo_{0.8}Re_{0.2}$	2–10	
$Co_{0.02-0.10}Nb_3Rh_{0.98-0.90}$	2.28–1.90	A15
$Co_xNi_{1-x}Si_2$	1.4 (max.)	C1
$Co_{0.5}Rh_{0.5}Si_2$	2.5	
$Co_xRh_{1-x}Si_2$	3.65 (max.)	
$Co_{\sim 0.3}So_{\sim 0.7}$	~0.35	

Substance	T_c/K	Crystal structure type
$Co_4Sc_5Si_{10}$	5.0	tP38 ($Co_4Sc_5Si_{10}$)
$CoSi_2$	1.40; 1.22	C1
Co_xSn_yYb	2.5	cP40
Co_3Th_7	1.83	$D10_2$
Co_xTi_{1-x}	2.8 (max.)	Co in α-Ti
Co_xTi_{1-x}	3.8 (max.)	Co in β-Ti
$CoTi_2$	3.44	$E9_3$
CoTi	0.71	A2
CoU	1.7	B2, distorted
CoU_6	2.29	$D2_c$
$Co_{0.28}Y_{0.72}$	0.34	
CoY_3	<0.34	
$CoZr_2$	6.3	C16
$Co_{0.1}Zr_{0.9}$	3.9	A3
$Cr_{0.6}Ir_{0.4}$	0.4	H.C.P.
$Cr_{0.65}Ir_{0.35}$	0.59	H.C.P.
$Cr_{0.7}Ir_{0.3}$	0.76	H.C.P.
$Cr_{0.72}Ir_{0.28}$	0.83	
Cr_3Ir	0.45	A15
$Cr_{0-0.1}Nb_{1-0.9}$	4.6–9.2	A2
$Cr_{0.80}Os_{0.20}$	2.5	Cubic
Cr_3Os	4.68	A15-cP8 (Cr_3Si)
Cr_xRe_{1-x}	1.2–5.2	
$Cr_{0.4}Re_{0.6}$	2.15	$D8_b$
$Cr_{0.8-0.6}Rh_{0.2-0.4}$	0.5–1.10	A3
Cr_3Rh	0.3	A15-cP8
Cr_3Ru (annealed)	3.3	A15
Cr_2Ru	2.02	$D8_b$
Cr_3Ru_2	2.10	$D8_b$-tP30 (CrFe)
$Cr_{0.1-0.5}Ru_{0.9-0.5}$	0.34–1.65	A3
Cr_xTi_{1-x}	3.6 (max.)	Cr in α-Ti
Cr_xTi_{1-x}	4.2 (max.)	Cr in β-Ti
$Cr_{0.1}Ti_{0.3}V_{0.6}$	5.6	
$Cr_{0.0175}U_{0.9825}$	0.75	β-phase
$Cs_{0.32}O_3W$	1.12	Hexagonal
$Cu_{0.15}In_{0.85}$ (film)	3.75	
$Cu_{0.04-0.08}In_{0.94-0.92}$	4.4	
CuLa	5.85	
$Cu_2Mo_6O_2S_6$	9	hR15 (Mo_6PbS_8)
$Cu_2Mo_6Se_8$	5.9	Same
Cu_xPb_{1-x}	5.7–7.7	
CuS	1.62	B18
CuS_2	1.48–1.53	C18
CuSSe	1.5–2.0	C18
$CuSe_2$	2.3–2.43	C18
CuSeTe	1.6–2.0	C18
Cu_xSn_{1-x}	3.2–3.7	
Cu_xSn_{1-x} (film, made at 10 K)	3.6–7	
Cu_xSn_{1-x} (film, made at 300 K)	2.8–3.7	
$CuTe_2$	<1.25–1.3	C18
$CuTh_2$	3.49	C16
$Cu_{0-0.027}V$	3.9–5.3	A2
CuY	0.33	B2-cP2 (CsCl)
Cu_xZn_{1-x}	0.5–0.845	
$DyMo_6S_8$	2.1	hR15
Er_xLa_{1-x}	1.4–6.3	
$ErMo_6S_8$	2.2	hR15
$ErMo_6Se_8$	6.2	hR15
$Fe_3Lu_2Si_5$	6.1	tP40 ($Fe_3Sc_2Si_5$)
$Fe_{0-0.04}Mo_{0.8}Re_{0.2}$	1–10	
$Fe_{0.05}Ni_{0.05}Zr_{0.90}$	~3.9	

Substance	T_c/K	Crystal structure type
Fe_3Re_2	6.55	$D8_b$-tP30 (FeCr)
$Fe_3Sc_2Si_5$	4.52	tP40
Fe_3Si_5Tm	1.3	Same
$Fe_3Si_5Y_2$	2.4	Same
Fe_3Th_7	1.86	D10
Fe_xTi_{1-x}	3.2 (max.)	Fe in α-Ti
Fe_xTi_{1-x}	3.7 (max.)	Fe in β-Ti
$Fe_xTi_{0.6}V_{1-x}$	6.8 (max.)	
FeU_6	3.86	$D2_c$
$Fe_{0.1}Zr_{0.9}$	1.0	A3
$Ga_{0.5}Ge_{0.5}Nb_3$	7.3	A15
Ga_2Ge_2U	0.87	B2-cP2
$GaHf_2$	0.21	C16-tI12 (Al_2Cu)
$GaLa_3$	5.84	
Ga_3Lu	2.3	B2-cP2
Ga_2Mo	9.5	
$GaMo_3$	0.76	A15
GaN (black)	5.85	B4
$Ga_{0.7}Pt_{0.3}$	2.9	C1
GaPt	1.74	B20
GaSb (120 kbar, 77 K, annealed)	4.24	A5
GaSb (unannealed)	~5.9	
$Ga_{0-1}Sn_{1-0}$ (quenched)	3.47–4.18	
$Ga_{0-1}Sn_{1-0}$ (annealed)	2.6–3.85	
GaTe	0.17	mC24 (GaTe)
Ga5V2	3.55	Tetragonal (Mn_2Hg_5)
$GaV_{4.5}$	9.15	
Ga_3Zr	1.38	
Ga_3Zr_5	3.8	D8b-hP16 (Mn_5Si_3)
Gd_xLa_{1-x}	< 1.0–5.5	
$GdMo_6S_8$	3.5	hR15
$GdMo_6Se_8$	5.6	hR15
$Gd_xOs_2Y_{1-x}$	1.4–4.7	
$Gd_xRu_2Th_{1-x}$	3.6 (max.)	C15
$Ge_{10}As_4Y_5$	9.06	tP38 ($C0_4Sc_5Si_{10}$)
GeIr	4.7	B31
GeIrLa	1.64	tI12 (LaPtSi)
$Ge_{10}Ir_4Lu_5$	2.60	tP38
$Ge_{10}Ir_4Y_5$	2.62	tP38
Ge_2La	1.49; 2.2	Orthorhombic, distorted (Mn_2Hg_5)
GeLaPt	3.53	tI12
$Ge_{13}Lu_3Os_4$	3.6	cP40 ($Pr_3Rh_2Sn_{13}$)
$Ge_{10}Lu_5Rh_4$	2.79	tP38
$Ge_{13}Lu_3Ru_4$	2.3	cP40
$GeMo_3$	1.43	A15
$GeNb_2$	1.9	
$Ge_{0.29}Nb_{0.71}$	6	A15
GePt	0.40	B31
Ge_3Rh_5	2.12	Orthorhombic, related to $InNi_2$
GeRh	0.96	B31-oP8 (MnP)
$Ge_{13}Rh_4Sc_3$	1.9	c P40
$Ge_{10}Rh_4Y_5$	1.35	tP38
$Ge_{13}Ru_4Y_3$	1.7	cP40
Ge_2So	1.3	
$GeTa_3$	8.0	A15-cP8 (Cr_3Si)
Ge_3Te_4 ($n = 1.06 \times 10^{22}$)	1.55–1.80	Rhombohedral
Ge_xTe_{1-x} ($n = 8.5–64 \times 10^{20}$)	0.07–0.41	R1
GeV_3	6.01	A15
Ge_2Y	3.80	C_c
$Ge_{1.62}Y$	2.4	

Substance	T_c/K	Crystal structure type
Ge_2Zr	0.30	oC12 ($ZrSi_2$)
$GeZr_3$	0.4	$L1_2$-tP32 (Ti_3P)
$H_{0.33}Nb_{0.67}$	7.28	B.C.C.
$H_{0.1}Nb_{0.9}$	7.38	Same
$H_{0.05}Nb_{0.95}$	7.83	Same
$H_{0.12}Ta_{0.88}$	2.81	B.C.C.
$H_{0.08}Ta_{0.92}$	3.26	Same
$H_{0.04}Ta_{0.96}$	3.62	Same
HfIrSi	3.50	C37-cP12 (Co_2Si)
$HfMo_2$	0.05	hP24 (Ni_2Mn)
$HfN_{0.989}$	6.6	B1
$Hf_{0-0.5}Nb_{1-0.5}$	8.3–9.5	A2
$Hf_{0.75}Nb_{0.25}$	> 4.2	
$HfOs_2$	2.69	C14
HfOsP	6.1	C22-hP9 (Fe_2P)
HfPRu	9.9	Same
$HfRe_2$	4.80	C14
$Hf_{0.14}Re_{0.86}$	5.86	A12
$Hf_{0.99-0.96}Rh_{0.01-0.04}$	0.85–1.51	
$Hf_{0-0.55}Ta_{1-0.45}$	4.4–6.5	A2
HfV_2	8.9–9.6	C15
Hg_xIn_{1-x}	3.14–4.55	
HgIn	3.81	
Hg_2K	1.20	Orthorhombic
Hg_3K	3.18	
Hg_4K	3.27	
Hg_8K	3.42	
Hg_3Li	1.7	Hexagonal
$HgMg_3$	0.17	hP8 (Na_3As)
Hg_2Mg	4.0	tI6 ($MoSi_2$)
Hg_3Mg_5	0.48	$D8_b$-hP16 (Mn_5Si_3)
Hg_2Na	1.62	Hexagonal
Hg_4Na	3.05	
Hg_xPb_{1-x}	4.14–7.26	
HgSn	4.2	
Hg_xTl_{1-x}	2.30–4.19	
Hg_5Tl_2	3.86	
Ho_xLa_{1-x}	1.3–6.3	
$Ho_{1.2}Mo_6Se_8$	6.1	$D10_2$-hR12 (Be_3Nb)
$In_{1-0.86}Mg_{0-0.14}$	3.395–3.363	
$In_2Mo_6Te_6$	2.6	hR15 (Mo_6PbS_8)
$InNb_3$ (high pressure and temp.)	4–8; 9.2	A15
$In_{0.5}Nb_3Zr_{0.5}$	6.4	
$In_{0.11}O_3W$	< 1.25–2.8	Hexagonal
$In_{0.95-0.85}Pb_{0.05-0.15}$	3.6–5.05	
$In_{0.98-0.91}Pb_{0.02-0.09}$	3.45–4.2	
InPb	6.65	
InPd	0.7	B2
InSb (quenched from 170 kbar into liquid N_2)	4.8	Like A5
InSb	2.1	B4
$(InSb)_{0.95-0.10}Sn_{0.05-0.90}$ (various heat treatments)	3.8–5.1	
$(InSb)_{0-0.07}Sn_{1-0.93}$	3.67–3.74	
In_3Sn	~5.5	
In_xSn_{1-x}	3.4–7.3	
$In_{0.82-1}Te$ ($n = 0.83–1.71 \times 10^{22}$)	1.02–3.45	B1
$In_{1.000}Te_{1.002}$	3.5–3.7	B1
In_3Te_4 ($n = 4.7 \times 10^{21}$)	1.15–1.25	Rhombohedral
In_xTl_{1-x}	2.7–3.374	
$In_{0.8}Tl_{0.2}$	3.223	
$In_{0.62}Tl_{0.38}$	2.760	
$In_{0.78-0.69}Tl_{0.22-0.31}$	3.18–3.32	Tetragonal
$In_{0.69-0.62}Tl_{0.31-0.38}$	2.98–3.3	F.C.C.
Ir_2La	0.48	C15
Ir_3La	2.32	$D10_2$
Ir_3La_7	2.24	$D10_2$
Ir_5La	2.13	
$IrLaSi_2$	2.03	oC16 ($CeNiSi_2$)
$IrLaSi_3$	2.7	tI10 ($BaNiSn_3$)
Ir_2Lu	2.47	C15
Ir_3Lu	2.89	C15
$Ir_4Lu_5Si_{10}$	3.9	tP38 ($Co_4Sc_5Si_{10}$)
IrMo	< 1.0	A3
$IrMo_3$	9.6	A15
$IrMo_3$	6.8	$D8_b$
$IrNb_3$	1.9	A15
$Ir_{0.4}Nb_{0.6}$	9.8	$D8_b$
$Ir_{0.37}Nb_{0.63}$	2.32	$D8_b$
IrNb	7.9	$D8_b$
$Ir_{1.15}Nb_{0.85}$	4.6	oP12 (IrTa)
$Ir_{0.02}Nb_3Rh_{0.98}$	2.43	A15
$Ir_{0.05}Nb_3Rh_{0.95}$	2.38	A15
$Ir_{0.287}O_{0.14}Ti_{0.573}$	5.5	$E9_3$
$Ir_{0.265}O_{0.035}Ti_{0.65}$	2.30	$E9_3$
Ir_xOs_{1-x}	0.3–0.98	
$Ir_{1.5}Os_{0.5}$	2.4	C14
IrOsY	2.6	C15
IrSiY	2.70	C37-oP12 (Co_2Si)
IrSiZr	2.04	Same
Ir_2Sc	2.07	C15
$Ir_{2.5}Sc$	2.46	C15
$Ir_4Sc_5Si_{10}$	8.46	tP38
Ir_2Si_2Th	2.14	tI10
$IrSi_3Th$	1.75	tI10
IrSiTh	6.50	tI12 (LaPtSi)
Ir_2Si_2Y	2.60	tI10 (Al4Ba)
$Ir_4Si_{10}Y_5$	3.10	tP38
$Ir_3Si_5Y_2$	2.83	oI40
$IrSn_2$	0.65–0.78	C1
Ir_2Sr	5.70	C15
Ir_7Ta_{13}	1.2	$D8_b$-tP30 (FeCr)
$Ir_{0.5}Te_{0.5}$	~3	
$IrTe_3$	1.18	C2
IrTh	< 0.37	B_f
Ir_2Th	6.50	C15
Ir_3Th	4.71	
Ir_3Th_7	1.52	$D10_2$
Ir_5Th	3.93	$D2_d$
$IrTi_3$	5.40	A15
IrV_2	1.39	A15
IrW_3	3.82	
$Ir_{0.28}W_{0.72}$	4.49	
Ir_2Y	2.18; 1.38	C15
$Ir_{0.69}Y_{0.31}$	1.98; 1.44	C15
$Ir_{0.70}Y_{0.30}$	2.16	C15
Ir_2Y_3	1.61	
Ir_3Y	3.50	$D10_2$-hR13 (Be_3Nb)
Ir_xY_{1-x}	0.3–3.7	
Ir_2Zr	4.10	C15
$Ir_{0.1}Zr_{0.9}$	5.5	A3
$K_2Mo_{15}S_{19}$	3.32	hR15
$K_{0.27-0.31}O_3W$	0.50	Hexagonal

Substance	T_c/K	Crystal structure type
$K_{0.40-0.57}O_3W$	1.5	Tetragonal
$La_{0.55}Lu_{0.45}$	2.2	Hexagonal, La type
$La_{0.8}Lu_{0.2}$	3.4	Same
$LaMg_2$	1.05	C15
$LaMo_6S_8$	7.1	hR15
LaN	1.35	
$LaOs_2$	6.5	C15
$LaPt_2$	0.46	C15
$La_{0.28}Pt_{0.72}$	0.54	C15
LaPtSi	3.48	tI12
$LaRh_3$	2.60	
$LaRh_5$	1.62	
La_7Rh_3	2.58	$D10_2$
$LaRhSi_2$	3.42	oC16 ($CeNiSi_2$)
$La_2Rh_3Si_5$	4.45	oI40 ($Co_3Si_5U_2$)
$LaRhSi_3$	2.7	tI10 ($BaNiSn_3$)
$LaRh_2Si_2$	3.90	tI10 (Al_4Ba)
$LaRu_2$	1.63	C15
La_3S_4	6.5	$D7_3$
La_3Se_4	8.6	$D7_3$
$LaSi_2$	2.3	C_c
La_xY_{1-x}	1.7–5.4	
LaZn	1.04	B2
$Li_2Mo_6S_8$	4.2	hR15
LiPb	7.2	
$LuOs_2$	3.49	C14
$Lu_{0.275}Rh_{0.725}$	1.27	C15
$LuRh_5$	0.49	
$Lu_5Rh_4Si_{10}$	3.95	tP38 ($Co_4So_5Si_{10}$)
$LuRu_2$	0.86	C14
$Mg_{1.14}Mo_{6.6}S_8$	3.5	hR15
Mg2Nb	5.6	
$Mg_{\sim0.47}Tl_{\sim0.53}$	2.75	B2
MgZn	0.9	A3-oP4 (AuCd)
Mn_xTi_{1-x}	2.3 (max.)	Mn in -Ti
Mn_xTi_{1-x}	1.1–3.0	Mn in -Ti
MnU_6	2.32	$D2_c$
Mo_2N	5.0	F.C.C.
$Mo_6Na_2S_8$	8.6	hR15
Mo_xNb_{1-x}	0.016–9.2	
$Mo_{5.25}Nb_{0.75}Se_8$	6.2	hR15
Mo_6NdSa_8	8.2	hR15
Mo_3Os	7.2	A15
$Mo_{0.62}Cs_{0.38}$	5.65	$D8_b$
Mo_3P	5.31	DO_e
$Mo_6Pb_{1.2}Se_8$	6.75	hR15
$Mo_{0.5}Pd_{0.5}$	3.52	A3
Mo_6PrSe_8	9.2	hR15
MoRe	7.8	$D8_b$-tP30
$MoRe_3$	9.25; 9.89	A12
Mo_xRe_{1-x}	1.2–12.2	
$Mo_{0.42}Re_{0.58}$	6.35	$D8_b$
MoRh	1.97	A3
Mo_xRh_{1-x}	1.5–8.2	B.C.C.
MoRu	9.5–10.5	A3
$Mo_{0.61}Ru_{0.39}$	7.18	$D8_b$
$Mo_{0.2}Ru_{0.8}$	1.66	A3
Mo_3Ru_2	7.0	$D8_b$-tP30
$Mo_4Ru_2Te_8$	1.7	hR15
Mo_6S_8	1.85	hR15
Mo_6S_8Sc	3.6	hR15

Substance	T_c/K	Crystal structure type
$Mo_6S_8Sm_{1.2}$	2.9	hR15
Mo_6S_8Tb	2.0	hR15
Mo_6S_8Tl	8.7	hR15
$Mo_6S_8Tm_{1.2}$	2.1	hR15
$Mo_6S_8Y_{1.2}$	3.0	hR15
Mo_6S_8Yb	9.2	hR15
$Mo_{6.6}S_8Zn_{11}$	3.6	hR15
Mo_3Sb_4	2.1	
Mo_6Se_8	6.3	hR15
$Mo_6Se_8Sm_{1.2}$	6.8	hR15
$Mo_6Se_8Sn_{1.2}$	6.8	hR15
Mo_6Se_8Tb	5.7	hR15
Mo_3Se_3Tl	4.0	hP14
$Mo_6Se_8Tm_{1.2}$	6.3	hR15
Mo_6Se_8Yb	6.2	hR15
Mo_3Si	1.30	A15
$MoSi_{0.7}$	1.34	
Mo_xSiV_{3-x}	4.54–16.0	A15
$Mo_{5.25}Ta_{0.75}Te_8$	1.7	hR15
Mo_6Te_8	1.7	hR15
$Mo_{0.16}Ti_{0.84}$	4.18; 4.25	
$Mo_{0.913}Ti_{0.087}$	2.95	
$Mo_{0.04}Ti_{0.96}$	2.0	Cubic
$Mo_{0.025}Ti_{0.975}$	1.8	
Mo_xU_{1-x}	0.7–2.1	
Mo_xV_{1-x}	0–~5.3	
Mo_2Zr	4.25–4.75	C15
NNb (film)	6–9	B1
$N_xO_yTi_z$	2.9–5.6	Cubic
$N_xO_yV_z$	5.8–8.2	Cubic
$N_{0.34}Re$	4–5	F.C.C.
NTa (film)	4.84	B1
$N_{0.6-0.987}Ti$	< 1.17–5.8	B1
$N_{0.82-0.99}V$	2.9–7.9	B1
NZr	9.8	B1
$N_{0.906-0.984}Zr$	3.0–9.5	B1
$Na_{0.28-0.35}O_3W$	0.56	Tetragonal
$Na_{0.28}Pb_{0.72}$	7.2	
NbO	1.25	
$NbOs_2$	2.52	A12
Nb_3Os	1.05	A15
$Nb_{0.6}Os_{0.4}$	1.89; 1.78	$D8_b$
$Nb_3Os_{0.02-0.10}Rh_{0.98-0.90}$	2.42–2.30	A15
Nb_3P	1.8	$L1_2$tP32 (Ti_3P)
NbPRh	4.08	C37-oP12 (Co_2Si)
$Nb_{0.6}Pd_{0.4}$	1.60	$D8_f$ plus cubic
$Nb_3Pd_{0.02-0.10}Rh_{0.92-0.90}$	2.49–2.55	A15
$Nb_{0.62}Pt_{0.38}$	4.21	$D8_b$
Nb_5Pt_3	3.73	$D8_b$
$Nb_3Pt_{0.02-0.98}Rh_{0.98-0.02}$	2.52–9.6	A15
$NbRe_3$	5.27	$D8_b$-tP30 (FeCr)
$Nb_{0.38-0.18}Re_{0.62-0.82}$	2.43–9.70	A15
NbRe	3.8	$D8_b$-tP30
NbReSi	5.1	oI36 (FeTiSi)
Nb_3Rh	2.64	A15
$Nb_{0.6}Rh_{0.40}$	4.21	$D8_b$ plus other
$Nb_{0.9}Rh_{1.1}$	3.07	A3-oP4 (AuCd)
$Nb_3Rh_{0.98-0.90}Ru_{0.02-0.10}$	2.42–2.44	A15
Nb_xRu_{1-x}	1.2–4.8	
NbRuSi	2.65	oI36
NbS_2	6.1–6.3	Hexagonal, $NbSe_2$ type

Substance	T_c/K	Crystal structure type
NbS_2	5.0–5.5	Hexagonal, three-layer type
Nb_3Sb	0.2	$L1_2$-tP32 (Ti_3P)
$Nb_3Sb_{0-0.7}Sn_{1-0.3}$	6.8–18	A15
$NbSe_2$	5.15–5.62	Hexagonal
$Nb_{1-1.05}Se_2$	2.2–7.0	Same
Nb_3Se_4	2.0	hP14
Nb_3Si	1.5	$L1_2$
Nb_3SiSnV_3	4.0	
$NbSn_2$	2.60	Orthorhombic
Nb_6Sn_5	2.8	oI44 (Sn_5Ti_6)
NbSnTaV	6.2	A15
NbSnV2	5.5	A15
Nb_2SnV	9.8	A15
Nb_xTa_{1-x}	4.4–9.2	A2
Nb_3Te_4	1.8	hP14
Nb_xTi_{1-x}	0.6–9.8	
$Nb_{0.6}Ti_{0.4}$	9.8	
Nb_xU_{1-x}	1.95 (max.)	
$Nb_{0.88}V_{0.12}$	5.7	A2
$Nb_{0.5}V_{1.5}Zr$	4.3	C15-hP12 ($MgZn_2$)
$Ni_{0.3}Th_{0.7}$	1.98	$D10_2$
$NiZr_2$	1.52	
$Ni_{0.1}Zr_{0.9}$	1.5	A3
$O_3Rb_{0.27-0.29}W$	1.98	Hexagonal
OSn	3.81	tP4 (PbO)
O_3SrTi (n = 1.7–12.0 × 10^{19})	0.12–0.37	
O_3SrTi (n = 10^{18}–10^{21})	0.05–0.47	
O_3SrTi (n = 10^{20})	0.47	
$O_3Sr_{0.08}W$	2–4	Hexagonal
OTi	0.58	
$O_3Tl_{0.30}W$	2.0–2.14	Hexagonal
OV_3Zr_3	7.5	$E9_3$
OW_3 (film)	3.35; 1.1	A15
OsPti	1.2	C22-hP9 (Fe_2P)
OsPZr	7.4	Same
OsReY	2.0	C14
Os2Sc	4.6	C14
OsTa	1.95	A12
Os_3Th_7	1.51	$D10_2$
Os_xW_{1-x}	0.9–4.1	
OsW_3	~3	
Os_2Y	4.7	C14
Os_2Zr	3.0	C14
Os_xZr_{1-x}	1.5–5.6	
PPb	7.8	
OsW_2	3.81	$D8_b$-tP30 (FeCr)
$PPd_{3.0-3.2}$	<0.35–0.7	DO_{11}
P_3Pd_7 (high temperature)	1.0	Rhombohedral
P_3Pd_7 (low temperature)	0.70	Complex
PRh	1.22	
PRh_2	1.3	C1
P_4Rh_5	1.22	oP28 ($CaFe_2O_4$)
PRhTa	4.41	C37-oP12 (Co_2Si)
PRhZr	1.55	Same
PRuTi	1.3	C22-hP9 (Fe_2P)
PRuZr	3.46	C37-oP12
PW_3	2.26	DO_e
Pb_2Pd	2.95	C16
Pb_4Pt	2.80	Related to C16
Pb_2Rh	2.66	C16
PbSb	6.6	
PbTe (plus 0.1 w/o Pb)	5.19	
PbTe (plus 0.1 w/o Te)	5.24–5.27	
$PbTl_{0.27}$	6.43	
$PbTl_{0.17}$	6.73	
$PbTl_{0.12}$	6.88	
$PbTl_{0.075}$	6.98	
$PbTl_{0.04}$	7.06	
$Pb_{1-0.26}Tl_{0-0.74}$	7.20–3.68	
$PbTl_2$	3.75–4.1	
Pb_3Zr_5	4.60	$D8_8$
$PbZr_3$	0.76	A15
$Pd_{0.9}Pt_{0.1}Te_2$	1.65	C6
$Pd_{0.05}Ru_{0.05}Zr_{0.9}$	~9	
$Pd_{2.2}S$ (quenched)	1.63	Cubic
$PdSb_2$	1.25	C2
PdSb	1.5	$B8_1$
PdSbSe	1.0	C2
PdSbTe	1.2	C2
Pd_4Se	0.42	Tetragonal
$Pd_{6-7}Se$	0.66	Like Pd_4Te
$Pd_{2.8}Se$	2.3	
Pd_xSe_{1-x}	2.5 (max.)	
PdSi	0.93	B31
PdSn	0.41	B31
$PdSn_2$	3.34	
Pd_2Sn	0.41	C37
Pd_3Sn	0.47–0.64	$B8_2$
Pd_2SnTm	1.77	DO_3-cF16 (BiF_3)
Pd_2SnY	4.92	Same
Pd_2SnYb	1.79	Same
PdTe	2.3; 3.85	$B8_1$
$PdTe_{1.02-1.08}$	2.56–1.88	$B8_1$
$PdTe_2$	1.69	C6
$PdTe_{2.1}$	1.89	C6
$PdTe_{2.3}$	1.85	C6
$Pd_{1.1}Te$	4.07	$B8_1$
Pd_3Te	0.76	cI2 (W)
$PdTh_2$	0.85	C16
$Pd_{0.1}Zr_{0.9}$	7.5	A3
PtSb	2.1	$B8_1$
PtSi	0.88	B31
PtSn	0.37	$B8_1$
$PtSn_4$	2.38	C16-oC20 ($PdSn_4$)
Pt_3Ta_7	1.5	$D8_b$-tP30
$PtTa_3$	0.4	A15-cP8 (Cr_3Si)
PtTe	0.59	Orthorhombic
PtTh	0.44	B_f
Pt_3Th_7	0.98	$D10_2$
Pt_5Th	3.13	
$PtTi_3$	0.58	A15
$Pt_{0.02}U_{0.98}$	0.87	β-phase
$PtV_{2.5}$	1.36	A15
PtV_3	2.87–3.20	A15
$PtV_{3.5}$	1.26	A15
$Pt_{0.5}W_{0.5}$	1.45	A1
Pt_xW_{1-x}	0.4–2.7	
Pt_2Y_3	0.90	
Pt_2Y	1.57; 1.70	C15
Pt_3Y_7	0.82	$D10_2$
PtZr	3.0	A3

Solids

Substance	T_c/K	Crystal structure type
Re_2Sc	4.2	C15-hP12 ($MgZn_2$)
$Re_{24}Sc_5$	2.2	A12-cI58 (Mg)
ReSiTa	4.4	oI36 (FeTiSi)
$Re_3Si_5Y_2$	1.76	tP40 ($Fe_3Sc_2Si_5$)
Re_3Ta_2	1.4	$D8_b$-tP30 (FeCr)
$Re_{0.64}Ta_{0.36}$	1.46	A12
Re_3Ta	6.78	A12-cI58 (Mn)
$Re_{24}Ti_5$	6.60	A12
Re_xTi_{1-x}	6.6 (max.)	
$Re_{0.76}V_{0.24}$	4.52	$D8_b$
Re_3V	6.26	$D8_b$-tP30
$Re_{0.92}V_{0.08}$	6.8	A3
$Re_{0.6}W_{0.4}$	6.0	
$Re_{0.5}W_{0.5}$	5.12	$D8_b$
$Re_{13}W_{12}$	5.2	$D8_b$-tP30
Re_3W	9.0	A12-cI58
Re_2Y	1.83	C14
Re_2Zr	5.9	C14
Re_3Zr	7.40	A12-cI58
Re_6Zr	7.40	Same
$Rh_{17}S_{15}$	5.8	Cubic
$Rh_{\sim0.24}Sc_{\sim0.76}$	0.88; 0.92	
$Rh_4Sc_5Si_{10}$	8.54	tP38
$Rh_4Sc_3Sn_{13}$	4.5	cP40
Rh_xSe_{1-x}	6.0 (max.)	
$RhSi_3Th$	1.76	tI10
$Rh_{0.86}Sc_{1.04}Th$	6.45	tI12
Rh_2Si_2Y	3.11	tI10
$Rh_3Si_5Y_2$	2.70	oI40
$Rh_4Sn_{13}Sr_3$	4.3	cP40
Rh_xSn_yTh	1.9	cI2 (W)
Rh_xSn_yTm	2.3	cP40
$Rh_4Sn_{13}Y_3$	3.2	cP40
Rh_2Sr	6.2	C15
$Rh_{0.4}Ta_{0.6}$	2.35	$D8_b$
$RhTe_2$	1.51	C2
$Rh_{0.67}Te_{0.33}$	0.49	
Rh_xTe_{1-x}	1.51 (max.)	
RhTh	0.36	B_f
Rh_3Th_7	2.15	$D10_2$
Rh_5Th	1.07	
Rh_xTi_{1-x}	2.25–3.95	
$Rh_{0.02}U_{0.98}$	0.96	
RhV_3	0.38	A15
RhW	~3.4	A3
RhY_3	0.65	
Rh_2Y_3	1.48	
Rh_3Y	1.07	C15
Rh_5Y	0.56	
Rh_3Y_7	0.32	hP20 (Fe_3Th_7)
$Rh_{0.005}Zr$ (annealed)	5.8	
$Rh_{0-0.45}Zr_{1-0.55}$	2.1–10.8	
$Rh_{0.1}Zr_{0.9}$	9.0	H.C.P.
Ru_2Sc	1.67	C14
RuSiTa	3.15	oI36
Ru_3Si_2Th	3.98	hP12
Ru_3Si_2Y	3.51	hP12
$Ru_{1.1}Sn_{3.1}Y$	1.3	cP40
Ru_2Th	3.56	C15
RuTi	1.07	B2
$Ru_{0.05}Ti_{0.95}$	2.5	
$Ru_{0.1}Ti_{0.9}$	3.5	
$Ru_xTi_{0.6}V_y$	6.6 (max.)	
Ru_3U	0.15	$L1_2$-cP4
$Ru_{0.45}V_{0.55}$	4.0	B2
RuW	7.5	A3
Ru_2Y	1.52	C14
Ru_2Zr	1.84	C14
$Ru_{0.1}Zr_{0.9}$	5.7	A3
STh	0.5	B1-cF8 (NaCl)
SbSn	1.30–1.42	B1 or distorted
$SbTa_3$	0.72	A15-cP8 (Cr_3Si)
$SbTi_3$	5.8	Same
Sb_2Ti_7	5.2	
$Sb_{0.01-0.03}V_{0.99-0.97}$	3.76–2.63	A2
SbV_3	0.80	A15
SeTh	1.7	B1-cF8
$SiMo_3$	1.4	A15-cP8
Si_2Th	3.2	C_c, α-phase
Si_2Th	2.4	C32, β-phase
$SiV_{2.7}Ru_{0.3}$	2.9	A15
Si_2W_3	2.8; 2.84	
$SiZr_3$	0.5	$L1_2$-tP32 (Ti_3P)
$Sn_{0.174-0.104}Ta_{0.826-0.896}$	6.5–< 4.2	A15
$SnTa_3$	8.35	A15, highly ordered
$SnTa_3$	6.2	A15, partially ordered
$SnTaV_2$	2.8	A15
$SnTa_2V$	3.7	A15
Sn_xTe_{1-x} (n = 10.5–20 × 10^{20})	0.07–0.22	B1
Sn_3Th	3.33	$L1_2$-cP4
$SnTi_3$	5.80	A15-cP8
Sn_xTl_{1-x}	2.37–5.2	
SnV_3	3.8	A15
$Sn_{0.02-0.057}V_{0.98-0.943}$	2.87–~1.6	A2
$SnZr_3$	0.92	A15-cP8
$Ta_{0.025}Ti_{0.975}$	1.3	Hexagonal
$Ta_{0.05}Ti_{0.95}$	2.9	Hexagonal
$Ta_{0.05-0.75}V_{0.95-0.25}$	4.30–2.65	A2
$Ta_{0.8-1}W_{0.2-0}$	1.2–4.4	A2
$Tc_{0.1-0.4}W_{0.9-0.6}$	1.25–7.18	Cubic
$Tc_{0.50}W_{0.50}$	7.52	α plus
$Tc_{0.60}W_{0.40}$	7.88	plus α
Tc_6Zr	9.7	A12
TeY	1.02	B1-cF8
$ThTl_3$	0.87	$L1_2$-cP4
$Th_{0-0.55}Y_{1-0.45}$	1.2–1.8	
$Ti_{0.70}V_{0.30}$	6.14	Cubic
Ti_xV_{1-x}	0.2–7.5	
$Ti_{0.5}Zr_{0.5}$ (annealed)	1.23	
$Ti_{0.5}Zr_{0.5}$ (quenched)	2.0	
Tl_3Y	1.52	$L1_2$-cP4
V_2Zr	8.80	C15
$V_{0.26}Zr_{0.74}$	5.9	
W_2Zr	2.16	C15
YZn	0.33	B2-cP2 (CsCl)

B. Superconductors with T_c > 10 K

Substance	T_c/K	Crystal structure type
Al_2CMo_3	10.0	A13
$Al_{0.5}Ge_{0.5}Nb$	12.6	A15
$Al_{\sim0.8}Ge_{\sim0.2}Nb_3$	20.7	A15
$AlNb_3$	18.0	A15 (Cr_3Si)
$AlNb_3$	12.0	(FeCr)
$A1_xNb_{1-x}$	<4.2–13.5	$D8_b$

Substance	T_c/K	Crystal structure type
$A1_xNb_{1-x}$	12–17.5	A15
$A1_{0.27}Nb_{0.73-0.48}V_{0-0.25}$	14.5–17.5	A15
$A1\,Nb_xV_{1-x}$	4.4–13.5	
$A1_{0.1}Si_{0.9}V_3$	14.05	
$A1V_3$	11.8	A15 (Cr_3Si)
$AuNb_3$	11.5	A15
$Au_{0-0.3}Nb_{1-0.7}$	1.1–11.0	
$Au_{0.02-0.98}Nb_3Rh_{0.98-0.02}$	2.53–10.9	A15
$AuNb_{3(1-x)}V_{3x}$	1.5–11.0	A15
$B_{0.03}C_{0.51}Mo_{0.47}$	12.5	
B_4LuRh_4	11.7	(B_4CeCo_4)
B_2LuRu	10	
B_4Rh_4Y	11.3	(B_4CeCo_4)
$B_{0.1}Si_{0.9}V_3$	15.8	A15
$BaBi_{0.2}O_3Pb_{0.8}$	13.2	
$Ba_2CaCu_2O_8T1_2$	120	
$Ba_2Cu_3LaO_6$	80	
$Ba_2Cu_3O_7Tm$	101	
$Ba_2Cu_3O_7Y$	90	
$(Ba,La)_2CuO_4$	36	A15 (K_2NiF_4)
$Bi_2CaCu_2O_8Sr_2$	110	
$Br_2Mo_6S_6$	13.8	(Mo_6PbS_8)
C_3La	11.0	(C_3Pu_2)
CMo	14.3	B1 (NaC1)
CMo_2	12.2	o**
$C_{0.5}Mo_xNb_{1-x}$	10.8–12.5	B1
CMo_xTi_{1-x}	10.2(max)	B1
$CMo_{0.83}Ti_{0.17}$	10.2	B1
$C_{0-0.38}N_{1-0.62}Ta$	10.0–11.3	
CNb (whiskers)	7.5–10.5	
CNb	11.5	B1
$C_{0.7-1.0}Nb_{0.3-0}$	6–11	B1
CNb_xTa_{1-x}	8.2–13.9	
$CNb_{0.6-0.9}W_{0.4-0.1}$	12.5–11.6	B1
$C_{0.1}Si_{0.9}V_3$	16.4	A15
CTa	10.3	B1
$CTa_{1-0.4}W_{0-0.6}$	8.5–10.5	B1
$C_{0.66}Th_{0.13}Y_{0.21}$	17	(C_3Pu_2)
C_3Y_2	11.5	(C_3Pu_2)
CW	10	B1
$(Ca,La)_2CuO_4$	18	(K_2NiF_4)
$Cu(La,Sr)_2O_4$	39	
$Cu_{1.8}Mo_6S_8$	10.8	(Mo_6PbS_8)
$Cr_{0.3}SiV_{2.7}$	11.3	A15
$GaNb_3$	14.5	A15 (Cr_3Si)
$Ga_xNb_3Sn_{1-x}$	14–18.37	A15
GaV_3	16.8	A15
$GaV_{2.1-3.5}$	6.3–14.45	A15
$GeNb_3$	23.2	A15
$GeNb_3$ (quenched)	6–17	A15
$Ge_xNb_3Sn_{1-x}$	17.6–18.0	A15
$Ge_{0.5}Nb_3Sn_{0.5}$	11.3	
$Ge_{0.1}Si_{0.9}V_3$	14.0	A15
GeV_3	11	A15
$InLa_3$	9.83; 10.4	LI_2 ($AuCu_3$)
$In_{0-0.3}Nb_3Sn_{1-0.7}$	18.0–18.19	A15
InV_3	13.9	A15
$Ir_{0.4}Nb_{0.6}$	10	(FeCr)
$LaMo_6Se_8$	11.4	(Mo_6PbS_8)
LiO_4Ti_2	13.7	($A1_2MgO_4$)
MgB_2	39.0±0.5	C32
MoN	12; 14.8	h*
Mo_3Os	12.7	A15
$Mo_6Pb_{0.9}S_{7.5}$	15.2	(Mo_6PbS_8)
Mo_3Re	10.0; 15	A15
Mo_xRe_{1-x}	1.2–12.2	
$Mo_{0.52}Re_{0.48}$	11.1	
$Mo_{0.57}Re_{0.43}$	14.0	
$Mo_{\sim 0.60}Re_{0.395}$	10.6	
MoRu	9.5–10.5	A3
Mo_3Ru	10.6	A15
Mo_6Se_8T1	12.2	(Mo_6PbS_8)
$Mo_{0.3}SiV_{2.7}$	11.7	A15
Mn_3Si	12.5	A15
Mo_3Tc	15	A15
$Mo_{0.3}Tc_{0.7}$	12.0	A15
Mo_xTc_{1-x}	10.8–15.8	
$MoTc_3$	15.8	
NNb (whiskers)	10–14.5	
NNb (diffusion wires)	16.10	
$N_{0.988}Nb$	14.9; 17.3	B1
$N_{0.824-0.988}Nb$	14.4–15.3	B1
$N_{0.7-0.795}Nb$	11.3–12.9	
NNb_xO_y	13.5–17.0	B1
NNb_xO_y	6.0–11	
$N_{100-42w/o}Nb_{0-58w/o}Ti$	15–16.8	
$N_{100-75w/o}Nb_{0-25w/o}Zr$	12.5–16.35	
NNb_xZr_{1-x}	9.8–13.8	B1
$N_{0.93}Nb_{0.85}Zr_{0.15}$	13.8	B1
NTa	12–14	B1
NZr	10.7	B1
Nb_3Pt	10.9	A15
$Nb_{0.18}Re_{0.82}$	10	(Mn)
Nb_3Si	19	A15
$Nb_{0.3}SiV_{2.7}$	12.8	A15
Nb_3Sn	18.05	A15
$Nb_{0.8}Sn_{0.2}$	18.18; 18.5	A15
Nb_xSn_{1-x} (film)	2.6–18.5	o*
Nb_3Sn_2	16.6	t*
$NbSnTa_2$	10.8	A15
Nb_2SnTa	16.4	A15
$Nb_{2.5}SnTa_{0.5}$	17.6	A15
$Nb_{2.75}SnTa_{0.25}$	17.8	A15
$Nb_{3x}SnTa_{3(1-x)}$	6.0–18.0	
$Nb_2SnTa_{0.5}V_{0.5}$	12.2	A15
$NbTc_3$	10.5	A12
$Nb_{0.75}Zr_{0.25}$	10.8	
$Nb_{0.66}Zr_{0.33}$	10.8	
$PbTa_3$	17	A15
$RhTa_3$	10	A15
$RhZr_2$	10.8; 11.3	C16 ($A1_2Cu$)
$Rh_{0-0.45}Zr_{1-0.55}$	2.1–10.8	
$SiTi_{0.3}V_{2.7}$	10.9	A15
SiV_3	17.1	A15
$SiV_{2.7}Zr_{0.3}$	13.2	A15

* *n* denotes current carriers concentration in cm^{-3}.

[a] Extrapolated.

Solids

TABLE 5. Critical Field Data

Substance	H_o/Oe	Substance	H_o/Oe	Substance	H_o/Oe
Ag_2F	2.5	$CoSi_2$	105	$NbSn_2$	620
Ag_7NO_{11}	57	$Cr_{0.1}Ti_{0.3}V_{0.6}$	1360	$PbTl_{0.27}$	756
Al_2CMo_3	1700	$In_{1-0.86}Mg_{0-0.14}$	272.4–259.2	$PbTl_{0.17}$	796
$BaBi_3$	740	InSb	1100	$PbTl_{0.12}$	849
Bi_2Pt	10	In_xTl_{1-x}	252–284	$PbTl_{0.075}$	880
Bi_3Sr	530	$In_{0.8}Tl_{0.2}$	252	$PbTl_{0.04}$	864
Bi_5Tl_3	>400	$Mg_{0.47}Tl_{0.53}$	220		
CdSn	>266	$Mo_{0.16}Ti_{0.84}$	<985		

TABLE 6. High Critical Magnetic-Field Superconductive Compounds and Alloys

Substance	T_c/K	H_{c1}/kOe	H_{c2}/kOe	H_{c3}/kOe	T_{obs}/K[a]
Al_2CMo_3	9.8–10.2	0.091	156		1.2
$AlNb_3$		0.375			
$Ba_xO_3Sr_{1-x}Ti$	<0.1–0.55	0.0039 max.			
$Bi_{0.5}Cd_{0.1}Pb_{0.27}Sn_{0.13}$			>24		3.06
Bi_xPb_{1-x}	7.35–8.4	0.122 max.	30 max.		4.2
$Bi_{0.56}Pb_{0.44}$	8.8		15		4.2
$Bi_{7.5w/o}Pb_{92.5w/o}$[b]			2.32		
$Bi_{0.099}Pb_{0.901}$		0.29	2.8		
$Bi_{0.02}Pb_{0.98}$		0.46	0.73		
$Bi_{0.53}Pb_{0.32}Sn_{0.16}$			>25		3.06
$Bi_{1-0.93}Sn_{0-0.07}$			0–0.032		3.7
Bi_5Tl_3	6.4		>5.6		3.35
C_8K (excess K)	0.55		0.160 (H⊥c)		0.32
			0.730 (H|c)		0.32
C_8K	0.39		0.025 (H⊥c)		0.32
			0.250 (H⊥c)		0.32
$C_{0.44}Mo_{0.56}$	12.5–13.5	0.087	98.5		1.2
CNb	8–10	0.12	16.9		4.2
$CNb_{0.4}Ta_{0.6}$	10–13.6	0.19	14.1		1.2
CTa	9–11.4	0.22	4.6		1.2
$Ca_xO_3Sr_{1-x}Ti$	<0.1–0.55	0.002–0.004			
$Cd_{0.1}Hg_{0.9}$ (by weight)		0.23	0.34		2.04
$Cd_{0.05}Hg_{0.95}$		0.28	0.31		2.16
$Cr_{0.10}Ti_{0.30}V_{0.60}$	5.6	0.071	84.4		0
GaN	5.85	0.725			4.2
Ga_xNb_{1-x}			>28		4.2
GaSb (annealed)	4.24		2.64		3.5
$GaV_{1.95}$	5.3		73[e]		
$GaV_{2.1-3.5}$	6.3–14.45		230–300[d]		0
GaV_3		0.4	350[c]		0
			500[d]		
$GaV_{4.5}$	9.15		121[c]		0
Hf^xNb^y			>52–>102		1.2
Hf^xTa^y			>28–>86		1.2
$Hg_{0.05}Pb_{0.95}$		0.235	2.3		
$Hg_{0.101}Pb_{0.899}$		0.23	4.3		4.2
$Hg_{0.15}Pb_{0.85}$	6.75		>13		2.93
$In_{0.98}Pb_{0.02}$	3.45	0.1		0.12	2.76
$In_{0.96}Pb_{0.04}$	3.68	0.1	0.12	0.25	2.94
$In_{0.94}Pb_{0.06}$	3.90	0.095	0.18	0.35	3.12
$In_{0.913}Pb_{0.087}$	4.2	~10.17	0.55	2.65	
$In_{0.316}Pb_{0.684}$		0.155	3.7		4.2
$In_{0.17}Pb_{0.83}$			2.8	5.5	4.2
$In_{1.000}Te_{1.002}$	3.5–3.7		1.2[c]		0
$In_{0.95}Tl_{0.05}$		0.263	0.263		3.3
$In_{0.90}Tl_{0.10}$		0.257	0.257		3.25
$In_{0.83}Tl_{0.17}$		0.242	0.39		3.21
$In_{0.75}Tl_{0.25}$		0.216	0.50		3.16

Substance	T_c/K	H_{c1}/kOe	H_{c2}/kOe	H_{c3}/kOe	T_{obs}/K[a]
LaN	1.35	0.45			0.76
La_3S_4	6.5	≈0.15	>25		1.3
La_3Se_4	8.6	≈0.2	>25		1.25
$Mo_{0.52}Re_{0.48}$	11.1		14–21	22–33	4.2
			18–28	37–43	1.3
$Mo_{0.6}Re_{0.395}$	10.6		14–20	20–37	4.2
			19–26		
				26–37	1.3
$Mo_{0.5}Ti_{0.5}$			75[c]		0
$Mo_{0.16}Ti_{0.84}$	4.18	0.028	98.7[c]		0
$Mo_{0.16}Ti_{0.84}$			36–38		3.0
$Mo_{0.913}Ti_{0.087}$	2.95	0.060	15		4.2
$Mo_{0.1-0.3}U_{0.9-0.7}$	1.85–2.06		>25		
$Mo_{0.17}Zr_{0.83}$			30		
$N_{(12.8\ w/o)}Nb$	15.2		>9.5		13.2
NNb (wires)	16.1		153[c]		0
			132		4.2
			95		8
			53		12
NNb_xO_{1-x}	13.5–17.0		38		
NNb_xZr_{1-x}	9.8–13.8		4- >130		4.2
$N_{0.93}Nb_{0.85}Zr_{0.15}$	13.8		>130		4.2
$Na_{0.086}Pb_{0.914}$		0.19	6.0		
$Na_{0.016}Pb_{0.984}$		0.28	2.05		
Nb	9.15		2.020		1.4
			1.710		4.2
Nb		0.4–1.1	3–5.5		4.2
Nb (unstrained)		1.1–1.8	3.40	6–9.1	4.2
Nb (strained)		1.25–1.92	3.44	6.0–8.7	4.2
Nb (cold-drawn wire)		2.48	4.10	≈10	4.2
Nb (film)			>25		4.2
NbSc			>30		
Nb_3Sn		0.170	221		4.2
			70		14.15
			54		15
			34		16
			17		17
$Nb_{0.1}Ta_{0.9}$		0.084	0.154		4.195
$Nb_{0.2}Ta_{0.8}$			10		4.2
$Nb_{0.65-0.73}Ta_{0.02-0.10}Zr_{0.25}$			>70–>90		4.2
Nb_xTi_{1-x}			148 max.		1.2
			120 max.		4.2
$Nb_{0.222}U_{0.778}$		1.98	23		1.2
Nb_xZr_{1-x}			127 max.		1.2
			94 max.		4.2
O_3SrTi	0.43	0.0049[c]	0.504[c]		0
O_3SrTi	0.33	0.00195[c]	0.420[c]		0
$PbSb_{1\ w/o}$(quenched)			>1.5		4.2
$PbSb_{1\ w/o}$(annealed)			>0.7		4.2
$PbSb_{2.8\ w/o}$(quenched)			>2.3		4.2
$PbSb_{2.8\ w/o}$(annealed)			>0.7		4.2
$Pb_{0.871}Sn_{0.129}$		0.45	1.1		
$Pb_{0.965}Sn_{0.035}$		0.53	0.56		
$Pb_{1-0.26}Tl_{0-0.74}$	7.20–3.68		2–6.9[c]		0
$PbTl_{0.17}$	6.73		4.5[c]		0
$Re_{0.26}W_{0.74}$			>30		
$Sb_{0.93}Sn_{0.07}$			0.12		3.7
SiV_3	17.0	0.55	156[e]		
Sn_xTe_{1-x}		0.00043–0.00236	0.005–0.0775		0.012–0.079
Ta (99.95%)		0.425	1.850		1.3
		0.325	1.425		2.27

Substance	T_c/K	H_{c1}/kOe	H_{c2}/kOe	H_{c3}/kOe	T_{obs}/K[a]
		0.275	1.175		2.66
		0.090	0.375		3.72
$Ta_{0.5}Nb_{0.5}$			3.55		4.2
$Ta_{0.65-0}Ti_{0.35-1}$	4.4–7.8		>14–138		1.2
$Ta_{0.5}Ti_{0.5}$			138		1.2
Te	3.3	0.25[c]			0
Tc_xW_{1-x}	5.75–7.88		8–44		4.2
Ti				2.7	4.2
$Ti_{0.75}V_{0.25}$	5.3	0.029[c]	199[c]		0
$Ti_{0.775}V_{0.225}$	4.7	0.024[c]	172[c]		0
$Ti_{0.615}V_{0.385}$	7.07	0.050	34		4.2
$Ti_{0.516}V_{0.484}$	7.20	0.062	28		4.2
$Ti_{0.415}V_{0.585}$	7.49	0.078	25		4.2
$Ti_{0.12}V_{0.88}$			17.3	28.1	4.2
$Ti_{0.09}V_{0.91}$			14.3	16.4	4.2
$Ti_{0.06}V_{0.94}$			8.2	12.7	4.2
$Ti_{0.03}V_{0.97}$			3.8	6.8	4.2
Ti_xV_{1-x}			108 max.		1.2
V	5.31	0.8	3.4		1.79
		0.75	3.15		2
		0.45	2.2		3
		0.30	1.2		4
$V_{0.26}Zr_{0.74}$	≈5.9	0.238			1.05
		0.227			1.78
		0.185			3.04
		0.165			3.5
W (film)	1.7–4.1		>34		1

[a] Temperature of critical field measurement.
[b] w/o denotes weight percent.
[c] Extrapolated.
[d] Linear extrapolation.
[e] Parabolic extrapolation.

Solids

References

1. B. W. Roberts, in *Superconductive Materials and Some of Their Properties. Progress in Cryogenics*, Vol. IV, 1964, pp. 160–231.
2. B. W. Roberts, Superconductive Materials and Some of Their Properties, NBS Technical Notes 408 and 482, U.S. Government Printing Office, 1966 and 1969; B. W. Roberts, *J. Phys. Chem. Ref. Data*, 5, 581, 1976.
3. B. W. Roberts, Properties of Selected Superconductive Materials, 1978 Supplement, NBS Technical Note 983, 1978.
4. T. Claeson, *Phys. Rev.*, 147, 340, 1966.
5. C. J. Raub, W. H. Zachariasen, T. H. Geballe, and B. T. Matthias, *J. Phys. Chem. Solids*, 24, 1093, 1963.
6. T. H. Geballe, B. T. Matthias, V. B. Compton, E. Corenzwit, G. W. Hull, Jr., and L. D. Longinotti, *Phys. Rev.*, 1A, 119, 1965.
7. C. J. Raub, V. B. Compton, T. H. Geballe, B. T. Matthias, J. P. Maita, and G. W. Hull, Jr., *J. Phys. Chem. Solids*, 26, 2051, 1965.
8. R. D. Blaugher, J. K. Hulm, and P. N. Yocom, *J. Phys. Chem. Solids*, 26, 2037, 1965.
9. T. Claeson and H. L. Luo, *J. Phys. Chem. Solids*, 27, 1081, 1966.
10. S. C. Ng and B. N. Brockhouse, *Solid State Comm.*, 5, 79, 1967.
11. O. I. Shulishova and I. A. Shcherbak, *Izv. AN SSSR, Neorg. Materials*, 3, 1495, 1967.
12. T. F. Smith and H. L. Luo, *J. Phys. Chem. Solids*, 28, 569, 1967.
13. A. C. Lawson, *J. Less-Common Metals*, 23, 103, 1971.
14. R. Chevrel, M. Sergent, and J. Prigent, *J. Solid State Chem.*, 3, 515, 1971.
15. M. Marezio, P. D. Dernier, J. P. Remeika, and B. T. Matthias, *Mat. Res. Bull.*, 8, 657, 1973.
16. J. K. Hulm and R. D. Blaugher in *Superconductivity in d- and f-Band Metals*, D. H. Douglass, Ed., American Institute of Physics, 4, 1, 1972.
17. R. N. Shelton, A. C. Lawson, and D. C. Johnston, *Mat. Res. Bull.*, 10, 297, 1975.
18. H. D. Wiesinger, *Phys. Status Sol.*, 41A, 465, 1977.
19. O. Fisher, *Applied Phys.*, 16, 1, 1978.
20. D. C. Johnston, *Solid State Comm.*, 24, 699, 1977.
21. H. C. Ku and R. H. Shelton, *Mat. Res. Bull.*, 15, 1441, 1980.
22. H. Barz, *Mat. Res. Bull.*, 15, 1489, 1980.
23. G. P. Espinosa, A. S. Cooper, H. Barz, and J. P. Remeika, *Mat. Res. Bull.*, 15, 1635, 1980.
24. E. M. Savitskii, V. V. Baron, Yu. V. Efimov, M. I. Bychkova, and L. F. Myzenkova, in *Superconducting Materials*, Plenum Press, 1981, p. 107.
25. R. Fluckiger and R. Baillif, in Topics in *Current Physics*, O. Fischer and M. B. Maple, Eds., Springer Verlag, 34, 113, 1982.
26. R. N. Shelton, in *Superconductivity in d- and f-Band Metals*, W. Buckel and W. Weber, Eds., Kernforschungszentrum, Karlsruhe, 1982, p. 123.
27. D. C. Johnston and H. F. Braun, *Topics in Current Phys.*, 32, 11, 1982.
28. R. Chevrel and M. Sergent, *Topics in Current Phys.*, 32, 25, 1982.
29. G. P. Espinosa, A. S. Cooper, and H. Barz, *Mat. Res. Bull.*, 17, 963, 1982.
30. R. Muller, R. N. Shelton, J. W. Richardson, Jr., and R. A. Jacobson, *J. Less-Comm. Met.*, 92, 177, 1983.
31. You-Xian Zhao and Shou-An He, in *High Pressure in Science and Technology*, North Holland, 22, 51, 1983.
32. You-Xian Zhao and Shou-An He, *Solid State Comm.*, 24, 699, 1983.
33. G. P. Meisner and H. C. Ku, *Appl. Phys.*, A31, 201, 1983.
34. R. J. Cava, D. W. Murphy, and S. M. Zahurak, *J. Electrochem. Soc.*, 130, 2345, 1983.
35. R. N. Shelton, *J. Less-Comm. Met.*, 94, 69, 1983.
36. B. Chevalier, P. Lejay, B. Lloret, Wang Xian–Zhong, J. Etourneau, and P. Hagenmuller, *Annales de Chemie*, 9, 191, 1984.
37. G. Venturini, M. Meot-Meyer, E. McRae, J. F. Mareche, and B. Rogues, *Mat. Res. Bull.*, 19, 1647, 1984.
38. J. M. Tarascon, F. G. DiSalvo, D. W. Murphy, G. Hull, and J. V. Waszczak, *Phys. Rev.*, 29B, 172, 1984.
39. G. V. Subba and G. Balakrishnan, *Bull. Mat. Sci.*, 6, 283, 1984.

40. B. Batlog, *Physica*, 126B, 275, 1984.
41. M. J. Johnson, Ames Lab (USA) Report IS-T-1140, 1984.
42. I. M. Chapnik, *J. Mat. Sci. Lett.*, 4, 370, 1985.
43. W. Rong-Yao, L. Q-Guang, and Z. Xiao, *Phys. Status Sol.*, 90A, 763, 1985.
44. W. Xian-Zhong, B. Chevalier, J. Etourneau, and P. Hagenmuller, *Mat. Res. Bull.*, 20, 517, 1985.
45. H. R. Ott, F. Hulliger, H. Rudigier, and Z. Fisk, *Phys. Rev.*, 31B, 1329, 1985.
46. P. Villars and L. D. Calver, *Pearson's Handbook of Crystallographic Data for Intermetallic Phases*, Vol. 1–3, ASM, 1985.
47. G. V. Subba Rao, K. Wagner, G. Balakhrishnan, J. Jakani, W. Paulus, and R. Scollhorn, *Bull. Mat. Sci.*, 7, 215, 1985.
48. J. G. Bednorz and K. A. Muller, *Zs. Physik*, B64, 189, 1986.
49. W. Rong-Yao, *Phys. Status Sol.*, 94A, 445, 1986.
50. H. D. Yang, R. N. Shelton, and H. F. Braun, *Phys. Rev.*, 33B, 5062, 1986.
51. G. Venturini, M. Kanta, E. McRae, J. F. Mareche, B. Malaman, and B. Roques, *Mat. Res. Bull.*, 21, 1203, 1986.
52. W. Rong-Yao, *J. Mat. Sci. Lett.*, 5, 87, 1986.
53. M. K. Wu, J. R. Ashburn, C. J. Torng, P. H. Hor, R. L. Meng, L. Gao, Z. J. Huang, Y. Q. Wang, and C. W. Chu, *Phys. Rev. Lett.*, 58, 908, 1987.
54. R. J. Cava, R. B. Van Dover, B. Batlog, and E. A. Rietman, *Phys. Rev. Lett.*, 58, 408, 1987.
55. L. C. Porter, T. J. Thorn, U. Geiser, A. Umezawa, H. H. Wang, W. K. Kwok, H-C. I. Kao, M. R. Monaghan, G. W. Crabtree, K. D. Carlson, and J. M. Williams, *Inorg. Chem.*, 26, 1645, 1987.
56. A. M. Kini, U. Geiser, H-C. I. Kao, K. D. Carlson, H. H. Wang, M. R. Monaghan, and K. M. Williams, *Inorg. Chem.*, 26, 1834, 1987.
57. T. Penney, S. von Molnar, D. Kaiser, F. Holtzberg, and A. W. Kleinsasser, *Phys. Rev.*, B38, 2918, 1988.
58. Y. K. Tao, J. S. Swinnea, A. Manthiram, J. S. Kim, J. B. Goodenoug, and H. Steinfink, *J. Mat. Res.*, 3, 248, 1988.
59. G. G. Peterson, B. R. Weinberger, L. Lynds, and H. A. Krasinski, *J. Mat. Res.*, 3, 605, 1988.
60. J. B. Torrance, Y. Tokura, A. Nazzai, and S. S. P. Parkin, *Phys. Rev. Lett.*, 60, 542, 1988.
61. K. Kourtakis, M. Robbins, P. K. Gallagher, and T. Teifel, *J. Mat. Res.*, 4, 1289, 1989.
62. J. C. Phillips, *Physics of High-T_c Superconductors*, Academic Press, 1989, p. 336.
63. Shui Wai Lin and L. I. Berger, *Rev. Sci. Instrum.*, 60, 507, 1989.
64. M. Tinkham, *Introduction to Superconductivity*, McGraw–Hill, New York, 1975.
65. O. Fischer and M.B. Maple, Eds., *Topics in Current Physics*, Volume 32: Superconductivity in Ternary Compounds I; Volume 34: Superconductivity in Ternary Compounds II, Springer–Verlag, Berlin, 1982.
66. K. J. Dunn and F. P. Bundy, *Phys. Rev.*, B25, 194, 1982.
67. A. Barone and G. Paterno, *Physics and Applications of the Josephson Effect*, Wiley, New York, 1982.
68. D. H. Douglass, Ed., *Superconductivity in d- and f-Band Metals*, Plenum Press, New York, 1976.
69. D. M. Ginsberg, Ed., *Physical Properties of High Temperature Superconductors*, (Volume II, 1990; Volume III, 1992; Volume V, 1996), World Scientific, Singapore.
70. T. Ishiguro and K. Yamji, *Organic Superconductors*, Springer-Verlag, Berlin, 1990.
71. Sh. Okada, K. Shimizu, T. C. Kobayashi, K. Amaya, and Sh. Endo., *J. Phys. Soc. Jpn.*, 65, 1924, 1996.
72. A. Bourdillon and N. X. Tan Bourdillon, *High Temperature Superconductors: Processing and Science*, Academic Press, 1994.
73. J. M. Williams, J. R. Ferraro, R. J. Thorn, K. Carlson, U. Geiser, H. H. Wang, A. M. Kini, and M.-H. Whangbo, *Organic Superconductors (Including Fullerenes): Synthesis, Structure, Properties, and Theory*, Prentice–Hall, 1992.
74. J. Nagamatsu, N. Nakagawa, T. Muranaka, Y. Zenitani, and J. Akimitsu, *Nature (London)*, 410, 63, 2001.
75. Y. Boguslavsky, G. K. Perkins, X. Qi, L. F. Cohen, and A. D. Caplin, *Nature (London)*, 410, 563, 2001.
76. B. Q. Fu, Y. Feng, G. Yan, Y. Zhao, A. K. Pradhan, C. H. Cheng, P. Ji, X. H. Liu, C. F. Liu, L. Zhou, and K. F. Yau, *J. Appl. Phys.*, 92, 7341, 2002.

HIGH-TEMPERATURE SUPERCONDUCTORS

C. N. R. Rao and A. K. Raychaudhuri

The following tables give properties of a number of high-temperature superconductors. Table 1 lists the crystal structure (space group and lattice constants) and the critical transition temperature T_c for the more important high-temperature superconductors studied so far. Table 2 gives the energy gap, critical current density, and penetration depth in the superconducting state. Table 3 gives electrical and thermal properties of some of these materials in the normal state. The tables were prepared in November 1992 and updated in November 1994.

References

1. Ginsburg, D. M., Ed., *Physical Properties of High-Temperature Superconductors*, Vols. I–III, World Scientific, Singapore, 1989–1992.
2. Rao, C. N. R., Ed., *Chemistry of High-Temperature Superconductors*, World Scientific, Singapore, 1991.
3. Shackelford, J. F., *The CRC Materials Science and Engineering Handbook*, CRC Press, Boca Raton, FL, 1992, 98–99 and 122–123.
4. Kaldis, E., Ed., *Materials and Crystallographic Aspects of HT$_c$-Superconductivity*, Kluwer Academic Publ., Dordrecht, The Netherlands, 1992.
5. Malik, S. K. and Shah, S. S., Ed., *Physical and Material Properties of High Temperature Superconductors*, Nova Science Publ., Commack, NY, 1994.
6. Chmaissem, O. et al., *Physica*, C230, 231–238, 1994.
7. Antipov, E. V. et al., *Physica*, C215, 1–10, 1993.

TABLE 1. Structural Parameters and Approximate T_c Values of High-Temperature Superconductors

Material	Structure	T_c/K (maximum value)
$La_2CuO_{4+\delta}$	Bmab; a = 5.355, b = 5.401, c = 13.15 Å	39
$La_{2-x}Sr_x(Ba_x)CuO_4$	I4/mmm; a = 3.779, c = 13.23 Å	35
$La_2Ca_{1-x}Sr_xCu_2O_6$	I4/mmm; a = 3.825, c = 19.42 Å	60
$YBa_2Cu_3O_7$	Pmmm; a = 3.821, b = 3.885, c = 11.676 Å	93
$YBa_2Cu_4O_8$	Ammm; a = 3.84, b = 3.87, c = 27.24 Å	80
$Y_2Ba_4Cu_7O_{15}$	Ammm; a = 3.851, b = 3.869, c = 50.29 Å	93
$Bi_2Sr_2CuO_6$	Amaa; a = 5.362, b = 5.374, c = 24.622 Å	10
$Bi_2CaSr_2Cu_2O_8$	A_2aa; a = 5.409, b = 5.420, c = 30.93 Å	92
$Bi_2Ca_2Sr_2Cu_3O_{10}$	A_2aa; a = 5.39, b = 5.40, c = 37 Å	110
$Bi_2Sr_2(Ln_{1-x}Ce_x)_2Cu_2O_{10}$	P4/mmm; a = 3.888, c = 17.28 Å	25
$Tl_2Ba_2CuO_6$	A_2aa; a = 5.468, b = 5.472, c = 23.238 Å; I4/mmm; a = 3.866, c = 23.239 Å	92
$Tl_2CaBa_2Cu_2O_8$	I4/mmm; a = 3.855, c = 29.318 Å	119
$Tl_2Ca_2Ba_2Cu_3O_{10}$	I4/mmm; a = 3.85, c = 35.9 Å	128
$Tl(BaLa)CuO_5$	P4/mmm; a = 3.83, c = 9.55 Å	40
$Tl(SrLa)CuO_5$	P4/mmm; a = 3.7, c = 9 Å	40
$(Tl_{0.5}Pb_{0.5})Sr_2CuO_5$	P4/mmm; a = 3.738, c = 9.01 Å	40
$TlCaBa_2Cu_2O_7$	P4/mmm; a = 3.856, c = 12.754 Å	103
$(Tl_{0.5}Pb_{0.5})CaSr_2Cu_2O_7$	P4/mmm; a = 3.80, c = 12.05 Å	90
$TlSr_2Y_{0.5}Ca_{0.5}Cu_2O_7$	P4/mmm; a = 3.80, c = 12.10 Å	90
$TlCa_2Ba_2Cu_3O_8$	P4/mmm; a = 3.853, c = 15.913 Å	110
$(Tl_{0.5}Pb_{0.5})Sr_2Ca_2Cu_3O_9$	P4/mmm; a = 3.81, c = 15.23 Å	120
$TlBa_2(La_{1-x}Ce_x)_2Cu_2O_9$	I4/mmm; a = 3.8, c = 29.5 Å	40
$Pb_2Sr_2La_{0.5}Ca_{0.5}Cu_3O_8$	Cmmm; a = 5.435, b = 5.463, c = 15.817 Å	70
$Pb_2(Sr,La)_2Cu_2O_6$	$P22_12$; a = 5.333, b = 5.421, c = 12.609 Å	32
$(Pb,Cu)Sr_2(La,Ca)Cu_2O_7$	P4/mmm; a = 3.820, c = 11.826 Å	50
$(Pb,Cu)(Sr,Eu)(Eu,Ce)Cu_2O_x$	I4/mmm; a = 3.837, c = 29.01 Å	25
$Nd_{2-x}Ce_xCuO_4$	I4/mmm; a = 3.95, c = 12.07 Å	30
$Ca_{1-x}Sr_xCuO_2$	P4/mmm; a = 3.902, c = 3.35 Å	110
$Sr_{1-x}Nd_xCuO_2$	P4/mmm; a = 3.942, c = 3.393 Å	40
$Ba_{0.6}K_{0.4}BiO_3$	Pm3m; a = 4.287 Å	31
Rb_2CsC_{60}	a = 14.493 Å	31
$NdBa_2Cu_3O_7$	Pmmm; a = 3.878, b = 3.913, c = 11.753	58
$SmBaSrCu_3O_7$	I4/mmm; a = 3.854, c = 11.62	84
$EuBaSrCu_3O_7$	I4/mmm; a = 3.845, c = 11.59	88
$GdBaSrCu_3O_7$	I4/mmm; a = 3.849, c = 11.53	86
$DyBaSrCu_3O_7$	Pmmm; a = 3.802, b = 3.850, c = 11.56	90
$HoBaSrCu_3O_7$	Pmmm; a = 3.794, b = 3.849, c = 11.55	87
$ErBaSrCu_3O_7$ (multiphase)	Pmmm; a = 3.787, b = 3.846, c = 11.54	82
$TmBaSrCu_3O_7$ (multiphase)	Pmmm; a = 3.784, b = 3.849, c = 11.55	88
$YBaSrCu_3O_7$	Pmmm; a = 3.803, b = 3.842, c = 11.54	84
$HgBa_2CuO_4$	I4/mmm; a = 3.878, c = 9.507	94
$HgBa_2CaCu_2O_6$ (annealed in O_2)	I4/mmm; a = 3.862, c = 12.705	127
$HgBa_2Ca_2Cu_3O_8$	Pmmm; a = 3.85, c = 15.85	133
HgBa2Ca3Cu4O10	Pmmm; a = 3.854, c = 19.008	126

TABLE 2. Superconducting Properties

J_c (0): Critical current density extrapolated to 0 K
λ_{ab}: Penetration depth in *a*-*b* plane
k_B: Boltzmann constant

Material	Form	Energy gap (Δ) $2\Delta_{pp}/k_BT_c$*	Energy gap (Δ) $2\Delta_{fit}/k_BT_c$†	$10^{-6} \times J_c$ (0)/A cm^{-2}	λ_{ab}/Å
$YBa_2Cu_3O_7$	Single Crystal	5–6	4–5	30 (film)	1400
$Bi_2Sr_2CaCu_2O_8$	Single Crystal	8–9	5.5–6.5	2	2700
$Tl_2Ba_2CaCu_2O_8$	Ceramic	6–7	4–6	10 (film, 80 K)	2000
$La_{2-x}Sr_xCuO_4$, $x = 0.15$	Ceramic	7–9	4–6		
$Nd_{2-x}Ce_xCuO_4$	Ceramic	8	4–5	0.2 (film)	

* Obtained from peak to peak value.
† Obtained from fit to BCS-type relation.

TABLE 3. Normal State Properties

ρ_{ab}: Resistivity in the *a*-*b* plane
ρ_c: Resistivity along the *c* axis
+ve: ρ_c has positive temperature coefficient of resistivity
−ve: ρ_c has negative temperature coefficient of resistivity
n_H: Hall density
k: Thermal conductivity
in plane: Along *a*-*b* plane
out of plane: Perpendicular to *a*-*b* plane

Material	Form	ρ_{ab}/μΩ cm 300 K	ρ_{ab}/μΩ cm 100 K	ρ_c/mΩ cm 300 K	$d\rho_c/dT$	$10^{-21} \times n_H$/cm^{-3} 300 K	$10^{-21} \times n_H$/cm^{-3} 100 K	*k*/(mW/cm K) at 300 K in plane	*k*/(mW/cm K) at 300 K out of plane
$YBa_2Cu_3O_7$	Single crystal	110	35	5	+ve	11–16	4–6	120	3
	Film	200–300	60–100			5–9	2–3		
$YBa_2Cu_4O_8$	Single crystal	75	20	10	−ve	14			
	Film	100–200	20–50			22	17		
$Bi_2Sr_2CuO_6$	Single crystal	300	150	5000	−ve	6	5		
$Bi_2Sr_2CaCu_2O_8$	Single crystal	150	50	>1000	−ve	4	3	60	8
$Tl_2Ba_2CuO_6$	Single crystal	300–400	50–75	200–300	+ve	3.1	2.5		
$Tl_2Ba_2Ca_2Cu_3O_{10}$	Ceramic	***	**				≈ 2*		
$La_{2-x}Sr_xCuO_4$, $x = 0.12$	Single crystal	900	350	200	+ve for $T > 225$ K	2.5			
$La_{2-x}Sr_xCuO_4$, $x = 0.20$	Single crystal	400	200	80	+ve for $T > 150$ K	10		50 (for $x = 0.04$)	20
	Film	400	160			8.4	6.3		
$Nd_{2-x}Ce_xCuO_4$, $x = 0.17$	Single crystal	500	275			53	17		
$x = 0.15$	Film	140–180	35			32	11	250 (for $x = 0.15$)	

* At 200 K
** ρ ~0.4 mΩ cm at 120 K
*** ρ ~1.5 mΩ cm at 300 K

ORGANIC SUPERCONDUCTORS

H. P. R. Frederikse

Although the vast majority of organic compounds are insulators, a small number of organic solids show considerable electrical conductivity. Some of these materials appear to be superconductors. The superconducting organics fall primarily into two groups: those containing fulvalenes (pentagonal rings containing sulfur or selenium) and those based on fullerenes, involving the nearly spherical cluster C_{60}.

The transition temperatures T_c of the fulvalene derivatives are shown in Table 1. The abbreviations of the various molecular groups are listed in Table 2 and their chemical structures are depicted in Figure 1. Most of the T_c's are between 1 and 12 K. Several of the compounds only show superconductivity under pressure.

The fullerenes are A_3C_{60} compounds, where A represents a single or a combination of alkali atoms. The C_{60} cluster is shown in Figure 2a, while Figure 2b illustrates how the alkali atoms fit into the A_3C_{60} molecule to form the A15 crystallographic structure. Their superconducting transition temperatures range from 8 to 31.3 K (see Table 3).

References

1. Ishigura, T. and Yamaji, K., *Organic Superconductors*, Springer-Verlag, Berlin, 1990.
2. Williams, J. M. et al., *Organic Superconductors (Including Fullerenes)*, Prentice Hall, Englewood Cliffs, N.J., 1992.
3. *The Fullerenes*, Ed.: Krato, H. W., Fisher, J. E., and Cox, D. E., Pergamon Press, Oxford, 1993.
4. Schluter, M. et al., in *The Fullerenes* (Ref. 3), p. 303.

TABLE 1. Critical Pressure and Maximum Critical Temperature of Organic Superconductors

Material	P_c/kbar	T_c/K	Material	P_c/kbar	T_c/K
$(TMTSF)_2PF_6$	6.5	1.2	β-$(ET)_2IBr_2$	0	2.8
$(TMTSF)_2AsF_6$	9	1.3	β-$(ET)_2AuI_2$	0	4.8
$(TMTSF)_2SbF_6$	11	0.4	$(ET)_4Hg_{2.89}Cl_8$	0	4.2
$(TMTSF)_2TaF_6$	12	1.4	$(ET)_4Hg_{2.89}Br_8$	12	1.8
$(TMTSF)_2ClO_4$	0	1.4	$(ET)_3Cl_2(H_2O)_2$	16	2
$(TMTSF)_2ReO_4$	9.5	1.3	κ-$(ET)_2Cu(NCS)_2$	0	10.4
$(TMTSF)_2FSO_3$	5	3	κ-$(d\text{-}ET)_2Cu(NCS)_2$	0	11.4
$(ET)_4(ReO_4)_2$	4.5	2	$(DMET)_2Au(CN)_2$	1.5	0.9
β_L-$(ET)_2I_3$	0	1.4	$(DMET)_2AuI_2$	5	0.6
β_H-$(ET)_2I_3$	0	8.1	$(DMET)_2AuBr_2$	0	1.9
γ-$(ET)_3I_{2.5}$	0	2.5	$(DMET)_2AuCl_2$	0	0.9
ε-$(ET)_2I_3(I_8)_{0.5}$	0	2.5	$(DMET)_2I_3$	0	0.6
α-$(ET)_2I_3I_2$-doped	0	3.3	$(DMET)_2IBr_2$	0	0.7
α_t-$(ET)_2I_3$	0	8	$(MDT\text{-}TTF)_2AuI_2$	0	3.5
ε→β-$(ET)_2I_3$[a]	0	6	$TTF[Ni(dmit)_2]_2$	2	1.6[b]
θ-$(ET)_2I_3$	0	3.6	$TTF[Pd(dmit)_2]_2$	20	6.5
κ-$(ET)_2I_3$	0	3.6	$(CH_3)_4N[Ni(dmit)_2]_2$	7	5

[a] Converted from ε-type to β-type by thermal treatment.
[b] For 7 kbar.
From Ishigura, T. and Yamaji, K., *Organic Superconductors*, Springer-Verlag, Berlin, 1990. With permission.

TABLE 2. List of Symbols and Abbreviations

TTF	tetrathiafulvalene
TMTSF	tetramethyltetraselenafulvalene
BEDT-TTF or "ET"	bis(ethylenedithio)tetrathiafulvalene
MDT-TTF	methylenedithiotetrathiafulvalene
DMET	[dimethyl(ethylenedithio)diselenadithiafulvalene]
dmit	4,5-dimercapto-1,3-dithiole-2-thione
T_c	transition temperature to superconducting state
P_c	minimum pressure required for superconducting transition

FIGURE 2. (a) C_{60} cluster placed in a fcc lattice. Each crystal axis crosses a double bond shared by two hexagons. (b) A hypothetical A_3C_{60} with the A15 structure. The structure can be seen to be an ordered defect structure of A_6C_{60}.

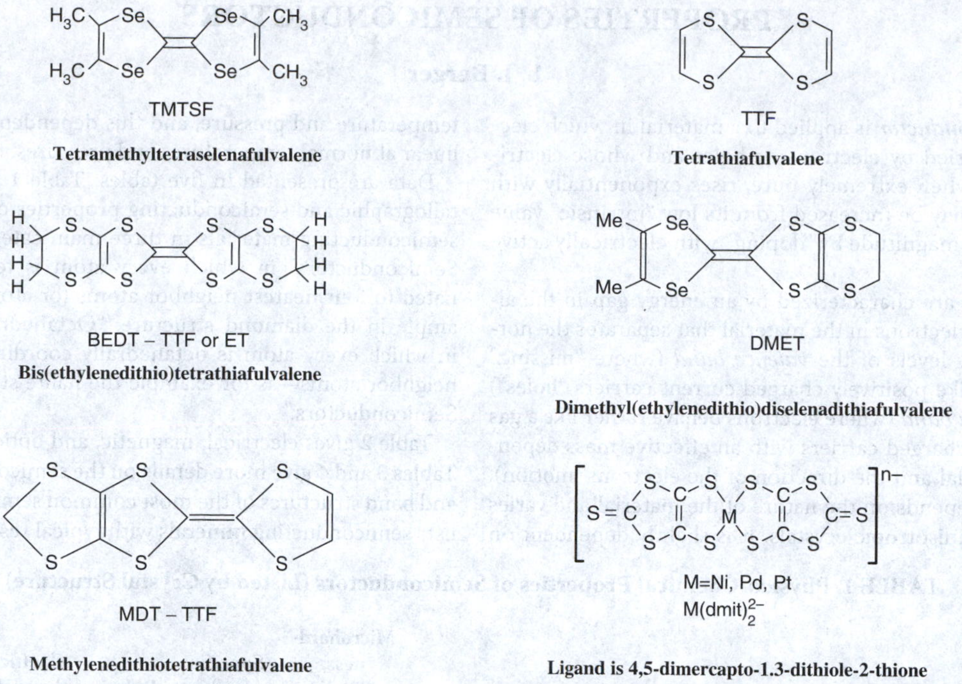

FIGURE 1. Structures of various donor molecules and acceptor species.

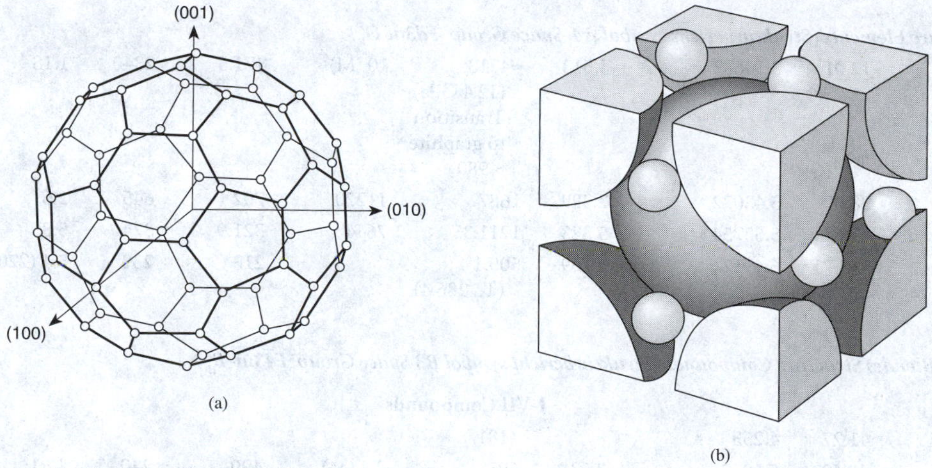

FIGURE 2. (a) C_{60} cluster placed in a fcc lattice. Each crystal axis crosses a double bond shared by two hexagons. (b) A hypothetical A_3C_{60} with the A15 structure. The structure can be seen to be an ordered defect structure of A_6C_{60}.

TABLE 3. Unit Cell and T_c for FCC A_3C_{60} Crystals

	Lattice parameter(s) (Å)	T_c/K
$Na_2Rb_{0.5}Cs_{0.5}C_{60}$	14.148(3)	8.0
Na_2CsC_{60} No. 1[a]	14.132(2)	10.5
Na_2CsC_{60} No. 2[a]	14.176(9)	14.0
K_3C_{60}	14.253(3)	19.3
K_2RbC_{60}	14.299(2)	21.8
Rb_2KC_{60} No. 1[a]	14.336(1)	24.4
Rb_2KC_{60} No. 2[a]	14.364(5)	26.4
Rb_3C_{60}	14.436(2)	29.4
Rb_2CsC_{60}	14.493(2)	31.3

[a] Samples labeled No. 1 and No. 2 have the same nominal composition.

From Schluter, M. et al., *The Fullerenes*, Krato, H. W., Fisher, J. E., and Cox, D. E., Eds., Pergamon Press, Oxford, 1993. With permission.

PROPERTIES OF SEMICONDUCTORS

L. I. Berger

The term *semiconductor* is applied to a material in which electric current is carried by electrons or holes and whose electrical conductivity, when extremely pure, rises exponentially with temperature and may be increased from its low "intrinsic" value by many orders of magnitude by "doping" with electrically active impurities.

Semiconductors are characterized by an energy gap in the allowed energies of electrons in the material that separates the normally filled energy levels of the *valence band* (where "missing" electrons behave like positively charged current carriers "holes") and the *conduction band* (where electrons behave rather like a gas of free negatively charged carriers with an effective mass dependent on the material and the direction of the electrons' motion). This energy gap depends on the nature of the material and varies with direction in anisotropic crystals. It is slightly dependent on temperature and pressure, and this dependence is usually almost linear at normal temperatures and pressures.

Data are presented in five tables. Table 1 lists the main crystallographic and semiconducting properties of a large number of semiconducting materials in three main categories: "Tetrahedral Semiconductors" in which every atom is tetrahedrally coordinated to four nearest neighbor atoms (or atomic sites) as for example in the diamond structure; "Octahedral Semiconductors" in which every atom is octahedrally coordinated to six nearest neighbor atoms—as for example the halite structure; and "Other Semiconductors."

Table 2 gives electrical, magnetic, and optical properties, while Tables 3 and 4 give more details on the semiconducting properties and band structures of the most common semiconductors. Table 5 lists semiconducting minerals with typical resistivity ranges.

TABLE 1. Physico-Chemical Properties of Semiconductors (Listed by Crystal Structure)

Substance	Molecular weight	Average atomic weight	Lattice parameters (Å, room temp.)	Density (g/cm³)	Melting point (K)	Microhardness, N/mm² (M-Mohs Scale)	Specific heat, J/kg·K (300 K)	Debye temp. (K)	Coefficient of thermal linear expansion [10^{-6} K^{-1} (300K)]	Thermal conductivity [mW/cm·K (300K)]
1.1. Tetrahedral (Adamantine) Semiconductors										
1.1.1. Diamond Structure Elements (Strukturbericht symbol A4, Space Group Fd3m-O_h^7)										
C (Diamond)		12.01	3.56683	3.513	≈4713 (12.4 GPa) Transition to graphite > 980	10 (M)	471.5	2340	1.18	9900(I) 23200(IIA) 13600(IIB)
Si		28.09	5.43072	2.329	1687	11270	702	645	2.6	1240
Ge		72.64	5.65754	5.323	1211.35	7644	321.9	374	5.8	640
α-Sn		118.71	6.4912	5.769	505.1 (Tr. 286.4)		213	230	5.4 (220 K)	
1.1.2. Sphalerite (Zinc Blende) Structure Compounds (Strukturbericht symbol B3 Space Group F$\bar{4}$3m-T_d^2)										
I-VII Compounds										
CuF	82.54	41.27	4.255		1181					
CuCl	98.99	49.49	5.4057	3.53	695	2.3 (M)	490	240	12.1	8.4
CuBr	143.45	71.73	5.6905	4.98	770	2.5 (M)	381	207	15.4	12.5
CuI	190.45	95.23	6.60427	5.63	878	192	276	181	19.2	16.8
AgBr	187.77	93.89		6.473	>1570 (Tr. 410)	2.5 (M)	270			
AgI	234.77	117.39	6.502	5.67	831	2.5 (M)	232	134	−2.5	4.2
II-VI Compounds										
BeS	41.08	20.54	4.865	2.36	dec.					
BeSe	87.97	43.99	5.139	4.315						
BeTe	136.61	68.31	5.626	5.090						
BePo	(2318)	(109)	5.838	7.3						
ZnO	81.39	40.69	4.63	5.675	2248	5.0 (M)	494	416	2.9	234
ZnS	97.46	48.72	5.4093	4.079	2100 (Tr. 1295)	1780	472	530	6.36	251
ZnSe	144.34	72.17	5.6676	5.42	1790	1350	339	400	7.2	140
ZnTe	192.99	96.5	6.101	6.34	1568	900	264	223	8.19	108
ZnPo	(274)	(137)	6.309							

Substance	Molecular weight	Average atomic weight	Lattice parameters (Å, room temp.)		Density (g/cm³)	Melting point (K)	Microhardness, N/mm² (M-Mohs Scale)	Specific heat, J/kg·K (300 K)	Debye temp. (K)	Coefficient of thermal linear expansion [10^{-6} K^{-1} (300K)]	Thermal conductivity [mW/cm·K (300K)]
CdS	144.48	72.24	5.832		4.826	1750	1250	330	219	4.7	200
CdSe	191.37	95.68	6.05		5.674	1512	1300	255	181	3.8	90
CdTe	240.01	120.00	6.477		5.86	1365	600	205	200	4.9	58.5
CdPo	(321)	(161)	6.665								
HgS	232.66	116.33	5.8517		7.73	1820	3 (M)	210			
HgSe	279.55	139.78	6.084		8.25	1070	2.5 (M)	178	151	5.46	10
HgTe	328.19	164.10	6.4623		8.17	943	300	164	242	4.6	20
III-V Compounds											
BN	24.82	12.41	3.615		3.49	3239	10 (M)	793	≈1900		200
BP(L.T.)	41.78	20.87	4.538		2.9	1398 (dec)	37000		≈980		
BAs	85.73	42.87	4.777			≈2300	19000		≈625		
AlP	57.95	28.98	5.451		2.42	≈2100	5.5 (M)		588		920
AlAs	101.90	50.95	5.6622		3.81	2013	5000		417	3.5	840
AlSb	148.74	74.37	6.1355		4.218	1330	4000		292	4.2	600
GaP	100.70	50.35	5.4495		4.13	1750	9450		446	5.3	752
GaAs	144.64	72.32	5.65315		5.316	1510	7500		344	5.4	560
GaSb	191.48	95.74	6.0954		5.619	980	4480	320	265	6.1	270
InP	145.79	72.90	5.86875		4.787	1330	4100		321	4.6	800
InAs	189.74	94.87	6.05838		5.66	1215	3300	268	249	4.7	290
InSb	236.58	118.29	6.47877		5.775	798	2200	144	202	4.7	160
Other Sphalerite Structure Compounds											
MnS	87.00	43.5	5.011								
MnSe	133.90	66.95	5.82								
β-SiC (3-C SiC)	40.10	20.1	4.348		3.21	3070				2.9	4.9
Ga_2Se_3	376.32	75.26	5.429		4.92	1020	3160			8.9	50
Ga_2Te_3	522.24	104.45	5.899		5.75	1063	2370				47
In_2Te_3(H.T.)	608.44	121.7	6.173		5.8	940	1660				69
$MgGeP_2$	158.84	39.71	5.652								
$ZnSnP_2$	246.00	61.5	5.65			1200					
$ZnSnAs_2$(H.T.)	333.90	82.38	5.851		5.53	1050					76
$ZnSnSb_2$	427.56	106.89	6.281		5.67	870	2500				76

1.1.3. Wurtzite (Zincite) Structure Compounds (Strukturbericht symbol B4, Space Group P 6_3 mc-C^4_{6v})

I-VII Compounds											
CuCl	99.0	49.5	3.91	6.42		703					
CuBr	143.45	71.73	4.06	6.66		770					
CuI	190.45	95.23	4.31	7.09							
AgI	234.77	117.40	4.580	7.494							
II-VI Compounds											
BeO	25.01	12.51	2.698	4.380		2800					
MgTe	151.9	76.0	4.54	7.39	3.85	≈2800					
ZnO	81.37	40.69	3.24950	5.2069	5.66	2250					600
ZnS	97.43	48.72	3.8140	6.2576	4.1	2100					460
ZnTe	192.99	46.50	4.27	6.99		1568					
CdS	144.48	72.23	4.1348	6.7490	4.82	1748					401
CdSe	191.37	95.68	4.299	7.010	5.66	1512					316
CdTe	240.01	120.00	4.57	7.47							
III-V Compounds											
BP(H.T.)	41.79	20.90	3.562	5.900							
AlN	40.99	20.50	3.111	4.978	3.26	≈2500					823
GaN	83.73	41.87	3.190	5.189	6.10	1500					656

Solids

Substance	Molecular weight	Average atomic weight	Lattice parameters (Å, room temp.)		Density (g/cm³)	Melting point (K)	Microhardness, N/mm² (M-Mohs Scale)	Specific heat, J/kg·K (300 K)	Debye temp. (K)	Coefficient of thermal linear expansion [10^{-6} K^{-1} (300K)]	Thermal conductivity [mW/cm·K (300K)]
InN	128.83	64.42	3.533	5.693	6.88	1200					556
Other Wurtzite Structure Compounds											
MnS	87.00	43.5	3.985	6.45	3.248						
MnSe	133.90	66.95	4.12	6.72							
SiC	40.10	20.1	3.076	5.048							
MnTe	182.54	91.27	4.078	6.701							
Al_2S_3	150.14	30.03	3.579	5.829	2.55	1400					
Al_2Se_3	290.84	58.17	3.890	6.30	3.91	1250					

1.1.4. Chalcopyrite Structure Compounds (Strukturbericht symbol $E1_1$, Space Group $I\bar{4}2d$-D_{2d}^{12})

Substance	Molecular weight	Average atomic weight	Lattice parameters (Å, room temp.)		Density (g/cm³)	Melting point (K)	Microhardness, N/mm² (M-Mohs Scale)	Specific heat, J/kg·K (300 K)	Debye temp. (K)	Coefficient of thermal linear expansion [10^{-6} K^{-1} (300K)]	Thermal conductivity [mW/cm·K (300K)]
I-III-VI_2 Compounds											
$CuAlS_2$	154.65	38.66	5.323	10.44	3.47	2500					
$CuAlSe_2$	248.45	62.11	5.617	10.92	4.70	2260					
$CuAlTe_2$	345.73	86.43	5.976	11.80	5.50	2550					
$CuGaS_2$	197.39	49.53	5.360	10.49	4.35	2300					
$CuGaSe_2$	291.19	72.80	5.618	11.01	5.56	1970	4200		275	5.4	42
$CuGaTe_2$	388.47	97.12	6.013	11.93	5.99	2400	3500			6.9	27
$CuInS_2$	242.49	60.62	5.528	11.08	4.75	1400	2550				
$CuInSe_2$	336.29	84.07	5.785	11.56	5.77	1600	2050			6.6	37
$CuInTe_2$	433.57	108.39	6.179	12.365	6.10	1660	400		195	7.1	49
$CuTlS_2$	322.05	83.01	5.580	11.17	6.32						
$CuTlSe_2$(L.T.)	425.85	106.46	5.844	11.65	7.11	900					
$CuFeS_2$	183.51	45.88	5.29	10.32	4.088	1135					
$CuFeSe_2$	277.31	69.33				850					
$CuLaS_2$	266.58	66.65	5.65	10.86							
$AgAlS_2$	198.97	49.74	5.707	10.28	3.94						
$AgAlSe_2$	292.77	73.19	5.968	10.77	5.07	1220					
$AgAlTe_2$	390.05	97.51	6.309	11.85	6.18	1000					
$AgGaS_2$	241.71	60.43	5.755	10.28	4.72						
$AgGaSe_2$	335.51	83.88	5.985	10.90	5.84	1120	4400				
$AgGaTe_2$	432.79	108.2	6.301	11.96	6.05	990	1800		212		10
$AgInS_2$(L.T.)	286.87	71.70	5.828	11.19	5.00		2250				
$AgInSe_2$	380.61	95.15	6.102	11.69	5.81	1053	1850				30
$AgInTe_2$	477.89	119.47	6.42	12.59	6.12	965				9.49, 0.69	
$AgFeS_2$	227.83	56.96	5.66	10.30	4.53						
II-IV-V_2 Compounds											
$ZnSiP_2$	155.40	38.85	5.400	10.441	3.39	1640	1100				
$ZnGeP_2$	199.90	49.98	5.465	10.771	4.17	1295	8100				180
$ZnSnP_2$	246.00	61.5					6500				
$CdSiP_2$	202.43	50.61	5.678	10.431	4.00	≈1470	10500		282		
$CdGeP_2$	246.94	61.74	5.741	10.775	4.48	1049	5650				110
$CdSnP_2$	243.03	73.26	5.900	11.518		840	5000		195		140
$ZnSiAs_2$	242.20	60.55	5.61	10.88	4.70	1311	9200				
$ZnGeAs_2$	287.80	71.95	5.672	11.153	5.32	1150	6800		263		110
$ZnSnAs_2$	333.90	83.48	5.8515	11.704	5.53	1048	4550		271		150
$CdSiAs_2$	290.34	72.58	5.884	10.882		>1120	6850				
$CdGeAs_2$	334.83	83.71	5.9427	11.2172	5.60	938	4700				48
$CdSnAs_2$	380.93	95.23	6.0944	11.9182	5.72	880	3450				40

1.1.5. Other Ternary Semiconductors with Tetrahedral Coordination

Substance	Molecular weight	Average atomic weight	Lattice parameters (Å, room temp.)		Density (g/cm³)	Melting point (K)	Microhardness, N/mm² (M-Mohs Scale)	Specific heat, J/kg·K (300 K)	Debye temp. (K)	Coefficient of thermal linear expansion [10^{-6} K^{-1} (300K)]	Thermal conductivity [mW/cm·K (300K)]
I_2-IV-VI_3 Compounds											
Cu_2SiS_3(H.T.)	251.36	41.89	3.684	6.004	3.81	1200					23

Substance	Molecular weight	Average atomic weight	Lattice parameters (Å, room temp.)		Density (g/cm³)	Melting point (K)	Microhardness, N/mm² (M-Mohs Scale)	Specific heat, J/kg·K (300 K)	Debye temp. (K)	Coefficient of thermal linear expansion [10^{-6} K^{-1} (300K)]	Thermal conductivity [mW/cm·K (300K)]
Cu_2SiS_3(L.T.)			5.290	10.156	3.63						
Cu_2SiTe_3	537.98	89.66	5.93		5.47						
Cu_2GeS_3(H.T.)	295.88	49.31	5.317		4.45	1210	4550	510	254	7.2	12
Cu_2GeS_3(L.T.)			5.327	5.215	4.46						
Cu_2GeSe_3	436.56	72.76	5.589	5.485	5.57	1030	3840	340	168	8.4	24
Cu_2GeTe_3	582.51	97.09	5.958	5.935	5.92		2890				130
Cu_2SnS_3	341.98	57.00	5.436		5.02	1110	2770	440	214	7.8	28
$CuSnSe_3$	482.66	80.44	5.687		5.94	960	2510	310	148	8.9	35
Cu_2SnTe_3	628.61	104.77	6.048		6.51	680	1970				144
Ag_2GeSe_3	525.21	87.54				810					
Ag_2SnSe_3	571.31	95.22									
Ag_2GeTe_3	671.13	111.86				600					
Ag_2SnTe_3	717.23	119.54									
I_3-V-VI_4-Compounds											
Cu_3PS_4	349.85	40.73	7.44	6.19							
Cu_3AsS_4	393.79	49.22	6.43	6.14	4.37	931				3.2	30.2
Cu_3AsSe_4	581.37	72.67	5.570	10.957	5.61	733			169	9.5	19
Cu_3SbS_4	440.64	55.08	5.38	16.76	4.90	830					
Cu_3SbSe_4	628.22	78.53	5.654	11.256	6.0	700			131	12.4	14.6
I-IV_2-V_3 Compounds											
$CuSi_2P_3$	212.64	35.44	5.25								
$CuGe_2P_3$	301.65	50.28	5.375		4.318	1113	8500	429	8.21	37.6	
$AgGe_2P_3$	345.97	57.66				1015	6150				

1.1.6. "Defect Chalcopyrite" Structure Compounds (Strukturbericht Symbol E3, Space Group I $\bar{4}$-S_4^2)

Substance	Molecular weight	Average atomic weight	Lattice parameters (Å, room temp.)		Density (g/cm³)	Melting point (K)
$ZnAl_2Se_4$	435.18	62.17	5.503	10.90	4.37	
$ZnAl_2Te_4$(?)	629.74	84.96	5.904	12.05	4.95	
$ZnGa_2S_4$(?)	333.06	47.58	5.274	10.44	3.80	
$ZnGa_2Se_4$(?)	520.66	74.38	5.496	10.99	5.21	
$ZnGa_2Te_4$(?)	715.22	102.17	5.937	11.87	5.67	
$ZnIn_2Se_4$	610.86	87.27	5.711	11.42	5.44	1250
$ZnIn_2Te_4$	805.42	115.06	6.122	12.24	5.83	1075
$CdAl_2S_4$	294.61	42.09	5.564	10.32	3.06	
$CdAl_2Se_4$	482.21	68.89	5.747	10.68	4.54	
$CdAl_2Te_4$(?)	676.77	97.68	6.011	12.21	5.10	
$CdGa_2S_4$	380.09	54.30	5.577	10.08	4.03	
$CdGa_2Se_4$	567.69	81.10	5.743	10.73	5.32	
$CdGa_2Te_4$	762.25	108.89	6.093	11.81	5.77	
$CdIn_2Te_4$	852.45	121.78	6.205	12.41	5.9	1060
$HgAl_2S_4$	382.79	54.68	5.488	10.26	4.11	
$HgAl_2Se_4$	570.39	82.48	5.708	10.74	5.05	
$HgAl_2Te_4$(?)	764.48	109.28	6.004	12.11	5.81	
$HgGa_2S_4$	468.27	66.90	5.507	10.23	5.00	
$HgGa_2Se_4$	655.87	93.70	5.715	10.78	6.18	
$HgIn_2Se_4$	746.07	106.58	5.764	11.80	6.3	1100
$HgIn_2Te_4$(?)	940.63	134.38	6.186	12.37	6.3	980

1.1.7. Other Adamantine Compounds

Substance	Molecular weight	Average atomic weight	Lattice parameters (Å, room temp.)		Density (g/cm³)	Melting point (K)
α–SiC	40.10	20.10	3.0817	15.12	3.21	3070
$Hg_5Ga_2Te_8$	2163.19	144.21	6.235			
$Hg_5In_2Te_8$	2253.39	150.23	6.328			
$CdIn_2Se_4$	657.89	93.98	a = c = 5.823			

Substance	Molecular weight	Average atomic weight	Lattice parameters (Å, room temp.)	Density (g/cm³)	Melting point (K)	Microhardness, N/mm² (M-Mohs Scale)	Specific heat, J/kg·K (300 K)	Debye temp. (K)	Coefficient of thermal linear expansion [10^{-6} K^{-1} (300K)]	Thermal conductivity [mW/cm·K (300K)]
1.2. Octahedral Semiconductors										
1.2.1. Halite Structure Semiconductors (Strukturbericht Symbol B1, Space Group $Fm3m\text{-}O_h^5$)										
GeTe	200.21	100.10	5.98	6.14						
SnSe	197.67	98.83	6.020		1133					
SnTe	246.31	123.15	6.313	6.45	1080 (max)					91
PbS	239.3	119.63	5.9362	7.61	1390					23
PbSe	286.2	143.08	6.1243	8.15	1340					17
PbTe	334.8	167.4	6.454	8.16	1180					23
1.2.2. Selected Other Binary Halites										
BiSe	287.94	143.97	5.99	7.98	880					
BiTe	336.58	168.29	6.47							
EuSe	230.92	115.46	6.191		2300					2.4
GdSe	236.21	118.11	5.771		2400					
NiO	74.69	37.35	4.1684	6.6	2260					
CdO	128.41	64.21	4.6953		1700					7
SrS	119.69	59.84	6.0199	3.643	3000					
1.3. Other Semiconductors										
1.3.1. Antifluorite Structure Compounds ($Fm3m\text{-}O_h^5$)										
Mg_2Si	76.70	25.57	6.338	1.88	1375				11.5	
Mg_2Ge	121.22	40.4	6.380	3.08	1388				15.0	
Mg_2Sn	167.32	55.77	6.765	3.53	1051				9.9	92
Mg_2Pb	225.81	85.27	6.836	5.1	823				10.0	
1.3.2. Tetradymite Structure Compounds ($\bar{R}3m\text{-}D_{3d}^5$)										
Sb_2Te_3	626.3	125.26	4.25 30.3	6.44	895					
Bi_2Se_3	654.84	130.97	4.14 28.7	7.51	979	167				24
Bi_2Te_3	800.76	160.15	4.38 30.45	7.73	858	155	16			30
1.3.3. Skutterudite Structure Compounds ($Im3\text{-}T_h^5$)										
CoP_3	151.85	37.96	7.7073		>1270					
$CoAs_3$	286.70	71.65	8.2060	6.73	1230					
$CoSb_3$	424.18	106.05	9.0385		1123			307		50
$NiAs_3$	283.45	70.86	8.330	6.43						
RhP_3	195.83	48.96	7.9951		>1470					
$RhAs_3$	327.67	81.92	8.4427		>1270					100
$RhSb_3$	468.16	117.04	9.2322		1170					
IrP_3	285.14	71.29	8.0151	7.36	>1470					
$IrAs_3$	416.98	104.25	8.4673	9.12	>1470					90
$IrSb_3$	557.47	139.37	9.2533	9.35	1170			303		
1.3.4. Selected Multinary Compounds										
$AgSbSe_2$	387.54	96.88	5.786	6.60	910					10.5
$AgSbTe_2$ (or $Ag_{19}Sb_{29}Te_{52}$)	484.82	121.2	6.078	7.12	830					86
$AgBiS_2$(H.T.)	380.97	95.24	5.648							
$AgBiSe_2$(H.T.)	474.77	118.69	5.82							
$AgBiTe_2$(H.T.)	572.05	143.01	6.155							
Cu_2CdSnS_4	486.43	60.80	5.586	10.83						
1.3.5. Some Elemental Semiconductors										
B		10.81	4.91 12.6	2.34	2348	9.5 (M)	1277	1370	8.3	600

Substance	Molecular weight	Average atomic weight	Lattice parameters (Å, room temp.)		Density (g/cm³)	Melting point (K)	Microhardness, N/mm² (M-Mohs Scale)	Specific heat, J/kg·K (300 K)	Debye temp. (K)	Coefficient of thermal linear expansion [10^{-6} K^{-1} (300K)]	Thermal conductivity [mW/cm·K (300K)]
Se(gray)		78.96	4.36	4.95	4.81	493	350	292.6		(‖C) 17.89 (⊥C) 74.09	(‖C) 45.2 (⊥C) 13.1
Te		127.60	4.45	5.91	6.23	723		196.5		16.8	(‖C) 33.8 (⊥C) 19.7

TABLE 2. Basic Thermodynamic, Electrical, and Magnetic Properties of Semiconductors (Listed by Crystal Structure)

Substance	Heat of formation [kJ/mol (300K)]	Volume compressibility (10^{-10}m²/N)	Static dielectric constant	Atomic magnetic susceptibility (10^{-6} cgs)	Index of refraction	Minimum room temperature energy gap (eV)	Mobility (room temp.) (cm²/V·s) Electrons	Holes	Optical transition	Breakdown voltage kV/mm	Remarks
2.1. Adamantine Semiconductors											
2.1.1. Diamond Structure Elements (Strukturbericht symbol A4, Space Group Fd 3m-O_h^7)											
C	714.4	18	5.7	−5.88	2.419 (589 nm)	5.4	1800	1400	i*	500	
Si	324	0.306	11.9	−3.9	3.49 (589 nm)	1.12	1900	500	i	30	
Ge	291	0.768	16	−0.12	3.99 (589 nm)	0.67	3800	1820	i		
α-Sn	267.5		24		2.75 (589 nm)	0.0; 0.8	2500	2400			
2.1.2. Sphalerite (Zinc Blende) Structure Compounds (Strukturbericht symbol B3 Space Group F$\bar{4}$3m-T_d^2)											
I-VII Compounds											
CuF											
CuCl	481	0.26	7.9		1.93	3.17			d		Nantokite
CuBr	481	0.26	7.9		2.12	2.91			d		
Cul	439	0.27	6.5		2.346	2.95			d		Marshite
AgBr	486		12.4		2.253	2.50	4000		i		Bromirite
AgI	389	0.41	10		2.22	2.22	30		d		Miersite
II-VI Compounds											
BeS					4.17				i		
BeSe					3.61				i		
BeTe					1.45		20		d		
BePo											
ZnO											See 2.1.3.
ZnS	477		8.9	−9.9	2.356	3.54	180	5(400°C)	d		See also 2.1.3.
ZnSe	422		9.2		2.89	2.58	540	28	d		
ZnTe	376		10.4		3.56	2.26	340	100	d		
ZnP											
CdS											See 2.1.3.
CdSe											See 2.1.3.
CdTe	339		7.2		2.50	1.44	1200	50	d		
CdPo											
HgS					2.85		250		d		Metacinnabarite
HgSe	247					2.10 (α)	20000	≈1.5	s		Tiemannite
HgTe	242					−0.06	25000	350	s		Coloradoite
III-V Compounds											
BN	815					4.6					Borazone

Solids

Substance	Heat of formation [kJ/mol (300K)]	Volume compressibility ($10^{-10}m^2/N$)	Static dielectric constant	Atomic magnetic susceptibility (10^{-6} cgs)	Index of refraction	Minimum room temperature energy gap (eV)	Mobility (room temp.) ($cm^2/V\cdot s$) Electrons	Holes	Optical transition	Breakdown voltage kV/mm	Remarks
BP(L.T.)						≈2.1	500	70			Ignites 470K
BAs						≈1.5					
AlP						2.45	80		i		
AlAs	627		10.9			2.16	1200	420	i		
AlSb	585	0.571	11		3.2	1.60	200–400	550	i		
GaP	635	0.110	11.1	−13.8	3.2	2.24	300	150	i		
GaAs	535	0.771	13.2	−16.2	3.30	1.35	8800	400	d		
GaSb	493	0.457	15.7	−14.2	3.8	0.67	4000	1400	d		
InP	560	0.735	12.4	−22.8	3.1	1.27	4600	150	d		
InAs	477	0.549	14.6	−27.7	3.5	0.36	33000	460	d		
InSb	447	0.442	17.7	−32.9	3.96	0.163	78000	750	d		
* i = indirect, d = direct, s = semimetal.											
Other Sphalerite Structure Compounds											
MnS											See also 2.1.3.
MnSe											See also 2.1.3.
β-SiC					2.697	2.3	4000				
Ga_2Te_3	271			−13.5		1.35	50				
In_2Te_3 (H.T.)	198			−13.6		1.04	50				
$MgGeP_2$											El–T[d12]
$ZnSnP_2$						2.1					Same
$ZnSnAs_2$ (H.T.)						≈0.7					Same
$ZnSnSb_2$						0.4					Same
2.1.3. Wurtzite (Zincite) Structure Compounds (Strukturbericht symbol B4, Space Group $P6_3mc$-C_{6v}^4)											
I-VII Compounds											
CuCl											
CuBr											
CuI											
AgI						2.63					Iodargirite
II-VI Compounds											
BeO											
MgTe											
ZnO	−350					3.2	180				
ZnS	−206					3.67					
ZnTe	−163										
CdS				8.45; 9.12	2.32	2.42	350	40	d		Greenockide
CdSe						1.74	900	50	d		Cadmoselite
CdTe						1.50	650				
III-V Compounds											
BP(H.T.)											
AlN						6.02					
GaN						3.34					
InN						2.0					
Other Wurtzite Structure Compounds											
MnS											
MnSe											
SiC					2.654						

Substance	Heat of formation [kJ/mol (300K)]	Volume compressibility ($10^{-10}m^2/N$)	Static dielectric constant	Atomic magnetic susceptibility (10^{-6} cgs)	Index of refraction	Minimum room temperature energy gap (eV)	Mobility (room temp.) ($cm^2/V{\cdot}s$)		Optical transition	Breakdown voltage kV/mm	Remarks
							Electrons	Holes			
MnTe						≈1.0					
Al_2S_3	426					4.1					
Al_2Se_3	367					3.1					

2.1.4. Chalcopyrite Structure Compounds (Strukturbericht symbol E11, Space Group $I\bar{4}2d$-D_{2d}^{12})

Substance	Heat of formation [kJ/mol (300K)]	Volume compressibility ($10^{-10}m^2/N$)	Static dielectric constant	Atomic magnetic susceptibility (10^{-6} cgs)	Index of refraction	Minimum room temperature energy gap (eV)	Mobility (room temp.) ($cm^2/V{\cdot}s$)		Optical transition	Breakdown voltage kV/mm	Remarks
							Electrons	Holes			
I-III-VI_2 Compounds											
$CuAlS_2$		0.106				2.5					
$CuAlSe_2$						2.67					
$CuAlTe_2$						0.88					
$CuCaS_2$		0.106				2.38					
$CuGaSe_2$		0.141				0.96, 1.63					
$CuGaTe_2$		0.227				0.82, 1.0					
$CuInS_2$		0.141				1.2					
$CuInSe_2$		0.187				0.86, 0.92					
$CuInTe_2$		0.278				0.95					
$CuTlS_2$											
$CuTlSe_2$ (L.T.)						1.07					
$CuFeS_2$						0.53					Chalcopyrite
$CuFeSe_2$						0.16					
$CuLaS_2$											
$AgAlS_2$											
$AgAlSe_2$						0.7					
$AgAlTe_2$						0.56					
$AgGaS_2$		0.150				1.66					
$AgGaSe_2$		0.182				1.1					
$AgGaTe_2$		0.280				1.9					
$AgInS_2$ (L.T.)		0.185				1.18					
$AgInSe_2$		0.238				0.96, 0.52					
$AgInTe_2$		0.338									
$AgFeS_2$											
II-IV-V_2 Compounds											
$ZnSiP_2$	312					2.3	1000				
$ZnGeP_2$	293					2.2					
$ZnSnP_2$	275					1.45					
$CdSiP_2$		0.103				2.2	1000				
$CdGeP_2$	289					1.8					
$CdSnP_2$	270					1.5					
$ZnSiAs_2$	290					1.7		50			
$ZnGeAs_2$	271			−14.4		0.85					
$ZnSnAs_2$	252			−18.4		0.65		300			Disorders at 910 K
$CdSiAs_2$		0.143				1.6					
$CdGeAs_2$	266			−23.4		0.53	70	25			Disorders at 903 K
$CdSnAs_2$	247		13.7	−21.5		0.26	22000	250			

2.1.5. Other Ternary Semiconductors with Tetrahedral Coordination

Substance	Heat of formation [kJ/mol (300K)]	Volume compressibility ($10^{-10}m^2/N$)	Static dielectric constant	Atomic magnetic susceptibility (10^{-6} cgs)	Index of refraction	Minimum room temperature energy gap (eV)	Mobility (room temp.) ($cm^2/V{\cdot}s$)		Optical transition	Breakdown voltage kV/mm	Remarks
							Electrons	Holes			
II_2-IV-VI_3 Compounds											
Cu_2SiS_3 (H.T.)											Wurtzite

Substance	Heat of formation [kJ/mol (300K)]	Volume compressibility ($10^{-10}m^2/N$)	Static dielectric constant	Atomic magnetic susceptibility (10^{-6} cgs)	Index of refraction	Minimum room temperature energy gap (eV)	Mobility (room temp.) ($cm^2/V{\cdot}s$) Electrons	Holes	Optical transition	Breakdown voltage kV/mm	Remarks
Cu_2SiS_3 (L.T.)											Tetragonal
Cu_2SiTe_3											Cubic
Cu_2GeS_3 (H.T.)				−18.7							Cubic
Cu_2GeS_3 (L.T.)							360				Tetragonal
Cu_2GeSe_3	211.5			−21.3		0.94	238				Same
Cu_2GeTe_3	190.2			−23.4							Same
Cu_2SnS_3				−18.2		0.91	405				Cubic
$CuSnSe_3$				−21.0		0.66	870				Cubic
Cu_2SnTe_3				−28.4							Cubic
Ag_2GeSe_3				−29.6		0.91 (77K)					
Ag_2SnSe_3				−29.5		0.81					
Ag_2GeTe_3				−31.4		0.25					
Ag_2SnTe_3				−31.0		0.08					
II_3-V-VI_4 Compounds											
Cu_3PS_4											Enargite
Cu_3AsS_4	269.6			−15.8		1.24					
Cu_3AsSe_4	161.3			−13.1		0.88					Famatinite
Cu_3SbS_4				−8.3		0.74					Famatinite
Cu_3SbSe_4	127.1			−20.5		0.31					
II-IV_2-V_3 Compounds											
$CuSi_2P_3$											El
$CuGe_2P_3$		0.12				0.90					El
$AgGe_2P_3$											
2.1.6. "Defect Chalcopyrite" Structure Compounds (Strukturbericht symbol E3, Space Group $I\bar{4}$-S_4^2)											
$ZnAl_2Se_4$											
$ZnAl_2Te_4$ (?)											
$ZnGa_2S_4$ (?)						≈3.4					
$ZnGa_2Se_4$ (?)						≈2.2					
$ZnGa_2Te_4$ (?)						1.35					
$ZnIn_2Se_4$	206					1.82	35				
$ZnIn_2Te_4$	198					1.2					
$CdAl_2S_4$											
$CdAl_2Se_4$											
$CdAl_2Te_4$ (?)											
$CdGa_2S_4$	256					3.44	60				
$CdGa_2Se_4$	216					2.43	33				
$CdGa_2Te_4$											
$CdIn_2Te_4$	195					(1.26 or 0.9)	4000				
$HgAl_2S_4$											
$HgAl_2Se_4$											
$HgAl_2Te_4$ (?)											
$HgGa_2S_4$	249					2.84					
$HgGa_2Se_4$	204					1.95	400				

Substance	Heat of formation [kJ/mol (300K)]	Volume compressibility (10^{-10}m²/N)	Static dielectric constant	Atomic magnetic susceptibility (10^{-6} cgs)	Index of refraction	Minimum room temperature energy gap (eV)	Mobility (room temp.) (cm²/V·s) Electrons	Holes	Optical transition	Breakdown voltage kV/mm	Remarks
$HgIn_2Se_4$	196					0.6	290				
$HgIn_2Te_4$ (?)	188					0.86	200				
2.1.7. Other Adamantine Compounds											
α–SiC			10.2	−6.4	2.67	2.86	400				6H structure
$Hg_5Ga_2Te_8$											B3 with superlattice
$Hg_5In_2Te_8$						0.7	2000				B3 with superlattice
$CdIn_2Se_4$						1.55					
2.2. Octahedral Semiconductors											
2.2.1. Halite Structure Semiconductors (Strukturbericht symbol B1, Space Group Fm3m-O_h^5)											
GeTe											
SnSe											
SnTe											
PbS	435					0.5	600	600			
PbSe	393		161			0.37	1000	900			
PbTe	393		280			0.26	1600	600			Altaite
			360			0.25					
2.2.2. Selected Other Binary Halites											
BiSe											
BiTe						0.4					
EuSe											
GdSe						1.8	4				
NiO						2.0 or 3.7	100				
CdO	531					2.5					
SrS						4.1					
2.3. Other Semiconductors											
2.3.1. Antifluorite Structure Compounds (Fm3m-O_h^5)											
Mg_2Si	79.08					0.77	405	70			
Mg_2Ge						0.74	520	110			
Mg_2Sn	76.57					0.36	320	260			
Mg_2Pb	52.72					0.1					
2.3.2. Tetradymite Structure Compounds (R$\bar{3}$-D_{3d}^5)											
Sb_2Te_3						0.3		360			
Bi_2Se_3						0.35	600				
Bi_2Te_3						0.21	1140	680			R3m (166)
2.3.3. Skutterudite Structure Compounds (Im3-T_h^5)											
CoP_3						0.43					
$CoAs_3$						0.69		~4000			
$CoSb_3$						0.63	70	~3000			
RhP_3								700			
$RhAs_3$						0.85		~3000			
$RhSb_3$						0.80		~7000			
$IrSb_3$						1.18		1500			
2.3.4. Selected Multinary Compounds											
$AgSbSe_2$						0.58					

Solids

Substance	Heat of formation [kJ/mol (300K)]	Volume compressibility (10^{-10}m²/N)	Static dielectric constant	Atomic magnetic susceptibility (10^{-6} cgs)	Index of refraction	Minimum room temperature energy gap (eV)	Mobility (room temp.) (cm²/V·s) Electrons	Holes	Optical transition	Breakdown voltage kV/mm	Remarks
$AgSbTe_2$ (or $Ag_{19}Sb_{29}Te_{52}$)						0.7, 0.27					
$AgBiS_2$ (H.T.)											
$AgBiSe_2$ (H.T.)											
$AgBiTe_2$ (H.T.)											
Cu_2CdSnS_4						1.16	<2				
2.3.5. Some Elemental Semiconductors											
B	397.1			−6.7	3.4	1.55	10				
Se (gray)			6.6 (0.1 GHz)	−22.1	2.5	1.5		5			$P3_121$(152)
Te				−39.5	3.3	0.33	1700	1200			Same

TABLE 3. Semiconducting Properties of Selected Materials

Substance	Minimum energy gap (eV) R.T.	0 K	10^4 dE/dT eV K^{-1}	dE/dT meV kbar^{-1}	Electron effective mass m_{da} (m_o)	Electron mobility and temperature dependence μ_a (cm² V^{-1} s^{-1})	$-x$	Hole effective mass m_{dp} (m_a)	Hole mobility and temperature dependence μ_p (cm² V^{-1} s^{-1})	$-x$
Elements										
Si	1.110	1.169	−2.8	−1.41	1.026	1900	2.6	0.056	500	2.3
Ge	0.664	0.744	−3.7	5.1	0.0823	3800	1.66	0.0438	1820	2.33
α-Sn	0.08	0.094	−0.5		0.0236	2500	1.65	0.195	z	2.0
Se	2.11	2.48								
Te	0.335				0.08	1100		0.19	560	
III-V Compounds										
AlAs	2.2	2.3				1200			420	
AlSb	1.6	1.7	−3.5	−1.6	0.09	200	1.5	0.4	500	1.8
GaP	2.272	2.350	−3.7	10.5	0.35	300	1.5	0.5	150	1.5
GaAs	1.441	1.579	−3.9	11.3	0.068	9000	1.0	0.5	500	2.1
GaSb	0.70	0.812	−3.7	14.5	0.050	5000	2.0	0.23	1400	0.9
InP	1.34	1.4236	−2.9	9.1	0.067	5000	2.0		200	2.4
InAs	0.356	0.418	−3.4	10.0	0.022	33 000	1.2	0.41	460	2.3
InSb	0.180	0.235	−2.8	15.7	0.014	78 000	1.6	0.4	750	2.1
II-VI Compounds										
ZnO	3.2	3.4376	−9.5	0.6	0.38	180	1.5			
ZnS	3.80	3.91	−4.7	−5.8		180			5(400 °C)	
ZnSe	2.713	2.820	−4.5	0.7		540			28	
ZnTe	2.26	2.391	−5.2	8.3		340			100	
CdO	1.20		−6		0.1	120				
CdS	2.485	2.585	−4.1	4.5	0.165	400		0.8		
CdSe	1.751	1.841	−3.6	5.0	0.13	650	1.0	0.6		
CdTe	1.43	1.606	−5.4	8	0.14	1200		0.35	50	
HgSe	−0.061				0.030	20 000	2.0			
HgTe	−0.141	−0.3025			0.017	25 000		0.5	350	
Halite Structure Compounds										
PbS	0.41	0.286	4		0.16	800		0.1	1000	2.2
PbSe	0.278	0.145	4		0.3	1500		0.34	1500	2.2

Substance	Minimum energy gap (eV) R.T.	Minimum energy gap (eV) 0 K	$10^4\ dE/dT$ eV K^{-1}	dE/dT meV kbar^{-1}	Electron effective mass m_{da} (m_o)	Electron mobility and temperature dependence μ_a (cm^2 V^{-1} s^{-1})	Electron mobility and temperature dependence $-x$	Hole effective mass m_{dp} (m_a)	Hole mobility and temperature dependence μ_p (cm^2 V^{-1} s^{-1})	Hole mobility and temperature dependence $-x$
PbTe	0.310	0.187	4	−7	0.21	1600		0.14	750	2.2
Others										
ZnSb	0.50	0.56			0.15	10				1.5
CdSb	0.459	0.57	−5.4		0.15	300			2000	1.5
Bi_2S_3	1.3	1.45				200			1100	
Bi_2Se_3	0.160					600			675	
Bi_2Te_3	0.13		−0.95		0.58	1200	1.68	1.07	510	1.95
Mg_2Si		0.77	−6.4		0.46	400	2.5		70	
Mg_2Ge	0.54	0.74	−9			280	2		110	
Mg_2Sn	0.18	0.36	−3.5		0.37	320			260	
Mg_2Sb_2		0.32				20			82	
Zn_3As_2	0.93					10	1.1		10	
Cd_3As_2	0.55				0.046	100 000	0.88			
GaSe	2.021	2.1275	−3.8						20	
GaTe	1.694	1.799	−3.6			14				
InSe	1.172	1.32				900				
TlSe	0.745		−3.9		0.3	30		0.6	20	1.5
$CdSnAs_2$	0.26				0.05	25 000	1.7			
Ga_2Te_2	1.22	1.55	−4.8							
α-In_2Te_2	0.92	1.2			0.7				50	1.1
β-In_2Te_2	1.0								5	
$Hg_5In_2Te_8$	0.5								11 000	
SnO_2	3.47	3.596							78	

TABLE 4. Band Properties of Semiconductors

Substance	Band curvature effective mass (Expressed as fraction of free electron mass) Heavy holes	Light holes	"Split-off" band holes	Energy separation of "split-off" band (eV)	Measured (light) hole mobility (cm^2/V·s)
4.1. Data on Valence Bands of Semiconductors (Room Temperature)					
4.1.1. Semiconductors with Valence Band Maximum at the Center of the Brillouin Zone ("Γ")					
Si	0.52	0.16	0.25	0.044	500
Ge	0.34	0.043	0.08	0.3	1820
Sn	0.3				2400
AlAs					
AlSb	0.4			0.7	550
GaP				0.13	100
GaAs	0.8	0.12	0.20	0.34	400
GaSb	0.23	0.06		0.7	1400
InP				0.21	150
InAs	0.41	0.025	0.083	0.43	460
InSb	0.4	0.015		0.85	750
CdTe	0.35				50
HgTe	0.5				350

Substance	Number of equivalent valleys and direction	Band curvature effective masses Longitudinal m_L	Transverse m_T	Anistrophy K = m_L/m_T	Measured (light) hole mobility (cm^2/V·s)
4.1.2. Semiconductors with Multiple Band Maxima					
PbSe	4 "L" [111]	0.095	0.047	2.0	1500
PbTe	4 "L" [111]	0.27	0.02	10	750
Bi_2Te_3	6	0.207	~0.045	4.5	515

Substance	Energy gap (eV)	Effective mass (m_o)	Mobility (cm²/V·s)	Comments

4.2. Data on Conduction Bands of Semiconductors (Room Temperature Data)

4.2.1. Single Valley Semiconductors

Substance	Energy gap (eV)	Effective mass (m_o)	Mobility (cm²/V·s)	Comments
GaAs	1.35	0.067	8500	3(or 6?) equivalent [100] valleys 0.36 eV above this maximum with a mobility of ~50.
InP	1.27	0.067	5000	3(or 6?) equivalent [100] valleys 0.4 eV above this minimum.
InAs	0.36	0.022	33,000	Equivalent valleys ~1.0 eV above this minimum.
InSb	0.165	0.014	78,000	
CdTe	1.44	0.11	1000	4(or 8?) equivalent [111] valleys 0.51 eV above this minimum.

Substance	Energy gap	Number of equivalent valleys and direction	Band curvature effective mass: Longitudinal m_L	Band curvature effective mass: Transverse m_T	Anisotropy $K = m_L/m_T$
4.2.2. Multivalley Semiconductors					
Si	1.107	6 in [100] "Δ"	0.00	0.192	4.7
Ge	0.67	4 in [111] at "L"	1.588	0.0815	19.5
GaSb	0.67	as Ge (?)	~1.0	~0.2	~5
PbSe	0.26	4 in [111] at "L"	0.085	0.05	1.7
PbTe	0.25	4 in [111] at "L"	0.21	0.029	5.5
Bi_2Te_3	0.13	6			~0.05

TABLE 5. Resistivity of Semiconducting Minerals

Mineral	ρ (ohm · m)
Diamond (C)	2.7
Sulfides	
Argentite, Ag_2S	1.5 to 2.0 × 10⁻³
Bismuthinite, Bi_2S_3	3 to 570
Bornite, $Fe_2S_3 \cdot nCu_2S$	1.6 to 6000 × 10⁻⁶
Chalcocite, Cu_2S	80 to 100 × 10⁻⁶
Chalcopyrite, $Fe_2S_3 \cdot Cu_2S$	150 to 9000 × 10⁻⁶
Covellite, CuS	0.30 to 83 × 10⁻⁶
Galena, PbS	6.8 × 10⁻⁶ to 9.0 × 10⁻²
Haverite, MnS_2	10 to 20
Marcasite, FeS_2	1 to 150 × 10⁻³
Metacinnabarite, HgS	2 × 10⁻⁶ to 1 × 10⁻³
Millerite, NiS	2 to 4 × 10⁻⁷
Mineral	ρ (ohm m)
Molybdenite, MoS_2	0.12 to 7.5
Cobaltite, CoAsS	6.5 to 130 × 10⁻³
Enargite, Cu_3AsS_4	0.2 to 40 × 10⁻³
Gersdorfite, NiAsS	1 to 160 × 10⁻⁶
Glaucodote, (Co, Fe)AsS	5 to 100 × 10⁻⁶
Antimonide	
Dyscrasite, Ag_3Sb	0.12 to 1.2 × 10⁻⁶
Arsenides	
Allemonite, $SbAs_3$	70 to 60,000
Lollingite, $FeAs_2$	2 to 270 × 10⁻⁶
Nicollite, NiAs	0.1 to 2 × 10⁻⁶
Skutterudite, $CoAs_3$	1 to 400 × 10⁻⁶
Smaltite, $CoAs_2$	1 to 12 × 10⁻⁶
Tellurides	
Altaite, PbTe	20 to 200 × 10⁻⁶
Calavarite, $AuTe_2$	6 to 12 × 10⁻⁶
Coloradoite, HgTe	4 to 100 × 10⁻⁶

Mineral	ρ (ohm · m)
Pentlandite, $(Fe, Ni)_4S_4$	1 to 11 × 10⁻⁶
Pyrrhotite, Fe_7S_4	2 to 160 × 10⁻⁶
Pyrite, FeS_2	1.2 to 600 × 10⁻³
Sphalerite, ZnS	2.7 × 10⁻³ to 1.2 × 10⁴
Antimony-sulfur compounds	
Berthierite, $FeSb_2S_4$	0.0083 to 2.0
Boulangerite, $Pb_5Sb_4S_{11}$	2 × 10³ to 4 × 10⁴
Cylindrite, $Pb_3Sn_4Sb_2S_{14}$	2.5 to 60
Franckeite, $Pb_5Sn_3Sb_2S_{14}$	1.2 to 4
Hauchecornite, $Ni_4(Bi, Sb)_2S_{14}$	1 to 83 × 10⁻⁶
Jamesonite, $Pb_4FeSb_6S_{14}$	0.020 to 0.15
Tetrahedrite, Cu_3SbS_3	0.30 to 30,000
Arsenic-sulfur compounds	
Mineral	ρ (ohm m)
Arsenopyrite, FeAsS	20 to 300 × 10⁻⁶
Hessite, Ag_2Te	4 to 100 × 10⁻⁶
Nagyagite, $Pb_6Au(S,Te)_{14}$	20 to 80 × 10⁻⁶
Sylvanite, $AgAuTe_4$	4 to 20 × 10⁻⁶
Oxides	
Braunite, Mn_2O_3	0.16 to 1.0
Cassiterite, SnO_2	4.5 × 10⁻⁴ to 10,000
Cuprite, Cu_2O	10 to 50
Hollandite, $(Ba, Na, K) Mn_8O_{16}$	2 to 100 × 10⁻³
Ilmenite, $FeTiO_3$	0.001 to 4
Magnetite, Fe_3O_4	52 × 10⁻⁶
Manganite, MnO OH	0.018 to 0.5
Melaconite, CuO	6000
Psilomelane, $BaMn_9O_{18} \cdot 2H_2O$	0.04 to 6000
Pyrolusite, MnO_2	0.007 to 30
Rutile, TiO_2	29 to 910
Uraninite, UO_2	1.5 to 200

References

1. Beer, A. C., *Galvanomagnetic Effects in Semiconductors*, Academic Press, New York, 1963.
2. Goryunova, N. A., *The Chemistry of Diamond-Like Semiconductors*, The MIT Press, Cambridge, MA, 1965.
3. Abrikosov, N. Kh., Bankina, V. F., Poretskaya, L. E., Shelimova, L. E., and Skudnova, E.V., *Semiconducting II-VI, IV-VI, and V-VI Compounds*, Plenum Press, New York, 1969.
4. Berger, L. I. and Prochukhan, V. D., *Ternary Diamond-Like Semiconductors*, Cons. Bureau/Plenum Press, New York, 1969.
5. Shay, J. L. and Wernick, J. H., *Ternary Chalcopyrite Semiconductors: Growth, Electronic Properties, and Applications*, Pergammon Press, 1975.
6. Bergman, R., *Thermal Conductivity in Solids*, Clarendon, Oxford, 1976.
7. Handbook of Semiconductors, Vol. 1, Moss, T. S. and Paul, W., Eds., *Band Theory and Transport Properties*; Vol. 2, Moss, T. S. and Balkanski, M., Eds., *Optical Properties of Solids*; Vol. 3, Moss, T. S. and Keller, S. P., Eds., *Materials Properties and Preparation*, North Holland Publ. Co., Amsterdam, 1980.
8. Böer, K. W., *Survey of Semiconductor Physics*, Van Nostrand Reinhold, 1990.
9. Rowe, D. M., Ed., *CRC Handbook of Thermoelectrics*, CRC Press, Boca Raton, FL, 1995.
10. Berger, L. I., *Semiconductor Materials*, CRC Press, Boca Raton, FL, 1997.
11. Glazov, V. M., Chizhevskaya, S.N., and Glagoleva, N.N., *Liquid Semiconductors, Plenum Press*, New York, 1969.
12. Phillips, J. C., *Bonds and Bands in Semiconductors*, Academic Press, New York, 1973.
13. Harrison, W. A., *Electronic Structure and the Properties of Solids*, Freeman Publ. House, San Francisco, 1980.
14. Balkanski, M., Ed., *Optical Properties of Solids*, North-Holland, Amsterdam, 1980.
15. *Landolt-Börnstein. Numerical Data and Functional Relationships in Science and Technology, New Series, Group III: Crystal and Solid State Physics*, Hellwege, K.-H. and Madelung, O., Eds., Volumes 17 and 22, Springer Verlag, Berlin, 1984 (and further).
16. Shklovskii, B. L. and Efros, A. L., *Electronic Processes in Doped Semiconductors*, Springer Verlag, Berlin, 1984.
17. Cohen, M. L. and Chelikowsky, J. R., *Electronic Structure and Optical Properties of Semiconductors*, Springer Verlag, New York, 1988.
18. Glass, J. T., Messier, R. F., and Fujimori, N., Eds., *Diamond, Silicon Carbide, and Related Wide Bandgap Semiconductors*, MRS Symposia Proc. 1652, Mater. Res. Soc., Pittsburgh, 1990.
19. Palik, E., Ed., *Handbook of Optical Constants of Solids II*, Academic Press, New York, 1991.
20. Reed, M., Ed., *Semiconductors and Semimetals*, Volume 35, Academic Press, Boston, 1992.
21. Haug, H. and Koch, S. W., *Quantum Theory of the Optical and Electronic Properties of Semiconductors*, 2nd Edition, World Scientific, Singapore, 1993.
22. Lockwood, D. J., Ed., *Proc. 22nd Intl. Conf. on the Physics of Semiconductors, Vancouver, 1994*, World Scientific, Singapore, 1994.
23. Morelli, D. T., Caillat, T., Fleurial, J.-P., Borschchevsky, A., Vandersande, J., Chen, B., and Uher, C., *Phys. Rev.*, B51, 9622, 1995.
24. Caillat, T., Borshchevsky, A., and Fleurial, J.-P., *J. Appl. Phys.*, 80, 4442, 1996.
25. Fleurial, J.-P.,Caillat, T., and Borshchevsky, A., *Proc. XVI Intl. Conf. Thermoelectrics*, Dresden, Germany, August 26–29, 1997 (in print).
26. Borshchevsky, A. et al., U.S. Patents 5,610,366 (March 1977) and 5,831,286 (March 1998).
27. Jarrendahl, K. and Davis, R. F., *Semiconductors and Semimetals*, Vol. 52, Y.S. Park, Ed., 1998, pp. 1–20.
28. Bettini, M., *Solid State Comm.*, 13, 599, 1973.
29. Chen, A. and Sher, A., *Semiconductor Alloys, Physics and Material Engineering*, Plenum Press, New York, 1995.
30. Holloway, P. H. and McGuire, G. E., Eds., *Handbook of Compound Semiconductors*, Noyes Publ., Park Ridge, NJ, 1995.
31. Madelung, O., Ed., *Semiconductors: Group IV Elements and III-V Compounds (Data in Science and Technology)*, Springer-Verlag, Berlin, Heidelberg, 1991.
32. Madelung, O., Ed., *Semiconductors: Other than Group IV Elements and III-V Compounds (Data in Science and Technology)*, Springer-Verlag, Berlin, Heidelberg, 1992.
33. Levinshtein, M., Rumyantsev, S., and Shur, M., Eds., Handbook Series on Semiconductor Parameters. Vol. 1, *Si, Ge, C (Diamond), GaAs, GaP, GaSb, InAs, InP, InSb*, World Scientific, Singapore, 1996.
34. Levinshtein, M., Rumyantsev, S., and Shur, M., Eds., Handbook Series on Semiconductor Parameters. Vol. 2, *Ternary and Quaternary III-V Compounds*, World Scientific, Singapore, 1996.
35. *Physical Properties of Semiconductors*, NSM Archive:http://www.ioffe.ru/SVA/NSM/Semicond/

SELECTED PROPERTIES OF SEMICONDUCTOR SOLID SOLUTIONS

L. I. Berger

Alloy system	Limits of solubility	Energy gap in eV (300 K)	Remarks, references
		Adamantine Semiconductors IV-IV	
Si_xGe_{1-x}	0≤x≤1	0.8941+0.0421x+0.1691x^2 0.7596+1.0860x+0.3306x^2	Transition Γ - X [Ref.1] Trans. Γ - L [Ref. 1]
		Adamantine Semiconductors III-V/III-V	
Common Anion			
$Al_xGa_{1-x}N$	0≤x≤1		Wurtzite Structure [Refs. 2 and 3]
$Al_xGa_{1-x}P$	0≤x≤0.5	2.28+0.16x	[Ref. 2]
$Al_xIn_{1-x}P$	0≤x≤0.44	at Γ: 134+2.23x; at X: 2.24+0.18x	[Ref. 2]
$Al_xGa_{1-x}As$	0≤x≤0.5	1.42=0.75x [Ref.3]; 1.424+1.429x-0.14x^2 [Ref.4]	
$Al_xIn_{1-x}As$	0≤x≤1	at Γ: 0.37+1.91x+0.74x^2; at X: 1.8+0.4x	[Refs. 2 and 6]
$Al_xGa_{1-x}Sb$	0≤x<≤1	0.73+1.10x+0.47x^2	Trans. Γ_{8v}- Γ_{6c} [Ref. 2]
$Al_xIn_{1-x}Sb$	0≤x≤1		[Ref. 6]
$Ga_xIn_{1-x}N$	0≤x≤1	1.950+1.487x-1.000x(1-x)	Wurtzite [Refs. 8 and 10]
$Ga_xIn_{1-x}P$	0≤x≤1		[Ref. 2]
$Ga_xIn_{1-x}As$	0≤x≤1	0.360+0.629x+0.436x^2	[Ref. 5]
$Ga_xIn_{1-x}Sb$	0<x<1	0.235+0.1653x+0.413x^2	[Ref. 2, see also Ref. 9]
Common Cation			
GaN_xAs_{1-x}	0≤x≤0.05	1.42-9.9x	[Ref. 2]
GaP_xAs_{1-x}	0<x<1	2.270-0.846x	[Ref. 2]
GaP_xAs_{1-x}	0≤x≤0.05	1.515+1.172x+0.186x^2 1.9715+0.144x+0.211x^2	(at 2 K, Γ–Γ) [Ref. 7] [Ref. 2]
$GaAs_xSb_{1-x}$	0≤x≤0.45, 0.6≤x≤1	1.43-1.9x+1.2x^2	[Ref. 5]
InP_xAs_{1-x}	0<x<1	0.356+0.675x+0.32x^2	[Ref. 2]
		Adamantine Binary Semiconductors II-VI/II-VI [Refs. 3 and 6]	
Common Anion			
$Zn_xCd_{1-x}S$	0≤x≤1		Wurtzite Structure
$Zn_xHg_{1-x}S$	0≤x≤1		
$Cd_xHg_{1-x}S$	0≤x≤1		Wurtzite Structure at x<0.6
$Zn_xCd_{1-x}Se$	0.7≤x≤1		
$Zn_xHg_{1-x}Se$	0≤x≤1		
$Cd_xHg_{1-x}Se$	0≤x≤0.7 and 0.75≤x*≤1		x*- Wurtzite Structure
$Zn_xCd_{1-x}Te$	0≤x≤1		
$Zn_xHg_{1-x}Te$	0≤x≤1		
$Cd_xHg_{1-x}Te$	0≤x≤1		
Common Cation			
ZnS_xSe_{1-x}	0≤x≤1		
ZnS_xTe_{1-x}	0≤x≤0.1 and 0.9≤x*≤1		x*- Wurtzite Structure
$ZnSe_xTe_{1-x}$	0≤x≤1		
CdS_xSe_{1-x}	0≤x≤1		Wurtzite Structure
CdS_xTe_{1-x}	0≤x≤0.25 and 0.8≤x*≤1		x*- Wurtzite Structure
$CdSe_xTe_{1-x}$	0≤x≤0.4 and 0.6≤x*≤1		x*- Wurtzite Structure
HgS_xSe_{1-x}	0≤x≤1		
HgS_xTe_{1-x}	0≤x≤1		
$HgSe_xTe_{1-x}$	0≤x≤1		
		Quaternary Adamantine Semiconductors II-VI/III-V [Ref. 6]	
$(ZnS)_x(AlP)_{1-x}$	0.99≤x≤1		
$(ZnSe)_x(GaAs)_{1-x}$	0≤x<1		
$(CdTe)_x(InAs)_{1-x}$	0<x≤0.2 and 0.7≤x≤1		
$(CdTe)_x(AlSb)_{1-x}$	0≤x≤1		
$(HgTe)_x(InAs)_{1-x}$	0≤x≤1		

Alloy system	Limits of solubility	Energy gap in eV (300 K)	Remarks, references
		Quaternary Adamantine Semiconductors III_x-III_{1-x}-V_y - V_{1-y}	
$Ga_xIn_{1-x}As_yP_{1-y}$	$0 \le x \le 1$, $0 \le x \le 1$	$1.35+0.668x-1.068y+0.758x^2+0.078y^2-0.069xy-0.322x^2y+0.03xy^2$	[Refs. 2 and 6]
		Quaternary Adamantine Semiconductors III_{1-x-y}-III_x-III_y-V	
$Al_xGa_yIn_{1-x-y}Sb$	$0 \le x \le 1$, $0 \le y \le 1$	$0.095+1.76x+0.28y+0.345(x^2+y^2)+0.085(1-x-y)^2+xy(1-x-y)(23-28y)$	[Refs. 2 and 6]

References

1. Krishnamurti, S., Sher, A., and Chen, A. *Appl. Phys. Lett.* 47, 160, 1985.
2. Madelung, O., Ed., *Semiconductors Group IV Elements and III-V Compounds*, Springer, 1991; *Semiconductors Other Than Group IV Elements and III-V Compounds*, Springer, 1992.
3. Goryunova, N. A., Kesamanly, F. P., and Nasledov, D. N., *Semiconductors and Semimetals*, Vol. 4, 1968, 413.
4. El Allali, M., Sorensen, C. B., Veje, E., and Tideman-Peterson, P., *Phys. Rev. B* 48, 4398, 1993.
5. Nahorny, R. E., Pollack, M. A., Johnson, W. D., and Barns, R. L. *Appl. Phys. Lett.* 33, 695, 1978.
6. Goryunova, N. A., *Multicomponent Diamond-Like Semiconductors*, Sov. Radio, Moscow, 1968 (in Russian).
7. Capizzi, M., Modesti, S., Martelli, F., and Frova, A., *Solid State Comm.* 39, 333, 1981.
8. Nakamura, S., Pearton, S., and Fasol, G., *The Blue Laser Diode*, 2nd Ed., Springer, 2000.
9. Roth, A. P., Keeler, W. J., and Fortin, E. *Canad. J. Phys.* 58, 560, 1980.
10. Wu, J., Walukiewicz, Yu, K. M., Ager, J. W., Haller, E. E., Lu, H., and Schaff, W. J., *Appl. Phys. Lett.* 80, 4741, 2002.

PROPERTIES OF ORGANIC SEMICONDUCTORS

L. I. Berger

Substance	Energy gap E/eV	Electrical resistivity[a] ρ/Ω cm	Mobility μ/cm^2 V^{-1} s^{-1}	Sign of majority carriers	Temperature range °C	Ref.
Metal-Free Molecular Crystals						
3-Acetylamino-*N*-methylphthalimide	3.46				67 to 100	1
3-Acetylamino-*N*-phenylphthalimide	3.50				54 to 124	1
4-Amino-*N*-cyclomethylphthalimide	2.90				73 to 100	1
4-Aminophthalimide	2.78				123 to 151	1
Acridine	3.90					1
Anthanthrene	1.67	$1.5 \cdot 10^{19}$			40 to 105	1
Anthanthrene	0.84	$1.5 \cdot 10^{19}$ (15°C)				2
Anthanthrone	1.70	$7.7 \cdot 10^{18}$			20 to 150	1
Anthracene	0.83	$1.3 \cdot 10^{14}$ (15°C)				2
Anthracene	2.50	$1.5 \cdot 10^{11}$	2.3	+	20 to 130	1
Anthracene	3.88 to 4.1	$>10^{15}$	1.74(n), 2.07(p)	+ & –		4
1,2-Benzanthracene	1.04	10^{16} (30°C)				2
Benzanthrone	3.12	$1.6 \cdot 10^{16}$				1
Benzene (liquid)	0.41					2
Benzene (amorphous)	0.84	10^{15}			–14 to 5	1
Benzene (cryst.)	7		2	–	–23	4
Benzimidazole	3.0 to 4.0	$5 \cdot 10^{13}$			84 to 144	1
Benzophenone	3.34				–23 to 14	1
Benzo[*f*]quinoline	2.77				30 to 50	1
Benzo[*h*]quinoline	2.72					1
3-Benzoylamino-*N*-methylphthalimide	3.28				84 to 112	1
Benzpentacene	1.72				0 to 150	1
Biphenyl	1.46	$1.7 \cdot 10^{15}$ (50°C)				2
Biphenyl	1.45				20 to 70	1
o-Chloranil	3.0	10^{15}				1
p-Chloranil	0.61					1
Chlorpromazine	2.1	10^{12} (32°C)		+ & –	32 to 80	1
Chrysene	1.1	$4 \cdot 10^{19}$ (15°C)				2
Chrysene	2.20	$4 \cdot 10^{19}$			25 to 90	1
Circumanthracene	1.8	$6 \cdot 10^{12}$				1
Coronene	1.7	$1.7 \cdot 10^{17}$			60 to 80	1
Coronene	0.85	$1.7 \cdot 10^{17}$ (15°C)				2
Cyananthrone	0.20	$1.2 \cdot 10^{7}$			30 to 125	1
1,6-Diaminopyrene	0.6	10^{8}				
Dibenzpentacene	1.50				0 to 150	1
Dinaphthopyrene	1.60				25 to 90	1
1,8-Diphenyl-1,3,5,7-octatetraene	1.7				72 to 191	1
Diphenylpentacene	1.60				0 to 150	1
4,4′-Diphenylstilbene	1.56				160 to 280	1
4,4′-Diphenylstilbene	0.80					2
Ferrocene	1.22	10^{14}		+		1
Flavanthrone	0.70	$1.4 \cdot 10^{11}$				1
Fluorene	2.7					2
Fluorene	1.4					2
Fluoridine	1.6	$6 \cdot 10^{13}$		+	20 to 140	1
Hexacene	0.57	$3.8 \cdot 10^{10}$ (50°C)				2
Hexacene	1.3					1
Hexamethylbenzene	1.86			+	20 to 140	1
3-Hydroxy-*N*-methylphthalimide	3.80				60 to 91	1
Imidazole	2.6	10^{11}			28 to 68	1

Substance	Energy gap E/eV	Electrical resistivity[a] ρ/Ω cm	Mobility μ/cm[2] V^{-1} s^{-1}	Sign of majority carriers	Temperature range °C	Ref.
Indanthrazine	0.66	$1.4 \cdot 10^{15}$			30 to 125	1
Indanthrone	0.64	$7.5 \cdot 10^{14}$			30 to 125	1
Indanthrone (black)	0.56	$2.5 \cdot 10^{8}$			30 to 125	1
Mesitylene (liquid)	0.19					2
Mesonaphthodianthracene	0.6	$4.0 \cdot 10^{18}$ (15°C)				2
Mesonaphthodianthrene	1.48				45 to 250	1
Mesonaphthodianthrone	0.86				5 to 110	1
3-Methoxy-*N*-methyl-phthalimide	3.18				54 to 78	1
Naphthacene	1.7	$1 \cdot 10^{15}$				1
Naphthalene	3.5	10^{14}			27 to 47	1
Naphthalene	1.15	$2.8 \cdot 10^{14}$ (50°C)				2
Naphthalene	4.9 to 5.1		0.64(n), 1.50(p)	+ & −		4
m-Naphthodianthrene	1.20	$4 \cdot 10^{18}$			40 to 150	1
m-Naphthodianthrone	1.30	$1.5 \cdot 10^{18}$			40 to 150	1
β-Naphthol	2.36	$2 \cdot 10^{5}$			60 to 110	1
β-Naphthoquinoline	2.77					1
1-Naphthylamine	2.2				25 to 42	1
1-Naphthylamine picrate	2.7				28 to 98	1
2-Naphthylphenyl sulphone	3.5				67 to 102	1
1-Nitronaphthalene	2.5				25 to 44	1
Ovalene	1.13	$2.3 \cdot 10^{15}$				1
Pentacene	0.58	$2.4 \cdot 10^{9}$ (50°C)				2
Pentacene	1.5	$6 \cdot 10^{13}$			20 to 140	1
Perylene	2.1	$4.1 \cdot 10^{13}$			40 to 100	1
Perylene	3.10		5.53(n), 87.4(p)	+ & −	−213	4
Phenanthrene	1.15	$1.3 \cdot 10^{14}$			12 to 72	1
Phenanthrene	0.65					2
1,10-Phenanthroline	2.73				50 to 90	1
Phenazine	2.1	$7 \cdot 10^{14}$ (100°C)			98 to 143	1
Phenazine	1.1			−		4
Phenothiazine	1.6	10^{11}			50 to 150	1
Phenothiazine			2.45(n), 0.02(p)	+ & −		4
Phenylanthranilic acid	3.30				87 to 119	1
4-Phenylstilbene	1.74				140 to 220	1
4-Phenylstilbene	0.86					2
Phosphonitrilic chloride trimer	1.68	10^{15}				1
Phthalocyanine, PcH_2	1.66	10^{13}	0.1 to 0.4	+	26 to 350	1
Phthalocyanine, PcH_2	2	10^{7}	1.2(n), 1.1(p)	+ & −	100	4
Pyranthrene	1.11	$1 \cdot 10^{15}$				1
Pyranthrene	0.51	$4.5 \cdot 10^{16}$ (15°C)				2
Pyranthrone	1.06	$3.9 \cdot 10^{15}$			40 to 150	1
Pyrene	2.02	$5 \cdot 10^{17}$		−		1
Pyrene			0.50	+		4
5,6-*N*-Pyridine-1,9-benzanthrone	1.60					2
p-Quaterphenyl	0.89	$1.0 \cdot 10^{15}$ (50°C)				2
Quaterrylene	0.6	10^{5}		−		1
p-Quinquiphenyl	0.91	$2.0 \cdot 10^{15}$ (50°C)				2
α-Resorcin	2.10	$2 \cdot 10^{16}$			30 to 94	1
β-Resorcin	3.27	$2 \cdot 10^{18}$			30 to 94	1
Salanil	4.1	10^{4}			20 to 40	1
p-Sexiphenyl	0.91	$7.0 \cdot 10^{14}$ (50°C)				2
cis-Stilbene	2.4			+	at 20	1
trans-Stilbene	1.80		2.4	+	70 to 120	1
trans-Stilbene	0.91					2
trans-Stilbene	1.4			+		4
o-Terphenyl		$3 \cdot 10^{-5}$		+		1
m-Terphenyl			10^{-5}	+		1
p-Terphenyl	0.6	10^{14} (25°C)				2
p-Terphenyl	1.2		0.025	+		1

Substance	Energy gap E/eV	Electrical resistivity[a] ρ/Ω cm	Mobility μ/cm^2 V^{-1} s^{-1}	Sign of majority carriers	Temperature range °C	Ref.
p-Terphenyl			1.2(n), 0.80(p)	+ & −		4
Tetracene	0.66	$3.2 \cdot 10^{12}$ (50°C)				2
Tetracene	1.7					1
Tetracene	3.4		0.85	+		4
1,1,10,10-Tetracyanodecapentaene	2.24	10^{13}			>68	1
1,1,6,6-Tetracyanohexatriene	1.54	10^{14}		−		1
Tetracyanoethylene			0.26(max)	+		4
7,7,8,8-Tetracyanoquinodimethane			0.65	−		4
1,1,8,8-Tetracyanooctatetraene	1.42	10^{12}		−		1
Tetraphenylpentacene	1.62				0 to 150	1
Tetrathiotetracene	0.46	10^4				1
Triphenodioxazine	1.65	$5 \cdot 10^{14}$		−	20 to 140	1
Triphenyldiamine			$2 \cdot 10^{-2}$		at 20	1
Violanthrene	0.85	$2.1 \cdot 10^{14}$			40 to 105	1
Isoviolanthrene	0.82	$8.4 \cdot 10^{13}$			40 to 150	1
Violanthrone	0.78	$2.3 \cdot 10^{10}$			40 to 150	1
Isoviolanthrone	0.76	$5.7 \cdot 10^9$			40 to 150	1
o-Xylene (liquid)	0.45					2
m-Xylene (liquid)	0.41					2
p-Xylene (liquid)	0.41					2
Long-Chain Compounds and Polymers						
Acrylic acid-amylproparylaniline copolymers		10^9–10^{10}				3
Acrylic acid-methylproparylaniline copolymers		10^9–10^{10}				3
Acrylic acid-octylproparylaniline copolymers		10^9–10^{10}				3
Anthrone polymers		0.28 ≥100 at 1.8 kbar ≥2 at 33 kbar				3
$[CH(AsF_5)_{0.1}]_x$		0.0005		+		3
$[CH \cdot I_{0.22}]_x$	1.9	*trans* 10^5, *cis* 10^9				3
1,6-Diacetylenes (cyclopolymerized)		10^{10}–10^{14}				3
Ionene elastomers		$2.7 \cdot 10^7$ to $2.2 \cdot 10^8$			−80 to 60	3
1,3,4-Oxydiazole polymers	0.81	$3 \cdot 10^{12}$			20 to 140	3
Oxypyrrole polymer films		0.125				3
Phenylformaldehyde polymeric pyrolysates						3
a) Pyrolysis Temperature 600°C		27	0.0014	−		
b) 1200°C		0.0044	7.84	+		
Phenylthiocyanate polymers	0.5 to 0.8	10^5–10^8				3
Polyacetylene (undoped)		10^{10}				3
Polyacetylene (I_2 doped)		0.04				3
Polyacetylene (*cis*-rich, undoped)		10^7				3
trans-Polyacetylene (I_2 doped, 0.22 mole %)				+		3
Polyacrylonitrile (heat treated 700°C)			0.01	−	−100 to 100	3
Poly-5,5′-biisatyl	air 0.84	air $2.6 \cdot 10^9$		+	20 to 140	3
thiophene-indophene	vacuum 1.0	vacuum $3.1 \cdot 10^9$				
Poly bis(amino)-phosphazenes	1.75	$1.8 \cdot 10^{11}$			20 to 180	3
Poly-5,5′-diisatylmetane-thiophene-indophenine	0.45	$7.3 \cdot 10^4$		+	20 to 140	3
Polyethylene	2.74				20 to 70	3
Polyethylene (low density)	0.17	$4 \cdot 10^9$			above T_g	3
Polyimide	2.84					3
Polymalonitrile	1.72			−		3
Poly(metalphthalocyanines) :Cu	0.12	$7 \cdot 10^6$				3
:Fe	0.15	$1.1 \cdot 10^6$				3
:Ni	0.46	100				3
:Sb		$3.1 \cdot 10^6$				3
:Zn	0.12	$5.3 \cdot 10^3$				3
Poly-*N*-methylpyrrole		$2 \cdot 10^6$				3
Polyoxypyrrole (black)	0.044	1790			−173 to 27	3
Polyphthalocyanines	0.01	7 to 58				3
Polypyrrole	0.01				−193 to 250	3

Substance	Energy gap E/eV	Electrical resistivity[a] ρ/Ω cm	Mobility μ/cm^2 V^{-1} s^{-1}	Sign of majority carriers	Temperature range °C	Ref.
Polypyrroline II	1.74					3
Polyselenomethylene	0.7 to 2.62	>10[13]			20 to 120	3
Poly(2-vinylpyridine):I_2 (1:2)	0.12	1000			−73 to 27	3
PVC (commercial)	2.84–3.04				$T<T_g$	3
PVC (commercial)	1.24–1.96				$T>T_g$	3
PVC (pure)	1.0±0.1				0 to 30	3
Salicylal-*N*-alkyliminate-Cu	1.62	$1.7 \cdot 10^{14}$				4
TTF-acetylacetonate polymers		$1.6 \cdot 10^4$				3
TTF-metal polymers		$1.6 \cdot 10^4$				3

References

1. F. Gutman and L. E. Lyons, *Organic Semiconductors*, John Wiley & Sons, New York, 1967.
2. Y. Okamoto and W. Brenner, *Organic Semiconductors*, Reinhold Publ. Corp., New York, 1964.
3. F. Gutman, H. Keyzer, L. E. Lyons, and R. B. Somoano, *Organic Semiconductors, Part B*, R. E. Krieger Publ. Co., Melbourne, FL, 1983.
4. L. I. Berger, *Semiconductor Materials*, CRC Press, Boca Raton, FL, 1997.

DIFFUSION DATA FOR SEMICONDUCTORS

B. L. Sharma

The diffusion coefficient D in many semiconductors may be expressed by an Arrhenius-type relation

$$D = D_o \exp(-Q/kT)$$

where D_o is a frequency factor, Q is the activation energy for diffusion, k is the Boltzmann constant, and T is the absolute temperature. This table lists D_o and Q for various diffusants in common semiconductors.

Abbreviations used in the table are

AES – Auger Electron Spectroscopy
DLTS – Deep Level Transient Spectroscopy
SEM – Scanning Electron Microscopy
SIMS – Secondary Ion Mass Spectrometry
$D(c)$ – Concentration Dependent Diffusion Coefficient
D_{max} – Maximum Diffusion Coefficient
(f) – Fast Diffusion Component
(i) – Interstitial Diffusion Component
(s) – Slow Diffusion Component
(‖) – Parallel to c Direction
(⊥) – Perpendicular to c Direction

Semiconductor	Diffusant	Frequency factor, D_o (cm²/s)	Activation energy, Q(eV)	Temperature range (°C)	Method of measurement	Ref.
Si	H	6×10^{-1}	1.03	120–1207	Electrical and SIMS	1
	Li	2.5×10^{-3}	0.65	25–1350	Electrical	2
	Na	1.65×10^{-3}	0.72	530–800	Electrical and flame photometry	3
	K	1.1×10^{-3}	0.76	740–800	Electrical and flame photometry	3
	Cu	4×10^{-2}	1.0	800–1100	Radioactive	4
		4.7×10^{-3}	0.43 (i)	300–700	Radioactive	5
	Ag	2×10^{-3}	1.6	1100–1350	Radioactive	6
	Au	2.4×10^{-4}	0.39 (i)	700–1300	Radioactive	7
		2.75×10^{-3}	2.05 (s)			
	Be	($D \sim 10^{-7}$)	–	1050	Electrical	8
	Ca	($D \sim 6 \times 10^{-14}$)	–	1100	Electrical and SIMS	1
	Zn	1×10^{-1}	1.4	980–1270	Electrical	9
	B	2.46	3.59	1100–1250	Electrical	10
		2.4×10^{1}	3.87	840–1250	Electrical	11
	Al	1.38	3.41	1119–1390	Electrical	12
		1.8	3.2	1025–1175	Electrical	13
	Ga	3.74×10^{-1}	3.39	1143–1393	Electrical	12
		6×10^{1}	3.89	900–1050	Radioactive	14
	In	7.85×10^{-1}	3.63	1180–1389	Electrical	12
		1.94×10^{1}	3.86	1150–1242	Radioactive	15
	Tl	1.37	3.7	1244–1338	Electrical	12
		1.65×10^{1}	3.9	1105–1360	Electrical	16
	Sc	8×10^{-2}	3.2	1100–1250	Radioactive	1
	Ce	($D \sim 3.9 \times 10^{-13}$)	–	1050	SIMS	1
	Pr	2.5×10^{-7}	1.74	1100–1280	Electrical	1
	Pm	7.5×10^{-9}	1.2 (s)	730–1270	Radioactive	1
		4.2×10^{-12}	0.13 (f)			
	Er	2×10^{-3}	2.9	1100–1250	Radioactive	1
	Tm	8×10^{-3}	3.0	1100–1280	Radioactive	1
	Yb	2.8×10^{-5}	0.95	947–1097	Neutron activation	1
	Ti	1.45×10^{-2}	1.79	950–1200	DLTS	17
	C	3.3×10^{-1}	2.92	1070–1400	Radioactive	18
	Si (self)	1.54×10^{2}	4.65	855–1175	SIMS	19
		1.6×10^{3}	4.77	1200–1400	Radioactive	20
	Ge	3.5×10^{-1}	3.92	855–1000	Radioactive	21
		2.5×10^{3}	4.97	1030–1302	Radioactive	21
		7.55×10^{3}	5.08	1100–1300	SIMS	22
	Sn	3.2×10^{1}	4.25	1050–1294	Neutron activation	23
	N	2.7×10^{-3}	2.8	800–1200	Out Diffusion; SIMS	1

Semiconductor	Diffusant	Frequency factor, D_o (cm²/s)	Activation energy, Q(eV)	Temperature range (°C)	Method of measurement	Ref.
	P	2.02×10^{1}	3.87	1100–1250	Electrical	10
		1.1	3.4	900–1200	Radioactive	24
		7.4×10^{-2}	3.3	1130–1405	Electrical	25
	As	6.0×10^{1}	4.2	950–1350	Radioactive	26
		6.55×10^{-2}	3.44	1167–1394	Electrical	27
		2.29×10^{1}	4.1	900–1250	Electrical	28
	Sb	1.29×10^{1}	3.98	1190–1398	Radioactive	29
		2.14×10^{-1}	3.65	1190–1405	Electrical	27
	Bi	1.03×10^{3}	4.64	1220–1380	Electrical	16
		1.08	3.85	1190–1394	Electrical	27
	Cr	1×10^{-2}	1	1100–1250	Radioactive	30
	Mo	($D \sim 2 \times 10^{-10}$)	–	1000	DLTS	1
	W	($D \sim 10^{-12}$)	–	1100	DLTS	1
	O	7×10^{-2}	2.44	700–1250	SIMS	31
		1.4×10^{-1}	2.53	700–1160	SIMS	32
	S	5.95×10^{-3}	1.83	975–1200	Radioactive	33
	Se	9.5×10^{-1}	2.6	1050–1250	Electrical	34
	Te	5×10^{-1}	3.34	900–1250	SIMS	1
	Mn	6.9×10^{-4}	0.63	900–1200	Radioactive	35
	Fe	1.3×10^{-3}	0.68	30–1250	Radioactive	36
	Co	2×10^{-3}	0.69	700–1300	Radioactive	37
	Ni	2×10^{-3}	0.47	800–1300	Radioactive	38
	Ru	($D \sim 5 \times 10^{-7}$ – 5×10^{-6})	–	1000–1280	Electrical	1
	Rh	($D \sim 10^{-6}$–10^{-4})	–	1000–1200	Electrical	39
	Pd	2.95×10^{-4}	0.22 (i)	702–1320	Nuclear Activation	1
	Pt	1.5×10^{2}	2.22	800–1000	Electrical	1
	Os	($D \sim 2 \times 10^{-6}$)	–	1280	Electrical	40
	Ir	4.2×10^{-2}	1.3	950–1250	Electrical	41
Ge	Li	1.3×10^{-3}	0.46	350–800	Electrical	42
		9.1×10^{-3}	0.57	800–500	Electrical	43
	Na	3.95×10^{-1}	2.03	700–850	Radioactive	44
	Cu	1.9×10^{-4}	0.18 (i)	750–900	Radioactive	45
		4×10^{-2}	0.99 (s)	600–700		
		4×10^{-3}	0.33 (i)	350–750	Radioactive	5
	Ag	4.4×10^{-2}	1.0 (i)	700–900	Radioactive	46, 47
		4×10^{-2}	2.23 (s)	800–900	Radioactive	48
	Au	2.25×10^{2}	2.5	600–900	Radioactive	49
	Be	5×10^{-1}	2.5	720–900	Electrical	50
	Mg	($D \sim 8 \times 10^{-9}$)	–	900	Electrical	1
	Zn	5	2.7	600–900	Radioactive and electrical	51
	Cd	1.75×10^{9}	4.4	760–915	Radioactive	52
	B	1.8×10^{9}	4.55	600–900	Electrical	51
	Al	1.0×10^{3}	3.45	554–905	SIMS	53
		$\sim 1.6 \times 10^{2}$	~3.24	750–850	Electrical	54
	Ga	1.4×10^{2}	3.35	554–916	SIMS	55
		3.4×10^{1}	3.1	600–900	Electrical	51
	In	1.8×10^{4}	3.67	554–919	SIMS	56
		3.3×10^{1}	3.02	700–855	Radioactive	57
	Tl	1.7×10^{3}	3.4	800–930	Radioactive	58
	Si	2.4×10^{-1}	2.9	650–900	(γ) resonance	59
	Ge (self)	2.48×10^{1}	3.14	549–891	Radioactive	60
		7.8	2.95	766–928	Radioactive	61
	Sn	1.7×10^{-2}	1.9	–	Radioactive	45
	P	3.3	2.5	600–900	Electrical	51
	As	2.1	2.39	700–900	Electrical	62
	Sb	3.2	2.41	700–855	Radioactive	57
		1.0×10^{1}	2.5	600–900	Radioactive and electrical	51

Semiconductor	Diffusant	Frequency factor, D_o (cm²/s)	Activation energy, Q(eV)	Temperature range (°C)	Method of measurement	Ref.
	Bi	3.3	2.57	650–850	–	63
	O	4×10^{-1}	2.08	–	Optical	64
	S	($D \sim 10^{-9}$)	–	920	–	65
	Se	($D \sim 10^{-10}$)	–	920	–	65
	Te	5.6	2.43	750–900	Radioactive	66
	Fe	1.3×10^{-1}	1.08	750–900	Radioactive	67
	Co	1.6×10^{-1}	1.12	750–850	Radioactive	47
	Ni	8×10^{-1}	0.9	670–900	Electrical	68
GaAs	Li	5.3×10^{-1}	1.0	250–500	Electrical and chemical	69
	Cu	3×10^{-2}	0.53	100–500	Radioactive	69
		6×10^{-2}	0.98	450–750	Ultrasonic	69
		1.5×10^{-3}	0.6	800–1000	Radioactive	69
	Ag	4×10^{-4}	0.8	500–1150	Radioactive	69
	Au	1×10^{-3}	1.0	740–1025	Radioactive	69
	Be	7.3×10^{-6}	1.2	800–990	Electrical	69
	Mg	4×10^{-5}	1.22	800–1200	Electrical	69
	Zn	1.5×10^{1}	2.49	600–980	Radioactive	69
		2.5×10^{-1}	3.0	750–1000	Radioactive	69
	Cd	1.3×10^{-3}	2.2	800–1100	Radioactive	69
		5×10^{-2}	2.43	868–1149	Radioactive	69
	Hg	($D \sim 5 \times 10^{-14}$)	–	1100	Radioactive	69
	Al	($D \sim 4 \times 10^{-18}$–10^{-14})	4.3	850–1100	AES	70
	Ga (self)	4×10^{-5}	2.6	1025–1100	Radioactive	69
		1×10^{7}	5.6	1125–1230	Radioactive	69
	In	($D \sim 7 \times 10^{-11}$)	–	1000	Radioactive	69
	C	($D \sim 1.04 \times 10^{-16}$)	–	825	SIMS	69
	Si	1.1×10^{-1}	2.5	850–1050	SIMS	69
	Ge	1.6×10^{-5}	2.06	650–850	SIMS	69
	Sn	6×10^{-4}	2.5	1060–1200	Radioactive	69
		1×10^{-5}	2	800–1000	Radioactive	69
	P	($D \sim 10^{-12}$–10^{-10})	2.9	800–1150	Reflectance measurements	69
	As (self)	7×10^{-1}	3.2	–	Radioactive	69
	Cr	2.04×10^{-6}	0.83 (f)	750–1000	SIMS	69
			1.7 (s)	700–900		
		7.9×10^{-3}	2.2	800–1100	Chemical analysis	69
	O	2×10^{-3}	1.1	700–900	Mass spectroscopy	69
	S	1.85×10^{-2}	2.6	1000–1300	Radioactive	69
		1.1×10^{1}	2.95	750–900	Electrical	69
	Se	3×10^{3}	4.16	1025–1200	Radioactive	69
	Te	1.5×10^{-1}	3.5	1000–1150	Radioactive	69
	Mn	6.5×10^{-1}	2.49	850–1100	Radioactive	69
	Fe	4.2×10^{-2}	1.8	850–1150	Radioactive	69
		2.2×10^{-3}	2.32	750–1050	Radioactive	69
	Co	5×10^{2}	2.5	800–1000	Radioactive	69
		1.2×10^{-1}	2.64	750–1050	Radioactive	69
	Tm	2.3×10^{-16}	1.0	800–1000	Radioactive	69
GaSb	Li	2.3×10^{-4}	1.9 (s)	527–657	Electrical and flame photometry	69
		1.2×10^{-1}	0.7 (f)	277–657		
	Cu	4.7×10^{-3}	0.9	470–650	Radioactive	69
	Zn	($D \sim 2 \times 10^{-13}$ – 1×10^{-11})	2	510–600	Radioactive	69
	Cd	1.5×10^{-6}	0.72	640–800	Electrical	69
	Ga (self)	3.2×10^{3}	3.15	658–700	Radioactive	69
	In	1.2×10^{-7}	0.53	320–650	Radioactive	69
	Sn	2.4×10^{-5}	0.8	320–650	Radioactive	69
		1.3×10^{-5}	1.1	500–650	Radioactive	69
	Sb (self)	3.4×10^{4}	3.45	658–700	Radioactive	69
	Se	($D \sim 2.4 \times 10^{-13}$ – 1.37×10^{-11})	–	400–500	Radioactive	69

Semiconductor	Diffusant	Frequency factor, D_o (cm²/s)	Activation energy, Q(eV)	Temperature range (°C)	Method of measurement	Ref.
	Te	3.8×10^{-4}	1.20	320–650	Radioactive	69
	Fe	5×10^{-2}	1.9 (I)	500–650	Radioactive	69
		5×10^{2}	2.3 (II)	500–650		
GaP	Ag	–	–	1000–1300	Radioactive	69
	Au	8	2.5 (I)	1050–1250	Radioactive	69
		20	2.4 (II)	1100–1250	Diffusion (I) A face and (II) B face	
	Be	($D_{max} \sim 2.4 \times 10^{-9}$– 8.5×10^{-8})	–	900–1000	Atomic absorption analysis	69
	Mg	5×10^{-5}	1.4	700–1050	Electrical	69
	Zn	1.0	2.1	700–1300	Radioactive	69
	Ge	–	–	900–1000	Radioactive	69
	Cr	6.2×10^{-4}	1.2	900–1130	Radioactive; ESR	69
	S	3.2×10^{3}	4.7	1120–1305	Radioactive	69
	Mn	2.1×10^{9}	4.7	T < 950	Radioactive; ESR	69
		1.1×10^{-6}	0.9	950–1130		
	Fe	1.6×10^{-1}	2.3	980–1180	Radioactive	69
	Co	2.8×10^{-3}	2.9	850–1100	Radioactive	69
InP	Cu	3.8×10^{-3}	0.69	600–900	Radioactive	69
	Ag	3.6×10^{-4}	0.59	500–900	Radioactive	69
	Au	1.32×10^{-5}	0.48	600–820	Radioactive	69
		1.37×10^{-4}	0.73	600–900	Radioactive	69
	Zn	1.6×10^{-8}	0.3	750–900	Electrical	69
		($D \sim 2 \times 10^{-9}$– 4×10^{-8})	–	700–900	Radioactive	69
	Cd	1.8	1.9	700–900	Radioactive	69
		1.1×10^{-7}	0.72	700–900	Electrical	69
		($D \sim 7 \times 10^{-13}$– 2×10^{-10})	–	450–650	Electrical	69
	In (self)	1×10^{5}	3.85	830–990	Radioactive	69
	Sn	($D \sim 3 \times 10^{-8}$)	–	550	Etching and cathodo-luminescence	69
	P (self)	7×10^{10}	5.65	900–1000	Radioactive	69
	Cr	–	–	600–900	Radioactive	69
	S	3.6×10^{-4}	1.94	585–708	Electrical	69
	Se	($D \sim 2 \times 10^{-8}$)	–	550	Cathodoluminescence	69
	Mn	–	2.9	650–750	SIMS	69
	Fe	3	2	600–950	Radioactive	69
		6.8×10^{5}	3.4	600–700	SIMS	69
	Co	9×10^{-1}	1.8	600–950	Radioactive	69
InAs	Cu	3.6×10^{-3}	0.52	342–875	Radioactive	69
		2.2×10^{-2}	0.54	525–890	Radioactive	69
	Ag	7.3×10^{-4}	0.26	450–900	Radioactive	69
	Au	5.8×10^{-3}	0.65	600–900	Radioactive	69
	Mg	1.98×10^{-6}	1.17	600–900	Electrical	69
	Zn	4.2×10^{-3}	0.96	600–900	Radioactive	69
		3.11×10^{-3}	1.17	600–900	Electrical	69
	Cd	7.4×10^{-4}	1.15	650–900	Radioactive	69
	Hg	1.45×10^{-5}	1.32	650–850	Radioactive	69
	In (self)	6×10^{5}	4.0	740–900	Radioactive	69
	Ge	3.74×10^{-6}	1.17	600–900	Electrical	69
	Sn	1.49×10^{-6}	1.17	600–900	Electrical	69
	As (self)	3×10^{7}	4.45	740–900	Radioactive	69
	S	6.78	2.2	600–900	Electrical	69
	Se	12.6	2.2	600–900	Electrical	69
	Te	3.43×10^{-5}	1.28	600–900	Electrical	69
InSb	Li	7×10^{-4}	0.28	0–210	Electrical	69
	Cu	9×10^{-4}	1.08	200–500	Radioactive	69
		3×10^{-5}	0.37	230–490	Radioactive	69
	Ag	1×10^{-7}	0.25	440–510	Radioactive	69
	Au	7×10^{-4}	0.32	140–510	Radioactive	69
	Zn	5×10^{-1}	1.35	362–508	Radioactive	69

Semiconductor	Diffusant	Frequency factor, D_o (cm²/s)	Activation energy, Q(eV)	Temperature range (°C)	Method of measurement	Ref.
		–	1.5	355–455	SIMS	69
	Cd	1×10^{-5}	1.1	250–500	Radioactive	69
		1.3×10^{-4}	1.2	360–500	Electrical	69
	Hg	4×10^{-6}	1.17	425–500	Radioactive	69
	In (self)	6×10^{-7}	1.45	400–500	Radioactive	69
		1.8×10^{13}	4.3	475–517	Radioactive	69
	Sn	5.5×10^{-8}	0.75	390–512	Radioactive	69
	Pb	$(D \sim 2.7 \times 10^{-15})$	–	500	Radioactive	71
	Sb (self)	5.35×10^{-4}	1.91	400–500	Radioactive	69
		3.1×10^{13}	4.3	475–517	Radioactive	69
	S	9×10^{-2}	1.4	360–500	Electrical	69
	Se	1.6	1.87	380–500	Electrical	69
	Te	1.7×10^{-7}	0.57	300–500	Radioactive	69
	Fe	1×10^{-7}	0.25	440–510	Radioactive	69
	Co	2.7×10^{-11}	0.39	420–500	Radioactive	69
AlAs	Ga	$(D \sim 2 \times 10^{-18}– 10^{-15})$	3.6	850–1100	AES	70
	Zn	$(D \sim 9 \times 10^{-11})$	–	557	SEM	69
AlSb	Cu	3.5×10^{-3}	0.36	150–500	Radioactive	69
	Zn	3.3×10^{-1}	1.93	660–860	Radioactive	69
	Cd	$D(c) \sim 4 \times 10^{-12}– 3 \times 10^{-10}$	–	900	Radioactive	69
	Al (self)	2	1.88	570–620	X-ray	69
	Sb (self)	1	1.7	570–620	X-ray	69
ZnS	Cu	2.6×10^{-3}	0.79	470–750	Radioactive	69
		4.3×10^{-4}	0.64	250–1200	Electroluminescence	69
		9.75×10^{-3}	1.04	400–800	Luminescence	69
	Au	1.75×10^{-4}	1.16	500–800	Radioactive	69
	Zn (self)	3×10^{-4}	1.5	$925<T<940$	Radioactive	69
		1.5×10^{4}	3.26	$940<T<1030$		
		1×10^{16}	6.5	$1030<T<1075$		
	Cd	$(D \sim 10^{-10})$	–	1100	Luminescence	72
	Al	5.69×10^{-4}	1.28	800–1000	Luminescence	69
	In	3×10^{1}	2.2	750–1000	Radioactive	69
	S (self)	2.16×10^{4}	3.15	600–800	Radioactive	69
		8×10^{-5}	2.2	740–1100	Radioactive	69
	Se	$(D \sim 5 \times 10^{-13})$	–	1070	X-ray microprobe	69
	Mn	2.3×10^{3}	2.46	500–800	Radioactive	69
ZnSe	Li	2.66×10^{-6}	0.49	950–980	Electrical	69
	Cu	1×10^{-4}	0.66	400–800	Luminescence	69
		1.7×10^{-5}	0.56	200–570	Radioactive	69
	Ag	2.2×10^{-2}	1.18	400–800	Luminescence	69
	Zn (self)	9.8	3.0	760–1150	Radioactive	69
	Cd	6.39×10^{-4}	1.87	700–950	Photoluminescence	69
	Al	2.3×10^{-2}	1.8	800–1100	Luminescence	69
	Ga	1.81×10^{2}	3.0	900–1100	Luminescence	69
		–	1.3	700–850	Electron probe	69
	In	$(D \sim 2 \times 10^{-12})$	–	940	–	69
	S	$(D \sim 8 \times 10^{-12})$	–	1060	X-ray microprobe	69
	Se (self)	1.3×10^{1}	2.5	860–1020	Radioactive	69
		2.3×10^{-1}	2.7	1000–1050	Radioactive	69
	Ni	$(D \sim 1.5 \times 10^{-8}– 1.7 \times 10^{-7})$	–	740–910	Luminescence	69
ZnTe	Li	2.9×10^{-2}	1.22 (s)	400–700	Nuclear and chemical analysis	69
		1.7×10^{-4}	0.78 (f)			
	Zn (self)	2.34	2.56	760–860	Radioactive	69
		1.4×10^{1}	2.69	667–1077	Radioactive	69
	Al	–	2.0	700–1000	Electrical and optical	69
	In	4	1.96	1100–1300	Radioactive	69
	Te (self)	2×10^{4}	3.8	727–977	Radioactive	69
CdS	Li	3×10^{-6}	0.68	610–960	Microhardness	69
	Na	$(D \sim 3 \times 10^{-7})$	–	800	Radioactive	69

Semiconductor	Diffusant	Frequency factor, D_o (cm²/s)	Activation energy, Q(eV)	Temperature range (°C)	Method of measurement	Ref.
	Cu	1.5×10^{-3}	0.76	400–700	Radioactive	69
		1.2×10^{-2}	1.05	300–700	Ultrasonic	69
		8×10^{-5}	0.72	20–200	Electrical	69
	Ag	2.5×10^{1}	1.2 (s)	300–500	Radioactive	69
		2.4×10^{-1}	0.8 (f)			
	Au	2×10^{2}	1.8	500–800	Radioactive	69
	Zn	1.27×10^{-9}	0.86 (s)	720–1000	Radioactive	69
		1.22×10^{-8}	0.66 (f)			
	Cd (self)	3.4	2.0	700–1100	Radioactive	69
	Ga	–	–	667–967	Optical and microprobe	69
	In	6×10^{1}	2.3 (∥)	650–930	Radioactive, optical and microprobe	69
		1×10^{1}	2.03 (⊥)			
	P	6.5×10^{-4}	1.6	800–1100	Radioactive	69
	S (self)	1.6×10^{-2}	2.05	800–900	Radioactive	69
		–	2.4	750–1050	Radioactive	69
	Se	$(D \sim 1.2 \times 10^{-9})$	–	900	Radioactive	69
	Te	1.3×10^{-7}	10.4	700–1000	Radioactive	69
	Cl	$(D \sim 3 \times 10^{-10})$	–	800	Electrical	69
	I	$(D \sim 5 \times 10^{-12})$	–	1000	Radioactive	69
	Ni	6.75×10^{-3}	10.9	570–900	Luminescence	69
	Yb	$(D \sim 1.3 \times 10^{-9})$	–	960	Photoluminescence	69
CdSe	Ag	2×10^{-4}	0.53	22–400	Ultrasonic	69
	Cd (self)	1.6×10^{-3}	1.5	700–1000	Radioactive	69
		6.3×10^{-2}	1.25 (I)	600–900	Radioactive; (I) saturated Cd and (II) saturated Se pressure	69
		4.12×10^{-2}	2.18 (II)	600–900		
	P	$(D \sim 5.3 \times 10^{-12} – 6 \times 10^{-11})$	–	900–1000	Radioactive	69
	Se (self)	2.6×10^{3}	1.55	700–1000	Radioactive; saturated Se pressure	69
CdTe	Li	$(D \sim 1.5 \times 10^{-10})$	–	300	Ion microprobe	69
	Cu	3.7×10^{-4}	0.67	97–300	Radioactive	69
		8.2×10^{-8}	0.64	290–350	Ion backscattering	69
	Ag	–	–	700–800	Electrical and photo-luminescence	69
	Au	6.7×10^{1}	2.0	600–1000	Radioactive	69
	Cd (self)	1.26	2.07	700–1000	Radioactive	69
		3.26×10^{2}	2.67 (I)	650–900	Radioactive; (I) saturated Cd and (II) saturated Te pressure	69
		1.58×10^{1}	2.44 (II)			
	In	8×10^{-2}	1.61	650–1000	Radioactive	69
		1.17×10^{2}	2.21 (I)	500–850	Radioactive; (I) saturated Cd and (II) saturated Te pressure	69
		6.48×10^{-4}	1.15 (II)			
	Sn	8.3×10^{-2}	2.2	700–925	Radioactive	69
	P	$(D \sim 1.2 \times 10^{-10})$	–	900	Radioactive	69
	As	–	–	850	–	69
	O	5.6×10^{-9}	1.22	200–650	Mass spectrometry	69
		6.0×10^{-10}	0.29	650–900		
	Se	1.7×10^{-4}	1.35	700–1000	Radioactive	69
	Te (self)	8.54×10^{-7}	1.42 (I)	600–900	Radioactive; (I) saturated Cd and (II) saturated Te pressure	69
		1.66×10^{-4}	1.38 (II)	500–800		
	Cl	7.1×10^{-2}	1.6	520–800	Radioactive	69
	Fe	$(D \sim 4 \times 10^{-8})$	0.77	900	Radioactive	69

Semiconductor	Diffusant	Frequency factor, D_o (cm²/s)	Activation energy, Q(eV)	Temperature range (°C)	Method of measurement	Ref.
HgSe	Sb	6.3×10^{-5}	0.85	540–630	Radioactive	69
	Se (self)	–	–	200–400	Radioactive	69
HgTe	Ag	6×10^{-4}	0.8	250–350	Radioactive	69
	Zn	5×10^{-8}	0.6	250–350	Radioactive	69
	Cd	3.1×10^{-4}	0.66	250–350	Radioactive	69
	Hg (self)	2×10^{-8}	0.6	200–350	Radioactive	69
	In	6×10^{-6}	0.9	200–300	Radioactive	69
	Sn	1.72×10^{-6}	0.66 (s)	200–300	Radioactive	69
		1.8×10^{-3}	0.80 (f)			
	Te (self)	10^{-6}	1.4	200–400	Radioactive	69
	Mn	1.5×10^{-4}	1.3	250–350	Radioactive	69
PbS	Cu	4.6×10^{-4}	0.36	150–450	Electrical	69
		5×10^{-3}	0.31	100–400	Electrical	69
	Pb (self)	8.6×10^{-5}	1.52	500–800	Radioactive	69
	S (self)	6.8×10^{-5}	1.38	500–750	Radioactive	69
	Ni	1.78×10^{1}	0.95	200–500	Electrical	69
PbSe	Na	1.5×10^{1}	1.74 (s)	400–850	Radioactive	69
		5.6×10^{-6}	0.4 (f)			
	Cu	2×10^{-5}	0.31	93–520	Radioactive	69
	Ag	7.4×10^{-4}	0.35	400–850	Radioactive	69
	Pb (self)	4.98×10^{-6}	0.83	400–800	Radioactive	69
	Sb	3.4×10^{-1}	2.0	650–850	Radioactive	69
	Se (self)	2.1×10^{-5}	1.2	650–850	Radioactive	69
	Cl	1.6×10^{-8}	0.45	400–850	Radioactive	69
	Ni	($D \sim 1 \times 10^{-10}$)	–	700	Radioactive	69
PbTe	Na	1.7×10^{-1}	1.91	600–850	Radioactive	69
	Sn	3.1×10^{-2}	1.56	500–800	Radioactive	69
	Pb (self)	2.9×10^{-5}	0.6	250–500	Radioactive	69
	Sb	4.9×10^{-2}	1.54	500–800	Radioactive	69
	Te	2.7×10^{-6}	0.75	500–800	Radioactive	69
	Cl	($D > 2.3 \times 10^{-10}$)	–	700	Radioactive	69
	Ni	($D > 1 \times 10^{-6}$)	–	700	Radioactive	69

References

1. N. A. Stolwijk and H. Bracht, in *Diffusion in Semiconductors and Non-Metallic Solids*, D. L. Beke, Ed., Springer-Verlag, Berlin, 1998, 2-1.
2. E. M. Pell, *Phys. Rev.*, 119, 1960; 119, 1014, 1960.
3. L. Svob, *Solid State Electron*, 10, 991, 1967.
4. B. I. Boltaks and I. I. Sosinov, *Zh. Tekh. Fiz.*, 28, 3, 1958.
5. R. N. Hall and J. N. Racette, *J. Appl. Phys.*, 35, 379, 1964.
6. B. I. Boltaks and Hsueh Shih-Yin, *Sov. Phys. Solid State*, 2, 2383, 1961.
7. W. R. Wilcox and T. J. LaChapelle, *J. Appl. Phys.*, 35, 240, 1964.
8. E. A. Taft and R. O. Carlson, *J. Electrochem. Soc.*, 117, 711, 1970.
9. R. Sh. Malkovich and N. A. Alimbarashvili, *Sov. Phys. Solid State*, 4, 1725, 1963.
10. R. N. Ghoshtagore, *Solid State Electron*, 15, 1113, 1972.
11. C. Hill, *Semiconductor Silicon 1981*, H. R. Huff, R. J. Kreiger, and Y. Takeishi, Eds., p. 988, *Electrochem. Soc.*, 1981.
12. R. N. Ghoshtagore, *Phys. Rev. B*, 3, 2507, 1971.
13. W. Rosnowski, *J. Electrochem. Soc.*, 125, 957, 1978.
14. J. S. Makris and B. J. Masters, *J. Appl. Phys.*, 42, 3750, 1971.
15. M. F. Millea, *J. Phys. Chem. Solids*, 27, 315, 1965 (refer Reference 2).
16. C. S. Fuller and J. A. Ditzenberger, *J. Appl. Phys.*, 27, 544, 1956.
17. S. Hocine and D. Mathiot, *Appl. Phys. Lett.*, 53, 1269, 1988.
18. R. C. Newman and J. Wakefield, *J. Phys. Chem. Solids*, 19, 230, 1961.
19. L. Kalinowski and R. Seguin, *Appl. Phys. Lett.*, 35, 211, 1979; *Appl. Phys. Lett.*, 36, 171, 1980.
20. R. F. Peart, *Phys. Stat. Sol.*, 15, K 119, 1966.
21. G. Hettich, H. Mehrer and K. Maler, *Inst. Phys. Conf. Ser.*, 46, 500, 1979.
22. M. Ogina, Y. Oana and M. Watanabe, *Phys. Stat. Sol. (a)*, 72, 535, 1982.
23. T. H. Yeh, S. M. Hu, and R. H. Kastl, *Appl. Phys.*, 39, 4266, 1968.
24. I. Franz and W. Langheinrich, *Solid State Electron*, 14, 835, 1971.
25. R. N. Ghoshtagore, *Hys. Rev. B*, 3, 389, 1971.
26. B. J. Masters and J. M. Fairfield, *J. Appl. Phys.*, 40, 2390, 1969.
27. R. N. Goshtagore, *Phys. Rev. B*, 3, 397, 1971.
28. R. S. Fair and J. C. C. Tsai, *J. Electrochem. Soc.*, 122, 1689, 1975.
29. J. J. Rohan, N. E. Pickering, and J. Kennedy, *J. Electrochem. Soc.*, 106, 705, 1969.
30. W. Wuerker, K. Roy, and J. Hesse, *Matsr. Res. Bull.*, 9, 971, 1974.
31. J. C. Mikkelsen, Jr., *Appl. Phys. Lett.*, 40, 336, 1982.
32. S. Tang Lee and D. Nicols, *Appl. Phys. Lett.*, 47, 1001, 1985.
33. P. L. Gruzin, S. V. Zemskii, A. D. Bullkin, and N. M. Makarov, *Sov. Phys. Sem.*, 7, 1241, 1974.
34. N. S. Zhdanovich and Yu. I. Kozlov, *Svoistva Legir, Poluprovodn.*, V. S. Zemskov, Ed., Nauka, Moscow, 1977, 115–120; *Fiz Tekh. Poluprovod.*, 9, 1594, 1975.
35. D. Gilles, W. Bergholze, and W. Schroeter, *J. Appl. Phys.*, 59, 3590, 1986.
36. E. R. Weber, *Appl. Phys.* A, 30, 1, 1983.
37. E. R. Weber, Properties of Silicon, EMIS Datareviews Ser. No. 4, INSPEC Publications, 1988, 409–451.
38. M. K. Bakhadyrkhanov, S. Zainabidinov, and A. Khamidov, *Sov. Phys. Sem.*, 14, 243, 1980.
39. S. A. Azimov, M. S. Yunosov, F. K. Khatamkulov, and G. Nasyrov, *Poluprovod.*, N. Kh. Abrikosov and V. S. Zemskov, Eds., Nauka, Moscow, 1975, 21–23.

40. S. A. Azimov, M. S. Yunosov, G. Nurkuziev, and F. R. Karimov, *Sov. Phys. Sem.*, 12, 981, 1978.
41. S. A. Azimov, B. V. Umarov, and M. S. Yunusov, *Sov. Phys. Sem.*, 10, 842, 1976.
42. C. S. Fuller and J. A. Ditzenberger, *Phys. Rev.*, 91, 193, 1953.
43. B. Pratt and F. Friedman, *J. Appl. Phys.*, 37, 1893, 1966.
44. M. Stojic, V. Spiric, and D. Kostoski, *Inst. Phys. Conf. Ser.*, 31, 304, 1976.
45. B. I. Boltaks, *Diffusion in Semiconductors*, Inforsearch, London, 1963, 162.
46. A. A. Bugai, V. E. Kosenko, and E. G. Miselyuk, *Zh. Tekh. Fiz.*, 27, 67, 1957.
47. L. Y. Wei, *J. Phys. Chem. Solids*, 18, 162, 1961.
48. V. E. Kosenko, *Sov. Phys. Solid State*, 4, 42, 1962.
49. W. C. Dunlap, Jr., *Phys. Rev.*, 97, 614, 1955
50. Yu. I. Belyaev and V. A. Zhidkov, *Sov. Phys. Solid State*, 3, 133, 1961.
51. W. C. Dunlap, Jr. *Phys. Rev.*, 94, 1531, 1954.
52. V. E. Kosenko, *Sov. Phys. Solid State*, 1, 1481, 1960.
53. P. Dorner, W. Gust, A. Lodding, H. Odelius, B. Predel, and U. Roll, *Acta Metall.*, 30, 941, 1982.
54. W. Meer and D. Pommerrening, *Z. Agnew. Phys.*, 23, 369, 1967.
55. U. Sodervall, H. Odelius, A. Lodding, U. Roll, B. Predel, W. Gust, and P. Dorner, *Phil. Mag. A*, 54, 539, 1986.
56. P. Dorner, W. Gust, A. Lodding, H. Odelius, B. Predel, and U. Roll, *Z. Metalkd.*, 73, 325, 1982.
57. P. V. Pavlov, *Sov. Phys. Solid State*, 8, 2377, 1967.
58. V. I. Tagirov and A. A. Kuliev, *Sov. Phys. Solid State*, 4, 196, 1962.
59. J. Raisanen, J. Hirvonen, and A. Anttila, *Solid State Electron.*, 24, 333, 1981.
60. C. Vogel, G. Hettich, and H. Mehrer, *J. Phys. C.*, 16, 6197, 1983.
61. H. Letaw, Jr., W. M. Portnoy, and L. Slifkin, *Phys. Rev.*, 102, 363, 1956.
62. W. Bosenberg, *Z. Naturforsch.*, 10a, 285, 1955.
63. V. M. Glazov and V. S. Zemskov, Physicochemical Principles of Semiconductor Doping, Israel Program for Scientific Translation, Jerusalem, 1968.
64. J. W. Corbett, R. S. McDonald, and G. D. Watkins, *J. Phys. Chem. Solids*, 25, 873, 1964.
65. W. W. Tyler, *J. Phys. Chem. Solids*, 8, 59, 1959.
66. V. D. Ignatkov and V. E. Kosenko, *Sov. Phys. Solid State*, 4, 1193, 1962.
67. A. A. Bugal, V. E. Kosenko, and E. G. Miseluk, *Zh. Tekh. Fiz.*, 27, 210, 1957.
68. F. van der Maesen and J. A. Brenkman, *Phillips Res. Rep.*, 9, 255, 1954.
69. M. B. Dutt and B. L. Sharma, in *Diffusion in Semiconductors and Non-Metallic Solids*, D. L. Beke, Ed., Springer-Verlag, Berlin, 1998, 3-1.
70. L. L. Chang and A. Koma, *Appl. Physics Lett.*, 29, 138, 1976.
71. D. L. Kendall, *Semiconductors and Semimetals*, Vol. 4, R. K. Willardson and A. C. Beer, Eds., Academic, 1968, 255.
72. H. J. Biter and F. Williams, *J. Luminescence*, 3, 395, 1971.

PROPERTIES OF MAGNETIC MATERIALS

H. P. R. Frederikse

Glossary of Symbols

Quantity	Symbol	SI Units	emu Units
Magnetic field	H	A m^{-1}	Oe (oersted)
Magnetic induction	B	T (tesla)	G (gauss)
Magnetization	M	A m^{-1}	emu cm^{-3}
Spontaneous magnetization	M_s	A m^{-1}	emu cm^{-3}
Saturation magnetization	M_0	A m^{-1}	emu cm^{-3}
Magnetic flux	Φ	Wb (weber)	maxwell
Magnetic moment	m, μ	A m^2	erg/G
Coercive field	H_c	A m^{-1}	Oe
Remanence	B_r	T	G
Saturation magnetic polarization	J_s	T	G
Magnetic susceptibility	χ		
Magnetic permeability	μ	H m^{-1} (henry/meter)	
Magnetic permeability of free space	μ_0	H m^{-1}	
Saturation magnetostriction	λ ($\Delta l/l$)		
Curie temperature	T_C	K	K
Néel temperature	T_N	K	K

Magnetic moment $\mu = \gamma hJ = g\mu_B J$

where

γ = gyromagnetic ratio
J = angular momentum
g = spectroscopic splitting factor (~2)
μ_B = Bohr magneton = $9.2741 \cdot 10^{-24}$
J/T = $9.2741 \cdot 10^{-21}$ erg/G
Earth's magnetic field H = 56 A m^{-1} = 0.7 Oe
For iron: $M_0 = 1.7 \cdot 10^6$ A m^{-1}; $B_r = 0.8 \cdot 10^6$ A m^{-1}
1 Oe = $(1000/4\pi)$ A m^{-1}; 1 G = 10^{-4} T; 1 emu cm^{-3} = 10^3 A m^{-1}
1 Maxwell = 10^{-8} Wb
$\mu_0 = 4\pi \cdot 10^{-7}$ H m^{-1}

Relation between Magnetic Induction and Magnetic Field

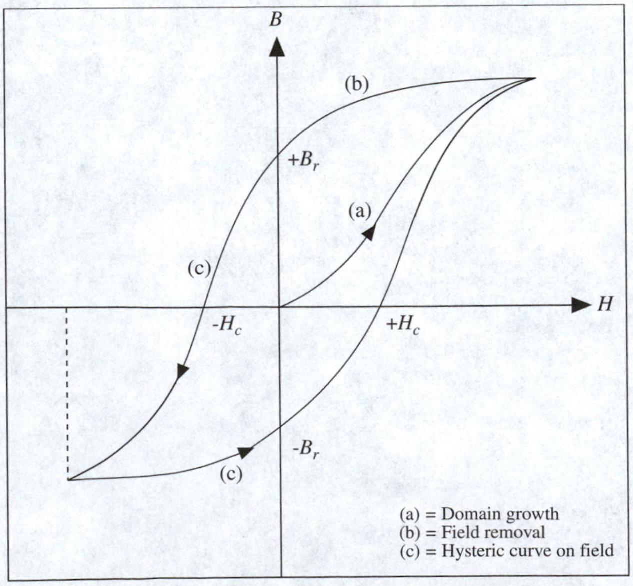

FIGURE 1. Typical curve representing the dependence of magnetic induction B on magnetic field H for a ferromagnetic material. When H is first applied, B follows curve (a) as the favorably oriented magnetic domains grow. This curve flattens as saturation is approached. When H is then reduced, B follows curve (b), but retains a finite value (the remanence B_r) at H = 0. In order to demagnetize the material, a negative field $-H_c$ (where H_c is called the coercive field or coercivity) must be applied. As H is further decreased and then increased to complete the cycle (curve c), a hysteresis loop is obtained. The area within this loop is a measure of the energy loss per cycle for a unit volume of the material.

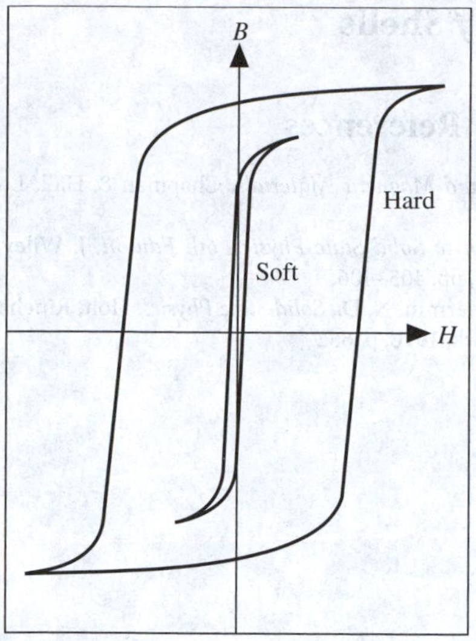

FIGURE 2. Schematic curve illustrating the B vs. H dependence for hard and soft magnetic materials. Hard materials have a larger remanence and coercive field, and a correspondingly large hysteresis loss.

Reference

Ralls, K. M., Courtney, T. H., and Wulff, J., *Introduction to Materials Science and Engineering*, J. Wiley & Sons, New York, 1976, p. 577, 582. With permission.

Magnetic Susceptibility of the Elements

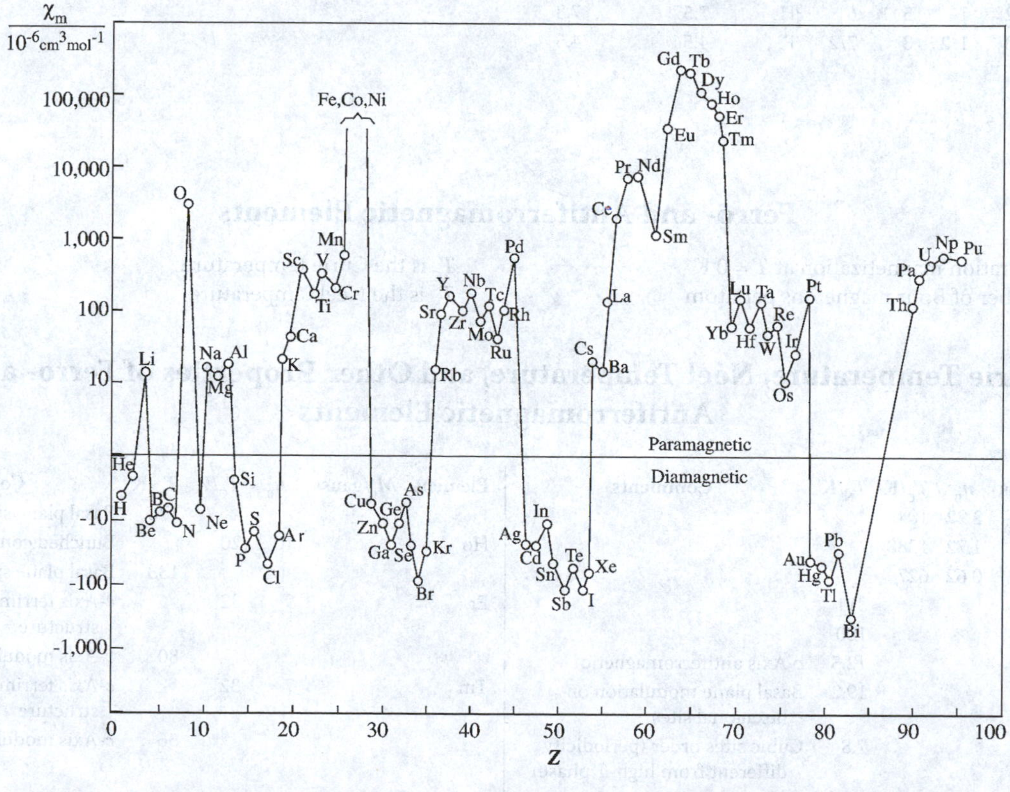

FIGURE 3. Molar susceptibility of the elements at room temperature (cgs units of 10^{-6} cm^3/mol). Values are not available for Z = 9, 61, and 84–89; Fe, Co, and Ni (Z = 26–28) are ferromagnetic. Data taken from the table "Magnetic Susceptibility of the Elements and Inorganic Compounds" in Section 4.

Reference

Gray, D. E., Ed., *American Institute of Physics Handbook, Third Edition*, McGraw-Hill, New York, 1972, pp. 5–224. With permission.

Ground State of Ions with Partly Filled *d* or *f* Shells

Z	Element	n	S	L	J	Gr. state	p_{calc}	$p_{calc}{}^{b}$	p_{meas}
22	Ti^{+3}	1	1/2	2	3/2	$^2D_{3/2}$	1.73^a	1.55	1.8
23	V^{+4}	1	1/2	2	3/2	$^2D_{3/2}$	1.73^a	1.55	1.8
23	V^{+3}	2	1	3	2	$^3F_2$	2.83^a	1.63	2.8
23	V^{+2}	3	3/2	3	3/2	$^4F_{3/2}$	3.87^a	0.77	3.8
24	Cr^{+3}	3	3/2	3	3/2	$^4F_{3/2}$	3.87^a	0.77	3.7
25	Mn^{+4}	3	3/2	3	3/2	$^4F_{3/2}$	3.87^a	0.77	4.0
24	Cr^{+2}	4	2	2	0	$^5D_0$	4.90^a	0	4.9
25	Mn^{+3}	4	2	2	0	$^5D_0$	4.90^a	0	5.0
25	Mn^{+2}	5	5/2	0	5/2	$^6S_{5/2}$	5.92^a	5.92	5.9
26	Fe^{+3}	5	5/2	0	5/2	$^6S_{5/2}$	5.92^a	5.92	5.9
26	Fe^{+2}	6	2	2	4	$^5D_4$	4.90^a	6.70	5.4
27	Co^{+2}	7	3/2	3	9/2	$^4F_{9/2}$	3.87^a	6.54	4.8
28	Ni^{+2}	8	1	3	4	$^3F_4$	2.83^a	5.59	3.2
29	Cu^{+2}	9	1/2	2	5/2	$^2D_{5/2}$	1.73^a	3.55	1.9
58	Ce^{+3}	1	1/2	3	5/2	$^2F_{5/2}$	2.54^c		2.4
59	Pr^{+3}	2	1	5	4	$^3H_4$	3.58^c		3.5
60	Nd^{+3}	3	3/2	6	9/2	$^4I_{9/2}$	3.62^c		3.5
61	Pm^{+3}	4	2	6	4	$^5I_4$	2.68^c		
62	Sm^{+3}	5	5/2	5	5/2	$^6H_{5/2}$	0.84^c		1.5
63	Eu^{+3}	6	3	3	0	$^7F_0$	0.0^c		3.4
64	Gd^{+3}	7	7/2	0	7/2	$^8S_{7/2}$	7.94^c		8.0
65	Tb^{+3}	8	3	3	6	$^7F_6$	9.72^c		9.5
66	Dy^{+3}	9	5/2	5	15/2	$^6H_{15/2}$	10.63^c		10.6
67	Ho^{+3}	10	2	6	8	$^5I_8$	10.60^c		10.4
68	Er^{+3}	11	3/2	6	15/2	$^4I_{15/2}$	9.59^c		9.5
69	Tm^{+3}	12	1	5	6	$^3H_6$	7.57^c		7.3
70	Yb^{+3}	13	1/2	3	7/2	$^2F_{7/2}$	4.54^c		4.5

[a] $p_{calc} = 2[S(S+1)]^{1/2}$
[b] $p_{calc} = 2[J(J+1)]^{1/2}$
[c] $p_{calc} = g[S(S+1)]^{1/2}$

References

1. Jiles, D., *Magnetism and Magnetic Materials*, Chapman & Hall, London, 1991, p. 243.
2. Kittel, C., *Introduction to Solid State Physics, 6th Edition*, J. Wiley & Sons, New York, 1986, pp. 405–406.
3. Ashcroft, N. W. and Mermin, N. D., *Solid State Physics*, Holt, Rinehart, and Winston, New York, 1976, p. 652.

Ferro- and Antiferromagnetic Elements

M_0 is the saturation magnetization at $T = 0$ K
n_B is the number of Bohr magnetons per atom
T_C is the Curie temperature
T_N is the Néel temperature

Curie Temperature, Néel Temperature, and Other Properties of Ferro- and Antiferromagnetic Elements

Element	M_0/gauss	n_B	T_C/K	T_N/K	Comments
Fe	22020	2.22	1043		
Co	18170	1.72	1388		
Ni	6410	0.62	627		
Cr				311	
Mn				100	
Ce				12.5	*c*-Axis antiferromagnetic
Nd				19.2	Basal plane modulation on hexagonal sites
				7.8	Cubic sites order (periodicity different from high-*T* phase)
Sm				106	Ordering on hexagonal sites
				13.8	Cubic site order
Eu				90.5	Spiral along cube axis
Gd	24880	7	293		
Tb		9	220		Basal plane ferromagnet
				230.2	Basal plane spiral
Dy		10	87		Basal plane ferromagnet
				176	Basal plane spiral
Ho		10	20		Bunched cone structure
				133	Basal plane spiral
Er		9	32		*c*-Axis ferrimagnetic cone structure
				80	*c*-Axis modulated structure
Tm		7	32		*c*-Axis ferrimagnetic cone structure
				56	*c*-Axis modulated structure

References

1. Ashcroft, N. W., and Mermin, N. D., *Solid State Physics*, Holt, Rinehart, and Winston, New York, 1976, p.652.
2. Gschneidner, K. A., and Eyring, L., *Handbook on the Physics and Chemistry of Rare Earths*, North Holland Publishing Co., Amsterdam, 1978.

Selected Ferromagnetic Compounds

M_0 is the saturation magnetization at $T = 293$ K
T_C is the Curie temperature

Saturation Magnetization and Curie Temperature of Selected Ferromagnetic Compounds

Compound	Formula	M_0/gauss	T_C/K	Crystal system
Manganese boride (MnB)	MnB	152	578	orthorh(FeB)
Manganese arsenide [MnAs]	MnAs	670	318	hex(FeB)
Manganese bismuthide [MnBi]	MnBi	620	630	hex(FeB)
Manganese antimonide (MnSb)	MnSb	710	587	hex(FeB)
Manganese nitride [Mn_4N]	Mn_4N	183	743	
Manganese silicide	MnSi		34	cub(FeSi)
Chromium monotelluride	CrTe	247	339	hex(NiAs)
Chromium(III) bromide	$CrBr_3$	270	37	hex(BiI_3)
Chromium(III) iodide	CrI_3		68	hex(BiI_3)
Chromium(IV) oxide	CrO_2	515	386	tetr(TiO_2)
Europium(II) oxide	EuO	1910*	77	cub
Europium(II) sulfide	EuS	1184*	16.5	cub
Gadolinium(III) chloride	$GdCl_3$	550*	2.2	orthorh
Iron boride (FeB)	FeB		598	orthorh
Iron boride (Fe_2B)	Fe_2B		1043	tetr ($CuAl_2$)
Iron beryllide [$FeBe_5$]	$FeBe_5$		75	cub($MgCu_2$)
Iron carbide	Fe_3C		483	orthorh
Iron phosphide (FeP)	FeP		215	orthorh (MnP)

* At $T = 0$ K

References

1. Kittel, C., *Introduction to Solid State Physics, 6th Edition*, J. Wiley & Sons, New York, 1986.
2. Ashcroft, N. W., and Mermin, N. D., *Solid State Physics*, Holt, Rinehart, and Winston, New York, 1976.

Magnetic Properties of High-Permeability Metals and Alloys (Soft)

μ_i is the initial permeability
μ_m is the maximum permeability
H_c is the coercive force
J_s is the saturation polarization
W_H is the hysteresis loss per cycle
T_C is the Curie temperature

Properties of High-Permeability Metals and Alloys (Soft)

Material	Composition (mass %)	μ_i/μ_0	μ_m/μ_0	H_c/A m^{-1}	J_s/T	W_H/J m^{-3}	T_C/K
Iron	Commercial 99Fe	200	6000	70	2.16	500	1043
Iron	Pure 99.9Fe	25000	350000	0.8	2.16	60	1043
Silicon-iron	96Fe-4Si	500	7000	40	1.95	50–150	1008
Silicon-iron (110) [001]	97Fe-3Si	9000	40000	12	2.01	35–140	1015
Silicon-iron {100} <100>	97Fe-3Si		100000	6	2.01		1015
Mild steel	Fe-0.1C-0.1Si-0.4Mn	800	1100	200			
Hypernik	50Fe-50Ni	4000	70000	4	1.60	22	753
Deltamax {100} <100>	50Fe-50Ni	500	200000	16	1.55		773
Isoperm {100} <100>	50Fe-50Ni	90	100	480	1.60		
78 Permalloy	78Ni-22Fe	4000	100000	4	1.05	50	651
Supermalloy	79Ni-16Fe-5Mo	100000	1000000	0.15	0.79	2	673
Mumetal	77Ni-16Fe-5Cu-2Cr	20000	100000	4	0.75	20	673
Hyperco	64Fe-35Co-0.5Cr	650	10000	80	2.42	300	1243
Permendur	50Fe-50Co	500	6000	160	2.46	1200	1253
2V-Permendur	49Fe-49Co-2V	800	4000	160	2.45	600	1253

Material	Composition (mass %)	μ_i/μ_0	μ_m/μ_0	H_c/A m^{-1}	J_s/T	W_H/J m^{-3}	T_C/K
Supermendur	49Fe-49Co-2V		60000	16	2.40	1150	1253
25Perminvar	45Ni-30Fe-25Co	400	2000	100	1.55		
7Perminvar	70Ni-23Fe-7Co	850	4000	50	1.25		
Perminvar (magnet. annealed)	43Ni-34Fe-23Co		400000	2.4	1.50		
Alfenol (or Alperm)	84Fe-16Al	3000	55000	3.2	0.8		723
Alfer	87Fe-13Al	700	3700	53	1.20		673
Aluminum-Iron	96.5Fe-3.5Al	500	19000	24	1.90		
Sendust	85Fe-10Si-5Al	36000	120000	1.6	0.89		753

References

1. McCurrie, R. A., *Structure and Properties of Ferromagnetic Materials*, Academic Press, London, 1994, p. 42.
2. Gray, D. E., Ed., *American Institute of Physics Handbook, Third Edition*, McGraw Hill, New York, 1972, p. 5–224.

Solids

Applications of High-Permeability Materials

Applications	Requirements
Power applications	
Distribution and power transformers	Low core losses, high permeability, high saturation magnetic polarization
High-quality motors and generators, stators and armatures, switched-mode power supplies	
Instrument transformers	
Audiofrequency transformers	Low core losses, high permeability, high magnetic polarization
Pulse transformers	High permeability
Cores for inductor coils	
Audiofrequency	Low hysteresis, high permeability
Carrier frequency	Very low hysteresis and eddy current loss
Radiofrequency	High permeability at low fields
Miscellaneous	
Relays, switches, earth leakage circuit	High permeability, low remanence, low coercivity
Magnetic shielding	Low core loss for AC applications
Magnetic recording heads	High initial permeability, low or zero remanence
Magnetic amplifiers, saturable reactors, saturable transformers, transformer cores	Rectangular hysteresis loops, low hysteresis loss
Magnetic shunts for temperature compensation in magnetic circuits	Low Curie temperature, appropriate decrease in permeability with increase in temperature
Electromagnets in indicating instruments, fire detection, quartz watches, electromechanical devices	High permeability, high saturation magnetic polarization
Magnetic yokes in permanent magnet devices, such as lifting and holding magnets, loudspeakers	High permeability, high saturation magnetic polarization

Reference

McCurrie, R. A., *Structure and Properties of Ferromagnetic Materials*, Academic Press, London, 1994. With permission.

Saturation Magnetostriction of Selected Materials

The tabulated parameter λ_s is related to the fractional change in length $\Delta l/l$ by $\Delta l/l = (3/2)\lambda_s(\cos^2\theta - 1/3)$, where θ is the angle of rotation.

Material	$\lambda_s \times 10^6$
Iron	-7
Fe - 3.2% Si	9
Nickel	-33
Cobalt	-62
45 Permalloy, 45% Ni - 55% Fe	27
Permalloy, 82% Ni - 18% Fe	0
Permendur, 49% Co - 49% Fe - 2% V	70
Alfer, 87% Fe - 13% Al	30

Material	$\lambda_s \times 10^6$
Magnetite, Fe_3O_4	40
Cobalt ferrite, $CoFe_2O_4$	-110
$SmFe_2$	-1560
$TbFe_2$	1753
$Tb_{0.3}Dy_{0.7}Fe_{1.93}$ (Terfenol D)	2000
$Fe_{66}Co_{18}B_{15}Si$ (amorphous)	35
$Co_{72}Fe_3B_6A_{13}$ (amorphous)	0

Reference

McCurrie, R.A., *Structure and Properties of Ferromagnetic Materials*, Academic Press, London, 1994, p. 91; additional data provided by A.E. Clark, Adelphi, MD.

Properties of Various Permanent Magnetic Materials (Hard)

B_r is the remanence
$_BH_c$ is the flux coercivity
$_iH_c$ is the intrinsic coercivity
$(BH)_{max}$ is the maximum energy product
T_C is the Curie temperature
T_{max} is the maximum operating temperature

Composition	B_r/T	$_BH_c$/10^3A m^{-1}	$_iH_c$/10^3A m^{-1}	$(BH)_{max}$/kJ m^{-3}	T_C/°C	T_{max}/°C
Alnico1 20Ni;12Al;5Co	0.72		35	25		
Alnico2 17Ni;10Al;12.5Co;6Cu	0.72		40-50	13-14		
Alnico3 24-30Ni;12-14Al;0-3Cu	0.5–0.6		40-54	10		
Alnico4 21-28Ni;11-13Al;3-5Co;2-4Cu	0.55–0.75		36-56	11-12		
Alnico5 14Ni;8Al;24Co;3Cu	1.25	53	54	40	850	520
Alnico6 16Ni;8Al;24Co;3Cu;2Ti	1.05		75	52		
Alnico8 15Ni;7Al;35Co;4Cu;5Ti	0.83	1.6	160	45		
Alnico9 15Ni;7Al;35Co;4Cu;5Ti	1.10	1.45	1.45	75	850	520
Alnico12 13.5Ni;8Al;24.5Co;2Nb	1.20		64	76.8		
$BaFe_{12}O_{19}$ (Ferroxdur)	0.4	1.6	192	29	450	400
$SrFe_{12}O_{19}$	0.4	2.95	3.3	30	450	400
$LaCo_5$	0.91			164	567	
$CeCo_5$	0.77			117	380	
$PrCo_5$	1.20			286	620	
$NdCo_5$	1.22			295	637	
$SmCo_5$	1.00	7.9	696	196	700	250
$Sm(Co_{0.76}Fe_{0.10}Cu_{0.14})_{6.8}$	1.04	4.8	5	212	800	300
$Sm(Co_{0.65}Fe_{0.28}Cu_{0.05}Zr_{0.02})_{7.7}$	1.2	10	16	264	800	300
$Nd_2Fe_{14}B$ sintered	1.22	8.4	1120	280	300	100
Fe;52Co;14V (Vicalloy II)	1.0	42		28	700	500
Fe;24Cr;15Co;3Mo (anisotropic)	1.54	67		76	630	500
Fe;28Cr;10.5Co (Chromindur II)	0.98	32		16	630	500
Fe;23Cr;15Co;3V;2Ti	1.35	4		44	630	500
Cu;20Ni;20Fe (Cunife)	0.55	4		12	410	350
Cu;21Ni;29Fe (Cunico)	0.34	0.5		8		
Pt;23Co	0.64	4		76	480	350
Mn;29.5Al;0.5C (anisotropic)	0.61	2.16	2.4	56	300	120

References

1. McCurrie, R. A., *Structure and Properties of Ferromagnetic Materials*, Academic Press, London, 1994, p. 204.
2. Gray, D. E., Ed., *American Institute of Physics Handbook, Third Edition*, McGraw Hill, New York, 1972, p. 5–165.
3. Jiles, D., *Magnetism and Magnetic Materials*, Chapman & Hall, London, 1991.

Selected Ferrites

J_s is the saturation magnetic polarization
T_C is the Curie temperature
ΔH is the line width

Properties and Applications of Ferrites

Material	J_s/T	T_C/°C	ΔH/kA m^{-1}	Applications
Spinels				
γ-Fe_2O_3	0.52	575		
Fe_3O_4	0.60	585		
$NiFe_2O_4$	0.34	575	350	Microwave devices
$MgFe_2O_4$	0.14	440	70	
$NiZnFe_2O_4$	0.50	375	120	Transformer cores
$MnFe_2O_4$	0.50	300	50	Microwave devices
$NiCoFe_2O_4$	0.31	590	140	Microwave devices
$NiCoAlFe_2O_4$	0.15	450	330	Microwave devices

Material	J_s/T	T_C/°C	ΔH/kA m^{-1}	Applications
$NiAl_{0.35}Fe_{1.65}O_4$	0.12	430	67	Microwave devices
$NiAlFe_2O_4$	0.05	1860	32	Microwave devices
$Mg_{0.9}Mn_{0.1}Fe_2O_4$	0.25	290	56	Microwave devices
$Ni_{0.5}Zn_{0.5}Al_{0.8}Fe_{1.2}O_4$	0.14		17	Microwave devices
$CuFe_2O_4$	0.17	455		Electromechanical transducers
$CoFe_2O_4$	0.53	520		
$LiFe_5O_8$	0.39	670		Microwave devices
Garnets				
$Y_3Fe_5O_{12}$	0.178	280	55	Microwave devices
$Y_3Fe_5O_{12}$ (single crys.)	0.178	292	0.5	Microwave devices
$(Y,Al)_3Fe_5O_{12}$	0.12	250	80	Microwave devices
$(Y,Gd)_3Fe_5O_{12}$	0.06	250	150	Microwave devices
$Sm_3Fe_5O_{12}$	0.170	305		Microwave devices
$Eu_3Fe_5O_{12}$	0.116	293		Microwave devices
$GdFe_5O_{12}$	0.017	291		Microwave devices
Hexagonal crystals				
$BaFe_{12}O_{19}$	0.45	430	1.5	Permanent magnets
$Ba_3Co_2Fe_{24}O_{41}$	0.34	470	12	Microwave devices
$Ba_2Zn_2Fe_{12}O_{22}$	0.28	130	25	Microwave devices
$Ba_3Co_{1.35}Zn_{0.65}Fe_{24}O_{41}$		390	16	Microwave devices
$Ba_2Ni_2Fe_{12}O_{22}$	0.16	500	8	Microwave devices
$SrFe_{12}O_{19}$	0.4	450		Permanent magnets

Reference

McCurrie, R. A., *Structure and Properties of Ferromagnetic Materials*, Academic Press, London, 1994.

Spinel Structure (AB_2O_4)

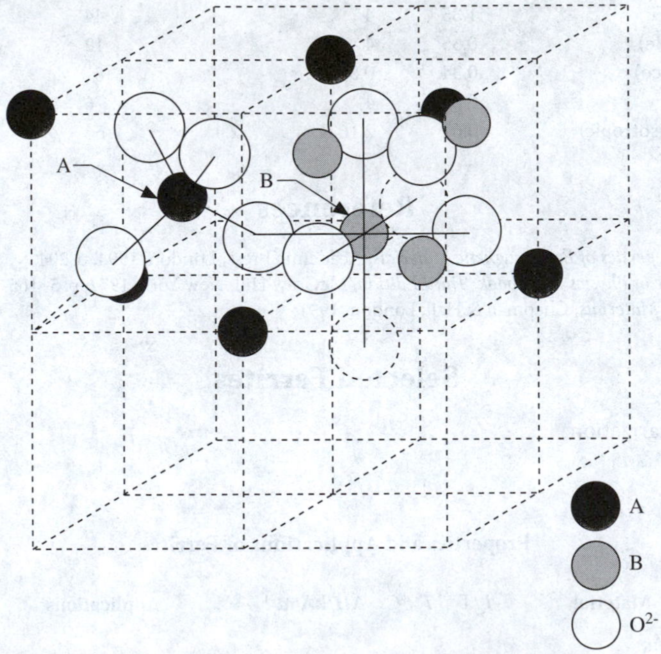

FIGURE 4. Arrangement of metal ions in the two octants A and B, showing tetrahedrally (A) and octahedrally (B) coordinated sites. (Reprinted from McCurrie, R.A., *Ferromagnetic Materials*, Academic Press, London, 1994. With permission.)

Crystal Structure and Néel Temperature of Selected Antiferromagnetic Solids

Material	Structure	T_N/K
Binary oxides		
MnO	cub(fcc)	122
FeO	cub(fcc)	198
CoO	cub(fcc)	291
NiO	cub(fcc)	525
α-Mn_2O_3	cub	90
CuO	monocl	230
UO_2	cub	30.8
Er_2O_3	cub	3.4
Gd_2O_3	cub	1.6
Perovskites		
$LaCrO_3$	orth	282
$LaMnO_3$	orth	100
$LaFeO_3$	orth	750
$NdCrO_3$	orth	224
$NdFeO_3$	orth	760
$YbCrO_3$	orth	118
$CaMnO_3$	cub	110
$EuTiO_3$	cub	5.3
$YCrO_3$	orth	141
$BiFeO_3$	cub*	673
$KCoF_3$	cub	125
$KMnF_3$	cub*	88.3
$KFeF_3$	cub	115
$KNiF_3$	cub	275
$NaMnF_3$	cub*	60
$NaNiF_3$	orth	149
$RbMnF_3$	cub	82
Spinels		
Co_3O_4	cub	40
$NiCr_2O_4$	tetr	65
$ZnCr_2O_4$	cub	15
$ZnFe_2O_4$	cub	9
$GeFe_2O_4$	cub	10
MgV_2O_4	cub	45
$MnGa_2O_4$	cub	33
NiAs and related structures		
Cr_2As	orth	300
CrSb	hex	714[a]
CrSe	hex	300
MnTe	hex	321[b]
NiS	hex	263
CrS	monocl	460
Rutile and related structures		
CoF_2	tetr	38
CrF_2	monocl	53
FeF_2	tetr	79
MnF_2	tetr	67
NiF_2	tetr	83
$CrCl_2$	orth	20
MnO_2	tetr	84
FeOF	tetr	315
Corundum and related structures		
Cr_2O_3	rhomb	318
α-Fe_2O_3	rhomb	948
$FeTiO_3$	rhomb	68
$MnTiO_3$	rhomb	41
$CoTiO_3$	rhomb	38
VF_3 and related structures		
CoF_3	rhomb	460
CrF_3	rhomb	80
FeF_3	rhomb	394
MnF_3	monocl	43
MoF_3	rhomb	185
Miscellaneous		
K_2NiF_4	tetr	97
MnI_2	hex	3.4
$CoUO_4$	orth	12
$CaMn_2O_4$	orth	225
CrN	cub*	273
CeC_2	tetr	33
$FeSn_2$	hex	373
Mn_2P	hex	103

* Distorted
[a] Range of 705 - 723 K
[b] Range of 320 - 323 K

References

1. Gray, D. E., Ed., *American Institute of Physics Handbook, Third Edition*, McGraw-Hill, New York, 1972, pp. 5–168 to 5–183.
2. Kittel, C., *Introduction to Solid State Physics, 6th Edition*, J. Wiley & Sons, New York, 1986.
3. Ashcroft, N. W., and Mermin, N. D., *Solid State Physics*, Holt, Rinehart, and Winston, New York, 1976, p. 697.

ORGANIC MAGNETS

J. S. Miller

Magnetic ordering, e.g., ferromagnetism, like superconductivity, is a property of a solid, not of an individual molecule or ion, and very rarely occurs for organic compounds. In contrast to superconductivity, where all electron spins pair to form a perfect diamagnetic material, magnetic ordering requires unpaired electron spins; hence, superconductivity and ferromagnetism are mutually exclusive.

The vast majority of organic compounds are diamagnetic (i.e., all electron spins are paired), and a relative few possess unpaired electrons (designated by an arrow, ↑) and are paramagnetic (PM), i.e., they are oriented in random directions. A few organic solids, however, exhibit strong magnetic behavior and magnetically order as ferromagnets (FO) with all spins aligned in the same direction. In some cases the spins align in the opposite direction and compensate to form an antiferromagnet (AF). In some cases these spins are not opposed to each other and do not compensate and lead to a canted antiferromagnet or weak ferromagnet (WF). If the number of spins that align in one direction differs from the number of spins that align in the opposite direction, the spins cannot compensate and a ferrimagnet (FI) results. Metamagnets (MM) are antiferromagnets in which all the spins become aligned like a ferromagnet in an applied magnetic field. Above the ordering or critical temperature, T_c, all magnets are paramagnets (PM). Organic magnets all possess electron spins in *p*-orbitals, but these may be in conjunction with metal ion-based spins.

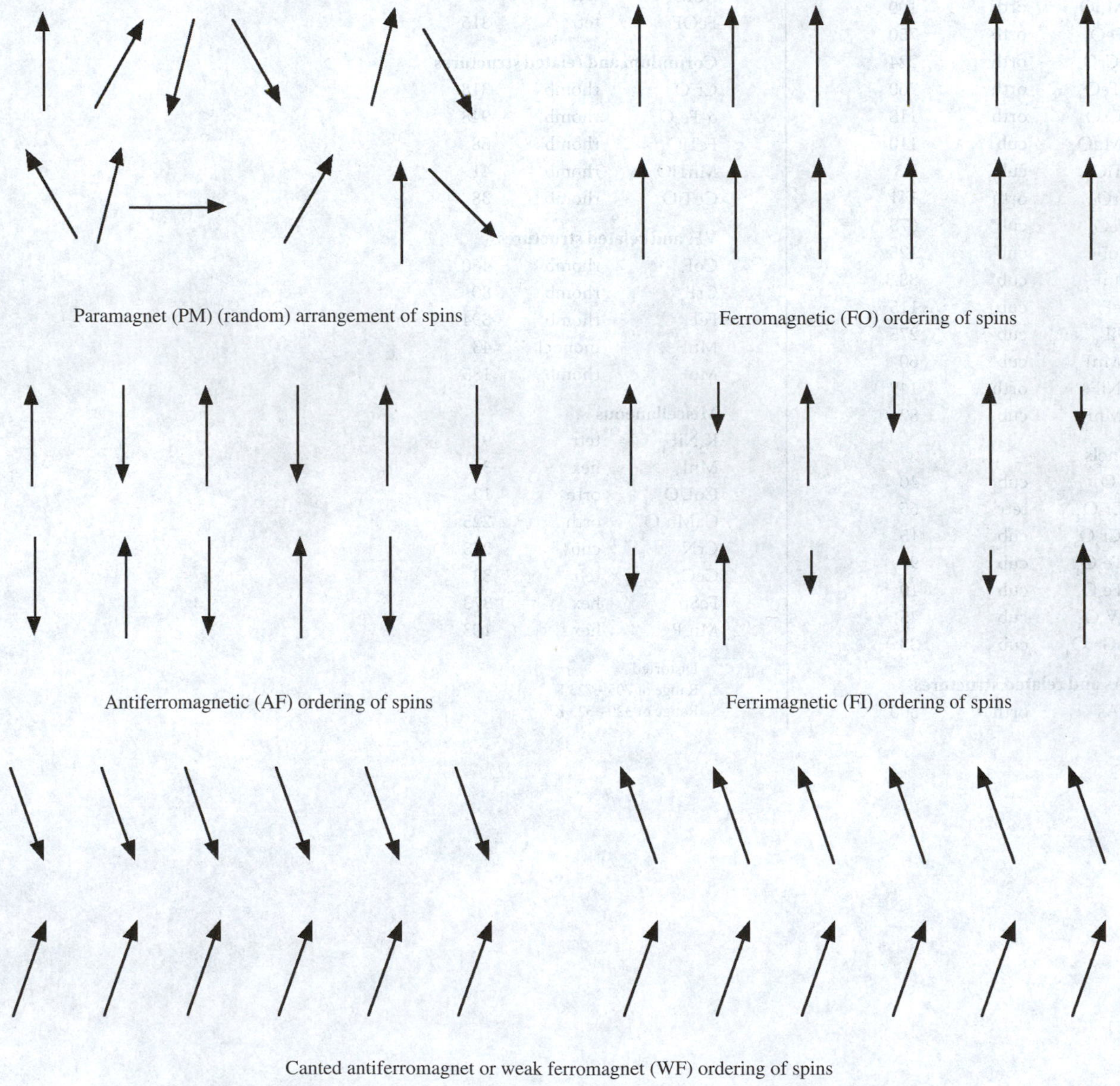

FIGURE 1. Schematic illustration of the different types of magnetic behavior.

Summary of the Critical Temperature, T_c, Saturation Magnetization, M_s, Coercive Field, H_{cr}, and Remanent Magnetization, M_r, for Selected Organic-Based Magnets

Magnet	Type	T_c/K	M_s/A m^{-1}	H_{cr}/T	M_r/A m^{-1}
α-1,3,5,7-Tetramethyl-2,6-diazaadamantane-*N,N'*-doxyl	FO	1.48	48,300	<0.00001	—
β-2-(4'-Nitrophenyl)-4,4,5,5-tetramethyl-4,5-dihydro-1H-imidazol-1-oxyl-3-*N*-oxide	FO	0.6	22,300	0.00008	<200
$\{Fe^{III}[C_5(CH_3)_5]_2\}$[TCNE]	FO	4.8	37,600	0.10	2,300
$\{Mn^{III}[C_5(CH_3)_5]_2\}$[TCNE]	FO	8.8	58,200	0.12	3,700
$\{Cr^{III}[C_5(CH_3)_5]_2\}$[TCNE]	FO	3.65	46,300	—	—
α-$\{Fe^{III}[C_5(CH_3)_5]_2\}$[TCNQ]	MM	2.55	34,200	—	—
β-$\{Fe^{III}[C_5(CH_3)_5]_2\}$[TCNQ]	FO	3.0	21,600	—	—
Tanol subarate	MM	0.38	20,700	—	—
$NCC_6F_4CN_2S_2$	WF	35.5	45	0.00009	—
$Mn^{II}(hfac)_2NITC_2H_5$	FI	7.8	39,400	0.03	27,600
$Mn^{II}(hfac)_2NIT(i\text{-}C_3H_8)$	FI	7.6	42,400	<0.0005	<420
$[Mn(hfac)_2]_3[\{ON[C_6H_3(t\text{-}C(CH_3)_3]_2NO]_2\}$	FI	46	24,400	—	—
$[MnTPP][TCNE]{\cdot}2C_6H_5CH_3$	FI	13	18,400	2.4	10,300
$V[TCNE]_x{\cdot}yCH_2Cl_2$ ($x \sim 2$; $y \sim 0.5$)	FI	~400	28,200	0.0015 - 0.006	1,650
$Mn[TCNE]_x{\cdot}yCH_2Cl_2$ ($x \sim 2$; $y \sim 0.5$)	FI	75	52,000	0.002	270
$Fe[TCNE]_x{\cdot}yCH_2Cl_2$ ($x \sim 2$; $y \sim 0.5$)	FI	97	46,300	0.23	3
$Co[TCNE]_x{\cdot}yCH_2Cl_2$ ($x \sim 2$; $y \sim 0.5$)	FI	44	22,000	0.65	—

List of Symbols and Abbreviations

M_s	Saturation magnetization at 2 K	hfac	Hexafluoroacetonate
H_{cr}	Coercive Field	NIT	Nitronyl nitroxide
T_c	Critical Temperature	FO	Ferromagnet
M_r	Remanent magnetization at 2 K	FI	Ferrimagnet
TCNE	Tetracyanoethylene	MM	Metamagnet
TCNQ	7,7,8,8-Tetracyano-*p*-quinodimethane	WF	Weak ferromagnet

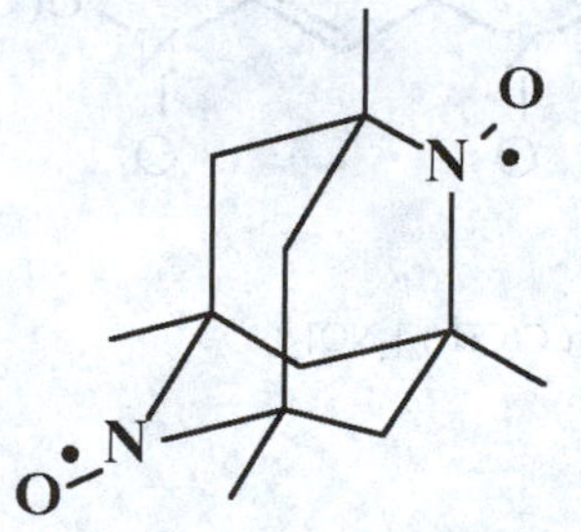

1,3,5,7-Tetramethyl-2,6-diazaadamantane-N,N'-doxyl

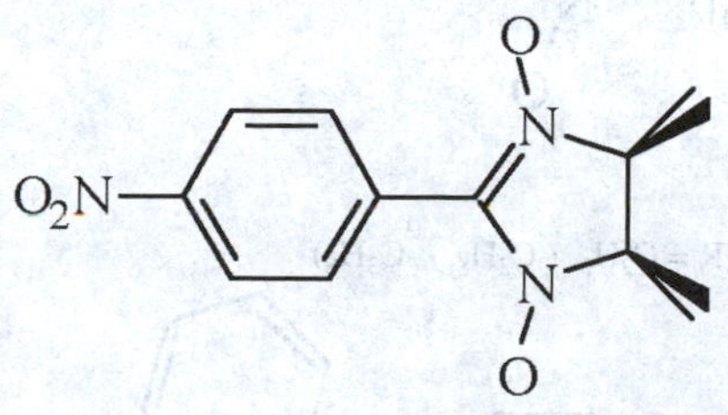

2-(4'-Nitrophyenyl)-4,4,5,5-tetramethyl-4,5-dihydro-1H-imidazol-1-oxyl-3-*N*-oxide

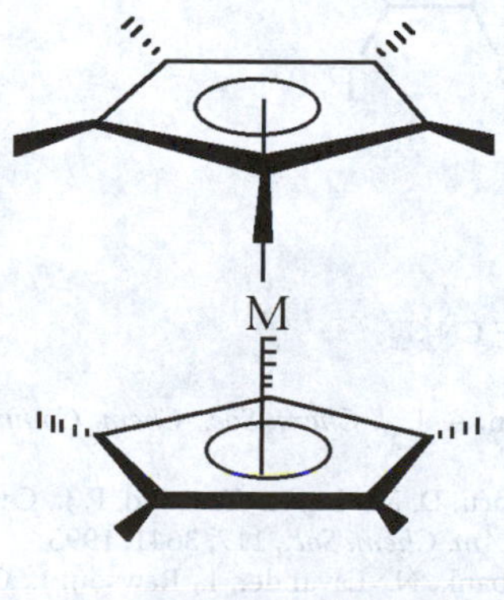

$M[C_5(CH_3)_5]_2$(M = Cr, Mn, Fe)

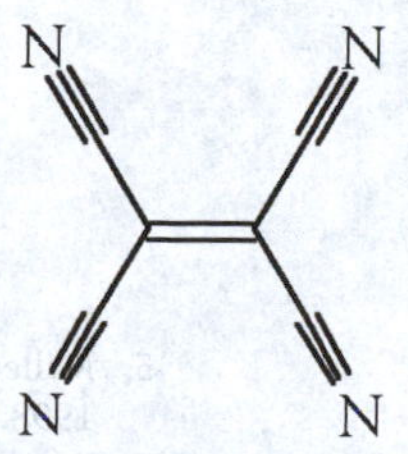

TCNE

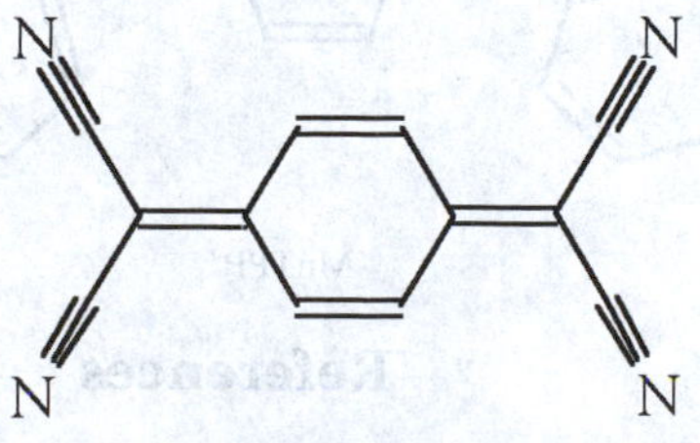

TCNQ

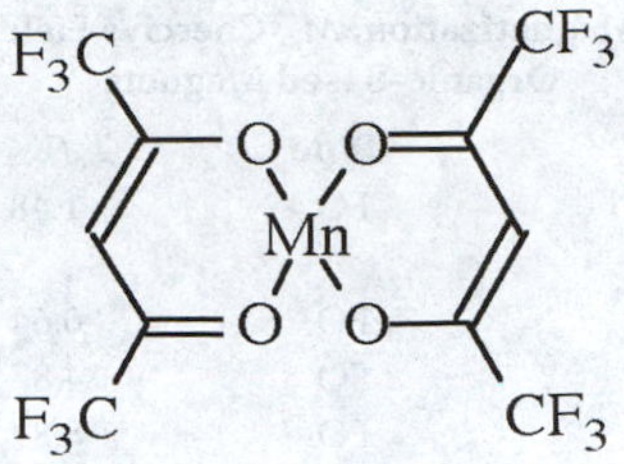

Mn(hfac)$_2$

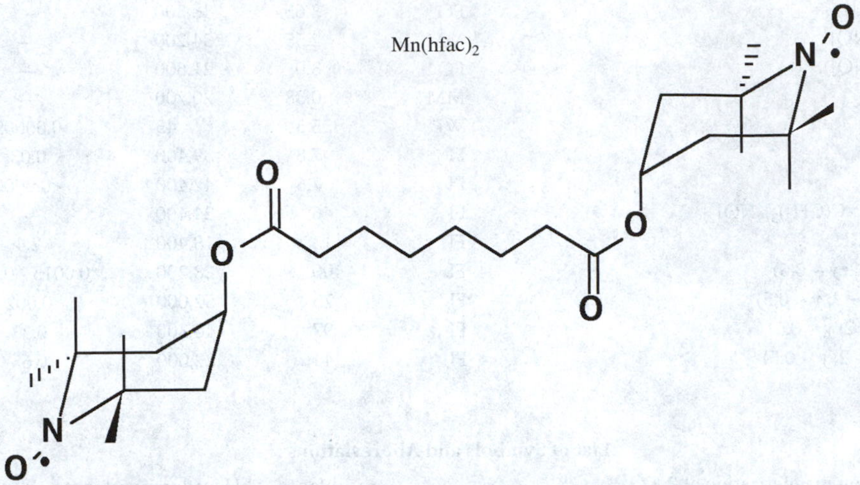

Tanol subarate

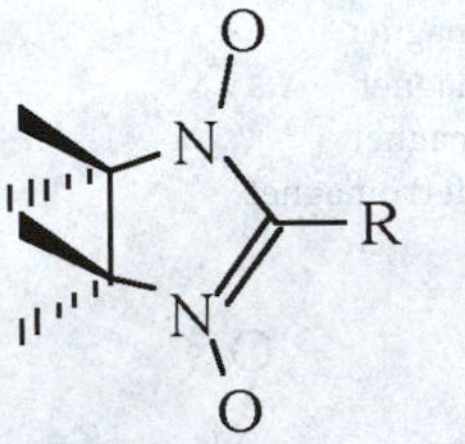

NITR (R = C_2H_5, *i*-C_3H_8, *n*-C_3H_8)

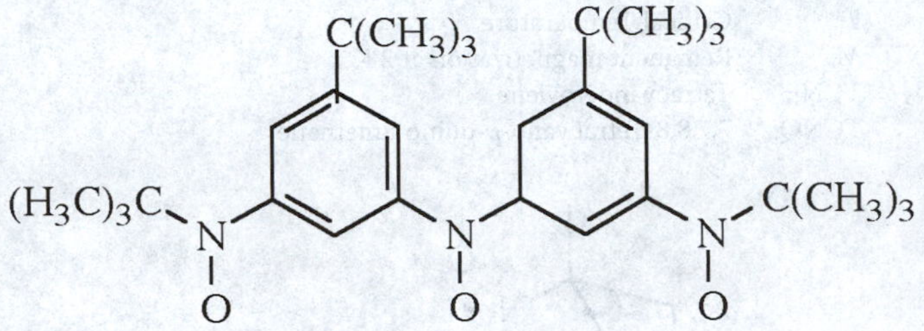

{$ON[C_6H_3(t\text{-}C(CH_3)_3]_2NO]_2$}

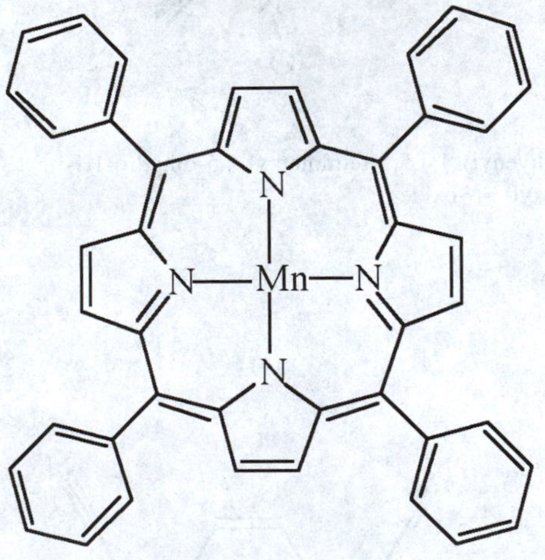

MnTPP

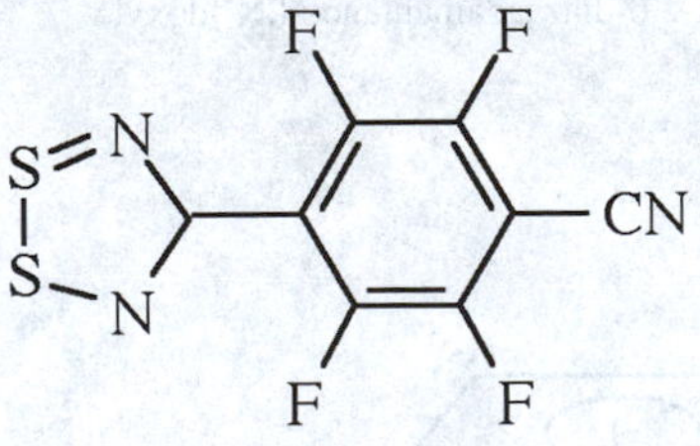

$NCC_6F_4CN_2S_2$

References

1. Miller, J. S. and Epstein, A. J., *Angew. Chem. Internat. Ed.*, 33, 385, 1994.
2. Chiarelli, R., Rassat, A., Dromzee, Y., Jeannin, Y., Novak, M. A., and Tholence, J. L., *Phys. Scrip.*, T49, 706, 1993.
3. Kinoshita, M., *Jap. J. Appl. Phys.*, 33, 5718, 1994.
4. Gatteschi, D., *Adv. Mat.*, 6, 635, 1994.
5. Miller, J. S. and Epstein, A. J., *J. Chem. Soc., Chem. Commun.*, 1319, 1998.
6. Broderick, W. E., Eichorn, D. M., Lu, X., Toscano, P. J., Owens, S. M. and Hoffman, B. M., *J. Am. Chem. Soc.*, 117, 3641, 1995.
7. Banister, A. J., Bricklebank, N., Lavander, I., Rawson, J., Gregory, C. I., Tanner, B. K., Clegg, W. J., Elsegood, M. R., and Palacio, F., *Angew. Chem. Internat. Ed.*, 35, 2533, 1996.

ELECTRON INELASTIC MEAN FREE PATHS

Cedric J. Powell

The inelastic mean free path (IMFP) of electrons impinging on a solid surface is defined as the average of distances, measured along the trajectories, that electrons with a given energy travel between inelastic collisions in the substance. It is an important parameter in analyzing results from surface characterization techniques such as Auger electron spectroscopy, x-ray photoelectron spectroscopy, low-energy electron diffraction, and others. IMFPs can be measured by the elastic-peak electron spectroscopy technique and other methods, and they can be calculated from optical data. A detailed analysis of the experimental and theoretical considerations in obtaining reliable IMPF values can be found in Refs. 4 and 5.

Table 1 below gives recommended IMFP values for 41 elemental solids in the energy range 50 eV to 30 000 eV. All values in Table 1 are taken from Ref. 1. Table 2 gives IMFP values for several inorganic compounds and organic materials in the range 50 eV to 2 000 eV. The entries in Table 1 are listed by atomic number, while substances in Table 2 are listed in alphabetical order by name, with inorganic compounds preceding the organic materials. IMFP values are given in Ångström units (1 Å = 10^{-10} m).

References

1. Tanuma, S., Powell, C. J., and Penn, D. R., *Surf. Interface Anal.* (in press); doi 10.1002/sia.3522.
2. Tanuma, S., Powell, C. J., and Penn, D. R., *Surf. Interface Anal.* 17, 927, 1991.
3. Tanuma, S., Powell, C. J., and Penn, D. R., *Surf. Interface Anal.* 21, 165, 1994.
4. Powell, C. J., and Jablonski, A., *J. Phys. Chem. Ref. Data* 28, 19, 1999.
5. Powell, C. J., and Jablonski, A., *Nucl. Instr. Meth. Phys. Res. A* 601, 54, 2009.

TABLE 1. Electron Inelastic Mean Free Paths of Elemental Solids in Å (10^{-10} m)

	Electron Energy in eV															
Element	**50**	**100**	**250**	**500**	**750**	**1000**	**1250**	**1500**	**1750**	**2000**	**3000**	**5000**	**10000**	**15000**	**20000**	**30000**
Li	4.6	7.0	13.3	22.7	31.4	39.7	47.8	55.7	63.4	71.0	100.4	156.2	286.6	410.1	529.6	760.6
Be	3.5	4.3	7.2	11.7	15.9	19.9	23.8	27.6	31.2	34.9	48.8	75.3	136.7	194.7	250.6	358.4
C-graphite	4.8	4.4	6.8	10.8	14.6	18.2	21.7	25.0	28.3	31.6	44.0	67.6	122.2	173.7	223.3	318.9
C-diamond	6.9	4.7	6.4	9.9	13.2	16.4	19.4	22.4	25.3	28.1	39.0	59.6	107.3	152.1	195.3	278.4
C-glassy	5.8	6.3	10.0	16.0	21.7	27.0	32.1	37.1	42.0	46.8	65.4	100.4	181.9	258.6	332.7	475.4
Na	5.0	7.6	13.8	23.1	31.8	40.0	48.0	55.8	63.5	71.0	100.1	155.1	283.4	404.7	522.0	748.5
Mg	4.0	5.4	9.4	15.3	20.7	25.9	30.9	35.8	40.6	45.3	63.5	97.9	177.7	253.0	325.7	465.9
Al	3.5	4.6	7.9	12.7	17.1	21.3	25.4	29.4	33.3	37.2	52.0	79.9	144.8	206.0	265.1	378.8
Si	4.1	5.3	9.0	14.6	19.7	24.5	29.2	33.7	38.2	42.6	59.5	91.5	165.7	235.7	303.1	433.2
K	7.1	10.0	18.3	31.4	43.5	54.8	65.7	76.4	86.8	97.1	136.8	212.0	387.5	553.5	714.0	1024.0
Sc	4.6	4.9	7.4	11.9	16.2	20.2	24.1	27.9	31.5	35.1	49.0	75.2	135.9	193.1	248.2	354.3
Ti	4.2	4.4	6.6	10.4	14.1	17.5	20.9	24.1	27.3	30.4	42.3	64.7	116.8	165.8	213.0	303.8
V	4.6	4.7	6.7	10.3	13.8	17.1	20.3	23.4	26.4	29.4	40.7	62.1	111.7	158.3	203.1	289.3
Cr	4.5	4.4	6.4	9.9	13.2	16.3	19.4	22.3	25.2	28.0	38.8	59.2	106.4	150.7	193.4	275.6
Fe	4.3	4.6	6.7	10.2	13.5	16.7	19.7	22.7	25.7	28.5	39.5	60.1	107.8	152.7	195.8	278.8
Co	5.1	4.6	6.1	9.0	11.8	14.5	17.2	19.8	22.3	24.8	34.2	51.9	92.9	131.4	168.3	239.5
Ni	4.9	4.6	6.2	9.2	12.1	14.9	17.6	20.2	22.8	25.3	35.0	53.1	95.0	134.3	172.0	244.7
Cu	5.0	5.0	7.0	10.4	13.6	16.7	19.6	22.6	25.4	28.2	39.0	59.2	105.7	149.5	191.5	272.4
Ge	4.0	5.0	8.2	12.9	17.2	21.2	25.2	29.0	32.8	36.5	50.7	77.5	139.5	197.9	254.2	362.5
Y	5.1	5.5	8.8	14.4	19.4	24.1	28.6	33.0	37.3	41.5	57.8	88.6	159.9	227.0	291.7	416.2
Nb	5.7	5.3	7.3	11.4	15.2	18.8	22.2	25.5	28.8	31.9	44.1	67.1	120.4	170.3	218.3	310.7
Mo	5.0	4.4	6.1	9.5	12.7	15.6	18.4	21.2	23.8	26.5	36.6	55.6	99.6	141.0	180.7	257.1
Ru	4.9	4.1	5.4	8.3	11.1	13.7	16.1	18.5	20.8	23.1	31.9	48.5	86.7	122.6	157.1	223.5
Rh	4.8	4.0	5.2	7.9	10.6	13.0	15.4	17.7	19.9	22.1	30.5	46.2	82.7	117.0	149.9	213.2
Pd	4.8	4.0	5.3	8.1	10.8	13.4	15.8	18.2	20.5	22.7	31.3	47.6	85.2	120.4	154.3	219.5
Ag	6.1	4.9	5.9	8.7	11.6	14.2	16.8	19.2	21.6	24.0	33.0	50.0	89.3	126.2	161.6	229.7
In	4.7	5.4	8.0	12.3	16.5	20.6	24.4	28.1	31.8	35.3	48.9	74.5	133.8	189.5	243.1	346.3
Sn	5.8	6.4	9.2	13.8	18.4	22.8	27.0	31.1	35.1	38.9	53.8	81.8	146.5	207.3	265.8	378.2
Cs	6.3	9.3	17.4	29.3	40.3	50.8	61.0	71.0	80.7	90.2	126.8	196.1	357.6	510.4	657.9	942.8
Gd	4.1	4.6	7.4	11.8	15.7	19.5	23.2	26.8	30.3	33.7	46.9	71.6	128.8	182.6	234.4	334.1
Tb	4.1	4.3	6.5	10.2	13.7	16.9	20.1	23.2	26.2	29.2	40.6	61.9	111.3	157.8	202.5	288.6
Dy	4.5	4.6	6.8	10.6	14.1	17.5	20.8	23.9	27.0	30.0	41.7	63.5	114.0	161.5	207.2	295.1
Hf	5.5	5.5	7.5	11.3	14.9	18.2	21.5	24.6	27.7	30.7	42.3	64.2	114.4	161.5	206.7	293.8
Ta	4.9	4.5	6.1	9.2	12.1	14.8	17.4	19.9	22.4	24.8	34.2	51.7	92.2	130.0	166.5	236.4
W	5.4	4.4	5.8	8.7	11.3	13.9	16.3	18.6	20.9	23.2	31.8	48.1	85.5	120.5	154.1	218.8

Element	Electron Energy in eV 50	100	250	500	750	1000	1250	1500	1750	2000	3000	5000	10000	15000	20000	30000
Re	5.4	4.2	5.4	8.0	10.5	12.8	15.0	17.2	19.3	21.3	29.3	44.3	78.7	111.0	142.0	201.5
Os	5.6	4.4	5.6	8.4	10.9	13.3	15.6	17.8	20.0	22.1	30.3	45.7	81.1	114.3	146.2	207.3
Ir	5.3	4.2	5.4	8.0	10.4	12.7	15.0	17.1	19.2	21.2	29.1	43.9	78.1	110.1	140.8	199.7
Pt	5.0	4.2	5.5	8.2	10.7	13.1	15.4	17.6	19.8	21.9	30.0	45.4	80.7	113.8	145.5	206.6
Au	5.1	4.3	5.6	8.4	11.0	13.5	15.9	18.2	20.5	22.7	31.1	47.1	83.9	118.4	151.5	215.1
Bi	4.9	5.5	8.0	12.3	16.4	20.2	23.9	27.5	31.0	34.4	47.5	72.1	129.3	182.8	234.4	333.6

TABLE 2. Electron Inelastic Mean Free Paths of Other Materials in Å (10^{-10} m)

Substance	Formula	Electron Energy in eV 50	100	150	200	300	400	600	800	1000	1200	1400	1600	1800	2000	Ref.
Gallium phosphide	GaP	5.6	6.5	7.8	9.0	11.4	13.7	18.1	22.3	26.3	30.2	34.1	37.8	41.5	45.2	2
Indium phosphide	InP	4.8	4.9	5.6	6.4	8.1	9.7	12.8	15.7	18.7	21.4	24.2	26.8	29.4	32.0	2
Lead(II) sulfide	PbS	4.8	5.6	6.7	7.8	10.0	12.1	16.1	19.8	23.6	27.1	30.6	33.9	37.2	40.5	2
Lead(II) telluride	PbTe	4.3	5.5	6.6	7.7	9.8	11.9	15.8	19.6	23.4	26.9	30.3	33.7	37.0	40.2	2
Potassium chloride	KCl	7.5	7.8	9.3	10.9	14.2	17.3	23.2	28.7	34.0	39.2	44.2	49.1	54.0	58.8	2
Silicon carbide	SiC	4.7	4.9	5.8	6.8	8.7	10.5	13.9	17.1	20.3	23.3	26.3	29.2	32.1	35.0	2
Silicon dioxide (vitreous)	SiO_2	8.0	7.7	8.8	10.0	12.6	15.2	20.0	24.7	29.3	33.7	38.0	42.2	46.4	50.5	2
Zinc sulfide	ZnS	5.8	6.5	7.7	8.9	11.3	13.6	17.9	22.0	26.0	29.8	33.6	37.3	40.9	44.5	2
Adenine	$C_5H_5N_5$	6.4	6.6	7.8	9.2	11.8	14.4	19.2	24.1	28.6	33.1	37.4	41.6	45.8	49.9	3
Bovine plasma albumin		7.3	7.2	8.5	9.9	12.7	15.4	20.7	25.8	30.8	35.6	40.2	44.8	49.4	53.8	3
β-Carotene	$C_{40}H_{56}$	6.4	7.0	8.5	10.0	13.0	15.9	21.4	26.9	32.0	37.0	41.9	46.6	51.3	56.0	3
Deoxyribonucleic acid (DNA)		7.3	7.3	8.5	9.8	12.6	15.4	20.7	25.9	30.8	35.6	40.3	44.9	49.4	53.8	3
1,6-Diphenyl-1,3,5-hexatriene	$C_{18}H_{16}$	6.4	7.0	8.4	9.9	12.9	15.8	21.3	26.7	31.8	36.8	41.7	46.4	51.1	55.7	3
Guanine	$C_5H_5N_5O$	6.2	6.2	7.2	8.4	10.8	13.1	17.5	21.8	25.9	29.9	33.8	37.6	41.4	45.1	3
Hexacosane	$C_{26}H_{54}$	7.0	7.6	9.2	10.9	14.1	17.2	23.2	29.2	34.7	40.1	45.4	50.6	55.7	60.7	3
Kapton		7.0	6.8	7.9	9.2	11.7	14.2	19.0	23.7	28.2	32.5	36.7	40.9	44.9	49.0	3
Polyacetylene		5.3	5.7	6.8	7.9	10.2	12.5	16.9	21.1	25.1	29.0	32.8	36.5	40.2	43.8	3
Poly(butene-1-sulfone)		7.1	7.2	8.5	9.9	12.7	15.4	20.7	25.8	30.6	35.3	39.9	44.4	48.8	53.2	3
Polyethylene		6.9	7.2	8.6	10.1	13.0	15.9	21.4	26.8	31.8	36.8	41.6	46.3	51.0	55.6	3
Poly(methyl methacrylate)		7.8	7.9	9.3	10.8	13.9	16.9	22.7	28.3	33.7	38.8	43.9	48.9	53.8	58.6	3
Polystyrene		6.9	7.3	8.7	10.2	13.2	16.1	21.6	27.1	32.2	37.2	42.1	46.9	51.6	56.2	3
Poly(2-vinylpyridine)		6.9	7.3	8.7	10.3	13.3	16.2	21.8	27.3	32.5	37.5	42.4	47.3	52.0	56.7	3

ELECTRON STOPPING POWERS

Cedric J. Powell

Numerical data are given for mass collision electron stopping powers in 41 elemental solids for energies between 100 eV and 30 keV. These stopping powers were determined with an algorithm that utilizes experimental optical data for each solid.

The stopping power for electrons and other charged particles in matter is often needed in calculations of electron transport in a medium, particularly in radiation physics and in descriptions of signal generation in analytical techniques such as electron-probe microanalysis and Auger electron spectroscopy. The stopping power is defined as the average rate at which the charged particles lose energy at any point along their trajectories. For electrons, it is customary to separate the total stopping power into two components, the collision stopping power due to inelastic-scattering events of the electrons in a medium and the radiative stopping power due to the emission of bremsstrahlung in the electric field of the atomic nucleus and atomic electrons (Ref. 1). For electron energies less than 30 keV, the radiative stopping power is less than 1% of the collision stopping power (Ref. 1), and is neglected in the numerical data given here.

Numerical data for collision and radiative stopping powers at electron energies between 10 keV and 1 GeV have been published for materials of interest in radiation physics and dosimetry (Ref. 1). Similar data can also be obtained from a website of the National Institute of Standards and Technology (Ref. 2). The collision stopping powers were calculated from the theory of Bethe (Refs. 3 and 4) and recommended values of the one material-dependent parameter, the mean excitation energy (Ref. 1). While the Bethe theory is expected to be valid for electron energies much larger than the largest K-shell binding energy of atoms in the particular material, the Bethe stopping-power equation is frequently utilized to calculate stopping powers for energies of 10 keV and above (Refs. 1 and 2). Detailed analyses of the Bethe stopping-power theory have been published (Refs. 1 and 4).

The Table below gives mass collision stopping powers for 41 elemental solids at energies between 100 eV and 30 keV (Ref. 5). The mass collision stopping power is the collision stopping power divided by the mass density of the solid. The mass collision stopping powers in the Table were determined by interpolation with a clamped cubic spline from the published data (Ref. 5), which had been calculated with an algorithm that utilizes experimental optical data for each solid. Comparisons with collision stopping powers from the Bethe stopping-power equation showed root-mean-square differences of 9.1% and 8.7% for energies of 9.897 keV and 29.733 keV, respectively (Ref. 5). This level of agreement was considered satisfactory on account of uncertainties of the algorithm and optical data used for the calculations as well as uncertainties of the mean excitation energies used with the Bethe equation. The mass collision stopping powers in the Table are given in units of MeV cm^2/g for a range of relativistic kinetic energies, T. The elemental solids are listed in order of atomic number.

References

1. *Stopping Powers for Electrons and Positrons,* ICRU Report 37 (International Commission on Radiation Units and Measurements, Bethesda, MD, 1984).
2. Berger, M. J., Coursey, J. S., Zucker, M. A., and Chang, J., *Stopping-Power and Range Tables for Electrons, Positrons, and Helium Ions,* Version 1.2.3, (http://www.nist.gov/pml/data/star/index.cfm), 2005.
3. Bethe, H., *Ann. Physik* 5, 325, 1930.
4. Inokuti, M., *Rev. Mod. Phys.* 43, 297, 1971.
5. Shinotsuka, H., Tanuma, S., Powell, C. J., and Penn, D. R., *Nucl. Instr. Methods Phys. Res. B* 270, 75, 2012.

Values of Mass Collision Electron Stopping Powers in MeV cm^2/g for the Indicated Elemental Solids and Relativistic Electron Kinetic Energies T in eV

	100	200	300	400	500	1000	1500	2000	3000	4000	5000	10000	15000	20000	30000
Li	336.6	263.5	233.4	207.0	185.3	122.0	92.2	74.7	55.0	44.1	37.0	21.3	15.4	12.2	8.86
Be	342.2	257.8	210.0	183.5	165.7	113.5	87.4	71.6	53.3	42.9	36.2	21.0	15.2	12.1	8.82
C-glassy	251.3	218.8	183.9	158.4	139.8	92.5	73.9	62.4	48.1	39.4	33.6	20.0	14.7	11.8	8.60
C-graphite	343.5	296.1	246.2	210.6	184.6	118.9	92.1	76.2	57.5	46.7	39.5	23.2	16.9	13.5	9.84
C-diamond	262.2	249.6	212.4	183.6	162.0	106.1	83.0	69.0	52.4	42.7	36.2	21.4	15.6	12.5	9.12
Na	161.5	160.4	153.7	143.3	132.6	94.4	73.5	60.7	45.7	37.1	31.5	19.0	14.0	11.3	8.30
Mg	203.8	175.9	165.9	154.5	143.2	103.5	81.6	67.8	51.3	41.7	35.4	21.3	15.7	12.6	9.32
Al	187.6	142.8	127.1	118.1	110.7	83.8	67.5	56.7	43.4	35.5	30.2	18.2	13.5	10.9	8.08
Si	199.4	148.9	125.7	115.9	109.2	84.8	68.6	57.8	44.3	36.3	30.9	18.6	13.8	11.1	8.23
K	189.9	156.2	127.5	108.8	95.8	64.1	52.4	45.2	35.8	29.8	25.7	15.7	11.7	9.42	7.01
Sc	204.5	204.0	171.7	148.1	130.7	85.4	66.0	55.5	43.2	35.8	30.7	18.7	13.8	11.1	8.20
Ti	147.7	153.7	132.6	116.2	103.6	69.3	53.8	45.2	35.5	29.7	25.6	15.8	11.7	9.46	7.01
V	101.1	119.0	112.1	102.0	92.8	64.5	50.6	42.5	33.5	28.2	24.4	15.2	11.3	9.15	6.79
Cr	97.4	104.1	96.9	88.5	81.0	57.4	45.3	38.1	30.1	25.5	22.3	14.0	10.5	8.51	6.34
Fe	80.4	86.2	83.7	78.7	73.7	54.8	43.9	37.0	28.9	24.4	21.3	13.6	10.2	8.31	6.20
Co	82.0	97.7	96.4	90.8	84.7	62.3	49.5	41.5	32.1	26.8	23.3	14.7	11.0	8.96	6.67
Ni	77.4	90.3	91.5	87.6	82.7	62.2	49.9	42.0	32.5	27.0	23.5	14.9	11.2	9.13	6.82
Cu	68.1	76.4	76.5	73.9	70.7	55.5	45.3	38.5	30.0	25.0	21.8	13.9	10.5	8.60	6.43
Ge	90.4	79.3	73.7	69.8	66.5	53.6	44.7	38.4	30.2	25.1	21.7	13.8	10.5	8.55	6.41
Y	117.2	106.8	88.0	75.6	67.1	48.4	40.5	35.2	28.2	23.7	20.5	12.8	9.66	7.92	5.97

	100	200	300	400	500	1000	1500	2000	3000	4000	5000	10000	15000	20000	30000
Nb	70.8	86.3	75.1	65.5	58.5	41.5	34.8	30.5	24.8	21.1	18.4	11.7	8.84	7.27	5.51
Mo	74.1	87.1	76.2	66.5	59.2	41.5	34.8	30.6	25.0	21.3	18.6	11.8	8.95	7.36	5.57
Ru	68.9	83.9	76.6	67.6	60.4	41.6	34.1	29.9	24.5	21.0	18.4	11.7	8.88	7.28	5.52
Rh	70.1	87.0	82.0	72.9	65.3	44.9	36.3	31.7	25.9	22.1	19.4	12.4	9.35	7.66	5.80
Pd	69.8	88.2	82.8	73.3	65.5	44.6	35.7	30.9	25.3	21.6	19.0	12.2	9.19	7.53	5.70
Ag	66.9	94.9	95.5	87.0	78.7	54.7	44.3	38.5	31.1	26.2	22.8	14.3	10.7	8.71	6.54
In	62.6	74.7	75.4	69.7	63.4	43.7	34.7	29.5	23.9	20.5	18.1	11.7	8.88	7.27	5.50
Sn	56.3	68.5	71.5	67.5	62.0	43.3	34.4	29.2	23.5	20.2	17.8	11.6	8.79	7.20	5.44
Cs	80.3	63.5	59.6	58.1	54.1	37.8	29.8	25.1	19.8	16.9	14.9	9.90	7.60	6.26	4.74
Gd	72.2	66.5	58.9	53.4	49.8	37.8	30.7	26.2	20.6	17.3	15.0	10.0	7.74	6.42	4.89
Tb	85.4	85.3	75.1	67.2	61.5	45.1	36.3	30.7	24.0	20.0	17.3	11.2	8.58	7.07	5.35
Dy	75.7	78.4	70.3	63.6	58.6	43.9	35.6	30.3	23.8	19.8	17.2	11.1	8.52	7.02	5.32
Hf	42.6	51.0	48.0	44.7	41.9	32.9	27.7	24.2	19.6	16.6	14.5	9.48	7.39	6.15	4.72
Ta	43.3	51.8	49.0	45.4	42.5	33.2	28.0	24.4	19.8	16.9	14.8	9.58	7.44	6.20	4.75
W	41.5	49.3	46.7	43.3	40.5	31.8	26.9	23.7	19.3	16.5	14.4	9.40	7.31	6.09	4.68
Re	42.3	50.5	47.7	44.0	40.9	31.7	26.8	23.5	19.2	16.4	14.4	9.35	7.27	6.06	4.65
Os	36.9	44.9	43.0	39.8	37.1	29.1	24.8	21.9	18.0	15.4	13.6	8.88	6.91	5.77	4.44
Ir	39.3	47.0	44.6	40.9	37.9	29.4	25.1	22.2	18.3	15.7	13.9	9.07	7.04	5.88	4.53
Pt	40.4	47.1	44.8	41.2	38.1	29.4	25.0	22.1	18.3	15.7	13.9	9.08	7.04	5.88	4.53
Au	41.7	52.1	49.7	45.5	41.7	31.1	26.0	22.8	18.7	16.0	14.1	9.21	7.12	5.94	4.58
Bi	47.8	56.8	56.1	51.9	47.5	34.2	28.2	24.6	20.2	17.4	15.3	10.0	7.65	6.34	4.87

ELECTRON WORK FUNCTION OF THE ELEMENTS

The electron work function Φ is a measure of the minimum energy required to extract an electron from the surface of a solid. It is defined more precisely as the energy difference between the state in which an electron has been removed to a distance from the surface of a single crystal face that is large enough that the image force is negligible but small compared to the distance to any other face (typically about 10^{-4} cm) and the state in which the electron is in the bulk solid. In general, Φ differs for each face of a monocrystalline sample.

Since Φ is dependent on the cleanliness of the surface, measurements reported in the literature often cover a considerable range. This table contains selected values for the electron work function of the elements which may be regarded as typical values for a reasonably clean surface. The method of measurement is indicated for each value. The following abbreviations appear:

TE – Thermionic emission
PE – Photoelectric effect
FE – Field emission
CPD – Contact potential difference
polycr – Polycrystalline sample
amorp – Amorphous sample

Values in parentheses are only approximate.

References

1. Hölzl, J., and Schulte, F. K., Work Functions of Metals, in *Solid Surface Physics*, Höhler, G., Ed., Springer-Verlag, Berlin, 1979.
2. Riviere, J. C., Work Function: Measurements and Results, in *Solid State Surface Science, Vol.1*, Green, M., Ed., Decker, New York, 1969.
3. Michaelson, H. B., *J. Appl. Phys.*, 48, 4729, 1977.

Element	Plane	Φ/eV	Method
Ag	100	4.64	PE
	110	4.52	PE
	111	4.74	PE
Al	100	4.20	PE
	110	4.06	PE
	111	4.26	PE
As	polycr	(3.75)	PE
Au	100	5.47	PE
	110	5.37	PE
	111	5.31	PE
B	polycr	(4.45)	TH
Ba	polycr	2.52	TH
Be	polycr	4.98	PE
Bi	polycr	4.34	PE
C	polycr	(5.0)	CPD
Ca	polycr	2.87	PE
Cd	polycr	4.08	CPD
Ce	polycr	2.9	PE
Co	polycr	5.0	PE
Cr	polycr	4.5	PE
Cs	polycr	1.95	PE
Cu	100	5.10	FE
	110	4.48	PE
	111	4.94	PE
	112	4.53	PE
Eu	polycr	2.5	PE
Fe	100	4.67	PE
	111	4.81	PE
Ga	polycr	4.32	PE
Gd	polycr	2.90	CPD
Ge	polycr	5.0	CPD
Hf	polycr	3.9	PE
Hg	liquid	4.475	PE
In	polycr	4.09	PE
Ir	100	5.67	PE
	110	5.42	PE
	111	5.76	PE

Element	Plane	Φ/eV	Method
	210	5.00	PE
K	polycr	2.29	PE
La	polycr	3.5	PE
Li	polycr	2.93	FE
Lu	polycr	(3.3)	CPD
Mg	polycr	3.66	PE
Mn	polycr	4.1	PE
Mo	100	4.53	PE
	110	4.95	PE
	111	4.55	PE
	112	4.36	PE
	114	4.50	PE
	332	4.55	PE
Na	polycr	2.36	PE
Nb	001	4.02	TH
	110	4.87	TH
	111	4.36	TH
	112	4.63	TH
	113	4.29	TH
	116	3.95	TH
	310	4.18	TH
Nd	polycr	3.2	PE
Ni	100	5.22	PE
	110	5.04	PE
	111	5.35	PE
Os	polycr	5.93	PE
Pb	polycr	4.25	PE
Pd	polycr	5.22	PE
	111	5.6	PE
Pt	polycr	5.64	PE
	110	5.84	FE
	111	5.93	FE
	320	5.22	FE
	331	5.12	FE
Rb	polycr	2.261	PE
Re	polycr	4.72	TE
Rh	polycr	4.98	PE

Element	Plane	Φ/eV	Method
Ru	polycr	4.71	PE
Sb	amorp	4.55	
	100	4.7	
Sc	polycr	3.5	PE
Se	polycr	5.9	PE
Si	n	4.85	CPD
	p 100	(4.91)	CPD
	p 111	4.60	PE
Sm	polycr	2.7	PE
Sn	polycr	4.42	CPD
Sr	polycr	(2.59)	TH
Ta	polycr	4.25	TH
	100	4.15	TH
	110	4.80	TH
	111	4.00	TH
Tb	polycr	3.0	PE
Te	polycr	4.95	PE
Th	polycr	3.4	TH
Ti	polycr	4.33	PE
Tl	polycr	(3.84)	CPD
U	polycr	3.63	PE
	100	3.73	PE
	110	3.90	PE
	113	3.67	PE
V	polycr	4.3	PE
W	polycr	4.55	CPD
	100	4.63	FE
	110	5.22	FE
	111	4.45	FE
	113	4.46	FE
	116	4.32	TH
Y	polycr	3.1	PE
Zn	polycr	3.63	PE
	polycr	(4.9)	CPD
Zr	polycr	4.05	PE

SECONDARY ELECTRON EMISSION

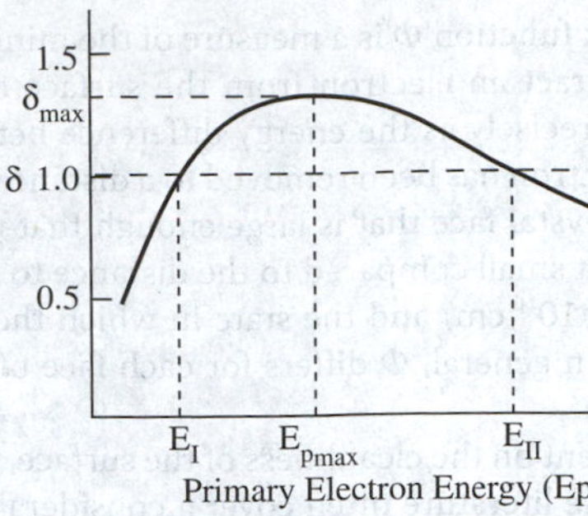

The secondary emission yield, or secondary emission ratio, δ, is the average number of secondary electrons emitted from a bombarded material for every incident primary electron. It is a function of the primary electron energy E_p. The maximum yield δ_{max} corresponds to a primary electron energy E_{pmax} (see figure). The two primary electron energies corresponding to a yield of unity are denoted by the first and second crossovers (E_I and E_{II}). An insulating target, or a conducting target that is electrically floating, will charge positively or negatively depending on the primary electron energy. For $E_I < E_p < E_{II}$, $\delta > 1$ and the surface charges positively provided there is a collector present that is positive with respect to the target. For $E_p < E_I$ or $E_p > E_{II}$, $\delta < 1$, and the surface charges negatively with respect to the potential of the source of primary electrons.

Secondary Electron Emission of the Elements

Element	δ_{max}	E_{pmax}/eV	E_I/eV	E_{II}/eV
Ag	1.5	800	200	>2000
Al	1.0	300	300	300
Au	1.4	800	150	>2000
B	1.2	150	50	600
Ba	0.8	400	None	None
Bi	1.2	550	None	None
Be	0.5	200	None	None
C (diamond)	2.8	750	None	>5000
C (graphite)	1.0	300	300	300
C (carbon black)	0.45	500	None	None
Cd	1.1	450	300	700
Co	1.2	600	200	None
Cs	0.7	400	None	None
Cu	1.3	600	200	1500
Fe	1.3	400	120	1400
Ga	1.55	500	75	None
Ge	1.15	500	150	900
Hg	1.3	600	350	>1200
K	0.7	200	None	None
Li	0.5	85	None	None

Element	δ_{max}	E_{pmax}/eV	E_I/eV	E_{II}/eV
Mg	0.95	300	None	None
Mo	1.25	375	150	1200
Na	0.82	300	None	None
Nb	1.2	375	150	1050
Ni	1.3	550	150	>1500
Pb	1.1	500	250	1000
Pd	1.3[a]	250[b]	120	None
Pt	1.8	700	350	3000
Rb	0.9	350	None	None
Sb	1.3	600	250	2000
Si	1.1	250	125	500
Sn	1.35	500	None	None
Ta	1.3	600	250	>2000
Th	1.1	800	None	None
Ti	0.9	280	None	None
Tl	1.7	650	70	>1500
W	1.4	650	250	>1500
Zr	1.1	350	None	None

[a] ≤1.3
[b] ≥250

Secondary Electron Emission of Inorganic Compounds

Material	δ_{max}	E_{pmax}/eV
Alkali halides		
CsCl	6.5	
KBr (crystal)	14	1800
KCl (crystal)	12	1600
KCl (layer)	7.5	1200
KI (crystal)	10	1600
KI (layer)	5.6	
LiF (crystal)	8.5	
LiF (layer)	5.6	700
NaBr (crystal)	24	1800
NaBr (layer)	6.3	
NaCl (crystal)	14	1200
NaCl (layer)	6.8	600
NaF (crystal)	14	1200
NaF (layer)	5.7	

Material	δ_{max}	E_{pmax}/eV
NaI (crystal)	19	1300
NaI (layer)	5.5	
RbCl (layer)	5.8	
Oxides		
Ag_2O	1.0	
Al_2O_3 (layer)	2–9	
BaO (layer)	2.3–4.8	400
BeO	3.4	2000
CaO	2.2	500
Cu_2O	1.2	400
MgO (crystal)	20–25	1500
MgO (layer)	3–15	400–1500
MoO_2	1.2	
SiO_2 (quartz)	2.1–4	400
SnO_2	3.2	640

Material	δ_{max}	E_{pmax}/eV
Sulfides		
MoS_2	1.1	
PbS	1.2	500
WS_2	1.0	
ZnS	1.8	350
Others		
BaF_2 (layer)	4.5	
CaF_2 (layer)	3.2	
$BiCs_3$	6	1000
BiCs	1.9	1000
GeCs	7	700
Rb_3Sb	7.1	450
$SbCs_3$	6	700
Mica	2.4	350
Glasses	2–3	300–450

OPTICAL PROPERTIES OF SELECTED ELEMENTS

J. H. Weaver and H. P. R. Frederikse

These tables list the index of refraction n, the extinction coefficient k, and the normal incidence reflection R ($\phi = 0$) as a function of photon energy E, which is expressed in electron volts (eV). To convert the energy in eV to the wavelength in µm, use $\lambda = 1.2398/E$. To compute the dielectric function $\tilde{\varepsilon} = \varepsilon_1 + i\varepsilon_2$ from the complex index of refraction $\tilde{N} = n + ik$, use $\varepsilon_1 = n^2 - k^2$ and $\varepsilon_2 = 2nk$.

The optical constants in these tables are abridged from three more extensive tabulations:

- *Optical Properties of Metals* (OPM), Volumes I and II, Physics Data, Nr. 18-1 and 18-2, J. H. Weaver, C. Krafka, D. W. Lynch, and E. E. Koch, Fachinformationzentrum, Karlsruhe, Germany.
- *Handbook of Optical Constants* (HOC), Vol. I, 1985, and Vol. II, 1991. E. D. Palik, Ed., Academic Press, Inc., London.
- *American Institute of Physics Handbook* (AIPH), 3rd Edition, D. E. Gray, Ed., McGraw-Hill, New York, 1972.

The first two of these major sources provide detailed comparisons of all optical data available in the literature at the time of the compilation. For critical applications the reader should refer to the original work. References for individual metals and semiconductors are listed at the end of the tables. Generally, tabulated values for the optical properties are accurate to better than 10%. Data in parentheses are extrapolated or interpolated values. For most elements the spectral range covered is from the far infrared (0.010 or 0.10 eV) to the far ultraviolet (10, 30 or 300 eV). The intervals between successive energies in the tables are chosen in such a way that the major spectral features are preserved.

Very small values of k are expressed in exponential notation, e.g., 1.23E-5 means 1.23×10^{-5}.

The following table is convenient for associating the energy entries in these tables with the corresponding wavelengths:

λ	E/eV	λ	E/eV
1 mm	0.00124	6000 Å	2.066
500 µm	0.00248	5000 Å	2.480
100 µm	0.01240	4000 Å	3.100
50 µm	0.02480	3000 Å	4.133
10 µm	0.12398	2000 Å	6.199
5 µm	0.24797	1000 Å	12.398
1 µm	1.240	400 Å	30.996

Solids

Energy (eV)	n	k	$R(\phi = 0)$
Aluminum[1]			
0.040	98.595	203.701	0.9923
0.050	74.997	172.199	0.9915
0.060	62.852	150.799	0.9906
0.070	53.790	135.500	0.9899
0.080	45.784	123.734	0.9895
0.090	39.651	114.102	0.9892
0.100	34.464	105.600	0.9889
0.125	24.965	89.250	0.9884
0.150	18.572	76.960	0.9882
0.175	14.274	66.930	0.9879
0.200	11.733	59.370	0.9873
0.250	8.586	48.235	0.9858
0.300	6.759	40.960	0.9844
0.350	5.438	35.599	0.9834
0.400	4.454	31.485	0.9826
0.500	3.072	25.581	0.9817
0.600	2.273	21.403	0.9806
0.700	1.770	18.328	0.9794
0.800	1.444	15.955	0.9778
0.900	1.264	14.021	0.9749
1.000	1.212	12.464	0.9697
1.100	1.201	11.181	0.9630
1.200	1.260	10.010	0.9521
1.300	1.468	8.949	0.9318
1.400	2.237	8.212	0.8852
1.500	2.745	8.309	0.8678
1.600	2.625	8.597	0.8794
1.700	2.143	8.573	0.8972
1.800	1.741	8.205	0.9069
1.900	1.488	7.821	0.9116
2.000	1.304	7.479	0.9148
2.200	1.018	6.846	0.9200
2.400	0.826	6.283	0.9228
2.600	0.695	5.800	0.9238
2.800	0.598	5.385	0.9242
3.000	0.523	5.024	0.9241
3.200	0.460	4.708	0.9243
3.400	0.407	4.426	0.9245
3.600	0.363	4.174	0.9246
3.800	0.326	3.946	0.9247
4.000	0.294	3.740	0.9248
4.200	0.267	3.552	0.9248
4.400	0.244	3.380	0.9249
4.600	0.223	3.222	0.9249
4.800	0.205	3.076	0.9249
5.000	0.190	2.942	0.9244
6.000	0.130	2.391	0.9257
6.500	0.110	2.173	0.9260
7.000	0.095	1.983	0.9262
7.500	0.082	1.814	0.9265
8.000	0.072	1.663	0.9269
8.500	0.063	1.527	0.9272
9.000	0.056	1.402	0.9277
9.500	0.049	1.286	0.9282
10.000	0.044	1.178	0.9286
10.500	0.040	1.076	0.9293
11.000	0.036	0.979	0.9298
11.500	0.033	0.883	0.9283
12.000	0.033	0.791	0.9224
12.500	0.034	0.700	0.9118
13.000	0.038	0.609	0.8960
13.500	0.041	0.517	0.8789
14.000	0.048	0.417	0.8486
14.200	0.053	0.373	0.8312
14.400	0.058	0.327	0.8102
14.600	0.067	0.273	0.7802
14.800	0.086	0.211	0.7202
15.000	0.125	0.153	0.6119
15.200	0.178	0.108	0.4903
15.400	0.234	0.184	0.3881
15.600	0.280	0.073	0.3182
15.800	0.318	0.065	0.2694
16.000	0.351	0.060	0.2326
16.200	0.380	0.055	0.2031
16.400	0.407	0.050	0.1789
16.750	0.448	0.045	0.1460
17.000	0.474	0.042	0.1278
17.250	0.498	0.040	0.1129
17.500	0.520	0.038	0.1005
17.750	0.540	0.036	0.0899
18.000	0.558	0.035	0.0809
18.500	0.591	0.032	0.0664
19.000	0.620	0.030	0.0554
19.500	0.646	0.028	0.0467
20.000	0.668	0.027	0.0398
20.500	0.689	0.025	0.0342
21.000	0.707	0.024	0.0296
21.500	0.724	0.023	0.0258
22.000	0.739	0.022	0.0226
22.500	0.753	0.021	0.0199
23.000	0.766	0.021	0.0177
23.500	0.778	0.020	0.0157
24.000	0.789	0.019	0.0140
24.500	0.799	0.018	0.0126
25.000	0.809	0.018	0.0113
25.500	0.817	0.017	0.0102
26.000	0.826	0.016	0.0092

Energy (eV)	n	k	R(φ = 0)
27.000	0.840	0.015	0.0076
28.000	0.854	0.014	0.0063
29.000	0.865	0.014	0.0053
30.000	0.876	0.013	0.0044
35.000	0.915	0.010	0.0020
40.000	0.940	0.008	0.0010
45.000	0.957	0.007	0.0005
50.000	0.969	0.006	0.0003
55.000	0.979	0.005	0.0001
60.000	0.987	0.004	0.0000
65.000	0.995	0.004	0.0000
70.000	1.006	0.004	0.0000
72.500	1.025	0.004	0.0002
75.000	1.011	0.024	0.0002
77.500	1.008	0.025	0.0002
80.000	1.007	0.024	0.0002
85.000	1.007	0.028	0.0002
90.000	1.005	0.031	0.0002
95.000	0.999	0.036	0.0003
100.000	0.991	0.030	0.0002
110.000	0.994	0.025	0.0002
120.000	0.991	0.024	0.0002
130.000	0.987	0.021	0.0001
140.000	0.989	0.016	0.0001
150.000	0.990	0.015	0.0001
160.000	0.989	0.014	0.0001
170.000	0.989	0.011	0.0001
180.000	0.990	0.010	0.0000
190.000	0.990	0.009	0.0000
200.000	0.991	0.007	0.0000
220.000	0.992	0.006	0.0000
240.000	0.993	0.005	0.0000
260.000	0.993	0.004	0.0000
280.000	0.994	0.003	0.0000
300.000	0.995	0.002	0.0000

Carbon (diamond)[2]

Energy (eV)	n	k	R(φ = 0)
0.06199	2.3741		0.166
0.06888	2.3741		0.166
0.07749	2.3745		0.166
0.08856	2.3750		0.166
0.1033	2.3757		0.166
0.1240	2.3765		0.166
0.1550	2.3772		0.166
0.1907		3.1 E-05	
0.2066	2.3779	5.7 E-05	0.166
0.22		1.21E-04	
0.23		2.36E-04	
0.24		3.82E-04	
0.25		5.21E-04	
0.26		2.96E-04	
0.27		4.39E-04	
0.28		2.75E-04	
0.29		7.82E-05	
0.30		1.32E-04	
0.31	2.3787	1.30E-04	0.167
0.32		1.11E-04	
0.33		2.99E-05	
0.34		1.89E-05	
0.35		2.11E-05	
0.36		2.47E-05	
0.37		2.80E-05	
0.38		3.11E-05	
0.39		3.67E-05	
0.40		3.58E-05	
0.41		3.25E-05	
0.4133	2.3795		0.167
0.42		2.94E-05	
0.43		2.87E-05	
0.44		3.14E-05	
0.45		3.62E-05	
0.46		3.22E-05	
0.47		1.57E-05	
0.48		6.17E-06	
0.4959	2.3801		0.167
0.6199	2.3813		0.167
0.8266	2.3837		0.167
1.240	2.3905		0.168
1.378	2.3934		0.169
1.459	2.3953		0.169
1.550	2.3975		0.169
1.653	2.4003		0.170
1.771	2.4036		0.170
1.889	2.4073		0.171
1.926	2.4084		0.171
2.066	2.4133		0.171
2.105	2.4147		0.172
2.271	2.4210		0.173
2.480	2.4299		0.174
2.650	2.4380		0.175
2.845		3.82E-07	
3.100	2.4627		0.178
3.434	2.4849		0.182
3.576	2.4955		0.183
3.961		8.97E-07	
4.160	2.5465		0.190
4.511		1.29E-06	
4.8187	2.6205	1.47E-06	0.200
5.00	2.6383		0.203
5.30		2.98E-06	
5.35		6.45E-06	
5.40		1.04E-05	
5.50		3.41E-05	
5.55		5.48E-04	
5.60	2.740	1.48E-03	0.216
5.80	2.780	5.02E-03	0.222
6.00	2.826	7.99E-03	0.228
6.10	2.852	8.62E-03	0.231
6.20	2.879	9.30E-03	0.235
6.30	2.910	9.74E-03	0.239
6.40	2.944	9.87E-03	0.243
6.50	2.985	1.10E-02	0.248
6.60	3.031	1.47E-02	0.254
6.70	3.085	2.20E-02	0.261
6.80	3.146	3.44E-02	0.268
6.90	3.220	5.24E-02	0.277
7.00	3.322	9.35E-02	0.289
7.10	3.444	0.210	0.304
7.15	3.464	0.307	0.308
7.20	3.437	0.388	0.307
7.30	3.376	0.473	0.303
7.40	3.335	0.515	0.300
7.50	3.321	0.533	0.299
7.60	3.306	0.592	0.300
7.80	3.276	0.659	0.300
8.00	3.251	0.712	0.300
8.25	3.232	0.765	0.301
8.50	3.228	0.806	0.303
8.75	3.247	0.855	0.308
9.00	3.272	0.910	0.314
9.25	3.308	0.978	0.322
9.50	3.348	1.055	0.331
9.75	3.398	1.147	0.342
10.00	3.453	1.258	0.355
10.25	3.514	1.403	0.371
10.50	3.565	1.581	0.389
10.75	3.600	1.813	0.411
11.00	3.582	2.078	0.434
11.25	3.507	2.380	0.460
11.50	3.346	2.693	0.488
11.75	3.090	2.986	0.518
12.00	2.736	3.228	0.551
12.20	2.383	3.354	0.580
12.40	1.983	3.382	0.610
12.60	1.532	3.265	0.641
12.80	1.312	2.953	0.627
13.00	1.223	2.722	0.604
13.50	1.129	2.379	0.557
14.00	1.070	2.178	0.526
14.50	1.018	2.034	0.504
15.00	0.972	1.929	0.489
15.50	0.917	1.845	0.482
16.00	0.861	1.767	0.477
16.50	0.805	1.692	0.474
17.00	0.753	1.619	0.471
17.50	0.707	1.546	0.467
18.00	0.665	1.476	0.463
18.50	0.626	1.408	0.459
19.00	0.589	1.341	0.455
19.50	0.557	1.273	0.449
20.00	0.527	1.203	0.442
21.00	0.487	1.052	0.413
22.00	0.518	0.888	0.330
23.00	0.597	0.850	0.270
24.00	0.586	0.829	0.268
25.00	0.562	0.787	0.265
26.00	0.538	0.736	0.260
27.00	0.516	0.679	0.252
28.00	0.501	0.616	0.239
29.00	0.494	0.552	0.221
30.00	0.493	0.490	0.201

Cesium (evaporated)[3]

Energy (eV)	n	k	R(φ = 0)
2.145	0.264	1.123	0.631
2.271	0.278	0.950	0.561
2.845	0.425	0.438	0.235
3.064	0.540	0.320	0.127
3.397	0.671	0.233	0.057
3.966	0.827	0.174	0.018
4.889	0.916	0.143	0.007

Energy (eV)	n	k	$R(\varphi = 0)$
Chromium[4]			
0.06	21.19	42.00	0.962
0.10	11.81	29.76	0.955
0.14	15.31	26.36	0.936
0.18	8.73	25.37	0.53
0.22	5.30	20.62	0.954
0.26	3.91	17.12	0.951
0.30	3.15	14.28	0.943
0.42	3.47	8.97	0.862
0.54	3.92	7.06	0.788
0.66	3.96	5.95	0.736
0.78	4.13	5.03	0.680
0.90	4.43	4.60	0.650
1.00	4.47	4.43	0.639
1.12	4.53	4.31	0.631
1.24	4.50	4.28	0.629
1.36	4.42	4.30	0.631
1.46	4.31	4.32	0.632
1.77	3.84	4.37	0.639
2.00	3.48	4.36	0.644
2.20	3.18	4.41	0.656
2.40	2.75	4.46	0.677
2.60	2.22	4.36	0.698
2.80	1.80	4.06	0.703
3.00	1.54	3.71	0.695
3.20	1.44	3.40	0.670
3.40	1.39	3.24	0.657
3.60	1.26	3.12	0.661
3.80	1.12	2.95	0.660
4.00	1.02	2.76	0.651
4.20	0.94	2.58	0.639
4.40	0.90	2.42	0.620
4.50	0.89	2.35	0.607
4.60	0.88	2.28	0.598
4.70	0.86	2.21	0.586
4.80	0.86	2.13	0.572
4.90	0.86	2.07	0.557
5.00	0.85	2.01	0.542
5.10	0.86	1.94	0.523
5.20	0.87	1.87	0.503
5.40	0.93	1.80	0.466
5.60	0.95	1.74	0.443
5.80	0.97	1.74	0.437
6.00	0.94	1.73	0.444
6.20	0.89	1.69	0.446
6.40	0.85	1.66	0.447
6.60	0.80	1.59	0.444
6.80	0.75	1.51	0.439
7.00	0.74	1.45	0.425
7.20	0.71	1.39	0.414
7.40	0.69	1.33	0.404
7.60	0.66	1.23	0.378
7.80	0.67	1.15	0.347
8.00	0.68	1.07	0.315
8.20	0.71	1.00	0.278
8.50	0.74	0.92	0.235
9.0	0.83	0.81	0.170
9.50	0.92	0.74	0.132
10.00	0.98	0.73	0.120
10.50	1.01	0.72	0.112

Energy (eV)	n	k	$R(\varphi = 0)$
11.00	1.05	0.69	0.103
11.50	1.09	0.69	0.100
12.00	1.13	0.70	0.101
12.50	1.15	0.73	0.108
13.00	1.15	0.77	0.119
13.50	1.12	0.80	0.128
14.00	1.09	0.82	0.135
14.50	1.03	0.82	0.142
15.00	1.00	0.82	0.143
15.50	0.96	0.80	0.141
16.00	0.92	0.77	0.139
16.50	0.31	0.75	0.134
17.00	0.90	0.73	0.132
17.50	0.88	0.72	0.130
18.00	0.87	0.70	0.129
18.50	0.84	0.69	0.130
19.00	0.82	0.68	0.131
20.00	0.77	0.64	0.130
20.5	0.76	0.63	0.129
21.0	0.74	0.58	0.121
21.5	0.72	0.55	0.116
22.0	0.71	0.52	0.112
22.5	0.70	0.50	0.109
23.0	0.69	0.48	0.105
23.5	0.68	0.45	0.101
24.0	0.68	0.43	0.096
24.5	0.67	0.39	0.089
25.0	0.68	0.36	0.080
25.5	0.68	0.33	0.072
26.0	0.70	0.31	0.063
26.5	0.71	0.28	0.055
27.0	0.72	0.26	0.048
27.5	0.73	0.25	0.043
28.0	0.75	0.23	0.037
29.0	0.77	0.22	0.032
30.0	0.78	0.21	0.030
Cobalt, single crystal, $\vec{E} \parallel \vec{c}$ [5]			
0.10	6.71	37.87	0.982
0.15	4.66	25.47	0.973
0.20	3.55	18.78	0.962
0.25	3.98	14.59	0.933
0.30	4.04	12.16	0.907
0.40	4.24	9.13	0.847
0.50	4.41	7.19	0.782
0.60	4.91	6.13	0.729
0.70	5.24	5.85	0.713
0.80	5.17	5.89	0.716
0.90	4.94	5.95	0.720
1.00	4.46	5.86	0.722
1.10	4.07	5.61	0.715
1.20	3.81	5.36	0.706
1.30	3.60	5.20	0.701
1.40	3.37	5.09	0.701
1.50	3.10	4.96	0.701
1.60	2.84	4.77	0.697
1.70	2.66	4.57	0.690
1.80	2.45	4.41	0.687
1.90	2.31	4.18	0.675
2.00	2.21	4.00	0.664
2.10	2.13	3.85	0.654

Energy (eV)	n	k	$R(\varphi = 0)$
2.20	2.07	3.70	0.642
2.30	2.01	3.59	0.634
2.40	1.95	3.49	0.627
2.50	1.88	3.40	0.622
2.60	1.81	3.32	0.618
2.70	1.73	3.24	0.615
2.80	1.66	3.13	0.607
2.90	1.61	3.05	0.600
3.00	1.55	2.96	0.594
3.20	1.46	2.80	0.579
3.40	1.38	2.64	0.563
3.60	1.31	2.48	0.544
3.80	1.28	2.33	0.519
4.00	1.26	2.20	0.495
4.20	1.25	2.10	0.471
4.40	1.24	2.01	0.452
4.60	1.24	1.94	0.435
4.80	1.23	1.88	0.423
5.00	1.22	1.83	0.411
5.20	1.21	1.79	0.403
5.40	1.19	1.77	0.399
5.60	1.16	1.75	0.400
5.80	1.10	1.73	0.406
6.00	1.03	1.68	0.407
6.20	0.97	1.62	0.401
6.40	0.94	1.53	0.386
6.60	0.91	1.46	0.368
6.80	0.91	1.38	0.345
7.00	0.91	1.32	0.326
7.00	0.91	1.26	0.305
7.40	0.92	1.21	0.286
7.60	0.93	1.17	0.269
7.80	0.94	1.13	0.253
8.00	0.95	1.09	0.239
Cobalt, single crystal, $\vec{E} \perp \vec{c}$ [5]			
0.10	5.83	32.36	0.979
0.15	4.24	21.37	0.965
0.20	3.87	15.53	0.042
0.30	4.34	10.01	0.865
0.40	4.66	7.39	0.785
0.50	5.17	5.75	0.709
0.60	5.77	5.17	0.682
0.70	6.15	5.20	0.685
0.80	6.08	5.61	0.702
0.90	5.57	5.93	0.715
1.00	4.83	5.94	0.721
1.10	4.31	5.60	0.711
1.20	4.02	5.34	0.701
1.30	3.78	5.16	0.694
1.40	3.55	5.05	0.692
1.50	3.26	4.93	0.692
1.60	3.03	4.74	0.687
1.70	2.83	4.60	0.684
1.80	2.61	4.45	0.683
1.90	2.41	4.27	0.677
2.00	2.25	4.09	0.670
2.10	2.13	3.89	0.659
2.20	2.04	3.72	0.646
2.30	1.99	3.56	0.632
2.40	1.95	3.44	0.620

Energy (eV)	n	k	$R(\phi = 0)$
2.50	1.90	3.34	0.611
2.60	1.86	3.26	0.605
2.70	1.79	3.19	0.602
2.80	1.72	3.11	0.596
2.90	1.66	3.03	0.591
3.00	1.60	2.94	0.586
3.20	1.50	2.78	0.571
3.40	1.42	2.62	0.553
3.60	1.36	2.47	0.533
3.80	1.33	2.33	0.511
4.00	1.31	2.21	0.488
4.20	1.28	2.12	0.471
4.40	1.27	2.03	0.452
4.60	1.26	1.95	0.435
4.80	1.25	1.90	0.423
5.00	1.24	1.84	0.411
5.20	1.22	1.80	0.403
5.40	1.21	1.78	0.399
5.60	1.17	1.76	0.400
5.80	1.11	1.74	0.406
6.00	1.04	1.69	0.407
6.20	0.98	1.62	0.401
6.40	0.94	1.54	0.386
6.60	0.92	1.46	0.368
6.80	0.91	1.38	0.345
7.00	0.91	1.32	0.326
7.20	0.91	1.26	0.305
7.40	0.92	1.21	0.285
7.60	0.93	1.17	0.269
7.80	0.94	1.13	0.253

Copper[6]

Energy (eV)	n	k	$R(\phi = 0)$
0.10	29.69	71.57	0.980
0.50	1.71	17.63	0.979
1.00	0.44	8.48	0.976
1.50	0.26	5.26	0.965
1.70	0.22	4.43	0.958
1.75	0.21	4.25	0.956
1.80	0.21	4.04	0.952
1.85	0.22	3.85	0.947
1.90	0.21	3.67	0.943
2.00	0.27	3.24	0.910
2.10	0.47	2.81	0.814
2.20	0.83	2.60	0.673
2.30	1.04	2.59	0.618
2.40	1.12	2.60	0.602
2.60	1.15	2.50	0.577
2.80	1.17	2.36	0.545
3.00	1.18	2.21	0.509
3.20	1.23	2.07	0.468
3.40	1.27	1.95	0.434
3.60	1.31	1.87	0.407
3.80	1.34	1.81	0.387
4.00	1.34	1.72	0.364
4.20	1.42	1.64	0.336
4.40	1.49	1.64	0.329
4.60	1.52	1.67	0.334
4.80	1.53	1.71	0.345
5.00	1.47	1.78	0.366
5.20	1.38	1.80	0.380
5.40	1.28	1.78	0.389

Energy (eV)	n	k	$R(\phi = 0)$
5.60	1.18	1.74	0.391
5.80	1.10	1.67	0.389
6.00	1.04	1.59	0.380
6.50	0.96	1.37	0.329
7.00	0.97	1.20	0.271
7.50	1.00	1.09	0.230
8.00	1.03	1.03	0.206
8.50	1.03	0.98	0.189
9.00	1.03	0.92	0.171
9.50	1.03	0.87	0.154
10.00	1.04	0.82	0.139
11.00	1.07	0.75	0.118
12.00	1.09	0.73	0.111
13.00	1.08	0.72	0.109
14.00	1.06	0.72	0.111
14.50	1.03	0.72	0.111
15.00	1.01	0.71	0.111
15.50	0.98	0.69	0.109
16.00	0.95	0.67	0.106
17.00	0.91	0.62	0.097
18.00	0.89	0.56	0.084
19.00	0.88	0.51	0.071
20.00	0.88	0.45	0.059
21.00	0.90	0.41	0.048
22.00	0.92	0.38	0.040
23.00	0.94	0.37	0.035
24.00	0.96	0.37	0.035
25.00	0.96	0.40	0.040
26.00	0.92	0.40	0.044
27.00	0.88	0.38	0.043
28.00	0.86	0.35	0.039
29.00	0.85	0.30	0.032
30.00	0.86	0.26	0.025
31.00	0.88	0.24	0.020
32.00	0.89	0.22	0.017
33.00	0.90	0.21	0.015
34.00	0.91	0.20	0.014
35.00	0.92	0.20	0.013
36.00	0.92	0.19	0.012
37.00	0.92	0.19	0.011
38.00	0.93	0.18	0.010
39.00	0.93	0.17	0.009
40.00	0.93	0.17	0.009
41.00	0.94	0.16	0.008
42.00	0.94	0.16	0.007
43.00	0.94	0.15	0.007
44.00	0.95	0.15	0.007
45.00	0.95	0.15	0.006
46.00	0.95	0.15	0.006
47.00	0.95	0.14	0.006
48.00	0.95	0.14	0.006
49.00	0.95	0.14	0.005
50.00	0.95	0.13	0.005
51.00	0.95	0.13	0.005
52.00	0.95	0.13	0.005
53.00	0.96	0.12	0.004
54.00	0.96	0.12	0.004
55.00	0.96	0.12	0.004
56.00	0.96	0.11	0.004
57.00	0.96	0.11	0.004

Energy (eV)	n	k	$R(\phi = 0)$
58.00	0.96	0.11	0.004
59.00	0.97	0.11	0.003
60.00	0.97	0.11	0.003
61.00	0.97	0.11	0.003
62.00	0.97	0.11	0.003
63.00	0.96	0.10	0.003
64.00	0.96	0.10	0.003
65.00	0.97	0.10	0.003
66.00	0.97	0.10	0.003
67.00	0.97	0.09	0.003
68.00	0.97	0.09	0.002
69.00	0.97	0.09	0.002
70.00	0.97	0.09	0.002
75.00	0.98	0.09	0.002
80.00	0.98	0.09	0.002
85.00	0.97	0.09	0.002
90.00	0.96	0.08	0.002

Gallium (liquid)[7]

Energy (eV)	n	k	$R(\phi = 0)$
1.425	2.40	9.20	0.900
1.550	2.09	8.50	0.898
1.771	1.65	7.60	0.898
2.066	1.25	6.60	0.897
2.480	0.89	5.60	0.898
3.100	0.59	4.50	0.896

Germanium, single crystal[8]

Energy (eV)	n	k	$R(\phi = 0)$
0.01240	(4.0065)	3.00E-03	0.361
0.01364	4.0063	2.40E-03	0.361
0.01488	(4.0060)	1.70E-03	0.361
0.01612	(4.0060)	1.55E-03	0.361
0.01736	(4.0060)	1.50E-03	0.361
0.01860		1.50E-03	
0.01984		1.60E-03	
0.02108		1.60E-03	
0.02232		1.55E-03	
0.02356		1.53E-03	
0.02480		1.50E-03	
0.02604		1.25E-03	
0.02728		8.50E-04	
0.02852		6.50E-04	
0.02976		7.00E-04	
0.03100	3.9827	8.50E-04	0.358
0.03224		1.55E-03	
0.03348		2.75E-03	
0.03472		3.55E-03	
0.03596	(3.9900)	3.05E-03	0.359
0.03720		2.75E-03	
0.03844		2.70E-03	
0.03968	(3.9930)	2.90E-03	0.359
0.04092		2.95E-03	
0.04215		3.20E-03	
0.04339		6.30E-03	
0.04463		3.40E-03	
0.04587	(3.9955)	2.50E-03	0.360
0.04711		2.10E-03	
0.04835		2.00E-03	
0.04959		8.00E-04	
0.05083		1.40E-03	
0.05207		1.35E-03	
0.05331		1.10E-03	
0.05455		8.00E-04	

Solids

Energy (eV)	n	k	$R(\phi = 0)$
0.05579		6.00E-04	
0.05703		9.0 E-04	
0.05827		6.5 E-04	
0.05951		4.6 E-04	
0.06075		4.0 E-04	
0.06199	3.9992	3.98E-04	0.360
0.06323		4.0 E-04	
0.06447		4.3 E-04	
0.06571		4.4 E-04	
0.06695	(4.0000)	4.3 E-04	0.360
0.06819		3.1 E-04	
0.06943		3.3 E-04	
0.07067		3.8 E-04	
0.07191		3.3 E-04	
0.07315		2.5 E-04	
0.07439		1.9 E-04	
0.07514		1.58E-04	
0.07749	4.0009	9.55E-05	0.360
0.07999	4.0011	1.71E-04	0.360
0.08266	4.0013	9.78E-05	0.360
0.08551	4.0015	5.77E-05	0.360
0.08920		3.98E-05	
0.09460		4.59E-05	
0.09840		3.51E-05	
0.1	4.0063	3.70E-05	0.361
0.2	4.0108		0.361
0.3	4.0246		0.362
0.4	4.0429		0.364
0.5	(4.074)		0.367
0.6	(4.104)	6.58E-07	0.370
0.7	4.180	1.27E-04	0.377
0.8	4.275	5.67E-03	0.385
0.9	4.285	7.45E-02	0.386
1.0	4.325	8.09E-02	0.390
1.1	4.385	0.103	0.395
1.2	4.420	0.123	0.398
1.3	4.495	0.167	0.405
1.4	4.560	0.190	0.411
1.5	4.635	0.298	0.418
1.6	4.763	0.345	0.428
1.7	4.897	0.401	0.439
1.8	5.067	0.500	0.453
1.9	5.380	0.540	0.475
2.0	5.588	0.933	0.495
2.1	5.748	1.634	0.523
2.2	5.283	2.049	0.516
2.3	5.062	2.318	0.519
2.4	4.610	2.455	0.508
2.5	4.340	2.384	0.492
2.6	4.180	2.309	0.480
2.7	4.082	2.240	0.471
2.8	4.035	2.181	0.464
2.9	4.037	2.140	0.461
3.0	4.082	2.145	0.463
3.1	4.141	2.215	0.471
3.2	4.157	2.340	0.482
3.3	4.128	2.469	0.490
3.4	4.070	2.579	0.497
3.5	4.020	2.667	0.502
3.6	3.985	2.759	0.509
3.7	3.958	2.863	0.517
3.8	3.936	2.986	0.527
3.9	3.920	3.137	0.539
4.0	3.905	3.336	0.556
4.1	3.869	3.614	0.579
4.2	3.745	4.009	0.612
4.3	3.338	4.507	0.659
4.4	2.516	4.669	0.705
4.5	1.953	4.297	0.713
4.6	1.720	3.960	0.702
4.7	1.586	3.709	0.690
4.8	1.498	3.509	0.677
4.9	1.435	3.342	0.664
5.0	1.394	3.197	0.650
5.1	1.370	3.073	0.636
5.2	1.364	2.973	0.622
5.3	1.371	2.897	0.609
5.4	1.383	2.854	0.600
5.5	1.380	2.842	0.598
5.6	1.360	2.846	0.602
5.7	1.293	2.163	0.479
5.8	1.209	2.873	0.632
5.9	1.108	2.813	0.641
6.0	1.30	2.34	0.517
6.5	1.10	2.05	0.489
7.0	1.00	1.80	0.448
7.5		1.60	
8.0	0.92	1.40	0.348
8.5	0.92	1.20	0.282
9.0	0.92	1.14	0.262
9.5		1.00	
10.0	0.93	0.86	0.167
20.0		0.237	
22.0		0.179	
24.0		0.144	
26.0		0.110	
28.0		0.0747	
30.0		0.1020	
32.0		0.0999	
34.0		0.0856	
36.0		0.0740	
38.0		0.0651	
40.0		0.0604	

Gold, electropolished, Au (110)[9]

Energy (eV)	n	k	$R(\phi = 0)$
0.10	8.17	82.83	0.995
0.20	2.13	41.73	0.995
0.30	0.99	27.82	0.995
0.40	0.59	20.83	0.995
0.50	0.39	16.61	0.994
0.60	0.28	13.78	0.994
0.70	0.22	11.75	0.994
0.80	0.18	10.21	0.993
0.90	0.15	9.01	0.993
1.00	0.13	8.03	0.992
1.20	0.10	6.54	0.991
1.40	0.08	5.44	0.989
1.60	0.08	4.56	0.986
1.80	0.09	3.82	0.979
2.00	0.13	3.16	0.953
2.10	0.18	2.84	0.925
2.20	0.24	2.54	0.880
2.40	0.50	1.86	0.647
2.50	0.82	1.59	0.438
2.60	1.24	1.54	0.331
2.70	1.43	1.72	0.356
2.80	1.46	1.77	0.368
2.90	1.50	1.79	0.368
3.00	1.54	1.80	0.369
3.10	1.54	1.81	0.371
3.20	1.54	1.80	0.368
3.30	1.55	1.78	0.362
3.40	1.56	1.76	0.356
3.50	1.58	1.73	0.349
3.60	1.62	1.73	0.346
3.70	1.64	1.75	0.351
3.80	1.63	1.79	0.360
3.90	1.59	1.81	0.366
4.00	1.55	1.81	0.369
4.10	1.51	1.79	0.368
4.20	1.48	1.78	0.367
4.30	1.45	1.77	0.368
4.40	1.41	1.76	0.370
4.50	1.35	1.74	0.370
4.60	1.30	1.69	0.364
4.70	1.27	1.64	0.354
4.80	1.25	1.59	0.344
4.90	1.23	1.54	0.332
5.00	1.22	1.49	0.319
5.20	1.21	1.40	0.295
5.40	1.21	1.33	0.275
5.60	1.21	1.27	0.256
5.80	1.21	1.20	0.236
6.00	1.22	1.14	0.218
6.20	1.24	1.09	0.203
6.40	1.25	1.05	0.190
6.60	1.27	1.01	0.177
6.80	1.30	0.97	0.167
7.00	1.34	0.95	0.162
7.20	1.36	0.95	0.161
7.40	1.38	0.96	0.164
7.60	1.38	0.98	0.169
7.80	1.35	0.99	0.171
8.00	1.31	0.96	0.165
8.20	1.30	0.92	0.155
8.40	1.30	0.89	0.147
8.60	1.31	0.88	0.144
8.80	1.31	0.86	0.140
9.00	1.30	0.83	0.133
9.20	1.31	0.81	0.126
9.40	1.33	0.78	0.122
9.60	1.36	0.78	0.121
9.80	1.37	0.79	0.124
10.00	1.37	0.80	0.126
10.20	1.36	0.80	0.127
10.40	1.35	0.80	0.125
10.60	1.34	0.79	0.123
10.80	1.34	0.77	0.120
11.00	1.34	0.76	0.116
11.20	1.34	0.74	0.113
11.40	1.35	0.73	0.111

Energy (eV)	n	k	$R(\phi = 0)$
11.60	1.36	0.72	0.109
11.80	1.38	0.71	0.108
12.00	1.39	0.71	0.109
12.40	1.44	0.73	0.115
12.80	1.45	0.79	0.127
13.20	1.42	0.84	0.137
13.60	1.37	0.86	0.140
14.00	1.33	0.86	0.140
14.40	1.29	0.86	0.139
14.80	1.26	0.84	0.135
15.20	1.24	0.83	0.132
15.60	1.22	0.81	0.127
16.00	1.21	0.79	0.123
16.40	1.20	0.78	0.119
16.80	1.19	0.76	0.116
17.20	1.19	0.75	0.114
17.60	1.19	0.74	0.111
18.00	1.19	0.74	0.109
18.40	1.19	0.73	0.109
18.80	1.20	0.74	0.110
19.20	1.21	0.76	0.116
19.60	1.21	0.80	0.125
20.00	1.18	0.83	0.133
20.40	1.14	0.85	0.141
20.80	1.10	0.87	0.149
21.20	1.05	0.88	0.156
21.60	1.00	0.88	0.162
22.00	0.94	0.86	0.164
22.40	0.89	0.83	0.163
22.80	0.85	0.79	0.157
23.20	0.82	0.75	0.149
23.60	0.80	0.70	0.138
24.00	0.80	0.66	0.125
24.40	0.80	0.62	0.113
24.80	0.80	0.58	0.101
25.20	0.82	0.56	0.090
25.60	0.83	0.54	0.084
26.00	0.84	0.52	0.079
26.40	0.85	0.51	0.074
26.80	0.85	0.50	0.071
27.20	0.86	0.49	0.068
27.60	0.86	0.49	0.065
28.00	0.87	0.48	0.063
28.40	0.88	0.48	0.062
28.80	0.88	0.48	0.062
29.20	0.88	0.48	0.062
29.60	0.87	0.48	0.064
30.00	0.86	0.48	0.064

Hafnium, single crystal, $\vec{E} \parallel \vec{c}$[10]

Energy (eV)	n	k	$R(\phi = 0)$
0.52	1.48	4.11	0.747
0.56	1.84	3.29	0.615
0.60	2.34	2.62	0.486
0.66	3.21	2.13	0.428
0.70	3.70	2.03	0.441
0.76	4.31	2.10	0.476
0.80	4.61	2.31	0.504
0.86	4.71	2.70	0.533
0.90	4.64	2.85	0.541
0.95	4.54	2.96	0.545
1.00	4.45	3.00	0.545
1.10	4.28	3.08	0.547
1.20	4.08	3.10	0.544
1.30	3.87	3.04	0.536
1.40	3.72	2.95	0.525
1.50	3.60	2.85	0.514
1.60	3.52	2.73	0.500
1.70	3.52	2.61	0.488
1.80	3.57	2.56	0.485
1.90	3.63	2.59	0.489
2.00	3.65	2.67	0.498
2.10	3.64	2.81	0.511
2.20	3.53	2.99	0.526
2.30	3.34	3.09	0.534
2.40	3.15	3.11	0.537
2.50	2.99	3.13	0.540
2.60	2.83	3.12	0.542
2.70	2.68	3.10	0.542
2.80	2.54	3.08	0.543
2.90	2.40	3.04	0.544
3.00	2.27	3.00	0.544
3.10	2.14	2.95	0.544
3.20	2.00	2.89	0.544
3.30	1.87	2.79	0.538
3.40	1.78	2.68	0.528
3.50	1.71	2.58	0.517
3.60	1.66	2.48	0.503
3.70	1.63	2.40	0.491
3.80	1.60	2.33	0.481
3.90	1.56	2.27	0.473
4.00	1.52	2.21	0.466
4.10	1.48	2.14	0.455
4.20	1.45	2.07	0.442
4.30	1.43	2.01	0.431
4.40	1.41	1.95	0.420
4.50	1.39	1.89	0.407
4.60	1.39	1.83	0.394
4.70	1.39	1.79	0.382
4.80	1.38	1.75	0.373
4.90	1.38	1.71	0.364
5.00	1.37	1.68	0.356
5.20	1.36	1.61	0.341
5.40	1.35	1.55	0.324
5.60	1.35	1.51	0.314
5.80	1.32	1.48	0.308
6.00	1.28	1.41	0.295
6.20	1.26	1.35	0.278
6.40	1.26	1.28	0.258
6.60	1.27	1.22	0.240
6.80	1.28	1.16	0.224
7.00	1.31	1.13	0.212
7.20	1.33	1.10	0.204
7.40	1.34	1.07	0.197
7.60	1.36	1.05	0.191
7.80	1.37	1.02	0.183
8.00	1.40	1.01	0.179
8.20	1.43	1.01	0.178
8.40	1.45	1.01	0.180
8.60	1.47	1.02	0.183
8.80	1.48	1.04	0.186
9.00	1.49	1.07	0.193
9.20	1.50	1.10	0.201
9.40	1.48	1.14	0.211
9.60	1.46	1.18	0.222
9.80	1.41	1.21	0.230
10.00	1.36	1.22	0.235
10.20	1.32	1.22	0.238
10.40	1.28	1.22	0.240
10.60	1.24	1.21	0.241
10.80	1.20	1.20	0.242
11.00	1.16	1.19	0.242
11.20	1.13	1.17	0.241
11.40	1.10	1.16	0.241
11.60	1.07	1.14	0.239
11.80	1.04	1.12	0.238
12.00	1.02	1.10	0.236
12.40	0.96	1.06	0.232
12.80	0.92	1.01	0.225
13.20	0.88	0.96	0.218
13.60	0.84	0.90	0.205
14.00	0.83	0.83	0.186
14.40	0.83	0.80	0.172
14.80	0.81	0.76	0.167
15.20	0.79	0.70	0.153
15.60	0.79	0.64	0.132
16.00	0.83	0.60	0.111
16.40	0.81	0.60	0.114
16.80	0.79	0.55	0.105
17.20	0.79	0.50	0.089
17.60	0.80	0.46	0.077
18.00	0.81	0.42	0.064
18.40	0.84	0.38	0.051
18.80	0.87	0.34	0.040
19.00	0.89	0.33	0.036
19.60	0.93	0.32	0.030
20.00	0.94	0.31	0.027
20.60	0.97	0.30	0.023
21.00	0.99	0.29	0.022
21.60	1.01	0.28	0.020
22.00	1.03	0.28	0.020
22.60	1.06	0.28	0.020
23.00	1.07	0.28	0.021
23.60	1.09	0.29	0.022
24.00	1.09	0.30	0.023
24.60	1.10	0.31	0.024

Hafnium, single crystal, $\vec{E} \perp \vec{c}$[10]

Energy (eV)	n	k	$R(\phi = 0)$
0.52	2.25	4.65	0.723
0.56	2.34	3.66	0.623
0.60	2.84	2.89	0.512
0.66	3.71	2.35	0.469
0.70	4.26	2.21	0.482
0.76	4.97	2.33	0.521
0.80	5.41	2.62	0.554
0.86	5.46	3.36	0.593
0.90	5.22	3.62	0.601
0.95	4.95	3.72	0.602
1.00	4.76	3.76	0.602
1.10	4.43	3.80	0.601
1.20	4.07	3.74	0.594
1.30	3.79	3.55	0.578
1.40	3.61	3.36	0.561

Energy (eV)	n	k	R(φ = 0)
1.50	3.55	3.13	0.540
1.60	3.58	3.01	0.529
1.70	3.63	2.98	0.526
1.80	3.66	3.02	0.530
1.90	3.63	3.14	0.541
2.00	3.51	3.26	0.551
2.10	3.35	3.33	0.558
2.20	3.18	3.36	0.563
2.30	2.99	3.39	0.568
2.40	2.78	3.35	0.569
2.50	2.65	3.26	0.562
2.60	2.54	3.22	0.560
2.70	2.42	3.17	0.559
2.80	2.31	3.13	0.558
2.90	2.20	3.08	0.558
3.00	2.08	3.05	0.561
3.10	1.94	2.98	0.560
3.20	1.83	2.88	0.555
3.30	1.74	2.78	0.547
3.40	1.68	2.69	0.538
3.50	1.62	2.61	0.529
3.60	1.57	2.52	0.519
3.70	1.53	2.45	0.510
3.80	1.49	2.38	0.501
3.90	1.45	2.32	0.493
4.00	1.41	2.25	0.484
4.10	1.38	2.18	0.474
4.20	1.35	2.11	0.462
4.30	1.33	2.05	0.451
4.40	1.31	1.99	0.438
4.50	1.30	1.93	0.427
4.60	1.29	1.88	0.415
4.70	1.28	1.82	0.402
4.80	1.28	1.77	0.389
4.90	1.27	1.73	0.379
5.00	1.27	1.69	0.367
5.20	1.27	1.62	0.349
5.40	1.27	1.57	0.335
5.60	1.26	1.52	0.322
5.80	1.24	1.48	0.313
6.00	1.21	1.42	0.302
6.20	1.19	1.36	0.285
6.40	1.18	1.29	0.265
6.60	1.19	1.22	0.244
6.80	1.21	1.18	0.230
7.00	1.22	1.14	0.217
7.20	1.23	1.10	0.206
7.40	1.26	1.06	0.194
7.60	1.28	1.04	0.187
7.80	1.30	1.02	0.180
8.00	1.33	1.00	0.174
8.20	1.35	0.99	0.173
8.40	1.38	0.99	0.173
8.60	1.40	1.00	0.174
8.80	1.42	1.02	0.178
9.00	1.43	1.04	0.184
9.20	1.45	1.08	0.193
9.40	1.43	1.12	0.204
9.60	1.40	1.16	0.214
9.80	1.37	1.19	0.223

Energy (eV)	n	k	R(φ = 0)
10.00	1.32	1.21	0.230
10.20	1.27	1.21	0.234
10.40	1.23	1.20	0.235
10.60	1.19	1.20	0.237
10.80	1.15	1.19	0.237
11.00	1.12	1.17	0.237
11.20	1.08	1.16	0.237
11.40	1.05	1.14	0.236
11.60	1.03	1.12	0.235
11.80	1.00	1.10	0.233
12.00	0.97	1.08	0.231
12.40	0.92	1.04	0.226
12.80	0.88	0.99	0.219
13.20	0.83	0.94	0.211
13.60	0.80	0.88	0.196
14.00	0.79	0.81	0.177
14.40	0.80	0.77	0.160
14.80	0.77	0.73	0.154
15.20	0.76	0.68	0.140
15.60	0.76	0.61	0.119
16.00	0.81	0.58	0.099
16.40	0.78	0.57	0.102
16.80	0.77	0.53	0.092
17.20	0.77	0.48	0.077
17.60	0.79	0.44	0.065
18.00	0.80	0.39	0.053
18.40	0.82	0.36	0.041
18.80	0.86	0.33	0.032
19.00	0.88	0.32	0.030
19.60	0.91	0.31	0.025
20.00	0.93	0.30	0.023
20.60	0.96	0.29	0.021
21.00	0.97	0.29	0.020
21.60	1.00	0.28	0.019
22.00	1.01	0.28	0.019
22.60	1.03	0.27	0.018
23.00	1.05	0.28	0.019
23.60	1.06	0.28	0.020
24.00	1.07	0.29	0.021
24.60	1.09	0.30	0.022
***Iridium*[11]**			
0.10	28.49	60.62	0.975
0.15	15.32	45.15	0.973
0.20	9.69	35.34	0.972
0.25	6.86	28.84	0.969
0.30	5.16	24.25	0.967
0.35	4.11	20.79	0.964
0.40	3.42	18.06	0.960
0.45	3.05	15.82	0.954
0.50	2.98	14.06	0.944
0.60	2.79	11.58	0.925
0.70	2.93	9.78	0.895
0.80	3.14	8.61	0.862
0.90	3.19	7.88	0.840
1.00	3.15	7.31	0.822
1.10	3.04	6.84	0.808
1.20	2.96	6.41	0.791
1.30	2.85	6.07	0.779
1.40	2.72	5.74	0.767
1.50	2.65	5.39	0.750

Energy (eV)	n	k	R(φ = 0)
1.60	2.68	5.08	0.728
1.70	2.69	4.92	0.716
1.80	2.64	4.81	0.710
1.90	2.57	4.68	0.704
2.00	2.50	4.57	0.699
2.10	2.40	4.48	0.697
2.20	2.29	4.38	0.695
2.30	2.18	4.26	0.692
2.40	2.07	4.14	0.689
2.50	1.98	4.00	0.682
2.60	1.91	3.86	0.673
2.70	1.85	3.73	0.665
2.80	1.81	3.61	0.655
2.90	1.77	3.51	0.646
3.00	1.73	3.43	0.640
3.20	1.62	3.26	0.629
3.40	1.53	3.05	0.610
3.60	1.52	2.81	0.573
3.80	1.61	2.69	0.541
4.00	1.64	2.68	0.535
4.20	1.58	2.71	0.549
4.40	1.45	2.68	0.561
4.60	1.31	2.60	0.567
4.80	1.18	2.49	0.570
5.00	1.10	2.35	0.559
5.20	1.04	2.22	0.543
5.40	1.00	2.09	0.522
5.60	0.98	1.98	0.499
5.80	0.96	1.86	0.474
6.00	0.95	1.78	0.454
6.20	0.94	1.68	0.427
6.40	0.94	1.59	0.401
6.60	0.94	1.50	0.375
6.80	0.95	1.42	0.345
7.00	0.97	1.34	0.318
7.20	0.99	1.27	0.290
7.40	1.02	1.20	0.262
7.60	1.03	1.14	0.241
7.80	1.08	1.06	0.208
8.00	1.13	1.03	0.191
8.20	1.18	1.00	0.179
8.40	1.22	0.98	0.171
8.60	1.26	0.96	0.164
8.80	1.29	0.95	0.160
9.00	1.33	0.94	0.157
9.20	1.36	0.95	0.159
9.40	1.39	0.95	0.161
9.60	1.42	0.97	0.163
9.80	1.44	0.99	0.169
10.00	1.45	1.01	0.175
10.20	1.45	1.04	0.182
10.40	1.44	1.07	0.187
10.60	1.43	1.09	0.193
10.80	1.41	1.12	0.200
11.00	1.38	1.13	0.206
11.20	1.34	1.14	0.208
11.40	1.31	1.13	0.208
11.60	1.28	1.12	0.206
11.80	1.25	1.10	0.203
12.00	1.24	1.08	0.199

Energy (eV)	*n*	*k*	*R*(ϕ = 0)
12.40	1.21	1.05	0.191
12.80	1.19	1.01	0.181
13.20	1.18	0.98	0.173
13.60	1.17	0.95	0.165
14.00	1.16	0.91	0.155
14.40	1.17	0.88	0.147
14.80	1.18	0.87	0.142
15.20	1.19	0.84	0.136
15.60	1.20	0.83	0.133
16.00	1.21	0.83	0.131
16.40	1.23	0.82	0.129
16.80	1.25	0.82	0.127
17.20	1.28	0.83	0.131
17.60	1.30	0.87	0.140
18.00	1.30	0.93	0.154
18.40	1.27	0.97	0.166
18.80	1.24	1.00	0.176
19.20	1.20	1.03	0.187
19.60	1.15	1.05	0.197
20.00	1.10	1.06	0.205
20.50	1.04	1.05	0.210
21.00	0.99	1.04	0.215
21.50	0.94	1.02	0.220
22.00	0.89	1.00	0.222
22.50	0.84	0.99	0.228
23.00	0.79	0.96	0.232
23.50	0.76	0.92	0.228
24.00	0.73	0.87	0.223
24.50	0.70	0.83	0.218
25.00	0.69	0.79	0.209
25.50	0.68	0.76	0.200
26.00	0.67	0.72	0.192
26.50	0.67	0.69	0.181
27.00	0.66	0.66	0.174
27.50	0.66	0.63	0.166
28.00	0.66	0.61	0.158
28.50	0.66	0.59	0.151
29.00	0.65	0.57	0.148
29.50	0.64	0.55	0.145
30.00	0.64	0.53	0.140
32.00	0.62	0.44	0.119
34.00	0.64	0.35	0.091
36.00	0.69	0.27	0.059
38.00	0.73	0.24	0.044
40.00	0.76	0.22	0.034

Iron[5]

Energy (eV)	*n*	*k*	*R*(ϕ = 0)
0.10	6.41	33.07	0.978
0.15	6.26	22.82	0.956
0.20	3.68	18.23	0.958
0.26	4.98	13.68	0.911
0.30	4.87	12.05	0.892
0.36	4.68	10.44	0.867
0.40	4.42	9.75	0.858
0.50	4.14	8.02	0.817
0.60	3.93	6.95	0.783
0.70	3.78	6.17	0.752
0.80	3.65	5.60	0.725
0.90	3.52	5.16	0.700
1.00	3.43	4.79	0.678
1.10	3.33	4.52	0.660
1.20	3.24	4.26	0.641
1.30	3.16	4.07	0.626
1.40	3.12	3.87	0.609
1.50	3.05	3.77	0.601
1.60	3.00	3.60	0.585
1.70	2.98	3.52	0.577
1.80	2.92	3.46	0.573
1.90	2.89	3.37	0.563
2.00	2.85	3.36	0.563
2.10	2.80	3.34	0.562
2.20	2.74	3.33	0.563
2.30	2.65	3.34	0.567
2.40	2.56	3.31	0.567
2.50	2.46	3.31	0.570
2.60	2.34	3.30	0.576
2.70	2.23	3.25	0.575
2.80	2.12	3.23	0.580
2.90	2.01	3.17	0.580
3.00	1.88	3.12	0.583
3.10	1.78	3.04	0.580
3.20	1.70	2.96	0.576
3.30	1.62	2.87	0.572
3.40	1.55	2.79	0.565
3.50	1.50	2.70	0.556
3.60	1.47	2.63	0.548
3.70	1.43	2.56	0.542
3.83	1.38	2.49	0.534
4.00	1.30	2.39	0.527
4.17	1.26	2.27	0.510
4.33	1.23	2.18	0.494
4.50	1.20	2.10	0.482
4.67	1.16	2.02	0.470
4.83	1.14	1.93	0.451
5.00	1.14	1.87	0.435
5.17	1.12	1.81	0.425
5.33	1.11	1.75	0.408
5.50	1.09	1.71	0.401
5.67	1.09	1.65	0.383
5.83	1.10	1.61	0.373
6.00	1.09	1.59	0.366
6.17	1.08	1.57	0.365
6.33	1.04	1.55	0.365
6.50	1.02	1.51	0.358
6.67	1.00	1.47	0.351
6.83	0.97	1.43	0.346
7.00	0.96	1.39	0.333
7.17	0.94	1.35	0.327
7.33	0.94	1.30	0.311
7.50	0.94	1.26	0.298
7.67	0.94	1.23	0.288
7.83	0.94	1.21	0.279
8.00	0.94	1.18	0.272
8.17	0.94	1.16	0.265
8.33	0.94	1.14	0.258
8.50	0.94	1.12	0.251
8.67	0.94	1.10	0.246
8.83	0.92	1.08	0.240
9.00	0.93	1.07	0.236
9.17	0.92	1.06	0.233
9.33	0.91	1.04	0.231
9.50	0.90	1.02	0.226
9.67	0.90	1.00	0.221
9.83	0.89	0.99	0.218
10.00	0.88	0.97	0.213
10.17	0.87	0.94	0.203
10.33	0.87	0.91	0.196
10.50	0.87	0.89	0.189
10.67	0.88	0.87	0.179
10.83	0.89	0.85	0.170
11.00	0.91	0.83	0.162
11.17	0.92	0.83	0.159
11.33	0.93	0.84	0.159
11.50	0.93	0.84	0.160
11.67	0.93	0.84	0.162
11.83	0.92	0.84	0.163
12.00	0.91	0.84	0.163
12.17	0.90	0.84	0.165
12.33	0.89	0.83	0.164
12.50	0.88	0.83	0.165
12.67	0.87	0.82	0.166
12.83	0.86	0.81	0.166
13.00	0.85	0.80	0.162
13.17	0.84	0.79	0.161
13.33	0.84	0.78	0.160
13.50	0.83	0.77	0.159
13.67	0.82	0.76	0.157
13.83	0.81	0.75	0.154
14.00	0.81	0.73	0.151
14.17	0.80	0.72	0.149
14.33	0.80	0.71	0.146
14.50	0.79	0.69	0.144
14.67	0.79	0.69	0.141
14.83	0.78	0.67	0.138
15.00	0.78	0.66	0.135
15.17	0.78	0.65	0.131
15.33	0.78	0.64	0.238
15.50	0.77	0.63	0.126
15.67	0.77	0.62	0.123
15.83	0.77	0.61	0.119
16.00	0.77	0.60	0.116
16.17	0.78	0.58	0.112
16.33	0.78	0.58	0.110
16.50	0.78	0.57	0.107
16.67	0.77	0.56	0.106
16.83	0.78	0.55	0.103
17.00	0.78	0.55	0.102
17.17	0.78	0.54	0.100
17.33	0.78	0.54	0.098
17.50	0.77	0.53	0.097
17.67	0.77	0.52	0.095
17.83	0.78	0.51	0.092
18.00	0.78	0.51	0.091
18.17	0.78	0.51	0.090
18.33	0.78	0.50	0.089
18.50	0.77	0.50	0.089
18.67	0.77	0.50	0.088
18.83	0.77	0.49	0.087
19.00	0.77	0.49	0.087
19.17	0.76	0.49	0.088
19.33	0.76	0.48	0.087

Energy (eV)	n	k	R(φ = 0)
19.50	0.75	0.47	0.086
19.67	0.75	0.47	0.085
19.83	0.75	0.46	0.084
20.00	0.74	0.45	0.083
20.17	0.74	0.44	0.081
20.33	0.74	0.44	0.081
20.50	0.74	0.42	0.080
20.67	0.73	0.43	0.079
20.83	0.73	0.42	0.078
21.00	0.73	0.41	0.077
21.17	0.72	0.40	0.076
21.33	0.72	0.39	0.074
21.50	0.72	0.38	0.073
21.67	0.72	0.38	0.071
21.83	0.72	0.37	0.070
22.00	0.72	0.36	0.068
22.17	0.71	0.35	0.067
22.33	0.72	0.34	0.064
22.50	0.72	0.34	0.063
22.67	0.72	0.33	0.062
22.83	0.72	0.32	0.059
23.00	0.72	0.31	0.058
23.17	0.72	0.30	0.056
23.33	0.72	0.29	0.054
23.50	0.73	0.28	0.050
23.67	0.73	0.28	0.049
23.83	0.74	0.27	0.047
24.00	0.74	0.27	0.045
24.17	0.74	0.26	0.044
24.33	0.74	0.26	0.043
24.50	0.74	0.25	0.042
24.67	0.75	0.25	0.040
24.83	0.75	0.24	0.039
25.00	0.75	0.24	0.038
26.00	0.76	0.21	0.031
27.00	0.78	0.18	0.026
28.00	0.79	0.16	0.021
29.00	0.81	0.14	0.017
30.00	0.82	0.13	0.014

Lithium[12]

Energy (eV)	n	k	R(φ = 0)
0.14	0.659	38.0	0.998
0.54	0.661	12.6	0.984
0.75	0.561	7.68	0.963
1.05	0.448	5.58	0.946
1.35	0.338	4.36	0.935
1.65	0.265	3.55	0.925
1.95	0.221	2.94	0.913
2.25	0.206	2.48	0.892
2.55	0.217	2.11	0.854
2.85	0.247	1.82	0.797
3.15	0.304	1.60	0.715
3.45	0.334	1.45	0.656
3.75	0.345	1.32	0.611
4.05	0.346	1.21	0.578
4.35	0.333	1.11	0.557
4.65	0.317	1.01	0.540
4.95	0.302	0.906	0.520
5.25	0.299	0.795	0.484
5.55	0.310	0.688	0.434
5.85	0.342	0.594	0.365

Energy (eV)	n	k	R(φ = 0)
6.15	0.376	0.522	0.306
6.45	0.408	0.460	0.256
6.75	0.440	0.407	0.214
7.05	0.466	0.364	0.183
7.35	0.492	0.320	0.155
7.65	0.517	0.282	0.131
7.95	0.545	0.246	0.109
8.25	0.572	0.214	0.091
8.55	0.601	0.189	0.075
8.85	0.624	0.163	0.063
9.15	0.657	0.144	0.050
9.45	0.680	0.130	0.042
9.75	0.708	0.119	0.034
10.1	0.726	0.108	0.029
10.4	0.743	0.102	0.025
10.6	0.753	0.080	0.022

Magnesium (evaporated)[13]

Energy (eV)	n	k	R(φ = 0)
2.145	0.48	3.71	0.880
2.270	0.57	3.47	0.843
2.522	0.53	2.92	0.805
2.845	0.52	2.65	0.777
3.064	0.52	2.05	0.681
5.167	0.10	1.60	0.894
5.636	0.15	1.50	0.832
6.200	0.20	1.40	0.765
6.889	0.25	1.30	0.693
7.750	0.20	1.20	0.722
8.857	0.15	0.95	0.730
10.335	0.25	0.40	0.419

Manganese[14]

Energy (eV)	n	k	R(φ = 0)
0.64	3.89	5.95	0.738
0.77	3.78	5.41	0.710
0.89	3.65	5.02	0.688
1.02	3.48	4.74	0.673
1.14	3.30	4.53	0.662
1.26	3.10	4.35	0.653
1.39	2.97	4.18	0.643
1.51	2.83	4.03	0.634
1.64	2.70	3.91	0.627
1.76	2.62	3.78	0.617
1.88	2.56	3.65	0.606
2.01	2.51	3.54	0.596
2.13	2.47	3.43	0.585
2.26	2.39	3.33	0.577
2.38	2.32	3.23	0.567
2.50	2.25	3.14	0.559
2.63	2.19	3.06	0.552
2.75	2.11	2.98	0.545
2.88	2.06	2.90	0.536
3.00	2.00	2.82	0.528
3.12	1.96	2.74	0.518
3.25	1.92	2.67	0.509
3.37	1.89	2.59	0.498
3.50	1.89	2.51	0.484
3.62	1.87	2.45	0.475
3.74	1.86	2.38	0.463
3.87	1.86	2.32	0.451
3.99	1.86	2.25	0.438
4.12	1.86	2.19	0.427

Energy (eV)	n	k	R(φ = 0)
4.24	1.85	2.14	0.417
4.36	1.85	2.08	0.406
4.49	1.86	2.03	0.395
4.61	1.85	1.99	0.388
4.74	1.84	1.94	0.378
4.86	1.83	1.91	0.372
4.98	1.82	1.86	0.362
5.11	1.82	1.82	0.354
5.23	1.81	1.79	0.348
5.36	1.78	1.76	0.342
5.48	1.74	1.73	0.337
5.60	1.73	1.70	0.331
5.73	1.72	1.67	0.325
5.85	1.70	1.64	0.319
5.98	1.67	1.61	0.313
6.10	1.63	1.58	0.307
6.22	1.62	1.55	0.301
6.35	1.59	1.52	0.295
6.47	1.55	1.50	0.292
6.60	1.48	1.47	0.288

Mercury (liquid)[15]

Energy (eV)	n	k	R(φ = 0)
0.2	13.99	14.27	0.869
0.3	11.37	11.95	0.846
0.4	9.741	10.65	0.830
0.5	8.528	9.805	0.818
0.6	7.574	9.195	0.808
0.8	6.086	8.312	0.796
1.0	4.962	7.643	0.789
1.2	4.050	7.082	0.786
1.4	3.324	6.558	0.785
1.6	2.746	6.054	0.783
1.8	2.284	5.582	0.782
2.0	1.910	5.150	0.782
2.2	1.620	4.751	0.780
2.4	1.384	4.407	0.779
2.6	1.186	4.090	0.779
2.8	1.027	3.802	0.779
3.0	0.898	3.538	0.777
3.2	0.798	3.294	0.773
3.4	0.713	3.074	0.770
3.6	0.644	2.860	0.763
3.8	0.589	2.665	0.755
4.0	0.542	2.502	0.749
4.2	0.507	2.341	0.738
4.4	0.477	2.195	0.727
4.6	0.452	2.058	0.715
4.8	0.431	1.929	0.701
5.0	0.414	1.806	0.685
5.2	0.401	1.687	0.666
5.4	0.394	1.569	0.642
5.6	0.386	1.454	0.617
5.7	0.386	1.396	0.601
5.8	0.386	1.341	0.585
5.9	0.385	1.287	0.569
6.0	0.386	1.232	0.551
6.1	0.388	1.176	0.531
6.2	0.390	1.118	0.510
6.3	0.399	1.058	0.481
6.4	0.412	1.002	0.450
6.5	0.428	0.949	0.418

Energy (eV)	n	k	R(φ = 0)
6.6	0.436	0.898	0.392
6.7	0.438	0.836	0.367
6.8	0.459	0.756	0.320
6.9	0.510	0.676	0.255
7.0	0.585	0.617	0.191
7.1	0.663	0.589	0.148
7.2	0.717	0.584	0.128
7.3	0.769	0.575	0.111
7.4	0.817	0.574	0.100
7.5	0.860	0.580	0.094
7.6	0.893	0.597	0.093
7.8	0.929	0.623	0.096
8.0	0.946	0.639	0.098
8.2	0.952	0.645	0.099
8.4	0.953	0.638	0.097
8.6	0.956	0.624	0.093
8.8	0.965	0.607	0.087
9.0	0.975	0.588	0.082
9.2	0.988	0.568	0.076
9.4	1.009	0.548	0.069
9.6	1.044	0.541	0.066
9.8	1.061	0.557	0.069
10.0	1.062	0.567	0.071
10.2	1.054	0.569	0.072
10.4	1.045	0.561	0.070
10.6	1.041	0.550	0.068
10.8	1.039	0.537	0.065
11.0	1.039	0.523	0.062
11.5	1.050	0.491	0.055
12.0	1.064	0.467	0.050
12.5	1.078	0.445	0.045
13.0	1.092	0.430	0.042
13.5	1.104	0.416	0.040
14.0	1.115	0.404	0.038
14.5	1.125	0.394	0.037
15.0	1.135	0.383	0.035
15.5	1.146	0.374	0.034
16.0	1.159	0.368	0.034
16.5	1.170	0.367	0.034
17.0	1.177	0.367	0.034
17.5	1.184	0.366	0.034
18.0	1.191	0.367	0.035
18.5	1.195	0.367	0.035
19.0	1.200	0.366	0.035
19.5	1.208	0.364	0.035

Molybdenum[16]

Energy (eV)	n	k	R(φ = 0)
0.10	18.53	68.51	0.985
0.15	8.78	47.54	0.985
0.20	5.10	35.99	0.985
0.25	3.36	28.75	0.984
0.30	2.44	23.80	0.983
0.34	2.00	20.84	0.982
0.38	1.70	18.44	0.980
0.42	1.57	16.50	0.978
0.46	1.46	14.91	0.975
0.50	1.37	13.55	0.971
0.54	1.35	12.36	0.966
0.58	1.34	11.34	0.960
0.62	1.38	10.44	0.952
0.66	1.43	9.67	0.942
0.70	1.48	8.99	0.932
0.74	1.51	8.38	0.921
0.78	1.60	7.83	0.906
0.82	1.64	7.35	0.892
0.86	1.70	6.89	0.876
9.90	1.74	6.48	0.859
1.00	1.94	5.58	0.805
1.10	2.15	4.85	0.743
1.20	2.44	4.22	0.671
1.30	2.77	3.74	0.608
1.40	3.15	3.40	0.562
1.50	3.53	3.30	0.550
1.60	3.77	3.41	0.562
1.70	3.84	3.51	0.570
1.80	3.81	3.58	0.576
1.90	3.74	3.58	0.576
2.00	3.68	3.52	0.571
2.10	3.68	3.45	0.565
2.20	3.76	3.41	0.562
2.30	3.79	3.61	0.578
2.40	3.59	3.78	0.594
2.50	3.36	3.73	0.591
2.60	3.22	3.61	0.582
2.70	3.13	3.51	0.573
2.80	3.08	3.42	0.565
2.90	3.05	3.33	0.566
3.00	3.04	3.27	0.550
3.10	3.03	3.21	0.544
3.20	3.05	3.18	0.540
3.30	3.06	3.18	0.540
3.40	3.06	3.19	0.541
3.50	3.06	3.21	0.543
3.60	3.05	3.23	0.546
3.70	3.04	3.27	0.550
3.80	3.04	3.31	0.554
3.90	3.04	3.40	0.564
4.00	3.01	3.51	0.576
4.20	2.77	3.77	0.610
4.40	2.39	3.88	0.640
4.60	2.06	3.84	0.658
4.80	1.75	3.76	0.678
5.00	1.46	3.62	0.695
5.20	1.22	3.42	0.706
5.40	1.07	3.20	0.706
5.60	0.96	2.99	0.700
5.80	0.89	2.80	0.688
6.00	0.85	2.64	0.674
6.20	0.81	2.50	0.660
6.40	0.79	2.36	0.641
6.60	0.78	2.24	0.619
6.80	0.78	2.13	0.592
7.00	0.80	2.04	0.568
7.20	0.81	1.98	0.548
7.40	0.81	1.95	0.542
7.60	0.75	1.90	0.552
7.80	0.71	1.81	0.542
8.00	0.69	1.73	0.530
8.20	0.67	1.65	0.512
8.40	0.66	1.57	0.495
8.60	0.65	1.49	0.475
8.80	0.65	1.41	0.450
9.00	0.65	1.33	0.420
9.20	0.67	1.25	0.385
9.40	0.69	1.19	0.355
9.60	0.71	1.12	0.320
9.80	0.74	1.05	0.285
10.00	0.77	0.99	0.250
10.20	0.81	0.93	0.217
10.40	0.86	0.88	0.188
10.60	0.91	0.83	0.162
10.80	0.98	0.79	0.138
11.00	1.05	0.77	0.125
11.20	1.12	0.78	0.123
11.40	1.18	0.80	0.125
11.60	1.23	0.85	0.135
11.80	1.25	0.89	0.145
12.00	1.26	0.92	0.154
12.40	1.25	0.98	0.168
12.80	1.23	1.00	0.178
13.20	1.20	1.02	0.185
13.60	1.17	1.02	0.187
14.00	1.15	1.01	0.185
14.40	1.13	1.00	0.182
14.80	1.13	0.99	0.179
15.00	1.14	0.99	0.179
15.60	1.15	1.01	0.184
16.00	1.14	1.04	0.194
16.60	1.10	1.10	0.216
17.00	1.04	1.12	0.233
17.60	0.94	1.14	0.257
18.00	0.87	1.12	0.270
18.60	0.77	1.08	0.283
19.00	0.71	1.02	0.284
19.60	0.66	0.94	0.275
20.00	0.64	0.89	0.264
20.60	0.62	0.81	0.245
21.00	0.61	0.77	0.234
21.60	0.61	0.71	0.215
22.00	0.60	0.69	0.207
22.60	0.59	0.63	0.195
23.00	0.58	0.60	0.185
23.60	0.58	0.53	0.166
24.00	0.58	0.49	0.151
24.60	0.60	0.43	0.124
25.00	0.62	0.39	0.106
25.60	0.66	0.35	0.085
26.00	0.68	0.33	0.072
26.50	0.71	0.31	0.060
27.00	0.73	0.29	0.050
27.50	0.76	0.28	0.041
28.00	0.79	0.27	0.036
28.50	0.81	0.26	0.031
29.00	0.83	0.26	0.028
29.50	0.86	0.26	0.025
30.00	0.88	0.26	0.023
31.00	0.92	0.29	0.024
32.00	0.92	0.32	0.030
33.00	0.90	0.33	0.032
34.00	0.91	0.34	0.034
35.00	0.87	0.37	0.043

Energy (eV)	n	k	R(φ = 0)
36.00	0.82	0.34	0.043
37.00	0.81	0.30	0.038
38.00	0.81	0.27	0.033
39.00	0.82	0.25	0.029
40.00	0.83	0.23	0.025

Nickel[17]

Energy (eV)	n	k	R(φ = 0)
0.10	9.54	45.82	0.983
0.15	5.45	30.56	0.978
0.20	4.12	22.48	0.969
0.25	4.25	17.68	0.950
0.30	4.19	15.05	0.934
0.35	4.03	13.05	0.918
0.40	3.84	11.43	0.900
0.50	4.03	9.64	0.864
0.60	3.84	8.35	0.835
0.70	3.59	7.48	0.813
0.80	3.38	6.82	0.794
0.90	3.18	6.23	0.774
1.00	3.06	5.74	0.753
1.10	2.97	5.38	0.734
1.20	2.85	5.10	0.721
1.30	2.74	4.85	0.708
1.40	2.65	4.63	0.695
1.50	2.53	4.47	0.688
1.60	2.43	4.31	0.679
1.70	2.28	4.18	0.677
1.80	2.14	4.01	0.670
1.90	2.02	3.82	0.659
2.00	1.92	3.65	0.649
2.10	1.85	3.48	0.634
2.20	1.80	3.33	0.620
2.30	1.75	3.19	0.605
2.40	1.71	3.06	0.590
2.50	1.67	2.93	0.575
2.60	1.65	2.81	0.557
2.70	1.64	2.71	0.542
2.80	1.63	2.61	0.525
2.90	1.62	2.52	0.509
3.00	1.61	2.44	0.495
3.10	1.61	2.36	0.480
3.20	1.61	2.30	0.467
3.30	1.61	2.23	0.454
3.40	1.62	2.17	0.441
3.50	1.63	2.11	0.428
3.60	1.64	2.07	0.416
3.70	1.66	2.02	0.405
3.80	1.69	1.99	0.397
3.90	1.72	1.98	0.393
4.00	1.73	1.98	0.392
4.20	1.74	2.01	0.396
4.40	1.71	2.06	0.409
4.60	1.63	2.09	0.421
4.80	1.53	2.11	0.435
5.00	1.40	2.10	0.449
5.20	1.27	2.04	0.454
5.40	1.16	1.94	0.449
5.60	1.09	1.83	0.435
5.80	1.04	1.73	0.417
6.20	1.00	1.54	0.371
6.40	1.01	1.46	0.345
6.60	1.01	1.40	0.325
6.80	1.02	1.35	0.308
7.00	1.03	1.30	0.291
7.20	1.03	1.27	0.282
7.40	1.03	1.24	0.273
7.60	1.02	1.22	0.265
7.80	1.01	1.18	0.256
8.00	1.01	1.15	0.248
8.20	1.00	1.13	0.242
8.40	0.99	1.11	0.235
8.60	0.98	1.08	0.228
8.80	0.97	1.05	0.220
9.00	0.97	1.01	0.211
9.20	0.96	0.99	0.203
9.40	0.95	0.96	0.194
9.60	0.95	0.93	0.185
9.80	0.95	0.89	0.175
10.00	0.95	0.87	0.166
10.20	0.95	0.83	0.155
10.40	0.95	0.80	0.145
10.60	0.97	0.76	0.129
10.80	0.99	0.75	0.123
11.00	1.01	0.73	0.115
11.25	1.04	0.72	0.111
11.50	1.05	0.71	0.109
11.75	1.07	0.71	0.108
12.00	1.07	0.71	0.108
12.25	1.07	0.71	0.107
12.50	1.08	0.71	0.106
12.75	1.08	0.71	0.106
13.00	1.08	0.71	0.105
13.25	1.08	0.71	0.105
13.50	1.07	0.70	0.105
13.75	1.07	0.70	0.105
14.00	1.07	0.71	0.106
14.25	1.06	0.70	0.106
14.50	1.05	0.70	0.106
14.75	1.04	0.70	0.107
15.00	1.03	0.70	0.107
15.25	1.02	0.69	0.106
15.50	1.01	0.69	0.105
15.75	1.00	0.68	0.104
16.00	0.99	0.67	0.103
16.50	0.98	0.66	0.101
17.00	0.96	0.64	0.098
17.50	0.94	0.63	0.096
18.00	0.92	0.61	0.092
18.50	0.91	0.58	0.087
19.00	0.90	0.56	0.082
19.50	0.90	0.54	0.077
20.00	0.89	0.51	0.071
20.50	0.89	0.49	0.066
21.00	0.90	0.47	0.061
21.50	0.91	0.46	0.057
22.00	0.91	0.45	0.055
22.50	0.91	0.44	0.053
23.00	0.92	0.44	0.051
23.50	0.91	0.44	0.052
24.00	0.90	0.43	0.051
24.50	0.90	0.43	0.051
25.00	0.89	0.42	0.050
26.00	0.88	0.39	0.046
27.00	0.87	0.37	0.042
28.00	0.87	0.35	0.040
29.00	0.86	0.34	0.037
30.00	0.86	0.32	0.034
35.00	0.86	0.24	0.022
40.00	0.87	0.18	0.014
45.00	0.88	0.13	0.008
50.00	0.92	0.10	0.004
60.00	0.96	0.08	0.002
65.00	0.98	0.09	0.002
68.00	0.96	0.12	0.004
70.00	0.94	0.11	0.004
75.00	0.94	0.09	0.003
80.00	0.94	0.07	0.002
90.00	0.94	0.06	0.002

Niobium[18]

Energy (eV)	n	k	R(φ = 0)
0.12	15.99	53.20	0.979
0.20	7.25	34.14	0.976
0.24	5.47	28.88	0.975
0.28	4.26	24.95	0.974
0.35	3.11	20.03	0.970
0.45	2.28	15.58	0.964
0.55	1.83	12.67	0.956
0.65	1.57	10.59	0.947
0.75	1.41	9.00	0.935
0.85	1.35	7.74	0.918
0.95	1.35	6.70	0.893
1.05	1.44	5.86	0.857
1.15	1.55	5.18	0.814
1.25	1.65	4.63	0.768
1.35	1.76	4.13	0.715
1.45	1.95	3.68	0.650
1.55	2.15	3.37	0.595
1.65	2.36	3.13	0.552
1.75	2.54	2.99	0.527
1.85	2.69	2.89	0.510
1.95	2.82	2.86	0.505
2.05	2.89	2.87	0.505
2.15	2.92	2.87	0.505
2.25	2.93	2.87	0.505
2.35	2.92	2.88	0.506
2.45	2.89	2.90	0.509
2.55	2.83	2.92	0.512
2.65	2.74	2.90	0.511
2.75	2.66	2.86	0.507
2.85	2.58	2.80	0.500
3.00	2.51	2.68	0.485
3.10	2.48	2.60	0.475
3.20	2.45	2.53	0.465
3.30	2.44	2.45	0.453
3.40	2.46	2.38	0.442
3.50	2.48	2.33	0.435
3.60	2.52	2.29	0.428
3.70	2.56	2.27	0.426
3.80	2.59	2.28	0.427
3.90	2.62	2.29	0.429
4.00	2.64	2.33	0.434
4.20	2.64	2.42	0.447

Energy (eV)	n	k	$R(\phi = 0)$
4.40	2.53	2.56	0.467
4.60	2.39	2.56	0.470
4.80	2.32	2.52	0.465
5.00	2.26	2.57	0.475
5.20	2.16	2.62	0.487
5.40	2.00	2.68	0.505
5.60	1.81	2.67	0.518
5.80	1.63	2.60	0.522
6.00	1.49	2.49	0.520
6.20	1.38	2.38	0.512
6.40	1.31	2.25	0.496
6.60	1.26	2.14	0.480
6.80	1.24	2.04	0.460
7.00	1.23	1.96	0.441
7.20	1.22	1.91	0.430
7.40	1.20	1.88	0.427
7.60	1.14	1.85	0.430
7.80	1.07	1.78	0.428
8.00	1.02	1.69	0.412
8.20	1.00	1.60	0.390
8.40	0.99	1.51	0.365
8.60	0.99	1.43	0.340
8.70	0.99	1.39	0.328
8.80	1.00	1.36	0.315
9.00	1.01	1.29	0.290
9.20	1.04	1.22	0.265
9.40	1.07	1.18	0.245
9.60	1.10	1.13	0.227
9.80	1.13	1.09	0.209
10.00	1.18	1.05	0.194
10.20	1.23	1.04	0.187
10.40	1.27	1.04	0.185
10.60	1.30	1.06	0.190
10.80	1.32	1.08	0.195
11.00	1.32	1.10	0.200
11.20	1.31	1.12	0.204
11.40	1.30	1.13	0.207
11.60	1.28	1.13	0.209
11.80	1.27	1.13	0.210
12.00	1.25	1.12	0.209
12.40	1.24	1.10	0.204
12.80	1.24	1.09	0.200
13.20	1.24	1.09	0.201
13.60	1.23	1.12	0.208
14.00	1.20	1.13	0.216
14.40	1.16	1.15	0.225
14.80	1.11	1.16	0.234
15.00	1.08	1.16	0.238
15.60	0.99	1.14	0.247
16.00	0.92	1.11	0.250
16.60	0.85	1.04	0.245
17.00	0.80	0.99	0.240
17.20	0.79	0.96	0.236
17.40	0.77	0.93	0.230
17.80	0.75	0.87	0.217
18.00	0.74	0.85	0.209
18.60	0.73	0.77	0.185
19.00	0.72	0.72	0.170
19.60	0.72	0.66	0.150
20.00	0.72	0.62	0.137

Energy (eV)	n	k	$R(\phi = 0)$
20.60	0.71	0.55	0.119
21.00	0.72	0.50	0.100
21.60	0.75	0.43	0.075
22.00	0.78	0.40	0.063
22.60	0.82	0.35	0.045
23.00	0.85	0.33	0.038
23.60	0.88	0.30	0.029
24.00	0.91	0.29	0.025
24.60	0.94	0.28	0.022
25.00	0.96	0.27	0.020
25.60	0.99	0.26	0.018
26.00	1.00	0.26	0.017
26.60	1.03	0.25	0.016
27.00	1.04	0.25	0.015
27.60	1.06	0.25	0.015
28.00	1.08	0.24	0.015
28.60	1.11	0.24	0.016
29.00	1.13	0.25	0.017
29.60	1.16	0.26	0.020
30.00	1.18	0.28	0.023
31.00	1.18	0.31	0.026
32.00	1.20	0.34	0.031
33.00	1.21	0.38	0.038
34.00	1.20	0.42	0.044
35.20	1.17	0.47	0.051
36.00	1.15	0.50	0.056
37.50	1.07	0.53	0.064
39.50	0.95	0.50	0.063
40.50	0.92	0.47	0.059

Osmium (Polycrystalline)[9]

Energy (eV)	n	k	$R(\phi = 0)$
0.10	4.08	50.23	0.994
0.15	2.90	33.60	0.990
0.20	2.44	25.11	0.985
0.25	2.35	19.99	0.977
0.30	2.23	16.54	0.969
0.35	2.33	14.06	0.955
0.40	2.45	12.32	0.940
0.45	2.43	11.02	0.927
0.50	2.41	9.97	0.913
0.55	2.33	9.12	0.901
0.60	2.21	8.37	0.890
0.65	2.11	7.68	0.877
0.70	2.02	7.04	0.862
0.75	2.00	6.46	0.842
0.80	2.00	5.95	0.820
0.85	2.01	5.51	0.796
0.90	2.03	5.10	0.769
0.95	2.05	4.74	0.742
1.00	2.09	4.41	0.712
1.10	2.15	3.84	0.651
1.20	2.16	3.35	0.592
1.30	2.25	2.77	0.506
1.40	2.49	2.23	0.419
1.50	2.84	1.80	0.369
1.60	3.36	1.62	0.379
1.70	3.70	1.75	0.411
1.80	3.78	1.83	0.423
1.90	3.81	1.75	0.418
2.00	3.98	1.60	0.418
2.10	4.26	1.54	0.432
2.20	4.58	1.62	0.457
2.30	4.84	1.76	0.479
2.40	5.10	2.01	0.506
2.50	5.28	2.38	0.532
2.60	5.36	2.82	0.557
2.70	5.30	3.29	0.580
2.80	5.07	3.78	0.603
2.90	4.65	4.18	0.624
3.00	4.05	4.40	0.639
3.20	3.29	3.96	0.614
3.40	2.93	3.79	0.607
3.60	2.75	3.45	0.577
3.80	2.73	3.32	0.562
4.00	2.71	3.34	0.565
4.20	2.53	3.44	0.584
4.40	2.24	3.44	0.599
4.60	2.01	3.31	0.598
4.80	1.88	3.19	0.592
5.00	1.74	3.12	0.596
5.20	1.58	3.00	0.597
5.40	1.46	2.88	0.593
5.60	1.36	2.77	0.589
5.80	1.27	2.65	0.582
6.00	1.20	2.54	0.575
6.20	1.13	2.44	0.571
6.40	1.06	2.33	0.562
6.60	1.01	2.21	0.548
6.80	0.97	2.11	0.532
7.00	0.95	2.00	0.514
7.20	0.92	1.91	0.497
7.40	0.91	1.81	0.476
7.60	0.90	1.72	0.451
7.80	0.90	1.63	0.426
8.00	0.91	1.55	0.400
8.20	0.91	1.48	0.375
8.40	0.94	1.40	0.344
8.60	0.96	1.34	0.319
8.80	0.98	1.29	0.296
9.00	1.01	1.24	0.274
9.20	1.04	1.19	0.255
9.40	1.08	1.16	0.238
9.60	1.10	1.14	0.229
9.80	1.13	1.11	0.217
10.00	1.16	1.10	0.209
10.20	1.19	1.08	0.203
10.30	1.20	1.08	0.201
10.40	1.22	1.08	0.200
10.50	1.23	1.09	0.201
10.60	1.24	1.10	0.203
10.80	1.25	1.11	0.206
11.00	1.24	1.13	0.213
11.20	1.23	1.14	0.217
11.40	1.19	1.15	0.223
11.60	1.17	1.12	0.216
11.80	1.16	1.10	0.211
12.00	1.15	1.08	0.205
12.40	1.14	1.03	0.191
12.80	1.15	1.01	0.183
13.20	1.16	0.98	0.174
13.60	1.17	0.97	0.170

Energy (eV)	n	k	$R(\phi = 0)$
14.00	1.17	0.96	0.169
14.40	1.16	0.94	0.165
14.80	1.16	0.91	0.156
15.20	1.17	0.89	0.148
15.60	1.20	0.86	0.140
16.00	1.25	0.87	0.140
16.40	1.28	0.90	0.147
16.80	1.28	0.94	0.157
17.20	1.27	0.97	0.167
17.60	1.26	1.01	0.178
18.00	1.23	1.04	0.189
18.40	1.19	1.08	0.200
18.80	1.14	1.10	0.210
19.20	1.10	1.10	0.219
19.60	1.05	1.11	0.227
20.00	0.96	1.10	0.239
20.40	0.93	1.09	0.240
20.80	0.89	1.05	0.240
21.20	0.86	1.02	0.237
21.60	0.83	0.99	0.235
22.00	0.80	0.96	0.230
22.40	0.78	0.93	0.226
22.80	0.77	0.90	0.220
23.20	0.75	0.88	0.217
23.60	0.75	0.86	0.211
24.00	0.73	0.84	0.209
24.40	0.72	0.82	0.207
24.80	0.70	0.80	0.205
25.20	0.69	0.77	0.202
25.60	0.67	0.75	0.199
26.00	0.66	0.72	0.195
26.40	0.65	0.69	0.189
26.80	0.63	0.66	0.183
27.20	0.65	0.62	0.165
28.00	0.64	0.59	0.156
28.40	0.64	0.57	0.148
28.80	0.65	0.55	0.140
29.20	0.65	0.53	0.134
29.60	0.65	0.51	0.128
30.00	0.65	0.49	0.121
31.00	0.65	0.45	0.111
32.00	0.66	0.41	0.095
33.00	0.68	0.37	0.079
34.00	0.70	0.34	0.068
35.00	0.72	0.31	0.057
36.00	0.74	0.29	0.048
37.00	0.77	0.27	0.040
38.00	0.79	0.26	0.035
39.00	0.81	0.26	0.031
40.00	0.84	0.26	0.026
***Palladium*[19]**			
0.10	4.13	54.15	0.994
0.15	3.13	35.82	0.990
0.20	3.07	26.59	0.983
0.26	3.11	20.15	0.971
0.30	3.56	17.27	0.955
0.36	3.98	14.41	0.932
0.40	4.27	13.27	0.916
0.46	4.27	12.11	0.902
0.50	4.10	11.44	0.896

Energy (eV)	n	k	$R(\phi = 0)$
0.56	3.92	10.49	0.883
0.60	3.80	9.96	0.876
0.72	3.51	8.70	0.854
0.80	3.35	8.06	0.840
1.00	2.99	6.89	0.811
1.10	2.81	6.46	0.800
1.20	2.65	6.10	0.790
1.30	2.50	5.78	0.781
1.40	2.34	5.50	0.774
1.50	2.17	5.22	0.767
1.60	2.08	4.95	0.755
1.70	2.00	4.72	0.745
1.80	1.92	4.54	0.737
1.90	1.82	4.35	0.729
2.00	1.75	4.18	0.721
2.10	1.67	4.03	0.714
2.20	1.60	3.88	0.707
2.30	1.53	3.75	0.700
2.40	1.47	3.61	0.693
2.50	1.41	3.48	0.685
2.60	1.37	3.36	0.676
2.70	1.32	3.25	0.668
2.80	1.29	3.13	0.658
2.90	1.26	3.03	0.648
3.00	1.23	2.94	0.639
3.10	1.20	2.85	0.630
3.20	1.17	2.77	0.622
3.30	1.14	2.68	0.613
3.40	1.12	2.60	0.602
3.50	1.10	2.52	0.591
3.60	1.08	2.45	0.581
3.70	1.07	2.38	0.570
3.80	1.06	2.31	0.558
3.90	1.05	2.25	0.547
4.00	1.03	2.19	0.537
4.20	1.04	2.09	0.510
4.40	1.03	2.01	0.493
4.60	1.03	1.94	0.476
4.80	1.01	1.90	0.470
5.00	0.96	1.86	0.472
5.20	0.90	1.79	0.474
5.40	0.85	1.70	0.463
5.60	0.81	1.62	0.449
5.80	0.78	1.54	0.437
6.00	0.76	1.45	0.418
6.20	0.74	1.37	0.397
6.40	0.73	1.29	0.375
6.60	0.72	1.21	0.350
6.80	0.73	1.13	0.316
7.00	0.73	1.05	0.287
7.20	0.75	0.98	0.255
7.40	0.77	0.91	0.223
7.60	0.79	0.85	0.195
7.80	0.83	0.78	0.163
8.00	0.88	0.73	0.133
8.20	0.94	0.70	0.117
8.40	0.96	0.70	0.114
8.60	1.00	0.65	0.097
8.80	1.04	0.65	0.094
9.00	1.07	0.64	0.090

Energy (eV)	n	k	$R(\phi = 0)$
9.50	1.12	0.65	0.089
10.00	1.14	0.65	0.088
10.50	1.16	0.65	0.087
11.00	1.18	0.64	0.086
11.50	1.19	0.65	0.087
12.00	1.20	0.66	0.089
12.50	1.19	0.67	0.091
13.00	1.18	0.67	0.091
13.50	1.18	0.67	0.092
14.00	1.17	0.67	0.093
14.50	1.15	0.68	0.095
15.00	1.13	0.69	0.098
15.50	1.10	0.68	0.096
16.00	1.08	0.66	0.092
16.50	1.06	0.63	0.086
17.00	1.07	0.61	0.081
17.50	1.06	0.61	0.080
18.00	1.07	0.59	0.077
18.50	1.07	0.59	0.077
19.00	1.08	0.59	0.077
19.50	1.08	0.61	0.080
20.00	1.07	0.65	0.090
20.50	1.03	0.67	0.098
21.00	0.99	0.67	0.103
21.50	0.95	0.66	0.103
22.00	0.91	0.64	0.103
22.50	0.88	0.62	0.101
23.00	0.86	0.59	0.097
23.50	0.85	0.56	0.091
24.00	0.84	0.54	0.086
25.00	0.81	0.51	0.084
26.40	0.80	0.43	0.066
27.80	0.81	0.38	0.052
29.20	0.82	0.35	0.046
***Platinum*[20]**			
0.10	13.21	44.72	0.976
0.15	8.18	31.16	0.969
0.20	5.90	23.95	0.962
0.25	4.70	19.40	0.954
0.30	3.92	16.16	0.945
0.35	3.28	13.66	0.936
0.40	2.81	11.38	0.922
0.45	3.03	9.31	0.882
0.50	3.91	7.71	0.813
0.55	4.58	7.14	0.777
0.60	5.13	6.75	0.753
0.65	5.52	6.66	0.746
0.70	5.71	6.83	0.751
0.75	5.57	7.02	0.759
0.80	5.31	7.04	0.762
0.85	5.05	6.98	0.763
0.90	4.77	6.91	0.765
0.95	4.50	6.77	0.763
1.00	4.25	6.62	0.762
1.10	3.86	6.24	0.753
1.20	3.55	5.92	0.746
1.30	3.29	5.61	0.736
1.40	3.10	5.32	0.725
1.50	2.92	5.07	0.716
1.60	2.76	4.84	0.706

Energy (eV)	n	k	$R(\varphi = 0)$
1.70	2.63	4.64	0.697
1.80	2.51	4.43	0.686
1.90	2.38	4.26	0.678
2.00	2.30	4.07	0.664
2.10	2.23	3.92	0.654
2.20	2.17	3.77	0.642
2.30	2.10	3.67	0.636
2.40	2.03	3.54	0.626
2.50	1.96	3.42	0.616
2.60	1.91	3.30	0.605
2.70	1.87	3.20	0.595
2.80	1.83	3.10	0.585
2.90	1.79	3.01	0.575
3.00	1.75	2.92	0.565
3.20	1.68	2.76	0.546
3.40	1.63	2.62	0.527
3.60	1.58	2.48	0.507
3.80	1.53	2.37	0.491
4.00	1.49	2.25	0.472
4.20	1.45	2.14	0.452
4.40	1.43	2.04	0.432
4.60	1.39	1.95	0.415
4.80	1.38	1.85	0.392
5.00	1.36	1.76	0.372
5.20	1.36	1.67	0.350
5.40	1.36	1.61	0.332
5.60	1.36	1.54	0.315
5.80	1.36	1.47	0.295
6.00	1.38	1.40	0.276
6.20	1.39	1.35	0.261
6.40	1.42	1.29	0.246
6.60	1.45	1.26	0.236
6.80	1.48	1.24	0.231
7.00	1.50	1.24	0.230
7.20	1.50	1.25	0.231
7.40	1.49	1.23	0.228
7.60	1.48	1.22	0.225
7.80	1.48	1.20	0.221
8.00	1.47	1.18	0.216
8.20	1.47	1.17	0.212
8.40	1.47	1.15	0.209
8.60	1.47	1.14	0.205
8.80	1.47	1.13	0.202
9.00	1.48	1.12	0.200
9.20	1.49	1.11	0.198
9.40	1.49	1.12	0.200
9.60	1.49	1.13	0.203
9.80	1.48	1.15	0.207
10.00	1.46	1.15	0.209
10.20	1.43	1.16	0.211
10.40	1.40	1.15	0.210
10.60	1.37	1.14	0.207
10.80	1.35	1.12	0.203
11.00	1.33	1.10	0.199
11.20	1.31	1.08	0.194
11.40	1.30	1.06	0.188
11.60	1.29	1.04	0.183
11.80	1.29	1.01	0.177
12.00	1.29	1.00	0.173
12.40	1.29	0.97	0.165

Energy (eV)	n	k	$R(\varphi = 0)$
12.80	1.29	0.94	0.158
13.20	1.31	0.93	0.155
13.60	1.31	0.93	0.155
14.00	1.31	0.93	0.155
14.40	1.30	0.93	0.156
14.80	1.27	0.93	0.157
15.20	1.27	0.93	0.155
15.60	1.25	0.92	0.151
16.00	1.24	0.89	0.146
16.50	1.24	0.87	0.142
17.00	1.25	0.86	0.138
17.50	1.27	0.85	0.135
18.00	1.31	0.88	0.142
18.50	1.30	0.94	0.157
19.00	1.28	0.99	0.171
19.50	1.23	1.03	0.184
20.00	1.18	1.06	0.197
20.50	1.11	1.09	0.212
21.00	1.03	1.10	0.226
21.50	0.94	1.08	0.238
22.00	0.87	1.04	0.240
22.50	0.81	0.98	0.235
23.00	0.77	0.92	0.226
23.50	0.75	0.87	0.213
24.00	0.74	0.82	0.201
24.50	0.73	0.77	0.187
25.00	0.73	0.73	0.174
25.50	0.73	0.70	0.162
26.00	0.74	0.67	0.150
26.50	0.74	0.65	0.142
27.00	0.74	0.63	0.136
27.50	0.74	0.62	0.130
28.00	0.75	0.60	0.125
28.50	0.75	0.59	0.121
29.00	0.75	0.58	0.118
29.50	0.74	0.58	0.120
30.00	0.73	0.58	0.124

Potassium[21]

Energy (eV)	n	k	$R(\varphi = 0)$
0.55	0.139	7.10	0.989
0.58	0.119	6.72	0.990
0.63	0.106	6.32	0.990
0.67	0.091	5.79	0.990
0.73	0.079	5.30	0.989
0.81	0.066	4.75	0.989
0.92	0.056	4.19	0.988
1.05	0.044	3.58	0.987
1.23	0.040	3.04	0.985
1.44	0.040	2.56	0.979
1.65	0.044	2.19	0.970
1.87	0.050	1.84	0.955
2.07	0.053	1.62	0.943
2.27	0.049	1.43	0.938
2.45	0.046	1.28	0.933
2.64	0.043	1.14	0.928
2.82	0.043	1.02	0.919
2.95	0.041	0.898	0.913
3.06	0.041	0.799	0.905
3.40	0.052	0.549	0.852
3.71	0.089	0.288	0.719
3.97	0.287	0.091	0.310

Energy (eV)	n	k	$R(\varphi = 0)$
4.00	0.34	0.08	0.245
4.065	0.38	0.07	0.204
4.133	0.41	0.07	0.177
4.203	0.45	0.06	0.145
4.275	0.48	0.06	0.125
4.350	0.52	0.05	0.101
4.428	0.55	0.05	0.085
4.509	0.58	0.05	0.072
4.592	0.61	0.05	0.060
4.679	0.64	0.04	0.049
4.769	0.66	0.04	0.043
4.862	0.68	0.04	0.037
4.959	0.70	0.04	0.032
5.061	0.72	0.04	0.027
5.166	0.74	0.04	0.023
5.276	0.76	0.04	0.019
5.391	0.78	0.04	0.016
5.510	0.79	0.05	0.015
5.637	0.81	0.05	0.012
5.767	0.83	0.05	0.009
6.048	0.85	0.05	0.007
6.199	0.87	0.05	0.006
6.358	0.88	0.05	0.005
6.526	0.90	0.06	0.004
6.702	0.91	0.06	0.003
6.888	0.92	0.06	0.003
7.085	0.92	0.06	0.003
7.293	0.93	0.06	0.002
7.514	0.93	0.06	0.002
7.749	0.94	0.06	0.002
7.999	0.94	0.06	0.002
8.260	0.94	0.06	0.002
8.551	0.94	0.06	0.002
8.856	0.94	0.05	0.002
9.184	0.94	0.05	0.002
9.537	0.94	0.04	0.001
9.919	0.94	0.04	0.001
10.33	0.94	0.03	0.001
11.0		0.03	
12.0		0.028	

Rhenium, single crystal, $\vec{E} \parallel \vec{c}$[9]

Energy (eV)	n	k	$R(\varphi = 0)$
0.10	6.06	51.03	0.991
0.15	4.66	33.96	0.984
0.20	4.16	25.36	0.975
0.25	4.03	20.10	0.962
0.30	4.37	16.69	0.943
0.35	4.50	14.53	0.925
0.40	4.53	12.96	0.909
0.45	4.53	11.78	0.893
0.50	4.53	10.88	0.878
0.55	4.50	10.26	0.867
0.60	4.29	9.75	0.861
0.65	4.07	9.35	0.856
0.70	3.80	8.94	0.853
0.75	3.48	8.55	0.850
0.80	3.21	8.10	0.846
0.85	2.96	7.68	0.841
0.90	2.73	7.24	0.835
0.95	2.56	6.79	0.826
1.00	2.45	6.36	0.813

Energy (eV)	n	k	$R(\phi = 0)$
1.10	2.38	5.61	0.778
1.20	2.35	5.02	0.742
1.30	2.39	4.54	0.702
1.40	2.44	4.13	0.662
1.50	2.50	3.79	0.624
1.60	2.59	3.49	0.587
1.70	2.70	3.27	0.557
1.80	2.82	3.10	0.535
1.90	2.90	3.00	0.520
2.00	2.97	2.91	0.510
2.10	3.03	2.86	0.504
2.20	3.06	2.84	0.501
2.30	3.07	2.82	0.499
2.40	3.06	2.81	0.498
2.50	3.02	2.80	0.497
2.60	2.96	2.77	0.493
2.70	2.89	2.68	0.482
2.80	2.89	2.57	0.468
2.90	2.99	2.47	0.457
3.00	3.11	2.57	0.470
3.20	2.90	2.68	0.482
3.40	2.83	2.50	0.459
3.60	2.93	2.48	0.457
3.80	2.86	2.56	0.467
4.00	2.81	2.51	0.460
4.20	2.86	2.55	0.466
4.40	2.81	2.74	0.489
4.60	2.56	2.83	0.504
4.80	2.41	2.71	0.493
5.00	2.39	2.68	0.488
5.20	2.34	2.75	0.500
5.40	2.20	2.81	0.515
5.60	2.02	2.84	0.530
5.80	1.83	2.80	0.538
6.00	1.65	2.71	0.541
6.20	1.54	2.59	0.532
6.40	1.45	2.50	0.526
6.80	1.32	2.31	0.508
7.00	1.26	2.23	0.500
7.20	1.20	2.15	0.493
7.40	1.16	2.06	0.480
7.60	1.12	1.99	0.470
7.80	1.08	1.89	0.454
8.00	1.05	1.80	0.435
8.20	1.05	1.71	0.411
8.40	1.05	1.62	0.386
8.60	1.06	1.55	0.360
8.80	1.09	1.48	0.336
9.00	1.11	1.43	0.317
9.20	1.13	1.39	0.301
9.40	1.16	1.34	0.281
9.60	1.18	1.32	0.274
9.80	1.20	1.29	0.264
10.00	1.23	1.26	0.252
10.20	1.25	1.25	0.246
10.40	1.28	1.25	0.242
10.60	1.29	1.25	0.242
10.80	1.30	1.26	0.244
11.00	1.30	1.27	0.247
11.20	1.29	1.28	0.249

Energy (eV)	n	k	$R(\phi = 0)$
11.40	1.28	1.28	0.252
11.60	1.26	1.28	0.252
11.80	1.24	1.26	0.249
12.00	1.23	1.24	0.244
12.40	1.22	1.21	0.237
12.80	1.21	1.18	0.230
13.20	1.22	1.16	0.222
13.60	1.22	1.13	0.215
14.00	1.24	1.12	0.209
14.40	1.27	1.11	0.204
14.80	1.29	1.15	0.213
15.20	1.29	1.19	0.225
15.60	1.26	1.22	0.236
16.00	1.23	1.25	0.248
16.40	1.19	1.27	0.259
16.80	1.14	1.29	0.269
17.00	1.12	1.30	0.275
17.40	1.07	1.30	0.286
18.00	0.99	1.30	0.300
18.40	0.93	1.29	0.311
18.80	0.87	1.28	0.321
19.20	0.81	1.25	0.330
19.60	0.77	1.21	0.332
20.00	0.73	1.18	0.333
20.40	0.70	1.14	0.332
20.80	0.67	1.11	0.332
21.20	0.64	1.08	0.334
21.60	0.61	1.04	0.335
22.00	0.58	1.01	0.340
22.40	0.55	0.97	0.341
22.80	0.53	0.93	0.338
23.20	0.51	0.89	0.334
23.60	0.50	0.85	0.329
24.00	0.48	0.80	0.319
24.40	0.48	0.76	0.207
24.80	0.47	0.72	0.296
25.20	0.47	0.68	0.282
25.60	0.47	0.65	0.270
26.00	0.47	0.61	0.255
26.40	0.48	0.57	0.240
26.80	0.48	0.54	0.225
27.20	0.49	0.51	0.208
27.60	0.50	0.48	0.193
28.00	0.51	0.45	0.176
29.00	0.54	0.39	0.145
30.00	0.57	0.33	0.114
31.00	0.62	0.29	0.086
32.00	0.66	0.26	0.065
33.00	0.68	0.24	0.054
34.00	0.72	0.21	0.041
35.00	0.76	0.20	0.031
36.00	0.79	0.20	0.025
37.00	0.82	0.19	0.021
38.00	0.85	0.20	0.018
39.00	0.89	0.21	0.016
40.00	0.88	0.26	0.022
42.00	0.88	0.26	0.022
44.00	0.89	0.29	0.026
46.00	0.85	0.32	0.035
48.00	0.82	0.30	0.036

Energy (eV)	n	k	$R(\phi = 0)$
50.00	0.80	0.30	0.038
52.00	0.78	0.30	0.044
54.00	0.72	0.30	0.055
56.00	0.66	0.24	0.061
58.00	0.65	0.16	0.055
Rhenium, single crystal, $\vec{E} \perp \vec{c}$[9]			
0.10	4.25	42.83	0.991
0.15	3.28	28.08	0.984
0.20	3.28	20.66	0.971
0.25	3.47	16.27	0.951
0.30	3.73	13.44	0.926
0.35	3.93	11.54	0.900
0.40	3.99	10.15	0.875
0.45	4.17	9.03	0.846
0.50	4.34	8.26	0.821
0.55	4.45	7.73	0.801
0.60	4.53	7.40	0.788
0.65	4.44	7.26	0.784
0.70	4.13	7.09	0.784
0.75	3.77	6.75	0.779
0.80	3.55	6.32	0.766
0.85	3.39	5.95	0.752
0.90	3.26	5.61	0.737
0.95	3.17	5.27	0.719
1.00	3.09	4.96	0.701
1.10	3.05	4.39	0.658
1.20	3.08	3.89	0.613
1.30	3.20	3.56	0.578
1.40	3.23	3.38	0.559
1.50	3.23	3.12	0.532
1.60	3.29	2.88	0.507
1.70	3.38	2.72	0.491
1.80	3.47	2.59	0.480
1.90	3.54	2.50	0.473
2.00	3.63	2.43	0.469
2.10	3.74	2.40	0.470
2.20	3.83	2.38	0.472
2.30	3.93	2.44	0.481
2.40	4.00	2.55	0.492
2.50	4.01	2.70	0.505
2.60	3.90	2.84	0.514
2.70	3.74	2.92	0.517
2.80	3.57	2.88	0.511
2.90	3.49	2.75	0.497
3.00	3.53	2.71	0.493
3.20	3.55	2.84	0.506
3.40	3.34	2.88	0.508
3.60	3.25	2.83	0.501
3.80	3.24	2.84	0.502
4.00	3.19	2.94	0.513
4.20	3.05	3.06	0.526
4.40	2.88	3.15	0.539
4.60	2.67	3.18	0.548
4.80	2.44	3.17	0.554
5.00	2.25	3.12	0.556
5.20	2.10	3.04	0.555
5.40	1.96	2.96	0.553
5.60	1.84	2.88	0.551
5.80	1.73	2.81	0.549
6.00	1.61	2.74	0.549

Energy (eV)	n	k	$R(\phi = 0)$
6.20	1.51	2.64	0.545
6.40	1.42	2.56	0.541
6.80	1.28	2.37	0.526
7.00	1.22	2.28	0.517
7.20	1.16	2.19	0.508
7.40	1.12	2.08	0.493
7.60	1.12	1.98	0.468
7.80	1.08	1.93	0.463
8.00	1.05	1.83	0.443
8.20	1.05	1.74	0.418
8.40	1.05	1.66	0.397
8.60	1.06	1.58	0.372
8.80	1.07	1.52	0.351
9.00	1.09	1.46	0.327
9.20	1.11	1.41	0.309
9.40	1.14	1.36	0.290
9.60	1.17	1.31	0.273
9.80	1.20	1.27	0.258
10.00	1.24	1.24	0.244
10.20	1.29	1.22	0.234
10.40	1.33	1.23	0.233
10.60	1.36	1.25	0.238
10.80	1.38	1.28	0.245
11.00	1.37	1.31	0.253
11.20	1.36	1.33	0.259
11.40	1.33	1.34	0.264
11.60	1.31	1.34	0.266
11.80	1.28	1.33	0.266
12.00	1.26	1.32	0.264
12.40	1.23	1.29	0.257
12.80	1.22	1.26	0.251
13.20	1.20	1.23	0.245
13.60	1.19	1.20	0.236
14.00	1.20	1.16	0.225
14.40	1.22	1.13	0.214
14.80	1.27	1.12	0.207
15.20	1.31	1.17	0.218
15.60	1.31	1.23	0.234
16.00	1.28	1.28	0.251
16.40	1.24	1.33	0.270
16.80	1.17	1.37	0.288
17.00	1.14	1.38	0.297
17.40	1.06	1.39	0.314
18.00	0.95	1.38	0.334
18.40	0.88	1.36	0.346
18.80	0.82	1.33	0.355
19.20	0.76	1.29	0.360
19.60	0.72	1.25	0.363
20.00	0.67	1.21	0.369
20.40	0.64	1.15	0.364
20.80	0.61	1.10	0.357
21.20	0.60	1.06	0.349
21.60	0.58	1.02	0.342
22.00	0.57	0.98	0.336
22.40	0.56	0.95	0.328
22.80	0.55	0.92	0.325
23.20	0.53	0.89	0.322
23.60	0.52	0.85	0.317
24.00	0.50	0.82	0.314
24.40	0.49	0.79	0.309
24.80	0.48	0.75	0.303
25.20	0.47	0.72	0.295
25.60	0.47	0.68	0.286
26.00	0.46	0.64	0.276
26.40	0.46	0.61	0.263
26.80	0.46	0.57	0.249
27.20	0.47	0.53	0.231
27.60	0.48	0.50	0.216
28.00	0.49	0.47	0.198
29.00	0.51	0.41	0.164
30.00	0.55	0.34	0.129
31.00	0.59	0.29	0.097
32.00	0.64	0.26	0.072
33.00	0.67	0.24	0.060
34.00	0.70	0.22	0.047
35.00	0.74	0.20	0.036
36.00	0.77	0.19	0.029
37.00	0.80	0.19	0.023
38.00	0.84	0.19	0.018
39.00	0.88	0.21	0.016
40.00	0.87	0.25	0.023
42.00	0.87	0.25	0.023
44.00	0.88	0.28	0.026
46.00	0.84	0.31	0.035
48.00	0.82	0.30	0.036
50.00	0.80	0.30	0.039
52.00	0.77	0.30	0.044
54.00	0.71	0.29	0.055
56.00	0.66	0.23	0.061
58.00	0.64	0.16	0.055

Rhodium[11]

Energy (eV)	n	k	$R(\phi = 0)$
0.10	18.48	69.43	0.986
0.20	8.66	37.46	0.977
0.30	5.85	25.94	0.967
0.40	4.74	19.80	0.955
0.50	4.20	16.07	0.941
0.60	3.87	13.51	0.925
0.70	3.67	11.72	0.908
0.80	3.63	10.34	0.887
0.90	3.62	9.36	0.867
1.00	3.71	8.67	0.848
1.10	3.67	8.26	0.837
1.20	3.51	7.94	0.832
1.30	3.26	7.63	0.829
1.40	3.01	7.31	0.827
1.50	2.78	6.97	0.823
1.60	2.60	6.64	0.818
1.70	2.42	6.33	0.813
1.80	2.30	6.02	0.805
1.90	2.20	5.76	0.798
2.00	2.12	5.51	0.789
2.10	2.05	5.30	0.780
2.20	2.00	5.11	0.772
2.30	1.94	4.94	0.765
2.40	1.90	4.78	0.756
2.50	1.88	4.65	0.748
2.60	1.85	4.55	0.743
2.70	1.80	4.49	0.742
2.90	1.63	4.36	0.748
3.00	1.53	4.29	0.753
3.10	1.41	4.20	0.760
3.20	1.30	4.09	0.764
3.30	1.20	3.97	0.767
3.40	1.11	3.84	0.769
3.50	1.04	3.71	0.768
3.60	0.99	3.58	0.764
3.70	0.95	3.45	0.759
3.80	0.91	3.34	0.753
3.90	0.88	3.23	0.747
4.00	0.86	3.12	0.739
4.20	0.83	2.94	0.722
4.40	0.80	2.76	0.706
4.60	0.78	2.60	0.684
4.80	0.79	2.46	0.659
5.00	0.79	2.34	0.635
5.20	0.79	2.23	0.613
5.40	0.80	2.14	0.591
5.60	0.80	2.06	0.573
5.80	0.79	2.00	0.561
6.00	0.76	1.93	0.556
6.20	0.73	1.85	0.544
6.40	0.70	1.77	0.534
6.60	0.68	1.69	0.518
6.80	0.67	1.60	0.498
7.00	0.66	1.52	0.476
7.20	0.66	1.43	0.452
7.40	0.66	1.35	0.423
7.60	0.67	1.27	0.394
7.80	0.68	1.20	0.363
8.00	0.69	1.12	0.329
8.20	0.71	1.04	0.288
8.40	0.74	0.97	0.252
8.60	0.78	0.89	0.212
8.80	0.83	0.83	0.179
9.00	0.88	0.77	0.148
9.20	0.95	0.73	0.125
9.40	1.01	0.71	0.110
9.60	1.07	0.69	0.102
9.80	1.12	0.69	0.098
10.00	1.17	0.69	0.098
10.60	1.26	0.73	0.106
11.00	1.29	0.76	0.113
11.60	1.32	0.80	0.124
12.00	1.32	0.82	0.127
12.60	1.32	0.82	0.129
13.00	1.32	0.83	0.131
13.60	1.32	0.85	0.134
14.00	1.32	0.86	0.138
14.60	1.30	0.89	0.144
15.00	1.28	0.90	0.147
15.60	1.25	0.90	0.147
16.00	1.24	0.89	0.147
16.50	1.23	0.88	0.145
17.00	1.22	0.88	0.144
17.50	1.22	0.87	0.143
18.00	1.23	0.88	0.145
18.50	1.25	0.92	0.155
19.00	1.24	0.98	0.172
19.50	1.18	1.05	0.193
20.00	1.10	1.09	0.213

Energy (eV)	n	k	R(φ = 0)
20.50	1.00	1.09	0.230
21.00	0.91	1.05	0.234
21.50	0.86	1.00	0.228
22.00	0.83	0.95	0.219
22.50	0.81	0.92	0.214
23.00	0.79	0.90	0.213
23.50	0.75	0.87	0.214
24.00	0.73	0.84	0.210
24.50	0.70	0.81	0.208
25.00	0.69	0.77	0.202
25.50	0.67	0.74	0.195
26.00	0.66	0.70	0.188
26.50	0.65	0.66	0.176
27.00	0.65	0.64	0.168
27.50	0.65	0.61	0.159
28.00	0.65	0.59	0.152
29.00	0.65	0.54	0.137
30.00	0.66	0.51	0.127
31.00	0.64	0.49	0.127
32.00	0.61	0.44	0.126
33.00	0.60	0.37	0.110
34.00	0.65	0.30	0.074
35.00	0.69	0.28	0.058
36.00	0.73	0.27	0.049
37.00	0.74	0.28	0.047
38.00	0.74	0.27	0.045
39.00	0.75	0.25	0.041

Ruthenium, single crystal, $\vec{E} \parallel \vec{c}$[5]

Energy (eV)	n	k	R(φ = 0)
0.10	11.50	51.38	0.984
0.20	5.93	27.14	0.970
0.30	4.33	18.50	0.953
0.40	3.60	13.97	0.933
0.50	3.18	11.04	0.909
0.60	3.28	8.89	0.865
0.70	3.62	7.73	0.822
0.80	3.42	7.02	0.801
0.90	3.25	6.12	0.766
1.00	3.39	5.33	0.715
1.10	3.66	4.83	0.675
1.20	3.84	4.57	0.654
1.30	3.94	4.38	0.638
1.40	4.02	4.19	0.624
1.50	4.16	4.07	0.614
1.60	4.33	4.08	0.615
1.70	4.42	4.21	0.624
1.80	4.40	4.38	0.636
1.90	4.29	4.61	0.651
2.00	4.04	4.81	0.667
2.10	3.69	4.90	0.679
2.20	3.35	4.82	0.683
2.30	3.09	4.70	0.681
2.40	2.89	4.55	0.677
2.50	2.74	4.40	0.671
2.60	2.64	4.25	0.663
2.70	2.58	4.14	0.656
2.80	2.54	4.05	0.650
2.90	2.48	4.03	0.650
3.00	2.38	4.03	0.656
3.10	2.26	4.00	0.661
3.20	2.13	3.96	0.666
3.30	2.00	3.91	0.671
3.40	1.87	3.83	0.673
3.50	1.76	3.74	0.674
3.60	1.66	3.65	0.675
3.70	1.57	3.55	0.673
3.80	1.49	3.45	0.672
3.90	1.42	3.35	0.668
4.00	1.37	3.24	0.661
4.20	1.29	3.08	0.649
4.40	1.22	2.93	0.639
4.60	1.16	2.79	0.628
4.80	1.11	2.67	0.617
5.00	1.06	2.56	0.607
5.20	1.01	2.46	0.600
5.40	0.95	2.35	0.593
5.60	0.92	2.23	0.576
5.80	0.90	2.14	0.559
6.00	0.88	2.05	0.545
6.20	0.87	1.98	0.531
6.40	0.84	1.91	0.521
6.60	0.82	1.84	0.510
6.80	0.79	1.77	0.500
7.00	0.76	1.69	0.489
7.20	0.75	1.61	0.472
7.40	0.73	1.54	0.455
7.60	0.73	1.46	0.433
7.80	0.73	1.39	0.411
8.00	0.72	1.33	0.391
8.20	0.72	1.26	0.366
8.40	0.73	1.20	0.342
8.60	0.74	1.14	0.318
8.80	0.74	1.08	0.295
9.00	0.75	1.02	0.267
9.20	0.77	0.97	0.243
9.40	0.79	0.91	0.217
9.60	0.82	0.86	0.190
9.80	0.85	0.81	0.167
10.00	0.88	0.76	0.144
10.20	0.92	0.72	0.125
10.40	0.96	0.69	0.110
10.60	1.01	0.67	0.100
10.80	1.05	0.66	0.094
11.00	1.09	0.65	0.090
11.20	1.12	0.65	0.088
11.40	1.15	0.65	0.087
11.60	1.18	0.65	0.088
11.80	1.21	0.66	0.090
12.00	1.23	0.67	0.092
12.40	1.26	0.69	0.098
12.80	1.27	0.72	0.104
13.20	1.28	0.74	0.108
13.60	1.28	0.75	0.111
14.00	1.28	0.76	0.114
14.40	1.27	0.76	0.114
14.80	1.27	0.76	0.114
15.00	1.27	0.76	0.114
15.60	1.28	0.77	0.115
16.00	1.30	0.78	0.118
16.50	1.32	0.80	0.123
17.00	1.34	0.85	0.136
17.50	1.32	0.93	0.155
18.00	1.26	0.99	0.173
18.50	1.18	1.02	0.185
19.00	1.11	1.02	0.192
19.50	1.05	1.02	0.199
20.00	0.99	1.02	0.208
20.50	0.92	0.99	0.212
21.00	0.86	0.94	0.209
21.50	0.83	0.90	0.203
22.00	0.81	0.86	0.193
23.00	0.77	0.79	0.182
24.00	0.74	0.74	0.171
25.00	0.71	0.69	0.163
26.00	0.68	0.63	0.154
27.00	0.67	0.57	0.140
28.00	0.66	0.51	0.124
29.00	0.67	0.46	0.107
30.00	0.67	0.43	0.097
31.00	0.67	0.37	0.084
32.00	0.69	0.33	0.070
33.00	0.71	0.30	0.058
34.00	0.73	0.27	0.048
35.00	0.75	0.25	0.039
36.00	0.77	0.24	0.035
37.00	0.79	0.23	0.039
38.00	0.80	0.22	0.027
39.00	0.82	0.22	0.024
40.00	0.83	0.22	0.022

Ruthenium, single crystal, $\vec{E} \perp \vec{c}$[5]

Energy (eV)	n	k	R(φ = 0)
0.10	11.85	50.81	0.983
0.20	6.68	27.18	0.966
0.30	4.94	18.92	0.950
0.40	3.90	14.51	0.933
0.50	3.27	11.63	0.915
0.60	2.98	9.54	0.888
0.70	2.82	7.99	0.856
0.80	2.73	6.71	0.815
0.90	2.82	5.54	0.751
1.00	3.17	4.59	0.670
1.10	3.69	3.91	0.604
1.20	4.28	3.66	0.585
1.30	4.66	3.72	0.593
1.40	4.86	3.79	0.601
1.50	4.99	3.89	0.609
1.60	5.08	4.03	0.618
1.70	5.12	4.22	0.629
1.80	5.10	4.45	0.642
1.90	4.96	4.78	0.660
2.00	4.61	5.06	0.677
2.10	4.21	5.09	0.682
2.20	3.94	5.00	0.681
2.30	3.69	4.97	0.684
2.40	3.44	4.88	0.684
2.50	3.27	4.77	0.681
2.60	3.14	4.66	0.677
2.70	3.06	4.59	0.674
2.80	2.99	4.59	0.676
2.90	2.87	4.64	0.686
3.00	2.64	4.69	0.701
3.10	2.40	4.64	0.710

Solids

Energy (eV)	n	k	R(φ = 0)
3.20	2.18	4.55	0.717
3.30	2.00	4.43	0.721
3.40	1.84	4.30	0.723
3.50	1.71	4.16	0.723
3.60	1.60	4.03	0.722
3.70	1.50	3.90	0.721
3.80	1.41	3.77	0.718
3.90	1.35	3.64	0.713
4.00	1.29	3.53	0.707
4.20	1.21	3.31	0.694
4.40	1.16	3.13	0.679
4.60	1.13	2.97	0.662
4.80	1.09	2.86	0.652
5.00	1.03	2.75	0.648
5.20	0.97	2.64	0.643
5.40	0.91	2.52	0.635
5.60	0.88	2.40	0.622
5.80	0.86	2.29	0.605
6.00	0.84	2.20	0.591
6.20	0.82	2.11	0.576
6.40	0.81	2.04	0.564
6.60	0.78	1.97	0.556
6.80	0.76	1.89	0.545
7.00	0.73	1.82	0.538
7.20	0.70	1.75	0.527
7.40	0.68	1.67	0.513
7.60	0.67	1.59	0.496
7.80	0.66	1.51	0.476
8.00	0.66	1.44	0.454
8.20	0.65	1.36	0.430
8.40	0.66	1.29	0.403
8.60	0.66	1.22	0.378
8.80	0.68	1.15	0.346
9.00	0.69	1.09	0.317
9.20	0.70	1.02	0.286
9.40	0.73	0.95	0.251
9.60	0.77	0.89	0.216
9.80	0.82	0.84	0.185
10.00	0.86	0.81	0.163
10.20	0.90	0.77	0.143
10.40	0.94	0.74	0.127
10.60	0.99	0.72	0.115
10.80	1.04	0.71	0.108
11.00	1.08	0.70	0.104
11.20	1.11	0.70	0.102
11.40	1.14	0.70	0.101
11.60	1.17	0.71	0.102
11.80	1.20	0.72	0.104
12.00	1.22	0.73	0.107
12.40	1.25	0.76	0.113
12.80	1.26	0.78	0.118
13.20	1.27	0.81	0.124
13.60	1.27	0.83	0.129
14.00	1.26	0.84	0.132
14.40	1.25	0.84	0.132
14.80	1.25	0.84	0.133
15.00	1.25	0.84	0.133
15.60	1.25	0.85	0.134
16.00	1.27	0.85	0.134
16.50	1.28	0.89	0.145
17.00	1.28	0.94	0.158
17.50	1.25	1.00	0.175
18.00	1.19	1.04	0.190
18.50	1.12	1.05	0.200
19.00	1.07	1.05	0.205
19.50	1.02	1.04	0.212
20.00	0.97	1.04	0.219
20.50	0.91	1.03	0.228
21.00	0.85	1.01	0.234
21.50	0.80	0.97	0.234
22.00	0.77	0.94	0.233
23.00	0.71	0.87	0.229
24.00	0.67	0.79	0.218
25.00	0.64	0.73	0.205
26.00	0.61	0.66	0.194
27.00	0.60	0.59	0.177
28.00	0.60	0.53	0.155
29.00	0.61	0.48	0.134
30.00	0.62	0.45	0.123
31.00	0.61	0.40	0.114
32.00	0.63	0.34	0.093
33.00	0.65	0.31	0.077
34.00	0.67	0.28	0.065
35.00	0.70	0.26	0.054
36.00	0.72	0.25	0.047
37.00	0.73	0.23	0.041
38.00	0.75	0.22	0.035
39.00	0.77	0.22	0.031
40.00	0.79	0.22	0.028

Selenium, single crystal, $\vec{E} \parallel \hat{c}$[22]

Energy (eV)	n	k	R(φ = 0)
0.01364	2.914	0.248	0.242
0.01488	3.175	9.95E-02	0.272
0.01612	3.263	2.13E-03	0.282
0.01736	3.306	3.81E-02	0.287
0.01860	3.330	7.04E-03	0.290
0.01984	3.346	4.23E-02	0.291
0.02108	3.358	3.40E-03	0.293
0.02232	3.366	5.31E-02	0.294
0.02356	3.372	1.96E-03	0.294
0.02480	3.377	2.39E-02	0.295
0.02604	3.380		0.295
0.02728		1.16E-02	
0.02976		7.96E-03	
0.03224		8.57E-03	
0.03472		2.70E-02	
0.03720	3.397	1.72E-02	0.297
0.04463		1.13E-02	
0.04959	3.403	2.79E-03	0.298
0.05703		1.56E-03	
0.06199	3.405	1.35E-03	0.298
0.06819		5.79E-04	
0.07439	3.407	4.44E-04	0.298
0.08059		4.41E-04	
0.08679	3.408	4.32E-04	0.298
0.09299		2.44E-04	
0.09919	3.409	3.23E-04	0.299
0.1116	3.409	2.87E-04	0.299
0.1240	3.410	2.71E-04	0.299
0.2480	3.417	2.67E-04	0.299
0.3720	3.427	1.90E-04	0.301
0.4959	3.442	1.41E-04	0.302
0.6199	3.462	1.12E-04	0.304
0.7439	3.486	9.42E-05	0.307
0.8679	3.516	8.07E-05	0.310
0.9919	3.551	7.11E-05	0.314
1.116	3.592	6.37E-05	0.319
1.240	3.640	5.81E-05	0.324
1.50		1.33E-04	
1.60		1.59E-04	
1.70		6.27E-04	
1.80	4.46	2.20E-02	0.402
2.0	4.79	0.76	0.438
2.2	4.49	1.19	0.431
2.4	4.28	1.21	0.417
2.6	4.40	1.32	0.430
2.8	4.59	1.70	0.462
3.0	4.44	2.29	0.490
3.2	3.92	2.59	0.493
3.4	3.69	2.76	0.502
3.6	3.39	3.01	0.521
3.8	(3.00)		
4.0	(2.65)		
4.2	(2.30)		
4.5	1.92	2.78	0.528
5.0	1.50	2.31	0.482
6.0	1.57	1.49	0.288
7.0	1.84	1.45	0.276
8.0	1.35	1.68	0.353
9.0	1.35	1.64	0.342
10.0	0.92	1.07	0.238
12.0	1.00	1.10	0.232
14.0	0.81	0.91	0.211
16.0	0.65	0.61	0.160
18.0	0.65	0.48	0.120
20.0	0.69	0.36	0.076
22.0	0.81	0.25	0.030
24.0	0.91	0.18	0.011
26.0	0.86	0.15	0.012
28.0	0.85	0.13	0.011
30.0	0.87	0.11	0.008

Selenium, single crystal, $\vec{E} \perp \hat{c}$[22]

Energy (eV)	n	k	R(φ = 0)
0.01364	2.854	0.0239	0.231
0.01488	2.932	0.0325	0.241
0.01612	3.140	0.1750	0.269
0.01736	2.959	1.3300	0.321
0.01860	2.111	0.2550	0.133
0.01984	2.356	0.0746	0.164
0.02108	2.462	0.0276	0.178
0.02232	2.502	0.0442	0.184
0.02356	2.543	0.0097	0.190
0.02480	2.550	0.0239	0.191
0.02604	2.582		0.195
0.02728	2.600	0.0101	0.198
0.02976	2.576	9.95E-03	0.194
0.03224	2.598	1.16E-02	0.197
0.03472	2.607	1.68E-02	0.199
0.03720	2.613	1.54E-02	0.199
0.04463		1.17E-02	
0.04959	2.627	3.58E-03	0.201
0.05703		8.65E-04	

Energy (eV)	n	k	R(φ = 0)
0.06199	2.632	2.07E-03	0.202
0.06819		2.89E-04	
0.07439	2.635	1.59E-04	0.202
0.08059		1.35E-04	
0.08679	2.636	1.42E-04	0.202
0.09299		1.04E-04	
0.09919	2.637	8.95E-05	0.203
0.1116	2.638	8.84E-05	0.203
0.1240	2.639	8.51E-05	0.203
0.2480	2.645	5.97E-05	0.204
0.3720	2.652	5.44E-05	0.205
0.4959	2.654	4.58E-05	0.205
0.6199	2.675	3.82E-05	0.208
0.7439	2.692	3.32E-05	0.210
0.8679	2.713	2.96E-05	0.213
0.9919	2.739	2.69E-05	0.216
1.116	2.772	2.48E-05	0.221
1.240	2.816	2.31E-05	0.226
1.50		7.37E-05	
1.60		8.63E-05	
1.70		3.60E-04	
1.80	3.32	0.11	0.289
2.00	3.38	0.65	0.310
2.20	3.07	0.73	0.282
2.40	2.93	0.61	0.259
2.60	3.00	0.53	0.263
2.80	3.12	0.58	0.279
3.00	3.30	0.70	0.305
3.20	3.35	1.01	0.328
3.40	3.22	1.24	0.334
3.60	3.06	1.47	0.344
3.80	2.84	1.66	0.351
4.00	2.51	1.81	0.356
4.20	2.18	1.83	0.352
4.50	1.75	1.94	0.382
5.00	1.25	1.50	0.316
6.00	1.32	0.73	0.107
7.00	1.62	0.61	0.105
8.00	1.81	0.69	0.135
9.00	1.66	1.02	0.182
10.00	1.72	0.95	0.171
12.00	1.25	1.02	0.181
14.00	0.98	0.92	0.178
16.00	0.68	0.96	0.274
18.00	0.61	0.65	0.191
20.00	0.73	0.48	0.094
22.00	0.78	0.39	0.060
24.00	0.78	0.32	0.046
26.00	0.78	0.26	0.036
28.00	0.80	0.19	0.023
30.00	0.79	0.14	0.020

Silicon, single crystal[23]

Energy (eV)	n	k	R(φ = 0)
0.01240	3.4185	2.90E-04	0.300
0.01488	3.4190	2.30E-04	0.300
0.01736	3.4192	1.90E-04	0.300
0.01984	3.4195	1.70E-04	0.300
0.02480	3.4197		0.300
0.03100	3.4199		0.300
0.04092	3.4200		0.300
0.04463		1.08E-04	
0.04959	3.4201	9.15E-05	0.300
0.05703		1.56E-04	
0.06199	3.4204	2.86E-04	0.300
0.06943		3.84E-04	
0.07439		7.16E-04	
0.08059	(3.4207)	1.52E-04	0.300
0.08679		1.02E-04	
0.09299		2.59E-04	
0.09919		1.77E-04	
0.1054		1.53E-04	
0.1116		2.02E-04	
0.1178		1.22E-04	
0.1240	3.4215	6.76E-05	0.300
0.1364		5.49E-05	
0.1488		2.41E-05	
0.1612		2.49E-05	
0.1736	(3.4230)	1.68E-05	0.300
0.1798		2.45E-05	
0.1860		2.66E-06	
0.1922		1.74E-06	
0.1984		8.46E-07	
0.2046		5.64E-07	
0.2108	(3.4244)	4.17E-07	0.300
0.2170		4.05E-07	
0.2232		3.94E-07	
0.2294		3.26E-07	
0.2356		2.97E-07	
0.2418		2.82E-07	
0.2480	3.4261	1.99E-07	0.300
0.3100	3.4294		0.301
0.3626	3.4327		0.301
0.4568	3.4393	2.50E-09	0.302
0.6199	3.4490		0.303
0.8093	3.4784		0.306
1.033	3.5193		0.311
1.1	(3.5341)	1.30E-05	0.312
1.2		1.80E-04	
1.3		2.26E-03	
1.4		7.75E-03	
1.5	3.673	5.00E-03	0.327
1.6	3.714	8.00E-03	0.331
1.7	3.752	1.00E-02	0.335
1.8	3.796	0.013	0.340
1.9	3.847	0.016	0.345
2.0	3.906	0.022	0.351
2.1	3.969	0.030	0.357
2.2	4.042	0.032	0.364
2.3	4.123	0.048	0.372
2.4	4.215	0.060	0.380
2.5	4.320	0.073	0.390
2.6	4.442	0.090	0.400
2.7	4.583	0.130	0.412
2.8	4.753	0.163	0.426
2.9	4.961	0.203	0.442
3.0	5.222	0.269	0.461
3.1	5.570	0.387	0.486
3.2	6.062	0.630	0.518
3.3	6.709	1.321	0.561
3.4	6.522	2.705	0.592
3.5	5.610	3.014	0.575
3.6	5.296	2.987	0.564
3.7	5.156	3.058	0.563
3.8	5.065	3.182	0.568
3.9	5.016	3.346	0.577
4.0	5.010	3.587	0.591
4.1	5.020	3.979	0.614
4.2	4.888	4.639	0.652
4.3	4.086	5.395	0.703
4.4	3.120	5.344	0.726
4.5	2.451	5.082	0.740
4.6	1.988	4.678	0.742
4.7	1.764	4.278	0.728
4.8	1.658	3.979	0.710
4.9	1.597	3.749	0.693
5.0	1.570	3.565	0.675
5.1	1.571	3.429	0.658
5.2	1.589	3.354	0.646
5.3	1.579	3.353	0.647
5.4	1.471	3.366	0.663
5.5	1.340	3.302	0.673
5.6	1.247	3.206	0.675
5.7	1.180	3.112	0.673
5.8	1.133	3.045	0.672
5.9	1.083	2.982	0.673
6.0	1.010	2.909	0.677
6.5	0.847	2.73	0.688
7.0	0.682	2.45	0.691
7.5	0.563	2.21	0.693
8.0	0.478	2.00	0.691
8.5	0.414	1.82	0.688
9.0	0.367	1.66	0.683
9.5	0.332	1.51	0.672
10.0	0.306	1.38	0.661
12.0	0.257	0.963	0.590
14.0	0.275	0.641	0.460
16.0	0.345	0.394	0.297
18.0	0.455	0.219	0.159
20.0	0.567	0.0835	0.079
22.14	0.675	0.0405	0.038
24.31	0.752	0.0243	0.020
26.38	0.803	0.0178	0.012
28.18	0.834	0.0152	0.008
30.24	0.860	0.0138	0.006
31.79	0.877	0.0132	0.004
34.44	0.899	0.0121	0.003
36.47	0.913	0.0113	0.002
38.75	0.925	0.0104	0.002
40.00	0.930	0.0100	0.001

Silver[6]

Energy (eV)	n	k	R(φ = 0)
0.10	9.91	90.27	0.995
0.20	2.84	45.70	0.995
0.30	1.41	30.51	0.994
0.40	0.91	22.89	0.993
0.50	0.67	18.32	0.992
1.00	0.28	9.03	0.987
1.50	0.27	5.79	0.969
2.00	0.27	4.18	0.944
2.50	0.24	3.09	0.914
3.00	0.23	2.27	0.864
3.25	0.23	1.86	0.816

Energy (eV)	n	k	$R(\phi = 0)$
3.50	0.21	1.42	0.756
3.60	0.23	1.13	0.671
3.70	0.30	0.77	0.475
3.77	0.53	0.40	0.154
3.80	0.73	0.30	0.053
3.90	1.30	0.36	0.040
4.00	1.61	0.60	0.103
4.10	1.73	0.85	0.153
4.20	1.75	1.06	0.194
4.30	1.73	1.13	0.208
4.50	1.69	1.28	0.238
4.75	1.61	1.34	0.252
5.00	1.55	1.36	0.257
5.50	1.45	1.34	0.257
6.00	1.34	1.28	0.246
6.50	1.25	1.18	0.225
7.00	1.18	1.06	0.196
7.50	1.14	0.91	0.157
8.00	1.16	0.75	0.114
9.00	1.33	0.56	0.074
10.00	1.46	0.56	0.082
11.00	1.52	0.56	0.088
12.00	1.61	0.59	0.100
13.00	1.66	0.64	0.112
14.00	1.72	0.78	0.141
14.50	1.64	0.88	0.152
15.00	1.56	0.92	0.156
16.00	1.42	0.91	0.151
17.00	1.33	0.86	0.139
18.00	1.28	0.80	0.124
19.00	1.27	0.75	0.111
20.00	1.29	0.71	0.103
21.00	1.35	0.75	0.112
21.50	1.37	0.80	0.124
22.00	1.34	0.87	0.141
22.50	1.26	0.93	0.157
23.00	1.17	0.94	0.163
23.50	1.10	0.93	0.165
24.00	1.04	0.90	0.165
24.50	0.99	0.87	0.160
25.00	0.95	0.83	0.154
25.50	0.91	0.78	0.144
26.00	0.90	0.74	0.133
26.50	0.89	0.69	0.121
27.00	0.89	0.65	0.109
27.50	0.89	0.62	0.099
28.00	0.90	0.59	0.090
28.50	0.91	0.57	0.084
29.00	0.92	0.56	0.079
30.00	0.93	0.54	0.074
31.00	0.93	0.53	0.072
32.00	0.92	0.53	0.072
33.00	0.90	0.51	0.071
34.00	0.88	0.49	0.067
35.00	0.86	0.45	0.061
36.00	0.89	0.44	0.055
38.00	0.89	0.39	0.043
40.00	0.90	0.37	0.039
42.00	0.90	0.35	0.036
44.00	0.90	0.33	0.033

Energy (eV)	n	k	$R(\phi = 0)$
46.00	0.90	0.32	0.031
48.00	0.89	0.31	0.030
50.00	0.88	0.29	0.027
52.00	0.89	0.28	0.024
54.00	0.88	0.17	0.024
56.00	0.87	0.26	0.024
58.00	0.87	0.24	0.021
60.00	0.87	0.22	0.018
62.00	0.88	0.21	0.016
64.00	0.88	0.21	0.016
66.00	0.88	0.21	0.016
68.00	0.87	0.21	0.017
70.00	0.83	0.20	0.021
72.00	0.85	0.18	0.016
74.00	0.85	0.17	0.014
76.00	0.85	0.16	0.013
78.00	0.85	0.15	0.013
80.00	0.85	0.14	0.012
85.00	0.85	0.11	0.011
90.00	0.85	0.08	0.009
95.00	0.86	0.06	0.007
100.00	0.87	0.04	0.005

Sodium[24]

Energy (eV)	n	k	$R(\phi = 0)$
0.55	0.262	9.97	0.990
0.58	0.241	9.45	0.989
0.63	0.207	8.80	0.990
0.67	0.175	8.09	0.990
0.73	0.147	7.42	0.990
0.81	0.123	6.67	0.989
0.92	0.099	5.82	0.989
1.05	0.078	5.11	0.989
1.23	0.064	4.35	0.987
1.44	0.053	3.72	0.986
1.65	0.050	3.22	0.983
1.87	0.049	2.76	0.978
2.07	0.053	2.48	0.971
2.27	0.059	2.23	0.961
2.45	0.063	2.07	0.953
2.64	0.066	1.88	0.943
2.82	0.068	1.76	0.936
2.95	0.068	1.63	0.928
3.06	0.069	1.54	0.921
3.20	0.065	1.47	0.921
3.40	0.061	1.33	0.916
3.71	0.055	1.13	0.908
3.97	0.049	1.01	0.908
6.199	0.390		0.193
6.358	0.454		0.141
6.526	0.485		0.120
6.702	0.533		0.093
6.888	0.574		0.073
7.130	0.616		0.056
7.328	0.641		0.048
7.583	0.674		0.038
7.847	0.700		0.031
8.015	0.710		0.029
8.634	0.762		0.018
9.143	0.800		0.012
9.709	0.819		0.010
10.20	0.843		0.007

Energy (eV)	n	k	$R(\phi = 0)$
11.08	0.870		0.005
11.83	0.887		0.004
12.73	0.907		0.002
13.05	0.913		0.002
13.42	0.914		0.002
13.73	0.917		0.002
14.07	0.922		0.002
14.83	0.934		0.001
15.05	0.936		0.001
15.46	0.942		0.001
16.21	0.948		0.001
18.10	0.964		0.000
21.12	0.979		0.000
25.51	0.993		0.000
26.95	1.00		0.000
27.68	1.01		0.000
28.37	1.01		0.000
29.52	1.02		0.000

Tantalum[16]

Energy (eV)	n	k	$R(\phi = 0)$
0.10	10.14	66.39	0.984
0.15	9.45	46.41	0.9834
0.20	5.77	35.46	0.982
0.26	3.67	27.53	0.981
0.30	2.87	23.90	0.980
0.38	2.03	18.87	0.978
0.50	1.37	14.26	0.974
0.58	1.15	12.19	0.970
0.70	0.96	9.92	0.962
0.78	0.89	8.77	0.956
0.90	0.84	7.38	0.942
1.00	0.89	6.47	0.992
1.10	0.93	5.75	0.899
1.20	0.98	5.14	0.872
1.30	1.00	4.62	0.842
1.40	1.04	4.15	0.805
1.50	1.09	3.73	0.762
1.60	1.15	3.33	0.707
1.70	1.24	2.95	0.640
1.80	1.35	2.60	0.560
1.90	1.57	2.24	0.460
2.00	1.83	1.99	0.388
2.10	2.10	1.84	0.354
2.20	2.36	1.81	0.351
2.30	2.56	1.86	0.365
2.40	2.68	1.92	0.378
2.50	2.75	1.98	0.388
2.60	2.80	2.02	0.395
2.70	2.84	2.08	0.405
2.80	2.85	2.14	0.412
2.90	2.84	2.20	0.420
3.00	2.81	2.24	0.425
3.20	2.73	2.31	0.432
3.40	2.61	2.33	0.435
3.60	2.49	2.30	0.430
3.80	2.40	2.22	0.418
4.00	2.36	2.14	0.406
4.20	2.35	2.06	0.392
4.40	2.39	2.01	0.384
4.60	2.45	2.00	0.384
4.80	2.53	2.06	0.394

Solids

Energy (eV)	n	k	$R(\phi = 0)$
5.00	2.58	2.20	0.416
5.20	2.52	2.44	0.450
5.40	2.31	2.61	0.480
5.60	2.06	2.67	0.501
5.80	1.83	2.63	0.510
6.00	1.63	2.56	0.515
6.20	1.48	2.45	0.512
6.40	1.37	2.33	0.504
6.60	1.29	2.22	0.492
6.80	1.23	2.11	0.478
7.00	1.18	2.01	0.462
7.20	1.15	1.91	0.445
7.40	1.13	1.82	0.425
7.60	1.12	1.75	0.406
7.80	1.11	1.68	0.390
8.00	1.11	1.61	0.370
8.20	1.12	1.55	0.350
8.40	1.13	1.50	0.332
8.60	1.14	1.45	0.317
8.80	1.17	1.41	0.301
9.00	1.19	1.40	0.294
9.20	1.21	1.38	0.289
9.40	1.21	1.38	0.287
9.60	1.21	1.38	0.285
9.80	1.21	1.37	0.285
10.00	1.20	1.37	0.286
10.20	1.19	1.37	0.286
10.40	1.18	1.37	0.287
10.60	1.16	1.36	0.288
10.80	1.15	1.36	0.289
11.00	1.13	1.35	0.290
11.20	1.11	1.35	0.292
11.40	1.09	1.34	0.293
11.60	1.07	1.33	0.294
11.80	1.05	1.32	0.295
12.00	1.02	1.31	0.296
12.20	1.00	1.29	0.295
12.40	0.98	1.28	0.294
12.60	0.96	1.26	0.292
12.80	0.94	1.24	0.289
13.00	0.93	1.22	0.286
13.60	0.91	1.16	0.272
14.00	0.90	1.15	0.272
14.60	0.85	1.15	0.285
15.00	0.80	1.13	0.293
15.60	0.72	1.08	0.301
16.00	0.68	1.04	0.304
16.60	0.63	0.97	0.301
17.00	0.60	0.92	0.296
17.60	0.60	0.92	0.296
18.00	0.55	0.79	0.274
18.60	0.53	0.71	0.254
19.00	0.53	0.65	0.236
19.60	0.53	0.57	0.207
20.00	0.54	0.52	0.185
20.60	0.55	0.44	0.153
21.00	0.57	0.39	0.127
21.60	0.64	0.34	0.089
22.00	0.64	0.32	0.081
22.60	0.69	0.27	0.058
23.00	0.73	0.24	0.043
23.60	0.80	0.26	0.033
24.00	0.80	0.26	0.034
24.60	0.82	0.25	0.029
25.00	0.83	0.25	0.026
25.60	0.86	0.24	0.022
26.00	0.88	0.25	0.022
26.60	0.87	0.26	0.023
27.00	0.87	0.25	0.022
27.60	0.89	0.23	0.019
28.00	0.90	0.23	0.017
28.60	0.91	0.22	0.015
29.00	0.92	0.22	0.014
29.60	0.94	0.22	0.014
30.00	0.95	0.22	0.014
31.00	0.97	0.23	0.014
32.00	0.98	0.24	0.015
33.00	0.98	0.25	0.015
34.00	0.99	0.25	0.016
35.00	0.99	0.26	0.017
36.00	0.99	0.27	0.018
37.00	0.99	0.28	0.019
38.00	0.98	0.28	0.021
39.00	0.97	0.29	0.022
40.00	0.95	0.29	0.023

***Tellurium*, $\vec{E} \parallel \hat{c}$**[25]

Energy (eV)	n	k	$R(\phi = 0)$
0.01364	4.82	0.118	0.431
0.01488	5.26	0.0505	0.463
0.01612	5.47	0.0278	0.477
0.01736	5.59	0.0174	0.485
0.01860		0.0796	
0.01984		0.0696	
0.02108		0.0749	
0.02232		0.1900	
0.02356		0.2220	
0.02480		0.0716	
0.02604		0.0682	
0.02728		0.0832	
0.02976		0.0149	
0.03224		2.14E-03	
0.03472		1.71E-02	
0.03720	5.94	3.71E-03	0.507
0.03968		2.44E-03	
0.04339	5.96	1.59E-03	0.508
0.04711		7.85E-04	
0.05083		7.38E-04	
0.05579		3.89E-04	
0.06199	5.98	3.09E-04	0.509
0.07439		2.52E-04	
0.08679		2.96E-04	
0.09919		3.68E-04	
0.12400	6.246	3.34E-04	0.524
0.15500	6.253		0.525
0.20660	6.286		0.526
0.24800	6.316	7.48E-05	0.528
0.31	6.372	1.18E-05	0.531
0.35		4.93E-04	
0.41		6.74E-03	
0.5	6.53	2.30E-02	0.539
0.6	6.71	7.50E-02	0.549
0.7	7.00	0.24	0.563
0.8	7.23	0.48	0.574
0.9	7.48	0.94	0.589
1.0	7.70	1.56	0.606
1.2	6.99	2.22	0.593
1.4	7.11	2.46	0.604
1.6	6.75	2.91	0.606
1.8	6.89	3.70	0.637
2.0	4.67	4.67	0.654
2.2	4.94	5.16	0.681
2.4	3.94	5.08	0.686
2.6	3.25	4.77	0.681
2.8	2.73	4.42	0.674
3.0	2.30	4.16	0.674
3.5	1.69	3.44	0.646
4.0	1.33	2.64	0.571
4.5	1.32	1.96	0.428
5.0	1.63	1.60	0.312
5.5	1.72	1.57	0.302
6.0	1.73	1.45	0.276
6.5	1.78	1.36	0.257
7.0	1.83	1.36	0.257
7.5	1.72	1.51	0.289
8.0	1.54	1.37	0.260
8.5	1.55	1.23	0.226
9.0	0.99	0.93	0.179
9.5	1.47	1.25	0.233
10.0	0.86	0.86	0.181
11.0	0.80	0.77	0.165
12.0	0.79	0.76	0.164
14.0	0.67	0.59	0.146
16.0	0.59	0.49	0.147
18.0	0.48	0.31	0.160
20.0	0.74	0.20	0.035
22.0	0.83	0.18	0.018
24.0	0.85	0.15	0.013
26.0	0.87	0.12	0.009
28.0	0.89	0.090	0.006
30.0	0.90	0.045	0.003

***Tellurium*, $\vec{E} \perp \hat{c}$**[25]

Energy (eV)	n	k	$R(\phi = 0)$
0.01364	2.61	0.2980	0.204
0.01488	3.65	0.0894	0.325
0.01612	4.10	0.0535	0.370
0.01736	4.63	0.4990	0.420
0.01860		0.1170	
0.01984		0.0343	
0.02108	(4.42)	0.0421	0.398
0.02232		0.1060	
0.02356		0.0880	
0.02480		0.0458	
0.02604		0.0928	
0.02728		0.0886	
0.02976		0.0232	
0.03224		3.06E-03	
0.03472		1.25E-02	
0.03720	4.71	2.65E-03	0.422
0.03968		1.89E-03	
0.04339	4.74	1.41E-03	0.425
0.04711		8.38E-04	
0.05083		6.79E-04	

Energy (eV)	n	k	$R(\phi = 0)$
0.05579		1.59E-04	
0.06199	4.77	1.16E-04	0.427
0.07439		7.23E-05	
0.08679		5.34E-05	
0.09919		4.28E-05	
0.1240	4.796	3.18E-05	0.429
0.1550	4.809		0.430
0.2066	4.838		0.432
0.2480	4.864	2.19E-05	0.434
0.31	4.929	3.18E-05	0.439
0.35		7.89E-02	
0.41		0.149	
0.5	4.90		0.437
0.6	4.93		0.439
0.7	4.95	0.11	0.441
0.8	5.10	0.13	0.452
0.9	5.22	0.22	0.461
1.0	5.35	0.45	0.472
1.2	5.17	0.63	0.462
1.4	5.56	0.63	0.488
1.6	5.88	1.15	0.517
1.8	6.10	1.80	0.545
2.0	5.94	2.69	0.571
2.2	5.10	3.61	0.594
2.4	4.24	3.77	0.593
2.6	3.57	3.75	0.591
2.8	3.03	3.63	0.588
3.0	2.51	3.39	0.578
3.5	1.72	2.70	0.532
4.0	1.32	2.01	0.440
4.5	1.28	1.28	0.251
5.0	1.47	0.82	0.132
5.5	1.74	0.51	0.104
6.0	1.94	0.39	0.118
6.5	2.19	0.32	0.148
7.0	2.48	0.40	0.192
7.5	2.60	0.69	0.226
8.0	2.59	0.91	0.245
8.5	2.39	1.00	0.235
9.0	1.11	1.24	0.259
9.5	2.08	1.11	0.224
10.0	0.99	1.04	0.215
11.0	0.84	1.01	0.237
12.0	0.87	0.87	0.182
14.0	0.59	0.87	0.282
16.0	0.64	0.55	0.144
18.0	0.52	0.41	0.161
20.0	0.50	0.38	0.165
22.0	0.56	0.29	0.110
24.0	0.54	0.25	0.113
26.0	0.50	0.20	0.127
28.0	0.48	0.17	0.135
30.0	0.46	0.088	0.140

Titanium (Polycrystalline)[14]

Energy (eV)	n	k	$R(\phi = 0)$
0.10	5.03	23.38	0.965
0.15	3.00	15.72	0.954
0.20	2.12	11.34	0.939
0.25	2.05	8.10	0.890
0.30	6.39	9.94	0.833
0.35	2.74	6.21	0.792
0.40	2.49	4.68	0.708
0.45	3.35	3.25	0.545
0.50	4.43	3.22	0.555
0.60	4.71	3.77	0.597
0.70	4.38	3.89	0.603
0.80	4.04	3.82	0.596
0.90	3.80	3.65	0.582
1.00	3.62	3.52	0.570
1.10	3.47	3.40	0.560
1.20	3.35	3.30	0.550
1.30	3.28	3.25	0.546
1.40	3.17	3.28	0.549
1.50	2.98	3.32	0.557
1.60	2.74	3.30	0.559
1.70	2.54	3.23	0.557
1.80	2.36	3.11	0.550
1.90	2.22	2.99	0.540
2.00	2.11	2.88	0.530
2.10	2.01	2.77	0.520
2.20	1.92	2.67	0.509
2.30	1.86	2.56	0.495
2.40	1.81	2.47	0.483
2.50	1.78	2.39	0.471
2.60	1.75	2.34	0.462
2.70	1.71	2.29	0.456
2.80	1.68	2.25	0.451
2.90	1.63	2.21	0.447
3.00	1.59	2.17	0.444
3.10	1.55	2.15	0.442
3.20	1.50	2.12	0.442
3.30	1.44	2.09	0.442
3.40	1.37	2.06	0.443
3.50	1.30	2.01	0.443
3.60	1.24	1.96	0.441
3.70	1.17	1.90	0.436
3.80	1.11	1.83	0.430
3.85	1.08	1.78	0.423
3.90	1.06	1.73	0.413
4.00	1.04	1.62	0.389
4.20	1.05	1.45	0.333
4.40	1.13	1.33	0.284
4.60	1.17	1.29	0.265
4.80	1.21	1.23	0.244
5.00	1.24	1.21	0.236
5.20	1.27	1.20	0.228
5.40	1.17	1.16	0.228
5.60	1.24	1.21	0.234
5.80	1.21	1.22	0.241
6.00	1.15	1.21	0.244
6.20	1.11	1.18	0.240
6.40	1.08	1.14	0.232
6.60	1.04	1.06	0.212
6.80	1.05	1.02	0.198
7.00	1.06	0.97	0.182
7.20	1.07	0.95	0.175
7.40	1.11	0.94	0.167
7.60	1.09	0.92	0.165
7.80	1.11	0.93	0.165
8.00	1.10	0.94	0.169
8.20	1.10	0.95	0.171
8.40	1.08	0.95	0.175
8.60	1.04	0.96	0.181
8.80	1.02	0.95	0.181
9.00	1.00	0.94	0.182
9.20	0.97	0.93	0.182
9.40	0.95	0.91	0.181
9.60	0.94	0.90	0.179
9.80	0.91	0.88	0.179
10.00	0.89	0.88	0.180
10.20	0.86	0.85	0.178
10.40	0.85	0.83	0.175
10.60	0.81	0.79	0.167
10.80	0.80	0.76	0.162
11.00	0.79	0.72	0.152
11.20	0.81	0.69	0.139
11.40	0.81	0.69	0.139
11.60	0.79	0.68	0.139
11.80	0.78	0.67	0.137
12.00	0.77	0.65	0.132
12.80	0.76	0.55	0.106
13.20	0.76	0.52	0.097
13.60	0.76	0.48	0.087
14.00	0.77	0.45	0.077
14.40	0.77	0.42	0.069
14.80	0.79	0.38	0.058
15.20	0.79	0.36	0.052
15.60	0.79	0.32	0.045
16.00	0.83	0.31	0.037
16.40	0.84	0.28	0.030
16.80	0.87	0.27	0.025
17.20	0.90	0.25	0.020
17.60	0.93	0.25	0.017
18.00	0.94	0.24	0.165
18.40	0.94	0.23	0.017
18.80	0.95	0.24	0.016
19.20	0.96	0.25	0.016
19.60	0.97	0.25	0.017
20.00	0.98	0.27	0.018
20.40	0.98	0.27	0.019
20.60	1.00	0.29	0.020
21.20	0.99	0.31	0.023
21.60	0.99	0.31	0.024
22.00	0.98	0.32	0.025
22.40	0.98	0.33	0.027
22.80	0.97	0.33	0.028
23.20	0.96	0.34	0.030
23.60	0.95	0.35	0.031
24.00	0.92	0.35	0.033
24.5	0.91	0.34	0.032
25.0	0.91	0.33	0.032
25.5	0.89	0.33	0.032
26.0	0.89	0.33	0.032
26.5	0.88	0.32	0.032
27.0	0.86	0.31	0.032
27.5	0.85	0.30	0.033
28.0	0.84	0.29	0.033
28.5	0.82	0.26	0.029
29.0	0.83	0.25	0.027
30.0	0.84	0.22	0.022

Energy (eV)	*n*	*k*	$R(\phi = 0)$
Tungsten[27]			
0.10	14.06	54.71	0.983
0.20	3.87	28.30	0.981
0.25	2.56	22.44	0.980
0.30	1.83	18.32	0.979
0.34	1.71	15.71	0.973
0.38	1.86	13.88	0.963
0.42	1.92	12.63	0.954
0.46	1.69	11.59	0.952
0.50	1.40	10.52	0.952
0.54	1.23	9.45	0.948
0.58	1.17	8.44	0.938
0.62	1.28	7.52	0.917
0.66	1.45	6.78	0.888
0.70	1.59	6.13	0.856
0.74	1.83	5.52	0.810
0.78	2.12	5.00	0.759
0.82	2.36	4.61	0.710
0.86	2.92	4.37	0.661
0.90	3.11	4.44	0.660
0.94	3.15	4.43	0.658
0.98	3.15	4.36	0.653
1.00	3.14	4.32	0.649
1.10	3.05	4.04	0.627
1.20	3.00	3.64	0.590
1.30	3.12	3.24	0.545
1.40	3.29	2.96	0.515
1.50	3.48	2.79	0.500
1.60	3.67	2.68	0.494
1.70	3.84	2.79	0.507
1.80	3.82	2.91	0.518
1.90	3.70	2.94	0.518
2.00	3.60	2.89	0.512
2.10	3.54	2.84	0.506
2.20	3.49	2.76	0.497
2.30	3.49	2.72	0.494
2.40	3.45	2.72	0.493
2.50	3.38	2.68	0.487
2.60	3.34	2.62	0.480
2.70	3.31	2.55	0.472
2.80	3.31	2.49	0.466
2.90	3.32	2.45	0.461
3.00	3.35	2.42	0.459
3.10	3.39	2.41	0.460
3.20	3.43	2.45	0.465
3.30	3.45	2.55	0.476
3.40	3.39	2.66	0.485
3.50	3.24	2.70	0.488
3.60	3.13	2.67	0.482
3.70	3.05	2.62	0.476
3.80	2.99	2.56	0.468
3.90	2.96	2.50	0.460
4.00	2.95	2.43	0.451
4.20	3.02	2.33	0.440
4.40	3.13	2.32	0.442
4.60	3.24	2.41	0.455
4.80	3.33	2.57	0.475
5.00	3.40	2.85	0.505
5.20	3.27	3.27	0.548
5.40	2.92	3.58	0.586
5.60	2.43	3.70	0.618
5.80	2.00	3.61	0.637
6.00	1.70	3.42	0.643
6.20	1.47	3.24	0.646
6.40	1.32	3.04	0.640
6.60	1.21	2.87	0.631
6.80	1.12	2.70	0.619
7.00	1.06	2.56	0.607
7.20	1.01	2.43	0.593
7.40	0.98	2.30	0.573
7.60	0.95	2.18	0.556
7.80	0.93	2.06	0.533
8.00	0.94	1.95	0.505
8.20	0.94	1.86	0.481
8.40	0.96	1.76	0.449
8.60	0.99	1.70	0.422
8.80	1.01	1.65	0.401
9.00	1.01	1.60	0.388
9.20	1.02	1.55	0.369
9.40	1.03	1.50	0.352
9.60	1.05	1.44	0.329
9.80	1.09	1.38	0.307
10.00	1.13	1.34	0.287
10.20	1.19	1.33	0.274
10.40	1.24	1.34	0.270
10.60	1.27	1.36	0.274
10.80	1.29	1.39	0.282
11.00	1.28	1.42	0.290
11.20	1.27	1.44	0.297
11.40	1.25	1.46	0.305
11.60	1.22	1.48	0.313
11.80	1.20	1.48	0.318
12.00	1.16	1.48	0.323
12.40	1.10	1.47	0.329
12.80	1.04	1.44	0.333
13.20	0.98	1.40	0.332
13.60	0.94	1.35	0.325
14.00	0.91	1.28	0.312
14.40	0.90	1.23	0.296
14.80	0.90	1.17	0.276
15.20	0.93	1.13	0.255
15.60	0.97	1.12	0.246
16.00	0.98	1.14	0.249
16.40	0.97	1.17	0.260
16.80	0.94	1.19	0.273
17.20	0.90	1.21	0.289
17.60	0.85	1.21	0.304
18.00	0.80	1.20	0.317
18.40	0.74	1.18	0.330
18.80	0.69	1.15	0.340
19.20	0.64	1.11	0.347
19.60	0.60	1.07	0.353
20.00	0.56	1.02	0.354
20.40	0.54	0.97	0.350
20.80	0.52	0.92	0.342
21.20	0.50	0.87	0.331
21.60	0.50	0.82	0.318
22.00	0.49	0.77	0.303
22.40	0.49	0.73	0.287
22.80	0.49	0.69	0.272
23.20	0.49	0.66	0.263
23.60	0.48	0.62	0.252
24.00	0.49	0.57	0.234
24.40	0.50	0.53	0.213
24.80	0.51	0.49	0.191
25.20	0.53	0.46	0.171
25.60	0.55	0.43	0.150
26.00	0.57	0.40	0.132
26.40	0.59	0.38	0.117
26.80	0.61	0.37	0.105
27.00	0.62	0.36	0.099
27.50	0.64	0.34	0.085
28.00	0.67	0.32	0.073
28.50	0.69	0.31	0.065
29.00	0.71	0.30	0.057
29.50	0.73	0.30	0.052
30.00	0.75	0.29	0.047
31.00	0.78	0.29	0.042
32.00	0.79	0.29	0.040
33.00	0.82	0.28	0.033
34.00	0.84	0.29	0.032
35.00	0.85	0.31	0.033
36.00	0.85	0.32	0.036
37.00	0.84	0.33	0.039
38.00	0.83	0.33	0.040
39.00	0.81	0.33	0.042
40.00	0.80	0.33	0.045
Vanadium[9]			
0.10	12.83	45.89	0.978
0.20	3.90	24.30	0.975
0.28	2.13	17.35	0.973
0.36	1.54	13.32	0.966
0.44	1.28	10.74	0.957
0.52	1.16	8.93	0.945
0.60	1.10	7.59	0.929
0.68	1.07	6.54	0.909
0.76	1.08	5.67	0.882
0.80	1.10	5.30	0.864
0.90	1.18	4.50	0.811
1.00	1.34	3.80	0.730
1.10	1.60	3.26	0.632
1.20	1.93	2.88	0.543
1.30	2.25	2.71	0.498
1.40	2.48	2.72	0.491
1.50	2.57	2.79	0.499
1.60	2.57	2.84	0.507
1.70	2.52	2.88	0.512
1.80	2.45	2.88	0.515
1.90	2.36	2.85	0.514
2.00	2.34	2.81	0.509
2.10	2.31	2.78	0.506
2.20	2.28	2.80	0.510
2.30	2.23	2.83	0.516
2.40	2.15	2.88	0.528
2.50	2.02	2.91	0.540
2.60	1.89	2.92	0.552
2.70	1.74	2.89	0.561
2.80	1.61	2.85	0.569
2.90	1.48	2.80	0.577

Energy (eV)	n	k	R(φ = 0)
3.00	1.36	2.73	0.582
3.20	1.16	2.55	0.585
3.40	0.99	2.37	0.586
3.60	0.87	2.17	0.575
3.80	0.80	1.96	0.547
4.00	0.78	1.76	0.503
4.20	0.80	1.60	0.449
4.40	0.83	1.47	0.400
4.60	0.87	1.38	0.355
4.80	0.90	1.31	0.326
5.00	0.91	1.26	0.304
5.25	0.93	1.18	0.271
5.50	0.94	1.14	0.258
5.75	0.96	1.09	0.235
6.00	0.98	1.06	0.223
6.25	0.97	1.02	0.212
6.50	0.97	0.98	0.199
6.75	0.97	0.94	0.185
7.00	0.98	0.91	0.175
7.33	0.97	0.89	0.170
7.66	0.98	0.87	0.162
8.00	0.98	0.85	0.155
8.33	0.98	0.81	0.146
8.66	0.98	0.81	0.145
9.00	0.96	0.79	0.142
9.50	0.94	0.77	0.136
10.00	0.91	0.74	0.133
10.50	0.89	0.71	0.126
11.00	0.87	0.65	0.112
11.50	0.88	0.58	0.091
12.00	0.90	0.58	0.089
12.50	0.89	0.57	0.086
13.00	0.88	0.55	0.082
13.50	0.87	0.53	0.079
14.00	0.86	0.51	0.075
14.50	0.86	0.49	0.070
15.00	0.86	0.47	0.065
15.50	0.86	0.46	0.062
16.00	0.85	0.45	0.061
16.50	0.84	0.43	0.059
17.00	0.84	0.41	0.056
17.50	0.83	0.40	0.054
18.00	0.82	0.38	0.051
18.50	0.82	0.37	0.048
19.00	0.82	0.35	0.045
19.50	0.82	0.34	0.043
20.00	0.81	0.32	0.041
20.50	0.81	0.31	0.038
21.00	0.81	0.29	0.036
21.50	0.81	0.28	0.033
22.00	0.81	0.27	0.032
22.50	0.81	0.25	0.029
23.00	0.82	0.24	0.027
23.50	0.82	0.23	0.025
24.00	0.82	0.22	0.024
24.50	0.83	0.21	0.022
25.00	0.83	0.20	0.020
25.50	0.83	0.19	0.019
26.00	0.83	0.18	0.018
26.50	0.84	0.17	0.016
27.00	0.84	0.16	0.015
27.50	0.85	0.16	0.014
28.00	0.85	0.15	0.013
28.50	0.86	0.14	0.012
29.00	0.86	0.14	0.011
29.50	0.86	0.13	0.010
30.00	0.87	0.13	0.009
31.00	0.88	0.12	0.008
32.00	0.90	0.11	0.007
33.00	0.90	0.10	0.005
34.00	0.91	0.10	0.005
35.00	0.92	0.09	0.004
36.00	0.94	0.10	0.004
37.00	0.94	0.10	0.004
38.00	0.95	0.11	0.004
39.00	0.95	0.12	0.004
40.00	0.95	0.13	0.005
***Zinc, $\vec{E} \parallel \hat{c}$*[28]**			
0.7514	1.9241	7.5619	0.883
0.827	1.7921	6.9973	0.874
0.866	1.5571	6.7753	0.881
0.952	1.4824	6.2296	0.868
0.992	1.5762	5.8843	0.847
1.033	1.5407	5.3192	0.823
1.078	1.5853	4.9013	0.793
1.127	1.7768	4.5307	0.748
1.181	1.9808	4.2004	0.701
1.240	2.8821	3.4766	0.575
1.305	3.2039	3.0042	0.520
1.377	2.9459	3.5761	0.584
1.459	3.2523	4.2447	0.640
1.550	3.8086	4.6212	0.657
1.653	3.7577	4.6239	0.659
1.722	3.5908	4.4614	0.650
1.823	3.4234	4.3232	0.642
1.937	3.0132	3.9974	0.624
1.984	1.8562	3.9706	0.690
2.066	1.4856	4.0555	0.737
2.094	1.2525	3.9961	0.762
2.119	1.0017	3.8683	0.789
2.275	0.7737	3.9129	0.832
2.445	0.6395	3.4013	0.821
2.666	0.4430	3.1379	0.851
2.917	0.3589	2.8140	0.853
3.220	0.3069	2.5088	0.847
3.594	0.2737	2.1737	0.828
4.065	0.2510	1.8528	0.799
4.678	0.2354	1.6357	0.776
***Zinc, $\vec{E} \perp \hat{c}$*[28]**			
0.751	1.4469	7.4158	0.905
0.827	1.4744	6.9688	0.892
0.866	1.3628	6.6886	0.892
0.952	1.3165	6.2212	0.881
0.992	1.3835	5.8910	0.863
1.033	1.2889	5.4001	0.850
1.078	1.3095	4.9025	0.822
1.127	1.6897	4.4062	0.746
1.181	1.9701	4.0176	0.684
1.240	2.8717	3.2873	0.555
1.305	3.3991	2.7684	0.497
1.377	3.1807	3.4709	0.569
1.459	3.5064	4.1994	0.630
1.550	4.1241	4.7768	0.664
1.653	4.0269	4.8027	0.667
1.722	3.9369	4.6356	0.657
1.823	3.7549	4.3042	0.635
1.937	3.4512	4.1942	0.631
1.984	3.2515	4.2980	0.644
2.066	2.0802	4.7231	0.738
2.094	1.7084	4.7923	0.774
2.119	1.3329	4.4751	0.791
2.275	0.9725	4.2879	0.825
2.455	0.7568	3.7627	0.824
2.666	0.5470	3.4277	0.845
2.917	0.4774	3.0476	0.834
3.220	0.3911	2.7463	0.835
3.594	0.3147	2.3041	0.821
4.065	0.3013	2.0077	0.789
4.678	0.2806	1.7997	0.770
***Zirconium (Polycrystalline)*[28]**			
0.10	6.18	1.76	0.300
0.15	3.37	1.30	0.123
0.20	2.34	1.08	0.058
0.26	2.24	1.06	0.052
0.30	2.59	1.14	0.073
0.36	3.17	1.26	0.110
0.40	3.09	1.24	0.105
0.46	3.36	1.30	0.123
0.50	4.13	1.44	0.175
0.56	5.01	1.58	0.231
0.60	5.18	1.61	0.242
0.70	4.54	1.51	0.202
0.80	4.03	1.42	0.168
0.90	3.74	1.37	0.149
0.96	3.69	1.36	0.145
1.00	3.66	1.35	0.143
1.10	3.65	1.35	0.142
1.20	3.53	1.33	0.134
1.30	3.25	1.27	0.116
1.40	3.10	1.25	0.106
1.50	3.02	1.23	0.100
1.60	2.88	1.20	0.091
1.70	2.68	1.16	0.078
1.80	2.49	1.12	0.067
2.00	2.14	1.03	0.047
2.10	1.99	1.00	0.040
2.20	1.87	0.97	0.034
2.30	1.78	0.94	0.030
2.40	1.71	0.92	0.027
2.50	1.62	0.90	0.024
2.60	1.54	0.88	0.022
2.70	1.46	0.86	0.019
2.80	1.40	0.84	0.018
2.90	1.34	0.82	0.016
3.00	1.30	0.81	0.016
3.10	1.26	0.80	0.015
3.30	1.19	0.77	0.014
3.40	1.16	0.76	0.013
3.50	1.13	0.75	0.013

Energy (eV)	n	k	$R(\phi = 0)$	Energy (eV)	n	k	$R(\phi = 0)$	Energy (eV)	n	k	$R(\phi = 0)$
3.60	1.10	0.74	0.013	9.00	1.65	0.91	0.025	17.20	1.09	0.74	0.013
3.70	1.07	0.73	0.013	9.20	1.63	0.90	0.025	17.60	1.13	0.75	0.013
3.80	1.04	0.72	0.012	9.40	1.60	0.89	0.024	18.00	1.17	0.76	0.014
3.90	1.01	0.71	0.012	9.60	1.57	0.89	0.023	18.40	1.21	0.78	0.014
4.00	0.98	0.70	0.012	9.80	1.52	0.87	0.021	18.80	1.24	0.79	0.014
4.20	0.94	0.68	0.013	10.00	1.47	0.86	0.020	19.20	1.27	0.80	0.015
4.40	0.89	0.67	0.013	10.20	1.42	0.84	0.018	19.60	1.29	0.80	0.015
4.60	0.85	0.65	0.014	10.40	1.35	0.82	0.016	20.00	1.30	0.81	0.015
4.80	0.81	0.64	0.014	10.50	1.32	0.81	0.016	20.60	1.29	0.80	0.015
5.00	0.78	0.63	0.015	10.60	1.28	0.80	0.015	21.00	1.27	0.80	0.015
5.20	0.77	0.62	0.016	10.80	1.23	0.78	0.014	21.60	1.23	0.78	0.014
5.40	0.77	0.62	0.016	11.00	1.19	0.77	0.014	22.00	1.20	0.77	0.014
5.60	0.80	0.63	0.014	11.20	1.16	0.76	0.013	22.60	1.15	0.76	0.013
5.80	0.87	0.66	0.013	11.40	1.13	0.75	0.013	23.00	1.12	0.75	0.013
6.00	1.00	0.71	0.012	11.60	1.11	0.74	0.013	23.60	1.08	0.73	0.013
6.20	1.11	0.75	0.013	11.80	1.09	0.74	0.013	24.00	1.05	0.73	0.013
6.40	1.23	0.78	0.014	12.00	1.08	0.73	0.013	24.60	1.02	0.71	0.012
6.60	1.33	0.81	0.016	12.40	1.05	0.72	0.012	25.00	1.00	0.71	0.012
6.80	1.42	0.84	0.018	12.80	1.01	0.71	0.012	25.60	0.97	0.69	0.012
7.00	1.49	0.86	0.020	13.20	0.98	0.70	0.012	26.00	0.95	0.69	0.013
7.20	1.54	0.88	0.022	13.60	0.95	0.69	0.013	26.60	0.91	0.67	0.013
7.40	1.58	0.89	0.023	14.00	0.92	0.68	0.013	27.00	0.88	0.66	0.013
7.60	1.61	0.90	0.024	14.40	0.89	0.67	0.013	27.60	0.84	0.65	0.014
7.80	1.63	0.90	0.025	14.80	0.90	0.67	0.013	28.00	0.83	0.64	0.014
8.00	1.66	0.91	0.026	15.20	0.92	0.68	0.013	28.60	0.82	0.64	0.014
8.20	1.67	0.91	0.026	15.60	0.95	0.69	0.013	29.00	0.81	0.64	0.014
8.40	1.68	0.92	0.026	16.00	0.98	0.70	0.012	29.60	0.82	0.64	0.014
8.60	1.68	0.92	0.026	16.40	1.01	0.71	0.012	30.00	0.82	0.64	0.014
8.80	1.66	0.91	0.026	16.80	1.04	0.72	0.012				

References

1. Shiles, E., Sasaki, T., Inokuti, M., and Smith, D. Y., *Phys. Rev. Sect. B*, 22, 1612, 1980.
2. Edwards, D. F., and Philipp, H. R., in *HOC-I*, p. 665.
3. Ives, H. E., and Briggs, N. B., *J. Opt. Soc. Am.*, 27, 395, 1937.
4. Bos, L. W., and Lynch, D. W., *Phys. Rev. Sect. B*, 2, 4567, 1970.
5. Weaver, J. H., Colavita, E., Lynch, D. W., and Rosei, R., *Phys. Rev. Sect. B*, 19, 3850, 1979.
6. Hagemann, H. J., Gudat, W., and Kunz, C., *J. Opt. Soc. Am.*, 65, 742, 1975.
7. Schulz, L. G., *J. Opt. Soc. Am.*, 47, 64, 1957.
8. Potter, R. F., in *HOC-I*, p. 465.
9. Olson, C. G., Lynch, D. W., and Weaver, J. H., unpublished.
10. Lynch, D. W., Olson, C. G., and Weaver, J. H., unpublished.
11. Weaver, J. H., Olson, C. G., and Lynch, D. W., *Phys. Rev. Sect. B*, 15, 4115, 1977.
12. Lynch, D. W., and Hunter, W. R., in *HOC-II*, p. 345.
13. Priol, M. A., Daudé, A., and Robin, S., *Compt. Rend.*, 264, 935, 1967.
14. Johnson, P. B., and Christy, R. W., *Phys. Rev. Sect. B*, 9, 5056, 1974.
15. Arakawn, E. T., and Inagaki, T., in *HOC-II*, p. 461.
16. Weaver, J. H., Lynch, D. W., and Olson, D. G., *Phys. Rev. Sect. B*, 10, 501, 1973.
17. Lynch, D. W., Rosei, R., and Weaver, J. H., *Solid State Commun.*, 9, 2195, 1971.
18. Weaver, J. H., Lynch, D. W., and Olson, C. G., *Phys. Rev. Sect. B*, 7, 4311, 1973.
19. Weaver, J. H., and Benbow, R. L., *Phys. Rev. Sect. B*, 12, 3509, 1975.
20. Weaver, J. H., *Phys. Rev., Sect. B*, 11, 1416, 1975.
21. Lynch, D. W., and Hunter, W. R., in *HOC-II*, p. 364.
22. Palik, E. D., in *HOC-II*, p. 691.
23. Edwards, D. F., in *HOC-I*, p. 547.
24. Lynch, D. W., and Hunter, W. R., in *HOC-II*, p. 354.
25. Palik, E. D., in *HOC-II*, p. 709.
26. Lynch, D. W., Olson, C. G., and Weaver, J. H., *Phys. Rev. Sect. B*, 11, 3671, 1975.
27. Weaver, J. H., Lynch, D. W., and Olson, C. G., *Phys. Rev. Sect. B*, 12, 1293, 1975.
28. Lanham, A. P., and Terherne, D. M., *Proc. Phys. Soc.*, 83, 1059, 1964.

OPTICAL PROPERTIES OF SELECTED INORGANIC AND ORGANIC SOLIDS

L. I. Berger

Optical properties of materials are closely related to their dielectric properties. The complex dielectric function (relative permittivity) of a material is equal to

$$\varepsilon(\omega) = \varepsilon'(\omega) - j\varepsilon''(\omega),$$

where $\varepsilon'(\omega)$ and $\varepsilon''(\omega)$ are its real and imaginary parts, respectively, and ω is the angular frequency of the applied electric field. For a non-absorbing medium, the index of refraction is $n = (\varepsilon\mu)^{1/2}$, where μ is the relative magnetic permeability of the medium (material); in the majority of dielectrics, $\mu \cong 1$.

For many applications, the most important optical properties of materials are the index of refraction, the extinction coefficient, k, and the reflectivity, R. The common index of refraction of a material is equal to the ratio of the phase velocity of propagation of an electromagnetic wave of a given frequency in vacuum to that in the material. Hence, $n \geqq 1$. The optical properties of highly conductive materials like metals and semiconductors (at photon energy range above the energy gap) differ from those of optically transparent media. Free electrons absorb the incident electromagnetic wave in a thin surface layer (a few hundred nanometers thick) and then release the absorbed energy in the form of secondary waves reflected from the surface. Thus, the light reflection becomes very strong; for example, highly conductive sodium reflects 99.8% of the incident wave (at 589 nm). Introduction of the effective index of refraction, $n_{eff} = (\varepsilon')^{1/2} = n - jk$, where $\varepsilon' = \varepsilon - j\delta/\omega\,\varepsilon_o$, δ is the electrical conductivity of the material in S/m, and $\varepsilon_o = 8.8542\cdot10^{-12}$ F/m is the permittivity of vacuum, allows one to apply the expressions of the optics of transparent media to the conductive materials. It is clear that the effective index of refraction may be smaller than 1. For example, $n = 0.05$ for pure sodium and $n = 0.18$ for pure silver (at 589.3 nm). At very high photon energies, the quantum effects, such as the internal photoeffect, start playing a greater role, and the optical properties of these materials become similar to those of insulators (low reflectance, existence of Brewster's angle, etc.).

The extinction coefficient characterizes absorption of the electromagnetic wave energy in the process of propagation of a wave through a material. The wave intensity, I, after it passes a distance x in an isotropic medium is equal to

$$I = I_0\exp(-\alpha x),$$

where I_0 is the intensity at $x = 0$ and α is called the absorption coefficient. For many applications, the extinction coefficient, k, which is equal to

$$k = \alpha\frac{\lambda}{4\pi},$$

where λ is the wavelength of the wave in the medium, is more commonly used for characterization of the electromagnetic losses in materials.

Reflection of an electromagnetic wave from the interface between two media depends on the media indices of refraction and on the angle of incidence. It is characterized by the reflectivity, which is equal to the ratio of the intensity of the wave reflected back into the first medium to the intensity of the wave approaching the interface. For polarized light and two non-absorbing media,

$$R = \frac{(N_1 - N_2)^2}{(N_1 + N_2)^2},$$

where $N_1 = n_1/\cos\theta_1$ and $N_2 = n_2/\cos\theta_2$ for the wave polarized in the plane of incidence, and $N_1 = n_1\cos\theta_1$ and $N_2 = n_2\cos\theta_2$ for the wave polarized normal to the plane of incidence; θ_1 and θ_2 are the angles between the normal to the interface in the point of incidence and the directions of the beams in the first and second medium, respectively. The reflectivity at normal incidence in this case is

$$R = [(n_1 - n_2)/(n_1 + n_2)]^2$$

For any two opaque (absorbing) media, the normal incidence reflectivity is

$$R = \frac{(n_1 - n_2)^2 + k_2^2}{(n_1 + n_2)^2 + k_2^2}.$$

In the majority of experiments, the first medium is air ($n \approx 1$), and hence,

$$R = \frac{(1-n)^2 + k^2}{(1+n)^2 + k^2}.$$

The data on n and k in the following table are abridged from the sources listed in the references. The reflectivity at normal incidence, R, has been calculated from the last equation. For convenience, the energy E, wavenumber $\bar{\nu}$, and wavelength λ are given for the incidence radiation.

E/eV	$\bar{\nu}$/cm^{-1}	λ/μm	n	n_a	n_c	k	k_a	k_c	R	R_a	R_c
Crystalline Arsenic Selenide (As_2Se_3) [Ref. 1]*											
2.194	17700	0.565					0.30				
2.168	17480	0.572					0.25				
2.141	17270	0.579					0.20				
2.123	17120	0.584					0.17				
2.098	16920	0.591					0.13				
2.094	16890	0.592						0.26			
2.091	16860	0.593						0.26			
2.073	16720	0.598					0.10	0.23			
2.060	16610	0.602						0.20			
2.049	16530	0.605					0.079	0.17			

E/eV	$\bar{\nu}$/cm^{-1}	λ/µm	n	n_a	n_c	k	k_a	k_c	R	R_a	R_c
2.036	16420	0.609						0.15			
2.023	16310	0.613						0.12			
2.013	16230	0.616					0.050				
2.009	16210	0.617						0.097			
2.000	16130	0.620						0.082			
1.987	16030	0.624						0.063			
1.977	15940	0.627					0.031				
1.974	15920	0.628						0.051			
1.962	15820	0.632						0.038			
1.953	15750	0.635						0.030			
1.949	15720	0.636					0.020				
1.937	15630	0.640						0.022			
1.925	15530	0.644						0.017			
1.922	15500	0.645					0.012				
1.905	15360	0.651					$8.6\cdot10^{-3}$				
1.893	15270	0.655					6.4				
1.881	15170	0.659					5.2				
1.859	14990	0.667					3.1				
1.848	14900	0.671						$1.7\cdot10^{-3}$			
1.845	14880	0.672					2.0				
1.842	14860	0.673						$1.2\cdot10^{-3}$			
1.831	14770	0.677					$1.3\ 10^{-3}$	$9.0\cdot10^{-4}$			
1.826	14730	0.679						6.4			
1.821	14680	0.681						4.7			
1.818	14660	0.682					$8.6\ 10^{-4}$				
1.815	14640	0.683						3.4			
1.807	14580	0.686					5.5				
1.802	14530	0.688					4.1				
0.06199	500.0	20.0		3.2	2.9		$1.7\cdot10^{-3}$	$1.8\cdot10^{-3}$		0.27	0.24
0.05904	476.2	21.0		3.1	2.9		$2.1\cdot10^{-3}$	$2.2\cdot10^{-3}$		0.26	0.24
0.05636	454.5	22.0		3.1	2.9		$2.5\cdot10^{-3}$	$2.6\cdot10^{-3}$		0.26	0.24
0.05391	434.8	23.0		3.1	2.9		$3.0\cdot10^{-3}$	$3.1\cdot10^{-3}$			
0.04592	370.4	27.0		3.0	2.8		$6.3\cdot10^{-3}$	$6.4\cdot10^{-3}$		0.25	0.22
0.04428	357.1	28.0		3.0	2.8		$7.6\cdot10^{-3}$	$7.7\cdot10^{-3}$		0.25	0.22
0.04275	344.8	29.0		3.0	2.8		0.0092	0.0093		0.25	0.22
0.04133	333.3	30.0		3.0	2.7		0.011	0.011		0.25	0.21
0.03542	285.7	35.0		2.7	2.5			0.037	0.034	0.21	0.18
0.03100	250.0	40.0		1.9	1.7			0.38	1.0	0.19	0.18
0.03061	247.0	40.5		2.0	2.6			0.33	0.95	0.12	0.25
0.03024	244.0	41.0		1.7	2.4			0.41	0.46	0.088	0.18
0.02883	232.6	43.0		1.2	1.3			2.2	0.94	0.50	0.16
0.02850	229.9	43.5		1.6	1.2			2.8	1.4	0.56	0.29
0.02818	227.3	44.0		2.3	1.2		3.3	2.0		0.58	0.48
0.02755	222.2	45.0		4.2	2.0		2.5	3.3		0.50	0.60
0.02480	200.0	50.0		6.5	4.0		3.6	0.26		0.62	0.36
0.02254	181.8	55.0		4.5	3.5		0.17	0.10		0.40	0.31
0.02066	166.7	60.0		4.0	3.2		0.089	0.10		0.36	0.27
0.01907	153.8	65.0		3.8	3.1		0.097	0.16		0.34	0.26
0.01771	142.9	70.0		3.6	3.0		0.19	0.30		0.32	0.25
0.01653	133.3	75.0		3.7	3.0		0.41	0.44		0.34	0.26
0.01550	125.0	80.0		3.8	3.1		0.29	0.40		0.34	0.27
0.01459	117.6	85.0		3.6	2.9		0.20	0.34		0.32	0.24
0.01378	111.1	90.0		3.2	2.6		0.43	0.49		0.28	0.21
0.01305	105.3	95.0		4.7	3.0		1.5	1.5		0.46	0.34
0.01240	100.0	100.0		4.4	2.7		0.22	0.81		0.40	0.25
0.01181	95.24	105.0		4.2	3.0		0.094	3.9		0.38	0.62
0.01127	90.91	110.0		4.1	5.3		0.059	0.70		0.37	0.47
0.01033	83.33	120.0		3.9	4.2		0.034	0.13		0.35	0.38
0.009537	76.92	130.0		3.9	4.0		0.024	0.069		0.35	0.36
0.008856	71.43	140.0		3.9	3.8		0.019	0.048		0.35	0.34
0.007749	63.50	160.0		3.8	3.7		0.014	0.032		0.34	0.33

E/eV	$\bar{\nu}$/cm^{-1}	λ/μm	n	n_a	n_c	k	k_a	k_c	R	R_a	R_c
0.006888	55.55	180.0		3.8	3.7		0.011	0.024		0.34	0.33
0.006199	50.0	200.0		3.8	3.6		0.0091	0.019		0.34	0.32

* Indices a and c relate to the radiation electric field parallel to the a and c axes of the crystal, respectively.

Vitreous Arsenic Selenide (As_2Se_3) [Ref. 1]

E/eV	$\bar{\nu}$/cm^{-1}	λ/μm	n	k	R
2.056	16580	0.603		0.12	
2.026	16340	0.612		0.11	
2.006	16180	0.618		0.099	
1.990	16050	0.623		9.0	
1.925	15530	0.644		5.6	
1.826	14730	0.679		1.4	
1.810	14600	0.685		0.012	
1.794	14470	0.691		0.0089	
1.771	14290	0.700		6.2	
1.715	13830	0.723		2.6	
1.701	13720	0.729		0.0022	
1.647	13280	0.753		0.00046	
1.629	13140	0.761	3.07	4.0	0.62
1.596	12870	0.777	3.06	2.7	0.49
1.579	12740	0.785	3.05	1.9	0.39
1.562	12590	0.794	3.05	0.00013	0.26
1.544	12450	0.803	3.04	0.000094	0.25
1.529	12330	0.811	3.03	6.3	0.78
1.512	12200	0.820	3.03	4.2	0.64
1.494	12050	0.830	3.02	2.8	0.50
1.476	11910	0.840	3.01	1.8	0.38
1.378	11110	0.90	2.98		
1.240	10000	1.00	2.93		
1.127	9091	1.10	2.90		
1.051	8475	1.18	2.89		
1.033	8333	1.20	2.88		
0.2555	1980	5.05		$1.6 \cdot 10^{-7}$	
0.2380	1919	5.21		$9.9 \cdot 10^{-8}$	
0.2344	1890	5.29		$1.1 \cdot 10^{-7}$	
0.1345	1085	9.22		4.4	
0.1339	1080	9.26		3.7	
0.1333	1075	9.30		4.4	
0.1308	1055	9.48		4.5	
0.1215	980	10.20		8.9	
0.1203	970	10.31		$9.9 \cdot 10^{-7}$	
0.1196	965	10.36		$1.0 \cdot 10^{-6}$	
0.1178	950	10.53		1.1	
0.1116	900	11.11		1.8	
0.1004	810	12.35		4.9	
0.09919	800	12.50		$7.0 \cdot 10^{-6}$	
0.09795	790	12.66		$1.0 \cdot 10^{-5}$	
0.09671	780	12.82		1.5	
0.09299	750	13.33		3.7	
0.08555	690	14.49		6.9	
0.08431	680	14.71		5.9	
0.08059	650	15.38		6.1	
0.07811	630	15.87		6.3	
0.07687	620	16.13		7.7	
0.07563	610	16.39		7.8	
0.07439	600	16.67		$9.3 \cdot 10^{-5}$	
0.07315	590	16.95	2.8	$1.2 \cdot 10^{-4}$	0.22
0.07191	580	17.24	2.8	1.4	0.32
0.07067	570	17.54	2.8	1.8	0.37
0.06943	560	17.86	2.8	2.8	0.50
0.06633	535	18.69	2.8	5.2	0.73
0.06571	530	18.87	2.8	$7.2 \cdot 10^{-4}$	0.22
0.06509	525	19.05	2.8	$1.2 \cdot 10^{-3}$	0.22

Solids

E/eV	$\bar{\nu}$/cm^{-1}	λ/μm	n	n_a	n_c	k	k_a	k_c	R	R_a	R_c
0.06447	520	19.23	2.8			1.7			0.35		
0.06075	490	20.41	2.7			4.9			0.71		
0.06024	485.9	20.58	2.7			5.2			0.73		
0.05331	430	23.26	2.7			1.4			0.31		
0.05269	425	23.53	2.7			$1.1 \cdot 10^{-3}$			0.21		
0.05207	420	23.81	2.7			$8.5 \cdot 10^{-4}$			0.21		
0.05145	415	24.10	2.7			7.3			0.84		
0.05083	410	24.39	2.7			8.3			0.87		
0.05021	405	24.69	2.7			$9.4 \cdot 10^{-4}$			0.21		
0.04959	400	25.0	2.7			$1.2 \cdot 10^{-3}$			0.21		
0.04862	392.2	25.5	2.6			1.6			0.33		
0.04679	377.4	26.5	2.6			5.0			0.73		
0.04592	370.4	27.0	2.6			$8.0 \cdot 10^{-3}$			0.20		
0.04509	363.6	27.5	2.6			$1.2 \cdot 10^{-2}$			0.20		
0.04428	357.1	28.0	2.6			1.7			0.34		
0.03875	312.5	32.0	2.5			8.2			0.87		
0.03815	307.7	32.5	2.5			$9.3 \cdot 10^{-3}$			0.18		
0.03757	303.0	33.0	2.4			0.11			0.17		
0.02988	241.0	41.5	2.2			0.89			0.20		
0.02952	238.1	42.0	2.2			1.0			0.22		
0.02725	219.8	45.5	3.2			1.8			0.39		
0.02362	190.5	52.5	3.6			0.30			0.32		
0.01937	156.2	64.0	3.2			0.10			0.27		
0.01922	155.0	64.5	3.2			$9.6 \cdot 10^{-2}$			0.27		
0.01907	153.8	65.0	3.2			9.4			0.88		
0.01734	139.9	71.5	3.1			8.7			0.87		
0.01653	133.3	75.0	3.1			9.4			0.88		
0.01642	132.5	75.5	3.1			0.096			0.26		
0.01494	120.5	83.0	3.0			0.15			0.25		
0.01246	100.5	99.5	3.2			0.60			0.26		
0.007606	61.35	163.0	3.3			0.12			0.29		
0.006199	50.00	200.0	3.2								
0.004592	37.04	270.0	3.1			0.072			0.26		
0.002799	22.57	443.0	3.0			4.5			0.67		
0.001826	14.73	679.0	3.0			2.8			0.50		
0.001273	10.27	974.0	3.0			2.1			0.41		
0.0006491	5.236	1910.0	3.0			$1.1 \cdot 10^{-2}$			0.25		
0.0004376	3.530	2833.0	3.0			$7.5 \cdot 10^{-3}$			0.25		
0.0002903	2.341	4271.0	3.0			5.0			0.71		
0.0001716	1.384	7224.0	3.0			3.1			0.53		
0.00009047	0.7297	13704	3.0			$1.6 \cdot 10^{-3}$			0.25		
0.00005621	0.4534	22056	3.0			$9.9 \cdot 10^{-4}$			0.25		
0.00002774	0.2237	44699	3.0			5.2			0.72		
0.00001439	0.1161	86153	3.0			2.6			0.47		
Vitreous Arsenic Sulfide (As_2S_3) - [Ref. 2]											
4.959	40000	0.2500	2.48			1.21			0.27		
3.100	25000	0.40	3.09			0.34			0.27		
2.48	20000	0.4999	2.83			0.013			0.23		
1.879	15150	0.66	2.59			$1.7 \cdot 10^{-6}$			0.20		
1.240	10000	1.0	2.48			$2.4 \cdot 10^{-7}$			0.18		
0.6199	5000	2.0	2.43						0.17		
0.3100	2500	4.0	2.41						0.17		
0.2480	2000	5.0	2.41						0.17		
0.1736	1400	7.143	2.40			$7.4 \cdot 10^{-7}$			0.17		
0.1240	1000	10.00	2.38			$1.3 \cdot 10^{-4}$			0.17		
0.09299	750	13.33	2.35			$3.0 \cdot 10^{-3}$			0.16		
0.07439	600	16.67	2.31			$4.6 \cdot 10^{-4}$			0.16		
0.04959	400.0	25.0	1.79			0.2			0.085		
0.03757	303.0	33.0	3.59			1.4			0.38		
0.03100	250.0	40.0	2.98			0.15			0.25		
0.02480	200.0	50	2.66			0.11			0.21		

E/eV	$\bar{\nu}$/cm^{-1}	λ/μm	n	n_a	n_c	k	k_a	k_c	R	R_a	R_c
0.02066	166.7	60	2.64			0.57			0.22		
0.01771	142.9	70	2.99			0.17			0.25		
0.01550	125.0	80	2.89			0.14			0.24		
0.01378	111.1	90	2.84			0.12			0.23		
0.01240	100	100	2.81			0.10			0.23		
0.008183	66	152	2.76			0.072			0.22		
0.004029	32.5	308	2.74			0.044			0.22		
0.002418	19.5	513	2.74			0.031			0.22		
0.001984	16	625	2.74			0.025			0.22		
0.001048	8.45	1180	2.73			$8.8 \cdot 10^{-3}$			0.22		
0.0001033	0.833	12000	2.73			$1.3 \cdot 10^{-3}$			0.22		
$4.129 \cdot 10^{-12}$	$3.33\ 10^{-8}$	$3 \cdot 10^{11}$	2.73						0.22		
						Cadmium Telluride (CdTe) - [Ref. 3]					
4.9	39520	0.2530	2.48			2.04			0.39		
4.1	33070	0.3024	2.33			1.59			0.32		
3.9	31460	0.3179	2.57			1.90			0.37		
3.5	28230	0.3542	2.89			1.52			0.34		
3.1	25000	0.4000	3.43			1.02			0.34		
3.0	24200	0.4133	3.37			0.861			0.32		
2.755	22220	0.45	3.080			0.485			0.27		
2.75	22180	0.4509	3.23			0.636			0.29		
2.610	21050	0.475	3.045								
2.5	20160	0.4959	3.14			0.525			0.28		
2.25	18150	0.5510	3.05			0.411			0.26		
1.771	14290	0.70	2.861			0.210			0.23		
1.512	12200	0.82	2.880			0.040			0.23		
1.50	12100	0.8266	2.98			0.319			0.25		
1.475	11900	0.840	2.905			0.00134			0.24		
1.47	11860	0.8434				0.000671					
1.465	11820	0.8463				3.37					
1.46	11780	0.8492				1.89					
1.459	11760	0.850	2.948						0.24		
1.455	11740	0.8521				$1.08 \cdot 10^{-4}$					
1.45	11690	0.8551	2.9565			$5.10 \cdot 10^{-5}$			0.24		
1.445	11650	0.8580				2.73					
1.442	11630	0.860	2.952						0.24		
1.44	11610	0.8610	2.9479			1.37			0.32		
1.43	11530	0.8670	2.9402						0.24		
1.30	10490	0.9537	2.8720						0.23		
1.24	10000	1.0	2.840						0.23		
1.20	9679	1.033	2.8353						0.23		
1.10	8872	1.127	2.8050						0.23		
1.00	8065	1.240	2.7793						0.22		
0.90	7259	1.378	2.7537						0.22		
0.80	6452	1.550	2.7384						0.22		
0.70	5646	1.771	2.7223						0.21		
0.60	4839	2.066	2.7086						0.21		
0.50	4033	2.480	2.6972						0.21		
0.40	3226	3.100	2.6878						0.21		
0.30	2420	4.133	2.6800						0.21		
0.20	1613	6.199	2.6722						0.21		
0.10	806.5	12.40	2.6535						0.20		
0.09	725.9	13.78	2.6482						0.20		
0.06819	550	18.18	2.623						0.20		
0.0573	462	21.6				$3.8 \cdot 10^{-6}$					
0.05	403.3	24.80	2.5801						0.19		
0.0469	378	26.5				$8.0 \cdot 10^{-5}$					
0.04592	370.3	27				$9.88 \cdot 10^{-5}$					
0.04133	333.3	30	2.55916			$2.86 \cdot 10^{-4}$			0.19		
0.04092	330	30.30	2.531			3.34			0.57		
0.03720	300	33.33	2.494			4.97			0.73		

E/eV	$\bar{\nu}$/cm^{-1}	λ/μm	n	n_a	n_c	k	k_a	k_c	R	R_a	R_c
0.03647	294.1	34.00				8.93					
0.03596	290	34.48	2.478			$5.77\cdot10^{-3}$			0.18		
0.03493	281.7	35.5				7.91					
0.03472	280	35.71	2.459			6.76			0.83		
0.03100	250	40	2.378			$1.18\cdot10^{-2}$			0.17		
0.02917	235.3	42.5				6.93					
0.02852	230	43.48	2.289			1.87			0.36		
0.02728	220	45.45	2.224			$2.47\cdot10^{-2}$			0.14		
0.02604	210	47.62	2.137			$3.4\cdot10^{-2}$			0.13		
0.02480	200	50.00	2.013			$4.97\cdot10^{-2}$			0.11		
0.02384	192.3	52.0				6.21					
0.01798	145	68.97	1.8			5.2			0.79		
0.01736	140	71.43	6.778			4.50			0.66		
0.01550	125	80.0	4.598			0.294			0.41		
0.01364	110	90.91	3.868			$9.47\cdot10^{-2}$			0.35		
0.01240	100	100	3.649			$5.68\cdot10^{-2}$			0.32		
0.009919	80	125	3.415			0.0262			0.30		
0.008679	70	142.9	3.348			0.0189			0.29		
0.007439	60	166.7	3.299			1.39			0.35		
0.006199	50	200	3.263			1.03			0.32		
0.004959	40	250	3.236			$7.52\cdot10^{-3}$			0.28		
0.003720	30	333.3	3.217						0.28		
0.023015	18.563		538.71			3.2096			0.28		
0.001550	12.50	800				6.18					
Gallium Arsenide (GaAs) - [Ref. 4]											
155		0.007999				0.0181					
145		0.008551				0.0203					
130		0.009537				0.0224					
110		0.01127				0.0278					
90		0.01378				0.0323					
70		0.01771				0.0376					
40		0.03100				0.0426					
23		0.05391	1.037			0.228					
7.0		0.1771	1.063			1.838					
6.0	48390	0.2066	1.264			2.472			0.61		
5.00	40330	0.2480	2.273			4.084			0.67		
4.00	32260	0.3100	3.601			1.920			0.42		
3.00	24200	0.4133	4.509			1.948			0.47		
2.50	20160	0.4959	4.333			0.441			0.39		
2.00	16130	0.6199	3.878			0.211			0.35		
1.80	14520	0.8888	3.785			0.151			0.34		
1.60	12900	0.7749	3.700			0.091			0.33		
1.50	12100	0.8266	3.666			0.080			0.33		
1.40	11290	0.8856	3.6140			$1.69\cdot10^{-3}$			0.32		
1.20	9679	1.033	3.4920						0.31		
1.00	8065	1.240	3.4232						0.30		
0.80	6452	1.550	3.3737						0.29		
0.50	4033	2.480	3.3240						0.29		
0.25	2016	4.959	3.2978						0.29		
0.15	1210	8.266	3.2831						0.28		
0.100	806.5	12.40	3.2597			$4.93\cdot10^{-6}$			0.28		
0.090	725.9	13.78	3.2493			$1.64\cdot10^{-5}$			0.28		
0.070	564.6	17.71	3.2081			$2.32\cdot10^{-4}$			0.28		
0.060	483.9	20.66	3.1609			$3.45\cdot10^{-3}$			0.27		
0.0495	399.2	25.05	3.058			$2.07\cdot10^{-3}$			0.26		
0.03968	320	31.25	2.495			$2.43\cdot10^{-2}$			0.18		
0.03496	282	35.46	0.307			$294\cdot10^{-2}$					
0.02976	240	41.67	4.57			$4.26\cdot10^{-2}$			0.41		
0.02066	166.7	60	3.77			$3.89\cdot10^{-3}$			0.34		
0.01550	125	80	3.681			$1.84\cdot10^{-3}$			0.33		
0.008266	66.67	150	3.62			$2.14\cdot10^{-3}$			0.32		

E/eV	$\bar{\nu}$/cm^{-1}	λ/μm	n	n_a	n_c	k	k_a	k_c	R	R_a	R_c
0.002480	20	500	3.607			$1.3\cdot10^{-3}$			0.32		
0.001240	10	1000	3.606						0.32		
Gallium Phosphide (GaP) - [Ref. 5]											
154.0		0.00805				$1.7\cdot10^{-2}$					
110.0		0.0113				$2.15\cdot10^{-2}$					
100.0		0.0124				$215\cdot10^{-2}$					
80.0		0.0155				$3.0\cdot10^{-2}$					
50.0		0.0248				$4.7\cdot10^{-2}$					
27.0		0.0459				$9.3\cdot10^{-2}$					
25.0		0.0496				0.122					
20.0		0.0620				0.180					
15.0		0.0826	0.748			0.628					
5.5	44360	0.2254	1.543			3.556			0.68		
4.68	37750	0.2649	4.181			2.634			0.50		
3.50	28230	0.3542	5.050			0.819			0.46		
3.00	24200	0.4133	4.081			0.224			0.37		
2.78	22420	0.4460	3.904			0.103			0.35		
2.621	21140	0.473	3.73			$6.37\cdot10^{-3}$			0.33		
2.480	20000	0.500	3.590			$2.47\cdot10^{-3}$			0.32		
2.18	17580	0.5687	3.411			$2.8\cdot10^{-7}$			0.30		
2.000	16130	0.62	3.3254						0.29		
1.6	12900	0.7749	3.209						0.28		
1.240	10000	1.0	3.1192						0.26		
0.6888	5556	1.8	3.0439						0.26		
0.4769	3846	2.6	3.0271						0.25		
0.1907	1538	6.5	2.995			$4.29\cdot10^{-4}$			0.25		
0.1550	1250	8.0	2.984						0.25		
0.1240	1000	10	2.964						0.25		
0.06199	500	20	2.615			$7.16\cdot10^{-3}$			0.20		
0.03100	250	40	3.594			$1.81\cdot10^{-2}$			0.32		
0.02480	200	50	3.461			$5.77\cdot10^{-3}$			0.30		
0.01727	139.27	71.80	3.3922			$4.34\cdot10^{-3}$			0.30		
0.01168	94.21	106.1	3.3621			$4.26\cdot10^{-3}$			0.29		
0.006199	50.00	200	3.3447			$1.3\cdot10^{-4}$			0.29		
0.004133	33.33	300	3.3413						0.29		
0.001240	10.00	1000	3.3319						0.29		
Indium Antimonide (InSb) - [Ref. 6]											
155		0.007999				$4.77\cdot10^{-3}$					
60		0.02066				$7.30\cdot10^{-2}$					
25		0.04959	1.15			.015					
24		0.05166	1.15			0.18					
15		0.08266	0.97			0.230					
10		0.1240	0.74			0.88					
5.00	40330	0.2480	1.307			2.441			0.53		
4.50	36290	0.2755	1.443			2.894			0.60		
4.00	32260	0.3100	2.632			3.694			0.61		
3.34	26940	0.3712	3.528			2.280			0.45		
2.84	22910	0.4366	3.340			2.021			0.45		
1.80	14520	0.6888	4.909			1.396			0.47		
1.50	12100	0.8266	4.418			0.643			0.41		
0.6	4839	2.066	4.03						0.36		
0.2480	2000	5.0	4.14			$9.1\cdot10^{-2}$			0.37		
0.1907	1538	6.5	4.30			$6.3\cdot10^{-2}$			0.39		
0.1653	1333	7.5	4.18			$2.7\cdot10^{-2}$			0.38		
0.06199	500	20.00	3.869			$2.0\cdot10^{-3}$			0.35		
0.03100	250	40.00	2.98			$2.6\cdot10^{-3}$			0.25		
0.02480	200	50.00	2.22			0.165			0.14		
0.02244	181	55.25	3.05			7.59			0.84		
0.02207	178	56.18	9.61			4.20			0.70		
0.02033	164	60.98	4.94			0.140			0.44		

E/eV	$\bar{\nu}$/cm^{-1}	λ/µm	n	n_a	n_c	k	k_a	k_c	R	R_a	R_c
0.01054	85	117.6	2.12			0.423			0.14		
0.005579	45	222.2	1.02			5.59			0.88		
0.001860	15	666.7	6.03			17.9			0.93		
0.001240	10	1000	10.7			24.0			0.94		
Indium Arsenide (InAs) - [Ref. 7]											
25		0.04959				1.139			0.168		
20		0.06199				1.125			0.225		
15		0.08266				0.894			0.336		
10		0.1240				0.835			1.071		
6	48390	0.2066	1.434			2.112			0.45		
5.0	40330	0.2480	1.524			2.871			0.58		
4.0	32260	0.3100	3.313			1.799			0.39		
3.5	28230	0.3542	3.008			1.754			0.37		
3.0	24200	0.4133	3.197			2.034			0.41		
2.5	20160	0.4959	4.364			1.786			0.45		
2.44	19680	0.5081	4.489			1.446			0.44		
1.86	15000	0.6666	3.889			0.554			0.36		
1.8	14520	0.6888	3.851			0.530			0.35		
1.7	13710	0.7293	3.798			0.493			0.35		
1.6	12900	0.7749	3.755			0.463			0.34		
1.5	12100	0.8266	3.714			0.432			0.34		
1.2	9679	1.033	3.613						0.32		
1.0	8065	1.240	3.548						0.31		
0.6	4839	2.066				0.161					
0.35	2823	3.542	3.608			$9.58 \cdot 10^{-3}$			0.32		
0.32	2581	3.875	3.512			$1.23 \cdot 10^{-4}$			0.31		
0.20	1613	6.199	3.427						0.30		
0.1240	1000	10.00	3.402						0.30		
0.06199	500	20.00	3.334						0.29		
0.04959	400	25.00	3.264						0.28		
0.04339	350	28.57	3.182			$5.46 \cdot 10^{-3}$			0.27		
0.03720	300	33.33	2.988						0.25		
0.03100	250	40.00	1.970			$6.37 \cdot 10^{-2}$			0.11		
0.02765	222	44.84	5.90			6.53			0.74		
0.02480	200	50.00	6.91			0.30			0.56		
0.01984	160	62.50	5.27			0.41			0.47		
0.01860	150	66.67	5.27			0.51			0.47		
0.01736	140	71.43	3.99			$1.1 \cdot 10^{-2}$			0.36		
0.01488	120	83.33	3.91			$6.6 \cdot 10^{-3}$			0.35		
0.01240	100	100.0	3.85			$4.3 \cdot 10^{-3}$			0.35		
0.009919	80	125.0	3.817						0.34		
0.007439	60	166.7	3.793						0.34		
0.004959	40	250.0	3.778						0.34		
0.002480	20	500	3.769						0.37		
0.001240	10	1000	3.766						0.34		
Indium Phosphide (InP) - [Ref. 8]											
20		0.06199	0.793			0.494					
15		0.08266	0.695			0.574					
10		0.1240	0.806			1.154					
5.5	44360	0.2254	1.426			2.562			0.79		
5.0	40330	0.2480	2.131			3.495			0.61		
4.0	32260	0.3100	3.141			1.730			0.38		
3.0	24200	0.4133	4.395			1.247			0.43		
2.0	16130	0.6199	3.549			0.317			0.32		
1.5	12100	0.8266	3.456			0.203			0.31		
1.25	10085	0.9915	3.324						0.29		
1.00	8068	1.239	3.220						0.28		
0.50	4034	2.479	3.114						0.26		
0.30	2420	4.131	3.089						0.26		
0.10	806.8	12.39	3.012						0.25		

E/eV	$\bar{\nu}$/cm^{-1}	λ/μm	n	n_a	n_c	k	k_a	k_c	R	R_a	R_c
0.075	605.1	16.53	2.932						0.24		
0.060	484.1	20.66	2.780			$1.46 \cdot 10^{-2}$			0.22		
0.050	403.4	24.79	2.429			$3.35 \cdot 10^{-2}$			0.17		
0.03992	322	31.06	0.307			3.57					
0.03496	282	35.46	3.89			0.282			0.35		
0.03100	250	40.00	4.27			$3.0 \cdot 10^{-2}$			0.39		
0.02728	220	45.45	3.93			$1.3 \cdot 10^{-2}$			0.35		
0.02480	200	50.0	3.81			$8.7 \cdot 10^{-3}$			0.34		
0.02418	195	51.28	3.19						0.27		
0.02232	180	55.56	3.19						0.27		
0.01860	150	66.67	3.65						0.32		
0.01240	100	100	3.57						0.32		
0.009919	80	125.0	3.551						0.31		
0.007439	60	166.7	3.538						0.31		
0.004959	40	250.0	3.529						0.31		
0.002480	20	500	3.523						0.31		
0.001240	10	1000.0	3.522						0.31		

Lead Selenide (PbSe) - [Ref. 9]

E/eV	$\bar{\nu}$/cm^{-1}	λ/μm	n	n_a	n_c	k	k_a	k_c	R	R_a	R_c
14.5		0.08551	0.72			0.20					
10		0.1240	0.68			0.50					
5	40330	0.2480	0.54			1.2					
2.0	16130	0.6199	3.65			2.9			0.51		
1.65	13310	0.7514	4.51			1.73			0.46		
1.5	12100	0.8266	4.64			2.64			0.52		
1.0	8065	1.240	4.65			1.1			0.44		
0.75	6049	1.653				0.269					
0.62	5001	2.000	4.59			0.770			0.42		
0.48	3871	2.583	4.90						0.44		
0.40	3226	3.100	4.91						0.44		
0.32	2581	3.875	4.98			0.173			0.44		
0.20	1613	6.199	4.82						0.43		
0.1190	960	10.42	4.74			$1.20 \cdot 10^{-3}$			0.42		
0.09919	800	12.50	4.72			$2.09 \cdot 10^{-3}$			0.42		
0.07935	640	15.63	4.68			$4.12 \cdot 10^{-3}$			0.42		
0.05951	480	20.83	4.59			$1.00 \cdot 10^{-2}$			0.41		
0.04959	400	25.00	4.49			$1.77 \cdot 10^{-2}$			0.40		
0.03968	320	31.25	4.31			$3.62 \cdot 10^{-2}$			0.39		
0.02976	240	41.67	3.89			$9.61 \cdot 10^{-2}$			0.24		
0.01984	160	62.50	2.34			0.56			0.18		
0.009919	80	125.0	1.73			7.38			0.88		
0.007935	64	156.3	2.91			10.1			0.90		
0.004959	40	250.0	11.2			14.6			0.88		
0.002480	20	500.0	12.6			12.2					
0.001736	14	714.3	14.1			16.6					
0.001240	10	1000	17.4			21.1					

Lead Sulfide (PbS) - [Ref. 10]

E/eV	$\bar{\nu}$/cm^{-1}	λ/μm	n	n_a	n_c	k	k_a	k_c	R	R_a	R_c
150		0.008266				$3.86 \cdot 10^{-3}$					
125		0.009919				$5.59 \cdot 10^{-3}$					
100		0.01240				$1.54 \cdot 10^{-2}$					
80		0.01550				$2.88 \cdot 10^{-2}$					
60		0.02066				$6.17 \cdot 10^{-2}$					
25		0.04959	0.845			0.171					
18.0		0.06888	0.846			0.294					
14.0		0.08856	0.651			0.665					
10.0		0.1240	0.879			1.050					
4.95	39920	0.2505	1.52			2.10			0.43		
4.0	32260	0.3100	1.73			2.83			0.55		
3.00	24200	0.4133	3.88			3.00			0.53		
2.90	23390	0.4275	4.12			2.70			0.51		
2.75	22180	0.4509	4.25			2.33			0.48		

E/eV	$\bar{\nu}$/cm^{-1}	λ/μm	n	n_a	n_c	k	k_a	k_c	R	R_a	R_c
2.55	20570	0.4862	4.35			2.00			0.47		
2.00	16130	0.6199	4.29			1.48			0.43		
1.60	12910	0.7749	4.62			0.94			0.43		
1.24	10000	1.00	4.43			0.597			0.41		
1.03	8333	1.2	4.30			0.458			0.39		
0.650	5263	1.9	4.24			0.318			0.39		
0.496	4000	2.5	4.30			0.235			0.39		
0.400	3226	3.1	4.30			$2.27\cdot10^{-2}$			0.39		
0.3100	2500	4.0	4.16			$6.38\cdot10^{-4}$			0.38		
0.2480	2000	5	4.115			$9.25\cdot10^{-4}$			0.37		
0.1240	1000	10	4.01			$6.32\cdot10^{-3}$			0.36		
0.1033	833.3	12	3.90			$1.14\cdot10^{-2}$			0.35		
0.08059	650	15.38	3.90						0.35		
0.06819	550	18.18	3.81						0.34		
0.04959	400	25.00	3.53						0.31		
0.03720	300	33.33	2.99						0.25		
0.02480	200.0	50	0.514			1.59					
0.01378	111.1	90	1.175			8.48			0.94		
0.01240	100.0	100	1.79			10.51			0.94		
0.008856	71.43	140	17.41			17.94			0.89		
0.006199	50.0	200	16.27			2.20			0.79		
0.003100	25.00	400	12.96			0.495			0.73		
0.001653	13.33	750	12.44			0.228			0.72		
0.001240	10.00	1000	12.35			0.167			0.72		
0.0006199	5.000	2000	12.27			0.0815			0.72		

Lead Telluride (PbTe) - [Ref. 11]

E/eV	$\bar{\nu}$/cm^{-1}	λ/μm	n	n_a	n_c	k	k_a	k_c	R	R_a	R_c
150		0.008266				$2.37\cdot10^{-3}$					
125		0.009919				$9.71\cdot10^{-3}$					
100		0.01240				$4.39\cdot10^{-2}$					
75		0.01653				$6.43\cdot10^{-2}$					
50		0.02480				$6.87\cdot10^{-2}$					
30		0.04133				$7.77\cdot10^{-2}$					
15		0.08266	0.72			0.17					
10		0.1240	0.66			0.60					
7.5		0.1653	0.8			0.92					
5.0	40330	0.2480	0.72			1.0					
3.0	24200	0.4133	1.0			2.2					
2.5	20160	0.4959	1.35			2.86			0.61		
1.5	12100	0.8266	3.8			3.1			0.53		
1.0	8065	1.240	4.55			2.2			0.49		
0.80	6452	1.550	6.25			0.71			0.53		
0.60	4839	2.066	6.10			0.521			0.52		
0.40	3226	3.100	6.075			0.331			0.52		
0.30	2420	4.133	5.95			$3.55\cdot10^{-2}$			0.51		
0.20	1613	6.199	5.77						0.50		
0.15	1210	8.266	5.76						0.50		
0.1017	820	12.20	5.47			$9.16\cdot10^{-3}$			0.48		
0.08927	720	13.89	5.38			$1.37\cdot10^{-2}$			0.47		
0.06943	560	17.86	5.13			$3.06\cdot10^{-2}$			0.45		
0.04959	400	25.00	4.50			$9.6\cdot10^{-2}$			0.40		
0.03968	320	31.25	3.58			0.23			0.32		
0.02976	240	41.67	1.01			1.9					
0.009919	80	125.0	2.95			16.6			0.96		
0.007439	60	166.7	4.9			22.5			0.96		
0.006199	50	200.0	6.9			27.2			0.97		
0.004959	40	250.0	11.6			34.8			0.97		
0.003720	30	333.3	27.7			35.7			0.95		
0.002480	20	500.0	27.6			39.1			0.95		
0.001240	10	1000	45.1			57.8			0.97		

Lithium Fluoride (LiF) - [Ref. 12]

E/eV	$\bar{\nu}$/cm^{-1}	λ/μm	n	n_a	n_c	k	k_a	k_c	R	R_a	R_c
2000		$6.199 \cdot 10^{-4}$	0.9999347			$4.33 \cdot 10^{-6}$					
1496		$8.287 \cdot 10^{-4}$	0.999883			$1.28 \cdot 10^{-5}$					
1016		$1.220 \cdot 10^{-3}$	0.999757			$5.18 \cdot 10^{-5}$					
725		$1.710 \cdot 10^{-3}$	0.999643			$1.62 \cdot 10^{-4}$					
504		$2.460 \cdot 10^{-3}$	0.999162			$4.96 \cdot 10^{-5}$					
303		$4.092 \cdot 10^{-3}$	0.99752			$3.12 \cdot 10^{-4}$					
250		$4.959 \cdot 10^{-3}$	0.99632			$6.17 \cdot 10^{-5}$					
200		$6.199 \cdot 10^{-3}$				$2.12 \cdot 10^{-3}$					
150		$8.265 \cdot 10^{-3}$	0.9899			$3.54 \cdot 10^{-3}$					
100		$1.240 \cdot 10^{-2}$	0.9801			$1.32 \cdot 10^{-2}$					
75		$1.653 \cdot 10^{-2}$				$2.63 \cdot 10^{-2}$					
50		$2.480 \cdot 10^{-2}$				$7.89 \cdot 10^{-2}$					
25		$4.959 \cdot 10^{-2}$	0.558			0.521					
20		$6.199 \cdot 10^{-2}$	1.20			0.58			0.10		
15.1		$8.211 \cdot 10^{-2}$	1.08			0.68			0.10		
13		$9.537 \cdot 10^{-2}$	1.04			1.64					
12.0		0.1033	2.28			0.11			0.15		
11.0		0.1127	1.77			$8.07 \cdot 10^{-7}$			0.08		
10.00		0.12398	1.606			$7.70 \cdot 10^{-7}$			0.05		
9		0.1375	1.53						0.04		
7		0.1771	1.46								
4.959	40000	0.250	1.4189						0.03		
4.000	32260	0.31	1.4073						0.03		
2.952	23810	0.42	1.3978						0.03		
2.000	16130	0.62	1.3915						0.03		
0.9919	8000	1.25	1.3851								
0.7999	6452	1.55	1.3858						0.03		
0.4959	4000	2.5	1.3731						0.02		
0.4000	3226	3.1	1.3650								
0.3100	2500	4.0	1.3493								
0.2480	2000	5.0	1.3266			$1.8 \cdot 10^{-6}$			0.02		
0.2000	1613	6.2	1.2912								
0.1698	1370	7.3	1.2499								
0.1494	1205	8.3	1.2036								
0.1240	1000	10.0	1.1005			$2.6 \cdot 10^{-3}$					
0.1127	909.1	11.0	1.0208			$8.0 \cdot 10^{-3}$					
0.1033	833.3	12.0				$1.9 \cdot 10^{-2}$					
0.09537	769.2	13.0				$3.7 \cdot 10^{-2}$					
0.08679	700	14.29	0.508			$7.74 \cdot 10^{-2}$					
0.07439	600	16.67	0.124			0.804					
0.06199	500	20.00	0.306			1.47			0.68		
0.05579	450	22.22	0.191			1.88			0.85		
0.04959	400	25.00	0.208			2.71			0.91		
0.03720	300	33.33	8.76			3.91			0.68		
0.03100	250	40.00	4.64			0.287			0.42		
0.02480	200	50.00	3.69			0.102			0.33		
0.01240	100.0	100	3.067			0.106			0.26		
0.06199	50.0	200	3.067			$4.0 \cdot 10^{-2}$			0.26		
0.04959	40.00	250	3.067			$2.2 \cdot 10^{-2}$			0.26		
0.02480	20.00	500	3.067			$6.3 \cdot 10^{-3}$					
0.01378	11.11	900				$3.1 \cdot 10^{-3}$					
$4.798\ 10^{-4}$	3.870	2584	3.023			$1.19 \cdot 10^{-3}$			0.25		
$1.464\ 10^{-4}$	1.181	8469	3.023			$6.20 \cdot 10^{-4}$			0.25		
$4.053\ 10^{-5}$	0.3269	30590	3.023			$2.63 \cdot 10^{-4}$			0.25		
$1.861\ 10^{-7}$	$1.501 \cdot 10^{-3}$	$6.662 \cdot 10^{6}$	3.018			$1.6 \cdot 10^{-5}$					
$3.718\ 10^{-8}$	$2.999 \cdot 10^{-4}$	$3.335 \cdot 10^{7}$	3.018			$1.6 \cdot 10^{-5}$					

Potassium Chloride (KCl) - [Ref. 13]

E/eV	$\bar{\nu}$/cm^{-1}	λ/μm	n	n_a	n_c	k	k_a	k_c	R	R_a	R_c
2860.3		$4.3347 \cdot 10^{-4}$				$3.93 \cdot 10^{-6}$					
2855.3		$4.3423 \cdot 10^{-4}$				$3.39 \cdot 10^{-6}$					
2849.3		$4.3514 \cdot 10^{-4}$				$4.61 \cdot 10^{-6}$					
2835.8		$4.3721 \cdot 10^{-4}$				$5.85 \cdot 10^{-6}$					

E/eV	$\bar{\nu}$/cm^{-1}	λ/μm	n	n_a	n_c	k	k_a	k_c	R	R_a	R_c
2832.3		$4.3775 \cdot 10^{-4}$				$5.85 \cdot 10^{-6}$					
2829.8		$4.3814 \cdot 10^{-4}$				$1.57 \cdot 10^{-6}$					
2828.3		$4.3837 \cdot 10^{-4}$				$4.19 \cdot 10^{-7}$					
219		$5.661 \cdot 10^{-3}$				$1.82 \cdot 10^{-3}$					
215		$5.767 \cdot 10^{-3}$				$1.84 \cdot 10^{-3}$					
212.5		$5.834 \cdot 10^{-3}$				$2.19 \cdot 10^{-3}$					
211		$5.876 \cdot 10^{-3}$				$1.82 \cdot 10^{-3}$					
185.1		$6.7 \cdot 10^{-3}$	0.99874						$1.01\ 10^{-3}$		
109.7		$1.13 \cdot 10^{-2}$	0.99578						$4.22\ 10^{-3}$		
43		0.02883	0.96			$3.0 \cdot 10^{-2}$					
40		0.03179	0.925			$1.8 \cdot 10^{-2}$					
29.9		0.04147	0.756			0.145					
20.1		0.06168	0.910			0.495					
15.1		0.08211	0.965			0.344					
10.0		0.1240	1.16			0.38			0.035		
9.0		0.1378	1.99			0.50			0.13		
8.0		0.1550	1.15			0.46			0.048		
7.0		0.1771	2.0			$8.46 \cdot 10^{-7}$			0.11		
6.199	50000	0.20	1.71739						0.070		
4.959	40000	0.25	1.58972								
3.999	32260	0.31	1.54005								
2.952	23810	0.42	1.50701								
2.695	21740	0.46	1.50115						0.040		
2.616	21100	0.474				$7.6 \cdot 10^{-11}$					
2.384	19230	0.52	1.49501								
2.066	16670	0.60	1.48969						0.039		
1.550	12500	0.80	1.48291						0.038		
1.033	8333	1.2	1.47813						0.037		
0.5166	4167	2.4	1.47464						0.037		
0.2480	2000	5.0	1.47048						0.036		
0.2000	1.613	6.2	1.46796						0.036		
0.1512	1220	8.2	1.46260						0.035		
0.09999	806.5	12.4	1.44611						0.033		
0.07560	609.8	16.4	1.42295						0.030		
0.04959	400.0	25.0	1.34059			$6.57 \cdot 10^{-4}$			0.021		
0.03999	322.6	31.0	1.2431						0.012		
0.02976	240	41.67	0.85			0.16					
0.02728	220	45.45	0.53			0.35					
0.02232	180	55.56	0.31			1.05					
0.01860	150	66.67	0.44			4.0					
0.01612	130	76.92	4.1			0.32			0.37		
0.01240	100	100.0	2.7			0.11			0.21		
0.008679	70	142.9	2.4			$9.2 \cdot 10^{-2}$			0.17		
0.006199	50	200.0	2.2						0.14		
0.001240	10.00	1000				$9.0 \cdot 10^{-3}$					
0.0006199	5.000	2000				$3.7 \cdot 10^{-3}$					
0.0004133	3.333	3000				$2.0 \cdot 10^{-3}$					
					Silicon Dioxide (Glass) - [Ref. 14]						
2000		$6.199 \cdot 10^{-4}$	0.99993			$1.503 \cdot 10^{-5}$					
1860		$6.665 \cdot 10^{-4}$	0.99991			$1.936 \cdot 10^{-5}$					
1609		$7.705 \cdot 10^{-4}$	0.99989			$9.941 \cdot 10^{-6}$					
1496		$8.287 \cdot 10^{-4}$	0.99987			$1.308 \cdot 10^{-5}$					
1204		$1.030 \cdot 10^{-3}$	0.99980			$2.916 \cdot 10^{-5}$					
1093		$1.134 \cdot 10^{-3}$	0.99975			$4.155 \cdot 10^{-5}$					
1016		$1.220 \cdot 10^{-3}$	0.99971			$5.423 \cdot 10^{-5}$					
798		$1.554 \cdot 10^{-3}$	0.99954			$1.289 \cdot 10^{-4}$					
597		$2.077 \cdot 10^{-3}$	0.99917			$3.560 \cdot 10^{-4}$					
396		$3.131 \cdot 10^{-3}$	0.99812			$4.04 \cdot 10^{-4}$					
303		$4.092 \cdot 10^{-3}$	0.99678			$9.91 \cdot 10^{-4}$					
201		$6.168 \cdot 10^{-3}$	0.99269			$3.63 \cdot 10^{-3}$					
151.2		$8.2 \cdot 10^{-3}$	0.9871			$7.3 \cdot 10^{-3}$					

Solids

E/eV	$\bar{\nu}$/cm^{-1}	λ/μm	n	n_a	n_c	k	k_a	k_c	R	R_a	R_c
99.99		$1.24 \cdot 10^{-2}$	0.9813			$7.0 \cdot 10^{-3}$					
49.59		$2.50 \cdot 10^{-2}$	0.9164			$6.5 \cdot 10^{-2}$					
40.00		$3.10 \cdot 10^{-2}$	0.907			$9.2 \cdot 10^{-2}$					
31.00		$4.00 \cdot 10^{-2}$	0.851			0.156					
25.00		0.04959	0.733			0.325					
20.00		0.06199	0.859			0.585					
15.00		0.08266	1.168			0.711			0.10		
13.00		0.09537	1.368			0.747			0.11		
11.00		0.1127	1.739			0.569			0.11		
10.00		0.1240	2.330			0.323			0.17		
9.00		0.1378	1.904			$1.89 \cdot 10^{-2}$			0.097		
7.00		0.1771	1.600						0.053		
6.00	48390	0.2066	1.543						0.046		
4.9939	40278.4	0.248272	1.50841						0.041		
4.1034	33096.1	0.302150	1.48719						0.038		
3.0640	24712.3	0.404656	1.46961						0.036		
2.5504	20570.5	0.486133	1.46313						0.035		
2.4379	19662.5	0.508582	1.46187						0.035		
2.2705	18312.5	0.546074	1.46008						0.035		
2.1489	17332.3	0.576959	1.45885						0.035		
2.1411	17269.2	0.579065	1.45877						0.035		
2.1102	17019.5	0.587561	1.45847						0.035		
2.1041	16970.4	0.589262	1.45841						0.035		
1.9257	15531.6	0.643847	1.45671						0.035		
1.8892	15237.6	0.656272	1.45637						0.035		
1.8566	14974.2	0.667815	1.45608						0.034		
1.7549	14153.9	0.706519	1.45515						0.034		
1.4550	11735.6	0.852111	1.45248						0.034		
1.0985	8860.06	1.12866	1.44888						0.034		
0.60243	4858.9	2.0581	1.43722						0.032		
0.35354	2851.4	3.5070	1.40568						0.028		
0.2976	2400	4.176	1.383			$1.07 \cdot 10^{-4}$			0.026		
0.2728	2200	4.545	1.365			$2.56 \cdot 10^{-4}$			0.024		
0.2480	2000	5.000	1.342			$3.98 \cdot 10^{-3}$			0.021		
0.2232	1800	5.556	1.306			$5.63 \cdot 10^{-3}$					
0.1984	1600	6.250	1.239			$6.52 \cdot 10^{-3}$					
0.1736	1400	7.143	1.053			$1.06 \cdot 10^{-2}$					
0.1674	1350	7.407	0.9488			$1.48 \cdot 10^{-2}$					
0.1612	1300	7.692	0.7719			$3.72 \cdot 10^{-2}$					
0.1500	1210	8.265	0.4530			0.704			0.30		
0.1401	1130	8.850	0.3563			1.53			0.66		
0.1302	1050	9.524	2.760			1.65			0.35		
0.1209	975	10.26	2.448			0.231			0.18		
0.1091	880	11.36	1.784			$7.75 \cdot 10^{-2}$			0.079		
0.09919	800	12.50	1.753			0.343			0.089		
0.08989	725	13.79	1.698			0.175			0.071		
0.06943	560	17.86	1.337			0.298			0.036		
0.06199	500	20.00	0.6616						0.882		
0.04959	400	25.0	2.739			0.397			0.23		
0.03720	300	33.33	2.210			$6.7 \cdot 10^{-2}$			0.14		
0.01240	100	100.0	1.967			$1.59 \cdot 10^{-2}$			0.11		
0.007439	60	166.7	1.959			$8.62 \cdot 10^{-3}$			0.11		
0.002480	20	500.0	1.955			$7.96 \cdot 10^{-3}$			0.10		

Silicon Monoxide (Noncrystalline) - [Ref. 15]

E/eV	$\bar{\nu}$/cm^{-1}	λ/μm	n	n_a	n_c	k	k_a	k_c	R	R_a	R_c
25		0.04959	0.8690			0.2717					
20		0.06199	0.8853			0.4919					
17.5		0.07085	0.9825			0.5961					
15		0.08266	1.132			0.6651			0.092		
12.5		0.09919	1.283			0.6523			0.090		
10		0.1240	1.378			0.6843			0.10		
7.5		0.1653	1.593			0.7473			0.12		

E/eV	$\bar{\nu}$/cm^{-1}	λ/μm	n	n_a	n_c	k	k_a	k_c	R	R_a	R_c
5	40330	0.2480	2.001			0.6052			0.15		
4	32260	0.3100	2.141			0.4006			0.15		
3	24200	0.4133	2.116			0.1211			0.13		
2.8	22580	0.4428	2.085			0.08374			0.12		
2.6	20970	0.4769	2.053			0.05544			0.12		
2.4	19360	0.5166	2.021			0.03533			0.11		
2.2	17740	0.5636	1.994			0.02153			0.11		
2	16130	0.6199	1.969			0.01175			0.11		
1.8	14520	0.6888	1.948			0.00523			0.10		
1.6	12900	0.7749	1.929			0.00151			0.10		
1.240	10000	1.000	1.87						0.092		
0.6199	5000	2.000	1.84						0.087		
0.3100	2500	4.000	1.80						0.082		
0.2480	2000	5.000	1.75						0.074		
0.2066	1667	6.000	1.70						0.067		
0.1771	1492	7.000	1.60						0.053		
0.1653	1333	7.500	1.42								
0.1459	1176	8.500	0.90			0.18					
0.1305	1053	9.500	1.20			1.20			0.024		
0.1240	1000	10.00	2.00			1.38			0.27		
0.1181	952.4	10.50	2.85			0.90			0.27		
0.1153	930.2	10.75	2.86			0.58			0.25		
0.1127	909.1	11.00	2.82			0.40			0.24		
0.1078	869.6	11.50	2.50			0.20			0.19		
0.1033	833.3	12.00	2.13			0.14			0.13		
0.09537	769.2	13.00	2.04			0.20			0.12		
0.08856	714.3	14.00	2.01			0.30			0.12		
Noncrystalline Silicon Nitride (Si_3N_4) - [Ref. 16]											
24		0.05166	0.655			0.420			0.28		
23		0.05391	0.625			0.481			0.22		
22		0.05636	0.611			0.560			0.16		
21		0.05904	0.617			0.647			0.19		
20		0.06199	0.635			0.743			0.21		
19		0.06526	0.676			0.841			0.23		
18		0.06888	0.735			0.936			0.26		
17		0.07293	0.810			1.03			0.25		
16		0.07749	0.902			1.11			0.26		
15		0.08266	1.001			1.18			0.26		
14		0.08856	1.111			1.26			0.26		
13		0.09537	1.247			1.35			0.27		
12	96790	0.1033	1.417			1.43			0.28		
11	88720	0.1127	1.657			1.52			0.29		
10.5	84690	0.1181	1.827			1.53			0.29		
10	80650	0.1240	2.000			1.49			0.29		
9.5	76620	0.1305	2.162			1.44			0.28		
9	72590	0.1378	2.326			1.32			0.27		
8	64520	0.1550	2.651			0.962			0.26		
7	56460	0.1771	2.752			0.493			0.23		
6	48390	0.2066	2.541			0.102			0.19		
5	40330	0.2480	2.278			$4.9 \cdot 10^{-3}$			0.15		
4.75	38310	0.2610	2.234			$1.2 \cdot 10^{-3}$			0.15		
4.5	36290	0.2755	2.198			$2.2 \cdot 10^{-4}$			0.14		
4	32260	0.3100	2.141						0.13		
3.5	28230	0.3542	2.099						0.13		
3	24200	0.4133	2.066						0.12		
2.5	20160	0.4959	2.041						0.12		
2	16130	0.6199	2.022						0.11		
1.5	12100	0.8266	2.008						0.11		
1	8065	1.240	1.998						0.11		

Sodium Chloride (NaCl) - [Ref. 17]

E/eV	$\bar{\nu}$/cm^{-1}	λ/μm	n	n_a	n_c	k	k_a	k_c	R	R_a	R_c
209.5		$5.918 \cdot 10^{-3}$				$2.54 \cdot 10^{-3}$					
206		$6.019 \cdot 10^{-3}$				$2.62 \cdot 10^{-3}$					
203		$6.107 \cdot 10^{-3}$				$2.08 \cdot 10^{-3}$					
200		$6.199 \cdot 10^{-3}$				$1.92 \cdot 10^{-3}$					
26.0		0.04769	0.83			0.15			0.015		
25.0		0.04959	0.83			0.18			0.018		
22.0		0.05636	0.83			0.31			0.057		
20.0		0.06199	0.88			0.34			0.036		
18.0		0.06888	0.89			0.33			0.033		
16.1		0.07700	0.74			0.45			0.084		
14.0		0.08856	0.98			0.89			0.17		
12.0		0.1033	1.22			0.79			0.12		
10.0		0.1240	1.55			0.71			0.12		
8.00		0.1550	1.38			1.10			0.20		
6.00	48390	0.2066	1.75						0.074		
5.00	40330	0.2480	1.65						0.060		
2.952	23810	0.42	1.56324						0.048		
2.480	20000	0.50	1.55157						0.047		
2.214	17860	0.56	1.54613						0.046		
2.000	16130	0.62	1.54228						0.045		
1.771	14290	0.70	1.53865						0.045		
1.675	13510	0.74	1.53728						0.045		
1.550	12500	0.80	1.53560						0.045		
1.240	10000	1.00	1.53200						0.044		
1.033	8333	1.2	1.53000						0.044		
0.6888	5556	1.8	1.52712						0.043		
0.4959	4000	2.5	1.52531						0.043		
0.4000	3226	3.1	1.52395						0.043		
0.3263	2632	3.8	1.52226			$(1.8 \pm 0.2) \cdot 10^{-9}$			0.043		
0.2952	2381	4.2	1.52121						0.043		
0.2755	2222	4.5	1.52036						0.043		
0.2480	2000	5.0	1.51883						0.042		
0.1240	1000	10.0	1.49473						0.039		
0.1033	833.3	12.0	1.48000						0.037		
0.08856	714.3	14.0	1.46188						0.035		
0.07749	625.0	16.0	1.4399						0.033		
0.06888	555.5	18.0	1.41364						0.029		
0.06199	500.0	20.0	1.3822						0.026		
0.04959	400	25.0	1.27			$3.5 \cdot 10^{-3}$			0.014		
0.04215	340	29.41	1.12			$1.7 \cdot 10^{-2}$			0.0032		
0.03720	300	33.33	0.85			0.85			0.18		
0.03410	275	36.36	0.59			0.22			0.084		
0.03286	265	37.74	0.42			0.50			0.26		
0.03224	260	38.46	0.45			0.45			0.22		
0.02480	200	50.00	0.14			1.99			0.89		
0.02108	170	58.82	1.35			6.03			0.87		
0.01984	160	62.50	6.92			2.14			0.59		
0.01922	155	64.52	5.50			0.87			0.49		
0.01860	150	66.67	4.52			0.380			0.41		
0.01736	140	71.43	3.72			0.219			0.33		
0.01612	130	76.92	3.31			0.135			0.29		
0.01488	120	83.33	3.02			0.110			0.25		
0.01240	100	100.0	2.74			0.087			0.22		
0.009919	80	125.0	2.57			0.077			0.19		
0.07439	60	166.7	2.48			0.055			0.18		
0.04959	40	250.00	2.44			0.041			0.18		
0.002480	20	500.0	2.43			0.024			0.17		
0.001240	10	1000	2.43			0.006			0.17		
0.001033	8.333	1200				$8.8 \cdot 10^{-3}$					
0.0006888	5.556	1800				$5.4 \cdot 10^{-3}$					
0.0006199	5.000	2000	2.43						0.17		

E/eV	$\bar{\nu}$/cm^{-1}	λ/μm	n	n_a	n_c	k	k_a	k_c	R	R_a	R_c
0.0004959	4.000	2500				$4.4\cdot10^{-3}$					
0.0004797	3.869	2584	2.43			$2.1\cdot10^{-3}$			0.17		
0.0003875	3.125	3200				$3.3\cdot10^{-3}$					
0.0001464	1.181	8469	2.43			$5.8\cdot10^{-4}$			0.17		
0.00004053	0.3269	30590	2.43			$2.5\cdot10^{-4}$					
Cubic Zinc Sulfide (ZnS) - [Ref. 18]											
2000		$6.199\cdot10^{-4}$	0.999904			$1.76\cdot10^{-5}$					
1204		$1.030\cdot10^{-3}$	0.999777			$1.00\cdot10^{-4}$					
1016		$1.220\cdot10^{-3}$	0.999838			$3.61\cdot10^{-5}$					
901		$1.376\cdot10^{-3}$	0.999647			$5.42\cdot10^{-5}$					
798		$1.554\cdot10^{-3}$	0.999520			$8.28\cdot10^{-5}$					
707		$1.754\cdot10^{-3}$	0.999372			$1.25\cdot10^{-4}$					
597		$2.077\cdot10^{-3}$	0.999160			$2.19\cdot10^{-4}$					
377		$9.50\cdot10^{-3}$	0.99789			$9.50\cdot10^{-4}$					
201		$6.168\cdot10^{-3}$	0.99553			$4.82\cdot10^{-3}$					
100		$1.240\cdot10^{-2}$	0.99061			$1.17\cdot10^{-2}$					
61.99		$2.000\cdot10^{-2}$	0.964			$3.32\cdot10^{-2}$			$6.2\ 10^{-4}$		
41.33		$3.000\cdot10^{-2}$	0.941			$5.10\cdot10^{-2}$					
31.00		$4.000\cdot10^{-2}$	0.847			$9.95\cdot10^{-2}$					
24.80		$5.000\cdot10^{-2}$	0.796			0.171			$2.2\ 10^{-2}$		
17.71		$7.000\cdot10^{-2}$	0.747			0.431			$7.7\ 10^{-2}$		
13.78		$9.000\cdot10^{-2}$	0.758			0.824			0.20		
12.40		0.1000	0.862			0.876			0.19		
9.919		0.125	1.02			1.36			0.31		
8.266		0.150	1.41			1.47			0.29		
6.199		0.200	2.32			1.62			0.32		
6.00	48390	0.2066	2.24			1.65			0.59		
4.00	32260	0.3100	2.70			0.44			0.22		
3.00	24200	0.4133	2.54			$4\cdot10^{-2}$			0.19		
2.50	20160	0.4959	2.42			$3\cdot10^{-2}$			0.17		
2.30	18550	0.5391	2.3950						0.17		
2.00	16130	0.6199	2.3576						0.16		
1.75	14110	0.7085	2.3319						0.16		
1.55	12500	0.7999	2.3146			$3.50\cdot10^{-6}$			0.16		
1.40	11290	0.8856	2.3033						0.16		
1.240	10000	1.000	2.2907			$3.02\cdot10^{-6}$			0.15		
1.00	8065	1.240	2.2795						0.15		
0.80	6452	1.550	2.2706						0.15		
0.6199	5000	2.000	2.2631			$6.2\cdot10^{-6}$			0.15		
0.45	3629	2.755	2.2587						0.15		
0.30	2420	4.133	2.2529						0.15		
0.20	1613	6.199	2.2443						0.15		
0.1550	1250	8.0	2.2213			$4.5\cdot10^{-6}$			0.14		
0.1240	1000	10.00	2.1986			$8.8\cdot10^{-6}$			0.14		
0.100	806.5	12.4	2.1969						0.14		
0.09	725.9	13.78	2.1793						0.14		
0.07999	645.2	15.5	2.1518			$3.82\cdot10^{-3}$			0.14		
0.07	564.6	17.71	2.1040						0.13		
0.06075	490	20.41	2.03			$8.0\cdot10^{-3}$			0.12		
0.05	403.3	24.80	1.6866						0.065		
0.03546	286	34.97	3.29			$8.3\cdot10^{-2}$			0.28		
0.03472	280	35.71	9.54			$5.2\cdot10^{-2}$			0.66		
0.02480	200	50.00	3.48			$3.1\cdot10^{-2}$			0.31		
0.01240	100	100.0	3.06			$5.8\cdot10^{-3}$			0.26		
0.004955	40	250.0	2.903			$6.2\cdot10^{-3}$			0.24		
0.004339	35	285.7	2.899			$7.0\cdot10^{-3}$			0.24		
0.003720	30	333.3	2.896						0.24		
0.003100	25	400.0	2.894						0.24		
0.002480	20	500.0	2.892						0.24		
0.001860	15	666.7	2.890						0.24		

Solids

E/eV	$\bar{\nu}$/cm^{-1}	λ/μm	n	n_a	n_c	k	k_a	k_c	R	R_a	R_c
Polytetrafluoroethylene (Teflon) - [Ref. 19] (Total diffuse hemispherical reflectance)											
4.960	40000	0.250							0.970		
4.769	38462	0.260							0.972		
4.593	37037	0.270							0.975		
4.426	35714	0.280							0.978		
4.276	34483	0.290							0.980		
4.133	33333	0.300							0.983		
4.000	32258	0.310							0.986		
3.875	31250	0.320							0.988		
3.758	30303	0.330							0.990		
3.647	29412	0.340							0.991		
3.543	28571	0.350							0.992		
3.444	27778	0.360							0.992		
3.351	27027	0.370							0.993		
2.255	18182	0.550							0.993		
2.067	16667	0.600							0.992		
1.378	11111	0.900							0.992		
1.305	10526	0.950							0.991		
1.078	8696	1.150							0.991		
1.033	8333	1.200							0.990		
0.9920	8000	1.250							0.990		
0.9538	7692	1.300							0.989		
0.9185	7407	1.350							0.988		
0.8857	7143	1.400							0.988		
0.8552	6897	1.450							0.989		
0.8267	6667	1.500							0.989		
0.8000	6452	1.550							0.988		
0.7750	6250	1.600							0.988		
0.7515	6061	1.650							0.987		
0.7294	5882	1.700							0.986		
0.7086	5714	1.750							0.986		
0.6889	5556	1.800							0.985		
0.6703	5405	1.850							0.980		
0.6526	5263	1.900							0.978		
0.6359	51282	1.950							0.978		
0.6200	5000	2.000							0.970		
0.6049	4878	2.050							0.959		
0.5905	4762	2.100							0.951		
0.5767	4651	2.150							0.946		
0.5636	4545	2.200							0.966		
0.5511	44444	2.250							0.965		
0.5487	44247	2.260							0.964		
0.5439	4386	2.280							0.963		
0.5415	4367	2.290							0.961		
0.5368	4329	2.310							0.959		
0.5345	4310	2.320							0.957		
0.5322	4292	2.330							0.956		
0.5299	4274	2.340							0.954		
0.5277	4255	2.350							0.951		
0.5232	4219	2.370							0.950		
0.5188	4184	2.390							0.949		
0.5167	4167	2.400							0.947		
0.5061	4082	2.450							0.946		
0.4960	4000	2.500							0.945		

References

1. Arsenic Selenide
   D. J. Treacy in *Handbook of Optical Constants of Solids*, E. D. Palik, Editor, Academic Press, 1985, p. 623. (Hereafter abbreviated as *HOCS*.)
   R. Zallen, R. E. Drews, R. L. Emerald, and M. L. Slade, *Phys. Rev. Lett.* 26, 1564 (1971).
   R. Zallen, M. L. Slade, and A. T. Ward, *Phys. Rev.* B 3, 4257 (1971).
   U. Strom and P. C. Taylor, *Phys. Rev.* B 16, 5512 (1977).
   G. Lucovsky, *Phys. Rev.* B 6, 1480 (1972).
   C. T. Moynihan, P. B. Macedo, M. S. Maklad, R. K. Mohr, and R. E. Howard, *J. Non-Cryst. Solids*, 17, 369 (1975).
   Y. Ohmachi, *J. Opt. Soc. Am.* 63, 630 (1973).
2. Arsenic Sulfide
   D. J. Treacy in *HOCS*, 1985, p. 641.
   P. A. Young, *J. Phys.* C 4, 93 (1971).
   W. S. Rodny, I. H. Malitson, and T. A. King, *J. Opt. Soc. Am.* 48, 633 (1958).
   R. Zallen, R.E. Drew, R. L. Emerald, and M.L. Slade, *Phys. Rev. Lett.* 26, 1564 (1971).
   M. S. Maklad, R. K. Mohr, R. E. Howard, P. B. Macedo, and C. T. Moynihan, *Solid State Commun.* 15, 855 (1974).
   P. B. Klein, P. C. Taylor, and D. J. Treacy, *Phys. Rev.* B16, 4511 (1977).
   G. Lucovsky, *Phys. Rev.* B 6, 1480 (1972).
3. Cadmium Telluride
   E. D. Palik in *HOCS*, 1985, p. 409.
   D. T. F. Marple and H. Ehrenreich, *Phys. Lett.* 8, 87 (1962).
   T. H. Myers, S. W. Edwards, and J. F. Schetzina, *J. Appl. Phys.* 52, 4231 (1981).
   D. T. F. Marple, *Phys. Rev.* 150, 728 (1966).
   A. N. Pikhtin and A. D. Yas'kov, *Sov. Phys. Semicond.* 12, 622 (1978).
   L. S. Ladd, *Infrared Phys.* 6, 145 (1966).
   J. E. Harvey and W. L. Wolfe, *J. Opt. Soc. Am.* 65, 1267 (1975).
   A. Manabe, A. Mitsuishi, and H. Yoshinaga, *Jpn. J. Appl. Phys.* 6, 593 (1967).
   A. Manabe, A. Mitsuishi, H. Oshinaga, Y. Ueda, and H. Sei, *Technol. Rep. Osaka Univ. Jpn.* 17, 263 (1967).
   J. R. Birch and D. K. Murrey, *Infrared Phys.* 18, 283 (1978).
4. Gallium Arsenide
   E. D. Palik in *HOCS*, 1985, p. 429.
   M. Cardona, W. Gudat, B. Sonntag, and P. Y. Yu, in *Proc. Intl. Conf. Phys. Semicond.*, 10th. Cambridge, 1970, p. 208. US Atom. Energy Commission, Oak Ridge, TN, 1970.
   H. R. Philipp and H. Ehrenreich, *Phys. Rev.* 129, 1550 (1963).
   J. B. Theeten, D. E. Aspnes, and R. P. H. Chang, *J. Appl. Phys.* 49, 6097 (1978).
   H. C. Casey, D. D. Sell, and K. W. Wecht, *J. Appl. Phys.* 46, 250 (1975).
   A. H. Kachare, W. G. Spitzer, F. K. Euler, and A. Kahan, *J. Appl. Phys.* 45, 2938 (1974).
   R. T. Holm, J. W. Gibson, and E. D. Palik, *J. Appl. Phys.* 48, 212 (1977).
   W. Cochran, S. J. Fray, F. A. Johnson, J. E. Quarrington, and N. Williams, *J. Appl. Phys. Suppl.* 32, 2102 (1961).
   C. P. Christensen, R. Joiner, S. K. T. Nieh, and W. H. Steier, *J. Appl. Phys.* 45, 4957 (1974).
   R. H. Stolen, *Phys. Rev.* B 11, 767 (1975); *Appl. Phys. Lett.* 15, 74 (1969).
5. Gallium Phosphide
   A. Borghesi and G. Guizzetti in *HOCS*, 1985, p. 445.
   M. Cardona, W. Gudat, B. Sonntag, and P. Y. Yu, *Proc. Intl. Conf. Phys. Semicond.* Cambridge, 1970, p. 208. US Atom. Energy Commission, Oak Ridge, TN, 1970.
   M. Cardona, W. Gudat, E. E. Koch, M. Skibowski, B. Sonntag, and P. Yu. *Phys. Rev. Lett.* 25, 659 (1970).
   S. E. Stokowski and D. D. Sell, *Phys. Rev.* B 5, 1636 (1972).
   S. A. Abagyan, G. A. Ivanov, Y. E. Shanurin, and V. I. Amosov, *Sov. Phys. Semicond.* 5, 889 (1971).
   P. G. Dean, G. Kaminsky, and R. B. Zetterstorm, *J. Appl. Phys.* 38, 3551 (1967).
   D. E. Aspnes and A. A. Studna, *Phys. Rev.* B 27, 985 (1983).
6. Indium Antimonide
   R. T. Holm in *HOCS*, 1985, p. 491.
   M. Cardona, W. Gudat, B. Sonntag, and P. Y. Yu, *Proc. Int. Conf. Phys. Semicond.*, 10th. Cambridge, 1970, p. 208. US Atom. Comm., Oak Ridge, TN, 1970.
   H. R. Philipp and H. Ehrenreich, *Phys. Rev.* 129, 1550 (1963).
   D. E. Aspnes and A. A. Studna, *Phys. Rev.* B 27, 985 (1983).
   T. S. Moss, S. D. Smith, and T. D. F. Hawkins, *Proc. Phys. Soc. London* 70B, 776 (1957).
   H. Yoshinaga and R. A. Oetjen, *Phys. Rev.* 101, 526 (1956).
   R. B. Sanderson, *J. Phys. Chem. Solids* 26, 803 (1965).
7. Indium Arsenide
   E. D. Palick and R. T. Holm in *HOCS*, 1985, p. 479.
   H. R. Philipp and H. Ehrenreich, *Phys. Rev.* 129, 1550 (1963).
   B. O. Seraphin and H. E. Bennett in *Semiconductors and Semimetals* (R. K. Willardson and A. C. Beer, Eds.), vol. 3, Academic, 1967, p. 499.
   D. E. Aspnes and A. A. Studna, *Phys. Rev.* B 27, 985 (1983).
   J. R. Dixon and J. M. Ellis, *Phys. Rev.* 123, 1560 (1961).
   A. Memon, T. J. Parker, and J. R. Birch, *Proc. SPIE*, 289, 20 (1981).
8. Indium Phosphide
   O. J. Glembocki and H. Piller in *HOCS*, 1985, p. 503.
   M. Cardona, *J. Appl. Phys.* 32, 958 (1961); 36, 2181 (1965).
   D. E. Aspnes and A. A. Studna, *Phys. Rev.* B 27, 985 (1983).
   G. D. Pettit and W. J. Turner, *J. Appl. Phys.* 36, 2081 (1965).
   R. Newman, *Phys. Rev.* 111, 1518 (1958).
   W. N. Reynolds, M. T. Lilburne, and R. M. Dell, *Proc. Phys. Soc. London* 71, 416 (1958).
   H. Jamshidi and T. J. Parker, Int. Meet. Infrared Mm. Waves, 7th., Marseilles, 1983.
9. Lead Selenide
   G. Bauer and H. Krenn in *HOCS*, 1985, p. 517.
   M. Cardona and D. L. Greenaway, *Phys. Rev.* A 133, 1685 (1964).
   T. S. Moss, *Optical Properties of Semiconductors*, Butterworth, 1959, p. 189.
   J. N. Zemel, J. D. Jensen, and R. B. Schoolar, *Phys. Rev.* A 140, 330 (1965).
   W. W. Scanlon, *J. Phys. Chem. Solids*, 8, 423 (1959).
   K. V. Vyatkin and A. P. Shotov, *Sov. Phys. Semicond.* 14, 785 (1980); Fiz. Tekh. Poluprovodn. 14, 1331 (1980).
10. Lead Sulfide
    G. Guizzetti and A. Borghesi in *HOCS*, 1985, p. 525.
    M. Cardona and R. Haensel, *Phys. Rev.* B 1, 2605 (1970).
    M. Cardona and D. L. Greenaway, *Phys. Rev.* A 133, 1685 (1964).
    M. Cardona, C. M. Penchina, E. E. Koch, and P. Y. Yu, *Phys. Status Solidi* B 53, 327 (1972).
    P. R. Wessel, *Phys. Rev.* 153, 836 (1967).
    C. E. Rossi and W. Paul, *J. Appl. Phys.* 38, 1803 (1967).
    J. N. Zemel, J. D. Jensen, and R. B. Schoolar, *Phys. Rev.* A 140, 330 (1965).
11. Lead Telluride
    G. Bauer and H. Krenn in *HOCS*, 1985, p. 535.
    M. Cardona and R. Haensel, *Phys. Rev.* B 1, 2605 (1970).
    M. Cardona and D. L. Greenaway, *Phys. Rev.* 133, A1685 (1964).
    D. M. Korn and R. Braunstein, *Phys. Rev.* B 5, 4837 (1972).
    W. W. Scanlon, *J. Phys. Chem. Solids* 8, 423 (1959).
    J. N. Zemel, J. D. Jensen, and R. B. Schoolar, *Phys. Rev.* 140, A330 (1965).
12. Lithium Fluoride
    E. D. Palik and W. R. Hunter in *HOCS*, 1985, p. 675.
    B. L. Henke, P. Lee, T. J. Tanaka, R. L. Shimabukuro, and B. K. Fujikawa, *Low Energy X-ray Diagnostics-1981* (D. T. Attwood and B. L. Henke, Eds.), AIP Conf. Proc. No. 75, 1981.
    A. P. Lukirskii, E. P. Savinov, O. A. Ershov, and Y. F. Shepelev, *Opt. Spektrosk.* 16, 168 (1964); 16, 310 (1964).
    F. C. Brown, C. Gahwiller, A. B. Kunz, and N. O. Lipari, *Phys. Rev. Lett.* 25, 927 (1970).
    A. Milgram and M. P. Givens, *Phys. Rev.* 125, 1506 (1962).
    T. Tomiki and T. Miyata, *J. Phys. Soc. Jpn.* 27, 658 (1969).
    A. Kachare, G. Andermann, and L. R. Brantley, *J. Phys. Chem. Solids* 33, 467 (1972).
13. Potassium Chloride
    E. D. Palik in *HOCS*, 1985, p. 703.
    O. Aita, I. Nagakura, and T. Sagawa, *J. Phys. Soc. Jpn.* 30, 1414 (1971).

A. P. Lukirskii, E. P. Savinov, O. A. Ershov, and Y. F. Shepelev, *Opt. Spectrosc.* 16, 168 (1964); *Opt. Spektrosk.* 16, 310 (1964).
T. Tomika, *J. Phys. Soc. Jpn.* 22, 463 (1967).
M. Antinori, A. Balzarotti, and M. Piacentini, *Phys. Rev.* B 7, 1541 (1973).
H. H. Li, *J. Phys. Chem. Ref. Data* 5, 329 (1976).
S. D. Allen and J. A. Harrington, *Appl. Opt.* 17, 1679 (1978).
K. W. Johnson and E. E. Bell, *Phys. Rev.* 139A, 1295 (1965).

14. Silicon Dioxide
H. R. Philipp in *HOCS*, 1985, p. 749.
J. Rife and J. Osantowski, *J. Opt. Soc. Am.* 70, 1513 (1980).
B. L. Henke, P. Lee, T. J. Tanaka, R. L. Shimabukuro, and B. K. Fujikawa, *Low Energy X-ray Diagnostics-1981* (D. T. Attwood and B. L. Henke, Eds.), AIP Conf. Proc. No. 75, 1981.
H. R. Philipp, *Solid State Commun.* 4, 73 (1966); *J. Phys. Chem. Solids*, 32, 1935 (1971).
P. L. Lamy, *Appl. Opt.* 16, 2212 (1977).
H. R. Philipp, *J. Appl. Phys.* 50 1053 (1979).
D. G. Drummond, *Proc. Roy. Soc. London*, 153, 328 (1935).

15. Silicon Monoxide
H. R. Philipp in *HOCS*, 1985, p. 765.
H. R. Philipp, *J. Phys. Chem. Solids*, 32, 1935 (1971).
G. Hass and C. D. Salzberg, *J. Opt. Soc. Am.* 44, 181 (1954).
E. Cremer, T. Kraus, and E. Ritter, *Zs. Electrochem.* 62, 939 (1958).
A. P. Bradford, G. Hass, M. McFarland, and E. Ritter, *Appl. Opt.* 4, 971 (1965).

16. Silicon Nitride
H. R. Philipp in *HOCS*, 1985, p. 771.
H. R. Philipp, *J. Electrochem. Soc.* 120, 295 (1973).
J. B. Theeten, D. E. Aspnes, F. Simondet, M. Erman, and P. C. Mürau, *J. Appl. Phys.* 52, 6788 (1981).
J. Bauer, *Phys. Status Solidi*, A 39, 411 (1977).

17. Sodium Chloride
J. E. Eldridge and E. D. Palik in *HOCS*, p. 775.
J. A. Harrington, C. J. Duthler, F. W. Patten, and M. Hass, *Solid State Commun.* 18, 1043 (1976).
T. Miyata and T. Tomiki, *J. Phys. Soc. Jpn.* 24, 1286 (1968); ibid., 22, 209 (1967).
D. M. Roessler and W. C. Walker, *J. Opt. Soc. Am.* 58, 279 (1968).
D. M. Roessler and W. C. Walker, *Phys. Rev.* 166, 599 (1968).
S. Allen and J. A. Harrington, *Appl. Opt.* 17, 1679 (1978).
O. Aita, I. Nagakura, and T. Sagawa, *J. Phys. Soc. Jpn.* 30, 1414 (1971).

18. Zinc Sulfide
E. D. Palik and A. Addamiano in *HOCS*, 1985, p. 597.
B. L. Henke, P. L. Lee, T. J. Tanaka, R. L. Shimabukuro, and B. F. Fujikawa, *Low Energy X-ray Diagnostics-1981* (D. T. Attwood and B. L. Henke, Eds.), AIP Conf. Proc. No. 75, 1981.
M. Cardona and G. Harbeke, *Phys. Rev.* 137, A1467 (1965).
Eastman Kodak, Publ. No. U-72, Rochester, New York (1981).
C. A. Klein and R. N. Donadio, *J. Appl. Phys.* 51, 797 (1980).
T. Deutsch, *Proc. Int. Conf. Phys. Semicond.*, 6th Exeter 1962, p. 505. The Inst. of Physics and the Physical Soc., London, 1962.
A. Manabe, A. Mitsuishi, and H. Yoshinaga, *Jpn. J. Appl. Phys.* 6, 593 (1967).
W. W. Piper, D. T. F. Marple, and P. D. Johnson, *Phys. Rev.* 110, 323 (1958).

19. Polytetrafluoroethylene
J. W. L. Thomas (NIST), Private communication.
NIST Certificate, STM 2044.
P. Y. Barnes, E. A. Early, and A. C. Parr, *NIST Special Publ. 250-48*, NIST Measurement Services: Spectral Reflectance.
Diffuse Reflectance Coatings and Materials Sections, Labsphere Catalog, 1996.
A. Arecchi and C. Ryder (Labsphere, North Sutten, NJ), private communication.

Solids

ELASTO-OPTIC, ELECTRO-OPTIC, AND MAGNETO-OPTIC CONSTANTS

When a crystal is subjected to a stress field, an electric field, or a magnetic field, the resulting optical effects are in general dependent on the orientation of these fields with respect to the crystal axes. It is useful, therefore, to express the optical properties in terms of the refractive index ellipsoid (or indicatrix):

$$\frac{x^2}{n_x^2}+\frac{y^2}{n_y^2}+\frac{z^2}{n_z^2}=1$$

or

$$\sum_{ij} B_{ij}x_i y_j = 1\,(i,j=1,2,3)$$

where

$$B_{ij}=\left[\frac{1}{\varepsilon}\right]_{ij}=\left[\frac{1}{n^2}\right]_{ij}$$

ε is the dielectric constant or permeability; the quantity B_{ij} is called impermeability.

A crystal exposed to a *stress* **S** will show a change of its impermeability. The photoelastic (or elasto-optic) constants, P_{ijkl}, are defined by

$$\Delta\left[\frac{1}{\varepsilon}\right]_{ij}=\Delta\left[\frac{1}{n^2}\right]_{ij}=\sum_{kl} P_{ijkl}S_{kl}$$

where n is the refractive index and S_{kl} are the strain tensor elements; the P_{ijkl} are the elements of a 4th rank tensor.

When a crystal is subjected to an *electric field* **E**, two possible changes of the refractive index may occur depending on the symmetry of the crystal.

1. All materials, including isotropic solids and polar liquids, show an electro-optic birefringence (Kerr effect) which is proportional to the square of the electric field, E:

$$\Delta\left[\frac{1}{n^2}\right]_{ij}=\sum_k K_{ijkl}E_kE_l=\sum_{k,l=1,2,3} g_{ijkl}p_kp_l$$

where E_k and E_l are the components of the electric field and p_k and p_l the electric polarizations. The coefficients K_{ijkl} are the quadratic electro-optic coefficients, while the constants g_{ijkl} are known as the Kerr constants.

2. The other electro-optic effect only occurs in the 20 piezoelectric crystal classes (no center of symmetry). This effect is known as the Pockels effect. The optical impermeability changes linearly with the static field

$$\Delta\left[\frac{1}{n^2}\right]_{ij}=\sum_k r_{ij,k}E_k$$

The coefficients $r_{ij,k}$ have the name (linear) electro-optic coefficients.

The values of the electro-optic coefficients depend on the boundary conditions. If the superscripts T and S denote, respectively, the conditions of zero stress (free) and zero strain (clamped) one finds:

$$r_{ij}^{\mathrm{T}}=r_{ij}^{\mathrm{S}}+q_{ik}^{\mathrm{E}}e_{jk}=r_{ij}^{\mathrm{S}}+P_{ik}^{\mathrm{E}}d_{jk}$$

where $e_{jk}=(\partial T_k/\partial E_j)_S$ and $d_{jk}=(\partial S_k/\partial E_j)_T$ are the appropriate piezoelectric coefficients.

The interaction between a *magnetic field* and a light wave propagating in a solid or in a liquid gives rise to a rotation of the plane of polarization. This effect is known as *Faraday rotation*. It results from a difference in propagation velocity for left and right circular polarized light.

The Faraday rotation, θ_F, is linearly proportional to the magnetic field H:

$$\theta_F = VlH$$

where l is the light path length and V is the *Verdet* constant (minutes/oersted·cm).

For ferromagnetic, ferrimagnetic, and antiferromagnetic materials the magnetic field in the above expression is replaced by the magnetization M and the magneto-optic coefficient in this case is known as the Kund constant K:

Specific Faraday rotation $F = KM$

In the tables below the *Faraday rotation* is listed at the saturation magnetization per unit length, together with the absorption coefficient α, the temperature T, the critical temperature T_C (or T_N), and the wavelength of the measurement.

In the tables that follow, the properties are presented in groups:

- Elasto-optic coefficients (photoelastic constants)
- Linear electro-optic coefficients (Pockels constants)
- Quadratic electro-optic coefficients (Kerr constants)
- Magneto-optic coefficients:
  - Verdet constants
  - Faraday rotation parameters

Within each group, materials are classified by crystal system or physical state. References are given at the end of each group of tables.

ELASTO-OPTIC COEFFICIENTS (PHOTOELASTIC CONSTANTS)

Elasto-Optic Coefficients: Cubic Crystals (43m, 432, m3m)

Name	Synonym	Formula	λ/μm	p_{11}	p_{12}	p_{44}	p_{11}-p_{12}	Ref.
Ammonium chloride	Sal ammoniac	NH_4Cl	0.589	0.142	0.245	0.042	-0.103	9
Cadmium telluride		CdTe	1.06	-0.152	-0.017	-0.057	-0.135	10
Calcium fluoride	Fluorite	CaF_2	0.60[a]	0.038	0.226	0.0254	-0.183	11
Carbon (diamond)	Diamond	C	0.565[b]	-0.278	0.123	-0.161	-0.385	13
Copper(I) bromide	Cuprous bromide	CuBr	0.633	0.072	0.195	-0.083	-0.123	12
Copper(I) chloride	Nantokite	CuCl	0.633	0.120	0.250	-0.082	-0.130	12
Copper(I) iodide	Marshite	CuI	0.633	0.032	0.151	-0.068	-0.119	12
Gallium arsenide		GaAs	1.15	-0.165	-0.140	-0.072	-0.025	15
Gallium phosphide		GaP	0.633	-0.151	-0.082	-0.074	-0.069	15
Germanium		Ge	3.39	-0.151	-0.128	-0.072	-0.023	14
KRS-5		Tl(Br,I)	0.633	-0.140	0.149	-0.0725	-0.289	18,20
KRS-6		Tl(Br,Cl)	0.633	-0.451	-0.337	-0.164	-0.114	19,20
Lithium chloride		LiCl	0.589			-0.0177	-0.0407	3
Lithium fluoride		LiF	0.589	0.02	0.13	-0.045	-0.11	5
Potassium bromide		KBr	0.589	0.212	0.165	-0.022	0.047	5
Potassium chloride	Sylvite	KCl	0.633	0.22	0.16	-0.025	0.06	4
Potassium fluoride		KF	0.546	0.26	0.20	-0.029	0.06	1
Potassium iodide		KI	0.590	0.212	0.171		0.041	6
Rubidium bromide		RbBr	0.589	0.293	0.185	-0.034	0.108	7,8
Rubidium chloride		RbCl	0.589	0.288	0.172	-0.041	0.116	7,8
Rubidium iodide		RbI	0.589	0.262	0.167	-0.023	0.095	7,8
Sodium bromide		NaBr	0.589	0.148	0.184	-0.0036	-0.035	1
Sodium chloride	Halite	NaCl	0.589	0.115	0.159	-0.011	-0.042	2
Sodium fluoride	Villiaumite	NaF	0.633	0.08	0.20	-0.03	-0.12	1
Sodium iodide		NaI	0.589			0.0048	-0.0141	3
Strontium fluoride		SrF_2	0.633	0.080	0.269	0.0185	-0.189	16
Strontium titanate		$SrTiO_3$	0.633	0.15	0.095	0.072		17
Zinc sulfide (sphalerite)	Sphalerite	ZnS	0.633	0.091	-0.01	0.075	0.101	15

[a] 0.55 to 0.65 μm
[b] 0.540 to 0.589 μm

Elasto-Optic Coefficients: Cubic Crystals (23, m3)

Name	Formula	λ/μm	p_{11}	p_{12}	p_{44}	p_{13}	Ref.
Barium nitrate	$Ba(NO_3)_2$	0.589		[a]	-0.0205	[b]	13
Lead(II) nitrate	$Pb(NO_3)_2$	0.589	0.162	0.24	-0.0198	0.20	24,25
Sodium bromate	$NaBrO_3$	0.589	0.185	0.218	-0.0139	0.213	26
Sodium chlorate	$NaClO_3$	0.589	0.162	0.24	-0.0198	0.20	26
Strontium nitrate	$Sr(NO_3)_2$	0.41	0.178	0.362	-0.014	0.316	27

[a] p_{11}–p_{22} = 0.992
[b] p_{11}–p_{13} = 0.713

Elasto-Optic Coefficients: Hexagonal Crystals (mmc, 6mm)

Name	Formula	λ/μm	p_{11}	p_{12}	p_{13}	p_{31}	p_{33}	p_{44}	Ref.
Beryllium aluminum metasilicate	$Be_3Al_2(SiO_3)_6$	0.589	0.0099	0.175	0.191	0.313	0.023	-0.152	28
Cadmium sulfide	CdS	0.633	-0.142	-0.066	-0.057	-0.041	-0.20	-0.099	15,2
Zinc oxide	ZnO	0.633	±0.222	±0.099	-0.111	±0.088	-0.235	0.0585	30
Zinc sulfide (wurtzite)	ZnS	0.633	-0.115	0.017	0.025	0.0271	-0.13	-0.0627	31

Elasto-Optic Coefficients: Trigonal Crystals (3m, 32, $\bar{3}$m)

Name	Formula	λ/μm	p_{11}	p_{12}	p_{13}	p_{14}	p_{31}	p_{33}	p_{41}	p_{44}	Ref.
Calcium carbonate (calcite)	$CaCO_3$	0.514	0.062	0.147	0.186	-0.011	0.241	0.139	-0.036	-0.058	33
Cinnabar	HgS	0.633			±0.445			±0.115			36
Lithium niobate	$LiNbO_3$	0.633	±0.034	±0.072	±0.139	±0.066	±0.178	±0.060	±0.154	±0.300	15,34
Lithium tantalate	$LiTaO_3$	0.633	-0.081	0.081	0.093	-0.026	0.089	-0.044	-0.085	0.028	15,35

Name	Formula	λ/μm	p_{11}	p_{12}	p_{13}	p_{14}	p_{31}	p_{33}	p_{41}	p_{44}	Ref.
Proustite	Ag_3AsS_3	0.633	±0.10	±0.19	±0.22		±0.24	±0.20			38
Quartz (α)	SiO_2	0.589	0.16	0.27	0.27	-0.030	0.29	0.10	-0.047	-0.079	37
Sapphire	Al_2O_3	0.644	-0.23	-0.03	0.02	0.00	-0.04	-0.20	0.01	-0.10	15,32
Sodium nitrite	$NaNO_2$	0.633		±0.21	±0.215	±0.027	±0.25		0.055	-0.06	39
Tellurium	Te	10.6	0.155	0.130							15

Elasto-Optic Coefficients: Tetragonal Crystals (4/mmm, $\bar{4}$2m, 422)

Name	Synonym	Formula	λ/μm	p_{11}	p_{12}	p_{13}	p_{31}	p_{33}	p_{44}	p_{66}	Ref.
Ammonium dihydrogen phosphate	ADP	$NH_4H_2PO_4$	0.589	0.319	0.277	0.169	0.197	0.167	-0.058	-0.091	40
Barium strontium niobate	Barium niobium strontium oxide	$Sr_{0.75}Ba_{0.25}Nb_2O_6$	0.633	0.16	0.10	0.08	0.11	0.47			46
Barium strontium niobate	Barium niobium strontium oxide	$Sr_{0.5}Ba_{0.5}Nb_2O_6$	0.633	0.06	0.08	0.17	0.09	0.23			46
Barium titanate	Barium metatitanate	$BaTiO_3$	0.633	0.425							41
Calomel	Mercury(I) chloride	Hg_2Cl_2	0.633	±0.551	±0.440	±0.256	±0.137	±0.010		±0.047	44
Cesium dihydrogen arsenate	CDA	CsH_2AsO_4	0.633	0.267	0.225	0.200	0.195	0.227			42
Magnesium fluoride	Sellaite	MgF_2	0.546						±0.0776	±0.0488	43
Potassium dihydrogen phosphate	KDP	KH_2PO_4	0.589	0.287	0.282	0.174	0.241	0.122	-0.019	-0.064	45
Rubidium dihydrogen arsenate	RDA	RbH_2AsO_4	0.633	0.227	0.239	0.200	0.205	0.182			41
Rubidium dihydrogen phosphate	RDP	RbH_2PO_4	0.633	0.273	0.240	0.218	0.210	0.208			41
Tellurium dioxide	Tellurite	TeO_2	0.633	0.0074	0.187	0.340	0.090	0.240	-0.17	-0.046	47
Titanium(IV) oxide (rutile)	Rutile	TiO_2	0.633	0.017	0.143	-0.139	-0.080	-0.057	-0.009	-0.060	48

Elasto-Optic Coefficients: Tetragonal Crystals (4, $\bar{4}$, 4/m)

Name	Synonym	Formula	λ/μm	p_{11}	p_{12}	p_{13}	p_{16}	p_{31}	p_{33}	p_{44}	p_{45}	p_{61}	p_{66}	Ref.
Cadmium molybdate		$CdMoO_4$	0.633	0.12	0.10	0.13		0.11	0.18					49
Lead(II) molybdate	Wulfenite	$PbMoO_4$	0.633	0.24	0.24	0.255	0.017	0.175	0.300	0.067	-0.01	0.013	0.05	52
Sodium bismuth molybdate		$NaBi(MoO_4)_2$	0.633	0.243	0.205	0.25		0.21	0.29					

Elasto-Optic Coefficients: Orthorhombic Crystals (222, m22, mmm)

Name	Formula	λ/μm	p_{11}	p_{12}	p_{13}	p_{21}	p_{22}	p_{23}	p_{31}	p_{32}	p_{33}	p_{44}	p_{55}	p_{66}	Ref.
Ammonium chlorate	NH_4ClO_3	0.633		0.24	0.18	0.23		0.20	0.19	0.18	±0.02	<±0.02		±0.04	51
Ammonium sulfate	$(NH_4)_2SO_4$	0.633	0.26	0.19	±0.260	±0.230	±0.27	±0.254	0.20	±0.26	0.26	0.015	±0.0015	0.012	52
Potassium sodium tartrate tetrahydrate	$KNaC_4H_4O_6{\cdot}4H_2O$	0.589	0.35	0.41	0.42	0.37	0.28	0.34	0.36	0.35	0.36	-0.030	0.0046	-0.025	53
Iodic acid	HIO_3	0.633	0.302	0.496	0.339	0.263	0.412	0.304	0.251	0.345	0.336	0.084	-0.030	0.098	54
Sulfur (rhombic)	S	0.633	0.324	0.307	0.268	0.272	0.301	0.310	0.203	0.232	0.270	0.143	0.019	0.118	54
Barite	$BaSO_4$	0.589	0.21	0.25	0.16	0.34	0.24	0.19	0.28	0.22	0.31	0.002	-0.012	0.037	55
Topaz	$Al_2(SiO_4)(F,OH)_2$		-0.085	0.069	0.052	0.095	-0.120	0.065	0.095	0.085	-0.083	-0.095	-0.031	0.098	28

Elasto-Optic Coefficients: Monoclinic Crystals (2, m, 2/m)

Name	Synonym	Formula	λ/μm	Coefficients		
Taurine	2-Aminoethanesulfonic acid	$C_2H_7NO_3S$	0.589	$p_{11} = 0.313$	$p_{25} = -0.0025$	$p_{51} = -0.014$
Taurine	2-Aminoethanesulfonic acid	$C_2H_7NO_3S$	0.589	$p_{12} = 0.251$	$p_{31} = 0.362$	$p_{52} = 0.006$
Taurine	2-Aminoethanesulfonic acid	$C_2H_7NO_3S$	0.589	$p_{13} = 0.270$	$p_{32} = 0.275$	$p_{53} = 0.0048$
Taurine	2-Aminoethanesulfonic acid	$C_2H_7NO_3S$	0.589	$p_{15} = -0.10$	$p_{33} = 0.308$	$p_{55} = 0.047$
Taurine	2-Aminoethanesulfonic acid	$C_2H_7NO_3S$	0.589	$p_{21} = 0.281$	$p_{35} = -0.003$	$p_{64} = 0.0024$
Taurine	2-Aminoethanesulfonic acid	$C_2H_7NO_3S$	0.589	$p_{22} = 0.252$	$p_{44} = 0.0025$	$p_{66} = 0.0028$
Taurine	2-Aminoethanesulfonic acid	$C_2H_7NO_3S$	0.589	$p_{23} = 0.272$	$p_{46} = -0.0056$	

Elasto-Optic Coefficients: Garnets

Name	Abbrev.	Formula	λ/μm	p_{11}	p_{12}	p_{44}	$p_{11} - p_{12}$	Ref.
Gadolinium gallium oxide	GGG	$Gd_3Ga_5O_{12}$	0.514	-0.086	-0.027	-0.078	-0.059	23
Yttrium iron oxide	YIG	$Y_3Fe_5O_{12}$	1.15	0.025	0.073	0.041		15
Yttrium gallium oxide	YGG	$Y_3Ga_5O_{12}$	0.633	0.091	0.019	0.079		17
Yttrium aluminum oxide	YAG	$Y_3Al_5O_{12}$	0.633	-0.029	0.0091	-0.0615	-0.038	15

Elasto-Optic Coefficients: Isotropic Materials

Name	Formula	λ/μm	p_{11}	p_{12}	p_{44}	Ref.
Fused silica	SiO_2	0.633	0.121	0.270	-0.075	15
Water	H_2O	0.633	±0.31	±0.31		15
Polystyrene		0.633	±0.30	±0.31		25
Lucite		0.633	±0.30	0.28		25
Orpiment	As_2S_3-glass	1.15	0.308	0.299	0.0045	15
Tellurium oxide	TeO_2-glass	0.633	0.257	0.241	0.0079	56
Laser glasses	LGS-247-2	0.488	±0.168	±0.230		57
	LGS-250-3		±0.135	±0.198		
	LGS-1		±0.214	±0.250		
	KGSS-1621		±0.205	±0.239		
Dense flint glasses (examples)	LaSF	0.633	0.088	0.147	-0.030	58
	SF_4		0.215	0.243	-0.014	
	U10502		0.172	0.179	-0.004	
	$TaFd_7$		0.099	0.138	-0.020	

Elasto-Optic Coefficients: Rare Gases

Name	Formula	λ/μm	p_{11}	p_{12}	p_{44}	$p_{11} - p_{12}$	Ref.
Neon[a]	Ne	0.488	0.157	0.168	0.004	-0.011	21
Argon[b]	Ar	0.488	0.256	0.302	0.015	-0.046	22
Krypton[c]	Kr	0.488	0.34	0.34	0.037	0	21
Xenon[d]	Xe	0.488	0.284	0.370	0.029	-0.086	22

[a] At 24.3 K
[b] At 82.3 K
[c] At 115.6 K
[d] At 160.5 K

References

A. Narasimhamurty, T. S., *Photoelastic and Electro-Optic Properties of Crystals*, Plenum Press, New York, 1981, pp. 290–293.
B. Weber, M. J., Ed., *CRC Handbook of Laser Science and Technology*, Volume IV, Part 2, CRC Press, Boca Raton, FL, 1986, pp. 324–331.
1. Petterson, H. E., *J. Opt. Soc. Am.*, 63, 1243, 197
2. Burstein, E. and Smith, P. L., *Phys. Rev.*, 74, 229, 1948.
3. Pakhnev, A. V., et al., *Sov. Phys. J. (transl.)*, 18, 1662, 197
4. Feldman, A., Horovitz, D., and Waxler, R. M., *Appl. Opt.*, 16, 2925, 1977.
5. Iyengar, K. S., *Nature (London)*, 176, 1119, 1955.
6. Bansigir, K. G. and Iyengar, K. S., *Acta Crystallogr.*, 14, 727, 1961.
7. Pakhev, A. V., et al., *Sov. Phys. J. (transl.)*, 20, 648, 1975.
8. Bansigir, K. G., *Acta Crystallogr.*, 23, 505, 1967.
9. Krishna Rao, K. V. and Krishna Murty, V. G., *Ind. J. Phys.*, 41, 150, 1967.
10. Weil, R. and Sun, M. J., *Proc. Int. Symp. CdTe (Detectors)*, Strasbourg Centre de Rech. Nucl., 1971, XIX-1 to 6, 1972.
11. Schmidt, E. D. D. and Vedam, K., *J. Phys. Chem. Solids*, 27, 1563, 1966.
12. Biegelsen, D. K., et al., *Phys. Rev. B*, 14, 3578, 1976.
13. Helwege, K. H., *Landolt-Börnstein, New Series Group III*, Vol. II, Springer-Verlag, Berlin, 1979.
14. Feldman, A., Waxler, R. M., and Horovitz, D., *J. Appl. Phys.*, 49, 2589, 1978.
15. Dixon, R. W., *J. Appl. Phys.*, 38, 5149, 1967.
16. Shabin, O. V., et al., *Sov. Phys. Sol. State (transl.)*, 13, 3141, 1972.
17. Reintjes, J. and Schultz, M. B., *J. Appl. Phys.*, 39, 5254, 1968.
18. Rivoallan, L. and Favre, F., *Opt. Commun.*, 8, 404, 1973.
19. Rivoallan, L. and Favre, F., *Opt. Commun.*, 11, 296, 1974.
20. Afanasev, I. I., et al., *Sov. J. Opt. Technol.*, 46, 663, 1979.
21. Rand, S. C., et al., *Phys. Rev. B*, 19, 4205, 1979.
22. Sipe, J. E., *Can J. Phys.*, 56, 199, 1978.
23. Christyi, I. L., et al., *Sov. Phys. Sol. State (transl.)*, 17, 922, 1975.
24. Narasimhamurty, T. S., *Curr. Sci. (India)*, 23, 149, 1954.
25. Smith, T. M. and Korpel, A., *IEEE J. Quant. Electron.*, QE-1, 283, 1965.
26. Narasimhamurty, T. S., *Proc. Ind. Acad. Sci.*, A40, 164, 1954.
27. Rabman, A., *Bhagarantam Commem. Vol.*, Bangalore Print. and Publ., 173, 1969.
28. Eppendahl, R., *Ann. Phys. (IV)*, 61, 591, 1920.
29. Laurenti, J. P. and Rouzeyre, M., *J. Appl. Phys.*, 52, 6484, 1981.
30. Sasaki, H., et al., *J. Appl. Phys.*, 47, 2046, 1976.
31. Uchida, N. and Saito, S., *J. Appl. Phys.*, 43, 971, 1972.
32. Waxler, R. M. and Farabaugh, E. M., *J. Res. Natl. Bur. Stand.*, A74, 215, 1970.
33. Nelson, D. F., Lazay, P. D., and Lax, M., *Phys. Rev.*, B6, 3109, 1972.
34. O'Brien, R. J., Rosasco, G. J., and Weber, A., *J. Opt. Soc. Am.*, 60, 716, 1970.
35. Avakyants, L. P., et al., *Sov. Phys.*, 18, 1242, 1976.
36. Sapriel, J., *Appl. Phys. Lett.*, 19, 533, 1971.
37. Narasimhamurty, T. S., *J. Opt. Soc. Am.*, 59, 682, 1969.
38. Zubrinov, I. I., et al., *Sov. Phys. Sol. State (transl.)*, 15, 1921, 1974.
39. Kachalov, O. V. and Shpilko, I. O., *Sov. Phys. JETP (transl.)*, 35, 957, 1972.
40. Narasimhamurty, T. S., et al., *J. Mater. Sci.*, 8, 577, 1973.
41. Tada, K. and Kikuchi, K., *Jpn. J. Appl. Phys.*, 19, 1311, 1980.
42. Aleksandrov, K. S., et al., *Sov. Phys. Sol. State (transl.)*, 19, 1090, 1977.
43. Afanasev, I. I., et al., *Sov. Phys. Sol. State (transl.)*, 17, 2006, 1975.
44. Silvestrova, I. M., et al., *Sov. Phys. Cryst. (transl.)*, 20, 649, 1975.
45. Veerabhadra Rao, K. and Narasimhamurty, T. S., *J. Mater. Sci.*, 10, 1019, 1975.
46. Venturini, E. L., et al., *J. Appl. Phys.*, 40, 1622, 1969.

47. Vehida, N. and Ohmachi, Y., *J. Appl. Phys.*, 40, 4692, 1969.
48. Grimsditch, M. H. and Ramdus, A. K., *Phys. Rev. B*, 22, 4094, 1980.
49. Schinke, D. P. and Viehman,W., unpublished data.
50. Coquin, G. A., et al., *J. Appl. Phys.*, 42, 2162, 1971.
51. Vasquez, F., et al., *J. Phys. Chem. Solids*, 37, 451, 1976.
52. Luspin, Y. and Hauret, G., *C.R.Ac. Sci. Paris*, B274, 995, 1972.
53. Narasimhamurty, T. S., *Phys. Rev.*, 186, 945, 1969.
54. Haussühl, S. and Weber, H. J., *Z. Kristall.*, 132, 266, 1970.
55. Vedam, K., *Proc. Ind. Ac. Sci.*, A34, 161, 1951.
56. Yano, T., Fukumoto, A., and Watanabe, A., *J. Appl. Phys.*, 42, 3674, 1971.
57. Manenkov, A. A. and Ritus, A. I., *Sov. J. Quant. Electr.*, 8, 78, 1978.
58. Eschler, H. and Weidinger, F., *J. Appl. Phys.*, 46, 65, 1975.

LINEAR ELECTRO-OPTIC COEFFICIENTS

Linear Electro-Optic Coefficients: Cubic Crystals ($\bar{4}3m$)

Name	Synonym	Formula	λ/μm	r_{41}/pm V^{-1}
Copper(I) bromide	Cuprous bromide	CuBr	0.525	0.85
Copper(I) chloride	Nantokite	CuCl	0.633	3.6
Copper(I) iodide	Marshite	CuI	0.55	-5.0
Eulytite	BSO	$Bi_4(SiO_4)_3$	0.63	0.54
Bismuth germanate	BGO	$Bi_4(GeO_4)_3$	0.63	1.0
Gallium arsenide		GaAs	10.6	1.6
Gallium phosphide		GaP	0.56	-1.07
Hexamethylenetetramine	Methenamine	$C_6H_{12}N_4$	0.633	0.78
Sphalerite	Zinc blende	ZnS	0.65	2.1
Zinc selenide	Stilleite	ZnSe	0.546	2.0
Zinc telluride		ZnTe	3.41	4.2
Cadmium telluride		CdTe	3.39	6.8

Linear Electro-Optic Coefficients: Cubic Crystals (23)

Name	Synonym	Formula	λ/μm	r_{41}/pm V^{-1}
Ammonium chloride	Sal ammoniac	NH_4Cl		1.5
Ammonium cadmium sulfate	Ammonium cadmium langbeinite	$(NH_4)_2Cd_2(SO_4)_3$	0.546	0.70
Ammonium manganese sulfate	Ammonium manganese langbeinite	$(NH_4)Mn_2(SO_4)_3$	0.546	0.53
Thallium cadmium sulfate	Thallium cadmium langbeinite	$Tl_2Cd_2(SO_4)_3$	0.546	0.37
Potassium magnesium sulfate	Potassium magnesium langbeinite	$K_2Mg_2(SO_4)_3$	0.546	0.40
Bismuth monogermanite	Bismuth germanium oxide	$6Bi_2O_3{\cdot}GeO_2$		3.3
Bismuth monosilicate		$6Bi_2O_3{\cdot}SiO_2$		3.3
Sodium chlorate		$NaClO_3$	0.589	0.4
Sodium uranyl acetate		$NaUO_2(CH_3COO)_3$	0.546	0.87
Tris(2-aminoethyl)amine hydrobromide	Tren hydrobromide	$(NH_2CH_2CH_2)_3N{\cdot}3HBr$		1.5
Tris(2-aminoethyl)amine hydrochloride	Tren hydrochloride	$(NH_2CH_2CH_2)_3N{\cdot}3HCl$		1.7

Linear Electro-Optic Coefficients: Tetragonal Crystals ($\bar{4}2m$)

Name	Synonym	Formula	T_{tran}/K	r_{41}/pm V^{-1}	r_{63}/pm V^{-1}
Ammonium dihydrogen phosphate	ADP	$NH_4H_2PO_4$	148	24.5	-8.5
Ammonium dideuterium phosphate	ADDP	$NH_4D_2PO_4$	242		11.9
Ammonium dihydrogen arsenate	ADA	$NH_4H_2AsO_4$			9.2
Cesium dihydrogen arsenate	CDA	CsH_2AsO_4	143		18.6
Cesium dideuterium arsenate	CDDA	CsD_2AsO_4	212		36.6
Potassium dihydrogen phosphate	KDP	KH_2PO_4	123	8.6	-10.5
Potassium dideuterium phosphate	KDDP	KD_2PO_4	222	8.8	23.8
Potassium dihydrogen arsenate	KDA	KH_2AsO_4	97	12.5	10.9
Potassium dideuterium arsenate	KDDA	KD_2AsO_4	162		18.2
Rubidium dihydrogen phosphate	RDP	RbH_2PO_4	147		15.5
Rubidium dihydrogen arsenate	RDA	RbH_2AsO_4	110		13.0
Rubidium dideuterium arsenate	RDDA	RbD_2AsO_4	178		21.4

Linear Electro-Optic Coefficients: Tetragonal Crystals (4mm)

Name	Formula	T_{tran}/K	r_{13}/pm V^{-1}	r_{33}/pm V^{-1}	r_{51}/pm V^{-1}
Barium titanate	$BaTiO_3$	406	8	28	
Potassium lithium niobate	$K_3Li_2Nb_5O_{15}$	693	8.9	5.9	
Lead titanate	$PbTiO_3$	765	13.8	5.9	
Strontium barium niobate (SBN75)	$Sr_{0.75}Ba_{0.25}Nb_2O_6$	330	6.7	1340	42
Strontium barium niobate (SBN46)	$Sr_{0.46}Ba_{0.54}Nb_2O_6$	602	180[a]	35	

[a] Approximate

Linear Electro-Optic Coefficients: Hexagonal Crystals (6mm)

Name	Formula	r_{13}/pm V^{-1}	r_{33}/pm V^{-1}	r_{42}/pm V^{-1}	r_{51}/pm V^{-1}
Greenockite	CdS	3.1	2.9	2.0	3.7
Greenockite (const. strain)	CdS	1.1	2.4		
Wurzite	ZnS	0.9	1.8		
Zincite	ZnO	-1.4	2.6		

Linear Electro-Optic Coefficients: Hexagonal Crystals (6)

Name	Formula	r_{13}/pm V^{-1}	r_{33}/pm V^{-1}	r_{42}/pm V^{-1}	r_{51}/pm V^{-1}
Lithium iodate	$LiIO_3$	4.1	6.4	1.4	3.3
Lithium potassium sulfate	$LiKSO_4$	$r_{13}-r_{33}$= 1.6			

Linear Electro-Optic Coefficients: Trigonal Crystals (3m)

Name	Formula	T_{tran}/K	r_{13}/pm V^{-1}	r_{22}/pm V^{-1}	r_{33}/pm V^{-1}	r_{42}/pm V^{-1}
Cesium nitrate	$CsNO_3$	425		0.43		
Lithium niobate	$LiNbO_3$	1483	8.6	7.0	30.8	28
Lithium tantalate	$LiTaO_3$	890	8.4		30.5	
Lithium sodium sulfate	$LiNaSO_4$			0.02[a]		
Tourmaline	$Na(Mg,Fe,Mn,Li,Al)_3Al_6Si_6O_{18}(BO_3)_3$			0.3		

[a] Upper limit

Linear Electro-Optic Coefficients: Trigonal Crystals (32)

Name	Formula	T_{tran}/K	r_{11}/pm V^{-1}	r_{41}/pm V^{-1}
Cesium tartrate	$Cs_2C_4H_4O_6$		1.0	
Cinnabar	HgS	659	3.1	1.5
Potassium dithionate	$K_2S_2O_6$		0.26	
Strontium dithionate	$SrS_2O_6{\cdot}4H_2O$		0.1	
Quartz	SiO_2	1140	-0.47	0.2
Selenium	Se	398	2.5	

Linear Electro-Optic Coefficients: Orthorhombic Crystals (222)

Name	Formula	T_{tran}/K	r_{41}/pm V^{-1}	r_{52}/pm V^{-1}	r_{63}/pm V^{-1}
Ammonium oxalate	$(NH_4)_2C_2O_4{\cdot}4H_2O$		230	330	250
Rochelle salt	$KNaC_4H_4O_6{\cdot}4H_2O$	T_u= 297; T_l= 255	-2.0	-1.7	0.32

Linear Electro-Optic Coefficients: Orthorhombic Crystals (mm2)

Name	Formula	T_{trans}/K	r_{13}/pm V^{-1}	r_{23}/pm V^{-1}	r_{33}/pm V^{-1}	r_{42}/pm V^{-1}	r_{51}/pm V^{-1}
Barium sodium niobate (BSN)	Ba_2NaNbO_{15}	833	15	13	48	92	90
Potassium niobate	$KNbO_3$	476	28	1.3	64	380	105

Linear Electro-Optic Coefficients: Monoclinic Crystals (2)

Name	Formula	T_{trans}/K	r_{22}/pm V^{-1}	r_{32}/pm V^{-1}
Calcium pyroniobate	$Ca_2Nb_2O_7$		0.33	13.7
Triglycine sulfate (TGS)	$(NH_2CH_2COOH)_3{\cdot}H_2SO_4$	322	7.2	13.6

References

1. Narasimhamurty, T. S., *Photoelastic and Electro-Optic Properties of Crystals*, Plenum Press, New York, 1981, pp. 405–407.
2. Weber, M. J., Ed., *CRC Handbook of Laser Science and Technology*, Vol. IV, CRC Press, Boca Raton, FL, 1986, pp. 258–278.

QUADRATIC ELECTRO-OPTIC COEFFICIENTS

Kerr Constants of Ferroelectric Crystals

Name	Formula	T_{tran}/K	λ/μm	$g_{11}/10^{10}$ esu	$g_{12}/10^{10}$ esu	$(g_{11}-g_{12})/10^{10}$ esu	$g_{44}/10^{10}$ esu
Barium sodium niobate	$Ba_{0.8}Na_{0.4}Nb_2O_6$	833		1.55	0.44	1.11	
Barium titanate	$BaTiO_3$	406	0.633	1.33	-0.11	1.44	
Lithium niobate	$LiNbO_3$	1483		0.94	0.25	0.7	0.6
Lithium tantalate	$LiTaO_3$	938		1.0	0.17	0.8	0.7
Potassium tantalate	$KTaO_3$	13	0.633			1.77	1.33
Potassium tantalate niobate	$KTa_{0.65}Nb_{0.35}O_3$	330	0.633	1.50	-0.42	1.92	1.63
Strontium titanate	$SrTiO_3$		0.633			1.56	

Kerr Constant K (at 589 nm and Room Temperature); Static Dielectric Constant ε; Melting Point t_m; and Normal Boiling Point t_b of Selected Liquids

Name	Mol. form.	$K/10^{-7}$ esu	ε	t_m/°C	t_b/°C
Acetone	C_3H_6O	16.3	21.0	-94.8	56.1
Acetophenone	C_8H_8O	66.6	17.44[e]	19.7	202.3
Benzaldehyde	C_7H_6O	80.8	17.85[b]	-26	179.05
Benzenesulfonyl chloride	$C_6H_5ClO_2S$	89.9	28.90	14.5	247
Benzyl alcohol	C_7H_8O	-15.4	11.92[c]	-15.3	205.8
2-Butanone	C_4H_8O	13.6	18.56	-86.67	79.6
Carbon disulfide	CS_2	3.23	2.63	-111.5	46.3
Carvone, (±)-	$C_{10}H_{14}O$	23.6	11.2	[f]	230
1-(3-Chlorophenyl)ethanone	C_8H_7ClO	69.1			
4-Chlorotoluene	C_7H_7Cl	23.0	6.25	7.5	162.4
m-Cresol	C_7H_8O	21.2	12.44[d]	11.8	202.27
o-Dichlorobenzene	$C_6H_4Cl_2$	42.6	10.12	-16.7	180
Ethyl *trans*-3-amino-2-butenoate	$C_6H_{11}NO_2$	31.0		33.9	210
Ethyl benzoylacetate	$C_{11}H_{12}O_3$	16.0	13.50	[f]	270
Ethyl cyanoacetate	$C_5H_7NO_2$	38.8	31.6	-22.5	205
Ethyl salicylate	$C_9H_{10}O_3$	19.6	8.48	1.3	231.5
Nitrobenzene	$C_6H_5NO_2$	326	35.6	5.7	210.8
2-Nitrotoluene	$C_7H_7NO_2$	174	26.26	-10	222.3
3-Nitrotoluene	$C_7H_7NO_2$	177	24.95	15.5	232
4-Nitrotoluene	$C_7H_7NO_2$	222	22.2	51.6	238.3
Paraldehyde	$C_6H_{12}O_3$	-23.0	14.7[a]	12.6	124
Pyridine	C_5H_5N	20.4	13.26	-42	115.23
Quinoline	C_9H_7N	15.0	9.16	-14.78	237.16
Water	H_2O	4.0	80.10	0.00	100.0

[a] 12.0 at 108 - 109 Hz frequency
[b] 14.1 at 108 - 109 Hz frequency
[c] 10.8 at 108 - 109 Hz frequency
[d] 5.0 at 108 - 109 Hz frequency
[e] 15.8 at 108 - 109 Hz frequency
[f] <0

References

1. Narasimhamurty, T. S., *Photoelastic and Electro-Optic Properties of Crystals*, Plenum Press, New York, 1981, p. 408.
2. Gray, D. E., Ed., *AIP Handbook of Physics*, McGraw-Hill, New York, 1972, p. 6–241.

MAGNETO–OPTIC CONSTANTS

Verdet Constants of Non-Magnetic Crystals[a]

Material	T/K	λ/nm	n	V/min Oe^{-1} cm^{-1}
Al_2O_3	300	546.1	1.771	0.0240
	300	589.3	1.768	0.0210
$BaTaO_3$	403	427		0.95
	403	496		0.38
	403	620		0.18
	403	826		0.072
$Bi_4Ge_3O_{12}$	300	442	2.077	0.289
	300	632.8	2.048	0.099
	300	1064	2.031	0.026
C (diamond)	300	589.3	2.417	0.0233

Material	T/K	λ/nm	n	V/min Oe^{-1} cm^{-1}
$CaCO_3$	300	589.3	1.658	0.019
CaF_2	300	589.3	1.434	0.0088
$Cd_{0.55}Mn_{0.45}Te$	300	632.8		6.87
CuCl	300	546.1	1.93	0.20
GaSe	298	632.8		0.80
$KAl(SO_4)_2{\cdot}12H_2O$	300	589.3	1.456	0.0124
KBr	300	546.1	1.564	0.0500
	300	589.3	1.560	0.0425
KCl	300	589.3	1.490	0.0275
KI	300	546.1	1.673	0.083
	300	589.3	1.666	0.070
$KTaO_3$	296	352		0.44
	296	413		0.19
	296	496		0.096
	296	620		0.051
	296	826		0.022
LaF_3	300	325	1.639	0.054
(H‖c)	300	442	1.615	0.028
	300	632.8	1.601	0.012
	300	1064	1.592	0.006
$MgAl_2O_4$	300	589.3	1.718	0.021
$NH_4AlSO_4{\cdot}12H_2O$	300	589.3	1.459	0.0128
NH_4Br	300	589.3	1.711	0.0504
NH_4Cl	300	546.1		0.0410
	300	589.3	1.643	0.0362
NaBr	300	546.1		0.0621
NaCl	300	546.1		0.0410
	300	589.3	1.544	0.0345
$NaClO_3$	300	546.1		0.0105
	300	589.3	1.515	0.0081
$NiSO_4{\cdot}6H_2O$	297	546.1		0.0256
	297	589.3	1.511	0.0221
SiO_2	300	546.1	1.546	0.0195
	300	589.3	1.544	0.0166
$SrTiO_3$	298	413	2.627	0.78
	298	496		0.31
	298	620		0.14
	298	826		0.066
ZnS	300	546.1		0.287
	300	589.3	2.368	0.226
ZnSe	300	476	2.826	1.50
	300	496	2.759	1.04
	300	514	2.721	0.839
	300	587	2.627	0.529
	300	632.8	2.592	0.406

[a] Ref. 1. V is the Verdet constant; n is the refractive index; and λ is the wavelength. "min" is minutes of angle.

Verdet Constants of Rare-Earth Aluminum Garnets at Various Wavelengths[a]

Material	T/K	λ = 405 nm	450 nm	480 nm	520 nm	546 nm	578 nm	635 nm	670 nm
$Tb_2Al_5O_{12}$	300	-2.266	-1.565	-1.290	-1.039	-0.912	-0.787	-0.620	-0.542
	77		-102.16	-83.45	-3.425	-3.051	-2.603	-2.008	-1.815
	4.2				-64.80	-58.35	-53.77	48.39	-45.15
	1.45		-200.95	-172.52	-139.28	-125.07	-111.27	97.47	-93.42
$Dy_3Al_5O_{12}$	300	-1.241	-0.942	-0.803	-0.667	-0.592	-0.518	-0.411	-0.359
$Ho_3Al_5O_{12}$	300	-0.709	-0.320	-0.260	-0.335	-0.304	-0.299		-0.206
$Er_3Al_5O_{12}$	300	-0.189	-0.240	-0.154	-0.162	-0.157	-0.145	-0.105	-0.089
$Tm_3Al_5O_{12}$	300	0.151	0.103	0.093	0.076	0.069	0.059	0.048	
$Yb_3Al_5O_{12}$	298	0.287	0.215	0.186	0.140	0.133	0.116	0.094	
	77	0.718	0.540	0.481	0.393	0.342	0.302	0.239	

[a] Ref. 1. The absorption coefficient α for these materials ranges from 0.2 to 0.6 cm^{-1} at 300 K.

Verdet Constants for KDP-Type Crystals[a]

Name	Formula	V/min Oe^{-1} cm^{-1}
Potassium dihydrogen phosphate	KH_2PO_4	0.0124
Potassium dideuterium phosphate	$KH_{0.3}D_{1.7}PO_4$	0.145
Ammonium dihydrogen phosphate	$NH_4H_2PO_4$	0.138
Potassium dihydrogen arsenate	KH_2AsO_4	0.238
Potassium dideuterium arsenate	$KH_{0.1}D_{1.9}AsO_4$	0.245
Ammonium dihydrogen arsenate	$NH_4H_2AsO_4$	0.244

[a] Ref. 1. Measurements refer to T = 298 K and λ = 632.8 nm, with $\mathbf{k}$ ‖ [001].

Verdet Constants of Gases[a]

Gas		$(n_D - 1) \times 10^3$	$10^6 \times V$/min Oe^{-1} cm^{-1}
Air		0.293	6.27
Ammonia	NH_3	0.376	19.0
Argon	Ar	2.81	9.36
Butane	C_4H_{10}		44.0
Carbon dioxide	CO_2	0.45	9.39
Carbon monoxide	CO	0.34	11.0
Chlorine	Cl_2	0.773	31.9
Helium	He	0.036	0.40
Hydrogen	H_2		6.29
Hydrogen chloride	HCl	0.447	21.5
Hydrogen sulfide	H_2S	0.63	41.5
Methane	CH_4	0.444	17.4
Nitric oxide	NO	0.297	-58
Nitrogen	N_2	0.297	6.46
Oxygen	O_2	0.272	5.69

[a] Ref. 2. Values refer to T = 0 °C and P = 101.325 kPa (760 mmHg); n_D is the refractive index at a wavelength of 589 nm.

Verdet Constants of Liquids[a]

Liquid		λ/nm	t/°C	$10^2 \times V$/min Oe^{-1} cm^{-1}	n_D
Acetone	$(CH_3)_2CO$	589[b]	20.0	1.116	1.3585
Antimony(V) chloride	$SbCl_5$	578	18	7.45	1.601[e]
Benzene	C_6H_6	589[b]	15.0	3.00	1.5005
Carbon disulfide	CS_2	589	20	4.255	1.6255
Chlorobenzene	C_6H_5Cl	589	15	2.92	1.5246
Methanol	CH_3OH	589	18.7	0.958	1.3289
Nitrobenzene	$C_6H_5NO_2$	589	15	2.17	1.5523
Phosphoric acid	H_3PO_4	578	97.4	1.35	
Phosphorus (white)	P	589	33	13.3	
Sulfur	S	589	114	8.1	1.929[c]
Tetrachloromethane	CCl_4	589[b]	25.1	1.60	1.463[d]
Titanium(IV) chloride	$TiCl_4$	578	17	-1.65	1.61
Titanium(IV) bromide	$TiBr_4$	578	46	-5.3	
Toluene	$C_6H_5CH_3$	589[b]	15.0	2.71	1.4950
Tribromomethane	$CHBr_3$	589	17.9	3.13	1.5960
Water	H_2O	589	20	1.309	1.3328
Water-d_2	D_2O	589	19.7	1.257	1.3384

[a] Ref. 2. n_D is the refractive index at a wavelength of 589 nm and a temperature of 20 °C, unless otherwise indicated. *V* is the Verdet constant.
[b] 578-589 nm
[c] At 110 °C
[d] At 15 °C
[e] At 14 °C

Verdet Constants of Rare-Earth Paramagnetic Crystals[a]

Rare Earth	Host	T/K	λ/nm	n	V/min Oe^{-1} cm^{-1}
Ce^{3+}(30%)	CaF_2	300	325	1.516	-0.956
		300	442	1.502	-0.297
		300	633	1.494	-0.111
		300	1064	1.489	-0.035
Ce^{3+}	CeF_3	300	442	1.613	-1.05
		300	633	1.598	-0.406
		77	633		-1.418
		300	1064	1.589	-0.113
Pr^{3+}(5%)	CaF_2	300	266	1.471	-0.172
		300	325	1.461	-0.0818
		300	442	1.451	-0.0089
		300	633	1.445	-0.0168
		300	1064	1.441	-0.0045
Nd^{3+}(2.9%)	CaF_2	4.2	426		-0.19
Nd^{3+}	NdF_3	300	442	1.60	-0.553
		290	633	1.59	-0.209
		77	633		-0.755
		300	1064	1.58	-0.097
Eu^{3+}(3%)	CaF_2	4.2	430		29
		4.2	440		22
Eu^{2+}	EuF_2	300	450		-4.5
		300	500		-2.6
		300	550		-1.6
		300	600		-1.1
		300	650		-0.8
		300	1064		-0.19
Tb^{3+}	KTb_3F_{10}	300	325	1.531	-2.174
		300	442	1.518	-0.933
		300	633	1.510	-0.386
		77	633		-1.94
		300	1064	1.505	-0.114
Tb^{3+}	$LiTbF_4$	300	325	1.493	-1.9
		300	442	1.481	-0.98
		300	633	1.473	-0.44

Solids

Rare Earth	Host	T/K	λ/nm	n	V/min Oe^{-1} cm^{-1}
		300	1064	1.469	-0.13
Tb^{3+}	$Tb_3Ga_5O_{12}$	300	500	1.989	-0.749
		300	570	1.981	-0.581
		300	633	1.976	-0.461
		300	830	1.967	-0.21
		300	1060	1.954	-0.12

[a] Ref. 1. n is the refractive index, and V is the Verdet constant at the wavelength and temperature indicated.

Verdet Constants of Paramagnetic Glasses[a]

R	x	λ= 405 nm	λ= 436 nm	λ= 480 nm	λ= 500 nm	λ= 520 nm	λ= 546 nm	λ= 578 nm	λ= 600 nm	λ= 635 nm	λ= 670 nm
Rare-earth phosphate glasses of composition $R_2O_3{\cdot}xP_2O_5$, where x is given in the second column											
La		0.037	0.030	0.024	0.022	0.020	0.018	0.015	-0.014	0.013	–
Ce	2.67	-0.672	0.510	-0.366	-0.326	-0.287	-0.253	-0.217	-0.197	-0.173	-0.150
Pr	3.09	-0.447	-0.332	-0.283	-0.261	-0.236	-0.208	-0.182	-0.170	-0.150	-0.132
Nd	2.92	-0.250	-0.209	-0.167	-0.155	-0.136	-0.134	-0.094	-0.080	-0.080	-0.071
Sm	2.87	0.026	0.024	0.020	0.020	0.017	0.015	0.014	0.012	0.011	0.010
Eu	2.93	-0.025	-0.017	-0.010	-0.006	-0.006	0.005	-0.004	-0.003	-0.002	-0.002
Gd	3.01	0.018	0.015	0.014	0.012	0.012	0.011	0.011	0.010	0.009	0.009
Tb	2.94	-0.560	-0.458	-0.357	-0.323	-0.295	-0.261	-0.226	-0.206	-0.190	-0.164
Dy	2.51	-0.540	-0.453	-0.359	-0.331	-0.301	0.268	-0.237	-0.217	-0.197	-0.173
Ho	2.94	-0.299	-0.313	-0.156	-0.153	-0.138	-0.138	-0.119	-0.110	-0.098	-0.084
Er	3.01	-0.139	-0.121	-0.100	-0.111	-0.095	-0.062	-0.060	-0.057	-0.051	-0.044
Tm	2.79	0.019	0.013	0.012	0.009	0.008	0.006	0.005	0.004	0.004	0.007
Yb	3.01	0.087	0.072	0.056	0.050	0.045	0.041	0.036	0.032	0.029	0.024
Rare-earth borate glasses of composition: for La and Pr: $R_2O_3{\cdot}xP_2O_5$; for Tb–Pr and Dy–Pr: $R_2O_3{\cdot}xB_2O_3$; and for other elements: $R_2O_3{\cdot}0.85La_3O_3{\cdot}xB_2O_3$.											
La	3.04	0.043	0.036	0.029	0.026	0.023	0.022	0.019	0.018	0.016	0.014
Pr-La	5.44	-0.380	-0.307	-0.230	-0.220	-0.201	-0.178	-0.153	-0.146	-0.128	-0.110
Nd-La	5.41	-0.180	-0.147	-0.120	-0.111	-0.096	-0.094	-0.100	-0.059	-0.056	-0.046
Sm-La	4.97	0.032	0.030	0.025	0.024	0.022	0.019	0.017	0.016	0.014	0.012
Eu-La	4.69	-0.081	-0.060	-0.038	-0.033	-0.029	-0.024	0.019	-0.016	0.014	-0.012
Gd-La	4.71	0.032	0.026	0.024	0.022	0.021	0.020	0.018	0.017	0.015	0.013
Tb-La	4.73	-0.512	-0.419	-0.319	-0.288	-0.262	-0.234	-0.205	-0.186	-0.167	-0.142
Dy-La	4.88	-0.436	-0.361	-0.299	-0.273	-0.246	-0.220	-0.193	-0.177	-0.159	-0.138
Ho-La	4.36	-0.269	-0.252	-0.123	-0.131	-0.112	-0.128	-0.104	-0.096		-0.074
Er-La	4.50	-0.093	-0.078	-0.068	-0.082		-0.045	-0.042	-0.040	-0.035	-0.034
Tm-La	4.75	0.060	0.046	0.039	0.034	0.031	0.026	0.023	0.021	0.018	0.016
Yb-La	8.58	0.115	0.094	0.073	0.066	0.060	0.054	0.046	0.043	0.037	0.033
Tb-Pr	4.99	-0.940	-0.786	-0.560	-0.536	-0.489	-0.436	-0.380	-0.348	-0.306	-0.265
Dy-Pr	4.63	-0.850			-0.497	-0.465	-0.413	-0.358	-0.332	-0.290	-0.252
Pr	2.56	-0.843	-0.646	-0.471	-0.480	-0.432	-0.390	-0.334	-0.317	-0.271	-0.243

[a] Ref. 1. Values of V at room temperature for the indicated wavelengths.

Verdet Constants of Diamagnetic Glasses[1]

The Verdet constant V in units of min/Oe cm is given at room temperature for the wavelengths indicated.

Verdet Constants in min/Oe cm at the Indicated Wavelengths (Part 1)

Glass type	Composition (wt. %)	λ= 325 nm	λ= 442 nm	λ= 633 nm	λ= 1064 nm
SiO_2	100% SiO_2			0.013	
B_2O_3	100% B_2O_3			0.010	
CdO	47.5% CdO, 52.5% P_2O_5	0.079	0.033	0.022	
ZnO	36.4% ZnO, 63.6% P_2O_5	0.072	0.044	0.020	
TeO_2	88.9% TeO_2, 11.1% P_2O_5		0.196	0.076	0.022
ZrF_4	63.1% ZrF_4, 14.9% BaF_2, 7.2% LaF_3, 1.9% AlF_3, 9.1% PbF_2, 3.8% LiF			0.011	

Verdet Constants in min/Oe cm at the Indicated Wavelengths (Part 2)

Glass type	Composition (wt. %)	λ = 700 nm	λ = 853 nm	λ = 1060 nm
Bi_2O_3	95% Bi_2O_3, 5% B_2O_3	0.086	0.051	0.033
PbO	95% PbO, 5% B_2O_3	0.093	0.061	0.031
	82% PbO, 18% SiO_2	0.077	0.045	0.027
	50% PbO, 15% K_2O, 35% SiO_2	0.032	0.020	0.011
Tl_2O	95% Tl_2O, 5% B_2O_3	0.092	0.061	0.032
	82% Tl_2O, 18% SiO_2	0.100	0.067	0.043
	50% Tl_2O, 15% K_2O, 35% SiO_2	0.036	0.022	0.012
SnO	76% SnO, 13% B_2O_3, 11% SiO_2	0.071	0.046	0.026
TeO_3	75% TeO_2, 25% Sb_2O_3	0.076	0.052	0.032
	80% TeO_2, 20% $ZnCl_2$	0.073	0.046	0.025
	84% TeO_2, 16% BaO	0.056	0.041	0.029
	70% TeO_2, 30% WO_3	0.052	0.035	0.022
	20% TeO_2, 80% PbO	0.128	0.075	0.048
Sb_2O_3	25% Sb_2O_3, 75% TeO_2	0.076	0.050	0.032
	75% Sb_2O_3, 75% Cs_2O, 5% Al_2O_3	0.074	0.044	0.025
	75% Sb_2O_3, 10% Cs_2O, 10% Rb_2O, 5% Al_2O_3	0.078	0.052	0.030

Verdet Constants of Commercial Glasses[a]

Glass type	ρ g/cm³	n_D	λ= 365.0 nm	λ= 404.7 nm	λ= 435.8 nm	λ= 546.1 nm	λ= 578.0 nm
BSC	2.49	1.5096	0.0499	0.0392	0.0333	0.02034	0.01798
HC	2.53	1.5189	0.0561	0.0440	0.0372	0.0225	0.01995
LBC	2.87	1.5406	0.0609	0.0477	0.0403	0.0245	0.0216
LF	3.23	1.5785	0.1143	0.0850	0.0693	0.0394	0.0344
BLF	3.48	1.6047	0.1112	0.0832	0.0685	0.0393	0.0344
DBC	3.56	1.6122	0.0662	0.0517	0.0435	0.0261	0.0231
DF	3.63	1.6203	0.1473	0.1076	0.0872	0.0485	0.0423
EDF	3.9	1.6533	0.1725	0.1248	0.1007	0.0556	0.0483

[a] Ref. 1. This table gives the density, ρ, refractive index at 589 nm, n_D, and Verdet constant, V,in min/Oe cm for the wavelengths indicated. The data refer to room temperature.

Glass Composition in wt %

Glass type	SiO_2	B_2O_3	K_2O	CaO	Al_2O_3	As_2O_3	Na_2O	BaO	ZnO	PbO
BSC	69.6	6.7	20.5	2.9	0.3	0.1				
HC	72.0		10.1	11.4	0.3	0.2	6.1			
LBC	57.1	1.8	13.7	0.3	0.2	0.1		26.9		
LF	52.5		9.5	0.3	0.2	0.1				37.6
BLF	45.2		7.8			0.4		16.0	8.3	22.2
DBC	36.2	7.7	0.2	0.2	3.5	0.7		44.6	6.7	
DF	46.3		1.1	0.3	0.2	0.1	5.0			47.0
EDF	40.6		7.5	0.2	0.2	0.2	0.1			51.5

References

1. Weber, M. J., *CRC Handbook of Laser Science and Technology*, Vol. IV, Part 2, CRC Press, Boca Raton, FL, 1988, pp. 299–310.
2. Gray, D. E., Ed., *American Institute of Physics Handbook, Third Edition*, McGraw-Hill, New York, 1972, p. 6–230.

FARADAY ROTATION

Faraday Rotation in Ferro-, Ferri-, and Antiferromagnetic Solids

Material	Formula	T_c/K	T/K	λ/nm	$4\pi M_s$/gauss	F/deg cm^{-1}	α/cm^{-1}	$(2F/\alpha)$/deg
Iron	Fe	1043	300	500	21,800	4.4×10^5	6.5×10^5	1.4
Iron			300	1000		6.5×10^5	5.0×10^5	2.6
Iron			300	1500		7×10^5	4.2×10^5	3.3
Iron			300	2000		7×10^5	3.5×10^5	4.0
Cobalt	Co	1390	300	500	18,200	2.9×10^5		
Cobalt			300	1000		5.5×10^5	6.1×10^5	1.8
Cobalt			300	1500		5.5×10^5	4.5×10^5	2.4
Cobalt			300	2000		5.5×10^5	3.6×10^5	2.7
Nickel	Ni	633	300	500	6,400	0.8×10^5		
Nickel			300	1000		2.6×10^5	5.8×10^5	0.9
Nickel			300	1500		1.5×10^5	4.8×10^5	0.6
Nickel			300	2000		1×10^5	4.1×10^5	0.25
Permalloy	Ni/Fe = 82/18	803	300	500	10,700	1.2×10^5	6×10^5	0.4
Permalloy	Ni/Fe = 100/0	633	300	632.8	6,000	1.2×10^5	7.05×10^5	0.34
Permalloy	Ni/Fe = 80/20		300	632.8	10,800	2.2×10^5	7.10×10^5	0.62
Permalloy	Ni/Fe = 60/40		300	632.8	14,900	2.9×10^5	7.54×10^5	0.77
Permalloy	Ni/Fe = 40/60		300	632.8	14,400	2.2×10^5	8.17×10^5	0.54
Permalloy	Ni/Fe = 20/80		300	632.8	19,400	3.3×10^5	8.10×10^5	0.81
Permalloy	Ni/Fe = 0/100	1043	300	632.8	21,600	3.5×10^5	8.13×10^5	0.86
Manganese bismuthide [MnBi]	MnBi		300	450	7,700	4.2×10^5	6.1×10^5	1.4
Manganese bismuthide [MnBi]			300	900		7.5×10^5	4.2×10^5	3.6
Manganese arsenide [MnAs]	MnAs	313	300	500		0.44×10^5	5.0×10^5	0.174
Manganese arsenide [MnAs]			300	900		0.62×10^5	4.4×10^5	0.28
Chromium monotelluride	CrTe	334	300	550	1015	0.5×10^5	2.0×10^5	0.5
Chromium monotelluride			300	900		0.4×10^5	1.2×10^5	0.7
Iron rhodium [FeRh]	FeRh	333	348	700		0.9×10^5	3.3×10^5	0.56
Yttrium iron oxide	$Y_3Fe_5O_{12}$ (YIG)	560	300	555	2500	2400	1500	3.2
Yttrium iron oxide			300	625		1250	1400	1.8
Yttrium iron oxide			300	770		750	450	3.3
Yttrium iron oxide			300	1500[f]		175	<0.06	$>3 \times 10^3$
Gadolinium iron garnet	$Gd_3Fe_5O_{12}$ (GdIG)	T_n = 564	300	500	7300	-2000	6000	0.6
Gadolinium iron garnet		T = 286	300	600		-1050	900	2.3
Gadolinium iron garnet			300	800		-300	100	6.0
Gadolinium iron garnet			300	1000		-80	70	2.3
Nickel ferrite	$NiFe_2O_4$	858	300	286	3350	2.0×10^4	5.9×10^4	0.7
Nickel ferrite			300	500		-1.0×10^4	10×10^4	0.2
Nickel ferrite			300	1500		-120	38	6
Nickel ferrite			300	3000		+75	15	10
Nickel ferrite			300	5000		+110	32	7
Cobalt(II) diiron tetroxide	$CoFe_2O_4$	793	300	286	4930	2.75×10^4	12×10^4	0.5
Cobalt(II) diiron tetroxide			300	400		3.6×10^4	17×10^4	0.4
Cobalt(II) diiron tetroxide			300	660		-2.5×10^4	6×10^4	0.8
Magnesium ferrite	$MgFe_2O_4$	593-713[e]	300	2500	1450[e]	-60	100	1
Magnesium ferrite			300	4000		0	12	0
Magnesium ferrite			300	6000		+35	6	11
Lithium ferrite [$Li_{0.5}Fe_{2.5}O_4$]	$Li_{0.5}Fe_{2.5}O_4$	863–953[e]	300	1500	3240[e]	-440	150	6
Lithium ferrite [$Li_{0.5}Fe_{2.5}O_4$]			300	3000	to 3900	+10	85	0.2
Lithium ferrite [$Li_{0.5}Fe_{2.5}O_4$]			300	5000		+110	44	5
Lithium ferrite [$Li_{0.5}Fe_{2.5}O_4$]			300	7000		+135	80	3
Barium ferrite	$BaFe_{12}O_{19}$	723	300	2000		-50	-38	3
Barium ferrite			300	3000		+75	20	7.5
Barium ferrite			300	5000		+150	20	15
Barium ferrite			300	7000		+165	22	15
Barium zinc hexaferrite [$Ba_2Zn_2Fe_{12}O_{19}$]	$Ba_2Zn_2Fe_{12}O_{19}$		300	5000		90	120	1.5
Barium zinc hexaferrite [$Ba_2Zn_2Fe_{12}O_{19}$]			300	7000		75	65	2.0
Rubidium nickel fluoride [$RbNiF_3$]	$RbNiF_3$	220	77	450[a]	1250	360	35	20
Rubidium nickel fluoride [$RbNiF_3$]			77	600[a]		70	10	14
Rubidium nickel fluoride [$RbNiF_3$]			77	800[a]		310	70	9

Material	Formula	T_c/K	T/K	λ/nm	$4\pi M_s$/gauss	F/deg cm^{-1}	α/cm^{-1}	$(2F/\alpha)$/deg
Rubidium nickel fluoride [$RbNiF_3$]			77	1000[a]		75	25	6
Rubidium nickel cobalt fluoride [$RbNi_{0.75}Co_{0.25}F_3$]	$RbNi_{0.75}Co_{0.25}F_3$	109	77	600[a]		180	9	40
Rubidium iron fluoride [$RbFeF_3$]	$RbFeF_3$	102	82	300[c]	–	3400	7	900
Rubidium iron fluoride [$RbFeF_3$]			82	400[c]		1600	3	1100
Rubidium iron fluoride [$RbFeF_3$]			82	600[c]		620	1.5	830
Rubidium iron fluoride [$RbFeF_3$]			82	800[c]		300	2.5	240
Iron(III) fluoride	FeF_3	365	300	349[d]	40	670	14	95
Iron(III) fluoride			300	522.5[d]	at 300 K	180	4.4	82
Chromium(III) chloride	$CrCl_3$	16.8	1.5	410	3880	2000	200	20
Chromium(III) chloride			1.5	450		-500	300	3
Chromium(III) chloride			1.5	590		-1000	70	30
Chromium(III) bromide	$CrBr_3$	32.5	1.5	478	3390	3×10^5	3×10^3	200
Chromium(III) bromide			1.5	500		1.6×10^5	1.4×10^4	23
Chromium(III) iodide	CrI_3	68	1.5	970	2690	1.1×10^5	6.3×10^3	35
Chromium(III) iodide			1.5	1000		0.8×10^5	3×10^3	53
Iron(III) borate	$FeBO_3$	348	300	500	115	3200	140	45
Iron(III) borate			300	700	at 300 K	450	38	24
Europium(II) oxide	EuO	69	5	1100	23700	-1.0×10^5	0.5×10^4	40
Europium(II) oxide			5	700		5×10^5	9.7×10^4	10
Europium(II) oxide			5	500		0.5×10^5	7.8×10^4	1.3
Europium(II) oxide			20	2500		3×10^4	>0.5	-105
Europium(II) oxide			20	10600		660	>1.0	1300
Europium(II) sulfide	EuS	16.3	6	825		-1.6×10^5	0	
Europium(II) sulfide			6	690		-9.6×10^5	3.3×10^4	58
Europium(II) sulfide			6	563		$+5.5 \times 10^5$	1.2×10^5	9.2
Europium(II) selenide	EuSe	7.0	4.2	750	13,200	1.45×10^5	80	3600
Europium(II) selenide			4.2	800		0.95×10^5	60	3170

[a] Measured along the C-axis (magnetic hard axis).
[b] Measured along the C-axis (magnetic easy axis).
[c] Measured along the C-axis ([100]-direction at room temperature).
[d] Strong natural bifringence interferes with the Faraday effect.
[e] Depends on heat treatment.
[f] 5000 to 1500 nm

Reference

1. Weber, M. J., Ed., *CRC Handbook of Laser Science and Technology*, Vol. IV, Part 2, CRC Press, Boca Raton, FL, 1988, pp. 288–296.

NONLINEAR OPTICAL CONSTANTS

H. P. R. Frederikse

The relation between the polarization density P of a dielectric medium and the electric field E is linear when E is small, but becomes nonlinear as E acquires values comparable with interatomic electric fields (10^5 to 10^8 V/cm). Under these conditions the relation between P and E can be expanded in a Taylor's series

$$P = \varepsilon_0 \chi^{(1)} E + 2\chi^{(2)} E^2 + 4\chi^{(3)} E^3 + \ldots \qquad (1)$$

where ε_o is the permittivity of free space, while $\chi^{(1)}$ is the linear and $\chi^{(2)}$, $\chi^{(3)}$, etc., the nonlinear optical susceptibilities.

If we consider two optical fields, the first $E_j^{\omega 1}$ (along the j-direction at frequency ω_1) and the second $E_k^{\omega 2}$ (along the k-direction at frequency ω_2) one can write the second term of the Taylor's series as follows

$$P_i(\omega_1 \omega_2) = 2\chi_{ijk}^{\omega_3=\omega_1\pm\omega_2} E_j^{\omega_1} E_k^{\omega_2}$$

When $\omega_1 \neq \omega_2$ the (parametric) mixing of the two fields gives rise to two new polarizations at the frequencies $\omega_3 = \omega_1 + \omega_2$ and $\omega_3' = \omega_1 - \omega_2$. When the two frequencies are equal, $\omega_1 = \omega_2 = \omega$, the result is Second Harmonic Generation (SHG): $\chi_{ijk}(2\omega, \omega, \omega)$, while equal and opposite frequencies, $\omega_1 = \omega$ and $\omega_2 = -\omega$ lead to Optical Rectification (OR): $\chi_{ijk}(0, \omega, -\omega)$. In the SHG case the following convention is adopted: the second order nonlinear coefficient d is equal to one half of the second order nonlinear susceptibility

$$d_{ijk} = 1/2\chi^{(2)}$$

Because of the symmetry of the indices j and k one can replace these two by a single index (subscript) m. Consequently the notation for the SHG nonlinear coefficient in reduced form is d_{im} where m takes the values 1 to 6. Only noncentrosymmetric crystals can possess a nonvanishing d_{ijk} tensor (third rank). The unit of the SHG coefficients is m/V (in the MKSQ/SI system).

In centrosymmetric media the dominant nonlinearity is of the third order. This effect is represented by the third term in the Taylor's series (Equation 1); it is the result of the interaction of a number of optical fields (one to three) producing a new frequency $\omega_4 = \omega_1 + \omega_2 + \omega_3$. The third order polarization is given by

$$P_j(\omega_1 \omega_2 \omega_3) = g_4 \chi_{jklm} E_k^{\omega_1} E_1^{\omega_2} E_m^{\omega_3}$$

Third Harmonic Generation (THG) is achieved when $\omega_1 = \omega_2 = \omega_3 = \omega$. In this case the constant $g_4 = 1/4$. The third order nonlinear coefficient C is related to the third order susceptibility as follows:

$$C_{jklm} = 1/4\chi_{jklm}$$

This coefficient is a fourth rank tensor. In the THG case the matrices must be invariant under permutation of the indices k, l, and m; as a result the notation for the third order nonlinear coefficient can be simplified to C_{jn}. The unit of C_{jn} is $m^2 \cdot V^{-2}$ (in the MKSQ/SI system).

Applications of second order nonlinear optical materials include the generation of higher (up to sixth) optical harmonics, the mixing of monochromatic waves to generate sum or difference frequencies (frequency conversion), the use of two monochromatic waves to amplify a third wave (parametric amplification) and the addition of feedback to such an amplifier to create an oscillation (parametric oscillation).

Third order nonlinear optical materials are used for THG, self-focusing, four wave mixing, optical amplification, and optical conjugation. Many of these effects — as well as the variation and modulation of optical propagation caused by mechanical, electric, and magnetic fields (see the preceding table on "Elasto-Optic, Electro-Optic, and Magneto-Optic Constants") are used in the areas of optical communication, optical computing, and optical imaging.

References

1. *Handbook of Laser Science and Technology*, Vol. 111, Part 1; Weber, M. J. Ed., CRC Press, Boca Raton, FL, 1986.
2. Dmitriev, V.G., Gurzadyan, G.G., and Nikogosyan, D., *Handbook of Nonlinear Optical Crystals*, Springer-Verlag, Berlin, 1991.
3. Shen, Y.R., *The Principles of Nonlinear Optics*, John Wiley, New York, 1984.
4. Yariv, A., *Quantum Electronics*, 3rd edition, John Wiley, New York, 1988.
5. Bloembergen, N., *Nonlinear Optics*, W.A. Benjamin, New York, 1965.
6. Zernike F. and Midwinter, J.E., *Applied Nonlinear Optics*, John Wiley, New York, 1973.
7. Hopf, F.A. and Stegeman, G.I., *Applied Classical Electrodynamics*, Volume 2: Nonlinear Optics, John Wiley, New York, 1986.
8. *Nonlinear Optical Properties of Organic Molecules and Crystals*, Chemla, D. S., and Zyss, J., Eds., Academic Press, Orlando, FL, 1987.
9. *Optical Phase Conjugation*, Fisher, R. A., Ed., Academic Press, New York, 1983.
10. Zyss, J., *Molecular Nonlinear Optics: Materials, Devices and Physics*, Academic Press, Boston, 1994.
11. Nonlinear Optics, 5 articles in *Physics Today, (Am. Inst. of Phys.)*, Vol. 47, No. 5, May, 1994.

Second Harmonic Generation Coefficients of Nonlinear Optical Crystals*

Material	Description	Symmetry class	d_{im}×10^{12}/m V^{-1}	λ/μm
Aluminum phosphate	$AlPO_4$	32	$d_{11} = 0.35 \pm 0.03$	1.058
Ammonium dihydrogen phosphate	$NH_4H_2PO_4$ (ADP)	$\bar{4}2$ m	$d_{36} = 0.53$	1.064
Ammonium dihydrogen phosphate			$d_{36} = 0.85$	0.694
Barium borate (β)	β-BaB_2O_4 [BBO]	3 m	$d_{22} = 2.22 \pm 0.09$	1.06
Barium borate (β)			$d_{31} = 0.16 \pm 0.08$	1.06
Barium sodium niobate	$Ba_2NaNb_5O_{15}$ [Banana]	mm 2	$d_{33} = -17.6 \pm 1.28$	1.064
Barium sodium niobate			$d_{31} = -12.8 \pm 1.28$	1.064
Barium strontium niobate	$BaSrNb_5O_{15}$	4 mm	$d_{33} = 11.3 \pm 3.3$	1.064

Material	Description	Symmetry class	d_{im}×10[12]/m V[-1]	λ/μm
Barium strontium niobate			$d_{31} = 4.31 \pm 1.32$	1.064
Barium strontium niobate			$d_{15} = 5.98 \pm 2$	1.064
Barium titanate	$BaTiO_3$	4 mm	$d_{33} = 6.8 \pm 1.0$	1.064
Barium titanate			$d_{31} = 15.7 \pm 1.8$	1.064
Barium titanate			$d_{15} = 17.0 \pm 1.8$	1.064
Benzil	$(C_6H_5CO)_2$	32	$d_{11} = 3.6 \pm 0.5$	1.064
Bismuth germanate	$Bi_4Ge_3O_{12}$	$\bar{4}$3 m	$d_{14} = 1.28$	1.064
Cadmium germanium arsenide	$CdGeAs_2$	$\bar{4}$2 m	$d_{36} = 351 \pm 105$	10.6
Cadmium selenide	CdSe	6 mm	$d_{33} = 54.5 \pm 12.6$	10.6
Cadmium selenide			$d_{31} = -26.8 \pm 2.7$	10.6
Cadmium sulfide	CdS	6 mm	$d_{33} = 25.8 \pm 1.6$	1.058
Cadmium sulfide			$d_{31} = -13.1 \pm 0.8$	1.058
Cadmium sulfide			$d_{15} = 14.4 \pm 0.8$	1.058
Cadmium telluride	CdTe	$\bar{4}$3 m	$d_{14} = 167.6 \pm 63$	10.6
1,3-Dinitrobenzene	$C_6H_4(NO_2)_2$ [MDB]	mm 2	$d_{33} = 0.74$	1.064
1,3-Dinitrobenzene			$d_{32} = 2.7$	1.064
1,3-Dinitrobenzene			$d_{31} = 1.78$	1.064
Gadolinium(III) molybdate	$Gd_2(MoO_4)_3$	mm 2	$d_{33} = -0.044 \pm 0.008$	1.064
Gadolinium(III) molybdate			$d_{32} = +2.42 \pm 0.36$	1.064
Gadolinium(III) molybdate			$d_{31} = -2.49 \pm 0.37$	1.064
Gallium arsenide	GaAs	$\bar{4}$3 m	$d_{14} = 134.1 \pm 42$	10.6
Gallium phosphide	GaP	$\bar{4}$3 m	$d_{14} = 71.8 \pm 12.3$	1.058
Hexamethylenetetramine	$N_4(CH_2)_6$	$\bar{4}$3 m	$d_{14} = 4.1$	1.06
Indium arsenide	InAs	$\bar{4}$3 m	$d_{14} = 364 \pm 47$	1.058
Indium arsenide			$d_{14} = 210$	10.6
Iodic acid	α-HIO_3	222	$d_{36} = 5.15 \pm 0.16$	1.064
Lead(II) titanate	$PbTiO_3$	4 mm	$d_{33} = 7.5 \pm 1.2$	1.064
Lead(II) titanate			$d_{31} = 37.6 \pm 5.6$	1.064
Lead(II) titanate			$d_{15} = 33.3 \pm 5$	1.064
Lithium iodate	$LiIO_3$	6	$d_{33} = -7.02$	1.06
Lithium iodate			$d_{31} = -5.53 \pm 0.3$	1.064
Lithium niobate	$LiNbO_3$	3 m	$d_{33} = 34.4$	1.06
Lithium niobate			$d_{31} = -5.95$	1.06
Lithium niobate			$d_{22} = 2.76$	1.06
Lithium tantalate	$LiTaO_3$	3 m	$d_{33} = -16.4 \pm 2$	1.058
Lithium tantalate			$d_{31} = -1.07 \pm 0.2$	1.058
Lithium tantalate			$d_{22} = +1.76 \pm 0.2$	1.058
Mercury(II) sulfide (red)	HgS	32	$d_{11} = 50.3 \pm 17$	10.6
Methyl-(2,4-dinitrophenyl)amino propanoate	$C_{10}H_{12}N_3O_6$ [MAP]	2	$d_{23} = 10.67 \pm 1.3$	1.064
Methyl-(2,4-dinitrophenyl)amino propanoate			$d_{22} = 11.7 \pm 1.3$	1.064
Methyl-(2,4-dinitrophenyl)amino propanoate			$d_{21} = 2.35 \pm 0.5$	1.064
Methyl-(2,4-dinitrophenyl)amino propanoate			$d_{25} = -0.35 \pm 0.3$	1.064
3-Nitroaniline	$NO_2C_6H_4NH_2$	mm 2	$d_{33} = 13.12 \pm 1.28$	1.064
3-Nitroaniline			$d_{32} = 1.02 \pm 0.22$	1.064
3-Nitroaniline			$d_{31} = 12.48 \pm 1.28$	1.064
Potassium sodium barium niobate	$K_{0.8}Na_{0.2}Ba_2Nb_5O_{15}$	4 mm	$d_{31} = 13.6 \pm 1.6$	1.064
Potassium dideuterium phosphate	KD_2PO_4 (KDDP)	$\bar{4}$2 m	$d_{36} = 0.38 \pm 0.016$	1.058
Potassium dideuterium phosphate			$d_{36} = 0.34 \pm 0.06$	0.694
Potassium dideuterium phosphate			$d_{14} = 0.37$	1.058
Potassium dihydrogen arsenate	KH_2AsO_4 (KDA)	$\bar{4}$2 m	$d_{36} = 0.43 \pm 0.025$	1.06
Potassium dihydrogen arsenate			$d_{36} = 0.39 \pm 0.4$	0.694
Potassium dihydrogen phosphate	KH_2PO_4 (KDP)	$\bar{4}$2 m	$d_{36} = 0.44$	1.064
Potassium dihydrogen phosphate			$d_{36} = 0.47 \pm 0.07$	0.694
Potassium lithium niobate	$K_3Li_2Nb_5O_{15}$	4 mm	$d_{33} = 11.2 \pm 1.6$	1.064
Potassium lithium niobate			$d_{31} = 6.18 \pm 1.28$	1.064
Potassium lithium niobate			$d_{15} = 5.45 \pm 0.54$	1.064
Potassium niobate	$KNbO_3$	mm 2	$d_{33} = -19.58 \pm 1.03$	1.064
Potassium niobate			$d_{32} = +11.34 \pm 1.03$	1.064
Potassium niobate			$d_{31} = -12.88 \pm 1.03$	1.064
Potassium titanyl phosphate	$KTiOPO_4$ [KTP]	mm 2	$d_{33} = 13.7$	1.06

Material	Description	Symmetry class	d_{im}×10¹²/m V⁻¹	λ/μm
Potassium titanyl phosphate			d_{32} = ± 5.0	1.06
Potassium titanyl phosphate			d_{31} = ± 6.5	1.06
Pyrargyrite	Ag_3SbS_3	3 m	d_{31} = 12.6 ± 4	10.6
Pyrargyrite			d_{22} = 13.4 ± 4	10.6
Selenium (gray)	Se	32	d_{11} = 97 ± 25	10.6
Silicon dioxide (α-quartz)	SiO_2 (quartz)	32	d_{11} = 0.335	1.064
Silver selenogallate	$AgGaSe_2$	$\bar{4}$2 m	d_{36} = 37.4 ± 6.0	10.6
Silver thioarsenate	Ag_3AsS_3 [proustite]	3 m	d_{31} = 11.3 ± 2.5	10.6
Silver thioarsenate			d_{22} = 18.0 ± 2.5	10.6
Silver thiogallate	$AgGaS_2$	$\bar{4}$2 m	d_{36} = 18 ± 2.7	10.6
Tellurium	Te	32	d_{11} = 650 ± 30	10.6
Triglycine sulfate	$(NH_2CH_2COOH)_3H_2SO_4$ [TGS]	2	d_{23} = 0.32	0.694
Urea	$(NH_2)_2CO$	$\bar{4}$2 m	d_{36} = 1.3	1.06
Zinc oxide	ZnO	6 mm	d_{33} = −5.86 ± 0.16	1.058
Zinc oxide			d_{31} = 1.76 ± 0.16	1.058
Zinc oxide			d_{15} = 1.93 ± 0.16	1.058
Zinc selenide	ZnSe	$\bar{4}$3 m	d_{14} = 78.4 ± 29.3	10.6
Zinc selenide			d_{36} = 26.6 ± 1.7	1.058
Zinc sulfide (sphalerite)	β-ZnS	$\bar{4}$3 m	d_{14} = 30.6 ± 8.4	10.6
Zinc sulfide (sphalerite)			d_{36} = 20.7 ± 1.3	1.058
Zinc sulfide (wurtzite)	α-ZnS	6 mm	d_{33} = 11.37 ± 0.07	1.058
Zinc sulfide (wurtzite)			d_{33} = 37.3 ± 12.6	10.6
Zinc sulfide (wurtzite)			d_{31} = −18.9 ± 6.3	10.6
Zinc sulfide (wurtzite)			d_{15} = 21.37 ± 8.4	10.6
Zinc telluride	ZnTe	$\bar{4}$3 m	d_{14} = 92.2 ± 33.5	10.6
Zinc telluride			d_{14} = 83.2 ± 8.4	1.058
Zinc telluride			d_{36} = 89.6 ± 5.7	1.058

* These data are taken from References 1 and 2.

Third Harmonic Generation Coefficients of Some Nonlinear Optical Materials*

Material	Description	NLO process	C_{jm}×10²⁰/m² V⁻²	λ/μm
Ammonium dihydrogen phosphate	$NH_4H_2PO_4$ [ADP]	(−3ω,ω,ω,ω)	C_{11} = 0.0104	1.06
Ammonium dihydrogen phosphate			C_{18} = 0.0098	1.06
Benzene	C_6H_6	(−3ω,ω,ω,ω)	C_{11} = 0.0184 ± 0.0042	1.89
Cadmium germanium arsenide	$CdGeAs_2$	(−3ω,ω,ω,ω)	C_{11} = 182 ± 84	10.6
Cadmium germanium arsenide	p-type: 5 × 10¹⁶ cm⁻³		C_{16} = 175	10.6
Cadmium germanium arsenide			C_{18} = −35	10.6
β-Carotene	$C_{40}H_{56}$	(−3ω,ω,ω,ω)	C_{11} = 0.263 ± 0.08	1.89
Gallium arsenide	GaAs high-resistivity	(−3ω,ω,ω,−ω)	C_{11} = 62 ± 31	1.06
Germanium	Ge	(−3ω,ω,ω,−ω)	C_{11} = 23.5 ± 12	1.06
Lithium iodate	$LiIO_3$	(−3ω,ω,ω,−ω)	C_{12} = 0.2285	1.06
Lithium iodate			C_{35} = 6.66 ± 1	1.06
Potassium bromide	KBr	(−3ω,ω,ω,−ω)	C_{11} = 0.0392	1.06
Potassium bromide			C_{18}/C_{11} = 0.3667	1.06
Potassium chloride	KCl	(−3ω,ω,ω,−ω)	C_{11} = 0.0168	1.06
Potassium chloride			C_{18}/C_{11} = 0.28	1.06
Potassium dihydrogen phosphate	KH_2PO_4 [KDP]	(−3ω,ω,ω,−ω)	$C_{11}−3C_{18}$ = 0.04	1.06
Silicon	Si p-type: 10¹⁴ cm⁻³	(−3ω,ω,ω,−ω)	C_{11} = 82.8 ± 25	1.06
Sodium chloride	NaCl	(−3,ω,ω,ω,−ω)	C_{11} = 0.0168	1.06
Sodium chloride			C_{18}/C_{11} = 0.4133	1.06
Sodium fluoride	NaF	(−3ω,ω,ω,−ω)	C_{11} = 0.0035	1.06

* These data are taken from Reference 1.

PHASE DIAGRAMS

H. P. R. Frederikse

A phase is a structurally homogeneous portion of matter. Regardless of the number of chemical constituents of a gas, there is only one vapor phase. This is true also for the liquid form of a pure substance, although a mixture of several liquid substances may exist as one or several phases, depending on the interactions among the substances. On the other hand a pure solid may exist in several phases at different temperatures and pressures because of differences in crystal structure (Ref. 1). At the phase transition temperature, T_{tr}, the chemical composition of the solid remains the same, but a change in the physical properties often will take place. Such changes are found in ferroelectric crystals (example $BaTiO_3$) that develop a spontaneous polarization below T_{tr}, in superconductors (example Pb) that lose all electrical resistance below the transition point, and in many other classes of solids.

In quite a few cases it is difficult to bring about the phase transition, and the high- (or low-) temperature phase persists in its metastable form. Many liquids remain in the liquid state for shorter or longer periods of time when cooled below the melting point (supercooling). However, often the slightest disturbance will cause solidification. Persistence of the high temperature phase in solid–solid transitions is usually of much longer duration. An example of this behavior is found in white tin; although gray tin is the thermodynamically stable form below T_{tr} (286.4 K), the metal remains in its undercooled, white tin state all the way to $T = 0$ K, and crystals of gray tin are very difficult to produce.

A *phase diagram* is a map that indicates the areas of stability of the various phases as a function of external conditions (temperature and pressure). Pure materials, such as mercury, helium, water, and methyl alcohol are considered one-component systems and they have *unary* phase diagrams. The equilibrium phases in two-component systems are presented in *binary* phase diagrams. Because many important materials consist of three, four, and more components, many attempts have been made to deduce their multicomponent phase diagrams. However, the vast majority of systems with three or more components are very complex, and no overall maps of the phase relationships have been worked out.

It has been shown during the last 20 to 25 years that very useful partial phase diagrams of complex systems can be obtained by means of thermodynamic modeling (Refs. 2, 3). Especially for complicated, multicomponent alloy systems the CALPHAD method has proved to be a successful approach for producing valuable portions of very intricate phase diagrams (Ref. 4). With this method thermodynamic descriptions of the free energy functions of various phases are obtained that are consistent with existing (binary) phase diagram information and other thermodynamic data. Extrapolation methods are then used to extend the thermodynamic functions into a ternary system. Comparison of the results of this procedure with available experimental data is then used to fine-tune the phase diagram and add ternary interaction functions if necessary. In principle this approximation strategy can be extended to four, five, and more component systems.

The 23 phase diagrams shown below present the reader with examples of some important types of single and multicomponent systems, especially for ceramics and metal alloys. This makes it possible to draw attention to certain features like the kinetic aspects of phase transitions (see Figure 22, which presents a time–temperature–transformation, or TTT, diagram for the precipitation of α-phase particles from the β-phase in a Ti-Mo alloy; Ref. 1, pp. 358–360). The general references listed below and the references to individual figures contain phase diagrams for many additional systems.

General References

1. Ralls, K. M., Courtney, T. H., and Wulff, J., *Introduction to Materials Science and Engineering*, Chapters 16 and 17, John Wiley & Sons, New York, 1976.
2. Kaufman, L., and Bernstein, H., *Computer Calculation of Phase Diagrams*, Academic Press, New York, 1970.
3. Kattner, U. R., Boettinger, W. J. B., and Coriell, S. R., *Z. Metallkd.*, 87, 9, 1996.
4. Dinsdale, A. T., Ed., *CALPHAD*, Vol. 1–20, Pergamon Press, Oxford, 1977–1996 and continuing.
5. Baker, H., Ed., *ASM Handbook, Volume 3: Alloy Phase Diagrams*, ASM International, Materials Park, OH, 1992.
6. Massalski, T. B., Ed., *Binary Alloy Phase Diagrams, Second Edition*, ASM International, Materials Park, OH, 1990.
7. Roth. R. S., Ed., *Phase Diagrams for Ceramists*, Vol. I (1964) to Volume XI (1995), American Ceramic Society, Waterville, OH.

References to Individual Phase Diagrams

Figure 1. Carbon: Reference 7, Vol. X (1994), Figure 8930. Reprinted with permission.
Figure 2. Si-Ge : Ref. 5, p. 2.231. Reprinted with permission.
Figure 3. H_2O (ice): See figure.
Figure 4. SiO_2: Reference 7, Vol. XI (1995), Figure 9174. Reprinted with permission.
Figure 5. Fe-O: Darken, L.S., and Gurry, R.W., *J. Am. Chem. Soc.*, 68, 798, 1946. Reprinted with permission.
Figure 6. Ti-O: Reference 5, p. 2.324. Reprinted with permission.
Figure 7. BaO-TiO_2: Reference 7, Vol. III (1975), Figure 4302. Reprinted with permission.
Figure 8. MgO-Al_2O_3: Reference 7, Vol. XI (1995), Figure 9239. Reprinted with permission.
Figure 9. Y_2O_3-ZrO_2: Reference 7, Vol. XI (1995), Figure 9348. Reprinted with permission.
Figure 10. Si-N-Al-O (Sialon): Reference 7, Vol. X (1994), Figure 8759. Reprinted with permission.
Figure 11. PbO-ZrO_2-TiO_2 (PZT): Reference 7, Vol. III (1975), Figure 4587. Reprinted with permission.
Figure 12. Al-Si-Ca-O: Reference 7 (1964), Vol. I, Figure 630. Reprinted with permission.
Figure 13. Y-Ba-Cu-O: Whitler, J.D., and Roth, R.S., *Phase Diagrams for High T_c Superconductors*, Figure S-082, American Ceramic Society, Waterville, OH, 1990. Reprinted with permission.
Figure 14. Al-Cu: Reference 5,p. 2.44. Reprinted with permission.
Figure 15. Fe-C: Ralls, K.M., Courtney, T.H., and Wulff, J., *Introduction to Materials Science and Engineering*, Figure 16.13, John Wiley & Sons, New York, 1976. Reprinted with permission.
Figure 16. Fe-Cr: Reference 5, p. 2.152. Reprinted with permission.
Figure 17. Cu-Sn: Reference 5, p. 2.178. Reprinted with permission.
Figure 18. Cu-Ni: Reference 5, p. 2.173. Reprinted with permission.
Figure 19. Pb-Sn (solder): Reference 5, p. 2.335. Reprinted with permission.
Figure 20. Cu-Zn (brass): Subramanian, P.R., Chakrabarti, D.J., and Laughlin, D.E., Eds., *Phase Diagrams of Binary Copper Alloys*, p. 487, ASM International, Materials Park, OH, 1994. Reprinted with permission.
Figure 21. Co-Sm: Reference 5, p. 2.148. Reprinted with permission.
Figure 22. Ti-Mo: Reference 5, p. 2.296; Reference 1, p. 359. Reprinted with permission.
Figure 23. Fe-Cr-Ni: Reference 5, Figure 48. Reprinted with permission.

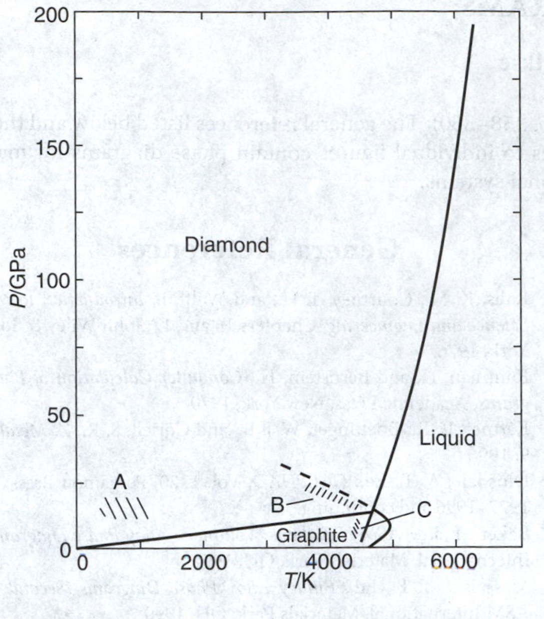

FIGURE 1. Phase diagram of carbon. (A) Martensitic transition: hex graphite → hex diamond. (B) Fast graphite-to-diamond transition. (C) Fast diamond-to-graphite transition.

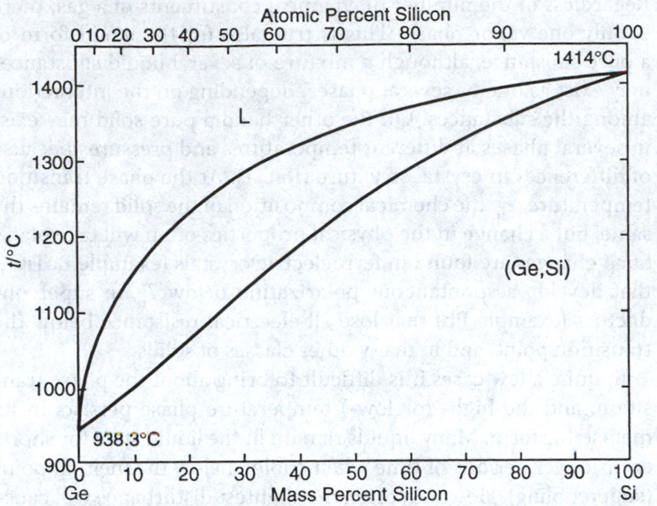

FIGURE 2. Si-Ge system.

Table 2. Si-Ge System

Phase	Composition, mass % Si	Pearson symbol	Space group
(Ge,Si)	0 to 100	*cF*8	$Fd\bar{3}m$
High-pressure phases			
GeII		*tI*4	$I4_1/amd$
SiII		*tI*4	$I4_1/amd$

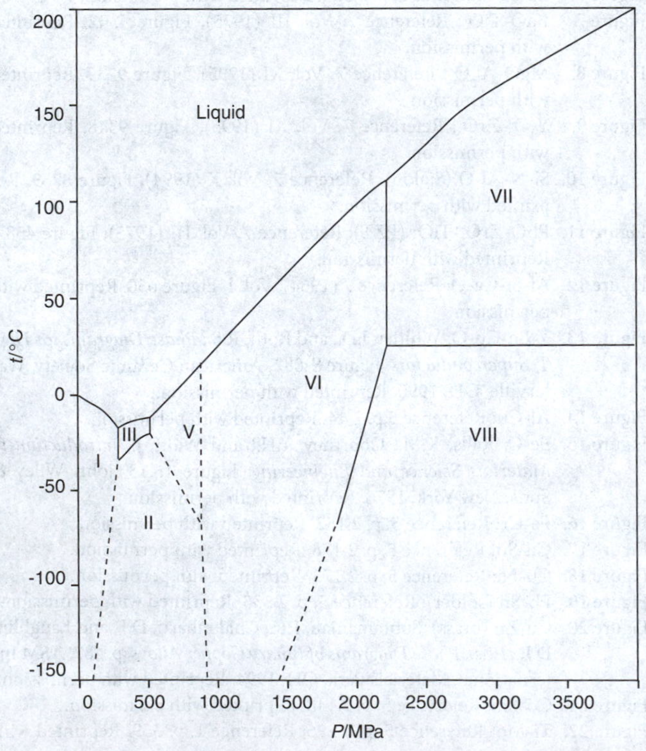

FIGURE 3. Diagram of the principal phases of ice. Solid lines are measured boundaries between stable phases; dotted lines are extrapolated. Ice IV is a metastable phase that exists in the region of ice V. Ice IX exists in the region below −100 °C and pressures in the range 200–400 MPa. Ice X exists at pressures above 44 GPa. See Table 1 for the coordinates of the triple points, where liquid water is in equilibrium with two adjacent solid phases.

Table 3. Crystal Structure, Density, and Transition Temperatures for the Phases of Ice (Cell Parameters in Å)

Phase	Crystal system	Cell parameters	Z	n	ρ/g cm^{-3}	Triple points
Ih	Hexagonal	a = 4.513; c = 7352	4	4	0.93	I-III: –21.99 °C, 209.9 MPa
Ic	Cubic	a = 6.35	8	4	0.94	
II	Rhombohedral	a = 7.78; α = 113.1°	12	4	1.18	
III	Tetragonal	a = 6.73; c = 6.83	12	4	1.15	III-V: –16.99 °C, 350.1 MPa
IV	Rhombohedral	a = 7.60; α = 70.1°	16	4	1.27	
V	Monoclinic	a = 9.22; b = 7.54; c = 10.35; β = 109.2°	28	4	1.24	V-VI: 0.16 °C, 632.4 MPa
VI	Tetragonal	a = 6.27; c = 5.79	10	4	1.31	VI-VII: 82 °C, 2216 MPa
VII	Cubic	a = 3.41	2	8	1.56	
VIII	Tetragonal	a = 4.80; c = 6.99	8	8	1.56	
IX	Tetragonal	a = 6.73; c = 6.83	12	4	1.16	
X	Cubic	a = 2.83	2	8	2.51	

References

1. Wagner, W., Saul, A., and Pruss, A., *J. Phys. Chem. Ref. Data*, 23, 515, 1994.
2. Lerner, R.G. and Trigg, G.L., Eds., *Encyclopedia of Physics,* VCH Publishers, New York, 1990.
3. Donnay, J.D.H. and Ondik, H.M, *Crystal Data Determinative Tables, Third Edition, Volume 2, Inorganic Compounds,* Joint Committee on Powder Diffraction Standards, Swarthmore, PA, 1973.
4. Hobbs, P.V., *Ice Physics,* Oxford University Press, Oxford, 1974.
5. Glasser, L., *J. Chem. Edu.*, 81, 414, 2004.

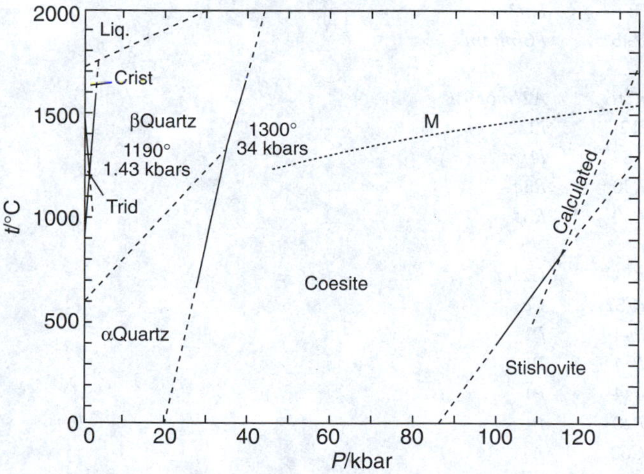

FIGURE 4. SiO_2 system. Crist = cristobalite; Trid = tridymite.

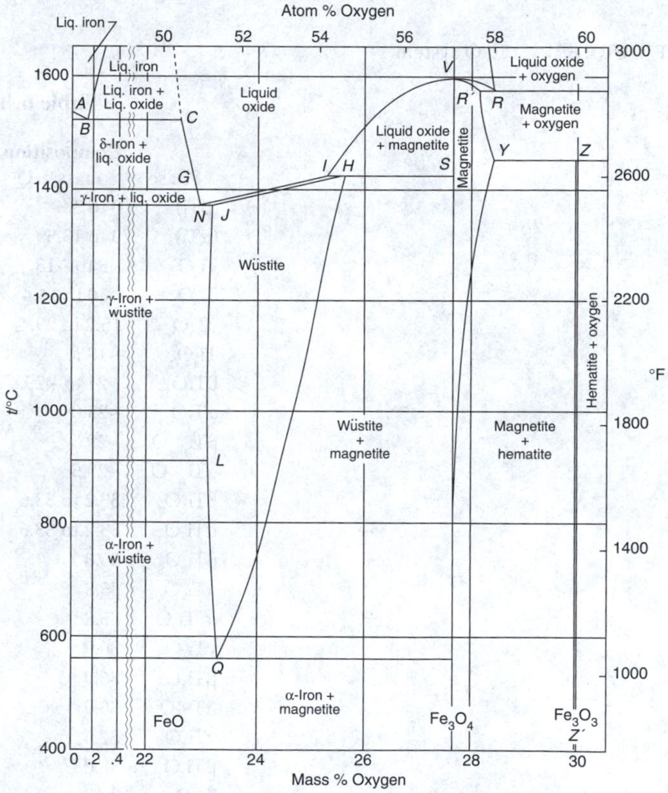

FIGURE 5. Fe-O system.

Table 5. Fe-O System

Point	t/°C	Mass % O	p_{CO_2}/p_{CO}	p_{O_2}/atm	Point	t/°C	Mass % O	p_{CO_2}/p_{CO}	p_{O_2}/atm	Point	t/°C	Mass % O	p_{CO_2}/p_{CO}	p_{O_2}/atm
A	1539				*J*	1371	23.16	0.282		*S*	1424	27.64	16.2	
B	1528	0.16	0.209		*L*	911[a]	23.10	0.447		*V*	1597	27.64		0.0575
C	1528	22.60	0.209		*N*	1371	22.91	0.282		*Y*	1457	28.36		1
G	1400[a]	22.84	0.263		*Q*	560	23.26	1.05		*Z*	1457	30.04		1
H	1424	25.60	16.2		*R*	1583	28.30		1	*Z′*		30.6		
I	1424	25.31	16.2		*R′*	1583	28.07		1					

[a] Values for pure iron.

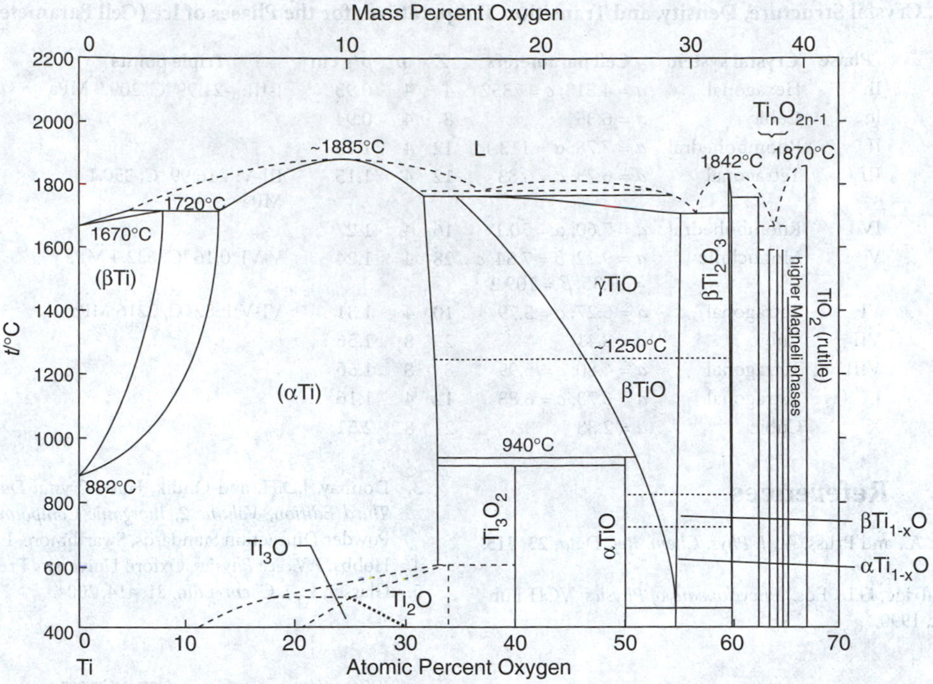

FIGURE 6. Ti-O system.

Table 6. Ti-O System

Phase	Composition, mass % O	Pearson symbol	Space group
(βTi)	0 to 3	*cI*2	$Im\bar{3}m$
(αTi)	0 to 13.5	*hP*2	*P*63/*mmc*
Ti_3O	~8 to ~13	*hP*~16	$P\bar{3}c$
Ti_2O	~10 to 14.4	*hP*3	$P\bar{3}m1$
γTiO	15.2 to 29.4	*cF*8	$Fm\bar{3}m$
Ti_3O_2	~18	*hP*~5	*P*6/*mmm*
βTiO	~24 to ~29.4	*c***	
αTiO	~25.0	*mC*16	*A*2/*morB**/*
$\beta Ti_{1-x}O$	~29.5	*oI*12	*I*222
$\alpha Ti_{1-x}O$	~29.5	*tI*18	*I*4/*m*
βTi_2O_3	33.2 to 33.6	*hR*30	$R\bar{3}c$
αTi_2O_3	33.2 to 33.6	*hR*30	$R\bar{3}c$
βTi_3O_5	35.8	*m***	
αTi_3O_5	35.8	*mC*32	*C*2/*m*
$\alpha' Ti_3O_5$	35.8	*mC*32	*Cc*
γTi_4O_7	36.9	*aP*44	$P\bar{1}$
βTi_4O_7	36.9	*aP*44	$P\bar{1}$
$\alpha Ti4O_7$	36.9	*aP*44	$P\bar{1}$
γTi_5O_9	37.6	*aP*28	$P\bar{1}$
βTi_6O_{11}	38.0	*aC*68	$A\bar{1}$
Ti_7O_{13}	38.3	*aP*40	$P\bar{1}$
Ti_8O_{15}	38.5	*aC*92	$A\bar{1}$
Ti_9O_{17}	38.7	*aP*52	$P\bar{1}$
Rutile TiO_2	40.1	*tP*6	$P4_2/mnm$
Metastable phases			
Anatase		*tI*12	$I4_1/amd$
Brookite		*oP*24	*Pbca*
High-pressure phases			
TiO_2-II		*oP*12	*Pbcn*
TiO_2-III		*hP*~48	

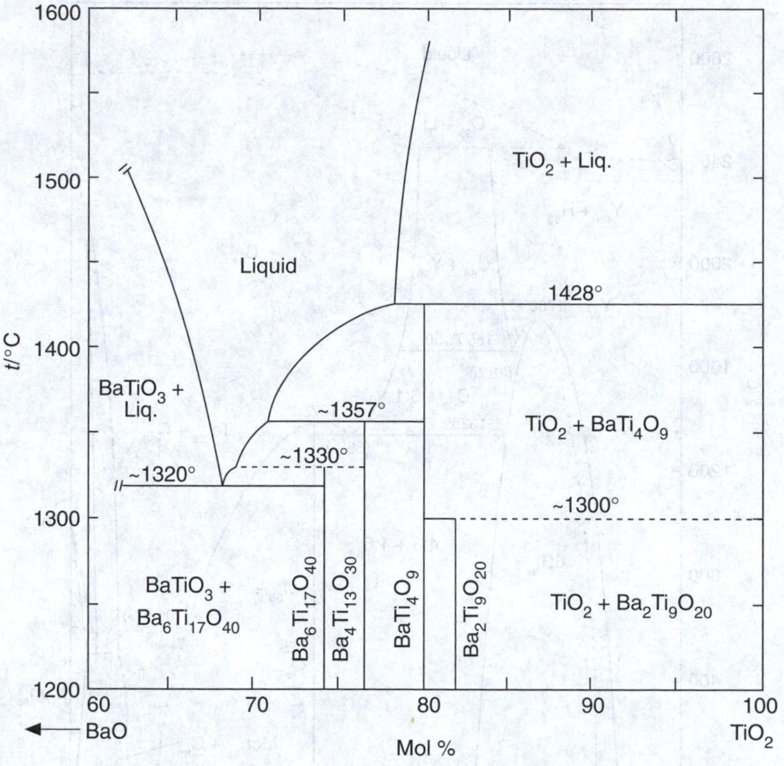

FIGURE 7. BaO-TiO_2 system.

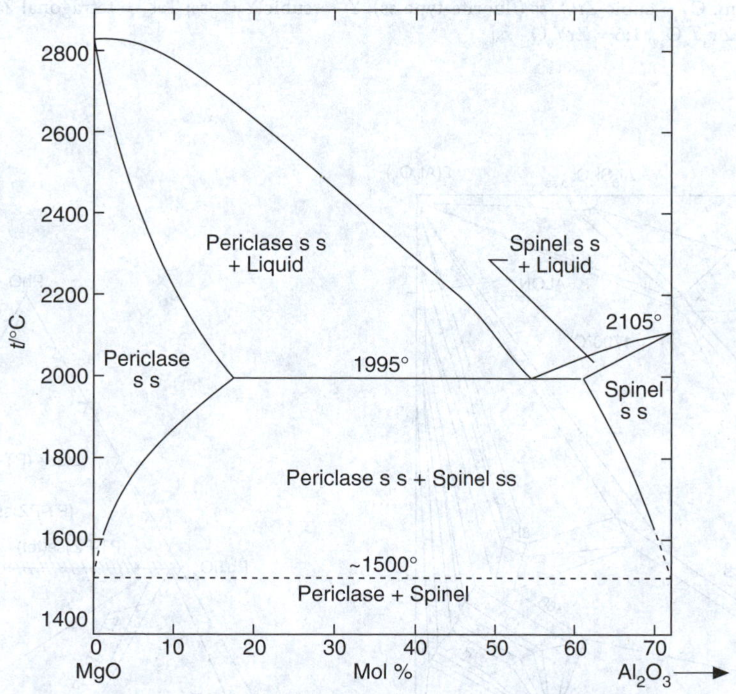

FIGURE 8. MgO-Al_2O_3 system.

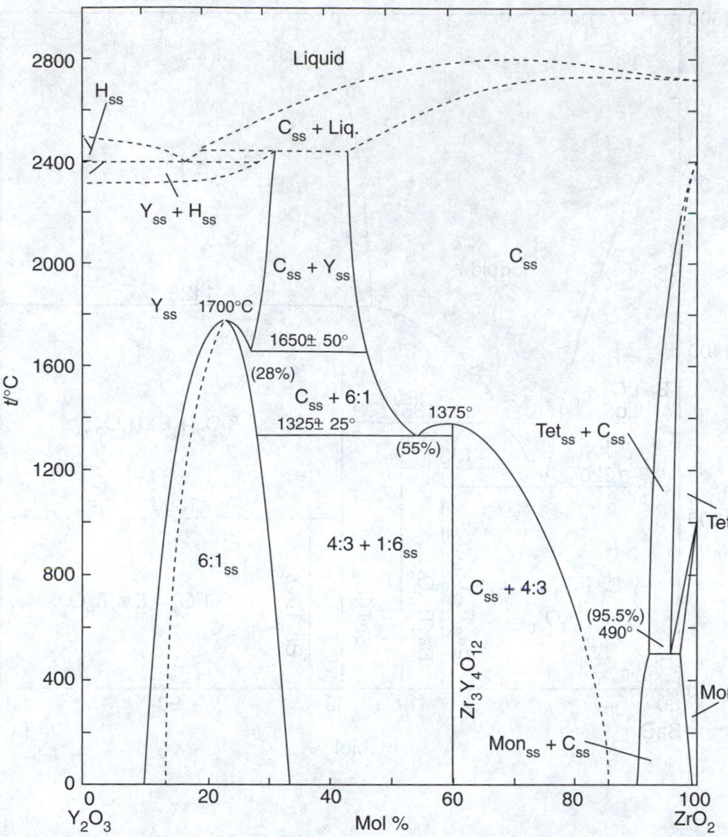

FIGURE 9. Y_2O_3-ZrO_2 system. C_{ss} = cubic ZrO_2 ss (fluorite-type ss); Y_{ss} = cubic Y_2O_3 ss; Tet_{ss} = tetragonal ZrO_2 ss; Mon_{ss} = monoclinic ZrO_2 ss; H_{ss} = hexagonal Y_2O_3 ss; 3:4 = $Zr_3Y_4O_{12}$; 1:6 = ZrY_6O_{11} ss.

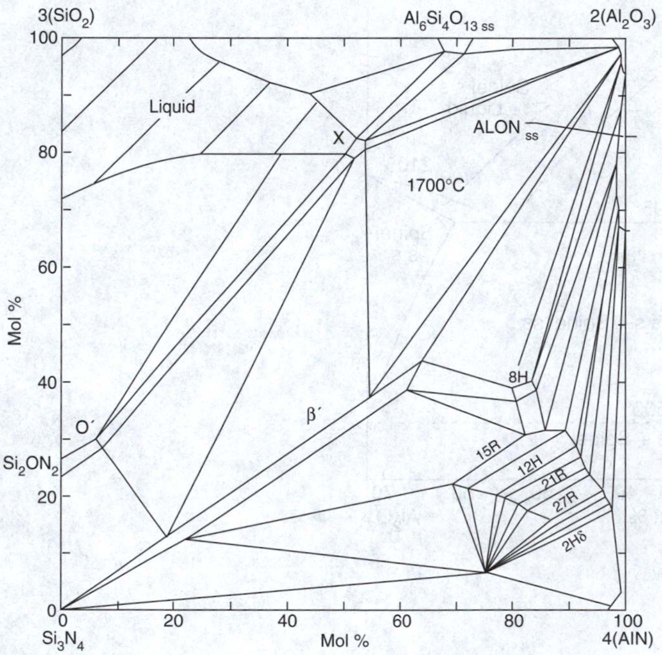

FIGURE 10. 3(SiO_2)-Si_3N_4-4(AlN)-2(Al_2O_3) system. "Behavior" diagram at 1700 °C. The labels 8H, 15R, 12H, 21R, 27R, $2H^{\delta}$ indicate defect AlN polytypes. β' = 3-sialon ($Si_{6-x}Al_xO_xN_{8-x}$); O' = sialon of Si_2ON_2 type; X = $SiAlO_2N$ ("nitrogen mullite"). ALON ss = aluminum oxynitride ss extending from approximately Al_7O_9N to $Al3O_3N$.

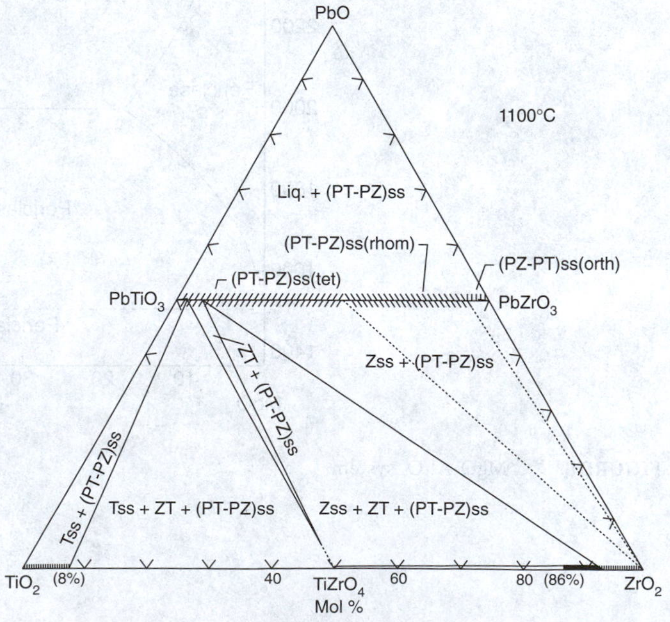

FIGURE 11. PbO-ZrO_2-TiO_2 (PZT) system, subsolidus at 1100 °C. *P* = PbO; *T* = TiO_2; *Z* = ZrO_2.

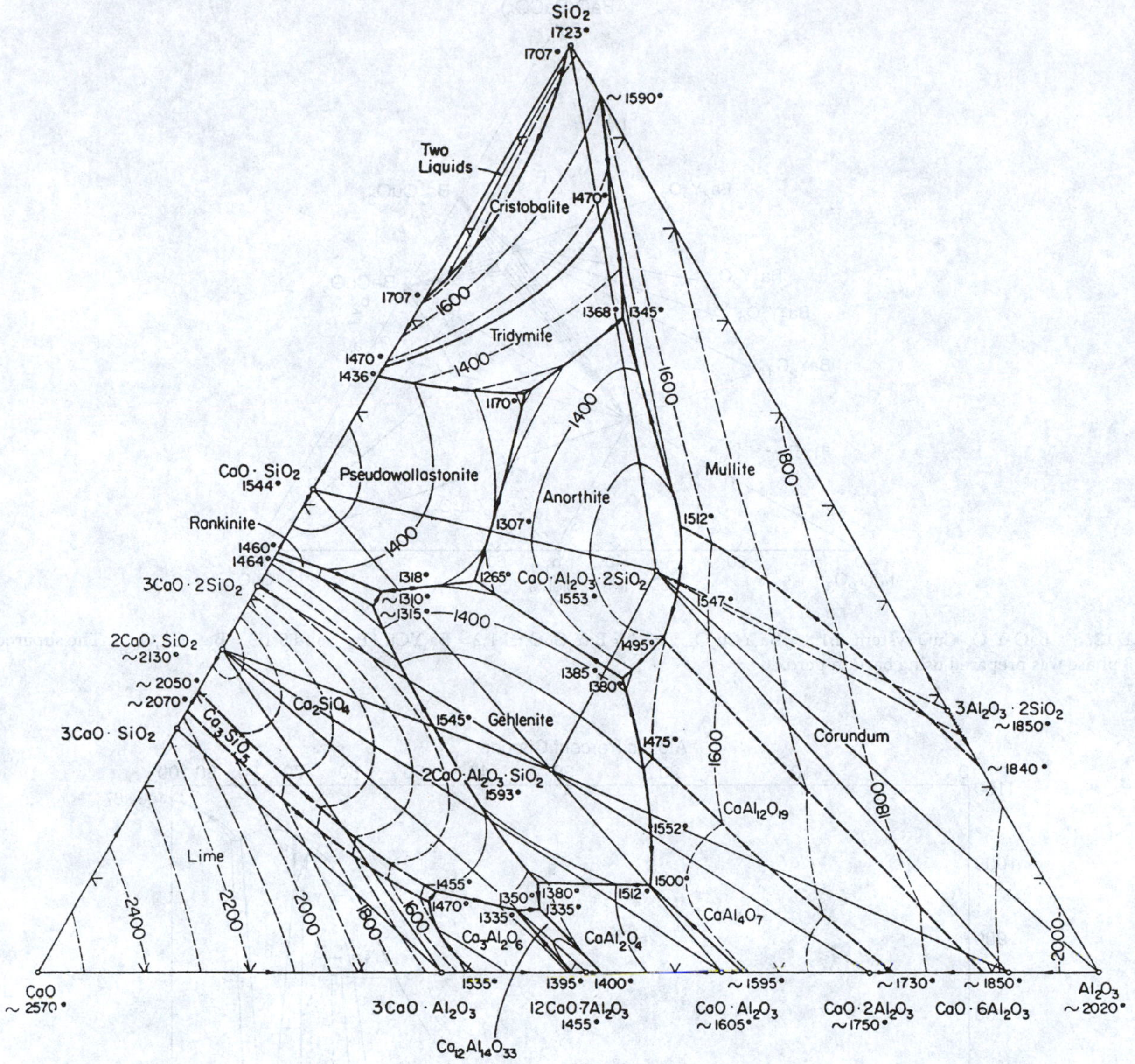

FIGURE 12. CaO-Al_2O_3-SiO_2 system (temperatures in °C).

Table 12. CaO-Al_2O_3-SiO_2 System

Notation	Oxide formula
Cristobalite/Tridymite	SiO_2
Pseudowollastonite	$CaO \cdot SiO_2$
Rankinite	$3CaO \cdot 2SiO_2$
Lime	CaO
Corundum	Al_2O_3
Mullite	$3Al_2O_3 \cdot 2SiO_2$
Anorthite	$CaO \cdot Al_2O_3 \cdot 2SiO_2$
Gehlenite	$2CaO \cdot Al_2O_3 \cdot SiO_2$

Temperatures up to approximately 1550 °C are on the Geophysical Laboratory Scale; those above 1550 °C are on the 1948 International Scale.

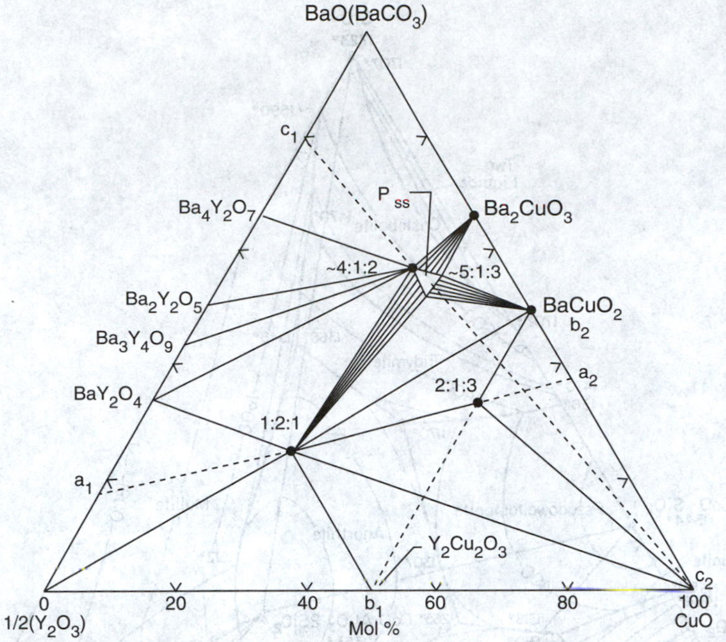

FIGURE 13. BaO-Y_2O_3-CuO system. 2:1:3 = $Ba_2YCu_3O_{7-x}$; 1:2:1 = BaY_2CuO_5; 4:1:2 = $Ba_4YCu_2O_{7.5+x}$; and 5:1:3 = $Ba_5YCu_3O_{9.5+x}$. The superconducting 2:1:3 phase was prepared using barium peroxide.

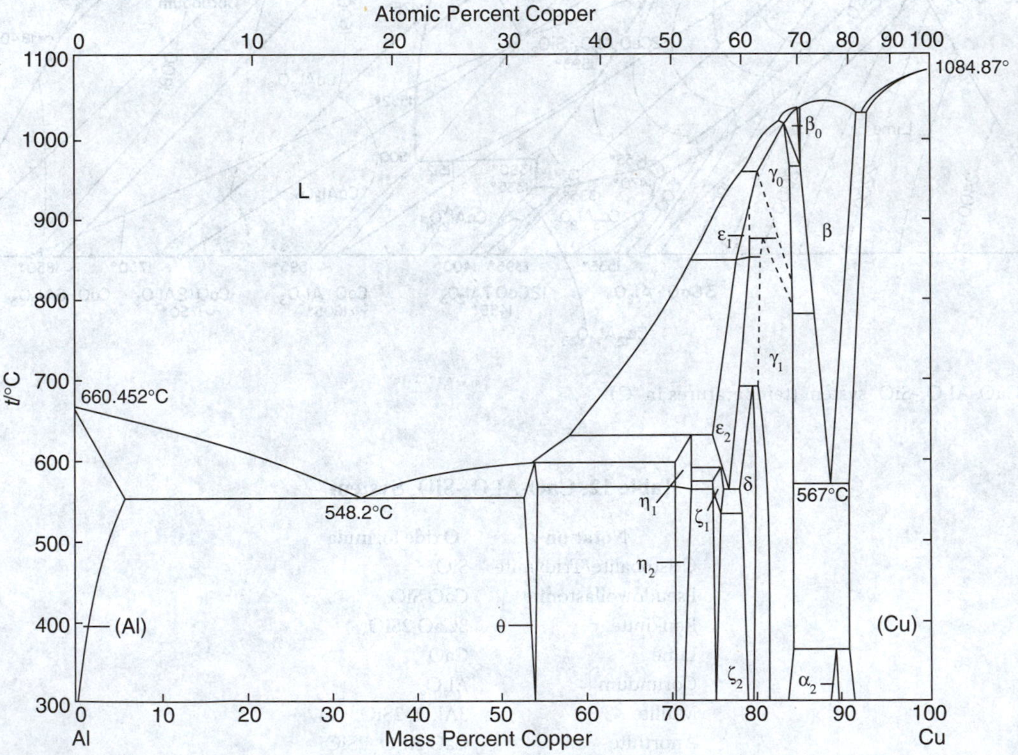

FIGURE 14. Al-Cu system.

Table 14. Al-Cu System

Phase	Composition, mass % Cu	Pearson symbol	Space group
(Al)	0 to 5.65	*cF4*	$Fm\bar{3}m$
θ	52.5 to 53.7	*tI12*	*I4/mcm*
η_1	70.0 to 72.2	*oP16* or*oC16*	*Pbanor Cmmm*
η_2	70.0 to 72.1	*mC20*	*C2/m*
ζ_1	74.4 to 77.8	*hP42*	*P6/mmm*
ζ_2	74.4 to 75.2	[a]	
ε_1	77.5 to 79.4	[b]	
ε_2	72.2 to 78.7	*hP4*	*P63/mmc*
δ	77.4 to 78.3	[c]	$R\bar{3}m$
γ_0	77.8 to 84	[d]	
γ_1	79.7 to 84	*cP52*	$P\bar{4}3m$
β_0	83.1 to 84.7	[d]	
β	85.0 to 91.5	*cI2*	$Im\bar{3}m$
α_2	88.5 to 89	[e]	
(Cu)	90.6 to 100	*cF4*	$Fm\bar{3}m$
Metastable phases			
θ´		*tP6*	
β´		*cF16*	$Fm\bar{3}m$
Al_3Cu_2	61 to 70	*hp5*	$P\bar{3}m1$

[a] Monoclinic?
[b] Cubic?
[c] Rhombohedral
[d] Unknown
[e] $D0_{22}$-type long-period superlattice

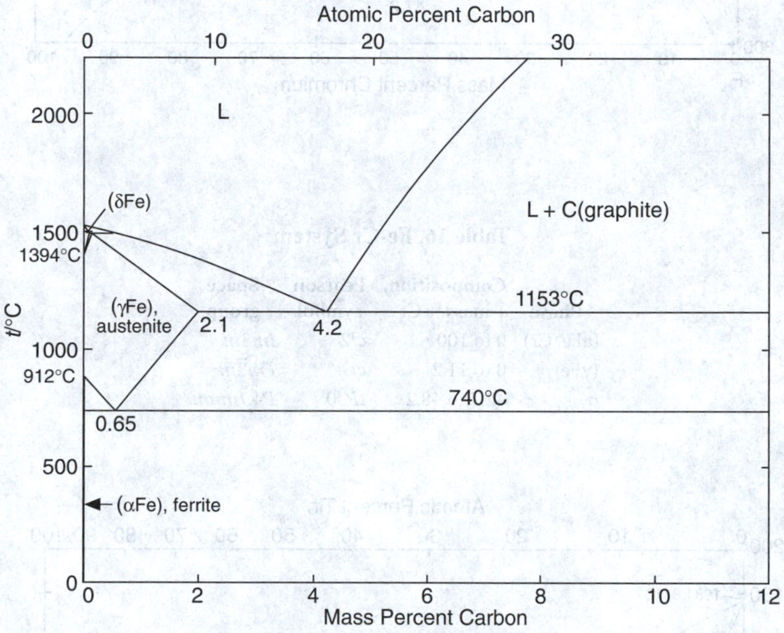

FIGURE 15. Fe-C system.

Table 15. Fe-C System

Phase	Composition, mass % C	Pearson symbol	Space group
(δFe)	0 to 0.09	*cI2*	$Im\bar{3}m$
(γFe)	0 to 2.1	*cF4*	$Fm\bar{3}m$
(αFe)	0 to 0.021	*cI2*	$Im\bar{3}m$
(C)	100	*hP4*	$P6_3/mmc$
Metastable/high-pressure phases			
(εFe)	0	*hP2*	$P6_3/mmc$
Martensite	< 2.1	*tI4*	*I4/mmm*
Fe_4C	5.1	*cP5*	$P\bar{4}3m$
Fe_3C (θ)	6.7	*oP16*	*Pnma*
Fe_5C_2 (χ)	7.9	*mC28*	*C2/c*
Fe_7C_3	8.4	*hP20*	$P6_3mc$
Fe_7C_3	8.4	*oP40*	*Pnma*
Fe_2C (η)	9.7	*oP6*	*Pnnm*
Fe_2C (ε)	9.7	*hP**	$P6_322$
Fe_2C	9.7	*hP**	$P\bar{3}m1$
(C)	100	*cF8*	$Fd\bar{3}m$

Solids

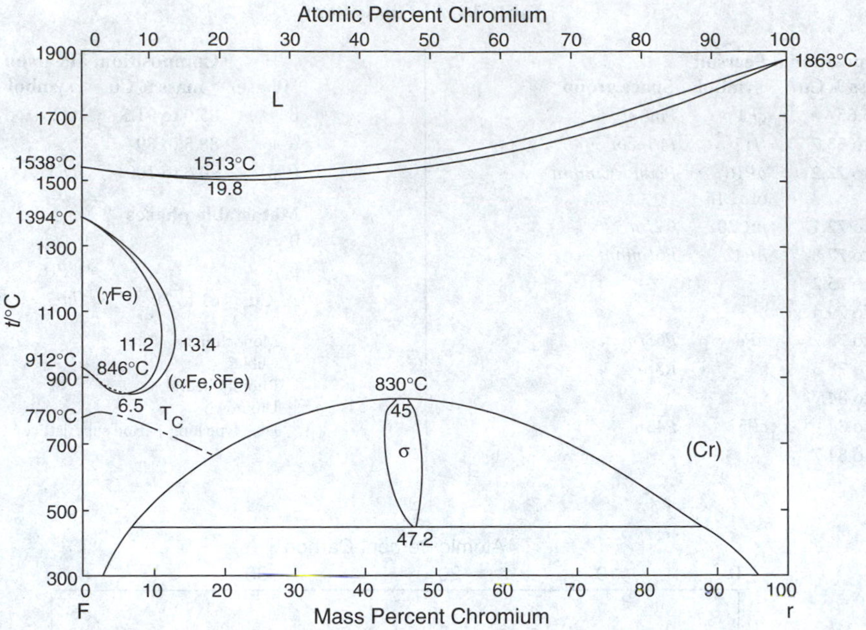

FIGURE 16. Fe-Cr system.

Table 16. Fe-Cr System

Phase	Composition, mass % Cr	Pearson symbol	Space group
(aFe, Cr)	0 to 100	*cI*2	$Im\bar{3}m$
(γFe)	0 to 11.2	*cF*4	$Fm\bar{3}m$
σ	42.7 to 48.2	*tP*30	$P4_2/mnm$

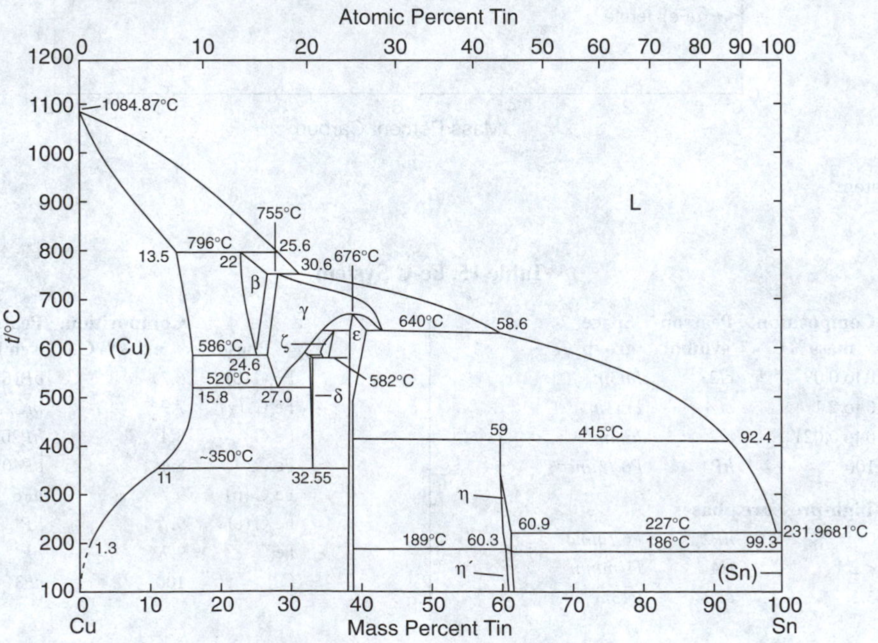

FIGURE 17. Cu-Sn system.

Table 17. Cu-Sn System

Phase	Composition, mass % Sn	Pearson symbol	Space group
α	0 to 15.8	*cF*4	*Fm*$\bar{3}$*m*
β	22.0 to 27.0	*cI*2	*Im*$\bar{3}$*m*
γ	25.5 to 41.5	*cF*16	*Fm*$\bar{3}$*m*
δ	32 to 33	*cF*416	*F*$\bar{4}$3*m*
ζ	32.2 to 35.2	*hP*26	*P*6_3
ε	27.7 to 39.5	*oC*80	*Cmcm*
η	59.0 to 60.9	*hP*4	*P*6_3/*mmc*
η′	44.8 to 60.9	[a]	
(βSn)	~100	*tI*4	*I*4_1/*amd*
(αSn)	100	*cF*8	*Fd*$\bar{3}$*m*

[a] Hexagonal; superlattice based on NiAs-type structure.

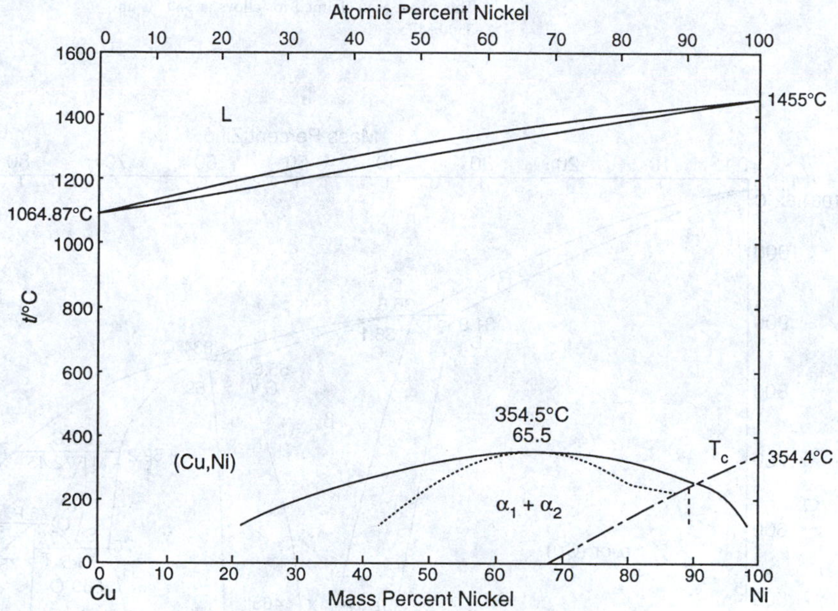

FIGURE 18. Cu-Ni system.

Table 18. Cu-Ni System

Phase	Composition, mass % Ni	Pearson symbol	Space group
(Cu, Ni) (above 354.5 °C)	0 to 100	*cF*4	*Fm*$\bar{3}$*m*

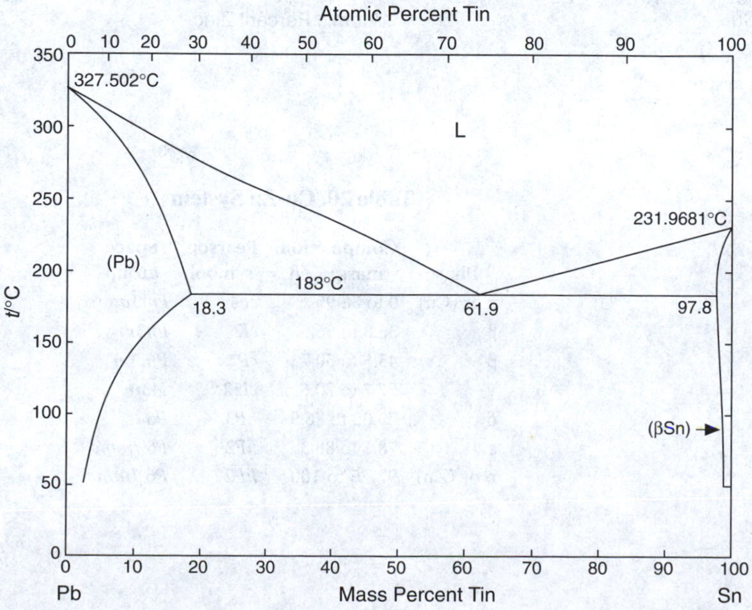

FIGURE 19. Pb-Sn system.

Table 19. Pb-Sn System

Phase	Composition, mass % Sn	Pearson symbol	Space group
(Pb)	0 to 18.3	*cF*4	$Fm\bar{3}m$
(βSn)	97.8 to 100	*tI*4	$I4_1/amd$
(αSn)	100	*cF*8	$Fd\bar{3}m$
High-pressure phases			
ε[a]	52 to 74	*hP*1	$P6/mmm$
ε′[b]	52	*hP*2	$P6_3/mmc$

[a] From phase diagram calculated at 2500 MPa.
[b] This phase was claimed for alloys at 350 °C and 5500 MPa.

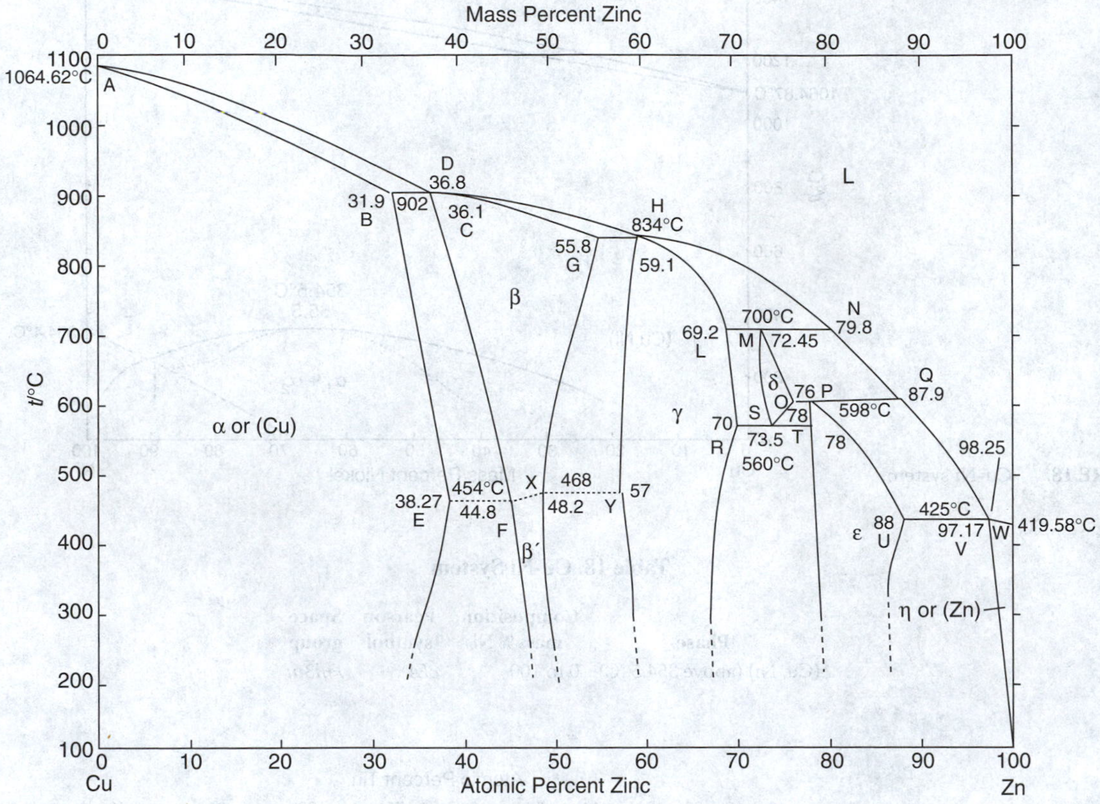

FIGURE 20. Cu-Zn system.

Table 20. Cu-Zn System

Phase	Composition, mass % Zn	Pearson symbol	Space group
α or (Cu)	0 to 38.95	*cF*4	$Fm\bar{3}m$
β	36.8 to 56.5	*cI*2	$Im\bar{3}m$
β′	45.5 to 50.7	*cP*2	$Pm\bar{3}m$
γ	57.7 to 70.6	*cI*52	$I\bar{4}3m$
δ	73.02 to 76.5	*hP*3	$P\bar{6}$
ε	78.5 to 88.3	*hP*2	$P6_3/mmc$
η or (Zn)	97.25 to 100	*hP*2	$P6_3/mmc$

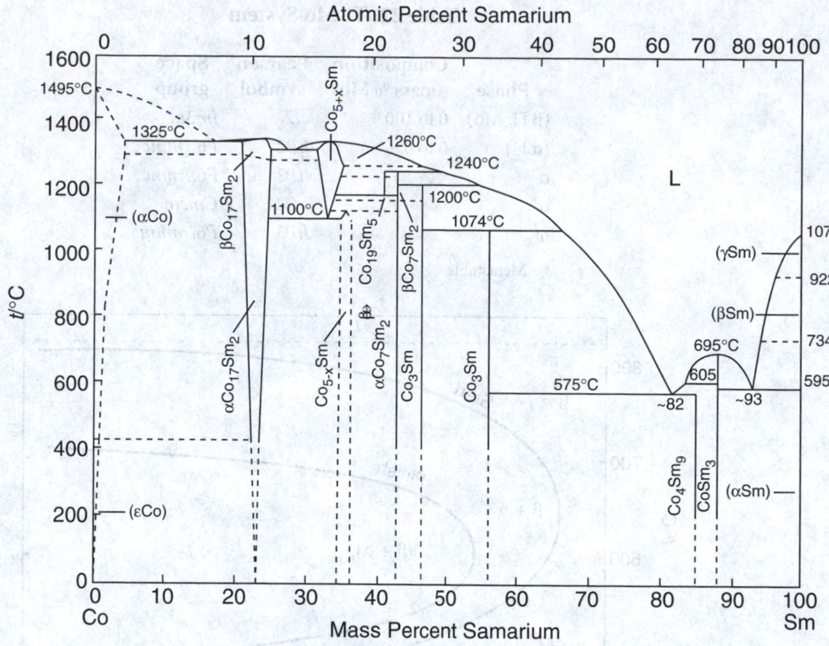

FIGURE 21. Co-Sm system.

Table 21. Cu-Sm System

Phase	Composition, mass % Sm	Pearson symbol	Space group
(αCo)	0 to ~3.7	*cF*4	$Fm\bar{3}m$
(εCo)	~0	*hP*2	$P6_3/mmc$
$\beta Co_{17}Sm_2$	~23.0	*hP*38	$P6_3/mmc$
$\alpha Co_{17}Sm_2$	~23.0	*hR*19	$R\bar{3}m$
		*hP*8	$P6/mmm$
$Co_{5+x}Sm$	~33 to 34		
$Co_{5-x}Sm$	~34 to 35		
$Co_{19}Sm_5$	~40.1	*hR*24	$R\bar{3}m$
		*hP*48	$P6_3/mmc$
αCo_7Sm_2	~42.1	*hR*18	$R\bar{3}m$
βCo_7Sm_2	~42.1	*hP*36	$P6_3/mmc$
Co_3Sm	46	*hR*12	$R\bar{3}m$
Co_2Sm	56.0	*hR*4	$R\bar{3}m$
		*cF*24	$Fd\bar{3}m$
Co_4Sm_9	~85.1	*o***	
$CoSm_3$	88	*oP*16	*Pnma*
(γSm)	~100	*cI*2	$Im\bar{3}m$
(βSm)	~100	*hP*2	$P6_3/mmc$
(αSm)	~100	*hR*3	$R\bar{3}m$
Other reported phases			
Co_5Sm	~33.8	*hP*6	$P6/mmm$
Co_2Sm_5	~86.4	*mC*28	$C2/c$

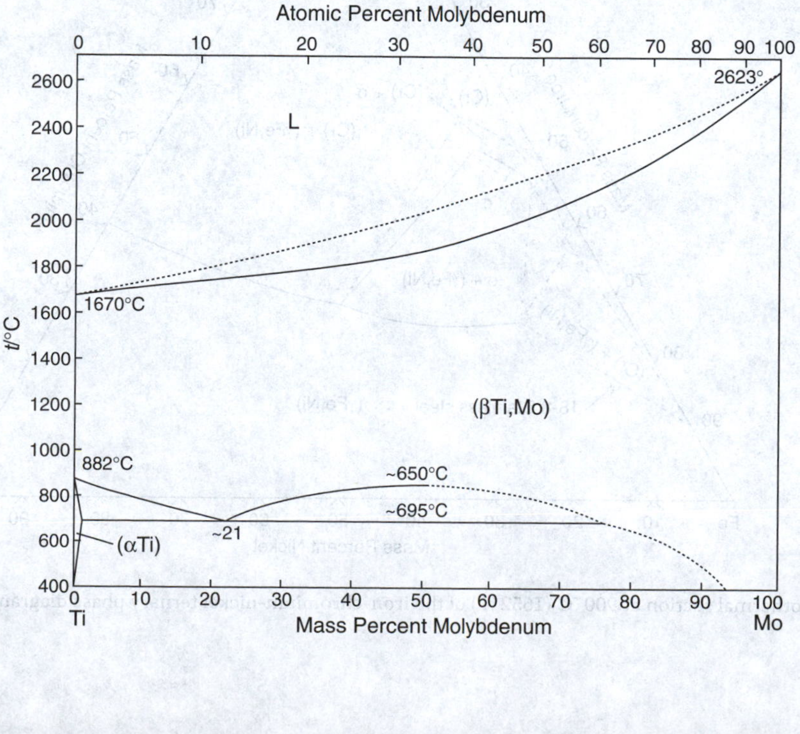

FIGURE 22. Ti-Mo system.

Table 22. Ti-Mo System

Phase	Composition, mass % Mo	Pearson symbol	Space group
(βTi, Mo)	0 to 100	*cI2*	$Im\bar{3}m$
(αTi)	0 to 0.8	*hP2*	$P6_3/mmc$
α′	[a]	*hP2*	$P6_3/mmc$
α″	[a]	*oC4*	*Cmcm*
ω	[a]	*hP3*	*P6/mmm*

[a] Metastable

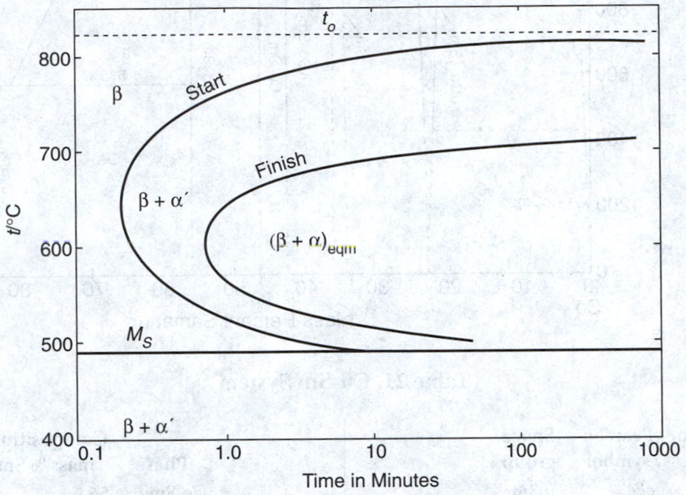

FIGURE 22a. Experimental time–temperature–transformation (TTT) diagram for Ti-Mo. The start and finish times of the isothermal precipitation reaction vary with temperature as a result of the temperature dependence of the nucleation and growth processes. Precipitation is complete, at any temperature, when the equilibrium fraction of α is established in accordance with the lever rule. The solid horizontal line represents the athermal (or nonthermally activated) martensitic transformation that occurs when the β phase is quenched.

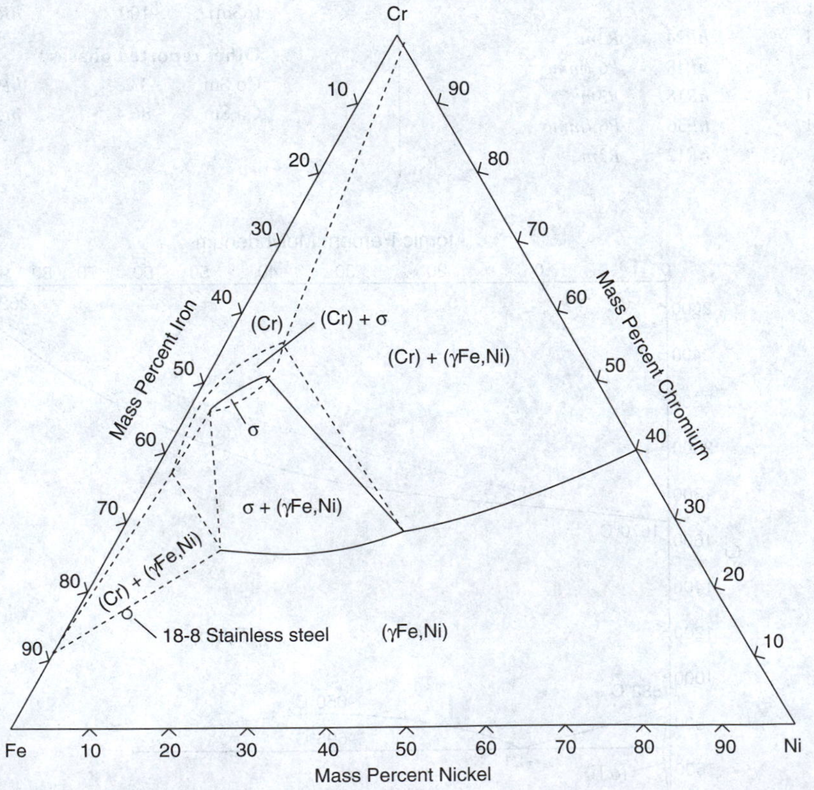

FIGURE 23. The isothermal section at 900 °C (1652 °F) of the iron-chromium-nickel ternary phase diagram, showing the nominal composition of 18-8 stainless steel.

PROPERTIES OF SELECTED MATERIALS AT CRYOGENIC TEMPERATURES

Peter E. Bradley and Ray Radebaugh

The design of systems for operation at cryogenic temperatures requires the use of material properties at these low temperatures. The properties at cryogenic temperatures can be much different than the room-temperature values. In addition, some properties can be strong functions of temperature. Property data at cryogenic temperatures are not easy to find. Many measurements were made at the National Institute of Standards and Technology (NIST) and other laboratories about 50 years ago. Some of the results were published in reports that are now out of print, which makes the results unavailable to most researchers. To correct that problem, NIST initiated a program to critically evaluate cryogenic material properties and to fit the available data for temperatures in the range of about 4 K to 300 K. The parameters for the fit, as well as a graph of the curve, are available on the website www.cryogenics.nist.gov. The properties available include thermal conductivity, specific heat, linear thermal expansion relative to 293 K, thermal expansion coefficient, and Young's modulus. Not all properties are available for all materials. The materials currently in the database are ones commonly used in the construction of cryogenic hardware.

The tables presented here as a function of temperature are the calculated values using the equations given on the website. In general, the equations fit a single set of data to within about 1 % to 2 %, but often several sets of data are used in determining the best fit, in which case deviations can be significantly higher, as much as 5 %. The website specifies the deviation of the fit relative to the experimental data for each property and each material. Uncertainties in the experimental data usually are in the range of 2 % to 5 %, and variations from sample to sample can also lead to similar uncertainties, especially in thermal conductivity. Some well-characterized materials, such as silicon, are used for standard reference materials. Thus, uncertainties in the experimental data for the thermal expansion coefficient of silicon are usually less than 0.2 %, and the standard deviation of the fit to the data is less than about 0.2 % over most of the temperature range. All values refer to ambient pressure, i.e., pressure in the neighborhood of 101.3 kPa.

Copper referred to here is of very high purity 99.99% (4N or better) and may be considered oxygen-free (sometimes referred to as OFHC – oxygen free high conductivity). Values are given with respect to the RRR (Residual Resistivity Ratio) which correlates the thermal resistivity and electrical resistivity as the impurity effect and is primarily additive in resistivity. Higher RRR values indicate higher purity and lower electrical and thermal resistance leading to higher thermal conductivity. Standard high-purity copper such as grade 101 or 102 has an RRR value of approximately 100. Higher values may be obtained with considerable effort at minimizing trace impurities by special annealing techniques that can achieve an RRR of about 1000 or greater in some special instances. Specially obtained high RRR value copper is often used only when very low temperatures (<40 K) and necessarily high thermal conduction at low temperatures are required.

Ti 15-3-3-3 has a nominal composition of 15% V, 3% Cr, 3% Sn, 3% Al, balance Ti. For the specific measurements documented by Canavan and Tuttle (Ref. 29), the exact composition is 14.88% V, 3.13% Cr, 2.88% Sn, 3.01% Al, balance Ti. The composition for brass is 65% Cu, 32% Zn, 3% Pb which is free machining. The composition for BeCu is 2% Be, 0.3% Co, balance Cu.

References

1. Mann, D., Ed., *LNG Materials and Fluids*, First Edition, Cryogenics Division, National Bureau of Standards, Boulder, CO, 1977. [Al, Invar, FeNi, Polystyrene, Polyurethane, PVC, Stainless steel, Be, G-10]
2. Veres, H. M., Ed., *Thermal Properties Database for Materials at Cryogenic Temperatures*, Vol. 1. [Al, G-10, Nylon, Teflon]
3. Touloukian, Y. S., *Recommended Values of the Thermophysical Properties of Eight Alloys, Major Constituents and their Oxides*, Purdue University, West Lafayette, IN, 1965. [Al, Stainless steel]
4. Touloukian, Y. S., Powell, R. W., Ho, C. Y., and Klemens, P. G., *The TPRC Data Series: Vol. 1, Thermal Conductivity-Metallic Elements and Alloys*, Shackelford, J., and Alexander, W., Eds., New York, Washington, 1970. [Be, Mo]
5. Johnson, V. J., Ed., A Compendium of the Properties of Materials at Low Temperature (Phase I), *Part II: Properties of Solids, Wadd Technical Report 60-56*, National Bureau of Standards, Boulder, CO, 1960. [BeCu, Pb, Pt, Nylon, Be, In, Al, Cu]
6. Berman, R., Foster, E. L., and Rosenberg, H. M., *Brit. J. Appl. Phys.* 6, 181, 1955. [BeCu]
7. Powell, R.L., Rogers, W.M., and Roder, H.M., Thermal Conductivities of Copper and Copper Alloys, *Adv. Cryog. Eng.* 2, 166, 1956. [Brass, Cu]
8. Simon, N. J., Drexler, E. S., and Reed, R. P., Properties of Copper and Copper Alloys at Cryogenic Temperature, *NIST Monograph 177*, 1992. [Brass, Cu, Phosphor-bronze, BeCu]
9. He, G. H., Wang, B. Q., Guo, X. N., Yang, F., Guo, J. D., and Zhou, B. L., Investigation of Thermal Expansion Measurement of Brass Strip H62 after High Current Density Electropulsing by the CCD Technique, *Mater. Sci. Eng., A* 292, 183, 2000. [Brass]
10. Hust, J. G., Thermal Conductivity of Glass Fiber/Epoxy Composite Support Bands for Cryogenic Dewars, Phase II, *NBSIR 84-3003*, National Bureau of Standards, Boulder, CO, 1984. [G-10]
11. Child, G., Erics, L. J., and Powell, R. L., Thermal Conductivity of Solids at Room Temperature and Below, *NBS Monograph 131*, 1973. [G-10]
12. Mechanical, Thermal, Electrical, and Magnetic Properties of Structural Materials, *Wadd Technical Report: Handbook on Materials for Superconducting Machinery Metals and Ceramics*, Information Center, Battelle, Columbus Laboratories, 1974 (with 1975 and 1977 Supplements). [Inconel]
13. Hust, J. G., Low-Temperature Thermal Conductivity of Two Fibre-Epoxy Composites, *Cryogenics* 15, 126, 1975. [Kevlar]
14. Foster, W. G., Naes, L. G., and Barnes, C.B., Thermal Conductivity Measurements of Fiberglass/Epoxy Structural Tubes from 4 K to 320 K, *AIAAS Paper 75-711*, American Institute of Aeronautics and Astronautics (A1AA), 10th Thermophysics Conference, Denver, CO, 1975. [Kevlar]
15. Harris, J. P., Yates, B., Batchelor, J., and Garrington, P. J., *J. Mater. Sci.* 17, 2925, 1982. [Kevlar]
16. Ventura, G., and Martelli, V., Thermal Conductivity of Kevlar 49 between 7 and 290 K, *Cryogenics* 49, 735, 2009. [Kevlar]
17. Ventura, G., and Martelli, V., Very Low Temperature Thermal Conductivity of Kevlar 49, *Cryogenics* 49, 376, 2008. [Kevlar]
18. Poulaert, B., Chieliens, J. C., Vandehande, C., and Legras, R., Temperature Variation of the Thermal Conductivity of Kevlar, *Polym. Commun.*, vol. 26, 1985 (digitized data). [Kevlar]
19. Hartwig, G., and Knaak, S., Fibre-Epoxy Composites at Low Temperatures, *Cryogenics* 24, 11, 1984. [Kevlar]
20. Shackelford, J. F., and Alexander, W., eds., *CRC Materials Science and Engineering Handbook*, 3rd Edition, CRC Press, Boca Raton, FL, 2001. [Mo]
21. Choy, C. L., and Grieg, D., The Low Temperature Thermal Conductivity of a Semi-crystalline Polymer, Polyethylene Terephthalate, *J. Phys. C: Solid State Phys.* 8, 3121, 1975. [Mylar]
22. Rule, D.L., Smith, D.R., and Sparks, L.L., Thermal Conductivity of a Polyimide Film Between 4.2 and 300 K, With and Without Alumina Particles as Filler, *NISTIR 3948*, 1990. [Kapton]

23. Touloukian, Y. S., and Ho, C. Y., Eds., *Thermophysical Properties of Selected Aerospace Materials, Part II: Thermophysical Properties of Seven Materials*, Plenum Press, New York, 1976. [Stainless steel]
24. Hust, J. G., and Sparks, L. L., Thermal Conductivity of Austentic Stainless Steel, SRM 735, From 5 to 280 K, *NBS Special Publication 260-35*, 1972. [Stainless steel 735]
25. Hust, J. G., and Giarratano, P. J., Thermal Conductivity and Electrical Resistivity Standard Reference Materials: Austentic Stainless Steel, SRM 735 and 798, From 4 to 1200 K, *NBS Special Publication 260-46*, 1975. [Stainless steel 735]
26. Ventura, G. Bianchini, G., Gottardi, E., Peroni, I., and Peruzzi, A., Thermal Expansion and Thermal Conductivity of Torlon at Low Temperatures, *Cryogenics* 39, 481, 1999. [Torlon]
27. Barucci, M., Olivieri, E., Pasca, E., Risegari, L., and Ventura, G., Thermal Conductivity of Torlon between 4.2 and 300 K, *Cryogenics* 45, 295, 2005. [Torlon]
28. Ziegler, W. T., Mullins, J. C., and Hwa, S. C. P., Specific Heat and Thermal Conductivity of Four Commercial Titanium Alloys From 20 to 300 K. *Adv. Cryog. Eng.* 8, 268, 1963. [Ti-6Al-4V]
29. Canavan, E. R., and Tuttle, J. G., Thermal Conductivity and Specific Heat Measurements of Candidate Structural Materials for the JWST Optical Bench, *Adv. Cryog. Eng.* 52, 233, 2006. [Ti 15-3-3-3]
30. Bunting, J. G., Ashworth, T., and Steeple, H., The Specific Heat of Apiezon N Grease, *Cryogenics* 9, 385, 1969. [Apiezon N Grease]
31. Bevolo, A. J., Heat Capacity of Apiezon N Grease from 1 to 50 K, *Cryogenics* 14, 661, 1974. [Apiezon N Grease]
32. Wun, M., and Phillips, N. E., Low Temperature Specific Heat of Apiezon N Grease, *Cryogenics* 15, 36, 1975. [Apiezon N Grease]
33. Kreitman, M. M., Ashworth, T., and Rechowicz, M., A Correlation Between Thermal Conductance and Specific Heat Anomalies and the Glass Temperature of Apiezon N and T Greases, *Cryogenics* 12, 32, 1972. [Apiezon N Grease]
34. Schnelle, W., Engelhardt, J., and Gemlin, E., Specific Heat Capacity of Apiezon N High Vacuum Grease and of Duran Borosilicate Glass, *Cryogenics* 39, 271, 1999. [Apiezon N Grease]
35. Touloukian, Y.S., and Buyco, E.H., *The TPRC Data Series: Vol. 4, Specific Heat-Metallic Elements and Alloys*, New York, Washington, 1970. [Be]
36. Touloukian, Y. S., Kirby, R. K, Taylor, R. E., and Desai, P. D, *The TPRC Data Series: Vol. 12, Thermal Expansion-Metallic Elements and Alloys*, Plenum Press, New York, 1970. [Be, Mo, Ti-6Al-4V]
37. Corruccini, R. J., and Gniewek, J. J., Thermal Expansion of Technical Solids at Low Temperatures, *NBS Monograph 29*, National Bureau of Standards, Boulder, CO, 1961. [BeCu, Nylon, Stainless steel, Teflon]
38. Reed, R. P., and Clark, A. F., *Materials at Low Temperatures*, American Society of Metals, Metals Park, OH, 1983. [G-10, NiTi, Nylon]
39. Arp, V., Wilson, J. H., Winrich, L., and Sikora, P., Thermal Expansion of Some Engineering Materials from 20 to 293 K, *Cryogenics* 2, 230, 1962. [Sapphire, Al]
40. Apostolescu, D. E., Gaal, P. S., and Chapman, A. S., A Proposed High Temperature Thermal Expansion Reference Material, Standard Reference Material, pp. 637-646. [Sapphire]
41. Swenson, C. A., Linear Thermal Expansivity (1.5-300 K) and Heat Capacity (1.2-90 K) of Stycast 2850FT, *Rev. Sci. Instrum.* 68, 1312, 1997. [Stycast epoxy]
42. Taylor, C. T., Notcutt, M., Wong, E. K., Mann, A. G., and Blair, D. G., Measurement of the Thermal Expansion Coefficient of an All-Sapphire Optical Cavity, *IEEE Trans. Instrum. Meas.* 46, 183, 1977. [Sapphire]
43. Taylor, C. T., Notcutt, M., Wong, E. K., Mann, A. G., and Blair, D. G, Measurement of the Thermal Expansion Coefficient of a Cryogenic All-Sapphire Optical Cavity, *Opt. Commun.* 131, 311, 1996. [Sapphire]
44. Lyon, K. G., Salinger, G. L., Swenson, C. A., and White, G. K., Linear Thermal Expansion Measurements on Silicon from 6 to 340 K, *J. Appl. Phys.* 48, 865, 1977. [Silicon]
45. Karlmann, P. B., Klein, K. J., Halverson, P. G., Peters, R. D., Levine, M. B., van Buren, D., and Dudik, M. J., Linear Thermal Expansion Measurements of Single Crystal Silicon for Validation of Interferometer Based Cryogenic Dilatometer, *Adv. Cryog. Eng.* 52, 35, 2006. [Silicon]
46. Roberts, R. B., Thermal Expansion Reference Data: Silicon 300-850 K, *J. Phys. D: Appl. Phys.* 14, L163, 1981. [Silicon]
47. White, G. K., and Minges, M. L., *Int. J. Thermophys.* 18, 1269, 1997. [Silicon]
48. Swenson, C. A., Recommended Values for the Thermal Expansivity of Silicon from 0 to 1000 K, *J. Phys. Chem. Ref. Data* 12, 179, 1983. [Silicon]

Thermal Conductivity λ in W m^{-1} K^{-1}

T/K	Al 1100	Al 3003	Al 5083	Al 6061	Al 6063	Be	Be-Cu	Brass	Cu[a] RRR= 50	Cu[a] RRR= 100	Cu[a] RRR= 300	G-10[b] (norm-dir)	G-10[b] (warp-dir)	Inconel-718	Invar (Fe-36Ni)	Kevlar-49 Composite	Kevlar-49 Fiber
1						181.4	0.3161									0.002234	0.003775
2						355.0	0.8999									0.008663	0.01457
3						551.1	1.402									0.01853	0.03077
4	54.11	10.81	3.295	5.347	34.36	741.4	1.879		320.4	642.3		0.07229	0.07322	0.4624	0.2419	0.03089	0.05042
6	83.26	16.77	4.982	8.268	51.64	1082	2.850	3.143	466.8	931.7	2810	0.09112	0.09693	0.8644	0.4001	0.06021	0.09448
8	113.5	22.81	6.685	11.23	69.70	1410	3.877	4.400	622.3	1239	3636	0.1023	0.1179	1.199	0.5630	0.09263	0.1405
10	141.8	28.94	8.427	14.20	86.51	1754	4.955	5.702	778.1	1540	4320	0.1122	0.1361	1.519	0.7339	0.1261	0.1865
12	170.1	35.15	10.19	17.15	103.5	2120	6.069	7.028	927.3	1814	4829	0.1217	0.1518	1.832	0.9114	0.1595	0.2320
14	199.1	41.35	11.97	20.05	121.1	2502	7.207	8.362	1064	2045	5147	0.1309	0.1654	2.136	1.094	0.1920	0.2770
16	228.0	47.49	13.73	22.91	139.2	2887	8.357	9.696	1185	2226	5276	0.1398	0.1775	2.426	1.279	0.2234	0.3213
18	256.2	53.51	15.48	25.70	157.5	3262	9.511	11.02	1287	2352	5234	0.1483	0.1883	2.699	1.465	0.2533	0.3649
20	282.6	59.37	17.21	28.43	175.4	3611	10.66	12.33	1368	2423	5052	0.1565	0.1982	2.956	1.652	0.2819	0.4077
30	371.7	85.21	25.43	41.10	246.3	4685	16.27	18.56	1444	2143	3257	0.1912	0.2393	3.998	2.570	0.4059	0.6122
40	389.5	104.7	32.89	52.23	276.6	4571	21.48	24.12	1163	1485	1833	0.2176	0.2740	4.744	3.437	0.5080	0.8031
50	369.2	118.7	39.66	62.05	277.7	3823	26.20	29.02	863.6	1005	1130	0.2382	0.3061	5.316	4.247	0.5976	0.9841
60	338.0	128.6	45.85	70.76	265.6	2968	30.42	33.36	670.0	741.2	801.8	0.2553	0.3366	5.784	5.005	0.6794	1.157
70	308.1	135.6	51.55	78.55	250.0	2243	34.08	37.26	561.1	603.6	638.5	0.2702	0.3660	6.182	5.717	0.7555	1.323
80	283.3	140.6	56.81	85.56	235.4	1693	37.15	40.84	500.3	529.3	551.0	0.2837	0.3942	6.532	6.386	0.8269	1.482
90	264.1	144.4	61.71	91.91	223.2	1296	39.60	44.21	465.1	487.0	501.0	0.2965	0.4214	6.842	7.016	0.8943	1.634
100	249.7	147.4	66.26	97.70	213.8	1014	41.41	47.45	443.9	461.5	471.1	0.3089	0.4477	7.119	7.611	0.9579	1.779
120	231.5	151.9	74.52	107.9	201.8	670.0	43.12		421.8	434.8	440.6	0.3335	0.4978	7.587	8.702	1.075	2.050
140	222.3	155.5	81.80	116.5	196.6	489.9			411.6	422.1	427.6	0.3586	0.5450	7.962	9.671	1.180	2.297
160	218.2	158.9	88.26	123.9	195.5	390.7			406.0	415.0	421.2	0.3846	0.5899	8.264	10.524	1.274	2.523
180	216.4	162.2	94.04	130.4	196.6	333.3			402.6	410.3	417.5	0.4117	0.6328	8.510	11.270	1.359	2.729
200	215.5	165.5	99.24	136.0	198.6	298.0			400.1	407.0	414.6	0.4399	0.6741	8.720	11.915	1.436	2.917
220	214.8	168.6	104.0	141.0	200.5	274.2			398.2	404.2	411.8	0.4692	0.7141	8.912	12.464	1.507	3.090
240	213.9	171.5	108.3	145.3	201.9	255.5			396.5	401.9	408.8	0.4995	0.7530	9.101	12.925	1.571	3.250

T/K	Al 1100	Al 3003	Al 5083	Al 6061	Al 6063	Be	Be-Cu	Brass	Cu[a] RRR= 50	Cu[a] RRR= 100	Cu[a] RRR= 300	G-10[b] (norm-dir)	G-10[b] (warp-dir)	Inconel-718	Invar (Fe-36Ni)	Kevlar-49 Composite	Kevlar-49 Fiber
260	213.0	174.1	112.2	149.1	202.3	237.6			395.0	399.9	405.5	0.5307	0.7909	9.301	13.303	1.630	3.398
280	212.1	176.2	115.9	152.4	201.8	217.9			393.6	398.0	401.8	0.5626	0.8278	9.528	13.604	1.685	3.536
300	211.8	177.8	119.3	155.3	200.5	194.9			392.4	396.3	397.9	0.5951	0.8636	9.793	13.835	1.735	3.664

T/K	Pb	Mo	Fe-2.25 Ni	Fe-3.5 Ni	Fe-5.0 Ni	Fe-9.0 Ni	Pt	Nylon	Kapton	Mylar[c]	PS[d] 12.66 kg m-3	PS[d] 32.04 kg m-3	PS[d] 49.98 kg m-3	PS[d] 99.96 kg m-3
1														
2										0.03092				
3										0.03610				
4	2682	56.74	1.119	0.9289	0.7754	0.6256	921.7	0.01245	0.01079	0.03764				0.009431
6	700.1	90.04	1.787	1.455	1.223	0.9814	1210	0.01983	0.01309	0.03978			0.004852	0.008950
8	313.3	118.4	2.460	1.999	1.679	1.349	1329	0.02880	0.01802	0.04334			0.004842	0.009014
10	177.1	146.3	3.137	2.535	2.138	1.716	1240	0.03902	0.02345	0.04791			0.004862	0.009055
12	118.3	175.4	3.809	3.064	2.599	2.080	1053	0.05014	0.02883	0.05295			0.004902	0.009098
14	89.02	205.3	4.470	3.588	3.061	2.443	856.0	0.06185	0.03397	0.05816			0.004956	0.009155
16	72.82	234.7	5.121	4.107	3.523	2.804	687.0	0.07388	0.03884	0.06337			0.005018	0.009226
18	63.12	262.2	5.760	4.624	3.983	3.163	553.5	0.08602	0.04345	0.06849			0.005087	0.009307
20	56.95	286.9	6.390	5.139	4.442	3.521	451.5	0.09811	0.04782	0.07349			0.005158	0.009393
30	45.29	353.6	9.423	7.678	6.691	5.293	209.9	0.1543	0.06704	0.09633		0.003803	0.005533	0.009833
40	41.97	345.0	12.28	10.11	8.825	7.022	137.0	0.1999	0.08319	0.1158		0.005073	0.005905	0.01026
50	40.10	307.3	14.96	12.40	10.82	8.688	108.3	0.2352	0.09706	0.1319		0.006153	0.006272	0.01071
60	38.73	267.3	17.42	14.52	12.68	10.28	94.57	0.2619	0.1090	0.1443		0.007232	0.006644	0.01121
70	37.70	233.7	19.67	16.47	14.39	11.79	87.02	0.2821	0.1192	0.1526		0.008311	0.007031	0.01177
80	36.98	207.9	21.71	18.25	15.98	13.21	82.32	0.2974	0.1280	0.1567		0.009382	0.007437	0.01239
90	36.51	188.6	23.56	19.89	17.44	14.55	79.09	0.3091	0.1355		0.01265	0.01045	0.007868	0.01306
100	36.23	174.5	25.23	21.38	18.79	15.81	76.69	0.3179	0.1419		0.01362	0.01151	0.008328	0.01378
120	36.03	156.8	28.14	24.01	21.19	18.09	73.37	0.3298	0.1522		0.01567	0.01368	0.009344	0.01535
140	36.02	147.9	30.56	26.23	23.24	20.07	71.39	0.3367	0.1599		0.01790	0.01594	0.01050	0.01705
160	35.96	143.7	32.62	28.10	24.98	21.77	70.43	0.3405	0.1659		0.02032	0.01829	0.01180	0.01890
180	35.78	142.0	34.36	29.69	26.47	23.23	70.27	0.3424	0.1708		0.02293	0.02072	0.01325	0.02089
200	35.46	141.4	35.84	31.01	27.72	24.46	70.65	0.3431	0.1749		0.02576	0.02320	0.01486	0.02309
220	35.10	141.2	37.06	32.10	28.78	25.49	71.25	0.3428	0.1786		0.02881	0.02572	0.01662	0.02557
240	34.80	140.9	38.01	32.98	29.64	26.32	71.72	0.3420	0.1820		0.03210	0.02826	0.01854	0.02845
260	34.70	140.4	38.70	33.65	30.35	26.98	71.65	0.3406	0.1853		0.03561	0.03082	0.02059	0.03185
280	34.95	139.6	39.09	34.13	30.90	27.48	70.66	0.3389	0.1886		0.03934	0.03343	0.02278	0.03599
300		138.6	39.17	34.43	31.32	27.83	68.42	0.3368	0.1919		0.04323	0.03615	0.02509	0.04112

T/K	PU[e] 31.88 kg m-3	PU[e] 32.04 kg m-3	PU[e] 49.02 kg m-3	PU[e] 64.08 kg m-3	PVC[f] 20.02 kg m-3	PVC[f] 56.06 kg m-3	Stainless Steel 304	304L	310	316	735	Teflon	Torlon 4203	Ti-6Al-4V	Ti-15V-3Cr-3Al-3Sn
1											0.4028		0.004330		
2											0.5637		0.008974		0.05608
3											0.7972		0.01145		0.08043
4							0.2724	0.2724	0.2409	0.2724	1.052	0.04599	0.01280		0.1010
6							0.4653	0.4653	0.3960	0.4653	1.315	0.06624	0.01410		0.1450
8							0.6770	0.6770	0.5654	0.6770	1.583	0.08244	0.01467		0.1904
10							0.9039	0.9039	0.7457	0.9039	1.854	0.09545	0.01477		0.2354
12							1.143	1.143	0.9348	1.143	2.126	0.1066	0.01871		0.2796
14							1.391	1.391	1.131	1.391	2.399	0.1166	0.03501		0.3235
16							1.647	1.647	1.331	1.647	3.729	0.1258	0.04760		0.3673
18							1.906	1.906	1.534	1.906	4.937	0.1343	0.05696		0.4114
20							2.169	2.169	1.739	2.169	5.985	0.1422	0.06377	0.8426	0.4559
30			0.01133				3.469	3.469	2.741	3.469	6.872	0.1738	0.07797	1.656	0.6871
40			0.01238				4.670	4.670	3.648	4.670	7.618	0.1953	0.08344	1.905	0.9301
50			0.01352				5.730	5.730	4.430	5.730	8.247	0.2099	0.08884	2.411	1.177
60			0.01465	0.007880			6.647	6.647	5.093	6.647	8.783	0.2203	0.09491	2.917	1.420
70			0.01575	0.008571			7.435	7.435	5.655	7.435	9.245	0.2280	0.1014	3.291	1.655
80	0.01018		0.01684	0.009551			8.114	8.114	6.135	8.114	9.650	0.2341	0.1079	3.531	1.881
90	0.01115		0.01792	0.01065	0.01652		8.705	8.705	6.551	8.705	10.01	0.2391	0.1111	3.685	2.095
100	0.01212	0.01206	0.01900	0.01179	0.01680	0.02844	9.224	9.224	6.917	9.224	10.33	0.2434	0.1143	3.804	2.300
120	0.01399	0.01465	0.02117	0.01407	0.01799	0.02952	10.101	10.10	7.544	10.10	10.91	0.2505	0.1264	4.061	2.687
140	0.01574	0.01736	0.02336	0.01624	0.01931	0.03223	10.834	10.83	8.081	10.83	11.42	0.2562	0.1374	4.412	3.054
160	0.01734	0.02005	0.02554	0.01827	0.02072	0.03503	11.479	11.48	8.568	11.48	11.89	0.2608	0.1477	4.846	3.412
180	0.01881	0.02276	0.02765	0.02015	0.02232	0.03762	12.071	12.07	9.027	12.07	12.34	0.2644	0.1579	5.311	3.770
200	0.02013	0.02550	0.02964	0.02184	0.02416	0.04015	12.633	12.63	9.473	12.63	12.77	0.2672	0.1684	5.750	4.133
220	0.02125	0.02829	0.03144	0.02331	0.02628	0.04291	13.178	13.18	9.912	13.18	13.19	0.2692	0.1800	6.129	4.505

T/K	PU[e] 31.88 kg m^{-3}	PU[e] 32.04 kg m^{-3}	PU[e] 49.02 kg m^{-3}	PU[e] 64.08 kg m^{-3}	PVC[f] 20.02 kg m^{-3}	PVC[f] 56.06 kg m^{-3}	Stainless Steel 304	304L	310	316	735	Teflon	Torlon 4203	Ti-6Al-4V	Ti-15V-3Cr-3Al-3Sn
240	0.02210	0.03112	0.03297	0.02452	0.02879	0.04609	13.714	13.71	10.35	13.71	13.61	0.2705	0.1930	6.447	4.885
260	0.02255	0.03401	0.03418	0.02541	0.03184	0.04972	14.247	14.25	10.78	14.25	14.01	0.2715	0.2083	6.740	5.270
280	0.02248	0.03702	0.03500	0.02595	0.03574	0.05366	14.779	14.78	11.21	14.78	14.22	0.2722	0.2263	7.075	5.658
300	0.02181	0.04027	0.03540	0.02608	0.04093	0.05756	15.309	15.31	11.63	15.31	15.01	0.2728	0.2480	7.555	6.041

[a] Oxygen-free high conductivity (OFHC) copper
[b] G-10 is a glass epoxy composite laminate material
[c] Polyethylene terephthalate
[d] Polystyrene
[e] Polyurethane
[f] Polyvinylchloride

Specific Heat at Constant Pressure C_p in J kg^{-1} K^{-1}

T/K	Al 3003	Al 5083	Al 6061	Apiezon-N Grease	Be	Cu (OFHC)	G-10[a]	In	Invar (Fe-36Ni)	Pb (normal)	Pb (superconducting)	Fe-2.25Ni	Fe-3.5Ni	Fe-5.0Ni	Fe-9.0Ni	Pt
1				0.02400							2.570					0.02435
2				0.2118							11.78					0.05529
3				0.8050							32.71					0.09667
4	0.2920	0.292	0.2920	2.031		0.09942	2.016	0.9463	0.9749		75.27					0.1524
6	0.6047	0.605	0.6047	6.761		0.2303	5.463	3.505	1.475		297.0					0.3300
8	1.048	1.048	1.048	14.38		0.4639	10.03	8.387	2.179		764.6					0.6349
10	1.573	1.573	1.573	24.32		0.8558	15.36	15.28	3.130	13.69	1539					1.115
12	2.268	2.268	2.268	35.98	0.5348	1.470	21.20	23.56	4.334	21.43						1.814
14	3.235	3.235	3.235	48.91	0.7169	2.375	27.41	32.62	5.814	29.51						2.770
16	4.581	4.581	4.581	62.76	0.9528	3.640	33.85	42.06	7.621	37.50						4.009
18	6.416	6.416	6.416	77.31	1.249	5.327	40.46	51.61	9.800	45.14						5.544
20	8.854	8.854	8.854	92.37	1.607	7.491	47.17	61.08	12.34	52.25						7.376
30	33.45	33.45	33.45	171.8	4.413	26.40	81.31	104.5		79.06						20.35
40	81.96	81.96	81.96	252.8	9.544	57.63	115.4	138.6		94.45						36.78
50	148.8	148.8	148.8	331.4	18.58	95.84	149.1	163.3		103.4		84.17	84.17	84.17	84.17	52.95
60	223.6	223.6	223.6	405.6	33.66	135.2	182.5	179.9		109.0		90.95	90.95	90.95	90.95	66.88
70	298.3	298.3	298.3	474.9	57.08	171.8	215.9	190.5		112.7		124.8	124.8	124.8	124.8	78.06
80	368.7	368.7	368.7	539.1	90.79	203.8	249.4	196.9		115.3		162.6	162.6	162.6	162.6	86.73
90	433.3	433.3	433.3	598.9	135.9	230.9	283.0	200.7		117.2		196.7	196.7	196.7	196.7	93.40
100	492.2	492.2	492.2	654.9	192.7	253.5	316.9	203.1		118.7		225.6	225.6	225.6	225.6	98.57
120	594.1	594.1	594.1	759.2	336.8	287.6	385.2	206.0		120.9		273.1	273.1	273.1	273.1	106.0
140	677.6	677.6	677.6	858.9	510.4	311.6	454.4	209.1		122.5		312.9	312.9	312.9	312.9	111.4
160	744.5	744.5	744.5	960.6	700.1	329.4	524.3	213.5		123.9		346.9	346.9	346.9	346.9	115.7
180	796.4	796.4	796.4	1070	897.4	343.4	594.4	219.0		125.2		375.1	375.1	375.1	375.1	119.5
200	835.2	835.2	835.2	1194	1097	355.0	664.4	225.0		126.3		397.6	397.6	397.6	397.6	122.9
220	863.7	863.7	863.7	1339	1296	364.7	733.9	230.6		127.4		415.8	415.8	415.8	415.8	125.7
240	885.4	885.4	885.4	1510	1488	372.6	802.5	234.5		128.4		431.0	431.0	431.0	431.0	127.7
260	904.6	904.6	904.6	1717	1661	378.6	869.8	235.7		129.2		443.8	443.8	443.8	443.8	128.8
280	925.8	925.8	925.8	1968	1800	382.5	935.3	233.1		129.8		453.5	453.5	453.5	453.5	128.7
300	953.9	953.9	953.9	2276	1883	384.0	998.7			130.1		457.7	457.7	457.7	457.7	

T/K	Nylon	Kapton	PS[b] (99.96 kg m^{-3})	PS[b] (9.93 kg m^{-3})	PS[b] (6.07 kg m^{-3})	PU[c] (49.02 kg m^{-3})	PU[c] (389.25 kg m^{-3})	PVC[d]	Stainless Steel 304	304L	310	316	Teflon
1													
2													
3													
4	1.560	0.7916							1.849	1.674	2.164	1.977	2.232
6	4.986	2.708							2.945	2.557	3.166	2.963	3.690
8	11.01	6.306							4.041	3.525	4.160	4.047	9.104
10	20.00	11.68						9.335	5.163	4.571	5.226	5.203	18.02
12	31.89	18.70						13.41	6.394	5.740	6.350	6.494	29.27
14	46.34	27.10						17.57	7.806	7.004	7.524	7.958	41.50
16	62.87	36.61						21.52	9.456	8.399	8.780	9.615	53.77
18	80.98	46.96						25.25	11.39	10.02	10.18	11.49	65.60
20	100.2	57.92				42.26		28.80	13.65	11.88	11.78	13.61	76.79
30	201.4	116.2				70.85		46.12	30.70		25.79	29.22	124.3
40	297.2	172.9				111.9		64.72	58.22		55.98	57.55	164.4
50	383.5	224.3				158.3		84.66	94.57		98.21	100.2	202.1
60	461.8	270.4				207.2		105.4	135.7		134.2	141.8	239.0

T/K	Nylon	Kapton	PS[b] (99.96 kg m^{-3})	PS[b] (9.93 kg m^{-3})	PS[b] (6.07 kg m^{-3})	PU[c] (49.02 kg m^{-3})	PU[c] (389.25 kg m^{-3})	PVC[d]	Stainless Steel 304	Stainless Steel 304L	Stainless Steel 310	Stainless Steel 316	Teflon
70	534.2	311.7				257.8		126.8	177.4		167.7	179.9	275.6
80	602.2	348.9				310.0		148.9	216.1		197.7	214.7	312.0
90	667.0	382.8				363.9		171.6	250.2		224.1	245.8	348.4
100	729.0	413.7	441.8	521.4	472.7	419.6	555.6	195.3	279.0		247.5	273.0	384.6
120	846.2	468.4	540.4	583.5	535.8	535.9	635.4	246.2	322.3		287.1	317.3	456.7
140	955.6	515.7	629.9	643.5	602.8	657.9	715.9	302.5	351.7		320.2	351.9	528.4
160	1058	557.1	719.3	705.6	670.8	783.7	796.3	365.2	373.6		348.8	380.3	599.6
180	1155	593.8	809.1	769.5	740.2	910.5	877.5	435.3	392.2		373.7	404.5	670.0
200	1247	626.8	898.5	834.6	811.4	1035	959.0	514.3	410.0		395.0	424.8	738.9
220	1335	656.6	987.3	900.6	884.2	1155	1039	604.1	428.0		413.2	441.6	805.8
240	1421	683.9	1076	968.1	958.5	1268	1117	708.2	446.0		428.9	455.1	869.5
260	1507	709.1	1164	1038	1035	1370	1193	831.5	463.5		443.0	466.5	929.2
280	1594	732.5	1251	1111	1115	1460	1269	981.0	479.1		457.0	477.4	983.6
300	1684	754.5	1337	1189	1201	1537	1351	1167	491.3		472.8	490.2	1032

[a] G-10 is a glass epoxy composite laminate material
[b] Polystyrene
[c] Polyurethane
[d] Polyvinylchloride

Linear Thermal Expansion ($10^5 \times$ Fractional Change in Length Relative to 293 K)

T/K	Al 3003	Al 5083	Al 6061	Be (a-axis)	Be (c-axis)	Be (polycrystalline)	Be-Cu	G-10[a] (norm-dir)	G-10[a] (warp-dir)	Inconel-718	Invar	Mo	Nb-Ti
4	-413.8		-413.8								-40		
6	-414.3		-414.3								-40		
8	-414.6		-414.6								-40		
10	-414.9		-414.9					-714.6			-40		
12	-415.2		-415.2					-713.3	-244.0		-40		
14	-415.3		-415.3					-712.1	-243.4		-40		
16	-415.4		-415.4					-710.7	-242.8		-40		
18	-415.5	-415.5	-415.5					-709.3	-242.2		-40		
20	-415.4	-415.4	-415.4					-707.9	-241.6	-238.9	-40	-92.4	-188.2
30	-414.3	-414.3	-414.3				-316.4	-699.7	-238.0	-238.4	-40	-92.46	-187.1
40	-411.5	-411.5	-411.5				-314.6	-690.1	-233.9	-236.9	-40	-92.04	-184.9
50	-407.3	-407.3	-407.3				-311.3	-679.0	-229.3	-234.5	-40	-91.21	-181.6
60	-401.6	-401.6	-401.6				-306.7	-666.5	-224.2	-231.2	-40	-89.98	-177.4
70	-394.5	-394.5	-394.5				-300.8	-652.6	-218.5	-227.0	-40	-88.39	-172.5
80	-386.1	-386.1	-386.1	-0.1480	-0.0970	-0.1313	-293.8	-637.2	-212.4	-222.1	-40.21	-86.45	-166.9
90	-376.4	-376.4	-376.4	-0.1468	-0.0967	-0.1302	-285.8	-620.5	-205.8	-216.3	-38.20	-84.19	-160.7
100	-365.4	-365.4	-365.4	-0.1450	-0.0960	-0.1287	-276.8	-602.4	-198.8	-209.9	-36.13	-81.64	-154.1
120	-340.2	-340.2	-340.2	-0.1395	-0.0935	-0.1240	-256.2	-562.1	-183.5	-194.9	-31.84	-75.74	-147.1
140	-310.8	-310.8	-310.8	-0.1316	-0.0892	-0.1172	-232.9	-516.4	-166.6	-177.6	-27.45	-68.91	-139.8
160	-277.8	-277.8	-277.8	-0.1213	-0.0832	-0.1083	-207.3	-465.4	-148.4	-158.2	-23.04	-61.31	-132.3
180	-241.5	-241.5	-241.5	-0.1088	-0.0755	-0.0973	-179.9	-409.3	-129.0	-137.0	-18.70	-53.08	-124.7
200	-202.6	-202.6	-202.6	-0.0942	-0.0660	-0.0844	-150.9	-348.2	-108.5	-114.4	-14.53	-44.34	-117.0
220	-161.5	-161.5	-161.5	-0.0774	-0.0547	-0.0696	-120.6	-282.1	-87.16	-90.64	-10.62	-35.22	-109.2
240	-118.6	-118.6	-118.6	-0.0588	-0.0419	-0.0529	-89.16	-211.2	-65.01	-66.17	-7.062	-25.80	-101.3
260	-74.40	-74.40	-74.40	-0.0382	-0.0273	-0.0345	-56.57	-135.6	-42.27	-41.26	-3.941	-16.18	-93.46
280	-29.45	-29.45	-29.45	-0.0160	-0.0113	-0.0144	-22.76	-55.33	-19.08	-16.25	-1.352	-6.431	-85.61
300	15.81	15.81	15.81	0.0079	0.0063	0.0073	12.43	29.44	4.398	8.536	0.6163	3.390	-77.77

T/K	Fe-2.25Ni	Fe-3.5Ni	Fe-5.0Ni	Fe-9.0Ni	Nylon	PS[b] (51.42 kg m^{-3})	PS[b] (102.2 kg m^{-3})	PU[c] (32.04 kg m^{-3})	PU[c] (64.07 kg m^{-3})	PVC[d]	Sapphire	Stainless Steel 304	Stainless Steel 304L	Stainless Steel 310	Stainless Steel 316	Stycast 2850 FT[e]	Teflon	Ti-6Al-4V
4	-210.5	-210.5	-210.5	-210.5	-1389	-1697	-1742			-1025		-297.0	-297.0	-297.0	-297.0	-44.57	-2127	-171.9
6	-210.6	-210.6	-210.6	-210.6	-1389	-1698	-1738			-1020		-297.6	-297.6	-297.6	-297.6	-44.53	-2127	-172.3
8	-210.5	-210.5	-210.5	-210.5	-1388	-1698	-1735			-1016		-298.1	-298.1	-298.1	-298.1	-44.48	-2127	-172.5
10	-210.5	-210.5	-210.5	-210.5	-1388	-1697	-1732			-1011		-298.6	-298.6	-298.6	-298.6	-44.43	-2127	-172.8
12	-210.4	-210.4	-210.4	-210.4	-1387	-1696	-1730			-1006		-299.0	-299.0	-299.0	-299.0	-44.38	-2126	-173.0
14	-210.2	-210.2	-210.2	-210.2	-1386	-1694	-1727			-1001	-79.13	-299.4	-299.4	-299.4	-299.4	-44.32	-2125	-173.2
16	-210.1	-210.1	-210.1	-210.1	-1384	-1691	-1724			-996.6	-79.16	-299.6	-299.6	-299.6	-299.6	-44.25	-2123	-173.4
18	-209.9	-209.9	-209.9	-209.9	-1383	-1688	-1721			-991.7	-79.19	-299.8	-299.8	-299.8	-299.8	-44.18	-2121	-173.5
20	-209.6	-209.6	-209.6	-209.6	-1381	-1685	-1718	-808.3	-1050	-986.7	-79.21	-300.0	-300.0	-300.0	-300.0	-44.11	-2119	-173.6
30	-207.9	-207.9	-207.9	-207.9	-1369	-1660	-1702	-803.6	-1036	-961.0	-79.29	-299.7	-299.7	-299.7	-299.7	-43.68	-2102	-173.4
40	-205.4	-205.4	-205.4	-205.4	-1352	-1624	-1679	-795.4	-1019	-934.0	-79.25	-297.9	-297.9	-297.9	-297.9	-43.15	-2077	-172.5
50	-202.2	-202.2	-202.2	-202.2	-1331	-1579	-1647	-784.1	-998.9	-905.9	-79.07	-294.7	-294.7	-294.7	-294.7	-42.53	-2047	-170.7

												Stainless Steel						
T/K	Fe-2.25Ni	Fe-3.5Ni	Fe-5.0Ni	Fe-9.0Ni	Nylon	PS[b] (51.42 kg m-3)	PS[b] (102.2 kg m-3)	PU[c] (32.04 kg m-3)	PU[c] (64.07 kg m-3)	PVC[d]	Sapphire	304	304L	310	316	Stycast 2850 FT[e]	Teflon	Ti-6Al-4V
60	-198.3	-198.3	-198.3	-198.3	-1306	-1525	-1605	-769.9	-975.6	-876.8	-78.73	-290.2	-290.2	-290.2	-290.2	-41.80	-2012	-168.2
70	-193.8	-193.8	-193.8	-193.8	-1278	-1464	-1551	-752.9	-949.5	-846.9	-78.21	-284.5	-284.5	-284.5	-284.5	-40.99	-1972	-165.0
80	-188.8	-188.8	-188.8	-188.8	-1246	-1398	-1489	-733.5	-921.0	-816.2	-77.49	-277.7	-277.7	-277.7	-277.7	-40.08	-1930	-161.1
90	-183.3	-183.3	-183.3	-183.3	-1211	-1329	-1419	-711.8	-890.1	-784.8	-76.54	-269.9	-269.9	-269.9	-269.9	-39.08	-1886	-156.6
100	-177.3	-177.3	-177.3	-177.3	-1173	-1257	-1346	-688.1	-857.3	-752.7	-75.36	-261.2	-261.2	-261.2	-261.2	-37.99	-1839	-151.6
120	-164.1	-164.1	-164.1	-164.1	-1090	-1117	-1203	-635.1	-786.3	-686.9	-72.23	-241.3	-241.3	-241.3	-241.3	-35.55	-1742	-139.9
140	-149.6	-149.6	-149.6	-149.6	-997.9	-979.1	-1059	-575.9	-709.3	-619.1	-68.01	-218.6	-218.6	-218.6	-218.6	-32.76	-1639	-126.4
160	-133.8	-133.8	-133.8	-133.8	-897.2	-842.6	-916.7	-511.6	-627.6	-549.2	-62.65	-193.7	-193.7	-193.7	-193.7	-29.65	-1529	-111.4
180	-117.0	-117.0	-117.0	-117.0	-788.5	-706.8	-777.2	-443.1	-542.0	-477.0	-56.12	-167.1	-167.1	-167.1	-167.1	-26.20	-1406	-95.39
200	-99.05	-99.05	-99.05	-99.05	-671.9	-571.7	-640.6	-371.2	-453.1	-402.2	-48.44	-139.1	-139.1	-139.1	-139.1	-22.43	-1265	-78.55
220	-80.06	-80.06	-80.06	-80.06	-546.8	-438.1	-506.0	-296.3	-361.3	-324.1	-39.65	-110.2	-110.2	-110.2	-110.2	-18.32	-1095	-61.28
240	-59.90	-59.90	-59.90	-59.90	-412.3	-307.4	-372.2	-218.9	-266.8	-241.7	-29.84	-80.50	-80.50	-80.50	-80.50	-13.87	-883.3	-43.92
260	-38.39	-38.39	-38.39	-38.39	-266.9	-182.1	-236.6	-138.9	-169.3	-154.0	-19.10	-50.35	-50.35	-50.35	-50.35	-9.082	-615.6	-26.82
280	-15.27	-15.27	-15.27	-15.27	-109.0	-65.04	-96.21	-56.39	-68.57	-59.6	-7.589	-19.87	-19.87	-19.87	-19.87	-3.936	-273.8	-10.30
300	9.770	9.770	9.770	9.770	64.02	39.90	53.03	28.96	36.16	43.09	4.523	10.86	10.86	10.86	10.86	1.579	162.2	5.285

[a] G-10 is a glass epoxy composite laminate material
[b] Polystyrene
[c] Polyurethane
[d] Polyvinylchloride
[e] Epoxy

Coefficient of Thermal Expansion α

T/K	Cu (OFHC) $\alpha/10^{-6}$ K^{-1}	Sapphire $\alpha/10^{-6}$ K^{-1}	Si[a] $\alpha/10^{-8}$ K^{-1}	Phosphor-bronze $\alpha/10^{-6}$ K^{-1}
0.1			0.00000004800	
0.2			0.0000003840	
0.4			0.000003072	
0.6			0.00001037	
0.8			0.00002458	
1			0.00000600	
2			0.00004800	
3			0.0003840	
4	0.002281	0.0001060	0.001296	
6	0.009244	0.0003781	0.003072	
8	0.01893	0.0008953	0.01037	
10	0.03338	0.0016535	0.02458	0.06000
12	0.05522	0.0027127	0.04800	0.1048
14	0.08720	0.0041822	0.08294	0.1657
16	0.1321	0.0061990	0.1331	0.2444
18	0.1927	0.0089156	0.1103	0.3421
20	0.2715	0.01249	-0.04541	0.4599
30	0.9888	0.04771	-0.3407	1.352
40	2.229	0.1152	-5.330	2.667
50	3.790	0.2093	-16.61	4.215
60	5.426	0.3323	-29.23	5.807
70	6.968	0.5154	-40.33	7.309
80	8.335	0.8437	-46.61	8.653
90	9.506		-47.18	9.815
100	10.49		-42.69	10.80
120	12.02		-21.48	12.32
140	13.13		11.13	13.39
160	13.97		49.33	14.17
180	14.65		87.86	14.78
200	15.21		123.6	15.29
220	15.67		156.3	15.73
240	16.05		185.1	16.12
260	16.33		210.8	16.48
280	16.51		233.1	16.78
300	16.58		252.6	17.04

[a] Note unit differs from other columns

Young's Modulus *E* in GPa

								Stainless Steel		
T/K	Al 5083	Al 6061	Invar	Fe-2.25Ni	Fe-3.5Ni	Fe-5.0Ni	Fe-9.0Ni	304	310	316
2	80.85	77.73	141.6	217.7	215.4	209.2	205.4			
4	80.87	77.75	141.7	217.7	215.4	209.2	205.4	210.1	206.7	
6	80.89	77.76	141.7	217.7	215.4	209.3	205.4	210.2	206.7	
8	80.90	77.78	141.7	217.7	215.4	209.3	205.5	210.2	206.7	207.9
10	80.91	77.79	141.7	217.7	215.4	209.3	205.5	210.2	206.7	207.8
12	80.92	77.80	141.7	217.7	215.4	209.4	205.5	210.2	206.7	207.8
14	80.92	77.80	141.7	217.7	215.4	209.4	205.5	210.1	206.7	207.9
16	80.93	77.81	141.7	217.7	215.4	209.4	205.5	210.1	206.7	207.9
18	80.93	77.81	141.7	217.7	215.4	209.4	205.5	210.1	206.6	208.0
20	80.93	77.81	141.7	217.7	215.4	209.4	205.5	210.1	206.6	208.1
30	80.90	77.78	141.6	217.7	215.4	209.5	205.6	210.4	206.5	208.8
40	80.82	77.71	141.4	217.6	215.3	209.5	205.5	211.7	206.3	209.3
50	80.70	77.60	141.1	217.4	215.1	209.5	205.4	213.4	206.0	209.6
60	80.54	77.46	140.8	217.2	214.9	209.4	205.2	214.6	205.7	209.5
70	80.34	77.28	140.5	217.0	214.6	209.3	205.0	214.5	205.3	209.3
80	80.11	77.08	140.1	216.7	214.3	209.1	204.8	214.4	204.9	209.0
90	79.85	76.86	139.9	216.4	214.0	208.9	204.5	214.1	204.5	208.7
100	79.56	76.61	139.7	216.0	213.6	208.7	204.2	213.7	204.0	208.2
120	78.92	76.07	139.5	215.2	212.8	208.1	203.5	212.8	202.9	207.2
140	78.21	75.48	139.7	214.3	211.9	207.3	202.7	211.5	201.7	206.0
160	77.45	74.86	140.3	213.3	211.0	206.4	201.8	210.1	200.4	204.7
180	76.65	74.21	141.3	212.3	210.0	205.4	200.9	208.7	199.1	203.3
200	75.82	73.56	142.8	211.2	208.9	204.3	199.9	207.2	197.7	201.8
220	74.98	72.89	144.6	210.2	207.9	203.1	198.9	205.6	196.3	200.3
240	74.14	72.22	146.7	209.1	206.8	201.9	197.9	204.1	194.8	198.8
260	73.30	71.54	148.8	208.0	205.7	200.6	196.9	202.4	193.4	197.3
280	72.46	70.83	150.8	206.9	204.7	199.4	195.9	200.7	191.9	195.7
300	71.62	70.10	152.4	205.8	203.6	198.2	194.9	198.8	190.3	194.1

HEAT CAPACITY OF SELECTED SOLIDS

This table gives the molar heat capacity at constant pressure of representative metals, semiconductors, and other crystalline solids as a function of temperature in the range 200 to 600 K.

References

1. Chase, M. W., et al., *JANAF Thermochemical Tables, 3rd ed., J. Phys. Chem. Ref. Data,* 14, Suppl. 1, 1985.
2. Garvin, D., Parker, V. B., and White, H. J., *CODATA Thermodynamic Tables,* Hemisphere Press, New York, 1987.
3. DIPPR Database of Pure Compound Properties, Design Institute for Physical Properties Data, American Institute of Chemical Engineers, New York, 1987.

Name	Formula	C_p(200 K)/ J mol^{-1} K^{-1}	C_p(250 K)/ J mol^{-1} K^{-1}	C_p(300 K)/ J mol^{-1} K^{-1}	C_p(350 K)/ J mol^{-1} K^{-1}	C_p(400 K)/ J mol^{-1} K^{-1}	C_p(500 K)/ J mol^{-1} K^{-1}	C_p(600 K)/ J mol^{-1} K^{-1}
Aluminum	Al	21.33	23.08	24.25	25.11	25.78	26.84	27.89
Aluminum oxide (α)	Al_2O_3	51.12	67.05	79.45	88.91	96.14	106.17	112.55
Anthracene	$C_{14}H_{10}$	138.6	173.9	210.7	248.8	288.4		
Benzoic acid	C_6H_5COOH	102.7	123.5	147.4	172.0			
Beryllium	Be	9.98	13.58	16.46	18.53	19.95	21.94	23.34
Biphenyl	$C_{12}H_{10}$	131.0	162.5	197.2				
Boron	B	5.99	8.82	11.40	13.65	15.69	18.72	20.78
Calcium	Ca	24.54	25.41	25.94	26.32	26.87	28.49	30.38
Calcium carbonate (calcite)	$CaCO_3$	66.50	75.66	83.82	91.51	96.97	104.52	109.86
Calcium oxide	CaO	33.64	38.59	42.18	45.07	46.98	49.33	50.72
Carbon (graphite)	C	5.01	6.82	8.58	10.24	11.81	14.62	16.84
Cesium chloride	CsCl	50.13	51.34	52.48	53.58	54.68	56.90	59.10
Chromium	Cr	19.86	22.30	23.47	24.39	25.23	26.63	27.72
Cobalt	Co	22.23	23.98	24.83	25.68	26.53	28.20	29.66
Copper	Cu	22.63	23.77	24.48	24.95	25.33	25.91	26.48
Copper(II) oxide	CuO	34.80		42.41	44.95	46.78	49.19	50.83
Copper(II) sulfate	$CuSO_4$	77.01	89.25	99.25	107.65	114.93	127.19	136.31
Germanium	Ge			23.25	23.85	24.31	24.96	25.45
Gold	Au			25.41	25.37	25.51	26.06	26.65
Hexachlorobenzene	C_6Cl_6	162.7	183.6	202.4				
Iodine	I	51.57	53.24	54.51	58.60			
Iron	Fe	21.59	23.74	25.15	26.28	27.39	29.70	32.05
Lead	Pb	25.87	26.36	26.85	27.30	27.72	28.55	29.40
Lithium	Li	21.57	23.42	24.64	25.96	27.60	29.28	
Lithium chloride	LiCl	43.35	46.08	48.10	49.66	50.97	53.34	55.59
Magnesium	Mg	22.72	24.02	24.90	25.57	26.14	27.17	28.18
Magnesium oxide	MgO			37.38	40.59	42.77	45.56	47.30
Manganese	Mn	23.05	24.95	26.35	27.52	28.53	30.29	31.90
Naphthalene	$C_{10}H_8$	105.8	134.1	167.8	204.1			
Potassium	K	27.00	28.01	29.60				
Potassium chloride	KCl	48.44	50.10	51.37	52.31	53.08	54.71	56.35
Silicon	Si	15.64	18.22	20.04	21.28	22.14	23.33	24.15
Silicon dioxide (α-quartz)	SiO_2	32.64	39.21	44.77	49.47	53.43	59.64	64.42
Silver	Ag			25.36	25.55	25.79	26.36	26.99
Sodium	Na	22.45	27.01	28.20	30.14			
Sodium chloride	NaCl	46.89	48.85	50.21	51.25	52.14	53.96	55.81
Tantalum	Ta	24.08	24.86	25.31	25.60	25.84	26.35	26.84
Titanium	Ti	22.37	24.07	25.28	26.17	26.86	27.88	28.60
Tungsten	W	22.49	23.69	24.30	24.65	24.92	25.36	25.79
Vanadium	V	21.88	23.70	24.93	25.68	26.23	26.94	27.49
Zinc	Zn	24.05	25.02	25.45	25.88	26.35	27.39	28.59
Zirconium	Zr	23.87	24.69	25.22	25.61	25.93	26.56	27.28

THERMAL AND PHYSICAL PROPERTIES OF PURE METALS

This table gives the following properties for the metallic elements under ambient conditions:

t_m: Melting point in °C
t_b: Normal boiling point in °C, at a pressure of 101.325 kPa (760 Torr)
$\Delta_{fus}H$: Enthalpy of fusion at the melting point in J g^{-1}
ρ: Density at 25 °C in g cm^{-3}
α: Coefficient of linear expansion at 25 °C in K^{-1} (the quantity listed is $10^6 \times \alpha$)
c_p: Specific heat capacity at constant pressure at 25 °C in J g^{-1} K^{-1}
λ: Thermal conductivity at 27 °C in W m^{-1} K^{-1}
ρ_{elec}: Electrical conductivity in μΩ cm

References

1. Dinsdale, A. T., *CALPHAD* 15, 317, 1991 [melting point, enthalpy of fusion].
2. Touloukian, Y. S., *Thermophysical Properties of Matter*, Vol. 12, Thermal Expansion, IFI/Plenum, New York, 1975 [coefficient of expansion, density].
3. Ho, C. Y., Powell, R. W., and Liley, P. E., *J. Phys. Chem. Ref. Data* 3, Suppl. 1, 1974 [thermal conductivity].
4. Cox, J. D., Wagman, D. D., and Medvedev, V. A., *CODATA Key Values for Thermodynamics*, Hemisphere Publishing Corp., New York, 1989 [heat capacity].
5. Glushko, V. P., Ed., *Thermal Constants of Substances*, VINITI, Moscow, 1965-1981 [enthalpy of fusion, heat capacity].
6. Wagman, D. D., Evans, W. H., Parker, V. B., Schumm, R. H., Halow, I., Bailey, S. M., Churney, K. L., and Nuttall, R. L., *The NBS Tables of Chemical Thermodynamic Properties, J. Phys. Chem. Ref. Data* 11, Suppl. 2, 1982 [heat capacity].
7. Gschneidner, K. A., *Bull. Alloy Phase Diagrams* 11, 216–224, 1990 [various properties of the rare earth metals].
8. Hellwege, K. H., Ed., *Landolt Börnstein, Numerical Values and Functions in Physics, Chemistry, Astronomy, Geophysics, and Technology*, Vol. 2, Part 1, Mechanical-Thermal Properties of State, 1971 [density].
9. *Physical Encyclopedic Dictionary*, Vol. 1–5, Encyclopedy Publishing House, Moscow, 1960–66.

Name	Formula	Mol. wt.	t_m/°C	t_b/°C	$\Delta_{fus}H$/ J g^{-1}	ρ/g cm^{-3}	$10^6 \times \alpha$/ K^{-1}	c_p/ J g^{-1} K^{-1}	λ/ W m^{-1} K^{-1}	ρ_{elec}/ μΩ cm
Actinium	Ac	227.0	1050	≈3200	52.9	10		0.120		
Aluminum	Al	26.982	660.323	2519	396.9	2.70	23.1	0.897	237	2.709
Antimony (gray)	Sb	121.76	630.628	1587	162.5	6.68	11.0	0.207	24.3	39
Barium	Ba	137.327	727	≈1845	51.85	3.62	20.6	0.204	18.4	34.0
Beryllium	Be	9.012	1287	2468	876.1	1.85	11.3	1.825	200	3.70
Bismuth	Bi	208.98	271.402	1564	53.144	9.79	13.4	0.122	7.87	107
Cadmium	Cd	112.411	321.069	767	55.2	8.69	30.8	0.231	96.8	6.8
Calcium	Ca	40.078	842	1484	213.1	1.54	22.3	0.647	200	3.42
Cerium	Ce	140.116	799	3443	38.97	6.770	6.3	0.192	11.3	82.8
Cesium	Cs	132.905	28.5	671	15.7	1.873	97	0.242	35.9	20.8
Chromium	Cr	51.996	1907	2671	403.9	7.15	4.9	0.449	93.7	12.6
Cobalt	Co	58.933	1495	2927	274.9	8.86	13.0	0.421	100	5.6
Copper	Cu	63.546	1084.62	2560	208.7	8.96	16.5	0.385	401	1.712
Dysprosium	Dy	162.5	1412	2567	69.85	8.55	9.9	0.173	10.7	92.6
Erbium	Er	167.259	1529	2868	118.98	9.07	12.2	0.168	14.5	86.0
Europium	Eu	151.964	822	1529	60.61	5.24	35.0	0.182	13.9	90.0
Gadolinium	Gd	157.25	1313	3273	61.49	7.90	9.4	0.235	10.5	131
Gallium	Ga	69.723	29.7646	2229	80.10	5.91	18	0.373	40.6	13.6
Gold	Au	196.967	1064.18	2836	63.72	19.3	14.2	0.129	317	2.255
Hafnium	Hf	178.49	2233	4600	152.39	13.3	5.9	0.144	23.0	33.7
Holmium	Ho	164.93	1472	2700	71.30	8.80	11.2	0.165	16.2	81.4
Indium	In	114.818	156.5985	2027	28.66	7.31	32.1	0.233	81.6	8.0
Iridium	Ir	192.217	2446	4428	213.93	22.562[a]	6.4	0.131	147	4.7
Iron	Fe	55.845	1538	2861	247.3	7.87	11.8	0.449	80.2	9.87
Lanthanum	La	138.905	920	3464	44.64	6.15	12.1	0.195	13.4	61.5
Lead	Pb	207.2	327.462	1749	23.04	11.3	28.9	0.130	35.3	21.1
Lithium	Li	6.94	180.50	1342	432	0.534	46	3.582	84.7	9.47
Lutetium	Lu	174.967	1663	3402	106.59	9.84	9.9	0.154	16.4	58.2
Magnesium	Mg	24.305	650	1090	348.9	1.74	24.8	1.023	156	4.48
Manganese	Mn	54.938	1246	2061	235.0	7.3	21.7	0.479	7.82	144
Mercury	Hg	200.59	-38.8290	356.619	11.44	13.5336	60.4	0.140	8.514	96.1
Molybdenum	Mo	95.96	2622	4639	390.6	10.2	4.8	0.251	138	5.47
Neodymium	Nd	144.242	1016	3074	49.50	7.01	9.6	0.190	16.5	64.3
Neptunium	Np	237.0	644	≈3902	13.50	20.2			6.3	
Nickel	Ni	58.693	1455	2913	297.8	8.90	13.4	0.444	90.7	7.12
Niobium	Nb	92.906	2477	4741	323	8.57	7.3	0.265	53.7	15.2

Name	Formula	Mol. wt.	t_m/°C	t_b/°C	$\Delta_{fus}H$/ J g^{-1}	ρ/g cm^{-3}	$10^6 \times \alpha$/ K^{-1}	c_p/ J g^{-1} K^{-1}	λ/ W m^{-1} K^{-1}	ρ_{elec}/ μΩ cm
Osmium	Os	190.23	3033	5008	304.11	22.587[a]	5.1	0.130	87.6	8.1
Palladium	Pd	106.42	1554.8	2963	157.3	12.0	11.8	0.244	71.8	10.73
Platinum	Pt	195.084	1768.2	3825	113.669	21.5	8.8	0.133	71.6	10.7
Plutonium	Pu	244.0	640	3228	11.57	19.7	46.7		6.74	
Polonium	Po	209.0	254	962	47.8	9.20	23.5		20	40
Potassium	K	39.098	63.5	759	59.72	0.89	83.3	0.757	102.4	7.39
Praseodymium	Pr	140.908	931	3520	48.90	6.77	6.7	0.193	12.5	70.0
Promethium	Pm	145.0	1042	≈3000		7.26	11	0.19	15	75
Protactinium	Pa	231.036	1572		53.41	15.4				17.7
Radium	Ra	226.0	696		34.1	5				
Rhenium	Re	186.207	3185	5590	183.02	20.8	6.2	0.137	47.9	17.2
Rhodium	Rh	102.906	1963	3695	258.4	12.4	8.2	0.243	150	4.3
Rubidium	Rb	85.468	39.30	688	25.6	1.53		0.363	58.2	13.1
Ruthenium	Ru	101.07	2333	4147	381.8	12.1	6.4	0.238	117	7.1
Samarium	Sm	150.36	1072	1794	57.33	7.52	12.7	0.196	13.3	94.0
Scandium	Sc	44.956	1541	2836	313.6	2.99	10.2	0.568	15.8	56.2
Silver	Ag	107.868	961.78	2162	104.8	10.5	18.9	0.235	429	1.617
Sodium	Na	22.99	97.794	882.940	113.1	0.97	71	1.228	141	4.88
Strontium	Sr	87.62	777	1377	84.8	2.64	22.5	0.306	35.3	13.4
Tantalum	Ta	180.948	3017	5455	202.10	16.4	6.3	0.140	57.5	13.4
Technetium	Tc	98.0	2157	4262	339.7	11			50.6	
Terbium	Tb	158.925	1359	3230	63.87	8.23	10.3	0.182	11.1	115
Thallium	Tl	204.38	304	1473	20.27	11.8	29.9	0.129	46.1	15
Thorium	Th	232.038	1750	4785	59.52	11.7	11.0	0.118	54.0	14.7
Thulium	Tm	168.934	1545	1950	99.68	9.32	13.3	0.160	16.9	67.6
Tin (white)	Sn	118.71	231.928	2586	60.2	7.287	22.0	0.227	66.6	11.5
Titanium	Ti	47.867	1670	3287	295.6	4.506	8.6	0.524	21.9	39
Tungsten	W	183.84	3414	5555	284.54	19.3	4.5	0.132	174	5.39
Uranium	U	238.029	1135	4131	38.40	19.1	13.9	0.116	27.6	28
Vanadium	V	50.942	1910	3407	422	6.0	8.4	0.489	30.7	20.1
Ytterbium	Yb	173.054	824	1196	44.26	6.90	26.3	0.155	38.5	25.0
Yttrium	Y	88.906	1522	3345	128.1	4.47	10.6	0.298	17.2	59.6
Zinc	Zn	65.38	419.527	907	108.11	7.134	30.2	0.388	116	6.01
Zirconium	Zr	91.224	1854	4406	230.2	6.52	5.7	0.278	22.7	42.9

[a] At 20 °C

Solids

THERMOPHYSICAL PROPERTIES OF STAINLESS STEEL 310

Stainless steel is used in a wide variety of applications, especially at high temperatures. This table gives properties of a reference standard Stainless Steel 310 certified by the U.K. National Physical Laboratory. The properties are:

a: thermal diffusivity in $mm^2 s^{-1}$
c_p: specific heat capacity at constant pressure in $J g^{-1} K^{-1}$
ρ: density in $g cm^{-3}$
k: thermal conductivity in $W m^{-1} K^{-1}$

With kind permission from Springer Science+Business Media: *International Journal of Thermophysics,* 28, 674, 2007, Table II.

Reference

Blumm, J., Lindemann, A. Niedrig, B., and Campbell, R., *Int. J. Thermophys.* 28, 674, 2007.

t/°C	a/$mm^2 s^{-1}$	c_p/$J g^{-1} K^{-1}$	ρ/$g cm^{-3}$	k/$W m^{-1} K^{-1}$
−125	3.170	0.376	7.878	9.39
−100	3.130	0.411	7.871	10.12
−75	3.145	0.435	7.863	10.76
−50	3.170	0.451	7.855	11.23
−25	3.210	0.464	7.846	11.69
0	3.256	0.475	7.838	12.12
25	3.352	0.483	7.829	12.67
50	3.439	0.490	7.820	13.18
101	3.611	0.501	7.801	14.11
150	3.763	0.512	7.782	14.99
200	3.917	0.518	7.762	15.75
250	4.075	0.525	7.742	16.56
300	4.205	0.533	7.722	17.31
350	4.331	0.541	7.701	18.04
400	4.455	0.548	7.681	18.75
450	4.571	0.555	7.660	19.43
500	4.686	0.562	7.639	20.12
550	4.806	0.570	7.618	20.86
600	4.920	0.595	7.596	22.24
651	5.058	0.601	7.574	23.02
701	5.179	0.607	7.551	23.74
750	5.207	0.611	7.529	23.95
800	5.288	0.617	7.506	24.49
850	5.404	0.624	7.483	25.23
901	5.506	0.633	7.460	26.00
950	5.618	0.645	7.436	26.94
1000	5.707	0.655	7.411	27.70

THERMAL CONDUCTIVITY OF METALS AND SEMICONDUCTORS AS A FUNCTION OF TEMPERATURE

This table gives the temperature dependence of the thermal conductivity of several metals and of carbon, germanium, and silicon. For graphite, separate entries are given for the thermal conductivity parallel (‖) and perpendicular (⊥) to the layer planes. The thermal conductivity of all these materials is very sensitive to impurities at low temperatures, especially below 100 K. Therefore, the values given here should be regarded as typical values for a highly purified specimen; the thermal conductivity of different specimens can vary by more than an order of magnitude in the low-temperature range. See Reference 2 for details.

References

1. Ho, C. Y., Powell, R. W., and Liley, P. E., *J. Phys. Chem. Ref. Data*, 1, 279, 1972.
2. White, G. K., and Minges, M. L., *Thermophysical Properties of Some Key Solids*, CODATA Bulletin No. 59, 1985.

Thermal Conductivity in W/cm K

				Carbon (C)						
				Diamond (type)			Pyrolytic graphite			
T/K	Ag	Al	Au	I	IIa	IIb	‖	⊥	Cr	Cu
1	39.4	41.1	5.46						0.402*	42.2
2	78.3	81.8	10.9	0.0138*	0.033*	0.0200*			0.803	84.0
3	115	121	16.1	0.0461	0.111	0.0676			1.20	125
4	147	157	20.9	0.108	0.261	0.160			1.60	162
5	172	188	25.2	0.206	0.494	0.307			2.00	195
6	187	213	28.5	0.344	0.820	0.510			2.39	222
7	193	229	30.9	0.523	1.24	0.778			2.27	239
8	190	237	32.3	0.762	1.77	1.12			3.14	248
9	181	239	32.7	1.05	2.41	1.53			3.50	249
10	168	235	32.4	1.40	3.17	2.03	0.811	0.0116	3.85	243
15	96.0	176	24.6	3.96	8.65	5.66			5.24	171
20	51.0	117	15.8	7.87	16.8	11.2	4.20	0.0397	5.93	108
30	19.3	49.5	7.55	18.8	38.9	26.5	9.86	0.0786	5.49	44.5
40	10.5	24.0	5.15	29.4	65.9	44.0	16.4	0.120	4.25	21.7
50	7.0	13.5	4.21	35.3	92.1	59.1	23.1	0.152	3.17	12.5
60	5.5	8.5	3.74	37.4	112	67.5	29.8	0.173	2.48	8.29
70	4.97	5.85	3.48	36.9	119	69.1	36.6	0.181	2.07	6.47
80	4.71	4.32	3.32	35.1	117	65.7	42.8	0.181	1.84	5.57
90	4.60	3.42	3.28	32.7	109	60.0	47.5	0.176	1.69	5.08
100	4.50	3.02	3.27	30.0	100	54.2	49.7	0.168	1.59	4.82
150	4.32	2.48	3.25	19.5	60.2	32.5	45.1	0.125	1.29	4.29
200	4.30	2.37	3.23	14.1	40.3	22.6	32.3	0.0923	1.11	4.13
250	4.29	2.35	3.21	11.0	29.7	17.0	24.4	0.0711	1.00	4.06
300	4.29	2.37	3.17	8.95	23.0	13.5	19.5	0.0570	0.937	4.01
350	4.27	2.40	3.14	7.55*	18.5*	11.1*	16.2	0.0477	0.929	3.96
400	4.25	2.40	3.11	6.5*	15.4*	9.32*	13.9	0.0409	0.909	3.93
500	4.19	2.36	3.04				10.8	0.0322	0.860	3.86
600	4.12	2.31	2.98				8.92	0.0268	0.807	3.79
800	3.96	2.18	2.84				6.67	0.0201	0.713	3.66
1000	3.79		2.70				5.34	0.0160	0.654	3.52
1200	3.61*		2.55				4.48	0.0134	0.619	3.39
1400							3.84	0.0116	0.588	
1600							3.33	0.0100	0.556	
1800							2.93	0.00895	0.526*	
2000							2.62	0.00807	0.494*	

T/K	Fe	Ge[a]	Mg	Ni	Pb	Pt	Si[a]	Sn	Ti	W
1	1.71	0.274	9.86	2.17	27.9	2.31	0.0693*	183	0.0144*	14.4
2	3.42	2.06	19.6	4.34	44.6	4.60	0.454	323	0.0288*	28.7
3	5.11	5.35	29.0	6.49	35.8	6.79	1.38	297	0.0432	42.8
4	6.77	8.77	37.6	8.59	22.2	8.8	2.97	181	0.0575	56.3
5	8.39	11.6	45.0	10.6	13.8	10.5	5.27	117	0.0719	68.7
6	9.93	13.9	50.8	12.5	8.10	11.8	8.23	76	0.0863	79.5
7	11.4	15.5	54.7	14.2	4.86	12.6	11.7	52	0.101	88.0
8	12.7	16.6	56.7	15.8	3.20	12.9	15.5	36	0.115	93.8
9	13.9	17.3	57.0	17.1	2.30	12.8	19.5	26	0.129	96.8
10	14.8	17.7	55.8	18.1	1.78	12.3	23.3	19.3	0.143	97.1
15	17.0	17.3	41.1	19.5	0.845	8.41	41.6	6.3	0.212	72.0
20	15.4	14.9	27.2	16.5	0.591	4.95	49.8	3.2	0.275	40.5
30	10.0	10.8	12.9	9.56	0.477	2.15	48.1	1.79	0.365	14.4
40	6.23	7.98	7.19	5.82	0.451	1.39	35.3	1.33	0.390	6.92
50	4.05	6.15	4.65	4.00	0.436	1.09	26.8	1.15	0.374	4.27
60	2.85	4.87	3.27	3.08	0.425	0.947	21.1	1.04	0.355	3.14
70	2.16	3.93	2.49	2.50	0.416	0.862	16.8	0.96	0.340	2.58
80	1.75	3.25	2.02	2.10	0.409	0.815	13.4	0.915	0.326	2.29
90	1.50	2.70	1.78	1.83	0.403	0.789	10.8	0.880	0.315	2.17
100	1.34	2.32	1.69	1.64	0.397	0.775	8.84	0.853	0.305	2.08
150	1.04	1.32	1.61	1.22	0.379	0.740	4.09	0.779	0.270	1.92
200	0.94	0.968	1.59	1.07	0.367	0.726	2.64	0.733	0.245	1.85
250	0.865	0.749	1.57	0.975	0.360	0.718	1.91	0.696	0.229	1.80
300	0.802	0.599	1.56	0.907	0.353	0.716	1.48	0.666	0.219	1.74
350	0.744	0.495	1.55	0.850	0.347	0.717	1.19	0.642	0.210	1.67
400	0.695	0.432	1.53	0.802	0.340	0.718	0.989	0.622	0.204	1.59
500	0.613	0.338	1.51	0.722	0.328	0.723	0.762	0.596	0.197	1.46
600	0.547	0.273	1.49	0.656	0.314	0.732	0.619		0.194	1.37
800	0.433	0.198	1.46*	0.676		0.756	0.422		0.197	1.25
1000	0.323	0.174		0.718		0.787	0.312		0.207	1.18
1200	0.283	0.174		0.762		0.826	0.257		0.220	1.12
1400	0.312			0.804		0.871	0.235		0.236	1.08
1600	0.330					0.919	0.221		0.253	1.04
1800	0.345*					0.961			0.270*	1.01
2000						0.994*				0.98

[a] Values below 300 K are typical values.

* Extrapolated.

Solids

THERMAL CONDUCTIVITY OF ALLOYS AS A FUNCTION OF TEMPERATURE

This table lists the thermal conductivity of selected alloys at various temperatures. The indicated compositions refer to weight percent. Since the thermal conductivity is sensitive to exact composition and processing history, especially at low temperatures, these values should be considered approximate.

References

1. Powell, R. L., and Childs, G. E., in *American Institute of Physics Handbook, 3rd Edition*, Gray, D. E., Ed., McGraw-Hill, New York, 1972.
2. Ho, C. Y., et al., *J. Phys. Chem. Ref. Data*, 7, 959, 1978.

Thermal Conductivity in W/m K

Alloy	λ(4 K)/ W m^{-1} K^{-1}	λ(20 K)/ W m^{-1} K^{-1}	λ(77 K)/ W m^{-1} K^{-1}	λ(194 K)/ W m^{-1} K^{-1}	λ(273 K)/ W m^{-1} K^{-1}	λ(373 K)/ W m^{-1} K^{-1}	λ(573 K)/ W m^{-1} K^{-1}	λ(973 K)/ W m^{-1} K^{-1}
Aluminum:								
1100	50	240	270	220	220			
2024	3.2	17	56	95	130			
3003	11	58	140	150	160			
5052	4.8	25	77	120	140			
5083, 5086	3	17	55	95	120			
Duralumin	5.5	30	91	140	160	180		
Bismuth:								
Rose metal		5.5	8.3	14	16			
Wood's metal	4	17	23					
Copper:								
electrolytic tough pitch	330	1300	550	400	390	380	370	350
free cutting, leaded	200	800	460	380	380			
phosphorus, deoxidized	7.5	42	120	190	220			
brass, leaded	2.3	12	39	70	120			
bronze, 68% Cu; 32% Zn	2.3	16	48	92	110			
beryllium	2	17	36	70	90	113	172	
german silver	0.75	7.5	17	20	23	25	30	40
silicon bronze A		3.4	11	23	30			
manganin	0.48	3.2	14	17	22			
constantan	0.9	8.6	17	19	22			
Gold:								
colbalt thermocouple	1.2	8.6	20					
65% Au; 35% Ag		12	24		61	89		
Indium:								
85.5% In; 14.5% Pb	1.9	7.8	24	41				
Iron								
commercial pure iron	15	72	106	82	76	66	54	34
plain carbon steel (AISI 1020)	13	20	58	65	65			
plain carbon steel (AISI 1095)		8.5	31	41	45			
3% Ni; 0.7% Cr; 0.6% Mo		6	22		33	35	36	30
4% Si					20	24	28	26
stainless steel	0.3	2	8	13	14	16	19	25
27% Ni; 15% Cr		1.7	55		11	12	16	21
Lead:								
60% Pb; 40% Sn (soft solder)		28	44					
64.35% Pb; 35.65% In	0.8	3.26	9.1		20.2			
Nickel:								
80% Ni; 20% Cr					12	14	17	23
contracid	0.2	2	7.3	9.5	13			
inconel	0.5	4.2	12.5	13	15	16	19	26
monel	0.9	7.1	15	20	21	24	30	43
Platinum:								
90% Pt; 10% Ir					31	31.4		
90% Pt; 10% Rh					30.1	30.5		
Silver:								
silver solder		12	34	58				
normal Ag thermocouple	48	230	310					
Tin								
60% Sn; 40% Pb	16	55	51					
Titanium:								
5.5% Al; 2.5% Sn; 0.2% Fe		1.8	4.3	6.4	7.8	8.4	10.8	
4.7% Mn; 3.99% Al; 0.14% C		1.7	4.5	6.5	8.5			

THERMAL CONDUCTIVITY OF CRYSTALLINE DIELECTRICS

This table lists the thermal conductivity, *k*, of a number of crystalline dielectrics, including some that find use as optical materials. Values are given at temperatures for which data are available.

Reference

Powell, R. L., and Childs, G. E., in *American Institute of Physics Handbook, 3rd Edition*, Gray, D. E., Ed., McGraw-Hill, New York, 1972.

Thermal Conductivity *k* at Various Temperatures

Material	Formula or other information	*T*/K	*k*/ W m^{-1} K^{-1}
Aluminum oxide (α)	Al_2O_3 (sintered)	4.2	0.5
		20	23
		77	150
		194	48
		273	35
		373	26
		973	8
Aluminum sulfide	As_2S_3 (glass)	283	0.16
		323	0.21
		373	0.27
Ammonium chloride	NH_4Cl	77	17
		194	23
		230	38
		273	27
Ammonium dihydrogen phosphate	∥ to optic axis	315	0.71
		339	0.71
	⊥ to optic axis	313	1.26
		342	1.34
Argon	Ar	8	6.0
		10	3.7
		20	1.4
		77	0.31
Barium fluoride	BaF_2	225	20
		260	13.4
		305	10.9
		370	10.5
Barium titanate	$BaTiO_3$	5	4.2
		30	24.0
		40	25.0
		100	12.0
		250	4.8
		300	6.2
Beryl	Al, Be silicate	315	6.4
Beryllium oxide	BeO	4.2	0.3
		20	16
		77	270
		373	210
		573	120
		1273	29
Bismuth telluride	Bi_2Te_3	80	6.4
		204	2.8
		303	3.6
		370	4.6
Boron nitride	BN	1047	36.2
		1475	22.7
		1928	21.9
		2111	18.5
Cadmium telluride	CdTe	160	7.0
		297	3.6
		422	2.9
Calcium carbonate (calcite)	∥ to *c* axis	83	25
		273	5.5
	⊥ to *c* axis	83	17
		194	6.5
		273	4.6
		373	3.6
Calcium fluoride	CaF_2 (fluorite)	83	39
		223	18
		273	10
		323	9.2
		373	9
Calcium tungstate	$CaWO_4$ (scheelite)	422	11.3
Carbon (diamond)	Type 1	4.2	13
		20	800
		77	3550
		194	1450
		273	1000
Cesium bromide	CsBr	223	1.2
		273	0.94
		323	0.81
		373	0.77
Cesium iodide	CsI	223	1.4
		273	1.2
		323	1
		373	0.95
Copper(I) oxide	Cu_2O (cuprite)	102	3.74
		163	7.76
		299	5.58
		360	4.86
Garnet	Al,Fe silicate	315	35.8
		358	35.4
		377	35.6
Glass	phoenix	4.2	0.095
		20	0.13
		77	0.37
	plastic perspex	4.2	0.058
		20	0.074
	pyrex	77	0.44
		194	0.88
		273	1
Helium	^3He (high pressure)	0.6	25
		1	2
		1.5	0.57
		2	0.21
	^4He (high pressure)	0.5	42
		0.8	120
		1	24
		2	0.18

Material	Formula or other information	*T*/K	*k*/ W m^{-1} K^{-1}
Hydrogen	H_2 (para + 0.5% ortho)	2.5	100
		3	150
		4	200
		6	30
		10	3
Ice	H_2O	173	3.5
		223	2.8
		273	2.2
Iodine	I_2	300	0.45
		325	0.42
		350	0.4
Iron(II,III) oxide	Fe_3O_4 (magnetite)	4.5	27.4
		20.5	293.0
		126.5	7.4
		304	7.0
Krypton	Kr	4.2	0.48
		10	1.7
		20	1.2
		77	0.36
Lanthanum fluoride	LaF_3	78	7.8
		197	5.0
		274	5.4
Lithium fluoride	LiF	4.2	620
		20	1800
		77	150
Manganese(II) oxide	MnO	4.2	0.25
		40	55
		120	8
		573	3.5
Neon	Ne	2	3.0
		3	4.6
		4.2	4.2
		10	0.8
		20	0.3
Nickel(II) oxide	NiO	4.2	5.9
		40	400
		194	82
Potassium bromide	KBr	2	150
		4.2	360
		100	12
		273	5
		323	4.8
		373	4.8
Potassium chloride	KCl	4.2	500
		25	140
		80	35
		194	10
		273	7.0
		323	6.5
		373	6.3
Potassium iodide	KI	4.2	700
		80	13
		194	4.6
		273	3.1
Sapphire	36° to *c* axis	4.2	110
		20	3500
		35	6000
		77	1100
	⊥ to *c* axis	373	2.6
		523	3.9
		773	5.8
Silicon dioxide (α-quartz)	∥ to *c* axis	20	720
		194	20
		273	12
	⊥ to *c* axis	20	370
		194	10
		273	6.8
Silicon dioxide (vitreous)	SiO_2 (fused silica)	4.2	0.25
		20	0.7
		77	0.8
		194	1.2
		273	1.4
		373	1.6
		673	1.8
Silver(I) chloride	AgCl	223	1.3
		273	1.2
		323	1.1
		373	1.1
Sodium chloride	NaCl	4.2	440
		20	300
		77	30
		273	6.4
		323	5.6
		373	5.4
Sodium fluoride	NaF	5	1100
		50	250
		100	90
Spinel	$MgO{\cdot}Al_2O_3$	373	13
		773	8.5
Strontium titanate	$SrTiO_3$	5	2.4
		30	21.0
		40	19.2
		100	18.5
		250	12.5
		300	11.2
Thallium(I) bromide	TlBr	316	0.59
Thallium(I) chloride	TlCl	311	0.75
	∥ to optic axis	4.2	200
		20	1000
		273	13
	⊥ to optic axis	4.2	160
		20	690
		273	9
Topaz	∥ to *c* axis	315	17.7
		358	15.6
		417	13.3
Tourmaline	∥ to *c* axis	398	2.9
		540	3.2
		723	3.5

THERMAL CONDUCTIVITY OF CERAMICS AND OTHER INSULATING MATERIALS

Thermal conductivity values λ for ceramics, refractory oxides, and miscellaneous insulating materials are given here. The thermal conductivity refers to samples with the density indicated in the second column. Since most of these materials are highly variable, the values should only be considered as a rough guide.

References

1. Powell, R. L., and Childs, G. E., in *American Institute of Physics Handbook, 3rd Edition*, Gray, D. E., Ed., McGraw-Hill, New York, 1972.
2. Perry, R. H., and Green, D., *Perry's Chemical Engineers' Handbook, Sixth Edition*, McGraw-Hill, New York, 1984.

Thermal Conductivity at Specified Values of Density and Temperature

Material	ρ/g cm^{-3}	t/°C	λ/W m^{-1} K^{-1}
Alumina (Al_2O_3)	3.8	100	30
	3.8	400	13
	3.8	1300	6
	3.8	1800	7.4
	3.5	100	17
	3.5	800	7.6
Al_2O_3 + MgO		100	15
		400	10
		1000	5.6
Asbestos	0.4	-100	0.07
	0.4	0	0.09
	0.4	100	0.10
Asbestos + 85% MgO	0.3	30	0.08
Asphalt	2.1	20	0.06
Beryllia (BeO)	2.8	100	210
	2.8	400	90
	2.8	1000	20
	2.8	1800	15
	1.85	50	64
	1.85	200	40
	1.85	600	23
Brick, dry	1.54	0	0.04
Brick, refractory, alosite		1000	1.3
Brick, refractory, aluminous	1.99	400	1.2
	1.99	1000	1.3
Brick, refractory, diatomaceous	0.77	100	0.2
	0.77	500	0.24
	0.4	100	0.08
	0.4	500	0.1
Brick, refractory, fireclay	2	400	1
	2	1000	1.2
Brick, refractory, silicon carbide	2	200	2
	2	600	2.4
Brick, refractory, vermiculite	0.77	200	0.26
	0.77	600	0.31
Calcium oxide		100	16
		400	9
		1000	7.5
Cement mortar	2	90	0.55
Charcoal	0.2	20	0.055
Coal	1.35	20	0.26
Concrete	1.6	0	0.8
Cork	0.05	0	0.03
	0.05	100	0.04
	0.35	0	0.06
	0.35	100	0.08
Cotton wool	0.08	30	0.04
Diatomite	0.2	0	0.05
	0.2	400	0.09
	0.5	0	0.09
	0.5	400	0.16
Ebonite	1.2	0	0.16
Felt, flax	0.2	30	0.05
	0.3	30	0.04
Fuller's earth	0.53	30	0.1
Glass wool	0.2	20[a]	0.005
	0.2	50	0.04
	0.2	100	0.05
	0.2	300	0.08
Graphite, 100 mesh	0.48	40	0.18
Graphite, 20-40 mesh	0.7	40	1.29
Linoleum cork	0.54	20	0.08
Magnesia (MgO)		100	36
		400	18
		1200	5.8
		1700	9.2
MgO + SiO_2		100	5.3
		400	3.5
		1500	2.3
Mica, muscovite		100	0.72
		300	0.65
		600	0.69
Mica, phlogopite		100	0.66
Mica, Canadian		300	0.19
		600	0.2
Micanite		30	0.3
Mineral wool	0.15	30	0.04
Perlite, expanded	0.1	20[a]	0.002
Plastics, bakelite	1.3	20	1.4
Plastics, celluloid	1.4	30	0.02
Plastics, polystyrene foam	0.05	20[a]	0.033
Plastics, mylar foil	0.05	20[a]	0.0001
Plastics, nylon		-253	0.10
		-193	0.23
		25	0.30
Plastics, polytetrafluoroethylene		-253	0.13
		-193	0.16
		25	0.26
		230	2.5
Plastics, urethane foam	0.07	20	0.06
Porcelain		90	1
Rock, basalt		20	2
Rock, chalk		20	0.92
Rock, granite	2.8	20	2.2
Rock, limestone	2	20	1
Rock, sandstone	2.2	20	1.3
Rock, slate, ⊥		95	1.4
Rock, slate, ǁ		95	2.5

Material	ρ/g cm^{-3}	t/°C	λ/W m^{-1} K^{-1}
Rubber, sponge	0.2	20	0.05
Rubber, 92 percent		25	0.16
Sand, dry	1.5	20	0.33
Sawdust	0.2	30	0.06
Shellac		20	0.23
Silica aerogel	0.1	20[a]	0.003
Snow	0.25	0	0.16
Steel wool	0.1	55	0.09
Thoria (ThO_2)		100	10
		400	5.8
		1500	2.4
Titanium dioxide		100	6.5
		400	3.8
		1200	3.3
Uranium dioxide		100	9.8
		400	5.5
		1000	3.4
Wood, balsa, ⊥	0.11	30	0.04

Material	ρ/g cm^{-3}	t/°C	λ/W m^{-1} K^{-1}
Wood, fir, ⊥	0.54	20	0.14
Wood, fir, ‖	0.54	20	0.35
Wood, oak		20	0.16
Wood, plywood		20	0.11
Wood, pine, ⊥	0.45	60	0.11
Wood, pine, ‖	0.45	60	0.26
Wood, walnut, ⊥	0.65	20	0.14
Wool	0.09	30	0.04
Zinc oxide		200	17
		800	5.3
Zirconia (ZrO_2)		100	2
		400	2
		1500	2.5
Zirconia + silica		200	5.6
		600	4.6
		1500	3.7

[a] 200 to 20 °C

THERMAL CONDUCTIVITY OF GLASSES

This table gives the composition of various types of glasses and the thermal conductivity k as a function of temperature. Because of the variability of glasses, the data should be regarded as only approximate.

Composition and Thermal Conductivity of Glasses

Type of glass	wt% SiO_2	Other oxides	wt% Other oxides	t/°C	k/W m^{-1} K^{-1}
Vitreous silica	100			-150	0.85
				-100	1.05
				-50	1.20
				0	1.30
				50	1.40
				100	1.50
Vycor glass	96	B_2O_3	3	-100	1.00
				0	1.25
				100	1.40
Pyrex type chemically resistant borosilicate glasses	80–81	B_2O_3	12–13	-100	0.90
		Na_2O	4	0	1.10
		Al_2O_3	2	100	1.25
Crown glasses					
Borosilicate crown glasses	60–65	B_2O_3	15–20	-100	0.65–0.75
				0	0.90–0.95
				100	1.00–1.05
	65–70	B_2O_3	10–15	-100	0.75–0.80
				0	0.95–1.00
				100	1.05–1.15
	70–75	B_3O_3	5–10	-100	0.80–0.85
				0	1.05–1.10
				100	1.15–1.20
Zinc crown glasses (i)	55–65	ZnO	5–15	-100	0.88–0.92
		Remainder:		0	1.10–1.15
		B_2O_3, Al_2O_3		100	1.15–1.25
		ZnO	5–15	-100	0.60–0.70
		Remainder:		0	0.70–0.90
		Na_2O, K_2O		100	0.85–0.95
		ZnO	15–25	-100	0.88–0.92
		Remainder:		0	1.10–1.15
		B_2O_3, Al_2O_3		100	1.15–1.20
		ZnO	15–25	-100	0.65–0.80
		Remainder:		0	0.85–0.95
		Na_2O, K_2O		100	0.90–1.05
Zinc crown glasses (ii)	65–75	ZnO	5–15	-100	0.88–0.92
		Remainder:		0	1.15–1.15
		B_2O_3, Al_2O_3		100	1.20–1.30
		ZnO	5–15	-100	0.70–0.85
		Remainder:		0	0.90–1.05
		Na_2O, K_2O		100	1.00–1.15

Type of glass	wt% SiO_2	Other oxides	wt% Other oxides	t/°C	k/W m^{-1} K^{-1}
		ZnO	15–25	-100	0.90–0.95
		Remainder:		0	1.15–1.15
		B_2O_3, Al_2O_3		100	1.20–1.25
		ZnO	15–25	-100	0.65–0.85
		Remainder:		0	0.85–1.00
		Na_2O, K_2O		100	1.05–1.20
Barium crown glasses	31	B_2O_3	12	-100	0.55
		Al_2O_3	8	0	0.70
		BaO	48	100	0.80
	41	B_2O_3	6	-100	0.60
		Al_2O_3	2	0	0.75
		ZnO	8	100	0.85
		BaO	43		
	47	B_2O_3	4	-100	0.65
		Na_2O	1	0	0.75
		K_2O	7	100	0.90
		ZnO	8		
		BaO	32		
	65	B_2O_3	2	-100	0.70
		Na_2O	5	0	0.90
		K_2O	15	100	1.00
		ZnO	2		
		BaO	10		
Borate glasses					
Borate flint glass	9	B_2O_3	36	-100	0.55
		Na_2O	1	0	0.65
		K_2O	2	100	0.80
		PbO	36		
		Al_2O_3	10		
		ZnO	6		
Borate flint glass	0	B_2O_3	56	-100	0.50
		Al_2O_3	12	0	0.65
		PbO	32	100	0.85
Borate flint glass	0	B_2O_3	43	-100	0.40
		Al_2O_3	5	0	0.55
		PbO	52	100	0.70
Borate glass	4	B_2O_3	55	-100	0.65
		Al_2O_3	14	0	0.80
		PbO	11	100	0.90
		K_2O	4		
		ZnO	12		
Borate crown glass	0	B_2O_3	64	-100	0.50
		Na_2O	8	0	0.65
		K_2O	3	100	0.85
		BaO	4		
		PbO	3		
		Al_2O_3	18		
Light borate crown glass	0	B_2O_3	69	-100	0.55
		Na_2O	8	0	0.70
		BaO	5	100	0.90

Type of glass	wt% SiO_2	Other oxides	wt% Other oxides	t/°C	k/W m⁻¹ K⁻¹
		Al_2O_3	18		
Zinc borate glass	0	B_2O_3	40	-100	0.65
		ZnO	60	0	0.75
				100	0.85
Phosphate crown glasses					
Potash phosphate glass	0	P_2O_5	70	0	0.75
		B_2O_3	3	100	0.85
		K_2O	12		
		Al_2O_3	10		
		MgO	4		
Baryta phosphate glass	0	P_2O_5	60	45	0.75
		B_2O_3	3		
		Al_2O_3	8		
		BaO	28		
Soda-lime glasses	75	Na_2O	17	-100	0.75
		CaO	8	0	0.95
				100	1.10
	75	Na_2O	12	-100	0.90
		CaO	13	0	1.10
				100	1.15
	72	Na_2O	15	-100	0.80
		CaO	11	0	1.00
		Al_2O_3	2	100	1.15
	65	Na_2O	25	-100	0.65
		CaO	10	0	0.85
				100	0.95
	65	Na_2O	15	-100	0.85
		CaO	20	0	1.00
				100	1.10
	60	Na_2O	20	-100	0.75
		CaO	20	0	0.90
				100	1.00
Other crown glasses					
Crown glass	75	Na_2O	9	-100	0.80
		K_2O	11	0	1.00
		CaO	5	100	1.10
High dispersion crown glass	68	Na_2O	16	-100	0.65
		ZnO	3	0	0.85
		PbO	13	100	1.00
Miscellaneous flint glasses					
(i) Silicate flint glasses	65	PbO	25	-100	0.65–0.70
Light flint glass		Others	10	0	0.88–0.92
				100	1.00–1.05
Light flint glass	55	PbO	35	-100	0.60–0.65
		Others	10	0	0.75–0.85
				100	0.88–0.92
Ordinary flint glass	45	PbO	45	-100	0.50–0.60
		Others	10	0	0.65–0.75

Type of glass	wt% SiO_2	Other oxides	wt% Other oxides	t/°C	k/W m^{-1} K^{-1}
				100	0.80–0.85
Heavy flint glass	35	PbO	60	-100	0.45–0.50
		Others	5	0	0.60–0.65
				100	0.70–0.75
Very heavy flint glass	25	PbO	73	-100	0.40–0.45
		Others	2	0	0.55–0.60
				100	0.63–0.67
Very heavy flint glass	20	PbO	80	-100	0.40
				0	0.50
				100	0.60
(ii) Borosilicate flint glass	33	B_2O_3	31	-100	0.65
		PbO	25	0	0.85
		Al_2O_3	7	100	0.95
		K_2O	3		
		Na_2O	1		
(iii) Barium flint glass	50	BaO	24	-100	0.60
		PbO	6	0	0.70
		K_2O	8	100	0.85
		Na_2O	3		
		ZnO	8		
		Sb_2O_3	1		
Other glasses					
Potassium glass	59	K_2O	33	50	0.88–0.92
		CaO	8		
Iron glasses	63	Fe_2O_3	10	-100	0.80
		Na_2O	17	0	0.95
		MgO	4	100	1.05
		CaO	3		
		Al_2O_3	2		
	67	Fe_2O_3	15	0	0.88–0.92
		Na_2O_3	18	100	1.00–1.05
	62	Fe_2O_3	20	0	0.85–0.90
		Na_2O	18	100	0.95–1.00
Rock glasses					
Obsidian				0	1.35
Artificial diabase				100	1.25

Solids

THERMOELECTRIC PROPERTIES OF METALS AND SEMICONDUCTORS

Lev I. Berger

There are three thermoelectric phenomena that result from correlation between propagation of heat through a conductor and displacement of the current carriers in the conductor. The Seebeck effect (Ref. 1) consists of formation of an electric current in an electrical circuit formed by two dissimilar conductors if the contacts between the conductors are held at different temperatures. A reverse phenomenon, the Peltier effect (Ref. 2), consists of formation of a temperature difference between the contacts in a circuit of this type if an electric current is created in the circuit by an external current source to which the circuit is connected. W. Thomson (Lord Kelvin), who explained both effects (Refs. 3,4), predicted and experimentally confirmed the existence of another thermoelectric phenomenon, named the Thomson effect, which consists of absorption or release of heat in a uniform conductor with a current passing through it when a temperature gradient (positive or negative) is present along the current direction.

The electromotive force, ΔU, which creates the Seebeck current in the circuit, is the algebraic sum of the emf's created in each of the conductors, and is proportional to the temperature difference, ΔT, between the electrical contact points: $\Delta U = \Delta U_1 + \Delta U_2 = \alpha_1 \Delta T + \alpha_2 \Delta T$. The coefficient of proportionality, α, called the Seebeck coefficient or thermoelectric power or thermal electromotive force (thermal emf), of each of the two materials depends on the electrical properties and temperature of the material. The Peltier effect is measured by the amount of heat, ΔQ, released or absorbed in a unit of time (in addition to the Joule heat) at a contact of two dissimilar conductors with electric current ΔI passing through the contact: $\Delta Q = \Pi \cdot \Delta I$. Thomson showed that $\Pi = \alpha T$. The Thomson effect's heat, dQ, released or absorbed in a unit of time along a part of a conductor of length dx is proportional to the current magnitude I, the temperature gradient along the conductor $\partial T/\partial x$, and the increment dx: $dQ = \tau I(\partial T/\partial x)dx$. Thomson showed that the magnitude of the coefficient of proportionality, τ, later named the Thomson coefficient, depends on only the properties of the conductor and the ambient temperature and correlates with the other thermoelectric parameters of a material through the equation $\tau = T(\partial \alpha/\partial T)$.

Another thermoelectric phenomenon, called the Bridgman effect or the internal Peltier effect (Ref. 5), occurs when an electric current passes through an anisotropic crystal, resulting in absorption or liberation of heat because of non-uniformity in current distribution.

In view of the correlations between α, Π, and τ, we need only to present data for one of these parameters, namely, thermal emf α and its dependence on temperature. These values are presented below, first for metals and then for semiconductors. In accordance with modern theory of solids, thermal emf in semiconductors is up to three or even four orders of magnitude higher than that in metals (Ref. 9).

The accuracy of the data presented in this table is dependent on a number of factors. The thermal emf of a material is sensitive to negligibly small amounts of impurities in the material, which may be below the limits of sensitivity of the chemical analysis; to orientation of crystal grains in a sample of the material, and to thermal processing of the material.

References

1. Seebeck, T. J., *Abhand. Deut. Akad. Wiss. Berlin*, 265–373, 1822.
2. Peltier, J. C. A, *Ann. Chem.*, LVI, 371–387, 1834.
3. Thomson, W., *Proc. Roy. Soc. Edinburgh*, 91–98, 1851.
4. Thomson, W., *Math. and Phys. Papers, Cambridge*, 1, 558, 1882; 2, 306, 1882.
5. Bridgman, P. W., *Proc. Natl. Acad. Sci. USA*, 13(2), 46–50, 1927; *Phys. Rev.*30, 911–921 (1927).
6. Blatt, F. J., *Thermoelectric Power of Metals*. Plenum Press, NY 1976.
7. Foiles, C. L., Thermopower of Pure Metals and Dilute Alloys, in *Landolt–Bornstein. Numerical Data and Functional Relationships in Science and Technology. New Series.* Group III, v. 15, Metals. Springer-Verlag, NY, 1985.
8. Burkov, A. T., and Vedernikov, M. V., in *CRC Handbook of Thermoelectrics*, D. M. Rowe, Ed., CRC Press, Boca Raton, FL, 1995, pp. 387–399.
9. Ioffe, A. F., *Semiconductor Thermoelements and Thermoelectric Cooling*, Infosearch Ltd., 1957.
10. Berger, L. I., and Prochuchan, V. D., *Ternary Diamond-Like Semiconductors*, Cons. Bureau, Plenum Press, New York, 1969.
11. Rowe, D. M., Ed., *Thermoelectrics Handbook Macro to Nano*, Taylor & Francis, Boca Raton, 2006.
12. Berger, L. I., *Semiconductor Materials*, CRC Press, Boca Raton, FL, 1996.
13. Glazov, V. M., Tshizhevskaya, S. N., and Glagoleva, N. N., *Liquid Semiconductors*, Nauka Publ. House, Moscow, 1967.
14. Shay, J. L., and Wernick, J. H., *Ternary Chalcopyrite Semiconductors: Growth, Electronic Properties and Applications*, Pergamon Press, New York, 1975.
15. Heikes, R. R., and Ure, R. W., *Thermoelectricity: Science and Engineering*, Interscience Publ., New York, 1961.
16. Goland, A. N., and Ewald, A. W., *Phys. Rev.* 104, 948 (1956).
17. Tauc J., *Photo and Thermoelectric Effects in Semiconductors*, Pergamon, New York, 1962.
18. Dugdale, J. S., *The Electrical Properties of Metals and Alloys*, Edward Arnold, London, 1977.
19. Rowe, D. M., Ed., *CRC Handbook of Thermoelectrics*, CRC Press, Boca Raton, FL, 1994.

Thermoelectric Properties of Elemental Metals

	Thermal emf α(*T*) in μV/K at Temperature *T*				
	100 K	300 K	500 K	1000 K	1500 K
Ag	0.73	1.51	2.82	7.95	
Al	−2.2	−1.66	1.96		
Au	0.82	1.94	2.86	3.85	
Ba	−4	12.1	28.5		
Be	−2.5	1.7	2.7	7.9	
Ca	1.05	10.3	17.1		
Cd	−0.05	2.55			
Ce	13.6	6.2	5.2	−4.8	
Co	−8.43	−30.8	−44.8	−35.9	−7.8
Cr	5	21.8	16.6	17.9	5.7
Cs		−0.9			
Cu	1.19	1.83	2.83	5.36	
Dy	−4.1	−1.8	0.9	2.3	
Er	−3.8	−0.1	1.9	4.2	
Eu	5.3	24.5	46		
Fe	11.6	15	3	0.4	
Ga	0.5				
Gd	−4.6	−1.6	−0.5	−0.8	
Hf	0	5.5	5.7	−0.5	
Ho	−6.7	−1.6	1.4	2.8	

	Thermal emf α(T) in μV/K at Temperature T				
	100 K	300 K	500 K	1000 K	1500 K
In	0.56	1.68			
Ir	1.42	0.86	−0.1	−2.7	−5.7
K	−5.2	−13.7			
La	0.1	1.7	2	−1.7	
Li	4.3				
Lu	−6.9	−4.3	−2.6	0	
Mg	−2.1	−1.46			
Mn	−2.5	−9.8	−8.4	−1.5	
Mo	0.1	5.6	11.4	17.4	13.7
Na	−2.6	−6.3			
Nb	1.05	−0.44	−1.1	0.45	3.2
Nd	−4	−2.3	0	−1.2	
Ni	−8.5	−19.5	−25.8	−29.9	
Np	8.9	−3.1			
Os	−3.2	−4.4	−4.7	−6.3	−8.5
Pb	−0.58	−1.05	−1.5		
Pd	1.1	−10.7	−16.3	−32.3	−46.4
Pt	−4.1	−5.3	−7.9	−8.2	
Pu	12				
Rb	−3.6	−10			

	Thermal emf α(T) in μV/K at Temperature T				
	100 K	300 K	500 K	1000 K	1500 K
Re	−1.4	−5.9	−5.9	−1.9	1.8
Rh	0.8	0.6	0.5	−1.5	
Ru	0.3	−1.4	−1.8	−4.2	−7.5
Sc	−14.3	−19	−17.5	−5.4	10.2
Sm	0.7	1.2	0.6	−3	
Sn	−0.04	−1			
Sr	−3	1.1	4.2		
Ta	0.7	−1.9	−2.3	1.6	7.2
Tb	−1.6	−1	0.3	0.6	
Th	0.6	−3.2	−9.2	−14.3	−10.4
Ti	−2	9.1	5.3	−3.1	−0.5
Tl	0.6	0.3	−1.5		
Tm	−1.3	1.9	2.7	2.2	
U	3	7.1	11	16.7	
V	2.9	0.23	1.1	4.6	
W	−4.4	0.9	9	19.8	21.3
Y	−5.1	−0.7	0.3	2.9	6.6
Yb	5.1	30	20.3	12.3	
Zn	0.7	2.4			
Zr	4.4	8.9	4.6	−3	1.1

Thermoelectric Properties of Selected Semiconductors; Values near Room Temperature Unless Otherwise Indicated

Material	α/μV K⁻¹	Material	α/μV K⁻¹	Material	α/μV K⁻¹
		Elemental Semiconductors			
B	600 (500 K)	n-Si	300	p-Si	−500
n-Ge	600	p-Ge	−830	α-Sn	−40 (250 K)
		I-VI Compounds			
Cu_2S	327	Cu_2Se	135	Cu_2Te	40
Ag_2Te	120				
		II-VI Compounds			
ZnO	300	CdS	700	ZnSe	55
CdSe	200				
		III-V Compounds			
GaN	70	GaP	1200	InP	−400
AlAs	70	n-GaAs	380	p-GaAs	−310
InAs	200	AlSb	500	n-GaSb	250
p-GaSb	−55	n-InSb	240	p-InSb	200
		V-VI Compounds			
Sb_2Te_3	110	n-Bi_2Te_3	224	p-Bi_2Te_3	−227
		I-III-VI Compounds			
$CuAlS_2$	50	$AgInSe_2$	−370	$CuTlTe_2$	80
CuGaSe	40	$AgTlSe_2$	800	$AgAlTe_2$	321
$CuInSe_2$	340	$CuGaTe_2$	340	$AgGaTe_2$	950
$CuTlSe_2$	−5	$CuInTe_2$	260	$AgInTe_2$	298
$AgGaSe_2$	90	$CuTlTe_2$	80		
		I-IV-VI Compounds			
Cu_2GeS_3	300	Cu_2GeSe_3	100	Cu_2GeTe_3	10
Cu_2SnS_3	600	Cu_2SnSe_3	250	Cu_2SnTe_3	30
		I-V-VI Compounds			
Cu_3AsS_4	130	Cu_3AsSe_4	120	Cu_3SbSe_4	200
		II-IV-V Compounds			
$ZnGeP_2$	1200	$ZnSiAs_2$	1100	$CdGeAs_2$	190
$CdSnAs_2$	600				

FERMI ENERGY AND RELATED PROPERTIES OF METALS

Lev I. Berger

In the classical Drude theory of metals, the Maxwell-Boltzmann velocity distribution of electrons is used. It states that the number of electrons per unit volume with velocities in the range of $d\bar{v}$ about any magnitude $\bar{v}$ at temperature T is

$$f_B(\bar{v})d\bar{v} = n\left(\frac{m}{2\pi k_B T}\right)\exp\left(-\frac{mv^2}{2k_B T}\right)d\bar{v}$$

where n is the total number of conduction electrons in a unit volume of a metal, m is the free electron mass, and k_B is the Boltzmann constant. In an attempt to explain a substantial discrepancy between the experimental data on the specific heat of metals and the values calculated on the basis of the Drude model, Sommerfeld suggested a model of the metal in which the Pauli exclusion principle is applied to free electrons. In this case, the Maxwell-Boltzmann distribution is replaced by the Fermi-Dirac distribution:

$$f(\bar{v})d\bar{v} = 2\left(\frac{m}{h}\right)^3 d\bar{v}\left\{\exp\left[\left(\frac{mv^2}{2} - k_B T_0\right)\Big/ k_B T\right] + 1\right\}^{-1}$$

Here h is the Planck constant and T_0 is a characteristic temperature which is determined by the normalization condition

$$n = \int d\bar{v}\cdot f(\bar{v})$$

The magnitude of T_0 is quite high; usually, $T_0 > 10^4$ K. So, at common temperatures ($T < 10^3$ K), the free electron density of a metal is much smaller than in the case of the Maxwell-Boltzmann distribution. This allows us to explain why the experimental data on specific heat for metals are close to those for insulators.

The maximum kinetic energy the electrons of a metal may possess at $T = 0$ K is called the Fermi energy, e.g.,

$$E_F = \frac{\hbar^2 k_F^2}{2m} = \left(\frac{e^2}{2k_B}\right)(k_F r_B)^2$$

where k_F is the Fermi momentum or the Fermi wave vector

$$k_F = (3\pi^2 n)^{1/3}$$

e is the electron charge, and r_B is the Bohr radius

$$r_B = \hbar^2/me^2 = 0.529\cdot 10^{-10}\ \text{m}$$

Another, more common expression for the Fermi energy is

$$E_F = \tfrac{1}{2} m v_F^2$$

where $v_F = \hbar k_F/m$ is the Fermi velocity which can be expressed using the concept of the electron radius, r_s. It is equal to radius of a sphere occupied by one free electron. If the total volume of a metal sample is V and the number of conduction electrons in this volume is N, then the volume per electron is equal to

$$\frac{V}{N} = \frac{1}{n} = \frac{4}{3}\pi r_S^3$$

and

$$r_S = \left(\frac{3}{4\pi n}\right)^{1/3}$$

The following table contains information pertinent to the Sommerfeld model for some metals. The magnitudes of T_0 are calculated using the expression

$$T_0 = \frac{E_F}{k_B} = \frac{58.2\cdot 10^4}{(r_S/r_B)^2}\ \text{K}$$

Fermi Energy and Other Ground-State Properties of the Electron Gas in Metals[d]

Metal	Name	Valency	$n/10^{28}$ m^{-3}	r_S/pm	r_S/r_B	E_F/eV	$T_0/10^4$ K	$k_F/10^{10}$ m^{-1}	$v_F/10^6$ m s^{-1}
Li[a]	Lithium	1	4.70	172	3.25	4.74	5.51	1.12	1.29
Na[b]	Sodium	1	2.65	208	3.93	3.24	3.77	0.92	1.07
K[b]	Potassium	1	1.40	257	4.86	2.12	2.46	0.75	0.86
Rb[b]	Rubidium	1	1.15	275	5.20	1.85	2.15	0.70	0.81
Cs[b]	Cesium	1	0.91	298	5.62	1.59	1.84	0.65	0.75
Cu	Copper	1	8.47	141	2.67	7.00	8.16	1.36	1.57
Ag	Silver	1	5.86	160	3.02	5.49	6.38	1.20	1.39
Au	Gold	1	5.90	159	3.01	5.53	6.42	1.21	1.40
Be	Beryllium	2	24.7	99	1.87	14.3	16.6	1.94	2.25
Mg	Magnesium	2	8.61	141	2.66	7.08	8.23	1.36	1.58
Ca	Calcium	2	4.61	173	3.27	4.69	5.44	1.11	1.28
Sr	Strontium	2	3.55	189	3.57	3.93	4.57	1.02	1.18
Ba	Barium	2	3.15	196	3.71	3.84	4.23	0.98	1.13
Nb	Niobium	1	5.56	163	3.07	5.32	6.18	1.18	1.37
Fe	Iron	2	17.0	112	2.12	11.1	13.0	1.71	1.98
Mn[c]	Manganese	2	16.5	113	2.14	10.9	12.7	1.70	1.96
Zn	Zinc	2	13.2	122	2.30	9.47	11.0	1.58	1.83
Cd	Cadmium	2	9.27	137	2.59	7.47	8.88	1.40	1.62

Metal	Name	Valency	$n/10^{28}\ m^{-3}$	r_S/pm	r_S/r_B	E_F/eV	$T_0/10^4$ K	$k_F/10^{10}\ m^{-1}$	$v_F/10^6\ m\ s^{-1}$
Hg[a]	Mercury	2	8.65	140	2.65	7.13	8.29	1.37	1.58
Al	Aluminum	3	18.1	110	2.07	11.7	13.6	1.75	2.03
Ga	Gallium	3	15.4	118	2.19	10.4	12.1	1.66	1.92
In	Indium	3	11.5	127	2.41	8.63	10.0	1.51	1.74
Tl	Thallium	3	10.5	131	2.48	8.15	9.46	1.46	1.69
Sn	Tin (white)	4	14.8	117	2.22	10.2	11.8	1.64	1.90
Pb	Lead	4	13.2	122	2.30	9.47	11.0	1.58	1.83
Bi	Bismuth	5	14.1	119	2.25	9.90	11.5	1.61	1.87
Sb	Antimony	5	16.5	113	2.14	10.9	12.7	1.70	1.96

[a] At 78 K.
[b] At 5 K.
[c] α-phase.
[d] The data in the table are for atmospheric pressure and room temperature unless otherwise noted.

References

1. Drude, P., *Ann. Physik*, 1, 566, 1900; ibid., 3, 369, 1900.
2. Sommerfeld, A. and Bethe, H., *Handbuch der Physik*, Chapter 3, Springer, 1933.
3. Wyckoff, R. W. G., *Crystal Structures*, 2nd. ed., Interscience, 1963.
4. Ashcroft, N. W. and Mermin, N. D., *Solid State Physics*, Holt, Rinehart and Winston, 1976.

PROPERTIES OF COMMERCIAL METALS AND ALLOYS

This table gives typical values of mechanical, thermal, and electrical properties of several common commercial metals and alloys. Values refer to ambient temperature (0 to 25 °C). All values should be regarded as typical, since these properties are dependent on the particular type of alloy, heat treatment, and other factors. Values for individual specimens can vary widely.

References

1. *ASM Metals Reference Book, Second Edition*, American Society for Metals, Metals Park, OH, 1983.
2. Lynch, C. T., *CRC Practical Handbook of Materials Science*, CRC Press, Boca Raton, FL, 1989.
3. Shackelford, J. F., and Alexander, W., *CRC Materials Science and Engineering Handbook*, CRC Press, Boca Raton, FL, 1991.

Common name	Thermal conductivity W/cm K	Density g/cm³	Coeff. of linear expansion 10^{-6}/°C	Electrical resistivity μΩ cm	Modulus of elasticity GPa	Tensile strength MPa	Approx. melting point °C
Ingot iron	0.7	7.86	11.7	9.7	205	–	1540
Plain carbon steel AISI-SAE 1020	0.52	7.86	11.7	18	205	450	1515
Stainless steel type 304	0.15	7.9	17.3	72	195	550	1425
Cast gray iron	0.47	7.2	10.5	67	90	180	1175
Malleable iron		7.3	12	30	170	345	1230
Hastelloy C	0.12	8.94	11.3	125	200	780	1350
Inconel	0.15	8.25	11.5	103	200	800	1370
Aluminum alloy 3003, rolled	1.9	2.73	23.2	3.7	70	110	650
Aluminum alloy 2014, annealed	1.9	2.8	23.0	3.4	70	185	650
Aluminum alloy 360	1.5	2.64	21.0	7.5	70	325	565
Copper, electrolytic (ETP)	3.9	8.94	16.5	1.7	120	300	1080
Yellow brass (high brass)	1.2	8.47	20.3	6.4	100	300-800	930
Aluminum bronze	0.7	7.8	16.4	12	120	400-600	1050
Beryllium copper 25	0.8	8.23	17.8	7	130	500-1400	925
Cupronickel 30%	0.3	8.94	16.2		150	400-600	1200
Red brass, 85%	1.6	8.75	18.7	11	90	300-700	1000
Chemical lead	0.35	11.34	29.3	21	13	17	327
Antimonial lead (hard lead)	0.3	10.9	26.5	23	20	47	290
Solder 50-50	0.5	8.89	23.4	15	–	42	215
Magnesium alloy AZ31B	1.0	1.77	26	9	45	260	620
Monel	0.3	8.84	14.0	58	180	545	1330
Nickel (commercial)	0.9	8.89	13.3	10	200	460	1440
Cupronickel 55-45 (constantan)	0.2	8.9	18.8	49	160	–	1260
Titanium (commercial)	1.8	4.5	8.5	43	110	330-500	1670
Zinc (commercial)	1.1	7.14	32.5	6	–	130	419
Zirconium (commercial)	0.2	6.5	5.85	41	95	450	1855

HARDNESS OF MINERALS AND CERAMICS

There are several hardness scales for describing the resistance of a material to indentation or scratching. This table lists a number of common materials in order of increasing hardness. Values are given, when available, on three different hardness scales: the original Mohs Scale (range 1 to 10); the modified Mohs Scale (range 1 to 15), and the Knoop Hardness Scale. In the last case, a load of 100 g is assumed.

Reference

Shackelford, J. F. and Alexander, W., *CRC Materials Science and Engineering Handbook*, CRC Press, Boca Raton, FL, 1991.

Solids

Hardness of Minerals and Ceramics

Name	Synonym	Formula	Mohs	Modified Mohs	Knoop
Carbon (graphite)	Graphite	C	0.5		
Talc		$3MgO{\cdot}4SiO_2{\cdot}H_2O$	1	1	
Alabaster		$CaSO_4{\cdot}2H_2O$	1.7		
Gypsum		$CaSO_4{\cdot}2H_2O$	2	2	32
Halite	Rock salt	NaCl	2		
Stibnite	Antimonite	Sb_2S_3	2.0		
Galena		PbS	2.5		
Mica		$(K,H)Al_3(SiO_4)_3$	2.8		
Calcite		$CaCO_3$	3	3	135
Barite		$BaSO_4$	3.3		
Marble			3.5		
Aragonite		$CaCO_3$	3.5		
Dolomite		$CaMg(CO_3)_2$	3.5		
Fluorite		CaF_2	4	4	163
Magnesium oxide	Magnesia	MgO	5		370
Apatite		$Ca_5(PO_4)_3(OH,F,Cl)$	5	5	430
Opal		$SiO_2{\cdot}nH_2O$	5		
Feldspar		$K_2O{\cdot}Al_2O{\cdot}6SiO_2$	6	6	560
Augite		$(Ca,Mg,Fe,Ti,Al)_2(Si,Al)_2O_6$	6		
Hematite		Fe_2O_3	6		750
Magnetite		Fe_3O_4	6		
Rutile		TiO_2	6.2		
Pyrite		FeS_2	6.3		
Agate		SiO_2	6.5		
Uranium(IV) oxide	Uranium dioxide	UO_2	6.7		600
Silicon dioxide (vitreous)	Fused silica	SiO_2		7	
Quartz (α)		SiO_2	7	8	820
Flint			7		
Silicon		Si	7		
Andalusite		Al_2OSiO_4	7.5		
Zircon		$ZrSiO_4$	7.5		
Zirconium(IV) oxide	Zirconia	ZrO_2			1200
Aluminum nitride		AlN			1225
Beryl		$Be_3Al_2(SiO_3)_6$	7.8		
Beryllium oxide	Beryllia	BeO			1300
Topaz		$Al_2(SiO_4)(F,OH)_2$	8	9	1340
Garnet		$Al_2O_3{\cdot}3FeO{\cdot}3SiO_2$		10	1360
Emery		Al_2O_3 (impure)	8		
Zirconium nitride		ZrN	8[a]		1510
Zirconium boride		ZrB_2			1560
Titanium nitride		TiN	9		1770
Zirconia (fused)		ZrO_2		11	
Tantalum carbide		TaC			1800
Tungsten carbide		WC			1880
Corundum		Al_2O_3	9		2025
Zirconium carbide		ZrC			2150
Alumina (fused)		Al_2O_3		12	

Name	Synonym	Formula	Mohs	Modified Mohs	Knoop
Beryllium carbide		Be_2C			2400
Titanium carbide		TiC			2470
Carborundum		SiC	9.3	13	2500
Aluminum monoboride		AlB			2500
Tantalum boride [TaB_2]	Tantalum diboride	TaB_2			2600
Boron carbide	Norbide	B_4C		14	2800
Boron		B	9.5		
Titanium boride		TiB_2			2850
Carbon (diamond)	Diamond	C	10	15	7000

[a] Can be higher than 8

Section 13
Polymer Properties

ABBREVIATIONS USED IN POLYMER SCIENCE AND TECHNOLOGY

ABA	triblock copolymers; acrylonitrile-butadiene acrylate
ABS	copolymer of acrylonitrile, butadiene, and styrene
ACS	acrylonitrile-chlorinated polyethylene styrene terpolymer
AIBN	2,2′-azobisisobutyronitrile
AMA	acrylate maleic anhydride terpolymer
AMMA	acrylate-methyl methacrylate copolymer
AN	acrylonitrile
AP	ethylene-propylene copolymers
APO	amorphous polyolefin
AS	acrylonitrile styrene copolymer
ASA	acrylonitrile-styrene-acrylonitrile block
ATR	attenuated total reflectance spectroscopy
AU	polyurethane
BMC	bulk molding compound
BMI	bis maleimide
BPO	benzoyl peroxide
CA	cellulose acetate
CAB	cellulose acetate butyrate
CAP	cellulose acetate proprionate
CAR	carbon fiber
CED	cohesive energy density
CFRP	carbon reinforced plastics
CMC	carboxymethylcellulose
CN	cellulose nitrate
COC	cycloolefin copolymer
COP	copolyester thermoplastic elastomer
CPE	chloronated polyethylene
CPVC	chlorinated poly(vinyl chloride)
CR	neoprene
CTA	cellulose triacetate
CTFE	chlorotrifluoroethylene
C_s	chain transfer constant
DAIP	diallyl isophthalate plasticizer
DAP	dially phthalate plasticizer
DNA	deoxyribonucleic acid
DP	degree of polymerization
DRS	dynamic reflectance spectroscopy
DS	degree of substitution
EAA	ethylene acrylic acid copolymer
EC	ethyl cellulose
ECTFE	ethylene-chlorotrifluoroethylene copolymer
EEA	ethylene-ethyl acetate copolymer
EGG	Einstein-Guth-Gold equation
EMAC	ethylene-methyl acrylate copolymer
EnBA	ethylene butyl acetate
EP	epoxy resin
EPDM	poly(ethylene-*co*-propylene) crosslinked
EPM	ethylene-propylene copolymer
EPR	ethylene propylene rubber
EPS	expanded polystyrene
ET	thiokol
ETFE	ethylene tetrafluoroethylene polymer
EU	polyether polyurethane
EVA	ethylene-vinyl acetate copolymer
EVOH	ethylene-vinyl alcohol copolymer
FEP	fluorinated ethylene propylene
FRP	fibrous glass reinforced polyester; fiber reinforced plastic

G	molar attraction constant
GF	glass reinforced
GRS	poly(butadiene-*co*-styrene)
HDPE	high-density polyethylene
HIPS	high-impact polystyrene
HMC	high strength molding compound
HMWHDPE	high-molecular-weight high-density polyethylene
I	ionomer
IIR	butyl rubber
IPN	interpenetrating polymer network
K	constant in Mark-Houwink equation
LC	liquid crystal
LCP	liquid crystal polymer
LDPE	low-density polyethylene
LLDPE	linear low-density polyethylene
LPE	linear polyethylene
MA	maleic anhydride
MABS	methyl methacrylate ABS copolymer
MBS	methyl methacrylate butadiene styrene terpolymer
MDPE	medium density polyethylene
MDI	methylene diphenylisocyanate
MF	melamine-formaldehyde resin
MP	melamine phenolic
MWD	molecular weight distribution
M_n	number-average molecular weight
M_v	viscosity-average molecular weight
M_w	weight-average molecular weight
M_z	*Z*-average molecular weight
NBR	poly(butadiene-co-acrylonitrile); nitrile butadiene rubber
NR	natural rubber
OSA	olefin modified styrene acrylonitrile
P	phenolic
PA	polyamide; nylon
PA6	polyamide 6, nylon6
PA11	polyamide 11, nylon11
PA12	polyamide 12, nylon12
PA46	polyamide 46, nylon46
PA66	polyamide 66, nylon6-6
PA66/6T	polyamide 66/6T
PA610	polyamide 610, nylon6-10
PA612	polyamide 612, nylon6-12
PA666	polyamide 666
PAA	poly(acrylic acid)
PAEK	polyaryletherketone
PAI	polyamide-imide
PAK	polyester alkyd
PAL	polyanaline
PAN	polyacrylonitrile
PARA	polyaryl amide
PAS	polyarylsulfone
PB	polybutylene
PBAN	polybutylene-acrylonitrile copolymer
PBD	polybutadine
PBI	polybenzimidazole
PBN	poly(butylene napthalate)
PBS	polybutadiene-styrene copolymer
PBT	poly(butylene terephthalate)
PC	polycarbonate

PC/ABS	polycarbonate/acrylonitrile butadiene styrene blend
PCB	polychlorinated biphenyl
PCCE	poly(cyclohexylene dimethylene cyclohexanedicarboxylate), glycol and acid comonomer
PCL	polycaprolactone
PCT	poly(cyclohexylene terephthalate)
PCTA	poly(cyclohexylene dimethylene terephthalate) copolyester
PCTFE	polychlorotrifluoroethylene
PCTG	poly(cyclohexylene dimethylene terephthalate) copolyester
PCT-G	glycol modified polycyclohexyl terephthallate
PE	polyethylene
PEBA	polyether block amide or polyester block amide
PEEK	poly(ether ether ketone)
PEG	poly(ethylene glycol)
PEI	polyetherimide
PEK	polyetherketone
PEKEKK	polyetherketone etherketone ketone
PEKK	polyetherketoneketone
PEN	poly(ethylene napthalene)
PEO	poly(ethylene oxide)
PES	polyethersulfone
PET	poly(ethylene terephthalate)
PET-G	glycol modified poly(ethylene terephthalate)
PEX	crosslinked polyethylene
PF	phenol-formaldehyde resin
PFA	perfluoroalkoxy
PI	polyimide, polyisoprene
PIB	polyisobutylene
PIR	polyisocyanurate
PK	polyketone
PLGA	poly(lactic-*co*-glycolic acid)
PMAN	polymethactylonitrile
PMMA	poly(methyl methacrylate)
PMP	polymethylpentene
PMS	polymethylstyrene
PNF	poly(phosphonitrilic fluorides)
PO	polyolefin
POM	polyoxymethylene, polyformaldehyde, acetals
PP	polypropylene
PPA	polyphthalamide
PPC	chlorinated polypropylene, polyphthalate carbonate
PPE	poly(phenylene ether)
PPI	polymeric polyisocyanate
PPO	poly(phenylene oxide)
PPOX	poly(propylene oxide)
PPS	poly(phenylene sulfide)
PPSU	poly(phenylene sulfone)
PPT	poly(propylene terephthalate)
PS	polystyrene
PS-b-PI	polystyrene/polyisoprene block copolymer
PSO, PSU	polysulfone
PTFE	polytetrafluoroethylene, Teflon
PTME	poly(tetramethylene terephthalate)
PTMT	poly(tetramethylene terephthalate)
PU	polyurethane
PUR	polyurethane rubber
PVA	poly(vinyl alcohol); sometimes poly(vinyl acetate)
PVAc	poly(vinyl acetate)
PVB	poly(vinyl butyral)
PVC	poly(vinyl chloride)
PVCA	copolymer of vinyl chloride and vinyl acetate
PVDA	polyvinylidene acetate
PVDC	poly(vinylidene chloride)
PVDF	poly(vinylidene fluoride)
PVF	poly(vinyl fluoride)
PVK	poly(vinyl carbazole)
PVOH	poly(vinyl alcohol)
PVP	poly(vinyl pyrrolidone)
RIM	reaction injection molding
RNA	ribonucleic acid
ROMP	ring opening metathesis polymerization
ROP	ring opening polymerization
S	radius of gyration
SAN	poly(styrene-*co*-acrylonitrile)
SB	styrene butadiene copolymer
SBR	poly(butadiene-*co*-styrene) elastomer
SBS	styrene butadiene styrene block copolymer
SEBS	styrene ethylene butylene styrene block copolymer
SI	silicon
SIS	styrene isoprene styrene block copolymer
SMA	poly(styrene-*co*-maleic anhydride)
SMC	sheet molding compound
SMMA	styrene methyl methacrylate copolymer
SMS	styrene/α-methyl styrene
SN	sulfur nitride
SR	synthetic rubber
SRP	styrene-rubber plastics
SVA	styrene vinyl acrylonitrile
TDI	toluenediisocyanate
TEO	thermoplastic elastic olefin
TGA, TG	thermal gravimetric analysis
TMC	thick molding compound
TMMV	threshold molecular weight value
TPA	polyamide thermoplastic elastomer
TPC	copolyester thermoplastic elastomer
TPE	thermoplastic elastomer
TPE-O, TPO	thermoplastic elastomer - olefinic
TPE-S, TPS	thermoplastic elastomer - styrenic
TPU	thermoplastic urethane
TPX	poly-4-methylpentene
TVO	thermoplastic vulcanites
T_c	ceiling temperature; cloud-point temperature
T_g	glass transition temperature
T_m	melting point temperature
UF	urea-formaldehyde resin
UHMWPE	ultrahigh molecular weight polyethylene
ULDPE	ultra low-density polyethylene
ULPE	ultra linear polyethylene
UP, UPE	unsaturated polyester (thermoset)
VA	vinyl acetate
VAE	vinyl acetate ethylene
VLDPE	very low-density polyethylene
WLF	Williams-Landel-Ferry equation
WS	polyurethane
XLPE	crosslinked polyethylene
XPS	expandable polystyrene

PHYSICAL PROPERTIES OF SELECTED POLYMERS

The physical properties of polymers are important parameters in determining their behavior and performance in a wide range of applications. This table lists some examples of general representative physical properties (including mechanical properties) of representative polymeric compounds. For glass transition temperatures of selected polymers, see pages 13-10 through 13-16 in this section. Some of the properties in this table are defined as follows:

The **heat deflection temperature** (HDT), or heat distortion temperature, is the temperature at which a polymer or plastic sample deforms under a specified load (normally either 0.455 MPa or 1.82 MPa).

The **crystalline melting point** is the temperature (or temperature range) at which a crystalline solid changes its state from solid to liquid. Although the phrase would suggest a specific temperature, most crystalline compounds actually melt over a range of a few degrees or less.

The **coefficient of linear thermal expansion** is the fractional change in length per °C change in temperature at constant pressure.

The **compressive strength** of a material is the maximum uniaxial compressive stress (compressive force per unit area) reached when the material fails completely on being subjected to a load that pushes it together.

The **tensile strength** is a measure of the ability of a material to withstand pulling stresses. It is defined as the stress (stretching force per unit area) required to break a specimen. Polymers are approximately 20 % stronger in compression than in tension.

The **flexural strength**, or cross-breaking strength, of a material is a measure of the bending strength or stiffness of a specimen expressed as the stress required to break a specimen by exerting a torque on it.

The **impact strength** is a measure of the energy needed to break a sample. The term toughness is sometimes used to describe the impact strength of a material. The notched izod impact test is a single point test that measures the resistance of a material to impact from a swinging pendulum. Izod impact is defined as the kinetic energy needed to initiate fracture and continue the fracture until the specimen is broken. Izod specimens are notched to prevent deformation of the specimen upon impact. This test can be used as a quick and easy quality control check to determine if a material meets specific impact properties or to compare materials for general toughness.

The **ultimate elongation** is a measure of how far a material will stretch before breaking, expressed as a percentage of its original length.

The properties of the following polymers are presented in this table:

PET	poly(ethylene terephthalate)
PBT	poly(butylene terephthalate)
PC	polycarbonate
Nylon 6,6	poly(iminoadipoyliminohexamethylene)
Nylon 6	poly[imino(1-oxohexamethylene)]
PPO	poly(phenylene ether)
POM	polyoxymethylene
LDPE	low-density polyethylene
HDPE	high-density polyethylene
UHMWPE	ultrahigh molecular weight polyethylene
iPP	isotactic polypropylene
ABS	copolymer of acrylonitrile, butadiene, and styrene (extrusion grade)
PTFE	polytetrafluoroethylene, Teflon
PCTFE	polymonochlorotrifluoroethylene
PVDF	poly(vinylidene fluoride)
PVF	poly(vinyl fluoride)
PVC (rigid)	poly(vinyl chloride)
PVC (plasterized)	poly(vinyl chloride)
PMMA	poly(methyl methacrylate)

The assistance of Charles E. Carraher, Jr. in providing these data is gratefully acknowledged.

Reference

Carraher, Jr., C. E., *Seymour/Carraher's Polymer Chemistry,* 7th Ed., CRC Press, Taylor & Francis Group, Boca Raton, FL, 2008.

	PET	**PBT**	**PC**	**Nylon 6,6**	**Nylon 6**	**PPO**
Heat deflection temperature at 1820 kPa (°C)	100	65	130	75	80	100
Maximum resistance to continuous heat (°C)	100	60	115	120	125	80
Crystalline melting point (°C)	—	—	225	265	225	215
Coefficient of linear expansion (10^{-5}/°C)	6.5	7.0	6.8	8.0	8.0	5.0
Compressive strength (kPa)	8.6×10^4	7.5×10^4	8.6×10^4	1×10^5	9.7×10^4	9.6×10^4
Flexural strength (kPa)	1.1×10^5	9.6×10^4	9.3×10^4	1×10^5	9.7×10^4	8.9×10^4
Impact strength (Izod: cm N/cm of notch)	26	53	530	80	160	270
Tensile strength (kPa)	6.2×10^4	5.5×10^4	7.2×10^4	8.3×10^4	6.2×10^4	5.5×10^4
Ultimate elongation (%)	100	100	110	30	—	50
Density (g cm^{-3})	1.35	1.35	1.2	1.2	1.15	1.1

	POM	LDPE	HDPE	UHMWPE	iPP	ABS
Heat deflection temperature at 1820 kPa (°C)	125	40	50	85	55	90
Maximum resistance to continuous heat (°C)	100	40	80	80	100	90
Crystalline melting point (°C)	180	—	—	—	—	—
Coefficient of linear expansion (10^{-5}/°C)	10.0	10	12	12	9	9.5
Compressive strength (kPa)	1.1×10^5	—	3×10^4	—	—	4.8×10^4
Flexural strength (kPa)	9.7×10^4	—	—	—	5×10^4	6.2×10^4
Impact strength (Izod: cm N/cm of notch)	80	No break	30	No break	27	320
Tensile strength (kPa)	6.9×10^4	5×10^3	2×10^4	6×10^4	3.5×10^4	3.4×10^4
Ultimate elongation (%)	30	—	—		100	60
Density (g cm^{-3})	1.4	0.91	0.96	0.93	0.90	1.0

	PTFE	PCTFE	PVDF	PVF	Rigid PVC	Plasticized PVC	PMMA
Heat deflection temperature at 1820 kPa (°C)	100	100	80	90	75	—	95
Maximum resistance to continuous heat (°C)	250	200	150	125	60	35	75
Crystalline melting point (°C)	—	—	—	—	170	—	
Coefficient of linear expansion (10^{-5}/°C)	10	14	8.5	10	6	12	7.0
Compressive strength (kPa)	2.7×10^4	3.8×10^4	—	—	6.8×10^4	6×10^3	1×10^5
Flexural strength (kPa)	—	6×10^4	—	—	9×10^4	—	9.6×10^4
Impact strength (Izod: cm N/cm of notch)	160	130	—	—	27	—	21
Tensile strength (kPa)	2.4×10^4	3.4×10^4	5.5×10^4	—	4.4×10^4	1×10^4	6.5×10^4
Ultimate elongation (%)	200	100	200	—	50	200	4
Density (g cm^{-3})	2.16	2.1	1.76	1.4	1.4	1.3	1.2

NOMENCLATURE FOR ORGANIC POLYMERS

Robert B. Fox and Edward S. Wilks

Organic polymers have traditionally been named on the basis of the monomer used, a hypothetical monomer or a semi-systematic structure. Alternatively, they may be named in the same way as organic compounds, i.e., on the basis of a structure as drawn. The former method, often called "source-based nomenclature" or "monomer-based nomenclature," sometimes results in ambiguity and multiple names for a single material. The latter method, termed "structure-based nomenclature," generates a sometimes cumbersome unique name for a given polymer, independent of its source. Within their limitations, both types of names are acceptable and well-documented.[1] The use of stereochemical descriptors with both types of polymer nomenclature has been published.[2]

Traditional Polymer Names

Monomer-Based Names

"Polystyrene" is the name of a homopolymer made from the single monomer styrene. When the name of a monomer comprises two or more words, the name should be enclosed in parentheses, as in "poly(methyl methacrylate)" or "poly(4-bromostyrene)" to identify the monomer more clearly. This method can result in several names for a given polymer: thus, "poly(ethylene glycol)," "poly(ethylene oxide)," and "poly(oxirane)" describe the same polymer. Sometimes, the name of a hypothetical monomer is used, as in "poly(vinyl alcohol)." Even though a name like "polyethylene" covers a multitude of materials, the system does provide understandable names when a single monomer is involved in the synthesis of a single polymer. When one monomer can yield more than one polymer, e.g., 1,3-butadiene or acrolein, some sort of structural notation must be used to identify the product, and one is not far from a formal structure-based name.

Copolymers, Block Polymers, and Graft Polymers. When more than one monomer is involved, monomer-based names are more complex. Some common polymers have been given names based on an apparent structure, as with "poly(ethylene terephthalate)." A better system has been approved by the IUPAC.[1] With this method, the arrangement of the monomeric units is introduced through use of an italicized connective placed between the names of the monomers. For monomer names represented by A, B, and C, the various types of arrangements are shown in Table 1.

Table 2 contains examples of common or semi-systematic names of copolymers. The systematic names of comonomers may also be used; thus, the polyacrylonitrile-*block*-polybutadiene-*block*-polystyrene polymer in Table 2 may also be named poly(prop-2-enenitrile)-*block*-polybuta-1,3-diene-*block*-poly(ethenylbenzene). IUPAC does not require alphabetized names of comonomers within a polymer name; many names are thus possible for some copolymers.

These connectives may be used in combination and with small, non-repeating (i.e. non-polymeric) junction units; see, for example, Table 2, line 8. A long dash may be used in place of the connective *-block-*; thus, in Table 2, the polymers of lines 7 and 8 may also be written as shown on lines 9 and 10.

IUPAC also recommends an alternative scheme for naming copolymers that comprises use of "copoly" as a prefix followed by the names of the comonomers, a solidus (an oblique stroke) to separate comonomer names, and the addition before "copoly" of any applicable connectives listed in Table 2 except *-co-*.

Table 3 gives the same examples shown in Table 2 but with the alternative format. Comonomer names need not be parenthesized.

TABLE 1. IUPAC Source-Based Copolymer Classification

No.	Copolymer type	Connective	Example
1	Unspecified or unknown	*-co-*	poly(A-*co*-B)
2	Random (obeys Bernoullian distribution)	*-ran-*	poly(A-*ran*-B)
3	Statistical (obeys known statistical laws)	*-stat-*	poly(A-*stat*-B)
4	Alternating (for two monomeric units)	*-alt-*	poly(A-*alt*-B)
5	Periodic (ordered sequence for 2 or more monomeric units)	*-per-*	poly(A-*per*-B-per-C)
6	Block (linear block arrangement)	*-block-*	polyA-*block*-polyB
7	Graft (side chains connected to main chains)	*-graft-*	polyA-*graft*-polyB

TABLE 2. Examples of Source-Based Copolymer Nomenclature

No.	Copolymer name
1	poly(propene-*co*-methacrylonitrile)
2	poly[(acrylic acid)-*ran*-(ethyl acrylate)]
3	poly(butene-*stat*-ethylene-*stat*-styrene)
4	poly[(sebacic acid)-*alt*-butanediol]
5	poly[(ethylene oxide)-*per*-(ethylene oxide)-*per*-tetrahydrofuran]
6	polyisoprene-*graft*-poly(methacrylic acid)
7	polyacrylonitrile-*block*-polybutadiene-*block*-polystyrene
8	polystyrene-*block*-dimethylsilylene-*block*-polybutadiene
9	polyacrylonitrile—polybutadiene—polystyrene
10	polystyrene—dimethylsilylene—polybutadiene

TABLE 3. Examples of Source-Based Copolymer Nomenclature (Alternative Format)

No.	Polymer name
1	copoly(propene/methacrylonitrile)
2	*ran*-copoly(acrylic acid/ethyl acrylate)
3	*stat*-copoly(butene/ethylene/styrene)
4	*alt*-copoly(sebacic acid/butanediol)
5	*block*-copoly(acrylonitrile/butadiene/styrene)
6	*per*-copoly(ethylene oxide/ethylene oxide/tetrahydrofuran)
7	*graft*-copoly(isoprene/methacrylic acid)

Source-based nomenclature for nonlinear macromolecules and macromolecular assemblies is covered by a 1997 IUPAC document.[11] The types of polymers in these classes, together with their connectives, are given in Table 4; the terms shown may be used as connectives, prefixes, or both to designate the features present.

TABLE 4. Connectives for Nonlinear Macromolecules and Macromolecular Assemblies

No.	Type	Connective
1	Branched (type unspecified)	branch
2	Branched with branch point of functionality f	f-branch
3	Comb	comb
4	Cross-link	ι (Greek iota)
5	Cyclic	cyclo
6	Interpenetrating polymer network	ipn
7	Long-chain branched	l-branch
8	Network	net
9	Polymer blend	blend
10	Polymer-polymer complex	compl
11	Semi-interpenetrating polymer network	sipn
12	Short-chain branched	sh-branch
13	Star	star
14	Star with f arms	f-star

Nonlinear polymers are named by using the italicized connective as a *prefix* to the source-based name of the polymer component or components to which the prefix applies; some examples are listed in Table 5.

TABLE 5. Nonlinear Macromolecules

No.	Polymer name	Polymer structural features
1	poly(methacrylic acid)-*comb*-polyacrylonitrile	Comb polymer with a poly(methacrylic acid) backbone and polyacrylonitrile side chains
2	*comb*-poly[ethylene-*stat*-(vinyl chloride)]	Comb polymer with unspecified backbone composition and statistical ethylene/vinyl chloride copolymer side chains
3	polybutadiene-*comb*-(polyethylene; polypropene)	Comb polymer with butadiene backbone and side chains of polyethylene and polypropene
4	*star*-(polyA; polyB; polyC; polyD; polyE)	Star polymer with arms derived from monomers A, B, C, D, and E, respectively
5	*star*-(polyA-*block*-polyB-*block*-polyC)	Star polymer with every arm comprising a tri-block segment derived from comonomers A, B, and C
6	*star*-poly(propylene oxide)	A star polymer prepared from propylene oxide
7	5-*star*-poly(propylene oxide)	A 5-arm star polymer prepared from propylene oxide
8	*star*-(polyacrylonitrile; polypropylene) (M_r 10000: 25000)	A star polymer containing polyacrylonitrile arms of MW 10000 and polypropylene arms of MW 25000

Macromolecular assemblies held together by forces other than covalent bonds are named by inserting the appropriate italicized connective between names of individual components; Table 6 gives examples.

TABLE 6. Examples of Polymer Blends and Nets

No.	Polymer name
1	polyethylene-*blend*-polypropene
2	poly(methacrylic acid)-*blend*-poly(ethyl acrylate)
3	*net*-poly(4-methylstyrene-ι-divinylbenzene)
4	*net*-poly[styrene-*alt*-(maleic anhydride)]-ι-(polyethylene glycol; polypropylene glycol)
5	*net*-poly(ethyl methacrylate)-*sipn*-polyethylene
6	[*net*-poly(butadiene-*stat*-styrene)]-*ipn*-[*net*-poly(4-methylstyrene-ι-divinylbenzene)]

Structure-Based Polymer Nomenclature

Regular Single-Strand Polymers

Structure-based nomenclature has been approved by the IUPAC[4] and is currently being updated; it is used by *Chemical Abstracts*.[5] Monomer names are not used. To the extent that a polymer chain can be described by a repeating unit in the chain, it can be named "poly(repeating unit)." For regular single-strand polymers, "repeating unit" is a bivalent group; for regular double-strand (ladder and spiro) polymers, "repeating unit" is usually a tetravalent group.[9]

Since there are usually many possible repeating units in a given chain, it is necessary to select one, called the "constitutional repeating unit" (CRU) to provide a unique and unambiguous name, "poly(CRU)," where "CRU" is a recitation of the names of successive units as one proceeds through the CRU from left to right. For this purpose, a portion of the main chain structure that includes at least two repeating sequences is written out. These sequences will typically be composed of bivalent subunits such as $-CH_2-$, -O-, and groups from ring systems, each of which can be named by the usual nomenclature rules.[6,7]

Where a chain is simply one long sequence comprising repetition of a single subunit, that subunit is itself the CRU, as in "poly(methylene)" or "poly(1,4-phenylene)." In chains having more than one kind of subunit, a seniority system is used to determine the beginning of the CRU and the direction in which to move along the main chain atoms (following the shortest path in rings) to complete the CRU. Determination of the first, most senior, subunit is based on a descending order of seniority: (1) heterocyclic rings, (2) hetero atoms, (3) carbocyclic rings, and lowest, (4) acyclic carbon chains.

Within each of these classes, there is a further order of seniority that follows the usual rules of nomenclature.

Heterocycles: A nitrogen-containing ring system is senior to a ring system not containing nitrogen.[4,9] Further descending order of seniority is determined by:

(i) the highest number of rings in the ring system
(ii) the largest individual ring in the ring system
(iii) the largest number of hetero atoms
(iv) the greatest variety of hetero atoms

Hetero atoms: The senior bivalent subunit is the one nearest the top right-hand corner of the Periodic Table; the order of seniority is: O, S, Se, Te, N, P, As, Sb, Bi, Si, Ge, Sn, Pb, B, Hg.

Carbocycles: Seniority[4] is determined by:

(i) the highest number of rings in the ring system
(ii) the largest individual ring in the ring system
(iii) degree of ring saturation; an unsaturated ring is senior to a saturated ring of the same size

Carbon chains: Descending order of seniority is determined by:

(i) chain length (longer is senior to shorter)
(ii) highest degree of unsaturation
(iii) number of substituents (higher number is senior to lower number)
(iv) ascending order of locants
(v) alphabetical order of names of substituent groups

Among equivalent ring systems, preference is given to the one having lowest locants for the free valences in the subunit, and among otherwise identical ring systems, the one having least hydrogenation is senior. Lowest locants in unsaturated chains are also given preference. Lowest locants for substituents are the final determinant of seniority.

Direction within the repeating unit depends upon the shortest path, which is determined by counting main chain atoms, both cyclic and acyclic, from the most senior subunit to another subunit of the same kind or to a subunit next lower in seniority. When identification and orientation of the CRU have been accomplished, the CRU is named by writing, in sequence, the names of the largest possible subunits within the CRU from left to right. For example, the main chain of the polymer traditionally named "poly(ethylene terephthalate)" has the structure shown in Figure 1.

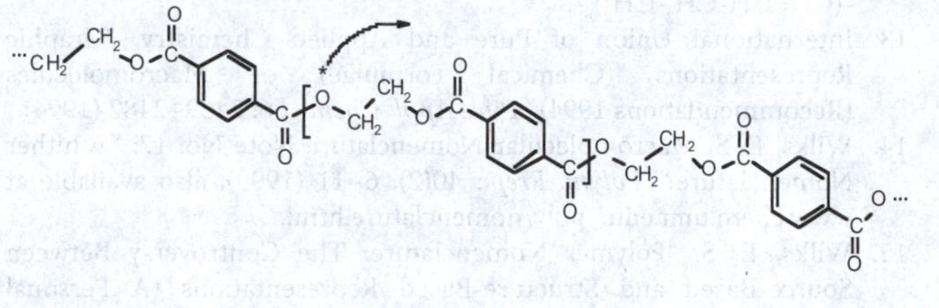

FIGURE 1. Structure-based name: poly(oxyethyleneoxyterephthaloyl); traditional name: poly(ethylene terephthalate).

The CRU in Figure 1 is enclosed in brackets and read from left to right. It is selected because (1) either backbone oxygen atom qualifies as the "most senior subunit," (2) the shortest path length from either -O- to the other -O- is via the ethylene subunit. Orientation of the CRU is thus defined by (1) beginning at the -O- marked with an asterisk, and (2) reading in the direction of the arrow. The structure-based name of this polymer is therefore "poly(oxyethyl eneoxyterephthaloyl)," not much longer than the traditional name and much more adaptable to the complexities of substitution. As organic nomenclature evolves, more systematic names may be used for subunits, e.g., "ethane-1,2-diyl" instead of "ethylene." IUPAC still prefers "ethylene" for the $-CH_2-CH_2-$ unit, however, but also accepts "ethane-1,2-diyl."

Structure-based nomenclature can also be used when the CRU backbone has no carbon atoms. An example is the polymer traditionally named "poly(dimethylsiloxane)," which on the basis of structure would be named "poly(oxydimethylsilylene)" or "poly(oxydimethylsilanediyl)." This nomenclature method has also been applied to inorganic and coordination polymers[8] and to double-strand (ladder and spiro) organic polymers.[9]

Irregular Single-Strand Polymers

Polymers that cannot be described by the repetition of a single CRU or comprise units not all connected identically in a directional sense can also be named on a structure basis.[10] These include copolymers, block and graft polymers, and star polymers. They are given names of the type "poly(A/B/C...)," where A, B, C, etc. are the names of the component constitutional units, the number of which are minimized. The constitutional units may include regular or irregular blocks as well as atoms or atomic groupings, and each is named by the method described above or by the rules of organic nomenclature.

The solidus denotes an unspecified arrangement of the units within the main chain.[10] For example, a statistical copolymer derived from styrene and vinyl chloride with the monomeric units joined head-to-tail is named "poly(l-chloroethylene/l-phenylethylene)." A polymer obtained by 1,4-polymerization and both head-to-head and head-to-tail 1,2-polymerization of 1,3-butadiene would be named "poly(but-1-ene-l,4-diyl/l-vinylethylene/2-vinylethylene)."[12] In graphic representations of these polymers, shown in Figure 2, the hyphens or dashes at each end of each CRU depiction are shown *completely within* the enclosing parentheses; this indicates that they are not necessarily the terminal bonds of the macromolecule.

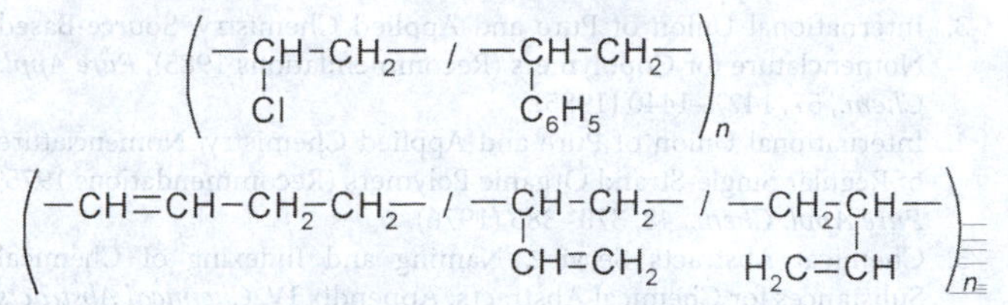

FIGURE 2. Graphic representations of copolymers.

A long hyphen is used to separate components in names of block polymers, as in "poly(A)—poly(B)—poly(C)," or "poly(A)—X—poly(B)" in which X is a non-polymeric junction unit, e.g., dimethylsilylene.

In graphic representations of these polymers, the blocks are shown connected when the bonding is known (Figure 3, for example); when the bonding between the blocks is unknown, the blocks are separated by solidi and are shown *completely within* the outer set of enclosing parentheses (Figure 4, for example).[10,13]

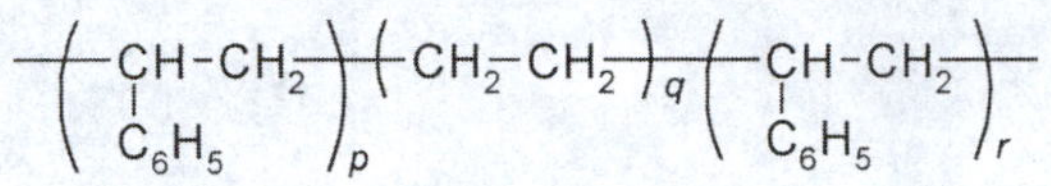

FIGURE 3. polystyrene—polyethylene—polystyrene.

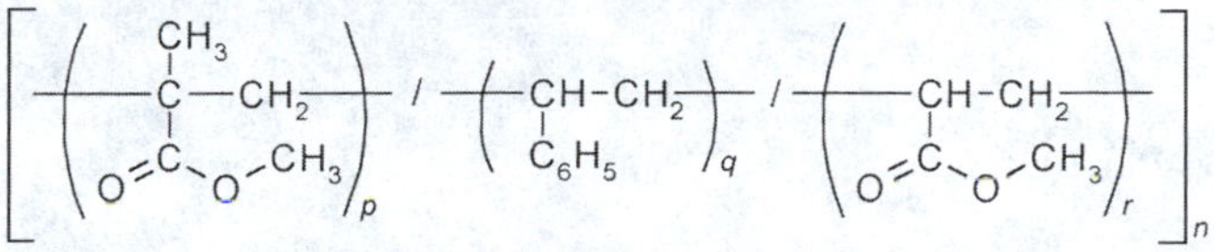

FIGURE 4. poly[poly(methyl methacrylate)—polystyrene—poly(methyl acrylate)].

Graft polymers are named in the same way as a substituted polymer but without the ending "yl" for the grafted chain; the name of a regular polymer, comprising Z units in which some have grafts of "poly(A)," is "poly[Z/poly(A)Z]." Star polymers are treated as a central unit with substituent blocks, as in "tetrakis(polymethylene) silane."[10,13]

Other Nomenclature Articles and Publications

In addition to the *Chemical Abstracts* and IUPAC documents cited above and listed below, other articles on polymer nomenclature are available. A 1999 article lists significant documents on polymer nomenclature published during the last 50 years in books, encyclopedias, and journals by *Chemical Abstracts*, IUPAC, and individual authors.[14] A comprehensive review of source-based and structure-based nomenclature for all of the major classes of polymers,[15] and a short tutorial on the correct identification, orientation, and naming of most commonly encountered constitutional repeating units were both published in 2000.[16]

References and Notes

1. International Union of Pure and Applied Chemistry, *Compendium of Macromolecular Nomenclature*, Blackwell Scientific Publications, Oxford, 1991.
2. International Union of Pure and Applied Chemistry, Stereochemical Definitions and Notations Relating to Polymers (Recommendations 1980), *Pure Appl. Chem.*, 53, 733–752 (1981).
3. International Union of Pure and Applied Chemistry, Source-Based Nomenclature for Copolymers (Recommendations 1985), *Pure Appl. Chem.*, 57, 1427–1440 (1985).
4. International Union of Pure and Applied Chemistry, Nomenclature of Regular Single-Strand Organic Polymers (Recommendations 1975, *Pure Appl. Chem.*, 48, 373–385 (1976).
5. Chemical Abstracts Service, Naming and Indexing of Chemical Substances for Chemical Abstracts, Appendix IV, *Chemical Abstracts 1999 Index Guide*.
6. International Union of Pure and Applied Chemistry, *A Guide to IUPAC Nomenclature of Organic Compounds* (1993), Blackwell Scientific Publications, Oxford, 1993.
7. International Union of Pure and Applied Chemistry, *Nomenclature of Organic Chemistry, Sections A, B, C, D, E, F, and H*, Pergamon Press, Oxford, 1979.
8. International Union of Pure and Applied Chemistry, Nomenclature of Regular Double-Strand and Quasi-Single-Strand Inorganic and Coordination Polymers (Recommendations 1984), *Pure Appl. Chem.*, 57, 149–168 (1985).
9. International Union of Pure and Applied Chemistry, Nomenclature of Regular Double-Strand (Ladder and Spiro) Organic Polymers (Recommendations 1993), *Pure Appl. Chem.*, 65, 1561–1580 (1993).
10. International Union of Pure and Applied Chemistry, Structure-Based Nomenclature for Irregular Single-Strand Organic Polymers (Recommendations 1994), *Pure Appl. Chem.*, 66, 873–889 (1994).
11. International Union of Pure and Applied Chemistry, "Source-Based Nomenclature for Non-Linear Macromolecules and Macromolecular Assemblies (Recommendations 1997)." *Pure Appl. Chem.*, 69, 2511–2521 (1997).
12. Poly(1,3-butadiene) obtained by polymerization of 1,3-butadiene in the so-called 1,4- mode is frequently drawn incorrectly in publications as $-(CH_2-CH=CH-CH_2)_n-$; the double bond should be assigned the lowest locant possible, i.e., the structure should be drawn as $-(CH=CH-CH_2-CH_2)_n-$.
13. International Union of Pure and Applied Chemistry, "Graphic Representations (Chemical Formulae) of Macromolecules (Recommendations 1994)." *Pure Appl. Chem.*, 66, 2469–2482 (1994).
14. Wilks, E. S. Macromolecular Nomenclature Note No. 17: "Whither Nomenclature?" *Polym. Prepr.* 40(2), 6–11 (1999); also available at www.chem.umr.edu/˜poly/nomenclature.html.
15. Wilks, E. S. "Polymer Nomenclature: The Controversy Between Source-Based and Structure-Based Representations (A Personal Perspective)." *Prog. Polym. Sci.* 25, 9–100 (2000).
16. Wilks, E. S. Macromolecular Nomenclature Note No. 18: "SRUs: Using the Rules." *Polym. Prepr.* 41(1), 6a–11a (2000); also available at www.chem.umr.edu/˜poly/nomenclature.html; a .pdf format version is also available.

SOLVENTS FOR COMMON POLYMERS

Abbreviations:
- HC: hydrocarbons
- MEK: methyl ethyl ketone
- THF: tetrahydrofuran
- DMF: dimethylformamide
- DMSO: dimethylsulfoxide

Polyethylene (HDPE)	HC and halogenated HC
Polypropylene (atactic)	HC and halogenated HC
Polybutadiene	HC, THF, ketones
Polystyrene	ethylbenzene, $CHCl_3$, CCl_4, THF, MEK
Polyacrylates	aromatic HC, chlorinated HC, THF, esters, ketones
Polymethacrylates	aromatic HC, chlorinated HC, THF, esters, MEK
Polyacrylamide	water
Poly(vinyl ethers)	halogenated HC, MEK, butanol
Poly(vinyl alcohol)	glycols (hot), DMF
Poly(vinyl acetate)	aromatic HC, chlorinated HC, THF, esters, DMF
Poly(vinyl chloride)	THF, DMF, DMSO
Poly(vinylidene chloride)	THF (hot), dioxane, DMF
Poly(vinyl fluoride)	DMF, DMSO (hot)
Polyacrylonitrile	DMF, DMSO
Poly(oxyethylene)	aromatic HC, $CHCl_3$, alcohols, esters, DMF
Poly(2,6-dimethylphenylene oxide)	aromatic HC, halogenated HC
Poly(ethylene terephthalate)	phenol, DMSO (hot)
Polyurethanes (linear)	aromatic HC, THF, DMF
Polyureas	phenol, formic acid
Polysiloxanes	HC, THF, DMF
Poly[bis(2,2,2-trifluoroethoxy)-phosphazene]	THF, ketones, ethyl acetate

GLASS TRANSITION TEMPERATURE FOR SELECTED POLYMERS

Robert B. Fox

Polymer names are based on the IUPAC structure-based nomenclature system described in the table "Naming Organic Polymers." Within each category, names are listed in alphabetical order. Source-based and trivial names are also given (in italics) for the most common polymers. The table does not include polymers for which T_g is not clearly defined because of variability of structure or because of reactions taking place near the glass transition.

All values of T_g cited in this table have been determined by differential scanning calorimetry (DSC) except those values indicated by:

(D)	dynamic method
(Dil)	dilatometry
(M)	mechanical method

Polymer name	Glass transition temperature (T_g/K)
ACYCLIC CARBON CHAINS	
Polyalkadienes	
Poly(alkenylene) *Polyalkadiene* –[CH=CHCH$_2$CH$_2$]–	
Poly(*cis*-1-butenylene)	171
cis-1,3-polybutadiene [PBD]	
Poly(*trans*-1-butenylene)	215
trans-1,3-polybutadiene [PBD]	
Poly(1-chloro-*cis*-1-butenylene)	253
cis-1,3-polychloroprene	
Poly(1-chloro-*trans*-1-butenylene)	233
trans-1,3-polychloroprene	
Poly(1-methyl-*cis*-1-butenylene)	200
cis-1,3-polyisoprene	
Poly(1-methyl-*trans*-1-butenylene)	207
trans-1,3-polyisoprene	
Poly(1,4,4-trifluoro-1-butenylene)	238
Polyalkenes	
Poly(alkylethylene) *Poly(alkylethylene)* -[RCHCH$_2$]-	
Poly(1-benzylethylene)	333
Poly(1-butylethylene)	223
Poly(1-cyclohexylethylene) (atactic)	393
Poly(1-cyclohexylethylene) (isotactic)	406 (D)
Poly(1,1-dimethylethylene)	200
Polyisobutylene [PIB]	
Poly(ethylene)	148
Poly(methylene)	155
Poly(1-phenethylethylene)	283
Poly(propylene) (isotactic)	272
Poly(propylene) (syndiotactic)	ca. 265
Poly[1-(2-pyridyl)ethylene]	377
Poly[1-(4-pyridyl)ethylene]	415
Poly(1-vinylethylene)	273
Polyacrylics	
Poly[1-(alkoxycarbonyl)ethylene] *Poly(alkyl acrylate)* –[(ROCO)CHCH$_2$]–	
Poly[1-(benzyloxycarbonyl)ethylene]	279
Poly[1-(butoxycarbonyl)ethylene]	219 (M)
Poly(butyl acrylate) [PBA]	
Poly[1-(*sec*-butoxycarbonyl)ethylene]	251
Poly[1-(butoxycarbonyl)-1-cyanoethylene]	358
Poly[1-(butylcarbamoyl)ethylene]	319 (M)
Poly(1-carbamoylethylene)	438
Polyacrylamide [PAM]	
Poly(1-carboxyethylene)	379

Polymer name	Glass transition temperature (T_g/K)
Poly(acrylic acid) [PAA]	
Poly[1-(2-chlorophenoxycarbonyl)ethylene]	326
Poly[1-(4-chlorophenoxycarbonyl)ethylene]	331
Poly[1-(4-cyanobenzyloxycarbonyl)ethylene]	317
Poly[1-(2-cyanoethoxycarbonyl)ethylene]	277
Poly[1-(cyanomethoxycarbonyl)ethylene)]	433 Dil
Poly[1-(4-cyanophenoxycarbonyl)ethylene]	363
Poly[1-(cyclohexyloxycarbonyl)ethylene]	292
Poly[1-(2,4-dichlorophenoxycarbonyl)ethylene]	333
Poly[1-(dimethylcarbamoyl)ethylene]	362
Poly[1-(ethoxycarbonyl)ethylene]	249
Poly(ethyl acrylate) [PEA]	
Poly[1-(ethoxycarbonyl)-1-fluoroethylene]	316
Poly[1-(2-ethoxycarbonylphenoxycarbonyl)ethylene]	303
Poly[1-(3-ethoxycarbonylphenoxycarbonyl)ethylene]	297
Poly[1-(4-ethoxycarbonylphenoxycarbonyl)ethylene]	310
Poly[1-(2-ethoxyethoxycarbonyl)ethylene]	223
Poly[1-(3-ethoxypropoxycarbonyl)ethylene]	218
Poly[1-(isopropoxycarbonyl)ethylene]	267–270
Poly[1-(methoxycarbonyl)ethylene]	283
Poly(methyl acrylate) [PMA]	
Poly[1-(2-methoxycarbonylphenoxycarbonyl)ethylene]	319
Poly[1-(3-methoxycarbonylphenoxycarbonyl)ethylene]	311
Poly[1-(4-methoxycarbonylphenoxycarbonyl)ethylene]	340
Poly[1-(2-methoxyethoxycarbonyl)ethylene]	223
Poly[1-(4-methoxyphenoxycarbonyl)ethylene]	324
Poly[1-(3-methoxypropoxycarbonyl)ethylene]	198
Poly[1-(2-naphthyloxycarbonyl)ethylene]	358
Poly[1-(pentachlorophenoxycarbonyl)ethylene]	420
Poly[1-(phenethoxycarbonyl)ethylene]	270
Poly[1-(phenoxycarbonyl)ethylene]	330
Poly[1-(*m*-tolyloxycarbonyl)ethylene]	298
Poly[1-(*o*-tolyloxycarbonyl)ethylene]	325
Poly[1-(*p*-tolyloxycarbonyl)ethylene]	316
Poly[1-(2,2,2-trifluorethoxycarbonyl)ethylene]	263
Polymethacrylics	
Poly[1-(alkoxycarbonyl)-1-methylethylene] *Poly(alkyl methacrylate)* $-[(ROCO)(Me)CCH_2]-$	
Poly[1-(benzyloxycarbonyl)-1-methylethylene]	327
Poly[1-(2-bromoethoxycarbonyl)-1-methylethylene]	325
Poly[(1-(butoxycarbonyl)-1-methylethylene]	293
Poly(butyl methacrylate) [PBMA]	
Poly[1-(*sec*-butoxycarbonyl)-1-methylethylene]	333
Poly[1-(*tert*-butoxycarbonyl)-1-methylethylene)]	391
Poly[1-(2-chloroethoxycarbonyl)-1-methylethylene]	ca 315
Poly[1-(2-cyanoethoxycarbonyl)-1-methylethylene]	364
Poly[1-(4-cyanophenoxycarbonyl)-1-methylethylene]	428
Poly[1-(cyclohexyloxycarbonyl)-1-methylethylene] (atactic)	356
Poly[1-(cyclohexyloxycarbonyl)-1-methylethylene)] (isotactic)	324
Poly[1-(dimethylaminoethoxycarbonyl)-1-methylethylene]	292
Poly[1-(ethoxycarbonyl)-1-ethylethylene]	300
Poly[1-(ethoxycarbonyl)-1-methylethylene] (atactic) *Poly(ethyl methacrylate)* [PEMA]	338
Poly[1-(ethoxycarbonyl)-1-methylethylene] (isotactic)	285
Poly[1-(ethoxycarbonyl)-1-methylethylene)] (syndiotactic)	339
Poly[1-(hexyloxycarbonyl)-1-methylethylene]	268
Poly[1-(isobutoxycarbonyl)-1-methylethylene]	326
Poly[1-(isopropoxycarbonyl)-1-methylethylene]	354
Poly[1-(methoxycarbonyl)-1-methylethylene] (atactic) *Poly(methyl methacrylate)* [PMMA]	378
Poly[1-(methoxycarbonyl)-1-methylethylene)] (isotactic)	311
Poly[1-(methoxycarbonyl)-1-methylethylene)] (syndiotactic)	378
Poly[1-(4-methoxycarbonylphenoxy)-1-methylethylene]	379

Polymers

Polymer name	Glass transition temperature (T_g/K)
Poly[1-(methoxycarbonyl)-1-phenylethylene)] (atactic)	391
Poly[1-(methoxycarbonyl)-1-phenylethylene)] (isotactic)	397
Poly[1-methyl-1-(phenethoxycarbonyl)ethylene]	299
Poly[1-methyl-1-(phenoxycarbonyl)ethylene]	383
Polyvinyl ethers, alcohols, and ketones	
Poly(1-alkoxyethylene) *Poly(alkyl vinyl ether)* –[ROCHCH$_2$]–	
Poly(1-hydroxyethylene) *Poly(vinyl alcohol)* –[HOCHCH$_2$]–	
Poly(1-alkanoylethylene) *Poly(alkyl vinyl ketone)* –[RCOCHCH$_2$]–	
Poly(1-butoxyethylene)	218
Poly(1-*sec*-butoxyethylene)	253
Poly(1-*tert*-butoxyethylene)	361
Poly[1-(butylthio)ethylene]	253
Poly(1-ethoxyethylene)	230
Poly[1-(4-ethylbenzoyl)ethylene]	325
Poly(1-hydroxyethylene)	358 (D)
Poly(vinyl alcohol) [PVA]	
Poly(hydroxymethylene)	407
Poly(1-isopropoxyethylene)	270
Poly[1-(4-methoxybenzoyl)ethylene]	319 (M)
Poly(1-methoxyethylene)	242
Poly(methyl vinyl ether) [PMVE]	
Poly[1-(methylthio)ethylene]	272
Poly(1-propoxyethylene)	224
Poly[1-(trifluoromethoxy)trifluoroethylene]	268
Polyvinyl halides and nitriles	
Poly(1-haloethylene) *Poly(vinyl halide)* –[XCHCH$_2$]–	
Poly(1-cyanoethylene) *Poly(acrylonitrile)* –[NCCHCH$_2$]–	
Poly(1-chloroethylene)	354
Poly(vinyl chloride) [PVC]	
Poly(chlorotrifluoroethylene)	373
Poly(1-cyanoethylene)	370
Polyacrylonitrile [PAN]	
Poly(1-cyano-1-methylethylene)	393
Polymethacrylonitrile	
Poly(1,1-dichloroethylene)	255
Poly(vinylidene chloride)	
Poly(1,1-difluoroethylene)	ca 233
Poly(vinylidene fluoride)	
Poly(1-fluoroethylene)	314 (M)
Poly(vinyl fluoride)	
Poly(1-hexafluoropropylene)	425
Poly[1-(2-iodoethyl)ethylene]	343
Poly(tetrafluoroethylene)	(160)
Poly[1-(trifluoromethyl)ethylene]	300
Polyvinyl esters	
Poly[1-(alkanoyloxy)ethylene] *Poly(vinyl alkanoate)* –[RCOOCHCH$_2$]–	
Poly(1-acetoxyethylene)	305
Poly(vinyl acetate) [PVAc]	
Poly[1-(benzoyloxy)ethylene]	344
Poly[1-(4-bromobenzoyloxy)ethylene]	365
Poly[1-(2-chlorobenzoyloxy)ethylene]	335
Poly[1-(3-chlorobenzoyloxy)ethylene]	338
Poly[1-(4-chlorobenzoyloxy)ethylene]	357
Poly[1-(cyclohexanoyloxy)ethylene]	349 (M)
Poly[1-(4-ethoxybenzoyloxy)ethylene]	343
Poly[1-(4-ethylbenzoyloxy)ethylene]	326

Polymer name	Glass transition temperature (T_g/K)
Poly[1-(4-isopropylbenzoyloxy)ethylene]	342
Poly[1-(2-methoxybenzoyloxy)ethylene]	338
Poly[1-(3-methoxybenzoyloxy)ethylene]	ca 317
Poly[1-(4-methoxybenzoyloxy)ethylene]	360
Poly[1-(4-methylbenzoyloxy)ethylene]	343
Poly[1-(4-nitrobenzoyloxy)ethylene]	395
Poly[1-(propionoyloxy)ethylene]	283 (M)
Polystyrenes	
Poly(1-phenylethylene) *Polystyrene* $–[C_6H_5CHCH_2]–$	
Poly[1-(4-acetylphenyl)ethylene]	389 (M)
Poly[1-(4-benzoylphenyl)ethylene]	371 (M)
Poly[1-(4-bromophenyl)ethylene]	391
Poly[1-(4-butoxyphenyl)ethylene]	ca 320 (M)
Poly[1-(4-butoxycarbonylphenyl)ethylene]	349 (M)
Pol[(1-(4-butylphenyl)ethylene]	279
Poly[1-(4-carboxyphenyl)ethylene]	386 (M)
Poly[1-(2-chlorophenyl)ethylene]	392
Poly[1-(3-chlorophenyl)ethylene]	363
Poly[1-(4-chlorophenyl)ethylene]	383
Poly[1-(2,4-dichlorophenyl)ethylene]	406
Poly[1-(2,5-dichlorophenyl)ethylene]	379
Poly[1-(2,6-dichlorophenyl)ethylene]	440
Poly[1-(3,4-dichlorophenyl)ethylene]	401
Poly[1-(2,4-dimethylphenyl)ethylene]	385
Poly[1-(4-(dimethylamino)phenyl)ethylene]	398 (M)
Poly[1-(4-ethoxyphenyl)ethylene]	ca 359 (M)
Poly[1-(4-ethoxycarbonylphenyl)ethylene]	367 (M)
Poly[1-(4-fluorophenyl)ethylene]	368
Poly[1-(4-iodophenyl)ethylene]	429
Poly[1-(4-methoxyphenyl)ethylene]	386
Poly[1-(4-methoxycarbonylphenyl)ethylene]	386 (M)
Poly(1-methyl-1-phenylethylene)	373
Poly(α-methylstyrene)	
Poly[1-(2-(methylamino)phenyl)ethylene]	462 (M)
Poly(1-phenylethylene)	373
Polystyrene [PS]	
Poly[1-(4-propoxyphenyl)ethylene]	343 (M)
Poly[1-(4-propoxycarbonylphenyl)ethylene]	365 (M)
Poly(1-*o*-tolylethylene)	409
CHAINS WITH CARBOCYCLIC UNITS	
Poly(arylenealkylene) $–[–Ar–(CH_2)_n]–$	
Poly[1-(2-bromo-1,4-phenylene)ethylene]	353 (M)
Poly[1-(2-chloro-1,4-phenylene)ethylene]	343 (M)
Poly[1-(2-cyano-1,4-phenylene)ethylene]	363 (M)
Poly[1-(2,5-dimethyl-1,4-phenylene)ethylene]	373 (M)
Poly[1-(2-ethyl-1,4-phenylene)ethylene]	298 (M)
Poly[1-(1,4-naphthylene)ethylene]	433 (M)
Poly[1-(1,4-phenylene)ethylene]	ca 353 (M)
CHAINS WITH HETEROATOM UNITS	
Main chain oxide units	
Poly(oxyalkylene) *Poly(alkylene oxide)* $–[O(CH_2)_n]–$	
Poly[oxy(1,1-bis(chloromethyl)trimethylene)]	265
Poly[oxy(1-(bromomethyl)ethylene)]	259
Poly[oxy(1-(butoxymethyl)ethylene)]	194
Poly[oxy(1-butylethylene)]	203
Poly[oxy(1-*tert*-butylethylene)]	308
Poly[oxy(1-(chloromethyl)ethylene)]	251

Polymer name	Glass transition temperature (T_g/K)
Poly(epichlorohydrin)	
Poly[oxy(2,6-dimethoxy-1,4-phenylene)]	440
Poly[oxy(1,1-dimethylethylene)]	264
Poly[oxy(2,6-dimethyl-1,4-phenylene)]	482
Poly[oxy(2,6-diphenyl-1,4-phenylene)]	493
Poly[oxy(1-ethylethylene)]	203
Poly(oxyethylidene)	243
Polyacetaldehyde	
Poly[oxy(1-(methoxymethyl)ethylene)]	211
Poly[oxy(2-methyl-6-phenyl-1,4-phenylene)]	428
Poly[oxy(1-methyltrimethylene)]	223 (D)
Poly[oxy(2-methyltrimethylene)]	218
Poly(oxy-1,4-phenylene)	358
Poly(phenylene oxide) [PPO]	
Poly[oxy(1-phenylethylene)]	313
Poly(oxytetramethylene)	189
Poly(tetrahydrofuran) [PTMO]	
Poly(oxytrimethylene)	195
Main-chain ester or anhydride units	
Poly(oxyalkyleneoxyalkanedioyl) *Poly(alkylene alkanedioate)*–[O(CH$_2$)$_m$OCO(CH$_2$)$_n$CO]–	
Poly(oxyadipoyloxydecamethylene)	217
Poly(oxyadipoyloxy-1,4-phenyleneisopropylidene-1,4-phenylene)	341
Poly(oxycarbonyloxy-1,4-phenylene-isopropylidene-1,4-phenylene)	422
Bisphenol A polycarbonate	
Poly(oxycarbonylpentamethylene)	213
Poly(oxycarbonyl-1,4-phenylenemethylene-1,4-phenylene)	395
Poly(oxycarbonyl-1,4-phenyleneisopropylidene-1,4-phenylene)	333
Poly[oxy(2,6-dimethyl-1,4-phenyleneisopropylidene-3,5-dimethyl-1,4-phenylene)oxysebacoyl]	318
Poly(oxyethylenecarbonyl-1,4-cyclohexylenecarbonyl) (trans)	291
Poly(oxyethyleneoxycarbonyl-1,4-naphthylenecarbonyl)	337
Poly(oxyethyleneoxycarbonyl-1,5-naphthylenecarbonyl)	344
Poly(oxyethyleneoxycarbonyl-2,6-naphthylenecarbonyl)	386
Poly(oxyethyleneoxycarbonyl-2,7-naphthylenecarbonyl)	392
Poly(oxyethyleneoxyterephthaloyl)	342
Poly(ethylene terephthalate) [PET]	
Poly(oxyisophthaloyl)	403 (D)
Poly(oxy(1-oxo-2,2-dimethyltrimethylene))	263
Poly(pivalolactone)	
Poly(oxy-1,4-phenyleneisopropylidene-1,4-phenyleneoxysebacoyl)	280
Poly(oxy-1,4-phenyleneoxy-1,4-phenyleneoxy-carbonyl-1,4-phenylene) [PEEK]	416
Poly(oxypropyleneoxyterephthaloyl)	341
Poly[oxyterephthaloyloxy(2,6-dimethyl-1,4-phenyleneisopropylidene-3,5-dimethyl-1,4-(D)phenylene)]	498
Poly(oxyterephthaloyloxyoctamethylene)	318 (D)
Poly(oxyterephthaloyloxy-1,4-phenyleneisopropylidene-1,4-phenylene)	478
Poly(bisphenol A terephthalate)	
Poly(oxytetramethyleneoxyterephthaloyl)	323
Poly(butylene terephthalate) [PBT]	
Main-chain amide units	
Poly(iminoalkyleneiminoalkanedioyl) *Poly(alkylene alkanediamide)*–[NH(CH$_2$)$_m$NHCO(CH$_2$)$_n$CO]–	
Poly(iminoadipoyliminodecamethylene)	313
Nylon 10,6	
Poly(iminoadipoyliminohexamethylene)	ca 323
Nylon 6,6	
Poly(iminoadipoyliminooctamethylene)	318
Nylon 8,6	
Poly[iminoadipoyliminotrimethylene(methylimino)trimethylene]	278
Poly(iminocarbonyl-1,4-cyclohexylenemethylene)	466
Poly[iminocarbonyl-1,4-phenylene(2-oxoethylene)iminohexamethylene]	377
Poly(iminoethylene-1,4-phenyleneethyleneiminosebacoyl)	378 (D)

Polymers

Polymer name	Glass transition temperature (T_g/K)
Poly(iminohexamethyleneiminoazelaoyl)	331
Nylon 6,9	
Poly(iminohexamethyleneiminododecanedioyl)	319
Nylon 6, 12	
Poly(iminohexamethyleneiminopimeloyl)	331
Nylon 6,7	
Poly(iminohexamethyleneiminosebacoyl)	323
Nylon 6,10	
Poly(iminohexamethyleneiminosuberoyl)	330
Nylon 6,8	
Poly(iminoisophthaloylimino-4,4′-biphenylylene)	558
Poly(iminoisophthaloyliminohexamethylene)	390
Poly(iminoisophthaloyliminomethylene-1,4-cyclohexylenemethylene)	481
Poly(iminoisophthaloyliminomethylene-1,3-phenylenemethylene)	438 (M)
Poly[iminomethylene(2,5-dimethyl-1,4-phenylene)methyleneiminosuberoyl]	351
Poly(imino-1,5-naphthyleneiminoisophthaloyl)	598
Poly(imino-1,5-naphthyleneiminoterephthaloyl)	578
Poly(iminooctamethyleneiminodecanedioyl)	333
Nylon 8,10	
Poly(iminooxalyliminohexamethylene)	430
Nylon 6,2	
Poly[imino(1-oxohexamethylene)]	326
Nylon 6	
Poly[imino(1-oxodecamethylene)]	315
Nylon 10	
Poly[imino(1-oxoheptamethylene)]	325
Nylon 7	
Poly[imino(1-oxo-3-methyltrimethylene]	369
Poly[imino(1-oxononamethylene)]	319
Nylon 9	
Poly[imino(1-oxooctamethylene)]	323
Nylon 8	
Poly[imino(1-oxotrimethylene)]	384
Nylon 3	
Poly(iminopentamethyleneiminoadipoyl)	318
Nylon 5,6	
Poly[iminopentamethyleneiminocarbonyl-1,4-phenylene(2-oxoethylene)]	376
Poly(imino-1,3-phenyleneiminoisophthaloyl)	553 (M)
Poly(imino-1,4-phenyleneiminoterephthaloyl)	618
Poly(iminopimeloyliminoheptamethylene)	328
Nylon 7,7	
Poly(iminoterephthaloylimino-4,4′-biphenylylene)	613
Poly(iminotetramethyleneiminoadipoyl)	316
Nylon 4,6	
Poly[iminotetramethyleneiminocarbonyl-1,4-phenylene(2-oxoethylene)]	357
Poly(iminotrimethyleneiminoadipoyliminotrimethylene)	307
Poly[iminotrimethyleneiminocarbonyl-1,4-phenylene(2-oxoethylene)]	382
Poly(oxy-1,4-phenyleneiminoterephthaloyl-imino-1,4-phenylene)	613
Poly(sulfonylimino-1,4-phenyleneiminoadipoylimino-1,4-phenylene)	467
Main-chain urethane units	
Poly(oxyalkyleneoxycarbonyliminoalkyleneiminocarbonyl)–$[O(CH_2)_mOCONH(CH_2)_nNHCO]$–	
Poly(oxyethyleneoxycarbonyliminohexamethyleneiminocarbonyl)	329
Poly[oxyethyleneoxycarbonylimino(6-methyl-1,3-phenylene)iminocarbonyl]	325
Poly(oxyethyleneoxycarbonylimino-1,4-phenylenemethylene-1,4-phenyleneiminocarbonyl)	412
Poly(oxyhexamethyleneoxycarbonyliminohexamethyleneiminocarbonyl)	332
Poly[oxyhexamethyleneoxycarbonylimino(6-methyl-1,3-phenylene)iminocarbonyl]	305
Poly(oxyhexamethyleneoxycarbonylimino-1,4-phenylenemethylene-1,4-phenyleneiminocarbonyl)	364
Poly(oxyoctamethyleneoxycarbonyliminohexamethyleneiminocarbonyl)	331
Poly[oxyoctamethyleneoxycarbonylimino(6-methyl-1,3-phenylene)iminocarbonyl]	337
Poly(oxyoctamethyleneoxycarbonylimino-1,4-phenylenemethylene-1,4-phenyleneiminocarbonyl)	352

Polymer name	Glass transition temperature (T_g/K)
Poly(oxytetramethyleneoxycarbonyliminohexamethyleneiminocarbonyl)	332
Poly[oxytetramethyleneoxycarbonylimino(6-methyl-1,3-phenylene)iminocarbonyl]	315
Poly(oxytetramethyleneoxycarbonylimino-1,4-phenylenemethylene-1,4-phenyleneiminocarbonyl)	382
Main-chain siloxanes	
Poly[oxy(dialkylsilylene)] *Poly(dialkylsiloxane)* –[O(R$_2$Si)]–	
Poly[oxy(dimethylsilylene)]	148
Poly(dimethylsiloxane) [PDMS]	
Poly[oxy(dimethylsilylene)oxy-1,4-phenylene]	363 (M)
Poly[oxy(dimethylsilylene)oxy-1,4-phenyleneisopropylidene-1,4-phenylene]	318 (M)
Poly[oxy(diphenylsilylene)]	238
Poly(diphenylsiloxane)	
Poly[oxy(diphenylsilylene)-1,3-phenylene]	ca 331
Poly[oxy((methyl)phenylsilylene)]	187
Poly[oxy((methyl)-3,3,3-trifluoropropylsilylene]	<193
Main-chain sulfur-containing units	
Poly(dithioethylene)	223
Poly(dithiomethylene-1,4-phenylenemethylene)	296
Poly(oxy-4,4′-biphenylylene-1,4-phenylenesulfonyl-1,4-phenylene)	503 (M)
Poly(oxycarbonyloxy-1,4-phenylenethio-1,4-phenylene)	ca 383
Poly(oxyethylenedithioethylene)	220 (M)
Poly[oxy(2-hydroxytrimethylene)oxy-1,4-phenylenesulfonyl-1,4-phenylene]	428
Poly(oxymethyleneoxyethylenedithioethylene)	214
Poly(oxy-1,4-phenylenesulfinyl-1,4-phenyleneoxy-1,4-phenylenecarbonyl-1,4-phenylene)	478 (M)
Poly(oxy-1,4-phenylenesulfinyl-1,4-phenyleneoxy-1,4-phenyleneisopropylidene-1,4-phenylene)	438 (M)
Poly(oxy-1,4-phenylenesulfonyl-1,4-phenylene)	487
Poly(oxy-1,4-phenylenesulfonyl-4,4′-biphenylylenesulfonyl-1,4-phenylene)	533
Poly[oxy-1,4-phenylenesulfonyl-1,4-phenyleneoxy(2,6-dimethyl-1,4-phenylene)isopropylidene (3,5-dimethyl-1,4-phenylene)]	508 (M)
Poly(oxy-1,4-phenylenesulfonyl-1,4-phenyleneoxy-1,4-phenylenecarbonyl-1,4-phenylene)	478 (M)
Poly[oxy-1,4-phenylenesulfonyl-1,4-phenyleneoxy-1,4-phenylene(hexafluoroisopropylidene)1,4-phenylene]	478 (M)
Poly(oxy-1,4-phenylenesulfonyl-1,4-phenyleneoxy-1,4-phenyleneisopropylidene-1,4-phenylene)	449
Poly(oxy-1,4-phenylenesulfonyl-1,4-phenyleneoxy-1.4-phenylenemethylene-1,4-phenylene)	453 (M)
Poly(oxy-1,4-phenylenesulfonyl-1,4-phenyleneoxy-1.4-phenylenethio-1,4-phenylene)	448 (M)
Poly(oxy-1,4-phenylenesulfonyl-1,4-phenyleneoxyterephthaloyl)	522
Poly(oxytetramethylenedithiotetramethylene)	197
Poly(sulfonyl-1,2-cyclohexylene)	401
Poly(sulfonyl-1,3-cyclohexylene)	381
Poly(sulfonyl-1,4-phenylenemethylene-1,4-phenylene)	497
Poly(thio-1,3-cyclohexylene)	221
Poly[thio(difluoromethylene)]	155
Poly(thioethylene)	223
Poly[thio(1-ethylethylene]	218
Poly[thio(1-methyl-3-oxotrimethylene)]	285
Poly[thio(1-methyltrimethylene)]	214
Pol[(thio(1-oxohexamethylene)]	292
Poly(thio-1,4-phenylene)	370
Poly(thiopropylene)	226
Main-chain heterocyclic units	
Poly(1,3-dioxa-4,6-cyclohexylenemethylene)	378
Poly(vinyl formal)	
Poly[(2,6-dioxopiperidine-1,4-diyl)trimethylene]	363
Poly[(2-methyl-1,3-dioxa-4,6-cyclohexylene)methylene]	355
Poly(vinyl acetal)	
Poly(1,4-piperazinediylcarbonyloxyethyleneoxycarbonyl)	333
Poly(1,4-piperazinediylisophthaloyl)	465 (M)
Poly[(2-propyl-1,3-dioxa-4,6-cyclohexylene)methylene]	322
Poly(vinyl butyral)	
Poly(3,6-pyridazinediyloxy-1,4-phenyleneisopropylidene-1,4-phenyleneoxy)	453 (M)
Poly(2,5-pyridinediylcarbonyliminohexamethyleneiminocarbonyl)	322

DIELECTRIC CONSTANT OF SELECTED POLYMERS

This table lists typical values of the dielectric constant (more properly called relative permittivity) of some important polymers. Values are given for frequencies of 1 kHz, 1 MHz, and 1 GHz; in most cases the dielectric constant at frequencies below 1 kHz does not differ significantly from the value at 1 kHz. Since the dielectric constant of a polymeric material can vary with density, degree of crystallinity, and other details of a particular sample, the values given here should be regarded as only typical or average values.

References

1. Gray, D. E., Ed., *American Institute of Physics Handbook, Third Edition*, p. 5-132, McGraw Hill, New York, 1972.
2. Anderson, H. L., Ed., *A Physicist's Desk Reference*, American Institute of Physics, New York, 1989.
3. Brandrup, J., and Immergut, E. H., *Polymer Handbook, Third Edition*, John Wiley & Sons, New York, 1989.

Name	*t*/°C	1 kHz	1 MHz	1 GHz
Polyacrylonitrile	25	5.5	4.2	
Polyamides (nylons)	25	3.50	3.14	2.8
	84	11	4.4	2.8
Polybutadiene	25	2.5		
Polycarbonate	23	2.92	2.8	
Polychloroprene (neoprene)	25	6.6	6.3	4.2
Polychlorotrifluoroethylene	23	2.65	2.46	2.39
Polyethylene	23	2.3		
Poly(ethylene terephthalate) (Mylar)	23	3.25	3.0	2.8
Polyisoprene (natural rubber)	27	2.6	2.5	2.4
Poly(methyl methacrylate)	27	3.12	2.76	2.6
	80	3.80	2.7	2.6
Polyoxymethylene (polyformaldehyde)	25	3.8		
Poly(phenylene oxide)	23	2.59	2.59	
Polypropylene	25	2.3	2.3	2.3
Polystyrene	25	2.6	2.6	2.6
Polysulfones	25	3.13	2.10	
Polytetrafluoroethylene (teflon)	25	2.1	2.1	2.1
Poly(vinyl acetate)	50		3.5	
	150		8.3	
Poly(vinyl chloride)	25	3.39	2.9	2.8
	100	5.3	3.3	2.7
Poly(vinylidene chloride)	23	4.6	3.2	2.7
Poly(vinylidene fluoride)	23	12.2	8.9	4.7

SECOND VIRIAL COEFFICIENTS OF POLYMER SOLUTIONS

Christian Wohlfarth

Second virial coefficients characterize the thermodynamic behavior of a dilute polymer solution. They are usually defined via the concentration dependence of the osmotic pressure, π, of a polymer solution:

$$\frac{\pi}{c_B} = RT\left[\frac{1}{M_n} + A_2 c_B + A_3 c_B^2 + \ldots\right] \quad (1)$$

where:

A_2, A_3	second, third osmotic virial coefficient
c_B	(g cm^{-3}) concentration of polymer (B)
M_n	number-average relative molar mass of the polymer
R	gas constant
T	(measuring) temperature

However, most experimental data are measured by light scattering. Scattering methods enable the determination of A_2 via the common relation:

$$\frac{Kc_B}{R(q)} = \frac{1}{M_w P_z(q)} + 2A_2Q(q)c_B + \ldots \quad (2)$$

where:

K	a constant that summarizes the optical parameters of a scattering experiment
M_w	mass-average relative molar mass of the polymer
$P_z(q)$	z-average of the scattering function
q	scattering vector $q = \frac{4\pi}{\lambda}\sin\frac{\theta}{2}$
$Q(q)$	function for the q-dependence of A_2
$R(q)$	excess intensity of the scattered beam at the value q
λ	wavelength
θ	scattering angle

In the dilute concentration region, the virial equation is usually truncated after the second virial coefficient which leads to a linear relationship. A linearized relation over a wider concentration range can be constructed if the Stockmayer-Casassa relation between A_2 and A_3 is applied:

$$A_3 M_n = \left(\frac{A_2 M_n}{2}\right)^2 \quad (3)$$

The values of second virial coefficients depend on the chosen polymer-solvent pair, on temperature and pressure, and on molar mass. In good solvents, a scaling relation can be applied for the molar mass dependence:

$$A_2 = \alpha M_w^{-\beta} \quad (4)$$

The constants α and β for a number of polymer-solvent pairs are given in the table below. They were newly fitted to experimental data by a nonlinear least-squares fitting procedure. Experimental data were taken from a recent table of experimental A_2-values in Ref. 1. The unit of A_2 (used to determine the constants α and β) is cm^3 mol g^{-2}. The used unit for the molar mass is g mol^{-1}.

Reference

1. Wohlfarth, C., *Thermodynamic Properties of Polymer Solutions*, in Landolt-Börnstein, New Series, Group VIII, Volume 6D, Lechner, M.D. (ed.), Springer Verlag, Berlin, Heidelberg, 2010.

Polymer	Solvent	T/K	α	β
Amylose	dimethylsulfoxide	298.15	0.0086	0.235
Amylose tris(*N*-phenyl carbamate)	2-propanone	290.15	0.0157	0.417
Cellulose tris(*N*-phenyl carbamate)	2-propanone	290.15	0.0023	0.177
Cellulose tris(*N*-phenyl carbamate)	tetrahydrofuran	298.15	0.0021	0.116
Dextran	water	298.15	0.0353	0.394
Dextran (acid-hydrolyzed)	water	298.15	0.0516	0.361
2-(Diethylamino)ethyl dextran	water	293.15	0.129	0.503
Hydroxypropyl cellulose	ethanol	298.15	0.108	0.371
Poly(*N*-acryloylmorpholine)	*N,N*-dimethylformamide	298.15	0.0084	0.295
Polybutadiene	cyclohexane	298.15	0.0112	0.204
Polybutadiene	tetrahydrofuran	296.15	0.0235	0.257
Poly(butyl methacrylate)	2-butanone	298.15	0.0099	0.285
Poly(4-*tert*-butylstyrene)	cyclohexane	298.15	0.0377	0.362
Poly(4-*tert*-butylstyrene)	toluene	310.15	0.0129	0.293
Polycarbonate-bisphenol-A	1,2-dichloroethane	298.15	0.0372	0.319
Polycarbonate-bisphenol-A	dichloromethane	298.15	0.0124	0.205
Polycarbonate-bisphenol-A	tetrahydrofuran	298.15	0.0426	0.333
Polycarbonate-bisphenol-A	trichloromethane	298.15	0.016	0.227
Poly(chloroprene)	butyl acetate	298.15	0.0041	0.229
Poly(chloroprene)	tetrachloromethane	298.15	0.0083	0.237
Poly(2-chlorostyrene)	toluene	298.15	0.0036	0.219
Poly(2-chlorostyrene)	toluene	308.15	0.0065	0.246
Poly(3-chlorostyrene)	toluene	308.15	0.0232	0.319
Poly(4-chlorostyrene)	2-butanone	294.15	0.0041	0.262

Polymer	Solvent	T/K	α	β
Poly(4-chlorostyrene)	toluene	303.15	0.0018	0.241
Poly(cyclobutyl methacrylate)	tetrahydrofuran	297.15	0.0114	0.271
Poly(cyclododecyl methacrylate)	cyclohexane	298.15	0.0029	0.242
Poly(cyclohexyl methacrylate)	cyclohexane	298.15	0.0071	0.291
Poly(cyclooctyl methacrylate)	cyclohexane	298.15	0.0056	0.271
Poly(cyclopentyl methacrylate)	tetrahydrofuran	297.15	0.0046	0.210
Poly(decyl methacrylate)	toluene	310.15	0.0389	0.398
Poly(2,5-dichlorostyrene)	1,4-dioxane	293.15	0.0044	0.295
Poly(dihexylsilylene)	tetrahydrofuran	298.15	0.0025	0.212
Poly(diisopropyl fumarate)	tetrahydrofuran	303.15	0.0044	0.197
Poly(*N*,*N*-dimethylacrylamide)	methanol	298.15	0.0168	0.266
Poly(*N*,*N*-dimethylacrylamide)	water	298.15	0.0298	0.365
Poly(3,5-dimethyl-1-methacryloylpyrazole)	trichloromethane	293.15	0.0228	0.378
Poly[*N*-(1,1-dimethyl-3-oxobutyl)acrylamide]	2-butanone	298.15	0.0134	0.310
Poly(2,6-dimethyl-1,4-phenylene ether)	toluene	298.15	0.0116	0.205
Poly(dimethylsiloxane)	toluene	298.15	0.0231	0.333
Poly(2,2-diphenylethyl methacrylate)	tetrahydrofuran	298.15	0.0052	0.275
Poly(diphenylmethyl methacrylate)	toluene	296.15	0.0027	0.256
Poly(docosyl methacrylate)	toluene	310.15	0.0445	0.423
Poly(dodecyl acrylate)	butyl acetate	296.15	0.0040	0.289
Poly(2-ethoxyethyl methacrylate)	2-butanone	298.15	0.0328	0.349
Poly(ethyl acrylate)	tetrahydrofuran	298.15	0.0320	0.328
Polyethylene (LDPE)	1-chloronaphthalene	398.15	0.109	0.419
Polyethylene (LDPE)	1,4-dimethylbenzene	354.15	0.0677	0.361
Polyethylene (LDPE)	tetrahydronaphthalene	354.65	0.113	0.443
Poly(ethylene-*alt*-propylene)	toluene	310.15	0.0451	0.333
Polyethylenimine	water	308.15	0.0813	0.490
Poly(ethyl methacrylate)	2-butanone	296.15	0.0575	0.392
Poly(ethyl methacrylate)	ethyl acetate	308.15	0.0069	0.265
Poly(3-fluorostyrene)	2-butanone	298.15	0.0542	0.397
Poly(4-fluorostyrene)	2-butanone	298.15	0.0164	0.280
Poly(hexadecyl methacrylate)	n-heptane	298.15	0.0196	0.328
Poly(hexyl acrylate)	tetrahydrofuran	298.15	0.0111	0.272
Poly(hexyl methacrylate)	2-butanone	296.15	0.0046	0.236
Poly(1-indanyl methacrylate)	tetrahydrofuran	298.15	0.0028	0.173
Poly(isobornyl methacrylate)	tetrahydrofuran	298.15	0.0042	0.244
Poly(isobutylene)	cyclohexane	298.15	0.0095	0.224
Poly(isobutylene)	n-heptane	298.15	0.0145	0.296
Poly(isobutylene)	n-hexane	297.15	0.0018	0.155
Poly(isobutylene)	trichloromethane	298.15	0.0034	0.204
Poly(isobutylene)	2,2,4-trimethylpentane	298.15	0.0051	0.219
Poly(isoprene) (70% *cis*, 23% *trans*)	cyclohexane	296.15	0.0175	0.260
Poly(isoprene) (*trans*)	cyclohexane	298.15	0.0064	0.179
Poly(isoprene) (3-arm star)	cyclohexane	296.15	0.0234	0.279
Poly(isoprene) (4-arm star)	cyclohexane	296.15	0.0172	0.249
Poly(isoprene) (8-arm star)	cyclohexane	296.15	0.0198	0.285
Poly(isoprene) (12-arm star)	cyclohexane	296.15	0.0108	0.257
Poly(isoprene) (18-arm star)	cyclohexane	296.15	0.0247	0.330
Poly(isoprene) (hydrogenated)	toluene	310.15	0.0455	0.333
Poly(*N*-isopropylacrylamide)	methanol	298.15	0.0085	0.250
Poly(*N*-isopropylacrylamide)	water	293.15	0.0028	0.195
Poly(isopropyl acrylate) (atactic)	bromobenzene	333.15	0.0172	0.285
Poly(isopropyl acrylate) (isotactic)	bromobenzene	333.15	0.0118	0.263
Poly(*N*-isopropylmethacrylamide)	water	293.15	0.0235	0.355
Poly(2-methoxyethyl methacrylate)	2-butanone	298.15	0.0525	0.397
Poly(methyl 2-butylacrylate)	2-butanone	303.15	0.0036	0.223
Poly(methyl 2-ethylacrylate)	2-butanone	303.15	0.0068	0.251
Poly(methyl methacrylate)	2-butanone	298.15	0.0052	0.254
Poly(methyl methacrylate)	1,4-dioxane	298.15	0.0164	0.295
Poly(methyl methacrylate)	nitroethane	298.15	0.0072	0.263

Polymer	Solvent	*T*/K	*α*	*β*
Poly(methyl methacrylate)	2-propanone	298.15	0.0035	0.235
Poly(methyl methacrylate)	tetrahydrofuran	298.15	0.0119	0.281
Poly(methyl methacrylate)	trichloromethane	298.15	0.0129	0.246
Poly(methylphenylsiloxane)	cyclohexane	298.15	0.0023	0.216
Poly(α-methylstyrene)	4-*tert*-butyltoluene	298.15	0.0027	0.239
Poly(α-methylstyrene)	1-chorobutane	298.15	0.0025	0.250
Poly(α-methylstyrene)	toluene	298.15	0.0124	0.298
Poly(2-methylstyrene)	toluene	303.15	0.0268	0.349
Poly(4-methylstyrene)	toluene	303.15	0.0145	0.289
Poly(2-methyl-5-vinylpyridine)	2-butanone	298.15	0.0151	0.335
Poly(2-methyl-5-vinyltetrazole)	*N,N*-dimethylformamide	293.15	0.0062	0.251
Poly(octadecyl methacrylate)	toluene	310.15	0.0106	0.312
Poly(1-octene)	bromobenzene	298.15	0.0051	0.222
Poly(1-octene)	n-heptane	293.15	0.0473	0.330
1,4-*trans*-Poly(1,3-pentadiene)	toluene	300.75	0.0377	0.405
trans-1,5-Polypentenamer	n-hexane	298.15	0.0048	0.233
Poly(1-pentene)	toluene	303.15	0.0554	0.387
Poly(phenyl methacrylate)	2-butanone	298.15	0.0105	0.328
Poly(1-phenyl-1-propyne)	toluene	298.15	0.025	0.325
Poly(phenylsilesquioxane)	toluene	310.15	0.0046	0.288
Poly(propyl acrylate)	tetrahydrofuran	298.15	0.0094	0.251
Polypropylene (atactic)	benzene	298.15	0.0108	0.311
Polypropylene (isotactic)	1-chloronaphthalene	418.15	0.0025	0.132
Polypropylene (isotactic)	tetrahydronaphthalene	408.15	0.0045	0.118
Poly(propylene sulfide)	benzene	293.15	0.0814	0.493
Polystyrene	benzene	298.15	0.0113	0.259
Polystyrene (4-arm star)	benzene	298.15	0.0092	0.258
Polystyrene (6-arm star)	benzene	298.15	0.0093	0.268
Polystyrene	bromobenzene	293.15	0.0142	0.282
Polystyrene	2-butanone	298.15	0.0028	0.233
Polystyrene (cyclic)	2-butanone	298.15	0.0207	0.395
Polystyrene	4-*tert*-butyltoluene	323.15	0.0035	0.221
Polystyrene	1,2-dichloroethane	308.15	0.0086	0.262
Polystyrene	ethylbenzene	298.15	0.0139	0.289
Polystyrene	tetrachloromethane	298.15	0.0084	0.246
Polystyrene	tetrahydrofuran	298.15	0.0129	0.268
Polystyrene	toluene	298.15	0.0118	0.264
Polystyrene	toluene	303.15	0.0076	0.241
Polystyrene	toluene	308.15	0.0102	0.267
Polystyrene (cyclic)	toluene	298.15	0.0144	0.292
Polystyrene (3-arm star)	toluene	293.15	0.0042	0.196
Polystyrene (12-arm star)	toluene	293.15	0.0282	0.369
Poly(tetrahydro-2H-pyran-2-yl methacrylate)	toluene	310.15	0.0182	0.362
Poly[4-(1,1,3,3-tetramethylbutyl)-phenyl methacrylate]	toluene	298.15	0.054	0.420
Poly(tridecyl methacrylate)	toluene	310.15	0.0112	0.325
Poly(vinyl acetate)	2-butanone	303.15	0.0117	0.269
Poly(vinyl acetate)	2-propanone	303.15	0.0132	0.268
Poly(*N*-vinylcarbazole)	benzene	298.15	0.0012	0.197
Poly(*N*-vinylcarbazole)	benzene	303.15	0.0014	0.193
Poly(*N*-vinylcarbazole)	benzene	308.15	0.0019	0.200
Poly(*N*-vinylcarbazole)	benzene	313.15	0.0025	0.208
Poly(*N*-vinylcarbazole)	1,4-dioxane	310.15	0.0038	0.261
Poly(vinyl chloride)	tetrahydrofuran	298.15	0.0096	0.157
Poly(vinyl phenylcarbamate)	1,4-dioxane	293.15	0.0130	0.307
Poly(2-vinylpyridine-1-oxide)	2-propanol	298.15	0.0332	0.365
Poly(1-vinyl-2-pyrrolidinone)	ethanol	298.15	0.0482	0.392
Poly(1-vinyl-2-pyrrolidinone)	methanol	298.15	0.0384	0.369
Poly(1-vinyl-2-pyrrolidinone)	1-propanol	298.15	0.0409	0.385
Pullulan	water	298.15	0.0092	0.291

PRESSURE–VOLUME–TEMPERATURE RELATIONSHIPS FOR POLYMER MELTS

Christian Wohlfarth

Numerous theoretical equations of state for polymer liquids have been developed. These, at the minimum, have to provide accurate fitting functions to experimental data. However, for the purpose of this table, the empirical Tait equation along with a polynomial expression for the zero pressure isobar is used. This equation is able to represent the experimental data for the melt state within the limits of experimental errors, i.e., the maximum deviations between measured and calculated specific volumes are about 0.001-0.002 cm^3/g.

The general form of the Tait equation is:

$$V(P,T) = V(0,T)\{1 - C\ \ln[1 + P/B(T)]\} \quad (1)$$

where the coefficient C is usually taken to be a universal constant equal to 0.0894. T is the absolute temperature in K and P the pressure in MPa. The volume V is the specific volume in cm^3/g. The Tait parameter $B(T)$ has the very simple meaning that it is inversely proportional to the compressibility κ at constant temperature and zero pressure:

$$\kappa(0,T) = -[1/V(0,T)](dV/dP) = C/B(T) \quad (2)$$

The $B(T)$ function is usually given by:

$$B(T) = B_0 \exp[-B_1(T\text{-}273.15)] \quad (3)$$

but, sometimes a polynomial expression is used:

$$B(T) = b_0 + b_1(T\text{-}273.15) + b_2(T\text{-}273.15)^2 \quad (4)$$

The zero-pressure isobar $V(0,T)$ is usually given by:

$$V(0,T) = A_0 + A_1(T\text{-}273.15) + A_2(T\text{-}273.15)^2 \quad (5)$$

where A_0, A_1, A_2 are specific constants for a given polymer (the expression T-273.15 is used because fitting to the zero-pressure isobar is usually done in terms of Celsius temperature). Other forms for $V(0,T)$ are also found in the literature, such as

$$V(0,T) = A_3 \exp[A_4(T\text{-}273.15)] \quad (6)$$

or

$$V(0,T) = A_5 \exp(A_6 T^{1.5}) \quad (7)$$

where A_3 and A_4 or A_5 and A_6 are again specific constants for a given polymer.

The Tait equation is particularly useful to calculate derivative quantities, such as the isothermal compressibility and the thermal expansivity coefficients. The isothermal compressibility $\kappa(P,T)$ is derived from equation (1) as:

$$\kappa(P,T) = -(1/V)(dV/dP) = 1/\{[P + B(T)][1/C - \ln(1 + P/B(T))]\} \quad (8)$$

and the thermal expansivity $\alpha(P,T)$ as:

$$\alpha(P,T) = (1/V)(dV/dT) = \alpha(0,T) - PB_1\kappa(P,T) \quad (9)$$

where $\alpha(0,T)$ represents the thermal expansivity at zero (atmospheric) pressure and is calculated from any suitable fit for the zero-pressure volume, such as equations (5) through (7) above.

Because polymer melt PVT-behavior depends only slightly on polymer molar mass above the oligomeric region, usually no information is given in the original literature for the average molar mass of the polymers.

Table 1 summarizes the polymers or copolymers considered here and the experimental ranges of pressure and temperature over which data are available. In Table 2 the Tait-equation functions, with parameters obtained from the fit, are given for 90 polymer or copolymer melts.

References

1. Zoller, P., *J. Appl. Polym. Sci.*, 23, 1051–1056, 1979.
2. Starkweather, H. W., Jones, G. A., and Zoller, P., *J. Polym. Sci., Pt. B Polym. Phys.*, 26, 257–266,1988.
3. Fakhreddine, Y. A., and Zoller, P., *J. Polym. Sci., Pt. B Polym. Phys.*, 29, 1141–1146, 1991.
4. Rodgers, P. A., *J. Appl. Polym. Sci.*, 48, 1061–1080, 1993.
5. Rodgers, P. A., *J. Appl. Polym. Sci.*, 48, 2075–2083, 1993.
6. Yi, Y. X., and Zoller, P., *J. Polym. Sci., Pt. B Polym. Phys.*, 31, 779–788, 1993.
7. Callaghan, T. A., and Paul, D. R., *Macromolecules*, 26, 2439–2450, 1993.
8. Wang, Y. Z., Hsieh, K. H., Chen, L. W.,and Tseng, H. C., *J. Appl. Polym. Sci.*, 53, 1191–1201, 1994.
9. Privalko, V. P., Arbuzova, A. P., Korskanov, V. V., and Zagdanskaya, N. E., *Polym. Intern.*, 35, 161–169, 1994.
10. Sachdev, V. K., Yashi, U., and Jain, R. K., *J. Polym. Sci., Pt. B Polym. Phys.*, 36, 841–850, 1998.

TABLE 1. Names of the Polymers, Abbreviation Used, and Range of Experimental Data Applied in the Determination of the Equation Constants

Polymer	Symbol	T/K	P/MPa	Ref.
Ethylene/propylene copolymer (50 wt%)	EP50	413-523	0.1-63	4
Ethylene/vinyl acetate copolymer				
18 wt% vinyl acetate	EVA18	385-491	0.1-177	4
25 wt% vinyl acetate	EVA25	367-506	0.1-177	4
28 wt% vinyl acetate	EVA28	367-508	0.1-177	4
40 wt% vinyl acetate	EVA40	348-508	0.1-177	4
Polyamide-6	PA6	509-569	0.1-196	4
Polyamide-11	PA11	478-542	0.1-200	5
Polyamide-66	PA66	519-571	0.1-196	4
cis-1,4-Polybutadiene	cPBD	277-328	0.1-284	4
Polybutadiene, 8% 1,2-content	PBD-8	298-473	0.1-200	6
Polybutadiene, 24% 1,2-content	PBD-24	298-473	0.1-200	6
Polybutadiene, 40% 1,2-content	PBD-40	298-473	0.1-200	6
Polybutadiene, 50% 1,2-content	PBD-50	298-473	0.1-200	6
Polybutadiene, 87% 1,2-content	PBD-87	298-473	0.1-200	6
Poly(1-butene), isotactic	iPB	406-519	0.1-196	4
Poly(butyl methacrylate)	PnBMA	307-473	0.1-200	4
Poly(butylene terephthalate)	PBT	508-576	0.1-200	3
Poly(ε-caprolactone)	PCL	373-421	0.1-200	4
Polycarbonate-bisphenol-A	PC	424-613	0.1-177	4
Polycarbonate-bisphenol-chloral	BCPC	428-557	0.1-200	4
Polycarbonate-hexafluorobisphenol-A	HFPC	432-553	0.1-200	4
Polycarbonate-tetramethylbisphenol-A	TMPC	491-563	0.1-160	4
Poly(cyclohexyl methacrylate)	PcHMA	396-471	0.1-200	4
Poly(2,5-dimethylphenylene oxide)	PPO	473-593	0.1-177	4
Poly(dimethyl siloxane)	PDMS	298-343	0.1-100	4
Poly(dimethyl siloxane) M_n = 1000	PDMS-10	304-420	0.1-250	10
Poly(dimethyl siloxane) M_n = 4000	PDMS-40	298-418	0.1-250	10
Poly(dimethyl siloxane) M_n = 6000	PDMS-60	291-423	0.1-250	10
Poly(epichlorohydrin)	PECH	333-413	0.1-200	4
Poly(ether ether ketone)	PEEK	619-671	0.1-200	4
Poly(ethyl acrylate)	PEA	310-490	0.1-196	4
Poly(ethyl methacrylate)	PEMA	386-434	0.1-196	4
Polyethylene, high density	HDPE	413-476	0.1-196	4
Polyethylene, linear	LPE	415-473	0.1-200	4
Polyethylene, linear, high MW	HMLPE	410-473	0.1-200	4
Polyethylene, branched	BPE	398-471	0.1-200	4
Polyethylene, low density	LDPE	394-448	0.1-196	4
Polyethylene, low density, type A	LDPE-A	385-498	0.1-196	1
Polyethylene, low density, type B	LDPE-B	385-498	0.1-196	1
Polyethylene, low density, type C	LDPE-C	385-498	0.1-196	1
Poly(ethylene oxide)	PEO	361-497	0.1- 68	4
Poly(ethylene terephthalate)	PET	547-615	0.1-196	4
Poly(4-hexylstyrene)	P4HS	303-403	30-100	4
Polyisobutylene	PIB	326-383	0.1-100	4
Polyisoprene, 8% 3,4-content	PI-8	298-473	0.1-200	6
Polyisoprene, 14% 3,4-content	PI-14	298-473	0.1-200	6
Polyisoprene, 41% 3,4-content	PI-41	298-473	0.1-200	6
Polyisoprene, 56% 3,4-content	PI-56	298-473	0.1-200	6
Poly(methyl acrylate)	PMA	310-493	0.1-196	4
Poly(methyl methacrylate)	PMMA	387-432	0.1-200	4
Poly(4-methyl-1-pentene)	P4MP	514-592	0.1-196	4
Poly(α-methylstyrene)	PαMS	473-533	0.1-170	7
Poly(*o*-methylstyrene)	PoMS	412-471	0.1-180	4
Polyoxymethylene	POM	463-493	0.1-196	2
Phenoxy[a]	PH	341-573	0.1-177	4
Polysulfone[b]	PSF	475-644	0.1-196	4
Polyarylate[c]	PAr	450-583	0.1-177	4
Polypropylene, atactic	aPP	353-393	0.1-100	4

Polymers

Polymer	Symbol	*T*/K	*P*/MPa	Ref.
Polypropylene, isotactic	iPP	443-570	0.1-196	4
Polystyrene	PS	388-469	0.1-200	4
Poly(tetrafluoroethylene)	PTFE	603-645	0.1- 39	4
Poly(tetrahydrofuran)	PTHF	335-439	0.1- 78	4
Poly(vinyl acetate)	PVAc	308-373	0.1- 80	4
Poly(vinyl chloride)	PVC	373-423	0.1-200	4
Poly(vinyl methyl ether)	PVME	303-471	0.1-200	4
Poly(vinylidene fluoride)	PVdF	451-521	0.1-200	5
Styrene/acrylonitrile copolymer				
2.7 wt% acrylonitrile	SAN3	378-539	0.1-200	4
5.7 wt% acrylonitrile	SAN6	370-540	0.1-200	4
15.3 wt% acrylonitrile	SAN15	405-531	0.1-200	4
18.0 wt% acrylonitrile	SAN18	377-528	0.1-200	4
40 wt% acrylonitrile	SAN40	373-543	0.1-200	4
70 wt% acrylonitrile	SAN70	373-544	0.1-200	4
Styrene/butadiene copolymer				
10 wt% styrene	SBR10	393-533	0.1-196	8
23.5 wt% styrene	SBR23	393-533	0.1-196	8
60 wt% styrene	SBR60	393-533	0.1-196	8
85 wt% styrene	SBR85	393-533	0.1-196	8
Styrene/methyl methacrylate copolymer				
20 wt% methyl methacrylate	SMMA20	383-543	0.1-200	4
60 wt% methyl methacrylate	SMMA60	383-543	0.1-200	4
N-Vinylcarbazole/4-ethylstyrene copolymer				
50 mol% ethylstyrene	VCES50	393-443	30-100	9
N-Vinylcarbazole/4-hexylstyrene copolymer				
80 mol% hexylstyrene	VCHS80	313-423	30-100	9
67 mol% hexylstyrene	VCHS67	333-423	30-100	9
60 mol% hexylstyrene	VCHS60	383-453	30-100	9
50 mol% hexylstyrene	VCHS50	373-443	30-100	9
40 mol% hexylstyrene	VCHS40	423-493	30-100	9
33 mol% hexylstyrene	VCHS33	463-523	30-100	9
20 mol% hexylstyrene	VCHS20	473-523	30-100	9
N-Vinylcarbazole/4-octylstyrene copolymer				
50 mol% octylstyrene	VCOS50	403-453	30-100	9
N-Vinylcarbazole/4-pentylstyrene copolymer				
50 mol% pentylstyrene	VCPS50	383-443	30-100	9

[a] Phenoxy = Poly(oxy-2-hydroxytrimethyleneoxy-1,4-phenyleneisopropylidene-1,4-phenylene)

[b] Polysulfone = Poly(oxy-1,4-phenylenesulfonyl-1,4-phenyleneoxy-1,4-phenyleneisopropylidene-1,4-phenylene)

[c] Polyarylate = Poly(oxyterephthaloyl/isophthaloyl T/I=50/50)oxy-1,4-phenyleneisopropylidene-1,4-phenylene

Polymers

TABLE 2. Tait Equation Parameter Functions for Polymer Melts

Polymer	$V(0,T)$/cm^{3}g^{-1}	$B(T)$/MPa
EP50	$1.2291 + 5.799 \cdot 10^{-5}(T-273.15) + 1.964 \cdot 10^{-6}(T-273.15)^2$	$487.0 \exp[-8.103 \cdot 10^{-3}(T-273.15)]$
EVA18	$1.02391 \exp(2.173 \cdot 10^{-5}T^{1.5})$	$188.2 \exp[-4.537 \cdot 10^{-3}(T-273.15)]$
EVA25	$1.00416 \exp(2.244 \cdot 10^{-5}T^{1.5})$	$184.4 \exp[-4.734 \cdot 10^{-3}(T-273.15)]$
EVA28	$1.00832 \exp(2.241 \cdot 10^{-5}T^{1.5})$	$183.5 \exp[-4.457 \cdot 10^{-3}(T-273.15)]$
EVA40	$1.06332 \exp(2.288 \cdot 10^{-5}T^{1.5})$	$205.1 \exp[-4.989 \cdot 10^{-3}(T-273.15)]$
PA6	$0.7597 \exp[4.701 \cdot 10^{-4}(T-273.15)]$	$376.7 \exp[-4.660 \cdot 10^{-3}(T-273.15)]$
PA11	$0.9581 \exp[6.664 \cdot 10^{-4}(T-273.15)]$	$254.7 \exp[-4.178 \cdot 10^{-3}(T-273.15)]$
PA66	$0.7657 \exp[6.600 \cdot 10^{-4}(T-273.15)]$	$316.4 \exp[-5.040 \cdot 10^{-3}(T-273.15)]$
cPBD	$1.0970 \exp[6.600 \cdot 10^{-4}(T-273.15)]$	$177.7 \exp[-3.593 \cdot 10^{-3}(T-273.15)]$
PBD-8	$1.1004 + 6.718 \cdot 10^{-4}(T-273.15) + 6.584 \cdot 10^{-7}(T-273.15)^2$	$200.0 \exp[-4.606 \cdot 10^{-3}(T-273.15)]$
PBD-24	$1.1049 + 6.489 \cdot 10^{-4}(T-273.15) + 7.099 \cdot 10^{-7}(T-273.15)^2$	$193.0 \exp[-4.519 \cdot 10^{-3}(T-273.15)]$
PBD-40	$1.1013 + 6.593 \cdot 10^{-4}(T-273.15) + 5.776 \cdot 10^{-7}(T-273.15)^2$	$188.0 \exp[-4.437 \cdot 10^{-3}(T-273.15)]$
PBD-50	$1.1037 + 5.955 \cdot 10^{-4}(T-273.15) + 7.789 \cdot 10^{-7}(T-273.15)^2$	$183.0 \exp[-4.425 \cdot 10^{-3}(T-273.15)]$
PBD-87	$1.1094 + 6.729 \cdot 10^{-4}(T-273.15) + 4.470 \cdot 10^{-7}(T-273.15)^2$	$175.0 \exp[-4.538 \cdot 10^{-3}(T-273.15)]$
iPB	$1.1417 \exp[6.751 \cdot 10^{-4}(T-273.15)]$	$167.5 \exp[-4.533 \cdot 10^{-3}(T-273.15)]$
PnBMA	$0.9341 + 5.5254 \cdot 10^{-4}(T-273.15) + 6.5803 \cdot 10^{-6}(T-273.15)^2 + 1.5691 \cdot 10^{-10}(T-273.15)^3$	$226.7 \exp[-5.344 \cdot 10^{-3}(T-273.15)]$
PBT	$0.9640 - 1.017 \cdot 10^{-3}(T-273.15) + 3.065 \cdot 10^{-6}(T-273.15)^2$	$263.0 \exp[-3.444 \cdot 10^{-3}(T-273.15)]$
PCL	$0.9049 \exp[6.392 \cdot 10^{-4}(T-273.15)]$	$189.0 \exp[-3.931 \cdot 10^{-3}(T-273.15)]$
PC	$0.73565 \exp(1.859 \cdot 10^{-5}T^{1.5})$	$310.0 \exp[-4.078 \cdot 10^{-3}(T-273.15)]$
BCPC	$0.6737 + 3.634 \cdot 10^{-4}(T-273.15) + 2.370 \cdot 10^{-7}(T-273.15)^2$	$363.4 \exp[-4.921 \cdot 10^{-3}(T-273.15)]$
HFPC	$0.6111 + 4.898 \cdot 10^{-4}(T-273.15) + 1.730 \cdot 10^{-7}(T-273.15)^2$	$236.6 \exp[-5.156 \cdot 10^{-3}(T-273.15)]$
TMPC	$0.8497 + 5.073 \cdot 10^{-4}(T-273.15) + 3.832 \cdot 10^{-7}(T-273.15)^2$	$231.4 \exp[-4.242 \cdot 10^{-3}(T-273.15)]$
PcHMA	$0.8793 + 4.0504 \cdot 10^{-4}(T-273.15) + 7.774 \cdot 10^{-7}(T-273.15)^2 - 7.7534 \cdot 10^{-10}(T-273.15)^3$	$295.2 \exp[-5.220 \cdot 10^{-3}(T-273.15)]$
PPO	$0.78075 \exp(2.151 \cdot 10^{-5}T^{1.5})$	$227.8 \exp[-4.290 \cdot 10^{-3}(T-273.15)]$
PDMS	$1.0079 \exp[9.121 \cdot 10^{-4}(T-273.15)]$	$89.4 \exp[-5.701 \cdot 10^{-3}(T-273.15)]$
PDMS-10	$0.8343 + 5.991 \cdot 10^{-4}(T-273.15) + 5.734 \cdot 10^{-7}(T-273.15)^2$	$542.63 \exp[-6.69 \cdot 10^{-3}(T-273.15)]$
PDMS-40	$0.8018 + 7.072 \cdot 10^{-4}(T-273.15) + 3.635 \cdot 10^{-7}(T-273.15)^2$	$482.73 \exp[-6.09 \cdot 10^{-3}(T-273.15)]$
PDMS-60	$0.8146 + 5.578 \cdot 10^{-4}(T-273.15) + 5.774 \cdot 10^{-7}(T-273.15)^2$	$482.73 \exp[-6.09 \cdot 10^{-3}(T-273.15)]$
PECH	$0.7216 \exp[5.825 \cdot 10^{-4}(T-273.15)]$	$238.3 \exp[-4.171 \cdot 10^{-3}(T-273.15)]$
PEEK	$0.7158 \exp[6.690 \cdot 10^{-4}(T-273.15)]$	$388.0 \exp[-4.124 \cdot 10^{-3}(T-273.15)]$
PEA	$0.8756 \exp[7.241 \cdot 10^{-4}(T-273.15)]$	$193.2 \exp[-4.839 \cdot 10^{-3}(T-273.15)]$
PEMA	$0.8614 \exp[7.468 \cdot 10^{-4}(T-273.15)]$	$260.9 \exp[-5.356 \cdot 10^{-3}(T-273.15)]$
HDPE	$1.1595 + 8.0394 \cdot 10^{-4}(T-273.15)$	$179.9 \exp[-4.739 \cdot 10^{-3}(T-273.15)]$
LPE	$0.9172 \exp[7.806 \cdot 10^{-4}(T-273.15)]$	$176.7 \exp[-4.661 \cdot 10^{-3}(T-273.15)]$
HMLPE	$0.8992 \exp[8.502 \cdot 10^{-4}(T-273.15)]$	$168.3 \exp[-4.292 \cdot 10^{-3}(T-273.15)]$
BPE	$0.9399 \exp[7.341 \cdot 10^{-4}(T-273.15)]$	$177.1 \exp[-4.699 \cdot 10^{-3}(T-273.15)]$
LDPE	$1.1944 + 2.841 \cdot 10^{-4}(T-273.15) + 1.872 \cdot 10^{-6}(T-273.15)^2$	$202.2 \exp[-5.243 \cdot 10^{-3}(T-273.15)]$
LDPE-A	$1.1484 \exp[6.950 \cdot 10^{-4}(T-273.15)]$	$192.9 \exp[-4.701 \cdot 10^{-3}(T-273.15)]$
LDPE-B	$1.1524 \exp[6.700 \cdot 10^{-4}(T-273.15)]$	$196.6 \exp[-4.601 \cdot 10^{-3}(T-273.15)]$
LDPE-C	$1.1516 \exp[6.730 \cdot 10^{-4}(T-273.15)]$	$186.7 \exp[-4.391 \cdot 10^{-3}(T-273.15)]$
PEO	$0.8766 \exp[7.087 \cdot 10^{-4}(T-273.15)]$	$207.7 \exp[-3.947 \cdot 10^{-3}(T-273.15)]$
PET	$0.6883 + 5.90 \cdot 10^{-4}(T-273.15)$	$369.7 \exp[-4.150 \cdot 10^{-3}(T-273.15)]$
P4HS	$0.8251 + 6.77 \cdot 10^{-4}T$	$103.1 \exp[-2.417 \cdot 10^{-3}(T-273.15)]$
PIB	$1.0750 \exp[5.651 \cdot 10^{-4}(T-273.15)]$	$200.3 \exp[-4.329 \cdot 10^{-3}(T-273.15)]$
PI-8	$1.1030 + 6.488 \cdot 10^{-4}(T-273.15) + 5.125 \cdot 10^{-7}(T-273.15)^2$	$188.0 \exp[-4.541 \cdot 10^{-3}(T-273.15)]$
PI-14	$1.0943 + 6.293 \cdot 10^{-4}(T-273.15) + 6.231 \cdot 10^{-7}(T-273.15)^2$	$202.0 \exp[-4.653 \cdot 10^{-3}(T-273.15)]$
PI-41	$1.0951 + 6.188 \cdot 10^{-4}(T-273.15) + 6.629 \cdot 10^{-7}(T-273.15)^2$	$199.0 \exp[-4.622 \cdot 10^{-3}(T-273.15)]$
PI-56	$1.0957 + 6.655 \cdot 10^{-4}(T-273.15) + 5.661 \cdot 10^{-7}(T-273.15)^2$	$200.0 \exp[-4.644 \cdot 10^{-3}(T-273.15)]$
PMA	$0.8365 \exp[6.795 \cdot 10^{-4}(T-273.15)]$	$235.8 \exp[-4.493 \cdot 10^{-3}(T-273.15)]$
PMMA	$0.8254 + 2.8383 \cdot 10^{-4}(T-273.15) + 7.792 \cdot 10^{-7}(T-273.15)^2$	$287.5 \exp[-4.146 \cdot 10^{-3}(T-273.15)]$
P4MP	$1.4075 - 9.095 \cdot 10^{-4}(T-273.15) + 3.497 \cdot 10^{-6}(T-273.15)^2$	$37.67 + 0.2134(T-273.15)] - 7.0445 \cdot 10^{-4}(T-273.15)^2$
PαMS	$0.89365 + 3.4864 \cdot 10^{-4}(T-273.15) + 5.0184 \cdot 10^{-7}(T-273.15)^2$	$297.7 \exp[-4.074 \cdot 10^{-3}(T-273.15)]$
PoMS	$0.9396 \exp[5.306 \cdot 10^{-4}(T-273.15)]$	$261.9 \exp[-4.114 \cdot 10^{-3}(T-273.15)]$
POM	$0.7484 \exp[6.770 \cdot 10^{-4}(T-273.15)]$	$305.6 \exp[-4.326 \cdot 10^{-3}(T-273.15)]$
PH	$0.76644 \exp(1.921 \cdot 10^{-5}T^{1.5})$	$359.9 \exp[-4.378 \cdot 10^{-3}(T-273.15)]$
PSF	$0.7644 + 3.419 \cdot 10^{-4}(T-273.15) + 3.126 \cdot 10^{-7}(T-273.15)^2$	$365.9 \exp[-3.757 \cdot 10^{-3}(T-273.15)]$
PAr	$0.73381 \exp(1.626 \cdot 10^{-5}T^{1.5})$	$296.9 \exp[-3.375 \cdot 10^{-3}(T-273.15)]$
aPP	$1.1841 - 1.091 \cdot 10^{-4}(T-273.15) + 5.286 \cdot 10^{-6}(T-273.15)^2$	$162.1 \exp[-6.604 \cdot 10^{-3}(T-273.15)]$
iPP	$1.1606 \exp[6.700 \cdot 10^{-4}(T-273.15)]$	$149.1 \exp[-4.177 \cdot 10^{-3}(T-273.15)]$

Polymer	$V(0,T)$/cm^{3}g^{-1}	$B(T)$/MPa
PS	$0.9287 \exp[5.131 \cdot 10^{-4}(T-273.15)]$	$216.9 \exp[-3.319 \cdot 10^{-3}(T-273.15)]$
PTFE	$0.3200 + 9.5862 \cdot 10^{-4}(T-273.15)$	$425.2 \exp[-9.380 \cdot 10^{-3}(T-273.15)]$
PTHF	$1.0043 \exp[6.691 \cdot 10^{-4}(T-273.15)]$	$178.6 \exp[-4.223 \cdot 10^{-3}(T-273.15)]$
PVAc	$0.82496 + 5.820 \cdot 10^{-4}(T-273.15) + 2.940 \cdot 10^{-7}(T-273.15)^2$	$204.9 \exp[-4.346 \cdot 10^{-3}(T-273.15)]$
PVC	$0.7196 + 5.581 \cdot 10^{-5}(T-273.15) + 1.468 \cdot 10^{-6}(T-273.15)^2$	$294.2 \exp[-5.321 \cdot 10^{-3}(T-273.15)]$
PVME	$0.9585 \exp[6.653 \cdot 10^{-4}(T-273.15)]$	$215.8 \exp[-4.588 \cdot 10^{-3}(T-273.15)]$
PVdF	$0.5790 \exp[8.051 \cdot 10^{-4}(T-273.15)]$	$244.0 \exp[-5.210 \cdot 10^{-3}(T-273.15)]$
SAN3	$0.9233 + 3.936 \cdot 10^{-4}(T-273.15) + 5.685 \cdot 10^{-7}(T-273.15)^2$	$239.8 \exp[-4.376 \cdot 10^{-3}(T-273.15)]$
SAN6	$0.9211 + 4.370 \cdot 10^{-4}(T-273.15) + 5.846 \cdot 10^{-7}(T-273.15)^2$	$226.9 \exp[-4.286 \cdot 10^{-3}(T-273.15)]$
SAN15	$0.9044 + 4.207 \cdot 10^{-4}(T-273.15) + 4.077 \cdot 10^{-7}(T-273.15)^2$	$238.4 \exp[-3.943 \cdot 10^{-3}(T-273.15)]$
SAN18	$0.9016 + 4.036 \cdot 10^{-4}(T-273.15) + 4.206 \cdot 10^{-7}(T-273.15)^2$	$240.4 \exp[-3.858 \cdot 10^{-3}(T-273.15)]$
SAN40	$0.8871 + 3.406 \cdot 10^{-4}(T-273.15) + 4.938 \cdot 10^{-7}(T-273.15)^2$	$289.3 \exp[-4.431 \cdot 10^{-3}(T-273.15)]$
SAN70	$0.8528 + 3.616 \cdot 10^{-4}(T-273.15) + 2.634 \cdot 10^{-7}(T-273.15)^2$	$335.4 \exp[-3.923 \cdot 10^{-3}(T-273.15)]$
SBR10	$0.9053 \exp(2.437 \cdot 10^{-5}T^{1.5})$	$530.3 \exp[-3.99 \cdot 10^{-3}(T-273.15)]$
SBR23	$0.8986 \exp(2.317 \cdot 10^{-5}T^{1.5})$	$551.6 \exp[-4.17 \cdot 10^{-3}(T-273.15)]$
SBR60	$0.8812 \exp(2.031 \cdot 10^{-5}T^{1.5})$	$486.0 \exp[-4.34 \cdot 10^{-3}(T-273.15)]$
SBR85	$0.8704 \exp(1.846 \cdot 10^{-5}T^{1.5})$	$356.7 \exp[-4.24 \cdot 10^{-3}(T-273.15)]$
SMMA20	$0.9063 + 3.570 \cdot 10^{-4}(T-273.15) + 6.532 \cdot 10^{-7}(T-273.15)^2$	$232.0 \exp[-4.143 \cdot 10^{-3}(T-273.15)]$
SMMA60	$0.8610 + 3.350 \cdot 10^{-4}(T-273.15) + 6.980 \cdot 10^{-7}(T-273.15)^2$	$261.0 \exp[-4.611 \cdot 10^{-3}(T-273.15)]$
VCES50	$0.6676 + 6.63 \cdot 10^{-4}T$	$5281.7 \exp[-9.264 \cdot 10^{-3}(T-273.15)]$
VCHS80	$0.7753 + 6.17 \cdot 10^{-4}T$	$247.6 \exp[-2.604 \cdot 10^{-3}(T-273.15)]$
VCHS67	$0.8028 + 6.50 \cdot 10^{-4}T$	$581.7 \exp[-4.553 \cdot 10^{-3}(T-273.15)]$
VCHS60	$0.8213 + 6.23 \cdot 10^{-4}T$	$229.1 \exp[-2.133 \cdot 10^{-3}(T-273.15)]$
VCHS50	$0.7827 + 5.05 \cdot 10^{-4}T$	$136.0 \exp[-1.083 \cdot 10^{-3}(T-273.15)]$
VCHS40	$0.7805 + 4.92 \cdot 10^{-4}T$	$155.0 \exp[-1.605 \cdot 10^{-3}(T-273.15)]$
VCHS33	$0.7710 + 4.86 \cdot 10^{-4}T$	$460.4 \exp[-3.453 \cdot 10^{-3}(T-273.15)]$
VCHS20	$0.6416 + 5.42 \cdot 10^{-4}T$	$489.8 \exp[-3.193 \cdot 10^{-3}(T-273.15)]$
VCOS50	$0.7081 + 7.40 \cdot 10^{-4}T$	$666.5 \exp[-4.503 \cdot 10^{-3}(T-273.15)]$
VCPS50	$0.7814 + 4.36 \cdot 10^{-4}T$	$880.1 \exp[-4.393 \cdot 10^{-3}(T-273.15)]$

UPPER CRITICAL (UCST) AND LOWER CRITICAL (LCST) SOLUTION TEMPERATURES OF BINARY POLYMER SOLUTIONS

Christian Wohlfarth

Liquid–liquid demixing in solutions of polymers in low molar mass solvents is not a rare phenomenon. Demixing depends on concentration, temperature, pressure, molar mass and molar mass distribution function of the polymer, chain branching and end groups of the polymer, the chemical nature of the solvent, isotope substitution in solvents or polymers, chemical composition of copolymers and its distributions, and other variables. Phase diagrams of polymer solutions can therefore show quite complicated behavior when they have to be considered in detail (see Ref. 1a).

Polymer solutions can undergo demixing when cooling a homogeneous solution as well as when heating such a solution. The corresponding cloud-point curves show a maximum (UCST behavior) or a minimum (LCST behavior). For common polymer solutions, the LCST region is at higher temperatures (in many cases near the critical temperature of the solvent) than the UCST region. The temperature range between both extrema provides the essential information where the one-phase region of a polymer solution can be found. In the case of monodisperse polymers the extrema are equal to the critical points. However, in the case of polydisperse polymers with distribution functions, these extrema are threshold temperatures whereas the critical point shifts to higher concentrations on the shoulder of the cloud-point curve. Usually, the critical concentration is much more strongly influenced than the critical temperature. Thus, the table below does not distinguish between threshold and critical temperatures.

UCST and LCST values depend somewhat on pressure. LCST values in the table are usually given at the vapor pressure of the solvent at this temperature. UCST values are measured in most cases at normal pressure; data at higher pressures are neglected here. The interested reader can find such information, for example, in Refs. 76, 84, 104, 157, 165, 177, 185–187, or 192.

However, UCST and LCST values of a given polymer/solvent pair depend strongly on the molar mass of the polymer. In the case of monodisperse polymers, this dependency can be described in good approximation by the so-called Shultz-Flory plot (see Refs. 6 and 8):

$$\frac{1}{T_{\text{crit}}} = \frac{1}{\theta}\left[1 + const.\left(\frac{1}{\sqrt{r}} - \frac{1}{2r}\right)\right] \qquad (1)$$

where r denotes the number of segments of a polymer (being proportional to the degree of polymerization or to the molar mass or molar volume of the polymer). Extrapolation to $r \to \infty$, i.e., to infinite molar mass, leads to the value of the θ-temperature. This θ-temperature is the highest temperature for UCST behavior or the lowest temperature for LCST behavior and a given polymer/solvent pair. In the case of polydisperse polymers, the segment number in equation (1) is to be replaced by its weight average, r_w (related to M_w). The constant in equation (1) reflects further thermodynamic properties of the given polymer/solvent pair, but should not depend on molar mass. A detailed discussion can be found in Ref. 1b.

The table gives the UCST and LCST values for selected polymer/solvent pairs at specified values of the molar mass, as measured by the number average M_n, mass average M_w, or viscosity average M_η. Further information on the molar mass dependence of UCST and LCST can be found in the references.

Polymer	M_n/g mol^{-1}	M_w/g mol^{-1}	M_η/g mol^{-1}	Solvent	UCST/K	LCST/K	Ref.
Acrylonitrile/butadiene copolymer							
(18% Acrylonitrile)			840000	Ethyl acetate		427	220
(26% Acrylonitrile)			1000000	Ethyl acetate		412	220
Butadiene/α-methylstyrene copolymer (10% α-Methylstyrene)			100000	Ethyl acetate	387	393	220
Carbon monoxide/ethylene copolymer (1:1, alternating)		1000000		1,1,1,3,3,3-Hexafluoro-2-propanol		453	159
Cellulose diacetate			120000	Benzyl alcohol	372		86
	59900	75500		2-Butanone	279.7	471.5	111
	59300			2-Propanone	216.2	438.2	42
Cellulose diacetate/styrene graft copolymer (77.4 wt% grafted polystyrene)		750000		N,N-Dimethylformamide	262	399	106
		750000		Tetrahydrofuran		363	106
Cellulose nitrate (13.3 wt% N)	unknown			2-Propanone	328	182	148
Cellulose triacetate			20000	Benzyl alcohol	322		86
		100500		2-Propanone	290.0	472.0	42
Cellulose tricaprylate	infinite			*N,N*-Dimethylformamide	413		5
	infinite			3-Phenyl-1-propanol	321		5
Decamethyltetrasiloxane	310.69			Tetradecafluorohexane	332.59		195
N,N-Dimethylacrylamide/2-butoxyethyl acrylate copolymer (50 wt% 2-butoxyethyl acrylate)				Water		<273.2	164
N,N-Dimethylacrylamide/butyl acrylate copolymer							
(15 wt% Butyl acrylate)				Water		346.2	164
(20 wt% Butyl acrylate)				Water		323.2	164
(30 wt% Butyl acrylate)				Water		294.2	164

Polymer	M_n/g mol^{-1}	M_w/g mol^{-1}	M_η/g mol^{-1}	Solvent	UCST/K	LCST/K	Ref.
(35 wt% Butyl acrylate)				Water		281.2	164
N,N-Dimethylacrylamide/2-ethoxyethyl acrylate copolymer							
(50 wt% 2-Ethoxyethyl acrylate)				Water		319.2	164
(75 wt% 2-Ethoxyethyl acrylate)				Water		285.2	164
N,N-Dimethylacrylamide/ethyl acrylate copolymer							
(25 wt% Ethyl acrylate)				Water		347.2	164
(30 wt% Ethyl acrylate)				Water		334.2	164
(50 wt% Ethyl acrylate)				Water		287.2	164
(55 wt% Ethyl acrylate)				Water		<273.2	164
N,N-Dimethylacrylamide/2-methoxyethyl acrylate copolymer							
(38 mol% 2-Methoxyethyl acrylate)				Water		353	184
(45 mol% 2-Methoxyethyl acrylate)				Water		333	184
(55 mol% 2-Methoxyethyl acrylate)				Water		315	184
(68 mol% 2-Methoxyethyl acrylate)				Water		305	184
(82 mol% 2-Methoxyethyl acrylate)				Water		288	184
(92 mol% 2-Methoxyethyl acrylate)				Water		283	184
N,N-Dimethylacrylamide/methyl acrylate copolymer							
(30 wt% Methyl acrylate)				Water		371.2	164
(40 wt% Methyl acrylate)				Water		338.2	164
(50 wt% Methyl acrylate)				Water		314.2	164
(55 wt% Methyl acrylate)				Water		294.2	164
(60 wt% Methyl acrylate)				Water		279.2	164
(70 wt% Methyl acrylate)				Water		<273.2	164
N,N-Dimethylacrylamide/propyl acrylate copolymer							
(20 wt% Propyl acrylate)				Water		353.2	164
(30 wt% Propyl acrylate)				Water		337.2	164
(40 wt% Propyl acrylate)				Water		294.2	164
(50 wt% Propyl acrylate)				Water		281.2	164
Dimethylsiloxane/methylphenylsiloxane	9100	41200		Anisole	291.45		198
copolymer (15 wt% methylphenylsiloxane)	9100	41200		2-Propanone	282.45		198
Ethylene/propylene copolymer			145000	Cyclohexane		534	101
(33 mol% ethylene)			145000	Cyclopentane		490	101
			145000	2,2-Dimethylbutane		428	101
			145000	2,3-Dimethylbutane		452	101
			145000	3,4-Dimethylhexane		541	101
			145000	2,2-Dimethylpentane		472	101
			145000	2,3-Dimethylpentane		500	101
			145000	2,4-Dimethylpentane		464	101
			145000	3-Ethylpentane		511	101
			145000	Heptane		502	101
			145000	Hexane		455	101
			145000	2-Methylbutane		396	101
			145000	Methylcyclohexane		558	101
			145000	Methylcyclopentane		512	101
			145000	2-Methylhexane		486	101
			145000	Nonane		558	101
			145000	Octane		528	101
			145000	Pentane		409	101
			145000	2,2,4,4-Tetramethylpentane		539	101
			145000	2,2,3-Trimethylbutane		500	101
			145000	2,2,4-Trimethylpentane		503	101
Ethylene/propylene copolymer	70000	140000		Hexane		436	127
(43 mol% ethylene)	70000	140000		2-Methylpentane		474	127
	70000	140000		Pentane		441	127
Ethylene/propylene copolymer			154000	2,2-Dimethylbutane		407	101
(53 mol% ethylene)			154000	2,3-Dimethylbutane		437	101
			154000	2,2-Dimethylpentane		453	101

Polymer	M_n/g mol^{-1}	M_w/g mol^{-1}	M_h/g mol^{-1}	Solvent	UCST/K	LCST/K	Ref.
			154000	2,3-Dimethylpentane		488	101
			154000	2,4-Dimethylpentane		445	101
			154000	3-Ethylpentane		500	101
			154000	Heptane		493	101
			154000	Hexane		443	101
			154000	Pentane		395	101
			154000	2,2,3-Trimethylbutane		488	101
			154000	2,3,4-Trimethylhexane		565	101
			154000	2,2,4-Trimethylpentane		484	101
Ethylene/propylene copolymer (63 mol% ethylene)			236000	Cyclohexane		526	101
			236000	Cyclopentane		481	101
			236000	2,3-Dimethylbutane		429	101
			236000	3,4-Dimethylhexane		530	101
			236000	2,2-Dimethylpentane		444	101
			236000	2,3-Dimethylpentane		482	101
			236000	2,4-Dimethylpentane		434	101
			236000	3-Ethylpentane		492	101
			236000	Heptane		485	101
			236000	Hexane		436	101
			236000	2-Methylbutane		348	101
			236000	Methylcyclopentane		498	101
			236000	Nonane		547	101
			236000	Octane		512	101
			236000	Pentane		387	101
			236000	2,2,4,4-Tetramethylpentane		528	101
			236000	2,2,3-Trimethylbutane		479	101
			236000	2,2,4-Trimethylpentane		479	101
Ethylene/propylene copolymer (75 mol% ethylene)			109000	2,2-Dimethylpentane		431	101
			109000	2,4-Dimethylpentane		425	101
			109000	Heptane		475	101
			109000	Hexane		427	101
			109000	Nonane		542	101
			109000	Octane		509	101
			109000	Pentane		378	101
			109000	2,2,4,4-Tetramethylpentane		523	101
			109000	2,2,4-Trimethylpentane		469	101
Ethylene/propylene copolymer (81 mol% ethylene)			195000	Cyclohexane		522	101
			195000	Cyclopentane		474	101
			195000	2,2-Dimethylbutane		381	101
			195000	2,3-Dimethylbutane		413	101
			195000	2,4-Dimethylhexane		478	101
			195000	2,5-Dimethylhexane		466	101
			195000	3,4-Dimethylhexane		522	101
			195000	2,2-Dimethylpentane		425	101
			195000	2,3-Dimethylpentane		471	101
			195000	2,4-Dimethylpentane		420	101
			195000	3-Ethylpentane		478	101
			195000	Heptane		468	101
			195000	Hexane		425	101
			195000	2-Methylbutane		327	101
			195000	Methylcyclohexane		541	101
			195000	Methylcyclopentane		493	101
			195000	2-Methylhexane		453	101
			195000	3-Methylhexane		459	101
			195000	Nonane		540	101
			195000	Octane		506	101
			195000	Pentane		370	101
			195000	2,2,4,4-Tetramethylpentane		519	101
			195000	2,2,3-Trimethylbutane		461	101
			195000	2,2,4-Trimethylpentane		460	101
Ethylene/vinyl acetate copolymer							

Polymer	M_n/g mol^{-1}	M_w/g mol^{-1}	M_h/g mol^{-1}	Solvent	UCST/K	LCST/K	Ref.
(2.3 wt% vinyl acetate)	52000	465000		Diphenyl ether	404.2		143
(4.0 wt% vinyl acetate)	47000	280000		Diphenyl ether	392.5		143
(7.1 wt% Vinyl acetate)	34000	460000		Diphenyl ether	378.2		143
(9.5 wt% Vinyl acetate)	53000	350000		Diphenyl ether	367.3		143
(9.7 wt% Vinyl acetate)	55000	490000		Diphenyl ether	370.8		143
(12.1 wt% Vinyl acetate)	66000	300000		Diphenyl ether	360.4		143
(42.6 mol% Vinyl acetate)	14800	41500		Methyl acetate	307.0		130
Ethylene/vinyl alcohol copolymer							
(87.2 mol% Vinyl alcohol)			infinite	Water	463.55	285.65	44
(88.9 mol% Vinyl alcohol)			infinite	Water	449.15	290.75	44
(91.0 mol% Vinyl alcohol)			infinite	Water	428.45	302.95	44
(94.1 mol% Vinyl alcohol)			infinite	Water	389.25	324.45	44
Ethylene oxide/propylene oxide copolymer							
(20.0 mol% Ethylene oxide)	3400			Water		303	211
(27.0 mol% Ethylene oxide)	3000			Water		309	210
(30.0 mol% Ethylene oxide)	5400			Water		313	211
(38.5 mol% Ethylene oxide)	5000			Water		309	210
(50.0 mol% Ethylene oxide)	3900			Water		323	211
(58.8 mol% Ethylene oxide)	3000			Water		326.65	210
(72.4 mol% Ethylene oxide)		36000		Water		333	153
(79.5 mol% Ethylene oxide)		30800		Water		345	153
(86.6 mol% Ethylene oxide)		30100		Water		355.5	153
Gutta Percha			194000	Propyl acetate	318.95		7
Hydroxypropylcellulose		75000		Water		318.45	43
		300000		Water		331.25	43
N-Isopropylacrylamide/acrylamide copolymer							
(15 mol% Acrylamide)		3100000		Water		315.15	172
(30 mol% Acrylamide)		4500000		Water		326.15	172
(45 mol% Acrylamide)		3900000		Water		347.15	172
N-Isopropylacrylamide/1-deoxy-1-methacrylamido-*D*-glucitol							
(12.9 mol% Glucitol)	78000	170000		Water		311.3	218
(13.7 mol% Glucitol)	51600	110000		Water		314.9	218
(14.0 mol% Glucitol)	145000	432000		Water		307.5	218
N-Isopropylacrylamide/*N*-isopropylmethacrylamide copolymer							
(10.56 mol% *N*-Isopropylmethacrylamide)	55300	177000		Water		307.15	212
(30.00 mol% *N*-Isopropylmethacrylamide)	28800	92000		Water		309.75	212
(39.99 mol% *N*-Isopropylmethacrylamide)	23100	74000		Water		311.05	212
(59.89 mol% *N*-Isopropylmethacrylamide)	23100	74000		Water		314.65	212
(79.81 mol% *N*-Isopropylmethacrylamide)	16600	53000		Water		317.35	212
(89.99 mol% *N*-Isopropylmethacrylamide)	14700	47000		Water		318.75	212
Methylcellulose (about 30 mol% methyl substitution)			70000	Water		324.75	47
Methylcellulose/hydroxypropylcellulose copolymer (25 mol% methyl, 8 mol% hydroxypropyl substitution)			80000	Water		340.15	63
Natural rubber		300000		Pentane		403	10
			74500	2-Pentanone	274.45		7
Phenol-formaldehyde resin (acetylated)				2-Ethoxyethanol	378.2		200
Poly(acrylic acid)		120000		Tetrahydrofuran		268.3	189
Poly[bis(2,3-dimethoxypropanoxy)phosphazene]	1070000	1500000		Water		317.15	183
Poly[bis(2-(2′-methoxyethoxy)ethoxy) phosphazene]	667000	1000000		Water		338.15	183
Poly[bis(2,3-bis(2-methoxyethoxy)propanoxy) phosphazene]	714000	1000000		Water		311.15	183
Poly[bis(2,3-bis(2-(2′-methoxyethoxy)ethoxy) propanoxy)phosphazene]	1420000	1700000		Water		322.65	183
Poly[bis(2,3-bis(2-(2′-(2″-dimethoxyethoxy) ethoxy)ethoxy)propanoxy)phosphazene]	857000	1200000		Water		334.65	183
Poly(1-butene) (atactic)	infinite			Anisole	359.4		11
	infinite			Toluene	356.2		28

Polymer	M_n/g mol^{-1}	M_w/g mol^{-1}	M_h/g mol^{-1}	Solvent	UCST/K	LCST/K	Ref.
Poly(1-butene) (isotactic)	infinite			Anisole	362.3		11
			530000	Cyclopentane		498	102
			530000	2,2-Dimethylbutane		444	102
			530000	2,5-Dimethylhexane		519	102
			530000	3,4-Dimethylhexane		559	102
			530000	2,3-Dimethylpentane		517	102
			530000	2,4-Dimethylpentane		480	102
			530000	3-Ethylpentane		523	102
			530000	Heptane		509	102
	infinite			Hexane		464	102
			530000	2-Methylbutane		416	102
	infinite			Nonane		564	102
			530000	Octane		540	102
	infinite			Pentane		421	102
			530000	2,2,3-Trimethylbutane		507	102
Poly(butyl methacrylate)	278000	470000		1-Butanol	287.15		132
	278000	470000		Decane	357.25		132
	278000	470000		Ethanol	315.25		132
	278000	470000		Heptane	342.55		132
	278000	470000		Octane	345.80		132
	278000	470000		1-Pentanol	286.30		132
	278000	470000		2-Propanol	294.90		132
	278000	470000		2,2,4-Trimethylpentane	347.50		132
Poly(2-chlorostyrene)	infinite			Benzene		298	40
Poly(4-chlorostyrene)	infinite			Benzene	274.0		22
	infinite			2-(Butoxyethoxy)ethanol	323.25		46
	infinite			Butyl acetate	502.4		22
	infinite			*tert*-Butyl acetate	338.55		46
	infinite			Chlorobenzene	128.8		22
	infinite			2-(Ethoxyethoxy)ethanol	300.95		46
	infinite			Ethyl acetate	613.2		22
	infinite			Ethylbenzene	283.2		22
	infinite			Ethylbenzene	258.45		46
	infinite			Ethyl chloroacetate	271.35		46
	infinite			Isopropyl acetate		348.65	46
	infinite			Isopropylbenzene	332.15		46
	infinite			Isopropyl chloroacetate	264.95		46
	infinite			Methyl chloroacetate	337.75		46
	infinite			Propyl acetate		908.7	22
	infinite			Tetrachloroethene	317.55		46
	infinite			Tetrachloromethane	323.85		46
	infinite			Toluene	236.8		22
Poly(decyl methacrylate)	390000	468000		1-Butanol	304.85		113
	390000	468000		1-Pentanol	278.40		113
	220000	252000		2-Propanol	346.85		132
Polydimethylsiloxane (cyclic)	9810	10300		2,2-Dimethylpropane		433	133
	9810	10300		Tetramethylsilane		448	133, 171
Polydimethylsiloxane			626000	Butane		392.95	53
	infinite			Decane		603	30
	14750	16370		2,2-Dimethylpropane		428	133
	infinite			Dodecane		643	30
			626000	Ethane		259.65	53
			100000	Ethoxybenzene	341.99		108
	infinite			Heptane		528	30
	infinite			Hexadecane		708	30
	infinite			Hexane		493	30
	infinite			Octane		553	30
	infinite			Pentane		453	30
			203000	Propane		340.15	53
	14750	16370		Tetramethylsilane		443	133, 171
Poly(ethyl acrylate)			48000	1-Butanol	310.05		27

Polymer	M_n/g mol^{-1}	M_w/g mol^{-1}	M_η/g mol^{-1}	Solvent	UCST/K	LCST/K	Ref.
			48000	Ethanol	301.15		27
			380000	Methanol	287.25		27
			48000	1-Propanol	305.15		27
Polyethylene (branched)	8400	32000		Diphenyl ether	384.7		95, 98
	24000	123000		Diphenyl ether	396.7		95, 98
	65000	425000		Diphenyl ether	415.3		95, 98
Polyethylene (linear)			20000	Anisole	368.15		24
			20000	Benzyl acetate	459.65		24
			20000	Benzyl phenyl ether	437.15		24
			20000	Benzyl propionate	436.15		24
			50900	Biphenyl	383.55		25
			61100	Butyl acetate	448	497	70
			20000	4-*tert*-Butylphenol	466.15		24
			134000	Cyclohexane		518	101
			20000	Cyclohexanone	389.65		24
			134000	Cyclopentane		472	101
	36700	49300		Decane		563.75	91
			20000	1-Decanol	400.15		24
			20000	Dibenzyl ether	448.65		24
			134000	3,4-Dimethylhexane		515	101
			134000	2,2-Dimethylpentane		399	101
			134000	2,3-Dimethylpentane		463	101
			134000	2,4-Dimethylpentane		395	101
	12000	150000		Diphenyl ether	416.2		95, 98
			97200	Diphenylmethane	400.25		25
	60400	82600		Dodecane		610.85	91
		218000		1-Dodecanol	405.15		141
			134000	3-Ethylpentane		471	101
	36700	49300		Heptane		464.70	91
			20000	1-Heptanol	440.15		24
	36700	49300		Hexane		414.65	91
	7900	92000		1-Hexanol	458.15		154
			20000	2-Methoxynaphthalene	427.65		24
			20000	3-Methylbutyl acetate	407.15		24
			134000	Methylcyclohexane		537	101
			134000	Methylcyclopentane		488	101
	60400	82600		Nonane		531.90	91
			20000	1-Nonanol	431.15		24
			20000	4-Nonylphenol	410.15		24
	36700	49300		Octane		502.40	91
	7900	92000		1-Octanol	426.65		154
			20000	4-Octylphenol	424.65		24
			134000	Pentane		353	101
			20000	1-Pentanol	445.15		24
			175000	Pentyl acetate	421	528	70
			20000	4-*tert*-Pentylphenol	443.65		24
			20000	Phenetole	366.65		24
			134000	2,2,4,4-Tetramethylpentane		513	101
	60400	82600		Tridecane		639.30	91
			134000	2,2,3-Trimethylbutane		444	101
			134000	2,3,4-Trimethylhexane		545	101
			134000	2,2,4-Trimethylpentane		495	101
	97700	135900		Undecane		583.95	91
Poly(ethylene glycol)			8000	*tert*-Butyl acetate	321.2	464.2	83
			21200	*tert*-Butyl acetate	353.2	431.2	83
	6100	6200		Water		404.79	185
	10457	11615		Water		394.33	205
	40800	151000		Water		378.25	205
Poly(ethylene oxide)-b-poly[bis(methoxyethoxyethoxy)-phosphazene] block copolymer (about 67 mol% Ethylene oxide)	22000	31500		Water		338	222

Polymers

Polymer	M_n/g mol^{-1}	M_w/g mol^{-1}	M_h/g mol^{-1}	Solvent	UCST/K	LCST/K	Ref.
Poly(ethylene oxide)-b-poly(propylene oxide)-b-poly(ethylene oxide) triblock copolymer (about 30 mol% Ethylene oxide)		4400		Water		286.65	209
Polyethylethylene	48000	52000		Diphenyl ether	411.2		95, 98
Poly(*p*-hexylstyrene)	infinite			2-Butanone	302.6		135
Poly(2-hydroxyethyl methacrylate)			77400	1-Butanol	337.25		35
			233600	2-Butanol	287		35
			233600	2-Metyl-1-propanol	342		35
			77400	1,2,3-Propanetriol	345		35
			77400	1-Propanol	311		35
Polyisobutylene	infinite			Anisole	377		3
			72000	Benzene		540.5	39
			703000	Butane		264.75	53
	infinite			Cycloheptane		572	34
			1500000	Cyclohexane		412	10
	infinite			Cyclooctane		637	34
			1500000	Cyclopentane		344	10
	infinite			Decane		535	34
			1500000	2,2-Dimethylbutane		376	10
			1500000	2,3-Dimethylbutane		404	10
	infinite			2,2-Dimethylhexane		454	34
	infinite			2,4-Dimethylhexane		458	34
	infinite			2,5-Dimethylhexane		446	34
	infinite			3,4-Dimethylhexane		497	34
	infinite			2,2-Dimethylpentane		404	34
	infinite			2,3-Dimethylpentane		451	34
	infinite			2,4-Dimethylpentane		403	34
	infinite			3,3-Dimethylpentane		451	34
	infinite			Diphenyl ether	306		3
	infinite			Decane		585	30
	infinite			Dodecane		582	34
	infinite			Ethylbenzene	249		3
	infinite			Ethylcyclopentane		524	34
	infinite			Ethyl heptanoate	306		3
	infinite			Ethyl hexanoate	330		3
	infinite			3-Ethylpentane		458	34
	infinite			Heptane		442	34
			72000	Hexane		428.5	39
			6030	2-Methylbutane		357.85	53
	infinite			Methylcyclohexane		526	34
	infinite			Methylcyclopentane		478	34
	infinite			2-Methylheptane		466	34
	infinite			3-Methylheptane		478	34
	infinite			2-Methylhexane		426	34
	infinite			3-Methylhexane		446	34
	infinite			2-Methylpentane		376	34
	infinite			3-Methylpentane		405	34
			470	2-Methylpropane		387	10
			72000	Octane		506.0	39
			6030	Pentane		403.55	53
			72000	Pentane		373.5	39
	infinite			Phenetole	357		3
			470	Propane		358	10
	infinite			Propylcyclopentane		547	34
	infinite			Toluene	260		3
	infinite			2,2,3-Trimethylbutane		445	34
	infinite			2,2,4-Trimethylpentane		435	34
1,4-*cis*-Polyisoprene			780000	2,5-Dimethylhexane		474.15	140
			780000	3,4-Dimethylhexane		520.15	140
			780000	2,2-Dimethylpentane		445.15	140
			780000	2,3-Dimethylpentane		484.15	140
			780000	2,4-Dimethylpentane		442.15	140

Polymer	M_n/g mol^{-1}	M_w/g mol^{-1}	M_h/g mol^{-1}	Solvent	UCST/K	LCST/K	Ref.
			780000	3-Methylpentane		483.15	140
			780000	Heptane		488.15	140
			780000	Hexane		434.15	140
			780000	Nonane		541.15	140
			780000	Octane		509.15	140
			780000	2,2,4,4-Tetramethylpentane		518.15	140
			780000	2,3,4-Trimethylhexane		548.15	140
			780000	2,2,4-Trimethylpentane		471.15	140
1,4-*trans*-Polyisoprene			180000	2,5-Dimethylhexane		451.15	140
			180000	3,4-Dimethylhexane		521.15	140
			180000	2,2-Dimethylpentane		405.15	140
			180000	2,3-Dimethylpentane		460.15	140
			180000	2,4-Dimethylpentane		404.15	140
			180000	3-Methylpentane		473.15	140
			180000	Heptane		467.15	140
			180000	Hexane		407.15	140
			180000	Nonane		540.15	140
			180000	Octane		503.15	140
			180000	2,2,4,4-Tetramethylpentane		519.15	140
			180000	2,3,4-Trimethylhexane		548.15	140
Poly(*N*-isopropylacrylamide)	5400	14000		Water		307.45	146
	146000	530000		Water		305.85	146
Poly(*N*-isopropylacrylamide)-poly[(*N*-acetylimino)ethylene] block copolymer (80 wt% *N*-Isopropylacrylamide)	5500			Water		306.2	223
Poly(*N*-isopropylacrylamide)-poly[(*N*-acetylimino)ethylene] graft copolymer (75 wt% *N*-Isopropylacrylamide)	6030			Water		306.2	223
Poly(*N*-isopropylmethacrylamide)	6250	20000		Water		319.95	212
Poly(methyl methacrylate)			127000	Acetonitrile	267.15		16
			970000	Acetonitrile	303.15		16
			50000	1-Butanol	353.25		2
	infinite			2-Butanone		482	80
	infinite			1-Chlorobutane	320	463	80
			970000	2,2-Dimethyl 3-pentanone	301.55		16
			127000	2,4-Dimethyl-3-pentanone	280.15		16
	200000	264000		2-Ethoxyethanol	312.15		196
		77000		Ethyl acetate	290	533	190
			127000	2-Ethylbutanal	264.65		16
	infinite			3-Heptanone	307.7		126
			970000	4-Heptanone	299.95		16
	infinite			3-Hexanone		522	80
	infinite			Methyl acetate		451	80
			50000	1-Methyl-4-isopropylbenzene	400.15		2
			1400000	2-Octanone	321.15		16
	572400	595300		3-Octanone	329.88		166
	infinite			3-Pentanone		506	80
			50000	1-Propanol	349.95		2
	infinite			2-Propanone		439	80
	200000	264000		Tetra(ethylene glycol)	390.15		196
			400000	Toluene	225.35		2
			50000	Trichloromethane	231.15		2
	200000	264000		Tri(ethylene glycol)	407.15		196
Poly(methyl methacrylate) (isotactic)	infinite			Acetonitrile	301	461	80
	infinite			2-Butanone		464	80
	infinite			1-Chlorobutane	309	454	80
	infinite			4-Heptanone	319	522	80
	infinite			3-Hexanone	279	511	80
	infinite			Methyl acetate		441	80
	infinite			3-Pentanone		497	80
	infinite			2-Propanone		428	80

Polymer	M_n/g mol^{-1}	M_w/g mol^{-1}	M_η/g mol^{-1}	Solvent	UCST/K	LCST/K	Ref.
Poly(4-methyl-1-pentene) (isotactic)			152000	Butane		388	102
			152000	Cyclopentane		505	102
			152000	2,2-Dimethylbutane		462	102
			152000	2,2-Dimethylpentane		499	102
			152000	2,4-Dimethylpentane		499	102
	infinite			Diphenyl	467.8		62
	infinite			Diphenyl ether	483.2		62
	infinite			Diphenylmethane	449.8		62
			152000	3-Ethylpentane		532	102
			152000	Heptane		522	102
			152000	Hexane		487	102
			152000	2-Methylbutane		431	102
			152000	Nonane		579	102
			152000	Octane		553	102
			152000	Pentane		441	102
			152000	2,2,3-Trimethylbutane		521	102
Poly(α-methylstyrene)	58500	61400		Butyl acetate	262.05	457.15	181
	99100	113000		Cyclohexane	293.55		152
	26000	31200		Cyclopentane	276.7	435.95	181
		289000		*trans*-Decahydronaphthalene	273		181
	69500	76500		Hexyl acetate	285.05	508.15	181
	72000	75600		Methylcyclohexane	328.9		203
	58500	61400		Pentyl acetate	287.1	484.6	181
Poly(2-methyl-5-vinylpyridine)			600000	Butyl acetate	287.95		20
			263000	Ethyl butyrate	319.05		20
			335000	Ethyl propionate	293.55		20
			275000	3-Methylbutyl acetate	314.75		20
			335000	4-Methyl-2-pentanone	299.95		20
			170000	2-Methylpropyl acetate	312.35		20
			165000	Pentyl acetate	316.95		20
			284000	Propionitrile	262.35		20
			152000	Propyl acetate	282.65		20
			181000	Propyl propionate	312.15		20
			233000	Tetrahydronaphthalene	316.95		20
Poly(1-pentene) (isotactic)			4500000	Cyclopentane		502	102
			4500000	2,2-Dimethylbutane		457	102
			4500000	3,4-Dimethylhexane		>569	102
			4500000	2,2-Dimethylpentane		502	102
			4500000	2,3-Dimethylpentane		529	102
			4500000	2,4-Dimethylpentane		493	102
			4500000	3-Ethylpentane		537	102
			4500000	Heptane		522	102
			4500000	Hexane		482	102
			4500000	2-Methylbutane		422	102
			4500000	Octane		556	102
			4500000	Pentane		433	102
			4500000	2,2,4-Trimethylpentane		527	102
Polypropylene (atactic)	infinite			Diphenyl ether	426.5		9
	infinite			Diethyl ether		383	68
			242000	Heptane		511	101
	infinite			Hexane		441	68
			242000	2-Methylbutane		413	101
			242000	Methylcyclohexane		564	101
	infinite			Pentane		397	68
Polypropylene (isotactic)			28000	Benzyl phenyl ether	429.2		31
			28000	Benzyl propionate	405.2		31
			28000	1-Butanol	395.2		31
			28000	4-*tert*-Butylphenol	413.2		31
			242000	Cyclohexane		540	101
			242000	Cyclopentane		495	101
			28000	Dibenzyl ether	433.2		31

Polymer	M_n/g mol^{-1}	M_w/g mol^{-1}	M_h/g mol^{-1}	Solvent	UCST/K	LCST/K	Ref.
			242000	2,2-Dimethylbutane		441	101
			242000	2,3-Dimethylbutane		465	101
			242000	3,4-Dimethylhexane		553	101
			242000	2,2-Dimethylpentane		489	101
			242000	2,3-Dimethylpentane		513	101
			242000	2,4-Dimethylpentane		481	101
			28000	Diphenyl	388.2		31
			28000	Diphenyl ether	395.2		31
			28000	Diphenylmethane	389.7		31
			242000	3-Ethylpentane		520	101
			28000	4-Ethylphenol	457.2		31
			242000	Heptane		511	101
			242000	Hexane		470	101
			242000	2-Methylbutane		413	101
			28000	3-Methylbutyl benzyl ether	384.2		31
			242000	Methylcyclohexane		564	101
			242000	Methylcyclopentane		518	101
			28000	4-Methylphenol	479.2		31
			28000	2-Methyl-1-propanol	395.2		31
			242000	Nonane		571	101
			242000	Octane		542	101
			28000	4-Octylphenol	379.2		31
			28000	4-Isooctylphenol	383.2		31
			242000	Pentane		422	101
			242000	2,2,4,4-Tetramethylpentane		548	101
			242000	2,2,3-Trimethylbutane		511	101
			242000	2,3,4-Trimethylhexane		585	101
			242000	2,2,4-Trimethylpentane		510	101
Poly(propylene glycol)	1000			Hexane	288.15		88
	575			Water		318.2	65
Polystyrene	34900	37000		Benzene		538.7	61
			62600	Butanedioic acid dimethyl ester	335.15		2
	3700	4000		1-Butanol	383.45		154
	91700	97200		2-Butanone		448.8	61
	545500	600000		Butyl acetate		489	181
	104000	110000		*tert*-Butyl acetate	250.0	417.9	74
			62600	Butyl stearate	387.15		2
	18400	19200		1-Chlorododecane	274.65		154
	18400	19200		1-Chlorohexadecane	337.05		154
	18400	19200		1-Chlorooctadecane	365.55		154
	18400	19200		1-Chlorotetradecane	309.35		154
	46400	51000		Cyclodecane	278.9		128
	46400	51000		Cycloheptane	276.2		128
	34900	37000		Cyclohexane	285.6	510.9	60
			236000	Cyclohexanol	353.5		8
	46400	51000		Cyclooctane	275.2		128
	91700	97200		Cyclopentane	275.2	445.5	61
	91500	97000		*trans*-Decahydronaphthalene	281.95		81
		4800		Decane	360.95		154
	3700	4000		1-Decanol	375.15		154
			570000	Decyl acetate		650	64
	18700	19800		Diethyl ether	235.6	314.5	51
	187000	200000		Diethyl malonate	285.8	589.6	74
	47200	50000		Diethyl oxalate	280.05		131
	151000	160000		Dimethoxymethane		401.2	51
		240000		1,4-Dimethylcyclohexane	387	482	116
			62600	Dimethyl malonate	409.15		2
			62600	Dimethyl oxalate	453.15		2
	116000	123000		Dodecadeuterocyclohexane	298.10		224

Polymer	M_n/g mol^{-1}	M_w/g mol^{-1}	M_h/g mol^{-1}	Solvent	UCST/K	LCST/K	Ref.
		25000		Dodecadeuteromethyl-cyclopentane	310.07		180
		4800		Dodecane	368.65		154
	3700	4000		1-Dodecanol	379.75		154
	infinite			Dodecyl acetate	285.2		206
	104000	110000		Ethyl acetate	213.9	435.4	72
	104000	110000		Ethyl butanoate		490.8	74
	221000	239000		Ethylcyclohexane	330.52		18
	9440	10000		Ethyl formate	272	451	74
		900000		Bis(2-ethylhexyl) phthalate	283.05		136
	4530	4800		Heptane	359	477	112
	3700	4000		1-Dexadecanol	386.25		154
	5500	5770		1,1,1,3,3,3-Hexadeutero-2-propanone	270	436	157
	1920	2030		Hexane	318	470	112
			62600	Hexanoic acid	448.15		2
	3700	4000		1-Hexanol	372.15		154
			62600	3-Hexanol	396.65		2
			90000	Hexyl acetate		578	64
	104000	110000		Methyl acetate	284.2	415.7	72
	104000	110000		3-Methyl-1-butyl acetate	210.1	510.1	72
	91700	97200		Methylcyclohexane	321.8	505.9	60
	10750	11500		Methylcyclopentane	295	480	157
	104000	110000		2-Methyl-1-propyl acetate	210.4	468.5	72
		48000		Nitroethane	303.1		151
		4800		Octadecane	403.55		154
	3700	4000		1-Octadecanol	390.55		154
	4530	4800		Octane	353	527	112
	3700	4000		1-Octanol	372.35		154
			62600	1-Octene	355.15		2
		4800		Pentadecane	385.25		154
		1100		Pentane	292		137
	3700	4000		1-Pentanol	375.05		154
	219800	233000		Pentyl acetate		519	181
		100000		1-Phenyldecane	283.60		105
	5500	5770		2-Propanone	251	452	157
	12750	13500		Propionitrile	312		187
	104000	110000		Propyl acetate	183.7	469.0	72
	104000	110000		2-Propyl acetate	220.9	414.2	72
	3700	4000		1-Tetradecanol	383.25		154
	34900	37000		Toluene		567.2	60
			62600	Vinyl acetate	384.15		2
Polystyrene (three-arm star)		230000		Cyclohexane	297.1	496.8	93
Polystyrene (four-arm star)		155000		Cyclohexane	294.13		199
Poly(trimethylene oxide)		infinite		Cyclohexane	300		79
Poly(vinyl alcohol)		40000		Water		514	45
Poly(*N*-vinyl caprolactam)		150000		Water		306.45	217
Poly(vinyl chloride)	55000			Dibutyl phthalate	353		114
	55000			Tricresyl phosphate	383		114
			85000	Dimethyl phthalate	355		219
Poly(*N*-vinylisobutyramide)	66000	105600		Water		313.25	208
Poly(vinyl methyl ether)	46500	98600		Deuterium oxide		307.2	173
	83000	155000		Water		306.95	146
Poly(*N*-vinyl-*N*-propylacetamide)			30000	Water		313.5	176
Styrene/acrylonitrile copolymer							
(21.1 wt% acrylonitrile)	infinite			Toluene	325.4		52
(23.2 wt% Acrylonitrile)	infinite			Toluene	355.1		52
(25.0 wt% Acrylonitrile)	90000	147000		Toluene	313.15		198
(51.0 wt% Acrylonitrile)		347000		Ethyl acetate		344.15	107
Styrene/methyl methacrylate copolymer (52.0 mol% Styrene)		infinite		Cyclohexanol	334.65		38

Polymer	M_n/g mol^{-1}	M_w/g mol^{-1}	M_η/g mol^{-1}	Solvent	UCST/K	LCST/K	Ref.
Styrene/α-methylstyrene copolymer (20.0 mol% Styrene)	100000	114000		Butyl acetate	288.85	453.05	181
	100000	114000		Cyclohexane	285.85	484.85	181
	100000	114000		Cyclopentane	290.95	421.05	181
	100000	114000		*trans*-Decahydronaphthalene	264.15		181
	100000	114000		Hexyl acetate	288.55	514.15	181
	100000	114000		Pentyl acetate	303.15	480.65	181
Trifluoronitrosomethane/tetrafluoroethylene copolymer (1:1) alternating							
	infinite			1,1,2-Trichloro-1,2,2-trifluoroethane	301.6		12
N-Vinylacetamide/vinyl acetate copolymer							
(58 mol% Vinyl acetate)	30000	57000		Water		340.15	225
(63 mol% Vinyl acetate)	27000	48600		Water		323.15	225
(78 mol% Vinyl acetate)	26000	46800		Water		282.15	225
Vinyl alcohol/vinyl butyrate copolymer (7.5 mol% Butyralized PVA)	infinite			Water	408.0	298.25	121
N-Vinylcaprolactam/*N*-vinylamine copolymer (3.8 mol% Vinyl amine)			160000	Water		308.8	176
N-Vinylformamide/vinyl acetate copolymer							
(60 mol% Vinyl acetate)	24000	45600		Water		310.15	225
(66 mol% Vinyl acetate)	25000	47500		Water		291.15	225
(73 mol% Vinyl acetate)	23000	50600		Water		277.15	225

References

1a. Koningsveld, R., Stockmayer, W.H., and Nies, E., *Polymer Phase Diagrams*, Oxford University Press, Oxford, 2001.
1b. Kamide, K., *Thermodynamics of Polymer Solutions*, Elsevier, Amsterdam, 1990.
2. Jenckel, E. and Gorke, K., *Z. Naturforsch.*, 5a, 317, 556, 1950.
3. Fox, T.G. and Flory, P.J., *J. Amer. Chem. Soc.*, 73, 1909, 1951.
4. Fox, T.G. and Flory, P.J., *J. Amer. Chem. Soc.*, 73, 1915, 1951.
5. Mandelkern, L. and Flory, P.J., *J. Amer. Chem. Soc.*, 74, 2517, 1952.
6. Shultz, A.R. and Flory, P.J., *J. Amer. Chem. Soc.*, 74, 4760, 1952.
7. Wagner, H.L. and Flory, P.J., *J. Amer. Chem. Soc.*, 74, 195, 1952.
8. Shultz, A.R.and Flory, P.J., *J. Amer. Chem. Soc.*, 75, 3888, 1953.
9. Kinsinger, J.B. and Wessling, R.A., *J. Amer. Chem. Soc.*, 81, 2908, 1959.
10. Freeman, P.I. and Rowlinson, J.S., *Polymer*, 1, 20, 1960.
11. Krigbaum, W.R., Kurz, J.E., and Smith, P., *J. Phys. Chem.*, 65, 1984, 1961.
12. Morneau, G.A., Roth, P.I., and Shultz, A.R., *J. Polym. Sci.*, 55, 609, 1961.
13. Debye, P., Coll, H., and Woermann, D., *J. Chem. Phys.*, 32, 939, 1960.
14. Debye, P., Coll, H., and Woermann, D., *J. Chem. Phys.*, 33, 1746, 1960.
15. Debye, P., Chu, B., and Woermann, D., *J. Chem. Phys.*, 36, 1803, 1962.
16. Fox, T.G., *Polymer*, 3, 111, 1962.
17. Ham, J.S., Bolen, M.C., and Hughes, J.K., *J. Polym. Sci.*, 57, 25, 1962.
18. Debye, P., Woermann, D., and Chu, B., *J. Polym. Sci.: Part A*, 1, 255, 1963.
19. Allen, G. and Baker, C.H., *Polymer*, 6, 181, 1965.
20. Gechele, G.B., Crescentini, L., *J. Polym. Sci.: Part A*, 3, 3599, 1965.
21. Myrat, C.D. and Rowlinson, J.S., *Polymer*, 6, 645, 1965.
22. Kubo, K. and Ogino, K., *Sci. Pap. Coll. Art. Sci. Univ. Tokyo*, 16, 193, 1966.
23. Rehage, G., Moeller, D., and Ernst, O., *Makromol.Chem.*, 88, 232, 1965.
24. Nakajima, A., Fujiwara, H., and Hamada, F., *J. Polym. Sci.: Part A-2*, 4, 507, 1966.
25. Nakajima, A., Hamada, F., and Hayashi, S., *J. Polym. Sci.: Part C*, 15, 285, 1966.
26. Koningsveld, R., *Proefschrift Univ. Leiden*, Heerlen, 1967.
27. Llopis, J., Albert, A., and Usobinaga P., *Eur. Polym. J.*, 3, 259, 1967.
28. Moraglio, G., Gianotti, G., and Danusso, F., *Eur. Polym. J.*, 3, 251, 1967.
29. Orwoll, R.A. and Flory, P.J., *J. Amer. Chem. Soc.*, 89, 6822, 1967.
30. Patterson, D., Delmas, G., and Somcynsky, T., *Polymer*, 8, 503, 1967.
31. Nakajima, A. and Fujiwara, H., *J. Polym. Sci.: Part A-2*, 6, 723, 1968.
32. Rehage, G. and Koningsveld, R., *J. Polym. Sci.: Polym. Lett.*, 6, 421, 1968.
33. Andreeva, V. M., et al., *Vysokomol. Soedin., Ser. B*, 11, 555, 1969.
34. Bardin, J.-M. and Patterson, D., *Polymer*, 10, 247, 1969.
35. Dusek, K., *Coll. Czech. Chem. Commun.*, 34, 3309, 1969.
36. Delmas, G. and Patterson, D., *J. Polym. Sci.: Part C*, 30, 1, 1970.
37. Koningsveld, R., Kleintjens, L.A., and Shultz, A.R., *J. Polym. Sci.: Part A-2*, 8, 1261, 1970.
38. Kotaka, T., et al., *Polym. J.*, 1, 245, 1970.
39. Liddell, A.H. and Swinton, F.L., *Discuss. Faraday Soc.*, 49, 115, 1970.
40. Matsumura, K., *Polym. J.*, 1, 322, 1970.
41. Nakayama, H., *Bull. Chem. Soc. Japan*, 43, 1683, 1970.
42. Cowie, J.M.G., Maconnachie, A., and Ranson, R.J., *Macromolecules*, 4, 57, 1971.
43. Kagemoto, A. and Baba, Y., *Kobunshi Kagaku*, 28, 784, 1971.
44. Shibatani, K. and Oyanagi, Y., *Kobunshi Kagaku*, 28 (1971) 361-367.
45. Tager, A.A., et al., *Vysokomol. Soedin., Ser. A*, 13, 659, 1971.
46. Izumi, Y. and Miyake, Y., *Polym. J.*, 3, 647, 1972.
47. Kagemoto, A., Baba, Y., and Fujishiro, R., *Makromol. Chem.*, 154, 105, 1972.
48. Kennedy, J.W., Gordon, M., and Koningsveld, R., *J. Polym. Sci.: Part C*, 39, 43, 1972.
49. Lirova, B.I., et al., *Vysokomol. Soedin., Ser. B*, 14,265, 1972.
50. Nakayama, H., *Bull. Chem. Soc. Japan*, 45, 1371, 1972.
51. Siow, K.S., Delmas, G., and Patterson, D., *Macromolecules*, 5, 29, 1972.
52. Teramachi, S. and Fujikawa, T., *J. Macromol. Sci.-Chem. A*, 6, 1393, 1972.
53. Zeman, L., Biros, J., Delmas, G., and Patterson, D., *J. Phys. Chem.*, 76, 1206, 1972.
54. Zeman, L. and Patterson, D., *J. Phys. Chem.*, 76, 1214, 1972.
55. Baba, Y., Fujita, Y., and Kagemoto, A., *Makromol. Chem.*, 164, 349, 1973.
56. Candau, F., Strazielle, C., and Benoit, H., *Makromol. Chem.*, 170, 165, 1973.
57. Hamada, F., Fujisawa, K., and Nakajima, A., *Polym. J.*, 4, 316, 1973.
58. Kuwahara, N., Nakata, M., and Kaneko, M., *Polymer*, 14, 415, 1973.
59. Kuwahara, N., Kojima, J., and Kaneko, M., *J. Polym. Sci.: Polym. Phys. Ed.*, 11, 2307, 1973.

60. Saeki, S., Kuwahara, N., Konno, S., and Kaneko, M., *Macromolecules*, 6, 246, 1973.
61. Saeki, S., Kuwahara, N., Konno, S., and Kaneko, M., *Macromolecules*, 6, 589, 1973.
62. Tani, S., Hamada, F., and Nakajima, A., *Polym. J.*, 5, 86, 1973.
63. Baba, Y. and Kagemoto, A., *Kobunshi Ronbunshu*, 31, 446, 1974.
64. Bataille, P., *J. Chem. Eng. Data*, 19, 224, 1974.
65. Bessonov, Yu.S. and Tager, A.A., *Trud. Khim. Khim. Tekhnol.*, 1, 150, 1974.
66. Cowie, J.M.G. and McEwen, L.J., *J. Chem. Soc., Faraday Trans. I*, 70, 171, 1974.
67. Cowie, J.M.G. and McEwen, I.J., *Macromolecules*, 7, 291, 1974.
68. Cowie, J.M.G. and McEwen, I.J., *J. Polym. Sci.: Polym. Phys. Ed.*, 12, 441, 1974.
69. Derham, K.W., Goldsbrough, J., and Gordon, M., *Pure Appl. Chem.*, 38, 97, 1974.
70. Kuwahara, N., Saeki, S., Chiba, T., and Kaneko, M., *Polymer*, 15, 777, 1974.
71. Nakajima, A., et al., *Makromol. Chem.*, 175, 197, 1974.
72. Saeki, S., Konno, S., Kuwahara, N., Nakata, M., and Kaneko, M., *Macromolecules*, 7, 521, 1974.
73. Ver Strate, G. and Philippoff, W., *J. Polym. Sci.: Polym. Lett. Ed.*, 12, 267, 1974.
74. Konno, S., et al., *Macromolecules*, 8, 799, 1975.
75. Nakata, M., Kuwahara, N., and Kaneko, M., *J. Chem. Phys.*, 62, 4278, 1975.
76. Saeki, S., Kuwahara, N., Nakata, M., and Kaneko, M., *Polymer*, 16, 445, 1975.
77. Strazielle, C. and Benoit, H., *Macromolecules*, 8, 203, 1975.
78. Tager, A. A., et al., *Vysokomol. Soedin., Ser. B*, 17, 61, 1975.
79. Chiu, D.S., Takahashi, Y., and Mark, J.E., *Polymer*, 17, 670, 1976.
80. Cowie, J.M.G. and McEwen, I.J., *J. Chem. Soc., Faraday Trans. I*, 72, 526, 1976.
81. Nakata, M., et al., *J. Chem. Phys.*, 64, 1022, 1976.
82. Nose, T. and Tan, T.V., *J. Polym. Sci.: Polym. Lett. Ed.*, 14, 705, 1976.
83. Saeki, S., Kuwahara, N., Nakata, M., and Kaneko, M., *Polymer*, 17, 685, 1976.
84. Saeki, S., Kuwahara, N., and Kaneko, M., *Macromolecules*, 101, 1976.
85. Slagowski, E., Tsai, B., and McIntyre, D., *Macromolecules*, 9, 687, 1976.
86. Panina, N.I., Lozgacheva, V.P., and Aver'yanova, V.M., *Vysokomol. Soedin., Ser. B*, 19, 786,. 1977.
87. Rigler, J.K., Wolf, B.A., and Breitenbach, J.W., *Angew. Makromol. Chem.*, 57, 15, 1977.
88. Vshivkov, S.A., et al., *Prots. Studneobras. Polimern. Sistem.*, (2), 3, 1977.
89. Wolf, B.A. and Jend, R., *Makromol. Chem.*, 178, 1811, 1977.
90. Wolf, B.A. and Sezen, M.C., *Macromolecules*, 10, 1010, 1977.
91. Kodama, Y. and Swinton, F.L., *Brit. Polym. J.*, 10, 191, 1978.
92. Nakata, M., Dobashi, T., Kuwahara, N., Kaneko, M., and Chu, B., *Phys. Rev. A*, 18, 2683, 1978.
93. Cowie, J.M.G., Horta, A., McEwen, I.J., and Prochazka, K., *Polym. Bull.*, 1, 329, 1979.
94. Hamano, K., Kuwahara, N., and Kaneko, M., *Phys. Rev. A*, 20, 1135, 1979.
95. Kleintjens, L.A.L., *Ph.D. Thesis*, Univ. Essex, U.K., 1979.
96. Dobashi, T., Nakata, M., and Kaneko, M., *J. Chem. Phys.*, 72, 6685, 1980.
97. Irvine, P. and Gordon, M., *Macromolecules*, 13, 761, 1980.
98. Kleintjens, L.A., Koningsveld, R., and Gordon, M., *Macromolecules*, 13, 303, 1980.
99. Lang, J.C. and Morgan, R.D., *J. Chem. Phys.*, 73, 5849, 1980.
100. Richards, R.W., *Polymer*, 21, 715, 1980.
101. Charlet, G. and Delmas, G., *Polymer*, 22, 1181, 1981.
102. Charlet, G., Ducasse, R., and Delmas, G., *Polymer*, 22, 1190, 1981.
103. Hashizume, J., Teramoto, A., and Fujita, H., *J. Polym. Sci.: Polym. Phys. Ed.*, 19, 1405, 1981.
104. Wolf, B.A. and Geerissen, H., *Colloid Polym. Sci.*, 259, 1214, 1981.
105. Geerissen, H. and Wolf, B.A., *Makromol. Chem., Rapid Commun.*, 3, 17, 1982.
106. Goloborod'ko, V.I., Valatin, S.M., and Tashmukhamedov, I.P., *Uzb. Khim. Zh.*, (3), 33, 1982.
107. Mangalam, P. V. and Kalpagam, V., *J. Polym. Sci.: Polym. Phys. Ed.*, 20, 773, 1982.
108. Shinozaki, K., Abe, M., and Nose, T., *Polymer*, 23, 722, 1982.
109. Shinozaki, K., Van Tan, T., Saito, Y., and Nose, T., *Polymer*, 23, 728, 1982.
110. Suzuki, H., Kamide, K., and Saitoh, M., *Eur. Polym. J.*, 18, 123, 1982.
111. Suzuki, H., Muraoka, Y., Saitoh, M., and Kamide, K., *Brit. Polym. J.*, 14, 23, 1982.
112. Cowie, J.M.G. and McEwen, I.J., *Polymer*, 24, 1445, 1983.
113. Herold, F.K., Schulz, G.V., and Wolf, B.A., *Materials Chem. Phys.*, 8, 243, 1983.
114. Tager, A.A., et al., *Vysokomol. Soedin., Ser. A*, 25, 1444, 1983.
115. Corti, M., Minero, C., and Degiorgio, V., *J. Phys. Chem.*, 88, 309, 1984.
116. Cowie, J.M.G. and McEwen, I.J., *Polymer*, 25, 1107, 1984.
117. Dobashi, T., Nakata, M., and Kaneko, M., *J. Chem. Phys.*, 80, 948, 1984.
118. Florin, E., Kjellander, R., and Eriksson, J.C., *J. Chem. Soc., Faraday Trans. I*, 80, 2889, 1984.
119. Gilluck, M., *Dissertation*, TH Leuna-Merseburg, 1984.
120. Rangel-Nafaile, C., Metzner, A.B., and Wissbrun, K.F., *Macromolecules*, 17, 1187, 1984.
121. Shiomi, T., et al., *J. Polym. Sci.: Polym. Phys. Ed.*, 22, 1305, 1984.
122. Tsuyumoto, M., Einaga, Y., and Fujita, H., *Polym. J.*, 16, 229, 1984.
123. Varennes, S., Charlet, G., and Delmas, G., *Polym. Eng. Sci.*, 24, 98, 1984.
124. Hamano, K., Kuwahara, N., Koyama, T., and Harada, S., *Phys. Rev. A*, 32, 3168, 1985.
125. Kraemer, H. and Wolf, B.A., *Makromol. Chem., Rapid Commun.*, 6, 21, 1985.
126. Herold, F.K. and Wolf, B.A., *Mater. Chem. Phys.*, 14, 311, 1986.
127. Irani, C.A. and Cozewith, C., *J. Appl. Polym. Sci.*, 31, 1879, 1986.
128. Cowie, J.M.G. and McEwen, I.J., *Brit. Polym. J.*, 18, 387, 1986.
129. Krüger, B., *Dissertation*, TH Leuna-Merseburg, 1986.
130. Rätzsch, M.T., et al., *J. Macromol. Sci.-Chem. A*, 23, 1349, 1986.
131. Saeki, S., et al., *Macromolecules*, 19, 2353, 1986.
132. Sander, U. and Wolf, B.A., *Angew. Makromol. Chem.*, 139, 149, 1986.
133. Barbarin-Castillo, J.-M., et al., *Polym.Commun.*, 28, 212, 1987.
134. Gruner, K. and Greer, S.C., *Macromolecules*, 20, 2238, 1987.
135. Magarik, S.Ya., Filippov, A.P., and D'yakonova, N.V., *Vysokomol. Soedin., Ser. A*, 29, 698, 1987.
136. Rangel-Nafaile, C. and Munoz-Lara, J.J., *Chem. Eng. Commun.*, 53, 177, 1987.
137. Kiepen, F. and Borchard, W., *Macromolecules*, 21, 1784, 1988.
138. Schuster, R., *Diploma Paper*, TH Leuna-Merseburg, 1988.
139. Tveekrem, J.L., Greer, S.C., and Jacobs, D.T., *Macromolecules*, 21, 147, 1988.
140. Bohossian, T., Charlet, G., and Delmas, G., *Polymer*, 30, 1695, 1989.
141. Chiu, G. and Mandelkern, L., *Macromolecules*, 23, 5356, 1990.
142. Goedel, W.A., et al., *Ber. Bunsenges. Phys. Chem.*, 94, 17, 1990.
143. Van der Haegen, R. and Van Opstal, L., *Makromol. Chem.*, 191, 1871, 1990.
144. Iwai, Y., et al., *Sekiyu Gakkaishi*, 33, 117, 1990.
145. Raetzsch, M.T., Krueger, B., and Kehlen, H., *J. Macromol. Sci.-Chem. A*, 27, 683, 1990.
146. Schild, H.G. and Tirrell, D.A., *J. Phys. Chem.*, 94, 4352, 1990.
147. Stafford, S.G., Ploplis, A.C., and Jacobs, D.T., *Macromolecules*, 23, 470, 1990.
148. Akhmadeev, I.R., et al., *Vysokomol. Soedin., Ser. B*, 33, 543, 1991.
149. Bae, Y.C., Lambert, S.M., Soane, D.S., and Prausnitz, J.M., *Macromolecules*, 24, 4403, 1991.
150. Chu, B., Linliu, K., Xie, P., Ying, Q., Wang, Z., and Shook, J.W., *Rev. Sci. Instr.*, 62, 2252, 1991.
151. Kawate, K., Imagawa, I., and Nakata, M., *Polym. J.*, 23, 233, 1991.
152. Lee, K.D. and Lee, D.C., *Pollimo*, 15, 274, 1991.
153. Louai, A., Sarazin, D., Pollet, G., Francois, J., and Moreaux, F., *Polymer*, 32, 703, 1991.
154. Van Opstal, L., Koningsveld, R., and Kleintjens, L.A., *Macromolecules*, 24, 161, 1991.
155. Schubert, K.-V., Strey, R., and Kahlweit, M., *J. Colloid Interface Sci.*, 141, 21, 1991.
156. Shen, W., Smith, G.R., Knobler, C.M., and Scott, R.L., *J. Phys. Chem.*, 95, 3376, 1991.

157. Szydlowski, J. and Van Hook, W.A., *Macromolecules*, 24, 4883, 1991.
158. Tager, A.A., et al., *Vysokomol. Soedin., Ser. B*, 33, 572, 1991.
159. Wakker, A., *Polymer*, 32, 279, 1991.
160. Yokoyama, H., Takano, A., Okada, M., and Nose, T., *Polymer*, 32, 3218, 1991.
161. Heinrich, M. and Wolf, B.A., *Polymer*, 33, 1926, 1992.
162. Heinrich, M. and Wolf, B.A., *Macromolecules*, 25, 3817, 1992.
163. Lecointe, J.P., Pascault, J.P., Suspene, L., and Yang, Y.S., *Polymer*, 33, 3226, 1992.
164. Mueller, K.F., *Polymer*, 33, 3470, 1992.
165. Szydlowski, J., Rebelo, L., and Van HooK, W.A., *Rev. Sci. Instrum.*, 63, 1717, 1992.
166. Xia, K.-Q., Franck, C., and Widom, B., *J. Chem. Phys.*, 97, 1446, 1992.
167. Arnauts, J., Berghmans, H., and Koningsveld, R., *Makromol. Chem.*, 194, 77, 1993.
168. Iwai, Y., Shigematsu, Y., Furuya, T., Fukuda, H., Arai, Y., *Polym. Eng. Sci.*, 33, 480, 1993.
169. Wakker, A., Van Dijk, F., and Van Dijk, M.A., *Macromolecules*, 26, 5088, 1993.
170. Wells, P.A., de Loos, Th.W., and Kleintjens, L.A., *Fluid Phase Equil.*, 83, 383, 1993.
171. Barbarin-Castillo, J.-M. and McLure, I.A., *Polymer*, 35, 3075, 1994.
172. Mumick, P.S. and McCormick, C.L., *Polym. Eng. Sci.*, 34, 1419, 1994.
173. Okano, K., Takada, M., Kurita, K., and Furusaka, M., *Polymer*, 35, 2284, 1994.
174. Sato, H., Kuwahara, N., and Kubota, K., *Phys. Rev. E*, 50, 1752, 1994.
175. Song, S.-W. and Torkelson, J.M., *Macromolecules*, 27, 6389, 1994.
176. Tager, A.A., et al., *Colloid Polym. Sci.*, 272, 1234, 1994.
177. Vanhee, S., et al., *Makromol. Chem. Phys.*, 195, 759, 1994.
178. Haas, C.K. and Torkelson, J.M., *Phys. Rev. Lett.*, 75, 3134, 1995.
179. Ikier, C. and Klein, H., *Macromolecules*, 28, 1003, 1995.
180. Luszczyk, M., Rebelo, L.P.N., and Van Hook, W.A., *Macromolecules*, 28, 745, 1995.
181. Pfohl, O., Hino, T., and Prausnitz, J.M., *Polymer*, 36, 2065, 1995.
182. Vshivkov, S.A. and Safronov, A.P., *Vysokomol. Soedin., Ser. B*, 37, 1779, 1995.
183. Allcock, H.R. and Dudley, G.K., *Macromolecules*, 29, 1313, 1996.
184. El-Ejmi, A.A.S. and Huglin, M.B., *Polym. Int.* 39, 113, 1996.
185. Fischer, V., Borchard, W., and Karas, M., *J. Phys. Chem.*, 100, 15992, 1996.
186. Imre, A. and Van Hook, W.A., *J. Polym. Sci.: Part B: Polym. Sci.*, 34, 751, 1996.
187. Luszczyk, M. and Van Hook, W.A., *Macromolecules*, 29, 6612, 1996.
188. Rong, Z., Wang, H., Ying, X., and Hu, Y., *J. East China Univ. Sci. Technol.*, 22, 754, 1996.
189. Safronov, A.P., Tager, A.A., and Koroleva, E.V., *Vysokomol. Soedin., Ser. B*, 38, 900, 1996.
190. Vshivkov, S.A. and Rusinova, E.V., *Vysokomol. Soedin., Ser. A*, 38, 1746, 1996.
191. Xia, K.-Q., An, X.-Q., and Shen, W.-G., *J. Chem.Phys.*, 105, 6018, 1996.
192. Imre, A. and Van Hook, W.A., *J. Polym. Sci.: Part B: Polym. Phys.*, 35, 1251, 1997.
193. Kita, R., Dobashi, T., Yamamoto, T., Nakata, M., and Kamide, K., *Phys. Rev. E*, 55, 3159, 1997.
194. Li, M., Zhu, Z.-Q., and Mei, L.-H., *Biotechnol. Progr.*, 13, 105, 1997.
195. McLure, I.A., Mokhtari, A., and Bowers, J., *J. Chem. Soc., Faraday Trans.*, 93, 249, 1997.
196. Chalykh, A.E., Dement'eva, O.V., and Gerasimov, V.K., *Vysokomol. Soedin., Ser. A*, 40, 815, 1998.
197. Kubota, K., Kita, R., and Dobashi, T., *J. Chem. Phys.*, 109, 711, 1998.
198. Schneider, A., *Dissertation*, Johannes Gutenberg Universität Mainz. 1998.
199. Terao, K., et al., *Macromolecules*, 31, 6885, 1998.
200. Yamagishi, T.-A., et al., *Macromol. Chem. Phys.*, 199, 423, 1998.
201. Lau, A.C.W. and Wu, C., *Macromolecules*, 32, 581, 1999.
202. Nakata, M., Dobashi, T., Inakuma, Y.-I., and Yamamura, K., *J. Chem. Phys.*, 111, 6617, 1999.
203. Pruessner, M.D., Retzer, M.E., and Greer, S.C., *J. Chem. Eng. Data*, 44, 1419, 1999.
204. Shimofure, S., Kubota, K., Kita, R., and Dobashi, T., *J. Chem.Phys.*, 111, 4199, 1999.
205. Fischer, V. and Borchard, W., *J. Phys. Chem. B*, 104, 4463, 2000.
206. Imre, A., and Van Hook, W.A., *Macromolecules*, 33, 5308, 2000.
207. Koizumi, J., et al., *J. Phys. Soc. Japan*, 69, 2543, 2000.
208. Kunugi, S., Tada, T., Yamazaki, Y., Yamamoto, K., and Akashi, M., *Langmuir*, 16, 2042, 2000.
209. La Mesa, C., *J. Therm. Anal. Calorim.*, 61, 493, 2000.
210. Persson, J., et al., *Bioseparation*, 9, 105, 2000.
211. Persson, J., Kaul, A., and Tjerneld, F., *J. Chromatogr. B*, 743, 115, 2000.
212. Djokpe, E. and Vogt, W., *Macromol. Chem. Phys.*, 202, 750, 2001.
213. Kujawa, P. and Winnik, F.M., *Macromolecules*, 34, 4130, 2001.
214. Pendyala, K.S., Greer, S.C., and Jacobs, D.T., *J. Chem. Phys.*, 115, 9995, 2001.
215. Berlinova, I. V., Nedelcheva, A. N., Samchikov, V., and Ivanov, Ya., *Polymer*, 43, 7243, 2002.
216. Freitag, R. and Garret-Flaudy, F., *Langmuir*, 18, 3434, 2002.
217. Maeda, Y., Nakamura, T., and Ikeda, I., *Macromolecules*, 35, 217, 2002.
218. Rebelo, L.P.N., et al., J., *Macromolecules*, 35, 1887, 2002.
219. Safronov, A.P. and Somova, T.V., *Vysokomol. Soedin., Ser. A*, 44, 2014, 2002.
220. Vshivkov, S.A., Rusinova, E.V., and Gur'ev, A.A., *Vysokomol. Soedin., Ser. B*, 44, 504, 2002.
221. Zhou, C.-S., An, X.-Q., Xia, K.-Q., Yin, X.-L., and Shen, W.-G., *J. Chem. Phys.*, 117, 4557, 2002.
222. Chang, Y., Powell, E.S., Allcock, H.R., Park, S.M., and Kim, C., *Macromolecules*, 36, 2568, 2003.
223. David, G., et al., *Eur. Polym. J.*, 39, 1209, 2003.
224. Siporska, A., Szydlowski, J., and Rebelo, L.P.N., *Phys. Chem. Chem. Phys.*, 5, 2996, 2003.
225. Yamamoto, K., Serizawa, T., and Akashi, M., *Macromol. Chem. Phys.*, 204, 1027, 2003.

VAPOR PRESSURES (SOLVENT ACTIVITIES) FOR BINARY POLYMER SOLUTIONS

Christian Wohlfarth

The vapor pressure of a binary polymer solution is given by the activity of the solvent A, a_A. Solvent activities in polymer solutions are measured either by the isopiestic method applying a reference system whose solvent activity is precisely known or by determining the solvent partial pressure, P_A, and calculating the activity of the solvent by equation (1):

$$a_A = (P_A/P_A^s)\exp\frac{(B_{AA}-V_A^L)(P-P_A^s)}{RT} \quad (1)$$

where B_{AA} is the second virial coefficient, P_A^s is the saturation vapor pressure, and V_A^L is the molar volume of the pure solvent A at the measuring temperature T. The exponential term is neglected in quite a lot of original papers, however, and only the reduced vapor pressures are given (such data are indicated by an asterisk in the table below). Vapor pressures of polymer solutions have been measured since the 1940s, but the amount of experimental data for polymer solutions is still relatively small in comparison to low-molecular mixtures and solutions. The data scatter with respect to temperature, concentration, molar mass, and other polymer characterization variables. Furthermore, the concentration range for measuring vapor pressures in good thermodynamic quality is often limited to the polymer mass fraction range between 0.4 and 0.85. A recent review on methods for the measurement of vapor pressures/solvent activities of polymer solutions and on related problems is given in Ref. [1]. Experimental data have been collected in several books [2-6].

The table in this *Handbook* provides data for a number of polymer solutions as smoothed values over the complete range of solvent activities between 0 (polymer mass fraction = 1) and 1 (polymer mass fraction = 0). For this purpose, the data were selected from data books [4–6] as well as from a number of original sources [7–22] which are not included in these books. The appropriate data were smoothed. The final table provides then the polymer mass fractions at given fixed solvent activities between 0.1 and 0.9. Of course, the user must keep in mind that the activity vs. concentration range of the experimental data is sometimes smaller than the below given complete range, thus the smoothed data should be used with sufficient care.

Generally, vapor pressures or solvent activities of binary polymer solutions depend on molar mass. However, for high molecular weight polymers (well above the oligomer region), this molar-mass dependence can be neglected in many cases. Therefore, the table below presents only data for polymer solutions where the number average molar mass, M_n, is in the order of 10^5 g/mol or even higher, therefore, the molar mass is not specified. The temperature is stated, even though the temperature dependence of a_A is relatively small for the temperature ranges where most of the experimental data exist.

References

1. Wohlfarth, C., Methods for the measurement of solvent activity of polymer solutions, in *Handbook of Solvents*, Wypych, G., Ed., ChemTec Publishing, Toronto, 2000, 146.
2. Wen, H., Elbro, H. S., and Alessi, P., *Polymer Solution Data Collection.* I. Vapor-liquid equilibrium; II. Solvent activity coefficients at infinite dilution; III. Liquid-liquid equilibrium, Chemistry Data Series, Vol. 15, DECHEMA, Frankfurt am Main, 1992.
3. Danner, R. P. and High, M. S., *Handbook of Polymer Solution Thermodynamics*, American Institute of Chemical Engineers, New York, 1993.
4. Wohlfarth, C., *Vapour-Liquid Equilibrium Data of Binary Polymer Solutions: Physical Science Data*, 44, Elsevier, Amsterdam, 1994.
5. Wohlfarth, C., *CRC Handbook of Thermodynamic Data of Copolymer Solutions*, CRC Press, Boca Raton, FL, 2001.
6. Wohlfarth, C., *CRC Handbook of Thermodynamic Data of Aqueous Polymer Solutions*, CRC Press, Boca Raton, FL, 2003.
7. Wang, K., Hu, Y., and Wu, D. T., *J. Chem. Eng. Data*, 39, 916, 1994.
8. Choi, J. S., Tochigi, K., and Kojima, K., *Fluid Phase Equil.*, 111, 143, 1995.
9. Tochigi, K., Kurita, S., Ohashi, M., and Kojima, K., *Kagaku Kogaku Ronbunshu*, 23, 720, 1997.
10. Wong, H. C., Campbell, S. W., and Bhethanabotla, V. R., *Fluid Phase Equil.*, 139, 371, 1997.
11. Kim, J., Joung, K. C., Hwang, S., Huh, W., Lee, C. S., and Yoo, K.-P., *Korean J. Chem. Eng.*, 15, 199, 1998.
12. Kim, N. H., Kim, S.J., Won, Y. S., and Choi, J. S., *Korean J. Chem. Eng.*, 15, 141, 1998.
13. Feng, W., Wang, W., and Feng, Z., *J. Chem. Ind. Eng. (China)*, 49, 271, 1998.
14. French, R. N. and Koplos, G. J., *Fluid Phase Equil.*, 160, 879, 1999.
15. Striolo, A. and Praunsitz, J. M., *Polymer*, 41, 1109, 2000.
16. Fornasiero, F., Halim, M., and Prausnitz, J. M., *Macromolecules*, 33, 8435, 2000.
17. Wong, H. C., Campbell, S. W., and Bhethanabotla, V. R., *Fluid Phase Equil.*, 179, 181, 2001.
18. Wibawa, G., Takahashi, M., Sato, Y., Takishima, S., and Masuoka, H., *J. Chem. Eng. Data*, 47, 518, 2002.
19. Wibawa, G., Hatano, R., Sato, Y., Takishima, S., and Masuoka, H., *J. Chem. Eng. Data*, 47, 1022, 2002.
20. Pfohl, O., Riebesell, C., and Dohrn, R., *Fluid Phase Equil.*, 202, 289, 2002.
21. Jung, J. K., Joung, S. N., Shin, H.Y., Kim, S. Y., Yoo, K.-P., Huh, W., Lee, C. S., *Korean J. Chem. Eng.*, 19, 296, 2002.
22. Kang, S., Huang, Y., Fu, J., Liu, H., and Hu, Y., *J. Chem. Eng. Data*, 47, 788, 2002.

Solvent Activity a_A as Function of Temperature and Mass Fraction

Polymer/ solvents	a_A: T/K	0.1	0.2	0.3	0.4	0.5	0.6	0.7	0.8	0.9
					Mass Fraction of the Polymer					
Acrylonitrile/Styrene Copolymer (28 wt% Acrylonitrile)										
Benzene*)	343.15	0.982	0.962	0.940	0.915	0.886	0.851	0.809	0.753	0.670
1,2-Dimethylbenzene*)	398.15	0.983	0.964	0.942	0.918	0.890	0.857	0.817	0.764	0.685
1,3-Dimethylbenzene*)	398.15	0.983	0.965	0.944	0.921	0.893	0.861	0.821	0.769	0.690
1,4-Dimethylbenzene*)	398.15	0.983	0.964	0.942	0.918	0.890	0.857	0.817	0.763	0.684
Propylbenzene*)	398.15	0.987	0.972	0.955	0.935	0.913	0.885	0.851	0.804	0.732
Toluene*)	343.15	0.982	0.962	0.940	0.915	0.886	0.851	0.809	0.753	0.669

Polymer/ solvents	a_A: T/K	0.1	0.2	0.3	0.4	0.5	0.6	0.7	0.8	0.9
		Mass Fraction of the Polymer								
Butadiene/Styrene Copolymer (41 wt% Styrene)										
Benzene*)	343.15	0.968	0.934	0.896	0.853	0.805	0.748	0.680	0.591	0.461
Cyclohexane*)	343.15	0.978	0.953	0.925	0.893	0.856	0.811	0.754	0.678	0.556
Ethylbenzene*)	398.15	0.974	0.945	0.912	0.875	0.831	0.779	0.713	0.625	0.491
Mesitylene*)	398.15	0.977	0.950	0.921	0.887	0.847	0.799	0.738	0.656	0.526
Toluene*)	343.15	0.970	0.936	0.899	0.857	0.808	0.751	0.682	0.591	0.456
Cellulose Triacetate										
Dichloromethane	298.15	0.979	0.956	0.930	0.899	0.863	0.819	0.762	0.683	0.554
Trichloromethane	298.15	0.978	0.953	0.924	0.892	0.853	0.806	0.747	0.665	0.533
Dextran										
Water	313.15	0.988	0.975	0.960	0.942	0.921	0.894	0.860	0.810	0.725
Hydroxyethylcellulose										
Water	368.15	0.988	0.974	0.958	0.939	0.915	0.884	0.841	0.775	0.650
Hydroxypropylstarch										
Water	293.15	0.989	0.977	0.963	0.947	0.927	0.903	0.872	0.827	0.749
Nitrocellulose										
Ethyl acetate	293.15	0.938	0.885	0.835	0.786	0.737	0.685	0.627	0.560	0.471
Ethyl formate	293.15	0.958	0.916	0.873	0.828	0.780	0.728	0.668	0.595	0.494
Ethyl propionate	293.15	0.941	0.889	0.839	0.789	0.739	0.685	0.625	0.555	0.460
Methyl acetate	293.15	0.890	0.820	0.763	0.711	0.660	0.609	0.554	0.490	0.406
2-Propanone	293.15	0.922	0.861	0.807	0.756	0.706	0.653	0.596	0.530	0.443
Propyl acetate	293.15	0.937	0.881	0.827	0.775	0.722	0.665	0.602	0.528	0.426
Polybutadiene (random cis-trans-vinyl)										
Benzene	298.15	0.964	0.925	0.884	0.839	0.788	0.731	0.663	0.578	0.455
Cyclohexane	298.15	0.974	0.945	0.913	0.876	0.833	0.782	0.719	0.635	0.507
Dichloromethane	298.15	0.951	0.902	0.852	0.800	0.745	0.684	0.616	0.532	0.415
Hexane	298.15	0.984	0.965	0.943	0.916	0.881	0.837	0.775	0.683	0.534
Tetrachloromethane	298.15	0.932	0.865	0.799	0.731	0.660	0.585	0.503	0.409	0.288
Toluene	298.15	0.969	0.935	0.898	0.856	0.809	0.754	0.688	0.603	0.476
Trichloromethane	298.15	0.925	0.855	0.788	0.720	0.650	0.578	0.498	0.406	0.289
1,4-cis-Polybutadiene										
Benzene	298.15	0.966	0.930	0.890	0.846	0.796	0.738	0.668	0.580	0.450
Cyclohexane	298.15	0.977	0.951	0.922	0.888	0.849	0.803	0.747	0.677	0.581
Dichloromethane	298.15	0.948	0.898	0.848	0.796	0.742	0.683	0.616	0.536	0.424
Hexane	298.15	0.983	0.963	0.941	0.916	0.886	0.850	0.804	0.741	0.639
Tetrachloromethane	298.15	0.936	0.871	0.805	0.736	0.665	0.588	0.505	0.409	0.287
Toluene	298.15	0.969	0.936	0.900	0.860	0.815	0.763	0.701	0.622	0.506
Trichloromethane	298.15	0.915	0.840	0.770	0.702	0.634	0.562	0.485	0.396	0.283
Poly(butyl acrylate)										
Benzene	298.15	0.964	0.926	0.887	0.845	0.799	0.749	0.691	0.619	0.519
Dichloromethane	298.15	0.868	0.801	0.744	0.690	0.636	0.577	0.511	0.430	0.318
Tetrachloromethane	298.15	0.932	0.868	0.805	0.742	0.677	0.607	0.529	0.438	0.317
Toluene	298.15	0.967	0.932	0.893	0.849	0.801	0.744	0.676	0.590	0.463
Trichloromethane	298.15	0.901	0.811	0.733	0.662	0.595	0.529	0.459	0.381	0.282
Poly(butyl methacrylate)										
Benzene	313.15	0.971	0.939	0.902	0.861	0.813	0.756	0.685	0.592	0.453
1-Butanol	313.15	0.991	0.980	0.968	0.953	0.936	0.914	0.885	0.842	0.762
2-Butanol	313.15	0.992	0.982	0.969	0.953	0.933	0.906	0.869	0.815	0.719
2-Butanone	313.15	0.982	0.963	0.940	0.914	0.884	0.846	0.799	0.732	0.623
Butyl acetate*)	308.15	0.982	0.961	0.936	0.908	0.875	0.836	0.789	0.730	0.652
Cyclohexane	313.15	0.985	0.968	0.948	0.925	0.899	0.866	0.823	0.764	0.666
Cyclopentane	313.15	0.984	0.965	0.944	0.918	0.886	0.846	0.792	0.714	0.579
Diethyl ether*)	298.15	0.987	0.973	0.956	0.937	0.914	0.885	0.848	0.795	0.703
1,4-Dimethylbenzene	333.15	0.971	0.940	0.905	0.866	0.822	0.770	0.706	0.622	0.497
Ethylbenzene	333.15	0.969	0.935	0.899	0.859	0.815	0.764	0.704	0.627	0.517
Methyl acetate	313.15	0.984	0.965	0.944	0.920	0.891	0.856	0.811	0.748	0.645

Polymer/ solvents	a_A: T/K	0.1	0.2	0.3	0.4	0.5	0.6	0.7	0.8	0.9
		Mass Fraction of the Polymer								
2-Methyl-1-propanol	333.15	0.988	0.974	0.958	0.940	0.919	0.893	0.860	0.815	0.744
Octane	313.15	0.988	0.974	0.959	0.942	0.921	0.896	0.865	0.823	0.758
1-Propanol	333.15	0.990	0.980	0.967	0.952	0.934	0.911	0.881	0.834	0.746
2-Propanol	313.15	0.991	0.981	0.970	0.956	0.939	0.918	0.889	0.845	0.755
2-Propanone	313.15	0.989	0.976	0.961	0.944	0.921	0.892	0.850	0.783	0.647
Propyl acetate	313.15	0.980	0.957	0.932	0.903	0.870	0.830	0.780	0.714	0.612
Toluene	313.15	0.971	0.939	0.903	0.863	0.818	0.764	0.698	0.613	0.485
Poly(ε-caprolacton)										
Tetrachloromethane*)	338.15	0.956	0.910	0.864	0.815	0.762	0.704	0.637	0.554	0.438
Poly(dimethylsiloxane)										
Chlorodifluoromethane	298.15	0.976	0.950	0.921	0.888	0.850	0.805	0.750	0.677	0.565
Cyclohexane	303.15	0.979	0.955	0.928	0.898	0.863	0.822	0.770	0.702	0.596
Hexane	303.15	0.982	0.962	0.939	0.912	0.880	0.842	0.793	0.724	0.611
Pentane	308.15	0.982	0.962	0.940	0.913	0.881	0.842	0.791	0.720	0.600
Pentane	423.15	0.984	0.966	0.946	0.922	0.893	0.858	0.813	0.749	0.641
Poly(ethyl acrylate)										
Benzene	298.15	0.970	0.939	0.904	0.866	0.823	0.774	0.716	0.641	0.533
Dichloromethane	298.15	0.900	0.830	0.768	0.709	0.648	0.584	0.512	0.427	0.313
Tetrachloromethane	298.15	0.950	0.900	0.848	0.794	0.736	0.672	0.598	0.509	0.385
Toluene	298.15	0.972	0.942	0.910	0.874	0.833	0.786	0.730	0.659	0.555
Trichloromethane	298.15	0.866	0.776	0.701	0.632	0.566	0.499	0.428	0.349	0.248
Poly(ethylene oxide)										
Benzene	323.15	0.972	0.942	0.908	0.869	0.824	0.771	0.706	0.620	0.490
2-Butanone	353.15	0.981	0.959	0.934	0.902	0.863	0.813	0.746	0.651	0.503
Cyclohexane	353.15	0.989	0.976	0.960	0.943	0.921	0.893	0.855	0.798	0.688
Methanol	303.15	0.964	0.927	0.887	0.844	0.797	0.744	0.682	0.604	0.494
2-Propanone	353.15	0.979	0.947	0.896	0.815	0.719	0.625	0.532	0.434	0.315
Water	293.15	0.977	0.951	0.923	0.890	0.852	0.806	0.748	0.671	0.550
Poly(ethylenimine)										
Water	353.15	0.975	0.947	0.917	0.883	0.845	0.801	0.748	0.680	0.581
Poly(ethyl methacrylate)										
Benzene	298.15	0.970	0.938	0.903	0.864	0.821	0.771	0.712	0.637	0.529
Dichloromethane	298.15	0.912	0.838	0.769	0.703	0.636	0.567	0.491	0.404	0.292
Tetrachloromethane	298.15	0.935	0.873	0.812	0.750	0.686	0.616	0.540	0.449	0.328
Toluene	298.15	0.974	0.945	0.913	0.877	0.836	0.787	0.727	0.647	0.527
Trichloromethane	298.15	0.859	0.760	0.678	0.604	0.533	0.464	0.392	0.313	0.217
Polyisobutylene										
Benzene	313.15	0.984	0.965	0.945	0.921	0.892	0.858	0.813	0.751	0.645
Cyclohexane	313.15	0.976	0.950	0.921	0.888	0.850	0.805	0.749	0.676	0.563
Cyclopentane	313.15	0.977	0.952	0.924	0.892	0.855	0.812	0.758	0.687	0.579
1,4-Dimethylbenzene	313.15	0.979	0.955	0.929	0.899	0.863	0.821	0.767	0.694	0.579
2,2-Dimethylbutane	298.15	0.983	0.964	0.942	0.917	0.887	0.852	0.806	0.743	0.640
Ethylbenzene	313.15	0.979	0.955	0.927	0.895	0.857	0.810	0.750	0.668	0.535
Heptane	298.15	0.983	0.964	0.942	0.917	0.887	0.851	0.804	0.741	0.637
Hexane	298.15	0.980	0.959	0.934	0.906	0.873	0.834	0.784	0.715	0.606
Octane	298.15	0.983	0.963	0.940	0.914	0.883	0.845	0.797	0.729	0.617
Tetrachloromethane	298.15	0.962	0.921	0.877	0.829	0.776	0.715	0.643	0.552	0.423
Toluene	313.15	0.984	0.966	0.944	0.918	0.884	0.840	0.779	0.688	0.537
Trichloromethane	298.15	0.969	0.935	0.899	0.858	0.813	0.761	0.698	0.619	0.503
2,4,4-Trimethylpentane	298.15	0.981	0.961	0.937	0.911	0.879	0.842	0.794	0.730	0.628
1,4-cis-Polyisoprene										
Benzene	313.15	0.982	0.962	0.937	0.908	0.873	0.827	0.766	0.679	0.537
2-Butanone	353.15	0.986	0.970	0.953	0.933	0.910	0.883	0.850	0.808	0.746
Cyclohexane	313.15	0.978	0.954	0.928	0.899	0.865	0.825	0.778	0.716	0.625
Dichloromethane	298.15	0.969	0.935	0.898	0.857	0.811	0.757	0.693	0.610	0.488
1,4-Dimethylbenzene	313.15	0.977	0.951	0.923	0.892	0.857	0.816	0.767	0.704	0.613

Polymer/ solvents	a_A: T/K	0.1	0.2	0.3	0.4	0.5	0.6	0.7	0.8	0.9
		Mass Fraction of the Polymer								
Ethylbenzene	313.15	0.978	0.954	0.928	0.898	0.864	0.823	0.774	0.709	0.612
Methyl acetate	313.15	0.968	0.935	0.900	0.862	0.820	0.773	0.717	0.649	0.554
Octane	313.15	0.984	0.967	0.948	0.926	0.901	0.871	0.834	0.785	0.711
Propyl acetate	333.15	0.983	0.964	0.942	0.916	0.886	0.850	0.803	0.738	0.633
Tetrachloromethane	298.15	0.929	0.864	0.800	0.737	0.672	0.602	0.526	0.435	0.316
Toluene	313.15	0.978	0.954	0.927	0.898	0.865	0.827	0.782	0.725	0.645
Trichloromethane	298.15	0.930	0.867	0.807	0.747	0.685	0.620	0.547	0.462	0.346
Poly(methyl acrylate)										
Benzene	298.15	0.979	0.956	0.930	0.901	0.867	0.826	0.776	0.710	0.608
Dichloromethane	298.15	0.917	0.851	0.791	0.732	0.671	0.605	0.532	0.444	0.326
Tetrachloromethane	298.15	0.963	0.924	0.882	0.838	0.788	0.733	0.668	0.586	0.470
Toluene	298.15	0.981	0.960	0.936	0.909	0.878	0.840	0.792	0.727	0.626
Trichloromethane	298.15	0.912	0.830	0.753	0.678	0.603	0.527	0.446	0.357	0.248
Poly(methyl methacrylate)										
Benzene	298.15	0.982	0.961	0.938	0.912	0.881	0.843	0.795	0.729	0.622
2-Butanone*)	308.15	0.989	0.976	0.961	0.945	0.925	0.900	0.869	0.825	0.751
Cyclohexanone*)	323.15	0.978	0.954	0.928	0.899	0.866	0.827	0.781	0.723	0.640
Dichloromethane	298.15	0.939	0.882	0.825	0.766	0.704	0.637	0.560	0.468	0.343
Ethyl acetate*)	308.15	0.986	0.969	0.950	0.928	0.902	0.869	0.826	0.763	0.649
Toluene	298.15	0.981	0.959	0.935	0.908	0.877	0.841	0.795	0.736	0.646
Trichloromethane	298.15	0.924	0.848	0.771	0.694	0.616	0.536	0.451	0.358	0.246
Poly(α-methylstyrene)										
Cumene	338.15	0.984	0.965	0.944	0.918	0.887	0.848	0.796	0.721	0.593
α-Methylstyrene	338.15	0.978	0.954	0.927	0.896	0.859	0.816	0.761	0.687	0.570
Poly(propylene oxide)										
Benzene	333.15	0.967	0.932	0.893	0.850	0.801	0.744	0.675	0.588	0.460
Metvhanol	298.15	0.992	0.982	0.970	0.955	0.936	0.910	0.872	0.812	0.689
Polystyrene										
Benzene	333.15	0.978	0.953	0.924	0.891	0.852	0.804	0.742	0.657	0.521
2-Butanone*)	298.15	0.986	0.971	0.954	0.935	0.912	0.885	0.851	0.804	0.724
Cyclohexane	313.15	0.990	0.978	0.965	0.949	0.931	0.908	0.877	0.833	0.754
Cyclohexanone*)	313.15	0.970	0.937	0.900	0.858	0.810	0.753	0.684	0.593	0.459
Dichloromethane	298.15	0.949	0.899	0.849	0.797	0.743	0.684	0.617	0.536	0.423
1,3-Dimethylbenzene*)	323.15	0.980	0.956	0.926	0.891	0.846	0.791	0.723	0.638	0.524
1,4-Dimethylbenzene	423.15	0.974	0.944	0.911	0.872	0.826	0.770	0.698	0.601	0.452
Ethyl acetate*)	313.15	0.976	0.948	0.918	0.882	0.841	0.791	0.728	0.642	0.507
Hexane	423.15	0.980	0.958	0.933	0.904	0.869	0.827	0.772	0.697	0.574
2-Propanone	323.15	0.991	0.980	0.969	0.955	0.938	0.918	0.892	0.854	0.788
Propyl acetate	343.15	0.983	0.965	0.943	0.919	0.891	0.858	0.815	0.758	0.667
Tetrachloromethane	298.15	0.961	0.917	0.869	0.814	0.751	0.678	0.592	0.486	0.344
Toluene	313.15	0.981	0.959	0.933	0.901	0.861	0.809	0.738	0.638	0.481
Trichloromethane	298.15	0.949	0.898	0.847	0.793	0.736	0.675	0.604	0.519	0.400
Poly(tetramethylene glycol)										
Methanol	303.15	0.981	0.961	0.938	0.913	0.883	0.849	0.806	0.751	0.671
Poly(vinyl acetate)										
Benzene	313.15	0.985	0.967	0.945	0.919	0.886	0.844	0.784	0.696	0.548
1-Butanol	313.15	0.992	0.982	0.971	0.958	0.942	0.923	0.896	0.856	0.779
2-Butanol	313.15	0.987	0.972	0.956	0.937	0.915	0.889	0.856	0.813	0.747
2-Butanone	313.15	0.980	0.958	0.934	0.906	0.873	0.835	0.787	0.724	0.626
1,2-Dichloroethane*)	300.15	0.955	0.906	0.851	0.790	0.722	0.644	0.556	0.450	0.315
1,4-Dimethylbenzene	313.15	0.990	0.978	0.964	0.948	0.928	0.903	0.868	0.814	0.705
Ethylbenzene	313.15	0.990	0.979	0.966	0.950	0.932	0.910	0.880	0.836	0.759
Methanol	333.15	0.990	0.978	0.965	0.949	0.931	0.908	0.877	0.834	0.757
Methyl acetate	313.15	0.976	0.949	0.919	0.886	0.849	0.805	0.752	0.684	0.583
2-Methyl-1-propanol	353.15	0.984	0.966	0.946	0.924	0.899	0.868	0.832	0.784	0.715
1-Propanol	353.15	0.987	0.972	0.955	0.936	0.914	0.888	0.856	0.815	0.753
2-Propanol	353.15	0.988	0.974	0.958	0.940	0.919	0.894	0.863	0.820	0.754

Polymer/ solvents	a_A: T/K	0.1	0.2	0.3	0.4	0.5	0.6	0.7	0.8	0.9
		Mass Fraction of the Polymer								
2-Propanone	333.15	0.983	0.963	0.940	0.913	0.880	0.838	0.784	0.707	0.578
Propyl acetate	333.15	0.979	0.955	0.930	0.901	0.869	0.831	0.786	0.728	0.645
Tetrahydrofuran	323.15	0.973	0.943	0.911	0.874	0.831	0.781	0.720	0.640	0.519
Toluene	333.15	0.983	0.965	0.944	0.920	0.891	0.857	0.815	0.756	0.664
Poly(vinyl chloride)										
2-Butanone*)	313.15	0.976	0.949	0.920	0.887	0.849	0.804	0.749	0.676	0.566
Cyclohexanone*)	333.15	0.971	0.934	0.889	0.839	0.781	0.714	0.635	0.536	0.397
Poly(vinyl methyl ether)										
Benzene*)	298.15	0.969	0.935	0.897	0.855	0.807	0.751	0.683	0.596	0.466
Chlorobenzene*)	343.15	0.972	0.941	0.906	0.867	0.822	0.769	0.705	0.620	0.494
1,2-Dimethylbenzene*)	363.15	0.973	0.943	0.910	0.871	0.826	0.772	0.705	0.616	0.478
Ethylbenzene*)	343.15	0.978	0.954	0.927	0.895	0.857	0.811	0.753	0.672	0.542
Propylbenzene*)	373.15	0.977	0.951	0.923	0.890	0.852	0.808	0.752	0.678	0.563
Poly(4-vinylpyridine)										
Methanol	343.15	0.986	0.971	0.953	0.931	0.905	0.871	0.825	0.756	0.627
2-Propanol	343.15	0.989	0.977	0.964	0.948	0.928	0.904	0.872	0.826	0.743
Poly(1-vinyl-2-pyrrolidinone)										
Water	368.15	0.984	0.966	0.946	0.924	0.899	0.870	0.835	0.790	0.727
Starch (amorphous)										
Water	383.15	0.991	0.981	0.970	0.956	0.939	0.918	0.889	0.845	0.754
Styrene/Methyl methacrylate Copolymer (41.45 wt% Styrene)										
Benzene*)	308.15	0.982	0.963	0.940	0.913	0.881	0.841	0.789	0.716	0.590
Vinyl acetate/Vinyl chloride Copolymer (12 wt% Vinyl acetate)										
Benzene	398.15	0.976	0.949	0.918	0.883	0.841	0.791	0.728	0.643	0.509
Chlorobenzene	398.15	0.984	0.965	0.944	0.920	0.891	0.856	0.810	0.746	0.638
1,4-Dimethylbenzene	398.15	0.989	0.977	0.963	0.946	0.926	0.899	0.863	0.807	0.692
Ethylbenzene	398.15	0.989	0.976	0.961	0.944	0.924	0.899	0.866	0.818	0.735
Octane	398.15	0.992	0.982	0.971	0.958	0.942	0.922	0.893	0.847	0.739

* $a_A = P_A / P_A^S$

SPECIFIC ENTHALPIES OF SOLUTION OF POLYMERS AND COPOLYMERS

Christian Wohlfarth

Enthalpies of solution or mixing, expressed as the enthalpy change per unit mass of polymer, are given in the table at infinite dilution, i.e., a very small amount of polymer and a large excess of solvent were mixed isothermally to form a homogeneous solution. By thermodynamics, $\Delta_{sol}H_B^\infty$ or $\Delta_M H_B^\infty$ are obtained from the following derivatives:

$$\Delta_{sol}H_B^\infty = \lim_{m_B \to 0}\left(\partial \Delta_{sol}h / \partial m_B\right)_{P,T,m_{j\neq B}} \quad (1)$$

$$\Delta_M H_B^\infty = \lim_{m_B \to 0}\left(\partial \Delta_M h / \partial m_B\right)_{P,T,m_{j\neq B}} \quad (2)$$

with a unit of J/g. Thus, they are the partial specific enthalpies of solution or mixing of the polymer B at infinite dilution where $\Delta_{sol}h$ or $\Delta_M h$ is the extensive enthalpy of the solution or mixing process.

The state of the polymer before dissolution can significantly affect the enthalpy of solution. The dissolving of a semicrystalline polymer requires an additional amount of heat associated with the disordering of crystalline regions. Consequently, its enthalpy of solution is usually positive and depends on the degree of crystallinity of the given polymer sample. An amorphous polymer below its glass transition temperature, T_g (see the T_g-table of this section), often dissolves with the release of heat. The enthalpy of solution of a glassy polymer is additionally dependent to some extent on the thermal history of the glass-forming process. An amorphous polymer above T_g can show endothermic or exothermic dissolution behavior depending on the nature of the solvent and the interaction energies involved as is the case for any enthalpy of mixing. This enthalpy of mixing is then independent of any crystalline or glassy aspects of the polymer. It can be obtained without difficulties for liquid/molten polymers mixed with a solvent. Therefore, the enthalpies given in the table are either enthalpies of solution or enthalpies of mixing, depending on the state of the polymer.

The enthalpies depend on temperature and molar mass. The necessary molar mass information for a system is provided in the table (if available) by the corresponding number average, M_n, mass average, M_w, or viscosity average, M_η, values of the polymer as given in the original sources. Outside the oligomer range, specific enthalpies of solution or mixing do not remarkably depend on molar mass, however. More enthalpy data of polymer-solvent systems can be found in Ref. 106.

Polymer	M_n/ g/mol	M_w/ g/mol	M_η/ g/mol	Solvent	T/K	ΔH_B^∞/ J/g	Ref.
Acrylonitrile/butadiene copolymer							
(18 wt% Acrylonitrile)				Benzene	298.15	0.0	18
(26 wt% Acrylonitrile)				Benzene	298.15	−1.9	18
(40 wt% Acrylonitrile)				Benzene	298.15	−2.9	18
Acrylonitrile/isoprene copolymer							
(15 mol% Isoprene)				*N,N*-Dimethylformamide	323.15	−32	66
Acrylonitrile/vinyl chloride copolymer							
(13 wt% Acrylonitrile)				*N,N*-Dimethylformamide	293.15	−38	35
(13 wt% Acrylonitrile)				*N,N*-Dimethylformamide	308.15	−22	35
(13 wt% Acrylonitrile)				*N,N*-Dimethylformamide	323.15	−18	35
(13 wt% Acrylonitrile)				*N,N*-Dimethylformamide	338.15	−15	35
(13 wt% Acrylonitrile)				*N,N*-Dimethylformamide	353.15	−12	35
(29 wt% Acrylonitrile)				*N,N*-Dimethylformamide	295.15	−42	35
(29 wt% Acrylonitrile)				*N,N*-Dimethylformamide	308.15	−27	35
(29 wt% Acrylonitrile)				*N,N*-Dimethylformamide	323.15	−21	35
(29 wt% Acrylonitrile)				*N,N*-Dimethylformamide	338.15	−19	35
(29 wt% Acrylonitrile)				*N,N*-Dimethylformamide	353.15	−16	35
(40 wt% Acrylonitrile)				*N,N*-Dimethylformamide	295.15	−47	35
(40 wt% Acrylonitrile)				*N,N*-Dimethylformamide	308.15	−30	35
(40 wt% Acrylonitrile)				*N,N*-Dimethylformamide	323.15	−28	35
(40 wt% Acrylonitrile)				*N,N*-Dimethylformamide	338.15	−18	35
(40 wt% Acrylonitrile)				*N,N*-Dimethylformamide	353.15	−17	35
Benzylcellulose							
				Benzene	298.15	−11	8
				Cyclohexanone	298.15	−15	25
				Trichloromethane	298.15	−38	25
Bisphenol A-isophthaloyl chloride/terephthaloyl chloride							
(50/50 Iso/terephtaloyl chloride)				*N,N*-Dimethylacetamide	298.15	−56	69
(50/50 Iso/terephtaloyl chloride)				1,1,2,2-Tetrachloroethane	298.15	+72	69

Polymer	M_n/ g/mol	M_w/ g/mol	M_η/ g/mol	Solvent	T/K	ΔH_B^∞/ J/g	Ref.
Butadiene/styrene copolymer							
(10 wt% Styrene)				Benzene	293.65	4.9	17
(30 wt% Styrene)				Benzene	293.65	3.0	17
(30 wt% Styrene)				Benzene	298.15	3.0	25
(50 wt% Styrene)				Benzene	293.65	1.8	17
(60 wt% Styrene)				Benzene	293.65	0.0	17
(70 wt% Styrene)				Benzene	293.65	0.0	17
(75 wt% Styrene)				Benzene	298.15	1.5	7
(80 wt% Styrene)				Benzene	293.65	−0.6	17
(90 wt% Styrene)				Benzene	293.65	−4.9	17
Butyl methacrylate/isobutyl methacrylate copolymer (50 wt%/50 wt%)							
Glass		150000		Cyclohexanone	303.15	5.9	98
Liquid		150000		Cyclohexanone	303.15	14.0	98
Butyl methacrylate/methyl methacrylate copolymer (45 wt%/55 wt%)							
Glass	107000	250000		Cyclohexanone	304.15	−5.4	99
Liquid	107000	250000		Cyclohexanone	304.15	+9.1	99
Cellulose acetate							
(52.2 wt% Acetate)				Formic acid	298.15	−30	10
(55.8 wt% Acetate)				Formic acid	298.15	−44	10
(52.5 wt% Acetate)				Methyl acetate	298.15	−80	1
(48 wt% Acetate)				2-Propanone	298.15	−35	25
(52.2 wt% Acetate)				2-Propanone	298.15	−30	10
(55.8 wt% Acetate)				2-Propanone	298.15	−26	10
(56 wt% Acetate)				2-Propanone	298.15	−45	25
(56 wt% Acetate)				2-Propanone	298.15	−30	4
Cellulose triacetate							
				2-Propanone	298.15	−29	4
				Trichloromethane	298.15	−47	4
Dextran							
	8200	10400		Dimethylsulfoxide	298.15	−185	75
	75900	101000		Dimethylsulfoxide	298.15	−187	70
	75900	101000		1,2-Ethanediol	298.15	−98	70
	75900	101000		Formamide	298.15	−228	70
	8200	10400		Water	298.15	−140	75
	75900	101000		Water	298.15	−150	75
(amorph)				Water	298.15	−123	65
Ethylene/propylene copolymer							
(33 mol% Ethylene)				Cyclohexane	298.15	1.4	74
(63 mol% Ethylene)				Cyclohexane	298.15	8.1	74
(75 mol% Ethylene)				Cyclohexane	298.15	11.8	74
(33 mol% Ethylene)				Cyclooctane	298.15	1.2	74
(63 mol% Ethylene)				Cyclooctane	298.15	6.9	74
(75 mol% Ethylene)				Cyclooctane	298.15	8.6	74
(33 mol% Ethylene)				Cyclopentane	298.15	−3.5	74
(63 mol% Ethylene)				Cyclopentane	298.15	1.1	74
(33 mol% Ethylene)				*cis*-Decahydronaphthalene	298.15	−2.4	74
(63 mol% Ethylene)				*cis*-Decahydronaphthalene	298.15	2.4	74
(75 mol% Ethylene)				*cis*-Decahydronaphthalene	298.15	3.9	74
(33 mol% Ethylene)				*trans*-Decahydronaphthalene	298.15	−4.8	74
(63 mol% Ethylene)				*trans*-Decahydronaphthalene	298.15	−1.3	74
(75 mol% Ethylene)				*trans*-Decahydronaphthalene	298.15	−0.3	74
(63 mol% Ethylene)				3,3-Diethylpentane	298.15	−1.4	74
(75 mol% Ethylene)				3,3-Diethylpentane	298.15	<0.1	74
(63 mol% Ethylene)				2,2-Dimethylpentane	298.15	5.3	74

Polymer	M_n/g/mol	M_w/g/mol	M_η/g/mol	Solvent	T/K	ΔH_B^{∞}/J/g	Ref.
(75 mol% Ethylene)				2,2-Dimethylpentane	298.15	2.3	74
(63 mol% Ethylene)				2,3-Dimethylpentane	298.15	0.7	74
(75 mol% Ethylene)				2,3-Dimethylpentane	298.15	0.4	74
(33 mol% Ethylene)				2,4-Dimethylpentane	298.15	−1.2	74
(63 mol% Ethylene)				2,4-Dimethylpentane	298.15	3.0	74
(75 mol% Ethylene)				2,4-Dimethylpentane	298.15	0.2	74
(33 mol% Ethylene)				3,3-Dimethylpentane	298.15	−2.7	74
(63 mol% Ethylene)				3,3-Dimethylpentane	298.15	0.3	74
(33 mol% Ethylene)				Dodecane	298.15	−0.1	73
(63 mol% Ethylene)				Dodecane	298.15	0.8	73
(75 mol% Ethylene)				Dodecane	298.15	−4.0	73
(63 mol% Ethylene)				3-Ethylpentane	298.15	2.6	74
(75 mol% Ethylene)				3-Ethylpentane	298.15	−0.6	74
(33 mol% Ethylene)				2,2,4,4,6,8,8-Heptamethylnonane	298.15	−0.5	73
(63 mol% Ethylene)				2,2,4,4,6,8,8-Heptamethylnonane	298.15	2.2	73
(75 mol% Ethylene)				2,2,4,4,6,8,8-Heptamethylnonane	298.15	−0.9	73
(33 mol% Ethylene)				Hexadecane	298.15	0.7	73
(63 mol% Ethylene)				Hexadecane	298.15	−1.1	73
(75 mol% Ethylene)				Hexadecane	298.15	−4.6	73
(63 mol% Ethylene)				3-Methylhexane	298.15	0.7	74
(75 mol% Ethylene)				3-Methylhexane	298.15	1.7	74
(33 mol% Ethylene)				Octane	298.15	−1.6	73
(63 mol% Ethylene)				Octane	298.15	3.6	73
(75 mol% Ethylene)				Octane	298.15	0.3	73
(33 mol% Ethylene)				2,2,4,6,6-Pentamethylheptane	298.15	−0.3	73
(63 mol% Ethylene)				2,2,4,6,6-Pentamethylheptane	298.15	3.6	73
(75 mol% Ethylene)				2,2,4,6,6-Pentamethylheptane	298.15	0.0	73
(63 mol% Ethylene)				2,2,4,4-Tetramethylpentane	298.15	2.7	74
(75 mol% Ethylene)				2,2,4,4-Tetramethylpentane	298.15	3.1	74
(33 mol% Ethylene)				2,2,4-Trimethylpentane	298.15	−0.2	73
(63 mol% Ethylene)				2,2,4-Trimethylpentane	298.15	1.9	73
(75 mol% Ethylene)				2,2,4-Trimethylpentane	298.15	3.5	73
Ethylene/vinylacetate copolymer							
(85 wt% Vinyl acetate)				Cyclopentanone	298.15	−0.5	104
(70 wt% Vinyl acetate)		220000		Tetrahydrofuran	304.65	−1.3	93
Gelatine							
				Water	293.15	−92	29
				Water	323.15	−63	29
Guttapercha							
				Trichloromethane	303.15	47	22
Isobutyl methacrylate/methyl methacrylate copolymer (51 wt%/49 wt%)							
Glass		150000		Cyclohexanone	303.15	−11	98
Liquid		150000		Cyclohexanone	303.15	15	98
Natural rubber							
				Benzene	298.15	10	25
				Benzene	298.15	12	20
Nitrocellulose							
			16600	2-Butanone	298.15	−80	4
			23000	2-Butanone	298.15	−81	4
			40000	2-Butanone	298.15	−81	4
				Butyl acetate	293.15	−75	23
				Butyl acetate	298.15	−75	23
				Butyl acetate	298.15	−73	26
				Butyl acetate	303.15	−75	23

Polymer	M_n/ g/mol	M_w/ g/mol	M_η/ g/mol	Solvent	T/K	ΔH_B^∞/ J/g	Ref.
				Butyl acetate	308.15	−71	23
				Butyl acetate	313.15	−65	23
				Butyl acetate	313.15	−67	26
				Butyl acetate	318.15	−59	23
				Butyl acetate	323.15	−54	23
				Butyl acetate	328.15	−50	23
				Butyl acetate	333.15	−59	26
				Butyl acetate	343.15	−55	26
				Butyl acetate	353.15	−47	26
				Diethyl ether	295.15	−62	3
				Dibutyl phthalate	273.15	−45	26
				Dibutyl phthalate	298.15	−46	26
				Dibutyl phthalate	313.15	−46	26
				Dibutyl phthalate	333.15	−42	26
				Ethanol	295.15	−46	3
				Ethyl acetate	293.15	−76	23
				Ethyl acetate	298.15	−75	23
				Ethyl acetate	303.15	−69	23
				Ethyl acetate	308.15	−61	23
				Ethyl acetate	313.15	−54	23
				Ethyl acetate	318.15	−50	23
				Ethyl acetate	323.15	−50	23
				Ethyl acetate	328.15	−50	23
				Methanol	293.15	−69	23
				Methanol	298.15	−56	23
				Methanol	303.15	−50	23
				Methanol	308.15	−50	23
				Methanol	313.15	−50	23
				Methanol	318.15	−50	23
				Methanol	323.15	−50	23
				Methanol	328.15	−50	23
				2,4-Pentanedione	298.15	−74	4
				2-Pentanone	298.15	−64	4
				2-Propanone	273.15	−75	26
				2-Propanone	293.15	−75	23
				2-Propanone	298.15	−83	2
				2-Propanone	298.15	−68	4
				2-Propanone	298.15	−71	8
				2-Propanone	298.15	−74	23
				2-Propanone	298.15	−79	25
				2-Propanone	298.15	−75	26
				2-Propanone	303.15	−60	23
				2-Propanone	308.15	−51	23
				2-Propanone	313.15	−50	23
				2-Propanone	313.15	−65	26
				2-Propanone	318.15	−50	23
				2-Propanone	323.15	−50	23
				2-Propanone	323.15	−50	26
				2-Propanone	328.15	−50	23
				Pyridine	298.15	−106	2
				Tri(4-methylphenyl) phosphate	298.15	−16	26
				Tri(4-methylphenyl) phosphate	313.15	−28	26
				Tri(4-methylphenyl) phosphate	333.15	−41	26
				Tri(4-methylphenyl) phosphate	343.15	−44	26
				Tri(4-methylphenyl) phosphate	353.15	−47	26

Polymer	M_n/ g/mol	M_w/ g/mol	M_η/ g/mol	Solvent	T/K	ΔH_B^∞/ J/g	Ref.
Nylon-6 (unoriented)							
				Formic acid	295.15	−53	24
				Tricresol	323.55	−66	22
				Tricresol	345.55	−66	22
Poly(acrylonitrile)							
				Benzene	298.15	0.0	18
				N,N-Dimethylformamide	295.15	−23	35
				N,N-Dimethylformamide	298.15	−21	18
				N,N-Dimethylformamide	298.15	−43	42
				N,N-Dimethylformamide	308.15	−17	35
				N,N-Dimethylformamide	323.15	−13	35
				N,N-Dimethylformamide	323.15	−15	66
				N,N-Dimethylformamide	338.15	−10	35
				Dimethylsulfoxide	298.15	−70	42
Poly(γ-benzyl-L-glutamate)							
		160000		Dichloroacetic acid	303.15	−35	46
		160000		1,2-Dichloroethane	303.15	−1.6	46
Polybutadiene							
				Benzene	298.15	6.1	7
				Benzene	298.15	7.1	25
				Benzene	298.15	10.5	32
				2,2,4-Trimethylpentane	298.15	1.1	32
1,4-cis-Polybutadiene							
	low			Cyclohexane	298.15	5.4	74
	low			Cyclooctane	298.15	5.8	74
	low			Cyclopentane	298.15	<0.1	74
	low			*cis*-Decahydronaphthalene	298.15	4.2	74
	low			*trans*-Decahydronaphthalene	298.15	2.6	74
	low			3,3-Diethylpentane	298.15	5.2	74
	low			2,2-Dimethylpentane	298.15	4.1	74
	low			2,3-Dimethylpentane	298.15	4.5	74
	low			2,4-Dimethylpentane	298.15	3.2	74
	low			3,3-Dimethylpentane	298.15	3.2	74
	low			Dodecane	298.15	4.2	74
	low			3-Ethylpentane	298.15	3.7	74
	low			2,2,4,4,6,8,8-Heptamethylnonane	298.15	4.8	74
	low			Hexadecane	298.15	4.9	74
	low			3-Methylhexane	298.15	3.6	74
	low			Octane	298.15	4.3	74
	low			2,2,4,6,6-Pentamethylheptane	298.15	5.0	74
	low			2,2,4,4-Tetramethylpentane	298.15	5.8	74
	low			2,3,3,4-Tetramethylpentane	298.15	5.1	74
Poly(1-butene)							
			20000	Cyclohexane	298.15	1.0	74
			20000	Cyclooctane	298.15	1.8	74
			20000	Cyclopentane	298.15	−2.9	74
			20000	*cis*-Decahydronaphthalene	298.15	<0.1	74
			20000	*trans*-Decahydronaphthalene	298.15	−2.0	74
			20000	Decane	298.15	1.2	62
			20000	3,3-Diethylpentane	298.15	−2.6	74
			20000	2,2-Dimethylpentane	298.15	−4.0	74
			20000	2,3-Dimethylpentane	298.15	−2.8	74

Polymer	M_n/ g/mol	M_w/ g/mol	M_η/ g/mol	Solvent	*T*/K	ΔH_B^∞/ J/g	Ref.
			20000	2,4-Dimethylpentane	298.15	−2.3	74
			20000	3,3-Dimethylpentane	298.15	−2.2	74
			20000	Dodecane	298.15	2.1	74
			20000	3-Ethylpentane	298.15	−2.8	74
			20000	Heptane	298.15	0.0	73
			20000	Hexadecane	298.15	3.0	62
			20000	Hexane	298.15	−1.2	62
			20000	3-Methylhexane	298.15	−2.1	74
			20000	Nonane	298.15	0.9	73
			20000	Octane	298.15	0.4	73
			20000	2,2,4,6,6-Pentamethylheptane	298.15	0.6	73
			20000	Pentane	298.15	−2.6	62
			20000	Tetradecane	298.15	2.7	62
			20000	2,2,4,4-Tetramethylpentane	298.15	−1.4	74
			20000	2,3,3,4-Tetramethylpentane	298.15	−2.2	74
			20000	2,2,4-Trimethylpentane	298.15	−0.5	73
Poly(butyl acrylate)				2-Propanone	298.15	0.8	25
Poly(butyl methacrylate)							
Glass	91300	210000		Cyclohexanone	304.15	7.7	99
Liquid	91300	210000		Cyclohexanone	304.15	8.2	99
				2-Propanone	298.15	19.5	25
Polychloroprene				Benzene	298.15	0.5	7
Poly(2,6-dimethyl phenylene oxide)							
	17000	46400		1,2-Dichlorobenzene	303.05	55	89
Poly(dimethylsiloxane)							
	13000			Benzene	298.15	11.2	50
			20000	Benzene	298.15	13.5	61
			100000	Benzene	298.15	14.2	40
			170000	Bromocyclohexane	303.15	10.2	51
			80000	2-Butanone	293.15	14.4	77
	30900			2-Butanone	303.15	14.2	44
			170000	2-Butanone	303.15	14.7	51
			80000	2-Butanone	308.15	14.3	77
			80000	2-Butanone	323.15	14.3	77
			80000	Butyl acetate	298.15	6.1	41
			80000	Butyl propanoate	298.15	4.9	41
			100000	Chlorobenzene	298.15	7.5	40
	13000			Cyclohexane	298.15	3.0	50
			20000	Cyclohexane	298.15	5.2	74
			100000	Cyclohexane	298.15	5.2	40
			20000	Cyclooctane	298.15	6.8	74
			20000	Cyclopentane	298.15	1.0	74
			20000	*cis*-Decahydronapthalene	298.15	7.1	74
			20000	*trans*-Decahydronapthalene	298.15	4.3	74
			20000	Decamethyltetrasiloxane	297.65	0.45	37
			20000	Decane	298.15	3.8	37
			20000	Decane	298.15	3.9	61
			80000	Decane	298.15	3.8	41
			80000	Decyl acetate	298.15	4.5	41
			80000	Dibutyl ether	298.15	0.6	41
			80000	Diethoxymethane	298.15	1.8	41
			80000	Diethyl ether	298.15	−1.3	41

Polymer	M_n/ g/mol	M_w/ g/mol	M_η/ g/mol	Solvent	T/K	ΔH_B^∞/ J/g	Ref.
			20000	3,3-Diethylpentane	298.15	1.9	74
			80000	Dihexyl ether	298.15	3.0	41
			80000	1,2-Dimethoxyethane	298.15	12.2	41
			80000	Dimethoxymethane	298.15	7.4	41
	13000			1,2-Dimethylbenzene	298.15	4.3	50
	13000			1,3-Dimethylbenzene	298.15	3.0	50
	13000			1,4-Dimethylbenzene	298.15	3.2	50
			20000	1,4-Dimethylbenzene	298.15	4.2	61
			80000	2,6-Dimethyl-4-heptanone	293.15	6.1	77
			20000	2,2-Dimethylpentane	298.15	0.8	74
			20000	2,3-Dimethylpentane	298.15	1.4	74
			20000	2,4-Dimethylpentane	298.15	1.6	74
			20000	3,3-Dimethylpentane	298.15	0.5	74
			80000	Dipentyl ether	298.15	2.1	41
			80000	Dipropyl ether	298.15	−1.2	41
			20000	Dodecamethylpentasiloxane	297.65	−0.3	37
			20000	Dodecane	297.65	4.5	37
			20000	Dodecane	298.15	4.4	73
			80000	Dodecane	298.15	4.5	41
			80000	Ethyl acetate	298.15	12.7	41
			170000	Ethyl acetate	303.15	13.7	51
	13000			Ethylbenzene	298.15	6.4	50
			20000	Ethylbenzene	298.15	6.2	61
			80000	Ethyl butanoate	298.15	6.0	41
			80000	Ethyl decanoate	298.15	3.8	41
			80000	Ethyl dodecanoate	298.15	3.8	41
			80000	Ethyl heptanoate	298.15	4.1	41
			80000	Ethyl hexanoate	298.15	4.3	41
			80000	Ethyl nonanoate	298.15	3.7	41
			80000	Ethyl octanoate	298.15	3.8	41
			20000	3-Ethylpentane	298.15	0.6	74
			80000	Ethyl propanoate	298.15	8.0	41
			20000	2,2,4,4,6,8,8-Heptamethylnonane	298.15	3.5	74
	13000			Heptane	298.15	1.8	50
			20000	Heptane	298.15	1.9	73
			20000	Heptane	297.65	2.0	37
			20000	Heptane	298.15	2.0	61
			80000	Heptane	298.15	2.0	41
			100000	Heptane	298.15	2.1	40
			170000	Heptane	303.15	1.9	51
			80000	3-Heptanone	308.15	8.8	77
			80000	3-Heptanone	323.15	8.8	77
			20000	Hexadecane	297.65	5.5	37
			20000	Hexadecane	298.15	5.5	73
			20000	Hexamethyldisiloxane	298.15	−1.2	37
			20000	Hexamethyldisiloxane	298.15	−1.6	61
			170000	Hexamethyldisiloxane	303.15	−1.5	51
			20000	Hexane	297.65	0.7	37
			20000	Hexane	298.15	0.7	61
			80000	Hexane	298.15	0.7	41
			170000	Hexane	303.15	0.3	51
			80000	Hexyl acetate	298.15	5.0	41
	13000			Isopropylbenzene	298.15	4.1	50
			80000	Methyl butanoate	298.15	8.6	41
	13000			Methylcyclohexane	298.15	2.9	50

Polymer	M_n/ g/mol	M_w/ g/mol	M_η/ g/mol	Solvent	T/K	ΔH_B^∞/ J/g	Ref.
			100000	Methylcyclohexane	298.15	1.9	40
			80000	Methyl decanoate	298.15	4.8	41
			20000	3-Methylhexane	298.15	1.3	74
			80000	Methyl hexanoate	298.15	5.3	41
			80000	Methyl octanoate	298.15	5.0	41
			80000	4-Methyl-2-pentanone	293.15	9.9	77
			80000	4-Methyl-2-pentanone	308.15	9.0	77
			80000	Methyl propanoate	298.15	12.1	41
			20000	Nonane	297.65	3.4	37
			20000	Nonane	298.15	3.3	73
			80000	Nonane	298.15	3.4	41
			80000	Octamethylcyclotetrasiloxane	293.15	−0.4	78
			20000	Octamethyltrisiloxane	297.65	−0.6	37
			20000	Octamethyltrisiloxane	298.15	−0.8	61
			170000	Octamethyltrisiloxane	303.15	−1.0	51
			20000	Octane	297.65	2.6	37
			20000	Octane	298.15	2.4	73
			20000	Octane	298.15	2.6	61
			80000	Octane	298.15	2.6	41
			20000	2,2,4,6,6-Pentamethylheptane	298.15	2.7	73
			20000	Pentane	298.15	−0.9	37
			20000	Pentane	298.15	−0.9	61
			80000	Pentane	298.15	−1.0	41
			80000	Pentyl acetate	298.15	5.8	41
			80000	Pentyl propanoate	298.15	3.8	41
			80000	Propyl acetate	298.15	8.6	41
			170000	Propyl acetate	303.15	9.9	51
			80000	Propyl propanoate	298.15	5.6	41
			100000	Tetrachloromethane	298.15	2.4	40
			20000	Tetradecane	297.65	5.1	37
			20000	Tetradecane	298.15	5.1	61
			80000	Tetradecane	298.15	5.1	41
			20000	2,2,4,4-Tetramethylpentane	298.15	2.1	73
			20000	2,2,4,4-Tetramethylpentane	298.15	2.3	74
			20000	2,3,3,4-Tetramethylpentane	298.15	1.9	74
	13000			Toluene	298.15	5.5	50
			20000	Toluene	298.15	6.7	61
			20000	Tridecane	297.65	4.8	37
			80000	Tridecane	298.15	4.8	41
	13000			1,3,5-Trimethylbenzene	298.15	3.7	50
			20000	2,2,4-Trimethylpentane	298.15	1.4	73
			20000	Undecane	297.65	4.2	37
			80000	Undecane	298.15	4.3	41
Polyethylene							
Semicrystalline		65000		1-Chloronaphthalene	373.15	780	47
Semicrystalline		65000		1-Chloronaphthalene	383.15	980	47
Semicrystalline		65000		1-Chloronaphthalene	393.15	800	47
Liquid		65000		1-Chloronaphthalene	403.15	49	47
Semicrystalline		144000		1-Chloronaphthalene	383.15	920	47
Semicrystalline		144000		1-Chloronaphthalene	393.15	990	47
Semicrystalline		144000		1-Chloronaphthalene	403.15	690	47
Liquid		144000		1-Chloronaphthalene	413.15	67	47
Liquid		144000		1-Chloronaphthalene	423.15	85	47
Semicrystalline		670000		1-Chloronaphthalene	363.15	380	47
Semicrystalline		670000		1-Chloronaphthalene	373.15	430	47

Polymer	M_n/ g/mol	M_w/ g/mol	M_η/ g/mol	Solvent	T/K	ΔH_B^∞/ J/g	Ref.
Semicrystalline		670000		1-Chloronaphthalene	383.15	165	47
Liquid		670000		1-Chloronaphthalene	393.15	39	47
Liquid		670000		1-Chloronaphthalene	403.15	36	47
Semicrystalline		900000		1-Chloronaphthalene	391.80	245	105
Semicrystalline		900000		Cyclohexane	379.50	205	105
Semicrystalline		900000		Cyclopentane	380.00	190	105
Alkathene				Decahydronaphthalene	349.85	142	56
Rigidex-3				Decahydronaphthalene	366.65	180	56
Rigidex-50				Decahydronaphthalene	374.05	233	56
Semicrystalline		900000		Decahydronaphthalene	384.00	260	105
Semicrystalline				1,2-Dichloroethane	333.15	30	27
Semicrystalline				1,2-Dichloroethane	338.15	38	27
Semicrystalline				1,2-Dichloroethane	343.15	54	27
Semicrystalline				1,2-Dichloroethane	348.15	65	27
Semicrystalline			10000	1,4-Dimethylbenzene	354.15	139	5
Semicrystalline			11800	1,4-Dimethylbenzene	352.15	139	5
Semicrystalline			15600	1,4-Dimethylbenzene	353.65	154	5
Semicrystalline			15600	1,4-Dimethylbenzene	363.65	113	5
Semicrystalline			15600	1,4-Dimethylbenzene	368.65	104	5
Semicrystalline		900000		2,4-Dimethylpentane	393.00	230	105
Semicrystalline		900000		2,2,4,4,6,8,8-Heptamethylnonane	399.50	170	105
Semicrystalline		900000		Hexadecane	399.50	262	105
Semicrystalline		900000		2-Methylbutane	394.20	165	105
Semicrystalline		65000		1,2,3,4-Tetrahydronaphthalene	373.15	940	47
Semicrystalline		65000		1,2,3,4-Tetrahydronaphthalene	383.15	990	47
Semicrystalline		65000		1,2,3,4-Tetrahydronaphthalene	393.15	790	47
Liquid		65000		1,2,3,4-Tetrahydronaphthalene	403.15	58	47
Semicrystalline		84000		1,2,3,4-Tetrahydronaphthalene	373.15	835	47
Semicrystalline		130000		1,2,3,4-Tetrahydronaphthalene	353.15	630	47
Semicrystalline		130000		1,2,3,4-Tetrahydronaphthalene	373.15	520	47
Liquid		130000		1,2,3,4-Tetrahydronaphthalene	393.15	69	47
Semicrystalline		144000		1,2,3,4-Tetrahydronaphthalene	373.15	610	47
Semicrystalline		144000		1,2,3,4-Tetrahydronaphthalene	383.15	1200	47
Semicrystalline		144000		1,2,3,4-Tetrahydronaphthalene	393.15	1130	47
Semicrystalline		144000		1,2,3,4-Tetrahydronaphthalene	403.15	800	47
Liquid		144000		1,2,3,4-Tetrahydronaphthalene	413.15	136	47
Liquid		144000		1,2,3,4-Tetrahydronaphthalene	423.15	88	47
Semicrystalline		310000		1,2,3,4-Tetrahydronaphthalene	343.15	485	47
Semicrystalline		670000		1,2,3,4-Tetrahydronaphthalene	353.15	560	47
Semicrystalline		670000		1,2,3,4-Tetrahydronaphthalene	363.15	560	47
Semicrystalline		670000		1,2,3,4-Tetrahydronaphthalene	373.15	460	47
Semicrystalline		670000		1,2,3,4-Tetrahydronaphthalene	383.15	155	47
Liquid		670000		1,2,3,4-Tetrahydronaphthalene	393.15	67	47
Liquid		670000		1,2,3,4-Tetrahydronaphthalene	403.15	39	47
Semicrystalline			16000	Toluene	353.15	110	22
Semicrystalline			22000	Toluene	358.35	118	22
Semicrystalline			22000	Toluene	367.35	106	22
Semicrystalline		900000		1,2,4-Trichlorobenzene	386.50	255	105
Poly(ethylene glycol)							
	180			Benzene	303.15	110	71
	385			Benzene	303.15	60	71
	560			Benzene	303.15	40	71
	1050			Benzene	303.15	90	71
	1610			Benzene	303.15	140	71
	1940			Benzene	303.15	215	71

Polymer	M_n/ g/mol	M_w/ g/mol	M_η/ g/mol	Solvent	T/K	ΔH_B^∞/ J/g	Ref.
	3200			Benzene	303.15	195	71
	4330			Benzene	303.15	195	71
	5850			Benzene	303.15	190	71
	9950			Benzene	303.15	195	71
			43400	Benzene	303.15	190	71
	400	420		Trichloromethane	303.15	−79	95
	590	615		Trichloromethane	303.15	−88	95
	180			Water	303.15	−136	71
	200			Water	321.35	−125	83
	355			Water	303.15	−159	71
	400			Water	321.35	−150	83
	560			Water	303.15	−150	71
	990			Water	321.35	−101	83
	1050			Water	303.15	−106	71
	1460			Water	321.35	−137	83
	1610			Water	303.15	−6	71
	1940			Water	303.15	57	71
	3200			Water	303.15	58	71
	4330			Water	303.15	28	71
	5850			Water	303.15	39	71
	9950			Water	303.15	30	71
			14000	Water	303.15	7	55
			14000	Water	313.15	27	55
			20300	Water	303.15	45	71
			34500	Water	303.15	34	71
			43300	Water	303.15	+40	71
Poly(ethylene glycol) dimethyl ether							
	250			Tetrachloromethane	303.15	−12	95
	250			Tetrachloromethane	318.15	−7.6	95
	400			Tetrachloromethane	303.15	−12	95
	520	550		Tetrachloromethane	303.15	−12	95
	520	550		Tetrachloromethane	303.15	−7.6	95
	250			Trichloromethane	303.15	−184	95
	520	550		Trichloromethane	303.15	−135	95
Poly(ethylene glycol) monododecyl ether							
	230			Dodecane	302.15	42	68
	274			Dodecane	302.15	23	68
	318			Dodecane	302.15	34	68
	362			Dodecane	302.15	37	68
	406			Dodecane	302.15	42	68
Poly(ethylene glycol) monomethyl ether							
	353	377		Trichloromethane	303.15	−125	95
	550	580		Trichloromethane	303.15	−117	95
Poly(ethylene oxide)							
Semicrystalline	6000			Dichloromethane	303.15	+84	58
Liquid	6000			Dichloromethane	303.15	−160	58
Semicrystalline	6000			Trichloromethane	303.15	+52	58
Liquid	6000			Trichloromethane	303.15	−186	58
Quenched	1520	1720		Water	293.15	−403	64
Annealed	1520	1720		Water	293.15	−392	64
Quenched	1520	1720		Water	298.15	−180	64
Annealed	1520	1720		Water	298.15	−150	64
Quenched	1520	1720		Water	303.15	+68	64

Polymer	M_n/ g/mol	M_w/ g/mol	M_η/ g/mol	Solvent	T/K	ΔH_B^∞/ J/g	Ref.
Annealed	1520	1720		Water	303.15	+109	64
Liquid	6000			Water	303.15	−50	58
Quenched	6840	7525		Water	293.15	−28	64
Annealed	6840	7525		Water	293.15	+209	64
Quenched	6840	7525		Water	298.15	+241	64
Annealed	6840	7525		Water	298.15	+540	64
Quenched	16600	19600		Water	293.15	−160	64
Annealed	16600	19600		Water	293.15	−143	64
Quenched	16600	19600		Water	298.15	+59	64
Annealed	16600	19600		Water	298.15	+155	64
Quenched	16600	19600		Water	303.15	+353	64
Annealed	16600	19600		Water	303.15	+490	64
Semicrystalline		20000		Water	298.15	+10	84
Polyindene							
	765	1023		Anisole	299.15	2.1	102
	765	1023		Benzene	299.15	−0.04	102
	765	1023		Benzonitrile	299.15	−4.4	102
	765	1023		Bromobenzene	299.15	−3.9	102
	765	1023		2-Butanone	299.15	1.9	102
	765	1023		Chlorobenzene	299.15	−3.9	102
	765	1023		1-Chlorobutane	299.15	−4.0	102
	765	1023		1-Chloroheptane	299.15	1.5	102
	765	1023		Cyclohexane	299.15	15	102
	765	1023		*N,N*-dimethylaniline	299.15	−8.2	102
	765	1023		Ethyl acetate	299.15	4.2	102
	765	1023		Ethylbenzene	299.15	−1.4	102
	765	1023		Ethyl benzoate	299.15	−0.7	102
	765	1023		Nitrobenzene	299.15	4.6	102
	765	1023		1-Nitropropane	299.15	8.4	102
	765	1023		Pyridine	299.15	−6.7	102
	765	1023		1,1,2,2-Tetrachloroethane	299.15	−19	102
	765	1023		Tetrachloromethane	299.15	−2.5	102
	765	1023		1,1,1-Trichloroethane	299.15	−1.8	102
	765	1023		Trichloromethane	299.15	−20	102
Polyisobutylene							
	360		700	Benzene	298.15	30	67
	1000		2000	Benzene	298.15	25	67
	1300		2500	Benzene	298.15	23	67
			30000	Benzene	297.65	19	38
			30000	Benzene	298.15	19	40
	44700			Benzene	303.15	16	44
			48000	Benzene	303.4	19	72
			50000	Benzene	303.15	16	45
			72000	Benzene	300.15	19	53
			72000	Benzene	323.15	18	53
			72000	Benzene	343.15	16	53
			72000	Benzene	375.15	13	53
			72000	Benzene	394.15	9.2	53
			72000	Benzene	423.15	3.5	53
			72000	Benzene	437.15	−0.5	53
			72000	Benzene	453.15	−4.7	53
			90000	Benzene	298.15	6.7	32
			160000	Benzene	303.15	16	51
			560000	Benzene	298.15	18	34

Polymer	M_n/ g/mol	M_w/ g/mol	M_η/ g/mol	Solvent	T/K	ΔH_B^∞/ J/g	Ref.
				Benzene	298.15	6.8	7
				Benzene	298.15	6.8	10
			30000	Chlorobenzene	297.65	12	38
			30000	Chlorobenzene	298.15	13	40
			50000	Chlorobenzene	303.15	12	45
			160000	Chlorobenzene	303.15	12	51
			560000	Chlorobenzene	298.15	12	34
	360		700	Cyclohexane	298.15	3.8	67
	1000		2000	Cyclohexane	298.15	1.2	67
	1300		2500	Cyclohexane	298.15	1.1	67
			4500	Cyclohexane	298.15	−0.6	74
			30000	Cyclohexane	297.65	−0.7	38
			30000	Cyclohexane	298.15	−0.6	40
			50000	Cyclohexane	303.15	−0.7	45
			160000	Cyclohexane	303.15	−0.6	51
			1990000	Cyclohexane	298.15	−0.7	39
			4500	Cyclooctane	298.15	+0.3	74
			4500	Cyclopentane	298.15	−5.9	74
			4500	*cis*-Decahydronaphthalene	298.15	0.2	74
			4500	*trans*-Decahydronaphthalene	298.15	−0.8	74
			30000	Decane	297.65	−0.5	38
			50000	Decane	303.15	−0.5	45
			30000	Dibutyl ether	297.65	1.2	37
			30000	Diethyl ether	297.65	2.8	37
			30000	Diethyl ether	297.65	2.8	38
			4500	3,3-Diethylpentane	298.15	−1.4	74
			30000	Dihexyl ether	297.65	0.9	37
			4500	2,2-Dimethylpentane	298.15	−1.1	74
			4500	2,3-Dimethylpentane	298.15	−1.9	74
			4500	2,4-Dimethylpentane	298.15	−1.1	74
			4500	3,3-Dimethylpentane	298.15	−1.7	74
			30000	Dipentyl ether	297.65	1.0	37
			30000	Dipropyl ether	297.65	1.8	37
	360		700	Dodecane	298.15	1.9	67
	1000		2000	Dodecane	298.15	0.7	67
	1300		2500	Dodecane	298.15	0.5	67
			4500	Dodecane	298.15	0.2	73
			30000	Dodecane	297.65	−0.1	38
			30000	Dodecane	298.15	−0.1	40
			48000	Ethylbenzene	291.15	9.5	72
			48000	Ethylbenzene	343.15	3.6	72
			30000	Ethyl decanoate	297.65	3.0	37
			30000	Ethyl heptanoate	297.65	5.6	37
			30000	Ethyl hexadecanoate	297.65	1.3	37
			30000	Ethyl hexanoate	297.65	6.7	37
			30000	Ethyl nonanoate	297.65	3.7	37
			30000	Ethyl octanoate	297.65	4.6	37
			4500	3-Ethylpentane	298.15	−2.0	74
			30000	Ethyl tetradecanoate	297.65	1.8	37
			4500	2,2,4,4,6,8,8-Heptamethylnonane	298.15	−0.5	74
	360		700	Heptane	298.15	−0.5	67
	1000		2000	Heptane	298.15	−1.0	67
	1300		2500	Heptane	298.15	−1.4	67
			4500	Heptane	298.15	−1.7	73
			30000	Heptane	297.65	−1.8	38

Polymer	M_n/ g/mol	M_w/ g/mol	M_η/ g/mol	Solvent	T/K	ΔH_B^∞/ J/g	Ref.
			30000	Heptane	298.15	−2.0	40
			50000	Heptane	303.15	−1.8	45
			160000	Heptane	303.15	−1.6	51
				Heptane	298.15	−1.4	7
				Heptane	298.15	−1.4	10
	360		700	Hexadecane	298.15	4.5	67
	1000		2000	Hexadecane	298.15	2.1	67
	1300		2500	Hexadecane	298.15	1.0	67
			4500	Hexadecane	298.15	0.9	73
			30000	Hexadecane	297.65	0.04	38
			30000	Hexane	297.65	−2.5	38
			30000	Hexane	298.15	−2.6	40
			50000	Hexane	303.15	−2.5	45
			72000	Hexane	303.15	−1.8	53
			72000	Hexane	324.15	−2.3	53
			72000	Hexane	348.15	−2.9	53
			72000	Hexane	373.15	−3.7	53
			72000	Hexane	393.15	−5.3	53
			72000	Hexane	408.15	−6.7	53
			72000	Hexane	423.15	−9.0	53
			72000	Hexane	433.15	−9.9	53
			160000	Hexane	303.15	−2.5	51
			30000	2-Methylbutane	297.65	−3.1	38
			30000	Methylcyclohexane	297.65	−1.2	38
			50000	Methylcyclohexane	303.15	−1.2	45
			160000	Methylcyclohexane	303.15	−1.2	51
			4500	3-Methylhexane	298.15	−1.0	74
			30000	3-Methylpentane	297.65	−2.8	38
			30000	Nonane	297.65	−0.8	38
			4500	Nonane	298.15	−0.8	73
			4500	Octane	298.15	−1.1	73
			30000	Octane	297.65	−1.2	38
			72000	Octane	303.15	−0.3	53
			72000	Octane	324.15	−0.8	53
			72000	Octane	348.15	−0.9	53
			72000	Octane	373.15	−1.1	53
			72000	Octane	393.15	−1.3	53
			72000	Octane	423.15	−3.6	53
			4500	2,2,4,6,6-Pentamethylheptane	298.15	−0.1	73
	360		700	Pentane	298.15	−1.9	67
	1000		2000	Pentane	298.15	−2.9	67
	1300		2500	Pentane	298.15	−3.2	67
			30000	Pentane	297.65	−3.6	38
			72000	Pentane	303.15	−2.8	53
			72000	Pentane	333.15	−3.4	53
			72000	Pentane	352.15	−4.5	53
			72000	Pentane	365.15	−5.5	53
	360		700	Tetrachloromethane	298.15	5.9	67
	1000		2000	Tetrachloromethane	298.15	5.8	67
	1300		2500	Tetrachloromethane	298.15	5.0	67
			1990000	Tetrachloromethane	298.15	4.1	39
			30000	Tetradecane	297.65	0.0	38
			4500	2,2,4,4-Tetramethylpentane	298.15	−0.6	74
			4500	2,3,3,4-Tetramethylpentane	298.15	−2.3	74
			50000	Toluene	303.15	7.4	45

Polymer	M_n/ g/mol	M_w/ g/mol	M_η/ g/mol	Solvent	T/K	ΔH_B^∞/ J/g	Ref.
			160000	Toluene	303.15	7.4	51
			1990000	Toluene	298.15	8.8	39
				Toluene	298.15	1.8	7
				Toluene	298.15	1.8	10
			30000	Tridecane	297.65	−0.04	38
			4500	2,2,4-Trimethylpentane	298.15	−0.4	73
			30000	2,2,4-Trimethylpentane	297.65	−0.6	38
			1990000	2,2,4-Trimethylpentane	298.15	0.0	39
				2,2,4-Trimethylpentane	298.15	0.0	7
				2,2,4-Trimethylpentane	298.15	0.0	10
			30000	Undecane	297.65	−0.4	38
Poly(isobutyl methacrylate)							
Glass		260000		Cyclohexanone	303.15	−5.2	98
Liquid		260000		Cyclohexanone	303.15	13	98
Poly(methyl acrylate)							
				2-Propanone	298.15	0.0	25
Poly(methyl methacrylate)							
Glass	73900	170000		Cyclohexanone	304.15	−14	98
Liquid	73900	170000		Cyclohexanone	304.15	17	98
	1930			1,2-Dichloroethane	298.15	−20	32
	240000			1,2-Dichloroethane	298.15	−27	32
			53000	Ethylbenzene	298.15	−31	28
			180000	Ethylbenzene	298.15	−29	28
	28900	35900		4-Methyl-2-pentanone	303.15	−21	76
	93940	101000		4-Methyl-2-pentanone	303.15	−24	76
	137000	215000		4-Methyl-2-pentanone	303.15	−28	76
				2-Propanone	298.15	−30	25
	93940	101000		Toluene	303.15	−22	76
	689000	782000		Toluene	303.15	−24	76
			12000	Trichloromethane	298.15	−65	52
			54000	Trichloromethane	298.15	−80	52
			80000	Trichloromethane	298.15	−81	52
			100000	Trichloromethane	298.15	−84	52
			320000	Trichloromethane	298.15	−83	52
	93940	101000		Trichloromethane	303.15	−71	76
	689000	782000		Trichloromethane	303.15	−72	76
		2320000		Trichloromethane	303.15	−73	76
Poly(4-methyl-1-pentene)							
Semicrystalline			350000	Cyclohexane	303.15	30	79
Poly(α-methylstyrene)							
	1030	1180		Toluene	298.15	−7.1	43
		1430		Toluene	298.15	−30	43
	1820	2230		Toluene	298.15	−34	43
	1920			Toluene	298.15	−37	43
	2700	3300		Toluene	298.15	−39	43
	3280			Toluene	298.15	−46	43
	5260			Toluene	298.15	−46	43
	8600			Toluene	298.15	−45	43
	12200			Toluene	298.15	−46	43
		10500		Toluene	310.15	−8.4	92
		53000		Toluene	310.15	−13	92
		55000		Toluene	333.15	−16	96
		87000		Toluene	298.15	−17	90
		87000		Toluene	310.15	−16	92

Polymer	M_n/ g/mol	M_w/ g/mol	M_η/ g/mol	Solvent	T/K	ΔH_B^∞/ J/g	Ref.
		87000		Toluene	333.15	−11	90
Poly(2-methyl-5-vinyltetrazole)							
				Acetic acid	298.15	47	100
				Acetonitrile	298.15	14	100
				1,2-Dichloroethane	298.15	17	100
				N,N-Diethylacetamide	298.15	17	100
				N,N-Dimethylformamide	298.15	33	100
				Dimethylsulfoxide	298.15	10	100
				Formamide	298.15	12	100
				Formic acid	298.15	110	100
				Nitromethane	298.15	10	100
				Pyridine	298.15	16	100
Poly(octamethylene oxide)							
	7000			Benzene	298.15	20	64
	7000			Benzene	303.15	22	64
	7000			Benzene	308.15	25	64
Polypentenamer							
			50000	Cyclohexane	298.15	4.6	74
			50000	Cyclooctane	298.15	5.1	74
			50000	Cyclopentane	298.15	−2.3	74
			50000	*cis*-Decahydronaphthalene	298.15	2.6	74
			50000	*trans*-Decahydronaphthalene	298.15	<0.1	74
			50000	3,3-Diethylpentane	298.15	2.4	74
			50000	2,2-Dimethylpentane	298.15	3.3	74
			50000	2,3-Dimethylpentane	298.15	2.2	74
			50000	2,4-Dimethylpentane	298.15	3.3	74
			50000	3,3-Dimethylpentane	298.15	2.7	74
			50000	Dodecane	298.15	2.9	74
			50000	3-Ethylpentane	298.15	2.1	74
			50000	2,2,4,4,6,8,8-Heptamethylnonane	298.15	3.2	74
			50000	Hexadecane	298.15	2.6	74
			50000	3-Methylhexane	298.15	2.4	74
			50000	Octane	298.15	2.2	74
			50000	2,2,4,6,6-Pentamethylheptane	298.15	3.8	74
			50000	2,2,4,4-Tetramethylpentane	298.15	4.5	74
			50000	2,3,3,4-Tetramethylpentane	298.15	2.4	74
			50000	2,2,4-Trimethylpentane	298.15	4.3	74
Poly(m-phenyleneisophthalamide)							
Glass				*N,N*-Dimethylacetamide	298.15	−171	60
Semicrystalline				*N,N*-Dimethylacetamide	298.15	−128	60
Glass				*N,N*-Dimethylformamide	298.15	−149	60
Semicrystalline				*N,N*-Dimethylformamide	298.15	−125	60
Glass				1-Methyl-2-pyrrolidone	298.15	−177	60
Semicrystalline				1-Methyl-2-pyrrolidone	298.15	−118	60
Polypropylene (atactic)							
			18000	Benzene	298.15	31	80
			6000	Cyclohexane	298.15	2.3	74
			18000	Cyclohexane	298.15	3.9	80
			6000	Cyclooctane	298.15	3.0	74
			6000	Cyclopentane	298.15	−2.3	74
			6000	*cis*-Decahydronaphthalene	298.15	0.5	74
			6000	*trans*-Decahydronaphthalene	298.15	−2.4	74

Polymer	M_n/ g/mol	M_w/ g/mol	M_η/ g/mol	Solvent	T/K	ΔH_B^∞/ J/g	Ref.
			18000	Decane	298.15	3.1	80
			6000	3,3-Diethylpentane	298.15	−3.9	74
			18000	1,2-Dimethylbenzene	298.15	13	80
			18000	1,3-Dimethylbenzene	298.15	12	80
			18000	1,4-Dimethylbenzene	298.15	10	80
			6000	2,2-Dimethylpentane	298.15	−2.2	74
			6000	2,3-Dimethylpentane	298.15	−2.5	74
			6000	2,4-Dimethylpentane	298.15	−1.8	74
			6000	3,3-Dimethylpentane	298.15	−3.0	74
			6000	Dodecane	298.15	1.7	73
			18000	Ethylbenzene	298.15	14	80
			6000	3-Ethylpentane	298.15	−2.5	74
			6000	2,2,4,4,6,8,8-Heptamethylnonane	298.15	−0.7	73
			6000	Heptane	298.15	−1.6	73
			18000	Heptane	298.15	0.5	80
			6000	Hexadecane	298.15	2.3	73
			18000	Hexane	298.15	−1.4	80
			6000	3-Methylhexane	298.15	−1.8	74
			6000	Nonane	298.15	0.8	73
			18000	Nonane	298.15	2.4	80
			6000	Octane	298.15	−1.2	73
			18000	Octane	298.15	1.0	80
			6000	2,2,4,6,6-Pentamethylheptane	298.15	−0.2	73
			18000	Pentane	298.15	−4.7	80
			18000	Tetrachloromethane	298.15	6.6	80
			6000	2,2,4,4-Tetramethylpentane	298.15	−0.8	74
			6000	2,3,3,4-Tetramethylpentane	298.15	−3.1	74
			18000	Toluene	298.15	17	80
			18000	Trichloromethane	298.15	17	80
			6000	2,2,4-Trimethylpentane	298.15	−1.0	73
Polypropylene (isotactic)							
				1-Chloronaphthalene	383.15	26	47
				1-Chloronaphthalene	393.15	170	47
				1-Chloronaphthalene	403.15	245	47
				1-Chloronaphthalene	423.15	275	47
				1,2,3,4-Tetrahydronaphthalene	373.15	140	47
				1,2,3,4-Tetrahydronaphthalene	383.15	215	47
				1,2,3,4-Tetrahydronaphthalene	393.15	330	47
				1,2,3,4-Tetrahydronaphthalene	403.15	330	47
				1,2,3,4-Tetrahydronaphthalene	413.15	335	47
				1,2,3,4-Tetrahydronaphthalene	423.15	290	47
Poly(propylene glycol)							
	150			Benzene	321.35	200	103
	425			Benzene	321.35	80	103
	2025			Benzene	321.35	45	103
	150			Ethanol	321.35	40	103
	425			Ethanol	321.35	60	103
	2025			Ethanol	321.35	65	103
	396	412		Tetrachloromethane	303.15	4.7	95
	396	412		Tetrachloromethane	318.15	5.2	95
	1900			Tetrachloromethane	303.15	−8.2	95
	1900			Tetrachloromethane	318.15	11	95
	1900			Trichloromethane	303.15	−81	95
		400		Water	298.15	−165	97

Polymer	M_n/ g/mol	M_w/ g/mol	M_η/ g/mol	Solvent	T/K	ΔH_B^∞/ J/g	Ref.
	150			Water	321.35	−90	103
	425			Water	321.35	−95	103
Polystyrene							
		600		Benzene	298.15	−1.3	54
		600		Benzene	313.15	−2.5	54
		900		Benzene	291.15	−10	54
		900		Benzene	318.15	−5.8	54
		2000		Benzene	291.15	−16	54
		2000		Benzene	318.15	−6.8	54
		5000		Benzene	291.15	−23	54
		5000		Benzene	318.15	−12	54
		10300		Benzene	291.15	−26	54
		10300		Benzene	318.15	−18	54
			18000	Benzene	298.15	−4.1	19
	20000			Benzene	296.15	−15	9
			29000	Benzene	298.15	−5.0	19
			30000	Benzene	298.15	−7.5	19
			59000	Benzene	298.15	−13	19
			91000	Benzene	298.15	−15	19
		97200		Benzene	318.15	−21	54
			142000	Benzene	298.15	−17	19
			190000	Benzene	303.15	−18	51
	214000			Benzene	300.15	−16	85
			216000	Benzene	298.15	−18	19
			272000	Benzene	298.15	−21	19
			300000	Benzene	298.15	−21	12
				Benzene	298.15	−27	25
				Benzene	298.15	−10	7
	20000			Butyl acetate	296.15	−13	9
	20000			2-Butanone	296.15	−15	9
			142000	2-Butanone	296.15	−17	30
			190000	Butylbenzene	303.15	−14	51
	150000			Chlorobenzene	293.15	−32	49
			266000	Chlorobenzene	298.15	5.4	34
				Chlorobenzene	293.15	−39	21
	1260			Cyclohexane	298.15	10	16
	1910			Cyclohexane	298.15	5.4	16
	3160			Cyclohexane	298.15	−5.4	16
	3980			Cyclohexane	298.15	−6.9	16
	5630			Cyclohexane	298.15	−9.3	16
	9070			Cyclohexane	298.15	−11	16
	20000			Cyclohexane	296.15	2.5	9
			190000	Cyclohexane	303.15	−2.1	51
				Cyclohexane	293.15	−14	21
	22400			Cyclohexanone	298.15	−29	25
	20000			Cyclohexene	296.15	−9.4	9
			190000	Decahydronaphthalene	303.15	3.8	51
	110000	115000		1,2-Dichlorobenzene	303.05	26	89
	20000			1,2-Dimethylbenzene	296.15	−13	9
	20000			1,3-Dimethylbenzene	296.15	−12	9
			190000	1,3-Dimethylbenzene	303.15	−15	51
			190000	1,4-Dioxane	303.15	−12	51
	20000			Ethyl acetate	296.15	−11	9
			142000	Ethyl acetate	296.15	−13	30
			785	Ethylbenzene	298.15	0.0	14

Polymer	M_n/ g/mol	M_w/ g/mol	M_η/ g/mol	Solvent	T/K	ΔH_B^∞/ J/g	Ref.
			18000	Ethylbenzene	298.15	−3.8	14
			18000	Ethylbenzene	298.15	−3.9	19
			30000	Ethylbenzene	298.15	−5.7	19
			35000	Ethylbenzene	298.15	−6.5	19
			45000	Ethylbenzene	298.15	−8.4	19
			91000	Ethylbenzene	298.15	−11	19
			142000	Ethylbenzene	298.15	−13	19
			216000	Ethylbenzene	298.15	−17	19
	60000			Ethylbenzene	303.15	−22	57
	113000	122000		Ethylbenzene	306.65	−24	63
	113000	122000		Ethylbenzene	317.15	−19	63
	113000	122000		Ethylbenzene	337.15	−11	63
	113000	122000		Ethylbenzene	347.15	−6.4	63
	113000	122000		Ethylbenzene	350.65	−4.9	63
	113000	122000		Ethylbenzene	366.65	−2.3	63
	113000	122000		Ethylbenzene	367.15	−2.1	63
	113000	122000		Ethylbenzene	368.65	−2.6	63
	113000	122000		Ethylbenzene	372.15	−4.2	63
	113000	122000		Ethylbenzene	378.15	−4.6	63
	113000	122000		Ethylbenzene	385.15	−5.7	63
	150000			Ethylbenzene	293.15	−34	49
			190000	Ethylbenzene	303.15	−15	51
			272000	Ethylbenzene	298.15	−18	14
			272000	Ethylbenzene	298.15	−18	19
			413000	Ethylbenzene	298.15	−24	39
				Ethylbenzene	293.15	−34	21
				Ethylbenzene	298.15	−17	11
				Ethylbenzene	298.15	−30	28
	1260			2-Propanone	298.15	−0.6	16
	1910			2-Propanone	298.15	−7.7	16
	3160			2-Propanone	298.15	−16	16
	3980			2-Propanone	298.15	−17	16
	5630			2-Propanone	298.15	−19	16
	9070			2-Propanone	298.15	−21	16
	20000			2-Propanone	296.15	−11	9
			190000	Propylbenzene	303.15	−14	51
	20000			Styrene	296.15	−18	9
				Styrene	296.15	−35	6
			413000	Tetrachloromethane	298.15	−22	39
	600			Toluene	296.15	−2.1	16
	600			Toluene	309.15	−1.8	16
	600			Toluene	318.15	−1.5	16
	1260			Toluene	296.15	−11	16
	1260			Toluene	303.15	−8.0	16
	1260			Toluene	309.15	−5.9	16
	1260			Toluene	318.15	−3.4	16
	1260			Toluene	328.15	−2.3	16
	1260			Toluene	338.15	−1.9	16
	1260			Toluene	346.65	−1.3	16
	1910			Toluene	298.15	−16	16
	1910			Toluene	318.15	−6.7	16
	1910			Toluene	338.15	−3.4	16
	1910			Toluene	348.15	−2.5	16
	3160			Toluene	298.15	−23	16
	3980			Toluene	298.15	−24	16

Polymer	M_n/ g/mol	M_w/ g/mol	M_η/ g/mol	Solvent	T/K	ΔH_B^∞/ J/g	Ref.
	3980			Toluene	318.15	−17	16
	3980			Toluene	338.15	−10	16
	5630			Toluene	298.15	−26	16
	9070			Toluene	298.15	−28	16
	270000			Toluene	298.15	−33	16
		600		Toluene	298.15	−1.4	54
		600		Toluene	313.15	−3.2	54
		900		Toluene	291.15	−7.3	54
		900		Toluene	318.15	−6.6	54
		2000		Toluene	291.15	−11	54
		2000		Toluene	318.15	−7.2	54
		5000		Toluene	291.15	−21	54
		5000		Toluene	318.15	−11	54
		10300		Toluene	291.15	−24	54
		10300		Toluene	318.15	−15	54
		97200		Toluene	318.15	−17	54
		9000		Toluene	310.15	−9.2	92
	20000			Toluene	296.15	−17	9
	20400			Toluene	298.15	−8.2	90
	20400			Toluene	310.15	−8.4	92
	20400			Toluene	333.15	−6.4	90
	47000			Toluene	310.15	−5.0	92
	50000			Toluene	333.15	−6.8	96
	60000			Toluene	303.15	−21	57
	113000	122000		Toluene	304.15	−29	63
	113000	122000		Toluene	306.15	−27	63
	113000	122000		Toluene	306.65	−26	63
	113000	122000		Toluene	316.15	−23	63
	113000	122000		Toluene	333.15	−15	63
	113000	122000		Toluene	337.15	−12	63
	113000	122000		Toluene	346.15	−8.3	63
	113000	122000		Toluene	347.15	−7.8	63
	113000	122000		Toluene	348.15	−8.2	63
	113000	122000		Toluene	350.65	−6.5	63
	113000	122000		Toluene	359.15	−4.3	63
	113000	122000		Toluene	362.15	−2.7	63
	113000	122000		Toluene	369.15	−3.3	63
	113000	122000		Toluene	372.15	−2.8	63
	115000			Toluene	310.15	−5.0	92
	150000			Toluene	293.15	−34	49
			190000	Toluene	303.15	−18	51
	214000			Toluene	300.15	−19	85
			250000	Toluene	303.15	−18	51
				Toluene	293.15	−34	21
				Toluene	298.65	−39	27
				Toluene	308.15	−34	27
				Toluene	318.15	−30	27
				Toluene	333.15	−23	27
				Toluene	343.15	−13	27
				Toluene	353.15	−13	27
		600		Trichloromethane	298.15	−13	54
		600		Trichloromethane	313.15	−9.9	54
		900		Trichloromethane	291.15	−22	54
		900		Trichloromethane	313.15	−15	54
		2000		Trichloromethane	291.15	−28	54

Polymers

Polymer	M_n/ g/mol	M_w/ g/mol	M_η/ g/mol	Solvent	T/K	ΔH_B^∞/ J/g	Ref.
		2000		Trichloromethane	313.15	−16	54
		5000		Trichloromethane	291.15	−30	54
		5000		Trichloromethane	313.15	−18	54
		10300		Trichloromethane	291.15	−33	54
		10300		Trichloromethane	313.15	−23	54
	22400			Trichloromethane	298.15	−17	25
		97200		Trichloromethane	313.15	−25	54
	20000			1,3,5-Trimethylbenzene	296.15	−11	9
			190000	1,3,5-Trimethylbenzene	303.15	−13	51
Poly(tetramethylene oxide)							
	650			Benzene	313.15	4.0	86
	1000			Benzene	313.15	2.0	86
	2000			Benzene	313.15	1.1	86
	650			1,2-Dichloroethane	313.35	3.1	88
	2000			1,2-Dichloroethane	313.35	0.3	88
	650			1,2-Dimethylbenzene	313.15	5.9	87
	1000			1,2-Dimethylbenzene	313.15	1.8	87
	2000			1,2-Dimethylbenzene	313.15	0.9	87
	650			1,3-Dimethylbenzene	313.15	6.4	87
	1000			1,3-Dimethylbenzene	313.15	0.6	87
	2000			1,3-Dimethylbenzene	313.15	0.8	87
	650			1,4-Dimethylbenzene	313.15	4.3	87
	1000			1,4-Dimethylbenzene	313.15	1.8	87
	2000			1,4-Dimethylbenzene	313.15	0.7	87
	650			1,4-Dioxane	321.35	4.0	81
	1000			1,4-Dioxane	321.35	2.3	81
	2000			1,4-Dioxane	321.35	1.0	81
	650			Ethylbenzene	313.15	6.9	86
	1000			Ethylbenzene	313.15	3.4	86
	2000			Ethylbenzene	313.15	−0.05	86
	650			Propylbenzene	313.15	5.8	86
	1000			Propylbenzene	313.15	1.3	86
	2000			Propylbenzene	313.15	0.9	86
	650			Tetrachloromethane	313.15	3.3	88
	1000			Tetrachloromethane	313.15	1.4	88
	2000			Tetrachloromethane	321.35	0.7	82
	650			Toluene	313.15	4.3	86
	1000			Toluene	313.15	2.0	86
	2000			Toluene	313.15	0.9	86
	650			1,3,5-Trimethylbenzene	313.15	6.1	87
	1000			1,3,5-Trimethylbenzene	313.15	2.7	87
	2000			1,3,5-Trimethylbenzene	313.15	0.6	87
Poly(vinyl acetate)							
			140000	Benzene	298.15	2.3	13
			350000	2-Butanone	303.15	−1.7	51
			350000	Butyl acetate	303.15	1.0	51
			135000	Chlorobenzene	298.15	5.0	34
			350000	Ethyl acetate	303.15	−6.7	51
				Ethyl acetate	303.15	0.0	11
			26000	3-Heptanone	303.15	7.0	44
			350000	3-Heptanone	303.15	4.9	51
			140000	Methanol	298.15	−45	13
			350000	Methyl acetate	303.15	−9.7	51
			350000	2-Pentanone	303.15	0.0	51

Polymer	M_n/ g/mol	M_w/ g/mol	M_η/ g/mol	Solvent	T/K	ΔH_B^∞/ J/g	Ref.
			93000	2-Propanone	303.15	−0.4	32
			350000	2-Propanone	303.15	−3.9	51
				2-Propanone	303.15	−2.9	25
			350000	Propyl acetate	303.15	−2.7	51
		150000		Tetrahydrofuran	304.65	4.5	93
			140000	Trichloromethane	298.15	28	13
Poly(vinyl alcohol)							
				Ethanol	298.15	3.8	11
				Ethanol	298.15	9.6	31
	7260			Water	303.15	−34	15
	17000			Water	303.15	−18	32
	61600			Water	303.15	−41	15
				Water	303.15	−8.4	11
Poly(vinyl chloride)							
				Chlorobenzene	298.15	−17	36
Glass	23200			Cyclohexanone	303.15	−27	59
Liquid	23200			Cyclohexanone	303.15	−7.5	59
Glass	38700			Cyclohexanone	303.15	−29	59
Liquid	38700			Cyclohexanone	303.15	−6.6	59
Glass	53500			Cyclohexanone	303.15	−28	59
Liquid	53500			Cyclohexanone	303.15	−6.3	59
Glass	66700			Cyclohexanone	303.15	−29	59
Liquid	66700			Cyclohexanone	303.15	−6.1	59
Glass	136000			Cyclohexanone	303.15	−31	59
Liquid	136000			Cyclohexanone	303.15	−5.8	59
Glass	155400			Cyclohexanone	303.15	−32	59
Liquid	155400			Cyclohexanone	303.15	−5.8	59
	48000			Cyclopentanone	298.15	−28	104
				1,2-Dichloroethane	323.65	24	27
				1,2-Dichloroethane	328.15	34	27
				1,2-Dichloroethane	333.15	38	27
				1,2-Dichloroethane	368.15	44	27
				1,2-Dichloroethane	373.15	46	27
				1,2-Dichloroethane	378.15	46	27
				N,N-Dimethylformamide	293.15	−28	35
				N,N-Dimethylformamide	308.15	−19	35
				N,N-Dimethylformamide	323.15	−14	35
				N,N-Dimethylformamide	338.15	−7.5	35
				N,N-Dimethylformamide	353.15	2.4	35
Glass	23200			Tetrahydrofuran	303.15	−34	59
Liquid	23200			Tetrahydrofuran	303.15	−14	59
Glass	38700			Tetrahydrofuran	303.15	−35	59
Liquid	38700			Tetrahydrofuran	303.15	−14	59
Glass	53500			Tetrahydrofuran	303.15	−39	59
Liquid	53500			Tetrahydrofuran	303.15	−14	59
Glass	66700			Tetrahydrofuran	303.15	−36	59
Liquid	66700			Tetrahydrofuran	303.15	−14	59
Glass	136000			Tetrahydrofuran	303.15	−39	59
Liquid	136000			Tetrahydrofuran	303.15	−14	59
Glass	155400			Tetrahydrofuran	303.15	−39	59
Liquid	155400			Tetrahydrofuran	303.15	−14	59
Poly(1-vinyl-3,5-dimethyl-1,2,4-triazole)							
				N,N-Dimethylformamide	298.15	−28	94
				Water	298.15	−139	94

Polymer	M_n/ g/mol	M_w/ g/mol	M_η/ g/mol	Solvent	T/K	ΔH_B^∞/ J/g	Ref.
Poly(1-vinylimidazole)							
	20700			Acetic acid	298.15	−393	94
	20700			Butanoic acid	298.15	−322	94
	20700			*N,N*-Dimethylacetamide	298.15	−48	94
	20700			*N,N*-Dimethylformamide	298.15	−48	94
	20700			1-Methyl-2-pyrrolidinone	298.15	−54	91
	20700			Pentanoic acid	298.15	−325	94
	20700			Propanoic acid	298.15	−278	94
	20700			Water	298.15	−119	91
Poly(1-vinylpyrazole)							
	18900			Acetic acid	298.15	−88	94
	18900			Butanoic acid	298.15	−60	94
	18900			*N,N*-Dimethylacetamide	298.15	−26	94
	18900			*N,N*-Dimethylformamide	298.15	−28	94
	18900			Pentanoic acid	298.15	−52	94
	18900			Propanoic acid	298.15	−36	94
Poly(1-vinyl-2-pyrrolidone)							
			32000	Trichloromethane	298.15	−75	48
			32000	Water	298.15	−150	48
Poly(1-vinyl-1,2,4-triazole)							
	69500			Acetic acid	298.15	−85	94
	69500			Butanoic acid	298.15	−74	94
	69500			*N,N*-Dimethylacetamide	298.15	−49	94
	69500			*N,N*-Dimethylformamide	298.15	−47	94
	69500			1-Methyl-2-pyrrolidinone	298.15	−55	91
	69500			Pentanoic acid	298.15	−75	94
	69500			Propanoic acid	298.15	−72	94
	69500			Water	298.15	−68	91
Vinyl acetate/vinyl alcohol copolymer							
(9 wt% Vinyl acetate)				2-Propanone	298.15	6.3	33
(44 wt% Vinyl acetate)				2-Propanone	298.15	4.6	33
(57 wt% Vinyl acetate)				2-Propanone	298.15	0.0	33
(67 wt% Vinyl acetate)				2-Propanone	298.15	−1.3	33
(4.2 mol% Vinyl acetate)	7560			Water	303.15	−41	15
(4.3 mol% Vinyl acetate)	64300			Water	303.15	−49	15
(9.0 mol% Vinyl acetate)	66900			Water	303.15	−55	15
(10.3 mol% Vinyl acetate)	7970			Water	303.15	−41	15
(15.3 mol% Vinyl acetate)	8300			Water	303.15	−60	15
(15.4 mol% Vinyl acetate)	70700			Water	303.15	−65	15
(19.5 mol% Vinyl acetate)	73100			Water	303.15	−66	15
(22.1 mol% Vinyl acetate)	8800			Water	303.15	−60	15
(26.2 mol% Vinyl acetate)	77000			Water	303.15	−64	15
(30.6 mol% Vinyl acetate)	9370			Water	303.15	−53	15
(34.0 mol% Vinyl acetate)	81600			Water	303.15	−60	15
(34.7 mol% Vinyl acetate)	9670			Water	303.15	−44	15

Polymer	M_n/ g/mol	M_w/ g/mol	M_η/ g/mol	Solvent	T/K	ΔH_B^∞/ J/g	Ref.
Vinyl acetate/vinyl chloride copolymer (90 wt% Vinyl chloride)							
Glass	12400	26000		Cyclohexanone	304.15	−37	101
Liquid	12400	26000		Cyclohexanone	304.15	−16	101
Vinyl chloride/vinylidene chloride copolymer							
				Trichloromethane	297.15	−17	2

References

1. Liepatoff, S. and Preobagenskaja, S., *Kolloid Z. Z. Polym.*, 68, 324, 1934.
2. Kargin, V. and Papkov, S., *Acta Physicochim. URSS*, 3, 839, 1935.
3. Papkov, S. and Kargin, V., *Acta Physicochim. URSS*, 7, 667, 1937.
4. Tager, A. and Kargin, V., *Acta Physicochim. URSS*, 14, 713, 1941.
5. Raine, H.C., Richards, R.B., and Ryder, H., *Trans Faraday Soc.*, 41, 56, 1945.
6. Roberts, D.E., Walton, W.W., and Jessup, R.S., *J. Polym. Sci.*, 2, 420, 1947.
7. Tager, A. and Sanatina, V., *Kolloidn. Zhur.*, 12, 474, 1950.
8. Glikman, S.A. and Root, L.A., *Zh. Obshch. Khim.*, 21, 58, 1951.
9. Hellfritz, H., *Makromol. Chem.*, 7, 191, 1951.
10. Tager, A. and Vershkain, R., *Kolloidn. Zhur.*, 13, 123, 1951.
11. Tager, A.A., and Kargin, V.A., *Kolloidn. Zhur.*, 14, 367, 1952.
12. Tager, A.A. and Dombek, Zh.S., *Kolloidn. Zhur.*, 15, 69, 1953.
13. Daoust, H. and Rinfret, M., *Can. J. Chem.*, 32, 492, 1954.
14. Gatovskaya, T.V., Kargin, V.A., and Tager, A.A., *Zh. Fiz. Khim.*, 29, 883, 1955.
15. Oya, S., *Chem. High Polym. Japan*, 12, 122, 1955.
16. Schulz, G.V., Guenner, K. von, and Gerrens, H., *Z. Phys. Chem., N. F.*, 4, 192, 1955.
17. Tager, A.A., Kosova, L.K., Karlinskaya, D.Yu., and Yurina, I.A., *Kolloid. Zhur.*, 17, 315, 1955.
18. Tager, A.A. and Kosova, L.K., *Kolloid. Zhur.*, 17, 391, 1955.
19. Tager, A.A., Krivokorytova, R.V., and Khodorov, P.M., *Dokl. Akad. Nauk SSSR*, 100, 741, 1955.
20. Glikman, S.A. and Root, L.A., *Kolloidn. Zhur.*, 18, 523, 1956.
21. Jenckel, E. and Gorke, K., *Z. Elektrochem.*, 60, 579, 1956.
22. Lipatov, Yu.S., Kargin, V.A., and Slonimskii, G.L., *Zh. Fiz. Khim.*, 30, 1202, 1956.
23. Meerson, S.I. and Lipatov, S.M., *Kolloidn., Zh.*, 18, 447, 1956.
24. Mikhailov, N.V. and Fainberg, E.Z., *Kolloidn. Zhur.*, 18, 44, 1956.
25. Struminskii, G.V. and Slonimskii, G.L., *Zh. Fiz. Khim.*, 30, 1941, 1956.
26. Gal'perin, D.I. and Moseev, L.I., *Kolloidn. Zhur.*, 19, 167, 1957.
27. Akhmedov, K.S., *Uzb. Khim. Zh.* (1), 19, 1958.
28. Kargin, V.A. and Lipatov, Yu.S., *Zh. Fiz. Khim.*, 32, 326, 1958.
29. Meerson, S.I. and Lipatov, S.M., *Kolloidn. Zhur.*, 20, 353, 1958.
30. Tager, A.A. and Galkina, L.A., *Nauchn. Dokl. Vyssh. Shkol., Khim. Khim. Tekhnol.*, (2), 357, 1958.
31. Tager, A. A. and Kargin, V. A., *Zh. Fiz. Khim.*, 32, 1362, 1958.
32. Tager, A. A. and Kargin, V. A., *Zh. Fiz. Khim.*, 32, 2694, 1958.
33. Tager, A.A. and Iovleva, M., *Zh. Fiz. Khim.*, 32, 1774, 1958.
34. Horth, A., Patterson, D., and Rinfret, M., *J. Polym. Sci.*, 39. 189, 1959.
35. Zelikman, S.G. and Mikhailov, N.V., *Vysokomol. Soedin.*, 1, 1077, 1959.
36. Mueller, F.H. and Engelter, A., *Kolloid Z.*, 171, 152, 1960.
37. Delmas, G., Patterson, D., and Boehme, A., *Trans. Faraday Soc.*, 58, 2116, 1962.
38. Delmas, G., Patterson, D., and Somcynsky, T., *J. Polym. Sci.*, 57, 79, 1962.
39. Tager, A.A. and Podlesnyak, A.I., *Vysokomol. Soedin., Ser. A*, 5, 87, 1963.
40. Delmas, G., Patterson, D., and Bhattacharyya, S.N., *J. Phys. Chem.*, 68, 1468, 1964.
41. Patterson, D., *J. Polym. Sci.: Part A*, 2, 5177, 1964.
42. Zverev, M.P., Barash, A.N., and Zubov, P.I., *Vysokomol. Soedin., Ser. A*, 6, 1012, 1964.
43. Cottam, B.J., Cowie, J.M.G., and Bywater, S., *Makromol. Chem.*, 86, 116, 1965.
44. Bianchi, U., Pedemonte, E., and Rossi, C., *Makromol. Chem.*, 92, 114, 1966.
45. Cuniberti, C. and Bianchi, U., *Polymer*, 7, 151, 1966.
46. Giacommeti, G. and Turolla, A., *Z. Phys. Chem., N.F.*, 51, 108, 1966.
47. Schreiber, H.P. and Waldman, M.H., *J. Polym. Sci.: Part A-2*, 5, 555, 1967.
48. Goldfarb, J. and Rodriguez, S., *Makromol. Chem.*, 116, 96, 1968.
49. Maron, S.H. and Daniels, C.A., *J. Macromol. Sci.-Phys. B*, 2, 769, 1968.
50. Morimoto, S., *J. Polym. Sci.: Part A-1*, 6, 1547, 1968.
51. Bianchi, U., Cuniberti, C., Pedemonte, E., and Rossi, C., *J. Polym. Sci.: Part A-2*, 7, 855, 1969.
52. Gerth, Ch. and Mueller, F.H., *Kolloid-Z. Z. Polym.*, 241, 1071, 1970.
53. Liddell, A.H. and Swinton, F.L., *Discuss. Faraday Soc.*, 49, 115, 1970.
54. Morimoto, S., *Nippon Kagaku Zasshi*, 91, 31, 1970.
55. Nakayama, H., *Bull. Chem. Soc. Japan*, 43, 1683, 1970.
56. Blackadder, D.A. and Roberts, T.L., *Angew. Makromol. Chem.*, 27, 165, 1972.
57. Maron, S.H. and Filisko, F.E., *J. Macromol. Sci.-Phys. B*, 6, 57, 1972.
58. Maron, S.H. and Filisko, F.E., *J. Macromol. Sci.-Phys. B*, 6, 79, 1972.
59. Maron, S.H. and Filisko, F.E., *J. Macromol. Sci.-Phys. B*, 6, 413, 1972.
60. Sokolova, D.F., Sokolov, L.B., and Gerasimov, V.D., *Vysokomol. Soedin., Ser. B*, 14, 580, 1972.
61. Chahal, R.S., Kao, W.-P., and Patterson, D., *J. Chem. Soc., Faraday Trans. I*, 69, 1834, 1973.
62. Delmas, G. and Tancrede, P., *Eur. Polym. J.*, 9, 199, 1973.
63. Filisko, F.E., Raghava, R.S., and Yeh, G.S.Y., *J. Macromol. Sci.-Phys. B*, 10, 371, 1974.
64. Ikeda, M., Suga, H., and Seki, S., *Polymer*, 16, 634, 1975.
65. Kiselev, V.P., Shakhova, E.M., Fainberg, E.Z., Virnik, A.D, and Rogovin, Z.A., *Vysokomol. Soedin., Ser. B*, 18, 847, 1976.
66. Petrosyan, V.A., Gabrielyan, G.A., and Rogovin, Z.A., *Arm. Khim. Zhur.*, 29, 516, 1976.
67. Deshpande, D.D. and Prabhu, C.S., *Macromolecules*, 10, 433, 1977.
68. Miura, T. and Nakamura, M., *Bull. Chem. Soc. Japan*, 50, 2528, 1977.
69. Sokolova, D.F., Kudim, T.V., Sokolov, L.B., Zhegalova, N.I., and Zhuravlev, N.D., *Vysokomol. Soedin., Ser. B*, 20, 596, 1978.
70 Basedow, A.M. and Ebert, K.H., *J. Polym. Sci.: Polym. Symp.*, 66, 101, 1979.
71 Koller, J., *Dissertation*, TU München, 1979.
72. Lee, J.-O., Ono, M., Hamada, F., and Nakajima, A., *Polym. Bull.*, 1, 763, 1979.
73. Phuong-Nguyen, H. and Delmas, G., *Macromolecules*, 12, 740, 1979.
74. Phuong-Nguyen, H. and Delmas, G., *Macromolecules*, 12, 746, 1979.
75. Basedow, A.M., Ebert, K.H., and Feigenbutz, W., *Makromol. Chem.*, 181, 1071, 1980.
76. Graun, K., *Dissertation*, TU München, 1980.
77. Shiomi, T., Izumi, Z., Hamada, F., and Nakajima, A., *Macromolecules*, 13, 1149, 1980.
78. Shiomi, T., Kohra, Y., Hamada, F., and Nakajima, A., *Macromolecules*, 13, 1154, 1980.
79. Aharoni, S.M., Charlet, G., and Delmas, G., *Macromolecules*, 14, 1390, 1981.
80. Ochiai, H., Ohashi, T., Tadokoro, Y., and Murakami, I., *Polym. J.*, 14, 457, 1982.
81. Sharma, S.C., Mahajan, R., Sharma, V.K., and Lakhanpal, M.L., *Indian J. Chem.*, 21A, 682, 1982.
82. Sharma, S.C., Mahajan, R., Sharma, V.K., and Lakhanpal, M.L., *Indian J. Chem.*, 21A, 685, 1982.
83. Lakhanpal, M.L. and Parashar, R.N., *Indian J. Chem.*, 22A, 48, 1983.
84. Daoust, H. and St-Cyr, D., *Macromolecules*, 17, 596, 1984.

85. Aeleni, N., *Mater. Plast. (Bucharest)*, 22, 92, 1985.
86. Sharma, S.C. and Sharma, V.K., *Indian J. Chem.*, 24A, 292, 1985.
87. Sharma, S.C., Bhalla, S., and Sharma, V.K., *Indian J. Chem.*, 25A, 131, 1986.
88. Sharma, S.C., Syngal, M., and Sharma, V.K., *Indian J. Chem.*, 26A, 285, 1987.
89. Aukett, P.N. and Brown, C.S., *J. Therm. Anal.*, 33, 1079, 1988.
90. Lanzavecchia, L. and Pedemonte, E., *Thermochim. Acta*, 137, 123, 1988.
91. Tager, A.A., Safronov, A.P., Voit, V.V., Lopyrev, V.A., Ermakova, T.G., Tatarova, L.A., and Shagelaeva, N.S., *Vysokomol. Soedin., Ser. A*, 30, 2360, 1988.
92. Pedemonte, E. and Lanzavecchia, L., *Thermochim. Acta*, 162, 223, 1990.
93. Shiomi, T., Ishimatsu, H., Eguchi, T., and Imai, K., *Macromolecules*, 23, 4970, 1990.
94. Tager, A.A. and Safronov, A.P., *Vysokomol. Soedin., Ser. A*, 33, 67, 1991.
95. Zellner, H., *Dissertation*, TU München, 1993.
96. Brunacci, A., Pedemonte, E., Cowie, J.M.G., and McEwen, I. J., *Polymer*, 35, 2893, 1994.
97. Carlsson, M., Hallen, D., and Linse, P., *J. Chem. Soc. Faraday Trans.*, 91, 2081, 1995.
98. Sato, T., Tohyama, M., Suzuki, M., Shiomi, T., and Imai, K., *Macromolecules*, 29, 8231, 1996.
99. Shiomi, T., Tohyama, M., Endo, M., Sato, T., and Imai, K., *J. Polym. Sci.: Part B: Polym. Phys.*, 34, 2599, 1996.
100. Kizhnyaev, V.N., Gorkovenko, O.P., Bazhenov, D.N., and Smirnov, A.I., *Vysokomol. Soedin., Ser. A*, 39, 856, 1997.
101. Sato, T., Suzuki, M., Tohyama, M., Endo, M., Shiomi, T., and Imai, K., *Polym. J.*, 29, 417, 1997.
102. Vanderryn, J. and Zettlemoyer, A.C., *Ind. Eng. Chem., Chem. Eng. Data Ser.*, 2, 56, 1957.
103. Parashar, R. and Sharma, S.C., *Indian J. Chem.*, 27A, 1092, 1988.
104. Righetti, M.C., Cardelli, C., Scalari, M., Tombari, E., and Conti, G., *Polymer*, 43, 5035, 2002.
105. Phuong-Nguyen, H. and Delmas, G., *J. Solution Chem.*, 23, 249, 1994.
106. Wohlfarth, C., *CRC Handbook of Enthalpy Data of Polymer-Solvent Systems*, CRC Press, Boca Raton, 2006.

SOLUBILITY PARAMETERS OF SELECTED POLYMERS

Christian Wohlfarth

The concept of cohesive energy density and solubility parameter was introduced by Hildebrand:

$$\delta^2 = \frac{\Delta U_m^{LV}}{V_m} = \frac{\Delta H_m^{LV} - RT}{V_m} \quad (1)$$

V_m is the molar volume, ΔU_m^{LV} is the molar energy of vaporization, and ΔH_m^{LV} is the molar enthalpy of vaporization. Units for the solubility parameter are $(MPa)^{1/2} = (J/cm^3)^{1/2} = 0.4887(cal/cm^3)^{1/2}$. The energy of vaporization is not accessible for polymers, but cohesive energy density of polymers can be determined from *PVT*-data. However, common ways for determining polymer solubility parameters use thermodynamic properties of polymer solutions and their relations to excess enthalpy or excess Gibbs energy per unit volume. These excess quantities are related to the (square) difference between the solubility parameters of solvents and polymers, i.e., $(\delta_1 - \delta_2)^2$.

$$\frac{H^E}{V} \sim (\delta_1 - \delta_2)^2 \quad \text{or} \quad \frac{G^E}{V} \sim (\delta_1 - \delta_2)^2 \quad (2)$$

Often, the Flory-Huggins solvent-polymer interaction parameter is applied instead of H^E or G^E. There are some books (Refs. 1–3) giving details for such procedures as well as extensive tables of polymer solubility parameters from which the table below is extracted. Methods for calculating solubility parameters can be found in Refs. 4–7.

References

1. Barton, A.F.M., *CRC Handbook of Polymer-Liquid Interaction Parameters and Solubility Parameters*, CRC Press, Boca Raton, 1991.
2. Brandrup, J., Immergut, E.H., Grulke, E.A. (eds.), *Polymer Handbook*, 4th ed., J. Wiley & Sons, New York, 1999.
3. Wohlfarth, C., *Thermodynamic Properties of Polymer Solutions*, in Landolt-Börnstein, New Series, Group VIII, Vol. 6D, Lechner, M.D., (ed.), Springer Verlag, Berlin, Heidelberg, 2010.
4. Hildebrand, J.H., Prausnitz, J.M., Scott, R.L., *Regular and Related Solutions*, Van Nostrand Reinhold Co., New York, 1970.
5. [Van] Krevelen, D.W., *Properties of Polymers*, 3rd ed., Elsevier, Amsterdam, 1990.
6. Bicerano, J., *Prediction of Polymer Properties*, 3rd ed., CRC Press, Boca Raton, 2002.
7. Hansen, C.M., *Hansen Solubility Parameters: A User's Handbook*, 2nd ed., CRC Press, Boca Raton, 2007.

Polymer	T/K	$\delta/(J/cm^3)^{1/2}$
Benzyl cellulose	298	25.2
Butadiene/acrylonitrile copolymer		
(25 wt% acrylonitrile)	298	19.4
(30 wt% acrylonitrile)	298	19.2
(34 wt% acrylonitrile)	298	20.4
(39 wt% acrylonitrile)	298	21.35
Butadiene/styrene copolymer		
(6 wt% styrene)	298	16.5
(12.5 wt% styrene)	298	16.5
(15 wt% styrene)	298	17.5
(25 wt% styrene)	298	17.6
(40 wt% styrene)	298	17.8
Cellulose	298	26.0
Cellulose acetate	298	25.1
Cellulose diacetate	298	22.3
Cellulose nitrate	298	22.0
Cellulose triacetate	298	19.0
Dextran	298	38.6
Ethyl cellulose	298	21.1
Ethylene/1-octene copolymer		
(2.0 wt% 1-octene)	473	17.2
(7.5 wt% 1-octene)	473	16.7
(12.0 wt% 1-octene)	473	16.5
(25.0 wt% 1-octene)	473	16.5
(39.4 wt% 1-octene)	473	16.4
(55.0 wt% 1-octene)	473	16.3
(64.0 wt% 1-octene)	473	16.3
Ethylene/vinyl acetate copolymer		
(11 wt% vinyl acetate)	323	16.6
(16 wt% vinyl acetate)	323	17.1
(24 wt% vinyl acetate)	323	16.8
(37 wt% vinyl acetate)	323	17.0
(42 wt% vinyl acetate)	323	17.2
Gelatine	298	24.6
Hydroxypropyl cellulose	298	26.8
Natural rubber	298	16.6
Poly(acrylonitrile)	298	26.0
Polyamide 4	298	24.0
Polyamide 6	298	21.7
Polyamide 66	298	22.9
Polyamide 7	298	24.1
Polyamide 8	298	20.3
Polyamide 9	298	22.6
Polyamide 10	298	19.4
Polyamide 11	298	22.9
Polyamide 12	298	20.8
Poly(*p*-benzamide)	298	23.0
Poly(benzyl methacrylate)	298	15.3
Polybutadiene	298	16.6
1,2-Polybutadiene	298	16.5
1,4-*cis*-Polybutadiene	298	16.5
Poly(1-butene), *isotactic*	298	16.0
Poly(butyl acrylate)	298	18.0
Poly(butyl methacrylate)	298	17.9

Polymer	T/K	$\delta/(J/cm^3)^{1/2}$
Poly(2-butyl methacrylate)	413	14.7
Poly(ε-caprolactone)	298	19.9
Polycarbonate bisphenol-A	298	20.0
Polycarbonate hexafluorobisphenol-A	298	20.1
Poly(chloroprene)	298	17.6
Poly(4-chlorostyrene)	298	19.3
Poly(cyclohexyl methacrylate)	298	19.8
Poly(3,3-diethyloxetane)	298	16.2
Poly(1,1-dihydroperfluorooctyl acrylate)	298	18.0
Poly(1,1-dihydroperfluorooctyl methacrylate)	298	17.0
Poly(3,3-dimethyloxetane)	298	16.2
Poly(2,6-dimethyl-1,4-phenylene ether)	298	18.1
Poly(dimethylsiloxane)	298	15.4
Poly(1,3-dioxepane)	298	18.8
Poly(1,3-dioxolane)	298	20.7
Poly(dodecyl methacrylate)	298	16.8
Poly(epichlorohydrin)	298	16.2
Poly(ethoxyethyl methacrylate)	298	18.4
Poly(ethyl acrylate)	298	19.2
Polyethylene, branched	298	16.2
Polyethylene, linear	298	16.2
Poly(ethylene adipate)	298	19.8
Poly(ethylene carbonate)	298	22.3
Poly(ethylene glycol)	298	23.7
Poly(ethylene oxide)	298	20.5
Poly(ethylene terephthalate)	298	21.9
Poly(ethyl methacrylate)	298	18.3
Poly(hexyl methacrylate)	298	17.7
Poly(3-hexyl thiophene)	298	19.0
Poly(4-hydroxystyrene)	298	24.0
Poly(isobornyl acrylate)	298	16.8
Poly(isobornyl methacrylate)	298	17.0
Poly(isobutylene)	298	16.2
Poly(isobutyl methacrylate)	413	14.6
1,4-*cis*-Poly(isoprene)	298	16.5
Poly(*N*-isopropylacrylamide)	298	23.5
Poly(DL-lactic acid)	298	20.5
Poly(L-lactide)	298	19.3
Poly(methacrylonitrile)	298	21.0
Poly(methyl acrylate)	298	20.5
Poly(methyl methacrylate)	298	19.3
Poly(4-methyl-1-pentene)	298	15.3
Poly(2-methylpropene)	298	17.7
Poly(2-methylstyrene)	298	18.4
Poly(4-methylstyrene)	298	19.3
Poly(methylvinylsiloxane)	298	15.65
Polynorbornene	298	14.0
Poly(1-octene)	298	16.6
Poly(octyl methacrylate)	298	18.0
Poly(*p*-phenylene terephthalamide)	298	23.3
Poly(propyl acrylate)	298	18.4
Poly(propyl methacrylate)	298	16.0
Polypropylene, *atactic*	298	15.5
Polypropylene, *isotactic*	298	17.5
Polypropylene, *syndiotactic*	298	17.6
Poly(propylene glycol)	298	19.5
Poly(propylene oxide)	298	18.5
Poly(propyl methacrylate)	298	20.0
Polystyrene	298	19.0
Polysulfone	298	19.9
Poly(tetrafluoroethylene)	298	19.6
Poly(tetramethylene oxide)	298	16.8
Poly(thioethylene)	298	18.8
Poly(trimethylene sulfide)	298	23.0
Poly(vinyl acetate)	298	20.6
Poly(vinyl alcohol)	298	22.0
Poly(vinyl bromide)	298	19.4
Poly(vinyl butyl ether)	298	19.2
Poly(*N*-vinylcarbazole)	298	19.2
Poly(vinyl chloride)	298	19.6
Poly(vinyl ethyl ether)	298	19.5
Poly(vinylidene fluoride)	298	23.2
Poly(vinyl methyl ether)	298	21.0
Poly(vinyl phenyl ether)	298	20.2
Poly(vinyl propionate)	298	18.1
Poly(vinyl propyl ether)	298	19.3
Poly(1-vinyl-2-pyrrolidinone)	298	25.6
Vinyl acetate/vinyl alcohol copolymer		
(43.4 mol% vinyl acetate)	298	21.8
(60.9 mol% vinyl acetate)	298	21.4
(74.4 mol% vinyl acetate)	298	20.9
(94.8 mol% vinyl acetate)	298	20.2
Vinyl acetate/vinyl chloride copolymer		
(3 wt% vinyl acetate)	298	18.8
(10 wt% vinyl acetate)	298	17.3
(17 wt% vinyl acetate)	298	19.1

Section 14
Geophysics, Astronomy, and Acoustics

ASTRONOMICAL CONSTANTS

Victor Abalakin

The constants in this table are based originally on the set of constants adopted by the International Astronomical Union (IAU) in 1976. Updates have been made when new data were available (Ref. 4). All values are given in SI units; thus masses are expressed in kilograms and distances in meters.

The astronomical unit of time is a time interval of one day (1 d) equal to 86400 s. An interval of 36525 d is one Julian century (1 cy).

References

1. Seidelmann, P. K., *Explanatory Supplement to the Astronomical Almanac*, University Science Books, Mill Valley, CA, 1990.
2. Lang, K. R., *Astrophysical Data: Planets and Stars*, Springer-Verlag, New York, 1992.
3. *The Astronomical Almanac for the Year 2007*, U.S. Government Printing Office, Washington, and Her Majesty's Stationary Office, London, 2005.
4. International Astronomical Union, <www.iau.org/static/resolutions/IAU2012_English.pdf>, 2012.

Defining constants	
Astronomical unit	au = 149597870700 m (exact)
Speed of light	c = 299792458 m s^{-1} (exact)
Primary constants	
Light-time for unit distance (1 au)	τ_A = 499.004786 s
Equatorial radius of Earth	a_e = 6378140 m
Equatorial radius of Earth (IUGG value)	a_e = 6378136 m
Dynamical form-factor for Earth	J_2 = 0.001082636
Geocentric gravitational constant	$GE = 3.986004 \times 10^{14}$ m^3 s^{-2}
Constant of gravitation	$G = 6.67428 \times 10^{-11}$ m^3 kg^{-1} s^{-2}
Ratio of mass of moon to that of Earth	μ = 0.01230002 $1/\mu$ = 81.300587
General precession in longitude, per Julian century, at standard epoch J2000	ρ = 5028″.796
Obliquity of the ecliptic at standard epoch J2000	ε = 23°26′21″.448
Derived constants	
Constant of nutation at standard epoch J2000	N = 9″.2025
Solar parallax ($\pi_0 = \arcsin(a_e/\mathrm{ua})$)	π_0 = 8″.794143
Constant of aberration for standard epoch J2000	κ = 20″.49552
Flattening factor for the Earth	f = 1/298.256 = 0.00335282
Heliocentric gravitational constant ($GS = A^3k^2/D^2$)	$GS = 1.32712438 \times 10^{20}$ m^3 s^{-2}
Ratio of mass of sun to that of the Earth ($S/E) = (GS)/(GE)$)	S/E = 332946.0
Ratio of mass of sun to that of Earth + moon	$(S/E)/(1 + \mu)$ = 328900.56
Mass of the sun ($S = (GS)/G$)	$S = 1.98844 \times 10^{30}$ kg
Ratios of mass of sun to masses of the planets	
Mercury	6023600
Venus	408523.7
Earth + moon	328900.56
Mars	3098708
Jupiter	1047.349
Saturn	3497.898
Uranus	22902.98
Neptune	19412.24

Geo-Astro

PROPERTIES OF THE SOLAR SYSTEM

The following tables give various properties of the planets and characteristics of their orbits in the solar system. Certain properties of the sun and of the earth's moon are also included.

Explanations of the column headings:

Mass: mass of the planet in units of 10^{24} kg
Radius: radius at the equator in km
Density: mean density in g/cm^3
Flattening: degree of oblateness, defined as $(r_e\text{-}r_p)/r_e$, where r_e and r_p are the equatorial and polar radii, respectively
Potential coefficients J_n: coefficients in the spherical harmonic representation of the gravitational potential U by the equation $U(r, \phi) = (GM/r)\,[1 - \Sigma J_n(a/r)^n\, P_n\,(\sin \phi)]$, where G is the gravitational constant, r the distance from the center of the planet, a the radius of the planet, M the mass, ϕ the latitude, and P_n the Legendre polynomial of degree n
Gravity: acceleration due to gravity at the surface
Escape vel.: velocity needed at the surface of the planet to escape the gravitational pull
Dist. to sun: semi-major axis of the elliptical orbit in astronomical units (1 ua ≈ $1.496{\cdot}10^8$ km)
ε: eccentricity of the orbit
Ecliptic angle: angle between the planetary orbit and the plane of the earth's orbit around the sun
Inclin.: angle between the equatorial plane of the planet and the plane of the planetary orbit
Orbit period: period of revolution around the sun measured in years
Rotation period: period of rotation of the planet measured in hours. A negative value indicates retrograde rotation.
Albedo: ratio of the light reflected from the planet to the light incident on it
No. of satellites: Number of confirmed satellites; this includes satellites that have not been named
T_{sur}: mean temperature at the surface
P_{sur}: pressure of the atmosphere at the surface

The last five entries in the table are *dwarf planets* as defined by the International Astronomical Union. These are bodies in orbit around the sun that are massive enough to adopt a near-spherical shape as a result of their self-gravity, but are appreciably smaller than the major planets. *Plutoids* form a subset of the dwarf planets; their orbits are larger than that of Neptune (see Ref. 9). As of 2015, the IAU has recognized the names for four plutoids: Pluto, Eris, Haumea, and Makemake. Since Ceres has an orbit much smaller than neptune, it is classified as a dwarf planet but not a plutoid.

The following general information on the solar system is of interest:

Mass of the earth = M_e = $5.9736{\cdot}10^{24}$ kg
Total mass of planetary system = $2.669{\cdot}10^{27}$ kg = 447 M_e
Total angular momentum of planetary system = $3.148{\cdot}10^{43}$ kg m^2 s^{-1}
Total kinetic energy of the planets = $1.99{\cdot}10^{35}$ J
Total rotational energy of planets = $0.7{\cdot}10^{35}$ J

Properties of the sun:

Mass = $1.9884{\cdot}10^{30}$ kg = 332943 M_e
Radius = $6.9551{\cdot}10^8$ m
Surface area = $6.079{\cdot}10^{18}$ m^2
Volume = $1.409{\cdot}10^{27}$ m^3
Mean density = 1.411 g/cm^3
Gravity at surface = 27398 cm/s^2
Escape velocity at surface = $6.177{\cdot}10^5$ m/s
Effective temperature = 5780 K
Total radiant power emitted (luminosity) = $3.828{\cdot}10^{26}$ W (Ref. 8)
Surface flux of radiant energy = $6.322{\cdot}10^7$ W/m^2
Flux of radiant energy at the earth (Solar Constant) = 1360.8 W/m^2 (Ref. 11)

References

1. Planetary Fact Sheet, NASA Goddard Space Flight Center, <nssdc.gsfc.nasa.gov/planetary/factsheet>, July 2015.
2. The Planetary Society, <www.planetary.org/explore/topics/groups/our_solar_system/>.
3. Arnet, B., The Nine Planets, <www.nineplanets.org>.
4. Onasch, B., Our Solar System, <www.onasch.de/astro/>.
5. *The Astronomical Almanac for the Year 2007*, U.S. Government Printing Office, Washington, DC, 2005; available online at <asa.usno.navy.mil/>.
6. Lang, K. R., *Astrophysical Data: Planets and Stars*, Springer-Verlag, New York, 1992.
7. Cox, A. N., Allen's Astrophysical Quantities, Fourth Edition, Springer-Verlag, New York, 2000; this is a revision of Allen, C. W., Astrophysical Quantities, Third Edition, 1983.
8. K. A. Olive et al. (Particle Data Group), *Chin. Phys. C* 38(9), 090001, 2014.; section on astrophysical constants available at <pdg.lbl.gov/2014/reviews/astrorpp.pdf>.
9. IAU Press Release, <www.iau.org/public_press/news/release/iau0804>, 11 June 2008.
10. Seidelmann, P. K., Editor, *Explanatory Supplement to the Astronomical Almanac*, University Science Books, Mill Valley, CA, 1992.
11. Kopp, G., and Lean, J. L., *Geophys. Res. Lett.* 38 (1), 2011. DOI: 10.1029/2010GL045777.

Geo-Astro

Properties of the Planets. Part 1

Planet	Mass/ 10^{24} kg	Radius/ km	Density/ g cm^{-3}	Flattening	Potential coefficient $10^3 J_2$	Potential coefficient $10^6 J_3$	Potential coefficient $10^6 J_4$	Gravity cm/ s^2	Escape velocity km/s
Mercury	0.33022	2439.7	5.43	0.0000				370	4.25
Venus	4.8685	6051.8	5.24	0.000	0.027			887	10.36
Earth	5.9736	6378.14	5.52	0.00335364	1.08263	-2.54	-1.61	980	11.18
(Moon)	0.07349	1738.1	3.35	0.0012	0.2027			162	2.38
Mars	0.64185	3396.2	3.93	0.0064763	1.964	36		371	5.03
Jupiter	1898.6	71492	1.33	0.0648744	14.75	-580		2312	59.54
Saturn	568.46	60268	0.69	0.09796	16.45	-1000		896	35.5
Uranus	86.832	25559	1.27	0.02293	3.343			869	21.29
Neptune	102.43	24764	1.64	0.01708	3.41			1100	23.5
Dwarf planets									
Pluto	0.0125	1195	1.75	0				58	1.2
Eris	0.0166	1225						80[a]	
Haumea	0.0040	620	3.0[a]					44	0.84
Makemake	0.004[a]	750	2[a]					47[a]	0.84[a]
Ceres	0.000943	471	2.08					27	0.51

[a] Approximate value

Properties of the Planets. Part 2

Planet	Dist. to Sun/au	Eccentricity ε	Ecliptic angle	Inclin.	Orbit period/yr	Rotation period/hr	Albedo	No. of satellites
Mercury	0.38710	0.2056	7.00°	0.01°	0.2408467	1407.6	0.106	0
Venus	0.72333	0.0068	3.39°	177.36°	0.61519726	-5832.5	0.65	0
Earth	1.00000	0.0167	0°	23.45°	1.0000174	23.9345	0.367	1
Mars	1.52366	0.0934	1.85°	25.19°	1.8808476	24.6229	0.150	2
Jupiter	5.20336	0.0484	1.305°	3.12°	11.862615	9.9250	0.52	67
Saturn	9.53707	0.0542	2.48°	26.73°	29.447498	10.656	0.47	62
Uranus	19.19126	0.0472	0.77°	97.86°	84.016846	-17.24	0.51	27
Neptune	30.06896	0.0086	1.77°	28.32°	164.79132	16.11	0.41	14
Dwarf planets								
Pluto	39.48168	0.2488	17.14°	118°	247.92065	-153.29	0.6	5
Eris	67.7	0.44		44.19°	557		0.8	1
Haumea	43.13	0.195		28.22°	283.28	3.8		
Makemake	45.79	0.159		28.96°	309.88			
Ceres	2.77	0.08		10.59°	4.599	9.074	0.090	

Mean Surface Temperature and Pressure and Atmospheric Composition

Planet	T_{sur}/K	P_{sur}/bar	CO_2	N_2	O_2	H_2O	H_2	He	Ar	Ne	CO	CH_4
Mercury	440	$1 \cdot 10^{-15}$										
Venus	737	90	96.5%	3.5%	69 ppm	20 ppm		12 ppm	70 ppm	7 ppm	17 ppm	
Earth	288	1.014	0.038%	78.084%	20.946%	0 to 3%	0.55 ppm	5.24 ppm	0.934%	18.18 ppm	1 ppm	1.7 ppm
Mars	208	0.007	95.32%	2.7%	0.13%	0.021%			1.6%	2.5 ppm	0.08%	
Jupiter	163	>>1000				4 ppm	89.8%	10.2%				0.30%
Saturn	133	>>1000					96.3%	3.25%				0.45%
Uranus	78	>>1000					82.5%	15.2%				2.3%
Neptune	73						80%	19%				1%
Pluto	48	$3 \cdot 10^{-6}$										

SATELLITES OF THE PLANETS

This table gives characteristics of the known satellites of the planets. The parameters covered are:

- Orbital period in units of earth days. An R following the value indicates a retrograde motion.
- Distance from the planet, as measured by the semi-major axis of the orbit
- Eccentricity of the orbit
- Inclination of the satellite orbit with respect to the equator of the planet
- Mass of the satellite in kilograms
- Radius of the satellite in kilometers
- Geometric albedo, which is a measure of the fraction of incident sunlight reflected by the satellite.

Since this is a very active field of research, the Internet sites listed below should be consulted for the most recent data.

References

1. *Solar System Dynamics*, Jet Propulsion Laboratory, California Institute of Technology, <ssd.jpl.nasa.gov/?phys_data>, June 2008.
2. The Planetary Society, <www.planetary.org/explore/topics/groups/our_solar_system/>.
3. Arnet, B., *The Nine Planets*, <www.nineplanets.org>.
4. Onasch, B., *Our Solar System*, <www.onasch.de/astro/>.
5. Sheppard, S. S., *The Giant Satellite and Moon Page*, <www.dtm.ciw.edu/sheppard/satellites/>.
6. *Gazetteer of Planetary Nomenclature*, U. S. Geological Survey, <planetarynames.wr.usgs.gov/append7.html>.
7. Seidelmann, P. K., Editor, *Explanatory Supplement to the Astronomical Almanac*, University Science Books, Mill Valley, CA, 1992.
8. Lang, K. R., *Astrophysical Data: Planets and Stars*, Springer-Verlag, New York, 1992.
9. Allen, C. W., *Astrophysical Quantities, Second Edition*, Athlone Press, London, 1955.

Planet		Satellite	Orb. period/ Earth days	Distance/ 10^3 km	Eccentricity	Inclination	Mass/kg	Radius/km	Albedo
Earth		Moon	27.321661	384.400	0.054900489	18.28–28.58°	$7.3483 \cdot 10^{22}$	1737.5	0.12
Mars	I	Phobos	0.31891023	9.378	0.0151	1.0°	$1.06 \cdot 10^{16}$	13.5×10.8×9.4	0.07
	II	Deimos	1.2624407	23.460	0.0005	0.9–2.7°	$2.4 \cdot 10^{15}$	7.5×6.1×5.5	0.07
Jupiter	I	Io	1.769137786	421.8	0.0041	0.04°	$8.932 \cdot 10^{22}$	1821.6	0.63
	II	Europa	3.551181041	671.1	0.0101	0.47°	$4.8 \cdot 10^{22}$	1560.8	0.67
	III	Ganymede	7.15455296	1070.4	0.0015	0.21°	$1.4819 \cdot 10^{23}$	2631.2	0.43
	IV	Callisto	16.6890184	1882.7	0.007	0.51°	$1.0759 \cdot 10^{23}$	2410.3	0.17
	V	Amalthea	0.49817905	181.4	0.003	0.40°	$7.17 \cdot 10^{18}$	131×73×67	0.09
	VI	Himalia	250.5662	11460	0.162	27.63°	$9.56 \cdot 10^{18}$	85	0.04
	VII	Elara	259.6528	11737	0.217	24.77°	$7.77 \cdot 10^{17}$	40	0.04
	VIII	Pasiphae	743.63 R	23620	0.409	145°	$1.91 \cdot 10^{17}$	18	0.04
	IX	Sinope	758.90 R	23940	0.250	153°	$7.77 \cdot 10^{16}$	14	0.04
	X	Lysithea	259.20	11720	0.112	29.02°	$7.77 \cdot 10^{16}$	12	0.04
	XI	Carme	734.17 R	23400	0.253	164°	$9.56 \cdot 10^{16}$	15	0.04
	XII	Ananke	629.77 R	21280	0.244	147°	$3.82 \cdot 10^{16}$	10	0.04
	XIII	Leda	240.92	11170	0.164	26.07°	$5.68 \cdot 10^{15}$	5	0.04
	XIV	Thebe	0.6745	221.9	0.018	0.8°	$7.77 \cdot 10^{17}$	55×45	0.05
	XV	Adrastea	0.29826	129	0.0015		$1.91 \cdot 10^{16}$	13×10×8	0.10
	XVI	Metis	0.29478	128	0.0002		$9.56 \cdot 10^{16}$	20	0.06
	XVII	Callirrhoe	758.77	24100	0.283		$8.7 \cdot 10^{14}$	4	0.04
	XVIII	Themisto	130.02	7507	0.242		$6.9 \cdot 10^{14}$	4	0.04
	XIX	Megaclite	752.86	23810	0.425		$2.1 \cdot 10^{14}$	2.7	0.04
	XX	Taygete	732.41	23360	0.251		$1.6 \cdot 10^{14}$	2.5	0.04
	XXI	Chaldene	723.72	23180	0.238		$7.5 \cdot 10^{13}$	1.9	0.04
	XXII	Harpalyke	623.32	21110	0.227		$1.2 \cdot 10^{14}$	2.2	0.04
	XXIII	Kalyke	742.06	23580	0.243		$1.9 \cdot 10^{14}$	2.6	0.04
	XXIV	Iocaste	631.60	21270	0.218		$1.9 \cdot 10^{14}$	2.6	0.04
	XXV	Erinome	728.46	23280	0.270		$4.5 \cdot 10^{13}$	1.6	0.04
	XXVI	Isonoe	726.63	23220	0.261		$7.5 \cdot 10^{13}$	1.9	0.04
	XVII	Praxidike	625.39	21150	0.220		$4.3 \cdot 10^{14}$	3.4	0.04
	XXVIII	Autonoe	760.95	23039	0.334		$9.0 \cdot 10^{13}$	2.0	0.04
	XXIX	Thyone	627.21	20940	0.229		$9.0 \cdot 10^{13}$	2.0	0.04
	XXX	Hermippe	633.90	21131	0.210		$9.0 \cdot 10^{13}$	2.0	0.04
	XXXI	Aitne	730.18	23231	0.264		$4.5 \cdot 10^{13}$	1.5	0.04
	XXXII	Eurydome	717.33	22685	0.276		$4.5 \cdot 10^{13}$	1.5	0.04
	XXXIII	Euanthe	620.49	20721	0.232		$4.5 \cdot 10^{13}$	1.5	0.04
	XXXIV	Euporie	550.74	19302	0.144		$1.5 \cdot 10^{13}$	1	0.04
	XXXV	Orthosie	622.56	20721	0.281		$1.5 \cdot 10^{13}$	1	0.04

Planet		Satellite	Orb. period/ Earth days	Distance/ 10^3 km	Eccentricity	Inclination	Mass/kg	Radius/km	Albedo
	XXXVI	Sponde	748.34	23487	0.312		$1.5\cdot10^{13}$	1	0.04
	XXXVII	Kale	729.47	23217	0.260		$1.5\cdot10^{13}$	1	0.04
	XXXVIII	Pasithee	719.44	23096	0.267		$1.5\cdot10^{13}$	1	0.04
	XXXIX	Hegemone	739.6	23947	0.328		$4.5\cdot10^{13}$	1.5	
	XL	Mneme	620.0	21069	0.227		$1.5\cdot10^{13}$	1	
	XLI	Aoede	761.5	23981	0.432		$9.0\cdot10^{13}$	2.0	
	XLII	Thelxinoe	628.1	21162	0.221		$1.5\cdot10^{13}$	1	
	XLIII	Arche	723.9	22931	0.259		$4.5\cdot10^{13}$	1.5	
	XLIV	Kallichore	764.7	24043	0.264		$1.5\cdot10^{13}$	1	
	XLV	Helike	634.8	21263	0.156		$9.0\cdot10^{13}$	2.0	
	XLVI	Carpo	456.1	16989	0.430		$4.5\cdot10^{13}$	1.5	
	XLVII	Eukelade	746.4	23661	0.272		$9.0\cdot10^{13}$	2.0	
	XLVIII	Cyllene	737.8	24349	0.319		$1.5\cdot10^{13}$	1	
	XLIX	Kore	779.2	24543	0.325			1	
Saturn	I	Mimas	0.942421813	185.52	0.0202	1.53°	$3.75\cdot10^{19}$	196	0.5
	II	Enceladus	1.370217855	238.02	0.00452	1.86°	$6.50\cdot10^{19}$	250	1.0
	III	Tethys	1.887802160	294.66	0.00000	1.86°	$6.27\cdot10^{20}$	530	0.9
	IV	Dione	2.736914742	377.40	0.002230	0.02°	$1.10\cdot10^{21}$	560	0.7
	V	Rhea	4.517500436	527.04	0.00100	0.35°	$2.31\cdot10^{21}$	765	0.7
	VI	Titan	15.94542068	1221.83	0.029192	0.33°	$1.3455\cdot10^{23}$	2575	0.21
	VII	Hyperion	21.2766088	1481.1	0.104	0.43°	$1.59\cdot10^{19}$	205×130×110	0.3
	VIII	Iapetus	79.3301825	3561.3	0.02828	14.72°	$1.59\cdot10^{21}$	730	0.6
	IX	Phoebe	550.31 R	12952	0.16326	177°	$7.2\cdot10^{18}$	110	0.08
	X	Janus	0.6945	151.472	0.007	0.14°	$1.92\cdot10^{18}$	110×100×80	0.6
	XI	Epimetheus	0.6942	151.422	0.009	0.34°	$5.4\cdot10^{17}$	70×60×50	0.5
	XII	Helene	2.7369	377.40	0.005	0.0°	$2.5\cdot10^{16}$	18×16×15	0.6
	XIII	Telesto	1.8878	294.66			$7.2\cdot10^{15}$	17×14×13	1.0
	XIV	Calypso	1.8878	294.66			$3.6\cdot10^{15}$	17×11×11	0.7
	XV	Atlas	0.6019	137.670		0.3°	$1.1\cdot10^{16}$	20×10	0.4
	XVI	Prometheus	0.6130	139.353	0.0024	0.0°	$3.3\cdot10^{17}$	70×50×40	0.6
	XVII	Pandora	0.6285	141.70	0.0042	0.0°	$1.9\cdot10^{17}$	55×45×35	0.5
	XVIII	Pan	0.5750	133.583			$2.7\cdot10^{15}$	10	0.5
	XIX	Ymir	1315.14	23096	0.470		$4.9\cdot10^{15}$	8	0.06
	XX	Paaliaq	686.95	15199	0.364		$8.2\cdot10^{15}$	9.5	0.06
	XXI	Tarvos	926.23	18247	0.536		$2.7\cdot10^{15}$	6.5	0.06
	XXII	Ijiraq	451.42	11440	0.322		$1.2\cdot10^{15}$	5	0.06
	XXIII	Suttungr	1016.67	19463	0.114		$2.1\cdot10^{14}$	2.8	0.06
	XXIV	Kiviuq	449.22	11365	0.334		$3.3\cdot10^{15}$	7	0.06
	XXV	Mundilfari	952.77	18709	0.208		$2.1\cdot10^{14}$	2.8	0.06
	XXVI	Albiorix	783.45	16404	0.478		$2.1\cdot10^{16}$	13	0.06
	XXVII	Skathi	728.20	15647	0.270		$3.1\cdot10^{14}$	3.2	0.06
	XXVIII	Erriapus	871.19	17616	0.474		$7.6\cdot10^{14}$	4.3	0.06
	XXIX	Siarnaq	895.53	18160	0.295		$3.9\cdot10^{16}$	16	0.06
	XXX	Thrymr	1094.11	20382	0.470		$2.1\cdot10^{14}$	2.8	0.06
	XXXI	Narvi	1003.86	19007	0.431		$4.9\cdot10^{15}$	3.3	0.04
	XXXII	Methone	1.010	194			$1.65\cdot10^{13}$	1.5	
	XXXIII	Pallene	1.154	211			$3.92\cdot10^{13}$	2	
	XXXIV	Polydeuces	2.737	377.4				4	
	XXXV	Daphnis	0.594	136.5				3.5	
	XXXVI	Aegir	1117.52	20735				3.5	
	XXXVII	Bebhionn	834.84	17119				3	
	XXXVIII	Bergelmir	1005.74	19338				3	
	XXXIX	Bestla	1088.72	20129				3.5	
	XL	Farbauti	1085.55	20390				2.5	
	XLI	Fenrir	1260.35	22453				2	
	XLII	Fornjot	1494.20	25108				3	
	XLIII	Hati	1038.61	19856				3	
	XLIV	Hyrrokkin	931.86	18437				4	
	XLV	Kari	1230.97	22118				3.5	
	XLVI	Loge	1311.36	23065				3	

Planet		Satellite	Orb. period/ Earth days	Distance/ 10^3 km	Eccentricity	Inclination	Mass/kg	Radius/km	Albedo
	XLVII	Skoll	878.29	17665				3	
	XLVIII	Surtur	1297.36	22707				3	
	XLIX	Anthe		197.7				1	
	L	Jarnsaxa		18600				3	
	LI	Greip		18105				3	
	LII	Tarqeq		19720				3.5	
Uranus	I	Ariel	2.52037935	191.02	0.0034	0.3°	$1.35 \cdot 10^{21}$	579	0.39
	II	Umbriel	4.1441772	266.30	0.0050	0.36°	$1.17 \cdot 10^{21}$	584.7	0.21
	III	Titania	8.7058717	435.91	0.0022	0.14°	$3.52 \cdot 10^{21}$	788.9	0.27
	IV	Oberon	13.4632389	583.52	0.0008	0.10°	$3.01 \cdot 10^{21}$	761.4	0.23
	V	Miranda	1.41347925	129.39	0.0027	4.2°	$6.59 \cdot 10^{19}$	236	0.32
	VI	Cordelia	0.335033	49.77	0.0003	0.1°	$5.4 \cdot 10^{16}$	20.1	0.07
	VII	Ophelia	0.376409	53.79	0.0099	0.1°	$5.4 \cdot 10^{16}$	21.4	0.07
	VIII	Bianca	0.434577	59.17	0.0009	0.2°	$9.3 \cdot 10^{16}$	25.7	0.07
	IX	Cressida	0.463570	61.78	0.0004	0.0°	$3.4 \cdot 10^{17}$	39.8	0.07
	X	Desdemona	0.473651	62.68	0.0001	0.2°	$1.8 \cdot 10^{17}$	32.0	0.07
	XI	Juliet	0.493066	64.35	0.0007	0.1°	$5.6 \cdot 10^{17}$	46.8	0.07
	XII	Portia	0.513196	66.09	0.0001	0.1°	$1.7 \cdot 10^{18}$	67.6	0.07
	XIII	Rosalind	0.558459	69.94	0.0001	0.3°	$2.6 \cdot 10^{17}$	36.0	0.07
	XIV	Belinda	0.623525	75.26	0.0001	0.0°	$3.6 \cdot 10^{17}$	40.3	0.07
	XV	Puck	0.761832	86.01	0.0001	0.31°	$2.9 \cdot 10^{18}$	81.0	0.07
	XVI	Caliban	579.73	7231	0.1587		$7.4 \cdot 10^{17}$	49	0.07
	XVII	Sycorax	1288.30	12179	0.5224		$5.4 \cdot 10^{18}$	95	0.07
	XVIII	Prospero	1978.29	16256	0.4448		$2.1 \cdot 10^{16}$	15	0.07
	XIX	Setebos	2225.21	17418	0.5914		$2.1 \cdot 10^{16}$	15	0.07
	XX	Stephano	677.36	8004	0.2292		$6.0 \cdot 10^{15}$	10	0.07
	XXI	Trinculo	749.24	8504	0.2200		$7.5 \cdot 10^{14}$	5	0.04
	XXII	Francisco	266.56	4276	0.146		$1.3 \cdot 10^{15}$	11	
	XXIII	Margaret	1687.01	14345	0.661		$1.0 \cdot 10^{15}$	5.5	
	XXIV	Ferdinand	2887.21	20901	0.368		$1.3 \cdot 10^{15}$	6	
	XXV	Perdita	0.638	76.42	0.0		$4.0 \cdot 10^{17}$	40	
	XXVI	Mab	0.923	97.73	0.0		$4.0 \cdot 10^{15}$	8	
	XXVII	Cupid	0.613	74.8	0.0		$1.2 \cdot 10^{15}$	6	
Neptune	I	Triton	5.8768541 R	354.76	0.000016	157.345°	$2.147 \cdot 10^{22}$	1353.4	0.76
	II	Nereid	360.13619	5513.4	0.7512	27.6°	$3.1 \cdot 10^{19}$	170	0.15
	III	Naiad	0.294396	48.227	0.0003	4.74°	$1.3 \cdot 10^{17}$	33	0.07
	IV	Thalassa	0.311485	50.075	0.0002	0.21°	$3.5 \cdot 10^{17}$	41	0.09
	V	Despina	0.334655	52.526	0.0001	0.07°	$2.3 \cdot 10^{18}$	75	0.09
	VI	Galatea	0.428745	61.953	0.0001	0.05°	$2.7 \cdot 10^{18}$	88	0.08
	VII	Larissa	0.554654	73.548	0.0014	0.20°	$4.8 \cdot 10^{18}$	104×89	0.09
	VIII	Proteus	1.122315	117.647	0.0004	0.55°	$4.9 \cdot 10^{19}$	218×208×201	0.10
	IX	Halimede	1879.08	16611	0.2646				
	X	Psamathe	9074.30	48096	0.3809		$1.5 \cdot 10^{16}$	14	
	XI	Sao	2912.72	22228	0.1365				
	XII	Laomedeia	3171.33	23567	0.3969				
	XIII	Neso	9740.73	49285	0.5714				
Pluto*	I	Charon	6.387	17.536	0.0022	99°	$1.6 \cdot 10^{21}$	593	0.37
	II	Nix	24.86	48.708	0.0030		$5 \cdot 10^{16}$	22–65	
	III	Hydra	38.20	64.749	0.0051		$5 \cdot 10^{16}$	22–65	
Eris*	I	Dysnomia		30				100–200	

* In June 2008 the International Astronomical Union decided on the name *plutoid* for the category of transneptunian dwarf planets. Plutoids are celestial bodies in orbit around the sun at a semimajor axis greater than that of Neptune and sufficiently massive to adopt a near-spherical shape. See <www.iau.org/public_press/news/release/iau0804/>.

INTERSTELLAR MOLECULES

Frank J. Lovas, Anthony J. Remijan, and Lewis E. Snyder

A number of molecules have been detected in the interstellar medium, in circumstellar envelopes around evolved stars, and comae and tails of comets through observation of their microwave, infrared, or optical spectra. The following list gives the molecules and the particular isotopic species that have been reported so far. Molecules are listed by molecular formula in the Hill order.* All species not footnoted otherwise are observed in interstellar clouds, while some are also found in comets and circumstellar clouds. The list was last updated in October 2016 and lists 203 molecules (393 isotopic forms). See Ref. 1 for transition frequencies reported for a typical molecular cloud and Ref. 2 for a discussion of comet observations. Laboratory measurements and predicted frequencies may be found in Ref. 3.

*The molecular formula gives the number of atoms of each element present in the molecule, without regard to the way in which the atoms are bonded. The Hill order specifies that C (if present) comes first; H (if present) is second; and the other element symbols follow in alphabetical order.

References

1. F. J. Lovas, Recommended Rest Frequencies for Observed Interstellar Molecule Microwave Transitions - 2002 Revision, *J. Phys. Chem. Ref. Data* 33, 177-355 (2004); and last update appearing at <www.nist.gov/pml/data/micro/index.cfm>.
2. Anita L. Cochran, Anny-Chantal Levasseur-Regourd, Martin Cordiner, Edith Hadamcik, Jérémie Lasue, Adeline Gicquel, David G. Schleicher, Steven B. Charnley, Michael J. Mumma, Lucas Paganini, Dominique Bockelée-Morvan, Nicolas Biver, and Yi-Jehng Kuan, The Composition of Comets, *Space Science Reviews*, pp 1-38, First online: 25 July 2015: <link.springer.com/article/10.1007%2Fs11214-015-0183-6>.
3. Splatalogue: <www.splatalogue.net>.

Interstellar Molecules Observed to Date (October 2016)

Mol. form.	Name	Isotopic species*
AlCl	Aluminum monochloride	$Al^{35}Cl$[a]
		$Al^{37}Cl$[a]
AlF	Aluminum monofluoride	AlF[a]
AlHO	Aluminum hydroxide	$^{27}AlOH$[a]
AlO	Aluminum monoxide	AlO[a]
ArH^+	Argon monohydride ion	$^{36}ArH^+$
		$^{38}ArH^+$
CAlN	Aluminum isocyanide	AlNC[a]
CF^+	Fluoromethylidynium ion	CF^+
CH	Methylidyne	CH
		^{13}CH
CH^+	Methyliumylidene	CH^+
		$^{13}CH^+$
CHMgN	Magnesium hydride isocyanide	HMgNC[a]
CHN	Hydrogen cyanide	HCN
		$H^{13}CN$
		$HC^{15}N$
		$H^{13}C^{15}N$
		DCN
CHN	Hydrogen isocyanide	HNC
		$H^{15}NC$
		$HN^{13}C$
		DNC
		$D^{15}NC$
CHNO	Cyanic acid	HOCN
CHNO	Isocyanic acid	HNCO
		$HN^{13}CO$
		$HNC^{18}O$
		DNCO
CHNO	Fulminic acid	HCNO
CHNS	Thiocyanic acid	HSCN
CHNS	Isothiocyanic acid	HNCS
CHO	Oxomethyl	HCO
CHO^+	Oxomethylium	HCO^+
		$H^{13}CO^+$
		$HC^{17}O^+$
		$HC^{18}O^+$
		DCO^+
		$D^{13}CO^+$
CHO^+	Hydroxymethylidyne	HOC^+
		$HO^{13}C^+$
CHO_2^+	Hydroxyoxomethylium	$HOCO^+$
		$DOCO^+$
CHP	Phosphaethyne	HCP
CHS^+	Thiooxomethylium	HCS^+
		DCS^+
CH_2	Methylene	CH_2
CH_2N^+	Iminomethylium	$HCNH^+$
CH_2N	Methylene amidogen	H_2CN
CH_2NO^+	*N*-Protonated isocyanic acid	H_2NCO^+
CH_2N_2	Cyanamide	NH_2CN
		$NH_2{}^{13}CN$
CH_2N_2	Methane diimine	HNCNH
CH_2O	Formaldehyde	H_2CO
		$H_2{}^{13}CO$
		$H_2C^{18}O$
		HDCO
		D_2CO
CH_2O_2	Formic acid	HCOOH
		$H^{13}COOH$
		HCOOD
		DCOOH
CH_2S	Thioformaldehyde	H_2CS
		$H_2{}^{13}CS$
		$H_2C^{34}S$
		HDCS

Mol. form.	Name	Isotopic species*
		D_2CS
CH_3	Methyl	CH_3
CH_3N	Methanimine	CH_2NH
		$^{13}CH_2NH$
CH_3NO	Formamide	NH_2CHO
		$NH_2{}^{13}CHO$
		NH_2CDO
		s-NHDCHO
		a-NHDCHO
CH_3O	Methoxy	CH_3O
CH_3O^+	Hydroxy methylium ion	H_2COH^+
CH_4	Methane	CH_4
CH_4O	Methanol	CH_3OH
		$^{13}CH_3OH$
		$CH_3{}^{18}OH$
		CH_2DOH
		CH_3OD
		CHD_2OH
		CD_3OH
CH_4S	Methanethiol	CH_3SH
CH_5N	Methylamine	CH_3NH_2
CFeN	Iron cyanide	FeCN
CKN	Potassium cyanide	KCN
CMgN	Magnesium cyanide	MgCN[a]
CMgN	Magnesium isocyanide	$^{24}MgNC$[a]
		$^{25}MgNC$[a]
		$^{26}MgNC$[a]
CN	Cyanide radical	CN
		^{13}CN
		$C^{15}N$
CN^+	Cyanide cation	CN^+[b]
CN^-	Cyanide anion	CN^-[a]
CNNa	Sodium cyanide	NaCN[a]
CNSi	Silicon cyanide	SiCN[a]
CNSi	Silicon isocyanide	SiNC[a]
CN_2	Cyanoimidogen	NCN[b]
CO	Carbon monoxide	CO
		^{13}CO
		$C^{17}O$
		$C^{18}O$
		$^{13}C^{17}O$
		$^{13}C^{18}O$
CO^+	Carbon monoxide ion	CO^+
COS	Carbon oxysulfide	OCS
		$OC^{33}S$
		$OC^{34}S$
		$O^{13}CS$
		$O^{13}C^{34}S$
		^{18}OCS
CO_2	Carbon dioxide	CO_2
CO_2^+	Carbon dioxide ion	CO_2^+[b]
CP	Carbon phosphide	CP[a]
CS	Carbon monosulfide	CS
		$C^{33}S$
		$C^{34}S$
		$C^{36}S$
		^{13}CS
		$^{13}C^{34}S$
CSi	Silicon carbide	SiC
CSi_2	Disilylidynemethylene	SiCSi[a]
C_2	Dicarbon	C_2

Mol. form.	Name	Isotopic species*
C_2H	Ethynyl	C_2H
		^{13}CCH
		$C^{13}CH$
		C_2D
C_2HN	Cyanomethylene	HCCN
C_2HNO	Cyanoformaldehyde	CNCHO
$C_2HN_2^+$	Protonated cyanogen	$NCCNH^+$
C_2HO	Ketenyl radical	HCCO
C_2H_2	Acetylene	HCCH[a]
		$HC^{13}CH$[a]
C_2H_2N	Cyanomethyl	CH_2CN
$C_2H_2N_2$	*E*-Cyanomethanimine	*E*-HNCHCN
C_2H_2O	Ketene	H_2CCO
		$H_2{}^{13}CCO$
		$H_2C^{13}CO$
C_2H_3N	Acetonitrile	CH_3CN
		$^{13}CH_3CN$
		$CH_3{}^{13}CN$
		$CH_3C^{15}N$
		$^{13}CH_3{}^{13}CN$
		CH_2DCN
C_2H_3N	Isocyanomethane	CH_3NC
C_2H_3N	Ketenimine	CH_2CNH
C_2H_3NO	Methyl isocyanate	CH_3NCO
C_2H_4	Ethylene	H_2CCH_2
$C_2H_4N_2$	Aminoacetonitrile	NH_2CH_2CN
C_2H_4O	Acetaldehyde	CH_3CHO
C_2H_4O	Ethylene oxide	*c*-C_2H_4O
C_2H_4O	*anti*-Ethenol	*a*-CH_2CHOH
C_2H_4O	*syn*-Ethenol	*s*-CH_2CHOH
$C_2H_4O_2$	Methyl formate	CH_3OCHO
		$^{13}CH_3OCHO$
		$CH_3O^{13}CHO$
		CH_2DOCHO
		CH_3OCDO
		$CH_3{}^{18}OCHO$
		$CH_3OCH^{18}O$
$C_2H_4O_2$	Acetic acid	CH_3COOH
$C_2H_4O_2$	Glycolaldehyde	CH_2OHCHO
		$^{13}CH_2OHCHO$
		$CH_2OH^{13}CHO$
		CHDOHCHO
		CH_2ODCHO
		CH_2OHCDO
C_2H_5N	Ethanimine	CH_3CHNH
C_2H_5NO	Acetamide	CH_3CONH_2
C_2H_6	Ethane	CH_3CH_3[a]
C_2H_6O	*trans*-Ethanol	*t*-CH_3CH_2OH
		t-$^{13}CH_3CH_2OH$
		t-$CH_3{}^{13}CH_2OH$
C_2H_6O	*gauche*-Ethanol	*g*-CH_3CH_2OH
C_2H_6O	Dimethyl ether	CH_3OCH_3
		$CH_3O^{13}CH_3$
		CH_2DOCH_3
$C_2H_6O_2$	*aGg'*-Ethylene glycol	*aGg'*-$HOCH_2CH_2OH$
$C_2H_6O_2$	*gGg'*-Ethylene glycol	*gGg'*-$HOCH_2CH_2OH$
C_2N	Cyanomethylidyne	CCN[a]
C_2O	Oxoethenylidene	CCO
C_2P	Phosohaethenylidene	CCP[a]
C_2S	Thioxoethenylidene	CCS
		$CC^{34}S$

Geo-Astro

Mol. form.	Name	Isotopic species*
		^{13}CCS
		$C^{13}CS$
C_2Si	Silicon dicarbide	*c*-SiC_2
		c-$^{29}SiC_2$
		c-$^{30}SiC_2$
		c-$Si^{13}CC$
C_3	Tricarbon	C_3
C_3H	Cyclopropenylidyne	*c*-C_3H
		c-$CC^{13}CH$
C_3H	Propenylidyne	*l*-C_3H
C_3H^+	2-Propyn-1-ylium-1-ylidene	*l*-C_3H^+
C_3HN	Cyanoacetylene	HCCCN
		$H^{13}CCCN$
		$HC^{13}CCN$
		$HCC^{13}CN$
		$HCCC^{15}N$
		$H^{13}C^{13}CCN$
		$H^{13}CC^{13}CN$
		$HC^{13}C^{13}CN$
		DCCCN
C_3HN	Isocyanoacetylene	HCCNC
C_3HN	3-Imino-1,2-propadienylidene	HNCCC
C_3H_2	Cyclopropenylidene	*c*-C_3H_2
		c-$H^{13}CCCH$
		c-$HC^{13}CCH$
		c-C_3HD
		c-C_3D_2
C_3H_2	Propadienylidene	*l*-H_2CCC
		l-HDCCC
$C_3H_2N^+$	Protonated cyanoacetylene	$HCCCNH^+$
C_3H_2O	2-Propynal	HCCCHO
C_3H_2O	Cyclopropenone	*c*-C_3H_2O
C_3H_3N	Acrylonitrile (vinyl cyanide)	CH_2CHCN
		$^{13}CH_2CHCN$
		$CH_2{}^{13}CHCN$
C_3H_4	Propyne	CH_3CCH
		$CH_3C^{13}CH$
		$^{13}CH_3CCH$
		$CH_3{}^{13}CCH$
		CH_2DCCH
		CH_3CCD
C_3H_4O	Propenal	CH_2CHCHO
C_3H_5N	Propanenitrile (ethyl cyanide)	CH_3CH_2CN
		$^{13}CH_3CH_2CN$
		$CH_3{}^{13}CH_2CN$
		$CH_3CH_2{}^{13}CN$
		$^{13}CH_3{}^{13}CH_2CN$
		$^{13}CH_3CH_2{}^{13}CN$
		$CH_3{}^{13}CH_2{}^{13}CN$
		$CH_3CH_2C^{15}N$
		CH_2DCH_2CN
		CH_3CHDCN
C_3H_6	Propylene	CH_2CHCH_3
C_3H_6O	Acetone	$(CH_3)_2CO$
C_3H_6O	Propanal	CH_3CH_2CHO
C_3H_6O	Methyloxirane (propylene oxide)	$CH_3CH\_CH_2O\_$
$C_3H_6O_2$	Methyl acetate	CH_3COOCH_3
$C_3H_6O_2$	Ethyl formate	$HCOOCH_2CH_3$
C_3N	Cyanoethynyl	CCCN
		$^{13}CCCN$
		$C^{13}CCN$
		$CC^{13}CN$
C_3N^-	Cyanoethynyl anion	$CCCN^-$
C_3O	3-Oxo-1,2-propadienylidene	CCCO
C_3S	3-Thioxo-1,2-propadienylidene	CCCS
		$CCC^{34}S$
		$C^{13}CCS$
C_3Si	Silicon tricarbon	SiC_3[a]
C_4H	1,3-Butadiynyl radical	HCCCC
		$H^{13}CCCC$
		$HC^{13}CCC$
		$HCC^{13}CC$
		$HCCC^{13}C$
		DCCCC
C_4H^-	1,3-Butadiynyl anion	$HCCCC^-$
C_4HN	3-Cyano-1,2-propadienylidene	HCCCCN
C_4H_2	Butatrienylidene	H_2CCCC
C_4H_2	1,3-Butadiyne	HCCCCH[a]
C_4H_3N	2-Butynenitrile	CH_3CCCN
C_4H_3N	Cyanoallene	CH_2CCHCN
C_4H_7N	Isopropylcyanide	*i*-$CH_3CH(CN)CH_3$
C_4H_7N	Propyl cyanide	C_3H_7CN
C_4Si	Silicon tetracarbide	SiC_4[a]
		$SiCCC^{13}C$[a]
C_5	Pentacarbon	C_5[a]
C_5H	2,4-Pentadiynylidyne	HCCCCC
C_5HN	2,4-Pentadiynenitrile	HCCCCCN
		$H^{13}CCCCCN$
		$HC^{13}CCCCN$
		$HCC^{13}CCCN$
		$HCCC^{13}CCN$
		$HCCCC^{13}CN$
		DCCCCCN
C_5H_4	1,3-Pentadiyne	CH_3C_4H
C_5N	4-Cyano-1,3-butadiyn-1-yl	C_5N[a]
C_5N^-	2,4-Pentadiynenitrile anion	C_5N^-
C_5S	5-Thioxo-1,2,3,4-pentatetraen-1-ylidene	CCCCCS[a]
C_6H	1,3,5-Hexatriynyl	HCCCCCC
C_6H^-	1,3,5-Hexatriynyl anion	$HCCCCCC^-$
C_6H_2	1,3,5-Hexatriyne	HCCCCCCH
C_6H_2	1,2,3,4,5-hexapentaenylidene	$H_2CCCCCC$
C_6H_3N	Methylcyanodiacetylene	CH_3C_4CN
C_6H_6	Benzene	C_6H_6
C_7H	2,4,6-Heptatriynylidyne	HCCCCCCC
C_7HN	2,4,6-Heptatriynenitrile	HC_7N
C_7H_4	Methyltriacetylene	CH_3C_6H
C_8H	1,3,5,7-Octatetraynyl	HC_8
C_8H^-	1,3,5,7-Octatetraynyl anion	HC_8^-
C_9HN	2,4,6,8-Nonatetraynenitrile	HC_9N
$C_{11}HN$	2,4,6,8,10-Undecapentaynenitrile	$HC_{11}N$
C_{60}	Fullerene-C_{60}	C_{60}[a]
$C_{60}{}^+$	Fullerene-C_{60} cation	$C_{60}{}^+$
C_{70}	Fullerene-C_{70}	C_{70}[a]
ClH	Hydrogen chloride	$H^{35}Cl$
		$H^{37}Cl$
ClH^+	Hydrogen chloride ion	$H^{35}Cl^+$
$ClH_2{}^+$	Chloronium	$H_2{}^{35}Cl^+$
		$H_2{}^{37}Cl^+$
ClK	Potassium chloride	$K^{35}Cl$[a]
		$K^{37}Cl$[a]
ClNa	Sodium chloride	$Na^{35}Cl$[a]

Mol. form.	Name	Isotopic species*
		$Na^{37}Cl$[a]
FH	Hydrogen fluoride	HF
FeO	Iron monoxide	FeO
HLi	Lithium Hydride	^{7}LiH
HN	Imidogen	HN
		ND
HNO	Nitrosyl hydride	HNO
HN_2^+	Hydrodinotrogen(1+)	N_2H^+
		$^{15}NNH^+$
		$N^{15}NH^+$
		N_2D^+
HO	Hydroxyl	OH
		^{17}OH
		^{18}OH
HO^+	Oxoniumylidene	OH^+
HO_2	Hydroperoxo	HO_2
HS	Mercapto	SH
HS^+	Hydrosulfur(1+)	SH^+
H_2	Hydrogen	H_2
		HD
H_2N	Amidogen	NH_2
H_2O	Water	H_2O
		$H_2{}^{17}O$
		$H_2{}^{18}O$
		HDO
		$HD^{18}O$
		D_2O
H_2O^+	Oxoniumyl	H_2O^+
H_2O_2	Hydrogen peroxide	HOOH
H_2S	Hydrogen sulfide	H_2S
		$H_2{}^{34}S$
		HDS
		D_2S
H_3^+	Trihydrogen ion	H_3^+
		H_2D^+
		D_2H^+
H_3N	Ammonia	NH_3
		$^{15}NH_3$
		NH_2D
		$^{15}NH_2D$
		NHD_2
		ND_3
H_3O^+	Oxonium	H_3O^+
H_3P	Phosphine	PH_3[a]
H_4N^+	Ammonium-*d*	NH_3D^+
H_4Si	Silane	SiH_4[a]
NO	Nitric oxide	NO
NO^+	Nitric oxide cation	NO^+
NP	Phosphorus nitride	PN
NS	Nitrogen sulfide	NS
		$N^{34}S$
NSi	Silicon nitride	SiN
N_2	Nitrogen	N_2
N_2^+	Nitrogen ion	N_2^+[b]
N_2O	Nitrous oxide	N_2O
OP	Phosphorus monoxide	PO[a]
OS	Sulfur monoxide	SO
		^{34}SO
		^{33}SO
		$S^{17}O$
		$S^{18}O$
OS^+	Sulfur monoxide ion	SO^+
OSi	Silicon monoxide	SiO
		$Si^{17}O$
		$Si^{18}O$
		^{29}SiO
		^{30}SiO
OTi	Titanium monoxide	TiO[a]
O_2	Oxygen	O_2
O_2S	Sulfur dioxide	SO_2
		$^{33}SO_2$
		$^{34}SO_2$
		$OS^{17}O$
		$OS^{18}O$
O_2Ti	Titanium dioxide	TiO_2[a]
SSi	Silicon monosulfide	SiS
		$Si^{33}S$
		$Si^{34}S$
		^{29}SiS
		$^{29}Si^{34}S$
		^{30}SiS
		$^{30}Si^{34}S$
S_2	Sulfur	S_2[b]

* *l*- before the isotopic species indicates a linear configuration, while *c*- indicates a cyclic molecule.

_ Underscore in formula indicates a ring compound.

[a] Reported only in circumstellar clouds.

[b] Reported only in comets.

MASS, DIMENSIONS, AND OTHER PARAMETERS OF THE EARTH

This table is a collection of data on various properties of the Earth. Most of the values are given in SI units. Note that 1 au (astronomical unit) = 149,597,870 km.

References

1. Seidelmann, P. K., *Ed., Explanatory Supplement to the Astronomical Almanac*, University Science Books, Mill Valley, CA, 1992.
2. Lang, K. R., *Astrophysical Data: Planets and Stars*, Springer-Verlag, New York, 1992.

Quantity	Symbol	Value	Unit
Mass	M	$5.9723 \cdot 10^{27}$	g
Major orbital semi-axis	a_{orb}	1.000000	au
		$1.4959787 \cdot 10^{8}$	km
Distance from Sun at perihelion	r_{π}	0.9833	au
Distance from Sun at aphelion	r_{α}	1.0167	au
Moment of perihelion passage	T_{π}	Jan. 2, 4 h 52 min	
Moment of aphelion passage	T_{α}	July 4, 5 h 05 min	
Siderial rotation period around Sun	P_{orb}	$31.5581 \cdot 10^{6}$	s
		365.25636	d
Mean rotational velocity	U_{orb}	29.78	km/s
Mean equatorial radius	$\bar{a}$	6378.140	km
Mean polar compression (flattening factor)	α	1/298.257	
Difference in equatorial and polar semi-axes	$a - c$	21.385	km
Compression of meridian of major equatorial axis	α_a	1/295.2	
Compression of meridian of minor equatorial axis	α_b	1/298.0	
Equatorial compression	ε	1/30 000	
Difference in equatorial semi-axes	$a - b$	213	m
Difference in polar semi-axes	$c_N - c_S$	~70	m
Polar asymmetry	η	$\sim 1.10^{-5}$	
Mean acceleration of gravity at equator	g_e	9.78036	m/s²
Mean acceleration of gravity at poles	g_p	9.83208	m/s²
Difference in acceleration of gravity at pole and at equator	$g_p - g_e$	5.172	cm/s²
Mean acceleration of gravity for entire surface of terrestrial ellipsoid	g	9.7978	m/s²
Mean radius	R	6371.0	km
Area of surface	S	$5.10 \cdot 10^{8}$	km²
Volume	V	$1.0832 \cdot 10^{12}$	km³
Mean density	ρ	5.515	g/cm³
Siderial rotational period	P	86,164.09	s
Rotational angular velocity	ω	$7.292116 \cdot 10^{-5}$	rad/s
Mean equatorial rotational velocity	v	0.46512	km/s
Rotational angular momentum	L	$5.861 \cdot 10^{33}$	J s
Rotational energy	E	$2.137 \cdot 10^{29}$	J
Ratio of centrifugal force to force of gravity at equator	q_c	0.0034677 = 1/288	
Moment of inertia	I	$8.070 \cdot 10^{37}$	kg m²
Relative braking of Earth's rotation due to tidal friction	$\Delta\omega_e/\omega$	$-4.2 \cdot 10^{-8}$	century⁻¹
Relative secular acceleration of Earth's rotation	$\Delta\omega_i/\omega$	$+1.4 \cdot 10^{-8}$	century⁻¹
Not secular braking of Earth's rotation	$\Delta\omega/\omega$	$-2.8 \cdot 10^{-8}$	century⁻¹
Probable value of total energy of tectonic deformation of Earth	E_t	$\sim 1 \cdot 10^{23}$	J/century
Secular loss of heat of Earth through radiation into space	$\Delta' E_k$	$1 \cdot 10^{23}$	J/century
Portion of Earth's kinetic energy transformed into heat as a result of lunar and solar tides in the hydrosphere	$\Delta'' E_k$	$1.3 \cdot 10^{23}$	J/century
Differences in duration of days in March and August	ΔP	0.0025 (March-August)	s
Corresponding relative annual variation in Earth's rotational velocity	$\Delta^* \omega/\omega$	$2.9 \cdot 10^{-8}$ (Aug.-March)	
Presumed variation in Earth's radius between August and March	$\Delta^* R$	−9.2 (Aug.-March)	cm
Annual variation in level of world ocean	Δh_o	~10 (Sept.-March)	cm
Area of continents	S_C	$1.49 \cdot 10^{8}$	km²
		29.2	% of surface
Area of world ocean	S_O	$3.61 \cdot 10^{8}$	km²
		70.8	% of surface
Mean height of continents above sea level	h_C	875	m
Mean depth of world ocean	h_O	3794	m
Mean thickness of lithosphere within the limits of the continents	$h_{C.l.}$	35	km
Mean thickness of lithosphere within the limits of the ocean	$h_{O.l.}$	4.7	km
Mean rate of thickening of continental lithosphere	$\Delta h/\Delta t$	10–40	m/10⁶ y
Mean rate of horizontal extension of continental lithosphere	$\Delta l/\Delta t$	0.75–20	km/10⁶ y
Mass of crust	m_l	$2.36 \cdot 10^{22}$	kg
Mass of mantle		$4.05 \cdot 10^{24}$	kg
Amount of water released from the mantle and core in the course of geological time		$3.40 \cdot 10^{21}$	kg
Total reserve of water in the mantle		$2 \cdot 10^{23}$	kg
Present content of free and bound water in the Earth's lithosphere		$2.4 \cdot 10^{21}$	kg
Mass of hydrosphere	m_h	$1.664 \cdot 10^{21}$	kg
Amount of oxygen bound in the Earth's crust		$1.300 \cdot 10^{21}$	kg
Amount of free oxygen		$1.5 \cdot 10^{18}$	kg
Mass of atmosphere	m_a	$5.136 \cdot 10^{18}$	kg
Mass of biosphere	m_b	$1.148 \cdot 10^{16}$	kg
Mass of living matter in the biosphere		$3.6 \cdot 10^{14}$	kg
Density of living matter on dry land		0.1	g/cm²
Density of living matter in ocean		$15 \cdot 10^{-8}$	g/cm³
Age of the Earth		$4.55 \cdot 10^{9}$	y
Age of oldest rocks		$4.0 \cdot 10^{9}$	y
Age of most ancient fossils		$3.4 \cdot 10^{9}$	y

GEOLOGICAL TIME SCALE

Period or epoch	Beginning and end, in 10^6 years	Key events
Cenozoic era		
Quaternary		
Holocene	0–0.0115	
Pleistocene*	0.0115–1.81	Homo erectus breakout
Tertiary		
Pliocene*	1.81–5.3	Ape man fossils
Miocene	5.3–23	Origin of grass
Oligocene	23–34	Rise of cats, dogs, pigs
Eocene	34–56	Debut of hoofed mammals
Paleocene	56–65	Earliest primates
Mesozoic era		
Cretaceous	65–145	Demise of dinosaurs
Jurassic	145–200	First birds
Triassic	200–251	Appearance of dinosaurs
Paleozoic era		
Permian	251–299	Flowers, insect pollination
Carboniferous	299–359	First conifers
Devonian	359–416	First vertebrates ashore
Silurian	416–444	Spore-bearing plants
Ordovician	444–488	First animals ashore
Cambrian	488–542	Vertebrates appear
Pre-Cambrian		
Pre-Cambrian III (Proterozoic)	542–2500	First plants, jellyfish
Pre-Cambrian II (Archean)	2500–3850	Photosynthetic bacteria
Pre-Cambrian I (Hadean)	3850–4600	Earth formed 4600 million years ago

* Some authorities place the boundary between the Pleistocene and Pliocene at $2.6{\cdot}10^6$ years.

References

1. U.S. Geological Survey Geologic Names Committee, 2007, Divisions of geologic time—Major chronostratigraphic and geochronologic units: U.S. Geological Survey Fact Sheet 2007-3015. Available on the Internet at <pubs.usgs.gov/fs/2007/3015/index.html>.
2. Walker, J. D., and Geissman, J. W., compilers, 2009, Geologic Time Scale: Geological Society of America, <www.geosociety.org/science/timescale/timescl.pdf>.
3. Calder, N., *Timescale - An Atlas of the Fourth Dimension*, Viking Press, New York, 1983.

ACCELERATION DUE TO GRAVITY

The acceleration due to gravity is tabulated here as a function of latitude and height above the Earth's surface. Values were calculated from the expression

$$g/(\mathrm{m/s^2}) = 9.780356\,(1 + 0.0052885 \sin^2 \phi - 0.0000059 \sin^2 2\,\phi) - 0.003086\,H$$

where ϕ is the latitude and H is the height in kilometers.

ϕ	$H = 0$	$H = 1$ km	$H = 5$ km	$H = 10$ km
0	9.78036	9.77727	9.76493	9.74950
5	9.78075	9.77766	9.76532	9.74989
10	9.78191	9.77882	9.76648	9.75105
15	9.78381	9.78072	9.76838	9.75295
20	9.78638	9.78330	9.77095	9.75552
25	9.78956	9.78647	9.77413	9.75870
30	9.79324	9.79016	9.77781	9.76238
35	9.79732	9.79424	9.78189	9.76646
40	9.80167	9.79858	9.78624	9.77081
45	9.80616	9.80307	9.79073	9.77530
50	9.81065	9.80757	9.79522	9.77979
55	9.81501	9.81193	9.79958	9.78415
60	9.81911	9.81602	9.80368	9.78825
65	9.82281	9.81972	9.80738	9.79195
70	9.82601	9.82292	9.81058	9.79515
75	9.82860	9.82551	9.81317	9.79774
80	9.83051	9.82743	9.81508	9.79965
85	9.83168	9.82860	9.81625	9.80082
90	9.83208	9.82899	9.81665	9.80122

Reference

Jursa, A. S., Ed., *Handbook of Geophysics and the Space Environment*, 4th Ed., Air Force Geophysics Laboratory, 1985, pp. 14–17.

DENSITY, PRESSURE, AND GRAVITY AS A FUNCTION OF DEPTH WITHIN THE EARTH

This table gives the density ρ, pressure p, and acceleration due to gravity g as a function of depth below the Earth's surface, as calculated from the model of the structure of the Earth in Reference 1. The model assumes a radius of 6371 km for the Earth. The boundary between the crust and mantle (the Mohorovicic discontinuity) is taken as 21 km, while in reality it varies considerable with location.

References

1. Anderson, D. L., and Hart, R. S., *J. Geophys. Res.* 81, 1461, 1976.
2. Carmichael, R. S., *CRC Practical Handbook of Physical Properties of Rocks and Minerals*, CRC Press, Boca Raton, FL, 1989, p. 467.

Depth km	ρ g/cm³	p kbar	g cm/s²
Crust			
0	1.02	0	981
3	1.02	3	982
3	2.80	3	982
21	2.80	5	983
Mantle (solid)			
21	3.49	5	983
41	3.51	12	983
61	3.52	19	984
81	3.48	26	984
101	3.44	33	984
121	3.40	39	985
171	3.37	56	987
221	3.34	73	989
271	3.37	89	991
321	3.47	106	993
371	3.59	124	994
571	3.95	199	999
871	4.54	328	997
1171	4.67	466	992
1471	4.81	607	991
1771	4.96	752	994
2071	5.12	903	1002
2371	5.31	1061	1017
2671	5.45	1227	1042
2886	5.53	1352	1069
Outer core (liquid)			
2886	9.96	1352	1069
2971	10.09	1442	1050
3371	10.63	1858	953
3671	11.00	2154	874
4071	11.36	2520	760
4471	11.69	2844	641
4871	11.99	3116	517
5156	12.12	3281	427
Inner core (solid)			
5156	12.30	3281	427
5371	12.48	3385	355
5771	12.52	3529	218
6071	12.53	3592	122
6371	12.58	3617	0

OCEAN PRESSURE AS A FUNCTION OF DEPTH AND LATITUDE

The following table is based upon an ocean model which takes into account the equation of state of standard seawater and the dependence on latitude of the acceleration of gravity. The tabulated pressure value is the excess pressure over the ambient atmospheric pressure at the surface.

References

1. *International Oceanographic Tables, Volume 4*, Unesco Technical Papers in Marine Science No. 40, Unesco, Paris, 1987.
2. Saunders, P. M., and Fofonoff, N. P., *Deep-Sea Res.* 23, 109–111, 1976.

Pressure in MPa at the Specified Latitude

Depth (meters)	0°	15°	30°	45°	60°	75°	90°
0	0.0000	0.0000	0.0000	0.0000	0.0000	0.0000	0.0000
500	5.0338	5.0355	5.0404	5.0471	5.0537	5.0586	5.0605
1000	10.0796	10.0832	10.0930	10.1064	10.1198	10.1296	10.1333
1500	15.1376	15.1431	15.1577	15.1778	15.1980	15.2127	15.2182
2000	20.2076	20.2148	20.2344	20.2613	20.2882	20.3080	20.3153
2500	25.2895	25.2985	25.3231	25.3568	25.3905	25.4153	25.4244
3000	30.3831	30.3940	30.4236	30.4641	30.5047	30.5345	30.5453
3500	35.4886	35.5012	35.5358	35.5832	35.6307	35.6654	35.6782
4000	40.6056	40.6201	40.6598	40.7140	40.7683	40.8082	40.8229
4500	45.7342	45.7505	45.7952	45.8564	45.9176	45.9626	45.9791
5000	50.8742	50.8924	50.9421	51.0102	51.0785	51.1285	51.1469
5500	56.0255	56.0456	56.1004	56.1755	56.2508	56.3059	56.3262
6000	61.1882	61.2100	61.2700	61.3521	61.4344	61.4947	61.5168
6500	66.3619	66.3857	66.4508	66.5399	66.6292	66.6947	66.7187
7000	71.5467	71.5724	71.6427	71.7388	71.8352	71.9059	71.9318
7500	76.7426	76.7701	76.8456	76.9488	77.0523	77.1282	77.1560
8000	81.9493	81.9788	82.0594	82.1697	82.2804	82.3614	82.3911
8500	87.1669	87.1983	87.2841	87.4016	87.5193	87.6057	87.6373
9000	92.3950	92.4284	92.5194	92.6440	92.7689	92.8606	92.8941
9500	97.6346	97.6698	97.7661	97.8978	98.0300	98.1269	98.1624
10000	102.8800	102.9170	103.0185	103.1572	103.2961	103.3981	103.4355

PROPERTIES OF SEAWATER

In addition to the dependence on temperature and pressure, the physical properties of seawater vary with the concentration of the dissolved constituents. A convenient parameter for describing the composition is the salinity, S, which is defined in terms of the electrical conductivity of the seawater sample. The defining equation for the practical salinity is:

$$S = a_0 + a_1K^{1/2} + a_2K + a_3K^{3/2} + a_4K^2 + a_5K^{5/2},$$

where K is the ratio of the conductivity of the seawater sample at 15 °C and atmospheric pressure to the conductivity of a potassium chloride solution in which the mass fraction of KCl is 0.0324356, at the same temperature and pressure. The values of the coefficients are:

$$a_0 = 0.0080$$
$$a_1 = -0.1692$$
$$a_2 = 25.3851$$
$$a_3 = 14.0941$$
$$a_4 = -7.0261$$
$$a_5 = 2.7081$$
$$\Sigma a_i = 35.0000$$

Thus when $K = 1$, $S = 35$ exactly (S is normally quoted in units of ‰, i.e., parts per thousand). The value of S can be roughly equated with the mass of dissolved material in grams per kilogram of seawater. Salinity values in the open oceans at midlatitudes typically fall between 34 and 36.

It is customary in oceanography to define the pressure at a given point as the pressure due to the column of water between that point and the surface. Thus by convention $P = 0$ at the sea surface. To a good approximation the pressure in decibars (dbar) can be equated to the depth in meters. Thus at 45° latitude the pressure is 5000 dbar at 4902 m, 10000 dbar at 9700 m.

The first table below gives several properties of seawater as a function of temperature for a salinity of 35. The second and third give density and electrical conductivity as a function of salinity at several temperatures, and the fourth lists typical concentrations of the main constituents of seawater as a function of salinity. The final table gives the freezing point as a function of salinity and pressure.

References

1. *The Practical Salinity Scale 1978 and the International Equation of State of Seawater 1980,* Unesco Technical Papers in Marine Science No. 36, Unesco, Paris, 1981; sections No. 37, 38, 39, and 40 in this series give background papers and detailed tables.
2. Kennish, M. J., *CRC Practical Handbook of Marine Science*, CRC Press, Boca Raton, FL, 1989.
3. Poisson, A. *IEEE J. Ocean. Eng.* OE-5, 50, 1981.
4. Webster, F., in *AIP Physics Desk Reference*, E. R. Cohen, D. R. Lide, and G. L. Trigg, Eds., Springer-Verlag, New York, 2003.

Properties of Seawater

ρ = density in g/cm³ $\beta = (1/\rho)\,(d\rho/dS)$ = fractional change in density per unit change in salinity
$\alpha = -(1/\rho)\,(d\rho/dt)$ = fractional change in density per unit change in temperature (°C⁻¹)
κ = electrical conductivity in S/cm
η = viscosity in mPa s (equal to cP)
c_p = specific heat in J/kg °C
ν = speed of sound in m/s

Properties of Seawater as a Function of Temperature at Salinity S = 35 and Normal Atmospheric Pressure

t/°C	ρ/g cm^{-3}	$10^7{\cdot}\beta$	$10^7{\cdot}\alpha$/°C^{-1}	κ/S cm^{-1}	η/mPa s	c_p/J kg^{-1} °C^{-1}	ν/m s^{-1}
0	1.028106	7854	526	0.029048	1.892	3986.5	1449.1
5	1.027675	7717	1136	0.033468	1.610		
10	1.026952	7606	1668	0.038103	1.388	3986.3	1489.8
15	1.025973	7516	2141	0.042933	1.221		
20	1.024763	7444	2572	0.047934	1.085	3993.9	1521.5
25	1.023343	7385	2970	0.053088	0.966		
30	1.021729	7338	3341	0.058373	0.871	4000.7	1545.6
35	1.019934	7300	3687				
40		7270	4004			4003.5	1563.2

Density of Surface Seawater in g/cm³ as a Function of Temperature and Salinity

t/°C	S = 0	S = 5	S = 10	S = 15	S = 20	S = 25	S = 30	S = 35	S = 40
0	0.999843	1.003913	1.007955	1.011986	1.016014	1.020041	1.024072	1.028106	1.032147
5	0.999967	1.003949	1.007907	1.011858	1.015807	1.019758	1.023714	1.027675	1.031645
10	0.999702	1.003612	1.007501	1.011385	1.015269	1.019157	1.023051	1.026952	1.030862
15	0.999102	1.002952	1.006784	1.010613	1.014443	1.018279	1.022122	1.025973	1.029834
20	0.998206	1.002008	1.005793	1.009576	1.013362	1.017154	1.020954	1.024763	1.028583
25	0.997048	1.000809	1.004556	1.008301	1.012050	1.015806	1.019569	1.023343	1.027128
30	0.995651	0.999380	1.003095	1.006809	1.010527	1.014252	1.017985	1.021729	1.025483
35	0.994036	0.997740	1.001429	1.005118	1.008810	1.012509	1.016217	1.019934	1.023662
40	0.992220	0.995906	0.999575	1.003244	1.006915	1.010593	1.014278	1.017973	1.021679

Electrical Conductivity of Seawater in S/cm as a Function of Temperature and Salinity

t/°C	S = 5	S = 10	S = 15	S = 20	S = 25	S = 30	S = 35	S = 40
0	0.004808	0.009171	0.013357	0.017421	0.021385	0.025257	0.029048	0.032775
5	0.005570	0.010616	0.015441	0.020118	0.024674	0.029120	0.033468	0.037734
10	0.006370	0.012131	0.017627	0.022947	0.028123	0.033171	0.038103	0.042935
15	0.007204	0.013709	0.019905	0.025894	0.031716	0.037391	0.042933	0.048355
20	0.008068	0.015346	0.022267	0.028948	0.035438	0.041762	0.047934	0.053968
25	0.008960	0.017035	0.024703	0.032097	0.039276	0.046267	0.053088	0.059751
30	0.009877	0.018771	0.027204	0.035330	0.043213	0.050888	0.058373	0.065683

Composition of Seawater and Ionic Strength at Various Salinities (Ref. 2)

Constituent	Molality S = 30	Molality S = 35	Molality S = 40	g/kg of Seawater S = 30	g/kg of Seawater S = 35	g/kg of Seawater S = 40
Cl^-	0.482	0.562	0.650	16.58	19.33	22.36
Br^-	0.00074	0.00087	0.00100	0.057	0.067	0.078
F^-		0.00007			0.001	
SO_4^{2-}	0.0104	0.0114	0.0122	0.97	1.06	1.14
HCO_3^-	0.00131	0.00143	0.00100	0.078	0.085	0.059
$NaSO_4^-$	0.0085	0.0108	0.0139	0.98	1.25	1.60
KSO_4^-	0.00010	0.00012	0.00015	0.013	0.016	0.020
Na^+	0.405	0.472	0.544	9.03	10.53	12.13
K^+	0.00892	0.01039	0.01200	0.338	0.394	0.455
Mg^{2+}	0.0413	0.0483	0.0561	0.974	1.139	1.323
Ca^{2+}	0.00131	0.00143	0.00154	0.051	0.056	0.060
Sr^{2+}	0.00008	0.00009	0.00011	0.007	0.008	0.009
$MgHCO_3^+$	0.00028	0.00036	0.00045	0.023	0.030	0.037
$MgSO_4$	0.00498	0.00561	0.00614	0.582	0.655	0.717
$CaSO_4$	0.00102	0.00115	0.00126	0.135	0.152	0.166
$NaHCO_3$	0.00015	0.00020	0.00024	0.012	0.016	0.020
H_3BO_3	0.00032	0.00037	0.00042	0.019	0.022	0.025
Ionic strength	0.5736	0.6675	0.7701			

Freezing Point of Seawater in °C as a Function of Salinity and Pressure

P/dbar[a]	S = 0	S = 5	S = 10	S = 15	S = 20	S = 25	S = 30	S = 35	S = 40
0	0.000	−0.274	−0.542	−0.812	−1.083	−1.358	−1.638	−1.922	−2.212
50	−0.038	−0.311	−0.580	−0.849	−1.121	−1.396	−1.676	−1.960	−2.250
100	−0.075	−0.349	−0.618	−0.887	−1.159	−1.434	−1.713	−1.998	−2.287
500	−0.377	−0.650	−0.919	−1.188	−1.460	−1.735	−2.014	−2.299	−2.589

[a] Pressure above atmospheric

ABUNDANCE OF ELEMENTS IN THE EARTH'S CRUST AND IN THE SEA

This table gives the estimated abundance of the elements in the continental crust (in mg/kg, equivalent to parts per million by mass) and in seawater near the surface (in mg/L). Values represent the median of reported measurements. The concentrations of the less abundant elements may vary with location by several orders of magnitude.

References

1. Carmichael, R. S., Ed., *CRC Practical Handbook of Physical Properties of Rocks and Minerals,* CRC Press, Boca Raton, FL, 1989.
2. Bodek, I., et al., *Environmental Inorganic Chemistry,* Pergamon Press, New York, 1988.
3. Ronov, A. B., and Yaroshevsky, A. A., "Earth's Crust Geochemistry," in *Encyclopedia of Geochemistry and Environmental Sciences,* Fairbridge, R. W., Ed., Van Nostrand, New York, 1969.

	Abundance	
Element	Crust mg/kg	Sea mg/L
Ac	5.5×10^{-10}	
Ag	7.5×10^{-2}	4×10^{-5}
Al	8.23×10^{4}	2×10^{-3}
Ar	3.5	4.5×10^{-1}
As	1.8	3.7×10^{-3}
Au	4×10^{-3}	4×10^{-6}
B	1.0×10^{1}	4.44
Ba	4.25×10^{2}	1.3×10^{-2}
Be	2.8	5.6×10^{-6}
Bi	8.5×10^{-3}	2×10^{-5}
Br	2.4	6.73×10^{1}
C	2.00×10^{2}	2.8×10^{1}
Ca	4.15×10^{4}	4.12×10^{2}
Cd	1.5×10^{-1}	1.1×10^{-4}
Ce	6.65×10^{1}	1.2×10^{-6}
Cl	1.45×10^{2}	1.94×10^{4}
Co	2.5×10^{1}	2×10^{-5}
Cr	1.02×10^{2}	3×10^{-4}
Cs	3	3×10^{-4}
Cu	6.0×10^{1}	2.5×10^{-4}
Dy	5.2	9.1×10^{-7}
Er	3.5	8.7×10^{-7}
Eu	2.0	1.3×10^{-7}
F	5.85×10^{2}	1.3
Fe	5.63×10^{4}	2×10^{-3}
Ga	1.9×10^{1}	3×10^{-5}
Gd	6.2	7×10^{-7}
Ge	1.5	5×10^{-5}
H	1.40×10^{3}	1.08×10^{5}
He	8×10^{-3}	7×10^{-6}
Hf	3.0	7×10^{-6}
Hg	8.5×10^{-2}	3×10^{-5}
Ho	1.3	2.2×10^{-7}
I	4.5×10^{-1}	6×10^{-2}
In	2.5×10^{-1}	2×10^{-2}
Ir	1×10^{-3}	
K	2.09×10^{4}	3.99×10^{2}
Kr	1×10^{-4}	2.1×10^{-4}
La	3.9×10^{1}	3.4×10^{-6}
Li	2.0×10^{1}	1.8×10^{-1}
Lu	8×10^{-1}	1.5×10^{-7}
Mg	2.33×10^{4}	1.29×10^{3}
Mn	9.50×10^{2}	2×10^{-4}
Mo	1.2	1×10^{-2}
N	1.9×10^{1}	5×10^{-1}
Na	2.36×10^{4}	1.08×10^{4}
Nb	2.0×10^{1}	1×10^{-5}
Nd	4.15×10^{1}	2.8×10^{-6}
Ne	5×10^{-3}	1.2×10^{-4}
Ni	8.4×10^{1}	5.6×10^{-4}
O	4.61×10^{5}	8.57×10^{5}
Os	1.5×10^{-3}	
P	1.05×10^{3}	6×10^{-2}
Pa	1.4×10^{-6}	5×10^{-11}
Pb	1.4×10^{1}	3×10^{-5}
Pd	1.5×10^{-2}	
Po	2×10^{-10}	1.5×10^{-14}
Pr	9.2	6.4×10^{-7}
Pt	5×10^{-3}	
Ra	9×10^{-7}	8.9×10^{-11}
Rb	9.0×10^{1}	1.2×10^{-1}
Re	7×10^{-4}	4×10^{-6}
Rh	1×10^{-3}	
Rn	4×10^{-13}	6×10^{-16}
Ru	1×10^{-3}	7×10^{-7}
S	3.50×10^{2}	9.05×10^{2}
Sb	2×10^{-1}	2.4×10^{-4}
Sc	2.2×10^{1}	6×10^{-7}
Se	5×10^{-2}	2×10^{-4}
Si	2.82×10^{5}	2.2
Sm	7.05	4.5×10^{-7}
Sn	2.3	4×10^{-6}
Sr	3.70×10^{2}	7.9
Ta	2.0	2×10^{-6}
Tb	1.2	1.4×10^{-7}
Te	1×10^{-3}	
Th	9.6	1×10^{-6}
Ti	5.65×10^{3}	1×10^{-3}
Tl	8.5×10^{-1}	1.9×10^{-5}
Tm	5.2×10^{-1}	1.7×10^{-7}
U	2.7	3.2×10^{-3}
V	1.20×10^{2}	2.5×10^{-3}
W	1.25	1×10^{-4}
Xe	3×10^{-5}	5×10^{-5}
Y	3.3×10^{1}	1.3×10^{-5}
Yb	3.2	8.2×10^{-7}
Zn	7.0×10^{1}	4.9×10^{-3}
Zr	1.65×10^{2}	3×10^{-5}

SOLAR IRRADIANCE AT THE EARTH

The solar luminosity (total radiant power emitted by the Sun) is $3.828 \cdot 10^{26}$ W, of which about 1361 W m^{-2} (the solar irradiance or "solar constant") reaches the top of the Earth's atmosphere (Ref. 6). To a zeroth approximation, the sun can be considered a black body with an effective temperature of 5780 K, which implies a peak in the radiation at around 0.520 μm (5200 Å). The actual solar spectral emission is more complex, especially at ultraviolet and shorter wavelengths. The graph in Figure 1, which was taken from Ref. 1, summarizes the solar irradiance at the top of the atmosphere in the range 0.3 μm to 10 μm.

The solar irradiance undergoes both long-term and short-term variations. Figure 2, which is taken from Ref. 4, shows the short-term variation over the 1976-2010 period. At the time of those measurements the mean value of the solar constant was believed to be about 1366 W/m^2. The more recent measurements in Ref. 6 show that the mean value (which is more difficult to measure than the short-term variations) is closer to 1361 W/m^2. Thus the curve in Figure 1 should be shifted down by about 5 W/m^2.

References

1. Jursa, A. S., Ed., *Handbook of Geophysics and the Space Environment*, Air Force Geophysics Laboratory, 1985.
2. Pierce, A. K., and Allen, R. G., "The Solar Spectrum between 0.3 and 10 μm", in *The Solar Output and Its Variation*, White, O. R., Ed., Colorado Associated University Press, Boulder, CO, 1977.
3. Lang, K. R., *Astrophysical Data. Planets and Stars*, Springer-Verlag, New York, 1992.
4. Hansen, J. E., and Sato, M., <www.columbia.edu/~mhs119/Solar>, 2011.
5. Frohlich, C., and Lean, J., *Astron. Astrophys. Rev.* 12, 273, 2004.
6. Kopp, G., and Lean, J. L., *Geophys. Res. Lett.* 38 (1), 2011. DOI: 10.1029/2010GL045777.

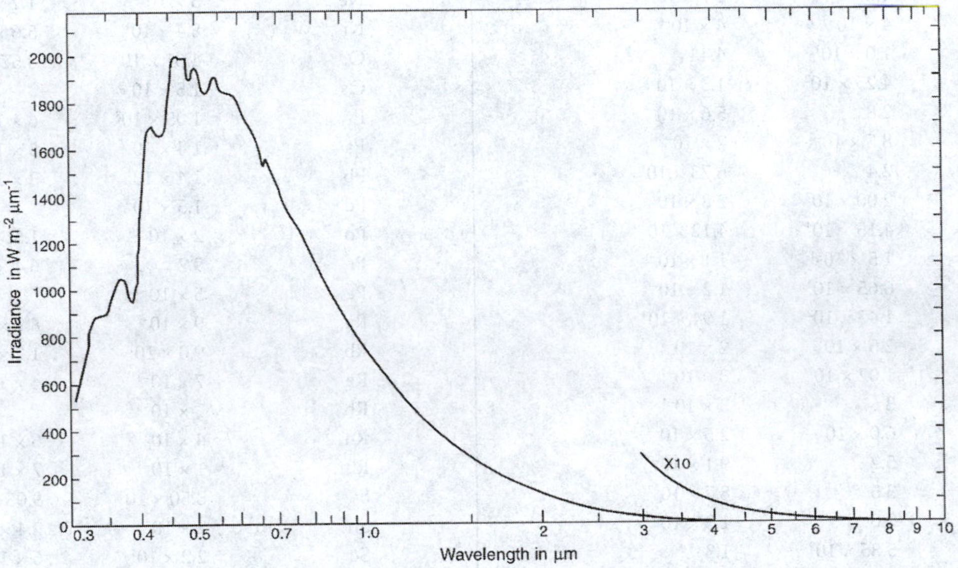

FIGURE 1. Wavelength dependence of solar irradiance.

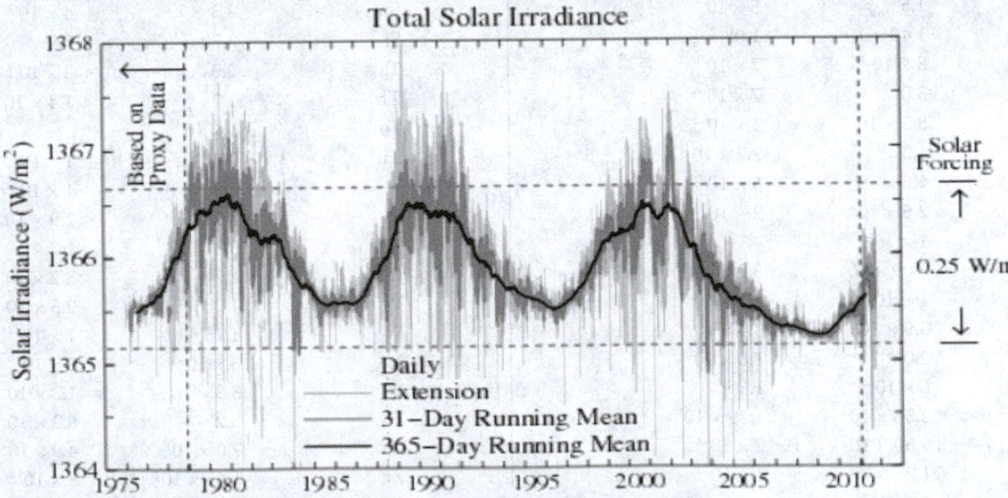

FIGURE 2. Variation of solar irradiance in the period 1976–2010.

U.S. STANDARD ATMOSPHERE (1976)

A Standard Atmosphere is a hypothetical vertical distribution of atmospheric temperature, pressure, and density that is roughly representative of year-round, midlatitude conditions. Typical uses are to serve as a basis for pressure altimeter calibrations, aircraft performance calculations, aircraft and rocket design, ballistic tables, meteorological diagrams, and various types of atmospheric modeling. The air is assumed to be dry and to obey the perfect gas law and the hydrostatic equation which, taken together, relate temperature, pressure, and density with vertical position. The atmosphere is considered to rotate with the Earth and to be an average over the diurnal cycle, the semiannual variation, and the range from active to quiet geomagnetic and sunspot conditions.

The U.S. Standard Atmosphere (1976) is an idealized, steady-state representation of mean annual conditions of the Earth's atmosphere from the surface to 1000 km at latitude 45° N, as it is assumed to exist during a period with moderate solar activity. The defining meteorological elements are sea-level temperature and pressure and a temperature-height profile to 1000 km. The 1976 Standard Atmosphere uses the following sea-level values that have been standard for many decades:

Temperature — 288.15 K (15 °C)
Pressure — 101325 Pa (1013.25 mbar, 760 mm of Hg, or 29.92 in. of Hg)
Density — 1225 g/m^3 (1.225 g/L)
Mean molar mass — 28.964 g/mol

The parameters included in this condensed version of the U.S. Standard Atmosphere are:

Z — Height (geometric) above mean sea level in meters
T — Temperature in kelvins
P — Pressure in pascals (1 Pa = 0.01 millibars)
ρ — Density in kilograms per cubic meter (1 kg/m^3 = 1 g/L)
n — Number density in molecules per cubic meter
ν — Mean collision frequency in collisions per second
l — Mean free path in meters
η — Absolute viscosity in pascal seconds (1 Pa s = 1000 cP)
k — Thermal conductivity in joules per meter second kelvin (W/m K)
v_s — Speed of sound in meters per second
g — Acceleration of gravity in meters per second square

The sea-level composition (percent by volume) is taken to be:

N_2 — 78.084%	He — 0.000524
O_2 — 20.9476	Kr — 0.000114
Ar — 0.934	Xe — 0.0000087
CO_2 — 0.0314	CH_4 — 0.0002
Ne — 0.001818	H_2 — 0.00005

The T and P columns for the troposphere and lower stratosphere were generated from the following formulas:

	T/K	P/Pa
$H \le 11000$ m	$288.15 - 0.0065\,H$	$101325(288.15/T)^{-5.25577}$
$11000 \text{ m} < H \le 20000$ m	216.65	$22632\,e^{-0.00015768832(H-11000)}$
$20000 \text{ m} < H \le 32000$ m	$216.65 + 0.0010(H-20000)$	$5474.87(216.65/T)^{34.16319}$

where $H = rZ/(r + Z)$ is the geopotential height in meters and r is the mean Earth radius at 45° N latitude, taken as 6356766 m. For altitudes up to 32 km, $\rho = 0.003483677(P/T)$ in the units used here. Formulas for the other quantities may be found in the references.

References

1. COESA, *U.S. Standard Atmosphere*, 1976, U.S. Government Printing Office, Washington, D.C., 1976.
2. Jursa, A. S., Ed., *Handbook of Geophysics and the Space Environment*, Air Force Geophysics Laboratory, 1985.

Z/m	T/K	P/Pa	ρ/kg m⁻³	n/m⁻³	ν/s⁻¹	l/m	η/Pa s	k/W m⁻¹ K⁻¹	ν_s/m s⁻¹	g/m s⁻²
–5000	320.68	1.778E+05	1.931	4.015E+25	1.151E+10	4.208E–08	1.942E–05	0.02788	359.0	9.822
–4500	317.42	1.685E+05	1.849	3.845E+25	1.096E+10	4.395E–08	1.927E–05	0.02763	357.2	9.830
–4000	314.17	1.596E+05	1.770	3.680E+25	1.044E+10	4.592E–08	1.912E–05	0.02738	355.3	9.819
–3500	310.91	1.511E+05	1.693	3.520E+25	9.933E+09	4.800E–08	1.897E–05	0.02713	353.5	9.818
–3000	307.66	1.430E+05	1.619	3.366E+25	9.448E+09	5.019E–08	1.882E–05	0.02688	351.6	9.816
–2500	304.41	1.352E+05	1.547	3.217E+25	8.982E+09	5.252E–08	1.867E–05	0.02663	349.8	9.814
–2000	301.15	1.278E+05	1.478	3.102E+25	8.623E+09	5.447E–08	1.852E–05	0.02638	347.9	9.813
–1500	297.90	1.207E+05	1.411	2.935E+25	8.106E+09	5.757E–08	1.836E–05	0.02613	346.0	9.811
–1000	294.65	1.139E+05	1.347	2.801E+25	7.693E+09	6.032E–08	1.821E–05	0.02587	344.1	9.810
–500	291.40	1.075E+05	1.285	2.672E+25	7.298E+09	6.324E–08	1.805E–05	0.02562	342.2	9.808
0	288.15	1.013E+05	1.225	2.547E+25	6.919E+09	6.633E–08	1.789E–05	0.02533	340.3	9.807
500	284.90	9.546E+04	1.167	2.427E+25	6.556E+09	6.961E–08	1.774E–05	0.02511	338.4	9.805
1000	281.65	8.988E+04	1.112	2.311E+25	6.208E+09	7.310E–08	1.758E–05	0.02485	336.4	9.804
1500	278.40	8.456E+04	1.058	2.200E+25	5.874E+09	7.680E–08	1.742E–05	0.02459	334.5	9.802
2000	275.15	7.950E+04	1.007	2.093E+25	5.555E+09	8.073E–08	1.726E–05	0.02433	332.5	9.801
2500	271.91	7.469E+04	0.957	1.990E+25	5.250E+09	8.491E–08	1.710E–05	0.02407	330.6	9.799
3000	268.66	7.012E+04	0.909	1.891E+25	4.959E+09	8.937E–08	1.694E–05	0.02381	328.6	9.797
3500	265.41	6.579E+04	0.863	1.795E+25	4.680E+09	9.411E–08	1.678E–05	0.02355	326.6	9.796
4000	262.17	6.166E+04	0.819	1.704E+25	4.414E+09	9.917E–08	1.661E–05	0.02329	324.6	9.794
4500	258.92	5.775E+04	0.777	1.616E+25	4.160E+09	1.046E–07	1.645E–05	0.02303	322.6	9.793
5000	255.68	5.405E+04	0.736	1.531E+25	3.918E+09	1.103E–07	1.628E–05	0.02277	320.6	9.791
5500	252.43	5.054E+04	0.697	1.450E+25	3.687E+09	1.165E–07	1.612E–05	0.02250	318.5	9.790
6000	249.19	4.722E+04	0.660	1.373E+25	3.467E+09	1.231E–07	1.595E–05	0.02224	316.5	9.788
6500	245.94	4.408E+04	0.664	1.299E+25	3.258E+09	1.302E–07	1.578E–05	0.02197	314.4	9.787
7000	242.70	4.111E+04	0.590	1.227E+25	3.058E+09	1.377E–07	1.561E–05	0.02170	312.3	9.785
7500	239.46	3.830E+04	0.557	1.159E+25	2.869E+09	1.458E–07	1.544E–05	0.02144	310.2	9.784
8000	236.22	3.565E+04	0.526	1.093E+25	2.689E+09	1.545E–07	1.527E–05	0.02117	308.1	9.782
8500	232.97	3.315E+04	0.496	1.031E+25	2.518E+09	1.639E–07	1.510E–05	0.02090	306.0	9.781
9000	229.73	3.080E+04	0.467	9.711E+24	2.356E+09	1.740E–07	1.493E–05	0.02063	303.9	9.779
9500	226.49	2.858E+04	0.440	9.141E+24	2.202E+09	1.848E–07	1.475E–05	0.02036	301.7	9.777
10000	223.25	2.650E+04	0.414	8.598E+24	2.056E+09	1.965E–07	1.458E–05	0.02009	299.5	9.776
10500	220.01	2.454E+04	0.389	8.079E+24	1.918E+09	2.091E–07	1.440E–05	0.01982	297.4	9.774
11000	216.77	2.270E+04	0.365	7.585E+24	1.787E+09	2.227E–07	1.422E–05	0.01954	295.2	9.773
11500	216.65	2.098E+04	0.337	7.016E+24	1.653E+09	2.408E–07	1.422E–05	0.01953	295.1	9.771
12000	216.65	1.940E+04	0.312	6.486E+24	1.528E+09	2.605E–07	1.422E–05	0.01953	295.1	9.770
12500	216.65	1.793E+04	0.288	5.996E+24	1.412E+09	2.818E–07	1.422E–05	0.01953	295.1	9.768
13000	216.65	1.658E+04	0.267	5.543E+24	1.306E+09	3.048E–07	1.422E–05	0.01953	295.1	9.767
13500	216.65	1.533E+04	0.246	5.124E+24	1.207E+09	3.297E–07	1.422E–05	0.01953	295.1	9.765
14000	216.65	1.417E+04	0.228	4.738E+24	1.116E+09	3.566E–07	1.422E–05	0.01953	295.1	9.764
14500	216.65	1.310E+04	0.211	4.380E+24	1.032E+09	3.857E–07	1.422E–05	0.01953	295.1	9.762
15000	216.65	1.211E+04	0.195	4.049E+24	9.538E+08	4.172E–07	1.422E–05	0.01953	295.1	9.761
16000	216.65	1.035E+04	0.166	3.461E+24	8.153E+08	4.881E–07	1.422E–05	0.01953	295.1	9.758
17000	216.65	8.850E+03	0.142	2.959E+24	6.969E+08	5.710E–07	1.422E–05	0.01953	295.1	9.754
18000	216.65	7.565E+03	0.122	2.529E+24	5.958E+08	6.680E–07	1.422E–05	0.01953	295.1	9.751
19000	216.65	6.467E+03	0.104	2.162E+24	5.093E+08	7.814E–07	1.422E–05	0.01953	295.1	9.748
20000	216.65	5.529E+03	8.891E–02	1.849E+24	4.354E+08	9.139E–07	1.422E–05	0.01953	295.1	9.745
21000	217.58	4.729E+03	7.572E–02	1.574E+24	3.716E+08	1.073E–06	1.427E–05	0.01961	295.1	9.742
22000	218.57	4.048E+03	6.451E–02	1.341E+24	3.173E+08	1.260E–06	1.432E–05	0.01970	296.4	9.739
23000	219.57	3.467E+03	5.501E–02	1.144E+24	2.712E+08	1.477E–06	1.438E–05	0.01978	297.1	9.736
24000	220.56	2.972E+03	4.694E–02	9.759E+23	2.319E+08	1.731E–06	1.443E–05	0.01986	297.7	9.733
25000	221.55	2.549E+03	4.008E–02	8.334E+23	1.985E+08	2.027E–06	1.448E–05	0.01995	298.4	9.730
26000	222.54	2.188E+03	3.426E–02	7.123E+23	1.700E+08	2.372E–06	1.454E–05	0.02003	299.1	9.727
27000	223.54	1.880E+03	2.930E–02	6.092E+23	1.458E+08	2.773E–06	1.459E–05	0.02011	299.7	9.724
28000	224.53	1.610E+03	2.508E–02	5.214E+23	1.250E+08	3.240E–06	1.465E–05	0.02020	300.4	9.721
29000	225.52	1.390E+03	2.148E–02	4.466E+23	1.073E+08	3.783E–06	1.470E–05	0.02028	301.1	9.718
30000	226.51	1.197E+03	1.841E–02	3.828E+23	9.219E+07	4.414E–06	1.475E–05	0.02036	301.7	9.715
31000	227.50	1.031E+03	1.579E–02	3.283E+23	7.925E+07	5.146E–06	1.481E–05	0.02044	302.4	9.712
32000	228.49	8.891E+02	1.356E–02	2.813E+23	6.818E+07	5.995E–06	1.486E–05	0.02053	303.0	9.709
33000	230.97	7.673E+02	1.157E–02	2.406E+23	5.852E+07	7.021E–06	1.499E–05	0.02073	304.7	9.706
34000	233.74	6.634E+02	9.887E–03	2.056E+23	5.030E+07	8.218E–06	1.514E–05	0.02096	306.5	9.703
35000	236.51	5.746E+02	8.463E–03	1.760E+23	4.331E+07	9.601E–06	1.529E–05	0.02119	308.3	9.700

Z/m	T/K	P/Pa	ρ/kg m^{-3}	n/m^{-3}	ν/s^{-1}	l/m	η/Pa s	k/W m^{-1} K^{-1}	v_s/m s^{-1}	g/m s^{-2}
36000	239.28	4.985E+02	7.258E−03	1.509E+23	3.736E+07	1.120E−05	1.543E−05	0.02142	310.1	9.697
38000	244.82	3.771E+02	5.367E−03	1.116E+23	2.794E+07	1.514E−05	1.572E−05	0.02188	313.7	9.690
40000	250.35	2.871E+02	3.996E−03	8.308E+22	2.104E+07	2.034E−05	1.601E−05	0.02233	317.2	9.684
42000	255.88	2.200E+02	2.995E−03	6.227E+22	1.594E+07	2.713E−05	1.629E−05	0.02278	320.7	9.678
44000	261.40	1.695E+02	2.259E−03	4.697E+22	1.215E+07	3.597E−05	1.657E−05	0.02323	324.1	9.672
46000	266.93	1.313E+02	1.714E−03	3.564E+22	9.318E+06	4.740E−05	1.685E−05	0.02376	327.5	9.666
48000	270.65	1.023E+02	1.317E−03	2.738E+22	7.208E+06	6.171E−05	1.704E−05	0.02397	329.8	9.660
50000	270.65	7.978E+01	1.027E−03	2.135E+22	5.620E+06	7.913E−05	1.703E−05	0.02397	329.8	9.654
52000	269.03	6.221E+01	8.056E−04	1.675E+22	4.397E+06	1.009E−04	1.696E−05	0.02384	328.8	9.648
54000	263.52	4.834E+01	6.390E−04	1.329E+22	3.452E+06	1.272E−04	1.660E−05	0.02340	325.4	9.642
56000	258.02	3.736E+01	5.045E−04	1.049E+22	2.696E+06	1.611E−04	1.640E−05	0.02296	322.0	9.636
58000	252.52	2.872E+01	3.963E−04	8.239E+21	2.095E+06	2.051E−04	1.612E−05	0.02251	318.6	9.632
60000	247.02	2.196E+01	3.097E−04	6.439E+21	1.620E+06	2.624E−04	1.584E−05	0.02206	315.1	9.624
65000	233.29	1.093E+01	1.632E−04	3.393E+21	8.294E+05	4.979E−04	1.512E−05	0.02093	306.2	9.609
70000	219.59	5.221	8.283E−05	1.722E+21	4.084E+05	9.810E−04	1.438E−05	0.01978	297.1	9.594
75000	208.40	2.388	3.992E−05	8.300E+20	1.918E+05	2.035E−03	1.376E−05	0.01883	289.4	9.579
80000	198.64	1.052	1.846E−05	3.838E+20	8.656E+04	4.402E−03	1.321E−05	0.01800	282.5	9.564
85000	188.89	4.457E−01	8.220E−06	1.709E+20	3.766E+04	9.886E−03	1.265E−05	0.01716	275.5	9.550
90000	186.87	1.836E−01	3.416E−06	7.116E+19	1.560E+04	2.370E−02				9.535
95000	188.42	7.597E−02	1.393E−06	2.920E+19	6.440E+03	5.790E−02				9.520
100000	195.08	3.201E−02	5.604E−07	1.189E+19	2.680E+03	1.420E−01				9.505
110000	240.00	7.104E−03	9.708E−08	2.144E+18	5.480E+02	7.880E−01				9.476
120000	360.00	2.538E−03	2.222E−08	5.107E+17	1.630E+02	3.310				9.447
130000	469.27	1.251E−03	8.152E−09	1.930E+17	7.100E+01	8.800				9.418
140000	559.63	7.203E−04	3.831E−09	9.322E+16	3.800E+01	1.800E+01				9.389
150000	634.39	4.542E−04	2.076E−09	5.186E+16	2.300E+01	3.300E+01				9.360
160000	696.29	3.040E−04	1.233E−09	3.162E+16	1.500E+01	5.300E+01				9.331
170000	747.57	2.121E−04	7.815E−10	2.055E+16	1.000E+01	8.200E+01				9.302
180000	790.07	1.527E−04	5.194E−10	1.400E+16	7.200	1.200E+02				9.274
190000	825.16	1.127E−04	3.581E−10	9.887E+15	5.200	1.700E+02				9.246
200000	854.56	8.474E−05	2.541E−10	7.182E+15	3.900	2.400E+02				9.218
220000	899.01	5.015E−05	1.367E−10	4.040E+15	2.300	4.200E+02				9.162
240000	929.73	3.106E−05	7.858E−11	2.420E+15	1.400	7.000E+02				9.106
260000	950.99	1.989E−05	4.742E−11	1.515E+15	9.300E−01	1.100E+03				9.051
280000	965.75	1.308E−05	2.971E−11	9.807E+14	6.100E−01	1.700E+03				8.997
300000	976.01	8.770E−06	1.916E−11	6.509E+14	4.200E−01	2.600E+03				8.943
320000	983.16	5.980E−06	1.264E−11	4.405E+14	2.900E−01	3.800E+03				8.889
340000	988.15	4.132E−06	8.503E−12	3.029E+14	2.000E−01	5.600E+03				8.836
360000	991.65	2.888E−06	5.805E−12	2.109E+14	1.400E−01	8.000E+03				8.784
380000	994.10	2.038E−06	4.013E−12	1.485E+14	1.000E−01	1.100E+04				8.732
400000	995.83	1.452E−06	2.803E−12	1.056E+14	7.200E−02	1.600E+04				8.680
450000	998.22	6.447E−07	1.184E−12	4.678E+13	3.300E−02	3.600E+04				8.553
500000	999.24	3.024E−07	5.215E−13	2.192E+13	1.600E−02	7.700E+04				8.429
550000	999.67	1.514E−07	2.384E−13	1.097E+13	8.400E−03	1.500E+05				8.307
600000	999.85	8.213E−08	1.137E−13	5.950E+12	4.800E−03	2.800E+05				8.188
650000	999.93	4.887E−08	5.712E−14	3.540E+12	3.100E−03	4.800E+05				8.072
700000	999.97	3.191E−08	3.070E−14	2.311E+12	2.200E−03	7.300E+05				7.958
750000	999.98	2.260E−08	1.788E−14	1.637E+12	1.700E−03	1.000E+06				7.846
800000	999.99	1.704E−08	1.136E−14	1.234E+12	1.400E−03	1.400E+06				7.737
850000	1000.00	1.342E−08	7.824E−15	9.717E+11	1.200E−03	1.700E+06				7.630
900000	1000.00	1.087E−08	5.759E−15	7.876E+11	1.000E−03	2.100E+06				7.525
950000	1000.00	8.982E−09	4.453E−15	6.505E+11	8.700E−04	2.600E+06				7.422
1000000	1000.00	7.514E−09	3.561E−15	5.442E+11	7.500E−04	3.100E+06				7.322

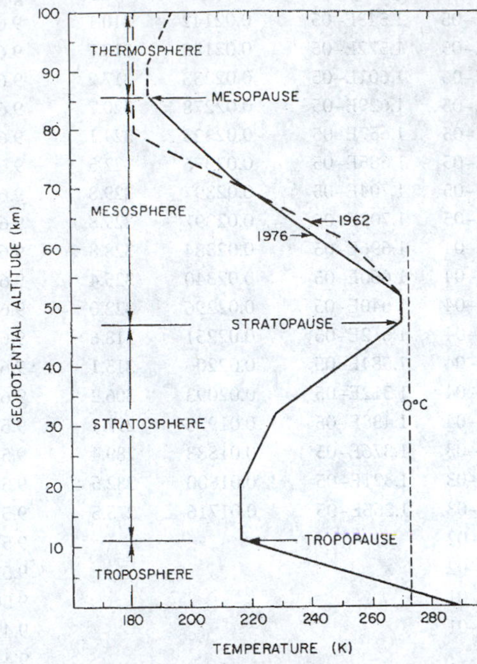

FIGURE 1. Temperature-height profile for U.S. Standard Atmosphere.

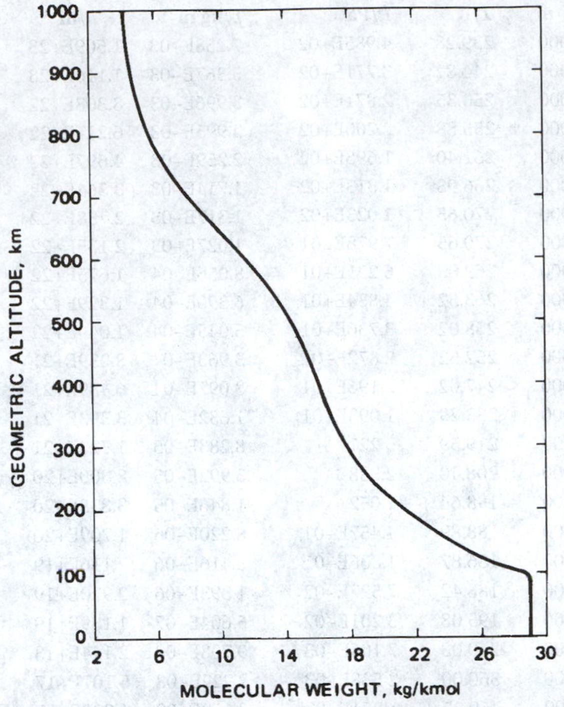

FIGURE 3. Mean molecular weight as a function of geometric altitude.

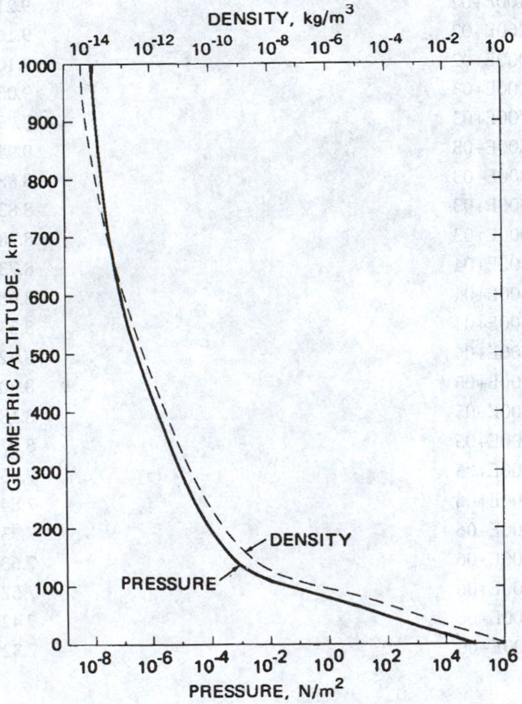

FIGURE 2. Total pressure and mass density as a function of geometric altitude.

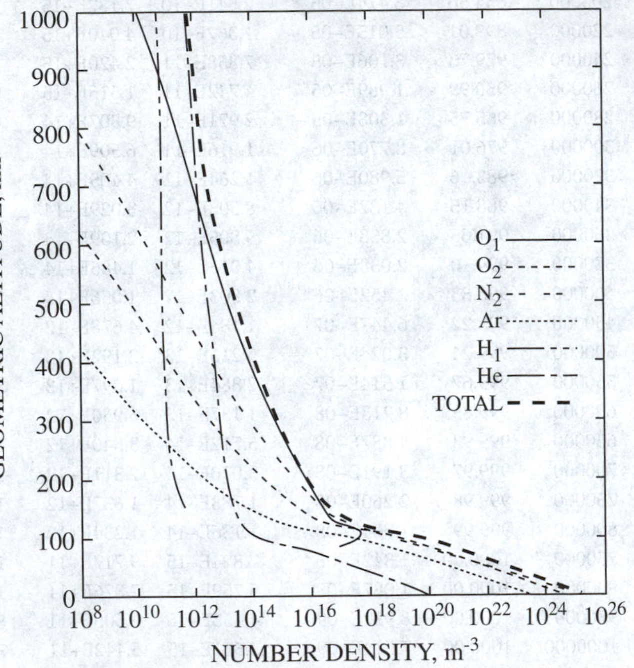

FIGURE 4. Number density of individual species and total number density as a function of geometric altitude.

Geo-Astro

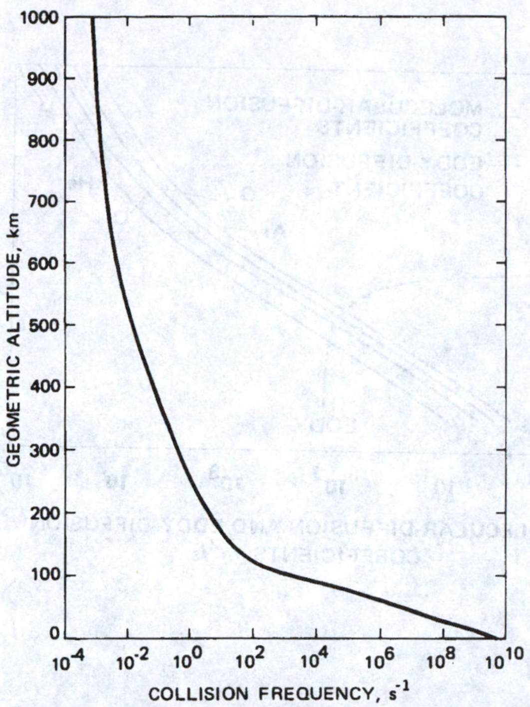

FIGURE 5. Collision frequency as a function of geometric altitude.

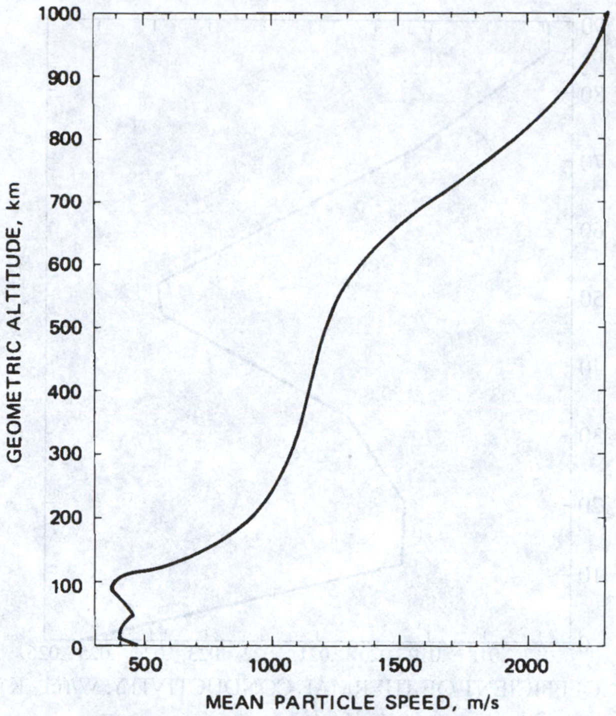

FIGURE 7. Mean air-particle speed as a function of geometric altitude.

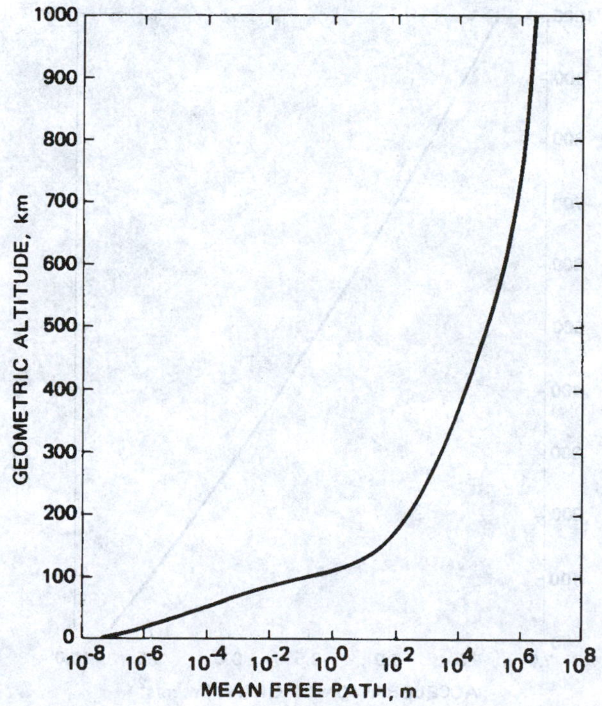

FIGURE 6. Mean free path as a function of geometric altitude.

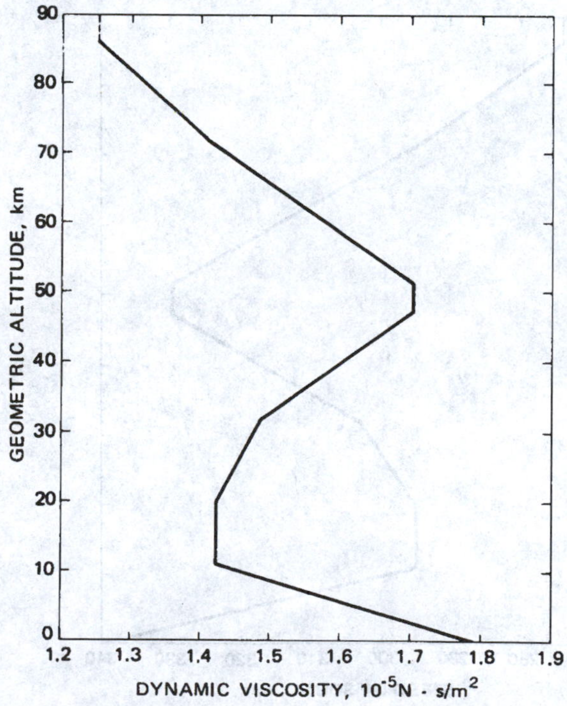

FIGURE 8. Dynamic viscosity as a function of geometric altitude.

Geo-Astro

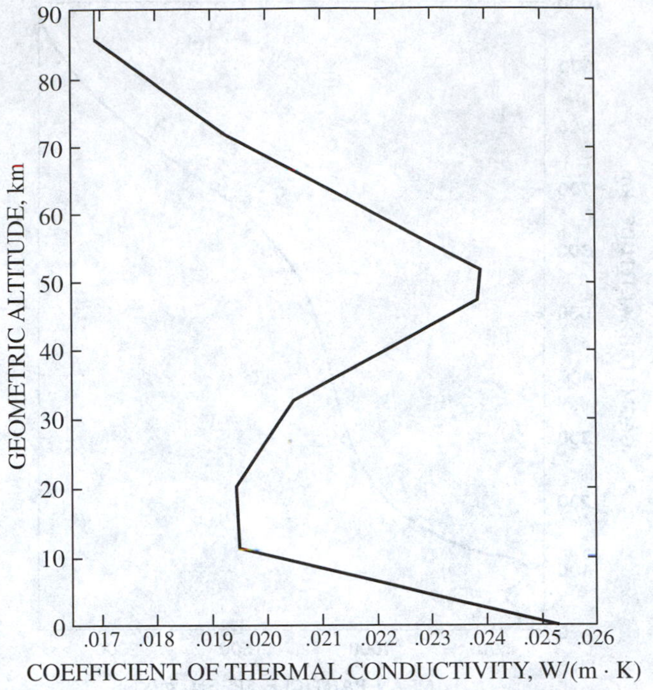

FIGURE 9. Coefficient of thermal conductivity as a function of geometric altitude.

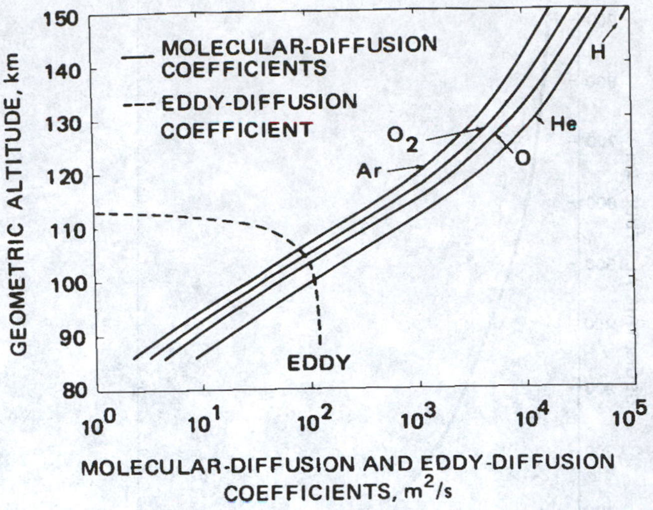

FIGURE 11. Molecular-diffusion and eddy-diffusion coefficients as a function of geometric altitude.

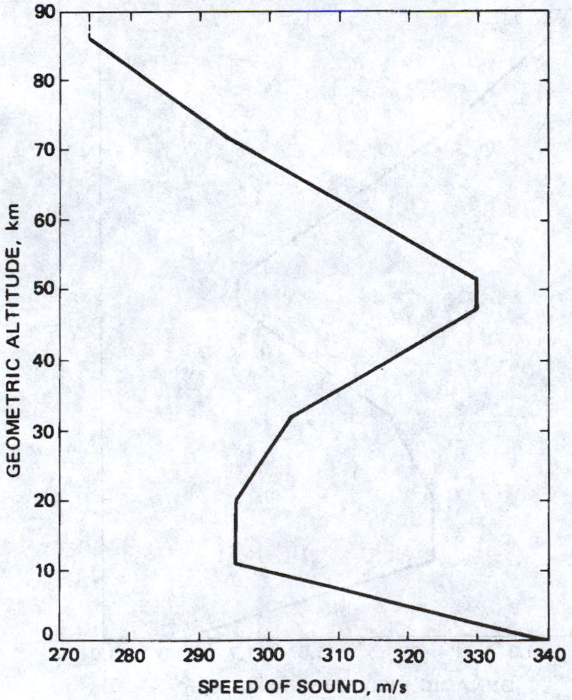

FIGURE 10. Speed of sound as a function of geometric altitude.

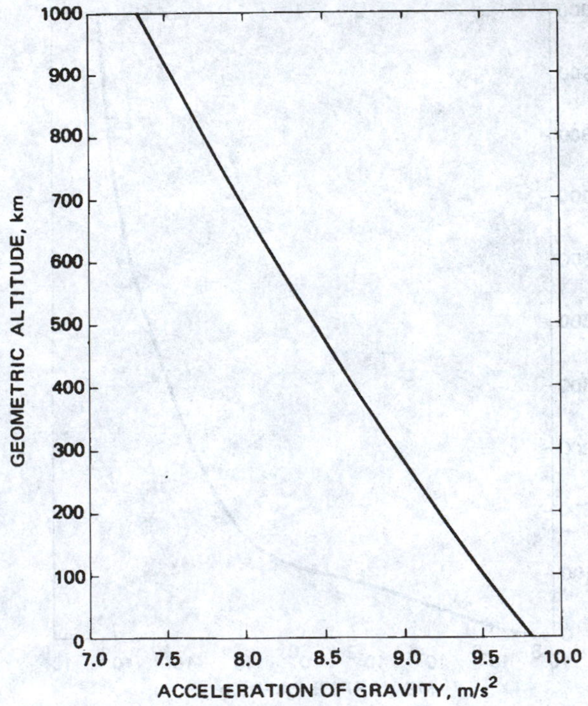

FIGURE 12. Acceleration of gravity as a function of geometric altitude.

GEOGRAPHICAL AND SEASONAL VARIATIONS IN SOLAR RADIATION

This table gives the amount of solar radiation reaching a unit area at the top of the Earth's atmosphere per day as a function of latitude and approximate date. It is based upon a solar constant (total energy per unit area at the Earth's average orbital distance) of 1373 W/m^2. Absorption of radiation by the atmosphere is not taken into consideration.

More recent measurements indicate the value of the solar constant is closer to 1361 W/m^2. Thus the values in this table should be reduced by about 1%.

Reference

List, R. J., *Smithsonian Meteorological Tables, Seventh Edition*, Smithsonian Institution Press, Washington, D.C., 1962.

Daily Solar Radiation in MJ/m^2

Lat.	Mar. 21	Apr. 13	May 6	May 29	Jun. 2	Jul. 15	Aug. 8	Aug. 31	Sep. 23	Oct. 16	Nov. 8	Nov. 30	Dec. 22	Jan. 13	Feb. 4	Feb. 26
90°		18.0	32.8	42.4	45.7	42.2	32.5	17.7								
80	6.6	18.0	32.3	41.8	45.0	41.6	32.0	17.7	6.5	0.3						0.3
70	13.0	22.3	31.8	39.9	43.0	39.7	31.5	22.0	12.9	5.5	1.0				1.0	5.6
60	19.0	27.0	34.4	39.7	41.6	39.4	34.0	26.7	18.8	11.6	6.2	3.1	2.1	3.1	6.2	11.7
50	24.4	31.1	36.8	40.7	42.0	40.5	36.5	30.8	24.1	17.6	12.1	8.7	7.5	8.7	12.3	17.8
40	29.1	34.3	38.6	41.3	42.1	41.1	38.3	33.9	28.7	23.1	18.2	14.8	13.5	14.9	18.4	23.5
30	32.9	36.7	39.4	41.1	41.4	40.8	39.1	36.3	32.5	28.2	23.9	20.9	19.8	21.0	24.1	28.4
20	35.7	38.0	39.2	39.7	39.7	39.5	38.9	37.5	35.3	32.3	29.1	26.6	25.7	26.7	29.3	32.7
10	37.4	38.1	37.9	37.4	37.1	37.2	37.6	37.7	37.0	35.5	33.5	31.8	31.1	31.9	33.8	35.9
0	38.0	37.1	35.5	34.1	33.5	34.0	35.2	36.6	37.6	37.6	36.9	36.1	35.8	36.3	37.3	38.0
-10	37.4	35.0	32.3	30.0	29.2	29.9	32.0	34.6	37.0	38.6	39.4	39.5	39.6	39.7	39.7	39.1
-20	35.7	31.8	28.0	25.2	24.1	25.1	27.8	31.5	35.3	38.5	40.7	42.0	42.4	42.2	41.1	39.0
-30	32.9	27.8	23.1	19.7	18.5	19.7	22.8	27.4	32.5	37.2	40.9	43.3	44.2	43.5	41.3	37.7
-40	29.1	22.8	17.5	14.0	12.6	13.9	17.4	22.6	28.7	34.8	40.1	43.6	45.0	43.8	40.5	35.2
-50	24.4	17.3	11.7	8.2	7.0	8.2	11.6	17.2	24.1	31.5	38.3	43.1	44.8	43.2	38.6	31.9
-60	19.0	11.4	5.9	2.9	2.0	2.9	5.9	11.3	18.8	27.3	35.7	41.9	44.4	42.1	36.0	27.7
-70	13.0	5.4	1.0				1.0	5.3	12.9	22.6	33.0	42.2	45.9	42.4	33.3	22.9
-80	6.6	0.3						0.3	6.5	18.2	33.5	44.2	48.1	44.4	33.8	18.4
-90										18.2	34.0	44.8	48.8	45.1	34.4	18.4

Geo-Astro

MAJOR WORLD EARTHQUAKES

The United States Geological Survey maintains a database of historic earthquakes throughout the world (Ref. 1). The table below is extracted from that database; it includes about 300 major earthquakes, based upon the magnitude and the degree of destruction. All recorded earthquakes of magnitude 7.5 or greater are listed, even if the fatalities are unknown or small. The death toll is often a rough estimate; in many cases the true toll could be much greater. More details on the exact location and degree of destruction can be found in Refs. 2 and 3.

The magnitude is given on the Richter scale, which was developed in 1935 by Charles F. Richter of the California Institute of Technology as a mathematical device to compare the size of earthquakes. The magnitude of an earthquake is measured by the logarithm of the amplitude of waves recorded by seismographs. Adjustments are included for the variation in the distance between the various seismographs and the epicenter of the earthquake. On the Richter scale, magnitude is expressed in whole numbers and decimal fractions, e.g., 6.3. Because of the logarithmic basis of the scale, each whole number increase in magnitude represents a tenfold increase in measured amplitude; as an estimate of energy, each whole number step in the magnitude scale corresponds to the release of about 31 times more energy than the amount associated with the preceding whole number value.

References

1. Earthquakes, <earthquake.usgs.gov/earthquakes/>
2. Most Destructive Known Earthquakes on Record in the World, <earthquake.usgs.gov/regional/world/most_destructive.php>.
3. Earthquakes with 1,000 or More Deaths since 1900, <earthquake.usgs.gov/regional/world/world_deaths.php>.

Major World Earthquakes, 856-2016

Date	Location	Magnitude	Fatalities
856/12/22	Damghan, Iran		200,000
893/03/23	Ardabil, Iran		150,000
1138/08/09	Aleppo, Syria		230,000
1268	Silicia, Asia Minor		60,000
1290/09/27	Chihli, China		100,000
1556/01/23	Shensi, China	8.0	830,000
1619/02/14	Trujillo, Peru	7.7	350
1667/11	Shemakha, Caucasia		80,000
1668/08/17	Anatolia, Turkey	8.0	
1687/10/20	Lima, Peru	8.5	
1693/01/11	Sicily, Italy	7.5	60,000
1700/01/26	Cascadia Subduction Zone (Oregon to British Columbia)	9.0	
1727/11/18	Tabriz, Iran		77,000
1730/07/08	Valparasio, Chile	8.7	
1755/11/01	Lisbon, Portugal	8.7	70,000
1783/02/04	Calabria, Italy		50,000
1787/05/02	Puerto Rico	8.0	
1811/12/16	New Madrid region, Missouri	8.1	
1812/02/07	New Madrid region, Missouri	8.0	
1812/03/26	Caracas, Venezuela	7.7	
1812/12/08	Southwest of San Bernardino County, California	6.9	40
1812/12/21	West of Ventura, California	7.1	1
1821/07/10	Camana, Peru	8.2	162
1835/02/20	Concepcion, Chile	8.2	500
1843/02/08	Leeward Islands	8.3	
1855/01/23	Wellington, New Zealand	8.0	4
1857/01/09	Fort Tejon, California	7.9	1
1868/04/03	Ka'u District, Island of Hawaii	7.9	77
1868/08/13	Arica, Peru (now Chile)	9.0	400
1872/03/26	Owens Valley, California	7.4	27
1877/05/10	Offshore Tarapaca, Chile	8.3	34
1886/09/01	Charleston, South Carolina	7.3	60
1887/05/03	Northern Sonora, Mexico	7.4	51
1891/10/27	Mino-Owari, Japan	8.0	
1892/02/24	Imperial Valley, California	7.8	
1896/06/15	Sanriku, Japan	8.5	28,000
1897/06/12	Assam, India	8.3	1,500
1899/09/04	Cape Yakataga, Alaska	7.9	
1899/09/10	Yakutat Bay, Alaska	8.0	
1900/10/09	Kodiak Island, Alaska	7.7	
1902/04/19	Quezaltenango and San Marcos, Guatemala	7.5	2,000
1902/12/16	Eastern Uzbekistan (Turkestan)	6.4	4,700
1903/04/28	Malazgirt, Turkey	7.0	3,500
1903/05/28	Gole, Turkey (Ottoman Empire)	5.8	1,000
1903/08/11	Southern Greece	8.3	
1905/04/04	Kangra, India	7.5	19,000
1905/07/09	Mongolia	8.4	
1905/09/08	Calabria, Italy	7.9	557
1906/01/31	Off the coast of Esmeraldas, Ecuador	8.8	1,000
1906/03/16	Chia-i, Taiwan	6.8	1,250
1906/04/18	San Francisco, California	7.8	3,000
1906/08/17	Valparaiso, Chile	8.2	20,000
1907/01/14	Kingston, Jamaica	6.5	800-1,000
1907/04/15	Guerrero, Mexico	7.7	
1907/10/21	Qaratog, Tajikistan, Russia	8.0	12,000
1908/12/12	Off the coast of Central Peru	8.2	
1908/12/28	Messina, Italy	7.2	72,000
1909/01/23	Silakhor, Iran (Persia)	7.3	5,000-6,000
1910/04/12	Taiwan region	7.6	
1911/01/03	Chong-Kemin, Kyrgyzstan	7.8	450
1911/02/18	Sarez, Tajikistan	7.4	90
1911/06/07	Off Guerrero, Mexico	7.7	45
1912/08/09	Murefte, Turkey (Ottoman Empire)	7.8	2,800
1914/10/03	Burdur, Turkey (Ottoman Empire)	7.0	4,000
1915/01/13	Avezzano, Italy	7.0	32,610
1917/01/20	Bali, Indonesia		1,500
1917/07/30	Daguan, Yunnan, China	7.5	1,800
1918/02/13	Nan'ao, Guangdong, (Kwangtung), China	7.3	1,000
1918/10/11	Mona Passage	7.5	116

Date	Location	Magnitude	Fatalities
1920/06/05	Taiwan region	8.0	
1920/12/16	Haiyuan, Ningxia, China	7.8	200,000
1922/11/11	Chile-Argentina border	8.5	100
1923/02/03	Kamchatka Peninsula	8.5	
1923/03/24	Near Luhuo, Sichuan, China	7.3	3,500
1923/05/25	Torbat-e Heydariyeh, Iran	5.7	2,200
1923/09/01	Kanto (Kwanto), Japan	7.9	142,800
1925/03/16	Yunnan, China	7.1	5,800
1927/03/07	Tango, Japan	7.6	3,020
1927/05/22	Tsinghai (Kansu), China	7.6	40,900
1928/12/01	Talca, Chile	7.6	225
1929/03/07	Fox Islands, Aleutian Islands, Alaska	7.8	
1929/05/01	Koppeh Dagh, Iran (Persia)	7.4	3,800
1930/05/06	Salmas, Iran (Persia)	7.2	2,500
1930/07/23	Irpinia, Italy	6.5	1,400
1931/01/15	Oaxaca, Mexico	7.8	114
1931/02/02	Hawke's Bay, New Zealand	7.9	256
1931/03/31	Managua, Nicaragua	6.0	2,500
1931/04/27	Zangezur Mountains, Armenia — Azerbaijan border	5.7	2,800
1931/08/10	Xinjiang, China	8.0	10,000
1932/06/03	Jalisco, Mexico	8.1	45
1932/06/18	Colima, Mexico	7.8	
1932/12/25	Gansu, China	7.6	275
1933/03/02	Sanriku, Japan	8.4	3,000
1933/03/11	Long Beach, California	6.4	115
1933/08/25	Sichuan, China	7.4	9,300
1934/01/15	Bihar, India — Nepal	8.1	10,700
1935/04/20	Taiwan (Formosa)	7.1	3,270
1935/05/30	Quetta, Pakistan	7.5	30,000
1935/07/16	Taiwan (Formosa)	6.5	2,740
1938/02/01	Banda Sea, Indonesia	8.5	
1938/11/10	Shumagin Islands, Alaska	8.2	
1939/01/25	Chillan, Chile	7.8	28,000
1939/12/26	Erzincan, Turkey	7.8	32,700
1940/05/19	Imperial Valley, California	7.1	9
1940/05/24	Callao, Peru	8.2	249
1940/11/10	Vrancea, Romania	7.3	1,000
1942/08/06	Guatemala	7.9	38
1942/08/24	Off the coast of central Peru	8.2	30
1942/11/26	Turkey	7.6	
1942/12/20	Erbaa, Turkey	7.3	1,100
1943/04/06	Illapel — Salamanca, Chile	8.2	25
1943/09/10	Tottori, Japan	7.4	1,190
1943/11/26	Ladik, Turkey	7.6	4,000
1944/01/15	San Juan, Argentina	7.4	8,000
1944/02/01	Gerede, Turkey	7.4	2,790
1944/12/07	Tonankai, Japan	8.1	998
1945/01/12	Mikawa, Japan	7.1	1,961
1945/11/27	Makran coast, Pakistan	8.0	4,000
1946/04/01	Unimak Island, Alaska	8.1	165
1946/05/31	Ustukran, Turkey	5.9	840-1,300
1946/08/04	Samana, Dominican Republic	8.0	100
1946/11/10	Ancash, Peru	7.3	1,400
1946/12/20	Nankaido, Japan	8.1	1,362
1947/11/01	Satipo, Peru	7.3	233
1948/05/11	Moquegua, Peru	7.4	70
1948/05/25	Sichuan, China	7.3	800
1948/06/28	Fukui, Japan	7.3	3,769
1948/10/05	Ashgabat, Turkmenistan	7.3	110,000
1949/04/13	Puget Sound, Washington	7.1	8
1949/07/10	Khait, Tajikistan	7.5	12,000
1949/08/05	Ambato, Ecuador	6.8	5,050
1949/08/22	Queen Charlotte Islands, British Columbia, Canada	8.1	
1950/08/15	Assam — Tibet	8.6	1,526
1951/08/02	Cosiguina, Nicaragua	5.8	1,000
1952/07/21	Kern County, California	7.3	12
1952/11/04	Kamchatka Peninsula	9.0	
1953/02/12	Torud, Iran	6.5	970
1953/03/18	Yenice-Gonen, Turkey	7.3	1,070
1953/08/12	Kefallinia, Greece	7.1	455
1953/12/12	Tumbes, Peru	7.4	7
1954/03/29	Spain	7.9	
1954/04/30	Greece	7.1	31
1954/09/09	Orleansville, Algeria	6.8	1,250
1957/03/09	Andreanof Islands, Alaska	8.6	
1957/04/25	Fethiye, Turkey	7.1	15
1957/05/26	Bolu Province, Turkey	7.1	66
1957/06/27	Stanovoy Mountains, Russia (USSR)	7.6	1,200
1957/07/02	Mazandaran, Iran	7.1	1,200
1957/07/28	Guerrero, Mexico	7.9	68
1957/12/04	Gobi-Altay, Mongolia	8.1	30
1957/12/13	Sahneh, Iran	7.1	1,130
1958/01/15	Arequipa, Peru	7.3	26
1958/07/10	Lituya Bay, Alaska	7.7	5
1958/11/06	Kuril Islands	8.3	
1959/04/26	Taiwan region	7.5	2
1959/08/18	Hebgen Lake, Montana	7.3	28
1960/01/13	Arequipa, Peru	7.5	57
1960/02/29	Agadir, Morocco	5.7	12,000-15,000
1960/05/21	Arauco Peninsula, Chile	7.9	
1960/05/22	Chile (off coast)	9.5	1,655
1962/05/11	Guerrero, Mexico	7.0	4
1962/05/19	Guerrero, Mexico	7.1	3
1962/09/01	Qazvin, Iran	7.1	12,225
1963/07/26	Skopje, Macedonia	6.0	1,100
1963/10/13	Kuril Islands	8.5	
1964/03/28	Prince William Sound, Alaska	9.2	128
1964/06/16	Niigata, Japan	7.5	26
1964/10/06	Western Turkey	7.0	36
1965/01/24	Sanana, Indonesia (Ceram Sea)	7.6	71
1965/02/04	Rat Islands, Alaska	8.7	
1965/03/14	Hindu Kush, Afghanistan	7.8	
1965/03/28	La Ligua, Chile	7.4	400
1965/03/31	Central Greece	7.1	6
1965/08/23	Oaxaca, Mexico	7.3	6
1966/03/07	Hebei, China	7.0	1,000
1966/03/22	Hebei, China	6.9	1,000
1966/08/19	Varto, Turkey	6.8	2,529
1966/10/17	Near the coast of Peru	8.1	125
1967/07/22	Mudurnu Valley, Turkey	7.3	173
1968/05/23	Inangahua, New Zealand	7.1	2
1968/08/02	Oaxaca, Mexico	7.1	18
1968/08/31	Dasht-e Bayaz, Iran	7.3	12,000
1969/02/28	Portugal-Morocco area	7.8	13
1969/07/25	Guangdong, China	5.9	3,000
1970/01/04	Yunnan Province, China	7.5	10,000
1970/03/28	Gediz, Turkey	6.9	1,086
1970/05/31	Chimbote, Peru	7.9	70,000

Date	Location	Magnitude	Fatalities
1970/07/31	Colombia	8.0	1
1971/02/09	San Fernando, California	6.6	65
1971/05/22	Eastern Turkey	6.9	1,000
1971/07/09	Valparaiso region, Chile	7.5	90
1972/01/25	Taiwan region	7.5	
1972/04/10	Southern Iran	7.1	5,054
1972/04/24	Taiwan region	7.2	4
1972/07/30	Sitka, Alaska	7.6	
1972/12/23	Nicaragua	6.2	5,000
1974/05/10	Near Zhaotong, China	6.8	20,000
1974/07/13	Panama-Colombia border region	7.3	11
1974/10/03	Near the coast of central Peru	8.1	78
1974/10/08	Leeward Islands	7.5	
1974/12/28	Northern Pakistan	6.2	5,300
1975/02/02	Near Islands, Alaska	7.6	
1975/02/04	Haicheng, China	7.0	2,000
1975/09/06	Diyarbakir Province, Turkey	6.7	2,300
1975/11/29	Kalapana, Hawaii	7.2	2
1976/02/04	Guatemala	7.5	23,000
1976/05/06	Northeastern Italy	6.5	1,000
1976/06/25	Papua, Indonesia	7.1	422
1976/07/27	Tangshan, China	7.5	255,000
1976/08/16	Mindanao, Philippines	7.9	8,000
1976/11/24	Turkey-Iran border region	7.3	5,000
1977/03/04	Romania	7.2	1,500
1978/09/16	Iran	7.8	15,000
1979/02/28	Mt. St. Elias, Alaska	7.5	
1980/10/10	El Asnam (formerly Orleansville), Algeria	7.7	5,000
1981/02/24	Greece	6.8	3,000
1981/06/11	Southern Iran	6.9	3,000
1981/07/28	Southern Iran	7.3	1,500
1982/12/13	Yemen	6.0	2,800
1983/10/30	Erzurum Province, Turkey	6.9	1,342
1985/03/03	Offshore Valparaiso, Chile	7.8	177
1985/09/19	Michoacan, Mexico	8.0	9,500
1986/05/07	Andreanof Islands, Alaska	7.9	
1986/10/10	El Salvador	5.5	1,000
1987/03/06	Colombia-Ecuador	7.0	1,000
1987/11/30	Gulf of Alaska	7.8	
1988/03/06	Gulf of Alaska	7.7	
1988/08/20	Nepal-India border region	6.8	1,000
1988/12/07	Spitak, Armenia	6.8	25,000
1989/10/18	Loma Prieta, California	6.9	63
1990/06/20	Western Iran	7.4	50,000
1990/07/16	Luzon, Philippine Islands	7.7	1,621
1991/04/22	Costa Rica	7.6	47
1991/10/19	Northern India	6.8	2,000
1992/09/02	Nicaragua	7.6	116
1992/12/12	Flores region, Indonesia	7.8	2,500
1993/08/08	South of the Mariana Islands	7.8	
1993/09/29	Latur-Killari, India	6.2	9,748
1994/01/17	Northridge, California	6.7	60
1994/06/09	Bolivia	8.2	10
1995/01/16	Kobe, Japan	6.9	5,502
1995/05/27	Sakhalin Island	7.1	1,989
1996/06/10	Andreanof Islands, Alaska	7.9	
1997/05/10	Northern Iran	7.3	1,567
1997/10/14	South of Fiji Islands	7.8	
1997/12/05	Near east coast of Kamchatka	7.8	
1998/01/04	Loyalty Islands region	7.5	
1998/02/04	Afghanistan-Tajikistan border region	5.9	2,323
1998/03/25	Balleny Islands region (off Antarctica)	8.1	
1998/05/03	Southeast of Taiwan	7.5	
1998/05/30	Afghanistan-Tajikistan border region	6.6	4,000
1998/07/17	Near north coast of New Guinea, Papua New Guinea	7.0	2,183
1999/01/25	Colombia	6.1	1,185
1999/08/17	Izmit, Turkey	7.6	17,118
1999/09/20	Taiwan	7.6	2,400
1999/09/30	Oaxaca, Mexico	7.5	
1999/11/12	Duzce, Turkey	7.2	894
2000/06/04	Southern Sumatera, Indonesia	7.9	103
2000/06/18	South Indian Ocean	7.9	
2000/11/16	New Ireland Region, Papua New Guinea	8.0	2
2001/01/01	Mindanao, Philippines	7.5	
2001/01/13	El Salvador	7.7	852
2001/01/26	Gujarat, India	7.6	20,085
2001/02/13	El Salvador	6.6	315
2001/06/23	Near the coast of Peru	8.4	138
2002/03/03	Hindu Kush region, Afghanistan	7.4	166
2002/03/05	Mindanao, Philippines	7.5	15
2002/03/25	Hindu Kush region, Afghanistan	6.1	1,000
2002/03/31	Taiwan region	7.1	5
2002/08/19	Fiji Islands	7.7	
2002/09/08	New Guinea, Papua New Guinea	7.6	
2002/10/10	Irian Jaya, Indonesia	7.6	8
2002/11/02	Northern Sumatera, Indonesia	7.4	3
2002/11/03	Denali Fault, Alaska	7.9	
2003/01/22	Offshore Colima, Mexico	7.6	29
2003/05/21	Northern Algeria	6.8	2,226
2003/05/26	Halmahera, Indonesia	7.0	1
2003/07/15	Carlsberg Ridge	7.6	
2003/08/04	Scotia Sea	7.6	
2003/09/25	Hokkaido, Japan region	8.3	
2003/09/27	Southwestern Siberia, Russia	7.3	3
2003/11/17	Rat Islands, Aleutian Islands, Alaska	7.8	
2003/12/26	Southeastern Iran	6.6	31,000
2004/02/05	Irian Jaya, Indonesia	7.0	37
2004/11/11	Kepulauan Alor, Indonesia	7.5	34
2004/11/26	Papua, Indonesia	7.1	32
2004/12/23	North of Macquarie Island, New Zealand	8.1	
2004/12/26	Sumatra-Andaman Islands	9.1	227,898
2005/03/28	Northern Sumatra, Indonesia	8.6	1,313
2005/06/13	Tarapaca, Chile	7.8	11
2005/09/09	New Ireland region, Papua New Guinea	7.6	
2005/09/26	Northern Peru	7.5	5
2005/10/08	Pakistan	7.6	86,000
2006/01/27	Banda Sea	7.6	
2006/02/22	Mozambique	7.0	4
2006/04/20	Koryakia, Russia	7.6	
2006/05/03	Tonga	8.0	
2006/05/26	Java, Indonesia	6.3	5,749
2006/07/17	South of Java, Indonesia	7.7	730

Date	Location	Magnitude	Fatalities
2006/11/15	Kuril Islands	8.3	
2006/12/26	Taiwan region	7.1	2
2007/01/13	East of the Kuril Islands	8.1	
2007/01/21	Molucca Sea	7.5	
2007/04/01	Solomon Islands	8.1	40
2007/08/08	Java, Indonesia	7.5	
2007/08/15	Near the coast of central Peru	8.0	514
2007/09/12	Southern Sumatra, Indonesia	8.5	25
2007/09/12	Kepulauan Mentawai region, Indonesia	7.9	
2007/09/28	Mariana Islands region	7.5	
2007/11/14	Antofagasta, Chile	7.7	2
2007/12/09	South of the Fiji Islands	7.8	
2008/05/12	Eastern Sichuan, China	7.9	87,587
2008/07/05	Sea of Okhotsk	7.7	
2009/01/03	Near the north coast of Papua, Indonesia	7.7	5
2009/03/19	Tonga region	7.6	
2009/04/06	Central Italy	6.3	295
2009/07/15	Off west coast of the South Island, New Zealand	7.8	
2009/08/10	Andaman Islands, India region	7.5	
2009/09/29	Samoa Islands region	8.1	192
2009/09/30	Southern Sumatra, Indonesia	7.5	1117
2009/10/07	Santa Cruz Islands	7.8	
2009/10/07	Vanuatu, Coral Sea	7.7	
2010/01/12	Haiti region	7.0	316,000
2010/02/27	Offshore Maule, Chile	8.8	547
2010/03/08	Eastern Turkey	6.1	42
2010/04/04	Baja California, Mexico	7.2	2
2010/04/13	Southern Qinghai, China	6.9	2,968
2010/07/23	Moro Gulf, Mindanao, Philippines	7.6	
2011/02/21	South Island of New Zealand	6.1	181
2011/03/11	Near the east coast of Honshu, Japan	9.0	20,896
2012/02/06	Negros-Cebu region, Philippines	6.7	113
2012/04/11	Off west coast of northern Sumatra	8.6[a]	10
2012/08/11	Northwestern Iran	6.4	306
2012/11/07	Offshore Guatamala	7.4	139
2013/01/05	Southeastern Alaska	7.5	
2013/02/06	Santa Cruz Islands	8.0	18
2013/04/06	Khash, Iran	7.7	35
2013/04/20	Linqiong, China	6.6	193
2013/05/24	Sea of Okhotsk	8.3	
2013/09/24	Awaran, Pakistan	7.7	825
2013/10/15	Sagbayan, Philippines	7.1	222
2013/11/17	Scotia Sea, Antarctica	7.8	
2014/04/03	Off west coast of northern Chile	7.7	6
2014/04/12	Solomon Islands	7.6	
2014/04/19	Panguna, Papua New Guinea	7.5	1
2014/06/23	Little Sitkin Island, Alaska	7.9	
2014/08/03	Wewnping, China	6.2	617
2015/04/25	E of Khudi, Nepal	7.8	9.000
2015/05/12	SE of Kodan, Nepal	7.3	218
2015/05/30	WNW of Chici-shima, Japan	7.8	0
2015/09/16	W of Illapel, Chile	8.3	19
2015/10/25	N of Alaqahdari-ye Kiran wa Munyan, Afghanistan	7.5	398
2016/02/06	SE of Yujing, Taiwan	6.4	117
2016/03/02	Southwest of Sumatra, Indonesia	7.8	
2016/04/16	SSE of Muisne, Ecuador	7.8	673
2016/07/29	SW of Agrihan, Northern Mariana Islands	7.7	
2016/08/19	South Georgia Island region	7.4	
2016/08/24	SE of Norcia, Italy	6.2	299
2016/11/13	NNE of Amberley, New Zealand	7.8	2
2016/12/08	Kirakira, Solomon Islands	7.8	
2016/12/17	E of Taron, Papua New Guinea	7.9	
2016/12/25	SW of Puerto Quellon, Chile	7.6	

[a] Initial shock of magnitude 8.2

WEATHER-RELATED SCALES

Saffir-Simpson Hurricane Scale

- **Tropical Storm**
  Winds 39–73 mph
- **Category 1 Hurricane** — winds 74–95 mph (64–82 knots); pressure greater than 980 mbar; storm surge 3–5 ft (1.0–1.7 m)
  No real damage to buildings. Damage to unanchored mobile homes. Some damage to poorly constructed signs. Also, some coastal flooding and minor pier damage.
  — Examples: Irene 1999 and Allison 1995
- **Category 2 Hurricane** — winds 96–110 mph (83–95 knots); pressure 979–965 mbar; storm surge 6–8 ft (1.8–2.6 m)
  Some damage to building roofs, doors and windows. Considerable damage to mobile homes. Flooding damages piers and small craft in unprotected moorings may break their moorings. Some trees blown down.
  — Examples: Bonnie 1998, Georges (FL & LA) 1998 and Gloria 1985
- **Category 3 Hurricane** — winds 111–130 mph (96–113 knots); pressure 964–945 mbar; storm surge 9–12 ft (2.7–3.8 m)
  Some structural damage to small residences and utility buildings. Large trees blown down. Mobile homes and poorly built signs destroyed. Flooding near the coast destroys smaller structures with larger structures damaged by floating debris. Terrain may be flooded well inland.
  — Examples: Keith 2000, Fran 1996, Opal 1995, Alicia 1983 and Betsy 1965
- **Category 4 Hurricane** — winds 131–155 mph (114–135 knots); pressure 944–920 mbar; storm surge 13–18 ft (3.9–5.6 m)
  More extensive curtainwall failures with some complete roof structure failure on small residences. Major erosion of beach areas. Terrain may be flooded well inland.
  — Examples: Hugo 1989 and Donna 1960
- **Category 5 Hurricane** — winds 156 mph and up (135+ knots); pressure less than 920 mbar; storm surge 19+ ft (5.7+ m)
  Complete roof failure on many residences and industrial buildings. Some complete building failures with small utility buildings blown over or away. Flooding causes major damage to lower floors of all structures near the shoreline. Massive evacuation of residential areas may be required.
  — Examples: Andrew (FL) 1992, Camille 1969 and Labor Day 1935

Fujita Scale for Tornado Damage

The original Fujita Scale was modified by NOAA in February 2007 and is now called the Enhanced Fujita Scale (EF). It is an operational scale based on the estimated speed of 3-second wind gusts, as indicated by typical damage levels. The table below describes the damage levels according to the original scale. In the enhanced scale, the damage is measured by a more elaborate set of criteria (see http://www.spc.noaa.gov/efscale/ef-scale.html).

Typical Damage (According to the Original Fujita Scale)

EF Number	3 s Gusts (mph)	
0	65–85	**Light damage.** Some damage to chimneys; branches broken off trees; shallow-rooted trees pushed over; sign boards damaged.
1	86–110	**Moderate damage.** Peels surface off roofs; mobile homes pushed off foundations or overturned; moving autos blown off roads.
2	111–135	**Considerable damage.** Roofs torn off frame houses; mobile homes demolished; boxcars overturned; large trees snapped or uprooted; light-object missiles generated; cars lifted off ground.
3	136–165	**Severe damage.** Roofs and some walls torn off well-constructed houses; trains overturned; most trees in forest uprooted; heavy cars lifted off the ground and thrown.
4	166–200	**Devastating damage.** Well-constructed houses leveled; structures with weak foundations blown away some distance; cars thrown and large missiles generated.
5	Over 200	**Incredible damage.** Strong frame houses leveled off foundations and swept away; automobile-sized missiles fly through the air in excess of 100 meters (109 yd); trees debarked; incredible phenomena will occur.

Beaufort Wind Scale

The Beaufort Wind Scale was devised by British Rear-Admiral Sir Francis Beaufort in 1805 based on observations of the effects of the wind.

Wind Effects on the Beaufort Scale

Force	Wind (knots)	WMO classification	On the water	On land
0	<1	Calm	Sea surface smooth and mirror-like	Calm, smoke rises vertically
1	1–3	Light Air	Scaly ripples, no foam crests	Smoke drift indicates wind direction, still wind vanes
2	4–6	Light Breeze	Small wavelets, crests glassy, no breaking	Wind felt on face, leaves rustle, vanes begin to move

Force	Wind (knots)	WMO classification	On the water	On land
3	7–10	Gentle Breeze	Large wavelets, crests begin to break, scattered whitecaps	Leaves and small twigs constantly moving, light flags extended
4	11–16	Moderate Breeze	Small waves 1–4 ft. becoming longer, numerous whitecaps	Dust, leaves, and loose paper lifted, small tree branches move
5	17–21	Fresh Breeze	Moderate waves 4–8 ft taking longer form, many whitecaps, some spray	Small trees in leaf begin to sway
6	22–27	Strong Breeze	Larger waves 8–13 ft, whitecaps common, more spray	Larger tree branches moving, whistling in wires
7	28–33	Near Gale	Sea heaps up, waves 13–20 ft, white foam streaks off breakers	Whole trees moving, resistance felt walking against wind
8	34–40	Gale	Moderately high (13–20 ft) waves of greater length, edges of crests begin to break into spindrift, foam blown in streaks	Whole trees in motion, resistance felt walking against wind
9	41–47	Strong Gale	High waves (20 ft), sea begins to roll, dense streaks of foam, spray may reduce visibility	Slight structural damage occurs, slate blows off roofs
10	48–55	Storm	Very high waves (20–30 ft) with overhanging crests, sea white with densely blown foam, heavy rolling, lowered visibility	Seldom experienced on land, trees broken or uprooted, "considerable structural damage"
11	56–63	Violent Storm	Exceptionally high (30–45 ft) waves, foam patches cover sea, visibility more reduced	
12	64+	Hurricane	Air filled with foam, waves over 45 ft, sea completely white with driving spray, visibility greatly reduced	

Wind Chill

The following chart prepared by the U.S. National Weather Service gives the temperature perceived by an average person as a function of the real air temperature and the wind speed. The current scale was adopted in 2001.

Temperature (°F)

Wind (mph) / Calm	40	35	30	25	20	15	10	5	0	-5	-10	-15	-20	-25	-30	-35	-40	-45
5	36	31	25	19	13	7	1	-5	-11	-16	-22	-28	-34	-40	-46	-52	-57	-63
10	34	27	21	15	9	3	-4	-10	-16	-22	-28	-35	-41	-47	-53	-59	-66	-72
15	32	25	19	13	6	0	-7	-13	-19	-26	-32	-39	-45	-51	-58	-64	-71	-77
20	30	24	17	11	4	-2	-9	-15	-22	-29	-35	-42	-48	-55	-61	-68	-74	-81
25	29	23	16	9	3	-4	-11	-17	-24	-31	-37	-44	-51	-58	-64	-71	-78	-84
30	28	22	15	8	1	-5	-12	-19	-26	-33	-39	-46	-53	-60	-67	-73	-80	-87
35	28	21	14	7	0	-7	-14	-21	-27	-34	-41	-48	-55	-62	-69	-76	-82	-89
40	27	20	13	6	-1	-8	-15	-22	-29	-36	-43	-50	-57	-64	-71	-78	-84	-91
45	26	19	12	5	-2	-9	-16	-23	-30	-37	-44	-51	-58	-65	-72	-79	-86	-93
50	26	19	12	4	-3	-10	-17	-24	-31	-38	-45	-52	-60	-67	-74	-81	-88	-95
55	25	18	11	4	-3	-11	-18	-25	-32	-39	-46	-54	-61	-68	-75	-82	-89	-97
60	25	17	10	3	-4	-11	-19	-26	-33	-40	-48	-55	-62	-69	-76	-84	-91	-98

Frostbite Times: 30 minutes, 10 minutes, 5 minutes

$$\text{Wind Chill } (^\circ F) = 35.74 + 0.6215T - 35.75(V^{0.16}) + 0.4275T(V^{0.16})$$

Where, T = Air Temperature (°F) V = Wind Speed (mph)

Effective 11/01/01

Reference

National Oceanic and Atmospheric Administration, http://www.noaa.gov.

Geo-Astro

INFRARED ABSORPTION BY THE EARTH'S ATMOSPHERE

Several constituents of the Earth's atmosphere absorb infrared radiation. At ground level the strongest absorbers are H_2O and CO_2, but 30 to 40 other compounds can make significant contributions. The centers of the most important absorption bands are listed below:

Molecule	Vibrational mode	Band center in cm^{-1}
H_2O	Bend	1595
H_2O	Symmetric O-H stretch	3657
H_2O	Antisymmetric O-H stretch	3756
CO_2	Bend	667
CO_2	Antisymmetric C-O stretch	2349
O_3	Bend	701
O_3	Antisymmetric O-O stretch	1042
O_3	Symmetric O-O stretch	1103
N_2O	Bend	589
N_2O	N-O stretch	1285
N_2O	N-N stretch	2224
CO	C-O stretch	2143
CH_4	Degenerate deformation	1306
CH_4	Degenerate stretch	3019

The HITRAN Molecular Spectroscopy Database (Refs. 1 and 2) is a compilation of wavenumbers and intensities of more than 1.7 million spectral lines of atmospheric constituents. It is a valuable resource for calculating transmission of the atmosphere, radiative energy transfer, and other phenomena. The graph below, which was supplied by Walter J. Lafferty (Ref. 3), gives the transmittance of the atmosphere for one set of conditions.

References

1. Rothman, L. S., et al., *J. Quant. Spectros. Radiat. Transfer* 82, 5, 2003; *ibid.*, to be published, 2005.
2. HITRAN Molecular Spectroscopy Database, <http://cfa-www.Harvard.edu/HITRAN/hitrandata04/>.
3. Lafferty, W. J., Some Aspects of High Resolution Molecular Spectroscopy, in *Lectures on Molecular Physics*, Institute for the Structure of Matter, Centro de Fisica Miguel A. Catalan, Madrid, 1997.

Transmittance of U.S. Standard Atmosphere at Ground Level for a Path of 1 km at 296 K

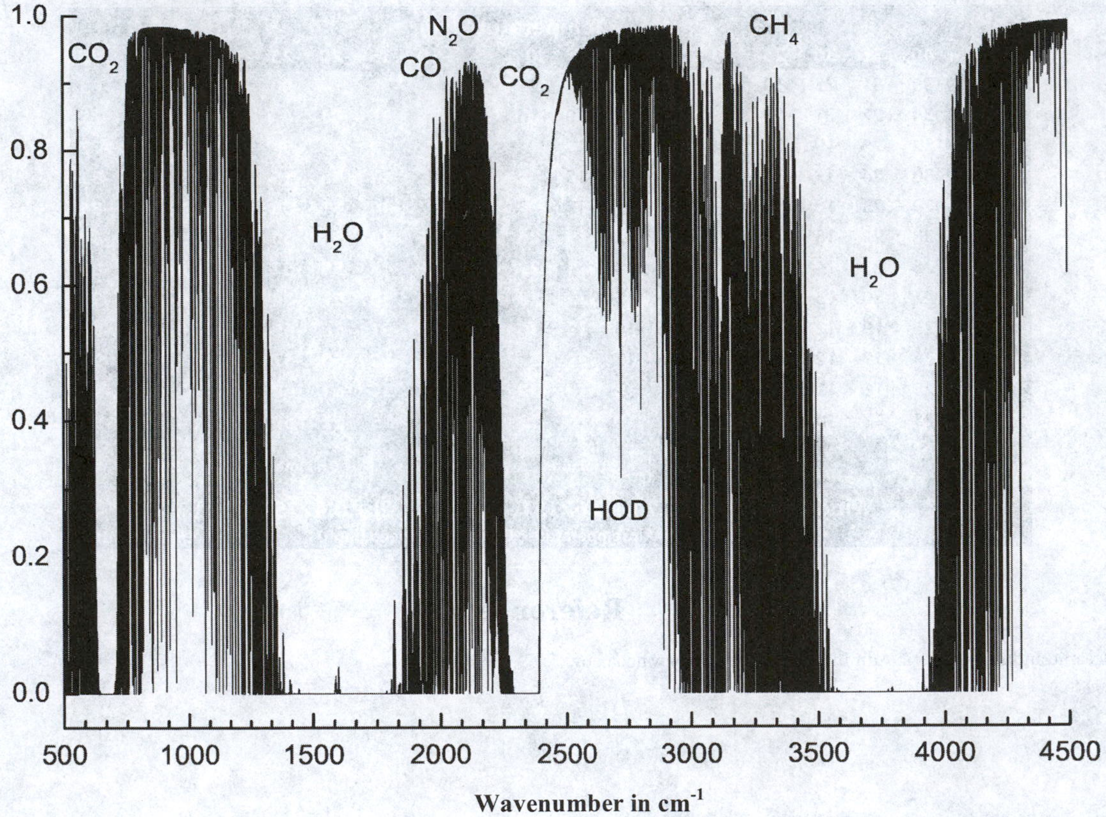

Geo-Astro

ATMOSPHERIC CONCENTRATION OF CARBON DIOXIDE, 1959-2016

The data summarized in this table were taken at the Mauna Loa Observatory in Hawaii and represent averages over each year. The concentration of CO_2 is given in parts per million (ppm) by volume (i.e., μmol/mol of dry air). Data from other measurement sites, as well as monthly data from Mauna Loa, may be found in Ref. 3.

The 2016 value from Ref. 2 is preliminary as of Jan. 11, 2017, and may change slightly with further analysis. The standard deviation of the concentration values is estimated to be 0.12 ppm.

The first graph (Ref. 1) illustrates the seasonal variation of CO_2 concentration and the steady increase over the last 55 years. The second graph (Ref. 4) summarizes the growth in global emissions of CO_2 into the atmosphere as a result of burning of fossil fuels.

References

1. Trends in Atmospheric Carbon Dioxide, <http:www.esrl.noaa.gov/gmd/ccgg/trends/co2_data_mlo.html>.
2. Tans, P., and Keeling, R., NOAA-ESRL Data, <ftp://aftp.cmdl.noaa.gov/products/trends/co2/co2_mm_mlo.txt>, Jan. 11, 2017.
3. Keeling, C. D., Piper, S. C., Bollenbacher, A. F., and Walker, S. J., Atmospheric carbon dioxide record from Mauna Loa, Carbon Dioxide Information Analysis Center, Oak Ridge National Laboratory, U.S. Department of Energy, Oak Ridge, TN, <cdiac.ornl.gov/trends/co2/sio-mlo.html>, February 2009.
4. Carbon Dioxide Information Analysis Center, Oak Ridge National Laboratory, U.S. Department of Energy, Oak Ridge, TN, <cdiac.onrl.gov>.
5. CO_2 Earth, <co2.earth>.

Year	Mean CO_2 Conc. in ppm	Year	Mean CO_2 Conc. in ppm	Year	Mean CO_2 Conc. in ppm	Year	Mean CO_2 Conc. in ppm
1959	315.97	1974	330.18	1989	353.12	2004	377.52
1960	316.91	1975	331.11	1990	354.39	2005	379.80
1961	317.64	1976	332.04	1991	355.61	2006	381.90
1962	318.45	1977	333.83	1992	356.45	2007	383.79
1963	318.99	1978	335.40	1993	357.10	2008	385.60
1964	319.62	1979	336.84	1994	358.83	2009	387.43
1965	320.04	1980	338.75	1995	360.82	2010	389.90
1966	321.38	1981	340.11	1996	362.61	2011	391.65
1967	322.16	1982	341.45	1997	363.73	2012	393.85
1968	323.04	1983	343.05	1998	366.70	2013	396.52
1969	324.62	1984	344.65	1999	368.38	2014	398.65
1970	325.68	1985	346.12	2000	369.55	2015	400.83
1971	326.32	1986	347.42	2001	371.14	2016	404.21
1972	327.45	1987	349.18	2002	373.28		
1973	329.68	1988	351.57	2003	375.80		

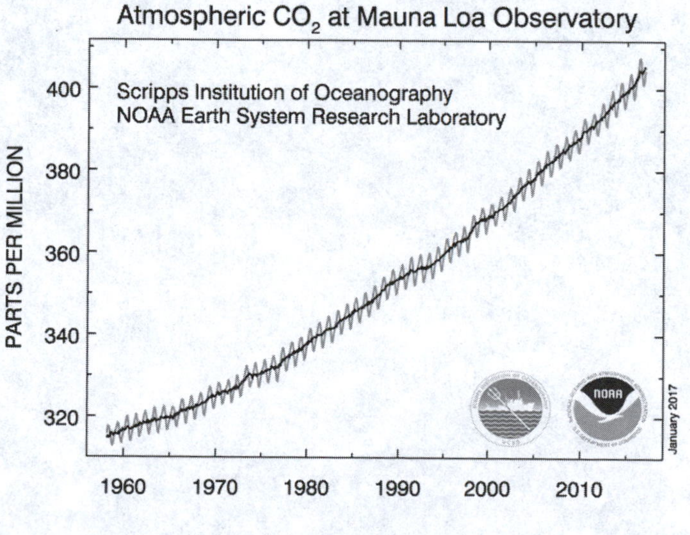

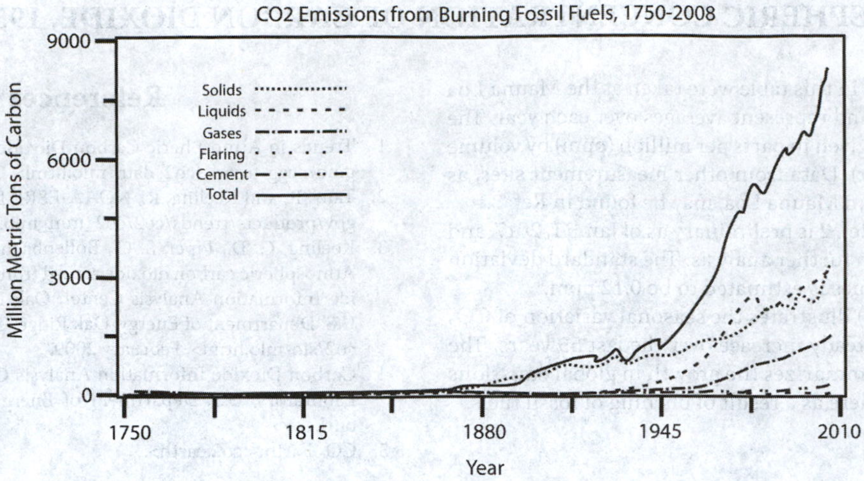
CO2 Emissions from Burning Fossil Fuels, 1750-2008
Million Metric Tons of Carbon
9000
6000
3000
0
1750
1815
1880
1945
2010
Year
Solids
Liquids
Gases
Flaring
Cement
Total

GLOBAL TEMPERATURE TREND, 1880-2016

This table and graph summarize the trend in annual mean global surface temperature from 1880 to 2016. The values were calculated from the global mean temperature anomalies in Ref. 2 by assuming an absolute global mean of 14.00 °C, which is the best estimate for the 1951–1980 period. The 95% confidence interval for comparing the annual mean temperature values for recent years is 0.05 °C. The values given here are for the annual mean land-ocean temperature.

Ref. 1 gives links to more extensive data sets that cover individual months, seasons, and zonal regions.

References

1. NASA Goddard Institute for Space Studies, <data.giss.nasa.gov/gistemp>.
2. Annual Mean Land-Ocean Temperature Index in 0.01 Degrees Celsius: Selected Zonal Means, <data.giss.nasa.gov/gistemp/tabledata_v3/ZonAnn.Ts+dSST.txt>, January 19, 2017.

Mean Global Temperature 1880-2016

Year	Mean t/°C	Year	Mean t/°C	Year	Mean t/°C	Year	Mean t/°C	Year	Mean t/°C
1880	13.80	1908	13.57	1936	13.85	1964	13.80	1992	14.23
1881	13.88	1909	13.53	1937	13.97	1965	13.90	1993	14.24
1882	13.90	1910	13.58	1938	13.97	1966	13.95	1994	14.32
1883	13.79	1911	13.56	1939	13.97	1967	13.98	1995	14.46
1884	13.72	1912	13.65	1940	14.08	1968	13.93	1996	14.35
1885	13.68	1913	13.66	1941	14.12	1969	14.07	1997	14.48
1886	13.69	1914	13.84	1942	14.09	1970	14.02	1998	14.64
1887	13.67	1915	13.89	1943	14.13	1971	13.91	1999	14.42
1888	13.80	1916	13.66	1944	14.25	1972	14.01	2000	14.42
1889	13.88	1917	13.60	1945	14.12	1973	14.15	2001	14.55
1890	13.63	1918	13.74	1946	13.96	1974	13.93	2002	14.63
1891	13.76	1919	13.78	1947	13.95	1975	13.98	2003	14.62
1892	13.73	1920	13.73	1948	13.91	1976	13.89	2004	14.55
1893	13.70	1921	13.79	1949	13.91	1977	14.18	2005	14.69
1894	13.69	1922	13.72	1950	13.82	1978	14.07	2006	14.63
1895	13.79	1923	13.76	1951	13.93	1979	14.17	2007	14.66
1896	13.85	1924	13.72	1952	14.01	1980	14.27	2008	14.54
1897	13.89	1925	13.79	1953	14.08	1981	14.33	2009	14.64
1898	13.72	1926	13.90	1954	13.87	1982	14.13	2010	14.71
1899	13.84	1927	13.79	1955	13.85	1983	14.30	2011	14.60
1900	13.91	1928	13.79	1956	13.80	1984	14.15	2012	14.63
1901	13.85	1929	13.64	1957	14.04	1985	14.12	2013	14.65
1902	13.73	1930	13.85	1958	14.07	1986	14.19	2014	14.74
1903	13.65	1931	13.91	1959	14.03	1987	14.33	2015	14.87
1904	13.56	1932	13.83	1960	13.98	1988	14.41	2016	14.99
1905	13.72	1933	13.71	1961	14.05	1989	14.29		
1906	13.77	1934	13.86	1962	14.03	1990	14.44		
1907	13.60	1935	13.80	1963	14.06	1991	14.43		

Annual Mean Global Temperature 1880-2016

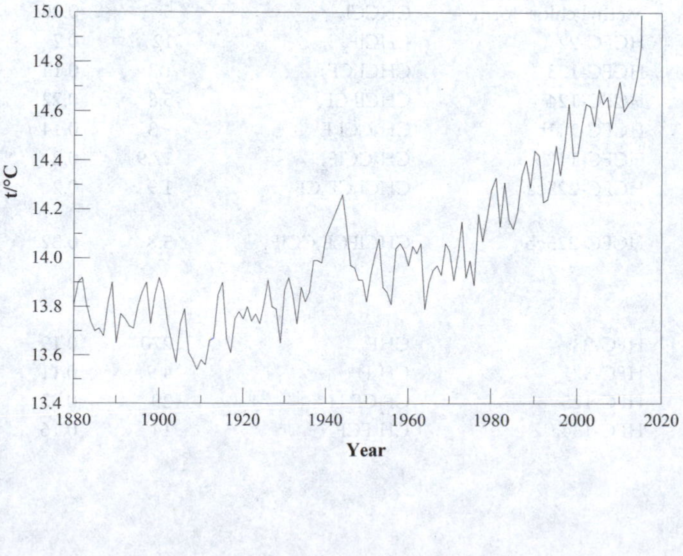

Geo-Astro

GLOBAL WARMING POTENTIAL OF GREENHOUSE GASES

The Global Warming Potential (GWP) of a gas is a measure of the degree, relative to carbon dioxide, to which the presence of that gas in the atmosphere will contribute to a long-term increase in global temperature. The calculation of the GWP for a given gas takes into account the efficiency of the gas in absorbing solar radiation (primarily determined by the infrared spectrum of the compound) and the time the compound will remain in the atmosphere before it is removed by natural processes. Thus if a pulse of 1 kg of the gas is emitted to the atmosphere at the same time as a pulse of 1 kg of CO_2, the GWP compares the warming effect of the gas relative to CO_2 over various time horizons.

This table, which is taken from the 2007 report of the Intergovernmental Panel of Climate Change (IPCC), gives the lifetime in years and the radiative efficiency in watts per square meter for a concentration of one part per billion for the major compounds identified in the Kyoto Protocol as contributing to global climate change. Radiative efficiency is a measure of the radiative forcing that influences the energy balance in the Earth–atmosphere system. The last four columns of the table give the Global Warming Potential, first as estimated in 1995 for a 100 year time horizon, and then as estimated with improved data in 2007 for 20-, 100-, and 500-year horizons. The calculation of a GWP involves a number of assumptions, and other measures have been proposed (see Reference).

The list of compounds includes those identified in the Montreal Protocol as contributing to ozone depletion, since these compounds also contribute to global warming. It also includes compounds used or proposed as replacements for the ozone-depleting compounds but which still have global warming potential.

Reference

Forster, P., V. Ramaswamy, P. Artaxo, T. Berntsen, R. Betts, D. W. Fahey, J. Haywood, J. Lean, D.C. Lowe, G. Myhre, J. Nganga, R. Prinn, G. Raga, M. Schulz, and R. Van Dorland, 2007: Changes in Atmospheric Constituents and in Radiative Forcing. In: *Climate Change 2007: The Physical Science Basis. Contribution of* Working *Group I to the Fourth Assessment Report of the Intergovernmental Panel on Climate Change* [Solomon, S., D. Sin, M. Manning, Z. Chen, M. Marquis, K. B. Averyt, M. Tignor and H.L. Miller (eds.)], Cambridge University Press, Cambridge, United Kingdom and New York, NY, USA. Available on the Internet at <http://ipcc-wg1.ucar.edu/wg1/wg1-report.html>.

Lifetime, Radiative Efficiency, and Global Warming Potential (GWP) of Greenhouse Gases

Compound	Synonym/code	Formula	Lifetime (years)	Radiative efficiency/ $W\ m^{2}\ ppb^{-1}$	GWP[a] 100 yr est.	GWP[b] 20 yr est.	GWP[b] 100 yr est.	GWP[b] 500 yr est.
Natural atmospheric constituents								
Carbon dioxide		CO_2		0.000014	1	1	1	1
Methane		CH_4	12	0.00037	21	72	25	7.6
Nitrous oxide		N_2O	114	0.00303	310	289	298	153
Substances controlled by the Montreal Protocol								
Trichlorofluoromethane	CFC-11	CCl_3F	45	0.25	3800	6730	4750	1620
Dichlorodifluoromethane	CFC-12	CF_2Cl_2	100	0.32	8100	11000	10900	5200
Chlorotrifluoromethane	CFC-13	CF_3Cl	640	0.25		10800	14400	16400
1,1,2-Trichloro-1,2,2-trifluoroethane	CFC-113	$C_2Cl_3F_3$	85	0.3	4800	6540	6130	2700
1,2-Dichloro-1,1,2,2-tetrafluoroethane	CFC-114	$C_2Cl_2F_4$	300	0.31		8040	10000	8730
Chloropentafluoroethane	CFC-115	CF_3CF_2Cl	1700	0.18		5310	7370	9990
Bromotrifluoromethane	Halon-1301	CF_3Br	65	0.32	5400	8480	7140	2760
Bromochlorodifluoromethane	Halon-1211	CF_2BrCl	16	0.3		4750	1890	575
1,2-Dibromotetrafluoroethane	Halon-2402	CF_2BrCF_2Br	20	0.33		3680	1640	503
Tetrachloromethane	Carbontetrachloride	CCl_4	26	0.13	1400	2700	1400	435
Bromomethane	Methyl bromide	CH_3Br	0.7	0.01		17	5	1
1,1,1-Trichloroethane	Methyl chloroform	$CHCCl_3$	5	0.06		506	146	45
Chlorodifluoromethane	HCFC-22	$CHClF_2$	12	0.2	1500	5160	1810	549
2,2-Dichloro-1,1,1-trifluoroethane	HCFC-123	$CHCl_2CF_3$	1.3	0.14	90	273	77	24
1-Chloro-1,2,2,2-tetrafluoroethane	HCFC-124	$CHClFCF_3$	5.8	0.22	470	2070	609	185
1,1-Dichloro-1-fluoroethane	HCFC-141b	CH_3CCl_2F	9.3	0.14		2250	725	220
1-Chloro-1,1-difluoroethane	HCFC-142b	CH_3CClF_2	17.9	0.2	1800	5490	2310	705
3,3-Dichloro-1,1,1,2,2-pentafluoropropane	HCFC-225ca	$CHCl_2CF_2CF_3$	1.9	0.2		429	122	37
1,3-Dichloro-1,1,2,2,3-pentafluoropropane	HCFC-225cb	$CHClFCF_2CClF_2$	5.8	0.32		2030	595	181
Hydrofluorocarbons								
Trifluoromethane	HFC-23	CHF_3	270	0.19	11700	12000	14800	12200
Difluoromethane	HFC-32	CH_2F_2	4.9	0.11	650	2330	675	205
Pentafluoroethane	HFC-125	CF_3CF_2H	29	0.23	2800	6350	3500	1100
1,1,1,2-Tetrafluoroethane	HFC-134a	CH_2FCF_3	14	0.16	1300	3830	1430	435

Compound	Synonym/code	Formula	Lifetime (years)	Radiative efficiency/ W m² ppb⁻¹	GWP[a] 100 yr est.	GWP[b] 20 yr est.	GWP[b] 100 yr est.	GWP[b] 500 yr est.
1,1,1-Trifluoroethane	HFC-143a	CH_3CF_3	52	0.13	3800	5890	4470	1590
1,1-Difluoroethane	HFC-152a	CH_3CHF_2	1.4	0.09	140	437	124	38
1,1,1,2,3,3,3-Heptafluoropropane	HFC-227ea	CF_3CHFCF_3	34.2	0.26	2900	5310	3220	1040
1,1,1,3,3,3-Hexafluoropropane	HFC-236fa	$CF_3CH_2CF_3$	240	0.28	6300	8100	9810	7660
1,1,1,3,3-Pentafluoropropane	HFC-245fa	$CHF_2CH_2CF_3$	7.6	0.28		3380	1030	314
1,1,1,3,3-Pentafluorobutane	HFC-365mfc	$CH_3CF_2CH_2CF_3$	8.6	0.21		2520	794	241
1,1,1,2,3,4,4,5,5,5-Decafluoropentane	HFC-43-10mee	$CF_3CHFCHFCF_2CF_3$	15.9	0.4	1300	4140	1640	500
Perfluorinated compounds								
Sulfur hexafluoride		SF_6	3200	0.52	23900	16300	22800	32600
Nitrogen trifluoride		NF_3	740	0.21		12300	17200	20700
Tetrafluoromethane	PFC-14	CF_4	50000	0.10	6500	5210	7390	11200
Hexafluoroethane	PFC-116	CF_3CF_3	10000	0.26	9200	8630	12200	18200
Perfluoropropane	PFC-218	$CF_3CF_2CF_3$	2600	0.26	7000	6310	8830	12500
Perfluorocyclobutane	PFC-318	$c\text{-}C_4F_8$	3200	0.32	8700	7310	10300	14700
Perfluorobutane	PFC-3-1-10	C_4F_{10}	2600	0.33	7000	6330	8860	12500
Perfluoropentane	PFC-4-1-12	C_5F_{12}	4100	0.41		6510	9160	13300
Perfluorohexane	PFC-5-1-14	C_6F_{14}	3200	0.49	7400	6600	9300	13300
Perfluorodecalin	PFC-9-1-18	$C_{10}F_{18}$	>1,000	0.56		>5500	>7500	>9500
(Trifluoromethyl)sulfur pentafluoride		SF_5CF_3	800	0.57		13200	17700	21200
Fluorinated ethers								
Trifluoromethyl difluoromethyl ether	HFE-125	CF_3OCF_2H	136	0.44		13800	14900	8490
Bis(difluoromethyl) ether	HFE-134	$(CF_2H)_2O$	26	0.45		12200	6320	1960
Methyl trifluoromethyl ether	HFE-143a	CH_3OCF_3	4.3	0.27		2630	756	230
2-Chloro-2-(difluoromethoxy)-1,1,1-trifluoroethane	HCFE-235da2	$CHF_2OCHClCF_3$	2.6	0.38		1230	350	106
Methyl 1,1,2,2-tetrafluoroethyl ether	HFE-245cb2	$CH_3OCF_2CHF_2$	5.1	0.32		2440	708	215
2-(Difluoromethoxy)-1,1,1-trifluoroethane	HFE-245fa2	$CHF_2OCH_2CF_3$	4.9	0.31		2280	659	200
Methyl pentafluoroethyl ether	HFE-254cb2	$CH_3OCF_2CHF_2$	2.6	0.28		1260	359	109
Perfluoropropyl methyl ether	HFE-347mcc3	$CH_3OCF_2CF_2CF_3$	5.2	0.34		1980	575	175
1,1,2,2-Tetrafluoroethyl 1,1,1-trifluoroethyl ether	HFE-347pcf2	$CF_2HCF_2OCH_3CF_3$	7.1	0.25		1900	580	175
1-Methoxy-1,1,2,2,3,3-hexafluoropropane	HFE-356pcc3	$CH_3OCF_2CF_2CHF_2$	0.33	0.93		386	110	33
Methyl nonafluorobutyl ether	HFE-449sl (HFE-7100)	$C_4F_9OCH_3$	3.8	0.31		1040	297	90
1-Ethoxy-1,1,2,2,3,3,4,4,4-nonafluorobutane	HFE-569sf2 (HFE-7200)	$C_4F_9OC_2H_5$	0.77	0.3		207	59	18
1-(Difluoromethoxy)-2-[(difluoromethoxy) difluoromethoxy]-1,1,2,2-tetrafluoroethane	HFE-43-10pccc124 (H-Galden 1040x)	$CHF_2OCF_2OC_2F_4OCHF_2$	6.3	1.37		6320	1870	569
Bis(difluoromethoxy)difluoromethane	HFE-236ca12 (HG-10)	$CHF_2OCF_2OCHF_2$	12.1	0.66		8000	2800	860
1,2-Bis(difluoromethoxy)-1,1,2,2-tetrafluoroethane	HFE-338pcc13 (HG-01)	$CHF_2OCF_2CF_2OCHF_2$	6.2	0.87		5100	1500	460
Perfluoropolymethylisopropyl ether	PFPMIE	$CF_3OCF(CF_3)CF_2OCF_2OCF_3$	800	0.65		7620	10300	12400
Other compounds - Direct effects								
Dimethyl ether	Methyl ether	CH_3OCH_3	0.015	0.02		1	1	<<1
Dichloromethane	Methylene chloride	CH_2Cl_2	0.38	0.03		31	8.7	2.7
Chloromethane	Methyl chloride	CH_3Cl	1	0.01		45	13	4

[a] Estimated in 1995.
[b] Estimated in 2007.

ATMOSPHERIC ELECTRICITY

Hans Dolezalek, Hannes Tammet, John Latham, and Martin A. Uman

I. SURVEY AND GLOBAL CIRCUIT

Hans Dolezalek

The science of atmospheric electricity originated in 1752 by an experimental proof of a related earlier hypothesis (that lightning is an electrical event). In spite of a large effort, in part by such eminent physicists as Coulomb, Lord Kelvin, and many others, an overall, proven theory able to generate models with sufficient resolution is not yet available. Generally accepted and encompassing textbooks are now more than 20 years old. The voluminous proceedings of the, so far, nine international atmospheric electricity conferences (1954 to 1992) give much valuable detail and demonstrate impressive progress, as do a number of less comprehensive textbooks published in the last 20 years, but a general theory as indicated above is not yet created. Only now, are certain related measuring techniques and mathematical possibilities emerging.

Applications to practical purposes do exist in the field of lightning research (including the electromagnetic radiation emanating from lightning) by the establishment of lightning-location networks and by the now developing possibility of detecting electrified clouds that pose hazards to aircraft. Application of atmospheric electricity to other parts of meteorology seems to be promising but so far has seldom been instituted. Because some atmospheric electric signals propagate around the Earth and because of the existence of a global circuit, applications for the monitoring of global change processes and conditions are now being proposed. Significant secular changes in the global circuit would indicate a change in the global climate; the availability of many old data (about a span of 100 years) could help detect a long-term trend.

The concept of the "global circuit" is based on the theory of the global spherical capacitor: both the solid (and liquid) Earth as one electrode, and the high atmospheric layers (about the ionosphere) as the other, are by orders of magnitude more electrically conductive than the atmosphere between them. According to the "classical picture of atmospheric electricity," this capacitor is continuously charged by the common action of all thunderstorms to a d.c. voltage difference of several hundred kilovolts, the Earth being negative. The much smaller but still existing conductivity of the atmosphere allows a current flowing from the ionosphere to the ground, integrated for all sink areas of the whole Earth, of the order of 1.5 kA. In this way, a global circuit is created with many generators and sink-areas both interspaced and distributed over the whole globe, all connected to two nodes: ionosphere and ground. Within the scope of the global circuit, for each location, the current density (order of several pA/m^2) is determined by the voltage difference between ionosphere and ground (which is the same for all locations but varying in time) and the columnar resistance reaching from the ground up to the ionosphere (in the order of 10^{17} Ω m^2).

Natural processes, especially meteorological processes and some human activity, which produce or move electric charges ("space charges") or affect the ion distribution, constitute local generators and thereby "local circuits," horziontally and/or in parallel or antiparallel to the local part of the global circuit. In many cases, the local currents are much stronger than the global ones, making the measurement of the global current at a given location and/or during a period of time very difficult or, often, impossible. The strongest local circuits usually occur with certain weather conditions (precipitation, fog, high wind, blown-up dust or snow, heavy cloudiness) that make measurement of the global circuit impossible everywhere; but even in their absence local generators exist in varying magnitudes and of different characters. The separation of the local and global shares in the measured values of current density is a central problem of the science of atmospheric electricity. Aerological measurements are of high value in this regard.

The above description is within the "classical picture" of atmospheric electricity, a group of hypotheses to explain the electrification of the atmosphere. It is probably fundamentally correct but certainly not complete; it has not yet been confirmed by systems of measurements resulting in no inner contradictions. In particular, extraterrestrial influences must be permitted; their general significance is still under debate.

Within this "classical picture" a kind of electric standard atmosphere may be constructed as shown in Table 1.

Values with a star, *, are rough average values from measurement. A star in parentheses, (*), points to a typical value from one or a few measurements. All other values have been calculated from starred values, under the assumption that at 2 km 50% and at 12 km 90% of the columnar resistance is reached. Voltage drop along one of the partial columns can be calculated by subtracting the value for the lower column from that of the upper one. Columnar resistances, conductances, and capacitances are valid for that particular part of the column which is indicated at the left. Capacitances are calculated with the formula for plate capacitors, and this fact must be considered also for the time constants for columns.

According to measurements, *U*, the potential difference between 0 m and 65 km may vary by a factor of approximately 2. The total columnar resistance, R_c, is estimated to vary up to a factor of 3, the variation being due to either reduction of conductivity in the exchange layer (about lowest 2 km of this table) or to the presence of high mountains; in both cases the variation is caused in the troposphere. Smaller variations in the stratosphere and mesosphere are being discussed because of aerosols there. The air–Earth current density in fair weather varies by a factor of 3 to 6 accordingly. Conductivity near the ground varies by a factor of about 3 but only decreasing; increase of conductivity due to extraordinary radioactivity is a singular event. The field strength near the ground varies as a consequence of variations of air–Earth current density and conductivity from about ⅓ to about 10 times of the value quoted in the table. Conductivity near the ground shows a diurnal and an annual variation which depends strongly on the locality: air–Earth current density shows a diurnal and annual variation because the Earth–ionosphere potential difference undergoes such variations, and also because the columnar resistance is supposed to have a diurnal and probably an annual variation.

Conductivities and air–earth current densities on high mountains are greater than at sea level by factors of up to 10. Conductivity decreases when atmospheric humidity increases. Values for space charges are not quoted because measurements are too few to allow calculation of average values. Values of parameters over the oceans are still rather uncertain.

Theoretically, in fair weather conditions, Ohm's law must be fulfilled for the electric field, the conduction current density, and the electrical conductivity of the atmosphere. Deviations point to

TABLE 1. Electrical Parameters of the Clear (Fair Weather) Atmosphere, Pertinent to the Classical Picture of Atm. Electricity (Electric Standard Atmosphere)

Part of atmosphere for which the values are calculated (elements are in free, cloudless atmosphere)	Currents, *I*, in A; and current densities, *i*, in A/m²	Potential differences, *U*, in V; field strength *E* in V/m; *U* = 0 at sea level	Resistances, *R*, in Ω; columnar resistances, R_c, in Ω m² and resistivities, *ρ*, in Ω m	Conductances, *G*, in Ω⁻¹; columnar conductances G_c, in Ω⁻¹ m⁻²; total conductivities, *γ*, in Ω⁻¹ m⁻¹	Capacitances, *C*, in F; columnar capacitances, C_c, in F m⁻² and capacitivities, ε, in F m⁻¹	Time constants *τ*, in seconds
Volume element at about sea level, 1 m³	$i = 3 \times 10^{-12*}$	$E_0 = 1.2 \times 10^{2*}$	$\rho_0 = 4 \times 10^{13}$	$\gamma_0 = 2.5 \times 10^{-14}$	$\varepsilon_0 = 8.9 \times 10^{-12*}$	$\tau_0 = 3.6\times 10^{2}$
Lower column of 1 m² cross section from sea level to 2 km height	$i = 3 \times 10^{-12}$	At upper end: $U_1 = 1.8 \times 10^{5}$	$R_{c1} = 6 \times 10^{16}$	$G_{c1} = 1.7 \times 10^{-17}$	$C_{c1} = 4.4 \times 10^{-15}$	$\tau_{c1} = 2.6 \times 10^{2}$
Volume element at about 2 km height, 1 m³	$i = 3 \times 10^{-12}$	$E_2 = 6.6 \times 10^{1}$	$\rho_2 = 2.2 \times 10^{13(*)}$	$\gamma_2 = 4.5 \times 10^{-14}$	$\varepsilon_2 = 8.9 \times 10^{-12*}$	$\tau_2 = 2 \times 10^{2}$
Center column of 1 m² cross section from 2 to 12 km	$i = 3 \times 10^{-12}$	At upper end: $U_m = 3.15 \times 10^{5}$	$R_{cm} = 4.5 \times 10^{16}$	$G_{cm} = 5 \times 10^{-17}$	$C_{cm} = 8.8 \times 10^{-16}$	$\tau_{cm} = 1.8 \times 10^{1}$
Volume element at about 12 km height, 1 m³	$i = 3 \times 10^{-12}$	$E_{12} = 3.9 \times 10^{0}$	$\rho_{12} = 1.3 \times 10^{12(*)}$	$\gamma_{12} = 7.7 \times 10^{-13}$	$\varepsilon_{12} = 8.9 \times 10^{-12}$	$\tau_{12} = 1.2 \times 10^{1}$
Upper column of 1 m² cross section from 12 to 65 km height	$i = 3 \times 10^{-12}$	At upper end: $U_u = 3.5 \times 10^{5}$	$R_{cu} = 1.5 \times 10^{16}$	$G_{cu} = 2.5 \times 10^{-17}$	$C_{cu} = 1.67 \times 10^{-16}$	$\tau_{cm} = 6.7 \times 10^{0}$
Whole column of 1 m² cross section from 0 to 65 km height	$i = 3 \times 10^{-12}$	At upper end: $U = 3.5 \times 10^{5}$	$R_c = 1.2 \times 10^{17}$	$G_c = 8.3 \times 10^{-18}$	$C_c = 1.36 \times 10^{-16}$	$\tau_c = 1.64 \times 10^{1}$
Total spherical capacitor area: 5 ×10¹⁴m²	$i = 1.5 \times 10^{3}$	$U = 3.5 \times 10^{5*}$	$R = 2.4 \times 10^{2}$	$G = 4.2 \times 10^{-3}$	$C = 6.8 \times 10^{-2}$	$\tau = 1.64 \times 10^{1}$

Note: All currents and fields listed are part of the global circuit, i.e., circuits of local generators are not included. Values are subject to variations due to latitude and altitude of the point of observation above sea level, locality with respect to sources of disturbances, meteorological and climatological factors, and man-made changes. For more explanations, see text.

shortcomings in the applied measuring techniques. Data which are representative for a large area (in the extreme, "globally representative data," i.e., data on the global circuit), can be obtained on the ground only by stations on an open plane and only if local generators are either small or constant or are independently measured. Certain measurements with instrumented aircraft provide globally representative information valid for the period of the actual measurement.

II. AIR IONS

Hannes Tammet

The term "air ions" signifies all airborne particles that are the carriers of the electrical current in the air and have drift velocities determined by the electric field.

The probability of electrical dissociation of molecules in the atmospheric air under thermodynamic equilibrium is near to zero. The average ionization at the ground level over the ocean is $2{\cdot}10^{6}$ ion pairs $m^{-3}\,s^{-1}$. This ionization is produced mainly by cosmic rays. Over the continents the ionizing radiation from soil and from radioactive substances in the air each add about $4{\cdot}10^{6}\ m^{-3}\,s^{-1}$. The total average ionization rate of $10^{7}\ m^{3}\,s^{-1}$ is equivalent to 17 μR/h, which is a customary expression of the background level of the ionizing radiations. The ionization rate over the ground varies in space due to the radioactivity of soil, and in time depending on the exchange of air between the atmosphere and radon-containing soil. Radioactive pollution increases the ionization rate. A temporary increase of about 10 times was registered in Sweden after the Chernobyl accident in 1986. The emission of ^{85}Krfrom nuclear power plants can noticeably increase the global ionization rate in the 21st century. The ionization rate decreases with altitude near the ground and increases at higher altitudes up to 15 km, where it has a maximum of about $5{\cdot}10^{7}\ m^{-3}\,s^{-1}$. Solar X-ray and extreme UV radiation cause a new increase at altitudes over 60 km.

Local sources of air ions are point discharges in strong electric fields, fluidization of charged drops from waves, etc.

The enhanced chemical activity of an ion results in a chain of ion-molecule reactions with the colliding neutrals, and, in the first microsecond of the life of an air ion, a charged molecular cluster called the *cluster ion* is formed. According to theoretical calculations in the air free from exotic trace gases the following cluster ions should be dominant:

$$NO_3^-{\bullet}(HNO_3){\bullet}H_2O,\ NO_2^-{\bullet}(H_2O)_2,\ NO_3^-{\bullet}H_2O,\ O_2^-{\bullet}(H_2O)_4,\ O_2^-{\bullet}(H_2O)_5,$$
$$H_3O^+{\bullet}(H_2O)_6,\ NH_4^+{\bullet}(H_2O)_2,\ NH_4^+{\bullet}(H_2O),\ H_3O^+{\bullet}(H_2O)_5,\ NH_4^+{\bullet}NH_3$$

A measurable parameter of air ions is the electrical mobility *k*, characterizing the drift velocity in the unit electric field. The mobility is inversely proportional to the density of air, and the results of measurements are as a rule reduced to normal conditions. According to mobility the air ions are called fast or small or light ions with mobility $k > 5{\cdot}10^{-5}\,m^2\,V^{-1}\,s^{1}$, intermediate ions, and slow or large or heavy ions with mobility $k > 10^{-6}\,m^2\,V^{-1}\,s^{1}$. The boundary between intermediate and slow ions is conventional.

Cluster ions are fast ions. The masses of cluster ions may be measured with mass spectrometers, but the possible ion-molecule reactions during the passage of the air through nozzles to the vacuum chamber complicate the measurement. Mass and mobility of cluster ions are highly correlated. The experimental results[5] can be expressed by the empirical formula

$$m \approx \frac{850\ \text{u}}{[0.3 + k/(10^{-4}\,\text{m}^2\text{V}^{-1}\text{s}^{-1})]^3}$$

where u is the unified atomic mass unit.

The value of the transport cross-section of a cluster ion is needed to calculate its mobility according to the kinetic theory of Chapman and Enskog. The theoretical estimation of transport cross-sections is rough and cannot be used to identify the chemical structures of cluster ions. Mass spectrometry is the main technique of identification of cluster ions.[2]

Märk and Castleman[4] presented an overview of over 1000 publications on the experimental studies of cluster ions. Most of them present information about ions of millisecond age range. The low concentration makes it difficult to get detailed information about masses and mobilities of the natural atmospheric ions at ground level. The results of a 1-year continuous measurement[6] are as follows:

	+ ions	– ions	unit
Average mobility	1.36	1.56	10^{-4} m² V⁻¹ s⁻¹
The corresponding mass	190	130	u
The corresponding diameter	0.69	0.61	nm
The average concentration	400	360	10^6 m⁻³
The corresponding conductivity	8.7	9.0	fS

The distribution of tropospheric cluster ions according to the mobility and estimated mass is depicted in Figure 1.

The problems and results of direct mass spectrometry of natural cluster ions are analyzed by Eisele[2] for ground level and by Meyerott, Reagan and Joiner[5] for stratospheric measurements. Air ions in the high atmosphere are a subject of ionospheric physics.

During its lifetime (about 1 min), a cluster ion at ground level collides with nearly 10^{12} molecules. Thus the cluster ions are able to concentrate trace gases of very low concentration if they have an extra high electron or proton affinity. For example, Eisele[2] demonstrated that a considerable fraction of positive atmospheric cluster ions in the unpolluted atmosphere at ground level probably consists of a molecule derived from pyridine. The concentration of these constituents is estimated to be about 10^{-12}. Therefore, air-ion mass and mobility spectrometry is considered as a promising technique for trace analysis in the air. Mass and mobility spectrometry of millisecond-age air ions has been developed as a technique of chemical analysis known as "plasma chromatography."[1] The sensitivity of the detection grows with the age of the cluster ions measured.

The mechanisms of annihilation of cluster ions are ion-ion recombination (on the average 3%) and sedimentation on aerosol particles (on the average 97% of cluster ions at ground level). The result of the combination of a cluster ion and neutral particle is a charged particle called an *aerosol ion*. In conditions of detailed thermodynamic equilibrium the probability that a spherical particle of diameter d carries q elementary charges is calculated from the Boltzmann distribution:

$$p_q(d) = (2\pi d/d_0)^{1/2} \exp(-q^2 d_0/2d)$$

where d_0 = 115 nm (at 18 °C). The supposition about the detailed equilibrium is an approximation and the formula is not valid for particles less than d_0. On the basis of numerical calculations by Hoppel and Frick[3] the following charge probabilities can be derived:

d	3	10	30	100	300	1000	3000	nm
P_0	98	90	70	42	24	14	8	%
$P_{-1}+P_1$	2	10	30	48	41	25	15	%
$P_{-2}+P_2$	0	0	0	10	23	21	14	%
$P_{q>2}$	0	0	0	0	12	40	63	%
k_1	15000	1900	250	28	5.1	1.11	0.33	10^{-9}m²V⁻¹s⁻¹

The last line of the table presents the mobility of a particle carrying one elementary charge. The distribution of the atmospheric aerosol ions over mobility is demonstrated in Figure 2.

Although the concentration of aerosol in continental air at ground level is an order of magnitude higher than the concentration of cluster ions, the mobilities of aerosol ions are so small that their percentage in air conductivity is less than 1%.

A specific class of aerosol ions is condensed aerosol ions produced as a result of the condensation of gaseous matter on the cluster ions. In aerosol physics the process is called ion-induced nucleation; it is considered as one of the processes of gas-to-particle conversion. The condensed aerosol ions have an inherent charge. Their sizes and mobilities are between the sizes and mobilities of cluster ions and of ordinary aerosol ions. Water and standard constituents of atmospheric air are not able to condense on the cluster ions in the real atmosphere. Thus the concentration of condensed aerosol ions depends on the trace constituents in the air and is very low in unpolluted air. Knowledge about condensed aerosol ions is poor because of measurement difficulties.

References

1. Carr, T. W., Ed., *Plasma Chromatography,* Plenum Press, New York and London, XII + 259 pp., 1984.
2. Eisele, F. L., Identification of Tropospheric Ions, *J. Geophys. Res.*, vol. 91, no. D7, pp. 7897–7906, 1986.
3. Hoppel, W. A., and Frick, G. M., The Nonequilibrium Character of the Aerosol Charge Distributions Produced by Neutralizers, *Aerosol Sci. Technol.*, vol. 12, no. 3, pp. 471–496, 1990.
4. Mark, T. D., and Castleman, A. W., Experimental Studies on Cluster Ions, in *Advances in Atomic and Molecular Physics*, vol. 20, pp. 65–172, Academic Press, 1985.

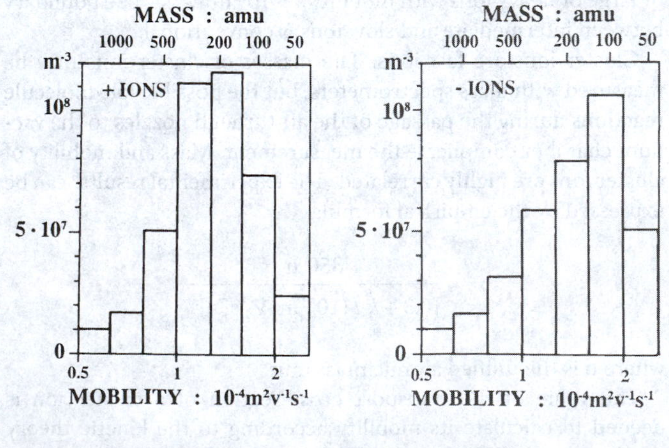

FIGURE 1. Average mobility and mass spectra of natural tropospheric cluster ions. Concentrations of the mobility fractions were measured in a rural site every 5 min over 1 year.[6] Ion mass is estimated according to the above empirical formula.

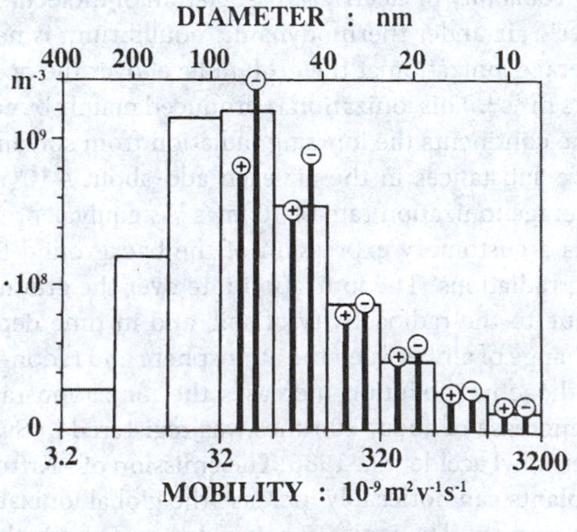

FIGURE 2. Mobility and size spectra of tropospheric aerosol ions.[6] The wide bars mark the fraction concentrations theoretically estimated on the basis of the standard size distribution of tropospheric aerosol. The pin bars with head + and – mark average values of positive and negative aerosol ion fraction concentrations measured in a rural site every 5 min during 4 months.

5. Meyerott, R. E., Reagan, J. B., and Joiner, R. G., The Mobility and Concentration of Ions and the Ionic Conductivity in the Lower Stratosphere, *J. Geophys. Res.*, vol. 85, no. A3, pp. 1273–1278, 1980.
6. Salm, J., Tammet, H., Iher, H., and Hõrrak, U., Atmospheric Electrical Measurements in Tahkuse, Estonia (in Russian), in *Voprosy Atmosfernogo Elektrichestva*, pp. 168–175, Gidrometeoizdat, Leningrad, 1990.

III. THUNDERSTORM ELECTRICITY

John Latham

The development of improved radar techniques and instruments for in-cloud electrical and physical measurements, coupled with a much clearer recognition by the research community that establishment of the mechanism or mechanisms responsible for electric field development in thunderclouds, culminating in lightning, is inextricably linked to the concomitant dynamical and microphysical evolution of the clouds, has led to significant progress over the past decade.

Field studies indicate that in most thunderclouds the electrical development is associated with the process of glaciation, which can occur in a variety of incompletely understood ways. In the absence of ice, field growth is slow, individual hydrometeor charges are low, and lightning is produced only rarely. Precipitation — in the solid form, as graupel — also appears to be a necessary ingredient for significant electrification, as does significant convective activity and mixing between the clouds and their environments, via entrainment.

Increasingly, the view is being accepted that charge transfer leading to field-growth is largely a consequence of rebounding collisions between graupel pellets and smaller vapor-grown ice crystals, followed by the separation under gravity of these two types of hydrometeor. These collisions occur predominantly within the temperature range −15 to −30 °C, and for significant charge transfer need to occur in the presence of supercooled cloud droplets.

The field evidence is inconsistent with an inductive mechanism, and extensive laboratory studies indicate that the principal charging mechanism is non-inductive and associated — in ways yet to be identified — with differences in surface characteristics of the interacting hydrometeors.

Laboratory studies indicate that the two most favored sites for corona emission leading to the lightning discharge are the tips of ephemeral liquid filaments, produced during the glancing collisions of supercooled raindrops, and protuberances on large ice crystals or graupel pellets. The relative importance of these alternatives will depend on the hydrometeor characteristics and the temperature in the regions of strongest fields; these features are themselves dependent on air-mass characteristics and climatological considerations.

A recently identified but unresolved question is why, in continental Northern Hemisphere thunderclouds at least, the sign of the charge brought to ground by lightning is predominantly negative in summer but more evenly balanced in winter.

IV. LIGHTNING

Martin A. Uman

From both ground-based weather-station data and satellite measurements, it has been estimated that there are about 100 lightning discharges, both cloud and ground flashes, over the whole Earth each second; representing an average global lightning flash density of about 6 $km^{-2}\,yr^{-1}$. Most of this lightning occurs over the Earth's land masses. For example, in central Florida, where thunderstorms occur about 90 days/yr, the flash density for discharges to earth is about 15 $km^{-2}\,yr^{-1}$. Some tropical areas of the Earth have thunderstorms up to 300 days/yr.

Lightning can be defined as a transient, high-current electric discharge whose path length is measured in kilometers and whose most common source is the electric charge separated in the ordinary thunderstorm or cumulonimbus cloud. Well over half of all lightning discharges occur totally within individual thunderstorm clouds and are referred to as intracloud discharges. Cloud-to-ground lightning, however, has been studied more extensively than any other lightning form because of its visibility and its more practical interest. Cloud-to-cloud and cloud-to-air discharges are less common than intracloud or cloud-to-ground lightning.

Lightning between the cloud and Earth can be categorized in terms of the direction of motion, upward or downward, and the sign of the charge, positive or negative, of the developing discharge (called a *leader*) that initiates the overall event. Over 90% of the worldwide cloud-to-ground discharges is initiated in the thundercloud by downward-moving negatively charged leaders and subsequently results in the lowering of negative charge to Earth. Cloud-to-ground lightning can also be initiated by downward-moving positive leaders, less than 10% of the worldwide cloud-to-ground lightning being of this type although the exact percentage is a function of season and latitude. Lightning between cloud and ground can also be initiated by leaders that develop upward from the Earth. These upward-initiated discharges are relatively rare, may be of either polarity, and generally occur from mountaintops and tall man-made structures.

We discuss next the most common type of cloud-to-ground lightning. A negative cloud-to-ground discharge or *flash* has an overall duration of some tenths of a second and is made up of various components, among which are typically three or four high-current pulses called *strokes*. Each stroke lasts about a millisecond, the separation time between strokes being typically several tens of milliseconds. Such lightning often appears to "flicker" because the human eye can just resolve the individual light pulse associated with each stroke. A drawing of the components of a negative cloud-to-ground flash is found in Figure 3. Some values for salient parameters are found in Table 1. The negatively charged *stepped leader* initiates the first stroke in a flash by propagating from cloud to ground through virgin air in a series of discrete steps. Photographically observed leader steps in clear air are typically 1 μs in duration and tens of meters in length, with a pause time between steps of about 50 μs. A fully developed stepped leader lowers up to 10 or more coulombs of negative cloud charge toward ground in tens of milliseconds with an average downward speed of about 2×10^5 m/s. The average leader current is in the 100 to 1000 A range. The steps have pulse currents of at least 1 kA. Associated with these currents are electric- and magnetic-field pulses with widths of about 1 μs or less and risetimes of about 0.1 μs or less. The stepped leader, during its trip toward ground, branches in a downward direction, resulting in the characteristic downward-branched geometrical structure commonly observed. The electric potential of the bottom of the negatively charged leader channel with respect to ground has a magnitude in excess of 10^7 V. As the leader tip nears ground, the electric field at sharp objects on the ground or at irregularities of the ground itself exceeds the breakdown value of air, and one or more upward-moving discharges (often called upward leaders) are initiated from those points, thus beginning the *attachment process*. An understanding of the physics of the attachment process is central to an understanding of the operation of lightning protection of ground-based objects and the effects of lightning on humans and animals, since it is the attach-

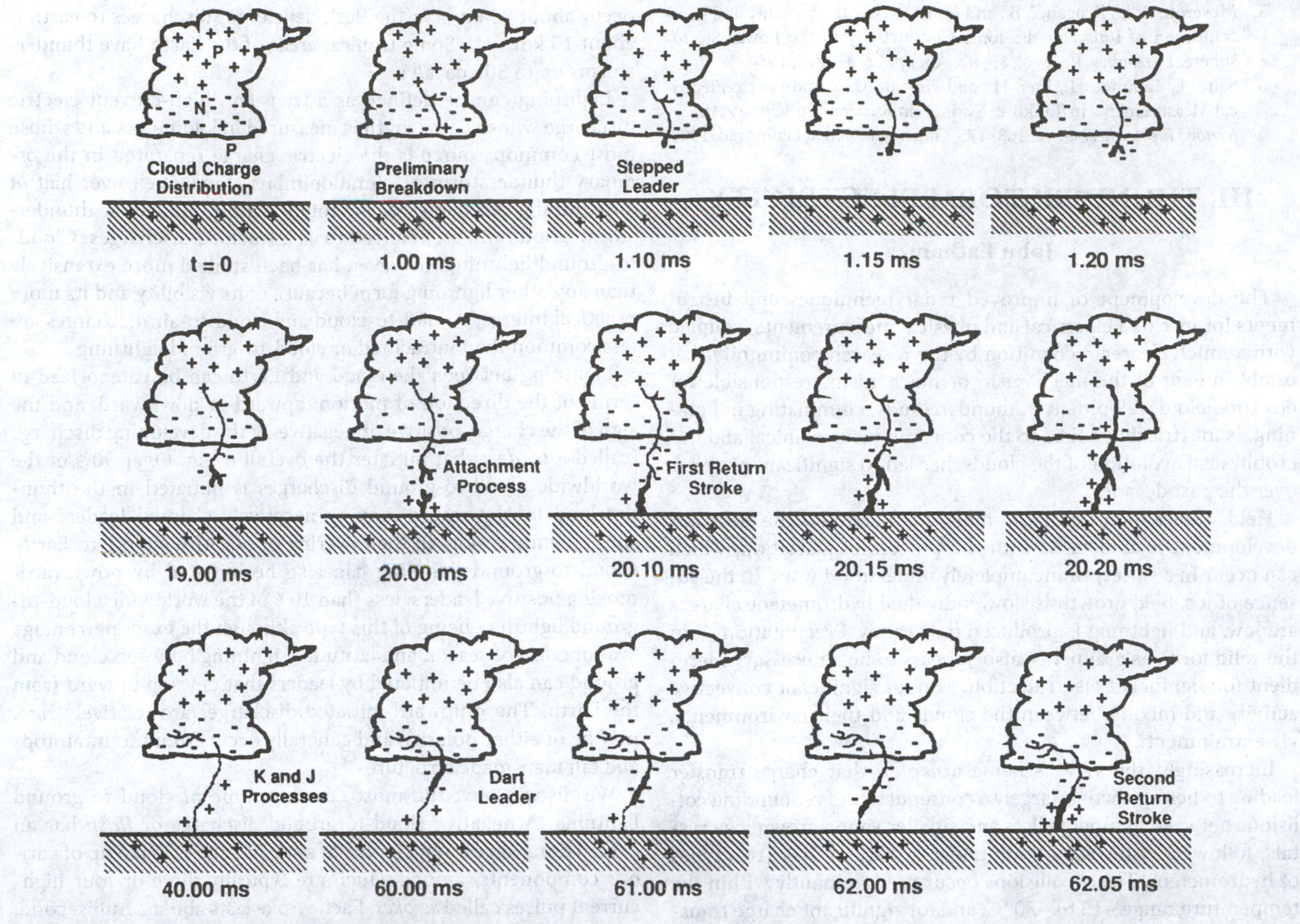

FIGURE 3. Sequence of steps in cloud-to-ground lightning.

ment process that determines where the lightning connects to objects on the ground and the value of the early currents which flow. When one of the upward-moving discharges from the ground (or from a lightning rod or an individual) contacts the tip of the downward-moving stepped leader, typically some tens of meters above the ground, the leader tip is effectively connected to ground potential. The negatively charged leader channel is then discharged to earth when a ground potential wave, referred to as the first *return stroke*, propagates continuously up the leader path. The upward speed of a return stroke near the ground is typically near one third the speed of light, and the speed decreases with height. The first return stroke produces a peak current near ground of typically 30 kA, with a time from zero to peak of a few microseconds. Currents measured at the ground fall to half of the peak value in about 50 μs, and currents of the order of hundreds of amperes may flow for times of a few milliseconds up to several hundred milliseconds. The longer-lasting currents are known as *continuing currents*. The rapid release of return stroke energy heats the leader channel to a temperature near 30,000 K and creates a high-pressure channel which expands and generates the shock waves that eventually become thunder, as further discussed later. The return stroke effectively lowers to ground the charge originally deposited onto the stepped-leader channel and additionally initiates the lowering of other charges that may be available to the top of its channel. First return-stroke electric fields exhibit a micro second scale rise to peak with a typical peak value of 5 V/m, normalized to a distance of 100 km by an inverse distance relationship. Roughly half of the field rise to peak, the so-called "fast transition," takes place in tenths of a microsecond, an observation that can only be made if the field propagation is over a highly conducting surface such as salt water.

After the first return-stroke current has ceased to flow, the flash, including charge motion in the cloud, may end. The lightning is then called a single-stroke flash. On the other hand, if additional charge is made available to the top of the channel, a continuous or *dart leader* may propagate down the residual first-stroke channel at a typical speed of about 1×10^7 m/s. The dart leader lowers a charge of the order of 1 C by virtue of a current of about 1 kA. The dart leader then initiates the second (or any subsequent) return stroke. Subsequent return-stroke currents generally have faster zero-to-peak rise times than do first-stroke currents, but similar maximum rates of change, about 100 kA/μs. Some leaders begin as dart leaders, but toward the end of their trip toward ground become stepped leaders. These leaders are known as *dart-stepped leaders* and may have different ground termination points (and separate upward leaders) from the first stroke. Most often the dart-stepped leaders are associated with the second stroke of the flash. Nearly half of all flashes exhibit more than one termination point on ground with the distance between separate terminations being up to several kilometers. Subsequent return-stroke radiated electric and magnetic fields are similar to, but usually a factor of two or so smaller than first return-stroke fields. About one third of all multiple-stroke flashes has at least one subsequent stroke that is larger than the first stroke.

Cloud-to-ground flashes that lower positive charge, though not common, are of considerable practical interest because their peak currents and total charge transfer can be much larger than for the more common negative ground flash. The largest recorded peak currents, those in the 200- to 300-kA range, are due to the return strokes of positive lightning. Such positive flashes to ground are initiated by downward-moving leaders that do not exhibit the distinct steps of their negative counterparts. Rather, they show a luminosity that is more or less continuous but modulated in intensity. Positive flashes are generally composed of a single stroke followed by a period of continuing current. Positive flashes are probably initiated from the upper positive charge in the thundercloud charge dipole when that cloud charge is horizontally separated from the negative charge beneath it, the source of the usual negative cloud-to-ground lightning. Positive flashes are relatively common in winter thunderstorms (snow storms), which produce few flashes overall, and are relatively uncommon in summer thunderstorms. The fraction of positive lightning in summer thunderstorms apparently increases with increasing latitude and with increasing height of the ground above sea level.

Distant lightning return stroke fields are often referred to as sferics (called "atmospherics" in the older literature). The peak in the sferics frequency spectrum is near 5 kHz due to the bipolar or ringing nature of the distant return-stroke electromagnetic signal and to the effects of propagation.

Thunder, the acoustic radiation associated with lightning, is sometimes divided into the categories "audible," sounds that one can hear, and "infrasonic," below a few tens of hertz, a frequency range that is inaudible. This division is made because it is thought that the mechanisms that produce audible and infrasonic thunder are different. Audible thunder is thought to be due to the expansion of a rapidly heated return stroke channel, as noted earlier, whereas infrasonic thunder is thought to be associated with the conversion to sound of the energy stored in the electrostatic field of the thundercloud when lightning rapidly reduces that cloud field.

The technology of artificially initiating lightning by firing upward small rocket trailing grounded wire of a few hundred meters length has been well-developed during the past decade. Such "triggered" flashes are similar to natural upward-initiated discharges from tall structure. They often contain subsequent strokes which, when they occur, are similar to the subsequent strokes in natural lightning. These triggered subsequent strokes have been the subject of considerable recent research.

Also in the past 10 years or so sophisticated lightning locating equipment has been installed throughout the world. For example, all ground flashes in the U.S. are now centrally monitored for research, for better overall weather prediction, and for hazard warning for aviation, electric utilities and other lightning-sensitive facilities.

Information on lightning physics can be found in M. A. Uman, *The Lightning Discharge*, Academic Press, San Diego, 1987; on lightning death and injury in *Medical Aspects of Lightning Injury*, C. Andrews, M. A. Cooper, M. Darveniza, and D. Mackerras, Eds., CRC Press, Boca Raton, FL, 1992. Ground flash location information for the U.S., in real time or archived, is available from Geomet Data Service of Tucson, AZ, which is also a source of the names of providers of those data in other countries.

Table 2 has data for cloud-to-ground lightning discharges bringing negative charge to earth. The values listed are intended to convey a rough feeling for the various physical parameters of lightning. No great accuracy is claimed since the results of different investigators are often not in good agreement. These values may, in fact, depend on the particular environment in which the lightning discharge is generated. The choice of some of the entries in the table is arbitrary.

TABLE 2. Data for Cloud-to-Ground Lightning Discharges

	Minimum[a]	Representative values	Maximum[a]
Stepped leader			
Length of step, m	3	50	200
Time interval between steps, μs	30	50	125
Average speed of propagation of stepped leader, m/s[b]	1.0×10^5	2.0×10^5	3.0×10^6
Charged deposited on stepped-leader channel, coulombs	3	5	20
Dart leader			
Speed of propagation, m/s[b]	1.0×10^6	1.0×10^7	2.4×10^7
Charged deposited on dart-leader channel, coulombs	0.2	1	6
Return stroke[c]			
Speed of propagation, m/s[b]	2.0×10^7	1.0×10^8	2.0×10^8
Maximum current rate of increase, kA/μs	<1	100	400
Time to peak current, μs	<1	2	30
Peak current, kA	2	30	200
Time to half of peak current, μs	10	50	250
Charge transferred excluding continuing current, coulombs	0.02	3	20
Channel length, km	2	5	15
Lightning flash			
Number of strokes per flash	1	4	26
Time interval between strokes in absence of continuing current, ms	3	60	100
Time duration of flash, s	10^{-2}	0.5	2
Charge transferred including continuing current, coulombs	3	30	200

[a] The words maximum and minimum are used in the sense that most measured values fall between these limits.

[b] Speeds of propagation are generally determined from photographic data and are "two-dimensional." Since many lightning flashes are not vertical, values stated are probably slight underestimates of actual values.

[c] First return strokes have longer times to current peak and generally larger charge transfer than do subsequent return strokes.

Adapted from Uman, M.A., *Lightning*, Dover Paperbook, New York, 1986, and Uman, M.A., *The Lightning Discharge*, Academic Press, San Diego, 1987.

Geo-Astro

SPEED OF SOUND IN VARIOUS MEDIA

The speed of sound in various solids, liquids, and gases is given in these tables. While only a single parameter v is needed for liquids and gases, sound propagation in isotropic solids is characterized by three velocity parameters. For a solid of infinite extent (or of finite extent if all dimensions are much larger than a wavelength), there are two relevant quantities.

v_l: velocity of longitudinal waves
v_s: velocity of shear waves

For a cylindrical rod with diameter much smaller than a wavelength,

v_{ext}: velocity of extensional waves along the rod. (Torsional waves in the rod are propagated at the same speed as sheer waves in an infinite solid.)

Table 1 lists values for a variety of solid materials. Table 2 covers liquids and gases; values for cryogenic liquids are given at the normal boiling point. Table 3 gives the speed of sound in pure water and in seawater of salinity S = 3.5% as a function of temperature. All values are in meters per second and are given for normal atmospheric pressure.

References

1. Gray, D. E., Ed., *American Institute of Physics Handbook, Third Edition*, McGraw-Hill, New York, 1972.
2. Anderson, H.L., Ed., *A Physicist's Desk Reference*, American Institute of Physics, New York, 1989.
3. 6. Mason, W. P., *Physical Acoustics and the Properties of Solids*, D. Van Nostrand Co., Princeton, N.J., 1958.
4. Landolt-Börnstein, *Numerical Data and Functional Relationships in Science and Technology, New Series, II/5, Molecular Acoustics*, Springer-Verlag, Heidelberg, 1967.
5. Lemmon, E.W., Huber, M.L., and McLinden, M.O., *NIST Standard Reference Database 23: Reference Fluid Thermodynamic and Transport Properties-REFPROP*, Version 9.0, National Institute of Standards and Technology, Standard Reference Data Program, Gaithersburg, MD, 2010.
6. International Association for the Properties of Water and Steam (IAPWS), *Release on the IAPWS Formulation 2008 for the Thermodynamic Properties of Seawater* (2008), available from http://www.iapws.org.
7. Feistel, R., A Gibbs Function for Seawater Thermodynamics for −6 to 80 °C and Salinity up to 120 g kg^{-1}, *Deep-Sea Res. I* 55, 1639-1671 (2008).
8. Millero, F.J., Feistel, R., Wright, D.G., and McDougall, T.J., The Composition of Standard Seawater and the Definition of the Reference-Composition Salinity Scale, *Deep-Sea Res. I* 55, 50-72 (2008).

TABLE 1. Speed of Sound in Solids at Room Temperature

Name	v_l/m s^{-1}	v_s/m s^{-1}	v_{ext}/m s^{-1}
Metals			
Aluminum, rolled	6420	3040	5000
Beryllium	12890	8880	12870
Brass (70 Cu, 30 Zn)	4700	2110	3480
Constantan	5177	2625	4270
Copper, annealed	4760	2325	3810
Copper, rolled	5010	2270	3750
Duralumin 17S	6320	3130	5150
Gold, hard-drawn	3240	1200	2030
Iron, cast	4994	2809	4480
Iron, electrolytic	5950	3240	5120
Iron, Armco	5960	3240	5200
Lead, annealed	2160	700	1190
Lead, rolled	1960	690	1210
Magnesium, annealed	5770	3050	4940
Molybdenum	6250	3350	5400
Monel metal	5350	2720	4400
Nickel	6040	3000	4900
Platinum	3260	1730	2800
Silver	3650	1610	2680
Steel (1% C)	5940	3220	5180
Steel, 347 Stainless	5790	3100	5000
Steel, K9	5940	3250	5250
Tin, rolled	3320	1670	2730
Titanium	6070	3125	5090
Tungsten, annealed	5220	2890	4620
Tungsten, drawn	5410	2640	4320
Zinc, rolled	4210	2440	3850
Other materials			
Fused silica	5968	3764	5760
Glass, heavy silicate flint	3980	2380	3720
Glass, light borate crown	5100	2840	4540
Glass, pyrex	5640	3280	5170
Lucite	2680	1100	1840
Nylon 6-6	2620	1070	1800
Polyethylene	1950	540	920
Polystyrene	2350	1120	1840
Rubber, butyl	1830		
Rubber, gum	1550		
Rubber, neoprene	1600		
Tungsten carbide	6655	3980	6220

TABLE 2. Speed of Sound in Liquids and Gases

Name	Synonym	t/°C	v/m s^{-1}
Liquids			
Acetone	2-Propanone	20	1187
Argon		-185.8	838
Benzene	[6]Annulene	25	1302
Bromobenzene	Phenyl bromide	20	1169
Butane		-0.5	1038
1-Butanol	Butyl alcohol	20	1262
Carbon disulfide	Carbon bisulfide	25	1140
Chlorobenzene	Phenyl chloride	20	1311
Cyclohexane	Hexahydrobenzene	19	1299
1-Decene		20	1250
Diethyl ether	Ethyl ether	25	951
Ethane		-88.6	1316
Ethanol	Ethyl alcohol	20	1159
Ethylene	Ethene	-103.8	1307
Ethylene glycol	Ethylene glycol	25	1639
Fluorobenzene	Phenyl fluoride	20	1183
Glycerol	1,2,3-Propanetriol	25	1904
Helium		-268.9	177
Heptane		20	1149
1-Heptene		20	1128
Hexane		20	1078
Hydrogen		-252.8	1119
Iodobenzene	Phenyl iodide	20	1114
Mercury	Quicksilver	25	1450
Methane		-161.5	1338
Methanol	Methyl alcohol	20	1116
Nitrobenzene		25	1463
Nitrogen		-195.8	851
1-Nonene	1-Nonylene	20	1218
Octane		20	1166
1-Octene	Caprylene	20	1184
Oxygen		-183.0	904
1-Pentadecene		20	1351
Pentane		20	1036
Propane	LPG	-42.1	1161
1-Propanol	Propyl alcohol	20	1223
Tetrachloromethane	Carbon tetrachloride	25	930
Trichloromethane	Chloroform	25	987
1-Undecene		20	1275
Water		25	1497
Water (sea, S = 3.5%)		25	1534
Gases at 1 atm			
Air, dry		25	346
Ammonia	R-717	0	414
Argon		27	323
Carbon monoxide	Carbon oxide	0	337
Carbon dioxide	Carbonic anhydride	0	258
Chlorine		0	205
Deuterium	Deuterium	0	888
Ethane		27	312
Ethylene	Ethene	27	331
Helium		0	973
Hydrogen		27	1320
Hydrogen bromide	Hydrobromic acid	0	200
Hydrogen chloride	Hydrochloric acid	0	294
Hydrogen iodide	Hydroiodic acid	0	157
Hydrogen sulfide		0	295
Methane		27	450
Neon		0	433
Nitric oxide		10	330
Nitrogen		27	353
Nitrous oxide		0	256
Oxygen		27	330
Sulfur dioxide		0	209
Water (steam)		100	472

TABLE 3. Speed of Sound in m/s in Water and Reference Seawater (Practical Salinity S = 35; Absolute Salinity S_A = 35.16504 g kg^{-1}) at Different Temperatures

t/°C	Water	Seawater
0	1402.4	1449.0
10	1447.3	1489.8
20	1482.3	1521.5
25	1496.7	1534.3
30	1509.2	1545.5
40	1528.9	1563.4
50	1542.6	
60	1551.0	
70	1554.7	
80	1554.4	
90	1550.4	
100	1543.5	

ATTENUATION AND SPEED OF SOUND IN AIR AS A FUNCTION OF HUMIDITY AND FREQUENCY

This table gives the attenuation and speed of sound as a function of frequency at various values of relative humidity. All values refer to still air at 20 °C.

References

1. Tables of Absorption and Velocity of Sound in Still Air at 68 °F (20 °C), AD-738576, National Technical Information Service, Springfield, VA.
2. Evans, L. B., Bass, H. E., and Sutherland, L. C., *J. Acoust. Soc. Am.*, 51, 1565, 1972.

Frequency (Hz)	Attenuation (dB/km)	Speed (m/s)
Relative humidity 0%		
20	0.51	343.477
40	1.07	343.514
50	1.26	343.525
63	1.43	343.536
100	1.67	343.550
200	1.84	343.559
400	1.96	343.561
630	2.11	343.562
800	2.27	343.562
1250	2.82	343.562
2000	4.14	343.562
4000	8.84	343.564
6300	14.89	343.565
10000	26.28	343.566
12500	35.81	343.566
16000	52.15	343.567
20000	75.37	343.567
40000	267.01	343.567
63000	644.66	343.567
80000	1032.14	343.567
Relative humidity 30%		
20	0.03	343.807
40	0.11	343.808
50	0.17	343.810
63	0.25	343.810
100	0.50	343.814
200	1.01	343.821
400	1.59	343.826
630	2.24	343.827
800	2.85	343.828
1250	5.09	343.828
2000	10.93	343.829
4000	38.89	343.831
6300	90.61	343.836
10000	204.98	343.846
12500	294.08	343.854
16000	422.51	343.865
20000	563.66	343.877
40000	1110.97	343.911
63000	1639.47	343.924
80000	2083.08	343.929

Frequency (Hz)	Attenuation (dB/km)	Speed (m/s)
Relative humidity 60%		
20	0.02	344.182
40	0.06	344.183
50	0.09	344.183
63	0.15	344.184
100	0.34	344.185
200	0.99	344.190
400	1.94	344.197
630	2.57	344.200
800	2.94	344.201
1250	4.01	344.202
2000	6.55	344.203
4000	18.73	344.204
6300	42.51	344.204
10000	101.84	344.206
12500	155.67	344.208
16000	247.78	344.211
20000	373.78	344.215
40000	1195.37	344.238
63000	2220.64	344.262
80000	2951.71	344.274
Relative humidity 100%		
20	0.01	344.685
40	0.04	344.685
50	0.06	344.685
63	0.09	344.685
100	0.22	344.686
200	0.77	344.689
400	2.02	344.695
630	3.05	344.699
800	3.57	344.701
1250	4.59	344.704
2000	6.29	344.705
4000	13.58	344.706
6300	27.72	344.706
10000	63.49	344.706
12500	96.63	344.707
16000	154.90	344.708
20000	237.93	344.709
40000	884.28	344.718
63000	1973.62	344.731
80000	2913.01	344.742

SPEED OF SOUND IN DRY AIR

Eric W. Lemmon

These values were calculated from the equation of state for dry air (average molecular weight 28.96) treated as a real gas. Values refer to standard atmospheric pressure. The speed of sound varies only slightly with pressure; at two atmospheres and −100 °C the value decreases by 0.16%, while at two atmospheres and 80 °C the speed increases by 0.05%. For additional values, see the table in Section 6 labeled "Thermophysical Properties of Air."

Reference

Lemmon, E.W., Jacobsen, R.T, Penoncello, S.G., and Friend, D.G., Thermodynamic Properties of Air and Mixtures of Nitrogen, Argon, and Oxygen from 60 to 2000 K at Pressures to 2000 MPa, *J. Phys. Chem. Ref. Data* 29, 331, 2000.

t/°C	v_s/m s^{-1}	*t*/°C	v_s/m s^{-1}	*t*/°C	v_s/m s^{-1}
−100	263.5	−30	312.7	40	354.9
−95	267.3	−25	315.9	45	357.7
−90	271.1	−20	319.1	50	360.4
−85	274.8	−15	322.2	55	363.2
−80	278.5	−10	325.4	60	365.9
−75	282.1	−5	328.4	65	368.7
−70	285.7	0	331.5	70	371.3
−65	289.2	5	334.5	75	374.0
−60	292.7	10	337.5	80	376.7
−55	296.1	15	340.5	85	379.3
−50	299.5	20	343.4	90	381.9
−45	302.9	25	346.3	95	384.5
−40	306.2	30	349.2	100	387.0
−35	309.5	35	352.0		

MUSICAL SCALES

Equal Tempered Chromatic Scale

A_4 = 440 Hz

International Concert Pitch

Note	Frequency in Hz	Note	Frequency in Hz	Note	Frequency in Hz	Note	Frequency in Hz
C_0	16.35	C_2	65.41	C_4	261.63	C_6	1046.50
$C\#_0$	17.32	$C\#_2$	69.30	$C\#_4$	277.18	$C\#_6$	1108.73
D_0	18.35	D_2	73.42	D_4	293.66	D_6	1174.66
$D\#_0$	19.45	$D\#_2$	77.78	$D\#_4$	311.13	$D\#_6$	1244.51
E_0	20.60	E_2	82.41	E_4	329.63	E_6	1318.51
F_0	21.83	F_2	87.31	F_4	349.23	F_6	1396.91
$F\#_0$	23.12	$F\#_2$	92.50	$F\#_4$	369.99	$F\#_6$	1479.98
G_0	24.50	G_2	98.00	G_4	392.00	G_6	1567.98
$G\#_0$	25.96	$G\#_2$	103.83	$G\#_4$	415.30	$G\#_6$	1661.22
A_0	27.50	A_2	110.00	A_4	440.00	A_6	1760.00
$A\#_0$	29.14	$A\#_2$	116.54	$A\#_4$	466.16	$A\#_6$	1864.66
B_0	30.87	B_2	123.47	B_4	493.88	B_6	1975.53
C_1	32.70	C_3	130.81	C_5	523.25	C_7	2093.00
$C\#_1$	34.65	$C\#_3$	138.59	$C\#_5$	554.37	$C\#_7$	2217.46
D_1	36.71	D_3	146.83	D_5	587.33	D_7	2349.32
$D\#_1$	38.89	$D\#_3$	155.56	$D\#_5$	622.25	$D\#_7$	2489.02
E_1	41.20	E_3	164.81	E_5	659.26	E_7	2637.02
F_1	43.65	F_3	174.61	F_5	698.46	F_7	2793.83
$F\#_1$	46.25	$F\#_3$	185.00	$F\#_5$	739.99	$F\#_7$	2959.96
G_1	49.00	G_3	196.00	G_5	783.99	G_7	3135.96
$G\#_1$	51.91	$G\#_3$	207.65	$G\#_5$	830.61	$G\#_7$	3322.44
A_1	55.00	A_3	220.00	A_5	880.00	A_7	3520.00
$A\#_1$	58.27	$A\#_3$	233.08	$A\#_5$	932.33	$A\#_7$	3729.31
B_1	61.74	B_3	246.94	B_5	987.77	B_7	3951.07
						C_8	4186.01

Scientific or Just Scale

C_4 = 256 Hz

Note	Frequency in Hz	Note	Frequency in Hz	Note	Frequency in Hz	Note	Frequency in Hz
C_0	16	C_2	64	C_4	256	C_6	1024
D_0	18	D_2	72	D_4	288	D_6	1152
E_0	20	E_2	80	E_4	320	E_6	1280
F_0	21.33	F_2	85.33	F_4	341.33	F_6	1365.33
G_0	24	G_2	96	G_4	384	G_6	1536
A_0	26.67	A_2	106.67	A_4	426.67	A_6	1706.67
B_0	30	B_2	120	B_4	480	B_6	1920
C_1	32	C_3	128	C_5	512	C_7	2048
D_1	36	D_3	144	D_5	576	D_7	2304
E_1	40	E_3	160	E_5	640	E_7	2560
F_1	42.67	F_3	170.67	F_5	682.67	F_7	2730.67
G_1	48	G_3	192	G_5	768	G_7	3072
A_1	53.33	A_3	213.33	A_5	853.33	A_7	3413.33
B_1	60	B_3	240	B_5	960	B_7	3840
						C_8	4096

CHARACTERISTICS OF HUMAN HEARING

The human ear is sensitive to sound waves with frequencies in the range from a few hertz to almost 20 kHz. Auditory response is usually expressed in terms of the *loudness level* of a sound, which is a measure of the sound pressure. The reference level, which is given in the unit *phon,* is a pure tone of frequency 1000 Hz with sound pressure of 20 μPa (in cgs units, $2 \cdot 10^{-4}$ dyn/cm^2); loudness level is usually expressed in decibels (dB) relative to this reference level. If a normal observer perceives an arbitrary sound to be equally loud as this reference sound, the sound is said to have the loudness level of the reference. The sensitivity of the typical human ear ranges from about 0 dB, the threshold loudness level, to about 140 dB, the level at which pain sets in. The minimum detectable level thus represents a sound wave of pressure 20 μPa and intensity (power density) 10^{-16} W/cm^2.

The following figure illustrates the frequency dependence of the threshold for an average young adult.

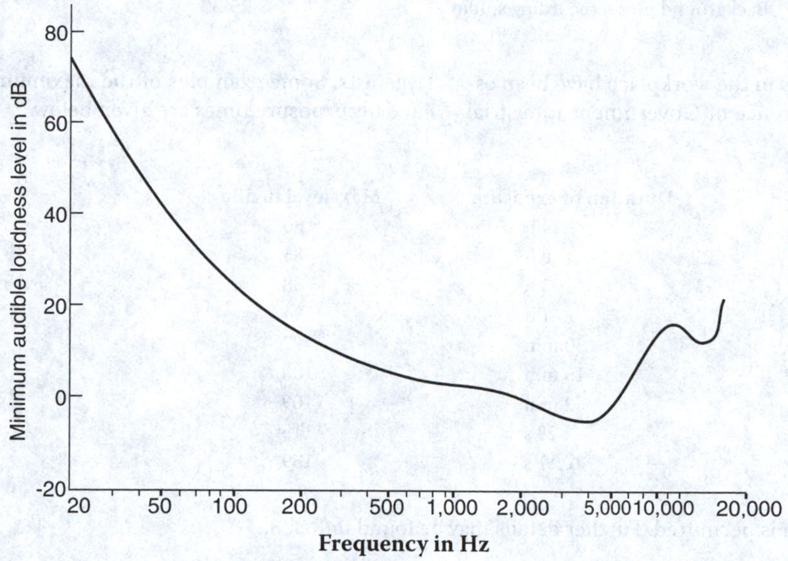

The relation between loudness level and frequency for a typical person is expressed by the following table:

Sound pressure level in dB relative to 20 μPa	Frequency in Hz 125	500	1000	4000	8000	10000
10			10	18		
20		16	20	28	11	
30	4	27	30	37	21	17
40	17	39	40	45	30	26
50	34	52	50	54	38	35
60	52	65	60	64	47	44
70	70	76	70	73	56	54
80	86	86	80	83	66	64
90	98	96	90	94	77	74
100	108	105	100	106	88	86

Thus, a 10,000 Hz tone at a pressure level of 50 dB seems equally loud as a 1000 Hz tone at a pressure of 35 dB.

The term *noise* refers to any unwanted sound, either a pure tone or a mixture of frequencies. Since the sensitivity of the ear is frequency dependent, as illustrated by the above table, noise level is expressed in a frequency-weighted scale, known as A-weighting. Decibel readings on this scale are designated as dBa. Typical noise levels from various sources are illustrated in this table:

Geo-Astro

Source	Noise level in dBa
Rocket engine	200
Jet aircraft engine	160
Light aircraft, cruising	140
Tractor, 150 hp	115
Electric motor, 100 hp at 2600 rpm	105
Pneumatic drill	100
Subway train	90
Vacuum cleaner	85
Heavy automobile traffic	75
Conversational speech	65
Whispered speech	40
Background noise, recording studio	25-30

Recommended noise thresholds in the workplace have been established by the American Conference of Government Industrial Hygenists. Some examples of the maximum safe levels for different daily exposure times are given below.

Duration of exposure	Max. level in dBa
24 h	80
8 h	85
4 h	88
1 h	94
30 min	97
15 min	100
2 min	109
28 s	115
0.11 s	139

No exposure greater than 140 dBa is permitted. Further details may be found in Ref. 3.

References

1. Anderson, H. L., Ed., *A Physicist's Desk Reference*, American Institute of Physics, New York, 1989, chap. 2.
2. Gray, D. E., Ed., *American Institute of Physics Handbook, Third Edition*, McGraw Hill, New York, 1972, chap. 3.
3. *Threshold Limit Values for Chemical Substances and Physical Agents; Biological Exposure Indices*, 2008 Edition, American Conference of Governmental Industrial Hygienists, 1330 Kemper Meadow Drive, Cincinnati, OH 45240-1634; <www.acgih.org>.

Section 15
Practical Laboratory Data

STANDARD ITS-90 THERMOCOUPLE TABLES

The Instrument Society of America (ISA) has assigned standard letter designations to a number of thermocouple types having specified emf-temperature relations. These designations and the approximate metal compositions that meet the required relations, as well as the useful temperature ranges, are given below:

Type B	(Pt + 30% Rh) vs. (Pt + 6% Rh)	0 to 1820 °C
Type E	(Ni + 10% Cr) vs. (Cu + 43% Ni)	–270 to 1000 °C
Type J	Fe vs. (Cu + 43% Ni)	–210 to 1200 °C
Type K	(Ni + 10% Cr) vs. (Ni + 2% Al + 2% Mn + 1% Si)	–270 to 1372 °C
Type N	(Ni + 14% Cr + 1.5% Si) vs. (Ni + 4.5% Si + 0.1% Mg)	–270 to 1300 °C
Type R	(Pt + 13% Rh) vs. Pt	–50 to 1768 °C
Type S	(Pt + 10% Rh) vs. Pt	–50 to 1768 °C
Type T	Cu vs. (Cu + 43% Ni)	–270 to 400 °C

The compositions are given in weight percent, and the positive leg is listed first. It should be emphasized that the standard letter designations do not imply a precise composition but rather that the specified emf-temperature relation is satisfied.

The first set of tables below lists, for each thermocouple type, the emf as a function of temperature on the International Temperature Scale of 1990 (ITS-90). The coefficients in the equation used to generate the table are also given. The second set of tables gives the inverse relationships, i.e., the coefficients in the polynomial equation that expresses the temperature as a function of thermocouple emf. The accuracy of these equations is also stated.

Further details and tables at closer intervals may be found in Ref. 1.

References

1. Burns, G. W., Seroger, M. G., Strouse, G. F., Croarkin, M. C., and Guthrie, W. F., *Temperature-Electromotive Force Reference Functions and Tables for the Letter-Designated Thermocouple Types Based on the ITS-90*, Natl. Inst. Stand. Tech. (U.S.) Monogr. 175, 1993.
2. Schooley, J. F., *Thermometry*, CRC Press, Boca Raton, FL, 1986.

Type B Thermocouples: emf-Temperature (°C) Reference Table and Equations

Thermocouple emf as a Function of Temperature in Degrees Celsius (ITS-90)

emf in Millivolts — **Reference junctions at 0 °C**

°C	0	10	20	30	40	50	60	70	80	90	100
0	0.000	–0.002	–0.003	–0.002	–0.000	0.002	0.006	0.011	0.017	0.025	0.033
100	0.033	0.043	0.053	0.065	0.078	0.092	0.107	0.123	0.141	0.159	0.178
200	0.178	0.199	0.220	0.243	0.267	0.291	0.317	0.344	0.372	0.401	0.431
300	0.431	0.462	0.494	0.527	0.561	0.596	0.632	0.669	0.707	0.746	0.787
400	0.787	0.828	0.870	0.913	0.957	1.002	1.048	1.095	1.143	1.192	1.242
500	1.242	1.293	1.344	1.397	1.451	1.505	1.561	1.617	1.675	1.733	1.792
600	1.792	1.852	1.913	1.975	2.037	2.101	2.165	2.230	2.296	2.363	2.431
700	2.431	2.499	2.569	2.639	2.710	2.782	2.854	2.928	3.002	3.078	3.154
800	3.154	3.230	3.308	3.386	3.466	3.546	3.626	3.708	3.790	3.873	3.957
900	3.957	4.041	4.127	4.213	4.299	4.387	4.475	4.564	4.653	4.743	4.834
1000	4.834	4.926	5.018	5.111	5.205	5.299	5.394	5.489	5.585	5.682	5.780
1100	5.780	5.878	5.976	6.075	6.175	6.276	6.377	6.478	6.580	6.683	6.786
1200	6.786	6.890	6.995	7.100	7.205	7.311	7.417	7.524	7.632	7.740	7.848
1300	7.848	7.957	8.066	8.176	8.286	8.397	8.508	8.620	8.731	8.844	8.956
1400	8.956	9.069	9.182	9.296	9.410	9.524	9.639	9.753	9.868	9.984	10.099
1500	10.099	10.215	10.331	10.447	10.563	10.679	10.796	10.913	11.029	11.146	11.263
1600	11.263	11.380	11.497	11.614	11.731	11.848	11.965	12.082	12.199	12.316	12.433
1700	12.433	12.549	12.666	12.782	12.898	13.014	13.130	13.246	13.361	13.476	13.591
1800	13.591	13.706	13.820								

NOTE—Temperature ranges and coefficients of equations used to compute the above table: The equations are of the form: $E = c_0 + c_1t + c_2t^2 + c_3t^3 + \ldots c_nt^n$, where E is the emf in millivolts, t is the temperature in degrees Celsius (ITS- 90), and c_0, c_1, c_2, c_3, etc., are the coefficients. These coefficients are extracted from Ref. 1.

Lab Data

		0 °C to 630.615 °C	630.615 °C to 1820 °C
c_0	=	0.000 000 000 0	−3.893 816 862 1 …
c_1	=	− 2.465 081 834 6 × 10^{-4}	2.857 174 747 0 × 10^{-2}
c_2	=	5.904 042 117 1 × 10^{-6}	−8.488 510 478 5 × 10^{-5}
c_3	=	− 1.325 793 163 6 × 10^{-9}	1.578 528 016 4 × 10^{-7}
c_4	=	1.566 829 190 1 × 10^{-12}	−1.683 534 486 4 × 10^{-10}
c_5	=	− 1.694 452 924 0 × 10^{-15}	1.110 979 401 3 × 10^{-13}
c_6	=	6.299 034 709 4 × 10^{-19}	−4.451 543 103 3 × 10^{-17}
c_7	=	…………	9.897 564 082 1 × 10^{-21}
c_8	=	…………	−9.379 133 028 9 × 10^{-25}

Type E Thermocouples: emf-Temperature (°C) Reference Table and Equations

Thermocouple emf as a Function of Temperature in Degrees Celsius (ITS-90)

emf in Millivolts — Reference junctions at 0 °C

°C	0	−10	−20	−30	−40	−50	−60	−70	−80	−90	−100
−200	−8.825	−9.063	−9.274	−9.455	−9.604	−9.718	−9.797	−9.835			
−100	−5.237	−5.681	−6.107	−6.516	−6.907	−7.279	−7.632	−7.963	−8.273	−8.561	−8.825
0	0.000	−0.582	−1.152	−1.709	−2.255	−2.787	−3.306	−3.811	−4.302	−4.777	−5.237
°C	**0**	**10**	**20**	**30**	**40**	**50**	**60**	**70**	**80**	**90**	**100**
0	0.000	0.591	1.192	1.801	2.420	3.048	3.685	4.330	4.985	5.648	6.319
100	6.319	6.998	7.685	8.379	9.081	9.789	10.503	11.224	11.951	12.684	13.421
200	13.421	14.164	14.912	15.664	16.420	17.181	17.945	18.713	19.484	20.259	21.036
300	21.036	21.817	22.600	23.386	24.174	24.964	25.757	26.552	27.348	28.146	28.946
400	28.946	29.747	30.550	31.354	32.159	32.965	33.772	34.579	35.387	36.196	37.005
500	37.005	37.815	38.624	39.434	40.243	41.053	41.862	42.671	43.479	44.286	45.093
600	45.093	45.900	46.705	47.509	48.313	49.116	49.917	50.718	51.517	52.315	53.112
700	53.112	53.908	54.703	55.497	56.289	57.080	57.870	58.659	59.446	60.232	61.017
800	61.017	61.801	62.583	63.364	64.144	64.922	65.698	66.473	67.246	68.017	68.787
900	68.787	69.554	70.319	71.082	71.844	72.603	73.360	74.115	74.869	75.621	76.373
1000	76.373										

NOTE—Temperature ranges and coefficients of equations used to compute the above table: The equations are of the form: $E = c_0 + c_1t + c_2t^2 + c_3t^3 + \ldots c_nt^n$, where E is the emf in millivolts, t is the temperature in degrees Celsius (ITS- 90), and c_0, c_1, c_2, c_3, etc., are the coefficients. These coefficients are extracted from Ref. 1.

		−270 to 0 °C	0 °C to 1000 °C
c_0	=	0.000 000 000 0 ...	0.000 000 000 0 ...
c_1	=	5.866 550 870 8 × 10^{-2}	5.866 550 871 0 × 10^{-2}
c_2	=	4.541 097 712 4 × 10^{-5}	4.503 227 558 2 × 10^{-5}
c_3	=	−7.799 804 868 6 × 10^{-7}	2.890 840 721 2 × 10^{-8}
c_4	=	−2.580 016 084 3 × 10^{-8}	−3.305 689 665 2 × 10^{-10}
c_5	=	−5.945 258 305 7 × 10^{-10}	6.502 440 327 0 × 10^{-13}
c_6	=	−9.321 405 866 7 × 10^{-12}	−1.919 749 550 4 × 10^{-16}
c_7	=	−1.028 760 553 4 × 10^{-13}	−1.253 660 049 7 × 10^{-18}
c_8	=	−8.037 012 362 1 × 10^{-16}	2.148 921 756 9 × 10^{-21}
c_9	=	−4.397 949 739 1 × 10^{-18}	−1.438 804 178 2 × 10^{-24}
c_{10}	=	−1.641 477 635 5 × 10^{-20}	3.596 089 948 1 × 10^{-28}
c_{11}	=	−3.967 361 951 6 × 10^{-23}	…………
c_{12}	=	−5.582 732 872 1 × 10^{-26}	…………
c_{13}	=	−3.465 784 201 3 × 10^{-29}	…………

Type J Thermocouples: emf-Temperature (°C) Reference Table and Equations

Thermocouple emf as a Function of Temperature in Degrees Celsius (ITS-90)

emf in Millivolts — Reference junctions at 0 °C

°C	0	−10	−20	−30	−40	−50	−60	−70	−80	−90	−100
−200	−7.890	−8.095									
−100	−4.633	−5.037	−5.426	−5.801	−6.159	−6.500	−6.821	−7.123	−7.403	−7.659	−7.890
0	0.000	−0.501	−0.995	−1.482	−1.961	−2.431	−2.893	−3.344	−3.786	−4.215	−4.633

emf in Millivolts										Reference junctions at 0 °C
°C	0	10	20	30	40	50	60	70	80	90	100
0	0.000	0.507	1.019	1.537	2.059	2.585	3.116	3.650	4.187	4.726	5.269
100	5.269	5.814	6.360	6.909	7.459	8.010	8.562	9.115	9.669	10.224	10.779
200	10.779	11.334	11.889	12.445	13.000	13.555	14.110	14.665	15.219	15.773	16.327
300	16.327	16.881	17.434	17.986	18.538	19.090	19.642	20.194	20.745	21.297	21.848
400	21.848	22.400	22.952	23.504	24.057	24.610	25.164	25.720	26.276	26.834	27.393
500	27.393	27.953	28.516	29.080	29.647	30.216	30.788	31.362	31.939	32.519	33.102
600	33.102	33.689	34.279	34.873	35.470	36.071	36.675	37.284	37.896	38.512	39.132
700	39.132	39.755	40.382	41.012	41.645	42.281	42.919	43.559	44.203	44.848	45.494
800	45.494	46.141	46.786	47.431	48.074	48.715	49.353	49.989	50.622	51.251	51.877
900	51.877	52.500	53.119	53.735	54.347	54.956	55.561	56.164	56.763	57.360	57.953
1000	57.953	58.545	59.134	59.721	60.307	60.890	61.473	62.054	62.634	63.214	63.792
1100	63.792	64.370	64.948	65.525	66.102	66.679	67.255	67.831	68.406	68.980	69.553
1200	69.553										

NOTE—Temperature ranges and coefficients of equations used to compute the above table: The equations are of the form: $E = c_0 + c_1t + c_2t^2 + c_3t^3 + \ldots c_nt^n$, where E is the emf in millivolts, t is the temperature in degrees Celsius (ITS-90), and c_0, c_1, c_2, c_3, etc., are the coefficients. These coefficients are extracted from Ref. 1.

		−260 °C to 760 °C	760 °C to 1200 °C
c_0	=	0.000 000 000 0 …	$2.964\,562\,568\,1 \times 10^{2}$
c_1	=	$5.038\,118\,781\,5 \times 10^{-2}$	−1.497 612 778 6 …
c_2	=	$3.047\,583\,693\,0 \times 10^{-5}$	$3.178\,710\,392\,4 \times 10^{-3}$
c_3	=	$-8.568\,106\,572\,0 \times 10^{-8}$	$-3.184\,768\,670\,1 \times 10^{-6}$
c_4	=	$1.322\,819\,529\,5 \times 10^{-10}$	$1.572\,081\,900\,4 \times 10^{-9}$
c_5	=	$-1.705\,295\,833\,7 \times 10^{-13}$	$-3.069\,136\,905\,6 \times 10^{-13}$
c_6	=	$2.094\,809\,069\,7 \times 10^{-16}$	…………
c_7	=	$-1.253\,839\,533\,6 \times 10^{-19}$	…………
c_8	=	$1.563\,172\,569\,7 \times 10^{-23}$	…………

Type K Thermocouples: emf-Temperature (°C) Reference Table and Equations

Thermocouple emf as a Function of Temperature in Degrees Celsius (ITS-90)

emf in Millivolts										Reference junctions at 0 °C
°C	0	−10	−20	−30	−40	−50	−60	−70	−80	−90	−100
−200	−5.891	−6.035	−6.158	−6.262	−6.344	−6.404	−6.441	−6.458			
−100	−3.554	−3.852	−4.138	−4.411	−4.669	−4.913	−5.141	−5.354	−5.550	−5.730	−5.891
0	0.000	−0.392	−0.778	−1.156	−1.527	−1.889	−2.243	−2.587	−2.920	−3.243	−3.554
°C	0	10	20	30	40	50	60	70	80	90	100
0	0.000	0.397	0.798	1.203	1.612	2.023	2.436	2.851	3.267	3.682	4.096
100	4.096	4.509	4.920	5.328	5.735	6.138	6.540	6.941	7.340	7.739	8.138
200	8.138	8.539	8.940	9.343	9.747	10.153	10.561	10.971	11.382	11.795	12.209
300	12.209	12.624	13.040	13.457	13.874	14.293	14.713	15.133	15.554	15.975	16.397
400	16.397	16.820	17.243	17.667	18.091	18.516	18.941	19.366	19.792	20.218	20.644
500	20.644	21.071	21.497	21.924	22.350	22.776	23.203	23.629	24.055	24.480	24.905
600	24.905	25.330	25.755	26.179	26.602	27.025	27.447	27.869	28.289	28.710	29.129
700	29.129	29.548	29.965	30.382	30.798	31.213	31.628	32.041	32.453	32.865	33.275
800	33.275	33.685	34.093	34.501	34.908	35.313	35.718	36.121	36.524	36.925	37.326
900	37.326	37.725	38.124	38.522	38.918	39.314	39.708	40.101	40.494	40.885	41.276
1000	41.276	41.665	42.053	42.440	42.826	43.211	43.595	43.978	44.359	44.740	45.119
1100	45.119	45.497	45.873	46.249	46.623	46.995	47.367	47.737	48.105	48.473	48.838
1200	48.838	49.202	49.565	49.926	50.286	50.644	51.000	51.355	51.708	52.060	52.410
1300	52.410	52.759	53.106	53.451	53.795	54.138	54.479	54.819			

NOTE—Temperature ranges and coefficients of equations used to compute the above table: The equations are of the form $E = c_0 + c_1t + c_2t^2 + c_3t^3 + \ldots c_nt^n$, where E is the emf in millivolts, t is the temperature in degrees Celsius (ITS-90), and c_0, c_1, c_2, c_3, etc., are the coefficients. In the 0 °C to 1372 °C range there is also an exponential term that must be evaluated and added to the equation. The exponential term is of the form $c_0e^{c_1(t-126.9686)^2}$, where t is the temperature in °C, e is the natural logarithm base, and c_0 and c_1 are the coefficients. These coefficients are extracted from Ref. 1.

		−270 °C to 0 °C	0 °C to 1372 °C	0 °C to 1372 °C (Exponential term)
c_0	=	0.000 000 000 0	$-1.760\,041\,368\,6 \times 10^{-2}$	$1.185\,976 \times 10^{-1}$
c_1	=	$3.945\,012\,802\,5 \times 10^{-2}$	$3.892\,120\,497\,5 \times 10^{-2}$	$-1.183\,432 \times 10^{-4}$
c_2	=	$2.362\,237\,359\,8 \times 10^{-5}$	$1.855\,877\,003\,2 \times 10^{-5}$	....
c_3	=	$-3.285\,890\,678\,4 \times 10^{-7}$	$-9.945\,759\,287\,4 \times 10^{-8}$	....
c_4	=	$-4.990\,482\,877\,7 \times 10^{-9}$	$3.184\,094\,571\,9 \times 10^{-10}$	....
c_5	=	$-6.750\,905\,917\,3 \times 10^{-11}$	$-5.607\,284\,488\,9 \times 10^{-13}$	....
c_6	=	$-5.741\,032\,742\,8 \times 10^{-13}$	$5.607\,505\,905\,9 \times 10^{-16}$	....
c_7	=	$-3.108\,887\,289\,4 \times 10^{-15}$	$-3.202\,072\,000\,3 \times 10^{-19}$	....
c_8	=	$-1.045\,160\,936\,5 \times 10^{-17}$	$9.715\,114\,715\,2 \times 10^{-23}$	....
c_9	=	$-1.988\,926\,687\,8 \times 10^{-20}$	$-1.210\,472\,127\,5 \times 10^{-26}$	....
c_{10}	=	$-1.632\,269\,748\,6 \times 10^{-23}$	......	....

Type N Thermocouples: emf-Temperature (°C) Reference Table and Equations

Thermocouple emf as a Function of Temperature in Degrees Celsius (ITS-90)

emf in Millivolts **Reference junctions at 0 °C**

°C	0	−10	−20	−30	−40	−50	−60	−70	−80	−90	−100
−200	3.990	−4.083	−4.162	−4.226	−4.277	−4.313	−4.336	−4.345			
−100	−2.407	−2.612	2.808	2.994	3.171	3.336	−3.491	3.634	−3.766	−3.884	−3.990
0	0.000	−0.260	−0.518	−0.772	−1.023	−1.269	−1.509	1.744	−1.972	−2.193	−2.407
°C	**0**	**10**	**20**	**30**	**40**	**50**	**60**	**70**	**80**	**90**	**100**
0	0.000	0.261	0.525	0.793	1.065	1.340	1.619	1.902	2.189	2.480	2.774
100	2.774	3.072	3.374	3.680	3.989	4.302	4.618	4.937	5.259	5.585	5.913
200	5.913	6.245	6.579	6.916	7.255	7.597	7.941	8.288	8.637	8.988	9.341
300	9.341	9.696	10.054	10.413	10.774	11.136	11.501	11.867	12.234	12.603	12.974
400	12.974	13.346	13.719	14.094	14.469	14.846	15.225	15.604	15.984	16.366	16.748
500	16.748	17.131	17.515	17.900	18.286	18.672	19.059	19.447	19.835	20.224	20.613
600	20.613	21.003	21.393	21.784	22.175	22.566	22.958	23.350	23.742	24.134	24.527
700	24.527	24.919	25.312	25.705	26.098	26.491	26.883	27.276	27.669	28.062	28.455
800	28.455	28.847	29.239	29.632	30.024	30.416	30.807	31.199	31.590	31.981	32.371
900	32.371	32.761	33.151	33.541	33.930	34.319	34.707	35.095	35.482	35.869	36.256
1000	36.256	36.641	37.027	37.411	37.795	38.179	38.562	38.944	39.326	39.706	40.087
1100	40.087	40.466	40.845	41.223	41.600	41.976	42.352	42.727	43.101	43.474	43.846
1200	43.846	44.218	44.588	44.958	45.326	45.694	46.060	46.425	46.789	47.152	47.513
1300	47.513										

NOTE—Temperature ranges and coefficients of equations used to compute the above table: The equations are of the form $E = c_0 + c_1t + c_2t^2 + c_3t^3 + \ldots c_nt^n$, where E is the emf in millivolts, t is the temperature in degrees Celsius (ITS-90), and c_0, c_1, c_2, c_3, etc., are the coefficients. These coefficients are extracted from Ref. 1.

		−270 °C to 0 °C	0 °C to 1300 °C
c_0	=	0.000 000 000 0 ...	0.000 000 000 0...
c_1	=	$2.615\,910\,596\,2 \times 10^{-2}$	$2.592\,939\,460\,1 \times 10^{-2}$
c_2	=	$1.095\,748\,422\,8 \times 10^{-5}$	$1.571\,014\,188\,0 \times 10^{-5}$
c_3	=	$-9.384\,111\,155\,4 \times 10^{-8}$	$4.382\,562\,723\,7 \times 10^{-8}$
c_4	=	$-4.641\,203\,975\,9 \times 10^{-11}$	$-2.526\,116\,979\,4 \times 10^{-10}$
c_5	=	$-2.630\,335\,771\,6 \times 10^{-12}$	$6.431\,181\,933\,9 \times 10^{-13}$
c_6	=	$-2.265\,343\,800\,3 \times 10^{-14}$	$-1.006\,347\,151\,9 \times 10^{-15}$
c_7	=	$-7.608\,930\,079\,1 \times 10^{-17}$	$9.974\,533\,899\,2 \times 10^{-19}$
c_8	=	$-9.341\,966\,783\,5 \times 10^{-20}$	$-6.086\,324\,560\,7 \times 10^{-22}$
c_9	=	....	$2.084\,922\,933\,9 \times 10^{-25}$
c_{10}	=	....	$-3.068\,219\,615\,1 \times 10^{-29}$

Type R Thermocouples: emf-Temperature (°C) Reference Table and Equations

Thermocouple emf as a Function of Temperature in Degrees Celsius (ITS-90)

emf in Millivolts — Reference junctions at 0 °C

°C	0	−10	−20	−30	−40	−50	−60	−70	−80	−90	−100
0	0.000	−0.051	−0.100	−0.145	−0.188	−0.226					
°C	**0**	**10**	**20**	**30**	**40**	**50**	**60**	**70**	**80**	**90**	**100**
0	0.000	0.054	0.111	0.171	0.232	0.296	0.363	0.431	0.501	0.573	0.647
100	0.647	0.723	0.800	0.879	0.959	1.041	1.124	1.208	1.294	1.381	1.469
200	1.469	1.558	1.648	1.739	1.831	1.923	2.017	2.112	2.207	2.304	2.401
300	2.401	2.498	2.597	2.696	2.796	2.896	2.997	3.099	3.201	3.304	3.408
400	3.408	3.512	3.616	3.721	3.827	3.933	4.040	4.147	4.255	4.363	4.471
500	4.471	4.580	4.690	4.800	4.910	5.021	5.133	5.245	5.357	5.470	5.583
600	5.583	5.697	5.812	5.926	6.041	6.157	6.273	6.390	6.507	6.625	6.743
700	6.743	6.861	6.980	7.100	7.220	7.340	7.461	7.583	7.705	7.827	7.950
800	7.950	8.073	8.197	8.321	8.446	8.571	8.697	8.823	8.950	9.077	9.205
900	9.205	9.333	9.461	9.590	9.720	9.850	9.980	10.111	10.242	10.374	10.506
1000	10.506	10.638	10.771	10.905	11.039	11.173	11.307	11.442	11.578	11.714	11.850
1100	11.850	11.986	12.123	12.260	12.397	12.535	12.673	12.812	12.950	13.089	13.228
1200	13.228	13.367	13.507	13.646	13.786	13.926	14.066	14.207	14.347	14.488	14.629
1300	14.629	14.770	14.911	15.052	15.193	15.334	15.475	15.616	15.758	15.899	16.040
1400	16.040	16.181	16.323	16.464	16.605	16.746	16.887	17.028	17.169	17.310	17.451
1500	17.451	17.591	17.732	17.872	18.012	18.152	18.292	18.431	18.571	18.710	18.849
1600	18.849	18.988	19.126	19.264	19.402	19.540	19.677	19.814	19.951	20.087	20.222
1700	20.222	20.356	20.488	20.620	20.749	20.877	21.003				

NOTE—Temperature ranges and coefficients of equations used to compute the above table: The equations are of the form $E = c_0 + c_1t + c_2t^2 + c_3t^3 + \ldots c_nt^n$, where E is the emf in millivolts, t is the temperature in degrees Celsius (ITS-90), and c_0, c_1, c_2, c_3, etc., are the coefficients. These coefficients are extracted from Ref. 1.

		−50 °C to 1064.18 °C	1064.18 °C to 1664.5 °C	1664.5 °C to 1768.1 °C
c_0	=	0.000 000 000 00 ...	2.951 579 253 16 ...	$1.522\ 321\ 182\ 09 \times 10^{2}$
c_1	=	$5.289\ 617\ 297\ 65 \times 10^{-3}$	$-2.520\ 612\ 513\ 32 \times 10^{-3}$	$-2.688\ 198\ 885\ 45 \times 10^{-1}$
c_2	=	$1.391\ 665\ 897\ 82 \times 10^{-5}$	$1.595\ 645\ 018\ 65 \times 10^{-5}$	$1.712\ 802\ 804\ 71 \times 10^{-4}$
c_3	=	$-2.388\ 556\ 930\ 17 \times 10^{-8}$	$-7.640\ 859\ 475\ 76 \times 10^{-9}$	$-3.458\ 957\ 064\ 53 \times 10^{-8}$
c_4	=	$3.569\ 160\ 010\ 63 \times 10^{-11}$	$2.053\ 052\ 910\ 24 \times 10^{-12}$	$-9.346\ 339\ 710\ 46 \times 10^{-15}$
c_5	=	$-4.623\ 476\ 662\ 98 \times 10^{-14}$	$-2.933\ 596\ 681\ 73 \times 10^{-16}$	....
c_6	=	$5.007\ 774\ 410\ 34 \times 10^{-17}$	....	....
c_7	=	$-3.731\ 058\ 861\ 91 \times 10^{-20}$	....	....
c_8	=	$1.577\ 164\ 823\ 67 \times 10^{-23}$	....	....
c_9	=	$-2.810\ 386\ 252\ 51 \times 10^{-27}$	....	....

Type S Thermocouples: emf-Temperature (°C) Reference Table and Equations

Thermocouple emf as a Function of Temperature in Degrees Celsius (ITS-90)

emf in Millivolts — Reference junctions at 0 °C

°C	0	−10	−20	−30	−40	−50	−60	−70	−80	−90	−100
0	0.000	−0.053	−0.103	−0.150	−0.194	−0.236					
°C	**0**	**10**	**20**	**30**	**40**	**50**	**60**	**70**	**80**	**90**	**100**
0	0.000	0.055	0.113	0.173	0.235	0.299	0.365	0.433	0.502	0.573	0.646
100	0.646	0.720	0.795	0.872	0.950	1.029	1.110	1.191	1.273	1.357	1.441
200	1.441	1.526	1.612	1.698	1.786	1.874	1.962	2.052	2.141	2.232	2.323
300	2.323	2.415	2.507	2.599	2.692	2.786	2.880	2.974	3.069	3.164	3.259
400	3.259	3.355	3.451	3.548	3.645	3.742	3.840	3.938	4.036	4.134	4.233
500	4.233	4.332	4.432	4.532	4.632	4.732	4.833	4.934	5.035	5.137	5.239
600	5.239	5.341	5.443	5.546	5.649	5.753	5.857	5.961	6.065	6.170	6.275
700	6.275	6.381	6.486	6.593	6.699	6.806	6.913	7.020	7.128	7.236	7.345
800	7.345	7.454	7.563	7.673	7.783	7.893	8.003	8.114	8.226	8.337	8.449
900	8.449	8.562	8.674	8.787	8.900	9.014	9.128	9.242	9.357	9.472	9.587
1000	9.587	9.703	9.819	9.935	10.051	10.168	10.285	10.403	10.520	10.638	10.757
1100	10.757	10.875	10.994	11.113	11.232	11.351	11.471	11.590	11.710	11.830	11.951
1200	11.951	12.071	12.191	12.312	12.433	12.554	12.675	12.796	12.917	13.038	13.159

Type S Thermocouples: emf-Temperature (°C) Reference Table and Equations

Thermocouple emf as a Function of Temperature in Degrees Celsius (ITS-90)

emf in Millivolts **Reference junctions at 0 °C**

°C	0	−10	−20	−30	−40	−50	−60	−70	−80	−90	−100
1300	13.159	13.280	13.402	13.523	13.644	13.766	13.887	14.009	14.130	14.251	14.373
1400	14.373	14.494	14.615	14.736	14.857	14.978	15.099	15.220	15.341	15.461	15.582
1500	15.582	15.702	15.822	15.942	16.062	16.182	16.301	16.420	16.539	16.658	16.777
1600	16.777	16.895	17.013	17.131	17.249	17.366	17.483	17.600	17.717	17.832	17.947
1700	17.947	18.061	18.174	18.285	18.395	18.503	18.609				

NOTE—Temperature ranges and coefficients of equations used to compute the above table: The equations are of the form $E = c_0 + c_1t + c_2t^2 + c_3t^3 + \ldots c_nt^n$, where E is the emf in millivolts, t is the temperature in degrees Celsius (ITS-90), and c_0, c_1, c_2, c_3, etc., are the coefficients. These coefficients are extracted from Ref. 1.

		−50 °C to 1064.18 °C	1064.18 °C to 1664.5 °C	1664.5 °C to 1768.1 °C
c_0	=	0.000 000 000 00 ...	1.329 004 440 85 ...	$1.466\,282\,326\,36 \times 10^{2}$
c_1	=	$5.403\,133\,086\,31 \times 10^{-3}$	$3.345\,093\,113\,44 \times 10^{-3}$	$-2.584\,305\,167\,52 \times 10^{-1}$
c_2	=	$1.259\,342\,897\,40 \times 10^{-5}$	$6.548\,051\,928\,18 \times 10^{-6}$	$1.636\,935\,746\,41 \times 10^{-4}$
c_3	=	$-2.324\,779\,686\,89 \times 10^{-8}$	$-1.648\,562\,592\,09 \times 10^{-9}$	$-3.304\,390\,469\,87 \times 10^{-8}$
c_4	=	$3.220\,288\,230\,36 \times 10^{-11}$	$1.299\,896\,051\,74 \times 10^{-14}$	$-9.432\,236\,906\,12 \times 10^{-15}$
c_5	=	$-3.314\,651\,963\,89 \times 10^{-14}$	....	....
c_6	=	$2.557\,442\,517\,86 \times 10^{-17}$	....	....
c_7	=	$-1.250\,688\,713\,93 \times 10^{-20}$	....	....
c_8	=	$2.714\,431\,761\,45 \times 10^{-24}$	....	....

Type T Thermocouples: emf-Temperature (°C) Reference Table and Equations

Thermocouple emf as a Function of Temperature in Degrees Celsius (ITS-90)

emf in Millivolts **Reference junctions at 0 °C**

°C	0	−10	−20	−30	−40	−50	−60	−70	−80	−90	−100
−200	−5.603	−5.753	−5.888	−6.007	−6.105	−6.180	−6.232	−6.258			
−100	−3.379	−3.657	−3.923	−4.177	−4.419	−4.648	−4.865	−5.070	−5.261	−5.439	−5.603
0	0.000	−0.383	−0.757	−1.121	−1.475	−1.819	−2.153	−2.476	−2.788	−3.089	−3.379
°C	**0**	**10**	**20**	**30**	**40**	**50**	**60**	**70**	**80**	**90**	**100**
0	0.000	0.391	0.790	1.196	1.612	2.036	2.468	2.909	3.358	3.814	4.279
100	4.279	4.750	5.228	5.714	6.206	6.704	7.209	7.720	8.237	8.759	9.288
200	9.288	9.822	10.362	10.907	11.458	12.013	12.574	13.139	13.709	14.283	14.862
300	14.862	15.445	16.032	16.624	17.219	17.819	18.422	19.030	19.641	20.255	20.872
400	20.872										

NOTE—Temperature ranges and coefficients of equations used to compute the above table: The equations are of the form $E = c_0 + c_1t + c_2t^2 + c_3t^3 + \ldots c_nt^n$, where E is the emf in millivolts, t is the temperature in degrees Celsius (ITS-90), and c_0, c_1, c_2, c_3, etc., are the coefficients. These coefficients are extracted from Ref. 1.

		270 °C to 0 °C	0 °C to 400 °C
c_0	=	0.000 000 000 0 ...	0.000 000 000 0 ...
c_1	=	$3.874\,810\,636\,4 \times 10^{-2}$	$3.874\,810\,636\,4 \times 10^{-2}$
c_2	=	$4.419\,443\,434\,7 \times 10^{-5}$	$3.329\,222\,788\,0 \times 10^{-5}$
c_3	=	$1.184\,432\,310\,5 \times 10^{-7}$	$2.061\,824\,340\,4 \times 10^{-7}$
c_4	=	$2.003\,297\,355\,4 \times 10^{-8}$	$-2.188\,225\,684\,6 \times 10^{-9}$
c_5	=	$9.013\,801\,955\,9 \times 10^{-10}$	$1.099\,688\,092\,8 \times 10^{-11}$
c_6	=	$2.265\,115\,659\,3 \times 10^{-11}$	$-3.081\,575\,877\,2 \times 10^{-14}$
c_7	=	$3.607\,115\,420\,5 \times 10^{-13}$	$4.547\,913\,529\,0 \times 10^{-17}$
c_8	=	$3.849\,393\,988\,3 \times 10^{-15}$	$-2.751\,290\,167\,3 \times 10^{-20}$
c_9	=	$2.821\,352\,192\,5 \times 10^{-17}$	....
c_{10}	=	$1.425\,159\,477\,9 \times 10^{-19}$	....
c_{11}	=	$4.876\,866\,228\,6 \times 10^{-22}$	....
c_{12}	=	$1.079\,553\,927\,0 \times 10^{-24}$	....
c_{13}	=	$1.394\,502\,706\,2 \times 10^{-27}$	....
c_{14}	=	$7.979\,515\,392\,7 \times 10^{-31}$	....

Type B Thermocouples: Coefficients (c_i) of Polynomials for the Computation of Temperatures in °C as a Function of the Thermocouple emf in Various Temperature and emf Ranges

Temperature range:	250 °C to 700 °C	700 °C to 1820 °C
emf range:	0.291 mV to 2.431 mV	2.431 mV to 13.820 mV
c_0 =	$9.842\,332\,1 \times 10^{1}$	$2.131\,507\,1 \times 10^{2}$
c_1 =	$6.997\,150\,0 \times 10^{2}$	$2.851\,050\,4 \times 10^{2}$
c_2 =	$-8.476\,530\,4 \times 10^{2}$	$-5.274\,288\,7 \times 10^{1}$
c_3 =	$1.005\,264\,4 \times 10^{3}$	9.916 080 4 ...
c_4 =	$-8.334\,595\,2 \times 10^{2}$	−1.296 530 3 ...
c_5 =	$4.550\,854\,2 \times 10^{2}$	$1.119\,587\,0 \times 10^{-1}$
c_6 =	$-1.552\,303\,7 \times 10^{2}$	$-6.062\,519\,9 \times 10^{-3}$
c_7 =	$2.988\,675\,0 \times 10^{1}$	$1.866\,169\,6 \times 10^{-4}$
c_8 =	−2.474 286 0 ...	$-2.487\,858\,5 \times 10^{-6}$

NOTE—The above coefficients are extracted from Ref. 1 and are for an expression of the form shown in Section 10.3.2. They yield approximate values of temperature that agree within ±0.03 °C with the values given in Table 10.2.

Type E Thermocouples: Coefficients (c_i) of Polynomials for the Computation of Temperatures in °C as a Function of the Thermocouple emf in Various Temperature and emf Ranges

Temperature range:	−200 °C TO 0 °C	0 °C to 1000 °C
emf range:	−8.825 mV to 0.0 mV	0.0 mV to 76.373 mV
c_0 =	0.000 000 0 ...	0.000 000 0 ...
c_1 =	$1.697\,728\,8 \times 10^{1}$	$1.705\,703\,5 \times 10^{1}$
c_2 =	$-4.351\,497\,0 \times 10^{-1}$	$-2.330\,175\,9 \times 10^{-1}$
c_3 =	$-1.585\,969\,7 \times 10^{-1}$	$6.543\,558\,5 \times 10^{-3}$
c_4 =	$-9.250\,287\,1 \times 10^{-2}$	$-7.356\,274\,9 \times 10^{-5}$
c_5 =	$-2.608\,431\,4 \times 10^{-2}$	$-1.789\,600 \times 10^{-6}$
c_6 =	$-4.136\,019\,9 \times 10^{-3}$	$8.403\,616\,5 \times 10^{-8}$
c_7 =	$-3.403\,403\,0 \times 10^{-4}$	$-1.373\,587\,9 \times 10^{-9}$
c_8 =	$-1.156\,489\,0 \times 10^{-5}$	$1.062\,982\,3 \times 10^{-11}$
c_9 =	....	$-3.244\,708\,7 \times 10^{-14}$

NOTE—The above coefficients are extracted from Ref. 1 and are for an expression of the form shown in Section 10.3.2. They yield approximate values of temperature that agree within ±0.02 °C with the values given in Table 10.4

Type J Thermocouples: Coefficients (c_i) of Polynomials for the Computation of Temperatures in °C as a Function of the Thermocouple emf in Various Temperature and emf Ranges

Temperature range:	−210 °C to 0 °C	0 °C to 760 °C	760 °C to 1200 °C
emf range:	−8.095 mV to 0.0 mV	0.0 mV to 42.919 mV	42.919 mV to 69.553 mV
c_0 =	0.000 000 0...	0.000 000 ...	$-3.113\,581\,87 \times 10^{3}$
c_1 =	$1.952\,826\,8 \times 10^{1}$	$1.978\,425 \times 10^{1}$	$3.005\,436\,84 \times 10^{2}$
c_2 =	−1.228 618 5 ...	$-2.001\,204 \times 10^{-1}$	−9.947 732 30 ...
c_3 =	−1.075 217 8 ...	$1.036\,969 \times 10^{-2}$	$1.702\,766\,30 \times 10^{-1}$
c_4 =	$-5.908\,693\,3 \times 10^{-1}$	$-2.549\,687 \times 10^{-4}$	$1.430\,334\,68 \times 10^{-3}$
c_5 =	$-1.725\,671\,3 \times 10^{-1}$	$3.585\,153 \times 10^{-6}$	$4.438\,860\,84 \times 10^{-6}$
c_6 =	$-2.813\,151\,3 \times 10^{-2}$	$-5.344\,285 \times 10^{-8}$	....
c_7 =	$-2.396\,337\,0 \times 10^{-3}$	$5.099\,890 \times 10^{-10}$	....
c_8 =	$-8.382\,332\,1 \times 10^{-5}$	....	....

NOTE—The above coefficients are extracted from Ref. 1 and are for an expression of the form shown in Section 10.3.2. They yield approximate values of temperature that agree within ± 0.5 °C with the values given in Table 10.6.

Lab Data

Type K Thermocouples: Coefficients (c_i) of Polynomials for the Computation of Temperatures in °C as a Function of the Thermocouple emf in Various Temperature and emf Ranges

Temperature range:	− 200 °C to 0 °C	0 °C to 500 °C	500 °C to 1372 °C
emf range:	− 5.891 mV to 0.0 mV	0.0 mV to 20.644 mV	20.644 mV to 54.886 mV
c_0 =	0.000 000 0 …	0.000 000 …	$-1.318\,058 \times 10^{2}$
c_1 =	$2.517\,346\,2 \times 10^{1}$	$2.508\,355 \times 10^{1}$	$4.830\,222 \times 10^{1}$
c_2 =	−1.166 287 8 …	$7.860\,106 \times 10^{-2}$	−1.646 031 …
c_3 =	− 1.083 363 8 …	$-2.503\,131 \times 10^{-1}$	$5.464\,731 \times 10^{-2}$
c_4 =	$-8.977\,354\,0 \times 10^{-1}$	$8.315\,270 \times 10^{-2}$	$-9.650\,715 \times 10^{-4}$
c_5 =	$-3.734\,237\,7 \times 10^{-1}$	$-1.228\,034 \times 10^{-2}$	$8.802\,193 \times 10^{-6}$
c_6 =	$-8.663\,264\,3 \times 10^{-2}$	$9.804\,036 \times 10^{-4}$	$3.110\,810 \times 10^{-8}$
c_7 =	$-1.045\,059\,8 \times 10^{-2}$	$-4.413\,030 \times 10^{-5}$	…….
c_8 =	$-5.192\,057\,7 \times 10^{-4}$	$1.057\,734 \times 10^{-6}$	…….
c_9 =	…….	$-1.052\,755 \times 10^{-8}$	…….

NOTE—The above coefficients are extracted from Ref. 1 and are for an expression of the form shown in Section 10.3.2. They yield approximate values of temperature that agree within ±0.05 °C with the values given in Table 10.8.

Type N Thermocouples: Coefficients (c_i) of Polynomials for the Computation of Temperatures in °C as a Function of the Thermocouple emf in Various Temperature and emf Ranges

Temperature range:	−200 °C to 0 °C	0 °C to 600 °C	600 °C to 1300 °C
emf range:	−3.990 mV to 0.0 mV	0.0 mV to 20.613 mV	20.613 mV to 47.513 mV
c_0 =	0.000 000 0 …	0.000 00 …	$1.972\,485 \times 10^{1}$
c_1 =	$3.843\,684\,7 \times 10^{1}$	$3.868\,96 \times 10^{1}$	$3.300\,943 \times 10^{1}$
c_2 =	1.101 048 5 …	−1.082 67 …	$-3.915\,159 \times 10^{-1}$
c_3 =	5.222 931 2 …	$4.702\,05 \times 10^{-2}$	$9.855\,391 \times 10^{-3}$
c_4 =	7.206 052 5 …	$-2.121\,69 \times 10^{-6}$	$-1.274\,371 \times 10^{-4}$
c_5 =	5.848 858 6 …	$-1.172\,72 \times 10^{-4}$	$7.767\,022 \times 10^{-7}$
c_6 =	2.775 491 6 …	$5.392\,80 \times 10^{-6}$	………
c_7 =	$7.707\,516\,6 \times 10^{-1}$	$-7.981\,56 \times 10^{-8}$	………
c_8 =	$1.158\,266\,5 \times 10^{-1}$	………	………
c_9 =	$7.313\,886\,8 \times 10^{-3}$	………	………

NOTE—The above coefficients are extracted from Ref. 1 and are for an expression of the form shown in Section 10.3.2. They yield approximate values of temperature that agree within ± 0.04 °C with the values given in Table 10.10.

Type R Thermocouples: Coefficients (c_i) of Polynomials for the Computation of Temperatures in °C as a Function of the Thermocouple emf in Various Temperature and emf Ranges

Temperature range:	− 50 °C to 250 °C	250 °C to 1200 °C	1064 °C to 1664.5 °C	1664.5 °C to 1768.1 °C
emf range:	− 0.226 mV to 1.923 mV	1.923 mV to 13.228 mV	11.361 mV to 19.739 mV	19.739 mV to 21.103 mV
c_0 =	0.000 000 0 …	$1.334\,584\,505 \times 10^{1}$	$-8.199\,599\,416 \times 10^{1}$	$3.406\,177\,836 \times 10^{4}$
c_1 =	$1.889\,138\,0 \times 10^{2}$	$1.472\,644\,573 \times 10^{2}$	$1.553\,962\,042 \times 10^{2}$	$-7.023\,729\,171 \times 10^{3}$
c_2 =	$-9.383\,529\,0 \times 10^{1}$	$-1.844\,024\,844 \times 10^{1}$	−8.342 197 663	$5.582\,903\,813 \times 10^{2}$
c_3 =	$1.306\,861\,9 \times 10^{2}$	4.031 129 726 …	$4.279\,433\,549 \times 10^{-1}$	$-1.952\,394\,635 \times 10^{1}$
c_4 =	$-2.270\,358\,0 \times 10^{2}$	$-6.249\,428\,360 \times 10^{-1}$	$-1.191\,577\,910 \times 10^{-2}$	$2.560\,740\,231 \times 10^{-1}$
c_5 =	$3.514\,565\,9 \times 10^{2}$	$6.468\,412\,046 \times 10^{-2}$	$1.492\,290\,091 \times 10^{-4}$	………
c_6 =	$-3.895\,390\,0 \times 10^{2}$	$-4.458\,750\,426 \times 10^{-3}$	………	………
c_7 =	$2.823\,947\,1 \times 10^{2}$	$1.994\,710\,149 \times 10^{-4}$	………	………
c_8 =	$-1.260\,728\,1 \times 10^{2}$	$-5.313\,401\,790 \times 10^{-6}$	………	………
c_9 =	$3.135\,361\,1 \times 10^{1}$	$6.481\,976\,217 \times 10^{-8}$	………	………
c_{10} =	−3.318 776 9 …	………	………	………

NOTE—The above coefficients are extracted from Ref. 1 and are for an expression of the form shown in Section 10.3.2. They yield approximate values of temperature that agree within ±0.02 °C with the values given in Table 10.12.

Type S Thermocouples: Coefficients (c_i) of Polynomials for the Computation of Temperatures in °C as a Function of the Thermocouple emf in Various Temperature and emf Ranges

Temperature range:	−50 °C to 250 °C	250 °C to 1200 °C	1064 °C to 1664.5 °C	1664.5 °C to 1768.1 °C
emf range:	−0.235 mV to 1.874 mV	1.874 mV to 11.950 mV	10.332 mV to 17.536 mV	17.536 mV to 18.693 mV
c_0 =	0.000 000 00 . . .	$1.291\ 507\ 177 \times 10^{1}$	$-8.087\ 801\ 117 \times 10^{1}$	$5.333\ 875\ 126 \times 10^{4}$
c_1 =	$1.849\ 494\ 60 \times 10^{2}$	$1.466\ 298\ 863 \times 10^{2}$	$1.621\ 573\ 104 \times 10^{2}$	$-1.235\ 892\ 298 \times 10^{4}$
c_2 =	$-8.005\ 040\ 62 \times 10^{1}$	$-1.534\ 713\ 402 \times 10^{1}$	−8.536 869 453 . . .	$1.092\ 657\ 613 \times 10^{3}$
c_3 =	$1.022\ 374\ 30 \times 10^{2}$	3.145 945 973 . . .	$4.719\ 686\ 976 \times 10^{-1}$	$-4.265\ 693\ 686 \times 10^{1}$
c_4 =	$-1.522\ 485\ 92 \times 10^{2}$	$-4.163\ 257\ 839 \times 10^{-1}$	$-1.441\ 693\ 666 \times 10^{-2}$	$6.247\ 205\ 420 \times 10^{-1}$
c_5 =	$1.888\ 213\ 43 \times 10^{2}$	$3.187\ 963\ 771 \times 10^{-2}$	$2.081\ 618\ 890 \times 10^{-4}$	....
c_6 =	$-1.590\ 859\ 41 \times 10^{2}$	$-1.291\ 637\ 500 \times 10^{-3}$	....	....
c_7 =	$8.230\ 278\ 80 \times 10^{1}$	$2.183\ 475\ 087 \times 10^{-5}$	....	....
c_8 =	$-2.341\ 819\ 44 \times 10^{1}$	$-1.447\ 379\ 511 \times 10^{-7}$	....	....
c_9 =	2.797 862 60 . . .	$8.211\ 272\ 125 \times 10^{-9}$	....	....

NOTE—The above coefficients are extracted from Ref. 1 and are for an expression of the form shown in Section 10.3.2. They yield approximate values of temperature that agree within ± 0.02 °C with the values given in Table 10.14.

Type T Thermocouples: Coefficients (c_i) of Polynomials for the Computation of Temperatures in °C as a Function of the Thermocouple emf in Various Temperature and emf Ranges

Temperature range:	−200 °C to 0 °C	0 °C to 400 °C
emf range:	−5.603 mV to 0.0 mV	0.0 mV to 20.872 mV
c_0 =	0.000 000 0 . . .	0.000 000 . . .
c_1 =	$2.594\ 919\ 2 \times 10^{1}$	$2.592\ 800 \times 10^{1}$
c_2 =	$-2.131\ 696\ 7 \times 10^{-1}$	$-7.602\ 961 \times 10^{-1}$
c_3 =	$7.901\ 869\ 2 \times 10^{-1}$	$4.637\ 791 \times 10^{-2}$
c_4 =	$4.252\ 777\ 7 \times 10^{-1}$	$-2.165\ 394 \times 10^{-3}$
c_5 =	$1.330\ 447\ 3 \times 10^{-1}$	$6.048\ 144 \times 10^{-5}$
c_6 =	$2.024\ 144\ 6 \times 10^{-2}$	$-7.293\ 422 \times 10^{-7}$
c_7 =	$1.266\ 817\ 1 \times 10^{-3}$	....

NOTE—The above coefficients are extracted from Ref. 1 and are for an expression of the form shown in Section 10.3.2. They yield approximate values of temperature that agree within ± 0.04 °C with the values given in Table 10.16.

REFERENCE POINTS ON THE ITS-90 TEMPERATURE SCALE

The International Temperature Scale of 1990 is described in Section 1 of this *Handbook*, where the defining fixed points are listed. The Consultative Committee on Thermometry (CCT) of the International Committee on Weights and Measures (CIPM), which oversees the temperature scale, has recommended a number of secondary reference points whose values have been accurately determined with respect to the primary fixed points. The most accurate of these, referred to as "first quality points", satisfy several criteria involving purity of the material, reproducibility, and documentation of the measurements. The CCT also lists "second quality points" that do not yet satisfy all the criteria but are still useful. These secondary reference points help fill in the gaps between the primary fixed points.

The table below describes these secondary reference points, along with the primary points. The values are listed on both the Kelvin and Celsius scales, together with an estimate of uncertainty. Full details are given in the reference.

The entries are listed in order of increasing temperature.

Reference

Bedford, R. E., Bonnier, G., Maas, H., and Pavese, F., *Metrologia* 33, 133, 1996.

Substance	Type of transition	*T*/K	*t*/°C	Uncert.	Note
Zinc	Superconductive transition	0.8500	−272.3000	0.0030	1st Quality
Aluminum	Superconductive transition	1.1810	−271.9690	0.0025	1st Quality
Helium	Superfluid transition (^{4}He)	2.1768	−270.9732	0.0001	1st Quality
Indium	Superconductive transition	3.4145	−269.7355	0.0025	1st Quality
Lead	Superconductive transition	7.1997	−265.9503	0.0025	1st Quality
Niobium	Superconductive transition	9.288	−263.862	0.0025	1st Quality
Hydrogen	Triple point (equilibrium H_2)	13.8033	−259.3467	0	Primary
Hydrogen	Triple point (normal H_2)	13.952	−259.198	0.002	2nd Quality
Hydrogen-d_2	Triple point (equilibrium D_2)	18.689	−254.461	0.001	1st Quality
Hydrogen-d_2	Triple point (normal D_2)	18.724	−254.426	0.001	1st Quality
Hydrogen	Boiling point (normal H_2)	20.388	−252.762	0.002	2nd Quality
Oxygen	α-β transition	23.868	−249.282	0.005	2nd Quality
Neon	Triple point (^{20}Ne)	24.541	−248.609	0.001	1st Quality
Neon	Triple point (normal Ne)	24.5561	−248.5939	0	Primary
Neon	Boiling point	27.097	−246.053	0.001	1st Quality
Nitrogen	α-β transition	35.614	−237.536	0.006	2nd Quality
Oxygen	β-γ transition	43.796	−229.354	0.001	2nd Quality
Oxygen	Triple point	54.3584	−218.7916	0	Primary
Nitrogen	Triple point	63.151	−209.999	0.001	1st Quality
Nitrogen	Boiling point	77.352	−195.798	0.002	1st Quality
Argon	Triple point	83.8058	−189.3442	0	Primary
Argon	Boiling point	87.303	−185.847	0.001	1st Quality
Oxygen	Condensation point	90.197	−182.953	0.001	1st Quality
Methane	Triple point	90.694	−182.456	0.001	1st Quality
Krypton	Triple point	115.775	−157.375	0.001	2nd Quality
Xenon	Triple point	161.405	−111.745	0.001	1st Quality
Carbon dioxide	Sublimation point	194.686	−78.464	0.003	2nd Quality
Carbon dioxide	Triple point	216.592	−56.558	0.001	1st Quality
Sulfur hexafluoride	Triple point	223.554	−49.596	0.005	2nd Quality
Mercury	Triple point	234.3156	−38.8344	0	Primary
Mercury	Freezing point	234.321	−38.829	0.0005	1st Quality
Water	Ice point	273.15	0	0	1st Quality
Water	Triple point	273.16	0.01	0	Primary
Gallium/20% indium	Eutectic melting point	288.8	15.65	0.001	2nd Quality
Gallium/8% tin	Eutectic melting point	293.626	20.476	0.002	2nd Quality
Diphenyl ether	Triple point	300.014	26.864	0.001	2nd Quality
Gallium	Melting point	302.9146	29.7646	0	Primary
Gallium	Triple point	302.9166	29.7666	0.0001	1st Quality
Ethylene carbonate	Triple point	309.465	36.315	0.001	2nd Quality
Succinonitrile	Triple point	331.215	58.065	0.002	2nd Quality
Sodium	Freezing point	370.944	97.794	0.005	2nd Quality
Water	Boiling point	373.124	99.974	0.001	1st Quality
Benzoic acid	Triple point	395.486	122.336	0.002	2nd Quality
Benzoic acid	Freezing point	395.502	122.352	0.007	2nd Quality

Substance	Type of transition	*T*/K	*t*/°C	Uncert.	Note
Indium	Triple point	429.7436	156.5936	0.0002	1st Quality
Indium	Freezing point	429.7485	156.5985	0	Primary
Tin (white)	Freezing point	505.078	231.928	0	Primary
Bismuth	Freezing point	544.552	271.402	0.001	1st Quality
Cadmium	Freezing point	594.219	321.069	0.001	1st Quality
Lead	Freezing point	600.612	327.462	0.001	1st Quality
Mercury	Boiling point	629.811	356.619	0.004	2nd Quality
Zinc	Freezing point	692.677	419.527	0	Primary
Sulfur	Boiling point	717.674	444.614	0.002	2nd Quality
Copper/66.9% aluminum	Eutectic melting point	821.308	548.158	0.01	2nd Quality
Silver/30% aluminum	Eutectic melting point	840.957	567.807	0.002	2nd Quality
Antimony	Freezing point	903.778	630.628	0.001	1st Quality
Aluminum	Freezing point	933.473	660.323	0	Primary
Copper/71.9% silver	Eutectic melting point	1052.78	779.63	0.05	1st Quality
Sodium chloride	Freezing point	1075.168	802.018	0.011	2nd Quality
Sodium	Boiling point	1156.090	882.940	0.005	2nd Quality
Silver	Freezing point	1234.93	961.78	0	Primary
Gold	Freezing point	1337.33	1064.18	0	Primary
Copper	Freezing point	1357.77	1084.62	0	Primary
Nickel	Freezing point	1728	1455	1	2nd Quality
Cobalt	Freezing point	1768	1495	3	2nd Quality
Iron	Freezing point	1811	1538	3	2nd Quality
Palladium	Freezing point	1828.0	1554.8	0.1	1st Quality
Titanium	Melting point	1943	1670	2	2nd Quality
Platinum	Freezing point	2041.3	1768.2	0.4	1st Quality
Zirconium	Melting point	2127	1854	8	2nd Quality
Rhodium	Freezing point	2236	1963	3	1st Quality
Aluminum oxide (α)	Melting point	2326	2053	2	2nd Quality
Ruthenium	Melting point	2606	2333	10	2nd Quality
Iridium	Freezing point	2719	2446	6	1st Quality
Molybdenum	Melting point	2895	2622	4	1st Quality
Tungsten	Melting point	3687	3414	7	1st Quality

RELATIVE SENSITIVITY OF BAYARD-ALPERT IONIZATION GAUGES TO VARIOUS GASES

Paul Redhead

Bayard-Alpert hot-cathode ionization gauges are widely used for pressure measurements in the 10^{-3} Torr to 10^{-10} Torr pressure range. The ion current I_+ in a hot-cathode ionization gauge is given by $I_+ = KI_eP$. The gauge constant is $K = (I_+/I_e)(1/P)$, where I_e is the electron current, and P the pressure. The sensitivity is given by $S = KI_e = I_+/P$. The constant K is independent of pressure below about 10^{-3} Torr, but it depends on the gas that is present. This table gives the sensitivity relative to nitrogen for several different gases.

Relative sensitivities for different Bayard-Alpert ionization gauges may differ by as much as ±15% as a result of differences in applied voltages, electron current, and electrode structure. The table below presents the average of the measurements of 12 experimenters on Bayard-Alpert ionization gauges in various gases.

References

1. Hollanda, R., *J. Vac. Sci. Technol.*, 10, 1133, 1973.
2. Nakayama, K., and Hojo, H., *Jap. J. Appl. Phys.*, Suppl. 2, part 1, p. 113, 1974.
3. Tilford, C. R., *J. Vac. Sci. Technol. A*, 1, 152, 1983.
4. Tilford, C. R., in *Physical Methods of Chemistry, Vol. 6, Determination of Thermodynamic Properties*, B. W. Rossiter and R. C. Baetzoid, Eds., pp. 101-173, John Wiley, New York, 1992.

Gas	Formula	Relative sensitivity $S/S(N_2)$
Nitrogen	N_2	**1.00**
Acetylene	HC≡CH	0.61
Allene	$CH_2=C=CH_2$	1.3
Argon	Ar	1.4
Benzene	C_6H_6	3.8
Butane	C_4H_{10}	4.3
Carbon dioxide	CO_2	1.4
Carbon monoxide	CO	1.0
Ethane	C_2H_6	2.6
Ethylene	$CH_2=CH_2$	1.3
Helium	He	0.18
Hydrogen	H_2	0.43
Krypton	Kr	1.9
Mercury	Hg	3.5
Methane	CH_4	1.6
Neon	Ne	0.31
Oxygen	O_2	0.96
Propane	C_3H_8	3.5
Propene	$CH_3CH=CH_2$	1.8
Propyne	$CH_3C\equiv CH$	1.4
Sulfur hexafluoride	SF_6	2.3
Water	H_2O	0.93
Xenon	Xe	2.7

LABORATORY SOLVENTS AND OTHER LIQUID REAGENTS

This table summarizes the properties of 574 liquids that are commonly used in the laboratory as solvents or chemical reagents. The properties tabulated are:

- M_r: Molecular weight
- mp: Melting point in °C
- bp: Normal boiling point in °C
- ρ: Density in g/cm³ at the temperature in °C indicated by the superscript
- ε: Dielectric constant (permittivity) at the temperature in °C indicated by the superscript (otherwise at ambient temperature)
- μ: Dipole moment in Debye units
- vp: Vapor pressure at 25 °C in kPa (1 kPa = 7.50 mmHg)
- FP: Flash point in °C
- Fl. lim: Flammable (explosive) limits in air in percent by volume
- IT: Autoignition temperature in °C
- TWA and STEL: Threshold limits for allowable airborne concentration in parts per million by volume at 25 °C and atmospheric pressure. TWA refers to a 8-hour time-weighted average and STEL to a short-term limit. More details on threshold limits can be found in Section 16.

Data on refractive index, specific heat capacity, and viscosity is included in the Internet version of the *Handbook*. Data on the temperature dependence of viscosity, dielectric constant, and vapor pressure can be found in the pertinent tables in Section 6.

References

1. Lide, D. R., *Handbook of Organic Solvents*, CRC Press, Boca Raton, FL, 1994.
2. Lide, D. R., and Kehiaian, H. V., *Handbook of Thermophysical and Thermochemical Data*, CRC Press, Boca Raton, FL, 1994.
3. Riddick, J. A., Bunger, W. B., and Sakano, T. K., *Organic Solvents, Fourth Edition*, John Wiley & Sons, New York, 1986.
4. *Fire Protection Guide to Hazardous Materials, 11th Edition*, National Fire Protection Association, Quincy, MA, 1994.
5. Urben, P. G., Ed., *Bretherick's Handbook of Reactive Chemical Hazards, 5th Edition*, Butterworth-Heinemann, Oxford, 1995.
6. *2015 TLV's and BEI's*, American Conference of Governmental Industrial Hygienists, Cincinnati, OH, 2015.

Physical Properties of Common Liquid Reagents

Name	Mol. form.	M_r	mp/°C	bp/°C	ρ/g cm⁻³	ε	μ/D	Vap. pres. (25 °C)/kPa	FP/°C	Flam. limits (%)	IT/°C	TWA/ ppm	STEL/ ppm
Acetaldehyde	C_2H_4O	44.052	-123.4	20.8	0.7834^{18}	21.0^{18}	2.750	120	-39	4.0-60%	175		25
Acetic acid	$C_2H_4O_2$	60.052	17	117.9	1.0510^{20}	6.20^{20}	1.70	2.07	39	4.0-19.9%	463	10	15
Acetic anhydride	$C_4H_6O_3$	102.089	-73.4	139.5	1.082^{20}	22.45^{20}	≈2.8	0.680	49	2.7-10.3%	316	1	3
Acetone	C_3H_6O	58.079	-94.9	56.08	0.7902^{20}	21.01^{20}	2.88	30.8	-20	2.5-12.8%	465	250	500
Acetone cyanohydrin	C_4H_7NO	85.105	-19	180	0.932^{19}				74	2.2-12.0%	688		
Acetonitrile	C_2H_3N	41.052	-44	81.6	0.7825^{20}	36.64^{20}	3.92519	11.9	6	3.0-16.0%	524	20	
Acetophenone	C_8H_8O	120.149	19.4	202.1	1.0281^{20}	17.44^{25}	3.02	0.049	77		570	10	
Acetyl bromide	C_2H_3BrO	122.948	-96.5	74	1.6625^{16}			16.2					
Acetyl chloride	C_2H_3ClO	78.497	-112.7	51	1.1051^{20}	15.8^{22}	2.72	38.4	4		390		
Acrolein	C_3H_4O	56.063	-87.8	52.3	0.840^{20}		[a]	36.2	-26	2.8-31%	220		0.1
Acrylic acid	$C_3H_4O_2$	72.063	13.56	142	1.0511^{20}			0.53	50	2.4-8.0%	438	2	
Acrylonitrile	C_3H_3N	53.063	-83.51	77.2	0.8007^{25}	33.0^{20}	3.92	14.1	0	3.0-17.0%	481	2	
Allyl alcohol	C_3H_6O	58.079	-129	96.9	0.8540^{20}	19.7^{20}	1.60	3.14	21	2.5-18.0%	378	0.5	
Allylamine	C_3H_7N	57.095	-88.2	54	0.758^{20}		≈1.2	33.1	-29	2.2-22%	374		
2-Amino-2-methyl-1-propanol	$C_4H_{11}NO$	89.136	25.5	163.8	0.934^{20}				67				
3-Amino-1-propanol	C_3H_9NO	75.109	12.1	185	0.9824^{26}				80				
Aniline	C_6H_7N	93.127	-6.0	184.1	1.0250^{20}	7.06^{20}	1.13	0.090	70	1.3-11%	615	2	
Anisole	C_7H_8O	108.138	-37.3	153.6	0.9940^{20}	4.30^{21}	1.38	0.472	52		475		
Antimony(V) chloride	Cl_5Sb	299.025	4	140 dec	2.34	3.222^{20}							
Antimony(V) fluoride	F_5Sb	216.752	8.3	141	3.10								
Arsenic(III) chloride	$AsCl_3$	181.281	-16	130	2.150		1.59	5.38					
Benzaldehyde	C_7H_6O	106.122	-57.12	178.7	1.0401^{25}	17.85^{20}	3.0	0.169	63		192		
Benzene	C_6H_6	78.112	5.538	80.08	0.8788^{20}	2.2825^{20}	0	12.7	-11	1.2-7.8%	498	0.5	2.5
Benzeneacetonitrile	C_8H_7N	117.149	-22.1	232	1.0205^{15}	17.87^{26}	3.5	0.012	113				
Benzeneethanamine	$C_8H_{11}N$	121.180	<0	204	0.9640^{25}								
Benzeneethanol	$C_8H_{10}O$	122.164	-19	220	1.0202^{20}	12.31^{20}		0.01	96				
Benzenemethanethiol	C_7H_8S	124.204	-30	199	1.058^{20}	4.705^{25}							
Benzenesulfonyl chloride	$C_6H_5ClO_2S$	176.621	14.5	252	1.3470^{15}	28.90^{50}		0.008					
Benzenethiol	C_6H_6S	110.177	-14.87	169.1	1.0775^{20}	4.26^{30}	1.23	0.26				0.1	
Benzonitrile	C_7H_5N	103.122	-12.82	191	1.0093^{15}	25.9^{20}	4.18	0.11					
Benzoyl chloride	C_7H_5ClO	140.567	-0.5	201	1.2120^{20}	23.0^{20}		0.084	72				0.5
Benzyl acetate	$C_9H_{10}O_2$	150.174	-51.5	215	1.0550^{20}	5.34^{30}	1.22	0.022	90		460	10	
Benzyl alcohol	C_7H_8O	108.138	-15.5	205.3	1.0419^{24}	11.916^{30}	1.71	0.015	93		436		
Benzylamine	C_7H_9N	107.153		185	0.9813^{20}	5.18^{20}		0.096					
2,2'-Bioxirane	$C_4H_6O_2$	86.090	2.0	144	1.113^{20}								

Lab Data

Name	Mol. form.	M_r	mp/°C	bp/°C	ρ/g cm^{-3}	ε	μ/D	Vap. pres. (25 °C)/kPa	FP/°C	Flam. limits (%)	IT/°C	TWA/ ppm	STEL/ ppm
Bis(2-aminoethyl)amine	$C_4H_{13}N_3$	103.166	-39	206.5	0.9569^{20}	12.62^{20}	1.89	0.03	98	2-6.7%	358	1	
N,N'-Bis(2-aminoethyl)-1,2-ethanediamine	$C_6H_{18}N_4$	146.234	12	266.5		10.76^{20}							
Bis(2-chloroethyl) ether	$C_4H_8Cl_2O$	143.012	-46.9	178	1.22^{20}	21.20^{20}	2.58	0.143	55	>2.7%	369	5	10
Bis(chloromethyl) ether	$C_2H_4Cl_2O$	114.958	-41.5	104	1.323^{15}	3.51^{20}						0.001	
Bis(2-ethylhexyl) phthalate	$C_{24}H_{38}O_4$	390.557	-55	384	0.981^{25}		2.84	0.00000005	218				
Bis(2-hydroxyethyl) sulfide	$C_4H_{10}O_2S$	122.186	-10.2	282	1.1793^{25}	28.61^{20}		0.08	160		298		
Boron tribromide	BBr_3	250.523	-46	91.3	2.6	2.58^{0}	0						0.7
Boron trichloride	BCl_3	117.170	-107.3	12.5	0.004789^{25} gas		0	156					0.7
Bromine	Br_2	159.808	-7.2	58.8	3.1028	3.1484^{25}		28.2				0.1	0.2
Bromobenzene	C_6H_5Br	157.008	-30.74	155.9	1.4950^{20}	5.45^{20}	1.70	0.556	51		565		
1-Bromobutane	C_4H_9Br	137.018	-112.5	101.4	1.2758^{20}	7.315^{10}	2.08	5.26	18	2.6-6.6%	265		
2-Bromobutane, (±)-	C_4H_9Br	137.018	-112.6	91	1.2585^{20}	8.64^{25}	2.23	9.32	21				
Bromochloromethane	CH_2BrCl	129.384	-87.9	67.9	1.9344^{20}		1.66	19.5				200	
Bromodichloromethane	$CHBrCl_2$	163.829	-56.0	90	1.980^{20}								
Bromoethane	C_2H_5Br	108.965	-118.4	38.2	1.4604^{20}	9.01^{25}	2.04	62.5		6.8-8.0%	511	5	
Bromoethene	C_2H_3Br	106.949	-139.5	16	1.4933^{20}		1.42	141		9-15%	530	0.5	
2-Bromo-2-methylpropane	C_4H_9Br	137.018	-16.8	73.3	1.4278^{20}	10.98^{20}	2.17	17.7					
1-Bromopentane	$C_5H_{11}Br$	151.045	-88.0	126	1.2182^{20}	6.31^{26}	2.20	1.68	32				
1-Bromopropane	C_3H_7Br	122.992	-110.1	70.8	1.3537^{20}	8.09^{20}	2.18	18.6			490	0.1	
2-Bromopropane	C_3H_7Br	122.992	-88.9	59.34	1.3140^{20}	9.46^{20}	2.21	28.9					
3-Bromopropene	C_3H_5Br	120.976	-119.3	70.1	1.398^{20}	7.0^{20}	≈1.9	18.6	-1	4.4-7.3%	295	0.1	0.2
2-Bromotoluene	C_7H_7Br	171.035	-27.5	182	1.4232^{20}	4.641^{20}		0.17	79				
Bromotrichloromethane	$CBrCl_3$	198.274	-5.6	103	2.012^{25}	2.405^{20}		5.35					
Butanal	C_4H_8O	72.106	-96.86	74.8	0.8016^{20}	13.45^{25}	2.72	15.7	-22	1.9-12.5%	218		
1,3-Butanediol	$C_4H_{10}O_2$	90.121	-77	208.2	1.0053^{20}	28.8^{25}		0.008	121		395		
1,4-Butanediol	$C_4H_{10}O_2$	90.121	20.43	229.5	1.0171^{20}	31.9^{25}	2.58	0.002	121				
2,3-Butanediol	$C_4H_{10}O_2$	90.121	7	178	1.0033^{20}			0.02			402		
2,3-Butanedione	$C_4H_6O_2$	86.090	-1.2	87.5	0.9808^{18}	4.04^{25}		7.45	27			0.01	0.02
Butanenitrile	C_4H_7N	69.106	-111.76	117.6	0.7936^{20}	24.83^{20}	≈3.8	2.55	24	>1.6%	501		
1-Butanethiol	$C_4H_{10}S$	90.187	-115.66	98.4	0.8416^{20}	5.204^{15}	1.53	6.07	2			0.5	
2-Butanethiol	$C_4H_{10}S$	90.187	-165	85.0	0.8295^{20}	5.645^{15}		10.8	-23				
Butanoic acid	$C_4H_8O_2$	88.106	-5.12	163.7	0.9528^{25}	2.98^{14}	1.65	0.221	72	2.0-10.0%	443		
Butanoic anhydride	$C_8H_{14}O_3$	158.195	-75.0	195	0.9668^{20}	12.8^{20}		0.07	54	0.9-5.8%	279		
1-Butanol	$C_4H_{10}O$	74.121	-88.60	117.6	0.8148^{20}	17.84^{20}	1.66	0.86	37	1.4-11.2%	343	20	
2-Butanol	$C_4H_{10}O$	74.121	-88.44	99.4	0.8063^{20}	17.26^{20}		2.32	24	1.7-9.8%	405	100	
2-Butanone	C_4H_8O	72.106	-86.67	79.6	0.7999^{25}	18.56^{20}	2.779	12.6	-9	1.4-11.4%	404	200	300
trans-2-Butenal	C_4H_6O	70.090	-76.6	102.2	0.8516^{20}		3.67	4.92	13	2.1-15.5%	232	0.3	
cis-2-Butenoic acid	$C_4H_6O_2$	86.090	15	169	1.0267^{20}			0.06					
2-Butoxyethanol	$C_6H_{14}O_2$	118.174	-74.8	171	0.9015^{20}		2.08	0.15	69	4-13%	238	20	
Butyl acetate	$C_6H_{12}O_2$	116.158	-77.0	126.0	0.8825^{20}	5.07^{20}	1.87	1.66	22	1.7-7.6%	425	50	150
sec-Butyl acetate	$C_6H_{12}O_2$	116.158	-98.9	108	0.8748^{20}	5.135^{20}	1.87		31	1.7-9.8%		50	150
Butyl acrylate	$C_7H_{12}O_2$	128.169	-63.6	146.6	0.8898^{20}	5.25^{28}		0.731	29	1.7-9.9%	292	2	
Butylamine	$C_4H_{11}N$	73.137	-49	77.0	0.7417^{20}	4.71^{20}	≈1.0	12.2	-12	1.7-9.8%	312		5
sec-Butylamine	$C_4H_{11}N$	73.137	-104.5	62.71	0.7246^{20}		1.28		-9				
tert-Butylamine	$C_4H_{11}N$	73.137	-66.92	44.02	0.6958^{20}		1.29	48.4	-9	1.7-8.9%	380		
Butylbenzene	$C_{10}H_{14}$	134.218	-87.81	183.3	0.8601^{20}	2.359^{20}		0.150	71	0.8-5.8%	410		
tert-Butylbenzene	$C_{10}H_{14}$	134.218	-57.84	169.1	0.8665^{20}	2.359^{20}	≈0.83	0.280	60	0.7-5.7%	450		
Butyl benzoate	$C_{11}H_{14}O_2$	178.228	-22.4	249	1.000^{20}	5.52^{30}		0.005	107				
tert-Butyl ethyl ether	$C_6H_{14}O$	102.174	-94.0	72.7	0.736^{25}			16.5				25	
tert-Butyl hydroperoxide	$C_4H_{10}O_2$	90.121	6	89 dec	0.8960^{20}				27				
1-*tert*-Butyl-4-methylbenzene	$C_{11}H_{16}$	148.245	-52.49	193	0.8612^{20}			0.09	68			1	
Butyl vinyl ether	$C_6H_{12}O$	100.158	-92	94	0.7888^{20}		1.25	6.65	-9		255		
γ-Butyrolactone	$C_4H_6O_2$	86.090	-43.36	204.6	1.1296^{20}	39.0^{20}	4.27	0.43	98				
Carbon disulfide	CS_2	76.141	-111.7	46.2	1.2632^{20}	2.6320^{20}	0	48.2	-30	1.3-50.0%	90	1	
Chloroacetaldehyde	C_2H_3ClO	78.497	-16.3	87	1.19								1
Chloroacetone	C_3H_5ClO	92.524	-44.5	116	1.15^{20}			2					1
Chloroacetyl chloride	$C_2H_2Cl_2O$	112.942	-21.7	106.0	1.4202^{20}		2.23	3.33				0.05	0.15
2-Chloroaniline	C_6H_6ClN	127.572	-2.3	209	1.2114^{22}	13.40^{20}	1.77	0.034					
3-Chloroaniline	C_6H_6ClN	127.572	-10.3	230	1.2161^{20}	13.3^{20}		0.0156			705		
Chlorobenzene	C_6H_5Cl	112.557	-45.2	131.6	1.1058^{20}	5.6895^{20}	1.69	1.6	28	1.3-9.6%	593	10	
2-Chloro-1,3-butadiene	C_4H_5Cl	88.536	-130	59	0.956^{20}	4.914^{20}		29.5	-20	4.0-20.0%		10	
1-Chlorobutane	C_4H_9Cl	92.567	-123.1	78.4	0.8857^{20}	7.276^{20}	2.05	13.7	-12	1.9-10.1%	240		
2-Chlorobutane	C_4H_9Cl	92.567	-131.3	71	0.8732^{20}	8.564^{20}	2.04	21.0	-10				
Chlorocyclohexane	$C_6H_{11}Cl$	118.604	-45	142.6	1.000^{20}	7.9505^{30}	[a]	1.0	32				
Chlorodibromomethane	$CHBr_2Cl$	208.280	-20	120	2.451^{20}								
Chloroethane	C_2H_5Cl	64.514	-138.4	12.3	0.9239^{0}	9.45^{20}	2.05	160	-50	3.8-15.4%	519	100	

Name	Mol. form.	M_r	mp/°C	bp/°C	ρ/g cm^{-3}	ε	μ/D	Vap. pres. (25 °C)/kPa	FP/°C	Flam. limits (%)	IT/°C	TWA/ ppm	STEL/ ppm
2-Chloroethanol	C_2H_5ClO	80.513	-68	126	1.2019^{20}	25.80^{20}	1.78	1.2	60	4.9-15.9%	425		1
2-Chloroethyl vinyl ether	C_4H_7ClO	106.551	-70	108	1.0495^{20}				27				
(Chloromethyl)benzene	C_7H_7Cl	126.584	-39.4	174	1.1004^{20}	6.854^{20}	1.82	0.164	67	>1.1%	585	1	
Chloromethyl methyl ether	C_2H_5ClO	80.513	-103.5	59	1.063^{10}			24.9					
1-Chloro-2-methylpropane	C_4H_9Cl	92.567	-130.3	69	0.8773^{20}	7.027^{20}	2.00	19.9	-6	2.0-8.7%			
2-Chloro-2-methylpropane	C_4H_9Cl	92.567	-25.60	50.9	0.8420^{20}	9.663^{20}	2.13	42.7	0				
1-Chloronaphthalene	$C_{10}H_7Cl$	162.616	-6.0	259	1.1880^{25}	5.04^{25}	1.57	0.003	121		>558		
1-Chlorooctane	$C_8H_{17}Cl$	148.674	-57.8	183	0.8734^{20}	5.05^{25}	2.00	0.11	70				
1-Chloropentane	$C_5H_{11}Cl$	106.594	-99.0	107.9	0.8820^{20}	6.654^{20}	2.16	4.36	13	1.6-8.6%	260		
2-Chlorophenol	C_6H_5ClO	128.556	8	173.4	1.2634^{20}	7.40^{23}		0.308	64				
1-Chloropropane	C_3H_7Cl	78.541	-122.9	46.2	0.8899^{20}	8.588^{20}	2.05	45.8	<-18	2.6-11.1%	520		
2-Chloropropane	C_3H_7Cl	78.541	-117.1	35.0	0.8617^{20}		2.17	68.9	-32	2.8-10.7%	593		
3-Chloro-1,2-propanediol	$C_3H_7ClO_2$	110.540		221	1.325^{18}	31.0^{20}							
3-Chloropropanenitrile	C_3H_4ClN	89.524	-51.4	175.5	1.1573^{20}				76				
2-Chloropropene	C_3H_5Cl	76.525	-137.4	23	0.9017^{20}	8.92^{26}	1.647	110	-37	4.5-16%			
3-Chloropropene	C_3H_5Cl	76.525	-136	44.8	0.9376^{20}	8.2^{20}	1.94	48.9	-32	2.9-11.1%	485	1	2
Chlorosulfonic acid	$ClHO_3S$	116.524	-80	152	1.75			0.42					
2-Chlorotoluene	C_7H_7Cl	126.584	-35.9	158.8	1.0825^{20}	4.721^{20}	1.56	0.482				50	
4-Chlorotoluene	C_7H_7Cl	126.584	7.4	161.8	1.0697^{20}	6.25^{20}	2.21	0.4					
Chromium(VI) dichloride dioxide	Cl_2CrO_2	154.901	-96.5	117	1.91							0.025	
trans-Cinnamaldehyde	C_9H_8O	132.159	-7.5	246	1.0497^{20}	17.72^{33}		0.005					
o-Cresol	C_7H_8O	108.138	31.0	191.0	1.0327^{35}	6.76^{25}	1.45	0.041	81	>1.4%	599		
m-Cresol	C_7H_8O	108.138	12.2	202.2	1.0339^{20}	12.44^{25}	1.48	0.019	86	>1.1%	558		
p-Cresol	C_7H_8O	108.138	34.77	201.9	1.0185^{40}	13.05^{25}	1.48	0.017	86	>1.1%	558		
Cyanogen chloride	CClN	61.471	-6.55	13	1.186^{20}		2.8331						0.3
Cyclobutane	C_4H_8	56.107	-90.7	12.5	0.7038^{0}		0	157	<10	>1.8%			
Cyclohexane	C_6H_{12}	84.159	6.7	80.7	0.7786^{20}	2.0243^{20}		13.0	-20	1.3-8%	245	100	
Cyclohexanol	$C_6H_{12}O$	100.158	26	160.9	0.9624^{20}	16.40^{20}		0.10	68	1-9%	300	50	
Cyclohexanone	$C_6H_{10}O$	98.142	-27.93	155.4	0.9478^{20}	16.1^{20}	3.246	0.53	44	1.1-9.4%	420	20	50
Cyclohexene	C_6H_{10}	82.143	-103.5	82.9	0.8110^{20}	2.2176^{20}	a	11.8	-12	>1.2%	310	300	
Cyclohexylamine	$C_6H_{13}N$	99.174	-17.7	133.6	0.8191^{20}	4.547^{20}	1.26	1.20	31	1.9-9.4%	293	10	
1,3-Cyclopentadiene	C_5H_6	66.102	-96.54	41	0.8021^{20}		0.419	58.5				75	
Cyclopentane	C_5H_{10}	70.133	-93.4	49.2	0.7457^{20}	1.9687^{20}		42.3	-25	>1.5%	361	600	
Cyclopentanol	$C_5H_{10}O$	86.132	-17	140.4	0.9488^{20}	18.5^{15}		0.294	51				
Cyclopentanone	C_5H_8O	84.117	-51.70	130.5	0.9487^{20}	13.58^{25}	≈3.3	1.55	26				
cis-Decahydronaphthalene	$C_{10}H_{18}$	138.250	-42.9	195.8	0.8965^{20}	2.219^{20}		0.10					
trans-Decahydronaphthalene	$C_{10}H_{18}$	138.250	-30.35	187.3	0.8659^{25}	2.184^{20}		0.164	54	0.7-5.4%	255		
Decamethylcyclopentasiloxane	$C_{10}H_{30}O_5Si_5$	370.770	-37.0	213	0.9593^{20}	2.50^{20}		0.02					
Decanal	$C_{10}H_{20}O$	156.265	-3.9	212	0.830^{15}			0.02					
Decane	$C_{10}H_{22}$	142.282	-29.61	174.1	0.7303^{20}	1.9853^{20}		0.170	51	0.8-5.4%	210		
Decanoic acid	$C_{10}H_{20}O_2$	172.265	31.39	270	0.8858^{40}								
1-Decanol	$C_{10}H_{22}O$	158.281	7	229	0.8294^{20}	7.93^{20}		0.009	82		288		
1-Decene	$C_{10}H_{20}$	140.266	-66.21	171	0.7408^{20}	2.136^{20}		0.210	<55		235		
Diacetone alcohol	$C_6H_{12}O_2$	116.158	-47	167.9	0.9387^{20}	18.2^{25}	3.24	0.224	58	1.8-6.9%	643	50	
Dibenzyl ether	$C_{14}H_{14}O$	198.260	1.8	298	1.0428^{20}	3.821^{20}			135				
Dibromodifluoromethane	CBr_2F_2	209.816	-110.1	22.79		2.939^{0}	0.66	110				100	
1,2-Dibromoethane	$C_2H_4Br_2$	187.861	9.8	131.3	2.1683^{25}	4.9612^{20}	1.19	1.55					
Dibromomethane	CH_2Br_2	173.835	-52.1	97.0	2.4969^{20}	7.77^{10}	1.43	6.12					
1,2-Dibromotetrafluoroethane	$C_2Br_2F_4$	259.823	-110.1	47.1	2.149^{25}	2.34^{25}		43.4					
Dibutylamine	$C_8H_{19}N$	129.244	-61.8	162	0.7670^{20}	2.765^{20}	0.98	0.34	47	1.1-6%			
Dibutyl ether	$C_8H_{18}O$	130.228	-96	141.6	0.7684^{20}	3.0830^{20}	1.17	0.898	25	1.5-7.6%	194		
Di-*tert*-butyl peroxide	$C_8H_{18}O_2$	146.228	-40	110.0	0.704^{20}			3.43	18				
Dibutyl phthalate	$C_{16}H_{22}O_4$	278.344	-35	338	1.0465^{20}	6.58^{20}	2.82		157	>0.5%	402		
Dibutyl sebacate	$C_{18}H_{34}O_4$	314.461	-9.2	356	0.9405^{15}	4.54^{20}	2.48		178	>0.4%	365		
Dibutyl sulfide	$C_8H_{18}S$	146.294	-74.97	168	0.8386^{20}	4.29^{25}	1.61	0.09	76				
Dichloroacetic acid	$C_2H_2Cl_2O_2$	128.942	12	193	1.5634^{20}	8.33^{20}		0.03				0.5	
o-Dichlorobenzene	$C_6H_4Cl_2$	147.002	-17.0	180.2	1.3059^{20}	10.12^{20}	2.50	0.18	66	2.2-9.2%	648	25	50
m-Dichlorobenzene	$C_6H_4Cl_2$	147.002	-24.8	172	1.2884^{20}	5.02^{20}	1.72	0.252	72				
trans-1,4-Dichloro-2-butene	$C_4H_6Cl_2$	124.997	3	155.4	1.183^{25}							0.005	
Dichlorodimethylsilane	$C_2H_6Cl_2Si$	129.061	-16	70.5	1.064^{25}			18.9	<21	3.4-9.5%			
1,1-Dichloroethane	$C_2H_4Cl_2$	98.959	-96.93	56.3	1.1757^{20}	10.10^{25}	2.06	30.5	-17	5.4-11.4%	458	100	
1,2-Dichloroethane	$C_2H_4Cl_2$	98.959	-35.6	83.4	1.2454^{25}	10.42^{20}	1.83	10.6	13	6.2-16%	413	10	
1,1-Dichloroethene	$C_2H_2Cl_2$	96.943	-122.5	31.6	1.213^{20}	4.60^{20}	1.34	80.0	-28	6.5-15.5%	570	5	
cis-1,2-Dichloroethene	$C_2H_2Cl_2$	96.943	-80.0	60	1.2837^{20}	9.20^{25}	1.90	26.8	6	3-15%	460	200	
trans-1,2-Dichloroethene	$C_2H_2Cl_2$	96.943	-49.8	47.64	1.2565^{20}	2.14^{20}	0	44.2	2	6-13%	460	200	
Dichloromethane	CH_2Cl_2	84.933	-94.9	39.8	1.3232^{20}	8.93^{25}	1.60	58.2		13-23%	556	50	
(Dichloromethyl)benzene	$C_7H_6Cl_2$	161.029	-17.0	205	1.26^{25}	6.9^{20}	2.07	0.06					

Lab Data

Name	Mol. form.	M_r	mp/°C	bp/°C	ρ/g cm^{-3}	ε	μ/D	Vap. pres. (25 °C)/kPa	FP/°C	Flam. limits (%)	IT/°C	TWA/ ppm	STEL/ ppm
1,1-Dichloropropane	$C_3H_6Cl_2$	112.986		88.4	1.1321^{20}			9.09					
1,2-Dichloropropane, (±)-	$C_3H_6Cl_2$	112.986	-100.53	96.4	1.1560^{20}	8.37^{20}	1.85	6.62	21	3.4-14.5%	557	10	
1,3-Dichloropropane	$C_3H_6Cl_2$	112.986	-99.5	120.8	1.1785^{25}	10.27^{30}	2.08	2.44					
2,3-Dichloropropene	$C_3H_4Cl_2$	110.970	10	93.0	1.211^{20}				15	2.6-7.8%			
2,4-Dichlorotoluene	$C_7H_6Cl_2$	161.029	-13.5	200	1.2476^{20}	5.68^{28}	1.70	0.055					
Dicyclohexylamine	$C_{12}H_{23}N$	181.318	-0.1	251	0.9123^{20}			0.003	>99				
Diethanolamine	$C_4H_{11}NO_2$	105.136	27.9	271.2	1.0966^{20}	25.75^{20}	2.8	<0.01	172	2-13%	662		
1,1-Diethoxyethane	$C_6H_{14}O_2$	118.174	-106.1	102	0.8254^{20}		1.38	3.68	-21	1.6-10.4%	230		
1,2-Diethoxyethane	$C_6H_{14}O_2$	118.174	-74.0	120.6	0.8351^{25}	3.90^{20}		4.33	27		205		
Diethylamine	$C_4H_{11}N$	73.137	-50	55.4	0.7056^{20}	3.680^{20}	0.92	30.1	-23	1.8-10.1%	312	5	15
N,N-Diethylaniline	$C_{10}H_{15}N$	149.233	-21.3	216	0.9307^{20}	5.15^{30}		0.025	85		630		
o-Diethylbenzene	$C_{10}H_{14}$	134.218	-31.4	183.4	0.8800^{20}	2.594^{20}		0.13	57		395		
m-Diethylbenzene	$C_{10}H_{14}$	134.218	-83.9	181.1	0.8602^{20}	2.369^{20}		0.14	56		450		
p-Diethylbenzene	$C_{10}H_{14}$	134.218	-43.3	184	0.8620^{20}	2.259^{20}		0.13	55	0.7-6.0%	430		
Diethyl carbonate	$C_5H_{10}O_3$	118.131	-43	125.9	0.9692^{25}	2.820^{24}	1.10	1.63	25				
Diethylene glycol	$C_4H_{10}O_3$	106.120	-10.3	245.5	1.1197^{15}	31.82^{20}	2.31	0.001	124	2-17%	224		
Diethylene glycol diethyl ether	$C_8H_{18}O_3$	162.227	-44.3	185	0.9063^{20}			0.10	82				
Diethylene glycol dimethyl ether	$C_6H_{14}O_3$	134.173	-64.0	162	0.9434^{20}	7.23^{25}	1.97	0.315	67				
Diethylene glycol monobutyl ether	$C_8H_{18}O_3$	162.227	-68	232	0.9553^{20}			0.0032				10	
Diethylene glycol monoethyl ether	$C_6H_{14}O_3$	134.173		202	0.9885^{20}		1.6	0.017	96				
Diethylene glycol monoethyl ether acetate	$C_8H_{16}O_4$	176.211	-25	218	1.0096^{20}		1.8	0.029	110		425		
Diethylene glycol monomethyl ether	$C_5H_{12}O_3$	120.147		194	1.035^{20}		1.6	0.024	96	1.38-22.7%	240		
Diethyl ether	$C_4H_{10}O$	74.121	-116.22	34.4	0.7135^{20}	4.2666^{20}	1.098	71.7	-45	1.9-36.0%	180	400	500
Diethyl maleate	$C_8H_{12}O_4$	172.179	-8.8	222	1.0662^{20}	7.560^{25}		0.015	121		350		
Diethyl malonate	$C_7H_{12}O_4$	160.168	-50	200	1.0551^{20}	7.550^{31}	2.54	0.048	93				
Diethyl oxalate	$C_6H_{10}O_4$	146.141	-40.6	186	1.0785^{20}	8.266^{20}	2.49	0.030	76				
Diethyl phthalate	$C_{12}H_{14}O_4$	222.237	-40.5	298	1.232^{14}	7.86^{20}		0.002	161	>0.7%	457		
Diethyl succinate	$C_8H_{14}O_4$	174.195	-21.6	217	1.0402^{20}	6.098^{20}		0.15	90				
Diethyl sulfate	$C_4H_{10}O_4S$	154.185	-26.0	208	1.172^{25}	29.2^{20}		0.05	104		436		
Diethyl sulfide	$C_4H_{10}S$	90.187	-103.9	92.1	0.8362^{20}	5.723^{25}	[a]	7.78					
Diiodomethane	CH_2I_2	267.836	6.0	182	3.3211^{20}	5.32^{25}	1.08	0.172					
Diiodosilane	H_2I_2Si	283.911	-1	150									
Diisobutylamine	$C_8H_{19}N$	129.244	-73.5	139.6				0.972	29				
Diisopentyl ether	$C_{10}H_{22}O$	158.281		172	0.7777^{20}	2.817^{20}	1.23	0.210					
Diisopropylamine	$C_6H_{15}N$	101.190	-61	84	0.7153^{20}		1.15	10.7	-1	1.1-7.1%	316	5	
Diisopropyl ether	$C_6H_{14}O$	102.174	-85.37	68.4	0.7192^{25}	3.805^{30}	1.13	19.9	-28	1.4-7.9%	443	250	310
1,2-Dimethoxyethane	$C_4H_{10}O_2$	90.121	-69.0	85.0	0.8637^{25}	7.30^{24}		9.93	-2		202		
Dimethoxymethane	$C_3H_8O_2$	76.095	-105.11	42.3	0.8593^{20}	2.644^{20}	0.74	53.1	-32	2.2-13.8%	237	1000	
Dimethylacetal	$C_4H_{10}O_2$	90.121	-113.2	63	0.8501^{20}			22.9					
N,N-Dimethylacetamide	C_4H_9NO	87.120	-19	165.9	0.9372^{25}	38.85^{21}	3.7	0.075	70	1.8-11.5%	490	10	
2,3-Dimethylaniline	$C_8H_{11}N$	121.180	3	223	0.9931^{20}				97	>1.0%		0.5	
2,6-Dimethylaniline	$C_8H_{11}N$	121.180	11.0	211	0.9842^{20}		1.63	0.45	96			0.5	
N,N-Dimethylaniline	$C_8H_{11}N$	121.180	2.1	193	0.9562^{20}	4.90^{25}	1.68	0.107	63		371	5	10
2,2-Dimethylbutane	C_6H_{14}	86.175	-99.0	49.7	0.6444^{25}	1.869^{20}		42.5	-48	1.2-7.0%	405	500	1000
2,3-Dimethylbutane	C_6H_{14}	86.175	-128.1	58.0	0.6616^{20}	1.889^{20}		31.3	-29	1.2-7.0%	405	500	1000
3,3-Dimethyl-2-butanone	$C_6H_{12}O$	100.158	-51.40	106.1	0.7229^{25}	12.73^{20}		4.27					
Dimethylcarbamic chloride	C_3H_6ClNO	107.539	-33	154	1.168^{25}								
Dimethyl disulfide	$C_2H_6S_2$	94.199	-84.67	109.72	1.0625^{20}	9.6^{25}	1.85	3.82	24				
N,N-Dimethylethanolamine	$C_4H_{11}NO$	89.136	-65	130.7	0.8866^{20}			0.9					
N,N-Dimethylformamide	C_3H_7NO	73.094	-60.3	152.8	0.9445^{25}	38.25^{20}	3.82	0.439	58	2.2-15.2%	445	10	
2,6-Dimethyl-4-heptanone	$C_9H_{18}O$	142.238	-46.0	157	0.8062^{20}	9.91^{20}	2.66	0.23	49	0.8-7.1%	396	25	
1,1-Dimethylhydrazine	$C_2H_8N_2$	60.098	-57.15	62.4	0.791^{22}			20.9	-15	2-95%	249	0.01	
Dimethyl phthalate	$C_{10}H_{10}O_4$	194.184	1.03	282.7	1.1905^{20}	8.66^{20}		0.001	146	>0.9%	490		
2,6-Dimethylpyridine	C_7H_9N	107.153	-6.12	144.0	0.9226^{20}	7.33^{20}	1.66	0.746					
Dimethyl sulfate	$C_2H_6O_4S$	126.132	-31.8	186	1.3322^{20}	55.0^{25}		0.13	83		188	0.1	
Dimethyl sulfide	C_2H_6S	62.134	-98.26	37.32	0.8483^{20}	6.70^{21}	1.554	64.4	-37	2.2-19.7%	206	10	
Dimethyl sulfoxide	C_2H_6OS	78.133	18.52	191.9	1.1010^{25}	47.24^{20}	3.96	0.084	95	2.6-42%	215		
1,4-Dioxane	$C_4H_8O_2$	88.106	11.75	101.2	1.0337^{20}	2.2189^{20}	0	4.95	12	2.0-22%	180	20	
1,3-Dioxolane	$C_3H_6O_2$	74.079	-97.21	75.3	1.060^{20}		1.19	14.6	2			20	
Dipentyl ether	$C_{10}H_{22}O$	158.281	-69.2	187	0.7833^{20}	2.798^{25}	1.20	0.13	57		170		
Dipropylamine	$C_6H_{15}N$	101.190	-63	107.5	0.7400^{20}	2.923^{20}	1.03	3.21	17		299		
Dipropylene glycol monomethyl ether	$C_7H_{16}O_3$	148.200	-80	203	0.95							100	150
Dipropyl ether	$C_6H_{14}O$	102.174	-114.8	90.1	0.7466^{20}	3.38^{24}	1.21	8.35	21	1.3-7.0%	188		

Lab Data

Name	Mol. form.	M_r	mp/°C	bp/°C	ρ/g cm^{-3}	ε	μ/D	Vap. pres. (25 °C)/kPa	FP/°C	Flam. limits (%)	IT/°C	TWA/ ppm	STEL/ ppm
Dodecane	$C_{12}H_{26}$	170.334	-9.55	216.3	0.7495^{20}	2.0120^{20}		0.016	74	>0.6%	203		
1-Dodecanol	$C_{12}H_{26}O$	186.333	24.2	264.1	0.8309^{24}	5.82^{30}		0.000016	127		275		
1-Dodecene	$C_{12}H_{24}$	168.319	-35.19	213.4	0.7584^{20}	2.152^{20}		0.019	79				
Epichlorohydrin	C_3H_5ClO	92.524	-26	111.99	1.1812^{20}	22.6^{20}	1.8	2.2	31	3.8-21.0%	411	0.5	
1,2-Epoxybutane	C_4H_8O	72.106	-150	63.4	0.8297^{20}		1.891	31.7	-22	1.7-19%	439		
1,2-Epoxy-4-(epoxyethyl) cyclohexane	$C_8H_{12}O_2$	140.180	<-55	227	1.0966^{20}							0.1	
1,2-Ethanediamine	$C_2H_8N_2$	60.098	11.14	116.9	0.8979^{20}	13.82^{20}	1.99	1.62	40	2.5-12.0%	385	10	
1,2-Ethanediol	$C_2H_6O_2$	62.068	-13	197.5	1.1135^{20}	41.4^{20}	2.36	0.01	111	3.2-22%	398		
1,2-Ethanediol, diacetate	$C_6H_{10}O_4$	146.141	-31	184	1.1043^{20}	7.7^{17}	2.34	0.030	88	1.6-8.4%	482		
1,2-Ethanediol, dinitrate	$C_2H_4N_2O_6$	152.062	-22.5	199	1.4918^{20}	28.26^{20}		0.009				0.05	
1,2-Ethanedithiol	$C_2H_6S_2$	94.199	-41.2	144	1.234^{20}	7.26^{20}	2.03						
Ethanethiol	C_2H_6S	62.134	-147.89	35.0	0.8315^{25}	6.667^{25}	[a]	70.3	-17	2.8-18.0%	300	0.5	
Ethanol	C_2H_6O	46.068	-114.14	78.24	0.7893^{20}	25.3^{20}	1.69	7.87	13	3.3-19%	363		1000
Ethanolamine	C_2H_7NO	61.083	10.4	170.3	1.0180^{20}	31.94^{20}	2.27	0.05	86	3.0-23.5%	410	3	6
4-Ethoxyaniline	$C_8H_{11}NO$	137.179	4.7	254	1.0652^{16}	7.43^{25}		0.0007	116				
Ethoxybenzene	$C_8H_{10}O$	122.164	-29.6	169.8	0.9651^{20}	4.216^{20}	1.45	0.204	63				
2-Ethoxyethanol	$C_4H_{10}O_2$	90.121	-70	134.7	0.9253^{25}	13.38^{25}	2.08	0.71	43	3-18%	235	5	
2-Ethoxyethyl acetate	$C_6H_{12}O_3$	132.157	-61.7	156.6	0.9740^{20}	7.567^{30}	2.25	0.24	56	2-8%	379	5	
Ethyl acetate	$C_4H_8O_2$	88.106	-83.8	77.1	0.9006^{20}	6.0814^{20}	1.78	12.6	-4	2.0-11.5%	426	400	
Ethyl acetoacetate	$C_6H_{10}O_3$	130.141	-45	180	1.0368^{10}	14.0^{20}		0.095	57	1.4-9.5%	295		
Ethyl acrylate	$C_5H_8O_2$	100.117	-71	98.9	0.9234^{20}	6.05^{30}	1.96	5.14	10	1.4-14%	372	5	15
Ethylamine	C_2H_7N	45.084	-81	16.6	0.689^{15}	8.7^{0}	1.22	141	-16	3.5-14%	385	5	15
N-Ethylaniline	$C_8H_{11}N$	121.180	-63.4	204	0.9625^{20}	5.87^{20}		0.039	85				
Ethylbenzene	C_8H_{10}	106.165	-94.95	136.2	0.8668^{20}	2.4463^{20}	0.59	1.28	21	0.8-6.7%	432	20	
Ethyl benzoate	$C_9H_{10}O_2$	150.174	-34.5	212.5	1.0415^{25}	6.20^{20}	2.00	0.04	88		490		
Ethyl butanoate	$C_6H_{12}O_2$	116.158	-97	121.1	0.8735^{25}	5.18^{28}	1.74	2.01	24		463		
2-Ethyl-1-butanol	$C_6H_{14}O$	102.174	<-15	155	0.8326^{20}	6.19^{89}		0.206	57				
Ethyl chloroacetate	$C_4H_7ClO_2$	122.551	-21	144	1.1585^{20}			0.640	64				
Ethyl chloroformate	$C_3H_5ClO_2$	108.524	-80.6	91	1.1352^{20}	9.736^{36}			16		500		
Ethyl cyanoacetate	$C_5H_7NO_2$	113.116	-26.1	216	1.0654^{20}	31.62^{-10}	2.17	0.003	110				
Ethyleneimine	C_2H_5N	43.068	-78.0	54	0.832^{25}	18.3^{25}	1.90	28.9	-11	3.3-54.8%	320	0.05	0.1
Ethyl formate	$C_3H_6O_2$	74.079	-79.6	54.09	0.9218^{20}	8.57^{15}	1.93	32.3	-20	2.8-16.0%	455		100
2-Ethylhexanal	$C_8H_{16}O$	128.212	<-100	161	0.8540^{20}				44	0.85-7.2%	190		
2-Ethyl-1,3-hexanediol	$C_8H_{18}O_2$	146.228	-40	243	0.9325^{22}	18.73^{20}			127		360		
2-Ethyl-1-hexanol	$C_8H_{18}O$	130.228	-70	186.2	0.8319^{25}	7.58^{25}	1.74	0.019	73	0.88-9.7%	231		
2-Ethylhexyl acetate	$C_{10}H_{20}O_2$	172.265	-80	200	0.8718^{20}		1.8	0.09	71	0.76-8.14%	268		
Ethyl lactate	$C_5H_{10}O_3$	118.131	-26	151	1.0328^{20}	15.4^{30}	2.4		46	>1.5%	400		
Ethyl 3-methylbutanoate	$C_7H_{14}O_2$	130.185	-99.3	135	0.8656^{20}	4.71^{20}		1.07					
Ethyl 2-methylpropanoate	$C_6H_{12}O_2$	116.158	-97.8	111	0.868^{20}			3.25	13				
Ethyl nitrite	$C_2H_5NO_2$	75.067		17.5	0.899^{15}			135	-35	4.0-50%	90		
Ethyl propanoate	$C_5H_{10}O_2$	102.132	-73.6	98.9	0.8895^{20}	5.76^{20}	1.74	4.97	12	1.9-11%	440		
Ethyl silicate	$C_8H_{20}O_4Si$	208.329	-82.2	168	0.9320^{20}	2.50^{20}		1.17	52			10	
Eucalyptol	$C_{10}H_{18}O$	154.249	1.4	176	0.9267^{20}	4.57^{25}		0.260	48				
Fluorobenzene	C_6H_5F	96.102	-42.18	84.7	1.0225^{20}	5.465^{20}	1.60	10.4	-15				
Fluorosulfonic acid	FHO_3S	100.069	-89	163	1.726			0.08					
Formamide	CH_3NO	45.041	2.57	217	1.1334^{20}	111.0^{20}	3.73	0.01	154			10	
Formic acid	CH_2O_2	46.026	8.3	101	1.220^{20}	51.1^{25}	1.425	5.75	50	18-57%	434	5	10
Furan	C_4H_4O	68.074	-85.58	31.3	0.9514^{20}	2.94^{25}	0.66	80.0	-36	2.3-14.3%			
Furfural	$C_5H_4O_2$	96.085	-38.3	161.5	1.1594^{20}	42.1^{20}	3.54	0.29	60	2.1-19.3%	316	2	
Furfuryl alcohol	$C_5H_6O_2$	98.101	-14.5	168	1.1296^{20}	16.85^{25}	1.92	0.097	75	1.8-16.3%	491	10	15
Germanium(IV) chloride	Cl_4Ge	214.45	-51.50	86.55	1.88	2.463^{0}	0						
Glycerol	$C_3H_8O_3$	92.094	18.2	289	1.2613^{20}	46.53^{20}	2.56	<0.01	199	3-19%	370		
Glycerol triacetate	$C_9H_{14}O_6$	218.203	-78	259	1.1583^{20}	7.11^{20}		<0.01	138	>1.0%	433		
Glycerol trioleate	$C_{57}H_{104}O_6$	885.432	5.3		0.915^{15}	3.109^{20}							
Heptanal	$C_7H_{14}O$	114.185	-43.94	153	0.8132^{25}	9.07^{22}		0.46					
Heptane	C_7H_{16}	100.202	-90.549	98.38	0.6837^{20}	1.9209^{20}		6.09	-4	1.05-6.7%	204	400	500
Heptanoic acid	$C_7H_{14}O_2$	130.185	-7.17	222	0.9181^{20}	3.04^{15}		0.001			275		
1-Heptanol	$C_7H_{16}O$	116.201	-33.2	178	0.8219^{20}	11.75^{20}		0.0044					
2-Heptanone	$C_7H_{14}O$	114.185	-34.7	151.0	0.8111^{20}	11.95^{20}	2.59	0.49	39	1.1-7.9%	393	50	
3-Heptanone	$C_7H_{14}O$	114.185	-37.2	146	0.8183^{20}	12.7^{20}	2.78	0.5	46			50	75
4-Heptanone	$C_7H_{14}O$	114.185	-32.1	144	0.8174^{20}	12.60^{20}		0.164	49			50	
1-Heptene	C_7H_{14}	98.186	-118.83	94	0.6970^{20}	2.092^{20}		7.52	-1		260		
Hexachloro-1,3-butadiene	C_4Cl_6	260.761	-21	216	1.556^{25}	2.55^{20}		0.13			610	0.02	
Hexachloro-1,3-cyclopentadiene	C_5Cl_6	272.772	-9	239	1.7019^{25}							0.01	
Hexafluorobenzene	C_6F_6	186.054	5.10	80.2	1.6175^{20}	2.029^{25}	0	11.3					
Hexamethyldisiloxane	$C_6H_{18}OSi_2$	162.377	-68.2	100.5	0.7638^{20}	2.179^{20}		5.57					

Name	Mol. form.	M_r	mp/°C	bp/°C	ρ/g cm^{-3}	ε	μ/D	Vap. pres. (25 °C)/kPa	FP/°C	Flam. limits (%)	IT/°C	TWA/ ppm	STEL/ ppm
Hexamethylphosphoric triamide	$C_6H_{18}N_3OP$	179.200	7.2	235	1.03^{20}	31.3^{20}	5.5						
Hexanal	$C_6H_{12}O$	100.158	-58.2	129.6	0.8335^{20}			1.48	32				
Hexane	C_6H_{14}	86.175	-95.27	68.72	0.6593^{20}	1.8865^{20}		20.2	-22	1.1-7.5%	225	50	
Hexanedinitrile	$C_6H_8N_2$	108.141	2.2	295	0.9676^{20}			<0.01	93	>1.0%	550	2	
Hexanoic acid	$C_6H_{12}O_2$	116.158	-4.1	204.9	0.9274^{20}	2.600^{25}	1.13	0.005	102		380		
1-Hexanol	$C_6H_{14}O$	102.174	-46.4	156.9	0.8195^{20}	13.03^{20}		0.11	63		290		
2-Hexanone	$C_6H_{12}O$	100.158	-55.45	127.6	0.8113^{20}	14.56^{20}	2.66	1.54	25	1-8%	423	5	10
1-Hexene	C_6H_{12}	84.159	-139.76	63.4	0.6685^{25}	2.077^{21}		24.8	-26	1.2-6.9%	253	50	
Hexyl acetate	$C_8H_{16}O_2$	144.212	-61.0	171.1	0.8779^{15}	4.42^{20}		0.185	45				
Hydrazine	H_4N_2	32.045	1.54	113.55	1.0036	51.7^{25}	1.75	1.91	38	5-100%		0.01	
Hydrazoic acid	HN_3	43.028	-80	35.7			1.70	68.2					0.11
Hydrogen cyanide	CHN	27.026	-13.28	25.63	0.6876^{20}	114.9^{20}	2.985188	98.8	-18	6-40%	538		4.7
Hydrogen peroxide	H_2O_2	34.015	-0.43	150.2	1.44	74.6^{17}	1.573	0.26				1	
3-Hydroxypropanenitrile	C_3H_5NO	71.078	-46	218	1.0404^{25}		[a]	0.010	129				
Indan	C_9H_{10}	118.175	-51.34	177.8	0.9639^{20}			0.2					
Indene	C_9H_8	116.160	-1.45	182.5	0.9960^{25}			0.220				5	
Iodine bromide	BrI	206.808	40	116 dec	4.3		0.726						
Iodine chloride	ClI	162.357	27.38	97.0 dec	3.24		1.24	3.59					
Iodobenzene	C_6H_5I	204.008	-30.7	188.5	1.8308^{20}	4.59^{20}	1.70	0.133					
1-Iodobutane	C_4H_9I	184.018	-103.5	130	1.6154^{20}	6.27^{20}	1.93	1.85					
Iodoethane	C_2H_5I	155.965	-111.0	72	1.9357^{20}	7.82^{20}	1.976	18.2					
Iodomethane	CH_3I	141.939	-66	42.4	2.2789^{20}	6.97^{20}	1.6406	53.9				2	
1-Iodopropane	C_3H_7I	169.992	-101.4	102	1.7489^{20}	7.07^{20}	2.04	5.75					
2-Iodopropane	C_3H_7I	169.992	-90.4	89	1.7042^{20}	8.19^{25}	1.95	9.36					
Iron pentacarbonyl	C_5FeO_5	195.896	-20	103	1.5^{20}	2.602^{20}		4				0.1	0.2
Isobutanal	C_4H_8O	72.106	-72.1	64.1	0.7891^{20}		[a]	23.0	-18	1.6-10.6%	196		
Isobutyl acetate	$C_6H_{12}O_2$	116.158	-97.1	116.9	0.8712^{20}	5.068^{20}	1.86	2.39	18	1.3-10.5%	421	150	
Isobutyl acrylate	$C_7H_{12}O_2$	128.169	-61	137.0	0.8896^{20}				30		427		
Isobutylamine	$C_4H_{11}N$	73.137	-86	68.8	0.724^{25}		1.27	19.0	-9	2-12%	378		
Isobutylbenzene	$C_{10}H_{14}$	134.218	-51.6	172.7	0.8532^{20}	2.318^{20}		0.257	55	0.8-6.0%	427		
Isobutyl formate	$C_5H_{10}O_2$	102.132	-95.5	98.4	0.8776^{20}	6.41^{20}	1.88	5.34	5	2-9%	320		
Isobutyl isobutanoate	$C_8H_{16}O_2$	144.212	-80.6	148	0.8542^{20}		1.9	0.552	38	0.96-7.59%	432		
Isopentane	C_5H_{12}	72.149	-159.8	27.83	0.6201^{20}	1.845^{20}	0.13	91.7	-51	1.4-7.6%	420	600	
Isopentyl acetate	$C_7H_{14}O_2$	130.185	-78.5	141.6	0.876^{15}	4.72^{20}	1.86	0.728	25	1.0-7.5%	360	50	100
Isophorone	$C_9H_{14}O$	138.206	-8.1	214.8	0.9255^{20}			0.06	84	0.8-3.8%	460		5
Isopropenyl acetate	$C_5H_8O_2$	100.117	-92.9	97.4	0.9090^{20}			6.02	26		432		
Isopropenylbenzene	C_9H_{10}	118.175	-22.36	165.4	0.9106^{20}	2.28^{20}		0.40	54	1.9-6.1%	574	10	
Isopropyl acetate	$C_5H_{10}O_2$	102.132	-73.4	88.6	0.8662^{25}			7.88	2	1.8-8%	460	100	200
Isopropylamine	C_3H_9N	59.110	-95.119	31.8	0.6891^{20}	5.6268^{20}	1.19	78.0	-37		402	5	10
Isopropylbenzene	C_9H_{12}	120.191	-96.01	152.4	0.8615^{20}	2.381^{20}	≈0.79	0.61	36	0.9-6.5%	424	50	
Isopropylbenzene hydroperoxide	$C_9H_{12}O_2$	152.190		153 expl	1.03^{20}			0.004	175				
1-Isopropyl-2-methylbenzene	$C_{10}H_{14}$	134.218	-71.5	178	0.8766^{20}			0.2					
1-Isopropyl-3-methylbenzene	$C_{10}H_{14}$	134.218	-63.8	175	0.8610^{20}			0.22					
1-Isopropyl-4-methylbenzene	$C_{10}H_{14}$	134.218	-68.1	177	0.8573^{20}	2.2322^{25}		0.19	47	0.7-5.6%	436		
Isoquinoline	C_9H_7N	129.159	26.46	243.2	1.0910^{30}	11.0^{25}	2.73	0.007					
d-Limonene	$C_{10}H_{16}$	136.234	-74.0	177.6	0.8411^{20}	2.3746^{25}		0.277	45	0.7-6.1%	237		
l-Limonene	$C_{10}H_{16}$	136.234		178	0.843^{20}	2.3738^{25}		0.254					
Mesityl oxide	$C_6H_{10}O$	98.142	-52.8	129.7	0.8653^{20}	15.6^{0}	2.79	1.47	31	1.4-7.2%	344	15	25
Methacrylic acid	$C_4H_6O_2$	86.090	14.6	160	1.0153^{20}		1.65	0.12	77	1.6-8.8%	68	20	
Methanol	CH_4O	32.042	-97.5	64.5	0.7909^{20}	33.0^{20}	1.70	16.9	11	6.0-36%	464	200	250
2-Methoxyaniline	C_7H_9NO	123.152	6.2	221	1.0923^{20}	5.230^{30}		0.013	118				
4-Methoxybenzaldehyde	$C_8H_8O_2$	136.149	0	255	1.119^{15}	22.0^{30}		0.004					
2-Methoxyethanol	$C_3H_8O_2$	76.095	-85.1	124.3	0.9647^{20}	17.2^{25}	[a]	1.31	39	1.8-14%	285	0.1	
2-Methoxyethyl acetate	$C_5H_{10}O_3$	118.131	-70	142	1.0074^{19}		2.13	0.67	49	1.5-12.3%	392	0.1	
Methyl acetate	$C_3H_6O_2$	74.079	-98.2	56.7	0.9346^{20}	7.07^{15}	1.72	28.8	-10	3.1-16%	454	200	250
Methyl acrylate	$C_4H_6O_2$	86.090	-75.6	80.1	0.9535^{20}	7.03^{30}	1.77	11.0	-3	2.8-25%	468	2	
2-Methylacrylonitrile	C_4H_5N	67.090	-35.8	90	0.8001^{20}		3.69	8.26	1	2-6.8%		1	
2-Methylaniline	C_7H_9N	107.153	-14.41	200.0	0.9984^{20}	6.138^{25}	1.60	0.043	85		482	2	
3-Methylaniline	C_7H_9N	107.153	-30.8	203.3	0.9889^{20}	5.816^{25}	1.45	0.036				2	
N-Methylaniline	C_7H_9N	107.153	-57	197	0.9859^{20}	5.96^{20}		0.05			533	0.5	
Methyl benzoate	$C_8H_8O_2$	136.149	-12.35	199	1.0837^{25}	6.642^{30}	1.94	0.052	83				
2-Methyl-1,3-butadiene	C_5H_8	68.118	-146.1	34.0	0.679^{20}	2.098^{20}	0.25	73.4	-54	1.5-8.9%	395		
Methyl butanoate	$C_5H_{10}O_2$	102.132	-85.8	101.9	0.8984^{20}	5.48^{28}		4.30	14				
3-Methylbutanoic acid	$C_5H_{10}O_2$	102.132	-29.6	176.5	0.931^{20}		0.63	0.067			416		
3-Methyl-1-butanol	$C_5H_{12}O$	88.148	-117.2	130.8	0.8104^{20}	15.63^{20}		0.315	43	1.2-9.0%	350	100	125
2-Methyl-2-butanol	$C_5H_{12}O$	88.148	-8.7	102.4	0.8096^{20}	5.78^{25}	1.82	2.19	19	1.2-9.0%	437		
3-Methyl-2-butanol, (±)-	$C_5H_{12}O$	88.148		113.7	0.8180^{20}	12.1^{25}		1.20	38				

Name	Mol. form.	M_r	mp/°C	bp/°C	ρ/g cm^{-3}	ε	μ/D	Vap. pres. (25 °C)/kPa	FP/°C	Flam. limits (%)	IT/°C	TWA/ ppm	STEL/ ppm
3-Methyl-2-butanone	$C_5H_{10}O$	86.132	-93.13	94.2	0.8051^{20}	10.37^{20}		6.99				20	
2-Methyl-1-butene	C_5H_{10}	70.133	-137.53	31.1	0.6504^{20}	2.180^{20}		81.4	-20				
2-Methyl-2-butene	C_5H_{10}	70.133	-133.72	38.5	0.6623^{20}	1.979^{23}		62.1	-20				
Methyl *tert*-butyl ether	$C_5H_{12}O$	88.148	-108.6	55.1	0.7353^{25}			33.6				50	
Methyl chloroacetate	$C_3H_5ClO_2$	108.524	-32.3	130	1.236^{20}	12.0^{20}		1.0	57	7.5-18.5%			
Methylcyclohexane	C_7H_{14}	98.186	-126.6	100.9	0.7694^{20}	2.024^{20}		6.18	-4	1.2-6.7%	250	400	
Methylcyclopentane	C_6H_{12}	84.159	-142.419	71.8	0.7486^{20}	1.9853^{20}		18.3	-29	1.0-8.35%	258		
N-Methylformamide	C_2H_5NO	59.067	-2.5	186	1.011^{19}	189.0^{20}	3.83	0.03					
Methyl formate	$C_2H_4O_2$	60.052	-99.7	31.6	0.9739^{20}	9.20^{15}	1.77	78.1	-19	4.5-23%	449	50	100
5-Methyl-2-hexanone	$C_7H_{14}O$	114.185		139	0.888^{20}	13.53^{20}		0.691	36	1.0-8.2%	191	20	50
Methylhydrazine	CH_6N_2	46.072	-52.3	83				6.61	-8	2.5-92%	194	0.01	
Methyl isocyanate	C_2H_3NO	57.051	-45	38.3	0.9588^{20}	21.75^{16}	≈2.8	57.7	-7	5.3-26%	534	0.02	0.06
Methyl lactate, (±)-	$C_4H_8O_3$	104.105		144.8	1.0928^{20}			0.62	49	>2.2%	385		
Methyl methacrylate	$C_5H_8O_2$	100.117	-47.55	100.6	0.9377^{25}	6.32^{30}	1.67	5.10	10	1.7-8.2%		50	100
1-Methylnaphthalene	$C_{11}H_{10}$	142.197	-30.43	244.4	1.0202^{20}	2.915^{20}		0.009			529	0.5	
Methyloxirane	C_3H_6O	58.079	-111.9	35	0.859^{0}		2.01	71.7	-37	3.1-27.5%	449	2	
2-Methylpentane	C_6H_{14}	86.175	-153.60	60.21	0.650^{25}	1.886^{20}		28.2	<-29	1.0-7.0%	264	500	1000
3-Methylpentane	C_6H_{14}	86.175	-162.89	63.3	0.6598^{25}	1.886^{20}		25.3	-7	1.2-7.0%	278	500	1000
2-Methyl-2,4-pentanediol	$C_6H_{14}O_2$	118.174	-50	197.9	0.923^{15}	25.86^{20}	2.9	<0.01	102	1-9%	306		25
2-Methyl-1-pentanol	$C_6H_{14}O$	102.174		157	0.8263^{20}			0.236	54	1.1-9.65%	310		
4-Methyl-2-pentanol	$C_6H_{14}O$	102.174	-90	132.0	0.8075^{20}			0.698	41	1.0-5.5%		25	40
4-Methyl-2-pentanone	$C_6H_{12}O$	100.158	-85	115.7	0.7965^{25}	13.11^{20}		2.64	18	1.2-8.0%	448	20	75
2-Methylpropanenitrile	C_4H_7N	69.106	-71.5	102	0.7704^{20}	24.42^{20}	4.29		8		482		
2-Methyl-2-propanethiol	$C_4H_{10}S$	90.187	1.27	64.2	0.7943^{25}	5.475^{20}	1.66	24.2	<-29				
Methyl propanoate	$C_4H_8O_2$	88.106	-87.5	78.6	0.9150^{20}	6.200^{20}		11.5	-2	2.5-13%	469		
2-Methylpropanoic acid	$C_4H_8O_2$	88.106	-46	154.4	0.9681^{20}	2.58^{20}	1.08	0.17	56	2.0-9.2%	481		
2-Methyl-1-propanol	$C_4H_{10}O$	74.121	-101.96	107.84	0.8018^{20}	17.93^{20}	1.64	1.39	28	1.7-10.6%	415	50	
2-Methyl-2-propanol	$C_4H_{10}O$	74.121	25.81	82.3	0.7887^{20}	12.47^{25}	1.66	5.52	11	2.4-8.0%	478	100	
2-Methylpyridine	C_6H_7N	93.127	-66.65	129.4	0.9443^{20}	10.18^{20}	1.85	1.5	39		538		
3-Methylpyridine	C_6H_7N	93.127	-18.1	144.1	0.9566^{20}	11.10^{30}	2.40	0.795					
4-Methylpyridine	C_6H_7N	93.127	3.68	145.3	0.9548^{20}	12.2^{20}	2.70	0.759	57				
N-Methyl-2-pyrrolidinone	C_5H_9NO	99.131	-24.0	204.2	1.0230^{25}	32.55^{20}	4.1	0.04	96	1-10%	346		
Methyl salicylate	$C_8H_8O_3$	152.148	-8.5	222.6	1.181^{25}	8.80^{41}	2.47	0.015	96		454		
4-Methylstyrene	C_9H_{10}	118.175	-37.8	172	0.9173^{25}			0.245	53	0.8-11.0%	538	50	100
Morpholine	C_4H_9NO	87.120	-4.8	128.2	1.0005^{20}	7.42^{25}	1.55	1.34	37	1.4-11.2%	290	20	
β-Myrcene	$C_{10}H_{16}$	136.234		171	0.8013^{15}	2.3^{25}		0.280					
Nickel carbonyl [$Ni(CO)_4$]	C_4NiO_4	170.734	-19.3	43 (exp 60)	1.31^{25}								0.05
L-Nicotine	$C_{10}H_{14}N_2$	162.231	-79	246	1.0097^{20}	8.937^{20}							
Nitric acid	HNO_3	63.013	-41.6	83	1.5129^{20}		2.17	8.34				2	4
2-Nitroanisole	$C_7H_7NO_3$	153.136	9.4	272	1.2540^{20}	45.75^{20}	5.0	0.002					
Nitrobenzene	$C_6H_5NO_2$	123.110	5.65	210.7	1.2037^{20}	35.6^{20}	4.22	0.03	88	>1.8%	482	1	
Nitroethane	$C_2H_5NO_2$	75.067	-89.42	114.1	1.0448^{25}	29.11^{15}	3.23	2.79	28	3.4-17%	414	100	
Nitromethane	CH_3NO_2	61.041	-28.7	101.19	1.1371^{20}	37.27^{20}	3.46	4.79	35	>7.3%	418	20	
1-Nitropropane	$C_3H_7NO_2$	89.094	-104.3	131.2	0.9961^{25}	24.70^{15}	3.66	1.36	36	>2.2%	421	25	
2-Nitropropane	$C_3H_7NO_2$	89.094	-91.3	120.2	0.9821^{25}	26.74^{15}	3.73	2.3	24	2.6-11.0%	428	10	
N-Nitrosodiethylamine	$C_4H_{10}N_2O$	102.134		172	0.9422^{20}								
N-Nitrosodimethylamine	$C_2H_6N_2O$	74.081		146	1.0048^{20}			0.73					
2-Nitrotoluene	$C_7H_7NO_2$	137.137	-3.6	220.9	1.1611^{19}	26.26^{20}		0.0014	106			2	
3-Nitrotoluene	$C_7H_7NO_2$	137.137	15.9	232.1	1.1581^{20}	24.95^{30}		0.03	106			2	
Nonane	C_9H_{20}	128.255	-53.47	150.8	0.7179^{20}	1.9722^{20}		0.570	31	0.8-2.9%	205	200	
Nonanoic acid	$C_9H_{18}O_2$	158.238	12.38	256	0.9052^{20}	2.475^{22}	0.79	0.00005					
1-Nonanol	$C_9H_{20}O$	144.254	-5.0	213.7	0.8280^{20}	8.83^{20}		0.00050			260		
1-Nonene	C_9H_{18}	126.239	-81.24	146.9	0.7253^{25}	2.180^{20}		0.714	26				
4-Nonylphenol	$C_{15}H_{24}O$	220.351	42	317	0.950^{20}								
cis,cis-9,12-Octadecadienoic acid	$C_{18}H_{32}O_2$	280.446	-6.9		0.9022^{20}	2.754^{20}							
cis-9-Octadecenoic acid	$C_{18}H_{34}O_2$	282.462	14	360	0.8935^{20}	2.336^{20}	1.18	0.000001	189		363		
Octane	C_8H_{18}	114.229	-56.73	125.62	0.7022^{20}	1.948^{20}		1.86	13	1.0-6.5%	206	300	375
Octanoic acid	$C_8H_{16}O_2$	144.212	16.51	240	0.9106^{20}	2.85^{15}	1.15	0.0002					
1-Octanol	$C_8H_{18}O$	130.228	-14.7	194.7	0.8262^{25}	10.30^{20}	1.76	0.01	81		270		
2-Octanol	$C_8H_{18}O$	130.228	-31.6	179	0.8193^{20}	8.13^{20}	1.71		88		265		
2-Octanone	$C_8H_{16}O$	128.212	-20.31	173	0.820^{20}	9.51^{20}	2.70	0.12	52				
1-Octene	C_8H_{16}	112.213	-101.66	121.3	0.7149^{20}	2.113^{20}		2.30	21		230		
Oxetane	C_3H_6O	58.079	-97	47.6	0.8930^{25}		1.94						
2-Oxetanone	$C_3H_4O_2$	72.063	-33.283	161	1.1460^{20}		4.18	0.3	74	>2.9%		0.5	
Oxirane	C_2H_4O	44.052	-112.46	10.4	0.8821^{10}	12.42^{20}	1.89	175	-20	3.0-100%	429	1	

Name	Mol. form.	M_r	mp/°C	bp/°C	ρ/g cm^{-3}	ε	μ/D	Vap. pres. (25 °C)/kPa	FP/°C	Flam. limits (%)	IT/°C	TWA/ ppm	STEL/ ppm
Oxiranemethanol, (±)-	$C_3H_6O_2$	74.079	-45	156	1.1143^{25}							2	
Paraldehyde	$C_6H_{12}O_3$	132.157	12	124	0.9943^{20}		1.43	1.6	36	>1.3%	238		
Parathion	$C_{10}H_{14}NO_5PS$	291.261	6.1	375	1.2681^{20}								
Pentachloroethane	C_2HCl_5	202.294	-29.0	161	1.6796^{20}	3.716^{25}	0.92	0.478					
cis-1,3-Pentadiene	C_5H_8	68.118	-140.81	44.0	0.6910^{20}	2.319^{25}	0.500	50.6					
trans-1,3-Pentadiene	C_5H_8	68.118	-87.5	42.0	0.6710^{25}		0.585	54.7					
Pentanal	$C_5H_{10}O$	86.132	-81.5	103	0.8095^{20}	10.00^{20}		4.58	12		222	50	
Pentane	C_5H_{12}	72.149	-129.67	36.06	0.6262^{20}	1.8371^{20}		68.3	-40	1.4-8.0%	260	1000	
Pentanedial	$C_5H_8O_2$	100.117	-14	176									0.05
1,5-Pentanediol	$C_5H_{12}O_2$	104.148	-20	241	0.9914^{20}	26.2^{20}	2.5	0.001	129		335		
2,4-Pentanedione	$C_5H_8O_2$	100.117	-18.3	140.7	0.9721^{25}	26.524^{30}	2.78	1.02	34		340		
1-Pentanethiol	$C_5H_{12}S$	104.214	-75.69	126.6	0.850^{20}	4.847^{20}		1.83	18				
Pentanoic acid	$C_5H_{10}O_2$	102.132	-33.63	186.1	0.9389^{20}	2.661^{21}	1.61	0.024	96		400		
1-Pentanol	$C_5H_{12}O$	88.148	-77.58	137.6	0.8144^{20}	15.13^{25}	1.7	0.259	33	1.2-10.0%	300		
2-Pentanol	$C_5H_{12}O$	88.148	-73	119.1	0.8094^{20}	13.71^{25}	1.66	0.804	34	1.2-9.0%	343		
3-Pentanol	$C_5H_{12}O$	88.148	-69.9	123	0.8203^{20}	13.35^{25}	1.64	1.10	41	1.2-9.0%	435		
2-Pentanone	$C_5H_{10}O$	86.132	-76.83	102.2	0.809^{20}	15.45^{20}	2.70	4.97	7	1.5-8.2%	452		150
3-Pentanone	$C_5H_{10}O$	86.132	-38.98	101.9	0.8098^{25}	17.00^{20}	2.82	4.72	13	>1.6%	450	200	300
1-Pentene	C_5H_{10}	70.133	-165.13	30.0	0.6405^{20}	2.011^{20}	≈0.5	85.0	-18	1.5-8.7%	275		
cis-2-Pentene	C_5H_{10}	70.133	-151.35	36.9	0.6556^{20}			66.0	<-20				
trans-2-Pentene	C_5H_{10}	70.133	-140.20	36.3	0.6431^{25}			67.4	<-20				
Pentyl acetate	$C_7H_{14}O_2$	130.185	-70.9	149.4	0.8756^{20}	4.79^{20}	1.75	0.60	16	1.1-7.5%	360	50	100
Pentylamine	$C_5H_{13}N$	87.164	-51	104.7	0.7544^{20}	4.27^{20}		4.00	-1	2.2-22%			
Perchloric acid	$ClHO_4$	100.459	-112	≈90 dec	1.77								
Peroxyacetic acid	$C_2H_4O_3$	76.051	-0.2	110	1.226^{15}			1.93	41				0.4
Phenol	C_6H_6O	94.111	40.89	181.8	1.0545^{45}	12.40^{30}	1.224	0.055	79	1.8-8.6%	715	5	
2-Phenoxyethanol	$C_8H_{10}O_2$	138.164	12	246	1.102^{22}			0.001	121				
Phenylhydrazine	$C_6H_8N_2$	108.141	20	244	1.0986^{20}	7.15^{20}		0.003	88			0.1	
1-Phenyl-2-propylamine, (±)-	$C_9H_{13}N$	135.206		198	0.9306^{25}			0.06	<100				
Phosphinic acid	H_3O_2P	65.997	26.5	130	1.49								
Phosphoric acid	H_3O_4P	97.995	42.4	407									
Phosphorothioc trichloride	Cl_3PS	169.398	-36.2	125	1.635	4.94^{25}							
Phosphorus(III) bromide	Br_3P	270.686	-41.5	173.2	2.8			0.38					
Phosphorus(III) chloride	Cl_3P	137.333	-93	76	1.574	3.498^{17}	0.56	16.1				0.2	0.5
Phosphoryl chloride	Cl_3OP	153.332	1.18	105.5	1.645	14.1^{20}	2.54	4.97				0.1	
α-Pinene	$C_{10}H_{16}$	136.234	-74	156.3	0.8539^{25}	2.1787^{25}		0.64	33		255	20	
β-Pinene	$C_{10}H_{16}$	136.234	-50.0	165.8	0.860^{25}	2.4970^{25}		0.61	38		275	20	
Piperidine	$C_5H_{11}N$	85.148	-11.05	106.19	0.8606^{20}	4.33^{20}	1.19	4.28	16	1-10%			
Propanal	C_3H_6O	58.079	-80	48.0	0.8657^{25}	18.5^{17}	2.72	42.2	-30	2.6-17%	207	20	
1,2-Propanediol	$C_3H_8O_2$	76.095	-60	187.3	1.0361^{20}	27.5^{30}	2.25	0.02	99	2.6-12.5%	371		
1,3-Propanediol	$C_3H_8O_2$	76.095	-27.6	214.7	1.0538^{20}	35.1^{20}	2.55	0.007			400		
Propanenitrile	C_3H_5N	55.079	-93	97.3	0.7818^{20}	29.7^{20}	4.05	6.14	2	3.1-14%	512		
Propanoic acid	$C_3H_6O_2$	74.079	-20.5	141.5	0.9882^{25}	3.44^{25}	1.75	0.553	52	2.9-12.1%	465	10	
Propanoic anhydride	$C_6H_{10}O_3$	130.141	-45.0	168	1.0110^{20}	18.30^{20}		0.45	63	1.3-9.5%	285		
1-Propanol	C_3H_8O	60.095	-124.39	97.04	0.8048^{20}	20.8^{20}	[a]	2.76	23	2.2-13.7%	412	100	400
2-Propanol	C_3H_8O	60.095	-87.91	82.21	0.7855^{20}	20.18^{20}	[a]	6.02	12	2.0-12.7%	399	200	400
Propargyl alcohol	C_3H_4O	56.063	-51.8	113	0.9478^{20}	20.8^{20}	1.13		36			1	
Propyl acetate	$C_5H_{10}O_2$	102.132	-93	101.0	0.8885^{20}	5.62^{20}	1.78	4.49	13	1.7-8%	450	200	250
Propylamine	C_3H_9N	59.110	-84.78	47.21	0.7173^{20}	5.08^{23}	1.17	42.1	-37	2.0-10.4%	318		
Propylbenzene	C_9H_{12}	120.191	-99.52	159.2	0.8619^{20}	2.370^{20}		0.45	30	0.8-6.0%	450		
Propyl butanoate	$C_7H_{14}O_2$	130.185	-95.2	144	0.8730^{20}	4.3^{20}		0.618	37				
Propylene carbonate	$C_4H_6O_3$	102.089	-48.8	241.6	1.2047^{20}	66.14^{20}	4.9	0.05	135				
Propyl formate	$C_4H_8O_2$	88.106	-92.9	80.6	0.9053^{20}	6.92^{30}	1.89	10.9	-3		455		
Propyl propanoate	$C_6H_{12}O_2$	116.158	-75.9	122.2	0.8755^{25}	5.249^{20}		1.88	79				
Pyridine	C_5H_5N	79.101	-41.63	115.2	0.9819^{20}	13.260^{20}	2.215	2.76	20	1.8-12.4%	482	1	
Pyrrole	C_4H_5N	67.090	-23.39	129.74	0.9698^{20}	8.00^{20}	1.767	1.10	39				
Pyrrolidine	C_4H_9N	71.121	-57.79	86.6	0.8586^{20}	8.30^{20}	1.57	8.40	3				
2-Pyrrolidone	C_4H_7NO	85.105	25.92	251.2	1.120^{20}	28.18^{25}	3.5		129				
Quinoline	C_9H_7N	129.159	-14.78	237.1	1.0977^{15}	9.16^{20}	2.29	0.011			480		
Safrole	$C_{10}H_{10}O_2$	162.185	11.2	235	1.1000^{20}			0.01	100				
Salicylaldehyde	$C_7H_6O_2$	122.122	-7	208	1.1674^{20}	18.35^{20}	2.86	0.075	78				
Selenium chloride	Cl_2Se_2	228.83	-85	127 dec	2.774								
Selenium oxychloride	Cl_2OSe	165.87	8.5	177	2.44	46.2^{20}		0.02					
Selenium oxyfluoride	F_2OSe	132.96	15	125	2.8			0.56					
Styrene	C_8H_8	104.150	-30.65	145.3	0.9016^{25}	2.4737^{20}	0.123	0.81	31	0.9-6.8%	490	20	40
Sulfolane	$C_4H_8O_2S$	120.171	28.45	286	1.2723^{18}		4.8	<0.01	177				
Sulfur chloride [ClSSCl]	Cl_2S_2	135.036	-77	137	1.69	4.79^{15}		1.27					1

Lab Data

Name	Mol. form.	M_r	mp/°C	bp/°C	ρ/g cm^{-3}	ε	μ/D	Vap. pres. (25 °C)/kPa	FP/°C	Flam. limits (%)	IT/°C	TWA/ ppm	STEL/ ppm
Sulfur dichloride	Cl_2S	102.971	-122	59.6	1.62	2.915^{25}	0.36	17.9					
Sulfuric acid	H_2O_4S	98.079	10.31	337	1.8305^{20}								
Sulfuryl chloride	Cl_2O_2S	134.970	-51	69.4	1.680	9.1^{20}	1.81	18.7					
α-Terpinene	$C_{10}H_{16}$	136.234		174	0.8375^{19}	2.4526^{25}							
1,1,2,2-Tetrabromoethane	$C_2H_2Br_4$	345.653	0	248	2.9655^{20}	6.72^{30}	1.38	0.003			335	0.1	
Tetrabromosilane	Br_4Si	347.702	5.39	154	2.8		0						
1,1,2,2-Tetrachloro-1,2-difluoroethane	$C_2Cl_4F_2$	203.830	26.54	92.83	1.5951^{50}	2.52^{35}		7.51				50	
1,1,1,2-Tetrachloroethane	$C_2H_2Cl_4$	167.849	-70.2	130.2	1.5406^{20}	9.22^{-66}		1.6	47	5-12%			
1,1,2,2-Tetrachloroethane	$C_2H_2Cl_4$	167.849	-42.4	146.0	1.5953^{20}	8.50^{20}	1.32	0.622	62	20-54%		1	
Tetrachloroethene	C_2Cl_4	165.833	-22.2	121.2	1.6230^{20}	2.268^{30}		2.42	45			25	100
Tetrachloromethane	CCl_4	153.823	-22.8	76.7	1.5940^{20}	2.2379^{20}		15.2				5	10
Tetrachlorosilane	Cl_4Si	169.898	-68.74	57.65	1.5	2.248^{0}	0	31.3					
Tetradecane	$C_{14}H_{30}$	198.388	5.87	253.5	0.7596^{20}	2.0343^{20}		0.002	112	>0.5%	200		
Tetraethylene glycol	$C_8H_{18}O_5$	194.226	-9.4	315	1.1285^{15}	20.44^{20}		0.000001	182				
Tetrafluoroboric acid	BF_4H	87.813		130 dec	≈1.8								
Tetrahydrofuran	C_4H_8O	72.106	-108.38	66.0	0.8833^{25}	7.52^{22}	1.75	21.6	-14	2-11.8%	321	50	100
Tetrahydrofurfuryl alcohol	$C_5H_{10}O_2$	102.132	<-80	176.3	1.0524^{20}	13.48^{30}	2.1	0.100	75	1.5-9.7%	282		
1,2,3,4-Tetrahydronaphthalene	$C_{10}H_{12}$	132.202	-35.76	207.2	0.9645^{25}	2.771^{25}		0.05	71	0.8-5.0%	385		
Tetrahydropyran	$C_5H_{10}O$	86.132	-49.1	88.0	0.8814^{20}	5.66^{20}	[a]	9.54	-20				
Tetrahydrothiophene	C_4H_8S	88.172	-96.13	121.1	0.9987^{20}		1.90	2.45					
Tetramethylsilane	$C_4H_{12}Si$	88.224	-99.063	26.7	0.648^{19}	1.921^{20}	0	94.2					
Tetramethylurea	$C_5H_{12}N_2O$	116.161	-2.67	177.1	0.9687^{20}	23.10^{20}	3.5	0.138	77				
Tetranitromethane	CN_4O_8	196.033	13.9	125.6	1.6380^{20}	2.317^{20}	0	1.13				0.005	
Thionyl bromide	Br_2OS	207.872	-50	140		9.06^{20}		0.84					
Thionyl chloride	Cl_2OS	118.970	-101	75.6	1.631	8.675^{25}	1.45	16.0					0.2
Thiophene	C_4H_4S	84.140	-38.12	84.1	1.0649^{20}	2.739^{20}	0.55	10.6	-1				
Tin(IV) chloride	Cl_4Sn	260.522	-34.07	114.15	2.234	3.014^{0}	0						
Titanium(IV) chloride	Cl_4Ti	189.679	-24.12	136.45	1.73	2.843^{-16}							
Toluene	C_7H_8	92.139	-95.0	110.60	0.8668^{20}	2.379^{23}	0.375	3.79	4	1.1-7.1%	480	20	
Toluene-2,4-diisocyanate	$C_9H_6N_2O_2$	174.156	20.5	251	1.2244^{20}	8.433^{20}		0.003	127	0.9-9.5%		0.001	0.005
Tribromomethane	$CHBr_3$	252.731	8.69	149.2	2.8788^{25}	4.404^{10}	0.99	0.726	83			0.5	
Tributylamine	$C_{12}H_{27}N$	185.349	-70	207	0.7770^{20}	2.340^{20}	0.78	0.01	63	1-5%			
Tributyl borate	$C_{12}H_{27}BO_3$	230.151	<-70	233.8	0.8567^{20}	2.23^{20}	0.77		93				
Tributyrin	$C_{15}H_{26}O_6$	302.363	-75	307.5	1.0350^{20}	5.72^{10}			180	>0.5%	407		
Trichloroacetaldehyde	C_2HCl_3O	147.387	-57.5	98	1.512^{20}	6.8^{25}		6.66					
1,2,4-Trichlorobenzene	$C_6H_3Cl_3$	181.447	17.0	213.5	1.459^{25}			0.057	105	2.5-6.6%	571		5
1,1,1-Trichloroethane	$C_2H_3Cl_3$	133.404	-30	74.02	1.3390^{20}	7.243^{20}	1.755	16.5	-1	8-10.5%	500	350	450
1,1,2-Trichloroethane	$C_2H_3Cl_3$	133.404	-36.3	113	1.4397^{20}	7.1937^{25}	1.4	3.1	32	6-28%	460	10	
Trichloroethene	C_2HCl_3	131.388	-84.7	86.8	1.4642^{20}	3.390^{28}	0.8	9.91	32	8-10.5%	420	10	25
Trichloroethylsilane	$C_2H_5Cl_3Si$	163.506	-105.6	98.7	1.2373^{20}		2.04	6.29	22				
Trichlorofluoromethane	CCl_3F	137.368	-110.44	23.7	1.4879^{20}	3.00^{20}	0.46	106					1000
Trichloromethane	$CHCl_3$	119.378	-63.3	61.2	1.4890^{20}	4.8069^{20}	1.04	26.2				10	
(Trichloromethyl)benzene	$C_7H_5Cl_3$	195.474	-17.0	221	1.3723^{20}		2.03	0.35	127		211		0.1
Trichloromethylsilane	CH_3Cl_3Si	149.480	-75.77	66	1.273^{20}		1.91	22.5	-9	7.6->20%	>404		
Trichloronitromethane	CCl_3NO_2	164.376	-69.4	112	1.6558^{20}	7.319^{20}		3.18				0.1	
1,2,3-Trichloropropane	$C_3H_5Cl_3$	147.431	-13.8	158	1.3889^{20}	7.5^{20}		0.492	71	3.2-12.6%		0.005	
Trichlorosilane	Cl_3HSi	135.453	-128.2	33	1.331		0.86		-50		104		
1,1,2-Trichloro-1,2,2-trifluoroethane	$C_2Cl_3F_3$	187.375		47.6	1.5635^{25}	2.41^{25}		44.8				1000	1250
Tri-*o*-cresyl phosphate	$C_{21}H_{21}O_4P$	368.363	11	410	1.1955^{20}	6.7^{25}	2.87	0.0000002	225		385		
Tridecane	$C_{13}H_{28}$	184.361	-5.35	235.4	0.7564^{20}	2.0213^{20}		0.005	79				
1-Tridecene	$C_{13}H_{26}$	182.345	-23.07	232.8	0.7658^{20}	2.139^{20}		0.0047	79				
Triethanolamine	$C_6H_{15}NO_3$	149.188	21.5	350	1.1242^{20}		3.57	<0.01	179	1-10%			
Triethylamine	$C_6H_{15}N$	101.190	-114.7	88.8	0.7275^{20}	2.418^{20}	0.66	7.70	-7	1.2-8.0%	249	0.5	1
Triethylene glycol	$C_6H_{14}O_4$	150.173	-9.4	288.6	1.1274^{15}	23.69^{20}		0.0002	177	0.9-9.2%	371		
Triethylene glycol dimethyl ether	$C_8H_{18}O_4$	178.227	-43.8	218	0.986^{20}	7.62^{25}			111				
Triethyl phosphate	$C_6H_{15}O_4P$	182.154	-56.4	216	1.0695^{20}	13.20^{25}	3.12		115		454		
Trifluoroacetic acid	$C_2HF_3O_2$	114.023	-15.2	72	1.5351^{25}	8.42^{20}	2.28	15.1					
(Trifluoromethyl)benzene	$C_7H_5F_3$	146.110	-28.99	102.0	1.1884^{20}	9.22^{25}	2.86	5.14	12				
1,2,3-Trimethylbenzene	C_9H_{12}	120.191	-25.32	176.0	0.8944^{20}	2.656^{20}		0.20	44	0.8-6.6%	470	25	
1,2,4-Trimethylbenzene	C_9H_{12}	120.191	-43.8	169.4	0.8758^{20}	2.377^{20}		0.30	44	0.9-6.4%	500	25	
1,3,5-Trimethylbenzene	C_9H_{12}	120.191	-44.69	164.7	0.8615^{25}	2.279^{20}	0	0.33	50	1-5%	559	25	
Trimethyl borate	$C_3H_9BO_3$	103.912	-29.3	67.4	0.915^{25}	2.2762^{20}		17.2	-8				
Trimethylchlorosilane	C_3H_9ClSi	108.642	-55.17	57.6	0.856^{25}	10.21^{0}		30.7	-28		395		
2,2,4-Trimethylpentane	C_8H_{18}	114.229	-107.36	99.2	0.6919^{20}	1.943^{20}		6.50	-12		418	300	
2,3,3-Trimethylpentane	C_8H_{18}	114.229	-101.2	114.7	0.7262^{20}	1.9780^{20}		3.60	<21		425	300	

Name	Mol. form.	M_r	mp/°C	bp/°C	ρ/g cm^{-3}	ε	μ/D	Vap. pres. (25 °C)/kPa	FP/°C	Flam. limits (%)	IT/°C	TWA/ ppm	STEL/ ppm
Trimethyl phosphate	$C_3H_9O_4P$	140.074	-46	197.2	1.2144^{20}	20.6^{20}	3.18	0.11	107				
2,4,6-Trimethylpyridine	$C_8H_{11}N$	121.180	-44.3	170	0.9166^{22}	7.807^{25}	2.05	4.1					
Trinitroglycerol	$C_3H_5N_3O_9$	227.087	12.8	218 exp	1.5931^{20}	19.25^{20}		0.00005			270	0.05	
Undecane	$C_{11}H_{24}$	156.309	-25.54	195.9	0.7402^{20}	1.9972^{20}		0.05	69				
Vanadium(IV) chloride	Cl_4V	192.754	-28	151	1.816	3.05^{25}							
Vanadyl trichloride	Cl_3OV	173.300	-79	127	1.829	3.4^{25}							
Vinyl acetate	$C_4H_6O_2$	86.090	-100.2	72.6	0.9256^{25}		1.79	15.4	-8	2.6-13.4%	402	10	15
4-Vinylcyclohexene	C_8H_{12}	108.181	-108.9	130	0.8299^{20}			1.87	16		269	0.1	
Water	H_2O	18.015	0.00	99.974	0.9970^{25}	80.1^{20}	1.8546	3.17					
o-Xylene	C_8H_{10}	106.165	-25.16	144.4	0.8802^{20}	2.562^{20}	0.640	0.88	32	0.9-6.7%	463	100	150
m-Xylene	C_8H_{10}	106.165	-47.85	139.1	0.8641^{20}	2.359^{20}		1.13	27	1.1-7.0%	527	100	150
p-Xylene	C_8H_{10}	106.165	13.3	138.3	0.8610^{20}	2.2735^{20}	0	1.19	27	1.1-7.0%	528	100	150
2,4-Xylenol	$C_8H_{10}O$	122.164	25	210.94	0.9650^{20}	5.060^{30}	1.4	0.022					

[a] See the table "Dipole Moments" in Section 9 for values for individual conformers.

Lab Data

MISCIBILITY OF ORGANIC SOLVENTS

The chart below gives qualitative information on the miscibility of pairs of organic liquids. Two liquids are considered *miscible* if mixing equal volumes produces a single liquid phase. If two phases separate, they are considered *immiscible.* An entry of *partially misc.* indicates two phases whose volumes differ appreciably, suggesting a partial miscibility of the components. The entry *reaction* indicates a chemical reaction between the components. All data refer to room temperature.

References

1. Drury, J. S., *Ind. Eng. Chem.* 44, 2744, 1959.
2. Jackson, W. M., and Drury, J. S., *Ind. Eng. Chem.* 51, 1491, 1959.

Miscibility of Organic Solvents

Name	Acetone	Benzaldehyde	Benzene	Butyl acetate	Butyl alcohol	Carbon tetrachloride	2-Chloroethanol	Chloroform	*o*-Cresol
Acetone	-	miscible	miscible	miscible	miscible	miscible	miscible	miscible	
2-Amino-2-methyl-1-propanol	miscible	miscible	miscible	miscible	miscible	miscible			
Benzaldehyde	miscible	-	miscible	miscible	miscible	miscible			
Benzene	miscible	miscible	-	miscible	miscible	miscible	miscible	miscible	
Benzenemethanethiol	miscible		miscible		miscible	miscible			
Benzonitrile	miscible		miscible		miscible	miscible			
Benzothiazole	miscible		miscible		miscible	miscible			
Benzyl alcohol	miscible	miscible	miscible	miscible	miscible	miscible			
Bis(2-aminoethyl)amine	miscible		miscible				reaction	miscible	
N,N'-Bis(2-aminoethyl)-1,2-ethanediamine	miscible		miscible				miscible	miscible	
2-Bromoethyl acetate	miscible		partially misc.					miscible	
1,3-Butanediol	miscible		immiscible					miscible	miscible
2,3-Butanediol	miscible		partially misc.					miscible	miscible
1-Butanol	miscible	miscible	miscible	miscible	-	miscible			
2-Butoxyethanol	miscible		miscible					miscible	miscible
Butyl acetate	miscible	miscible	miscible	-	miscible	miscible			
2-Chloroethanol	miscible		miscible				-		miscible
3-Chloro-1,2-propanediol	miscible		immiscible					miscible	miscible
trans-Cinnamaldehyde	miscible		miscible				miscible	miscible	
o-Cresol							miscible		-
Diacetone alcohol	miscible	miscible	partially misc.	miscible	miscible	partially misc.			
Dibenzyl ether	miscible		miscible				miscible	miscible	
Dibutylamine							reaction		
Dibutyl carbonate	miscible		miscible					miscible	
Dibutyl ether	miscible	miscible	miscible	miscible	miscible	miscible			
Diethanolamine	miscible	immiscible	immiscible	immiscible	miscible	immiscible			
Diethylene glycol dibutyl ether	miscible		miscible					miscible	miscible
Diethylene glycol diethyl ether	miscible		miscible					miscible	miscible
Diethylene glycol monobutyl ether	miscible		miscible					miscible	miscible
Diethylene glycol monoethyl ether	miscible		miscible					miscible	miscible
Diethylene glycol monomethyl ether	miscible		miscible					miscible	
Diethyl ether	miscible	miscible	miscible	miscible	miscible	miscible	miscible	miscible	
N,N-Diethylformamide	miscible		miscible				miscible	miscible	
Dihexyl ether	miscible		miscible				miscible	miscible	
Diisopentyl sulfide	miscible		miscible		miscible	miscible			
Diisopropylamine	miscible		miscible				reaction	miscible	
N,N-Dimethylaniline	miscible	miscible	miscible	miscible	miscible	miscible			
Dimethyl disulfide	miscible		miscible		miscible	miscible			
2,6-Dimethyl-4-heptanone	miscible		miscible				miscible	miscible	
Dipentylamine	miscible		miscible					miscible	
N,N-Dipropylaniline	miscible		miscible		miscible	miscible	miscible	miscible	
Dipropylene glycol	miscible		miscible				miscible	miscible	miscible
1,2-Ethanediol	miscible	partially misc.	immiscible	partially misc.	miscible	immiscible		partially misc.	miscible
Ethanol	miscible	miscible	miscible	miscible	miscible	miscible		miscible	
2-Ethoxyaniline	miscible		miscible				miscible	miscible	
2-Ethoxyethanol	miscible		miscible					miscible	miscible
Ethyl benzoate	miscible	miscible	miscible	miscible	miscible	miscible		miscible	
2-Ethylbutanoic acid	miscible		miscible				miscible	miscible	
Ethyl chloroacetate	miscible		miscible				miscible	miscible	
Ethyl *trans*-cinnamate	miscible		miscible					miscible	
2-Ethyl-1-hexanol	miscible	miscible	miscible	miscible	miscible	miscible			

Name	Acetone	Benzaldehyde	Benzene	Butyl acetate	Butyl alcohol	Carbon tetrachloride	2-Chloroethanol	Chloroform	*o*-Cresol
Ethyl phenylacetate	miscible		miscible				miscible	miscible	
Ethyl thiocyanate	miscible		miscible		miscible	miscible			
Formamide	miscible	miscible	immiscible	immiscible	miscible	immiscible			
Furfuryl alcohol	miscible	miscible	miscible	miscible	miscible	miscible			
Glycerol	immiscible	partially misc.	immiscible	immiscible	miscible	immiscible		immiscible	miscible
1-Heptadecanol	miscible		miscible					miscible	
3-Heptanol, (*S*)-	miscible		miscible				miscible	miscible	
Heptyl acetate	miscible		miscible				miscible	miscible	
Hexanedinitrile	miscible		miscible		miscible	immiscible			
Hexanenitrile	miscible		miscible				miscible	miscible	
Isopentyl acetate	miscible		miscible				miscible	miscible	
4-Methoxybenzaldehyde							miscible		
2-Methoxyethanol	miscible		miscible						miscible
3-Methyl-1-butanol	miscible	miscible	miscible	miscible	miscible	miscible			
3-Methyl-2-butanone	miscible		miscible				miscible	miscible	
4-Methylpentanoic acid	miscible		miscible					miscible	
4-Methyl-2-pentanone	miscible	miscible	miscible	miscible	miscible	miscible			
2-Methyl-1-propanethiol	miscible		miscible		miscible	miscible			
Nitromethane	miscible	miscible	immiscible	miscible	miscible	miscible			
1-Octanol	miscible	miscible	miscible	miscible	miscible	miscible			
1,2-Propanediol	miscible		immiscible					miscible	miscible
1,3-Propanediol	miscible	miscible	immiscible	partially misc.	miscible	immiscible		miscible	miscible
Pyridine	miscible	miscible	miscible	miscible	miscible	miscible	miscible		
Tetrachloromethane	miscible	miscible	miscible	miscible	miscible	-			
1-Tetradecanol	miscible		miscible				miscible	miscible	
Tributyl phosphate	miscible		miscible				miscible	miscible	
Trichloromethane	miscible		miscible				miscible	-	
Triethylene glycol	miscible		partially misc.					miscible	miscible
Triethyl phosphate	miscible		miscible					miscible	miscible
2,6,8-Trimethyl-4-nonanone	miscible		miscible				miscible	miscible	

Name	Diethyl ether	*N,N*-Dimethylaniline	Dipentylamine	Ethyl alcohol	Ethylene glycol	Ethylene glycol monoethyl ether	Formamide	Furfuryl alcohol
Acetone	miscible	miscible	miscible	miscible	miscible	miscible	miscible	miscible
2-Amino-2-methyl-1-propanol	miscible	miscible		miscible	miscible		miscible	miscible
Benzaldehyde	miscible	miscible		miscible	partially misc.		miscible	miscible
Benzene	miscible	miscible	miscible	miscible	immiscible	miscible	immiscible	miscible
Benzenemethanethiol	miscible	miscible		miscible	immiscible		immiscible	miscible
Benzonitrile	miscible	miscible		miscible	immiscible		immiscible	miscible
Benzothiazole	miscible	miscible		miscible	miscible		immiscible	miscible
Benzyl alcohol	miscible	miscible		miscible	miscible		miscible	miscible
Bis(2-aminoethyl)amine			immiscible	miscible	miscible	miscible		
N,N'-Bis(2-aminoethyl)-1,2-ethanediamine	miscible		immiscible	miscible	miscible	miscible		
2-Bromoethyl acetate			reaction	miscible				
1,3-Butanediol	partially misc.		miscible	miscible				
2,3-Butanediol	miscible		miscible	miscible				
1-Butanol	miscible	miscible		miscible	miscible		miscible	miscible
2-Butoxyethanol	miscible		miscible	miscible				
Butyl acetate	miscible	miscible		miscible	partially misc.		immiscible	miscible
2-Chloroethanol	miscible		miscible					
3-Chloro-1,2-propanediol	miscible		reaction	miscible				
trans-Cinnamaldehyde			miscible	miscible	immiscible	miscible		
o-Cresol					miscible	miscible		
Diacetone alcohol	miscible	miscible		miscible	miscible		miscible	miscible
Dibenzyl ether			miscible		immiscible	miscible		
Dibutylamine					miscible	miscible		
Dibutyl carbonate			miscible	miscible				
Dibutyl ether	miscible	miscible		miscible	immiscible		immiscible	miscible
Diethanolamine	immiscible	partially misc.		miscible	miscible		miscible	miscible
Diethylene glycol dibutyl ether	miscible		reaction	miscible				
Diethylene glycol diethyl ether	miscible		miscible	miscible				
Diethylene glycol monobutyl ether	miscible		miscible	miscible				
Diethylene glycol monoethyl ether	miscible		miscible	miscible				
Diethylene glycol monomethyl ether	miscible		miscible	miscible				
Diethyl ether	-	miscible	miscible	miscible	immiscible	miscible	immiscible	miscible
N,N-Diethylformamide			reaction	miscible	miscible	miscible		

Name	Diethyl ether	*N,N*-Dimethylaniline	Dipentylamine	Ethyl alcohol	Ethylene glycol	Ethylene glycol monoethyl ether	Formamide	Furfuryl alcohol
Dihexyl ether			miscible	miscible	immiscible	miscible		
Diisopentyl sulfide	miscible	miscible		miscible	immiscible		immiscible	immiscible
Diisopropylamine			miscible	miscible	miscible	miscible		
N,N-Dimethylaniline	miscible	-		miscible	immiscible		immiscible	miscible
Dimethyl disulfide	miscible	miscible		miscible	immiscible		immiscible	miscible
2,6-Dimethyl-4-heptanone			miscible	miscible	immiscible	miscible		
Dipentylamine	miscible		-	miscible	partially misc.	miscible		
N,N-Dipropylaniline	miscible	miscible	miscible	miscible	immiscible	miscible	immiscible	miscible
Dipropylene glycol	miscible			miscible				
1,2-Ethanediol	immiscible	immiscible	partially misc.	miscible	-		miscible	miscible
Ethanol	miscible	miscible	miscible	-	miscible	miscible	miscible	miscible
2-Ethoxyaniline			miscible	miscible	miscible	miscible		
2-Ethoxyethanol	miscible		miscible	miscible		-		
Ethyl benzoate	miscible	miscible	miscible	miscible	immiscible		immiscible	miscible
2-Ethylbutanoic acid			reaction	miscible	miscible	miscible		
Ethyl chloroacetate			miscible	miscible	immiscible	miscible		
Ethyl *trans*-cinnamate			miscible	miscible	immiscible	miscible		
2-Ethyl-1-hexanol	miscible	miscible		miscible	miscible		immiscible	miscible
Ethyl phenylacetate			miscible		immiscible	miscible		
Ethyl thiocyanate	miscible	miscible		miscible	immiscible		immiscible	miscible
Formamide	immiscible	immiscible		miscible	miscible		-	miscible
Furfuryl alcohol	miscible	miscible		miscible	miscible		miscible	-
Glycerol	immiscible	immiscible	partially misc.	miscible	miscible		miscible	miscible
1-Heptadecanol			miscible	miscible				
3-Heptanol, (*S*)-			miscible	miscible	miscible	miscible		
Heptyl acetate			miscible		immiscible	miscible		
Hexanedinitrile	immiscible	miscible		miscible	immiscible		miscible	miscible
Hexanenitrile			miscible	miscible	immiscible	miscible		
Isopentyl acetate			miscible	miscible	immiscible	miscible		
4-Methoxybenzaldehyde					immiscible	miscible		
2-Methoxyethanol	miscible		miscible	miscible				
3-Methyl-1-butanol	miscible	miscible		miscible	miscible		miscible	miscible
3-Methyl-2-butanone	miscible		miscible	miscible	immiscible	miscible		
4-Methylpentanoic acid			miscible	miscible	miscible	miscible		
4-Methyl-2-pentanone	miscible	miscible		miscible	immiscible		partially misc.	miscible
2-Methyl-1-propanethiol	miscible	miscible		miscible	immiscible		immiscible	miscible
Nitromethane	miscible	miscible		miscible	immiscible		miscible	miscible
1-Octanol	miscible	miscible		miscible	miscible		immiscible	miscible
1,2 Propanediol	partially misc.		miscible	miscible				
1,3-Propanediol	immiscible	immiscible	miscible	miscible	miscible		miscible	miscible
Pyridine	miscible	miscible		miscible	miscible	miscible	miscible	miscible
Tetrachloromethane	miscible	miscible		miscible	immiscible		immiscible	miscible
1-Tetradecanol			miscible	miscible	immiscible	miscible		
Tributyl phosphate			miscible	miscible	partially misc.	miscible		
Trichloromethane	miscible		miscible	miscible	partially misc.	miscible		
Triethylene glycol	immiscible		partially misc.	miscible				
Triethyl phosphate	miscible		miscible	miscible				
2,6,8-Trimethyl-4-nonanone			miscible	miscible	immiscible	miscible		

Name	Glycerol	Methyl isopropyl ketone	Nitromethane	1-Octanol	1,3-Propanediol	Pyridine	Triethylenetetramine	Triethyl phosphate
Acetone	immiscible	miscible	miscible	miscible	miscible	miscible	miscible	miscible
2-Amino-2-methyl-1-propanol	miscible		miscible	miscible	miscible	miscible		
Benzaldehyde	partially misc.		miscible	miscible	miscible	miscible		
Benzene	immiscible	miscible	immiscible	miscible	immiscible	miscible	miscible	miscible
Benzenemethanethiol	immiscible				immiscible	miscible		
Benzonitrile	immiscible				immiscible	miscible		
Benzothiazole	immiscible				miscible	miscible		
Benzyl alcohol	miscible		miscible	miscible	miscible	miscible		
Bis(2-aminoethyl)amine	miscible	reaction					miscible	miscible
N,N'-Bis(2-aminoethyl)-1,2-ethanediamine	miscible	reaction			miscible		-	miscible
2-Bromoethyl acetate		miscible					reaction	
1,3-Butanediol		miscible				miscible	miscible	
2,3-Butanediol		miscible				miscible	miscible	
1-Butanol	miscible		miscible	miscible	miscible	miscible		
2-Butoxyethanol		miscible				miscible	miscible	

Name	Glycerol	Methyl isopropyl ketone	Nitromethane	1-Octanol	1,3-Propanediol	Pyridine	Triethylenetetramine	Triethyl phosphate
Butyl acetate	immiscible		miscible	miscible	partially misc.	miscible		
2-Chloroethanol		miscible				miscible	miscible	miscible
3-Chloro-1,2-propanediol		miscible				miscible	reaction	
trans-Cinnamaldehyde	immiscible	miscible			immiscible		reaction	miscible
o-Cresol	miscible				miscible			miscible
Diacetone alcohol	immiscible		miscible	miscible	miscible	miscible		
Dibenzyl ether	immiscible	miscible					miscible	miscible
Dibutylamine	partially misc.							miscible
Dibutyl carbonate		miscible					immiscible	
Dibutyl ether	immiscible		immiscible	miscible	immiscible	miscible		
Diethanolamine	miscible		immiscible	miscible	miscible	miscible		
Diethylene glycol dibutyl ether		miscible				miscible		
Diethylene glycol diethyl ether		miscible				miscible	miscible	
Diethylene glycol monobutyl ether		miscible				miscible	miscible	
Diethylene glycol monoethyl ether		miscible				miscible	miscible	
Diethylene glycol monomethyl ether		miscible				miscible	miscible	
Diethyl ether	immiscible	miscible	miscible	miscible	immiscible	miscible	miscible	miscible
N,N-Diethylformamide	miscible	miscible			miscible		reaction	miscible
Dihexyl ether	immiscible	miscible			immiscible		immiscible	
Diisopentyl sulfide	immiscible				immiscible	miscible		
Diisopropylamine	miscible	miscible			miscible		miscible	miscible
N,N-Dimethylaniline	immiscible		miscible	miscible	immiscible	miscible		
Dimethyl disulfide	immiscible				reaction	miscible		
2,6-Dimethyl-4-heptanone	immiscible	miscible			immiscible		miscible	miscible
Dipentylamine	partially misc.	miscible			miscible		immiscible	miscible
N,N-Dipropylaniline	immiscible	miscible			immiscible	miscible	miscible	miscible
Dipropylene glycol		miscible				miscible	miscible	
1,2-Ethanediol	miscible	immiscible	immiscible	miscible	miscible	miscible	miscible	
Ethanol	miscible	miscible	miscible	miscible	miscible	miscible	miscible	miscible
2-Ethoxyaniline		miscible			miscible		miscible	
2-Ethoxyethanol		miscible				miscible	miscible	
Ethyl benzoate	immiscible	miscible	miscible	miscible	partially misc.	miscible	miscible	
2-Ethylbutanoic acid	immiscible	miscible			miscible		reaction	miscible
Ethyl chloroacetate	immiscible	miscible			immiscible		reaction	miscible
Ethyl *trans*-cinnamate	immiscible	miscible			immiscible		miscible	miscible
2-Ethyl-1-hexanol	immiscible		immiscible	miscible	miscible	miscible		
Ethyl phenylacetate	immiscible	miscible			immiscible		miscible	miscible
Ethyl thiocyanate	immiscible				immiscible	miscible		
Formamide	miscible		miscible	immiscible	miscible	miscible		
Furfuryl alcohol	miscible		miscible	miscible	miscible	miscible		
Glycerol	-	immiscible	immiscible	immiscible	miscible	miscible	miscible	
1-Heptadecanol		miscible					miscible	
3-Heptanol, (*S*)-	immiscible				miscible		miscible	miscible
Heptyl acetate	immiscible	miscible			immiscible		reaction	miscible
Hexanedinitrile	immiscible				miscible	miscible		
Hexanenitrile	immiscible	miscible			immiscible		miscible	miscible
Isopentyl acetate	immiscible	miscible			immiscible		miscible	miscible
4-Methoxybenzaldehyde	immiscible				immiscible			miscible
2-Methoxyethanol						miscible	miscible	
3-Methyl-1-butanol	immiscible		miscible	miscible	miscible	miscible		
3-Methyl-2-butanone	immiscible	-			miscible		reaction	miscible
4-Methylpentanoic acid	immiscible	miscible			miscible		reaction	miscible
4-Methyl-2-pentanone	immiscible		miscible	miscible	immiscible	miscible		
2-Methyl-1-propanethiol	immiscible				reaction	miscible		
Nitromethane	immiscible		-	partially misc.	immiscible	miscible		
1-Octanol	immiscible		partially misc.	-	miscible	miscible		
1,2-Propanediol		miscible				miscible	miscible	
1,3-Propanediol	miscible	miscible	immiscible	miscible	-	miscible	miscible	
Pyridine	miscible		miscible	miscible	miscible	-		miscible
Tetrachloromethane	immiscible		miscible	miscible	immiscible	miscible		
1-Tetradecanol	immiscible	miscible			partially misc.		miscible	miscible
Tributyl phosphate	immiscible	miscible			miscible		miscible	miscible
Trichloromethane	immiscible	miscible			miscible		miscible	miscible
Triethylene glycol		miscible				miscible	miscible	
Triethyl phosphate		miscible				miscible	miscible	-
2,6,8-Trimethyl-4-nonanone	immiscible	miscible			immiscible		immiscible	miscible

Lab Data

DENSITY OF SOLVENTS AS A FUNCTION OF TEMPERATURE

The table below lists the densities of several common solvents over the temperature range from 0 °C to 100 °C. The values have been calculated from the REFPROP program where equations are available or from the Rackett equation (as indicated with an asterisk) using parameters in Ref. 1. Density values refer to the liquid at its saturation vapor pressure; thus, entries for temperatures above the normal boiling point are for pressures greater than atmospheric pressure.

References

1. Lide, D. R. and Kehiaian, H. V., *CRC Handbook of Thermophysical and Thermochemical Data*, CRC Press, Boca Raton, FL, 1994, pp. 49-59.
2. Lemmon, E. W., Huber, M. L., and McLinden, M. O., *NIST Standard and Reference Database 23: Reference Fluid Thermodynamic and Transport Properties-REFPROP*, Version 9.1, National Institute of Standards and Technology, Standard Reference Data Program, Gaithersburg, Maryland, 2013.

	Density in g/mL										
Solvent	**0 °C**	**10 °C**	**20 °C**	**30 °C**	**40 °C**	**50 °C**	**60 °C**	**70 °C**	**80 °C**	**90 °C**	**100 °C**
Acetic acid*			1.0510	1.0380	1.0250	1.0120	0.9993	0.9861	0.9728	0.9592	0.9454
Acetone	0.8121	0.8012	0.7902	0.7790	0.7677	0.7561	0.7443	0.7322	0.7198	0.7070	0.6938
Acetonitrile*			0.7825	0.7707	0.7591	0.7473	0.7353	0.7231	0.7106	0.6980	0.6851
Aniline*	1.0410	1.0330	1.0250	1.0160	1.0080	1.0000	0.9909	0.9823	0.9735	0.9646	0.9557
Benzene		0.8894	0.8788	0.8681	0.8574	0.8466	0.8357	0.8247	0.8135	0.8022	0.7907
1-Butanol	0.8347	0.8246	0.8148	0.8051	0.7955	0.7860	0.7766	0.7671	0.7577	0.7481	0.7385
Butylamine*	0.7606	0.7512	0.7417	0.7320	0.7221	0.7120	0.7017	0.6911	0.6803	0.6693	0.6579
Carbon disulfide*	1.2900	1.2770	1.2630	1.2480	1.2340						
Chlorobenzene*	1.1270	1.1160	1.1060	1.0960	1.0850	1.0740	1.0640	1.0530	1.0420	1.0300	1.0190
Cyclohexane		0.7880	0.7786	0.7691	0.7595	0.7499	0.7401	0.7303	0.7203	0.7101	0.6997
Decane	0.7459	0.7381	0.7303	0.7226	0.7148	0.7070	0.6992	0.6913	0.6834	0.6754	0.6674
1-Decanol*			0.8294	0.8229	0.8162	0.8093	0.8024	0.7955	0.7884	0.7813	0.7740
Dichloromethane	1.3632	1.3433	1.3232	1.3031	1.2828	1.2624	1.2417	1.2207	1.1995	1.1777	1.1555
Diethyl ether	0.7358	0.7248	0.7135	0.7020	0.6901	0.6778	0.6652	0.6521	0.6384	0.6242	0.6093
N,N-Dimethylaniline*		0.9638	0.9562	0.9483	0.9401	0.9318	0.9234	0.9150	0.9064	0.8978	0.8890
Ethanol	0.8063	0.7979	0.7893	0.7807	0.7720	0.7631	0.7540	0.7446	0.7348	0.7247	0.7140
Ethyl acetate*	0.9245	0.9126	0.9006	0.8884	0.8759	0.8632	0.8503	0.8370	0.8234	0.8095	0.7952
Ethylbenzene	0.8842	0.8755	0.8668	0.8581	0.8493	0.8405	0.8316	0.8226	0.8135	0.8044	0.7951
Ethyl formate*	0.9472	0.9346	0.9218	0.9087	0.8954	0.8818	0.8678	0.8535	0.8389	0.8238	0.8082
Ethyl propanoate*	0.9113	0.9005	0.8895	0.8784	0.8671	0.8556	0.8439	0.8319	0.8197	0.8072	0.7944
Heptane	0.7004	0.6921	0.6837	0.6753	0.6667	0.6581	0.6493	0.6404	0.6314	0.6221	0.6127
Hexane	0.6773	0.6684	0.6593	0.6501	0.6407	0.6313	0.6216	0.6117	0.6016	0.5913	0.5806
1-Hexanol*	0.8359	0.8278	0.8195	0.8111	0.8027	0.7941	0.7854	0.7766	0.7676	0.7585	0.7492
Isopropylbenzene*	0.8769	0.8696	0.8615	0.8533	0.8450	0.8366	0.8280	0.8194	0.8106	0.8017	0.7927
Methanol	0.8096	0.8003	0.7909	0.7815	0.7721	0.7625	0.7528	0.7428	0.7326	0.7220	0.7109
Methyl acetate*	0.9606	0.9478	0.9346	0.9211	0.9074	0.8933	0.8790	0.8643	0.8491	0.8336	0.8176
N-Methylaniline*	1.0010	0.9933	0.9859	0.9785	0.9709	0.9633	0.9556	0.9478	0.9399	0.9319	0.9239
Methylcyclohexane	0.7865	0.7780	0.7694	0.7607	0.7520	0.7433	0.7344	0.7255	0.7164	0.7072	0.6979
Methyl formate*	1.0030	0.9887	0.9739	0.9588	0.9433	0.9275	0.9112	0.8945	0.8772	0.8594	0.8409
Methyl propanoate*	0.9383	0.9268	0.9150	0.9030	0.8907	0.8783	0.8656	0.8526	0.8393	0.8257	0.8117
Nitromethane*			1.1390	1.1250	1.1110	1.0970	1.0830	1.0690	1.0550	1.0400	1.0260
Nonane	0.7337	0.7258	0.7179	0.7101	0.7021	0.6942	0.6862	0.6781	0.6700	0.6617	0.6534
Octane	0.7182	0.7102	0.7022	0.6941	0.6860	0.6779	0.6696	0.6613	0.6528	0.6442	0.6354
Pentanoic acid*	0.9563	0.9476	0.9389	0.9301	0.9211	0.9121	0.9029	0.8937	0.8843	0.8748	0.8652
1-Propanol*	0.8252	0.8151	0.8048	0.7943	0.7837	0.7729	0.7619	0.7506	0.7391	0.7273	0.7152
2-Propanol	0.8003	0.7931	0.7855	0.7774	0.7689	0.7599	0.7504	0.7402	0.7295	0.7182	0.7062
Propyl acetate*	0.9101	0.8994	0.8885	0.8775	0.8662	0.8548	0.8432	0.8313	0.8192	0.8069	0.7942
Propylbenzene*	0.8779	0.8700	0.8619	0.8538	0.8456	0.8373	0.8289	0.8204	0.8117	0.8030	0.7943
Propyl formate*	0.9275	0.9166	0.9053	0.8938	0.8821	0.8702	0.8581	0.8457	0.8330	0.8201	0.8068
Tetrachloromethane*	1.6290	1.6110	1.5930	1.5750	1.5570	1.5380	1.5180	1.4990	1.4790	1.4580	1.4370
Toluene	0.8854	0.8761	0.8668	0.8575	0.8481	0.8387	0.8291	0.8195	0.8098	0.7999	0.7899
Trichloromethane*	1.5240	1.5070	1.4890	1.4710	1.4520	1.4330	1.4140	1.3940			
2,2,4-Trimethylpentane	0.7082	0.7001	0.6919	0.6837	0.6753	0.6668	0.6583	0.6496	0.6407	0.6316	0.6224
o-Xylene	0.8969	0.8886	0.8802	0.8717	0.8633	0.8547	0.8461	0.8375	0.8287	0.8199	0.8109
m-Xylene	0.8811	0.8727	0.8641	0.8555	0.8469	0.8381	0.8293	0.8204	0.8113	0.8022	0.7929
p-Xylene			0.8610	0.8523	0.8436	0.8348	0.8260	0.8171	0.8081	0.7990	0.7898

* Calculated from the Rackett equation (Ref. 1).

DEPENDENCE OF BOILING POINT ON PRESSURE

The normal boiling point of a liquid is defined as the temperature at which the vapor pressure reaches standard atmospheric pressure, 101.325 kPa. Calculated values of the change in boiling point with small changes in pressure, $\Delta t/\Delta p$, are given in this table in units of both °C/kPa and °C/mmHg. The values from Ref. 1 were calculated from the representation of the vapor pressure by the Antoine Equation,

$$\ln p/p_0 = A_1 - A_2/(T + A_3)$$

where p is the vapor pressure in kPa, p_0 is 101.325 kPa, T is the absolute temperature, and A_1, A_2, and A_3 are constants, taken from Ref. 1. The correction to the boiling point is generally accurate to 0.1 °C to 0.2 °C as long as the pressure is within 10% of standard atmospheric pressure. Values from Ref. 2 were calculated from the REFPROP program, which employs a more sophisticated equation.

A slightly less accurate estimate of $\Delta t/\Delta p$ may be obtained from the Clausius-Clapeyron equation, with the assumption that the change in volume upon vaporization equals the ideal-gas volume of the vapor. This leads to the equation,

$$\Delta t/\Delta p = RT_b^2/[p_0\Delta_{vap}H(T_b)]$$

where R is the molar gas constant, T_b is the normal boiling point temperature (absolute), and $\Delta_{vap}H(T_b)$ is the molar enthalpy of vaporization at the normal boiling point. Values of the last quantity may be obtained from the table "Enthalpy of Vaporization" in Section 6.

References

1. Lide, D. R., and Kehiaian, H. V., *CRC Handbook of Thermophysical and Thermochemical Data*, CRC Press, Boca Raton, FL, 1994, pp. 49-59.
2. Lemmon, E.W., Huber, M.L., and McLinden, M.O., *NIST Standard Reference Database 23: Reference Fluid Thermodynamic and Transport Properties-REFPROP*, Version 9.1, National Institute of Standards and Technology, Standard Reference Data Program, Gaithersburg, MD, 2013.

Name	bp/°C	Δt(°C)/Δp(kPa)	Δt(°C)/Δp(mmHg)	Ref.
Acetaldehyde	20.8	0.261	0.0348	1
Acetic acid	117.9	0.324	0.0432	1
Acetic anhydride	139.5	0.343	0.0457	1
Acetone	56.08	0.289	0.0385	2
Acetonitrile	81.6	0.316	0.0421	1
Ammonia	-33.33	0.197	0.0262	2
Aniline	184.1	0.378	0.0504	1
Anisole	153.6	0.367	0.0489	1
Benzaldehyde	178.7	0.392	0.0523	1
Benzene	80.08	0.321	0.0427	2
Bromine	58.8	0.300	0.0400	1
1,3-Butadiene	-4.6	0.254	0.0339	1
Butane	-0.5	0.260	0.0346	2
1-Butanol	117.6	0.286	0.0381	2
2-Butanone	79.6	0.309	0.0412	1
Carbon disulfide	46.2	0.304	0.0405	1
Chlorine	-34.04	0.224	0.0299	2
Chlorobenzene	131.6	0.365	0.0487	1
1-Chlorobutane	78.4	0.321	0.0428	1
Chloroethane	12.3	0.262	0.0349	1
Chloroethene	-13.8	0.241	0.0321	1
o-Cresol	191.0	0.381	0.0508	1
m-Cresol	202.2	0.372	0.0496	1
p-Cresol	201.9	0.372	0.0496	1
Cyclohexane	80.7	0.329	0.0438	2
Cyclohexanol	160.9	0.344	0.0459	1
Cyclohexanone	155.4	0.382	0.0509	1
Decane	174.1	0.388	0.0518	2
Dibutyl ether	141.6	0.363	0.0484	1
Dichloromethane	39.8	0.282	0.0377	2
Diethyl ether	34.4	0.278	0.0370	2
Dimethyl sulfoxide	191.9	0.379	0.0505	1
1,4-Dioxane	101.2	0.321	0.0428	1
Dipropyl ether	90.1	0.326	0.0435	1
1,2-Ethanediol	197.5	0.333	0.0444	2
Ethanol	78.24	0.250	0.0334	2
Ethyl acetate	77.1	0.300	0.0400	1
Heptane	98.38	0.336	0.0448	2
Hexafluorobenzene	80.2	0.305	0.0407	1
Hexane	68.72	0.314	0.0419	2
1-Hexanol	156.9	0.318	0.0424	1
Hydrogen fluoride	20	0.276	0.0368	1
Iodomethane	42.4	0.291	0.0388	1
Isobutane	-11.7	0.252	0.0336	2
Isopropylbenzene	152.4	0.381	0.0508	1
Methanol	64.5	0.251	0.0334	2
Methyl acetate	56.7	0.282	0.0376	1
N-Methylaniline	197	0.396	0.0528	1
N-Methylformamide	186	0.371	0.0495	1
Methyl formate	31.6	0.582	0.0776	1
Nitrobenzene	210.7	0.418	0.0557	1
Nitromethane	101.19	0.320	0.0427	1
1-Nonanol	213.7	0.346	0.0461	1
Pentane	36.06	0.289	0.0386	2
1-Pentanol	137.6	0.296	0.0395	1
Phenol	181.8	0.349	0.0465	1
Propane	-42.11	0.224	0.0298	2
1-Propanol	97.04	0.261	0.0348	1
2-Propanol	82.21	0.247	0.0329	2
Pyridine	115.2	0.340	0.0453	1
Pyrrole	129.74	0.330	0.0440	1
Pyrrolidine	86.6	0.309	0.0412	1
Styrene	145.3	0.369	0.0492	1
Sulfur dioxide	-10.02	0.221	0.0294	2
Tetrachloroethene	121.2	0.354	0.0472	1
Tetrachloromethane	76.7	0.325	0.0433	1
Toluene	110.60	0.348	0.0463	2
Trichloroethene	86.8	0.330	0.0440	1
Trichloromethane	61.2	0.302	0.0403	1
Trimethylamine	2.8	0.248	0.0331	1
Water	99.974	0.277	0.0369	2
o-Xylene	144.4	0.373	0.0497	2
m-Xylene	139.1	0.368	0.0490	2
p-Xylene	138.3	0.369	0.0492	2

EBULLIOSCOPIC CONSTANTS FOR CALCULATION OF BOILING POINT ELEVATION

The boiling point T_b of a dilute solution of a non-volatile, non-dissociating solute is elevated relative to that of the pure solvent. If the solution is ideal (i.e., follows Raoult's Law), the amount of elevation depends only on the number of particles of solute present. Hence the change in boiling point ΔT_b can be expressed as

$$\Delta T_b = E_b m_2$$

where m_2 is the molality (moles of solute per kilogram of solvent) and E_b is the Ebullioscopic Constant, a characteristic property of the solvent. The Ebullioscopic Constant may be calculated from the relation

$$E_b = RT_b^2 M/\Delta_{vap}H$$

where R is the molar gas constant, T_b is the normal boiling point temperature (absolute) of the solvent, M the molar mass of the solvent, and $\Delta_{vap}H$ the molar enthalpy (heat) of vaporization of the solvent at its normal boiling point.

This table lists E_b values for some common solvents, as calculated from data in the table "Enthalpy of Vaporization" in Section 6.

Name	t_b/°C	E_b/K kg mol^{-1}
Acetic acid	117.9	3.22
Acetone	56.08	1.80
Acetonitrile	81.6	1.44
Aniline	184.1	3.82
Anisole	153.6	4.20
Benzaldehyde	178.7	4.24
Benzene	80.08	2.64
1-Butanol	117.6	2.17
Carbon disulfide	46.2	2.42
Chlorobenzene	131.6	4.36
1-Chlorobutane	78.4	3.13
Cyclohexane	80.7	2.92
Cyclohexanol	160.9	3.5
Decane	174.1	6.10
Dichloromethane	39.8	2.42
Diethyl ether	34.4	2.20
Dimethyl sulfoxide	191.9	3.22
1,4-Dioxane	101.2	3.01
1,2-Ethanediol	197.5	2.26
Ethanol	78.24	1.23
Ethyl acetate	77.1	2.82
Heptane	98.38	3.62
Hexane	68.72	2.90
Iodomethane	42.4	4.31
Methanol	64.5	0.86
Methyl acetate	56.7	2.21
N-Methylaniline	197	4.3
N-Methylformamide	186	2.2
Nitrobenzene	210.7	5.2
Nitromethane	101.19	2.09
1-Octanol	194.7	5.06
Phenol	181.8	3.54
1-Propanol	97.04	1.66
2-Propanol	82.21	1.58
Pyridine	115.2	2.83
Pyrrole	129.74	2.33
Pyrrolidine	86.6	2.32
Tetrachloroethene	121.2	6.18
Tetrachloromethane	76.7	5.26
Toluene	110.60	3.40
Trichloroethene	86.8	4.52
Trichloromethane	61.2	3.80
Water	99.974	0.513
o-Xylene	144.4	4.25

CRYOSCOPIC CONSTANTS FOR CALCULATION OF FREEZING POINT DEPRESSION

The freezing point T_f of a dilute solution of a non-volatile, non-dissociating solute is depressed relative to that of the pure solvent. If the solution is ideal (i.e., follows Raoult's Law), this lowering is a function only of the number of particles of solute present. Thus the absolute value of the lowering of freezing point ΔT_f can be expressed as

$$\Delta T_f = E_f m_2$$

where m_2 is the molality (moles of solute per kilogram of solvent) and E_f is the Cryoscopic Constant, a characteristic property of the solvent. The Cryoscopic Constant may be calculated from the relation

$$E_f = RT_f^2 M/\Delta_{fus}H$$

where R is the molar gas constant, T_f is the freezing point temperature (absolute) of the solvent, M the molar mass of the solvent, and $\Delta_{fus}H$ the molar enthalpy (heat) of fusion of the solvent.

This table lists cryscopic constants for selected substances, as calculated from data in the table "Enthalpy of Fusion" in Section 6.

Name	t_f/°C	E_f/ K kg mol^{-1}
Acetamide	80.16	3.92
Acetic acid	17	3.63
Acetophenone	19.4	5.16
Aniline	-6.0	5.23
Benzene	5.538	5.07
Benzonitrile	-12.82	5.35
Benzophenone	48.0	8.58
Camphor, (+)	178.7	37.8
1-Chloronaphthalene	-6.0	7.68
o-Cresol	31.0	5.92
m-Cresol	12.2	7.76
p-Cresol	34.77	7.20
Cyclohexane	6.7	20.8
Cyclohexanol	26	42.2
cis-Decahydronaphthalene	-42.9	6.42
trans-Decahydronaphthalene	-30.35	4.70
Dibenzyl ether	1.8	6.17
p-Dichlorobenzene	53.1	7.57
Diethanolamine	27.9	3.16
Dimethyl sulfoxide	18.52	3.85
1,4-Dioxane	11.75	4.63
Diphenylamine	53.2	8.38
1,2-Ethanediol	-13	3.11
Formamide	2.57	4.25
Formic acid	8.3	2.38
Glycerol	18.2	3.56
4-Methylaniline	43.3	4.91
Methylcyclohexane	-126.6	2.60
Naphthalene	80.22	7.45
Nitrobenzene	5.65	6.87
Phenol	40.89	6.84
Pyridine	-41.63	4.26
Quinoline	-14.78	6.73
Succinonitrile	57.985	19.3
1,1,2,2-Tetrabromoethane	0	21.4
1,1,2,2-Tetrachloro-1,2-difluoroethane	26.54	41.0
Toluene	-95.0	3.55
Tribromomethane	8.69	15.0
Water	0.00	1.86
p-Xylene	13.3	4.31

Lab Data

FREEZING POINT LOWERING BY ELECTROLYTES IN AQUEOUS SOLUTION

Reference

Forsythe, W. E., *Smithsonian Physical Tables, Ninth Edition*, Smithsonian Institution, Washington, DC, 1956.

Lowering of Freezing Point of Water (in °C) as a Function of Molality *m* (in mol/kg)

Compound	0.05 m	0.10 m	0.25 m	0.50 m	0.75 m	1.00 m	1.50 m	2.00 m	2.50 m	3.00 m
$CaCl_2$	0.25	0.49	1.27	2.66	4.28	6.35	10.78	15.27	20.42	28.08
$CuSO_4$	0.13	0.23	0.47	0.96						
HCl	0.18	0.36	0.90	1.86	2.90	4.02	6.63	9.94		
HNO_3	0.18	0.35	0.88	1.80	2.78	3.80	5.98	8.34	10.95	13.92
H_2SO_4	0.20	0.39	0.96	1.95	3.04	4.28	7.35	11.35	16.32	
KBr	0.18	0.36	0.92	1.78						
KCl	0.17	0.35	0.86	1.68	2.49	3.29	4.88	6.50	8.14	9.77
KNO_3	0.17	0.33	0.78	1.47	2.11	2.66				
K_2SO_4	0.23	0.43	1.01	1.87						
LiCl	0.18	0.35	0.88	1.80	2.78					
$MgSO_4$	0.13	0.24	0.55	1.01	1.50	2.08	3.41			
NH_4Cl	0.17	0.34	0.85	1.70	2.55					
NaCl	0.18	0.35	0.85	1.68	2.60					
$NaNO_3$	0.18	0.36	0.80	1.62	2.63	3.10				

CORRECTION OF BAROMETER READINGS TO 0 °C TEMPERATURE

The following corrections are used to reduce the reading of a mercury barometer with a brass scale to 0 °C. The number in the table should be subtracted from the observed height of the mercury column to give the true pressure in mmHg (1 mmHg = 133.322 Pa). The table is calculated from the formula

$$\Delta h = -0.0001634\, ht/(1+0.0001818\, t),$$

where h is the observed column height in mm and t the Celsius temperature. This relation is based on thermal expansion coefficients of $181.8 \cdot 10^{-6}\ °C^{-1}$ for mercury and $18.4 \cdot 10^{-6}\ °C^{-1}$ for brass.

	Observed Height in mm																		
t/°C	**620**	**630**	**640**	**650**	**660**	**670**	**680**	**690**	**700**	**710**	**720**	**730**	**740**	**750**	**760**	**770**	**780**	**790**	**800**
0	0.00	0.00	0.00	0.00	0.00	0.00	0.00	0.00	0.00	0.00	0.00	0.00	0.00	0.00	0.00	0.00	0.00	0.00	0.00
1	0.10	0.10	0.10	0.11	0.11	0.11	0.11	0.11	0.11	0.12	0.12	0.12	0.12	0.12	0.12	0.13	0.13	0.13	0.13
2	0.20	0.21	0.21	0.21	0.22	0.22	0.22	0.23	0.23	0.23	0.24	0.24	0.24	0.25	0.25	0.25	0.25	0.26	0.26
3	0.30	0.31	0.31	0.32	0.32	0.33	0.33	0.34	0.34	0.35	0.35	0.36	0.36	0.37	0.37	0.38	0.38	0.39	0.39
4	0.40	0.41	0.42	0.42	0.43	0.44	0.44	0.45	0.46	0.46	0.47	0.48	0.48	0.49	0.50	0.50	0.51	0.52	0.52
5	0.51	0.51	0.52	0.53	0.54	0.55	0.56	0.56	0.57	0.58	0.59	0.60	0.60	0.61	0.62	0.63	0.64	0.64	0.65
6	0.61	0.62	0.63	0.64	0.65	0.66	0.67	0.68	0.69	0.70	0.71	0.71	0.72	0.73	0.74	0.75	0.76	0.77	0.78
7	0.71	0.72	0.73	0.74	0.75	0.77	0.78	0.79	0.80	0.81	0.82	0.83	0.85	0.86	0.87	0.88	0.89	0.90	0.91
8	0.81	0.82	0.84	0.85	0.86	0.87	0.89	0.90	0.91	0.93	0.94	0.95	0.97	0.98	0.99	1.01	1.02	1.03	1.04
9	0.91	0.92	0.94	0.95	0.97	0.98	1.00	1.01	1.03	1.04	1.06	1.07	1.09	1.10	1.12	1.13	1.15	1.16	1.17
10	1.01	1.03	1.04	1.06	1.08	1.09	1.11	1.13	1.14	1.16	1.17	1.19	1.21	1.22	1.24	1.26	1.27	1.29	1.30
11	1.11	1.13	1.15	1.17	1.18	1.20	1.22	1.24	1.26	1.27	1.29	1.31	1.33	1.35	1.36	1.38	1.40	1.42	1.44
12	1.21	1.23	1.25	1.27	1.29	1.31	1.33	1.35	1.37	1.39	1.41	1.43	1.45	1.47	1.49	1.51	1.53	1.55	1.57
13	1.31	1.34	1.36	1.38	1.40	1.42	1.44	1.46	1.48	1.50	1.53	1.55	1.57	1.59	1.61	1.63	1.65	1.67	1.70
14	1.41	1.44	1.46	1.48	1.51	1.53	1.55	1.57	1.60	1.62	1.64	1.67	1.69	1.71	1.73	1.76	1.78	1.80	1.83
15	1.52	1.54	1.56	1.59	1.61	1.64	1.66	1.69	1.71	1.74	1.76	1.78	1.81	1.83	1.86	1.88	1.91	1.93	1.96
16	1.62	1.64	1.67	1.69	1.72	1.75	1.77	1.80	1.82	1.85	1.88	1.90	1.93	1.96	1.98	2.01	2.03	2.06	2.09
17	1.72	1.74	1.77	1.80	1.83	1.86	1.88	1.91	1.94	1.97	1.99	2.02	2.05	2.08	2.10	2.13	2.16	2.19	2.22
18	1.82	1.85	1.88	1.91	1.93	1.96	1.99	2.02	2.05	2.08	2.11	2.14	2.17	2.20	2.23	2.26	2.29	2.32	2.35
19	1.92	1.95	1.98	2.01	2.04	2.07	2.10	2.13	2.17	2.20	2.23	2.26	2.29	2.32	2.35	2.38	2.41	2.44	2.48
20	2.02	2.05	2.08	2.12	2.15	2.18	2.21	2.25	2.28	2.31	2.34	2.38	2.41	2.44	2.47	2.51	2.54	2.57	2.60
21	2.12	2.15	2.19	2.22	2.26	2.29	2.32	2.36	2.39	2.43	2.46	2.50	2.53	2.56	2.60	2.63	2.67	2.70	2.73
22	2.22	2.26	2.29	2.33	2.36	2.40	2.43	2.47	2.51	2.54	2.58	2.61	2.65	2.69	2.72	2.76	2.79	2.83	2.86
23	2.32	2.36	2.40	2.43	2.47	2.51	2.54	2.58	2.62	2.66	2.69	2.73	2.77	2.81	2.84	2.88	2.92	2.96	2.99
24	2.42	2.46	2.50	2.54	2.58	2.62	2.66	2.69	2.73	2.77	2.81	2.85	2.89	2.93	2.97	3.01	3.05	3.08	3.12
25	2.52	2.56	2.60	2.64	2.68	2.72	2.77	2.81	2.85	2.89	2.93	2.97	3.01	3.05	3.09	3.13	3.17	3.21	3.25
26	2.62	2.66	2.71	2.75	2.79	2.83	2.88	2.92	2.96	3.00	3.04	3.09	3.13	3.17	3.21	3.26	3.30	3.34	3.38
27	2.72	2.77	2.81	2.85	2.90	2.94	2.99	3.03	3.07	3.12	3.16	3.20	3.25	3.29	3.34	3.38	3.42	3.47	3.51
28	2.82	2.87	2.91	2.96	3.00	3.05	3.10	3.14	3.19	3.23	3.28	3.32	3.37	3.41	3.46	3.51	3.55	3.60	3.64
29	2.92	2.97	3.02	3.06	3.11	3.16	3.21	3.25	3.30	3.35	3.39	3.44	3.49	3.54	3.58	3.63	3.68	3.72	3.77
30	3.02	3.07	3.12	3.17	3.22	3.27	3.32	3.36	3.41	3.46	3.51	3.56	3.61	3.66	3.71	3.75	3.80	3.85	3.90
31	3.12	3.17	3.22	3.27	3.32	3.37	3.43	3.48	3.53	3.58	3.63	3.68	3.73	3.78	3.83	3.88	3.93	3.98	4.03
32	3.22	3.28	3.33	3.38	3.43	3.48	3.54	3.59	3.64	3.69	3.74	3.79	3.85	3.90	3.95	4.00	4.05	4.11	4.16
33	3.32	3.38	3.43	3.48	3.54	3.59	3.64	3.70	3.75	3.81	3.86	3.91	3.97	4.02	4.07	4.13	4.18	4.23	4.29
34	3.42	3.48	3.53	3.59	3.64	3.70	3.75	3.81	3.87	3.92	3.98	4.03	4.09	4.14	4.20	4.25	4.31	4.36	4.42
35	3.52	3.58	3.64	3.69	3.75	3.81	3.86	3.92	3.98	4.03	4.09	4.15	4.21	4.26	4.32	4.38	4.43	4.49	4.55
36	3.62	3.68	3.74	3.80	3.86	3.92	3.97	4.03	4.09	4.15	4.21	4.27	4.32	4.38	4.44	4.50	4.56	4.62	4.68
37	3.72	3.78	3.84	3.90	3.96	4.02	4.08	4.14	4.20	4.26	4.32	4.38	4.44	4.50	4.56	4.62	4.68	4.74	4.80
38	3.82	3.88	3.95	4.01	4.07	4.13	4.19	4.25	4.32	4.38	4.44	4.50	4.56	4.62	4.69	4.75	4.81	4.87	4.93
39	3.92	3.99	4.05	4.11	4.18	4.24	4.30	4.37	4.43	4.49	4.56	4.62	4.68	4.75	4.81	4.87	4.94	5.00	5.06
40	4.02	4.09	4.15	4.22	4.28	4.35	4.41	4.48	4.54	4.61	4.67	4.74	4.80	4.87	4.93	5.00	5.06	5.13	5.19

DETERMINATION OF RELATIVE HUMIDITY FROM DEW POINT

The relative humidity of a water vapor–air mixture is defined as 100 times the partial pressure of water divided by the saturation vapor pressure of water at the same temperature. The relative humidity may be determined from the dew point t_{dew}, which is the temperature at which liquid water first condenses when the mixture is cooled from an initial temperature t. This table gives relative humidity as a function of the dew point depression $t - t_{dew}$ for several values of the dew point. Values are calculated from the vapor pressure table in Section 6.

Relative Humidity as a Function of Ambient Temperature t and Dew Point t_{dew}

$(t-t_{dew})$/°C	t_{dew} = −10 °C	t_{dew} = 0 °C	t_{dew} = 10 °C	t_{dew} = 20 °C	t_{dew} = 30 °C
0.0	100	100	100	100	100
0.2	99	99	99	99	99
0.4	97	97	97	98	98
0.6	95	96	96	96	97
0.9	94	94	95	95	96
1.0	92	93	94	94	94
1.2	91	92	92	93	93
1.4	90	90	91	92	92
1.6	88	89	90	91	91
1.8	87	88	89	90	90
2.0	86	87	88	88	89
2.2	84	85	86	87	89
2.4	83	84	85	86	87
2.6	82	83	84	85	86
2.8	80	82	83	84	85
3.0	79	81	82	83	84
3.2	78	80	81	82	83
3.4	77	79	80	81	82
3.6	76	77	79	80	82
3.8	75	76	78	79	81
4.0	73	75	77	78	80
4.2	72	74	76	77	79
4.4	71	73	75	77	78
4.6	70	72	74	76	77
4.8	69	71	73	75	76
5.0	69	70	72	74	75
5.2	67	69	71	73	75
5.4	66	68	70	72	74
5.6	65	67	69	71	73
5.9	64	66	69	70	72
6.0	63	66	68	70	71
6.2	62	65	67	69	71
6.4	61	64	66	68	70
6.6	60	63	65	67	69
6.8	60	62	64	66	68
7.0	59	61	63	66	68
7.2	58	60	63	65	67
7.4	57	60	62	64	66
7.6	56	59	61	63	65
7.8	55	58	60	63	65
8.0	54	57	60	62	64
8.2	54	56	59	61	63
8.4	53	56	58	60	63
8.6	52	55	57	60	62
8.8	51	54	57	59	61
9.0	51	53	56	58	61
9.2	50	53	55	58	60
9.4	49	52	55	57	59
9.6	48	51	54	56	59
9.8	48	51	53	56	58
10.0	47	50	53	55	57
10.5	45	48	51	54	56
11.0	44	47	49	52	55
11.5	42	45	48	51	53
12.0	41	44	47	49	52
12.5	39	42	45	48	50
13.0	38	41	44	46	49
13.5	37	40	43	45	48
14.0	35	38	41	44	47
14.5	34	37	40	43	45
15.0	33	36	39	42	44
15.5	32	35	38	40	
16.0	31	34	37	39	
16.5	30	33	36	38	
17.0	29	32	35	37	
17.5	28	31	34	36	
18.0	27	30	33	35	
18.5	26	29	32	34	
19.0	25	28	31	33	
19.5	24	27	30	33	
20.0	24	26	29	32	
21.0	22	25	27	30	
22.0	21	23	26	29	
23.0	19	22	24	27	
24.0	18	21	23	26	
25.0	17	19	22	24	
26.0	16	18	21	23	
27.0	15	17	20	22	
28.0	14	16	19	21	
29.0	13	15	18	20	
30.0	12	14	17	19	

DETERMINATION OF RELATIVE HUMIDITY FROM WET AND DRY BULB TEMPERATURES

Relative humidity may be determined by comparing temperature readings of wet and dry bulb thermometers. The following table, extracted from more extensive U.S. National Weather Service tables, gives the relative humidity as a function of air temperature t_d (dry bulb) and the difference $t_d - t_w$ between dry and wet bulb temperatures. The data assume a pressure near normal atmospheric pressure and an instrumental configuration with forced ventilation.

	$(t_d - t_w)$/°C											
t_d/°C	**0.5**	**1.0**	**1.5**	**2.0**	**2.5**	**3.0**	**3.5**	**4.0**	**4.5**	**5.0**	**5.5**	**6.0**
–10	83	67	51	35	19							
–8	86	71	57	43	29	15						
–6	88	74	61	49	37	25	8					
–4	89	77	66	55	44	33	23	12				
–2	90	79	69	60	50	40	31	22	12			
0	91	81	72	64	55	46	38	29	21	13	5	
2	91	84	76	68	60	52	44	37	29	22	14	7
4	92	85	78	71	63	57	49	43	36	29	22	16
6	93	86	79	73	66	60	54	48	41	35	29	24
8	93	87	81	75	69	63	57	51	46	40	35	29
10	94	88	82	77	71	66	60	55	50	44	39	34
12	94	89	83	78	73	68	63	58	53	48	43	39
14	95	90	85	79	75	70	65	60	56	51	47	42
16	95	90	85	81	76	71	67	63	58	54	50	46
18	95	91	86	82	77	73	69	65	61	57	53	49
20	96	91	87	83	78	74	70	66	63	59	55	51
22	96	92	87	83	80	76	72	68	64	61	57	54
24	96	92	88	84	80	77	73	69	66	62	59	56
26	96	92	88	85	81	78	74	71	67	64	61	58
28	96	93	89	85	82	78	75	72	69	65	62	59
30	96	93	89	86	83	79	76	73	70	67	64	61
35	97	94	90	87	84	81	78	75	72	69	67	64
40	97	94	91	88	85	82	80	77	74	72	69	67

	$(t_d - t_w)$/°C											
t_d/°C	**6.5**	**7.0**	**7.5**	**8.0**	**8.5**	**9.0**	**10.0**	**11.0**	**12.0**	**13.0**	**14.0**	**15.0**
4	9											
6	17	11	5									
8	24	19	14	8								
10	29	24	20	15	10	6						
12	34	29	25	21	16	12	5					
14	38	34	30	26	22	18	10					
16	42	38	34	30	26	23	15	8				
18	45	41	38	34	30	27	20	14	7			
20	48	44	41	37	34	31	24	18	12	6		
22	50	47	44	40	37	34	28	22	17	11	6	
24	53	49	46	43	40	37	31	26	20	15	10	5
26	54	51	49	46	43	40	34	29	24	19	14	10
28	56	53	51	48	45	42	37	32	27	22	18	13
30	58	55	52	50	47	44	39	35	30	25	21	17
32	60	57	54	51	49	46	41	37	32	28	24	20
34	61	58	56	53	51	48	43	39	35	30	26	23
36	62	59	57	54	52	50	45	41	37	33	29	25
38	63	61	58	56	54	51	47	43	39	35	31	27
40	64	62	59	57	54	53	48	44	40	36	33	29

CONSTANT HUMIDITY SOLUTIONS

Anthony Wexler

An excess of a water soluble salt in contact with its saturated solution and contained within an enclosed space produces a constant relative humidity and water vapor pressure according to

$$RH = A\exp(B/T)$$

where RH is the percent relative humidity (generally accurate to ±2 %), T is the temperature in kelvin, and the constants A and B and the range of valid temperatures are given in the table below. The vapor pressure, p, can be calculated from

$$p = (RH/100) \times p_0$$

where p_0 is the vapor pressure of pure water at temperature T as given in the table in Section 6 titled "Vapor Pressure of Water from 0 to 370 °C".

References

1. Wexler, A. S. and Seinfeld, J. H., *Atmospheric Environment*, 25A, 2731, 1991.
2. Greenspan, L., *J. Res. National Bureau of Standards*, 81A, 89, 1977.
3. Broul, et al., *Solubility of Inorganic Two-Component Systems*, Elsevier, New York, 1981.
4. Wagman, D. D. et al., *J. Phys. Chem. Ref. Data, Vol. 11, Suppl.* 2, 1982.

Compound	Temperature range (°C)	RH (25 °C)	A	B
$NaOH \cdot H_2O$	15–60	6	5.48	27
$LiBr \cdot 2H_2O$	10–30	6	0.23	996
$ZnBr_2 \cdot 2H_2O$	5–30	8	1.69	455
$KOH \cdot 2H_2O$	5–30	9	0.014	1924
$LiCl \cdot H_2O$	20–65	11	14.53	–75
$CaBr_2 \cdot 6H_2O$	11–22	16	0.17	1360
$LiI \cdot 3H_2O$	15–65	18	0.15	1424
$CaCl_2 \cdot 6H_2O$	15–25	29	0.11	1653
$MgCl_2 \cdot 6H_2O$	5–45	33	29.26	34
$NaI \cdot 2H_2O$	5–45	38	3.62	702
$Ca(NO_3)_2 \cdot 4H_2O$	10–30	51	1.89	981
$Mg(NO_3)_2 \cdot 6H_2O$	5–35	53	25.28	220
$NaBr \cdot 2H_2O$	0–35	58	20.49	308
NH_4NO_3	10–40	62	3.54	853
KI	5–30	69	29.35	254
$SrCl_2 \cdot 6H_2O$	5–30	71	31.58	241
$NaNO_3$	10–40	74	26.94	302
NaCl	10–40	75	69.20	25
NH_4Cl	10–40	79	35.67	235
KBr	5–25	81	40.98	203
$(NH_4)_2SO_4$	10–40	81	62.06	79
KCl	5–25	84	49.38	159
$Sr(NO_3)_2 \cdot 4H_2O$	5–25	85	28.34	328
$BaCl_2 \cdot 2H_2O$	5–25	90	69.99	75
CsI	5–25	91	70.77	75
KNO_3	0–50	92	43.22	225
K_2SO_4	10–50	97	86.75	34

Lab Data

STANDARD SALT SOLUTIONS FOR HUMIDITY CALIBRATION

Saturated aqueous solutions of inorganic salts are convenient secondary standards for calibration of instruments for measurement of relative humidity. The International Union of Pure and Applied Chemistry has recommended salt solutions for calibrations in the range of 10% to 90% relative humidity, and the American Society for Testing and Materials has published similar standards. The data in this table are taken from the IUPAC recommendations, except for K_2CO_3 and K_2SO_4, which are ASTM recommendations.

Details on the preparation and use of these standards may be found in References 1 and 2. Data for other salts are given in Reference 3.

References

1. Marsh, K. N., Ed., *Recommended Reference Materials for the Realization of Physicochemical Properties*, Blackwell Scientific Publications, Oxford, 1987, pp. 157–162.
2. *Standard Practice for Maintaining Constant Relative Humidity by Means of Aqueous Solutions*, ASTM Standard E 104-85, Reapproved 1991.
3. Greenspan, L., *J. Res. Nat. Bur. Stand.* 81A, 89, 1977.

	Relative Humidity in %						
t/°C	LiCl	$MgCl_2$	K_2CO_3	$Mg(NO_3)_2$	NaCl	KCl	K_2SO_4
0		33.66 ± 0.33	43.1 ± 0.7	60.35 ± 0.55	75.51 ± 0.34	88.61 ± 0.53	98.8 ± 2.1
5		33.60 ± 0.28	43.1 ± 0.5	58.86 ± 0.43	75.65 ± 0.27	87.67 ± 0.45	98.5 ± 0.9
10		33.47 ± 0.24	43.1 ± 0.4	57.36 ± 0.33	75.67 ± 0.22	86.77 ± 0.39	98.2 ± 0.8
15		33.30 ± 0.21	43.2 ± 0.3	55.87 ± 0.27	75.61 ± 0.18	85.92 ± 0.33	97.9 ± 0.6
20	11.31 ± 0.31	33.07 ± 0.18	43.2 ± 0.3	54.38 ± 0.23	75.47 ± 0.14	85.11 ± 0.29	97.6 ± 0.5
25	11.30 ± 0.27	32.78 ± 0.16	43.2 ± 0.4	52.89 ± 0.22	75.29 ± 0.12	84.34 ± 0.26	97.3 ± 0.5
30	11.28 ± 0.24	32.44 ± 0.14	43.2 ± 0.5	51.40 ± 0.24	75.09 ± 0.11	83.62 ± 0.25	97.0 ± 0.4
35	11.25 ± 0.22	32.05 ± 0.13		49.91 ± 0.29	74.87 ± 0.12	82.95 ± 0.25	96.7 ± 0.4
40	11.21 ± 0.21	31.60 ± 0.13		48.42 ± 0.37		82.32 ± 0.25	96.4 ± 0.4
45	11.16 ± 0.21	31.10 ± 0.13		46.93 ± 0.47		81.74 ± 0.28	96.1 ± 0.4
50	11.10 ± 0.22	30.54 ± 0.14		45.44 ± 0.60		81.20 ± 0.31	95.8 ± 0.5
55	11.03 ± 0.23	29.93 ± 0.16				80.70 ± 0.35	
60	10.95 ± 0.26	29.26 ± 0.18				80.25 ± 0.41	
65	10.86 ± 0.29	28.54 ± 0.21				79.85 ± 0.48	
70	10.75 ± 0.33	27.77 ± 0.25				79.49 ± 0.57	
75	10.64 ± 0.38	26.94 ± 0.29				79.17 ± 0.66	
80	10.51 ± 0.44	26.05 ± 0.34				78.90 ± 0.77	

LOW-TEMPERATURE BATHS FOR MAINTAINING CONSTANT TEMPERATURE

A liquid–solid slurry is a convenient means of maintaining a constant temperature environment below room temperature. The following is a list of readily available organic liquids suitable for this purpose, arranged in order of their melting (freezing) points t_m. The normal boiling points t_b are also given.

Compound	t_m/°C	t_b/°C
Isopentane (2-Methylbutane)	−159.9	27.8
Methylcyclopentane	−142.5	71.8
3-Chloropropene (Allyl chloride)	−134.5	45.1
Pentane	−129.7	36.0
Allyl alcohol	−129	97.0
Ethanol	−114.1	78.2
Carbon disulfide	−111.5	46
Isobutyl alcohol	−108	107.8
Toluene	−94.9	110.6
Acetone	−94.8	56.0
Ethyl acetate	−83.6	77.1
Dry ice + acetone	−78	
p-Cymene	−68.9	177.1
Trichloromethane (Chloroform)	−63.6	61.1
N-Methylaniline	−57	196.2
Chlorobenzene	−45.2	131.7
Anisole	−37.5	153.7
Bromobenzene	−30.6	156.0
Tetrachloromethane (Carbon tetrachloride)	−23	76.8
Benzonitrile	−12.7	191.1

METALS AND ALLOYS WITH LOW MELTING TEMPERATURE

L. I. Berger

Metal or alloy system	Composition, % * Weight	Atomic	Melting temperature (°C)	Comments	Ref.
Hg	100	100	−38.829		
Cs–K	77.0–23.0	50.0–50.0	−37.5	Eutectic (?)	1
Cs–Na	94.5–5.5	75.0–25.0	−30.0	Eutectic	2
K–Na	76.7–23.3	65.9–34.1	−12.65	Eutectic	3
Na–Rb	8.0–92.0	24.4–75.6	−5	Eutectic	4
Ga–In–Sn	62.5–21.5–16.0	73.6–15.3–11.1	11	Eutectic	5
Ga–Sn–Zn	82.0–12.0–6.0	86.0–7.3–6.7	17	Eutectic	5
Cs	100	100	28.44		
Ga	100	100	29.7646		
K–Rb	32.0–68.0	50–50	33	Eutectic	4
Bi–Cd–In–Pb–Sn	44.7–5.3–19.1–22.6–8.3	35.1–8.2–27.3–17.9–11.5	46.7	Eutectic	6
Bi–In–Pb–Sn	49.5–21.3–17.6–11.6	39.2–30.7–14.0–16.2	58.2	Eutectic	6
Bi–In–Sn	32.5–51.0–16.5	21.1–60.1–18.8	60.5	Eutectic	7
K	100	100	63.38		
Bi–Cd–Pb–Sn	50.0–12.5–25.0–12.5	41.5–19.3–21.0–18.2	70	Wood's alloy	6
Bi–In	33.0–67.0	21.3–78.7	72	Eutectic	8
Bi–Cd–Pb	51.6–8.2–40.2	48.1–14.2–37.7	91.5	Eutectic	6
Bi–Pb–Sn	52.5–32.0–15.5	46.8–28.7–24.5	95	Eutectic	6
Na	100	100	97.8		
Bi–Cd–Sn	54.0–20.0–26.0	39.4–27.2–33.4	102.5	Eutectic	6
In–Sn	51.8–48.2	52.6–47.4	119	Eutectic	9
Cd–In	25.3–74.7	25.7–74.3	120	Eutectic	10
Bi–Pb	55.5–44.5	55.3–44.7	124	Eutectic	11
Bi–Sn–Zn	56.0–40.0–4.0	40.2–50.6–9.2	130	Eutectic	6, 7
Bi–Sn	70–30	57.0–43.0	138.5	Eutectic	6, 12
Bi–Cd	60.3–39.7	45.0–55.0	145.5	Eutectic	13, 14
In	100	100	156.6		
Li	100	100	180.5		
Pb–Sn	38.1–61.9	26.1–73.9	183	Eutectic	6,15
Bi–Tl	48.0–52.0	47.5–52.5	185	Eutectic	13
Sn–Zn	91.0–9.0	85.0–15.0	198	Eutectic	14
Sb–Sn	8.0–92.0	7.8–92.2	199	White Metal	16
Au–Pb	14.6–85.4	15.2–84.8	212	Eutectic	17
Ag–Sn	3.5–96.5	3.8–96.2	221	Eutectic	13,18
Bi–Pb–Sb–Sn	48.0–28.5–9.0–14.5	40.8–24.5–13.1–21.6	226	Matrix Alloy	6
Cu–Sn	0.75–99.25	1.3–98.7	227	Eutectic	13, 19
Sn	100	100	231.928		

* The useful expressions for correlations between the atomic and weight concentrations of an alloy components are:

$$f\left(\mathrm{a},A_k\right)=\frac{f\left(\mathrm{w},A_k\right)}{M_k\sum_{i=1}^{N}\frac{f\left(\mathrm{w},A_i\right)}{M_i}} \quad \text{and} \quad f\left(\mathrm{w},A_k\right)=\frac{M_k\cdot f\left(\mathrm{a},A_k\right)}{\sum_{i=1}^{N}M_i\cdot f\left(\mathrm{a},A_i\right)} \quad (i=1,\ldots,k,\ldots,N)$$

where $f(\mathrm{a}, A_i)$ and $f(\mathrm{w}, A_i)$ are the atomic and weight concentrations of component A_i, respectively, and M_i is the atomic weight of this component.

References

1. Zintle, E., and Hauke, W., *Z. Electrochem.* 44, 104, 1938.
2. Rinck, E., *Compt. Rend.* 199, 1217, 1934.
3. Krier, C. A., Craign, R. S., and Wallace, W. E., *J. Phys. Chem.* 61, 522, 1957.
4. Goria, C., *Gazz. Chim. Ital.* 65, 865, 1935.
5. Baker, H., Ed., *ASM Handbook, Volume 3: Alloy Phase Diagrams*, ASM Intl., Materials Park, OH, 1992.
6. Sedlacek, V., *Non-Ferrous Metals and Alloys*, Elsevier, New York, 1986.
7. Villars, P., Prince, A., Okamoto, H., Eds., *Handbook of Ternary Alloy Phase Diagrams*, ASM Intl., 1994.
8. Palatnik, L. S., Kosevich, V. M., and Tyrina, L. V., *Phys. Metals Metallog. (USSR)* 11, 75, 1961.
9. Neumann, T., and Alpout, O., *J. Less-Common Metals* 6, 108, 1964.
10. Neumann, T., and Predel, B., *Z. Metallk.* 50, 309, 1959.
11. Roy, P., Orr, R. L., and Hultgren, R., *J. Phys. Chem.* 64, 1034, 1960.
12. Dobovicek, B., and Smajic, N., *Rudarsko–Met. Zbornik* 4, 353, 1962.
13. Massalski, T. B., Okamoto, H., Subramanian, P. R., and Kacprzak, L., Eds., *Binary Alloy Phase Diagrams*, 2nd Ed., ASM Intl. 1990.
14. Dobovicek, B., and Straus, B., *Rudarsko–Met. Zbornik* 3, 273, 1960.
15. Schurmann, E., and Gilhaus, F. J., *Arch. Eisenhuettenw.* 32, 867, 1961.
16. Rosenblatt, G. M., and Birchenall, C. E., *Trans. AIME* 224, 481, 1962.
17. Evans, D. S., and Prince, A., in *Alloy Phase Diagrams*, MRS Simposia Proc., Vol. 19, North Holland, 1983, p. 383.
18. Umanskiy, M. M., *Zh. Fiz. Khim.* 14, 846, 1940.
19. Homer, C. E., and Plummer, H., *J. Inst. Met.* 64, 169, 1939.

WIRE TABLES

The resistance per unit length of wires of various metals is tabulated here. Values were calculated from resistivity values in the tables "Electrical Resistivity of Pure Metals" and "Electrical Resistivity of Selected Alloys", which appear in Section 12. In practice, resistance may vary because of differing heat treatments and metal composition. The values in the table refer to 20 °C, but values at other temperatures may be calculated from the following resistivity data.

Resistivity in 10^{-8} Ω m at the Indicated Temperature

Metal	0 °C	20 °C	25 °C	100 °C
Aluminum	2.417	2.650	2.709	3.56
Brass (70% Cu, 30% Zn)	5.87	6.08	6.13	6.91
Constantan (60% Cu, 40% Ni)	45.43	45.38	45.35	45.11
Copper	1.543	1.678	1.712	2.22
Nichrome (79% Ni, 21% Cr)	107.3	107.5	107.6	108.3
Platinum	9.6	10.5	10.7	13.6
Silver	1.467	1.587	1.617	2.07
Tungsten	4.82	5.28	5.39	7.18

Resistance per Unit Length at 20 °C in Ω/m

AWG Gauge[a]	Diameter (mm)	Aluminum	Brass	Constantan	Copper	Nichrome	Platinum	Silver	Tungsten
0	8.252	0.000495	0.00114	0.00848	0.000314	0.0201	0.00196	0.000297	0.00099
2	6.543	0.000788	0.00181	0.0135	0.000499	0.0320	0.00312	0.000472	0.00157
4	5.189	0.00125	0.00287	0.0214	0.000793	0.0508	0.00496	0.000750	0.00250
6	4.115	0.00199	0.00457	0.0341	0.00126	0.0808	0.00789	0.00119	0.00397
8	3.264	0.00317	0.00727	0.0542	0.00200	0.128	0.0125	0.00190	0.00631
10	2.588	0.00504	0.0115	0.0863	0.00319	0.204	0.0200	0.00302	0.0100
12	2.053	0.00800	0.0184	0.137	0.00507	0.325	0.0317	0.00479	0.0159
14	1.628	0.0127	0.0292	0.218	0.00806	0.516	0.0504	0.00762	0.0254
16	1.291	0.0202	0.0464	0.347	0.0128	0.821	0.0802	0.0121	0.0403
18	1.024	0.0322	0.0738	0.551	0.0204	1.30	0.127	0.0193	0.0641
20	0.8118	0.0512	0.117	0.877	0.0324	2.08	0.203	0.0307	0.102
22	0.6439	0.0814	0.187	1.39	0.0515	3.30	0.322	0.0487	0.162
24	0.5105	0.129	0.297	2.22	0.0820	5.25	0.513	0.0775	0.258
26	0.4049	0.206	0.472	3.52	0.130	8.35	0.815	0.123	0.410
28	0.3211	0.327	0.751	5.60	0.207	13.3	1.30	0.196	0.652
30	0.2548	0.520	1.19	8.90	0.329	21.1	2.06	0.311	1.03
32	0.2019	0.828	1.90	14.2	0.524	33.6	3.28	0.496	1.65
34	0.1601	1.32	3.02	22.5	0.833	53.4	5.22	0.788	2.62
36	0.1270	2.09	4.80	35.8	1.32	84.9	8.29	1.25	4.17
38	0.1007	3.33	7.63	57.0	2.11	135	13.2	1.99	6.63
40	0.07988	5.29	12.1	90.5	3.35	214	20.9	3.17	10.5

[a] Often called Brown & Sharpe Gauge.

STANDARD FITTINGS FOR COMPRESSED GAS CYLINDERS

Thomas J. Bruno and Paris D. N. Svoronos

The following table presents a partial list of gases and the Compressed Gas Association (CGA) fittings that are required to use those gases when they are stored in, and dispensed from, compressed gas cylinders (Refs. 1 and 2). These fittings are matched to gas cylinder content for safe operation, and substitution and adaptation should be avoided.

References

1. CGA Pamphlet V-1-87, American Canadian and Compressed Gas Association Standard for Compressed Gas Cylinder Valve Outlet and Inlet Connections, ANSI-B57.1; CSA B96, 1987.
2. Bruno, T. J., and Svoronos, P. D. N., *CRC Handbook of Basic Tables for Chemical Analysis,* 3rd. Ed., CRC Press, Boca Raton, FL, 2011.

Gas	Fitting
Acetylene	510
Air	346
Carbon dioxide	320
Carbon monoxide	350
Chlorine	660
Ethane	350
Ethylene	350
Ethylene oxide	510
Helium	580
Hydrogen	350
Hydrogen chloride	330
Methane	350
Neon	580
Nitrogen	580
Nitrous oxide	326
Oxygen	540
Sulfur dioxide	660
Sulfur hexafluoride	590
Xenon	580

The following graphic shows the geometry and dimensions of common CGA fittings for compressed gas cylinders (Ref. 1).

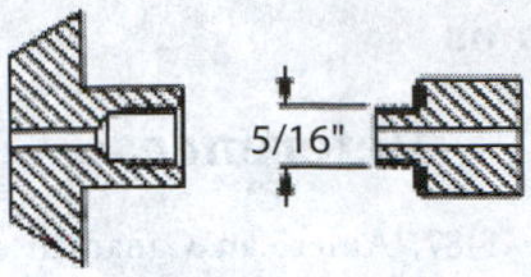

Connection 110 - Lecture bottle outlet for corrosive gases - 5/16" - 32 RH INT., with gasket

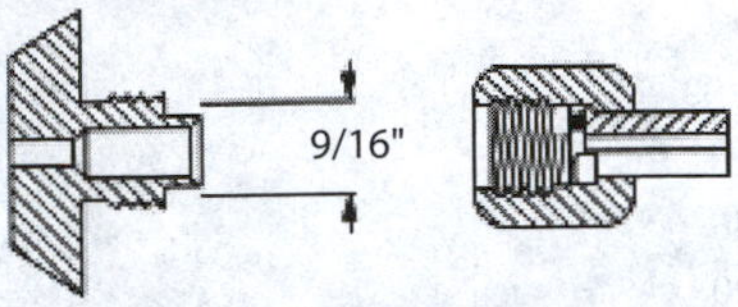

Connection 170 - Lecture bottle outlet for non-corrosive gases 9/16" - 18 RH EXT. and 5/16" - 32 RH INT., with gasket

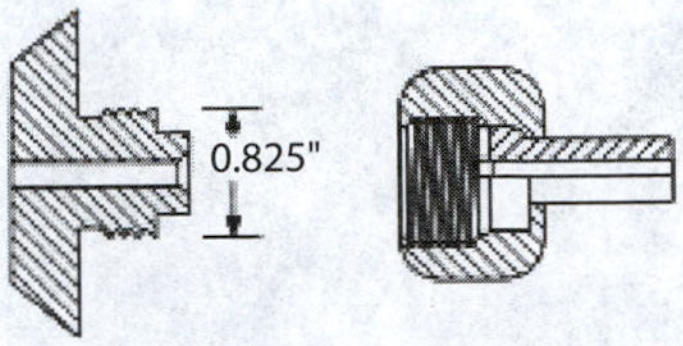

Connection 320 - 0.825" - 14 RH EXT., with gasket

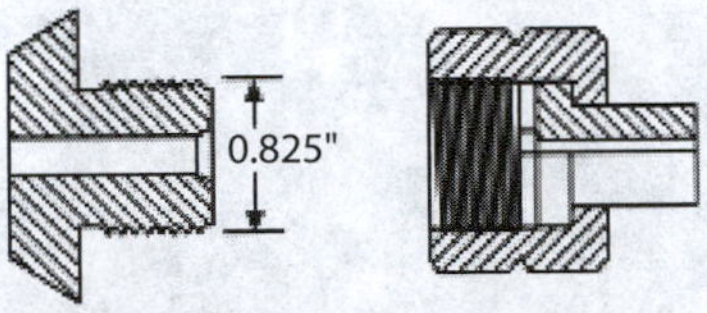

Connection 330 - 0.825" - 14 LH EXT., with gasket

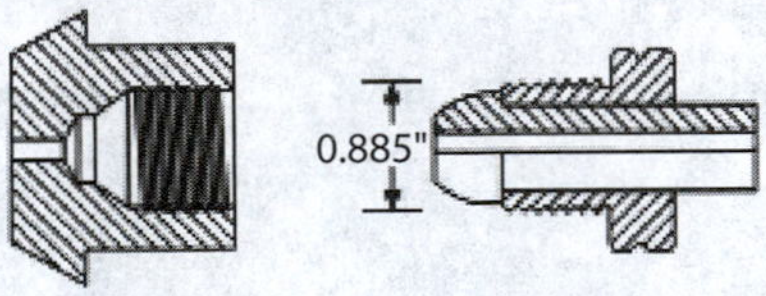

Connection 510 - 0.885" - 14 LH INT.

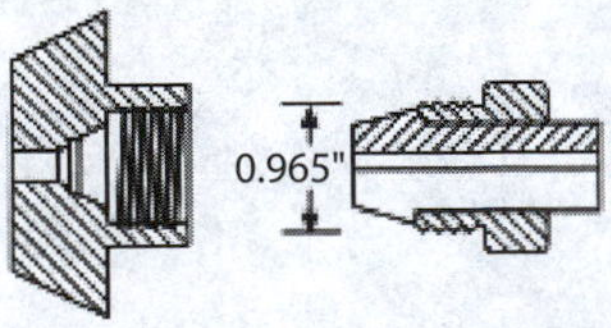

Connection 580 - 0.965" - 14 RH INT.

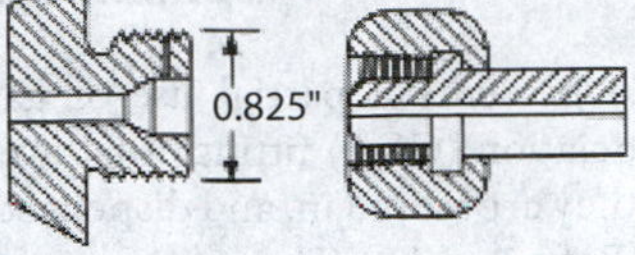

Connection 326 - 0.825" - 14 RH EXT.

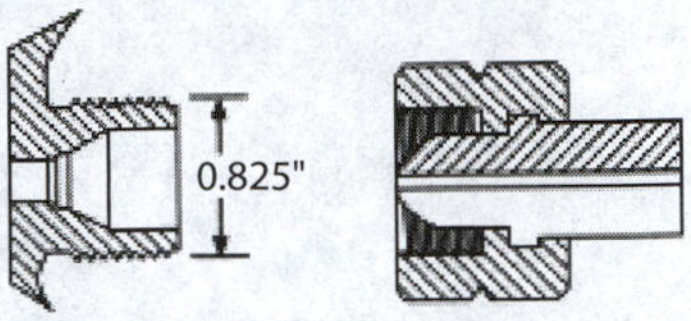

Connection 350 - 0.825" - 14 LH EXT.

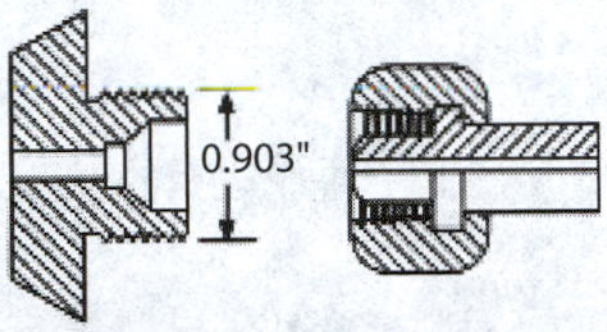

Connection 540 - 0.903" - 14 RH EXT.

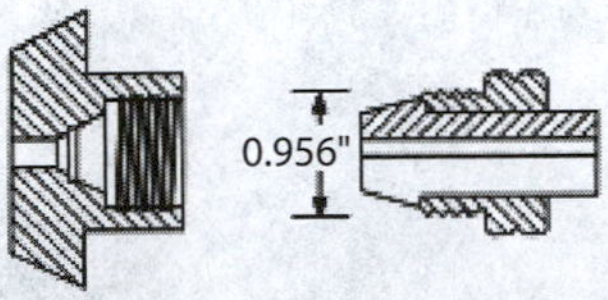

Connection 590 - 0.956" - 14 LH INT.

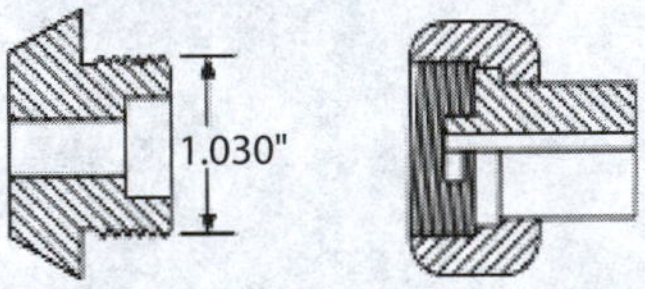

Connection 660 - 1.030" - 14 RH EXT., with gasket

PLUG AND OUTLET CONFIGURATIONS FOR COMMON LABORATORY DEVICES

Thomas J. Bruno and Paris D. N. Svoronos

The following schematic diagrams show typical plug (P) and outlet receptacle (R) configurations used on common laboratory instruments and devices (Refs. 1 and 2). These figures will assist in identifying which circuits and capacities will be needed to operate different pieces of equipment. Good laboratory practice requires the use of the appropriate plug-outlet combination, without alteration.

References

1. Plugs, Receptacles, and Connectors of the Pin and Sleeve Type for Hazardous Locations, National Electrical Manufacturer Association, Standard FB 11, 2000.
2. Bruno, T. J., and Svoronos, P. D. N., *CRC Handbook of Basic Tables for Chemical Analysis*, 3rd Ed., CRC Press, Boca Raton, FL, 2011.

2 pole, 2 wire

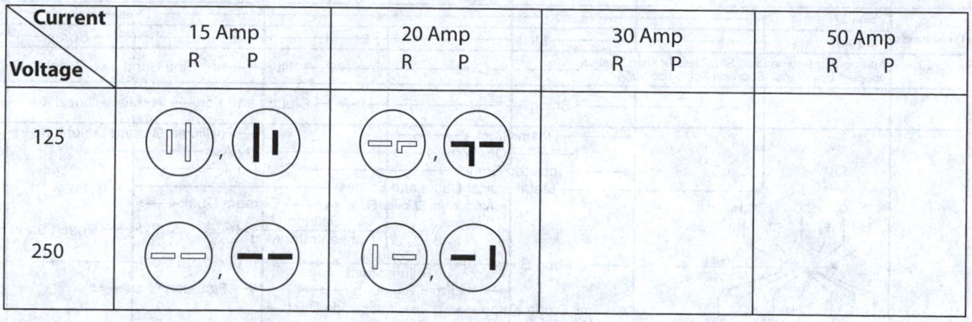

2 pole, 3 wire (grounding)

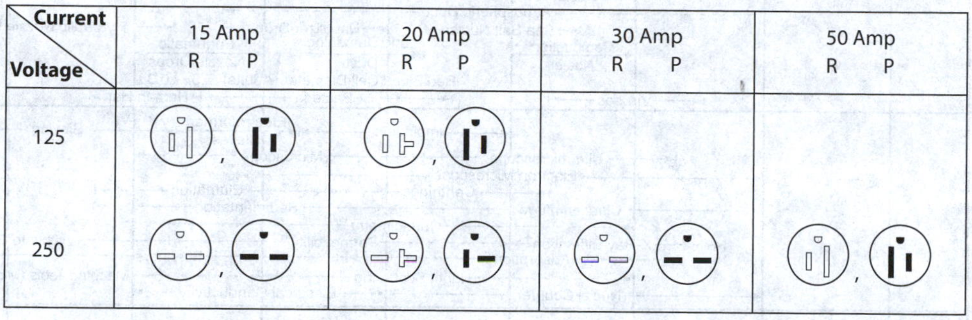

CHARACTERISTICS OF PARTICLES AND PARTICLE DISPERSOIDS

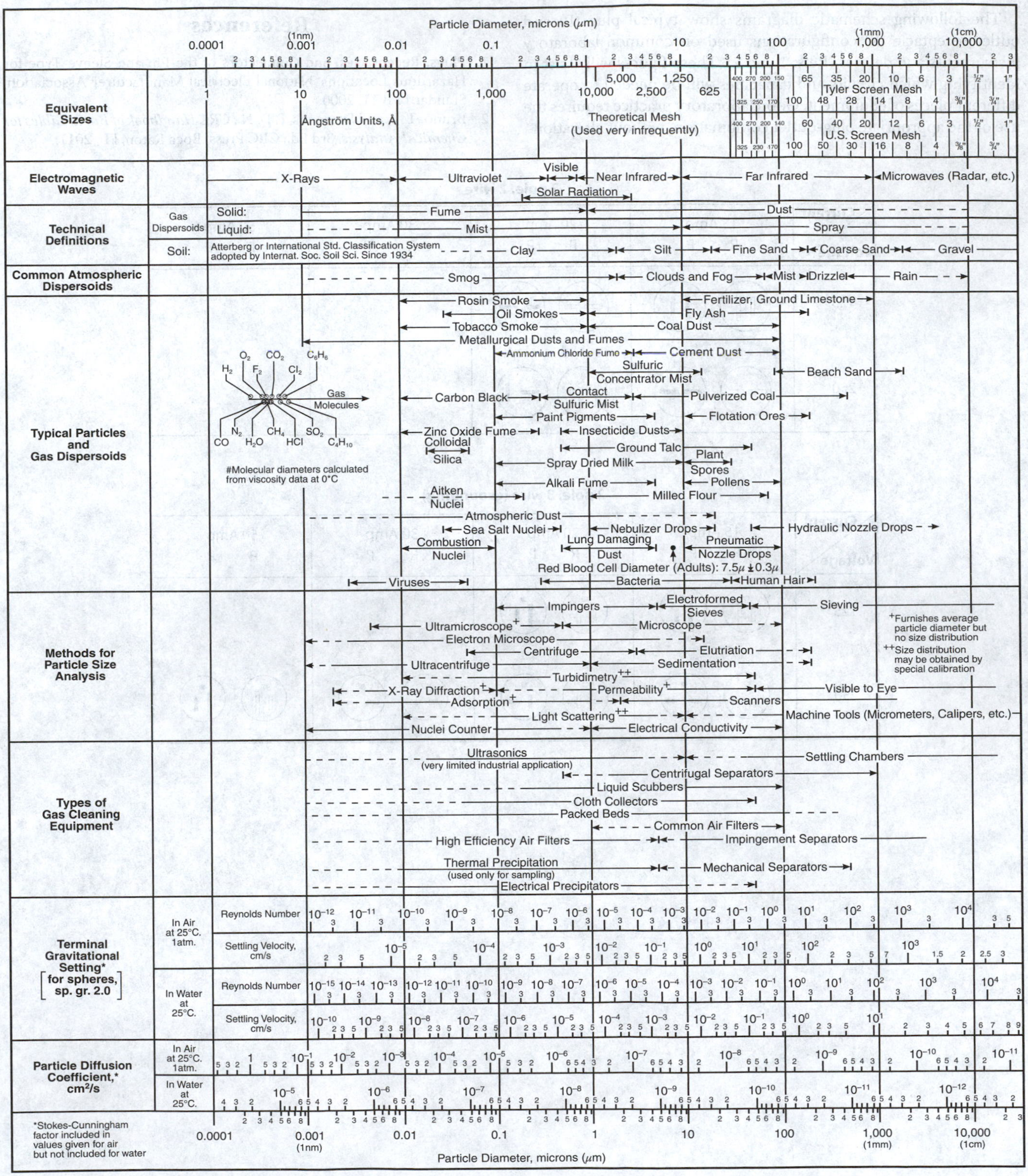

From C. E. Lapple, *Stanford Research Institute Journal*, Vol. 5, p. 95 (Third Quarter) 1961.

DENSITY OF VARIOUS SOLIDS

This table gives the range of density for miscellaneous solid materials whose characteristics depend on the source or method of preparation. The entries for different types of wood refer to seasoned wood.

References

1. Forsythe, W. E., *Smithsonian Physical Tables, Ninth Edition*, Smithsonian Institution, Washington, D.C., 1956.
2. Kaye, G. W. C., and Laby, T. H., *Tables of Physical and Chemical Constants, 16th Edition*, Longman, London, 1995.
3. Brandrup, J., and Immergut, E. H., *Polymer Handbook, Third Edition*, John Wiley & Sons, New York, 1989.

Density Ranges for Solid Materials

Material	ρ/g cm^{-3}	Material	ρ/g cm^{-3}	Material	ρ/g cm^{-3}
Agate	2.5 - 2.7	Graphite	2.30 - 2.72	Solder	8.7 - 9.4
Alabaster, carbonate	2.69 - 2.78	Gum arabic	1.3 - 1.4	Starch	1.53
Alabaster, sulfate	2.26 - 2.32	Gypsum	2.31 - 2.33	Steel, stainless	7.8
Albite	2.62 - 2.65	Halite	2.18	Sugar	1.59
Amber	1.06 - 1.11	Hematite	4.9 - 5.3	Talc	2.7 - 2.8
Amphiboles	2.9 - 3.2	Hornblende	3.0	Tallow, beef	0.94
Anorthite	2.74 - 2.76	Ice	0.917	Tar	1.02
Asbestos	2.0 - 2.8	Iron, cast	7.0 - 7.4	Topaz	3.5 - 3.6
Asbestos slate	1.8	Ivory	1.83 - 1.92	Tourmaline	3.0 - 3.2
Asphalt	1.1 - 1.5	Kaolin	2.6	Tungsten carbide	14.0 - 15.0
Basalt	2.4 - 3.1	Leather, dry	0.86	Wax, sealing	1.8
Beeswax	0.96 - 0.97	Lime, slaked	1.3 - 1.4	Wood, alder	0.42 - 0.68
Beryl	2.69 - 2.70	Limestone	2.68 - 2.76	Wood, apple	0.66 - 0.84
Biotite	2.7 - 3.1	Linoleum	1.18	Wood, ash	0.65 - 0.85
Bone	1.7 - 2.0	Magnetite	4.9 - 5.2	Wood, balsa	0.11 - 0.14
Brasses	8.44 - 8.75	Malachite	3.7 - 4.1	Wood, bamboo	0.31 - 0.40
Brick	1.4 - 2.2	Marble	2.60 - 2.84	Wood, basswood	0.32 - 0.59
Bronzes	8.74 - 8.89	Meerschaum	0.99 - 1.28	Wood, beech	0.70 - 0.90
Butter	0.86 - 0.87	Mica	2.6 - 3.2	Wood, birch	0.51 - 0.77
Calamine	4.1 - 4.5	Muscovite	2.76 - 3.00	Wood, blue gum	1.00
Calcspar	2.6 - 2.8	Ochre	3.5	Wood, box	0.95 - 1.16
Camphor	0.99	Opal	2.2	Wood, butternut	0.38
Cardboard	0.69	Paper	0.7 - 1.15	Wood, cedar	0.49 - 0.57
Celluloid	1.4	Paraffin	0.87 - 0.91	Wood, cherry	0.70 - 0.90
Cement, set	2.7 - 3.0	Peat blocks	0.84	Wood, dogwood	0.76
Chalk	1.9 - 2.8	Pitch	1.07	Wood, ebony	1.11 - 1.33
Charcoal, oak	0.57	Polyamides	1.15 - 1.25	Wood, elm	0.54 - 0.60
Charcoal, pine	0.28 - 0.44	Polyethylene	0.91 - 0.96	Wood, hickory	0.60 - 0.93
Cinnabar	8.12	Poly(methyl methacrylate)	1.19	Wood, holly	0.76
Clay	1.8 - 2.6	Polypropylene	0.90 - 0.94	Wood, juniper	0.56
Coal, anthracite	1.4 - 1.8	Polystyrene	1.06 - 1.12	Wood, larch	0.50 - 0.56
Coal, bituminous	1.2 - 1.5	Polytetrafluoroethylene	2.16 - 2.30	Wood, locust	0.67 - 0.71
Coke	1.0 - 1.7	Poly(vinyl acetate)	1.19	Wood, logwood	0.91
Copal	1.04 - 1.14	Poly(vinyl chloride)	1.39 - 1.42	Wood, mahogany	0.66 - 0.85
Cork	0.22 - 0.26	Porcelain	2.3 - 2.5	Wood, maple	0.62 - 0.75
Corundum	3.9 - 4.0	Porphyry	2.6 - 2.9	Wood, oak	0.60 - 0.90
Diamond	3.51	Pyrite	4.95 - 5.10	Wood, pear	0.61 - 0.73
Dolomite	2.84	Quartz	2.65	Wood, pine, pitch	0.83 - 0.85
Ebonite	1.15	Resin	1.07	Wood, pine, white	0.35 - 0.50
Emery	4.0	Rubber, hard	1.19	Wood, pine, yellow	0.37 - 0.60
Epidote	3.25 - 3.50	Rubber, soft	1.1	Wood, plum	0.66 - 0.78
Feldspar	2.55 - 2.75	Rubber, pure gum	0.91 - 0.93	Wood, poplar	0.35 - 0.50
Flint	2.63	Rubber, Neoprene	1.23 - 1.25	Wood, satinwood	0.95
Fluorite	3.18	Sandstone	2.14 - 2.36	Wood, spruce	0.48 - 0.70
Galena	7.3 - 7.6	Serpentine	2.50 - 2.65	Wood, sycamore	0.40 - 0.60
Garnet	3.15 - 4.30	Silica, fused	2.21	Wood, teak, Indian	0.66 - 0.98
Gelatin	1.27	Silicon carbide	3.16	Wood, walnut	0.64 - 0.70
Glass, common	2.4 - 2.8	Slag	2.0 - 3.9	Wood, water gum	1.00
Glass, lead	3 - 4	Slate	2.6 - 3.3	Wood, willow	0.40 - 0.60
Glass, Pyrex	2.23	Soapstone	2.6 - 2.8	Wood's metal	9.70
Granite	2.64 - 2.76				

DENSITY OF SULFURIC ACID

This table gives the density of aqueous sulfuric acid solutions as a function of concentration (in mass percent of H_2SO_4) and temperature.

Reference

Washburn, E. W., Ed., *International Critical Tables of Numerical Data of Physics, Chemistry, and Technology*, Vol. 3, p. 56, McGraw-Hill, New York, 1926–1932.

	Density in g/mL										
Mass %	0 °C	10 °C	15 °C	20 °C	25 °C	30 °C	40 °C	50 °C	60 °C	80 °C	100 °C
1	1.0074	1.0068	1.0060	1.0051	1.0038	1.0022	0.9986	0.9944	0.9895	0.9779	0.9645
2	1.0147	1.0138	1.0129	1.0118	1.0104	1.0087	1.0050	1.0006	0.9956	0.9839	0.9705
3	1.0219	1.0206	1.0197	1.0184	1.0169	1.0152	1.0113	1.0067	1.0017	0.9900	0.9766
4	1.0291	1.0275	1.0264	1.0250	1.0234	1.0216	1.0176	1.0129	1.0078	0.9961	0.9827
5	1.0364	1.0344	1.0332	1.0317	1.0300	1.0281	1.0240	1.0192	1.0140	1.0022	0.9888
6	1.0437	1.0414	1.0400	1.0385	1.0367	1.0347	1.0305	1.0256	1.0203	1.0084	0.9950
7	1.0511	1.0485	1.0469	1.0453	1.0434	1.0414	1.0371	1.0321	1.0266	1.0146	1.0013
8	1.0585	1.0556	1.0539	1.0522	1.0502	1.0481	1.0437	1.0386	1.0330	1.0209	1.0076
9	1.0660	1.0628	1.0610	1.0591	1.0571	1.0549	1.0503	1.0451	1.0395	1.0273	1.0140
10	1.0735	1.0700	1.0681	1.0661	1.0640	1.0617	1.0570	1.0517	1.0460	1.0338	1.0204
12	1.0886	1.0846	1.0825	1.0802	1.0780	1.0756	1.0705	1.0651	1.0593	1.0469	1.0335
14	1.1039	1.0994	1.0971	1.0947	1.0922	1.0897	1.0844	1.0788	1.0729	1.0603	1.0469
16	1.1194	1.1145	1.1120	1.1094	1.1067	1.1040	1.0985	1.0927	1.0868	1.0740	1.0605
18	1.1351	1.1298	1.1271	1.1243	1.1215	1.1187	1.1129	1.1070	1.1009	1.0879	1.0744
20	1.1510	1.1453	1.1424	1.1394	1.1365	1.1335	1.1275	1.1215	1.1153	1.1021	1.0885
22	1.1670	1.1609	1.1579	1.1548	1.1517	1.1486	1.1424	1.1362	1.1299	1.1166	1.1029
24	1.1832	1.1768	1.1736	1.1704	1.1672	1.1640	1.1576	1.1512	1.1448	1.1313	1.1176
26	1.1996	1.1929	1.1896	1.1862	1.1829	1.1796	1.1730	1.1665	1.1599	1.1463	1.1325
28	1.2160	1.2091	1.2057	1.2023	1.1989	1.1955	1.1887	1.1820	1.1753	1.1616	1.1476
30	1.2326	1.2255	1.2220	1.2185	1.2150	1.2115	1.2046	1.1977	1.1909	1.1771	1.1630
32	1.2493	1.2421	1.2385	1.2349	1.2314	1.2278	1.2207	1.2137	1.2068	1.1928	1.1787
34	1.2661	1.2588	1.2552	1.2515	1.2479	1.2443	1.2371	1.2300	1.2229	1.2088	1.1946
36	1.2831	1.2757	1.2720	1.2684	1.2647	1.2610	1.2538	1.2466	1.2394	1.2251	1.2109
38	1.3004	1.2929	1.2891	1.2855	1.2818	1.2780	1.2707	1.2635	1.2561	1.2418	1.2276
40	1.3179	1.3103	1.3065	1.3028	1.2991	1.2953	1.2880	1.2806	1.2732	1.2589	1.2446
42	1.3357	1.3280	1.3242	1.3205	1.3167	1.3129	1.3055	1.2981	1.2907	1.2762	1.2619
44	1.3538	1.3461	1.3423	1.3384	1.3346	1.3308	1.3234	1.3160	1.3086	1.2939	1.2796
46	1.3724	1.3646	1.3608	1.3569	1.3530	1.3492	1.3417	1.3343	1.3269	1.3120	1.2976
48	1.3915	1.3835	1.3797	1.3758	1.3719	1.3680	1.3604	1.3528	1.3455	1.3305	1.3159
50	1.4110	1.4029	1.3990	1.3951	1.3911	1.3872	1.3795	1.3719	1.3644	1.3494	1.3348
52	1.4310	1.4228	1.4188	1.4148	1.4109	1.4069	1.3991	1.3914	1.3837	1.3687	1.3540
54	1.4515	1.4431	1.4391	1.4350	1.4310	1.4270	1.4191	1.4113	1.4036	1.3884	1.3735
56	1.4724	1.4640	1.4598	1.4557	1.4516	1.4475	1.4396	1.4317	1.4239	1.4085	1.3934
58	1.4937	1.4852	1.4809	1.4768	1.4726	1.4685	1.4604	1.4524	1.4446	1.4290	1.4137
60	1.5154	1.5067	1.5024	1.4983	1.4940	1.4898	1.4816	1.4735	1.4656	1.4497	1.4344
62	1.5375	1.5287	1.5243	1.5200	1.5157	1.5115	1.5031	1.4950	1.4869	1.4708	1.4554
64	1.5600	1.5510	1.5465	1.5421	1.5378	1.5335	1.5250	1.5167	1.5086	1.4923	1.4766
66	1.5828	1.5736	1.5691	1.5646	1.5602	1.5558	1.5472	1.5388	1.5305	1.5140	1.4981
68	1.6059	1.5965	1.5920	1.5874	1.5829	1.5785	1.5697	1.5611	1.5528	1.5359	1.5198
70	1.6293	1.6198	1.6151	1.6105	1.6059	1.6014	1.5925	1.5838	1.5753	1.5582	1.5417
72	1.6529	1.6433	1.6385	1.6338	1.6292	1.6246	1.6155	1.6067	1.5981	1.5806	1.5637
74	1.6768	1.6670	1.6622	1.6574	1.6526	1.6480	1.6387	1.6297	1.6209	1.6031	1.5857
76	1.7008	1.6908	1.6858	1.6810	1.6761	1.6713	1.6619	1.6526	1.6435	1.6252	1.6074
78	1.7247	1.7144	1.7093	1.7043	1.6994	1.6944	1.6847	1.6751	1.6657	1.6469	1.6286
80	1.7482	1.7376	1.7323	1.7272	1.7221	1.7170	1.7069	1.6971	1.6873	1.6680	1.6493
82	1.7709	1.7599	1.7544	1.7491	1.7437	1.7385	1.7281	1.7180	1.7080	1.6882	1.6692
84	1.7916	1.7804	1.7748	1.7693	1.7639	1.7585	1.7479	1.7375	1.7274	1.7072	1.6878
86	1.8095	1.7983	1.7927	1.7872	1.7818	1.7763	1.7657	1.7552	1.7449	1.7245	1.7050
88	1.8243	1.8132	1.8077	1.8022	1.7968	1.7914	1.7809	1.7705	1.7602	1.7397	1.7202
90	1.8361	1.8252	1.8198	1.8144	1.8091	1.8038	1.7933	1.7829	1.7729	1.7525	1.7331
91	1.8410	1.8302	1.8248	1.8195	1.8142	1.8090	1.7986	1.7883	1.7783	1.7581	1.7388
92	1.8453	1.8346	1.8293	1.8240	1.8188	1.8136	1.8033	1.7932	1.7832	1.7633	1.7439

Mass %	Density in g/mL										
	0 °C	10 °C	15 °C	20 °C	25 °C	30 °C	40 °C	50 °C	60 °C	80 °C	100 °C
93	1.8490	1.8384	1.8331	1.8279	1.8227	1.8176	1.8074	1.7974	1.7876	1.7681	1.7485
94	1.8520	1.8415	1.8363	1.8312	1.8260	1.8210	1.8109	1.8011	1.7914		
95	1.8544	1.8439	1.8388	1.8337	1.8286	1.8236	1.8137	1.8040	1.7944		
96	1.8560	1.8457	1.8406	1.8355	1.8305	1.8255	1.8157	1.8060	1.7965		
97	1.8569	1.8466	1.8414	1.8364	1.8314	1.8264	1.8166	1.8071	1.7977		
98	1.8567	1.8463	1.8411	1.8361	1.8310	1.8261	1.8163	1.8068	1.7976		
99	1.8551	1.8445	1.8393	1.8342	1.8292	1.8242	1.8145	1.8050	1.7958		
100	1.8517	1.8409	1.8357	1.8305	1.8255	1.8205	1.8107	1.8013	1.7922		

Lab Data

DENSITY OF ETHANOL–WATER MIXTURES

This table gives the density of mixtures of ethanol and water as a function of composition and temperature. The composition is specified in weight percent of ethanol, i.e., mass of ethanol per 100 g of solution. Values from the reference have been converted to true densities.

Reference

Washburn, E. W., Ed., *International Critical Tables of Numerical Data of Physics, Chemistry, and Technology*, Vol. 3, McGraw-Hill, New York, 1926–1932.

Weight %	Density in g/cm³						
Ethanol	10 °C	15 °C	20 °C	25 °C	30 °C	35 °C	40 °C
0	0.99970	0.99910	0.99820	0.99705	0.99565	0.99403	0.99222
5	0.99095	0.99029	0.98935	0.98814	0.98667	0.98498	0.98308
10	0.98390	0.98301	0.98184	0.98040	0.97872	0.97682	0.97472
15	0.97797	0.97666	0.97511	0.97331	0.97130	0.96908	0.96667
20	0.97249	0.97065	0.96861	0.96636	0.96392	0.96131	0.95853
25	0.96662	0.96421	0.96165	0.95892	0.95604	0.95303	0.94988
30	0.95974	0.95683	0.95379	0.95064	0.94738	0.94400	0.94052
35	0.95159	0.94829	0.94491	0.94143	0.93787	0.93422	0.93048
40	0.94235	0.93879	0.93515	0.93145	0.92767	0.92382	0.91989
45	0.93223	0.92849	0.92469	0.92082	0.91689	0.91288	0.90881
50	0.92159	0.91773	0.91381	0.90982	0.90577	0.90165	0.89747
55	0.91052	0.90656	0.90255	0.89847	0.89434	0.89013	0.88586
60	0.89924	0.89520	0.89110	0.88696	0.88275	0.87848	0.87414
65	0.88771	0.88361	0.87945	0.87524	0.87097	0.86664	0.86224
70	0.87599	0.87184	0.86763	0.86337	0.85905	0.85467	0.85022
75	0.86405	0.85985	0.85561	0.85131	0.84695	0.84254	0.83806
80	0.85194	0.84769	0.84341	0.83908	0.83470	0.83027	0.82576
85	0.83948	0.83522	0.83093	0.82658	0.82218	0.81772	0.81320
90	0.82652	0.82225	0.81795	0.81360	0.80920	0.80476	0.80026
95	0.81276	0.80850	0.80422	0.79989	0.79553	0.79112	0.78668
100	0.79782	0.79358	0.78932	0.78504	0.78073	0.77639	0.77201

DIELECTRIC STRENGTH OF INSULATING MATERIALS

L. I. Berger

The loss of the dielectric properties by a sample of a gaseous, liquid, or solid insulator as a result of application to the sample of an electric field* greater than a certain critical magnitude is called *dielectric breakdown*. The critical magnitude of electric field at which the breakdown of a material takes place is called the *dielectric strength* of the material (or *breakdown voltage*). The dielectric strength of a material depends on the specimen thickness (as a rule, thin films have greater dielectric strength than that of thicker samples of a material), the electrode shape**, the rate of the applied voltage increase, the shape of the voltage vs. time curve, and the medium surrounding the sample, e.g., air or other gas (or a liquid — for solid materials only).

Breakdown in Gases

The current carriers in gases are free electrons and ions generated by external radiation. The equilibrium concentration of these particles at normal pressure is about 10^3 cm^{-3}, and hence the electrical conductivity is very small, of the order of 10^{-16} – 10^{-15} S/cm. But in a strong electric field, these particles acquire kinetic energy along their free path, large enough to ionize the gas molecules. The new charged particles ionize more molecules; this avalanche-like process leads to formation between the electrodes of channels of conducting plasma (streamers), and the electrical resistance of the space between the electrodes decreases virtually to zero.

Because the dielectric strength (breakdown voltage) of gases strongly depends on the electrode geometry and surface condition and the gas pressure, it is generally accepted to present the data for a particular gas as a fraction of the dielectric strength of either nitrogen or sulfur hexafluoride measured at the same conditions. In Table 1, the data are presented in comparison with the dielectric strength of nitrogen, which is considered equal to 1.00. For convenience to the reader, a few average magnitudes of the dielectric strength of some gases are expressed in kilovolts per millimeter. The data in the table relate to the standard conditions, unless indicated otherwise.

Breakdown in Liquids

If a liquid is pure, the breakdown mechanism in it is similar to that in gases. If a liquid contains liquid impurities in the form of small drops with greater dielectric constant than that of the main liquid, the breakdown is the result of formation of ellipsoids from these drops by the electric field. In a strong enough electric field, these ellipsoids merge and form a high-conductivity channel between the electrodes. The current increases the temperature in the channel, liquid boils, and the current along the steam canal leads to breakdown. Formation of a conductive channel (bridge) between the electrodes is observed also in liquids with solid impurities. If a liquid contains gas impurities in the form of small bubbles, breakdown is the result of heating of the liquid in strong electric fields. In the locations with the highest current density, the liquid boils, the size of the gas bubbles increases, they merge and form gaseous channels between the electrodes, and the breakdown medium is again the gas plasma.

Breakdown in Solids

It is known that the current in solid insulators does not obey Ohm's law in strong electric fields. The current density increases almost exponentially with the electric field, and at a certain field magnitude it jumps to very high magnitudes at which a specimen of a material is destroyed. The two known kinds of electric breakdown are thermal and electrical breakdowns. The former is the result of material heating by the electric current. Destruction of a sample of a material happens when, at a certain voltage, the amount of heat produced by the curresnt exceeds the heat release through the sample surface; the breakdown voltage in this case is proportional to the square root of the ratio of the thermal conductivity and electrical conductivity of the material. A semi-empirical expression for dependence of the breakdown voltage, V_B, on the physical properties and geometry of a sample of a solid material for the one-dimensional case is

$$V_B = [A\rho\kappa/a\phi(d)]^{1/2}$$

where A is a numerical constant related to the system of units used, ρ and κ are the volume resistivity and thermal conductivity of the sample material, a is a constant related to the chemical bond nature and crystal structure of the sample material, and $\phi(d)$ is a function of the sample geometry, first of all, thickness, d (see, e.g., Ref. R6). In the majority of materials, $\phi(d)$ increases with d, hence, the magnitude of V_B is greater in the thinner samples of a particular material.

The electrical breakdown results from the tunneling of the charge carriers from electrodes or from the valence band or from the impurity levels into the conduction band, or by the impact ionization. The tunnel effect breakdown happens mainly in thin layers, e.g., in thin p-n junctions. Otherwise, the impact ionization mechanism dominates. For this mechanism, the dielectric strength of an insulator can be estimated using Boltzmann's kinetic equation for electrons in a crystal.

In the following tables, the dielectric strength values are for room temperature and normal atmospheric pressure, unless indicated otherwise.

* The unit of electric field in the SI system is newton per coulomb or volt per meter.

** For example, the U.S. standard ASTM D149 is based on use of symmetrical electrodes, while per U.K. standard BS2918 one electrode is a plane and the other is a rod with the axis normal to the plane.

TABLE 1. Dielectric Strength of Gases

Material	Dielectric strength*	Ref.
Nitrogen, N_2	1.00	
Hydrogen, H_2	0.50	1,2
Helium, He	0.15	1
Oxygen, O_2	0.92	2
Air	0.97	6
Air[b]	3.0[a]	3
Air	0.4-0.7[a]	4
Air	1.40[a]	5
Neon, Ne	0.25	1
	0.16	2
Argon, Ar	0.18	2
Chlorine, Cl_2	1.55	1
Carbon monoxide, CO	1.02	1
	1.05	2
Carbon dioxide, CO_2	0.88	1
	0.82	2
	0.84	6
Nitrous oxide, N_2O	1.24	2
Sulfur dioxide, SO_2	2.63	2
	2.68	6
Sulfur monochloride, S_2Cl_2[c]	1.02	1
Thionyl fluoride, SOF_2	2.50	1
Sulfur hexafluoride, SF_6	2.50	1
	2.63	2
Sulfur hexafluoride, SF_6	8.50[a]	7
	9.8	8
Perchloryl fluoride, ClO_3F	2.73	1
Tetrachloromethane, CCl_4	6.33	1
	6.21	2
Tetrafluoromethane, CF_4	1.01	1
Methane, CH_4	1.00	1
	1.13	2
Bromotrifluoromethane, CF_3Br	1.35	1
	1.97	2
Bromomethane, CH_3Br	0.71	2
Chloromethane, CH_3Cl	1.29	2
Iodomethane, CH_3I	3.02	2
Iodomethane, CH_3I[d]	2.20	7
Dichloromethane, CH_2Cl_2	1.92	2
Dichlorodifluoromethane, CCl_2F_2	2.42	1
	2.63	2,6
Chlorotrifluoromethane, $CClF_3$	1.43	1
	1.53	2
Trichlorofluoromethane, CCl_3F	3.50	1
	4.53	2
Trichloromethane, $CHCl_3$	4.2	1
	4.39	2
Methylamine, CH_3NH_2	0.81	1
Difluoromethane, CH_2F_2	0.79	2
Trifluoromethane, CHF_3	0.71	2
Bromochlorodifluoromethane, CF_2ClBr	3.84	2
Chlorodifluoromethane, $CHClF_2$	1.40	1
	1.11	2
Dichlorofluoromethane, $CHCl_2F$	1.33	1
	2.61	2
Chlorofluoromethane, CH_2ClF	1.03	1
Hexafluoroethane, C_2F_6	1.82	1
	2.55	2
Ethyne (Acetylene), C_2H_2	1.10	1
	1.11	2
Chloropentafluoroethane, C_2ClF_5	2.3	1
	3.0	6
Dichlorotetrafluoroethane, $C_2Cl_2F_4$	2.52	1
Chlorotrifluoroethylene, C_2ClF_3	1.82	2
1,1,1-Trichloro-2,2,2-trifluoroethane	6.55	2
1,1,2-Trichloro-1,2,2-trifluoroethane	6.05	2
Chloroethane, C_2H_5Cl	1.00	1
1,1-Dichloroethane	2.66	2
Trifluoroacetonitrile, CF_3CN	3.5	1
Acetonitrile, CH_3CN	2.11	2
Dimethylamine, $(CH_3)_2NH$	1.04	1
Ethylamine, $C_2H_5NH_2$	1.01	1
Ethylene oxide (oxirane), CH_3CHO	1.01	1
Perfluoropropene, C_3F_6	2.55	2
Octafluoropropane, C_3F_8	2.19	1
	2.47	2
3,3,3-Trifluoro-1-propene, CH_2CHCF_3	2.11	2
Pentafluoroisocyanoethane, C_2F_5NC	4.5	1
1,1,1,4,4,4-Hexafluoro-2-butyne, CF_3CCCF_3	5.84	2
Octafluorocyclobutane, C_4F_8	3.34	2
1,1,1,2,3,4,4,4-Octafluoro-2-butene	2.8	1
Decafluorobutane, C_4F_{10}	3.08	1
Perfluorobutanenitrile, C_3F_7CN	5.5	1
Perfluoro-2-methyl-1,3-butadiene, C_5F_8	5.5	1
Hexafluorobenzene, C_6F_6	2.11	2
Perfluorocyclohexane, C_6F_{12}, (saturated vapor)	6.18	2

* Relative to nitrogen, unless units of kV/mm are indicated.
[a] In kV/mm
[b] Flat electrodes
[c] At 12.5 torr
[d] At 370 torr

TABLE 2. Dielectric Strength of Liquids

Material	Dielectric strength kV/mm	Ref.
Helium, He, liquid, 4.2 K	10	9
Static	10	11
Dynamic	5	11
	23	12
Nitrogen, N_2, liquid, 77 K		
Coaxial cylinder electrodes	20	10
Sphere to plane electrodes	60	10
Water, H_2O, distilled	65-70	13
Carbon tetrachloride, CCl_4	5.5	14
	16.0	15
Hexane, C_6H_{14}	42.0	16
Two 2.54 cm diameter spherical electrodes, 50.8 μm space	156	17,18
Cyclohexane, C_6H_{12}	42-48	16
2-Methylpentane, C_6H_{14}	149	17,18
2,2-Dimethylbutane, C_6H_{14}	133	17,18
2,3-Dimethylbutane, C_6H_{14}	138	17,18
Benzene, C_6H_6	163	17,18
Chlorobenzene, C_6H_5Cl	7.1	14
	18.8	15
2,2,4-Trimethylpentane, C_8H_{18}	140	17,18
Phenylxylylethane	23.6	19
Heptane, C_7H_{16}	166	17,18
2,4-Dimethylpentane, C_7H_{16}	133	17,18
Toluene, $C_6H_5CH_3$	199	17,18
	46	16
	12.0	14
	20.4	15
Octane, C_8H_{18}	16.6	14
	20.4	15
	179	17,18
Ethylbenzene, C_8H_{10}	226	17,18
Propylbenzene, C_9H_{12}	250	17,18
Isopropylbenzene, C_9H_{12}	238	17,18
Decane, $C_{10}H_{22}$	192	17,18
Synthetic Paraffin Mixture Synfluid 2cSt PAO	29.5	37
Butylbenzene, $C_{10}H_{14}$	275	17,18
Isobutylbenzene, $C_{10}H_{14}$	222	17,18
Silicone oils—polydimethylsiloxanes, $(CH_3)_3Si\text{-}O\text{-}[Si(CH_3)_2]_x\text{-}O\text{-}Si(CH_3)_3$		
Polydimethylsiloxane silicone fluid	15.4	20
Dimethyl silicone	24.0	21,22
Phenylmethyl silicone	23.2	22
Silicone oil, Basilone M50	10-15	23
Mineral insulating oils	11.8	6
Polybutene oil for capacitors	13.8	6
Transformer dielectric liquid	28-30	6
Isopropylbiphenyl capacitor oil	23.6	6
Transformer oil	110.7	24
Transformer oil Agip ITE 360	9-12.6	23
Perfluorinated hydrocarbons		
Fluorinert FC 6001	8.0	23
Fluorinert FC 77	10.7	23
Perfluorinated polyethers		
Galden XAD (Mol. wt. 800)	10.5	23
Galden D40 (Mol. wt. 2000)	10.2	23
Castor oil	65	25

TABLE 3. Dielectric Strength of Solids

Material	Dielectric strength kV/mm	Ref
Sodium chloride, NaCl, crystalline	150	26
Potassium bromide, KBr, crystalline	80	26
Ceramics		
Alumina (99.9% Al_2O_3)	13.4	6,27a
Aluminum silicate, Al_2SiO_5	5.9	6
Berillia (99% BeO)	13.8	6,27b
Boron nitride, BN	37.4	6
Cordierite, $Mg_2Al_4Si_5O_{18}$	7.9	6,27c
Forsterite, Mg_2SiO_4	9.8	28
Porcelain	35-160	26
Steatite, $Mg_3Si_4O_{11}\cdot H_2O$	9.1-15.4	6
Titanates of Mg, Ca, Sr, Ba, and Pb	20-120	3
Barium titanate, glass bonded	>30	36
Zirconia, ZrO_2	11.4	29
Glasses		
Fused silica, SiO_2	470-670	26
Alkali-silicate glass	200	26
Standard window glass	9.8-13.8	28
Micas		
Muscovite, ruby, natural	118	6
Phlogopite, amber, natural	118	6
Fluorophlogopite, synthetic	118	6
Glass-bonded mica	14.0-15.7	6
Thermoplastic Polymers		
Polypropylene	23.6	6
Amide polymer nylon 6/6, dry	23.6	6
Polyamide-imide copolymer	22.8	6
Modified polyphenylene oxide	21.7	6
Polystyrene	19.7	6
Polymethyl methacrylate	19.7	6
Polyetherimide	18.9	6
Amide polymer nylon 11(dry)	16.7	6
Polysulfone	16.7	6
Styrene-acrylonitrile copolymer	16.7	6
Acrylonitrile-butadiene-styrene	16.7	6
Polyethersulfone	15.7	6
Polybutylene terephthalate	15.7	6
Polystyrene-butadiene copolymer	15.7	6
Acetal homopolymer	15.0	6
Acetal copolymer	15.0	6
Polyphenylene sulfide	15.0	6
Polycarbonate	15.0	6
Acetal homopolymer resin (molding resin)	15.0	6
Acetal copolymer resin	15.0	6
Thermosetting Molding Compounds		
Glass-filled allyl (Type GDI-30 per MIL-M-14G)	15.7	6
Glass-filled epoxy, electrical grade	15.4	6

Material	Dielectric strength kV/mm	Ref
Glass-filled phenolic (Type GPI-100 per MIL-M-14G)	15.0	6
Glass-filled alkyd/polyester (Type MAI-60 per MIL-M-14G)	14.8	6
Glass-filled melamine (Type MMI-30 per MIL-M-14G)	13.4	6
Extrusion Compounds for High-Temperature Insulation		
Polytetrafluoroethylene	19.7	6
Perfluoroalkoxy polymer	21.7	6
Fluorinated ethylene-propylene copolymer	19.7	6
Ethylene-tetrafluoroethylene copolymer	15.7	6
Polyvinylidene fluoride	10.2	6
Ethylene-chlorotrifluoroethylene copolymer	19.3	6
Polychlorotrifluoroethylene	19.7	6
Extrusion Compounds for Low-Temperature Insulation		
Polyvinyl chloride		
Flexible	11.8-15.7	30
Rigid	13.8-19.7	30
Polyethylene	18.9	28
Polyethylene, low-density	21.7	6
	300	31
Polyethylene, high-density	19.7	6
Polypropylene/polyethylene copolymer	23.6	6
Embedding Compounds		
Basic epoxy resin: bisphenol A/epichlorohydrin polycondensate	19.7	6
Cycloaliphatic epoxy: alicyclic diepoxy carboxylate	19.7	6
Polyetherketone	18.9	30
Polyurethanes		
Two-component, polyol-cured	25.4	6
Two-part solventless, polybutylene-based	24.0	6
Silicones		
Clear two-part heat curing electrical grade silicone embedding resin	21.7	6
Red insulating enamel (MIL-E-22118)		
Dry	47.2	6
Wet	11.8	6
Enamels		
Red enamel, fast cure		
Standard conditions	78.7	6
Immersion conditions	47.2	6
Black enamel		
Standard conditions	70.9	6
Immersion conditions	47.2	6
Varnishes		
Vacuum-pressure impregnated baking type solventless polyester varnish		
Rigid, two-part	70.9	6
Semiflexible high-bond thixotropic	78.7	6
Rigid high-bond high-flash freon-resistant	68.9	6
Baking type epoxy varnish		
Solventless, rigid, low viscosity, one-part	90.6	6
Solventless, semiflexible, one-part	82.7	6
Solventless, semirigid, chemical resistant, low dielectric constant	106.3	6
Solvable, for hermetic electric motors	181.1	6
Polyurethane coating		
Clear conformal, fast cure		
Standard conditions	78.7	6
Immersion conditions	47.2	6
Insulating Films and Tapes		
Low-density polyethylene film (40 μm thick)	300	31
Poly-*p*-xylylene film	410-590	32
Aromatic polymer films		
Kapton H (Du Pont)	389-430	33
Ultem (GE Plastic and Roem AG)	437-565	33
Hostaphan (Hoechst AG)	338-447	33
Amorphous Stabar K2000 (ICI film)	404-422	33
Stabar S100 (ICI film)	353-452	33
Polyetherimide film (26 μm)	486	34
Parylene N/D (poly-*p*-xylylene/polydichloro-*p*-xylylene) 25 μm film	275	6
Cellulose acetate film	157	6
Cellulose triacetate film	157	6
Polytetrafluoroethylene film	87-173	6
Perfluoroalkoxy film	157-197	6
Fluorinated ethylene-propylene copolymer film	197	6
Ethylene-tetrafluoroethylene film	197	6
Ethylene-chlorotrifluoroethylene copolymer film	197	6
Polychlorotrifluoroethylene film	118-153.5	6
High-voltage rubber insulating tape	28	6
Composites		
Isophthalic polyester (vinyl toluene monomer) filled with		
Calcium carbonate, $CaCO_3$	15.0	38
Gypsum, $CaSO_4$	14.4	38
Alumina trihydrate	15.4	38
Clay	14.4	38
BPA fumarate polyester (vinyl toluene monomer) filled with		
Calcium carbonate	6.1	38
Gypsum	5.9	38
Alumina trihydrate	11.8	38
Clay	12.6	38
Polysulfone resin—30% glass fiber	16.5-18.7	38
Polyamid resin (Nylon 66)—30% carbon fiber		
Polyimide thermoset resin, glass reinforced	12.0	39
Polyester resin (thermoplastic)—40% glass fiber	20.0	38
Epoxy resin (diglycidyl ether of bisphenol A), glass reinforced	16.0	40
Various Insulators		
Rubber, natural	100-215	26
Butyl rubber	23.6	6
Neoprene	15.7-27.6	6
Silicone rubber	26-36	6
Room-temperature vulcanized silicone rubber	9.2-10.9	35
Ureas (from carbamide to tetraphenylurea)	11.8-15.7	28
Dielectric papers		
Aramid paper, calendered	28.7	6
Aramid paper, uncalendered	12.2	6
Aramid with Mica	39.4	6

References

1. Vijh, A. K., *IEEE Trans.*, EI-12, 313, 1997.
2. Brand, K. P., *IEEE Trans.*, EI-17, 451, 1982.
3. *Encyclopedic Dictionary in Physics*, Vedensky, B. A. and Vul, B. M., Eds., Vol. 4, Soviet Encyclopedia Publishing House, Moscow, 1965.
4. Kubuki, M., Yoshimoto, R., Yoshizumi, K., Tsuru, S., and Hara, M., *IEEE Trans.*, DEI-4, 92, 1997.
5. Al-Arainy, A. A., Malik, N. H., and Qureshi, M. I., *IEEE Trans.*, DEI-1, 305, 1994.
6. Shugg, W. T., *Handbook of Electrical and Electronic Insulating Materials*, Van Nostrand Reinhold, New York, 1986.
7. Devins, J. C., *IEEE Trans.*, EI-15, 81, 1980.
8. Xu, X., Jayaram, S., and Boggs, S. A., *IEEE Trans.*, DEI-3, 836, 1996.
9. Okubo, H., Wakita, M., Chigusa, S., Nayakawa, N., and Hikita, M., *IEEE Trans.*, DEI-4, 120, 1997.
10. Hayakawa, H., Sakakibara, H., Goshima, H., Hikita, M., and Okubo, H., *IEEE Trans.*, DEI-4, 127, 1997.
11. Okubo, H., Wakita, M., Chigusa, S., Hayakawa, N., and Hikita, M., *IEEE Trans.*, DEI-4, 220, 1997.
12. Von Hippel, A. R., *Dielectric Materials and Applications*, MIT Press, Cambridge, MA, 1954.
13. Jones, H. M. and Kunhards, E. E., *IEEE Trans.*, DEI-1, 1016, 1994.
14. Nitta, Y. and Ayhara, Y., *IEEE Trans.*, EI-11, 91, 1976.
15. Gallagher, T. J., *IEEE Trans.*, EI-12, 249, 1977.
16. Wong, P. P. and Forster, E. O., in *Dielectric Materials. Measurements and Applications*, IEE Conf. Publ. 177, 1, 1979.
17. Kao, K. C. *IEEE Trans.*, EI-11, 121, 1976.
18. Sharbaugh, A. H., Crowe, R. W., and Cox, E. B., *J. Appl. Phys.*, 27, 806, 1956.
19. Miller, R. L., Mandelcorn, L., and Mercier, G. E., in *Proc. Intl. Conf. on Properties and Applications of Dielectric Materials*, Xian, China, June 24-28, 1985; cited in Ref. 6, p. 492.
20. Hakim, R. M., Oliver, R. G., and St-Onge, H., *IEEE Trans.*, EI-12, 360, 1977.
21. Hosticka, C., *IEEE Trans.*, 389, 1977.
22. Yasufuku, S., Umemura, T., and Ishioka, Y., *IEEE Trans.*, EI-12, 402, 1977.
23. Forster, E. O., Yamashita, H., Mazzetti, C., Pompini, M., Caroli, L., and Patrissi, S., *IEEE Trans.*, DEI-1, 440, 1994.
24. Bell, W. R., *IEEE Trans.*, 281, 1977.
25. Ramu, T. C. and Narayana Rao, Y., in *Dielectric Materials. Measurements and Applications*, IEE Conf. Publ. 177, 37.
26. Skanavi, G. I., *Fizika Dielektrikov; Oblast Silnykh Polei* (Physics of Dielectrics; Strong Fields). Gos. Izd. Fiz. Mat. Nauk (State Publ. House for Phys. and Math. Scis.), Moscow, 1958.
27. Kleiner, R. N., in *Practical Handbook of Materials Science*, Lynch, C. T., Ed., CRC Press, 1989; 27a: p. 304; 27b: p.300; 27c: p. 316.
28. *Materials Selector Guide. Materials and Methods*, Reinhold Publ., New York, 1973.
29. Flinn, R. A. and Trojan, P. K., *Engineering Materials and Their Applications*, 2nd ed., Houghton Mifflin, 1981, p. 614.
30. Lynch, C. T., Ed., *Practical Handbook of Materials Science*, CRC Press, Boca Raton, FL, 1989.
31. Suzuki, H., Mukai, S., Ohki, Y., Nakamichi, Y., and Ajiki, K., *IEEE Trans.*, DEI-4, 238, 1997.
32. Mori, T., Matsuoka, T., and Muzitani, T., *IEEE Trans.*, DEI-1, 71, 1994.
33. Bjellheim, P. and Helgee, B., *IEEE Trans.*, DEI-1, 89, 1994.
34. Zheng, J. P., Cygan, P. J., and Jow, T. R., *IEEE Trans.*, DEI-3, 144, 1996.
35. Danukas, M. G., *IEEE Trans.*, DEI-1, 1196, 1994.
36. Burn, I. and Smithe, D. H., *J. Mater. Sci.*, 7, 339, 1972.
37. Hope, K.D., Chevron Chemical, Private Communication.
38. *Engineering Materials Handbook*, Vol. 1, Composites, C.A. Dostal, Ed., ASM Intl., 1987.
39. 1985 Materials Selector, *Mater. Eng.*, (12) 1984.
40. *Modern Plastics Encyclopedia*, McGraw-Hill, v. 62 (No. 10A) 1985–1986.
41. *Review Literature on the Subject*
42. R1. Kuffel, E. and Zaengl, W. S., *HV Engineering Fundamentals*, Pergamon, 1989.
43. R2. Kok, J. A., *Electrical Breakdown of Insulating Liquids*, Phillips Tech. Library, Cleaver-Hum, London, 1961.
44. R3. Gallagher, T. J., *Simple Dielectric Liquids*, Clarendon, Oxford, 1975.
45. R4. Meek, J. M. and Craggs, J. D., Eds., *Electric Breakdown in Gases*, John Wiley & Sons, 1976.
46. R5. Von Hippel, A. R., *Dielectric Materials and Applications*, MIT Press, Cambridge, MA, 1954.
47. R6. O'Dwyer, J. J. *The Theory of Dielectric Breakdown of Solids*, Clarendon Press, 1964.

COEFFICIENT OF FRICTION

The coefficient of friction between two surfaces is the ratio of the force required to move one over the other to the force pressing the two together. Thus if F is the minimum force needed to move one surface over the other, and W is the force pressing the surfaces together, the coefficient of friction μ is given by $\mu = F/W$. A greater force is generally needed to initiate movement from rest than to continue the motion once sliding has started. Thus the static coefficient of friction μ(static) is usually larger than the sliding or kinetic coefficient μ(sliding).

This table gives characteristic values of both the static and sliding coefficients of friction for a number of material combinations. In each case Material 1 is moving over the surface of Material 2. The type of lubrication or any other special condition is indicated in the third column. All values refer to room temperature unless otherwise indicated. It should be emphasized that the coefficient of friction is very sensitive to the condition of the surface, so that these values represent only a rough guide.

References

1. Minshall, H., in *CRC Handbook of Chemistry and Physics, 73rd Edition*, Lide, D. R., Ed., CRC Press, Boca Raton, FL, 1992.
2. Fuller, D. D., in *American Institute of Physics Handbook, 3rd Edition*, Gray, D. E., Ed., McGraw-Hill, New York, 1972.

Material 1	Material 2	Conditions	μ(static)	μ(sliding)
Metals				
Hard steel	Hard steel	Dry	0.78	0.42
Hard steel	Hard steel	Castor oil	0.15	0.081
Hard steel	Hard steel	Steric acid	0.005	0.029
Hard steel	Hard steel	Lard	0.11	0.084
Hard steel	Hard steel	Light mineral oil	0.23	
Hard steel	Hard steel	Graphite		0.058
Hard steel	Graphite	Dry	0.21	
Mild steel	Mild steel	Dry	0.74	0.57
Mild steel	Mild steel	Oleic acid		0.09
Mild steel	Phosphor bronze	Dry		0.34
Mild steel	Cast iron	Dry		0.23
Mild steel	Lead	Dry	0.95	0.95
Mild steel	Lead	Mineral oil	0.5	0.3
Mild steel	Brass	Dry	0.35	
Cast iron	Cast iron	Dry	1.10	0.15
Aluminum	Aluminum	Dry	1.05	1.4
Aluminum	Mild steel	Dry	0.61	0.47
Brass	Mild steel	Dry	0.51	0.44
Brass	Mild steel	Castor oil	0.11	
Brass	Cast iron	Dry		0.30
Bronze	Cast iron	Dry		0.22
Cadmium	Mild steel	Dry		0.46
Copper	Copper	Dry	1.6	
Copper	Mild steel	Dry	0.53	0.36
Copper	Mild steel	Oleic acid		0.18
Copper	Cast iron	Dry	1.05	0.29
Copper	Glass	Dry	0.68	0.53
Lead	Cast iron	Dry		0.43
Magnesium	Magnesium	Dry	0.6	
Magnesium	Mild steel	Dry		0.42
Magnesium	Cast iron	Dry		0.25
Nickel	Nickel	Dry	1.10	0.53
Nickel	Mild steel	Dry		0.64
Tin	Cast iron	Dry		0.32
Zinc	Cast iron	Dry	0.85	0.21
Nonmetals				
Diamond	Diamond	Dry	0.1	
Diamond	Metals	Dry	0.12	
Garnet	Mild steel	Dry		0.39
Glass	Glass	Dry	0.94	0.4
Glass	Nickel	Dry	0.78	0.56
Graphite	Graphite	Dry	0.1	
Mica	Mica	Freshly cleaved	1.0	
Nylon	Nylon	Dry	0.2	
Nylon	Steel	Dry	0.40	
Polyethylene	Polyethylene	Dry	0.2	
Polyethylene	Steel	Dry	0.2	
Polystyrene	Polystyrene	Dry	0.5	
Polystyrene	Steel	Dry	0.3	
Sapphire	Sapphire	Dry	0.2	
Teflon	Teflon	Dry	0.04	0.04
Teflon	Steel	Dry	0.04	0.04
Tungsten carbide	Tungsten carbide	Dry, room temp.	0.17	
Tungsten carbide	Tungsten carbide	Dry, 1000 °C	0.45	
Tungsten carbide	Tungsten carbide	Dry, 1600 °C	1.8	
Tungsten carbide	Tungsten carbide	Oleic acid	0.12	
Tungsten carbide	Graphite	Dry	0.15	
Tungsten carbide	Steel	Dry	0.5	
Tungsten carbide	Steel	Oleic acid	0.08	
Miscellaneous Materials				
Cotton	Cotton	Threads	0.3	
Leather	Cast iron	Dry	0.6	0.56
Leather	Oak	Parallel	0.61	0.52
Oak	Oak	Parallel	0.62	0.48
Oak	Oak	Perpendicular	0.54	0.32
Silk	Silk	Clean	0.25	
Wood	Wood	Dry	0.35	
Wood	Wood	Wet	0.2	
Wood	Brick	Dry	0.6	
Wood	Leather	Dry	0.35	
Various Materials on Ice and Snow				
Ice	Ice	Clean, 0 °C	0.1	0.02
Ice	Ice	Clean, -12 °C	0.3	0.035
Ice	Ice	Clean, -80 °C	0.5	0.09
Aluminum	Snow	Wet, 0 °C	0.4	
Aluminum	Snow	Dry, 0 °C	0.35	
Brass	Ice	Clean, 0 °C		0.02
Brass	Ice	Clean, -80 °C		0.15
Nylon	Snow	Wet, 0 °C	0.4	
Nylon	Snow	Dry, -10 °C	0.3	
Teflon	Snow	Wet, 0 °C	0.05	
Teflon	Snow	Dry, 0 °C	0.02	
Wax, ski	Snow	Wet, 0 °C	0.1	
Wax, ski	Snow	Dry, 0 °C	0.04	
Wax, ski	Snow	Dry, -10 °C	0.2	

FLAME TEMPERATURES

This table gives the adiabatic flame temperatures for stoichemetric mixtures of various fuels and oxidizers. The temperatures are calculated from thermodynamic and transport properties under ideal adiabatic conditions, using methods described in the reference.

Reference

Fristrom, R. M., *Flame Structures and Processes*, Oxford University Press, New York, 1995.

Adiabatic Flame Temperature in K for Various Fuel-Oxidizer Combinations

	Oxidizer							
Fuel	Air	O_2	F_2	Cl_2	N_2O	NO	ClO_2	O_3
Organic liquids and gases								
Acetaldehyde	2288							
Acetone	2253							
Acetylene	2607							
Benzene	2363							
Butane	2248							
Carbon disulfide	2257							
Cyanogen	2596	4855						5207
Cyclohexane	2250							
Cyclopropane	2370							
Decane	2286							
Ethane	2244							
Ethanol	2238							
Ethylene	2375							
Hexane	2238							
Methane	2236							
Methanol	2222							
Oxirane	2177							
Pentane	2250							
Propane	2250							
Toluene	2344							
Solids								
Aluminum		4005						
Lithium		2711						
Phosphorus (white)		3242						
Zirconium		4278						
Other								
Ammonia		2845						
Carbon monoxide	1388							
Diborane		3350						
Hydrazine		3037						
Hydrogen	2169	3000	4006	2493	2965	3127	3098	
Hydrogen sulfide	2091	3414						
Phosphine		3139						
Silane		3043						

ALLOCATION OF FREQUENCIES IN THE RADIO SPECTRUM

In the United States the National Telecommunications and Information Administration (NTIA) has responsibility for assigning each portion of the radio spectrum (9 kHz to 300 GHz) for different uses. These assignments must be compatible with the rules of the International Telecommunications Union (ITU), to which the United States is bound by treaty. The current assignments are given in a wall chart (Reference 1) and may also be found on the NTIA web site (Reference 2). The list below summarizes the broad features of the spectrum allocation, with particular attention to those sections of scientific interest. The references should be consulted for details of the allocations in the frequency bands listed here, which in some cases are quite complex.

References

1. *United States Frequency Allocations*, 1996 Spectrum Wall Chart, Stock No. 003-000-00652-2, U. S. Government Printing Office, P. O. Box 371954, Pittsburgh, PA 15250-7954.
2. http://www.ntia.doc.gov/files/ntia/publications/2003-allochrt.pdf.

Frequency range	Allocation
9–19.95 kHz	Maritime communication, navigation
19.95–20.05 kHz	Standard frequency and time signal (also at 60 kHz and 2.5, 5, 10, 15, 20, 25 MHz)
20.05–535 kHz	Maritime and aeronautical communication, navigation
535–1605 kHz	AM radio broadcasting
1605–3500 kHz	Mobile communication and navigation, amateur radio (1800–1900 kHz)
3.5–4.0 MHz	Amateur radio
4.0–5.95 MHz	Mobile communication
5.95–13.36 MHz	Mobile communication, amateur, short-wave broadcasting
13.36–13.41 MHz	Radioastronomy
13.41–25.55 MHz	Mobile communication, amateur, short-wave broadcasting
25.55–25.67 MHz	Radioastronomy
25.67–37.5 MHz	Mobile communication, amateur, short-wave broadcasting
37.5–38.25 MHz	Radioastronomy
38.25–50.0 MHz	Mobile communication
50.0–54.0 MHz	Amateur
54.0–72.0 MHz	TV channels 2–4
72.0–73.0 MHz	Mobile communication
73.0–74.6 MHz	Radioastronomy
74.6–76.0 MHz	Mobile communication
76.0–88.0 MHz	TV channels 5–6
88.0–108.0 MHz	FM radio broadcasting
108.0–118.0 MHz	Aeronautical navigation
118.0–174.0 MHz	Mobile communication, space research, meteorological satellites
174.0–216.0 MHz	TV channels 7–13
216.0–400.05 MHz	Mobile communication
400.05–400.15 MHz	Standard frequency and time satellite (also 20 and 25 GHz)
400.15–406.1 MHz	Meteorological aids (radiosonde)
406.1–410.0 MHz	Radioastronomy
410.0–470.0 MHz	Mobile communication, amateur
470.0–512.0 MHz	TV channels 14–20
512.0–608.0 MHz	TV channels 21–36
608.0–614.0 MHz	Radioastronomy
614.0–806.0 MHz	TV channels 38–69
806–1400 MHz	Mobile communication, navigation
1400–1427 MHz	Radioastronomy, space research
1427–1660 MHz	Various navigation and satellite applications
1660–1710 MHz	Radioastronomy, space research, meteorology
1710–2655 MHz	Various navigation and satellite applications
2655–2700 MHz	Radioastronomy, space research
2.7–4.99 GHz	Various navigation and satellite applications
4.99–5.0 GHz	Radioastronomy, space research
5.0–10.6 GHz	Various navigation and satellite applications
10.6–10.7 GHz	Radioastronomy, space research
10.7–15.35 GHz	Various navigation and satellite applications
15.35–15.4 GHz	Radioastronomy, space research
15.4–22.21 GHz	Various navigation and satellite applications

Frequency range	Allocation
22.21–22.5 GHz	Radioastronomy, space research
22.25–23.6 GHz	Various navigation and satellite applications
23.6–24.0 GHz	Radioastronomy, space research
24.0–31.3 GHz	Various navigation and satellite applications
31.3–31.8 GHz	Radioastronomy, space research
31.8–42.5 GHz	Various navigation and satellite applications
42.5–43.5 GHz	Radioastronomy
43.5–51.4 GHz	Various navigation and satellite applications
51.4–54.25 GHz	Radioastronomy, space research
54.25–58.2 GHz	Space research
58.2–59.0 GHz	Radioastronomy, space research
59.0–64.0 GHz	Satellite applications
64.0–65.0 GHz	Radioastronomy, space research
65.0–72.77 GHz	Various navigation and satellite applications
72.77–72.91 GHz	Radioastronomy, space research
72.91–86.0 GHz	Various navigation and satellite applications
86.0–92.0 GHz	Radioastronomy, space research
92.0–105.0 GHz	Various navigation and satellite applications
105.0–116.0 GHz	Radioastronomy, space research
116.0–164.0 GHz	Various navigation and satellite applications
164.0–168.0 GHz	Radioastronomy, space research
168.0–182.0 GHz	Various navigation and satellite applications
182.0–185.0 GHz	Radioastronomy, space research
185.0–217.0 GHz	Various navigation and satellite applications
217.0–231.0 GHz	Radioastronomy, space research
231.0–265.0 GHz	Various navigation and satellite applications
265.0–275.0 GHz	Radioastronomy
275.0–300.0 GHz	Mobile communications

Section 16
Health and Safety Information

ABBREVIATIONS USED IN THE ASSESSMENT AND PRESENTATION OF LABORATORY HAZARDS

Thomas J. Bruno and Paris D. N. Svoronos

The following abbreviations are commonly encountered in presentations of laboratory and industrial hazards. The reader is urged to consult the references (Refs. 1 and 2) for additional information.

References

1. Furr, A.K., ed., *CRC Handbook of Laboratory Safety*, 5th ed., CRC Press, Boca Raton, FL, 2000.
2. Bruno, T. J., and Svoronos, P. D. N., *CRC Handbook of Basic Tables for Chemical Analysis*, 3rd. Ed., CRC Press, Boca Raton, FL, 2011.

CC: Closed Cup; method for the measurement of the flash point. With this method, sample vapors are not allowed to escape as they can with the open cup method. Because of this, flash points measured with the CC method are usually a few degrees lower than those measured with the OC. The choice between CC and OC is dependent on the (usually ASTM) standard method chosen for the test.

COC: Cleveland Open Cup, see Open Cup

IDLH: Immediately Dangerous to Life and Health; the maximum concentration of chemical contaminants, normally expressed as parts per million (ppm, mass/mass), from which one could escape within 30 minutes without a respirator, and without experiencing any escape impairing (severe eye irritation) or irreversible health effects. Set by NIOSH. Note that this term is also used to describe electrical hazards.

LEL: Lower Explosion Limit; the minimum concentration of a chemical in air at which detonation can occur.

LFL: Lower Flammability Limit; the minimum concentration of a chemical in air at which flame propagation occurs.

MSDS: Material Safety Data Sheet; a (legal) document that must accompany any supplied chemical that provides information on chemical content, physical properties, hazards, and treatment of hazards. The MSDS should be considered only a minimal source of information, and cannot replace additional information available in other, more comprehensive sources.

NOEL: No Observed Effect Level; the maximum dose of a chemical at which no signs of harm are observed. This term can also be used to describe electrical hazards.

OC: Open Cup; also called Cleveland Open Cup. This refers to the test method for determining the flash point of common compounds. It consists of a brass, aluminum or stainless steel cup, a heater base to heat the cup, a thermometer in a fixture and a test flame applicator. The flash point is the lowest temperature at which a material will form a flammable mixture with air above its surface. The lower the flash point, the easier it is to ignite.

PEL: Permissible Exposure Level; an exposure limit that is published and enforced by OSHA as a legal standard. The PEL may be expressed as a time-weighted-average (TWA) exposure limit (for an 8 hour workday), a 15 minute short-term exposure limit (STEL), or a ceiling (C, or CEIL, or TLV-C).

RTECS: Registry of Toxic Effects of Chemical Substances; a database maintained by the National Institute of Occupational Safety and Health (NIOSH). The goal of the database is to include data on all known toxic substances, along with the concentration at which toxicity is known to occur. There are approximately 140 000 compounds listed.

STEL: Short-Term Exposure Level; an exposure limit for a short-term, 15 minute, exposure that cannot be exceeded during the workday, enforced by OSHA as a legal standard. Short-term exposures below the STEL level generally will not cause irritation, chronic or reversible tissue damage, or narcosis.

REL: Recommended Exposure Level; average concentration limit recommended for up to a 10 hour workday during a 40 hour workweek, by NIOSH.

TLV: Threshold Limit Value; guidelines suggested by the American Conference of Governmental Industrial Hygienists to assist industrial hygienists with limiting hazards of chemical exposures in the workplace.

TLV-C: Threshold Limit Ceiling Value; an exposure limit which should not be exceeded under any circumstances.

TWA: Time-weighted average concentration for a conventional 8 hour workday and a 40 hour workweek. It is the concentration to which it is believed possible that nearly all workers can be exposed without adverse health effects.

UEL: Upper Explosion Limit; the maximum concentration of a chemical in air at which detonation can occur.

UFL: Upper Flammability Limit; the maximum concentration of a chemical in air at which flame propagation can occur.

WEEL: Workplace Environmental Exposure Limit; set by the American Industrial Hygiene Association (AIHA).

Some abbreviations that are sometimes used on material safety data sheets, and in other sources, are ambiguous. The most common meanings of some of these vague abbreviations are provided below, but the reader is cautioned that these are only suggestions:

CC: Closed Cup; method for the measurement of the flash point. With this method, sample vapors are not allowed to escape as they can with the open cup method. Because of this, flash points measured with the CC method are usually a few degrees lower than those measured with the OC. The choice between CC and OC is dependent on the (usually ASTM) standard method chosen for the test.

COC: Cleveland Open Cup, see Open Cup

IDLH: Immediately Dangerous to Life and Health; the maximum concentration of chemical contaminants, normally expressed as parts per million (ppm, mass/mass), from which one could escape within 30 minutes without a respirator, and without experiencing any escape impairing (severe eye irritation) or irreversible health effects. Set by NIOSH. Note that this term is also used to describe electrical hazards.

LEL: Lower Explosion Limit; the minimum concentration of a chemical in air at which detonation can occur.

LFL: Lower Flammability Limit; the minimum concentration of a chemical in air at which flame propagation occurs.

MSDS: Material Safety Data Sheet; a (legal) document that must accompany any supplied chemical that provides information on chemical content, physical properties, hazards, and treatment of hazards. The MSDS should be considered only a minimal source of information, and cannot replace additional information available in other, more comprehensive sources.

NOEL: No Observed Effect Level; the maximum dose of a chemical at which no signs of harm are observed. This term can also be used to describe electrical hazards.

OC: Open Cup; also called Cleveland Open Cup. This refers to the test method for determining the flash point of common compounds. It consists of a brass, aluminum or stainless steel cup, a heater base to heat the cup, a thermometer in a fixture and a test flame applicator. The flash point is the lowest temperature at which a material will form a flammable mixture with air above its surface. The lower the flash point, the easier it is to ignite.

PEL: Permissible Exposure Level; an exposure limit that is published and enforced by OSHA as a legal standard. The PEL may be expressed as a time-weighted-average (TWA) exposure limit (for an 8 hour workday), a 15 minute short-term exposure limit (STEL), or a ceiling (C, or CEIL, or TLV-C).

RTECS: Registry of Toxic Effects of Chemical Substances; a database maintained by the National Institute of Occupational Safety and Health (NIOSH). The goal of the database is to include data on all known toxic substances, along with the concentration at which toxicity is known to occur. There are approximately 140 000 compounds listed.

STEL: Short-Term Exposure Level; an exposure limit for a short-term, 15 minute, exposure that cannot be exceeded during the workday, enforced by OSHA as a legal standard. Short-term exposures below the STEL level generally will not cause irritation, chronic or reversible tissue damage, or narcosis.

REL: Recommended Exposure Level; average concentration limit recommended for up to a 10 hour workday during a 40 hour workweek, by NIOSH.

TLV: Threshold Limit Value; guidelines suggested by the American Conference of Governmental Industrial Hygienists to assist industrial hygienists with limiting hazards of chemical exposures in the workplace.

TLV-C: Threshold Limit Ceiling Value; an exposure limit which should not be exceeded under any circumstances.

TWA: Time-weighted average concentration for a conventional 8 hour workday and a 40 hour workweek. It is the concentration to which it is believed possible that nearly all workers can be exposed without adverse health effects.

UEL: Upper Explosion Limit; the maximum concentration of a chemical in air at which detonation can occur.

UFL: Upper Flammability Limit; the maximum concentration of a chemical in air at which flame propagation can occur.

WEEL: Workplace Environmental Exposure Limit; set by the American Industrial Hygiene Association (AIHA).

EST: Established; estimated

MST: Mist

N/A, NA: Not applicable

ND: None determined; not determined

NE: None established; not established

NEGL: Negligible

NF: None found; not found

N/K, NK: Not known

N/P, NP: Not provided

SKN: Skin

TS: Trade secret

UKN: Unknown

INCOMPATIBLE CHEMICALS

The term "incompatible chemicals" refers to chemicals that can react with each other

- Violently
- With evolution of substantial heat
- To produce flammable products
- To produce toxic products

Good laboratory safety practice requires that incompatible chemicals be stored, transported, and disposed of in ways that will prevent their coming together in the event of an accident. Tables 1 and 2 give some basic guidelines for the safe handling of acids, bases, reactive metals, and other chemicals. Neither of these tables is exhaustive, and additional information on incompatible chemicals can be found in the following references.

1. Urben, P. G., Ed., *Bretherick's Handbook of Reactive Chemical Hazards*, 5th Ed., Butterworth-Heinemann, Oxford, 1995.
2. Luxon, S. G., Ed., *Hazards in the Chemical Laboratory*, 5th Ed., Royal Society of Chemistry, Cambridge, 1992.
3. *Fire Protection Guide to Hazardous Materials*, 11th Ed., National Fire Protection Association, Quincy, MA, 1994.

TABLE 1. General Classes of Incompatible Chemicals

A	B
Acids	Bases, reactive metals
Oxidizing agents[a]	Reducing agents[a]
Chlorates	Ammonia, anhydrous and aqueous
Chromates	Carbon
Chromium trioxide	Metals
Dichromates	Metal hydrides
Halogens	Nitrites
Halogenating agents	Organic compounds
Hydrogen peroxide	Phosphorus
Nitric acid	Silicon
Nitrates	Sulfur
Perchlorates	
Peroxides	
Permanganates	
Persulfates	

[a] The examples of oxidizing and reducing agents are illustrative of common laboratory chemicals; they are not intended to be exhaustive.

TABLE 2. Examples of Incompatible Chemicals

Chemical	Is incompatible with
Acetic acid	Chromic acid, nitric acid, hydroxyl compounds, ethylene glycol, perchloric acid, peroxides, permanaganates
Acetylene	Chlorine, bromine, copper, fluorine, silver, mercury
Acetone	Concentrated nitric and sulfuric acid mixtures
Alkali and alkaline earth metals (such as powdered aluminum or magnesium, calcium, lithium, sodium, potassium)	Water, carbon tetrachloride or other chlorinated hydrocarbons, carbon dioxide, halogens
Ammonia (anhydrous)	Mercury (in manometers, for example), chlorine, calcium hypochlorite, iodine, bromine, hydrofluoric acid (anhydrous)
Ammonium nitrate	Acids, powdered metals, flammable liquids, chlorates, nitrites, sulfur, finely divided organic or combustible materials
Aniline	Nitric acid, hydrogen peroxide
Arsenical materials	Any reducing agent
Azides	Acids
Bromine	See Chlorine
Calcium oxide	Water
Carbon (activated)	Calcium hypochlorite, all oxidizing agents
Carbon tetrachloride	Sodium
Chlorates	Ammonium salts, acids, powdered metals, sulfur, finely divided organic or combustible materials
Chromic acid and chromium troixide	Acetic acid, naphthalene, camphor, glycerol, alcohol, flammable liquids in general
Chlorine	Ammonia, acetylene, butadiene, butane, methane, propane (or other petroleum gases), hydrogen, sodium carbide, benzene, finely divided metals, turpentine
Chlorine dioxide	Ammonia, methane, phosphine, hydrogen sulfide
Copper	Acetylene, hydrogen peroxide

Safety

Chemical	Is incompatible with
Cumene hydroperoxide	Acids (organic or inorganic)
Cyanides	Acids
Flammable liquids	Ammonium nitrate, chromic acid, hydrogen peroxide, nitric acid, sodium peroxide, halogens
Fluorine	Everything
Hydrocarbons (such as butane, propane, benzene)	Fluorine, chlorine, bromine, chromic acid, sodium peroxide
Hydrocyanic acid	Nitric acid, alkali
Hydrofluoric acid (anhydrous)	Ammonia (aqueous or anhydrous)
Hydrogen peroxide	Copper, chromium, iron, most metals or their salts, alcohols, acetone, organic materials, aniline, nitro-methane, combustible materials
Hydrogen sulfide	Fuming nitric acid, oxidizing gases
Hypochlorites	Acids, activated carbon
Iodine	Acetylene, ammonia (aqueous or anhydrous), hydrogen
Mercury	Acetylene, fulminic acid, ammonia
Nitrates	Sulfuric acid
Nitric acid (concentrated)	Acetic acid, aniline, chromic acid, hydrocyanic acid, hydrogen sulfide, flammable liquids, flammable gases, copper, brass, any heavy metals
Nitrites	Acids
Nitroparaffins	Inorganic bases, amines
Oxalic acid	Silver, mercury
Oxygen	Oils, grease, hydrogen, flammable liquids, solids, or gases
Perchloric acid	Acetic anhydride, bismuth and its alloys, alcohol, paper, wood, grease, oils
Peroxides, organic	Acids (organic or mineral), avoid friction, store cold
Phosphorus (white)	Air, oxygen, alkalis, reducing agents
Potassium	Carbon tetrachloride, carbon dioxide, water
Potassium chlorate	Sulfuric and other acids
Potassium perchlorate (see also chlorates)	Sulfuric and other acids
Potassium permanganate	Glycerol, ethylene glycol, benzaldehyde, surfuric acid
Selenides	Reducing agents
Silver	Acetylene, oxalic acid, tartartic acid, ammonium compounds, fulminic acid
Sodium	Carbon tetrachloride, carbon dioxide, water
Sodium nitrite	Ammonium nitrate and other ammonium salts
Sodium peroxide	Ethyl or methyl alcohol, glacial acetic acid, acetic anhydride, benzaldehyde, carbon disulfide, glycerin, ethylene glycol, ethyl acetate, methyl acetate, furfural
Sulfides	Acids
Sulfuric acid	Potassium chlorate, potassium perchlorate, potassium permanganate (similar compounds of light metals, such as sodium, lithium)
Tellurides	Reducing agents

EXPLOSION (SHOCK) HAZARDS

Table 1 lists some common classes of laboratory chemicals that have potential for producing a violent explosion when subjected to shock or friction. These chemicals should never be disposed of as such, but should be handled by procedures given in *Prudent Practices for Disposal of Chemicals from Laboratories*, National Academy Press, 1983, chapters 6 and 7. Additional information on these, as well as on some less common classes of explosives, can be found in L. Bretherick, *Handbook of Reactive Chemical Hazards*, 5th ed., Butterworth-Heinemann, Oxford, 1995.

Table 2 lists some illustrative combinations of common laboratory reagents that can produce explosions when they are brought together or that form reaction products that can explode without any apparent external initiating action. This list is not exhaustive, and additional information on potentially explosive reagent combinations can be found in *Manual of Hazardous Chemical Reactions, A Compilation of Chemical Reactions Reported to be Potentially Hazardous*, National Fire Protection Association, NFPA 491M, 1991, NFPA, Quincy, MA.

TABLE 1. Shock-Sensitive Compounds

Acetylenic compounds, especially polyacetylenes, haloacetylenes, and heavy metal salts of acetylenes (copper, silver, and mercury salts are particularly sensitive)
Acyl nitrates
Alkyl nitrates, particularly polyol nitrates such as nitrocellulose and nitroglycerine
Alkyl and acyl nitrites
Alkyl perchlorates
Amminemetal oxosalts: metal compounds with coordinated ammonia, hydrazine, or similar nitrogenous donors and ionic perchlorate, nitrate, permanganate, or other oxidizing group
Azides, including metal, nonmetal, and organic azides
Chlorite salts of metals, such as $AgClO_2$ and $Hg(ClO_2)_2$
Diazo compounds such as CH_2N_2
Diazonium slats, when dry
Fulminates (silver fulminate, AgCNO, can form in the reaction mixture from the Tollens' test for aldehydes if it is allowed to stand for some time; this can be prevented by adding dilute nitric acid to the test mixture as soon as the test has been completed)
Hydrogen peroxide becomes increasingly treacherous as the concentration rises above 30%, forming explosive mixtures with organic materials and decomposing violently in the presence of traces of transition metals
N–Halogen compounds such as difluoroamino compounds and halogen azides
N–Nitro compounds such as *N*–nitromethylamine, nitrourea, nitroguanidine, and nitric amide
Oxo salts of nitrogenous bases: perchlorates, dichromates, nitrates, iodates, chlorites, chlorates, and permanganates of ammonia, amines, hydroxylamine, guanidine, etc.
Perchlorate salts. Most metal, nonmetal, and amine perchlorates can be detonated and may undergo violent reaction in contact with combustible materials
Peroxides and hydroperoxides, organic (see Chapter 6, Section II.P)
Peroxides (solid) that crystallize from or are left from evaporation of peroxidizable solvents (see Chapter 6 and Appendix I)
Peroxides, transition–metal salts
Picrates, especially salts of transition and heavy metals, such as Ni, Pb, Hg, Cu, and Zn; picric acid is explosive but is less sensitive to shock or friction than its metal salts and is relatively safe as a water–wet paste (see Chapter 7)
Polynitroalkyl compounds such as tetranitromethane and dinitroacetonitrile
Polynitroaromatic compounds, especially polynitro hydrocarbons, phenols, and amines

TABLE 2. Potentially Explosive Combinations of Some Common Reagents

Acetone + chloroform in the presence of base
Acetylene + copper, silver, mercury, or their salts
Ammonia (including aqueous solutions) + Cl_2, Br_2, or I_2
Carbon disulfide + sodium azide
Chlorine + an alcohol
Chloroform or carbon tetrachloride + powdered Al or Mg
Decolorizing carbon + an oxidizing agent
Diethyl ether + chlorine (including a chlorine atmosphere)
Dimethyl sulfoxide + an acyl halide, $SOCl_2$ or $POCl_3$
Dimethyl sulfoxide + CrO_3
Ethanol + calcium hypochlorite
Ethanol + silver nitrate
Nitric acid + acetic anhydride or acetic acid
Picric acid + a heavy–metal salt, such as of Pb, Hg, or Ag
Silver oxide + ammonia + ethanol
Sodium + a chlorinated hydrocarbon
Sodium hypochlorite + an amine

WATER-REACTIVE CHEMICALS

This table lists some common laboratory chemicals that react violently with water and that should always be stored and handled so that they do not come into contact with liquid water or water vapor. Procedures for decomposing laboratory quantities are given in *Prudent Practices for Disposal of Chemicals from Laboratories*, chapter 6; the pertinent section of that chapter is given in parentheses.

Water-Reactive Chemicals

Alkali metals (III.D)
Alkali metal hydrides (III.C.2)
Alkali metal amides (III.C.7)
Metal alkyls, such as lithium alkyls and aluminum alkyls (IV.A)
Grignard reagents (IV.A)
Halides of nonmetals, such as BCl_3, BF_3, PCl_3, PCl_5, $SiCl_4$, S_2Cl_2 (III.F)
Inorganic acid halides, such as $POCl_3$, $SOCl_2$, SO_2Cl_2 (III.F)
Anhydrous metal halides, such as $AlCl_3$, $TiCl_4$, $ZrCl_4$, $SnCl_4$ (III.E)
Phosphorus pentoxide (III.I)
Calcium carbide (IV.E)
Organic acid halides and anhydrides of low molecular weight (II.J)

TESTING REQUIREMENTS FOR PEROXIDIZABLE COMPOUNDS

Thomas J. Bruno and Paris D. N. Svoronos

Because some compounds form peroxides more easily or faster than others, prudent practices require testing the supply on hand in the laboratory on a periodic basis. The following list provides guidelines on test scheduling (Refs. 1 and 2). The peroxide hazard of the compounds listed in Group 1 is on the basis of time in storage. The compounds in Group 2 present a peroxide hazard primarily due to concentration, mainly by evaporation of the liquid. The compounds listed in Group 3 are hazardous because of the potential of peroxide-initiated polymerization. When stored as liquids, the peroxide formation may increase, and therefore these compounds should be treated as Group 1 peroxidizable compounds.

Group 1 Test every 3 months

Divinylacetylene
Isopropyl ether
Potassium
Sodium amide
Vinylidene chloride

Group 2 Test every 6 months

Acetal
Cumene
Cyclohexene
Diacetylene
Dicyclopentadiene
Diethyl ether
Dimethyl ether
1,4-Dioxane
Ethylene glycol dimethyl ether (glyme)
Methylacetylene
Methyl isobutyl ketone
Methylcyclopentane
Tetrahydrofuran
Tetrahydronaphthalene (tetralin)
Vinyl ethers

Group 3 Test every 12 months

Acrylic acid
Acrylonitrile
Butadiene
Chloroprene
Chlorotrifluoroethene
Methyl methacrylate
Styrene
Tetrafluoroethylene
Vinyl acetate
Vinyl acetylene
Vinyl chloride
Vinyl pyridine

References

1. Ringen, S., Environmental Health and Safety Manual - Chemical Safety, Sec. 4-50, University of Wyoming, June, 2000.
2. Bruno, T. J., and Svoronos, P. D. N., *CRC Handbook of Basic Tables for Chemical Analysis*, 3rd. Ed., CRC Press, Boca Raton, FL, 2011.

TESTS FOR THE PRESENCE OF PEROXIDES

Thomas J. Bruno and Paris D. N. Svoronos

Peroxides may be detected qualitatively with one of the following test procedures (Refs. 1 and 2):

References

1. Gordon, A. J., and Ford, R. A., *The Chemist's Companion*, John Wiley and Sons, New York, 1972.
2. Bruno, T. J., and Svoronos, P. D. N., *CRC Handbook of Basic Tables for Chemical Analysis*, 3rd. Ed., CRC Press, Boca Raton, FL, 2011.

Ferrithiocyanate Test:

Reagent Preparation, in sequence:

Add 9 g $FeSO_4{\bullet}7H_2O$ to 50 mL 18 percent (vol/vol) $HCl_{(aq)}$
add 1 mg to 3 mg granular Zn
add 5 g NaSCN

After the red color fades, add an additional 12 g NaSCN, decant leaving unreacted Zn.
Upon mixing this reagent with a peroxide-containing liquid, the colorless solution will produce a red color, the result of the conversion of ferrothiocyanide to ferrithiocyanide. This test is very sensitive, and can be used to detect peroxides at a concentration of 0.001 percent (mass/mass).

Potassium Iodide Test:

Reagent Preparation:
Make a 10 percent (mass/mass) solution of KI in water.
Upon mixing this reagent with a peroxide-containing liquid, a yellow color will appear within one minute.

Acidic Iodide Test:

Reagent Preparation:
To 1 mL of glacial acetic acid, add 100 mg KI or NaI.
Upon mixing this reagent with an equal volume of a peroxide-containing liquid, a yellow coloration will appear. The color will appear dark or even brown if the peroxide concentration is very high.

Perchromate Test:

Reagent Preparation:
Dissolve 1 mg of $Na_2Cr_2O_7$ in 1 mL of water, add a drop of dilute H_2SO_4 (aq).
Upon mixing this reagent with a peroxide-containing liquid, a blue color will develop in the organic layer indicating the formation of the perchromate ion.

PYROPHORIC COMPOUNDS – COMPOUNDS THAT ARE REACTIVE WITH AIR

Thomas J. Bruno and Paris D. N. Svoronos

The following listing provides the classes of compounds, with some examples, that can undergo spontaneous reaction upon exposure to air (Refs. 1 to 4). In some cases the reaction is vigorous, while in others, the reaction is more subdued, or will only occur if other conditions (such as temperature, humidity, or a reactive surface) are present. The reader is advised to check the literature for more specific information.

References

1. Bretherick, L., Urben, P. G., and Pitt, M. J., *Bretherick's Handbook of Reactive Chemical Hazards: An Indexed Guide to Published Data*, 6th Ed., Butterworths, London, 1999.
2. Pyrophoric Materials, www.safety.science.tamu.edu/pyrophorics.html, Texas A&M University, 2003.
3. Virginia Commonwealth University Office of Environmental Health & Safety, Chemical/Biological Safety Section, Pyrophoric Chemicals: Safe Work Practices Information Page, http://www.vcu.edu/oehs/chemical/biosafe/pyrophorics.pdf.
4. Bruno, T. J., and Svoronos, P. D. N., *CRC Handbook of Basic Tables for Chemical Analysis*, 3rd. Ed., CRC Press, Boca Raton, FL, 2011.

Alkali metals* (sodium, potassium, potassium/sodium alloy, lithium/tin alloys)
Alkylaluminum derivatives (diethylaluminum hydride)
Alkylated metal alkoxides (diethylethoxyaluminum)
Alkylboranes
Alkylhaloboranes (bromodimethyl borane)
Alkylhalophosphines
Alkylhalosilanes
Alkyl metals
Alkyl non-metal hydrides
Boranes (diborane)
Carbonyl metals (pentacarbonyl iron, octacarbonyl dicobalt, nickel carbonyl)
Complex acetylides
Complex hydrides (diethylaluminum hydride)
Finely divided metals* (calcium, zirconium)
Grignard reagents
Haloacetylene derivatives
Hexamethylnitrato dialluminum salts
Metal hydrides (germane, sodium hydride, lithium aluminum hydride)
Non-metal hydrides
Some nonmetal (organic) halides (dichloro(methyl)silane)
Spent hydrogenation catalysts (can be especially hazardous because of adsorbed hydrogen; for example, Raney nickel)
White phosphorus

*Note that the reactivity depends on the particle size and the ease at which oxides are formed on the metal surface.

FLAMMABILITY HAZARDS OF COMMON SOLVENTS

Thomas J. Bruno and Paris D. N. Svoronos

The following table lists relevant data regarding the flammability of common organic solvents. The codes for entries in the last column are given in the footnotes.

References

1. Turner, C. F., and McCreery, J. W., *The Chemistry of Fire and Hazardous Materials*, Allyn and Bacon, Boston, 1981.
2. *Fire Protection Guide to Hazardous Materials, 11th Edition*, National Fire Protection Association, Quincy, MA, 1994.

Flammability Hazards of Common Solvents

Solvent	Formula	Density/g cm^{-3}	Boiling point/°C	Flash point/°C	Autoignition point/°C	How to extinguish fires
Acetaldehyde	CH_3CHO	0.7834^{18}	20.8	-39	175	a,b,c
Acetone	$(CH_3)_2CO$	0.7902^{20}	56.08	-20	465	a,b
Acetonitrile	CH_3CN	0.7825^{20}	81.6	6	524	a,c,d
Acrolein	$CH_2{=}CHCHO$	0.840^{20}	52.3	-26	220	a,b,c
Acrylonitrile	$CH_2{=}CHCN$	0.8007^{25}	77.2	0	481	a,c,d
Allylamine	$C_3H_5NH_2$	0.758^{20}	54	-29	374	a,b
Aniline	$C_6H_5NH_2$	1.0250^{20}	184.1	70	615	a,b,c; use masks
Anisole	C_7H_8O	0.9940^{20}	153.6	52	475	a,b,c
Benzaldehyde	C_6H_5CHO	1.0401^{25}	178.7	63	192	a,b,c
Benzene	C_6H_6	0.8788^{20}	80.08	-11	498	a,b,c
Butanal	C_4H_8O	0.8016^{20}	74.8	-22	218	a,b,c
2,3-Butanedione	$C_4H_6O_2$	0.9808^{18}	87.5	27		a,b,c
1-Butanol	C_4H_9OH	0.8148^{20}	117.6	37	343	a,b,c
2-Butanone	C_4H_8O	0.7999^{25}	79.6	-9	404	a,b,c
2-Butanone peroxide	$C_8H_{16}O_4$		63 exp			a,b
trans-2-Butenal	C_4H_6O	0.8516^{20}	102.2	13	232	a,b,c
tert-Butyl peroxyacetate	$C_6H_{12}O_3$	>1.0	expl	<27		a,b,c
tert-Butyl peroxybenzoate	$C_{11}H_{14}O_3$	1.021^{25}	118 expl	>88	8	b,c
Carbon disulfide	CS_2	1.2632^{20}	46.2	-30	90	b,d; use masks
Cyclohexanone	$C_6H_{10}O$	0.9478^{20}	155.4	44	420	a,b,c
Diethanolamine	$C_4H_{11}NO_2$	1.0966^{20}	271.2	172	662	b,c
Diethylene glycol diethyl ether	$C_8H_{18}O_3$	0.9063^{20}	185	82		a,e
Diethyl ether	$(C_2H_5)_2O$	0.7135^{20}	34.4	-45	180	a,b,e
Diisopropyl ether	$C_6H_{14}O$	0.7192^{25}	68.4	-28	443	a,b
Dimethyl sulfate	$C_2H_6O_4S$	1.3322^{20}	186	83	188	a,b,c,d
Dimethyl sulfide	$(CH_3)_2S$	0.8483^{20}	37.32	-37	206	b,c
1,4-Dioxane	$C_4H_8O_2$	1.0337^{20}	101.2	12	180	a,b,c
1,2-Ethanediamine	$C_2H_8N_2$	0.8979^{20}	116.9	40	385	a,b,c
1,2-Ethanediol	$(CH_2OH)_2$	1.1135^{20}	197.5	111	398	a,b,c,d
Ethanol	C_2H_5OH	0.7893^{20}	78.24	13	363	a,b,c
Ethylamine	$C_2H_5NH_2$	0.689^{15}	16.6	-16	385	a,b,c
Formaldehyde	HCHO	0.815^{-20}	-19.1	85	424	a,b,c
Furfural	$C_5H_4O_2$	1.1594^{20}	161.5	60	316	a,b,c,d
Furfuryl alcohol	$C_5H_6O_2$	1.1296^{20}	168	75	491	a,b,c
Gasoline[f]		≈0.8	38-204	-43	280	a,b,c
Hexylamine	$C_6H_{13}NH_2$	0.7660^{20}	132	29		a,b
Isopropylbenzene hydroperoxide	$C_9H_{12}O_2$	1.03^{20}	153 expl	175		a,b,c
Kerosene[f]		<1.0	150-300	38-72	210	a,b,c
Methanol	CH_3OH	0.7909^{20}	64.5	11	464	a,b
Methylamine	CH_3NH_2	0.656^{25} (p>1 atm)	-6.4	0	430	a,b,c
N-Methylaniline	C_7H_9N	0.9859^{20}	197		533	a,b
Naphtha[f]		<1.0	100-160[h]	0-30[h]	230[h]	a,b,c
Paraldehyde	$C_6H_{12}O_3$	0.9943^{20}	124	36	238	a,b,c
2,4-Pentanedione	$CH_3COCH_2COCH_3$	0.9721^{25}	140.7	34	340	a,b,c
1-Pentanethiol	$C_5H_{11}SH$	0.850^{20}	126.6	18		a,b
2-Pentanone	$C_5H_{10}O$	0.809^{20}	102.2	7	452	a,b,c
2-Pentanone	$C_5H_{10}O$	0.809^{20}	102.2	7	452	

Solvent	Formula	Density/g cm^{-3}	Boiling point/°C	Flash point/°C	Autoignition point/°C	How to extinguish fires
3-Pentanone	$C_5H_{10}O$	0.8098^{25}	101.9	13	450	a,b,c
3-Pentenenitrile	C_5H_7N	0.83^{i}	14.5	40		b,c,d
Peroxyacetic acid	$C_2H_4O_3$	1.226^{15}	110	41		a,b,c
Petroleum ether[f]		≈0.6	35-60	<-18	288	a,b,c
Propanal	C_2H_5CHO	0.8657^{25}	48.0	-30	207	a,b,c
1,3-Propanediol	$C_3H_8O_2$	1.0538^{20}	214.7		400	a,b
2-Propanol	$CH_3CHOHCH_3$	0.7855^{20}	82.21	12	399	b,c
Propylamine	$C_3H_7NH_2$	0.7173^{20}	47.21	-37	318	a,b,c
Sulfur dichloride	SCl_2	1.62	59.6			b,c
Sulfuryl chloride	SO_2Cl_2	1.680	69.4			a,b,c
Tetrahydrofuran	C_4H_8O	0.8833^{25}	66.0	-14	321	a,b,c
Thionyl chloride	$SOCl_2$	1.631	75.6			
Toluene	$C_6H_5CH_3$	0.8668^{20}	110.60	4	480	a,b,c
Triethanolamine	$C_6H_{15}NO_3$	1.1242^{20}	350	179		b,c,d
Triethylamine	$(C_2H_5)_3N$	0.7275^{20}	88.8	-7	249	a,b,c
o-Xylene	$C_6H_4(CH_3)_2$	0.8802^{20}	144.4	32	463	a,b,c
m-Xylene	$C_6H_4(CH_3)_2$	0.8641^{20}	139.1	27	527	a,b,c
p-Xylene	$C_6H_4(CH_3)_2$	0.8610^{20}	138.3	27	528	a,b,c

[a] Extinguish using foam
[b] Extinguish using carbon dioxide
[c] Extinguish using dry chemical (ABC)
[d] Extinguish using water
[e] Extinguish using halons
[f] Mixture of C_5 to C_{10} hydrocarbons, mostly aliphatic (C_5 to C_6 for petroleum ether)
[g] Aqueous solution
[h] Varies widely with manufacturer

Safety

SELECTION OF LABORATORY GLOVES

Thomas J. Bruno and Paris D. N. Svoronos

The following table provides guidance in the selection of protective gloves for laboratory use (Refs. 1–5). If protection from more than one class of chemical is required, double gloving should be considered.

References

1. Garrod, A. N., Martinez, M., and Pearson, J., *Ann. Occup. Hyg.* 43, 543, 1999.
2. Garrod, A. N., Phillips, A. M., and Pemberton, J. A., *Ann. Occup. Hyg.* 45, 55, 2001.
3. Mockelsen, R. L., and Hall, R. C., *Am. Ind. Hyg. Assoc. J.* 48, 941, 1987.
4. OSHA, Federal Register, Vol. 59, No. 66, 29 CFR 1910, 16334-16364, 1994.
5. Bruno, T. J., and Svoronos, P. D. N., *CRC Handbook of Basic Tables for Chemical Analysis,* 3rd. Ed., CRC Press, Boca Raton, FL, 2011.

Glove material	Resistant to:
Viton	PCBs, chlorinated solvents, aromatic solvents
Viton/Butyl	acetone, toluene, aromatics, aliphatic hydrocarbons, chlorinated solvents, ketones, amines and aldehydes
SilverShield and 4H (PE/EVAL)	morpholine, vinyl chloride, acetone, ethyl ether, many toxic solvents and caustics
Barrier	wide range of chlorinated solvents, aromatic acids
PVA	ketones, aromatics, chlorinated solvents, , xylene, MIBK, trichloroethylene; DO NOT USE WITH WATER/AQUEOUS SOLUTIONS
Butyl	aldehydes, ketones, esters, alcohols, most inorganic acids, caustics, dioxane
Neoprene	oils, grease, petroleum-based solvents, detergents, acids, caustics, alcohols, solvents
PVC	acids, caustics, solvents, solvents, grease, oil
Nitrile	oils, fats, acids, caustics, alcohols
Latex	body fluids, blood, acids, alcohols, alkalis
Vinyl	body fluids, blood, acids, alcohols, alkalis
Rubber	organic acids, some mineral acids, caustics, alcohols; not recommended for aromatic solvents, chlorinated solvents

SELECTION OF PROTECTIVE LABORATORY GARMENTS

Thomas J. Bruno and Paris D. N. Svoronos

The following table provides guidance in the selection of special protective garments that are used in the laboratory for specific tasks (Refs. 1 and 2).

References

1. Mount Sinai School of Medicine Personal Protective Equipment Guide, www.mssm.edu/biosafety/policies, 2013.
2. Bruno, T. J., and Svoronos, P. D. N., *CRC Handbook of Basic Tables for Chemical Analysis,* 3rd. Ed., CRC Press, Boca Raton, FL, 2011.

Material	Type of garment	Common use
Cotton/Natural Fiber/Blends	Coveralls; lab coats; sleeve protectors; aprons	For dry dusts, particulates and aerosols
Tyvek	Coveralls; lab coats; sleeve protectors; aprons; hoods	For dry dusts and aerosols
Saranax/Tyvek SL	Coveralls; lab coats; sleeve protectors; aprons; hoods; level B suits	Aerosols; liquids; solvents
Polyethylene	Barrier gowns; aprons	Body fluids
Polypropylene	Clean room suits; coveralls; lab coats	For dry dusts; non-toxic particulates
Polyethylene/Tyvek (QC)	Coveralls; aprons; lab coats; shoe covers	Moisture; solvents
Polypropylene	Coveralls; lab coats; shoe covers; caps; clean room suits	Non-toxic particulates; dry dusts
Tychem BR; Tychem TK	Full level A and level B suits	Highly toxic particulates; dry dusts
CPF	Full level A and level B suits; splash suits	Highly toxic chemicals; gases; aerosols
PVC	Full level A suits	Highly toxic chemicals; gases; aerosols

SELECTION OF RESPIRATOR CARTRIDGES AND FILTERS

Thomas J. Bruno and Paris D. N. Svoronos

Respirators are sometimes desirable or required when performing certain tasks in the laboratory, provided adequate ventilation prevents asphyxiation or oxygen deficiency conditions from developing. Oxygen deficiency conditions require the use of a self-contained breathing apparatus or a hose mask with blower. There is a standardized color code system used by all manufacturers for the specification and selection of the cartridges and filters that are used with respirators (Ref. 1). The following table provides guidance in the selection of the proper cartridge using the color code. We note that the use of respirators in the laboratory is subject to regulation and management. In the United States, for example, before a respirator can be used, a medical evaluation must be performed to establish that the user has sufficient lung capacity and that the use of the respirator would not cause health problems. This evaluation must be performed by a physician or nurse, and if performed by a nurse, the evaluation must be approved and signed by a physician. The respirator must be fitted properly by an industrial hygienist or other trained personnel, and that fitting procedure must adhere to an established protocol with approved equipment. Finally, training by an industrial hygienist or other appropriate personnel must be provided.

Reference

1. Bruno, T. J., and Svoronos, P. D. N., *CRC Handbook of Basic Tables for Chemical Analysis*, 3rd. Ed., CRC Press, Boca Raton, FL, 2011.

Color code	Application
Gray	Organic vapors, ammonia, methylamine, chlorine, hydrogen chloride, and sulfur dioxide or hydrogen sulfide (for escape only) or hydrogen fluoride or formaldehyde.
Black	Organic vapors, not to exceed regulatory standards.
Yellow	Organic vapors, chlorine, chlorine dioxide, hydrogen chloride, hydrogen fluoride, sulfur dioxide, or hydrogen sulfide (for escape only).
White	Chlorine, hydrogen chloride, hydrogen chloride, hydrogen fluoride, sulfur dioxide, or hydrogen sulfide (for escape only).
Green	Ammonia and methylamine.
Orange	Mercury and/or chlorine.
Purple	Solid and liquid aerosols and mists.
Purple + gray	Organic vapors, ammonia, methylamine, chlorine, hydrogen chloride, and sulfur dioxide or hydrogen sulfide (for escape only) or hydrogen fluoride or formaldehyde; solid and liquid aerosols and mists.
Purple + black	Organic vapors, and solid and liquid aerosols and mists.
Purple + yellow	Organic vapors, chlorine, chlorine dioxide, hydrogen chloride, hydrogen fluoride, sulfur dioxide, or hydrogen sulfide (for escape only); solid and liquid aerosols and mists.
Purple + white	Chlorine, hydrogen chloride, hydrogen chloride, hydrogen fluoride, sulfur dioxide, or hydrogen sulfide (for escape only); solid and liquid aerosols and mists.
Purple + green	Ammonia, methylamine, and solid and liquid aerosols and mists.

In addition to the cartridges specified in the table, particulate filters are available that can be used alone or in combination with those cartridges specified.

MATERIALS COMPATIBLE WITH AND RESISTANT TO 72 PERCENT PERCHLORIC ACID

Thomas J. Bruno and Paris D. N. Svoronos

Perchloric acid is used in the preparation of ion pairing agents for HPLC (see Section 8 dealing with this topic), and must be handled with great care since it can be a very powerful oxidizing agent. Cold perchloric acid at a concentration of 70 percent (mass/mass) or less is not considered a very strong oxidizing agent. At concentrations of 73 percent or higher, or at lower concentrations but at higher temperatures, perchloric acid is a powerful oxidant. The following table provides some guidance in handling this material in the laboratory (Ref. 1).

Reference

1. Bruno, T. J., and Svoronos, P. D. N., *CRC Handbook of Basic Tables for Chemical Analysis, 3rd Edition*, CRC Press, Boca Raton, FL, 2011.

Material	Comments
Elastomers	
Gum Rubber	Each batch must be tested to determine compatibility
Vitons	Slight swelling only
Metals and alloys	
Tantalum	Excellent
Titanium (chemically pure grade)	Excellent
Zirconium	Excellent
Columbium (Niobium)	Excellent
Hastelloy	Slight corrosion rate
Plastics	Adequate
Polyvinyl chloride	
Teflon	
Polyethylene	
Polypropylene	
Kel-F	
Vinylidine fluoride	
Saran	
Epoxies	
Others	Adequate
Glass	
Glass-lined steel	
Alumina	
Fluorolube	

Incompatible

Plastics

Polyamide (nylon)
Modacrylic ester, Dynel (35–85%) acrylonitrile
Polyester (dacron)
Bakelite
Lucite
Micarta
Cellulose-based lacquers

Metals

Copper
Copper alloys (brass, bronze, etc.) for very shock-sensitive perchlorate salts
Aluminum (dissolves at room temperature)
High nickel alloys (dissolves)

Others

Cotton
Wood
Glycerin-lead oxide (letharge)

Safety

PROTECTIVE CLOTHING LEVELS

Thomas J. Bruno and Paris D. N. Svoronos

In the United States, OSHA defines various levels of protective clothing, and sets parameters that govern their use with chemical spills and in environments where chemical exposure is a possibility. A summary of the definitions is provided below (Refs. 1 and 2).

References

1. OSHA Technical Manual, Section VIII, Chapter 1, Chemical Protective Clothing, 2003.
2. Bruno, T. J., and Svoronos, P. D. N., *CRC Handbook of Basic Tables for Chemical Analysis*, 3rd. Ed., CRC Press, Boca Raton, FL, 2011.

Level A:

Vapor protective suit (meets NFPA 1991), pressure-demand, full-face SCBA, inner chemical-resistant gloves, chemical-resistant safety boots, two-way radio communication.

Protection Provided: Highest available level of respiratory, skin, and eye protection from solid, liquid and gaseous chemicals.

Used When: The chemical(s) have been identified and have high level of hazards to respiratory system, skin and eyes; substances are present with known or suspected skin toxicity or carcinogenity; operations must be conducted in confined or poorly ventilated areas.

Limitations: Protective clothing must resist permeation by the chemical or mixtures present.

Level B:

Liquid splash-protective suit (meets NFPA 1992), pressure-demand, full-facepiece SCBA, inner chemical-resistant gloves, chemical-resistant safety boots, two-way radio communications.

Protection Provided: Provides same level of respiratory protection as Level A, but somewhat less skin protection. Liquid splash protection is provided, but not protection against chemical vapors or gases.

Used When: The chemical(s) have been identified but do not require a high level of skin protection; the primary hazards associated with site entry are from liquid and not vapor contact.

Limitations: Protective clothing items must resist penetration by the chemicals or mixtures present.

Level C:

Support function protective garment (meets NFPA 1993), full-facepiece, air-purifying, canister-equipped respirator, chemical resistant gloves and safety boots, two-way communications system.

Protection Provided: The same level of skin protection as Level B, but a lower level of respiratory protection; liquid splash protection but no protection to chemical vapors or gases.

Used When: Contact with site chemical(s) will not affect the skin; air contaminants have been identified and concentrations measured; a canister is available which can remove the contaminant; the site and its hazards have been completely characterized.

Limitations: Protective clothing items must resist penetration by the chemical or mixtures present; chemical airborne concentration must be less than IDLH levels; the atmosphere must contain at least 19.5% oxygen.

Not Acceptable for Chemical Emergency Response.

Level D:

Coveralls, safety boots/shoes, safety glasses or chemical splash goggles.

Protection Provided: No respiratory protection, minimal skin protection.

Used When: The atmosphere contains no known hazard; work functions preclude splashes, immersion, potential for inhalation, or direct contact with hazard chemicals.

Limitations: The atmosphere must contain at least 19.5% oxygen.

Not Acceptable for Chemical Emergency Response.

Optional items may be added to each level of protective clothing. Options include items from higher levels of protection, as well as hardhats, hearing protection, outer gloves, a cooling system, etc.

CHEMICAL FUME HOODS AND BIOLOGICAL SAFETY CABINETS

Thomas J. Bruno

Engineered safety equipment is preferred over the reliance on personal protective equipment, and the fume hood is one such engineered safety device that is nearly ubiquitous in laboratories. Laboratories concerned with biological specimens (microbes, spores) also commonly are equipped with biological safety cabinets. The following section provides basic information on the function and application of these devices. The purpose here is not to provide design or installation instructions, since most users will find this equipment already installed in their work spaces. Rather, this information is to allow optimal use to be made of the installation that is preexisting.

Types of Chemical Fume Hoods

Most of the chemical fume hoods considered here consist of a cabinet or enclosure set at waist level (above a table or storage cabinet) that is connected to a blower located above the hood or external to the hood through a duct system. The cabinet has an open side (or sides) to allow a user to perform work within. A movable transparent sash separates the user from the work. Most chemical fume hoods have a sill that functions as an airfoil at the work surface below the sash. The connection to the blower might be by use of a v-belt, or it may be direct drive. This allows provision of a smooth flow of air with minimal turbulence. In some installations, axially mounted blowers are used, especially if multiple hoods are ducted into a common blower. Baffles located in the rear of the cabinet provide control of the air flow patterns, and can usually be adjusted to provide the best air flow around the experiment or procedure being performed. Many chemical fume hoods are equipped with air flow indicators, low flow monitors and alarms, and differential pressure sensors to allow the user to operate safely. The major types of chemical fume hoods include the standard/conventional, walk-in, bypass, variable air volume, auxiliary air, or ductless types. Additional types include snorkels and canopies that are portable. Each type must be understood to be operated most efficiently within specifications (see the section below on safe operation).

Standard or Conventional:

The standard chemical fume hood utilizes a constant speed motor, and for this reason the volume of air drawn into the hood will change with movement of the sash position. As the sash is lowered, the velocity of the air drawn into the hood will increase.

Bypass:

The bypass chemical fume hood is very similar to the standard/conventional hood except that as the sash is lowered, a vent is opened above the sash to allow additional air flow into the hood. This prevents a large increase in velocity in the working area inside the hood.

Variable Air Volume:

The variable volume chemical fume hood controls the volume of air drawn into the hood as a function of sash position, while maintaining the face velocity of the air at a constant rate, within the specifications required. These types of chemical fume hoods are more energy efficient than the standard or bypass hoods because they minimize costs incurred by laboratory heating and cooling.

Auxiliary Air:

The auxiliary air chemical fume hood includes an additional blower that injects air into or at the face of the hood, providing additional flow inside the enclosed cabinet. These types of hoods are rarely installed in renovations or new construction, but may be encountered in older laboratories. They are less desirable than the standard/conventional, bypass or variable volume types because they require a great deal of energy to operate (although the early designs featured the addition of an auxiliary air stream that was not air conditioned). These devices are mechanically more complex than other types, and consequently more prone to maintenance problems.

Walk-In Hood:

The walk-in hood is a chemical fume hood that is mounted directly on the laboratory floor or a slightly raised chemical resistance platform. It is used for the ventilation of larger pieces of equipment, with the advantage that these pieces of equipment can be wheeled in and out of the walk-in hood. The walk-in hood typically uses two separate sashes.

Ductless Hoods:

This type of chemical fume hood does not duct the air flow to outside the laboratory, but rather the air flow is returned to the room or interstitial space after passing through a means to remove contaminants. The contaminants may be removed by HEPA filters, activated carbon cartridges, adsorbents, or catalyst beds. The means of contaminant removal must be inspected and serviced at regular intervals.

Laminar Flow Hood (Clean Bench):

Related at least in principle to ductless hoods are laminar flow hoods, sometimes called clean hoods. These are devices intended to protect the work being performed from particulates in the air, which is accomplished by bathing the work area with HEPA filtered air either blown at low velocity over the work area or blown from the bottom of the hood as an air curtain. Only approximately 10 % of the air flow is through the face of the hood. These are intended to protect the work or samples inside the hood, not primarily the user. These units should not be used in place of chemical fume hoods, rather they are used to protect the work from dust or pollen.

Snorkels and Canopies:

Snorkels are flexible ducts routed from a blower duct that can be placed temporarily atop or near an experiment to provide some measure of protection. A canopy is similar, but it incorporates an additional bell shaped collector that might be suspended above an experiment, but for the same purpose as that for the snorkel.

Chemical Fume Hood Operations

While the operation of chemical fume hoods is straightforward, safe operating practices must be observed. The face velocities should be optimized at between 80 and 120 fpm. Face velocities in excess of 125 fpm can cause turbulent flow and allow outflow of contaminants from the hood and potentially expose the user to hazards. Face velocities are checked periodically by facilities managers. Ideally, chemical fume hoods are located in low traffic areas of the laboratory, away from entry doors. Safe operation of chemical fume hoods requires observance of the following practices, divided into primary (applicable to all installations) and secondary (applicable on a case by case basis):

Primary Guidelines:

- Before using the chemical fume hood, ensure that it is in working order. A simple air flow test can be done with a laboratory wipe, or one can observe the flow monitor if one is present. Be aware of clattering sounds that might indicate a broken belt, or screeching sounds that might indicate a failed or failing bearing.
- The motor of the chemical fume hood should be running at all times, except for maintenance. If a switch is located inside the laboratory, it should be equipped with a lockout to protect maintenance staff.
- Users should perform all work at least six inches inside the plane of the sash
- The user should avoid placing his/her head inside the chemical fume hood, beyond the plane of the sash.
- Laboratory occupants should avoid traffic in front of the chemical fume hood to minimize turbulence.
- Users should avoid rapid movements in front of the chemical fume hood; this includes rapidly raising and lowering the sash.
- Keep the sash closed down as far as possible.
- The baffles of the chemical fume hood should be free of obstruction.
- If a heating device such as a hot plate is being used in the hood, the interior air flow can be dramatically changed by convection. In those instances, the lower baffles of the hood should be closed or minimized and the middle baffles fully opened to accommodate the convection.
- Equipment located inside the cabinet that can potentially disrupt the air flow should be raised above the level of the lower baffles by a small shelf or blocks.
- Hoses and power cords that must be run into the chemical fume hood from the outside should be run through the airfoil beneath the sash.

Secondary Guidelines:

- Some chemical fume hoods have a small cup sink located inside the cabinet. Water should be run into this sink periodically to maintain it free of obstruction and to keep the P-trap full and thereby prevent sewer gas back up.
- It is imperative to prevent the discharge of chemicals into the cup sink.
- Some chemical fume hoods have compressed air lines that allow the use of air operated equipment such as vortex tubes (for heating and cooling). When using vortex tubes inside a fume hood, the sash should be fully closed because additional air flow inside the hood is present.
- In a power outage, lower the sash to within an inch of the fully closed position.
- Evaporations and digestions with perchloric acid must only be done in a specifically designed and designated perchloric acid chemical fume hood. The same considerations apply for the use of radionuclides and infectious agents.

Types of Biological Safety Cabinets (BSC):

Biological safety cabinets, as distinct from chemical fume hoods, are enclosed ventilated cabinets intended to provide both a clean and a safe working environment for aerosols and biological hazards. All exhaust air is HEPA-filtered as it exits the biosafety cabinet, removing harmful bacteria and viruses. The Centers for Disease Control (U.S.) lists three classes of biological safety cabinets, Classes I, II, and III.

Class I cabinets are open front negative pressure cabinets that provide personnel and environmental protection but do not provide protection to the sample or media (the product) being used in the cabinet. There is air flow into the cabinet (at a face velocity of 75 fpm) that can potentially cause sample contamination. Class I cabinets are often used to enclose specific equipment (centrifuges, harvesting equipment or fermenters) or ongoing procedures (cultures) that potentially generate aerosols. BSCs of this class are either ducted (connected to the building exhaust system) or unducted (recirculating HEPA-filtered exhaust back into the laboratory, provided there is an interlock with the building exhaust system). Some Class I cabinets are used for animal cage changing and these typically require frequent HEPA filter changes due to odoriferous compounds saturating the filter.

A Class II biological safety cabinet provides protection for both the worker and the sample or product, making it suitable as a sterile compartment for cell culture. This type of cabinet is the most versatile and most common, with face velocities similar to those of Class I. There are four types of Class II cabinets, the main features of which are discussed below. Note that there are additional differences among the types in Class II (types A1, A2, B1 and B2), primarily concerning the geometry of the air flows and placement of HEPA filters.

Type A1 (formerly Type A) does not have to be duct vented (although it is possible to connect to building ventilation systems by use of canopy exhaust connections), which makes it suitable for use in laboratories inaccessible to ductwork. This cabinet can be used for use of low to moderate hazard agents that do not include volatile toxic chemicals and volatile radionuclides. The supply air is HEPA filtered to present the sample or media being used with a particulate free air stream with a face velocity of at least 75 fps. This type of BSC cannot be used for volatile and toxic compounds and solvents because small quantities of these materials can quickly load the filter. Type A2 differs from Type A1 in that protection of the operator and the environment is only afforded if the exhaust line is canopy vented to the building exhaust. The face velocities of these units are at least 100 fps.

Type B1 cabinets must be hard vented, with 50 % of the air exhausted from the cabinet while 50 % can be recirculated back into the room. This cabinet may be used with etiologic agents and traces of volatile and toxic chemicals and radionuclides required as an adjunct to microbiological studies (if the work is done in the directly exhausted portion of the cabinet). The air intake velocity of the B1 type is specified to be 100 fps. Type B2 cabinets are must be 100 % exhausted through a dedicated duct. The air intake velocity is specified to be 100 fps. This cabinet may be used with etiologic agents treated with toxic chemicals and radionuclides required as an adjunct to microbiological studies.

The Class III cabinet is designed for highly infectious microbial agents. It is entirely gas tight, with a non-operating view window (cannot be opened). Access to the interior is through a dunk tank accessible through the floor of the cabinet or through a double door system. Both the supply and exhaust gas streams pass through HEPA filters. Heavy duty rubber gloves are used for manipulations in the interior of the cabinet.

GAS CYLINDER SAFETY AND STAMPED MARKINGS

Thomas J. Bruno and Paris D. N. Svoronos

The graphic below describes the permanent, stamped markings that are used on high-pressure gas cylinders commonly found in laboratories (Ref. 1). Note that individual jurisdictions and institutions have requirements for marking the cylinder contents as well. These requirements are in addition to the stamped markings, which pertain to the cylinder itself rather than to the fill contents (Ref. 2).

There are four fields of markings on cylinders that are used in the United States, labeled 1-4 on the figure.

Field 1 – Cylinder Specifications:

DOT stands for the United States Department of Transportation, the agency that regulates the transport and specification of gas cylinders in the United States. The next entry, for example, 3AA, is the specification for the type and material of the cylinder. The most common cylinders are 3A, 3AA, 3AX, 3AAX, 3T, and 3AL. All but the last refer to steel cylinders, while 3AL refers to aluminum. The individual specifications differ mainly in chemical composition of the steel, and the gases that are approved for containment and transport. The 3T deals with large bundles of tube trailer cylinders.

The next entry in this field is the service pressure, in psig.

Field 2 – Serial Number:

This is a unique number assigned by the manufacturer.

Field 3 – Identifying Symbol:

The manufacturer identifying symbol historically can be a series of letters or a unique graphical symbol. In recent years, the DOT has standardized this identification with the "M" number, for example, M1004. This is a number issued by DOT that identifies the cylinder manufacturer.

Field 4 – Manufacturing Data:

The data of manufacture is provided as a month and year. With this date is the inspector's official mark, for example, H. In recent years, this letter has been replaced with an IA number, for example IA02, pertaining to an independent agency that is approved by DOT as an inspector. If "+" is present, the cylinder qualifies for an overfill of 10 percent in service pressure. If "★" is present, the cylinder qualifies for a 10 year rather than a 5 year retest interval.

In some cases, the original insignia of the inspector that performed the first hydrostatic test is found below Field 4. Also stamped on the cylinder will be the retest dates. A cylinder must have a current (that is, within 5 or 10 years) test stamp. On the collar of the cylinder, the owner of the cylinder may be stamped.

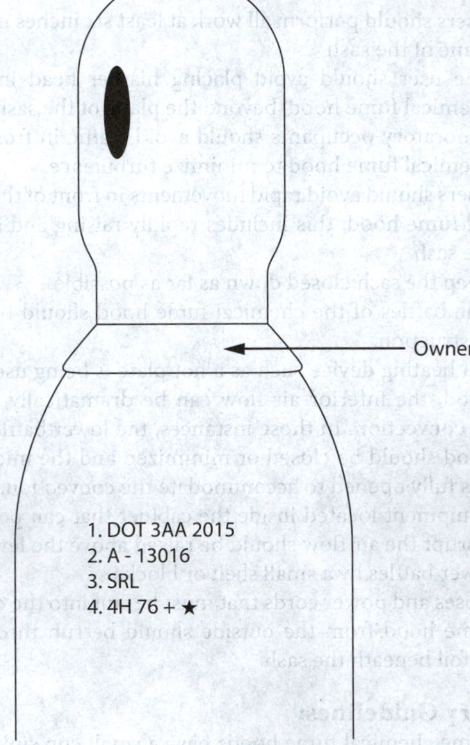

References

1. Hazardous Materials: Requirements for Maintenance, Requalification, Repair and Use of DOT Specification Cylinders, 49 CFR Parts 107, 171, 172, 173, 177, 178, 179, and 180; [Docket No. RSPA-01-10373 (HM-220D)]RIN 2137-AD58, August 8, 2002.
2. Bruno, T. J., and Svoronos, P. D. N., *CRC Handbook of Basic Tables for Chemical Analysis*, 3rd. Ed., CRC Press, Boca Raton, FL, 2011.

FLAMMABILITY OF CHEMICAL SUBSTANCES

This table gives properties related to the flammability of about 900 chemical substances. The properties listed are:

t_b: Normal boiling point in °C at 101.325 kPa pressure (sp indicates sublimation point).

FP: lash point, which is the minimum temperature at which the vapor pressure of a liquid is sufficient to form an ignitable mixture with air near the surface of the liquid. Flash point is not an intrinsic physical property but depends on the conditions of measurement (see Ref. 1).

Flam. limits: Flammable limits (often called explosive limits), which specify the range of concentration of the vapor in air (in percent by volume) for which a flame can propagate. Below the lower flammable limit, the gas mixture is too lean to burn; above the upper flammable limit, the mixture is too rich. Values refer to ambient temperature and pressure and are dependent on the precise test conditions. A ? indicates that one of the limits is not known.

IT: Ignition temperature (sometimes called autoignition temperature), which is the minimum temperature required for self-sustained combustion in air in the absence of an external ignition source. As in the case of flash point, the value depends on specified test conditions.

VP: Vapor pressure at 25 °C.

Even in cases where very careful measurements of flash point have been replicated in several laboratories, observed values can differ by 3 to 6 °C (Ref. 4). For more typical measurements, larger uncertainties should be assumed in both flash points and autoignition temperatures. The absence of a flash point entry in this table does not mean that the substance is nonflammable, but only that no reliable value is available.

Compounds are listed by name.

References

1. *Fire Protection Guide to Hazardous Materials, 11th Edition*, National Fire Protection Association, Quincy, MA, 1994.
2. Urben, P. G., Ed., *Bretherick's Handbook of Reactive Chemical Hazards, 5th Edition*, Butterworth-Heinemann, Oxford, 1995.
3. Daubert, T. E., Danner, R. P., Sibul, H. M., and Stebbins, C. C., *Physical and Thermodynamic Properties of Pure Compounds: Data Compilation*, extant 1994 (core with 4 supplements), Taylor & Francis, Bristol, PA.
4. *Report of Investigation: Flash Point Reference Materials*, National Institute of Standards and Technology, Standard Reference Materials Program, Gaithersburg, MD, 1995.

Flash Point, Flammability Limits, and Ignition Temperature of Selected Substances

Name	Mol. form.	t_b/°C	FP/°C	Flam. limits	IT/°C	VP (25 °C)/kPa
Acetaldehyde	C_2H_4O	20.8	-39	4.0-60%	175	120
Acetanilide	C_8H_9NO	292	169		530	
Acetic acid	$C_2H_4O_2$	117.9	39	4.0-19.9%	463	2.07
Acetic anhydride	$C_4H_6O_3$	139.5	49	2.7-10.3%	316	0.680
Acetoacetanilide	$C_{10}H_{11}NO_2$		185			
Acetone	C_3H_6O	56.08	-20	2.5-12.8%	465	30.8
Acetone cyanohydrin	C_4H_7NO	180	74	2.2-12.0%	688	
Acetonitrile	C_2H_3N	81.6	6	3.0-16.0%	524	11.9
Acetophenone	C_8H_8O	202.1	77		570	0.049
Acetyl chloride	C_2H_3ClO	51	4		390	38.4
Acetylene	C_2H_2	-84.7 sp		2.5-100%	305	
N-Acetylethanolamine	$C_4H_9NO_2$		179		460	
4-Acetylmorpholine	$C_6H_{11}NO_2$		113			
Acrolein	C_3H_4O	52.3	-26	2.8-31%	220	36.2
Acrylic acid	$C_3H_4O_2$	142	50	2.4-8.0%	438	0.53
Acrylonitrile	C_3H_3N	77.2	0	3.0-17.0%	481	14.1
Allyl acetate	$C_5H_8O_2$	104	22		374	
Allyl alcohol	C_3H_6O	96.9	21	2.5-18.0%	378	3.14
Allylamine	C_3H_7N	54	-29	2.2-22%	374	33.1
Allyl hexanoate	$C_9H_{16}O_2$	186	66			
Allyltrichlorosilane	$C_3H_5Cl_3Si$	118	35			
2-Aminobiphenyl	$C_{12}H_{11}N$	298.3			450	
2-Amino-1-butanol, (±)-	$C_4H_{11}NO$	178	74			
2-Amino-2-methyl-1-propanol	$C_4H_{11}NO$	163.8	67			
3-Amino-1-propanol	C_3H_9NO	185	80			
1-Amino-2-propanol	C_3H_9NO	141	77		374	0.165
Ammonia	H_3N	-33.33		16-25%		
Aniline	C_6H_7N	184.1	70	1.3-11%	615	0.090
Aniline hydrochloride	C_6H_8ClN		193			

Name	Mol. form.	t_b/°C	FP/°C	Flam. limits	IT/°C	VP (25 °C)/kPa
Anisole	C_7H_8O	153.6	52		475	0.472
Anthracene	$C_{14}H_{10}$	341.3	121	>0.6%	540	
9,10-Anthracenedione	$C_{14}H_8O_2$	377	185			
Benzaldehyde	C_7H_6O	178.7	63		192	0.169
Benzene	C_6H_6	80.08	-11	1.2-7.8%	498	12.7
Benzeneacetaldehyde	C_8H_8O	202	71			0.052
Benzeneacetic acid	$C_8H_8O_2$	268	>100			
Benzeneacetonitrile	C_8H_7N	232	113			0.012
1,2-Benzenediamine	$C_6H_8N_2$	257	156	>1.5%		
Benzeneethanol	$C_8H_{10}O$	220	96			0.01
Benzoic acid	$C_7H_6O_2$	250.2	121		570	
Benzoyl chloride	C_7H_5ClO	201	72			0.084
Benzyl acetate	$C_9H_{10}O_2$	215	90		460	0.022
Benzyl alcohol	C_7H_8O	205.3	93		436	0.015
Benzyl benzoate	$C_{14}H_{12}O_2$	321.3	148		480	
Benzyl methyl ether	$C_8H_{10}O$	175	135			
2-(Benzyloxy)ethanol	$C_9H_{12}O_2$	256	129		352	
Benzyl salicylate	$C_{14}H_{12}O_3$	320	>100			
Biphenyl	$C_{12}H_{10}$	255.2	113	0.6-5.8%	540	
Bis(2-aminoethyl)amine	$C_4H_{13}N_3$	206.5	98	2-6.7%	358	0.03
1,2-Bis(2-chloroethoxy)ethane	$C_6H_{12}Cl_2O_2$	214	121			
Bis(2-chloroethyl) ether	$C_4H_8Cl_2O$	178	55	>2.7%	369	0.143
Bis(2-ethylhexyl)amine	$C_{16}H_{35}N$		132			
Bis(2-ethylhexyl) azelate	$C_{25}H_{48}O_4$		227	>0.3%	374	
Bis(2-ethylhexyl) hexanedioate	$C_{22}H_{42}O_4$		206	>0.4%	377	
Bis(2-ethylhexyl) phthalate	$C_{24}H_{38}O_4$	384	218			0.00000005
Bis(2-hydroxyethyl) sulfide	$C_4H_{10}O_2S$	282	160		298	0.08
Borneol, (±)-	$C_{10}H_{18}O$	213	66			
Bromobenzene	C_6H_5Br	155.9	51		565	0.556
4-Bromobiphenyl	$C_{12}H_9Br$	309	144			
1-Bromobutane	C_4H_9Br	101.4	18	2.6-6.6%	265	5.26
2-Bromobutane, (±)-	C_4H_9Br	91	21			9.32
1-Bromo-2-butene	C_4H_7Br	98		4.6-12.0%		
1-Bromododecane	$C_{12}H_{25}Br$	275	144			
Bromoethane	C_2H_5Br	38.2		6.8-8.0%	511	62.5
Bromoethene	C_2H_3Br	16		9-15%	530	141
Bromomethane	CH_3Br	3.4		10-16%	537	217
1-Bromopentane	$C_5H_{11}Br$	126	32			1.68
1-Bromopropane	C_3H_7Br	70.8			490	18.6
3-Bromopropene	C_3H_5Br	70.1	-1	4.4-7.3%	295	18.6
3-Bromo-1-propyne	C_3H_3Br	73	10	>3.0%	324	
Bromosilane	BrH_3Si	1.9	<0		≈20	
2-Bromotoluene	C_7H_7Br	182	79			0.17
4-Bromotoluene	C_7H_7Br	184	85			
1,3-Butadiene	C_4H_6	-4.6		2.0-12.0%	420	
Butanal	C_4H_8O	74.8	-22	1.9-12.5%	218	15.7
Butanal oxime	C_4H_9NO	152	58			
Butane	C_4H_{10}	-0.5	-60	1.9-8.5%	287	242
1,2-Butanediol, (±)-	$C_4H_{10}O_2$	196.42	40			
1,3-Butanediol	$C_4H_{10}O_2$	208.2	121		395	0.008
1,4-Butanediol	$C_4H_{10}O_2$	229.5	121			0.002
2,3-Butanediol	$C_4H_{10}O_2$	178			402	0.02
2,3-Butanedione	$C_4H_6O_2$	87.5	27			7.45
Butanenitrile	C_4H_7N	117.6	24	>1.6%	501	2.55
1-Butanethiol	$C_4H_{10}S$	98.4	2			6.07
2-Butanethiol	$C_4H_{10}S$	85.0	-23			10.8
Butanoic acid	$C_4H_8O_2$	163.7	72	2.0-10.0%	443	0.221
Butanoic anhydride	$C_8H_{14}O_3$	195	54	0.9-5.8%	279	0.07
1-Butanol	$C_4H_{10}O$	117.6	37	1.4-11.2%	343	0.86
2-Butanol	$C_4H_{10}O$	99.4	24	1.7-9.8%	405	2.32
2-Butanone	C_4H_8O	79.6	-9	1.4-11.4%	404	12.6
2-Butanone oxime	C_4H_9NO	151.5	≈70			0.396
trans-2-Butenal	C_4H_6O	102.2	13	2.1-15.5%	232	4.92

Name	Mol. form.	t_b/°C	FP/°C	Flam. limits	IT/°C	VP (25 °C)/kPa
1-Butene	C_4H_8	-6.3		1.6-10.0%	385	296
cis-2-Butene	C_4H_8	3.72		1.7-9.0%	325	214
trans-2-Butene	C_4H_8	0.88		1.8-9.7%	324	234
trans-2-Butenenitrile	C_4H_5N	120	16			
2-Buten-1-ol (unspecified isomer)	C_4H_8O	122	27	4.2-35.3%	349	
3-Buten-2-one	C_4H_6O	81	-7	2.1-15.6%	491	12.0
1-Buten-3-yne	C_4H_4	6.0		21-100%		
2-Butoxyethanol	$C_6H_{14}O_2$	171	69	4-13%	238	0.15
N-Butylacetamide	$C_6H_{13}NO$	229	116			
Butyl acetate	$C_6H_{12}O_2$	126.0	22	1.7-7.6%	425	1.66
sec-Butyl acetate	$C_6H_{12}O_2$	108	31	1.7-9.8%		
Butyl acetoacetate	$C_8H_{14}O_3$		85			
Butyl acrylate	$C_7H_{12}O_2$	146.6	29	1.7-9.9%	292	0.731
Butylamine	$C_4H_{11}N$	77.0	-12	1.7-9.8%	312	12.2
sec-Butylamine	$C_4H_{11}N$	62.71	-9			
tert-Butylamine	$C_4H_{11}N$	44.02	-9	1.7-8.9%	380	48.4
N-*tert*-Butylaminoethyl methacrylate	$C_{10}H_{19}NO_2$		96			
N-Butylaniline	$C_{10}H_{15}N$	254	107			
Butylbenzene	$C_{10}H_{14}$	183.3	71	0.8-5.8%	410	0.150
sec-Butylbenzene, (±)-	$C_{10}H_{14}$	173.3	52	0.8-6.9%	418	
tert-Butylbenzene	$C_{10}H_{14}$	169.1	60	0.7-5.7%	450	0.280
4-*tert*-Butyl-1,2-benzenediol	$C_{10}H_{14}O_2$	286	130			
Butyl benzoate	$C_{11}H_{14}O_2$	249	107			0.005
2-Butyl-1,1'-biphenyl	$C_{16}H_{18}$	292	>100		430	
Butyl butanoate	$C_8H_{16}O_2$	164.95	53			
Butyl citrate	$C_{18}H_{32}O_7$		157		368	
Butylcyclohexane	$C_{10}H_{20}$	180.9			246	
tert-Butylcyclohexane	$C_{10}H_{20}$	171.6			342	
Butylcyclohexylamine	$C_{10}H_{21}N$		93			
Butylcyclopentane	C_9H_{18}	156			250	
Butylethylamine	$C_6H_{15}N$	104.8	18			3.19
Butyl ethyl ether	$C_6H_{14}O$	89	4			7.46
Butyl formate	$C_5H_{10}O_2$	106.1	18	1.7-8.2%	322	3.53
tert-Butyl hydroperoxide	$C_4H_{10}O_2$	89 dec	27			
Butyl methacrylate	$C_8H_{14}O_2$	163.7	52			
Butylmethylamine	$C_5H_{13}N$	91	13			7.28
1-*tert*-Butyl 4-methylbenzene	$C_{11}H_{16}$	193	68			0.09
Butyl 3-methylbutanoate	$C_9H_{18}O_2$		53			
4-*tert*-Butyl-2-methylphenol	$C_{11}H_{16}O$	256	118			
1-Butylnaphthalene	$C_{14}H_{16}$	288	360			
Butyl nitrate	$C_4H_9NO_3$	133	36			
2-Butyl-1-octanol	$C_{12}H_{26}O$	248	110			
Butyl oleate	$C_{22}H_{42}O_2$		180			
N-Butyl-*N*-phenylacetamide	$C_{12}H_{17}NO$	281	141			
Butyl phenyl ether	$C_{10}H_{14}O$	210	82			
Butyl propanoate	$C_7H_{14}O_2$	145.1	32		426	
Butyl stearate	$C_{22}H_{44}O_2$	343	160		355	
Butyltrichlorosilane	$C_4H_9Cl_3Si$	148.5	54			
Butyl vinyl ether	$C_6H_{12}O$	94	-9		255	6.65
2-Butyne	C_4H_6	27.1	-31	>1.4%		94.3
γ-Butyrolactone	$C_4H_6O_2$	204.6	98			0.43
Camphor, (+)	$C_{10}H_{16}O$	209	66	0.6-3.5%	466	0.032
Caprolactam	$C_6H_{11}NO$	270.8	125			0.0003
Carbon disulfide	CS_2	46.2	-30	1.3-50.0%	90	48.2
Carbon monoxide	CO	-191.51		12.5-74%	609	
Carbon oxysulfide	COS	-50.2		12-29%		
α-Chloroacetophenone	C_8H_7ClO	247	118			0.001
4-Chlorobenzaldehyde	C_7H_5ClO	213.5	88			
Chlorobenzene	C_6H_5Cl	131.6	28	1.3-9.6%	593	1.6
2-Chloro-1,3-butadiene	C_4H_5Cl	59	-20	4.0-20.0%		29.5
1-Chlorobutane	C_4H_9Cl	78.4	-12	1.9-10.1%	240	13.7
2-Chlorobutane	C_4H_9Cl	71	-10			21.0
2-Chloro-1-butene	C_4H_7Cl	58	-19	2.3-9.3%		

Safety

Name	Mol. form.	t_b/°C	FP/°C	Flam. limits	IT/°C	VP (25 °C)/kPa
Chlorocyclohexane	$C_6H_{11}Cl$	142.6	32			1.0
1-Chloro-1,1-difluoroethane	$C_2H_3ClF_2$	-9.12		6-18%	632	351
1-Chloro-2,4-dinitrobenzene	$C_6H_3ClN_2O_4$	315	194	2.0-22%		
Chloroethane	C_2H_5Cl	12.3	-50	3.8-15.4%	519	160
2-Chloroethanol	C_2H_5ClO	126	60	4.9-15.9%	425	1.2
Chloroethene	C_2H_3Cl	-13.8	-78	3.6-33.0%	472	355
1-Chloro-4-ethylbenzene	C_8H_9Cl	184.4	64			
2-Chloroethyl vinyl ether	C_4H_7ClO	108	27			
1-Chlorohexane	$C_6H_{13}Cl$	135.0	35			1.25
Chloromethane	CH_3Cl	-24.1		8.1-17.4%	632	574
(Chloromethyl)benzene	C_7H_7Cl	174	67	>1.1%	585	0.164
1-Chloro-3-methylbutane	$C_5H_{11}Cl$	99	<21	1.5-7.4%		
2-Chloro-2-methylbutane	$C_5H_{11}Cl$	85		1.5-7.4%	345	11.5
3-(Chloromethyl)heptane	$C_8H_{17}Cl$	171	60			
1-Chloro-2-methylpropane	C_4H_9Cl	69	-6	2.0-8.7%		19.9
2-Chloro-2-methylpropane	C_4H_9Cl	50.9	0			42.7
3-Chloro-2-methylpropene	C_4H_7Cl	72	-12	3.2-8.1%		16.9
1-Chloronaphthalene	$C_{10}H_7Cl$	259	121		>558	0.003
1-Chloro-4-nitrobenzene	$C_6H_4ClNO_2$	238	127			0.003
1-Chloro-1-nitroethane	$C_2H_4ClNO_2$	124.5	56			
1-Chloro-1-nitropropane	$C_3H_6ClNO_2$	142	62			
2-Chloro-2-nitropropane	$C_3H_6ClNO_2$	142	57			
1-Chloro-4-nitro-2-(trifluoromethyl)benzene	$C_7H_3ClF_3NO_2$	232	135			
1-Chlorooctane	$C_8H_{17}Cl$	183	70			0.11
1-Chloropentane	$C_5H_{11}Cl$	107.9	13	1.6-8.6%	260	4.36
2-Chlorophenol	C_6H_5ClO	173.4	64			0.308
4-Chlorophenol	C_6H_5ClO	219	121			
1-Chloropropane	C_3H_7Cl	46.2	<-18	2.6-11.1%	520	45.8
2-Chloropropane	C_3H_7Cl	35.0	-32	2.8-10.7%	593	68.9
3-Chloropropanenitrile	C_3H_4ClN	175.5	76			
2-Chloropropanoic acid	$C_3H_5ClO_2$	185	107		500	
2-Chloro-1-propanol	C_3H_7ClO	133.5	52			
1-Chloro-2-propanol	C_3H_7ClO	124.4	52			
2-Chloropropene	C_3H_5Cl	23	-37	4.5-16%		110
3-Chloropropene	C_3H_5Cl	44.8	-32	2.9-11.1%	485	48.9
Chlorotrifluoroethene	C_2ClF_3	-28.3		8.4-16.0%		1293
1-Chloro-2-(trifluoromethyl)benzene	$C_7H_4ClF_3$	153	59			
o-Cresol	C_7H_8O	191.0	81	>1.4%	599	0.041
m-Cresol	C_7H_8O	202.2	86	>1.1%	558	0.019
p-Cresol	C_7H_8O	201.9	86	>1.1%	558	0.017
Cyanamide	CH_2N_2		141			
Cyanogen	C_2N_2	-21.1		6.6-32%		
Cyclobutane	C_4H_8	12.5	<10	>1.8%		157
1,5,9-Cyclododecatriene	$C_{12}H_{18}$	240	71			
Cycloheptane	C_7H_{14}	118.8	<21	1.1-6.7%		2.90
Cyclohexane	C_6H_{12}	80.7	-20	1.3-8%	245	13.0
1,4-Cyclohexanedimethanol	$C_8H_{16}O_2$	283	167		316	
Cyclohexanethiol	$C_6H_{12}S$	158.8	43			
Cyclohexanol	$C_6H_{12}O$	160.9	68	1-9%	300	0.10
Cyclohexanone	$C_6H_{10}O$	155.4	44	1.1-9.4%	420	0.53
Cyclohexene	C_6H_{10}	82.9	-12	>1.2%	310	11.8
3-Cyclohexene-1-carboxaldehyde	$C_7H_{10}O$	164	57			
Cyclohexyl acetate	$C_8H_{14}O_2$	174	58		335	
Cyclohexylamine	$C_6H_{13}N$	133.6	31	1.9-9.4%	293	1.20
Cyclohexylbenzene	$C_{12}H_{16}$	239	99			
Cyclohexyl formate	$C_7H_{12}O_2$	162	51			
cis,cis-1,5-Cyclooctadiene	C_8H_{12}	149	35			
Cyclopentane	C_5H_{10}	49.2	-25	>1.5%	361	42.3
Cyclopentanol	$C_5H_{10}O$	140.4	51			0.294
Cyclopentanone	C_5H_8O	130.5	26			1.55
Cyclopentene	C_5H_8	44.2	-29		395	50.7
Cyclopropane	C_3H_6	-31		2.4-10.4%	498	
trans-Decahydronaphthalene	$C_{10}H_{18}$	187.3	54	0.7-5.4%	255	0.164

Name	Mol. form.	t_b/°C	FP/°C	Flam. limits	IT/°C	VP (25 °C)/kPa
Decane	$C_{10}H_{22}$	174.1	51	0.8-5.4%	210	0.170
1-Decanol	$C_{10}H_{22}O$	229	82		288	0.009
1-Decene	$C_{10}H_{20}$	171	<55		235	0.210
Decylamine	$C_{10}H_{23}N$	217	99			
Decylbenzene	$C_{16}H_{26}$	298	107			
1-Decylnaphthalene	$C_{20}H_{28}$	379	177			
Diacetone alcohol	$C_6H_{12}O_2$	167.9	58	1.8-6.9%	643	0.224
3,3-Diacetoxy-1-propene	$C_7H_{10}O_4$	176	82			
Diallyl ether	$C_6H_{10}O$	94	-7			
4,4'-Diaminodiphenylmethane	$C_{13}H_{14}N_2$	379	220			
Dibenzyl ether	$C_{14}H_{14}O$	298	135			
Diborane	B_2H_6	-92.49	-90	1-98%	≈40	
1,2-Dibutoxyethane	$C_{10}H_{22}O_2$	198	85			
Dibutylamine	$C_8H_{19}N$	162	47	1.1-6%		0.34
N,N-Dibutylaniline	$C_{14}H_{23}N$	274.8	110			
Dibutyl ether	$C_8H_{18}O$	141.6	25	1.5-7.6%	194	0.898
Dibutyl maleate	$C_{12}H_{20}O_4$	280	141			0.002
2,6-Di-*tert*-butyl-4-methylphenol	$C_{15}H_{24}O$	265	127			0.0004
Dibutyl oxalate	$C_{10}H_{18}O_4$	244	104			
Di-*tert*-butyl peroxide	$C_8H_{18}O_2$	110.0	18			3.43
Dibutyl phthalate	$C_{16}H_{22}O_4$	338	157	>0.5%	402	
Dibutyl sebacate	$C_{18}H_{34}O_4$	356	178	>0.4%	365	
Dibutyl sulfide	$C_8H_{18}S$	168	76			0.09
Dibutyl tartrate	$C_{12}H_{22}O_6$	320	91		284	
Dichloroacetyl chloride	C_2HCl_3O	108	66			
3,4-Dichloroaniline	$C_6H_5Cl_2N$	273.0	166			
o-Dichlorobenzene	$C_6H_4Cl_2$	180.2	66	2.2-9.2%	648	0.18
m-Dichlorobenzene	$C_6H_4Cl_2$	172	72			0.252
p-Dichlorobenzene	$C_6H_4Cl_2$	173.9	66			0.235
1,2-Dichlorobutane	$C_4H_8Cl_2$	123.9			275	2.82
1,4-Dichlorobutane	$C_4H_8Cl_2$	155	52			
Dichlorodimethylsilane	$C_2H_6Cl_2Si$	70.5	<21	3.4-9.5%		18.9
Dichlorodiphenylsilane	$C_{12}H_{10}Cl_2Si$	304	142			
1,1-Dichloroethane	$C_2H_4Cl_2$	56.3	-17	5.4-11.4%	458	30.5
1,2-Dichloroethane	$C_2H_4Cl_2$	83.4	13	6.2-16%	413	10.6
1,1-Dichloroethene	$C_2H_2Cl_2$	31.6	-28	6.5-15.5%	570	80.0
cis-1,2-Dichloroethene	$C_2H_2Cl_2$	60	6	3-15%	460	26.8
trans-1,2-Dichloroethene	$C_2H_2Cl_2$	47.64	2	6-13%	460	44.2
Dichloromethane	CH_2Cl_2	39.8		13-23%	556	58.2
Dichloromethylsilane	CH_4Cl_2Si	40.9	-9	6.0-55%	316	20.1
1,1-Dichloro-1-nitroethane	$C_2H_3Cl_2NO_2$	123.5	76			
1,1-Dichloro-1-nitropropane	$C_3H_5Cl_2NO_2$	145	66			
1,5-Dichloropentane	$C_5H_{10}Cl_2$	182.9	>27			
2,4-Dichlorophenol	$C_6H_4Cl_2O$	210	114			
1,2-Dichloropropane, (±)-	$C_3H_6Cl_2$	96.4	21	3.4-14.5%	557	6.62
1,3-Dichloro-2-propanol	$C_3H_6Cl_2O$	171	74			0.125
2,3-Dichloropropene	$C_3H_4Cl_2$	93.0	15	2.6-7.8%		
Dichlorosilane	Cl_2H_2Si	8.3		4.1-99%	36	
Dicyclohexylamine	$C_{12}H_{23}N$	251	>99			0.003
Diethanolamine	$C_4H_{11}NO_2$	271.2	172	2-13%	662	<0.01
1,1-Diethoxyethane	$C_6H_{14}O_2$	102	-21	1.6-10.4%	230	3.68
1,2-Diethoxyethane	$C_6H_{14}O_2$	120.6	27		205	4.33
Diethylamine	$C_4H_{11}N$	55.4	-23	1.8-10.1%	312	30.1
N,N-Diethylaniline	$C_{10}H_{15}N$	216	85		630	0.025
o-Diethylbenzene	$C_{10}H_{14}$	183.4	57		395	0.13
m-Diethylbenzene	$C_{10}H_{14}$	181.1	56		450	0.14
p-Diethylbenzene	$C_{10}H_{14}$	184	55	0.7-6.0%	430	0.13
N,N'-Diethylcarbanilide	$C_{17}H_{20}N_2O$		150			
Diethyl carbonate	$C_5H_{10}O_3$	125.9	25			1.63
Diethylene glycol	$C_4H_{10}O_3$	245.5	124	2-17%	224	0.001
Diethylene glycol diacetate	$C_8H_{14}O_5$	200	135			
Diethylene glycol dibutyl ether	$C_{12}H_{26}O_3$	255	118		310	0.005
Diethylene glycol diethyl ether	$C_8H_{18}O_3$	185	82			0.10

Name	Mol. form.	t_b/°C	FP/°C	Flam. limits	IT/°C	VP (25 °C)/kPa
Diethylene glycol dimethyl ether	$C_6H_{14}O_3$	162	67			0.315
Diethylene glycol monoethyl ether	$C_6H_{14}O_3$	202	96			0.017
Diethylene glycol monoethyl ether acetate	$C_8H_{16}O_4$	218	110		425	0.029
Diethylene glycol monomethyl ether	$C_5H_{12}O_3$	194	96	1.38-22.7%	240	0.024
N,*N*-Diethyl-1,2-ethanediamine	$C_6H_{16}N_2$	144	46			
Diethyl ether	$C_4H_{10}O$	34.4	-45	1.9-36.0%	180	71.7
Diethyl fumarate	$C_8H_{12}O_4$	214	104			
Diethyl maleate	$C_8H_{12}O_4$	222	121		350	0.015
Diethyl malonate	$C_7H_{12}O_4$	200	93			0.048
1,3-Diethyl-5-methylbenzene	$C_{11}H_{16}$	205			455	0.074
Diethyl oxalate	$C_6H_{10}O_4$	186	76			0.030
3,3-Diethylpentane	C_9H_{20}	146.2		0.7-5.7%	290	
Diethyl phthalate	$C_{12}H_{14}O_4$	298	161	>0.7%	457	0.002
N,*N*-Diethyl-1,3-propanediamine	$C_7H_{18}N_2$	165	59			
Diethyl selenide	$C_4H_{10}Se$	108		>2.5%		
Diethyl succinate	$C_8H_{14}O_4$	217	90			0.15
Diethyl sulfate	$C_4H_{10}O_4S$	208	104		436	0.05
Diethyl DL-tartrate	$C_8H_{14}O_6$	281	93			
Diethyl terephthalate	$C_{12}H_{14}O_4$	303	117			
1,1-Difluoroethene	$C_2H_2F_2$	-85.5		5.5-21.3%		
Digermane	Ge_2H_6	29			≈50	
3,4-Dihydro-2*H*-pyran	C_5H_8O	85.5	-18			11.4
Diisobutylamine	$C_8H_{19}N$	139.6	29			0.972
Diisopropanolamine	$C_6H_{15}NO_2$	250	127		374	
Diisopropylamine	$C_6H_{15}N$	84	-1	1.1-7.1%	316	10.7
Diisopropyl ether	$C_6H_{14}O$	68.4	-28	1.4-7.9%	443	19.9
Diketene	$C_4H_4O_2$	127.0	34			1.41
3,3'-Dimethoxybenzidine	$C_{14}H_{16}N_2O_2$		206			
1,2-Dimethoxyethane	$C_4H_{10}O_2$	85.0	-2		202	9.93
Dimethoxymethane	$C_3H_8O_2$	42.3	-32	2.2-13.8%	237	53.1
N,*N*-Dimethylacetamide	C_4H_9NO	165.9	70	1.8-11.5%	490	0.075
Dimethylamine	C_2H_7N	7.3	20	2.8-14.4%	400	203
3-(Dimethylamino)propanenitrile	$C_5H_{10}N_2$	177	65			
2,3-Dimethylaniline	$C_8H_{11}N$	223	97	>1.0%		
2,6-Dimethylaniline	$C_8H_{11}N$	211	96			0.45
N,*N*-Dimethylaniline	$C_8H_{11}N$	193	63		371	0.107
2,2-Dimethylbutane	C_6H_{14}	49.7	-48	1.2-7.0%	405	42.5
2,3-Dimethylbutane	C_6H_{14}	58.0	-29	1.2-7.0%	405	31.3
2,3-Dimethyl-1-butene	C_6H_{12}	55.59	<-20		360	33.6
2,3-Dimethyl-2-butene	C_6H_{12}	73.19	<-20		401	16.7
Dimethyl carbonate	$C_3H_6O_3$	90.11	19			
Dimethylcyanamide	$C_3H_6N_2$	162	71			
cis-1,2-Dimethylcyclohexane	C_8H_{16}	129.7	16		304	1.93
trans-1,2-Dimethylcyclohexane	C_8H_{16}	123.4	11		304	2.58
cis-1,4-Dimethylcyclohexane	C_8H_{16}	124.3	16			2.39
Dimethyl disulfide	$C_2H_6S_2$	109.72	24			3.82
Dimethyl ether	C_2H_6O	-24.8	-41	3.4-27.0%	350	
N,*N*-Dimethylformamide	C_3H_7NO	152.8	58	2.2-15.2%	445	0.439
2,5-Dimethylfuran	C_6H_8O	96	7			
2,6-Dimethyl-4-heptanone	$C_9H_{18}O$	157	49	0.8-7.1%	396	0.23
2,3-Dimethylhexane	C_8H_{18}	115.6	7		438	3.13
2,4-Dimethylhexane	C_8H_{18}	109.4	10			4.05
1,1-Dimethylhydrazine	$C_2H_8N_2$	62.4	-15	2-95%	249	20.9
Dimethyl isophthalate	$C_{10}H_{10}O_4$	285.1	138			
Dimethyl maleate	$C_6H_8O_4$	202	113			
2,6-Dimethylmorpholine	$C_6H_{13}NO$	147	44			
3,7-Dimethyl-6-octen-1-ol, (±)-	$C_{10}H_{20}O$	224	96			
2,3-Dimethylpentane	C_7H_{16}	89.8	-56	1.1-6.7%	335	9.18
2,4-Dimethylpentane	C_7H_{16}	80.4	-12			13.1
2,4-Dimethyl-3-pentanol	$C_7H_{16}O$	142	49			
Dimethyl phthalate	$C_{10}H_{10}O_4$	282.7	146	>0.9%	490	0.001
2,2-Dimethyl-1,3-propanediol	$C_5H_{12}O_2$	207	129		399	
2,2-Dimethyl-1-propanol	$C_5H_{12}O$	112	37			

Name	Mol. form.	t_b/°C	FP/°C	Flam. limits	IT/°C	VP (25 °C)/kPa
2,5-Dimethylpyrazine	$C_6H_8N_2$	152	64			
Dimethyl sebacate	$C_{12}H_{22}O_4$	289	145			
Dimethyl sulfate	$C_2H_6O_4S$	186	83		188	0.13
Dimethyl sulfide	C_2H_6S	37.32	-37	2.2-19.7%	206	64.4
Dimethyl sulfoxide	C_2H_6OS	191.9	95	2.6-42%	215	0.084
Dimethyl terephthalate	$C_{10}H_{10}O_4$	288	153		518	
2,4-Dinitroaniline	$C_6H_5N_3O_4$		224			
Dioctyl ether	$C_{16}H_{34}O$	289	>100		205	
1,4-Dioxane	$C_4H_8O_2$	101.2	12	2.0-22%	180	4.95
1,3-Dioxolane	$C_3H_6O_2$	75.3	2			14.6
Dipentene	$C_{10}H_{16}$	176	45		237	0.259
Dipentylamine	$C_{10}H_{23}N$	204	51			
Dipentyl ether	$C_{10}H_{22}O$	187	57		170	0.13
Dipentyl sulfide	$C_{10}H_{22}S$		85			
Diphenylamine	$C_{12}H_{11}N$	305.1	153		634	
1,3-Diphenyl-2-buten-1-one	$C_{16}H_{14}O$	342.5	177			
1,1-Diphenylethane	$C_{14}H_{14}$	285.8	>100		440	
Diphenyl ether	$C_{12}H_{10}O$	258.0	112	0.8-1.5%	618	
Diphenylmethane	$C_{13}H_{12}$	264.2	130		485	
Diphenyl phthalate	$C_{20}H_{14}O_4$		224			
Diphosphine	H_4P_2	63.5 dec			≈20	
Dipropylamine	$C_6H_{15}N$	107.5	17		299	3.21
Dipropyl ether	$C_6H_{14}O$	90.1	21	1.3-7.0%	188	8.35
Disilane	H_6Si_2	-14.8	-14		≈20	
Divinyl ether	C_4H_6O	28	<-30	1.7-27%	360	89.3
Dodecane	$C_{12}H_{26}$	216.3	74	>0.6%	203	0.016
1-Dodecanethiol	$C_{12}H_{26}S$	277	128			
1-Dodecanol	$C_{12}H_{26}O$	264.1	127		275	0.000016
1-Dodecene	$C_{12}H_{24}$	213.4	79			0.019
Dodecylbenzene	$C_{18}H_{30}$	329	140			
Eicosane	$C_{20}H_{42}$	344.1	>100		232	
Epichlorohydrin	C_3H_5ClO	111.99	31	3.8-21.0%	411	2.2
1,2-Epoxybutane	C_4H_8O	63.4	-22	1.7-19%	439	31.7
Ethane	C_2H_6	-88.6		3.0-12.5%	472	
1,2-Ethanediamine	$C_2H_8N_2$	116.9	40	2.5-12.0%	385	1.62
1,2-Ethanediol	$C_2H_6O_2$	197.5	111	3.2-22%	398	0.01
1,2-Ethanediol, diacetate	$C_6H_{10}O_4$	184	88	1.6-8.4%	482	0.030
1,2-Ethanediol, monoacetate	$C_4H_8O_3$	188	102			
Ethanethiol	C_2H_6S	35.0	-17	2.8-18.0%	300	70.3
Ethanol	C_2H_6O	78.24	13	3.3-19%	363	7.87
Ethanolamine	C_2H_7NO	170.3	86	3.0-23.5%	410	0.05
Ethoxyacetylene	C_4H_6O	50	<-7			
2-Ethoxyaniline	$C_8H_{11}NO$	233	115			
4-Ethoxyaniline	$C_8H_{11}NO$	254	116			0.0007
Ethoxybenzene	$C_8H_{10}O$	169.8	63			0.204
2-Ethoxyethanol	$C_4H_{10}O_2$	134.7	43	3-18%	235	0.71
2-Ethoxyethyl acetate	$C_6H_{12}O_3$	156.6	56	2-8%	379	0.24
3-Ethoxypropanal	$C_5H_{10}O_2$	135.2	38			
N-Ethylacetamide	C_4H_9NO	205	110			
Ethyl acetate	$C_4H_8O_2$	77.1	-4	2.0-11.5%	426	12.6
Ethyl acetoacetate	$C_6H_{10}O_3$	180	57	1.4-9.5%	295	0.095
Ethyl acrylate	$C_5H_8O_2$	98.9	10	1.4-14%	372	5.14
Ethylamine	C_2H_7N	16.6	-16	3.5-14%	385	141
N-Ethylaniline	$C_8H_{11}N$	204	85			0.039
Ethylbenzene	C_8H_{10}	136.2	21	0.8-6.7%	432	1.28
α-Ethylbenzenemethanol	$C_9H_{12}O$	219	100			
Ethyl benzoate	$C_9H_{10}O_2$	212.5	88		490	0.04
Ethyl benzoylacetate	$C_{11}H_{12}O_3$	267 dec	141			
Ethyl bromoacetate	$C_4H_7BrO_2$	159	48			
2-Ethylbutanal	$C_6H_{12}O$		21	1.2-7.7%		
Ethyl butanoate	$C_6H_{12}O_2$	121.1	24		463	2.01
2-Ethylbutanoic acid	$C_6H_{12}O_2$	193	99		400	
2-Ethyl-1-butanol	$C_6H_{14}O$	155	57			0.206

Name	Mol. form.	t_b/°C	FP/°C	Flam. limits	IT/°C	VP (25 °C)/kPa
2-Ethyl-1-butene	C_6H_{12}	64.7	<-20		315	23.4
Ethyl *trans*-2-butenoate	$C_6H_{10}O_2$	140	2			
2-Ethylbutyl acetate	$C_8H_{16}O_2$	161	54			
Ethyl *N*-butylcarbamate	$C_7H_{15}NO_2$	221	92			
Ethyl chloroacetate	$C_4H_7ClO_2$	144	64			0.640
Ethyl chloroformate	$C_3H_5ClO_2$	91	16		500	
Ethyl cyanoacetate	$C_5H_7NO_2$	216	110			0.003
Ethylcyclobutane	C_6H_{12}	70	-15	1.2-7.7%	210	
Ethylcyclohexane	C_8H_{16}	131.8	35	0.9-6.6%	238	1.71
Ethylcyclopentane	C_7H_{14}	103.5	<21	1.1-6.7%	260	5.32
Ethyl decanoate	$C_{12}H_{24}O_2$	242	>100			
3-Ethyl-2,4-dimethylpentane	C_9H_{20}	123	390			1.33
Ethylene	C_2H_4	-103.8		2.7-36%	450	
Ethylene carbonate	$C_3H_4O_3$	246	143			0.003
Ethyleneimine	C_2H_5N	54	-11	3.3-54.8%	320	28.9
Ethyl formate	$C_3H_6O_2$	54.09	-20	2.8-16.0%	455	32.3
2-Ethylhexanal	$C_8H_{16}O$	161	44	0.85-7.2%	190	
2-Ethyl-1,3-hexanediol	$C_8H_{18}O_2$	243	127		360	
Ethyl hexanoate	$C_8H_{16}O_2$	165	49			0.158
2-Ethyl-1-hexanol	$C_8H_{18}O$	186.2	73	0.88-9.7%	231	0.019
2-Ethylhexyl acetate	$C_{10}H_{20}O_2$	200	71	0.76-8.14%	268	0.09
2-Ethylhexyl acrylate	$C_{11}H_{20}O_2$		82		252	
2-Ethylhexylamine	$C_8H_{19}N$	172	60			
Ethyl lactate	$C_5H_{10}O_3$	151	46	>1.5%	400	
Ethyl methacrylate	$C_6H_{10}O_2$	116	20			2.62
Ethyl methyl ether	C_3H_8O	6	-37	2.0-10.1%	190	195
3-Ethyl-4-methylhexane	C_9H_{20}	140	24			
4-Ethyl-2-methylhexane	C_9H_{20}	134	<21	>0.7%	280	
3-Ethyl-2-methylpentane	C_8H_{18}	115.6	<21		460	3.19
Ethyl 2-methylpropanoate	$C_6H_{12}O_2$	111	13			3.25
N-Ethylmorpholine	$C_6H_{13}NO$	145	32			
1-Ethylnaphthalene	$C_{12}H_{12}$	258			480	
Ethyl nitrate	$C_2H_5NO_3$	89	10	>4%		8.56
Ethyl nitrite	$C_2H_5NO_2$	17.5	-35	4.0-50%	90	135
3-Ethyloctane	$C_{10}H_{22}$	165			230	
4-Ethyloctane	$C_{10}H_{22}$	163			229	
Ethyl octanoate	$C_{10}H_{20}O_2$	206	79			
4-Ethylphenol	$C_8H_{10}O$	217.97	104			
Ethyl phenylacetate	$C_{10}H_{12}O_2$	228	99			
5-Ethyl-2-picoline	$C_8H_{11}N$	178	68	1.1-6.6%		
Ethyl propanoate	$C_5H_{10}O_2$	98.9	12	1.9-11%	440	4.97
Ethyl propyl ether	$C_5H_{12}O$	63	<-20	1.7-9.0%		24.2
Ethyl silicate	$C_8H_{20}O_4Si$	168	52			1.17
2-Ethyltoluene	C_9H_{12}	165.1			440	
3-Ethyltoluene	C_9H_{12}	161.3			480	
4-Ethyltoluene	C_9H_{12}	162.0			475	
Ethyl *p*-toluenesulfonate	$C_9H_{12}O_3S$		158			
Ethyl vinyl ether	C_4H_8O	36	<-46	1.7-28%	202	68.8
Eucalyptol	$C_{10}H_{18}O$	176	48			0.260
Fluorobenzene	C_6H_5F	84.7	-15			10.4
Fluoroethene	C_2H_3F	-72		2.6-21.7%		
Formaldehyde	CH_2O	-19.1	85	7.0-73%	424	
Formamide	CH_3NO	217	154			0.01
Formic acid	CH_2O_2	101	50	18-57%	434	5.75
Furan	C_4H_4O	31.3	-36	2.3-14.3%		80.0
2-Furanmethanamine	C_5H_7NO	145.5	37			
Furfural	$C_5H_4O_2$	161.5	60	2.1-19.3%	316	0.29
Furfuryl alcohol	$C_5H_6O_2$	168	75	1.8-16.3%	491	0.097
Geraniol	$C_{10}H_{18}O$	229	>100			
Germane	GeH_4	-88.1			≈20	
Glycerol	$C_3H_8O_3$	289	199	3-19%	370	<0.01
Glycerol triacetate	$C_9H_{14}O_6$	259	138	>1.0%	433	<0.01
1-Heptadecanol	$C_{17}H_{36}O$	324	154			

Name	Mol. form.	t_b/°C	FP/°C	Flam. limits	IT/°C	VP (25 °C)/kPa
2-Heptadecanone	$C_{17}H_{34}O$	320	120			
Heptane	C_7H_{16}	98.38	-4	1.05-6.7%	204	6.09
2-Heptanol, (±)-	$C_7H_{16}O$	159	71			
3-Heptanol, (*S*)-	$C_7H_{16}O$	163	60			
2-Heptanone	$C_7H_{14}O$	151.0	39	1.1-7.9%	393	0.49
3-Heptanone	$C_7H_{14}O$	146	46			0.5
4-Heptanone	$C_7H_{14}O$	144	49			0.164
1-Heptene	C_7H_{14}	94	-1		260	7.52
trans-2-Heptene	C_7H_{14}	98	<0			6.56
Heptylamine	$C_7H_{17}N$	153	54			
Hexachloro-1,3-butadiene	C_4Cl_6	216			610	0.13
Hexadecane	$C_{16}H_{34}$	286.9	136		202	0.0002
1,4-Hexadiene (unspecified isomer)	C_6H_{10}	65	-21	2.0-6.1%		
1,5-Hexadien-3-yne	C_6H_6	85	<-20	>1.5%		10.6
Hexanal	$C_6H_{12}O$	129.6	32			1.48
Hexane	C_6H_{14}	68.72	-22	1.1-7.5%	225	20.2
Hexanedinitrile	$C_6H_8N_2$	295	93	>1.0%	550	<0.01
1,6-Hexanedioic acid	$C_6H_{10}O_4$	337.5	196		420	
2,5-Hexanediol	$C_6H_{14}O_2$	229	110			
2,5-Hexanedione	$C_6H_{10}O_2$	194	79		499	
Hexanedioyl dichloride	$C_6H_8Cl_2O_2$		72			
1,2,6-Hexanetriol	$C_6H_{14}O_3$		191			
Hexanoic acid	$C_6H_{12}O_2$	204.9	102		380	0.005
1-Hexanol	$C_6H_{14}O$	156.9	63		290	0.11
2-Hexanone	$C_6H_{12}O$	127.6	25	1-8%	423	1.54
3-Hexanone	$C_6H_{12}O$	123.5	35	1-8%		
1-Hexene	C_6H_{12}	63.4	-26	1.2-6.9%	253	24.8
cis-2-Hexene	C_6H_{12}	68.9	-21			20.0
cis-3-Hexen-1-ol	$C_6H_{12}O$	157	54			
Hexyl acetate	$C_8H_{16}O_2$	171.1	45			0.185
sec-Hexyl acetate	$C_8H_{16}O_2$	147.5	45			
Hexylamine	$C_6H_{15}N$	132	29			1.17
2-Hexyne	C_6H_{10}	84.3	-10			10.7
Hydrazine	H_4N_2	113.55	38	5-100%		1.91
Hydrogen	H_2	-252.879		4-74%		
Hydrogen cyanide	CHN	25.63	-18	6-40%	538	98.8
Hydrogen disulfide	H_2S_2	70.7	<22			
Hydrogen sulfide	H_2S	-59.55		4-44%	260	
Hydrogen telluride	H_2Te	-2			-50	
p-Hydroquinone	$C_6H_6O_2$	288	165		516	
2-Hydroxybenzoic acid	$C_7H_6O_3$		157	>1.1%	540	
2-Hydroxybiphenyl	$C_{12}H_{10}O$	281	124		530	
3-Hydroxybutanal	$C_4H_8O_2$		66		250	
3-Hydroxypropanenitrile	C_3H_5NO	218	129			0.010
Isobutanal	C_4H_8O	64.1	-18	1.6-10.6%	196	23.0
Isobutane	C_4H_{10}	-11.7	-87	1.8-8.4%	460	348
Isobutene	C_4H_8	-7.0		1.8-9.6%	465	300
Isobutyl acetate	$C_6H_{12}O_2$	116.9	18	1.3-10.5%	421	2.39
Isobutyl acrylate	$C_7H_{12}O_2$	137.0	30		427	
Isobutylamine	$C_4H_{11}N$	68.8	-9	2-12%	378	19.0
Isobutylbenzene	$C_{10}H_{14}$	172.7	55	0.8-6.0%	427	0.257
Isobutyl butanoate	$C_8H_{16}O_2$	157	50			0.500
Isobutylcyclohexane	$C_{10}H_{20}$	171.3			274	
Isobutyl formate	$C_5H_{10}O_2$	98.4	5	2-9%	320	5.34
Isobutyl isobutanoate	$C_8H_{16}O_2$	148	38	0.96-7.59%	432	0.552
Isobutyl vinyl ether	$C_6H_{12}O$	83	-9			9.30
Isopentane	C_5H_{12}	27.83	-51	1.4-7.6%	420	91.7
Isopentyl acetate	$C_7H_{14}O_2$	141.6	25	1.0-7.5%	360	0.728
Isopentyl butanoate	$C_9H_{18}O_2$	184.8	59			0.160
Isopentyl nitrite	$C_5H_{11}NO_2$	99			210	
Isophorone	$C_9H_{14}O$	214.8	84	0.8-3.8%	460	0.06
Isopropenyl acetate	$C_5H_8O_2$	97.4	26		432	6.02
Isopropenylbenzene	C_9H_{10}	165.4	54	1.9-6.1%	574	0.40

Name	Mol. form.	t_b/°C	FP/°C	Flam. limits	IT/°C	VP (25 °C)/kPa
2-Isopropoxyethanol	$C_5H_{12}O_2$	141.6	33			
Isopropyl acetate	$C_5H_{10}O_2$	88.6	2	1.8-8%	460	7.88
Isopropylamine	C_3H_9N	31.8	-37		402	78.0
Isopropylbenzene	C_9H_{12}	152.4	36	0.9-6.5%	424	0.61
Isopropyl benzoate	$C_{10}H_{12}O_2$	218	99			
Isopropylcyclohexane	C_9H_{18}	154.4			283	
Isopropyl formate	$C_4H_8O_2$	68	-6		485	18.2
1-Isopropyl-4-methylbenzene	$C_{10}H_{14}$	177	47	0.7-5.6%	436	0.19
d-Limonene	$C_{10}H_{16}$	177.6	45	0.7-6.1%	237	0.277
Linalol	$C_{10}H_{18}O$	198	71			
Maleic anhydride	$C_4H_2O_3$	202	102	1.4-7.1%	477	
2-Mercaptoethanol	C_2H_6OS	150.0	74			0.202
Mesityl oxide	$C_6H_{10}O$	129.7	31	1.4-7.2%	344	1.47
Methacrylic acid	$C_4H_6O_2$	160	77	1.6-8.8%	68	0.12
Methane	CH_4	-161.5		5.0-15.0%	537	
Methanethiol	CH_4S	6.0	-18	3.9-21.8%		202
Methanol	CH_4O	64.5	11	6.0-36%	464	16.9
2-Methoxyaniline	C_7H_9NO	221	118			0.013
2-Methoxybenzaldehyde	$C_8H_8O_2$	243	118			
2-Methoxyethanol	$C_3H_8O_2$	124.3	39	1.8-14%	285	1.31
2-Methoxyethyl acetate	$C_5H_{10}O_3$	142	49	1.5-12.3%	392	0.67
4-Methoxyphenol	$C_7H_8O_2$	253	132		421	
Methyl abietate	$C_{21}H_{32}O_2$		180			
4-Methylacetanilide	$C_9H_{11}NO$	307	168			
Methyl acetate	$C_3H_6O_2$	56.7	-10	3.1-16%	454	28.8
Methyl acetoacetate	$C_5H_8O_3$	168	77		280	0.241
4-Methylacetophenone	$C_9H_{10}O$	225	96			
Methyl acrylate	$C_4H_6O_2$	80.1	-3	2.8-25%	468	11.0
2-Methylacrylonitrile	C_4H_5N	90	1	2-6.8%		8.26
Methylamine	CH_5N	-6.4	0	4.9-20.7%	430	353
Methyl 2-aminobenzoate	$C_8H_9NO_2$	256	>100			0.002
2-Methylaniline	C_7H_9N	200.0	85		482	0.043
4-Methylaniline	C_7H_9N	201	87		482	1.74
4-Methylanisole	$C_8H_{10}O$	175	60			
α-Methylbenzenemethanol	$C_8H_{10}O$	205	93			
Methyl benzoate	$C_8H_8O_2$	199	83			0.052
α-Methylbenzylamine, (±)-	$C_8H_{11}N$	193	79			
2-Methylbiphenyl	$C_{13}H_{12}$	258	137		502	
2-Methyl-1,3-butadiene	C_5H_8	34.0	-54	1.5-8.9%	395	73.4
3-Methyl-2-butanethiol	$C_5H_{12}S$	109.8	3			
Methyl butanoate	$C_5H_{10}O_2$	101.9	14			4.30
3-Methylbutanoic acid	$C_5H_{10}O_2$	176.5			416	0.067
2-Methyl-1-butanol, (±)-	$C_5H_{12}O$	129.0	50		385	0.416
3-Methyl-1-butanol	$C_5H_{12}O$	130.8	43	1.2-9.0%	350	0.315
2-Methyl-2-butanol	$C_5H_{12}O$	102.4	19	1.2-9.0%	437	2.19
3-Methyl-2-butanol, (±)-	$C_5H_{12}O$	113.7	38			1.20
2-Methyl-1-butene	C_5H_{10}	31.1	-20			81.4
3-Methyl-1-butene	C_5H_{10}	20.1	-7	1.5-9.1%	365	120
2-Methyl-2-butene	C_5H_{10}	38.5	-20			62.1
3-Methyl-3-buten-2-one	C_5H_8O	97		1.8-9.0%		
2-Methyl-1-buten-3-yne	C_5H_6	33	<-7			
Methyl chloroacetate	$C_3H_5ClO_2$	130	57	7.5-18.5%		1.0
Methylcyclohexane	C_7H_{14}	100.9	-4	1.2-6.7%	250	6.18
cis-2-Methylcyclohexanol	$C_7H_{14}O$	165	65		296	
trans-2-Methylcyclohexanol, (±)-	$C_7H_{14}O$	168.4	65		296	
cis-3-Methylcyclohexanol, (±)-	$C_7H_{14}O$	168	70		295	
trans-3-Methylcyclohexanol, (±)-	$C_7H_{14}O$	167	70		295	
cis-4-Methylcyclohexanol	$C_7H_{14}O$	174	70		295	
trans-4-Methylcyclohexanol	$C_7H_{14}O$	175	70		295	
4-Methylcyclohexene	C_7H_{12}	103	-1			5.43
Methylcyclopentane	C_6H_{12}	71.8	-29	1.0-8.35%	258	18.3
2-Methyldecane	$C_{11}H_{24}$	189.2			225	
1-Methyl-2,4-dinitrobenzene	$C_7H_6N_2O_4$	300 dec	207			

Name	Mol. form.	t_b/°C	FP/°C	Flam. limits	IT/°C	VP (25 °C)/kPa
N-Methyl-2-ethanolamine	C_3H_9NO	159.24	74			
Methyl formate	$C_2H_4O_2$	31.6	-19	4.5-23%	449	78.1
3-Methylfuran	C_5H_6O	66	-30			
2-Methylhexane	C_7H_{16}	90.0	-1	1.0-6.0%	280	8.78
3-Methylhexane	C_7H_{16}	92	-4		280	
5-Methyl-2-hexanone	$C_7H_{14}O$	139	36	1.0-8.2%	191	0.691
Methylhydrazine	CH_6N_2	83	-8	2.5-92%	194	6.61
Methyl isocyanate	C_2H_3NO	38.3	-7	5.3-26%	534	57.7
Methyl lactate, (±)-	$C_4H_8O_3$	144.8	49	>2.2%	385	0.62
Methyl methacrylate	$C_5H_8O_2$	100.6	10	1.7-8.2%		5.10
4-Methylmorpholine	$C_5H_{11}NO$	121.4	24			20.4
1-Methylnaphthalene	$C_{11}H_{10}$	244.4			529	0.009
2-Methylnonane	$C_{10}H_{22}$	167			210	
Methyloxirane	C_3H_6O	35	-37	3.1-27.5%	449	71.7
cis-2-Methyl-1,3-pentadiene	C_6H_{10}	80	-12			
4-Methyl-1,3-pentadiene	C_6H_{10}	76	-34			
2-Methylpentanal	$C_6H_{12}O$	118	17		199	
2-Methylpentane	C_6H_{14}	60.21	<-29	1.0-7.0%	264	28.2
3-Methylpentane	C_6H_{14}	63.3	-7	1.2-7.0%	278	25.3
2-Methyl-2,4-pentanediol	$C_6H_{14}O_2$	197.9	102	1-9%	306	<0.01
2-Methylpentanoic acid, (±)-	$C_6H_{12}O_2$	195	107		378	
2-Methyl-1-pentanol	$C_6H_{14}O$	157	54	1.1-9.65%	310	0.236
4-Methyl-2-pentanol	$C_6H_{14}O$	132.0	41	1.0-5.5%		0.698
4-Methyl-2-pentanone	$C_6H_{12}O$	115.7	18	1.2-8.0%	448	2.64
2-Methyl-1-pentene	C_6H_{12}	62.1	-28		300	26.0
4-Methyl-1-pentene	C_6H_{12}	54	-7		300	36.1
2-Methyl-2-pentene	C_6H_{12}	67.3	<-7			21.0
4-Methyl-*cis*-2-pentene	C_6H_{12}	56.4	-32			32.5
4-Methyl-*trans*-2-pentene	C_6H_{12}	58.58	-29			29.7
Methyl 2-phenylacetate	$C_9H_{10}O_2$	215	91			
1-Methylpiperazine	$C_5H_{12}N_2$	135	42			
2-Methylpropanenitrile	C_4H_7N	102	8		482	
2-Methyl-1-propanethiol	$C_4H_{10}S$	88.5	2			9.30
2-Methyl-2-propanethiol	$C_4H_{10}S$	64.2	<-29			24.2
Methyl propanoate	$C_4H_8O_2$	78.6	-2	2.5-13%	469	11.5
2-Methylpropanoic acid	$C_4H_8O_2$	154.4	56	2.0-9.2%	481	0.17
2-Methylpropanoic anhydride	$C_8H_{14}O_3$	183	59	1.0-6.2%	329	
2-Methyl-1-propanol	$C_4H_{10}O$	107.84	28	1.7-10.6%	415	1.39
2-Methyl-2-propanol	$C_4H_{10}O$	82.3	11	2.4-8.0%	478	5.52
2-Methyl-2-propenol	C_4H_8O	114.5	33			
Methyl propyl ether	$C_4H_{10}O$	38.5	-20	2.0-14.8%		60.9
2-Methylpyrazine	$C_5H_6N_2$	129	50			
2-Methylpyridine	C_6H_7N	129.4	39		538	1.5
4-Methylpyridine	C_6H_7N	145.3	57			0.759
1-Methylpyrrole	C_5H_7N	112.7	16			
N-Methylpyrrolidine	$C_5H_{11}N$	206.1	-14			13.5
N-Methyl-2-pyrrolidinone	C_5H_9NO	204.2	96	1-10%	346	0.04
Methyl salicylate	$C_8H_8O_3$	222.6	96		454	0.015
Methyl octadecanoate	$C_{19}H_{38}O_2$	353	153			
2-Methylstyrene	C_9H_{10}	170	53	0.8-11.0%	538	0.245
3-Methylstyrene	C_9H_{10}	170	53	0.8-11.0%	538	0.236
4-Methylstyrene	C_9H_{10}	172	53	0.8-11.0%	538	0.245
2-Methyltetrahydrofuran	$C_5H_{10}O$	80	-11			12.6
Methyl vinyl ether	C_3H_6O	6			287	180
Morpholine	C_4H_9NO	128.2	37	1.4-11.2%	290	1.34
4-Morpholineethanol	$C_6H_{13}NO_2$	220	99			
4-Morpholinepropanamine	$C_7H_{16}N_2O$	220	104			
Naphthalene	$C_{10}H_8$	218.0	79	0.9-5.9%	526	0.011
2-Naphthol	$C_{10}H_8O$	286	153			
1-Naphthylamine	$C_{10}H_9N$	300.7	157			
Neopentane	C_5H_{12}	9.50	-65	1.4-7.5%	450	171
4-Nitroaniline	$C_6H_6N_2O_2$	328	199			
Nitrobenzene	$C_6H_5NO_2$	210.7	88	>1.8%	482	0.03

Safety

Name	Mol. form.	t_b/°C	FP/°C	Flam. limits	IT/°C	VP (25 °C)/kPa
Nitrocyclohexane	$C_6H_{11}NO_2$	205	88			
Nitroethane	$C_2H_5NO_2$	114.1	28	3.4-17%	414	2.79
Nitromethane	CH_3NO_2	101.19	35	>7.3%	418	4.79
1-Nitropropane	$C_3H_7NO_2$	131.2	36	>2.2%	421	1.36
2-Nitropropane	$C_3H_7NO_2$	120.2	24	2.6-11.0%	428	2.3
2-Nitrotoluene	$C_7H_7NO_2$	220.9	106			0.0014
3-Nitrotoluene	$C_7H_7NO_2$	232.1	106			0.03
4-Nitrotoluene	$C_7H_7NO_2$	238.66	106			0.0007
1-Nitro-3-(trifluoromethyl)benzene	$C_7H_4F_3NO_2$	202.8	103			
Nonadecane	$C_{19}H_{40}$	330	>100		230	
2-Nonadecanone	$C_{19}H_{38}O$		124			
Nonane	C_9H_{20}	150.8	31	0.8-2.9%	205	0.570
2-Nonanone	$C_9H_{18}O$	194	60	0.9-5.9%	360	
1-Nonene	C_9H_{18}	146.9	26			0.714
Nonyl acetate	$C_{11}H_{22}O_2$	225	68			
Nonylbenzene	$C_{15}H_{24}$	280	99			
2,5-Norbornadiene	C_7H_8	90	-21			8.97
Octadecane	$C_{18}H_{38}$	316	>100		227	
1-Octadecanol	$C_{18}H_{38}O$	351			450	
cis-9-Octadecenoic acid	$C_{18}H_{34}O_2$	360	189		363	0.000001
Octahydroindene	C_9H_{16}	164			296	
Octanal	$C_8H_{16}O$	174	52			0.321
Octane	C_8H_{18}	125.62	13	1.0-6.5%	206	1.86
1-Octanethiol	$C_8H_{18}S$	199	69			
1-Octanol	$C_8H_{18}O$	194.7	81		270	0.01
2-Octanol	$C_8H_{18}O$	179	88		265	
2-Octanone	$C_8H_{16}O$	173	52			0.12
Octanoyl chloride	$C_8H_{15}ClO$	195	82			
1-Octene	C_8H_{16}	121.3	21		230	2.30
Octylamine	$C_8H_{19}N$	178.6	60			
4-Octylphenyl salicylate	$C_{21}H_{26}O_3$		216		416	
1,4-Oxathiane	C_4H_8OS	149	42			
2-Oxetanone	$C_3H_4O_2$	161	74	>2.9%		0.3
Oxirane	C_2H_4O	10.4	-20	3.0-100%	429	175
Paraformaldehyde	$(CH_2O)x$		70	7.0-73%	300	1.40
Paraldehyde	$C_6H_{12}O_3$	124	36	>1.3%	238	1.6
Pentaborane(9)	B_5H_9	60.10	30	>0.4%	35	
Pentamethylbenzene	$C_{11}H_{16}$	232	93		427	
Pentanal	$C_5H_{10}O$	103	12		222	4.58
Pentane	C_5H_{12}	36.06	-40	1.4-8.0%	260	68.3
1,5-Pentanediol	$C_5H_{12}O_2$	241	129		335	0.001
2,4-Pentanedione	$C_5H_8O_2$	140.7	34		340	1.02
1-Pentanethiol	$C_5H_{12}S$	126.6	18			1.83
Pentanoic acid	$C_5H_{10}O_2$	186.1	96		400	0.024
1-Pentanol	$C_5H_{12}O$	137.6	33	1.2-10.0%	300	0.259
2-Pentanol	$C_5H_{12}O$	119.1	34	1.2-9.0%	343	0.804
3-Pentanol	$C_5H_{12}O$	123	41	1.2-9.0%	435	1.10
2-Pentanone	$C_5H_{10}O$	102.2	7	1.5-8.2%	452	4.97
3-Pentanone	$C_5H_{10}O$	101.9	13	>1.6%	450	4.72
1-Pentene	C_5H_{10}	30.0	-18	1.5-8.7%	275	85.0
cis-2-Pentene	C_5H_{10}	36.9	<-20			66.0
trans-2-Pentene	C_5H_{10}	36.3	<-20			67.4
Pentyl acetate	$C_7H_{14}O_2$	149.4	16	1.1-7.5%	360	0.60
sec-Pentyl acetate, (*R*)-	$C_7H_{14}O_2$	142	32			
Pentylamine	$C_5H_{13}N$	104.7	-1	2.2-22%		4.00
4-*tert*-Pentylaniline	$C_{11}H_{17}N$	260.5	102			
Pentylbenzene	$C_{11}H_{16}$	203	66			
Pentyl butanoate	$C_9H_{18}O_2$	180	57			
Pentylcyclohexane	$C_{11}H_{22}$	204			239	
Pentyl formate	$C_6H_{12}O_2$	126	26			
Pentyl lactate	$C_8H_{16}O_3$		79			
1-Pentylnaphthalene	$C_{15}H_{18}$	305	124			
Pentyl propanoate	$C_8H_{16}O_2$	166	41		378	

Name	Mol. form.	t_b/°C	FP/°C	Flam. limits	IT/°C	VP (25 °C)/kPa
Pentyl salicylate	$C_{12}H_{16}O_3$	270	132			
Pentyl stearate	$C_{23}H_{46}O_2$		185			
1-Pentyne	C_5H_8	39.9	<-20			58.1
Peroxyacetic acid	$C_2H_4O_3$	110	41			1.93
β-Phellandrene	$C_{10}H_{16}$	177	49			0.189
Phenanthrene	$C_{14}H_{10}$	338.4	171			
Phenol	C_6H_6O	181.8	79	1.8-8.6%	715	0.055
2-Phenoxyethanol	$C_8H_{10}O_2$	246	121			0.001
Phenyl acetate	$C_8H_8O_2$	195	80			
N-Phenyl-*N*,*N*-diethanolamine	$C_{10}H_{15}NO_2$		196	>0.7%	387	
N-Phenylethanolamine	$C_8H_{11}NO$	280	152			
Phenylhydrazine	$C_6H_8N_2$	244	88			0.003
Phenyloxirane	C_8H_8O	194.1	74		498	
1-Phenyl-1-propanone	$C_9H_{10}O$	217.4	99			
1-Phenyl-2-propylamine, (±)-	$C_9H_{13}N$	198	<100			0.06
Phorone	$C_9H_{14}O$	197	85			
Phosphine	H_3P	-87.75		>1.8%	100	
Phosphorus (white)	P	280.5			38	
Phthalic acid	$C_8H_6O_4$	dec	168			
Phthalic anhydride	$C_8H_4O_3$	285.3	152	1.7-10.5%	570	
α-Pinene	$C_{10}H_{16}$	156.3	33		255	0.64
β-Pinene	$C_{10}H_{16}$	165.8	38		275	0.61
Piperazine	$C_4H_{10}N_2$	148.63	81			
1-Piperazineethanamine	$C_6H_{15}N_3$	225	93			
1-Piperazineethanol	$C_6H_{14}N_2O$	259	124			
Piperidine	$C_5H_{11}N$	106.19	16	1-10%		4.28
Propanal	C_3H_6O	48.0	-30	2.6-17%	207	42.2
Propane	C_3H_8	-42.11	-104	2.1-9.5%	450	939
1,3-Propanediamine	$C_3H_{10}N_2$	139.2	24			
1,2-Propanediol	$C_3H_8O_2$	187.3	99	2.6-12.5%	371	0.02
1,3-Propanediol	$C_3H_8O_2$	214.7			400	0.007
Propanenitrile	C_3H_5N	97.3	2	3.1-14%	512	6.14
Propanoic acid	$C_3H_6O_2$	141.5	52	2.9-12.1%	465	0.553
Propanoic anhydride	$C_6H_{10}O_3$	168	63	1.3-9.5%	285	0.45
1-Propanol	C_3H_8O	97.04	23	2.2-13.7%	412	2.76
2-Propanol	C_3H_8O	82.21	12	2.0-12.7%	399	6.02
Propanoyl chloride	C_3H_5ClO	80	12			
Propargyl alcohol	C_3H_4O	113	36			
Propene	C_3H_6	-47.6		2.0-11.1%	455	
Propyl acetate	$C_5H_{10}O_2$	101.0	13	1.7-8%	450	4.49
Propylamine	C_3H_9N	47.21	-37	2.0-10.4%	318	42.1
Propylbenzene	C_9H_{12}	159.2	30	0.8-6.0%	450	0.45
Propyl butanoate	$C_7H_{14}O_2$	144	37			0.618
Propylcyclohexane	C_9H_{18}	156.7			248	
Propylcyclopentane	C_8H_{16}	130.9			269	1.64
Propylene carbonate	$C_4H_6O_3$	241.6	135			0.05
Propyl formate	$C_4H_8O_2$	80.6	-3		455	10.9
Propyl nitrate	$C_3H_7NO_3$	110	20	2-100%	175	3.12
Propyl propanoate	$C_6H_{12}O_2$	122.2	79			1.88
Propyne	C_3H_4	-23.2		2.1-12.5%		581
Pyridine	C_5H_5N	115.2	20	1.8-12.4%	482	2.76
Pyrocatechol	$C_6H_6O_2$	246	127			
Pyrrole	C_4H_5N	129.74	39			1.10
Pyrrolidine	C_4H_9N	86.6	3			8.40
2-Pyrrolidone	C_4H_7NO	251.2	129			
Quinoline	C_9H_7N	237.1			480	0.011
Resorcinol	$C_6H_6O_2$	280	127	>1.4%	608	
Safrole	$C_{10}H_{10}O_2$	235	100			0.01
Salicylaldehyde	$C_7H_6O_2$	208	78			0.075
Silane	H_4Si	-111.9	-112	>1.4%	≈20	
Stearic acid	$C_{18}H_{36}O_2$	371	196		395	
Styrene	C_8H_8	145.3	31	0.9-6.8%	490	0.81
Succinonitrile	$C_4H_4N_2$	266	132			0.001

Safety

Name	Mol. form.	t_b/°C	FP/°C	Flam. limits	IT/°C	VP (25 °C)/kPa
Sulfolane	$C_4H_8O_2S$	286	177			<0.01
L-Tartaric acid	$C_4H_6O_6$		210		425	
Terephthalic acid	$C_8H_6O_4$	300 sp	260		496	
o-Terphenyl	$C_{18}H_{14}$	337	163			
m-Terphenyl	$C_{18}H_{14}$	375	191			
α-Terpineol	$C_{10}H_{18}O$	218	90			
1,2,4,5-Tetrachlorobenzene	$C_6H_2Cl_4$	247	155			
Tetradecane	$C_{14}H_{30}$	253.5	112	>0.5%	200	0.002
1-Tetradecanol	$C_{14}H_{30}O$	295.8	141			
1-Tetradecene	$C_{14}H_{28}$	251.1	110		235	
Tetraethylene glycol	$C_8H_{18}O_5$	315	182			0.000001
Tetraethylene glycol dimethyl ether	$C_{10}H_{22}O_5$	274	141			
Tetraethylenepentamine	$C_8H_{23}N_5$	341.5	163		321	
Tetrafluoroethene	C_2F_4	-76		10.0-50.0%	200	
Tetrahydrofuran	C_4H_8O	66.0	-14	2-11.8%	321	21.6
Tetrahydrofurfuryl alcohol	$C_5H_{10}O_2$	176.3	75	1.5-9.7%	282	0.100
1,2,3,4-Tetrahydronaphthalene	$C_{10}H_{12}$	207.2	71	0.8-5.0%	385	0.05
Tetrahydropyran	$C_5H_{10}O$	88.0	-20			9.54
1,2,3,4-Tetramethylbenzene	$C_{10}H_{14}$	205	74		427	
1,2,3,5-Tetramethylbenzene	$C_{10}H_{14}$	198	71		427	
1,2,4,5-Tetramethylbenzene	$C_{10}H_{14}$	197	54			
2,2,3,3-Tetramethylpentane	C_9H_{20}	140.2	<21	0.8-4.9%	430	
2,2,3,4-Tetramethylpentane	C_9H_{20}	133.0	<21			
Tetramethylstannane	$C_4H_{12}Sn$	76.8	-12	>1.9%		14.6
Tetramethylurea	$C_5H_{12}N_2O$	177.1	77			0.138
Tetraphenylstannane	$C_{24}H_{20}Sn$	420	232			
Thiophene	C_4H_4S	84.1	-1			10.6
Toluene	C_7H_8	110.60	4	1.1-7.1%	480	3.79
Toluene-2,4-diisocyanate	$C_9H_6N_2O_2$	251	127	0.9-9.5%		0.003
p-Toluenesulfonic acid	$C_7H_8O_3S$		184			
Tribromosilane	Br_3HSi	109			≈20	
Tributylamine	$C_{12}H_{27}N$	207	63	1-5%		0.01
Tributyl borate	$C_{12}H_{27}BO_3$	233.8	93			
Tributyl phosphate	$C_{12}H_{27}O_4P$	289	146			
Tributyrin	$C_{15}H_{26}O_6$	307.5	180	>0.5%	407	
1,2,4-Trichlorobenzene	$C_6H_3Cl_3$	213.5	105	2.5-6.6%	571	0.057
1,1,1-Trichloroethane	$C_2H_3Cl_3$	74.02	-1	8-10.5%	500	16.5
1,1,2-Trichloroethane	$C_2H_3Cl_3$	113	32	6-28%	460	3.1
Trichloroethene	C_2HCl_3	86.8	32	8-10.5%	420	9.91
Trichloroethylsilane	$C_2H_5Cl_3Si$	98.7	22			6.29
(Trichloromethyl)benzene	$C_7H_5Cl_3$	221	127		211	0.35
Trichloromethylsilane	CH_3Cl_3Si	66	-9	7.6->20%	>404	22.5
Trichlorooctadecylsilane	$C_{18}H_{37}Cl_3Si$		89			
Trichloropentylsilane	$C_5H_{11}Cl_3Si$	172	63			
Trichlorophenylsilane	$C_6H_5Cl_3Si$	201	91			
1,2,3-Trichloropropane	$C_3H_5Cl_3$	158	71	3.2-12.6%		0.492
Trichloropropylsilane	$C_3H_7Cl_3Si$	123.5	37			
Trichlorosilane	Cl_3HSi	33	-50		104	
Trichlorovinylsilane	$C_2H_3Cl_3Si$	90.9	21			8.79
Tri-*o*-cresyl phosphate	$C_{21}H_{21}O_4P$	410	225		385	0.0000002
Tridecane	$C_{13}H_{28}$	235.4	79			0.005
1-Tridecanol	$C_{13}H_{28}O$	287	121			
2-Tridecanone	$C_{13}H_{26}O$	268	107			
1-Tridecene	$C_{13}H_{26}$	232.8	79			0.0047
Triethanolamine	$C_6H_{15}NO_3$	350	179	1-10%		<0.01
Triethylamine	$C_6H_{15}N$	88.8	-7	1.2-8.0%	249	7.70
Triethylene glycol	$C_6H_{14}O_4$	288.6	177	0.9-9.2%	371	0.0002
Triethylene glycol dimethyl ether	$C_8H_{18}O_4$	218	111			
Triethyl phosphate	$C_6H_{15}O_4P$	216	115		454	
(Trifluoromethyl)benzene	$C_7H_5F_3$	102.0	12			5.14
Triisopropanolamine	$C_9H_{21}NO_3$		160		320	
Triisopropyl borate	$C_9H_{21}BO_3$	139.6	28			
Trimethylamine	C_3H_9N	2.8	-5	2.0-11.6%	190	215

Name	Mol. form.	t_b/°C	FP/°C	Flam. limits	IT/°C	VP (25 °C)/kPa
1,2,3-Trimethylbenzene	C_9H_{12}	176.0	44	0.8-6.6%	470	0.20
1,2,4-Trimethylbenzene	C_9H_{12}	169.4	44	0.9-6.4%	500	0.30
1,3,5-Trimethylbenzene	C_9H_{12}	164.7	50	1-5%	559	0.33
Trimethyl borate	$C_3H_9BO_3$	67.4	-8			17.2
2,2,3-Trimethylbutane	C_7H_{16}	80.8	<0		412	28.5
2,3,3-Trimethyl-2-butanol	$C_7H_{16}O$	128.3	<0		375	
Trimethylchlorosilane	C_3H_9ClSi	57.6	-28		395	30.7
2,2,5-Trimethylhexane	C_9H_{20}	124	13			2.21
Trimethylolpropane	$C_6H_{14}O_3$		149			
2,2,3-Trimethylpentane	C_8H_{18}	109.8	<21		346	4.28
2,2,4-Trimethylpentane	C_8H_{18}	99.2	-12		418	6.50
2,3,3-Trimethylpentane	C_8H_{18}	114.7	<21		425	3.60
2,2,4-Trimethyl-1,3-pentanediol	$C_8H_{18}O_2$	230.1	113		346	
2,4,4-Trimethyl-1-pentene	C_8H_{16}	101.3	-5	0.8-4.8%	391	5.96
2,4,4-Trimethyl-2-pentene	C_8H_{16}	104.9	2		305	4.80
Trimethyl phosphate	$C_3H_9O_4P$	197.2	107			0.11
Trimethyl phosphite	$C_3H_9O_3P$	110	54			
Trinitroglycerol	$C_3H_5N_3O_9$	218 exp			270	0.00005
1,3,5-Trioxane	$C_3H_6O_3$	116	45	3.6-29%	414	
Tripentylamine	$C_{15}H_{33}N$	242.5	102			
Triphenylmethane	$C_{19}H_{16}$	359	>100			
Triphenyl phosphate	$C_{18}H_{15}O_4P$		220			
Triphenylphosphine	$C_{18}H_{15}P$		180			
Triphenyl phosphite	$C_{18}H_{15}O_3P$	360	218			
Tripropylamine	$C_9H_{21}N$	153	41			
Trisilane	H_8Si_3	52.9	<0		≈20	
Undecane	$C_{11}H_{24}$	195.9	69			0.05
2-Undecanol	$C_{11}H_{24}O$	231	113			
2-Undecanone	$C_{11}H_{22}O$	233.1	89			
Vinyl acetate	$C_4H_6O_2$	72.6	-8	2.6-13.4%	402	15.4
Vinyl butanoate	$C_6H_{10}O_2$	116.7	20	1.4-8.8%		
4-Vinylcyclohexene	C_8H_{12}	130	16		269	1.87
Vinyloxirane	C_4H_6O	68	<-50			
Vinyl propanoate	$C_5H_8O_2$	94.8	1			
o-Xylene	C_8H_{10}	144.4	32	0.9-6.7%	463	0.88
m-Xylene	C_8H_{10}	139.1	27	1.1-7.0%	527	1.13
p-Xylene	C_8H_{10}	138.3	27	1.1-7.0%	528	1.19

NANOMATERIAL SAFETY GUIDELINES

Thomas J. Bruno and Beverly L. Smith

The classification of nanomaterials includes nano-objects and nanoparticles; nano-objects are materials with at least one dimension (length, width, height, and/or diameter) that is between 1 nm (1×10^{-9} m) and 100 nm, and nanoparticles are materials in which all three dimensions are on this scale. [Note that the ASTM definition allows for two dimensions between 1 nm and 100 nm.] Beyond scale, nanomaterials can be classified as natural, incidental, and engineered, depending on origin. Natural nanomaterials include volcanic products, viruses, sea spray, and mineral aerosols, and are ubiquitous in nature at appreciable concentrations. Incidental nanomaterials include metal vapors produced during welding, sandblasting dust and other industrial effluents, cooking smoke, and diesel engine particulates. The environmental health and safety aspects of natural nanomaterials have received some study, and among the incidental nanomaterials, welding vapors and diesel fuel particulates have received extensive study. In recent years, however, there has been a great emphasis on engineered nanomaterials, and it is this class, which includes metal nanoparticles, nanorods, nanowires, nanotubes, Buckeyballs, nanocapsules, and quantum dots that are the main concern here. Study of the environmental health and safety risks of engineered nanomaterials remains an active area of research that is receiving increasing attention due to the widespread use of these materials in numerous applications ranging from medicine to energy storage. While much is still unknown regarding the fate and toxicity of this class of materials, here, we provide some general guidelines for the safe handling of nanomaterials. We begin with some simplified definitions or terms used in nanotechnology, needed for understanding of these safety guidelines as well as those provided elsewhere.

Aerodynamic diameter: An indirect measure of particle diameter defined as the diameter of a sphere with a density of 1000 kg/m^3, having the same settling velocity of a particle of interest.

Agglomerate: A group of particles (which may include nanoparticles) held together in a loose cluster by weak forces that may include van der Waals forces, surface tension, and electrostatic forces. Agglomerates are often re-suspendable.

Aggregate: A heterogeneous particle held together with relatively strong forces such that the particle is not easily disassembled. Aggregates are typically not re-suspendable.

Buckeyballs: Spherical carbon (C60) fullerenes.

Fullerenes: Molecules composed entirely of carbon, usually in the form of a hollow sphere, ellipsoid, or tubes.

Graphene: A one atom thick sheet of carbon.

Multi-walled carbon nanotube: Multiple sheets of sheet graphene wrapped into a tube of nanoscale dimensions.

Nanoaerosol: A collection of nanomaterials suspended in a gas.

Nanocolloid: A nanomaterial suspended in a gel or other semi-solid substance.

Nanocomposite: A solid material composed of two or more nanomaterials having different physical characteristics.

Nanohydrosol: A nanomaterial suspended in a solution.

Nanotube: A seamless tube with a diameter on the order of nanometers.

Nanowire: A wire of dimensions on the order of nanometers.

Quantum dot: A nanomaterial or nanoparticle that confines the motion of conduction band electrons, valence band holes, or excitons (pairs of conduction band electrons and valence band holes) in all three spatial directions.

Single-walled carbon nanotube: A single-sheet graphene wrapped into a tube of nanoscale dimensions.

Ultrafine particle: A (usually) airborne particle with a diameter less than 100 nm.

Safety Issues and Exposure Routes

The unique safety issues posed by nanomaterials result from the potential of deep penetration into tissue, the potential of passing through the blood-brain barrier, and the possible ability to translocate between organs. Biological effects result from the size, shape, polarity/charge, adsorptive capacity, and surface composition, and the ability to bind biological proteins and receptors. Nanomaterials have a higher reactivity than the parent compounds, often having catalytic effects and often presenting greater flammability or explosion risks. For example, bulk elemental gold is considered inert, but gold nanoparticles below 5 nm are catalytic toward a number of oxidation reactions.

The most obvious exposure route of nanomaterials is respiratory; particles depositable in the air exchange region of the lungs are considered respirable. Ingestion can occur from unintentional hand-to-mouth transfer. Finally, nanoparticles can be absorbed through skin or cuts/abrasions to the skin.

Guidelines for Safe Handling of Nanomaterials

The safe handling of nanomaterials will generally follow the usual laboratory safety grid:

Elimination – A change in the experimental design to avoid the hazard

Substitution – The use of a surrogate of lower hazard

Engineering Controls – The use of enclosures, fume hoods, etc.

Administrative Controls – Adherence to standard procedures and protocols

Personal Protective Equipment (PPE) – The last line of safety, including gloves, clothing, respirators, etc.

Clearly, elimination and substitution are most useful for nanoparticles of incidental origin. Research with engineered nanoparticles must make use of engineering controls, administrative controls, and PPE.

The use of non-regenerating general ventilation systems, such as fume hoods and high efficiency particulate air (HEPA) dust collection systems, are critical to safe handling. Where possible, installation of ultralow particulate air (ULPA) filters should be used, since they are widely viewed as being more effective for engineered nanomaterials. The lab in which nanomaterials are handled should be under negative pressure relative to the surroundings (corridor, service galley, etc.). The entry on "Chemical Fume Hoods and Biological Safety Cabinets" in this section provides additional information on fume hood selection and operation.

Where possible, manipulations should be conducted in solution (or in a liquid phase) to minimize the potential of aerosol formation.

Solution phase nanomaterials should be handled wearing gloves. Gloves should be compatible with the solvent used to disperse the nanomaterials in solution. In general, nitrile gloves are recommended, and double gloving is advisable for heavy usage or prolonged usage. Gloves with cuffs, clothing with full sleeves to protect wrists, or a laboratory coat are recommended. Some glove materials may have reactivity with certain nanomaterials and this must be considered before selecting the glove material. Liquid or solution phase nanomaterial manipulations are best conducted in a fume hood or biological safety cabinet, especially when employing nanomaterials dispersed in open containers, in solvents with known health risks, or when higher risk activities such as sonication, agitation, and vortex mixing are involved. Manipulations conducted in closed containers need not be performed in a fume hood or biological safety cabinet; however, a closed container should be opened inside such an enclosure because aerosols could be released upon opening. Disposable bench covers should be used where spillage is possible. Any spillage should be cleaned up immediately. Contaminated gloves should be removed and replaced immediately.

Manipulation of dry nanomaterials must be performed in a fume hood or biological safety cabinet. Transport of dry nanomaterials from place to place in the lab must be done in closed containers.

For manipulations of air-sensitive nanomaterials, a glove box or glove bag is required.

Hand washing must be done after manipulations of nanomaterials. Work areas should be cleaned after completion of tasks. Adequate consideration should be given to tasks involved with the maintenance of equipment or instrumentation used in work on nanomaterials. Such maintenance should be done with the assumption of the presence of nanomaterials.

Waste Disposal

Though the fate and toxicity of nanomaterials remains largely unknown and is still an area of active investigation, nanomaterials and any by-products from their synthesis should be treated as potentially hazardous waste. Nanomaterials should be properly disposed of based on their nanomaterial and solvent compositions. Nanomaterials containing heavy metals should be treated accordingly and separately from other waste streams. When possible, it is advisable to collect or fully dissolve nanomaterials that are present in solution to limit the volume of waste generated. Often nanomaterials can be aggregated, or precipitated from solution with an anti-solvent and filtered off to be recycled or collected as solid waste. This is called "crashing out" in laboratory vernacular. Alternatively, adsorbents such as activated charcoal can often be used to remove certain types of nanomaterials from solution upon filtration and collection of the adsorbent following exposure to the nanomaterial-containing solution. Other types of nanomaterials such as metal oxides can often be dissolved completely with strong acids.

THRESHOLD LIMITS FOR AIRBORNE CONTAMINANTS

Several organizations recommend limits of exposure to airborne contaminants in the workplace. These include the Occupational Safety and Health Administration (OSHA), the National Institute for Occupational Safety and Health (NIOSH), and the non-governmental organization, American Conference of Governmental Industrial Hygienists (ACGIH). The threshold limit value (TLV) for a substance (also known as permissible exposure limit, PEL) is defined as the concentration level under which the majority of workers may be repeatedly exposed, day after day, without adverse effects. The TLV recommendations are given in two forms:

- Time-weighted average (TWA) concentration for a normal 8-hr workday and 40-hr workweek.
- Short-term exposure limit (STEL), which should not be exceeded for more than 15 min.

Both kinds of limits are specified for some substances.

The following table gives threshold limit values for a number of substances that may be encountered in the atmosphere of a chemical laboratory or industrial facility. All values refer to the concentration in air at 25 °C and normal atmospheric pressure. Data for gases are given in parts per million by volume (ppm). Values for liquids refer to mists or aerosols, and those for solids to dusts or fumes; both are stated in mass concentration units (mg/m^3). In some cases the Comments column includes the statement "Ceiling value", which indicates a ceiling limit that should not be exceeded even for very brief periods because of acute toxic effects of the substance. The notation "levels as low as possible" in the Comments column indicates such a high degree of hazard that no safe limit can be recommended. The expression "inhalable fraction" appearing in the Comments column indicates that the TLV values have been adjusted to reflect the influence of particle size on the respiratory hazard. See Ref. 1 for details.

The notation NIC 2016 indicates that changes will probably be made in 2017.

Substances are listed alphabetically by systematic name. The Comments column gives common synonyms (in brackets) and further information on the physical form of the substance and the basis to which the limit is referred. The Formula column gives the molecular formula in the Hill convention for organic compounds and the customary line formula for inorganic compounds. The TWA and STEL limits appear in the last two columns.

References

1. *2016 TLV's and BEI's*, American Conference of Governmental Industrial Hygienists, 1330 Kemper Meadow Drive, Cincinnati, OH 45240-1634, 2016 <www.acgih.org>.
2. *NIOSH Pocket Guide to Chemical Hazards*, U.S. Department of Health and Human Services, National Institute for Occupational Health and Safety, 2010 <www.cdc.gov/niosh/npg/>.
3. OSHA Standard Number 1910.1000, TABLE Z-1, Limits for Air Contaminants, U.S. Department of Labor, Occupational Safety and Health Administration, <www.osha.gov/pls/oshaweb/owadisp.show_document?p_table=STANDARDS&p_id=9992>.

Threshold Limits for Airborne Contaminants

Name	Comments	Formula	CAS Reg. No.	Time-weighted average	Short-term exposure limit
Abate	[Temephos]	$C_{16}H_{20}O_6P_2S_3$	3383-96-8	10 mg/m^3	
Acetaldehyde	[Ethanal] Ceiling value	CH_3CHO	75-07-0		25 ppm
Acetic acid	[Ethanoic acid]	CH_3COOH	64-19-7	10 ppm	15 ppm
Acetic anhydride	[Acetyl acetate]	$C_4H_6O_3$	108-24-7	1 ppm	3 ppm
Acetone	[2-Propanone]	$(CH_3)_2CO$	67-64-1	250 ppm	500 ppm
Acetone cyanohydrin	as CN. Ceiling value	C_4H_7NO	75-86-5		5 mg/m^3
Acetonitrile	[Methyl cyanide]	CH_3CN	75-05-8	20 ppm	
Acetophenone	[Methyl phenyl ketone]	C_8H_8O	98-86-2	10 ppm	
2-(Acetyloxy)benzoic acid	[Acetylsalicylic acid (Aspirin)]	$C_9H_8O_4$	50-78-2	5 mg/m^3	
Acrolein	[2-Propenal] Ceiling value	CH_2=CHCHO	107-02-8		0.1 ppm
Acrylamide	[2-Propenamide]	C_3H_5NO	79-06-1	0.03 mg/m^3	
Acrylic acid	[2-Propenoic acid]	$C_3H_4O_2$	79-10-7	2 ppm	
Acrylonitrile	[Propenenitrile]	CH_2=CHCN	107-13-1	2 ppm	
Alachlor	[Acetamide, 2-chloro-*N*-(2,6-diethylphenyl)-*N*-(methoxymethyl)-]	$C_{14}H_{20}ClNO_2$	15972-60-8	1 mg/m^3	
Aldrin	inhalable fraction & vapor	$C_{12}H_8Cl_6$	309-00-2	0.05 mg/m^3	
Allyl alcohol	[2-Propen-1-ol]	C_3H_6O	107-18-6	0.5 ppm	
Allyl glycidyl ether	[AGE]	$C_6H_{10}O_2$	106-92-3	1 ppm	
Allyl propyl disulfide		$C_6H_{12}S_2$	2179-59-1	0.5 ppm	
Aluminum	metal dust & insoluble compounds	Al	7429-90-5	1 mg/m^3	
4-Aminobiphenyl	[*p*-Biphenylamine]; levels as low as possible	$C_{12}H_{11}N$	92-67-1		
4-Amino-3,5,6-trichloro-2-pyridinecarboxlic acid	[Picloram]	$C_6H_3Cl_3N_2O_2$	1918-02-1	10 mg/m^3	
Ammonia		NH_3	7664-41-7	25 ppm	35 ppm
Ammonium chloride	[Sal ammoniac]; fume	NH_4Cl	12125-02-9	10 mg/m^3	20 mg/m^3
Ammonium perfluorooctanoate		$C_8H_4F_{15}NO_2$	3825-26-1	0.01 mg/m^3	
Ammonium sulfamate		$NH_4NH_2SO_3$	7773-06-0	10 mg/m^3	
Aniline	[Benzenamine]	$C_6H_5NH_2$	62-53-3	2 ppm	
Antimony	and compounds, as Sb	Sb	7440-36-0	0.5 mg/m^3	
Antimony(III) oxide (senarmontite)	[Senarmontite]; levels as low as possible	Sb_2O_3	1309-64-4		
Arsenic	and inorganic compounds, as As	As	7440-38-2	0.01 mg/m^3	

Name	Comments	Formula	CAS Reg. No.	Time-weighted average	Short-term exposure limit
Arsine	[Arsenic hydride]	AsH_3	7784-42-1	0.005 ppm	
Asbestos	all forms; limit is 0.1 fibers/mL		1332-21-4		
Asphalt	fume, as aerosol		8052-42-4	0.5 mg/m³	
Atrazine	[6-Chloro-*N*-ethyl-*N'*-(1-methylethyl)-1,3,5-triazine-2,4-diamine]	$C_8H_{14}ClN_5$	1912-24-9	2 mg/m³	
Azinphos-methyl	inhalable fraction & vapor	$C_{10}H_{12}N_3O_3PS_2$	86-50-0	0.2 mg/m³	
Barium	and soluble compounds, as Ba	Ba	7440-39-3	0.5 mg/m³	
Barium sulfate	[Barite]	$BaSO_4$	7727-43-7	5 mg/m³	
Benomyl	inhalable fraction	$C_{14}H_{18}N_4O_3$	17804-35-2	1 mg/m³	
Benz[*a*]anthracene	[1,2-Benzanthracene]; levels as low as possible	$C_{18}H_{12}$	56-55-3		
Benzene		C_6H_6	71-43-2	0.5 ppm	2.5 ppm
1,2-Benzenediamine	[*o*-Phenylenediamine]	$C_6H_8N_2$	95-54-5	0.1 mg/m³	
1,3-Benzenediamine	[*m*-Phenylenediamine]	$C_6H_8N_2$	108-45-2	0.1 mg/m³	
1,4-Benzenediamine	[*p*-Phenylenediamine]	$C_6H_8N_2$	106-50-3	0.1 mg/m³	
1,3-Benzenedimethanamine	[*m*-Xylene diamine] Ceiling value	$C_8H_{12}N_2$	1477-55-0		0.1 mg/m³
Benzenethiol	[Phenyl mercaptan]	C_6H_5SH	108-98-5	0.1 ppm	
p-Benzidine	[[1,1'-Biphenyl]-4,4'-diamine]; levels as low as possible	$C_{12}H_{12}N_2$	92-87-5		
Benzo[*b*]fluoranthene	[Benz[*e*]acephenanthrylene]; levels as low as possible	$C_{20}H_{12}$	205-99-2		
Benzo[*a*]pyrene	[2,3-Benzopyrene]; levels as low as possible	$C_{20}H_{12}$	50-32-8		
p-Benzoquinone	[2,5-Cyclohexadiene-1,4-dione]	$C_6H_4O_2$	106-51-4	0.1 ppm	
Benzoyl chloride	[Benzoic acid, chloride] Ceiling value	C_6H_5COCl	98-88-4		0.5 ppm
Benzoyl peroxide		$C_{14}H_{10}O_4$	94-36-0	5 mg/m³	
Benzyl acetate	[(Acetoxymethyl)benzene]	$C_9H_{10}O_2$	140-11-4	10 ppm	
Beryllium	and compounds, as Be	Be	7440-41-7	0.00005 mg/m³	
Biphenyl	[Diphenyl]	$C_{12}H_{10}$	92-52-4	0.2 ppm	
Bis(2-aminoethyl)amine	[Diethylenetriamine]	$C_4H_{13}N_3$	111-40-0	1 ppm	
Bis(2-chloroethyl) ether	[Dichloroethyl ether]	$C_4H_8Cl_2O$	111-44-4	5 ppm	10 ppm
Bis(chloromethyl) ether	[Chloromethyl ether]	$C_2H_4Cl_2O$	542-88-1	0.001 ppm	
Bis(2-dimethylaminoethyl) ether	[2,2'-Oxybis[*N,N*-dimethylethanamine (DMAEE)]	$C_8H_{20}N_2O$	3033-62-3	0.05 ppm	0.15 ppm
Bis(2-ethylhexyl) phthalate	[Di-*sec*-octyl phthalate (DEHP)]	$C_{24}H_{38}O_4$	117-81-7	5 mg/m³	
Bismuth telluride	[Tetradymite]	Bi_2Te_3	1304-82-1	10 mg/m³	
Boric acid	[Orthoboric acid]; and inorganic borate compounds	H_3BO_3	10043-35-3	2 mg/m³	6 mg/m³
Boron oxide	[Boric oxide]	B_2O_3	1303-86-2	10 mg/m³	
Boron tribromide	[Tribromoborane] Ceiling value.	BBr_3	10294-33-4		0.7 ppm
Boron trichloride	[Trichloroborane] Ceiling value	BCl_3	10294-34-5		0.7 ppm
Boron trifluoride	[Trifluoroborane] Ceiling value.	BF_3	7637-07-2	0.1 ppm	0.7 ppm
Bromacil	[5-Bromo-3-*sec*-butyl-6-methyluracil]	$C_9H_{13}BrN_2O_2$	314-40-9	10 mg/m³	
Bromine		Br_2	7726-95-6	0.1 ppm	0.2 ppm
Bromine pentafluoride		BrF_5	7789-30-2	0.1 ppm	
Bromochloromethane	[Halon 1011]	CH_2BrCl	74-97-5	200 ppm	
2-Bromo-2-chloro-1,1,1-trifluoroethane	[Halothane]	$C_2HBrClF_3$	151-67-7	50 ppm	
Bromoethane	[Ethyl bromide]	C_2H_5Br	74-96-4	5 ppm	
Bromoethene	[Vinyl bromide]	$CH_2{=}CHBr$	593-60-2	0.5 ppm	
Bromomethane	[Methyl bromide]	CH_3Br	74-83-9	1 ppm	
1-Bromopropane	[Propyl bromide]	C_3H_7Br	106-94-5	0.1 ppm	
3-Bromopropene	[Allyl bromide]	C_3H_5Br	106-95-6	0.1 ppm	0.2 ppm
Bromotrifluoromethane	[Halon-1301]	CF_3Br	75-63-8	1000 ppm	
1,3-Butadiene	[Divinyl]	C_4H_6	106-99-0	2 ppm	
Butane	both isomers, NIC 2016	C_4H_{10}	106-97-8		1000 ppm
2,3-Butanedione	[Diacetyl]	$C_4H_6O_2$	431-03-8	0.01 ppm	0.02 ppm
1-Butanethiol	[Butyl mercaptan]	$C_4H_{10}S$	109-79-5	0.5 ppm	
1-Butanol	[Butyl alcohol]	C_4H_9OH	71-36-3	20 ppm	
2-Butanol	[*sec*-Butyl alcohol]	$C_4H_{10}O$	78-92-2	100 ppm	
2-Butanone	[Methyl ethyl ketone (MEK)]	C_4H_8O	78-93-3	200 ppm	300 ppm
2-Butanone peroxide	[Methyl ethyl ketone peroxide] Ceiling value	$C_8H_{16}O_4$	1338-23-4		0.2 ppm
trans-2-Butenal	[*trans*-Crotonaldehyde] Ceiling value	C_4H_6O	123-73-9		0.3 ppm
1-Butene	[1-Butylene]	C_4H_8	106-98-9	250 ppm	
cis-2-Butene		C_4H_8	590-18-1	250 ppm	
trans-2-Butene		C_4H_8	624-64-6	250 ppm	
3-Buten-2-one	[Methyl vinyl ketone] Ceiling value	C_4H_6O	78-94-4		0.2 ppm
2-Butoxyethanol	[Ethylene glycol monobutyl ether (EGBE)]	$C_6H_{14}O_2$	111-76-2	20 ppm	
2-Butoxyethyl acetate	[Ethylene glycol monobutyl ether acetate (EGBEA)]	$C_8H_{16}O_3$	112-07-2	20 ppm	
Butyl acetate		$C_6H_{12}O_2$	123-86-4	50 ppm	150 ppm
sec-Butyl acetate	[1-Methylpropyl acetate]	$C_6H_{12}O_2$	105-46-4	50 ppm	150 ppm
tert-Butyl acetate	[1,1,-Dimethylethyl acetate]	$C_6H_{12}O_2$	540-88-5	50 ppm	150 ppm
Butyl acrylate	[Butyl 2-propenoate]	$C_7H_{12}O_2$	141-32-2	2 ppm	
Butylamine	[1-Butanamine] Ceiling value	$C_4H_{11}N$	109-73-9		5 ppm

Name	Comments	Formula	CAS Reg. No.	Time-weighted average	Short-term exposure limit
tert-Butyl chromate	as CrO_3. Ceiling value	$C_8H_{18}CrO_4$	1189-85-1		0.1 mg/m^3
tert-Butyl ethyl ether	[Ethyl *tert*-butyl ether (ETBE)]	$C_6H_{14}O$	637-92-3	25 ppm	
Butyl glycidyl ether	[BGE]	$C_7H_{14}O_2$	2426-08-6	3 ppm	
Butyl lactate		$C_7H_{14}O_3$	34451-18-8	5 ppm	
1-*tert*-Butyl-4-methylbenzene	[4-*tert*-Butyltoluene]	$C_{11}H_{16}$	98-51-1	1 ppm	
2-*sec*-Butylphenol		$C_{10}H_{14}O$	89-72-5	5 ppm	
Cadmium	metal	Cd	7440-43-9	0.01 mg/m^3	
Cadmium	compounds, as Cd	Cd	7440-43-9	0.002 mg/m^3	
Calcium chromate	as Cr	$CaCrO_4$	13765-19-0	0.001 mg/m^3	
Calcium cyanamide	[Calcium carbimide]	$CaCN_2$	156-62-7	0.5 mg/m^3	
Calcium hydroxide	[Portlandite]	$Ca(OH)_2$	1305-62-0	5 mg/m^3	
Calcium metasilicate	[Parawollastonite]; synthetic, nonfibrous.	$CaSiO_3$	1344-95-2	1 mg/m^3	
Calcium oxide	[Lime]	CaO	1305-78-8	2 mg/m^3	
Calcium sulfate	[Anhydrite]	$CaSO_4$	7778-18-9	10 mg/m^3	
Camphor, (+)	[1,7,7-Trimethylbicyclo[2.2.1]heptan-2-one, (1*R*)-]	$C_{10}H_{16}O$	464-49-3	2 ppm	3 ppm
Caprolactam	[6-Hexanelactam]	$C_6H_{11}NO$	105-60-2	5 mg/m^3	
Captafol	[Difolatan] NIC 2016	$C_{10}H_9Cl_4NO_2S$	2425-06-1	0.1 mg/m^3	
Captan	inhalable fraction	$C_9H_8Cl_3NO_2S$	133-06-2	5 mg/m^3	
Carbaryl		$C_{12}H_{11}NO_2$	63-25-2	0.5 mg/m^3	
Carbofuran	[7-Benzofuranol, 2,3-dihydro-2,2-dimethyl-, methylcarbamate]	$C_{12}H_{15}NO_3$	1563-66-2	0.1 mg/m^3	
Carbon (graphite)	[Graphite]; except fibers	C	7782-42-5	2 mg/m^3	
Carbon black	[Carbon (amorphous)] inhalable fraction	C	1333-86-4	3 mg/m^3	
Carbon dioxide	[Carbonic anhydride]	CO_2	124-38-9	5000 ppm	30,000 ppm
Carbon disulfide	[Carbon bisulfide]	CS_2	75-15-0	1 ppm	
Carbon monoxide	[Carbon oxide]	CO	630-08-0	25 ppm	
Carbon oxysulfide	[Carbonyl sulfide]	OCS	463-58-1	5 ppm	
Carbonyl chloride	[Phosgene]	$COCl_2$	75-44-5	0.1 ppm	
Carbonyl fluoride		COF_2	353-50-4	2 ppm	5 ppm
Cellulose			9004-34-6	10 mg/m^3	
Cesium hydroxide		CsOH	21351-79-1	2 mg/m^3	
Chlordane	[1,2,4,5,6,7,8,8-Octachloro-2,3,3a,4,7,7a-hexahydro-4,7-methano-1*H*-indene]	$C_{10}H_6Cl_8$	57-74-9	0.5 mg/m^3	
o-Chlorinated diphenyl oxide			31242-93-0	0.5 mg/m^3	
Chlorine		Cl_2	7782-50-5	0.5 ppm	1 ppm
Chlorine dioxide		ClO_2	10049-04-4	0.1 ppm	0.3 ppm
Chlorine trifluoride	Ceiling value	ClF_3	7790-91-2		0.1 ppm
Chloroacetaldehyde	[2-Chloro-1-ethanal] Ceiling value	C_2H_3ClO	107-20-0		1 ppm
Chloroacetic acid		$CH_2ClCOOH$	79-11-8	0.5 ppm	
Chloroacetone	Ceiling value	C_3H_5ClO	78-95-5		1 ppm
α-Chloroacetophenone	[ω-Chloroacetophenone]	C_8H_7ClO	532-27-4	0.05 ppm	
Chloroacetyl chloride		$C_2H_2Cl_2O$	79-04-9	0.05 ppm	0.15 ppm
Chlorobenzene	[Phenyl chloride]	C_6H_5Cl	108-90-7	10 ppm	
o-Chlorobenzylidene malononitrile	Ceiling value	$C_{10}H_5ClN_2$	2698-41-1		0.05 ppm
2-Chloro-1,3-butadiene	[Chloroprene] NIC 2016	C_4H_5Cl	126-99-8	10 ppm	
Chlorodifluoromethane	[HCFC-22]	$CHClF_2$	75-45-6	1000 ppm	
Chloroethane	[Ethyl chloride]	C_2H_5Cl	75-00-3	100 ppm	
2-Chloroethanol	[Ethylene chlorohydrin] Ceiling value	C_2H_5ClO	107-07-3		1 ppm
Chloroethene	[Vinyl chloride]	CH_2=CHCl	75-01-4	1 ppm	
Chloromethane	[Methyl chloride]	CH_3Cl	74-87-3	50 ppm	100 ppm
(Chloromethyl)benzene	[Benzyl chloride]	C_7H_7Cl	100-44-7	1 ppm	
Chloromethyl methyl ether	levels as low as possible	C_2H_5ClO	107-30-2		
1-Chloro-4-nitrobenzene	[*p*-Chloronitrobenzene]	$C_6H_4ClNO_2$	100-00-5	0.1 ppm	
1-Chloro-1-nitropropane		$C_3H_6ClNO_2$	600-25-9	2 ppm	
Chloropentafluoroethane	[CFC-115]	CF_3CF_2Cl	76-15-3	1000 ppm	
2-Chloropropanoic acid	[2-Chloropropionic acid]	$C_3H_5ClO_2$	598-78-7	0.1 ppm	
2-Chloro-1-propanol	[Propylene chlorohydrin]	C_3H_7ClO	78-89-7	1 ppm	
1-Chloro-2-propanol	[*sec*-Propylene chlorohydrin]	C_3H_7ClO	127-00-4	1 ppm	
3-Chloropropene	[Allyl chloride]	C_3H_5Cl	107-05-1	1 ppm	2 ppm
2-Chlorostyrene		C_8H_7Cl	2039-87-4	50 ppm	75 ppm
2-Chlorotoluene	[1-Chloro-2-methylbenzene]	C_7H_7Cl	95-49-8	50 ppm	
Chlorpyrifos	[Phosphorothioic acid, *O,O*-diethyl *O*-(3,5,6-trichloro-2-pyridinyl) ester]	$C_9H_{11}Cl_3NO_3PS$	2921-88-2	0.1 mg/m^3	
Chromium	metal	Cr	7440-47-3	0.5 mg/m^3	
Chromium	Cr(III) compounds, as Cr	Cr	7440-47-3	0.5 mg/m^3	
Chromium	soluble Cr(VI) compounds, as Cr	Cr	7440-47-3	0.05 mg/m^3	
Chromium	insoluble Cr(VI) compounds, as Cr	Cr	7440-47-3	0.01 mg/m^3	

Name	Comments	Formula	CAS Reg. No.	Time-weighted average	Short-term exposure limit
Chromium(VI) dichloride dioxide	[Chromyl chloride]	CrO_2Cl_2	14977-61-8	0.025 ppm	
Chrysene	[Benzo[a]phenanthrene]; levels as low as possible	$C_{18}H_{12}$	218-01-9		
Clopidol	inhable fraction & vapor	$C_7H_7Cl_2NO$	2971-90-6	3 mg/m^3	
Coal	dust, anthracite			0.4 mg/m^3	
Coal	dust, bituminous			0.9 mg/m^3	
Coal tar	volatiles, as aerosol		65996-93-2	0.2 mg/m^3	
Cobalt	metal & inorganic compounds, as Co	Co	7440-48-4	0.02 mg/m^3	
Cobalt	thoracic particulate matter in hard metals containing Co and tungsten carbide, as Co	Co	7440-48-4	0.005 mg/m^3	
Cobalt carbonyl	[Dicobalt octacarbonyl]; as Co	$Co_2(CO)_8$	10210-68-1	0.1 mg/m^3	
Cobalt hydrocarbonyl	[Tetracarbonylhydrocobalt]; as Co	C_4HCoO_4	16842-03-8	0.1 mg/m^3	
Copper	fume	Cu	7440-50-8	0.2 mg/m^3	
Copper	dusts & mists, as Cu	Cu	7440-50-8	1 mg/m^3	
Cotton	dust			0.1 mg/m^3	
Coumaphos		$C_{14}H_{16}ClO_5PS$	56-72-4	0.05 mg/m^3	
o-Cresol	[2-Methylphenol]; inhalable fraction & vapor	C_7H_8O	95-48-7	20 mg/m^3	
m-Cresol	[3-Methylphenol]; inhalable fraction & vapor	C_7H_8O	108-39-4	20 mg/m^3	
p-Cresol	[4-Methylphenol]; inhalable fraction & vapor	C_7H_8O	106-44-5	20 mg/m^3	
Crufomate		$C_{12}H_{19}ClNO_3P$	299-86-5	5 mg/m^3	
Cyanamide	[Cyanogenamide]	H_2NCN	420-04-2	2 mg/m^3	
Cyanide anion	cyanide salts, as CN. Ceiling value	CN^-	57-12-5		5 mg/m^3
Cyanogen	Ceiling value	C_2N_2	460-19-5		5 ppm
Cyanogen bromide	[Bromine cyanide] Ceiling value	BrCN	506-68-3		0.3 ppm
Cyanogen chloride	[Chlorine cyanide] Ceiling value	ClCN	506-77-4		0.3 ppm
Cyclohexane	[Hexahydrobenzene]	C_6H_{12}	110-82-7	100 ppm	
Cyclohexanol	[Cyclohexyl alcohol]	$C_6H_{12}O$	108-93-0	50 ppm	
Cyclohexanone	[Pimelic ketone]	$C_6H_{10}O$	108-94-1	20 ppm	50 ppm
Cyclohexene	[Tetrahydrobenzene]	C_6H_{10}	110-83-8	300 ppm	
Cyclohexylamine	[Cyclohexanamine]	$C_6H_{13}N$	108-91-8	10 ppm	
1,3-Cyclopentadiene	[Pyropentylene]	C_5H_6	542-92-7	75 ppm	
Cyclopentane	[Pentamethylene]	C_5H_{10}	287-92-3	600 ppm	
Cyhexatin	[Tricyclohexylhydroxystannane]	$C_{18}H_{34}OSn$	13121-70-5	5 mg/m^3	
Decaborane(14)		$B_{10}H_{14}$	17702-41-9	0.05 ppm	0.15 ppm
Demeton	[Systox]	$C_8H_{19}O_3PS_2$	8065-48-3	0.05 mg/m^3	
Demeton-*S*-methyl	[Phosphorothioic acid, *S*-[2-(ethylthio)ethyl] *O*,*O*-dimethyl ester]	$C_6H_{15}O_3PS_2$	919-86-8	0.05 mg/m^3	
Diacetone alcohol	[4-Hydroxy-4-methyl-2-pentanone]	$C_6H_{12}O_2$	123-42-2	50 ppm	
4,4'-Diaminodiphenylmethane	[4,4'-Methylenedianiline]	$C_{13}H_{14}N_2$	101-77-9	0.1 ppm	
Diazinon		$C_{12}H_{21}N_2O_3PS$	333-41-5	0.01 mg/m^3	
Diazomethane		$CH_2{=}N{\equiv}N$	334-88-3	0.2 ppm	
Diborane		B_2H_6	19287-45-7	0.1 ppm	
Dibromodifluoromethane		CBr_2F_2	75-61-6	100 ppm	
2-Dibutylaminoethanol		$C_{10}H_{23}NO$	102-81-8	0.5 ppm	
2,6-Di-*tert*-butyl-4-methylphenol	[Butylated hydroxytoluene (BHT)]	$C_{15}H_{24}O$	128-37-0	2 mg/m^3	
Dibutylphenyl phosphate		$C_{14}H_{23}O_4P$	2528-36-1	0.3 ppm	
Dibutyl phosphate	inhalable fraction & vapor	$C_8H_{19}O_4P$	107-66-4	5 mg/m^3	
Dibutyl phthalate	[Butyl phthalate]	$C_{16}H_{22}O_4$	84-74-2	5 mg/m^3	
Dichloroacetic acid		$CHCl_2COOH$	79-43-6	0.5 ppm	
Dichloroacetylene	Ceiling value	C_2Cl_2	7572-29-4		0.1 ppm
o-Dichlorobenzene	[1,2-Dichlorobenzene]	$C_6H_4Cl_2$	95-50-1	25 ppm	50 ppm
p-Dichlorobenzene	[1,4-Dichlorobenzene]	$C_6H_4Cl_2$	106-46-7	10 ppm	
3,3'-Dichloro-*p*-benzidine	[3,3'-Dichloro[1,1'-biphenyl]-4,4'-diamine]; levels as low as possible	$C_{12}H_{10}Cl_2N_2$	91-94-1		
cis-1,4-Dichloro-2-butene		$C_4H_6Cl_2$	1476-11-5	0.005 ppm	
trans-1,4-Dichloro-2-butene		$C_4H_6Cl_2$	110-57-6	0.005 ppm	
Dichlorodifluoromethane	[CFC-12]	CF_2Cl_2	75-71-8	1000 ppm	
1,3-Dichloro-5,5-dimethyl hydantoin		$C_5H_6Cl_2N_2O_2$	118-52-5	0.2 mg/m^3	0.4 mg/m^3
1,1-Dichloroethane	[Ethylidene dichloride]	CH_3CHCl_2	75-34-3	100 ppm	
1,2-Dichloroethane	[Ethylene dichloride]	$C_2H_4Cl_2$	107-06-2	10 ppm	
1,1-Dichloroethene	[Vinylidene chloride]	$CH_2{=}CCl_2$	75-35-4	5 ppm	
cis-1,2-Dichloroethene	[*cis*-1,2-Dichloroethylene]	$C_2H_2Cl_2$	156-59-2	200 ppm	
trans-1,2-Dichloroethene	[*trans*-1,2-Dichloroethylene]	$C_2H_2Cl_2$	156-60-5	200 ppm	
Dichlorofluoromethane	[Refrigerant 21]	$CHCl_2F$	75-43-4	10 ppm	
Dichloromethane	[Methylene chloride]	CH_2Cl_2	75-09-2	50 ppm	
1,1-Dichloro-1-nitroethane	[Ethide]	$C_2H_3Cl_2NO_2$	594-72-9	2 ppm	
(2,4-Dichlorophenoxy)acetic acid	[2,4-D]; inhalable fraction. NIC 2016	$C_8H_6Cl_2O_3$	94-75-7	10 mg/m^3	
1,2-Dichloropropane, (±)-	[Propylene dichloride]	$C_3H_6Cl_2$	78-87-5	10 ppm	

Name	Comments	Formula	CAS Reg. No.	Time-weighted average	Short-term exposure limit
2,2-Dichloropropanoic acid	[2,2-Dichloropropionic acid]	$C_3H_4Cl_2O_2$	75-99-0	5 mg/m³	
cis-1,3-Dichloropropene	[*cis*-1,3-Dichloropropylene]	$C_3H_4Cl_2$	10061-01-5	1 ppm	
trans-1,3-Dichloropropene	[*trans*-1,3-Dichloropropylene]	$C_3H_4Cl_2$	10061-02-6	1 ppm	
1,2-Dichloro-1,1,2,2-tetrafluoroethane	[CFC-114]	$C_2Cl_2F_4$	76-14-2	1000 ppm	
Dichlorvos	[Phosphoric acid, 2,2-dichloroethenyl dimethyl ester]	$C_4H_7Cl_2O_4P$	62-73-7	0.1 mg/m³	
Dicrotophos		$C_8H_{16}NO_5P$	141-66-2	0.05 mg/m³	
o-Dicyanobenzene	[*o*-Phthalodinitrile]; inhalable fraction or vapor	$C_8H_4N_2$	91-15-6	1 mg/m³	
m-Dicyanobenzene	[*m*-Phthalodinitrile]; inhalable fraction & vapor	$C_8H_4N_2$	626-17-5	5 mg/m³	
Dicyclopentadiene		$C_{10}H_{12}$	1755-01-7	5 ppm	
Dieldrin	inhalable fraction & vapor	$C_{12}H_8Cl_6O$	60-57-1	0.1 mg/m³	
Diesel fuel	as total hydrocarbons		68334-30-5	100 mg/m³	
Diethanolamine	[Bis(2-hydroxyethyl)amine]; inhalable fraction & vapor	$C_4H_{11}NO_2$	111-42-2	1 mg/m³	
Diethylamine	[*N*-Ethylethanamine]	$(C_2H_5)_2NH$	109-89-7	5 ppm	15 ppm
2-Diethylaminoethanol		$C_6H_{15}NO$	100-37-8	2 ppm	
Diethylene glycol monobutyl ether	[2-(2-Butoxyethoxy)ethanol]; inhalable fraction & vapor	$C_8H_{18}O_3$	112-34-5	10 ppm	
Diethyl ether	[Ethyl ether]	$(C_2H_5)_2O$	60-29-7	400 ppm	500 ppm
3,3-Diethylpentane	[Tetraethylmethane]	C_9H_{20}	1067-20-5	200 ppm	
Diethyl phthalate		$C_{12}H_{14}O_4$	84-66-2	5 mg/m³	
1,1-Difluoroethene	[Vinylidene fluoride]	$CH_2{=}CF_2$	75-38-7	500 ppm	
Diglycidyl ether	[Bis(2,3-epoxypropyl) ether (DGE)]	$C_6H_{10}O_3$	2238-07-5	0.1 ppm	
Diisopropylamine	[*N*-Isopropyl-2-propanamine]	$C_6H_{15}N$	108-18-9	5 ppm	
Diisopropyl ether	[Isopropyl ether]	$C_6H_{14}O$	108-20-3	250 ppm	310 ppm
Dimethoxymethane	[Methylal]	$C_3H_8O_2$	109-87-5	1000 ppm	
N,N-Dimethylacetamide	[*N,N*-Dimethylethanamide]	C_4H_9NO	127-19-5	10 ppm	
Dimethylamine	[*N*-Methylmethanamine]	$(CH_3)_2NH$	124-40-3	5 ppm	15 ppm
Dimethylaniline (unspecified isomer)	[Xylidine (unspecified isomer)]; all isomers	$C_8H_{11}N$	1300-73-8	0.5 ppm	
2,3-Dimethylaniline	[2,3-Xylidine]	$C_8H_{11}N$	87-59-2	0.5 ppm	
2,4-Dimethylaniline	[2,4-Xylidine]	$C_8H_{11}N$	95-68-1	0.5 ppm	
2,5-Dimethylaniline	[2,5-Xylidine]	$C_8H_{11}N$	95-78-3	0.5 ppm	
2,6-Dimethylaniline	[2,6-Xylidine]	$C_8H_{11}N$	87-62-7	0.5 ppm	
3,4-Dimethylaniline	[3,4-Xylidine]	$C_8H_{11}N$	95-64-7	0.5 ppm	
3,5-Dimethylaniline	[3,5-Xylidine]	$C_8H_{11}N$	108-69-0	0.5 ppm	
N,N-Dimethylaniline	[*N,N*-Dimethylbenzenamine]	$C_8H_{11}N$	121-69-7	5 ppm	10 ppm
2,2-Dimethylbutane	[Neohexane]	C_6H_{14}	75-83-2	500 ppm	1000 ppm
2,3-Dimethylbutane	[Diisopropyl]	C_6H_{14}	79-29-8	500 ppm	1000 ppm
Dimethylcarbamic chloride	[Dimethylcarbamoyl chloride]	C_3H_6ClNO	79-44-7	0.005 mg/m³	
Dimethyl disulfide	[Methyl disulfide]	$C_2H_6S_2$	624-92-0	0.5 mg/m³	
N,N-Dimethylformamide	[DMF]	C_3H_7NO	68-12-2	10 ppm	
2,2-Dimethylheptane		C_9H_{20}	1071-26-7	200 ppm	
2,3-Dimethylheptane		C_9H_{20}	3074-71-3	200 ppm	
2,4-Dimethylheptane		C_9H_{20}	2213-23-2	200 ppm	
2,5-Dimethylheptane		C_9H_{20}	2216-30-0	200 ppm	
2,6-Dimethylheptane		C_9H_{20}	1072-05-5	200 ppm	
3,3-Dimethylheptane		C_9H_{20}	4032-86-4	200 ppm	
3,4-Dimethylheptane		C_9H_{20}	922-28-1	200 ppm	
3,5-Dimethylheptane		C_9H_{20}	926-82-9	200 ppm	
4,4-Dimethylheptane		C_9H_{20}	1068-19-5	200 ppm	
2,6-Dimethyl-4-heptanone	[Diisobutyl ketone]	$C_9H_{18}O$	108-83-8	25 ppm	
2,2-Dimethylhexane		C_8H_{18}	590-73-8	300 ppm	
2,3-Dimethylhexane		C_8H_{18}	584-94-1	300 ppm	
2,4-Dimethylhexane		C_8H_{18}	589-43-5	300 ppm	
2,5-Dimethylhexane	[Biisobutyl]	C_8H_{18}	592-13-2	300 ppm	
3,3-Dimethylhexane		C_8H_{18}	563-16-6	300 ppm	
3,4-Dimethylhexane		C_8H_{18}	583-48-2	300 ppm	
1,1-Dimethylhydrazine	[UDMH]	$C_2H_8N_2$	57-14-7	0.01 ppm	
Dimethyl mercury	[Mercury dimethyl]	$Hg(CH_3)_2$	593-74-8	0.01 mg/m³	
trans-3,7-Dimethyl-2,6-octadienal	[Citral]; inhalable fraction & vapor	$C_{10}H_{16}O$	141-27-5	5 ppm	
Dimethyl phthalate	[Methyl phthalate]	$C_{10}H_{10}O_4$	131-11-3	5 mg/m³	
2,2-Dimethyl-1-propanol acetate		$C_7H_{14}O_2$	926-41-0	50 ppm	100 ppm
Dimethyl sulfate		$C_2H_6O_4S$	77-78-1	0.1 ppm	
Dimethyl sulfide	[2-Thiapropane]	$(CH_3)_2S$	75-18-3	10 ppm	
1,2-Dinitrobenzene	[*o*-Dinitrobenzene]	$C_6H_4N_2O_4$	528-29-0	0.15 ppm	
1,3-Dinitrobenzene	[*m*-Dinitrobenzene]	$C_6H_4N_2O_4$	99-65-0	0.15 ppm	
1,4-Dinitrobenzene	[*p*-Dinitrobenzene]	$C_6H_4N_2O_4$	100-25-4	0.15 ppm	
1,4-Dioxane	[1,4-Dioxacyclohexane]	$C_4H_8O_2$	123-91-1	20 ppm	
Dioxathion		$C_{12}H_{26}O_6P_2S_4$	78-34-2	0.1 mg/m³	
1,3-Dioxolane	[1,3-Dioxacyclopentane]	$C_3H_6O_2$	646-06-0	20 ppm	

Name	Comments	Formula	CAS Reg. No.	Time-weighted average	Short-term exposure limit
Diphenylamine	[*N*-Phenylbenzenamine]	$(C_6H_5)_2NH$	122-39-4	10 mg/m³	
Diphenyl ether	[Oxybisbenzene]	$(C_6H_5)_2O$	101-84-8	1 ppm	2 ppm
4,4'-Diphenylmethane diisocyanate	[Methylene diphenyl diisocyanate (MDI)]	$C_{15}H_{10}N_2O_2$	101-68-8	0.005 ppm	
Dipropylene glycol monomethyl ether	[1-(2-Methoxyisopropoxy)-2-propanol (DPGME)]	$C_7H_{16}O_3$	34590-94-8	100 ppm	150 ppm
Diquat		$C_{12}H_{12}N_2$	2764-72-9	0.5 mg/m³	
Disulfiram		$C_{10}H_{20}N_2S_4$	97-77-8	2 mg/m³	
Disulfoton	[Phosphorodithioic acid, *O,O*-diethyl *S*-[2-(ethylthio) ethyl] ester]	$C_8H_{19}O_2PS_3$	298-04-4	0.05 mg/m³	
Diuron		$C_9H_{10}Cl_2N_2O$	330-54-1	10 mg/m³	
o-Divinylbenzene	[1,2-Divinylbenzene]	$C_{10}H_{10}$	91-14-5	10 ppm	
m-Divinylbenzene	[1,3-Divinylbenzene]	$C_{10}H_{10}$	108-57-6	10 ppm	
p-Divinylbenzene	[1,4-Divinylbenzene]	$C_{10}H_{10}$	105-06-6	10 ppm	
1-Dodecanethiol	[Dodecyl mercaptan]	$C_{12}H_{26}S$	112-55-0	0.1 ppm	
Endosulfan	inhalable fraction & vapor	$C_9H_6Cl_6O_3S$	115-29-7	0.1 mg/m³	
Endrin		$C_{12}H_8Cl_6O$	72-20-8	0.1 mg/m³	
Enflurane		$C_3H_2ClF_5O$	13838-16-9	75 ppm	
Epichlorohydrin	[(Chloromethyl)oxirane]	C_3H_5ClO	13403-37-7	0.5 ppm	
1,2-Epoxy-4-(epoxyethyl)cyclohexane	[4-Vinyl-1-cyclohexene dioxide]	$C_8H_{12}O_2$	106-87-6	0.1 ppm	
1,2-Ethanediamine	[Ethylenediamine]	$C_2H_8N_2$	107-15-3	10 ppm	
1,2-Ethanediol	[Ethylene glycol] aerosol. Ceiling value. NIC 2016	$(CH_2OH)_2$	107-21-1		100 mg/m³
1,2-Ethanediol, dinitrate	[Ethylene glycol dinitrate (EGDN)]	$C_2H_4N_2O_6$	628-96-6	0.05 ppm	
Ethanethiol	[Ethyl mercaptan]	C_2H_5SH	75-08-1	0.5 ppm	
Ethanol	[Ethyl alcohol]	C_2H_5OH	64-17-5		1000 ppm
Ethanolamine	[Glycinol; 2-Aminoethanol]	C_2H_7NO	141-43-5	3 ppm	6 ppm
Ethion	[Phosphorodithioic acid, *S,S'*-methylene *O,O,O',O'*-tetraethyl ester]	$C_9H_{22}O_4P_2S_4$	563-12-2	0.05 mg/m³	
Ethoxydimethylsilane	[Dimethylethoxysilane]	$C_4H_{12}OSi$	14857-34-2	0.5 ppm	1.5 ppm
2-Ethoxyethanol	[Ethylene glycol monoethyl ether (EGEE)]	$C_4H_{10}O_2$	110-80-5	5 ppm	
2-Ethoxyethyl acetate	[Ethylene glycol monoethyl ether acetate (EGEEA)]	$C_6H_{12}O_3$	111-15-9	5 ppm	
Ethyl acetate		$C_4H_8O_2$	141-78-6	400 ppm	
Ethyl acrylate	[Ethyl propenoate]	$C_5H_8O_2$	140-88-5	5 ppm	15 ppm
Ethylamine	[Ethanamine]	$C_2H_5NH_2$	75-04-7	5 ppm	15 ppm
Ethylbenzene	[Phenylethane]	C_8H_{10}	100-41-4	20 ppm	
Ethyl 2-cyanoacrylate	[Ethyl 2-cyano-2-propenoate]	$C_6H_7NO_2$	7085-85-0	0.2 ppm	
3-Ethyl-2,2-dimethylpentane		C_9H_{20}	16747-32-3	200 ppm	
3-Ethyl-2,3-dimethylpentane		C_9H_{20}	16747-33-4	200 ppm	
3-Ethyl-2,4-dimethylpentane		C_9H_{20}	1068-87-7	200 ppm	
Ethylene	[Ethene]	$CH_2{=}CH_2$	74-85-1	200 ppm	
Ethyleneimine	[Aziridine]	C_2H_5N	151-56-4	0.05 ppm	0.1 ppm
Ethyl formate		$C_3H_6O_2$	109-94-4		100 ppm
3-Ethylheptane		C_9H_{20}	15869-80-4	200 ppm	
4-Ethylheptane		C_9H_{20}	2216-32-2	200 ppm	
3-Ethylhexane		C_8H_{18}	619-99-8	300 ppm	
2-Ethylhexanoic acid		$C_8H_{16}O_2$	149-57-5	5 mg/m³	
N-Ethyl-*N*-hydroxyethanamine	[*N,N*-Diethylhydroxylamine]	$C_4H_{11}NO$	3710-84-7	2 ppm	
5-Ethylidene-2-norbornene	[5-Ethylidenebicyclo[2.2.1]hept-2-ene]	C_9H_{12}	16219-75-3	2 ppm	4 ppm
Ethyl isocyanate		C_3H_5NO	109-90-0	0.02 ppm	0.06 ppm
3-Ethyl-2-methylhexane		C_9H_{20}	16789-46-1	200 ppm	
3-Ethyl-3-methylhexane		C_9H_{20}	3074-76-8	200 ppm	
3-Ethyl-4-methylhexane	[2,3-Diethylpentane]	C_9H_{20}	3074-77-9	200 ppm	
4-Ethyl-2-methylhexane		C_9H_{20}	3074-75-7	200 ppm	
3-Ethyl-2-methylpentane	[2-Methyl-3-ethylpentane]	C_8H_{18}	609-26-7	300 ppm	
3-Ethyl-3-methylpentane	[3-Methyl-3-ethylpentane]	C_8H_{18}	1067-08-9	300 ppm	
N-Ethylmorpholine		$C_6H_{13}NO$	100-74-3	5 ppm	
O-Ethyl *O*-*p*-nitrophenyl benzenethiophosphonate	[EPN]	$C_{14}H_{14}NO_4PS$	2104-64-5	0.1 mg/m³	
Ethyl silicate	[Tetraethoxysilane]	$Si(OC_2H_5)_4$	78-10-4	10 ppm	
Fenamiphos		$C_{13}H_{22}NO_3PS$	22224-92-6	0.05 mg/m³	
Fensulfothion	[Phosphorothioic acid, *O,O*-diethyl *O*-[4-(methylsulfinyl) phenyl] ester]	$C_{11}H_{17}O_4PS_2$	115-90-2	0.01 mg/m³	
Fenthion	[Phosphorothioic acid, *O,O*-dimethyl *O*-[3-methyl-4-(methylthio)phenyl] ester]	$C_{10}H_{15}O_3PS_2$	55-38-9	0.05 mg/m³	
Ferbam	[Iron, tris(dimethylcarbamodithioato-*S,S'*)-, (OC-6-11)-]; inhalable fraction	$C_9H_{18}FeN_3S_6$	14484-64-1	5 mg/m³	
Ferrocene	[Bis(cyclopentadienyl)iron]	$Fe(C_5H_5)_2$	102-54-5	10 mg/m³	
Ferrovanadium	dust		12604-58-9	1 mg/m³	3 mg/m³
Flour	dust, inhalable fraction			0.5 mg/m³	
Fluoride ion [F⁻]	fluoride salts, as F	F^-	16984-48-8	2.5 mg/m³	
Fluorine		F_2	7782-41-4	1 ppm	2 ppm

Name	Comments	Formula	CAS Reg. No.	Time-weighted average	Short-term exposure limit
Fluorine monoxide	[Oxygen difluoride] Ceiling value	F_2O	7783-41-7		0.05 ppm
Fluoroethene	[Vinyl fluoride]	CH_2=CHF	75-02-5	1 ppm	
Fonofos	[Phosphonodithioic acid, ethyl-, *O*-ethyl *S*-phenyl ester]	$C_{10}H_{15}OPS_2$	944-22-9	0.01 mg/m³	
Formaldehyde	[Methanal] Ceiling value. NIC 2016	HCHO	50-00-0		0.3 ppm
Formamide	[Methanamide]	$HCONH_2$	75-12-7	10 ppm	
Formic acid	[Methanoic acid]	HCOOH	64-18-6	5 ppm	10 ppm
Furfural	[2-Furaldehyde] NIC 2016	$C_5H_4O_2$	98-01-1	2 ppm	
Furfuryl alcohol	[2-Furanmethanol] NIC 2016	$C_5H_6O_2$	98-00-0	10 ppm	15 ppm
Gallium arsenide		GaAs	1303-00-0	0.0003 mg/m³	
Gasoline			8006-61-9	300 ppm	500 ppm
Germane	[Germanium tetrahydride]	GeH_4	7782-65-2	0.2 ppm	
Glyoxal	inhalable fraction	$H_2C_2O_2$	107-22-2	0.1 mg/m³	
Grain	dust			4 mg/m³	
Hafnium	metal & compounds, as Hf	Hf	7440-58-6	0.5 mg/m³	
Heptachlor		$C_{10}H_5Cl_7$	76-44-8	0.05 mg/m³	
Heptachlor epoxide		$C_{10}H_5Cl_7O$	1024-57-3	0.05 mg/m³	
Heptane	all isomers	C_7H_{16}	142-82-5	400 ppm	500 ppm
2-Heptanone	[Methyl pentyl ketone]	$C_7H_{14}O$	110-43-0	50 ppm	
3-Heptanone	[Ethyl butyl ketone]	$C_7H_{14}O$	106-35-4	50 ppm	75 ppm
4-Heptanone	[Dipropyl ketone]	$C_7H_{14}O$	123-19-3	50 ppm	
Hexachlorobenzene	[Perchlorobenzene]	C_6Cl_6	118-74-1	0.002 mg/m³	
Hexachloro-1,3-butadiene	[Perchlorobutadiene]	C_4Cl_6	87-68-3	0.02 ppm	
1,2,3,4,5,6-Hexachlorocyclohexane, (1α,2α,3β,4α,5α,6β)	[Lindane]	$C_6H_6Cl_6$	58-89-9	0.5 mg/m³	
Hexachloro-1,3-cyclopentadiene	[Perchlorocyclopentadiene]	C_5Cl_6	77-47-4	0.01 ppm	
Hexachloroethane	[Perchloroethane]	CCl_3CCl_3	67-72-1	1 ppm	
Hexachloronaphthalene (unspecified isomer)	all isomers	$C_{10}H_2Cl_6$	1335-87-1	0.2 mg/m³	
Hexahydro-1,3-isobenzofurandione	[Hexahydrophthalic anhydride] Ceiling value; all isomers	$C_8H_{10}O_3$	85-42-7		0.005 mg/m³
Hexahydro-1,3,5-trinitro-1,3,5-triazine	[Cyclonite]	$C_3H_6N_6O_6$	121-82-4	0.5 mg/m³	
Hexamethylene diisocyanate		$C_8H_{12}N_2O_2$	822-06-0	0.005 ppm	
Hexane		C_6H_{14}	110-54-3	50 ppm	
1,6-Hexanediamine	[Hexamethylenediamine]	$C_6H_{16}N_2$	124-09-4	0.5 ppm	
Hexanedinitrile	[Adiponitrile]	$C_6H_8N_2$	111-69-3	2 ppm	
1,6-Hexanedioic acid	[Adipic acid]	$C_6H_{10}O_4$	124-04-9	5 mg/m³	
2-Hexanone	[Butyl methyl ketone]	$C_6H_{12}O$	591-78-6	5 ppm	10 ppm
1-Hexene		C_6H_{12}	592-41-6	50 ppm	
sec-Hexyl acetate	[4-Methyl-2-pentyl acetate]	$C_8H_{16}O_2$	108-84-9	50 ppm	
Hydrazine		N_2H_4	302-01-2	0.01 ppm	
Hydrazoic acid	[Hydrogen azide] vapor. Ceiling value	HN_3	7782-79-8		0.11 ppm
Hydrogen bromide	[Hydrobromic acid] Ceiling value	HBr	10035-10-6		2 ppm
Hydrogen chloride	[Hydrochloric acid] Ceiling value	HCl	7647-01-0		2 ppm
Hydrogen cyanide	[Hydrocyanic acid] Ceiling value	HCN	74-90-8		4.7 ppm
Hydrogen fluoride	[Hydrofluoric acid] Ceiling value	HF	7664-39-3	0.5 ppm	2 ppm
Hydrogen peroxide		H_2O_2	7722-84-1	1 ppm	
Hydrogen selenide		H_2Se	7783-07-5	0.05 ppm	
Hydrogen sulfide		H_2S	7783-06-4	1 ppm	5 ppm
p-Hydroquinone	[1,4-Benzenediol]	$C_6H_6O_2$	123-31-9	1 mg/m³	
2-Hydroxypropyl acrylate		$C_6H_{10}O_3$	999-61-1	0.5 ppm	
Indene	[Indonaphthene]	C_9H_8	95-13-6	5 ppm	
Indium	metal & compounds, as In	In	7440-74-6	0.1 mg/m³	
Iodine	and volatile iodides. Ceiling value	I_2	7553-56-2	0.01 ppm	0.1 ppm
Iodomethane	[Methyl iodide]	CH_3I	74-88-4	2 ppm	
Iron ion [Fe^{+2}]	soluble ferrous salts, as Fe	Fe^{+2}	15438-31-0	1 mg/m³	
Iron ion [Fe^{+3}]	soluble ferric salts, as Fe	Fe^{+3}	20074-52-6	1 mg/m³	
Iron(III) oxide	[Hematite]; dust & fume, as Fe	Fe_2O_3	1309-37-1	5 mg/m³	
Iron pentacarbonyl	[Iron carbonyl [$Fe(CO)_5$]]; as Fe	$Fe(CO)_5$	13463-40-6	0.1 ppm	0.2 ppm
Isobutane	[2-Methylpropane]	C_4H_{10}	75-28-5		1000 ppm
Isobutene	[2-Methyl-1-propene]	$(CH_3)_2C$=CH_2	115-11-7	250 ppm	
Isobutyl acetate	[2-Methylpropyl acetate]	$C_6H_{12}O_2$	110-19-0	150 ppm	
Isobutyl nitrite	Ceiling value	$C_4H_9NO_2$	542-56-3		1 ppm
Isopentane	[2-Methylbutane]	C_5H_{12}	78-78-4	600 ppm	
Isopentyl acetate	[Isoamyl acetate]	$C_7H_{14}O_2$	123-92-2	50 ppm	100 ppm
Isophorone	[3,5,5-Trimethyl-2-cyclohexen-1-one] Ceiling value	$C_9H_{14}O$	78-59-1		5 ppm
Isophorone diisocyanate		$C_{12}H_{18}N_2O_2$	4098-71-9	0.005 ppm	
Isopropenylbenzene	[α-Methyl styrene]	C_9H_{10}	98-83-9	10 ppm	
2-Isopropoxyethanol		$C_5H_{12}O_2$	109-59-1	25 ppm	
Isopropyl acetate	[1-Methylethyl acetate]	$C_5H_{10}O_2$	108-21-4	100 ppm	200 ppm
Isopropylamine	[2-Propanamine]	C_3H_9N	75-31-0	5 ppm	10 ppm

Name	Comments	Formula	CAS Reg. No.	Time-weighted average	Short-term exposure limit
N-Isopropylaniline		$C_9H_{13}N$	768-52-5	2 ppm	
Isopropylbenzene	[Cumene]	C_9H_{12}	98-82-8	50 ppm	
Isopropyl glycidyl ether	[(1-Methylethoxy)methyloxirane (IGE)]	$C_6H_{12}O_2$	4016-14-2	50 ppm	75 ppm
Kaolin			1332-58-7	2 mg/m³	
Kerosene			8008-20-6	200 mg/m³	
Ketene		$CH_2{=}C{=}O$	463-51-4	0.5 ppm	1.5 ppm
Lead	metal & inorganic compounds, as Pb	Pb	7439-92-1	0.05 mg/m³	
Lead(II) arsenate		$Pb_3(AsO_4)_2$	3687-31-8	0.15 mg/m³	
Lead(II) chromate	[Crocoite]; as Pb	$PbCrO_4$	7758-97-6	0.05 mg/m³	
Lithium hydride	Ceiling value	LiH	7580-67-8		0.05 mg/m³
Magnesium oxide	[Magnesia]	MgO	1309-48-4	10 mg/m³	
Malathion		$C_{10}H_{19}O_6PS_2$	121-75-5	1 mg/m³	
Maleic anhydride	inhalable fraction & vapor	$C_4H_2O_3$	108-31-6	0.01 mg/m³	
Manganese	metal & inorganic compounds, as Mn. Inhalable fraction is 0.1 mg/m³; see Ref. 1.	Mn	7439-96-5	0.02 mg/m³ (respirable fraction)	
Manganese cyclopentadienyl tricarbonyl	as Mn	$C_8H_5MnO_3$	12079-65-1	0.1 mg/m³	
Manganese 2-methylcyclopentadienyl tricarbonyl	as Mn	$C_9H_7MnO_3$	12108-13-3	0.2 mg/m³	
Mercury	[Quicksilver]; metal & inorganic compounds, as Hg	Hg	7439-97-6	0.025 mg/m³	
Mercury	[Quicksilver]; alkyl compounds, as Hg	Hg	7439-97-6	0.01 mg/m³	0.03 mg/m³
Mercury	[Quicksilver]; aryl compounds, as Hg	Hg	7439-97-6	0.1 mg/m³	
Mesityl oxide	[Isobutenyl methyl ketone]	$C_6H_{10}O$	141-79-7	15 ppm	25 ppm
Methacrylic acid	[2-Methylpropenoic acid]	$C_4H_6O_2$	79-41-4	20 ppm	
Methanethiol	[Methyl mercaptan]	CH_3SH	74-93-1	0.5 ppm	
Methanol	[Methyl alcohol]	CH_3OH	67-56-1	200 ppm	250 ppm
Methomyl	[Acetimidic acid, *N*-[(methylcarbamoyl)oxy]thio-, methyl ester]	$C_5H_{10}N_2O_2S$	16752-77-5	0.2 mg/m³	
2-Methoxyaniline	[*o*-Anisidine]	C_7H_9NO	90-04-0	0.5 mg/m³	
4-Methoxyaniline	[*p*-Anisidine]	C_7H_9NO	104-94-9	0.5 mg/m³	
Methoxychlor		$C_{16}H_{15}Cl_3O_2$	72-43-5	10 mg/m³	
2-Methoxyethanol	[Ethylene glycol monomethyl ether (EGME)]	$C_3H_8O_2$	109-86-4	0.1 ppm	
2-Methoxyethyl acetate	[Ethylene glycol monomethyl ether acetate (EGMEA)]	$C_5H_{10}O_3$	110-49-6	0.1 ppm	
2-Methoxy-2-methylbutane	[Methyl *tert*-pentyl ether (TAME)]	$C_6H_{14}O$	994-05-8	20 ppm	
4-Methoxyphenol		$C_7H_8O_2$	150-76-5	5 mg/m³	
1-Methoxy-2-propanol	[1,2-Propylene glycol monomethyl ether (PGME)]	$C_4H_{10}O_2$	107-98-2	50 ppm	100 ppm
Methyl acetate		$C_3H_6O_2$	79-20-9	200 ppm	250 ppm
Methyl acrylate	[Methyl propenoate]	$C_4H_6O_2$	96-33-3	2 ppm	
2-Methylacrylonitrile	[2-Methylpropenenitrile]	C_4H_5N	126-98-7	1 ppm	
Methylamine	[Methanamine]	CH_3NH_2	74-89-5	5 ppm	15 ppm
2-Methylaniline	[*o*-Toluidine]	C_7H_9N	95-53-4	2 ppm	
3-Methylaniline	[*m*-Toluidine]	C_7H_9N	108-44-1	2 ppm	
4-Methylaniline	[*p*-Toluidine]	C_7H_9N	106-49-0	2 ppm	
N-Methylaniline	[*N*-Methylbenzenamine]	C_7H_9N	100-61-8	0.5 ppm	
3-Methyl-1-butanol	[Isopentyl alcohol (Isoamyl alcohol)]	$C_5H_{12}O$	123-51-3	100 ppm	125 ppm
2-Methyl-1-butanol acetate		$C_7H_{14}O_2$	624-41-9	50 ppm	100 ppm
3-Methyl-2-butanol acetate		$C_7H_{14}O_2$	5343-96-4	50 ppm	100 ppm
3-Methyl-2-butanone	[Methyl isopropyl ketone]	$C_5H_{10}O$	563-80-4	20 ppm	
Methyl *tert*-butyl ether	[*tert*-Butyl methyl ether (MTBE)]	$C_5H_{12}O$	1634-04-4	50 ppm	
Methyl 2-cyanoacrylate	[Mecrylate] NIC 2016	$C_5H_5NO_2$	137-05-3	0.2 ppm	
Methylcyclohexane		C_7H_{14}	108-87-2	400 ppm	
1-Methylcyclohexanol		$C_7H_{14}O$	590-67-0	50 ppm	
cis-2-Methylcyclohexanol		$C_7H_{14}O$	615-38-3	50 ppm	
trans-2-Methylcyclohexanol, (±)-		$C_7H_{14}O$	615-39-4	50 ppm	
cis-3-Methylcyclohexanol, (±)-		$C_7H_{14}O$	5454-79-5	50 ppm	
trans-3-Methylcyclohexanol, (±)-		$C_7H_{14}O$	7443-55-2	50 ppm	
cis-4-Methylcyclohexanol		$C_7H_{14}O$	7731-28-4	50 ppm	
trans-4-Methylcyclohexanol		$C_7H_{14}O$	7731-29-5	50 ppm	
2-Methylcyclohexanone, (±)-		$C_7H_{12}O$	24965-84-2	50 ppm	75 ppm
Methyl demeton		$C_6H_{15}O_3PS_2$	8022-00-2	0.05 mg/m³	
2-Methyl-3,5-dinitrobenzamide	[Dinitolmide]	$C_8H_7N_3O_5$	148-01-6	1 mg/m³	
1-Methyl-2,3-dinitrobenzene	[2,3-Dinitrotoluene]	$C_7H_6N_2O_4$	602-01-7	0.2 mg/m³	
1-Methyl-2,4-dinitrobenzene	[2,4-Dinitrotoluene]	$C_7H_6N_2O_4$	121-14-2	0.2 mg/m³	
1-Methyl-3,5-dinitrobenzene	[3,5-Dinitrotoluene]	$C_7H_6N_2O_4$	618-85-9	0.2 mg/m³	
2-Methyl-1,3-dinitrobenzene	[2,6-Dinitrotoluene]	$C_7H_6N_2O_4$	606-20-2	0.2 mg/m³	
2-Methyl-1,4-dinitrobenzene	[2,5-Dinitrotoluene]	$C_7H_6N_2O_4$	619-15-8	0.2 mg/m³	
4-Methyl-1,2-dinitrobenzene	[3,4-Dinitrotoluene]	$C_7H_6N_2O_4$	610-39-9	0.2 mg/m³	
2-Methyl-4,6-dinitrophenol	[4,6-Dinitro-*o*-cresol]	$C_7H_6N_2O_5$	534-52-1	0.2 mg/m³	

Name	Comments	Formula	CAS Reg. No.	Time-weighted average	Short-term exposure limit
4,4'-Methylenebis[2-chloroaniline]	[3,3'-Dichloro-4,4'-diaminodiphenylmethane (MBOCA)]	$C_{13}H_{12}Cl_2N_2$	101-14-4	0.01 ppm	
Methylenebis(4-cyclohexylisocyanate)		$C_{15}H_{22}N_2O_2$	5124-30-1	0.005 ppm	
Methyl formate		$C_2H_4O_2$	107-31-3	50 ppm	100 ppm
2-Methylheptane		C_8H_{18}	592-27-8	300 ppm	
3-Methylheptane		C_8H_{18}	589-81-1	300 ppm	
4-Methylheptane		C_8H_{18}	589-53-7	300 ppm	
6-Methyl-1-heptanol	[Isooctyl alcohol]	$C_8H_{18}O$	1653-40-3	50 ppm	
5-Methyl-3-heptanone	[Ethyl 2-methylbutyl ketone]	$C_8H_{16}O$	541-85-5	10 ppm	
5-Methyl-2-hexanone	[Methyl isopentyl ketone]	$C_7H_{14}O$	110-12-3	20 ppm	50 ppm
Methylhydrazine		CH_6N_2	60-34-4	0.01 ppm	
Methyl isocyanate		CH_3NCO	624-83-9	0.02 ppm	0.06 ppm
Methyl methacrylate	[Methyl 2-methyl-2-propenoate]	$C_5H_8O_2$	80-62-6	50 ppm	100 ppm
1-Methylnaphthalene		$C_{11}H_{10}$	90-12-0	0.5 ppm	
2-Methylnaphthalene		$C_{11}H_{10}$	91-57-6	0.5 ppm	
2-Methyl-5-nitroaniline	[5-Nitro-*o*-toluidine]	$C_7H_8N_2O_2$	99-55-8	1 mg/m³	
2-Methyloctane		C_9H_{20}	3221-61-2	200 ppm	
3-Methyloctane		C_9H_{20}	2216-33-3	200 ppm	
4-Methyloctane		C_9H_{20}	2216-34-4	200 ppm	
Methyloxirane	[1,2-Propylene oxide]	C_3H_6O	75-56-9	2 ppm	
Methyl parathion	inhalable fraction & vapor	$C_8H_{10}NO_5PS$	298-00-0	0.02 mg/m³	
2-Methylpentane	[Isohexane]	C_6H_{14}	107-83-5	500 ppm	1000 ppm
3-Methylpentane		C_6H_{14}	96-14-0	500 ppm	1000 ppm
2-Methyl-2,4-pentanediol	[Hexylene glycol] Ceiling value. NIC 2016	$C_6H_{14}O_2$	107-41-5		25 ppm
4-Methyl-2-pentanol	[Methyl isobutyl carbinol]	$C_6H_{14}O$	108-11-2	25 ppm	40 ppm
4-Methyl-2-pentanone	[Isobutyl methyl ketone]	$C_6H_{12}O$	108-10-1	20 ppm	75 ppm
2-Methyl-1-propanol	[Isobutyl alcohol]	$C_4H_{10}O$	78-83-1	50 ppm	
2-Methyl-2-propanol	[*tert*-Butyl alcohol]	$(CH_3)_3COH$	75-65-0	100 ppm	
Methyl silicate	[Tetramethoxysilane]	$Si(OCH_3)_4$	681-84-5	1 ppm	
2-Methylstyrene	[2-Vinyl toluene]	C_9H_{10}	611-15-4	50 ppm	100 ppm
3-Methylstyrene	[3-Vinyl toluene]	C_9H_{10}	100-80-1	50 ppm	100 ppm
4-Methylstyrene	[4-Vinyl toluene]	C_9H_{10}	622-97-9	50 ppm	100 ppm
N-Methyl-*N*,2,4,6-tetranitroaniline	[Tetryl]	$C_7H_5N_5O_8$	479-45-8	1.5 mg/m³	
Metribuzin		$C_8H_{14}N_4OS$	21087-64-9	5 mg/m³	
Mevinphos, *E*		$C_7H_{13}O_6P$	7786-34-7	0.01 ppm	
Mica		$(K,H)Al_3(SiO_4)_3$	12001-26-2	3 mg/m³	
Mineral oil	mist			5 mg/m³	10 mg/m³
Molybdenum	metal & insoluble compounds, as Mo	Mo	7439-98-7	10 mg/m³	
Molybdenum	soluble compounds, as Mo	Mo	7439-98-7	0.5 mg/m³	
Monocrotophos	[*trans*-Dimethyl 1-methyl-3-(methylamino)-3-oxo-1-propenyl phosphate]	$C_7H_{14}NO_5P$	6923-22-4	0.05 mg/m³	
Morpholine	[Tetrahydro-1,4-oxazine]	C_4H_9NO	110-91-8	20 ppm	
Naled	[1,2-Dibromo-2,2-dichloroethylphosphoric acid, dimethyl ester]	$C_4H_7Br_2Cl_2O_4P$	300-76-5	0.1 mg/m³	
Naphtha			8030-30-6	400 ppm	
Naphthalene		$C_{10}H_8$	91-20-3	10 ppm	
1-Naphthalenylthiourea	[ANTU]	$C_{11}H_{10}N_2S$	86-88-4	0.3 mg/m³	
2-Naphthylamine	[β-Naphthylamine]; levels as low as possible	$C_{10}H_9N$	91-59-8		
Neopentane	[2,2-Dimethylpropane]	$C(CH_3)_4$	463-82-1	600 ppm	
Nickel	metal	Ni	7440-02-0	1.5 mg/m³	
Nickel	soluble compounds, as Ni	Ni	7440-02-0	0.1 mg/m³	
Nickel	insoluble compounds, as Ni	Ni	7440-02-0	0.2 mg/m³	
Nickel carbonyl [$Ni(CO)_4$]	[Nickel tetracarbonyl] as Ni. Ceiling value	$Ni(CO)_4$	13463-39-3		0.05 ppm
Nickel subsulfide	[Heazlewoodite]; as Ni	Ni_3S_2	12035-72-2	0.1 mg/m³	
L-Nicotine	[3-(1-Methyl-2-pyrrolidinyl)pyridine, (*S*)-]	$C_{10}H_{14}N_2$	54-11-5	0.5 mg/m³	
Nitrapyrin	[Pyridine, 2-chloro-6-(trichloromethyl)-]	$C_6H_3Cl_4N$	1929-82-4	10 mg/m³	
Nitric acid		HNO_3	7697-37-2	2 ppm	4 ppm
Nitric oxide		NO	10102-43-9	25 ppm	
4-Nitroaniline		$C_6H_6N_2O_2$	100-01-6	3 mg/m³	
Nitrobenzene		$C_6H_5NO_2$	98-95-3	1 ppm	
4-Nitrobiphenyl	levels as low as possible	$C_{12}H_9NO_2$	92-93-3		
Nitroethane		$C_2H_5NO_2$	79-24-3	100 ppm	
Nitrogen dioxide		NO_2	10102-44-0	0.2 ppm	
Nitrogen trifluoride		NF_3	7783-54-2	10 ppm	
Nitromethane	[Nitrocarbol]	CH_3NO_2	75-52-5	20 ppm	
1-Nitropropane		$C_3H_7NO_2$	108-03-2	25 ppm	
2-Nitropropane	[Isonitropropane]	$C_3H_7NO_2$	79-46-9	10 ppm	
N-Nitrosodimethylamine	[Dimethylnitrosamine]; levels as low as possible	$C_2H_6N_2O$	62-75-9		

Name	Comments	Formula	CAS Reg. No.	Time-weighted average	Short-term exposure limit
2-Nitrotoluene	[1-Methyl-2-nitrobenzene]	$C_7H_7NO_2$	88-72-2	2 ppm	
3-Nitrotoluene	[1-Methyl-3-nitrobenzene]	$C_7H_7NO_2$	99-08-1	2 ppm	
4-Nitrotoluene	[1-Methyl-4-nitrobenzene]	$C_7H_7NO_2$	99-99-0	2 ppm	
Nitrous oxide		N_2O	10024-97-2	50 ppm	
3,3,4,4,5,5,6,6,6,-Nonafluoro-1-hexene	[Perfluorobutylethene]	$C_6H_3F_9$	19430-93-4	100 ppm	
Nonane	all isomers	C_9H_{20}	111-84-2	200 ppm	
Octachloronaphthalene	[Perchloronaphthalene]	$C_{10}Cl_8$	2234-13-1	0.1 mg/m^3	0.3 mg/m^3
Octane	all isomers	C_8H_{18}	111-65-9	300 ppm	375 ppm
Osmium(VIII) oxide	[Osmic acid (Osmium tetroxide)]	OsO_4	20816-12-0	0.0002 ppm	0.0006 ppm
Oxalic acid	and dihydrate	$H_2C_2O_4$	144-62-7	1 mg/m^3	2 mg/m^3
2-Oxetanone	[β-Propiolactone]	$C_3H_4O_2$	57-57-8	0.5 ppm	
Oxirane	[Ethylene oxide]	C_2H_4O	75-21-8	1 ppm	
Oxiranemethanol, (±)-	[Glycidol]	$C_3H_6O_2$	61915-27-3	2 ppm	
4,4'-Oxybis(benzenesulfonyl hydrazide)		$C_{12}H_{14}N_4O_5S_2$	80-51-3	0.1 mg/m^3	
Ozone	depends on workload	O_3	10028-15-6	0.1 ppm	
Paraquat		$C_{12}H_{14}N_2$	4685-14-7	0.5 mg/m^3	
Parathion		$C_{10}H_{14}NO_5PS$	56-38-2	0.05 mg/m^3	
Pentaborane(9)		B_5H_9	19624-22-7	0.005 ppm	0.015 ppm
Pentachloronaphthalene (unspecified isomer)	all isomers	$C_{10}H_3Cl_5$	1321-64-8	0.5 mg/m^3	
Pentachloronitrobenzene	[Quintozene]	$C_6Cl_5NO_2$	82-68-8	0.5 mg/m^3	
Pentachlorophenol	inhalable fraction & vapor	C_6HCl_5O	87-86-5	0.5 mg/m^3	1 mg/m^3
Pentaerythritol		$C_5H_{12}O_4$	115-77-5	10 mg/m^3	
Pentanal	[Valeraldehyde]	$C_5H_{10}O$	110-62-3	50 ppm	
Pentane	all isomers	C_5H_{12}	109-66-0	1000 ppm	
Pentanedial	[Glutaraldehyde] Ceiling value	$C_5H_8O_2$	111-30-8		0.05 ppm
3-Pentanol acetate		$C_7H_{14}O_2$	620-11-1	50 ppm	100 ppm
2-Pentanone	[Methyl propyl ketone]	$C_5H_{10}O$	107-87-9		150 ppm
3-Pentanone	[Diethyl ketone]	$C_5H_{10}O$	96-22-0	200 ppm	300 ppm
Pentyl acetate	[Amyl acetate]; all isomers	$C_7H_{14}O_2$	628-63-7	50 ppm	100 ppm
sec-Pentyl acetate, (*R*)-	[*sec*-Amyl acetate, (*R*)-]	$C_7H_{14}O_2$	54638-10-7	50 ppm	100 ppm
Perchloryl fluoride	[Chlorine trioxide fluoride]	ClO_3F	7616-94-6	3 ppm	6 ppm
Perfluoroacetone	[Hexafluoroacetone]	C_3F_6O	684-16-2	0.1 ppm	
Perfluoroisobutene	[Perfluoroisobutylene] Ceiling value	C_4F_8	382-21-8		0.01 ppm
Perfluoropropene	[Hexafluoropropene]	C_3F_6	116-15-4	0.1 ppm	
Peroxyacetic acid	[Peracetic acid] inhalable fraction & vapor	$C_2H_4O_3$	79-21-0		0.4 ppm
Phenol	[Hydroxybenzene]	C_6H_5OH	108-95-2	5 ppm	
10*H*-Phenothiazine	[Thiodiphenylamine]	$C_{12}H_9NS$	92-84-2	5 mg/m^3	
Phenyl glycidyl ether	[PGE]	$C_9H_{10}O_2$	122-60-1	0.1 ppm	
Phenylhydrazine		$C_6H_8N_2$	100-63-0	0.1 ppm	
Phenyl isocyanate		C_6H_5NCO	103-71-9	0.005 ppm	0.015 ppm
N-Phenyl-2-naphthalenamine	[*N*-Phenyl-β-naphthylamine]; levels as low as possible	$C_{16}H_{13}N$	135-88-6		
Phenylphosphine	[Monophenylphosphine] Ceiling value	$C_6H_5PH_2$	638-21-1		0.05 ppm
Phorate	[Phosphorodithioic acid, *O,O*-diethyl *S*-[(ethylthio) methyl] ester]	$C_7H_{17}O_2PS_3$	298-02-2	0.05 mg/m^3	0.2 mg/m^3
Phosphine	[Phosphorus hydride]	PH_3	7803-51-2	0.3 ppm	1 ppm
Phosphoric acid	[Orthophosphoric acid]	H_3PO_4	7664-38-2	1 mg/m^3	3 mg/m^3
Phosphorus (white)	[Yellow phosphorus]	P	7723-14-0	0.1 mg/m^3	
Phosphorus(III) chloride	[Phosphorus trichloride]	PCl_3	7719-12-2	0.2 ppm	0.5 ppm
Phosphorus(V) chloride	[Phosphorus pentachloride]	PCl_5	10026-13-8	0.1 ppm	
Phosphorus(V) sulfide	[Phosphorus pentasulfide]	P_2S_5	1314-80-3	1 mg/m^3	3 mg/m^3
Phosphoryl chloride	[Phosphorus oxychloride]	$POCl_3$	10025-87-3	0.1 ppm	
Phthalic anhydride	NIC 2016	$C_8H_4O_3$	85-44-9	1 ppm	
α-Pinene	[2-Pinene]	$C_{10}H_{16}$	80-56-8	20 ppm	
β-Pinene	[Nopinene]	$C_{10}H_{16}$	127-91-3	20 ppm	
Piperazine	and salts, as piperazine; inhalable fraction & vapor	$C_4H_{10}N_2$	110-85-0	0.03 ppm	
2-Pivaloyl-1,3-indandione	[Pindone]	$C_{14}H_{14}O_3$	83-26-1	0.1 mg/m^3	
Platinum	metal dust	Pt	7440-06-4	1 mg/m^3	
Platinum	soluble salts, as Pt	Pt	7440-06-4	0.002 mg/m^3	
Polychlorinated biphenyls (42% chlorine)	[PCBs]		53469-21-9	1 mg/m^3	
Polychlorinated biphenyls (54% chlorine)	[PCBs]		11097-69-1	0.5 mg/m^3	
Poly(vinyl chloride)	[PVC]		9002-86-2	1 mg/m^3	
Portland cement	respirable fraction; no asbestos		65997-15-1	1 mg/m^3	
Potassium hydroxide	Ceiling value	KOH	1310-58-3		2 mg/m^3
Propanal	[Propionaldehyde]	C_2H_5CHO	123-38-6	20 ppm	
1,3-Propane sultone	[1,2-Oxathiolane, 2,2-dioxide]; levels as low as possible	$C_3H_6O_3S$	1120-71-4		
Propanoic acid	[Propionic acid]	$C_3H_6O_2$	79-09-4	10 ppm	
1-Propanol	[Propyl alcohol]	$CH_3CH_2CH_2OH$	71-23-8	100 ppm	400 ppm

Name	Comments	Formula	CAS Reg. No.	Time-weighted average	Short-term exposure limit
2-Propanol	[Isopropyl alcohol]	$CH_3CHOHCH_3$	67-63-0	200 ppm	400 ppm
Propargyl alcohol	[3-Hydroxy-1-propyne (2-Propyn-1-ol)]	C_3H_4O	107-19-7	1 ppm	
Propene	[Propylene]	$CH_3CH{=}CH_2$	115-07-1	500 ppm	
Propoxur	[Phenol, 2-(1-methylethoxy)-, methylcarbamate] inhalable fraction & vapor	$C_{11}H_{15}NO_3$	114-26-1	0.5 mg/m³	
Propyl acetate		$C_5H_{10}O_2$	109-60-4	200 ppm	250 ppm
1,2-Propylene glycol dinitrate		$C_3H_6N_2O_6$	6423-43-4	0.05 ppm	
Propyleneimine	[2-Methylaziridine]	C_3H_7N	75-55-8	0.2 ppm	0.4 ppm
Propyl nitrate		$C_3H_7NO_3$	627-13-4	25 ppm	40 ppm
Propyne	[Methylacetylene] NIC 2016	$CH_3C{\equiv}CH$	74-99-7	1000 ppm	
Pyrethrin I	[Pyrethrum]	$C_{21}H_{28}O_3$	121-21-1	5 mg/m³	
2-Pyridinamine	[2-Aminopyridine]	$C_5H_6N_2$	504-29-0	0.5 ppm	
Pyridine	[Azine]	C_5H_5N	110-86-1	1 ppm	
Pyrocatechol	[1,2-Benzenediol (Catechol)]	$C_6H_6O_2$	120-80-9	5 ppm	
Resorcinol	[1,3-Benzenediol]	$C_6H_6O_2$	108-46-3	10 ppm	20 ppm
Rhodium	metal & insoluble compounds, as Rh	Rh	7440-16-6	1 mg/m³	
Rhodium	soluble compounds, as Rh	Rh	7440-16-6	0.01 mg/m³	
Ronnel		$C_8H_8Cl_3O_3PS$	299-84-3	5 mg/m³	
Rotenone		$C_{23}H_{22}O_6$	83-79-4	5 mg/m³	
Rubber	natural latex, as inhalable proteins			0.0001 mg/m³	
Selenium	element & compounds, as Se	Se	7782-49-2	0.2 mg/m³	
Selenium hexafluoride		SeF_6	7783-79-1	0.05 ppm	
Sesone	[Sodium 2-(2,4-dichlorophenoxy)ethyl sulfate]	$C_8H_7Cl_2NaO_5S$	136-78-7	10 mg/m³	
Silane	[Silicon tetrahydride]	SiH_4	7803-62-5	5 ppm	
Silicon carbide (hexagonal)	[Moissanite]; lower limits for fibrous SiC	SiC	409-21-2	10 mg/m³	
Silicon dioxide (α-quartz)	[Silica]; respirable fraction	SiO_2	14808-60-7	0.025 mg/m³	
Silicon dioxide (cristobalite)	[Cristobalite]; respirable fraction	SiO_2	14464-46-1	0.025 mg/m³	
Silver	metal	Ag	7440-22-4	0.1 mg/m³	
Silver	soluble compounds, as Ag	Ag	7440-22-4	0.01 mg/m³	
Simazine	inhalable particulate matter		122-34-9	0.5 mg/m³	
Sodium azide	[Smite] Ceiling value	NaN_3	26628-22-8		0.29 mg/m³
Sodium fluoroacetate		$C_2H_2FNaO_2$	62-74-8	0.05 mg/m³	
Sodium hydrogen sulfite	[Sodium bisulfite]	$NaHSO_3$	7631-90-5	5 mg/m³	
Sodium hydroxide	[Caustic soda] Ceiling value	NaOH	1310-73-2		2 mg/m³
Sodium metabisulfite	[Sodium pyrosulfite]	$Na_2S_2O_5$	7681-57-4	5 mg/m³	
Sodium tetraborate decahydrate	[Borax]	$Na_2B_4O_7{\cdot}10H_2O$	1303-96-4	2 mg/m³	6 mg/m³
Starch			9005-25-8	10 mg/m³	
Stibine	[Antimony(III) hydride]	SbH_3	7803-52-3	0.1 ppm	
Stoddard solvent			8052-41-3	100 ppm	
Strontium chromate	as Cr	$SrCrO_4$	7789-06-2	0.0005 mg/m³	
Strychnine		$C_{21}H_{22}N_2O_2$	57-24-9	0.15 mg/m³	
Styrene	[Vinylbenzene]	C_8H_8	100-42-5	20 ppm	40 ppm
Subtilisins	as crystalline active enzyme. Ceiling value		9014-01-1		0.00006 mg/m³
Sucrose		$C_{12}H_{22}O_{11}$	57-50-1	10 mg/m³	
Sulfometuron methyl		$C_{15}H_{16}N_4O_5S$	74222-97-2	5 mg/m³	
Sulfotep	[Tetraethyl thiodiphosphate (TEDP)]	$C_8H_{20}O_5P_2S_2$	3689-24-5	0.2 mg/m³	
Sulfur chloride [ClSSCl]	[Sulfur monochloride (Disulfur dichloride)] Ceiling value	S_2Cl_2	10025-67-9		1 ppm
Sulfur decafluoride	[Disulfur decafluoride (Sulfur pentafluoride)] Ceiling value	S_2F_{10}	5714-22-7		0.01 ppm
Sulfur dioxide		SO_2	7446-09-5		0.25 ppm
Sulfur hexafluoride		SF_6	2551-62-4	1000 ppm	
Sulfuric acid	[Oil of vitriol]	H_2SO_4	7664-93-9	0.2 mg/m³	
Sulfur tetrafluoride	Ceiling value	SF_4	7783-60-0		0.1 ppm
Sulfuryl fluoride		SO_2F_2	2699-79-8	5 ppm	10 ppm
Sulprofos	inhalable fraction & vapor	$C_{12}H_{19}O_2PS_3$	35400-43-2	0.1 mg/m³	
Talc	respirable fraction; no asbestos	$3MgO{\cdot}4SiO_2{\cdot}H_2O$	14807-96-6	2 mg/m³	
Tantalum	dust	Ta	7440-25-7	5 mg/m³	
Tantalum(V) oxide	[Tantalum pentoxide]; dust, as Ta	Ta_2O_5	1314-61-0	5 mg/m³	
Tellurium	and compounds, as Te (except H_2Te)	Te	13494-80-9	0.1 mg/m³	
Tellurium hexafluoride		TeF_6	7783-80-4	0.02 ppm	
Terbufos		$C_9H_{21}O_2PS_3$	13071-79-9	0.01 mg/m³	
Terephthalic acid	[1,4-Benzenedicarboxylic acid]	$C_8H_6O_4$	100-21-0	10 mg/m³	
o-Terphenyl	Ceiling value	$C_{18}H_{14}$	84-15-1		5 mg/m³
m-Terphenyl	Ceiling value	$C_{18}H_{14}$	92-06-8		5 mg/m³
p-Terphenyl		$C_{18}H_{14}$	92-94-4		5 mg/m³
1,1,2,2-Tetrabromoethane	[Acetylene tetrabromide]	$C_2H_2Br_4$	79-27-6	0.1 ppm	
Tetrabromomethane	[Carbon tetrabromide]	CBr_4	558-13-4	0.1 ppm	0.3 ppm

Name	Comments	Formula	CAS Reg. No.	Time-weighted average	Short-term exposure limit
1,1,1,2-Tetrachloro-2,2-difluoroethane	[Tetrachloro-1,1-difluoroethane]	$C_2Cl_4F_2$	76-11-9	100 ppm	
1,1,2,2-Tetrachloro-1,2-difluoroethane	[Tetrachloro-1,2-difluoroethane]	$C_2Cl_4F_2$	76-12-0	50 ppm	
1,1,2,2-Tetrachloroethane	[Acetylene tetrachloride]	$C_2H_2Cl_4$	79-34-5	1 ppm	
Tetrachloroethene	[Perchloroethylene]	C_2Cl_4	127-18-4	25 ppm	100 ppm
Tetrachloromethane	[Carbon tetrachloride]	CCl_4	56-23-5	5 ppm	10 ppm
1,2,3,4-Tetrachloronaphthalene		$C_{10}H_4Cl_4$	20020-02-4	2 mg/m³	
Tetraethyl lead	as Pb	$C_8H_{20}Pb$	78-00-2	0.1 mg/m³	
Tetraethyl pyrophosphate	[TEPP]	$C_8H_{20}O_7P_2$	107-49-3	0.01 mg/m³	
Tetrafluoroethene	[Tetrafluoroethylene]	$F_2C{=}CF_2$	116-14-3	2 ppm	
Tetrahydrofuran	[Tetramethylene oxide (Oxolane)]	C_4H_8O	109-99-9	50 ppm	100 ppm
Tetrakis(hydroxymethyl)phosphonium chloride	also the sulfate	$C_4H_{12}ClO_4P$	124-64-1	2 mg/m³	
2,2,3,3-Tetramethylbutane		C_8H_{18}	594-82-1	300 ppm	
Tetramethyl lead	as Pb	$C_4H_{12}Pb$	75-74-1	0.15 mg/m³	
2,2,3,3-Tetramethylpentane		C_9H_{20}	7154-79-2	200 ppm	
2,2,3,4-Tetramethylpentane		C_9H_{20}	1186-53-4	200 ppm	
2,2,4,4-Tetramethylpentane	[Di-*tert*-butylmethane]	C_9H_{20}	1070-87-7	200 ppm	
2,3,3,4-Tetramethylpentane		C_9H_{20}	16747-38-9	200 ppm	
Tetramethylsuccinonitrile	[Tetramethylbutanedinitrile]	$C_8H_{12}N_2$	3333-52-6	0.5 ppm	
Tetranitromethane		$C(NO_2)_4$	509-14-8	0.005 ppm	
Thallium	and compounds, as Tl	Tl	7440-28-0	0.02 mg/m³	
4,4′-Thiobis(6-*tert*-butyl-*m*-cresol)	[Bis(5-*tert*-butyl-4-hydroxy-2-methylphenyl) sulfide]; inhalable fraction	$C_{22}H_{30}O_2S$	96-69-5	1 mg/m³	
Thioglycolic acid	NIC 2016	$C_2H_4O_2S$	68-11-1	1 ppm	
Thionyl chloride	[Sulfinyl dichloride] Ceiling value	$SOCl_2$	7719-09-7		0.2 ppm
Thiram	inhalable fraction & vapor	$C_6H_{12}N_2S_4$	137-26-8	0.05 mg/m³	
Tin	metal	Sn	7440-31-5	2 mg/m³	
Tin	inorganic compounds, as Sn	Sn	7440-31-5	2 mg/m³	
Tin	organic compounds, as Sn	Sn	7440-31-5	0.1 mg/m³	0.2 mg/m³
Titanium(IV) oxide (rutile)	[Rutile (Titanium dioxide)]	TiO_2	1317-80-2	10 mg/m³	
Toluene	[Methylbenzene]	$C_6H_5CH_3$	108-88-3	20 ppm	
Toluene-2,4-diisocyanate	inhalable fraction and vapor	$C_9H_6N_2O_2$	584-84-9	0.001 ppm	0.005 ppm
Toluene-2,6-diisocyanate	inhalable fraction and vapor	$C_9H_6N_2O_2$	91-08-7	0.001 ppm	0.005 ppm
Toxaphene	[Polychlorocamphene]	$C_{10}H_{10}Cl_8$	8001-35-2	0.5 mg/m³	
1*H*-1,2,4-Triazol-3-amine	[Amitrole]	$C_2H_4N_4$	61-82-5	0.2 mg/m³	
Tribromomethane	[Bromoform]	$CHBr_3$	75-25-2	0.5 ppm	
Tributyl phosphate	[Butyl phosphate]; inhalable fraction & vapor	$C_{12}H_{27}O_4P$	126-73-8	5 mg/m³	
Trichlorfon	[2,2,2-Trichloro-1-hydroxyethylphosphonic acid, dimethyl ester]	$C_4H_8Cl_3O_4P$	52-68-6	1 mg/m³	
Trichloroacetic acid		CCl_3COOH	76-03-9	0.5 ppm	
1,2,4-Trichlorobenzene	Ceiling value	$C_6H_3Cl_3$	120-82-1		5 ppm
1,1,1-Trichloro-2,2-bis(4-chlorophenyl)ethane	[Dichlorodiphenyltrichloroethane (DDT)]	$C_{14}H_9Cl_5$	50-29-3	1 mg/m³	
1,1,1-Trichloroethane	[Methyl chloroform]	$CHCCl_3$	71-55-6	350 ppm	450 ppm
1,1,2-Trichloroethane	[Vinyl trichloride]	$C_2H_3Cl_3$	79-00-5	10 ppm	
Trichloroethene	[Trichloroethylene]	C_2HCl_3	79-01-6	10 ppm	25 ppm
Trichlorofluoromethane	[CFC-11] Ceiling value	CCl_3F	75-69-4		1000 ppm
Trichloromethane	[Chloroform]	$CHCl_3$	67-66-3	10 ppm	
Trichloromethanesulfenyl chloride	[Perchloromethyl mercaptan]	CCl_4S	594-42-3	0.1 ppm	
(Trichloromethyl)benzene	[Benzotrichloride] Ceiling value	$C_6H_5CCl_3$	98-07-7		0.1 ppm
Trichloronaphthalene (unspecified isomer)	all isomers	$C_{10}H_5Cl_3$	1321-65-9	5 mg/m³	
Trichloronitromethane	[Chloropicrin]	CCl_3NO_2	76-06-2	0.1 ppm	
2,4,5-Trichlorophenoxyacetic acid	[2,4,5-T]	$C_8H_5Cl_3O_3$	93-76-5	10 mg/m³	
1,2,3-Trichloropropane	[Allyl trichloride]	$C_3H_5Cl_3$	96-18-4	0.005 ppm	
1,1,2-Trichloro-1,2,2-trifluoroethane	[CFC-113]	$C_2Cl_3F_3$	76-13-1	1000 ppm	1250 ppm
Tri-*o*-cresyl phosphate	[Tri-*o*-tolyl phosphate]. inhalable fraction and vapor	$C_{21}H_{21}O_4P$	78-30-8	0.02 mg/m³	
Triethanolamine	[Tris(2-hydroxyethyl)amine]	$C_6H_{15}NO_3$	102-71-6	5 mg/m³	
Triethylamine	[*N,N*-Diethylethanamine]	$(C_2H_5)_3N$	121-44-8	0.5 ppm	1 ppm
1,3,5-Triglycidyl-*s*-triazinetrione	[1,3,5-Tris(oxiranemethyl)-1,3,5-triazine-2,4,6(1*H*,3*H*,5*H*)-trione]	$C_{12}H_{15}N_3O_6$	2451-62-9	0.05 mg/m³	
Triiodomethane	[Iodoform] NIC 2016	CHI_3	75-47-8	0.6 ppm	
Trimellitic anhydride	[1,2,4-Benzenetricarboxylic anhydride] Ceiling value	$C_9H_4O_5$	552-30-7	0.0005 mg/m³	0.002 mg/m³
Trimethylamine	[*N,N*-Dimethylmethanamine]	$(CH_3)_3N$	75-50-3	5 ppm	15 ppm
1,2,3-Trimethylbenzene	[Hemimellitene]	C_9H_{12}	526-73-8	25 ppm	
1,2,4-Trimethylbenzene	[Pseudocumene]	C_9H_{12}	95-63-6	25 ppm	
1,3,5-Trimethylbenzene	[Mesitylene]	C_9H_{12}	108-67-8	25 ppm	
2,2,3-Trimethylhexane		C_9H_{20}	16747-25-4	200 ppm	
2,2,4-Trimethylhexane		C_9H_{20}	16747-26-5	200 ppm	
2,2,5-Trimethylhexane		C_9H_{20}	3522-94-9	200 ppm	

Name	Comments	Formula	CAS Reg. No.	Time-weighted average	Short-term exposure limit
2,3,3-Trimethylhexane		C_9H_{20}	16747-28-7	200 ppm	
2,3,4-Trimethylhexane		C_9H_{20}	921-47-1	200 ppm	
2,3,5-Trimethylhexane		C_9H_{20}	1069-53-0	200 ppm	
2,4,4-Trimethylhexane		C_9H_{20}	16747-30-1	200 ppm	
3,3,4-Trimethylhexane		C_9H_{20}	16747-31-2	200 ppm	
2,2,3-Trimethylpentane	[2-*tert*-Butylbutane]	C_8H_{18}	564-02-3	300 ppm	
2,2,4-Trimethylpentane	[Isooctane]	C_8H_{18}	540-84-1	300 ppm	
2,3,3-Trimethylpentane		C_8H_{18}	560-21-4	300 ppm	
2,3,4-Trimethylpentane		C_8H_{18}	565-75-3	300 ppm	
Trimethyl phosphite		$C_3H_9O_3P$	121-45-9	2 ppm	
Trinitroglycerol	[Nitroglycerin]	$C_3H_5N_3O_9$	55-63-0	0.05 ppm	
2,4,6-Trinitrophenol	[Picric acid]	$C_6H_3N_3O_7$	88-89-1	0.1 mg/m³	
2,4,6-Trinitrotoluene	[2-Methyl-1,3,5-trinitrobenzene (TNT)]	$C_7H_5N_3O_6$	118-96-7	0.1 mg/m³	
Triphenyl phosphate		$C_{18}H_{15}O_4P$	115-86-6	3 mg/m³	
Tungsten	[Wolfram]; metal & insoluble compounds, as W	W	7440-33-7	5 mg/m³	10 mg/m³
Tungsten	[Wolfram]; soluble compounds, as W. NIC 2016	W	7440-33-7	1 mg/m³	3 mg/m³
Turpentine	and selected monoterpenes		8006-64-2	20 ppm	
Uranium	metal & compounds, as U	U	7440-61-1	0.2 mg/m³	0.6 mg/m³
Vanadium(V) oxide	[Vanadium pentoxide]; inhalable fraction, as V	V_2O_5	1314-62-1	0.05 mg/m³	
Vinyl acetate		$C_4H_6O_2$	108-05-4	10 ppm	15 ppm
4-Vinylcyclohexene		C_8H_{12}	100-40-3	0.1 ppm	
1-Vinyl-2-pyrrolidinone		C_6H_9NO	88-12-0	0.05 ppm	
Warfarin	[Coumadin]; inhalable particulate matter.	$C_{19}H_{16}O_4$	81-81-2	0.01 mg/m³	
Wood	dust, inhalable fraction (0.5 mg/m³ for western red cedar)			1 mg/m³	
o-Xylene	[1,2-Dimethylbenzene]	$C_6H_4(CH_3)_2$	95-47-6	100 ppm	150 ppm
m-Xylene	[1,3-Dimethylbenzene]	$C_6H_4(CH_3)_2$	108-38-3	100 ppm	150 ppm
p-Xylene	[1,4-Dimethylbenzene]	$C_6H_4(CH_3)_2$	106-42-3	100 ppm	150 ppm
Yttrium	metal & compounds, as Y	Y	7440-65-5	1 mg/m³	
Zinc chloride	fume	$ZnCl_2$	7646-85-7	1 mg/m³	2 mg/m³
Zinc chromate	as Cr	$ZnCrO_4$	13530-65-9	0.01 mg/m³	
Zinc oxide	[Zincite]	ZnO	1314-13-2	2 mg/m³	10 mg/m³
Zirconium	metal & compounds, as Zr	Zr	7440-67-7	5 mg/m³	10 mg/m³

LASER HAZARDS IN THE LABORATORY

Thomas J. Bruno and Paris D. N. Svoronos

Lasers are commonly used in the laboratory, although in many instruments, most lasers are embedded in instrumentation and are therefore shielded or protected by optical barriers and interlocks that, when functioning properly, prevent accidental exposure. Care must be exercised when performing maintenance or when changing samples in such instruments. In this section we provide basic information on laser safety and hazards (Refs. 1 to 3). This is by no means exhaustive nor is it meant to substitute for an understanding of the specific safety requirements of instrumentation, or applicable law or regulations. The special case of common laser pointers has received considerable attention recently and is treated separately (Ref. 4). We note that as of 2007, the general practice in the United States is to use the IEC definitions.

References

1. American National Standard for Safe Use of Lasers, American National Standards Institute, ANSI Z136.1, 2007
2. Safety of Laser Products – Part 1: Equipment Classification and Requirements, International Electrotechnical Commission, IEC 60825-1, 2nd Ed., 2007.
3. Bruno, T. J., and Svoronos, P. D. N., *CRC Handbook of Basic Tables for Chemical Analysis*, 3rd. Ed., CRC Press, Boca Raton, FL, 2011.
4. Hadler, J., and Dowell, M., Accurate, Inexpensive Testing of Laser Pointer Power for Safe Operation, *Meas. Sci. Technol.* 24, 1, 2013.

Classes of Lasers:

The following is a summary for the laser classes following the ANSI guidelines used in the United States:

Class I

Class I lasers are inherently safe with no possibility of eye damage under conditions of normal use. The safety can result from a low output power (in which case eye damage is impossible even after prolonged exposure), or due to an enclosure preventing user access to the laser beam during normal operation, such as in CD players, laser printers, surveying transits, or measurement instruments.

Class II

The blink reflex of the human eye will prevent eye damage, unless the person deliberately stares into the beam for an extended period. Thus, a Class II laser can cause some eye damage if this is done. Output power may be up to 1 mW. This class includes only lasers that emit visible light. Some laser pointers are in this category.

Class IIIa

Lasers in this class are mostly dangerous in combination with certain optical instruments that change the beam diameter or power density. Output power does not exceed 5 mW. Beam power density may not exceed 2.5 mW cm^{-2}. Many laser sights for firearms and some laser pointers are included in this category.

Class IIIb

Lasers in this class may cause damage if the beam enters the eye directly. This generally applies to lasers powered from 5 mW to 500 mW. Lasers in this category can cause permanent eye damage with exposures of 1/100th of a second or less depending on the strength of the laser. A diffuse reflection (on paper or from a matte surface) is generally not hazardous but a specular reflection from a highly reflective surface can be just as dangerous as direct exposure. Protective eyewear is recommended when direct beam viewing of Class IIIb lasers may occur. Lasers at the high power end of this class may also present a fire hazard and can lightly burn skin.

Class IV

Lasers in this class have output powers of more than 500 mW in the beam and may cause severe, permanent damage to eye or skin without being magnified by optics of eye or instrumentation. Diffuse reflections of the laser beam can be hazardous to skin or eye within the Nominal Hazard Zone. Many industrial, scientific, military, and medical lasers are in this category.

The following is a summary of the laser classes following the IEC guidelines:

Class 1

A Class 1 laser is safe under all conditions of normal use, with no known biological hazard present. This class includes high-power lasers within an enclosure that prevents exposure to the radiation and that cannot be opened without shutting down the laser. This typically requires an interlocking.

Class 1M

A Class 1M laser is safe for all conditions of normal use except when passed through magnifying optics such as microscopes, telescopes, or on optical benches. Class 1M lasers typically produce large-diameter beams, or beams that are divergent. The classification of a Class 1M laser must be changed if the emergent light is refocused.

Class 2

A Class 2 laser is safe for all conditions of normal use because the blink reflex will limit the exposure to no more than 0.25 seconds. It only applies to visible-light lasers (400 nm to 700 nm) limited to 1 mW continuous wave, or more if the emission time is less than 0.25 s or if the light is not spatially coherent. Intentional suppression of the blink reflex could lead to eye injury. Many laser pointers are Class 2.

Class 2M

A Class 2M laser is similar to a Class 2, but it is used in an instrument that may focus the beam. This laser is safe because of the blink reflex provided the beam is not viewed through optical instruments as described above for Class 1M.

Class 3R

A Class 3R laser is considered safe if handled carefully, with restricted beam viewing. These lasers can be hazardous to the human eye if the beam is viewed for extended periods of time or under fixated conditions. Continuous beam Class 3R lasers operating in the visible region are limited in power output to 5

Safety

mW. For other wavelengths and for pulsed lasers, other limits will apply.

Class 3B

A Class 3B laser is hazardous if the eye is exposed directly, but diffuse reflections such as from paper surfaces are not harmful. Continuous lasers in the wavelength range from 315 nm to the far infrared are limited in power output to 0.5 W. For pulsed lasers between 400 nm and 700 nm, the limit is 30 mJ. Other limits apply to other wavelengths and to short pulse lasers. Protective eyewear is typically required where direct viewing of a class 3B laser beam may occur. Class 3B lasers must be equipped with a key switch and a safety interlock.

Class 4

Class 4 lasers include all lasers with beam power greater than those covered in class 3B. By definition, a Class 4 laser can burn the skin, in addition to causing severe and permanent eye damage. This eye damage can result from of direct or diffuse beam viewing. These lasers may ignite combustible materials, and thus may represent a fire risk. Class 4 lasers must be equipped with a key switch and a safety interlock. Many industrial, scientific, military, and medical lasers are in this category.

Laser Pointers

Laser pointers, ubiquitous at meetings, shows, and in the classroom, deserve separate consideration because of recent work on actual observed power output. For purposes of classification into the levels discussed above, the typical laser pointer is classified as 3R. The human light aversion response can generally protect against 3R lasers; however, this response is less sensitive in the near infrared range (700 nm to 1400 nm). Thus, to prevent retinal burns, laser pointers must not emit hazardous levels of infrared, and must be Class 1 compliant in terms of accessible emission level (AEL) in that wavelength range. In a testing program undertaken at NIST, cited in Ref. 4, it was found that of the laser pointers randomly chosen and tested, all but two pointers failed to comply by more than 15 % of the specified AEL, and 48 % emitted more than twice the specified AEL at one or more specified wavelength. This indicates a risk of 3B exposure from these devices that are nominally classified as 3R.

GENERAL CHARACTERISTICS OF IONIZING RADIATION FOR THE PURPOSE OF PRACTICAL APPLICATION OF RADIATION PROTECTION

Thomas W. Grove and Thomas J. Bruno

The following table provides practical information to allow the design and implementation of radiation protection in laboratory and industrial environments. Additional information and details can be found in Refs. 1 to 3.

References

1. Johnson, T. E., and Birky, B. K., *Health Physics Radiological Health*, 4th Ed., Lippencott, Williams and Wilkins, Baltimore, MD, 2012.
2. Cember, H., and Johnson, T. E., *Introduction to Health Physics*, 4th Ed., McGraw Hill, New York, 2011.
3. 1990 Recommendations of the International Commission on Radiological Protection. ICRP Publication 60. Ann. ICRP 21 (1-3), 1990.

Type (symbol)	Physical properties	Range	Shielding	Biological hazards	Comments
Alpha particle (α)	Very large mass (2 protons, 2 neutrons, 0 electrons) +2 charge	Very short 3 cm to 6 cm (~ 1 inch to 2 inches) in air	Few centimeters of air Sheet of paper Dead (outer) layer of skin	Internally, the source of alpha radiation is in close contact with live body tissue. It can deposit large amounts of energy in a small amount of body tissue. Rarely an external hazard	Alpha particles of at least 7.5 MeV are required to penetrate the epidermis, the protective layer of skin, which is about 0.07 mm (70 μm) thick. The range of most particles of common emitters (4.5 MeV to 5.5 MeV) is 3 cm to 4 cm in air. Range in air = $0.322E^{3/2}$ cm, when E is expressed in MeV.
Beta particle (β)	Small mass -1 or +1 charge -1 charge particle is an electron, +1 charge particle is a positron	Short 6 cm to 600 cm (1 inch to 20 feet) in air	Low atomic number materials Plastic Glass Aluminum	Internal hazard Externally, may be hazardous to the skin and eyes	Beta particles of at least 70 keV are required to penetrate the epidermis. $R_{air} \approx 3.65$ m/MeV. The range of beta particles in material in g cm^{-2} (thickness in cm multiplied by the density in g cm^{-3}) is approximately half the maximum energy in MeV. ($R \approx E_{max}/2$). Dose rate (in rad/hr) at 1 cm from a beta point source; ≈ 300 rad/hr per Curie. Dose rate (in rad/hr) in a solution ≈ $2.12\bar{E}C/\rho$; where $\bar{E}$ = average energy in MeV, C = concentration in μCi cm^{-3}, and ρ = density of the solution in g cm^{-3}. The dose rate is about one-half this value at the surface. An aqueous solution of 1 Curie ^{32}P in a glass vial typically produces 3 mrad/hr at 1 meter from Bremsstrahlung. Shielding causes Bremsstrahlung radiation similar to x-rays and gamma rays; generally the higher the atomic number of a material, the more intense is the Bremsstrahlung.
Neutron (n or n^0)	Large mass No charge	Very far in air Easily travel several hundred meters High penetrating power due to lack of charge	High hydrogen content material Water Concrete Plastic	External whole body exposure hazard May be external and/or internal hazard Depends on whether source is inside or outside the body Energy dependent	Shielding can be provided by hydrogen-rich materials such as hydrocarbons, water, waxes, high water concrete. Can cause neutron activation, in which radio nuclides are formed. Particularly damaging to soft tissue such as the cornea.
X-Ray, Gamma ray (γ)	No mass No charge Electromagnetic wave Similar but place of origin and energy levels may differ	Very far in air Easily travel several hundred meters Very high penetrating power since it has no mass & no charge	High atomic number materials Depleted uranium Lead Steel Concrete Water	Whole body exposure hazard May be external and/or internal hazard Can be a skin and/or eye hazard Depends on whether source is inside or outside the body Energy dependent	Shielding requires large mass and density materials; lead or depleted uranium are commonly used. Protective clothing and other PPE can guard against ingestion or absorption but not against the radiation itself. Doubling the distance from a point source will result in a reduction of exposure (or dose) by a factor of four. Protective clothing and other PPE can effectively guard against ingestion or absorption of radioactive material but is not usually practical for protecting against x-rays or gamma rays.

RADIATION SAFETY UNITS

Thomas J. Bruno and Paris D. N. Svoronos

Ionizing radiation, consisting of x-rays, gamma rays, alpha particles, beta particles, and neutron particles, is measured and quantified in units of the radioactivity source and dose (Refs. 1 to 4). The radioactivity measures the strength of a source in terms of events of emission per second. Dose is a measure of the energy that is actually absorbed into matter.

References

1. Radiation - Quantities and Units of Ionizing Radiation, Canadian Centre for Occupational Health and Safety, OSH Answer List Series, 2008.
2. Radioactivity Units, Health Physics Society, <www.hps.org/>, 2013.
3. Furr, A. K., *CRC Handbook of Laboratory Safety*, 5th Ed., CRC Press, Boca Raton, FL, 2000.
4. Bruno, T. J., and Svoronos, P. D. N., *CRC Handbook of Basic Tables for Chemical Analysis*, 3rd. Ed., CRC Press, Boca Raton, FL, 2011.

Radioactivity:

In the SI system, the bequerel (Bq) has replaced the curie (Ci) as the accepted unit of radioactivity (or simply activity). One Bq is one event of radiation emission (such as a disintegration) per second. It is related to the older unit by:

$$1 \text{ Ci} = 3.7 \times 10^{10} \text{ Bq}$$

$$1 \text{ Ci} = 37 \text{ GBq} = 37\,000 \text{ MBq}$$

The following chart provides a practical guide between the two units:

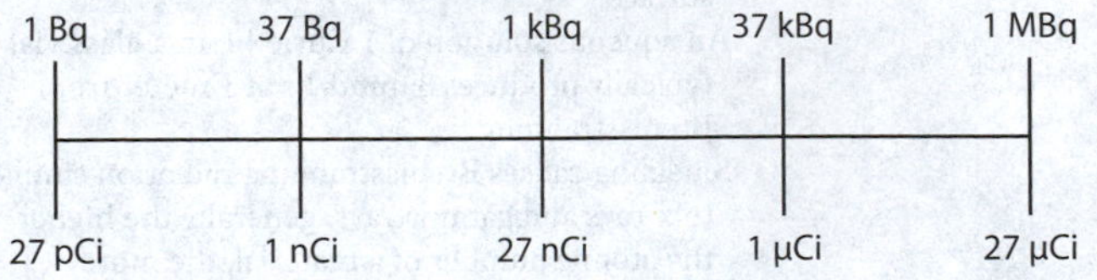

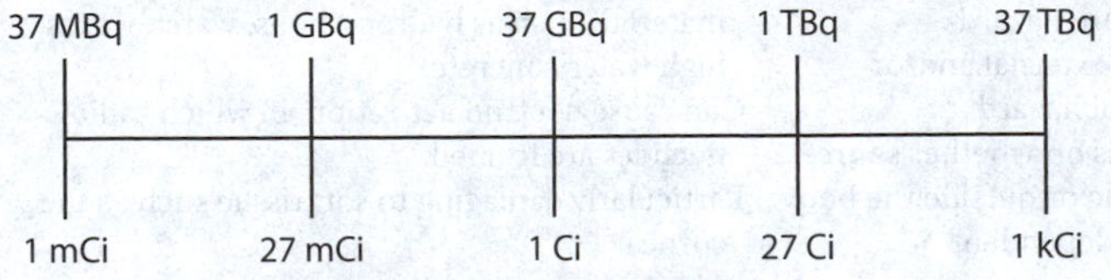

Class A radionuclides: $0.3 \text{ Bq/cm}^2 = 8.1 \text{ pCi/cm}^2$
Class B radionuclides: $3 \text{ Bq/cm}^2 = 81 \text{ pCi/cm}^2$
Class C radionuclides: $30 \text{ Bq/cm}^2 = 810 \text{ pCi/cm}^2$

Energy:

For ionizing radiation, the energy is measured in electron volts (eV), which is related to other energy quantities by:

$$1 \text{ eV} = 1.6021766208(98) \times 10^{-19} \text{ J}$$

Dose:

The older unit of dose, which is defined as the energy that is actually absorbed, is the radiation absorbed dose (RAD). The RAD was defined as the dose that would cause 0.01 J to be absorbed in 1 kg of matter (or 100 ergs per gram). The modern SI unit is the Gray (Gy):

$$100 \text{ RAD} = 1 \text{ Gy}$$

Equivalent Dose:

The equivalent dose (also called the dose equivalent or biological dose) describes the effect of radiation on human tissue, rather then the physical effects of the radiation alone. This quantity is expressed in Sieverts (Sv), and is found by multiplying the absorbed dose, in grays, by a dimensionless quality factor Q (which depends on the radiation type), and by another dimensionless factor N (the tissue weighting factor). Q is also called the Relative Biological Effectiveness (RBE). The factor N depends upon the part of the body irradiated, the time and volume over which the dose was spread, and the species of the subject.

The currently accepted, approximate Q factors are provided below:

Radiation type	***Q***
X-rays	1
Gamma rays	1
Beta particles	1
Thermal neutrons (<10 keV)	5
Fast neutrons (10 keV to 100 keV)	10
Fast neutrons (100 keV to 2 MeV)	20
Fast neutrons (2 MeV to 20 MeV)	10
Fast neutrons (>20 MeV)	5
Protons (>2 MeV)	5
Alpha particles	20
Other atomic nuclei	20

The currently accepted N factors for human body parts are provided below:

Body part	***N***
Gonads	0.20
Bone marrow	0.12
Colon	0.12
Lung	0.12
Stomach	0.12
Bladder	0.05
Brain	0.05
Breast	0.05
Kidney	0.05
Liver	0.05
Muscle	0.05
Esophagus	0.05
Pancreas	0.05
Small intestine	0.05

Body part	N
Spleen	0.05
Thyroid	0.05
Uterus	0.05
Bone surface	0.01
Skin	0.01

Relative to the effect on humans, the following N factors have been suggested for other organisms:

Organism	N
Viruses	0.03 to 0.0003
Bacteria	0.03 to 0.0003
Single cell organisms	0.03 to 0.0003
Insects	0.1 to 0.002
Mollusks	0.06 to 0.006
Plants	2 to 0.02
Fish	0.75 to 0.03
Amphibians	0.4 to 0.14
Reptiles	1 to 0.075
Birds	0.6 to 0.15
Humans (scale definition)	≡ 1

In terms of the older unit, rem (roentgen equivalent in man):

$$1 \text{ rem} = 0.01 \text{ Sv, assuming } Q = 1.$$

The following chart provides a practical guide between the two units:

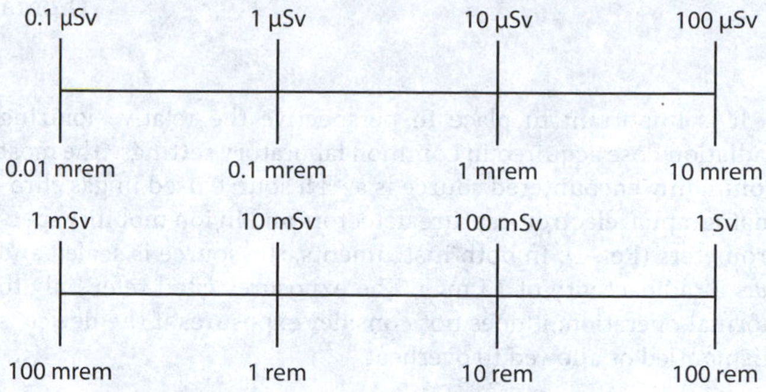

The approximate effects of full body dosages is summarized below:

Dose/Sv	Effect
1	Nausea
2–5	Hair loss, hemorrhage, death is possible
>3	Death is likely in 50 % of cases within 30 days
6	Death is likely in all cases

The relationship between nuclide half-lives elapsed and the remaining radioactivity is provided below:

Half-lives elapsed	Percent remaining
0	100
1	50
2	25
3	12.5
4	6.25
5	3.125

RELATIVE DOSE RANGES FROM IONIZING RADIATION

Thomas J. Bruno

It is important to place in perspective the relative ionizing radiation dose acquired in common laboratory settings. The most commonly encountered source is a ^{63}Ni source used in gas chromatographic electron capture detectors and in ion mobility spectrometers (Ref. 1). In both instruments, the source is sealed and has a radioactivity of 15 mCi. The exposures cited refer only to normal operation; it does not consider exposures if the device is dismantled or allowed to overheat.

Background

Natural background consists of the highly variable sum of all ubiquitous sources of ionizing radiation encountered on the planet (Ref. 2). Background in general can be divided into the following four major contributions:

Contribution	Average dose, mrem/ year, United States
Terrestrial contribution	21
Cosmic contribution	33
Airborne radon (and daughter) contribution	228
Internal consumption contribution	29
Total Natural Background	**311**

In the United States, the average natural background ionizing radiation level is 311 mrem/yr. This is variable due primarily to differences in altitude and primordial radionuclides and their daughters. For example, the averages in the United Kingdom and Finland are 200 mrem/yr and 700 mrem/yr, respectively. Higher levels are found at higher altitudes and regions with a larger radon budget. Within the United States, for example, the background in Denver, Colorado is approximately 450 mrem/yr, while in most of Florida, it is closer to 230 mrem/yr. The terrestrial contribution primarily arises from radionuclides of potassium, uranium, and thorium, and their daughters. The cosmic contribution arises primarily from muons, neutrons, and electrons, and varies with terrestrial magnetic field and altitude. The internal contribution results from consuming radionuclides of potassium and carbon in food and water. By far the largest contribution is from radon and radon daughters. The radon budget results from terrestrial sources of uranium (Ref. 3). Within the United States, the action level requiring indoor radon mitigation is reached when a measurement results in 4 pCi/L (150 Bq/m^3) or higher. This level of radioactivity results in a dose of between 300 mrem/yr and 700 mrem/yr, assuming 80% indoor occupancy. The range cited results from different dose conversion coefficients and dosimetric models used by different agencies (Ref. 4).

Typical Incremental Increases above Background

Exposure to ionizing radiation in the laboratory results in a dose level in excess of the background levels discussed above (Ref. 4). The following charts place in perspective the additional dose received above background for some common exposures. Since in Figure 1 the ranges are dwarfed by smoking tobacco use, Figure 2 is presented with this contribution removed. This is significant because the tobacco dose is specific to the lungs, and not the whole body. A more relevant comparison for Figure 1 would be obtained by multiplying the listed 8000 mrem by the tissue-weighting factor for the lungs, 0.12.

The high incremental level associated with tobacco use (1 mrem/hr while smoking) results from the accumulation of ^{210}Po (a radon daughter) on tobacco leaves. The incremental dose accrued by air travel is dependent on altitude, with higher levels associated with higher altitudes, and will range between 0.35 mrem/hr and 0.5 mrem/hr. The incremental dose due to medical imaging or radiation treatment can be misleading. For many individuals the dose can be close to zero, but in the case of radiation treatment it can be much higher. Indeed, patients given certain radiation treatments become incremental sources themselves, resulting in incremental dosages above background to attending medical personnel, caregivers, and the general public. The listing for clocks in both Figures 1 and 2 are for older (even antique) clocks that have dials coated with Ra/ZnS paint to provide illumination.

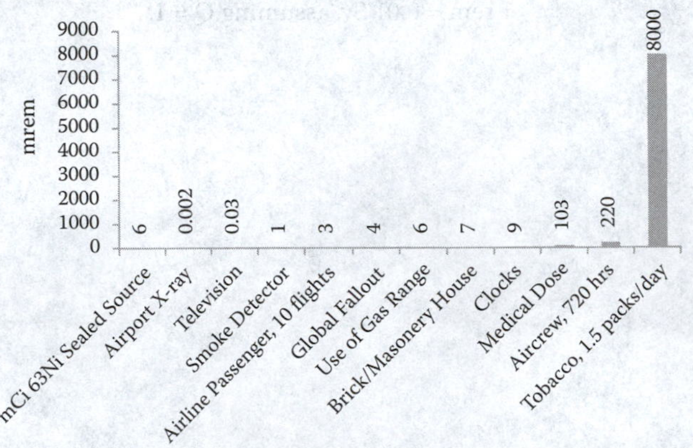

FIGURE 1 A comparison of increments to natural background levels, explicit for the 15 mCi ^{63}Ni sources used in electron capture detectors and ion mobility spectrometers.

FIGURE 2 The same comparison as in Figure 1, but with tobacco use removed. Thus, the data shown above are all whole body dosages.

Safety

References

1. Jeffery, P. G., and Kipping, P. J., *Gas Analysis by Gas Chromatography*, Pergamon Press, Oxford, 1972.
2. Metting, N. F., Ionizing Radiation Dose Ranges, Office of Science, U.S. Department of Energy, <www.lowdose.energy.gov>, 2010.
3. *NCRP Report 160*, Ionizing Radiation Exposure of the Population of the United States, Recommendations of the National Council on Radiation Protection and Measurements, Bethesda, MD, 2009.
4. Johnson, T. E., and Fellman, A., *Estimated Dose and Risk from 15mCi ^{63}Ni Sealed Source Type NR-348-D-111-B*, Report Prepared forHewlett Packard Co., CSI Radiation Safety, Gaithersburg, MD, 1999.

ANNUAL LIMITS ON INTAKES OF RADIONUCLIDES

K. F. Eckerman

The following table lists, for workers, the annual limits on oral and inhalation intakes (ALI) for selected radionuclides based on the occupational radiation protection guidance of the International Commission on Radiological Protection (Refs. 1 and 2). An intake of one ALI corresponds to an annual whole body dose of 0.02 Sv (2 rem).

The ALI is expressed in the SI unit of activity, the becquerel (Bq), and in the conventional unit, the microcurie (μCi); 1 μCi = $3.7 \cdot 10^4$ Bq for both inhalation intakes (inh) and oral intakes (orl). The chemical form of inhaled radionuclides is, in most instances, stated in terms of the rate of absorption to blood from the lungs and the fractional absorption from the small intestine. Types F, M, and S denote chemical forms that are absorbed from the lungs at rates characterized as fast, moderate, and slow, respectively. The time to absorb 90% of the deposited radionuclide, in the absence of radioactive decay, corresponds to about 10 minutes, 150 days, and 7000 days for Type F, M, and S compounds, respectively. Type F compounds can be considered to be more soluble than M or S, S being the most insoluble. Chemical form consideration for ingestion is specified by the fractional absorption from the small intestine, denoted as f_1. The f_1 values range from 10^{-5} to 1. Higher fractional absorption is associated with greater solubility of the compound.

The notation 1.1E+13 means $1.1 \cdot 10^{13}$.

References

1. *1990 Recommendations of the International Commission on Radiological Protection, ICRP Publication 60, Annals of the ICRP 21, (1–3)*, Pergamon Press, Oxford, 1991.
2. *Dose Coefficients for Intakes of Radionuclides by Workers, ICRP Publication 68, Annals of the ICRP*, 24(4), Pergamon Press, Oxford, 1995.

Annual Limits on Intakes of Radionuclides

	Physical half-life	Chemical form type/f_1 (inh)	ALI/Bq (inh)	ALI/μCi (inh)	Chemical form f_1 (orl)	ALI/Bq (orl)	ALI/μCi (orl)
^{3}H	12.3 y	HT gas	1.1E+13	3.0E+08	1.000	1.1E+13	3.0E+08
		HTO vapor	1.1E+09	3.0E+04			
^{11}C	0.340 h	CO	1.7E+10	4.5E+05	1.000	8.3E+08	2.3E+04
		CO_2	9.1E+09	2.5E+05			
		Organic compounds	6.2E+09	1.7E+05			
^{14}C	5730 y	CO	2.5E+10	6.8E+05	1.000	3.4E+07	9.3E+02
		CO_2	3.1E+09	8.3E+04			
		Organic compounds	3.4E+07	9.3E+02			
^{18}F	1.83 h	F 1.000	3.7E+08	1.0E+04	1.000	4.1E+08	1.1E+04
		M 1.000	2.2E+08	6.1E+03			
		S 1.000	2.2E+08	5.8E+03			
^{22}Na	2.60 y	F 1.000	1.0E+07	2.7E+02	1.000	6.3E+06	1.7E+02
^{24}Na	15.0 h	F 1.000	3.8E+07	1.0E+03	1.000	4.7E+07	1.3E+03
^{32}P	14.3 d	F 0.800	1.8E+07	4.9E+02	0.800	8.3E+06	2.3E+02
		M 0.800	6.9E+06	1.9E+02			
^{35}S	87.4 d	Inorganic compounds					
		F 0.800	2.5E+08	6.8E+03	0.800	1.4E+08	3.9E+03
		M 0.800	1.8E+07	4.9E+02	0.100	1.1E+08	2.8E+03
		Vapor	1.7E+08	4.5E+03			
		Organic compounds			1.000	2.6E+07	7.0E+02
^{42}K	12.4 h	F 1.000	1.0E+08	2.7E+03	1.000	4.7E+07	1.3E+03
^{43}K	22.6 h	F 1.000	7.7E+07	2.1E+03	1.000	8.0E+07	2.2E+03
^{45}Ca	163 d	M 0.300	8.7E+06	2.4E+02	0.300	2.6E+07	7.1E+02
^{47}Ca	4.53 d	M 0.300	9.5E+06	2.6E+02	0.300	1.3E+07	3.4E+02
^{51}Cr	27.7 d	F 0.100	6.7E+08	1.8E+04	0.100	5.3E+08	1.4E+04
		M 0.100	5.9E+08	1.6E+04	0.010	5.4E+08	1.5E+04
		S 0.100	5.6E+08	1.5E+04			
^{54}Mn	312 d	F 0.100	1.8E+07	4.9E+02	0.100	2.8E+07	7.6E+02
		M 0.100	1.7E+07	4.5E+02			
^{52}Fe	8.28 h	F 0.100	2.9E+07	7.8E+02	0.100	1.4E+07	3.9E+02
		M 0.100	2.1E+07	5.7E+02			
^{55}Fe	2.70 y	F 0.100	2.2E+07	5.9E+02	0.100	6.1E+07	1.6E+03
		M 0.100	6.1E+07	1.6E+03			
^{59}Fe	44.5 d	F 0.100	6.7E+06	1.8E+02	0.100	1.1E+07	3.0E+02
		M 0.100	6.3E+06	1.7E+02			
^{57}Co	271 d	M 0.100	5.1E+07	1.4E+03	0.100	9.5E+07	2.6E+03
		S 0.050	3.3E+07	9.0E+02	0.050	1.1E+08	2.8E+03
^{58}Co	70.8 d	M 0.100	1.4E+07	3.9E+02	0.100	2.7E+07	7.3E+02

	Physical half-life	Chemical form type/f_1 (inh)	ALI/Bq (inh)	ALI/μCi (inh)	Chemical form f_1 (orl)	ALI/Bq (orl)	ALI/μCi (orl)
		S 0.050	1.2E+07	3.2E+02	0.050	2.9E+07	7.7E+02
^{60}Co	5.27 y	M 0.100	2.8E+06	7.6E+01	0.100	5.9E+06	1.6E+02
		S 0.050	1.2E+06	3.2E+01	0.050	8.0E+06	2.2E+02
^{64}Cu	12.7 h	F 0.500	2.9E+08	7.9E+03	0.500	1.7E+08	4.5E+03
		M 0.500	1.3E+08	3.6E+03			
		S 0.500	1.3E+08	3.6E+03			
^{59}Ni	75000 y	F 0.050	9.1E+07	2.5E+03	0.050	3.2E+08	8.6E+03
		M 0.050	2.1E+08	5.8E+03			
		Vapor	2.4E+07	6.5E+02			
^{63}Ni	96.0 y	F 0.050	3.8E+07	1.0E+03	0.050	1.3E+08	3.6E+03
		M 0.050	6.5E+07	1.7E+03			
		Vapor	1.0E+07	2.7E+02			
^{65}Zn	244 d	S 0.500	7.1E+06	1.9E+02	0.500	5.1E+06	1.4E+02
^{67}Ga	3.26 d	F 0.001	1.8E+08	4.9E+03	0.001	1.1E+08	2.8E+03
		M 0.001	7.1E+07	1.9E+03			
^{68}Ga	1.13 h	F 0.001	4.1E+08	1.1E+04	0.001	2.0E+08	5.4E+03
		M 0.001	2.5E+08	6.7E+03			
^{68}Ge	288 d	F 1.000	2.4E+07	6.5E+02	1.000	1.5E+07	4.2E+02
		M 1.000	2.5E+06	6.8E+01			
^{75}Se	120 d	F 0.800	1.4E+07	3.9E+02	0.800	7.7E+06	2.1E+02
		M 0.800	1.2E+07	3.2E+02	0.050	4.9E+07	1.3E+03
^{79}Se	65000 y	F 0.800	1.3E+07	3.4E+02	0.800	6.9E+06	1.9E+02
		M 0.800	6.5E+06	1.7E+02	0.050	5.1E+07	1.4E+03
^{86}Rb	18.6 d	F 1.000	1.5E+07	4.2E+02	1.000	7.1E+06	1.9E+02
^{85}Sr	64.8 d	F 0.300	3.6E+07	9.7E+02	0.300	3.6E+07	9.7E+02
		S 0.010	3.1E+07	8.4E+02	0.010	6.1E+07	1.6E+03
^{87m}Sr	2.80 h	F 0.300	9.1E+08	2.5E+04	0.300	6.7E+08	1.8E+04
		S 0.010	5.7E+08	1.5E+04	0.010	6.1E+08	1.6E+04
^{89}Sr	50.5 d	F 0.300	1.4E+07	3.9E+02	0.300	7.7E+06	2.1E+02
		S 0.010	3.6E+06	9.7E+01	0.010	8.7E+06	2.4E+02
^{90}Sr	29.1 y	F 0.300	6.7E+05	1.8E+01	0.300	7.1E+05	1.9E+01
		S 0.010	2.6E+05	7.0E+00	0.010	7.4E+06	2.0E+02
^{99}Mo	2.75 d	F 0.800	5.6E+07	1.5E+03	0.800	2.7E+07	7.3E+02
		S 0.050	1.8E+07	4.9E+02	0.050	1.7E+07	4.5E+02
^{99m}Tc	6.02 h	F 0.800	1.0E+09	2.7E+04	0.800	9.1E+08	2.5E+04
		M 0.800	6.9E+08	1.9E+04			
^{99}Tc	213000 y	F 0.800	5.0E+07	1.4E+03	0.800	2.6E+07	6.9E+02
		M 0.800	6.3E+06	1.7E+02			
^{106}Ru	1.01 y	F 0.050	2.0E+06	5.5E+01	0.050	2.9E+06	7.7E+01
		M 0.050	1.2E+06	3.2E+01			
		S 0.050	5.7E+05	1.5E+01			
^{111}In	2.83 d	F 0.020	9.1E+07	2.5E+03	0.020	6.9E+07	1.9E+03
		M 0.020	6.5E+07	1.7E+03			
^{113m}In	1.66 h	F 0.020	1.1E+09	2.8E+04	0.020	7.1E+08	1.9E+04
		M 0.020	6.3E+08	1.7E+04			
^{113}Sn	115 d	F 0.020	2.5E+07	6.8E+02	0.020	2.7E+07	7.4E+02
		M 0.020	1.1E+07	2.8E+02			
^{123}I	13.2 h	F 1.000	1.8E+08	4.9E+03	1.000	9.5E+07	2.6E+03
		Vapor	9.5E+07	2.6E+03			
^{125}I	60.1 d	F 1.000	2.7E+06	7.4E+01	1.000	1.3E+06	3.6E+01
		Vapor	1.4E+06	3.9E+01			
^{129}I	$1.57 \cdot 10^7$ y	F 1.000	3.9E+05	1.1E+01	1.000	1.8E+05	4.9E+00
		Vapor	2.1E+05	5.6E+00			
^{131}I	8.04 d	F 1.000	1.8E+06	4.9E+01	1.000	9.1E+05	2.5E+01
		Vapor	1.0E+06	2.7E+01			
^{129}Cs	1.34 d	F 1.000	2.5E+08	6.7E+03	1.000	3.3E+08	9.0E+03
^{134}Cs	2.06 y	F 1.000	2.1E+06	5.6E+01	1.000	1.1E+06	2.8E+01
^{136}Cs	13.1 d	F 1.000	1.1E+07	2.8E+02	1.000	6.7E+06	1.8E+02
^{137}Cs	30.0 y	F 1.000	3.0E+06	8.1E+01	1.000	1.5E+06	4.2E+01
^{141}Ce	32.5 d	M 5.0E-04	7.4E+06	2.0E+02	5.0E-04	2.8E+07	7.6E+02
		S 5.0E-04	6.5E+06	1.7E+02			
^{144}Ce	284 d	M 5.0E-04	8.7E+05	2.4E+01	5.0E-04	3.8E+06	1.0E+02
		S 5.0E-04	6.9E+05	1.9E+01			

	Physical half-life	Chemical form type/f_1 (inh)	ALI/Bq (inh)	ALI/μCi (inh)	Chemical form f_1 (orl)	ALI/Bq (orl)	ALI/μCi (orl)
^{133}Ba	10.7 y	F 0.100	1.1E+07	3.0E+02	0.100	2.0E+07	5.4E+02
^{140}Ba	12.7 d	F 0.100	1.3E+07	3.4E+02	0.100	8.0E+06	2.2E+02
^{169}Yb	32.0 d	M 5.0E-04	9.5E+06	2.6E+02	5.0E-04	2.8E+07	7.6E+02
		S 5.0E-04	8.3E+06	2.3E+02			
^{198}Au	2.69 d	F 0.100	5.1E+07	1.4E+03	0.100	2.0E+07	5.4E+02
		M 0.100	2.0E+07	5.5E+02			
		S 0.100	1.8E+07	4.9E+02			
^{198m}Au	2.30 d	F 0.100	3.4E+07	9.2E+02	0.100	1.5E+07	4.2E+02
		M 0.100	1.0E+07	2.7E+02			
		S 0.100	1.1E+07	2.8E+02			
^{197}Hg	2.67 d	Inorganic compounds					
		F 0.400	2.4E+08	6.4E+03	1.000	2.0E+08	5.5E+03
					0.400	1.2E+08	3.2E+03
		Vapor	4.5E+06	1.2E+02			
		Organic compounds					
		F 0.020	2.0E+08	5.4E+03	0.020	8.7E+07	2.4E+03
		M 0.020	7.1E+07	1.9E+03			
^{203}Hg	46.6 d	Inorganic compounds					
		F 0.400	2.7E+07	7.2E+02	1.000	1.1E+07	2.8E+02
					0.400	1.8E+07	4.9E+02
		Vapor	2.9E+06	7.7E+01			
		Organic compounds					
		F 0.020	3.4E+07	9.2E+02	0.020	3.7E+07	1.0E+03
		M 0.020	1.1E+07	2.8E+02			
^{201}Tl	3.04 d	F 1.000	2.6E+08	7.1E+03	1.000	2.1E+08	5.7E+03
^{210}Pb	22.3 y	F 0.200	1.8E+04	4.9E-01	0.200	2.9E+04	7.9E-01
^{207}Bi	38.0 y	F 0.050	2.4E+07	6.4E+02	0.050	1.5E+07	4.2E+02
		M 0.050	6.3E+06	1.7E+02			
^{210}Po	138 d	F 0.100	2.8E+04	7.6E-01	0.100	8.3E+04	2.3E+00
		M 0.100	9.1E+03	2.5E-01			
^{224}Ra	3.66 d	M 0.200	8.3E+03	2.3E-01	0.200	3.1E+05	8.3E+00
^{226}Ra	1600 y	M 0.200	1.7E+03	4.5E-02	0.200	7.1E+04	1.9E+00
^{228}Ra	5.75 y	M 0.200	1.2E+04	3.2E-01	0.200	3.0E+04	8.1E-01
^{228}Th	1.91 y	M 5.0E-04	8.7E+02	2.4E-02	5.0E-04	2.9E+05	7.7E+00
		S 2.0E-04	6.3E+02	1.7E-02	2.0E-04	5.7E+05	1.5E+01
^{230}Th	77000 y	M 5.0E-04	7.1E+02	1.9E-02	5.0E-04	9.5E+04	2.6E+00
		S 2.0E-04	2.8E+03	7.5E-02	2.0E-04	2.3E+05	6.2E+00
^{232}Th	$1.40 \cdot 10^{10}$ y	M 5.0E-04	6.9E+02	1.9E-02	5.0E-04	9.1E+04	2.5E+00
		S 2.0E-04	1.7E+03	4.5E-02	2.0E-04	2.2E+05	5.9E+00
^{234}U	$2.44 \cdot 10^{5}$ y	F 0.020	3.1E+04	8.4E-01	0.020	4.1E+05	1.1E+01
		M 0.020	9.5E+03	2.6E-01	0.002	2.4E+06	6.5E+01
		S 0.002	2.9E+03	7.9E-02			
^{235}U	$7.04 \cdot 10^{8}$ y	F 0.020	3.3E+04	9.0E-01	0.020	4.3E+05	1.2E+01
		M 0.020	1.1E+04	3.0E-01	0.002	2.4E+06	6.5E+01
		S 0.002	3.3E+03	8.9E-02			
^{238}U	$4.47 \cdot 10^{9}$ y	F 0.020	3.4E+04	9.3E-01	0.020	4.5E+05	1.2E+01
		M 0.020	1.3E+04	3.4E-01	0.002	2.6E+06	7.1E+01
		S 0.002	3.5E+03	9.5E-02			
^{237}Np	$2.14 \cdot 10^{6}$ y	M 5.0E-04	1.3E+03	3.6E-02	5.0E-04	1.8E+05	4.9E+00
^{239}Np	2.36 d	M 5.0E-04	1.8E+07	4.9E+02	5.0E-04	2.5E+07	6.8E+02
^{238}Pu	87.7 y	M 5.0E-04	6.7E+02	1.8E-02	5.0E-04	8.7E+04	2.4E+00
		S 1.0E-05	1.8E+03	4.9E-02	1.0E-05	2.3E+06	6.1E+01
					1.0E-04	4.1E+05	1.1E+01
^{239}Pu	24100 y	M 5.0E-04	6.3E+02	1.7E-02	5.0E-04	8.0E+04	2.2E+00
		S 1.0E-05	2.4E+03	6.5E-02	1.0E-05	2.2E+06	6.0E+01
					1.0E-04	3.8E+05	1.0E+01
^{241}Pu	14.4 y	M 5.0E-04	3.4E+04	9.3E-01	5.0E-04	4.3E+06	1.2E+02
		S 1.0E-05	2.4E+05	6.4E+00	1.0E-05	1.8E+08	4.9E+03
					1.0E-04	2.1E+07	5.6E+02
^{241}Am	432 y	M 5.0E-04	7.4E+02	2.0E-02	5.0E-04	1.0E+05	2.7E+00
^{244}Cm	18.1 y	M 5.0E-04	1.2E+03	3.2E-02	5.0E-04	1.7E+05	4.5E+00
^{252}Cf	2.64 y	M 5.0E-04	1.5E+03	4.2E-02	5.0E-04	2.2E+05	6.0E+00

CHEMICAL CARCINOGENS

The following chemical substances are listed in the *14th Report on Carcinogens,* 2016, released by the National Institute of Environmental Health Sciences (NIEHS) under the National Toxicology Program (NTP). Substances are grouped in two classes:

- **Known to be human carcinogens:** There is sufficient evidence of cancer from studies in humans showing a cause-and-effect relationship between exposure to the substance and human cancer.
- **Reasonably anticipated to be human carcinogens:** There is limited evidence of cancer from studies in humans or sufficient evidence of cancer from studies in experimental animals showing a cause-and-effect relationship between exposure to the substance and cancer. Alternatively, a substance can be listed in this category if there is evidence that it is a member of a class of substances already listed in the *Report on Carcinogens* or causes biological effects known to lead to the development of cancer.

The NTP report also lists many poorly defined materials such as soots, tars, mineral oils, and coke oven emissions, as well as viruses, sunlight, ionizing radiation, etc. These carcinogenic agents are not included here.

The table is ordered by the substance name normally used in the *CRC Handbook of Chemistry and Physics.* In many cases the primary name given here is different from that used in the NTP report; however, names used in the NTP report appear in the Synonym or Comments columns. Additional information and acronyms for the compounds are given in the Comments column. The known carcinogens are listed first.

Extensive details on each substance are given in the reference.

Reference

14th Report on Carcinogens, http://ntp.niehs.nih.gov/pubhealth/roc/index-1.html.

Chemical Carcinogens (14th NTP Report, 2016)

Name	Synonym	Comments	CAS Reg. No.
Known to Be Human Carcinogens			
Aflatoxins			1402-68-2
4-Aminobiphenyl	*p*-Biphenylamine		92-67-1
Aristolochic acids			
Arsenic		and inorganic arsenic compounds	7440-38-2
Asbestos			1332-21-4
Azathioprine	6-[(1-Methyl-4-nitro-1*H*-imidazol-5-yl)thio]-1*H*-purine		446-86-6
Benzene	[6]Annulene		71-43-2
p-Benzidine	[1,1'-Biphenyl]-4,4'-diamine	includes dyes metabolized to benzidene	92-87-5
Beryllium	Glucinium	and beryllium compounds	7440-41-7
Bis(2-chloroethyl) sulfide	Mustard gas		505-60-2
Bis(chloromethyl) ether	Chloromethyl ether	and technical grade chloromethyl methyl ether	542-88-1
1,3-Butadiene	Divinyl		106-99-0
Cadmium		and cadmium compounds	7440-43-9
Chlorambucil			305-03-3
Chloroethene	Vinyl chloride		75-01-4
1-(2-Chloroethyl)-3-(4-methylcyclohexyl)-1-nitrosourea	Semustine	also known as MeCCNU	13909-09-6
Chloromethyl methyl ether			107-30-2
Chromium		hexavalent compounds only	7440-47-3
Cyclophosphamide	Cyclophosphane		50-18-0
Cyclosporin A	Cyclosporine		59865-13-3
trans-Diethylstilbestrol			56-53-1
Erionite			66733-21-9
Estrogens, steroidal			
N-(4-Ethoxyphenyl)acetamide	Phenacetin	in analgesic mixtures	62-44-2
Formaldehyde	Methanal		50-00-0
Melphalan	L-Phenylalanine, 4-[bis(2-chloroethyl)amino]-		148-82-3
Methoxsalen	9-Methoxy-7*H*-furo[3,2-*g*][1]benzopyran-7-one	with UV-A therapy (PUVA)	298-81-7
2-Methylaniline	*o*-Toluidine	also the hydrochloride	95-53-4
2-Naphthylamine	β-Naphthylamine		91-59-8
Nickel compounds		both inorganic and organic	
Oxirane	Ethylene oxide		75-21-8
Radon		source of ionizing radiation	10043-92-2

Name	Synonym	Comments	CAS Reg. No.
Silicon dioxide (α-quartz)	Silica	respirable size	14808-60-7
Silicon dioxide (tridymite)	Tridymite	respirable size	15468-32-3
Silicon dioxide (cristobalite)	Cristobalite	respirable size	14464-46-1
Sulfuric acid	Oil of vitriol	in strong acid mists	7664-93-9
Tamoxifen			10540-29-1
2,3,7,8-Tetrachlorodibenzo-*p*-dioxin	Dioxin	also known as TCDD	1746-01-6
Thorium(IV) oxide	Thoria	source of ionizing radiation	1314-20-1
Trichloroethene	Trichloroethylene	also known as TCE	79-01-6
Triethylenethiophosphoramide	Thiotepa		52-24-4
Reasonably Anticipated to Be Human Carcinogens			
Acetaldehyde	Ethanal		75-07-0
2-(Acetylamino)fluorene			53-96-3
Acrylamide	2-Propenamide		79-06-1
Acrylonitrile	Propenenitrile		107-13-1
4-Allyl-1,2-dimethoxybenzene	Methyleugenol		93-15-2
2-Amino-9,10-anthracenedione	2-Aminoanthraquinone		117-79-3
1-Amino-2,4-dibromo-9,10-anthracenedione	1-Amino-2,4-dibromoanthraquinone		81-49-2
2-Amino-3,4-dimethylimidazo[4,5-*f*]quinoline	Me-IQ		77094-11-2
2-Amino-3,8-dimethylimidazo[4,5-*f*]quinoxaline	MeIQx		77500-04-0
1-Amino-2-methyl-9,10-anthracenedione	1-Amino-2-methylanthraquinone		82-28-0
2-Amino-3-methyl-3*H*-imidazo(4,5-*f*)quinoline	IQ		76180-96-6
2-Amino-1-methyl-6-phenylimidazo[4,5-*b*]pyridine	PhIP		105650-23-5
Azacitidine	4-Amino-1-β-D-ribofuranosyl-1,3,5-triazine-2(1*H*)-one		320-67-2
Benz[*a*]anthracene	1,2-Benzanthracene		56-55-3
Benzo[*b*]fluoranthene	Benz[*e*]acephenanthrylene		205-99-2
Benzo[*j*]fluoranthene	Dibenzo[*a,jk*]fluorene		205-82-3
Benzo[*k*]fluoranthene	2,3,1',8'-Binaphthylene		207-08-9
Benzo[*a*]pyrene	2,3-Benzopyrene		50-32-8
2,2'-Bioxirane	Diepoxybutane		1464-53-5
2,2-Bis(bromomethyl)-1,3-propanediol	Pentaerythritol dibromide		3296-90-0
Bis(2-chloroethyl)methylamine hydrochloride	Nitrogen mustard hydrochloride	also known as Mechlorethamine	55-86-7
N,N'-Bis(2-chloroethyl)-*N*-nitrosourea	Carmustine	also known as BCNU	154-93-8
Bis[4-(dimethylamino)phenyl]methane	Michler's Base		101-61-1
1,3-Bis(2,3-epoxypropoxy)benzene	Diglycidyl resorcinol ether		101-90-6
Bis(2-ethylhexyl) phthalate	Di-*sec*-octyl phthalate	also known as DEHP	117-81-7
Bromodichloromethane			75-27-4
Bromoethene	Vinyl bromide		593-60-2
1-Bromopropane	Propyl bromide		106-94-5
1,4-Butanediol dimethylsulfonate	Busulfan		55-98-1
tert-Butyl-4-hydroxyanisole	Butylated hydroxyanisole		25013-16-5
Captafol	Difolatan		2425-06-1
Chloramphenicol			56-75-7
Chlorendic acid	1,4,5,6,7,7-Hexachloro-5-norbornene-2,3-dicarboxylic acid		115-28-6
Chlorinated paraffins (C_{12}, 60% Cl)			108171-26-2
4-Chloro-1,2-benzenediamine	4-Chloro-*o*-phenylenediamine		95-83-0
2-Chloro-1,3-butadiene	Chloroprene		126-99-8
1-(2-Chloroethyl)-3-cyclohexyl-1-nitrosourea	Lomustine	also known as CCNU and Belustine	13010-47-4
4-Chloro-2-methylaniline	*p*-Chloro-*o*-toluidine	also the hydrochloride	95-69-2
1-Chloro-2-methylpropene	Dimethylvinyl chloride		513-37-1
3-Chloro-2-methylpropene			563-47-3
Chlorozotocin	2-[[[(2-Chloroethyl)nitrosoamino]carbonyl]amino]-2-deoxy-D-glucose		54749-90-5

Name	Synonym	Comments	CAS Reg. No.
Cobalt		and Co compounds that release Co ions *in vivo*	7440-48-4
Cobalt-tungsten carbide	Co/WC	powders and hard metal	
Cupferron			135-20-6
Dacarbazine	5-(3,3-Dimethyl-1-triazenyl)-1*H*-imidazole-4-carboxamide		4342-03-4
Decabromobiphenyl			13654-09-6
2,4-Diaminoanisole sulfate	1,3-Benzenediamine, 4-methoxy, sulfate		39156-41-7
4,4'-Diaminodiphenyl ether	4,4'-Oxydianiline		101-80-4
4,4'-Diaminodiphenylmethane	4,4'-Methylenedianiline	also the dihydrochloride	101-77-9
4,4'-Diaminodiphenylmethane dihydrochloride	4,4'-Methylenedianiline dihydrochloride		13552-44-8
4,4'-Diaminodiphenyl sulfide	4,4'-Thiodianiline		139-65-1
cis-Diamminedichloroplatinum	Cisplatin		15663-27-1
Dibenz[*a,h*]acridine			226-36-8
Dibenz[*a,j*]acridine	7-Azadibenz[*a,j*]anthracene		224-42-0
Dibenz[*a,h*]anthracene	1,2:5,6-Dibenzanthracene		53-70-3
7*H*-Dibenzo[*c,g*]carbazole			194-59-2
Dibenzo[*a,e*]pyrene	Naphtho[1,2,3,4-*def*]chrysene		192-65-4
Dibenzo[*a,h*]pyrene	Dibenzo[*b,def*]chrysene		189-64-0
Dibenzo[*a,i*]pyrene	Benzo[*rst*]pentaphene		189-55-9
Dibenzo[*a,l*]pyrene	Dibenzo[*def,p*]chrysene		191-30-0
1,2-Dibromo-3-chloropropane			96-12-8
1,2-Dibromoethane	Ethylene dibromide		106-93-4
2,3-Dibromo-1-propanol	DBP		96-13-9
2,3-Dibromo-1-propanol, phosphate (3:1)	Tris(2,3-dibromopropyl) phosphate		126-72-7
p-Dichlorobenzene	1,4-Dichlorobenzene		106-46-7
3,3'-Dichloro-*p*-benzidine	3,3'-Dichloro[1,1'-biphenyl]-4,4'-diamine		91-94-1
3,3'-Dichloro-*p*-benzidine dihydrochloride	3,3'-Dichloro-[1,1'-biphenyl]-4,4'-diamine dihydrochloride		612-83-9
1,2-Dichloroethane	Ethylene dichloride		107-06-2
Dichloromethane	Methylene chloride		75-09-2
1,3-Dichloropropene (unspecified isomer)		technical grade	542-75-6
Diethyl sulfate			64-67-5
2,3-Dihydro-6-propyl-2-thioxo-4(1*H*)-pyrimidinone	Propylthiouracil	also known as PROP	51-52-5
1,8-Dihydroxy-9,10-anthracenedione	Danthron		117-10-2
3,3'-Dimethoxybenzidine	Dianisidine	and dyes metabolized to 3,3'-Dimethoxybenzidine	119-90-4
4-(Dimethylamino)azobenzene	Methyl Yellow		60-11-7
2',3-Dimethyl-4-aminoazobenzene	4-*o*-Tolylazo-*o*-toluidine	also known as *o*-Aminoazotoluene	97-56-3
Dimethylcarbamic chloride	Dimethylcarbamoyl chloride		79-44-7
1,1-Dimethylhydrazine	UDMH		57-14-7
Dimethyl sulfate			77-78-1
1,6-Dinitropyrene			42397-64-8
1,8-Dinitropyrene			42397-65-9
1,4-Dioxane	1,4-Dioxacyclohexane		123-91-1
1,2-Diphenylhydrazine	Hydrazobenzene		122-66-7
1,3-Diphenyl-1-triazene	Diazoaminobenzene		136-35-6
Disperse Blue No. 1	1,4,5,8-Tetraamino-9,10-anthracenedione		2475-45-8
Doxorubicin hydrochloride	Adriamycin		25316-40-9
Epichlorohydrin	(Chloromethyl)oxirane		13403-37-7
1,2-Epoxy-4-(epoxyethyl)cyclohexane	4-Vinyl-1-cyclohexene dioxide		106-87-6
Ethyl carbamate	Urethane		51-79-6
Ethyl methanesulfonate			62-50-0
N-Ethyl-*N*-nitrosourea	*N*-Nitroso-*N*-ethylurea	also known as ENU	759-73-9
Fluoroethene	Vinyl fluoride		75-02-5
Fuchsin	C.I. Basic Red 9, monohydrochloride		569-61-9
Furan	Oxacyclopentadiene		110-00-9
Glass wool fibers (inhalable)		certain types	
Hexabromobiphenyl (unspecified isomer)	Firemaster FF-1		67774-32-7
Hexachlorobenzene	Perchlorobenzene		118-74-1

Name	Synonym	Comments	CAS Reg. No.
1,2,3,4,5,6-Hexachlorocyclohexane, (1α,2α,3β,4α,5α,6β)	Lindane	also known as γ-Hexachlorocyclohexane	58-89-9
1,2,3,4,5,6-Hexachlorocyclohexane, (1α,2α,3β,4α,5β,6β)	α-Hexachlorocyclohexane		319-84-6
1,2,3,4,5,6-Hexachlorocyclohexane, (1α,2β,3α,4β,5α,6β)	β-Hexachlorocyclohexane		319-85-7
Hexachlorocyclohexane (unspecified isomer)		all isomers	608-73-1
Hexachloroethane	Perchloroethane		67-72-1
Hexamethylphosphoric triamide	Tris(dimethylamino)phosphine oxide	also known as HMPA	680-31-9
Hydrazine			302-01-2
Hydrazine sulfate			10034-93-2
2-Imidazolidinethione	Ethylene thiourea		96-45-7
Indeno[1,2,3-*cd*]pyrene	1,10-(1,2-Phenylene)pyrene		193-39-5
Isopropylbenzene	Cumene		98-82-8
Kepone	Chlordecone		143-50-0
Lead		and lead compounds	7439-92-1
2-Methoxyaniline	*o*-Anisidine	also the hydrochloride	90-04-0
2-Methoxy-5-methylaniline	5-Methyl-*o*-anisidine	also known as *p*-Cresidine	120-71-8
4-Methyl-1,3-benzenediamine	Toluene-2,4-diamine	also known as 2,4-Diaminotoluene	95-80-7
2-Methyl-1,3-butadiene	Isoprene		78-79-5
5-Methylchrysene		also known as 5-MC	3697-24-3
4,4'-Methylenebis[2-chloroaniline]	3,3'-Dichloro-4,4'-diaminodiphenylmethane	also known as MBOCA	101-14-4
Methyl methanesulfonate			66-27-3
N-Methyl-*N*'-nitro-*N*-nitrosoguanidine			70-25-7
N-Methyl-*N*-nitrosourea	*N*-Nitroso-*N*-methylurea		684-93-5
Methyloxirane	1,2-Propylene oxide		75-56-9
Metronidazole	2-Methyl-5-nitro-1*H*-imidazole-1-ethanol		443-48-1
Mirex	Hexachloropentadiene dimer		2385-85-5
Naphthalene			91-20-3
Nickel		metallic (nickel compounds are known carcinogens)	7440-02-0
Nitrilotriacetic acid	*N*,*N*-Bis(carboxymethyl)glycine		139-13-9
2-Nitroanisole	1-Methoxy-2-nitrobenzene		91-23-6
Nitrobenzene			98-95-3
6-Nitrochrysene			7496-02-8
Nitrofen	2,4-Dichloro-1-(4-nitrophenoxy)benzene		1836-75-5
Nitromethane	Nitrocarbol		75-52-5
2-Nitropropane	Isonitropropane		79-46-9
1-Nitropyrene			5522-43-0
4-Nitropyrene			57835-92-4
N-Nitrosodibutylamine	Dibutylnitrosamine		924-16-3
N-Nitrosodiethanolamine	2,2'-(Nitrosoimino)ethanol		1116-54-7
N-Nitrosodiethylamine	Diethylnitrosamine	also known as DEN	55-18-5
N-Nitrosodimethylamine	Dimethylnitrosamine	also known as DMN	62-75-9
4-(*N*-Nitrosomethylamino)-1-(3-pyridyl)-1-butanone	Ketone, 3-pyridyl-3-(*N*-methyl-*N*-nitrosamino) propyl	also known as NNK	64091-91-4
N-Nitroso-*N*-methylvinylamine	*N*-Methyl-*N*-nitrosoethenamine		4549-40-0
4-Nitrosomorpholine	*N*-Nitrosomorpholine		59-89-2
N-Nitrosonornicotine	*N*'-Nitroso-3-(2-pyrrolidinyl)pyridine		16543-55-8
N-Nitrosopiperidine	1-Nitrosopiperidine		100-75-4
N-Nitrosodipropylamine	*N*-Nitroso-*N*-propyl-1-propanamine		621-64-7
N-Nitrosopyrrolidine			930-55-2
N-Nitrososarcosine	*N*-Methyl-*N*-nitrosoglycine		13256-22-9
2-Nitrotoluene	1-Methyl-2-nitrobenzene		88-72-2
Norethisterone	19-Norpregn-4-en-20-yn-3-one,17-hydroxy-,(17α)-		68-22-4
Ochratoxin A			303-47-9
Octabromobiphenyl (unspecified isomer)			61288-13-9
2-Oxetanone	β-Propiolactone		57-57-8
Oxiranemethanol, (±)-	Glycidol		61915-27-3
Oxymetholone	Androstan-3-one,17-hydroxy-2-(hydroxymethylene)-17-methyl-		434-07-1

Safety

Name	Synonym	Comments	CAS Reg. No.
Pentachlorophenol		and synthesis byproducts	87-86-5
Phenazopyridine hydrochloride	3-(Phenylazo)-2,6-pyridinediamine, monohydrochloride		136-40-3
Phenolphthalein	3,3-Bis(4-hydroxyphenyl)-1(3*H*)-isobenzofuranone		77-09-8
Phenoxybenzamine hydrochloride			63-92-3
Phenyloxirane	Styrene-7,8-oxide		96-09-3
Phenytoin	5,5-Diphenyl-2,4-imidazolidinedione	also Phenytoin sodium; also known as Dilantin	57-41-0
Polybrominated biphenyls	PBBs		
Polychlorinated biphenyls	PCBs		1336-36-3
Procarbazine		also the hydrochloride	671-16-9
Progesterone	Pregn-4-ene-3,20-dione		57-83-0
1,3-Propane sultone	1,2-Oxathiolane, 2,2-dioxide		1120-71-4
Propyleneimine	2-Methylaziridine		75-55-8
Reserpine			50-55-5
Riddelline			23246-96-0
Safrole	5-(2-Propenyl)-1,3-benzodioxole		94-59-7
Selenium monosulfide			7446-34-6
Streptozotocin	D-Glucopyranose, 2-deoxy-2-[[(methylnitrosoamino)carbonyl]amino]-		18883-66-4
Styrene	Vinylbenzene		100-42-5
Sulfallate	*N,N*-Diethyldithiocarbamic acid, 2-chloroallyl ester		95-06-7
Tetrachloroethene	Perchloroethylene		127-18-4
Tetrachloromethane	Carbon tetrachloride		56-23-5
Tetrafluoroethene	Tetrafluoroethylene		116-14-3
N,N,N',N'-Tetramethyl-4,4'-diaminobenzophenone	Michler's ketone	also known as 4,4'-Methylenebis[*N,N*-dimethylaniline]	90-94-8
Tetranitromethane			509-14-8
Thioacetamide	Ethanethioamide		62-55-5
Thiourea	Thiocarbamide		62-56-6
o-Tolidine	3,3'-Dimethylbenzidine	and dyes metabolized to *o*-Tolidine	119-93-7
Toluene diisocyanate (unspecified isomer)		includes both 2,4- and 2,6- isomers; also known as TDI	26471-62-5
Toxaphene	Polychlorocamphene		8001-35-2
1*H*-1,2,4-Triazol-3-amine	Amitrole		61-82-5
1,1,1-Trichloro-2,2-bis(4-chlorophenyl) ethane	Dichlorodiphenyltrichloroethane (DDT)		50-29-3
Trichloromethane	Chloroform		67-66-3
(Trichloromethyl)benzene	Benzotrichloride		98-07-7
2,4,6-Trichlorophenol			88-06-2
1,2,3-Trichloropropane	Allyl trichloride		96-18-4

Appendix A
Sources of Physical and Chemical Data

Web site	Address	Comments
SABIO-Reaction Kinetics Database	sabio.villa-bosch.de/	Data on kinetics of biochemical reactions
Sigma-Aldrich	www.sigmaaldrich.com/	Chemical catalogs; includes some physical property data
Spectral Database for Organic Compounds	sdbs.db.aist.go.jp/sdbs/cgi-bin/cre_index.cgi?lang=eng	MS, NMR, IR, Raman, and ESR spectra; 32,000 compounds measured at AIST, Japan
SpecInfo	onlinelibrary.wiley.com/book/10.1002/9780471692294	IR, NMR, and mass spectra
Spectra Online	www.ftirsearch.com/	FTIR and Raman spectra
SPRESI-web	www.spresi.de/	Structures, reactions, and some physical properties
SpringerMaterials	www.springermaterials.com	The online version of *Landolt-Börnstein Tables* (C.6)
STN Easy	stneasy.cas.org	Chemical directory (and access to Chemical Abstracts databases)
STN Easy-Europe	stneasy.fiz-karlsruhe.de	European node of STN Easy
STN Easy-Japan	stneasy-japan.cas.org	Japanese node of STN Easy
Swissprot	bo.expasy.org/enzyme/	Enzyme nomenclature and related information
Thermodynamics of Enzyme-Catalyzed Reactions	xpdb.nist.gov/enzyme_thermodynamics/	Equilibrium constants of biochemical reactions
Thermodynamics Research Center	www.trc.nist.gov	See B.2
TOXNET	toxnet.nlm.nih.gov	Portal to HSDB and other databases on hazardous chemicals

Index

The most efficient way to use this index is to look for the pertinent *property* (e.g., vapor pressure, entropy), *process* (e.g., shielding, calibration), or *general concept* (e.g., units, radiation). Most primary entries are subdivided into several secondary entries, e.g., under heat capacity there are 24 secondary entries such as air, metals, water, etc. Primary entries will be found for certain *classes of substances*, such as alloys, elements, organic compounds, refrigerants, semiconductors, etc. Primary entries are also given for the individual chemical elements and for a few compounds such as water and carbon dioxide. However, only the most important tables are listed under these substances. Therefore, the user will find in most cases that it is best to look first for the property of interest, then examine the table or tables that are referenced.

Entries in boldface type are the titles of tables as they appear in the Table of Contents.

The reference given for each index term is the inclusive pages of the pertinent table (e.g., **8**-45 to 55). The introduction to each table describes the method of ordering the substances within that table.

The editor would be grateful for comments and suggestions on this index.

D

E

SOURCES OF PHYSICAL AND CHEMICAL DATA

In addition to the primary research journals, there are many useful sources of property data of the type contained in the *CRC Handbook of Chemistry and Physics*. A selected list of these is presented here, with emphasis on print and electronic sources whose contents have been subject to a reasonable level of quality control.

A. Data Journals

1. **Journal of Physical and Chemical Reference Data** — Published jointly by the National Institute of Standards and Technology and the American Institute of Physics, this quarterly journal contains compilations of evaluated data in chemistry, physics, and materials science. It is available in print and on the Internet. [scitation.aip.org/content/aip/journal/jpcrd/]
2. **Journal of Chemical and Engineering Data** — This bimonthly journal of the American Chemical Society publishes articles reporting original experimental measurements carried out under carefully controlled conditions. The main emphasis is on thermochemical and thermophysical properties. Review articles with evaluated data from the literature are also published. [pubs.acs.org/journals/jceaax/index.html]
3. **Journal of Chemical Thermodynamics** — This journal publishes original research papers that include highly accurate measurements of thermodynamic and thermophysical properties. [www.sciencedirect.com/science/journal/00219614]
4. **Atomic Data and Nuclear Data Tables** — This is a bimonthly journal containing compilations of data in atomic physics, nuclear physics, and related fields. [www.science direct.com/science/journal/aip/0092640X]
5. **Journal of Phase Equilibria and Diffusion** — This journal presents critically evaluated phase diagrams, kinetic properties, and related data on alloy systems. It is now published by Springer and is the successor to the previous ASM periodical *Bulletin of Alloy Phase Diagrams*. [www.springer.com/materials/journal/11669]

B. Data Centers

This section lists selected organizations that perform a continuing function of compiling and critically evaluating data in specific fields of science.

1. **National Institute of Standards and Technology** — Under its Standard Reference Data program, NIST supports a number of data centers in chemistry, physics, and materials science. Topics covered include thermodynamics, fluid properties, chemical kinetics, mass spectroscopy, atomic spectroscopy, fundamental physical constants, ceramics, and crystallography. Address: Office of Standard Reference Data, National Institute of Standards and Technology, Gaithersburg, MD 20899 [www.nist.gov/srd/].
2. **Thermodynamics Research Center** — Now located at the National Institute of Standards and Technology, TRC maintains an extensive archive of data covering thermodynamic, thermochemical, and transport properties of organic compounds and mixtures. Data are distributed in both print and electronic form. Address: Mail code 838.00, 325 Broadway, Boulder, CO 80305-3328 [www.trc.nist.gov].
3. **Design Institute for Physical Property Data** — Under the auspices of the American Institute of Chemical Engineers [www.aiche.org/dippr/], DIPPR offers evaluated data on industrially important chemical compounds. The largest project deals with physical, thermodynamic, and transport properties of pure compounds [aiche.org/dippr].
4. **Dortmund Data Bank** — Maintains extensive databases on thermodynamic and transport properties of pure compounds and mixtures of industrial interest. The data are distributed through DECHEMA, FIZ CHEMIE, and other outlets. Address: DDBST GmbH, Industriestr. 1, 26121 Oldenburg, Germany [www.ddbst.de].
5. **Cambridge Crystallographic Data Centre** — Maintains the Cambridge Structural Database of over 800,000 organic compounds. The data files and manipulation software are distributed in several ways. Address: 12 Union Rd., Cambridge CB2 1EZ, U.K. [www.ccdc.cam.ac.uk].
6. **FIZ Karlsruhe** — In addition to many bibliographic databases, FIZ Karlsruhe maintains the Inorganic Crystal Structure Database in collaboration with the National Institute of Standards and Technology. The ICSD contains the atomic coordinates and related data on over 50,000 inorganic crystals. Address: Fachinformationszentrum (FIZ) Karlsruhe, Hermann-von-Helmholtz-Platz 1, D-76344 Eggenstein-Leopoldshafen, Germany [www.fiz-karlsruhe.de/en/leistungen/kristallographie/kristallstrukturdepot.htm].
7. **International Centre for Diffraction Data** — Maintains and distributes the Powder Diffraction File (PDF), a file of over 500,000 x-ray powder diffraction patterns used for identification of crystalline materials. The ICDD also distributes the NIST Crystal Data file, which contains lattice parameters for over 235,000 inorganic, organic, metal, and mineral crystalline materials. Address: 12 Campus Blvd., Newton Square, PA 19073-3273 [www.icdd.com].
8. **Research Collaboratory for Structural Bioinformatics** — Maintains the Protein Data Bank (PDB), a file of 3-dimensional structures of proteins and other biological macromolecules. Address: Department of Chemistry and Chemical Biology, Rutgers University, 610 Taylor Road, Piscataway, NJ 08854-8087 [www.rcsb.org].
9. **Toth Information Systems** — Maintains the Metals Crystallographic Data File (CRYSTMET). Address: 2045 Quincy Ave., Gloucester, ON, Canada K1J 6B2 [www.tothcanada.com].
10. **Atomic Mass Data Center** — Collects and evaluates high-precision data on masses of individual isotopes and maintains a comprehensive database. Address: C.S.N.S.M. (IN2P3-CNRS), Batiment 108, F-91405 Orsay Campus, France [amdc.impcas.ac.cn]
11. **Particle Data Group** — International center for data of high-energy physics; maintains a database of properties of fundamental particles that is published in both print and electronic form. Address: MS 50-308, Lawrence Berkeley National Laboratory, Berkeley, CA 94720 [pdg.lbl.gov].

12. **National Nuclear Data Center** — Maintains databases on nuclear structure and reactions, including neutron cross sections. The NNDC is the U.S. node in an international network of nuclear data centers. Address: Brookhaven National Laboratory, Upton, NY 11973-5000 [www.nndc.bnl.gov].
13. **International Union of Pure and Applied Chemistry** — Address: PO Box 13757, Research Triangle Park, NC 27709-3757 [www.iupac.org]. IUPAC supports a number of long-term data projects, including these examples:
    a. **Solubility Data Project** — Carries out evaluation of all types of solubility data. The results are published in the Solubility Data Series, whose current outlet is the *Journal of Physical and Chemical Reference Data*. [srdata.nist.gov/solubility/]
    b. **Kinetic Data for Atmospheric Chemistry** — Maintains a comprehensive database on the kinetics of reactions important in the chemistry of the atmosphere. [www.iupac-kinetic.ch.cam.ac.uk/]
    c. **Stability Constants Database** — Collection of metal-ligand stability constants and associated software. [www.acadsoft.co.uk]

C. Major Multi-Volume Handbook Series

1. **CRC Chemical Dictionaries** — These originally appeared in print form as the *Dictionary of Organic Compounds*, *Dictionary of Natural Products*, etc. They are now published in electronic form and are available in DVD format [www.crcpress.com] and on the Internet [www.chemnetbase. com]. The consolidated version, called the *Combined Chemical Dictionary*, has data on more than 560,000 compounds spanning all branches of chemistry. The coverage includes physical properties, biological sources, hazard information, uses, and literature references.
2. **Properties of Organic Compounds** — Originally published in three editions as the *Handbook of Data on Organic Compounds*, it is now in electronic form as *Properties of Organic Compounds*. The database includes about 30,000 compounds; physical properties and spectral data (mass, infrared, Raman, ultraviolet, and NMR) are covered. It is offered as CD-ROM [www.crcpress.com] and by Web access [www.chemnetbase.com].
3. **Beilstein Handbook of Organic Chemistry** — The classic source of data on organic compounds, dating from the 19th century, *Beilstein* was converted to electronic form in the last decade of the 20th century. Over 8 million compounds and 10 million chemical reactions are now covered, with a broad range of physical properties as well as synthetic methods and ecological data. The database is now accessed through Reaxys [www.elsevier.com/solutions/reaxys].
4. **Gmelin Handbook of Inorganic and Organometallic Chemistry** — A subset of the information in the print series has been converted to electronic form and is now distributed in the same manner as *Beilstein*. In addition to the standard physical properties, the coverage includes a wide range of optical, magnetic, spectroscopic, thermal, and transport properties for about 1.4 million compounds [www.elsevier.com/solutions/reaxys].
5. **DECHEMA Chemical Data Series** — DECHEMA distributes the DETHERM database, which emphasizes data used in process design in the chemical industry, including thermodynamic and transport properties of about 20,000 pure compounds and 90,000 mixtures. Access is available through in-house databases and via the Internet [www.dechema.de].
6. **Landolt-Börnstein Numerical Data and Functional Relationships in Science and Technology** — Landolt-Börnstein covers a very broad range of data in physics, chemistry, crystallography, materials science, biophysics, astronomy, and geophysics. Hard-copy volumes are no longer published, but the entire collection is available online [www.springermaterials.com].

D. Selected Single-Volume Handbooks

The following handbooks offer broad coverage of high-quality data in a single volume. This list is only representative; an extensive listing of handbooks in all fields of science may be found in *Handbooks and Tables in Science and Technology, Third Edition* (Russell H. Powell, ed., Oryx Press, Westport, CT, 1994).

1. **American Institute of Physics Handbook** — Although an old book, it contains much data that are still useful, especially in acoustics, mechanics, optics, and solid state physics. (Dwight E. Gray, ed., McGraw-Hill, New York, 1972)
2. **Constants of Inorganic Substances** — This book presents physical constants, thermodynamic data, solubility, reactivity, and other information on over 3000 inorganic compounds. Since it draws heavily on Russian literature, it contains a great deal of data that do not make their way into most U.S. handbooks. (R. A. Lidin, L. L. Andreeva, and V. A. Molochko, Begell House, New York, 1995)
3. **Handbook of Chemistry and Physics** — Now in the 98th Edition, the *CRC Handbook* covers data from most branches of chemistry and physics. The annual revisions permit regular updating of the information. Also available on the Web [hbcpnetbase.com]. (John Rumble, ed., CRC Press, Boca Raton, FL, 2016)
4. **Handbook of Inorganic Compounds, Second Edition** — This book covers physical constants and other properties for about 3300 inorganic compounds. (Dale L. Perry, CRC Press, Boca Raton, FL, 2011)
5. **Handbook of Physical Properties of Liquids and Gases** — This is a valuable source of data on all types of fluids, ranging from liquid and gaseous hydrocarbons to molten metals and ionized gases. Detailed tables of physical, thermodynamic, and transport properties are given for temperatures from the cryogenic region to 6000 K. Western and Russian literature is covered. (N. B. Vargaftik, Y. K. Vinogradov, and V. S. Yargin, Begell House, New York, 1996)
6. **Handbook of Physical Quantities** — The range of coverage is somewhat similar to the *CRC Handbook of Chemistry and Physics*, but with a stronger emphasis on physics than on chemistry. Solid state physics, lasers, nuclear physics, geophysics, and astronomy receive considerable attention. (Igor S. Grigoriev and Evgenii Z. Meilikhov, eds., CRC Press, Boca Raton, FL, 1997)
7. **Kaye & Laby Tables of Physical and Chemical Constants** — Kaye & Laby dates from 1911, and the 16th Edition was prepared in 1995 by a committee of experts. The coverage extends to almost every field of physics and chemistry; data on a limited number of representative substances or materials are given for each topic. (Longman Group Limited, Harlow, Essex, U.K., 1995)

8. **Lange's Handbook of Chemistry** — Provides broad coverage of chemical data; last updated in 2005. Also available on the Web [www.knovel.com]. (James G. Speight, ed., McGraw-Hill, New York, 2005)
9. **Recommended Reference Materials for the Realization of Physicochemical Properties** — This IUPAC book emphasizes highly accurate data on substances and materials that can be used as calibration standards. It covers physical, thermal, optical, and electrical properties. (K. N. Marsh, ed., Blackwell Scientific Publications, Oxford, 1987)
10. **The Merck Index** — Now in its 15th Edition, *The Merck Index* is a widely used source of data on over 10,000 compounds, chosen particularly for their importance in biology, medicine, and ecology. A short monograph on each compound gives information on the synthesis and uses as well as physical and toxicological properties. (Maryadele J. O'Neil, ed., RSC Publishing, 2013)

E. Summary of Useful Web Sites for Physical and Chemical Properties

Most of the Web sites in the following list provide direct access to factual data on physical and chemical properties. However, the list also includes portals that link to different property databases or describe the procedure for gaining access to electronic sources of property data. There are also a few chemical directory sites that are useful for obtaining formulas, synonyms, and registry numbers for substances of interest.

Web site	Address	Comments
ACD/Labs Spectral Data	www.acdlabs.com/products/adh/	Infrared, Raman and NMR spectra collections from Coblentz Society and other sources
Advanced Chemistry Development	www.acdlabs.com	Chemical directory, with programs for estimating physical and spectral properties
ASM Alloy Center Database	mio.asminternational.org/ac/	Physical, electrical, thermal, and mechanical properties of alloys
American Mineralogist Crystal Structure Database	www.geo.arizona.edu/AMS/amcsd.php	Lattice constants of minerals
Atomic Mass Data Center	amdc.impcas.ac.cn	See B.10
Beilstein Database	www.elsevier.com/solutions/reaxys	Properties and reactions of organic compounds. See C.3
Biocatalysis/Biodegradation Database	umbbd.ethz.ch/	Biocatalytic reactions, biodegradation of chemical compounds
BioCyc	biocyc.org/	Metabolic pathways of microorganisms
BioInfo Bank	www.abren.net/bioinfo/	Portal to ProTherm (protein thermodynamics), ProNit (protein–nucleic acid interactions), biomolecule structures
Biological Macromolecule Crystallization Database	xpdb.nist.gov:8060/BMCD4/index.faces	Crystal data and crystallization conditions for proteins, nucleic acids, and complexes
BRENDA	www.brenda-enzymes.info/	Enzyme nomenclature and properties
Cambridge Structural Database	www.ccdc.cam.ac.uk	See B.5
Carbon Dioxide Information Center	cdiac.esd.ornl.gov/	Data on atmospheric carbon dioxide
Ceramic Properties Databases	ceramics.org/knowledge-center/ceramic-resource-center	Mechanical, thermal, and other properties of ceramic materials
ChemExper	www.chemexper.com/	Consolidated chemical catalogs from various suppliers; provides physicavl properties and safety data; links to molfiles and MSDS
Chemical Entities of Biological Interest (ChEBI)	www.ebi.ac.uk.chebi	Dictionary of molecules and fragments, with identifiers and structures
ChemIDplus	chem.sis.nlm.nih.gov/chemidplus/	Chemical directory
ChemIndustry	www.chemindustry.com/chemicals/	Chemical directory
CHEMnetBASE	www.chemnetbase.com	Portal to *CRC Chemical Dictionaries, Handbook of Chemistry and Physics, Properties of Organic Compounds*, etc.
ChemSpider	www.chemspider.com	Aggregation of chemical structures and other information from many public sources References to synthesis; limited property data
ChemSynthesis Chemical Database	www.chemsynthesis.com	References to syntheses; limited property data
CODATA Databases	www.codata.org/resources/databases/	Thermodynamic key values and fundamental constants
Comparative Toxicogenomics Database (CTD)	ctdbase.org/	Chemical – gene/protein interactions
Cool Prop	coolprop.sourceforge.net	Thermophysical properties of 110 pure fluids and pseudo-pure fluids including humid air of interest to the organic Rankine cycle community
CRC Combined Chemical Dictionary	www.chemnetbase.com/	See C.1
Crystallography Open Database (COD)	www.crystallography.net	Crystal data on 52,000 compounds
DECHEMA (DETHERM)	www.dechema.de/en/Databases.html	See C.5
DIPPR Pure Compound Database	www.aiche.org/dippr	See B.3
Dortmund Data Bank	www.ddbst.de	See B.4

Web site	Address	Comments
eMolecules	www.emolecules.com	Portal to databases of chemical suppliers
Enzyme Nomenclature Database	www.expasy.ch/enzyme/	IUBMB nomenclature for enzymes
European Bioinformatics Institute	www.ebi.ac.uk/Databases/	Nucleotide and protein sequences, protein structures, enzyme nomenclature and reactions
FDM Reference Spectra Databases	www.fdmspectra.com/	Infrared, Raman, and mass spectra
FIZ Karlsruhe — ICSD	www.fiz-karlsruhe.de	ICSD crystal structure databases
Fundamental Physical Constants	physics.nist.gov/cuu/constants	CODATA fundamental constants
Gmelin Database	www.elsevier.com/solutions/reaxys	Properties and reactions of inorganic compounds. See C.4
Handbook of Chemistry and Physics	hbcpnetbase.com	Web version of *CRC Handbook of Chemistry and Physics*
Hazardous Substances Data Bank	toxnet.nlm.nih.gov/cgi-bin/sis/htmlgen?HSDB	Physical and toxicological properties of chemicals of health or environmental importance
HITRAN Database	www.cfa.harvard.edu/hitran/	High-resolution spectroscopic data for constituents of the atmosphere; parameters for calculating atmospheric transmission
Human Metabolome Database	www.hmdb.ca	Chemical and biological data on small molecule metabolites in humans
Infotherm	www.infotherm.com	Physical and thermal properties of pure compounds and mixtures
International Centre for Diffraction Data	www.icdd.com	See B.7
International Spectroscopic Data Bank	www.is-db.org	All types of spectra, deposited by users. Access is free
Ionic Liquids Database (ILThermo)	ilthermo.boulder.nist.gov/	Thermodynamic and thermophysical properties of ionic liquids and mixtures
IUBMB	www.chem.qmul.ac.uk/iubmb/	Enzyme and nucleic acid nomenclature
IUCr Data Activities	www.iucr.org/resources/data	Portal to crystallographic databases
IUPAC Home Page	www.iupac.org/	See B.13
IUPAC Kinetics Data	iupac.pole-ether.fr/	See B.13.b
IUPAC Nomenclature Rules	www.chem.qmul.ac.uk/iupac/	Useful site for organic and biochemical nomenclature
IUPAC-NIST Solubility Database	srdata.nist.gov/solubility/	See B.13.a
Knovel.com	www.knovel.com	Portal to *Lange's Handbook*, *Perry's Chemical Engineers' Handbook*, etc.
Kyoto Encyclopedia of Genes and Genomes (KEGG)	www.genome.ad.jp/kegg/	Includes data on drugs and other biochemical compounds
Landolt-Börnstein Online	www.springermaterials.com	See C.6
Lipidat	www.lipidat.tcd.ie	Structures and thermodynamic properties of lipids; crystal polymorphic transitions
MatWeb	www.matweb.com	Thermal, electrical, and mechanical properties of engineering materials
Metals Crystallographic Data File	www.tothcanada.com	See B.9
NASA Chemical Kinetics Data	jpldataeval.jpl.nasa.gov	Kinetic and photochemical data for stratospheric modeling
National Center for Biotechnology Information	www.ncbi.nlm.nih.gov	Portal to GenBank and other sequence databases
National Nuclear Data Center	www.nndc.bnl.gov	See B.12
National Toxicology Program	ntp.niehs.nih.gov	Chemical health and safety data
NIST Atomic Spectra Database	www.nist.gov/pml/data/asd.cfm	Energy levels, wavelengths, and transition probabilities of atoms and atomic ions
NIST Ceramics Webbook	www.nist.gov/mml/mmsd/webbook.cfm	See B.1
NIST Chemistry Webbook	webbook.nist.gov	Broad range of physical, thermal, and spectral properties
NIST Data Gateway	srdata.nist.gov/gateway/	Portal to all NIST data systems; see B.1
NIST Physical Reference Data	www.nist.gov/pml/data	Atomic and molecular spectra, cross sections, x-ray attenuation, and dosimetry data
NLM Gateway	www.nlm.nih.gov/databases	Portal to all National Library of Medicine databases
NMR Shift DB	www.nmrshiftdb.org	NMR data submitted by users
Nucleic Acid Database	ndbserver.rutgers.edu/	Crystal structures of nucleic acids
Particle Data Group	pdg.lbl.gov	See B.11
Polymers — A Property Database	www.polymersdatabase.com/	Properties of commercial polymers
Powder Diffraction File	www.icdd.com	See B.7
Properties of Organic Compounds	www.chemnetbase.com/tours/poc/intro.jsf	See C.2
Protein Data Bank	www.rcsb.org	See B.8
PubChem	pubchem.ncbi.nlm.nih.gov/	Chemical directory with links to biological information
Reaxys	www.elsevier.com/solutions/reaxys	Access to Beilstein and Gmelin databases

Web site	Address	Comments
SABIO-Reaction Kinetics Database	sabio.villa-bosch.de/	Data on kinetics of biochemical reactions
Sigma-Aldrich	www.sigmaaldrich.com/	Chemical catalogs; includes some physical property data
Spectral Database for Organic Compounds	sdbs.db.aist.go.jp/sdbs/cgi-bin/cre_index.cgi?lang=eng	MS, NMR, IR, Raman, and ESR spectra; 32,000 compounds measured at AIST, Japan
SpecInfo	onlinelibrary.wiley.com/book/10.1002/9780471692294	IR, NMR, and mass spectra
Spectra Online	www.ftirsearch.com/	FTIR and Raman spectra
SPRESI-web	www.spresi.de/	Structures, reactions, and some physical properties
SpringerMaterials	www.springermaterials.com	The online version of *Landolt-Börnstein Tables* (C.6)
STN Easy	stneasy.cas.org	Chemical directory (and access to Chemical Abstracts databases)
STN Easy-Europe	stneasy.fiz-karlsruhe.de	European node of STN Easy
STN Easy-Japan	stneasy-japan.cas.org	Japanese node of STN Easy
Swissprot	bo.expasy.org/enzyme/	Enzyme nomenclature and related information
Thermodynamics of Enzyme-Catalyzed Reactions	xpdb.nist.gov/enzyme_thermodynamics/	Equilibrium constants of biochemical reactions
Thermodynamics Research Center	www.trc.nist.gov	See B.2
TOXNET	toxnet.nlm.nih.gov	Portal to HSDB and other databases on hazardous chemicals

Index

The most efficient way to use this index is to look for the pertinent *property* (e.g., vapor pressure, entropy), *process* (e.g., shielding, calibration), or *general concept* (e.g., units, radiation). Most primary entries are subdivided into several secondary entries, e.g., under heat capacity there are 24 secondary entries such as air, metals, water, etc. Primary entries will be found for certain *classes of substances*, such as alloys, elements, organic compounds, refrigerants, semiconductors, etc. Primary entries are also given for the individual chemical elements and for a few compounds such as water and carbon dioxide. However, only the most important tables are listed under these substances. Therefore, the user will find in most cases that it is best to look first for the property of interest, then examine the table or tables that are referenced.

Entries in boldface type are the titles of tables as they appear in the Table of Contents.

The reference given for each index term is the inclusive pages of the pertinent table (e.g., **8**-45 to 55). The introduction to each table describes the method of ordering the substances within that table.

The editor would be grateful for comments and suggestions on this index.

INDEX

A

C

D

F

I

J

K

N

O

T